AMERICAN
MEN AND WOMEN
OF SCIENCE

AMERICAN MEN AND WOMEN OF SCIENCE

13TH EDITION

Edited by the Jaques Cattell Press

Volume 1 A–C

R. R. BOWKER COMPANY
A Xerox Publishing Company
New York & London, 1976

Contents

Advisory Committee

Dr. Dael L. Wolfle, Chairman
Graduate School of Public Affairs
University of Washington

Dr. Randolph W. Bromery
Chancellor,
University of Massachusetts

Dr. Janet W. Brown
Program Head,
Office of Opportunities
in Science
American Association for the
Advancement of Science

Dr. Robert W. Cairns
Executive Director,
American Chemical Society

Dr. S. D. Cornell
Assistant to the President
National Academy of Sciences

Dr. Ruth M. Davis
Director,
Institute for Computer Science
and Technology
National Bureau of Standards

Dr. Carl D. Douglass
Deputy Director,
Division of Research Grants
National Institutes of Health

Dr. Richard G. Folsom
President Emeritus,
Rensselaer Polytechnic Institute

Dr. Robert E. Henze
Director,
Membership Division
American Chemical Society

Dr. Eugene L. Hess
Executive Director,
Federation of American Societies
for Experimental Biology

Dr. William C. Kelly
Executive Director,
Commission on Human Resources
National Research Council

Dr. Kenneth B. Raper
Department of Bacteriology
University of Wisconsin

Dr. A. L. Schawlow
Department of Physics
Stanford University

Dr. John F. Sherman
Vice President,
Association of American Medical
Colleges

Dr. Matthias Stelly
Executive Vice President,
American Society of Agronomy

Dr. John R. Whinnery
Department of Electrical Engineering
and Computer Sciences
University of California, Berkeley

Preface

The observance of an anniversary prompts reflection on the past, and it is natural to note here that the astounding growth in the dimension and stature of this 200 year old American nation has been influenced immeasurably by the achievements of her scientists. 1976 also marks the 70th anniversary of AMERICAN MEN & WOMEN OF SCIENCE as a chronicle of the lives and professional activities of those men and women most instrumental in affecting the shape and quality of science in America. The explosion in scientific activity over the past seven decades is clearly evident when one compares the 1906 edition, a single volume containing 4,000 entries, with the present edition, six volumes profiling nearly 110,000 men and women of importance in their fields.

This edition of AMERICAN MEN & WOMEN OF SCIENCE is a landmark in biographic achievement. The information contained in all seven volumes has been gathered, edited and compiled by the Jaques Cattell Press in the space of ten months. This is a radical, and beneficial, departure from the production of the 12th edition, which took three years to publish in its entirety. The acceleration was made possible by the use of a computerized printing method and more efficient production procedures.

The editors have not sacrificed quality in the interest of speed, however. The criteria were stringently applied in the selection of new entrants, all nominated by former biographees. The criteria follow.

> Achievement, by reason of experience and training, of a stature in scientific work equivalent to that associated with the doctoral degree, coupled with presently continued activity in such work;

> or

> Research activity of high quality in science as evidenced by publication in reputable scientific journals; or, for those whose work cannot be published because of governmental or industrial security, research activity of high quality in science as evidenced by the judgment of the individual's peers;

> or

> Attainment of a position of substantial responsibility requiring scientific training and experience to the extent described for (1) and (2).

All data forms submitted by biographees were carefully edited and proofread. Those whose information was appearing for the first time received a proof before publication. Information on scientists who did not supply current material is included only if verification of present professional activity was established by our researchers. A former biographee whose current status could not be verified is given a reference to the 12th edition if the probability exists of a continued activity in science. References are also given to scientists who have died since publication of the last edition. Omitted are the names of those previously listed as retired and those who have entered fields or activities not covered by the scope of the directory. More than 20,000 nominations for inclusion were received and over 12,000 of those nominated returned information and were selected for listing. This is a significant increase in both nominations and selections over any previous edition and gives fuller representation to the developing scientist.

Geographic and discipline indexes make up the seventh volume of the set. The discipline index has been rearranged and is organized with major subject headings, providing easier access to this information.

Certain disciplines previously included in the directory are not represented in this edition. Engineering and economics will appear in separate directories with expanded criteria, to enable even broader coverage of these areas. Others, including sociology, psychology and political science were omitted because they are fully covered by membership directories in each field.

Great appreciation is expressed to the AMERICAN MEN & WOMEN OF SCIENCE Advisory Committee for their guidance in the planning of the 13th edition. Their efforts have contributed to an unusually good response to our requests for information and nominations, which has enhanced the value of the publication. Also to be thanked are the many scientific societies that provided membership lists for the use of our researchers or that published announcements in their bulletins and journals.

The staff of the Jaques Cattell Press deserves the highest accolade for their sustained interest, devotion and good will through the many hours of learning and implementing the new procedures necessary to the successful completion of this book. The overwhelming workload was shared by temporary employees who performed with the greatest diligence and responsibility. The job could not have been completed without their fine help. Everyone involved with this project gave outstanding service, but the contributions of Alice Smith, Pauline Stump, Joyce Howell and Ila Martin cannot go without mention. Special acknowledgement is given to Fred Scott, former general manager of the Jaques Cattell Press, who was responsible for initiating the formation of the advisory committee and for overseeing the planning and early stages of work on the 13th edition.

Comments and suggestions are invited and should be addressed to The Editors, American Men & Women of Science, Jaques Cattell Press, P.O. Box 25001, Tempe, Arizona, 85282.

Renee Lautenbach, Supervising Editor
Anne Rhodes, Administrative Managing Editor

JAQUES CATTELL PRESS

Desmond Reaney, Manager Book Editorial

R.R. BOWKER COMPANY

October, 1976

Abbreviations

AAAS—American Association for the Advancement of Science
abnorm—abnormal
abstr—abstract(s)
acad—academic, academy
acct—account, accountant, accounting
acoust—acoustic(s), acoustical
ACTH—adrenocorticotrophic hormone
actg—acting
activ—activities, activity
addn—addition(s), additional
Add—Address
adj—adjunct, adjutant
adjust—adjustment
Adm—Admiral
admin—administration, administrative
adminr—administrator(s)
admis—admission(s)
adv—adviser(s), advisory
advan—advance(d), advancement
advert—advertisement, advertising
AEC—Atomic Energy Commission
aerodyn—aerodynamic(s)
aeronaut—aeronautic(s), aeronautical
aerophys—aerophysical, aerophysics
aesthet—aesthetic(s)
AFB—Air Force Base
affil—affiliate(s), affiliation
agr—agricultural, agriculture
agron—agronomic, agronomical, agronomy
agrost—agrostologic, agrostological, agrostology
agt—agent
AID—Agency for International Development
Ala—Alabama
allergol—allergological, allergology
alt—alternate
Alta—Alberta
Am—America, American
AMA—American Medical Association
anal—analysis, analytic, analytical
analog—analogue
anat—anatomic, anatomical, anatomy
anesthesiol—anesthesiology
angiol—angiology
Ann—Annal(s)
ann—annual
anthrop—anthropological, anthropology
anthropom—anthropometric, anthropometrical, anthropometry
antiq—antiquary, antiquities, antiquity
antiqn—antiquarian

apicult—apicultural, apiculture
APO—Army Post Office
app—appoint, appointed
appl—applied
appln—application
approx—approximate(ly)
Apr—April
apt—apartment(s)
aquacult—aquaculture
arbit—arbitration
arch—archives
archaeol—archaeological, archaeology
archit—architectural, architecture
Arg—Argentina, Argentine
Ariz—Arizona
Ark—Arkansas
artil—artillery
asn—association
assoc(s)—associate(s), associated
asst(s)—assistant(s), assistantship(s)
Assyriol—Assyriology
astrodyn—astrodynamics
astron—astronomical, astronomy
astronaut—astronautical, astronautics
astronr—astronomer
astrophys—astrophysical, astrophysics
attend—attendant, attending
atty—attorney
audiol—audiology
Aug—August
auth—author
AV—audiovisual
Ave—Avenue
avicult—avicultural, aviculture

b—born
bact—bacterial, bacteriologic, bacteriological, bacteriology
BC—British Columbia
bd—board
behav—behavior(al)
Belg—Belgian, Belgium
bibl—biblical
bibliog—bibliographic, bibliographical, bibliography
bibliogr—bibliographer
biochem—biochemical, biochemistry
biog—biographical, biography
biol—biological, biology
biomed—biomedical, biomedicine
biomet—biometric(s), biometrical, biometry
biophys—biophysical, biophysics

bk(s)—book(s)
bldg—building
Blvd—Boulevard
Bor—Borough
bot—botanical, botany
br—branch(es)
Brig—Brigadier
Brit—Britain, British
Bro(s)—Brother(s)
bryol—bryology
Bull—Bulletin
bur—bureau
bus—business
BWI—British West Indies

c—children
Calif—California
Can—Canada, Canadian
cand—candidate
Capt—Captain
cardiol—cardiology
cardiovasc—cardiovascular
cartog—cartographic, cartographical, cartography
cartogr—cartographer
Cath—Catholic
CEngr—Corps of Engineers
cent—central
Cent Am—Central America
cert—certificate(s), certification, certified
chap—chapter
chem—chemical(s), chemistry
chemother—chemotherapy
chmn—chairman
citricult—citriculture
class—classical
climat—climatological, climatology
clin(s)—clinic(s), clinical
cmndg—commanding
Co—Companies, Company
coauth—coauthor
co-dir—co-director
co-ed—co-editor
coeduc—coeducation, coeducational
col(s)—college(s), collegiate, colonel
collab—collaboration, collaborative
collabr—collaborator
Colo—Colorado
com—commerce, commercial
Comdr—Commander
commun—communicable, communication(s)
comn(s)—commission(s), commissioned

ABBREVIATIONS

comnr—commissioner
comp—comparative
compos—composition
comput—computation, computer(s), computing
comt(s)—committee(s)
conchol—conchology
conf—conference
cong—congress, congressional
Conn—Connecticut
conserv—conservation, conservatory
consol—consolidated, consolidation
const—constitution, constitutional
construct—construction, constructive
consult(s)—consult, consultant(s), consultant-
 ship(s), consultation, consulting
contemp—contemporary
contrib—contribute, contributing, contribu-
 tion(s)
contribr—contributor
conv—convention
coop—cooperating, cooperation, cooperative
coord—coordinate(d), coordinating, coordination
coordr—coordinator
corp—corporate, corporation(s)
corresp—correspondence, correspondent, corresponding
coun—council, counsel, counseling
counr—councilor, counselor
criminol—criminological, criminology
cryog—cryogenic(s)
crystallog—crystallographic, crystallograph-
 ical, crystallography
crystallogr—crystallographer
Ct—Court
Ctr—Center
cult—cultural, culture
cur—curator
curric—curriculum
cybernet—cybernetic(s)
cytol—cytological, cytology
CZ—Canal Zone
Czech—Czechoslovakia

DC—District of Columbia
Dec—December
Del—Delaware
deleg—delegate, delegation
delinq—delinquency, delinquent
dem—democrat(s), democratic
demog—demographic, demography
demogr—demographer
demonstr—demonstrator
dendrol—dendrologic, dendrological, dendrology
dent—dental, dentistry
dep—deputy
dept—department(al)
dermat—dermatologic, dermatological, dermatology
develop—developed, developing, development,
 developmental
diag—diagnosis, diagnostic
dialectol—dialectological, dialectology
dict—dictionaries, dictionary
Dig—Digest
dipl—diploma, diplomate
dir(s)—director(s), directories, directory
dis—disease(s), disorders
Diss Abstr—Dissertation Abstracts
dist—district
distrib—distributed, distribution, distributive
distribr—distributor(s)
div—division, divisional, divorced

DNA—deoxyribonucleic acid
doc—document(s), documentary, documentation
Dom—Dominion
Dr—Drive

e—east
ecol—ecological, ecology
econ(s)—economic(s), economical, economy
economet—econometric(s)
ECT—electroconvulsive or electroshock therapy
ed—edition(s), editor(s), editorial
ed bd—editorial board
educ—education, educational
educr—educator(s)
EEG—electroencephalogram, electroencephalographic, electroencephalography
Egyptol—Egyptology
EKG—electrocardiogram
elec—electric, electrical, electricity
electrochem—electrochemical, electrochemistry
electrophys—electrophysical, electrophysics
elem—elementary
embryol—embryologic, embryological, embryology
emer—emeriti, emeritus
employ—employment
encour—encouragement
encycl—encyclopedia
endocrinol—endocrinologic, endocrinology
eng—engineering
Eng—England, English
engr(s)—engineer(s)
enol—enology
Ens—Ensign
entom—entomological, entomology
environ—environment(s), environmental
enzym—enzymology
epidemiol—epidemiologic, epidemiological,
 epidemiology
equip—equipment
ESEA—Elementary & Secondary Education
 Act
espec—especially
estab—established, establishment(s)
ethnog—ethnographic, ethnographical, ethnography
ethnogr—ethnographer
ethnol—ethnologic, ethnological, ethnology
Europ—European
eval—evaluation
evangel—evangelical
eve—evening
exam—examination(s), examining
examr—examiner
except—exceptional
exec(s)—executive(s)
exeg—exegeses, exegesis, exegetic, exegetical
exhib(s)—exhibition(s), exhibit(s)
exp—experiment, experimental
exped(s)—expedition(s)
explor—exploration(s), exploratory
expos—exposition
exten—extension

fac—faculty
facil—facilities, facility
Feb—February
fed—federal
fedn—federation
fel(s)—fellow(s), fellowship(s)
fermentol—fermentology

fertil—fertility, fertilization
Fla—Florida
floricult—floricultural, floriculture
found—foundation
FPO—Fleet Post Office
Fr—French
Ft—Fort

Ga—Georgia
gastroenterol—gastroenterological, gastroenterology
gen—general
geneal—genealogical, genealogy
geod—geodesy, geodetic
geog—geographic, geographical, geography
geogr—geographer
geol—geologic, geological, geology
geom—geometric, geometrical, geometry
geomorphol—geomorphologic, geomorphology
geophys—geophysical, geophysics
Ger—German, Germanic, Germany
geriat—geriatric(s)
geront—gerontological, gerontology
glaciol—glaciology
gov—governing, governor(s)
govt—government, governmental
grad—graduate(d)
Gt Brit—Great Britain
guid—guidance
gym—gymnasium
gynec—gynecologic, gynecological, gynecology

handbk(s)—handbook(s)
helminth—helminthology
hemat—hematologic, hematological,
 hematology
herpet—herpetologic, herpetological, herpetology
Hisp—Hispanic, Hispania
hist—historic, historical, history
histol—histological, histology
HM—Her Majesty
hochsch—hochschule
homeop—homeopathic, homeopathy
hon(s)—honor(s), honorable, honorary
hort—horticultural, horticulture
hosp(s)—hospital(s), hospitalization
hq—headquarters
HumRRO—Human Resources Research Office
husb—husbandry
Hwy—Highway
hydraul—hydraulic(s)
hydrodyn—hydrodynamic(s)
hydrol—hydrologic, hydrological, hydrology
hyg—hygiene, hygienic(s)
hypn—hypnosis

ichthyol—ichthyological, ichthyology
Ill—Illinois
illum—illuminating, illumination
illus—illustrate, illustrated, illustration
illusr—illustrator
immunol—immunologic, immunological, immunology
Imp—Imperial
improv—improvement
Inc—Incorporated
in-chg—in charge
incl—include(s), including
Ind—Indiana
indust(s)—industrial, industries, industry
inf—infantry
info—information
inorg—inorganic

ins—insurance
inst(s)—institute(s), institution(s)
instnl—institutional(ized)
instr(s)—instruct, instruction, instructor(s)
instrnl—instructional
int—international
intel—intelligence
introd—introduction
invert—invertebrate
invest(s)—investigation(s)
investr—investigator
irrig—irrigation
Ital—Italian

J—Journal
Jan—January
Jct—Junction
jour—journal, journalism
jr—junior
jurisp—jurisprudence
juv—juvenile

Kans—Kansas
Ky—Kentucky

La—Louisiana
lab(s)—laboratories, laboratory
lang—language(s)
laryngol—laryngological, laryngology
lect—lecture(s)
lectr—lecturer(s)
legis—legislation, legislative, legislature
lett—letter(s)
lib—liberal
libr—libraries, library
librn—librarian
lic—license(d)
limnol—limnological, limnology
ling—linguistic(s), linguistical
lit—literary, literature
lithol—lithologic, lithological, lithology
Lt—Lieutenant
Ltd—Limited

m—married
mach—machine(s), machinery
mag—magazine(s)
maj—major
malacol—malacology
mammal—mammalogy
Man—Manitoba
Mar—March
Mariol—Mariology
Mass—Massachusetts
mat—material(s)
mat med—materia medica
math—mathematic(s), mathematical
Md—Maryland
mech—mechanic(s), mechanical
med—medical, medicinal, medicine
Mediter—Mediterranean
Mem—Memorial
mem—member(s), membership(s)
ment—mental(ly)
metab—metabolic, metabolism
metall—metallurgic, metallurgical, metallurgy
metallog—metallographic, metallography
metallogr—metallographer
metaphys—metaphysical, metaphysics
meteorol—meteorological, meteorology
metrol—metrological, metrology
metrop—metropolitan
Mex—Mexican, Mexico
mfg—manufacturing

mfr(s)—manufacture(s), manufacturer(s)
mgr—manager
mgt—management
Mich—Michigan
microbiol—microbiological, microbiology
micros—microscopic, microscopical, microscopy
mid—middle
mil—military
mineral—mineralogical, mineralogy
Minn—Minnesota
Miss—Mississippi
mkt—market, marketing
Mo—Missouri
mod—modern
monogr—monograph
Mont—Montana
morphol—morphological, morphology
Mt—Mount
mult—multiple
munic—municipal, municipalities
mus—museum(s)
musicol—musicological, musicology
mycol—mycologic, mycology

n—north
NASA—National Aeronautics & Space Administration
nat—national, naturalized
NATO—North Atlantic Treaty Organization
navig—navigation(al)
NB—New Brunswick
NC—North Carolina
NDak—North Dakota
NDEA—National Defense Education Act
Nebr—Nebraska
nematol—nematological, nematology
nerv—nervous
Neth—Netherlands
neurol—neurological, neurology
neuropath—neuropathological, neuropathology
neuropsychiat—neuropsychiatric, neuropsychiatry
neurosurg—neurosurgical, neurosurgery
Nev—Nevada
New Eng—New England
New York—New York City
Nfld—Newfoundland
NH—New Hampshire
NIH—National Institutes of Health
NIMH—National Institute of Mental Health
NJ—New Jersey
NMex—New Mexico
nonres—nonresident
norm—normal
Norweg—Norwegian
Nov—November
NS—Nova Scotia
NSF—National Science Foundation
NSW—New South Wales
numis—numismatic(s)
nutrit—nutrition, nutritional
NY—New York State
NZ—New Zealand

observ—observatories, observatory
obstet—obstetric(s), obstetrical
occas—occasional(ly)
occup—occupation, occupational
oceanog—oceanographic, oceanographical, oceanography
oceanogr—oceanographer
Oct—October
odontol—odontology

OEEC—Organization for European Economic Cooperation
off—office, official
Okla—Oklahoma
olericult—olericulture
oncol—oncologic, oncology
Ont—Ontario
oper(s)—operation(s), operational, operative
ophthal—ophthalmologic, ophthalmological, ophthalmology
optom—optometric, optometrical, optometry
ord—ordnance
Ore—Oregon
org—organic
orgn—organization(s), organizational
orient—oriental
ornith—ornithological, ornithology
orthod—orthodontia, orthodontic(s)
orthop—orthopedic(s)
osteop—osteopathic, osteopathy
otol—otological, otology
otolaryngol—otolaryngological, otolaryngology
otorhinol—otorhinologic, otorhinology

Pa—Pennsylvania
Pac—Pacific
paleobot—paleobotanical, paleobotany
paleont—paleontological, paleontology
Pan-Am—Pan-American
parasitol—parasitology
partic—participant, participating
path—pathologic, pathological, pathology
pedag—pedagogic(s), pedagogical, pedagogy
pediat—pediatric(s)
PEI—Prince Edward Islands
penol—penological, penology
periodont—periodontal, periodontic(s), periodontology
petrog—petrographic, petrographical, petrography
petrogr—petrographer
petrol—petroleum, petrologic, petrological, petrology
pharm—pharmacy
pharmaceut—pharmaceutic(s), pharmaceutical(s)
pharmacog—pharmacognosy
pharmacol—pharmacologic, pharmacological, pharmacology
phenomenol—phenomenologic(al), phenomenology
philol—philological, philology
philos—philosophic, philosophical, philosophy
photog—photographic, photography
photogeog—photogeographic, photogeography
photogr—photographer(s)
photogram—photogrammetric, photogramme-photogrammetry
photom—photometric, photometrical, photometry
phycol—phycology
phys—physical
physiog—physiographic, physiographical, physiography
physiol—physiological, physiology
Pkwy—Parkway
Pl—Place
polit—political, politics
polytech—polytechnic(al)
pomol—pomological, pomology
pontif—pontifical
pop—population
Port—Portugal, Portuguese
postgrad—postgraduate

ABBREVIATIONS

PQ—Province of Quebec
PR—Puerto Rico
pract—practice
practr—practitioner
prehist—prehistoric, prehistory
prep—preparation, preparative, preparatory
pres—president
Presby—Presbyterian
preserv—preservation
prev—prevention, preventive
prin—principal
prob(s)—problem(s)
proc—proceedings
proctol—proctologic, proctological, proctology
prod—product(s), production, productive
prof—professional, professor, professorial
Prof Exp—Professional Experience
prog(s)—program(s), programmed, programming
proj—project(s), projection(al), projective
prom—promotion
protozool—protozoology
prov—province, provincial
psychiat—psychiatric, psychiatry
psychoanal—psychoanalysis, psychoanalytic, psychoanalytical
psychol—psychological, psychology
psychomet—psychometric(s)
psychopath—psychopathologic, psychopathology
psychophys—psychophysical, psychophysics
psychophysiol—psychophysiological, psychophysiology
psychosom—psychosomatic(s)
psychother—psychotherapeutic(s), psychotherapy
Pt—Point
pub—public
publ—publication(s), publish(ed), publisher, publishing
pvt—private

Qm—Quartermaster
Qm Gen—Quartermaster General
qual—qualitative, quality
quant—quantitative
quart—quarterly

radiol—radiological, radiology
RAF—Royal Air Force
RAFVR—Royal Air Force Volunteer Reserve
RAMC—Royal Army Medical Corps
RAMCR—Royal Army Medical Corps Reserve
RAOC—Royal Army Ordnance Corps
RASC—Royal Army Service Corps
RASCR—Royal Army Service Corps Reserve
RCAF—Royal Canadian Air Force
RCAFR—Royal Canadian Air Force Reserve
RCAFVR—Royal Canadian Air Force Volunteer Reserve
RCAMC—Royal Canadian Army Medical Corps
RCAMCR—Royal Canadian Army Medical Corps Reserve
RCASC—Royal Canadian Army Service Corps
RCASCR—Royal Canadian Army Service Corps Reserve
RCEME—Royal Canadian Electrical & Mechanical Engineers
RCN—Royal Canadian Navy
RCNR—Royal Canadian Naval Reserve
RCNVR—Royal Canadian Naval Volunteer Reserve
Rd—Road

RD—Rural Delivery
rec—record(s), recording
redevelop—redevelopment
ref—reference(s)
refrig—refrigeration
regist—register(ed), registration
registr—registrar
regt—regiment(al)
rehab—rehabilitation
rel(s)—relation(s), relative
relig—religion, religious
REME—Royal Electrical & Mechanical Engineers
rep—represent, representative
repub—republic
req—requirements
res—research, reserve
rev—review, revised, revision
RFD—Rural Free Delivery
rhet—rhetoric, rhetorical
RI—Rhode Island
Rm—Room
RM—Royal Marines
RN—Royal Navy
RNA—ribonucleic acid
RNR—Royal Naval Reserve
RNVR—Royal Naval Volunteer Reserve
roentgenol—roentgenologic, roentgenological, roentgenology
RR—Railroad, Rural Route
rte—route
Russ—Russian
rwy—railway

s—south
SAfrica—South Africa
SAm—South America, South American
sanit—sanitary, sanitation
Sask—Saskatchewan
SC—South Carolina
Scand—Scandinavia(n)
sch(s)—school(s)
scholar—scholarship
sci—science(s), scientific
SDak—South Dakota
SEATO—Southeast Asia Treaty Organization
sec—secondary
sect—section
secy—secretary
seismog—seismograph, seismographic, seismography
seismogr—seismographer
seismol—seismological, seismology
sem—seminar, seminary
sen—senator, senatorial
Sept—September
ser—serial, series
serol—serologic, serological, serology
serv—service(s), serving
silvicult—silvicultural, silviculture
soc(s)—societies, society
soc sci—social science
sociol—sociologic, sociological, sociology
Span—Spanish
spec—special
specif—specification(s)
spectrog—spectrograph, spectrographic, spectrography
spectrogr—spectrographer
spectrophotom—spectrophotometer, spectrophotometric, spectrophotometry
spectros—spectroscopic, spectroscopy
speleol—speleological, speleology
Sq—Square

sr—senior
St—Saint, Street(s)
sta(s)—station(s)
stand—standard(s), standardization
statist—statistical, statistics
Ste—Sainte
steril—sterility
stomatol—stomatology
stratig—stratigraphic, stratigraphy
stratigr—stratigrapher
struct—structural, structure(s)
stud—student(ship)
subcomt—subcommittee
subj—subject
subsid—subsidiary
substa—substation
super—superior
suppl—supplement(s), supplemental, supplementary
supt—superintendent
supv—supervising, supervision
supvr—supervisor
supvry—supervisory
surg—surgery, surgical
surv—survey, surveying
survr—surveyor
Swed—Swedish
Switz—Switzerland
symp—symposia, symposium(s)
syphil—syphilology
syst(s)—system(s), systematic(s), systematical

taxon—taxonomic, taxonomy
tech—technical, technique(s)
technol—technologic(al), technology
tel—telegraph(y), telephone
temp—temporary
Tenn—Tennessee
Terr—Terrace
Tex—Texas
textbk(s)—textbook(s)
text ed—text edition
theol—theological, theology
theoret—theoretic(al)
ther—therapy
therapeut—therapeutic(s)
thermodyn—thermodynamic(s)
topog—topographic, topographical, topography
topogr—topographer
toxicol—toxicologic, toxicological, toxicology
trans—transactions
transl—translated, translation(s)
translr—translator(s)
transp—transport, transportation
treas—treasurer, treasury
treat—treatment
trop—tropical
tuberc—tuberculosis
TV—television
Twp—Township

UAR—United Arab Republic
UK—United Kingdom
UN—United Nations
undergrad—undergraduate
unemploy—unemployment
UNESCO—United Nations Educational Scientific & Cultural Organization
UNICEF—United Nations International Childrens Fund
univ(s)—universities, university
UNRRA—United Nations Relief & Rehabilitation Administration

UNRWA—United Nations Relief & Works Agency
urol—urologic, urological, urology
US—United States
USA—US Army
USAAF—US Army Air Force
USAAFR—US Army Air Force Reserve
USAF—US Air Force
USAFR—US Air Force Reserve
USAR—US Army Reserve
USCG—US Coast Guard
USCGR—US Coast Guard Reserve
USDA—US Department of Agriculture
USMC—US Marine Corps
USMCR—US Marine Corps Reserve
USN—US Navy
USNAF—US Naval Air Force
USNAFR—US Naval Air Force Reserve
USNR—US Naval Reserve
USPHS—US Public Health Service
USPHSR—US Public Health Service Reserve
USSR—Union of Soviet Socialist Republics
USWMC—US Women's Marine Corps

USWMCR—US Women's Marine Corps Reserve

Va—Virginia
var—various
veg—vegetable(s), vegetation
vent—ventilating, ventilation
vert—vertebrate
vet—veteran(s), veterinarian, veterinary
VI—Virgin Islands
vinicult—viniculture
virol—virological, virology
vis—visiting
voc—vocational
vocab—vocabulary
vol(s)—voluntary, volunteer(s), volume(s)
vpres—vice president
vs—versus
Vt—Vermont

w—west
WAC—Women's Army Corps

Wash—Washington
WAVES—Women Accepted for Voluntary Emergency Service
WHO—World Health Organization
WI—West Indies
wid—widow, widowed, widower
Wis—Wisconsin
WRCNS—Women's Royal Canadian Naval Service
WRNS—Women's Royal Naval Service
WVa—West Virginia
Wyo—Wyoming

yearbk(s)—yearbook(s)
YMCA—Young Men's Christian Association
YMHA—Young Men's Hebrew Association
yr(s)—year(s)
YWCA—Young Women's Christian Association
YWHA—Young Women's Hebrew Association

zool—zoological, zoology

AMERICAN
MEN AND WOMEN
OF SCIENCE

A

AABOE, ASGER (HARTVIG), b Copenhagen, Denmark, Apr 26, 22; m 50; c 4. MATHEMATICS. Educ: Brown Univ, PhD(hist math), 57. Prof Exp: Vis lectr math, Washington Univ, 47-48; adj, Birkeröd State Sch, Denmark, 49-52; from instr to assoc prof, Tufts Univ, 52-62; assoc prof, 62-67, PROF MATH & HIST SCI & MED, YALE UNIV, 67-, CHMN DEPT HIST SCI & MED, 68- Concurrent Pos: Vis assoc prof, Yale Univ, 61-62; Guggenheim fels, 64-66. Mem: Am Math Soc; Hist Sci Soc; Danish Math Soc. Res: History of mathematics and astronomy. Mailing Add: Dept of Hist of Sci & Med Yale Univ Box 2036 New Haven CT 06520

AAGAARD, GEORGE NELSON, b Minneapolis, Minn, Aug 16, 13; m 39; c 5. INTERNAL MEDICINE, CARDIOLOGY. Educ: Univ Minn, BS, 34, BMed, 36, MD, 37. Prof Exp: Instr internal med, Univ Minn, 42-45, asst prof, 45-48, assoc prof & dir postgrad med educ, 48-51; prof med & dean, Univ Tex Southwestern Med Sch Dallas, 52-54; dean sch med, 54-64, ombudsman, 69-70, PROF MED & HEAD DIV CLIN PHARMACOL, SCH MED, UNIV WASH, 64- Concurrent Pos: Fel internal med, Univ Minn, 41-42; mem health res facil adv coun, NIH; mem nat adv coun, Nat Heart Inst; mem bd dirs, Am Heart Asn; mem spec med adv group, Vet Admin, 70-74; chmn subcomt drug spotlight prog, Am Soc Clin Pharmacol & Therapeut. Mem: AMA; Am Heart Asn; Am Pub Health Asn; Am Col Clin Pharmacol & Chemother; NY Acad Sci. Res: Medical education; cardiovascular disease. Mailing Add: Sch of Med Univ of Wash Seattle WA 98195

AAGAARD, KNUT, b Brooklyn, NY, Feb 5, 39; m 63; c 2. PHYSICAL OCEANOGRAPHY. Educ: Oberlin Col, AB, 61; Univ Wash, MS, 64, PhD(phys oceanog), 66. Prof Exp: Res assoc phys oceanog, Univ Wash, 66; Nat Acad Sci-US Air Force Off Sci Res research assoc, Geophys Inst, Univ Bergen, 67; res asst prof, 68-73, RES ASSOC PROF PHYS OCEANOG, UNIV WASH, 73- Concurrent Pos: Mem overflow working group, Int Coun Explor of the Seas, 74- Mem: AAAS; fel Arctic Inst NAm. Res: Arctic oceanography, current measurements, general physical oceanography. Mailing Add: Dept of Oceanog Univ of Wash Seattle WA 98105

AAGARD, ROGER L, b East Chain, Minn, Aug 2, 31; m 60; c 3. PHYSICS, MATHEMATICS. Educ: Univ Minn, BA, 57, MS, 59. Prof Exp: Asst atmospheric physics, Univ Minn, 57-59; asst infrared physics, 59-63, sr res scientist optical physics, 63-67, PRIN RES SCIENTIST OPTICAL PHYSICS, HONEYWELL RES CTR, 67- Mem: Am Phys Soc; Optical Soc Am. Res: Thermal radiation sensors; semiconductor infrared detectors and dielectric lasers; atmospheric infrared radiation and optical properties of lasers; lasers and magneto-optic materials for computer information storage; integrated optics; thin flim optical components and systems. Mailing Add: Honeywell Corp Res Ctr 10701 Lyndale Ave Bloomington MN 55420

AALDERS, LEWIS ELDON, b Kentville, NS, July 19, 33; div. CYTOGENETICS. Educ: Acadia Univ, BSc, 52, MSc, 53; Cornell Univ, PhD(cytogenetics), 57. Prof Exp: HORTICULTURIST PLANT BREEDING, RES STA, CAN DEPT AGR, 53- Mem: Am Pomol Soc; Am Soc Hort Sci; Can Soc Hort Sci; Genetics Soc Can. Res: Cytogenetics and breeding of fruit crops; breeding of lowbush blueberry, strawberry and raspberry; culture of lowbush blueberry. Mailing Add: Res Sta Can Dept of Agr Kentville NS Can

AAMODT, R L, b Bacchus, Utah, Feb 14, 17; m 41; c 5. PHYSICS. Educ: Univ Utah, BS, 47; Univ Calif, PhD(physics), 51. Prof Exp: Res physicist, Radiation Lab, 47-51, MEM STAFF, LOS ALAMOS SCI LAB, UNIV CALIF, 51- Mem: Am Phys Soc. Res: Nuclear and surface physics. Mailing Add: Los Alamos Sci Lab Q22 PO Box 1663 Los Alamos NM 87545

AAMODT, RICHARD E, b St Paul, Minn, Nov 22, 36; m 60; c 3. PHYSICS. Educ: Univ Mich, BSE, 58, MS, 59, PhD(physics), 63. Prof Exp: Staff assoc plasma physics, Gen Atomic Div, Gen Dynamics Corp, 62-64, staff mem, 64-65; from asst prof to assoc prof physics, Univ Tex, Austin, 65-69; vis assoc prof astro-geophys, Univ Colo, 69-70, from assoc prof to prof, 70-75; SR STAFF SCIENTIST, SCI APPLN, INC, 75- Concurrent Pos: Consult, Lawrence Radiation Lab, Univ Calif, 66-; consult, Boeing Sci Res Lab, 66-70; vis staff mem, Los Alamos Sci Lab, 69-; res physicist, Naval Res Lab, 74. Res: Theoretical plasma physics; transport theory. Mailing Add: Sci Appl Inc 934 Pearl St Boulder CO 80302

AARON, CHARLES SIDNEY, b Jonesboro, La, Feb 17, 43; m 64; c 2. MOLECULAR GENETICS, MUTAGENESIS. Educ: La Tech Univ, BS, 68; La State Univ, Baton Rouge, PhD(biochem), 72. Prof Exp: Teacher chem, Episcopal High Sch, Baton Rouge, 71-72; RES ASSOC CHEM MUTAGENESIS, LA STATE UNIV, BATON ROUGE, 72- Mem: Am Chem Soc; Sigma Xi; Environ Mutagen Soc; Genetics Soc Am; AAAS. Res: Development of quantitative chemical mutagen dosimetry methods for assessing environmental hazard to man; theoretical-practical aspects of distinction of exposure to mutagen from dose-to-target; experimental determination of dose-to-

target vs genetic response correlation using Drosophila melanogaster. Mailing Add: Dept of Zool Life Sci Bldg La State Univ Baton Rouge LA 70802

AARON, HERBERT SAMUEL, b Minneapolis, Minn, Jan 12, 29; m 59; c 2. ORGANIC CHEMISTRY. Educ: Univ Minn, BA, 49; Univ Calif, Los Angeles, PhD(chem), 53. Prof Exp: Res chemist, Yerkes Res Lab, E I du Pont de Nemours & Co, 53-54; CHEMIST, CHEM RES LABS, EDGEWOOD ARSENAL, 56- Mem: Am Chem Soc. Res: Stereochemistry; conformational analysis; organo-phosphorus chemistry. Mailing Add: Chem Res Labs Edgewood Arsenal MD 21010

AARON, RONALD, b Philadelphia, Pa, June 14, 35; m 62; c 2. THEORETICAL PHYSICS. Educ: Temple Univ, AB, 56; Univ Pa, PhD(physics), 61. Prof Exp: Res assoc physics, Univ Md, 61-62; res assoc, Ind Univ, 62-63; physicist, Goddard Space Ctr, NASA, 63; asst prof physics, 63-72, PROF PHYSICS, NORTHEASTERN UNIV, 72- Concurrent Pos: Consult, Lawrence Radiation Lab, Univ Calif, Livermore, 64-67; vis staff mem, Los Alamos Sci Lab, NMex, 67- Mem: Am Phys Soc. Res: Theoretical high energy physics and scattering theory. Mailing Add: Dept of Physics Northeastern Univ Boston MA 02115

AARONOFF, BURTON ROBERT, b Brooklyn, NY, July 4, 35; m 66; c 2. ORGANIC CHEMISTRY. Educ: Brooklyn Col, BS, 56; Johns Hopkins Univ, MA, 59, PhD(chem), 62. Prof Exp: Res assoc chem, Lever Bros Co, 62-63; sr chemist, Merck, Sharp & Dohme Labs, 63-66; asst to vpres res & develop, Block Drug Co, 66-68, assoc res dir, 68-73; DIR EXPLOR RES, PFIZER CONSUMER OPERS, 73- Mem: Am Chem Soc; Am Pharmaceut Asn. Res: Organic and medicinal chemistry; proprietary drugs and toiletries; pharmaceutical dosage forms; research administration. Mailing Add: Pfizer Inc 100 Jefferson Rd Parsippany NJ 07054

AARONS, JULES, b New York, NY, Oct 3, 21; m 43; c 2. IONOSPHERIC PHYSICS. Educ: City Col, NY, BS, 42; Boston Univ, MA, 49; Univ Paris, PhD, 54. Prof Exp: Physicist, Air Force Cambridge Res Labs, 46-55, chief radio astron br, 55-72, CHIEF TRANS-IONOSPHERIC PROPAGATION BR & SR SCIENTIST, IONOSPHERIC PHYSICS LAB, AIR FORCE CAMBRIDGE RES LABS, 72- Concurrent Pos: US panel coordr adv group for aerospace res & develop, Electromagnetic Propagation Panel, 73-; secy comn III, US Sect Int Sci Radio Union. Mem: Fel Inst Elec & Electronics Engrs; Am Astron Soc; Am Geophys Union. Res: Upper atmosphere physics; ionospheric propagation, solar radio astronomy, solar terrestrial relationships. Mailing Add: Air Force Cambridge Res Labs Hanscom AFB Bedford MA 01731

AARONS, MELVIN WOLF, physics, see 12th edition

AARONSON, DAVID ANDREW, physics, mathematics, see 12th edition

AARONSON, SHELDON, b New York, NY, Oct 9, 22; m 48; c 4. MICROBIAL BIOCHEMISTRY. Educ: City Col New York, BS, 44; Univ Pa, MA, 48; NY Univ, PhD(protistology), 53. Prof Exp: Lectr biol, City Col New York, 48-53; from instr to assoc prof, 53-64, PROF BIOL, QUEENS COL, NY, 65- Concurrent Pos: Res assoc, Haskins Labs, Inc, 49-65, sr staff mem, 65-68; pres fel, Am Soc Microbiol, 57; NSF sci fac fel, 59-60. Mem: Fel AAAS; Am Soc Microbiol; Soc Protozool; Am Soc Cell Biol; Phycol Soc Am. Res: Biochemistry of microorganisms; biochemical phylogeny; cellular action of drugs; microbial ecology. Mailing Add: Dept of Biol Queens Col Flushing NY 11367

AASE, JAN KRISTIAN, b Stavanger, Norway, Nov 21, 35; US citizen; m 65; c 4. SOIL SCIENCE. Educ: Brigham Young Univ, BA, 58; Univ Minn, MS, 60; Colo State Univ, PhD(soil physics), 67. Prof Exp: RES SOIL SCIENTIST, NORTHERN PLAINS SOIL & WATER RES CTR, AGR RES SERV, USDA, 67- Mem: Am Soc Agron; Soil Sci Soc Am; Soil Conserv Soc Am. Res: Soil-plant-water relations; plant water use efficiency and microclimatological research on agronomic crops and native rangeland. Mailing Add: Agr Res Serv US Dept of Agr PO Box 1109 Sidney MT 59270

AASLESTAD, HALVOR GUNERIUS, b Birmingham, Ala, Sept 6, 37; m 60; c 4. VIROLOGY. Educ: La State Univ, BS, 59, PhD(microbiol), 65; Pa State Univ, 61. Prof Exp: Nat Acad Sci-Nat Res Coun fel, US Army Biol Ctr, 65-66, prin investr microbiol, 66-68; asst prof, Univ Ga, 68-70; asst prof, Wistar Inst, 70-73; sr scientist, Cancer Res Ctr, Nat Cancer Inst, Frederick, Md, 73-76; EXEC SECY DIV RES GRANTS, NIH, 76- Mem: Am Soc Microbiol. Res: Virology; virus and phage biochemistry and genetics. Mailing Add: Div of Res Grants Nat Inst of Health Bethesda MD 20014

ABADI, DJHANGUIR M, b Tehran, Iran, Sept 1, 29; US citizen; m 60; c 2. CLINICAL CHEMISTRY. Educ: Doane Col, BA, 52; Univ Nebr, MS, 55; Univ Wash, PhD(biochem), 58; Am Bd Clin Chem, dipl. Prof Exp: Biochemist, Swed Hosp, Seattle, Wash, 59-61; clin chemist, Good Samaritan Hosp, Portland, Ore, 61-64; CLIN CHEMIST, VET ADMIN HOSP, IOWA CITY, 64- Mem: Fel Am Asn Clin

Chemists. Res: Proteolytic enzymes; protein chemistry. Mailing Add: Vet Admin Hosp Iowa City IA 52240

ABADIE, STANLEY HERBERT, b New Orleans, La, Nov 13, 25; m 62; c 2. MEDICAL PARASITOLOGY, MEDICAL EDUCATION. Educ: Loyola Univ, La, BS, 55; La State Univ, New Orleans, MS, 58, PhD(med parasitol), 63. Prof Exp: Grad asst med microbiol & parasitol, 56-58, from instr to assoc prof med parasitol, 58-72, MEM GRAD FAC MED PARASITOL, LA STATE UNIV MED CTR, 60-, PROF MED PARASITOL & MICROBIOL, 72-, DEAN, SCH ALLIED HEALTH PROFESSIONS, 75- Concurrent Pos: La State Univ Interam Training Prog fel trop med, Costa Rica, 64; Schlieder Educ Found res grant Chagas' dis, 74-77; consult mycol, Touro Infirm, New Orleans, 63-64; scientist, Vis Staff, Charity Hosp, New Orleans, 63-; consult parasitol, Vet Admin Hosp, New Orleans, 70-; lectr parasitic dis, Sch Dent, La State Univ, New Orleans, 71-; contribr, Manual of Microbiol & Manual of Trop Med. Mem: AAAS; Am Soc Trop Med & Hyg; Am Soc Parasitol. Res: Hookworm disease; heartworm; filarial infections. Mailing Add: Dept Trop Med & Med Parasitol La State Univ Med Ctr New Orleans LA 70112

ABAIDOO, KODWO-JAMES R, b Saltpond, Ghana, Feb 8, 40; m 69; c 2. HEMATOLOGY, PHYSIOLOGY. Educ: Univ Freiburg, BSc, 62, cert anat, biochem & physiol, 62-66; Thomas Jefferson Univ, PhD(hemat, physiol), 72. Prof Exp: Asst physiol, Univ Hamburg, 66; instr gen physiol, 71-72, lectr physiol & pharmacol, 72-73, ASST PROF PHYSIOL & PHARMACOL, PHILADELPHIA COL OSTEOP MED, 73- Concurrent Pos: Fel Hemopath, Path Inst, Kiel, 73-74; consult pathophysiol, Metrop Hosp, 73. Mem: Am Asn Univ Prof. Res: Erythrokinetics of bone marrow; erythopoietin and its inhibition; enzyme-cytochemical reactions of normal and abnormal hemopoietic tissues. Mailing Add: Hematol Res Lab Evans Hall 418 Philadelphia Col of Osteop Med Philadelphia PA 19131

ABAJIAN, PAUL G, b Framingham, Mass, Aug 4, 41; m 63; c 3. PHYSICAL CHEMISTRY. Educ: Worcester Polytech Inst, BS, 63; Univ Vt, PhD(chem), 67. Prof Exp: Asst anal chem, Univ Vt, 63-67; asst prof chem, Southern Conn State Col, 67-69; ASST PROF CHEM, JOHNSON STATE COL, 69- Res: Solution calorimetry of transition metal compounds and also polarographic studies of rare earth metals. Mailing Add: Dept of Chem Johnson State Col Johnson VT 05656

ABARBANEL, ABRAHAM ROBERT, b Jersey City, NJ, Aug 8, 12; m 41; c 4. PHYSIOLOGY. Educ: Cornell Univ, AB, 32; Long Island Col Med, MD, 36. Prof Exp: Resident obstet & gynec, Morrisania Hosp, 39-40; resident gynec, Harlem Hosp, 41-42; instr, 45-48, asst clin prof, 48-51, ASST PROF OBSTET & GYNEC, LOMA LINDA UNIV, 52- Concurrent Pos: Res fel, Mt Sinai Hosp, 40-41; fel obstet & gynec, Sch Med, George Washington Univ, 42-44; med dir, Fertil Inst. Mem: Am Geriat Soc; AMA; Am Fertil Soc; Endocrine Soc; fel Int Col Surg. Res: Disturbances of menstruation; anovulation; biology of human cervical mucus; transvaginal pelvioscopy; myometcomy; psychosomatic aspects of gynecology. Mailing Add: Suite 9 6515 Wilshire Blvd Los Angeles CA 90048

ABARBANEL, HENRY D I, b Washington, DC, May 31, 43; m 65. THEORETICAL PHYSICS. Educ: Princeton Univ, PhD(physics), 66. Prof Exp: Res assoc physics, Princeton Univ, 66, vis fel, 66-67; res assoc, Stanford Linear Accelerator Ctr, 67-68; asst prof, Princeton Univ, 68-72; PHYSICIST, FERMI NAT ACCELERATOR LAB, 72- Concurrent Pos: NSF fel, Princeton Univ, 67-68, A P Sloan res fel, 69- Res: High energy collision phenomena of elementary particles. Mailing Add: Fermi Nat Accelerator Lab Batavia IL 60510

ABASHIAN, ALEXANDER, b Binghamton, NY, Mar 11, 30; m 57; c 3. NUCLEAR PHYSICS. Educ: Purdue Univ, BS, 52; Johns Hopkins Univ, PhD, 57. Prof Exp: Tech engr, Air Force Develop, Int Bus Mach, 52; jr instr physics, Johns Hopkins Univ, 52-55; jr res assoc high energy physics, Brookhaven Cosmotron, 55-57; instr physics, Univ Rochester, 57-59; res assoc, Radiation Lab, Univ Calif, 59-61; from asst prof to prof, Univ Ill, Urbana-Champaign, 61-74; PROG DIR ELEM PARTICLE PHYSICS, NSF, 72- Mem: AAAS; Am Phys Soc. Res: Pion interactions; nuclear polarization; invariance tests; weak interactions; resonances; muon scattering. Mailing Add: Physics Div Nat Sci Found Washington DC 20550

ABASHIAN, STEVEN, b Binghamton, NY, Mar 11, 25; m 54; c 4. ORGANIC CHEMISTRY. Educ: Rensselaer Polytech Inst, BS, 50; Johns Hopkins Univ, MA, 52, PhD(chem), 54. Prof Exp: Jr instr chem, Johns Hopkins Univ, 50-53; SR RES CHEM, E I DU PONT DE NEMOURS & CO, INC, 54- Mem: Am Chem Soc. Res: Polymers; polymerization processes; synthetic fibers; organo-metallics; organic syntheses. Mailing Add: 2200 Emerson Rd Kinston NC 28501

ABBAS, MIAN MOHAMMAD, b Pakistan, June 22, 33; US citizen; m 60; c 2. INFRARED ASTRONOMY, SPACE PHYSICS. Educ: Panjab Univ, BS, 55; Laval Univ, MS, 60; Univ RI, PhD, 67. Prof Exp: Lectr elec eng, WPakistan Eng Univ, 55-57; electronics engr, Electronics Lab, Canadair Ltd, 60-62; res fel, Univ RI, 64-68; asst prof, Ohio Univ, 68-70; assoc prof, Univ Ky, Lexington, 70-74; NAT ACAD SCI-SR RES ASSOC, GODDARD SPACE FLIGHT CTR, NASA, 74- Mem: Inst Elec & Electronics Engrs; Am Geophys Union. Res: Observation and analysis of astronomical infrared sources; planetary atmospheres; stratosphere. Mailing Add: Code 692 Lab Extraterr Physics Goddard Space Flight Ctr Greenbelt MD 20771

ABBASI, QAMAR ALI, b Amroha, India, July 8, 41; m 70; c 1. SYSTEMATIC ENTOMOLOGY. Educ: Aligarh Muslim Univ, India, BS, 59, MS, 62; Karachi Univ, Pakistan, PhD(zool), 75. Prof Exp: Res asst entom, Indian Agr Res Inst, New Delhi, 62-67; res fel, Dept Zool, Univ Karachi, 67-71, lectr zool, 71-73; RES ASSOC BIOL SCI, UNIV ILL, CHICAGO CIRCLE, 73- Res: Morphotaxonomy of super family Pentatomoidea with with reference to phylogeny of the group; biological control of insect pests and taxonomy of Chalcidoidea, Hymenoptera. Mailing Add: 2422 N Sawyer Ave Chicago IL 60647

ABBATE, FRANKLIN WILLIAM, organic chemistry, polymer chemistry, see 12th edition

ABBATIELLO, MICHAEL JAMES, b New York, NY, June 19, 21; m 45; c 1 ZOOLOGY, BIOLOGY. Educ: Hofstra Col, BA, 54, MA, 57; St John's Univ, NY, PhD(zool), 66. Prof Exp: Tech asst hort, 47-53, from jr instr to assoc prof biol, 53-66, PROF BIOL, STATE UNIV NY AGR & TECH COL FARMINGDALE, 66- Mem: AAAS; Entom Soc Am; Nat Asn Biol Teachers. Res: Biology and ecology of oribatid mites. Mailing Add: Dept of Biol State Univ of NY Agr & Tech Col Farmingdale NY 11735

ABBE, ERNST CLEVELAND, b Washington, DC, Aug 21, 05; m 30; c 2. BOTANY. Educ: Cornell Univ, BS, 28, MS, 30; Harvard Univ, AM, 31, PhD, 34. Prof Exp: Asst bot, Cornell Univ, 27-30 & Harvard Univ, 31-32; Nat Res Coun fel biol, Columbia, Harvard & Cornell Univs, 33-35; from instr to prof, 35-74, chmn dept, 44-47 & 62-67, EMER PROF BOT, UNIV MINN, MINNEAPOLIS, 74- Concurrent Pos: Botanist, Grenfell-Forbes exped, Labrador, 31; mem, Arnold Arboretum Exped, BC & Alta, Can, 32; leader, Univ Minn Expeds to Hudson Bay, 39, Southeast Asia, 59-60 & Mt Kinabalu, 62; res assoc & Guggenheim fel, Harvard Univ, 41-42; Fulbright prof, Univ Singapore, 61-62; vis investr, Forest Dept, Sarawak, 64. Mem: Bot Soc Am; Linnean Soc London. Res: Phylogeny of Amentiferae; phytogeography of Eastern Subarctic; morphogenesis in relation to genetical constitution. Mailing Add: Dept of Bot 220 Bio-Sci Ctr Univ of Minn St Paul MN 55108

ABBE, WINFIELD JONATHAN, b Cleveland, Ohio, Feb 27, 39; m 66; c 1. THEORETICAL HIGH ENERGY PHYSICS. Educ: Univ Calif, Berkeley, AB, 61; Calif State Col, Los Angeles, MS, 62; Univ Calif, Riverside, PhD(physics), 66. Prof Exp: Asst prof, 66-70, ASSOC PROF PHYSICS, UNIV GA, 70- Concurrent Pos: Res fel, Univ Mich, 66-67. Res: Elementary particle theory; S-matrix theory of strong interactions; Regge Pole theory; dispersion relations. Mailing Add: Dept of Physics Univ of Ga Athens GA 30601

ABBEY, ANTHONY ALFRED, b Brooklyn, NY, Apr 11, 17; m 39; c 2. BACTERIOLOGY. Prof Exp: Asst dept bact & immunol, LI Col Med, 36-39; asst biol control dept, Schering Corp, 39-43; bacteriologist, Antibiotics Res & Develop Div, Heyden Chem, 43-54; SUPVRY GROUP LEADER, MICROBIOL LAB, AGR DIV, AM CYANAMID CO, 54- Mem: AAAS; Am Chem Soc; Am Soc Microbiol. Res: Microbiological evaluation and application of antibiotics; chemotherapy; foods; bioassay and quality control. Mailing Add: 18 School St Milltown NJ 08850

ABBEY, HELEN, b Ann Arbor, Mich, Sept 1, 20. BIOSTATISTICS. Educ: Battle Creek Col, BS, 40; Univ Mich, MA, 42; Johns Hopkins Univ, ScD, 51. Prof Exp: From instr to assoc prof, 51-69, PROF BIOSTATIST, JOHNS HOPKINS UNIV, 69- Concurrent Pos: Mem epidemiol & dis control study sect, Div Res Grants, NIH, 72-76. Mem: Am Statist Asn; Pop Asn Am; Am Soc Human Genetics; Am Pub Health Asn; Biomet Soc. Res: Application of statistics in medicine; biological sciences; epidemiology; human genetics; demography. Mailing Add: Sch of Hyg & Pub Health Johns Hopkins Univ Baltimore MD 21205

ABBINK, HENRY C, physical chemistry, see 12th edition

ABBISS, JOSEPH WILLIAM, b Tividale, Eng, Oct 4, 16; US citizen; m 63. CLINICAL PATHOLOGY. Educ: Univ Birmingham, MD, 42. Prof Exp: Lectr path & bact, Univ Birmingham, 46-47; assoc prof, Dalhousie Univ, 47-50; pathologist & dir labs, Mem Hosp, 51-67; DIR LABS, ST FRANCIS HOSP, 67- Mem: Am Soc Clin Path; Col Am Path. Res: Cytology and hematology. Mailing Add: St Francis Hosp Eighth & DuPont Sts Wilmington DE 19805

ABBOT, CHARLES GREELEY, astrophysics, deceased

ABBOTT, AGATIN T, geomorphology, deceased

ABBOTT, ANDREW DOYLE, b Waterville, Wash, Nov 20, 13; c 3. PHYSICAL CHEMISTRY. Educ: Univ Wash, BS, 37, PhD(phys chem), 42. Prof Exp: Res chemist, Procter & Gamble Co, 42-46 & Calif Res Corp, 46-66; res chemist, 66-70, SR RES CHEMIST, CHEVRON RES CO, 70- Mem: Am Chem Soc. Res: Fatty oil chemistry; colloidal chemistry; petroleum chemistry. Mailing Add: Chevron Res Co 576 Standard Ave Richmond CA 94802

ABBOTT, BERNARD C, b Yeovil, Eng, Oct 13, 20; m 44; c 1. BIOPHYSICS. Educ: Univ London, BSc, 41 & 48, PhD(biophys), 50. Prof Exp: Lectr biophys, Univ Col, Univ London, 46-53; prin sci officer, Marine Biol Asn, UK, 53-57; assoc res zoologist, Univ Calif, Los Angeles, 57-58, assoc prof zool, 58-62; prof biophys & physiol, Univ Ill, Urbana, 62-68; PROF PHYSIOL, SCH MED, CHMN DEPT BIOL SCI & DIR, ALLAN HANCOCK FOUND, UNIV SOUTHERN CALIF, 68- Mem: Am Physiol Soc; Biophys Soc; Soc Gen Physiol; Am Soc Zoologists; NY Acad Sci. Res: Muscle physiology; thermodynamics of bioelectric phenomena; toxins present in poisonous red tides. Mailing Add: Allan Hancock Found Univ of Southern Calif Los Angeles CA 90007

ABBOTT, BETTY JANE, b Roanoke, Va, Oct 13, 31. CANCER. Educ: Radford Col, BS, 52; Va Polytech Inst, MS, 55. Prof Exp: Instr high schs, Va, 52-57, chmn dept biol, 53-57; asst biol, Va Polytech Inst, 57-59; biologist, Va Agr Res Sta, Agr Res Serv, USDA, 59-60; biologist, 60-67, HEAD SCREENING SECT, DRUG EVAL BR, DRUG RES & DEVELOP, NAT CANCER INST, 67- Honors & Awards: Sustained High Qual Performance Award, Nat Cancer Inst, 66 & 72. Mem: AAAS; Am Asn Cancer Res; NY Acad Sci. Res: Development and selection of new drugs for trial as potential cancer chemotherapeutic agents; ways to increase effectiveness of agents of established clinical usefulness. Mailing Add: Drug Eval Br Drug Res & Develop Nat Cancer Inst Bethesda MD 20014

ABBOTT, DONALD CLAYTON, b Lincoln, Kans, Nov 11, 23; m 45; c 3. BIOCHEMISTRY, CEREAL CHEMISTRY. Educ: Kans State Univ, BS, 49, MS, 51, PhD, 64. Prof Exp: Asst to head exp baking lab, Fleischman Labs, Standard Brands, Inc, NY, 51; cereal technologist, Soft Wheat Qual Lab, Ohio Agr Exp Sta, 51-54; PROF BIOCHEM, OKLA STATE UNIV, 54- Mem: Am Asn Cereal Chem; Am Chem Soc. Res: Baking quality of wheat varieties; immunological properties of wheat proteins; development of wheat proteins during maturation. Mailing Add: Dept of Biochem Okla State Univ Stillwater OK 74074

ABBOTT, DONALD PUTNAM, b Chicago, Ill, Oct 14, 20; m 43; c 1. INVERTEBRATE ZOOLOGY. Educ: Univ Hawaii, BA, 41; Univ Calif, MA, 48, PhD(zool), 50. Prof Exp: Instr educ, Univ Hawaii, 42-43, instr zool, 46; from instr to assoc prof biol, Stanford Univ, 50-64, PROF BIOL, STANFORD UNIV, 64-, ASSOC DIR HOPKINS MARINE STA, 66- Concurrent Pos: Mem & later leader, Pac Sci Bd Ifaluk Atoll Exped, W Carolines, 53; Sulu Sea Exped, 57; Galapagos Int Sci Proj, 64; chief scientist, Te Vega Exped Cruise 5, Indian Ocean, 64, Cruise 18, East Trop Pac, 68, Cruise 22-2, Queen Charlotte Strait, 70, mem Cruise 23-2, Hawaii, 71; asst dir, Hopkins Marine Sta, 62-66. Mem: Am Soc Zool; AAAS; Soc Syst Zool; Ecol Soc Am. Res: Budding and taxonomy of ascidians; functional anatomy, behavior and autecology of marine invertebrates. Mailing Add: Hopkins Marine Sta Stanford Univ Pacific Grove CA 93950

ABBOTT, EDWIN HUNT, b New York, NY, Dec 28, 41; m 64; c 2. INORGANIC CHEMISTRY. Educ: Tufts Univ, BS, 64; Tem A&M Univ, PhD(chem), 69. Prof Exp: Fel chem, Tex A&M Univ, 69-70; asst prof, 70-73, ASSOC PROF CHEM, HUNTER COL, 74- Concurrent Pos: Mem doctoral fac, City Univ New York, 70- Mem: Am Chem Soc; fel NY Acad Sci; Sigma Xi. Res: Coordination chemistry; homogeneous catalysis; bioinorganic chemistry; catalysts for water pollution abatement; nuclear magnetic resonance of metal chelate compounds; automation. Mailing Add: Dept of Chem Hunter Col 695 Park Ave New York NY 10021

ABBOTT, FRANK SIDNEY, b Ottawa, Ont, Apr 12, 28; m 63; c 3. ANIMAL PHYSIOLOGY. Educ: Carleton Univ, Ont, BSc, 53; Univ Toronto, MA, 57; Univ Alta, PhD(zool), 66. Prof Exp: Tech officer, Geol Surv Can, 53-56, admin officer,

Dept Northern Affairs & Nat Resources, 58-63; from asst prof to assoc prof biol, Sir George Williams Univ, 66-74; SCI PROC MGR, SCI CTR, DEPT SUPPLY & SERV, GOVT CAN, 74- Concurrent Pos: Nat Res Coun Can res grants-in-aid, 67-70. Mem: Can Soc Environ Biologists; AAAS; Am Inst Biol Sci; Can Soc Zoologists; Can Physiol Soc. Res: Physiology of melanophore response to background tint in Fundulus heteroclitus. Mailing Add: 33 Moorcroft Rd Ottawa ON Can

ABBOTT, HERSCHEL GEORGE, b Woodstock, Maine, May 4, 21; m 43. FORESTRY. Educ: Univ Maine, BS, 43; Harvard Univ, MF, 52, MA, 59. Prof Exp: Forester, Chadbourne Lumber Co, Maine, 46-50; from instr to assoc prof forestry, 53-70, PROF FORESTRY & WILDLIFE MGT, UNIV MASS, AMHERST, 70- Concurrent Pos: Vpres & treas, Rich Tree Farms & Forestry Corp, 64- Honors & Awards: Distinguished Serv Award, Soc Am Foresters, 75. Mem: Soc Am Foresters. Res: Silviculture. Mailing Add: Dept of Forestry & Wildlife Mgt Univ of Mass Amherst MA 01002

ABBOTT, ISABELLA AIONA, b Hana, Hawaii, June 20, 19; m 43; c 1. PHYCOLOGY, MARINE BIOLOGY. Educ: Univ Hawaii, AB, 41; Univ Mich, MS, 42; Univ Calif, Berkeley, PhD(bot), 50. Prof Exp: Lectr bot, Univ Hawaii, 43-46; res assoc & lectr phycol, 60-70, PROF BIOL, STANFORD UNIV, 71- Honors & Awards: Darbaker Award, Bot Soc Am, 69. Mem: Int Phycol Soc (treas, 67-71); Phycological Soc Am; Bot Soc Am; Brit Phycological Soc. Res: Life histories, development and systematics of marine red algae, particularly Pacific marine algae. Mailing Add: Hopkins Marine Sta Stanford Univ Pacific Grove CA 93950

ABBOTT, JOAN, b Newton, Mass, Aug 5, 32. DEVELOPMENTAL BIOLOGY. Educ: Conn Col, BA, 54; Wash Univ, MA, 57; Univ Pa, PhD(develop biol), 65. Prof Exp: Asst prof biol, Barnard Col, Columbia Univ, 65-67, asst prof dept biol sci, Columbia Univ, 67-72; ASSOC, MEM SLOAN-KETTERING CANCER CTR, 72-, ASST PROF BIOL, SLOAN-KETTERING DIV, GRAD SCH MED SCI, CORNELL UNIV, 73- Mem: AAAS; Am Soc Cell Biol; Soc Develop Biol; NY Acad Sci. Res: Definition of precursor cell compartments in murine lymphocyte entogeny using serological assays for cell surface markers; analysis in vitro of molecular events associated with sequential passage through these compartments. Mailing Add: Sloan-Kettering Inst Cancer Res 1275 York Ave New York NY 10021

ABBOTT, JOHN RICHARDS, b Geneva, NY, Dec 9, 32; m 54; c 4. ORGANIC CHEMISTRY. Educ: Hobart Col, BS, 55; Brown Univ, MS, 57. Prof Exp: Org chemist, 58-73, LAB HEAD ORG CHEM, COLOR PHOTOG DIV, RES LABS, EASTMAN KODAK CO, 73- Mem: Am Chem Soc. Res: Organic synthesis for color photographic systems. Mailing Add: Res Labs Eastman Kodak Co Kodak Park Rochester NY 14650

ABBOTT, LYNN DE FORREST, JR, b Ithaca, NY, Nov 23, 13; m 40; c 3. BIOCHEMISTRY. Educ: Wayne Univ, BS, 36, MS, 37; Univ Mich, PhD(biochem), 40. Prof Exp: Asst chem, Wayne Univ, 35-36, asst biochem, 36-37; asst, Univ Mich, 37-40; res assoc, 40-43, from asst prof to assoc prof, 43-48, PROF BIOCHEM, MED COL VA, 56-, CHMN DEPT, 62- Concurrent Pos: Ed, Va J Sci, 68-73. Mem: AAAS; Am Chem Soc; Am Soc Biol Chemists; Soc Exp Biol & Med; Am Soc Clin Nutrit. Res: Blood chemistry and amino acid metabolism; enzyme chemistry; isotopic nitrogen. Mailing Add: Dept of Biochem Med Col of Va Richmond VA 23298

ABBOTT, MARIE BOHRN-LAMBERT, b Buenos Aires, Arg, Dec 30, 22; US citizen; m 66. MARINE ZOOLOGY. Educ: Hunter Col, AB, 42; Columbia Univ, AM, 45; Univ Conn, PhD(zool), 71. Prof Exp: Instr geol & paleont, Mt Holyoke Col, 47-49; geologist, US Geol Surv, Washington, DC, 49-52; instr geol & paleont, Mt Holyoke Col, 55-56; instr geol, Smith Col, 56-57; spec consult biol & geol, Niskayuna Schs, NY, 64-66; lectr geol, Russell Sage Col, 66; CUR GRAY MUS, MARINE BIOL LAB, WOODS HOLE, 71- Concurrent Pos: Consult, Normandeau Assocs, NH, 72- Mem: Am Soc Zoologists; Geol Soc Am; Soc Syst Zoologists; Soc Econ Paleontologists & Mineralogists; Sigma Xi. Res: Autecology of Cheilostome Bryozoa; systematics of colonial animals, particularly Bryozoa; structure and diversity of marine epibenthic communities. Mailing Add: RD 2 Box 18 High St Coventry CT 06238

ABBOTT, MAXINE LANGFORD, b Carpenter, Tex, Oct 31, 15; m 39; c 2. PALEOBOTANY. Educ: Tex Tech Col, BS, 37, MS, 39, PhD(paleobot), 56. Prof Exp: Asst bot, Univ Cincinnati, 49-54, inst, 57-58; cur, J H Hoskins Mem Paleobot Collections, 58-64; instr paleobot, Univ Cincinnati, 64-66; assoc prof biol, Western State Col Colo, 66-69; PROF BIOL, SUL ROSS STATE UNIV, 69- Concurrent Pos: NSF grant, 59-64. Mem: AAAS; Int Orgn Paleobot; Int Union Biol Sci; Torrey Bot Club; Bot Soc Am. Res: Paleozoic paleobotany; compression floras of Pennsylvanian period. Mailing Add: Dept of Biol Sul Ross State Univ Alpine TX 79830

ABBOTT, MITCHEL THEODORE, b Los Angeles, Calif, June 6, 30; m 62; c 3. BIOCHEMISTRY. Educ: Univ Calif, Los Angeles, BS, 57, PhD(biochem), 63. Prof Exp: Fel biochem, Sch Med, NY Univ, 62-64; asst prof, 64-71, PROF BIOCHEM, SAN DIEGO STATE UNIV, 71- Concurrent Pos: Vis scientist, Roche Inst Molecular Biol, 72-73. Mem: Am Soc Biol Chem. Res: Intermediary metabolism of nucleic acids and of their components. Mailing Add: Dept of Chem San Diego State Univ San Diego CA 92182

ABBOTT, NORMAN JOHN, b Toronto, Ont, Nov 14, 18; m 44; c 3. TEXTILE PHYSICS. Educ: Univ Toronto, BA, 40. Prof Exp: Elec engr, Amalgamated Elec Corp, 40-41; spec instr armed forces dept physics, Univ Toronto, 41-43; tech writer, Radar Div, Res Enterprises Ltd, 43-45; fel textile dept, Ont Res Found, 45-47; asst dir appl physics, 57-69, ASSOC DIR, FABRIC RES LABS, 69- Mem: AAAS; Fiber Soc; Can Asn Physicists (secy, 45-46, treas, 47); fel Brit Textile Inst; Inst Textile Sci. Res: Mechanical properties of textile structures; behavior of textiles under extreme conditions; coated fabric mechanics; characterization, utilization of high temperature, high performance, nonflammable fibers; design of mechanical structures containing such fibers. Mailing Add: Fabric Res Labs 1000 Providence Hwy Dedham MA 02026

ABBOTT, OKRA JONES, b Ritner, Ky, Feb 11, 21; m 67. POULTRY NUTRITION, BIOCHEMISTRY. Educ: Berea Col, BS, 49; Univ Ky, MS, 52; Univ Wis, PhD(nutrit, biochem), 56. Prof Exp: Pub sch teacher, Ky, 49-50; asst poultry nutrit, Univ Ky, 51-52 & Univ Wis, 52-55; asst prof poultry prod tech, Univ Del, 55-60; assoc prof poultry sci, Univ Ky, 60-67; assoc prof, 67-73, PROF ANIMAL & VET SCI, W VA UNIV, 73- Concurrent Pos: Instr, US Armed Forces Inst, 54-55; vis sr scientist pharmacol, Univ Ky, 64-65; AID sr lectr, Makerere Univ, Uganda, 67-73. Mem: AAAS; Am Chem Soc; NY Acad Sci; Poultry Sci Asn; Am Soc Animal Sci. Res: Nutrition and metabolism in animals with emphasis on effects of intestinal microorganisms. Mailing Add: Div of Animal & Vet Sci W Va Univ Morgantown WV 26506

ABBOTT, PATRICK LEON, b San Diego, Calif, Sept 30, 40; m 61; c 3. SEDIMENTOLOGY. Educ: San Diego State Col, BS, 63; Univ Tex, Austin, MA, 65, PhD(geol), 73. Prof Exp: Lectr geol, 68-69, asst prof, 71-73, ASSOC PROF GEOL, SAN DIEGO STATE UNIV, 73- Mem: AAAS; Geol Soc Am; Soc Econ Paleontologists & Mineralogists; Am Asn Petrol Geologists. Res: Hydrogeology of carbonate rock terrane, particularly south-central Texas; Paleogene paleosols and paleoclimates, especially western North America; petrology and provenance of Mesozoic and Cenozoic sandstone and conglomerates, particularly Baja and Alta California. Mailing Add: Dept of Geol Sci San Diego State Univ San Diego CA 92182

ABBOTT, RICHARD LEIGH, organic chemistry, see 12th edition

ABBOTT, ROBERT CLASSIE, b Syracuse, NY, June 7, 26; m 62; c 2. COMPUTER SCIENCES, STATISTICAL ANALYSIS. Educ: Syracuse Univ, BEE, 50, MS, 58, PhD(physics), 59. Prof Exp: Elec engr, Wright Air Develop Ctr, Ohio, 50; elec engr, Navy Ord Div, Eastman Kodak Co, NY, 50-51; asst physics, Syracuse Univ, 51-59; res physicist, Cornell Aeronaut Lab, Inc, 59-61, prin physicist, 61-62, head solid state physics sect, 62-66; assoc prof fac eng & appl sci, State Univ NY Buffalo, 66-73; VPRES RES, MULTITECH RES & INVESTMENT CORP, 73- Mem: Am Phys Soc. Res: Surface physics and biophysics; ion-metal surface interactions; metallic adsorption systems; macromolecular information storage mechanisms, especially by field emission and field ion microscopy; digital signal processing for biological and economic systems. Mailing Add: Multitech Res & Investment Corp 5225 Sheridan Dr Buffalo NY 14221

ABBOTT, ROBERT TUCKER, b Watertown, Mass, Sept 28, 19; m 46; c 3. MALACOLOGY. Educ: Harvard Univ, BS, 46; George Washington Univ, PhD, 55. Prof Exp: From asst cur to assoc cur, Div Mollusks, US Nat Mus, 46-54; Pilsbry chair malacol, Acad Natural Sci, Philadelphia, 54-69; DUPONT CHAIR & ASST DIR, DEL MUS NATURAL HIST, 69- Concurrent Pos: Zoologist, Harvard-Archbold Exped, Melanesia & Polynesia, 40-41 & Pac Sci Bd, EAfrica, 51; assoc ed, Johnsonia, Harvard Univ, 42-69; ed, Indo-Pac Mollusca, 59- & Nautilus, 68-; trustee, Bermuda Biol Sta. Mem: Fel AAAS; Am Malacol Union; Soc Syst Zool. Res: Marine mollusks; medical malacology; popular conchology. Mailing Add: Del Mus of Natural Hist Box 3937 Greenville DE 19807

ABBOTT, ROBINSON SHEWELL, b Philadelphia, Pa, June 23, 26; m 56; c 4. ECOLOGY. Educ: Bucknell Univ, BS, 49; Cornell Univ, PhD(biol), 56. Prof Exp: Teacher pub sch, LI, 50-51; asst bot, Cornell Univ, 53; asst, Wiegand Herbarium, 53-54, econ bot, 55-56; instr bot, Smith Col, 56-61; from asst prof to assoc prof biol, 61-71, chmn div sci & math, 64-68, PROF BIOL, UNIV MINN, MORRIS, 71- Mem: Bot Soc Am; Ecol Soc Am; Phycol Soc Am; Int Phycol Soc. Res: General ecology, especially beach and salt marsh ecology; use of filamentous algae as ecological indicators; environmental effects of the growth and development of algae. Mailing Add: Dept of Biol Univ of Minn Morris MN 56267

ABBOTT, ROSE MARIE SAVELKOUL, b Chaska, Minn, Feb 7, 31; m 56; c 4. PLANT MORPHOLOGY. Educ: Univ Minn, BA, 53, MS, 55; Cornell Univ, PhD(plant morphol), 59. Prof Exp: Instr biol, Smith Col, 57-58; instr plant morphol, 64-, INSTR GENETICS, UNIV MINN, MORRIS, 66- Mem: Bot Soc Am. Res: Developmental anatomy of primary xylem elements; development of shoot apices; developmental anatomy of Berberidaceae. Mailing Add: Div of Sci & Math Univ of Minn Morris MN 56267

ABBOTT, ROYAL K, JR, organic chemistry, see 12th edition

ABBOTT, SETH R, b Brooklyn, NY, Mar 14, 44; m 68; c 2. CHEMICAL INSTRUMENTATION, SPECTROSCOPY. Educ: Brooklyn Col, BS, 64; Mass Inst Technol, PhD(chem), 69. Prof Exp: Res chemist anal & polymer chem, Memorex Corp, 69-72; SR RES CHEMIST LIQUID CHROMATOG, VARIAN ASSOCS, 72- Res: Liquid chromatography; emission and Raman spectroscopy; mass spectroscopy; electrochemistry. Mailing Add: Varian Assocs 2700 Mitchell Dr Walnut Creek CA 94598

ABBOTT, SUSAN, b Seattle, Wash, Feb 27, 45. ANTHROPOLOGY. Educ: Idaho State Univ, BS, 69; Univ NC, Chapel Hill, MA, 71, PhD(anthrop), 74. Prof Exp: ASST PROF ANTHROP & BEHAV SCI, UNIV KY, 74- Mem: Fel Am Anthrop Asn; Soc Cross Cult Res; Soc Med Anthrop. Res: Socialization and sex roles in cross-cultural perspective; modernization and culture change in East Africa; comparative social organizations; Kikuyu social organization. Mailing Add: Dept of Anthrop Univ of Ky Lexington KY 40506

ABBOTT, THOMAS B, b Pa, June 27, 24; m 45; c 2. SPEECH PATHOLOGY. Educ: Muskingum Col, BA, 43; Western Reserve Univ, MA, 48; Univ Fla, PhD, 57. Prof Exp: Asst prof speech & hearing, St Cloud State Col, 49-53; instr, Univ Fla, 54-57; dir speech ther prog, Los Angeles County Soc Crippled Children, 57-58; from assoc prof to prof speech path, Baylor Univ, 58-63, dir speech clin, 58-63; asst prof speech path & head speech & hearing clin, 63-66, ASSOC PROF SPEECH PATH & DIR DIV, UNIV FLA, 66- Concurrent Pos: Lectr, Univ Southern Calif, 57-58. Mem: Fel Am Speech & Hearing Asn; Speech Asn Am. Res: Stuttering; language. Mailing Add: 436 ASB Bldg Dept of Speech Univ of Fla Gainesville FL 32611

ABBOTT, THOMAS PAUL, b Peoria, Ill, Sept 26, 42; m 61; c 3. POLYMER CHEMISTRY. Educ: Univ Akron, BS, 64, MS, 68, PhD(polymer chem), 72. Prof Exp: Chemist polymer anal, Cent Res, Gen Tire & Rubber Co, 64-69; RES CHEMIST ELASTOMERS & PLASTICS, USDA, 71- Mem: Am Chem Soc. Res: Modifying starch to make elastomers, plastics and composites of modified starch with synthetic elastomers. Mailing Add: 1815 N University St Peoria IL 61604

ABBOTT, URSULA K, b Chilliwack, BC, July 30, 27; m 54; c 2. DEVELOPMENTAL GENETICS. Educ: Univ BC, BSA, 49, MSA, 50; Univ Calif, PhD(genetics), 55. Prof Exp: Asst poultry, Univ BC, 48-50; res asst poultry genetics, Univ Calif, Berkeley, 50-54; prin lab tech, 54-55, lectr & asst specialist, 55-56, instr & jr poultry geneticist, 56-57, from asst prof & asst geneticist to assoc prof & assoc geneticist, 57-67, PROF AVIAN SCI & GENETICIST, UNIV CALIF, DAVIS, 67- Concurrent Pos: Prog dir develop biol, NSF, 68-69. Honors & Awards: Hon prof, Univ BC, 67. Mem: Am Genetic Asn; Poultry Sci Asn; Am Physiol Soc; Soc Study Develop & Growth; NY Acad Sci. Res: Experimental embryology and teratology. Mailing Add: Dept of Avian Sci Univ of Calif Davis CA 95616

ABBOTT, WILLIAM HAROLD, b Atlanta, Ga, Mar 23, 44; m 69; c 1. MICROPALEONTOLOGY, OCEANOGRAPHY. Educ: Ga State Univ, Atlanta, BS, 69; Northeast La Univ, MS, 70; Univ SC, Columbia, PhD(geol), 72. Prof Exp: Asst prof geol & oceanog, Univ SC, Beaufort, 72-74; STAFF GEOLOGIST, SC STATE GEOL SURV, 74- Concurrent Pos: Geol Soc Am Penrose Bequest grant, 73; adj prof, Univ SC, Columbia & researcher, Belle W Baruch Marine Inst, 73-74. Mem: Geol Soc Am; Paleont Soc; Paleont Res Inst. Res: Stratigraphic relationships of marine Cenozoic sediments utilizing fossil diatoms, particularly southeastern United States coastal plain. Mailing Add: State Develop Bd Geol Div Harbison Forest Rd Columbia SC 29210

ABBOUD, FRANCOIS MITRY, b Cairo, Egypt, Jan 5, 31; US citizen; m 55; c 4. INTERNAL MEDICINE. Educ: Ain Shams Univ, Cairo, MB, BCh, 55; Am Bd Internal Med, dipl, 64. Prof Exp: Intern, Demerdash Govt Hosp, Cairo, 55; resident, Milwaukee County Hosp, Wis, 55-58; res assoc & instr med & cardiol, Marquette Univ, 58-60; from asst prof to assoc prof internal med, 61-68, PROF MED, UNIV IOWA, 68-, PROF PHYSIOL & BIOPHYS, 75-, DIR CARDIOVASC DIV, DEPT INTERNAL MED, 70-, DIR CARDIOVASC CTR, 75- Concurrent Pos: Am Heart Asn res fel, 58-60, adv res fel, 60-62; adv res fel cardiol, Cardiovasc Res Labs, Univ Iowa, 60-61; USPHS res career develop award, 62-71. Mem: Am Fedn Clin Res (pres, 71-72); Am Heart Asn; AMA; Am Physiol Soc; Soc Exp Biol & Med. Res: Cardiovascular physiology and pharmacology; peripheral circulation; vascular reactivity. Mailing Add: Dept of Internal Med Univ of Iowa Iowa City IA 52240

ABBRECHT, PETER H, b Toledo, Ohio, Nov 27, 30; m 57; c 2. PHYSIOLOGY, BIOENGINEERING. Educ: Purdue Univ, BS, 52; Univ Mich, MS, 53, PhD(chem eng), 57, MD, 62. Prof Exp: Sr chem engr, Minn Mining & Mfg Co, 56-58; intern med, Univ Calif Hosp, Los Angeles, 62-63; from asst prof to assoc prof physiol, 63-72, resident internal med, 70-72, PROF PHYSIOL & CHMN BIOENG PROG, UNIV MICH, ANN ARBOR, 72- Concurrent Pos: NIH career develop award, 69-; fel pulmonary dis, Med Ctr, Univ Mich, Ann Arbor, 74-75; instr, Wayne State Univ, 57-58. Mem: Am Physiol Soc; Am Soc Artificial Internal Organs; Soc Exp Biol & Med; Biomed Eng Soc. Res: Cardiopulmonary and renal physiology and modeling; hemoglobin regulation; use of systems theory in medical diagnosis and therapy. Mailing Add: Dept of Physiol Univ of Mich Ann Arbor MI 48104

ABDALI, SYED KAMAL, b Patna, India, Mar 10, 40; Pakistani citizen; m 65; c 2. COMPUTER SCIENCES. Educ: Am Univ Beirut, BE, 63; Univ Montreal, MSc, 68; Univ Wis-Madison, PhD(comput sci), 74. Prof Exp: Lectr elec eng, Govt Eng Col, Karachi, Pakistan, 63-64; jr engr, Tel Indust Pakistan, Haripur, 64-65; design engr, Int Tel & Tel, Montreal, 65-66; consult, All-Tech Assocs, Montreal, 66-67; teaching asst comput sci, Univ Wis, 68-71; asst prof, NY Univ, 71-73; ASST PROF MATH SCI, RENSSELAER POLYTECH INST, 73- Mem: Asn Comput Mach; Asn Symbolic Logic. Res: Semantics of programming languages; validation methods for programs and machine designs; combinatory logic. Mailing Add: Dept of Math Sci Rensselaer Polytech Inst Troy NY 12181

ABDEL-GAWAD, MONEM, b Edku, Egypt, Dec 5, 30; US citizen; m 60; c 3. GEOLOGY. Educ: Univ Alexandria, BSc, 51; Columbia Univ, MA, 58, PhD(geol, mineral), 60. Prof Exp: Anal & indust chemist, Egyptian Fertilizer & Chem Indust Co, 52-55; geologist, Sahara Petrol Corp, 55-56; asst prof mineral, Egypt Atomic Energy Estab, 60-65, assoc prof, 65-66; staff mem geochem, Am Univ Cairo, 66; sr scientist, Space Sci, Rockwell Int Sci Ctr, Thousand Oaks, Calif, 66-75; DIR MINERAL RESOURCES DIV, THE ADAR CORP, 75- Mem: Fel Geol Soc Am; Am Geophys Union. Res: Mineral resources development and exploration; application of remote sensing and space technology in geology, hydrology, soil surveys, with emphasis on developing countries. Mailing Add: Dir Mineral Resources Div Adar Corp 1156 15th St Suite 410 Washington DC 20005

ABDEL-LATIF, ATA A, b Beitunia, Palestine, Jan 22, 33; US citizen; m 57; c 4. BIOCHEMISTRY, NEUROCHEMISTRY. Educ: DePaul Univ, BS, 55, MS, 58; Ill Inst Technol, PhD(biochem, physiol), 63. Prof Exp: Control chemist, Ninol Chem Labs, 55-56; med res assoc, State of Ill Pediat Inst, 63-67; assoc prof, 67-74, PROF BIOCHEM, MED COL GA, 74- Concurrent Pos: Fel psychiat, Col Med, Univ Ill, 63-65; State of Ill Dept Ment Health fel, 63-67; NIH grants, 65-; asst prof, Ill Inst Technol, 65-67. Mem: AAAS; Am Soc Biol Chem; Am Physiol Soc; Am Chem Soc; Am Soc Neurochem. Res: Metabolism of synaptic nerve endings during development of the brain, especially phospholipids, transport of ions and metabolites and effect of drugs and neurohormones on those processes associated with the synapse; mechanism of action of adrenergic and cholinergic neurotransmitters at the molecular level in brain and iris muscle. Mailing Add: Dept of Cell & Molecular Biol Med Col of Ga Augusta GA 30902

ABDEL-MONEM, MAHMOUD MOHAMED, b Cairo, Egypt, June 7, 38; m 64. MEDICINAL CHEMISTRY. Educ: Cairo Univ, BS, 59; Univ Minn, Minneapolis, PhD(med chem), 66. Prof Exp: Instr plant chem, Cairo Univ, 60-61; head res, Nile Co Pharmaceut, Cairo, 65-68; res specialist med chem, Univ Minn, Minneapolis, 68-70; asst prof, Univ Ill Med Ctr, Chicago, 70-71; ASST PROF MED CHEM, COL PHARM, UNIV MINN, MINNEAPOLIS, 71- Concurrent Pos: Assoc ed, J Med Chem, 72-; consult, Zinpro Corp, Excelsior, Minn, 72- Res: Study of distribution and metabolism of drugs; structure activity correlation and chemical mechanisms in drug metabolism. Mailing Add: Col of Pharm Univ of Minn Minneapolis MN 55455

ABDOU, NABIH I, b Cairo, UAR, Oct 11, 34; m 67. MEDICINE, IMMUNOLOGY. Educ: Cairo Univ, MD, 57; Univ Pa, MSc, 67; McGill Univ, PhD(immunol), 69. Prof Exp: Asst prof med, Univ Pa, 69-75, ASSOC PROF MED, UNIV KANS, 75- Mem: Am Asn Immunol; Am Acad Allergy; Transplantation Soc; Am Fedn Clin Res. Res: Cellular immunology. Mailing Add: Med Ctr Univ of Kans Kansas City KS 66103

ABDUL-BAKI, AREF ASAD, b Apr 9, 32; US citizen; m 62; c 1. PLANT PHYSIOLOGY. Educ: Am Univ Beirut, BSc, 56; Univ Ill, Urbana, PhD(plant physiol), 64. Prof Exp: Asst prof plant physiol, Eastern Mich Univ, 64; res assoc, Univ Mich, Ann Arbor, 65-66 & Univ Calif, Santa Cruz, 66-67; res plant physiologist, 67-75, CHIEF SEED QUAL LAB, AGR RES SERV, USDA, 75- Honors & Awards: Nat Canners Asn Award, 72. Mem: Am Soc Plant Physiol; Am Soc Agron; Crop Sci Soc Am; Am Soc Hort Sci. Res: Physiology and biochemistry of agronomic seeds as they relate to marketing quality attributes such as vigor, deterioration, dormancy, hard seed coat and stress tolerance. Mailing Add: Seed Qual Lab Beltsville Agr Res Ctr-West Beltsville MD 20705

ABE, RONALD KURASO, b Hilo, Hawaii, Sept 17, 39. ANIMAL NUTRITION. Educ: Univ Hawaii, BS, 62; Kans State Univ, MS, 65, PhD(animal nutrit), 70. Prof Exp: ASSOC PROF ANIMAL SCI, FT VALLEY STATE COL, 68- Mem: Am Soc Animal Sci; Am Dairy Sci Asn. Res: Calf and ruminant nutrition. Mailing Add: Div of Agr Ft Valley State Col Ft Valley GA 31030

ABEGG, ROLAND, b Perth Amboy, NJ, Dec 14, 14; m 38; c 3. ZOOLOGY. Educ: Univ Mich, BA, 36; La State Univ, MS, 39, PhD(zool), 48. Prof Exp: Supt La State Univ quail farm, La Dept Wildlife & Fisheries, 37-42; analyst, Cities Serv Ref Corp, 43-45 & Esso Standard Oil, 45-46; assoc prof biol, Northwestern State Col La, 48-50; from assoc prof to prof biol, Southeastern La Col, 50-59; PROF ZOOL & HEAD DEPT, LA TECH UNIV, 59- Mem: Am Heart Asn. Res: Waste control and pollution in relation to game fishes; cytology and genetics. Mailing Add: Dept of Zool La Tech Univ Ruston LA 71270

ABEGG, VICTOR PAUL, b Torrance, Calif, Feb 14, 45. ORGANIC CHEMISTRY, HISTORY OF SCIENCE. Educ: Loyola Univ Chicago, AB, 67; Mass Inst Technol, PhD(chem), 70; Univ Toronto, MDiv, 73. Prof Exp: Teaching fel, 74-75, SESSIONAL LECTR CHEM, YORK UNIV, 75- Mem: Am Chem Soc. Res: An examination of new synthetic pathways to small-ring carbon compounds is continuing with emphasis on photochemical methods; these species are subsequently subjected to strong-acid-catalyzed rearrangement and product behavior determined. Mailing Add: 226 A St George St Toronto ON Can

ABEL, ALAN WILSON, b Wilkinsburg, Pa, Mar 7, 39. PHYSICAL CHEMISTRY. Educ: Univ Pittsburgh, BS, 61, PhD(chem), 67. Prof Exp: Asst ed, 67-69, assoc indexer, 69-70, SR ASSOC INDEXER, GEN SUBJ INDEXING DIV, CHEM ABSTR SERV, 70- Mem: Am Chem Soc. Res: Indexing of chemical literature; magnetic properties of metals, alloys and intermetallic compounds; gas-metal reactions. Mailing Add: 3015 Stadium Dr Apt 2 Columbus OH 43202

ABEL, ERNEST LAWRENCE, b Toronto, Ont, Feb 10, 43; m 71. PSYCHOPHARMACOLOGY. Educ: Univ Toronto, BA, 65, MA, 67, PhD(psychol), 71. Prof Exp: Psychologist, Dept Pub Health, City of Toronto, 68; SR RES SCIENTIST, RES INST ALCOHOLISM, NY STATE DEPT MENT HYG, 73- Concurrent Pos: Med Res Coun Can fel pharmacol, Univ NC, Chapel Hill, 71-73; United Way grant, 74-75. Res: Psychology; pharmacology. Mailing Add: Res Inst on Alcoholism 1021 Main St Buffalo NY 14203

ABEL, FRANCIS LEE, b Iowa City, Iowa, Apr 12, 31; m 54; c 3. CARDIOVASCULAR PHYSIOLOGY. Educ: Univ Kans, BA, 52; Harvard Med Sch, MD, 57; PhD(physiol), 60. Prof Exp: Intern pediat, Children's Hosp, Los Angeles, 60-61; res instr, Univ Wash, 61-62; from asst prof to prof physiol, Sch Med, Ind Univ, Indianapolis, 72-75; chmn dept physiol & pharmacol, 75-76, PROF PHYSIOL, SCH MED, UNIV SC, 75-, INTERIM DEAN, 76- Concurrent Pos: USPHS trainee physiol, Univ Wis, 59-60; NIH career develop award, 69-73; vis prof, Ind Univ-USAID at Jinnah Postgrad Med Ctr, Karachi, 64-65; vis prof, Univ Southern Calif, 70. Mem: AAAS; Biophys Soc; Inst Elec & Electronics Engrs; Am Physiol Soc; Am Fedn Clin Res. Res: Cardiovascular physiology; biophysical instrumentation; blood flow studies; reactions to hemorrhage; electro-magnetic flowmeters; shock; venous return; arterial pressure; regulation; cardiac function. Mailing Add: Sch of Med Univ of SC Columbia SC 29208

ABEL, JAMES EDWARD, b New York, NY, Apr 26, 15; m 50; c 1. X-RAY CRYSTALLOGRAPHY, EXPLOSIVES. Educ: Univ Wash, BS, 48; Stevens Inst Technol, MS, 55. Prof Exp: Insecticides chemist, Boyle-Midway Div, Am Home Prod, 43-46; EXPLOSIVES RES CHEMIST, FELTMAN RES LAB, SOLID STATE BR, PICATINNY ARSENAL, 49- Mem: Am Crystallog Asn; Am Chem Soc; Am Inst Chemists. Res: Propellants; primary and secondary explosives; crystal structure of explosive solids using x-ray diffraction. Mailing Add: 265 West Shore Trail Sparta NJ 07871

ABEL, JOHN H, JR, b Painesville, Ohio, Feb 25,37; m 58; c 3. CELL BIOLOGY. Educ: Col Wooster, BA, 59; Brown Univ, MA, 64; PhD(cell biol), 66. Prof Exp: Chmn dept high sch, Ohio, 59-61; teacher high sch, Ill, 61-62; instr cell biol, NY Med Col, 66-67; from asst prof to assoc prof, 67-75, PROF CELL BIOL, COLO STATE UNIV, 75- Concurrent Pos: Consult, Ciba Chem & Dye Co, 66-68; ed consult, McGraw-Hill Bk Co, 66-; guest prof, Univ Bonn, Fed Repub Ger, 74-75. Honors & Awards: Sr US Scientist Award for Res & Teaching, Alexander von Humboldt Stiftung, Fed Repub Ger, 73. Mem: Am Soc Cell Biol; Am Assoc Anatomists; NY Acad Sci; Soc Study Reproduction & Fertil; AAAS. Res: Endocrine and neuroendocrine control of ovarian and adrenal function; osmoregulation in euryhaline birds. Mailing Add: Dept of Physiol & Biophys Colo State Univ Ft Collins CO 80523

ABEL, ROBERT BERGER, b Providence, RI, July 21, 26; m 53; c 2. SCIENCE ADMINISTRATION. Educ: Brown Univ, BS, 47; George Washington Univ, MEA, 62; American Univ, PhD(polit sci), 72. Prof Exp: Chemist, Woods Hole Oceanog Inst, 47-50; oceanogr & head, Oceanog Surv Br, 50-55; asst to dir, Hydrographic Off, US Navy, 55-60, asst res coordr, Off Naval Res & exec secy, Inter-agency Comt Oceanog, 60-67; DIR NAT SEA GRANT PROG, NAT OCEANIC & ATMOSPHERIC ADMIN, 67- Concurrent Pos: Instr, US Naval Res Off Sch, 62-67; adj instr oceanog, Fairleigh Dickinson Univ; chmn US deleg UNESCO working group on educ & training in marine sci. Honors & Awards: Super Serv Award & Distinguished Serv Award, US Govt; Gold Medal, US Dept of Commerce, 73. Mem: Am Chem Soc; Am Geophys Union; Sigma Xi; Am Soc Oceanog; Marine Technol Soc (past vpres, pres elect, 70, pres, 74-75). Res: Oceanography; public administration. Mailing Add: 821 Braeburn Dr Oxon Hill MD 20022

ABEL, WILLIAM ROBERT, b Spencer, Wis, June 3, 27; m 50; c 1. MATHEMATICS. Educ: Morningside Col, BA, 49; Univ SDak, MA, 51; Univ Mo, PhD, 57. Prof Exp: Instr math, Univ Mo, 52-57; asst prof, Univ Nebr, 57-64; assoc prof, 64-67, PROF MATH, WESTERN WASH STATE COL, 67- Mem: Am Math Soc; Math Asn Am. Res: Geometry of metric arcs. Mailing Add: Dept of Math Western Wash State Col Bellingham WA 98225

ABEL, WILLIAM T, b Marion, Ind, Feb 16, 22; m 49; c 4. FUEL SCIENCE, PHYSICAL CHEMISTRY. Educ: Franklin Col, BA, 44; Univ Ill, Urbana, MS, 47. Prof Exp: Res asst chem, Ill Geol Surv, 44-47; res asst, Mound Lab, Ohio, 47-53; chemist, Dowell Inc, Okla, 53-56; chemist, Elec Auto-Lite Co, 56-58; supvy chemist, US Bur Mines, 58-60, chem res engr, 60-74; CHEM ENGR, ENERGY RES DEVELOP ADMIN, 74- Mem: AAAS; Am Chem Soc. Res: Corrosion inhibition; gas-solid reaction; dry processes for removal of pyrite from coal; coal liquefaction. Mailing Add: 564 Killarney Dr Morgantown WV 26505

ABELES, ANN LINDSTROM, b Tracy, Minn, Sept 23, 42; m 62; c 2. BIOPHYSICS. Educ: Univ Minn, BA, 63; Univ Md, PhD(biophys), 69. Prof Exp: Asst prof, 70-74, ASSOC PROF SCI, FREDERICK COMMUNITY COL, 75- Mem: Am Chem Soc. Res: Study of tree growth to determine the possible effects of air pollution from a generating system station on the growth of surrounding trees. Mailing Add: Dept of Biophys Frederick Community Col Frederick MD 21701

ABELES, BENJAMIN, b Vienna, Austria, June 23, 25; m 58; c 1. SOLID STATE PHYSICS. Educ: Prague Tech Univ, MS, 49; Hebrew Univ, Israel, PhD, 56. Prof Exp: Res tech, Meteorol Serv, Israel, 49-51; researcher, Weizmann Inst, Israel, 51-56; Radio Corp Am, 56-69, RESEARCHER, RCA LABS, 69- Mem: Am Phys Soc. Mailing Add: RCA Labs Princeton NJ 08540

ABELES, FRANCINE, b New York, NY, Oct 19, 35; m 57; c 3. MATHEMATICS. Educ: Barnard Col, Columbia Univ, AB, 57; Columbia Univ, MA, 59, EdD(math), 64. Prof Exp: Teacher high schs, NY & Ger, 57-63; from asst prof to assoc prof, 64-72, PROF MATH, KEAN COL NJ, 72- Mem: Am Math Soc; Math Asn Am. Res: Affine geometry; linear algebra; convex sets; history of mathematics. Mailing Add: 351 E 19th St New York NY 10003

ABELES, FRED BERNARD, botany, biochemistry, see 12th edition

ABELES, ROBERT HEINZ, b Vienna, Austria, Jan 14, 26; nat US; m 48.

BIOCHEMISTRY. Educ: Univ Chicago, MS, 50; Univ Colo, PhD(biochem), 55. Prof Exp: Asst dept pediat, Univ Chicago, 50-51; fel, Nat Found Infantile Paralysis, 55-56; fel chem, Harvard Univ, 56-57; asst prof chem, Ohio State Univ, 57-60; from asst prof to assoc prof biochem, Univ Mich, 60-64; assoc prof, 64-67, PROF BIOCHEM, BRANDEIS UNIV, 67-, CHMN GRAD DEPT, 72- Concurrent Pos: Ed, J Biol Chem, Am Soc Biol Chem, 67-72. Mem: Nat Acad Sci; Am Chem Soc; Am Soc Biol Chem; Am Acad Sci. Res: Mechanism of enzyme action; biological oxidations. Mailing Add: 415 Ward St Newton Centre MA 02159

ABELING, EDWIN JOHN, b Amsterdam, NY, Oct 4, 15; m 45; c 3. BACTERIOLOGY. Educ: Univ Ky, BS, 44. Prof Exp: Med technologist, Montgomery County Lab, NY, 39-41; med technologist, St Joseph's Hosp, Lexington, Ky, 41-44; bacteriologist, Beech-Nut Packing Co, NY, 44-47, head, San Jose Lab, Calif, 47-50, food lab, Canajoharie, NY, 50-54, assoc dir res, Beech-Nut Life Savers, 54-57, dir res & develop, Quality Control, 57-61; dir, 61-68, V PRES RES & DEVELOP, PETER PAUL, INC, 68- Concurrent Pos: Mem, Exec Prog Bus Admin, Grad Sch Bus, Columbia Univ, 60. Mem: Inst Food Technologists. Res: Development of new products; food, packaging and market research; quality control. Mailing Add: Ansonia Rd Woodbridge CT 06525

ABELL, CREED WILLS, b Charlottesville, Va, July 8, 34; m 56; c 2. BIOCHEMISTRY, ONCOLOGY. Educ: Va Mil Inst, BS, 56; Purdue Univ, MS, 58; Univ Wis, PhD(oncol), 62. Prof Exp: Chemist, Chem Sect, Carcinogenesis Studies Br, Nat Cancer Inst, 62-67; from assoc prof to prof biochem & molecular biol, Sch Med, Univ Okla, 67-72; PROF HUMAN BIOL CHEM & GENETICS, UNIV TEX MED BR GALVESTON, 72- Mem: Am Asn Cancer Res; Am Soc Biol Chemists; Sigma Xi. Res: Hydrocarbon carcinogenesis; alkylating agent carcinogenesis; protein biosynthesis and coding; alteration in coding by chemicals and carcinogens; induction of lung tumors in mice by alkylating agents; regulation of division of normal and leukemic lymphocytes and enzyme therapy of neoplasia. Mailing Add: Dept Human Biol Chem & Genetics Basic Sci Bldg Univ Tex Med Br Galveston TX 77550

ABELL, DANA LEROY, zoology, aquatic biology, see 12th edition

ABELL, GEORGE OGDEN, b Los Angeles, Calif, Mar 1, 27; m 51, 72; c 2. ASTRONOMY. Educ: Calif Inst Technol, BS, 51, MS, 52, PhD(astron), 57. Prof Exp: Observer, Nat Geog Soc-Palomar Observ Sky Surv, 53-56; from instr to assoc prof, 56-67, chmn dept, 68-75, PROF ASTRON, UNIV CALIF, LOS ANGELES, 67- Concurrent Pos: Planetarium lectr, Griffith Observ, Los Angeles, 53-61; guest investr, Hale Observ, 57- & Max Planck Inst Physics & Astrophys, 65-66. Mem: AAAS; Int Astron Union; Am Astron Soc. Res: Problems relating to organization, structure and distribution of clusters of galaxies; planetary nebulae. Mailing Add: Dept of Astron Univ of Calif Los Angeles CA 90024

ABELL, JARED, b Los Angeles, Calif, Sept 5, 28; m 50; c 3. ORGANIC CHEMISTRY. Educ: Calif Inst Technol, BS, 50; Univ Calif, Los Angeles, PhD(org chem), 54. Prof Exp: Sr res chemist, Chevron Res Co, 54-66; res chemist, Ortho Div, 66-68, MGR PROD RES & SERV, DIV RES & DEVELOP, ORTHO DIV, CHEVRON CHEM CO, 68- Mem: Am Chem Soc. Res: Pesticide formulation and agricultural chemicals in general. Mailing Add: Chevron Chem Co Ortho Div 940 Hensley St Richmond CA 94804

ABELL, LIESE LEWIS, b Frankfurt, Ger, Aug 17, 09; nat US. BIOCHEMISTRY. Educ: Univ Frankfurt, PhD, 35. Prof Exp: Chemist, Flower & Fifth Ave Hosp, New York Med Col, 40-42; chemist, Wyeth Inst Appl Biochem, Pa, 45-47; asst prof biochem, Columbia Univ, 47-68; sr res scientist, Bur Labs, Pub Health Labs, New York City Dept Health, 68-74. Concurrent Pos: Res fel, Columbia Univ, 38-40 & 42-45; mem coun arteriosclerosis, Am Heart Asn. Mem: AAAS; Am Heart Asn; Am Soc Biol Chemists. Res: Lipid metabolism in connection with arteriosclerosis. Mailing Add: 7 Peter Cooper Rd New York NY 10010

ABELL, MURRAY RICHARDSON, b Aylmer, Ont, Oct 14, 20; nat US; m 44; c 4. MEDICINE, PATHOLOGY. Educ: Univ Western Ont, MD, 44, PhD(path), 51; Am Bd Path, dipl; FRCPS(C). Prof Exp: From instr to assoc prof, 52-59, PROF PATH, UNIV MICH, ANN ARBOR, 59- Concurrent Pos: Trustee, Am Bd Path. Mem: Am Asn Path & Bact; Int Acad Path; Am Soc Clin Path. Res: Hepatic disease; mediastinal tumors; neoplasms of female reproductive tract; antigen-antibody reactions in tissues; testicular neoplasms. Mailing Add: Dept of Path Univ of Mich Ann Arbor MI 48104

ABELL, PAUL IRVING, b Pelham, Mass, July 24, 23; m 51; c 1. ORGANIC CHEMISTRY. Educ: Univ NH, BS, 48; Univ Wis, PhD(chem), 51. Prof Exp: From instr to assoc prof, 51-64, PROF ORG CHEM, UNIV RI, 64- Concurrent Pos: Petrol Res Fund Int Award, Univ Wales, 60-61 & Univ Bristol, 69-70; Fulbright lectr, UAR, 65-66; mem, Omo River Res Exped, 67 & Lake Rudolf Res Expeds, 68-74. Mem: Am Chem Soc; NY Acad Sci; Brit Chem Soc; Faraday Soc. Res: Stereochemistry and kinetics of free radical reactions; organic geochemistry; paleontology. Mailing Add: Dept of Chem Univ of RI Kingston RI 02881

ABELLA, ISAAC D, b Toronto, Ont, June 20, 34; m 66. EXPERIMENTAL PHYSICS, QUANTUM OPTICS. Educ: Univ Toronto, BA, 57; Columbia Univ, MA, 59, PhD(physics), 63. Prof Exp: Asst physics, Columbia Univ, 57-63, res assoc, 63-65; asst prof, 65-72, ASSOC PROF PHYSICS, UNIV CHICAGO, 72- Concurrent Pos: Mem laser apparatus comt, Am Inst Physics, 65, vis scientist, 66-71; consult, Mithras, Inc, Mass, 65-70 & Sanders Assoc, NH, 70-; mem laser adv comt, Ill Dept Pub Health, 68-72; vis fel, Joint Inst Lab Astrophys, Univ Colo, Boulder, 72-73. Mem: AAAS; Am Phys Soc; fel Optical Soc Am. Res: Experimental physics, especially non-linear optics, photon echoes, quantum beats, atomic physics and lasers. Mailing Add: Ryerson Phys Lab Univ of Chicago Chicago IL 60637

ABELMANN, WALTER H, b Frankfurt, Ger, May 16, 21; nat US; m 58; c 5. MEDICINE. Educ: Harvard Univ, AB, 43; Univ Rochester, MD, 46. Prof Exp: Asst, 51-53, instr, 53-55, assoc, 55-58, asst prof, 58-64, assoc clin prof, 64-69, assoc prof, 69-72, PROF MED, HARVARD MED SCH, 72- Concurrent Pos: Res fel, Thorndike Mem Lab, Boston City Hosp, 51-; Am Heart Asn res fel, 53-55; estab investr, Am Heart Asn, 55-60. Mem: AAAS; Am Fedn Clin Res; Am Heart Asn; Asn Univ Cardiol; Am Soc Clin Invest. Res: Clinical cardiovascular physiology. Mailing Add: 330 Brookline Ave Boston MA 02215

ABELS, LARRY L, b Freeport, Ill, Feb 16, 37; m 60; c 3. PHYSICS. Educ: Knox Col, Ill, BA, 59; Ohio State Univ, MSc, 61, PhD(physics), 65. Prof Exp: ASST PROF PHYSICS, UNIV ILL, CHICAGO CIRCLE, 65- Mem: Optical Soc Am. Res: Studies of infrared transmission through synthetic atmospheres; infrared line parameter determinations. Mailing Add: Dept of Physics Univ of Ill at Chicago Circle Chicago IL 60680

ABELSON, JOHN NORMAN, b Grand Coulee Dam, Wash, Oct 19, 38. MOLECULAR BIOLOGY, MOLECULAR GENETICS. Educ: Wash State Univ, BS, 60; Johns Hopkins Univ, PhD(biophys), 65. Prof Exp: Fel, Lab Molecular Biol, Cambridge Univ, Eng, 65-69; asst prof, 69-73, ASSOC PROF CHEM, UNIV CALIF, SAN DIEGO, 73- Mem: Am Soc Biol Chemists; Am Chem Soc. Res: Genetic control mechanisms; nucleotide sequences of DNA and RNA; RNA synthesis and post-transcriptional processing. Mailing Add: Dept of Chem B017 Univ of Calif La Jolla CA 92093

ABELSON, PHILIP HAUGE, b Tacoma, Wash, Apr 27, 13; m 36; c 1. PHYSICAL CHEMISTRY. Educ: Wash State Univ, BS, 33, MS, 35; Univ Calif, PhD(nuclear physics), 39. Hon Degrees: DSc, Yale Univ, 64, Southern Methodist Univ, 69; DHL, Univ Puget Sound, 68. Prof Exp: Asst physics, Wash State Univ, 33-35; asst, Univ Calif, 35-38, asst, Radiation Lab, 38-39; asst physicist, Carnegie Inst, 39-41; from assoc physicist to prin physicist, Naval Res Lab, 41-45; chmn biophys sect, Dept Terrestrial Magnetism, 46-53, dir geophys lab, 53-71, PRES, CARNEGIE INST, 71- Concurrent Pos: Mem, Nat Defense Res Comt, 40-42; chmn radiation cataract comt, Nat Res Coun, 49-57; mem biophysics & biophys chem study sect, NIH, 56-59, phys biol training grants comt, 58-60; plowshare adv comt, AEC, 59-63, gen adv comt, 60-63; sci counsr, Nat Inst Arthritis & Metab Dis, 60-63. Honors & Awards: Distinguished Civilian Serv Medal, US Navy, 45; Mod Med Award, 67; Mellon Inst Award, Carnegie-Mellon Univ, 70; Joseph Priestley Award, Dickinson Col, 73; Kalinga Prize, UNESCO, 73; Sci Achievement Award, AMA, 74. Mem: Nat Acad Sci; AAAS; Am Nuclear Soc; fel Am Phys Soc; fel Am Acad Arts & Sci. Res: Nuclear physics; radioactive tracers; fission products of uranium; characteristic x-rays emitted by radioactive substances; neptunium; separation of uranium isotopes; mechanisms of ion transport into living matter; biosynthesis in microorganisms; plasma volume expanders; petrology; paleobiochemistry; geochemistry. Mailing Add: Carnegie Inst of Washington 1530 P St NW Washington DC 20005

ABEND, PHILLIP GARY, b Washington, DC, Aug 9, 30; m 52; c 5. ORGANIC CHEMISTRY. Educ: George Washington Univ, BS, 51; Univ Pittsburgh, PhD(chem), 58. Prof Exp: Res chemist, Gulf Res & Develop Co, Pa, 58-59; sr res chemist, Armour & Co, Ill, 59-62 & B F Goodrich Res Ctr, 62-65; group leader, Harchem Div, Wallace & Tiernan, Inc, NJ, 65-68, Pennwalt Corp, 68-70 & Standard Chem Prod, Inc, 70-74; SUPVR ORG RES LAB, HENKEL INC, 74- Concurrent Pos: Lectr eve div, Fairleigh Dickinson Univ, 68- Mem: Am Chem Soc; Am Oil Chemists Soc; Soc Cosmetic Chemists. Res: Surfactants; fatty acids; cosmetic chemicals. Mailing Add: 1332 Taft Rd Teaneck NY 07666

ABERCROMBIE, JAY, b Akron, Ohio, Sept 10, 42; m 68. ENTOMOLOGY. Educ: Univ Akron, BS, 64; Cornell Univ, PhD(parasitol), 70. Prof Exp: Entomologist, Walter Reed Army Inst Res, 71-72; chief entomologist, Mil Asst Command, Vietnam, 72-73; entomologist, Walter Reed Army Inst Res, 73-75; ENTOMOLOGIST, US ARMY ENVIRON HYG AGENCY, 75- Concurrent Pos: Taxonomist, Med Entom Proj, Smithsonian Inst, 73- Mem: Entom Soc Am; Asn Trop Biol. Res: Taxonomy and biology of aquatic Diptera, especially Culicidae, Sciomyzidae and Simuliidae. Mailing Add: US Army Environ Hy Agency Regional Div North Ft George G Meade MD 20755

ABERCROMBIE, WARREN FULTON, b Leeds, Ala, June 8, 11; m 43; c 4. ZOOLOGY. Educ: Howard Col, AB, 31; NY Univ, MS, 33, PhD(entom), 35. Prof Exp: Asst biol, NY Univ, 31-35; asst prof, Howard Col, 35-41; asst prof & head dept, Erskine Col, 41-42; tech ed & actg chief info sect, Med Div, Army Chem Ctr, Md, 46-48; pub ed in-chg tech reports & libr sect, Commun Dis Ctr, USPHS, 48-56, training officer, Training Br, 56-57; training officer in-chg radiol defence sch, Off Civil Defense Mobilization, 58-59; dep chief, Training Br, Div Emergency Health Serv, USPHS, 59-74, actg chief, 65-67, EDUC SPECIALIST, PERSONNEL TRAINING & DEVELOP BR, BUR MED SERV, USPHS, 74- Mem: AAAS; Am Micros Soc; Am Asn Anat; Am Pub Health Asn. Res: Effects of iodine compounds on the thyroid gland; growth of Japanese beetle larvae; effects of chemicals on the cockroach. Mailing Add: Personnel Training & Develop Br Bur of Med Serv USPHS 6525 Belcrest Rd Hyattsville MD 20782

ABERE, JOSEPH FRANCIS, b New York, NY, Mar 30, 20; m 44; c 7. POLYMER CHEMISTRY. Educ: Queens Col, NY, BS, 41; Polytech Inst Brooklyn, MS, 43, PhD(high polymers), 48. Prof Exp: Tutor chem, Queens Col, NY, 43-44, instr, 46-47; res chemist, 48-53, head appl res sect, Cent Res, 53-57, proj mgr chem div, 57-62, assoc mgr tape res, 62-67, mgr pioneering res, Indust Spec Prod Dept, 67-72, tech mgr composites, 70-72, TECH MGR OVERSEAS OPERS, MINN MINING & MFG CO, 72- Mem: AAAS; Am Chem Soc; Am Inst Chemists; Am Inst Aeronaut & Astronaut. Res: High polymers; fluorine containing polymers; vulcanization; adhesives and adhesion; resins; plastics; chemical development; imine chemistry; composites; reinforced plastics. Mailing Add: Indust Spec Prod Dept 3M Co 2501 Hudson Rd St Paul MN 55119

ABERHART, DONALD JOHN, b St John, NB, Oct 20, 41. ORGANIC CHEMISTRY. Educ: Univ Western Ont, BSc, 63, PhD(org chem), 67. Prof Exp: NATO sci fel, Glasgow Univ, 67-68; fel, Worcester Found Exp Biol, 69-70; res assoc, Mass Inst Technol, 70-71; ASST PROF CHEM, CATH UNIV AM, 71- Mem: Am Chem Soc; NY Acad Sci. Res: Natural products chemistry; biosynthetic studies. Mailing Add: Dept of Chem Cath Univ of Am Washington DC 20064

ABERLE, DAVID FRIEND, b St Paul, Minn, Nov 23, 18; m 55; c 1. ETHNOLOGY. Educ: Harvard Univ, AB, 40, Columbia Univ, PhD(anthrop), 50. Prof Exp: Inst social anthrop, Harvard Univ, 47-50; vis assoc prof anthrop, Walter Hines Page Sch Int Rels, Johns Hopkins Univ, 50-52; from assoc prof to prof, Univ Mich, 52-60; Simon vis prof social anthrop, Univ Manchester, 60-61; prof anthrop & chmn dept, Brandeis Univ, 61-63; prof, Univ Ore, 63-67; PROF ANTHROP, UNIV BC, 67- Concurrent Pos: Res assoc, Harvard Sch Pub Health, Harvard Univ, 48-50; Soc Sci Res Coun grant, Navajo Reservation, Johns Hopkins Univ, 51; NIMH res grants, Navajo Reservation, Johns Hopkins Univ & Univ Mich, 51-53; fel, Ctr Advan Study Behav Sci, 55-56; mem behav sci study sect, NIMH, 57-60; hon res fel social anthrop, Univ Manchester, 61; NSF res grant, Navajo Reservation, Univ Ore, 65-72; Can Coun res grants, Univ BC, 67-; adj res assoc anthrop, Univ Ore, 67-72. Mem: Am Anthrop Asn; Am Sociol Asn; Royal Anthrop Inst Gt Brit & Ireland; Can Sociol & Anthrop Asn; Soc Appl Anthrop. Res: Cultural evolution and cultural ecology; development and underdevelopment; kinship, including lexical reconstruction of kinship terminology of protolanguages; religious movements; Athapaskan Indians and especially Navajo Indians. Mailing Add: Dept of Anthrop & Sociol Univ of BC Vancouver BC Can

ABERLE, ELTON D, b Sabetha, Kans, Aug 30, 40; m 65; c 2. MEAT SCIENCES, FOOD SCIENCE. Educ: Kans State Univ, BS, 62; Mich State Univ, MS, 65, PhD(food sci), 67. Prof Exp: From asst prof to assoc prof, 67-76, PROF ANIMAL SCI, PURDUE UNIV, 76- Concurrent Pos: Mem NC-91, Tech Comt Regional Res Proj, Agr Res Serv, USDA, 68-74, NC-131, 75-; assoc, Univ Minn, 75. Mem: Am Soc Animal Sci; Inst Food Tech; Am Meat Sci Asn. Res: Muscle biochemistry and physiology and muscle growth and differentiation in meat animals relating to use of muscle as food; adipose tissue growth in animals; meat processing. Mailing Add: Dept of Animal Sci Purdue Univ West Lafayette IN 47907

ABERLE, SOPHIE BLEDSOE, b Schenectady, NY, July 21, 99; m 40. NUTRITION. Educ: Stanford Univ, AB, 23, MA, 25, PhD(genetics), 27; Yale Univ, MD, 30. Prof Exp: Asst histol, Stanford Univ, 24-25, asst embryol & neurol, 25-26; instr anthrop, Inst Human Rels, Yale Univ, 27-30, Sterling fel, Sch Med, 30-31, instr, 30-34; assoc res, Carnegie Inst, 34-35; supt, Pueblo Indians, Bur Indian Affairs, US Dept Interior & secy, Southwest Supts Coun, US Indian Serv, 35-44; div med sci, Nat Res Coun, 44-49; spec res dir, Univ NMex, 49-54; chief nutrit, Bernalillo County Indian Hosp, 53-66; mem staff, Dept Psychiat, Med Sch, 66-69, MEM STAFF, LAW SCH, UNIV N MEX, 70- Concurrent Pos: Field worker among Pueblo Indians under grant from Comt Res Probs of Sex, 27-35; chief emergency med serv, NMex State Coun Nat Defence, 42-44; mem upper Rio Grande drainage basin comt, Nat Resources Planning Bd, 36-44; NMex Nutrit Comt, 40-; White House Conf Children in Democracy, 40-46; chmn bd dirs, Southwest Field Training Sch for Fed Serv, 37-41; mem, Defence Saving Comt, NMex, 41-44; chmn, Bernalillo County Hosp Comt, 49-50; mem Nat Sci Bd, 50-58, exec comt, 52-54; mus comt, Albuquerque Chamber of Commerce, 50-56; consult, Health Comt, All Pueblo Coun, 53-; mem comt maternal & infant mortality, NMex Med Soc, 54-; actg exec dir, Comn Rights, Liberties & Responsibilities of Am Indian, 57-63; dir surv Indian Educ, Bur Indian Affairs, 63-; bd dir, Bernalillo County Planned Parenthood, 64-; dir, YWCA, 64; mem bd dirs, Planned Parenthood, World Pop, 69-75; consult, Inst Math Studies Soc Sci on Indian Educ, Stanford Univ, 71-; consult, All Indian Pueblo Coun comput assisted instruction prog & soc sci, Sch Law, Univ NMex, 70-; consult bilingual/bicultural proj, Bernalillo Sch Dist Title VII, 75- Mem: AAAS; AMA; fel Soc Res Child Develop; fel Am Anthrop Asn; Am Asn Anat. Res: Anthropology; human nutrition. Mailing Add: Rte 3 Box 3030 Albuquerque NM 87120

ABERNATHIE, JAN W, b Anna, Ill, Oct 21, 29; m 52; c 4. HORTICULTURE. Educ: Univ Ill, BS, 51, MS, 52, PhD, 59. Prof Exp: Asst mgr, Caribbean Gardens, Fla, 56-58; asst prof ornamental hort, Univ Ky, 59-67; ORNAMENTALS HORTICULTURIST, VA TRUCK & ORNAMENTALS RES STA, 67- Mem: Am Soc Hort Sci. Res: Floricultural physiology, especially growth substances and effect on flowering. Mailing Add: Va Truck & Ornamentals Res Sta PO Box 2160 Norfolk VA 23501

ABERNATHY, CHARLES OWEN, b Brunswick, Ga, Nov 18, 41; m 72. PHARMACOLOGY. Educ: Asbury Col, AB, 64; Univ Ky, MS, 67; NC State Univ, PhD(physiol), 70. Prof Exp: Res asst entom, Univ Ky, 64-66; USPHS trainee toxicol, NC State Univ, 67-70; res entomologist, Univ Calif, Berkeley, 70-73; PHARMACOLOGIST, LIVER RES UNIT, VET ADMIN HOSP, WASHINGTON, DC, 73- Res: Effects of drugs and other compounds; endotoxin on liver function with the goal of establishing in vitro models to study hepatotoxicity. Mailing Add: Liver Res Unit 151 W Vet Admin Hosp 50 Irving St NW Washington DC 20042

ABERNATHY, HENRY HERMAN, b Vidalia, Ga, Oct 7, 13; m 41; c 4. POLYMER CHEMISTRY. Educ: Emory Univ, AB, 38, MS, 39. Prof Exp: Anal chemist, Jackson Lab, 39-40, rubber chemist, Rubber Lab, 40-52, tech mgr, Rubber Chem Div, 54-57, mgr mkt develop, Elastomer Chem Dept, 57-64, dir & mkt adv, Du Pont Far East, Inc, Showa Neoprene KK, Tokyo, 64-65, MGR PLANNING & ENG, ELASTOMER CHEM DEPT, E I DU PONT DE NEMOURS & CO, INC, 65- Mem: Am Chem Soc. Res: Synthetic rubber and latices. Mailing Add: 802 Princeton Rd Wilmington DE 19807

ABERNATHY, JAMES RALPH, b Dadeville, Ala, Jan 8, 26; m 55; c 4. BIOSTATISTICS. Educ: Samford Univ, BS, 51; Uni MSPH, 53, PhD(biostatist), 65. Prof Exp: Statistician, Jeffer Bd Health, Ala, 52-59; biostatistician, NC State Bd Health, 59-61; res assoc, 64-65, from asst prof to assoc prof, 65-75, PROF BIOSTATIST, UNIV NC, CHAPEL HILL, 75- Mem: Am Pub Health Asn; Am Statist Asn; Pop Asn Am. Res: Statistical methodology in perinatal mortality and morbidity; demography and life tables; population control programs. Mailing Add: Dept of Biostatist Univ of NC Chapel Hill NC 27514

ABERNATHY, RICHARD PAUL, b McCaysville, Ga, Mar 22, 32; m 57; c 3. NUTRITION. Educ: Berry Col, BSA, 52; Univ Ga, MS, 57; Cornell Univ, PhD(animal nutrit), 60. Prof Exp: Asst prof & nutritionist, Ga Exp Sta, 60-66; from assoc prof to prof nutrit, Va Polytech Inst & State Univ, 66-74; HEAD DEPT FOODS & NUTRIT, PURDUE UNIV, 74- Mem: Am Inst Nutrit; Am Home Econ Asn; Soc Nutrit Educ. Res: Metabolic studies and nutrient requirements of preadolescent children. Mailing Add: Dept Foods & Nutrit HADM Purdue Univ West Lafayette IN 47907

ABERNATHY, ROBERT O, b Dallas, Tex, Apr 16, 27; m 64. MATHEMATICS. Educ: Prairie View Agr & Mech Col, BS, 48; Univ Calif, Berkeley, MA, 54, PhD(math), 62. Prof Exp: Asst math, Univ Calif, Berkeley, 53-56; asst prof, Southern Univ, 58-59; asst, Univ Calif, Berkeley, 60-62; prof, Tenn State Univ, 62-69; chmn dept, 69-74, PROF MATH, SC STATE COL, 69- Mem: Am Math Soc; Math Asn Am; Soc Indust & Appl Math. Res: Functional analysis and partial differential equations. Mailing Add: Dept of Math SC State Col Box 1675 Orangeburg SC 29117

ABERNATHY, ROBERT SHIELDS, b Gastonia, NC, Nov 18, 23; m 49; c 5. MEDICINE, MICROBIOLOGY. Educ: Duke Univ, BS & MD, 49; Univ Minn, PhD, 57. Prof Exp: From intern to resident med, Univ Minn, 49-54, instr med & microbiol, 55-57; from asst prof to assoc prof, 57-67, PROF MED & MICROBIOL & HEAD DEPT MED, UNIV ARK, LITTLE ROCK, 67- Mem: Am Fedn Clin Res; Am Col Physicians. Res: Infectious disease. Mailing Add: Dept of Med Univ of Ark Med Ctr Little Rock AR 72201

ABERNETHY, JOHN LEO, b San Jose, Calif, Mar 6, 15. BIO-ORGANIC CHEMISTRY. Educ: Univ Calif, Los Angeles, BA, 36; Northwestern Univ, MS, 38, PhD(org chem), 40. Prof Exp: Instr chem, Univ Tex, El Paso, 40-41; asst prof, Univ Tex, Austin, 41-45 & Washington & Lee Univ, 47-48; assoc prof, Bowling Green State Univ, 48-49 & Univ SC, 49-50; from asst prof to assoc prof, Calif State Col Syst, 51-60, res assoc, Univ Calif, Davis & Univ Calif, Los Angeles, 60-69; ASST PROF CHEM, CALIF STATE POLYTECH UNIV, 69- Concurrent Pos: NSF grant, 60-61; NIH fel, 60-61; assoc prof, Claremont Men's Col, 61-62; Fulbright fel, San Marcos Univ, Lima, 62-63; Sigma Xi grants, 67-70, 71-73; Res Corp grant, 69-70. Mem: Am Chem Soc. Res: Papain-catalyzed reactions involving resolutions of racemic mixtures, partial asymmetric syntheses and other organic syntheses involving amino acids and amino-containing bases, as well as other papain-catalyzed reactions. Mailing Add: Dept of Chem Calif State Polytech Univ Pomona CA 91768

ABERNETHY, ROBERT BRUCE, engineering, statistics, see 12th edition

ABERNETHY, VIRGINIA DEANE, b Havana, Cuba, Oct 4, 34; US citizen; m 55, 65; c 4. MEDICAL ANTHROPOLOGY. Educ: Wellesley Col, BA, 55; Harvard Univ, MA, 68, PhD(anthrop), 70. Prof Exp: Fel social psychiat, Med Sch, Harvard Univ, 70-72, res assoc anthrop, Dept Psychiat, 72, assoc, 72-75; ASST PROF PSYCHIAT & DIR DIV HUMAN BEHAV, DEPT PSYCHIAT, SCH MED, VANDERBILT UNIV, 75- Mem: Am Anthrop Asn; AAAS; Pop Asn Am; Soc Med Anthrop. Res: Psychologic factors in unwanted pregnancy, especially among adolescents; socioeconomic determinants of birth rates and mortality; biological bases of human behavior. Mailing Add: Dept of Psychiat Sch of Med Vanderbilt Univ Nashville TN 37232

ABERS, ERNEST S, b San Mateo, Calif, Dec 31, 36; m 60; c 1. PHYSICS. Educ: Harvard Univ, AB, 58; Univ Calif, Berkeley, PhD(physics), 63. Prof Exp: Res fel physics, Calif Inst Technol, 63-64; vis scientist, Europ Orgn Nuclear Res, Switz, 64-65; from asst prof to assoc prof, 65-74, PROF PHYSICS, UNIV CALIF, LOS ANGELES, 74- Concurrent Pos: Visitor, Ctr Theoret Physics, Mass Inst Technol, 68-69; Alfred P Sloan Found fel, 68-70. Mem: Am Phys Soc. Res: Theoretical elementary particle physics. Mailing Add: Dept of Physics Univ of Calif Los Angeles CA 90024

ABERTH, OLIVER GEORGE, b Akron, Ohio, July 23, 29; m 58; c 3. MATHEMATICS. Educ: City Col New York, BS, 50; Mass Inst Technol, MS, 51; Univ Pa, PhD(math), 62. Prof Exp: Engr, Remington Rand Univac Div, Sperry Rand Corp, 51-53 & 56-58; instr math, Swarthmore Col, 60-62; asst prof, Univ Ill, 62-66; assoc prof, Rutgers Univ, 66-69; lectr, City Col New York, 69-70; PROF MATH, TEX A&M UNIV, 70- Mem: AAAS; Am Math Soc; Math Asn Am; Soc Indust & Appl Math. Res: Constructive analysis; geometry; tensor analysis. Mailing Add: Dept of Math Tex A&M Univ College Station TX 77843

ABERTH, WILLIAM H, b Los Angeles, Calif, Jan 4, 33; m 54; c 3. MASS SPECTROMETRY. Educ: City Col New York, BS, 54; Columbia Univ, MA, 57; NY Univ, PhD(physics), 63. Prof Exp: Lectr physics, City Col New York, 57-60; res fel, Stanford Res Inst, 63-65, physicist, 65-69; assoc prof physics, Sonoma State Col, 69-70; sr physicist, Stanford Res Inst, 70-72, asst mgr, Mass Spectrom Res Ctr, 72-75; res assoc, Space Sci Lab, Univ Calif, Berkeley, 75-76; JR FEL, LINUS PAULING INST SCI & MED, 76- Mem: Am Phys Soc; Am Soc Mass Spectrom. Res: Electron scattering; atomic beams; universal detectors; ionic and atomic collisions; ultra high vacuum techniques; mass spectrometer development. Mailing Add: Linus Pauling Inst Sci & Med 2700 Sand Hill Rd Menlo Park CA 94025

ABEY, ALBERT EDWARD, b Spokane, Wash, Aug 12, 35; m 57; c 2. SOLID MECHANICS. Educ: Wash State Univ, BS, 58; Univ Wash, MS, 60; Univ Ariz, PhD(physics), 64. Prof Exp: Res assoc physics, Univ Ariz, 64; res assoc, Advan Mat Res & Develop Lab, Pratt & Whitney Aircraft, 64-67; PHYSICIST, LAWRENCE RADIATION LAB, 67- Mem: Am Phys Soc. Res: High hydrostatic and nearly hydrostatic pressure studies of solid state physics, particularly ionic conductivity; elastic and plastic properties of materials; surface physics in connection with thermionics. Mailing Add: Lawrence Radiation Lab Box 808 Livermore CA 94550

ABHYANKAR, SHREERAM, b Ujjain, India, July 22, 30; m 58; c 1. MATHEMATICS. Educ: Inst Sci, India, BSc, 51; Harvard Univ, AM, 52, PhD(math), 54. Prof Exp: Res fel math, Harvard Univ, 54-55; res assoc, Columbia Univ, 55-56, vis asst prof, 56-57; asst prof, Cornell Univ, 57-58; vis asst prof, Princeton Univ, 58-59; assoc prof, Johns Hopkins Univ, 59-63; prof, 63-67, DISTINGUISHED PROF MATH, PURDUE UNIV, WEST LAFAYETTE, 67- Concurrent Pos: Sloan Found fel, 58- Mem: Am Math Soc. Res: Algebraic geometry; algebra; several complex variables; circuit theory. Mailing Add: Div of Math Sci Purdue Univ West Lafayette IN 47906

ABIAN, ALEXANDER, b Tabriz, Iran, Jan 1, 25; US citizen; m 59; c 3. PURE MATHEMATICS. Educ: Univ Tehran, BS, 46; Univ Chicago, MS, 54; Univ Cincinnati, PhD(math), 56. Prof Exp: Asst prof math, Univ Tenn, 56-57, Queens Col, 57-59 & Univ Pa, 59-62; assoc prof, Ohio State Univ, 62-67; PROF MATH, IOWA STATE UNIV, 67- Mem: Am Math Soc. Res: Theory of sets, analysis and mathematical logic. Mailing Add: Dept of Math Iowa State Univ Ames IA 50010

ABIKOFF, WILLIAM, b New York, NY, Aug 18, 44. MATHEMATICS. Educ: Polytech Inst Brooklyn, BS, 65, MS, 66, PhD(amth), 71. Prof Exp: Mem tech staff, Bell Tel Labs, Inc, 65-70; instr math, Columbia Univ, 70-71; res fel, Mittag-Leffler Inst, Swed Royal Acad Sci, 71-72; asst prof, Columbia Univ, 72-75; ASST PROF MATH, UNIV ILL, 75- Mem: Am Math Soc. Res: Complex analysis; hyperbolic geometry; Riemann surfaces. Mailing Add: Dept of Math Univ of Ill Urbana IL 61801

ABILDGAARD, CHARLES FREDERICK, b Winfield, Kans, Aug 10, 30; m; c 3. PEDIATRICS. Educ: Stanford Univ, AB, 52, MD, 55; Am Bd Pediat, dipl, 59. Prof Exp: Intern, Boston City Hosp, 54-55; resident pediat, Stanford Univ Hosps, 55-57; pediatrician, Coco Solo Hosp, Cristobal, CZ, 59-60; from asst prof to assoc prof pediat, Col Med, Univ Ill, 61-68; PROF PEDIAT, SCH MED, UNIV CALIF, DAVIS, 68- Concurrent Pos: USPHS trainee hemat, Children's Mem Hosp, Chicago, 60-61. Mem: Am Soc Hemat; Am Pediat Soc; Soc Pediat Res; Am Acad Pediat; Am Physiol Soc. Res: Pediatric hematology, thrombopoiesis, hemostasis, hemophilia. Mailing Add: Dept of Pediat Univ of Calif Sch of Med Davis CA 95616

ABILDSKOV, JUNIOR A, b Salen, Utah, Sept 22, 23; m 44; c 4. MEDICINE. Educ: Univ Utah, BA, 44, MD, 46. Prof Exp: Instr internal med, Tulane Univ, 51-54; cardiologist & chief serv, William Beaumont Army Hosp, 54-56; from asst prof to prof med, State Univ NY Upstate Med Ctr, 56-68; PROF MED, UNIV UTAH, 68- Concurrent Pos: Fel internal med, Tulane Univ, 47-51. Mem: AAAS; Am Soc Clin Invest; Am Fedn Clin Res; Soc Exp Biol & Med; Asn Am Physicians. Res: Clinical cardiology; cardiac physiology. Mailing Add: Univ of Utah Med Ctr 50 N Medical Dr Salt Lake City UT 84132

ABINANTI, FRANCIS RALPH, virology, see 12th edition

AB IORWERTH, HEFIN, b Denbigh, N Wales, June 7, 43; m 66. GEOPHYSICS. Educ: Univ Col Wales, BSc, 65; Univ Birmingham, MSc, 66, PhD(rock magnetism), 69. Prof Exp: GEOPHYSICIST SEISMIC EXPLOR, AMOCO CAN PETROL CO, LTD, 69- Mem: Am Geophys Union; Soc Explor Geophys. Res: Anisotropy of magnetic susceptibility and of saturation magnetization of rocks and its application to geological problems; digital analysis of seismic data. Mailing Add: Amoco Can Petrol Co Ltd 444 Seventh Ave SW Calgary AB Can

ABKOWITZ, MARTIN ARNOLD, b New York, NY, Feb 22, 36; m 57; c 2. EXPERIMENTAL SOLID STATE PHYSICS. Educ: City Col New York, BS, 57; Univ Rochester, MA, 59; Syracuse Univ, PhD(physics), 64. Prof Exp: Fel physics, Univ Pittsburgh, 64-65; scientist, 65-73, SR SCIENTIST, XEROX WEBSTER CORP RES LAB, 73-; ASSOC LECTR PHYSICS, UNIV ROCHESTER, 69- Mem: Am Phys Soc. Res: Magnetic resonance; dielectric spectroscopy; charge generation and transport in disordered molecular solids; polymer physics. Mailing Add: Xerox Webster Corp Res Lab 800 Phillips Rd Webster NY 14580

ABLARD, JAMES ELBERT, b Oshkosh, Wis, Dec 7, 10; m 42; c 3. CHEMISTRY. Educ: Univ Wis, BSc, 33; Carnegie Inst Technol, MS, 34, DS(phys chem), 37. Prof Exp: Instr chem, Carnegie Inst Technol, 33-38; phys chemist, B F Drankenfield Co,

Inc, Pa, 38-42; group leader, Nat Defense Res Comt, 42-45; dir res, Synthetic Rubber Polymerization, Gen Tire & Rubber Co, 45-49; consult, Explosives Div, Naval Ord Lab, Md, 49-53, dep chief explosives res dept, 53-55, chief, 55-57, prog chief, Chem & Explosives Area, 57-73; PRES, ABLARD ENTERPRISES, INC, 73- Mem: AAAS; Am Chem Soc. Res: Dielectric constant and conductance; resistant ceramic colors for glass; rubber latex compounding; synthetic rubber latex preparation; conductivity of triethyl amine solutions; explosives; propellants; high pressure-high temperature thermodynamics and hydrodynamics; electrochemistry; plastics; electron diffraction of vapors. Mailing Add: 14851 Amherst Ave Silver Spring MD 20902

ABLE, KENNETH PAUL, b Louisville, Ky, Feb 5, 44; m 67; c 1. BEHAVIORAL & COMMUNITY ECOLOGY. Educ: Univ Louisville, BS, 66, MS, 68; Univ Ga, PhD(zool), 71. Prof Exp: ASST PROF BIOL, STATE UNIV NY ALBANY, 71- Concurrent Pos: NSF res grant, 74. Mem: Animal Behav Soc; Am Soc Zoologists; Ecol Soc Am; Am Ornithologists Union; AAAS. Res: Migration and orientation of birds; structure of forest bird communities. Mailing Add: Dept of Biol Sci State Univ of NY Albany NY 12222

ABLER, RONALD FRANCIS, b Milwaukee, Wis, May 30, 39; m 60; c 2. GEOGRAPHY. Educ: Univ Minn, BA, 63, MA, 66, PhD(geog), 68. Prof Exp: Asst prof geog, 67-71, ASSOC PROF GEOG, PA STATE UNIV, UNIVERSITY PARK, 71- Concurrent Pos: NSF instnl grant, Pa State Univ, 70-71; vis prof geog, Univ BC, 71; vis prof, Univ Minn, Minneapolis, 72-74; assoc dir, Urban Goals Res Proj, Asn Am Geogrs, 72-74. Mem: AAAS; Asn Am Geogrs; Am Geog Soc; World Future Soc; Int Commun Asn. Res: Origin, evolution and diffusion of communications media; effects of communications media on human spatial behavior, with special attention to settlement patterns and urbanism. Mailing Add: 403 Deike Bldg Pa State Univ University Park PA 16802

ABLER, THOMAS STRUTHERS, b Saginaw, Mich, Aug 31, 41. ETHNOLOGY, SOCIAL ANTHROPOLOGY. Educ: Northwestern Univ, BA, 63; Univ Wis-Milwaukee, MS, 65; Univ Toronto, PhD(anthrop), 69. Prof Exp: Asst prof, 67-74, ASSOC PROF ANTHROP, UNIV WATERLOO, 74- Mem: Am Anthrop Asn; Royal Anthrop Inst Gt Brit & Ireland; Am Soc Ethnohist; Can Ethnol Soc. Res: North American Indians; Iroquois; social and political organization; ethnohistory; warfare. Mailing Add: Dept of Anthrop Univ of Waterloo Waterloo ON Can

ABLES, ERNEST D, b Hugo, Okla, Jan 13, 34; m 60. WILDLIFE ECOLOGY, ZOOLOGY. Educ: Okla State Univ, BS, 61; Univ Wis, MS, 64, PhD(wildlife ecol), 68. Prof Exp: From asst prof to assoc prof ecol, Tex A&M Univ, 67-73; PROF ECOL, UNIV IDAHO, 73-, ASSOC ACAD DEAN, COL FORESTRY, WILDLIFE & RANGE SCI, 74- Mem: Ecol Soc Am; Wildlife Soc; Am Soc Mammal; Am Inst Biol Sci; Wilderness Soc. Res: Home range and activity studies of red foxes by radio tracking; radio tracking studies of impala in Africa; ecology of exotic ungulates in Texas; infrared scanning for censusing. Mailing Add: Col Forestry Wildlife & Range Sci Univ of Idaho Moscow ID 83843

ABLES, HAROLD DWAYNE, b Hico, Tex, Feb 18, 38; m 59. ASTRONOMY. Educ: Univ Tex, Austin, BA, 61, PhD(astron), 68. Prof Exp: ASTRONR, NAVAL OBSERV, FLAGSTAFF STA, 64- Mem: Am Astron Soc; Int Astron Union. Res: Observational studies of the luminosity distributions in Magellanic type irregular galaxies and the low luminosity halos and the globular clusters around early type galaxies. Mailing Add: Naval Observ Flagstaff Sta PO Box 1149 Flagstaff AZ 86001

ABLON, JOAN, b Dallas, Tex, Oct 21, 34. ANTHROPOLOGY, MEDICAL ANTHROPOLOGY. Educ: Univ Tex, BA, 55; Univ Chicago, MA, 58, PhD(anthrop), 63. Prof Exp: Res anthropologist, Sch Criminol, Univ Calif, Berkeley, 63-66; res anthropologist, NMex State Div Ment Health, 66-67; asst prof, 68-74, RESIDENT ASSOC PROF MED ANTHROP, UNIV CALIF, SAN FRANCISCO, 74- Mem: Fel Am Anthrop Asn; fel Soc Appl Anthrop; Soc Med Anthrop. Res: Cultural values and mental health; family structure in alcoholism; urban anthropology; urban ethnic and racial minority groups; action and applied anthropology; middle class life style. Mailing Add: Dept of Psychiat Univ of Calif San Francisco CA 94122

ABLOW, CLARENCE MAURICE, b New York, NY, Nov 6, 19; m 45. MATHEMATICS. Educ: Univ Calif, Los Angeles, BA, 40, MA, 42; Brown Univ, PhD(appl math), 51. Prof Exp: Mathematician, Boeing Airplane Co, 51-55; MATHEMATICIAN, STANFORD RES INST, 55- Mem: Am Math Soc; Soc Indust & Appl Math; Math Asn Am. Res: Solution of differential equations. Mailing Add: Div of Eng Res Stanford Res Inst Menlo Park CA 94025

ABNEY, THOMAS SCOTT, b Galatia, Ill, July 25, 38; m 62; c 3. PLANT PATHOLOGY, MYCOLOGY. Educ: Southern Ill Univ, BS, 60, MS, 64; Iowa State Univ, PhD(plant path), 67. Prof Exp: Res assoc corn dis, Iowa State Univ, 64-66; RES PATHOLOGIST, CROPS RES DIV, AGR RES SERV, USDA, 67-; ASST PROF BOT & PLANT PATH, PURDUE UNIV, WEST LAFAYETTE, 67- Mem: Am Phytopath Soc. Res: Diseases of plants, particularly organisms that attack Midwest field crops; nature of resistance; physiology of parasitism. Mailing Add: Dept of Bot & Plant Path Purdue Univ West Lafayette IN 47906

ABODEELY, ROBERT ASSAD, b Worcester, Mass, Feb 18, 42; c 2. MICROBIOLOGY. Educ: Clark Univ, BA, 64; Univ Conn, MS, 66; Univ Miss, PhD(microbiol), 70; Rutgers Univ, MBA, 75. Prof Exp: Fel, St Jude Children's Hosp, 70-71; sr scientist microbiol, 71-75, MGR REGULATORY AFFAIRS, ETHICON, INC DIV, JOHNSON & JOHNSON, 75- Mem: Am Soc Microbiol; Tissue Cult Asn; Soc Indust Microbiol. Res: In vitro biologic screening of materials intended for human implant; sterilization; biological indicators. Mailing Add: Ethicon Inc Rte 22 Somerville NJ 08876

ABOLINS, MARIS ARVIDS, b Liepaja, Latvia, Feb 5, 38; US citizen; m 59; c 2. HIGH ENERGY PHYSICS. Educ: Univ Wash, BS, 60; Univ Calif, San Diego, MS, 62, PhD(physics), 65. Prof Exp: Res asst physics, Univ Calif, San Diego, 60-65; physicist, Lawrence Radiation Lab, Univ Calif, Berkeley, 65-68; assoc prof, 68-72, PROF PHYSICS, MICH STATE UNIV, 72- Mem: Am Phys Soc. Res: High energy physics research with counters and spark chambers; high energy instrumentation. Mailing Add: Dept of Physics Mich State Univ East Lansing MI 48823

ABOOD, LEO GEORGE, b Erie, Pa, Jan 15, 22; m 47; c 2. BIOCHEMISTRY. Educ: Ohio State Univ, BS, 43; Univ Chicago, PhD(pharmacol), 50. Prof Exp: Instr physiol, Univ Chicago, 50-52; from asst prof to prof neurophysiol & biochem, Col Med, Univ Ill, 52-65, dir res labs, Dept Psychiat, 56-65; PROF BIOCHEM, UNIV ROCHESTER, 65-, PROF BRAIN RES, 67- Concurrent Pos: Mem neurol sci res comt, Nat Inst Neurol Dis & Stroke. Mem: Am Physiol Soc; Am Chem Soc; Am Col Neuropsychopharmacol; Soc Neurosci; Int Soc Neurochem. Res: Chemistry and cellular physiology of nervous system. Mailing Add: Ctr for Brain Res Univ of Rochester Med Ctr Rochester NY 14642

ABOU-EL-SEOUD, MOHAMED OSMAN, b Cairo, Egypt, Apr 6, 21; m 49; c 4. MYCOLOGY. Educ: Cairo Univ, BSc, 43; Ain Shams Univ, Cairo, Dipl Educ, 59; Ohio State Univ, MSc, 62, PhD(mycol), 64. Prof Exp: Sec sch teacher, Egypt, 45-51; tech secy, Sci Div, Egyptian Ministry Educ, 51-54; asst prof bot, Teachers Col, Cairo, 54-60; res assoc, Southern Ill Univ, Carbondale, 64-65; prof, Ain Shams Univ, Cairo, 65-67; lectr, 68-69, asst prof, 69-71, ASSOC PROF NATURAL SCI, MICH STATE UNIV, 71- Mem: Am Inst Biol Sci; Soc Indust Microbiol. Res: Lipid and protein synthesis by fungi; light effect on sporulation and pigmentation of fungi. Mailing Add: Dept of Natural Sci Mich State Univ East Lansing MI 48824

ABOUL-ELA, MOHAMED MOHAMED, b Damanhur, Egypt, Mar 10, 18; US citizen; m 51; c 6. PLANT PHYSIOLOGY. Educ: Univ Cairo, BA, 40; Iowa State Univ, MS, 47, PhD(plant physiol), 50. Prof Exp: Teacher high sch, Damanhur, Egypt, 40-42; from instr to asst prof crops, Univ Alexandria, 42-55; assoc prof, Univ Bagdad, 55-56; assoc prof, Univ Alexandria, 56-62; fel plant sci, Tex A&M Univ, 62-65; asst prof, 65-67, ASSOC PROF BIOL, TEX WOMAN'S UNIV, 67- Mem: Consult, Agrarian Reform Ministry Egypt, 56-62. Mem: AAAS; Am Inst Biol Sci. Res: Crop production; cotton defoliation; harvest aids; plants recovery from radiation injury. Mailing Add: Dept of Biol Tex Woman's Univ Denton TX 76204

ABOUL-ENEIN, HASSAN YOUSSEFF, b Cairo, Egypt, Jan 12, 43; m 67; c 1. MEDICINAL CHEMISTRY, PHARMACOLOGY. Educ: Cairo Univ, BSc, 64; Univ Miss, MSc, 69, PhD(med chem), 71. Prof Exp: Instr org chem, Fac Pharm, Cairo Univ, 64-66; res asst med chem, Sch Pharm, Univ Miss, 66-71; NIH res fel pharmacol & drug metab, Med Ctr, Univ Ala, Birmingham, 71-73; USPHS res investr med chem & drug metab, Dept Pharmacol, Col Med, Univ Iowa, 73-75; SR RES CHEMIST, USV PHARMACEUT CORP, 75- Concurrent Pos: Contrib ed, Int Pharmaceut Asbtr, 72-; Sigma Xi grant-in-aid, 73. Honors & Awards: Mead Johnson Award, Mead Johnson & Co, 71. Mem: Am Pharmaceut Asn; Sigma Xi; NY Acad Sci. Res: Synthesis of biologically active compounds, their spectroscopic analysis, structure-activity-relationships studies and their correlation ot the pharmacological properties and drug-receptor interactions; drug metabolism. Mailing Add: USV Pharmaceut Corp Tuckahoe NY 10707

ABOU-SABE, MORAD A, b Cairo, Egypt, Apr 3, 37; m 60; c 2. MOLECULAR BIOLOGY. Educ: Univ Alexandria, BSc, 58; Univ Calif, Berkeley, MSc, 62; Univ Pittsburgh, PhD(bact), 65. Prof Exp: Instr agr, Univ Alexandria, 58-59; asst prof genetics, Cairo Univ, 66-68; asst prof, 68-74, ASSOC PROF BACT, RUTGERS UNIV, 74- Mem: Am Soc Microbiol; Genetics Soc Am; UAR Soc Microbiol. Res: Gene enzyme interactions; specific genetic control mechanism in bacteria; cyclic adenosine monophosphate regulation of cell growth and growth control. Mailing Add: Dept of Microbiol Rutgers Univ New Brunswick NJ 08903

ABPLANALP, HANS, b Zurich, Switz, Apr 22, 25; m 54; c 3. GENETICS. Educ: Inst Technol, Zurich, Ing Agron, 49; Wash State Col, MS, 51; Univ Calif, PhD(genetics), 55. Prof Exp: Res asst genetics, Wash State Col, 49-51 & Univ Calif, 51-53; asst specialist statist & genetics, Div Poultry Husb, 53-56, asst prof genetics, 56-69, PROF AVIAN SCI, UNIV & POULTRY GENETICIST, EXP STA, UNIV CALIF, DAVIS, 69- Mem: Genetics Soc Am; Am Genetic Asn; Soc Study Evolution. Res: Population genetics; application to breeding; basic research in underlying genetic mechanisms. Mailing Add: Div of Avian Sci Univ of Calif Davis CA 95616

ABPLANALP, PAUL LEROY, b Ridgway, Pa, Oct 6, 36; m 62; c 1. NEUROANATOMY. Educ: Pa State Univ, BS, 63; Mass Inst Technol, PhD(physiol psychol), 68. Prof Exp: ASST PROF BIOL, PA STATE UNIV, 68- Concurrent Pos: NIH res grant neuroanat, 69- Res: Comparative neuroanatomy of the visual system network; interconnections of the nucleus lateralis posterior and the cerebral cortex; visually guided behavior. Mailing Add: 321 Life Sci I Pa State Univ University Park PA 16802

ABRAHAM, BERNARD M, b Kansas City, Mo, Nov 21, 18; m 42; c 3. CHEMISTRY. Educ: Univ Chicago, BS, 40, PhD(chem), 46. Prof Exp: Jr assoc, Univ Chicago, 41-42 & Northwestern Univ, 42; jr assoc & jr chemist, Univ Chicago, 42-45; sr chemist, Monsanto Chem Co, 45; Universal Oil Prod fel, Northwestern Univ, 46-47; assoc chemist, 47-59, SR CHEMIST & GROUP LEADER, ARGONNE NAT LAB, 59- Concurrent Pos: Guggenheim Mem fel, 59. Mem: AAAS; Am Chem Soc; fel Am Phys Soc. Res: Thermodynamic and transport properties of liquid helium-3 and liquid helium-4; thermal and magnetic measurements as very low temperatures; suspension of inorganic compounds in liquid metals. Mailing Add: Argonne Nat Lab Bldg 200 9700 S Cass Ave Argonne IL 60439

ABRAHAM, DONALD JAMES, b Greensburg, Pa, Nov 19, 36; m 62; c 1. ORGANIC CHEMISTRY, MEDICINAL CHEMISTRY. Educ: Pa State Univ, BS, 58; Marshall Univ, MS, 59; Purdue Univ, PhD(org chem), 63. Prof Exp: Res assoc med chem & fel, Univ Va, 63-64; res asst prof, 64-67, assoc prof, 68-72, PROF MED CHEM, UNIV PITTSBURGH, 72-, HEAD DEPT, 68- Concurrent Pos: Mem preclin psychopharmacol study sect, NIMH, Alcohol & Drug Abuse Ment Health Admin, 73-77. Mem: Am Chem Soc. Res: Structure elucidation of biologically important compounds; organic synthesis; synthesis of medicinals; x-ray crystallography; molecular action of drug molecules; sickle cell and psychopharmacology research. Mailing Add: Sch of Pharm Univ of Pittsburgh Pittsburgh PA 15216

ABRAHAM, EDATHARA CHACKO, b Kottayam, India, m 63; c 3. BIOCHEMISTRY. Educ: Univ Kerala, BS, 59; Univ Louisville, PhD(biochem), 71. Prof Exp: Clin biochemist, Christian Med Col, Vellore, India, 59-62, res biochemist, 63-67; res biochemist renal biochem, Univ Louisville Sch Med, 67-69, grad res asst biochem, 69-71; ASST PROF BIOCHEM, MED COL GA, 74- Concurrent Pos: Res assoc, Univ Louisville Sch Med, 72-73 & Med Col Ga, 73-74; biochemist, Vet Admin Hosp, Augusta, Ga, 75- Mem: Am Soc Hematol; Am Asn Clin Chem; Am Soc Human Genetics; AAAS. Res: Physico-chemical and functional properties and the structure-function relationships of hemoglobins; Red Cell Metabolism; abnormal hemoglobins and hemoglobinopathies; gene activation in Eukaryotes; developing macro and micro chromatographic procedures. Mailing Add: Protein Chem Lab Sickle Ctr Med Col Ga Augusta GA 30902

ABRAHAM, FARID FADLOW, b Phoenix, Ariz, May 5, 37; m 59; c 4. PHYSICS. Educ: Univ Ariz, BS, 59, PhD(physics), 62. Prof Exp: Res assoc high energy physics, Univ Chicago, 62-64; sr physicist, Lawrence Radiation Lab, Univ Calif, 64-66; staff mem, IBM Sci Ctr, 66-71, proj leader air qual dynamics group, 71-72, PROJ LEADER, IBM RES LAB, 72- Mem: Am Phys Soc. Res: Nucleation theory; phase stability, thermodynamics of nonuniform systems, statistical physics; microphysics of clouds; air pollution diffusion models. Mailing Add: IBM Res Lab San Jose CA 95193

ABRAHAM, MARVIN MEYER, b New York, NY, Dec 8, 30; m 58; c 3. PHYSICS. Educ: City Col New York, BS, 53; Univ Calif, MA, 54, PhD(physics), 58. Prof Exp: Asst physics, Univ Calif, 54-58; Fulbright res fel, Clarendon Lab, Oxford Univ, 58-60; res physicist, Lawrence Radiation Lab, Univ Calif, Berkeley, 60-63; RES PHYSICIST, SOLID STATE DIV, OAK RIDGE NAT LAB, 63- Concurrent Pos: Pressed steel fel, Oxford Univ, 59-60; Int Atomic Energy Auth vis prof, San Carlos de Bariloche Inst

Physics, Arg, 62-63. Mem: Fel Am Phys Soc; Am Ceramic Soc. Res: Experimental solid state physics; electron paramagnetic resonance; nuclear magnetic resonance. Mailing Add: Solid State Div Oak Ridge Nat Lab Oak Ridge TN 37830

ABRAHAM, RALPH HERMAN, b Burlington, Vt, July 4, 36; m 58; c 2. MATHEMATICS. Educ: Univ Mich, BSE, 56, MS, 58, PhD(math), 60. Prof Exp: Lectr math, Univ Calif, Berkeley, 60-62; Off Naval Res res assoc, Columbia Univ, 62-63, asst prof, 63-64; asst prof, Princeton Univ, 64-68; ASSOC PROF NATURAL SCI I, UNIV CALIF, SANTA CRUZ, 68- Mem: Am Math Soc. Res: General relativity, qualitative theory of ordinary differential equations and mechanical systems; analysis on manifolds. Mailing Add: Natural Sci I Univ of Calif Santa Cruz CA 95064

ABRAHAM, SAMUEL, b New York, NY, Sept 23, 23; m 52; c 2. PHYSIOLOGY, BIOCHEMISTRY. Educ: City Col New York, BS, 48; Univ Calif, PhD(comp biochem), 53. Prof Exp: Asst chem, Syracuse Univ, 48; res asst physiol, 48-53, from asst res physiologist to res physiologist, 53-66, lectr, 66-73, ADJ PROF NUTRIT SCI, UNIV CALIF, BERKELEY, 73-; DIR RES, BRUCE LYON MEM RES LAB, CHILDREN'S HOSP MED CTR NORTHERN CALIF, 66- Concurrent Pos: Grants, Nat Cancer Inst, NASA, NSF, Nat Cystic Fibrosis Res Found, Life Ins Med Res Fund & Nat & Calif Div, Am Cancer Soc; lectr physiol chem, Univ Calif, Davis, 67-68; consult, Vet Admin Hosp, Martinez, Calif, Highland Gen Hosp, Oakland & NASA, Moffett Field. Mem: AAAS; Am Inst Nutrit; Soc Exp Biol & Med; Am Soc Biol Chemists; Am Asn Cancer Res. Res: Biochemical studies of metabolic pathways; carcinogenesis; enzymology; lipid and carbohydrate chemistry and metabolism in normal and diseased states. Mailing Add: Bruce Lyon Mem Res Lab Children's Hosp Med Ctr Oakland CA 94609

ABRAHAMS, ELIHU, b Port Henry, NY, Apr 3, 27; m 53; c 2. PHYSICS. Educ: Univ Calif, PhD(physics), 52. Prof Exp: Res assoc physics, Univ Calif, 52-53; res assoc, Univ Ill, 53-55, asst prof, 55-56; from asst prof to assoc prof, 56-64, PROF PHYSICS, RUTGERS UNIV, NEW BRUNSWICK, 64- Mem: Am Phys Soc. Res: Theoretical solid state physics; many body problem; metals. Mailing Add: Dept of Physics Rutgers Univ New Brunswick NJ 08903

ABRAHAMS, IRVING, b New York, NY, July 10, 13; m 40. MICROBIOLOGY, IMMUNOLOGY. Educ: City Col New York, BS, 34; Cornell Univ, PhD(microbiol), 52. Prof Exp: Lab technician, Div Lab & Res, NY State Dept Health, 38-43; head dept bact & viral testing, Lederle Labs, 46-48; instr bact & immunol, Med Col, Cornell Univ, 50-51, lectr microbiol & immunol, 52-57, asst prof, 57-65, clin asst prof microbiol, 65-67; res microbiologist, Nassau County Hosp Pulmonary Dis, Plainview, NY, 65-69, dir labs, 65-67; dir div microbiol, Meadowbrook Hosp, East Meadow, 67-69; dir labs & res, Plainview Div, Nassau County Med Ctr, 69-72; DIR LABS & RES, NASSAU COUNTY DEPT HEALTH, 72- Concurrent Pos: Clin assoc prof microbiol, Sch Allied Health Sci, State Univ NY Stony Brook, 72- Mem: AAAS; Am Asn Immunol; Am Soc Microbiol; NY Acad Sci; Am Pub Health Asn. Res: Immunology and pathogenesis of systemic mycosis; concomittant lymphoma infections; medical microbiology; tetanus toxoid responses; effects of antimicrobials in actinomycosis and tuberculosis; enteropathogenic Escherichia coli. Mailing Add: Div of Labs & Res Nassau County Dept of Health Hempstead NY 11550

ABRAHAMS, ROBERT ALAN, physical chemistry, atomic physics, see 12th edition

ABRAHAMS, SIDNEY CYRIL, b London, Eng, May 28, 24; US citizen; m 50; c 3. CRYSTALLOGRAPHY. Educ: Glasgow Univ, BSc, 46, PhD, 49, DSc, 57. Prof Exp: Mem staff, Div Indust Coop, Lab Insulation Res, Mass Inst Technol, 50-54; MEM TECH STAFF, BELL LABS, 57- Concurrent Pos: Guest scientist, Brookhaven Nat Lab, 57-; mem comn crystallog apparatus, Int Union Crystallog, 63-72, chmn, 72-75; mem, Nat Comt Crystallog, 66-, chmn, 70-72; mem comt chem crystallog, Div Chem & Chem Technol, Nat Res Coun, 67-, chmn, 68-76. Mem: Am Crystallog Asn (pres, 68); Am Phys Soc; Brit Chem Soc. Res: Crystal chemistry and solid state physics; ferroelectric, ferroelastic and magnetic structure in relation to properties; crystallographic instrumentation and automation. Mailing Add: Bell Labs Murray Hill NJ 07974

ABRAHAMS, VIVIAN CECIL, b London, Eng, Oct 19, 27; m 55; c 4. NEUROPHYSIOLOGY. Educ: Univ Edinburgh, BSc, 52, PhD(physiol), 55. Prof Exp: Asst lectr physiol, Univ Edinburgh, 52-54, lectr, 54-55; vis instr, Univ Pa, 55-56; Beit fel med res, Nat Inst Med Res, Univ London, 56-59, mem sci staff, Med Res Coun, 59-63; assoc prof, 63-67, PROF PHYSIOL, QUEEN'S UNIV (ONT), 67- Concurrent Pos: Med Res Coun vis scientist, Cerebral Functions Group, Dept Anat, Univ Col, Univ London, 70-71. Mem: Am Physiol Soc; Can Physiol Soc; Brit Physiol Soc; Brit Pharmacol Soc; Soc Neurosci. Res: Organization of sensory and effector systems in the brain stem. Mailing Add: Dept of Physiol Queens Univ Kingston ON Can

ABRAHAMSON, ADOLF AVRAHAM, b Berlin, Ger, Mar 14, 20; US citizen; m 51; c 1. THEORETICAL PHYSICS. Educ: NY Univ, BA, 50, MS, 54, PhD(physics), 60. Prof Exp: Tutor physics, 57-60, from instr to assoc prof, 60-74, PROF PHYSICS, CITY COL NEW YORK, 74- Concurrent Pos: Consult, Brookhaven Nat Lab, 60-; NASA res grant, 61-; nat scientist personnel registr, Am Inst Physics, 63; Israel Atomic Energy Comn fel, 64-65. Mem: Am Phys Soc; Am Asn Physics Teachers; NY Acad Sci. Res: Interatomic potentials at very small and intermediate separations; applications to radiation damage in solids; gas scattering; astrophysics; statistical model of the atom. Mailing Add: Dept of Physics City Col of New York New York NY 10031

ABRAHAMSON, DEAN EDWIN, b Hasty, Minn, Dec 21, 34; m 54; c 2. FORENSIC SCIENCE, ENVIRONMENTAL SCIENCES. Educ: Gustavus Adolphus Col, BS, 55; Univ Nebr, MA, 58; Univ Minn, Minneapolis, MD & PhD(anat), 67. Prof Exp: First asst physics, Royal Inst Technol, Sweden, 58; reactor physicist, Babcock & Wilcox Co, Va, 58-59; sr scientist, Honeywell Inc, Minn, 59-63; asst prof anat & lab med, 67-70, assoc prof anat, lab med & physics & dir ctr studies phys environ, 70-72, PROF PUB AFFAIRS & CHMN ALL-UNIV COUN ENVIRON QUAL, UNIV MINN, 72- Concurrent Pos: Mem adv bd, Energy Pol Proj, Ford Found, 72-74; mem bd dirs, Natural Resources Defense Coun, 72- Mem: Sci Inst for Pub Info; Fedn Am Scientists; AAAS. Res: Technology assessment; energy policy with emphasis on nuclear issues; environmental implications of energy policies. Mailing Add: Sch of Pub Affairs Univ of Minn 967 Soc Sci Bldg Minneapolis MN 55455

ABRAHAMSON, EARL ARTHUR, b Aberdeen, SDak, June 5, 24; m 48; c 3. ANALYTICAL CHEMISTRY. Educ: Univ NDak, BS, 48; Univ Kans, PhD(chem), 51. Prof Exp: Anal res chemist, 51-58, asst div supvr, Phys & Anal Div, 58-69, SUPVR CENT COMPUT TASK FORCE, CENT RES DEPT, E I DU PONT DE NEMOURS & CO, INC, 69- Mem: AAAS; Am Chem Soc; Soc Appl Spectros; Sigma Xi. Res: Analytical and physical chemistry; electrochemistry; on-line computer systems. Mailing Add: Cent Res Dept Exp Sta E I du Pont de Nemours & Co Wilmington DE 19898

ABRAHAMSON, EDWIN WILLIAM, b Medina, NY, Jan 6, 22; m 48; c 2. PHYSICAL CHEMISTRY, MOLECULAR BIOLOGY. Educ: Univ Buffalo, AB, 44; Columbia Univ, MA, 48; Syracuse Univ, PhD(phys chem), 52. Prof Exp: Res chemist, Manhattan Proj, Columbia Univ, 44-46; instr chem, Colgate Univ, 48-50; res assoc photochem, Mass Inst Technol, 52-53; res assoc, Syracuse Univ, 53-55; res assoc chem, State Univ NY Col Forestry, Syracuse Univ, 55-56, asst prof, 56-59; assoc profphys chem, Case Western Reserve Univ, 59-67, prof chem, 67-75; PROF CHEM, UNIV GUELPH, 75- Concurrent Pos: Grants, Res Corp, 58-59, NIH, 58-, AEC, 60-72 & NSF, 72-; sr vis dept theoret chem, Cambridge Univ, 64-65; vis prof, Yale Univ, 70; mem comt photobiol, Nat Res Coun, 71; hon ed, Photochem & Photobiol, 71-73; continuing vis prof biophys, Sch Med, Univ Ala, 72-; mem Int Comt Eye Res, 72-; mem comt on automation in med lab sci, Nat Inst Gen Med Sci, 78-80. Mem: Am Chem Soc; Am Soc Biol Chem; Am Phys Soc; fel Am Inst Chem. Res: Spectroscopic basis of photochemical reactions, especially biologically important pigments; chemistry of vision. Mailing Add: Dept of Chem Univ of Guelph Guelph ON Can

ABRAHAMSON, LILA, b Patterson, NJ, Dec 26, 27. PLANT BIOCHEMISTRY. Educ: Queens Col, NY, BS, 49; Brooklyn Col, AM, 55; Univ Mich, PhD(bot), 60. Prof Exp: Instr biol, Brooklyn Col, 53-54; from instr to asst prof, Cornell Col, 59-61; asst prof biol, Univ Chicago, 61-64, asst prof biol & bot, 64-67; ASSOC PROF BIOL, W VA UNIV, 67- Mem: AAAS; Bot Soc Am; Soc Plant Physiol. Res: Growth and development of plant tissue cultures, especially biochemisty of development; purification and characterization of homocysteine methyltransferase from plant seeds; methionone biosynthesis in higher plants. Mailing Add: Dept of Biol WVa Univ Morgantown WV 26506

ABRAHAMSON, SEYMOUR, b New York, NY, Nov 28, 27; m 53; c 1. GENETICS. Educ: Rutgers Univ, AB, 51; Ind Univ, PhD(genetics), 56. Prof Exp: USPHS fel, Univ Wis, 56-57; from instr to asst prof biol, Rutgers Univ, 57-60; from asst prof to prof zool, 60-69, PROF ZOOL & GENETICS, UNIV WIS-MADISON, 69- Concurrent Pos: Mem, UN Sci Comt Effect of Atomic Radiation, 65-66 & 70; USPHS fel, Calif Inst Technol, 69; Ford Found biol adv, Agrarian Univ, Peru; mem radiation protection adv comt, Wis State Bd Health; mem, Nat Comt Radiation Protection & Measurements, adv comt biol effect of ionizing radiation, Nat Acad Sci & comt I, Int Comn Radiation Protection. Mem: Radiation Res Soc; Genetics Soc Am; Environ Mutagen Soc. Res: Radiation genetics; induced mutation by radiation and other environmental mutagens. Mailing Add: Dept of Zool Univ of Wis Madison WI 53706

ABRAHAMSON, WARREN GENE, II, b Ludington, Mich, Mar 26, 47; m 69; c 1. PLANT ECOLOGY, POPULATION BIOLOGY. Educ: Univ Mich, Ann Arbor, BS, 69; Harvard Univ, AM, 71, PhD(biol), 73. Prof Exp: Res asst plant ecol, Univ Mich, 68-69; teaching asst biol, Harvard Univ, 69-72; tutor, 72-73; ASST PROF BIOL, BUCKNELL UNIV, 73- Mem: Ecol Soc Am; Soc Study Evolution; Am Inst Biol Sci; Sigma Xi. Res: Life history strategies of plants, vegetative and seed reproduction; environmental influences on life histories; effects of parasitism, insect, on life history parameters of goldenrods; fire ecology of south central Florida. Mailing Add: Dept of Biol Bucknell Univ Lewisburg PA 17837

ABRAM, DINAH, b Jerusalem, Israel, Nov 30, 28. MICROBIOLOGY, ELECTRON MICROSCOPY. Educ: Hebrew Univ, Israel, MSc, 52; Imp Col, dipl & Univ London, PhD(microbiol), 57. Prof Exp: Asst biochem, Hebrew Univ, Israel, 51-52, asst bact, Sch Med, 52-53, instr, 57; res assoc microbiol, biochem & electron micros, Purdue Univ, 59-69; ASST PROF MICROBIOL, SCH MED, UNIV PITTSBURGH, 69- Concurrent Pos: Nat Res Coun Can res fel microbiol, 57-59. Mem: Am Soc Microbiol. Res: Soil microbiology; physiology of extreme halphilic bacteria; structure-function relationship at the molecular and organelle levels and specialized structure of bacterial flagella; chemistry of flagellins; fine structure of bacterial cytoplasmic membrane. Mailing Add: Dept of Microbiol Univ of Pittsburgh Sch of Med Pittsburgh PA 15213

ABRAM, HARRY SHORE, b Roanoke, Va, Mar 25, 31; m 56; c 5. PSYCHIATRY, PSYCHOSOMATIC MEDICINE. Educ: Northwestern Univ, BS, 52; Univ Va, MD, 56. Prof Exp: From instr to assoc prof psychiat, Sch Med, Univ Va, 64-70, prog dir undergrad training psychiat, 65-70, co-dir renal dialysis unit, Univ Hosp, 65-70; PROF PSYCHIAT, SCH MED, VANDERBILT UNIV, 70- Concurrent Pos: Clin & res fel psychiat, Harvard Med Sch, 59-61; career teacher award, NIMH, 61-63; med consult, Borland-Coogan Assoc Inc, 73-; consult, Psychiat Educ Br, NIMH, 74- Mem: Am Psychiat Asn; Am Psychoanal Asn. Res: Psychological aspects of physical illness and medical progress; dialysis and transplantation. Mailing Add: Dept of Psychiat Vanderbilt Univ Med Ctr Nashville TN 37232

ABRAM, JAMES BAKER, JR, b Tulsa, Okla, Dec 5, 37; m 61; c 2. ZOOLOGY, PARASITOLOGY. Educ: Langston Univ, BS, 59; Okla State Univ, MS, 63, PhD(zool), 68. Prof Exp: Teacher high sch, Okla, 59-62; from instr to assoc prof biol, Md State Col, Princess Anne, 63-70; assoc prof, 70-73, PROF BIOL, HAMPTON INST, 73- Mem: Am Soc Parasitol; Am Soc Mammal. Res: Helminths of muskrats; helminths of small mammals; ecological influences upon parasites. Mailing Add: Dept of Biol Hampton Inst Hampton VA 23368

ABRAMITIS, WALTER WILLIAM, b Newark, NJ, Oct 31, 16; m 47; c 3. AGRICULTURAL CHEMISTRY. Educ: Rutgers Univ, BSc, 40, MSc, 42; Iowa State Univ, PhD(insecticides), 51. Prof Exp: Entomologist, Standard Oil Co, NJ, 40; sect head pesticide screening & develop, Res Div, Armour & Co, 51-61, head agr chem sect, Armour Indust Chem Co, 61-65, mgr agr chem res & develop, 65-71, MGR AGR CHEM RES & DEVELOP, ARMAK CO, 71- Mem: Entom Soc Am; Weed Sci Soc Am; Am Chem Soc; fel Am Inst Chem. Res: Screening and development program on insecticides; fungicides; herbicides; nematocides; plant growth regulators. Mailing Add: 8401 W 47th St McCook IL 60525

ABRAMOFF, PETER, b Montreal, Que, Dec 28, 27; nat US; m 53; c 6. ZOOLOGY, IMMUNOLOGY. Educ: Univ Western Ont, BA, 50; Univ Detroit, MSc, 52; Univ Wis, PhD(zool), 56. Asst, Univ Detroit, 50-52; from instr to assoc prof, 55-67, PROF BIOL, MARQUETTE UNIV, 67-, CHMN DEPT, 65- Mem: AAAS; Am Inst Biol Sci; Soc Exp Biol & Med; Reticuloendothelial Soc; Int Soc Exp Hemat. Res: Cellular and subcellular sites of antibody formation; hypersensitivity pneumonitis. Mailing Add: Dept of Biol Marquette Univ Milwaukee WI 53233

ABRAMOVICI, MIRON, b Bacau, Romania, May 19, 34; m 67; c 2. ORGANIC CHEMISTRY, PHYSICAL ORGANIC CHEMISTRY. Educ: Univ Sao Paulo, BS, 60; Northwestern Univ, MS, 68; Loyola Univ Chicago, PhD(chem), 70. Prof Exp: Res chemist, Uniao Oil Refinery Co, Brazil, 60-63; res chemist, Cent Res Lab, GAF Corp, Pa, 69-70; RES CHEMIST, CATALYTIC & COORD CHEM SECT, RES & DEVELOP, ENGELHARD INDUSTS DIV, ENGELHARD MINERALS & CHEM CORP, 70- Mem: Am Chem Soc; Catalysis Soc. Res: Preparative and synthetic organic chemistry; analytical instrumentation for organic compounds; homogeneous and heterogeneous catalysis in organic chemistry; catalysis; chemical and physical characterization of heterogeneous catalysts. Mailing Add: Catalytic & Coord Chem

Sect R & D Engelhard Industs Div Menlo Park Engelhard Minerals & Chem Corp Edison NJ 08817

ABRAMOVITCH, RUDOLPH ABRAHAM HAIM, b Alexandria, Egypt, July 19, 30; m 52. ORGANIC CHEMISTRY. Educ: Univ London, BSc, 50, PhD(chem), 53, DSc, 64. Prof Exp: Med Res Coun fel, Univ Exeter, 54; res chemist, Weizmann Inst Sci, Israel, 54-55; Imp Chem Indust res fel, King's Col, Univ London, 55-57; from asst prof to prof chem, Univ Sask, 57-67; prof, 67-70, RES PROF CHEM, UNIV ALA, 70- Concurrent Pos: Consult, Mead Johnson Res Ctr, 73- Mem: The Chem Soc; Am Chem Soc; Chem Inst Can; Int Soc Heterocyclic Chem. Res: Reactive intermediates; heterocyclic chemistry; stereochemistry; medicinal chemistry. Mailing Add: Dept of Chem Col of Arts & Sci Univ of Ala University AL 35486

ABRAMOWITZ, STANLEY, b Brooklyn, NY, Sept 18, 36; m 57; c 3. CHEMICAL PHYSICS. Educ: Brooklyn Col, BS, 57; Polytech Inst Brooklyn, PhD(phys chem), 63. Prof Exp: Asst chem, Polytech Inst Brooklyn, 57-59; Nat Acad Sci-Nat Res Coun fel, 62-64, PHYS CHEMIST, MOLECULAR SPECTROS, NAT BUR STANDARDS, 64- Mem: Am Phys Soc. Res: Molecular spectroscopy and structure. Mailing Add: Nat Bur of Standards Washington DC 20234

ABRAMS, ADOLPH, b Springfield, Mass, Apr 30, 19; m 51; c 3. BIOCHEMISTRY. Educ: Univ Wis, PhD(biochem), 49. Prof Exp: From asst prof to assoc prof, 54-65, PROF BIOCHEM, SCH MED, UNIV COLO, DENVER, 65-, ACTG CHMN DEPT, 73- Concurrent Pos: Nat Res Coun fel, Cancer Soc, Carlsberg Lab & Copenhagen Univ, 49-50; Lilly fel biochem, Calif Inst Technol, 50-54. Mem: Am Chem Soc; Am Soc Biol Chemists. Res: Physical chemistry and biosynthesis of proteins; intermediary metabolism of amino and nucleic acids; chemistry and enzymology of bacterial cell membranes; relations between membranes, nucleic acids and protein synthesis; membrane permeability and active transport; bacterial membrane ATPase. Mailing Add: Dept of Biochem Univ of Colo Sch Med Denver CO 80220

ABRAMS, ALBERT, b St Joseph, La, Sept 2, 21; m 54; c 2. PHYSICAL CHEMISTRY, PETROLEUM ENGINEERING. Educ: City Col NY, BS, 48; NY Univ, PhD(chem), 52. Prof Exp: Control chemist, Stauffer Chem Co, 48; chemist, 52-63, res chemist, 63-71, RES ASSOC, EXPLOR & PROD RES DIV, SHELL DEVELOP CO, 71- Mem: Am Chem Soc; Am Inst Mining, Metall & Petrol Eng. Res: Fluid flow; transport phenomena; surface chemistry; flow of fluids through porous media; oil formation impairment; oil production problems. Mailing Add: Shell Develop Co Bellaire Res Ctr PO Box 481 Houston TX 77001

ABRAMS, ALBERT MAURICE, b Los Angeles, Calif, June 7, 31; m 56; c 2. PATHOLOGY. Educ: Univ Calif, DDS, 57; Univ Southern Calif, MS, 61. Prof Exp: Instr periodont & oral path, 58-59, PROF PATH & CHMN DEPT, SCH DENT, UNIV SOUTHERN CALIF, 65- Concurrent Pos: Fel oral path, Armed Forces Inst Path, 62-64; mem attend staff, Dept Path, Los Angeles County Gen Hosp, 65-; consult, Long Beach Vet Hosp, 65- Mem: Am Acad Oral Path; fel Am Col Dent; Am Soc Clin Path; Am Asn Cancer Educ. Res: Clinical-pathological studies of odontogenic tumors, salivary gland diseases and fibro-osseous lesions. Mailing Add: Dept of Path Univ of Southern Calif Sch Dent Los Angeles CA 90007

ABRAMS, ARCHIE ADAM, b Cleveland, Ohio, Apr 10, 06; m 32; c 2. MEDICINE. Educ: Western Reserve Univ, AB, 27, MD, 31. Prof Exp: Asst obstet & gynec, Med Sch, Tufts Univ, 35-42; asst clin prof, 48-53, ASSOC PROF OBSTET & GYNEC, SCH MED, BOSTON UNIV, 53-, CHIEF OBSTET, UNIV HOSP, MED CTR, 66-, VIS GYNECOLOGIST, 70- Concurrent Pos: Teaching fel obstet & gynec, Med Sch, Tufts Univ, 33-35; chief obstet, Mass Mem Hosp & Salvation Army Unmarried Mothers Prog; res consult, Eli Lilly Co; consult, Boston City Hosp; mem bd dir, Planned Parenthood Pregnancy Coun Serv. Res: Obstetrics and gynecology. Mailing Add: Dept of Obstet & Gynec Boston Univ Sch of Med Boston MA 02215

ABRAMS, EDWARD, b Bridgeport, Conn, Sept 19, 12; m; c 1. RESEARCH ADMINISTRATION. Educ: City Col NY, BS, 33; George Washington Univ, MA, 41; Georgetown Univ, PhD(biochem), 46. Prof Exp: Lab aide, Bur Standards, 41-42, from jr technologist to asst technologist, 42-44, microbiologist, 44-46; res asst, Inst Textile Technol, 46-47, res assoc & proj leader, 47-48, res biologist & head biol div, 48-50; head textile sect, Southern Res Inst, 50-56; res adminr, 56-61, tech consult, 61-64, mgr tech adv serv, 64-72, DIR TECH DEVELOP, CHEMETRON CORP, 72- Mem: Am Chem Soc; fel Am Inst Chem; Fiber Soc. Res: Evaluation of new products, processes and acquisitions. Mailing Add: Chemetron Corp 111 E Wacker Dr Chicago IL 60601

ABRAMS, ELLIS, b Pittsburgh, Pa, Jan 24, 17; m 46; c 2. INDUSTRIAL CHEMISTRY. Educ: Univ Pittsburgh, BS, 38, PhD(chem), 42. Prof Exp: Asst anal chem, Univ Pittsburgh, 38-42; res chemist, Quaker Chem Prod Corp, 42-49, dir textile res, 49-54; supvr org textile & paper res, 54-58, asst mgr, 59-69, mgr res, 70-72, DIR RES, E F HOUGHTON & CO, 72- Mem: Am Chem Soc; Am Soc Lubrication Engrs. Res: Surface active agents; textile finishes; derivatives of m-diphenyl benzene; chemical defoamers; cellulose reactants; cutting and rolling oils; pollution control. Mailing Add: E F Houghton & Co Van Buren & Madison Aves Norristown PA 19401

ABRAMS, GERALD DAVID, b Detroit, Mich, Apr 27, 32; m 54; c 2. PATHOLOGY. Educ: Wayne State Univ, AB, 51; Univ Mich, MD, 55. Prof Exp: Instr path, Med Sch, Univ Mich, 59-60; asst chief dept exp path, Walter Reed Army Inst Res, 61-62; from asst prof to assoc prof, 63-69, PROF PATH, MED SCH, UNIV MICH, ANN ARBOR, 69- Concurrent Pos: Markle scholar acad med, 63-68; attend consult, Vet Admin Hosp, Ann Arbor, Mich, 63-; dep med exam, Washtenaw Co, Mich, 63- Mem: AAAS; Am Soc Exp Path; Am Asn Path & Bact; Int Acad Path. Res: Pathologic anatomy; experimental pathology; gnotobiotics; effects of normal flora on host resistance; comparative pathology. Mailing Add: Dept of Path Univ of Mich Med Sch Ann Arbor MI 48109

ABRAMS, GERALD STANLEY, b New York, NY, May 30, 41; m 63; c 2. HIGH ENERGY PHYSICS. Educ: Cornell Univ, BA, 62; Univ Md, PhD(physics), 67. Prof Exp: Res assoc physics, Univ Ill, Urbana, 67-69; PHYSICIST, LAWRENCE BERKELEY LAB, 69- Mem: Am Phys Soc. Res: Experimental high energy physics. Mailing Add: Bldg 50-A Lawrence Berkeley Lab Berkeley CA 94720

ABRAMS, HERBERT L, b New York, NY, Aug 16, 20; m 43; c 2. RADIOLOGY, PHYSIOLOGY. Educ: Cornell Univ, BA, 41; State Univ NY, MD, 46; Am Bd Radiol, cert. Prof Exp: Intern, Long Island Col Hosp, 46-47; resident, Montefiore Hosp, 47-48; resident radiol, Stanford Univ Hosp, 48-51; from instr to assoc prof, Sch Med, Stanford Univ, 51-61, actg exec dir, Dept Radiol, 61-62, dir, Div Diag Roentgenol, 61-67, prof radiol, 62-67; PHILIP H COOK PROF RADIOL & CHMN DEPT, HARVARD MED SCH, 67- Concurrent Pos: Nat Heart Inst spec res fel, Univ Lund, 60-; mem radiation study sect, NIH, 63-67, exec comt, Coop Study Renovascular Hypertension; radiologist-in-chief, Peter Bent Brigham Hosp, 67-; consult surv renovascular hypertension, Nat Heart Inst; consult US Naval Hosp & Vet Admin, Calif & coun pharmacol & chem, AMA; mem overseas adv comt, Tel Aviv Univ Med Sch; mem exec & adv comts, Inter-Soc Comn Heart Dis Resources. Mem: AAAS; Asn Univ Radiol; AMA; Am Col Radiol; Am Asn Cancer Res. Res: Cardiovascular radiology and physiology. Mailing Add: Dept of Radiol Harvard Med Sch Boston MA 02115

ABRAMS, IRVING MELVIN, b St Paul, Minn, Feb 25, 17; m 52; c 3. INDUSTRIAL CHEMISTRY. Educ: Univ Minn, BCh, 38, PhD(biochem), 42. Prof Exp: Res chemist, Econ Lab, Inc, St Paul, 38-39; Bristol-Myers fel, Stanford Univ, 42-44, Nat Adv Comt Aeronaut fel, 42-44; sr res chemist, Int Minerals & Chem Corp, Chicago, 44-47; lab dir, Chem Process Co, 47-55, assoc tech dir, 55-62; tech mgr, Ion Exchange Dept, 62-75, MGR TECH DEVELOP, FUNCTIONAL POLYMERS GROUP, PROCESS CHEM DIV, DIAMOND SHAMROCK CHEM CO, 75- Concurrent Pos: Instr human rels & effective speech, Simmons Inst, San Francisco & Los Angeles, 48-68; consult, Stanford Res Inst, 48; mem adv comt, Int Water Conf, 62- Mem: Am Chem Soc; Am Inst Mining, Metall & Petrol Eng; Am Soc Beet Sugar Technol; Am Soc Enologists. Res: Anomalous osmosis and collodion membranes; foaming phenomena; chemistry of sugar beets; ion exchange resins; adsorption phenomena; water treatment and wastewater renovation; counter-current ion exchange; condensate polishing. Mailing Add: PO Box 829 Redwood City CA 94064

ABRAMS, ISRAEL JACOB, b Syracuse, NY, June 11, 26. MATHEMATICAL STATISTICS. Educ: Univ Calif, BA, 50, MA, 52, PhD(math statist), 57. Prof Exp: Asst, Univ Calif, 50-57; math statistician, Space Tech Labs, 57-61; vpres technol, Mkt Res Corp Am, 61-69; PRES, DATANAMICS CORP AM, 69-; SR VPRES, MKT RES CORP AM, 71- Concurrent Pos: Lectr, Exten, Univ Calif, Los Angeles, 59-60; ed comput sect, J Mkt Res, 67-72. Mem: AAAS; Inst Math Statist; Am Statist Asn; Opers Res Soc Am; Soc Indust & Appl Math. Res: Marketing research; applied statistics; computer systems design; operations research. Mailing Add: Mkt Res Corp of Am 624 S Michigan Ave Chicago IL 60605

ABRAMS, JEROME SANFORD, b Cleveland, Ohio, Mar 24, 29; m 51; c 2. SURGERY. Educ: Western Reserve Univ, BS, 50; Ohio State Univ, MD, 57. Prof Exp: From instr to asst prof surg, Case Western Reserve Univ, 62-69; assoc prof, 69-71, PROF SURG, COL MED, UNIV VT, 71-, ASSOC DIR DEPT, 69- Concurrent Pos: Asst dir dept surg, Cleveland Metrop Gen Hosp, 67-69. Mem: AMA; Am Col Surg; Am Asn Surg of Trauma; Soc Surg Alimentary Tract; Am Burn Asn. Res: General surgery; shock; trauma; infection. Mailing Add: Dept of Surg Univ of Vt Col of Med Burlington VT 05401

ABRAMS, JOHN WERNER, b San Francisco, Calif, Dec 15, 13; m 44; c 2. HISTORY OF SCIENCE, OPERATIONAL RESEARCH. Educ: Univ Calif, AB, 33, PhD(statist astrophys), 39. Prof Exp: Instr astron, Exten, Univ Calif, 40; astronomer, US Naval Observ, 40-41; asst prof physics, Univ Man, 45-46; astron, Wesleyan Univ, 46-49; mem staff, Defence Res Bd, Ottawa, Can, 49-62, dep dir oper res group, 51-52, supt, 52-54, sci adv to chief air staff, Royal Can Air Force, 54-58, consult air defense, Supreme Allied Comdr, Europe, 58-61, chief oper res, Defence Res Bd, 61-62; assoc prof indust eng, 63-67, dir inst hist & philos sci & technol, 67-72, PROF INDUST ENG & HIST, UNIV TORONTO, 67-; CONSULT, 62- Concurrent Pos: Asst prof, Trinity Col (Conn), 47; vis prof, Sch Bus, Queen's Univ (Ont), 64-66; secy-gen Int Comt Coop Hist Technol, 74- Mem: Fel AAAS; Opers Res Soc Am; Hist Sci Soc; Can Opers Res Soc (vpres, 58-59, pres, 61-62); Can Soc Hist Sci (pres, 72-75). Res: Philosophical bases of science; history of science and technology. Mailing Add: Dept Indust Eng Univ of Toronto Toronto ON Can

ABRAMS, LLOYD, b New York, NY, June 9, 39; m 71. PHYSICAL CHEMISTRY. Educ: City Col NY, BChE, 61; Rutgers Univ, PhD(phys chem), 67. Prof Exp: Anal engr, Conn Aircraft Nuclear Engine Lab, Pratt & Whitney Aircraft, 61, exp engr, Adv Powers Systs, Conn, 61-62, Apollo Proj, 62-63; teaching asst chem, Rutgers Univ, 63-64; res assoc, Brookhaven Nat Lab, 66-68; res chemist, 68-72, SR RES CHEMIST, PIGMENTS DIV, EXP STA LAB, E I DU PONT DE NEMOURS & CO, 72- Mem: AAAS; Am Chem Soc; Am Ceramic Soc; Sigma Xi; NY Acad Sci. Res: Gas-solid reactions; adsorption studies; infrared spectroscopy of surfaces; radiation chemistry of adsorbed species; solid state chemistry; ceramics; basic studies. Mailing Add: Pigments Div Exp Sta Lab E I du Pont de Nemours & Co Wilmington DE 19898

ABRAMS, MARVIN COLIN, b Reno, Nev, Sept 6, 30; m 56; c 2. PHYSICAL CHEMISTRY. Educ: Univ Nev, BS, 52, MS, 54; Wash State Univ, PhD(chem), 58. Prof Exp: Res chemist, AEC, Univ Nev, 52-54; asst chem, Wash State Univ, 54-58; staff scientist chem & physics, 59-67, sr staff scientist, 67-70, CHIEF MFG TECHNOL, GEN DYNAMICS/POMONA, 70- Mem: Am Chem Soc; Combustion Inst. Res: Chemical mechanism and kinetics of combustion processes; spectroscopic studies of solids and reacting gaseous media. Mailing Add: Gen Dynamics Technol Dept 1675 W Mission Blvd Pomona CA 91766

ABRAMS, RICHARD, b Chicago, Ill, Sept 19, 17; m 47; c 4. BIOCHEMISTRY. Educ: Univ Chicago, SB, 38, PhD(chem), 41. Prof Exp: Res instr chem, Univ Chicago, 41-42, group leader, Manhattan Proj, 42-46; Donner Found fel, Karolinska Inst, Sweden, 46; asst prof biochem, Inst Radiobiol & Biophys, Univ Chicago, 47-51; assoc dir inst res, Montefiore Hosp, 51-58, dir res, 58-65; prof biochem & nutrit & head dept, Grad Sch Pub Health, 65-70, PROF BIOCHEM & CHMN DEPT, FAC ARTS & SCI, UNIV PITTSBURGH, 70- Mem: AAAS; Am Chem Soc; Am Soc Biol Chemists. Res: Isolation and properties of respiratory enzymes; metabolism of fission products and transuranic elements; tracer studies of nucleic acid metabolism. Mailing Add: Dept of Biochem Univ of Pittsburgh Pittsburgh PA 15213

ABRAMS, RICHARD LEE, b Cleveland, Ohio, Apr 20, 41; m 62; c 2. APPLIED PHYSICS. Educ: Cornell Univ, BEP, 64, PhD(appl physics), 68. Prof Exp: Mem tech staff laser physics, Bell Tel Labs, 68-71; sr staff physicist, 71-72, head electro-optics sect, 72-75, MGR OPTICAL PHYSICS DEPT, HUGHES RES LABS, 75- Mem: Am Phys Soc; Inst Elec & Electronics Eng. Res: Quantum electronics; atomic physics; infrared physics; laser communications; nonlinear optics; electro-optics. Mailing Add: Hughes Res Labs 3011 Malibu Cyn Rd Malibu CA 90265

ABRAMS, ROBERT JAY, b Chicago, Ill, June 16, 38; m 66; c 3. EXPERIMENTAL HIGH ENERGY PHYSICS. Educ: Ill Inst Technol, BS, 59; Univ Ill, Urbana, MS, 61, PhD(physics), 66. Prof Exp: Res assoc physics, Brookhaven Nat Lab, 66-68, asst physicist, 68-70; asst prof, 70-74, ASSOC PROF PHYSICS, UNIV ILL, CHICAGO CIRCLE, 74- Mem: Am Phys Soc. Res: High energy multiparticle reactions; K meson decays. Mailing Add: Dept of Physics Univ of Ill Box 4348 Chicago IL 60680

ABRAMS, ROBERT MARLOW, b Rome, NY, June 8, 31; m 57; c 4. PHYSIOLOGY. Educ: Hamilton Col, AB, 53; Univ Pa, MS & DDS, 60, PhD(physiol), 63. Prof Exp: Asst prof pub health & epidemiol, Yale Univ, 68-69; asst prof, 69-74, ASSOC PROF OBSTET & GYNEC, COL MED, UNIV FLA, 74- Concurrent Pos: USPHS career develop award, 63-68; asst fel, John B Pierce Found Lab, Med Sch, Yale Univ, 63-69; mem study sect, Human Embryol & Develop, NIH, 75. Mem: Am Physiol Soc; Soc

Gynecol Invest; Perinatal Res Soc. Res: Thermal physiology; reproduction. Mailing Add: Dept of Obstet & Gynec Univ of Fla Col of Med Gainesville FL 32601

ABRAMS, ROCHELLE B, chemistry, see 12th edition

ABRAMSKY, TESSA, b Poland, Mar 30, 21; US citizen; m 74. BIOCHEMISTRY. Educ: Hunter Col, BA, 42; Columbia Univ, PhD(biochem), 63. Prof Exp: Technician, Rockefeller Inst, 43-48; res assoc physiol, Sch Med, Georgetown Univ, 48-50; res worker biochem, Col Physicians & Surgeons, Columbia Univ, 52-63; asst res biochemist, Univ Calif, Berkeley, 63-65; res assoc, 65-67, ASST PROF BIOCHEM GENETICS, ROCKEFELLER UNIV, 67- Res: Biosynthesis of antibiotics; intermediary metabolism; biosynthesis of porphyrins; cellular physiology; biochemical basis of morphogenesis and development in microorganisms. Mailing Add: 41 Wayland Dr Verona NJ 07044

ABRAMSON, ALLAN LEWIS, b New York, NY, Nov 8, 40; m 64; c 2. OTOLARYNGOLOGY. Educ: Colgate Univ, BA, 62; State Univ NY Downstate Med Ctr, MD, 67. Prof Exp: Resident otolaryngol, Mt Sinai Sch Med, 69-72; dir div otolaryngol, Naval Hosp, Camp LeJeune, NC, 72-74; ASSOC PROF OTOLARYNGOL, STATE UNIV NY STONY BROOK, 74-; DIR OTOLARYNGOL, LONG ISLAND JEWISH-HILLSIDE MED CTR, 74- Mem: Deafness Res Found; Am Broncho-Esophagol Soc; Am Acad Ophthal & Otolaryngol; Am Acad Facial Plastic & Reconstruct Surg. Res: Molecular and antibiotic transfer across normal and infected sinus membranes; bone grafting into the frontal sinus cavity. Mailing Add: Dept of Otolaryngol LIJ-Hillside Med Ctr New Hyde Park NY 11040

ABRAMSON, ARTHUR SIMON, b Montreal, Que, June 4, 12; nat US; m 56; c 1. REHABILITATION MEDICINE. Educ: McGill Univ, BSc, 33, MD & CM, 37. Prof Exp: Asst chief phys med & rehab serv, Vet Admin Hosp, NY, 48-50, chief, 50-55; asst clin prof, Col Med, NY Univ-Bellevue Med Ctr, 50-52; clin prof phys med & rehab, New York Med Col, 52-55; PROF REHAB MED & CHMN DEPT, ALBERT EINSTEIN COL MED, 55- Concurrent Pos: Metrop res fel, Montefiore Hosp, 53-55; vis prof phys med & med dir phys ther div, Ithaca Col; Gross Mem lectr, 50; mem, Truman Comn Vet Med Serv, 50; consult, Kessler Inst Rehab, 51; vis physician, Bird S Coler Mem Hosp, 53-55; consult, Kingsbrook Jewish Hosp, 54; mem med & sci comt, Arthritis & Rheumatism Found, NY, 55; area med consult, Vet Admin Hosps, 55, 56 & 57; consult, Nat Surgeon Gen, Amvets, 55-56; vis physician & dir phys med & rehab, Bronx Munic Hosp Ctr, 55-; vchmn, Am Bd Phys Med & Rehab, 56-68; mem, President's Comt Employ Physically Handicapped, 56-; consult, USPHS Hosp, 57 & Misericordia Hosp, Bronx, 58-; mem, Gov Rockefeller's Coun Rehab, 59-70; mem nat adv coun voc rehab admin, Dept Health, Educ & Welfare, 62-66, consult emer, 69-; mem, Joint Comn Educ in Phys Med & Rehab, 63-70; Albee Mem lectr, 64; Gaffney Mem lectr, 66; Coulter Mem lectr, 68; Wise Mem lectr, 69; consult, Muscular Dystrophy Asns Am; assoc ed, Archives Phys Med & Rehab, 73-; Zeiter lectr, 75; vis prof, Can Asn Phys Med & Rehab, Vancouver & Calgary, 75. Honors & Awards: Award, Amvets, 54; President's Trophy, 55; Gold Key Award, Am Cong Rehab Med. Mem: Am Acad Phys Med & Rehab (pres, 71-72); fel Am Col Physicians; AMA; Am Cong Rehab Med; fel NY Acad Med. Res: Physical medicine rehabilitation; neurogenic bladder; therapeutic community. Mailing Add: Hawthorne Way Hartsdale NY 10530

ABRAMSON, CARL, microbiology, immunochemistry, see 12th edition

ABRAMSON, DAVID C, b Philadelphia, Pa, Mar 30, 38; m 60; c 3. PEDIATRICS, NEONATOLOGY. Educ: Muhlenberg Col, BS, 60; Georgetown Univ, MD(med, physiol), 66. Prof Exp: DIR NURSERIES, GEORGETOWN UNIV HOSP, 69- Concurrent Pos: Consult, Columbia Hosp Women, Providence Hosp & Arlington Hosp, 68-; dir nurseries, Fairfax County Hosp, 70-; prin investr, NIH, 68- Mem: Soc Obstet Anesthesia & Perinatology. Res: Placental transfer of iron; disseminated intravascular coagulation in the newborn; carbohydrate metabolism in the newborn; congenital heart disease. Mailing Add: 2910 Garfield NW Washington DC 20008

ABRAMSON, DAVID IRVIN, b New York, NY, Oct 14, 05; m 40; c 2. CARDIOVASCULAR DISEASES. Educ: Long Island Col Med, MD, 29; Am Bd Cardiovasc Dis, dipl. Prof Exp: Asst physiol, Long Island Col Med, 28-29; intern, Bushwick Hosp, Brooklyn, 29-30; instr physiol & pharm, Long Island Col Med, 30-36; dir cardiovasc res, May Inst Med Res, Univ Cincinnati, 37-42; from asst prof to prof med, 46-72, prof phys med & rehab & head dept, 55-72, EMER PROF MED, COL MED, UNIV ILL, 72- Concurrent Pos: Fels, Michael Reese Hosp, Chicago, 36 & Mt Sinai Hosp, New York, 36-37; dir dept electrocardiography, Jewish Hosp, Cincinnati, 38-42; instr, Sch Med, Univ Cincinnati, 38-42; attend physician, Hines Vet Hosp, Ill, 46-53, consult, 63-; attend physician, West Side Vet Hosp, Chicago, 46-53, consult, 63-; attend physician, Michael Reese Hosp, Chicago, 47- & Mt Sinai Hosp, 49- Mem: Fel Am Col Physicians; Am Soc Clin Invest; Soc Exp Biol & Med; Am Physiol Soc. Res: Physiology of peripheral circulation; peripheral vascular disorders; convalescence; rate of peripheral blood flow in man under normal and abnormal conditions and in various pathologic states, using the venous occlusion plethysmographic method. Mailing Add: 916 N Oak Park Ave Oak Park IL 60302

ABRAMSON, EDWARD, b Philadelphia, Pa, Mar 2, 33; m 54; c 2. OPTICS. Educ: Ursinus Col, BS, 54; Univ Pa, MS, 57, PhD(physics), 62. Prof Exp: Physicist, Cent Res Dept, 62-67, res physicist, Org Chem Dept, 67-71, RES PHYSICIST, PHOTO PROD DEPT, E I DU PONT DE NEMOURS & CO, 71- Mem Optical Soc Am. Res: Ultraviolet sources and optics; photochemistry; optical imaging systems; applications of lasers; electrooptics and fiber optics. Mailing Add: Berg Electronics Div E I Du Pont de Nemours New Cumberland PA 17070

ABRAMSON, FRED PAUL, physical chemistry, see 12th edition

ABRAMSON, FREDRIC DAVID, b Philadelphia, Pa, Nov 9, 41; m 64; c 1. COMMUNITY HEALTH, EPIDEMIOLOGY. Educ: Univ Pa, AB, 63; Univ Rochester, MS, 66; Univ Mich, PhD(genetics), 71. Prof Exp: Genetic counr birth defects, Wayne State Univ, 71-72; asst prof community med, Col Med, Univ Ky, 72-75; PRES, HEALTH MGT SYSTS, INC, 75- Concurrent Pos: Consult, Cedars Lebanon Hosp, Miami, Fla, 73- Mem: Am Pub Health Asn; Am Soc Human Genetics; Am Statist Asn; Behav Genetics Asn; Soc Epidemiol Res. Res: Nosocomial infections; spontaneous abortion; medical education; health care systems. Mailing Add: Health Mgt Systs Inc 1087 The Lane Lexington KY 40504

ABRAMSON, HANLEY N, b Detroit, Mich, June 10, 40; m 67; c 2. MEDICINAL CHEMISTRY. Educ: Wayne State Univ, BS, 62; Univ Mich, MS, 63, PhD(pharmaceut chem), 66. Prof Exp: NSF fel, Hebrew Univ Jerusalem, 66-67; asst prof, 67-73, ASSOC PROF PHARMACEUT CHEM, COL PHARM & ALLIED HEALTH PROFESSIONS, WAYNE STATE UNIV, 73- Mem: Am Pharmaceut Asn; Am Chem Soc; Am Heart Asn; AAAS; NY Acad Sci. Res: Synthesis of natural products and their analogs. Mailing Add: Col Pharm & Allied Health Prof Wayne State Univ Detroit MI 48202

ABRAMSON, IRWIN JEROME, b Philadelphia, Pa, Aug 3, 35; m 59; c 2. MICROBIOLOGY. Educ: Pa State Univ, BS, 57; Va Polytech Inst & State Univ, PhD(microbiol), 71; Nat Registry Microbiologists, regist, 73. Prof Exp: Chief microbiol, Philadelphia Labs, Inc, 63-67; staff fel & res microbiologist, Baltimore Cancer Res Ctr, Nat Cancer Inst, 71-73; SR MICROBIOLOGIST, BIOQUEST DIV, BECTON DICKINSON CORP, 73- Concurrent Pos: Consult anaerobe lab apparatus, Kontes Glass Co, 72-74. Honors & Awards: Recognition Award, USPHS, 73. Mem: Am Soc Microbiol. Res: Development, design and testing of media and devices for collection, transport, and isolation of anaerobic and aerobic microorganisms from a clinical specimen; investigation of methods to improve growth of anaerobes on commercial media; effect of antimicrobial agents on Treponemas. Mailing Add: 6503 Copper Ridge Dr Baltimore MD 21209

ABRAMSON, LEE RICHARD, b New York, NY, May 19, 33; m 54; c 2. MATHEMATICAL STATISTICS. Educ: Columbia Univ, AB, 54, PhD(math statist), 63. Prof Exp: Sr res mathematician, Electronics Res Lab, Columbia Univ, 57-63, prin res mathematician, Labs & lectr biostatist, Univ, 64-67, lectr, Sci Hons Prog, 64-67, res assoc, 65-67; mem res staff, Riverside Res Inst, 67-72; SR RES ANALYST, ENERGY RES & DEVELOP ADMIN, 72- Concurrent Pos: Lectr, Math Prog High Sch Students, City Univ New York, 70-; consult, Sandoz Pharmaceut, 69- Mem: AAAS; Inst Math Statist; Am Math Soc. Res: Operations research; sequential analysis; weapons systems evaluation. Mailing Add: ERDA Int Security Affairs Mail Sta C-111 Washington DC 20545

ABRAMSON, MORRIS BARNET, b New York, NY, July 8, 10; m 38; c 2. PHYSICAL BIOCHEMISTY, NEUROCHEMISTY. Educ: NY Univ, BS, 31, MS, 34, PhD(chem), 39. Prof Exp: Res assoc, 62-73, ASSOC PROF PHYS CHEM, DEPT NEUROL, ALBERT EINSTEIN COL MED, 73- Concurrent Pos: Adj assoc prof, NY Univ, 42-; Sir Ernest Oppenheimer fel, Cambridge Univ, 61 & Polytech Inst Brooklyn, 63; NIH fel, 63. Res: Physical chemistry of lipids; surface chemistry of particles in aqueous media; model membrane systems; physical chemistry of neurotransmitters and enzymes acting on these. Mailing Add: 53-35 Hollis Ct Blvd Flushing NY 11365

ABRAMSON, NORMAN JAY, b Los Angeles, Calif, May 4, 27; m 51; c 1. FISHERIES, BIOMETRICS. Educ: Univ Wash, BS, 55. Prof Exp: Marine biologist, Calif State Fisheries Lab, 55-57, head biomet anal sect, 57-70; consult, Food & Agr Orgn, UN, 70; surv biometrician, Nat Marine Fisheries Serv, 71-72; fishery biologist, Tiburon Fisheries Lab, 72-74, DIR SOUTHWEST FISHERIES CTR, TIBURON LAB, 74- Concurrent Pos: Consult, Fisheries Res Inst, Univ Wash, 69. Mem: Biomet Soc; Am Statist Asn; Am Fisheries Soc; Am Inst Fishery Res Biologists. Res: Population dynamics of marine fishes and related statistical problems. Mailing Add: Tiburon Lab 3150 Paradise Dr Tiburon CA 94920

ABRAMSON, STANLEY L, b Akron, Ohio, Mar 29, 25; m 49; c 3. PHYSICS, OPERATIONS RESEARCH. Educ: Univ Akron, BS, 47, MS, 48. Prof Exp: Instr physics, Univ Akron, 46-49; asst prof, Nebr Wesleyan Univ, 49-50; chief ballistics sect, US Army Chem Ctr, 50-53, mem opers res group, 53-55; sr engr, Air Arm Div, Westinghouse Elec Corp, 55-58; sr engr, Arma Div, Am Bosch Arma Corp, 58-60; adv engr, Res Lab, Westinghouse Elec Corp, 60-63, mgr tech utilization prog res & develop, Astronuclear Lab, 63-67, planning consult, Westinghouse Power Systs Planning, 67-69, CONSULT, WESTINGHOUSE NUCLEAR FUEL DIV, PITTSBURGH, WESTINGHOUSE ELEC CORP, 69- Mem: AAAS; Am Phys Soc; Am Nuclear Soc; Am Mgt Asn. Res: Ballistics and aerodynamics of various blunt configurations; operations research applied to chemical, biological and radiological warfare and to selection of new products and new business ventures. Mailing Add: 1190 Colgate Dr Monroeville Pa 15146

ABRASH, HENRY I, b Paterson, NJ, Sept 27, 35; m 62. ORGANIC CHEMISTRY, BIOCHEMISTRY. Educ: Harvard Univ, BA, 56; Calif Inst Technol, PhD(chem), 61. Prof Exp: Proj asst chem, Univ Wis, 60-61; from asst prof to assoc prof, 61-70, PROF CHEM, CALIF STATE UNIV, NORTHRIDGE, 70- Concurrent Pos: USPHS grant, 62-68; Hays-Fulbright travel grant, Carlsberg Labs, Copenhagen Univ, 68-69. Mem: Am Chem Soc; The Chem Soc. Res: Mechanism of enzyme action, particularly on hydrolytic enzymes; mechanism of hydrolytic reactions; proteins in solution. Mailing Add: Dept of Chem Calif State Univ Northridge CA 91324

ABRELL, JOHN WILLIAM, b Martinsburg, WVa, Aug 30, 36; m 67. BIOORGANIC CHEMISTRY. Educ: Duke Univ, BS, 58; Dartmouth Col, MA, 60; Wash Univ, PhD(chem), 65. Prof Exp: NIH fel, 65-68; res chemist, Nat Inst Arthritis & Metab Dis, NIH, 68-71; biochemist, Litton-Bionetics, Inc, 71-74; SCI ADMINR, DIV LUNG DIS, NAT HEART & LUNG INST, NIH, 74- Mem: Am Chem Soc; Sigma Xi; Am Asn Cancer Res; AAAS. Res: Enzymology; ribonuclease I for E coli; mechanism of action; nucleic acid metabolism, in particular DNA synthesis by the DNA polymerases found in RNA viruses; isolation and characterization of mammalian cellular DNA polymerases. Mailing Add: Div of Lung Dis WW Bldg Rm 6A07 Nat Heart & Lung Inst NIH Bethesda MD 20014

ABRUZZO, JOHN L, b West New York, NJ, Apr 27, 31; m; c 7. MEDICINE. Educ: St Peter's Col, NJ, BS, 53; Georgetown Univ, MD, 57. Prof Exp: Jr intern, Childrens Convalescent Hosp, DC, 56-57; med intern, Jersey City Med Ctr, NJ, 57-58; asst in med, Seton Hall Col Med, 58-60, instr, 63-64; asst prof, NJ Col Med, 64-67; from asst prof to assoc prof, 67-74, PROF MED, JEFFERSON MED COL, 74- Concurrent Pos: USPHS fel, Columbia-Presby Med Ctr, 60-61, vis fel, Col Physicians & Surgeons, 63-64; Arthritis Found fel, 63-66; NIH res grants, 64-70; from jr asst med resident to sr asst med resident, Jersey City Med Ctr, 58-60; spec investr, Arthritis Found, 66-67; attend physician, Thomas Jefferson Univ Hosp. Mem: Am Fedn Clin Res; AMA; fel Am Col Physicians; Am Rheumatism Asn. Mailing Add: Jefferson Med Col 1025 Walnut St Philadelphia PA 19107

ABSE, DAVID WILFRED, b Cardiff, Wales, Mar 15, 15; US citizen; c 2. PSYCHIATRY. Educ: Univ Wales, BSc, 35, MD, 48; Welsh Nat Sch Med, MB, BCh, 38; Univ London, DPM, 40. Prof Exp: Dep med supt, Monmouthshire Ment Hosp, Abergavenny, UK, 47-48; chief asst psychiatrist, Med Sch, Charing Cross Hosp, London, 49-51; clin dir, Dorothea Dix State Hosp, Raleigh, NC, 52-53; from assoc prof to prof psychiat, Univ NC, Chapel Hill, 53-62; PROF PSYCHIAT, SCH MED, UNIV VA, 62- Concurrent Pos: Lectr psychol, City Col, London, 50-51; lectr psychol & clin assoc prof psychiat, Univ NC, Chapel Hill, 52-53; teaching analyst, Washington Psychoanal Inst, 62- & Washington Sch Psychiat, 71- Mem: Fel Royal Soc Med; fel Am Psychiat Asn; Brit Med Asn; Am Psychoanal Asn; Int Psychoanal Asn. Res: Clinical evaluation of psychotropic drugs; psychiatric aspects of cancer; psychotherapy of schizophrenics. Mailing Add: Box 203 Dept of Psychiat Univ of Va Hosp Charlottesville VA 22901

ABSHIRE, CLAUDE JAMES, b Kaplan, La, Nov 19, 33; Can citizen. BIOCHEMISTRY, MICROBIOLOGY. Educ: Univ Southwestern La, BSc, 55; Univ Tex, Austin, MA, 57, PhD(org chem), 60. Prof Exp: Res assoc chem, Univ Ill, Urbana, 59-60; res chemist, fabrics and finishes dept, Du Pont Chem Co, Pa, 60-61;

res assoc chem, Mass Inst Technol, 61-62; fel biochem, Laval Univ, 62-64, from asst prof to assoc prof, 64-72; PROF CHEM, UNIV QUE, MONTREAL, 72- Concurrent Pos: Scholar Med Res Coun Can, 65-68. Res: Synthesis of new unnatural amino acids and the study of their metabolic action in microbial systems; food science research. Mailing Add: Dept of Chem Univ of Que Montreal PQ Can

ABSOLON, KAREL B, b Brno, Czech, Mar 21, 26; US citizen; m 54; c 5. THORACIC & CARDIOVASCULAR SURGERY. Educ: Masaryk Univ, BM, 48; Yale Univ, MD, 52; Univ Minn, MS & PhD(physiol), 63; Am Bd Surg, dipl; Am Bd Thoracic Surg, dipl; Am Bd Cardiovasc Dis, dipl. Prof Exp: Asst prof surg, Med Col, Univ Minn, 63-69; asst prof, Univ Tex Health Sci Ctr Dallas, 69-70; ASSOC PROF SURG, MED SCH, GEORGE WASHINGTON UNIV, 70-; CHMN DEPT SURG, WASHINGTON HOSP CTR, 70- Concurrent Pos: Am Cancer Soc fel, 54-56; USPHS trainee, 58-61, spec fel, 61-63; chg surg, Cardiopulmonary Inst, Methodist Hosp, Dallas, 69-70. Mem: Am Col Angiol; NY Acad Sci; Am Heart Asn; Am Col Surg; AMA. Res: Clinical and experimental surgery; experimental physiology and surgery; cardiovascular and pulmonary and hepatic physiology and surgery; history of surgical development. Mailing Add: Washington Hosp Ctr Dept of Surg 110 Irving St Washington DC 20010

ABSON, DEREK, b Wakefield, Eng, July 9, 29. PULP CHEMISTRY, PAPER CHEMISTRY. Educ: Univ Birmingham, Eng, BSc, 52, MSc, 61; Univ BC, PhD(chem), 65. Prof Exp: Sr scientist cellulose res, Columbia Cellulose Co Ltd, 65-68; SECT MGR FIBER RES, WEYERHAEUSER CO, 68- Mem: Tech Asn Pulp & Paper Indust; Am Chem Soc. Res: Management of a program of longer range research projects aimed at developing new and improved products and processes of future value to the pulp and paper industry. Mailing Add: Weyerhaeuser Co 3400 13th Ave SW Seattle WA 98134

ABT, HELMUT ARTHUR, b Helmstedt, Ger, May 26, 25; US citizen. ASTRONOMY. Educ: Northwestern Univ, BS, 46, MS, 48; Calif Inst Technol, PhD(astron), 52. Prof Exp: Jr res astronomer, Lick Observ, Univ Calif, 52-53; res assoc, Yerkes Observ, Univ Chicago, 53-55, asst prof, 55-59; assoc astronomer, 59-63, ASTRONOMER, KITT PEAK NAT OBSERV, 63- Concurrent Pos: Consult, NASA, 67; assoc ed, Astron J, 67-69; vpres comn 30, Int Astron Union, 67-70; managing ed, Astrophys J, 71- Mem: AAAS; Am Astron Soc; fel Royal Astron Soc. Res: Stellar spectroscopy and rotation; spectroscopic binaries. Mailing Add: Kitt Peak Nat Observ Box 26732 Tucson AZ 85726

ABTS, MARY LAVONNE, b Fountain City, Wis, Feb 17, 20. CHEMISTRY. Educ: Viterbo Col, BS, 51; St Louis Univ, PhD(chem), 57. Prof Exp: Parochial sch teacher, Wis, 38-51 & Iowa, 51-52; from teacher to assoc prof & chmn dept, 56-71, PROF CHEM, VITERBO COL, 71- Mem: Am Chem Soc. Res: Kinetics and mechanism of acid-catalyzed dehydration of carbinols. Mailing Add: Dept of Chem Viterbo Col 815 S Ninth St La Crosse WI 54601

ABU-ISA, ISMAT ALI, b Tarshiha, Palestine, Dec 12, 38; m 66; c 1. PHYSICAL CHEMISTRY, POLYMER CHEMISTRY. Educ: Am Univ Beirut, BSc, 59, MS, 61; Northwestern Univ, PhD(phys chem), 65. Prof Exp: Asst prof chem, St Dunstan's Univ, 64-67; assoc sr res chemist, 67-69, SR RES CHEMIST, GEN MOTORS TECH CTR, 69- Concurrent Pos: Nat Res Coun Can res awards, 65-67. Mem: AAAS; Am Chem Soc. Res: Physical chemistry of polymers, their thermodynamics properties, oxidation and surface characteristics; thermodynamics and kinetics of chelates and complexes; radiation chemistry of polymers. Mailing Add: Polymers Dept GM Res Labs GM Tech Ctr 12 Mile & Mound Rd Warren MI 48090

ABUL-HAJJ, JUSUF J, b Jerusalem, Palestine, Apr 3, 40; m 65; c 1. BIOCHEMISTRY, ORGANIC CHEMISTRY. Educ: Am Univ, Beirut, BSc, 62, MSc, 64; Univ Wis-Madison, PhD(pharmaceut biochem), 68. Prof Exp: Asst chem, Am Univ, Beirut, 62-64; asst pharmaceut biochem, Univ Wis-Madison, 64-68; ASST PROF PHARMACOG, UNIV MINN, MINNEAPOLIS, 68- Res: Mechanisms of enzymic reactions; mechanisms in steroid metabolism; biotransformation of steroids and alkaloids. Mailing Add: Col of Pharm Univ of Minn Minneapolis MN 55455

ABUNGU, CORNELIO OYOLA, b Kamagambo, Kenya, Jan 1, 35; m 73; c 1. MATHEMATICS. Educ: Jarvis Christian Col, BA, 67; ETex State Univ, MS, 69, EdD(math educ), 75. Prof Exp: Music librn, ETex State Univ, 67-69; instr math, 69-70, ASST PROF MATH, JARVIS CHRISTIAN COL, 70-, DIR INSTNL RES OFF, 74- & HEAD DEPT MATH & PHYSICS, 75- Mailing Add: Dept of Math & Physics Jarvis Christian Col Box 47 Hawkins TX 75765

ABUSHANAB, ELIE, b Damascus, Syria, Nov 6, 36; US citizen; m 65; c 3. PHARMACEUTICAL CHEMISTRY. Educ: Am Univ Beirut, BSc, 60; Univ Wis, MSc, 62, PhD(pharmaceut chem), 65. Prof Exp: Asst prof pharmaceut chem, Univ Md, Baltimore, 65-67; staff chemist, Chas Pfizer & Co, Inc, Conn, 67-70; asst prof, 70-73, ASSOC PROF MED CHEM, UNIV RI, 73- Mem: Am Chem Soc; Sigma Xi. Res: Chemistry of naturally occurring and synthetic 9-beta, 19 cyclosteroids; synthesis of 9, 19- abeo analogs of biologically active steroids; synthesis of novel nitrogen heterocycles as potential anti-tumor agents. Mailing Add: 113 Church St Peace Dale RI 02883

ABU-SHUMAYS, IBRAHIM KHALIL, b Jaffa, Palestine, Feb 11, 37; m 68. APPLIED MATHEMATICS, MATHEMATICAL PHYSICS. Educ: Am Univ Beirut, BS, 60; Harvard Univ, PhD(physics), 66. Prof Exp: Res asst, Harvard Univ, 60-61, teaching fel, 61-64, Off Naval Res asst, 64-66; res assoc appl math, Argonne Nat Lab, 66-69, asst mathematician, Appl Math Div, 69-72; vis assoc prof eng sci dept, Northwestern Univ, 72-73; SR & FEL MATHEMATICIAN, BETTIS ATOMIC POWER LAB, 73- Concurrent Pos: Res fel, Harvard Univ, 67. Mem: Soc Indust & Appl Math. Res: Numerical analysis; mathematical analysis; differential equations; approximation theory; reactor mathematics; reactor science and engineering. Mailing Add: Bettis Atomic Power Lab PO Box 79 West Mifflin PA 15122

ABU-ZAHRA, NADIA, b Guiza Cairo, Egypt. ANTHROPOLOGY. Educ: Cairo Univ, BA, 59; Oxford Univ, dipl anthrop, 62, BLitt, 64, DPhil(anthrop), 68. Prof Exp: Asst prof anthrop, Calgary Univ, 69-70; ASST PROF ANTHROP, UNIV BC, 71- Concurrent Pos: Fel Islamic studies, Grad Sch, Univ Toronto, 70-71. Mem: Can Sociol & Anthrop Asn; fel Royal Anthrop Inst. Res: Tunisia and North Africa as a whole; the Islamic world and civilization. Mailing Add: Dept of Anthrop Univ of BC Vancouver BC Can

ABU-ZEID, MOHYI ELDIN, b Mansoura, Egypt, Sept 10, 36; US citizen; m 69; c 2. MOLECULAR SPECTROSCOPY. Educ: Univ Tenn, MS, 65, PhD(physics), 68. Prof Exp: Consult physics, Oak Ridge Nat Lab, 68-69; asst prof, 69-72, ASSOC PROF PHYSICS, UNIV PR, MAYAGUEZ, 72- Concurrent Pos: Scientist physics, PR Nuclear Ctr, Mayaguez, 70-; NIH grant, 73; NSF grant, 74. Mem: Am Asn Physics Teachers; Sigma Xi. Res: Emission and decay-times of poly-atomic molecules of biological interest under laser, electron and x-ray excitation; high pressure and low temperature physics. Mailing Add: Dept of Physics Univ of PR Mayaguez PR 00708

ABUZZAHAB, FARUK S, SR, b Beirut, Lebanon, Oct 12, 32; m 62; c 4. PSYCHIATRY, PHARMACOLOGY. Educ: Am Univ Beirut, BS, 55, MD, 59; Univ Minn, Minneapolis, PhD, 68. Prof Exp: Pharmaceut Mfr Asn Found fac career develop award & asst prof clin pharmacol, 67-69, asst clin prof psychiat & pharmacol, 69-73, ASSOC CLIN PROF PSYCHIAT, PHARMACOL & FAMILY PRACT, UNIV MINN, 73-; PRES, CLIN PSYCHOPHARMACOL CONSULTS, P A, 73- Concurrent Pos: Consult, Northern Pines Ment Health Clin, Turtle Lake, Wis, 63-70; chmn pharm subcomt, Hastings State Hosp, Minn, 64-, chief psychiat, 70-; clin dir, Clin Psychopharmacol Serv, Univ Minn Hosp, Minneapolis, 67-69; mem subcomt on alcoholism & drug abuse, Minn State Med Soc, 71-; consult polydrug abuse proj, Multi-Resource Ctr, 74- Mem: AMA; Am Psychiat Asn; Am Soc Clin Pharmacol & Therapeut; Int Col Neuropsychopharmacol. Res: Clinical psychopharmacology research in the area of psychoactive drugs mainly in depression, schizophrenia and memory; neurochemical correlates of behavior and drug action; toxicity of lithium carbonate in animals and man. Mailing Add: Univ Minn Hosps Minneapolis MN 55455

ACCARDO, CARL ANTHONY, nuclear physics, atmospheric physics, see 12th edition

ACEDO, GREGORIA N, b Albay, Philippines, Apr 24, 36; m 65; c 1. PLANT GENETICS, PLANT PATHOLOGY. Educ: Univ Philippines, Los Banos, BSA, 59, MS, 65; Univ Mass, Amherst, PhD(plant path), 69. Prof Exp: Res asst plant path, Bur Plant Indust, Philippines, 59-62; res instr, Univ Philippines, 65-66; asst, Univ Mass, 66-69; fel plant path, 70-71, RES ASST PLANT GENETICS, UNIV MO-COLUMBIA, 73- Mem: Philippine Phytopath Soc. Res: Genetic aspect of disease resistance in plants; genetic control of flower differentiation; search for auxotrophs; tissue culture. Mailing Add: 803 Woodrow St Columbia MO 65201

ACERBO, SAMUEL NICHOLAS, b Port Chester, NY, Jan 19, 29. ORGANIC CHEMISTRY. Educ: Iona Col, BS, 55; Fordham Univ, MS, 57, PhD(org chem), 60. Prof Exp: From instr to assoc prof, 58-70, PROF CHEM, IONA COL, 70- Concurrent Pos: NIH res grant, 61-62. Mem: Am Chem Soc. Res: Trace quantities of boron produce profound effects on plant growth; the use of C14 on celi wall formation in plants grown in media of varying boron content. Mailing Add: Dept of Chem Iona Col North Ave New Rochelle NY 10801

ACETO, MARIO DOMENICO GIULIO, b Providence, RI, Oct 26, 30; m 59; c 2. PHARMACOLOGY. Educ: Univ RI, BS, 53; Univ Md, MS, 55; Univ Conn, PhD(pharmacol), 59. Prof Exp: From instr to asst prof pharmacol, Sch Pharm, Univ Pittsburgh, 58-62; assoc res biologist, Sterling Winthrop Res Inst, 62, res biologist & group leader, 63-66, sr res biologist & head sect pharmacol, 66-72, proj dir, 71-72; ASSOC PROF PHARMACOL, MED COL VA, 72- Mem: Fel Am Col Angiol; Am Soc Pharmacol & Exp Therapeut; Am Pharmaceut Asn. Res: Central nervous system pharmacology and drug dependence. Mailing Add: Dept of Pharmacol Med Col of Va Richmond VA 23298

ACHE, HANS JOACHIM, b Steinau, Ger, Jan 18, 31; m 63; c 1. RADIOCHEMISTRY, RADIATION CHEMISTRY. Educ: Univ Cologne, BS, 55, MS, 57, PhD(chem), 59. Prof Exp: Res assoc nuclear chem & radiochem, Inst Nuclear Chem, Univ Cologne, 59-62; res assoc radiochem, Brookhaven Nat Lab, 62-64, vis assoc chemist, 64-65; from asst prof to assoc prof, 65-71, PROF RADIOCHEM, VA POLYTECH INST & STATE UNIV, 71- Mem: Am Chem Soc; Am Nuclear Soc; Soc Ger Chem. Res: Hot atom chemistry; positronium chemistry; production of radioisotopes; radioactive labeling of organic compounds. Mailing Add: Dept of Chem Va Polytech Inst & State Univ Blacksburg VA 24061

ACHEE, FRANCES M, biochemistry, see 12th edition

ACHESON, CYRUS HAROLD, b Johannesburg, SAfrica, Jan 15, 13; m 44; c 2. GEOPHYSICS. Educ: Univ Toronto, BA, 37, MA, 38. Prof Exp: Gravity meter operator, Trop Oil Co Ltd, 38-40, gravity supvr, 40-43; gravity interpreter & supvr, Imp Oil Ltd, 43-46; gravity interpreter, Int Petrol Co, 46-47; sr gravity interpreter, Trop Oil Co, Ltd, 47-50; seismic comput, Imp Oil Co Ltd, 50-53; res geophysicist, 53-71; RES GEOPHYSICIST, PAN ARCTIC OILS CO, 71- Mem: Soc Explor Geophysicists; Can Soc Explor Geophysicists; Can Asn Petrol Geologists; Europ Asn Explor Geophysicists. Res: Improvement of methods if interpretation. Mailing Add: 435 49th Ave SW Calgary AB Can

ACHESON, DONALD THEODORE, b Waltham, Mass, June 18, 35; m 61; c 2. METEOROLOGY, INSTRUMENTATION. Educ: Harvard Univ, BA, 57; Univ Md, MS, 65, PhD(meteorol), 74. Prof Exp: Gen engr physics, Engr Res & Develop Labs, Dept Army, 57-61; gen engr physics, Nat Weather Serv, 61-72, GEN ENGR PHYSICS & METEOROL, CTR EXP DESIGN & DATA ANAL, NAT OCEANIC & ATMOSPHERIC ADMIN, 72- Mem: Am Meteorol Soc; AAAS. Res: Origin and propagation of instrumental errors and optimal processing of observations in meteorology. Mailing Add: Ctr for Exp Design & Data Anal 3300 Whitehaven St NW Washington DC 20234

ACHESON, GEORGE HAWKINS, b Pittsburgh, Pa, Feb 11, 12; m 35; c 3. MEDICINE. Educ: Harvard Univ, AB, 33, MD, 37. Prof Exp: Intern, Mass Gen Hosp, 38-39; instr physiol, Harvard Med Sch, 39-42, assoc pharmacol, 42-46, asst prof, 46-48; prof pharmacol, Col Med, Univ Cincinnati, 48-75; PRACT MED, 75- Concurrent Pos: Ed, Pharmacol Rev, 62-70; mem med bd, Grass Found. Mem: Int Brain Res Orgn; Am Soc Pharmacol & Exp Therapeut; Soc Exp Biol & Med. Mailing Add: 25 Quissett Ave Woods Hole MA 02543

ACHESON, WILLARD PHILLIPS, b Cairo, Egypt, Nov 24, 27; US citizen; m 49; c 3. PHYSICS, APPLIED MATHEMATICS. Educ: Westminster Col, Pa, BS, 48; Pa State Univ, MS, 50; Univ Pittsburgh, PhD(appl math), 61. Prof Exp: Teacher sci & Head dept, Assiut Col, Egypt, 50-53; fel struct clay prod, Mellon Inst, 53-60; asst prof physics, Muskingum Col, 60-62; sr res engr petrophys, 62-75, RES ASSOC ENERGY RESOURCES, GULF RES & DEVELOP CO, 75- Mem: Am Phys Soc; Am Asn Physics Teachers. Res: Plasticity; rheology; fluid flow; acoustics; petrophysics; enhanced recovery technology of petroleum and alternative fuels. Mailing Add: Gulf Res & Develop Co PO Drawer 2038 Pittsburgh PA 15230

ACHEY, FREDERICK AUGUSTUS, b Lancaster, Pa, Sept 10, 29; m 58; c 3. ANALYTICAL CHEMISTRY. Educ: Franklin & Marshall Col, BS, 51; Lehigh Univ, MS, 53, PhD, 56. Prof Exp: Asst, Lehigh Univ, 51-56; vpres, Serfass Corp, Pa, 56-60; supvr processes & temperature, 60-73, ASST SECT MGR, BETHLEHEM STEEL CORP, 73- Mem: Am Chem Soc (treas, 63); Instrument Soc Am. Res: Analytical instrumentation; x-ray spectroscopy. Mailing Add: 3345 Nazareth Pike Bethlehem PA 18017

ACHEY, PHILLIP M, b Lancaster, Pa, June 25, 39; m 64; c 4. RADIATION BIOPHYSICS. Educ: Franklin & Marshall Col, AB, 61; Pa State Univ, MS, 64, PhD(biophys), 66. Prof Exp: Fel, M D Anderson Hosp & Tumor Inst Houston, 66-67; asst prof biophys, 67-75, ASSOC PROF MICROBIOL, UNIV FLA, 75- Mem:

Biophys Soc; AAAS; Am Soc Microbiol; Radiation Res Soc; Sigma Xi. Res: Biological and physical action of radiation on cells and DNA; repair of radiation damage. Mailing Add: Radiation Biol Lab 403 Nuclear Sci Bldg Univ Fla Gainesville FL 32601

ACHHAMMER, BERNARD GEORGE, b Saugerties, NY, June 20, 20; m 46; c 2. POLYMER CHEMISTRY. Educ: Am Univ, BS, 43, MA, 56. Prof Exp: Chemist, Resinous Prod & Chem Co, Philadelphia, Pa, 43-44; polymer res, Nat Bur Standards, 46-52, asst chief org plastics sect, 52-61; POLYMER PROG MGR, MAT PROG, NASA HQ, 61- Honors & Awards: Silver Medal, US Dept Commerce, 60. Mem: Soc Plastics Eng (int pres, 67). Res: Chemical structure and degradation of high polymers; specifications and test methods for plastics; administration of research organizations; program management; flammability of polymers. Mailing Add: Mat Prog NASA Code RWM 600 Independence Ave SW Washington DC 20546

ACHILLES, ROBERT F, b Inman, Kans, July 5, 24; m 46; c 2. SPEECH PATHOLOGY, AUDIOLOGY. Educ: McPherson Col, BS, 49; Wichita State Univ, MA, 55, PhD(logopedics), 68. Prof Exp: Speech pathologist, Inst Logopedics, 49-52; dir speech & hearing, Children's Rehab Ctr, Md, 52-53; clin supvr, Inst Logopedics, 53-58, dir clin serv, 58-60; assoc prof, 63-68, PROF SPEECH, LAMAR UNIV, 68-, DIR SPEECH PATH, 63- Concurrent Pos: Lectr logopedics, Wichita State Univ, 55-57, instr, 57-60. Mem: Am Speech & Hearing Asn. Res: Cerebral palsy and speech; voice science. Mailing Add: Speech & Hearing Ctr Lamar Univ Box 10076 Beaumont TX 77705

ACHOLONU, ALEXANDER DOZIE, b Owerri, Nigeria, Nov 30, 32; m 67; c 5. ZOOLOGY. Educ: Howard Univ, BS, 58; Prairie View Agr & Mech Col, MS, 61; Colo State Univ, PhD(zool), 64. Prof Exp: Asst instr zool, Colo State Univ, 61-62; assoc prof, Southern Univ, 64-68, prof, 68-69; assoc prof, Inter Am Univ PR, San German, 69-70; prof zool & chmn dept biol, Inte Am Univ PR, San Juan, 70-72; prof & chmn res & develop, Cath Univ, PR, 72-73; actg chmn int & foreign area studies, 73-76, DEAN LIB STUDIES, STATE UNIV NY COL ONEONTA, 73- Concurrent Pos: Deleg, Int Cong Protozool, London, Eng, 65, Leningrad, USSR, 69 & Clermont Ferrand, France, 73; deleg, Int Cong Parasitol, DC, 70 & Munich, Ger, 74; guest lectr, Cath Univ, PR, 70; NSF inst grant; Sigma Xi grants-in-aid; Caribbean Inst & Study Ctr for Latin Am grant. Mem: Am Soc Parasitol; Soc Protozool; Wildlife Dis Asn; Am Micros Soc; Am Soc Trop Med & Hyg. Res: Hematozoa; taxonomy of helminth parasites; biology and life history of trematodes; human parasitism. Mailing Add: Dean of Lib Studies State Univ NY Col Oneonta NY 13820

ACHOR, WILLIAM THOMAS, b Birmingham, Ala, June 18, 29; m 58; c 2. PHYSICS. Educ: Ala Polytech Inst, BS, 52; Vanderbilt Univ, MS, 54, PhD, 58. Prof Exp: Instr physics, Western Reserve Univ, 57-59; mem tech staff, Radio Corp Am Labs, 59-62; asst prof physics, Earlham Col, 63-65; PROF PHYSICS & CHMN DEPT, WESTERN MD COL, 65- Mem: Am Phys Soc; Am Asn Physics Teachers. Res: Nuclear spectrometry; Auger effect; physical electronics; history of science; science education. Mailing Add: Dept of Physics Western Md Col Westminster MD 21157

ACHORN, PETER J, surface chemistry, colloid chemistry, see 12th edition

ACITELLI, MARIO A, organic chemistry, polymer chemistry, see 12th edition

ACKART, RICHARD JENKS, b Wilmington, Del, Mar 24, 15; m 41; c 3. PATHOLOGY. Educ: Wesleyan Univ, BA, 37; Univ Rochester, MD, 42; Columbia Univ, MS, 48. Prof Exp: Asst dir, Johns Hopkins Hosp, 47-51; dir, Va Hosp, 51-53; exec dir, Blue Cross & Blue Shield Asn, Richmond, 53-61; asst dir, Am Hosp Asn, 61-66; asst resident pathologist, NC Baptist Hosp, Winston-Salem, 66-70; PATHOLOGIST, CAPE FEAR MEM HOSP, 70- Mem: Am Soc Clin Path; Col Am Path. Mailing Add: 6302 Mallard Dr Wilmington NC 28401

ACKART, WATSON BOUDINOT, b New Brunswick, NJ, July 4, 20; m 44; c 2. INDUSTRIAL MICROBIOLOGY. Educ: Rutgers Univ, BS, 41, MS, 50, PhD(bact), 54. Prof Exp: Asst bact, Rutgers Univ, 41-42; assoc microbiologist, Merck & Co, Inc, 46-52; biochemist, 54-55, develop assoc, 55-62, proj scientist, 62-69, res scientist, 69-75, SPECIALIST GOVT LIAISON, UNION CARBIDE CORP, 75- Mem: Am Soc Microbiol; Inst Food Technol; Am Soc Testing Mat. Res: Food bacteriology and food preservation; food additives; biodegradation. Mailing Add: Union Carbide Corp River Rd Bound Brook NJ 08805

ACKELL, EDMUND FERRIS, b Danbury, Conn, Nov 29, 25; m 53; c 4. MEDICAL ADMINISTRATION, MAXILLOFACIAL SURGERY. Educ: Col Holy Cross, BS, 49; Tufts Univ, DMD, 53; Western Reserve Univ, MD, 62; Am Bd Oral Surg, dipl. Prof Exp: From asst prof to assoc prof oral surg & chmn dept, Western Reserve Univ, 57-66, asst prof oral path, 62-66; dean col dent, Univ Fla, 66-69, vpres health affairs, J Hillis Miller Health Ctr, 69-74; VPRES HEALTH AFFAIRS, UNIV SOUTHERN CALIF, 74- Concurrent Pos: Am Cancer Soc clin res fel, 55-56; consult, Vet Admin Hosp, Crilĕ, 63-; mem rev comt, Health Professions Educ Act, 64-; mem med adv bd, Vet Admin Hosp, Brecksville, 64- Mem: Am Dent Asn; AMA. Res: Dental education; education in the health professions. Mailing Add: Off of Vpres Health Affairs Univ of Southern Calif Los Angeles CA 90007

ACKER, DONALD STANLEY, organic chemistry, see 12th edition

ACKER, DUANE CALVIN, b Atlantic, Iowa, Mar 13, 31; m 52; c 2. ANIMAL NUTRITION. Educ: Iowa State Univ, BS, 52, MS, 53; Okla State Univ, PhD(animal nutrit), 57. Prof Exp: Asst animal nutrit, Iowa State Univ, 52-53; instr animal husb, Okla State Univ, 53-55; from instr to asst prof animal sci, Iowa State Univ, 55-58, assoc prof animal sci in-chg farm oper curriculum, 58-62; assoc dean agr & dir resident instr, Kans State Univ, 62-66, asst dir Agr Exp Sta, 65-66; dean agr & biol sci & dir agr exp sta, SDak State Univ, 66-74, dir coop exten serv, 71-74; vchancellor agr & natural resources, Univ Nebr-Lincoln, 74-75; PRES, KANS STATE UNIV, 75- Concurrent Pos: Mem adv comt to Iowa Legis Res Bur, Higher Educ, 60; exec comt, Great Plains Agr Coun, 66-67; US team for review of Marshall Plan Aid to WGer, 67; co-chmn, USDA, Exp Sta Task Force Qual of Environ, 67-68; mem, Nat Gov Conf Task Force, Rural-Urban Develop, 69-70; del, Orgn Econ Coop & Develop Conf Higher Educ in Agr, Paris, France, 70. Honors & Awards: Fed Land Bank Golden Anniversary Medal, Contrib to Agr, 67. Mem: AAAS; Am Soc Animal Sci. Res: Meat quality as influenced by nutrients, hormones, drugs; protein and amino acid requirements and metabolism. Mailing Add: 106 Anderson Hall Kans State Univ Manhattan KS 66506

ACKER, GEORGE GERALD, ecology, see 12th edition

ACKER, GERALDINE ENOD, b Fargo, NDak, June 7, 16. FOODS, NUTRITION. Educ: NDak Agr Col, BS, 37; Kans State Col, MS, 49. Prof Exp: High sch teacher, NDak, 37-40; asst state supvr prod proj, War Proj Admin, NDak, 40-41; supv & high sch teacher, NDak, 41-43; instr, Food Serv Sch, US Dept Army, 46-48; EXTEN SPECIALIST FOODS & NUTRIT, UNIV ILL, URBANA & USDA, 49-, PROF, 67- Res: Sorghum flour. Mailing Add: Home Econ Exten 533 Bevier Hall Univ of Ill Urbana IL 61801

ACKER, ROBERT FLINT, b Chicago, Ill, Aug 24, 20; m 48; c 4. MICROBIOLOGY. Educ: Ind Univ, BA, 42, MA, 48; Rutgers Univ, PhD(microbiol), 53; Am Bd Microbiol, dipl, 65. Prof Exp: Microbiologist, Merck & Co, Inc, 48-50; Eli Lilly fel mold metab, Purdue Univ, 53-54; asst prof bact, Iowa State Univ, 54-59; asst proj off, Microbiol Assocs, Inc, 59-60, chief qual control dept, 60-62, chief prod dept, 61-62; head microbiol br, Off Naval Res, 62-69; prof biol, asst dean fac res & dir off res coord, Northwestern Univ, 69-74; EXEC DIR, AM SOC MICROBIOL, 74- Mem: Am Soc Microbiol; Soc Indust Microbiol; Brit Soc Gen Microbiol. Res: Science administration; virology; microbial physiology; bacteriophage; lysozyme; antibiotics. Mailing Add: Am Soc for Microbiol 1913 I St NW Washington DC 20006

ACKER, THOMAS STEPHEN, invertebrate zoology, entomology, see 12th edition

ACKER, WILLIAM JAMES, b Junction City, Kans, June 8, 24; m 44; c 2. GEOGRAPHY, ECONOMICS. Educ: Purdue Univ, BS, 49; Univ Kans, MS, 55; Syracuse Univ, MA, 57, PhD(soc sci), 69. Prof Exp: Pres, Skyway Aviation Corp, 49-51; ASSOC PROF GEOG, ARIZ STATE UNIV, 70- Mem: Asn Am Geog; Nat Coun Geog Educ; Am Inst Urban & Regional Affairs. Res: Spatial analysis of economic aspects of urban growth. Mailing Add: Dept of Geog Ariz State Univ Tempe AZ 85281

ACKERMAN, BRUCE DAVID, b New York, NY, Mar 12, 34; m 57; c 2. PEDIATRICS. Educ: NY Univ, AB, 54; Univ Chicago, MD, 58. Prof Exp: Fel neonatology, Magee-Womens Hosp, Univ Pittsburgh, 63-65; from instr to assoc prof pediat, Univ Calif, Irvine, 65-72; assoc prof, State Univ NY Stony Brook, 72-74; assoc prof, Mt Sinai Sch Med, 74-75; ASSOC PROF PEDIAT, STATE UNIV NY DOWNSTATE MED CTR, 75- Mem: Soc Pediat Res; Am Acad Pediat. Res: Perinatal physiology; fetal, placental and neonatal blood flow; effect of narcotic drugs on the fetus and newborn. Mailing Add: Maimonides Med Ctr Brooklyn NY 11234

ACKERMAN, CLEMENS JOHN, b Chicago, Ill, Apr 24, 24; m 55. BIOCHEMISTRY, NUTRITION. Educ: Univ Ill, BS, 51; Ala Polytech Inst, MS, 52, PhD, 55. Prof Exp: Assoc prof, 55-69, PROF BIOCHEM & NUTRIT, VA POLYTECH INST & STATE UNIV, 69- Concurrent Pos: Consult & head biochem lab, Surv in Libya, NIH, 57. Res: Protein reactions with ethylene oxide; choline biosynthesis; thyroid function. Mailing Add: Dept of Biochem & Nutrit Va Polytech Inst & State Univ Blacksburg VA 24061

ACKERMAN, CLINTON CRAIG, cryogenics, solid state physics, see 12th edition

ACKERMAN, DONALD GODFREY, JR, b McAllen, Tex, Apr 8, 44. ANALYTICAL CHEMISTRY. Educ: Univ Wis-Madison, BS, 66; Univ Ariz, PhD(chem), 73. Prof Exp: MEM TECH STAFF CHEM, TRW SYSTS GROUP, 74- Mem: Am Chem Soc. Res: Physicochemical bases of gas chromatographic separations; application of chromatographic techniques to detection of extraterrestrial life; studies on new battery systems; trace analysis in environmental chemistry. Mailing Add: TRW Systs Group 1 Space Park O1/2020 Redondo Beach CA 90278

ACKERMAN, DONALD REID, human biology, anthropology, see 12th edition

ACKERMAN, EUGENE, b Brooklyn, NY, July 8, 20; m 43; c 3. COMPUTER SCIENCES. Educ: Swarthmore Col, BA, 41; Brown Univ, ScM, 43; Univ Wis, PhD(biophys), 49. Prof Exp: Asst, Brown Univ, 41-43; asst, Univ Wis, 46-47; res asst, Wis Alumni Res Found, 47-49; assoc biophys, Johnson Res Found, Univ Pa, 49-51; asst prof physics, Pa State Univ, 51-58, prof biophys, 58-60; assoc prof biophys, Mayo Grad Sch Med, 60-65, prof, 65-67, PROF BIOMED COMPUT & BIOMET, COL MED SCI, UNIV MINN, MINNEAPOLIS, 67-, DIR DIV HEALTH COMPUT SCI, 69- Concurrent Pos: Johnson Res Found fel, Univ Pa, 57-58; consult, Mayo Found, 60-67, dir comput facil, 63-65; mem var comts, NIH, 64- Mem: AAAS; Asn Comput Mach; NY Acad Sci; Soc Math Biol; Inst Elec & Electronics Eng. Res: Health computer sciences; biomathematics; compartmental analysis; bioacoustics; bioenergetics. Mailing Add: Box 511 Mayo Mem Bldg Univ of Minn Minneapolis MN 55455

ACKERMAN, GUSTAVE ADOLPH, JR, b Columbus, Ohio, m 56; c 2. ANATOMY, MEDICINE. Educ: Ohio State Univ, BA, 48, MSc, 49, MD & PhD(anat), 54. Prof Exp: Intern med, Salt Lake County Hosp, 54-55; from asst prof to assoc prof, 57-64, PROF ANAT, OHIO STATE UNIV, 64- Concurrent Pos: Lederle med fac award, 61-63. Honors & Awards: Borden Award, 54. Mem: Am Asn Anat; Histochem Soc; Am Soc Cell Biol; Am Soc Hemat. Res: Morphology and histochemistry of the cells of the blood and hemopoietic system; cellular differentiation and origin. Mailing Add: Dept of Anat Ohio State Univ Columbus OH 43210

ACKERMAN, HERVEY WINFIELD, JR, b Easton, Pa, Feb 28, 29; m 47, 64; c 6. ORGANIC CHEMISTRY. Educ: Lafayette Col, BS, 49; Yale Univ, PhD(org chem), 53. Prof Exp: Res chemist, Film Dept, E I du Pont de Nemours & Co, 52-54; res chemist, Nat Aniline Div, Allied Chem Corp, 54-62; res chemist, Air Prod & Chem, Inc, 62-64; res chemist, M & T Chem, Inc, 64-68; opers res, Am Can Co, 68-70, adminr mgt sci, 70-72; assoc, ROI Controls Corp, 72-75; CONSULT, CONSOLIDATED EDISON CO NY, INC, 75- Mem: Am Chem Soc. Res: Management sciences. Mailing Add: 22 Huckleberry Lane Ridgefield CT 06877

ACKERMAN, JAMES HOWARD, b Detroit, Mich, Oct 4, 28; m 54; c 2. MEDICINAL CHEMISTRY. Educ: Univ Mich, BS, 50; Univ Wis, PhD(org chem), 54. Prof Exp: Chemist, Mallinckrodt Chem Works, 54-58; SR RES CHEMIST & GROUP LEADER, STERLING-WINTHROP RES INST, 58- Mem: Am Chem Soc. Res: Radiopaques; steroids. Mailing Add: Sterling-Winthrop Res Inst Rensselaer NY 12144

ACKERMAN, JAMES L, b Elizabeth, NJ, Feb 2, 38; m 70. ORTHODONTICS. Educ: Univ Pa, DDS, 60; Harvard Univ, cert orthod, 62; Am Bd Orthod, dipl, 70. Prof Exp: Res fel orthod, Harvard Univ, 60-62; res assoc orthod, Nat Inst Dent Res, 62-64; dir res, Fairleigh Dickinson Univ, 64-66; assoc prof, 68-71, chmn dept, 68-73, assoc dean student affairs, 70-72, PROF ORTHOD, UNIV PA, 71-, PROF PEDIAT DENT & CHMN DEPT, 73- Concurrent Pos: Pvt practr, Westfield, NJ, 64-72; consult, Children's Specialized Hosp, Mountainside, NJ, 65-72 & Univ Ky, 66-68; sr dentist, Children's Hosp, Philadelphia, 73-; vis prof pediat dent, Harvard Univ, 74-75. Mem: AAAS; Am Dent Asn; Am Asn Orthod; Int Asn Dent Res; Am Soc Dent Children. Res: Skeletal morphogenesis at the cellular, tissue and gross levels; relationship of form and function in bone; growth and development. Mailing Add: Sch of Dent Med Univ of Pa Philadelphia PA 19104

ACKERMAN, JOSEPH FRANCIS, b Newport, Ky, Jan 29, 22; m 46; c 4. ORGANIC CHEMISTRY. Educ: Xavier Univ, Ohio, BS, 42; Univ Notre Dame, MS, 47, PhD(chem), 49. Prof Exp: Res chemist, High Polymers, B F Goodrich Co, Ohio, 49-

51; res chemist, Finishes Div, Interchem Corp, Ohio & NJ, 51-69; DIR PACKAGING RES & DEVELOP, INMONT CORP, 69- Mem: Am Chem Soc. Res: Resinous film formers. Mailing Add: Inmont Corp 1255 Broad St Clifton NJ 07015

ACKERMAN, LAUREN VEDDER, b Auburn, NY, Mar 12, 05; m 39; c 4. PATHOLOGY. Educ: Hamilton Col, AB, 27; Univ Rochester, MD, 32. Hon Degrees: DSc, Hamilton Col, 62. Prof Exp: From intern to resident med, Univ Calif, 32-35; asst path & bact, Univ Rochester, 35-36; resident path, Pondville Cancer Hosp, Mass, 36-37; pathologist, Rutland State Sanitarium, 37-39; instr med & path, Univ Calif, 39-40; pathologist, Ellis Fischel State Cancer Hosp, Mo, 40-48, med dir, 42-48; from asst prof path to prof surg path & path, 42-74, EMER PROF SURG PATH & PATH, SCH MED, WASH UNIV, 74-; PROF PATH, HEALTH SCI CTR, STATE UNIV NY STONY BROOK, 74- Concurrent Pos: Resident, Arroyo Sanitarium, Calif, 33-34. Mem: AMA; Asn Cancer Res; Am Asn Path & Bact; Am Soc Clin Path. Res: Pathologic anatomy of cancer; diagnosis, treatment and prognosis of cancer; surgical pathology. Mailing Add: Dept of Path State Univ of NY Stony Brook NY 11790

ACKERMAN, NEIL RICHARD, b Baltimore, Md, May 1, 43; m 64; c 1. PULMONARY DISEASES, IMMUNOLOGY. Educ: Univ Md, BA, 65, BS, 66, PhD(zool), 69. Prof Exp: Asst embryol & neurophysiol, Univ Md, 66-69; res assoc, Stanford Univ, 70-71; SR PHARMACOLOGIST, PFIZER, INC, 71- Concurrent Pos: Fel immunophysiol, Stanford Univ, 69-70. Mem: AAAS; Am Soc Zool; NY Acad Sci. Res: Developmental physiology; mammalian reproductive physiology; immunophysiology; pulmonary diseases; cell biology; dermatologic diseases. Mailing Add: Med Res Lab Pfizer Inc Groton CT 06340

ACKERMAN, NORMAN BERNARD, b New York, NY, Nov 27, 30; m 53; c 4. SURGERY, CANCER. Educ: Harvard Univ, AB, 52; Univ Pa, MD, 56; Univ Minn, PhD(surg), 64. Prof Exp: From asst prof to assoc prof surg, Sch Med, Boston Univ, 65-69; assoc clin prof surg, Univ Kans, 69-74; PROF SURG, STATE UNIV NY UPSTATE MED CTR, 74- Concurrent Pos: Nat Cancer Inst fel, Univ Minn, 60-63; Am Cancer Soc fac res assoc award, 66-69; attend surgeon, Univ Hosp, Boston & Boston Vet Admin Hosp, 65-69; consult, Providence Vet Admin Hosp, RI, 65-69; chmn dept surg, Menorah Med Ctr, Kansas City, 69-74; prof surg, Univ Mo-Kansas City, 72-74; consult, Syracuse Vet Hosp, 74- Mem: Am Col Surg; Soc Univ Surg; Am Soc Exp Path; Am Asn Cancer Res; Int Soc Surg. Res: Cancer vascularity; gastrointestinal physiology; carcinogenesis. Mailing Add: Dept of Surg State Univ of NY Upstate Med Ctr Syracuse NY 13210

ACKERMAN, ROBERT EDWIN, b Grand Rapids, Mich, May 21, 28; m 52; c 3. ANTHROPOLOGY, ETHNOGRAPHY. Educ: Univ Mich, BA, 50, MA, 51; Univ Pa, PhD(anthrop), 61. Prof Exp: Fel anthrop, Eastern Pa Psychiat Inst, Philadelphia, 59-61; from instr to assoc prof anthrop, 61-71, actg chmn dept, 71-72, PROF ANTHROP, WASH STATE UNIV, 71- Concurrent Pos: Grants, Arctic Inst NAm, Archeol Surv Kuskokwim Bay Area, Alaska, 62-63, Nat Park Serv, Archeol Surv Glacier Bay Nat Monument, 63-65, NSF, Archeol & Ethnol Invest Cape Newenham Region, 66-67 & Archeol Invest Icy Strait Region, 71-72. Mem: Fel AAAS; fel Arctic Inst NAm; fel Am Anthrop Asn; Soc Am Archeol; Soc Hist Archaeol. Res: Archeology and ethnology of the north Pacific Arctic and subarctic, including northeast Asia and northwest North America. Mailing Add: Lab of Anthrop Wash State Univ Pullman WA 99163

ACKERMAN, ROY ALAN, b Brooklyn, NY, Sept 9, 51; m 74. BIOENGINEERING. Educ: Polytechnic Inst Brooklyn, BS, 72; Mass Inst Technol, SM, 75. Prof Exp: Asst mgr chem eng, Arlee, Inc, 67-72; process engr bioeng, Corning Glass Works, 73; chief ed med sci, Linguistics Systems, Inc, 74-75; DIR RES & DEVELOP, TRI-FLO RES LABS, LTD, NEW YORK, 72- Concurrent Pos: Sr consult chem eng & bioeng, Appl Sci Through Res & Eng, Ltd, 75- Mem: Int Ozone Inst; Am Inst Chem Engrs; Am Chem Soc; Am Soc Artificial Internal Organs; Am Soc Microbiol. Res: Development of biosupport apparatus drawing upon chemical engineering principles, hemodialysis, aerosol therapy, and membrane oxygenators; development of recycling systems for water and waste utilizing microbial and physicochemical techniques. Mailing Add: 3135 Shore Rd Bellmore Harbor NY 11710

ACKERMAN, WILLIAM VAUGHN, b Sundance, Wyo, Jan 30, 44; m 70. GEOGRAPHY OF LATIN AMERICA, ECONOMIC GEOGRAPHY. Educ: Univ Wyo, BA, 66, MA, 68; Ohio State Univ, PhD(geog), 72. Prof Exp: ASST PROF GEOG & DEPT CHMN, CALIF STATE COL, SAN BERNARDINO, 72- Concurrent Pos: Vchmn, Joint Utilities Mgt Prog Adv Comt, 75- Mem: Asn Am Geog; Conf Latin Am Geog. Res: Analysis and development of regional economic models for manipulating the geographical aspects of regional development in less developed countries; Latin America, with special interest in Argentina; urban geography. Mailing Add: Dept of Geog Calif State Col San Bernardino CA 92407

ACKERMANN, GUENTER ROLF, b Fellbach, Ger, Mar 20, 24; US citizen; m 51; c 3. ORGANIC CHEMISTRY, PHARMACEUTICAL CHEMISTRY. Educ: Ursinus Col, BS, 51; Univ Md, MS, 54; Univ Mich, PhD(pharmaceut chem), 56. Prof Exp: Chemist, Synthetic Org Chem, Atlantic Refining Co, Pa, 56-62; SR RES CHEMIST, ROHM AND HAAS CO, PHILADELPHIA, 62- Mem: AAAS; Am Chem Soc. Res: Polymer synthesis; synthesis, spinning and testing of synthetic fibers; synthesis of coagulants and flocculents; development of ion-exchange resins; development of desalination processes; pollution control research; removal of pollutants from waste streams. Mailing Add: 616 Andover Rd Newtown Square PA 19073

ACKERMANN, HANS WOLFGANG, b Berlin, Ger, June 16, 36. MEDICAL MICROBIOLOGY. Educ: Free Univ Berlin, MD, 62. Prof Exp: Asst gynec, Rudolf Virchow Hosp, Berlin, Ger, 61, asst surg & internal med, 62-63; sci asst, Inst Hyg & Med Microbiol, Free Univ Berlin, 63-67; asst prof, 67-70, ASSOC PROF, FAC MED, LAVAL UNIV, 70- Concurrent Pos: Airlift Mem Found res fel, Pasteur Inst, Paris, France, 61-62; French Govt spec training fel med microbiol, 63-64. Mem: Can Soc Microbiol; French Soc Microbiol. Res: Bacteriophages, also their implications with bacterial and viral taxonomy; airborne fungi. Mailing Add: Dept of Microbiol Laval Univ Fac of Med Quebec PQ Can

ACKERMANN, MARTIN NICHOLAS, b Philadelphia, Pa, Feb 19, 41; m 63; c 2. INORGANIC CHEMISTRY. Educ: Carnegie-Mellon Univ, BS, 63; Univ Calif, Berkeley, PhD(chem), 66. Prof Exp: Asst prof, 66-72, ASSOC PROF CHEM, OBERLIN COL, 72- Concurrent Pos: Petrol Res Fund grant, 66-69, 69-72 & 74-76; Res Corp grant, 74-76. Mem: Am Chem Soc. Res: Transition metal organometallic synthesis; coordination chemistry of azo compounds; chemistry of diazenes. Mailing Add: Dept of Chem Oberlin Col Oberlin OH 44074

ACKERMANN, PHILIP GULICK, b Waterville, Minn, May 3, 09; m 45. BIOCHEMISTRY. Educ: Ore State Col, BS, 31; Johns Hopkins Univ, PhD(phys chem), 36. Prof Exp: Chief chemist, Cole Chem Co, 39-45; asst biochem, Sch Med, Washington Univ, 45-54, res assoc, 54-58, res asst prof, 58-64; BIOCHEMIST, DE PAUL HOSP, ST LOUIS, MO, 64- Concurrent Pos: Biochemist, St Louis Chronic Hosp, 47-64; biochem consult, Faith Hosp & Homer G Phillips Hosp. Mem: AAAS; Am Chem Soc; Am Asn Clin Chemists; NY Acad Sci. Res: Determination of molecular structure by electron diffraction; statistical mechanics; analysis of pharmaceutical preparations; biochemistry of aging in man; nutrition and hormonal changes in the aged; blood lipoproteins and atherosclerosis. Mailing Add: 6033 Childress Ave St Louis MO 63109

ACKERMANN, RAYMOND J, b Spearville, Kans, Apr 19, 30; m 55; c 3. HIGH TEMPERATURE CHEMISTRY. Educ: Univ Kans, AB, 52, PhD(chem), 55. Prof Exp: CHEMIST, ARGONNE NAT LAB, 55- Concurrent Pos: Expert thermodyn, Int Atomic Energy Agency, Vienna, 70. Mem: Am Chem Soc; Sigma Xi. Res: Phenomena occurring in high temperature systems; evaporation behavior; vapor pressure; gaseous species in equilibrium with condensed phases; high temperature x-ray diffraction; thermodynamics of lanthanide and actinide compounds. Mailing Add: Chem Div Argonne Nat Lab Argonne IL 60439

ACKERMANN, UWE, b Greifswald, Ger, Dec 24, 39; m 66. MAMMALIAN PHYSIOLOGY. Educ: Univ Toronto, BASc, 66, MASc, 68, PhD(physiol), 73. Prof Exp: Asst prof physiol, Fac Med, Mem Univ, Nfld, 73-76; ASST PROF PHYSIOL, FAC MED, UNIV TORONTO, 76- Res: Mechanisms of extracellular fluid volume regulation. Mailing Add: Dept of Physiol Fac of Med Univ of Toronto Toronto ON Can

ACKERS, GARY KEITH, b Dodge City, Kans, Oct 25, 39; m 59; c 3. PHYSICAL BIOCHEMISTRY. Educ: Harding Col, BS, 61; Johns Hopkins Univ, PhD(biochem), 64. Prof Exp: NIH fel physiol chem, Sch Med, Johns Hopkins Univ, 64-65, instr, 65-66; from asst prof to assoc prof, 66-72, PROF BIOCHEM, UNIV VA, 72- Concurrent Pos: Guggenheim fel, 72-73. Mem: AAAS; Biophys Soc. Res: Interacting systems of macromolecules; molecular sieve methods. Mailing Add: Dept of Biochem Univ of Va Charlottesville VA 22903

ACKLES, KENNETH NORMAN, b Hamilton, Ont, June 29, 35. PHYSIOLOGY. Educ: Queen's Univ, Ont, BSc, 59, MSc, 61; Univ Alta, PhD(physiol), 66. Prof Exp: Sci officer hyperbaric physiol, 67-70, HEAD PRESSURE PHYSIOL GROUP, DEFENCE RES ESTAB, TORONTO, 70- Concurrent Pos: Defence Res Bd Can fel aviation med, Royal Air Force, Eng, 66-67. Mem: Aerospace Med Asn; Undersea Med Soc. Res: Physiological measurement of inert gas narcosis in diving; bubble detection and hematological aspects of decompression sickness; physiological effects of exposure to hyperbaric environment; assessment of stress in operational environments; life support systems in high performance aircraft. Mailing Add: Defence & Civil Inst Environ Med PO Box 2000 Toronto ON Can

ACKLEY, WILLIAM BENTON, b Portis, Kans, June 5, 18; m 41; c 2. HORTICULTURE. Educ: Kans State Col, BS, 40, MS, 47; State Col Wash, PhD(hort), 53. Prof Exp: Tech asst, US Bur Census, Washington, DC, 40-43; asst, Kans State Col, 46-47; res asst, 47-48, horticulturist, 48-64, PROF HORT, WASH STATE UNIV, 48-, CHMN DEPT, 64- Mem: Am Soc Hort Sci; Am Soc Plant Physiologists. Res: Water relations of plants; physiological disorders of tree fruits; orchard pest control. Mailing Add: Dept of Hort Wash State Univ Pullman WA 99163

ACKMAN, ROBERT GEORGE, b Dorchester, NB, Sept 27, 27; m 57; c 2. ORGANIC CHEMISTRY. Educ: Univ Toronto, BA, 50; Dalhousie Univ, MS, 52; Univ London, PhD(org chem), 56; Imp Col, dipl(org chem), 56. Prof Exp: Asst scientist, Atlantic Fisheries Exp Sta, Halifax, NS, 52-53; assoc scientist, 56-62, group leader marine lipids, 62-70, PROG MGR, HALIFAX LAB, FISHERIES RES BD CAN, 70- Concurrent Pos: Mem comt fats & oils, Nat Res Coun Can, 62-66, chmn, 67- Mem: Fel Chem Inst Can; Am Oil Chemists Soc. Res: Analysis and utilization of marine lipids and oils; gas-liquid chromatography of fatty acids and derivatives. Mailing Add: Halifax Lab PO Box 429 Fisheries & Marine Environ Can Halifax NS Can

ACOSTA, DANIEL, JR, b El Paso, Tex, Mar 25, 45; m 73; c 1. PHARMACOLOGY. Educ: Univ Tex, Austin, BS, 68; Univ Kans, PhD(pharmacol), 74. Prof Exp: Res asst pharmacol, Sch Pharm, Univ Kans, 70-74; ASST PROF PHARMACOL, COL PHARM, UNIV TEX, AUSTIN, 74- Mem: AAAS; Tissue Cult Asn. Res: Cellular toxicology, particularly use of cell culture techniques to evaluate the effects of drugs and toxicants at the cellular and subcellular level. Mailing Add: Col of Pharm Univ of Tex Austin TX 78712

ACOSTA, PHYLLIS BROWN, b Postell, NC, Dec 27, 33; wid; c 3. FOODS, NUTRITION. Educ: Andrews Univ, BA, 55; Univ Iowa, MS, 57; Univ Calif, Los Angeles, MPH, 65, DPH, 69. Prof Exp: Teaching dietitian & instr, White Mem Hosp, Loma Linda Univ, 57-59, asst prof & asst dir, Sch Dietetics, 59-62, assoc prof, Sch & dir dietary serv, White Mem Med Ctr, 62-64; prof nutrit, Sch Home Econ & Sch Med Sci, Univ Nev, Reno, 70-74; assoc prof foods & nutrit, Univ Ga, 74-75; ASSOC PROF PEDIAT, SCH MED, UNIV NMEX, 75- Concurrent Pos: Co-investr, NIH Grant Study Phenylketonuria, 61-64; nutrition consult, Child Develop Clin, Children's Hosp Soc, Los Angeles, 62-70, nutritionist, 65-; consult nat collab study phenylketonuria, Maternal & Child Health Sect, Health Serv & Ment Health Admin, US Dept Health, Educ & Welfare; consult pediat, Div Med Genetics, Sch Med, Emory Univ, 75- Mem: Am Inst Nutrit; Am Dietetic Asn; Am Pub Health Asn; Soc Nutrit Educ. Res: Relationship of diet to mental retardation in phenyketonuria and galactosemia; iron nutriton and growth of infants. Mailing Add: Univ NMex Surge Bldg 2701 Frontier Ave NE Albuquerque NM 87131

ACOSTA, VIRGILIO, b Camaguey, Cuba, June 14, 16; US citizen; m 48; c 3. COSMIC RAY PHYSICS. Educ: Univ Havana, DSc(physics, math), 42. Prof Exp: Prof physics & math, Riemann Inst, Cuba, 40-55; asst dean fac sci & technol, Univ Villanueva, 55-60; res prof physics, Cath Univ Am, 60-65, prof, summers 64-69, part-time res prof, Keane Res Ctr, 65-69; ASSOC PROF SCI, US NAVAL ACAD, 65- Concurrent Pos: High sch prof, Cuba, 40-60; head dept physics, Univ Villanueva, 51-60; asst prof, Nat Inst Econ, Cuba, 58-60; Smith-Mundt Act grant physics, Cath Univ Am, 60. Mem: Am Phys Soc; Am Geophys Union; Am Asn Physics Teachers. Res: Neutrino detection; measurement of weak magnetic fields; demonstration equipment showing composition and decomposition of water. Mailing Add: Dept of Physics US Naval Acad Annapolis MD 21402

ACREE, TERRY EDWARD, b West Hamlin, WVa, Aug 6, 40; m 66. FOOD CHEMISTRY. Educ: Univ Calif, Berkeley, BA, 63; Cornell Univ, MS, 66, PhD(biochem), 68. Prof Exp: Asst prof, 68-74, ASSOC PROF BIOCHEM, NY STATE AGR EXP STA, 74- Concurrent Pos: Fulbright-Hays res scholar, Portugal, 72-73. Mem: Am Chem Soc; Inst Food Technol; AAAS; Sigma Xi. Res: Chemistry and biochemistry of the flavor components of fruit and vegetable foods. Mailing Add: NY State Agr Exp Sta Cornell Univ Geneva NY 14456

ACRIVOS, JUANA LUISA VIVO, b June 24, 28; US citizen; m 56. PHYSICAL CHEMISTRY. Educ: Univ Minn, PhD(phys chem, math), 56; Univ Havana, DSc(phys chem), 56. Prof Exp: Res assoc biophys, Hanasen Labs, Stanford Univ, 57-59; fel phys chem, Univ Calif, 59-62; fac res grant, 62, from asst prof to assoc prof,

62-74, PROF PHYS CHEM, SAN JOSE STATE UNIV, 74- Concurrent Pos: Res Corp grants, 64, 66 & 71; NSF res grants, 66, 68 & 70. Mem: Am Chem Soc; Am Phys Soc. Res: Molecular structure determinations of semiquinones and biologically active free radicals by means of ESR spectroscopy; alkali metal and alkali halide solutions in ammonia, amines and ethers. Mailing Add: Dept of Chem San Jose State Univ San Jose CA 95114

ACS, GEORGE, b Dunaszerdahely, Hungary, Aug 14, 23; m 48; c 2. BIOCHEMISTRY. Educ: Orvosegyetem, Budapest, MD, 50, PhD, 53. Prof Exp: Demonstr, Orvosivegytan, Szeged, 47-48; demonstr med chem, Orvosivegytan, Budapest, 48-50, from asst prof to assoc prof, 50-56; res assoc, Rockefeller Inst, 57-60; sr res chemist, NY Psychiat Inst, 60-61; asst prof, Columbia Univ, 61; MEM, INST MUSCLE DIS, 61-; PROF PEDIAT, MT SINAI SCH MED, 74-, PROF BIOCHEM, 75- Concurrent Pos: Res fel, Mass Gen Hosp, 56-57. Mem: Am Soc Cell Biol; Brit Biochem Soc. Res: Protein synthesis. Mailing Add: Dept of Pediat Mt Sinai Sch of Med New York NY 10029

ACTON, EDWARD MCINTOSH, b San Jose, Calif, May 30, 30. BIO-ORGANIC CHEMISTRY. Educ: Stanford Univ, BS, 51; Mass Inst Technol, PhD(org chem), 57. Prof Exp: Chemist, Merck & Co, NJ, 51-53; from org chemist to sr org chemist, 57-75, PROG MGR, STANFORD RES INST, 75- Mem: Am Chem Soc; The Chem Soc; AAAS. Res: Organic synthesis related to sugars, nucleosides, nitrogen heterocycles, alicyclics; study of reactions, neighboring group participation, tertiary amine oxide pyrolysis; stereochemistry; flavor chemistry; development of new artificial sweeteners, new anticancer agents. Mailing Add: 281 Arlington Way Menlo Park CA 94025

ACTON, LOREN WILBER, b Lewistown, Mont, Mar 7, 36; m 57; c 2. SOLAR PHYSICS. Educ: Mont State Col, BS, 59; Univ Colo, PhD(astro-geophys), 65. Prof Exp: Jr scientist, Hanford Atomic Prod Oper, Gen Elec Co, 59; physicist, Cent Radio Propagation Lab, Nat Bur Standards, Colo, 60-61; physicist, US Naval Res Lab, DC, 62; res asst, High Altitude Observ, Colo, 62-64; assoc res scientist x-ray astron, 64-66, res scientist, 66-70, dir, Lockheed Solar Observ, 68-71, staff scientist & group leader space astron, 70-75, SR STAFF SCIENTIST & SR MEM RES LAB, LOCKHEED MISSILES & SPACE CO, 75- Concurrent Pos: Assoc astronomer, Inst Astron, Univ Hawaii, 68-69; vis scientist, Group Phys Space Res, WGer, 71-72. Mem: AAAS; Am Astron Soc; Int Astron Union; Sigma Xi. Res: X-ray astronomy. Mailing Add: Lockheed Palo Alto Res Lab 3251 Hanover St Palo Alto CA 94304

ACTON, W PAUL, organic chemistry, see 12th edition

ACTOR, PAUL, b New York, NY, 1933; m; c 3. CHEMOTHERAPY. Educ: Hunter Col, AB, 55; Rutgers Univ, MA, 57, PhD(zool), 59. Prof Exp: Asst zool, Rutgers Univ, 55-59; sr res scientist, Squibb Inst Med Res, 59-63; sr microbiologist, 63-66, group leader microbiol, 66-68, asst dir microbiol, 68-73, ASSOC DIR MICROBIOL, SMITH KLINE & FRENCH LABS, 73- Mem: Fel Am Acad Microbiol; Am Soc Microbiol; Am Soc Parasitol; Am Soc Trop Med & Hyg; Infectious Dis Soc Am. Res: Chemotherapy and immunology of bacterial, fungal and parasitic infection. Mailing Add: 632 Pickering Lane Phoenixville PA 19460

ACZEL, JANOS D, b Budapest, Hungary, Dec 26, 24; m 46; c 2. MATHEMATICAL ANALYSIS, GEOMETRY. Educ: Eötvös Lorand Univ, Budapest, PhD(math), 47; Hungarian Acad Sci, Habil, 52, DSc(functional equations), 57. Prof Exp: Asst prof math, Univ Szeged, 48-50; assoc prof, Miskolc Inst Technol, 50-52; from assoc prof to prof, Debrecen Univ, 52-65; prof, 65-69, DISTINGUISHED PROF MATH, UNIV WATERLOO, 69- Concurrent Pos: Vis prof, Univ Fla, 63-64, Stanford Univ, 64, Univ Cologne, 65, Univ Giessen, 66, 70, Univ Bochum, 68, Nat Inst Math, Rome, 71 & Monash Univ, 72; NSF grant, 63-64; Nat Res Coun Can grant, 66- Honors & Awards: M Beke Award; Award, Hungarian Acad Sci, 62. Mem: Fel Royal Soc Can; Can Math Cong; Am Math Soc; Austrian Math Soc. Res: Functional equations; inequalities; mean values; geometric objects; generalized groups; webs; nomography; orthogonal polynomials; projective geometry; probability theory; statistical distributions; theory of measurement; theory of information. Mailing Add: Fac of Math Univ of Waterloo Waterloo ON Can

ACZEL, THOMAS, b Nagykanizsa, Hungary, Dec 18, 30; US citizen; m 62; c 4. ANALYTICAL CHEMISTRY. Educ: Univ Trieste, DSc(phys chem), 54. Prof Exp: Chemist, Aquila Co, Italy, 54-55; tech adv, Petroli Aquila Co, 55-58; from res chemist to sr res chemist, Humble Oil & Ref Co, 59-66; res specialist, 66-69, RES ASSOC, EXXON RES & ENG CO, 69- Honors & Awards: Bituminous Coal Res Award, Am Chem Soc. Mem: Am Chem Soc; Am Soc Testing & Mat; Am Soc Mass Spectrometry. Res: Mass spectrometry; gas chromatography; high resolution mass spectrometry; application to characterization of complex petroleum mixtures; automatic data acquisition and handling systems required by such analyses. Mailing Add: PO Box 4255 Baytown TX 77520

ADAIR, CHARLES ROY, agronomy, see 12th edition

ADAIR, DENNIS WILTON, b San Jose, Calif, Dec 7, 39; m; c 1. BIOPHARMACEUTICS. Educ: Univ Calif, San Francisco, PharmD, 64, PhD(pharm chem), 76. Prof Exp: SR SCIENTIST PHARMACEUT, CIBA-GEIGY PHARMACEUT DIV, 73- Mem: Sigma Xi; Am Pharmaceut Asn; Acad Pharmaceut Sci. Res: Gastrointestinal absorption of drugs; drug delivery system with in vivo-in vitro evaluation. Mailing Add: Ciba-Geigy Pharmaceut Div Summit NJ 07901

ADAIR, FRANK WILLIAM, b Philadelphia, Pa, May 30, 40; m 65. MICROBIOLOGY. Educ: Pa State Univ, BS, 63; Rutgers Univ, MS, 65, PhD(microbiol), 67. Prof Exp: Instr microbiol, Univ Hawaii, 67-68; mgr anal microbiol, 68-74, ASST DIR ANAL MICROBIOL & BIOL CONTROL, PHARMACEUT DIV, CIBA-GEIGY CORP, 74- Concurrent Pos: Lectr, Dept Bact & Pub Health, Wagner Col, 69-71. Mem: AAAS; Am Soc Microbiol. Res: Mechanism of action of antimicrobial agents; physiology, metabolism, nutrition and ultrastructure of antibiotic resistant bacteria. Mailing Add: Dept of Microbiol Ciba-Geigy Corp Summit NJ 07901

ADAIR, KENT THOMAS, b Allegan, Mich, Jan 21, 33; m 55; c 3. FORESTRY, MANAGEMENT SCIENCES. Educ: Colo State Univ, BS, 58, PhD(forest econ), 68; Ore State Univ, MF, 61. Prof Exp: Forester, US Forest Serv, 58-59; forester & consult, Bigley & Feiss, Foresters, Inc, 60-61; res forester, Colo State Univ, 61-63; asst prof forest econ, Univ Mont, 63-66; PROF FORESTRY, UNIV MO-COLUMBIA, 67- Mem: Soc Am Foresters; Forest Prod Res Soc. Res: Management, marketing and industrial organization. Mailing Add: Sch of Forestry Univ of Mo Columbia MO 65201

ADAIR, ROBERT KEMP, b Ft Wayne, Ind, Aug 14, 24; m 52; c 3. NUCLEAR PHYSICS. Educ: Univ Wis, PhD(physics), 51. Prof Exp: Instr physics, Univ Wis, 51-53; from assoc physicist to physicist, Brookhaven Nat Lab, 53-58; assoc prof, 58-61, chmn dept, 67-70, PROF PHYSICS, YALE UNIV, 61- Concurrent Pos: Guggenheim fel, 53; Ford Found fel, Europ Orgn Nuclear Res, 62. Mem: Nat Acad Sci; fel Am Phys Soc. Res: Nuclear and particle physics. Mailing Add: Dept of Physics Yale Univ New Haven CT 06520

ADAIR, SUZANNE FRANK, b Milwaukee, Wis, Aug 17, 41; m 66; c 1. BIOPHARMACEUTICS. Educ: Univ Wis, BS, 63; Univ Calif, San Francisco, PhD(pharmaceut chem), 76. Prof Exp: Actg asst prof pharm, Univ Calif, San Francisco, 74; ASST PROF PHARM, RUTGERS UNIV, BUSCH CAMPUS, 74- Mem: Sigma Xi; NY Acad Sci; Acad Pharmaceut Sci; AAAS. Res: Membrane transport and the in vitro absorption of drugs; effect of drugs and various agents on tissue electrical parameters; short-circuit current, potential difference, resistance. Mailing Add: Rutgers Univ Col of Pharm Busch Campus New Brunswick NJ 08903

ADAIR, THOMAS WEYMON, III, b Houston, Tex, Jan 23, 35; m 62. SOLID STATE PHYSICS, LOW TEMPERATURE PHYSICS. Educ: Tex A&M Univ, BS, 57, PhD(physics), 65; Rice Univ, MA, 60. Prof Exp: Res physicist, Prod Res Lab, Humble Oil & Refining Co, 59-62; asst prof, 66-72, asst to pres, 72-75, ASSOC PROF PHYSICS, TEX A&M UNIV, 72- Concurrent Pos: NSF fel, Kamerlingh Onnes Lab, Univ Leiden, 65-66. Mem: Am Phys Soc. Res: Magnetic properties of alkali halide single crystals; photoelectron spectroscopy. Mailing Add: 1008 Dominik College Station TX 77840

ADAIR, WINSTON LEE, JR, b Chicago, Ill, May 27, 44; m 72; c 1. BIOCHEMISTRY. Educ: Brown Univ, ScB, 66; Georgetown Univ, PhD(biochem), 72. Prof Exp: ASST PROF BIOCHEM, COL OF MED, UNIV S FLA, 75- Concurrent Pos: NIH fel, Dept Hematol-Oncol, Sch Med, Wash Univ, 73-75. Mem: Sigma Xi; Am Chem Soc; Soc Complex Carbohydrates. Res: Cell surface glycoproteins; membrane biochemistry; biosynthesis of long chain polyisoprenoid alcohols; interactions of plant lectins with cell membranes. Mailing Add: Dept of Biochem Col of Med Univ of SFla Tampa FL 33620

ADAM, DAVID PETER, b Berkeley, Calif, May 18, 41; m 67; c 2. PALYNOLOGY, CLIMATOLOGY. Educ: Harvard Univ, BA, 62; Univ Ariz, MS, 65, PhD(geochronology), 70. Prof Exp: Res assoc palynology, Univ Ariz, 66-69, teaching asst, 70, res assoc, DendrochronolDendrochronology, Lab Tree-ring Res, 70-71; RES ASSOC PALYNOLOGY, US GEOL SURV, 71- Mem: AAAS; Ecol Soc Am; Glaciol Soc; Brit Freshwater Biol Asn; Am Asn Quaternary Environ. Res: Theoretical palynology; causes of climatic change; computer applications in paleoecology; climatic history of California. Mailing Add: US Geol Surv 345 Middlefield Rd Menlo Park CA 94025

ADAM, FRANK CUTHBERT, b Chemainus, BC, June 21, 27; m 55, 63; c 3. PHYSICAL CHEMISTRY. Educ: Univ BC, 50, MSc, 53; Univ Wash, PhD(chem), 57. Prof Exp: Fel, Univ Wash, 57-58; ASSOC PROF CHEM, UNIV CALGARY, 58- Res: Study of molecular structure and interactions by using the techniques of visible and ultraviolet spectroscopy and paramagnetic and nuclear spin resonance. Mailing Add: Dept of Chem Univ of Calgary Calgary AB Can

ADAM, WALDEMAR, b Alexanderdorf, Ukraine, July 26, 37; US citizen; m 58; c 3. Educ: Univ Ill, Urbana, BSc, 58; Mass Inst Technol, PhD(org chem), 61. Prof Exp: Chemist, Monsanto Chem Co, 59; res assoc, Mass Inst Technol, 61; from asst prof to assoc prof, 61-70, PROF CHEM, UNIV PR, 70- Concurrent Pos: NSF fel, Univ Karlsruhe, 62; A P Sloan fel, 68-72; vis prof, Univ Zurich, Switz, 72-73; res fel, Guggenheim Found, 72-73; career fel, NIH, 75-80. Mem: Am Chem Soc; NY Acad Sci. Res: Mechanisms of free radical reactions; cyclic peroxides; bioluminescence; prostaglandin endoperoxides; diradicals. Mailing Add: Dept of Chem Univ of PR Rio Piedras PR 00931

ADAMACZAK, ROBERT L, b Buffalo, NY, Aug 16, 27; m 58; c 3. PHYSICAL CHEMISTRY. Educ: Univ Buffalo, BA, 51, MA, 55, PhD(chem), 56. Prof Exp: Res analyst propellant chem, Olin Mathieson Chem Corp, 51-53; res chemist fuel chem, Esso Res & Eng Co, Stand Oil Co NJ, 56-58; res phys chemist, Wright Air Develop Ctr, 59-60, BR FLUID & LUBRICANTS, RES & TECHNOL DIV, US AIR FORCE MAT LAB, 60- Concurrent Pos: Mem, Lubricants Comt, Res & Eng Div, US Dept Defense, 61- Mem: Am Chem Soc; Am Soc Mech Engrs; Am Soc Lubrication Engrs; fel Am Inst Chemists. Res: Energy transfer fluids; lubricants; conventional and propellant lubrication fuels; inorganic polymer chemistry; theory of friction and wear; properties of liquid metals and inorganics. Mailing Add: 7556 Beldale Ave Dayton OH 45424

ADAMCIK, JOE ALFRED, b Taylor, Tex, June 28, 30. ORGANIC CHEMISTRY. Educ: Univ Tex, BS, 51, MA, 54, Univ Ill, PhD(chem), 58. Prof Exp: Asst prof, 57-61, ASSOC PROF CHEM, TEX TECH UNIV, 61- Mem: AAAS; Am Chem Soc; Am Geophys Union; Geochem Soc; The Chem Soc. Res: Physical organic chemistry; cosmochemistry. Mailing Add: Dept of Chem Tex Tech Univ Lubbock TX 79409

ADAMEK, EDUARD GEORG, b Mistelbach, Austria, Apr 19, 25; Can citizen; m 54. ENVIRONMENTAL MANAGEMENT, AIR POLLUTION. Educ: Univ Vienna, BSc, 50; Innsbruck Univ, Drs, 52; McGill Univ, PhD(org chem), 59. Prof Exp: Res chemist, Starch & Chem Div, Ogilvie Flour Mills Co, Ltd, 52-58; sr develop chemist, Textile Fibres Div, Du Pont of Can, 59-60, from res chemist to sr res chemist, Cent Res Labs, 60-70; scientist, Air Mgt Br, Ont Dept Energy & Resources, 70-71; ORG LAB SUPVR, ONT MINISTRY ENVIRON, 71- Mem: AAAS; Am Chem Soc; Chem Inst Can; NY Acad Sci. Res: New organic products and processes; polymers; organic air pollutants research; waste product utilization; metallo-organic compounds; analytical methods development; carbohydrates, protein and amino acids. Mailing Add: Ont Ministry of the Environ 880 Bay St Toronto ON Can

ADAMKEWICZ, LAURA, genetics, population biology, see 12th edition

ADAMKIEWICZ, VINCENT WITOLD, b Poland, Nov 27, 24; m 54; c 5. PHYSIOLOGY, IMMUNOLOGY. Educ: Bristol Univ, BSc, 48 & 49, MSc, 50; Univ Montreal, PhD(endocrinol), 53. Prof Exp: Agr engr, Poland, 44; res assoc, Sch Vet Med, Bristol Univ, 50; assoc prof physiol, 56-67, PROF IMMUNOPHYSIOL, FAC MED, UNIV MONTREAL, 67- Mem: Am Physiol Soc; Can Physiol Soc; Asn French Speaking Physiol. Res: Hormone-drug interactions; physiology of immune responses. Mailing Add: Dept of Microbiol & Immunol Univ of Montreal Fac of Med Montreal PQ Can

ADAMO, JOSEPH ALBERT, b Jersey City, NJ, Oct 22, 38; m 64; c 2. NEMATOLOGY. Educ: Jersey City State Col, BA, 64; Fairleigh Dickinson Univ, MS, 67; Rutgers Univ, PhD(nematol), 75. Prof Exp: Instr bot, Fairleigh Dickinson Univ, 65-66; asst prof bot & zool, Jersey City State Col, 66-67; ASSOC PROF MICROBIOL, OCEAN COUNTY COL, 67- Mem: Soc Nematol; Am Phytopath Soc; Am Microbiol Soc. Res: Nematode behavior and physiology specifically related to the vertical migration, quiescence, and senescence of above-ground phytoparasitic forms. Mailing Add: Dept of Biol Ocean County Col Toms River NJ 08753

ADAMO, NORMA JEAN, neuroanatomy, deceased

ADAMS, ALBERT WHITTEN, b Ft Scott, Kans, Oct 9, 27; m 49; c 4. ANIMAL NUTRITION. Educ: Kans State Col, BS, 51, MS, 55; SDak State Univ, PhD(animal sci), 65. Prof Exp: Asst hatchery mgr, Swift & Co, 51-52, hatchery mgr, 52-54; asst poultry sci, Kans State Col, 54-55; from instr to asst prof, SDak State Univ, 55-60; mgr, Stant's Turkey Farms, 60-62; asst prof dairy & poultry sci, Univ, 62-66, assoc scientist, Agr Exp Sta, 66-72, ASSOC PROF DAIRY & POULTRY SCI, KANS STATE UNIV, 66-, RES POULTRY SCIENTIST, AGR EXP STA, 73- Mem: AAAS; Poultry Sci Asn; World Poultry Sci Asn. Res: Relationship between the effects of various nutritional and environmental factors on reproduction in chickens, turkeys and pheasants. Mailing Add: Dept of Poultry Sci Kans State Univ Manhattan KS 66502

ADAMS, ALDEN ROSS, b Edgartown, Mass, Feb 29, 24; m 50; c 2. ORGANIC CHEMISTRY, BIOCHEMISTRY. Educ: Harvard Univ, AB, 49; Univ Del, MS, 50, PhD(chem), 52. Prof Exp: Res chemist, Am Viscose Corp, 52-55, asst to vpres, 55-57, admin mgr, 57-59; res chemist, Sun Oil Co, 59-61, prod mgr, Avisun Corp, 59-61, proj mgr, 61-63, mgr mkt develop & planning, 63-64, mgr com develop div, 64-65, gen mkt mgr, 65-69; GEN MGR PLASTICS, AIR PROD & CHEM CO, 69- Mem: Am Chem Soc. Res: Condensation and addition polymerization; enzymes; synthetic organic chemistry. Mailing Add: Prod & Chem Co 656 E Swedesford Rd Wayne PA 19087

ADAMS, ALFRED BIRK, b Steubenville, Ohio, Jan 16, 34; m 60; c 1. BIOCHEMISTRY, MICROBIOLOGY. Educ: Bethany Col, WVa, BS, 55; Mich State Univ, MS, 59; Ohio State Univ, PhD(physiol chem), 66. Prof Exp: Res instr biochem-microbiol, Col Dent, Ohio State Univ, 59-64; res assoc biochem, Mich State Univ, 65-67, res assoc microbiol, 67-68; ASST PROF BIOCHEM, COL DENT, UNIV NEBR, LINCOLN, 68-, ASSOC PROF ORAL BIOL, 73- Mem: AAAS; Int Asn Dent Res; Am Asn Dent Schs; Am Soc Microbiol. Res: Specificity of DNases; antibacterial properties of saliva; molecular mechanisms of bacterial sporulation. Mailing Add: Col of Dent Univ of Nebr Lincoln NE 68503

ADAMS, ANDREW BORDEN, b Fall River, Mass, Dec 20, 19; m 43, 53; c 3. ANATOMY, RADIOLOGY. Educ: Harvard Univ, AB, 42; Temple Univ, MD, 45. Prof Exp: Intern med & surg, Abington Mem Hosp, 45-46; from asst instr to asst prof anat, Med Sch, Temple Univ, 48-54, resident radiol, Univ Hosp, 51-54; instr, Med Sch, Univ Rochester, 54; asst radiotherapist, Strong Mem Hosp, 54-56, asst radiologist, 56-58; CHIEF RADIOLOGIST, OUR LADY OF LOURDES MEM HOSP, 58- Res: Radiotherapy. Mailing Add: Our Lady of Lourdes Mem Hosp 169 Riverside Dr Binghamton NY 13905

ADAMS, ANGUS MACAULAY, b Toronto, Ont, Sept 23, 16; m 42; c 1. FOOD SCIENCE, ENOLOGY. Educ: Ont Agr Col, BSA, 38; McMaster Univ, MA, 48, PhD(bot), 52. Prof Exp: Bacteriologist, Kennedy Labs, Ltd, Ont, 38-40; asst bacteriologist, Div Appl Biol, Nat Res Coun Can, 40-41; res asst dept bot, McMaster Univ, 47-49; res asst, Hort Exp Sta, 49-54, res scientist microbiol, 54-67, res scientist, Hort Res Inst Ont, 67-70, CHIEF RES SCIENTIST & HEAD PROD LAB, HORT RES INST ONT, ONT DEPT AGR, 70- Mem: Soc Indust Microbiol; Agr Inst Can; Am Inst Biol Sic; Can Soc Oenol. Res: Food preservation; grape and other fruit fermentations; sporulation in yeasts; yeast physiology and taxonomy. Mailing Add: Prod Lab Hort Res Inst of Ont Vineland ON Can

ADAMS, ARTHUR CURTIS, b Robinson, Ill, Sept 3, 39; m 61; c 3. INORGANIC CHEMISTRY. Educ: Univ Ill, BS, 61; Univ Wis, PhD(inorg chem), 65. Prof Exp: Res assoc inorg chem, Univ Colo, 65-66; MEM TECH STAFF, BELL LABS, 66- Mem: AAAS; Am Chem Soc; Electrochem Soc. Res: Kinetics and mechanisms of inorganic reactions; chemical vapor deposition; plasma chemistry; thin film characterization. Mailing Add: Bell Labs Mountain Ave Murray Hill NJ 07974

ADAMS, BRUCE EDWARD, b Lock Haven, Pa, Nov 20, 16; m 45; c 1. PHYSICAL GEOGRAPHY, GEOGRAPHY OF EUROPE. Educ: Lock Haven State Col, BS, 41; Pa State Univ, MEd, 49, PhD(geog), 60. Prof Exp: Teacher geog & world hist, Canton High Sch, Pa, 41-42 & 45-49; teacher social studies, Roosevelt High Sch, Pa, 49-56; chmn dept geog, 61-68, PROF GEOG, BLOOMSBURG STATE COL, 56- Concurrent Pos: Dir, NDEA Inst Advan Study Geog, 65-66. Mem: Nat Coun Geog Educ; Asn Am Geogr. Res: Status of geographic education. Mailing Add: Dept of Geog & Earth Sci Bloomsburg State Col Bloomsburg PA 17815

ADAMS, BRUCE GORDON, b Seattle, Wash, Mar 31, 42; m 65. MICROBIOLOGY, MICROBIAL PHYSIOLOGY. Educ: Whitman Col, BA, 64; Ore State Univ, PhD(microbial physiol), 68. Prof Exp: NSF fel biol, Mass Inst Technol, 68-70; asst prof, 70-74, ASSOC PROF MICROBIOL, UNIV HAWAII, HONOLULU, 75- Mem: AAAS; Am Soc Microbiol; Brit Soc Gen Microbiol. Res: Function and physiological forms of ergosterol in Saccharomyces cerevisiae; physiology and metabolism of blood flagellates; catabolite repression of inducible soluble enzymes in Saccharomyces cerevisiae. Mailing Add: Dept of Microbiol 2538 The Mall Univ of Hawaii Snyder Hall 202 Honolulu HI 96822

ADAMS, BUDD B, geophysics, see 12th edition

ADAMS, CAROLINE LANDER, b Chicago, Ill, Apr 23, 03; m 35; c 1. CYTOLOGY. Educ: Ill Col, AB, 25; Univ Chicago, MS, 28; Univ Wis, PhD(mycol), 33. Prof Exp: Instr bot, Hood Col, 29-30; teaching & res asst, Univ Wis, 30-34, Am Asn Univ Women fel, 35-36; res asst, Cotton Div, USDA, 34-35; from instr to prof, 46-72, EMER PROF BOT, GEORGE WASHINGTON UNIV, 72-; ASSOC PROF, NAT CATHEDRAL SCH, 62- Concurrent Pos: Chief dept sci, Nat Cathedral Sch, 52-62. Mem: AAAS; Bot Soc Am. Res: Morphological development and cytology of gasteromycetes. Mailing Add: 242 N Granada St Arlington VA 22203

ADAMS, CHARLES HENRY, b Burdick, Kans, Nov 7, 18; m 43. ANIMAL SCIENCE, MEAT SCIENCE. Educ: Kans State Univ, BS, 41, MS, 42; Mich State Univ, PhD, 64. Prof Exp: Asst instr animal husb, Kans State Univ, 46-47; from asst prof to assoc prof, 47-70, PROF ANIMAL HUSB & MEM GRAD FAC, UNIV NEBR, LINCOLN, 70-, ASST DEAN, COL AGR, 73- Honors & Awards: Reciprocal Meat Conf Signal Serv Award, Am Meat Sci Asn, 67, Spec Recognition Award, 74. Mem: AAAS; Am Soc Animal Sci; Am Meat Sci Asn; Inst Food Technol; Am Inst Biol Sci. Res: Meat and meat products; beef and pork carcass evaluation; packaging of frozen meat and meat products; tenderness of meat. Mailing Add: Inst of Agr & Natural Resources Univ of Nebr Lincoln NE 68583

ADAMS, CHARLES REX, b Mt Vernon, Tex, Aug 19, 30; m 51, 67; c 4. PETROLEUM CHEMISTRY. Educ: ETex State Teachers Col, BA, 50; Rice Inst, MA, 52, PhD(chem), 54. Prof Exp: Res chemist, 54-64, on assignment to Shell Grundlagenforschung Gmb H, Siegburg bei, Bonn, WGer, 64-66, RES SUPVR, PETROL CHEM DEPT, SHELL DEVELOP CO, 66- Honors & Awards: Ipatieff Prize, Am Chem Soc, 68. Mem: Am Chem Soc; fel Am Inst Chem; Catalysis Soc. Res: Mechanisms of heterogeneous catalysis; structure of solid catalysts; structure of the solid state. Mailing Add: Shell Develop Co Westhollow Res Ctr PO Box 1380 Houston TX 77001

ADAMS, CLARK EDWARD, b Algona, Iowa, Mar 18, 42; m 62; c 3. MAMMALIAN ECOLOGY. Educ: Concordia Teachers Col, BS, 64; Univ Ore, MS, 66; Univ Nebr-Lincoln, PhD(zool), 73. Prof Exp: Instr biol, St John's Acad & Jr Col, 64-65; ASSOC PROF ECOL, CONCORDIA TEACHERS COL, 66- Mem: Am Soc Mammalogists; Wildlife Soc; Sigma Xi; Nat Asn Biol Teachers. Res: Measuring territoriality in fox squirrels; Sciurus niger rufiventer pre and post weaning and pre and post mating. Mailing Add: Dept of Biol Concordia Teachers Col Seward NE 68434

ADAMS, CLIFFORD, geology, see 12th edition

ADAMS, CLIFFORD LOWELL, b Knox Co, Ind, Jan 28, 15; m 43; c 2. PHYSICS. Educ: Ind State Teachers Col, BS, 40; Mo Sch Mines & Metals, MS, 50. Prof Exp: Teacher high sch, Ind, 44-46; instr physics, Morehead State Col, 42-44; asst prof, Tri-State Col, 46-48; instr, Mo Sch Mines & Metal, 48-50; assoc prof, Union Univ, 50-51; asst prof & assoc prof physics & elec eng, US Naval Acad, 51-56; assoc prof physics, Northeastern La State Col, 56-58; head dept physics, 58-66, assoc provost res, 73-74, PROF PHYSICS, OLD DOM UNIV, 58-, DIR RES ADMIN, 66-, ASST TO VPRES UNIV RELS, 74- Concurrent Pos: Mem, Nat Coun Univ Res Adminr; Nat Conf on Admin of Res. Mem: AAAS; Am Phys Soc; Am Asn Physics Teachers. Res: Energy levels of potassium chloride; optics. Mailing Add: 1325 Monterey Ave Norfolk VA 23508

ADAMS, CURTIS H, b DeKalb, Miss, Sept 12, 17; m 42; c 1. ENTOMOLOGY, ZOOLOGY. Educ: Miss State Univ, BS, 41, PhD(entom), 65; Henderson State Teachers Col, MSEd, 63. Prof Exp: Assoc prof mil sci, The Citadel, 51-54; prof, Henderson State Teachers Col, 61-63; head life sci sect & chmn biol fac, 68-73, ASSOC PROF BIOL, UNIV ALA, HUNTSVILLE, 65- Mem: Wildlife Soc; AAAS; Am Inst Biol Sci; Entom Soc Am. Res: Insecticide resistance in beneficial insect parasites. Mailing Add: Dept of Biol Univ of Ala PO Box 1247 Huntsville AL 35807

ADAMS, DALE W, b Pleasant Grove, Utah, June 21, 34; m 60; c 3. AGRICULTURAL ECONOMICS. Educ: Utah State Univ, BS, 56; Mich State Univ, MS, 61, PhD(agr econ), 64. Prof Exp: Asst prof agr econ, Univ Wis, 64-66; PROF AGR ECON, OHIO STATE UNIV, 66- Concurrent Pos: Staff economist, AID, 68-69; vis prof, Stanford Univ, 74-75. Mem: Am Agr Econ Asn. Res: Rural savings and credit activities in low income countries. Mailing Add: Dept of Agr Econ & Rural Sociol Ohio State Univ Columbus OH 43210

ADAMS, DANIEL OTIS, b Portland, Maine, Mar 14, 18; m 45; c 4. CHEMISTRY, PAPER TECHNOLOGY. Educ: Oberlin Col, AB, 39; Lawrence Col, MS, 41, PhD(pulp & paper), 43. Prof Exp: Res chemist, WVa Pulp & Paper Co, 43-44, res proj leader, 44-48; res chemist, Bird & Son, Inc, 48-50, chief chemist, Res Div, 50-55; res dir, WVa Pulp & Paper Co, 55-59, supt tech serv, Kraft Div, 59-64, MGR TECH SERV, WESTVACO CORP, 64- Mem: Am Tech Asn Pulp & Paper Indust. Res: Morphology of paper making fibers as related to their papermaking characteristics; air and water pollution abatement; resin solution saturation of paper; pulp and paper manufacture. Mailing Add: Westvaco Corp Box 5207 North Carleston SC 29406

ADAMS, DARIUS MAINARD, b Los Angeles, Calif, July 5, 44; m 66. FOREST ECONOMICS. Educ: Humboldt State Univ, BS, 66; Yale Univ, MFS, 68; Univ Calif, Berkeley, PhD(wildland resource sci), 73. Prof Exp: Asst prof forestry, Univ Wis-Madison, 71-74; ASST PROF FOREST MGT, ORE STATE UNIV, 75- Mem: Soc Am Foresters; Am Econ Asn. Res: Forest products, especially markets for public and private timber and logs; optimal management programs for uneven-aged stands and forests. Mailing Add: Sch of Forestry Ore State Univ Corvallis OR 97331

ADAMS, DAVID A, b Lakewood, Ohio, Nov 26, 31; m 55; c 1. RESOURCE MANAGEMENT. Educ: NC State Col, BS, 53, MS, 57, PhD(plant ecol), 62. Prof Exp: Waterfowl proj leader, Wildlife Resources Comn, NC, 57; chief park naturalist, NC State Parks, 57-59; asst, NC State Univ, 59-62; cur, NC State Mus, 62-63; comnr, NC Div Com & Sports Fisheries, 63-68; sr staff mem, Nat Coun Marine Resources & Eng Develop, 68-69; PRES, COASTAL ZONE RESOURCES CORP & VPRES, OCEAN DATA SYSTS, INC, 69- Concurrent Pos: Mem, Comn Marine Sci, Eng & Resources, 67-69; mem, Cape Fear Tech Inst Adv Comt & Res Adv Comt, US AID. Mem: Ecol Soc Am; Am Fisheries Soc. Res: Estuarine ecology; coastal fisheries management; mammal and bird distribution and ecology; salt marsh ecology; waterfowl management. Mailing Add: 4505 Franklin Ave Wilmington NC 28401

ADAMS, DAVID GEORGE, physical chemistry, organic chemistry, see 12th edition

ADAMS, DAVID LAWRENCE, b Brockton, Mass, Oct 10, 45; m 66; c 2. SCIENCE EDUCATION. Educ: Univ Mass, BS, 67; Univ Conn, MS, 69, PhD(org chem), 71. Prof Exp: Instr org chem, Pa State Univ, 71-72; INSTR CHEM & CHMN DIV SCI & ALLIED HEALTH, N SHORE COMMUNITY COL, 72- Concurrent Pos: Textbook consult, Wadsworth Publ Co, 73-, Prentice Hall Publ Co, 75- & Dickenson Publ Co, 75- Mem: Am Chem Soc; Sigma Xi. Res: Determination and quantitative evaluation of environmental pollutants as influenced by various other environmental parameters; synthesis of bicyclic turpines and acid catalyzed rearrangements of same. Mailing Add: Div of Sci & Allied Health N Shore Community Col Beverly MA 01915

ADAMS, DOLPH OLIVER, b Montezuma, Ga, Apr 12, 39; m 69. PATHOLOGY, IMMUNOLOGY. Educ: Duke Univ, AB, 60; Med Col Ga, MS & MD, 65; Univ NC, PhD(exp path), 72. Prof Exp: Chief path, US Army Joint Laser Safety Team, Frankford Arsenal, 70-72; asst prof, 72-76, ASSOC PROF PATH, MED CTR, DUKE UNIV, 76-, DIR AUTOPSY SERV, 72- Mem: Am Asn Pathologists & Bacteriologists; Am Soc Exp Path; Reticuloendothelial Soc. Res: Role of macrophages in cellular immunology and tumor biology. Mailing Add: Dept of Path Duke Univ Med Ctr Durham NC 27710

ADAMS, DONALD F, b Spokane, Wash, Oct 26, 19; m 51; c 3. ANALYTICAL CHEMISTRY. Educ: State Col Wash, BS, 41, MS, 42. Prof Exp: Teaching asst, State Col Wash, 41-42; chemist, Boeing Airplane Co, 42-45; CHEMIST, WASH STATE UNIV, 45-, DIR AIR POLLUTION TRAINING PROG & HEAD AIR POLLUTION RES SECT, COL ENG, 49- Mem: Am Chem Soc; Instrument Soc Am; Air Pollution Control Asn. Res: Fluoride analysis; automatic air pollution analysis instrumentation; gas chromatography of sulfur compounds; effects of fluorides on plants; human response to malodors; odor control in draft pulp mills. Mailing Add: Res Div Col of Eng Wash State Univ Pullman WA 99163

ADAMS, EARNEST DWIGHT, b Carrollton, Ga, Feb 16, 33; m 53; c 3. LOW TEMPERATURE PHYSICS. Educ: Berry Col, AB, 53; Emory Univ, MS, 54; Duke Univ, PhD, 60. Prof Exp: Res assoc physics, Stanford Univ, 60-62; from asst prof to assoc prof, 62-70, PROF PHYSICS, UNIV FLA, 70- Concurrent Pos: Vis prof,

Helsinki Univ Technol, 71. Mem: Fel Am Phys Soc; Am Asn Physics Teachers. Res: Low temperature physics, especially properties of helium-3 and helium-4. Mailing Add: Dept of Physics Univ of Fla Gainesville FL 32611

ADAMS, EDWARD NEUFVILLE, b Bogalusa, La, Sept 29, 22; m 44; c 3. INFORMATION SCIENCE. Educ: Southwestern at Memphis, BS, 43; Univ Wis, MS, 47, PhD(physics), 50. Prof Exp: From instr to asst prof physics, Inst Nuclear Studies, Univ Chicago, 50-54; mgr semiconductor res, Westinghouse Elec Co, 54-59; sr physicist, IBM Res, Int Bus Mach Corp, 59-64, dir eng sci, 60-63, dir syst & appln, 63-64, res dir comput assisted instruct, 64-70, RES STAFF MEM, T J WATSON CTR, IBM, 70- Concurrent Pos: Assoc physicist, Chicago Midway Labs, 51-54; lectr, Int Colloquium Edmonton-Banff, 57; adj prof, Carnegie Inst Technol, 57-58; vis prof physics, State Univ NY Stony Brook, 66-68; vis prof educ psychol, Grad Ctr, City Univ New York, 72-74. Mem: AAAS; fel Am Phys Soc; sr mem Inst Elec & Electronics Eng; Am Asn Physics Teachers. Res: Solid state electronics; applied optics; pattern recognition; computer administered instruction; information retrieval; educational technology. Mailing Add: Thomas J Watson Res Ctr IBM Corp PO Box 218 Yorktown Heights NY 10598

ADAMS, ELIJAH, b Buffalo, NY, Jan 14, 18; m 43; c 4. BIOCHEMISTRY. Educ: Johns Hopkins Univ, BA, 38; Univ Rochester, MD, 42. Prof Exp: Assoc prof pharmacol, Col Med, NY Univ, 55-58; prof pharmacol & dir dept, Sch Med, St Louis Univ, 58-63; PROF BIOCHEM & CHMN DEPT, SCH MED, UNIV MD AT BALTIMORE, 63- Mem: Am Chem Soc; Am Soc Biol Chem. Res: Enzymes; intermediary metabolism; amino acids; collagen. Mailing Add: Dept of Biochem Univ of Md Sch of Med Baltimore MD 21201

ADAMS, ELIZABETH, b Dalton, Mass, May 23, 06. ORGANIC CHEMISTRY. Educ: Middlebury Col, BS, 27; Smith Col, AM, 29; Pa State Col, PhD(org chem), 38. Prof Exp: Asst chem, Smith Col, 27-29; instr, Wells Col, 29-31; instr, Sweet Briar Col, 32-39; instr, Queens Col & City Col NY, 39-44; res chemist, Plastics Div, Gen Elec Co, 44-49; asst prof chem, Woman's Col, NC, 49-55; from asst prof to assoc prof, Queens Col NY, 55-71; RETIRED. Mem: Am Chem Soc. Res: Rearrangement of the propyl group during Friedel-Crafts reaction. Mailing Add: Dept of Chem Queens Col Flushing NY 11367

ADAMS, EMORY TEMPLE, JR, b Tampico, Mex, Sept 5, 28; US citizen. BIOPHYSICAL CHEMISTRY. Educ: Rice Univ, BA, 49; Baylor Col Med, MS, 52; Univ Wis-Madison, PhD(biochem), 62. Prof Exp: Res assoc chem, Univ Wis-Madison, 62-63; res chemist, Hercules Res Ctr, Del, 63-64; res assoc biol, Brookhaven Nat Lab, 64-65; staff fel, Nat Inst Arthritis & Metab Dis, 66; asst prof chem, Ill Inst Technol, 66-70; ASSOC PROF CHEM, TEX A&M UNIV, 70- Mem: Am Chem Soc; Biophys Soc; Am Soc Biol Chemists. Res: Analysis of self-associating proteins and mixed associations of biopolymers by sedimentation equilibrium and osmometry; molecular weights and molecular-weight distributions from sedimentation equilibrium experiments; sedimentation equilibrium theory. Mailing Add: Dept of Chem Tex A&M Univ College Station TX 77843

ADAMS, ERNEST CLARENCE, b Meridian, Miss, Mar 27, 25. BIOCHEMISTRY, IMMUNOCHEMISTRY. Educ: Miss State Col, BS, 48; Northwestern Univ, PhD(biochem), 50. Prof Exp: Asst prof chem, Miss State Col, 50-51; res biochemist, Miles-Ames Res Lab, 51-59, sr res scientist, Ames Res Lab, 59-73, sect head, 64-73, adminr immunochem opers, 70-73, STAFF RES SCIENTIST, AMES RES LAB, AMES CO DIV, MILES LABS, 73- Mem: AAAS; Am Chem Soc; NY Acad Sci; Asn Clin Scientists. Res: Carbohydrate, protein and drug metabolism; clinical diagnostics; organic syntheses; enzymatic methods of analyses; immunochemical assays; radioimmune assays. Mailing Add: Ames Res Lab 1127 Myrtle St Elkhart IN 46514

ADAMS, EUGENE VAN, virology, see 12th edition

ADAMS, FORREST HOOD, b Minneapolis, Minn, Sept 20, 19; m 43, 69; c 10. PEDIATRIC CARDIOLOGY. Educ: Univ Minn, BA, 41, MB, 43, MD, 44, MS, 46; Am Bd Pediat, dipl, 48, cert cardiol, 61. Prof Exp: Intern, Univ Minn Hosp, 43-44, from instr to asst prof pediat, 48-52; assoc prof, 52-58, actg chmn dept, 58-59 & 64-65, vchmn dept, 62-64, PROF PEDIAT, SCH MED, UNIV CALIF, LOS ANGELES, 58- Concurrent Pos: Nat Res fel pediat, Univ Minn Hosp, 48-49; asst dir heart sect, Crippled Children's Prog, St Paul, 49-50; physician chg pediat, Sister Elizabeth Kenny Inst, 49-50; assoc physician chg pediat, Minneapolis Gen Hosp, 49-50, chief, 50-52; dir pediat heart clins, Univ Minn Hosp & Variety Club Heart Hosp, 51-52; med dir, Marion Davies Children's Clin, 52-; mem staff, St John's Hosp & Harbor Gen Hosp, 52-; consult, State Bd Pub Health, Calif, 63-; pres, Sub-Specialty Bd Pediat Cardiol, 67, emer mem, 69-; mem adv bd, Inter-Soc Comn Heart Dis Resources, 68-, mem comt congenital heart dis, 68-71, mem exec comt, 68-71; consult, Off Surgeon Gen. Honors & Awards: Theodore & Susan Cummings Humanitarian Award, 64, 65, 66, 67, 71 & 72; Distinguished Fel Award, Am Col Cardiol, 74. Mem: Soc Pediat Res; Am Acad Pediat; Am Col Cardiol (vpres, 68-69, pres-elect, 70, pres, 71-72); Am Pediat Soc; Am Heart Asn. Res: Toxoplasmosis; congenital heart disease; rheumatic fever; pediatric cardiology. Mailing Add: Dept of Pediat Univ of Calif Sch of Med Los Angeles CA 90024

ADAMS, FRANK WILLIAM, b Billings, Mont, Feb 17, 25; m 47; c 7. ANIMAL NUTRITION. Educ: Mont State Col, BS, 48; Ore State Univ, MS, 50, PhD(zool, biochem), 65. Prof Exp: Instr, 53-65, ASST PROF AGR CHEM, ORE STATE UNIV, 65- Mem: AAAS; Am Chem Soc. Res: Mammalian embryology; normal and nutritionally caused abnormalities in placentae; trace element relationships in animal nutrition; animal fluorosis; nutrition in today's society. Mailing Add: Dept of Agr Chem Ore State Univ Corvallis OR 97331

ADAMS, FRANKLIN SCOTT, b Philadelphia, Pa, July 31, 29; m 51; c 4. BOTANY, MORPHOLOGY. Educ: Johnson State Col, BS, 56; Univ Pa, MS, 59; Univ NH, PhD(bot), 67. Prof Exp: Asst prof biol, Pa State Univ, 67-74; ASSOC PROF BOT, UNIV MICH, DEARBORN, 74- Mem: AAAS; Am Inst Biol Sci; Bot Soc Am. Res: Plant anatomy and morphology, especially the biology of the higher aquatic plants. Mailing Add: Dept of Natural Sci Univ of Mich 4901 Evergreen Dearborn MI 48128

ADAMS, FRED, b Marion, La, Mar 1, 21; m 43; c 2. SOIL SCIENCE. Educ: La State Univ, BS, 43, MS, 48; Univ Calif, PhD(soil sci), 51. Prof Exp: Prof soil sci, Am Univ, Beirut, 52-55; assoc prof, 55-66, PROF SOILS, AUBURN UNIV, 66- Mem: Am Soc Agron; Soil Sci Soc Am. Res: Chemistry and fertility of soils. Mailing Add: Dept of Agron & Soils Auburn Univ Auburn AL 36830

ADAMS, GABRIELLE H M, b Mateszalka, Hungary, Oct 29, 39; Can citizen; m 64; c 2. MOLECULAR BIOLOGY. Educ: McMaster Univ, BSc, 63; Carleton Univ, PhD(cell biol), 68. Prof Exp: Fel biochem, Nat Res Coun Can, 67-69; publ asst, Can J Bot, Nat Res Coun Can, 69-70; res assoc biochem, Nat Res Coun Can & Dept Biol, Carleton Univ, 70-75; PUBL SUPVR RES JOURNALS, NAT RES COUN CAN, 75- Mem: Can Soc Cell Biol. Res: Chemistry and function of nuclear chromatin proteins. Mailing Add: Rm 1157 100 Sussex Dr Ottawa ON Can

ADAMS, GAIL DAYTON, JR, b Cleveland, Ohio, Jan 27, 18; m 42, 66, 76; c 3. RADIOLOGICAL PHYSICS. Educ: Case Inst Technol, BS, 40; Univ Ill, MS, 42, PhD(physics), 43; Am Bd Radiol, dipl; Am Bd Health Physics, dipl. Prof Exp: Asst physics, Univ Ill, 40-43, physicist, 43-45, asst prof physics, 45-51; res physicist & assoc dir radiol lab, Med Sch, Univ Calif, San Francisco, 51-65, lectr radiol, 51-53, clin prof physics, 53-65; PROF RADIOL & RADIATION PHYSICS, HEALTH SCI CTR, UNIV OKLA, 65-, VCHMN DEPT RADIOL SCI, 70-, CAMPUS RADIATION SAFETY OFFICER, 71- Concurrent Pos: Consult, Vet Admin Hosp; mem staff, Univ Okla Hosp & Okla Childrens Mem Hosp; with AEC & Off Sci Res & Develop. Mem: Fel Am Phys Soc; Radiation Res Soc (secy-treas, 63-69); Am Asn Physicists in Med (pres, 58-60, ed, 73-); fel Am Col Radiol; NY Acad Sci. Res: Absorption of high energy quanta; magnet design and testing; radiography at 20-mev; effect of magnetic fields on biological systems; application of x-rays, particularly 70 mev x-rays to radiology and to radiation therapy. Mailing Add: Dept of Radiol Sci Univ of Okla Oklahoma City OK 73104

ADAMS, GEORGE BAKER, JR, b Kentfield, Calif, Feb 23, 19. PHYSICAL CHEMISTRY. Educ: Univ Calif, BS, 41, MS, 47; Ohio State Univ, PhD(phys chem), 51. Prof Exp: Asst chem, Univ Calif, 46-47; res assoc, Univ Ore, 53-55, from instr to asst prof, 55-58; sr scientist, 58-59, res scientist, 59-62, sr staff scientist, 62-69, SR MEM RES & LAB, LOCKHEED MISSILES & SPACE CO INC, 62-, CONSULT SCIENTIST, 70- Mem: Electrochem Soc; Int Soc Electrochem. Res: Electrochemical energy conversion; chemical thermodynamics; environmental chemistry; metallurgical chemistry; anodic oxide film formation kinetics. Mailing Add: Lockheed Res Lab 0/52-32 3251 Hanover St Palo Alto CA 94304

ADAMS, GORDON ALBERT, b Watford, Ont, Aug 5, 07; m 38. MICROBIAL BIOCHEMISTRY. Educ: Queen's Univ (Ont), BA, 28; Univ Western Ont, MSc, 30; Univ Chicago, PhD(biochem), 38. Prof Exp: Lectr biochem, Med Sch, Univ Western Ont, 30-39; res biochemist, Biochem Lab, Nat Res Coun Can, 39-72; vis prof microbiol, Colo State Univ, 73-74; RETIRED. Mem: Fel Chem Inst Can; fel Royal Soc Can. Res: Cellulose and polysaccharides starch; microbial polysaccharides; bacterial lipopolysaccharides. Mailing Add: 99 Front St W Strathroy ON Can

ADAMS, HARRY, b Ft Riley, Kans, June 13, 24; m 53; c 4. NUCLEAR PHYSICS. Educ: Kans State Univ, BS, 45, MS, 47; Univ Minn, PhD(physics), 62. Prof Exp: Instr physics, Pac Lutheran Univ, 47-52; assoc scientist optical eng, Aeronaut Labs, Univ Minn, 53-55; res assoc & vis prof physics, Fla State Univ, 60-62; PROF PHYSICS, PAC LUTHERAN UNIV, 62- Mem: Am Phys Soc. Res: Nuclear reactions and structure. Mailing Add: Dept of Physics Pac Lutheran Univ Tacoma WA 98447

ADAMS, HELEN ELIZABETH, b Cowra, Australia, Feb 21, 45; m 67. MATHEMATICS. Educ: Univ New Eng, Australia, BS, 66; Monash Univ, Australia, PhD(math), 71. Prof Exp: Sr tutor math, Univ Melbourne, 66-68; fel, Mt Holyoke Col, 72-73; ASST PROF MATH, SMITH COL, 73- Mem: Am Math Soc; Math Asn Am. Res: Commutative algebra. Mailing Add: Dept of Math Smith Col Northampton MA 01060

ADAMS, HENRY RICHARD, b Dallas, Tex, Apr 12, 42; m 65; c 2. VETERINARY PHARMACOLOGY. Educ: Tex A&M Univ, BS, 65, DVM, 66; Univ Pittsburgh, PhD(pharmacol), 72. Prof Exp: Res vet, US Army, Ft Detrick, Md, 66-68; res assoc pharmacol, Univ Pittsburgh, 68-72; asst prof, 72-75, ASSOC PROF PHARMACOL, SOUTHWESTERN MED SCH, UNIV TEX HEALTH SCI CTR DALLAS, 75- Mem: Am Vet Med Asn; Am Soc Pharmacol & Exp Therapeut; Am Soc Vet Physiol & Pharmacol. Res: Comparative cardiovascular pharmacology. Mailing Add: Univ of Tex Health Sci Ctr 5323 Harry Hines Blvd Dallas TX 75235

ADAMS, HERBERT JACK, b Hamburg, NY, July 11, 24; m 45; c 2. PHARMACOLOGY, PHYSIOLOGY. Educ: Univ Buffalo, BA, 58, PhD(pharmacol), 61. Prof Exp: Asst pharmacol, Sch Med, Univ Buffalo, 52-57; sr pharmacologist, SKF Labs, 61-67; SECT HEAD GEN PHARMACOL, ASTRA PHARMACEUT PROD, INC, 67- Mem: Am Pharmaceut Asn. Res: Absorption, distribution and excretion of drugs; neuromuscular blocking agents; antiparkinson agents; antiarrhythmic agents; local anesthetics. Mailing Add: Res & Develop Dept Astra Pharmaceut Prod Inc Framingham MA 01701

ADAMS, HERMAN RAY, b Abilene, Tex, Sept 15, 39; m 63; c 2. CLINICAL CHEMISTRY. Educ: Tex A&M Univ, BS, 62, MS, 65; Univ Tex Southwestern Med Sch Dallas, PhD(pharmacol), 72; Am Bd Clin Chem, cert, 76. Prof Exp: DIR, SPEC CHEM & TOXICOL LAB, SCOTT & WHITE CLIN, 72- Mem: Am Acad Forensic Sci; Am Asn Clin Chemists; Am Chem Soc; Sigma Xi. Mailing Add: Scott & White Clin Temple TX 76501

ADAMS, HOLYOKE PURINTON, b St Johsbury, Vt, Oct 30, 22; m 47; c 6. ANIMAL NUTRITION. Educ: Univ Maine, BS, 47; Univ Wis, MS, 48, PhD(dairy husb), 52. Prof Exp: Asst feed res, Eastern States Farmer's Exchange, 51-65; dairy specialist, Univ Nev, Reno, 65-71, assoc prof animal nutrit, 67-71; DAIRY SPECIALIST & ASSOC PROF ANIMAL NUTRIT, ORE STATE UNIV, 71- Res: Dairy husbandry; fat utilization by dairy cattle. Mailing Add: 120 Withycombe Hall Ore State Univ Corvallis OR 97331

ADAMS, HORACE, JR, organic chemistry, see 12th edition

ADAMS, J MACK, b Marfa, Tex, Aug 14, 33; m 52; c 2. COMPUTER SCIENCE. Educ: Tex Western Col, BS, 54; NMex State Univ, MS, 60, PhD(math), 63. Prof Exp: Assoc scientist, Bettis Atomic Power Div, Westinghouse Elec Corp, 54-56; mathematician, Flight Simulation Lab, White Sands Missile Range, NMex, 56-58, supvr mathematician, 58-60, mathematician, 61-62, res mathematician, Electronics Res & Develop Activity, 63, supvry mathematician, 63-64; assoc prof math, Tex Western Col, 64-65; PROF COMPUT SCI & HEAD DEPT, NMEX STATE UNIV, 65- Honors & Awards: US Govt Award, 64. Mem: Math Asn Am; Asn Comput Mach. Res: Programming languages; theory of algorithms; computer science education. Mailing Add: Dept of Comput Sci NMex State Univ Las Cruces NM 88001

ADAMS, JACK DONALD, b Gary, Ind, June 3, 19; m 47; c 2. GENETICS. Educ: Purdue Univ, BS, 50, MS, 52; Wash State Univ, PhD(radiation genetics), 56. Prof Exp: Asst bot, Purdue Univ, 50-52; exp aide radiation genetics, Wash State Univ, 52-53, asst, 53-56; asst prof agron, NMex State Univ, 56-59; ASSOC PROF BIOL, CENT MICH UNIV, 59- Mem: Genetics Soc Am. Res: Plant genetics; cytology; cytogenetics; radiation genetics. Mailing Add: Dept of Biol Cent Mich Univ Mt Pleasant MI 48858

ADAMS, JAMES HALL, JR, b Statesville, NC, Aug 7, 43; m 68. COSMIC RAY PHYSICS. Educ: NC State Univ, BS, 66, MS, 68, PhD(physics), 72. Prof Exp: Res asst cosmic ray physics, Johnson Space Ctr, 69-72; res assoc, Nat Res Coun, 72-74; RES PHYSICIST COSMIC RAY PHYSICS, NAVAL RES LAB, 74- Mem: Am Phys Soc; AAAS. Res: Measurements of cosmic ray, elemental and isotopic composition;

energy spectra; studies of heavy ion nuclear interactions. Mailing Add: Code 6603M Naval Res Lab 4555 Overlook Ave SW Washington DC 20375

ADAMS, JAMES MILLER, b Cleveland, Ohio, Sept 9, 24; m 60; c 7. BIOCHEMISTRY, ANALYTICAL CHEMISTRY. Educ: Case Inst Technol, BS, 45; Va Polytech Inst, BS, 48, MS, 49, PhD(biochem), 54. Prof Exp: Instr, Va Polytech Inst, 48-51; res chemist, Visking Corp, Ill, 54-55; chief chemist, Lab Vitamin Technol, 55-58; res chemist, Chemagro Corp, Mo, 58-63; res assoc drug residues & atherosclerosis, Merck Inst Therapeut Res, NJ, 63-67; sect leader phys & anal res dept, Merck, Sharp & Dohme Res Labs, 67-69; REGIONAL DIR LAB DIV, US CUSTOMS SERV, 69- Concurrent Pos: Instr, Univ Kansas City, 58-63 & Middlesex County Col, 67-69; safety officer, US Customs, Region III, 69-75. Mem: Am Chem Soc; Am Soc Microbiol; Soc Appl Spectros. Res: Saran coatings; large cellulose casings; vitamin assays; radiation sterilization of meat; organophosphate insecticide assay; drug residue assay; bacterial metabolism; atherosclerosis; narcotics detection; forensic analysis; chemical characterization of fruit juice; country of origin of petroleum. Mailing Add: 307 Sheffield Ct Joppa MD 21085

ADAMS, JAMES NORMAN, b Brooklyn, NY, Nov 4, 32; m 55; c 3. MICROBIOLOGY, GENETICS. Educ: Univ Ky, BS, 54; Univ Ga, PhD(bact), 61. Prof Exp: Bacteriologist, Dugway Proving Ground, Utah, 57-59; asst prof bact, Univ Ga, 62-63; from asst prof to prof microbiol, Sch Med, Univ SDak, 63-75, PROF & COORDR MED MICROBIOL, SCH MED, UNIV SC, 75- Concurrent Pos: USPHS fel, 61-62, grants, 62-63 & 64-; consult ed, Aquarium Jour, 61-; Univ Minn fel, 62; Am Soc Microbiol pres fel, 62; NSF res grant, 64-; Nat Inst Gen Med Sci career develop award, 66- Mem: AAAS; Am Soc Microbiol; Genetics Soc Am; Brit Soc Gen Microbiol. Res: Genetics, cytology and taxonomy of the Nocardiae; bacterial taxonomy; cytology and physiology of fungi; Actinomycetes genetics. Mailing Add: Dept of Biol Univ of SC Columbia SC 29208

ADAMS, JAMES RUSSELL, b Independence, Pa, May 31, 14; m 40; c 3. ZOOLOGY. Educ: McGill Univ, BSc, 36, MSc, 37, PhD(zool), 40. Prof Exp: Demonstr zool, McGill Univ, 36-37; lectr biol, Sir George Williams Col, 37-40, from asst prof to assoc prof, 40-46; assoc prof, 46-55, PROF ZOOL, UNIV BC, 55- Concurrent Pos: Nat Res Coun Can sr res fel, 62. Mem: AAAS; Am Soc Zoologists; Can Soc Zoologists; Wildlife Dis Asn. Res: Parasites of fishes and wildlife; host-parasite relations. Mailing Add: Dept of Zool Univ of BC Vancouver BC Can

ADAMS, JAMES WILLIAM, Conover, Wis, Oct 29, 21; c 2. INDUSTRIAL CHEMISTRY. Educ: Univ Wis, BS, 43. Prof Exp: Control chemist, US Rubber Co, WVa, 43, res group leader, 43-47, chief chemist, Conn, 48-51; sr scientist, Marathon Corp, 51-57, res assoc, Marathon Div, Am Can Co, 57-63, sr res assoc, 63-71, RES SUPVR II, AM CAN CO, 71- Mem: Am Chem Soc. Res: Grafting vinyl polymers on cellulose. Mailing Add: 2008 Clarberth St Schofield WI 54476

ADAMS, JANE N, b Marion, Ind, Aug 28, 29; m 54. MICROBIOLOGY, GENETICS. Educ: Earlham Col, AB, 51; Purdue Univ, MS, 54; Univ Ill, PhD(bact), 59. Prof Exp: Instr bact, Univ Ill, 59-61; asst prof, 61-69, ASSOC PROF MICROBIOL, SAN FRANCISCO STATE UNIV, 69- Res: Bacterial and viral genetics. Mailing Add: Dept of Biol San Francisco State Univ San Francisco CA 94132

ADAMS, JEAN BURNHAM, b Fredericton, NB, July 8, 13; m 41. INSECT PHYSIOLOGY. Educ: Univ NB, BA, 35; Univ Maine, MS, 36. Prof Exp: Entomologist, 36-46, officer in charge field crop insect lab, 46-58, RES SCIENTIST, RES BR, SCI SERV, CAN DEPT AGR, 58- Honors & Awards: Hons for contrib to the natural sci, Nat Mus Natural Sci, Ottawa, Can, 75. Mem: Can Soc Zool; Entom Soc Can. Res: Host-parasite relationship as expressed by aphids and their host plants; mechanism of host selection and exploitation; aphid enzymology and the application to pest management. Mailing Add: Res Sta Box 20280 Canada Agr Fredericton NB Can

ADAMS, JEAN RUTH, b Edgewater Park, NJ, Aug 17, 28. INSECT PATHOLOGY, ELECTRON MICROSCOPY. Educ: Rutgers Univ, BS, 50, PhD(entom), 62. Prof Exp: Technician, Rohm and Haas Co, 51-57; res fel zool, Univ Pa, 61-62; RES ENTOMOLOGIST, AGR RES CTR, USDA, 62- Mem: Electron Micros Soc Am; Entom Soc Am; Am Soc Cell Biol; NY Acad Sci. Res: Electron microscopic investigations on histopathology of pathogens of insects; virus invasion and replication of insect pathogens in insect tissues and in insect tissue culture cells. Mailing Add: USDA Insect Path Lab Bldg 011A Rm 214 Agr Res Ctr West Beltsville MD 20705

ADAMS, JERRY L, b Oklahoma City, Okla, June 1, 41. NUCLEAR PHYSICS. Educ: Mass Inst Technol, SB, 62; Okla State Univ, MS, 64; Fla State Univ, PhD(physics), 67. Prof Exp: Asst prof, 67-74, ASSOC PROF PHYSICS, OHIO UNIV, 74- Mem: Am Phys Soc. Res: Low energy theoretical nuclear physics. Mailing Add: Dept of Physics Ohio Univ Athens OH 45701

ADAMS, JOHN ALLAN STEWART, b Independence, Mo, Nov 1, 26; div; c 4. GEOCHEMISTRY. Educ: Univ Chicago, PhB, 46, BS, 48, MS, 49, PhD(geol), 51. Prof Exp: Proj assoc geochem, Dept Chem, Univ Wis, 51-54; from asst prof to assoc prof, 54-60, chmn dept, 65-71, PROF GEOL, RICE UNIV, 60- Concurrent Pos: Lectr, Univ Wis, 53 & Am Asn Petrol Geologists, 55; consult, Shell Develop Co, 54-58 & Humble Oil & Ref Co, 58-; exec ed, Geochimica et Cosmochimica Acta, 60-66; adj prof, Sch Pub Health, Univ Tex, 73- Mem: Am Chem Soc; fel Geol Soc Am; Geochem Soc; Am Asn Petrol Geologists. Res: Geochemistry of thorium and uranium; physical geology; mineralogy; petrology; analytical chemistry of rocks and minerals; geochronology; remote sensing and environmental management studies; lunar samples. Mailing Add: Dept of Geol Rice Univ Houston TX 77001

ADAMS, JOHN COLLINS, b New Albany, Ind, Sept 18, 38; m 60; c 2. MICROBIAL ECOLOGY, WATER POLLUTION. Educ: Purdue Univ, BS, 62; Iowa State Univ, MS, 64; Wash State Univ, PhD(bact), 69. Prof Exp: Asst prof microbiol, 69-75, ASSOC PROF MICROBIOL, UNIV WYO, 75- Mem: Am Soc Microbiol; Water Pollution Control Fedn; Soc Gen Microbiol; AAAS; Sigma Xi. Res: Microbiology of black water associated with oil shale; microbiology of coal strip mines in the western United States; microbiology of fresh water; microbiology of game meat; fluorescent and denitrifying bacteria. Mailing Add: Div of Microbiol & Vet Med Univ of Wyo Box 3354 Univ Sta Laramie WY 82071

ADAMS, JOHN EDGAR, b Curtis, Nebr, Jan 18, 18; m 42; c 3. SOIL PHYSICS. Educ: Univ Nebr, BSc, 42, MSc, 49; Iowa State Univ, PhD(soil physics), 56. Prof Exp: Soil scientist sedimentation & hydrol, Qual Water Br, US Geol Surv, 48-50; SOIL SCIENTIST SOUTHERN REGION, OKLA-TEX AREA, AGR RES SERV, USDA, 55- Mem: Am Soc Agron; Soil Sci Soc Am; Soil Conserv Soc Am; Sigma Xi. Res: Water intake and erodibility of heavy clay soils; management factors affecting evaporation and water use efficiency of row crops in dryland areas. Mailing Add: Blackland Conserv Res Ctr USDA Box 748 Temple TX 76501

ADAMS, JOHN EDWARD, b Milwaukee, Wis, June 10, 35; m 60; c 4. CULTURAL

GEOGRAPHY, GEOGRAPHY OF THE CARIBBEAN. Educ: Univ Minn, BA, 58, MA, 60, PhD(geog & anthrop), 70. Prof Exp: Instr geog, St Cloud State Col, 61-65; asst prof, 66-75, ASSOC PROF GEOG, UNIV MINN, DULUTH, 75- Res: Traditional and commercial fishing in the Caribbean; whaling, shipbuilding and maritime commerce in the Caribbean. Mailing Add: Dept of Geog Univ of Minn Duluth MN 55812

ADAMS, JOHN EDWIN, b Berkeley, Calif, Apr 18, 14; m 35; c 3. NEUROSURGERY. Educ: Univ Calif, AB, 35; Harvard Univ, MD, 39; Am Bd Neurol Surg, dipl. Prof Exp: Asst neuroanat, 47, from instr to assoc prof neurosurg, 48-65, res assoc med physics, 54-55, chmn div neurosurg, 57-65, prof neurol surg & chmn div, 65-68, GUGGENHEIM PROF NEUROL SURG, MED CTR, UNIV CALIF, SAN FRANCISCO, 68- Concurrent Pos: Attend neurosurgeon, Hosps, 50-; consult, Naval Hosp, 52- Mem: AAAS; Soc Univ Surg; Am Asn Neurol Surg; Am Col Surg; AMA. Res: Stereotactic surgery; cerebral metabolism and circulation; hypothermia. Mailing Add: Div of Neurol Surg Univ of Calif Med Ctr San Francisco CA 94122

ADAMS, JOHN EVI, b Durham, NC, May 23, 37; m 65; c 2. PSYCHIATRY. Educ: Swarthmore Col, BA, 59; Univ NC, 59-60; Cornell Univ, MD, 64. Prof Exp: Intern med, Vanderbilt Univ Hosp, Tenn, 64-65; resident psychiat, Stanford Univ Hosp, Calif, 65-68; spec asst to the dir, NIMH, 68-69, assoc dir, Div Manpower & Training, 69-70; asst prof psychiat, Sch Med, Stanford Univ, 70-74; PROF PSYCHIAT & CHMN DEPT, COL MED, UNIV FLA, 74- Concurrent Pos: USPHS res training fel, Dept Psychiat, Stanford Univ, 67-68; consult, Vet Admin Hosps, Palo Alto, Calif, 70-74 & Gainesville, Fla, 74-; mem training rev comt, Nat Inst Alcoholism & Alcohol Abuse, 71-, chmn, 74-75; chmn nat adv comt, Nat Ctr Alcohol Educ, 74-; mem, Fla Adv Coun Ment Health, 75- Mem: AAAS; Am Psychiat Asn (manpower comn, 70-73); Am Psychosom Soc. Res: Coping behavior in acute and transitional crisis; aggressive behavior and conflict resolution; ego psychology; alcoholism and aggression. Mailing Add: Dept of Psychiat Univ of Fla Col of Med Gainesville FL 32610

ADAMS, JOHN GEORGE, b Pittsburgh, Pa, Jan 31, 21; m 53; c 2. PHARMACOLOGY. Educ: Duquesne Univ, BS, 47; Univ Ill, MS, 52, PhD, 55. Prof Exp: Instr pharmacol & pharmacog, Sch Pharm, Duquesne Univ, 47-49, asst prof pharmacol, 52-55, prof & dean, 55-61; prof pharmacol, Univ Conn, 61-65; dir off sci activities, 65-68, VPRES OFF SCI & PROD RELS, PHARMACEUT MFRS ASN, 68- Mem: AAAS; Am Pharmaceut Asn; NY Acad Sci. Res: Relationship of molecular architecture to biological activity; biogenesis of neurohumoral agents in nerve impulse transmission and muscular contraction; drug enzymology; biochemical pharmacology. Mailing Add: 8107 Whites Ford Way Rockville MD 20854

ADAMS, JOHN HOWARD, b Portland, Maine, May 7, 39; m 65; c 3. ORGANIC CHEMISTRY. Educ: Union Col, BS, 61; Univ Colo, PhD(org chem), 65. Prof Exp: Asst chem, Univ Colo, 61-64; RES CHEMIST, CHEVRON RES CO, 65- Mem: Am Soc Lubrication Engrs. Res: Fluorocarbons; reaction of fluorohalocyclobutenes with nucleophiles; alkyl flavanoid synthesis; polymer antioxidant and stabilizers; thermal and photo oxidation of polyolefins; lubricating oil additives; solar radiation. Mailing Add: 9 Mt Rainier Ct San Rafael CA 94903

ADAMS, JOHN KENDAL, b Willimantic, Conn, June 14, 33; m 67. GEOLOGY. Educ: Univ Conn, BA, 55; Rutgers Univ, MS, 57, PhD(geol), 59. Prof Exp: Asst prof geol, Univ Del, 58-63; geologist, US Peace Corps, 63-65; ASSOC PROF GEOL, TEMPLE UNIV, 65- Mem: Geol Soc Am; Soc Econ Paleont & Mineral; Sigma Xi. Res: Environmental stratigraphy; engineering soils mapping; diagenesis. Mailing Add: Dept of Geol Temple Univ Philadelphia PA 19122

ADAMS, JOHN LESTER, b Manitou, Okla, Oct 9, 21; m 42; c 2. POULTRY HUSBANDRY. Educ: Okla Agr & Mech Col, BS, 43; Univ Wis, MS, 47, PhD(genetics, poultry husb), 50. Prof Exp: Instr poultry husb, Okla Agr & Mech Col, 46; teaching asst, Univ Wis, 46-48, instr, 48-50, from asst prof to assoc prof, 51-57; prof & chmn dept, 57-64, ASSOC DIR COOP AGR EXTEN SERV, UNIV NEBR, LINCOLN, 64-, DIR AGR EXTEN SERV, 65- Mem: Endocrine Soc; Am Genetic Asn; Soc Exp Biol & Med; Poultry Sci Asn. Res: Genetics, physiology and management of poultry; agricultural extension methods. Mailing Add: Univ of Nebr Agr Exten Serv East Campus Lincoln NE 68503

ADAMS, JOHN MILTON, b Minneapolis, Minn, June 7, 05; m. MEDICINE. Educ: Princeton Univ, BS, 29; Columbia Univ, MD, 33; Univ Minn, PhD(med), 42. Prof Exp: Chmn dept, 50-64, PROF PEDIAT, SCH MED, UNIV CALIF, LOS ANGELES, 64- Honors & Awards: Commonwealth Fund Award, 64. Mem: Am Pediat Soc; Am Soc Clin Invest; Soc Pediat Res; Infectious Dis Soc Am. Res: Pneumotropic viruses; influenza; tuberculosis; measles and encephalitis; multiple sclerosis; slow-virus infections. Mailing Add: Dept of Pediat Univ of Calif Med Ctr Los Angeles CA 90024

ADAMS, JOHN PLETCH, b Ashburn, Mo, Feb 2, 22; m 48; c 1. ORTHOPEDIC SURGERY. Educ: Univ Mo, BS, 43; Wash Univ, MD, 45; Am Bd Orthop Surg, dipl, 51. Prof Exp: PROF ORTHOP SURG, SCH MED, GEORGE WASHINGTON UNIV, 53-, CHMN DEPT, 69-, CHIEF ORTHOP SERV, UNIV HOSP, 53- Concurrent Pos: Am Orthop Asn travel fel, Gt Brit, 59; chief orthop serv & chief orthop consult, Crippled Children's Unit & chief hand clin, 53; consult orthop surg, Vet Hosp, Washington, DC, 53; consult, Vet Admin Hosp, WVa, Col Med, Howard Univ, Dept Child & Maternal Welfare, DC, Walter Reed Army Med Ctr, Hand Surg & Div Indian Health, USPHS & US Navy Med Ctr. Mem: Orthop Res Soc; AMA; Am Rheumatism Asn; fel Am Col Surg; Am Acad Cerebral Palsy; fel Am Acad Orthop Surg. Res: Children's orthopedics; reconstructive surgery of upper extremities; vascular problems in the extremities. Mailing Add: 2150 Pennsylvania Ave NW Washington DC 20037

ADAMS, JOHN QUINCY, b Louisville, Ky, July 3, 31; m 54. PHYSICAL CHEMISTRY. Educ: Purdue Univ, BS, 53; Univ Ill, MS, 55, PhD(phys chem), 61. Prof Exp: Assoc res chemist, Calif Res Corp, 61-64; from res chemist to sr res chemist, 64-75, SR RES ASSOC, CHEVRON RES CO, 75- Mem: Am Chem Soc. Res: Carbon and phosphorus; electron spectroscopy of surfaces; secondary ion mass spectroscopy. Mailing Add: Chevron Res Co 576 Standard Ave Richmond CA 94802

ADAMS, JOHN R, b Eureka, Calif, Jan 6, 18; m 42; c 4. MEDICINE, PSYCHIATRY. Educ: Northwestern Univ, 35-39; McGill Univ, MD & CM, 43. Prof Exp: Staff psychiatrist, Menninger Found, Kans, 49-54, dir admis, 52-54; chief psychiat, Passavant Mem Hosp, 54-72; PROF PSYCHIAT, MED SCH, NORTHWESTERN UNIV, CHICAGO, 54-, SR PSYCHIAT, INST PSYCHIAT, 75- Concurrent Pos: Consult, US Army Hosp, Ft Riley, Kans, 53-54; Vet Admin Res Hosp, Chicago, 54-61 & Ill State Psychiat Inst, 60- Mem: AMA; Am Psychiat Asn. Res: Psychiatry teaching; curriculum design; general hospital psychiatry. Mailing Add: 707 N Fairbanks Ct Chicago IL 60611

ADAMS, JOHN STEPHEN, b Minneapolis, Minn, Sept 7, 38; m 62; c 4. URBAN

GEOGRAPHY, URBAN ECONOMICS. Educ: St Thomas Col, BA, 60; Univ Minn, MA, 62, PhD(geog), 66. Prof Exp: Asst prof geog, Pa State Univ, 66-70; actg dir urban & regional planning prog, 71-74, assoc prof geog, 70-71, ASSOC PROF GEOG & PUB AFFAIRS, UNIV MINN, MINNEAPOLIS, 71-, CHMN URBAN STUDIES PROG, 74- Concurrent Pos: Mem, Comn Col Geog, 71-74; dir comp metrop anal proj, NSF-Asn Am Geog, 72- Mem: Asn Am Geog; Am Geog Soc (secy, 75-78). Res: Urban housing structures and housing markets; spatial organization of American urban areas; residential mobility of American urban households. Mailing Add: 414 Social Sci Univ of Minn Minneapolis MN 55455

ADAMS, JOHN WAGSTAFF, b New York, NY, Sept 17, 15; m 42; c 2. ECONOMIC GEOLOGY. Educ: Colo Sch Mines, GeolE, 41. Prof Exp: Stud engr, Empire Zinc Co, 42-43; asst res geologist, Pine Creek Unit, US Vanadium Corp, 46-47; geologist & engr, Am Zinc Lead & Smelting Co, 47; jr geologist, 43-45 & 47-48, assoc geologist, 48-51, GEOLOGIST, US GEOL SURV, 51- Concurrent Pos: Geologist-mineralogist, Colo Sch Mines Res Found, 58. Mem: Fel Mineral Soc Am; Soc Econ Geologists; Mineral Asn Can. Res: Geology and mineralogy of the rare-earth elements. Mailing Add: 705 Garland St Lakewood CO 80215

ADAMS, JOSEPH EDISON, b Middletown, NY, Nov 23, 03; m 29; c 2. BOTANY. Educ: Columbia Univ, PhG, 25, PhC, 27, AM, 32; Univ Mich, BS, 29; Univ Calif, PhD(bot), 35. Prof Exp: Instr mat med & bot, Columbia Univ, 25-32; from instr to prof, 35-69, EMER PROF BOT, UNIV NC, CHAPEL HILL, 69- Mem: Am Soc Plant Taxon. Res: Systematic botany; plant anatomy. Mailing Add: PO Box 333 Chapel Hill NC 27514

ADAMS, JULIAN PHILIP, b Windsor, Eng, Oct 30, 44. POPULATION GENETICS. Educ: Univ Wales, BSc, 65; Univ Calif, Davis, PhD(genetics), 69. Prof Exp: Fel genetics, Dept Pomol, Univ Calif, Davis, 69-70; asst prof bot, 70-75, ASSOC PROF BIOL, UNIV MICH, ANN ARBOR, 75- Mem: Genetics Soc Am; Brit Genetical Soc; Am Statist Asn; Am Soc Naturalists. Res: Population genetics of microorganisms; adaptive significance of diploidy; evolution at the level of the enzyme; human genetics. Mailing Add: Div of Biol Sci Univ of Mich Ann Arbor MI 48104

ADAMS, KENNETH ALLEN HARRY, b Melville, Sask, Nov 16, 34; m 60; c 2. ORGANIC CHEMISTRY. Educ: Univ Man, BSc, 55; McMaster Univ, MSc, 58; Univ Sask, PhD(chem), 61. Prof Exp: Res assoc chem, Ind Univ, 61-62; res & teaching fel, McMaster Univ, 62-63; asst prof, 63-68, ASSOC PROF CHEM, MT ALLISON UNIV, 68- Concurrent Pos: Nat Res Coun Can grants, 63-65. Mem: Chem Inst Can; The Chem Soc. Res: Synthesis and chemistry of nitrogenous heterocyclic compounds, including natural products; chemistry and synthetic application of nitrene intermediates. Mailing Add: Dept of Chem Mt Allison Univ Sackville NB Can

ADAMS, KENNETH HOWARD, b Elgin, Ill, July 7, 06; m 34, 67; c 1. ORGANIC CHEMISTRY. Educ: Univ Chicago, BS, 28, PhD(chem), 32. Prof Exp: Instr & cur chem, Univ Chicago, 33-40; assoc prof, Harris Teachers Col, Mo, 40-43; from asst prof to assoc prof, 43-55, PROF CHEM, ST LOUIS UNIV, 55- Mem: AAAS; Am Chem Soc. Res: Mechanism of carbinol dehydrations and rearrangements; acid catalysis in nonaqueous solvents; aromatic nucleophilic substitution. Mailing Add: Dept of Chem Box 8089 St Louis Univ Pierre Laclede Sta St Louis MO 63156

ADAMS, LEON MILTON, b Waco, Tex, Mar 11, 13; m 37; c 2. ORGANIC CHEMISTRY. Educ: Agr & Mech Col Tex, BS, 33, MS, 34; Univ Nebr, PhD(phys chem), 37. Prof Exp: Res fel, Mellon Inst, 37-38; res chemist, Pittsburgh Plate Glass Co, 38-43; asst chief chemist, Taylor Ref Corp, Tex, 43-46; group leader, Am Oil Co, 46-60; MGR ORG & POLYMER SECT, SOUTHWEST RES INST, 60- Mem: Am Chem Soc. Res: Encapsulation; polymers; organic synthesis; gelation; radiation grafting; reverse osmosis. Mailing Add: Southwest Res Inst 8500 Culebra Rd San Antonio TX 78228

ADAMS, LOWELL, vertebrate ecology, see 12th edition

ADAMS, MARK F, b Hoquiam, Wash, Aug 4, 14; m 42; c 4. ANALYTICAL CHEMISTRY, INDUSTRIAL CHEMISTRY. Educ: Wash State Univ, BS, 40, MS, 42, PhD(phys chem), 51; Nat Asn Corrosion Engrs, cert corrosion specialist, No. 843, 72. Prof Exp: Res chemist, State Dept Conserv & Develop, Wash, 42-45; from asst chemist to chemist, Div Indust Res, 45-69, head chem res, 51-69, head mat chem, Col Eng Res Div, 69-74, EMER PROF, COL ENG RES DIV, WASH STATE UNIV, 74-; CONSULT, 74- Concurrent Pos: Abstr ed, 62-; assoc consult, Norton Corrosion Ltd, Inc, 75- Mem: AAAS; Am Chem Soc; Nat Asn Corrosion Engrs. Res: Animal diseases; corrosion; water quality; wheat utilization; forensic investigations; solar power for electric energy. Mailing Add: PO Box 226 Pacific Beach WA 98571

ADAMS, MARTHA LOVELL, b Springfield, Ill, May 8, 25. ANALYTICAL CHEMISTRY. Educ: Col William & Mary, BS, 46; Univ Md, MS, 52. Prof Exp: Instr inorg chem, St Helena Exten, Col William & Mary, 46-48; res chemist coating & chem lab, Aberdeen Proving Ground, 51-74; RES CHEMIST, US ARMY MOBILITY EQUIP RES & DEVELOP CTR, FT BELVOIR, VA, 74- Mem: Am Chem Soc; fel Am Inst Chemists; Soc Appl Spectros. Res: Analytical methods for coating materials, especially the use of ultraviolet and infrared spectrophotometry. Mailing Add: 1612 Pickett Rd Lutherville MD 21093

ADAMS, MAURICE WAYNE, b Rosedale, Ind, June 23, 18; m 45; c 5. PLANT BREEDING. Educ: Purdue Univ, BSA, 41; Univ Wis, PhD, 49. Prof Exp: Prof agron, SDak State Col, 47-58, agronomist, Exp Sta, 47-58; assoc prof farm crops, 58-64, PROF CROP SCI, MICH STATE UNIV, 64- Res: Breeding behavior and genetics of legumes; genetic regulation of ion transport; genetic bases of yield structures and processes in edible dry beans. Mailing Add: Dept of Crop & Soil Sci Mich State Univ East Lansing MI 48824

ADAMS, MAX DAVID, b St Marys, WVa, July 25, 41; m 63; c 2. PHARMACOLOGY. Educ: WVa Univ, BS, 65; Purdue Univ, MS, 68, PhD(pharmacol), 71. Prof Exp: Instr, 71-72, ASST PROF PHARMACOL, MED COL VA, 72- Res: Cardiovascular pharmacology; hypertension and antihypertensive drugs. Mailing Add: Dept of Pharmacol Med Col Va Richmond VA 23298

ADAMS, MAX DWAIN, b Red Oak, Iowa, May 23, 26; m 47; c 3. INORGANIC CHEMISTRY. Educ: Tarkio Col, BA, 48; Okla State Univ, MS, 50; St Louis Univ, PhD(chem), 55. Prof Exp: Anal chemist, Mallinckrodt Chem Works, 50-52; res assoc, St Louis Univ, 52-55; chemist, Argonne Nat Lab, 55-73; PROPRIETOR, M D ADAMS ASSOCS, 73- Mem: AAAS; Am Chem Soc; Sigma Xi; Am Inst Chemists; Am Nuclear Soc. Res: High temperature chemistry; nuclear materials and fission product elements; molten salt chemistry; microanalytical chemistry; particle identification; trace contamination detection methods. Mailing Add: 313 Tamarack Ave Naperville IL 60540

ADAMS, MERLE VERNON, agronomy, plant physiology, see 12th edition

ADAMS, MICHAEL STUDEBAKER, b Houston, Tex, Aug 26, 38; m 64. PLANT ECOLOGY. Educ: Univ Calif, Davis, BS, 62, MS, 64; Univ Calif, Riverside, PhD(biol), 68. Prof Exp: NASA trainee, Univ Calif, Riverside, 65-68; asst prof, 68-72, ASSOC PROF BOT, UNIV WIS-MADISON, 72- Mem: AAAS; Ecol Soc Am; Am Inst Biol Sci; Am Soc Limnol & Oceanog; Int Asn Theoret & Appl Limnol. Res: Physiological ecology of aquatic macrophytes; analysis of littoral ecosystems. Mailing Add: Dept of Bot Univ of Wis Madison WI 53706

ADAMS, ORA R, veterinary surgery, see 12th edition

ADAMS, OTIS WILLIAM, b Chicago, Ill, Sept 22, 25; m 55; c 3. PHYSICAL ORGANIC CHEMISTRY. Educ: Univ Chicago, BS, 51; Ill Inst Technol, PhD, 61. Prof Exp: Chemist, Armour & Co, 51-52; res chemist, Armour Res Found, 52-62; sr chemist, Abbott Labs, 62-69; PROG DIR, NAT SCI FOUND, 69- Mem: AAAS; Am Chem Soc. Res: Quantum chemistry; chemical reactivity. Mailing Add: Chem Div NSF 1800 G St NW Washington DC 20550

ADAMS, P B, b Elmira, NY, Sept 4, 29; m 61; c 2. SURFACE CHEMISTRY. Educ: Hobart Col, BS, 51. Prof Exp: From jr chemist to sr chemist, 51-62, res chemist, 62-75, RES SUPVR, CORNING GLASS WORKS, 75- Mem: Am Chem Soc; fel Am Ceramic Soc; Am Soc Testing & Mat; fel Am Inst Chemists. Res: Silicate chemistry; emission spectroscopy of silicates; degradation processes in non-metallic materials; chemical durability of glass; surface chemistry of glass. Mailing Add: Corning Glass Works Corning NY 14830

ADAMS, PAUL LIEBER, b Broken Bow, Okla, Jan 22, 24; m 46; c 3. PSYCHIATRY, SOCIOLOGY. Educ: Centre Col, AB, 43; Columbia Univ, MA, 48, MD, 55; Am Bd Psychiat & Neurol, dipl, 63, cert child psychiat, 64. Prof Exp: Instr soc sci, Bennett Col, 47-51; from asst to instr psychiat, Duke Univ, 56-60; from asst prof to prof psychiat & pediat, Col Med, Univ Fla, 60-74, dir, Children's Ment Health Unit, 66-74; PROF PSYCHIAT & VCHMN GRAD EDUC, SCH MED, UNIV LOUISVILLE, 74- Concurrent Pos: Fel child psychiat, Duke Univ Hosp & Durham Child Guid Clin, 58-60; vis prof, NC Col, 57-60; dir, Cumberland County Guid Ctr, 57-60; consult, Jacksonville Hosp Educ Prog, 60-. Mem: Fel Am Psychiat Asn; Am Acad Child Psychiat; fel Am Col Psychiat; fel Am Orthopsychiat Asn; Acad Psychosom Med. Res: Child psychiatry. Mailing Add: Dept of Psychiat & Behav Sci Univ Louisville Health Sci Ctr Louisville KY 40201

ADAMS, PAUL LOUIS, b Memphis, Tenn, May 15, 48. MEDICAL PHYSICS. Educ: Rice Univ, BA, 70; Stanford Univ, MS, 72, PhD(appl physics), 75. Prof Exp: INSTR MED PHYSICS, DEPT DIAG RADIOL, UNIV TENN, 75- Res: Applications of computed tomography and radiological imaging in general. Mailing Add: 6115 Lake Hickory Memphis TN 38138

ADAMS, PETER B, b Concord, Mass, Apr 18, 36; m 58; c 2. PLANT PATHOLOGY. Educ: Univ Vt, BS, 59, MS, 61; Univ RI, PhD(plant path), 64. Prof Exp: Fel plant path, US Army Biol Lab, 74-75; PLANT PATHOLOGIST, SOILBORNE DIS LAB, AGR RES SERV, USDA, 75- Mem: AAAS; Am Phytopath Soc. Res: Ecology of soilborne pathogens and control of the diseases which they cause. Mailing Add: Soilborne Dis Lab Agr Res Ctr West Beltsville MD 20705

ADAMS, PETER DAVID, b Cardiff, Wales, Nov 18, 37; m 60; c 2. SOLID STATE PHYSICS. Educ: Univ Wales, BSc, 59, Univ London, dipl, Imp Col & PhD(physics), 64. Prof Exp: From asst physicist to assoc physicist, 64-69, ED, PHYS REV, BROOKHAVEN NAT LAB, 69- Mem: Am Phys Soc. Res: General theory of liquid metals and alloys, especially the experimental and theoretical analysis of the electrical and acoustic properties. Mailing Add: Phys Review Brookhaven Nat Lab Upton NY 11973

ADAMS, PHILIP DELMAR, b Wilkinsburg, Pa, Sept 30, 01; m 30; c 2. BIOCHEMISTRY. Educ: Pa State Col, BS, 25, MS, 30; Univ Cincinnati, PhD(biochem), 35. Prof Exp: Instr chem, Am Univ Beirut, 25-28; instr, Pa State Col, 28-30; res biochemist, Skin & Cancer Hosp, Philadelphia, Pa, 30-32; dir res & control labs, Andrew Jergens Co, 35-61, dir spec proj, 61-63; TECH REP, HILL TOP RES, INC, MIAMIVILLE, OHIO, 64- Mem: Am Chem Soc; Soc Cosmetic Chemists. Res: Chemistry of fungi, dermatology and cosmetics; biochemistry of skin; soap; oxygen uptake and composition of the skin of rats in vitamin G deficiency. Mailing Add: 2053 Sutton Ave Cincinnati OH 45230

ADAMS, PHILLIP, b Brooklyn, NY, June 2, 25; m 55; c 3. ORGANIC CHEMISTRY. Educ: Univ Md, BS, 45; Cornell Univ, PhD(chem), 50. Prof Exp: Lab asst org chem, Cornell Univ, 45-50; res chemist, Am Cyanamid Co, NJ, 50-51; asst dir res & develop, Berkeley Chem Co, 51-68, dir res & develop, Berkeley Chem Div & asst dir corp res & develop, Millmaster Onyx Corp, 68-69, mem exec comt, 69-75, MEM OPERATING COMT, MILLMASTER ONYX CORP, 75- Mem: Am Chem Soc; NY Acad Sci. Res: Custom synthesis in organic and inorganic chemistry. Mailing Add: Millmaster Onyx Corp 11 Summit Ave Berkeley Heights NJ 07922

ADAMS, PHILLIP A, b Los Angeles, Calif, Jan 13, 29. ENTOMOLOGY. Educ: Univ Calif, BS, 51; Harvard Univ, AM & PhD(biol), 58. Prof Exp: From instr to asst prof zool, Univ Calif, Santa Barbara, 58-63; from asst prof to assoc prof, 63-71, PROF ZOOL, CALIF STATE UNIV FULLERTON, 71- Concurrent Pos: Vis assoc prof biol & vis assoc cur insects, Yale Univ, 68-69. Mem: AAAS; Am Entom Soc. Res: Insect evolution; phylogeny and taxonomy of Neuroptera; Myrmeleontidae and Chrysopidae; insect thermo-regulation. Mailing Add: Dept of Biol Calif State Univ Fullerton CA 92634

ADAMS, PRESTON, b Madison, Fla, Jan 12, 30; m 63. BOTANY. Educ: Univ Ga, BS, 54, MS, 56; Harvard Univ, PhD(biol), 59. Prof Exp: Asst bot, Univ Ga, 54-56 & Gray Herbarium, Harvard Univ, 56-57; res assoc bot, Fla State Univ, 59-61; from asst prof to assoc prof, 61-73, PROF BOT, DePAUW UNIV, 74- Mem: Bot Soc Am; Am Soc Plant Taxon; Int Asn Plant Taxon. Res: Evolution of vascular plants; monographic studies of flowering plants, especially the family Guttiferae. Mailing Add: Dept of Bot DePauw Univ Greencastle IN 46135

ADAMS, RALPH NORMAN, b Atlantic City, NJ, Aug 26, 24; m 53; c 1. ANALYTICAL CHEMISTRY. Educ: Rutgers Univ, BS, 50; Princeton Univ, PhD(anal chem), 53. Prof Exp: Instr anal chem, Princeton Univ, 53-55; from asst prof to assoc prof, 55-63, PROF CHEM, UNIV KANS, 63- Mem: Am Chem Soc. Res: Electroanalytical and instrumental methods. Mailing Add: Dept of Chem Univ of Kans Lawrence KS 66044

ADAMS, RANDALL HENRY, b Milburn, Okla, Aug 24, 39; m 62; c 2. ENTOMOLOGY, PLANT SCIENCE. Educ: Okla State Univ, BS, 61, MS, 66, PhD(entom ecol), 72. Prof Exp: Res asst entom, Okla State Univ, 66-68; asst prof, NMex State Univ, 68-70, agron, 70-74; ASST PROF ENTOM & PLANT SCI, SOUTHERN STATE COL, ARK, 74- Concurrent Pos: Res consult entom & agron, Shell Chem Co, Dow Chem USA, PPG Chem Co, Thuron Industs & Cooper USA,

70- Mem: Sigma Xi; Entom Soc Am. Res: Evaluation of pesticides on animal and plant pests. Mailing Add: Dept of Agr Southern State Col Magnolia AR 71753

ADAMS, RAYMOND D, b Portland, Ore, Feb 13, 11; m 33; c 4. NEUROPATHOLOGY, NEUROLOGY. Educ: Harvard Univ, AM, 55; Univ Ghent & Univ Lausanne, MD; Univ Newcastle, DSc. Prof Exp: From asst prof to assoc clin prof neurol, 48-54, BULLARD PROF NEUROPATH, HARVARD UNIV, 54-; CHIEF NEUROL SERV, MASS GEN HOSP, 51- Concurrent Pos: Dir, Kennedy & Shriver Labs Study Ment Retardation & Neurol Dis of Children. Mem: Am Acad Arts & Sci; Am Neurol Asn; Am Asn Neuropath; Asn Res Nerv & Ment Dis; Am Soc Clin Invest. Res: Diseases of skeletal muscle; alcoholic, nutritional and metabolic diseases of the nervous system; developmental diseases of the nervous system; clinical research. Mailing Add: Neurol Serv Mass Gen Hosp Boston MA 02114

ADAMS, RICHARD AUGUST, biology, see 12th edition

ADAMS, RICHARD EDWARD WOOD, b Kansas City, Mo, July 17, 31; m 55; c 4. ANTHROPOLOGY, ARCHAEOLOGY. Educ: Univ NMex, BA, 53; Harvard Univ, MA, 60, PhD(anthrop), 63. Prof Exp: Archaeologist, Tikal Proj, Mus, Univ Pa, 58; archaeologist & ceramic analyst, Altar de Sacrificios Prog, Peabody Mus, Harvard Univ, 61-63; from asst prof to assoc prof anthrop, Univ Minn, Minneapolis, 63-72; PROF ANTHROP & DEAN HUMANITIES & SOC SCI, UNIV TEX, SAN ANTONIO, 72- Concurrent Pos: NSF grant, Univ Minn & Cotzal Valley, Guatemala, 65-66; Univ Minn Found grant, Becan ruins, Campeche, Mex, 70; field dir, Becan Proj Tulane Univ, Nat Geog Soc, 70; proj dir, Rio Bec Proj, Nat Geog Soc grant, 73. Mem: AAAS; Am Archaeol Soc (secy, 72-76); Am Anthrop Asn; Archaeol Inst Am; Int Cong Archaeolists. Res: Prehistory of Mesoamerica; cultural ecology; ceramics; culture process; Maya lowlands and highlands; ethno-history; method and theory of archaeology. Mailing Add: Dept of Anthrop Univ of Tex San Antonio TX 78228

ADAMS, RICHARD LINWOOD, b Dixfield, Maine, Sept 17, 28; m 53; c 2. POULTRY NUTRITION. Educ: Univ Maine, BS, 52; Purdue Univ, MS, 59, PhD(poultry nutrit), 61. Prof Exp: Fieldman, Swift & Co, Ore, 52-53 & C M T Co, Maine, 53-55; 4-H Club agent, York County Exten Serv, 55-57; POULTRY EXTEN SPECIALIST, PURDUE UNIV, 62- Res: Poultry nutrition. Mailing Add: Poultry Sci Bldg Purdue Univ Lafayette IN 47907

ADAMS, RICHARD MELVERNE, b Gary, Ind, Nov 1, 16; m 44; c 2. PHYSICAL CHEMISTRY. Educ: Univ Chicago, BS, 39, MS, 48; Ill Inst Technol, PhD(chem), 53. Prof Exp: Chemist, Portland Cement Asn, 40-42; res assoc, Off Sci Res & Develop Proj, Univ Chicago, 42-43, jr chemist, Metall Lab, Manhattan Dist, 43-46; asst group leader, Los Alamos Sci Lab, 46; lectr chem, Ind Univ, 46-49; assoc chemist, 49-59, sci asst to dir, 59-65, ASST LAB DIR, ARGONNE NAT LAB, 65- Mem: AAAS; Am Chem Soc; Am Phys Soc; Sigma Xi; Am Nuclear Soc; Korean Nuclear Soc. Res: Etherates of boron trifluoride; radiochemistry; fission products; kinetics of exchange reactions; fluorine chemistry; reactor safety. Mailing Add: Argonne Nat Lab 9700 S Cass Ave Argonne IL 60439

ADAMS, RICHARD NEWBOLD, b Ann Arbor, Mich, Aug 4, 24; m 51; c 3. SOCIAL ANTHROPOLOGY, CULTURAL ANTHROPOLOGY. Educ: Univ Mich, AB, 47; Yale Univ, MA, 49, PhD(anthrop), 51. Prof Exp: Ethnologist, Inst Social Anthrop, Smithsonian Inst, 50-51; specialist anthrop, US Dept State, Guatemala, 51-52; scientist, WHO, 53-56; prof anthrop, Mich State Univ, 56-62; asst dir, Inst Latin Am Studies, Univ Tex, Austin, 62-67, chmn dept anthrop, Univ, 64-67, PROF ANTHROP, UNIV TEX, AUSTIN, 62- Concurrent Pos: US Dept State grant anthrop, Guatemala, 53-56; vis prof anthrop, Univ Calif, Berkeley, 60-61; Ford Found res fel, Univ Tex & Guatemala, 64-67; mem, Joint Comt Latin Am, Soc Sci Res Coun, 66-69; prog adv, Soc Sci, Ford Found, Arg, 69-70 & prog adv & consult, Arg, Brazil, Paraguay & Peru, 70-; mem, Comt Int Nutrit Progs, 72-; Guggenheim fel, 73-74. Mem: AAAS (vpres, sect anthrop, 72-73); Am Anthrop Asn; Soc Appl Anthrop (pres, 62-63); Latin Am Studies Asn (pres, 68); Am Ethnol Soc. Res: Social anthropology of complex societies, especially Latin America; evolution of society; ecology and energy of society. Mailing Add: Dept of Anthrop Univ of Tex Austin TX 78712

ADAMS, RICHARD OWEN, b Garden Home, Ore, Mar 5, 33; m 64. SOLID STATE PHYSICS. Educ: Willamette Univ, BA, 55; Wash State Univ, PhD(physics), 64. Prof Exp: Sr develop specialist, 63-67, sr res physicist, 67-68, res mgr, 68-75, SR RES SPECIALIST, ROCKY FLATS PLANT, ROCKWELL INT, 75- Mem: Am Vacuum Soc; Am Phys Soc; Am Soc Nondestructive Testing. Res: Surface physics, particularly the interactions of gases with metal surfaces; acoustic emission, its sources and propagation through metals. Mailing Add: Rockwell Int Rocky Flats Plant PO Box 464 Golden CO 80401

ADAMS, RICHARD SANFORD, b Lewiston, Maine, May 5, 28; m 51; c 3. ANIMAL NUTRITION, DAIRY HUSBANDRY. Educ: Univ Maine, BS, 50; Univ Minn, PhD(dairy husb), 55. Prof Exp: Asst prof dairy sci, Pa State Univ, 54-57; res assoc ruminant nutrit, Gen Mills, Inc, 57-58; from asst prof to assoc prof, 58-65, PROF DAIRY SCI, PA STATE UNIV, 65- Concurrent Pos: Consult, US Feed Grains Coun, SEurope, 69 & Univ Hawaii, 71. Mem: Am Dairy Sci Asn; Am Soc Animal Sci. Res: Dairy cattle nutrition; forage evaluation; extension teaching methods. Mailing Add: 213 Borland Lab Pa State Univ University Park PA 16802

ADAMS, ROBERT ALTON, microbiology, finance, see 12th edition

ADAMS, ROBERT D, b Grand Rapids, Mich, Dec 10, 26; m 56; c 2. MATHEMATICS. Educ: Univ Minn, PhD(math), 60. Prof Exp: ASSOC PROF MATH, UNIV KANS, 60- Mem: Am Math Soc; Math Asn Am. Res: Application of functional analysis to partial differential equations. Mailing Add: Dept of Math Univ of Kans Lawrence KS 66045

ADAMS, ROBERT EVANS, b Sodus, NY, Nov 20, 24; m 46. PLANT PATHOLOGY. Educ: Cornell Univ, PhD(plant path), 53. Prof Exp: From asst prof to assoc prof, 53-72, PROF PLANT PATH, W VA UNIV, 72- Mem: Soc Nematol; Am Phytopath Soc. Res: Deciduous fruit tree diseases; nematology. Mailing Add: 1292 Fairlawns Ave Morgantown WV 26505

ADAMS, ROBERT JOHN, b Solon, Iowa, May 24, 15; m 41; c 2. ELECTRONICS. Educ: Univ Iowa, BA, 36; Univ Wis, PhD(physics), 41. Prof Exp: Physicist, Corning Glass Works, NY, 40-42; head antenna sect, 42-54, HEAD SEARCH RADAR BR, NAVAL RES LAB, 54- Concurrent Pos: Assoc ed, Trans Antennas, Inst Elec & Electronic Engrs. Mem: AAAS; Am Geophys Union; Am Phys Soc; Inst Elec & Electronic Engrs; Sigma Xi. Res: Radar antennas and systems; radio physics. Mailing Add: Code 5330 Naval Res Lab Washington DC 20375

ADAMS, ROBERT MCCORMICK, b Chicago, Ill, July 23, 26; m 53; c 1. ANTHROPOLOGY, HISTORICAL GEOGRAPHY. Educ: Univ Chicago, PHB, 47, MA, 52, PhD(anthrop), 56. Prof Exp: Instr anthrop, Univ Chicago, 55-57, from asst prof to assoc prof, Univ & Oriental Inst, 57-63, dir, Oriental Inst, 62-68, dean, Div Soc Sci, 70-74, prof, Univ & Oriental Inst, 63-75, HAROLD SWIFT DISTINGUISHED SERV PROF, UNIV CHICAGO & ORIENTAL INST, 75- Concurrent Pos: Lewis Henry Morgan lectr, Univ Rochester, 65; annual prof, Am Sch Oriental Res, Baghdad, 66-67, resident dir, 68-69; mem div behav sci, Nat Res Coun, 67-, chmn, 72-, chmn, Assembly Behav & Soc Sci, 73- Mem: Nat Acad Sci; fel Am Anthrop Asn; Ger Archaeol Inst; fel Am Acad Arts & Sci; Am Philos Soc. Res: Ecologically oriented study of historic patterns of land use settlement and urbanization; substantive fieldwork principally in Middle East, especially Iraq; comparative interest in other centers of early civilization. Mailing Add: Oriental Inst Univ Chicago 1155 E 58th St Chicago IL 60637

ADAMS, ROBERT PHILLIP, b Denison, Tex, Dec 8, 39; m 62; c 2. EVOLUTIONARY BIOLOGY, TAXONOMY. Educ: Univ Tex, Austin, BA, 62, PhD(biol sci), 69. Prof Exp: Systs analyst comput prog, Control Data Corp, 62-64; farmer, 64-66; asst prof bot & biol, 69-74, ASSOC PROF BOT & BIOL, COLO STATE UNIV, 74- Concurrent Pos: Prin investr, NSF grant, 70-72. Mem: Phytochem Soc NAm; Bot Soc Am; Am Soc Plant Taxon; Int Asn Plant Taxon; Soc Syst Zool. Res: Chemosystematics; numerical taxonomy; computerized mapping of plant distributions; numerical approaches to population problems; seasonal variations in terpenoid production; terpene metabolism; gas-liquid chromatography. Mailing Add: Dept of Bot Colo State Univ Ft Collins CO 80521

ADAMS, ROBERT TRAIN, organic chemistry, see 12th edition

ADAMS, ROBERT W, b New York, NY, July 8, 34; m 57; c 3. GEOLOGY. Educ: Univ Rochester, BA, 56; Johns Hopkins Univ, PhD(geol), 64. Prof Exp: Explor geologist, Shell Oil Co, 64-67; from asst prof to assoc prof geol, 67-74, PROF GEOL, STATE UNIV NY COL BROCKPORT, 74- Mem: Geol Soc Am; Soc Econ Paleontologists & Mineralogists; Am Asn Petrol Geologists; Nat Asn Geol Teachers. Res: Sedimentology and sedimentary petrology. Mailing Add: Dept of Geol & Earth Sci State Univ of NY Col Brockport NY 14420

ADAMS, ROBERT WALKER, JR, b Ashburn, Ga, Nov 9, 20; m 45; c 4. MEDICINE. Educ: Vanderbilt Univ, MD, 48. Prof Exp: Asst prof, 55-70, ASSOC PROF PSYCHIAT, MED SCH, VANDERBILT UNIV, 71-; CLIN ASST PROF, MEHARRY MED SCH, 55- Mem: AMA; Am Psychiat Asn. Res: Clinical psychiatry; teaching. Mailing Add: Dept of Psychiat Vanderbilt Univ Sch of Med Nashville TN 37232

ADAMS, ROGER FRANCIS, physical chemistry, polymer chemistry, see 12th edition

ADAMS, ROGER JAMES, b Rio, Wis, Aug 27, 36; m 58; c 3. PHYSICAL CHEMISTRY. Educ: Univ Wis, BS, 58; Iowa State Univ, PhD(phys chem), 61. Prof Exp: Sr chemist, 61-67, res supvr, 67-68, tech mgr res & develop, 68-74, LAB MGR MED PROD, MINN MINING & MFG CO, 74- Mem: Am Chem Soc. Res: Chemical kinetics; surface chemistry of metals; electrochemistry; displacement plating, electroplating, vapor plating; applications of tagged tracers in surface chemistry. Mailing Add: Minn Mining & Mfg Co 3M Ctr Bldg 230-3 St Paul MN 55101

ADAMS, ROWLAND KEEDY, JR, organic chemistry, see 12th edition

ADAMS, ROY MELVILLE, b Cheung Chau, Hong Kong, Oct 19, 19; m 46; c 4. CHEMISTRY. Educ: Sterling Col, AB, 40; Univ Kans, AM, 42, PhD(chem), 49. Prof Exp: From instr to assoc prof, 46-58, PROF CHEM & CHMN DEPT, GENEVA COL, 58- Concurrent Pos: Chief dept chem, Callery Chem Co, 52-53, res coordr, 53-58; titular mem inorg nomenclature comn, Int Union Pure & Appl Chem, 68- Mem: Am Chem Soc; Am Sci Affil; Brit Chem Soc; Am Inst Physics. Res: Boron; inorganic nomenclature. Mailing Add: Dept of Chem Geneva Col Beaver Falls PA 15010

ADAMS, RUSSELL BLAIR, b Enderlin, SDak, Jan 1, 26; m 71; c 2. GEOGRAPHY, STATISTICS. Educ: Univ Minn, BBA, 49, BS, 52, MA, 55, PhD(geol), 69. Prof Exp: Teacher math & social studies, Hill City High Sch, Minn, 52-53; instr social sci & teaching asst geog, Univ Minn, 53-57; programmer comput sci, Univac, Minn, 57-58; asst dir transp, Minn Hwy Dept, 58-61; asst dir urban res, Upper Midwest Econ Study, 61-64; from instr geog to to asst prof geog, 64-72, ASSOC PROF GEOG, UNIV MINN, MINNEAPOLIS, 72- Mem: AAAS; Asn Am Geog; Am Statist Asn; Am Geog Soc; Int Studies Asn. Res: Economic-quantitative geography; transportation; Union of Soviet Socialist Republics; Eastern Europe. Mailing Add: Dept of Geog Univ of Minn Minneapolis MN 55455

ADAMS, RUSSELL S, JR, b Kincaid, Kans, Mar 9, 26; m 59; c 1. SOIL CHEMISTRY. Educ: Kans State Univ, BS, 58, MS, 59; Univ Ill, PhD(soil biochem), 62. Prof Exp: Asst agron, Univ Ill, 61-62; res assoc, 62-63, from asst prof to assoc prof, 63-71, PROF SOILS, UNIV MINN, ST PAUL, 71- Mem: Am Soc Agron; Am Chem Soc; Soil Sci Soc Am; Weed Sci Soc Am; Int Soil Sci Soc. Res: Contamination of soils by petroleum hydrocarbons; fixation of ammonium ions by soil-forming minerals; the fate of pesticide residues in soils; soybean nutrition; disposal of solid wastes in soils. Mailing Add: Dept of Soil Sci Univ of Minn Inst of Agr St Paul MN 55108

ADAMS, SAM, b Walthall, Miss, Mar 14, 16; m 44; c 1. SCIENCE EDUCATION. Educ: Delta State Teachers Col, BS, 36; La State Univ, MA, 40, PhD(sci educ), 51. Prof Exp: Teacher high sch, Miss, 36-38; instr, Univ Ala, 40-42, Univ exten supvr, 42; civilian instr radio, US Air Force, SDak, 42-43; elec supply foreman, Tenn Eastman Corp, Oak Ridge, 43-44; physicist and spectroscopist, Carbide & Carbon Chem Co, 46-49; assoc prof physics, McNeese State Col, 51-54; assoc prof sci educ, 54-65, PROF SCI EDUC, LA STATE UNIV, BATON ROUGE, 65-, ASSOC DEAN ACAD AFFAIRS, 64- Concurrent Pos: Teacher high sch, Ala, 40-42. Mem: Nat Sci Teachers Asn. Res: Emission spectra of uranium isotopes; science education. Mailing Add: Dept of Educ La State Univ Baton Rouge LA 70803

ADAMS, STUART LYLE, b Louisville, Ky, July 30, 16; m 40; c 1. BACTERIOLOGY. Educ: Centre Col, AB, 38; Univ Louisville, MS, 42; State Col Wash, PhD(bact), 49. Prof Exp: Instr biol, Centre Col, 38-39; bacteriologist, Res Dept, 39-43, head fermentation div, 43-45, 46-47, asst dir res, 49-50, dir res, 50-61, TECH DIR, JOS E SEAGRAM & SONS, INC, 61- Mem: AAAS; Am Soc Microbiol; Am Chem Soc; Am Soc Testing & Mat. Res: Yeast fermentation; cereal grains; microbial enzymes; enzymatic hydrolysis of starch; 2-3 butylene glycol. Mailing Add: Jos E Seagram & Sons Inc 375 Park Ave New York NY 10022

ADAMS, TERRANCE STURGIS, b Los Angeles, Calif, Dec 14, 38; m 63. INSECT PHYSIOLOGY. Educ: Calif State Col Los Angeles, BA, 61; Univ Calif, Riverside, PhD(entom), 66. Prof Exp: RES ENTOMOLOGIST, METAB LAB, AGR RES SERV, USDA, 66- Mem: Entom Soc Am; Am Zool Soc. Res: Factors affecting insect reproduction; endocrine control of reproduction in dipterous insects. Mailing Add: Metab Lab Agr Res Serv USDA Univ Sta Fargo ND 58102

ADAMS, THOMAS, b Buffalo, NY, Oct 3, 30; m 53; c 2. PHYSIOLOGY. Educ: Univ Md, BS, 52; Purdue Univ, MS, 55; Univ Wash, PhD(physiol), 63. Prof Exp: Chief thermal sect, Civil Aeromed Inst, Fed Aviation Agency, Okla, 63-66; PROF PHYSIOL, MICH STATE UNIV, 66- Mem: AAAS; Am Physiol Soc; Soc Exp Biol & Med; Am Asn Univ Prof. Res: Body temperature regulation; neurophysiology of temperature regulation; mechanisms of thermal acclimatization; sweat gland physiology. Mailing Add: Dept of Physiol Mich State Univ East Lansing MI 48824

ADAMS, THOMAS C, b San Francisco, Calif, Nov 4, 18; m 57; c 2. FOREST ECONOMICS, INTERNATIONAL ECONOMICS. Educ: Univ Calif, Berkeley, BS, 40, AB, 41; Univ Mich, MA, 51, PhD(forestry), 52. Prof Exp: Forest economist, Pac Northwest Forest & Range Exp Sta, US Forest Serv, 52-55; asst prof forestry, Ore State Univ, 55-57; PRIN ECONOMIST, PAC NORTHWEST FOREST & RANGE EXP STA, US FOREST SERV, 57- Concurrent Pos: Consult, Stanford Res Inst, Ore, 55. Mem: Am Econ Asn; Soc Am Foresters; Am Agr Econ Asn; Regional Sci Asn. Res: Marketing, market development and international trade of forest products; economic analysis of new logging systems, thinning operations and forest residue reduction. Mailing Add: Pac Northwest Forest & Range Exp Sta US Forest Serv PO Box 3141 Portland OR 97208

ADAMS, WADE J, b Calhoun Co, Mich, Jan 9, 40; m 65; c 1. STRUCTURAL CHEMISTRY, CHEMICAL PHYSICS. Educ: Western Mich Univ, BS, 65; Univ Mich, MS, 68, PhD(phys chem), 69. Prof Exp: Fel, Univ Mich, 71-72; asst prof chem, Macalester Col, 72-73; ASST PROF CHEM, WESTERN MICH UNIV, 73- Mem: Am Chem Soc. Res: Experimental studies of the structure and force field of molecules by gaseous electron diffraction; molecular mechanics calculations of physical and chemical properties. Mailing Add: Dept of Chem Western Mich Univ Kalamazoo MI 49001

ADAMS, WALTER CHURCH, b Newtown, Pa, Aug 22, 36; m 62; c 2. ENDOCRINOLOGY, PHYSIOLOGY. Educ: Drew Univ, AB, 58; Rutgers Univ, MS, 62, PhD(endocrinol), 63. Prof Exp: Res scientist, NJ Bur Res Neurol & Psychiat, 63-67; asst prof, 67-71, ASSOC PROF BIOL SCI, KENT STATE UNIV, 71- Mem: AAAS; Am Physiol Soc; Am Soc Zoologists; Brit Soc Endocrinol. Res: Interrelationships of hypothalamic control of gonadotrophin secretion and ovarian physiology with respect to abnormal ovarian responses; comparative reproductive physiology. Mailing Add: Dept of Biol Sci Kent State Univ Kent OH 44242

ADAMS, WILLIAM ALFRED, b Toronto, Ont, June 15, 41; m 64; c 1. GLACIOLOGY. Educ: McMaster Univ, BSc, 63; Univ Ottawa, PhD(chem), 68. Prof Exp: Fel, Div Appl Chem, Nat Res Coun Can, 67-69; res scientist, Hydrol Sci Div, Inland Waters Br, 69-71, res scientist, Water Qual Br, 71-75, RES SCIENTIST, GLACIOL DIV, INLAND WATERS DIRECTORATE, ENVIRON CAN, 75- Concurrent Pos: Ed, Electrochem Soc, 75-78. Mem: Am Chem Soc; Chem Inst Can; Spectros Soc Can (pres, 75-76); Electrochem Soc. Res: Effects of pressure and temperature on the interactions of ions and polar molecules in solution using spectroscopic and electrochemical methods; impact of oil spills on Arctic aquatic ecosystems; deep well disposal. Mailing Add: Glaciol Div Environ Can 562 Booth St Ottawa ON Can

ADAMS, WILLIAM CURTIS, b Ft Smith, Ark, Mar 5, 24; m 55; c 2. PEDIATRICS. Educ: Hamilton Col, AB, 45; Temple Univ, MD, 47. Prof Exp: Resident, Sch Med, Temple Univ, 49-50, 52-54; from instr to asst prof child health, Sch Med, Univ Louisville, 54-60; assoc prof pediat, Univ Miami, 60-73, prof pediat & family med, 73-75. Concurrent Pos: Fel, Univ Ill, 48-49; dir pediat clin, Louisville Gen Hosp, 54-60, chief, 55-60; med dir, Variety Childrens Hosp, Miami, 60-64; consult, Food & Drug Admin; spec consult, Accident Prev Study Sect, NIH. Mem: Am Asn Poison Control Ctrs; AMA; NY Acad Sci; Am Acad Pediat. Res: Infectious and chest diseases; accident prevention; tuberculosis; mental retardation; accidental poisoning; emergency medicine. Mailing Add: 4820 San Amaro Dr Coral Gables FL 33146

ADAMS, WILLIAM ELIAS, surgery, medicine, deceased

ADAMS, WILLIAM HENRY, b Baltimore, Md, Dec 21, 33; m 58; c 1. THEORETICAL CHEMISTRY. Educ: Johns Hopkins Univ, AB, 55; Univ Chicago, SM, 56, PhD(chem phys), 60. Prof Exp: NSF fel quantum chem group, Univ Uppsala, 60-62; asst prof chem, Pa State Univ, 62-66; from asst prof to assoc prof, 66-75, PROF CHEM, RUTGERS UNIV, 75- Concurrent Pos: Vis prof, Tech Univ, Munich, 70-71. Mem: Am Phys Soc. Res: Quantum theory, particularly interactions between many electron systems. Mailing Add: Sch of Chem Rutgers State Univ New Brunswich NJ 08903

ADAMS, WILLIAM HENSLEY, b Nashville, Tenn, Aug 14, 29; m 51; c 3. ECOLOGY. Educ: Univ Tenn, AB, 51; La State Univ, MS, 56; Auburn Univ, PhD(zool), 59. Prof Exp: Asst, Auburn Univ, 56-59; sr game biologist, Tenn Game & Fish Comn, 59-60; prof biol & chmn div & dept, Tenn Wesleyan Col, 60-64; prof biol & dean col arts & sci, Tenn Tech Univ, 64-66; PROG MGR, NAT SCI FOUND, 66- Concurrent Pos: Sigma Xi res grant-in-aid, 59; mem, NSF insts radiation biol, Oak Ridge Inst Nuclear Studies, 61, comp anat, Harvard Univ, 62 & marine biol, Duke Univ, 63; NSF res partic, Highlands Biol Sta, 61; mem vis scientist prog, NSF-Tenn Acad Sci, 62-66; dir coop col-sch sci prog, NSF, 63-65; mem ed adv bd, Fed Notes, 73- Mem: AAAS; Am Soc Mammal (honorarium, 59); Am Ornith Union; Ecol Soc Am; Cooper Ornith Soc; Wildlife Soc. Res: Vertebrate ecology at the population level; improvement of higher education. Mailing Add: 4004 Moss Dr Annandale VA 22003

ADAMS, WILLIAM L, b Clay Center, Kans, May 23, 29; m 53; c 4. GEOLOGY. Educ: Univ Kans, BS, 51; Univ Calif, Los Angeles, MA, 56. Prof Exp: Geologist, Pan Am Petrol Corp, 56-64, dist geologist, 64-66, explor info coordr, 66-68, div geologist, 68-70; div explor mgr, 70-75, GEN MGR EXPLOR, AMOCO PROD CO, 75- Mem: Am Asn Petrol Geologists; Geol Soc Am. Res: Petroleum geology and oil exploration management. Mailing Add: 546 Timber Lane Lake Forest IL 60045

ADAMS, WILLIAM MANSFIELD, b Kissimmee, Fla, Feb 19, 32; m 55; c 3. GEOPHYSICS. Educ: Univ Chicago, BA, 51; Univ Calif, BA, 53; St Louis Univ, MS, 55, PhD, 57; Univ Santa Clara, MBA, 64. Prof Exp: Geophys trainee, Stanolind Oil & Gas Co, 53; computer, Western Geophys Co, 54; seismologist, Geotech Corp, 57-58; physicist, Lawrence Radiation Lab, Univ Calif, 59-62; pres, Planetary Sci, Inc, 62-64; PROF, UNIV & GEOPHYSICIST, INST GEOPHYS, UNIV HAWAII, 64-, Concurrent Pos: Dir tsunami res, Univ Hawaii, 67-72; vis fel, Coop Inst Res Environ Sci, Univ Colo, 70-71; vis prof geophys, Ind Univ, 75-76. Mem: Seismol Soc Am; Geol Soc Am; Soc Explor Geophys; Am Geophys Union. Res: Direct exploration of the interior of the earth; indexing and retrieval of scientific literature; seismic holography; tsunamis; systems for alerting society to natural hazards; hydrogeophysics. Mailing Add: Univ of Hawaii Inst Geophys Tsunami Div Honolulu HI 96822

ADAMS, WILLIAM PETER, b Ellesmere Port, Eng, Apr 17, 36; m 60; c 4. GEOGRAPHY, GLACIOLOGY. Educ: Univ Sheffield, BA, 58; McGill Univ, MSc, 62, PhD(geog, glaciol), 66. Prof Exp: Dir geog, McGill Sub-Arctic Res Lab, 63-66; vis fel, Univ Grenoble, 66-67; sr lectr, St Luke's Col, Eng, 67-68; assoc prof geog, 68-74, PROF GEOG, TRENT UNIV, 74-, CHMN DEPT, 68- Mem: Glaciol Soc; Arctic Inst NAm; Can Asn Geog. Res: Physical geography; hydrology. Mailing Add: Dept of Geog Trent Univ Peterborough ON Can

ADAMS, WILLIAM S, b Sodus, NY, May 28, 19; m 47; c 4. METABOLISM. Educ: Univ Rochester, MD, 43. Prof Exp: Intern, Strong Mem Hosp, Rochester, NY, 44; asst resident med, Univ Rochester, 44-45, instr, 47-48; resident, Vet Admin Ctr Hosp, Los Angeles, 48-49, admin sect chief, asst resident & physician full grade, 49-50; asst clin prof, 48-49, from asst prof to assoc prof, 49-59, vchmn dept, 67, PROF MED, SCH MED, UNIV CALIF, LOS ANGELES, 59- Concurrent Pos: Res fel, Univ Rochester, 46-47; consult, Atomic Bomb Casualty Comn. Mem: Asn Am Physicians; Am Col Physicians. Res: Metabolic aspects of malignant disease. Mailing Add: Dept of Med Univ of Calif Sch of Med Los Angeles CA 90024

ADAMS, WILLIAM WELLS, b Redlands, Calif, July 23, 37; m 64; c 2. NUMBER THEORY. Educ: Univ Calif, Los Angeles, AB, 59; Columbia Univ, PhD(math), 64. Prof Exp: From instr to asst prof math, Univ Calif, Berkeley, 64-69; assoc prof, 69-71, PROF MATH, UNIV MD, COLLEGE PARK, 71- Concurrent Pos: NSF fel, Inst Advan Study, Princeton Univ, 66-67. Mem: Am Math Soc; Math Asn Am. Res: Number theory with emphasis on Diophantine approximations. Mailing Add: Dept of Math Univ of Md College Park MD 20742

ADAMS, WILLIAM YEWDALE, b Los Angeles, Calif, Aug 6, 27; m 55; c 2. CULTURAL ANTHROPOLOGY. Educ: Univ Calif, AB, 48; Univ Ariz, PhD(anthrop), 58. Prof Exp: Sr archaeologist, Glen Canyon Proj, Mus Northern Ariz, 57-59; prog specialist archaeol, UNESCO & Sudan Antiq Serv, 59-66; assoc prof anthrop, Univ Ky, 66-71, PROF ANTHROP, UNIV KY, 71- Mem: Fel AAAS; fel Am Anthrop Asn; fel Soc Appl Anthrop; Am Ethnol Soc; Am Soc Ethnohist. Res: Cultural dynamics and culture change; theory of culture history. Mailing Add: Dept of Anthrop Univ of Ky Lexington KY 40506

ADAMS, WRIGHT (ROWE), b Sheridan, Ill, June 14, 03; m 27; c 2. CARDIOLOGY. Educ: Univ Ill, BS, 26, MD, 29; Am Bd Internal Med, dipl. Prof Exp: Intern, St Lukes Hosp, Chicago, 29-30; vol path, Cook County Hosp, 30; asst resident med, Billings Hosp, 30-31; clin asst, 31-34, from instr to prof, 34-70, assoc dean div biol sci, 47-49 & 61-67, chmn dept med, 49-61, EMER PROF MED, UNIV CHICAGO, 70- Concurrent Pos: Chmn, Am Bd Internal Med, 61-63; exec dir, Ill Regional Med Prog, 67-70. Mem: Am Physiol Soc; Am Soc Clin Invest; Asn Am Physicians; Am Col Physicians; Am Heart Asn. Res: Electrocardiography; circulatory and work physiology. Mailing Add: Box 298 Gulf Shores AL 36542

ADAMSKI, ROBERT J, b Newark, NJ, Mar 26, 35; m 56; c 3. MEDICINAL CHEMISTRY. Educ: Univ Wis-Madison, BS, 60; Univ Iowa, PhD(pharmaceut chem), 64. Prof Exp: Fel, Ciba Pharmaceut Co, NJ, 64-66; res assoc, Inst Microbiol, Rutgers Univ, 66-67; sr chemist, Carter-Wallace, Inc, 67-68; MGR ALLERGY/SURG SERV, ALCON LABS, INC, 68- Mem: Am Chem Soc; Am Pharmaceut Asn; NY Acad Sci; Am Inst Chemists. Res: Chemical modification of known and structural elucidation of unknown antimicrobial materials; design and synthesis of agents having potential medicinal value as antiglaucomics, anti-inflammatories, antibacterial agents. Mailing Add: Alcon Labs Inc Box 1959 Ft Worth TX 76109

ADAMS-MAYNE, MABELLE ELAINE, b Wichita Falls, Tex, May 13, 26; m; c 3. CLINICAL CHEMISTRY, PROTEIN CHEMISTRY. Educ: Baylor Univ, BS, 46, PhD(biochem), 63; Tulane Univ, MS, 52. Prof Exp: Chemist, Southern Regional Lab, USDA, 46-52, Brit Gelatine & Glue Res Assoc, Eng, 52-54, Courtaulds Ala, Ltd, 54 & Samuel Roberts Noble Found, Okla, 55-59; asst biochem, Col Med, Baylor Univ, 59-61; Robert A Welch fel, M D Anderson Hosp & Tumor Inst, Univ Tex, 63-65, asst biochemist res clin path sect, hosp & inst & asst prof biomed sci, Univ, 65-71; lab dir, Nat Health Labs, Inc, 71-72; CLIN CHEMIST, DEPT PATH, TEX CHILDREN'S HOSP, 72- Concurrent Pos: William Osler lectureship, Philadelphia Gen Hosp, 67; asst prof, Baylor Col Med, 72- Mem: AAAS; Am Chem Soc; Am Asn Cancer Res; fel Am Asn Clin Chemists; NY Acad Sci. Res: Protein physical chemistry; immunoassay and radioimmunoassay; structure determinations on myeloma proteins, Bence-Jones proteins, immunoglobulins and tumor antigens; clinical chemistry methods development; pediatric hormone levels; pediatric normal range determinations. Mailing Add: 3310 Aberdeen Way Houston TX 77025

ADAMSON, ARTHUR WILSON, b Shanghai, China, Aug 15, 19; m 42; c 3. PHYSICAL CHEMISTRY. Educ: Univ Calif, BS, 40; Univ Chicago, PhD(phys chem), 44. Prof Exp: Asst metall lab, Univ Chicago, 42-43; res assoc, Clinton Labs, Tenn, 43-46; from asst prof to assoc prof, 46-52, chmn dept, 72-75, PROF CHEM, UNIV SOUTHERN CALIF, 52- Concurrent Pos: NSF sr fel, 62-63; Unilever prof, Bristol Univ, 65-66; Australian Acad Sci sr fel, 69; Foster lectr, State Univ NY Buffalo, 70; Venable lectr, Univ NC, 75. Honors & Awards: Assocs Res Award, Univ Southern Calif, 70. Mem: AAAS; fel Am Inst Chemists; Am Chem Soc; Sigma Xi. Res: Coordination chemistry; photochemistry of coordination compounds; surface chemistry. Mailing Add: Dept of Chem Univ of Southern Calif Los Angeles CA 90007

ADAMSON, IAN YOUNG RADCLIFFE, b Wishaw, Scotland, Mar 16, 41; m 67; c 2. EXPERIMENTAL PATHOLOGY. Educ: Univ Glasgow, BSc, 63, PhD(chem), 66. Prof Exp: From lectr to asst prof path, 67-75, ASSOC PROF PATH, UNIV MANITOBA, 75- Mem: Am Soc Exp Path; Am Soc Cell Biol. Res: Experimentally induced lung injury and repair, with emphasis on changes in ultrastructure, turnover and uptake of radioactive metabolic precursors in the different cell types. Mailing Add: Dept of Path Univ of Man 770 Bannatyne Ave Winnipeg MB Can

ADAMSON, JOHN DOUGLAS, b St Boniface, Man, Mar 1, 32; m 57; c 6. PSYCHIATRY, PSYCHOPHYSIOLOGY. Educ: Univ Man, MD, 56; Royal Col Physicians & Surgeons Can, cert psychiat, 61. Prof Exp: Asst psychiat, Sch Med & Dent, Univ Rochester, 59-60, instr, 60-61; lectr, 61-63, asst prof, 63-65, ASSOC PROF PSYCHIAT, FAC MED, UNIV MAN, 65- Concurrent Pos: Mem subcomt rehab, Dept Nat Health & Welfare Can, 63-65 & mem subcomt ment health res, 65-68; coordr psychiat res, Dept Health & Social Develop, Govt Prov Man, 65-; dir psychiat clin invest, Health Sci Gen Ctr, 65-; chmn prof & tech adv comt, Alcoholism Found Man, 65-; chmn, Man Ment Health Res Found, 74- Mem: Am Psychosom Soc; Soc Psychophysiol Res; Can Med Asn; Can Psychiat Asn. Res: Biology and psychopathology of schizophrenia; clinical and biological studies on alcoholism; psychophysiology of emotion; sleep. Mailing Add: Health Sci Ctr 700 William Ave Winnipeg MB Can

ADAMSON, LUCILE FRANCES, b Chetopa, Kans, Nov 10, 26. NUTRITIONAL BIOCHEMISTRY, ENVIRONMENTAL SCIENCES. Educ: Kans State Col, BS, 48; Univ Iowa, MS, 50; Univ Calif, PhD(biochem), 56. Prof Exp: Asst nutrit, Univ Ill, 50-52; asst nutritionist, Univ Hawaii, 56-60; asst prof pediat & biochem, Univ Mo, 60-64; chief biochemist, Thorndike Mem Lab, Harvard Med Sch, 64-70; res fel biochem, Monash Univ, Australia, 70-72; staff scientist, Environ Defense Fund, Washington,

DC, 72-74; PROF & CHMN PROG MACROENVIRON STUDIES, HOWARD UNIV, 74- Mem: AAAS; Am Inst Nutrit. Res: Chemical and nutritional aspects of environmental problems. Mailing Add: Sch of Human Ecol Howard Univ Washington DC 20059

ADAMSON, RICHARD H, b Council Bluffs, Iowa, Aug 9, 37; m 63. PHARMACOLOGY, BIOCHEMISTRY. Educ: Drake Univ, BA, 57; Univ Iowa, MS, 59, PhD(pharmacol), 61; George Washington Univ, MA, 68. Prof Exp: Asst, Col Med, Univ Iowa, 57-58; sr investr, Lab Chem Pharmacol, 63-69, head pharmacol & exp therapeut sect, 69-73, CHIEF LAB CHEM PHARMACOL, NAT CANCER INST, 73- Concurrent Pos: Lectr, Col Med, George Washington Univ, 63-70; Fulbright vis scientist, Dept Biochem, St Mary's Hosp Med Sch, Univ London, 65-66. Mem: AAAS; Am Asn Cancer Res; Am Soc Pharmacol & Exp Therapeut; NY Acad Sci; Sigma Xi. Res: Drug metabolism; cancer chemotherapy; antibiotics; toxicology; marine biology; comparative pharmacology; anesthesiology; science and government. Mailing Add: Lab of Chem Pharmacol Nat Cancer Inst Bethesda MD 20014

ADAMSON, WILLIAM CHARLES, b Turin, Ga, Nov 17, 38; m 66; c 1. PLANT BREEDING. Educ: Univ Ga, BSA, 61, MS, 64; Auburn Univ, PhD(agron), 69. Prof Exp: RES PLANT GENETICIST, AGR RES SERV, USDA, 69- Mem: AAAS; Am Soc Agron; Crop Sci Soc Am; Am Genetic Asn. Res: Root-knot resistance in Sericea lespedeza, kenaf and roselle; kenaf production. Mailing Add: Rte 4 Box 433 Savannah GA 31405

ADAMSONS, KARLIS, JR, b Riga, Latvia, Oct 30, 26; US citizen; m 68. OBSTETRICS & GYNECOLOGY. Educ: Univ Göttingen, MD, 52; Columbia Univ, PhD(pharmacol), 56. Prof Exp: Intern, St Vincents Hosp, NY, 52-53; asst resident obstet & gynec, Columbia-Presby Med Ctr, 56-58, from resident to chief resident, 59-61, asst prof, 61-65, assoc obstetrician & gynecologist, 65-69; from assoc prof to prof, Col Physicians & Surgeons, Columbia Univ, 65-70; prof obstet, gynec & pharmacol, Mt Sinai Sch Med, City Univ New York, 70-75; PROF OBSTET & GYNEC & CHMN DEPT, BROWN UNIV, 75-; OBSTETRICIAN & GYNECOLOGIST IN CHIEF, WOMEN & INFANTS HOSP OF RI, 75- Concurrent Pos: Macy fel, Nuffield Inst Med Res, Oxford Univ, Eng, 58-59; Wellcome sr res fel, 61-; vis scientist, NIH, 62, consult, Nat Inst Child Health & Human Develop, 66-71; vis prof obstet & gynec, Univ PR, 63; consult, Coun Drugs, AMA, 63; mem adv coun, Food & Drug Admin, 65-70; Fulbright prof, Univ Uruguay, 69. Res: Perinatal physiology; biochemistry and physiology of neurohypophysical hormones. Mailing Add: Women & Infants Hosp of RI 50 Maude St Providence RI 02906

ADAMS SMITH, WILLIAM NELSON, b Wellington, NZ, Dec 7, 29; m 58; c 4. MEDICINE, NEUROENDOCRINOLOGY. Educ: Univ NZ, MB, ChB, 57; Oxford Univ, MA, 64, DPhil(neuroendocrinol), 66; Univ Otago, MD, 70. Prof Exp: From asst lectr to lectr anat, Univ Otago, 60-62; lectr & tutor, Oxford Univ, 63-66; from asst prof to assoc prof, 66-69, prof human biol & dean, Sch Allied Health Sci, Med Univ SC, 69-72; PROF & VPRES HEALTH AFFAIRS, UNIV SC, 72- Concurrent Pos: Nuffield Dom demonstr, Oxford Univ, 62-65, Beit Mem res fel, 65-66; Roe E Remington Mem lectr, SC Acad Sci, 68- Honors & Awards: Prizewinner Rolleston Mem, Oxford & Cambridge Univs, 66. Mem: AAAS; Am Asn Anat; Am Soc Zool; Am Soc Study Reproduction; Brit Soc Endocrinol. Res: Teratology of developing heart; hormonal influences on sexual development of the brain. Mailing Add: Div of Health Affairs Univ of SC Columbia SC 29208

ADAWI, IBRAHIM (HASAN), b Palestine, Apr 18, 30; US citizen; m 56; c 4. THEORETICAL SOLID STATE PHYSICS. Educ: Wash Univ, BS, 53; Cornell Univ, PhD(eng physics), 57. Prof Exp: Teacher math & physics, Terra Santa Sch, Syria, 49-51; physicist, Radio Corp Am Labs, 56-60; res consult, Battelle Mem Inst, 60-68; PROF PHYSICS, UNIV MO-ROLLA, 68- Concurrent Pos: Adj prof elec eng, Ohio State Univ, 65-68. Mem: Am Phys Soc. Res: Penetration of charged particles and radiation in matter; transport theory. Mailing Add: Dept of Physics Univ of Mo Rolla MO 65401

ADCOCK, JAMES LUTHER, b Crane, Tex, June 26, 43; m 67; c 1. FLUORINE CHEMISTRY, INORGANIC CHEMISTRY. Educ: Univ Tex, Austin, BS, 66, PhD(chem), 71. Prof Exp: Res asst fluorine chem, Mass Inst Technol, 71-74; ASST PROF CHEM, UNIV TENN, KNOXVILLE, 74- Mem: Am Chem Soc; Sigma Xi. Res: Synthetic organofluorine and organometallic chemistry; elemental fluorine as a synthetic reagent, preparation of fluorinated organometallic monomers and polymers, inorganic ring systems, boron containing polymers. Mailing Add: Dept of Chem Univ of Tenn Knoxville TN 37916

ADCOCK, LOUIS HENRY, b Durham, NC, Mar 28, 29; m 60; c 3. ANALYTICAL CHEMISTRY. Educ: Duke Univ, BS, 51, MA, 53; La State Univ, PhD, 70. Prof Exp: Asst prof, 56-69, ASSOC PROF PHYS SCI & CHEM, UNIV NC, WILMINGTON, 69- Mem: AAAS; Am Chem Soc; fel Am Inst Chemists. Res: Chromatography; analysis of trace elements; pesticides in marine environment. Mailing Add: Dept of Chem Univ of NC Wilmington NC 28401

ADCOCK, WILLIS ALFRED, b St Johns, Que, Nov 25, 22; nat US; m 43; c. 4. CHEMISTRY. Educ: Hobart Col, BS, 43; Brown Univ, PhD(chem), 48. Prof Exp: Chemist, Stanolind Oil & Gas Co, 48-53; res group supvr, Tex Instruments, 53-55, dir device res, 55-58, mgr res & eng dept, 58-61, mgr qual assurance dept, 60-61, mgr integrated circuit dept, 61-64; tech dir, Sperry Semiconducotr Div, Sperry Rand Corp, 64-65; mgr advan planning dept, Semiconductor Div, 65-67, vpres tech develop, Components Group, 67-69, vpres strategic planning, Corp Develop, 69-72, ASST VPRES & TECH DIR CONSUMER PROD, TEX INSTRUMENTS INC, 72- Mem: Am Chem Soc; Inst Elec & Electronics Engrs. Res: Semiconductors. Mailing Add: Tex Instruments Inc Mail Sta 974 PO Box 5012 Dallas TX 75222

ADDA, LIONEL PAUL, b Allentown, Pa, Jan 17, 22. SOLID STATE SCIENCE. Educ: Lehigh Univ, BS, 49, MS, 50, PhD(physics), 62. Prof Exp: Engr fed tel telecommun labs, Int Tel & Tel Co, 50-54; MEM TECH STAFF, BELL TEL LABS, 54- Res: Measurement of eletrical and infrared properties; crystalline perfection and impurities in semiconductors; role of imperfections on device behavior; insulator properties and metal-insulator-semiconductor systems. Mailing Add: Bell Tel Labs Allentown PA 18103

ADDAMIANO, ARRIGO, b Molfetta, Italy, Feb 16, 23; US citizen; m 52; c 4. INORGANIC CHEMISTRY, PHYSICAL CHEMISTRY. Educ: Univ Rome, PhD(chem), 44, PhD(physics), 46. Prof Exp: Asst prof gen & inorg chem, Univ Rome, 44-48; Ital Bd Educ scholar res x-ray crystallog, Res Assoc Dept X-ray, Oxford Univ, 48-50; asst prof gen & inorg chem, Univ Rome, 51-53; Fulbright fel & res assoc x-ray & crystal structure lab, Pa State Univ, 53-54; phys chemist, Gen Elec Co, 54-69; RES CHEMIST, US NAVAL RES LAB, 69- Concurrent Pos: Nat Res Coun fel, Rome, 51-53. Mem: Am Chem Soc; Am Crystallog Asn; Electrochem Soc. Res: Crystal structure determination; crystal physics; single crystal growth; inorganic syntheses; phosphors; semiconductors. Mailing Add: Code 5214A US Naval Res Lab Washington DC 20390

ADDANKI, SOMASUNDARAM, b Lakkavaram, India, Mar 7, 32; m 54; c 3. CLINICAL CHEMISTRY, BIOCHEMISTRY. Educ: Madras Univ, BVS, 57; Ohio State Univ, MSc, 62, PhD(nutrit biochem), 64; Am Bd Clin Chem, cert, 70. Prof Exp: From instr to asst prof, 64-70, ASSOC PROF PEDIAT & PHYSIOL CHEM, OHIO STATE UNIV, 70-; DIR LABS, CLIN STUDY CTR, COLUMBUS CHILDREN'S HOSP, 66- Concurrent Pos: Consult, Pathologists, Inc, Licking Mem Hosp, Newark, Ohio, 73- Mem: Fel Am Asn Clin Chem; Am Fedn Clin Res; Brit Biochem Soc. Res: Metabolic pediatric chemistry; radio-immuno assays-ketotic hypoglycemia; catecholamine chemistry and metabolism; endocrinology. Mailing Add: Dept of Pediat Ohio State Univ Columbus OH 43205

ADDELMAN, SIDNEY, b Ottawa, Ont, Nov 7, 32; m 56; c 4. EXPERIMENTAL STATISTICS. Educ: Carleton Univ, Ont, BA, 54; Univ Del, MA, 56; Iowa State Univ, PhD(statist), 60. Prof Exp: Statistician, Res Triangle Inst, 61-67; assoc prof, 67-72, PROF STATIST, STATE UNIV NY BUFFALO, 72- Concurrent Pos: Adj asst prof, NC State Univ, 62-65, adj assoc prof, 65-67. Mem: Fel Am Statist Asn; Biomet Soc. Res: Design and analysis of experiments. Mailing Add: State Univ of NY at Buffalo Dept of Statist 4230 Ridge Lea Rd Amherst NY 14226

ADDICOTT, FREDRICK TAYLOR, b Oakland, Calif, Nov 16, 12; m 35; c 4. PLANT PHYSIOLOGY. Educ: Stanford Univ, AB, 34; Calif Inst Technol, PhD(plant physiol), 39. Prof Exp: Asst, Stanford Univ, 34-37; teaching fel, Calif Inst Technol, 37-39; from instr to asst prof bot, Santa Barbara State Col, 39-46; assoc physiologist, Spec Guayule Res Proj, USDA, Calif, 43-44; from asst prof to assoc prof bot, Univ Calif, Los Angeles, 46-54; prof bot, 54-61, prof agron, 61-72, PROF BOT, UNIV CALIF, DAVIS, 72- Concurrent Pos: Sigma Xi grant-in-aid, 40; Fulbright res scholar, Vicotria Univ, Wellington, NZ, 57; vis prof, Univ Adelaide, 66 & Univ Natal, 70. Mem: Int Soc Hort Sci; Am Soc Plant Physiol; Bot Soc Am; Can Soc Plant Physiol; Phytochem Soc NAm. Res: Morphology, physiology and ecology of abscission; abscisic acid; physiology of plant hormones. Mailing Add: Dept of Bot Univ of Calif Davis CA 95616

ADDICOTT, JOHN FREDRICK, b Santa Barbara, Calif, Dec 28, 44; m 66; c 2. ECOLOGY. Educ: Univ Calif, Davis, AB, 67; Univ Mich, MSc, 68, PhD(zool), 72. Prof Exp: Res asst prof biol, Univ Utah, 72-73; ASST PROF ZOOL, UNIV ALTA, 73- Mem: Ecol Soc Am; Soc Study of Evol. Res: Competition in patchy environments; effects of predation on prey community structure. Mailing Add: Dept of Zool Univ of Alta Edmonton AB Can

ADDICOTT, WARREN O, b Fresno, Calif, Feb 17, 30; m 55; c 2. PALEONTOLOGY. Educ: Pomona Col, AB, 51; Stanford Univ, MA, 52; Univ Calif, PhD(paleont), 56. Prof Exp: Asst, Univ Calif, 52-54; geologist, Gen Petrol Corp, 54-62, explor coordr, 58-60, dist geologist, 60-62; geologist, 62-70, RES GEOLOGIST, US GEOL SURV, 70- Concurrent Pos: Consult prof, Stanford Univ, 71- Mem: AAAS; Paleont Soc (secy, 71-); Geol Soc Am; Am Asn Petrol Geol; Paleont Res Inst. Res: Cenozoic marine invertebrate paleontology and stratigraphy; molluscan zoogeography; paleoclimatology; molluscan taxonomy. Mailing Add: Paleont & Stratig US Geol Surv 345 Middlefield Rd Menlo Park CA 94025

ADDINALL, CARL RUPERT, b Dewsbury, Eng, Dec 1, 90; US citizen; m 47. ORGANIC CHEMISTRY. Educ: Harvard Univ, BS, 25, AM, 26, PhD(org chem), 30. Prof Exp: Asst chem, Radcliffe Col, 26-27; instr, Harvard Univ, 28-30, res chemist, 30-34; res chemist, Merck & Co, Inc, NJ, 34-44, asst dir res & develop, 44-48, dir, Tech Info Dept, 48-51, foreign sci adv, 51-55, consult, 56-62; CONSULT CHEM, 62- Mem: Fel AAAS; fel Am Chem Soc; fel NY Acad Sci; fel Am Inst Chemists. Res: Nitrogen ring compounds; alkaloids; chemical literature; foreign scientific developments. Mailing Add: 746 Belvidere Ave Westfield NJ 07090

ADDINK, SYLVAN, b Sheldon, Iowa, Dec 23, 41; m 62; c 2. WEED SCIENCE. Educ: Ore State Univ, BS, 69; NDak State Univ, MS, 72, PhD(agron), 73. Prof Exp: Plant sci rep, 73-74, REGIONAL RES REP, ELI LILLY & CO, 74- Mem: Weed Sci Soc Am. Res: Control of pests that are harmful to crops, animals, mankind and his environment. Mailing Add: 1011 Victoria Dr Nesbit MS 38651

ADDIS, FRANK WILLIAM, solid state physics, see 12th edition

ADDIS, PAUL BRADLEY, b Honolulu, Hawaii, Feb 13, 41; m 67; c 1. FOOD SCIENCE. Educ: Wash State Univ, BS, 62; Purdue Univ, PhD(food sci), 67. Prof Exp: Fel, Max Planck Soc, Ger, 67; from asst prof to assoc prof food sci, 67-75, PROF FOOD SCI & NUTRIT, UNIV MINN, ST PAUL, 75- Concurrent Pos: Fulbright travel grant, 67; consult, Totino's Finer Foods, Inc & Brother Redi-Roast Prod, Inc, Minn, 70-72; researcher, Muscle Biol Lab, Univ Calif, Davis, 71. Mem: AAAS; Am Chem Soc; Inst Food Technologists; Am Soc Animal Sci. Res: Physiology and biochemistry of post-mortem change in muscle. Mailing Add: Dept of Food Sci & Nutrit Univ of Minn St Paul MN 55108

ADDISON, ANTHONY WILLIAM, b Sydney, NSW, June 24, 46; m 72. BIOINORGANIC CHEMISTRY. Educ: Univ New South Wales, BSc, 68; Univ Kent, PhD(chem), 71. Prof Exp: Res asst chem, Univ Kent, 68-70; res fel, Northwestern Univ, 70-72; ASST PROF CHEM, UNIV BC, 72- Mem: Fel The Chem Soc; Am Chem Soc; Chem Inst Can. Res: Chemistry of metalloproteins and of synthetic models for their properties particularly oxygen transport and redox chemistry. Mailing Add: Dept of Chem Univ of BC Vancouver BC Can

ADDISON, JOHN RUNDLE, b Montreal, Que, Mar 26, 34. LOW TEMPERATURE PHYSICS. Educ: McGill Univ, BSc, 55, BEng, 58, MSc, 62, PhD(physics), 67. Prof Exp: Lectr physics, McGill Univ, 62-65; asst prof, Sir George Williams Univ, 65-67; ASST PROF PHYSICS, McGILL UNIV, 67- Res: Electrical properties of sea ice and of pure ice at low temperatures. Mailing Add: Dept of Physics McGill Univ Montreal PQ Can

ADDISON, JOHN WEST, JR, b Washington, DC, Apr 2, 30; m 55; c 4. MATHEMATICAL LOGIC. Educ: Princeton Univ, AB, 51; Univ Wis, MS, 53, PhD(math), 55. Prof Exp: Instr math, Univ Mich, 54-56; mem, Inst Advan Study, 56-57; asst prof math, Univ Mich, 57-62; assoc prof, 62-68, chmn group in logic & methodology of sci, 63-65, PROF MATH & CHMN DEPT, UNIV CALIF, BERKELEY, 68- Concurrent Pos: NSF fel, 56-57; mem math inst, Polish Acad Sci, 57; lectr, Univ Calif, Berkeley, 59-60. Mem: Fel AAAS; Am Math Soc; Math Asn Am; Asn Symbolic Logic. Res: Theory of definability, including recursive function theory and descriptive set theory; foundations of mathematics; theory of models; axiomatic set theory. Mailing Add: Dept of Math Univ of Calif Berkeley CA 94720

ADDISON, LESLIE MANDEVILLE, b Los Angeles, Calif, May 12, 22; m 46; c 2. ORGANIC CHEMISTRY. Educ: Millsaps Col, BS, 41; Univ NC, MA, 43; Purdue Univ, PhD(org chem), 50. Prof Exp: Res chemist, Esso Labs, Stand Oil Develop Co, 43 & 46-47, sect head, 57-61; vpres & gen mgr, Micro-Tek Instruments, Inc, 61-66; pres, Systs Design Corp, 66-68; TECH DIR, SCI SYSTS CORP, 68- Mem: Am Chem Soc. Res: Heterogeneous catalysis; petroleum refining; gas chromatography;

instrumentation; environmental control; solar simulation; atmospheric pollution. Mailing Add: Sci Systs Corp 9020 S Choctaw Dr Baton Rouge LA 70815

ADDISS, RICHARD ROBERT, JR, b Jersey City, NJ, May 13, 29; m 52; c 1. SOLID STATE PHYSICS. Educ: Rensselaer Polytech Inst, 51; Cornell Univ, MS, 56, PhD(physics), 58. Prof Exp: Asst mech, heat, sound, elec & magnetism, Cornell Univ, 51-56, asst oxidation metals, 56-58; mem tech staff evaporated films, Radio Corp Am Labs, 58-65; mem sci staff photoelectronic mat, Itek Corp, 65-75; MGR RES & DEVELOP, SOLAR POWER CORP, 75- Concurrent Pos: Mem steering comt, Int Conf Thin Films, Mass, 69. Mem: Am Phys Soc; Am Vacuum Soc. Res: Photoelectronic processes in crystals; properties of solids relating to defect structure; structure and properties of surfaces and thin film; photovoltaic devices. Mailing Add: Solar Power Corp 23 North Ave Wakefield MA 01880

ADDLEMAN, ALBERT DUANE, genetics, physiology, see 12th edition

ADDOR, ROGER WILLIAMS, organic chemistry, see 12th edition

ADDUCI, JERRY M, b Rochester, NY, June 5, 34; m 58; c 2. ORGANIC CHEMISTRY. Educ: Univ Rochester, BS, 57; Univ Pa, PhD(org chem), 63. Prof Exp: Lab asst chem, Eastman Kodak Co, 52-53; instr math, Rochester Inst Technol, 58; lab technician, Strong Mem Hosp, Rochester, 58-59; res chemist, Fabrics & Finishes Dept, E I du Pont de Nemours & Co, Del, 63-65; res assoc & fel, Univ Md, 65-66, instr chem, 66; ASST PROF CHEM, ROCHESTER INST TECHNOL, 66- Mem: Am Chem Soc. Res: Reactions mechanisms; polymer synthesis. Mailing Add: One Lomb Memorial Dr Rochester NY 14623

ADDY, JOHN KEITH, b Sheffield, Eng, June 30, 37; m 63; c 2. PHYSICAL ORGANIC CHEMISTRY. Educ: Univ London, BSc, 58; Univ Southampton, PhD(phys org chem), 62. Prof Exp: Res assoc phys chem, Univ Ore, 61-62; vis lectr chem, Northeast Essex Tech Col, Eng, 62-63; lectr phys chem, John Dalton Col, 63-66; from asst prof to assoc prof, 66-74, PROF CHEM, WAGNER COL, 74- Mem: The Chem Soc; fel Royal Inst Chem; Am Asn Univ Prof. Res: Reaction kinetics, especially mechanisms of epoxide ring fission and allylic rearrangements; plasticization and structure of polymers, especially nylon. Mailing Add: Dept of Chem Wagner Col Staten Island NY 10301

ADEL, ARTHUR, b Brooklyn, NY, Nov 22, 08; m 35. ASTROPHYSICS. Educ: Univ Mich, AB, 31, PhD(physics), 33. Prof Exp: Res assoc, Lowell Observ, 33-35; fel, Johns Hopkins Univ, 35-36; mem staff, Lowell Observ, 36-42; asst prof physics, Univ Mich, 42-46, asst prof astron, McMath Hulbert Observ, 46-48; PROF PHYSICS & DIR RES OBSERV, NORTHERN ARIZ UNIV, 48- Mem: AAAS; fel Am Phys Soc; Am Astron Soc; Am Meteorol Soc. Res: Atmospheres of the solar system; infrared spectra and molecular structure; far infrared spectroscopy of the solar and terrestrial atmospheres; composition and temperatures of upper atmosphere; infrared spectroscopy of atmospheric minor constituents; discovered atmospheric nitrous oxide, atmospheric heavy water and the 20-micron window. Mailing Add: Atmospheric Res Lab Northern Ariz Univ Flagstaff AZ 86001

ADELBERG, ARNOLD M, b Brooklyn, NY, Mar 17, 36; m 62; c 2. GEOMETRY, ALGEBRA. Educ: Columbia Univ, BA, 56; Princeton Univ, MA, 62. Prof Exp: Instr math, Columbia Univ, 59-62; from instr to asst prof, 62-68, chmn dept math, 69-71, chmn sci div, 71-73, chmn fac, 74-76, ASSOC PROF MATH, GRINNELL COL, 68- Concurrent Pos: Hon res fel, Harvard Univ, 68-69. Mem: Am Math Soc; Math Asn Am; Am Asn Univ Prof. Res: Algebraic geometry, especially rationality questions of algebraic groups; Koszul resolutions; Bezout's theorem; elementary metric geometry; homological algebra. Mailing Add: Dept of Math Grinnell Col Grinnell IA 50112

ADELBERG, EDWARD ALLEN, b Cedarhurst, NY, Dec 6, 20; m 42; c 3. GENETICS. Educ: Yale Univ, BS, 42, MS, 47, PhD(microbiol), 49. Prof Exp: From instr to prof bact, Univ Calif, Berkeley, 49-60, chmn dept, 57-61; prof microbiol, 61-74, chmn dept, 61-64, dir div biol sci, 64-69, chmn dept microbiol, 70-72, PROF HUMAN GENETICS, YALE UNIV, 74- Concurrent Pos: Guggenheim fels, Pasteur Inst, Paris, 56-57 & Nat Ctr Sci Res, 65-66; consult, Chem Res Div, Eli Lilly & Co, 59-67; ed, J Bact, 64-67; ed-in-chief, Bact Rev, 67-70; mem genetics training comt, Nat Inst Gen Med Sci, 70-73; consult, Genetics Br, NSF, 71-74. Mem: Nat Acad Sci; fel Am Acad Arts & Sci; fel Am Acad Microbiol; Am Soc Microbiol. Res: Molecular genetics; microbial genetics; genetics of cultured mammalian cells. Mailing Add: Dept of Human Genetics Yale Univ Med Sch New Haven CT 06510

ADELBERGER, ERIC GEORGE, b Bryn Mawr, Pa, June 26, 38; m 61; c 1. EXPERIMENTAL NUCLEAR PHYSICS. Educ: Calif Inst Technol, BS, 60, PhD(physics), 67. Prof Exp: Res fel physics, Calif Inst Technol, 67-68; res assoc, Stanford Univ, 68-69; asst prof, Princeton Univ, 69-71; from asst prof to assoc prof, 71-75, PROF PHYSICS, UNIV WASH, 75- Concurrent Pos: Mem adv panel physics, NSF, 73-; fac adv nuclear physics, Max Planck Soc, Ger, 75- Mem: Am Phys Soc. Res: Experimental studies of fundamental symmetries in nuclei; nuclear structure. Mailing Add: Dept of Physics Univ of Wash Seattle WA 98195

ADELBERGER, REXFORD E, b Cleveland, Ohio, Mar 30, 40; m 66; c 1. NUCLEAR PHYSICS. Educ: Col William & Mary, BSc, 61; Univ Rochester, PhD(physics), 67. Prof Exp: Asst prof physics, State Univ NY Col Geneseo, 67-73; PROF PHYSICS & CHMN DEPT, GUILFORD COL, 73- Mem: Am Asn Physics Teachers; Am Phys Soc. Res: Experimental study of the few nucleon problem at moderate energies. Mailing Add: Dept of Physics Guilford Col Greensboro NC 27410

ADELMAN, ALBERT H, b New York, NY, Dec 17, 30; m 52; c 2. PHYSICAL CHEMISTRY. Educ: Brooklyn Col, BS, 51; Polytech Inst Brooklyn, PhD(chem), 56. Prof Exp: Assoc res scientist, NY Univ, 56-60; sr scientist, 60-70, assoc chief, Chem Dept, 70-72, sect mgr org chem, 72-75, MGR, CHEM DEPT, COLUMBUS DIV, BATTELLE MEM INST, 75- Concurrent Pos: NSF fel, 56-58; instr, City Col New York, 59- Mem: AAAS; Am Phys Soc; Am Chem Soc. Res: Charge-transfer photochemistry; stepwise excitation of luminescence; solar energy utilization employing physical and biological systems. Mailing Add: Battelle Mem Inst Dept Chem 505 King Ave Columbus OH 43201

ADELMAN, FRANK LOUIS, nuclear physics, see 12th edition

ADELMAN, FRED, b Malden, Mass, May 1, 30; m 52; c 2. ANTHROPOLOGY. Educ: Yale Univ, BA, 51; Univ Pa, MA, 54, PhD(anthrop), 61. Prof Exp: Asst prof anthrop, Univ Pittsburgh, 59-62; assoc prof, 62-74, PROF ANTHROP, CHATHAM COL, 74- Concurrent Pos: Lectr, Chatham Col, 61-62; NDEA foreign lang fel, 64-65. Mem: AAAS; fel Am Anthrop Asn. Res: Culture change; social structures; peoples of Central and Northern Asia. Mailing Add: Dept of Anthrop Chatham Col Pittsburgh PA 15232

ADELMAN, RICHARD CHARLES, b Newark, NJ, Mar 10, 40; m 63; c 2. GERONTOLOGY, BIOCHEMISTRY. Educ: Kenyon Col, AB, 62; Temple Univ, MA, 65, PhD(biochem), 67. Prof Exp: Asst prof, 69-73, ASSOC PROF BIOCHEM, FELS RES INST, MED SCH, TEMPLE UNIV, 73- Concurrent Pos: Am Cancer Soc fel, Albert Einstein Col Med, 67-69; Am Can Soc res grant aging, Fels Res Inst, Med Sch, Temple Univ, 70-73; NIH res grants, 70-; mem study sect pathobiol chem, NIH, 75-78; mem nat adv coun geriat res educ & clin ctrs, Vet Admin, 75-78; chmn, Gordon Res Conf Biol of Aging, 76. Mem: AAAS; fel Geront Soc (secy-treas biol, 72-, vpres-elect); Am Soc Biol Chemists; Am Chem Soc. Res: Biology of aging; hormonal regulation of enzyme activity. Mailing Add: Fels Res Inst Temple Univ Sch of Med Philadelphia PA 19140

ADELMAN, ROBERT LEONARD, b Chicago, Ill, May 20, 19. ORGANIC CHEMISTRY. Educ: Univ Chicago, BS, 41, PhD(org chem), 45. Prof Exp: Res chemist, Nat Defense Res Comt & Off Sci Res & Develop, Univ Chicago, 42-45; from res chemist to sr chemist, 45-61, staff scientist, 61-66, res assoc, Indust Chem Dept, 72-74, RES ASSOC, PLASTICS DEPT, E I DU PONT DE NEMOURS & CO, INC, 74- Mem: Am Chem Soc; Sigma Xi. Res: Reaction mechanism; synthesis and characterization of monomers, polymers, adhesives, protective coatings, structural plastics, plated plastics, polymers in electroplating and electronics. Mailing Add: Plastics Dept Exp Sta E I du Pont de Nemours & Co Wilmington DE 19898

ADELMAN, WILLIAM JOSEPH, JR, b Mt Vernon, NY, Jan 29, 28; m 51; c 3. PHYSIOLOGY, BIOPHYSICS. Educ: Fordham Univ, BS, 50; Univ Vt, MS, 52; Univ Rochester, PhD(physiol), 55. Prof Exp: Asst physiol & biophys, Univ Vt, 50-52; aviation physiologist, US Air Force Sch Aviation Med, 55-56; from instr to asst prof physiol, Sch Med, Univ Buffalo, 56-59; physiologist, Biophys Lab, Nat Inst Neurol Dis & Blindness, 59-62; from assoc prof to prof physiol, Sch Med, Univ Md, Baltimore, 62-71; CHIEF LAB BIOPHYS, NAT INST NEUROL COMMUN DIS & STROKE, 71- Concurrent Pos: Nat Inst Neurol Dis & Stroke spec fel, Marine Biol Lab, Woods Hole, 69-70. Mem: Fel AAAS; Am Physiol Soc; Biophys Soc; Soc Gen Physiol; Soc Neurosci. Res: Neurophysiology and electrobiology; role of ions in membrane phenomena in nerve; modeling of neural behavior. Mailing Add: Nat Inst of Neurolog & Communicative Dis & Stroke NIH Bethesda MD 20014

ADELMANN, HOWARD BERNHARDT, b Buffalo, NY, May 8, 98. ZOOLOGY. Educ: Cornell Univ, AB, 20, AM, 22, PhD(histol, embryol), 24. Hon Degrees: ScD, Ohio State Univ, 62. Prof Exp: Asst histol & embryol, 19-21, from instr to prof, 21-74, chmn dept zool, 44-59, EMER PROF HISTOL & EMBRYOL, CORNELL UNIV, 74- Concurrent Pos: Nat Res fel, 27-28. Honors & Awards: William H Welch Medal, Am Asn Hist Med, 67; Pfizer Award, Hist Sci Soc, 67. Mem: Am Soc Zoologists; Am Asn Anatomists; Am Asn Hist Med; Hist Sci Soc; fel Int Inst Embryol. Res: Embryology; morphology and development of the neural tube and cranial ganglia in the rat, prechordal mesoderm of the chick and salamander, and eyes of the salamander; cyclopia and mesodermal somites; history of embryology. Mailing Add: Stimson Hall Cornell Univ Ithaca NY 14850

ADELSBERGER, LUCIE, immunology, deceased

ADELSON, BERNARD HENRY, b Tampa, Fla, Mar 16, 20; m 50; c 3. MEDICINE, CHEMISTRY. Educ: Northwestern Univ, BS, 41, PhD(chem), 46, BM, 50, MD, 51. Prof Exp: Asst chem, 42-44, res assoc, 44-46, instr, 46-47, lectr, 47, clin asst med, 54-57, from instr to asst prof med, 58-73, ASSOC PROF CLIN MED, NORTHWESTERN UNIV, CHICAGO, 73- Concurrent Pos: Resident, Evanston Hosp, Ill, 51-53, dir artificial kidney unit & chief nephrology; resident, Cook County Hosp, 53-54. Mem: AAAS; AMA; Royal Soc Med; Am Soc Nephrology. Res: Organic chemistry; synthesis of 5-substituted quinolines; cardiac outputs in myocardial infarcts and medical shock; extracorporeal hemodialysis. Mailing Add: 595 Lincoln Ave Glencoe IL 60022

ADELSON, DAVID E, b Chicago, Ill, Mar 21, 12; m 39; c 2. CHEMISTRY. Educ: Univ Fla, BS, 32, MS, 33, PhD(chem), 35. Prof Exp: Asst chem, Univ Fla, 30-32, 33-34, instr, 34-35; Fritzche fel org chem, Columbia Univ, 35-36, Nat Res Found fel chem, 36-37; res chemist, Shell Develop Co, 37-51; CONSULT CHEMIST, 51- Mem: Am Chem Soc. Res: Derivatives of piperazine; retene; synthetic organic chemistry; synthetic resins; chemical addition agents for lubricants. Mailing Add: 4348 San Pablo Ave Emeryville CA 94608

ADELSON, HAROLD ELY, b New York, NY, May 5, 31. RESOURCE MANAGEMENT. Educ: City Col New York, BS, 53; Univ Calif, MA, 54, PhD(physics), 59. Prof Exp: Asst physics, Univ Calif, 53-55, asst nuclear physics, Radiation Lab, 55-59; sr res physicist, Convair Astronaut, 59-61, sr staff scientist, Gen Dynamics/Astronaut, 61-65; mgr environ res satellite dept, TRW Systs Group, 65-67, from asst dir to dir res appln lab, 67-71, mgr Viking Lander biol & meteorol instruments prog, 71-75, MGR APPL TECHNOL DIV DESIGN REV OFF, TRW SYSTS, INC, 75- Mem: Am Phys Soc; Am Geophys Union; Am Inst Aeronaut & Astronaut. Res: Spacecraft technology; space instrumentation. Mailing Add: Apt 1902 2170 Century Park E Los Angeles CA 90067

ADELSON, LESTER, b Chelsea, Mass, Aug 20, 14; m 42; c 2. PATHOLOGY. Educ: Harvard Univ, AB, 35; Tufts Univ, MD, 39. Prof Exp: From asst prof legal med to assoc prof forensic path, 53-69, PROF FORENSIC PATH, SCH MED, CASE WESTERN RESERVE UNIV, 69-; PATHOLOGIST & CHIEF DEP CORONER, CUYAHOGA COUNTY, 50- Concurrent Pos: Res fel path & legal med, Harvard Med Sch, 49-50. Mem: AMA; Am Soc Clin Path. Mailing Add: Cuyahoga County Coroner's Off 2121 Adelbert Rd Cleveland OH 44106

ADELSON, LIONEL MORTON, b Holyoke, Mass, Dec 29, 28; m 51. MARINE ECOLOGY, HISTOLOGY. Educ: Am Int Col, BA, 50; Drexel Inst Technol, MS, 57. Prof Exp: Res assoc neurol, 54-58, from asst prof to assoc prof biol, 58-62, chmn dept, 59-66, PROF BIOL, DEL VALLEY COL, 62-, CHMN DIV SCI, 66-, ASSOC DEAN, 73- Concurrent Pos: Grants, Nat Multiple Sclerosis Soc, 54-56, NIH, 56-58. Mem: AAAS; Am Chem Soc; Am Soc Microbiol. Res: Experimental in vitro demyelination; oyster bacteriology. Mailing Add: 300 Sandy Knoll Dr Doylestown PA 18901

ADELSTEIN, GILBERT WILLIAM, medicinal chemistry, organic chemistry, see 12th edition

ADELSTEIN, PETER Z, b Montreal, Que, Sept 1, 24; nat US; m 47; c 3. PHYSICAL CHEMISTRY. Educ: McGill Univ, BE, 46, PhD(chem), 49. Prof Exp: Chemist, 49-63, GROUP LEADER, EASTMAN KODAK CO, 58- Concurrent Pos: Chmn PHI-3, Am Nat Standards Inst, 67-; chmn WG-5, Int Standards Inst, TC/42, 73- Mem: Am Chem Soc; Am Soc Testing & Mat; Nat Fire Protection Asn; Soc Photog Sci & Eng. Res: Behavior of high polymers in solution; physical properties of high polymer materials in solid states; physical behavior of photographic film; archival stability of photographic materials. Mailing Add: 149 Warrington Dr Rochester NY 14618

ADELSTEIN, STANELY JAMES, b New York, NY, Jan 24, 28; m 57; c 2.

RADIATION BIOPHYSICS, NUCLEAR MEDICINE. Educ: Mass Inst Technol, BS & MS, 49, PhD(biophys), 57; Harvard Univ, MD, 53. Prof Exp: House officer, Peter Bent Brigham Hosp, Boston, 53-54, sr asst resident, 57-58, chief resident physician, 59-60; assoc anat, 63-65, from asst prof to assoc prof radiol, 65-72, PROF RADIOL, HARVARD MED SCH, 72- Concurrent Pos: Moseley traveling fel, Harvard Med Sch, 58-59, HA&C Christian fel, 59-60, Med Found fel, 60-63, P H Cook fel radiol, 60-68; USPHS career develop award, 65-68; assoc med & radiol, Peter Bent Brigham Hosp, 63-68, dir nuclear med, 68-; vis fel, Johns Hopkins Med Inst, 68; chief nuclear med, Children's Hosp Med Ctr, Boston, 70-; mem, Am Bd Nuclear Med. Mem: Biophys Soc; Radiation Res Soc; Am Soc Cell Biol; Soc Nuclear Med. Res: Cellular and molecular radiation biology. Mailing Add: Harvard Med Sch Boston MA 02115

ADEM, JULIAN, b Tuxpan, Mex, Jan 8, 24; m 58; c 2. METEOROLOGY, APPLIED MATHEMATICS. Educ: Brown Univ, PhD, 53. Prof Exp: Prof appl math, Nat Univ Mex, 54-55; fel, Int Meteorol Inst, Univ Stockholm, 55-56; prof geophys & asst dir, Inst Geophys, Nat Univ Mex, 57-59, prof & dir, 59-65; res meteorologist, Nat Meteorol Ctr, Nat Weather Serv, Nat Oceanic & Atmospheric Admin, 65-71; PROF GEOPHS & DIR INST GEOPHYS, NAT UNIV MEX, 71- Concurrent Pos: Vchmn Mex comt, Int Geophys Year, 57-59; founder & ed, Geofisica Internacional, 60-; vis prof, Univ Hamburg, 61-62; consult, Nat Oceanic & Atmospheric Admin, 63-65. Mem: Am Geophys Union; Am Meteorol Soc; Mex Geophys Union (pres, 60-); Mex Math Soc; Mex Acad Sci. Res: Elastic wave propagation; motion of atmospheric vortices; development of thermodynamical model for long-range forecasting; heat budget of the atmosphere; air-sea interaction. Mailing Add: Inst of Geophys Nat Univ Mex Mex 20 D F Mexico

ADEN, DAVID PAUL, b Quincy, Ill, Feb 23, 46; m 71. IMMUNOBIOLOGY. Educ: Quincy Col, BS, 64; Mont State Univ, MS, 70, PhD(microbiol), 73. Prof Exp: Res investr, 73-76; RES ASSOC, WISTAR INST ANAT & BIOL, 76- Concurrent Pos: Nat Cancer Inst fel, 74-76. Mem: Am Soc Microbiol. Res: Cell surface antigens coded for by specific human chromosomes; chromosomes involves in tumorgenicity. Mailing Add: Wistar Inst of Anat & Biol 36th & Spruce Sts Philadelphia PA 19146

ADENSTEDT, ROLF KARL, b Dessau, Ger, Nov 2, 41; US citizen; m 66; c 2. MATHEMATICS, MATHEMATICAL STATISTICS. Educ: Brown Univ, ScB, 63, PhD(math), 67. Prof Exp: Asst prof appl math, Brown Univ, 67-69; asst prof, 69-74, ASSOC PROF APPL MATH, LEHIGH UNIV, 74- Mem: Inst Math Statist. Res: Time series analysis; regression analysis. Mailing Add: Ctr Appl Math Lehigh Univ Bethlehem PA 18015

ADER, OLIN BLAIR, b Amantha, NC, Jan 30, 05; m 67. APPLIED MATHEMATICS. Educ: Duke Univ, AB, 26, AM, 28; Univ Ky, PhD(math), 37. Prof Exp: Teacher soc & math, Brevard Inst, 28-30; prof math & physics, Lindsey Wilson Jr Col, 32-35; instr math, Southern Methodist Univ, 37-39; asst prof, Wofford Col, 39-42; instr, Univ Ariz, 42-44; prof, Southern Ariz Sch Boys, 45-46; assoc prof, Miss Southern Col, 47-50 & Fenster Ranch Sch, Tucson, Ariz, 50-51; from asst prof to prof, 51-74, EMER PROF MATH, NMEX STATE UNIV, 74- Mem: Am Math Soc. Res: Convex regions of metric spaces; mechanics of a conical spring. Mailing Add: Dept of Math NMex State Univ University Park NM 88001

ADES, HARLOW WHITING, b Rockford, Ill, Dec 31, 11; m 33, 63; c 4. EXPERIMENTAL NEUROLOGY. Educ: Univ Ill, BS, 34, MS, 35, PhD, 38. Prof Exp: Asst zool & Psychol, Univ Ill, 34-38; asst physiol optics, Johns Hopkins Univ, 38-39; from instr to asst prof anat, Sch Med, Emory Univ, 39-47, prof & chmn dept, 47-54; prof, Southwestern Med Sch, Univ Tex, 54-57; head neurol sci div, US Naval Aerospace Med Inst, Fla, 57-65; PROF ELEC ENG, PHYSIOL, BIOPHYS & PSYCHOL, UNIV ILL, URBANA-CHAMPAIGN, 65- Concurrent Pos: Annual res prof, Neurol Inst, Northwestern Univ, 47-48; consult electroencephalog, Vet Admin Hosps, Atlanta, Ga, 49-54 & McKinney, Tex, 54-57; consult aviation physiol, US Navy, 55 & Biophys Res Lab, Univ Ill, 55-65; mem, Navy-Air Force Proj Biol Effects Noise, 53 & Armed Forces-Nat Res Coun-NASA comt hearing, bioacoustics & biomech, 55; mem exec comt res adv comt biotech & human res, NASA, 65-67; mem exec coun comt hearing, bioacoustics & biomech, Nat Acad Sci-Nat Res Coun, 74-75. Honors & Awards: SKYLAB Achievement Award, NASA, 74. Mem: Am Physiol Soc; Am Asn Anat; Am Neurol Asn; Aerospace Med Asn; Acoustical Soc Am. Res: Structure and function of the central auditory pathway and cortical projection area; funtion of the cerebral cortex in learning; visual projection system; non-auditory effects of high intensity noise; extrapyramidal system; electronmicroscopical studies of the effect of noise on the inner ear; effects of aging in the ear and auditory system. Mailing Add: Dept of Elec Eng Univ of Ill Urbana IL 61801

ADEY, WILLIAM ROSS, b Adelaide, Australia, Jan 31, 22; US citizen; m 70; c 3. NEUROPHYSIOLOGY. Educ: Univ Adelaide, BS & MB, 43, MD, 49. Prof Exp: Lectr, sr lectr & reader anat, Univ Adelaide, 46-53; asst prof, Univ Calif, Los Angeles, 53-54; sr lectr, Univ Melbourne, 54-57; PROF ANAT & PHYSIOL, UNIV CALIF, LOS ANGELES, 57-, DIR ENVIRON LAB, BRAIN RES INST, 75- Concurrent Pos: Nuffield Found Dom traveling fel, 50; Rockefeller Found grants, 51, 55; Royal Soc London & Nuffield Found traveling bursary, 56-57; mem space sci panel, President's Sci Adv Comt, 66-, biol & med sci panel, 69-72; mem electromagnetic radiation coun, Exec Off of President, 69- Honors & Awards: Herrick Award, Am Asn Anat, 63. Mem: AAAS; Am Physiol Soc; Am Asn Anat; fel Am Acad Arts & Sci; fel Inst Elec & Electronics Eng. Res: Neuroanatomy and neurophysiology of brain functions in behavioral mechanisms; medical electronics and bioinstrumentation. Mailing Add: Dept of Anat Univ of Calif Los Angeles CA 90024

ADHIKARI, P K, b Nagpur, India, May 18, 28; m 66. INTERNAL MEDICINE. Educ: Univ Calcutta, MB, BS, 52; FRCP(C), 65. Prof Exp: Resident med, Univ Vt, 55-57; res assoc, Pulmonary Function Lab, Wayne State Univ, 60-62; res assoc, 62-64, ASST PROF PEDIAT, UNIV MAN, 65- Concurrent Pos: Fel cardio-respiration, Univ Vt, 57-59. Res: Respiratory physiology and diseases. Mailing Add: 400-309 Hargrave St Winnipeg MB Can

ADIARTE, ARTHUR LARDIZABAL, b San Nicolas, Philippines, Oct 27, 43; m 72; c 1. BIOPHYSICAL CHEMISTRY. Educ: Univ Philippines, BS, 63; Univ Pittsburgh, PhD(biophys), 72. Prof Exp: Instr physics math, Univ Philippines, 63-66; teaching asst physics, Univ Pittsburgh, 66-67, res asst biophys, 67-72; fel, Univ Regensburg, 72-74; FEL, LAB BIOPHYS CHEM, DEPT CHEM, UNIV MINN, 75- Mem: Biophys Soc; AAAS. Res: Protein structure and function, thermodynamics and kinetics of biological processes; the role of water in protein systems and enthalpy-entropy compensation phenomena. Mailing Add: Lab for Biophys Chem Dept of Chem Univ of Minn Minneapolis MN 55455

ADIBI, SIAMAK A, b Tehran, Iran, Mar 17, 32; US citizen; m 63; c 3. GASTROENTEROLOGY, NUTRITION. Educ: Johns Hopkins Univ, BA, 55; Jefferson Med Col, MD, 59; Mass Inst Technol, PhD(nutrit biochem, metab), 67. Prof Exp: Intern, Jefferson Med Col Hosp, 59-60; rotating med resident, Joslin Clin, Lahey Clin & New Eng Deaconess Hosp, 61-62; sr resident & chief resident, New Eng Deaconess Hosp, 62-63, asst in med, 63-66; CHIEF GASTROINTESTINAL & HUMAN NUTRIT UNIT, MONTEFIORE HOSP, 66-; PROF MED, UNIV PITTSBURGH, 74- Concurrent Pos: Asst in med, Peter Bent Brigham Hosp, 63-66; assoc, Dept Nutrit & Food Sci, Mass Inst Technol, 63-66. Mem: AAAS; Am Inst Nutrit; Am Fedn Clin Res; Am Soc Clin Invest; Am Physiol Soc. Res: Intestinal fate of the products of protein digestion such as amino acids and dipeptides in humans; amino acid metabolism in altered nutritional and hormonal states. Mailing Add: Gastrointestinal & Human Nutrit Montefiore Hosp Pittsburgh PA 15213

ADICKES, H WAYNE, b Cuero, Tex, Sept 6, 40. ORGANIC CHEMISTRY. Educ: Stephen F Austin State Col, BS, 62; Tex Christian Univ, PhD(chem), 68. Prof Exp: Instr chem, Arlington State Col, 62-65; fel, La State Univ, New Orleans, 68-69 & Yale Univ, 69-70; sr res chemist, 70-75, MGR IMAGING PAPERS, PLASTIC TECHNOL & CONVERTED PROD, ST REGIS PAPER CO, 75- Mem: AAAS; Am Chem Soc; Soc Photog Scientists & Engrs. Res: Investigation of five-membered heterocyclic arynes; development of new synthetic techniques; cyclopropanone chemistry; new routes to beta-lactams; investigation and synthesis of organic photo-conductors. Mailing Add: St Regis Paper Co W Nyack Rd West Nyack NY 10994

ADICOFF, ARNOLD, polymer chemistry, physical chemistry, see 12th edition

ADIN, ANTHONY, b Sheffield, Eng, Nov 11, 42; m 69; c 2. PHYSICAL INORGANIC CHEMISTRY. Educ: Leeds Univ, BSc, 64, PhD(chem), 67. Prof Exp: Fel, Brookhaven Nat Lab, 67-69 & Ames Lab, AEC, 69-71; SR RES CHEMIST, EASTMAN KODAK CO, 71- Res: Application of inorganic and organometallic chemistry to unconventional photography. Mailing Add: Eastman Kodak Co Bldg 82A Kodak Park Rochester NY 14615

ADINOFF, BERNARD, b Toronto, Ont, Mar 9, 19; US citizen; m 48; c 3. CHEMISTRY. Educ: Univ Chicago, BS, 39, PhD(phys chem), 43. Prof Exp: Res asst, Univ Chicago, 43-44; chemist, Dayton Rubber Co, 44-59; chief chem engr, Fruehauf Trailer Co, 59-60; CHIEF CHEM ENGR AUTOMOTIVE OPERS, ROCKWELL INT, 60- Mem: Am Chem Soc; Nat Asn Corrosion Eng; Soc Automotive Engrs; Am Soc Testing & Mat; Am Soc Lubrication Engrs (vpres, 72-74). Res: Lubrication; corrosion; plastics; adhesives; rubber; paints; latex foam; urethane foam; industrial fabrics. Mailing Add: 18817 Goldwin Ave Southfield MI 48075

ADINOLFI, ANTHONY M, b New Haven, Conn, May 5, 39; m 63; c 2. NEUROCYTOLOGY. Educ: Yale Univ, BA, 60; Columbia Univ, MA, 65, PhD(anat), 67. Prof Exp: Asst prof anat, 69-74, ASST PROF ANAT & PSYCHIAT, SCH MED, UNIV CALIF, LOS ANGELES, 74- Concurrent Pos: Fel anat, Med Ctr, Univ Calif, Los Angeles, 67-69; chief neurocytol res lab, Vet Admin Hosp, 69-74. Mem: Am Asn Anat; Electron Micros Soc Am; Soc Neurosci. Res: Electron microscopic analysis of synaptic organization in the central nervous system. Mailing Add: Dept of Anat Univ of Calif Med Ctr Los Angeles CA 90024

ADISESH, SETTY RAVANAPPA, b Y N Hosakote, Mysore, India, June 17, 26; m 52. PHYSICAL CHEMISTRY, INORGANIC CHEMISTRY. Educ: Univ Mysore, BSc, 52, MSc, 56; Kent State Univ, PhD(chem), 65. Prof Exp: Lectr chem, Univ Mysore, 52-60; from asst prof to assoc prof, 65-72, PROF CHEM, ST LEO COL, 72- Concurrent Pos: Res assoc, Kent State Univ, 66; res appointment, Tex A&M Univ, 68 & Ga Inst Technol, 69; NSF res grant, 69. Mem: Am Chem Soc. Res: X-ray diffraction; voltammetry; kinetic and reaction mechanisms. Mailing Add: Dept of Chem St Leo Col St Leo FL 33574

ADISMAN, I KENNETH, b New York, NY, Aug 3, 19; m 57; c 2. PROSTHODONTICS. Educ: Univ Buffalo, DDS, 40; NY Univ, MS, 60. Prof Exp: PROF MAXILLOFACIAL PROSTHODONT, BROOKDALE DENT CTR, COL DENT, NY UNIV, 71- Concurrent Pos: USPHS fel cancer control, NY Univ, 66-70; dir training prog, Nat Cancer Inst, 74-76; consult, Vet Admin Hosp, 69-; ed, Acad Maxillofacial Prosthetics, 69-; assoc ed, J Prosthetic Dent, 69-; consult, Clemson Univ, 72-; pres-elect, Greater NY Acad Prosthodont, 74- Honors & Awards: Ackerman Award, Am Acad Maxillofacial Prosthetics, 73. Mem: Fel Am Col Dent; fel Am Col Prosthodont; Int Asn Dent Res; Acad Dent Prosthetics; Am Cleft Palate Asn. Res: Cleft palate prosthetics; maxillofacial prosthetics; bio-materials. Mailing Add: 100 Central Park S New York NY 10019

ADKINS, BENJAMIN JEFFERSON, b DuQuoin, Ill, Jan 28, 32; div; c 2. BIOCHEMISTRY. Educ: Southern Ill Univ, BA, 56, MA, 58; Purdue Univ, PhD(chem), 64. Prof Exp: Fel, Purdue Univ, 64-65, Roswell Park Mem Inst, 65 & Univ Calif, San Francisco, 65-67; sr res assoc biochem, Palo Alto Med Res Found, 67-70; fel, Inst Humane Studies, 70-71; sr resident res assoc, Nat Res Coun, Ames Res Ctr, NASA, 71-73; staff scientist, Anal Develop Assocs Corp, 73-74; DIR, B JEFFERSON ADKINS & ASSOCS, 74- Res: Protein chemistry; immunobiology; stimulation of the cellular immune response system of the host. Mailing Add: 835 Webster St Palo Alto CA 94301

ADKINS, ELIZABETH KOCHER, b Washington, DC, July 12, 45. BEHAVIORAL BIOLOGY, REPRODUCTIVE ENDOCRINOLOGY. Educ: Univ Md, BS, 67; Univ Pa, PhD(psychol), 71. Prof Exp: Asst prof psychol, State Univ NY Col Cortland, 74-75; ASST PROF PSYCHOL, NEUROBIOL & BEHAV, CORNELL UNIV, 75- Mem: AAAS; Animal Behav Soc; Am Ornithologists Union; Am Inst Biol Sci; Int Soc Study Aggression. Res: Role of sex hormones in the development and maintenance of social behavior in vertebrates, especially reproductive behavior. Mailing Add: Dept of Psychol Uris Hall Cornell Univ Ithaca NY 14850

ADKINS, JOHN EARL, JR, b Lynchburg, Va, June 20, 37; m 64; c 2. ANALYTICAL CHEMISTRY. Educ: Davidson Col, BS, 58; Univ Tenn, PhD(anal chem), 63. Prof Exp: Res chemist, Plastics Dept, 63-68, div chemist, Tech Dept, 68-72, SR CHEMIST, TECH DEPT, E I DU PONT DE NEMOURS & CO, INC, 72- Mem: Am Chem Soc. Res: Atomic absorption and infrared spectrophotometry; gas chromatography, ambient air quality monitoring and analysis. Mailing Add: E I du Pont de Nemours & Co Tech Dept Box 1089 Orange TX 77631

ADKINS, JOHN NATHANIEL, b Spokane, Wash, July 23, 11; m 41. GEOPHYSICS. Educ: Univ Calif, AB, 36, PhD(seismol), 39. Prof Exp: Nat Res Coun fel, Mass Inst Technol, 39-41, asst prof geophys, 46-48; mem staff, War Res Div, Columbia Univ, 41-45; supvr, Antenna Sect, Airborne Instruments Lab, Inc, NY, 45-46; head geophys br, Off Naval Res, 48-49, dir earth sci div, 49-59, asst chief scientist, 59-72; RETIRED. Concurrent Pos: Mem, Nat Res Coun, 49-62, chmn div earth sci, 58-60. Mem: Geol Soc Am. Res: Earthquake travel times; crustal structure of the earth. Mailing Add: The Green Weems VA 22576

ADKINS, JULIA ELIZABETH, b Wayne, WVa, June 26, 10. MATHEMATICS. Educ: Marshall Col, AB, 30; Ohio State Univ, MA, 43, PhD(math educ), 56. Prof Exp: Pub sch teacher & prin, WVa, 30-50; instr sci, Marshall Col, 50-52; math, Ohio State Univ, 52-53, supvr student teachers sec math, 53-56; asst prof math, State Col Iowa, 56-57; from asst prof to assoc prof, 57-65, PROF MATH, CENT MICH UNIV, 65-

Res: History and teaching of mathematics. Mailing Add: Dept of Math Cent Mich Univ Mount Pleasant MI 48858

ADKINS, LARRY RAY, physics, see 12th edition

ADKINS, RONALD JAMES, b Bremerton, Wash, June 28, 32; m 55; c 2. NEUROPHYSIOLOGY. Educ: Univ Wash, BS, 54, MS, 57, PhD(physiol, psychol), 65. Prof Exp: Res asst neurophysiol, Univ Wash, 57-59; instr psychol & sociol, Lower Columbia Col, 59-61; behavioral trainee physiol & psychol, Univ Wash, 61-65; instr physiol, NY Med Col, 65-66; asst prof zool & physiol, 66-71, ASSOC PROF ZOOL & PHYSIOL, WASH STATE UNIV, 71- Mem: AAAS; assoc Am Physiol Soc. Res: Physiology of behavior; interaction among descending and ascending systems within the sensory apparatus, currently emphasizing the pyramidal tract and the dorsal column-medial lemniscal system in cats. Mailing Add: Dept of Zool Wash State Univ Pullman WA 99163

ADKINS, RUTHERFORD HAMLET, b Alexandria, Va, Nov 21, 24; m 69; c 3. NUCLEAR PHYSICS. Educ: Va State Col, BS, 47; Howard Univ, MS, 50; Cath Univ Am, PhD(physics), 55. Prof Exp: Instr physics, Va State Col, 49-51; res assoc, Cath Univ Am, 52-54; assoc prof, Va State Col, 54-58; prof, Tenn State Univ, 58-62; assoc dean, 68-70, PROF PHYSICS & MATH, FISK UNIV, 62-, VPRES, 70- Mem: Am Phys Soc; Am Asn Physics Teachers. Res: Theoretical studies of structure of light nuclei revealed by low energy experiments. Mailing Add: Fisk Univ Nashville TN 37203

ADKINS, THEODORE ROOSEVELT, JR, b San Antonio, Tex, Dec 26, 30; m 49; c 3. ENTOMOLOGY. Educ: Auburn Univ, BS, 52, MS, 54, PhD(entom), 58. Prof Exp: Asst entom, Auburn Univ, 56-57; asst prof entom & zool & asst entomologist, 57-61, assoc prof entom & zool, 61-68, PROF ENTOM & ZOOL, CLEMSON UNIV, 68- Mem: Entom Soc Am; Am Mosquito Control Asn. Res: Medical-veterinary entomology. Mailing Add: Dept Entom & Econ Zool Clemson Univ Clemson SC 29631

ADKINSON, BURTON WILBUR, b Everson, Wash, Mar 5, 09; m 42; c 2. INFORMATION SCIENCE. Educ: Univ Wash, BA, 36, MA, 39; Clark Univ, PhD(geog), 42. Prof Exp: Teacher pub schs, Wash, 29-39; instr geog, Univ Wash, 42; from asst to assoc regional asst, Dept of State, 42-43; res assoc & dep dir, Dept of Interior, 43-44; asst chief, Map Intel Sect, Off Strategic Serv, 44-45; asst chief & chief map div, Libr Cong, 45-48, from asst dir to dir ref dept, 48-57; head off sci info serv, NSF, 57-71; dir, Am Geog Soc, 71-73; RETIRED. Concurrent Pos: Mem, US Nat Comt, Int Geog Union, 71-73; mem adv comt libr, arch & doc, UNESCO, 71-72. Mem: AAAS; Spec Libr Asn (pres, 59-60); Am Soc Info Sci; Asn Am Geog (secy, 54-57); Am Geog Soc; Int Fedn Doc (pres, 63-65). Mailing Add: 5907 Welborn Dr Washington DC 20016

ADKISSON, CURTIS SAMUEL, b Little Rock, Ark, Feb 2, 42; m 69; c 1. ORNITHOLOGY. Educ: Oberlin Col, BA, 65; Miami Univ, MA, 67; Univ Mich, PhD(zool), 72. Prof Exp: Teaching fel zool, Univ Mich, 68-72; ASST PROF ZOOL, VA POLYTECH INST & STATE UNIV, 72-; CUR ANIMAL COLLECTIONS, ROCKY MOUNTAIN BIOL LAB, 73- Mem: Am Ornith Union; Soc Study of Evol; AAAS. Res: Dialects in bird vocalizations; interspecific imitation in bird song; systematics and behavior of cardueline finches; habitat utilization in North American birds. Mailing Add: Dept of Biol Va Polytech Inst & State Univ Blacksburg VA 24061

ADKISSON, PERRY LEE, b Blytheville, Ark, Mar 11, 29; m 56; c 1. ECONOMIC ENTOMOLOGY. Educ: Univ Ark, BS, 50, MS, 54; Kans State Col, PhD(entom), 56. Prof Exp: Asst prof entom, Univ Mo, 56-57; asst prof, 58-63, PROF ENTOM, TEX A&M UNIV, 63-, HEAD DEPT, 68- Concurrent Pos: Fel, Harvard Univ, 63-64; consult, Int Atomic Energy Agency, Vienna; chmn sci adv panel agr chem, Gov of Tex, 70-71; mem panel of experts on integrated pest control, Food & Agr Orgn, Rome, Italy, 70-74; consult, Hazardous Mat Adv Comt, US Environ Protection Agency, 71; mem exec comt study prob pest mgt, Environ Sci Bd, Nat Acad Sci- Nat Acad Eng, 72-; mem, Struct Pest Control Bd Tex, 72-; mem comt biol pest species, Nat Res Coun, 73-; mem ed bd, Ann Rev Entom, 74-; mem, US Insect Control Deleg to People's Repub of China, 75; mem UN-Food & Agr Orgn consult on pesticides in agr & pub health, Rome, Italy, 75; mem, US Directorate to UNESCO Man & Biosphere Prog, 75- Honors & Awards: J Everett Bussart Award, Entom Soc Am, 67. Mem: AAAS; Entom Soc Am (pres, 74); Am Inst Biol Sci; Sigma Xi; Int Orgn Biol Control. Res: Basic and applied research on cotton insects; insect photoperiodism. Mailing Add: Dept of Entom Tex A&M Univ College Station TX 77843

ADLDINGER, HANS KARL, b Munich, WGer. VETERINARY VIROLOGY, CANCER. Educ: Univ Munich, dipl vet med, 61, Dr med vet, 62; Cornell Univ, PhD(vet microbiol), 71. Prof Exp: Res assoc, Bavarian State Vaccination Inst, Munich, 61-63; sr res assoc virol, Inst Comp Trop Med, Univ Munich, 64-65; sr res assoc, NY State Vet Col, Cornell Univ, 66-68; res assoc virol, Albert Einstein Med Ctr, Philadelphia, 71; ASSOC PROF MICROBIOL, COL VET MED, UNIV MO, 72- Concurrent Pos: Nat Cancer Inst res grant, 75. Res: Virology; immunology; pathogenesis of animal virus diseases; mechanisms of oncogenesis in Herpes-virus-induced cancer; cell-mediated immunity and pathogenesis in Marek's disease of chickens. Mailing Add: Dept of Microbiol Col of Vet Med Univ Mo Columbia MO 65201

ADLER, ALAN DAVID, b Nyack, NY, Oct 5, 31; m 54; c 3. MOLECULAR BIOLOGY, BIOPHYSICAL CHEMISTRY. Educ: Univ Rochester, BS, 53; Univ Pa, PhD(chem), 60. Prof Exp: Asst prof molecular biol, Univ Pa, 61-67; staff scientist & assoc prof chem sci & chmn div, New Eng Inst, 67-74; ASSOC PROF CHEM, WESTERN CONN STATE COL, 74- Mem: AAAS; Am Chem Soc; Am Phys Soc; Am Inst Chemists; NY Acad Sci. Res: Physical, chemical and biological aspects of porphyrin materials; thermodynamics of electrolytic solutions; physical chemistry. Mailing Add: Dept of Chem Western Conn State Col Danbury CT 06810

ADLER, ALEXANDRA, b Vienna, Austria, Sept 24, 01; nat US; m 59. NEUROLOGY, PSYCHIATRY. Educ: Univ Vienna, MD, 26. Prof Exp: Intern, resident & vis physician, Univ Vienna Hosp, 26-34; res fel, asst & instr neurol, Harvard Univ, 35-44; from asst to assoc clin prof neurol, 46-69, CLIN PROF PSYCHIAT, 69-, VIS PHYSICIAN, BELLEVUE & UNIV HOSPS, 46- Concurrent Pos: Vis physician, Goldwater Mem Hosp, 46-56; psychiatrist, Dept Correction, New York City, 48-72; med dir, Alfred Adler Ment Hyg Clin, 54- Mem: AMA; Am Psychiat Asn; Asn Res Nerv & Ment Dis; Am Acad Neurol; Int Asn Individual Psychol (past pres). Res: Psychotherapy; general psychiatry; organic mental syndromes; psycho-pharmacol-therapy. Mailing Add: 30 Park Ave New York NY 10016

ADLER, ALFRED, b Ger, Feb 21, 30; US citizen; m 54; c 3. MATHEMATICS. Educ: Mass Inst Technol, BS, 52; Univ Calif, Los Angeles, PhD(math), 56. Prof Exp: Instr math, Princeton Univ, 56-58; lectr, Mass Inst Technol, 58-60; asst prof, Rutgers Univ, 60-61; vis prof, Univ Bonn, 61-63; from assoc prof to prof, Purdue Univ, 63-67; PROF MATH, STATE UNIV NY STONY BROOK, 67- Res: Geometry of complex manifolds. Mailing Add: Dept of Math State Univ of NY Stony Brook NY 11794

ADLER, ALICE JOAN, b Jersey City, NJ, Dec 8, 35; m 58; c 3. BIOPHYSICAL CHEMISTRY. Educ: Columbia Univ, AB, 56; Harvard Univ, PhD(phys chem), 61. Prof Exp: Fel biochem, Children's Cancer Res Found, Boston, 61-62; fel biol, Mass Inst Technol, 62-64; fel biochem, Oxford Univ, 64-65; SR RES ASSOC CHEM, BRANDEIS UNIV, 65- Mem: AAAS; Am Chem Soc; Biophys Soc; Am Soc Biol Chem. Res: Physical studies of biological macromolecules and model compounds, optical rotatory properties and kinetics of nucleic acids, polynucleotides, polypeptides and proteins; nucleic acid and protein interactions; chromatin structure. Mailing Add: Dept of Biochem Brandeis Univ Waltham MA 02154

ADLER, BEATRIZ RAQUEL, b Havana, Cuba, Oct 29, 29; US citizen; m 51. BIOLOGY. Educ: Univ Havana, Dr nat sci, 53. Prof Exp: Parochial sch teacher, Cuba, 48-61 & Pa, 61-64; from instr to assoc prof biol & phys sci, 64-73, PROF BIOL, CALDWELL COL, 73- Res: Geographical survey and methods of fishery of early Cuban Indians. Mailing Add: Dept of Biol Caldwell Col Caldwell NJ 07006

ADLER, CARL GEORGE, b Buffalo, NY, Oct 3, 39; m 63; c 2. THEORETICAL PHYSICS. Educ: Univ Notre Dame, BS, 61, PhD(physics), 66. Prof Exp: ASSOC PROF PHYSICS, E CAROLINA UNIV, 65- Concurrent Pos: Assoc dir, Eastern Carolina Coop Physics. Mem: Am Asn Physics Teachers. Res: Theoretical nuclear structure; quantum mechanics; physics pedagogy. Mailing Add: Dept of Physics E Carolina Univ Greenville NC 27834

ADLER, DAVID, b Bronx, NY, Apr 13, 35; m 58; c 3. SOLID STATE PHYSICS. Educ: Rensselaer Polytech Inst, BS, 56; Harvard Univ, AM, 58, PhD(physics), 64. Prof Exp: Res assoc theoret physics, UK Atomic Energy Res Estab, Harwell, 64-65; res assoc, 65-67, from asst prof to assoc prof elec eng, 67-75, PROF ELEC ENG, MASS INST TECHNOL, 75- Concurrent Pos: Mem ed bd, J Nonmetals & Semiconductors, 72-; mem comt basic res, Nat Res Coun, 73-; chmn solar photovoltaic panel, Mass Inst Technol, 75. Mem: Fel Am Phys Soc; Am Vacuum Soc; Inst Elec & Electronics Engrs. Res: Electrical and optical properties of low-mobility and amorphous semiconductors; nonmetal-metal transitions; magnetic semiconductors; electronic phase transitions; correlations in narrow energy bands; transition-metal oxides; solar cells; amorphous semiconductor devices. Mailing Add: Rm 13-3050 Mass Inst of Technol Cambridge MA 02139

ADLER, ERIC, b Vienna, Austria, Oct 20, 37; US citizen; m 58; c 1. SOLID STATE PHYSICS. Educ: City Col New York, BS, 59; Columbia Univ, PhD(physics), 64. Prof Exp: Res asst physics, Watson Lab, Columbia Univ, 64-65; asst prof, City Col New York, 65-68; STAFF PHYSICIST, IBM CORP, 68- Res: Theoretical solid state physics; electron-phonon interaction; optical properties of semiconductors. Mailing Add: IBM Corp Dept 384 Box A Creek Rd Essex Junction VT 05452

ADLER, FELIX T, b Zurich, Switz, Jan 5, 15; nat US; m 59. MATHEMATICAL PHYSICS. Educ: Univ Zurich, MS, 36, PhD(theoret physics), 38. Prof Exp: Res physics, Nuclear Chem Lab, Col France, 39-40 & Inst Advan Study, Princeton, 41-42; vis asst prof physics, Univ Wis, 42-45; sci officer, Can Nat Res Coun, Brit Ministry of Supply, 43-46; assoc prof physics, Univ Colo, 46; asst prof, Univ Wis, 46-50; assoc prof, Carnegie Inst Technol, 50-57; PROF PHYSICS & NUCLEAR ENG, UNIV ILL, URBANA-CHAMPAIGN, 58- Concurrent Pos: Consult, Nuclear Develop Assocs, 52-53, Oak Ridge Nat Lab, 55-57, 66-68, Radiation Lab, Univ Calif, Livermore, 60 & Los Alamos Sci Lab, Univ Calif, 61-; consult & staff mem, Gen Atomics, 56-61; mem & chmn adv comt, Divs Reactor Eng & Reactor Physics, Argonne Nat Lab, 59-69. Mem: Fel Am Phys Soc; fel Am Nuclear Soc; Am Asn Physics Teachers. Res: Neutron physics; reactor physics; reactor safety; stability of plasmas under intense EM radiation. Mailing Add: Dept of Physics Univ of Ill Urbana IL 61801

ADLER, FRANK LEO, b Graz, Austria, Aug 17, 22; nat US; m 49; c 2. BACTERIOLOGY, IMMUNOLOGY. Educ: City Col New York, BS, 47; Univ Ky, MS, 49; Wash Univ, PhD(bact, immunol), 52. Prof Exp: Asst bact & immunol, Sch Med, Wash Univ, 49-52; mem assoc staff, Peter Bent Brigham Hosp, Boston, 52-54; res assoc bact & immunol, Harvard Med Sch, 53-54; MEM, PUB HEALTH RES INST OF CITY OF NEW YORK, INC, 54- Mem: Am Asn Immunol. Res: Immune reactions. Mailing Add: Pub Health Res Inst of New York 455 First Ave New York NY 10016

ADLER, FRED PETER, applied physics, see 12th edition

ADLER, FREDERICK E W, b Wood, Wis, Apr 8, 15; m 71; c 5. FOOD SCIENCE. Educ: Marquette Univ, BS, 36, MS, 38. Prof Exp: Asst chem, Marquette Univ, 36-38; city jr water chemist, Milwaukee, 38; state conserv chemist, Wis, 39-43; res chemist, Red Star Yeast Co, 46-50, asst dir prod, 50-54, plant mgr, 54-58, dir prod, 58-64, DIR QUAL CONTROL, UNIVERSAL FOODS CORP, 64- Mem: AAAS; fel Am Inst Chemists; Am Chem Soc; Wildlife Soc. Res: Determination of particle size by sedimentation; wax constants analysis; wildlife food investigation; amino acid and vitamin studies on yeast; protein hydrolysates; cytochrome c; molasses filtration. Mailing Add: Universal Foods Corp 433 E Michigan St Milwaukee WI 53201

ADLER, GEORGE, b Ger, Nov 25, 20; US citizen; m 49; c 2. SOLID STATE CHEMISTRY, PHYSICAL ORGANIC CHEMISTRY. Educ: City Col New York, BS, 48; Brooklyn Col, MA, 52. Prof Exp: Asst lectr, Brooklyn Col, 49-51; rubber technologist, Mat Lab, NY Naval Shipyard, 50-56; CHEMIST, BROOKHAVEN NAT LAB, 56- Mem: AAAS; Am Chem Soc; Am Crystallog Asn; Int Soc Magnetic Resonance. Res: Solid state polymerization; electron spin resonance; organic solid state chemistry; colloids and monolayers; gas-solid reactions; photo and radiation chemistry of organic solids. Mailing Add: Dept of Appl Sci Brookhaven Nat Lab Upton NY 11973

ADLER, HANS HENRY, b Bremen, Ger, May 17, 23; US citizen; m 40; c 5. GEOLOGY. Educ: City Col New York, BS, 47; Columbia Univ, MA, 49, PhD(geol), 62. Prof Exp: Mineralogist, NMex Bur Mines, 50-51; chief, Mineral Lab, Div Raw Mat, US AEC, 51-54, staff res geologist, 54-74, STAFF RES GEOLOGIST, DIV RAW MAT, ENERGY RES & DEVELOP ADMIN, 74- Mem: Geol Soc Am; Mineral Soc Am; Soc Econ Geol. Res: Uranium geology and geochemistry; infrared absorption spectroscopy; differential thermal analysis. Mailing Add: Div of Raw Mat Energy Res & Develop Admin Washington DC 20461

ADLER, HENRY ELLIOT, b New York, NY, Apr 2, 17; m 41; c 6. VETERINARY PATHOLOGY. Educ: Wash State Univ, BS, 43, DVM, 46; Univ Calif, PhD(comp path), 55. Prof Exp: Poultry pathologist, Dept Agr, State Wash, 46-47; animal pathologist, Territory Hawaii, 47-49; clin & poultry pathologist, Wash State Col, 49-53; assoc prof poultry path, 53-64, PROF VET MED, UNIV CALIF, DAVIS, 64- Concurrent Pos: NIH fel, 61. Honors & Awards: Newman Award, Int Poultry Asn, Gt Brit; Nat Turkey Fedn Res Award, 60; Corn Prod Co Res Award, 70. Mem: Am

Vet Med Asn; Poultry Sci Asn. Res: Mycoplasma; infections of birds and mammals. Mailing Add: Dept of Vet Med Univ of Calif Davis CA 95616

ADLER, HOWARD IRVING, b New York, NY, July 1, 31; m 53; c 4. BACTERIOLOGY. Educ: Cornell Univ, BS, 53, MS, 55, PhD, 56. Prof Exp: Res assoc, 56-57, biologist, 57-69, dir biol div, 69-75, SR STAFF MEM BIOL DIV, OAK RIDGE NAT LAB, 75- Concurrent Pos: Sr staff mem, Inst Energy Anal, Oak Ridge Assoc Univs, 75-76. Mem: Am Soc Microbiol; Radiation Res Soc; Genetics Soc Am. Res: Mechanisms of radiation damage in microorganisms and control of cell division; bacterial mitogenesis and cytology. Mailing Add: Biol Div Oak Ridge Nat Lab Box Y Oak Ridge TN 37830

ADLER, IRVING LARRY, b Philadelphia, Pa, Sept 12, 43; m 66; c 2. ORGANIC CHEMISTRY, ANALYTICAL CHEMISTRY. Educ: Temple Univ, BA, 64; Pa State Univ, PhD(org chem), 68. Prof Exp: SR CHEMIST, ROHM AND HAAS CO, 68- Mem: Am Chem Soc. Res: Organolithium chemistry; pesticide residue analysis; isolation and identification of metabolites; gas chromatography. Mailing Add: 11017 Greiner Rd Philadelphia PA 19116

ADLER, ISIDORE, physical chemistry, see 12th edition

ADLER, JOHN G, b Budapest, Hungary, Sept 29, 35; Can citizen; m 54; c 3. SOLID STATE PHYSICS. Educ: Univ BC, BSc, 59; Univ Alta, MSc, 61, PhD(physics), 63. Prof Exp: Geophysicist, Imp Oil Ltd, 57-59; asst prof physics, Dalhousie Univ, 63-64; res fel, Case Western Reserve Univ, 64-65, from asst prof to assoc prof, 65-72, PROF PHYSICS, UNIV ALTA, 72- Mem: Am Phys Soc. Res: Electron tunneling; superconductivity; low temperature physics. Mailing Add: Dept of Physics Univ of Alta Edmonton AB Can

ADLER, JOHN HENRY, b Brooklyn, NY. BIOLOGICAL CHEMISTRY. Educ: Univ Md, College Park, BS, 70, MS, 73, PhD(plant physiol), 75. Prof Exp: Fel, 75-76, RES ASSOC BIOCHEM, DREXEL UNIV, 76- Mem: Am Chem Soc; Am Soc Plant Physiologists; AAAS; Sigma Xi. Res: Examination of the role of sterols and phospholipids in the form and function of biological membranes; the ontogenetic and phylogenetic implications of different lipid structures. Mailing Add: Dept of Biol Sci Drexel Univ Philadelphia PA 19104

ADLER, JULIUS, b Edelfingen, Ger, Apr 30, 30; nat US; m 63; c 2. BIOCHEMISTRY, GENETICS. Educ: Harvard Univ, AB, 52; Univ Wis, MS, 54, PhD(biochem), 57. Prof Exp: Fel microbiol, Wash Univ, 57-59; fel biochem, Stanford Univ, 59-60; from asst prof to assoc prof, 60-66, PROF BIOCHEM & GENETICS, UNIV WIS-MADISON, 66- Mem: Am Chem Soc; Am Soc Biol Chemists. Res: Biochemistry and genetics of behavior, especially in microorganisms. Mailing Add: Dept of Biochem Univ of Wis Madison WI 53706

ADLER, KRAIG (KERR), b Lima Ohio, Dec 6, 40; m 67; c 1. ANIMAL BEHAVIOR, EVOLUTION. Educ: Ohio Wesleyan Univ, 62; Univ Mich, MS, 65, PhD(zool), 68. Prof Exp: Asst prof, Univ Notre Dame, 68-72; ASSOC PROF BIOL, CORNELL UNIV, 72-, CHMN NEUROBIOL & BEHAV SCI DIV, 76- Concurrent Pos: Ed, Jour, Soc Study Amphibians & Reptiles, 58-63, miscellaneous publ, 61-; temporary cur, Mus Zool, Univ Mich, 65; co-ed, Int J Interdisciplinary Cycle Res, Neth, 70-; leader, Univ Notre Dame Exped Mex, 69. Mem: Soc Syst Zool; Am Soc Ichthyologists & Herpetologists; Soc Study Evolution; Animal Behav Soc; Soc Study Amphibians & Reptiles. Res: Photoreception, orientation, navigation and circadian rhythms of vertebrates; evolution, systematics and zoogeography of amphibians and reptiles; paleo- and archeozoology. Mailing Add: Langmuir Lab Cornell Univ Ithaca NY 14853

ADLER, KURT ALFRED, b Vienna, Austria, Feb 25, 05; US citizen; m; c 1. PSYCHIATRY. Educ: Univ Vienna, PhD(physics), 35; Long Island Col Med, MD, 41. Prof Exp: Chief psychiat serv, US Army, 42-46; DEAN, ALFRED ADLER INST, NEW YORK, 52-, PSYCHIAT CONSULT, ADVAN CTR PSYCHOTHER, 58-, DEAN, ADVAN INST ANAL PSYCHOTHER, 69- Concurrent Pos: Consult psychiatrist, Youth-House, New York, 47-48; emer psychiatrist, Lenox Hill Hosp, New York, 47-; consult, Alfred Adler Inst Minn, 70-; dir, Bowie State Col, 75. Honors & Awards: Physicians Recognition Award, AMA, 70. Mem: AMA; fel Am Psychiat Asn; Am Soc Adlerian Psychol; Int Asn Individual Psychol (pres, 63-70); fel Asn Advan Psychother. Mailing Add: 30 E 60th St New York NY 10022

ADLER, LASZLO, b Debrecen, Hungary, Dec 3, 32; US citizen; m 63; c 2. PHYSICAL ACOUSTICS. Educ: Mich State Univ, MS, 61; Univ Tenn, Knoxville, PhD(physics), 69. Prof Exp: Assoc prof physics, Gen Motors Inst, Univ Mich, 60-64; instr, 66-69, res assoc, 69-71, ASST PROF PHYSICS, UNIV TENN, KNOXVILLE, 71- Concurrent Pos: Consult, Metal & Ceramic Div, Oak Ridge Nat Lab, Tenn, 69- Mem: Acoust Soc Am. Res: Non-linear mechanisms of ultrasonic waves in liquids, especially harmonic and fractional harmonics generation and correlation to liquid structure; defects in metals by interference of a multifrequency pulse echo; physics of non-destructive testing. Mailing Add: Dept of Physics Univ of Tenn Knoxville TN 37916

ADLER, MARTIN WILLIAM, b Philadelphia, Pa, Oct 30, 29; m 53; c 2. PHARMACOLOGY. Educ: NY Univ, BA, 49; Brooklyn Col Pharm, BS, 53; Columbia Univ, MS, 57; Albert Einstein Col Med, PhD(pharmacol), 60. Prof Exp: From instr to assoc prof, 60-73, PROF PHARMACOL, SCH MED, TEMPLE UNIV, 73- Concurrent Pos: NIMH grant, 61-69 & 70-; consult coun drugs, AMA, 62-63. Mem: AAAS; Soc Neurosci; Am Soc Pharmacol & Exp Therapeut; Am Col Neuropsychopharmacol; Int Asn Study Pain. Res: Neuropharmacology; psychopharmacology; effect of brain damage on responsiveness to drugs; relationship of sensory input to brain excitability; narcotic receptors; brain lesions and narcotic dependence. Mailing Add: Dept of Pharmacol Temple Univ Sch of Med Philadelphia PA 19140

ADLER, NORMAN, b Brooklyn, NY, June 19, 28. ANALYTICAL CHEMISTRY, PHYSICAL CHEMISTRY. Educ: Brooklyn Col, BS, 49; Polytech Inst Brooklyn, PhD(anal chem), 54. Prof Exp: Res analyst, Control Div, Merck & Co, Inc, 53 & 55, chief instrumental anal, 56-64; consult, Arthur D Little, Inc, 64-69; dir radiopharmaceut res & develop, New Eng Nuclear Corp, 70-71, div mgr, Radiopharm Div, 72-75; TECH DIR, CLIN ASSAYS, INC, 75- Mem: Am Chem Soc (pres anal sect, 71-72); Soc Nuclear Med. Res: Pharmaceutical and trace analysis; gas and liquid chromatography; isotope generators; diagnostic parenteral radiopharmaceuticals; in vitro radioactive clinical test kits. Mailing Add: Clin Assays Inc 237 Binney St Cambridge MA 02142

ADLER, RICHARD, b Buffalo, NY, July 2, 22; m 70; c 5. SURGERY. Educ: Univ Buffalo, MD, 45; Univ Colo, MS, 55; Am Bd Surg, cert, 51; Bd Thoracic Surg, cert, 54. Prof Exp: Asst physiol, Sch Med, Univ Buffalo, 49; instr surg, Med Sch, Univ Mich, 50-52; hon registr, Brompton Hosp, London, Eng, 52-53; from asst prof to assoc prof, 55-67, PROF SURG, SCH MED, STATE UNIV NY BUFFALO, 67-; DIR THORACIC SURG & ATTEND SURGEON, BUFFALO GEN HOSP, 57- Concurrent Pos: Markle scholar, 57-; attend thoracic surgeon, Vet Hosp, Buffalo, 55-; asst attend surgeon, Buffalo Children's Hosp, 56-; consult, Roswell Park Mem Inst, 57-; vis prof surg, Nat Defense Med Ctr, Taipei, Taiwan, 68-69; assoc consult, E J Meyer Mem Hosp; consult thoracic surg, Millard Fillmore, Brooks Mem & Tri-County Hosps. Mem: AAAS; Am Col Surg; Am Asn Thoracic Surg; Soc Univ Surg; Am Col Chest Physicians. Res: General and thoracic surgery. Mailing Add: Univ Cardiothoracic Res Prog State Univ of NY Sch of Med Buffalo NY 14214

ADLER, ROBERT, b Vienna, Austria, Dec 4, 13; nat US; m 46. PHYSICS, ELECTRONICS. Educ: Univ Vienna, PhD(physics), 37. Prof Exp: Asst to patent attorney, Vienna, Austria, 37-38; in chg lab, Sci Acoust, Ltd, London, 39-40 & Assoc Res, Inc, Chicago, 40-41; res engr, 41-52, assoc dir res, 52-63, VPRES RES, ZENITH RADIO CORP, 63- Mem: Nat Acad Eng; AAAS; fel Inst Elec & Electronics Engrs. Res: Electron beam parametric amplifiers; opto-acoustic interaction devices; surface wave amplifiers; video disc recording and playback. Mailing Add: Zenith Radio Corp 6001 W Dickens Ave Chicago IL 60639

ADLER, ROBERT FREDERICK, b West Reading, Pa, Jan 19, 44; m 72. METEOROLOGY. Educ: Pa State Univ, BS, 65, MS, 67; Colo State Univ, PhD(atmospheric sci), 74. Prof Exp: Res meteorologist, Navy Weather Res Fac, Dept of Defense, 67-71; METEOROLOGIST, GODDARD SPACE FLIGHT CTR, NASA, 74- Concurrent Pos: Res assoc meteorol, Colo State Univ, 74. Mem: Am Meteorol Soc. Res: Application of meteorological satellite data to detect and predict severe storms. Mailing Add: Code 911 Goddard Space Flight Ctr Greenbelt MD 20771

ADLER, ROBERT GARBER, b Upland, Calif, July 11, 29; m 63; c 3. ANALYTICAL CHEMISTRY. Educ: Calif Inst Technol, BSc, 51; Univ Southern Calif, MSc, 55; Univ Calif, Riverside, PhD(inorg chem), 68. Prof Exp: Sr chemist, Whittier Res Lab, Am Potash & Chem Corp, 57-63; chemist, Sci Ctr, NAm Aviation, Inc, Calif, 63-64; asst prof chem, Bethel Col, Kans, 68-71; QUAL CONTROL COORDR, OCCUP SAFETY & HEALTH ADMIN, US DEPT LABOR, UTAH, 72- Mem: Am Chem Soc. Res: Boron hydrides, especially preparations, properties and improvement of analytical chemical techniques. Mailing Add: Occup Safety & Health Admin 390 Wakara Way Salt Lake City UT 84108

ADLER, RONALD JOHN, b Pittsburgh, Pa, Apr 17, 37. THEORETICAL PHYSICS. Educ: Carnegie Inst Technol, BS, 59; Stanford Univ, PhD(physics), 65. Prof Exp: Res asst prof physics, Univ Wash, 64-66; res assoc, Univ Colo, 66-67; asst prof, Va Polytech Inst, 67-70; assoc prof, Am Univ, 70-73; PROF PHYSICS, FED UNIV PERNAMBUCO, BRAZIL, 73- Mem: Am Phys Soc; fel Brit Interplanetary Soc. Res: High energy theory; two nucleon problem; general relativity. Mailing Add: Dept of Physics Univ Fed de Pernambuco Recife Brazil

ADLER, ROY LEE, b Newark, NJ, Feb 22, 31; m 53; c 2. MATHEMATICS. Educ: Yale Univ, BS, 52; Columbia Univ, AM, 54; Yale Univ, PhD(math), 61. Prof Exp: Jr engr, Nat Union Radio, 52 & Bendix Aviation Corp, 53-54; RES STAFF MEM MATH, INT BUS MACH CORP, 60- Concurrent Pos: Adj prof, Columbia Univ, 63-65 & Yeshiva Univ, 65-66; vis prof, Stanford Univ, 70-71. Mem: Am Math Soc; Math Asn Am; Sigma Xi. Res: Ergodic theory; classification and structure problem of measure preserving transformations. Mailing Add: IBM Corp Watson Res Ctr Box 218 Yorktown Heights NY 10598

ADLER, SEYMOUR JACOB, b New York, NY, May 12, 18; m 47; c 3. ANALYTICAL CHEMISTRY. Educ: Cooper Union, BChE, 41; Columbia Univ, MA, 50, PhD(chem), 54. Prof Exp: Anal chemist, Columbia Mineral Beneficiation Lab, 53-56, Brookhaven Nat Lab, 56-57 & Radio Corp Am, NJ, 57-69; ASST PROF CHEM, TRENTON STATE COL, 70- Mem: AAAS; Am Chem Soc; The Chem Soc. Res: Trace analysis; spectrophotometry; radio-chemistry; emission spectrography; solvent extraction; ion exchange. Mailing Add: Dept of Chem Trenton State Col Trenton NJ 08625

ADLER, SOLOMON STANLEY, b New York, NY, May 26, 45; m 67; c 3. HEMATOLOGY, ONCOLOGY. Educ: City Col New York, BS, 66; Albert Einstein Col Med, MD, 70. Prof Exp: Instr med, 73-75, ASST PROF MED, RUSH MED COL, 75-, CHIEF, SPECIAL HEMAT LAB, RUSH-PRESBYTERIAN-ST LUKE'S MED CTR, 74-, ASST ATTENDING PHYSICIAN, PRESBYTERIAN-ST LUKE'S HOSP, 75- Concurrent Pos: Adj attending physician, Presbyterian-St Luke's Hosp, 74-75; NIH res fel, Nat Cancer Inst, 74-77. Res: Hemopoietic stem cell proliferation, differentiation and immunological characteristics; myeloproliferative disorders, especially the animal models of myelofibrosis and the myelogenous leukemia of the RFM mouse; extramedullary hemopoiesis, benign and malignant. Mailing Add: 1753 W Congress Pkwy Chicago IL 60612

ADLER, STEPHEN FRED, b Berlin, Ger, Sept 27, 30; nat US; m 52; c 2. INDUSTRIAL CHEMISTRY. Educ: Roosevelt Univ, BS, 51; Northwestern Univ, MS, 53, PhD(inorg chem), 54. Prof Exp: Res chemist, Am Cyanamid Co, 54-60, group leader, 60-69; sect mgr, 68-70, MGR CHEM RES DEPT, STAUFFER CHEM CO, 70- Mem: Am Chem Soc. Res: Product development of laundry, dry cleaning and food ingredients; dentifrice polishing agents; industrial catalysis. Mailing Add: Chem Res Dept Stauffer Chem Co Livingston St Dobbs Ferry NY 10522

ADLER, STEPHEN L, b New York, NY, Nov 30, 39; m 62; c 3. THEORETICAL HIGH ENERGY PHYSICS. Educ: Harvard Univ, AB, 61; Princeton Univ, PhD(physics), 64. Prof Exp: Jr fel, Soc Fels, Harvard Univ, 64-66; mem, 66-69, PROF THEORET PHYSICS, INST ADVAN STUDY, PRINCETON, 69- Mem: Nat Acad Sci; fel Am Phys Soc; fel Am Acad Arts & Sci. Res: Theoretical research on problems in elementary particle physics. Mailing Add: 9 Veblen Circle Princeton NJ 08540

ADLER, STEPHEN MILLER, b Brooklyn, NY, June 6, 35; m 59; c 2. ASTRONOMY. Educ: Univ Mich, BS, 57, MS, 59, PhD(astron), 66. Prof Exp: From instr to asst prof astron, Mt Holyoke Col, 64-70; asst prof, 70-74, ASSOC PROF ASTRON, WELLESLEY COL, 74- Mem: AAAS; Am Astron Soc. Res: Solar physics; sunspots; velocity fields in sunspots; correlations between changes in sunspot structure and the occurrence of solar flares. Mailing Add: Whitin Observ Wellesley Col Wellesley MA 02181

ADLER, VICTOR EUGENE, b New York, NY, Jan 18, 24; m 65; c 2. ENTOMOLOGY. Educ: Memphis State Univ, BS, 50; Kans State Univ, MS, 56. Prof Exp: Biol aide plant pest control, 56-57, entomologist, 57-61, res entomologist, Grain & Forage Insects Res Br, Agr Res Ctr, 61-65, Pesticide Chem Res Br, 65-66 & Pesticide Chem Res Div, 66-72, RES ENTOMOLOGIST, AGR ENVIRON QUAL INST, BIOL ACTIVE NATURAL PROD LAB, AGR RES CTR, USDA, 72- Mem: Entom Soc Am. Res: Safer chemical and practical applications of insecticide for controlling insect pests; study of insect attractant and repellents electrophysiologically. Mailing Add: Agr Environ Qual Inst Rm 108 Bldg 476 Agr Res Ctr East Beltsville MD 20705

ADLERZ, WARREN CLIFFORD, b Worcester, Mass, Apr 5, 28; m 52; c 1. ENTOMOLOGY. Educ: Univ Mass, BS, 53; Ore State Col, MS, 55, PhD(entom), 58. Prof Exp: Asst entom, 58-66, assoc entomologist, 66-74, ENTOMOLOGIST WATERMELON & GRAPE INVESTS LAB, AGR RES CTR, UNIV FLA, 74- Mem: Entom Soc Am. Res: Insect pollination; insect transmission of plant diseases. Mailing Add: Inst Food & Agr Sci Agr Res Ctr Box 388 Univ of Fla Leesburg FL 32748

ADMAN, ELINOR THOMSON, b New York, NY, Jan 3, 41; c 2. BIOLOGICAL STRUCTURE. Educ: Col Wooster, BA, 62; Brandeis Univ, MA, 64, PhD(phys chem), 67. Prof Exp: Sr fel crystallog, 67-71, RES ASSOC CRYSTALLOG, DEPT BIOL STRUCT, UNIV WASH, 71- Mem: AAAS; Am Crystallog Asn. Res: Determination and refinement of protein structures, iron-sulfur proteins. Mailing Add: Dept of Biol Struct SM-20 Univ of Wash Seattle WA 98195

ADMAN, RAYMOND LANCE, b Columbus, Ohio, Apr 24, 40; m 63; c 2. BIOCHEMISTRY. Educ: Ohio Univ, BS, 61; Brandeis Univ, PhD(biochem), 66; Miami Univ, MD, 73. Prof Exp: NIH fel, Dept Chem, Harvard Univ, 66-67; NIH fel, Dept Pediat, Univ Wash, 67-69, fel, Dept Genetics, 69-71; Am Cancer Soc scholar, Univ Miami, 72-73; intern, Dept Med, 73-74, RESIDENT, DEPT LAB MED, UNIV WASH, 74- Res: Endocrinology. Mailing Add: Dept of Lab Med Univ of Wash Seattle WA 98195

ADNEY, JOSEPH ELLIOTT, JR, b DeLand, Fla, Aug 20, 23; m 52; c 2. MATHEMATICS. Educ: Stetson Univ, BS, 44; Ohio State Univ, MA, 49, PhD(math), 54. Prof Exp: Assoc, Res Found, Ohio State Univ, 52-54, instr math, 54-55; asst prof, Purdue Univ, 55-64; assoc prof, 64-69, PROF MATH, MICH STATE UNIV, 69-, CHMN DEPT, 74- Concurrent Pos: Consult, US Air Force, 57, 59. Mem: Am Math Soc. Res: Abstract algebra; ground theory. Mailing Add: Dept of Math Mich State Univ East Lansing MI 48823

ADOLPH, ALAN ROBERT, b New York, NY, Feb 5, 32; m 57; c 3. NEUROPHYSIOLOGY. Educ: Rensselaer Polytech Inst, BEE, 53; Mass Inst Technol, SM, 57; Rockefeller Inst, PhD(biol sci), 63. Prof Exp: Sr scientist, Bolt, Beranek & Newman, Inc, 63-64; HEAD NEUROSCI LAB, EYE RES INST OF RETINA FOUND, 64- Concurrent Pos: Consult, Lockheed Missile & Space Co Labs, 58-59 & Stanford Res Inst, 62. Mem: Am Physiol Soc; Soc Neurosci; Asn Res Vision & Ophthal. Res: Neurophysiology and pharmacology of the retina; electrophysiology of sensory neurons; marine neurobiology; mathematical biophysics; visual behavior; psychophysiology of vision. Mailing Add: Eye Res Inst Retina Found Neurosci Lab 20 Staniford St Boston MA 02114

ADOLPH, HORST GUENTER, b Pforzheim, Ger, Nov 27, 32; m 60; c 2. ORGANIC CHEMISTRY. Educ: Univ Tübingen, Dr rer nat, 59. Prof Exp: Res asst, Inst Appl Chem & asst to ed, Houben-Weyl, Univ Tübingen, 61; RES CHEMIST, US NAVAL SURFACE WEAPONS CTR, 61- Mem: Am Chem Soc. Res: Synthetic organic chemistry; reaction mechanisms; spectroscopy; chemistry and physical chemistry of explosives and propellants. Mailing Add: 13008 Ivy Dr Beltsville MD 20705

ADOLPH, ROBERT J, b Chicago, Ill, May 12, 27; m 58; c 3. INTERNAL MEDICINE. Educ: Univ Ill, MD, 52; Am Bd Internal Med, cert, 63; Am Bd Cardiovasc Dis, dipl. Prof Exp: Asst prof med, Univ Ill, 59-60; from asst prof to assoc prof, 62-70, PROF MED, COL MED, UNIV CINCINNATI, 70- Concurrent Pos: Am Heart Asn res fel, Col Med, Univ Ill, 56-58; NIH res fel physiol & biophys, Sch Med, Univ Wash, 60-62; mem staff, Dept Internal Med, Cardiac Res Lab, Cincinnati Gen Hosp, 62-, dir lab, 70-; mem med adv bd, Sect on Circulation & Coun Clin Cardiol, Am Heart Asn. Mem: Am Fedn Clin Res; Am Col Cardiol. Res: Cardiovascular physiology and investigation; myocardial contractility; effect of digitalis on myocardial electrolytes; high out-put states; cardiac pacing; biomedical engineering developments. Mailing Add: Cardiac Res Lab H3 Cincinnati Gen Hosp Cincinnati OH 45229

ADOMAITIS, VYTAUTAS ALBIN, b Worcester, Mass, Apr 6, 26; m 56; c 3. WATER CHEMISTRY. Educ: Holy Cross Col, BS, 47, MS, 48. Prof Exp: Res chemist, Sherwin-Williams Co, Chicago, Ill, 48-49; asst prof, NEng Col Pharm, 49-51; exploitation engr, Shell Oil Co, Calif, 53-56; develop chemist, Norda Essential Oil & Chem Co, NJ, 56-57; RES CHEMIST, US FISH & WILDLIFE SERV, 57-, RES CHEMIST, NORTHERN PRAIRIE WILDLIFE RES CTR, 65- Concurrent Pos: Mem nat tech adv comt on water qual, Secy of Interior. Mem: Ecol Soc Am; Am Chem Soc; Am Water Resources Asn. Res: Organic pigments; effect of chemicals on geological strata; evaluation of rodent repellents; biochemistry of pesticides in wildlife; chemical literature of Lithuania; history and philosophy of science; ecological and environmental chemistry. Mailing Add: Chem Lab US Fish & Wildlife Serv North Prairie Wildlife Res Ctr Jamestown ND 58401

ADOMIAN, GEORGE, b Buffalo, NY, Mar 21, 22; m 56; c 2. APPLIED MATHEMATICS, SYSTEMS THEORY. Educ: Univ Mich, BS, 44, MS, 48; Univ Calif, Los Angeles, PhD(theoret physics), 63. Prof Exp: Res engr, Gen Elec Co, 48-49; instr math, Wayne State Univ, 51-52; res assoc, Univ Mich, 52-53; sr scientist, Hughes Aircraft Co, 53-64; prof math & res prof eng, Pa State Univ, 64-66; BARROW DISTINGUISHED PROF MATH, UNIV GA, 66- Concurrent Pos: Consult, Nat Acad Sci, 71- Mem: Fel AAAS; Am Phys Soc; Am Math Soc; Inst Elec & Electronics Engrs; Soc Indust & Appl Math. Res: Stochastic differential equations and stochastic operator theory; theoretical physics applications. Mailing Add: Dept of Math Univ of Ga Athens GA 30602

ADORNO, DAVID SAMUEL, b New Britain, Conn, June 17, 27; m 49; c 4. MATHEMATICS, STATISTICS. Educ: Univ Tex, BA, 53; Pa State Univ, MA, 55; Harvard Univ, PhD(statist), 63. Prof Exp: Sr res mathematician, Jet Propulsion Labs, Calif Inst Technol, 59-62; res staff mem, Int Bus Mach Corp, 62-64; assoc prof opers res, NY Univ, 64-65; Fulbright prof statist, Madrid, 65-66; assoc prof math, Univ Hartford, 66-69; chmn dept, 69-71, PROF MATH, ITHACA COL, 69- Concurrent Pos: Foreign mem, Spain's Higher Res Coun, Madrid & Opers Res Inst Spain, 66- Mem: Math Asn Am; Am Statist Asn. Res: Mathematical and statistical theory of control systems; applied spectral analysis. Mailing Add: Dept of Math Ithaca College Ithaca NY 14850

ADOVASIO, JAMES MICHAEL, b Youngstown, Ohio, Feb 17, 44; m 70. ANTHROPOLOGY. Educ: Univ Ariz, BA, 65; Univ Utah, PhD(anthrop), 70. Prof Exp: From instr to asst prof anthrop, Youngstown State Univ, 66-71; res assoc, Smithsonian Inst, 71-72; ASST PROF ANTHROP, UNIV PITTSBURGH, 72-, DIR ARCHAEOL RES PROG, 74- Concurrent Pos: Smithsonian Inst res fel, 71-72; Nat Geog Soc res grant, Cyprus, 72; res assoc, Smithsonian Inst, 74-; NSF grant, 75- Res: Primitive technology, including lithics, basketry and textiles; post Pleistocene adaptation to arid lands, particularly Great Basin, American Southwest, Texas, northern Mexico and Cyprus; early man in North America. Mailing Add: Dept of Anthrop Univ of Pittsburgh Pittsburgh PA 15260

ADRIAN, ALAN PATRICK, b Kaukauna, Wis, Feb 13, 16; m 42; c 2. CHEMISTRY. Educ: Lawrence Col, AB, 38, MS, 40, PhD(paper chem & technol), 42. Prof Exp: Develop chemist, Detroit Sulphite Pulp & Paper Co, 45-51, oper mgr, 51-58; chief bond papers & allied prods, Res & Develop, Kimberly Clark Corp, 58-60, dir bus paper mfg, 60-62, prod mgr, Neenah Paper Div, 62-70, OPERS & DEVELOP MGR, NEENAH PAPER DIV, KIMBERLY CLARK CORP, 70- Mem: Am Chem Soc; Am Tech Asn Pulp & Paper Indust. Res: Chemistry and technology of pulp and paper; evaluating the optical constants of pigments in paper; resin application army and chart papers. Mailing Add: Neenah Paper Div Kimberly Clark Corp Neenah WI 54956

ADRIAN, ERLE KEYS, JR, b Temple, Tex, Apr 18, 36; m 62; c 3. ANATOMY. Educ: Rice Inst, BA, 58; Univ Tex, Galveston, MA, 61, PhD(anat), 67; Harvard Med Sch, MD, 63. Prof Exp: Res asst anat, Med Br, Univ Tex, Galveston, 59-60, from instr to assoc prof, 63-69; assoc prof, 69-74, PROF ANAT, UNIV TEX HEALTH SCI CTR SAN ANTONIO, 74- Mem: Am Asn Anat. Res: Neuropathology; gerontology. Mailing Add: Dept of Anat Univ of Tex Health Sci Ctr San Antonio TX 78284

ADRIAN, FRANK JOHN, b Brooklyn, NY, Oct 7, 29; m 69; c 2. PHYSICAL CHEMISTRY. Educ: Cath Univ Am, AB, 51; Cornell Univ, PhD(phys chem), 55. Prof Exp: PRIN PROF STAFF CHEMIST, APPL PHYSICS LAB, JOHNS HOPKINS UNIV, 55- Concurrent Pos: Assoc ed, J Chem Physics, 75-78. Mem: Fel Am Phys Soc. Res: Magnetic and nuclear quadrupole resonance; structure of molecules, free radicals and paramagnetic imperfections in solids; photochemistry; chemically induced magnetic polarization. Mailing Add: Appl Physics Lab Johns Hopkins Univ Laurel MD 20810

ADRIANI, JOHN, b Bridgeport, Conn, Dec 2, 08; m 36, 53; c 1. SURGERY, ANESTHESIOLOGY. Educ: Columbia Univ, AB, 30, MD, 34; Am Bd Anesthesiol, dipl, 40. Prof Exp: Intern, French Hosp, New York, 34-36; resident, Bellevue Hosp, 36-37; instr surg, NY Univ, 39-41; dir blood bank, Charity Hosp, New Orleans, 41-70, asst dir hosp, 60-63; asst prof, 47-51, PROF SURG & ANESTHESIOL, TULANE UNIV, 51-; DIR DEPT ANESTHESIOL, CHARITY HOSP, 41-, ASSOC DIR HOSP, 63- Concurrent Pos: Fel surg, NY Univ, 37-39, physiol, 38-39; from asst prof to assoc prof anesthesiol, Univ New Orleans, 41-55, clin prof surg & prof pharmacol, 55-, clin prof oral surg, 71-; asst prof, Loyola Univ, La, 45-57, prof 57-71; consult, Vet Admin Hosp, 49-, USPHS Hosp, Hotel Dieu, Ochsner Found Hosp & US Food & Drug Admin; ed, Am Lect Series Anesthesiol; mem subcomt anesthesia & comt surg, Nat Res Coun; mem adv comt, Div Invest Drugs, Food & Drug Admin; mem rev comt, US Pharmacopoeia, 50-70, 74-; mem bd dirs, Am Bd Anesthesiol, 60-, pres, 68-69, chmn exam comt; mem coun on drugs, AMA, 64-, chmn, 68-71. Honors & Awards: Award, Am Soc Anesthesiol, 49; Distinguished Serv Award, Int Anesthesia Res Soc, 57, 72; Ralph M Waters Medal & Award Anesthesiol, 68. Mem: Soc Exp Biol & Med; fel Am Col Anesthesiol; AMA; Am Heart Asn. Res: Pharmacology. Mailing Add: Charity Hosp 1532 Tulane Ave New Orleans LA 70140

ADROUNIE, V HARRY, b Battle Creek, Mich, Apr 29, 15; m 43; c 2. ENVIRONMENTAL MANAGEMENT. Educ: St Ambrose Col, BS, 40, BA, 59; Air Univ, dipl mil mgt, 57; Am Bd Indust Hyg, dipl, 62; Am Int-Soc Acad, cert sanit dipl, 67. Prof Exp: Chief labs & asst chief prev med, Med Serv Corps, US Air Force, 50-52, chief planning & test br, Field Test & Meteorol Div, 52-53, mem staff biol & chem spec weapons br & bioenviron specialist, Off Inspector Gen, 53-56, environ med officer, Off Surgeon Gen, 57-61, comdr, Detachment 10, 1st Aeromed Transport Group, 61-63; vis lectr environ health, Sch Pub Health, Am Univ Beirut, 63-64, vis assoc prof & actg chmn dept, 64-66; dep comdr, 1st Aeromed Evacuation Group, Pope AFB, 66-68; tech dir, ARA Environ Serv, 68-70; dir, Div Environ Sanit, Chester County Health Dept, PA, 70-75; ENVIRON CONSULT, 75- Concurrent Pos: US Air Force environ health rep, Nat Acad Sci, 57-61; mem, Nat Cong Environ Health, 59-60; US Air Force rep, US Interdept Comt Nutrit for Nat Defense, 59-61; mem, President's Coun Youth Fitness, 60; Nat Environ Health Asn del, Nat Health Coun, 61; consult health mobilization prog, USPHS, 61-62 & UN Relief & Works Agency Educ Div, 64-66; chmn, Bd Registr Sanitarians in Pa, 70-; mem bd dirs, Chester Co Pa Water Resources Auth, 70-, chmn, 74-; mem, Chester Co Bd Health, 75- Honors & Awards: Cert Achievement, USArmy, 50; Commendation Medal, Dept of Air Force, 57, Oak Leaf, 68, Legion of Merit, 67; Mangold Award, 63. Mem: Environ Health Asn; fel Am Pub Health Asn; fel Royal Soc Health; NY Acad Sci; Am Acad Health Admin. Res: Environmental health through the media of research, teaching and direction; beneficial control of the environment. Mailing Add: Reeds Rd R D 2 Downingtown PA 19335

ADROUNY, GEORGE ADOUR (KUYUMJIAN), b Kilis, Turkey, Apr 1, 16; US citizen; m 45; c 3. BIOCHEMISTRY. Educ: Am Univ Beirut, BA, 34, PhC, 40; Emory Univ, PhD(biochem), 54. Prof Exp: Instr & res assoc biochem, Emory Univ, 54-57; from asst prof to assoc prof, 57-72, PROF BIOCHEM, SCH MED, TULANE UNIV, 72- Mem: AAAS; Am Chem Soc. Res: Isolation and characterization of fire ant venom; effects of nutritional and hormonal factors on the carbohydrate metabolism of cardiac and skeletal muscles; the glycogen storing effect of somatotrophin. Mailing Add: Dept of Biochem Tulane Univ Sch of Med New Orleans LA 70112

ADUSS, HOWARD, b Brooklyn, NY, Mar 31, 32; m 53; c 4. ORTHODONTICS. Educ: Purdue Univ, BS, 54; Northwestern Univ, DDS, 57; Univ Rochester, MS, 62, Eastman Dent Ctr, cert orthod, 62. Prof Exp: PROF ORTHOD, CTR FOR CRANIOFACIAL ANOMALIES & DEPT ORTHOD, UNIV ILL, 66- Concurrent Pos: Prof orthod, Rush Univ. Mem: Am Dent Asn; Am Asn Orthod; Am Cleft Palate Asn (pres, 75-76); fel Am Col Dent; Int Asn Dent Res. Res: Craniofacial malformations, particularly cleft lip and palate; craniofacial growth and development in syndromes. Mailing Add: 237 Lakeside Pl Highland Park IL 60035

ADVANI, SHYAM BHOJRAJ, b Karachi, Pakistan, Aug 16, 35; US citizen; m 63; c 5. MEDICINAL CHEMISTRY, AGRICULTURAL CHEMISTRY. Educ: Univ Bombay, MS, 59; Univ Utah, BS, 62; Univ Miss, PhD(med chem), 67. Prof Exp: Teaching asst med chem & biochem, Univ Utah, 60-62; fel med chem, Univ Miss, 67; RES CHEMIST, PENNWALT CORP, 67- Mem: Am Chem Soc. Res: Cardiovascular medicine; encapsulation and synthesis of various medicinal and pesticidal agents. Mailing Add: Pennwalt Corp 900 First Ave King of Prussia PA 19406

AEGERTER, ERNEST E, b Randolph, Nebr, Jan 4, 06. MEDICINE. Educ: Yankton Col, AB, 28; Univ SDak, BS, 30; Univ Pa, MD, 32. Prof Exp: Prof, 45-70, EMER PROF PATH, SCH MED, TEMPLE UNIV, 72- Mem: Am Soc Clin Path; Am Asn Path & Bact; fel Col Am Path; hon mem Am Acad Orthop Surg. Res: Orthopedic pathology. Mailing Add: RD 1 Chalfont PA 18914

AEPLI, OTTO THEODORE, b Bryn Mawr, Pa, July 20, 12; m 41; c 1. RESEARCH ADMINISTRATION. Educ: Univ Pa, BS, 38; Temple Univ, AM, 46. Prof Exp: Anal chemist, Graselli Chem Co, Philadelphia, 31-38; res chemist, Attapulgus Clay Co, 38-44; res chemist, Pennsalt Chem Corp, Mich, 44-50, chief chemist, Wyandotte Plant, 50-57, head anal sect, 57-60; asst tech dir, Magnus Prod Corp, 60-63; res assoc, 63-64, RES SUPVR SPECIALTY PROD, BASF-WYANDOTTE CORP, 64- Concurrent

Pos: Lab instr, Temple Univ, 46-47. Mem: Am Chem Soc; Nat Environ Health Asn; Am Soc Brewing Chem; Inst Food Technologists. Res: Specialty chemicals; analytical and physical chemical properties of inorganic fluorine compounds; catalysis and adsorption; analysis and manufacture of insecticides; redox indicator silicate chemistry; spectrographic analysis. Mailing Add: 13329 Veronica Dr Southgate MI 48195

AEPPLI, ALFRED, b Zurich, Switz, Nov 8, 28; nat US; m 55; c 2. MATHEMATICS. Educ: Swiss Fed Inst Technol, dipl, 51, Dr math, 56. Prof Exp: Asst math, Swiss Fed Inst Technol, 52-57; from instr to asst prof, Cornell Univ, 57-62; assoc prof, 62-66, PROF MATH, UNIV MINN, MINNEAPOLIS, 66- Concurrent Pos: Guest prof, Univ Heidelberg, 74-75. Mem: Math Asn Am; Am Math Soc; Swiss Math Soc; Soc Indust & Appl Math. Res: Topology; differential geometry; complex variables. Mailing Add: Sch of Math Univ of Minn Minneapolis MN 55455

AFANADOR, ARTHUR JOSEPH, b Tampa, Fla, Nov 2, 42; m 65; c 2. OPTOMETRY, PHYSIOLOGICAL OPTICS. Educ: Southern Col Optom, BS & OD, 65; Univ Calif, Berkeley, PhD(physiol optics), 72. Prof Exp: ASST PROF OPTOM, IND UNIV, BLOOMINGTON, 72- Mem: Fel Am Acad Optom. Res: Neurophysiology of the retina; single cell recording of goldfish ganglion cells. Mailing Add: Div of Optom Ind Univ Bloomington IN 47401

AFFENS, WILBUR ALLEN, b New York, NY, Jan 9, 18; m 45; c 2. CHEMISTRY. Educ: City Col New York, BS, 37; Polytech Inst Brooklyn, MS, 46; Georgetown Univ, PhD(phys chem), 60. Prof Exp: Chemist, Rubatex Prod Inc, NY, 37-38; chemist, Maltbie Chem Co, NJ, 38-42; asst chemist, Corps Engrs, US War Dept, 42-45; assoc chemist, USDA, 45-53, phys chemist, Entom Res Br, Agr Res Serv, 53-56; RES CHEMIST, US NAVAL RES LAB, 56- Mem: Am Chem Soc; Combustion Inst; Am Soc Testing & Mat. Res: Hydrocarbon chemistry; combustion; ignition; flammability; petroleum; fuels. Mailing Add: 8302 26th Pl Adelphi MD 20783

AFFRONTI, LEWIS FRANCIS, b Rochester, NY, Aug 12, 28; m 56; c 4. MICROBIOLOGY, IMMUNOLOGY. Educ: Univ Buffalo, BA, 50, MA, 51; Duke Univ, PhD(microbiol), 58. Prof Exp: Asst, Univ Buffalo, 50-51; res assoc, Vet Admin Hosp, 51-52; med entomologist, US Air Force, 52-54; res assoc, Roswell Park Mem Inst, 54; dir vaccine lab, Duke Univ Hosp, 56; res assoc tuberc, Henry Phipps Inst, Univ Pa, 57-58; from sr asst scientist to scientist, USPHS, 58-62; from asst prof to assoc prof, 62-72, PROF MICROBIOL, SCH MED, GEORGE WASHINGTON UNIV, 72-, CHMN DEPT, 73- Concurrent Pos: Spec consult, Vet Admin Hosp, Wilmington, Del, 66-; consult, Vet Admin Ctr, Martinsburg, WVa; US rep, Int Meeting Stand & Uses of Tuberculins, WHO, Geneva, Switz, 66; USPHS spec res fel, Inst Superiore Sanita, Italy, 69-70; US-Japanese res in tuberc grant, 69-; WHO Exchange of Res Workers fel, Univ Goteborg, 70; Nat Tuberc Asn travel fel, 71; mem med adv bd, Wilmington VA Hosp Ctr, 72- Mem: Am Pub Health Asn; Nat Tuberc Asn; Am Soc Microbiol; Soc Exp Biol & Med; Am Asn Immunol. Res: Antigens of mycobacteria; immunochemistry; bacterial physiology; isolation and characterization of bacterial tumor isolates; cellular mechanisms in delayed allergy. Mailing Add: Dept of Microbiol George Washington Univ Washington DC 20006

AFGHAN, BADERUDDIN KHAN, b Shikarpur, Pakistan, Dec 12, 40; Can citizen; m 69; c 2. ANALYTICAL CHEMISTRY, ENVIRONMENTAL CHEMISTRY. Educ: Sind Univ, Pakistan, BSc, 62, Hons, 63; Univ London, DIC, 64, PhD(anal chem), 66. Prof Exp: Fel org reagents for fluorometric analysis, Dalhousie Univ, 66-68; res assoc solution chem, Univ Montreal, 68-69; res scientist anal methods develop, Dept Energy Mines & Resources, Fed Govt, Ottawa, Ont, 69-72; res scientist environ anal chem, 72-75, RES MGR ANAL METHODS RES, CAN CENTRE FOR INLAND WATERS, DEPT ENVIRON, 75- Mem: Fel Chem Inst Can; Spectros Soc Can; Am Soc Testing & Mat. Res: Modern polarographic and related electroanalytical techniques; high speed liquid chromatography; atomic and molecular absorption as well as fluorescence spectroscopy; trace analysis; environmental analytical chemistry. Mailing Add: Anal Methods Res Sect Can Centre for Inland Waters PO Box 5050 867 Lakeshore Rd Burlington ON Can

AFGHANI, HISHAM T, b Jerusalem, Israel, Feb 12, 34; US citizen; m 61; c 1. ANALYTICAL CHEMISTRY. Educ: Ottawa Univ, BSc, 58; Univ Mo-Columbia, MSc, 62. Prof Exp: Asst chemist, Midwest Res Inst, 58-59; chemist, Thompson-Hayward Chem Co, 59-60 & Paniplus Div, Continental Baking Co, 60-61; chemist, Luzier, Inc, 63-65, dir res & develop, 65-68, vpres res & develop, 68-74; VPRES RES & DEVELOP, BRISTOL-MYERS CO, KANSAS CITY, 74- Mem: AAAS; Am Chem Soc; Am Oil Chem Soc; Soc Cosmetic Chem; fel Am Inst Chemists. Res: Development of cosmetics. Mailing Add: Bristol-Myers Co 4007 Pennsylvania Kansas City MO 64111

AFONSKY, DIMITRI ALEKSANDROVICH, b Ufa, Russia, Mar 30, 10; wid; c 1. PATHOLOGY. Educ: Harbin Dent Sch, China, LDS, 32; West China Union Univ, DDS, 47; Univ Rochester, PhD(path), 54. Prof Exp: Instr oral path & surg, Harbin Dent Sch, China, 34-37; head dept dent, Dent Clin, Yale-in-China Med Col & Hosp, 37-39, lectr, 37-43, asst prof dent, 47-49; res assoc, Eastman Dent Clin, Rochester, 54-55; in charge div dent med, Colgate Palmolive Co, 55-58; asst prof dent & dent res, Univ Rochester, 58-61; PROF DENT PATH, SCH DENT MED, TUFTS UNIV, 61-, RES PROF ORAL PATH, 68- Concurrent Pos: Secy med sect, Int Red Cross Ctr, China, 40-47; assoc res specialist, Bur Biol Res, Rutgers Univ, 57; res assoc, Sch of Dent, Univ Ala, 58; consult, Off Naval Res, 59. Honors & Awards: Cert of Merit, Am Bd of Dent Med, 53. Mem: AAAS; Int Asn Dent Res; NY Acad Sci. Res: Oral pathology. Mailing Add: Dept of Oral Path Tufts Univ Sch of Dent Med Boston MA 02155

AFONSO, ADRIANO, b Goa, India, Mar 5, 35; m 66; c 1. ORGANIC CHEMISTRY. Educ: Univ Bombay, MS, 57; Univ Wis, PhD(pharmaceut chem), 61. Prof Exp: Res asst, Univ Wis, 58; res assoc, Dept Chem, Ind Univ, Bloomington, 61-63; from chemist to sr chemist, 63-70, PRIN CHEMIST, NATURAL PROD DEPT, SCHERING CORP, 70- Honors & Awards: Lunsford Richardson Award, Schering Corp, 60. Mem: Am Chem Soc. Res: Synthetic organic chemistry; structural studies on veratrum alkaloids; syntheses of diterpene resin acids; chemical modifications of steroids for biological activity; peptide structure elucidations. Mailing Add: 10 Woodmere Rd West Caldwell NJ 07006

AFT, HARVEY, b Chicago, Ill, June 17, 29; m 53; c 4. ORGANIC CHEMISTRY. Educ: Univ Southern Calif, AB, 50; Univ Puget Sound, MS, 52; Ore State Univ, PhD(chem), 62. Prof Exp: Res chemist, Forest Prod Res Lab, 57-61; asst prof forest prod chem, Forest Res Lab, Ore State Univ, 61-65; asst prof chem, Mich Technol Univ, 65-66; assoc prof, Ore State Univ, 66-69; assoc prof, 69-72, PROF CHEM & CHMN DEPT, UNIV MAINE, FARMINGTON, 72- Mem: AAAS; Am Chem Soc; Brit Chem Soc. Res: Chemistry of phenolic and oxygen containing heterocyclic compounds and natural products chemistry. Mailing Add: Dept of Chem Univ Maine Farmington ME 04938

AFTANDILIAN, VICTOR DANIEL, inorganic chemistry, see 12th edition

AFTERGOOD, LILLA, b Krakow, Poland, Jan 10, 25; nat US; m 49; c 3. BIOCHEMISTRY, NUTRITION. Educ: Univ Paris, Sorbonne, Lic en Sc, 48; Univ Southern Calif, MS, 51, PhD(biochem), 56. Prof Exp: Asst biochem, Univ Southern Calif, 50-56, res assoc, 56-62; RES BIOCHEMIST, SCH PUB HEALTH, UNIV CALIF, LOS ANGELES, 62- Concurrent Pos: Fel Coun Arteriosclerosis, Am Heart Asn. Honors & Awards: Bond Award, Am Oil Chemists Soc, 74. Mem: AAAS; Am Inst Nutrit; Am Oil Chemists Soc. Res: Essential fatty acids; cholesterol, their metabolic interrelationship; effect of sex hormones on lipid metabolism; metabolic effects of oral contraceptives; vitamin E; nutrition and cancer. Mailing Add: 8056 El Manor Ave Los Angeles CA 90045

AFTERGUT, SIEGFRIED, b Frankfurt am Main, Ger, Jan 7, 27; nat US; m 51; c 2. ORGANIC CHEMISTRY. Educ: Syracuse Univ, BA, 50, MS, 54, PhD(org chem), 56. Prof Exp: Asst instr chem, Syracuse Univ, 51-56; proj engr, Advan Tech Labs, 56-66, mgr photorecording mat, 66-73, MGR DISPLAY MAT, RES & DEVELOP CTR, GEN ELEC CO, 73- Mem: AAAS; Am Chem Soc; Soc Photog Scientists & Engrs; Soc Info Display. Res: Mannich reaction; aromatic silanes and ethers; electrical properties of organic compounds; photoplastic recording; organic photoconductors; liquid crystals. Mailing Add: 1063 Nott St Schenectady NY 12308

AGALIDES, EUGENE, b Braila, Romania, Sept 24, 12; US citizen; m 53. BIOPHYSICS. Educ: Acad High Indust & Com Sci, Romania, PhD, 39; Bucharest Acad Tech Sci, DE, 54. Prof Exp: Asst mgr, Stan Rizescu Mfg Co, 32-41; pres, Gen Automobile Co, 41-48; chief, Stand Sect, Philips Radio Co, 48-51; sect chief, Stand Tel Co, 51-52; head elec sect, Iprochim Inst Res & Develop, 52-53, chief engr, Ipromin & Iceprom Res & Develop Inst, 53-56; proj engr, Mason Labs, Inc, 58; head biophys commun lab, Res Dept, Gen Dynamics/Electronics, 58-68, sr staff scientist, Gen Dynamics/Pomona, 68-74; SR STAFF SCIENTIST APPL PHYSICS, SCI APPLICATIONS INC, 74- Concurrent Pos: Mem, Nat Comt Elaboration Elec Stand in Bucharest, 48-56; prof, State Univ Romania, 53-56; mem organizing comt, Rochester Conf Data Acquisition & Processing Biol & Med, 61- Mem: AAAS; Sigma Xi; sr mem Inst Elec & Electronic Engrs; NY Acad Sci. Res: Cybernetics in biology; engineering communication; information processes in living organisms; nervous system; transmitting and receiving ability of coded signals of electric fishes. Mailing Add: Sci Applications Inc PO Box 2351 La Jolla CA 92037

AGAN, RAYMOND JOHN, b Knoxville, Iowa, July 8, 19; m 39; c 2. AGRICULTURE. Educ: Iowa State Univ, BS, 40, MS, 50; Univ Mo, EdD(agr), 55. Prof Exp: Asst agr educ, Univ Mo, 54-55; from instr to asst prof, Ore State Univ, 55-58; from assoc prof to prof, Kans State Univ, 58-71; PROF AGR EDUC & CHMN DEPT VOC EDUC, SAM HOUSTON STATE UNIV, 71- Concurrent Pos: Consult, Univ Costa Rica, 63-64; UNESCO consult, Colombian projs & ODECA/OCE PLAN, Cent Am, 65-70. Mem: Asn Teachers Educators Agr (vpres, 63-64). Res: Agricultural occupations and training needs of Oregon Indian tribes; agricultural needs in Central and South America. Mailing Add: Dept of Voc Educ Sam Houston State Univ Huntsville TX 77340

AGAR, MICHAEL HENRY, b Evanston, Ill, May 7, 45. ETHNOGRAPHY, ANTHROPOLOGICAL LINGUISTICS. Educ: Stanford Univ, AB, 67; Univ Calif, Berkeley, PhD(anthrop), 71. Prof Exp: Cult anthropologist, Clin Res Ctr, Nat Inst Ment Health, 68-70; asst prof anthrop, Univ Hawaii, 71-73; chief sociocult sect, Bur Res, Narcotic Addition Control Comn, 73-75; ASSOC PROF ANTHROP, UNIV HOUSTON, 75- Concurrent Pos: Res consult, Waikiki Drug Clin, 71-73; vis asst prof, Univ Calif, Berkeley, 73; consult, Nat Inst Drug Abuse, 73-; adj prof, Sch Pub Health, Univ Tex, 76- Mem: Fel Am Anthrop Asn; Soc Appl Anthrop. Res: Anthropology of communication; mathematical anthropology; ethnography of India and urban America. Mailing Add: Dept Anthrop Univ of Houston Houston TX 77004

AGARWAL, ARUN KUMAR, b Lucknow, India, Mar 17, 44; m 67; c 2. PURE MATHEMATICS. Educ: Lucknow Univ, BSc, 63, MSc & PhD(math), 67. Prof Exp: Coun Sci Indust Res sr res fel, New Delhi, India, 67-68; fel, WVa Univ, 68-69; ASSOC PROF MATH, GRAMBLING STATE UNIV, 69- Concurrent Pos: Reviewer, Math Rev, 70- & Zentrablatt für Math, 70- Mem: Am Math Soc. Res: Entire functions of a single and several complex variables; special functions. Mailing Add: Box 679 Dept of Math Sci Grambling State Univ Grambling LA 71245

AGARWAL, KAILASH C, b Uttarpradesh, India. BIOCHEMISTRY. Educ: Allahabad Univ, India, MSc, 63; Agra Univ, India, PhD(sci), 67. Prof Exp: Lectr biochem, Kanpur Med Col, 66-68; res assoc bio-med, Brown Univ, 68-71, instr, 71-74, ASST PROF BIO-MED, BROWN UNIV, 75- Res: Nucleotide metabolism in formed elements of human blood, such as erythrocytes, lymphocytes and platelets. Mailing Add: Bio Med Div Brown Univ Providence RI 02912

AGARWAL, RAM PRAKASH, b Mondona, India, June 2, 39; m 59; c 5. BIOCHEMICAL PHARMACOLOGY. Educ: Univ Lucknow, BSc, 56, MSc, 59; Banaras Hindu Univ, PhD(biochem), 67. Prof Exp: Demonstr-instr chem, Indian Sch Mines, 59-63; demonstr biochem, Banaras Hindu Univ, 63-64, lectr biochem-med chem, 64-66; instr & res assoc biochem pharmacol, Brown Univ, 69-71; enxymologist biochem, RI Hosp, 70-72; ASST PROF BIOCHEM PHARMACOL, BROWN UNIV, 71- Concurrent Pos: Res fel, K G Med Col, Univ Lucknow, 62-63. Mem: AAAS. Res: Localization, isolation, kinetics and mechanism of action of enzymes; metabolism of purines and purine antimetabolites; cancer chemotherapy; metabolism in red blood cells. Mailing Add: Div of Biol & Med Sci Brown Univ Providence RI 02912

AGARWAL, SOM PRAKASH, b Aligarh, India, Apr 6, 30; US citizen; m 55; c 3. COSMIC RAY PHYSICS, NUCLEAR PHYSICS. Educ: Agra Univ, BSc, 49; Aligarh Muslim Univ, MSc, 51; Temple Univ, PhD(physics), 62. Prof Exp: Lectr physics, MMH Col, India, 51-52; assoc prof, Univ Tex, El Paso, 62-64; physicist, Braddock, Dunn & McDonald, Inc, 64-65; asst prof, 65-66, ASSOC PROF PHYSICS, UNIV MINN, MORRIS, 66- Concurrent Pos: Consult, Braddock, Dunn & McDonald, Inc, 63-64. Mem: Am Phys Soc; Am Geophys Union; Am Asn Physics Teachers. Res: Cosmic rays; interplanetary particles and fields; nuclear reactions; elementary particles; radiation effects and electronic physics. Mailing Add: Dept of Physics Univ of Minn Morris MN 56267

AGARWAL, SURESH KUMAR, b Pilibhit, India, June 22, 32; m 50; c 4. RADIOLOGICAL PHYSICS. Educ: Agra Univ, BS, 50; Lucknow Univ, MS, 52, PhD(physics), 65; Am Bd Radiol, cert radiol physics, 72. Prof Exp: Physicist, GM & Assoc Hosps, India, 52-58; lectr radiol physics, Lucknow Univ, 58-69; asst prof, 69-74, ASSOC PROF RADIOL PHYSICS, UNIV VA, 74- Mem: Am Asn Physicists Med; Am Col Radiol; Health Physics Soc; Soc Nuclear Med; AAAS. Res: Radiation treatment planning of irregular fields and its computerization; quality control in diagnostic radiology; radio-bioassay program, its guidelines and implementation. Mailing Add: Radiol Physics Div Univ of Va Sch of Med Box 249 Charlottesville VA 22901

AGATSTON, ROBERT STEPHEN, b New York, NY, Apr 20, 23; m 57. GEOLOGY. Educ: Ohio State Univ, BS, 43; Columbia Univ, MA, 47, PhD(geol), 52. Prof Exp:

Div res geologist, Atlantic Refining Co, 48-67, DIR GEOL SCI, ATLANTIC RICHFIELD CO, 67- Mem: Am Asn Petrol Geologists. Res: Pennsylvanian of Wyoming. Mailing Add: Atlantic Richfield Co Box 2819 Dallas TX 75221

AGEE, ERNEST MASON, b Richmond, Ky, Oct 2, 42; m 63. DYNAMIC METEOROLOGY, FLUID MECHANICS. Educ: Eastern Ky Univ, BS, 64; Univ Mo-Columbia, MS, 66, PhD(atmospheric sci), 68. Prof Exp: Asst prof, 68-72, ASSOC PROF GEOSCI, PURDUE UNIV, 72- Concurrent Pos: NSF res grant, 70- & Nat Oceanic & Atmospheric Admin, 72-; Purdue Univ Sci Rep, Univ Corp Atmospheric Res, Boulder, Colo, 76-80. Mem: Am Meteorol Soc; Meteorol Soc Japan. Res: Thunderstorms and tornadoes; mesoscale cellular convection; theoretical and laboratory study of convective patterns and vortex features; field study of convective clouds in the atmosphere. Mailing Add: Dept of Geosci Purdue Univ West Lafayette IN 47906

AGEE, HERNDON ROYCE, b Cottonburg, Ky, Dec 21, 33; m 53; c 2. INSECT PHYSIOLOGY. Educ: Berea Col, BA, 58; Univ Minn, MS, 60; Tufts Univ, PhD, 68. Prof Exp: Asst entom, Univ Minn, 58-60; entomologist, Cotton Insect Br, Agr Res Serv, USDA, 60-75, RES ENTOMOLOGIST INSECT ATTRACTANTS, BEHAV & BASIC BIOL RES LAB, AGR RES SERV, USDA, 75- Mem: Entom Soc Am; Am Soc Zoologists. Res: Electrophysiological and behavioral studies of the sensory systems of insects; visual and acoustic responses; quality control. Mailing Add: Behav & Basic Biol Res Lab Agr Res Serv USDA Box 14565 Gainesville FL 32604

AGENBROAD, LARRY DELMAR, b Nampa, Idaho, Apr 3, 33; m 55; c 2. GEOLOGY. Educ: Univ Ariz, BS, 59, MS, 62, PhD(geol), 67, MA, 70. Prof Exp: Explor geophysicist, Pan Am Petrol Corp, 60-62; geologist AEC shoal event, Nev Bur Mines, 62-63; teaching asst geol, Univ Ariz, 63-65 , teaching asst geol & anthrop, 65-67; asst prof, 67-70, ASSOC PROF EARTH SCI, CHADRON STATE COL, 70- Concurrent Pos: Dir, Hudson-Meng Paleo-Indian Bison Kill, 70-75; asst dir, Lehner Ranch, 74-75; dir, Hot Springs Mammoth Site, 74-75. Mem: AAAS; Soc Am Archaeol; Am Asn Quaternary Environ; Nat Asn Geol Teachers; Am Geol Inst. Res: Quaternary geology and its relation to early man in the New World; present paleohydrology and ground water resources; archaeology, particularly prehistory. Mailing Add: Dept of Earth Sci Chadron State Col Chadron NE 69337

AGER, JOHN WINFRID, JR, b Birmingham, Ala, Nov 2, 26; m 56; c 2. CHEMISTRY. Educ: Harvard Univ, BA, 49; Univ NC, MA, 51; Oxford Univ, PhD(org chem), 55. Prof Exp: Res chemist, Olin Mathieson Chem Corp, NY, 56-60; RES CHEMIST, FMC CORP, 60- Mem: Am Chem Soc. Res: Boron based high energy fuels; synthesis in the field of natural products; process development of commercial chemicals; agricultural chemicals. Mailing Add: FMC Corp Box 8 Princeton NJ 08540

AGERSBORG, HELMER PARELI KJERSCHOW, JR, b Decatur, Ill, Dec 2, 28; m 52; c 3. PHYSIOLOGY, TOXICOLOGY. Educ: Southern Ill Univ, AB, 54; Univ Tenn, PhD, 57. Prof Exp: Asst physiol, Univ Tenn, 54-57, instr clin physiol & obstet & gynec & dir obstet & gynec, 57-58; clin physiologist, Med Div, 58-61, mgr toxicol & comp pharmacol, Res Div, 61-69, ASSOC DIR, RES DIV, WYETH LABS, INC, 69- Mem: AAAS; NY Acad Sci; Am Physiol Soc; Am Soc Zool; Soc Toxicol. Res: Metabolic disease; cardiovascular physiology; ion and fluid dynamics; drug safety evaluation. Mailing Add: Res Div Wyeth Labs Inc Philadelphia PA 19101

AGETT, ALBERT HENRY, b Corning, NY, Aug 20, 14; m 38; c 4. CHEMISTRY. Educ: Mich State Col, BS, 36, MS, 38, PhD(org chem), 40. Prof Exp: Develop chemist, Tenn Eastman Corp, 40-44, supt dye dept, 44-57, asst to div supt, Org Chem Div, 57-58, asst to div supt, Filber Prod Div, 58-59, asst div supt, 59-63; DIR RES & DEVELOP, ASG INDUSTS, INC, 63- Mem: Am Chem Soc. Res: Organic chemistry; dyes; triethyl phosphate; air oxidation. Mailing Add: ASG Industs Inc PO Box 929 Kingsport TN 37662

AGGARWAL, ROSHAN LAL, b Salala, India, Feb 15, 37; m 58; c 2. QUANTUM OPTICS. Educ: Punjab Univ, India, BSc, 57, MSc, 58; Purdue Univ, PhD(physics), 65. Prof Exp: Res assoc physics, Purdue Univ, 65; staff mem, 65-71, proj leader quantum optics, 71-74, GROUP LEADER QUANTUM OPTICS & PLASMA PHYSICS, FRANCIS BITTER NAT MAGNET LAB, MASS INST TECHNOL & SR RES SCIENTIST, PHYSICS DEPT, 75- Mem: Am Phys Soc; Sigma Xi; Am Optical Soc. Res: Near and far infrared spectroscopy; modulation magnetospectroscopy; lasers and nonlinear optics; light scattering in solids. Mailing Add: F B Nat Magnet Lab Bldg NW 14 Mass Inst of Technol Cambridge MA 02139

AGGARWAL, SUNDAR LAL, b Jullundur, India, Oct 15, 22; nat US; m 48; c 3. POLYMER CHEMISTRY. Educ: Punjab Univ, India, BSc, 42, MSc, 43; Cornell Univ, PhD(phys chem), 49. Prof Exp: Lectr anal chem, Govt Col, India, 43-45; res assoc polymer chem, Cornell Univ, 49-50; sci officer, Nat Chem Lab, India, 50-52; sr scientist, Film Res Sect, Olin Industs, 52-54, group leader, 54-55, sect chief, Phys Properties & Polymer Sect, 55-57; head chem physics res, 57-62, mgr basic res, 62-68, mat res & tech serv, 68-75, V PRES & DIR, RES DIV, GEN TIRE & RUBBER CO, 75- Concurrent Pos: Chmn, Gordon Res Conf Elastomers, 69; assoc ed, Int J Polymer Mat; chmn, Symp Block Polymers. Mem: Fel AAAS; Am Chem Soc; Am Phys Soc; Soc Rheol; NY Acad Sci. Res: Structure and properties of block polymers and elastomers; dynamic mechanical and strength properties of polymers; electron microscopy and molecular spectroscopy methods for polymer structure; polymer physics of composite materials. Mailing Add: Res & Develop Ctr Gen Tire & Rubber Co PO Box 1829 Akron OH 44329

AGGARWALA, BHAGWAN D, b India, Mar 9, 31. APPLIED MATHEMATICS. Educ: Punjab Univ, India, BA, 50, MA, 52, PhD(math), 59. Prof Exp: Instr math, Indian Inst Technol, Kharagpur, 53-55; instr, Carnegie Inst Technol, 55-57; asst prof mech, Rensselaer Polytech Inst, 57-60; asst prof math, McGill Univ, 60-66; assoc prof, 66-71, PROF MATH, STATIST & COMPUT SCI, UNIV CALGARY, 71- Mem: Can Math Cong; Soc Indust & Appl Math; Am Acad Mech. Res: Stress analysis of elastic and viscoelastic materials; heat transfer. Mailing Add: Dept of Math Univ of Calgary Calgary AB Can

AGHAJANIAN, GEORGE KEVORK, b Beirut, Lebanon, Apr 14, 32; m 59; c 4. NEUROPHARMACOLOGY. Educ: Cornell Univ, AB, 54; Yale Univ, MD, 58. Prof Exp: Intern, Jackson Mem Hosp, 58-59; resident, 59-63, from asst prof psychiat to assoc prof psychiat & pharmacol, 65-74, PROF PSYCHIAT & PHARMACOL, SCH MED, YALE UNIV, 74- Concurrent Pos: NIMH fel, 61-63, career develop award, 65-75. Mem: Am Soc Pharmacol & Exp Therapeut; Psychiat Res Soc; NY Acad Sci; fel Am Col Neuropsychopharmacol; Soc Neurosci. Res: Brain monoamines and LSD. Mailing Add: Dept of Psychiat Yale Univ Sch of Med New Haven CT 06519

AGIN, DANIEL PIERRE, b New York, NY, May 19, 30; m 56; c 2. NEUROPHYSIOLOGY, BIOPHYSICS. Educ: City Col New York, BA, 53; Univ Rochester, PhD(psychol), 61. Prof Exp: NIH fel, 61-62; from instr to asst prof, 62-67, actg chmn dept, 68-69, ASSOC PROF PHYSIOL, UNIV CHICAGO, 67- Concurrent Pos: Vis prof, Max Planck Inst Biophys, 69-70. Res: Membrane biophysics; biophysical pharmacology. Mailing Add: Dept Physiol & Pharmacol Univ Chicago 5801 Ellis Ave Chicago IL 60637

AGIN, GARY PAUL, b Kansas City, Mo, Dec 22, 40. NUCLEAR PHYSICS. Educ: Univ Kans, BS, 63; Kans State Univ, MS, 67, PhD(physics), 68. Prof Exp: ASST PROF PHYSICS, MICH TECHNOL UNIV, 68- Mem: AAAS; Am Asn Physics Teachers; Am Phys Soc. Res: Low energy nuclear physics; beta and gamma ray spectroscopy; computer applications. Mailing Add: Dept of Physics Mich Technol Univ Houghton MI 49931

AGINS, BARNETT ROBERT, b New York, NY, May 19, 22; m 45; c 2. MATHEMATICS. Educ: NY Univ, BEE, 52, MEE, 56; Stanford Univ, MSc, 61. Prof Exp: US Air Force, 50-, commun off Air Defense Command, 50-54, proj engr electronic syst, 56-59, chief appl math div, Off Sci Res, 61-67, asst to dir, Courant Inst Math Sci, 67-69, PROG DIR APPL MATH & STATIST, NSF, 69- Concurrent Pos: Prof lectr, Am Univ, 62-; assoc ed, J Optimization Theory & Applns, 67-; assoc ed, Comput & Math with Applns, 74- Mem: Inst Elec & Electronics Engrs; Soc Indust & Appl Math. Res: Sophisticated electronic warfare techniques; mathematics with respect to ordinary differential equations, especially those of celestial mechanics. Mailing Add: Nat Sci Found Washington DC 20550

AGINSKY, BERNARD WILLARD, b New York, NY, Feb 10, 05; m 29. ANTHROPOLOGY, CULTURAL ANTHROPOLOGY. Educ: NY Univ, BS, 31, MA, 32; Columbia Univ, PhD(anthrop), 34. Prof Exp: Res assoc anthrop, Columbia Univ, 34-36; res assoc, Univ Calif, Berkeley, 36; res assoc, Columbia Univ, 37-38; instr anthrop & sociol, NY Univ, 38-41; lectr, Hunter Col, 46; prof, City Col New York, 46-65, chmn dept, 47-52; DIR, INST FOR WORLD UNDERSTANDING OF PEOPLES, CULT & LANG, 65- Concurrent Pos: Columbia Univ-Soc Sci Res Coun fel, 34-36; vis prof, Yencheng Univ, China, 36-37; Rockefeller Found-Soc Sci Res Coun fel, 39; dir soc sci field lab, NY Univ, 39-41 & Syracuse Univ, 47-48; Wenner-Gren Found fel, 46-50; US policy bd adv, Nat Indian Inst for 2nd Inter-Am Conf on Indian Life, Peru, 48; Am Jewish comt rep, White House Conf on Youth Probs, 61; TV ed & course lectr, Cooper Union & Sta WUHF, 62; Res Inst for Study of Man fel methodology, Dolphin-Seaquarium, Key Biscayne, Fla, 64-65; Sperry Hutchison lectr, Univ Miami, 65; Inst for World Understanding of Peoples, Cult & Lang fel, 66. Mem: Fel AAAS; Am Acad Polit & Soc Sci; fel Am Anthrop Asn; Am Ethnol Soc (pres, 49); Soc Appl Anthrop. Res: The establishment of a universally applicable methodology for the comparative study and understanding of the components and the differential developments of the language, physical characteristics, and culture of all populations and species. Mailing Add: Inst for World Understanding 939 Coast Blvd 19DE La Jolla CA 92037

AGINSKY, ETHEL G, b Scranton, Pa, Sept 24, 10; m 29. ANTHROPOLOGY. Educ: NY Univ, BA, 32; Columbia Univ, MA, 33, PhD(anthrop), 34. Prof Exp: Res assoc ling, Columbia Univ, 34-36 & field res, 36-37; instr sociol anthrop, 39-47, from asst prof to assoc prof, 48-61, prof anthrop, 61-66, EMER PROF ANTHROP, HUNTER COL, 66-; ASSOC DIR SOCIAL SCI LAB, NY UNIV, 39-; ASSOC DIR, INST WORLD UNDERSTANDING PEOPLE, CULT & LANG, 65- Concurrent Pos: Inst Pac Rels fel, Univ Calif, 39; Viking Fund grants, 46, 47. Mem: AAAS; Am Ethnol Soc; Ling Soc Am; Am Anthrop Asn (treas, 47-); Am Sociol Asn. Res: Acculturation; social anthropology. Mailing Add: Inst of World Understanding 939 Coast Blvd 19DE La Jolla CA 92037

AGNELLO, EUGENE JOSEPH, b Rochester, NY, Oct 3, 19; m 59; c 5. ORGANIC CHEMISTRY. Educ: State Univ NY Albany, AB, 41; Univ Rochester, PhD(chem), 50. Prof Exp: Teacher pub sch, NY, 41-46; asst, Univ Rochester, 46-50; fel, Univ Ill, 50-51; res chemist, Chas Pfizer & Co, Inc, 51-63; asst prof chem, Waynesburg Col, 63-64; from asst prof to assoc prof, 64-73, PROF CHEM, HOFSTRA UNIV, 73- Mem: Am Chem Soc. Res: Steroids; medicinal chemistry; natural products. Mailing Add: Dept of Chem Hofstra Univ Hempstead NY 11550

AGNEW, ALLEN FRANCIS, b Ogden, Ill, Aug 24, 18; m 46; c 4. GEOLOGY. Educ: Univ Ill, AB, 40, MS, 42; Stanford Univ, PhD(geol), 49. Prof Exp: Asst geologist, Ill State Geol Surv, 39-42; geologist, US Geol Surv, Wis, 42-45 & Water Resources Div, Iowa, 45-47; asst prof geol, Univ Ala, 48-49; from assoc prof to prof, Univ SDak, 55-63; prof geol & dir water resources res ctr, Ind Univ, Bloomington, 63-69; prof geol & dir water res ctr, Wash State Univ, 69-74; SR SPECIALIST MINING, CONG RES SERV, LIBR CONG, 74- Concurrent Pos: Geologist, US Geol Surv, Wis, 48-55; geologist, SDak Geol Surv, 55-57, state geologist, 57-63. Mem: AAAS; Asn Eng Geol; Geol Soc Am; Soc Econ Geologists; Soc Environ Geochem Health. Res: Lead-zinc deposits; geology of South Dakota; surface mining hydrology; mineral deposits and resources. Mailing Add: 11906 Escalante Ct Reston VA 22091

AGNEW, HAROLD MELVIN, b Denver, Colo, Mar 28, 21; m 42; c 2. PHYSICS. Educ: Univ Denver, AB, 42; Univ Chicago, MS, 48, PhD(physics), 49. Prof Exp: Physicist, Manhattan Dist, Univ Calif, Los Alamos, NMex, 42-46; Nat Res fel, Physics Div, Los Alamos Sci Lab, 49-50, asst to tech assoc dir, 51-53 & Theoret Div, 54-56, leader alt weapons div, 56-61; sci adv to Supreme Allied Comdr, Europe, 61-64; leader weapons div, 64-70, DIR, LOS ALAMOS SCI LAB, 70- Concurrent Pos: Mem, US Air Force Sci Adv Bd, 57-68; chmn, US Army Combat Develop Command Sci Adv Group, 65-66; mem, President's Sci Adv Comt, 65-73; mem, Defense Sci Bd, 66-70; chmn, US Army Sci Adv Panel, 66-70, mem, 70-74; mem, NASA Aerospace Safety Adv Panel, 68-74; chmn gen adv comt, US Arms Control & Disarmament Agency Coun on Foreign Rels, 74-, mem, 75- Honors & Awards: Ernest Orlando Lawrence Award, AEC, 66. Mem: Fel AAAS; Am Phys Soc. Res: Neutron physics; light particle reactions; particle accelerators. Mailing Add: Los Alamos Sci Lab PO Box 1663 Los Alamos NM 87545

AGNEW, JEANNE LE CAINE, b Port Arthur, Ont, May 3, 17; US citizen; m 42; c 5. MATHEMATICS. Educ: Queen's Univ, Ont, BA, 37, MA, 38; Harvard Univ, PhD(math), 41. Prof Exp: Instr math, Smith Col, 41-42; res physicist, Nat Res Coun Can, 42-45; instr math, Cambridge Jr Col, 46-47; from asst prof to assoc prof, 56-69, PROF MATH, OKLA STATE UNIV, 69- Concurrent Pos: Vis assoc prof, Ga State Col, 66-67. Mem: Am Math Soc; Math Asn Am. Res: Applications of undergraduate level mathematics to industrial problems. Mailing Add: Dept of Math Okla State Univ Stillwater OK 74074

AGNEW, LESLIE ROBERT CORBET, b Newcastle-on-Tyne, Eng, Nov 18, 23. HISTORY OF MEDICINE, EXPERIMENTAL PATHOLOGY. Educ: Glasgow Univ, MB, ChB, 46, MD, 50; Harvard Univ, AM, 57. Prof Exp: Mem staff path, Rowett Res Inst, Scotland, 47-49; res assoc prof, Univ Fla, 53-55; assoc nutrit, Sch Pub Health, Harvard Univ, 55-56, resident tutor, Harvard Col, 55-57; from assoc prof to prof hist med & chmn dept, Med Ctr, Univ Kans, 57-65; sr lectr, 65-66, ASSOC PROF MED HIST, SCH MED, UNIV CALIF, LOS ANGELES, 66- Concurrent Pos: Trent Mem lectr, Duke Univ, 62; Shuman Mem lectr, Univ Calif, Los Angeles, 64; Davis lectr, Univ Ill, 65; mem fac med hist, Worshipful Co of Apothecaries; mem study sect hist life sci, NIH, 62-67. Mem: Am Asn Hist Med; Hist Sci Soc; NY Acad

Sci; corresp mem Int Acad Hist Med; Path Soc Gt Brit & Ireland. Res: Eighteenth and nineteenth century British and American medicine; vitamin deficiency states; hormones and cancer. Mailing Add: Div of Med Hist Univ of Calif Sch of Med Los Angeles CA 90024

AGNEW, LEWIS EDGAR, JR, b Glendale, Mo, June 20, 26; m 52; c 4. PHYSICS. Educ: Univ Mo-Rolla, BS, 50; Univ Calif, Berkeley, MA, 56, PhD(physics), 60. Prof Exp: Res asst, Los Alamos Sci Lab, 50-53, staff mem, 53-54; res asst, Lawrence Radiation Lab, Univ Calif, 56-59; staff mem, Lab, 59-69, GROUP LEADER, LAMPF ACCELERATOR PROJ, LOS ALAMOS SCI LAB, 70- Concurrent Pos: Head physics sect, Div Res & Labs, Int Atomic Energy Agency, Vienna, Austria, 66-68. Mem: Fel AAAS; Am Phys Soc. Res: Medium energy physics; plasma spectroscopy. Mailing Add: Los Alamos Sci Lab PO Box 1663 Los Alamos NM 87544

AGNEW, ROBERT MORSON, b Cardigan, PEI, Nov 20, 32. IMMUNOBIOLOGY. Educ: Dalhousie Univ, BSc, 53, MSc, 55; Cambridge Univ, PhD(bact), 59. Prof Exp: Lectr bact, 58-64, asst prof biol, 64-69, ASSOC PROF BIOL, UNIV REGINA, 69- Concurrent Pos: Vis mem staff, Rheumatic Dis Unit, Univ Toronto, 70-71. Mem: AAAS; NY Acad Sci; Can Soc Microbiol. Res: Immunology; medical bacteriology; immune reactions in rheumatoid arthritis; tumor immunology. Mailing Add: Dept of Biol Univ of Regina Regina SK Can

AGNEW, WILLIAM FINLEY, b Greenville, SC, Aug 28, 25; m 58; c 2. PHYSIOLOGY. Educ: Wheaton Col, AB, 49; Univ Ill, Urbana, MS, 54; Univ Southern Calif, PhD(physiol), 64. Prof Exp: Res asst physiol, Baxter Labs, 52-53; res asst biol, Calif Inst Technol, 55; SR INVESTR, INST MED RES, HUNTINGTON MEM HOSP, 63- Concurrent Pos: Laband neurosurg res fel physiol, Inst Med Res, Huntington Mem Hosp, 55-63; NIH res grant, 61-63; Nat Inst Neurol Dis & Stroke fel, Univ Copenhagen, 65-66, res grant, 69-71; res asst, Univ Southern Calif, 57-60. Mem: Teratology Soc; Am Physiol Soc; Soc Exp Biol & Med; Soc Neurosci. Res: Cerebral circulation; blood-brain cerebrospinal fluid barriers and electron microscopy of the central nervous system. Mailing Add: Inst of Appl Med Res Huntington Mem Hosp Pasadena CA 91105

AGNIHOTRI, RAM K, b Kanpur, India, Oct 15, 33. CHEMISTRY. Educ: Agra Univ, BS, 54, MS, 56; Purdue Univ, PhD, 63. Prof Exp: Lectr chem, India, 57-58; res fel & asst, Purdue Univ, 58-63; res chemist, E I du Pont de Nemours & Co, 63-67; staff chemist, Int Bus Mach Corp, 67-69, proj chemist, 69-70, develop chemist & mgr photolithographic mat, 70-71, ADV ENGR, IBM CORP, 71- Honors & Awards: Achievement Award, IBM Corp, 71. Mem: Am Chem Soc. Res: Photostabilizers; polymers; resins; organometallic chemistry; optical rotatory power; vinyl fluoride coatings; semiconductor personalization; semiconductor packaging. Mailing Add: D692 B330-135 IBM Corp Hopewell Junction NY 12533

AGOCS, WILLIAM BAILEY, b South Bethlehem, Pa, Nov 28, 11; m 37. PHYSICS. Educ: Lehigh Univ, BS, 34, MS, 44, PhD(physics), 46. Prof Exp: Geophysicist & party chief, Gulf Res & Develop Co, 34-40; asst physics, Lehigh Univ, 40-43, asst prof, 46-48; geophysicist, Seismograph Serv Corp, Okla, 47; chief geophysicist, Venezuela Atlantic Ref Co, 44-45 & 48-50; from assoc prof to prof geophys & physics, Univ Tulsa, 50-53, head dept geophys, 52-53; dir geophys, Aero Serv Corp, 53-60; consult, UN Spec Fund, 60-61; consult, Govt Saudi Arabia, 61-63; PROF PHYSICS, KUTZTOWN STATE COL, 64-, CHMN DEPT PHYS SCI, 67- Concurrent Pos: Consult, Bethlehem Steel Co, 48-50 & Develop & Res Corp, Ivory Coast, 60-64. Mem: Soc Explor Geophys; Am Phys Soc; Inst Elec & Electronics Engrs; Am Geophys Union; Venezuelan Soc Geol. Res: Total magnetic fields interpretation; thermal conductivity of metals; potential fields theory; airborne scintillometer gamma radiation study; seismic wave propagation in solids and fluids; method of determining time break on deep sea seismic records and sea bottom slope. Mailing Add: Dept of Phys Sci Kutztown State Col Kutztown PA 19530

AGOGINO, GEORGE ALLAN, b West Palm Beach, Fla, Nov 18, 20; m 51; c 2. ANTHROPOLOGY, SOCIAL PSYCHOLOGY. Educ: Univ NMex, BA, 49, MA, 51; Syracuse Univ, PhD(anthrop), 58. Hon Degrees: PhD, Rome Inst Arts & Sci, Italy, 60. Prof Exp: Asst prof, Nasson Col, 54-56, Univ SDak, 59-60 & Univ Wyo, 60-62; assoc prof, Baylor Univ, 63-64; assoc prof, 64-66, PROF ANTHROP, EASTERN NMEX UNIV, 66-, CHMN DEPT, 63-, DIR SPEC PROG ANTHROP, 75-, DIR PALEO-INDIAN INST & MUS, 63-, DIR MILES MUS, 69- Concurrent Pos: Wenner-Gren Found grant anthrop, Harvard Univ, 62-63; dir, Blackwater Draw Mus, 69- Mem: Fel AAAS; fel Am Anthrop Asn; Am Soc Archaeol; Soc Clin & Exp Hypnosis; fel Inter-Am Inst Indian Affairs. Res: Primitive religion; Paleo-Indians; Mexico. Mailing Add: Dept of Anthrop Eastern NMex Univ Portales NM 88130

AGOSIN, MOISES, b Marseille, France, Dec 1, 22; m 48; c 3. BIOCHEMISTRY, PARASITOLOGY. Educ: Univ Chile, MD, 48. Prof Exp: From asst prof to assoc prof parasitol, Sch Med, Univ Chile, 48-57, prof biochem, 57-61 & chem, 61-68; vis prof, 68-69, RES PROF ZOOL, UNIV GA, 69- Concurrent Pos: Rockefeller Found fel, NIH, 52-54, res assoc, 54-55, res grants, 58-74; Rockefeller Found res grant, 56; vis prof, Univ Calif, Berkeley, 61 & Univ London, 65. Mem: Fel Am Acad Microbiol; Am Chem Soc; Biochem Soc. Res: Biochemistry of parasitic organisms; biochemistry of insecticide resistance. Mailing Add: Dept of Zool Univ of Ga Athens GA 30601

AGOSTA, WILLIAM CARLETON, b Dallas, Tex, Jan 1, 33; m 58; c 2. ORGANIC CHEMISTRY. Educ: Rice Univ, BA, 54; Harvard Univ, AM, 55, PhD(chem), 57. Prof Exp: Nat Res Coun fel org chem, Dyson Perrins Lab, Oxford Univ, 57-58; Pfizer fel, Univ Ill, 58-59; asst prof chem, Univ Calif, 59-61; sci liaison officer, US Naval Force Europe, 61-63; from asst prof to assoc prof, 63-74, PROF CHEM, ROCKEFELLER UNIV, 74- Concurrent Pos: Alfred P Sloan Found fel, 69-71. Mem: Am Chem Soc; The Chem Soc; Am Soc Photobiol. Res: Photochemistry; synthesis; pheromone chemistry. Mailing Add: Rockefeller Univ New York NY 10021

AGOSTINO, DOMENICO, biology, see 12th edition

AGOSTON, MAX KARL b Stockerau, Austria, Mar 25, 41; US citizen. MATHEMATICS. Educ: Reed Col, BA, 62; Yale Univ, MA, 64, PhD(math), 67. Prof Exp: Lectr math, Wesleyan Univ, 66-67, asst prof, 67-75. Concurrent Pos: Vis prof math, Univ Heidelberg, 70-71; vis fel, Univ Auckland, 73-74. Mem: Am Math Soc; Math Asn Am. Res: Differential topology; imbedding problems of manifolds; algebraic topology. Mailing Add: 24 Hawthorn Dr Atherton CA 94025

AGRANOFF, BERNARD WILLIAM, b Detroit, Mich, June 6, 26; m 57; c 2. BIOCHEMISTRY. Educ: Wayne State Univ, MD, 50; Univ Mich, BS, 54. Prof Exp: Intern, Robert Packer Hosp, 50-51; NSF fel, Mass Inst Technol, 51-52; from asst officer chg to officer chg, Dept Chem, US Naval Med Sch, 52-54; biochemist, Nat Inst Neurol Dis & Blindness, NIH, 54-60; assoc prof biochem, res biochemist & chief sect biochem, 61-65, PROF BIOCHEM, MENT HEALTH RES INST, UNIV MICH, ANN ARBOR, 65- Concurrent Pos: Mem staff, Max Planck Inst Cell Chem, 58-59; adv ed, Advan in Lipid Res, 62-; fel comt biochem & nutrit, NIH, 64-67, mem study sect neurol A, 67-71; mem adv comn fundamental res, Nat Multiple Sclerosis Soc, 70-73; chmn panel biochem, Nat Comn Multiple Sclerosis, 73; vis scientist, Med Res Coun, Mill Hill, 74-75; ed, J Biol Chem, J Neurochem & Brain Res. Mem: Asn Res Nerv & Ment Dis; fel Am Psychol Asn; Soc Neurosci; Am Soc Neurochem (pres, 73-75); fel Am Col Neuropsychopharmacol. Res: Biochemistry of lipids; neurochemistry; biochemical correlates of behavior. Mailing Add: Neurosci Lab Bldg Univ of Mich Ann Arbor MI 48104

AGRAS, WILLIAM STEWART, b London, Eng, May 17, 29; Can citizen; m 55; c 2. MEDICINE, EXPERIMENTAL PSYCHIATRY. Educ: Univ London, MB, BS, 55. Prof Exp: Demonstr psychiat, McGill Univ, 60-61; instr, Col Med, Univ Vt, 61-62, from asst prof to assoc prof, 62-69; prof & chmn dept, Med Ctr, Univ Miss, 69-73; PROF PSYCHIAT, SCH MED, STANFORD UNIV, 73- Concurrent Pos: Ed, J Appl Behav Anal, 74-; fel, Ctr Advan Study Behav Sci, 76-77. Mem: Am Psychiat Asn; Psychiat Res Soc; Soc Exp Anal Behav. Res: Application of techniques derived from learning theory to the investigation of neuroses; research in behavioral medicine. Mailing Add: Dept of Psychiat Stanford Univ Sch of Med Stanford CA 94305

AGRAWAL, JAGDISH CHANDRA, b Mathura, India, July 29, 38; m 58; c 4. APPLIED MATHEMATICS. Educ: Agra Univ, BSc, 57, MSc, 60; Univ Windsor, MSc, 66; Purdue Univ, PhD(appl math), 69. Prof Exp: Lectr math, Vaish Col, Shamli, India, 60-62; from sr res asst to lectr, Indian Inst Technol, Bombay, 62-65; instr, Univ Windsor, 66 & Purdue Univ, 66-69; PROF MATH, CALIFORNIA STATE COL, PA, 69- Concurrent Pos: Consult, Int Bus Mach Sci Ctr, 66. Mem: Am Math Soc; Soc Indust & Appl Math; Math Asn Am; London Math Soc; Edinburgh Math Soc. Res: Magnetohydrodynamics; fluid mechanics; supersonic flow and shock waves; nonlinear stability theory; differential equations. Mailing Add: Dept of Math California State Col California PA 15419

AGRAWAL, KRISHNA CHANDRA, b Calcutta, India, Mar 15, 37; m 60; c 3. MEDICINAL CHEMISTRY, PHARMACOLOGY. Educ: Andhra Univ, India, BS, 59, MS, 60; Univ Fla, PhD(pharmaceut chem), 65. Prof Exp: Res assoc, 66-69, instr, 69-70, asst prof pharmacol, 70-76, ASSOC PROF SCH MED, YALE UNIV, 76- Concurrent Pos: NIH fel grant, Univ Fla, 65-66. Mem: Am Soc Pharmacol & Exp Therapeut; Am Asn Cancer Res; Am Chem Soc. Res: Design and synthesis of chemical agents for cancer chemotherapy; studies of structure-activity relationship and biochemical mechanisms involved in cell death. Mailing Add: Dept of Pharmacol Yale Univ Sch of Med New Haven CT 06510

AGRAWALA, ASHOK KUMAR, b Meerut, India, June 28, 43. COMPUTER SCIENCE. Educ: Indian Inst Sci, Bangalore Univ, BE, 63, ME, 65; Harvard Univ, AM, 70, PhD(appl math), 70. Prof Exp: Prin engr, Data Systs Div, Honeywell Inc, 70-71; ASST PROF COMPUT SCI, UNIV MD, COLLEGE PARK, 71- Concurrent Pos: Vpres, LNK Corp, 71- Mem: Inst Elec & Electronic Engrs; Asn Comput Mach; Simulation Coun Inc; Pattern Recognition Soc; Sigma Xi. Res: Analysis, modelling, measurement and evaluation of computer systems, their design and architecture. Mailing Add: Dept of Comput Sci Univ of Md College Park MD 20742

AGRE, COURTLAND LEVERNE, b Boyd, Minn, Sept 11, 13; m 46; c 6. CHEMISTRY. Educ: Univ Minn, BChE, 34, PhD(org chem), 37. Prof Exp: Res chemist, E I du Pont de Nemours & Co, Inc, 37-40 & Minn Mining & Mfg Co, St Paul, 41-46; prof chem, St Olaf Col, 46-58; PROF CHEM, AUGSBURG COL, 59- Concurrent Pos: NSF fac fel, Univ Calif, 58-59; consult, Minn Mining & Mfg Co, 46- Mem: AAAS; Am Chem Soc. Res: Synthetic resins; organic chemicals; organo-silicon compounds; use of benzoin and diacetyl as catalysts for light activated polymerization of unsaturated compounds; preparation of polyvinyl ketals; preparation of nitroesters. Mailing Add: Dept of Chem Augsburg Col Minneapolis MN 55404

AGRE, KARL, b Feb 24, 32; US citizen; m 55; c 3. CLINICAL PHARMACOLOGY, PEDIATRICS. Educ: Villanova Col, BS, 52; Hahnemann Med Col, MS, 54, PhD(pharmacol), 56; Duke Univ, MD, 59. Prof Exp: Instr pharmacol, Sch Med, Duke Univ, 56-59; intern-resident pediat, Bronx Munic Hosp Ctr, 59-62, clin instr, 64-66; dir clin pharmacol, Bristol Labs, Inc, 66-75; MEM STAFF, SEARLE LABS, 75- Concurrent Pos: Clin instr, State Univ NY Upstate Med Ctr, 66-75; pvt med pract, NY, 64-66. Mem: Fel Am Acad Pediat. Mailing Add: Searle Labs Skokie IL 60076

AGRESTA, JOSEPH, b Long Island City, NY, June 13, 29; m 53; c 7. PHYSICS, APPLIED MATHEMATICS. Educ: Cooper Union, BEE, 50; NY Univ, MS, 52, PhD(physics), 58. Prof Exp: Asst physics, NY Univ, 50-53; physicist, Curtiss-Wright Corp, 53-56; res assoc physics, NY Univ, 56-58; physicist, United Nuclear Corp, 58-62; physicist, Union Carbide Res Inst, 62-67, PHYSICIST, UNION CARBIDE CORP, 68- Concurrent Pos: Lectr, City Col New York, 58-62; adj asst prof, NY Univ, 59-62. Mem: Am Asn Physics Teachers. Res: Radiation transport; nuclear reactor theory; computing machine methods; hydrodynamics; operations research; systems analysis; simulation. Mailing Add: 92 Oak Hill Rd Chappaqua NY 10514

AGRESTI, DAVID GEORGE, b Washington, DC, Aug 8, 38; m 69. NUCLEAR SPECTROSCOPY, MATERIALS SCIENCE. Educ: Ohio State Univ, BSc, 59; Calif Inst Technol, MS, 62, PhD(physics), 67. Prof Exp: Asst prof physics, Calif State Col, Los Angeles, 67-69; asst prof, 69-74, ASSOC PROF PHYSICS & CHMN DEPT, UNIV ALA, BIRMINGHAM, 74- Mem: Am Phys Soc. Res: Nuclear and solid state properties by means of Mössbauer effect experiments. Mailing Add: Dept of Physics Univ of Ala Birmingham AL 35233

AGRESTI, WILLIAM W, b Erie, Pa, Oct 19, 46. COMPUTER SCIENCES. Educ: Case Inst Technol, BS, 68; NY Univ, MS, 71, PhD(comput sci), 73. Prof Exp: Systs analyst oper res, Lord Corp, 68-69; ASST PROF INDUST & SYSTS ENG, UNIV MICH-DEARBORN, 73-, DIR COMPUT & INFO SCI, 75- Concurrent Pos: Asst dir res, Traffic Safety Systs Proj, Hwy Safety Comn, PR, 70. Mem: Asn Comput Mach; Opers Res Soc Am; Am Inst Indust Engrs. Res: Software engineering, especially programming language design and compiler optimization; simulation and dynamic programming. Mailing Add: Univ of Mich 4901 Evergreen Rd Dearborn MI 48128

AGRIOS, GEORGE NICHOLAS, b Galarinos, Greece, Jan 16, 36; US citizen; m 62; c 2. PLANT PATHOLOGY, VIROLOGY. Educ: Univ Thessaloniki, BS, 57; Iowa State Univ, PhD(plant path & genetics), 60. Prof Exp: Asst prof, 63-68, ASSOC PROF PLANT PATH, UNIV MASS, AMHERST, 69- Res: Plant viruses, particularly fruit trees and their transmission and identification; physiological effects of viruses on host plants; methods for virus detection. Mailing Add: Dept of Plant Path Univ of Mass Amherst MA 01002

AGRON, SAM LAZRUS, b Russia, Nov 27, 20; nat US; m 44; c 2. GEOLOGY. Educ: Northwestern Univ, BS, 41; Johns Hopkins Univ, PhD(geol), 49. Prof Exp: Instr geol, Brown Univ, 49-51; from asst prof to assoc prof geol, 51-62, dir, NSF Earth Sci Inst, 64-67, PROF GEOL, RUTGERS UNIV, NEWARK, 62- Mem: AAAS; Am Geophys Union; Geol Soc Am; Nat Asn Geol Teachers; Int Asn Planetology. Res: Structural, economic and environmental geology; structural petrology and petrofabrics; petrology; astrogeology. Mailing Add: Dept of Geol Rutgers Univ Newark NJ 07102

AGRUSS, BERNARD, electrochemistry, see 12th edition

AGUAYO, CARLOS G, b Havana, Cuba, Dec 19, 99; m 31; c 1. MALACOLOGY. Educ: Univ Havana, ScD(zool), 25. Prof Exp: Asst prof biol, Univ Havana, 25-33, prof zool, 33-58; prof zool, 58-59, PROF BIOL, UNIV PR, 60- Concurrent Pos: Guggenheim Mem Found fel, 31-33; ed, Caribbean J Sci, 63- Mem: Am Malacol Union; Soc Study Evolution (vpres, 48); Soc Syst Zool. Res: Antillean mollusks; systematic malacology; zoogeography. Mailing Add: Dept of Biol Univ of PR Col Sta Mayaguez PR 00708

AGUIAR, ADAM MARTIN, b Newark, NJ, Aug 11, 29; m 64. ORGANIC CHEMISTRY. Educ: Fairleigh Dickinson Univ, BS, 55; Columbia Univ, MA, 57, PhD(chem), 59. Prof Exp: Org chemist, Otto B May, 47-55; asst, Columbia Univ, 55-59; asst prof chem, Fairleigh Dickinson Univ, 59-63; from asst prof to prof, Tulane Univ, 63-72; dean grad progs & res, William Paterson Col, 72-73; PROF CHEM, FAIRLEIGH DICKINSON UNIV, 73- Concurrent Pos: Fel, NIH, 59; sabbatical, Europe, 69-70; hon res prof, Biekbeck Col, Univ London, 70; res specialist, Rutgers Univ, Newark, 73-; consult, Cargill Industs, Cedar Grove NJ, 74- Mem: AAAS; Am Chem Soc; The Chem Soc; fel Am Inst Chemists; Am Asn Consult Chemists & Engrs. Res: Organo-phosphorus chemistry; medicinal chemistry; biochemistry and physiology. Mailing Add: Dept of Chem Fairleigh Dickinson Univ Madison NJ 07940

AGUIAR, ARMANDO JOSEPH, b Eldoret, Kenya, May 25, 27; m 55; c 7. PHYSICAL PHARMACY. Educ: Univ Poona, India, BSc, 52; St Louis Col Pharm, BS & MS, 55; Univ Wis, PhD, 59. Prof Exp: Asst, Sch Pharm, Univ Wis, 55-57; from assoc res pharmacist, to res pharmacist, 59-65, sr res scientist, 65-68, assoc lab dir, 68-70; DIR PHARMACEUT RES & DEVELOP, PFIZER INC, 70- Mem: AAAS; Am Chem Soc; Am Pharmaceut Asn; fel Acad Pharmaceut Sci. Res: Application of physical chemistry to pharmaceutical problems; pharmaceutical research and development. Mailing Add: Med Res Labs Pfizer Inc Groton CT 06340

AGUIAR, HARRIET GREENBERG, organic chemistry, see 12th edition

AGUILERA, FRANCISCO ENRIQUE, b Philadelphia, Pa, May 1, 43; m 67. CULTURAL ANTHROPOLOGY. Educ: Univ Pa, BA, 65, MA, 68, PhD(cult anthrop), 72. Prof Exp: ASST PROF ANTHROP, BOSTON UNIV, 70- Mem: Am Anthrop Asn; Royal Anthrop Inst Gt Brit & Ireland; Am Ethnol Soc. Res: Latin America and Iberia; social organization of rural agrarian peoples; community social organization; ritual process; acculturation; ethnohistory. Mailing Add: Dept of Anthrop Boston Univ Col of Lib Arts Boston MA 02215

AGUILO, ADOLFO, b Buenos Aires, Arg, Sept 6, 28; m; c 4. INDUSTRIAL CHEMISTRY. Educ: Univ Buenos Aires, MS, 53, PhD(chem), 55. Prof Exp: Asst res chemist, Steel Factory, Arg, 50-51; res chemist, Arg Air Force, 51-53; group leader, Arg AEC, 53-60; from res chemist to sr res chemist, 60-66, res assoc, 66-70, GROUP LEADER, CELANESE CHEM CO, 70- Concurrent Pos: Lab supvr, Univ Buenos Aires, 55-58, assoc prof, 58-60. Mem: Am Chem Soc. Res: Organometallics; solvent extraction applied to metal recovery; analytical and inorganic chemistry related with beryllium and uranium technology; catalysts; coordination compounds in catalysis; liquid phase oxidation of olefins. Mailing Add: Celanese Chem Co PO Box 9077 Corpus Christi TX 78408

AH, HYONG-SUN, b Suwon, Korea, May 27, 31; US citizen; m 58; c 3. VETERINARY PARASITOLOGY, ACAROLOGY. Educ: Seoul Nat Univ, DVM, 55; Univ Ga, PhD(med entom), 68. Prof Exp: Vet res assoc, Diag & Res Lab, Tifton, 69-71, ASST PROF PARASITOL, COL VET MED, UNIV GA, 71- Mem: Am Soc Parasitologists; Acarological Soc Am; Int Filariasis Asn; World Fedn Parasitologists. Res: Experimental filariasis in the areas of pathology, chemotherapy, immunology and host-parasite relationships. Mailing Add: Dept of Parasitol Col of Vet Med Univ of Ga Athens GA 30602

AHARONI, SHAUL MOSHE, b Tel Aviv, Israel, Dec 3, 33; US citizen; m 56; c 2. POLYMER CHEMISTRY. Educ: Univ Wis-Eau Claire, BS, 67; Case Western Reserve Univ, MS, 69, PhD(polymer sci), 72. Prof Exp: Asst chief chemist, Nat Presto Industs, Wis, 66-67, consult, 68-69; sr chemist, Gould Labs, Gould Inc, Ohio, 71-73; SR RES CHEMIST, ALLIED CHEM CORP, 73- Mem: Am Chem Soc; Am Phys Soc. Res: Structure property relationships in polymers, especially between geometrical parameters and thermomechanical performance of linear polymers; organization of the amorphous state, packing density and free volume. Mailing Add: Chem Res Lab Allied Chem Corp PO Box 1021R Morristown NJ 07960

AHEARN, DONALD G, b Grove City, Pa, Feb 1, 34; m 59; c 3. MYCOLOGY, MARINE MICROBIOLOGY. Educ: Mt Union Col, BS, 57; Univ Miami, MS, 59, PhD(microbiol), 64. Prof Exp: Instr microbiol & marine biol, Univ Miami, 63-64, asst prof, Sch Med & Inst Marine Sci, 64-66; asst prof microbiol, 67-68, ASSOC PROF MICROBIOL, GA STATE UNIV, 68-, DEAN, GRAD DIV, SCH ARTS & SCI, 70- Concurrent Pos: Mem, Int Oceanog Found. Mem: Am Soc Microbiol; Soc Indust Microbiol; Mycol Soc Am; Int Soc Human & Animal Mycol. Res: Ecology, physiology and systematics of fungi, chiefly yeasts, in aquatic habitats; epidemology of yeast-like fungi pathogenic to man. Mailing Add: Dept Biol Sch Arts & Sci Ga State Univ Atlanta GA 30303

AHEARN, GREGORY ALLEN, b Cambridge, Mass, Nov 28, 43; m 67. COMPARATIVE PHYSIOLOGY. Educ: Univ Calif, Los Angeles, BA, 65; Univ Hawaii, MS, 67; Ariz State Univ, PhD(zool), 70. Prof Exp: Teaching asst gen zool, Ariz State Univ, 67-68, teaching assoc, 68-69; res fel, Zoophysiol Lab A, August Krogh Inst, Univ Copenhagen, 70-72; ASST MARINE BIOLOGIST, HAWAII INST MARINE BIOL, 72- Concurrent Pos: Fel, Danish Res Coun, 70; Instnl grant, NSF, 75. Mem: Am Soc Zoologists; NY Acad Sci; Am Physiol Soc. Res: Membrane transport, epithelial biology, osmotic and ionic regulation, water balance, nutritional physiology, thermoregulation, environmental physiology. Mailing Add: Hawaii Inst Marine Biol PO Box 1346 Kaneohe HI 96744

AHEARN, JAMES JOSEPH, JR, b Beverly, Mass, May 21, 43. ANALYTICAL CHEMISTRY. Educ: Boston Col, BS, 65, PhD(chem), 69. Prof Exp: Res chemist, Org Chem Dept, Res Div, E I du Pont de Nemours & Co, 69-70; SCIENTIST ANAL CHEM, POLAROID CORP, 70- Mem: Am Chem Soc. Res: Chromatographic procedures for dyes, dye intermediates and photographic chemicals. Mailing Add: Polaroid Corp 600 Main St 1F Cambridge MA 02139

AHEARN, JAYNE NEWTON, b Sheboygan, Wis, Dec 30, 42; m 67. DEVELOPMENTAL GENETICS. Educ: Univ Wis, Madison, BS, 66; Ariz State Univ, PhD(zool), 73. Prof Exp: Jr researcher genetics, 72-75, ASST PROF GENETICS, UNIV HAWAII, 76- Mem: Am Soc Zool. Res: Gene control of gametogenesis; origins and genetic bases of reproductive interspecific isolating mechanisms; maternal effects; Drosophila developmental genetics. Mailing Add: Dept of Genetics Univ of Hawaii 1960 East West Rd Honolulu HI 96822

AHEARN, MICHAEL JOHN, b Jacksonville, Tex, June 22, 36; m 64. CELL BIOLOGY. Educ: Univ Tex, BA, 58, MA, 62, PhD(zool), 65. Prof Exp: Lectr zool, Univ Tex, 64-65; ASSOC BIOLOGIST, UNIV TEX M D ANDERSON HOSP & TUMOR INST, TEX MED CTR, 65- Mem: Int Soc Exp Hemat; Am Asn Cancer Res; Am Soc Cell Biol; Electron Micros Soc Am. Res: Hematology, especially cellular alterations induced by chemotherapeutic agents; electron microscopy; time-lapse cinematography; cytological techniques. Mailing Add: M D Anderson Hosp & Tumor Inst Tex Med Ctr 6723 Bertner Ave Houston TX 77025

AHEARNE, JOHN FRANCIS, b New Britain, Conn, June 14, 34; m 56; c 5. RESOURCE MANAGEMENT. Educ: Cornell Univ, BEngPhys, 57, MS, 58; Princeton Univ, MA, 63, PhD(plasma phys), 66. Prof Exp: Proj off nuclear weapons effects, Weapons Lab, US Air Force, Kirtland AFB, NMex, 59-61; from instr to asst prof physics, US Air Force Acad, 64-69; syst analyst, Off Asst Secy, Devense for Systs Anal, 69-70, dir tactical air prog, 70-72, dep asst secy defense, Off Asst Secy Defense for Prog Anal & Eval, 72-74, PRIN DEPT ASST SECY DEFENSE, OFF ASST SECY DEFENSE FOR MANPOWER & RESERVE AFFAIRS, 75- Mem: Am Phys Soc. Res: Kinetic theory of plasmas; electromagnetic theory; quantum mechanics. Mailing Add: Off Asst Secy Defense Systs Anal Pentagon Washington DC 20330

AHERN, FRANCIS JOSEPH, b New York, NY, June 21, 44; m 70; c 2. APPLIED PHYSICS. Educ: Cornell Univ, AB, 66; Univ Md, PhD(astron), 72. Prof Exp: Fel astrophys, Univ Toronto, 72-74; RES SCIENTIST APPL PHYSICS, CAN CTR REMOTE SENSING, 75- Mem: Am Astron Soc; Astron Soc Pac. Res: Research into methods of remote sensing of the earth for environmental and resource management with emphasis on spectrometry in visual and near infrared regions. Mailing Add: Can Ctr for Remote Sensing 2464 Sheffield Rd Ottawa ON Can

AHERNE, FRANCIS XAVIER, b Dublin, Ireland, June 14, 36; m 68; c 2. ANIMAL NUTRITION. Educ: Univ Col, Dublin, BSc, 59, MSc, 62; Iowa State Univ, PhD(animal nutrit), 67. Prof Exp: Teaching asst agr, Univ Col, Dublin, 59-61; res vis animal nutrit, Nat Res Sta, France, 61-63; lectr, Univ Col, Dublin, 62-63; asst, Iowa State Univ, 63-67; lectr, Univ Col, Dublin, 68-71; fel, Iowa State Univ, 71-72; ASSOC PROF ANIMAL SCI, UNIV ALTA, 72- Mem: Am Soc Animal Sci; Can Soc Animal Sci. Res: Arthrosis in growing pigs; baby pig mortality and performance; evaluation of the nutritive value of feedstuff available for animal nutrition. Mailing Add: 11634 77th Ave Edmonton AB Can

AHL, ALWYNELLE S, b Leesville, La, Mar 18, 41; m 63; c 1. ZOOLOGY, MAMMALOGY. Educ: Centenary Col, BS, 61; Univ Wyo, MS, 63, PhD(zool), 67. Prof Exp: Res asst biochem, Univ Wyo, 65-67; ASST PROF NATURAL SCI, MICH STATE UNIV, 67- Mem: AAAS; Am Soc Mammalogists; Res: Comparative physiology and biochemistry; physiological and biochemical studies of mammals in their natural environment; altitude physiology; biochemical systematics of mammals. Mailing Add: Dept of Natural Sci Univ Col Mich State Univ East Lansing MI 48823

AHLBERG, HENRY DAVID, b Boston, Mass, Oct 9, 39; m 62; c 2. BIOLOGY. Educ: NPark Col, AB, 61; Boston Univ, PhD(biol), 69. Prof Exp: Asst prof biol, Northeastern Univ, 67-76; ASSOC PROF BIOL, AM INT COL, 76- Mem: Wildlife Soc; Am Soc Zoologists; Am Soc Mammalogists. Res: Geographic variation of North American porcupine; seasonal variation of genital systems in rodents; aging in natural populations of rodents. Mailing Add: Dept of Biol Am Int Col Springfield MA 01109

AHLBERG, JOHN HAROLD, b Middletown, Conn, Dec 10, 27. APPLIED MATHEMATICS, NUMERICAL ANALYSIS. Educ: Yale Univ, BA, 50, MA, 54, PhD(math), 56; Wesleyan Univ, MA, 52. Prof Exp: Chief, Math Anal, United Aircraft Res Labs, 56-68; PROF APPL MATH, BROWN UNIV, 68- Mem: Am Math Soc; Math Asn Am; Soc Indust & Appl Math. Res: Application of splines to the numerical solution of boundary and initial value problems and their application to the representation of curves and surfaces. Mailing Add: Deepwood Dr Amston CT 06231

AHLBORN, BOYE, b Kampen, Ger, July 16, 33; m 61; c 3. PLASMA PHYSICS. Educ: Univ Kiel, dipl physics, 60; Munich Tech Univ, Dr rer nat, 64. Prof Exp: Sci asst, Inst Plasma Physics, Garching, Ger, 62-64; instr, 64-65, from asst prof to assoc prof, 65-72, PROF PLASMA PHYSICS, UNIV BC, 72- Mem: Can Asn Physicists; Ger Phys Soc. Res: Plasma flow with heat sources—arcs, detonations, radiation gas dynamics; shock waves; gas dynamical and chemical lasers. Mailing Add: Dept of Physics Univ of BC Vancouver BC Can

AHLBRANDT, CALVIN DALE, b Scottsbluff, Nebr, Aug 13, 40; m 61; c 3. MATHEMATICS. Educ: Univ Wyo, BS, 62; Univ Okla, MA, 65, PhD(math), 68. Prof Exp: ASSOC PROF MATH, UNIV MO-COLUMBIA, 68- Mem: Am Math Soc; Math Asn Am; Soc Indust & Appl Math. Res: Study of boundary value problems for systems of ordinary differential equations. Mailing Add: Dept of Math Univ of Mo Columbia MO 65201

AHLBRANDT, THOMAS STUART, b Torrington, Wyo, May 31, 48; m 69; c 2. GEOLOGY, SEDIMENTOLOGY. Educ: Univ Wyo, BA, 69, PhD(geol), 73. Prof Exp: Sr res geologist, Esso Prod Res Co, Tex, 73-74; RES GEOLOGIST, BR OIL & GAS RESOURCES, US GEOL SURV, 74- Concurrent Pos: NSF fel, 70, 71 & 72. Mem: Soc Econ Paleontologists & Mineralogists. Res: Modern and ancient Eolian deposits; development of equipment and techniques to recognize and describe dune and interdune deposits and practical applications of such work. Mailing Add: US Geol Surv Mail Stop 934 Denver Fed Ctr Denver CO 80225

AHLBRECHT, ARTHUR H, chemistry, see 12th edition

AHLER, STANLEY ALBERT, b Florence, Ala, Sept 10, 43; m 73. ANTHROPOLOGY. Educ: Univ Tenn, Knoxville, BS, 67; Univ Mo, Columbia Univ, MA, 70, PhD(anthrop), 75. Prof Exp: ASSOC CUR ANTHROP, ILL STATE MUS, 73- Concurrent Pos: Co-prin investr, Completion Archeol Invest at Rodgers Shelter, Mo, an archeol site on Nat Regis of His Places, 75-78. Mem: Am Anthrop Asn; Soc Am Archaeol; AAAS; Am Quaternary Asn; Plains Conf Anthrop. Res: Functional, technological and formal analysis of prehistoric stone tools and assemblages; the application of computerized techniques to the analysis of New and Old World archeological assemblages. Mailing Add: Quaternary Studies Ctr Ill State Mus Springfield IL 62706

AHLERS, GUENTER, b Bremen, Ger, May 28, 34; US citizen. PHYSICAL CHEMISTRY, SOLID STATE PHYSICS. Educ: Univ Calif, Riverside, BA, 59; Univ Calif, Berkeley, PhD(phys chem), 63. Prof Exp: Chemist silicate chem, Riverside Cement Co, 56-58; MEM TECH STAFF SOLID STATE PHYSICS, BELL TEL LABS, 63- Mem: AAAS; fel Am Phys Soc. Res: Thermodynamic properties of solidified gases; heat capacities; critical phenomena; liquid helium; transport properties. Mailing Add: Bell Tel Labs Murray Hill NJ 07974

AHLFELD, CHARLES EDWARD, b Aug 9, 40; m 62; c 3. REACTOR PHYSICS. Educ: Univ Fla, BS, 62; Fla State Univ, MS, 64, PhD(physics), 68. Prof Exp: RES PHYSICIST REACTOR PHYSICS, SAVANNAH RIVER LAB, E I DU PONT DE NEMOURS & CO, INC, 67- Mem: Am Phys Soc; Am Nuclear Soc. Res: Experimental and theoretical studies of static and kinetic nuclear reactor behavior; cross sections and nuclear data pertinent to reactor design; application of minicomputers to reactor experimentation. Mailing Add: E I du Pont de Nemours & Co Savannah River Lab Bldg 777M Aiken SC 29801

AHLFORS, LARS VALERIAN, b Helsinki, Finland, Apr 18, 07; nat US; m 33; c 3. MATHEMATICS. Educ: Univ Helsinki, PhD(math), 30. Hon Degrees: AM, Harvard Univ, 38; LLD, Boston Col, 51. Prof Exp: Adj math, Univ Helsinki, 33-36; asst prof, Harvard Univ, 36-38; prof, Univ Helsinki, 38-44 & Univ Zurich, 45-46; PROF MATH, HARVARD UNIV, 46- Concurrent Pos: Rockefeller fel, Paris, 32. Honors & Awards: Field's Medal, 36. Mem: Nat Acad Sci; Am Math Soc; Swedish Royal Soc. Res: Theory of functions of a complex variable; conformal mapping; Riemann surfaces. Mailing Add: Dept of Math Harvard Univ Cambridge MA 02138

AHLGREN, CLIFFORD ELMER, b Toimi, Minn, Apr 22, 22; m 54; c 2. FORESTRY. Educ: Univ Minn, BS, 48, MS, 53. Prof Exp: Forester, Iron Range Resources & Rehab, 48; res forester, 48-52, DIR, QUETICO-SUPERIOR WILDERNESS RES CTR, 52- Mem: AAAS; Soc Am Foresters; Am Forestry Asn; Ecol Soc Am. Res: Vegetational succession following natural disturbances in wilderness areas; field grafting and breeding of northern coniferous species; ecological effect of fire on northern coniferous forests. Mailing Add: Quetico-Superior Wilderness Res Ctr 215 W Oxford St Duluth MN 55803

AHLGREN, GEORGE E, b Cloquet, Minn, Dec 20, 31; m 61; c 2. PLANT PHYSIOLOGY. Educ: Univ Minn, BS, 59, MS, 62, PhD(agr, plant physiol), 66. Prof Exp: Asst prof, 66-71, ASSOC PROF BIOL, UNIV MINN, DULUTH, 71- Mem: Am Soc Plant Physiol. Res: Absorption and translocation of mineral ions and other substances by the plant. Mailing Add: Dept of Biol Univ of Minn Duluth MN 55812

AHLGREN, HENRY LAWRENCE, b Wyoming, Minn, Oct 3, 08; m 36; c 2. AGRONOMY. Educ: Univ Wis, BS, 31, MS, 33, PhD(agron), 35. Prof Exp: Asst agron & soils, 29-35, from instr to prof agron, 35-74, chmn dept, 49-52, assoc dir agr exten, 52-67, asst chancellor, Univ Exten, 66-67, vchancellor, 67-69, chancellor, 69-74, dir coop exten serv, 69-70, EMER PROF AGRON & EMER CHANCELLOR UNIV EXTEN, UNIV WIS-MADISON, 74- Concurrent Pos: Traveling fel, Europe, 36; dep undersecy rural develop, USDA, 70. Honors & Awards: Distinguished Serv Award, USDA, 59. Mem: Fel Am Soc Agron; fel Royal Swed Acad Agr & Forestry. Res: Pasture improvement, particularly the fertilization, management and ecological aspects; effect of various fertilizers, cutting treatments and irrigation on yield of forage and chemical composition of rhizomes of Kentucky bluegrass. Mailing Add: 24 Park Pl Madison WI 53705

AHLGREN, ISABEL FULTON, b Viroqua, Wis, Feb 11, 24; m 54; c 2. BOTANY. Educ: DePauw Univ, AB, 46; Ind Univ, PhD(bot), 50. Prof Exp: Asst bot, Ind Univ, 46-49, instr, 50; instr, Wheaton Col, Mass, 50-52 & Wellesley Col, 52-54; RES ASSOC, QUETICO-SUPERIOR WILDERNESS RES CTR, 55- Concurrent Pos: Lectr, Univ Minn, Duluth, 55, 62-73. Mem: Bot Soc Am. Res: Boreal forest ecology. Mailing Add: Quetico-Superior Wilderness Res Ctr 215 W Oxford St Duluth MN 55803

AHLQUIST, RAYMOND PERRY, b Missoula, Mont, July 26, 14; m 39. PHARMACOLOGY. Educ: Univ Wash, BS, 35, MS, 37, PhD(pharmacol), 40. Prof Exp: Asst prof pharmacol & pharmacog, SDak State Col, 40-44, pharmacologist, Exp Sta, 40-44; from asst prof to prof pharmacol & head dept, 44-63, assoc dean, 63-70, PROF PHARMACOL & HEAD DEPT, MED COL GA, 70- Mem: Am Soc Pharmacol & Exp Therapeut; NY Acad Sci; Am Physiol Soc; Am Col Clin Pharmacol; Am Pharmaceut Asn. Res: Pharmacology of sympathomimetic and adrenergic blocking agents; clinical pharmacology. Mailing Add: Dept of Pharmacol Med Col of Ga Augusta GA 30902

AHLRICHS, JAMES LLOYD, b Palmer, Iowa, Sept 13, 28; m 52; c 4. SOIL MINERALOGY. Educ: Univ Iowa, BS, 50, MS, 55; Purdue Univ, PhD(soil chem), 61. Prof Exp: Teacher, Pierson Pub Sch, 50-51; from instr to assoc prof, 57-68, PROF SOIL CHEM, PURDUE UNIV, WEST LAFAYETTE, 68- Concurrent Pos: Vis scientist, Macaulay Inst Soil Sci, Scotland, 66-67; sr Fulbright lectr, Coun Sci Invest & Autonomous Univ Madrid, 73-74. Mem: Am Chem Soc; Soil Sci Soc Am; Soil Conserv Soc Am; Am Soc Agron; Clay Minerals Soc. Res: Clay mineralogy and reactions at colloidal surfaces. Mailing Add: Dept of Agron Purdue Univ West Lafayette IN 47907

AHLSCHWEDE, WILLIAM T, b Lincoln, Nebr, Jan 31, 42; m 64; c 2. ANIMAL BREEDING. Educ: Univ Nebr, BS, 64; NC State Univ, MS, 67, PhD(animal sci), 70. Prof Exp: Instr animal sci, NC State Univ, 67-69; res assoc med genetics, Univ Wis, 69-70; asst prof, 70-75, ASSOC PROF ANIMAL SCI, UNIV NEBR, 75- Mem: Am Soc Animal Sci. Res: Genetic and maternal effects in swine and beef cattle. Mailing Add: Dept of Animal Sci Col Agr Univ Nebr Lincoln NE 68503

AHLSTROM, ELBERT HALVOR, b Sharon, Pa, Feb 15, 10; m 54. BIOLOGY. Educ: Marietta Col, AB, 30; Ohio State Univ, MA, 33, PhD, 34. Prof Exp: Aquatic biologist, Div Conserv, Ohio Univ, 30; RES BIOLOGIST, FISH & WILDLIFE SERV, STANFORD UNIV, 35, 39-; ADMINR, FISHERY-OCEANOG CTR, NAT MARINE FISHERIES SERV, 64-, SR SCIENTIST, CALIF CURRENT RESOURCES LAB, 67- Concurrent Pos: Res biologist, Fish & Wildlife Serv, La Jolla, Calif, 49-54, asst chief, SPac Fishery Invest, 54-59, lab dir, 59-64; dir, Calif Current Resources Lab, 64-67; res assoc, Scripps Inst Oceanog, 51-67, adj prof, 67- Honors & Awards: Gold Medal, US Dept Com, 73; Outstanding Achievement Award, Am Inst Fishery Res Biologists, 75. Mem: AAAS; Am Soc Ichthyologists & Herpetologists; Am Micros Soc; Am Soc Limnol & Oceanog. Res: Life history and ecology of pelagic marine fishes; taxonomy rotifera and fishes. Mailing Add: 2475 Chatsworth Blvd San Diego CA 92106

AHLUWALIA, BALWANT SINGH, b India, Mar 12, 32; m 62; c 1. ENDOCRINOLOGY. Educ: Univ Minn, MS, 59, PhD(reproduction), 62. Prof Exp: Scientist, Worcester Found Exp Biol, 67-68; scientist, Nat Inst Arthritis & Metab Dis, 68-70; asst prof, 70-72, ASSOC PROF OBSTET & GYNEC, COL MED, HOWARD UNIV, 72- Concurrent Pos: Fel physiol, Hormel Inst, Univ Minn, 63-67. Mem: Am Vet Med Asn; Endocrine Soc; Am Inst Nutrit; Am Oil Chemists Soc. Res: Lipid metabolism in the reproductive organs with respect to fertility and sterility in animals. Mailing Add: Dept of Obstet & Gynec Howard Univ Washington DC 20001

AHLUWALIA, DALJIT SINGH, b Sialkot, India, Sept 5, 32; m 60; c 4. APPLIED MATHEMATICS. Educ: Punjab Univ, India, BA, 52, MA, 55; Ind Univ, Bloomington, MS, 65, PhD(appl math), 65. Prof Exp: Lectr math, R G Col, Phagwara, Punjab, 55-57, Khalsa Col, Bombay, 57-62 & Univ Bombay, 59-62; part

time teaching assoc, Ind Univ, Bloomington, 62-65, asst prof, 65-66; vis mem & adj asst prof, 66-68, asst prof, 68-69, ASSOC PROF MATH, COURANT INST MATH SCI, NY UNIV, 69- Concurrent Pos: Prof math, Univ SFla, Tampa, 72-74. Mem: Am Math Soc; Soc Indust & Appl Math; Tensor Soc; Int Sci Radio Union. Res: Plastic flow and fracture in solids; uniform theories of diffraction for the edges and convex bodies; study of wave propagation in elastic media and acoustics. Mailing Add: Courant Inst Math Sci NY Univ 251 Mercer St New York NY 10012

AHLUWALIA, GURJIT SINGH, b Burewala, India, Nov 11, 32; m 67. BIOCHEMISTRY. Educ: Govt Col, India, BS, 52; Punjab Univ, India, BS, 54, MS, 56; Cornell Univ, PhD(biochem), 67. Prof Exp: Teaching asst org chem, Punjab Univ, India, 55-56; res asst biochem, Malaria Inst India, 56-58, res officer, 58-62; res asst biochem, Cornell Univ, 62-67; res assoc, St Jude Childrens Res Hosp, Memphis, Tenn, 67-68; proj scientist med diag opers, Xerox Corp, Calif, 68-70; asst dir clin biochem, Calbiochem, 70-71; CLIN CHEMIST, BINGHAMTON GEN HOSP, 71- Mem: AAAS; Am Chem Soc; Am Asn Clin Chemists; NY Acad Sci; Fedn Am Scientists. Res: Enzymology; metabolism; diagnostic reagents; automated blood analyses; clinical chemistry; diagnostic kits; quality control; radioimmunoassay. Mailing Add: Binghamton Gen Hosp Mitchell Ave Binghamton NY 13903

AHLUWALIA, HARJIT SINGH, b Bombay, India, May 13, 34; m 64; c 2. COSMIC RAY PHYSICS, HIGH ENERGY ASTROPHYSICS. Educ: Panjab Univ, India, BSc, 53, MSc, 54; Gujarat Univ, India, PhD(physics), 60. Prof Exp: Sr res asst cosmic rays, Phys Res Lab, Navrangpura, Gujarat, India, 54-62; tech asst expert, UNESCO, France, 62; res assoc cosmic rays & geomagnetism, Southwest Ctr Advan Studies, Tex, 63-64; prof physics, Int Atomic Energy Agency, Austria & sci dir, Lab Cosmic Physics, Univ La Paz, 65-67; prof physics, Pan Am Union, Washington, DC, 67; assoc prof, 68-73, PROF PHYSICS, UNIV N MEX, 73- Concurrent Pos: NASA fel, 63-64; chmn, Bolivian Space Res Comt, 65-67; prin investr, US Air Force res proj, 65-68, NSF res proj, 68- & Sandia Corp res proj, Albuquerque, 69-71; Bolivian nat rep, Comt Space Res & Int Union Pure & Appl Physics, 66-67; corresp mem, Cosmic Ray Comn, Int Union Pure & Appl Physics, 66-69. Mem: AAAS; Am Geophys Union; Inst Elec & Electronics Engrs; Am Meteorol Soc; Am Phys Soc. Res: Cosmic rays; geomagnetism; nuclear electronics; space physics; plasma physics; astrophysics. Mailing Add: Dept of Physics & Astron Univ of NMex Albuquerque NM 87131

AHMAD, IQBAL, b Sangroor, India, Aug 1, 25; m 51; c 4. PHYSICAL CHEMISTRY, METALLURGY. Educ: Univ Panjab, WPakistan, BSc, 45, MSc, 50; Univ London, PhD(phys chem, extractive metall) & DIC, 63; FRIC; cert chem, Royal Inst Chemists, 74. Prof Exp: Res scholar chem, Fazli-Omer Res Inst, Qadian, 45-47; tech asst, Govt Indust Res Inst, Lahore, 47-49; res officer chem & explosives, 50-56; res officer & head sect ore dressing & metall, WRegional Labs, Pakistan Coun Sci & Indust Res, 56-59; res fel phys chem, Rensselaer Polytech Inst, 63-64; chief phys chem lab, 64-70; GROUP LEADER, BENET WEAPONS LAB, US ARMY WATERVLIET ARSENAL, 70- Mem: Am Chem Soc; Am Ceramic Soc; Am Inst Mining, Metall & Petrol Eng. Res: Extractive metallurgy and high temperature physical chemistry including phase equilibria and thermodynamics; growth of high strength whisker crystals and high strength filament reinforced metal composites; chemical vapor deposition of refractory metals and alloys; development of high temperature materials for armaments and gas turbines. Mailing Add: Benet Weapons Lab Watervliet Arsenal Watervliet NY 12189

AHMAD, IRSHAD, b Azamgarh, India, Nov 1, 39; m 69; c 2. NUCLEAR CHEMISTRY. Educ: Univ Punjab, Pakistan, BSc, 58; Univ Peshawar, MSc, 62; Univ Pac, MS, 65; Univ Calif, Berkeley, PhD(chem), 66. Prof Exp: Demonstr chem, Edwardes Col, Peshawar, 58-60; res fel, Lawrence Radiation Lab, Univ Calif, Berkeley, 66; res assoc, 66-68, asst chemist, 68-70, CHEMIST, ARGONNE NAT LAB, 70- Mem: Am Phys Soc; fel Am Inst Chemists; Am Chem Soc. Res: Synthesis of new actinide isotopes; nuclear structure studies of transuranium nuclei by high-resolution alpha, beta and gamma spectroscopy and nuclear reactions. Mailing Add: Chem Div Argonne Nat Lab Argonne IL 60439

AHMAD, NAZIR, b WPakistan, Feb 22, 36; US citizen; m 55; c 1. HISTOLOGY, REPRODUCTIVE ENDOCRINOLOGY. Educ: San Francisco State Col, BA, 55; Univ Calif, Berkeley, MA, 59; Univ Calif, San Francisco, PhD(anat), 68. Prof Exp: Asst prof histol, Sch Med, Georgetown Univ, 68-69; ASSOC PROF HISTOL, SCH MED, UNIV SOUTHERN CALIF, 69- Mem: AAAS; Endocrine Soc; Soc Study Reproduction; Am Asn Anat. Res: Reproductive endocrinology, utilizing pituitary as well as ovarian hormones essential for the maintenance of gestation and lactation; maintenance of spermatogenesis; human histology. Mailing Add: Dept of Anat Univ of Southern Calif Los Angeles CA 90033

AHMAD, SHAIR, b Kabul, Afghanistan, June 19, 34; m 58; c 3. MATHEMATICS. Educ: Univ Utah, BS, 60, MS, 62; Case Western Reserve Univ, PhD(math), 68. Prof Exp: Asst math, Univ Utah, 60-62; instr, SDak State Univ, 62-64; asst, Case Western Reserve Univ, 64-65, instr, 66-68; asst prof, Univ NDak, 65-66; ASSOC PROF, OKLA STATE UNIV, 68- Mem: Am Math Soc; Math Asn Am. Res: Dynamical systems; algebra; differential equations. Mailing Add: Dept of Math & Statist Okla State Univ Stillwater OK 74074

AHMADJIAN, VERNON, b Whitinsville, Mass, May 19, 30; m 56; c 3. BOTANY. Educ: Clark Univ, AB, 52, MA, 56; Harvard Univ, PhD, 59. Prof Exp: Lab asst, Clark Univ, 54-56; lab asst, Harvard Univ, 56, asst, Farlow Herbarium, 56-58; from asst prof to prof bot, Clark Univ, 59-68; prof, Univ Mass, 68-69; assoc dean grad sch, 69-71, PROF BOT & COORDR RES, CLARK UNIV, 69-, DEAN GRAD SCH, 71- Concurrent Pos: Vis prof, Univ Calif, Berkeley, 65-66. Honors & Awards: New York Bot Garden Award, 68; Antarctic Medal. Mem: AAAS. Res: Cryptogamic botany; mycology; lichenology. Mailing Add: Dept of Biol Clark Univ Worcester MA 01610

AHMADZADEH, AKBAR, b Isfahan, Iran, Sept 2, 32; US citizen; m 61; c 2. PHYSICS. Educ: Univ Calif, Berkeley, BA, 56, PhD(physics), 64. Prof Exp: Res assoc physics, Univ Calif, San Diego, 64-66; asst prof, 66-68, ASSOC PROF PHYSICS, ARIZ STATE UNIV, 68- Concurrent Pos: AEC fel, 64-66. Res: Theoretical nuclear physics; high energy physics; scattering theory; symmetries. Mailing Add: Dept of Physics Ariz State Univ Tempe AZ 85281

AHMAN, MOID UDDIN, b Agra, India, Aug 18, 27; m; c 2. GROUNDWATER HYDROLOGY, GEOPHYSICS. Educ: Agra Univ, BSc, 44, LLB, 47; NMex Inst Mining & Technol, MS, 61; Univ London, PhD(geol, seismol), 66. Prof Exp: Asst seismologist, Geophys Inst Quetta, Pakistan, 49-58; head dept physics, Battersea Grammar Sch, London, Eng, 61-67; geophysicist, Food & Agr Orgn; UN & consult to Govt of Kuwait, 67-69; assoc prof hydrol & geophys, 69-74, PROF HYDROL & GEOPHYS, OHIO UNIV, 74- Mem: Am Geophys Union; Soc Explor Geophys; Europ Asn Explor Geophys; Am Water Resources Asn. Res: Exploration of groundwater in arid regions; acid mine drainage control, development of a hydrological approach to control acid discharge from strip and drift mines and techniques to map acid producing areas by thermal mapping. Mailing Add: Dept of Geol Rm 416 Porter Hall Ohio Univ Athens, OH 45701

AHMED, ASAD, b Saharanpur, India, Nov 7, 39; Can citizen. MOLECULAR GENETICS. Educ: Aligarh Muslim Univ, India, BSc, 56, MSc, 58, PhD(plant path), 61; Yale Univ, PhD(biochem genetics), 64. Prof Exp: Fulbright scholar, Yale Univ, 60-64, univ fel, 61-62, Wadsworth fel, 62-63, Sterling fel, 63-64; res assoc, Inst Molecular Biol, Univ Ore, 64-65; res scientist biochem, Dept Nat Health & Welfare, Can, 66-67; asst prof, 67-70, ASSOC PROF MOLECULAR GENETICS, UNIV ALTA, 70- Mem: Genetics Soc Am; Am Soc Microbiol. Res: Repression of methionine biosynthesis in Escherichia coli; insertion mutations in Escherichia coli; organization of histidine-3 region of Neurospora. Mailing Add: Dept of Genetics Univ of Alta Edmonton AB Can

AHMED, ESAM MAHMOUD, b Cairo, UAR, Oct 7, 25; US citizen; m 56, 67; c 4. FOOD SCIENCE. Educ: Univ Cairo, BS, 45; Univ Alexandria, MS, 53; Univ Md, PhD(hort), 57. Prof Exp: From instr to asst prof veg crops physiol, Univ Alexandria, 45-59; res assoc hort physiol, Univ Md, 59-64; from asst prof to assoc prof, 64-75, PROF FOOD SCI, UNIV FLA, 75- Concurrent Pos: Joseph H Gourley res award, 64. Mem: Am Soc Host Sci; Inst Food Technologists. Res: Food quality; psychophysical aspects of foods; food color and texture measurements; food irradiation; post-harvest physiology. Mailing Add: Dept of Food Sci Univ of Fla Gainesville FL 32601

AHMED, ISMAIL YOUSEF, b Silwad, Jordan, Sept 27, 39. INORGANIC CHEMISTRY. Educ: Cairo Univ, BS, 60; Pa State Univ, MS, 65; Southern Ill Univ, Carbondale, PhD(chem), 68. Prof Exp: High sch teacher, Jordan, 60-64; lectr chem, Southern Ill Univ Carbondale, 68-69; asst prof, 69-73, ASSOC PROF CHEM, UNIV MISS, 73- Mem: Am Chem Soc. Res: Lewis acid-Lewis base interactions; interactions of metal halides with compounds containing N-S bond; characterization of solute species in nonaqueous solvents; solvation and rates of solvent exchange by nuclear magnetic resonance. Mailing Add: Dept of Chem Univ of Miss University MS 38677

AHMED, KHALIL, b Lahore, Pakistan, Nov 30, 34; US citizen; m 69. BIOCHEMISTRY. Educ: Univ Panjab, Pakistan, BSc, 54, MSc, 55; McGill Univ, PhD(biochem), 60. Prof Exp: Res chemist, WRegional Labs, Pakistan, 55-57; res asst biochem, Montreal Gen Hosp & Res Inst, McGill Univ, 57-60; res fel, Wistar Inst, Univ Pa, 60-61, res assoc, 61-63; asst prof metab res, Chicago Med Sch, 63-67; sr staff mem & res biochemist, Lab Pharmacol, Baltimore Cancer Res Ctr, Nat Cancer Inst, 67-72; RES BIOCHEMIST & CHIEF TOXICOL RES LAB, VET ADMIN HOSP, MINNEAPOLIS, 72-; ASSOC PROF LAB MED & PATH, UNIV MINN, MINNEAPOLIS, 73- Concurrent Pos: Vis lectr, Chicago Med Sch, 68-69. Mem: AAAS; Am Soc Biol Chemists; Am Chem Soc; Am Soc Pharmacol & Exp Therapeut; Endocrine Soc. Res: Enzymic mechanism of drug action; biochemistry of ion transport; biochemistry of androgen action in male sex glands; neurochemistry; phosphoproteins; biochemistry of gene action. Mailing Add: Toxicol Res Lab Vet Admin Hosp Minneapolis MN 55417

AHMED, SAIYED I, b Desna, Pakistan, Jan 5, 41; m 64; c 2. MICROBIOLOGY, BIOCHEMICAL GENETICS. Educ: Univ Karachi, BSc, 60; Univ Frankfurt, PhD(microbiol, biochem, bot), 63. Prof Exp: Sr res off microbiol & fermentation res, Pakistan Coun Sci & Indust Res, 64; Nat Res Coun Can fel biochem genetics, Univ Man, 64-66; NIH sr fel, Yale Univ, 66-69, res assoc, 69-70; vis scientist, Sch Med, 70-73, RES ASSOC, DEPT OCEANOG, UNIV WASH, 73- Mem: Am Soc Microbiol; Am Soc Plant Physiologists. Res: Mutation genetics in serratia and E coli; control mechanisms and regulations; enzymology; mutagenicity; pollution and environmental studies; marine bacteria and marine phytoplankton metabolism. Mailing Add: Dept Oceanog WB-10 Univ of Wash Seattle WA 98195

AHN, BIRUTA V EIMANIS, plant pathology, botany, see 12th edition

AHNERT, FRANK, b Wittgensdorf, Ger, Dec 12, 27. PHYSICAL GEOGRAPHY, GEOGRAPHY OF AMERICA. Educ: Univ Heidelberg, PhD(geog), 53. Prof Exp: From asst prof to assoc prof geog, 56-66, PROF GEOG, UNIV MD, COLLEGE PARK, 66- Concurrent Pos: Ger Res Asn-Nat Res Coun fel, Univ Md, 54-56; corresp mem comn study slope evolution, Int Geog Union, 64-68; vis assoc prof, Pa State Univ, 66; corresp mem comn study present-day geomorphol processes, Int Geog Union, 68- Mem: AAAS; Asn Am Geog; fel Am Geog Soc. Res: Geomorphology; general physical geography; regional geography of North America and Europe. Mailing Add: Dept of Geog Univ of Md College Park MD 20742

AHO, AARO E, b Ladysmith, BC, June 20, 25; m 49; c 2. PETROLOGY, ECONOMIC GEOLOGY. Educ: Univ BC, BA & BASc, 49; Univ Calif, Berkeley, PhD(geol sci), 54. Prof Exp: Explor mgr, White Pass & Yukon Corp, 53-57; consult, various mining co, 57-64; PRES, DYNASTY EXPLOR LTD, 64- & Atlas Explor Ltd, 65- Concurrent Pos: Instr, Ore State Univ, 53-54; mem senate, Univ BC, 69. Mem: Geol Soc Am; Geol Asn Can; Can Inst Mining & Metall. Res: Igneous petrology of nickelferrous ultrabasics; regional structural environment of mineral deposits in Yukon. Mailing Add: 330-355 Burrard St Vancouver BC Can

AHO, PAUL E, b Worcester, Mass, June 29, 34; m 57; c 3. PLANT PATHOLOGY. Educ: Univ Mass, BS, 56; Yale Univ, MF, 57; Ore State Univ, PhD(plant pathology), 76. Prof Exp: PLANT PATHOLOGIST, FORESTRY SCI LAB, FOREST SERV, USDA, 57- Mem: Soc Am Foresters; Am Phytopath Soc. Res: Heart rots of western conifers; dwarf mistletoes; development of methods for estimating or predicting the extent of damage. Mailing Add: Forestry Sci Lab Forest Serv USDA 3200 Jefferson Way Corvallis OR 97331

AHO, WILLIAM A, b Bessemer, Mich, Jan 29, 18; m 42; c 2. POULTRY SCIENCE. Educ: Mich State Col, BS, 42, MS, 48. Prof Exp: Exten poultryman, Mich State Col, 48-51; assoc prof poultry sci, 52-72, PROF POULTRY SCI, COL AGR, UNIV CONN, 72- Mem: Poultry Sci Asn. Mailing Add: Dept of Animal Industs Univ of Conn Col of Agr Storrs CT 06280

AHRAMJIAN, LEO, b Boston, Mass, Aug 24, 30; m 59; c 2. ORGANIC CHEMISTRY. Educ: Univ Calif, Berkeley, BS, MS, 54; Northwestern Univ, PhD(chem), 59. Prof Exp: RES CHEMIST, ELASTOMER CHEM DEPT, E I DU PONT DE NEMOURS & CO, 59- Mem: Am Chem Soc. Res: Adhesives and coatings research and development; isocyanate and carbodiimide chemistry; synthesis and properties of polyurethanes; elastomers for vibration damping and vibration isolation; general elastomer technology. Mailing Add: 2122 Valley Rd Wilmington DE 19810

AHRENKIEL, RICHARD K, b Springfield, Ill, Jan 7, 36; m 56; c 1. SOLID STATE PHYSICS. Educ: Univ Ill, BS, 59, PhD(physics), 64. Prof Exp: Res physicist, 64-70, RES ASSOC, RES LABS, EASTMAN KODAK CO, 70- Mem: Am Phys Soc. Res: Transport phenomena in high impedance semiconductors, particularly the silver and alkali halides and studies of defect centers; optical properties of magnetic semiconductors. Mailing Add: Res Labs Eastman Kodak Co Rochester NY 14650

AHRENS, EDWARD HAMBLIN, JR, b Chicago, Ill, May 21, 15; m 40; c 3. MEDICINE. Educ: Harvard Univ, BS, 37, MD, 41. Prof Exp: Intern, Babies Hosp, 42-43, chief resident, 51-52; asst, Rockefeller Inst, 46-49; assoc, 52-58, assoc prof, 58-60, PROF BIOCHEM & MED, ROCKEFELLER UNIV, 60- Concurrent Pos: NSF sr fel, 48-59; Nat Res Coun sr fel, 49-50; Nat Found Infantile Paralysis sr fel, 50-52; mem metab study sect, USPHS, 56-61, chmn, 59-61; mem adv bd, J Lipid Res, 58-74, ed, 63-69; mem bd sci counsr, Nat Heart Inst, 63-67, chmn diet-heart rev panel, 67-68; mem sci adv comt, New Eng Regional Primate Ctr, 63-69; pres, Lipid Res, Inc, 63-74; mem sci adv coun, 65; mem, Stouffer Prize Selection Comt, 66-69; mem tech adv comt, Inst Human Nutrit, Columbia Univ, 66-71; mem gen clin res ctrs comt, NIH, 70-74; master prof, Cornell Univ Med Col, 70-74; mem sci adv comt, Hirschl Trust, 72-74; mem bd dirs & exec comt, Regional Plan Asn, New York, 73-; mem adv comt, Ernst Klenk Found, 75- Mem: Nat Acad Sci; Inst Med Nat Acad Sci; Soc Pediat Res; Asn Am Physicians; Soc Exp Biol & Med. Res: Biochemistry of lipids and clinical investigation in field of lipid metabolism. Mailing Add: Rockefeller Univ York Ave & 66th St New York NY 10021

AHRENS, FRANKLIN ALFRED, b Leigh, Nebr, Apr 27, 36; m 60; c 4. PHARMACOLOGY, TOXICOLOGY. Educ: Kans State Univ, BS & DVM, 59; Cornell Univ, MS, 65, PhD(pharmacol), 68. Prof Exp: From asst prof to assoc prof, 68-75, PROF PHARMACOL, IOWA STATE UNIV, 75- Mem: NY Acad Sci; Sigma Xi. Res: Toxicity of chelating agents and lead; enteric disease and drug absorption. Mailing Add: Dept of Vet Physiol & Pharmacol Iowa State Univ Ames IA 50010

AHRENS, JOHN FREDERICK, b Bellmore, NY, Nov 21, 29; m 52; c 4. WEED SCIENCE. Educ: Univ Ga, BS, 54; Iowa State Univ, MS, 55, PhD(plant physiol), 57. Prof Exp: From asst plant physiologist to assoc plant physiologist, 57-70, PHYSIOLOGIST, CONN AGR EXP STA, 70- Mem: Am Soc Plant Physiol; Weed Sci Soc Am; Int Plant Propagators Soc. Res: Weed control; residual effects of herbicides in soils; plant growth; inactivation of herbicide residues in soil. Mailing Add: Valley Lab Conn Agr Exp Sta PO Box 248 Windsor CT 06095

AHRENS, RICHARD AUGUST, b Manitowoc, Wis, Sept 18, 36; m 61; c 3. NUTRITION, PHYSIOLOGY. Educ: Univ Wis, BS, 58; Univ Calif, Davis, PhD(nutrit), 63. Prof Exp: Res physiologist human nutrit, Agr Res Serv, USDA, 63-66; assoc prof, 66-75, PROF FOOD & NUTRIT, UNIV MD, COLLEGE PARK, 75- Mem: Am Inst Nutrit; Am Home Econ Asn; NY Acad Sci; Soc Nutrit Educ; Nutrit Today Soc. Res: Diet and atherosclerosis; enzymes and energy metabolism; dietary carbohydrate and the effects of physical activity on health; sucrose and high blood pressure. Mailing Add: Dept of Food Nutrit & Instnl Admin Univ of Md College Park MD 20742

AHRENS, ROLLAND WILLIAM, b Clarkson, Nebr, Sept 28, 33; m 63. PHYSICAL CHEMISTRY. Educ: Univ Nebr, BSc, 54, MS, 55; Univ Wis, PhD(phys chem), 59. Prof Exp: Chemist, Separations Div, Savannah River Lab, 59-64, chemist, Seaford Textile Fibers Plant, 65-67, ANAL RES SUPVR, TEXTILE FIBERS DEPT, E I DU PONT DE NEMOURS & CO, 67- Concurrent Pos: Traveling lectr, Oak Ridge Inst Nuclear Studies, 63- Mem: Am Chem Soc. Res: Radiation chemistry of aqueous solutions, ion exchange, and liquid ammonia; recovery of transplutonium elements by solvent extraction; nylon carpet new products development. Mailing Add: 901 Short Ln Seaford DE 19973

AHRENS, RUDOLF MARTIN (TINO), b St Louis, Mo, Feb 6, 28; m 53; c 2. PHYSICS. Educ: Wash Univ, PhD, 52. Prof Exp: Sr nuclear engr, Convair, 52-54; staff specialist, Lockheed Aircraft Corp, 54-56, dept mgr nuclear anal, 56-57; assoc prof, 57-66, PROF THEORET PHYSICS, GA INST TECHNOL, 66- Concurrent Pos: Vis prof, Max-Planck Inst Physics & Astrophys, 59-60 & Univ SC, 67-68; consult & vpres, Advan Res Corp Ind. Res: Beta-decay; reactors; shielding and hazards; elementary particles; symmetries; relativity. Mailing Add: Dept of Physics Ga Inst of Technol Atlanta GA 30332

AHRENS, THOMAS J, b Frankfurt, Ger, Apr 25, 36; US citizen; m 58; c 2. GEOPHYSICS, HIGH-PRESSURE PHYSICS. Educ: Mass Inst Technol, BS, 57; Calif Inst Technol, MS, 58; Rensselaer Polytech Inst, 62. Prof Exp: Intermediate explor geophysicist, Pan Am Petrol Corp, 58-59; asst geophys, Rensselaer Polytech Inst, 62; geophysicist, Poulter Res Labs, Stanford Res Inst, 62-66, head geophys group, Shock Wave Physics Div, 66-67; ASSOC PROF GEOPHYS, CALIF INST TECHNOL, 67- Concurrent Pos: Assoc ed, Rev Sci Instruments & J Geophys Res, 71-74; mem earth sci adv panel, NSF, 72-75. Mem: AAAS; Am Geophys Union; Soc Explor Geophys; Am Phys Soc. Res: Shock and ultrasonic wave propagation in solids; exploration seismology; high-pressure physics of the earth's interior; impact effects on planetary surfaces. Mailing Add: Div Geol & Planetary Sci Calif Inst of Technol Pasadena CA 91125

AHRING, ROBERT M, b Lincoln, Kans, Oct 4, 28; m 51; c 4. AGRONOMY, SOILS. Educ: Okla State Univ, BS, 52, MS, 58; Univ Nebr, PhD, 72. Prof Exp: Instr, 52-56, res asst, 56-57, asst prof, 57-75, ASSOC PROF AGRON, OKLA STATE UNIV, 75-; RES AGRONOMIST, FORAGE & RANGE, SOUTHERN REGION, AGR RES SERV, USDA, 57- Mem: Crop Sci Soc Am; Am Soc Agron; Am Soc Range Mgt. Res: Seed production and technology; native and introduced grasses and legumes; crop physiology. Mailing Add: Dept of Agron Okla State Univ Stillwater OK 74075

AHSAN, S REZA, b Monghyr, India, Mar 31, 37; m 57; c 2. GEOGRAPHY. Educ: Aligarh Muslim Univ, India, BSc, 54, MSc & LLB, 56; Univ Fla, PhD(geog, geol, civil eng), 63. Prof Exp: Lectr geog, Univ Dacca, 56-58; asst prof, Western Ky Col, 62-63 & Haile Selassie Univ, 63-67; assoc prof, Wis State Univ, 67-68; assoc prof, 68-70, PROF GEOG, WESTERN KY UNIV, 70- Concurrent Pos: Lectr & consult, Dacca & Addis Ababa; fac fel, Johnson Space Ctr, NASA, 74-75. Mem: AAAS; Asn Am Geographers; Am Soc Photogram. Res: Cartography; agricultural geography; food; space photo uses. Mailing Add: 108 Springhill Rd Bowling Green KY 42101

AHSHAPANEK, DON COLESTO, b Milton, Del, Apr 29, 32; m 62; c 3. PLANT ECOLOGY. Educ: Cent State Univ, Okla, BS, 56; Univ Okla, MS, 59, PhD(bot), 62. Prof Exp: Asst prof biol, Kans State Teachers Col, 62-67, assoc prof, 67-71; instr biol, 71-73, CHMN DIV NATIVE AM CULT, HASKELL AM INDIAN JR COL, 73- Mem: Ecol Soc Am. Res: Ecological research in causes of plant succession in grasslands with studies in phenology, plant inhibition and the inter-relations between micro-organisms and plant roots in the rhizosphere and rhizoplane. Mailing Add: Dept of Biol Haskell Am Indian Jr Col Lawrence KS 66044

AHUJA, JAGAN N, b Rawalpindi, WPakistan, Nov 12, 35; m 60; c 3. BIOCHEMISTRY, CLINICAL CHEMISTRY. Educ: Agra Univ, BSc, 54, MSc, 56; Mich State Univ, PhD(biochem), 61. Prof Exp: NIH res asst, Mich State Univ, 58-61; instr biochem, Sch Med, Univ Miami, 61-64; reader, Maulana Azad Med Col, India, 66; NIH sr fel clin chem, Univ Hosp, Seattle, Wash, 66-68; CHIEF CLIN CHEM, KAISER FOUND HOSP, 68- Mem: AAAS; Am Asn Clin Chemists. Res: Effects of cytotoxic alkylating agents on the metabolism of pyridine nucleotides in ascites tumor cells; techniques for the isolation and measurement of nucleotides; measurements of protein bound iodine and serum thyroxine. Mailing Add: Kaiser Found Hosp Cent Lab 3772 Howe St Oakland CA 94611

AHUJA, JAGDISH C, b Rawalpindi, India, Dec 24, 27; m 55; c 2. STATISTICS, MATHEMATICS. Educ: Banaras Hindu Univ, BA, 53, MA, 55; Univ BC, PhD(math), 63. Prof Exp: Teacher high sch, Nairobi, Kenya, 55-56; teacher math, Dept Educ, Tanganyika, 56-58; lab instr statist, Dept Econ, Univ BC, 59-61, lectr, 61-63; asst prof math, Univ Alta, 63-66; assoc prof, 66-69, PROF MATH, PORTLAND STATE UNIV, 69- Mem: Inst Math Statist; Am Statist Asn. Res: Distribution theory; regression analysis; estimation and statistical inference. Mailing Add: Dept of Math Portland State Univ Portland OR 97207

AHUJA, SATINDER, b Jehlum, WPakistan, Sept 11, 33; m 62; c 2. PHARMACEUTICAL CHEMISTRY. Educ: Banaras Univ, BPharm, 55, MPharm, 56; Philadelphia Col Pharm, PhD(pharmaceut anal chem), 64. Prof Exp: Asst prof pharmaceut chem & pharm, Univ Nagpur, 57-58; teaching asst pharmaceut anal chem, Philadelphia Col Pharm, 58-64; assay develop chemist, Lederle Labs, NY, 64-66; res assoc pharmaceut anal chem, Geigy Chem Corp, Ardsley, 66-69, group leader, Ciba-Geigy Corp, 69-73, SR STAFF SCIENTIST, CIBA-GEIGY CORP, 73- Concurrent Pos: Ed, Progress in Anal Chem, 73. Mem: Am Chem Soc; Acad Pharmaceut Sci. Res: Various modes of chromatography; kinetic studies; stability-indicating methods; dosage variation and availability of drug products; automation and computerization. Mailing Add: Ciba-Geigy Corp Old Mill Rd Suffern NY 10901

AICHELE, MURIT DEAN, b Freewater, Ore, July 17, 28; m 52; c 1. PLANT PATHOLOGY. Educ: Wash State Col, BS, 52. Prof Exp: PLANT PATHOLOGIST, WASH STATE DEPT AGR, 55- Mem: Am Phytopath Soc; Am Soc Hort Sci. Res: Pomology; ornamental horticulture; stone and pome fruit viruses. Mailing Add: Irrig Agr Res & Exten Ctr Wash State Dept of Agr Prosser WA 99350

AIELLO, EDWARD LAWRENCE, b Flushing, NY, June 12, 28; m 53; c 3. PHARMACOLOGY. Educ: St Peters Col, BS, 53; Columbia Univ, MA, 54, PhD(zool), 60. Prof Exp: Instr anat & physiol, Flint Jr Col, 57-58; instr pharmacol, NY Med Col, Flower & Fifth Ave Hosps, 58-60; from asst prof to assoc prof, 60-73, PROF BIOL SCI, FORDHAM UNIV, 73- Mem: AAAS; Am Soc Zoologists; Am Soc Pharmacol & Exp Therapeut. Res: Invertebrate neuropharmacology; ciliary movement; mucociliary transport. Mailing Add: Dept of Biol Sci Fordham Univ Bronx NY 10458

AIGNER, JEAN STEPHANIE, b Los Angeles, Calif, May 5, 43. ANTHROPOLOGY. Educ: Univ Wis, AB, 64, MA, 66, PhD(anthrop), 69. Prof Exp: Res asst anthrop, Univ Wis, 68-69; asst prof, 69-73, actg head dept anthrop, 73-74, actg head dept biocult anthrop, 74-75, ASSOC PROF ANTHROP, UNIV CONN, 73- Concurrent Pos: NSF res grants & Conn Res Found grant, Univ Conn, 69-73. Mem: Fel Am Anthrop Asn; Soc Am Archaeol; Am Quaternary Asn; Far-Eastern Prehist Asn. Res: Human adaptation and cultural variation in the Aleutian Islands and Arctic Pleistocene Chinese faunal evolution and hominid and cultural developments; Asiatic-New World continuum; anthropological perspectives on women. Mailing Add: U-176 Biocult Anthrop Univ of Conn Storrs CT 06268

AIKAWA, JERRY KAZUO, b Stockton, Calif, Aug 24, 21; m 44; c 1. CLINICAL MEDICINE. Educ: Univ Calif, AB, 42; Wake Forest Col, MD, 45. Prof Exp: From intern to asst resident, NC Baptist Hosp, 45-57; instr med, Wake Forest Col, 51-53; Am Heart Asn estab investr, 53-58; from asst prof to assoc prof med, Sch Med, 53-67, dir technol, 67-69, PROF MED, SCH MED, UNIV COLO MED CTR, DENVER, 67-, DIR LAB SERV, 60-, DIR ALLIED HEALTH PROG, 69- Concurrent Pos: Nat Res Coun fel, Med Sch, Univ Calif, 48-49; Nat Res Coun & AEC fel, Duke Univ & Bowman Gray Sch Med, Wake Forest Col, 49-51, Am Heart Asn res fel, Wake Forest Col, 51-53. Res: Cell physiology; electrolyte metabolism; computer applications in medicine; allied health training programs. Mailing Add: 619 S Poplar Way Denver CO 80222

AIKEN, CHARLES S, b Thyatira, Miss, Aug 20, 38; m 70. CULTURAL GEOGRAPHY, ECONOMIC GEOGRAPHY. Educ: Memphis State Univ, BS, 60; Univ Ga, MA, 62, PhD(geog), 69. Prof Exp: Instr geog, Memphis State Univ, 62-65; asst prof, Univ Tenn, Knoxville, 69-72, ASSOC PROF GEOG, UNIV TENN, KNOXVILLE, 72- Mem: Asn Am Geog; Am Geog Soc. Res: Changing geography of the American South; rural America. Mailing Add: Dept of Geog Univ of Tenn Knoxville TN 37916

AIKEN, DAVID EDWIN, invertebrate endocrinology, physiological ecology, see 12th edition

AIKEN, JOHN M, b Ottawa, Kans, Jan 11, 23; m 47; c 4. VETERINARY MEDICINE. Educ: Kans State Univ, BS & DVM, 49; Univ Nebr, Lincoln, MS, 63. Prof Exp: Gen practitioner, Nebr, 49-59; vpres res, Develop & Prof Serv, Baldwin Labs, Inc, 59-61; res assoc vet immunol, Univ Nebr, Lincoln, 64. Res: Fluorescent antibody techniques in viral pathogenesis studies; neonatal immunology. Mailing Add: Leverage Tools, Inc 300 E Ninth St Hastings NE 68901

AIKEN, ROBERT BASCOM, b Troy, NY, Oct 7, 09; m 38; c 2. PUBLIC HEALTH. Educ: Univ Vt, PhB, 31, MS, 33, MD, 37; Harvard Univ, MPH, 48. Prof Exp: Pvt pract, 38-39; dir indust hyg div, 39-41, dir commun dis div, 41-46, secy & exec officer, 46-49, COMNR HEALTH, VT STATE DEPT HEALTH, 49-; ASSOC PROF PREV MED, UNIV VT, 41- Concurrent Pos: Med consult, Dept Educ, Vt; health adv, Gov & Human Serv Agency, State Vt, 73-74; exec dir, Vt PSRO, Inc, 74- Res: Preventive medicine. Mailing Add: VT State Dept Health 115 Colchester Ave Burlington VT 05401

AIKEN, ROBERT MCLEAN, b Springfield, Ill, May 24, 41; m 66. COMPUTER SCIENCE. Educ: Northwestern Univ, BS, 63, MS, 65, PhD(indust eng), 68. Prof Exp: Asst prof, 68-75, ASSOC PROF COMPUT SCI, UNIV TENN, KNOXVILLE, 75- Concurrent Pos: Vis prof, Univ Info Ctr, Univ Geneva, Switz, 75-76. Mem: AAAS; Asn Comput Mach; Sigma Xi. Res: Application of formal language theory to the design of programming languages and the construction of efficient compilers; design and implementation of computer managed instruction curriculum. Mailing Add: Dept Comput Sci Univ of Tenn Knoxville TN 37916

AIKEN, WILLIAM HAMBLEN, chemistry, deceased

AIKENS, CLYDE MELVIN, b Ogden, Utah, July 13, 38; m 63; c 2. ANTHROPOLOGY, ARCHAEOLOGY. Educ: Univ Utah, BA, 60; Univ Chicago, MA, 62, PhD(anthrop), 66. Prof Exp: Staff archeologist, Univ Utah, 62-63, cur, Mus Anthrop, 63-66; asst prof, Univ Nev, Reno, 66-68; asst prof, 68-71, ASSOC PROF ANTHROP, UNIV ORE, 71- Concurrent Pos: NSF res grant, Univ Ore, 70-72; NSF sci fac fel, Kyoto Univ, 71-72; ed, Univ Ore Anthrop Papers, 71- Mem: AAAS; fel Am Anthrop Asn; Soc Am Archaeol; Inst Oriental Studies. Res: Archeology of North America and Japan; culture history and cultural ecology. Mailing Add: Dept of Anthrop Univ of Ore Eugene OR 97403

AIKENS, DAVID ANDREW, b Boston, Mass, Apr 27, 32; m 58; c 6. ELECTROANALYTICAL CHEMISTRY. Educ: Northeastern Univ, BS, 54; Mass Inst Technol, PhD, 60. Prof Exp: Fel, Univ NC, 60-62; from asst prof to assoc prof, 62-69, PROF CHEM, RENSSELAER POLYTECH INST, 69- Mem: Am Chem Soc. Res: Electrode reaction mechanisms; ion-selective electrodes; coordination chemistry. Mailing Add: Dept of Chem Rensselaer Polytech Inst Troy NY 12181

AIKIN, A M, nuclear chemistry, see 12th edition

AIKIN, ARTHUR COLDREN, JR, physics, space science, see 12th edition

AIKMAN, EDWARD PERCY, b Montreal, Que, Jan 24, 11; US citizen; m 38; c 2. PHYSICS. Educ: McGill Univ, BSc, 32, MSc, 33, PhD(physics), 35. Prof Exp: From physicist to lab mgr, Gen Chem Div, Allied Chem Corp, 35-51, from gen mgr to exec vpres, Nichols Chem-Allied Chem Can, 51-64, tech dir nat aniline div, Indust Chem Div, Allied Chem Corp, 64-68; PRES, BLAIK CONSULTANTS INC, 68- Mem: Am Phys Soc; Am Chem Soc. Res: Raman effect of concentrated hydrogen peroxide and liquid ozone; stark effect anomalies in hydrogen and deuterium; spectroscopy; manufacture of chemicals; physical measurements; microscopy; high voltage phenomena. Mailing Add: Woodland Dr Centerbrook CT 06409

AIKMAN, GEORGE CHRISTOPHER LAWRENCE, b Ottawa, Ont, Nov 11, 43; m 72; c 1. ASTROPHYSICS. Educ: Bishop's Univ, BSc, 65; Univ Toronto, MSc, 68. Prof Exp: RES ASST ASTROPHYS, NAT RES COUN CAN, 68- Mem: Can Astron Soc; Am Astron Soc; Astron Soc Pac; Royal Astron Soc Can. Res: The origin of abundance anomalies in chemically peculair stars of the upper main sequence; galactic structure; cometary spectroscopy. Mailing Add: Herzberg Inst Nat Res Coun Can 5071 W Saanich Rd Victoria BC Can

AIKMAN, JOHN MULVANEY, botany, deceased

AILION, DAVID CHARLES, b London, Eng, Mar 21, 37; US citizen; m 64. SOLID STATE PHYSICS. Educ: Oberlin Col, AB, 56; Univ Ill, MS, 58, PhD(physics), 64. Prof Exp: Asst prof, 64-70, ASSOC PROF PHYSICS, UNIV UTAH, 70- Concurrent Pos: Prin investr, NSF grants; vis assoc, Calif Inst Technol, 71-72; NIH spec fel, 71-72. Honors & Awards: Inventor's Incentive Award, Univ Utah, 69. Mem: Am Phys Soc. Res: Nuclear magnetic resonance studies of atomic and molecular motions in solids; electronics; biophysics. Mailing Add: Dept of Physics Univ of Utah Salt Lake City UT 84112

AILMAN, DAVID EDGAR, organic chemistry, see 12th edition

AINBINDER, ZARAH, b Brooklyn, NY, July 12, 37; m 62; c 2. ORGANIC CHEMISTRY. Educ: Brooklyn Col, BS, 59; Univ Chicago, PhD(chem), 63. Prof Exp: Res chemist, 63-69, SR RES CHEMIST, E I DU PONT DE NEMOURS & CO, INC, 69- Mem: Am Chem Soc. Res: Polymer intermediates; catalysis; petro chemicals. Mailing Add: Exp Sta Lab Bldg 336 E I du Pont de Nemours & Co Wilmington DE 19898

AINES, PHILIP DEANE, b Lancaster, Pa, Mar 29, 25; m 51; c 2. NUTRITION. Educ: Rutgers Univ, BS, 49; Cornell Univ, MNS, 51, PhD(nutrit), 54. Prof Exp: Asst biochem, Cornell Univ, 53-54; res chemist, Procter & Gamble Co, 54-55; head prod res dept, Buckeye Cellulose Corp, 56-58, res chemist, Procter & Gamble Co, 59, Head basic develop, Food Prod Div, 60-65, from assoc dir to dir Food Prod Develop Div, 65-73; VPRES RES & ENG, PILLSBURY CO, 73- Mem: AAAS; Am Chem Soc. Res: Protein nutrition and metabolism; food research and development; animal nutrition. Mailing Add: Pillsbury Co Res Labs 311 Second St SE Minneapolis MN 55414

AINIS, HERMAN, b Pittsburgh, Pa, Jan 10, 24; m 46; c 3. BACTERIOLOGY, IMMUNOLOGY. Educ: Univ Calif, Los Angeles, BS, 48; Univ Southern Calif, MS, 56, PhD, 57. Prof Exp: Asst med, Univ Southern Calif, 54-57; immunologist, Jewish Nat Home Asthmatic Children & Children's Asthma Res Inst, 60-64; immunologist, Hektoen Inst Med Res Cook County, 64-71; dir immunol lab, Dept Hemat, 71-74, ASSOC DIR DIV IMMUNOL, COOK COUNTY HOSP, 74- Concurrent Pos: USPHS fel chem & chem eng, Calif Inst Technol, 57-59 & career develop award, Immunochem Lab, Hektoen Inst Med Res Cook County, 66-71; asst prof microbiol, Med Ctr, Univ Ill, 66-73; consult, Dept Exp Immunol, Nat Jewish Hosp, Denver, 63-65. Mem: AAAS; Am Soc Cell Biol; Tissue Cult Asn; Am Soc Microbiol. Res: Induction and production of antibodies in tissue culture; transplantation immunity. Mailing Add: Div of Immunol Cook County Hosp 627 S Wood St Chicago IL 60612

AINLEY, DAVID GEORGE, b Bridgeport, Conn, Apr 3, 46. MARINE ECOLOGY, ORNITHOLOGY. Educ: Dickinson Col, Pa, BS, 68; Johns Hopkins Univ, PhD(ecol), 71. Prof Exp: Res mem antarctic ecol, Johns Hopkins Univ, 68-70; BIOLOGIST MARINE ORNITH, POINT REYES BIRD OBSERV, BOLINAS, 71- Honors & Awards: Polar Res Medal, US Antarctic Res Prog, 74-76. Mem: Am Ornithologists' Union; Cooper Ornith Soc; Wilson Ornith Soc. Res: Marine ecology of seabirds in California, the Gulf of California and Antarctica; trophic relationships of seabirds in marine communities. Mailing Add: Point Reyes Bird Observ Bolinas CA 94924

AINSLIE, HARRY ROBERT, b Hartwick, NY, Dec 2, 23; m 47; c 5. ANIMAL NUTRITION. Educ: Kans State Univ, BS, 49, MS, 50, PhD(animal nutrit), 65. Prof Exp: From asst prof to assoc prof, 50-66, PROF DAIRY HUSB, CORNELL UNIV, 66-, EXTEN LEADER DEPT ANIMAL SCI, 69- Concurrent Pos: Supt off testing, NY, 54-66; vis prof exten educ, Univ Philippines, 66-67; mem northeast subcomt, Nat Dairy Herd Improv Coord Group, 70. Mem: Am Diary Sci Asn. Res: Dairy husbandry extension; educational programs in dairy cattle nutrition; breeding, management and production testing; organization of inservice training programs for extension personnel. Mailing Add: Dept of Animal Sci Cornell Univ Ithaca NY 14850

AINSLIE, JOHN DURHAM, b Guatemala, May 22, 24; US citizen; m 44; c 2. PSYCHIATRY. Educ: Univ Calif, BA, 44, MD, 46; Am Bd Psychiat & Neurol, dipl, 58. Prof Exp: Res assoc epidemiol, Sch Pub Health, Univ Mich, 50-51; intern, USPHS Hosp, Staten Island, NY, 51-52, mem staff internal med & psychiat, USPHS Hosp, Ft Worth, Tex, 52-53; resident psychiat, USPHS, Ft Worth & Lexington, Univ Cincinnati Med Sch & Univ Louisville Child Guid Clin, 53-56; staff psychiatrist, chief admis & dep chief psychiat rehab serv, USPHS Hosp, Lexington, Ky, 56-57; dep chief, Vet Admin Ment Hyg Clin, Miami, Fla, 57-58; from asst prof to assoc prof psychiat, Col Med, Univ Fla, 58-65; dir res & educ, Kings View Community Ment Health Ctr, 65-72; dir, Fla Ment Health Inst, Tampa, 72-74; MED DIR, MERCED COUNTY MENT HEALTH SERV, 74- Concurrent Pos: Res fel epidemiol, Sch Pub Health, Univ Mich, 47-50; consult Vet Admin Hosp, Lake City, Fla & Alachua County Health Dept, 62-65; staff psychiatrist, Kings View Hosp, 65-70, med dir, Visalia Outpatient & Day Treatment Ctr, 70-72; assoc prof, Fresno State Col, 68-69; assoc clin prof, Col Med, Univ Southern Calif, 68-72. Mem: AAAS; AMA; Asn Advan Psychother; Am Group Psychother Asn; Am Pub Health Asn. Res: Psychiatric clinical teaching; individual and group psychotherapy; psychopharmacology; virology;

epidemiology. Mailing Add: Merced County Ment Health Serv 480 E 13th St Merced CA 95340

AINSWORTH, CAMERON, b Alta, Can, Nov 13, 20; US citizen; M 48; c 1. ORGANIC CHEMISTRY. Educ: Univ Alta, BS, 45; Univ Rochester, PhD(chem), 49. Prof Exp: Instr chem, Univ Colo, 49-51; sr org chemist, Eli Lilly & Co, 51-66; assoc prof chem, Colo State Univ, 66-71; PROF CHEM, SAN FRANCISCO STATE UNIV, 72- Mem: AAAS; Am Chem Soc. Res: Synthetic organic compounds of nitrogen related to the natural products histamine and serotonin; Raney nickel nitrogen-nitrogen cleavage; thermal transformations, mesoionic compounds, acyloin reaction and ketene acetal chemistry. Mailing Add: Dept of Chem San Francisco State Univ San Francisco CA 94132

AINSWORTH, EARL JOHN, b Indianapolis, Ind, May 18, 33; m 60; c 3. TOXICOLOGY, RADIOBIOLOGY. Educ: Butler Univ, AB, 55; Brown Univ, ScM, 57, PhD(biol), 59. Prof Exp: Trainee, Nat Cancer Inst, 57-59; resident res assoc, Argonne Nat Lab, 59-61; sr investr cellular radiobiol br, US Naval Radiol Defense Lab, 61-64, actg head, 64-65, head mammalian radiobiol sect, 65-69, prin investr, 69; biologist, Div Biol & Med Res, 69-73, GROUP LEADER, NEUTRON & GAMMA RADIATION TOXICITY PROG, ARGONNE NAT LAB, 73- Concurrent Pos: Mem comt genetic stand, Nat Res Coun, 75. Honors & Awards: Gold Medal Sci Achievement, US Naval Radiol Defense Lab, 69. Mem: Radiation Res Soc; Sigma Xi; Int Soc Exp Hemat. Res: Radiation toxicology; life shortening neoplasia and late functional injury to the hematopoietic system. Mailing Add: Div of Biol & Med Res Argonne Nat Lab D-202 B 117 Argonne IL 60439

AINSWORTH, LOUIS, b Preston, Eng, June 4, 37; m 59; c 2. REPRODUCTIVE ENDOCRINOLOGY. Educ: Univ Leeds, BSc, 59; McGill Univ, MSc, 61, PhD(steroid biochem), 64. Prof Exp: Res assoc obstet & gynec, Case Western Reserve Univ, 64-65, res fel, 65-67, instr reproductive biol, 67-69; res scientist, 69-75, SR RES SCIENTIST, RES BR, CAN DEPT AGR, 75- Mem: Endocrine Soc; Soc Study Reproduction. Res: Mammalian reproduction; hormonal control of reproductive cycles; mechanism of ovulation; development of procedures for increasing reproductive efficiency in domestic animals. Mailing Add: Reproductive Physiology Sect Animal Res Inst Ottawa ON Can

AINSWORTH, OSCAR RICHARD, b Vicksburg, Miss, July, 28, 22; m 47. MATHEMATICS. Educ: Univ Miss, BA & MA, 46; Univ Calif, PhD(math), 51. Prof Exp: From asst prof to assoc prof, 50-60, PROF MATH, UNIV ALA, 60-; MATHEMATICIAN, REDSTONE ARSENAL, NASA, 53- 53- Mem: Am Math Soc; Sigma Xi. Res: Special functions; elasticity; missile guidance systems. Mailing Add: Univ of Ala PO Box 1416 University AL 35486

AIRD, ROBERT BURNS, b Provo, Utah, Nov 5, 03; m 35; c 4. NEUROLOGY. Educ: Cornell Univ, AB, 26; Harvard Univ, MD, 30; Am Bd Psychiat & Neurol, dipl, 44. Prof Exp: Intern surg, Strong Mem Hosp, Rochester, NY, 30-31, asst resident, 31-32; res fel, 32-35, from instr to asst prof surg, 35-46, from assoc prof to prof neurol, 46-71, chmn dept, 47-66, EMER PROF NEUROL, SCH MED, UNIV CALIF, SAN FRANCISCO, 71- Concurrent Pos: Physician-in-chg, Electroencephalog & Electromyography Labs, 40-71; vis neurologist, Univs Calif Hosp & San Francisco Gen Hosp, 47-71; attend neurologist, Langley Porter Neuropsychiat Inst, 47-71; chmn dean's comt, Vet Admin Hosp, San Francisco, 50-52; mem study sect, Nat Inst Neurol Dis & Blindness, 51-54 & 64-68, spec consult & chmn panel eval clin ther, 63-67; vis lectr, Swed Med Training Ctrs, Gothenburg, Lund, Stockholm & Uppsala, 53; Fulbright res fel, Univ Aix Marseille, 57; trustee, Deep Springs Col, 59-71, dir, 60-66, chmn bd trustees, 67-71; chmn, Lennox Trust Fund, 61-69; neurol consult, Letterman Gen Hosp, Vet Admin Hosp plus others. Honors & Awards: Order of Hipolito, Peru, 63; Royer & Lennox Awards, 70; Hope Chest Award, Nat Multiple Sclerosis Soc, 75. Mem: AMA; Am Neurol Asn (vpres, 55 & 71); Am Acad Neurol; Am Electroencephalog Soc (pres, 53); Pan-Am Med Asn (vpres, 66). Res: Epilepsy; electroencephalography; human neurophysiology; blood-brain barrier and cerebrovascular permeability. Mailing Add: Dept of Neurol Univ of Calif Sch of Med San Francisco CA 94143

AIREE, SHAKTI KUMAR, b Hoshiarpur, India, Nov 12, 34; m 63; c 3. PHYSICAL CHEMISTRY, BIOCHEMISTRY. Educ: Panjab Univ, BSc, 56, MSc, 58; Okla State Univ, PhD(chem), 67. Prof Exp: asst prof, 66-69, ASSOC PROF CHEM, UNIV TENN, MARTIN, 69- Mem: Am Chem Soc. Res: Hydroxysteroid dehydrogenases; mercepto disulfide reactivity and enzymatic activity. Mailing Add: Dept of Chem Univ of Tenn Martin TN 38237

AISEN, PHILIP, b New York, NY, Mar 28, 29; m 51; c 3. BIOCHEMISTRY, MEDICINE. Educ: Columbia Univ, AB, 49, MD, 53. Prof Exp: Assoc med, 60-63, from asst prof to assoc prof, 63-73, PROF BIOPHYS & MED, ALBERT EINSTEIN COL MED, 73-, ACTG CHMN DEPT BIOPHYS, 70- Concurrent Pos: Am Cancer Soc fel, 54-55; Nat Res Coun fel, 57-58; Guggenheim fel, Univ Gothenburg, 67-68; mem staff, Watson Lab, IBM Corp, 64-70. Mem: Am Soc Clin Invest; Am Soc Biol Chemists; NY Acad Sci; Am Fedn Clin Res. Res: Protein chemistry and function; physics of metal proteins; biophysics. Mailing Add: Dept of Biophys Albert Einstein Col of Med Bronx NY 10461

AISENBERG, ALAN CLIFFORD, b New York, NY, Dec 7, 26; m 52; c 2. ONCOLOGY. Educ: Harvard Univ, SB, 45, MD, 50; Univ Wis, PhD, 56. Prof Exp: Instr, 57-59, assoc 59-62, asst prof, 62-69, ASSOC PROF MED, HARVARD MED SCH, 69- Concurrent Pos: Clin asst, Mass Gen Hosp, 57-61, asst physician, 61-69, assoc physician, 69- Mem: Am Fedn Clin Res; Am Asn Cancer Res; Am Col Physicians; Am Asn Immunol; Am Soc Clin Oncol. Res: Hodgkin's disease; tumor metabolism. Mailing Add: Huntington Mem Lab Mass Gen Hosp Boston MA 02114

AISENBERG, SOL, b New York, NY, Aug 26, 28; m 56; c 3. PHYSICS. Educ: Brooklyn Col, BS, 51; Mass Inst Technol, PhD(physics), 57. Prof Exp: Res staff mem, Res Div, Raytheon Co, 57-61, sr res scientist, 61-64; sr scientist, Space Sci, Inc, 64-65, actg head physics dept, 65-69, vpres, 68-70, gen mgr, 69-70, GEN MGR, SPACE SCI DIV, WHITTAKER CORP, 70-, CHIEF RES ENGR, 74- Mem: AAAS; Am Phys Soc; Am Vacuum Soc; Am Inst Aeronaut & Astronaut; Asn Adv Med Instrumentation. Res: Plasma diagnostics; gaseous discharges; thin films; lasers; energy conversion; electron emission; plasma accelerators; ultra high vacuum; physical electronics; atmospheric physics; electro-optics; solid state; medical instrumentation; diagnostic instrumentation. Mailing Add: Whittaker Corp Space Sci Div 335 Bear Hill Rd Waltham MA 02154

AISSEN, MICHAEL ISRAEL, b Istanbul, Turkey, Jan 16, 21; nat US; m 44; c 3. MATHEMATICS. Educ: City Col New York, BS, 47; Stanford Univ, PhD(math), 51. Prof Exp: Instr & res assoc math, Univ Pa, 49-51; res scientist, Radiation Lab, Johns Hopkins Univ, 51-61; from assoc prof to prof math, Fordham Univ, 61-70; PROF MATH & CHMN DEPT, NEWARK CAMPUS, RUTGERS UNIV, 70- Mem: Am Math Soc; Math Asn Am; Soc Indust & Appl Math; NY Acad Sci. Res: Zeroes of polynomials; completion processes; combinatorics. Mailing Add: Dept of Math Rutgers Univ Newark Campus Newark NJ 07102

AIST, JAMES ROBERT, b Cheverly, Md, Feb 20, 45; m 67; c 2. PLANT PATHOLOGY, PLANT CYTOLOGY. Educ: Univ Ark, BS, 66, MS, 68; Univ Wis, PhD(plant path), 71. Prof Exp: NATO fel, Swiss Fed Inst Technol, 71-72; ASST PROF PLANT PATH, CORNELL UNIV, 72- Mem: Biol Stain Comn; Am Phytopath Soc; Am Inst Biol Sci. Res: Time course, experimental and cytochemical studies of the role of a plant cell response, papilla formation, during fungal attack; nuclear studies of plant pathogenic fungi. Mailing Add: Dept of Plant Path Cornell Univ Ithaca NY 14853

AISTON, STEWART SAMUEL, b Africa, May 8, 19; nat US; m 42; c 5. BACTERIOLOGY. Educ: Guilford Col, BSc, 48. Prof Exp: Supvr blood plasma processing, 41-43, serum albumin processing & filtration, 43-44, typhus, Japanese b & influenza prod, 44-48 & all human viral & rickettsial prod, 48-55, supt human & vet viral & rickettsial prod, 55-59 & modified live virus poliomyelitics vaccine for clin trial, 59-61, adminr viral & rickettsial prod, 61-69, ADMINR CELL CULT VACCINE PROD, LEDERLE LABS, AM CYANAMID CO, 69- Res: Viral and rickettsial production and research. Mailing Add: Lederle Labs Am Cyanamid Co Pearl River NY 10965

AITA, JOHN ANDREW, b Council Bluffs, Iowa, June 20, 14; m 37; c 2. NEUROLOGY, PSYCHIATRY. Educ: Univ Iowa, MD, 37; Univ Minn, PhD(psychiat), 44. Prof Exp: Intern, US Marine Hosp, Calif, 37-38; asst instr, Sch Med, Yale Univ, 38-40; instr, Mayo Found, Univ Minn, 40-44; assoc prof, 46-69, PROF NEUROL & PSYCHIAT, COL MED, UNIV NEBR MED CTR, OMAHA, 69- Concurrent Pos: Res physician & ment hygienist, Yale Univ, 38-40; first asst, Mayo Clin, 43-44. Mem: Am Psychiat Asn; Am Acad Neurol. Res: Brain injury. Mailing Add: 105 S 49th St Omaha NE 68132

AITKEN, ALFRED H, b New York, NY, June 22, 25; m 50; c 1. THEORETICAL PHYSICS. Educ: Lehigh Univ, BS, 49; Ind Univ, MS, 50, PhD(physics), 55. Prof Exp: PHYSICIST, US NAVAL RES LAB, 54- Mem: Am Phys Soc. Mailing Add: Spec Proj Orgn US Naval Res Lab Washington DC 20375

AITKEN, DONALD W, JR, b Hilo, Hawaii, Mar 14, 36; m 58; c 2. PHYSICS, ASTROPHYSICS. Educ: Dartmouth Col, AB, 58; Stanford Univ, MS, 61, PhD(physics), 63. Prof Exp: Res assoc physics, Stanford Univ, 63-70; PROF ENVIRON STUDIES, CALIF STATE UNIV, SAN JOSE, 70- Mem: AAAS. Res: X-ray astronomy; interaction mechanisms of x-rays in solids; nucleon structure from high energy electron scattering experiments. Mailing Add: Dept of Environ Sci Calif State Univ San Jose CA 95114

AITKEN, JAMES HENRY, b Glasgow, Scotland, Mar, 25, 17; Can citizen; m 55; c 3. NUCLEAR PHYSICS. Educ: Univ Edinburgh, BSc, 52; Univ Toronto, PhD, 69. Prof Exp: Asst lectr physics, Univ Edinburgh, 53-56; physicist, Comput Devices Can, Ltd, 56-57; physicist res off, Nat Res Coun Can, 57-65; res assoc, Univ Toronto, 65-70; analyst, Hydro-elec Power Comn Ont, 70-73; CHIEF HEALTH PHYSICS, MINISTRY HEALTH, ONT, 73- Mem: Can Asn Physicists. Res: Methods of nuclear radiation detection and measurement; reactor physics; reactor safety; health physics. Mailing Add: 22 Churchill Ave Willowdale ON Can

AITKEN, JANET MORA, b Millers Falls, Mass, Dec 23, 16. GEOLOGY. Educ: Smith Col, AB, 39; Johns Hopkins Univ, AM, 41, PhD(geol), 48. Prof Exp: Librn, Geol Dept, Johns Hopkins Univ, 41-42; asst instr geol, 42-43, from instr to prof, 42-72, head dept geol & geog, 63-72, EMER PROF GEOL, UNIV CONN, 72- Mem: Geol Soc Am; Am Geophys Union. Res: Petrology and structure of metamorphic rocks of northeast Connecticut. Mailing Add: Dept of Geol & Geog Univ of Conn Storrs CT 06268

AITKEN, JOHN MALCOLM, b Staten Island, NY, Jan 1, 45; m 68; c 2. SOLID STATE PHYSICS. Educ: Fordham Univ, BS, 66; Rensselaer Polytech Inst, MS & PhD(physics), 73. Prof Exp: Assoc solid state physics, Rensselaer Polytech Inst, 73-74; RES STAFF MEM, T J WATSON RES CTR, IBM CORP, 74- Res: Metal-oxide-semiconductor device physics; electron and hole trapping in silicon dioxide. Mailing Add: T J Watson Res Ctr PO Box 218 IBM Corp Yorktown Heights NY 10598

AITKEN, THOMAS HENRY GARDINER, b Porterville, Calif, Aug 31, 12; m 48; c 2. MEDICAL ENTOMOLOGY, PARASITOLOGY. Educ: Univ Calif, BS, 35, PhD(med entom), 40. Prof Exp: Asst entom & parasitol, Univ Calif, 36-40; mem field staff, Rockefeller Found, 46-71; SR RES ASSOC EPIDEMIOL, YALE ARBOVIRUS RES UNIT, SCH MED, YALE UNIV, 71- Honors & Awards: Gold Medal, Govt of Sardinia, 51. Mem: Am Soc Trop Med & Hyg; Am Mosquito Control Asn; Entom Soc Am. Res: Bloodsucking diptera, particularly mosquitoes; malaria and arthropod-borne viruses. Mailing Add: Yale Arbovirus Res Unit Dept Epid & Pub Health Yale Univ Sch of Med New Haven CT 06510

AIVAZIAN, GARABED HAGPOP, b Turkey, Dec 11, 12; US citizen; m 39; c 2. PSYCHIATRY. Educ: Am Univ Beirut, MD, 35. Prof Exp: Intern internal med, Am Univ Beirut Hosp, 35-37; resident psychiat, Am Univ Beirut & Lebanon Hosp Ment & Nerv Dis, 38-41; clin asst, Am Univ Beirut, 41-52, clin asst prof, 52-54; from asst prof to assoc prof, 54-62, from actg chmn dept to chmn dept, 63-74, PROF PSYCHIAT, COL MED, UNIV TENN, MEMPHIS, 62-, DIR RESIDENCY TRAINING PROG, 61- Concurrent Pos: Consult, Am Univ Beirut Hosp, 41-54, John Gaston Hosp, Memphis, Tenn, 55-, Gailor Ment Health Ctr, 56-, Tenn Psychiat Hosp & Inst, 62-, Baptist Mem Hosp, 63-, William F Bowld Hosp, 65-, Vet Admin Hosp, 67- & Methodist Hosp, 74-; Rockefeller Found fel psychiat, Col Med, Univ Tenn, Payne Whitney Clin & Med Col, Cornell Univ, 48. Mem: AAAS; AMA; Am Psychiat Asn. Res: Medical education and psychopharmaco-therapy; clinical evaluation of psychotropic drugs. Mailing Add: Dept of Psychiat Univ of Tenn Memphis TN 38103

AIZLEY, PAUL, b Boston, Mass, Feb 16, 36; m 69; c 3. MATHEMATICS. Educ: Harvard Univ, AB, 57; Univ Ariz, MS, 59; Ariz State Univ, PhD(math), 69. Prof Exp: Instr math, Tufts Univ, 59-61; vis asst prof, Ariz State Univ, 67-68; asst prof, 68-70, ASSOC PROF MATH, UNIV NEV, LAS VEGAS, 70-, ADMIN ASST TO PRES, 74- Mem: Am Math Soc; Math Asn Am. Res: Convolution algebras. Mailing Add: Dept of Math Univ of Nev Las Vegas NV 89109

AJAMI, ALFRED MICHEL, b Caracas, Venezuela, May 18, 48; US citizen. PESTICIDE CHEMISTRY, PHARMACEUTICAL CHEMISTRY. Educ: Harvard Univ, AB, 70, AM, 72, PhD(biol), 73. Prof Exp: VPRES, ECO-CONTROL, INC, 71- Concurrent Pos: Consult, Environ Protection Agency, 74. Mem: Am Chem Soc; Sigma Xi; AAAS; Am Soc Zoologists. Res: Development of selective narrow spectrum pesticides, pharmaceuticals and formulation processes for their controlled release. Mailing Add: Eco-Control Inc PO Box 305 Cambridge MA 02142

AJAX, ERNEST THEODORE, b Salt Lake City, Utah, Oct 11, 26; m 50; c 4.

CLINICAL NEUROLOGY. Educ: Univ Utah, BS, 49, MD, 51. Prof Exp: Instr psychiat, 58-65, from instr to assoc prof neurol, 59-71, instr med, 61-69, ASST PROF PSYCHIAT, COL MED, UNIV UTAH, 65-, PROF NEUROL, 71-, ASST NEUROLOGIST, UNIV HOSP, 65- Concurrent Pos: Asst chief neurol serv, Vet Admin Hosp, Salt Lake City, 57-62, chief, 62-; assoc neurologist, Salt Lake County Gen Hosp, 59-65. Mem: Am Acad Neurol; Am Electroencephalog Soc; AMA. Res: Electroencephalography; disorders of language. Mailing Add: Col of Med Univ of Utah Salt Lake City UT 84112

AJELLO, LIBERO, b New York, NY, Jan 19, 16; m 42; c 1. MEDICAL MYCOLOGY. Educ: Columbia Univ, AB, 38, MA, 40, PhD(mycol), 48. Prof Exp: Med mycologist, Off Sci Res & Develop, Ft Benning, Ga, 42-43; med mycologist, Johns Hopkins Hosp, 42-46; scientist dir chg mycol div, 47-67, DIR MYCOL DIV, LAB BUR, CTR DIS CONTROL, USPHS, 67- Mem: AAAS; Am Soc Microbiol; Mycol Soc Am; Brit Mycol Soc; Int Soc Human & Animal Mycol. Res: Fungi pathogenic to man. Mailing Add: Mycol Div Lab Bur USPHS Ctr for Dis Control Atlanta GA 30333

AJEMIAN, MARTIN, b Harpoot, Armenia, May 15, 1907; nat US; m 46; c 1. ANATOMY. Educ: Boston Univ, BS, 37; Univ Ark, MA, 51; Georgetown Univ, PhD(anat), 53. Prof Exp: Teacher, Pub Sch, Mass, 41-42; instr biol, Ark Polytech Col, 48-50; instr anat, Sch Med & Dent Sch, Georgetown Univ, 50-53; prof & head dept, South Col Optom, 53-54; prof biol, Ark State Univ, 54-55; assoc prof anat, Univ Miss, 55-74; RETIRED. Res: Degeneration and regeneration of peripheral nerves. Mailing Add: Rte 3 Box 199B Carthage MS 39051

AJL, SAMUEL JACOB, b Poland, Nov 15, 23; nat US; m 46; c 3. MICROBIOLOGY. Educ: Brooklyn Col, BA, 45; Iowa State Col, PhD(physiol bact), 49; Dropsie Univ, DHL, 68. Prof Exp: Asst prof bact & immunol, Sch Med, Wash Univ, 49-52; chief microbiol chem sect, Dept Bact, Army Med Serv Grad Sch, Walter Reed Army Med Ctr, 52-54, asst chief, 54-58; prog dir metab biol, NSF, 59-60; dir res, Albert Einstein Med Ctr, 60-73; VPRES RES, NAT FOUND MARCH DIMES, 73- Concurrent Pos: NSF spec fel, Hebrew Univ, Jerusalem & Oxford Univ, Eng, 68; consult, NIH & NSF; prof biol & microbiol, Sch Med, Temple Univ, 60-73. Mem: Fel AAAS; Am Soc Biol Chem; Am Soc Microbiol; Am Acad Microbiol; fel NY Acad Sci. Res: Respiratory mechanisms of microorganisms; energy transfer reactions; microbial toxins; purification, mechanism of action and immunology of microbial toxins. Mailing Add: Nat Found for March of Dimes Box 2000 White Plains NY 10602

AJMONE-MARSAN, COSIMO, b Cossato, Italy, Jan 2, 18; nat US; m 43; c 2. NEUROPHYSIOLOGY. Educ: Univ Turin, MD, 42. Prof Exp: Asst neurol, Clin Nerv Dis, Italy, 41-47, chief EEG & neurophysiol, 48-49; Rockefeller fel, Montreal Neurol Inst, McGill Univ, 50-51, lectr, Inst, 52-53; CHIEF NEUROSCI BR, NAT INST NEUROL COMMUNICATIVE DIS & STROKE, 54- Concurrent Pos: Clin instr, George Washington Univ, 55-58, assoc, 58-64, asst prof neurol, 65- Mem: Am Epilepsy Soc (pres, 73); Am Acad Neurol; Am Physiol Soc; Int Brain Res Orgn; Am EEG Soc (pres, 62-63). Res: Electroencephalography; physiology of cerebral cortex and thalamus; clinical electroencephalography; epilepsy; clinical neurophysiology. Mailing Add: Neurosci Br Nat Inst Neurol Com Dis & Stroke Bethesda MD 20014

AJZENBERG-SELOVE, FAY, b Berlin, Ger, Feb 13, 26; m 55. NUCLEAR PHYSICS. Educ: Univ Mich, BSE, 46; Univ Wis, MS, 49, PhD(physics), 52. Prof Exp: From asst prof to assoc prof physics, Boston Univ, 52-57; from asst prof to prof, Haverford Col, 57-70, actg chmn dept, 60-61 & 67-69, comnr, Comn Col Physics, 68-70; res prof, 70-73, PROF PHYSICS, UNIV PA, 73- Concurrent Pos: Fel & mem staff, Mass Inst Technol, 52-53; fel, Calif Inst Technol, 52, 54; lectr, Smith Col, 52-53; Smith-Mundt fel, US Dept State, 55; vis asst prof, Columbia Univ, 55; vis prof, Nat Univ Mex, 55; vis assoc physicist, Brookhaven Nat Lab, 56; lectr, Univ Pa, 57; Guggenheim fel, Lawrence Radiation Lab, Univ Calif, 65-66; consult, Calif Inst Technol, 70-72. Exec secy comt physics in cols, Am Inst Physics, 62-65, mem adv comt vis scientist prog, 63-65 & adv comt manpower, 64-68; exec secy ad hoc panel nuclear data compilation, Nat Acad Sci-Nat Res Coun, 71-75; mem comn nuclear physics, Int Union Pure & Appl Physics, 72- Mem: Fel AAAS; fel Am Phys Soc. Res: Neutron spectra; nuclear structure. Mailing Add: Dept of Physics Univ of Pa Philadelphia PA 19174

AKAGI, JAMES MASUJI, b Seattle, Wash, Dec 23, 27; m 60; c 2. MICROBIAL PHYSIOLOGY. Educ: Univ Ill, BS, 51; Univ Kans, MA, 55, PhD, 59. Prof Exp: Instr bact, Univ Kans, 58-59; fel microbiol, Sch Med, Western Reserve Univ, 59-61; from asst prof to assoc prof, 61-67, PROF MICROBIOL, UNIV KANS, 67- Mem: Am Soc Microbiol. Res: Biochemistry. Mailing Add: Dept of Microbiol Univ of Kans Lawrence KS 66045

AKAMATSU, YASUYUKI, b Hyogoken, Japan, Apr 18, 28; m 60; c 3. PATHOLOGY, ONCOLOGY. Educ: Nara Med Col, Japan, MD, 50; Osaka Univ, PhD(path), 57; Am Bd Path, dipl & cert anat path, 60. Prof Exp: Intern med, Nissei-Hosp, Osaka, Japan, 50-51; resident path, Osaka Univ, 51-57; Rockefeller Found fel, Univ Rochester, 58; Am Cancer Soc fel, 59; jr res pathologist, Univ Calif, Los Angeles, 60; instr, Inst Cancer Res, Osaka Univ, 61-68; ASST PROF PATH, MED COL GA, 68- Concurrent Pos: Physician, Atomic Bomb Casualty Comt, Japan, 55-57. Mem: AAAS; Int Acad Path; Am Asn Cancer Res; Am Asn Path & Bact; fel Am Soc Clin Pathologists. Res: Pathology of spontaneous tumors in inbred strains of mice; lathyrism; carcinogenecity of actinomycin and mitomycin; resistant tumor cells to actinomycin; hepatomas of mice; chemical carcinogenesis and amyloidosis; experimental oncology. Mailing Add: Dept of Path Med Col of Ga Augusta GA 30902

AKAMINE, ERNEST KISEI, b Hilo, Hawaii, May 10, 12; m 38; c 3. PLANT PHYSIOLOGY. Educ: Univ Hawaii, BS, 35, MS, 41. Prof Exp: Asst agr, 35-36, agron, 36-38, plant physiol, 38-42, assoc, 42-44, jr plant physiologist, 52-62, assoc plant physiologist & assoc prof plant physiol, 62-68, PLANT PHYSIOLOGIST & PROF PLANT PHYSIOL, EXP STA, UNIV HAWAII, 68- Concurrent Pos: Mem sci invests, Ryukyu Islands Prog, Nat Res Coun, 52; cousult, Nat Lexicog Bd, Ltd, 53-; tech consult, New Wonder World Encycl, 57; vis consult, US Civil Admin Ryukyu Islands, US Dept Army, 62. Mem: AAAS; Am Soc Plant Physiol; Am Soc Hort Sci; Bot Soc Am. Res: Commodity shipment studies; postharvest physiology. Mailing Add: Dept of Bot Univ of Hawaii 3190 Maile Way Honolulu HI 96822

AKASAKI, TAKEO, b Long Beach, Calif, Dec 19, 36. ALGEBRAIC TOPOLOGY. Educ: Univ Calif, Los Angeles, BA, 58, MA, 61, PhD(math), 64. Prof Exp: Actg asst prof math & asst res mathematician, Air Force Off Sci Res grant, Univ Calif, Los Angeles, 64; asst prof math, Rutgers Univ, 64-66; asst prof, 66-72, ASSOC PROF MATH, UNIV CALIF, IRVINE, 72- Mem: Am Math Soc. Res: Homotopy theory of absolute neighborhood retracts; integral group rings; projective modules. Mailing Add: Dept of Math Univ of Calif Irvine CA 92664

AKASOFU, SYUN-ICHI, b Nagano-ken, Japan, Dec 4, 30; m 61. AERONOMY. Educ: Tohoku Univ, BS, 53, MS, 57; Univ Alaska, PhD(geophys), 61. Prof Exp: Asst geophys, Nagasaki Univ, 53-55; asst geophys, 58-61, res geophysicist, 61-62, assoc prof, 62-64, PROF GEOPHYS, UNIV ALASKA, FAIRBANKS, 64- Concurrent Pos: Assoc ed, J Geomagnetism & Geoelec, 72 & J Geophys Res, 72-74. Mem: Am Geophys Union; Soc Terrestrial Magnetism & Elec Japan. Res: Geomagnetic storms; aurora polaris; physics of the magnetosphere; solar-terrestrial relationships. Mailing Add: Geophys Inst Univ of Alaska Fairbanks AK 99701

AKAWIE, RICHARD ISIDORE, b New York, NY, June 11, 23; m 45; c 3. ORGANIC CHEMISTRY, POLYMER CHEMISTRY. Educ: Univ Calif, Los Angeles, AB, 42, AM, 43, PhD(chem), 47. Prof Exp: Asst chem, Univ Calif, Los Angeles, 42-44, 46-47; org res chemist, Gasparcolor, Inc, 47-49; org chemist, US Vet Admin, 49-56; sr chemist, Atomics Int, 56-61; MEM TECH STAFF, HUGHES AIRCRAFT CO, 61- Mem: Am Chem Soc; Sigma Xi. Res: Synthesis of labeled compounds; pharmaceuticals; organic dyes; Grignard reagents; antimalarials; radiation chemistry; silicones; fluorinated compounds; heat-resistant polymers; thermal control coatings. Mailing Add: 12301 Deerbrook Lane Los Angeles CA 90049

AKBAR, ABULFATAH MAKSOOD, b Hyderabad, India. ENDOCRINOLOGY. Educ: Osmania Med Col, India, DVM, 59; Brigham Young Univ, MS, 68; Ore State Univ, PhD(physiol), 71. Prof Exp: Asst res officer vaccine prod & qual control, Biol Res Inst, India, 60-65; fel reproductive endocrinol, Ann Arbor, Mich, 71-72 & Colo State Univ, 72-73; DIR RADIO-IMMUNOASSAY LAB, DIV ENDOCRINOL, COOK COUNTY HOSP, CHICAGO, 73- Mem: Soc Study Reproduction, India; AAAS; Sigma Xi. Res: Elucidation of physiological and biochemical factors involved in the regression of corpus luteum in primates. Mailing Add: Div of Endocrinol Cook County Hosp 1825 W Harrison Chicago IL 60612

AKELEY, DAVID FRANCIS, b Presque Isle, Maine, Apr 13, 28; m 56; c 2. POLYMER CHEMISTRY. Educ: Univ Maine, BS, 49; Univ Wis, PhD(chem), 53. Prof Exp: PATENT CHEMIST, TEXTILE FIBERS DEPT, E I DU PONT DE NEMOURS & CO, 63- Res: Patent law; textile fibers. Mailing Add: 1607 Shadybrook Rd Wilmington DE 19803

AKELEY, ROBERT VINTON, vegetable crops, see 12th edition

AKER, FRANKLIN DAVID, b Norristown, Pa, Apr 26, 43; m 68; c 2. NEUROANATOMY, ANATOMY. Educ: Gettysburg Col, BA, 65; Temple Univ, PhD(anat), 70. Prof Exp: ASST PROF ANAT SCI, SCH DENT, TEMPLE UNIV, 70- Mem: Am Asn Anat. Res: Peripheral and central nervous system in conjunction with fluorescence microscopy and ultra-freezing and drying techniques; development of hard tissues of head region. Mailing Add: Dept of Anat Sci Temple Univ Sch of Dent Philadelphia PA 19140

AKERA, TAI, b Tokyo, Japan, July 13, 32; m 62; c 3. PHARMACOLOGY. Educ: Keio Univ, Japan, MD, 58, PhD(pharmacol), 65. Prof Exp: Intern med, Keio Univ Hosp, 58-59, asst pharmacol, Sch Med, 59-62, instr, 64-66, asst prof, 66-67 & 70-71; from asst prof to assoc prof, 67-74, PROF PHARMACOL, MICH STATE UNIV, 74- Mem: Am Soc Pharmacol & Exp Therapeut. Res: Biochemical and biophysical aspects of the action of narcotics, tranquilizers and cardiac glycosides. Mailing Add: Dept of Pharmacol Mich State Univ Col of Human Med East Lansing MI 48824

AKERBOOM, JACK, b Bridgeton, NJ, Apr 8, 28; m 56; c 2. FOOD TECHNOLOGY. Educ: Lehigh Univ, BS, 49. Prof Exp: Chemist qual control, Best Foods Inc, 49-53, food technologist prod develop, 53-60, dir qual assurance, Knorr Prod, Best Foods Div, CPC Int, 60-62, dir res & develop, Corn Prod, Food Technol Inst, 62-70, ASST DIR FOOD TECHNOL, BEST FOODS DIV, CPC INT, 70- Res: New food product development; vegetables, pasta, shelf stable foods, frozen chicken products, beverages, and egg products. Mailing Add: Best Food Res Ctr PO Box 1543 Union NJ 07083

AKERLOF, CARL W, b New Haven, Conn, Mar 5, 38; m 65; c 2. ELEMENTARY PARTICLE PHYSICS. Educ: Yale Univ, BA, 60; Cornell Univ, PhD(physics), 67. Prof Exp: Res assoc, 66-68, asst prof, 68-72, ASSOC PROF PHYSICS, UNIV MICH, ANN ARBOR, 72- Mem: Am Phys Soc. Res: Strong and electromagnetic interactions of elementary particles. Mailing Add: Dept of Physics Univ of Mich Ann Arbor MI 48104

AKERS, LAWRENCE KEITH, b Ashburn, Ga, Apr 16, 19; m 46; c 4. PHYSICS. Educ: Univ Ga, BS, 49, MS, 50; Vanderbilt Univ, PhD, 55. Prof Exp: Scientist, Oak Ridge Inst Nuclear Studies, 54-55, sr scientist, 55-58, res scientist & actg chmn univ rels div, 58-59; head training unit, Int Atomic Energy Agency, 59-60; prin scientist, Spec Training Div, Oak Rdige Inst Nuclear Studies, 61-64, asst chmn, 64-65, CHMN SPEC TRAINING DIV, OAK RIDGE ASSOC UNIVS, 65- Concurrent Pos: Consult, Cordoba Nat Univ; mem steering comt tech-physics proj, Am Inst Physics. Mem: Am Phys Soc. Res: Infrared and Raman spectra of corrosive compounds; nuclear spectroscopy. Mailing Add: 108 S Purdue Ave Oak Ridge TN 37830

AKERS, ROBERT PRESTON, b Andover, Maine, Mar 7, 17; m 43; c 2. PUBLIC HEALTH ADMINISTRATION. Educ: Bates Col, BS, 39; Boston Univ, MA, 42, PhD, 51. Prof Exp: Asst prof biol, Univ Bridgeport, 46-49; res asst, Grad Sch, Boston Univ, 49-51; res assoc, Nat Heart Inst, 51-57, sci adminr res grants, 57-64, adminr, NIH Latin Am Off, Brazil, 64-66, ADMINR, POLICY & PROCEDURES OFF, OFF OF DIR, NIH, 66- Mem: AAAS; Am Physiol Soc. Mailing Add: Off of Dir NIH 9000 Rockville Pike Bethesda MD 20014

AKERS, SHELDON BUCKINGHAM, JR, b Washington, DC, Oct 22, 26; m 53; c 4. MATHEMATICS. Educ: Univ Md, BS, 48, MA, 52. Prof Exp: Electronics scientist, Nat Bur Stand, Washington, DC, 48-50; radio engr, US Coast Guard Hq, 50-53; comput engr, Nat Bur Stand, 53-54; mathematician, Avion Div, ACF Industs, Inc, Va, 54-56; MATHEMATICIAN & COMPUT SCIENTIST, ELECTRONICS LAB, GEN ELEC CO, 56- Concurrent Pos: Adj prof, Syracuse Univ, 75- Mem: Fel Inst Elec & Electronics Engrs; Soc Indust & Appl Math; Sigma Xi; Am Math Soc. Res: Switching circuit theory; design and application of digital computers; operations research; design automation; combinatorial analysis; graph theory. Mailing Add: 110 Treeland Circle Syracuse NY 13219

AKERS, THOMAS GILBERT, b Oakland, Calif, Jan 5, 28; m 54; c 3. VIROLOGY, MICROBIOLOGY. Educ: Univ Calif, Berkeley, BS, 50, MPH, 56; Mich State Univ, PhD(virol), 63. Prof Exp: US Navy, 50-, pub health officer, Formosa, 51-54, res virologist, Naval Biomed Res Lab, Univ Calif, Berkeley, 54-56, res virologist, Cairo, Egypt, 56-60, RES VIROLOGIST, NAVAL BIOMED RES LAB, UNIV CALIF, BERKELEY, 63- Mem: Am Pub Health Asn; Soc Exp Biol & Med; Am Asn Immunol; Am Soc Trop Med & Hyg; NY Acad Sci. Res: Public health microbiology; biophysical properties of virus nucleic acids; epidemiology of respiratory and enteric virus; cytochemical techniques. Mailing Add: Naval Biomed Res Lab Univ of Calif Sch of Pub Health Berkeley CA 94720

AKERS, THOMAS KENNY, b Brooklyn, NY, Jan 16, 31; m 56; c 4. COMPARATIVE PHYSIOLOGY, ENVIRONMENTAL PHYSIOLOGY. Educ: DePaul Univ, BS, 56; Loyola Univ Chicago, MS, 59, PhD(physiol), 61. Prof Exp: Instr physiol, Norweg Am

Sch Nursing, 58-59; instr, St Anthony of Padua, 59-60; from instr to asst prof pharmacol, Stritch Sch Med, Loyola Univ Chicago, 61-65; asst prof, 65-69, ASSOC PROF PHYSIOL, SCH MED, UNIV NDAK, 69- Concurrent Pos: Res assoc, Inst Study Mind, Drug & Behav, 63-65; dir, High Pressure Life Lab, 74- Honors & Awards: Peck Fund Res Award, 62. Mem: Am Physiol Soc; Aerospace Med Asn; Undersea Med Soc. Res: Hyperbaric studies; oxygen toxicity; adrenergic function; bioengineering. Mailing Add: Dept of Physiol Univ of NDak Sch of Med Grand Forks ND 58201

AKERS, WILLIAM ALEXANDER, b Ashland, Ky, Sept 16, 26; m 50; c 6. DERMATOLOGY. Educ: Univ Louisville, MD, 51. Prof Exp: Intern pediat, Louisville Gen Hosp, 51-52; resident, Children's Hosp, Louisville, 52-53; staff psychiatrist, Cent State Hosp, Lakeland, Ky, 53-54; MedCorps, US Army, 54-, staff pediatrician, Walter Reed Gen Hosp, DC, 54-55, resident dermat, Brooke Gen Hosp, Ft Sam Houston, Tex, 56-59, chief dermat sect, 121 Evacuation Hosp, Korea, 59-60, chief outpatient & dermat, US Army Hosp, Ft Campbell, Ky, 60-62, chief dermat serv, Fitzsimons Gen Hosp, Denver, 62-67, CHIEF DEPT DERMAT RES, LETTERMAN ARMY INST RES, US ARMY, 67-, DEP COMMANDING OFFICER MED-SURG RES SUPV, 73- Concurrent Pos: Asst clin prof, Med Sch, Univ Colo, 62-68, Med Ctr, Univ Calif, 67- & Med Sch, Stanford Univ, 70-; consult, Food & Drug Admin, US Dept Health, Educ & Welfare, 73- Honors & Awards: Cert Achievement & A Prefix, US Army Med Dept, 68; James Clarke White Award, 75. Mem: AAAS; fel Am Acad Dermat; Asn Mil Surg US. Res: Man's skin response to environmental assaults; friction blisters; insect repellents; cutaneous fungal bacterial infections, experimental and natural; contact dermatitis; urticaria; genodermatoses. Mailing Add: Letterman Army Inst Res San Francisco CA 94129

AKESON, WALTER ROY, b Chappell, Nebr, Nov 2, 37; m 59; c 3. BIOCHEMISTRY, AGRONOMY. Educ: Univ Nebr, BS, 59, MS, 61; Univ Wis, PhD(biochem), 66. Prof Exp: Asst prof agron, Univ Nebr, Lincoln, 65-69; SR PLANT PHYSIOLOGIST, AGR RES CTR, GREAT WESTERN SUGAR CO, 69- Mem: AAAS; Am Soc Agron; Am Soc Plant Physiol; Am Soc Sugar Beet Technol. Res: Metabolism of aromatic compounds; extraction and evaluation of leaf protein concentrates; nature of biochemical resistance of plants to disease and insects; sugar beet storage losses. Mailing Add: Agr Res Ctr Great Western Sugar Co Sugarmill Rd Longmont CO 80501

AKESON, WAYNE HENRY, b Sioux City, Iowa, May 5, 28; m 51, 69; c 4. ORTHOPEDIC SURGERY. Educ: Univ Chicago, MD, 53; Am Bd Orthop Surg, dipl. Prof Exp: Instr orthop surg, Univ Chicago, 57-58; asst prof, Sch Med, Univ Wash, 61-70; PROF ORTHOP & HEAD DIV, UNIV CALIF, SAN DIEGO, 70- Concurrent Pos: Consult, Vet Admin. Honors & Awards: Nicholas Andry Award, 66; Kappa Delta Award, 68. Mem: Am Orthop Asn; Orthop Res Soc; Am Acad Orthop Surg. Res: Collagen and mucopolysaccharides and their interaction; biochemical and biophysical basis of joint stiffness; metabolism of intercellular substances of articular cartilage and of arthroplastic surfaces. Mailing Add: Div of Orthop Univ of Calif Sch of Med San Diego CA 92103

AKI, KEIITI, b Yokohama, Japan, Mar 3, 30; m 56; c 2. SEISMOLOGY. Educ: Univ Tokyo, BS, 52, PhD(geophys), 58. Prof Exp: Fulbright res fel geophys, Calif Inst Technol, 58-60; res fel seismol, Earthquake Res Inst, Univ Tokyo, 60-62; vis assoc prof geophys, Calif Inst Technol, 62-63; assoc prof seismol, Earthquake Res Inst, Univ Tokyo, 63-65; PROF GEOPHYS, MASS INST TECHNOL, 66- Concurrent Pos: Lectr, Int Inst Seismol & Earthquake Eng, UNESCO, 60-65; geophysicist, Nat Ctr Earthquake Res, US Geol Surv, 67-; vis scientist, Royal Norweg Res Coun; consult, Sandia Labs, 76-79. Mem: Am Geophys Union; Seismol Soc Am; Soc Explor Geophys; Royal Astron Soc; fel Am Acad Arts & Sci, 73. Res: Geophysical research directed toward predicting, preventing and controlling earthquake hazard; seismic wave propagation; earthquake statistcs; structure of earth's crust and upper mantle; thermal processes in the earth; geothermal energy source exploration. Mailing Add: Rm 54-526 Mass Inst of Technol Cambridge MA 02139

AKIN, WALLACE ELMUS, b Murphysboro, Ill, May 18, 23; m 48; c 2. PHYSICAL GEOGRAPHY. Educ: Southern Ill Univ, BA, 48; Ind Univ, MA, 49; Northwestern Univ, PhD(geog), 52. Prof Exp: Field team chief, Rural Land Classification Prog PR, Commonwealth Dept Agr & Com, 50-51; instr geog, Univ Ill, Chicago, 52-53; PROF GEOG & CHMN DEPT, DRAKE UNIV, 53- Concurrent Pos: Fulbright res grant, Univ Copenhagen, 61-62. Mem: Asn Am Geogr; Royal Danish Geog Soc; Arctic Inst NAm; Soil Conserv Soc Am; Am Geog Soc. Res: The role of physical geography in resource inventory and management. Mailing Add: Dept of Geog & Geol Drake Univ Des Moines IA 50311

AKINS, DANIEL L, b Miami, Fla, July 8, 41; m 63; c 1. PHYSICAL CHEMISTRY. Educ: Howard Univ, BS, 63; Univ Calif, Berkeley, PhD(chem), 68. Prof Exp: Res assoc chem, Inst Molecular Biophys, Fla State Univ, 68-69, asst prof, 69-70, ASST PROF CHEM, UNIV S FLA, 70- Res: Energy transfer in molecular collisions; intermolecular potential functions determination through study of the pressure dependence of absorption and fluorescence. Mailing Add: Dept of Chem Univ of SFla Tampa FL 33620

AKINS, ERVIN LORAINE, b Warsaw, Ohio, Mar 26, 28; m 53; c 2. PHYSIOLOGY, ENDOCRINOLOGY. Educ: Ohio State Univ, BSc, 54, MSc, 57, DVM, 62; Okla State Univ, PhD(reproductive physiol), 68. Prof Exp: Dairy herd mgr, Ohio Agr Exp Sta, 54-56; res asst genetics, Agr Res Serv, USDA, 56-58; instr ambulatory clin, Col Vet Med, Ohio State Univ, 62-64; NIH trainee grant physiol, Okla State Univ, 64-65, fel, 65-68; asst prof vet med, Sch Vet Sci & Med, Purdue Univ, 68-74; MEM FAC, DEPT VET PHYSIOL, TEX A&M UNIV, 74- Mem: Am Vet Med Asn; Soc Study Reproduction. Res: Domestic animal reproductive physiology and endocrinology, especially of equine female; temporal relationships between anterior pituitary corpora lutea and endometrium during estrous cycle. Mailing Add: Dept of Vet Physiol Tex A&M Univ College Station TX 77843

AKINS, GLENN JOHN, b Denver, Colo, Aug 21, 42; m 65; c 2. RESOURCE MANAGEMENT. Educ: Univ Minn, BA, 68; Western Wash State Col, MS, 70. Prof Exp: Sr planner environ mgt, Mid-Valley Coun Govts, 71-72; chief planner resource mgt, Ore Coastal Conserv & Develop Comn, 72-75; consult resource mgt, 75; DIR RESOURCE MGT, ALASKA COASTAL MGT PROG, 75- Mem: Am Inst Planners. Res: Use of remote sensing for land use and environmental management; identification and management of coastal wetlands. Mailing Add: PO Box 4780-7 Juneau AK 99803

AKINS, VIRGINIA, b Lafayette Co, Wis. BIOLOGY. Educ: Univ Wis, BA, 37, PhD(bot), 40. Prof Exp: Asst bot, Univ Wis, 37-40; agent forage crops & dis, USDA, Wis, 42-43; asst forest prod technologist, Forest Prod Lab, US Forest Serv, 43-47; from teacher to assoc prof, 47-71, PROF BIOL, UNIV WIS-RIVER FALLS, 71- Mem: AAAS. Res: Cytology of flagellates and grasses; detection and effect of gelatinous fibers in wood; cytological study of Carteria crucifera. Mailing Add: Dept of Biol Univ Wis River Falls WI 54022

AKLONIS, JOHN JOSEPH, b Elizabeth, NJ, Sept 28, 40; m 66. PHYSICAL CHEMISTRY, POLYMER CHEMISTRY. Educ: Rutgers Univ, BS, 62; Princeton Univ, MA, 64, PhD(phys chem), 65. Prof Exp: Fel polymer physics, Princeton Univ, 65-66; asst prof, 66-70, ASSOC PROF CHEM, UNIV SOUTHERN CALIF, 70- Concurrent Pos: Fulbright fel polymer chem, Cath Univ Louvain, 66-67; res assoc, Macromolecule Res Ctr, Strasbourg, France, 74-75. Mem: Am Chem Soc. Res: Physics and physical chemistry of high polymeric systems. Mailing Add: Dept of Chem Univ of Southern Calif Los Angeles CA 90007

AKRE, ROGER DAVID, b Grand Rapids, Minn, Mar 27, 37; m 56; c 1. ENTOMOLOGY. Educ: Univ Minn, Duluth, BS, 60; Kans State Univ, MS, 62, PhD(entom), 64. Prof Exp: From asst prof to assoc prof, 64-74, PROF ENTOM, WASH STATE UNIV, 74- Honors & Awards: R M Wade Award Excellence in Teaching, 69. Mem: Entom Soc Am; Animal Behav Soc; Entom Soc Can. Res: Biology and behavior of arthropods associated with neotropical army ants; behavior of vespine wasps. Mailing Add: Dept of Entom Wash State Univ Pullman WA 99163

AKS, STANLEY ØLAF, b Brooklyn, NY, Mar 6, 35; m 61. MATHEMATICAL PHYSICS. Educ: Polytech Inst Brooklyn, BS, 56; Univ Md, PhD(physics), 64. Prof Exp: Res assoc math, Inst Fluid Dynamics & Appl Math, Univ Md, 64-69; ASSOC PROF PHYSICS, UNIV ILL, CHICAGO CIRCLE, 69- Mem: Am Phys Soc; Am Math Soc; Italian Phys Soc. Res: Analytical properties of partial differential equations; several complex variable theory; nonlinear differential equations particularly as they apply to the quantum theory. Mailing Add: Dept of Physics Box 4348 Univ of Ill at Chicago Circle Chicago IL 60680

AKSELRAD, ALINE, b Wilno, Poland, Apr 14, 35; Can citizen; m 62; c 1. EXPERIMENTAL SOLID STATE PHYSICS, MAGNETISM. Educ: McGill Univ, BSc, 57; Univ Toronto, MSc, 59, PhD(solid state spectros), 61. Prof Exp: Asst physics, Univ Toronto, 59-61; grant & mem tech staff, Weizmann Inst Sci, 61-62; MEM TECH STAFF, RCA LABS, 62- Honors & Awards: Achievement Awards, RCA Labs, 69 & 73. Mem: Am Phys Soc. Res: Electronic structure and properties of solids; magnetic insulators and their properties; propagation and applications of acoustic waves; solar energy conversion. Mailing Add: RCA Labs Princeton NJ 08540

AKSNES, KAARE, b Kvan, Norway, Mar 25, 38; m 59; c 3. ASTRONOMY. Educ: Univ Bergen, Norway, BS, 60; Univ Oslo, MS, 63; Yale Univ PhD(astron), 69. Prof Exp: Res asst astron, Inst Theoret Astrophys, Univ Oslo, 61-63, 64-65; comput programmer, Norweg Defense Res Estab, 63-64; mathematician, Smithsonian Astrophys Observ, 65-67; res asst, Yale Univ Observ, 67-69; sr engr space res, Jet Propulsion Lab, Calif Inst Technol, 69-71; CELESTIAL MECHANICIAN, SMITHSONIAN ASTROPHYS OBSERV, 71- Concurrent Pos: Assoc, Harvard Univ Observ, 71-; vis scientist, Tokyo Astron Observ, 76. Mem: Int Astron Union; Royal Astron Soc Eng; Am Astron Soc. Res: Celestial mechanics; dynamical studies of satellites, minor planets and comets. Mailing Add: Smithsonian Astrophys Observ 60 Garden St Cambridge MA 02138

AKST, IRVING BERNARD, physics, see 12th edition

AKTIPIS, STELIOS, b Athens, Greece, July 14, 35; US citizen; m 65. BIOCHEMISTRY, BIOPHYSICS. Educ: Nat Univ Athens, dipl, 59; Brown Univ, ScM, 62; Brandeis Univ, PhD(phys org chem), 65. Prof Exp: Res fel biochem, Harvard Med Sch, 65-66; from asst prof to assoc prof, 66-74, PROF BIOCHEM, STRITCH SCH MED, LOYOLA UNIV CHICAGO, 74-, ASST CHMN DEPT, 70- Mem: AAAS; Am Soc Biol Chemists; Am Chem Soc. Res: Optical rotatory dispersion and circular dichroism of macromolecules; interaction of nucleic acids with intercalating molecules; inhibition of RNA polymerase by template inactivators. Mailing Add: Stritch Sch of Med Loyola Univ of Chicago Med Ctr Maywood IL 60153

AKUTSU, TETSUZO, b Japan, Aug 20, 22; m 50; c 2. MEDICINE. Educ: Nagoya Univ, MD, 47. Prof Exp: Instr, Nagoya Univ, 57; res fel, Cleveland Clin Found, 57-60, mem staff, 60-64; asst prof surg, State Univ NY Downstate Med Ctr & dir div surg res, Maimonides Hosp Brooklyn, 64-67; from assoc prof to prof surg, Med Ctr, Univ Miss, 69-74; ASSOC DIR CARDIOVASC SURG RES LABS, TEX HEART INST, 74- Mem: Soc Cryobiol; Am Heart Asn. Res: Cardiovascular research; extracorporeal circulation; artificial organs and heart. Mailing Add: Cardiovasc Surg Res Labs Tex Heart Inst PO Box 20269 Houston TX 77025

ALABRAN, DAVID MAX, organic chemistry, see 12th edition

ALADJEM, FREDERICK, b Vienna, Austria, Feb 8, 21; nat US; m 57; c 3. IMMUNOCHEMISTRY, BIOPHYSICS. Educ: Univ Calif, AB, 44, PhD(biophys), 54. Prof Exp: Physicist, Radiation Lab, Univ Calif, 50-54; physicist, Nat Microbiol Inst, USPHS fel & res fel div chem, Calif Inst Technol, 54-57; from instr to assoc prof med microbiol, 56-65, PROF MED MICROBIOL, SCH MED, UNIV SOUTHERN CALIF, 65- Concurrent Pos: Consult path, Los Angeles County Gen Hosp, 57- Mem: AAAS; Am Asn Immunol; Am Soc Microbiol; Biophys Soc. Res: Antigen-antibody reaction; immunochemistry of lipoproteins; chemical aspects of allergic reactions. Mailing Add: Dept of Microbiol Univ of Southern Calif Sch Med Los Angeles CA 90033

AL-AIDROOS, KAREN MESSING, b Springfield, Mass, Feb 2, 43. MICROBIAL GENETICS. Educ: Harvard Univ, BA, 63; McGill Univ, MSc, 70, PhD(biol), 75. Prof Exp: Res asst biochem, Jewish Gen Hosp, Montreal, 70-71; NIH FEL GENETICS, BOYCE THOMPSON INST PLANT RES, 75- Mem: Am Soc Microbiol. Res: Genetic manipulation of Metarrhizium anisopliae, a fungal pathogen of mosquitoes. Mailing Add: Boyce Thompson Inst Plant Res Yonkers NY 10701

ALAIMO, ROBERT J, b Rochester, NY, Mar 2, 40; m 63; c 3. ORGANIC CHEMISTRY. Educ: Univ Miami, BS, 61; Cornell Univ, PhD(org chem), 65. Prof Exp: SR RES CHEMIST, NORWICH PHARMACAL CO, 65- Mem: Am Chem Soc; The Chem Soc. Res: Synthetic organic chemistry involving the synthesis of medicinal agents. Mailing Add: 29 Hillview Dr Norwich NY 13815

ALAKA, MIKHAIL A, b Baghdad, Iraq, June 26, 13; US citizen; m 54; c 3. METEOROLOGY, MATHEMATICS. Educ: Am Univ, Beirut, BA, 46; Univ Chicago, SM, 49, PhD(meteorol), 55. Prof Exp: Forecaster, Iraq Meteorol Serv, 37-49, dep dir, 49-50; asst meteorol, Univ Chicago, 50-54; tech officer, World Meteorol Orgn, Switz, 54-60 res meteorologist, Nat Hurricane Res Proj, Miami, Fla, 60-64, supvry res meteorologist, Tech Develop Lab, 64-66, CHIEF SPEC PROJS BR, US WEATHER BUR, 67- Mem: Am Meteorol Soc; Royal Meteorol Soc. Res: Tropical meteorology; severe storm research. Mailing Add: 8515 Madison St New Carrollton MD 20784

ALAM, ASHRAF UL, b Dacca, Pakistan, Jan 15, 36; m 63. BIOCHEMISTRY. Educ: Univ Dacca, BSc, 59, MSc, 61; Tex A&M Univ, PhD(biochem), 67. Prof Exp: Supvr anal & qual control, Karnafuli Paper Mills Ltd, Pakistan, 54-58; asst biochemist, Univ

Dacca, 61-63; res assoc biochem, Baylor Col Med, 67-69; sr lectr, 69-73, ASSOC PROF BIOCHEM, UNIV DACCA, 73- Mem: Am Chem Soc; Nutrit Soc Bangladesh (gen secy, 72-). Res: Analytical and biochemical aspects of plant pigments and alkaloids; biochemical changes in germinating seedlings, lipids and lipolytics enzymes; nutritional parameters in developing countries; urea cycle enzymes. Mailing Add: Dept of Biochem Univ of Dacca Dacca 2 Bangladesh

ALAM, MOHAMMED ASHRAFUL, b Rajshahi, Bangladesh, Jan 1, 32; m 65; c 2. ANALYTICAL CHEMISTRY, PHYSICAL CHEMISTRY. Educ: Rajshahi Govt Col, BSc, 51; Univ Dacca, MSc, 53; La State Univ, PhD(chem), 62. Prof Exp: Asst chemist, Jute Res Inst, Pakistan, 54-58; res assoc, Tulane Univ, 62; sr lectr phys chem, Univ Dacca, 64-65; assoc prof chem, Elizabeth City State Col, 65-68; reader phys chem, Islamabad, 68-69; ASSOC PROF CHEM, ELIZABETH CITY STATE UNIV, 69- Mem: Am Chem Soc; Sigma Xi. Res: Physicochemical studies of proteins and enzymes; hydrogen bonding studies; metal ion interaction with nucleic acid and polynucleotides. Mailing Add: Dept of Chem Elizabeth City State Univ Elizabeth City NC 27909

ALAM, SYED QAMAR, b Jagrawan, India, June 22, 32; US citizen. NUTRITIONAL BIOCHEMISTRY, DENTAL RESEARCH. Educ: Univ Punjab, WPakistan, BSc, 52; Univ Calif, Berkeley, MS, 61; Mass Inst Technol, PhD(nutrit biochem), 65. Prof Exp: Assoc prof nutrit, Fla A&M Univ, 65-67; res assoc nutrit & oral sci, Mass Inst Technol, 67-71; asst prof, 71-74, ASSOC PROF BIOCHEM, LA STATE UNIV MED CTR, NEW ORLEANS, 74- Mem: Am Inst Nutrit; Int Asn Dent Res; Am Asn Dent Res. Res: Effects of nutrition on the composition of oral tissues and their resistance to disease; lipids in salivary glands, saliva and teeth. Mailing Add: Dept of Biochem Sch of Dent La State Univ Med Ctr New Orleans LA 70119

ALANEN, JACK DAVID, b Painesville, Ohio, Aug 27, 38; m 69; c 1. COMPUTER SCIENCES. Educ: Case Inst Technol, BS, 60, MS, 62; Yale Univ, MS, 66, MPH, 67, PhD(statist), 72. Prof Exp: Mgr opers comput, Comput Ctr, Yale Univ, 63-64; asst prof statist, Univ Conn, 67-68; staff syst consult comput, CHI Corp, 69-70; mem staff, Comput Dept, Math Ctr, Univ Amsterdam, 70-73; chmn, Comput Ctr Dept, Univ Nairobi, 73-74; assoc prof comput sci, State Univ NY Buffalo, 74-75; SR STAFF MEM, DEPT COMPUT SCI, UNIV UTRECHT, NETH, 75- Mem: Asn Comput Mach; Inst Math Statist; Am Statist Asn; Math Asn Am; Sigma Xi. Res: Computer science education; programming systems and languages; statistical data analysis; software engineering. Mailing Add: Frederik Hendrikstraat 112 Utrecht Netherlands

ALAOGLU, LEONIDAS, b Red Deer, Alta, Mar 19, 14; US citizen; m 47; c 3. MATHEMATICS. Educ: Univ Chicago, SB, 36, SM, 37, PhD, 38. Prof Exp: Instr math, Pa State Univ, 38-39; Peirce instr, Harvard Univ, 39-42; instr, Purdue Univ, 42-44; STAFF ENGR & SR SCIENTIST OPERS RES, LOCKHEED AIRCRAFT CORP, 53- Concurrent Pos: Mem, Weapons Systs Eval Group, 50-53. Res: Operations research. Mailing Add: Systs Anal Dept Lockheed Calif Co Burbank CA 91520

ALARIE, ALBERT, b Ste Rose, Que, June 5, 13; m 38; c 2. SOIL MICROBIOLOGY. Educ: Ste Anne de la Pocatiere, BSA, 38, BSc, 41; McGill Univ, MSc, 43, PhD(microbiol), 45. Prof Exp: Soil survr, Dept Agr, Que, 38-45, soil microbiologist, 45-47, head dept soil microbiol, 57-62; PROF MICROBIOL, LAVAL UNIV, 62- Concurrent Pos: Assoc prof, Ste Anne de la Pocatiere, 45-62. Mem: Am Soc Microbiol; Can Soc Microbiol. Res: Cellulose decomposing bacteria in soils; symbiotic fixation of nitrogen by Rhizobium bacteria; soil organic matter; composting of organic refuses. Mailing Add: Fac of Agr Laval Univ Ste Foy PQ Can

ALARIE, YVES, b Montreal, Que, Mar 18, 39; m 60; c 4. PHYSIOLOGY. Educ: Univ Montreal, BS, 60, MS, 61, PhD(physiol), 63. Prof Exp: Physiologist, Hazleton Labs, Inc, 63-70; PROF OCCUP HEALTH, UNIV PITTSBURGH, 70- Concurrent Pos: Nat Inst Environ Health Sci spec fel, 70. Honors & Awards: Achievement Award, Soc Toxicol, 71, F R Blood Award, 74. Mem: AAAS; Soc Toxicol. Res: Respiratory physiology, effects of air pollutants on respiratory system; sensory irritation; chemoreceptors. Mailing Add: Grad Sch of Pub Health Univ of Pittsburgh Pittsburgh PA 15261

AL-ASKARI, SALAH, b Baghdad, Iraq, Nov 16, 27; m 56; c 2. IMMUNOLOGY, UROLOGY. Educ: Royal Col Med, Baghdad, MB, ChB, 51; NY Univ, MSc, 59. Prof Exp: Instr, 58, univ fel transplantation immunol, 60, USPHS fel, 61, instr urol, 61-63, from asst prof to assoc prof, 63-70, PROF UROL, NY UNIV, 70- Concurrent Pos: Co-investr, res contract transplantation immunol, NY Health Res Coun, 61-64, career scientist award, 64- Mem: Harvey Soc; Soc Exp Biol & Med; Am Asn Immunol; Transplantation Soc; Am Urol Asn. Res: Ileal segments in genitourinary surgery; homograft rejection mechanism and transplantation antigens. Mailing Add: Dept of Urol NY Univ Med Ctr New York NY 10016

ALAUPOVIC, PETAR, b Prague, Czech, Aug 3, 23; m 47; c 1. BIOCHEMISTRY, ORGANIC CHEMISTRY. Educ: Univ Zagreb, ChemE, 48, PhD(chem), 56. Prof Exp: Researcher, Pharmaceut Res Lab, Chem Corp, Prague, 48-49; researcher, Org Lab, Inst Indust Res, Yugoslavia, 49-50; asst, Agr Fac, Univ Zagreb, 51-54, asst, Chem Inst Med Fac, 54-56; res biochemist, Univ Ill, 57-60; MEM CARDIOVASC SECT, OKLA MED RES FOUND, 60-, HEAD LIPOPROTEIN LAB, 72-; PROF RES BIOCHEM, SCH MED, UNIV OKLA, 60- Concurrent Pos: NIH grants, 61-68; assoc ed, Lipids, 74- Mem: AAAS; Am Chem Soc; Am Heart Asn. Res: Chemistry of naturally occurring macromolecular lipid compounds such as serum and tissue lipoproteins and bacterial endotoxins; biochemistry of red cell membranes; isolation and characterization of tissue lipases. Mailing Add: Lipoprotein Lab Okla Med Res Found Oklahoma City OK 73104

ALAVI, YOUSEF, b Iran, Mar 19, 29. MATHEMATICS. Educ: Mich State Univ, BS, 52, MS, 55, PhD(math), 58. Prof Exp: From asst prof to assoc prof, 58-68, PROF MATH, WESTERN MICH UNIV, 68- Mem: Am Math Soc; Soc Indust & Appl Math; Math Asn Am. Res: Special functions; graph theory. Mailing Add: Dept of Math Western Mich Univ Kalamazoo MI 49007

AL-AWQATI, QAIS, b Baghdad, Iraq, Aug 18, 39. PHYSIOLOGY, BIOPHYSICS. Educ: Univ Baghdad, Iraq, MB, ChB, 62; Am Bd Internal Med, cert, 72; Am Bd Nephrology, cert, 72. Prof Exp: Fel med, Johns Hopkins Med Sch, 67-70; fel med, Harvard Med Sch, Mass Gen Hosp, 70-71, instr, 71-73; ASST PROF MED, COL MED, UNIV IOWA, 73- Concurrent Pos: NIH res award, 76- Mem: Soc Gen Physiologists; Am Physiol Soc; Am Fedn Clin Res; Am Soc Nephrology. Res: Mechanisms of water and ion transport across biological membranes; particularly the energetics of active transport. Mailing Add: Univ Hosp Col of Med Univ of Iowa Iowa City IA 52242

ALBACH, RICHARD ALLEN, b Chicago, Ill, Mar 3, 31; m 53; c 7. CELL PHYSIOLOGY, MICROBIOLOGY. Educ: Univ Ill, BS, 56, MS, 58; Northwestern Univ, PhD(microbiol), 63. Prof Exp: Res assoc microbiol, Lutheran Gen Hosp, 63-67; assoc, Inst Med Res, 67-68, from asst prof to assoc prof, 68-73, PROF MICROBIOL, CHICAGO MED SCH, 73- Concurrent Pos: Co-prin investr, Nat Inst Allergy & Infectious Dis Grant, 65-68 & 68-73, prin investr, 73-76. Honors & Awards: Bd Trustees Res Award, Chicago Med Sch, 68. Mem: AAAS; Am Soc Microbiol; Soc Protozool; NY Acad Sci. Res: Aspects of DNA and RNA metabolism; nutritional requirements; metabolism of entamoeba histolytica. Mailing Add: 2020 Ogden Chicago IL 60612

ALBACH, ROGER FRED, b Chicago, Ill, Mar 25, 32; m 57; c 3. ORGANIC CHEMISTRY. Educ: Fresno State Col, BS, 57; Univ Calif, Davis, MS, 60, PhD(chem), 64. Prof Exp: Range aide, Agr Res Serv, USDA, 54-55; field inspector, Calif Spray Chem Corp, 56; lab technician chem, Fresno State Col, 57; RES CHEMIST, FOOD CROPS UTILIZATION RES LAB, AGR RES SERV, USDA, 63- Mem: Am Chem Soc; Phytochem Soc NAm. Res: Phytochemistry; natural products; food chemistry. Mailing Add: Food Crops Utilization Res Lab Southern Region PO Box 388 Weslaco TX 78596

ALBAN, EVAN KENNETH, b Columbus, Ohio, July 5, 13; m 40. HORTICULTURE. Educ: Denison Univ, BA, 36; Ohio State Univ, MSc, 43, PhD(hort), 45. Prof Exp: Asst hort, 42-45, from instr to assoc prof, 46-59, asst, Res Found, 44-45, PROF HORT, OHIO STATE UNIV, 59- Concurrent Pos: Instr, Denison Univ, 45; asst, Ohio Exp Sta, 43-46. Mem: AAAS; Am Inst Biol Sci; Am Soc Plant Physiol; Am Soc Hort Sci; Weed Sci Soc Am. Res: Vegetable crop physiology; chemical weed control; post-harvest physiology; industrial and non-crop land vegetation control. Mailing Add: Dept of Hort Ohio State Univ Columbus OH 43210

ALBANESE, ANTHONY AUGUST, b New York, NY, Feb 12, 08; m 33. BIOCHEMISTRY. Educ: NY Univ, BS, 30; Columbia Univ, PhD(biochem), 40; Am Bd Nutrit, dipl, 51. Prof Exp: Res asst path, Col Physicians & Surgeons, Columbia Univ, 33-40; res assoc, Div Chem, NIH, 40-41; assoc pediat & res, Johns Hopkins Univ, 41-45; from asst prof to assoc prof pediat biochem, Col Med, NY Univ, 45-49; DIR GERIAT NUTRIT LAB, MIRIAM OSBORN MEM HOME, 50-; DIR NUTRIT & METAB RES DIV, BURKE REHAB CTR, 59- Concurrent Pos: Corn Indust fel, NIH, 40-41; responsible investr, Off Sci Res & Develop, 41-45, Off Naval Res, 45-72 & Air Res & Develop Command, 60-65; dir res, St Luke's Convalescent Hosp, 49-60; guest lectr, Univ Brazil, 59, Univ Tokyo & Univ Osaka, 60 & Univ Istanbul, 70; assoc ed, NY State J Med, 59-; ed-in-chief, Nutrit Reports Int, 70-; mem, President's Sci Adv Comt Toxicol Info Prog. Mem: Fel Am Inst Chemists; Am Chem Soc; Harvey Soc; Am Soc Biol Chemists; Soc Exp Biol & Med. Res: Nutritional effects of steroids; fat metabolism; metabolic effects of enzymes; amino acid chemistry; biological value of various proteins in the human; electrolytic method for determination of basic amino acids in proteins; carbohydrate and cholesterol metabolisms. Mailing Add: Burke Rehab Ctr 785 Mamaroneck Ave White Plains NY 10605

ALBATS, PAUL, b Latvia, Dec 31, 41; US citizen; m 69; c 2. COSMIC RAY PHYSICS. Educ: Univ Chicago, BS, 64; Cornell Univ, PhD(physics), 71. Prof Exp: From res assoc to sr res assoc physics, 69-73, ASST PROF PHYSICS, CASE WESTERN RESERVE UNIV, 74- Mem: Am Phys Soc; Sigma Xi. Res: Experimental high energy astrophysics gamma ray astronomy; experiments to measure atmospheric high energy neutron flux; experimental search for solar neutrons. Mailing Add: Dept of Physics Case Western Reserve Univ Cleveland OH 44106

ALBAUGH, A HENRY, b Chicago, Ill, Aug 21, 22; m 50; c 4. MATHEMATICS, STATISTICS. Educ: Mich State Univ, BS, 48; Univ Mich, MA, 50. Prof Exp: Teacher high schs, Mich, 52-57; asst prof, 57-60, head dept, 57-75, ASSOC PROF MATH, HILLSDALE COL, 60- Mem: Nat Coun Teachers Math. Res: Educational games; thermocycling. Mailing Add: Dept of Math Hillsdale Col Hillsdale MI 49242

ALBAUM, HARRY GREGORY, b Odessa, Russia, Feb 9, 10; nat US; m 36; c 1. BIOCHEMISTRY. Educ: Brooklyn Col, AB, 32; NY Univ, MS, 34; Columbia Univ, PhD(biol), 38. Prof Exp: Tutor, 34-38, from instr to assoc prof, 38-53, PROF BIOL, BROOKLYN COL, 53-, DIR DIV GRAD STUDIES & DEAN GRAD STUDIES, 66-, DEAN FACULTIES, 69- Concurrent Pos: Res grants, AAAS, Am Philos Soc, Am Cancer Soc, USPHS, US Air Force; Nat Res fel, Univ Wis, 42-43; res assoc, Univ Chicago, 43-44; exec officer, PhD prog biol, City Univ New York, assoc dean, 66-; biochemist, Edgewood Arsenal, Md, 44-45. Mem: Soc Exp Biol & Med; Am Soc Biol Chem; Soc Gen Physiol; fel NY Acad Sci. Res: Enzymes in tissue trauma; serum enzymes in cancer. Mailing Add: Dept of Biol Brooklyn Col Bedford Ave & Ave H Brooklyn NY 11210

ALBAUM, MELVIN, b New York, NY, July 13, 36; div; c 1. GEOGRAPHY. Educ: Hunter Col, BA, 60; Univ Wis, MS, 64; Ohio State Univ, PhD(geog), 69. Prof Exp: Instr geog, Ohio State Univ, 64-68; asst prof, Univ Ky, 68-71; ASSOC PROF GEOG, UNIV COLO, BOULDER, 71- Concurrent Pos: Consult structure soc sci, NSF & admin training prog urban educ, Off Educ, Univ Colo, 71-; consult, Short Course Prog Ecosysts Mgt, US Forest Serv Mgt, 71- & Human Sci Proj, Biol Sci Curric Study, 72-73; Nat Inst Ment Health grant, 73-77. Mem: Pop Asn Am; Asn Am Geog. Res: Human fertility behavior; migration processes; social problems in a spatial context; population dynamics training; regionalization problems. Mailing Add: Dept of Geog Univ of Colo Boulder CO 80302

ALBEE, ARDEN LEROY, b Port Huron, Mich, May 28, 28; m 53; c 4. GEOLOGY. Educ: Harvard Univ, AB, 50, MA, 51, PhD(geol), 57. Prof Exp: Geologist, US Geol Surv, 50-59; from vis asst prof to assoc prof, 59-66, PROF GEOL, CALIF INST TECHNOL, 66- Concurrent Pos: Chmn, Lunar Sci Rev Panel, 73- Mem: Geol Soc Am; Mineral Soc Am. Res: Metamorphic petrology; electron microprobe; lunar rock investigations; seismo-tectonic siting. Mailing Add: Div of Geol & Planetary Sci Calif Inst Technol Pasadena CA 91125

ALBEE, HOWARD FRANKLIN, b Fruitland, Idaho, Feb 1, 15; m 43; c 3. GEOLOGY. Educ: Idaho State Col, BS, 49. Prof Exp: Geologist, Simplot Fertilizer Co, 47-50; geologist, 51-55, field geologist, Off Minerals Explor, 55-59 & 60-68, dist geologist, Conserv Div, 68-75, CONSULT GEOLOGIST, CONSERV DIV, US GEOL SURV, 75- Mem: Geol Soc Am; Am Inst Prof Geologists. Res: Triassic stratigraphy of Colorado Plateau; conglomerates of Triassic rocks on Colorado Plateau; exploration geology; areal geologic mapping. Mailing Add: Conserv Div US Geol Surv 8422 Fed Bldg Salt Lake City UT 84138

ALBEN, JAMES O, b Seattle, Wash, July 13, 30; m 54; c 3. BIOCHEMISTRY. Educ: Reed Col, BA, 51; Univ Ore, MS, 57, PhD(biochem), 59. Prof Exp: USPHS res fel physiol chem, Sch Med, Johns Hopkins Univ, 59-62; from asst prof to assoc prof, 62-73, PROF PHYSIOL, OHIO STATE UNIV, 73- Concurrent Pos: NIH grant, 64- Mem: Am Chem Soc; NY Acad Sci; Coblentz Soc; Brit Chem Soc; Am Soc Biol Chemists. Res: Molecular spectroscopy and structure of heme proteins; Fourier transform infrared interferometry; hemoglobin structure and function; oxygen binding and transport; physical inorganic biochemistry of metal proteins; mechanisms of metal-enzyme catalysis. Mailing Add: Dept of Physiol Chem Ohio State Univ Columbus OH 43210

ALBEN, RICHARD (SAMUEL), b Brooklyn, NY, July 12, 44; m 71. SOLID STATE PHYSICS. Educ: Harvard Univ, AM, 65, PhD(physics), 67. Prof Exp: Physicist, US Naval Ord Lab, 64-65; US-Japan Coop Sci grant magnetic resonance, Osaka Univ, 67-68; asst prof eng, appl sci & solid state physics, 68-73, ASSOC PROF ENG & APPL SCI, YALE UNIV, 73- Concurrent Pos: US-France exchange fel, Nat Ctr Sci Res Magnetic Lab, Grenoble, 74. Mem: Am Phys Soc. Res: Theory of magnetic impurities in garnets; theory of magnetoelastic effect; antiferromagnetic resonance; theory of liquid crystals; structure and excitations of amorphous materials. Mailing Add: Becton Ctr Yale Univ New Haven CT 06520

ALBER, HERBERT KARL, b Klagenfurt, Austria, Apr 4, 04; nat US; m 35; c 2. MICROCHEMISTRY. Hon Degrees: Ing Ch, Graz Tech Univ, 26, Dr Tech Sc, 28. Prof Exp: Asst chem, Graz Tech Univ, 26-28, asst inorg & anal chem, 28-29 & asst chem, 29-31; univ fel & teaching fel, Wash Sq Col, NY Univ, 34, from asst to instr anal & org microanal, 34-36; chief microanalyst, Biochem Res Found, Franklin Inst, 36-40; res technologist, A H Thomas Co, 41-49, Dir Res, 49-73; RETIRED. Concurrent Pos: Lectr, Univ Stockholm, 32 & Univ Berne, 32 & 33. Mem: Fel AAAS; Am Chem Soc; fel Am Inst Chem; Am Soc Test & Mat; Am Microchem Soc. Res: General, analytical and applied microanalysis; microanalysis equipment; instrumentation; pollution; standardization glassware; commercial standards; worldwide history of microchemistry. Mailing Add: 36 South Brookside Rd Springfield PA 19064

ALBERDA, WILLIS JOHN, b Bozeman, Mont, Feb 7, 36; m; c 2. MATHEMATICAL STATISTICS. Educ: Calvin Col, AB, 59; Mont State Univ, MS, 63, Nat Defense Educ Act fel & PhD(math statist), 64. Prof Exp: PROF MATH, DORDT COL, 64- Mem: Inst Math Statist; Am Statist Asn; Biomet Soc; Math Asn Am; Am Sci Affil. Res: Central limit theorems in probabilistic models; design and analysis of sampling procedures and population models. Mailing Add: Dept of Math Dordt Col Fourth Ave NE Sioux Center IA 51250

ALBERDING, HERBERT, b Chicago, Ill, Jan 14, 11; m 38; c 1. EARTH SCIENCES. Educ: Northwestern Univ, BS, 33; Univ Ariz, PhD(geol), 38. Prof Exp: Geologist, Nat Develop Co Philippines, 38-40, Anaconda Copper Mining Co, 41-45, Tex Petrol Co, Colombia, 45-47 & Phillips Petrol Co, Venezuela, 47-57; mgr explor, Signal Oil & Gas of Venezuela, 58-59; PROF GEOL & HEAD DEPT EARTH SCI, SIMPSON COL, 60- Concurrent Pos: Deleg, Int Geol Cong, Mex, 56; Am Asn Petrol Geologists dist rep, Caracas, Venezuela, 57-59. Mem: Fel Geol Soc Am; Venezuelan Geol, Mining & Petrol Asn (pres, 57-58). Res: Venezuelan stratigraphy; tectonics of Caribbean area; dynamics of wrench-faulting; development of sedimentary basins; isostasy in Glacier National Park; coal deposits of the Philippines. Mailing Add: Dept of Earth Sci Simpson Col Indianola IA 50125

ALBERGOTTI, JESSE CLIFTON, b Columbia, SC, Jan 4, 37. EXPERIMENTAL NUCLEAR PHYSICS. Educ: Wheaton Col, BS, 58; Univ NC, PhD(physics), 63. Prof Exp: Asst prof physics, Davidson Col, 62-64; ASST PROF PHYSICS, UNIV SAN FRANCISCO, 64-, CHMN DEPT, 68- Mem: Am Phys Soc; Am Asn Physics Teachers. Mailing Add: Dept of Physics Univ of San Francisco San Francisco CA 94117

ALBERINO, LOUIS MICHAEL, polymer chemistry, polymer physics, see 12th edition

ALBERNAZ, JOSE GERALDO, b Januaria, Brazil, Dec 3, 23; m 50; c 5. NEUROANATOMY, NEUROSURGERY. Educ: Univ Minas Gerais, BS, 40, MD, 46, MS, 55; Univ Guanabara, PhD(neurosurg), 58. Prof Exp: Prof neurol & actg head dept, Univ Minas Gerais, 56-58, asst prof neuropath, 58-59; Rockefeller Found res fel & travel grant, Sch Med, Univ Wash, 59-60; prof neurol & actg head dept, Univ Minas Gerais, 61-62, prof neurol & chmn dept, 62-66, prof neurol & neurol surg & head dept, 66-68; prof neuroanat, Med Col Ohio, 68-72; MEM STAFF, DEPT NEUROL & NEUROL SURG, FREDERICK C SMITH CLIN, 72- Concurrent Pos: Exec comt mem, Int Cong Neurol Surg, Denmark, 66; mem staff, Community Mem Hosp, Marion, Ohio & mem courtesy staff, Marion Gen Hosp. Mem: Am Asn Anat; cor mem Am Asn Neurol Surg; fel Am Col Surg; Pan Am Asn Anat (secy, 70-); Brazilian Acad Neurol. Res: Significance of supraspinal control of the gamma system; structure and function of cerebral arteries; causes of reactions to spinal puncture and pneumoencephalography. Mailing Add: F C Smith Clin Dept Neurol Surg 1040 Delaware Ave Marion OH 43302

ALBERS, EDWIN WOLF, b Schenectady, NY, July 29, 30; m 57; c 3. PHYSICAL CHEMISTRY. Educ: Clarkson Col Technol, BS, 52; Rensselaer Polytech Inst, PhD(phys chem), 62. Prof Exp: Chemist, Houston Refinery, Shell Oil Co, 52-55; asst chem, Rensselaer Polytech Inst, 55-57 & 58-60, instr, 60-61, from res asst to res assoc, 61-65; phys chemist, Inst Gas Technol, Ill Inst Technol Res Inst Ctr, 65-67; sr res scientist, 67-69, SUPVR CATALYTIC PREP, DAVISON CHEM DIV, W R GRACE & CO, 69- Mem: Am Chem Soc; Am Phys Soc. Res: Adsorption; gas solid systems; low temperature physics and chemistry; chemical kinetics related to the upper atmosphere; interstellar space research; zeolite and zeolitic promoted catalysts for petroleum and petrochemical applications. Mailing Add: Davison Chem Div W R Grace & Co Washington Res Ctr Clarksville MD 21029

ALBERS, HENRY, b Andover, Mass, Nov 17, 25; m 50; c 3. ASTRONOMY. Educ: Harvard Univ, AB, 50; Univ Minn, MA, 52; Case Inst Technol, PhD(astron), 56. Prof Exp: Instr astron, Univ Minn, 53-55 & Case Inst Technol, 55-56; asst prof math & astron, Butler Univ, 56-58; from asst prof to assoc prof, 58-68, PROF ASTRON, VASSAR COL, 68- Concurrent Pos: Vis lectr, Macalester Col, 54-55; NSF sci fac fel, 66. Mem: Am Astron Soc; Sigma Xi; AAAS. Res: Objective prism spectroscopy; galactic structure. Mailing Add: Vassar Col Observ Poughkeepsie NY 12601

ALBERS, JAMES RAY, b Tacoma, Wash, Sept 4, 34; m 61; c 2. THEORETICAL PHYSICS. Educ: Wash State Univ, BS, 56; George Washington Univ, MS, 58; Univ Wash, PhD(physics), 62. Prof Exp: Physicist, Nat Bur Standards, 57-58; lectr physics, Seattle Univ, 63-66, from asst prof to assoc prof, 66-71; assoc prof, 71-74, PROF PHYSICS, HUXLEY COL ENVIRON STUDIES, WESTERN WASH STATE COL, 74-, VPROVOST INSTRUCT & RES, 74- Mem: AAAS; Am Phys Soc; Am Meteorol Soc. Res: Field theory as it pertains to the elementary particle mass spectrum; atomic physics; cascade showers; air resources. Mailing Add: Dept of Physics Huxley Col Western Wash State Col Bellingham WA 98225

ALBERS, JOHN P, b Terry, Mont, May 25, 19; m 42; c 3. GEOLOGY. Educ: Carleton Col, AB, 40; Univ Minn, MS, 42; Stanford Univ, PhD(geol), 58. Prof Exp: Geologist, 42-45 & 46-70, ASSOC CHIEF GEOLOGIST, US GEOL SURV, 70- Mem: Geol Soc Am; Soc Econ Geol; Mineral Soc Am; Geochem Soc. Res: Mineral deposits; structural geology. Mailing Add: US Geol Surv Sunrise Valley Dr Reston VA 22070

ALBERS, ROBERT JAY, b Byron Center, Mich, Sept 24, 37; m 60; c 1. BIOPHYSICAL CHEMISTRY. Educ: Calvin Col, AB, 59; Univ Conn, PhD(chem), 62. Prof Exp: NSF fel chem, Univ Leiden, 62-63; res assoc, Fla State Univ, 63-66; asst prof chem, Northern Ill Univ, 67-74; MEM FAC, DEPT CHEM, WILLIAM RAINEY HARPER COL, 74- Mem: AAAS; Brit Chem Soc. Res: Chemical kinetics; free radical chemistry; protein conformations; block peptide polymers. Mailing Add: Dept of Chem William Rainey Harper Col Algonquin & Roselle Palatine IL 60067

ALBERS, ROBERT WAYNE, b Hebron, Nebr, Aug 5, 29. NEUROCHEMISTRY. Educ: Univ Nebr, BS, 50; Wash Univ, PhD(pharmacol), 54. Prof Exp: Biochemist, Lab Neuroanat Sci, 54-61, HEAD SECT ENZYMES, LAB NEUROCHEM, NIH, 61- Concurrent Pos: Prof lectr, George Washington Univ, 64-65. Mem: Am Soc Biol Chem; AAAS; Am Soc Neurochem; Int Soc Neurochem; Fedn Am Sci. Res: Biochemistry of the nervous system; relation of enzyme systems to physiological mechanisms. Mailing Add: Lab of Neurochem Nat Inst of Health Bethesda MD 20014

ALBERS, VERNON MARTIN, physics, deceased

ALBERS, WALTER ANTHONY, JR, b McKeesport, Pa, July 19, 30; m 52; c 5. RESEARCH ADMINISTRATION, SOLID STATE PHYSICS. Educ: Wayne State Univ, BS, 52, MS, 54, PhD(physics), 59. Prof Exp: Mem tech staff, Bell Tel Labs, 54-55; physicist, Res Labs, Bendix Corp, 55-57; res assoc, Wayne State Univ, 57-59; sr physicist, Res Labs, Bendix Corp, 59-62; supvry physicist, 62-73, HEAD SOCIETAL ANAL DEPT, RES LABS, GEN MOTORS CORP, WARREN, 73- Mem: AAAS; Am Phys Soc. Res: Surface and chemical physics; optics; quantitative social sciences. Mailing Add: 21168 Centerfarm Lane Northville MI 48167

ALBERSHEIM, PETER, b New York, NY, Mar 30, 34; m 58; c 3. PLANT BIOCHEMISTRY. Educ: Cornell Univ, BS, 56; Calif Inst Technol, PhD(biochem), 59. Prof Exp: Fel biochem, Calif Inst Technol, 59; NSF fel, Swiss Fed Inst Technol, 59-60; from instr to asst prof biol, Harvard Univ, 60-64; assoc prof biochem, 64-67, PROF BIOCHEM, UNIV COLO, 67-, PROF MOLECULAR BIOL, 70- Concurrent Pos: Fac fel Univ Colo, 70-71 & 75-76. Honors & Awards: Charles A Shull Award, Am Soc Plant Physiologists, 73. Mem: AAAS; Am Chem Soc; Am Soc Plant Physiol; Am Soc Biol Chem; Am Phytopath Soc. Res: Biochemistry of polysaccharides; function and nature of plant cell walls; mechanisms underlying disease resistance in plants and virulence in pathogens; mechanism of host selection by nitrogen-fixing bacteria. Mailing Add: Dept of Chem Univ of Colo Boulder CO 80302

ALBERS-SCHÖNBERG, GEORG, b Berlin, Ger, Dec 2, 29; m 69. ORGANIC CHEMISTRY. Educ: Swiss Fed Inst Technol, BS, 54; Univ Zurich, PhD(org chem), 62. Prof Exp: Res assoc org chem, Mass Inst Technol, 62-63 & mass spectrometry, 64-65; RES FEL BIOPHYS, MERCK INST THERAPEUT RES, 65- Mem: Am Chem Soc; NY Acad Sci; Swiss Chem Soc; Ger Chem Soc; fel Am Inst Chem. Res: Structure determination of natural products, mass spectrometry; relation between molecular structures and biological activities. Mailing Add: 30 Tyson Lane Princeton NJ 08540

ALBERT, ABRAHAM ADRIAN, mathematics, deceased

ALBERT, ALEXANDER, b Spring Valley, NY, Mar 25, 11. ENDOCRINOLOGY. Educ: St Stephens Col, Columbia, BA, 32; Harvard Univ, MA, 33, PhD(zool), 35; Harvard Med Sch, MD, 43. Hon Degrees: ScD, Bard Col, 61. Prof Exp: Res assoc, Harvard Univ, 35-45; res assoc, Mayo Clin, 46-47; from asst prof to assoc prof, 47-53, PROF PHYSIOL, MAYO GRAD SCH MED, UNIV MINN, ROCHESTER, 53- Concurrent Pos: Res fel, Mass Gen Hosp & Harvard Med Sch, 45-46; mem Marine Biol Corp, 35; intern, Beth Israel Hosp, Boston, 43-44, resident, 45; head endocrinol lab, Mayo Clin, 47-; mem, Laurentian Hormone Conf, 50-, sr adv comt, 60-; ed-in-chief, J Clin Endocrinol & Metab, 57-63; mem & actg chmn multidisciplinary study sect comt gerontol, NIH, 57, mem endocrinol study sect, 57-60, chmn, 60-63, mem pituitary hormone distrib comt, 60, chmn subcomt standards, 63, mem career awards study sect, 63-; mem ment health policy adv comt, Minn, 58-62; Mossman lectr, Col Med, Univ Nebr, 61; dir, Med Res Inst Worcester, Ind, 63- Honors & Awards: Merit Award, Goiter Asn, 55; Distinguished Serv Award, Am Thyroid Asn, 63. Mem: Am Soc Zool; Am Asn Anat; Am Thyroid Asn (pres, 60-61); Endocrine Soc; Am Soc Clin Invest. Res: Physiology. Mailing Add: Mayo Grad Sch of Med Univ of Minn Rochester MN 55902

ALBERT, ANTHONY HAROLD, b Los Angeles, Calif, Nov 24, 40; m 69. PHARMACEUTICAL CHEMISTRY. Educ: Occidental Col, BA, 63; San Diego State Univ, MS, 65; Ariz State Univ, PhD(org chem), 71. Prof Exp: Res assoc med chem, Nucleic Acid Res Inst, 69-71, res chemist, 71-73, HEAD PROCESS RES, PROD DEVELOP, ICN PHARMACEUT, 73- Mem: Am Chem Soc. Res: Development of industrial processes for pharmaceutical and chemical products. Mailing Add: ICN Pharmaceut 2727 Campus Dr Irvine CA 92715

ALBERT, ARTHUR EDWARD, b New York, NY, Nov 6, 35; m 59; c 3. MATHEMATICAL STATISTICS, APPLIED MATHEMATICS. Educ: Mass Inst Technol, BS, 56; Stanford Univ, MS, 57, PhD(math statist), 59. Prof Exp: NSF fel math & statist, Inst Math Sci, Stockholm, 59-60; asst prof, Columbia Univ, 60-61; res assoc elec eng, Mass Inst Technol, 61-62; sr scientist, Arcon Corp, 62-70; PROF MATH, BOSTON UNIV, 70- Mem: Inst Math Statist; Soc Indust & Appl Math. Res: Statistical decision theory; stochastic processes; time series analysis; pattern recognition; statistical estimation procedures; design of experiments. Mailing Add: Dept of Math Boston Univ Boston MA 02215

ALBERT, CHARLES GERALD, b Piper City, Ill, Apr 3, 12; m 37; c 5. CHEMISTRY. Educ: Johns Hopkins Univ, PhD(chem), 35. Prof Exp: Res chemist, Edgar Bros Co, 35-41, tech dir, 41-54; from asst dir res to dir res, 54-65, vpres res, Minerals & Chem Div, 65-74, VPRES RES, MINERALS & CHEM DIV & RES & DEVELOP DIV, ENGELHARD MINERALS & CHEM CORP, 74- Mem: Am Chem Soc; Tech Asn Pulp & Paper Indust; Soc Rheol. Res: Loading and coating of paper; refining of nonmetallic minerals; applications of fillers; beneficiation of nonmetallic minerals; particle size analysis; paint pigments and extenders; contact and percolation absorbents; activated absorbents; petroleum cracking catalysts and drilling fluids; agricultural carriers and diluents; pharmaceutical formulations. Mailing Add: Res & Develop Div Engelhard Minerals & Chem Corp Middlesex & Essex Turnpike Menlo Park NJ 08817

ALBERT, ERNEST NARINDER, b Gujarat, WPakistan, July 21, 37; m 60; c 4. ANATOMY, CELL BIOLOGY. Educ: High Point Col, BS, 58; Univ Pittsburgh, MS, 63; Georgetown Univ, PhD(anat), 65. Prof Exp: Instr gen & oral histol, Sch Med & Dent, Georgetown Univ, 65-66; res assoc, Univ Calif, Los Angeles, 66-67; asst prof histol & neuroanat, Jefferson Med Col, 67-68; asst prof, 68-72, ASSOC PROF HISTOL & NEUROANAT, SCH MED, GEORGE WASHINGTON UNIV, 72- Mem: Am Asn Anat; Electron Micros Soc Am; Am Heart Asn. Res: Subcellular effects of contraceptive steroids on blood vessels; development of electron dense stains; ultrastructural changes in tissue from exposure to microwaves. Mailing Add: Dept of Anat George Washington Univ Washington DC 20005

ALBERT, ETHEL MARY, b New Britain, Conn, Mar 28, 18. ANTHROPOLOGY. Educ: Brooklyn Col, AB, 42; Columbia Univ, AM, 47; Univ Wis, PhD(anthrop, philos), 49. Prof Exp: Instr philos, Brooklyn Col, 46-47, Univ Wis, 47-49 & Syracuse Univ, 49-52; res assoc, Lab Social Rels, Harvard Univ, 53-57; fel, Ctr Advan Study Behav Sci, 57-58; prof speech, Univ Calif, Berkeley, 58-66; PROF ANTHROP & SPEECH, NORTHWESTERN UNIV, EVANSTON, 66- Concurrent Pos: Ford Found Overseas Prog res fel, 55-57; asst dir, Educ Resources in Anthrop Proj, 60-61; Soc Sci Res Coun fel, 62; NSF sr fel, 65-66. Mem: Am Anthrop Asn; Am Philos Asn; African Studies Asn; Philos Sci Asn. Res: Ethnophilosophy; philosophy of social science; value theory; cultural logics; semiotics. Mailing Add: Dept of Anthrop Northwestern Univ Evanston IL 60201

ALBERT, EUGENE, b New York, NY, Jan 9, 30; m 60; c 1. MATHEMATICS. Educ: Brooklyn Col, BA, 50, MA, 51; Univ Va, PhD(math), 61. Prof Exp: Instr math, Brooklyn Col, 53-56; engr, Gen Elec Corp, NY, 56-57; asst prof math, Union Col, NY, 57-58; instr, Univ Va, 58-61; John Wesley Young res instr, Dartmouth Col, 61-63; asst prof, Univ Calif, Davis, 63-67; ASSOC PROF MATH, CALIF STATE COL, LONG BEACH, 67- Mem: Am Math Soc; Math Asn Am. Res: Probability theory; Markov chains. Mailing Add: Dept of Math Calif State Col Long Beach CA 90804

ALBERT, HAROLD MARCUS, b Russellville, Ala, Mar 5, 19; m 47; c 2. CARDIOVASCULAR SURGERY. Educ: Univ Chicago & Tulane Univ, BS, 40; Tulane Univ, MD, 44. Prof Exp: Intern, Touro Infirmary, 44-45, resident surg, 45-47; sr fel, Ochsner Found, 47-48; sr surgeon, Huey P Long Mem Hosp, 48-49; clin instr, 49-56, from asst prof to assoc prof, 56-73, PROF SURG, SCH MED, LA STATE UNIV, NEW ORLEANS, 73- Concurrent Pos: Preceptorship, Huey P Long Mem Hosp, 49-51; La Heart Asn fel, Northwestern Univ, 51-52; mem staff, Charleston Naval Hosp & Portsmouth Naval Hosp, 54-56; mem cardiovasc surg coun, Am Heart Asn; sr vis surgeon, Charity Hosp La, New Orleans; sr assoc, Touro Infirmary; active staff mem, Hotel Dieu, New Orleans, Methodist Hosp & West Jefferson Hosp, Marrero. Mem: Am Col Surg; AMA; Am Col Chest Physicians; Am Soc Artificial Internal Organs; Am Col Cardiol. Res: Congenital and acquired heart disease; shock. Mailing Add: Dept of Surg La State Univ Sch of Med New Orleans LA 70112

ALBERT, HARRISON BERNARD, b Oakland, Calif, Dec 12, 36; m 59; c 4. CHEMICAL INSTRUMENTATION, COMPUTER SCIENCES. Educ: Univ Calif, Berkeley, BS, 59; Univ Colo, Boulder, PhD(phys chem), 69. Prof Exp: Mem res staff, Thomas J Watson Res Ctr, Int Bus Mach Corp, NY, 69-70; ASST PROF CHEM, UNIV COLO, BOULDER, 70-, RES ASSOC, 74- Concurrent Pos: NSF fel, 63-66. Mem: AAAS; Am Chem Soc. Res: Analytical chemistry instrumentation and computing science; gas liquid chromatography and disc gel chromatography data acquisition hardware; algorithms for storage compression, deconvolutions, displays and removing noise and drift; innovative microcalorimeter configurations and electronics. Mailing Add: Dept of Chem Univ of Colo Boulder CO 80302

ALBERT, JERRY DAVID, b Milwaukee, Wis, June 6, 37; m 61; c 1. CLINICAL BIOCHEMISTRY. Educ: Occidental Col, BA, 59; Iowa State Univ, PhD(biochem), 64. Prof Exp: Sr scientist appl res labs, Aeronutronic Div, Philco-Ford Corp, 64-67; fel prebiotic chem, Salk Inst Biol Studies, 67-68; staff res assoc IV, Univ Hosp, San Diego County, 68-73; RES BIOCHEMIST, MERCY HOSP & MED CTR, 73- Mem: Am Chem Soc. Res: Steroid hormone metabolism; mechanism of action analysis. Mailing Add: 5202 Cobb Place San Diego CA 92117

ALBERT, LUKE SAMUEL, b Palmyra, Pa, Feb 19, 27; m 50; c 2. PLANT PHYSIOLOGY. Educ: Lebanon Valley Col, BS, 50; Rutgers Univ, MS, 52, PhD(plant physiol), 58. Prof Exp: Res plant physiologist, Biol Res Lab, Ft Detrick, Md, 53-55; res asst, Rutgers Univ, 55-58; plant physiologist, Am Cyanamid Co, 58-60; from asst prof to assoc prof, 60-70, PROF BOT, UNIV RI, 70- Mem: AAAS; Am Soc Plant Physiol; Ecol Soc Am; Bot Soc Am; Soc Gen Systs Res. Res: Physiological ecology; boron physiology; growth and development; systems research. Mailing Add: Dept of Bot Univ of RI Kingston RI 02881

ALBERT, MARY ROBERTS FORBES (DAY), b Manchester, NH, Mar 2, 26; m 55; c 2. CYTOLOGY, BIOCHEMISTRY. Educ: Univ NH, BS, 48; Bryn Mawr Col, MA, 50; Brown Univ, PhD, 55. Prof Exp: Asst cytol, Brown Univ, 51-55; res fel med, Mass Gen Hosp, 55-61; instr, Northeastern Univ, 61-62 & Wellesley Col, 63-64; asst prof biol sci, Newton Col Sacred Heart, 64-75; DIR, BIOL LABS, BOSTON COL, 75- Res: Irradiation effects on mitosis in the rat and mouse; abnormal mitosis in Hela cells. Mailing Add: 56 Chapin Rd Newton MA 02158

ALBERT, OSCAR J, b July 11, 12; Can citizen; m 42; c 3. BACTERIOLOGY, FOOD CHEMISTRY. Educ: Laval Univ, BS, 35; Iowa State Univ, MS, 39, PhD(bact chem), 43. Prof Exp: Vpres & dir res, Mil-Ko Prod, 56-64; VPRES & DIR RES, RICH PROD CORP, 64- Mem: Can Inst Food Technol; Inst Food Technologists. Mailing Add: Rich Prod Corp PO Box 245 Buffalo NY 14240

ALBERT, PAUL JOSEPH, b Edmunston, Can, Mar 11, 46. SENSORY PHYSIOLOGY. Educ: Univ NB, BSc, 68, PhD(biol), 72. Prof Exp: Fel, Univ Sask, 72-73; ASST PROF BIOL, CONCORDIA UNIV, 73- Mem: Can Soc Zool; Entom Soc Can. Res: Structure and physiology of insect sense organs. Mailing Add: Dept of Biol Concordia Univ Loyola Campus Montreal PQ Can

ALBERT, RICHARD DAVID, b Elmira, NY, Aug 9, 22; m 46; c 4. NUCLEAR PHYSICS. Educ: Univ Mich, BA, 43; Columbia Univ, AM, 46, PhD(physics), 51. Prof Exp: Asst, Columbia Univ, 46-49; sr scientist, Westinghouse Elec Corp, 49-51; res assoc, Knolls Atomic Power Lab, 51-55; sr physicist, Lawrence Radiation Lab, Univ Calif, 55-65, res physicist, Space Sci Lab, Univ Calif, 66-70; physicist, Terradynamics, Inc, 70-71; PRES, PHYSICS ANAL LABS, INC, 71- Concurrent Pos: Consult, Space Sci Lab, Univ Calif, 70-; NASA grant. Mem: AAAS; Am Phys Soc; Am Geophys Union. Res: Scintillation spectrometry; measurement of neutron capture resonances; design of scintillation detectors for use with neutron time-of-flight spectrometer; beta ray spectroscopy; p,n and alpha,n cross sections; p,n and d,p spectra and angular distributions; nuclear temperatures; neutron cross-sections using space nuclear explosions; neutron and charged particle space experiments. Mailing Add: 317 Hartford Rd Danville CA 94526

ALBERT, ROY ERNEST, b New York, NY, Jan 11, 24; m 45; c 4. ENVIRONMENTAL MEDICINE, PULMONARY PHYSIOLOGY. Educ: Columbia Univ, AB, 43; NY Univ, MD, 46. Prof Exp: Med officer, NY Opers Off, AEC, 52-54, chief med br, Div Biol & Med, 54-56; asst prof med, Sch Med, George Washington Univ, 56-59; assoc prof, 59-66, PROF ENVIRON MED, MED CTR, NY UNIV, 66- Concurrent Pos: USPHS fel, 49-51. Mem: Radiation Res Soc; Am Indust Hyg Asn; Am Asn Cancer Res. Res: Radiation and chemical carcinogenesis; aerosol deposition and clearance; cancer and environmental toxicants epidemiology. Mailing Add: Inst of Environ Med NY Univ Med Ctr New York NY 10016

ALBERT, SAMUEL, b Montreal, Que, Aug 15, 15; m 46; c 1. ONCOLOGY. Educ: McGill Univ, BA, 36, MD, 40, PhD(anat), 48. Prof Exp: Demonstr histol, McGill Univ, 41-42; res assoc, Nat Cancer Inst Can, 48; INSTR PATH, WAYNE STATE UNIV, 48-; CHIEF LAB HUMAN ECOL, DEPT EPIDEMIOL, MICH CANCER FOUND, 74- Mem: Am Asn Anat; Am Asn Cancer Res; NY Acad Sci. Res: Experimental morphology and oncology; normal and neoplastic growth; hematopoietic system; epidemiology. Mailing Add: Mich Cancer Found 110 E Warren Ave Detroit MI 48201

ALBERT, WACO W, b Ainsworth, Nebr, Oct 22, 21; m 47. ANIMAL NUTRITION. Educ: Univ Nebr, BS, 48; Univ Okla, MS, 49; Univ Ill, PhD(animal nutrit), 54. Prof Exp: Instr animal husb, Agr & Mech Col, Tex, 49-51; asst prof, 54-70, ASSOC PROF ANIMAL SCI, UNIV ILL, URBANA-CHAMPAIGN, 70- Res: Sulfur requirements of growing-fattening lambs; use of urea in lamb rations; drylotting beef cows plus utilization of husklage and stalklage for beef cows; energy requirements of light horses; maintenance; growth; work and locations. Mailing Add: 104 Stock Pavilion Urbana IL 61801

ALBERT, WILLIAM CHARLES, chemistry, see 12th edition

ALBERTA, MARY, pure mathematics, statistics, see 12th edition

ALBERTE, RANDALL SHELDON, b Newark, NJ, June 7, 47. PLANT PHYSIOLOGY, BIOCHEMISTRY. Educ: Gettysburg Col, BA, 69; Duke Univ, PhD(bot), 73. Prof Exp: Res assoc, 73-75, NSF ENERGY-RELATED FEL BIOCHEM, UNIV CALIF, LOS ANGELES, 75- Mem: Am Soc Plant Physiologists; Soc Exp Biol; Bot Soc Am; AAAS; Am Inst Biol Sci. Res: Photobiology and adaptive physiology of photosynthesis; chloroplast development; organization of chlorophyll in vivo; membrane biochemistry; biochemistry and physiology of water stress and of the control of development and differentiation. Mailing Add: Dept of Biol Univ of Calif Los Angeles CA 90024

ALBERTI, PETER W R M, b Aug 23, 34; m 61; c 3. OTOLARYNGOLOGY. Educ: Univ Durham, MB, BS, 57; Washington Univ, PhD(anat), 65; FRCS, 65; FRCPS(C), 68. Prof Exp: Demonstr anat, Med Sch, Univ Durham, 58-59, first asst otolaryngol, 64-67; instr anat, Emory Univ, 60-61; clin teacher, 67-68, asst prof, 68-70, ASSOC PROF OTOLARYNGOL, FAC MED, UNIV TORONTO, 70- Mem: AAAS; Am Asn Anat; Can Otolaryngol Soc; Royal Soc Med. Res: Anatomy of middle ear; teaching techniques; hearing testing. Mailing Add: 600 University Ave Toronto ON Can

ALBERTS, ARNOLD A, b Davis, SDak, May 28, 06; m 31; c 2. ORGANIC CHEMISTRY. Educ: Univ SDak, AB, 28; Univ Okla, MS, 30; Ohio State Univ, PhD(org chem), 34. Prof Exp: Asst gen & org chem, Univ Okla, 28-30; asst Ohio State Univ, 30-31, asst gen chem, 31-34, instr chem, 35-36, res asst metall eng exp sta, 35-37, investr indust probs, 36-37; asst prof org & phys chem, Otterbein Col, 37; from instr to asst prof chem, Washington & Jefferson Col, 37-42; from asst prof to assoc prof, Purdue Univ, 43-52; res chemist, Western Co, 52-56; PROF CHEM & HEAD DEPT, HASTINGS COL, 56- Concurrent Pos: Admin asst, Res Found, Purdue Univ, 46-49, supvr chem, Exten Ctr, 47-52. Mem: Am Chem Soc. Res: Corrosion and corrosion inhibitors. Mailing Add: Dept of Chem Hastings Col Hastings NE 68901

ALBERTS, BRUCE M, b Chicago, Ill, Apr 14, 38; m 60; c 3. MOLECULAR BIOLOGY. Educ: Harvard Univ, AB, 60, PhD(biophys), 65. Prof Exp: NSF res fel, 65-66; asst prof chem, 66-71, assoc prof biochem sci, 71-73, PROF BIOCHEM SCI, PRINCETON UNIV, 73- Honors & Awards: Eli Lilly Award, Am Chem Soc, 72; US Steel Award, Nat Acad Sci, 75. Mem: Am Soc Biol Chem; Am Chem Soc. Res: Molecular genetics; DNA replication; eukaryotic gene control. Mailing Add: Dept of Biochem Sci Frick Chem Lab Princeton Univ Princeton NJ 08540

ALBERTS, GENE S, b Brookings, SDak, Sept 20, 37; m 59. ANALYTICAL CHEMISTRY. Educ: Wash State Univ, BS, 59; Univ Wis, PhD(chem), 63. Prof Exp: Staff chemist, Systs Develop Div, Int Bus Mach Corp, 63-66, res staff mem, T J Watson Res Lab, 66-67, develop chemist, Component Div, IBM Corp, 67-69, sr eng, 69-72, SR ENG, SYSTS PROD DIV, IBM CORP, 72- Res: Electroanalytical and electrokinetic chemistry; magnetic film deposition; semi conductor process engineering. Mailing Add: Systs Prod Div IBM Corp PO Box A Essex Junction VT 05452

ALBERTS, WALTER WATSON, b Los Angeles, Calif, Dec 31, 29; m 59; c 2. NEUROSCIENCES, MEDICAL ADMINISTRATION. Educ: Univ Calif, AB, 51, PhD(biophys), 56. Prof Exp: Res physiologist, Med Ctr, Univ Calif, San Francisco, 55-56; biophysicist, Mt Zion Hosp & Med Ctr, 56-72; grants assoc, NIH, 72-73, spec asst to assoc dir, C & FR, 73-74, head res contracts sect, 74-75, ASST DIR CONTRACTS RES PROGS, EXTRAMURAL ACTIV PROG, NAT INST NEUROL & COMMUNICATIVE DIS & STROKE, NIH, 75-Concurrent Pos: Consult, Donner Lab, Univ Calif, Berkeley, 59-64; res career prog award, Nat Inst Neurol Dis & Blindness, 63-68; lectr dept physiol, Med Ctr, Univ Calif, San Francisco, 69. Mem: Am Physiol Soc; fel AAAS; Biophys Soc; Inst Elec & Electronics Eng; Soc Neurosci. Res: Neurophysiology; biological and medical physics, particularly the central nervous system of man. Mailing Add: Nat Inst of Neurolog & Communicative Dis & Stroke NIH Bethesda MD 20014

ALBERTSON, CLARENCE ELMO, chemistry, see 12th edition

ALBERTSON, JAMES STANISLAUS, b Los Angeles, Calif, May 26, 27. THEORETICAL PHYSICS. Educ: St Louis Univ, BA, 52; Harvard Univ, PhD(physics), 58. Prof Exp: From instr to assoc prof physics, Loyola Univ Los Angeles, 62-67, chmn dept, 65-68; prof physics & acad vpres, Univ Santa Clara, 68-73; dir anal studies, 73-74, ASST VPRES ACAD AFFAIRS, UNIV CALIF, BERKELEY, 74- Concurrent Pos: NSF fac fel, Stanford Univ, 67-68. Res: Quantum-mechanical measurement theory. Mailing Add: Off of the Pres Univ of Calif Berkeley CA 94720

ALBERTSON, JOHN NEWMAN, JR, b New Haven, Conn, Jan 18, 33; m 55; c 7. MICROBIOLOGY, LABORATORY MANAGEMENT. Educ: Univ Conn, AB, 54; Hahnemann Med Col, MS, 64; Am Bd Microbiol, dipl. Prof Exp: US ARMY, 54-, comndg officer med dispensary, Munich, Ger, 56-57, asst chief dept immunol & serol, Med Lab, Landstuhl, Ger, 57-59, chief clin path & bacteriologist, Valley Forge Gen Hosp, Pa, 59-62, chief clin microbiol, Hahnemann Med Col, 62-64, chief bact div, First Med Lab, Ft George G Meade, Md, 64-68, chief med & biol sci br, Off Chief Res & Develop, Hq Dept Army, Washington, DC, 68-70, exec res, 70-71, comndg officer, 9th Med Lab, Repub of Vietnam, 71-72, exec officer, Walter Reed Army Inst Res, Washington, DC, 72-74, EXEC OFFICER, ARMED FORCES INST PATH, US ARMY, WASHINGTON, DC, 75- Concurrent Pos: Consult, Europ Command & partic coop study on Treponemal pallidum immobilization & other tests, WHO, Ger, 57-59. Mem: Fel AAAS; Am Soc Microbiol; Am Acad Microbiol; Sigma Xi. Res: Chemotherapy and bacteriology of tuberculosis; diagnostic microbiology. Mailing Add: 5226 Ferndale St Springfield VA 22151

ALBERTSON, MICHAEL OWEN, b Philadelphia, Pa, June 24, 46; m 69; c 1. COMBINATORIAL MATHEMATICS. Educ: Mich State Univ, BS, 66; Univ Pa, PhD(math), 71. Prof Exp: Asst prof math, Swarthmore Col, 72-73; ASST PROF MATH, SMITH COL, 73- Concurrent Pos: Prin investr, Cottrell grant, Res Corp, 74. Mem: Am Math Soc. Res: Independent sets in graphs-four color problem; topological graph theory. Mailing Add: Dept of Math Smith Col Northampton MA 01060

ALBERTSON, NOEL FREDERICK, b New Haven, Conn, Oct 29, 15; m 42; c 3. MEDICINAL CHEMISTRY. Educ: Polytech Inst Brooklyn, BS 36, MS, 38; Ohio State Univ, PhD, 41. Prof Exp: Jr chemist, Merck & Co, 38-39; asst, Ohio State Univ, 39-41, res assoc & fel, Univ Res Found, 41-43; group leader, Winthrop Chem Co, 43-46; group leader, 46-64, sect head, 64-66, ASST DIR CHEM DIV, STERLING-WINTHROP RES INST, 66- Mem: Am Chem Soc. Res: Molecular addition compounds of sulphur dioxide; production of sulphur from natural gas; synthesis of amino acids, peptides, piperidines and benzomorphans. Mailing Add: Chem Div Sterling-Winthrop Res Inst Rensselaer NY 12144

ALBERTY, ROBERT ARNOLD, b Winfield, Kans, June 21, 21; m 44; c 3. CHEMISTRY. Educ: Univ Nebr, BS, 43, MS, 44; Univ Wis, PhD(phys chem), 47. Prof Exp: Asst, Univ Wis-Madison, 44-46, from instr to prof phys chem, 47-62, assoc dean letters & sci, 62-63, dean grad sch, 63-67; DEAN SCH SCI, MASS INST TECHNOL, 67- Concurrent Pos: Guggenheim fel, Calif Inst Technol, 50-51; chmn, Comn Human Resources, Nat Res Coun, 74- Honors & Awards: Eli Lilly Award, 56. Mem: Nat Acad Sci; Am Acad Arts & Sci; Am Chem Soc. Res: Enzyme kinetics; chemical kinetics and exchange; theory of electrophoresis; fast reactions in solutions. Mailing Add: Rm 6-215 Sch of Sci Mass Inst of Technol Cambridge MA 02139

ALBERTY, RONNIE LEE, b Mt Vernon, Mo, Feb 4, 37; m 1; c 3. METEOROLOGY. Educ: Univ Mo-Columbia, BS, 63, MS, 65, PhD(atmospheric sci), 67. Prof Exp: From asst prof to assoc prof meteorol, Naval Postgrad Sch, 67-72; chief storm modeling, 72-74, CHIEF METEOROL RES, NAT SEVERE STORMS LAB, 74- Concurrent Pos: Adj assoc prof meteorol, Univ Okla, 73- Mem: Am Meteorol Soc; Sigma Xi. Res: Dynamic meteorology; cloud physics; severe storms research. Mailing Add: Nat Severe Storms Lab 1313 Halley Circle Norman OK 73069

ALBIN, ROBERT CUSTER, b Beaver City, Okla, May 19, 39; m 60; c 2. ANIMAL NUTRITION. Educ: Tex Technol Col, BS, 61, MS, 62; Univ Nebr, PhD(animal nutrit), 65. Prof Exp: Asst prof animal husb, 64-67, assoc prof, 67-73, PROF ANIMAL SCI, TEX TECH UNIV, 73- Mem: Am Soc Animal Sci. Res: Ruminant nutrition; beef cattle nutrition and management. Mailing Add: Dept of Animal Sci Tex Tech Univ Lubbock TX 79409

ALBINAK, MARVIN JOSEPH, b Detroit, Mich, June 21, 28; m 61; c 3. INORGANIC CHEMISTRY. Educ: Univ Detroit, AB, 49, MS, 52; Wayne State Univ, PhD(inorg chem), 59. Prof Exp: Chemist explor res, Ethyl Corp, 52-54; from instr to asst prof chem, Univ Detroit, 54-61; sr res scientist, Elec Autolite, 61-62; res chemist, Owens-Ill Glass Co, 62-65; from asst prof to assoc prof chem, Wheeling Col, 65-68; PROF & CHMN DIV SCI & MATH, ESSEX COMMUNITY COL, 68- Mem: AAAS; Am Chem Soc; Am Ceramic Soc; Brit Soc Glass Technol. Res: Inorganic chemistry of glass; synthesis and resolution of coordination compounds; absorption and fluorescence of inorganics; science for the layman; science and public policy. Mailing Add: 717 Hillen Rd Towson MD 21204

ALBISSER, ANTHONY MICHAEL, b Johannesburg, Africa, Sept 5, 41; m 64; c 3. BIOMEDICAL ENGINEERING. Educ: McGill Univ, BEng, 64; Univ Toronto, MASc, 66, PhD(biomed eng), 68. Prof Exp: Assoc dir med eng, 68-71, dir, 71-75, INVESTR BIOMED RES, HOSP FOR SICK CHILDREN, 75- Concurrent Pos: Spec lectr elec eng, Univ Toronto, 68-71, asst prof, 72-75, adj prof, 75-; subcomt mem, Nat Comn Diabetes, 75; co-ed-in-chief, Med Progress Technol, 75. Mem: Can Med Biol Eng Soc; Am Diabetes Asn; Asn Prof Eng Ont; Asn Advan Med Instrumentation. Res: Study of the pathophysiology of diabetes and how this relates to the development of an artificial endocrine pancreas; the application of technology and modern materials to the solution of the orthotic problems of disabled children. Mailing Add: Div of Biomed Res Res Inst Hosp for Sick Children Toronto ON Can

ALBRECHT, ALBERTA MARIE, b Reading, Pa, June 30, 30. MICROBIAL PHYSIOLOGY, CANCER. Educ: Seton Hill Col, BA, 51; Fordham Univ, MS, 52; Rutgers Univ, PhD(microbiol), 61. Prof Exp: Biologist, Res Div, Am Cyanamid, 52-58; res fel, 61-63, res assoc, 63-65, assoc, 65-75, ASSOC MEM, SLOAN-KETTERING INST CANCER RES, 75- Concurrent Pos: Instr microbiol, Sloan-Kettering Div, Grad Sch Med Sci, Cornell Univ, 64-68, asst prof, 68- Mem: Am Soc Microbiol; AAAS; Am Asn Cancer Res; Sigma Xi; NY Acad Sci. Res: Interests in folic acid metabolism and regulatory mechanisms in folate auxotrophs, folate and antifolate-transforming bacteria, microbial enzymes in therapy of neoplasia, nutrition of malignant cells. Mailing Add: Sloan-Kettering Inst Cancer Res 145 Boston Post Rd Rye NY 10580

ALBRECHT, ANDREAS CHRISTOPHER, b Berkeley, Calif, June 3, 27; m 51; c 4. QUANTUM CHEMISTRY, SOLID STATE CHEMISTRY. Educ: Univ Calif, BS, 50; Univ Wash, PhD(chem), 54. Prof Exp: Res assoc, Mass Inst Technol, 54-56; from instr to assoc prof, 56-65, PROF CHEM, CORNELL UNIV, 65- Concurrent Pos: Consult, Eastman Kodak, 66-; partic, US-USSR Cult Exchange, 63-64 & US-USSR Acad Sci Exchange Prog, 74; NSF sci fac fel, 70-71; mem adv comt on USSR & Eastern Europe, Nat Acad Sci, 70-73; vis prof, Univ Calif, Santa Cruz, 71. Mem: Fedn Am Scientists. Res: Electronic, vibronic and Raman spectroscopy; organic solid state. Mailing Add: Dept of Chem Cornell Univ Ithaca NY 14850

ALBRECHT, FREDERICK XAVIER, b New York, NY, Oct 25, 43. PHYSICAL ORGANIC CHEMISTRY. Educ: State Univ NY Albany, BS, 67; State Univ NY Buffalo, PhD(chem), 72. Prof Exp: Asst prof chem, State Univ NY Col Canton, 69-70; fel chem, Univ Wis-Madison, 72-74; RES CHEMIST, EASTMAN KODAK CO, 74- Mem: Am Chem Soc; Sigma Xi. Res: Mechanistic study of organic photoconductive materials for use in electrophotography. Mailing Add: Eastman Kodak Co Res Labs Bldg 82 Kodak Park Rochester NY 14650

ALBRECHT, GUSTAV, physical chemistry, see 12th edition

ALBRECHT, HERBERT RICHARD, b Kenosha, Wis, Nov 14, 09; m 36; c 2. AGRONOMY. Educ: Univ Wis, BS, 32, MS, 33, PhD(plant genetics), 38. Hon Degrees: DAgr, Purdue Univ, 62; DSc, NDak State Univ, 72. Prof Exp: Asst genetics, Univ Wis, 31-36; from asst prof to assoc prof agron, Auburn Univ, 36-44, from asst agronomist to assoc agronomist exp sta, 36-44; assoc prof agron, Purdue Univ, 44-45, asst chief agron, 45-47; prof & head dept, Pa State Univ, 47-52, dir coop agr & home econ exten serv, 53-61; pres, NDak State Univ, 61-68; dir, Int Inst Trop Agr, Ibadan, Nigeria, 68-75; RETIRED. Mem: Fel AAAS; Crop Sci Soc Am (pres, 53); fel Am Soc Agron; Sigma Xi. Res: Legume and turf grass breeding; inoculation and fertilizer studies with legumes; crop rotation; insect resistance in legumes; inheritance of resistance to bacterial wilt in alfalfa. Mailing Add: Rte 1 Box 61 U Clarksville VA 23927

ALBRECHT, NORMAN EDWARD, b St Paul, Minn, Mar 1, 22; m 46; c 4. MATHEMATICS. Educ: Hamline Univ, BS, 43; Univ Minn, MA, 47. Prof Exp: Instr math, Bethel Col, Minn, 42-44; from instr to asst prof, Hamline Univ, 46-54; mathematician, Remington Rand Univac, 54-58; MATH ANALYST, INVESTORS DIVERSIFIED SERV, 58- Mem: Math Asn Am; Opers Res Soc Am; Inst Mgt Sci. Mailing Add: 3069 W Owasso Blvd St Paul MN 55112

ALBRECHT, PAUL, b Presov, Czech, Jan 27, 25; m 53; c 2. VIROLOGY, PATHOLOGY. Educ: Comenius Univ Bratislava, MUDr, 49; Czech Acad Sci, CSc, 60. Prof Exp: Asst prof path, Med Fac, Comenius Univ Bratislava, 50-54; from res assoc to chief lab, Inst Virol, Czech Acad Sci, 55-65; res assoc microbiol, Nat Inst Neurol Dis & Blindness, 65-68; res virologist, Div Biol Standards, NIH, 68-72; RES VIROLOGIST, BUR BIOLOGICS, FOOD & DRUG ADMIN, 72- Concurrent Pos: Czech Acad Sci Award, 58-63. Honors & Awards: Czech Med Asn Prize, 64. Mem: Am Soc Microbiol; Soc Exp Biol & Med; NY Acad Sci. Res: Immunogenesis of viral infections; slow virus infections; development and control of vaccines. Mailing Add: Nat Insts of Health Bldg 29A Bethesda MD 20014

ALBRECHT, ROBERT H, b Ohio, Ill, May 1, 16; m 43; c 3. CHEMISTRY. Educ: NCent Col, BS, 37; NDak State Univ, MS, 41. Prof Exp: Chemist, 41-70, dir chem coatings lab, 70-75, TECH MGR BROWN LABEL LINES, SHERWIN WILLIAMS CO, 75- Mem: Am Chem Soc. Res: Protective coatings; lacquers. Mailing Add: AW Stendel Tech Ctr 549 E 115th St Chicago IL 60628

ALBRECHT, ROBERT MICHAEL, b Green Island, NY, Feb 25, 17. EPIDEMIOLOGY. Educ: Col of the Holy Cross, AB, 38; Albany Med Col, MD, 42; Harvard Univ, MPH, 49. Prof Exp: Pub health physician, NY State Health Dept, 47-65, dir, Off Epidemiol, 54-65; dir, Psychiat Epidemiol Field Sta, Col Physicians & Surgeons, Columbia Univ, 65-67, dir, Air Pollution Epidemiol Res Unit, Sch Pub Health, 67-70, assoc prof epidemiol, 65-70; asst comnr health, Ariz State Health Dept, 70-71; from vis scientist to med officer, Bur Radiol Health, Food & Drug Admin, 71-73; SCI RES & WRITING, 73- Concurrent Pos: Med officer, Epidemiol Studies Sect, WHO, 55-56. Mem: Am Epidemiol Soc; Soc Occup & Environ Health; Soc Epidemiol Res; Int Epidemiol Asn; Air Pollution Control Asn. Res: Adverse effects of prolonged exposure to environmental agents on health, especially air pollution, ionizing and non-ionizing radiation; methodology of studying delayed adverse effects of environmental agents. Mailing Add: 5 Myrtle Ave Troy NY 12180

ALBRECHT, STEVEN HAROLD, b Milwaukee, Wis, Feb 5, 39; m 61; c 2. PHYSICAL CHEMISTRY. Educ: St Olaf Col, BA, 61; NDak State Univ, PhD(phys chem), 66. Prof Exp: Res chemist graphite, Air Reduction, Inc, 66-69; asst prof phys chem, NDak State Univ, 69-70; asst prof chem, Huron Col, 70-72, chmn div natural sci, 72-73; RES SCIENTIST CHEM, BETZ LABS, INC, 73- Mem: Am Chem Soc. Res: Industrial water technology; chemistry of coal; powder x-ray diffraction; crystal growth; futures research. Mailing Add: Betz Labs Inc Trevose PA 19047

ALBRECHT, THOMAS WYMAN, biochemistry, see 12th edition

ALBRECHT, WALTER AUGUST, JR, mathematics, see 12th edition

ALBRECHT, WILLIAM LIND, b Marietta, Ohio, July 4, 37; m 63; c 2. MEDICINAL CHEMISTRY. Educ: Ohio State Univ, BS, 61; Univ Kans, PhD(med chem), 65. Prof Exp: Res org chemist, 66-70, SECT HEAD, MERRELL-NAT LABS DIV, RICHARDSON-MERRELL INC, 70- Concurrent Pos: Adj assoc prof, Col Pharm, Univ Cincinnati. Mem: AAAS; Am Pharmaceut Asn; Am Chem Soc. Res: Non-steroidal antiinflammatory agents; antiviral agents; immuno-regulators; antisecretory agents. Mailing Add: Org Chem Dept Merrell-Nat Labs 110 E Amity Rd Cincinnati OH 45215

ALBRECHT, WILLIAM LLOYD, b Aurora, Ill, May 9, 33; m 56; c 4. PHYSICAL CHEMISTRY. Educ: Oberlin Col, AB, 55; Univ Wis, PhD(phys chem), 60. Prof Exp: Sr res chemist, 60-66, group leader, 66-69, mgr water treat chem res, 69-70, tech dir, 70-71, mgr technol & mkt, 71-75, MGR RES, PULP & PAPER CHEM NALCO CHEM CO, 75- Mem: Am Chem Soc. Res: Radiochemistry and tracer techniques; colloid chemistry; paper process chemicals. Mailing Add: Nalco Chem Co 6216 W 66th Pl Chicago IL 60638

ALBRECHT, WILLIAM MELVIN, b Hungerford, Pa, Feb 18, 26; m 48; c 8. PHYSICAL CHEMISTRY, MATERIALS ENGINEERING. Educ: Lebanon Valley Col, BS, 48; Univ Cincinnati, MS, 50. Prof Exp: Prin chemist physics div, Battelle Mem Inst, 50-57, asst div chief ferrous metall div, 57-60; staff chemist, 60-62, adv chemist, 60-66, develop engr & mgr advan technol, 66-73, SR ENGR & MGR, INT BUS MACH CORP, 73- Mem: Am Chem Soc; Electrochem Soc; Am Soc Metals; fel Am Inst Chem; AAAS. Res: Analysis, diffusion, kinetics, thermodynamics and sorption in gas metal systems; electrochemistry; surface chemistry; research and development of new materials for electronic packaging. Mailing Add: Int Bus Mach Corp PO Box 6 Glendale Lab Endicott NY 13760

ALBRECHT-BUEHLER, GUENTER WILHELM, b Berlin, WGer, Mar 8, 42; m 67; c 2. CELL BIOLOGY. Educ: Univ Munich, BSc, 63, dipl physics, 67, PhD(physics), 71. Prof Exp: Investr elec physiol, Ger Radiol Soc, 67-70; fel cell biol, Friedrich Miescher Inst, Basel, Switz, 70-73; fel, Univ Fla, 73-74; staff investr cell biol, 74-75, HEAD CELL BIOL DEPT, COLD SPRING HARBOR LAB, 75- Concurrent Pos: Training fel, Int Agency for Res of Cancer, Lyon, France, 73-74. Res: Motility phenomena in the surface of animal cells and their relation to cell communication and transformation. Mailing Add: Box 100 Cold Spring Harbor NY 11724

ALBRECHTSEN, RULON S, b Emery, Utah, Mar 12, 33; m 59; c 5. GENETICS, PLANT BREEDING. Educ: Utah State Univ, BS, 56, MS, 57; Purdue Univ, PhD(genetics), 65. Prof Exp: Res agronomist, Agr Res Serv, USDA, 57-59; res asst agron, Purdue Univ, 59-63; from asst prof to assoc prof, SDak State Univ, 63-69; PROF AGRON, UTAH STATE UNIV, 69- Mem: Am Soc Agron. Res: Genetics and breeding of barley, safflower, birdsfoot trefoil, oats, flax, rye, and wheat. Mailing Add: Dept of Plant Sci Utah State Univ Logan UT 84321

ALBREGTS, EARL EUGENE, b Earl Park, Ind, May 30, 29; m 49; c 3. SOIL CHEMISTRY. Educ: Purdue Univ, BS, 64, PhD(soils), 68. Prof Exp: Asst prof, 67-73, ASSOC PROF SOIL CHEM, UNIV FLA, 73- Mem: Am Soc Agron; Am Soc Hort Sci. Res: Soil fertility; plant nutrition. Mailing Add: RR 2 PO Box 629 Dover FL 33527

ALBRIDGE, ROYAL, b Lima, Ohio, Jan 20, 33; m 57; c 2. NUCLEAR PHYSICS, ATOMIC PHYSICS. Educ: Ohio State Univ, BS, 55; Univ Calif, Berkeley, PhD(nuclear physics), 60. Prof Exp: From asst prof to assoc prof, 61-73, PROF PHYSICS, VANDERBILT UNIV, 73- Mem: AAAS; Am Phys Soc. Res:

Photoelectron, beta and gamma ray spectroscopy. Mailing Add: Dept of Physics Vanderbilt Univ Nashville TN 37203

ALBRIGHT, BRUCE CALVIN, b Kansas City, Mo, July 30, 46; m 69; c 2. NEUROSCIENCES. Educ: Univ Md, BS, 69; Med Col Va, Va Commonwealth Univ, MS, 72, PhD(anat), 74. Prof Exp: ASST PROF ANAT, SCH MED, UNIV N DAK, 74- Mem: Am Asn Anatomists; Sigma Xi. Res: Comparative prosimian neuroanatomy. Mailing Add: Dept of Anat Sch of Med Univ of NDak Grand Forks ND 58201

ALBRIGHT, CARL HOWARD, b Allentown, Pa, June 1, 33; m 70. PARTICLE PHYSICS. Educ: Lehigh Univ, BS, 55; Princeton Univ, PhD(physics), 60. Prof Exp: Instr physics, Princeton Univ, 59-61; res assoc, Northwestern Univ, 61-62, asst prof, 62-68; assoc prof, 68-71, PROF PHYSICS, NORTHERN ILL UNIV, 71- Concurrent Pos: NSF grant, 65-68. Mem: Am Phys Soc. Res: Theoretical particle physics; weak and electromagnetic interactions. Mailing Add: Dept of Physics Northern Ill Univ DeKalb IL 60115

ALBRIGHT, CHARLES HARRISON, JR, analytical chemistry, see 12th edition

ALBRIGHT, DON ALAN, physical chemistry, see 12th edition

ALBRIGHT, EDWIN C, b Iowa City, Iowa, Oct 8, 15; m 40; c 4. MEDICINE. Educ: Univ Iowa, BA, 36; Harvard Univ, MD, 40; Am Bd Internal Med, dipl. Prof Exp: Intern med, Mass Gen Hosp, 40-42; res fel, 46-47, resident, 47-48, instr, 48-49, from asst prof to prof, 49-75, EMER PROF MED, UNIV WIS-MADISON, 75- Concurrent Pos: Consult, Vet Admin Hosp, Wis, 51- Mem: AAAS; Endocrine Soc; Am Thyroid Asn; Am Col Physicians; Am Soc Clin Invest. Res: Thyroid physiology and disease. Mailing Add: 3901 Euclid Ave Madison WI 53711

ALBRIGHT, FRED RONALD, b Talmage, Pa, Feb 16, 44; m 66; c 1. BIOCHEMISTRY. Educ: Muhlenberg Col, BS, 66; Univ Ill, PhD(biochem), 70. Prof Exp: Fel, Johns Hopkins Univ, 70-72; ASST TECH DRI, LANCASTER LABS, INC, 72- Concurrent Pos: Damon Runyon Mem Fund Cancer Res fel, 70-71; Am Cancer Soc fel, 71-72. Mem: Am Chem Soc; Am Oil Chemists Soc; Inst Food Technologists. Res: Analytical method development for detection of trace elements, pesticides, nutrients, and toxins in agricultural products and our environment; effect of these materials on plant and animal life. Mailing Add: Lancaster Labs Inc 2425 New Holland Pike Lancaster PA 17601

ALBRIGHT, JACK LAWRENCE, b San Francisco, Calif, Mar 14, 30; m 57; c 2. ANIMAL SCIENCE. Educ: Calif Polytech State Univ, BS, 52; Wash State Univ, MS, 54, PhD(animal sci), 58. Prof Exp: Actg instr dairy sci, Wash State Univ, 54; from instr to asst prof dairy husb, Calif Polytech State Univ, 55-59; asst prof dairy sci, Univ Ill, 59-63; assoc prof, 63-66, PROF ANIMAL SCI, PURDUE UNIV, WEST LAFAYETTE, 66- Concurrent Pos: Partic, Int Dairy Cong, Copenhagen, 62, Munich, 66 & Sydney, 70; Fulbright sr res fel, Ruakura Animal Res Sta, Hamilton, NZ, 70-71. Mem: Fel AAAS; Animal Behav Soc; Am Dairy Sci Asn; Am Soc Animal Sci. Res: Analysis and measurement of management; life cycle management, housing and behavior; animal behavior; bovine physiology; dairy herd health. Mailing Add: Dept of Animal Sci Smith Hall Purdue Univ West Lafayette IN 47906

ALBRIGHT, JAMES ANDREW, b Amsterdam, NY, Jan 31, 45; m 64; c 5. ORGANIC CHEMISTRY. Educ: State Univ NY, BS, 66; Clark Univ, PhD(chem), 69. Prof Exp: Res chemist pesticides, Air Prod & Chem Inc, 69-70; res chemist, Glidden-Durkee Div SCM Corp, 70-72; GROUP LEADER FLAME RETARDANTS, MICH CHEM CORP, 72- Mem: Am Chem Soc. Res: Synthesis of novel heterocycles and the stereo chemical studies of cyclic organophosphorous compounds. Mailing Add: Mich Chem Corp Tech Ctr 1975 Green Rd Ann Arbor MI 48105

ALBRIGHT, JAMES CURTICE, b Madison, Wis, Sept 8, 29; m 51; c 3. PHYSICS, MATHEMATICS. Educ: Univ Wichita, BA, 50; Univ Okla, MS, 52, PhD(physics), 56. Prof Exp: Res engr, 55-57, sr res engr, 57-61, res group leader, 61-64, RES ASSOC FORMATION EVAL, CONTINENTAL OIL CO, 64- Mem: Soc Petrol Eng; Soc Prof Well Log Anal. Res: Oil well logging; electrokinetics; computer programming; Raman and infrared spectroscopy; physics. Mailing Add: Continental Oil Co PO Box 1267 Ponca City OK 74601

ALBRIGHT, JAY DONALD, b Lancaster Co, Pa, Apr 28, 33; m 54; c 3. ORGANIC CHEMISTRY. Educ: Elizabethtown Col, BS, 55; Univ Ill, PhD(chem), 58. Prof Exp: Asst org chem, Univ Ill, 55-57; from res chemist to sr res chemist, 58-74, GROUP LEADER, LEDERLE LABS, AM CYANAMID CO, 74- Mem: Am Chem Soc; NY Acad Sci; The Chem Soc. Res: Synthetic organic and medicinal chemistry; heterocycles; alkaloids. Mailing Add: 5 Clifford Ct Nanuet NY 10954

ALBRIGHT, JOHN RUPP, b Wilkes-Barre, Pa, June 10, 37; m 60; c 3. PHYSICS. Educ: Susquehanna Univ, AB, 59; Univ Wis, MS, 61, PhD, 64. Prof Exp: Asst prof, 63-70, ASSOC PROF PHYSICS, FLA STATE UNIV, 70- Concurrent Pos: Consult, Fermi Nat Accelerator Lab, 75- Mem: Am Phys Soc; Am Asn Physics Teachers. Res: Elementary particles; digital computer programming. Mailing Add: Dept of Physics Fla State Univ Tallahassee FL 32306

ALBRIGHT, JOHN T, b Fairmont, WVa, Mar 1, 17; m 49; c 2. BIOLOGY, DENTISTRY. Educ: WVa Univ, AB, 41; St Louis Univ, DDS, 46; Boston Univ, PhD(biol), 64. Prof Exp: Pvt pract, 47-50; sr asst & dent surgeon, Nat Inst Dent Res, 51-53; assoc prof, 64-69, PROF BIOL, GRAD SCH, BOSTON UNIV, 69-; CLIN ASSOC ORAL PATH, SCH DENT MED, HARVARD UNIV, 61- Concurrent Pos: Res assoc oral path, Sch Dent Med, Harvard Univ, 54-61. Mem: NY Acad Sci; Int Asn Dent Res. Res: Electron microscopic studies in teeth, oral mucosa, keratinization, formation and resorption of bone, blood vessels and connective tissue; aging. Mailing Add: Dept of Biol Boston Univ Boston MA 02215

ALBRIGHT, JOSEPH FINLEY, b New Tazewell, Tenn, May 9, 27; m 51; c 2. ZOOLOGY. Educ: Southwestern Univ, Tex, BS, 49; Ind Univ, PhD(zool), 56. Prof Exp: Res assoc biol, Oak Ridge Nat Lab, 56-57; NIH fel, Nat Cancer Inst, 57-58; asst prof surg, Med Col Va, 58-61; biologist, Oak Ridge Nat Lab, 61-70; sr immunologist, Smith, Kline & French Labs, 70-71; PROF LIFE SCI, IND STATE UNIV, TERRE HAUTE, 71-, PROF MICROBIOL, TERRE HAUTE CTR MED EUDC, 73- Mem: AAAS; Soc Study Develop Biol; Am Asn Immunologists. Res: Immunology; embryology; cellular physiology. Mailing Add: Dept of Life Sci Ind State Univ Terre Haute IN 47809

ALBRIGHT, LAWRENCE JOHN, b Owen Sound, Ont, July 22, 41. MICROBIOLOGY, OCEANOGRAPHY. Educ: McGill Univ, BSc, 63; Ore State Univ, MS, 65, PhD(marine microbiol), 67. Prof Exp: Asst prof, 67-72, ASSOC PROF MARINE MICROBIOL, SIMON FRASER UNIV, 72- Concurrent Pos: Grants, Nat Res Coun Can, 67- & Can Dept Environ, 71- Mem: Can Soc Microbiol; Am Soc Microbiol. Res: Physiology and biochemistry of marine microbes, microbe pollutant interactions in natural waters and sediments. Mailing Add: Dept of Biol Sci Simon Fraser Univ Burnaby BC Can

ALBRIGHT, RAYMOND GERARD, b Detroit, Mich, Apr 1, 26. ANATOMY, ZOOLOGY. Educ: Loyola Univ, Ill, BA, 49, MS, 53, PhD(anat), 55; WBaden Col, STL, 59. Prof Exp: From instr to asst prof, 60-66, chmn dept, 62-66, ASSOC PROF BIOL, UNIV DETROIT, 66- Mem: Am Soc Zoologists; Am Soc Ichthyologists & Herpetologists. Res: Functional morphology of chordates, especially reptiles. Mailing Add: Dept of Biol Univ of Detroit Detroit MI 48221

ALBRIGHT, ROBERT LEE, b Leola, Pa, Jan 28, 32; m 59; c 2. ORGANIC CHEMISTRY. Educ: Elizabethtown Col, BS, 54; Univ Ill, PhD(org chem), 58. Prof Exp: Asst, Univ Ill, 54-56; res chemist, 58-70, SR CHEMIST & CONSULT, ROHM AND HAAS CO, 70- Concurrent Pos: Instr, Elizabethtown Col, 59; lectr, Holy Family Col, 63. Mem: Am Chem Soc. Res: Reaction mechanisms; anchimerically assisted reactions; reactions of iodonium salts and organophosphorus compounds; polymerization, especially mechanism of formation of macrostructure and microstructure; polymer morphology; mechanism of reactions of polymers; synthesis and properties of ion exchange polymers. Mailing Add: 36 Autumn Rd Churchville PA 18966

ALBRIGO, LEO GENE, b Palmdale, Calif, Aug 24, 40; m 59; c 3. HORTICULTURE, PLANT PHYSIOLOGY. Educ: Univ Calif, Davis, BS, 62, MS, 64; Rutgers Univ, PhD(hort), 68. Prof Exp: From res asst to res assoc pomol, Rutgers Univ, 64-68; ASST HORTICULTURIST, AGR RES & EDUC CTR, UNIV FLA, 68- Mem: Am Hort Soc. Res: Fresh fruit quality; environmental influences on fruit development. Mailing Add: Agr Res & Educ Ctr PO Box 1088 Lake Alfred FL 33850

ALBRINK, MARGARET JORALEMON, b Bisbee, Ariz, Jan 6, 20; m 44; c 3. INTERNAL MEDICINE. Educ: Radcliffe Col, BA, 41; Yale Univ, MD, 46, MPH, 51. Prof Exp: Asst med, Yale Univ, 46-47, from instr internal med to asst prof med, 51-61; assoc prof, 61-66, PROF MED, SCH MED, W VA UNIV, 66- Concurrent Pos: Intern, New Haven Hosp, 46-47; estab investr, Am Heart Asn, 58-63. Mem: Am Soc Clin Invest; Am Fedn Clin Res; Am Soc Clin Nutrit. Res: Serum lipids in metabolic diseases. Mailing Add: Sch of Med WVa Univ Morgantown WV 26506

ALBRINK, WILHELM STOCKMAN, b Napoleon, Ohio, Aug 22, 15; m 44; c 3. PATHOLOGY. Educ: Oberlin Col, AB, 37; Yale Univ, PhD(zool), 41, MD, 47. Prof Exp: Asst zool, Yale Univ, 37-40, asst path, Off Sci Res & Develop Proj, Sch Med, 42-45, asst, Univ, 47-49, from instr to assoc prof, 51-61; chmn dept, 61-69, PROF PATH, MED CTR, W VA UNIV, 61- Concurrent Pos: Intern, New Haven Hosp, Conn, 47-48, asst resident, 48-49; Am Cancer Soc fel, 49-51. Mem: AAAS; Am Asn Path & Bact; Am Asn Cancer Res; NY Acad Sci; Am Soc Exp Path. Res: Electromotive force of living tissues; pathological physiology of war gases; synthetic membrane behavior; biology of neoplasia; pathology of anthrax; localization of particulate matter in the lung. Mailing Add: Dept of Path WVa Univ Med Ctr Morgantown WV 26506

ALBRITTON, CLAUDE CARROLL, JR, b Corsicana, Tex, Apr 7, 13; m 44; c 3. GEOLOGY. Educ: Southern Methodist Univ, BS & AB, 33; Harvard Univ, MA, 34, PhD(geol), 36. Prof Exp: Lab instr geol, 30-33, from instr to prof, 36-62, chmn dept, 47-51, dean col arts & sci, 52-57, grad sch, 57-59 & grad sch humanities & sci, 59-71, dir grad res ctr, 61-65 & sci info inst, 65-71, vprovost univ, 71-73, W B HAMILTON PROF GEOL, SOUTHERN METHODIST UNIV, 62-, VPRES INST STUDY EARTH & MAN, 68-, DEAN LIBR, 73- Concurrent Pos: From asst geologist to geologist, US Geol Surv, 42-50; assoc ed, Field & Lab, 52-59 & Tex J Sci, 52-; ed jour, Grad Res Ctr, 59-64; mem subcomt doc, US Nat Comt, Int Union Geol Sci, 62-65 & mem exec comt, US Nat Comt, Int Comt for Hist of Geol Sci, 74-; trustee, Ft Burgwin Res Ctr, 64-74; Rosenbach fel bibliog, Univ Pa, 70. Mem: AAAS; fel Geol Soc Am; Paleont Soc (treas, 56-63); Am Asn Petrol Geologists; Soc Econ Paleont & Mineral. Res: Geology of early man sites in Egypt and Ethiopia; philosophy and history of geology. Mailing Add: Box 153 Southern Methodist Univ Dallas TX 75275

ALBRO, LEWIS PEARSON, b White Plains, NY, May 23, 22; m 48; c 5. ANALYTICAL CHEMISTRY. Educ: Rensselaer Polytech Inst, BChE, 46; Wesleyan Univ, Am, 49; Univ Del, PhD(chem), 52. Prof Exp: Mem staff, Burroughs Wellcome & Co, 46-47, ORDWES, Wesleyan Univ, 47-49 & Biochem Res Inst, 49-51; chemist, Sterling Winthrop Res Inst, 52-69; PRES, INSTRANAL LAB, INC, 69- Mem: Am Chem Soc; AAAS. Res: Organic microanalysis including design of instrumentation involving electronics as well as chemistry. Mailing Add: Instranal Lab Inc 273 Columbia Turnpike PO Box 70 Rensselaer NY 12144

ALBRO, PHILLIP WILLIAM, b Geneva, NY, Aug 24, 39; m 65; c 4. BIOCHEMISTRY. Educ: Univ Rochester, BA, 61; St Louis Univ, PhD(biochem), 68. Prof Exp: Chemist, US Army Biol Labs, Md, 64-65; BIOCHEMIST, NAT INST ENVIRON HEALTH SCI, 68- Concurrent Pos: Adj asst prof, Duke Univ, 75- Mem: AAAS; Am Chem Soc; Am Oil Chemists Soc. Res: Analytical biochemistry, especially of microbial metabolites; intermediary metabolism of simple liquids. Mailing Add: Nat Inst of Environ Health Sci Box 12233 Research Triangle Park NC 27709

ALBU, EVELYN D, b Mt Vernon, NY, Sept 25, 38. MEDICAL EDUCATION. Educ: Ohio Univ, BS, 60; Georgetown Univ Med & Dent Sch, PhD(microbiol), 73. Prof Exp: Microbiol biochemist res, Squibb Inst Med Res, 60-64, med writer, 64-65; clin res assoc int clin res, Schering Corp, 65-69; teaching asst microbiol, Georgetown Univ, 69-73; ASSOC DIR MED EDUC, SCHERING CORP, 73- Mem: Am Soc Microbiol. Res: Clinical use of antibiotics, resistance development, epidemiology. Mailing Add: Prof Serv Dept Schering Corp Galloping Hill Rd Kenilworth NJ 07033

ALBURGER, DAVID ELMER, b Philadelphia, Pa, Oct 6, 20; m 45; c 4. PHYSICS. Educ: Swarthmore Col, BA, 42; Yale Univ, MS, 47, PhD(physics), 48. Prof Exp: Proj engr, Naval Res Lab, 42-45; lab asst physics, Yale Univ, 45-47; PHYSICIST, BROOKHAVEN NAT LAB, 48- Concurrent Pos: NSF fel, Nobel Inst Physics, Sweden, 52-53. Mem: Fel Am Phys Soc. Res: Radioactivity and van de Graaff accelerator research on nuclear energy levels; gamma-ray pair spectrometer design. Mailing Add: Brookhaven Nat Lab Upton NY 11973

ALBURN, HARVEY EUGENE, b Youngstown, Ohio, Feb 23, 15; m 46; c 2. BIOCHEMISTRY. Educ: Western Reserve Univ, AB, 37, AM, 38, PhD(biochem), 45. Prof Exp: Res chemist, Ohio, 43-44, Pa, 44-46, head biochem dept, 46-59, mgr biochem sect, 59-74, ASSOC RES DIR BIOCHEM & FERMENTATION SCI, WYETH LABS, 74- Mem: Am Chem Soc; Am Soc Microbiol. Res: Peptide hormones; diabetes; blood platelets; antibiotics biosynthesis; antitumor and antiviral agents. Mailing Add: Wyeth Labs Box 8299 Philadelphia PA 19101

ALCALA, JOSE RAMON, b Ponce, PR, May 1, 40; m 64. ANATOMY. Educ: Univ Mo-Columbia, BA, 64, MA, 66; Univ Ill Med Ctr, PhD(anat), 72. Prof Exp: Instr biol, Mt Union Col, 67-70; ASST PROF ANAT, SCH MED, WAYNE STATE UNIV, 72- Mem: AAAS; Am Asn Anatomists. Res: Biochemistry and immunochemistry of lens plasma membranes. Mailing Add: Dept of Anat Wayne State Univ Sch of Med Detroit MI 48201

ALCARAZ, ERNEST CHARLES, b Coronado, Calif, Dec 4, 35; m 60; c 1. LASER, ATMOSPHERIC PHYSICS. Educ: San Diego State Col, BA, 58, MS, 60; Wayne State Univ, PhD(physics), 68. Prof Exp: Instr physics, San Diego State Col, 60-62; from teaching asst to instr, Wayne State Univ, 62-68; res physicist, Signature & Propagation Lab, US Army Ballistic Res Labs, 68-75; SCIENTISt, SCI APPLN, INC, 75- Mem: Optical Soc Am. Res: Properties of liquid helium; coherent light propagation through the near earth atmosphere; micrometeorological structure of the near earth atmosphere; field test evaluation of laser/eo systems. Mailing Add: Sci Appln Inc 2361 S Jefferson Davis Hwy Arlington VA 22202

ALCOCK, NORMAN ZINKAN, b Edmonton, Alta, May 29, 18; m 48; c 4. NUCLEAR PHYSICS, ELECTRICAL ENGINEERING. Educ: Queen's Univ, Ont, BSc, 40; Calif Inst Technol, MS, 41; McGill Univ, PhD(physics), 49. Prof Exp: Jr res engr radar antenna, Nat Res Coun Can, 41-45; asst res physicist cyclotron design, McGill Univ, 45-46; asst res physicist neutron diffraction, Atomic Energy Can, 47-50; vpres & dir, Isotope Prod, Ltd, 50-57; gen mgr isotope prod div, Can Curtiss-Wright Ltd, 58-59, dir gen eng, 59; DIR GEN ENG, CAN PEACE RES INST, 60-, PRES, 61- Mem: Am Phys Soc; Inst Elec & Electronics Engrs; Can Asn Physicists. Res: Development and design in the physical sciences; social science. Mailing Add: 244 Lakewood Dr Oakville ON Can

ALCORN, GORDON DEE, b Olympia, Wash, Apr 6, 07; m 35; c 1. BOTANY, ORNITHOLOGY. Educ: Col Puget Sound, BS, 30; Univ Wash, Seattle, MS, 33, PhD(bot), 35. Prof Exp: Asst prof biol, Col Puget Sound, 32-33; asst bot, Univ Wash, Seattle, 34-35; asst prof, Univ Idaho, 35-37; vpres & head dept biol, Grays Harbor Col, 37-43, pres, 45-46; assoc prof biol, 46-47, dir grad studies, 70-72, PROF BIOL, UNIV PUGET SOUND, 47-, CHMN DEPT, 51-, PROF, SUMMER SCH, 30-, EMER DIR GRAD STUDIES, 72- Concurrent Pos: Dir & cur, Puget Sound Mus Natural Hist, 51-73; ed, Murrelet. Mem: Cooper Ornith Soc; Am Ornith Union. Res: Air sacs in birds; salt glands. Mailing Add: Dept of Biol Univ of Puget Sound Tacoma WA 98416

ALCORN, STANLEY MARCUS, b Modesto, Calif, June 18, 26; m 49; c 4. PLANT PATHOLOGY. Educ: Univ Calif, PhD(plant path), 54. Prof Exp: Res asst plant path, Univ Calif, 49-53, Merck & Co fel, 54-55; plant pathologist, USDA, 55-63; assoc prof, 63-65, PROF PLANT PATH, UNIV ARIZ, 65- Mem: Fel AAAS; Am Soc Microbiol; Am Phytopath Soc; Am Soc Hort Sci. Res: Verticillium albo-atrum; Erwinia carnegieana. Mailing Add: Dept of Plant Path Univ of Ariz Tucson AZ 85721

ALDEN, RICHARD ALLEN, b Brockton, Mass, May 17, 35. X-RAY CRYSTALLOGRAPHY. Educ: Mass Inst Technol, BS, 56; Univ Wash, PhD(org chem), 62. Prof Exp: Res chemist, Nat Cash Register Co, 56-58; chemist, 62-63, NIH fel, 63-65, assoc specialist, 65-70, asst res chemist, 70-71, ASSOC RES CHEMIST, UNIV CALIF, SAN DIEGO, 71- Concurrent Pos: NIH career develop award, 70-74. Mem: AAAS; Am Crystallog Asn, Sigma Xi. Res: X-ray crystallography of biological macro-molecules; computer programming. Mailing Add: Dept of Chem B-017 Univ of Calif San Diego La Jolla CA 92093

ALDEN, ROLAND HERRICK, b Champaign, Ill, Feb 4, 14; m 37; c 3. ANATOMY. Educ: Stanford Univ, AB, 36; Yale Univ, PhD(zool), 41. Prof Exp: Instr zool, Yale Univ, 41-42; mem fac, Div Anat, 49-51, chief, 51-61, assoc dean grad sch med sci, 60-68, chancellor pro tem, 70, DEAN COL BASIC MED SCI, MED UNITS, UNIV TENN, MEMPHIS, 61-, DEAN GRAD SCH MED SCI, 68- Concurrent Pos: Mem anat comt, Nat Bd Med Examrs, 59-62; secy-treas, Tenn Bd Basic Sci Examrs, 63-; spec consult, Anat Sci Training Comt, USPHS, 60-64; mem comt grad educ, Nat Asn State Univ & Land-Grant Cols, 68- Mem: Am Asn Anat (pres, 69-70); Am Physiol Soc; Soc Exp Biol & Med; Am Soc Zoologists. Res: Mammalian reproduction; experimental embryology; implantation. Mailing Add: Univ of Tenn Ctr Health Sci Memphis TN 38103

ALDER, BERNI JULIAN, b Duisburg, Ger, Sept 9, 25; nat US; m 56; c 3. CHEMICAL PHYSICS, STATISTICAL MECHANICS. Educ: Univ Calif, BS, 47, MS, 48; Calif Inst Technol, PhD(chem), 51. Prof Exp: Instr chem, Univ Calif, Berkeley, 51-54, THEORET PHYSICIST, LAWRENCE LIVERMORE LAB, UNIV CALIF, BERKELEY, 55- Concurrent Pos: Guggenheim fel, 54-55; NSF fel, 63-64; Van der Waals vis prof, Univ Amsterdam, 70-71. Mem: Nat Acad Sci; Am Chem Soc; Am Phys Soc. Res: Statistical mechanical behavior of systems, theoretical and computational. Mailing Add: 1245 Contra Costa Dr El Cerrito CA 94530

ALDER, EDWIN FRANCIS, b Hugo, Okla, Sept 1, 27; m 51; c 4. PLANT PHYSIOLOGY. Educ: Univ Okla, BS, 51, PhD(plant sci), 56; Univ Chicago, MS, 52. Prof Exp: Instr sci, Ark State Col, 53; from asst to instr plant sci, Univ Okla, 54-55; instr bot, Univ Ark, 55-57; sr plant physiologist, 57-59, asst head agr res, 59-61, head plant sci res, 61-65, dir, 65-66, agr res, 66-69, vpres agr res & develop, 69-73, VPRES LILLY RES LAB DIV, ELI LILLY & CO, 73- Mem: AAAS; Am Soc Plant Physiol; Weed Sci Soc Am; Scand Soc Plant Physiol. Res: Pesticide screening and development; weed control; plant growth regulators. Mailing Add: 10140 E Troy Ave Indianapolis IN 46239

ALDER, HENRY LUDWIG, b Duisburg, Ger, Mar 26, 22; nat US; m 63; c 1. MATHEMATICS, STATISTICS. Educ: Univ Calif, Berkeley, AB, 42, PhD(math), 47. Prof Exp: Asst math, Univ Calif, Berkeley, 42-43, assoc math & jr instr meteorol, 43-44. instr math, 47-48; from instr to assoc prof, 48-65, PROF MATH, UNIV CALIF, DAVIS, 65- Honors & Awards: Lester R Ford Award, Math Asn Am, 70; Cert of Merit, Nat Coun Teachers Math, 75. Mem: Am Math Soc; Math Asn Am (secy, 60-74, pres-elect, 76, pres, 77-78); Inst Math Statist. Res: Number theory; existence and nonexistence of certain identities in the theory of partitions and compositions. Mailing Add: Dept of Math Univ of Calif Davis CA 95616

ALDERFER, RONALD GODSHALL, b Harleysville, Pa, July 14, 43; m 65; c 2. ECOLOGY. Educ: Wash Univ, AB, 65, PhD(biol), 69. Prof Exp: Asst prof biol, Univ Chicago, 69-75; CHIEF ECOLOGIST, HARLAND BARTHOLOMEW & ASSOCS, 75- Mem: AAAS; Ecol Soc Am; Am Inst Biol Sci; Am Soc Naturalists. Res: Environmental regulation of physiological processes; mechanisms of physiological adaptation; applied ecology. Mailing Add: Harland Bartholomew & Assocs 165 N Meramec Ave St Louis MO 63105

ALDERFER, RUSSELL BRUNNER, b Lansdale, Pa, Sept 27, 13; m 41 c 2. SOILS. Educ: Pa State Col, BS, 36, MS, 40, PhD(soil technol), 47. Prof Exp: Soil technologist, Soil Conserv Serv, USDA, 36-43; instr soil technol & coop agent, Pa State Univ, 43-47, from asst prof to prof soil technol, 47-54; prof & chmn dept, 54-62, RES PROF SOILS, RUTGERS UNIV, NEW BRUNSWICK, 62- Mem: Fel AAAS; Soil Sci Soc Am; Am Soc Agron; Soil Conserv Soc Am; Int Soc Soil Sci. Res: Soil physics; soil, water and waste management. Mailing Add: Dept of Soils & Crops Rutgers Univ New Brunswick NJ 08903

ALDERMAN, DE FOREST CHARLES, b Morgantown, WVa, June 22, 14; m 38; c 3. HORTICULTURE. Educ: Univ Minn, BS, 37, PhD(hort), 47; Iowa State Col, MS, 38. Prof Exp: County agr agent, Exten Serv, Mich State Col, 42-43, asst prof hort, 43-44; asst prof hort, Univ La, 45-48; mgr, La Fruit Growers Asn, 48-50; assoc prof hort, Agr & Mech Col Tex, 50-52 & Univ WVa, 52-56; agriculturist, Calif Exten Serv, Univ Calif, Davis, 56-70; PUB SERV SPECIALIST PLANT SCI, UNIV CALIF, BERKELEY, 70- Res: Food processing; harvest handling and marketing of produce; fruit studies; physiological and nutritional studies with apples, prunes, strawberries, cherries and apricots. Mailing Add:

ALDERMAN, JOHN FREMONT, organic chemistry, see 12th edition

ALDERMAN, LOUIS CLEVELAND, JR, b Douglas, Ga, Aug 12, 24; m 52; c 4. BIOLOGY, PARASITOLOGY. Educ: Emory Univ, AB, 46; Univ Ga, MS, 49; Auburn Univ, EdD, 59. Prof Exp: Asst biol, Univ Ga, 48-49, instr, Rome Ctr, 49-50, dir & asst prof, Savannah Ctr, 50-51, Rome Ctr, 51-56 & Columbus Ctr, 56-59; asst prof, Henderson Col, Univ Ky, 59-64; PROF BIOL & PRES, MID GA COL, 64- Concurrent Pos: Asst educ admin, Auburn Univ, 58-59. Mem: AAAS. Res: In vitro cultivation of avian malaria parasite; Plasmodium cathemerium. Mailing Add: Old Chester Rd Cochran GA 31014

ALDERMAN, MICHAEL HARRIS, b New Haven, Conn, Mar 26, 36; m 68; c 4. PUBLIC HEALTH. Educ: Harvard Col, BA, 58; Yale Univ, MD, 62. Prof Exp: Fel human genetics, New York Hosp, 67-68; chief renal serv, Lincoln Hosp, 68-70; asst prof, 70-76, ASSOC PROF MED & PUB HEALTH, MED SCH, CORNELL UNIV, 76- Concurrent Pos: Coordr pub health, Rural Health Proj, Jamaica, WI, 71-; mem comt mother & child health, Jamaican Govt, 71-; mem NY comt, US Civil Rights Comn, 72-76; WHO traveling fel, 73; assoc ed, Milbank Mem Quart Health & Soc, 73-; contrib ed, Sci Yearbk, 73-; mem bd dirs, Int Med & Res Found, 74-; physician to outpatients, New York Hosp, 75-; chmn adv comt hypertension, New York City Health Dept, 75- Mem: Fel Am Col Physicians; Am Heart Asn; fel Am Soc Clin Nutrit; Am Soc Nephrol; Am Soc Trop Med & Hyg. Res: Developing and establishing methods of health care delivery; malnutrition in Jamaica and detection and treatment of hypertension in the United States. Mailing Add: Cornell Univ Med Ctr 1300 York Ave New York NY 10021

ALDINGER, EARL EDWARD, b Charleston, SC, July 11, 29; m 62; c 1. PHARMACOLOGY. Educ: Col Charleston, BS, 50; Med Col SC, MS, 59, PhD(pharmacol), 61. Prof Exp: Assoc pharmacol, Med Col SC, 57-61, instr, 61-62; asst prof, 62-67, ASSOC PROF PHARMACOL, SCH MED, TULANE UNIV, 67- Concurrent Pos: SC Heart Asn fel, 61-62. Mem: Am Soc Pharmacol & Exp Therapeut. Res: Cardiovascular pharmacology and physiology. Mailing Add: Dept of Pharmacol Tulane Univ Sch of Med New Orleans LA 70118

ALDON, EARL F, b Chicago, Ill, Mar 5, 30; m 57; c 2. FORESTRY. Educ: Univ Mich, BSF, 52, MS, 53. Prof Exp: Forester, 53-56, RES FORESTER, ROCKY MOUNTAIN FOREST & RANGE EXP STA, US FOREST SERV, 56- Mem: AAAS; Soc Am Foresters. Res: Watershed rehabiliation. Mailing Add: Rocky Mountain Exp Sta 5423 Fed Bldg 517 Gold Ave SW Albuquerque NM 87101

AL-DOORY, YOUSEF, b Baghdad, Iraq, 1924. MEDICAL MYCOLOGY. Educ: Univ Baghdad, BS, 45; Univ Tex, MA, 51; La State Univ, PhD(bot, plant path), 54; Duke Univ, dipl, 60. Prof Exp: Teacher high sch, Baghdad, 45-49; teaching & res, Univ Baghdad, 54-58; res assoc med mycol, Univ Okla, 58-59, res assoc microbiol, Sch Med, 59-60, fel med mycol, 60-61; fel, NY State Dept Health, 61-62; sr res scientist in-chg diag med mycol lab, New York City Dept Health, 62-64; chmn dept mycol, Southwest Found Res & Educ, 64-70; asst prof epidemiol & environ health, 70-75, ASSOC PROF PATH, SCH MED, GEORGE WASHINGTON UNIV, 75- Concurrent Pos: Consult, Santa Rosa Med Ctr, San Antonio, Tex, 66-; chief microbiol lab med, George Washington Univ Hosp, 75- Mem: Mycol Soc Am; Am Soc Microbiol; Int Soc Human & Animal Mycol; Med Mycol Soc of the Americas; fel Am Acad Microbiol. Res: Pathogenics dematiaceous fungi; application of fluorescent antibody procedures; lipid of H capsulatum; mycoflora of subhuman primates; soil and air; electron microscopy studies of host-parasite in mycotic diseases. Mailing Add: Dept of Path George Washington Univ Sch Med Washington DC 20037

ALDOUS, DUANE LEO, b Albuquerque, NMex, Nov 2, 30; m 55; c 5. ORGANIC CHEMISTRY, BIOCHEMISTRY. Educ: Univ NMex, BS, 53, PhD(chem), 61. Prof Exp: Fel, Univ NMex, 61-62; res chemist, E I du Pont de Nemours & Co, NC, 62-68; asst prof, 68-71, ASSOC PROF PHARMACEUT CHEM, XAVIER UNIV LA, 71-, DEAN COL PHARM, 74- Concurrent Pos: Consult, Inst Serv to Educ, 75- Mem: Am Chem Soc; Sigma Xi; Am Asn Cols Pharm; Int Soc Heterocyclic Chem. Res: Synthetic organic and heterocyclic chemistry; synthesis of potential antitumor compounds; formulation of cosmetics. Mailing Add: 7325 Palmetto St New Orleans LA 70125

ALDOUS, JOHN GRAY, b Bristol, Eng, Nov 16, 16; m 44; c 3. ENZYMOLOGY. Educ: Univ BC, BA, 39, MA, 41; Univ Toronto, PhD(gen physiol), 45. Prof Exp: Asst gen physiol & biol, Univ BC, 39-41; demonstr gen physiol, Univ Toronto, 41-45; from asst prof to assoc prof pharmacol, 45-50, PROF PHARMACOL, DALHOUSIE UNIV, 51-, HEAD DEPT, 74- Concurrent Pos: Nuffield travel grantee, Univ Cambridge, 58. Mem: Can Physiol Soc; Pharmacol Soc Can (pres, 70-71). Res: Cell metabolism in relation to toxicology. Mailing Add: Dept of Pharmacol Dalhousie Univ Halifax NS Can

ALDRED, J PHILLIP, b Noblesville, Ind, Oct 11, 35; m 59; c 3. PHYSIOLOGY, PHARMACOLOGY. Educ: Purdue Univ, BS, 57, PhD(animal physiol), 62; Univ Ill, MS, 59. Prof Exp: RES PHYSIOLOGIST, PHARMACOL DEPT, ARMOUR PHARMACEUT CO, 62- Mem: Am Soc Pharmacol & Exp Therapeut; Soc Exp Biol & Med. Res: Reproductive physiology; endocrinology. Mailing Add: Armour Pharmaceut Co PO Box 511 Kankakee IL 61443

ALDRETE, JORGE ANTONIO, b Mexico City, Mex, Feb 28, 37; c 3. ANESTHESIOLOGY. Educ: Nat Univ Mex, BS, 53, MD, 60; Univ Colo, MS, 67. Prof Exp: Med dir surg care, Jackson Mem Hosp, Miami, Fla, 71; prof anesthesiol & chmn dept, Sch Med, Univ Louisville, 71-75; PROF ANESTHESIA & CHMN DEPT, MED CTR, UNIV COLO, DENVER, 75- Concurrent Pos: Chmn drug & therapeut comt, Vet Admin Hosp, Denver, Colo, 69-70 & Jackson Mem Hosp, 70-71; attend anesthesiol, Jackson Mem Hosp, 71; chief anesthesiol, Louisville Gen Hosp, 71-75, Children's Hosp, 72-75, Vet Admin Hosp, 72-, Norton Mem Infirmary, 72-75, Kosair Crippled Children Hosp, 72-75 & Community Hosp, 74-75; chmn intensive care comt, Kosair Crippled Children Hosp, 74-75. Mem: Am Soc Anesthesiologists. Mailing Add: Dept of Anesthesia Univ of Colo Med Ctr Denver CO 80220

ALDRICH, CLARENCE KNIGHT, b Chicago, Ill, Apr 12, 14; m 42; c 4. PSYCHIATRY. Educ: Wesleyan Univ, BA, 35; Northwestern Univ, MD, 40. Prof Exp: With USPHS, 40-46; asst prof neuropsychiat, Univ Wis, 46-47; from asst prof to assoc prof psychiat, Sch Med, Univ Minn, 47-55; prof, Sch Med, Univ Chicago, 55-

69, chmn dept, 55-64; prof & chmn dept, NJ Col Med & Dent, 69-73; PROF PSYCHIAT & DIR BLUE RIDGE COMMUNITY MENT HEALTH CTR, SCH MED, UNIV VA, 73- Concurrent Pos: Vis prof, Univ Edinburgh, 63-64; Erskine fel, Univ Canterbury, 71. Mem: Am Psychiat Asn; Am Orthopsychiat Asn. Res: Psychiatric education; community mental health. Mailing Add: Dept of Psychiat Univ of Va Sch of Med Charlottesville VA 22903

ALDRICH, DANIEL GASKILL, JR, b Northwood, NH, July 12, 18; m 41; c 3. SOIL FERTILITY. Educ: RI State Col, BS, 39; Univ Ariz, MS, 41; Univ Wis, PhD(soils), 43. Hon Degrees: ScD, RI State Col, 60. Prof Exp: Jr chemist, Citrus Exp Sta, Univ Calif, Riverside, 43, from asst chemist to chemist, 46-55; prof soils, Univ Calif, Davis, 55-62; CHANCELLOR, UNIV CALIF, IRVINE, 62- Concurrent Pos: Dean agr, Univ Calif, Berkeley, 59-62; mem agr adv comt, W K Kellogg Found, 61-63; mem agr educ policy comt, Nat Acad Sci, 61-65 & comn on educ in agr & nat resources, 65-67; mem adv comt, Carnegie Study on Agr Educ, 63-64; mem sci adv comt, Subpanel on Res & Educ, World Food Supply Panel, 66; trustee, Agron Sci Found, 67- Mem: Fel AAAS; fel Am Soc Agron; Am Soc Hort Sci; Soil Sci Soc Am; Soil Conserv Soc Am. Res: Soil management; land use. Mailing Add: Univ of Calif Irvine CA 92664

ALDRICH, DAVID VIRGIL, bJamestown, NY, Oct 22, 28. MARINE ECOLOGY. Educ: Kenyon Col, AB, 50; Rice Inst, MA, 52, PhD(parasitol), 54. Prof Exp: Physiologist, US Fish & Wildlife Serv, 56-66; ASSOC PROF BIOL & OCEANOG, 66-, TEX A&M UNIV, 66-, WILDLIFE & FISHERY SCI, 70-, MARINE SCI, 72- Mem: AAAS; Am Soc Limnol & Oceanog; Am Soc Zool; Marine Biol Asn UK; World Mariculture Soc. Res: Ecology and behavior of penaeid shrimp, dinoflagellates and parasites; mariculture and ecological impact at estuarine power plants. Mailing Add: Dept of Marine Sci Tex A&M Univ Bldg 311 Ft Crockett Galveston TX 77550

ALDRICH, FRANK THATCHER, b Dallas, Tex, June 21, 41; m 62; c 2. GEOGRAPHY. Educ: Univ Tex, Austin, BA, 64; Ore State Univ, MS, 66, PhD(geog), 72. Prof Exp: Instr geog, Ore State Univ, 66-67; ASST PROF GEOG, ARIZ STATE UNIV, 69- Concurrent Pos: Dir comput mapping, Mountain W Res Inc, 74- Mem: Asn Am Geogr; Ecol Soc Am; Am Soc Photogram; Asn Pac Coast Geogr; Nat Coun Geog Educ. Res: Computer mapping and graphics; land use; field methodology; remote sensing; biogeography and cartography. Mailing Add: Dept of Geog Ariz State Univ Tempe AZ 85281

ALDRICH, FRANKLIN DALTON, b Detroit, Mich, Jan 25, 29; m 52; c 3. TOXICOLOGY, INTERNAL MEDICINE. Educ: Mich State Univ, BS, 50; Ore State Univ, MA, 53, PhD(plant physiol), 54; Western Reserve Univ, MD, 62. Prof Exp: Plant physiologist, US Dept Agr, Denver, Colo, 56-58; intern, State Univ Iowa Hosps, 62-63; resident, Vet Admin Hosp, Denver, 63-64; staff physician, Upjohn Co, Mich, 64-65; prin investr, Colo Community Study Pesticides, Colo Dept Health, Greeley, 66-69; from resident to chief resident med, Lemuel Shattuck Hosp, Boston, 69-71; ASST MED DIR, MED DEPT & PHYSICIAN-IN-CHG, ENVIRON MED SERV, MASS INST TECHNOL, 71- Concurrent Pos: Res fel, Sch Med, Western Reserve Univ, 58-62; fel med, Univ Colo, Denver, 65-66; Mead Johnson scholar med, 70-71. Mem: Am Acad Clin Toxicol. Res: Environmental and clinical toxicology; agricultural medicine; pesticides; chemical human ecology. Mailing Add: Rm 20B-238 Mass Inst Technol Cambridge MA 02139

ALDRICH, FREDERICK ALLEN, b Butler, NJ, May 1, 27; m 52. INVERTEBRATE ZOOLOGY. Educ: Drew Univ, BA, 49; Rutgers Univ, MS, 53, PhD(zool), 54. Prof Exp: Asst zool, Drew Univ, 47-49; teaching asst, Rutgers Univ, 49-54; asst cur limnol, Acad Nat Sci, Philadelphia, 54-57, assoc cur, Div Estuarine Sci & in chg marine invert, 58-61; assoc prof biol, 61-63, head dept, 63-65, dir marine sci res lab, 65-70, PROF BIOL, MEM UNIV NFLD, 65-, DEAN GRAD STUDIES, 70- Concurrent Pos: Mem bd dirs, Huntsman Marine Lab, St Andrews, NB. Mem: Fel AAAS; Am Soc Limnol & Oceanog; Nat Shellfisheries Asn; Can Soc Zool; fel Zool Soc London. Res: Marine invertebrate zoology; functional morphology of marine invertebrates, especially the Asteroidae and Decapoda Cephalopodae; invertebrate ecology. Mailing Add: Mem Univ of Nfld St Johns NF Can

ALDRICH, HENRY CARL, b Beaumont, Tex, Feb 17, 41; m 62; c 2. BOTANY, MYCOLOGY. Educ: Univ Tex, BA, 63, PhD(bot), 66. Prof Exp: Asst prof, 66-71, ASSOC PROF BOT, UNIV FLA, 71- Mem: Mycol Soc Am; Bot Soc Am; Am Soc Cell Biol. Res: Ultrastructure of myxomycetes, algae, fungi; membranes, freeze-etching. Mailing Add: Dept of Bot Bartram Hall Univ of Fla Gainesville FL 32611

ALDRICH, JOHN WARREN, b Providence, RI, Feb 23, 06; m 33; c 2. ORNITHOLOGY. Educ: Brown Univ, PhB, 28; Western Reserve Univ, MA, 33, PhD(biol), 37. Prof Exp: Gen curatorial asst, Buffalo Mus Sci, 28-29; biol asst, Cleveland Mus Natural His, 30-37, cur ornith, 38-40; ornithologist, Fish & Wildlife Serv, US Dept Interior, 41-47, in charge sect distribution birds & mammals, 47-58, res staff specialist, 58-73; res assoc, Smithsonian Inst, 73; RETIRED. Honors & Awards: Distinguished Serv Award, US Dept Interior, 67. Mem: AAAS; fel Ornith Union; Wilson Ornith Soc; Cooper Ornith Soc; Am Ornithologists Union (pres, 68-70). Res: Taxonomy, geographic distribution, ecology, population and systematics of birds. Mailing Add: 6324 Lakeview Dr Falls Church VA 22041

ALDRICH, LEWIS EUGENE, JR, b Baker, Ore, May 22, 22. INVERTEBRATE ZOOLOGY, PARASITOLOGY. Educ: Linfield Col, BA, 50; Ore State Univ, MS, 54, PhD(invert zool), 60. Prof Exp: Instr gen zool, Portland State Col, 60-62; asst prof invert zool & parasitol, Tex Technol Col, 62-63; assoc prof biol & head dept, St Francis Col, Maine, 63-68, chmn div natural sci, 64-68; ASSOC PROF BIOL & CHMN DEPT, SEATTLE UNIV, 68- Concurrent Pos: NSF fel, Ariz State Univ, 71. Mem: Am Soc Parasitol; Am Inst Biol Sci. Res: Monogenetic and digenetic trematodes of fishes. Mailing Add: Dept of Biol Seattle Univ 12th & E Columbia Seattle WA 98122

ALDRICH, LYMAN THOMAS, b Hopkins, Minn, June 28, 17; m 41; c 2. GEOPHYSICS. Educ: Univ Minn, PhD(physics), 48. Prof Exp: Asst physicist, Naval Ord Lab, 40-45, assoc physicist, 45; asst & res assoc, Univ Minn, 45-48; asst prof physics, Univ Mo, 48-50; mem staff, 50-65, from asst dir to assoc dir, 65-74, ACTG DIR, CARNEGIE INST DEPT TERRESTRIAL MAGNETISM, 74- Concurrent Pos: Vis prof, Kyoto Univ, 62. Mem: AAAS; Seismol Soc Am; Am Phys Soc; Am Geophys Union (gen secy, 74-). Res: Properties of earth's crust and upper mantle from electrical conductivity measurements and use of controlled and natural seismic sources in South American Andes. Mailing Add: Carnegie Inst 5241 Broad Branch Rd NW Washington DC 20015

ALDRICH, PAUL E, b Springfield, Ohio, June 29, 28; m 63. ORGANIC CHEMISTRY. Educ: Mass Inst Technol, BS, 52; Univ Wis, PhD(chem), 58. Prof Exp: Res org chemist, Merck & Co, 52-54; RES ORG CHEMIST, E I DU PONT DE NEMOURS & CO, 58- Mem: Am Chem Soc. Res: Natural products; fluorocarbons; medicinal agents. Mailing Add: E I du Pont de Nemours & Co Wilmington DE 19898

ALDRICH, PAUL HARWOOD, organic chemistry, see 12th edition

ALDRICH, RICHARD JOHN, b Fairgrove, Mich, Apr 16, 25; m 42; c 3. AGRONOMY. Educ: Mich State Col, BS, 48; Ohio State Univ, PhD(agron), 50. Prof Exp: Agronomist & coordr weed control res, Northeastern Region, USDA, 50-57, asst dir, Mich Agr Exp Sta, 57-64; assoc dir agr exp sta, 64-67, ASSOC DEAN COL AGR, UNIV MO, COLUMBIA, 67- Concurrent Pos: Mem gov bd, Agr Res Inst, Nat Acad Sci-Nat Res Coun, 64-67 & 67-70, pres elect, 74; pres, 74-75; chmn, N Cent Agr Exp State Dir Asn, 70-71, Nat Soybean Res Coord Comt, 72-76 & Div Agr, Nat Asn State Univs & Land Grant Cols, 74-75. Mem: AAAS. Mailing Add: 117 W Burnam Rd Columbia MO 65202

ALDRICH, ROBERT ANDERSON, b Evanston, Ill, Dec 13, 17; m 40; c 3. PEDIATRICS. Educ: Amherst Col, BA, 39; Northwestern Univ, MD, 44. Prof Exp: Instr pediat, Grad Sch, Univ Minn, 50; asst prof, Sch Med, Univ Ore, 51-53, assoc prof pediat & res assoc biochem, 53-56; prof & exec officer pediat, Sch Med, Univ Wash, 56-62; dir, Nat Inst Child Health & Human Develop, 62-64; prof pediat, Sch Med, Univ Wash, 64-70; vpres health affairs, 70-74, PROF PEDIAT & PREV MED, UNIV COLO, DENVER, 70- Concurrent Pos: Mem training comt, NIH, 59; vchmn, President's Comt Ment Retardation, 66-71; chief pediat, King County Hosp, Seattle; pediatrician-in-chief, Univ Wash Hosp; assoc chief med serv, Children's Orthop Hosp; consult, Madigan Army Hosp, Tacoma. Mem: AAAS; fel Am Psychiat Asn; Soc Pediat Res; Am Pediat Soc. Am Pub Health Asn. Res: Biochemistry of prophyrins; chemistry of bilirubin; mechanisms of heme synthesis; inborn errors of metabolism. Mailing Add: Dept of Prev Med Univ of Colo Med Ctr Denver CO 80220

ALDRICH, ROBERT CLEMENT, b Madison, Wis, Dec 3, 22; m 49; c 2. FOREST MENSURATION. Educ: State Univ NY Col Forestry, Syracuse Univ, BSF, 44, MF, 48. Prof Exp: Jr forester, Va Forest Serv, 48; forester, 48-54, from res forester to prin res forester, 54-74, PROJ LEADER REMOTE SENSING RES WORK UNIT, PAC SOUTHWEST FOREST & RANGE EXP STA, US FOREST SERV, 74- Mem: Soc Am Foresters; Am Soc Photogram. Res: Aerial survey and remote sensing techniques; remote sensing applications in forest resource inventories, including use of aircraft, satellite and ground data in multi-stage sampling designs. Mailing Add: 217 Conifer Lane Walnut Creek CA 94598

ALDRICH, SAMUEL ROY, b Fairgrove, Mich, June 12, 17; m 40; c 1. AGRONOMY. Educ: Mich State Col, BS, 38; Ohio State Univ, PhD(agron), 42. Prof Exp: Asst agron, Ohio State Univ, 38-42, instr, 42; from asst prof to assoc prof, Exten, Cornell Univ, 42-52, prof field crops, 52-57; prof soils exten in agron, 57-73, ASST DIR ILL AGR EXP STA, UNIV ILL, URBANA-CHAMPAIGN, 73- Concurrent Pos: Mem, Ill Pollution Control Bd, 70-71. Honors & Awards: Agron Educ Award, Am Soc Agron, 65. Mem: AAAS; fel Am Soc Agron; Soil Sci Soc Am. Res: Corn maturity; culture of corn and small grains, land use, rotations; soil fertility; environmental problems. Mailing Add: 210 Mumford Hall Univ of Ill Urbana IL 61801

ALDRICH, WILLARD WALKER, b Philadelphia, Pa, Nov 9, 01; m 27. HORTICULTURE. Educ: Johns Hopkins Univ, BS, 23; Univ Md, MS, 26, PhD(hort), 30. Prof Exp: Orchard mgr, Am Fruit Growers, Inc, Md, 26-29; from asst horticulturist to prin pomologist, Bur Plant Indust, Agr Res Serv, USDA, 29-46; tech adv, Am Fruit Growers, Inc, 46-51; mgr, Wahiawa Plantation Div, Dole Co, 52-66; AGR CONSULT, 67- Mem: Am Soc Hort Sci; Am Soc Plant Physiol; Soil Sci Soc Am. Res: Deciduous and subtropical fruit production; soil moisture-plant relationships; physiology of nutrient intake and of leaf functioning; effect of soil moisture, pruning, leaf area and mineral nutrients upon fruit set and quality; chemical induction of flowering of pineapple. Mailing Add: 406 Rogue Valley Manor Medford OR 97501

ALDRIDGE, FREDERICK THOMAS, physical chemistry, see 12th edition

ALDRIDGE, JACK P, III, physics, see 12th edition

ALDRIDGE, KEITH DOUGLAS, b Toronto, Ont, Jan 23, 41; m 71; c 1. GEOPHYSICS. Educ: Univ Toronto, BASc, 62; Mass Inst Technol, PhD(geophys), 67. Prof Exp: Sci officer psychophys, Defence Res Estab, Toronto, 67-68; NATO fel geophys, Brit Meteorol Off, 69-71; res assoc, Univ Alta, 71-75; LECTR GEOPHYS, UNIV B C, 75- Concurrent Pos: Consult, US Army, 74-75. Mem: Sigma Xi. Res: Laboratory experiments in rotating fluids; hydrodynamics of the earth's fluid outer core. Mailing Add: Dept of Geophys & Astron Univ of B C Vancouver BC Can

ALDRIDGE, MARY HENNEN, b Ark, Jan 11, 19; m 41; c 1. ORGANIC CHEMISTRY, BIOCHEMISTRY. Educ: Univ Ga, BS, 39; Duke Univ, MA, 41; Georgetown Univ, PhD(chem), 54. Prof Exp: Chemist, E I du Pont de Nemours & Co, NY, 41-47; asst prof chem, Univ Md, 47-55; assoc prof, 55-63, PROF CHEM, AM UNIV, 63- Concurrent Pos: Res grants, Eve Star, 60, US Army Med Res & Develop Command, 61-64, 66-69, NIH, 67-68 & Water Resources Res Ctr, 75-76. Mem: Am Chem Soc; AAAS; Am Asn Univ Prof. Res: Structure-activity relationships; trace organics in drinking water. Mailing Add: 2930 45th St NW Washington DC 20016

ALDRIDGE, WILLIAM GORDON, b Gladstone, NJ, Apr 28, 34; m 55; c 5. ANATOMY. Educ: Rutgers Univ, AB, 60; Univ Rochester, NIH fel, 60-62, PhD(anat), 62. Prof Exp: From instr to asst prof, 63-71, ASSOC PROF ANAT & RADIATION BIOL & DIR MULTIDISCIPLINE LABS, UNIV ROCHESTER, 71- Mem: AAAS; Histochem Soc; Am Soc Cell Biologists; Am Asn Anatomists. Res: Human anatomy; biology of nucleic acids and correlation of biochemical and electron morphological observations. Mailing Add: Multidiscipline Labs Univ of Rochester Med Ctr Rochester NY 14642

ALEEM, M I HUSSIAN, b Lyallpur, WPakistan, Jan 2, 24; m 64; c 1. MICROBIAL BIOCHEMISTRY. Educ: Univ Panjab, WPakistan, BSc, 45, MSc, 50; State Univ Groningen, dipl, 53; Cornell Univ, PhD(microbiol), 59. Prof Exp: Asst bacteriologist, Punjab Govt, Pakistan, 51-52; res assoc biochem, Johns Hopkins Univ, 58-61; asst prof microbiol, Univ Man, 61-64; biochemist, Res Inst Advan Studies, 64-66; assoc prof microbiol, 66-69, PROF MICROBIOL, UNIV KY, 69- Concurrent Pos: Brit Coun vis prof, Oxford & Bristol Univs, 63. Mem: Am Soc Microbiol. Res: Energy conversions in autotrophic bacteria. Mailing Add: Dept of Microbiol Univ of Ky Lexington KY 40506

ALEGNANI, WILLIAM CHARLES, b Springfield, Ill, Jan 18, 24; m 44; c 5. BACTERIOLOGY. Educ: Wayne Univ, BA, 48; Mich State Col, MS, 50, PhD(bact), 52. Prof Exp: Teaching asst, Mich State Col, 48-52; from assoc res microbiologist to sr res microbiologist, 52-65, sr control microbiologist, 65-68, MGR MICROBIOL CONTROL, PARKE, DAVIS & CO, 68- Mem: Am Soc Microbiol; Am Asn Contamination Control. Res: Antiseptics and disinfectants; vitamin analysis; microbiological assays; antibiotic assays; environmental control. Mailing Add: Parke, Davis & Co Joseph Campus at River Detroit MI 48232

ALEKMAN, STANLEY L, b New York, NY, Mar 21, 38; m 61; c 3. PHYSICAL

ORGANIC CHEMISTRY. Educ: City Col New York, BA, 62; Univ Del, PhD(phys org chem), 68. Prof Exp: Res chemist, US Army Ballistics Res Lab, 63-64 & Atlas Chem Co, Inc, 65; RES CHEMIST, E I DU PONT DE NEMOURS & CO, INC, 68- Mem: AAAS; Sigma Xi; Am Chem Soc; The Chem Soc; NY Acad Sci. Res: Kinetics of chromic acid oxidation; carbonium ion structure; kinetics of polycondensation reactions; kinetics of fast reactions. Mailing Add: Org Chem Dept E I du Pont de Nemours & Co Deepwater NJ 08023

ALEKSIUK, MICHAEL, b Grassland, Alta, Dec 6, 42; m 68; c 2. ENVIRONMENTAL PHYSIOLOGY. Educ: Univ Alta, BSc, 65; Univ BC, PhD(zool), 68. Prof Exp: Asst prof zool, 68-73, ASSOC PROF ZOOL, UNIV MAN, 73- Mem: Can Soc Zoologists; Ecol Soc Am; Am Soc Mammalogists; Am Soc Ichthyologists & Herpetologists. Res: Energetics of reptiles and mammals; behavioural and physiological adaptation of reptiles and mammals to the environment. Mailing Add: Dept of Zool Univ of Man Winnipeg MB Can

ALEO, JOSEPH JOHN, b Wilkes-Barre, Pa, Oct 8, 25; m 49; c 2. PATHOLOGY. Educ: Bucknell Univ, BS, 48; Temple Univ, DDS, 53; Univ Rochester, PhD(path), 65. Prof Exp: Asst biochem, Hahnemann Med Sch, 48-49; asst dent surgeon, USPHS, 53-54; pvt pract, 54-60; USPHS fel path, Univ Rochester, 60-65; chmn dept, 65-70, ASST DEAN ADVAN EDUC & RES, SCH DENT, TEMPLE UNIV, 70- Concurrent Pos: Consult, NIH, 69- Mem: AAAS; Am Dent Asn; Am Soc Exp Path; Tissue Cult Asn; Int Asn Dent Res. Res: Connective tissue diseases; experimental carcinogenesis; tissue culture; periodontal diseases. Mailing Add: Div of Advan Educ Temple Univ Sch of Dent Philadelphia PA 19140

ALEX, JACK FRANKLIN, b Rutland, Sask, Aug 20, 28; m 57; c 3. PLANT TAXONOMY, ECOLOGY. Educ: Univ Sask, BSA, 50, MSc, 52; Wash State Univ, PhD(bot), 59. Prof Exp: Asst plant ecol, Univ Sask, 50-52; field asst, Div Bot, Can Dept Agr, 52; asst bot, State Col Wash, 52-54; lectr agr bot, Univ Ceylon, 54-56; res assoc, Univ Sask, 57-58; weed ecologist, Plant Res Inst, Can Dept Agr, 58-62, ecologist, Regina Res Sta, 62-68; ASSOC PROF TAXON, UNIV GUELPH, 68- Mem: Ecol Soc Am; Weed Sci Soc Am; Am Inst Biol Sci; Agr Inst Can; Can Bot Asn. Res: Ecological investigations on weeds and native vegetation; taxonomy and ecology of weedy species. Mailing Add: Dept of Environ Biol Univ of Guelph Guelph ON Can

ALEXAKOS, LOUIS GEORGE, physical chemistry, see 12th edition

ALEXANDER, A ALLAN, b Hudson, Mass, July 19, 28; m 52; c 1. VERTEBRATE ANATOMY, TAXONOMY. Educ: Univ Mass, BS, 50; Springfield Col, MS, 47; State Univ NY, Buffalo, PhD(biol), 66. Prof Exp: Asst biol, Springfield Col, 54-55, asst physiol, 55-57; asst biol, State Univ NY, Buffalo, 58-60; from instr to asst prof, 61-68, ASSOC PROF BIOL, CANISIUS COL, 68-, CHMN DEPT, 71- Mem: Am Soc Zool; Soc Study Evolution; Am Soc Ichthyol & Herpet; Asn Study Animal Behav. Res: Herpetological taxonomy and comparative herpetological morphology and development. Mailing Add: Dept of Biol Canisius Col Buffalo NY 14208

ALEXANDER, AARON D, b New York, NY, Jan 14, 17; m 41; c 3. MEDICAL MICROBIOLOGY. Educ: City Col New York, BS, 38; George Washington Univ, MS, 53, PhD(microbiol), 61; Am Bd Microbiol, dipl, 65. Prof Exp: Bacteriologist antibiotics, Food & Drug Admin, 46-49; res bacteriologist div vet med, Walter Reed Army Inst Res, 49-53, chief res sect dept vet bact, 53-60, from asst chief to chief dept vet microbiol, 60-74; PROF MICROBIOL & CHMN DEPT, CHICAGO COL OSTEOP MED, 74- Concurrent Pos: Mem expert comt leptospirosis, WHO-Food Agr Argn, 57- mem leptospira subcomt, Int Comt Bact Nomenclature & Taxon, 58-; chief, Leptospirosis Ref Lab, 60-74. Mem: Am Soc Microbiol; Am Soc Exp Biol; Am Asn Immunol; Wildlife Dis Asn. Res: Microbial zoonoses; leptospirosis, melioidosis, listeriosis; microbial diseases of wildlife. Mailing Add: Dept of Microbiol Chicago Col of Osteop Med Chicago IL 60615

ALEXANDER, ALEX G, plant physiology, horticulture, see 12th edition

ALEXANDER, ALEXANDRE EMIL, b Leipzig, Ger, Feb 8, 05; m 39. MINERALOGY, CERAMICS. Educ: Cornell Univ, BS, 29, PhD(mineral), 33. Prof Exp: Mineralogist and petrographer, Univ NC, 30-31; petrographer, Spencer Lens Co, 34-35; ceramic petrographer, Elec Auto-Lite Co, 36-38; indust fel, Mellon Inst, 38-40; ceramic engr, B G Corp, 40-45; dir, Gem Trade Lab, Inc, 45-48; asst treas & mem bd dir, Tiffany & Co, 48-56; PRES, A E ALEXANDER RES CO, 56- Concurrent Pos: Asst, Woods Hole Oceanog Inst, 32; dir, Bur Natural Pearl Info, 40-45; res assoc, Buffalo Mus Sci, 45-63; div exec, Zale Corp, New York & Dallas, 66-; merchandising exec, Fortunoff Silver Sales, Inc, Westbury, NY, 69-70; consult engr, Champion Spark Plug Co, Toledo, 45-47, Horizons, Inc, Princeton, 47-48; Swiss Jewel Co, Switz, 47-49; jewelry consult, Designcraft Jewel Industs, New York, 75; Am ed, Int Diamond Ann, 73-74. Mem: Fel AAAS; Am Soc Metals; Am Chem Soc; Am Ceramic Soc; fel Am Inst Chem. Res: Gemology; petrography; ceramic engineering. Mailing Add: 155 E 47th St Apt 8D New York NY 10017

ALEXANDER, ALLEN LEANDER, b Statesville, NC, Feb 12, 10; m 41; c 2. CHEMISTRY. Educ: Univ NC, BS, 31, MS, 32, PhD(chem), 36. Prof Exp: Teaching fel chem, Univ NC, 31-32 & 33-36; res chemist, Sherwin-Williams Co, Ohio, 36-38; chemist, US Naval Res Lab, 38-40, head org chem br, 40-72; RETIRED. Mem: Am Chem Soc; Nat Asn Corrosion Engrs; Fedn Socs Paint Technol. Res: Paints, varnishes, lacquers; fungicides; antifouling and temperature indicating paints; chemical compounds obtained from destructive distillation of tobacco; wood preservation; corrosion and biodegradation in tropical environments. Mailing Add: 4216 Sleepy Hollow Rd Annandale VA 22003

ALEXANDER, ARCHIBALD FERGUSON, b Minneapolis, Minn, Oct 13, 28; m 53; c 4. VETERINARY PATHOLOGY. Educ: Univ Minn, BS & DVM, 51; Colo State Univ, MS, 58, PhD(animal path), 62. Prof Exp: Chief animal supply, Dugway Proving Ground, Utah, 51-56; from instr to assoc prof, 56-65, PROF VET PATH, COLO STATE UNIV, 65-, HEAD DEPT PATH, 66- Concurrent Pos: Nat Heart Inst spec fel, Glasgow, 63-64. Mem: AAAS; Am Vet Med Asn; Am Col Vet Path; Int Acad Path. Res: Experimental veterinary medicine; cardiovascular-pulmonary pathology in relation to high altitude acclimatization. Mailing Add: Dept of Path Colo State Univ Ft Collins CO 80521

ALEXANDER, BENJAMIN, b Boston, Mass, Mar 20, 09; m 37; c 3. MEDICINE. Educ: Harvard Univ, AB, 30, MD, 34. Prof Exp: House officer med, Beth Israel Hosp, 35-36; asst med, Harvard Med Sch, 39-41, instr, 41-46, assoc, 46-49, from asst prof to assoc prof, 49-66; SR INVESTR COAGULATION, NEW YORK BLOOD CTR, 66- Concurrent Pos: Teaching fel biochem, Harvard Med Sch, 36-38; Moseley fel, Dunn Inst, Cambridge & Carlsberg Lab, Copenhagen, 37-38; Commonwealth Fund fel, Weizmann Inst, 63-64; from jr vis physician to vis physician, Beth Israel Hosp, 39-67, assoc med res, 39-41, assoc dir med serv, 51-63; clin prof, Cornell Univ, 66-; attend physician, New York Hosp, 66-; assoc attend physician, Mem Hosp, 66-; consult ed, Transfusion. Mem: AAAS; Am Acad Arts & Sci; AMA; Am Soc Biol Chem; Am Soc Clin Invest. Res: Blood coagulation; hemorrhagic disease; hematology. Mailing Add: New York Blood Ctr 310 E 67th St New York NY 10021

ALEXANDER, BENJAMIN H, b Roberta, Ga, Oct 18, 21; m 48; c 2. ORGANIC CHEMISTRY. Educ: Univ Cincinnati, BA, 43; Bradley Univ, MS, 50; Georgetown Univ, PhD(chem), 57. Prof Exp: Technician, Cincinnati Chem Works, 44-45; chemist, Agr Res Serv, USDA, Ill, 45-54, res chem, Md, 54-62; chemist, Walter Reed Army Inst Res, 62-67; health scientist admin, NIH, 67-68, spec asst to dir for disadvantaged, Nat Ctr Health Serv Res & Develop, Health Serv & Ment Health Admin, USPHS, 68-69, adminr new health career projs & dep equal employ off, 69-70, prog officer, Health Care Orgn & Resources Div, 70-74; PRES, CHICAGO STATE UNIV, 74- Concurrent Pos: Adj prof, Am Univ, 58-; lectr, Grad Sch, USDA, 60-68; mem, Comt Women Higher Educ, Am Coun Educ, 75-77. Honors & Awards: Cert Achievement, Am Chem Soc. Mem: Am Chem Soc. Res: Syntheses of organophosphorus compounds for medicinal purposes; syntheses of pesticide chemicals; preparation of useful compounds from agricultural wastes. Mailing Add: Off of the Pres Chicago State Univ Chicago IL 60628

ALEXANDER, C ALEX, b Kumbanad, India, Mar 1, 35; nat US; m 60. MEDICAL ADMINISTRATION. Educ: Univ Madras, MB, 58; Johns Hopkins Univ, MPH, 64, DrPH, 66; Am Bd Prev Med, dipl, 67. Prof Exp: Asst prof pub health admin, Sch Hyg & Pub Health, Johns Hopkins Univ, 66-70, asst prof int health, 67-70, assoc prof pub health & int health, 70-72; assoc prof social & prev med, Sch Med, Univ Md, Baltimore City, 72-75; ASST DEAN & CLIN PROF COMMUNITY MED, SCH MED, WRIGHT STATE UNIV, 75-; CHIEF STAFF, VET ADMIN CTR, DAYTON, OHIO, 75- Concurrent Pos: Consult, Hosp Health Serv Res, USPHS, Baltimore & US Social Security Admin, 74. Honors & Awards: Distinguished Health & Community Serv Award, Community Health Coun Md, Inc, 74. Mem: Fel Am Pub Health Asn; Soc Int Develop; fel Am Col Prev Med. Res: Preventive medicine; epidemiology; health services administra- tion; international health; health planning; public health administration and population control. Mailing Add: Vet Admin Ctr Dayton OH 45428

ALEXANDER, CARL STUART, b New York, NY, Dec 15, 22; c 3. MEDICINE. Educ: Univ Wis, BA, 44; Columbia Univ, MD, 48; Univ Minn, PhD(med), 61. Prof Exp: Staff physician, Pulmonary Dis Serv, 56-57, clin investr, Res Div, 57-60, chief cardiovasc sect, 60-72, STAFF PHYSICIAN, CARDIOVASC SECT, VET ADMIN HOSP, 60-; PROF MED, UNIV MINN, MINNEAPOLIS, 61- Mem: Am Fedn Clin Res; Soc Exp Biol & Med; Am Physiol Soc; fel Am Col Physicians; fel Am Col Cardiol. Res: Cardiac research; antidiuretic hormone. Mailing Add: Cardiovasc Sect Vet Admin Hosp Minneapolis MN 55417

ALEXANDER, CHARLES EDWARD, JR, b Port Washington, NY, Nov 25, 30; m 60; c 3. MEDICINE, PUBLIC HEALTH. Educ: Yale Univ, AB, 51; Univ Pa, MD, 55; Johns Hopkins Univ, MPH, 59; DrPH(chronic dis), 64; Am Bd Prev Med, dipl, 64. Prof Exp: Intern med, Geisinger Med Ctr, Danville, Pa, 55-56; Med Corps, US Navy, 56-, med officer, 56-57, clinician, Naval Air Sta, Alameda, Calif, 57-58, resident pub health, Johns Hopkins Univ, 58-60, asst officer in chg, Prev Med Unit, 60-62, head venereal dis & tuberc control sects, Bur Med & Surg, 62-64, head commun dis br, 63-65, officer in chg, Prev Med Univ 7, 65-67, officer in chg, Navy Prev Med Unit, Danang, Vietnam, 67-68, prev med officer, US Mil Assistance Command, Thailand, 68-69, dep surgeon, 69-71, dir prev med div, Bur Med & Surg, 71-75, DIR OCCUP & PREV MED DIV, BUR MED & SURG, NAVY DEPT, 75- Mem: AMA; Am Pub Health Asn; Am Soc Trop Med & Hyg; fel Am Col Prev Med; Asn Mil Surg US. Res: Communicable diseases; venereal disease; tuberculosis; military preventive medicine; public health administration; epidemiology; tropical medicine; menogococcal meningitis. Mailing Add: BUMED-55 Navy Dept Washington DC 20372

ALEXANDER, CHARLES STEVENSON, b Santiago, Chile, Dec 26, 16; US citizen; m 43. GEOGRAPHY, GEOMORPHOLOGY. Educ: Univ Calif, Berkeley, BA, 47, MA, 50, PhD(geog), 55. Prof Exp: Instr geog, San Francisco State Col, 51-52; from instr to assoc prof, 53-68, PROF GEOG, UNIV ILL, URBANA, 68- Mem: Asn Am Geogr; Am Geog Soc; fel Geol Soc Am; Sigma Xi; AAAS. Res: Fluvial and coastal geomorphology; paleosols of the Caribbean Islands. Mailing Add: Dept of Geog Univ of Ill Urbana IL 61801

ALEXANDER, CHARLES WILLIAM, b Olathe, Kans, June 2, 31; m 53; c 2. AGRONOMY Educ: Kans State Col, BS, 53, MS, 54; NC State Col, PhD(agron), 57. Prof Exp: Res agronomist, Humid Pasture Mgt, 57-62, agt adminr, 62-69 & budget officer, 60-72, asst area dir, 72-75, AREA DIR, AGR RES SERV, USDA, COLUMBIA, MO, 75- Mem: Crop Sci Soc Am; Am Soc Agron; Am Soc Pub Admin. Res: Plant physiology; forage crop ecology; agricultural research administration; public administration. Mailing Add: Agr Res Serv USDA 800 N Providence Rd Columbia MO 65201

ALEXANDER, CHESTER, JR, b Tarboro, NC, Nov 6, 37; m 61; c 2. PHYSICS. Educ: Davidson Col, BS, 60; Emory Univ, MS, 62; Duke Univ, PhD(physics), 68. Prof Exp: Instr physics, Emory Univ, 61-62; physicist, Feltman Labs, Picatinny Arsenal, 62-64; asst prof physics, 68-73, ASSOC PROF PHYSICS, UNIV ALA, 73- Res: Electron spin resonance; microwave spectroscopy. Mailing Add: Dept of Physics Univ of Ala Box 1921 University AL 35486

ALEXANDER, CLAUDE GORDON, b San Diego, Calif, Sept 15, 24; m 54; c 3. ZOOLOGY. Educ: Ore State Col, BS, 48, MS, 50; Univ Calif, Los Angeles, PhD, 55. Prof Exp: From asst prof to assoc prof, 55-69, PROF BIOL, SAN FRANCISCO STATE UNIV, 69- Mem: Am Soc Parasitol; Am Soc Zoologists. Res: Helminth parasites comparative physiology of Elasmobranch fishes. Mailing Add: Dept of Biol San Francisco State Univ San Francisco CA 94132

ALEXANDER, DENTON EUGENE, b Potomac, Ill, Dec 18, 17; m 43. PLANT BREEDING, CYTOGENETICS. Educ: Ill State Norm Univ, BS, 41; Univ Ill, BS, 41, PhD(agron), 50. Prof Exp: Instr aircraft engine mechs, Army Air Force, 41-44; prod supvr Manhattan proj, Oak Ridge Nat Lab, 44-47; fel bot, 50-51, from instr & asst prof to assoc prof, 51-63, PROF PLANT BREEDING, UNIV ILL, URBANA-CHAMPAIGN, 63- Concurrent Pos: Hybrid maize expert, Food & Agr Orgn, Yugoslavia, 57; Ford Found consult, Latin Am, 65-67; mem bed dir, Funk Bros Seed Co, 69-74. Honors & Awards: Crop Sci Award, Am Soc Agron, 70. Mem: Am Soc Agron; Genetics Soc Am; Soviet Acad Agr Sci. Res: Breeding and genetics of oil and protein in maize. Mailing Add: Dept of Agron Univ of Ill Urbana IL 61801

ALEXANDER, EARL BETSON, (JR), soil science, see 12th edition

ALEXANDER, EARL GLYNN, b Nacogdoches, Tex, Dec 9, 36; m 57; c 2. RESOURCE MANAGEMENT, FOOD SCIENCE. Educ: Baylor Univ, BS, 58; Brown Univ, PhD(phys chem), 63. Prof Exp: Asst, Brown Univ, 58-59, 60-61 & 62; mem tech staff, Semiconductor Res & Develop Labs, Tex Instruments, Inc, 63-69, asst lab dir, 69-73, mgr strategic planning, Components Group, 73-74; DIR RESOURCES DEVELOP, YOUTH WITH A MISSION, 75- Mem: Electrochem Soc. Res:

Heterogeneous catalysis by transition metals and alloys; solar cells; semiconductor processing; high voltage p-n junction formation; thermal printing; agricultural economics; conservation; population studies. Mailing Add: PO Box 3097 Kailua-Kona HI 96740

ALEXANDER, EBEN, JR, b Knoxville, Tenn, Sept 14, 13; m 42; c 4. NEUROSURGERY. Educ: Univ NC, AB, 35; Harvard Med Sch, MD, 39; Am Bd Neurol Surg, dipl, 49. Prof Exp: Intern surg, Peter Bent Brigham Hosp, Boston, Mass, 39-41; asst resident surg & neurosurg, Children's Hosp, 41-42; from asst resident to resident neurosurg, Peter Bent Brigham Hosp, Boston, 46-48; from asst prof to assoc prof, 49-60, PROF NEUROSURG, BOWMAN GRAY SCH MED, WAKE FOREST UNIV, 49-; CHIEF PROF SERV, NC BAPTIST HOSP, 53- Concurrent Pos: Res fel surg, Harvard Med Sch, 46; fel neurosurg, Children's Hosp, 46; hon fel, Toronto Gen Hosp, Can, 48; mem bd sci coun, Nat Inst Neurol Dis & Blindness, 61-64, neurol sci res training comt, 62-66; mem prog proj grant comt, Nat Inst Neurol Dis & Stroke, 67-71; sr ed, J Neurosurg, 69-70; mem brain tumor study group, Nat Cancer Inst. Honors & Awards: Physician of Year Award, NC Gov. Mem: Am Asn Neurol Soc (treas, Harvey Cushing Soc, 59-62, secy, 62-65, pres-elect, 65, pres, 66); Am Acad Neurol Surg (secy-treas, 53-57, vpres, 62-63); Neurol Soc Am (vpres, 55); Soc Neurol Surg (pres, 72-73); fel Am Col Surg. Res: Hydrocephalus; fracture-dislocation of cervical spine; cerebrovascular disease, particularly intracranial aneurysm. Mailing Add: Bowman Gray Sch of Med Wake Forest Univ Winston-Salem NC 27103

ALEXANDER, EDWARD CLEVE, b Knoxville, Tenn, Nov 20, 43. ORGANIC CHEMISTRY. Educ: City Col New York, BS, 65; State Univ NY, Buffalo, PhD(org chem), 69. Prof Exp: Fel, Iowa State Univ, 69-70; ASST PROF CHEM, UNIV CALIF, SAN DIEGO, 70- Mem: AAAS; Am Chem Soc; NY Acad Sci. Res: Organic photochemistry; highly strained ring systems; small ring heterocyclic chemistry. Mailing Add: Dept of Chem Univ of Calif San Diego PO Box 109 La Jolla CA 92037

ALEXANDER, EDWARD LAWSON, b Lewiston, Maine, Oct 22, 25; m 49; c 4. NUCLEAR CHEMISTRY, RADIATION CHEMISTRY. Educ: Univ Maine, BS, 50, MS, 51; Vanderbilt Univ, PhD(radiation chem), 55. Prof Exp: Res assoc, Knolls Atomic Power Lab, Gen Elec Co, 55-57; assoc prof chem, Ga Inst Technol, 57-58; mgr radiol sci, Indust Reactor Labs, Inc, 58-62; prof radiation sci & dir radiation sci ctr, Rutgers Univ, 62-67; coordr grad studies & head dept radiol sci, Lowell Technol Inst, 67-68; DEAN GRAD SCH, UNIV LOWELL, 68- Mem: Am Chem Soc; Am Nuclear Soc; Health Phys Soc. Res: Radiation and health physics. Mailing Add: 80 Chestnut St Andover MA 01810

ALEXANDER, EDWARD RUSSELL, b Chicago, Ill, June 15, 28; m 51; c 2. EPIDEMIOLOGY. Educ: Univ Chicago, PhB, 48, SB, 50, MD, 53. Prof Exp: Intern, Cincinnati Gen Hosp, 53-54; asst resident pediat, Univ Chicago, 54-55, 57-58, instr & chief resident, 58-59; asst chief surveillance sect, Commun Dis Ctr, USPHS, 55-56, chief, 56-57 & 59-60; from asst prof to prof prev med & pediat, 61-70, chmn dept epidemiol, 70-75, PROF EPIDEMIOL, SCH PUB HEALTH, UNIV WASH, 70- Concurrent Pos: Markel scholar med sci, 62-67; consult & guest investr, US Naval Med Res Unit, 64-66, Bur Med consult, 68-; mem, Armed Forces Epidemiol Bd, 74- Mem: Fel Am Acad Pediat; fel Am Pub Health Asn; Am Epidemiol Soc; Soc Pediat Res; Infectious Dis Soc Am. Res: Infectious disease and cancer epidemiology. Mailing Add: Dept of Epidemiol Univ of Wash Sch Pub Health Seattle WA 98195

ALEXANDER, EMMIT CALVIN, JR, b Lawton, Okla, July 4, 43; m 66; c 2. GEOCHRONOLOGY. Educ: Okla State Univ, BS, 66; Univ Mo-Rolla, PhD(chem), 70. Prof Exp: Asst res chemist, Univ Calif, Berkeley, 70-73; ASST PROF RARE GAS ISOTOPES, UNIV MINN, MINNEAPOLIS, 73- Concurrent Pos: Prin investr & co-investr, Lunar Sample Anal Prog, NASA, 70- Mem: Am Geophys Union; AAAS; Geochem Soc. Res: Rare gas isotope studies of lunar, meteoritic and terrestrial samples, Ar-Ar dating; origin and evolution of terrestrial planets and their atmospheres. Mailing Add: Dept of Geol & Geophys Univ of Minn Minneapolis MN 55455

ALEXANDER, ERNEST JOHN, b St Albans, Vt, Nov 29, 23; m 48; c 2. ORGANIC CHEMISTRY. Educ: Dartmouth Col, AB, 48, MA, 50; Rensselaer Polytech Inst, PhD(chem), 56. Prof Exp: RES CHEMIST-GROUP LEADER, STERLING-WINTHROP RES INST, 50- Mem: Am Chem Soc. Res: Medicinal chemistry. Mailing Add: Sterling-Winthrop Res Inst Rensselaer NY 12144

ALEXANDER, FORREST DOYLE, b Trenton, Tenn, Oct 27, 27; m 51; c 3. MATHEMATICS. Educ: Union Univ, Tenn, BS, 50; George Peabody Col, MA, 55, PhD(math), 61. Prof Exp: Teacher high schs, Tenn, 51-56; instr, 56-59, from asst prof to assoc prof, 61-67, PROF MATH, STEPHEN F AUSTIN STATE UNIV, 67-, ADMIN ASST, 68- Concurrent Pos: NSF grant, 61-62, lectr, NSF Coop Col-Sch Sci Prog, Stephen F Austin State Univ, 70-71; consult, Math Staff, ETex Baptist Col & Nacogdoches County Pub Schs, 63-64. Mem: Math Asn Am. Res: Elementary modern mathematics; abstract algebra; geometry; analysis. Mailing Add: Dept of Math Box 3040 Stephen F Austin State Univ Nacogdoches TX 75961

ALEXANDER, FRANK CREIGHTON, JR, b Aspinwall, Pa, Nov 30, 18; m 42; c 3. PHYSICS. Educ: Carnegie Inst Technol, 42. Prof Exp: Physicist, Gulf Res & Develop Co, 47-56; head, Electronic Lab, Res & Develop Div, Am Viscose Corp, 56-58, instrumentation group, 58-62, HEAD APPL PHYSICS & INSTRUMENTATION, RES & DEVELOP LAB, FMC CORP, 62- Concurrent Pos: Instr, Carnegie Inst Technol, 50-51. Mem: Am Phys Soc; Inst Elec & Electronics Engrs. Res: Design and development of electronic instrumentation; electrical properties of organic materials; digital computer application. Mailing Add: Res & Develop Bldg FMC Corp Marcus Hook PA 19061

ALEXANDER, FRED, b NJ, Sept 17, 17; m 50; c 4. INTERNAL MEDICINE. Educ: St John's Col, Md, AB, 37; Univ Md, MD, 41. Prof Exp: Intern med, Univ Pittsburgh Hosps, 42-43; asst resident pediat, Children's Hosp, Philadelphia, 43-45; resident med, Jefferson Hosp, 45-46; resident cardiol, Mass Gen Hosp, 46-50; sr scientist med, Los Alamos Sci Lab, 50-52; ASST PROF MED, UNIV PA, 52-; DIR CLIN LAB, SMITH KLINE & FRENCH LABS, 62- Concurrent Pos: Consult, Pa Mutual Ins Co, 50-51. Mem: Am Col Physicians; Sigma Xi; AMA; Am Heart Asn; Am Col Cardiol. Res: Initial human pharmacological trials of potential therapeutic agents; cardiovascular diseases, hypertension and renal pathology; chemotherapy; metabolism; psychopharmacological entities and diseases of nervous system. Mailing Add: Hosp of the Univ of Pa 34th & Spruce St Philadelphia PA 19107

ALEXANDER, GEORGE JAY, b Paris, France, June 27, 25; nat US; m 58. BIOLOGICAL CHEMISTRY. Educ: Hobart Col, BS, 49; Rutgers Univ, PhD(microbiol & chem), 53. Prof Exp: Res assoc, Worcester Found Exp Biol, 53-58; assoc, 58-60, ASST PROF DEPT BIOCHEM, COL PHYSICIANS & SURGEONS, COLUMBIA UNIV, 60- Concurrent Pos: Assoc, NY State Psychiat Inst, 58-60; assoc res scientist, Neurotoxicol Res Unit, NY State Dept Ment Hyg, 66- Mem: Am Chem Soc; Am Soc Biol Chem; Soc Exp Biol & Med. Res: Cholesterol synthesis: transmethylation; brain metabolism; neurotoxicology. Mailing Add: 722 W 168th St New York NY 10032

ALEXANDER, GUY B, b Ogden, Utah, May 31, 18; m 41; c 2. CHEMISTRY, METALS. Educ: Univ Utah, BS, 41, MS, 42; Univ Wis, PhD(inorg chem), 47. Prof Exp: Res chemist, Manhattan Proj, Monsanto Chem Co, Ohio, 44-45; instr chem, Univ Utah, 46-47; res chemist, E I du Pont de Nemours & Co, 47-50, res supvr, 50-68; mgr chem res, Fansteel Inc, Md, 68-69, dir res, Vr-Wesson Div, Ill, 69-70 & San Fernando Labs, 70-73, DIR RES & LAB MGR, FANSTEEL RES CTR, 73- Concurrent Pos: Mem adv bd, Chem Mag, 72-75; adj prof, Univ Utah, 73- Mem: Am Chem Soc. Res: Colloid chemistry; sintered carbides; arc reactions; hot pressing; carbide powder processes; cermets; metals for high temperatures. Mailing Add: Fansteel Inc 540 Arapeen Dr Res Park Salt Lake City UT 84108

ALEXANDER, HERMAN DAVIS, b Sweetwater, Tex, Dec 19, 19; m 46; c 2. NUTRITION, BIOCHEMISTRY. Educ: Auburn Univ, BS, 50, MS, 52, PhD(nutrit, biochem), 55. Prof Exp: Asst nutrit & biochem, 50-55, asst prof nutrit, 55-60, from asst prof to assoc prof physiol, 60-66, ASSOC PROF PHARMACOL, AUBURN UNIV, 66- Mem: Am Chem Soc. Res: Nutrition, effects of various nutritional deficiencies. Mailing Add: Dept of Physiol & Pharmacol Auburn Univ Auburn AL 36830

ALEXANDER, HOWARD WRIGHT, b Toronto, Ont, June 19, 11; m 42; c 5. MATHEMATICS, STATISTICS. Educ: Univ Toronto, BA, 33, MA, 34; Princeton Univ, PhD(math), 39. Prof Exp: Instr math, Lehigh Univ, 37-40 & Fenn Col, 40-41; prof math & physics, Adrian Col, 46-47, prof math, 47-52; assoc prof, 52-60, PROF MATH, EARLHAM COL, 60- Concurrent Pos: Prof, Univ Col, Nairobi, Kenya, 65-67; vis lectr, Harvard Univ, 70-71. Mem: Math Asn Am; Am Statist Asn. Res: Differential geometry; analysis of variance; computers in the teaching of statistics. Mailing Add: 111 SW G St Richmond IN 47374

ALEXANDER, JAMES CHARLES, physical chemistry, see 12th edition

ALEXANDER, JAMES CRAIG, b Ont, Feb 12, 26; m 53; c 2. BIOCHEMISTRY. Educ: Univ Toronto, BSA, 49, MSA, 51; Univ Wis, PhD(biochem), 54. Prof Exp: Res biochemist, Procter & Gamble Co, 54-66; assoc prof biochem, 66-69, actg chmn dept nutrit, 71-72, PROF BIOCHEM, UNIV GUELPH, 69- Mem: Fel AAAS; Am Inst Nutrit; Poultry Sci Asn; Nutrit Soc Can; Can Inst Food Sci & Technol. Res: Nutritional, biochemical and physiological studies on the effect of lipids on biological systems. Mailing Add: Dept of Nutrit Univ of Guelph Guelph ON Can

ALEXANDER, JAMES CREW, b Zanesville, Ohio, Mar 22, 42. MATHEMATICS. Educ: Johns Hopkins Univ, BA, 64, PhD(math), 68. Prof Exp: Instr math, Johns Hopkins Univ, 67-69; res assoc, 69-70, ASST PROF MATH, UNIV MD, COLLEGE PARK, 70- Mem: Am Math Soc. Res: Algebraic topology. Mailing Add: Dept of Math Univ of Md College Park MD 20742

ALEXANDER, JAMES E, b Niagara Falls, NY, Jan 23, 27; m 57; c 1. OCEANOGRAPHY. Educ: Niagara Univ, BS, 54; Univ Miami, MS, 57, PhD(chem oceanog), 64. Prof Exp: Res aide fisheries technol, Univ Miami, 55-57, instr, 57-59, instr chem, oceanog, 59-64; from instr to assoc prof chem, Fordham Univ, 64-70; sr res scientist, NY Ocean Sci Lab, 70-75; PROG MGR, STATE UNIV SYST FLA, INST OCEANOG, 75- Concurrent Pos: Asst to dir, Louis Colder Conserv & Ecol Study Ctr, 67-; actg dir, NY Ocean Sci Lab, 69-70. Mem: AAAS; Am Soc Limnol & Oceanog; Am Chem Soc; Ecol Soc Am; Marine Biol Asn UK. Res: Chemical and biological oceanography; biological ecology of traces of metals in the marine environment. Mailing Add: 830 First St St Petersburg FL 33701

ALEXANDER, JAMES ERNEST, b Wilmington, Del, Nov 5, 43; m 67; c 2. ORGANIC POLYMER CHEMISTRY, RUBBER CHEMISTRY. Educ: Col Wooster, BA, 65; NDak State Univ, PhD(org chem), 70. Prof Exp: Chemist, Plastics Dept, 70-72 & Elastomer Chem Dept, 72-75, TECH REP FLUOROCARBON ELASTOMERS, E I DU PONT DE NEMOURS & CO, INC, 75- Mem: Am Chem Soc. Res: Compounding, reinforcement, rheology, processing and vulcanization chemistry of fluorocarbon elastomers. Mailing Add: Elastomer Chem Dept E I du Pont de Nemours & Co Wilmington DE 19808

ALEXANDER, JAMES KERMOTT, b Evanston, Ill, Dec 25, 20; m 45; c 3. INTERNAL MEDICINE. Educ: Amherst Col, AB, 42; Harvard Univ, MD, 46; Am Bd Internal Med, dipl, 55. Prof Exp: Intern, First Med Div, Bellevue Hosp, NY, 46-47; chief radiol sect, US Army Hosp, Guam, 47-49; resident internal med, Bellevue Hosp, 50-51; asst med, Columbia Univ, 53-54; from asst prof to assoc prof, 54-60, PROF MED, CHIEF CARDIAC SECT & DIR CARDIOPULMONARY LAB, BAYLOR COL MED, 60- Concurrent Pos: Res fel, Cardiopulmonary Lab, Bellevue Hosp, NY, 49-50; Am Heart Asn res fels, Dept Physiol, Harvard Med Sch, 51-52 & Cardiopulmonary Lab, Presby Hosp, 52-53; dir cardiac lab, Ben Taub Gen Hosp, 54-; asst attend physician, Presby Hosp, 53-54; assoc internal med, Methodist Hosp, 55. Mem: AMA; Am Heart Asn; Am Fedn Clin Res; Am Physiol Soc; Am Soc Clin Invest. Res: Cardiopulmonary function in a variety of chronic lung and circulatory diseases. Mailing Add: Baylor Col of Med Houston TX 77025

ALEXANDER, JAMES KING, b Hysham, Mont, Jan 9, 28; m 55; c 3. MICROBIOLOGY, BIOCHEMISTRY. Educ: Univ Mont, BA, 50; Mont State Univ, MS, 54, PhD(bact), 59. Prof Exp: Instr bact, NDak State Univ, 54-55; res assoc microbiol, 58-60, from instr to assoc prof, 60-72, PROF BIOL CHEM, HAHNEMANN MED COL, 72-, PROF MICROBIOL, 73- Concurrent Pos: Vis prof biol, Mass Inst Technol, 73. Mem: AAAS; Am Soc Biol Chemists; Am Soc Microbiol. Res: Carbohydrate metabolism of microorganisms; regulation of enzyme synthesis; cellulose and hexose catabolism; oligosaccharide synthesis; microbial phosphorylases. Mailing Add: Dept of Biol Chem Hahnemann Med Col Philadelphia PA 19102

ALEXANDER, JAMES WESLEY, b El Dorado, Kans, May 23, 34; m 65; c 3. SURGERY, IMMUNOLOGY. Educ: Univ Tex, MD, 57; Univ Cincinnati, ScD(surg), 64; Am Bd Surg & Bd Thoracic Surg, cert, 65. Prof Exp: Chief trauma study br, Surg Res Unit, Ft Sam Houston, Tex, 65-66; from asst prof to assoc prof surg, 66-75, PROF SURG, MED CTR, UNIV CINCINNATI, 75-, DIR TRANSPLANTATION DIV, 67-, ASST DIR SURG RES LABS, 68-, ACTG DIR BLOOD CTR, 72- Concurrent Pos: Ad hoc consult, NIH, 67- Mem: AAAS; fel Am Col Surg; Am Burn Asn; Asn Acad Surg; Am Surg Asn. Res: Infections; host defense mechanisms; transplantation; burn injury. Mailing Add: Dept of Surg Univ of Cincinnati Med Ctr Cincinnati OH 45219

ALEXANDER, JOHN J, b Indianapolis, Ind, Apr 13, 40. INORGANIC CHEMISTRY, ORGANOMETALLIC CHEMISTRY. Educ: Columbia Col, AB, 62; Columbia Univ, MA, 63, NSF fel & PhD(chem), 67. Prof Exp: Res assoc chem, Ohio State Univ, 67-69, fel, 67-68; asst prof, 70-74, ASSOC PROF CHEM, UNIV CINCINNATI, 74- Mem: Am Chem Soc. Res: Electronic structures of transition metal complexes:

synthetic organometallic chemistry. Mailing Add: Dept of Chem Univ of Cincinnati Cincinnati OH 45221

ALEXANDER, JOHN MACMILLAN, JR, b Columbia, Mo, Aug 17, 31; m 53; c 4. PHYSICAL CHEMISTRY, NUCLEAR SCIENCE. Educ: Davidson Col, BS, 53; Mass Inst Technol, PhD(phys chem), 56. Prof Exp: Res assoc chem, Mass Inst Technol, 56-57; chemist, Lawrence Radiation Lab, Univ Calif, Berkeley, 57-63; assoc prof, 63-68, chmn dept, 70-72, PROF CHEM, STATE UNIV NY STONY BROOK, 68- Concurrent Pos: Sloan fel, 64-66; chmn, Gordon Res Conf Nuclear Chem, 66; Guggenheim fel, 69-70. Res: Nuclear chemistry and reactions; stopping of heavy atoms. Mailing Add: 14 Highwood Rd Setauket NY 11785

ALEXANDER, JOHN WILLIAM, b Milwaukee, Wis, Oct 3, 12; m 36; c 3. ANALYTICAL CHEMISTRY. Educ: Univ Wis, BS, 35, PhD(anal & inorg chem), 41. Prof Exp: Instr chem, Univ Exten, Univ Wis, 35-37; anal chemist, 41-43, group leader instrumental anal, 43-59, dir, Anal Serv, 59-63, Anal & Phys Chem, 63-69 & Anal Res, 69-75, SR SCIENTIST, HARSHAW CHEM CO, 75- Mem: Am Chem Soc; Am Soc Test & Mat. Res: Spectrographic analysis of inorganic chemicals and ceramic materials; ultraviolet spectrophotometry; infrared spectroscopy; preparation of ultra pure inorganic materials, preparation of catalysts. Mailing Add: Harshaw Chem Co 1945 E 97th St Cleveland OH 44106

ALEXANDER, KLIEM, b Lebanon, Tenn, July 8, 01; m 35; c 2. CHEMISTRY. Educ: Mid Tenn State Col, BS, 27; Peabody Col, MA, 28; Univ Iowa, PhD(org chem), 38. Prof Exp: High sch prin & teacher, SC, 25-27; instr sci, Ala State Normal Sch, Troy, 28-29; instr chem & biol, WTenn State Col, 29-35; asst chem, Univ Iowa, 35-39; assoc prof, WTex State Col, 39-43; asst prof org chem, Ala Polytech Inst, 43 & Univ Fla, 43-45; res chemist, Northern Regional Res Lab, USDA, 45-50; res chemist, Ord Dept, Rock Island Arsenal, US Dept Army, 51; asst prof chem, NMex State Col, 51-53; supvr chem res, Masonite Corp, 54-59; prof chem, 59-74, EMER PROF CHEM, FLORENCE STATE UNIV, 74- Concurrent Pos: Vis prof, Univ Tenn, 60. Mem: AAAS; Am Chem Soc; fel Am Inst Chemists. Res: Organic chemistry; acyl derivatives of o-amino-phenol; furan derivatives; catalytic hydrogenation; polyhydroxy compounds; polymer and plastics intermediates. Mailing Add: 1006 Willingham Rd Florence AL 35630

ALEXANDER, LEO, b Vienna, Austria, Oct 11, 05; nat US; m 36; c 3. PSYCHIATRY, NEUROLOGY. Educ: Univ Vienna, MD, 29. Prof Exp: Instr neurol, Harvard Univ, 34-41; assoc prof neuropsychiat, Duke Univ, 41-46; asst clin prof, 46-73, LECTR PSYCHIAT, COL MED, TUFTS UNIV, 73-; CO-DIR, CHANDLER HOVEY UNIT RES & TREATMENT MULTIPLE SCLEROSIS, BROOKS HOSP, 61- Concurrent Pos: Lectr, Peiping Union Med Col, 33; assoc, Boston State Hosp, 35-41, assoc dir res, 46-48, dir neurobiol unit, Div Psychiat Res, 48-64; vis psychiatrist, Washington Hosp, Boston, 38-41; consult, US Secy War, Nuremberg War Crimes Trial, Ger, 46-47. Mem: AMA; Am Psychiat Asn; Am Neurol Asn; Am Soc Med Psychiat (vpres, 63-64, pres, Electroshock Res Asn, 51); Soc Biol Psychiat (pres, 71-72). Res: Psychophysiology; conditional reflexes; endocrine effects on the central nervous system; prognosis and treatment of multiple sclerosis; psychiatric diagnosis and treatment; hypnosis. Mailing Add: 29 Forest Ave West Newton MA 02165

ALEXANDER, LEROY ELBERT, b West Bend, Wis, Nov 7, 10; m 40; c 2. STRUCTURAL CHEMISTRY, CRYSTALLOGRAPHY. Educ: Wis State Teachers Col, BE, 37; Univ Minn, MS, 42, PhD(phys chem), 43. Prof Exp: Asst, Univ Minn, 38-43; res chemist, Gen Elec Co, Mass, 43-46; sr fel, Chem Physics Group, Carnegie-Mellon Univ, 46-47, prof chem, Mellon Inst, 67-76; RETIRED. Concurrent Pos: NSF sr fel, Technol Univ Delft, 62-63. Mem: Am Chem Soc; Am Crystallog Asn (secy, 58-60); AAAS. Res: X-ray crystallography; x-ray diffraction theory, techniques and instrumentation; polymer structure. Mailing Add: 263 Franklin Dr Pittsburgh PA 15241

ALEXANDER, LESLIE LUTHER, b Kingston, Jamaica, Oct 10, 17; US citizen; m 51; c 5. MEDICINE, RADIOLOGY. Educ: NY Univ, AB, 47, AM, 48; Howard Univ, MD, 52. Prof Exp: From instr to assoc prof, 56-69, PROF RADIOL, COL MED, STATE UNIV NY DOWNSTATE MED CTR, 69- Concurrent Pos: Nat Med Fels, Inc fel, 54-56; consult, Brooklyn Vet Hosp, 62-, Cath Med Ctr Brooklyn & Queens, 67- & Bur Health Prof Educ & Manpower Training, NIH, 70-; dir radiation ther, North Shore Hosp, Manhasset, NY. Mem: AMA; Nat Med Asn; fel Am Col Radiol; Radiol Soc NAm; Am Radium Soc. Res: Radiology, including radiation therapy, nuclear medicine, radiobiology and cancer research. Mailing Add: Col of Med State Univ NY Downstate Med Ctr Brooklyn NY 11203

ALEXANDER, LEWIS MCELWAIN, b Summit, NJ, June 15, 21; m 50; c 2. GEOGRAPHY. Educ: Middlebury Col, AB, 42; Clark Univ, MA, 48, PhD(geog), 49. Prof Exp: Instr geog, Hunter Col, 49-50; from asst prof to assoc prof, Harpur Col, State Univ NY, Binghamton, 50-60; PROF GEOG & CHMN DEPT, UNIV RI, 60-, DIR, MASTER MARINE AFFAIRS PROG, 68- Concurrent Pos: Off Naval Res grant, Univ London, 58-59; consult, Bur Intel & Res, US State Dept, 63- & Nat Coun Marine Resources & Eng Develop, 70-71; dir, Law of Sea Inst, 65-73; Mershon Soc Sci fel, Ohio State Univ, 66-67; dep dir, President's Comn Marine Sci, Eng & Resources, 67-68; mem, Int Marine Sci Affairs Panel, Nat Acad Sci & Interagency Task Force on Law of Sea, 73- Mem: Asn Am Geog; Marine Technol Soc; Am Soc Int Law. Res: Geography and the law of the sea; the nation-state and its boundaries; coastal zone management. Mailing Add: Dept of Geog Univ of RI Kingston RI 02881

ALEXANDER, LLOYD EPHRAIM, b Salem, Va, Aug 17, 02; m 34; c 2. EMBRYOLOGY. Educ: Univ Mich, AB, 27, AM, 28; Univ Rochester, PhD(embryol), 36. Prof Exp: From instr to assoc prof biol, Fisk Univ, 30-48; prof & head dept, 49-72, EMER PROF BIOL & HEAD DEPT, KY STATE COL, 73- Res: Experimental embryology; grafting and transplanting tissues in vertebrates; production of lenses in chick embryos; capacities of optic tissues of Gallus domesticus for induction and regeneration. Mailing Add: 2908 Virginia Ave Louisville KY 40211

ALEXANDER, MADELINE J, b San Francisco, Calif, May 13, 21; m 51; c 3. MATHEMATICS, RESOURCE MANAGEMENT. Educ: Stanford Univ, AB & MA, 43, PhD(math), 46. Prof Exp: Instr math, Purdue Univ, 46-52; lectr, Univ Del, 55-57; mem tech staff, Rocketdyne Div, NAm Rockwell Co, 62-70; sr math specialist, Aerojet Electrosysts Co, 70-72; SECT HEAD, TRW SYSTS INC, 72- Mem: Am Math Soc; Am Statist Asn. Res: Design and analysis of experiments; parametric and nonparametric sensitivity methods; mathematical modelling; data analysis; systems analysis; design of control system for resource management; development of real time software. Mailing Add: 19380 Halsted St Northridge CA 91324

ALEXANDER, MARTIN, b Newark, NJ, Feb 4, 30; m 51; c 2. SOIL MICROBIOLOGY, MICROBIAL ECOLOGY. Educ: Rutgers Univ, BS, 51; Univ Wis, MS, 53, PhD(bact), 55. Prof Exp: From asst prof to assoc prof, 55-64, PROF SOIL MICROBIOL, CORNELL UNIV, 64- Concurrent Pos: Consult to var pvt industs, int & nat agencies. Mem: Fel AAAS; fel Am Soc Agron; fel Am Acad Microbiol; Am Soc Microbiol. Res: Nitrogen transformations in soil and water; nitrification; metabolism of aromatic compounds; pesticide decomposition; biochemical ecology; environmental pollution. Mailing Add: 708 Bradfield Hall Cornell Univ Ithaca NY 14850

ALEXANDER, MARTIN DALE, b Grants, NMex, Nov 27, 38; m 68; c 1. INORGANIC CHEMISTRY. Educ: NMex State Univ, BSc, 60; Ohio State Univ, PhD(inorg chem), 64. Prof Exp: From asst prof to assoc prof, 64-74, PROF CHEM, NMEX STATE UNIV, 74- Mem: Am Chem Soc; fel Brit Chem Soc. Res: Mechanisms of reactions of coordination compounds of transition metals; synthesis of transition metal coordination compounds. Mailing Add: Dept of Chem NMex State Univ Las Cruces NM 88001

ALEXANDER, MARY, LOUISE, b Ennis, Tex, Jan 15, 26. GENETICS. Educ: Univ Tex, BA, 47, MA, 49, PhD(zool, genetics), 51. Prof Exp: Fel, Oak Ridge Nat Lab, AEC, 51-52; res assoc zool, Genetics Found, Univ Tex, 52-55, instr, 54, asst prof, Med Sch & assoc biologist, M D Anderson Hosp & Tumor Inst, 56-62, res scientist, Genetics Found, 62-67; assoc prof, 67-70, PROF BIOL, SOUTHWEST TEX STATE COL, 70- Concurrent Pos: Consult, Brookhaven Nat Lab, 55; res partic, Oak Ridge Inst Nuclear Studies, 56-; NIH fel, Univ Scotland, 60-62. Mem: Genetics Soc Am; Soc Study Evolution; Am Soc Naturalists; Am Soc Human Genetics; Radiation Res Soc. Res: Genetics of Drosophila; evolution and population genetics; radiation and chemical genetics. Mailing Add: Dept of Biol Southwest Tex State Col San Marcos TX 78666

ALEXANDER, MAURICE MYRON, b South Onondaga, NY, Dec 18, 17; m 43; c 3. VERTEBRATE ECOLOGY. Educ: NY State Col Forestry, BS, 40, PhD(wildlife mgt), 50; Univ Conn, MS, 42. Prof Exp: Res asst, State Bd Fisheries & Game, Conn, 45-46; instr forestry & wildlife mgt, Univ Conn, 46-47; teaching fel, 47-49, from instr to assoc prof, 49-65, PROF FOREST ZOOL & CHMN DEPT, STATE UNIV NY COL ENVIRON SCI & FORESTRY, 65- Mem: Wildlife Soc; Am Soc Mammalogists; Ecol Soc Am. Res: Animal ecology; aging techniques; furbearer management; marsh ecology. Mailing Add: Dept Forest Zool State Univ NY Col of Environ Sci & Forestry Syracuse NY 13210

ALEXANDER, MICHAEL NORMAN, b Washington, DC, Mar 27, 41; m 64; c 2. SOLID STATE PHYSICS, CHEMICAL PHYSICS. Educ: Harvard Univ, AB, 62; Cornell Univ, PhD(physics), 67. Prof Exp: Res asst physics, Cornell Univ, 64-67; RES PHYSICIST, MAT SCI DIV, ARMY MAT & MECH RES CTR, 67- Mem: AAAS; Am Phys Soc. Res: Electronic structure of metals, alloys and semiconductors; motions in molecular and polymeric solids, especially application of nuclear magnetic resonance. Mailing Add: Mat Sci Div Army Mat & Mech Res Ctr Watertown MA 02172

ALEXANDER, MILLARD HENRY, b Boston, Mass, Feb 17, 43; m 66; c 1. THEORETICAL CHEMISTRY. Educ: Harvard Col, BA, 64; Univ Paris, PhD(chem), 67. Prof Exp: Res fel chem, Harvard Univ, 67-71; asst prof molecular physics, 71-73, asst prof chem, 73-75, ASSOC PROF CHEM, UNIV MD, COLLEGE PARK, 75- Mem: Am Chem Soc; Am Phys Soc. Res: Theoretical study of rotationally and vibrationally inelastic collisions between atoms and molecules. Mailing Add: Dept of Chem Univ of Md College Park MD 20742

ALEXANDER, MORRIS WILBURN, b Westminster, SC, Oct 10, 26. AGRONOMY. Educ: Clemson Col, BS, 51; NC State Col, MS, 59. Prof Exp: ASST PROF AGRON, TIDEWATER RES & CONTINUING EDUC CTR, VA POLYTECH INST & STATE UNIV, 54- Mem: Am Soc Agron; assoc mem Sigma Xi. Res: General agronomic research and extension; soybeans. Mailing Add: PO Box 7217 Holland Sta Suffolk VA 23437

ALEXANDER, NANCY J, b Cleveland, Ohio, Dec 1, 39; c 2. REPRODUCTIVE PHYSIOLOGY, IMMUNOLOGY. Educ: Miami Univ, BS, 60, MA, 61; Univ Wis, PhD(entom), 65. Prof Exp: Asst prof zool, Miami Univ, 66-67; fel electron micros, 67-69, asst scientist, 69-72, ASSOC SCIENTIST REPRODUCTIVE PHYSIOL, ORE REGIONAL PRIMATE CTR, 72-; ASSOC PROF OBSTET, GYNEC & UROL & DIR INFERTILITY SERV, UNIV ORE HEALTH SCI CTR, 75- Concurrent Pos: Mem animal resources adv comt, NIH, 73-77. Mem: Am Soc Cell Biol; Am Fertility Soc; Am Soc Andrology (treas, 75); Soc Study Reproduction; AAAS. Res: Hormonal effects on epididymal structure and function; semen banking, infertility and immunoreproduction; immunological effects of vasectomy. Mailing Add: Dept of Reproductive Physiol Ore Regional Primate Ctr Beaverton OR 97005

ALEXANDER, NATALIE, b Los Angeles, Calif, Aug 31, 26; m 50; c 1. MEDICAL PHYSIOLOGY. Educ: Univ Southern Calif, MS, 48, PhD(med physiol), 52. Prof Exp: From instr to asst prof physiol, 50-58, res assoc med physiol, 58-70, asst prof med, 70-72, ASSOC PROF MED, SCH MED, UNIV SOUTHERN CALIF, 72- Concurrent Pos: USPHS career develop grant, 60-; mem staff, Los Angeles County Hosp, 55; mem nat adv comt, Nat Heart & Lung Inst, 73-75. Mem: AAAS; Am Physiol Soc; Am Heart Asn. Res: Circulatory physiology and hypertension. Mailing Add: Clin Pharmacol Sect Univ of Southern Calif Sch of Med Los Angeles CA 90033

ALEXANDER, NICHOLAS MICHAEL, b Boise, Idaho, June 30, 25; m 53; c 3. BIOCHEMISTRY. Educ: Univ Calif, AB, 50, PhD(biochem), 55. Prof Exp: Lectr biochem, Yale Univ, 56-64, asst prof, 64-65, assoc prof, Sch Med, 65-70; assoc adj prof, 70-74, PROF PATH, UNIV CALIF, SAN DIEGO, 74-, ASSOC DIR CLIN CHEM, UNIV HOSP, 70- Concurrent Pos: Prin scientist biochem, Vet Admin Hosp, 55- Honors & Awards: Van Meter Award, Am Thyroid Asn, 60. Mem: Am Chem Soc; Am Thyroid Asn; Am Soc Biol Chem; Endocrine Soc; Am Asn Clin Chemists; AAAS. Res: Iodine and amino acid metabolism; thyroid hormone biosynthesis; fatty acid and protein metabolism; clinical chemistry, thyroid hormones and liver regeneration. Mailing Add: Univ Hosp Univ of Calif 225 W Dickinson St San Diego CA 92103

ALEXANDER, PAUL MARION, b Akron, Ohio, Aug 21, 27; m 55; c 3. PLANT PATHOLOGY. Educ: Calif State Polytech Col, BS, 53; Ohio State Univ, MSc, 55, PhD(bot, plant path), 58. Prof Exp: Asst plant pathologist, Clemson Univ, 58-66, from asst prof to assoc prof hort, 66-69; agronomist, Green Sect, US Golf Asn, 69-70; dir educ, Golf Course Supt Asn Am, 70-73; staff vpres agron, Sea Pines Co, 73-74; CHIEF AGRONOMIST & TURF MGR, GOLTRA, INC, WINSTON-SALEM, NC, 75- Concurrent Pos: Proj leader, SC Turfgrass Res Proj, 58-69. Mem: Am Soc Agron; Soil Sci Soc Am; Crop Sci Soc Am; Am Inst Biol Sci; Am Phytopath Soc. Res: Diseases of turf grasses and ornamental plants; phytonematology; turf fungicides; weed control; soil amendments; irrigation practices; turf insects. Mailing Add: PO Box 1654 Clemson SC 29631

ALEXANDER, PETER, b New York, NY, Feb 14, 35; m 56; c 2. NUCLEAR PHYSICS. Educ: Mass Inst Technol, BS, 56; Purdue Univ, PhD(physics), 60. Prof Exp: Res assoc physics, Purdue Univ, 60-61; res fel, Calif Inst Technol, 61-64; physicist, Brookhaven Nat Lab, 64-65; mgr, Physics Dept, Teledyne isotopes, 65-72; head advan technol br, US Naval Intel Support Ctr, 72-74; DIR ADVAN TECHNOL

DIV, BENDIX FIELD ENG CORP, 74- Mem: Am Phys Soc; Am Nuclear Soc; Inst Elec & Electronic Engr. Res: Sensor development; beta and gamma spectroscopy; fission yields; geophysical tools for uranium exploration; semiconducting devices; radiation effects. Mailing Add: Bendix Field Eng Corp Box 1569 Grand Junction CO 81501

ALEXANDER, RALPH WILLIAM, b Schley, Ohio, Mar 14, 11; m 37; c 5. MEDICINE. Educ: Marietta Col, AB, 32; Univ Rochester, 36; Am Bd Internal Med, dipl. Prof Exp: Rotating intern, Jefferson Hosp, Philadelphia, Pa, 36-38; resident, William Pepper Lab Clin Med, Hosp Univ Pa, 38-39, student health serv, 39-40, staff physician, 40-46; from asst prof to assoc prof clin & prev med, 46-61, attend physician clin & infirmary, 53-61, dep dir & attend physician, Gannett Clin & Safe Infirm, 61-69, actg dir, Dept Univ Health Serv, 69-71, PROF CLIN MED, DEPT UNIV HEALTH SERV, CORNELL UNIV, 61-, DEP DIR DEPT, 71- Concurrent Pos: Ed, Student Med, 52-62; ed, J Am Col Health Asn, 62-73. Mem: AMA; Am Col Physicians; fel Am Col Health Asn. Res: Internal medicine; university students health. Mailing Add: Gannett Med Clin 10 Central Ave Ithaca NY 14853

ALEXANDER, RALPH WILLIAM, JR, b Philadelphia, Pa, May 17, 41; m 65; c 2. SOLID STATE PHYSICS, SPECTROSCOPY. Educ: Wesleyan Univ, BA, 63; Cornell Univ, PhD(physics), 68. Prof Exp: Ger Res Asn fel, Univ Freiburg, 68-69; fel, 70, ASST PROF PHYSICS, UNIV MO-ROLLA, 70- Mem: Am Phys Soc. Res: Spectroscopy of solid surfaces; catalysis. Mailing Add: Dept of Physics Univ of Mo Rolla MO 65401

ALEXANDER, RENEE R, b Leipzig, Ger, Jan 23, 32; US citizen; m 51; c 2. MOLECULAR BIOLOGY, BIOCHEMISTRY. Educ: Univ Wis, BS, 54, MS, 55; Cornell Univ, PhD(microbiol), 58. Prof Exp: Asst bact, Univ Wis, 54-55; asst bact, Cornell Univ, 55-58, res assoc microbiol, 58, phys biol, 62-65 & biochem & molecular biol, 65-70; asst prof genetics, State Univ NY Col Cortland, 70-71; LECTR BIOCHEM, CORNELL UNIV, 71- Mem: Am Soc Microbiol. Res: Genetic studies of Escherichia coli, especially radiation resistance; radioactive contamination of the food chain; chromosome mapping and study of regulation of leucine biosynthesis in Salmonella typhimurium. Mailing Add: 301 Winthrop Dr Ithaca NY 14850

ALEXANDER, RICHARD COVAL, fluid mechanics, physics, see 12th edition

ALEXANDER, RICHARD DALE, b White Heath, Ill, Nov 18, 29; m 50; c 2. ZOOLOGY. Educ: Ill State Univ, BSc, 50; Ohio State Univ, MSc, 51, PhD(entom), 56. Prof Exp: Res assoc, Rockefeller Found, 56-57; from instr to assoc prof, 57-69, PROF ZOOL, UNIV MICH, ANN ARBOR, 69-, CUR INSECTS, MUS ZOOL, 57- Honors & Awards: Daniel Giraud Elliot Medal, 71; Newcomb Cleveland Prize, AAAS, 61. Mem: Nat Acad Sci; Soc Study Evolution; Soc Syst Zool; Entom Soc Am; Ecol Soc Am. Res; Acoustical communication in insects; systematics of Orthoptera and Cicadidae; evolution of insect and social behavior. Mailing Add: Mus of Zool Univ of Mich Ann Arbor MI 48104

ALEXANDER, RICHARD RAYMOND, b Covington, Ky, Feb 2, 46; m 72. PALEOECOLOGY. Educ: Univ Cincinnati, BS, 68; Ind Univ, MA, 70, PhD(geol), 72. Prof Exp: ASST PROF GEOL, UTAH STATE UNIV, 72- Mem: Soc Econ Paleont & Mineral; Paleont Soc; Paleont Asn; Sigma Xi; Int Paleont Union. Res: Paleoautoecology of brachiopods and bivalves, particularly demographic and morphologic adaptations to sedimentologic influences and intraspecific-interspecific competition, and secondly, the relationship of functional morphology to generic longevity. Mailing Add: Dept of Geol Utah State Univ Logan UT 84322

ALEXANDER, ROBERT ALLEN, b Webster County, Miss, Oct 2, 32; m 51; c 3. ANIMAL HUSBANDRY. Educ: Univ Southwestern La, BS, 54; Univ Fla, MSA, 59, PhD(animal husb & nutrit), 61. Prof Exp: Mem staff, Dr Hornsby's Animal Clin, 47-50; salesman, Realsilk Hosiery Mills, 50-54; herdsman & mgr, Com Ranch, 54; asst, Univ Fla, 57-61; asst prof agr, 61-63, assoc prof in charge animal sci sect pre-vet & dir lab farms, 63-64, PROF AGR & CHMN DEPT, MID TENN STATE UNIV, 64- Concurrent Pos: Consult farm & livestock mgt. Mem: Am Dairy Sci Asn; Am Soc Animal Sci. Res: Animal nutrition; biochemistry and statistics. Mailing Add: Dept of Agr Mid Tenn State Univ Murfreesboro TN 37130

ALEXANDER, ROBERT BENJAMIN, b West, Tex, Oct 5, 17; m 40; c 1. PHYSICAL CHEMISTRY. Educ: Baylor Univ, BS, 45, MA, 46; Agr & Mech Col Tex, PhD(chem), 57. Prof Exp: Instr chem, Baylor Univ, 46-51; from instr to asst prof, 51-59, ASSOC PROF CHEM, TEX A&M UNIV, 59- Mem: Am Chem Soc. Res: Electrochemistry; corrosion inhibition; measurement of corrosion rates. Mailing Add: Dept of Chem Tex A&M Univ College Station TX 78363

ALEXANDER, ROBERT HOUSTON, b Griswold, Iowa, Apr 20, 27; m 55, 73; c 1. GEOGRAPHY. Educ: Iowa State Univ, BS, 50; Calif Inst Technol, MS, 53. Prof Exp: Geologist, Shell Oil Co, Calif, 53-56; geogr, Off Naval Res, 60-67; GEOGR, GEOG PROG, US GEOL SURV, 67- Mem: Asn Am Geogr; Am Soc Photogram; Am Geog Soc. Res: Geographic applications of remote sensors; earth's surface energy budget; utilization of land and natural resources. Mailing Add: Geog Prog US Geol Surv Reston VA 22092

ALEXANDER, ROBERT SPENCE, b Melrose, Mass, June 14, 17; m 42; c 5. PHYSIOLOGY. Educ: Amherst Col, AB, 38, MA, 40; Princeton Univ, PhD(gen physiol), 42. Prof Exp: Asst physiol, Princeton Univ, 40-42; instr, Med Col, Cornell Univ, 42-45; sr instr, Sch Med, Western Reserve Univ, 45-47, from asst prof to assoc prof, 47-53; assoc prof, Med Col, Univ Ga, 53-55; chmn dept, 55-72, PROF PHYSIOL, ALBANY MED COL, 55- Concurrent Pos: Res consult, Vet Admin Hosp, 48-52. Mem: AAAS; Am Physiol Soc; Harvey Soc. Res: Dynamics of arterial pulses; regulation of venous system; smooth muscle tone. Mailing Add: Dept of Physiol Albany Med Col Albany NY 12208

ALEXANDER, ROBERT STANLEY, b Topeka, Kans, Jan 5, 09; m 41; c 2. ASTRONOMY, PHYSICS. Educ: Washburn Col, BS, 36; Univ Kans, MA, 38; Univ Pa, PhD(astron), 41. Prof Exp: Asst, Flower Observ, Univ Pa, 38-39; from instr to assoc prof physics & astron, 40-50, PROF PHYSICS & ASTRON & HEAD DEPT, WASHBURN UNIV, 51- Concurrent Pos: Field engr, Western Elec Co, 45. Mem: AAAS; Am Astron Soc; Am Phys Soc; Am Asn Physics Teachers. Res: Astronomical photographic and photoelectric photometry; eclipsing variable stars. Mailing Add: Dept of Physics & Astron Washburn Univ Topeka KS 66621

ALEXANDER, ROBERT W, b North Creek, NY, July 11, 26; m 54; c 4. PATHOLOGY. Educ: Union Col, NY, BS, 49; Univ Ill, MD, 55. Prof Exp: Dir dept surg path, Presby-St Lukes Hosp, 60-67; DIR LABS, COPLEY MEM HOSP, AURORA, 67- Concurrent Pos: Med consult & dir, Nalline Testing Ctr, Div Narcotic Control, Ill Dept Pub Safety, 57-; asst prof path, Univ Ill Col Med, 60-71. Mem: Int Acad Path. Res: Surgical pathology. Mailing Add: Copley Mem Hosp Lincoln & Weston Ave Aurora IL 60507

ALEXANDER, SAMUEL CRAIGHEAD, JR, b Upper Darby, Pa, May 3, 30; m 51; c 3. ANESTHESIOLOGY, PHARMACOLOGY. Educ: Davidson Col, BS, 51; Univ Pa, MD, 55. Prof Exp: Intern, Philadelphia Gen Hosp, 55-56; sr asst surgeon, USPHS, 56-58; instr pharmacol, Univ Pa, 58-60, instr anesthesiol, 60-63, assoc, 63-65, asst prof, 65-70; prof & head dept, Univ Conn, 70-72; PROF ANESTHESIOL & CHMN DEPT, MED SCH, UNIV WIS-MADISON, 72- Concurrent Pos: Pharmaceut Mfrs fel clin pharmacol, 57-58; USPHS career develop award, 65-70; consult, Philadelphia Vet Admin Hosp, 64-68 & Madison Vet Admin Hosp, 72-; vis scientist clin physiol, Bispebjerg Hosp, Copenhagen, Denmark, 68-69. Mem: Am Soc Pharmacol & Exp Therapeut; Am Soc Anesthesiol; Asn Univ Anesthetists. Res: Respiratory control; effects of anesthetics on brain metabolism. Mailing Add: Dept of Anesthesiol Univ of Wis Med Sch Madison WI 53706

ALEXANDER, STUART DAVID, b Brooklyn, NY, Nov 18, 38; m 62; c 2. PULP & PAPER TECHNOLOGY. Educ: Cornell Univ, BChE, 60; State Univ NY Col Environ Sci & Forestry, PhD(pulp & paper technol), 66. Prof Exp: Tech serv engr, Papermill, Westvaco Corp, NY, 60-62; sr develop engr, 66-74, STAFF SPECIALIST-PAPERMAKING FIELD SERV, TECH CTR, ST REGIS PAPER CO, 74- Mem: Tech Asn Pulp & Paper Indust; Am Chem Soc. Res: Paper grade development based on paper and fiber rheology and chemistry; improving papermaking processes and manufacturing technology by means of applied mechanics, fluid dynamics, time series analysis and instrumentation development. Mailing Add: St Regis Paper Co Tech Ctr Rte 59A West Nyack NY 10994

ALEXANDER, SYDENHAM BENONI, b Charlotte, NC, May 28, 19; m 44; c 4. INTERNAL MEDICINE. Educ: Univ NC, AB, 41; Med Col Va, MD, 44; Am Bd Internal Med, dipl, 51. Prof Exp: Asst physician & instr med, Univ NC, 46-47; asst, Med Col Va, 47-48, instr, 48-49; assoc physician & instr med, Univ NC, Chapel Hill, 49-52, assoc physician, 52-56, clin asst prof med, Div Health Affairs, 52-66, asst adminr, 56-66; PROF MED & PREV MED & DIR STUDENT HEALTH SERV, UNIV ALA, TUSCALOOSA, 66- Concurrent Pos: Vis scientist, USPHS, 62-63; consult, Vet Admin Hosp, Tuscaloosa, 73- Mem: Am Col Physicians; Am Col Health Asn (vpres, 69-70). Res: Medical education. Mailing Add: Univ of Ala PO Box Y University AL 35486

ALEXANDER, TAYLOR RICHARD, b Hope, Ark, May 27, 15; m 43; c 2. BOTANY. Educ: Ouachita Col, AB, 36; Univ Chicago, MS, 38, PhD(physiol), 41. Prof Exp: From instr to assoc prof, 40-46, chmn dept, 47-65, PROF BOT, UNIV MIAMI, 47- Mem: AAAS; Ecol Soc Am; Bot Soc Am; Am Inst Biol Sci. Res: Plant nutrition in subtropical conditions; minor element deficiencies and subtropical ecology. Mailing Add: Dept of Biol Univ of Miami Coral Gables FL 33124

ALEXANDER, THOMAS GOODWIN, b Washington, DC, Sept 6, 28; m 51; c 3. ANALYTICAL CHEMISTRY. Educ: Univ Md, BS, 50; George Washington Univ, MS, 57. Prof Exp: Chem analyst, NY State Dept Agr, Cornell Univ, 50; anal chemist dairy prod lab, USDA, DC, 52-56; res anal chemist div pharmaceut chem, 56-67, supvry chemist div pharmaceut sci, 67-70, SUPVRY CHEMIST NAT CTR ANTIBIOTIC ANAL, FOOD & DRUG ADMIN, 70- Mem: Am Chem Soc; fel Asn Off Anal Chem. Res: Elemental analysis milk and milk products; complete analysis of ergot alkaloids in pharmaceuticals; chromatographic methods; development of nuclear magnetic resonance methods for pharmaceuticals. Mailing Add: 407 Hurtt Pl Oxon Hill MD 20022

ALEXANDER, THOMAS KENNEDY, b Vancouver, BC, Mar 6, 31; m 55; c 3. NUCLEAR PHYSICS. Educ: Univ BC, BA, 53, MSc, 55; Univ Alta, PhD(physics), 64. Prof Exp: Res officer reactor physics, 55-56, electronics, 56-61, RES OFFICER NUCLEAR PHYSICS, ATOMIC ENERGY CAN LTD, 64- Mem: Can Asn Physicists; Am Phys Soc. Res: Nuclear spectroscopy; measurements of lifetimes, moments and reactions with heavy ions. Mailing Add: Physics Div Atomic Energy of Can Ltd Chalk River ON Can

ALEXANDER, VERA, b Budapest, Hungary, Oct 26, 32; m 53, 67; c 2. AQUATIC ECOLOGY. Educ: Univ Wis, BA, 55, MS, 62; Univ Alaska, PhD(marine sci), 65. Prof Exp: Sr res asst, 63-65, from asst prof to assoc prof, 65-74, PROF MARINE SCI, UNIV ALASKA, FAIRBANKS, 74- Concurrent Pos: Chmn Alpha Helix Rev Comt, Univ Nat Oceanog Lab Syst, 75-; mem oceanog panel, NSF, 75- Mem: Sigma Xi; Am Soc Limnol & Oceanog; Nat Oceanog Found; Brit Freshwater Biol Asn. Res: Primary production processes in marine and freshwater systems. Mailing Add: Inst of Marine Sci Univ of Alaska Fairbanks AK 99701

ALEXANDER-JACKSON, ELEANOR GERTRUDE, b New York, NY, July 1, 04; m 34; c 1. BACTERIOLOGY. Educ: Wellesley Col, BA, 25; Columbia Univ, MA, 28; NY Univ, PhD(bact), 34. Prof Exp: Res asst bact, Univ Hosp, Univ Mich, 28-30; lab instr bact & immunol, Col Med, NY Univ, 34-36; spec res bacteriologist, NY State Dept Health, 36-39; res fel, Med Col, Cornell Univ, 41-50; bacteriologist, Cancer Res Lab, Newark, NJ, 50-54; guest investr, Inst Comp Med, Col Physicians & Surgeons, Columbia Univ, 61-65; CONSULT RES MICROBIOLOGIST, 65- Concurrent Pos: Sigma Delta Epsilon grant-in-aid award, 65. Honors & Awards: Morrison Prize, NY Acad Sci. Mem: AAAS; Am Soc Microbiol; assoc fel NY Acad Med; NY Acad Sci. Res: Forms of M-tuberculosis and M-leprae; microbiology of cancer; cultivation of M-leprae with production of disease in animals. Mailing Add: 390 Riverside Dr New York NY 10025

ALEXANDERSON, GERALD LEE, b Caldwell, Idaho, Nov 13, 33. MATHEMATICS. Educ: Univ Ore, BA, 55; Stanford Univ, MS, 58. Prof Exp: From instr to assoc prof, 58-72, PROF MATH, UNIV SANTA CLARA, 72-, CHMN DEPT, 67- Mem: Am Math Soc; Math Asn Am. Res: Combinatorial analysis. Mailing Add: Dept of Math Univ of Santa Clara Santa Clara CA 95053

ALEXANDROPOULOS, NIKOS G, b Stavros, Greece, Aug 7, 34; m 64; c 2. SOLID STATE PHYSICS, X-RAY ASTRONOMY. Educ: Nat Univ Athens, dipl, 60, DSc(physics), 64. Prof Exp: Asst physics, Nat Univ Athens for US Air Force, 61-63, Democritus Nuclear Res Ctr, Greece, 63-64 & Purdue Univ, 64-66; asst chem, Syracuse Univ, 66-67; asst prof physics, Polytech Inst Brooklyn, 67-72; resident vis, Bell Tell Lab, NJ, 71-72; mem tech staff, Aerospace Corp, 72-74, vis assoc prof sci, Rice Univ, 74-75; VIS ASSOC PROF PHYSICS, UNIV HOUSTON, 75- Mem: Am Phys Soc; Am Crystallog Asn; Am Astron Soc; Sigma Xi; NY Acad Sci. Res: X-ray spectroscopy, application to atomic molecular physics, x-ray astrophysics, plasma diagnostics and chemical analysis; x-ray scattering, including small angle, nonlinear and it applications to crystallography, liquid metal and statistical physics; hydrogen in metals. Mailing Add: 9721 Kempwood Dr Houston TX 77080

ALEXEFF, IGOR, b Pittsburgh, Pa, Jan 5, 31; m 54; c 2. NUCLEAR PHYSICS. Educ: Harvard Univ, BA, 52; Univ Wis, MS, 55, PhD(physics), 59. Prof Exp: Res engr nuclear physics, Westinghouse Res Labs, 52-53; NSF fel, Univ Zurich, 59-60; group leader turbulent heating, basic physics & levitated multipole, Oak Ridge Nat Lab, 60-71; PROF ELEC ENG, UNIV TENN, KNOXVILLE, 71- Concurrent Pos: Vis prof elec eng, Univ Tenn, 67-71; consult, Oak Ridge Nat Lab & Space Inst, Tenn; chmn,

Gordon Res Conf, 74; organizer, First Int Inst Elec & Electronics Engr Conf Plasma Sci, Tenn, 74. Mem: Sr mem Inst Elec & Electronics Engr; fel Am Phys Soc. Res: Experimental plasma physics; controlled thermonuclear fusion; low energy nuclear physics; ultra-high vacuum techniques. Mailing Add: Dept of Elec Eng Univ of Tenn Knoxville TN 37916

ALEXOPOULOS, CONSTANTINE JOHN, b Chicago, Ill, Mar 17, 07; m 39. MYCOLOGY. Educ: Univ Ill, BS, 27, MS, 28, PhD(mycol), 32. Prof Exp: Asst bot, Univ Ill, 28-30 & 32-34, instr, 34-35; from instr to asst prof biol, Kent State Univ, 35-38; plant pathologist, Inst Chimie et d'Agr, NCanel, Piraeus, Greece, 38-39; from asst prof to assoc prof biol, Kent State Univ, 39-43; assoc field technician, US Rubber Develop Corp, Brazil, 43, field technician, 43-44; agr rehab officer, UNRRA, Greece, 44-46, dep dir div agr & fish, 46-47; from assoc prof to prof bot & plant path, Mich State Univ, 47-56; prof bot & head dept, Univ Iowa, 56-62; actg chmn dept, 64-65, PROF BOT, UNIV TEX, AUSTIN, 62- Concurrent Pos: Fulbright res scholar, Nat Univ Athens, 54-55. Honors & Awards: Cert of Merit, Bot Soc Am, 67. Mem: AAAS; Bot Soc Am (pres, 63); Mycol Soc Am (secy-treas, 54-56, vpres, 57, pres elect, 58, pres, 59); Am Soc Plant Taxon; Int Mycol Asn (pres, 71-). Res: Mycoflora of Greece; taxonomy, morphology and laboratory cultivation of myxomycetes; experimental approach to the taxonomy of the myxomycetes; floristic studies of tropical myxomycetes. Mailing Add: 917 Calithea Rd Austin TX 78746

ALEY, THOMAS JOHN, b Steubenville, Ohio, Sept 8, 38; c 2. GROUNDWATER HYDROLOGY. Educ: Univ Calif, Berkeley, BS, 60, MS, 62. Prof Exp: Chief hydrologist, Toups Eng, Inc, 64-65; hydrologist, Mark Twain Nat Forest, US Forest Serv, 66-73; DIR GOUNDWATER HYDROL, OZARK UNDERGROUND LAB, 73- Concurrent Pos: Dir, Ozark Underground Lab, 66-73; consult, Time-Life Bks, 73, Nat Park Serv, 73, 75, Ark Dept Planning, 74, Corps Engrs, US Army, 74-75, US Forest Serv, 74-75 & Mo Clean Water Comn, 74-75. Honors & Awards: Dill Award, Nat Speleol Soc, 73. Mem: Nat Speleol Soc. Res: Hydrology of karst terrains with emphasis on interrelationships of surface and subsurface hydrology; conceptual and predictive hydrologic modeling of spring systems in karst regions. Mailing Add: Ozark Underground Lab 1025 S Roanoke Springfield MO 65807

ALF, CAROL JEAN, b Hamilton, Ohio, Nov 7, 48. STATISTICS, PURE MATHEMATICS. Educ: Bowling Green State Univ, BS, 70, MA, 72, PhD(math), 75. Prof Exp: ASST PROF MATH STATIST, WRIGHT STATE UNIV, 75- Mem: Am Math Soc; Inst Math Statist; Am Statist Asn; Asn Women in Math. Res: Probability in abstract spaces; probability limit theorems. Mailing Add: Dept of Math Wright State Univ Dayton OH 45431

ALF, RAYMOND MANFRED, b Canton, China, Dec 10, 05; US citizen; m 30; c 2. VERTEBRATE PALEONTOLOGY. Educ: Doane Col, AB, 28; Univ Colo, MA, 39. Hon Degrees: DSc, Claremont Grad Sch, 71; DH, Lewis & Clark Col, 71. Prof Exp: High sch teacher, Nebr, 28-29; MASTER BIOL & DEAN FAC, WEBB SCH, 29- Concurrent Pos: Res fel vert paleont, Harvard Univ, 61-62. Mem: Geol Soc Am; Soc Vert Paleont. Res: Vertebrate paleontology of the Mojave Desert, California; Pre-Cambrian fossils of the bass formation of Grand Canyon, Arizona. Mailing Add: Webb Sch of Calif 1175 W Baseline Rd Claremont CA 91711

ALFANO, MICHAEL CHARLES, b Newark, NJ, Aug 8, 47; m 69; c 2. PERIODONTOLOGY, NUTRITION. Educ: Col Med & Dent NJ, DMD, 71; Mass Inst Technol, PhD(biochem), 74. Prof Exp: CHMN GRAD ORAL BIOL & ASST PROF PERIODONT & ORAL MED, FAIRLEIGH DICKINSON UNIV, 74- Concurrent Pos: Dent consult, Inst Nutrit Cent Am & Panama, 72; lectr, Sch Dent Med, Tufts Univ, 73-74; Nat Inst Dent Res grant, Fairleigh Dickinson Univ, 74- Mem: AAAS; Int Asn Dent Res; NY Acad Sci; Am Acad Periodont; Am Acad Oral Med. Res: Role of nutrition in the etiology of oral diseases; interactions of nutrition and infection; nutritional modulation of epithelial permeability. Mailing Add: Dept of Periodont Sch of Dent Fairleigh Dickinson Univ Hackensack NJ 07601

ALFEIS, FRANZ JUERGEN, b Hamburg, Ger, Feb 22, 09; US citizen; m 43. CHEMISTRY. Educ: Munich Tech Univ, BSc, 31; Brunswick Tech Inst, MSc, 33, DrIng(chem technol), 35. Prof Exp: Res chemist, Rubberoid Co, Ill, 35-38, Ga, 38-46; works man, Bldg Prod, Celotex Corp, Ill, 46-52, NJ, 52-58, Mich, 58-62; VPRES & TECH DIR CHEM SPECIALTIES & BLDG PROD, UPSON CO, 63- Mem: AAAS; Am Soc Testing & Mat; Fedn Socs Paint Technol; Am Chem Soc; Am Tech Asn Pulp & Paper Indust. Res: Asphalt, coal-tar and cellulose fibre products; polymer adhesives and coatings; dimensional stabilization of cellulose fibre products. Mailing Add: Upson Co 72 Stevens St Lockport NY 14094

ALFERT, MAX, b Vienna, Austria, Apr 23, 21; nat US; m 45; c 1. CYTOCHEMISTRY. Educ: Wagner Col, BS, 47; Columbia Univ, MA, 48, PhD(zool), 51. Prof Exp: From instr to assoc prof, 50-63, PROF ZOOL, UNIV CALIF, BERKELEY, 63- Concurrent Pos: Guggenheim fel, 56. Mem: Am Soc Zoologists; Soc Gen Physiol; Int Soc Cell Biol; Am Soc Cell Biol. Res: Quantitative cytochemical studies of nuclear composition; development of cytochemical techniques. Mailing Add: Dept of Zool Univ of Calif Berkeley CA 94720

ALFIERI, CHARLES C, b Groton, Conn, July 19, 22; m 53; c 2. ORGANIC CHEMISTRY, INORGANIC CHEMISTRY. Educ: Brown Univ, ScB, 44; Purdue Univ, MS, 48, PhD(org chem), 50. Prof Exp: Jr chemist, Shell Oil Co, 43-44; asst chemist, Ansco Div, Gen Aniline & Film Corp, 44, 46-47; sr res chemist, Cent Res Labs, US Rubber Co, 50-52, group leader explosives & agr chem res, Naugatuck Chem Div, 52-58; HEAD CHEM RES SECT, ELKTON DIV, THIOKOL CHEM CORP, 58-; COUNR, TECH SERV DIV, UNIV DEL, 71- Mem: Am Chem Soc. Res: Use of high energy chemicals for solid rocket propellants and high explosives; synthesis and application of new liquid polymers. Mailing Add: 101 Hullihen Ct Newark DE 19711

ALFIERI, GAETANO T, b Brooklyn, NY, Feb 7, 26; c 1. PHYSICAL CHEMISTRY. Educ: Fordham Univ, BS, 50; NY Univ, MS, 55; Polytech Inst Brooklyn, PhD(chem), 69. Prof Exp: From instr to assoc prof, 53-69, PROF CHEM, NEW YORK CITY COMMUNITY COL, 69-, CHMN DEPT, 70- Mem: AAAS; Am Chem Soc. Res: High pressure chemistry; pressure dependence of magnetic phase transitions. Mailing Add: Dept of Chem New York City Community Col Brooklyn NY 11201

ALFIERI, SALVATORE A, JR, plant pathology, see 12th edition

ALFIN-SLATER, ROSLYN BERNICE (MRS GRANT G SLATER), b New York, NY, July 28, 16; m 48. BIOCHEMISTRY. Educ: Brooklyn Col, AB, 36; Columbia Univ, AM, 42, PhD(chem), 46. Prof Exp: Asst chem, Brooklyn Col, 38-43, tutor, 43; asst instr, Columbia Univ, 43-45; res chemist enzymes, Takamine Labs, 46-47; res fel cancer, Sloan-Kettering Inst, 47-48; res assoc cholesterol metab & biochem, Med Sch, Univ Southern Calif, 48-52, vis asst prof, 52-56, vis assoc prof, 56-59; assoc prof nutrit, 59-65, PROF NUTRIT & BIOL CHEM, SCH PUB HEALTH, UNIV CALIF, LOS ANGELES, 65- Concurrent Pos: Instr, Col Dent, NY Univ, 45-46; instr, Fairleigh Dickinson Jr Col, 47 & Brooklyn Col, 46-48; mem nutrit study sect, NIH, 68-72; fel coun arteriosclerosis, Am Heart Asn. Honors & Awards: Osborne Mendel Award, 70. Mem: AAAS; Am Soc Biol Chemists; Soc Exp Biol & Med; Am Inst Nutrit; fel Am Pub Health Asn. Res: Biochemical and nutritional aspects of lipid metabolism; cholesterol and essential fatty acids; saturated and unsaturated fats; carbohydrate-lipid interrelationships; essential fatty acids and vitamin E; experimental atherosclerosis; studies with aflatoxin. Mailing Add: 986 Somera Rd Los Angeles CA 90024

ALFKE, DOROTHY, b Brooklyn, NY, Nov 30, 19. SCIENCE EDUCATION. Educ: Cornell Univ, BS, 41, MS, 47, PhD(bot, ornith), 52. Prof Exp: Teacher high schs, NY, 41-46; instr sci, Oneonta State Col, 47-50; assoc prof, 52-74, PROF SCI EDUC, PA STATE UNIV, 74- Concurrent Pos: Consult, Better Light, Better Sight Bur, NY, 56-57; teacher, Int Sch, Thailand & guest lectr, Chulalongkorn, Bangkok, 61-62. Mem: Nat Asn Res Sci Teaching; Nat Sci Teachers Asn. Res: Science for elementary children and improved techniques for teaching science to non-science majors. Mailing Add: 142 Chambers Bldg Pa State Univ University Park PA 16802

ALFORD, BETTY BOHON, b St Louis, Mo, June 9, 32; m 58; c 2. NUTRITION. Educ: Tex Woman's Univ, BS, 54, MA, 56, PhD(nutrit), 65. Prof Exp: Instr nutrit & home econ, Baylor Univ, 56-57; instr foods & nutrit, 58-63, res assoc nutrit, 63-66, ASST PROF NUTRIT, TEX WOMAN'S UNIV, 66- Mem: Am Dietetic Asn; Am Pub Health Asn. Res: Metabolism during recumbancy of healthy young men; nutrition of mentally retarded children; effectiveness of nutrition education for low income families. Mailing Add: Box 23564 Tex Woman's Univ Denton TX 76204

ALFORD, BOBBY R, b Dallas, Tex, May 30, 32; m 53; c 3. OTOLARYNGOLOGY. Educ: Baylor Col Med, MD, 56. Prof Exp: From asst prof to assoc prof, 62-66, PROF OTOLARYNGOL, BAYLOR COL MED, 66-, CHMN DEPT, 67- Concurrent Pos: NIH fel, Sch Med, Johns Hopkins Univ, 61-62; consult to Surgeon Gen, US Army, 64-70 & USPHS, 65-68 & 70-; chief ed, Arch Otolaryngol, AMA, 70- Honors & Awards: Herman Johnson Award, Baylor Col Med, 56. Mem: Am Otol Soc; Am Laryngol, Rhinol & Otol Soc; Soc Univ Otolaryngol (secy, 65-69); Am Col Surg; Am Acad Ophthal & Otolaryngol. Res: Otology; otophysiology; otopathology. Mailing Add: Dept of Otolaryngol Baylor Col of Med Houston TX 77025

ALFORD, CHARLES AARON, JR, b Birmingham, Ala, Dec 8, 28; m 62; c 2. VIROLOGY. Educ: Univ Ala, BS, 51; Med Col Ala, MD, 55. Prof Exp: From intern to chief resident pediat, Med Col Ala, 55-58; pediatrician in chief, Sta Naval Hosp, Sasebo, Japan, 58-60; instr pediat, Med Col Ala, 60-62; Nat Inst Allergy & Infectious Dis spec fel virol, Dept Trop Pub Health, Sch Pub Health, Harvard Univ & res fel med, Children's Hosp Med Ctr, Boston, 62-65; from asst prof to assoc prof pediat, 65-67, MEYER PROF PEDIAT, SCH MED, UNIV ALA, BIRMINGHAM, 67-, ASSOC PROF CLIN PATH, 69-, PROF MICROBIOL, 73- Mem: Soc Pediat Res; Am Acad Pediat; Am Fedn Clin Res; NY Acad Sci. Res: Pediatric virology and immunology in the study of congenital infections. Mailing Add: Dept of Pediat Univ of Ala Sch of Med Birmingham AL 35294

ALFORD, HARVEY EDWIN, b Ashtabula, Ohio, Aug 30, 24; m 47; c 2. ORGANIC CHEMISTRY. Educ: Hiram Col, BA, 47; Western Reserve Univ, MS, 52. Prof Exp: RES ASSOC, STANDARD OIL CO OHIO, 47- Mem: Am Chem Soc. Res: Petroleum processing; coke production; activated carbon production and use; waste treatment. Mailing Add: 4440 Warrensville Center Rd Cleveland OH 44128

ALFORD, JOHN ABRIGHT, b Yazoo City, Miss, May 12, 19; m 44; c 3. FOOD MICROBIOLOGY. Educ: Miss State Col, BS, 40; La State Univ, MS, 42; Univ Wis, PhD(bact), 49. Prof Exp: Asst prof bact, La State Univ, 49-52; assoc prof, Miss State Col, 52-57; microbiologist, 57-59, head microbiol invests, Meat Lab, 59-70, head cheese & butterfat invests, Dairy Prod Lab, 70-72, CHIEF DAIRY FOODS NUTRIT LAB, NUTRIT INST, BELTSVILLE AGR RES CTR, USDA, 72- Mem: Am Soc Microbiol; Inst Food Technol; Am Dairy Sci Asn; fel Am Acad Microbiol; Brit Soc Gen Microbiol. Res: Food and dairy microbiology; psychrophiles; microorganisms related to dairy products; food safety; lipolytic activity of microorganisms. Mailing Add: Dairy Foods Nutrit Lab Bldg 157 USDA Beltsville MD 20705

ALFORD, WILLIAM CURTIS, chemistry, see 12th edition

ALFORD, WILLIAM LUMPKIN, b Albertville, Ala, Oct 6, 24; m 48; c 5. NUCLEAR PHYSICS. Educ: Vanderbilt Univ, AB, 48; Calif Inst Technol, MS, 49, PhD(physics), 53. Prof Exp: From asst prof to assoc prof physics, Auburn Univ, 53-58; nuclear physicist, US Army Missile Command, 58-64; PROF PHYSICS, AUBURN UNIV, 64- Mem: Am Phys Soc. Mailing Add: Dept of Physics Auburn Univ Auburn AL 36830

ALFORD, WILLIAM PARKER, b London, Ont, Mar 22, 27; m 49; c 3. NUCLEAR PHYSICS. Educ: Univ Western Ont, BSc, 49; Princeton Univ, PhD(physics), 54. Prof Exp: Instr physics, Princeton Univ, 54-55; from instr to prof, Univ Rochester, 55-73; PROF PHYSICS & CHMN DEPT, UNIV WESTERN ONT, 73- Concurrent Pos: Vis prof, Univ Munich, 70-71. Mem: Fel Am Phys Soc; Am Asn Physics Teachers. Res: Nuclear reactions and spectroscopy. Mailing Add: Dept of Physics Univ of Western Ont London ON Can

ALFORD, WILLIAM R, mathematics, see 12th edition

ALFORS, JOHN THEODORE, b Reedley, Calif, Nov 24, 30; m 69; c 1. GEOLOGY. Educ: Univ Calif, Berkeley, BA, 52, MA, 56, PhD(geol), 59. GEOLOGIST, CALIF DIV MINES & GEOL, 60- Mem: Geol Soc Am; Am Asn Petrol Geol; Mineral Soc Am. Res: Mineralogy and petrology of glaucophane schists; geology and mineralogy of the barium silicate deposits of Fresno County, California; geochemical study of the Rocky Hill granodiorite, Tulare County, California; geology of Tiburon Peninsula, Marin County, California. Mailing Add: Resources Bldg Rm 118 Calif Div of Mines & Geol 1416 Ninth St Sacramento CA 95814

ALFRED, LOUIS CHARLES ROLAND, b Mauritius, June 2, 29; m 64; c 3. SOLID STATE PHYSICS, PHYSICAL METALLURGY. Educ: Univ Bombay, BSc, 52, MSc, 54; Univ Sheffield, PhD(theoret physics), 57. Prof Exp: Res assoc phys chem, Brandeis Univ, 57-59; res fel physics, Brookhaven Nat Lab, 59-61 & Atomic Energy Res Estab, Eng, 61-65; assoc physicist, Argonne Nat Lab, 65-70; ASSOC PROF PHYSICS, TRENT UNIV, 70- Res: Defect properties and structure of solids; electron states in solids; magnetic resonance; atomic structure; mathematical methods. Mailing Add: Dept of Physics Trent Univ Peterborough ON Can

ALFREY, CLARENCE P, JR, b Brownwood, Tex, May 25, 30; m 55; c 4. HEMATOLOGY, BIOENGINEERING. Educ: Baylor Univ, MD, 55; Univ Minn, PhD(med), 66. Prof Exp: From instr to assoc prof, 61-71, PROF MED, BAYLOR COL MED, 71-, CO-DIR HEMAT RES, 66- Concurrent Pos: Chief radioisotope & hemat serv, Vet Admin Hosp, Houston, 61-66, assoc chief of staff res & educ, 62-66; adj assoc prof bioeng, Rice Univ, 67-71, adj prof, 71- Mem: Am Soc Hemat; Am Fedn Clin Res; AMA. Res: Medical uses of radioisotopes; iron metabolism in man;

effects of physical forces on red blood cells; ultrasound in medical diagnosis. Mailing Add: Dept of Med Baylor Col of Med Houston TX 77025

ALFREY, TURNER, JR, b Siloam Springs, Ark, May 7, 18; m; c 2. POLYMER CHEMISTRY. Educ: Washington Univ, BS, 38, MS, 40; Polytech Inst Brooklyn, PhD(polymer chem), 43. Prof Exp: Res chemist, Monsanto Chem Co, Mass, 40 & 43-45; asst prof chem, Polytech Inst Brooklyn, 45-48, assoc prof polymer chem, 48-50, assoc dir, Polymer Res Inst, 49-50; group leader, 50-53, asst dir phys res lab, 53-56, dir polymer res lab, 56-60, RES SCIENTIST, DOW CHEM CO, 60- Honors & Awards: Co-holder Chaire Francqui Liege, Belg, 47. Mem: Am Chem Soc; Soc Rheol; Am Phys Soc. Res: Rheology; free radical reactions; mechanical behavior of high polymers; copolymerization. Mailing Add: Dow Chem Co 1702 Bldg Midland MI 48640

ALFRIEND, KYLE TERRY, theoretical mechanics, applied mechanics, see 12th edition

ALGARD, FRANKLIN THOMAS, b Montrose, Colo, Oct 28, 22; m 49; c 2. EMBRYOLOGY. Educ: San Jose State Col, BA, 47; Stanford Univ, PhD(biol), 51. Prof Exp: Teaching fel anat, Stanford Univ, 52-53, from instr to asst prof, 53-64; assoc prof, 64-71, PROF ANAT, UNIV VICTORIA, 72- Concurrent Pos: USPHS sr res fel cell physiol, Col Physicians & Surgeons, Columbia Univ, 56-57 & Strangeways Lab, Eng, 62-63; USPHS sr res fel cell path, Imp Cancer Res Fund, Eng, 72-73. Mem: Tissue Cult Asn; Can Asn Anat; Sigma Xi. Res: Growth and differentiation in vitro; hormone dependency in vitro; cancer. Mailing Add: Dept of Biol Univ of Victoria Victoria BC Can

ALGER, ELIZABETH A, b Plainfield, NJ, Apr 20, 39; m 71. INTERNAL MEDICINE, ANATOMY. Educ: Seton Hall Univ, MD, 64. Prof Exp: ASSOC PROF ANAT & ASST PROF MED, COL MED & DENT NJ, 66- Concurrent Pos: Lectr, St Barnabas Med Ctr, 67- Mem: AAAS; Am Asn Anatomists; Am Fedn Clin Res; Am Med Women's Asn; Soc Study Reprod. Res: Hormonal regulation of spermatogenesis; exogenous control of male fertility; computers in medicine and medical education. Mailing Add: Dept of Anat Col of Med & Dent of NJ Newark NJ 07103

ALGER, NELDA ELIZABETH, b Ithaca, NY, Dec 14, 23. ZOOLOGY. Educ: Univ Mich, BS, 45, MS, 47; NY Univ, PhD(zool), 62. Prof Exp: Instr zool, Highland Park Jr Col, 47-48; instr zool, Hunter Col, 49-50; lab instr parasitol, Columbia Univ, 50-51; from instr to asst prof prev med, NY Univ, 52-64; asst prof, 64-70, ASSOC PROF ZOOL, UNIV ILL, URBANA-CHAMPAIGN, 70- Mem: Am Soc Parasitol; Am Soc Trop Med & Hyg; Am Soc Zoologists; Soc Protozool; Tissue Cult Asn. Res: Malaria; trypanosomiasis; parasitic immunology. Mailing Add: Dept of Genetics & Develop Univ of Ill Urbana IL 61801

ALGER, TERRY DEAN b Royal, Utah, Jan 22, 40; m 61; c 2. PHYSICAL CHEMISTRY. Educ: Univ Utah, BS, 62, PhD(chem), 66. Prof Exp: Nat Sci Found fel, Univ Ill, 66-67; asst prof, 67-71, ASSOC PROF CHEM, UTAH STATE UNIV, 71-; ASST DIR ACAD AFFAIRS, UTAH STATE BD REGENTS, 72- Mem: Am Chem Soc. Res: Use of nuclear magnetic resonance techniques to determine molecular structure and interactions. Mailing Add: Utah State Bd of Regents 1201 University Club Bldg Salt Lake City UT 84111

ALGERMISSEN, SYLVESTER THEODORE, b St Louis, Mo, May 9, 32; m 68. GEOPHYSICS. Educ: Mo Sch Mines, BS, 53; Wash Univ, AM, 55, PhD(geophys), 57. Prof Exp: Proj engr, Sinclair Res Labs, Inc, 57-59; asst prof geophys, Univ Utah, 59-63; chief data anal & res br, Coast & Geod Surv, Environ Sci Serv Admin, 63-65, chief geophys res group, 65-71; dir seismol res group, Environ Res Labs, Nat Oceanic & Atmospheric Admin, 71-73; chief seismicity & risk anal br, 73-75, RES GEOPHYSICIST, US GEOL SURV, 75- Concurrent Pos: Mem sci & eng task force, Fed Reconstruct & Develop Comn for Alaska, 64; mem comt Alaska earthquake, Nat Acad Sci-Nat Res Coun, 64-; fel, Earthquake Eng Res Inst; mem US deleg, UNESCO Intergovt Conf Mitigation of Seismic Risk, 76. Mem: Fel AAAS; Soc Explor Geophys; Seismol Soc Am; Am Geophys Union. Res: Focal mechanisms of earthquakes; earthquake risk studies; engineering seismology; interpretation of geophysical data; structure of the earth. Mailing Add: US Geol Surv Stop 978 Box 25046 DFC Denver CO 80225

ALI, MAHAMED ASGAR, b Burdwan, India, Nov 1, 34; m 68. PHYSICAL CHEMISTRY, QUANTUM CHEMISTRY. Educ: Presidency Col, Calcutta, India, BSc, 54; Univ Col Sci, Calcutta, MSc, 56; Oxford Univ, DPhil(theoret chem), 60. Prof Exp: Fel, Quantum Chem Group, Royal Univ Uppsala, 60-61; Imp Chem Indust res fel, Univ Keele, 61-62; prof chem, Presidency Col, Calcutta, 62-65; lectr, Univ Sheffield, 65-66 & Univ York, 66-68; res fel, Battelle Mem Inst, 68-69; vis assoc prof, Vanderbilt Univ, 69; res fel metals & ceramics, Oak Ridge Nat Lab, 69; ASSOC PROF CHEM, HOWARD UNIV, 69- Mem: Am Phys Soc; fel Brit Inst Physics & Phys Soc; Am Chem Soc. Res: Study of atomic structure and spectra by quantum mechanical methods; theoretical study of molecular-electronic structure; potential scattering involving atoms and molecules; theoretical molecular spectroscopy. Mailing Add: Dept of Chem Howard Univ Washington DC 20001

ALI, MIR MASWOOD, b Patuakhali, Bangladesh, Mar 1, 29; Can citizen; m 62; c 5. MATHEMATICAL STATISTICS. Educ: Univ Dacca, Pakistan, BSc, 48, MSc, 50; Univ Mich, MS, 58; Univ Toronto, PhD, 61. Prof Exp: Lectr statist, Univ Dacca, Pakistan, 51-52; mem actuarial dept, Norwich Union Ins Soc, Eng, 52-56; group actuarial dept, Can Life Assurance Co, 56-57; from asst prof to assoc prof math, 61-66, PROF MATH, UNIV WESTERN ONT, 66- Mem: Inst Math Statist; Can Math Cong; fel Royal Statist Soc; Am Statist Asn; Statist Sci Asn Can. Res: Theory of estimation; statistical inference; foundation of probability theory. Mailing Add: Dept of Math Univ of Western Ont London ON Can

ALI, MOHAMED ATHER bBangalore, India, July 16, 32; nat Can. ZOOLOGY. Educ: Presidency Col, Madras, India, BSc, 52; Univ Madras, MSc, 54, DSc(physiol), 69; Univ BC, PhD(zool), 58; Dr es Sc Nat, Sorbonne, 71. Prof Exp: Demonstr zool, Loyola Col, Madras, 52, demonstr invert embryol, 52-54; demonstr comp anat, Univ BC, 54-56, demonstr physiol, 56-58; lectr physiol & histol, McGill Univ, 58-59; asst prof biol, Mem Univ, 59-61; from asst prof to assoc prof, 61-68, PROF BIOL, UNIV MONTREAL, 68- Concurrent Pos: Res fel, Queen's Univ, Ont, 59; Alexander von Humboldt sr fel, Max-Planck Inst Physiol Behav, Ger, 64-65; NATO investr, Mus Nat, Paris, 65; Brit Coun vis prof, UK, 65; chief scientist, Te Vega Exped 15, Stanford Univ, 67; vis investr, Harvard Univ, 68 & Carnegie-Mellon Univ, 69; vis prof, Univ Sao Paulo, Univ Rio de Janeiro & Rio Grande do Sul, Brazil, 71-74. Honors & Awards: Mem Int Comt Photobiol Res Award, Arctic Inst NAm, 58, 59. Mem: AAAS; Am Soc Ichthyologists & Herpetologists; Am Soc Zoologists; Can Soc Zoologists; Can Physiol Soc. Res: Sensory physiology; behavior; neurophysiology; histology; embryology; experimental biology and ecology. Mailing Add: Dept of Biol Univ of Montreal Montreal PQ Can

ALI, MONICA MCCARTHY, b Boston, Mass, Nov 22, 41; c 2. CHEMISTRY. Educ: Emmanuel Col, AB, 63; Georgetown Univ, MS, 66, PhD(org anal chem), 71. Prof Exp: Teacher chem, biol & gen sci, Gate of Heaven High Sch, South Boston, Mass, 63-64; org chemist drug synthesis, Arthur D Little, Inc, Mass, 66-67; teacher chem, Georgetown Visitation Prep Sch, Washington, DC, 71-72; lectr gen & org chem, George Mason Univ, 72-75; ASST PROF GEN & ORG CHEM, OXFORD COL, EMORY UNIV, 75- Concurrent Pos: Lectr org chem, Northern Va Community Col, 74-75. Mem: Am Chem Soc. Res: Formation of aziridinium ion intermediates in the reactions of beta chloro amines. Mailing Add: Dept of Chem Oxford Col of Emory Univ Oxford GA 30267

ALIBERT, VERNON F, physics, mathematics, see 12th edition

ALICATA, JOSEPH EVERETT, b Carlentini, Italy, Nov 5, 04; nat; m 29, 58; c 2. PARASITOLOGY. Educ: Grand Island Col, AB, 27; Northwestern Univ, AM, 28; George Washington Univ, PhD(zool), 34. Prof Exp: Jr zoologist, bur animal indust, USDA, 28-35; instr parasitol, grad sch, 35; assoc prof zool, 35-46, parasitologist, exp sta, 35-70, prof, 47-70, EMER PROF PARASITOL & EMER PARASITOLOGIST, UNIV HAWAII, 70- Concurrent Pos: NIH & USPHS res grants; parasitologist, Bd Health, Hawaii, 36-37; parasitol investr, health & sanit comn, Hawaiian Sugar Planters's Asn, 43; Fulbright scholar, 50-51; sr scientist, USPHS, 53-70; parasitologist, For Oper Admin Mission, Jordan, 53-54. Honors & Awards: Excellence in Res Award, Univ Hawaii, 70. Mem: Am Soc Parasitologists; Am Soc Trop Med & Hyg. Res: Taxonomy, morphology and life histories of nematodes, cestodes and trematodes; trichinosis; fasciola infections and control; immunity in parasitic diseases; protozoology; epidemiology of typhus; human, canine and murine leptospirosis; control of livestock parasites; murine angiostrongylosis in the Pacific Islands and South Asia and its relationship to human eosinophilic meningoencephalitis. Mailing Add: 2130 Hunnewell St Honolulu HI 96822

ALICH, AGNES AMELIA, b International Falls, Minn, June 6, 32. ORGANOMETALLIC CHEMISTRY. Educ: Marquette Univ, BS, 60, MS, 61; Northwestern Univ, Evanston, PhD(inorg chem), 72. Prof Exp: Instr chem, Col St Scholastica, 61-64; instr, Gerard High Sch, Phoenix, 64-67; ASSOC PROF CHEM, COL ST SCHOLASTICA, 67- Mem: Am Chem Soc; Sigma Xi; The Chem Soc. Res: Synthesis of low- and zero-valent compounds of transition metals, mainly organometallic carbonyls; trace metal analysis in natural products. Mailing Add: Dept of Chem Col of St Scholastica Duluth MN 55811

ALICINO, NICHOLAS J, b New York, NY, Apr 29, 21; m 44; c 6. ANIMAL SCIENCE. Educ: Fordham Univ, BS, 42. Prof Exp: Anal chemist, Calco Div, Am Cyanamid Co, 42-44; anal chem, Nopco Chem Co, 46-50, chief analyst, 50-58, dir anal chem, 58-70; mgr res fire chem div, Diamond Shamrock Chem Co, NJ, 70, MGR RES & DEVELOP, NUTRIT & ANIMAL HEALTH DIV, DIAMOND SHAMROCK CHEM CO, OHIO, 70- Mem: Am Chem Soc; Am Microchem Soc. Res: Analytical research; chemical method by means of functional groups; new drug research for animal diseases and growth promotion. Mailing Add: Diamond Shamrock Chem Co 1100 Superior Ave Cleveland AOH 44114

ALIFF, JOHN VINCENT, b Bluefield, WVa, June 6, 42; m 64; c 1. PARASITOLOGY, ENVIRONMENTAL SCIENCES. Educ: Marshall Univ, BS, 64, MS, 65; Univ Ky, PhD(biol), 73. Prof Exp: Asst prof biol, 68-75, coordr environ sci, 73, ASSOC PROF BIOL, GA COL, 75- Concurrent Pos: Mem sci adv comn, Ga State Dept Natural Resources, 74. Mem: Am Soc Parasitologists; Sigma Xi. Res: Taxonomy and life histories of digenetic trematodes and other fish parasites; environmental curricula. Mailing Add: Dept of Biol Ga Col Milledgeville GA 31061

ALIG, ROGER CASANOVA, b Indianapolis, Ind, Nov 7, 41; m 63; c 3. SOLID STATE PHYSICS. Educ: Wabash Col, BA, 63; Purdue Univ, MS, 65, PhD(physics), 67. Prof Exp: MEM TECH STAFF PHYSICS, RCA LABS, 67- Concurrent Pos: Vis prof, Sao Carlos Sch Eng, Brazil, 70-71. Mem: Am Phys Soc. Res: Transport phenomena in metals; color centers. Mailing Add: RCA Labs Princeton NJ 08540

ALIKHAN, MUHAMMAD AKHTAR, b Lyallpur, Pakistan, Apr 1, 32; Can citizen; m 61; c 3. ANIMAL PHYSIOLOGY, BIOCHEMISTRY. Educ: Univ Panjab, Pakistan, BSc, 52; Univ Leeds, dipl, 58; Marie Curie-Sklodowska Univ, Poland, MSc, 59, PhD(physiol), 61. Prof Exp: Lectr zool, Agr Col, Lyallpur, 52-57; assoc prof, WPakistan Agr Univ, 61-64; asst prof biol, Univ Calgary, 64-65 & Lethbridge Univ, 65-67; from asst prof to assoc prof, 67-74, PROF BIOL, LAURENTIAN UNIV, 74- Concurrent Pos: Polish Acad Sci res fel, 64, 68 & 69; Nat Res Coun fel physiol & travel grant, 66 & res grant, 66-; vis scientist, Czechoslovak Acad Sci, Prague, 72 & 73; vis prof, State Univ Gent, Belg, 73, Univ Agr, Lyallpur, Pakistan & Univ Mosul, Iraq, 74. Mem: Fel Royal Entom Soc London; Asn Sci Eng & Technol Community Can; Int Asn Ecologists; Asn Can Fr Advan Sci; Can Biochem Soc. Res: Biochemical adaptations in animals to their environments; biochemistry and physiological ecology of terrestrial Crustacea. Mailing Add: Dept of Biol Laurentain Univ, Sudbury ON Can

ALI-KHAN, SYED TAHIR, b Hyderabad, India, Aug 26, 34; m 60; c 4. PLANT BREEDING, GENETICS. Educ: Osmania Univ, India, BSc, 56; Okla State Univ, MS, 60; Tex A&M Col, PhD(plant breeding), 67. Prof Exp: Agr asst, Dept Agr, India, 56-57; instr agron, Osmania Univ, India, 57-58; asst, Okla State Univ, 59-60; asst prof bot & genetics, Agr Col, Osmania Univ, India, 60-63; asst, Tex A&M Univ, 63-66; asst prof biol, Morris Col, 66-67; RES SCIENTIST PLANT BREEDING, CAN DEPT AGR, 67- Mem: Am Soc Agron; Can Soc Agron; Genetics Soc Can. Res: Breeding and genetics of crop plants, including development of improved varieties of crop plants by genetic manipulation. Mailing Add: Can Dept of Agr Res Sta Box 3001 Morden MB Can

ALIN, JOHN SUEMPER, b LaMoure, NDak, Aug 14, 40; m 64; c 1. ALGEBRA. Educ: Concordia Col, BA, 63; Univ Nebr, MA, 65, PhD(math), 67. Prof Exp: Vis asst prof math, Univ Southern Calif, 67-68; asst prof, Univ Utah, 68-72; ASSOC PROF MATH, LINFIELD COL, 72- Concurrent Pos: Consult, India Prog, NSF, 70; adj prof math, Portland State Univ, 72- Mem: Am Math Soc; Math Asn Am. Mailing Add: Dept of Math Linfield Col McMinnville OR 97128

ALIPRANTIS, CHARALAMBOS DIONISIOS, b Cephalonia, Greece, May 12, 46; m 74. MATHEMATICAL ANALYSIS. Educ: Univ Athens, dipl math, 68; Calif Inst Technol, MS, 71, PhD(math), 73. Prof Exp: Lectr math, Occidental Col, 73-74, asst prof, 74-75; ASST PROF MATH, IND UNIV-PURDUE UNIV, INDIANAPOLIS, 75- Concurrent Pos: Res scientist, STD Res Corp, 73-74. Mem: Am Math Soc; Math Asn Am. Res: Functional analysis, particularly the theory of Riesz spaces. Mailing Add: Dept of Math Ind Univ-Purdue Univ Indianapolis IN 46205

ALIRE, RICHARD MARVIN bMogote, Colo, July 5, 32; m 57; c 2. CHEMISTRY. Educ: York Col, Nebr, BA, 54; Univ Nebr, Lincoln, MS, 56; Univ NMex, PhD(chem), 62. Prof Exp: Staff mem chem res, 61-69, SECT LEADER, LOS ALAMOS SCI LAB, 69- Mem: Am Chem Soc; fel Am Inst Chemists. Res: Molecular spectroscopy of rare earth chelates; kinetics of gas-solid reactions; application of microbalance techniques to studying gas-solid interactions; materials for controlled thermonuclear reactors. Mailing Add: 107 Ft Union White Rock Los Alamos NM 87544

ALIVISATOS, SPYRIDON GERASIMOS ANASTASIOS, b Cephallonia, Greece, Oct 20, 18; nat US; m 53; c 2. BIOCHEMISTRY. Educ: Nat Univ Athens, MD, 46; McGill Univ, MSc, 49, PhD(biochem), 51. Prof Exp: Instr pharmacol, Nat Univ Athens, 46-47; res assoc biochem, Inst Chem Piraeus, Greece, 47-48; res assoc, McGill Univ, 51-52; head dept biochem res, Mt Sinai Med Res Found, 55-62; res assoc path, 57-58, from res asst prof to res assoc prof, 58-62, prof enzym & dir div, 62-68, PROF BIOCHEM & CHMN DEPT, CHICAGO MED SCH, 68- Concurrent Pos: Nat Res Coun Can Merck fel, NY Univ, 52-53; Runyon Mem Fund fel, Rockefeller Inst, 53-55; prof biochem & chmn dept, Nat Univ Athens, 66-67. Mem: Am Biochem Soc; Am Chem Soc; Soc Exp Biol & Med; NY Acad Sci; Am Soc Pharmacol & Exp Therapeut. Res: Biochemistry and chemistry of pyridine coenzymes; oligonucleotide biochemistry; mechanism of action of biogenic amines; brain metabolism; purine biosynthetic processes at the dinucleotide level; incorporation of nucleotides in nucleic acids; histaminepyridine coenzyme interactions. Mailing Add: Dept of Biochem Chicago Med Sch Chicago IL 60612

ALKER, JULIUS, environmental geology, vertebrate paleontology, see 12th edition

ALKEZWEENY, ABDUL JABBAR, b Amarah, Iraq, Sept 1, 35. CLOUD PHYSICS, AIR POLLUTION. Educ: Univ Baghdad, BS, 58; Univ Calif, Santa Barbara, MA, 63; Univ Wash, PhD(atmospheric sci), 68. Prof Exp: Scientist, Meteorol Res Inc, 68-71; SR RES SCIENTIST, BATTELLE-PAC NORTHWEST LAB, 71- Mem: Am Meteorol Soc; Sigma Xi. Res: Physical and chemical transformations of pollutants in urban plume using airborne and ground instrumentation. Mailing Add: Battelle-Northwest Richland WA 99352

ALKIRE, WILLIAM HENRY, b Bremerton, Wash, June 6, 35; m 67. CULTURAL ANTHROPOLOGY, ETHNOLOGY. Educ: Univ Wash, BA, 57; Univ Hawaii, MA, 59; Univ Ill, PhD(anthrop), 65. Prof Exp: Res assoc anthrop, Bernice P Bishop Mus, Honolulu, 64-66; asst prof anthrop, San Francisco State Col, 67-68; lectr, Univ Malaya, 68-69; ASSOC PROF ANTHROP, UNIV VICTORIA, 70- Concurrent Pos: NSF res grant, Micronesia, Bishop Mus, Honolulu, 64-66; asst prof, Univ Hawaii, 66-67; univ res grant, Univ Victoria, 70-71; Can Coun grant, res Saipan, Mariana Islands, 71 & res grant, 75. Mem: Fel Am Anthrop Asn; Am Ethnol Soc; Can Sociol & Anthrop Asn; Asn Anthrop Micronesia; Asn Social Anthrop Oceania. Res: Ethnology of Micronesia; cultural ecology; interisland migration, central Caroline islands. Mailing Add: Dept of Anthrop & Sociol Univ of Victoria Victoria BC Can

ALKJAERSIG, NORMA KIRSTINE (MRS A P FLETCHER), b Ikast, Denmark, Dec 25, 21; nat US; m 61. BIOCHEMISTRY. Educ: Tech Univ Denmark, 49; Univ Copenhagen, PhD, 65. Prof Exp: Res assoc, Biol Inst, Carlsberg Found, 49-51 & Col Med, Wayne State Univ, 51-55; res asst, Jewish Hosp, 55-58; res asst prof, 58-69, RES ASSOC PROF MED, SCH MED, WASH UNIV, 69- Mem: Am Physiol Soc. Res: Fibrinolytic enzymes; blood clotting. Mailing Add: Dept of Internal Med Wash Univ Sch of Med St Louis MO 63110

ALKSNE, ALBERTA YEARIAN b Charleston, Ill, July 16, 05; m 31; c 2. MATHEMATICS, GEOPHYSICS. Educ: Stanford Univ, BA, 27. Prof Exp: Statistician, Supvr Shipbldg, US Dept Navy, 41-43; computer, Ames Aeronaut Lab, Nat Adv Comt Aeronaut, 43-49, res scientist, Ames Res Ctr, NASA, 49-70; RETIRED. Mem: AAAS. Res: Aeronautical research, particularly transonic flow theory; space science, particularly magnetohydrodynamics and theory of the solar wind; applied mathematics. Mailing Add: 4115 Amaranta Ave Palo Alto CA 94306

ALLABASHI, JOHN CHRISTO, organic chemistry, see 12th edition

ALLAIN, RONALD JOSEPH, organic chemistry, see 12th edition

ALLAIRE, FRANCIS RAYMOND b Fall River, Mass, Nov 18, 37; m 61; c 4. DAIRY SCIENCE. Educ: Univ Mass, BVA, 59, MS, 63; Cornell Univ, PhD(animal breeding), 65. Prof Exp: From fel to asst prof, 65-71, ASSOC PROF DAIRY SCI, OHIO STATE UNIV, 72- Mem: Am Soc Animal Sci; Am Dairy Sci Asn; Biomet Soc; AAAS. Res: Understanding and utilization of genetic variations in quantitative traits of animal populations, with emphasis on breeding and selection among animals serving as sources of food protein. Mailing Add: Dept Dairy Sci Ohio State Univ 625 Stadium Dr Columbus OH 43210

ALLAM, MARK WHITTIER, b Fernwood, Pa, Aug 17, 08; m 33; c 2. Educ: Univ Pa, VMD, 32. Prof Exp: From instr to prof vet surg & chmn dept, Sch Vet Med, 43-73, ASST VPRES HEALTH AFFAIRS, UNIV PA, 73- Concurrent Pos: Researcher, Harrison Dept Surg Res, Sch Med, Univ Pa & mem bd mgr, Wistar Inst, 47-, dean fac, Sch Vet Med, Univ Pa, 53-73; mem adv bd bur med, US Food & Drug Admin, 66-70 & adv bur vet med, 70- Mem: Am Vet Med Asn. Res: Peripheral nerve regeneration; developed transplantable carcinoma in dog. Mailing Add: Univ of Pa New Bolton Ctr Kennett Square PA 19348

AL-LAMI, FADHIL, b Baghdad, Iraq, Sept 6, 32; m 63; c 2. ANATOMY, PHYSIOLOGY. Educ: Univ Baghdad, BSc, 55; Ind Univ, MSc, 59, PhD(anat), 64. Prof Exp: Res assoc electron micros, Ind Univ, 66-67; researcher, Putnam Hosp, 67-71; ASST PROF IND UNIV, 74- Mem: Am Asn Anatomists. Res: Ultrastructure of the carotid body of normal anoxic mammals as well as the study of the effects of some cholenergic drugs on this organ. Mailing Add: Dept of Anat & Physiol Indiana Univ Bloomington IN 47401

ALLAN, BARRY DAVID b Steubenville, Ohio, Jan 20, 35; m 61; c 2. PHYSICAL CHEMISTRY, LASERS. Educ: Ariz State Univ, BS, 56; Univ Ala, MS, 64, PhD(chem), 68. Prof Exp: Proj officer propellants, Army Rocket & Guided Missile Agency, 56-58, from chemist to res chemist, 58-70, res staff, Chem Laser Br, 70-75, CHIEF HIGH ENERGY LASER DIRECTORATE, DEVICE/COMPONENTS BR, ARMY MISSILE COMMAND, 75- Concurrent Pos: Mem, Army Explosives & Propellant Res & Develop Liaison Group, 58-62; mem panel on liquid test methods, Chem Propellant Info Agency, 62-; mem Army Ad Hoc Comt Sensitivity of New Mat, 62-; reviewer, NSF, 74- Honors & Awards: Army Res & Develop Award, Dept of Army, 62. Mem: Am Chem Soc. Res: Combustion and oscillatory combustion; physico-chemical properties of high energy compounds; sensitivity and desensitization of high energy compounds; theoretical propellant performance analysis; physico-chemical and rheological analysis of thixotropic gels. Mailing Add: 7803 Michael Circle Huntsville AL 35802

ALLAN, BENJAMIN WILSON, b Baltimore, Md, Feb 10, 07; m 34; c 3. INORGANIC CHEMISTRY, PHYSICAL CHEMISTRY. Educ: Johns Hopkins Univ, PhD(phys chem), 32. Prof Exp: Instr chem, Johns Hopkins Univ, 32-33, res chemist, 34-35; dir res, Titanium Div, Glidden Co, 35-47 & Chem, Pigments & Metals Div, 47-57, asst coordr res, 57-62; tech adv, Am Potash & Chem Corp, 62-70, Kerr-McGee Chem Corp, Kerr-McGee Corp, 70-73; MEM STAFF, CREST CHEM CO INC, LOS ANGELES, CALIF, 73- Mem: Fel AAAS; Am Chem Soc; fel Am Inst Chem; Am Soc Testing & Mat. Res: Production of titanium dioxide pigments; heavy chemicals. Mailing Add: Crest Chem Co Inc 6920 S Stanford Ave Los Angeles CA 90001

ALLAN, DAVID WAYNE, b Mapleton, Utah, Sept 25, 36; m 59; c 6. PHYSICS, MATHEMATICS. Educ: Brigham Young Univ, BS, 60; Univ Colo, MS, 65. Prof Exp: Physicist, 60-68, ASST SECT CHIEF, TIME & FREQUENCY DIV, NAT BUR STANDARDS, DEPT COMMERCE, 68- Concurrent Pos: Mem, Int Radio Consultative Comt, 67-, US del, 68; Ital Govt vis prof grant, 69. Honors & Awards: Silver Medal Award & Sustained Super Performance Award, Dept Commerce, 68. Mem: Sci Res Soc Am. Res: Atomic time scale; classification of the statistical characteristics of atomic and molecular frequency standards. Mailing Add: Time & Frequency Div Nat Bur Standards Boulder CO 80302

ALLAN, FRANK DUANE, b Salt Lake City, Utah, May 19, 25; m 46; c 7. ANATOMY, EMBRYOLOGY. Educ: Univ Utah, BS, 47, MS, 49; La State Univ, PhD(anat), 54. Prof Exp: Asst anat, Univ Utah, 48-49; from asst prof to assoc prof, 54-68, PROF ANAT, SCH MED, GEORGE WASHINGTON UNIV, 68-, DIR AV SERV, 62-, DIR AV EDUC, 73- Concurrent Pos: Consult, Sibley Mem Hosp, 58-62, Wash Hosp Ctr, 60-64 & Columbia Hosp Women, 62-66; vis scientist, Carnegie Inst Washington, 62-64. Mem: AAAS; Am Asn Anatomists. Res: Early human embryology; cardiovascular physiology; electrophysiological studies on the development of tolerance to alcohol; anomalous arteries; innervation of the ducts arteriosus; anatomy of the fetus and newborn; audiovisual technology in medical education; gross and microscopic anatomy; neuroanatomy. Mailing Add: Dept of Anat George Washington Univ Sch Med Washington DC 20037

ALLAN, GEORGE GRAHAM b Glasgow, Scotland, Nov 21, 30; m 55; c 6. ORGANIC CHEMISTRY. Educ: Royal Col Sci & Technol, Scotland, dipl, 51, assoc, 52; Glasgow Univ, BSc, 52, PhD(org chem), 55; Univ Strathclyde, DSc(fiber & polymer sci), 70; FRIC. Prof Exp: Asst to Prof F S Spring, Royal Col Sci & Technol, Scotland, 55-56; res chemist electrochem, E I du Pont de Nemours, 56-62; sr scientist, pioneering res dept, Weyerhaeuser Co, 62-66; PROF FIBER SCI & POLYMER CHEM, DEPT CHEM ENG & COL FOREST RESOURCES, UNIV WASH, 66- Concurrent Pos: Head sci leather technologists, David Dale Tech Col, Glasgow & lectr, Paisley Tech Col, Scotland, 52-56. Mem: Am Chem Soc; Am Inst Chem Engrs; Tech Asn Pulp & Paper Indust. Res: Lignin and forest products chemistry; polymers; biologically active organic heterocycles; epoxidation; nonwovens; fiber surface modification; adhesion; controlled release pesticides; pollution control; marine polymers. Mailing Add: Univ of Wash AR-10 Seattle WA 98195

ALLAN, J DAVID, b London, Ont, Feb 5, 45; m 68. ECOLOGY. Educ: Univ BC, BSc, 66; Univ Mich, MS, 68, PhD(zool), 71. Prof Exp: Fel biol, Univ Chicago, 71-72; ASST PROF ZOOL, UNIV MD, COLLEGE PARK, 72- Mem: AAAS; Ecol Soc Am; Int Soc Theoret & Appl Limnol. Res: Ecology and aquatic biology; population dynamics and life histories; competition and predation; distributional patterns. Mailing Add: Dept of Zool Univ of Md College Park MD 20742

ALLAN, JOHN R, b Detroit, Mich, Feb 6, 37; m 63; c 4. PLANT PHYSIOLOGY. Educ: Univ Mich, BS, 59; Ind Univ, MA, 62; Univ Sask, PhD(biol, plant physiol), 67. Prof Exp: Teaching asst aquatic plants, Mich Biol Sta, 59; res asst plant physiol, Argonne Nat Lab, 61; teaching asst, Ind Univ, 61-62; res asst, Univ Sask, 62-64; RES SCIENTIST PLANT PHYSIOL, CAN AGR RES STA, 66- Mem: Phytochem Soc NAm; Weed Sci Soc Am. Res: Plant biochemistry; aquatic plant taxonomy and ecology. Mailing Add: Can Agr Res Sta Lethbridge AB Can

ALLAN, ROBERT EMERSON, b Morris, Ill, Jan 12, 31; c 4. AGRONOMY. Educ: Iowa State Col, BS, 52; Kans State Col, MS, 56, PhD(agron), 58. Prof Exp: Res asst, Kans State Col, 54-57; agronomist, 57-59, GENETICIST & RES LEADER WHEAT BREEDING & PROD, AGR RES SERV, USDA, WASH STATE UNIV, 59- Mem: Genetics Soc Am; Am Soc Agron; Am Genetic Asn. Res: Genetics and cytogenetics of wheat and its relatives; genetics of host, parasite relationships; nuclear and cytoplasmic interactions in Triticum species. Mailing Add: 209 Johnson Hall Dept of Agron Wash State Univ Pullman WA 99163

ALLAN, ROBERT K, b Hamilton Ont, Jan 25, 40; m 62; c 3. BIOCHEMISTRY. Educ: McMaster Univ, BSc, 62, PhD(biochem), 67. Prof Exp: Asst prof, 66-70, ASSOC PROF CHEM, YORK UNIV, 70- Mem: AAAS; Chem Inst Can. Res: Biochemistry of microorganisms; chloroplast mutagenesis; mechanism of action of chemical mutagens. Mailing Add: Dept of Chem York Univ 4700 Keele St Toronto ON Can

ALLAND, ALEXANDER, JR, b Newark, NJ, Sept 23, 31; m 56; c 2. ANTHROPOLOGY, CULTURAL ANTHROPOLOGY. Educ: Univ Wis, BS, 54; Univ Conn, MS, 58; Yale Univ, PhD(anthrop), 63. Prof Exp: Instr anthrop, Vassar Col, 62-63 & Hunter Col, 63-64; from asst prof to assoc prof, 64-72, PROF ANTHROP, COLUMBIA UNIV, 72- Mem: AAAS; Am Anthrop Asn (comt women, 70-72, comt ethics, 73-, chmn, 75-76); Royal Anthrop Inst; Am Ethnol Soc. Res: Biological and cultural evolution, including evolution of creativity and artistic behavior; primitive art; Africa; ecology. Mailing Add: Dept of Anthrop Columbia Univ New York NY 10027

ALLARA, DAVID LAWRENCE, b Vallejo, Calif, Nov 3, 37; m 68. PHYSICAL ORGANIC CHEMISTRY. Educ: Univ Calif, Berkeley, BS, 59; Univ Calif, Los Angeles, PhD(chem), 64. Prof Exp: NSFfel chem, Oxford Univ, 64-65; fel, Stanford Res Inst, 65-66, org chemist, 66-67; assoc prof chem, San Francisco State Col, 67-69; MEM TECH STAFF, BELL TEL LABS, 69- Mem: AAAS; Am Chem Soc; NY Acad Sci; The Chem Soc. Res: Reactions of organic compounds with molecular oxygen; kinetics of free radical processes; reactions at interfaces. Mailing Add: Bell Tel Labs 7F-212 Murray Hill NJ 07974

ALLARD, CLAUDE, b Montreal, Que, Aug 15, 22; m 46; c 3. BIOCHEMISTRY. Educ: Univ Montreal, PhD(chem), 48. Prof Exp: Res assoc, Montreal Cancer Inst, 50-59; chief dept biochem & epidemiol, Montreal Inst Cardiol, 59-67; dir dept labs, Montreal Heart Inst, 67-74; RES PROF DEPT PHYS EDUC, LAVAL UNIV, 74- Concurrent Pos: Res assoc prof, Univ Montreal, 58-75. Mem: Am Heart Asn; Can Biochem Soc; Can Cardiovasc Soc; Can Soc Clin Invest; Int Soc Cardiol. Res: Epidemiology of cardiovascular disease; nutrition; sports medicine; biochemistry of effort. Mailing Add: Dept of Phys Educ Fac of Educ Sci Laval Univ Quebec PQ Can

ALLARD, GILLES OLIVER, b Rougemont, Que, Dec 12, 27; m 52; c 3. GEOLOGY, ECONOMIC GEOLOGY. Educ: Univ Montreal, BA, 48, BS, 51; Queen's Univ, Ont, MA, 53; Johns Hopkins Univ, PhD(geol), 56. Prof Exp: Consult, Que Dept Natural Resources, 52-55; field mgr, Chibougamau Mining & Smelting, Que, 55-58; asst prof geol, Univ Va, 58-59; prof, Ctr Res & Develop Petrol-Petrobras, Salvador, Brazil, 59-64; vis lectr, Univ Calif, Riverside, 64-65; assoc prof , 65-69, actg head dept, 69-70, PROF GEOL, UNIV GA, 70- Concurrent Pos: Consult, Que Dept Natural Resources, 66- Mem: Fel Geol Soc Am; Mineral Soc Am; Can Inst Mineral & Metall; Mineral Asn Can; Brazilian Geol Soc. Res: Structural and economic geology of Chibougamau,

Quebec; discovery of Henderson Copper Mine, Chibougamau; Dore Lake layered complex; Propria geosyncline, Brazil; geologic link Brazil-Gabon and continental drift; vanadium deposits, Chibougamau. Mailing Add: Dept of Geol Univ of Ga Athens GA 30602

ALLARD, NONA MARY, b Minneapolis, Minn, Dec 23, 28. MATHEMATICS. Educ: Col St Catherine, BA, 50; Cath Univ Am, MA. 60, PhD(math), 67. Prof Exp: Teacher, Cromwell High Sch, Minn, 50-52, Alexander Ramsey High Sch, 53-54 & Visitation High Sch, Ill, 56-58; instr, 60-61, from instr to asst prof, 65-73, ASSOC PROF MATH, ROSARY COL, 74- Mem: Am Math Soc; Math Asn Am. Res: Functional analysis; convex topological algebras and other generalizations of Banach algebras. Mailing Add: Dept of Math Rosary Col 7900 W Division St River Forest IL 60305

ALLARD, ROBERT WAYNE, b Los Angeles, Calif, Sept 3, 19; m 44; c 5. GENETICS. Educ: Univ Calif, BS, 41; Univ Wis, PhD, 46. Prof Exp: From asst prof to prof agron, 46-64, PROF GENETICS, UNIV CALIF, DAVIS, 64-, CHMN DEPT, 67- Concurrent Pos: Guggenheim fels, 54-55 & 60-61; Fulbright sr res fel, 55. Mem: Nat Acad Sci; Genetics Soc am; Am Genetic Asn; Am Soc Naturalists; Am Soc Agron. Res: Population genetics. Mailing Add: Dept of Genetics Univ of Calif Davis CA 95616

ALLARD, ROMEO PAUL, b Franklin, NH, Nov 13, 06; m 35; c 3. CHEMISTRY. Educ: Univ Notre Dame, BS, 31, MS, 32, PhD(phys chem), 34. Prof Exp: Chmn dept, 35-75, EMER CHMN DEPT CHEM, LOYOLA UNIV, LOS ANGELES, 75- Concurrent Pos: Dean col sci, Loyola Univ, Los Angeles, 46-50. Mem: Am Chem Soc; fel Am Inst Chem. Res: Dipole moments; general and physical chemistry. Mailing Add: PO Box 185 Culver City CA 90230

ALLARD, WILLIAM KENNETH, b Lowell, Mass, Oct 29, 41; m 68; c 1. PURE MATHEMATICS. Educ: Villanova Univ, ScB, 63; Brown Univ, PhD(math), 68. Prof Exp: ASST PROF MATH, PRINCETON UNIV, 68- Concurrent Pos: Sloan Found fel, 70-72. Mem: Am Math Soc. Res: Application of geometric measure theoretic techniques to the study of elliptic variational problems. Mailing Add: Fine Hall Princeton Univ Princeton NJ 08540

ALLAS, RICHARD G, b Riga, Latvia, Sept 16, 34; US citizen; m 69. NUCLEAR PHYSICS. Educ: Wash Univ, AB, 54, PhD(physics), 61. Prof Exp: Res assoc physics, Argonne Nat Lab, 61-64; NUCLEAR PHYSICIST, US NAVAL RES LAB, 64- Mem: Am Phys Soc; Am Nuclear Soc. Res: Nuclear spectroscopy. Mailing Add: Code 6611 US Naval Res Lab Washington DC 20390

ALLAWAY, NORMAN C, b New York, NY, Aug 5, 22; m 47; c 3. BIOSTATISTICS. Educ: Brooklyn Col, BA, 42; Columbia Univ, MS, 55. Prof Exp: Jr statistician, New York City Dept Health, 46-51; biostatistician, NY State Dept Health, 51-55, sr biostatistician, 56-57, assoc biostatistician, 57-62, prin biostatistician, NY State Dept Ment Hyg, 62-67, ASST DIR BUR MEDICAID, NY STATE DEPT HEALTH, 67- Mem: Am Statist Asn; Am Pub Health Asn. Res: Medical care; effects of state and local medical care programs. Mailing Add: NY State Dept of Health ESP Tower Bldg Albany NY 12237

ALLAWAY, WILLIAM HUBERT, soil science, see 12th edition

ALLBRIGHT, CHARLES SIMAR, b Rock Rapids, Iowa, Sept 26, 17; m 40; c 2. ANALYTICAL CHEMISTRY. Educ: Iowa State Col, BS, 40. Prof Exp: Lab asst oil well core anal, Core Labs, Inc, 41; chemist, Anderson-Prichard Oil Corp, 41-50; shale oil anal, 50-56, petrol chemist, 56-70, RES CHEMIST, US BUR MINES, 70- Mem: Am Chem Soc; Am Inst Chem; Sigma Xi. Res: Composition of shale oils. Mailing Add: Energy Res & Develop Admin Box 3395 Univ Station Laramie WY 82071

ALLBRITTEN, FRANK F, JR, b Cunningham, Kans, Dec 1, 14; m 40; c 5. SURGERY. Educ: Univ Kans, AB, 35; Univ Pa, MD, 38; Am Bd Surg, dipl, 46; Am Bd Thoracic Surg, dipl, 51. Prof Exp: Asst surgeon, Jefferson Med Col, 47-52, assoc prof surg, 52-54, surg dir, Barton Mem Div, 47-54; prof surg, Univ Kans Med Ctr, Kansas City, 54-72, chmn dept, 54-70; RETIRED. Concurrent Pos: Adj, Pa Hosp, 50-54; surg consult hosps, 54- Mem: AAAS; Soc Clin Surg; AMA; Am Surg Asn; Am Asn Thoracic Surg; Soc Univ Surg; Am Col Surg. Mailing Add: PO Box 177 Cunningham KS 67035

ALLBRITTEN, HERBERT GRAVES, b Murray, Ky, Apr 26, 11; m 39; c 1. SOIL BIOCHEMISTRY, PLANT BIOCHEMISTRY. Educ: Murray State Col, BS, 31; Univ Ky, MS, 41; Pa State Col, PhD(soil chem & technol), 51. Prof Exp: High sch instr, Ky, 32-36; soils surv, Soil Conserv Serv, USDA, Southwest, 41-42; ord training, Picatinny Arsenal, Dover, NJ, 42; asst chief inspector ord, Ark Ord Plant, Little Rock, 42-43; asst agronomist, Ky Agr Exp Sta, 43-44, civilian instr ord, Ft Knox, Ky, 44-46; asst agronomist & exten agronomist, RI Agr Exp Sta, 47-50; assoc agronomist, SC Agr Exp Sta, 51-54, agronomist & dir soil testing lab, 54-59; prof soils & chem, Murray State Col, 59-63; assoc prof chem, 63-68, PROF CHEM, MEMPHIS STATE UNIV, 68- Mem: AAAS; fel Am Inst Chemists; Am Chem Soc; Am Soc Agron; Soil Sci Soc Am. Res: Soil fertility; plant nutrition; crop production; pesticides; chemical methods in soil testing. Mailing Add: Dept of Chem Memphis State Univ Memphis TN 38111

ALLCOCK, HARRY REX, b Loughborough, Eng, Apr 8, 32; m 59. INORGANIC CHEMISTRY, POLYMER CHEMISTRY. Educ: Univ London, BSc, 53, PhD(chem), 56. Prof Exp: Fel inorg chem, Purdue Univ, 56-57; res fel phys-org chem, Nat Res Coun Can, 58-60; res chemist, Stamford Res Labs, Am Cyanamid Co, 61-65, sr res chemist, 65-66; assoc prof, 66-70, PROF CHEM, PA STATE UNIV, 70- Mem: Am Chem Soc; The Chem Soc; Am Crystallographic Asn. Res: Inorganic polymer chemistry; ionic polymerization reactions; biomedical polymers; synthesis; reaction mechanisms; molecular structure studies; phosphorus-nitrogen, organophosphorus, organosilicon and other heteroatom systems. Mailing Add: Dept of Chem Pa State Univ University Park PA 16802

ALLDREDGE, GERALD PALMER, b Hereford, Tex, Oct 20, 35. PHYSICS, MATERIALS SCIENCE. Educ: Tex Technol Col, BA, 58, MS, 60; Mich State Univ, PhD(physics), 66. Prof Exp: From instr to asst prof physics, Southern Ill Univ, 64-68; res scientist, Univ Tex, Austin, 68-75; SR RES INVESTR, GRAD CTR MAT RES, UNIV MO-ROLLA, 75-, ASSOC PROF PHYSICS, 76- Concurrent Pos: Partic, Prof Activ Continuing Educ Prog, Argonne Nat Lab, 65; consult, Columbia Sci Res Inst, Tex, 69- Mem: AAAS; Am Phys Soc; Am Asn Physics Teachers; Am Chem Soc. Res: Theory of condensed matter; physics and chemistry of surfaces and interfaces; chemical physics; statistical mechanics; computational physics; mathematical physics; materials science. Mailing Add: Grad Ctr for Mat Res Univ of Mo Rolla MO 65401

ALLDREDGE, LEROY ROMNEY, b Mesa, Ariz, Feb 6, 17; m 40; c 7. GEOPHYSICS. Educ: Univ Ariz, MS, 40; Harvard Univ, ME, 52; Univ Md, PhD, 55. Prof Exp: Instr physics, Univ Ariz, 40-41; radio inspector, Fed Commun Comn, 41-44; radio engr, Dept Terrestrial Magnetism, Carnegie Inst Technol, 44-45; physicist & div chief, Res Dept, Naval Ord Lab, 45-55; analyst, Opers Res Off, Johns Hopkins Univ, 55-59; geophysicist, Coast & Geodetic Surv, US Dept Com, 59-65; from actg dir to dir, Inst Earth Sci, Environ Sci Serv Admin, 65-70, dir, Earth Sci Labs, Nat Oceanic & Atmospheric Admin Environ Res Labs, 70-73; RES GEOPHYSICIST, US GEOL SURV, DEPT INTERIOR, DENVER, 73- Mem: Am Geophys Union; Int Asn Geomag & Aeronomy. Res: Electricity and magnetism. Mailing Add: US Geol Surv Theoret & Appl Geophys Stop 964 PO Box 25046 Denver Fed Ctr Denver CO 80225

ALLDRIDGE, NORMAN ALFRED, b Eugene, Ore, July 28, 24; m 48; c 4. BOTANY. Educ: Utah State Agr Col, BS, 50; Univ Tex, PhD(bot), 56. Prof Exp: Instr bot, Univ Tex, 56-57; asst prof biol, 57-69, dir biol field sta, 61-69, exec officer dept biol, 64-69, ASSOC PROF BIOL, CASE WESTERN RESERVE UNIV, 69- Mem: AAAS; Am Soc Plant Physiol; Am Soc Cell Biol. Res: Plant growth, development and metabolism; physiology of dwarfism; nature, distribution and function of plant peroxidases; metabolic pathways involved in germination. Mailing Add: Dept of Biol Case Western Reserve Univ Cleveland OH 44106

ALLEE, BRIAN JAMES, b San Francisco, Calif, July 14, 39; m 62; c 2. FISHERIES. Educ: Univ Calif, BS, 65; Univ Wash, PhD(fisheries), 74. Prof Exp: Res asst fisheries, Univ Wash, 67-71; fisheries scientist, Quinault Indian Tribe, 71-73; TECH DIR AQUACULT, NEW BUS RES, WEYERHAEUSER CO, 73- Mem: Am Fisheries Soc; Am Inst Fisheries Res Biologists. Res: Aquaculture research both extensive and intensive; salmonid fish and freshwater shrimp are the subject species; utilization of warm water effluent in culture systems. Mailing Add: Weyerhaeuser Co 3400 13th Ave SW Seattle WA 98134

ALLEE, MARSHALL CRAIG, b Rome, Ga, July 18, 41; m 60; c 3. ENTOMOLOGY, ZOOLOGY. Educ: Shorter Col, Ga, BA, 63; Clemson Univ, MS, 65, PhD(entom), 68. Prof Exp: Instr biol, Clemson Univ, 66-67; ASST PROF BIOL, SHORTER COL, GA, 68-, DEAN MEN, 72- Mem: Entom Soc Am. Res: Response of the face fly, Musca autumnalis DeGeer, to various concentrations of aphotate and tepa; potential of Aleochara tristis Gravenhorst as a means of controlling the face fly. Mailing Add: Dept of Biol Shorter Col Rome GA 30161

ALLEE, PAUL ANDREW, b Indianapolis, Ind, May 18, 19; m 43; c 2. ATMOSPHERIC PHYSICS. Educ: Occidental Col, AB, 48, MA, 50. Prof Exp: Meteorologist, US Weather Bur, 51-52, PHYSICIST, ATMOSPHERIC PHYSICS & CHEM LAB, ENVIRON RES LAB, NAT OCEANIC & ATMOSPHERIC ADMIN, COLO, 52- Mem: Am Meteorol Soc; Sigma Xi; Weather Modification Asn. Res: Measurement of the concentration of atmospheric particulates, and identification of those active on cloud droplet condensation nuclei; ice crystal nuclei; identification of the particulate sources, sinks and effect on cloud physics. Mailing Add: Atmospheric Physics & Chem Lab Nat Oceanic & Atmospheric Admin Boulder CO 80302

ALLEGRE, CHARLES FREDERICK, b Osage City, Kans, Oct 6, 11; m 38. BIOLOGY. Educ: Kans State Teachers Col, BS, 36; Univ Iowa, MS, 41, PhD(zool), 47. Prof Exp: Assoc prof biol & actg head dept, Gustavus Adolphus Col, 47-50; from asst prof to assoc prof, 50-75, PROF BIOL, UNIV NORTHERN IOWA, 75- Mem: AAAS; assoc Am Soc Zool; Am Micros Soc. Res: Protozoology; fresh water and parasitic taxonomy; microbiology; life history of a gregarine parasitic in grasshoppers. Mailing Add: Dept of Biol Univ of Northern Iowa Cedar Falls IA 50613

ALLEMAN, RAY STARR, b Springville, Utah, Jan 22, 12; m 35; c 5. UNDERWATER ACOUSTICS, APPLIED STATISTICS. Educ: Brigham Young Univ, AB, 33, MA, 34; Johns Hopkins Univ, PhD(physics), 39. Prof Exp: Asst physics, Brigham Young Univ, 33-34; jr instr, Johns Hopkins Univ, 34-38, res physicist, Sch Med, 36-37; asst prof physics, St John's Col, Md, 38-41, tutor, 39-41; physicist div war res, Columbia Univ, 41-45 & Univ Calif, 45-46; head div, US Navy Electronics Lab, 46; asst prof physics & math, Brigham Young Univ, 46-48; physicist & sound br head, US Navy Electronics Lab, 49-53; prin physicist, Stone & Smith, Inc, 53-57; tech specialist, Aero-Jet Gen Corp, 57-68; staff scientist, Oceanics Div, Lockheed-Calif, Co, 68-71; physicist, Thompson Ramo Wooldridge, 72-73; PHYSICIST, EG&G WASHINGTON ANAL SERV CTR, 73- Mem: AAAS; fel Am Phys Soc; Acoustical Soc Am; Am Asn Physics Teachers. Res: Ultrasonics. Mailing Add: 11221 Wedge Dr Reston VA 22090

ALLEMAND, CHARLY D, b Soengeiliat, Banka, Sept 2, 24; m; c 1. APPLIED PHYSICS. Educ: Swiss Fed Inst Technol, dipl, 49; Univ Neuchatel, PhD(semiconductors), 54. Prof Exp: Asst solid state physics, Univ Neuchatel, 51-52, chief adv, 52-54; scientist, Fabrique Suisse de Ressorts d'Horlogerie, SA, 54-58, tech mgr metall, 58-60, gen mgr method, time & motion orgn prog, 60-61; mgr eng res, Jarrell-Ash Co, Switz, 62-66, staff scientist, Jarrell-Ash Div, 66-69, Fisher Res Labs, 69-73, SR STAFF SCIENTIST, JARRELL-ASH DIV, FISHER SCI CO, 73- Honors & Awards: Meggers Award, Soc Appl Spectros, 71. Mem: Optical Soc Am; NY Acad Sci; Soc Appl Spectros. Res: Underwater telephony; electronic conduction in dielectric crystals; high tensile strength alloys; automation; controlled electrolytic polishing; excitation sources for spectroscopy; geometrical and physical optics; light detection electronics; micro-analysis systems; automated computer controlled spectrometric analyser. Mailing Add: Fisher Res Labs 590 Lincoln St Waltham MA 02154

ALLEN, A D, b London, Eng, Aug 25, 19; m 40; c 4. INORGANIC CHEMISTRY. Educ: Univ London, BSc, 50, PhD(chem), 53. Prof Exp: Lectr chem, Univ Col, London, 52-57; res chemist, Int Nickel Co Can, 57-59; from asst prof to assoc prof chem, 59-64, from assoc dean to dean arts & sci, 64-74, PROF CHEM, UNIV TORONTO, 64- Concurrent Pos: Res grants, Nat Res Coun Can, 59- & Petrol Res Fund, 64-65. Mem: Brit Chem Soc; Royal Inst Chem; Chem Inst Can. Res: Transition metal compounds; factors affecting bonding and reactivity; reaction mechanisms and spectral data. Mailing Add: Dept of Chem Univ of Toronto Toronto ON Can

ALLEN, AGNES MORGAN, b Normal, Ill, July 17, 01. GEOGRAPHY. Educ: Ill State Univ, BEd, 24; Colo State Col, MA, 25; Clark Univ, MA, 34, PhD(geog), 37. Prof Exp: Teacher, pub schs, Oak Park & Normal, Ill, 19-21; training sch teacher, Nebr State Teachers Col, Chadron, 24-30; asst prof geog, Miss Southern Col, 30-33; asst prof educ geog, 34, prof sci, 48-72, dir Sch Math & Sci, 48-66, prof geog, 66-72, EMER PROF GEOG, NORTHERN ARIZ UNIV, 72- Mem: Fel AAAS; Asn Am Geogrs; Am Geog Soc. Mailing Add: 220 S Beaver St Flagstaff AZ 86001

ALLEN, ALEXANDER CHARLES, b Morristown, NJ, Dec 27, '33; m 60; c 3. PEDIATRICS, NEONATOLOGY. Educ: Haverford Col, BS, 55; McGill Univ, MD, CM, 59; Am Pediat, dipl, 70. Prof Exp: Intern med, Royal Victoria Hosp, Montreal, Que, 59-60; jr asst resident pediat, Montreal Children's Hosp, 60-61; jr asst resident internal med, Royal Victoria Hosp, 61-62; sr asst resident pediat, Montreal Children's Hosp, 62-63; res assoc neonatology, McGill Univ & Royal Victoria Hosp, Montreal, 63-66; ASST PROF PEDIAT, OBSTET & GYNEC, SCH MED, UNIV PITTSBURGH, 69-; DIR, INFANT REFERRAL CTR, MAGEE-WOMEN'S HOSP, 73- Mem: Fel Am Acad Pediat; Soc Comput Med; Soc Pediat Res. Res: Idiopathic respiratory distress syndrome of prematurity; renal acid excretion in sick and well newborn infants; neonatal infections. Mailing Add: Dept of Pediat Magee-Women's Hosp Pittsburgh PA 15213

ALLEN, ALFRED MARSTON, b Glendale, Ohio, Dec 22, 37. EPIDEMIOLOGY, PREVENTIVE MEDICINE. Educ: Princeton Univ, BA, 59; Univ Calif, San Francisco, MD, 63; Univ Calif, Berkeley, MPH, 70. Prof Exp: From field investr med res to mem staff, Walter Reed Army Inst Res, 68-73; MEM STAFF MED RES, LETTERMAN ARMY INST RES, 73- Mem: Fel Am Col Prev Med; AMA; Soc Epidemiol Res. Res: Epidemiology of skin diseases, especially common infections; epidemiology of hepatitis. Mailing Add: Dept of Dermat Res Letterman Army Inst of Res San Francisco CA 94129

ALLEN, ALICE STANDISH, geology, see 12th edition

ALLEN, ANNEKE S, b Dronrijp, Netherlands, Sept 14, 30; US citizen; m 57; c 3. PHYSICAL CHEMISTRY. Educ: Tulane Univ, PhD(phys chem), 55. Prof Exp: Instr chem, Tulane Univ, 54-55; chemist, Cent Labs, Dow Chem Co, Tex, 55-58; pres, Alchem, Fla, 59-63; asst prof, 64-71, ASSOC PROF CHEM, WICHITA STATE UNIV, 71- Concurrent Pos: Asst prof, Orlando Jr Col, 62-63. Mem: AAAS; Am Chem Soc; Sigma Xi. Res: Inorganic reactions in solution; bioinorganic reactions; electrochemistry. Mailing Add: Dept of Chem Wichita State Univ Wichita KS 67208

ALLEN, ANTON MARKERT, b Augusta, Ga, Feb 9, 31; m 51; c 4. VETERINARY PATHOLOGY. Educ: Univ Ga, DVM, 55; Univ Wis, PhD(path), 61. Prof Exp: Asst vet officer, Comp Path Sect, Nat Cancer Inst, 55-57; vet officer, Comp Path & Cytol, Sch Med, Univ Wis, 57-59; asst chief, 59-61, CHIEF COMP PATH SECT, VET RESOURCES BR, NIH, 61- Mem: AAAS; Int Primatol Soc; Am Vet Med Asn; Am Asn Path & Bact; Conf Res Workers Animal Dis. Res: Pathology of primate diseases; mitosis and binucleation of mast cells; pathogenesis of naturally occurring diseases of laboratory animals. Mailing Add: Vet Resources Br Comp Path Sect Bldg 28A NIH Bethesda MD 20014

ALLEN, ARCHIE C, b Ash, NC, Dec 23, 29; m 59; c 2. ZOOLOGY, GENETICS. Educ: Univ NC, BS, 55, MA, 58; Univ Pittsburgh, PhD(zool), 61. Prof Exp: NIH fel genetics, Univ Tex, 61-63; asst prof biol, 63-67, ASSOC PROF BIOL, TEX TECH UNIV, 67- Mem: AAAS; Genetics Soc Am. Res: Recombination, population and lethal studies of Drosophila melanogaster. Mailing Add: Dept of Biol Tex Tech Univ Lubbock TX 79409

ALLEN, ARNOLD ORAL, b Malcolm, Nebr, Aug 2, 29; m 61; c 1. MATHEMATICS. Educ: Univ Nebr, BA, 50; Univ Calif, Los Angeles, MA, 59, PhD(math), 62. Prof Exp: Asst physics, Univ Nebr, 50; comput programmer, IBM Corp, 56-58; asst math, Univ Calif, Los Angeles, 58-59; systs engr, 62-64, mathematician, 64-68, PhD col rels rep, 68-71, SR STAFF MEM SYSTS SCI INST, IBM CORP, 71- Concurrent Pos: Lectr, Univ Calif, Los Angeles, 65-71. Mem: Math Asn Am. Res: Applications of probability, statistics and queueing theory to analysis of computer systems. Mailing Add: IBM Systs Sci Inst 3550 Wilshire Blvd Los Angeles CA 90010

ALLEN, ARTHUR, b Philadelphia, Pa, Nov 20, 28; m 56; c 2. BIOCHEMISTRY. Educ: Temple Univ, BA, 50, MA, 53, PhD(chem), 56. Prof Exp: Chief div chem & toxicol, First US Army Med Lab, NY, 56-58; asst prof, 58-65, ASSOC PROF BIOCHEM, JEFFERSON MED COL, 65- Concurrent Pos: Lectr chem, Temple Univ, 60-67, adj assoc prof, 67-69. Mem: AAAS; Am Chem Soc. Res: Intermediary metabolism of carbohydrates and lipids. Mailing Add: Dept of Biochem Jefferson Med Col Philadelphia PA 19107

ALLEN, ARTHUR CHARLES, b Trenton, NJ, Dec 16, 10; wid. PATHOLOGY. Educ: Univ Calif, MD, 36. Prof Exp: Asst med examr, New York, 46-48; assoc pathologist, Mem Ctr Cancer, 48-57; prof path, Sch Med, Univ Miami, 57-61; DIR LABS, JEWISH HOSP BROOKLYN, 61-; CLIN PROF PATH, STATE UNIV NY DOWNSTATE MED CTR, 61- Concurrent Pos: Consult, Armed Forces Inst Path, 46-49, Bronx Vet Admin Hosp, 46-57, Coral Gables Vet Admin Hosp, 57-61, Hunterdon Med Ctr, 56-, Brooklyn Vet Admin Hosp, Phelps Mem Hosp, Kew Gardens Hosp & Huntington Hosp, Long Island, 69-; assoc prof, Sch Med, Cornell Univ, 53-56. Mem: AAAS; Am Soc Clin Path; Asn Am Med Cols; Am Asn Path & Bact; Col Am Path. Res: Diseases of kidney and skin. Mailing Add: 555 Prospect Pl Brooklyn NY 11238

ALLEN, ARTHUR DELOS, animal physiology, pharmacology, see 12th edition

ALLEN, ARTHUR (SILSBY), b Brewer, Maine, Mar 5, 34; m 58; c 2. PLANT PATHOLOGY, MYCOLOGY. Educ: Univ Maine, BS, 56, MS, 60; Mich State Univ, PhD(phytopath), 68. Prof Exp: Asst bot, Univ Maine, 58-60; asst phytopath, Mich State Univ, 60-63, res assoc & asst instr, 65-68; asst prof bot, Humboldt State Col, 63-65; ASSOC PROF PHYTOPATH, NORTHWESTERN STATE UNIV, 68- Concurrent Pos: Pres, La Agr Consults Asn, 75-76. Mem: Am Phytopath Soc. Res: Host-pathogen relations; disease resistance; epidemiology and control of pecan diseases. Mailing Add: Dept of Biol Sci Northwestern State Univ Natchitoches LA 71457

ALLEN, ARTHUR T, JR, b Darlington, SC, Sept 8, 17; m 40; c 1. GEOLOGY. Educ: Emory Univ, AB, 39; Univ Tenn, MS, 47; Univ Colo, PhD(geol), 50. Prof Exp: Mining geologist, Am Zinc Co, 40-46; from asst prof to assoc prof geol, 46-58, PROF GEOL, EMORY UNIV, 58-, CHMN DEPT, 64- Mem: Am Asn Petrol Geologists; Geol Soc Am. Res: Sedimentation; stratigraphy; paleontology. Mailing Add: Dept of Geol Emory Univ Atlanta GA 30307

ALLEN, ASHAEL LESTER, b Los Angeles, Calif, Sept 24, 23; m 43; c 11. ZOOLOGY. Educ: Univ Calif, Los Angeles, AB, 46, PhD(zool), 51. Prof Exp: Instr life sci, Orange Coast Col, 51-54; prof zool, 54-65, chmn dept, 65-69, DEAN COL BIOL & AGR SCI, BRIGHAM YOUNG UNIV, 69- Res: Radiation biology; embryology; genetics. Mailing Add: Col of Biol & Agr Sci Brigham Young Univ Provo UT 84602

ALLEN, AUGUSTINE OLIVER, b San Rafael, Calif, July 16, 10; m 38; c 5. PHYSICAL CHEMISTRY, RADIATION CHEMISTRY. Educ: Univ Calif, BS, 30; Harvard Univ, PhD(phys chem), 38. Prof Exp: Res chemist, Bell Tel Labs, NY, 30-31; fel, Harvard Univ, 31-35; res chemist, Ethyl Corp, Detroit, 37-43; group leader & assoc sect chief, metall lab, Univ Chicago, 43-46; prin chemist, Oak Ridge Nat Lab, 46-48; SR SCIENTIST, BROOKHAVEN NAT LAB, 48- Concurrent Pos: Vis scientist, Int Atomic Energy Agency, Greece, 65-66, Danish Atomic Energy Comn Lab, 70-71 & Interuniv Reactor Inst, Delft, Netherlands, 75. Mem: AAAS; Am Chem Soc; Radiation Res Soc (pres, 63-64). Res: Reaction kinetics; electronic properties of liquids. Mailing Add: Brookhaven Nat Lab Upton NY 11973

ALLEN, BONNIE L, b Hillsboro, Tex, Aug 11, 24. SOIL GENESIS, SOIL MINERALOGY. Educ: Tex Tech Univ, BS, 48; Mich State Univ, MS, 51, PhD(soil sci), 60. Prof Exp: Asst prof agron, Eastern NMex Univ, 52-57, assoc prof soils & geol, 57-59; assoc prof, 59-72, PROF SOILS, TEX TECH UNIV, 72- Concurrent Pos: Rockefeller Found grants, 61-64. Mem: Am Soc Agron; Soil Sci Soc Am; Am Chem Soc; Clay Minerals Soc; Int Soc Soil Sci. Res: Mineralogical transformations and micromorphological changes with soil development; soil-nutritive vegetation relationships. Mailing Add: Dept of Agron Tex Tech Univ Lubbock TX 79409

ALLEN, CHARLES EUGENE, b Burley, Idaho, Jan 25, 39; m 60; c 2. ANIMAL SCIENCE, BIOCHEMISTRY. Educ: Univ Idaho, BS, 61; Univ Wis, MS, 63, PhD(animal sci), 66. Prof Exp: NSF fel, Div Food Preservation, Commonwealth Sci & Indust Res Orgn, Australia, 66-67; from asst prof to assoc prof, 67-72, PROF ANIMAL SCI, UNIV MINN, ST PAUL, 72- Concurrent Pos: Consult, Am Cyanamid Co, 74- Mem: Am Soc Animal Sci; Am Meat Sci Asn; Inst Food Technol. Res: Lipid deposition and composition; mechanisms related to growth and development of muscle and adipose tissue. Mailing Add: Dept of Animal Sci Univ of Minn St Paul MN 55108

ALLEN, CHARLES FRANCIS HITCHCOCK, b Milford, NH, Aug 12, 95; m 20; c 2. ORGANIC CHEMISTRY. Educ: Boston Univ, AB, 19; Harvard Univ, AM & PhD(org chem), 24. Hon Degrees: DSc, McGill Univ, 37 & Boston Univ, 44. Prof Exp: Asst chem, Boston Univ, 20-21; instr, Tufts Col, 24-29; asst prof, McGill Univ, 29-37; asst supt in chg org & polymer res, Eastman Kodak Co, 37-61; PROF CHEM, ROCHESTER INST TECHNOL, 61- Mem: Am Chem Soc. Res: Photographic chemicals and dyes; delta ketonic nitriles; vinyl phenyl ketone; carbonyl bridge compounds; polynuclear systems; heterocyclic compounds; continuous reactors; Michael reaction. Mailing Add: Dept of Chem Rochester Inst of Technol Rochester NY 14623

ALLEN, CHARLES FREEMAN, b Berkeley, Calif, Feb 16, 28; m 50; c 3. ORGANIC CHEMISTRY, BIOCHEMISTRY. Educ: Univ Calif, BS, 48; Univ Wis, PhD(org chem), 52. Prof Exp: Res chemist, Univ Calif, 52-54, instr, 53-54; from asst prof to assoc prof chem, 54-66, PROF CHEM, POMONA COL, 66- Concurrent Pos: NSF fac fel, Univ Cologne, 61-62, Kettering Res Lab, Ohio & Scripps Inst Oceanog, 68-69. Mem: Am Chem Soc. Res: Lipid-protein interactions; biological membranes; photosynthesis. Mailing Add: Dept of Chem Pomona Col Claremont CA 91711

ALLEN, CHARLES MARSHALL, JR, b Cortland, NY, Sept 13, 38; m 63; c 2. BIOCHEMISTRY. Educ: Syracuse Univ, BS, 60; Brandeis Univ, PhD(biochem), 64. Prof Exp: USPHS fel chem, Harvard Univ, 64-67; asst prof, 67-73, ASSOC PROF BIOCHEM, UNIV FLA, 73- Mem: Am Chem Soc; Am Soc Biol Chemists. Res: Biosynthesis of long chain polyprenyl phosphates which function in glycosyl transfer reactions; biosynthesis of isoprenoid containing fungal metabolites derived from tryptophan containing cyclic dipeptides. Mailing Add: Dept of Biochem Univ of Fla Gainesville FL 32601

ALLEN, CHARLES MARVIN, b Mt Gilead, NC, July 31, 18; m 43. CYTOLOGY, INVERTEBRATE ZOOLOGY. Educ: Wake Forest Col, BS, 39, MA, 41; Duke Univ, PhD, 55. Prof Exp: From instr to assoc prof, 41-62, PROF BIOL, WAKE FOREST UNIV, 62-, DIR CONCERTS & LECTS, 74- Mem: AAAS. Res: Chromosome morphology and germ cell cycles; morphology and cytology of marine annelids. Mailing Add: Wake Forest Univ Dept of Biol Box 7211 Reynolds Sta Winston-Salem NC 27106

ALLEN, CHARLES ROBERT, b Bowling Green, Ky, June 26, 11; m 35; c 3. ANESTHESIOLOGY. Educ: Western Ky State Teachers Col, BS, 32, MA, 33; Univ Wis, PhD(physiol), 41, MD, 46. Prof Exp: Teacher, Ky, 35-38; instr physiol, Univ Wis, 40-42; from asst prof to assoc prof, 42-53, PROF ANESTHESIOL & CHMN DEPT, UNIV TEX MED BR GALVESTON, 53-; CHIEF ANESTHESIOL, JOHN SEALY HOSP, 53- Concurrent Pos: Attend anesthetist, John Sealy Hosp, 42-53. Mem: Am Physiol Soc. Res: Physiology and pharmacology of anesthetic drugs; protection from cyclopropane-adrenalin tachycardia by various drugs. Mailing Add: Dept of Anesthesiol Univ of Tex Med Br Galveston TX 77551

ALLEN, CHERYL, b Detroit, Mich, Jan 30, 39. BIOCHEMISTRY. Educ: Nazareth Col, BS, 69; Univ Ill Med Ctr, PhD(biochem), 74. Prof Exp: Sec teacher math & sci, Detroit Parochial Sch Syst, 59-68; teaching asst, Univ Ill Med Ctr, 69-73; ASST PROF BIOCHEM, NAZARETH COL, 74- Mem: Am Chem Soc; AAAS. Res: The role of serum proteins, especially the alpha-2 globulins, in immunity and host defense. Mailing Add: Dept of Biochem Nazareth Col Nazareth MI 49074

ALLEN, CHRISTOPHER WHITNEY, b Waterbury, Conn, Oct 19, 42; m 65; c 2. INORGANIC CHEMISTRY. Educ: Univ Conn, BA, 64; Univ Ill, MS, 66, PhD(inorg chem), 67. Prof Exp: Asst prof, 67-72, ASSOC PROF CHEM, UNIV VT, 72- Mem: Am Chem Soc. Res: Synthesis, structure and spectroscopic properties of non-metal and organometallic compounds. Mailing Add: 20 Grandview Ave Essex Junction VT 05452

ALLEN, CLARENCE RODERIC, b Palo Alto, Calif, Feb 15, 25. GEOLOGY, SEISMOLOGY. Educ: Reed Col, BA, 49; Calif Inst Technol, MS, 51, PhD(geol), 54. Prof Exp: Asst prof geol, Univ Minn, 54-55; from asst prof to assoc prof, 55-64, interim dir seismol lab, 65-67, actg chmn div geol sci, 67-68, chmn fac, 70-71, PROF GEOL & GEOPHYS, CALIF INST TECHNOL, 64- Concurrent Pos: Consult, Calif Dept Water Resources, US Geol Surv & UNESCO; mem adv panel earth sci, NSF, 65-68, chmn, 67-68, mem adv panel environ sci, 70-; mem, Calif State Mining & Geol Bd, 69-; mem task force earthquake hazard reduction, Off Sci & Technol, Exec Off President, 70; chmn panel on earthquake prediction, Nat Acad Sci, 74-; chmn, Calif State Mining & Geol Bd, 75- Honors & Awards: Gilbert Award, 60. Mem: Nat Acad Sci; Soc Explor Geophys; Geol Soc Am (pres, 73-74); Seismol Soc Am (pres, 75-76); fel Am Acad Arts & Sci. Res: Mechanics of faulting; San Andreas fault system; relation of seismicity to geologic structure; geophysical exploration of glaciers; tectonics of Southern California and Baja California; Circum-Pacific earthquakes and faulting; micro-earthquakes; geologic hazards. Mailing Add: Seismol Lab Calif Inst of Technol Pasadena CA 91125

ALLEN, CLIFFORD MARSDEN, b Winnipeg, Man, Dec 14, 27; m 60; c 3. GEOLOGY, GEOCHEMISTRY. Educ: Univ Man, BSc, 49, MSc, 51. Prof Exp: From asst prof to assoc prof, 54-75, PROF GEOL, MT ALLISON UNIV, 75- Mem: Geol Soc Am; Nat Asn Geol Teachers; fel Geol Asn Can; fel Geol Soc London. Res: Igneous and metamorphic petrology; volcanic geology of the Maritime Provinces of Canada. Mailing Add: Dept of Geol Mt Allison Univ Sackville NB Can

ALLEN, CLIFFORD V, b Akron, Iowa, Mar 25, 08; m 34. NUCLEAR MEDICINE. Educ: Univ Iowa, MD, 33. Prof Exp: Pvt pract, 34-51; chief isotopes & radiother, Vet Admin Hosp, Portland, Ore, 55-60; prof & dir radiother, 60-72, chmn dept radiation ther, 67-72, EMER PROF RADIOTHER, MED SCH, UNIV ORE, 72- Concurrent Pos: Mem staff, Vet Admin Hosp, Portland, 60-64. Mem: Soc Nuclear Med; fel Am Col Radiol; Radiol Soc NAm; AMA; Am Radium Soc. Res: Radiation therapy from conventional and isotope sources. Mailing Add: 14031 Lakeforest Dr Sun City AZ 85351

ALLEN, DAVID MITCHELL, b Sebree, Ky, July 15, 38; m 64; c 3. STATISTICS. Educ: Univ Ky, BS, 61; NC State Univ, MES, 66, PhD(statist), 68. Prof Exp: Asst prof, 67-72, ASSOC PROF STATIST, UNIV KY, 72- Concurrent Pos: Consult, Mead Johnson & Co, 68-; vis assoc prof, Cornell Univ, 74-75; assoc ed, Biometrics, 75- Mem: Biomet Soc; Inst Math Statist; Am Statist Asn. Res: Statistical computation; linear models; prediction. Mailing Add: Dept of Statist Univ of Ky Lexington KY 40506

ALLEN, DELORAN MATTHEW, b Cherryvale, Kans, Feb 28, 39; m 65. ANIMAL SCIENCE, MEAT SCIENCE. Educ: Kans State Univ, BSAgr, 61; Univ Idaho, MSAgr, 63; Mich State Univ, PhD(meat sci), 66. Prof Exp: Asst prof animal sci, 66-70, ASSOC PROF ANIMAL SCI & INDUST & ASSOC ANIMAL SCIENTIST, KANS STATE UNIV, 70- Mem: Am Soc Animal Sci; Am Meat Sci Asn. Res: Beef carcass composition and cutability. Mailing Add: Dept of Animal Sci & Indust Kans State Univ Manhattan KS 66506

ALLEN, DON LEE, b Burlington, NC, Mar 13, 34; m 58; c 3. DENTISTRY, PERIODONTOLOGY. Educ: Univ NC, DDS, 59; Univ Mich, MS, 64. Prof Exp: From instr to prof periodont, Sch Dent, Univ NC, Chapel Hill, 59-70, assoc dean, 69-70; assoc dean col dent, 70-74, PROF PERIODONT & DEAN COL DENT, UNIV FLA, 74- Concurrent Pos: Consult, USPHS Dent Asst Teaching Prog, Univ NC, 60-62 & Vet Admin Hosp, Gainesville, Fla. Mem: Fel Am Col Dent; fel Int Col Dent; Am Dent Asn; Am Acad Periodont; Int Asn Dent Res. Res: Clinical periodontics; stimulus for osteogenesis; tissue response in guinea pig to sterile and nonsterile calculus. Mailing Add: Univ of Fla Col of Dent Gainesville FL 32601

ALLEN, DONALD ORRIE, b Belding, Mich, Jan 11, 39; m 61; c 4. PHARMACOLOGY. Educ: Ferris State Col, BS, 62; Marquette Univ, PhD(pharmacol), 67. Prof Exp: Res assoc pharmacol, Sch Med, Ind Univ, Indianapolis, 67-68, from asst prof to assoc prof, 68-75; PROF & COORDR PHARMACOL, SCH MED, UNIV SC, 75- Mem: Am Soc Pharmacol & Exp Therapeut. Res: Autonomic pharmacology and drug interaction; biochemical pharmacology. Mailing Add: Dept of Pharmacol Univ of SC Sch of Med Columbia SC 29208

ALLEN, DONALD STEWART, b Saugus, Mass, Sept 9, 11; m 42; c 2. CHEMISTRY. Educ: Dartmouth Col, AB, 32, AM, 34; Yale Univ, PhD(chem), 43. Prof Exp: Instr chem, Dartmouth Col, 32-34; master sci & math, Tex Country Day Sch, 34-40; from asst instr to instr chem, Yale Univ, 41-43; asst prof chem, Bates Col, 43-45; prof chem & chmn div natural sci, State Teachers Col, New Paltz, 45-59; prof chem, State Univ NY Albany, 59-68, chmn dept, 59-63; PROF CHEM & DIR DIV SCI & MATH, EISENHOWER COL, 68- Concurrent Pos: Res scientist, Oak Ridge Inst Nuclear Studies, 59; consult, State Univ NY Ford Found Proj, Indonesia, 63-65; vis scholar hist & philos sci, Cambridge Univ, 74-75. Mem: Am Chem Soc; fel Am Inst Chemists. Res: Friedel-Crafts reaction; latex research; standard electrode potentials; world studies science. Mailing Add: Div Sci & Math Eisenhower Col Seneca Falls NY 13148

ALLEN, DOUGLAS CHARLES, b Brattleboro, Vt, Mar 8, 40; m 65; c 2. INSECT ECOLOGY. Educ: Univ Maine, BS, 62, MS, 65; Univ Mich, PhD(forest entom), 68. Prof Exp: ASSOC PROF FOREST INSECT ECOL, STATE UNIV NY COL ENVIRON SCI & FORESTRY, 68- Mem: Entom Soc Am; Soc Am Foresters; Entom Soc Can. Res: Population dynamics and bionomics of forest insects. Mailing Add: State Univ of NY Col of Environ Sci & Forestry Syracuse NY 13210

ALLEN, DUFF SHEDERIC, JR, b St Louis, Mo, Dec 8, 28; m 60; c 3. ORGANIC CHEMISTRY. Educ: Princeton Univ, BA, 49; Univ Wis, PhD(org chem), 60. Prof Exp: Res chemist, Monsanto Chem Co, 52-53; asst org chem, Univ Wis, 54-58, res assoc, 58-60; res chemist, Org Chem Div, 60-62, group leader, Lederle Labs, 63-65, HEAD DEPT ORG CHEM PROCESS RES, PROCESS & PREP RES SECT, LEDERLE LABS, AM CYANAMID CO, 65- Mem: Am Chem Soc; AAAS; NY Acad Sci. Res: Organic synthesis of steroids; heterocycles; natural products; medicinal chemicals. Mailing Add: 178 Byram Shore Rd Byram CT 10573

ALLEN, DURWARD LEON, b Uniondale, Ind, Oct 11, 10; m 35; c 3. VERTEBRATE ECOLOGY. Educ: Univ Mich, AB, 32; Mich State Univ, PhD(zool), 37. Hon Degrees: LHD, Northern Mich Univ, 71. Prof Exp: Res biologist, game div, Mich State Dept Conserv, 37-46; biologist in charge wildlife invest on agr lands, US Fish & Wildlife Serv, Washington, DC, 46-51, asst chief, bur wildlife res, 51-54; assoc prof wildlife mgt, 54-57, PROF WILDLIFE ECOL, PURDUE UNIV, 57- Concurrent Pos: Asst secy gen, Inter-Am Conf Renewable Natural Resources, 48; mem comn land use & wildlife, Nat Acad Sci, 66-69; mem adv bd nat parks, US Dept Interior, 66-72, chmn adv comt, Fish, Wildlife & Parks, 75- Honors & Awards: Awards, Wildlife Soc, 44 & 55, Leopold Award, 68. Mem: Hon mem Wildlife Soc (pres, 56-57); Ecol Soc Am; Am Soc Mammalogists. Res: Ecological and management studies on fox squirrel, skunk, rabbit, pheasant, moose and wolf; population biology of vertebrates; predator-prey relationships; human ecology; management of renewable natural resources; ecology of the wolf and its prey in Isle Royale Nat Park. Mailing Add: Dept Forestry & Natural Resources Purdue Univ Lafayette IN 47907

ALLEN, EDWARD FRANKLIN, b Denver, Colo, Aug 28, 07; m 30; c 2. PHYSICS. Educ: Acad New Church, BS, 28; Univ Pa, MA, 32. Prof Exp: Instr math & physics, 28-45, head dept math & phys sci, 52-69, PROF MATH, ACAD OF THE NEW CHURCH, 45-, PROF PHILOS & CHMN DEPT, 69- Concurrent Pos: Res physicist, Franklin Inst, 42-45; ed, New Philos, 61-70. Honors & Awards: Naval Ord Develop Award, 45; Glencairn Award, 58. Mem: AAAS; Am Asn Physics Teachers. Res: History and philosophy of science; instrumentation; specialization in the philosophy of Swedenborg. Mailing Add: Dept of Philos Acad of the New Church Bryn Athyn PA 19009

ALLEN, EDWARD HAROLD, biochemistry, see 12th edition

ALLEN, EDWARD SWITZER, b Kansas City, Mo, Dec 12, 87; m 15; c 3. MATHEMATICS. Educ: Harvard Univ, AB, 09, AM, 10, PhD(math), 14; Univ Rome, 11-13; Univ Göttingen, 12; Univ Berlin, 31-32. Prof Exp: Instr math, Dartmouth Col, 13-14, Brown Univ, 14-15 & Univ Mich, 15-19; asst prof math, WVa Univ, 19-21; assoc prof, 21-43, PROF MATH, IOWA STATE UNIV, 43- Concurrent Pos: Vis prof, Grinnell Col, 60-62, Cottey Col, 64-65 & Wartburg Col, 67-69. Mem: Am Math Soc; Math Asn Am; Ger Math Asn. Res: Algebraic geometry; probability; mathematics applied to chemistry. Mailing Add: Dept of Math Iowa State Univ Ames IA 50010

ALLEN, EMMA GATES, b Bridgeton, NJ, Dec 12, 20. MICROBIOLOGY, VIROLOGY. Educ: Philadelphia Col Pharm, BSc, 42; Univ Pa, MSc, 50, PhD(med microbiol), 52. Prof Exp: Bacteriologist, Charles E Hires, Co, Pa, 42-44; asst resident, Children's Hosp Philadelphia, 47-52; asst, Res Found, State Univ NY, 52-58; from asst prof to assoc prof biol, Long Island Univ, 58-63; asst prof, 63-67, ASSOC PROF MICROBIOL & IMMUNOL, STATE UNIV NY DOWNSTATE MED CTR, 67- Concurrent Pos: NIH spec res fel, 70-71. Mem: Am Soc Microbiol. Res: Viral enzymatic activity; host response to infectious agents. Mailing Add: 319 Main St Newport NJ 08345

ALLEN, EMORY RAWORTH, b Augusta, Ga, Jan 21, 35; m 59; c 3. ANATOMY, CELL BIOLOGY. Educ: Univ Md, BS, 59; Univ Pa, PhD(develop anat), 64. Prof Exp: Asst instr anat, Univ Pa, 62-65; from instr to asst prof anat & cell biol, Sch Med, Univ Pittsburgh, 65-71; ASSOC PROF ANAT, LA STATE UNIV MED CTR, NEW ORLEANS, 71- Concurrent Pos: Fel anat, Univ Pa, 64-65. Res: Ultrastructure of developing muscle; myosin synthesis. Mailing Add: Dept of Anat La State Univ Med Ctr New Orleans LA 70112

ALLEN, ERIC RAYMOND, b Teneriffe, Canary Isles, Apr 12, 32; m 61; c 2. PHYSICAL CHEMISTRY. Educ: Univ Liverpool, BSc, 56, PhD(phys chem), 59. Prof Exp: Fel gas kinetics, Nat Res Coun Can, 59-61; fel photochem, Univ Calif, Riverside, 61-63; res chemist, Nat Ctr Atmospheric Res, 63-68, prog scientist, 68-74; SR RES ASSOC, ATMOSPHERIC SCI RES CTR, STATE UNIV NY ALBANY, 74- Concurrent Pos: Vis assoc prof dept civil eng, Univ Tex, Austin, 71; mem, panel on carbon monoxide, Div Med Sci, Nat Acad Sci-Nat Res Coun, lectr, Environ Studies Prog, State Univ NY Albany & adj assoc prof innovative & interdisciplinary studies, 74- Mem: AAAS; Am Chem Soc; Am Geophys Union; Sci Res Soc Am; Brit Chem Soc. Res: Gas phase kinetics; atomic and free radical reactions; photochemistry; atmospheric and air pollution chemistry; water chemistry; solar energy storage systems. Mailing Add: Atmospheric Sci Res Ctr State Univ NY 1400 Washington Ave Albany NY 12222

ALLEN, ERNEST E, b Detroit, Mich, Sept 7, 33; m 57; c 3. MATHEMATICS EDUCATION. Educ: Wayne State Univ, BA, 55; Mich State Univ, BA, 57, MA, 59; Univ Detroit, MATM, 63; Univ Northern Colo, EdD(math educ), 70. Prof Exp: Pub sch teacher, 58-60; asst prof math educ, Eastern Mich Univ, 60-63; ASSOC PROF MATH, SOUTHERN COLO STATE COL, 63- Concurrent Pos: Dir, Colo Coun Teachers Math, 68-70, ed, 70- Res: Selected characteristics of junior high level mathematics teachers; methods of teaching mathemathematics. Mailing Add: Dept of Math Southern Colo State Col Pueblo CO 81005

ALLEN, ERNEST MASON, b Terrell, Tex, Dec 1, 04; m 28; c 3. PUBLIC HEALTH ADMINISTRATION. Educ: Emory Univ, PhB, 26, MA, 39, DSc, 56. Hon Degrees: LLD, Clemson Univ, 68. Prof Exp: Instr French, Jr Col Augusta, 26-41; proj mgr, Nat Youth Admin, 41-43; sr pub health rep, Div Venereal Dis, USPHS, 43-46; asst chief div res grants, NIH, 46-51, chief div res grant, 51-60, assoc dir, Insts, 60-63; grants policy officer, 63-68, dir off extramural prog, Off Asst Secy Health & Sci Affairs, 68-69, dir div grants admin policy, 69-70, dep asst secy grant admin policy, 70-73, ASSOC DIR NAT LIBR MED, USPHS, 73- Mem: AAAS; Asn Mil Surg US. Mailing Add: 8507 Hazelwood Dr Bethesda MD 20014

ALLEN, EUGENE (MURRAY), b Newark, NJ, Nov 7, 16; m 37; c 2. PHYSICAL CHEMISTRY, COLOR SCIENCE. Educ: Columbia Univ, BA, 38; Stevens Inst Technol, MS, 44; Rutgers Univ, PhD(chem), 52. Prof Exp: Res chemist, United Color & Pigment Co, 39-41 & E R Squibb & Sons, 41-42; anal res chemist, Picatinny Arsenal, 42-45; anal res chemist, Am Cyanamid Co, NJ, 45-56, group leader, 56-61, sr res scientist, 61-63, res assoc, 63-66, res fel, 66-67, sr res award, 58; PROF CHEM, LEHIGH UNIV, 67- Concurrent Pos: Consult, Int Comn Illum, 67-; mem US nat comt, Int Comn on Illumination, 67- Mem: Am Chem Soc; Optical Soc Am; Am Asn Textile Chem & Colorists; Inter Soc Color Coun. Res: Spectrophotometry; measurement and specification of color; colorimetry of fluorescent substances; color matching by digital computer; radiative transfer theory; thermodynamics. Mailing Add: Color Sci Lab Lehigh Univ Ctr for Surface Coatings & Res Bethlehem PA 18015

ALLEN, FRANK B, b Mt Vernon, Ill, Oct 21, 09; m 43; c 3. MATHEMATICS, PHYSICS. Educ: Southern Ill Univ, BEd, 29; Univ Iowa, MS, 34. Prof Exp: High sch teacher, Ill, 29-41; teacher math, Lyons Twp High Sch & Jr Col, 41-42, 46-56, chmn dept, 56-68; assoc prof, 68-74, chmn dept, 70-74, PROF MATH, ELMHURST COL, 74- Mem: Math Asn Am. Res: Application of elementary logic to the improvement of exposition in school mathematics. Mailing Add: Dept of Math Elmhurst Col Elmhurst IL 60126

ALLEN, FRANK JOSEPH, b Brooklyn, NY, Feb 8, 21; m 48; c 3. PHYSICS. Educ: Clarkson Col Technol, BS, 43; Columbia Univ, MA, 49. Prof Exp: Physicist, electronics & mech, penetration & fragmentation br, 48-51, mech, spec probs br, 51-55, physicist, chief theory sect & dep chief nuclear physics br, 55-63, actg chief, laser effects br, 63-64, CHIEF, LASER EFFECTS BR, BALLISTIC RES LABS, 64- Honors & Awards: Sustained Super Performance Award, 63. Mem: Am Phys Soc; Am Nuclear Soc. Res: Applied mechanics; elasticity; vibrations; nuclear weapon phenomenology; terminal ballistics including radiation shielding; neutron and gamma ray transport theory; laser effects; plasma physics. Mailing Add: Laser Effects Br Ballistic Res Labs Aberdeen Proving Ground MD 21005

ALLEN, FRANK LLUBERAS, b Ponce, PR, July 5, 25; m 51; c 2. INFORMATION SCIENCE. Educ: Univ Chicago, PhB & BS, 46, PhD(biophys), 53. Prof Exp: Res asst physics, Univ Chicago, 44-46; physicist, Los Alamos Sci Lab, NMex, 46-47; res assoc biophys, Univ Chicago, 53-54; sr staff, 55-65, sect head info systs, 65-72, VPRES INFO SYSTS, ARTHUR D LITTLE, INC, 72- Res: Operations research; business data processing; weapons systems evaluation; project management; computers and management information systems. Mailing Add: 35 Acorn Park Cambridge MA 02140

ALLEN, FRED ERNEST, b Everett, Mass, Dec 27, 10; m 33; c 8. VETERINARY MEDICINE. Educ: Univ NH, BS, 32; Ohio State Univ, DVM, 36. Prof Exp: Jr veterinarian, US Bur Animal Indust, 36-37; munic veterinarian, City of Columbus, Ohio, 37-40; PROF ANIMAL SCI & VETERINARIAN, AGR EXP STA, UNIV NH, 40- Mem: NY Acad Sci. Res: Bovine mastitis; immunization studies of staphylococcal type. Mailing Add: Kendall Hall Univ of NH Durham NH 03824

ALLEN, FRED HAROLD, JR, b Holyoke, Mass, Feb 23, 12; m 38; c 4. IMMUNOHEMATOLOGY. Educ: Amherst Col, AB, 34; Harvard Univ, MD, 38. Prof Exp: Assoc dir, Blood Grouping Lab Boston, Mass, 47-63; SR INVESTR, NEW YORK BLOOD CTR, 63- Concurrent Pos: Asst clin prof, Harvard Univ, 58-63; clin assoc prof, Med Col, Cornell Univ, 63; ed-in-chief, N & SAm Vox Sanguinis, 63- Mem: Soc Pediat Res; Am Soc Human Genetics. Res: Erythroblastosis fetalis; serology and genetics of human blood groups. Mailing Add: 310 E 67th St New York NY 10021

ALLEN, FRED WILLIAM, b Kansas City, Kans, Feb 27, 02; m 28; c 3. PHYSIOLOGY. Educ: Univ Kans, AB, 26, AM, 27, PhD(bact), 33. Prof Exp: From asst prof to assoc prof biol, Univ NMex, 29-39; supt, NMex Indust Sch for Boys, 39-

45; prof biol & head dept, Southwestern State Col, 47-67; prof microbiol, chmn dept & dir clin labs, 67-72, EMER PROF MICROBIOL, KANSAS CITY COL OSTEOP MED, 72- Mem: AAAS; Am Soc Parasitol. Res: Immunity; blood groups. Mailing Add: Rte 1 Forsyth MO 65653

ALLEN, FREDDIE LEWIS, b Jellico, Tenn. AGRONOMY. Educ: Tenn Technol Univ, BS, 70; Va Polytech Inst & State Univ, MS, 72; Univ Minn, PhD(plant breeding), 75. Prof Exp: ASST PROF AGRON, UNIV TENN, KNOXVILLE, 75- Mem: Am Soc Agron; Crop Sci Soc Am; Am Soybean Asn. Res: Soybean breeding and genetics with emphasis on morphological, anatomical and physiological characteristics as they affect yield. Mailing Add: Dept of Plant & Soil Sci Univ of Tenn Knoxville TN 37916

ALLEN, FREDERICK GRAHAM, b Boston, Mass, Feb 2, 23; m 49; c 3. SOLID STATE PHYSICS. Educ: Cornell Univ, BME, 44; Harvard Univ, PhD(appl physics), 56. Prof Exp: Mem tech staff, Bell Tel Res Labs, NJ, 55-65; dept head manned space flight exp prog, Bellcomm, Inc, Washington, DC, 67-69; PROF ELEC SCI & ENG & CHMN DEPT, UNIV CALIF, LOS ANGELES, 69- Mem: Fel Am Phys Soc; Inst Elec & Electronics Engrs. Res: Physics of semiconductor and metal surfaces; photoemission. Mailing Add: Sch of Eng & Appl Sci Univ Calif Los Angeles CA 90024

ALLEN, GARLAND EDWARD, III, b Louisville, Ky, Feb 13, 36; m 66. HISTORY OF SCIENCE, HISTORY OF BIOLOGY. Educ: Univ Louisville, AB, 57; Harvard Univ, AMT, 58, AM, 64, PhD(hist sci), 66. Prof Exp: Allston-Burr sr tutor & instr hist sci, Harvard Univ, 65-67; asst prof, 67-72, ASSOC PROF BIOL, WASH UNIV, 72- Concurrent Pos: Consult, Educ Res Coun, Ohio, 67; comnr, Comn Undergrad Educ Biol Sci, 65-68; mem div soc sci, NSF Panel, 68-70. Mem: AAAS; Hist Sci Soc; Am Asn Hist Med. Res: History of late 19th and early 20th century biology, especially genetics and evolution; history of eugenics in the 20th century. Mailing Add: Dept of Biol Wash Univ St Louis MO 63130

ALLEN, GARY CURTISS, b Stockton, Calif, July 18, 39; m 65; c 2. GEOCHEMISTRY, PETROLOGY. Educ: Stanford Univ, BS, 61; Rice Univ, MA, 63; Univ NC, PhD(geochem), 68. Prof Exp: Geologist, Va Div Mineral Resources, 66-68; asst prof earth sci, La State Univ, New Orleans, 68-72; ASSOC PROF EARTH SCI, UNIV NEW ORLEANS, 72- Concurrent Pos: Consult mineral, Tulane Univ Med Sch, 72-; consult several indust firms, 71- Mem: Am Chem Soc; Geochem Soc; Geol Soc Am; Sigma Xi; Mineral Soc Am. Res: Petrogenic geochemistry; mineralogy; analytical chemistry of major and trace elements in rocks and minerals; medical geochemistry; archeological geochemistry. Mailing Add: Dept of Earth Sci Univ of New Orleans Lake Front New Orleans LA 70122

ALLEN, GARY IRVING, b Lockport, NY, Apr 7, 42; m 64; c 2. NEUROPHYSIOLOGY. Educ: Cornell Univ, BS, 65; State Univ NY Buffalo, PhD(physiol), 69. Prof Exp: Fel physiol, 69-71, res asst prof, 71-74, ASST PROF PHYSIOL, STATE UNIV NY BUFFALO, 74-, ACTG DIR, LAB NEUROBIOL, 75- Mem: Am Physiol Soc; Soc Neurosci; Int Brain Res Orgn. Res: Role of cerebellum and cerebro-cerebellar loops in control of movement; analysis of capabilities and integrational properties of neurons in cerebro-cerebellar pathways using electrophysiological techniques in anesthetized cats and monkeys. Mailing Add: Dept of Physiol State Univ of NY Buffalo NY 14214

ALLEN, GARY WILLIAM, b Washington, DC, Mar 3, 44; m 73; c 1. PHYSICAL ORGANIC CHEMISTRY, PHOTOGRAPHIC CHEMISTRY. Educ: Univ Wash, BA, 67; Wesleyan Univ, PhD(chem), 71. Prof Exp: NIH res fel chem, Harvard Univ, 71-73; asst prof, Ariz State Univ, 73-75; SR RES CHEMIST, EASTMAN KODAK CO, 75- Mem: AAAS; Am Chem Soc; The Chem Soc; Sigma Xi. Res: Physical organic chemistry of the photographic process. Mailing Add: Eastman Kodak Res Labs Kodak Park Rochester NY 14650

ALLEN, GEORGE E, b Seminole, Tex, Mar 18, 32; m; c 5. ENTOMOLOGY, INVERTEBRATE PATHOLOGY. Educ: Stephen F Austin State Col, BS, 58; Tex A&M Univ, MS, 60; Miss State Univ, PhD(entom), 68. Prof Exp: Res asst entom, Tex A&M Univ, 58-60; instr microbiol & gen biol, Tex A&I Univ, 60-62; asst prof biol, Mid Tenn State Univ, 62-64; asst prof invert path & entom, Miss State Univ, 64-68; assoc prof microbiol, Fla Technol Univ, 68-70, prof biol sci & pre-prof coordr, 70-73, chmn dept, 70-71; PROF ENTOM, DEPT ENTOM & NEMATOL, UNIV FLA, 73- Mem: Soc Invert Path; Am Soc Microbiol; Entom Soc Am; Am Inst Biol Sic. Res: Virus diseases of insects, especially viruses of the Heliothis genus; their ecological and pathological relationships. Mailing Add: Dept of Entom & Nematol Univ of Fla Gainesville FL 32601

ALLEN, GEORGE OTIS, b Port Arthur, Tex, Jan 30, 28; m 56; c 4. PHYSIOLOGY. Educ: Univ Tex, BA, 51, MA, 54. Prof Exp: Res scientist physiol, Radiobiol Lab, Univ Tex, 52-56; physiologist, 57-64, asst scientist, 64-68, SCIENTIST, ORTHO RES FOUND, 68- Mem: NY Acad Sci; Soc Study Reprod. Res: Physiology of reproduction. Mailing Add: Ortho Res Found Raritan NJ 08869

ALLEN, GEORGE PERRY, b Frankfort, Ky, Dec 16, 41; m 72. MICROBIOLOGY. Educ: Georgetown Col, BS, 63; Univ Ky, PhD(microbiol), 75. Prof Exp: Lab technician microbiol, Ky State Dept Health, 64-66; lab technician, Univ Cincinnati, 66-68; lab technician, Univ Ky, 68-70; Am Cancer Soc fel, 75-77, RES ASST MICROBIOL, UNIV MISS MED CTR, 75- Mem: Am Soc Microbiol; Sigma Xi. Res: Molecular biology of herpes-virus replication. Mailing Add: Dept of Microbiol Univ of Miss Med Ctr Jackson MS 39216

ALLEN, GEORGE RODGER, JR, b Port Arthur, Tex, Nov 8, 29; m 55; c 2. MEDICINAL CHEMISTRY. Educ: Univ Tex, BS, 49, AM, 51, PhD(chem), 53. Prof Exp: Instr, Lamar Col, 51-52; assoc, Univ Tex, 52-53; res chemist, Calco Div, Am Cyanamid Co, 53-54, Lederle Labs, 55-58; res chemist, Mead Johnson & Co, 58-60; res chemist, 60-65, group leader, 65-74, dept head, 74-75, ASSOC DIR NEW PROD ACQUISITIONS, LEDERLE LABS, AM CYANAMID CO, 75- Mem: Am Chem Soc. Res: Inositols; steroids; antibiotics and related compounds; synthetic antibacterials and heterocyclic compounds. Mailing Add: Lederle Labs Am Cyanamid Co Pearl River NY 10965

ALLEN, GOERGE HERBERT, b Zurich, Switz, Aug 16, 23; m 55; c 3. FISHERIES. Educ: Univ Wyo, BS, 50; Univ Wash, PhD(fisheries), 56. Prof Exp: Fisheries biologist, Wyo Game & Fish Dept, 48-51; asst, Col Fisheries, Univ Wash, 52, assoc, 54, fisheries biologist, Fisheries Res Inst, 53-54, res instr oceanog, 56-57; fisheries biologist, Wash State Dept Fisheries, 55-56; prog leader, 67-75, PROF FISHERIES, HUMBOLDT STATE UNIV, 57- Mem: AAAS; Am Fisheries Soc; Am Soc Ichthyol & Herpet; Am Inst Biol Sci; Am Inst Fish Res Biol. Res: Ichthyology; biology of sewage fish ponds; rearing of salmonids in salt water ponds fertilized with sea water; utilization of domestic waste water to fertilize brackish water ponds for production of smolts of Pacific salmon and trout with application to commercial, private or public fish farming. Mailing Add: Sch of Natural Resources Humboldt State Univ Arcata CA 95521

ALLEN, GORDON, b Brookline, Mass, Oct 27, 19; m 49; c 4. HUMAN GENETICS. Educ: Harvard Univ, BS, 42; Columbia Univ, MD, 51. Prof Exp: Intern, USPHS Hosp, NY, 51-52; RES GENETICIST, NIMH, 52- Mem: AAAS; Am Soc Human Genetics; Soc Study Social Biol; Behav Genetics Asn; Am Soc Naturalists. Res: Biology of twins; twin research methods; behavior and population genetics. Mailing Add: NIH Bethesda MD 20014

ALLEN, GORDON AINSLIE, b London, Ont, May 24, 22; m 54; c 3. ENVIRONMENTAL MANAGEMENT, PULP CHEMISTRY. Educ: Univ Western Ont, BA, 43; Univ Rochester, PhD(chem), 49. Prof Exp: Demonstr chem, Univ Toronto, 43-44; develop chemist, L-Air Liquide Soc, Montreal, 44-46; res asst chem, Univ Rochester, 46-49; fel, Nat Res Coun Can, 49-50; chemist, Pulp & Paper Res Inst Can, 50-53; asst dir res, Fraser Co, Ltd, 53-58, spec asst admin dept, 58-62, coordr com develop, 62-65; dir res & develop, 65-74, DIR ENVIRON SERV, GREAT LAKES PAPER CO, LTD, 74- Mem: Am Chem Soc; Am Tech Asn Pulp & Paper Industs; Can Pulp & Paper Asn; Chem Inst Can. Res: Pulp and paper. Mailing Add: 622 Rosewood Crescent Thunder Bay ON Can

ALLEN, GUY FLETCHER, physical chemistry, deceased

ALLEN, HALSEY LEONARD, III, nuclear physics, see 12th edition

ALLEN, HAROLD ALFRED, physical chemistry, solid state physics, see 12th edition

ALLEN, HARRY CLAY, JR, b Saugus, Mass, Nov 26, 20; m 48; c 2. CHEMICAL PHYSICS. Educ: Northeastern Univ, BS, 48; Brown Univ, MS, 49; Univ Wash, PhD(phys chem), 51. Prof Exp: Atomic Energy Comn fel, Harvard Univ, 51-53; asst prof physics, Mich State Col, 53-54; physicist, radiometry sect, atomic & radiation physics div, Nat Bur Standards, 54-61, chief anal & inorg chem div, 61-63, chief inorg mat div, 63-65, dep dir, inst mat res, 65-66; asst dir minerals res, Bur Mines, US Dept Interior, 66-69; PROF CHEM & CHMN DEPT, CLARK UNIV, 69- Concurrent Pos: Vis prof, Univ Wash, 58; res fel theoret chem, Cambridge Univ, 59-60; vis lectr, Univ Md, 64. Honors & Awards: Except Serv Award & Gold Medal, US Dept Commerce, 64; Stratton Award, 65. Mem: Am Chem Soc; fel Am Phys Soc; fel Am Inst Chemists. Res: Solid state; molecular spectroscopy. Mailing Add: Dept of Chem Clark Univ Worcester MA 01610

ALLEN, HARRY PRINCE, b New York, NY, July 31, 38; m 62; c 2. MATHEMATICS. Educ: Brooklyn Col, BS, 60; Yale Univ, MA, 62, PhD(math), 65. Prof Exp: Actg instr math, Yale Univ, 64-65; instr, Mass Inst Technol, 65-67; NATO res fel, Math Inst, State Univ Utrecht, 67-68; asst prof, Rutgers Univ, 68-70; ASSOC PROF MATH, OHIO STATE UNIV, 70- Mem: Am Math Soc. Res: Nonassociative and lie algebras. Mailing Add: Dept of Math Ohio State Univ Columbus OH 43210

ALLEN, HARRY WILLIS, b Pelham, Mass, Aug 8, 92; m 19; c 2. ENTOMOLOGY. Educ: Univ Mass, BS, 13; Miss State Univ, MS, 22; Ohio State Univ, PhD, 26. Prof Exp: Sci asst, Bur Entom, USDA, 13-17, 19-20; from instr to assoc prof entom, Miss State Univ, 20-26, assoc entomologist, Exp Sta, 23-26; from assoc entomologist to entomologist, Bur Entom & Plant Quarantine, 26-53, entomologist, Entom Res Br, Agr Res Serv, 53-58, COLLABR, ENTOM RES BR, AGR RES SERV, USDA, 58- Concurrent Pos: Res assoc, Acad Natural Sci Philadelphia; NSF grant. Mem: AAAS; Entom Soc Am; Am Entom Soc (pres, 58-59). Res: Taxonomy of Tiphiinae. Mailing Add: USDA Res Labs Box 150 Moorestown NJ 08057

ALLEN, HENRY FREEMAN, b Boston, Mass, Nov 23, 16; m 41; c 3. OPHTHALMOLOGY. Educ: Harvard Univ, AB, 39, MD, 43. Prof Exp: From asst clin prof to clin prof ophthal, 47-70, dir postgrad prog, 53-68, chmn dept, 68-74, HENRY WILLARD WILLIAMS CLIN PROF OPHTHAL, HARVARD MED SCH, 70- Concurrent Pos: Fel ophthal, Harvard Med Sch, 46-47; asst chief ed, Arch Ophthal, AMA, 61-66, chief ed, 66-; consult, Binder-Schweitzer Amazonian Hosp, 62-; dir, Lancaster Course, 65-; chief ophthal, Mass Eye & Ear Infirmary, 68-73. Mem: AMA; Am Ophthal Soc; Am Acad Ophthal & Otolaryngol; Am Asn Ophthal (pres, 70). Res: Infectious diseases in ophthalmic practice; prevention of infection in eye surgery; hospital hygiene; preschool vision testing. Mailing Add: 243 Charles St Boston MA 02114

ALLEN, HENRY L, b Philadelphia, Pa, Apr 26, 45; m 73; c 1. VETERINARY PATHOLOGY. Educ: Pa State Univ, BA, 67; Univ Pa, VMD, 71. Prof Exp: From resident to instr vet path, Sch Vet Med, Univ Pa, 71-75; RES FEL PATH, MERCK SHARP & DOHME RES LABS, 75- Mem: Am Vet Med Asn. Res: Neoplastic diseases and clinical oncology in domestic animals; comparative aspects. Mailing Add: Merck Sharp & Dohme Res Labs 44-1 West Point PA 19486

ALLEN, HERBERT CLIFTON, JR, b Richmond, Va, Jan 7, 17; m 49; c 6. INTERNAL MEDICINE, NUCLEAR MEDICINE. Educ: Univ Richmond, BSc, 37; Med Col Va, MD, 41. Prof Exp: Intern, Philadelphia Gen Hosp, 41-42; resident path, Pa Hosp, 46-47; resident med, Med Col Va, 47-48; asst chief radioisotope sect, Dept Med & Surg, Vet Admin, Washington, DC, 48-49; chief metab serv & asst dir, Birmingham Vet Admin Hosp, Van Nuys, Calif, 49-50; asst dir radioisotope unit, Wadsworth Vet Admin Hosp, Los Angeles, 50-51; ASST PROF MED, BAYLOR COL MED, 51- Concurrent Pos: Dir radioisotope unit, Vet Admin Hosp, Houston, 51-55, actg dir, 55-57; dir dept nuclear med, Methodist Hosp, 52-, Hermann Hosp, 56- & Mem Baptist Hosp, 65-; mem, Radiation Adv Bd; secy, Tex State Dept Health, 61-65; pres, Atomic Energy Indust Labs Southwest, Inc & Atomic Food Processing Corp Am. Mem: AMA; Am Thyroid Asn; Am Heart Asn; Soc Nuclear Med; NY Acad Sci. Res: Diagnostic and therapeutic atomic medicine; diseases of thyroid; development of diagnostic instrumentation for nuclear medicine, particularly thyroid, brain and renal scanning. Mailing Add: 4010 Martinshire Dr Houston TX 77025

ALLEN, J FRANCES, b Arkville, NY, Apr 14, 16. ECOLOGY. Educ: Radford Col, BS, 38; Univ Md, MS, 48, PhD(zool), 52. Prof Exp: Teacher various high schs, 38-47; asst zool, Univ Md, 47-48, instr, 48-55, asst prof, 55-58; prof asst, Syst Biol Prog, NSF, 58-61, assoc prog dir, 61-67; chief br water qual requirements, Fed Water Pollution Control Admin, US Dept Interior, 67-68, asst dir biol sci, Div Water Qual Res, 68-70; chief biol sci br, Div Water Qual Res, 70-71, chief ecol effects, Processes & Effects Div, 71-73, STAFF SCIENTIST-ECOLOGIST, SCI ADV BD, US ENVIRON PROTECTION AGENCY, 73- Concurrent Pos: Biol Exam, NY State Dept Educ, 47. Mem: Fel AAAS; Am Soc Zool; Am Fisheries Soc; Soc Syst Zool; Am Malacol Union. Res: Fishery biology; shellfish; gastropod ecology and distribution; taxonomy of the gastropods; aquatic biology; pollution; biology; ecological effects. Mailing Add: 7507 23rd Ave Hyattsville MD 20783

ALLEN, JACK C, JR, b Dallas, Tex, Sept 17, 35; m 61. GEOLOGY. Educ: Southern Methodist Univ, BS, 58; Princeton Univ, MA, 60, PhD(geol), 61. Prof Exp: Asst prof geol, Wesleyan Univ, 61-63; from asst prof to assoc prof, 63-74, PROF GEOL,

BUCKNELL UNIV, 74- Mem: AAAS; Geol Soc Am; Mineral Soc Am. Res: Structure and petrology of granitic rocks; stability of amphibole in andesite and basalt. Mailing Add: Dept of Geol & Geog Bucknell Univ Lewisburg PA 17837

ALLEN, JAMES DURWOOD, b Commerce, Tex, Nov 22, 35; m 58. ORGANIC POLYMER CHEMISTRY. Educ: ETex State Col, BA, 58; Rice Univ, PhD(chem), 62. Prof Exp: Res chemist, Shell Oil Co, 62-63; NIH fel chem, Northwestern Univ, 63-64; res chemist, Phillips Petrol Co, 64-69; mem tech staff, NAm Rockwell Corp, 69-75; MGR RES & DEVELOP, ADVAN COMPOSITE DIV, FIBERITE CORP, 75- Mem: Am Chem Soc; Soc Aerospace Mat & Process Engrs. Res: Polymer chemistry as applied to graphite fiber reinforced advanced composites. Mailing Add: 1344 Conrad Dr Winona MN 55987

ALLEN, JAMES HARRILL, b Chattanooga, Tenn, Jan 31, 06; m 34; c 3. OPHTHALMOLOGY, BACTERIOLOGY. Educ: Univ Tenn, BA, 26; Univ Mich, MD, 30; Univ Iowa, MS, 38. Prof Exp: From intern to resident, Univ Iowa Hosps, 31-36, res assoc ophthal, Univ, 36-37, from instr to prof, 37-50; prof ophthal, Med Sch, Tulane Univ, 50-76, chmn dept, 52-68, assoc dean & med dir clins, 68-71; RETIRED. Concurrent Pos: Mem, Nat Coun Combat Blindness; consult, Nat Soc Prev Blindness, 50-, Eye Bank Sight Restoration, 54- & Surgeon Gen, US Air Force, 52-59; mem, Int Contact Lens Coun of Ophthalmologists. Mem: AAAS; AMA; Asn Res Vision & Ophthal; Am Asn Ophthal; Contact Lens Asn Ophthalmologists. Res: Staphlyococcic infections of the eye; inclusion blennorrhea; ocular complications of diabetes mellitus; ophthalmic surgical techniques. Mailing Add: 9104 Quince St New Orleans LA 70118

ALLEN, JAMES PAUL, b Providence, RI, Oct 13, 36; m 74. GEOGRAPHY. Educ: Amherst Col, BA, 58; Harvard Univ, MAT, 59; Syracuse Univ, PhD(geog), 70. Prof Exp: Asst prof, 69-72, ASSOC PROF GEOG, CALIF STATE UNIV, NORTHRIDGE, 72- Mem: Asn Am Geogr; Am Geog Soc. Res: Studies of migration in the United States, including recent immigration; trend of migration rates and ethnic settlement patterns in the United States. Mailing Add: Dept of Geog Calif State Univ Northridge CA 91324

ALLEN, JAMES R, JR, b Mars Hill, NC, Dec 13, 27; m 48; c 4. PATHOLOGY. Educ: Univ Tenn, BS, 50, MS, 51; Univ Ga, DVM, 55; Univ Wis, PhD(path), 61. Prof Exp: Asst nutrit, Univ Tenn, 50-51; dir field serv, Res Div, Cent Soya Co, Inc, 55-59; asst prof path, Sch Vet Med, Univ Ga, 61-62; asst prof, 62-70, PROF PATH, SCH MED, UNIV WIS-MADISON, 70- Concurrent Pos: NIH res grant, 63-; sr scientist & assoc, Wis Regional Primate Res Ctr, 62- Mem: AAAS; Soc Toxicol; Am Vet Med Asn; Am Soc Cell Biol; Int Acad Path. Res: Biochemical pathologic and ultramicroscopic changes in the liver and cardiovascular system from consumption of certain fats and alkaloids. Mailing Add: Dept of Path Univ of Wis Sch of Med Madison WI 53706

ALLEN, JAMES RICHARD, b Eau Claire, Wis, Mar 3, 41; m 72; c 1. GEOMORPHOLOGY. Educ: Wis State Univ-Eau Claire, BS, 65; Eastern Mich Univ, MA, 69; Rutgers Univ, PhD(geog), 73. Prof Exp: ASST PROF EARTH SCI, NORTHEASTERN UNIV, 72- Concurrent Pos: Consult, Marine Sci Ctr, Rutgers Univ, 75- Mem: Asn Am Geogr; Sigma Xi. Res: Analysis of beach dynamics that leads toward empirical modeling in coastal geomorphology. Mailing Add: Dept of Earth Sci Northeastern Univ Boston MA 02115

ALLEN, JAMES ROY, b Kingston, Ont, Aug 27, 26; m 58; c 2. THEORETICAL PHYSICS. Educ: Queen's Univ, Ont, BA, 47, MA, 49; Univ Manchester, PhD(theoret physics), 63. Prof Exp: Exhib of 1851 res scholar, 49-52; lectr physics, Queen's Univ, Ont, 53-55, asst prof, 56-61; res assoc theoret physics, Univ Manchester, 62-63; assoc prof physics, 64-70, PROF PHYSICS, QUEEN'S UNIV, ONT, 71- Mem: Am Phys Soc; Can Asn Physicists; fel Brit Inst Physics & Phys Soc. Res: Nuclear emulsions; cosmic rays; passage of fast charged particles through solids; photonuclear reactions; hydrodynamic and magnetohydrodynamic turbulence; quasi-stationary states and resonant reactions. Mailing Add: Dept of Physics Queen's Univ Kingston ON Can

ALLEN, JAMES SIRCOM, b Halifax, NS, Aug 11, 11; US citizen; m 36. NUCLEAR PHYSICS. Educ: Univ Cincinnati, BA, 33; Univ Chicago, PhD(physics), 37. Prof Exp: Instr physics, Kenyon Col, 36-37; asst, Univ Minn, 37-39; assoc prof, Kans State Col, 39-42; asst prof, Univ Chicago, 46-48; from assoc prof to prof physics, Univ Ill, Urbana-Champaign, 48-73; RETIRED. Concurrent Pos: Consult, Los Alamos Sci Lab, 46-; with AEC; with Off Sci Res & Develop, 44. Mem: Fel Am Phys Soc. Res: Application of electron-multiplier tubes as nuclear detectors; recoil nuclei from beta ray disintegrations; disintegration of beryllium by protons; electron-neutrino angular correlation experiments; nuclear 6reactions induced by protons, deuterons, helium 3 and helium 4 particles. Mailing Add: Box 41 Taos Canyon Rt Taos NM

ALLEN, JAMES WARD, b Livingston, Mont, May 4, 41; m 62; c 2. SOLID STATE PHYSICS. Educ: Stanford Univ, BS, 63, MS, 65, PhD(elec eng), 68. Prof Exp: Staff mem, Lincoln Lab, Mass Inst Technol, 68-73; STAFF MEM, XEROX PALO ALTO RES CTR, 73- Mem: Am Phys Soc. Res: Optical properties of transition metal and rare earth compounds, and of defects in semiconductors; magnetic anisotropy and magnetoelastic effects; insulator-to-metal transition materials. Mailing Add: Xerox Palo Alto Res Ctr 3333 Coyote Hill Rd Palo Alto CA 94304

ALLEN, JOE FRANK, b Hogansville, Ga, June 3, 34; m 55; c 2. INORGANIC CHEMISTRY, SCIENCE EDUCATION. Educ: Berry Col, AB, 55; Univ Miss, MS, 59; Ga Inst Technol, PhD(inorg chem), 63. Prof Exp: Instr chem, Ga Inst Technol, 57-61; radio chemist, Oak Ridge Nat Lab, 62-64; asst prof, 64-68, ASSOC PROF INORG CHEM, CLEMSON UNIV, 68- Concurrent Pos: Consult, mobile lab prog, Oak Ridge Assoc Univs, 66-; consult develop learning mat, Ulster Co Community Col & Environ Protection Agency, 72-75. Mem: AAAS; Am Chem Soc; The Chem Soc. Res: Chemistry of complex compounds; application of radioisotopes; chemical separations using ion exchange and solvent extraction; development of individualized instructional systems for science and technical training. Mailing Add: Dept of Chem Clemson Univ Clemson SC 29631

ALLEN, JOHN ED, b Danville, La, Nov 18, 37; m 60; c 2. MATHEMATICS. Educ: La Polytech Inst, BS, 58; Okla State Univ, MS, 60, PhD(math), 63. Prof Exp: Asst prof, 63-72, ASSOC PROF MATH, NORTH TEX STATE UNIV, 72- Concurrent Pos: NSF sci fac fel, Purdue Univ, 71-72. Mem: Math Asn Am; Am Math Soc. Res: Convexity; numerical analysis. Mailing Add: Dept of Math North Tex State Univ Denton TX 76203

ALLEN, JOHN ELIOT, b Seattle, Wash, Aug 12, 08; m 33; c 1. GEOLOGY. Educ: Univ Ore, BA, 31, MA, 32; Univ Calif, PhD(geol), 44. Prof Exp: Field geologist, Rustless Iron & Steel Co, Md, 35-38; field geologist, Ore Dept Geol & Mineral Indust, 38-39, chief geologist in chg field invests & assoc ed publ, 39-47; actg head dept geol, NMex Inst Mining & Technol, 49-51; chief, Navaho Mineral Serv, 51-54; econ geologist, Bur Mines, NMex, 54-56; prof geol & head dept earth sci, 56-74, EMER PROF GEOL, PORTLAND STATE UNIV, 74- Concurrent Pos: SEATO prof & chmn dept, Univ Peshawar, 63-64; prof geol, Whitman Col, 75; res geologist, Nev Bur Mines & Geol, Reno, 75-76. Honors & Awards: Niel Miner Award, Nat Asn Geol Teachers, 75. Mem: Fel Geol Soc Am; Soc Econ Geologists; Nat Asn Geol Teachers; Am Inst Prof Geologists. Res: Structural relations of chromite deposits; Oregon economic geology and areal geology of the Coos Bay and Wallowa Mountains; geology of San Juan Bautista Quadrangle, California and Northern Sacramento Mountains, Southwest San Juan Basin, New Mexico; glaciation in Northwest; volcanoes of Cascade Range. Mailing Add: Dept Earth Sci Box 751 Portland State Univ Portland OR 97207

ALLEN, JOHN KAY, b Rochdale, Eng, Mar 14, 36; m 67. POLYMER CHEMISTRY. Educ: Univ Birmingham, BS, 57, PhD(chem), 60. Prof Exp: Res asst chem, Univ Louisville, 60-62; res chemist, Shell Chem Co, Eng, 62-65; proj chemist, 65-68, sr proj chemist, 68-70, RES CHEMIST, AMOCO CHEM CORP, 70- Mem: Am Chem Soc. Res: Use of radioactive tracers in polymer chemistry; properties of ion-exchange resins; chemistry of polyether polyols and polyurethanes; preparation of fire resistant cellular plastics; plasticizer alcohols synthesis and aromatic acids synthesis. Mailing Add: Res & Develop Dept Amoco Chem Corp PO Box 400 Naperville IL 60540

ALLEN, JOHN LOGAN, b Laramie, Wyo, Dec 27, 41; m 64; c 2. HISTORICAL GEOGRAPHY, CULTURAL GEOGRAPHY. Educ: Univ Wyo, BA, 63, MA, 64; Clark Univ, PhD(geog), 69. Prof Exp: NSF dept develop grant, Clark Univ, 70-71; from instr to asst prof, 67-73, ASSOC PROF GEOG, UNIV CONN, 73- Mem: Asn Am Geogr; fel Am Geog Soc; Int Geog Union. Res: Man's impact on the environment; perception of the environment; exploration and the formation and transmission of geographical images and conceptions. Mailing Add: Dept of Geog Univ of Conn Storrs CT 06268

ALLEN, JOHN MORGAN, b Springfield, Mo, Sept 1, 24; m 51. CELL BIOLOGY. Educ: Drury Col, AB, 48; Univ Chicago, PhD(zool), 54. Hon Degrees: ScD, Drury Col, 70. Prof Exp: From instr to assoc prof, 52-64, chmn dept, 66-71, PROF ZOOL, UNIV MICH, ANN ARBOR, 64- Mem: Am Soc Zoologists; Histochem Soc (pres, 73); Am Soc Cell Biol. Res: Enzyme cytochemistry; biology of lysosomes and perosomes. Mailing Add: Dept of Zool Univ Mich Ann Arbor MI 48104

ALLEN, JOHN RYBOLT, b Indianapolis, Ind, Sept 14, 26; m 53. CHEMISTRY, BIOCHEMISTRY. Educ: Ball State Teachers Col, AB, 49; Univ Ill, Urbana, PhD(biochem), 54. Prof Exp: Res assoc biochem, Northwestern Univ, 53-56; asst prof, Col Med, Baylor Univ, 56-59; sr scientist, Warner-Lambert Pharm Co, NJ, 59-60; res assoc dent sch, Wash Univ, 60-62; prof chem & head dept, Union Col, Ky, 62-64; clin assoc clin chem, Univ Hosp, Case Western Reserve Univ, 64-65; asst prof path & radiol, Col Med, Ohio State Univ, 65-68; clin chemist, St John's Mercy Hosp, St Louis, Mo, 68-69, Decatur Mem Hosp, Ill, 69-70, San Diego Inst Path, 70 & San Bernardino County Hosp, 70-75; INSTR CHEM, PHOENIX COL, 75- Mem: Fel AAAS; fel Am Asn Clin Chem; Am Chem Soc; Acad Clin Lab Physicians & Scientists; fel Am Inst Chem. Res: Quality control and methods; creatine phosphokinase; vitamin E deficiency, lipid metabolism and structure. Mailing Add: 301-B 6131 N 16th St Phoenix AZ 85016

ALLEN, JOSEPH GARROTT, b Elkins, WVa, June 5, 12; m; c 5. MEDICINE. Educ: Washington Univ, AB, 34; Harvard Univ, MD, 38. Prof Exp: Smith fel, Univ Chicago, 40-43, instr surg, 43-44, res assoc, Manhattan Proj, 44-46, chief surg res, 46-47, from instr to prof surg, 46-59; PROF SURG, SCH MED, STANFORD UNIV, 59- Concurrent Pos: Mem panel blood coagulation, Nat Res Coun; mem surg study sect, NIH, 54-59, med sci training comt, 63-; chmn study sect cancer chemother, Am Cancer Soc, 60-63. Honors & Awards: Abel Prize, 48; Silver Medal, Am Roentgen Ray Soc, 48; Gold Medal, AMA, 48; Gross Surg Prize, Philadelphia Acad Surg, 55; Elliot Award, Am Asn Blood Banks, 56. Mem: AAAS; fel AMA; fel Am Col Surg; fel Am Physiol Soc; fel Soc Exp Biol & Med. Res: Surgery; ionizing irradiation injury; parenteral protein nutrition; coagulation of blood; posttransfusion hepatitis in relation to paid donors and the need for an all-volunteer national blood program. Mailing Add: Dept of Surg Stanford Univ Stanford CA 94305

ALLEN, JOSEPH HUNTER, b St Joseph, Mo, Aug 31, 25; m 56; c 1. RADIOLOGY. Educ: Westminster Col, 42-43; Washington Univ, MD, 48. Prof Exp: From asst prof to assoc prof, 58-69, PROF RADIOL, VANDERBILT UNIV, 69- Mem: AMA; Am Soc Neuroradiol; fel Am Col Radiol; Asn Univ Radiol. Res: Neuroradiology. Mailing Add: Dept of Radiol Vanderbilt Univ Nashville TN 37203

ALLEN, JOSEPH PERCIVAL, b Crawfordsville, Ind, June 27, 37; m 61. NUCLEAR PHYSICS, SPACE PHYSICS. Educ: DePauw Univ, BA, 59; Yale Univ, MS, 61, PhD(physics), 65. Prof Exp: Staff physicist, nuclear struct lab, Yale Univ, 65-66; res assoc nuclear physics lab, Univ Wash, 66-67; scientist-astronaut, 67-75, ASST ADMINR OFF LEGIS AFFAIRS, MANNED SPACECRAFT CTR, NASA, 75- Concurrent Pos: Guest res assoc, Brookhaven Nat Lab, 64-67; staff mem, President's Coun Int Econ Policy, 73. Mem: AAAS; Am Phys Soc; Am Astronaut Soc; Am Inst Aeronaut & Astronaut. Res: Experimental studies on detailed nuclear structure and nuclear reaction mechanisms using a variety of particle accelerators of beam energies 1 to 200 million electron volts. Mailing Add: Code CB NASA Manned Spacecraft Ctr Houston TX 77058

ALLEN, JULIUS CADDEN, b New London, Conn, Oct 18, 38; m 61; c 2. PHARMACOLOGY, BIOCHEMISTRY. Educ: Amherst Col, BA, 60; Univ Mass, MA, 62; Univ Alta, PhD(pharmacol), 67. Prof Exp: Asst prof pharmacol, 70-72, asst prof cell biophys, 72-73, ASSOC PROF CELL BIOPHYS, BAYLOR COL MED, 73- Concurrent Pos: Fel, Baylor Col Med, 67-69, Nat Heart & Lung Inst fel pharmacol, 69-70, Nat Heart & Lung Inst grant, 74-77; Am Heart Asn grant, 70-73; mem, Int Study Group Res Cardiac Metab, 72-; estab investr, Am Heart Asn, 74- Mem: Am Heart Asn; Am Soc Pharmacol & Exp Therapeut; Biophys Soc. Res: Cardiac glycoside interaction; biochemistry of smooth muscle membrane systems; biochemistry and biophysics of ion transport; biochemistry of cardiac ischemia. Mailing Add: Dept of Cell Biophys Baylor Col of Med Houston TX 77025

ALLEN, KENNETH RICHARD, b Cleveland, Ohio, Dec 11, 39; m 59; c 2. STATISTICAL MECHANICS. Educ: Ga Inst Technol, BS, 61, MS, 64, PhD(physics), 67. Prof Exp: Res physicist, United Aircraft Corp, Conn, 61-62; asst solid state physics, Ga Inst Technol, 62-63, physics, 63-65 & solid state physics & statist mech, 65-66, instr physics, 66; fel, Univ Fla, 67-68, asst prof, 68-73; RES PHYSICIST, NAVAL COASTAL SYSTS LAB, 74- Mem: Am Phys Soc; Sigma Xi. Res: Energy transport in solids; quantum fluids. Mailing Add: Naval Coastal Systs Lab Panama City FL 32401

ALLEN, KENNETH WILLIAM, b St Stephens, NB, June 20, 30; nat US; m 52; c 2. ZOOLOGY. Educ: Wheaton Col, BS, 52; Univ Maine MS, 56; Rice Univ, PhD, 59. Prof Exp: Fel biol, Rice Univ, 59-60; asst prof zool, Univ Calif, Los Angeles, 60-63; PROF ZOOL & OCEANOG & HEAD DEPT, UNIV MIANE, ORONO, 63-, ACTG DEAN COL ARTS & SCI, 74- Mem: Am Soc Zoologists; Am Soc Parasitol.

Res: Comparative biochemistry and parasitology. Mailing Add: Dept of Zool Univ of Maine Orono ME 04473

ALLEN, LANE, b Ellijay, Ga, Aug 14, 09; m 39, 60; c 3. ANATOMY. Educ: Univ Ga, BS, 30, MS, 32, MD, 39; Georgetown Univ, PhD(anat), 35. Prof Exp: Instr zool, Univ Ga, 30-31; instr anat, George Washington Univ, 32-33; instr, Georgetown Univ, 33-35; from asst prof to assoc prof anat, 35-42, prof gross anat, 42-64, PROF ANAT, MED COL GA, 64- Mem: Am Asn Anatomists; Am Physiol Soc; Pan Am Asn Anat. Res: Anatomy and physiology of the lymphatic system; physiology of ascites; anatomy of hernia. Mailing Add: Dept of Anat Med Col of Ga Augusta GA 30902

ALLEN, LAWRENCE HARVEY, b Ottawa, Ont, Mar 27, 43; m 68. PHYSICAL CHEMISTRY, COLLOID CHEMISTRY. Educ: Carleton Univ, BSc, 65; Clarkson Col Technol, PhD(phys chem), 70. Prof Exp: RES SCIENTIST COLLOID CHEM, PULP & PAPER RES INST CAN, 72- Concurrent Pos: Nat Res Coun Can fel, Inland Waters Directorate, Environ Can, 71-72. Mem: Am Chem Soc; Chem Inst Can; Can Pulp & Paper Asn. Res: Colloid and emulsion stability; precipitation phenomena; wood resins; chemistry of papermaking processes. Mailing Add: Pulp & Paper Res Inst of Can 570 St Johns Rd Pointe Claire PQ Can

ALLEN, LELAND CULLEN, b Cincinnati, Ohio, Dec 3, 26; m 60; c 3. CHEMISTRY. Educ: Univ Cincinnati, BS, 49, EE, 55; Mass Inst Technol, PhD(theoret physics), 57. Prof Exp: Fel, Mass Inst Technol, 57-59; NSF fel physics, Univ Calif, Berkeley, 59-60; from asst prof to assoc prof, 60-65, PROF CHEM, PRINCETON UNIV, 65- Concurrent Pos: NSF sr fel, Centre Mecanique Ondulatoire Appliquee, Paris, 67; Guggenheim fel, Math Inst, Oxford Univ, 67; NIH spec fel biochem sci, Princeton Univ, 72. Mem: AAAS; Am Phys Soc; Am Chem Soc. Res: Electronic structure theory, applications to inorganic, organic and biochemistry. Mailing Add: Dept of Chem Princeton Univ Princeton NJ 08540

ALLEN, LEROY RICHARD, b Garden Grove, Calif, Sept 14, 13; m 38; c 3. RESEARCH ADMINISTRATION, PUBLIC HEALTH. Educ: Univ Redlands, AB, 36; Univ Southern Calif, MD, 41; Univ Mich, MPH, 48; Am Bd Prev Med & Pub Health, dipl, 49. Prof Exp: Tuberc control officer, State Dept Pub Health, Md, 45-48; tuberc consult, Regional Off San Francisco, USPHS, 49-50; med dir, US Tech Coop Mission to Burma, 50-53, prof prev & social med, Vellore Med Col, South India, 54-56; med educ & pub health rep, Rockefeller Found, New Delhi, India & Quezon City, Philippines, 57-72; from dep dir to dir, Atomic Bomb Casualty Comn, Nat Acad Sci, 72-75, VCHMN, RADIATION EFFECTS RES FOUND, 75- Concurrent Pos: Consult, Indian Pub Health Asn, USPHS, 44-53; vis prof community med, Med Col, Univ Philippines, 68-71; on leave, consult, Atomic Bomb Casualty Comn, US Nat Acad Sci-Japanese Nat Inst Health, 71-72. Mem: Am Thoracic Soc; Am Pub Health Asn; hon mem Indian Pub Health Asn. Res: Medical education and care; problems related to population control. Mailing Add: Radiation Effects Res Found 5-2 Hijiyama Koen Hiroshima Japan

ALLEN, LEW, JR, b Miami, Fla, Sept 30, 25; m 49; c 5. PHYSICS. Educ: US Mil Acad, BS, 46; Univ Ill, MS, 52, PhD(physics), 54. Prof Exp: Mem staff, Los Alamos Sci Lab, 54-57; sci adv, Air Force Spec Weapons Ctr, 57-61, mem staff defense dir res & eng, 61-65, dep dir spec projs, Off of Secy of Air Force, 65-68, dir space systs, 68-70, dir spec projs, 71-73; DIR, NAT SECURITY AGENCY, 73- Mem: Am Phys Soc; Am Geophys Union. Res: Satellite systems. Mailing Add: 4526 Butler St Ft Meade MD 20755

ALLEN, LEWIS EDWIN, b Monroe, La, Aug 9, 37; m 64; c 3. ORGANIC CHEMISTRY. Educ: Queens Col, BS, 58; Syracuse Univ, PhD(org chem), 70. Prof Exp: Assoc prof chem, Fla A&M Univ, 63-70; sr chemist, 70-74, COORDR ENVIRON SERV, EASTMAN KODAK CO, 74- Mem: Am Chem Soc; Brit Chem Soc; Soc Photog Sci & Eng. Res: Physical organic chemistry; stereochemistry; organometallic compounds; photographic science; disposal of photographic waste solutions. Mailing Add: Res Lab Eastman Kodak Co Rochester NY 14650

ALLEN, LINDSAY HELEN, b Chippenham, Eng, July 24, 46; m 69. NUTRITION. Educ: Univ Nottingham, BSc, 67; Univ Calif, PhD(nutrit), 73. Prof Exp: Res asst nutrit, Med Res Coun, Cambridge Univ, 67-69, Univ Calif Davis, 69-73; fel nutrit, Univ Calif, Berkeley, 73-74; ASST PROF NUTRIT, UNIV CONN, 74- Mem: Brit Nutrit Soc; NY Acad Sci; Sigma Xi. Res: Interrelationships betwen dietary protein and mineral metabolism; assessment of nutritional status; nutrition and development. Mailing Add: Dept of Nutrit Sci Univ of Conn Storrs CT 06268

ALLEN, LINUS SCOTT, b Dallas, Tex, Dec 9, 35; m 54; c 2. APPLIED PHYSICS. Educ: Southern Methodist Univ, BS, 54, MS, 61, PhD(physics), 71. Prof Exp: Scientist reactor physics, Jet Propulsion Lab, Calif Inst Technol, 61-62; SR RES PHYSICIST APPL PHYSICS, MOBIL RES & DEVELOP CORP, 62- Mem: Am Nuclear Soc; Soc Prof Well Log Analysts. Res: Development of new or improved methods for locating and assessing petroleum and mineral occurances. Mailing Add: 10260 Newcombe Dallas TX 75228

ALLEN, LOIS BRENDA, b Martin, Ky, July 4, 39. VIROLOGY, CHEMOTHERAPY. Educ: Geogetown Col, Ky, BS, 61; Mich State Univ, MS, 63; Univ Mich, Ann Arbor, PhD(epidemiol), 71. Prof Exp: From virologist to sr virologist, 71-75, HEAD DEPT VIROL, ICN NUCLEIC ACID RES INST, 75- Mem: AAAS; Am Asn Immunol; Am Soc Microbiol. Res: Susceptibility and resistance factors of disease with emphasis on development of new antiviral drugs; development of Virazole and ara-HxMP. Mailing Add: ICN Nucleic Acid Res Inst 2727 Campus Dr Irvine CA 92664

ALLEN, LOUIS PINCKNEY JR, b Wash, DC, Mar 30, 13; m 36; c 3. METEOROLOGY, OCEANOGRAPHY. Educ: Wilson Teachers Col, BS, 36; Univ Md, MEd, 43. Prof Exp: Instr pub schs, Wash, DC, 36-42; instr meteorol & navig, Pa Cent Airlines, Wash, DC, 42-43; meteorologist, US Weather Bur, 46; meteorologist, oceanographer & head prog br, Hydrographic Off, US Dept Navy, 46-51, staff oceanographer, Off Naval Res, 51-53; PRES ALLEN WEATHER CORP, 53- Concurrent Pos: Mem, US Naval Exped, Arctic, 47; meteorologist, Nat Broadcasting Co, 48-50 & Columbia Broadcasting Co, 50-55; consult, Nat Res Coun & Nat Acad Sci, 54 & US Maritime Admin, 55; meteorologist, Am Broadcasting Co, 55-74 & CBS, 74- Mem: AAAS; Am Geophys Union; Am Oceanog Orgn; Marine Tech Soc; Am Meteorol Soc. Res: Weather and oceanographic forecasting; navigating with the weather; forecasting mathematically the trajectories of tropical storms; forecasting sea ice; operational climatology; pioneer and developer optimum ship weather routing. Mailing Add: 5207 Wisconsin Ave NW Washington DC 20015

ALLEN, M GEORGE, organic chemistry, see 12th edition

ALLEN, MARCIA KATZMAN, b New York, NY, Jan 10, 35; m 57; c 3. MICROBIAL GENETICS. Educ: Bryn Mawr Col, BA, 56; Stanford Univ, PhD(biol), 62. Prof Exp: NIH res fel, Mass Inst Technol, 63-65; instr molecular biol, Syntex Corp, 65-67; LECTR BIOL SCI, STANFORD UNIV, 67- Mem: Genetics Soc Am; Am Soc Microbiol. Res: Biochemical genetics; biochemistry; gene action. Mailing Add: Dept of Biol Sci Stanford Univ Stanford CA 94305

ALLEN, MARSHALL B, JR, b Long Beach, Miss, Oct 19, 27; m 57; c 3. NEUROSURGERY. Educ: Univ Miss, BA, 49; Harvard Med Sch, MD, 53. Prof Exp: Intern, Jefferson Med Col, 53-54; asst thoracic surg, Miss State Sanitorium, 54-55; gen pract med, Houston Hosp Inc, Miss, 55-56; resident gen surg, Sch Med, Univ Miss, 56-57, resident neurosurg, 57-61; chief serv neurosurg, Vet Admin Hosp, Jackson, Miss, 62-64; PROF NEUROSURG & CHIEF DIV, MED COL GA, 64- Concurrent Pos: Nat Inst Neurol Dis & Blindness fel neurophysiol, Hosp Henri-Rouselle, Paris, France, 61-62; consult, Vet Admin Hosp, Augusta & Ga State Penitentiary, 65- & Milledgeville State Hosp, 66- Mem: AMA; Cong Neurol Surgeons; Am Asn Neurol Surgeons; Soc Neurol Surgeons; Am Col Surgeons. Res: Neurophysiology; clinical neurosurgery. Mailing Add: Div of Neurosurg Med Col of Ga Augusta GA 30902

ALLEN, MARTIN, b New York, NY, Mar 26, 18; m 42; c 4. PHYSICAL CHEMISTRY. Educ: Brooklyn Col, AB, 38; Univ Minn, MS, 41, PhD(phys chem), 44. Prof Exp: Asst, Univ Minn, 40-43, instr phys chem, 43-45; res assoc, Allegany Ballistics Lab, Md, 45; sr technician, B F Goodrich Chem Co, Ohio, 45-47; assoc prof chem, Butler Univ, 47-56; assoc prof, 56-57, PROF CHEM, COL ST THOMAS, 57-, CHMN DEPT, 75; ABSTRACTOR, CHEM ABSTR SERV, 57- Mem: AAAS; Am Chem Soc; Fedn Am Scientists; The Chem Soc. Res: Thermodynamics; solutions of electrolytes; properties of high polymers; solubility of silver acetate in mixed solvents containing other electrolytes. Mailing Add: Dept of Chem Col of St Thomas St Paul MN 55105

ALLEN, MARVIN CARROL, b Searcy, Ark, Nov 30, 39; m 65; c 2. ANALYTICAL CHEMISTRY. Educ: Ark State Univ, BS, 61; Univ Ark, MS, 64, PhD(phys chem), 67. Prof Exp: RES SCIENTIST CHEM, CONTINENTAL OIL CO, 69- Mem: Am Chem Soc; Sigma Xi. Res: Separation and quantitation of organic compounds by thin layer chromatography and high performance liquid chromatography; characterization of polynuclear aromatic compounds in petroleum fractions by fluorescence spectrophotometry and/or chromatography. Mailing Add: Continental Oil Co 1000 S Pine Ponca City OK 74601

ALLEN, MARY A MENNES, b Owatonna, Minn, Apr 26, 38; m 61. MICROBIOLOGY. Educ: Univ Wis-Madison, BS, 60, MS, 61; Univ Calif, Berkeley, PhD(microbiol), 66. Prof Exp: USPHS trainee, Univ Calif, Berkeley, 62-63; from instr to asst prof, 68-75, ASSOC PROF BIOL SCI, WELLESLEY COL, 75- Concurrent Pos: Res assoc dept biol sci, Tufts Univ, 68; Brown-Hazen res grant, Res Corp, 69; NSF res grant, 71-; Cottrel Col sci grant, Res Corp, 73. Mem: AAAS; Phycol Soc Am; Am Soc Cell Biol; Am Soc Microbiol. Res: Fine structure and physiology of blue-green algae; growth and cell division of unicellular blue-green algae; nitrogen chlorosis and heterotrophy in blue-green algae; effect of changes in environment on blue-green algae. Mailing Add: Dept of Biol Sci Wellesley Col Wellesley MA 02181

ALLEN, MARY BELLE, microbiology, see 12th edition

ALLEN, MARY-MAURICE BELT, genetics, biochemistry, deceased

ALLEN, MATTHEW ARNOLD, b Edinburgh, Scotland, Apr 27, 30; US citizen; m 57; c 3. PHYSICS. Educ: Univ Edinburgh, BSc, 51; Stanford Univ, PhD(physics), 59. Prof Exp: Res assoc microwave physics, Hansen Labs, Stanford Univ, 55-61; mgr res tube div, Microwave Assocs, Inc, Mass, 61-65; MEM TECH STAFF, STANFORD LINEAR ACCELERATOR CTR, STANFORD UNIV, 65- Concurrent Pos: Consult, Microwave Assocs, Inc, 65- & Bechtel Corp, 70- Mem: Am Phys Soc; sr mem Inst Elec & Electronics Eng. Res: Microwave electronics; accelerator and plasma physics; high energy electron-positron storage ring research. Mailing Add: Stanford Linear Accelerator Ctr Stanford Univ Stanford CA 94305

ALLEN, MAX SCOTT, b Chanute, Kans, Jan 13, 11; m 37; c 3. INTERNAL MEDICINE. Educ: Univ Wichita, AB, 33, Univ Kans, MD, 37. Prof Exp: Assoc prof, 50-59, PROF MED, UNIV KANS MED CTR, KANSAS CITY, 59- Mem: AMA; fel Am Col Physicians; Am Asn Hist Med. Res: Clinical research in internal medicine. Mailing Add: Univ of Kans Med Ctr Kansas City KS 66103

ALLEN, MERRILL JAMES, b San Antonio, Tex, Aug 2, 18; m 42; c 2. OPTOMETRY. Educ: Ohio State Univ, BSc, 41, MSc, 42, PhD(physiol optics), 49. Prof Exp: Asst optom, Ohio State Univ, 41-43; physicist, Frankfort Arsenal, Pa, 43-44; asst prof optom, Ohio State Univ, 49-53; assoc prof, 53-57, PROF OPTOM, IND UNIV, BLOOMINGTON, 57- Concurrent Pos: Mem, Am Optom Found, 55-; Am Optom Found grant, Ind Univ, Bloomington, 61-67; visual consult to attys, 63-; mem adv panel automotive safety, US Gen Serv Admin, 66-69. Honors & Awards: Apollo Award, Am Optom Asn, 72. Mem: Am Acad Optom; Asn Res Vision & Ophthal. Res: Visibility problems in vehicle and highway design; development of techniques and equipment for therapy of amblyopia and strabismus; tonometer modification; translid bionocular trainer; night vision performance tester. Mailing Add: 1311 Valley Forge Rd Bloomington IN 47401

ALLEN, MILDRED, b Sharon, Miss, Mar 25, 94. MECHANICS. Educ: Vassar Col, AB, 16; Clark Univ, AM, 17, PhD(physics), 22. Prof Exp: Instr physics, Mt Holyoke Col, 18-20 & Wellesley Col, 22-23; from instr to asst prof, Mt Holyoke Col, 23-26; res fel, Bartol Res Found, Franklin Inst, 27-30; res instr, Oberlin Col, 30-31; from assoc prof to prof, 33-59, EMER PROF PHYSICS, MT HOLYOKE COL, 59- Concurrent Pos: Vis lectr physics, Vassar Col, 60 & Oberlin Col, 61. Mem: Fel Am Phys Soc; Am Asn Physics Teachers; fel AAAS; Sigma Xi. Res: Study of a heavy torsion pendulum, especially its unexplained variations in period which are periodic. Mailing Add: 41 Woodbridge Terr South Hadley MA 01075

ALLEN, MILLER SHANNON, JR, b White Gate, Va, Sept 23, 22; m 51; c 4. PATHOLOGY, SURGERY. Educ: Univ Va, BA, 43, MD, 45, MS, 50; Am Bd Surg, dipl, 54; Am Bd Path, dipl, 61. Prof Exp: Intern surg, Presby Hosp, New York, 45-46; asst resident, Univ Va Hosp, 48-50, chief resident & instr surg, 51-52; attend surgeon, Holston Valley Commun Hosp, Kingsport, Tenn, 52-57; from instr to assoc prof, 57-72, PROF PATH, SCH MED, UNIV VA, 72- Concurrent Pos: Am Cancer Soc fel path, 50-51. Mem: Int Acad Path. Res: Cancer; chronic lung disease. Mailing Add: Dept of Path Univ of Va Sch of Med Charlottesville VA 22903

ALLEN, MILTON JOEL, electrochemistry, see 12th edition

ALLEN, NEIL KEITH, b Elburn, Ill, July 24, 44; m 67. POULTRY NUTRITION, ANIMAL NUTRITION. Educ: Univ Ill, Urbana, BS, 67, MS, 68, PhD(animal sci), 71. Prof Exp: Res asst animal nutrit, Univ Ill, Urbana, 67-71; res assoc poultry nutrit, Cornell Univ, 71-73; formulation mgr, Dawe's Lab, Inc, 73-75; ASST PROF ANIMAL SCI, UNIV MINN, ST PAUL, 75- Mem: Poultry Sci Asn. Res: Protein and amino acid requirements of chickens as a function of age and production; mineral and vitamin needs for optimal calcium metabolism. Mailing Add: Dept of Animal Sci Univ of Minn St Paul MN 55108

ALLEN, NELSON, b Franklin, Ky, May 12, 05; m 37. PLASTICS CHEMISTRY. Educ: Centre Col, AB, 27; Univ Chicago, MS, 29; Princeton Univ, AM, 32, PhD(chem), 33. Prof Exp: Instr chem, Centre Col, 27-29, asst prof, 29-31; res chemist, E I du Pont de Nemours & Co, 33-38, res supt chem, 38-44, develop supt films, 44-50, mgr develop, 50-65, tech coordr packaging films, 65-70; CONSULT PACKAGING & PLASTIC FILMS, 70- Concurrent Pos: Adv container develop, Nat Res Coun; volunteer exec, Int Exec Serv Corps, 70- Honors & Awards: Prof Award, Packaging Inst, 65; Packaging Hall of Fame Award, Packaging Educ Found, 72. Mem: Am Chem Soc; Sigma Xi; League Int Food Educ. Res: Analytical methods; flexible films for packaging. Mailing Add: 509 St George Dr Georgian Terr Wilmington DE 19809

ALLEN, NORMAN, b Atlanta, Ga, Jan 7, 25; m 56; c 3. NEUROLOGY. Educ: Harvard Univ, MD, 49. Prof Exp: Med house officer, Peter Bent Brigham Hosp, 49-50; asst resident neurol, Boston City Hosp, 50, sr resident, 52-53; neurol consult, Medfield State Hosp, Mass, 54-55; from instr to asst prof neurol, Sch Med, Univ NC, Chapel Hill, 55-60, assoc prof neurol, anat & biochem, 60-65; PROF NEUROL & PHYSIOL CHEM & DIR DIV NEUROL, COL MED, OHIO STATE UNIV, 65- Concurrent Pos: Teaching fel, Harvard Med Sch, 52-53; NIH fels, Neurol Unit, Boston City Hosp, 53-54 & Res Lab, McLean Hosp, 54-55; USPHS res career develop award, 63; assoc attend physician, NC Mem Hosp, 50- Mem: Am Acad Neurol; Am Neurol Asn; Am Soc Neurochem; Soc Neurosci; Am Asn Neuropath. Res: Neurochemistry; developmental neurology; clinical neurology. Mailing Add: Div of Neurol Ohio State Univ Col of Med Columbus OH 43210

ALLEN, NORRIS ELLIOTT, b Baltimore, Md, Sept 15, 39; m 64; c 1. MICROBIOLOGY. Educ: Univ Md, BS, 64, MS, 66, PhD(microbiol), 69. Prof Exp: Res assoc biol, Univ Pa, 69-71; SR MICROBIOLOGIST, LILLY RES LABS, ELI LILLY & CO, 71- Mem: Am Soc Microbiol; AAAS; Sigma Xi. Res: Mechanisms of antibiotic action; inducible macrolide resistance; inhibitors of ribosome function. Mailing Add: Microbiol & Ferment Prod Res Lilly Res Labs Indianapolis IN 46206

ALLEN, OSCAR NELSON, b Corsicana, Tex, May 15, 05; m 30. BACTERIOLOGY. Educ: Univ Tex, AB & AM, 27; Univ Wis, PhD(agr bact), 30. Prof Exp: Asst agr bact, Univ Wis, 27-29, instr, 29-30; from asst prof to prof bact & plant path, Univ Hawaii, 30-45; head dept bact, Univ Md, 45-46; prof bact, 46-75, EMER PROF BACT, UNIV WIS, MADISON, 75- Concurrent Pos: Bacteriologist, Grad Sch Trop Agr, Univ Hawaii, 31-34, collab, Exp Sta, 32-39, actg head bot dept, 35-36 & 39-40, chmn dept, 40-42, bact, 42-45; coop soil bacteriologist, Pineapple Producers Coop Asn, 31-39; res bacteriologist, Rothamsted Exp Sta, Eng; sci adv yeast prod, Honolulu Plantation, 43-45; secy spec comt bact nomenclature, Int Bot Cong, Sweden, 50; mem, Int Cong Soil Sci, Netherlands, 50, mem orgn comt, Wis, 60; mem int comt bact nomenclature, Int Cong Microbiol, Brazil, 50, mem, Italy, 53, Soil Sci Soc Am del & prog chmn soil microbiol sect, USSR, 66, mem, Mexico City, 70; mem, Int Conf Rust Workers, Mex, 56; George Ives Haight traveling fel, from Univ Wis, Far East & SPac, 62; convenor, Int Soil Conf, NZ, 62; convenor biol colloquium, Ore State Univ, 64; mem judicial comn, Int Comt Nomenclature of Bacteria, 66; mem gen assembly & comt nitrogen fixation, Int Biol Prog, Paris, 66; consult, US Agency Int Develop Prog, Porto Alegre, Brazil, 67; mem int organizing comt, Int Conf Global Impacts of Appl Microbiol, Addis Ababa, 67; invitation mem, Northwest Europ Microbiol Group, Stockholm, 69. Mem: AAAS; Am Soc Microbiol; Am Phytopath Soc; fel Am Soc Agron; Soil Sci Soc Am. Res: Physiology of rhizobia; bacterial-plant symbiosis; structure of root nodules; bacteria of non-leguminous plants; aggregation of soil particles by microorganisms. Mailing Add: 116 Bact Bldg Dept of Bact Univ of Wis Madison WI 53706

ALLEN, PATTON TOLBERT, b Lockhart, Tex, Oct 7, 39; m 61; c 2. ANIMAL VIROLOGY. Educ: Southwest Tex State Col, BS, 62; Univ Tex, Austin, PhD(microbiol), 66. Prof Exp: ASST PROF VIROL & ASST VIROLOGIST, UNIV TEX M D ANDERSON HOSP & TUMOR INST, 71- Mem: AAAS; Am Soc Microbiol. Res: Viral interference in cell culture; biochemical and biophysical studies of arboviruses replication; production and assay of interferon inducers in mice; molecular biology of cancer viruses. Mailing Add: Dept of Virol Univ of Tex M D Anderson Hosp & Tumor Inst Houston TX 77025

ALLEN, PAUL, JR, b Litchfield, Conn, Sept 2, 98; m 26; c 2. ORGANIC CHEMISTRY. Educ: Harvard Univ, AB, 19, AM, 22, PhD(chem), 24. Hon Degrees: ME, Stevens Inst Technol, 55. Prof Exp: Instr chem, NY Univ, 23-24; res chemist, Marx & Rawolle, Brooklyn, 24-25 & Vacuum Oil Co, Paulsboro, 25-29; from asst prof to assoc prof chem, St John's Col, Md, 29-39; from assoc prof to prof, Lynchburg Col, 39-42; from asst prof to prof, 42-69, assoc dean grad studies, 52-53 & 68-73, EMER PROF CHEM, STEVENS INST TECHNOL, 73- Mem: Am Chem Soc; fel Am Inst Chemists. Res: Reactions of unsaturated keytones; aliphatic sulfinic acids and sulfur compounds. Mailing Add: Stevens Inst of Technol Hoboken NJ 07030

ALLEN, PAUL D, b Atlanta, Ga, Feb 24, 44; m 68; c 2. EXERICISE PHYSIOLOGY. Educ: Boston Univ, BA, 67, MD, 67; Univ Fla, PhD(physiol), 73. Prof Exp: Intern surg, Boston City Hosp, 67-68; resident surg, Univ Fla Hosp & Clin, 68-70, fel, 70-71, assoc, 71-72; RES INTERNIST EXERCISE PHYSIOL, US ARMY RES INST ENVIRON MED, 73- Concurrent Pos: Mem, Coun Thrombosis, Am Heart Asn, 69-72. Mem: Am Heart Asn; Am Col Sports Med. Res: Isolated muscle physiology; limiting factors in exercise; local control of blood flow. Mailing Add: USARIEM Exercise Physiol Div Natick MD 91760

ALLEN, PAUL JAMES, b Stockbridge, Mass, Sept 28, 14; m 43; c 3. PLANT PHYSIOLOGY. Educ: Harvard Univ, BA, 36; Univ Rochester, MS, 38; Univ Calif, PhD(plant physiol), 41. Prof Exp: Teaching asst biol, Univ Rochester, 36-38; teaching asst bot, Univ Calif, 38-41; instr, Univ Pa, 41-43; asst microbiologist, Eastern Regional Res Labs, Bur Agr & Indust Chem, USDA, 43; microbiologist, Emergency Rubber Proj, Calif, 43-46; from asst prof to assoc prof, 46-59, chmn dept, 65-70, PROF BOT, UNIV WIS, MADISON, 59-, PLANT PATH, 64- Concurrent Pos: Merck fel, Eng, 53-54; vis prof, Univ Calif, Berkeley, 60-61. Mem: Bot Soc Am; Am Phytopath Soc; Am Inst Biol Sci. Res: Physiology of plant parasitism, rusts and powdery mildews; spore germination; metabolism and growth of fungi. Mailing Add: Dept of Bot Univ of Wis Madison WI 53706

ALLEN, PETER HERBERT, forest ecology, see 12th edition

ALLEN, PETER ZACHARY, b New York, NY, July 21, 25. IMMUNOLOGY, IMMUNOCHEMISTRY. Educ: Columbia Col, AB, 49; Columbia Univ, PhD(microbiol), 58. Prof Exp: Nat Res Coun fel, Lister Inst, Eng, 58-59; NIH fel, 59-60; from asst prof to assoc prof, 60-71, PROF MICROBIOL, SCH MED, UNIV ROCHESTER, 71- Concurrent Pos: Adv ed, Immunochemistry, 71-75. Mem: Am Asn Immunologists; Biochem Soc. Res: Immunochemistry of polysaccharides and proteins. Mailing Add: Dept of Microbiol Univ of Rochester Sch of Med Rochester NY 14620

ALLEN, RALPH GORDON, physics, see 12th edition

ALLEN, RAYMOND A, b Lyman, Utah, Nov 7, 21; m 60; c 2. PATHOLOGY. Educ: Univ Utah, BS, 46; Univ Louisville, MD; Univ Minn, MS, 54. Prof Exp: Asst path & microbiol, Rockefeller Inst, 54-56; from asst prof to prof path, Univ Calif, Los Angeles, 56-68, assoc dir, Jules Stein Eye Inst, 66-68; pathologist, Good Samaritan Hosp, Phoenix, Ariz, 68-70; chief path & dir labs, St Dominic-Jackson Mem Hosp, 70-75; PATHOLOGIST, CENTINELA HOSP, INGLEWOOD, 75- Concurrent Pos: Attend physician, Los Angeles County Harbor Gen Hosp, 55-; consult, Vet Admin Ctr, Los Angeles, Calif, 54- Mem: AAAS; Am Soc Clin Path; Col Am Path; Int Acad Path. Res: Clinical pathology; hematology. Mailing Add: 1272 Via Landeta Palos Verdes Estates CA 90274

ALLEN, RAYMOND CLAYTON, b Barre, Mass, Jan 17, 07; m 34; c 2. FLORICULTURE. Educ: Mass State Col, BS, 31; Cornell Univ, PhD(floricult & plant physiol), 38. Prof Exp: From asst to asst prof floricult & ornamental hort, Cornell Univ, 32-43; exec secy & ed, Am Rose Soc, 43-53; DIR, KINGWOOD CTR, MANSFIELD, OHIO, 53- Honors & Awards: Silver Medal, Am Rose Soc, 58, Gold Hon Medal, 64, Gold Medal, 66; Silver Medal, Men's Garden Clubs Am, 61. Mem: Am Soc Hort Sci. Res: Rose culture; herbaceous plant materials. Mailing Add: Kingwood Ctr PO Box 966 Mansfield OH 41902

ALLEN, RAYMOND W, mathematics, statistics, see 12th edition

ALLEN, REX WAYNE, b Brocton, Ill, Oct 16, 08; m 33; c 2. ANIMAL PARASITOLOGY. Educ: Univ Chicago, BS, 38, MS, 47; Univ Wis, PhD(vet sci), 70. Prof Exp: STAFF MEM, VET SCI RES DIV, AGR RES SERV, USDA, 40-75, LAB DIR, 48-75; RETIRED. Concurrent Pos: Adv vet parasitol, Nat Univ Paraguay, 65-66, consult, 71-72. Mem: Fel AAAS; Am Soc Parasitol; Wildlife Dis Asn; Conf Res Workers Animal Dis. Res: Parasites and parasitic diseases of animals. Mailing Add: 3015 Fairway Dr Las Cruces NM 88001

ALLEN, RHESA MCCOY, JR, b Wash, DC, June 26, 16; m 42; c 2. GEOLOGY. Educ: Va Polytech Inst & State Univ, BS, 38; Univ Idaho, MS, 40; Cornell Univ, PhD(econ geol), 47. Prof Exp: Fel geol, Univ Idaho, 38-40; asst geol, Cornell Univ, 40-41, 45-47; asst prof, Rutgers Univ, 47-48; assoc prof, Va Polytech Inst & State Univ, 48-50; gen mgr & chief engr, French Coal Co & chief engr, Home Creek Smokeless Coal Co, 50-57; from assoc prof to prof geol, La Tech Univ, 57-75, dir div eng res, 65-73, assoc dean engr, 67-73; GEOL & ENGR CONSULT, 73- Concurrent Pos: Field asst, US Geol Surv, Idaho, 39; geologist, Ore Dept Geol, 46 & Calif Co, 47-48, 49; vpres, Peter White Coal Co, 50-; geologist, Va Div Mineral Res, 58-65. Mem: Fel Geol Soc Am; Am Inst Mining, Metall & Petrol Eng. Res: Geology of mineral deposits; Appalachian structure; geology of coal; hydrology. Mailing Add: Angel Fire Eagle Nest NM 87718

ALLEN, RICHARD BALLANTINE, b Springfield, Mass, Sept 18, 22; m 52; c 1. OPTICAL PHYSICS. Educ: Dartmouth Col, AB, 47; Harvard Univ, MA, 49; Brown Univ, PhD(physics), 55. Prof Exp: Instr physics, Worcester Polytech Inst, 47-48, 55, asst prof, 55-58; sr scientist, semiconductor div, Minneapolis Honeywell Regulator Co, 59; asst prof, 60-63, ASSOC PROF PHYSICS, UNIV HARTFORD, 63- Concurrent Pos: Consult, res labs, United Aircraft Corp, Conn, 62-63. Mem: Am Phys Soc; Am Asn Physics Teachers; Optical Soc Am. Res: Semiconductor devices and solid state diffusion; surface physics and high vacuum technique; holography. Mailing Add: Dept of Physics Univ Hartford 200 Bloomfield Ave West Hartford CT 06117

ALLEN, RICHARD CRENSHAW, JR, b Detroit, Mich, July 17, 33; m 56; c 2. APPLIED MATHEMATICS. Educ: Murray State Univ, BS, 59; Univ Mo-Columbia, MA, 60; Univ Colo, 64-66; Univ NMex, PhD(math), 68. Prof Exp: Mathematician, Martin-Marietta Corp, Colo, 60-64; ASST PROF MATH, UNIV NMEX, 68- Mem: Am Math Soc; Soc Indust & Appl Math. Res: Transport theory; analytical and numerical studies of partial differential integral equations arising from transport theory; numerical solution of two-point boundary value problems, linear and nonlinear; integral equations and ordinary differential equations. Mailing Add: Dept of Math & Statist Univ of NMex Albuquerque NM 87106

ALLEN, RICHARD DEAN, b Dallas Center, Iowa, Sept 20, 35; m 58; c 3. CELL BIOLOGY. Educ: Greenville Col, BA, 57; Univ Ill, MS, 60; Iowa State Univ, PhD(cell biol), 64. Prof Exp: Teacher sci & math, Niagara Christian Col, Ont, 57-59; instr biol & bot, Greenville Col, 60-61; asst prof biol, Messiah Col, 65-68; lectr biol & res assoc electron micros, Biol Labs, Harvard Univ, 68-69; assoc prof, 69-75, PROF MICROBIOL, PAC BIOMED RES CTR, UNIV HAWAII, MANOA, 75- Concurrent Pos: NIH res fel cell biol, Biol Labs, Harvard Univ, 64-65; vis prof, Univ Colo, Boulder, 75-76. Mem: AAAS; Soc Protozool; Am Soc Cell Biologists. Res: Fine structure, function and morphogenesis of cell organelles and organelle systems with emphasis on ciliated protozoans; studies on intracellular motion and membrane turnover. Mailing Add: Pac Biomed Res Ctr Univ of Hawaii at Manoa Honolulu HI 96822

ALLEN, RICHARD GUY, organic chemistry, physical chemistry, see 12th edition

ALLEN, RICHARD K, b Salt Lake City, Utah, Apr 21, 25. ENTOMOLOGY. Educ: Univ Utah, MS, 55, PhD(entom), 60. Prof Exp: Assoc investr, NSF grant & res assoc entom, Univ Utah, 60-62; lectr, Univ Calif, Los Angeles, 63; prin investr, NSF grants, Univ Utah, 63-64; assoc prof, 64-70, PROF INVERT ZOOL, CALIF STATE UNIV, LOS ANGELES, 70- Concurrent Pos: Prin investr, NSF grant, Calif State Univ, Los Angeles, 65-67; NSF grants, 68-72 & 73-77. Mem: Entom Soc Am; Entom Soc Can; Royal Entom Soc London. Res: Forest and aquatic entomology; scale insects; ephemeroptera; marine invertebrates of southern California. Mailing Add: Dept of Zool Calif State Univ 5151 State University Dr Los Angeles CA 90032

ALLEN, ROBERT CARTER, b Natick, Mass, Feb 5, 30; m 56; c 2. PATHOLOGY. Educ: Univ Vt, BS, 52, MS, 55; Va Polytech Inst, PhD(bact), 59. Prof Exp: Asst prof vet path, Va Polytech Inst, 55-60; NIH training grant, Jackson Mem Lab, 60-62; sr scientist radiation path, Oak Ridge Nat Lab, 62-68; sr scientist life sci, Ortec Inc, 68-73; ASSOC PROF PATH, MED UNIV SC, 73- Concurrent Pos: Consult biol div, Oak Ridge Nat Lab, 68-69; guest prof, Univ Heidelberg, 71-72 & 74. Mem: NY Acad Sci; Histochem Soc; Am Soc Clin Path. Res: Chronic obstructive lung disease; isozymes, proteinases and antiproteinases. Mailing Add: Dept of Path Med Univ of SC Charleston SC 29401

ALLEN, ROBERT DAY, b Providence, RI, Aug 28, 27; m 50, 70. CELL BIOLOGY, BIOPHYSICS. Educ: Brown Univ, BA, 49; Univ Pa, PhD, 53. Prof Exp: Fel, Nat Cancer Inst, 53-54; instr zool, Univ Mich, 54-56; asst prof biol, Princeton Univ, 56-61, assoc prof, 61-66; prof biol sci & chmn dept, State Univ NY Albany, 66-72; PROF BIOL SCI, DARTMOUTH COL, 72-, CHMN DEPT, 75- Concurrent Pos:

Guggenheim fel, 61, 66. Honors & Awards: Vis Prof Award, Japan Soc for the Promotion of Sci. Mem: Fel AAAS; Am Soc Zool; Biophys Soc; Fedn Am Sci; Am Soc Cell Biol. Res: Cell movement; cytoplasmic streaming and other motility phenomena; modulated light techniques in quantitative microscopy; cytoplasmic structure; ultrasensitive thermal and optical measurements. Mailing Add: Dept of Biol Sci Dartmouth Col Hanover NH 03755

ALLEN, ROBERT EDWARD, b Hampton, Iowa, Dec 27, 19; m 44; c 4. CHEMISTRY. Educ: Grinnell Col, AB, 41; Univ Ill, PhD(org chem), 44. Prof Exp: Asst chem, Univ Ill, 41-43; War Prod Bd, 43-45; res chemist, William S Merrell Co, 46-58; res chemist, 58-60, dir org res, 60-70, PATENT & LICENSING COORDR, CUTTER LABS, 70- Mem: Am Chem Soc. Res: Synthetic rubber; substituted styrene synthetics; medicinal chemistry. Mailing Add: Cutter Labs Berkeley CA 94710

ALLEN, ROBERT ERWIN, b Lufkin, Tex, Oct 9, 41; m 69. MEDICAL PHYSIOLOGY, HUMAN FACTORS ENGINEERING. Educ: Stephen F Austin State Col, BA, 63; Vanderbilt Univ, PhD(med physiol), 68. Prof Exp: Aerospace physiologist, 68-69, chief biomed res-develop, 69-72, sect chief, Biotechnol Br, 72-74, CHIEF BIOTECHNOL BR, GEORGE C MARSHALL SPACE FLIGHT CTR, NASA, 74- Mem: Aerospace Med Asn. Res: Exercise and aerospace physiology. Mailing Add: EH35 Marshall Space Flight Ctr Huntsville AL 35812

ALLEN, ROBERT HENRY, b Orange, Mass, July 25, 27; m 63; c 2. ORGANIC CHEMISTRY. Educ: Univ Conn, BA, 50; Univ Calif, PhD(org chem), 53. Prof Exp: Res chemist, Polymer Res Lab, Dow Chem Co, 53-62, sr res chemist, Plastics Dept Res Lab, 62-69; assoc prof chem, Tri-State Col, 69-74; PRES & CHMN, GULF RESOURCES & CHEM CORP, HOUSTON, TEX, 74- Mem: Sigma Xi. Res: Kinetics and mechanisms of alkylbenzene alkylation and isomerization; polymer chemistry; psychology of learning and growth. Mailing Add: Gulf Resources & Chem Corp 2125 Tenneco Bldg Houston TX 77002

ALLEN, ROBERT JOHN, JR, b Melrose, Mass, Apr 20, 12; m 36; c 2. AGRONOMY. Educ: Mass State Col, BS, 35; Univ Mass, MS, 49; Univ Md, PhD(agron), 51. Prof Exp: Florist, R J Allen, Sr, Mass, 35-42; ASST AGRONOMIST, AGR RES & EDUC CTR, UNIV FLA, 51- Concurrent Pos: Mem, Am Forage & Grassland Coun. Mem: Am Soc Agron. Res: Pasture grass management under grazing and mechanical harvesting; fertilization; preservation as silage; plant introduction of grasses and forage crops; dehydration and pelleting of forages; solar radiation and plant growth. Mailing Add: Univ of Fla Agr Res & Educ Ctr Drawer A Belle Glade FL 33430

ALLEN, ROBERT MAX, b Ill, June 5, 21; m 46; c 3. FORESTRY. Educ: Iowa State Univ, BS, 47, MS, 51; Duke Univ, PhD(tree physiol), 58. Prof Exp: Forester, Southern Forest Exp Sta, US Forest Serv, 48-57, plant physiologist, 57-66; Belle W Baruch prof forestry, 66-70, PROF FORESTRY & HEAD DEPT, CLEMSON UNIV, 70- Mem: Soc Am Foresters; Am Forestry Asn; Ecol Soc Am. Res: Tree physiology and genetics; growth of southern pine. Mailing Add: Dept of Forestry Clemson Univ Clemson SC 29631

ALLEN, ROBERT RAY, b Potwin, Kans, Sept 2, 20; m 44; c 5. ORGANIC CHEMISTRY. Educ: Kans State Col, BS, 47, MS, 48, PhD(Chem), 50. Prof Exp: Chemist, Armour & Co, 50-52; head chem res dept, Foods Div, 56-68, DIR EXPLOR RES, ANDERSON CLAYTON FOODS, ANDERSON, CLAYTON & CO, 68- Mem: Am Chem Soc; Am Oil Chem Soc (pres, 72-73). Res: Fats and oils. Mailing Add: W L Clayton Res Ctr Anderson Clayton & Co 3333 N Central Expressway Richardson TX 75080

ALLEN, ROBERT SCOTT, b Tabonia, Utah, Nov 13, 17; m 40; c 3. BIOCHEMISTRY. Educ: Brigham Young Univ, BS, 39, MS, 40; Iowa State Col, PhD(biochem), 49. Prof Exp: Technician biochem, Iowa State Univ, 40-43, instr chem, 46-47, res assoc, 47-49, from asst prof to prof chem & dairy husb, 49-60, prof biochem & dairy sci, 60-67; PROF BIOCHEM & HEAD DEPT, LA STATE UNIV, 67- Concurrent Pos: Spec consult, NIH, 60-64. Honors & Awards: Am Feed Mfg Award, 55. Mem: Am Inst Nutrit; Am Chem Soc; Am Dairy Sci Asn. Res: Lipid metabolism; carotene and vitamin A metabolism; forage preservation and utilization; ruminant nutrition; bloat syndrome; metabolism in marine organisms. Mailing Add: 256 Court St Baton Rouge LA 70810

ALLEN, ROBERT THOMAS, b Farmerville, La, Dec 14, 39; m 62. SYSTEMATIC ENTOMOLOGY. Educ: La State Univ, BS, 62, MS, 64; Univ Ill, PhD(entom), 68. Prof Exp: Asst prof, 67-71, ASSOC PROF ENTOM, UNIV ARK, FAYETTEVILLE, 71- Mem: Entom Soc Am; Soc Syst Zool; Soc Study Evolution. Res: Systematics and ecology of Coleoptera, especially the family Carabidae. Mailing Add: Dept of Entom Univ of Ark Fayetteville AR 72701

ALLEN, ROGER BAKER, b Portland, Maine, Mar 26, 29; m 54; c 3. PHYSICAL CHEMISTRY. Educ: Univ Idaho, BS, 51; Univ NH, MS, 57, PhD(chem), 59. Prof Exp: Res chemist, Electro- chem Syst, US Naval Ord Lab, Md, 59-61; RES CHEMIST, S D WARREN CO DIV, SCOTT PAPER CO, 61- Mem: Am Chem Soc; Tech Asn Pulp & Paper Indust. Res: Kinetic studies of electrochemical systems for ordnance hardware; specialty coatings for paper, lithographic and electroelectrophotographic. Mailing Add: Paper Coatings S D Warren Co Div Scott Paper Co Westbrook ME 04092

ALLEN, ROGER WILLIAMS, b Birmingham, Ala, Mar 29, 97; m 27; c 2. ORGANIC CHEMISTRY. Educ: Auburn Univ, BS, 18, MS, 19; Univ Mich, AM, 21; Columbia Univ, PhD(chem), 27. Prof Exp: Chemist, State Chem Lab, Ala, 19; chemist, bur entom & plant quarantine, USDA, 21; prof chem, Howard Col, 21-22, 23-26; chemist, Inecto, Inc & Marinello Co, New York, NY, 26-28; prof chem, 28-41, dean sch sci & lit, 41-67, EMER DEAN SCH SCI & LIT, AUBURN UNIV, 67- Res: Organic dyestuffs; electrolytic reduction of organic acids and other compounds. Mailing Add: Sch of Sci & Lit Auburn Univ Auburn AL 36830

ALLEN, ROLAND EMERY, b Houston, Tex, Sept 3, 41. SOLID STATE PHYSICS. Educ: Rice Univ, BA, 63; Univ Houston, BA, 65; Univ Tex, Austin, PhD(physics), 69. Prof Exp: Teacher high sch, Tex, 64-65; res assoc physics, Univ Tex, Austin, 69-70; ASST PROF PHYSICS, TEX A&M UNIV, 70- Mem: AAAS; Am Phys Soc; Am Asn Physics Teachers. Res: Surface physics. Mailing Add: Dept of Physics Tex A&M Univ College Station TX 77843

ALLEN, ROSS LORRAINE, b Newark, NJ, Feb 28, 05; m 27; c 2. PUBLIC HEALTH. Educ: Univ Mich, BS, 27, MS, 34, DrPH, 36. Prof Exp: Field rep, Sch Health Study, Am Child Health Asn, NY, 27-28; high sch teacher, NY, 28-35; asst ed, J Asn Health, Phys Educ & Recreation, 35-36; asst exec dir, Am Camping Asn, 36-39, exec dir, 39-41; asst supvr dept phys educ & athletics, Univ Mich, 41-45; dir educ, Div Health, Phys Educ & Recreation, Mich, 41-45; head dept health educ, State Univ NY Col Cortland, 45-54, prof, 45-74, dir educ, Div Health, Phys Educ & Recreation, 54-65, dean div grad studies & res, 65-74; RETIRED. Mem: Am Asn Health, Phys Educ & Reacreation; Am Sch Health Asn. Res: School health education; training of school health educators; weight deviation and health in a college freshman group. Mailing Add: 1222 Bell Dr Cortland NY 13045

ALLEN, ROSS MARVIN, b Oil City, Pa, Nov 2, 17; m 41; c 3. PLANT PATHOLOGY. Educ: Univ Ariz, BS, 50, PhD(plant path), 53. Prof Exp: Asst prof plant path, Univ Ariz, 53-54; plant pathologist, Abaca Proj, USDA, Inter-Am Inst Agr Sci, 54-57; from asst plant pathologist to plant pathologist, Yuma Br Agr Exp Sta, 57-67; PROF PLANT PATH & PLANT PATHOLOGIST, UNIV ARIZ, 67- Mem: AAAS; Am Phytopath Soc; Int Orgn Citrus Virol. Res: Diseases of citrus; citrus virology. Mailing Add: Dept of Plant Path Univ of Ariz Tucson AZ 85721

ALLEN, ROVELLE HARPER, b Aylesbury, Sask, Sept 5, 10; m 40; c 3. CLINICAL CHEMISTRY, SEROLOGY. Educ: Univ Sask, BS, 32, MSc, 34; Univ Wis, PhD(physiol chem), 40, MD, 41. Prof Exp: Med officer, Can Commun Dis Centre, Dept Nat Health & Welfare, Can, 43-75. Res: Methods and quality control. Mailing Add: 262 Second Ave Ottawa ON Can

ALLEN, SALLY LYMAN, b New York, NY, Aug 3, 26; m 51; c 1. GENETICS. Educ: Vassar Col, AB, 46; Univ Chicago, PhD(zool), 54. Prof Exp: Asst, R B Jackson Mem Lab, Bar Harbor, 46-48; res asst, 53-54, RES ASSOC ZOOL, UNIV MICH, 55-, PROF BOT, 70- Concurrent Pos: Assoc prof bot, Univ Mich, 67-70. Mem: Fel AAAS; Genetics Soc Am; Am Genetics Asn; Am Soc Zool; Soc Protozool. Res: Genetics of tumor transplantation; bacteriophage and protozoan genetics; factors affecting control of protein structure and function; phenotypic drift; genomic exclusion; DNA hybridization. Mailing Add: Dept of Bot 3121 Natural Sci Bldg Univ of Mich Ann Arbor MI 48104

ALLEN, SEWARD ELLERY, b Macedon, NY, Mar 21, 20; m 49; c 2. PLANT PHYSIOLOGY, SOIL CHEMISTRY. Educ: Cornell Univ, BS, 42; Univ Wis, MS, 47, PhD(bot), 50. Prof Exp: Res asst, Univ Wis, 47-50; plant physiologist, Tex Res Found, 50-52; agr chemist, Midwest Res Inst, 52-57; res chemist, US Rubber Co, 57-62; PLANT PHYSIOLOGIST, TENN VALLEY AUTH, 62- Mem: Am Soc Agron; Soil Sci Soc Am; Am Soc Plant Physiol. Res: Plant growth regulators; soil fertility; utilization of fertilizers by isotope technique; nutrition and physiology of Hevea rubber; agronomic evaluation and development of experimental plant nutrients. Mailing Add: 2934 Alexander St Florence AL 35632

ALLEN, STEPHEN IVES, b Holyoke, Mass, Dec 13, 14; m 43; c 2. MATHEMATICS. Educ: Amherst Col, BA, 37; Harvard Univ, MA, 46; Univ Pittsburgh, PhD(math, anal), 63. Prof Exp: Instr math & physics, Ruston Acad, Havana, Cuba, 38-39 & Wilbraham Acad, 39-40; instr & prin, Grahmam-Eckes Sch, Fla, 40-41 & 46-48; from instr to asst prof, 48-67, ASSOC PROF MATH, UNIV MASS, AMHERST, 67-, ASST DEAN COL ARTS & SCI, 73- Mem: Math Asn Am; Am Math Soc. Res: Algebraic functions; algebra. Mailing Add: Dept of Math Univ of Mass Amherst MA 01002

ALLEN, SUSAN DAVIS, b Jacksonville, Fla, Sept 13, 43; m 62; c 1. OPTICAL PHYSICS. Educ: Colo Col, BS, 66; Univ Southern Calif, PhD(chem physics), 71. Prof Exp: Res assoc chem physics, Univ Southern Calif, 71-73; MEM TECH STAFF CHEM PHYSICS, HUGHES RES LABS, 73- Mem: Am Phys Soc; Am Chem Soc; AAAS. Res: Low absorbing optics for high power lasers and optical damage; surface and bulk optical properties of solids in the infrared; new instrumentation to measure low optical absorption and surface properties. Mailing Add: Hughes Res Labs 3011 Malibu Canyon Rd Malibu CA 90265

ALLEN, SYDNEY HENRY GEORGE, b Cambridge, Mass, Aug 26, 29; m 52; c 2. BIOCHEMISTRY. Educ: Tufts Univ, BS, 51; Univ Mass, MS, 53; Purdue Univ, PhD, 57. Prof Exp: Instr bact, Univ Conn, 57-60; from asst prof to assoc prof biochem, 63-75, PROF BIOCHEM, ALBANY MED COL, 75- Concurrent Pos: Fel biochem, Sch Med, Western Reserve Univ, 60-63; Lederle med fac award, 65-68. Mem: AAAS; Am Soc Microbiol; Am Soc Biol Chemists. Res: Metabolic role of carbon dioxide; carboxylation and decarboxylation reactions; mechanism of enzyme action; relationship between vitamin and enzyme activity; fermentation pathways; membrane-bound enzymes and intestinal lipid absorption. Mailing Add: Dept of Biochem Albany Med Col Albany NY 12208

ALLEN, TED TIPTON, b McKenzie, Tenn, Mar 22, 32; m 58; c 2. VERTEBRATE ZOOLOGY. Educ: Murray State Col, BA, 54; Univ Wis, MS, 58; Univ Fla, PhD(zool), 62. Prof Exp: From instr to assoc prof, 61-74, PROF BIOL, JACKSONVILLE UNIV, 74-, HEAD DEPT, 63- Mem: Am Ornith Union; Am Inst Biol Sci; Wilson Ornith Soc. Res: Ornithology; ecology and mycology of birds. Mailing Add: Dept of Biol Jacksonville Univ Box 3 Jacksonville FL 32211

ALLEN, THOMAS CORT, JR, b Madison, Wis, Oct 28, 31; m 53; c 1. PLANT PATHOLOGY. Educ: Univ Wis, BS, 53; Univ Calif, PhD(plant path), 56. Prof Exp: Plant pathologist, Biol Warfare Labs, Ft Detrick, Md, 56-58; head plant path sect, Agr Res & Develop Labs, Stauffer Chem Co, Calif, 58-62; from asst prof to assoc prof, 62-73, PROF BOT & PLANT PATH, ORE STATE UNIV, 73- Concurrent Pos: NSF fel, 64; NATO sr fel sci, 70. Mem: Am Phytopath Soc; Am Inst Biol Sci; Electron Micros Soc Am; fel Royal Hort Soc. Res: Plant virology; ultrastructural pathology; biological control of plant viruses. Mailing Add: Dept of Bot & Plant Path Ore State Univ Corvallis OR 97331

ALLEN, THOMAS HUNTER, physiology, see 12th edition

ALLEN, THOMAS J, b Wortham, Tex, Aug 4, 25; m 46; c 3. RANGE SCIENCE, WEED SCIENCE. Educ: Tex A&M Univ, BS, 49, PhD(range sci), 69; Sul Ross State Univ, MS, 56. Prof Exp: Teacher voc agr, Marathon Pub Schs, Tex, 49-55 & Alpine Pub Schs, 55-57; range specialist, Tex A&M Univ, 57-64, range scientist, 64-67, asst prof range sci, 67-74; ASST PROF RES, TEX AGR RES & EXTEN CTR, 74- Concurrent Pos: Consult, Velsicol Chem Corp, 73. Mem: Soc Range Mgt; Am Soc Plant Physiologists; Sigma Xi; Weed Sci Soc Am. Res: Industrial vegetation management, reproduction and establishment of forage plants for rangeland use, and chemical control of herbaceous and woody plants on rangelands. Mailing Add: Tex Agr Res & Exten Ctr PO Box 1658 Vernon TX 76384

ALLEN, THOMAS LOFTON, b San Jose, Calif, Jan 20, 24; m 44; c 4. PHYSICAL CHEMISTRY. Educ: Univ Calif, BS, 44; Calif Inst Technol, PhD(chem), 49. Prof Exp: Asst chem, Calif Inst Technol, 46-47; asst prof chem, Univ Idaho, 48-49; lectr chem, Univ Calif, Davis, 49-51; res chemist, Calif Res Corp, 51-52; from asst prof to assoc prof, 52-63, PROF CHEM, UNIV CALIF, DAVIS, 63- Concurrent Pos: Vis res scholar, Ind Univ, 59-60, vis prof, 70-71; vis prof, Univ Nottingham, 71- Mem: Am Chem Soc; Am Phys Soc. Res: Chemical kinetics; thermodynamics; chemical bonding; molecular quantum mechanics. Mailing Add: Dept of Chem Univ of Calif Davis CA 95616

ALLEN, VERNON R, b Nashville, Tenn, Mar 25, 28; m 51; c 4. PHYSICAL

CHEMISTRY, POLYMER CHEMISTRY. Educ: Tenn Polytech Inst, BS, 52; Univ Akron, MS, 57, PhD(polymer chem), 60. Prof Exp: Analyst, Carbide & Carbon Chem Co, 53-54; instr gen chem, Univ Wis, Milwaukee Exten, 54-55; res fel polymer physics, Mellon Inst, 60-63; asst prof phys chem, 63-69, assoc prof chem, 69-74, PROF CHEM, TENN TECHNOL UNIV, 74- Mem: Am Chem Soc. Res: Rheology of polymeric system; physical properties of rubbery materials. Mailing Add: Dept of Chem Tenn Technol Univ Cookeville TN 38501

ALLEN, WAYNE ROBERT, b New Castle, Pa, Feb 28, 34; m 63; c 3. PLANT PATHOLOGY. Educ: Hiram Col, BA, 56; Cornell Univ, PhD(plant path), 62. Prof Exp: RES SCIENTIST, CAN DEPT AGR, 61- Mem: Am Phytopath Soc; Can Phytopath Soc. Res: Chemical and physical properties of plant viruses; diseases of fruit trees; nuclear stock production. Mailing Add: Res Sta Can Dept of Agr Vineland Station ON Can

ALLEN, WENDALL E, b Elizabethtown, Ky, Nov 16, 36; m 66. MICROBIOLOGY, MICROBIAL GENETICS. Educ: Vanderbilt Univ, BA, 58, MA, 61; Univ Ky, PhD(microbiol), 68. Prof Exp: Instr microbiol, Ala Col, 61-63; res assoc bact genetics, Univ Ky, 65-67; asst dean gen col, 70-75, ASST PROF BIOL, EAST CAROLINA UNIV, 67-, ASSOC DEAN GEN COL, 75- Mem: Am Soc Microbiol. Res: Genetics of staphylococci and their bacteriophage, with particular emphasis on hemolysins and nucleases; lysogeny and conversion in staphylococci, comparing phage relationships in Staphylococcus aureus and Staphylococcus epidermidis. Mailing Add: Dept of Biol East Carolina Univ Greenville NC 27834

ALLEN, WILLARD FINLAY, b London, Ont, May 15, 24; m 46; c 4. ANALYTICAL CHEMISTRY. Educ: Univ Western Ont, BSc, 45; Univ Toronto, MA, 46, PhD(chem), 50. Prof Exp: From asst prof to assoc prof, 48-71, PROF CHEM, UNIV ALTA, 71-, ASSOC VPRES, 70- Mem: Am Chem Soc. Res: Spectrophotometry; absorption spectroscopy; photochemistry. Mailing Add: 3-12 Univ Hall Univ Alta Edmonton AB Can

ALLEN, WILLARD M, b Macedon, NY, Nov 5, 04; m 27, 46; c 1. OBSTETRICS & GYNECOLOGY, ENDOCRINOLOGY. Educ: Hobart Col, BS, 26; Univ Rochester, MS, 29, MD, 32. Hon Degrees: ScD, Hobart Col, 40 & Univ Rochester, 57. Prof Exp: From instr to asst prof obstet & gynec, Sch Med & Dent, Univ Rochester, 36-40; prof & chmn dept, 40-71, EMER PROF OBSTET & GYNEC, SCH MED, WASHINGTON UNIV, 71-; PROF OBSTET & GYNEC, SCH MED, UNIV MD, BALTIMORE CITY, 71- Honors & Awards: Eli Lilly Award in Biochem, 35. Mem: AMA; Am Gynec Soc; Am Asn Obstet & Gynec; Endocrine Soc; Am Physiol Soc. Res: Female sex hormones and endocrinology. Mailing Add: Univ of Md Hosp 22 S Greene St Baltimore MD 21201

ALLEN, WILLARD ROSS, entomology, see 12th edition

ALLEN, WILLIAM E, JR, b Pensacola, Fla, Aug 14, 03; m 38. MEDICINE, RADIOLOGY. Educ: Howard Univ, BS, 27, MD, 30. Prof Exp: Med dir radiol, Homer G Phillips Hosp, 45-73; assoc clin prof, 73-75, CLIN PROF RADIOL, SCH MED, ST LOUIS UNIV, 75-, RADIOTHERAPIST, 70- Concurrent Pos: Consult, Secy War, 45-47; from instr to asst clin prof, Sch Med, St Louis Univ, 67-73. Honors & Awards: Nat Med Asn Distinguished Serv Award, 67; Am Col Radiol Gold Medal, 74. Mem: AAAS; Nat Med Asn (vpres, 62-63); fel Am Col Radiol; Am Soc Therapeut Radiol; Radiol Soc NAm. Res: Nuclear medicine; radiation therapy-carcinoma of the prostate. Mailing Add: 720 N Sarah St St Louis MO 63108

ALLEN, WILLIAM HUBERT, b Yakima, Wash, Oct 9, 16; m 42; c 4. PHYSICS. Educ: Univ Tex, BA, 51. Prof Exp: Chemist, Am Potash & Chem Co, 39-40; weather observer, US Weather Bur, Alaska, 40-41; sr instr meteor, Tech Sch Aviation, Brazil, 46-47; res & educ specialist, Arctic Desert Tropic Info Ctr, US Air Force Air Univ, 49-57; ed, US Nat Comt, Int Geophys Year, 57-58; specialist weapons systs eval div, Inst Defense Anal, Pentag, 58-59; chief, Spec Pub Br, NASA, 59-63, life sci prog mgr, 63-65, res scientist, Mission Anal Div, Ames Res Ctr, 65-69; SYSTS ANALYST, 69- Concurrent Pos: From asst prof to assoc prof, US Air Force Air Univ, 52-57; mem comt vision, Nat Acad Sci-Nat Res Coun; dir congressional inquiries, Nat Aeronautics & Space Admin, 72- Mem: AAAS; Optical Soc Am; cor mem Int Acad Astronaut; World Future Soc; Int Soc Technol Assessment. Res: Man-system integration; aerospace visual problems; simulation of aerospace environments; environmental impact of aerospace systems; technology assessment; lexicography of science and technology. Mailing Add: 5024 Garfield St NW Washington DC 20016

ALLEN, WILLIAM MERLE, b San Luis Obispo, Calif, Oct 9, 39; m 63; c 2. ORGANIC CHEMISTRY. Educ: Loma Linda Univ, BA, 61; Univ Md, College Park, PhD(org chem), 67. Prof Exp: Asst prof chem, Andrews Univ, 66-68; asst prof, 68-71, ASSOC PROF CHEM & CHMN DEPT, LOMA LINDA UNIV, LA SIERRA CAMPUS, 71- Mem: Am Chem Soc. Res: Synthetic and mechanistic studies of organic compounds. Mailing Add: Dept of Chem Loma Linda Ave La Sierra Campus Riverside CA 92505

ALLEN, WILLIAM PETER, b Buffalo, NY, Jan 15, 27; m 51; c 3. MEDICAL MICROBIOLOGY. Educ: Univ Buffalo, BA, 49, MA, 51; Univ Mich, PhD(med bact), 56. Prof Exp: Res asst immunol, Univ Mich, 54-55; microbiologist, US Army Biol Labs, 55-68; supvry microbiologist, US Dept Army, 68-71; res scientist, Delta Primate Ctr, Tulane Univ, 71-75; HEALTH SCIENTIST ADMINR, NAT INST ALLERGY & INFECTIOUS DIS, NIH, 75- Concurrent Pos: Mem, Am Comt Arthropod-Borne Viruses, Am Inst Biol Sci; adj assoc prof microbiol & immunol, Tulane Univ, 73-75. Honors & Awards: Res Award, Sci Res Soc Am, 62. Mem: AAAS; Am Soc Microbiol; Am Soc Trop Med & Hyg; Am Inst Biol Sci; Sigma Xi. Res: Natural and acquired immunity to microbial diseases; administrator and program manager of grants on viral diseases. Mailing Add: 5333 Westbard Ave Rm 706 Bethesda MD 20014

ALLEN, WILLIAM ROSS, b Sydney, Australia, Dec 8, 42; m 75. DEVELOPMENTAL BIOLOGY. Educ: Hawkesbury Agr Col, dipl agr, 62; Univ Calif, Davis, BS, 66, PhD(genetics), 71. Prof Exp: Fel, Dept Zool, Univ Calif, Berkeley, 70-72; ASST PROF BIOL, UNIV CALIF, RIVERSIDE, 72- Mem: Sigma Xi; Soc Develop Biol; Am Soc Zoologists. Res: Organization and function of nuclear RNA in sea urchin embryos. Mailing Add: Dept of Biol Univ of Calif Riverside CA 92502

ALLEN, WILLIAM WAYNE, agronomy, see 12th edition

ALLEN, WILLIAM WESTHEAD, b Santa Cruz, Calif, Oct 13, 21; m 54; c 2. ECONOMIC ENTOMOLOGY, ACAROLOGY. Educ: Univ Calif, BS, 43, PhD(entom), 52. Prof Exp: Assoc entomologist, 52-66, ENTOMOLOGIST, UNIV CALIF, BERKELEY, 66- Mem: Entom Soc Am; Sigma Xi; AAAS. Res: Insect pest management of agricultural crops with emphasis on economic injury levels and the use of selective pesticides; floriculture. Mailing Add: Berkeley CA

ALLENBACH, CHARLES ROBERT, b Buffalo, NY, Aug 3, 28; m 55; c 2. INORGANIC CHEMISTRY, PHYSICAL CHEMISTRY. Educ: Univ Buffalo, BA, 49, PhD(chem), 52. Prof Exp: Instr chem, Univ Buffalo, 51-52; sect mgr chem eng process develop div, Tech Dept, Metals Div, 52-63, sr develop chemist, Develop Dept, Consumer Prod Div, 63-66, sect mgr res & develop, Mining & Metals Div, 66-70, STAFF ENGR ENVIRON CONTROL, ENG DEPT, FERROALLOYS DIV, UNION CARBIDE CORP, 70-, MGR ENVIRON AFFAIRS, METALS DIV, 75- Mem: Am Chem Soc; Am Inst Mining, Metall & Petrol Eng; Am Soc Test Mat; Mfg Chem Asn. Res: Environmental control activities; new product development; powder technology; aerosol technology; metal chemicals and compounds and reactive metals. Mailing Add: Opers Serv Metals Div Union Carbide Corp Niagara Falls NY 14302

ALLENBY, CLEMENT W, b Franklin Centre, Que, Sept 7, 17; m 58; c 3. ORGANIC CHEMISTRY. Educ: McGill Univ, BSc, 39, PhD, 42; Univ Toronto, MA, 41. Prof Exp: Res chemist, Can Industs Ltd, 42-44, group leader, 44-54; mgr res lab, 54-60, mgr res & develop dept, 60-65, mgr films dept, 65-69, MGR RES DIV, DU PON OF CAN LTD, 69- Mailing Add: Res Div Du Pont Can Ltd PO Box 660 Montreal PQ Can

ALLENBY, RICHARD JOHN, JR, b Chicago, Ill, July 4, 23; m 46; c 3. GEOPHYSICS. Educ: Dartmouth Col, AB, 43, MA, 48; Univ Toronto, PhD(geophys), 52. Prof Exp: Instr physics, Dartmouth Col, 46-48; demonstr, Univ Toronto, 49-52; res seismologist, Calif Res Corp, Standard Oil Co, Calif, 52-55, lead geophysicist, 59-62; chief geophysicist, Richmond Petrol Ço, 55-59; staff scientist, Off Space Sci, NASA, 62-63, chief planetology, 63-64, dep dir, Manned Space Sci Div, 64-68, asst dir lunar sci, 68-75, RES GEOPHYSICIST, GODDARD SPACE FLIGHT CTR, 75- Honors & Awards: Apollo 8 Letter of Commendation; Apollo Achievement Award; NASA Medal Exceptional Sci Achievement, 71. Mem: AAAS; Am Geophys Union; Soc Explor Geophys; Europ Asn Explor Geophys. Res: Geophysics and geology of earth moon and planets; spacecraft scientific instrumentation; explorational petroleum geophysics and geology. Mailing Add: 6389 Lakeview Dr Falls Church VA 22041

ALLENDE, MANUEL FRANCISCO, b Spain, Nov 11, 18; nat US; m 52; c 2. MEDICINE. Educ: Univ Paris, BS, 36, MD, 42. Prof Exp: From asst clin prof to assoc clin prof, 48-72, CLIN PROF DERMAT, SCH MED, UNIV CALIF, SAN FRANCISCO, 72-; CHIEF SERV DERMAT, CHILDREN'S HOSP, 66- Honors & Awards: 1st Award, Am Acad Dermat, 52. Mem: Am Acad Dermat; AMA; Soc Invest Dermat; hon & corresp mem Fr, Brazilian, Ecuadorian & Cent Am Socs Dermat. Mailing Add: 490 Post St San Francisco CA 94102

ALLENDER, DAVID WILLIAM, b Mt Pleasant, Iowa, Nov 20, 47; m 70; c 2. THEORETICAL SOLID STATE PHYSICS. Educ: Univ Iowa, BA, 70; Univ Ill, MS, 71, PhD(physics), 75. Prof Exp: Res assoc physics, Brown Univ, 74-75; ASST PROF PHYSICS, KENT STATE UNIV, 75- Mem: Am Phys Soc. Res: Liquid crystal; phase transitions; superconductivity. Mailing Add: Dept of Physics Kent State Univ Kent OH 44242

ALLENDER, JAMES HARRY, b Flint, Mich, June 15, 48; m 69. PHYSICAL OCEANOGRAPHY. Educ: Univ Mich-Flint, AB, 70, MS, 72, PhD(oceanic sci), 75. Prof Exp: Physicist, DuPont Chem Co, 70-71; asst res oceanographer, Univ Mich, 74-75; RES ASSOC PHYS LIMNOL RES, ARGONNE NAT LAB, 75- Res: Physics of lake and ocean circulation, especially numerical modeling of Great Lakes circulation via stochastic and deterministic methods; analysis of observational data to test present theories in physical limnology. Mailing Add: EES Div Argonne Nat Lab Argonne IL 60439

ALLENDOERFER, CARL BARNETT, mathematics, deceased

ALLENDORF, FREDERICK WILLIAM, b Philadelphia, Pa, Apr 29, 47; m 68; c 1. POPULATION GENETICS. Educ: Pa State Univ, BS, 71; Univ Wash, MS, 73, PhD(fisheries, genetics), 75. Prof Exp: Lectr, Genetics Inst, Aarhus Univ, Denmark, 75-76; ASST PROF GENETICS, DEPT ZOOL, UNIV MONT, 76- Honors & Awards: Spec Achievement Award, Nat Oceanic & Atmospheric Agency, US Dept Commerce, 73, 75. Mem: Genetics Soc Am; Am Fisheries Soc. Res: Experimental and theoretical study of genetic variation in natural populations and the application of the principles of population genetics to the management of aquatic species. Mailing Add: Dept of Zool Univ of Mont Missoula MT 59801

ALLENSON, DOUGLAS ROGERS, physical chemistry, water chemistry, see 12th edition

ALLENSPACH, ALLAN LEROY, b Campbell, Minn, Sept 23, 35; m 57; c 3. DEVELOPMENTAL BIOLOGY. Educ: Sioux Falls Col, BS, 57; Iowa State Univ, MS, 59, PhD(embryol, bot), 61; Marine Biol Lab, cert, 61. Prof Exp: Asst prof biol, Albright Col, 61-66; ASSOC PROF ZOOL & PHYSIOL, MIAMI UNIV, 66- Concurrent Pos: Res consult, Bernville Biol Labs, 64; USPHS fel, 65. Mem: AAAS; Am Soc Zool; Am Soc Cell Biol; Soc Develop Biol; Am Inst Biol Sci. Res: Normal and experimental studies on organogenesis in the chick embryo. Mailing Add: Dept of Zool Miami Univ Oxford OH 45056

ALLENSTEIN, RICHARD VAN, b Lamont, Iowa, Aug 28, 30; m 55; c 3. ANALYTICAL CHEMISTRY. Educ: Univ Minn, BS, 52, MA, 57; Wash Univ, EdD, 61. Prof Exp: Dir NSF acad yr inst, Wash Univ, 62-66; asst prof chem, Univ Wis, Waukesha, 66-69; asst prof, 69-72, ASSOC PROF CHEM, NORTHERN MICH UNIV, 72- Mem: Fel AAAS; Am Chem Soc; Nat Asn Res Sci Teaching; Nat Sci Teachers Asn. Res: Measuring the effectiveness of science teaching and determining its correlates. Mailing Add: 61 W Elder Dr Marquette MI 49855

ALLENTOFF, NORMAN, b New York, NY, Nov 16, 23; m 56; c 4. PHOTOGRAPHIC CHEMISTRY. Educ: Univ Toronto, BA, 50, MA, 51, PhD(org chem), 56. Prof Exp: Res chemist, Chem Div, Can Dept Agr, 51-53, 55-58; sr chemist, Photog Chem Res Labs, 58-63, res assoc, 64-65, TECH ASSOC, FILM EMULSION DIV, EASTMAN KODAK CO, 65- Mem: Am Chem Soc. Res: Plant biochemistry; stereochemistry of the Grignard reaction; photographic chemistry. Mailing Add: 236 Paddy Hill Dr Rochester NY 14616

ALLER, HAROLD ERNEST, b New York, NY, May 13, 34; m 57; c 3. ENTOMOLOGY. Educ: City Col New York, BS, 55; Cornell Univ, PhD(entom), 62. Prof Exp: Res entomologist, Niagara Chem Div, FMC Corp, 62-67; GROUP LEADER INSECTICIDES & NEMATOCIDES, ROHM & HAAS CO, 67- Mem: Entom Soc Am; Am Mosquito Control Asn. Res: Chemical control; pesticides; toxicology; acaricides; molluscicides. Mailing Add: Res Labs Rohm & Haas Co Spring House PA 19477

ALLER, LAWRENCE HUGH, b Tacoma, Wash, Sept 24, 13; m 41; c 3. ASTROPHYSICS. Educ: Univ Calif, AB, 36; Harvard Univ, AM, 38, PhD(astron), 43. Prof Exp: Lectr astron, Tufts Col, 40; instr physics, Harvard Univ, 42-43, res assoc, Observ, 42-43, 47; physicist, Univ Calif, 43-45; asst prof astron, Ind Univ & res assoc, W J McDonald Observ, 45-48; assoc prof, Univ Mich, 48-54, prof, 54-62; PROF ASTRON, UNIV CALIF, LOS ANGELES, 62- Concurrent Pos: Pres stellar spectros comn, Int Astron Union, 58-64; vis prof, Australian Nat Univ, 60-61 & Univ Toronto, 61-62; NSF sr fel, Australia, 60-61 & 68-69; vis prof, Univ Sydney & Univ Tasmania, 68-69; res assoc, Commonwealth Sci & Indust Res Orgn, Australia, 68, 69 & 71. Mem: Fel Nat Acad Sci; fel AAAS; fel Am Acad Arts & Sci; Am Astron Soc. Res: Spectroscopic and theoretical studies of the gaseous nebulae and stellar atmospheres; transition probabilities for spectral lines; cosmic abundances of elements. Mailing Add: Dept of Astron Univ of Calif Los Angeles CA 90024

ALLER, MARGO FRIEDEL, b Springfield, Ill, Aug 27, 38; m 64. ASTRONOMY. Educ: Vassar Col, BA, 60; Univ Mich, MS, 64, PhD(astron), 69. Prof Exp: Mathematician programmer astron, Smithsonian Astrophys Observ, 60-62; teaching fel, 62-66, res asst, 66-68, teaching fel, 69, RES ASSOC ASTRON, UNIV MICH, 70- Mem: Sigma Xi; Am Astron Soc; Int Astron Union. Res: Primary research in microwave extra-galactic astronomy and space solar radio astronomy with the goal of finding theoretical models which can explain the observed properties of the objects studied. Mailing Add: Dept of Astron Physics Physics Astron Bldg Univ of Mich Ann Arbor MI 48109

ALLERHAND, ADAM, b Krakow, Poland, May 23, 37. PHYSICAL CHEMISTRY, BIOPHYSICAL CHEMISTRY. Educ: State Tech Univ, Chile, BS, 58; Princeton Univ, PhD(chem), 62. Prof Exp: Res assoc chem, Univ Ill, 62-63, instr phys chem, 63-64, res assoc, 64-65; asst prof chem, Johns Hopkins Univ, 65-67; from asst prof to assoc prof, 67-72, PROF CHEM, IND UNIV, BLOOMINGTON, 72- Mem: Am Chem Soc. Res: Nuclear magnetic resonance. Mailing Add: Dept of Chem Ind Univ Bloomington IN 47401

ALLERTON, SAMUEL E, b Three Rivers, Mich, Aug 21, 33; m 66; c 2. BIOPHYSICAL CHEMISTRY, MEDICAL SCIENCE. Educ: Kalamazoo Col, BA, 55; Harvard Univ, PhD(biochem), 62. Prof Exp: Res assoc biochem, Rockefeller Inst, 61-64; asst prof biophys, 65-67, asst prof biochem, 67-69, ASSOC PROF BIOCHEM, ALLAN HANCOCK FOUND, SCH DENT, UNIV SOUTHERN CALIF, 69- Mem: AAAS; Biophys Soc; Am Chem Soc; NY Acad Sci. Res: Isolation and characterization of proteins; studies on lipoproteins and glycoproteins; ultracentrifugation and electrophoresis; biochemical pathology; cell surface substances and Wilm'stumor; pediatric oncology. Mailing Add: Allan Hancock Found Sch of Dent Univ of Southern Calif Los Angeles CA 90007

ALLES, HAROLD GENE, b Greeley, Colo, Nov 23, 46; m 67; c 2. INSTRUMENTATION, COMMUNICATIONS ENGINEERING. Educ: Case Inst Technol, BS, 68; Univ Ore, PhD(physics), 72. Prof Exp: MEM TECH STAFF, BELL LABS, INC, 72- Res: Digital data systems and high speed signal processors and their application to communication systems, instrumentation problems and music synthesis. Mailing Add: Bell Labs Inc 600 Mountain Ave Murray Hill NJ 07974

ALLEVA, FREDERIC REMO, b Norristown, Pa, Oct 16, 33; m 62. PERINATAL BIOLOGY. Educ: Gettysburg Col, AB, 56; George Washington Univ, MPhil, 70, PhD(zool), 71. Prof Exp: Neuropharmacologist, Merck Sharp & Dohme, 59-66; PHARMACOLOGIST, FOOD & DRUG ADMIN, 66- Mem: AAAS; Sigma Xi; Endocrine Soc. Res: Assessing the effects of drugs, commonly used in pediatrics, on various parameters of development in neonatal rodents. Mailing Add: Food & Drug Admin Washington DC 20204

ALLEVA, JOHN J, b Norristown, Pa, Apr 17, 28; m 60; c 4. BIOLOGY. Educ: Univ Pa, AB, 50; Univ Mo, MS, 52; Harvard Univ, PhD(biol), 59. Prof Exp: Sr scientist biochem, Smith Kline & French Labs, 59-62; instr physiol, Albany Med Col, 62-63; RES BIOLOGIST & PHARMACOLOGIST, DIV DRUG BIOL, FOOD & DRUG ADMIN, 63- Mem: AAAS; Endocrine Soc; Soc Study Reproduction. Res: Reproductive endocrinology; effects of drugs on central mechanisms, including the biological clock; controlling ovulation in hamsters. Mailing Add: Div of Drug Biol Food & Drug Admin Washington DC 20204

ALLEWELL, NORMA MARY, b Hamilton, Ont. BIOLOGY. Educ: McMaster Univ, BSc, 65; Yale Univ, PhD, 69. Prof Exp: Fel, Nat Inst Arthritis & Metab Dis, 69-70; asst prof biochem, Polytech Inst Brooklyn, 70-73; ASST PROF BIOL, WESLEYAN UNIV, 73- Mem: Am Chem Soc. Res: Thermodynamics of macromolecular interactions; role of the cell surface in development. Mailing Add: Dept of Biol Wesleyan Univ Middletown CT 06457

ALLEY, CLYDE DUNN, JR, physical chemistry, see 12th edition

ALLEY, CURTIS J, b Calif, Aug 13, 18; m 43; c 4. VITICULTURE, ENOLOGY. Educ: Univ Calif, BS, 40, PhD, 51. Prof Exp: Specialist, 49-50. from jr specialist to assoc specialist, 52-67, SPECIALIST VITICULT & ENOL, UNIV CALIF, DAVIS, 67- Concurrent Pos: Peach breeder, Calif, 50-52. Mem: Am Soc Enol; Int Plant Propagators Soc. Res: Grapevine propagation studies, especially budding, grafting and benchgrafting; rooting of cuttings; wine variety adaptability; fruit maturity-wine quality. Mailing Add: Dept of Viticult & Enol Univ of Calif Davis CA 95616

ALLEY, EARL GIFFORD, b Corsica, SDak, Dec 10, 35; m 62; c 2. PESTICIDE CHEMISTRY, PHOTOCHEMISTRY. Educ: Miss State Univ, BS, 59, MS, 61; Univ Ill, Urbana, PhD(org chem), 68. Prof Exp: Chemist, Dow Chem Co, 67-69; DIR, RES & INDUST AGR SERV DIV, MISS STATE CHEM LAB, 70- Mem: Am Chem Soc; AAAS; Sigma Xi. Res: Photochemistry and degradation of agricultural chemicals; involvement of charge transfer and adsorption in these processes; analyses of trace metals and organic materials. Mailing Add: Box CR Miss State Chem Lab Mississippi State MS 39762

ALLEY, HAROLD PUGMIRE, b Cokeville, Wyo, Mar 26, 24; m 46; c 2. WEED SCIENCE. Educ: Univ Wyo, BS, 49, MS, 56; Colo State Univ, PhD(bot sci), 65. Prof Exp: Voc agr instr, La Grange Sch Dist, Wyo, 49-55; instr plant sci, 56-59, from asst prof to assoc prof, 59-68, PROF WEED CONTROL & EXTEN WEED SPECIALIST, UNIV WYO, 68- Mem: Am Soc Agron; Weed Sci Soc Am. Res: Weed physiology; morphological and physiological effects of herbicides; chemistry and mode of action of herbicides; control of undesirable plants. Mailing Add: Plant Sci Div Univ Wyo Laramie WY 82070

ALLEY, KEITH EDWARD, b Palm Springs, Calif, June 27, 43; m 66; c 1. NEUROBIOLOGY. Educ: Univ Ill, DDS, 68, PhD(anat), 72. Prof Exp: Fel anat, Univ Ill, 68-72; asst prof oral biol, Univ Iowa, 72-74; ASST PROF ANAT, CASE WESTERN RESERVE UNIV, 74- Concurrent Pos: Fel neurobiol, Univ Iowa, 72-74. Mem: Soc Neurosci; Am Asn Anatomists; Sigma Xi. Res: Development of the neural circuits that regulate motor activity in vertebrates and the environmental role in their formation. Mailing Add: Dept of Anat Case Western Reserve Univ Cleveland OH 44106

ALLEY, PEGGY WHITE, organic chemistry, see 12th edition

ALLEY, PHILLIP WAYNE b Chicago, Ill, Mar 11, 32; m 58; c 4. SOLID STATE PHYSICS. Educ: Lawrence Col, BS, 53; Rutgers Univ, PhD(physics), 58. Prof Exp: Asst physics, Rutgers Univ, 53-56, instr, 56-58; asst prof, Franklin & Marshall Col, 58-64; PROF PHYSICS, STATE UNIV NY COL GENESEO, 64- Mem: Am Phys Soc; Am Asn Physics Teachers; Sigma Xi. Res: Electrical conductivity in dilute metal alloys; teaching physics at undergraduate and graduate level. Mailing Add: Dept of Physics State Univ NY Col at Geneseo Geneseo NY 14454

ALLEY, REUBEN EDWARD, JR, b Petersburg, Va, July 16, 18; m 49; c 1. PHYSICS. Educ: Univ Richmond, BA, 38; Princeton Univ, EE, 40, PhD(elec eng), 49. Prof Exp: Instr physics, Univ Richmond, 40-42; staff mem, radiation lab, Mass Inst Technol, 42-43; instr elec eng, Princeton Univ, 43-44; assoc prof physics & chmn dept, Univ Richmond, 49-51; mem tech staff, Bell Tel Labs, 51-53; assoc prof physics & chmn dept, Univ Richmond, 53-55; assoc prof physics, Washington & Lee Univ, 55-57; mem tech staff, Bell Tel Labs, 57-59; sr proj engr, semiconductor-components div, Tex Instruments, Inc, 59-60; prof physics, Vassar Col, 60-62, chmn dept, 61-62; prof elec eng, Univ SC, 62-65; PROF ELEC ENG, US NAVAL ACAD, 65- Concurrent Pos: Bd of Gov, Am Inst Physics, 75-78. Mem: Am Phys Soc; Am Asn Physics Teachers. Res: Direct current controlled reactors; electrical analogy networks for solution of nonlinear differential equations; magnetic materials and high frequency magnetic properties; nuclear radiation effects. Mailing Add: 299 Halsey Rd Annapolis MD 21401

ALLEY, STARLING KESSLER, JR, b Crab Orchard, Tenn, Sept 20, 30; m 52; c 2. PETROLEUM CHEMISTRY. Educ: Berea Col, AB, 52; Univ Calif, Los Angeles, PhD(phys chem), 61. Prof Exp: Res chemist. Olin Mathieson Chem Corp, NY, 52-55, res chemist, high energy fuels div, Calif, 55-57; asst chem, Univ Calif, Los Angeles, 57-61; sr res chemist, 61-69, res assoc, 70-73, SUPVR CATALYST RES, UNION OIL CO CALIF, 73- Mem: Am Chem Soc. Res: High energy fuels; boron chemistry; thermodynamics of nonelectrolytes; nuclear magnetic resonance; hydrogen bonding; heterogeneous catalysis. Mailing Add: Res Dept Union Oil Co of Calif PO Box 76 Brea CA 92621

ALLGAIER, ROBERT STEPHEN, b Union City, NJ, Nov 29, 25; m 54; c 2. SOLID STATE PHYSICS. Educ: Columbia Univ, AB, 50, AM, 52; Univ Md, PhD(physics), 58. Prof Exp: Asst physics, Columbia Univ, 50-51; physicist, Naval Ord Lab, 51-73, chief, solid state div, 73-74; adminr mat technol prog, Off Chief of Naval Develop, 74-75; SR SCIENTIST, MAT DIV, NAVAL SURFACE WEAPONS CTR, SILVER SPRING, 75- Concurrent Pos: Vis scientist, Cavendish Lab, Cambridge Univ, 65-66; mem tech rev comt, Joint Serv, Electronics Prog, Dept of Defense, 72-; instr physics, Univ Md, 75- Honors & Awards: US Navy Meritorious Civilian Serv Award, 67. Mem: Am Phys Soc. Res: Transport properties of semiconductors and metals in liquid, solid, amorphous and crystalline forms. Mailing Add: 11034 Seven Hill Lane Potomac MD 20854

ALLGOOD, JOSEPH PATRICK, b Calvert, Ala, June 27, 27. PHYSIOLOGY. Educ: Univ Ala, BS, 49; Northwestern Univ, DDS, 53, MS, 60, PhD(physiol), 63. Prof Exp: Instr, 62-63, ASST PROF PHYSIOL, DENT SCH, NORTHWESTERN UNIV, CHICAGO, 63- Mem: AAAS; Int Asn Dent Res. Res: Masticatory efficiency; electromyography; temporomandibular joint disturbances; endocrinology. Mailing Add: Dept of Physiol Northwestern Univ Chicago IL 60611

ALLGOWER, EUGENE L, b Chicago, Ill, Aug 11, 35; m 58; c 1. MATHEMATICS. Educ: Ill Inst Technol, BS, 57, MS, 59, PhD(math), 64. Prof Exp: Instr math, Ill Inst Technol, 60-62 & Univ Ariz, 62-64; asst prof, Sacramento State Col, 64-65; assoc prof, Univ Tex, El Paso, 65-67; assoc prof, 67-74, PROF MATH, COLO STATE UNIV, 67- Concurrent Pos: Marathon Oil Co Indust fel, 65. Mem: Am Math Soc; Math Asn Am; Soc Indust & Appl Math. Res: Analysis; differential equations. Mailing Add: Dept of Math Colo State Univ Ft Collins CO 80521

ALLIGER, GLEN, b Leadville, Colo, Mar 13, 14; m 40; c 3. ORGANIC CHEMISTRY. Educ: State Col Wash, BS, 37; Univ Iowa, MS, 39, PhD(org chem), 41. Prof Exp: Asst, Univ Iowa, 37-41; res chemist, 41-42 & 45-53, head res compounding div, 53-59, asst dir, 59-63, res dir, 63-67, dir chem labs, 67-71, DIR RES, FIRESTONE TIRE & RUBBER CO, 71- Concurrent Pos: Firestone rep, Ind Res Inst. Mem: AAAS; Am Chem Soc. Res: Reactions of chloro ureas; synthetic, chlorinated and stereo rubbers; accelerators of vulcanization; research compounding; organic nitrogen compounds; synthetic rubbers; petrochemicals; textile fiber chemistry; tire physics and design. Mailing Add: Firestone Tire & Rubber Co 1200 Firestone Pkwy Akron OH 44317

ALLIN, EDGAR FRANCIS, b Edmonton, Alta, Jan 31, 39; m 67. ANATOMY. Educ: Univ Alta, MD, 63, BSc, 66. Prof Exp: Instr anat, Univ Alta, 61-62; intern, Vancouver Gen Hosp, 63-64; jr asst resident surg, Univ Hosp, Edmonton, Alta, 64-65; from proj specialist & teaching asst to instr, 67-69, ASST PROF ANAT, SCH MED, UNIV WIS-MADISON, 69- Mem: AAAS; Am Asn Phys Anthropologists; Soc Vert Paleont. Res: Adaptive myology, skeletal muscle and fiber type histochemistry; human biology; vertebrate paleontology, especially primates and advanced therapsids; human and comparative vertebrate morphology. Mailing Add: Dept of Anat Univ of Wis Sch of Med Madison WI 53706

ALLIN, ELIZABETH JOSEPHINE, b Blackwater, Ont, July 8, 05. PHYSICS. Educ: Univ Toronto, BA, 26, MA, 27, PhD(spectros), 31. Prof Exp: Asst demonstr physics, Univ Toronto, 26-27, demonstr & asst, 30-33; Royal Soc Can fel, Cambridge Univ, 33-34; lectr, 34-41, from asst prof to prof, 41-72, EMER PROF PHYSICS, UNIV TORONTO, 72- Mem: Can Asn Physicists. Res: Underwater spark spectra; hyperfine structure of spectral lines; x-ray; structure of alloys; spectroscopy; low temperature spectroscopy. Mailing Add: 36 Willowbank Toronto ON Can

ALLING, CHARLES CALVIN III, b Guthrie, Okla, Dec 27, 23; m 47; c 3. ORAL SURGERY, DENTISTRY. Educ: Ind Univ, AB, 43, DDS, 46; Univ Mich, MS, 54; Am Bd Oral Surg, dipl, 54. Prof Exp: Chief oral surg & training prog, 387th Sta & 11th Field Hosps, Ger, Dent Corps, US Army, 47-51, asst prof mil sci, Univ Mich, 51-54, chief oral surg, William Beaumont Gen Hosp, El Paso, Tex, 54-57, Letterman Gen Hosp, San Francisco, Calif, 57-59 & 121 Evacuation Hosp, Korea, 59-60, chief dent & oral surg & dir dent educ, Ireland Army Hosp, Ft Knox, Ky, 60-63, chief oral & maxillofacial res & proj officer mil dent, US Army Med Res & Develop Command, 63-66, chief oral surg serv, Walter Reed Gen Hosp, 66-68, chief prof serv, US Army Dent Corps, 68-69; PROF ORAL SURG & CHMN DEPT, SCH DENT, UNIV ALA, BIRMINGHAM, 69- Concurrent Pos: Lectr, Col Physicians & Surgeons, San Francisco, 57-59 & Georgetown Univ, 63-; consult, UN Command, Korea & Japan, 59-60; mil mem, Dent Study Sect, NIH, 63-; liaison mem, Dent Res Adv Comt, US Army, 63-; chief dent prof div, Off Surgeon Gen; with US Vet Admin & Provident Hosp, Baltimore. Mem: Am Dent Asn; Am Soc Oral Surgeons; fel Am Col Dent; Asn Mil Surgeons US; Int Asn Dent Res. Res: Oral and maxillofacial surgery, undergraduate, postgraduate and graduate education; clinical investigations; national research administration. Mailing Add: Dept of Oral Surg Univ of Ala Sch of Dent Birmingham AL 35233

ALLING, DAVID WHEELOCK, b Rochester, NY, July 5, 18; m 48; c 1. MATHEMATICAL STATISTICS, MEDICINE. Educ: Univ Rochester, BA, 40, MD, 48; Cornell Univ, PhD(statist), 59. Prof Exp: Intern med, Arnot-Ogden Hosp, 48-49; resident Hermann M Briggs Mem Hosp, 49-56; med officer statist, Nat Cancer Inst, 59-60, med officer statist, Nat Inst Allergy & Infectious Dis, 60-64, RES MATH STATISTICIAN, NAT INST ALLERGY & INFECTIOUS DIS, NIH, 64- Mem: Inst Math Statist; Am Statist Asn; Biomet Soc. Res: Partition of chi-squared variables; sequential tests of hypotheses; stochastic models of chronic diseases. Mailing Add: Bldg 10 Rm 11N118 NIH Bethesda MD 20014

ALLING, NORMAN LARRABEE, b Rochester, NY, Feb 8, 30; m 57; c 2. MATHEMATICS. Educ: Bard Col, BA, 52; Columbia Univ, MA, 54, PhD(math), 58. Prof Exp: Lectr math, Columbia Univ, 55-57; from asst prof to assoc prof, Purdue Univ, 57-65; assoc prof, 65-70, PROF MATH, UNIV ROCHESTER, 70- Concurrent Pos: NSF fel, Harvard Univ, 61-62; lectr, Mass Inst Technol, 62-64, NSF fel, 64-65; vis prof, math inst, Univ Würzburg, 71. Mem: Am Math Soc. Res: Analysis, algebra and cohomology of real and complex algebraic curves. Mailing Add: Dept of Math Univ of Rochester Rochester NY 14627

ALLINGER, NORMAN LOUIS, b Alameda, Calif, Apr 6, 28; m 52. ORGANIC CHEMISTRY. Educ: Univ Calif, BS, 51; Univ Calif, Los Angeles, PhD, 54. Prof Exp: Fel, Univ Calif, Los Angeles, 54-55, NSF fel, 55-56; from asst prof to prof chem, Wayne State Univ, 56-69; PROF CHEM, UNIV GA, 69- Mem: Am Chem Soc. Res: Mechanism of organic reactions; conformational analysis; organic quantum chemistry; physical properties of organic compounds; macro-rings. Mailing Add: Dept of Chem Univ of Ga Athens GA 30601

ALLINGHAM, JOHN WING, b Los Angeles, Calif, Jan 20, 20; m 55; c 2. GEOLOGY, GEOPHYSICS. Educ: Calif Inst Technol, BS, 48, MS, 54. Prof Exp: Geologist ground water br, 48-49, mineral deposits br, 49-55, geophys br, 55-69, environ geol br, 69-75, GEOLOGIST, ENVIRON IMOPACT ANAL PROG, US GEOL SURV, 75- Concurrent Pos: Staff scientist, NASA Planetology Prog, 71-73. Mem: Geol Soc Am; Soc Econ Geol; Soc Explor Geophys; Am Geophys Union; AAAS. Res: Investigations of copper, lead and zinc deposits in the central states; interpretation of aeromagnetic and gravity surveys within the continental United States; task force management; impact statement preparation. Mailing Add: US Geol Surv 760 Nat Ctr Reston VA 22092

ALLINGTON, WILLIAM B, b Lodgepole, Nebr, July 28, 12; m 34; c 2. PLANT PATHOLOGY. Educ: Univ Nebr, BS, 33, AM, 35; Univ Wis, PhD(plant path), 38. Prof Exp: Asst plant path, Univ Nebr, 33-35; asst plant path, Univ Wis, 35-38, instr hort, 38-43; agent, Div Tobacco Invests, Bur Plant Indust, USDA, 38-43, plant pathologist, Div Forage Crops & Dis, Bur Plant Indust, Soils & Agr Eng, 43-48; assoc plant pathologist, Univ Nebr, 48-52, prof plant path & bot, 52-68, chmn dept plant path, 49-62; MEM RES & DEVELOP STAFF, INSTRUMENTATION SPECIALTIES CO, 68- Mem: Am Inst Biol Sci; AAAS. Res: Bacterial diseases of plants; virus diseases of plants; soybean diseases. Mailing Add: ISCO 4700 Superior Lincoln NE 68504

ALLINSON, MORRIS JONATHAN CARL, b New Haven, Conn, Feb 20, 12; m 48; c 4. RADIOLOGY. Educ: Yale Univ, BS, 32; Boston Univ, PhD(biochem), 38; Univ Ark, MD, 45; Am Bd Radiol, dipl. Prof Exp: Res biochemist, Res Lab, Sharp & Dohme, 34-35; instr biochem, Sch Med, La State Univ, 39-40; instr physiol & pharmacol, Sch Med, Univ Ark, 40-45; intern, Grace-New Haven Hosp, 45-46; resident radiol, Hosp St Raphael's New Haven, 50-52 & Bridgeport Hosp, 52; resident radiation ther, Hosp, Joint Dis, 53, staff therapist, 53-57; practicing radiologist, 53-57; RADIOLOGIST, FRANKLIN HOSP, 57- Res: Metabolism and colorimetric methods of analysis of creatine and creatinine; action of bacterial enzymes on creatine and nicotinic acid analysis of inositol and penicillin. Mailing Add: Franklin Hosp Benton IL 62812

ALLIS, JOHN W, b Buffalo, NY, Apr 24, 39; m 65; c 2. PHYSICAL CHEMISTRY. Educ: Syracuse Univ, BS, 60; Univ Wis, PhD(phys chem), 65. Prof Exp: Res assoc protein chem, Georgetown Univ, 67-69; res chemist, USPHS, 70-71; RES CHEMIST, ENVIRON PROTECTION AGENCY, 71- Mem: Am Chem Soc; NY Acad Sci; Biophys Soc. Res: Physical chemistry of macromolecules; investigation by the methods of physical chemistry into the conformation and structure of proteins and their relation to chemical reactivity; effects of radiation. Mailing Add: Health Effects Res Lab Environ Protect Agency Research Triangle Park NC 27711

ALLIS, WILLAM PHELPS, b Menton, France, Nov 15, 01; US citizen; m 35; c 3. PLASMA PHYSICS, ELECTRON PHYSICS. Educ: Mass Inst Technol, BS, 23, MS, 24; Univ Nancy, DSc(high frequency resonance), 25. Hon Degrees: MA, Oxford Univ, 68. Prof Exp: Res assoc plasma physics, 25-29, from instr to prof, 31-67, EMER PROF & SR LECTR PLASMA PHYSICS, MASS INST TECHNOL, 67- Concurrent Pos: Vis prof, Harvard Univ, 58, Univ Tex, 60, St Catherine's Col, Oxford Univ, 68, Univ Paris, Sud, 69, 72 & 74, Mid East Tech Univ, Ankara, 70 & Univ S Fla, 71; chmn, Gaseous Electronics Conf, 49-62, hon chmn, 66-; consult, Los Alamos Sci Lab, 52-; asst secy gen sci affairs, NATO, France, 62-64; Fulbright sr lectr, Univ Innsbruck, 74-75. Honors & Awards: Legion of Honor, France, 68. Mem: Fel Am Phys Soc; Am Acad Arts & Sci (vpres, 61-62); fel Brit Inst Physics; Royal Soc Arts. Res: Free electrons in gases; ionized gases; electron distributions and processes in lasers. Mailing Add: 33 Reservoir St Cambridge MA 02138

ALLISON, BETZABE M, b Chuquicamata, Chile, Apr 14, 32; US citizen; m 56; c 1. CELL PHYSIOLOGY. Educ: Univ Calif, Berkeley, BA, 53; Univ Del, PhD(physiol), 66. Prof Exp: Instr biol, Hobbs Tutorial Col, Univ London, 56-57; instr anat & physiol, Del Hosp Sch Nursing, 58-59 & Univ Del, 61-62; asst prof biol, Lincoln Univ, Pa, 66-67; ASST PROF BIOL, MICH TECHNOL UNIV, 67- Res: Aging in biological systems; morphological and biochemical change in aging Tetrahymena pyriformis; aging and morphological change in human skin. Mailing Add: Dept of Biol Sci Mich Technol Univ Houghton MI 49931

ALLISON, DAVID C, b Monmouth, Ill, Feb 25, 31; m 54; c 3. PLANT GENETICS, CYTOLOGY. Educ: Univ Ill, BS, 56, MS, 57; Pa State Univ, PhD(genetics, breeding), 60. Prof Exp: Asst plant breeder, Univ Ariz, 61-62; from asst prof to assoc prof, 62-73, PROF BIOL, MONMOUTH COL, ILL, 73- Mem: Am Soc Agron; Crop Sci Soc Am. Res: Taxonomy and cytotaxonomy of the Gramineae; plant breeding. Mailing Add: Dept of Biol Monmouth Col Monmouth IL 61462

ALLISON, FLOYD ELVIN, physics, see 12th edition

ALLISON, FRED, physics, deceased

ALLISON, FRED, JR, b Abingdon, Va, Sept 8, 22; m 49; c 4. MEDICINE. Educ: Ala Polytech Inst, BS, 44; Vanderbilt Univ, MD, 46. Prof Exp: Instr microbiol, Sch Med, La State Univ, 50-51, instr med, 51-52; fel, Div Infectious Dis, Univ Wash, 52-53, instr med, 53-54, from instr to asst prof prev med, 54-55; from asst prof to prof med, Univ Miss, 55-68, chief div infectious dis, 55-68, assoc prof microbiol, 64-68; PROF MED & HEAD DEPT, MED CTR, UNIV NEW ORLEANS, 68- Concurrent Pos: Consult infectious dis, Jackson Vet Hosp, 55-67; vis investr, Rockefeller Univ, 66-67. Mem: AAAS; AMA; Am Fedn Clin Res; Am Col Physicians; Microcirc Soc. Res: Mechanisms relating to acute inflammatory reaction; clinical aspects of infectious diseases. Mailing Add: Univ of New Orleans Med Ctr 1542 Tulane Ave New Orleans LA 70112

ALLISON, JAMES, genetics, plant breeding, see 12th edition

ALLISON, JAMES LEROY, clinical chemistry, biophysics, see 12th edition

ALLISON, JEAN BATCHELOR, b Teague, Tex, Dec 16, 31; m 53; c 2. PHYSICAL CHEMISTRY. Educ: Rice Univ, BA, 53; Univ Houston, MS, 58, PhD(molecular spectros), 62. Prof Exp: Chemist, med res, Vet Admin Hosp, Houston, Tex, 53-57; mem staff, 62-72, RES CHEMIST, BELLAIRE RES LABS, TEXACO, INC, 72- Mem: Am Chem Soc; Sigma Xi. Res: Electronic and vibrational molecular spectroscopy of metalloporphyrins; development of inorganic and organic analytical chemistry techniques. Mailing Add: Texaco Inc PO Box 425 Bellaire TX 77402

ALLISON, JERRY ROBERT, b Union City, Pa, Apr 29, 45; m 67; c 2. ORGANIC CHEMISTRY. Educ: Bethany Col, BS, 67; Purdue Univ, MS, 69, PhD(org chem), 74. Prof Exp: Forensic chemist, US Army Criminal Invest Lab, 69-71; ASST PROF ORG CHEM, LAFAYETTE COL, 74- Mem: Am Chem Soc. Res: Synthetic and mechanistic organosulfur chemistry. Mailing Add: Dept of Chem Lafayette Col Easton PA 18042

ALLISON, JOHN ARTHUR CHARLES, b London, Eng, June 24, 27; US citizen; m 56; c 1. ORGANIC CHEMISTRY, INORGANIC CHEMISTRY. Educ: Cambridge Univ, BA, 48, MA & PhD(chem), 53. Prof Exp: Tech officer, Plastics Div, Imp Chem Industs Ltd, 51-54; res fel inorg chem, Harvard Univ, 54-55 & Univ Wash, 55-56; res chemist explosives dept exp sta, E I du Pont de Nemours & Co, 58-64; res assoc, Tidewater Oil Co, Pa, 64-66; res assoc, Hercules inc, 66-67; ASSOC PROF CHEM, MICH TECHNOL UNIV, 67- Concurrent Pos: Instr, eve col, Phila Col Textiles & Sci, 66-67. Mem: Am Chem Soc; Sigma Xi. Res: Organo- inorganic chemistry; organometallics; fluorine chemistry; petrochemicals. Mailing Add: Dept of Chem & Chem Eng Mich Technol Univ Houghton MI 49931

ALLISON, JOHN EVERETT, b Mont, Aug 16, 17; m 43; c 1. ANATOMY. Educ: Concordia Col, BA, 40; Univ Minn, MA, 47; Univ Iowa, PhD(zool), 52. Prof Exp: Instr biol, Drake Univ, 47-49; asst zool, Univ Iowa, 49-52; from instr to asst prof anat, Med Sch, St Louis Univ, 52-57; assoc prof, 57-69, PROF ANAT, COL MED, UNIV OKLA, 69- Mem: Am Asn Anatomists; Endocrine Soc. Res: Reproductive endocrinology. Mailing Add: Dept of Anat Sci Univ of Okla Health Sci Ctr Oklahoma City OK 73104

ALLISON, JOHN P b Beckenham, Kent, Eng, Feb 17, 36; US citizen; m 66; c 2. ORGANIC CHEMISTRY, POLYMER CHEMISTRY. Educ: Univ Birmingham, BSc, 58, PhD(chem), 61. Prof Exp: Fel chem, Univ Ariz, 61-63; sr res chemist, polymer dept, res labs, Gen Motors Tech Ctr, 63-66; staff mem, polymer chem, res div, Raychem Corp, 66-70; RES CHEMIST, KIMBERLY-CLARK CORP, 70- Mem: Am Chem Soc; The Chem Soc. Res: Synthetic polymer chemistry; fiber bonding; chemistry of natural polymers and polymer reactions. Mailing Add: Kimberly-Clark Corp W Off Bldg Neenah WI 54956

ALLISON, JOSEPH LEWIS, b Billings, Mont, Dec 13, 11; m 39; c 3. PLANT PATHOLOGY. Educ: Mont State Col, BS, 34; Wash State Col, MS, 36; Univ Minn, PhD(plant path), 40. Prof Exp: Asst, Univ Minn, 36-39; instr, La State Univ, 39-40; asst prof, Univ Wis, 40-46; prof, NC State Col, 49-57; dir, Farm Seed Res Corp, 57-68; forage res & prod specialist, Ferry-Morse Seed Co, 68-69; SUPT & PLANT PATHOLOGIST, IRRIGATED AGR RES & EXTEN CTR, WASH STATE UNIV, 69- Concurrent Pos: Plant pathologist, USDA, 40-57 & Food & Agr Orgn, UN, Iraq, 52-53. Mem: AAAS; Am Phytopath Soc; Am Soc Agron. Res: Diseases and breeding of forage legumes and grasses. Mailing Add: Irrigated Res & Exten Ctr Wash State Univ PO Box 30 Prosser WA 99350

ALLISON, LOWELL EDWARD, b Stryker, Ohio, Sept 19, 04; m 30; c 1. SOIL CHEMISTRY. Educ: Purdue Univ, BSA, 30; Univ Ill, MS, 33, PhD(chem), 42. Prof Exp: Res asst soil chem & microbiol, Agr Exp Sta, Ill, 30-36; asst prof soils, Purdue Univ, 37-42; res chemist, Monsanto Chem Co, Seattle, 42-44; soil scientist, US Salinity Lab, USDA, 44-69,soil scientist, Agr Res Serv, Okla, 69-71; RETIRED. Concurrent Pos: Consult training & salinity probs, Agency Int Develop, soil scientist, Uganda, 71-72. Mem: Am Soc Agron; Soil Sci Soc Am. Res: Problems relating to the reclamation and management of saline and alkali soils of western US. Mailing Add: 1151 5th Ave S Edmonds WA 98020

ALLISON, MARVIN J, b Schenectady, NY, Jan 6, 21; m 55; c 3. MICROBIOLOGY. Educ: Col William & Mary, BA, 42; Univ Pa, MS, 47, PhD(microbiol), 60. Prof Exp: Res assoc pub health, Univ Pa, 60-61; from asst prof to assoc prof, 61-70, PROF CLIN PATH, MED COL VA, 70- Mem: Am Soc Exp Path; fel Am Thoracic Soc; NY Acad Med. Res: Metabolism of the host's cells to resistance to infectious disease. Mailing Add: Med Col of Va 5th Floor Clin Bldg Richmond VA 23219

ALLISON, MILTON JAMES, b South Shore, SDak, May 10, 31; m 53; c 3. MICROBIOLOGY. Educ: SDak State Col, BS, 53, MS, 54; Univ Md, PhD, 61. Prof Exp: Instr bact, SDak State Col, 54-55; bacteriologist, Dairy Cattle Res Br, Agr Res Serv, 57-62, RES MICROBIOLOGIST, NAT ANIMAL DIS LAB, USDA, 62- Mem: Am Soc Microbiol; Am Soc Animal Sci; AAAS; Am Acad Microbiol. Res: Ecology of microorganisms; bacterial physiology and nutrition; biosynthesis of microbial amino acids and lipids; anaerobic bacteria of the mammalian digestive system. Mailing Add: Nat Animal Dis Lab USDA Box 70 Ames IA 50010

ALLISON, PATRICIA (LEE) (VAN BURGH), b Wash, DC, May 8, 23. PLANT PATHOLOGY. Educ: Univ Houston, BA & BS, 46; Ohio State Univ, MS, 48, PhD, 50. Prof Exp: Lab asst & fel biol, Univ Houston, 43-46; asst bot & plant path, Ohio State Univ, 46-48 & 49-50, plant path, 50, res assoc, 51-55; asst prof, Univ Minn, 57; lectr bot, Morris Arboretum, Univ Pa, 57-75, pathologist, 63-75, assoc pathologist, 57-63, ed Morris Arboretum Bull, 68-73, CONSULT MORRIS ARBORETUM, UNIV PA & BARNES FOUND ARBORETUM, 75- Concurrent Pos: Res asst, Univ Minn, 49; consult, Wynne S Eastman Prod, Tex, 56; prof, Univ Guayaquil, 63. Honors & Awards: Bronze Award Neographics, 72; Citation Consejo Directivo. Mem: Am Phytopath Soc; Mycol Soc Am; NAm Mycol Soc; Bot Soc Am. Res: Diseases of woody plants; mycology. Mailing Add: 908 Hunters Lane Oreland PA 19075

ALLISON, RICHARD C, b Seattle, Wash, Oct 19, 35; m 67; c 2. INVERTEBRATE

PALEONTOLOGY, GEOLOGY. Educ: Univ Wash, BS, 57, MS, 59; Univ Calif, Berkeley, PhD(paleont), 67. Prof Exp: Instr geol, Col San Mateo, 65-68; ASSOC PROF GEOL, UNIV ALASKA, 68- Mem: Paleont Soc; Int Paleont Union; Brit Palaeont Asn; Paleont Res Inst; Mex Geol Soc. Res: Cenozoic stratigraphy of southern Mexico; systematic paleontology of turritellid gastropods; Cenozoic invertebrate paleontology. Mailing Add: Dept of Geol Univ of Alaska College AK 99701

ALLISON, RICHARD GALL, b Hanover, Pa, Jan 28, 43; m 66; c 1. NUTRITIONAL BIOCHEMISTRY. Educ: Pa State Univ, BS, 64; Univ Calif, Davis, PhD(nutrit), 68. Prof Exp: Res biochemist, Dept Food Sci & Technol, Univ Calif, Davis, 69; biochemist, Walter Reed Army Inst Res, Walter Reed Army Med Ctr, 69-73; RES ASSOC LIFE SCI RES OFF, FEDN AM SOCS EXP BIOL, 74- Mem: AAAS; Am Chem Soc; Sigma Xi. Res: Biological significance in human nutrition of biochemically and biophysically demonstrable interactions of biological molecules with food substances, including vitamins, minerals, water, drugs and environmental factors. Mailing Add: Fedn Am Socs Exp Biol 9650 Rockville Pike Bethesda MD 20014

ALLISON, ROBERT DEAN, cardiopulmonary physiology, see 12th edition

ALLISON, TERRY C, b Robstown, Tex, Oct 14, 39; m 60; c 2. INVERTEBRATE ZOOLOGY. Educ: Tex A&I Univ, BS, 61; Tex A&M Univ, MS, 64, PhD(zool), 67. Prof Exp: Instr biol, Kilgore Col, 67-68; supvr invert & fish group, Brown & Root-Northrop, Manned Spacecraft Ctr, Tex, 68-70; ASSOC PROF BIOL, PAN AM UNIV, 70- Mem: AAAS; Am Soc Zool; Soc Invert Path. Res: Parasites of freshwater fishes, especially monogenetic trematodes; vertical migrations of marine copepods. Mailing Add: Dept of Biol Pan-Am Univ Edinburg TX 78539

ALLISON, WILLIAM EARL, b Claremore, Okla, Oct 2, 32; m 56; c 2. ENTOMOLOGY. Educ: Okla A&M Col, BS, 57; Okla State Univ, MS, 58; Tex A&M Univ, PhD(biochem), 63. Prof Exp: Res entomologist, Dow Chem Co, Seal Beach, Calif, 63-67, group leader entom, Midland, Mich, 67-71, group leader entom & nematol, Walnut Creek, 71-73, sr res entomologist, Dow Chem Co Japan Ltd, 74 & Dow Chem Co Pac Ltd, Malaysia, 75, SR DEVELOP SPECIALIST ENTOM, PLANT PATH & PLANT NUTRIT, DOW CHEM CO, WALNUT CREEK, 75- Mem: Entom Soc Am; AAAS. Res: Structure versus activity to search for new insecticides. Mailing Add: 724 Lisboa Ct Walnut Creek CA 94598

ALLISON, WILLIAM HUGH, b Harrison Twp, Pa, Nov 25, 34; m 58; c 3. MYCOLOGY. Educ: Pa State Univ, BS, 56, MS, 57, PhD(bot), 63. Prof Exp: Plant pathologist crops div, US Army Biol Labs, Md, 60-63; dir res com mushroom cultivation, Brandywine Mushroom Co Div, Borden Co, 63-66; mgr, Great Lakes Spawn Co, 66-68; ASST PROF BIOL, DEL VALLEY COL SCI & AGR, 68- Mem: AAAS; Mycol Soc Am; Bot Soc Am; Am Phytopath Soc. Res: General biology; commercial mushroom production; epidemiology of rice blast disease. Mailing Add: Dept of Biol Del Valley Col of Sci & Agr Doylestown PA 18901

ALLISON, WILLIAM S, b North Adams, Mass, June 16, 35; m 64; c 2. PROTEIN CHEMISTRY. Educ: Dartmouth Col, AB, 57, MA, 59; Brandeis Univ, PhD(biochem), 63. Prof Exp: USPHS fel, Lab Molecular Biol, Cambridge Univ, 64-65; res assoc, Brandeis Univ, 65-66, asst prof biochem, 66-69; asst prof chem, 69-73, ASSOC PROF CHEM, UNIV CALIF, SAN DIEGO, 73- Honors & Awards: Career Develop Award, NIH, 67-69. Res: Enzymology; comparative biochemistry; functional groups of enzymes; mechanism of enzyme action. Mailing Add: Dept of Chem Univ of Calif San Diego La Jolla CA 92093

ALLISTON, CHARLES WALTER, b Florence, Miss, May 27, 30; m 53; c 2. ENVIRONMENTAL PHYSIOLOGY, REPRODUCTIVE PHYSIOLOGY. Educ: Miss State Col, BS, 51, MS, 57; NC State Col, PhD(physiol), 60. Prof Exp: Res asst, Miss State Col, 55-56, res technician, 56-57; res asst, NC State Col, 57-60, asst prof zool, 60-65, assoc prof, 65-67; assoc prof, 67-75, PROF ANIMAL SCI, PURDUE UNIV, 75- Concurrent Pos: NIH res grant, 62-; Sigma Xi res award, 66; NSF grant, 74- Mem: Am Physiol Soc; Am Soc Animal Sci; Soc Study Reproduction; Brit Soc Study Fertil; Am Asn Lab Animal Sci. Res: Environmental and reproductive physiology, particularly influence of the physical environment upon reproductive efficiency of the mammalian female. Mailing Add: Dept of Animal Sci Lilly Hall Purdue Univ West Lafayette IN 47907

ALLMAN, JOHN MORGAN, b Columbus, Ohio, May 17, 43. NEUROPHYSIOLOGY. Educ: Univ Va, BA, 65; Univ Chicago, PhD(anthrop), 71. Prof Exp: Fel neurophysiol, Univ Wis, 70-73; res asst prof psychol, Vanderbilt Univ, 73-74; ASST PROF BIOL, CALIF INST TECHNOL, 74- Concurrent Pos: Sloan fel, 74. Mem: Soc Neurosci; Am Asn Anatomists; Asn Res Vision & Ophthal. Res: Functional organization of the visual system in primates; evolution of the brain in primates. Mailing Add: Div of Biol 216-76 Calif Inst Tech Beckman Lab Pasadena CA 91125

ALLMANN, DAVID WILLIAM, b Peru, Ind, May 20, 35; m 56; c 2. BIOCHEMISTRY. Educ: Ind Univ, BS, 58, PhD(biochem), 64. Prof Exp: Res asst bact, Ind Univ, 57, res asst biochem, 58-60; asst prof, Univ Wis-Madison, 66-70; ASSOC PROF BIOCHEM, SCH MED, IND UNIV-PURDUE UNIV, INDIANAPOLIS, 70-, ASSOC PROF DENT SCH, 73- Concurrent Pos: NIH fel biochem, Univ Wis-Madison, 64-66; res biochemist, Vet Admin Hosp, 70-72. Mem: Biophys Soc; Int Asn Dent Res; Am Soc Biol Chemists; fel Am Inst Chemists; Am Soc Cell Biologists. Res: Structure and function of mitochondria; effect of F-ions on adenylate cyclase and glucose metabolism in vivo. Mailing Add: Dept of Biochem Ind Univ Sch of Med Indianapolis IN 46202

ALLMARAS, RAYMOND RICHARD, b New Rockford, NDak, Sept 11, 26; m 52; c 6. SOIL SCIENCE. Educ: NDak State Univ, BS, 52; Univ Nebr, MS, 56; Iowa State Univ, PhD(soil sci), 60. Prof Exp: Soil scientist, agr res serv, USDA, Nebr, 52-56; asst, Iowa State Univ, 57-60; soil scientist, soil & water conserv res div, 60-72, RES LEADER & TECH ADV, COLUMBIA PLATEAU CONSERV RES CTR, AGR RES SERV, USDA, 72- Mem: Am Soc Agron; Soil Sci Soc Am; Biomet Soc; Am Statist Asn; AAAS. Res: Tillage, soil structure, plant rooting, soil and plant environment and plant-water relations. Mailing Add: Columbia Plateau Res Ctr Agr Res Serv USDA PO Box 370 Pendleton OR 97801

ALLMENDINGER, DAVIS FREDERICK b Wenatchee, Wash, July 19, 09; m 35; c 3. HORTICULTURE. Educ: State Col Wash, BS, 36, PhD(hort), 47; Ohio State Univ, MS, 38. Prof Exp: Asst tree fruit br exp sta, State Col Wash, 34-36; asst prof hort, NMex Col Agr & Mech Arts, 38-39; asst horticulturist, southwestern exp sta, Wash State Univ, 43-47, assoc horticulturist & supt, 47-51, horticulturist & supt, 51-53, horticulturist & supt, Western Wash Res & Exten Ctr, 53-75, vdir, Col Agr Res Ctr, 54-75; RETIRED. Mem: Am Soc Hort Sci. Res: Nutrition of fruits and vegetables; effect of fluorine on crops; variety testing and culture of fruits and vegetables. Mailing Add: Western Wash Res & Exten Ctr Puyallup WA 98371

ALLNATT, ALAN RICHARD, b Portsmouth, Eng, July 18, 33. THEORETICAL CHEMISTRY, PHYSICAL CHEMISTRY. Educ: Univ London, BSc, 56, PhD(phys chem), 59. Prof Exp: NATO fel chem, Univ Chicago, 59-61; lectr, Univ Manchester, 61-69; PROF CHEM, UNIV WESTERN ONT, 69- Concurrent Pos: Fel, Chem Inst Can, 74. Mem: The Chem Soc; fel Can Inst Chem. Res: Statistical mechanics of matter transport and thermodynamic properties of imperfect solids and simple liquids and of complex biochemical systems. Mailing Add: Dept of Chem Univ of Western Ont London ON Can

ALLRED, ALBERT LOUIS, b Mt Airy, NC, Sept 19, 31; m 58; c 3. INORGANIC CHEMISTRY. Educ: Univ NC, BS, 53; Harvard Univ, AM, 55, PhD(chem), 56. Prof Exp: Instr chem, 56-58, from asst prof to assoc prof, 58-69, assoc dean arts & sci, 70-74, PROF CHEM, NORTHWESTERN UNIV, 69- Concurrent Pos: Alfred P Sloan res fel, 63-65. Res: Nuclear and electron magnetic resonance; organometallic chemistry; synthetic inorganic chemistry. Mailing Add: Dept of Chem Northwestern Univ Evanston IL 60201

ALLRED, DORALD MERVIN, b Lehi, Utah, July 11, 23; m 52; c 5. ENTOMOLOGY, ECOLOGY. Educ: Brigham Young Univ, AB, 50, MA, 51; Univ Utah, PhD(entom), 54. Prof Exp: Field entomologist, State Exten Serv, Utah, 48-50; ranger-naturalist, US Nat Park Serv, 50-51; res fel entom, Brigham Young Univ, 51-53; instr biol, St Mary-of-the-Wasatch Acad, 53; instr biol, Univ Utah, 53-54, assoc ecologist & chief entom & arachnid sect, ecol res, 54-56; from asst prof to assoc prof, 56-66, PROF ZOOL, BRIGHAM YOUNG UNIV, 66- Res: Parasitic acarology; parasitology; medical entomology and ecology. Mailing Add: Dept of Zool Brigham Young Univ Provo UT 84602

ALLRED, EVAN LEIGH, b Deseret, Utah, May 22, 29; m 55; c 4. ORGANIC CHEMISTRY. Educ: Brigham Young Univ, BS, 51, MS, 56; Univ Calif, Los Angeles, PhD(chem), 59. Prof Exp: Res chemist, Phillips Petrol Co, 51-54; asst org chem, Brigham Young Univ, 54-55 & Univ Calif, Los Angeles, 55-59; instr chem, Univ Wash, 60-61; sr res chemist, Rohm and Haas Co, 61-63; from asst prof to assoc prof, 63-70, PROF CHEM, UNIV UTAH, 70- Concurrent Pos: NSF fel, Univ Colo, 59-60; David P Gardner fac fel, Univ Utah, 76. Mem: Am Chem Soc. Res: Physical-organic chemistry; factors affecting the reactivity of molecules; synthesis and reactivity of strained ring structures; reaction mechanisms. Mailing Add: Dept of Chem Univ of Utah Salt Lake City UT 84112

ALLRED, HARRY MILBURN, b Salt Lake City, Utah, July 26, 14; m 41; c 4. PHYSICS. Educ: Fresno State Col, AB, 35; Stanford Univ, MA, 48. Prof Exp: Physicist, Navy Dept, Calif & Va, 41-45; PHYSICIST, TEXACO, INC, 45- Mem: Am Phys Soc; Electron Micros Soc Am; Royal Micros Soc. Res: Measurement and analysis of the magnetic fields of ships; measurement of the magnetic susceptibility of various ferromagnetic powders; absorption of gases on solids; electron microscopy of catalysts, colloids and greases; petroleum products; x-ray diffraction analysis. Mailing Add: RD 1 Wappingers Falls NY 12590

ALLRED, JOHN B, b Oklahoma City, Okla, June 17, 34; m 54; c 3. BIOCHEMISTRY, NUTRITION. Educ: Okla State Univ, BS, 55; Wash State Univ, MS, 57; Univ Calif, Davis, PhD(biochem), 62. Prof Exp: Asst nutrit, Wash State Univ, 55-57; asst nutrit & biochem, Univ Calif, Davis, 58-60; from asst prof to assoc prof chem, Okla City Univ, 61-68, chmn dept, 62-63; DIR INST NUTRIT, OHIO STATE UNIV, 68-, ASSOC PROF FOOD SCI & NUTRIT, 74- Mem: AAAS; Am Chem Soc; Am Inst Nutrit; Brit Biochem Soc. Res: Intermediary metabolism; metabolic control mechanisms; enzymology; nutritional interest in energy, vitamins and minerals. Mailing Add: Inst of Nutrit & Food Technol Ohio State Univ 1314 Kinnear Rd Columbus OH 43212

ALLRED, JOHN CALDWELL, b Breckenridge, Tex, Apr 24, 26; m 50; c 3. PHYSICS. Educ: Tex Christian Univ, BA, 44; Univ Tex, MA, 48, PhD(physics), 50. Prof Exp: Asst, Los Alamos Sci Lab, 48-49, mem staff, 49-55; res scientist, Convair, 55-56; assoc prof, 56-61, assoc dean col arts & sci, 59-61, asst to pres, 61-62, vpres & dean faculties, 62-68, PROF PHYSICS, UNIV HOUSTON, 61- Mem: Fel Am Phys Soc; Am Nuclear Soc; Acoust Soc Am. Res: Light particle scattering and interactions; neutron scattering and spectra; fluid dynamics; reactor design architectural acoustics. Mailing Add: Dept of Physics Univ of Houston Houston TX 77004

ALLRED, KEITH REID, b Spring City, Utah, Feb 19, 25; m 45; c 6. CROP PRODUCTION, PLANT BIOCHEMISTRY. Educ: Brigham Young Univ, BS, 51; Cornell Univ, PhD(crop prod), 55. Prof Exp: Asst, Cornell Univ, 51-54; res assoc, Co-op Grange League Fedn Exchange, Inc, 54-57; from asst prof to assoc prof agron, 57-65, from assoc prof to prof plant sci, 65-70, PROF AGRON, UTAH STATE UNIV, 70-, ASSOC DEAN, 75- Mem: Am Soc Agron. Res: Forage crop physiology and production; study of relationship between dodder, parasitic plant, and alfalfa as host. Mailing Add: Int Progs UMC35 Utah State Univ Logan UT 84321

ALLRED, RAYMOND CHARLES, b Monett, Mo, Nov 8, 23; m 48; c 3. PETROLEUM MICROBIOLOGY, ENVIRONMENTAL SCIENCES. Educ: Okla State Univ, BS, 48; Univ Ky, MS, 51. Prof Exp: Res microbiologist, Phillips Petrol Co, 51-54 & Petrolite Corp, 54; supvr res scientist, Cent Res Div, Okla, 54-70, ASST DIR ENVIRON CONSERV, RES ENG DEPT, CONTINENTAL OIL CO, WASH, DC, 70- Mem: Am Soc Microbiol; Soc Indust Microbiol; Am Inst Mining, Metall & Petrol Eng; Am Petrol Inst. Res: Microbial problems in oil recovery; microbial corrosion; microbial oxidation of hydrocarbons and petroleum products; effect of oil on biological communities; environmental control of air and water discharges. Mailing Add: Continental Oil Co PO Box 1267 Ponca City OK 74601

ALLRED, RODNEY CHASE, b Lehi, Utah, June 5, 19; m 47; c 7. AGRONOMY. Educ: Brigham Young Univ, BS, 48; Kans State Col, 49; Univ Nebr, PhD(agron), 52. Prof Exp: Asst agron, Kans State Col, 48-49; asst agron, Univ Nebr, 49-51, asst prof forage crops, 52-55; from assoc prof to prof agron, Brigham Young Univ, 55-71; prof agron, Utah State Univ, US Agency Int Develop, La Paz, Bolivia, 71-73; PROF AGRON, BRIGHAM YOUNG UNIV, 73- Mem: Am Soc Agron. Res: Ecology and breeding of forage crops. Mailing Add: Dept of Agron Brigham Young Univ Provo UT 84601

ALLSBROOK, JANET S, topology, see 12th edition

ALLUM, FRANK RAYMOND, b Melbourne, Australia, Jan 31, 36; m 68. COSMIC RAY PHYSICS, SPACE PHYSICS. Educ: Univ Melbourne, BS, 55, MS, 58, PhD(nuclear physics), 63. Prof Exp: Sr demonstr, Dept Physics, Univ Melbourne, 61-63; res assoc auroral physics, Space Sci Dept, Rice Univ, 64-66; vis res assoc cosmic rays, 66, res assoc, 66-69, asst prof, 69-71, RES SCIENTIST, UNIV TEX, DALLAS, 71- Mem: Am Geophys Union; Am Phys Soc. Res: Propagation characteristics of low energy solar cosmic rays in the interplanetary medium using temporal, spectral and anisotropic data from the satellites, Explorers 34 and 41; solar-terrestrial relationships. Mailing Add: Univ of Tex at Dallas PO Box 688 Richardson TX 75080

ALM, ALVIN ARTHUR, b Albert Lea, Minn, July 30, 35; m 60; c 2. FORESTRY. Educ: Oniv Minn, BS, 61, MS, 65, PhD(forestry), 71. Prof Exp: Assoc forester, Forestry Consult Serv, Inc, 61-62; res asst forestry, Univ Minn, 63-65; appraiser real estate, Bur Pub Rds, 65-66; asst prof, 71-75, ASSOC PROF FORESTRY, UNIV MINN, 75- Mem: Soc Am Foresters; Sigma Xi. Res: Forest regeneration with emphasis on containerized seedlings; vegetative successional changes; silvicultural practices such as thinnings and harvesting; Christmas tree management; moisture and tree growth relationships. Mailing Add: Cloquet Forestry Ctr Univ of Minn Cloquet MN 55720

ALM, ROBERT M, b Princeton, Ill, Sept 19, 21; m 43; c 3. ORGANIC CHEMISTRY. Educ: Monmouth Col, BS, 43; Ohio State Univ, PhD(chem), 48. Prof Exp: Chemist, 48-53, group leader, 53-63, res assoc, 63-69, dir anal res, Naperville, 70-74, SR CONSULT CHEMIST, STANDARD OIL CO (IND), 74- Mem: Am Chem Soc. Res: Shale oil; synthetic fuels. Mailing Add: 927 Stoddard Ave Wheaton IL 60187

ALMAN, JOHN ERNEST, b Salina, Kans, July 3, 11; m 42; c 2. STATISTICS. Educ: Pomona Col, AB, 33; Claremont Col, AM, 39. Prof Exp: Instr math, Boston Univ, 47-50, dir statist & res serv, 50-61, DIR COMPUT CTR, BOSTON UNIV, 61- Mem: AAAS; Am Statist Asn; Asn Comput Mach; Psychomet Soc. Res: Statistical methods applied to the behavioral sciences; computing and data processing. Mailing Add: Comput Ctr Boston Univ 111 Cummington St Boston MA 02215

ALMAZAN, JAMES A, b San Antonio, Tex, Sept 7, 35; m 59; c 2. METEOROLOGY. Educ: Tex A&M Univ, BS, 64; Univ Tex, Austin, BA, 57, MS, 68, PhD(atmospheric sci), 70. Prof Exp: Meteorologist, Nat Climatic Ctr, 65-70, RES METEOROLOGIST, CTR EXP DESIGN & DATA ANAL, ENVIRON DATA SERV, NAT OCEANIC & ATMOSPHERIC ADMIN, 70- Concurrent Pos: Res assoc, Atmospheric Sci Group, Univ Tex, Austin, 68-69; mem, Boundary Layer Panel, Int Field Yr Great Lakes, 72-, Boundary Layer Working Group, US Global Atmospheric Res Prog Atlantic Trop Exp, 75- Honors & Awards: Spec Achievement Award, Nat Oceanic & Atmospheric Admin, US Dept Com, 73, 74 & 75. Mem: Am Meteorol Soc. Res: Boundary layer meteorology; air-sea and air-lake energy exchange processes; meteorological experiment design; systems analysis for data processing and validation. Mailing Add: Ctr Exp Design & Data Anal Page Bldg 2 3300 Whitehaven St NW Washington DC 20235

ALMEIDA, SILVERIO PEDRO, b Hudson, Mass, July 27, 33. ELECTROOPTICS, BIOPHYSICS. Educ: Clark Univ, BA, 57; Mass Inst Technol, MS, 59; Cambridge Univ, PhD(physics), 64. Prof Exp: Res asst physicist, Lawrence Radiation Lab, Univ Calif, Berkeley, 60-62; res asst physicist, Europ Orgn Nuclear Res, Geneva, Switz, 62-63, fel, 63-64; sr res assoc elem particle physics, Cavendish Lab, Cambridge Univ, 64-66; sr scientist, Aeronutronic-Philco Ford, Calif, 66-67; asst prof, 66-68, ASSOC PROF PHYSICS, VA POLYTECH INST, 68- Mem: Am Phys Soc; Am Optical Soc; Inst Elec & Electronics Eng, Comput Soc. Res: Coherent optics; holography; pattern recognition; image analysis; light scattering; elementary particles. Mailing Add: Dept of Physics Va Polytech Inst Blacksburg VA 24061

ALMGREN, FREDERICK JUSTIN, JR, b Birmingham, Ala, July 3, 33; m 58; c 2. MATHEMATICS. Educ: Princeton Univ, BSE, 55; Brown Univ, PhD(math), 62. Prof Exp: Instr math, Princeton Univ, 62-63; mem staff, Inst Advan Study, 63-65; from asst prof to assoc prof, 65-74, PROF ANAT, PRINCETON UNIV, 74- Res: Geometric measure theory. Mailing Add: Dept of Math Princeton Univ Princeton NJ 08540

ALMODOVAR, ISMAEL, b San German, PR, Apr 14, 32; US citizen; wid; c 2. INORGANIC CHEMISTRY, NUCLEAR CHEMISTRY. Educ: Univ PR, BS, 52; Carnegie Inst Technol, MS, 58, PhD(nuclear & inorg chem), 60. Prof Exp: Head nuclear sci & technol div, Univ PR, Rio Piedras, 60-62, nuclear sci & eng fel adv, 60-64, assoc prof chem, 61-70, dir neutron diffraction prog, 62-74, chmn dept chem, 67-74, prof chem & dean fac, 70-74; health scientist adminr, div res resources, NIH, Bethesda, Md, 74-75; COORDR SCI AFFAIRS, OFF OF PRES, UNIV PR SYST, 75- Concurrent Pos: NSF lectr, high schs, 60-; mem bd of examr, chem exam bd, Dept of State, San Juan, PR, 70-; consult MBS prog, Div Res Resources, NIH, 73 & 75- Mem: Am Chem Soc; Am Inst Chemists; PR Chemists Asn; PR Acad Arts & Sci; NY Acad Sci. Res: Solid state chemistry and physics; nuclear physics; geochemistry. Mailing Add: Off of Pres Univ PR Syst GPO Box 4984-6 San Juan PR 00936

ALMODOVAR, LUIS RAUL, algology, see 12th edition

ALMON, LOIS, bacteriology, botany, see 12th edition

ALMOND, HAROLD RUSSELL, JR, b Oakland, Calif, Dec 21, 34; m 56; c 2. ANALYTICAL CHEMISTRY, ORGANIC CHEMISTRY. Educ: Calif Inst Technol, BS, 56, PhD(org chem), 61. Prof Exp: Sr res chemist, 60-63, GROUP LEADER ANAL CHEM, McNEIL LABS, INC, JOHNSON & JOHNSON, 64- Mem: Am Chem Soc; Am Soc Mass Spectrometry. Res: Synthesis of amino acid derivatives and peptides for alpha-chymotrypsin kinetics; general analytical chemistry, especially structure determinations via nuclear magnetic resonance and mass spectrometry; quantitative structure-activity relationships. Mailing Add: McNeil Labs Inc Camp Hill Rd Ft Washington PA 19034

ALMOND, HY, b Chattanooga, Tenn, June 11, 14; m 49; c 2. INORGANIC CHEMISTRY. Educ: Univ Chicago, BS, 38. Prof Exp: Explosives supvr, Army Ord, 42-43; chemist, US Geol Surv, 46-57; PROCESS CONTROL SUPVR, AUTONETICS GROUP, ROCKWELL INT CORP, 57- Mem: Am Chem Soc; Geochem Soc. Res: Printed circuitry; geochemistry; analytical chemistry; general electroplating; thin films; thick films. Mailing Add: Autonetics Grp Rockwell Int Corp Dept 144 GA 25 Lab 3370 Mira Loma Anaheim CA 92803

ALMOND, PETER RICHARD, physics, see 12th edition

ALMQUIST, JOHN OLSON, b Holdrege, Nebr, Feb 10, 21; m 42; c 3. REPRODUCTIVE PHYSIOLOGY. Educ: Cornell Univ, BS, 42; Purdue Univ, MS, 44; Pa State Univ, PhD(dairy sci), 47. Prof Exp: Asst animal husb, Purdue Univ, 42-44; instr dairy sci, 44-47, assoc prof, 47-51, PROF DAIRY PHYSIOL, PA STATE UNIV, 51-, IN CHARGE DAIRY BREEDING RES CTR, 49- Concurrent Pos: Consult, Dept Agr, PR; vis prof reproductive physiol, Univ Hawaii, 69. Honors & Awards: Res Award, Glycerine Producers Asn, 59; Borden Res Award, Am Dairy Sci Asn, 63; Am Cyanamid Animal Physiol & Endocrinol Res Award, Am Soc Animal Sci, 74; Animal Prod Res Award, Spallanzani Italian Res Inst for Artificial Insemination, 72. Mem: Fel AAAS; Am Dairy Sci Asn; Am Soc Animal Sci; Brit Soc Study Fertility. Res: Reproductive physiology and artificial insemination of farm animals; semen physiology and preservation; sexual behavior and reproductive efficiency of bulls; cattle infertility. Mailing Add: Dairy Breeding Res Ctr Pa State Univ University Park PA 16802

ALMS, GREGORY RUSSELL, b Sycamore, Ill, July 8, 47; m 69; c 2. CHEMICAL PHYSICS. Educ: Monmouth Col, BA, 69; Stanford Univ, PhD(chem), 73. Prof Exp: Fel chem, Univ Ill, 73-75; ASST PROF CHEM, FORDHAM UNIV, 75- Res: Laser light scattering spectroscopy studies on orientational correlation in liquids. Mailing Add: Dept of Chem Fordham Univ Bronx NY 10458

ALMS, THOMAS H, b Orangefield, Tex, Nov 8, 31; m 54; c 2. MICROBIOLOGY, IMMUNOLOGY. Educ: Univ Houston, BS, 54; Univ Tex, MA, 59, PhD(microbiol), 66. Prof Exp: Supvr bact res lab, Plastic Surg Dept, Univ Tex Med Br Galveston, 59-62; asst prof, 66-68, ASSOC PROF MICROBIOL, UNIV MO-KANSAS CITY, 68-, CHMN DEPT, 74- Mem: Am Soc Microbiol. Res: Isolation and purification of protective antigens from Pseudomonas aeruginosa; immunological factors and peridontal disease; immunoglobulin A and caries prevention; role of cellular and humoral factors in antigen antibody reactions and protection. Mailing Add: Dept of Microbiol Univ of Mo Sch of Dent Kansas City MO 64108

ALMY, THOMAS PATTISON, b New York, NY, Jan 10, 15; m 43; c 3. INTERNAL MEDICINE. Educ: Cornell Univ, AB, 35, MD, 39. Honors & Awards: MA, Dartmouth Col, 70. Prof Exp: Asst med, Med Col, Cornell Univ, 40-42, from instr to asst prof, 42-48, assoc prof neoplastic dis, 48-54, from assoc prof to prof med, 54-68; Nathan Smith prof & chmn dept, 68-74, THIRD CENTURY PROF MED, DARTMOUTH MED SCH, 74- Concurrent Pos: Physician outpatients, NY Hosp, 43-48, from asst attend physician to attend physician, 48-68; asst attend physician, Mem Hosp, 48-68 & James Ewing Hosp, 50-68; vis physician & dir 2nd med div, Bellevue Hosp, 54-68; dir med, Dartmouth-Hitchcock Affiliated Hosps, 68-74; consult med, Vet Admin Hosp, NY, 54-68 & Vet Admin Ctr, White River Junction, Vt, 68- Mem: Asn Am Physicians; Am Soc Clin Invest; Am Gastroenterol Asn (pres, 64); Am Physiol Soc; fel Am Col Physicians (regent, 68-73). Res: Psychosomatic medicine; physiology of human gastrointestinal tract; methods of medical teaching. Mailing Add: Dept of Med Dartmouth Med Sch Hanover NH 03755

AL-NAKEEB, SHAHEEN MUSTAFA, b Kirkuk, Iraq, May 4, 39; m 64; c 2. VETERINARY PATHOLOGY, VETERINARY SURGERY. Educ: Univ Baghdad, Iraq, BVMS, 62; Iowa State Univ, MS, 66, PhD(vet path), 69. Prof Exp: Instr vet clin sci, Univ Baghdad, 63-64; instr vet surg, Iowa State Univ, 66-69; asst prof vet path & surg, 69-72, assoc prof lab animal sci, 69-75, ASSOC PROF VET PATH & SURG, STATE UNIV NY BUFFALO, 72- Concurrent Pos: Consult, Aquarium of Niagara Falls Inc, NY, 71- & Vet Admin Hosp, Buffalo, 72- Mem: Am Vet Med Asn; Am Asn Lab Animal Sci; AAAS; Am Microbiol Soc; Turkish Vet Med Asn. Res: Chronic implantable reservoirs for collection of biological fluids; oviductal fluid immunology and biochemistry; Wilson's disease in dogs. Mailing Add: 409 Farber Hall State Univ of NY Buffalo NY 14214

ALO, RICHARD ANTHONY, b Erie, Pa, Nov 24, 38; m 60; c 4. MATHEMATICS. Educ: Gannon Col, BA, 60; Pa State Univ, MA, 65, PhD(math), 65. Prof Exp: Engr systs test retrofit, defense projs div, Western Elec Co, 60-61; teaching asst math, Pa State Univ, 61-64, instr, 65-66; asst prof, 66-70, ASSOC PROF MATH, CARNEGIE-MELLON UNIV, 70- Concurrent Pos: Kanpur Indo-Am Prog math sci adv, Indian Inst Technol, 69-70; Nat Res Coun Italy vis fel, 74; vis prof math, Univ Parma, 75; Nat Acad Sci vis fel, 75. Mem: Am Math Soc; Math Asn Am; Ital Math Union; Soc Indust & Appl Math. Res: General topology; measure theory; function spaces; functional and numerical analyses. Mailing Add: Dept Math Carnegie-Mellon Univ 5000 Forbes Ave Pittsburgh PA 15213

ALONSO, CAROL TRAVIS, b Montreal, Que, Dec 5, 41; US citizen; m 69; c 2. NUCLEAR PHYSICS. Educ: Allegheny Col, BS, 63; Bryn Mawr Col, MS, 65; Mass Inst Technol, PhD(nuclear physics), 70. Prof Exp: Mem res staff heavy ion physics, Yale Univ, 70-72; mem res staff, Lawrence Radiation Lab, Univ Calif, Berkeley, 72-75, MEM RES STAFF NUCLEAR THEORY, LAWRENCE LIVERMORE LAB, UNIV CALIF, 75- Mem: Am Phys Soc. Res: Heavy ion reaction theory; liquid drop model; heavy ion fusion energy sources; hydrodynamics. Mailing Add: L-32 Lawrence Livermore Lab PO Box 808 Livermore CA 94550

ALONSO, MARCELO, b Havana, Cuba, Feb 6, 21; m 43; c 4. NUCLEAR PHYSICS, QUANTUM THEORY. Educ: Univ Havana, PhD(physics), 43. Prof Exp: Fel physics, Yale Univ, 43-44; prof physics, Univ Havana, 44-60, chmn dept, 56-60; lectr physics, Georgetown Univ, 61-71; dep dir, 60-73, DIR SCI & TECHNOL, ORGN AM STATES, 73- Concurrent Pos: Dir, Nuclear Energy Comn, Cuba, 55-60; tech dir, Nat Bank Econ Develop, Cuba, 58-60; hon prof, Univ Guadalajara, Mex, 71. Mem: Hon mem, Guatemala Acad Sci; Am Phys Soc; Am Asn Physics Teachers. Mailing Add: Orgn of Am States Washington DC 20006

ALPEN, EDWARD LEWIS, b San Francisco, Calif, May 14, 22; m 45; c 2. PHYSIOLOGY, RADIOBIOLOGY. Educ: Univ Calif, BS, 46, PhD(pharmaceut chem), 50. Prof Exp: Asst prof pharmacol, Sch Med, George Washington Univ, 50-51; investr & head thermal injury br, Biol & Med Sci Div, US Naval Radiol Defense Lab, San Francisco, 53-55, head biophys br, 55-58, head div, 58-69; assoc lab dir, 69-73, lab dir, Pac Northwest Labs, Battelle Mem Inst, 73-75; PROF MED PHYSICS & DIR DONNER LAB, UNIV CALIF, BERKELEY, 75- Concurrent Pos: NSF fel, Oxford Univ, 58-59; Guggenheim fel, Univ Paris, 65-66. Honors & Awards: Mem Award, Asn Mil Surg US, 61; Sci Medal, US Secy Navy, 62; Distinguished Civilian Serv Medal, US Dept Defense, 63. Mem: Asn Mil Surg US; Am Physiol Soc; Radiation Res Soc; Brit Soc Exp Biol; Royal Soc Med. Res: Radiation biology and biophysics; cellular kinetics; erythropoetic mechanisms; regulation of erythropoesis; environmental science. Mailing Add: Donner Lab Univ of Calif Berkeley CA 94720

ALPER, ALLEN MYRON, b New York, NY, Oct 23, 32; m 59; c 2. PETROLOGY, MINERALOGY. Educ: Brooklyn Col, BS, 54; Columbia Univ, PhD(petrol, mineral), 57. Prof Exp: Instr phys geol, Brooklyn Col, 56-57; sr mineralogist, Ceramic Res Lab, Corning Glass Works, 57-59, res mineralogist, 59-62, res mgr, 62-69; mgr chem & electronic mat, 69-70, mgr res & develop, 70-71, chief engr, 71-72, DIR RES & ENG, CHEM & METALL DIV, GTE SYLVANIA INC, 72- Concurrent Pos: Kemp Mem res grant, 54; lectr, St John's Seminary, NY, 58 & Elmira Col, 59-60. Mem: AAAS; Am Ceramic Soc; Am Chem Soc; Am Inst Mining, Metall & Petrol Eng; Geol Soc Am. Res: Superalloys; rhenium; tungsten; molybdenum; powder metallurgy; rare-earth oxides and metals; iodides; metallurgical extraction; photochemical machining technology; phosphors; crystal growth; ceramics; phase studies; crystal chemistry. Mailing Add: Chem & Metall Div GTE Sylvania Inc Towanda PA 18848

ALPER, CARL, b Hoboken, NJ, May 28, 20; m 49; c 4. CLINICAL BIOCHEMISTRY. Educ: Drew Univ, BA, 41; Tulane Univ, MS, 43, PhD(chem), 47. Prof Exp: Instr chem, Tulane Univ, 43-46; res assoc nutrit, E R Squibb & Sons, 47-48, res assoc immunochem develop, 48-49; from asst prof to assoc prof biochem, Hahnemann Med Col, 49-66, biochemist, Hosp, 62-66; assoc prof biochem & dif dept clin biochem, Temple Univ, 66-70; DIR PHILADELPHIA BR, BIO-SCI LABS, 70- Concurrent Pos: Adj assoc prof, Temple Univ, 70- Mem: Am Chem Soc; Am Soc Clin Path; Am Asn Clin Chem; Am Inst Nutrit; Soc Acad Clin Lab Physicians & Scientists. Res: Biochemistry of disease; protein nutrition and metabolism; body composition;

enzymology; immunochemistry. Mailing Add: Bio-Sci Lab 114-116 S 18th St Philadelphia PA 19103

ALPER, HOWARD, b Montreal, Que, Oct 17, 41; m 66; c 2. ORGANIC CHEMISTRY, ORGANOMETALLIC CHEMISTRY. Educ: Sir George Williams Univ, BSc, 63; McGill Univ, PhD(chem), 67. Prof Exp: NATO fel, Princeton Univ, 67-68; from asst prof to assoc prof chem, State Univ NY Binghamton, 68-74; ASSOC PROF CHEM, UNIV OTTAWA, 75- Mem: Am Chem Soc; Chem Inst Can; The Chem Soc. Res: Metal carbonyls as reagents in organic chemistry; mechanistic organometallic chemistry; synthesis of theoretically important organic molecules; ring-chain tautomerism. Mailing Add: Dept of Chem Univ of Ottawa Ottawa ON Can

ALPER, JOSEPH SETH, b Brooklyn, NY, Aug 2, 42; m 68. THEORETICAL CHEMISTRY. Educ: Harvard Univ, AB, 63; Yale Univ, PhD(chem), 68. Prof Exp: Fel chem, Mass Inst Technol, 68-69; ASSOC PROF CHEM, UNIV MASS, DORCHESTER, 69- Concurrent Pos: Petrol Res Fund-Am Chem Soc type G grant, 70-73. Mem: Am Phys Soc. Res: Applications of group theory to problems in atomic and molecular structure and spectra. Mailing Add: Dept of Chem Univ of Mass Dorchester MA 02125

ALPER, MILTON H, b Lynn, Mass, July 26, 30; m 54; c 3. ANESTHESIOLOGY, PHARMACOLOGY. Educ: Harvard Univ, AB, 50, MD, 54. Prof Exp: NIH training grant & res fel pharmacol, 61-62, instr anesthesia, 62-64, clin assoc, 64-69, ASSOC PROF ANESTHESIA, HARVARD MED SCH, 69-; ANESTHESIOLOGIST-IN-CHIEF, LYING-IN-CHIEF, LYING-IN-DIV, BOSTON HOSP WOMEN, 69- Concurrent Pos: From jr assoc to assoc, Peter Bent Brigham Hosp, 62-69; attend, West Roxbury Vet Admin Hosp, 63- Honors & Awards: Mead Johnson Training Award Anesthesiol, 59. Mem: Am Soc Anesthesiol; Am Soc Pharmacol & Exp Therapeut; Asn Univ Anesthetists. Res: Pharmacology of anesthetic drugs; obstetrical anesthesia. Mailing Add: Lying-In-Div Boston Hosp for Women Boston MA 02115

ALPER, ROBERT, b New York, NY, Sept 8, 33; m 70; c 4. BIOCHEMISTRY. Educ: Utica Col, BA, 59; State Univ NY Upstate Med Ctr, PhD(biochem), 69. Prof Exp: Res asst biochem, Vet Admin Hosp, Syracuse, NY, 60-64; res assoc, State Univ NY Buffalo, 68-69 & Fla State Univ, 69-71; res assoc med, Univ Pa Div, Philadelphia Gen Hosp, 71-72, ASST PROF BIOCHEM, SCH DENT MED, UNIV PA, 72- Mem: AAAS; Am Chem Soc. Res: Involvement of glycoproteins and mucopolysaccharides in atherosclerosis; structure of the amyloid fibril; identification and characterization of cell surface glycoproteins in neoplastic cells; biochemistry of basement membrane; collagen biochemistry. Mailing Add: Gen Clin Res Ctr Philadelphia Gen Hosp Philadelphia PA 19104

ALPERIN, HARVEY ALBERT, b New York, NY, Mar 13, 29; m 56; c 2. SOLID STATE PHYSICS. Educ: Rensselaer Polytech Inst, BS, 49; Univ Mich, MS, 50; Univ Conn, PhD(physics), 60. Prof Exp: Res assoc physics, Willow Run Res Ctr, Univ Mich, 50-53; physicist, Underwater Sound Lab, US Navy, 53-57, PHYSICIST, US NAVAL ORD LAB, 57- Concurrent Pos: Guest scientist, Brookhaven Nat Lab, 60-68; vis scientist, Weizmann Inst Sci, 70-71. Mem: AAAS; Am Phys Soc. Res: Neutron diffraction; magnetic materials; computer applications; hydrogen in metals. Mailing Add: US Naval Surface Weapons Ctr White Oak Silver Spring MD 20910

ALPERIN, JONATHAN L, b Boston, Mass, June 2, 37. MATHEMATICS. Educ: Harvard Univ, AB, 59; Princeton Univ, MA, 60, PhD(math), 61. Prof Exp: Inst math, Mass Inst Technol, 62-63; asst prof, 63-73, PROF MATH, UNIV CHICAGO, 73- Mem: Am Math Soc; Math Asn Am. Mailing Add: Dept of Math Univ of Chicago Chicago IL 60637

ALPERIN, RICHARD JUNIUS, b Philadelphia, Pa, Dec 16, 41. CYTOCHEMISTRY, DEVELOPMENTAL BIOLOGY. Educ: Univ Pa, AB, 59, PhD(biol), 69. Prof Exp: Asst prof biol, anat & physiol, 69-75, ASSOC PROF BIOL, COMMUNITY COL PHILADELPHIA, 75-, CHMN FAC SENATE, 74- Concurrent Pos: Fel, Prof I Gersh's Lab, Dept Animal Biol, Univ Pa, 69-74; consult, Biophys Hemat Unit to Dr S Srinnivasan, State Univ NY Downstate Med Sch, 70-71, Montgomery County Supt Schs, Pa, 71- & NASA, 75-; guest lectr electron micros, Sch Vet Med, Univ Pa, 71-74. Mem: Am Genetic Asn; Biol Photog Asn; Am Micros Soc; Am Soc Zool; Pattern Recognition Soc. Res: Hypoblast function in the chick embryo using microsurgical and ultraviolet light injury methods; changing submicroscopic distribution of nucleic acids during metaplasia and determination revealed cytochemically and recorded with electronmicroscope. Mailing Add: 842 Lombard St Philadelphia PA 19147

ALPERN, HERBERT P, b New York, NY, Oct 3, 40. PSYCHOPHARMACOLOGY, NEUROSCIENCES. Educ: City Col New York, BS, 63; Univ Ore, MA, 65; Univ Calif, Irvine, PhD(psychobiol), 68. Prof Exp: Nat Inst Gen Med Sci grant, 69-71, asst prof, 68-74, ASSOC PROF PSYCHOL, UNIV COLO, BOULDER, 74-, FAC FEL, INST BEHAV GENETICS, 69- Concurrent Pos: NIMH grant, 71-74. Mem: AAAS; Soc Neurosci; Int Brain Res Orgn. Mailing Add: Dept of Psychol Univ of Colo Boulder CO 80302

ALPERN, MATHEW, b Akron, Ohio, Sept 22, 20; m 51; c 4. VISUAL PHYSIOLOGY. Educ: Univ Fla, BME, 46; Ohio State Univ, PhD(physics), 50. Prof Exp: Res assoc physiol optics, 49-51; asst prof optom, Pac Univ, 51-55; from instr to asst prof, 55-58, assoc prof opthal & psychol, 58-63, PROF PHYSIOL OPTICS & PSYCHOL, VISION LAB, UNIV HOSP, UNIV MICH, 63- Concurrent Pos: Mem vis sci study sect, NIH, 70-74. Honors & Awards: Jonas S Friedenwald Award, Asn Res Vision & Opthal, 74. Mem: Optical Soc Am; Am Physiol Soc; Biophys Soc; Asn Res Vision Opthal; Am Psychol Asn. Res: Electrophysiology of the retina; visual contrast effects, scattering of light within the eye; visual interaction; intensity time relations of the visual stimulus; critical flicker frequency; factors influencing size of the pupil; accomodation-convergence relations; human visual pigments in normal and abnormal eyes. Mailing Add: Vision Lab Univ Hosp Univ Mich Ann Arbor MI 48104

ALPERS, JOSEPH BENJAMIN, b Salem, Mass, Aug 24, 25; m 56; c 2. BIOLOGICAL CHEMISTRY, METABOLISM. Educ: Yale Univ, AB, 49; Columbia Univ, MD, 53; NY Univ, PhD(biochem), 62. Prof Exp: Res fel med, Sloan-Kettering Inst, NY, 53-55; from intern to asst resident med, NY Hosp, 55-57; res fel biochem, New York City Pub Health Res Inst, 57-60, asst biochem, 60-61; assoc, 61-65, assoc prof, 66-69, ASSOC PROF BIOCHEM, HARVARD MED SCH, 69-; DIR CLIN LABS, CHILDREN'S HOSP, 71- Mem: Am Soc Biol Chemists. Res: Intermediary metabolism; enzymology of carbohydrates. Mailing Add: Children's Hosp 300 Longwood Ave Boston MA 02115

ALPERT, ARNOLD, pharmaceutical chemistry, biochemistry, see 12th edition

ALPERT, LEO, b Boston, Mass, Oct 31, 15; m 43; c 2. ENVIRONMENTAL SCIENCES, METEOROLOGY. Educ: Mass State Col, BS, 37; Clark Univ, MA, 39, PhD(climat), 46. Prof Exp: Staff weather officer, US Air Force, Panama, 41-43; res weather officer, Hqs Air Weather Serv, DC, 44-46; environ specialist, Engr Intel Div, Corps Engrs, 50-51; chief climat lab, Geophys Directorate, Air Force Cambridge Res Labs, 52; geogr, Engr Strategic Intel Div, Corps Engrs, US Army, 53-60, trop environ scientist, Environ Sci Div, Army Res Off, DC, 61-62, chief scientist, Tropic Test Ctr, Panama, CZ, 63-65, geographer, Environ Sci Div, Army Res Off, 65-69, meteorologist, 69-71, CHIEF ATMOSPHERIC SCI BR, ARMY RES OFF, RES TRIANGLE PARK, 71- Concurrent Pos: Consult climatologist, US, SAm & Cent Am, 52-63; del, Pan Am Inst Geog & Hist, DC, 52, Mex, 55; Nat Res Coun del, Int Geog Cong, Brazil, 56; chief engr, Field Party, Mex, 57; del, Comn Climat, DC, 57; Geophys Union del, Int Union Geod & Geophys, Finland, 60; del, Int Geog Cong, Sweden, 60; Nat Res Coun del, Pac Sci Cong, Hawaii, 61; Int Geog Union del, Comn Agr Climat, Can, 62; del, Int Union Conserv Nature & Natural Resources, Galapagos Island Symp, Ecuador, 64. Mem: Am Meteorol Soc; Am Geophys Union; Asn Am Geog. Res: Environmental research and testing of material; tropical environment; applied and tropical meteorology; weather modification; climate of the Galapagos Islands and eastern Tropical Pacific Ocean Area; pollution; environmental quality and degradation; climatic change. Mailing Add: 4122 Cobblestone Pl Durham NC 27707

ALPERT, LOUIS KATZ, b New York, NY, Aug 8, 07; m 32; c 1. MEDICINE. Educ: Yale Univ, BS, 28, MD, 32. Prof Exp: Asst path, Yale Univ, 32-33, asst med, 33-35; from asst to instr, Univ Chicago, 35-38; Nat Res Coun fel, Rockefeller Inst, NY, 38-39; instr, Johns Hopkins Univ, 39-43; from adj clin prof to clin prof, 48-56, prof, 56-74, EMER PROF MED, SCH MED, GEORGE WASHINGTON UNIV, 74- Mem: Am Fedn Clin Res; Am Asn Cancer Res; Endocrine Soc; Am Diabetes Asn; Am Col Physicians. Res: Cancer chemotherapy; endocrinology. Mailing Add: 4220 Van Ness St NW Washington DC 20016

ALPERT, MARSHALL B, physical chemistry, deceased

ALPERT, MORTON, b Brooklyn, NY, June 10, 24; m 52; c 2. CYTOCHEMISTRY, RESEARCH ADMINISTRATION. Educ: Univ La, BS, 47; Ohio State Univ, MS, 48; Univ Minn, PhD(anat), 52; Ind Univ, South Bend, MSBA, 72. Prof Exp: Asst anat, Ohio State Univ, 47-49, from instr to asst prof, 52-59; assoc prof, Sch Med, Ind Univ-Purdue Univ, Indianapolis, 59-68; dir clin res, 68-75, MGR ANAL CYTOL, AMES CO DIV, MILES LABS, INC, 75-; ASSOC DIR RES LAB, ELKHART GEN HOSP, IND, 68- Concurrent Pos: Asst, Univ Minn, 50-51; vis scientist, Karolinska Inst, Sweden, 63; expert, Off Sci & Tech Equip, Bus & Defense Serv Admin, US Dept Com, 67-68; vis assoc prof anat, Stritch Sch Med, Loyola Univ Chicago, 67-68; mem, Biol Stain Comn. Honors & Awards: Hofheimer Prize, Am Psychiat Asn, 61. Mem: Soc Exp Biol & Med; Histochem Soc; Am Asn Anat; Endocrine Soc; Am Soc Clin Pathologists. Res: Adrenal histophysiology; ceroid pigments; microquantitation of ascorbic acid in tissue sections; maturation of adrenal-pituitary axis, maturation of central nervous system as a function of early stress. Mailing Add: Ames Co Div Miles Labs Inc Elkhart IN 46514

ALPERT, NELSON LEIGH, b New Haven, Conn, June 14, 25; m 50; c 2. MEDICAL TECHNOLOGY, SPECTROSCOPY. Educ: Yale Univ, BS, 45; Mass Inst Technol, PhD(physics), 48. Prof Exp: Asst prof physics, Rutgers Univ, 48-52; physicist, White Develop Corp, 52-58; group leader, Perkin-Elmer Corp, 58-61, chief engr, Spectros Prod Develop, 61-67; dir instrument develop, Becton Dickinson & Co, 67-68; INDEPENDENT CONSULT, 68- Concurrent Pos: Consult, Dept Path, Hartford Hosp, 69- Mem: AAAS; Optical Soc Am; Soc Appl Spectros; Am Soc Test & Mat; Am Asn Clin Chem. Res: Analytical and clinical instruments, particularly spectroscopic, ultraviolet, visible, infrared absorption and emission, flame, fluorescence. Mailing Add: PO Box 3403 Ridgeway Sta Stamford CT 06905

ALPERT, NORMAN, b Philadelphia, Pa, May 5, 21; m 48; c 2. PETROLEUM TECHNOLOGY, PETROLEUM CHEMISTRY. Educ: Temple Univ, AB, 42, MA, 47; Purdue Univ, PhD(chem), 49. Prof Exp: Res chem, Publicker Industs, Inc, 42-45; group leader, Fuels Res, Texas Co, 49-59; MGR PROD RES DIV, EXXON RES & ENG CO, 59- Mem: Am Chem Soc; Soc Automotive Engrs. Res: Development of new, improved, lower cost petroleum fuels, lubricants, specialties; technical services, pollution abatement from use of petroleum products. Mailing Add: Exxon Res & Eng Co PO Box 51 Linden NJ 07036

ALPERT, NORMAN ROLAND, b Stamford, Conn, July 28, 22; m 52. PHYSIOLOGY. Educ: Wesleyan Univ, AB, 43; Columbia Univ, PhD(physiol), 51. Prof Exp: From asst to instr physiol, Columbia Univ, 38-53; from asst prof to prof, Col Med, Univ Ill, 53-66; PROF PHYSIOL & CHMN DEPT, COL MED, UNIV VT, 66- Mem: Am Physiol Soc; Biophys Soc; Harvey Soc; Am Soc Gen Physiol; Am Soc Exp Biol & Med. Res: Anaerobiosis in vivo; metabolism, respiration and circulation; thermodynamics, chemistry and mechanics of skeletal and cardiac muscle contraction. Mailing Add: Dept of Physiol & Biophys Univ of Vt Burlington VT 05401

ALPERT, SEYMOUR, b New York, NY, Apr 20, 18; m 41. MEDICINE. Educ: Columbia Univ, AB, 39; Long Island Col Med, MD, 43; Am Bd Anesthesiol, dipl, 49. Prof Exp: Intern, Beth Israel Hosp, 43-44; chief med officer anesthesiol, Gallinger Munic Hosp, 46-47; from instr to assoc prof, 48-61, assoc, Univ Hosp, 48-50, from asst dir to assoc dir, 61-69, PROF ANESTHESIOL, SCH MED, GEORGE WASHINGTON UNIV, 61-, VPRES DEVELOP, UNIV, 69- Concurrent Pos: Univ fel anesthesiol, Sch Med, George Washington Univ, 48; consult, Walter Reed Army Hosp, 48-, DC Gen Hosp, 48-69 & Mt Alto Vet Hosp, 59-69. Mem: Am Soc Anesthesiologists; AMA; fel Am Col Anesthesiologists; Asn Am Med Cols; Pan-Am Med Asn. Mailing Add: Off of VPres for Develop George Washington Univ Washington DC 20052

ALPERT, SEYMOUR SAMUEL, b Amsterdam, NY, Dec 21, 30; m 61; c 3. PHYSICS. Educ: Univ Calif, Berkeley, AB, 53, PhD(physics), 62. Prof Exp: Physicist, Lawrence Radiation Lab, 61-62; mem tech staff, Bell Tel Labs, Inc, 62-64; res assoc laser scattering, Columbia Univ, 64-66; from asst prof to assoc prof, 66-74, PROF PHYSICS, UNIV NMEX, 74- Mem: Am Phys Soc; Am Asn Physics Teachers. Res: Laser light scattering; optical spectroscopy of liquids; critical opalescence. Mailing Add: Dept of Physics & Astron Univ of NMex Albuquerque NM 87106

ALPHA, ANDREW GRAY, b Lethbridge, Alta, May 11, 12; nat US; m 36; c 3. PETROLEUM GEOLOGY. Educ: Univ NDak, BS, 34, MS, 35. Prof Exp: Asst geol, Univ NDak & NDak Geol Surv, 31-35; soil technologist & ground water geologist, Soil Conserv Serv, USDA, 35-40; ground water geologist, Grazing Serv, US Dept Interior, 40-43; petrol geologist, Gen Petrol Corp, 43-52; geologist & mgr, Rocky Mountain Div, Signal Oil & Gasoline Co, 53-58, asst chief geologist, 58-64; assoc div geologist, Mobil Oil Co, Socony Mobil Oil Co, Inc, 64-69, regional geol adv, Mobil Oil Corp, 69-70, SR GEOLOGIST GEOTHERMAL EXPLOR, MOBIL OIL CORP, 70- Concurrent Pos: Consult, Andrew G Alpha & Assocs, 77- Mem: Fel Am Geol Soc; Am Asn Petrol Geol; Marine Technol Soc; Asn Prof Geol Sci. Res: Geothermal stratigraphy; tectonics; glacial geology; geomorphology; ground water; hydrocarbons; soils; environmental geology. Mailing Add: 1101 Monaco Pkwy Denver CO 80220

ALPHER, RALPH ASHER, b Washington, DC, Feb 3, 21; m 42; c 2. PHYSICS. Educ: George Washington Univ, BS, 43, PhD(physics), 48. Physicist, appl physics lab, Johns Hopkins Univ, 44-55; PHYSICIST, CORP RES & DEVELOP, GEN ELEC CO, 55-

Concurrent Pos: Adj prof, Rensselaer Polytech Inst, 60-64. Honors & Awards: Magellanic Premium, Am Philos Soc, 75; Prix Georges Vander Linden, Royal Acad Sci, Lett & Fine Arts, Belg, 75. Mem: Fel Am Phys Soc; fel AAAS; Sigma Xi. Res: Theoretical physics; cosmology; physics of fluids; astrophysics. Mailing Add: Corp Res & Develop Gen Elec Co PO Box 8 Schenectady NY 12301

ALPHIN, REEVIS STANCIL, b Mt Olive, NC, Apr 21, 29; m 55; c 2. PHARMACOLOGY, BIOCHEMISTRY. Educ: Univ NC, BA, 51; Duke Univ, MA, 56; Med Col Va, PhD(pharmacol), 66. Prof Exp: Res asst physiol, Duke Univ, 53-56; pharmacologist, Lilly Res Labs, Eli Lilly & Co, 56-60; pharmacologist, 60-64, head sect gastroenterol, 64-73, ASSOC DIR PHARMACOL, A H ROBINS CO, INC, 73- Mem: AAAS; assoc Am Gastroenterol Asn; Am Physiol Soc; NY Acad Sci; Am Soc Pharmacol & Exp Therapeut. Res: Gastrointestinal tract; peptic ulcer; anorectic and antihidrotic agents. Mailing Add: A H Robins Co Inc Res Labs 1211 Sherwood Ave Richmond VA 23220

ALPINER, JEROME GERALD, b Massillon, Ohio, Apr 19, 32; m 55; c 3. AUDIOLOGY, SPEECH PATHOLOGY. Educ: Ohio Univ, BFA, 54, PhD, 61; Western Reserve Univ, MA, 59. Prof Exp: Speech & hearing therapist, Mich Pub Schs, 56-58; instr psychol & debate coach, Alpena Jr Col, 56-58; asst chief audiol, Vet Admin Hosp, Cleveland, 60-61; asst clin prof clin audiol, Western Reserve Univ, 61-62; assoc prof speech & dir audiol, Northern Ill Univ, 62-65; assoc prof speech & dir speech & hearing ctr, 65-69, PROF SPEECH PATH & AUDIOL & ACTG ASSOC DEAN COL ARTS & SCI, UNIV DENVER, 69- Concurrent Pos: Consult, Rehab Serv Admin, Dept Health, Educ & Welfare. Honors & Awards: Cert Clin Competence Audiol, Am Speech & Hearing Asn. Mem: AAAS; fel Am Speech & Hearing Asn; Speech Asn Am; Nat Asn Deaf; Acad Rehab Audiol (pres). Res: Clinical and rehabilitative audiology; psychology. Mailing Add: Dept of Speech Path & Audiol Univ of Denver Denver CO 80210.

ALRUTZ, ROBERT WILLARD, b Pittsburgh, Pa, Aug 20, 21; m 46; c 1. ECOLOGY. Educ: Univ Pittsburgh, BS, 43; Univ Ill, MS, 47, PhD, 51. Prof Exp: Instr biol, Univ Minn, 51-52; from asst prof to assoc prof, 52-64, PROF BIOL, DENISON UNIV, 64- DIR BIOL RESERVE, 69- Concurrent Pos: Chmn dept biol, Denison Univ, 64-65, chmn biol reserve, 65-69. Mem: AAAS; Ecol Soc Am; Am Inst Biol Sci; Nat Asn Biol Teachers; Conserv Educ Asn. Res: Distribution and ecology of odonata. Mailing Add: Dept of Biol Denison Univ Granville OH 43023

AL-SAADI, ABDUL A, b Iraq, Oct 20, 35; US citizen; m 61; c 3. CELL BIOLOGY, MICROSCOPIC ANATOMY. Educ: Univ Baghdad, BA, 55; Univ Kans, MA, 59; Univ Mich, PhD(zool), 63. Prof Exp: Res assoc radiation effects, Sch Med, Univ Mich, 62-66, asst prof cellular biol, 66-70; CHIEF CYTOGENETICS & ASSOC DIR RES, DEPT ANAT PATH, WILLIAM BEAUMONT HOSP, 70- Concurrent Pos: NIH grant, 62-67; Univ Mich Cancer Inst grants, 67, 68-69; Am Cancer Soc grant. Mem: AAAS; Am Soc Human Genetics; Am Thyroid Asn; Am Asn Cancer Res; Am Fedn Clin Res. Res: Cytogenetics in cancer research; clinical cytogenetics; ultrastructure changes in carcinogenesis; thyroid physiology in health and disease; radiation biology. Mailing Add: Dept of Anat Path William Beaumont Hosp Royal Oak MI 48072

ALSBERG, HENRY, b Ger, Oct 6, 21; US citizen; m 49; c 2. POLYMER CHEMISTRY, INDUSTRIAL CHEMISTRY. Educ: Univ Toronto, BASc, 47; Purdue Univ, MS, 49. Prof Exp: Group leader polymers, Koppers Co, 51-64; MGR RES & DEVELOP, RICHARDSON CO, 64- Mem: Am Chem Soc; Soc Plastic Engrs; Am Soc Testing & Mat. Res: Development of specialty suspension and emulsion polymers; pilot plant operations; plant process development. Mailing Add: Richardson Co 2700 Lake St Melrose Park IL 60160

ALSCHER, RUTH PAULA, b New York, NY, Sept 11, 21. CELL PHYSIOLOGY. Educ: Col New Rochelle, BA, 43; Fordham Univ, MS, 46, PhD(physiol), 51. Prof Exp: Res asst dept nutrit, Fleischmann Labs, Standard Brands, Inc, New York, 43-45; asst gen physiol, Grad Sch & Gen Sci Sch Educ, Fordham Univ, 45-46; instr, 46-74, PROF BIOL & CHMN DEPT, MANHATTANVILLE COL, 73- Concurrent Pos: Gen asst dept biol & genetics, Fordham Univ, 50-51; mem corp, Marine Biol Lab, Woods Hole, 52. Mem: Fel AAAS; Am Micros Soc; Am Soc Zool; NY Acad Sci. Res: Vitamin A toxicity; physiology of body cavity fluid of marine annelids; morphology and physiology of cells of body cavity fluid of marine annelids; proteolytic activity of cell structures of spinach leaves; chloroplastic structures of spinach leaf epidermis. Mailing Add: Dept of Biol Manhattanville Col Purchase NY 10577

ALSEVER, JOHN BELLOWS, b Syracuse, NY, Nov 24, 08; m 38; c 3. INTERNAL MEDICINE, IMMUNOHEMATOLOGY. Educ: Syracuse Univ, AB, 30; Harvard Univ, MD, 34; Am Bd Internal Med, dipl, 41. Prof Exp: Asst path, Syracuse Med Col, 34-35; intern & resident internal med, Peter Bent Brigham Hosp, 35-37; pvt pract, NY, 37-42; tech dir blood prog, US Off Civilian Defense, USPHS, 42-44, assoc med dir, Am Nat Red Cross, 44-46, dir training & prof standards, Hosp Div, 46-49, asst chief, 49-51, dir blood prog, Fed Civil Defense Admin, 51-54, dept dir, Health & Spec Weapons Defense, 53-54, med officer, Chronic Dis Div, 54-55; med dir blood bank, 55-70, vpres med affairs, 70-75, MED CONSULT, BLOOD SERV, 75- Concurrent Pos: Liaison officer to comt on blood, Nat Res Coun, 42-55; consult blood prog, NY State Health Dept, 44, Fed Civil Defense Admin, 55-60 & Ministry Health, Egypt, 56. Mem: AMA; Am Col Physicians; Am Soc Clin Path; Am Geriat Soc; Asn Mil Surg US. Res: Blood transfusion, especially blood banking. Mailing Add: PO Box 1030 Scottsdale AZ 85252

ALSMEYER, RICHARD HARVEY, b Sebring, Fla, Feb 4, 29; m 53; c 2. MEAT SCIENCE, BIOCHEMISTRY. Educ: Univ Fla, BSA, 52, MSA, 56, PhD(animal sci & nutrit), 60. Prof Exp: Res technician meat sci, Agr Exp Sta, Univ Fla, 56-57; res animal husbandman, Meat Qual Lab, 60-67, head stand group, Tech Serv Div, 67-71, sr staff officer, Prod Stands, Sci & Tech Serv, 71-75, DIR PROG EVAL & SPEC REPORTS, COOP STATE RES SERV, USDA, 75- Concurrent Pos: Res fel, Brit Meat Res Inst, Cambridge Univ, Eng, 66; mem adv bd marine safety & ecol, Int Maritime Comt; AM Meat Sci Asn rep, Nat Res Coun-Nat Acad Sci, 72-75. Mem: Am Meat Sci Asn; Inst Food Sci & Technol; Am Soc Animal Sci. Res: Research administration; directs agency budget development-program reviews at state universities and experiment stations; responsible for reports on research by states; maintains files on research projects by state experiment stations and forestry schools. Mailing Add: 8500 Wild Olive Dr Potomac MD 20854

ALSMEYER, WILLIAM LOUIS, b Sebring, Fla, Aug 17, 34; m 58; c 2. ANIMAL NUTRITION, BIOCHEMISTRY. Educ: Univ Fla, BSAgr, 56, MSAgr, 57; Univ Ill, Urbana, PhD(animal nutrit), 66. Prof Exp: Res fel animal nutrit, Chas Pfizer & Co, Ind, 57; asst prof animal sci, NC State Univ, 66-671; mgr granite diag div, Carolina Biol Supply Co, 71-72; MEM STAFF, HONNEGERS & CO, INC, 72- Mem: Am Soc Animal Sci; Poultry Sci Asn; Am Inst Biol Sci. Res: Nonruminant nutrition; amino acid metabolism of swine and poultry; interactions of amino acid metabolism; vitamin B6 metabolism and minerals as affected by chelating agents; digestive enzymes of the neonate. Mailing Add: Honnegers & Co Inc 201 W Locust Fairbury IL 61739

ALSMILLER, RUFARD G, JR, b Louisville, Ky, Nov 16, 27; m 52. PHYSICS, MATHEMATICS. Educ: Univ Louisville, BS, 49; Purdue Univ, MS, 52; Univ Kans, PhD(physics), 57. Prof Exp: Fire control systs analyst, Naval Ord Plant, Ind, 52-53; physicist, Allison Div, Gen Motors Corp, 56; assoc dir div, 69-73, GROUP LEADER HIGH & MEDIUM ENERGY SHIELDING, NEUTRON PHYSICS DIV, OAK RIDGE NAT LAB, 57- Mem: Am Phys Soc; fel Am Nuclear Soc. Mailing Add: Neutron Physics Div Oak Ridge Nat Lab PO Box X Oak Ridge TN 37830

ALSOP, DAVID W, b Lindsay, Ont, Nov 15, 39. CYTOLOGY, MORPHOLOGY. Educ: Cornell Univ, BS, 64, PhD(biol), 70. Prof Exp: ASST PROF BIOL, QUEENS COL, NY, 70- Mem: Entom Soc Am; Am Soc Zool. Res: Comparative studies on cockroaches. Mailing Add: Dept of Biol Queens Col Flushing NY 11367

ALSOP, JOHN HENRY, III, b Okmulgee, Okla, Oct 4, 24; m 51; c 2. ANALYTICAL CHEMISTRY. Educ: Okla State Univ, BS, 50; Univ Tex, PhD(anal chem), 57. Prof Exp: Chemist, Eagle-Picher Co, Okla, 50-51, Am Window Glass Co, 51-52 & Stanolind Oil & Gas Co, 56-58; sr chemist, Pan Am Petrol Corp, 58-59, Chemstrand Corp, Ala, 59-60 & Chemstrand Res Ctr, NC, 60-62; RES CHEMIST, RICHMOND RES LABS, TEXACO INC, 62- Mem: Am Chem Soc; Nat Asn Corrosion Engrs; Soc Appl Spectros. Res: Spectrophotometry; fluorometry; gas chromatography; infrared analysis of gases; analytical methods development in air and water conservation. Mailing Add: Richmond Res Labs Texaco Inc PO Box 3407 Richmond VA 23234

ALSOP, LEONARD E, b Oroville, Calif, Apr 10, 30; m 55; c 3. GEOPHYSICS. Educ: Columbia Univ, AB, 51, AM, 56, PhD(physics), 61. Prof Exp: Res scientist, Lamont Geol Observ, 59-65, adj assoc prof geol, 65-73, MEM ACAD STAFF, LAMONT-DOHERTY GEOL OBSERV, COLUMBIA UNIV, 73- Concurrent Pos: Mem staff, IBM Corp, 65-73. Mem: Am Geophys Union; Am Seismol Soc; Royal Astron Soc. Res: Free oscillations of the earth; use of a laser as a transducer for seismograph; tides of the solid earth. Mailing Add: Lamont-Doherty Geol Observ Columbia Univ Palisades NY 10964

ALSTADT, DON MARTIN, b Erie, Pa, July 29, 21. PHYSICS. Educ: Univ Pittsburgh, BS, 47. Prof Exp: Develop engr, 47-50, chief physicist, 50-52, head basic res, 52-56, mgr cent res, 56-61, vpres & gen mgr, 64-66, exec vpres & mem bd dirs, 66-68, VPRES, LORD MFG CO, 61-, GEN MGR, HUGHSON CHEM CO DIV, 58-, PRES, LORD CORP, 68-, VCHMN, LORD CORP, 75- Concurrent Pos: Consult, Carborundum Co, 47-48, Sci Specialties Corp, 51-52 & Transistor Prod Corp, 52-56; metrop chmn, Nat Alliance Businessmen, 69-; mem bd dirs, Lord Corp, Kiethley Instruments, Inc, Marine Bank; mem bd, Schs Management, Univ Pittsburgh, Case Western Reserve Univ & Univ Denver, 69-; mem bd adv, Case Inst Technol, Sch Eng, Tulane Univ & Mellon Inst, 70-; mem bd trustees, Polytech Inst NY, Halmot Hosp, Erie, Kolff Found, Cleveland & Villa Maria Col; mem corp, Conf Bd, NY. Mem: Am Chem Soc; Am Phys Soc; Electrochem Soc; fel Am Inst Chem; Inst Mgt Sci. Res: Adhesives and adhesion; dielectrics and rheology of materials; physics and chemistry of surfaces; frictional phenomena; bioengineering. Mailing Add: 228 Rosemont Ave Erie PA 16505

ALSTON, JIMMY ALBERT, b Temple, Tex, Mar 23, 42; m 67; c 2. PLANT BREEDING. Educ: Tex A&M Univ, BS, 64, MS, 67, PhD(plant breeding), 68. Prof Exp: Res asst, Tex A&M Univ, 64-68; supvr plant breeding dept, Seed Prod Div, Baker Castor Oil Co, Tex, 68-70; PLANT BREEDER, GEORGE W PARK SEED CO, INC, 70- Mem: Am Soc Agron; Crop Sci Soc Am. Res: Cyclic gametic selection as a breeding method in corn; inheritance of female characters in Ricinus communis; development of new floricultural varieties by selective breeding methods; devising more efficient methods of seed production. Mailing Add: Route 8 Belle Meade 206 Yosemite Greenwood SC 29646

ALSTON, PETER VAN, b Windsor, NC, Apr 27, 46; m 68; c 2. PHYSICAL ORGANIC CHEMISTRY, ANALYTICAL CHEMISTRY. Educ: Univ NC, BS, 68; Va Commonwealth Univ, PhD(chem), 74. Prof Exp: SR CHEMIST, E I DU PONT DE NEMOURS & CO, INC, 74- Concurrent Pos: Adj fac, J Sargeant Reynolds Community Col, 74- Mem: Am Chem Soc; Soc Appl Spectros. Res: Mechanism of cycloaddition reactions; use of perturbation molecular orbital theory in organic chemistry; use of fluorescence spectroscopy in polymer characterization. Mailing Add: 11136 Olympic Rd Richmond VA 23235

ALSTON, WILLIAM JEFFRIES, III, nuclear physics, see 12th edition

ALSTON-GARNJOST, MARGARET, b Ashtead, Eng, Jan 23, 29; m 66. Educ: Univ Liverpool, BSc, 51, Hons, 52, PhD(physics), 55. Prof Exp: Demonstr physics, Univ Liverpool, 54-56, Imp Chem Industs res fel, 56-58; physicist, 59-72, SR STAFF MEM, LAWRENCE BERKELEY LAB, UNIV CALIF, 72- Mem: Am Phys Soc. Res: High energy particle physics; experimental high energy physics. Mailing Add: Lawrence Berkeley Lab Univ of Calif Berkeley CA 94720

ALSUM, DONALD JAMES, b Randolph, Wis, June 27, 37; m 61; c 2. REPRODUCTIVE PHYSIOLOGY. Educ: Calvin Col, BA, 61; Purdue Univ, MS, 66; Univ Minn, PhD(animal physiol), 74. Prof Exp: Jr high sch teacher sci, Western Suburbs Sch, Ill, 61-63; sec sci teacher, Timothy Christian High Sch, Ill, 63-70; res asst physiol, Univ Minn, Minneapolis, 70-74; ASST PROF BIOL, ST MARY'S COL, 74- Mem: Soc Study Reproduction; Am Inst Biol Sci. Res: Testicular functions with particular interests in spermatozial maturation and levels of adenosinetriphosphate in certain regions of the male reproductive tracts of specific mammals, fish and insects. Mailing Add: Dept of Biol St Mary's Col Winona MN 55987

ALSUP, RICHARD GLENN, organic chemistry, see 12th edition

ALT, DAVID D, b St Louis, Mo, Sept 17, 33. GEOLOGY, GEOCHEMISTRY. Educ: Wash Univ, St Louis, AB, 55; Univ Minn, MS, 58; Univ Tex, PhD(geol), 61. Prof Exp: Sr res assoc chem, Univ Leeds, 61-62; asst prof geol, Univ Fla, 62-65; from asst prof to assoc prof, 65-73, PROF GEOL, UNIV MONT, 73- Mem: Geol Soc Am; Soc Econ Paleont & Mineral; Nat Asn Geol Teachers. Res: Emergent shorelines and coastal plain geomorphology; karst landscapes. Mailing Add: Dept of Geol Univ of Mont Missoula MT 59801

ALT, FRANZ LEOPOLD, b Vienna, Austria, Nov 30, 10; nat US; m 38; c 2. MATHEMATICS. Educ: Univ Vienna, PhD(math), 32. Prof Exp: Actuary, Assicurazioni Generali, Vienna, 35-38; from res prin to chief analyst, Economet Inst, NY, 38-43, asst dir res, 45-46; dep chief comput lab, Ballistics Res Lab, Ord Dept, US Army, 46-48; asst chief comput lab, Nat Bur Stand, 48-52, appl math div, 52-64, Off Stand Ref Data, 64-67; dep dir info div, Am Inst Physics, 67-73; RETIRED. Mem: Am Math Soc; Asn Comput Mach (pres, 50-52); Asn Comput Ling. Res: Numerical computation using high-speed computing devices; information retrieval; machine translation of languages; operations research; econometrics; statistical methods of business forecasting; foundations of geometry. Mailing Add: 245 Bennet Ave New York NY 10040

ALT, GERALD HORST, b Berlin, Ger, Jan 24, 25; m 59; c 2. ORGANIC CHEMISTRY. Educ: Univ London, BSc, 49, PhD(org chem), 54. Prof Exp: Res fel, Wayne State Univ, 54-56; res chemist, Monsanto Co, 56-64, res specialist, Agr Div, 64-67, SR RES SPECIALIST, MONSANTO AGR PROD CO, 67- Mem: Am Chem Soc; The Chem Soc. Res: Enamine chemistry; steroids; conformational analysis; reaction mechanisms; heterocycles. Mailing Add: Monsanto Agr Prod Co St Louis MO 63166

ALT, LESLIE L, b Arad, Rumania, Oct 6, 22; m 57. CHEMISTRY. Educ: Northwestern Univ, MS, 52. Prof Exp: Lab mgr, Diversey Corp, 51-59; scientist, 60-70, mgr cardiovasc studies, 70-74, MGR MED TRENDS STUDIES, GEN ELEC CO, 74- Mem: Am Chem Soc; Electrochem Soc. Res: Solid state chemistry of metals and semiconductors. Mailing Add: 12134 W Bel Mar Dr Franklin WI 53130

ALTAR, WILLIAM, b Vienna, Austria, Aug 27, 00; US citizen; m 37; c 2. THEORETICAL PHYSICS, ELECTRONICS. Educ: Univ Vienna, PhD(physics, math), 23. Prof Exp: Asst prof physics, Pa State Univ, 29-35; res fel chem physics, Princeton Univ, 36-37; instr elec eng, Robert Col, Istanbul, 37-40; instr, Case Inst Technol, 40-42; res physicist, Westinghouse Res Labs, 42-59; res specialist, US Dept Army, 50-51; res physicist, Ramo Wooldridge Corp & Space Technol Labs, Inc, 59-61; sr adv physicist, Aerospace Corp, 61-65; staff physicist & consult, Sound Div, Naval Res Labs, 66-71; PVT CONSULT, 71- Res: Sonar propagation in oceanic subsurface ducts; quantum mechanics; optics; chemical physics; microwave measurements and theory; error correcting codes; signal detection and processing. Mailing Add: 12255 Richwood Dr Los Angeles CA 90049

ALTARELLI, MASSIMO, b Rome, Italy, Feb 12, 48; m 73. THEORETICAL SOLID STATE PHYSICS. Educ: Univ Rome, Dr Physics, 69. Prof Exp: Res fel physics, Univ Rome, 70-71; res assoc, Univ Rochester, 71-73; res asst prof, 74-75, ASST PROF PHYSICS, UNIV ILL, URBANA, 75- Mem: Am Phys Soc. Res: Theory of the electronic structure and of the optical properties of perfect and imperfect solids. Mailing Add: Dept of Physics Univ of Ill Urbana IL 61801

ALTARES, TIMOTHY, JR, b Philadelphia, Pa, Dec 24, 29; m 53; c 3. PHYSICAL CHEMISTRY, POLYMER CHEMISTRY. Educ: Carnegie Inst Technol, BS, 52. Prof Exp: Res chemist, Ashland Oil & Refining Co, 54-56; res scientist, Callery Chem Co, 56-60; jr fel, polymer sect, Mellon Inst, 60-66; dir res, Pressure Chem Co, 66-69; SR RES SCIENTIST, ARCO/POLYMERS, INC, 69- Mem: AAAS; Am Chem Soc; fel Am Inst Chemists. Res: High energy fuels based on boron hydride chemistry; polymers, plastics and materials; correlation of properties with microstructure; polymers in general. Mailing Add: Res Dept Arco/Polymers Inc 440 College Park Dr Monroeville PA 15146

ALTENAU, ALAN GILES, b Cincinnati, Ohio, May 16, 38; m 74; c 1. ANALYTICAL CHEMISTRY. Educ: Univ Cincinnati, BS, 59; Purdue Univ, MS, 61, PhD(anal chem), 64. Prof Exp: Group leader, Anal Div, 66-69, mgr, 69-74, ASST DIR, CENT RES, FIRESTONE TIRE & RUBBER CO, 74- Mem: Am Chem Soc. Mailing Add: Cent Res Firestone Tire & Rubber Co Akron OH 44317

ALTENBERGER-SICZEK, ALDONA, b Warsaw, Poland, Jan 3, 44. THEORETICAL CHEMISTRY. Educ: Polytech Univ Warsaw, Poland, MS, 65; Univ Chicago, PhD(phys chem), 74. Prof Exp: Asst nuclear chem, Polytech Univ Warsaw, Poland, 65-67; RES ASSOC THEORET CHEM, ARGONNE NAT LAB, 74- Res: Quantum dynamics of molecular collisions; quantum mechanical close-coupled calculations of three body exchange reactions; theoretical study of promotion of chemical reactions by excitation of reactants; molecular structure calculations. Mailing Add: Argonne Nat Lab 9700 S Cass Ave Argonne IL 60439

ALTENBERN, ROBERT ALLEN, bacteriology, biochemistry, see 12th edition

ALTENBURG, LEWIS CONRAD, b Houston, Tex, May 12, 42; m 65; c 1. CYTOGENETICS. Educ: Rice Univ, BA, 65; Univ Tex Grad Sch Biomed Sci, MS, 68, PhD(biochem), 70. Prof Exp: ASST PROF MED GENETICS, UNIV TEX GRAD SCH BIOMED SCI, HOUSTON, 73- Concurrent Pos: Fel, Max Planck Inst Exp Med, Gottingen, WGer, 70-71; fel, Univ Tex Grad Sch Biomed Sci, 71-72; NIH fel, Dept Biol, Univ Tex M D Anderson Hosp & Tumor Inst, 71-73; res assoc med genetics, Univ Tex Grad Sch Biomed Sci, Houston, 73. Mem: Am Chem Soc; Am Soc Microbiol; Am Soc Cell Biol; Environ Mutagen Soc; Genetics Soc Am. Res: Molecular cytogenetics of human chromosomes. Mailing Add: Med Genetics Ctr Univ of Tex Grad Sch Biomed Sci Houston TX 77030

ALTER, ABRAHAM, b Brooklyn, NY, Apr 5, 05; m 30; c 2. CHEMISTRY. Educ: NY Univ, BS, 26; Columbia Univ, MA, 28. Prof Exp: Chemist, Am Cyanamid Co, 28-29; chemist, US Food & Drug Admin, NY, 29-35, Md, 35-44; asst chief, Wash DC Health Dept, 44-46; asst chief insecticides, USDA, 46-54; chemist, Brooklyn Navy Yard, 54-63 & Coney Island Hosp, NY, 63; chemist, New York City Dept Health, 63-73, RETIRED. Mem: Am Chem Soc. Res: Analytical insecticides; determination of trace elements; lead in blood, urine, paint chips and paint; spray residue; corrosion of metals. Mailing Add: Camden F125 West Palm Beach FL 33409

ALTER, HARVEY, b New York, NY, Sept 4, 32; m 57; c 2. PHYSICAL CHEMISTRY, ENVIRONMENTAL CHEMISTRY. Educ: City Univ New York, BS, 52; Univ Cincinnati, MS, 54, PhD(phys chem), 57. Prof Exp: Physicist plastics dept, Union Carbide Corp, 57-59; sr chemist, Harris Res Labs, Gillette Co Res Inst, 59-63, group leader, 64-65, assoc dir res, Toni Div, 65-66, tech dir res, 66-68, mgr, Harris Res Labs Dept Gillette Co Res Inst, 68-72, vpres, 72; DIR RES PROGS, NAT CTR FOR RESOURCE RECOVERY, INC, 72- Concurrent Pos: Lectr, City Col New York, 58-59; prof lectr, Am Univ, 74-; mem, US Air Force PBI Technol Rev Comt, 70, mem ad hoc adhesion Adv Comt, 71; chmn, Gordon Res Conf Sci Adhesion, 72. Mem: AAAS; Am Chem Soc. Res: Materials and energy recovery from wastes; science and public policy; chemistry of non-metallic materials. Mailing Add: Nat Ctr for Resource Recovery Inc 1211 Connecticut Ave Washington DC 20036

ALTER, HARVEY JAMES, b New York, NY, Sept 12, 35; m 65; c 1. INTERNAL MEDICINE, HEMATOLOGY. Educ: Univ Rochester, BA, 56, MD, 60. Prof Exp: From intern to resident med, Strong Mem Hosp, Univ Rochester, 60-61; clin assoc blood bank, NIH, 61-64; resident med, Seattle Univ Hosp, 64-65; from instr to asst prof med, Georgetown Univ Hosp, 66-69; SR INVESTR, NIH, 69-, CHIEF IMMUNOL SECT, BLOOD BANK, 73- Concurrent Pos: Univ fel hemat, Georgetown Univ Hosp, 65-66, clin assoc prof, 69- Honors & Awards: Superior Serv Award, Dept Health, Educ & Welfare, 74. Mem: Fedn Clin Res; Am Soc Hemat; Int Soc Hemat. Res: Studies relating to nature and significance of the hepatitis associated antigen. Mailing Add: 16705 George Washington Dr Rockville MD 20853

ALTER, HENRY WARD, b Taxila, Punjab, India, Dec 26, 23; US citizen; m44; c 3. PHYSICAL CHEMISTRY. Educ: Univ Calif, AB, 43, PhD(chem), 48. Prof Exp: Asst chem, Univ Calif, 42-44, 46-47, chemist, Manhattan Proj, 44-48; mgr separations chem & res assoc, Knolls Atomic Power Lab, Gen Elec Co, 48-57; tech specialist & mgr nuclear & inorg chem, Vallecitos Atomic Lab, 57-68; mgr nucleonics lab, Gen Elec Co, 68-71, mgr nuclear technol & appln oper, 71-74; PRES, TERRADEX CORP, 74- Mem: Am Chem Soc; Am Nuclear Soc; Atomic Indust Forum. Res: Nuclear energy; nuclear fuels; plutonium technology; nuclear chemistry; fuel reprocessing; uranium exploration. Mailing Add: 7 Kirkcrest Lane Alamo CA 94507

ALTER, JOHN EMANUEL, b Davenport, Iowa, June 25, 45. PHYSICAL CHEMISTRY. Educ: Univ Iowa, BS, 68; Cornell Univ, MS, 70, PhD(phys chem), 74. Prof Exp: RES SCIENTIST PHYS CHEM, MILES LABS, INC, 74- Mem: Am Chem Soc; AAAS. Res: Properties of immobilized enzymes; texturization of vegetable protein. Mailing Add: Miles Labs Inc 1127 Myrtle St Elkhart IN 46514

ALTER, MILTON, b Buffalo, NY, Nov 11, 29; m 52; c 5. NEUROLOGY, EPIDEMIOLOGY. Educ: State Univ NY Buffalo, BA, 51, MD, 55; Univ Minn, PhD, 64. Prof Exp: Mem staff epidemiol br, Nat Inst Neurol Dis & Blindness, 52-62; consult neurol, Vet Hosp, St Cloud & Ancker Hosp, St Paul, 64-65; consult neurologist, Vet Admin Hosp, Fargo, NDak & Faribault State Hosp, Minn, 65-67; assoc prof, 67-71, PROF NEUROL, MED SCH, UNIV MINN, MINNEAPOLIS, 71-; CHIEF NEUROL SERV, MINNEAPOLIS VET ADMIN HOSP, 67- Concurrent Pos: USPHS fel, 62-64; Minn Asn Retarded Children grant, 63-64; Off Int Res grant, NIH, 63-66; consult, Honeywell, Inc, 63-72; USPHS grant, 65-68; Multiple Sclerosis Soc grants, 68, 71 & 73; mem task force epidemiol of epilepsy, NIH, 70-73. Mem: Am Acad Neurol; AMA; Soc Epidemiol Res; fel Am Neurol Asn. Res: Epidemiology and genetics of neurological disorders, especially multiple sclerosis, myasthenia gravis and congenital malformations of the nervous system; dermatoglyphics of mental retardation. Mailing Add: Minneapolis Vet Admin Hosp 54th & 48th Ave S Minneapolis MN 55417

ALTER, RONALD, b New York, NY, Mar 27, 39; m 63; c 3. MATHEMATICS, COMPUTER SCIENCE. Educ: City Col New York, BS, 60; Univ Pa, MA, 62, PhD(math), 65. Prof Exp: Asst prof math, Univ Calif, Los Angeles, 64-67; res fel, Syst Develop Corp, 67-68, assoc res scientist, Res & Technol Div, 68-69; asst prof, 69-70, ASSOC PROF MATH & COMPUT SCI, UNIV KY, 70- Concurrent Pos: Math analyst, Litton Systs, Inc, 66; fac res fel, Oak Ridge Nat Lab, 75. Mem: Am Math Soc; Asn Comput Mach; Math Asn Am. Res: Analytic, algebraic and elementary number theory; combinatorial mathematics; graph theory; algebraic coding theory; computer applications to mathematics. Mailing Add: Dept of Comput Sci Univ Ky Lexington KY 40506

ALTERA, KENNETH P, b Chicago, Ill, June 30, 36; m 63; c 1. VETERINARY PATHOLOGY. Educ: Univ Ill, BS, 58, DVM, 60; Univ Calif, PhD(comp path), 65. Prof Exp: Asst pathologist, Univ Calif, 64-65; asst prof vet path, Colo State Univ, 65-71, assoc prof vet path & vet pathologist, 71-72; pathologist, United Med Labs, 72-73; PATHOLOGIST, SYNTEX CORP, 73- Mem: Am Col Vet Path. Res: Comparative and experimental oncology; drug safety evaluation. Mailing Add: 736 Casa Bonita Ct Los Altos CA 94022

ALTEVEER, ROBERT JAN GEORGE, b Rheden, Neth, June 5, 35; US citizen; m 74; c 2. PHYSIOLOGY, BIOPHYSICS. Educ: Acad Phys Educ, Amsterdam, BPE, 57; Springfield Col, MS, 58; Univ Minn, PhD(educ, physiol), 65. Prof Exp: Instr physiol hyg, Univ Minn, 62-65; lectr physiol, Ind Univ, Bloomington, 65, asst prof physiol & med, 65-67; asst prof, 67-72, ASSOC PROF PHYSIOL, HAHNEMANN MED COL, 72- Concurrent Pos: Adj asst prof, Univ Nev, Las Vegas, 66-67; vis prof bioeng, Rose Polytech Inst. Mem: AAAS. Res: Physiology of stress; environmental stress; biophysics of respiration and circulation. Mailing Add: Dept of Physiol Hahnemann Med Col Philadelphia PA 19102

ALTGELT, KLAUS H, b Ger, Jan 30, 27; m 51; c 2. PHYSICAL CHEMISTRY, PETROLEUM CHEMISTRY. Educ: Univ Mainz, BS, 52, MS, 55, PhD(phys chem), 58. Prof Exp: Instr phys chem, Inst Phys Chem, Univ Mainz, 54-59; res assoc phys chem, Mass Inst Technol, 59-61; res chemist, 61-65, sr res chemist, 65-68, SR RES ASSOC, CHEVRON RES CO, 68- Concurrent Pos: Fel, Univ Mainz, 58-59; pres, Ger Sch San Francisco, 68-71. Mem: AAAS; Am Chem Soc. Res: Physical chemical characterization of macromolecules such as natural rubber, collagen and asphaltenes in dilute solutions; chromatography, especially gel permeation chromatography; polymeric fuel and oil additives in petroleum chemistry; chemical structure of asphalt, heavy petroleum fractions, and coal. Mailing Add: Chevron Res Co 576 Standard Ave Richmond CA 94802

ALTHAUS, RALPH ELWOOD, b Bluffton, Ohio, May 27, 25; m 49; c 3. PLANT PATHOLOGY. Educ: Bluffton Col, BS, 49; Ohio State Univ, MS, 51. Prof Exp: Asst, Boyce-Thompson Inst, 51-53; prod develop rep, B F Goodrich Chem Co, 53-57 & Merck & Co, Inc, 57-60; prod develop rep, 60-68, MGR INT MKT DEVELOP DEPT, MONSANTO CO, 68- Res: Herbicides for agricultural and industrial vegetation control. Mailing Add: Monsanto Co 800 N Lindbergh Blvd St Louis MO 63166

ALTHAUSEN, DARRELL, chemistry, see 12th edition

ALTHOUSE, PAUL M, biochemistry, deceased

ALTHOUSE, VICTOR E, organic chemistry, see 12th edition

ALTHUIS, THOMAS HENRY, b Kalamazoo, Mich, June 21, 41; m 67; c 3. ORGANIC CHEMISTRY, MEDICINAL CHEMISTRY. Educ: Western Mich Univ, BS, 63, MA, 65; Mich State Univ, PhD(org chem), 68. Prof Exp: Res scientist, Med Chem Res Dept, 68-73, SR RES SCIENTIST, MED CHEM RES DEPT, PFIZER INC, 74- Mem: Am Chem Soc. Res: Synthesis of nitrogen heterocycles; natural product synthesis; asymmetric synthesis. Mailing Add: Pfizer Inc Med Chem Res Dept Eastern Point Rd Groton CT 06340

ALTICK, PHILIP LEWIS, b Los Angeles, Calif, June 27, 33; m 55; c 4. ATOMIC PHYSICS. Educ: Stanford Univ, BS, 55; Univ Calif, Berkeley, MA, 60, PhD(atomic physics), 63. Prof Exp: Res physicist, Lawrence Radiation Lab. Univ Calif, Berkeley, 63; from asst prof to assoc prof, 63-75, PROF PHYSICS, UNIV NEV, RENO, 75- Concurrent Pos: Consult, Lockheed Missile & Space Co, 63-; res assoc, Univ Chicago, 67-68; vis prof, Univ Trier-Kaiserslautern, 72-73. Mem: Am Phys Soc. Res: Theoretical study of atomic processes, especially the interaction of photons with atoms. Mailing Add: Dept of Physics Univ of Nev Reno NV 89507

ALTLAND, HENRY WOLF, b Takoma Park, Md, Feb 26, 45; m 68; c 1. ORGANIC CHEMISTRY. Educ: Gettysburg Col, AB, 67; Princeton Univ, PhD(org chem), 71. Prof Exp: NIH fel, Univ Va, 71-73; SR RES CHEMIST, EASTMAN KODAK RES LABS, 73- Mem: Am Chem Soc; Soc Photog Scientists & Engrs. Res: Synthesis and reactions of heterocyclic compounds, heterocyclic rearrangements; application of organometallic reactions to organic synthesis; novel photographic developers and silver-ion stabilizers. Mailing Add: Kodak Res Labs 1999 Lake Ave Rochester NY 14650

ALTLAND, PAUL DANIEL, b York, Pa, Apr 1, 13; m 44; c 3. ZOOLOGY, PHYSIOLOGY. Educ: Gettysburg Col, BS, 34; Duke Univ, AM, 36, PhD(cytol), 37. Prof Exp: Asst zool, Duke Univ, 34-36; from instr to asst prof biol, Gettysburg Col, 37-44; from asst physiologist to physiologist, 44-50, sr physiologist, 50-57, physiologist, 57-62, chief physiol sect, Lab Phys Biol, 62-72, CHIEF PHYSIOL SECT, LAB CHEM PHYSICS, NIH, USPHS, 72- Mem: Am Soc Zool; Soc Exp Biol & Med; Am Physiol Soc; Am Inst Biol Sci; Int Soc Biometeorol. Res: Physiology and pathology of animals exposed to simulated altitude, exercise and cold; blood formation in reptiles; longevity of erythrocytes in reptiles and birds; immunity at altitude; experimental endocarditis in animals. Mailing Add: Lab Chem Physics Nat Inst Arthritis USPHS Bethesda MD 20014

ALTMAN, ALBERT, b New York, NY, Dec 11, 32; m 55 & 68; c 2. THEORETICAL NUCLEAR PHYSICS, SOLID STATE PHYSICS. Educ: Brooklyn Col, BS, 54; Univ Md, MS, 58, PhD(nuclear physics), 62. Prof Exp: Sr physicist, Gen Dynamics Convair, 60-61; asst prof nuclear theory, Univ Va, 62 & Univ Md, 63-64; physicist, Naval Ord Lab, 64-66; PROF PHYSICS, LOWELL TECHNOL INST, 66- Mem: Am Phys Soc. Res: Nuclear reactions and nuclear structure; quantum mechanics and scattering theory; molecular physics. Mailing Add: Dept of Physics Lowell Technol Inst Lowell MA 01854

ALTMAN, ALLEN BURCHARD, b Berkeley, Calif, Nov 16, 42; m 64. GEOMETRY, ALGEBRA. Educ: Columbia Univ, PhD(math), 68. Prof Exp: Actg asst prof math, Univ Calif, San Diego, 68-69, asst prof, 69-75; PROF MATH, SIMON BOLIVAR UNIV, 75- Concurrent Pos: Fulbright fel, 71-72. Mem: AAAS; Am Math Soc; Math Asn Am. Mailing Add: Dept of Math Simon Bolivar Univ Caracas Venezuela

ALTMAN, CARL, physical chemistry, see 12th edition

ALTMAN, DAVID, b Paterson, NJ, Feb 13, 20; m 47; c 1. CHEMISTRY. Educ: Cornell Univ, AB, 40; Univ Calif, PhD(chem), 43. Prof Exp: Asst chem, Univ Calif, 40-43, res assoc, Radiation Lab, Manhattan Proj, 43-45, chief chemist, Jet Propulsion Lab, 45-56; mgr vehicle technol, Aeronutronic Systs, Inc, 56-59; VPRES CHEM SYSTS DIV, UNITED TECHNOL CORP, 59- Concurrent Pos: Consult, Dept Defense, US Air Force & NASA. Honors & Awards: Propulsion Award, Am Inst Aeronaut & Astronaut, 64. Mem: Am Chem Soc; Am Phys Soc; fel Am Inst Aeronaut & Astronaut; Aerospace Indust Asn Am. Res: High temperature; equilibrium and kinetics; thermodynamics; aerochemistry; surface chemistry; propulsion; missile systems. Mailing Add: Chem Systs Div United Technol Corp 1050 E Arques Sunnyvale CA 94088

ALTMAN, ISIDORE, b New York, NY, Sept 29, 12; m 35; c 3. APPLIED STATISTICS. Educ: City Col New York, BS, 32, MS, 33; Univ Pittsburgh, PhD(biostatist), 52. Prof Exp: Statistician, USPHS, 36-52; chief statistician, Comn Financing Hosp Care, 52-54; statist consult, United Cerebral Palsy, 54-55; from asst prof to prof biostatist, Grad Sch Pub Health, Univ Pittsburgh, 55-70; adj prof & sr assoc health servs res, Ctr Community Health & Med Care, Harvard Univ, 70-75; RES PROF BIOSTATIST, GRAD SCH PUB HEALTH, UNIV PITTSBURGH, 75- Concurrent Pos: Mem panel on health servs res, Comt on Study of Nat Needs for Biomed & Behav Res Personnel, Comn on Human Resources, Nat Res Coun. Mem: AAAS; Am Pub Health Asn. Res: Health services delivery systems; quality of medical care; health data processing. Mailing Add: Grad Sch Pub Health Univ of Pittsburgh Pittsburgh PA 15261

ALTMAN, JACK, b Antwerp, Belg, Feb 17, 23; nat US; m 47; c 4. PLANT PATHOLOGY. Educ: Rutgers Univ, BSc, 54, PhD(plant path), 57. Prof Exp: From asst prof & asst plant pathologists to assoc prof & assoc plant pathologist, 57-70, PROF BOT & PLANT PATH & PLANT PATHOLOGIST, EXP STA, COLO STATE UNIV, 70- Concurrent Pos: Tech rep, Regional Res Proj W-56, 58-; mem Colo Gov Comt Air Pollution, 63-; Colo State Rep, Fed Mold Metabolites, 64-; NSF awards, 63-65; Indust Res grants, 64-68; Dept Health, Educ & Welfare Air Pollution Proj res award, 67-69; pest control adv & plant pathologist, Khuzestan Water & Power Authority, Iran, 70-72. Mem: Am Phytopath Soc; Soc Nematologists; Sigma Xi; Am Soc Agron; Soil Sci Soc Am. Res: Associated nematode complex of root rot diseases of plants; use of antibiotics to control plant diseases; side effects of chlorinated hydrocarbons on nitrogenous transformations in soil; use of chemicals for Rhizoctonia control; turf diseases; changes in amino acids in soil following fumigation. Mailing Add: Dept of Bot & Plant Path Colo State Univ Ft Collins CO 80521

ALTMAN, JOSEPH, b Budapest, Hungary, Oct 7, 25; m 50; c 1. NEUROPSYCHOLOGY. Educ: NY Univ, PhD(physiol psychol), 59. Prof Exp: Asst neurophysiol, Mt Sinai Hosp, NY, 57-59; res fel neuroanat, Col Physicians & Surg, Columbia Univ, 59-60; res assoc physiol psychol, Sch Med, NY Univ, 60-61; res assoc psychophysiol lab, Mass Inst Technol, 61-62, assoc prof psychol, 62-68; PROF, DEPT BIOL SCI, PURDUE UNIV, 68-, PROF PSYCHOL SCI, 71- Mem: Soc Neurosci; Int Soc Develop Psychobiol. Res: Development of cerebellum and motor functions in normal and experimentally retarded animals. Mailing Add: Dept of Biol Sci Purdue Univ Lafayette IN 47907

ALTMAN, JOSEPH HENRY, b Cambridge, Mass, Oct 11, 21; m 48; c 2. PHYSICS. Educ: Mass Inst Technol, BS, 42. Prof Exp: Physicist, Res Labs, 46-59, res assoc, 59-70, head res labs, 70-75, SR LAB HEAD, EASTMAN KODAK CO, 75- Concurrent Pos: Part-time lectr, Univ Rochester, 68- Honors & Awards: Jour Award, Soc Motion Picture & TV Eng, 68. Mem: Fel Optical Soc Am; fel Soc Photog Sci & Eng. Res: Structures of optical and photographic images; microphotography; microminiaturization; properties of photographic materials; photographic sensitometry. Mailing Add: Res Labs Eastman Kodak Co Rochester NY 14650

ALTMAN, KURT ISON, b Breslau, Ger, Oct 13, 19; US citizen; m 45; c 3. BIOCHEMISTRY, RADIOBIOLOGY. Educ: Univ Chicago, BS, 41, MS, 42; Univ Rochester, PhD(biochem), 63. Prof Exp: From res assoc to instr biochem, Univ Chicago, 44-57; res assoc radiation biol, 47-65, head div radiation chem, Atomic Energy Proj, 61-65, asst prof, 65-68, ASSOC PROF BIOCHEM, UNIV ROCHESTER, 68-, ASSOC PROF EXP RADIOL, SCH MED & DENT, 65- Concurrent Pos: Vis sr scientist, Ctr Study Nuclear Energy, Europ AEC, Mol & Donk, Belg, 64-65; vis prof physiol chem, Fac Med, Univ Düsseldorf, 73. Mem: Am Soc Biol Chemists; Geront Soc. Res: Enzymology; intermediary metabolism; isotope technology; biosynthesis of porphyrins and bile pigments; biochemical effects of ionizing radiations; physiological and radiological aging; connective tissue metabolism; bio-organic aspects of organic chemistry; erythrocyte metabolism; splenic function in hemolysis. Mailing Add: Dept of Radiation Biol & Biophys Univ of Rochester Sch Med & Dent Rochester NY 14642

ALTMAN, LAWRENCE JAY, b Chicago, Ill, Nov 12, 41; m 65; c 1. ORGANIC CHEMISTRY. Educ: Calif Inst Technol, BS, 62; Columbia Univ, PhD(org chem), 65. Prof Exp: Air Force Off Sci Res fel photochem, Calif Inst Technol, 65-66; asst prof org chem, Stanford Univ, 66-72; ASSOC PROF CHEM, STATE UNIV NY STONY BROOK, 72- Concurrent Pos: Grants, Res Corp, 66-67 & 70-71, Am Chem Soc, 66-71, Lilly Res Found, 67-69 & NIH, 72-; A P Sloan Found fel, 71-75. Mem: AAAS; Am Chem Soc; Brit Chem Soc. Res: Biogenesis of sterols; small ring compounds; biosynthesis of natural products; tritium nuclear magnetic resonance spectroscopy; small ring chemistry; synthetic organic chemistry. Mailing Add: Dept of Chem State Univ NY Stony Brook NY 11794

ALTMAN, LEONARD CHARLES, b Fresno, Calif, Sept 1, 44; m 70. IMMUNOLOGY. Educ: Univ Pa, BA, 65; Harvard Med Sch, MD, 69. Prof Exp: From intern to resident med, Univ Wash Affil Hosps, 69-71, chief resident, Harborview Med Ctr, 74-75; ASST PROF MED, DEPT MED, DIV ALLERGY, UNIV WASH, 75- Concurrent Pos: From res assoc to sr res assoc, Lab Microbiol & Immunol, Nat Inst Dent Health, NIH, 71-74. Mem: Am Asn Immunologists; Am Acad Allergy; Am Bd Internal Med; Am Fedn Clin Res; Reticuloendothelial Soc. Res: Host defense mechanisms with emphasis on white cell function; cell mediated immune mechanisms with emphasis on the role of cell mediated immune function in allergic patients. Mailing Add: Div of Allergy Dept of Med Univ of Wash RM-13 Seattle WA 98195

ALTMAN, PHILIP LAWRENCE, b Kansas City, Mo, Jan 6, 24; m 46; c 2. INFORMATION SCIENCE, COMMUNICATION SCIENCE. Educ: Univ Southern Calif, AB, 48; Western Reserve Univ, MS, 49. Prof Exp: Technician histopath, Univ Kans Med Ctr, 51; biologist mycol, Kansas City Field Sta, USPHS, 52-54; biologist virol, NIH, 55-56; res analyst, Nat Acad Sci, 57-59; DIR OFF BIOL HANDBKS, FEDN AM SOCS EXP BIOL, 59- Concurrent Pos: Int Union Biol Sci del, Comt on Data for Sci & Technol, 70- Mem: AAAS; Am Soc Info Sci; Coun Biol Educ (chmn, 72-73); Am Inst Biol Sci. Res: Comparative vertebrate anatomy; histoplasmosis; virus crystallization; analysis, editing and compilation of data in biological and medical sciences. Mailing Add: 9206 Ewing Dr Bethesda MD 20034

ALTMAN, ROBERT LEON, b Brooklyn, NY, Apr 27, 31; m 58; c 2. PHYSICAL CHEMISTRY. Educ: NY Univ, BA, 52; Univ Southern Calif, PhD(protein denaturation), 59. Prof Exp: Sr res engr, Rocketdyne Div, NAm Aviation, Inc, 56-59; fel phys chem, Univ Calif, Berkeley, 59-60, chemist, Lawrence Radiation Lab, 60-62; asst prof, Calif State Univ, Hayward, 62-66; RES SCIENTIST, NASA AMES RES CTR, 66- Concurrent Pos: Vis asst prof, Univ Calif, Berkeley, 64. Mem: Am Chem Soc; Combustion Inst. Res: Protein chemistry; calculation of thermodynamic properties; high temperature chemistry; improvement of fire extinguishment by use of dry chemical powders. Mailing Add: NASA Ames Res Ctr Moffett Field CA 94035

ALTMAN, ROBERT M, entomology, see 12th edition

ALTMAN, SIDNEY, b Montreal, Que, May 7, 39. MOLECULAR BIOLOGY. Educ: Mass Inst Technol, BS, 60; Univ Colo, PhD(biophys), 67. Prof Exp: Teaching asst, Columbia Univ, 60-62; Damon Runyon Mem Fund Cancer Res fel molecular biol, Harvard Univ, 67-69; Anna Fuller Fund fel, Med Res Coun Lab Molecular Biol, 69-70, med res coun fel, 69-71; asst prof, 71-75, ASSOC PROF BIOL, YALE UNIV, 75- Concurrent Pos: Tutor biol, Radcliffe Col, 68-69; consult ed, Am Scientist, 73- Mem: AAAS. Res: Effects of acridines on T4 DNA replication; mutants; precursors of tRNA; RNA processing and ribonuclease function. Mailing Add: Dept of Biol Yale Univ New Haven CT 06520

ALTMANN, STUART ALLEN, b St Louis, Mo, June 8, 30; m 59; c 2. ANIMAL BEHAVIOR. Educ: Univ Calif, Los Angeles, BA, 53, MA, 54; Harvard Univ, PhD(biol), 60. Prof Exp: Biologist, NIH, 56-58; from asst prof to assoc prof zool, Univ Alta, 60-65; sociobiologist, Yerkes Regional Primate Res Ctr, 65-70; PROF BIOL & ANAT, UNIV CHICAGO, 70- Concurrent Pos: Res grants, Nat Res Coun Can, 61-63, NIMH, 62-64 & 70-71 & NSF, 63-65 & 69; mem primate conserv comt, Nat Res Coun; mem exp psychol res rev comt, NIMH. Mem: Soc Study Evolution; Ecol Soc Am; fel Animal Behav Soc; Am Soc Zool. Res: Social communication in animals; field studies of primate behavior; mathematical models of social behavior. Mailing Add: 1507 E 56th St Chicago IL 60637

ALTMILLER, DALE HENRY, b Shattuck, Okla, Sept 11, 40; m 64; c 2. CLINICAL CHEMISTRY, MEDICAL GENETICS. Educ: WTex State Univ, BS, 63; Univ Tex, PhD(genetics), 69. Prof Exp: Asst prof human genetics, Univ Tex Med Br, 69-75; ASST PROF PATH, UNIV OKLA COL MED, 75- Concurrent Pos: Res grants, NSF, 70-72 & Robert A Welch Found, 71-74; dir clin endocrinol serv lab, Okla Children's Mem Hosp, 75- Mem: Am Asn Clin Chem; Am Soc Human Genetics; Sigma Xi; AAAS. Res: Metabolism and function of hormones and development of new methods of hormone analysis. Mailing Add: Dept of Path PO Box 26901 Univ Okla Health Sci Ctr Oklahoma City OK 73125

ALTMILLER, HENRY, b Ft Smith, Ark, July 13, 41. PHYSICAL CHEMISTRY, RADIATION CHEMISTRY. Educ: Univ Notre Dame, BS, 64, PhD(chem), 69. Prof Exp: Teacher, Holy Cross High Sch, Tex, 64-65; asst prof, 69-73, ASSOC PROF CHEM, ST EDWARD'S UNIV, 73-, ACAD DEAN, 74- Mem: Am Chem Soc. Res: Energy transfer; production of excited species in photo and radiation chemistry. Mailing Add: Dept of Chem St Edward's Univ Austin TX 78704

ALTON, ALVIN JOHN, b Linden, Wis, Mar 31, 13; m 40; c 1. FOOD SCIENCE. Educ: Univ Wis, BS, 36. Prof Exp: Asst dairy indust, Univ Wis, 36-37; food chemist, Beatrice Creamery Co, Chicago, 36-48; dir prod res, Taylor Freezer, Wis, 48-49; dir prod res, 49-65, GEN MGR, BEATRICE FOODS CO, 65- Mem: Am Dairy Sci Asn; Inst Food Technol. Res: Dairy and allied food products. Mailing Add: Beatrice Foods Co PO Box 749 Beloit WI 53511

ALTON, DONALD ALVIN, b Chicago, Ill, Sept 28, 43; m 66; c 4. COMPUTER SCIENCES, MATHEMATICAL LOGIC. Educ: Rice Univ, BA, 65; Cornell Univ, PhD(math), 70. Prof Exp: Asst prof math, 70-71, asst prof comput sci, 71-74, ASSOC PROF COMPUT SCI, UNIV IOWA, 74- Concurrent Pos: NSF grants, 72- Mem: Asn Comput Mach; Am Math Soc; Asn Symbolic Logic. Res: Computational complexity; recursion-theoretic computational complexity in the spirit of work of Manuel Blum and others; design and analysis of efficient algorithms; recursion theory. Mailing Add: Dept of Comput Sci Univ of Iowa Iowa City IA 52242

ALTON, EARL ROBERT, b Oelwein, Iowa, Oct 28, 33; m 55; c 2. INORGANIC CHEMISTRY. Educ: St Olaf Col, BA, 55; Univ Mich, MS, 58, PhD(chem), 61. Prof Exp: From asst prof to assoc prof, 60-69, PROF CHEM, AUGSBURG COL, 69-, CHMN DEPT, 68- Concurrent Pos: NSF sci fac fel, 67-68. Mem: AAAS; Am Chem Soc; Sigma Xi. Res: Chemistry of phosphorus trifluoride, especially reactions with group II and III halides and with transition metal halides; chemistry of coordination compounds and of hydrides. Mailing Add: Dept of Chem Augsburg Col Minneapolis MN 55454

ALTON, ELAINE VIVIAN, b Watertown, NY, Aug 30, 25. MATHEMATICS EDUCATION. Educ: State Univ NY Albany, BA, 42; St Lawrence Univ, MEd, 51; Univ Mich, MA, 58; Mich State Univ, PhD(math educ), 65. Prof Exp: Teacher math, Cobleskill High Sch & Fultonville High Sch, 46-48; assoc prof math, Ferris State Col, 48-62; asst instr, Mich State Univ, 62-64; asst prof, 64-71, ASSOC PROF MATH EDUC, IND UNIV-PURDUE UNIV, INDIANAPOLIS, 71- Concurrent Pos: Mem, Adv Comt Math, State Ind, 68-; NSF proposal panelist, Coop Col-Sch Proj, 71. Mem: Nat Coun Teachers Math; Math Asn Am; Asn Teacher Educrs. Res: Methods and materials for teaching the metric system; commercial and teacher made materials for teaching concepts and computational skills to a wide range of abilities; changing teacher attitudes towards mathematics. Mailing Add: Dept of Math Sci Ind Univ-Purdue Univ Indianapolis IN 46205

ALTON, GERALD DODD, b Murray, Ky. EXPERIMENTAL ATOMIC PHYSICS, THEORETICAL ATOMIC PHYSICS. Educ: Murray State Univ, Ky, BS, 60; Univ Tenn, Knoxville, MS, 67, PhD(physics), 72. Prof Exp: Develop physicist atomic physics, Isotopes Div, 60-72, RES STAFF PHYSICIST EXP & THEORET ATOMIC PHYSICS, PHYSICS DIV, OAK RIDGE NAT LAB, 73- Concurrent Pos: Consult, GCA Corp, Sunnyvale, Calif, 74- Mem: Sigma Xi. Res: Negative and multiply charged positive ion source research and development; accelerating systems and theoretical and experimental atomic physics. Mailing Add: Physics Div Oak Ridge Nat Lab Oak Ridge TN 37830

ALTROCK, RICHARD CHARLES, b Omaha, Nebr, Dec 20, 40; m 63; c 2. SOLAR PHYSICS. Educ: Univ Nebr, BSc, 62; Univ Colo, PhD(astro-geophys), 68. Prof Exp: ASTROPHYSICIST, SACRAMENTO PEAK OBSERV, AIR FORCE CAMBRIDGE RES LAB, 67- Concurrent Pos: Travel grant, Australian Commonwealth Sci & Indust Res Orgn, 71-; vis res fel, Dept Appl Math, Univ Sydney, 71-72. Mem: AAAS; Am Astron Soc; Sci Res Soc Am; Astron Soc Pac; Int Astron Union. Res: Physical conditions in the quiet solar chromosphere and photosphere; analysis of solar atomic-line profiles; global variations in solar physical conditions. Mailing Add: Sacramento Peak Observ Sunspot NM 88349

ALTSCHER, SIEGFRIED, b Berlin, Ger, July 18, 22; US citizen; m 46; c 3. POLYMER CHEMISTRY, ORGANIC CHEMISTRY. Educ: Brooklyn Col, BA, 42; Polytech Inst Brooklyn, MS, 50, PhD(chem), 65. Prof Exp: Prod supvr fine chem, Wallace & Tiernan Prod Inc, 46-54; group leader res & develop, Nopco Chem Co, 54-66; MGR POLYMER/ANAL DEPT, EASTERN RES CTR, STAUFFER CHEM CO, 66- Mem: Am Chem Soc; Fedn Socs Paint Technol; Soc Plastics Engrs. Res: Development of commercial specialty and surfactant chemicals; basic studies of polymer and plastic properties to develop new compositions which satisfy newer commercial requirements of improved performance. Mailing Add: Eastern Res Ctr Stauffer Chem Co Dobbs Ferry NY 10522

ALTSCHUL, AARON MAYER, b Chicago, Ill, Mar 13, 14; m 37; c 2. NUTRITION. Educ: Univ Chicago, BS, 34, PhD(chem), 37. Hon Degrees: DSc, Tulane Univ, 68. Prof Exp: Instr chem, Univ Chicago, 37-41; biochemist, Southern Regional Res Lab, USDA, 41-52, head oil seed sect, Agr Res Serv, 52-58, chief chemist, Seed Protein Pioneering Res Lab, 58-67, spec asst int nutrit improv, 67-69, Off Secy, 69-71; PROF COMMUNITY MED & INT HEALTH, SCH MED, GEORGETOWN UNIV, 71-, CHMN DIV NUTRIT, 75- Concurrent Pos: Res consult, Tulane Univ, 43-58, consult prof, 58-64, prof, 64-66; mem, Select Comt Substances Gen Recognized As Safe, 72-73 & Fedn Am Socs Exp Biol Select Comt Flavor Eval Criteria, 75- Mem: Am Chem Soc; AAAS; Am Soc Biol Chem; Inst Food Technologists; Am Inst Nutrit. Res: Seed enzymes and proteins; food utilization of plant proteins; fortification of foods; community nutrition. Mailing Add: 700 New Hampshire Rd NW Washington DC 20037

ALTSCHUL, ROLF, b Duesseldorf, Ger, Jan 24, 18; nat US; m 55; c 2. PHYSICAL ORGANIC CHEMISTRY. Educ: Harvard Univ, AM, 39, PhD(chem), 41. Prof Exp: Fel, Harvard Univ, 41, Pittsburgh Plate Glass fel, 41-44; lectr, Bryn Mawr Col, 44-45; CHMN DEPT CHEM, SARAH LAWRENCE COL, 45- Concurrent Pos: Res assoc, Brandeis Univ, 54- Res: Kinetics and mechanisms of liquid phase reactions. Mailing Add: Dept of Chem Sarah Lawrence Col Bronxville NY 10708

ALTSCHUL, SIRI VON REIS, b Detroit, Mich, Feb 10, 31; m 63, 75; c 7. BOTANICAL TAXONOMY, ETHNOBOTANY. Educ: Univ Mich, AB, 53; Radcliffe Col, MA, 57, PhD(biol), 61. Prof Exp: HON CUR ETHNOBOT, NY BOT GARDEN & HON RES ASSOC ETHNOPHARMACOL, BOT MUS HARVARD UNIV, 74- Concurrent Pos: Nat Inst Ment Health res grants, Bot Mus, Harvard Univ, 64-66; Nat Libr Med grant, 69; trustee, Radcliffe Col, 70-; Rockefeller Found grant for herbarium search for plants affecting human fertility, NY Bot Garden, 75- Mem: AAAS; Am Antrhop Asn; Am Inst Biol Sci; Am Soc Pharmacog; Int Asn Plant Taxon. Res: Higher plants which are used by primitive societies for curing, magic or eating and which contain biodynamic agents potentially useful in modern medicine and nutrition. Mailing Add: 45 E 89th St New York NY 10028

ALTSCHULE, MARK DAVID, b New York, NY, July 16, 06; m 34. MEDICINE. Educ: City Col New York, BS, 27; Harvard Univ, MD, 33; Am Bd Internal Med, dipl, 42. Prof Exp: Technician, Montefiore Hosp, New York, 26-28; intern, Peter Bent Brigham Hosp, Boston, 32; intern med, Beth Israel Hosp, 32-34; res fel, Harvard Med Sch, 34-35, from asst to asst prof med, 35-52, from asst prof to prof clin med, 52-73; dir internal med & res clin physiol, 47-68, CONSULT CLIN PHYSIOLOGIST, McLEAN HOSP, BELMONT, MASS, 68-; HON CUR PRINTS & PHOTOG COLLECTIONS, FRANCIS A COUNTWAY LIBR MED, HARVARD MED SCH, 70-, VIS PROF MED, 73- Concurrent Pos: Resident med, Beth Israel Hosp, 34-36, assoc, 37-43 & 46-66, actg dir med res, 43-46, assoc vis physician, 38-46, vis physician, 46-66; consult internist, McLean Hosp, 45-47; attend physician, Vet Admin Hosp, Boston, 53-; assoc, Thorndike Lab, staff consult & vis physician, Boston City Hosp 55-74; ed-in-chief, Lippincott's Med Sci, 59-68 & Med Counterpoint, 69-73; consult, Childrens Hosp Med Ctr, Boston, 64-67, Naval Blood Res Lab, Chelsea, Mass, 68- & Yale-New Haven Hosp, Conn, 70-; lectr med, Sch Med, Yale Univ, 66- Mem: Emer mem Asn Am Physicians; Am Heart Asn; NY Acad Sci; Am Col Clin Pharmacol & Chemother; emer mem Am Soc Clin Invest. Res: Physiology of cardiac and pulmonary disease; physiology of mental disease; clinical significance of cardiac and respiratory adjustments in chronic anemia. Mailing Add: 23 Warwick Rd Brookline MA 02146

ALTSCHULER, HELMUT MARTIN, b Mannheim, Ger, Feb 13, 22; US citizen; m 42; c 3. ELECTROPHYSICS. Educ: Polytech Inst Brooklyn, BEE, 47, MEE, 49, PhD(electrophys), 63. Prof Exp: Res asst, Microwave Res Inst, Polytech Inst Brooklyn, 48-51, jr res assoc, 51-52, sr res assoc, Electrophys Dept, 52-63, res assoc prof, 63-64; chief radio standards eng div, 64-69, SR RES SCIENTIST, ELECTROMAGNETICS DIV, NAT BUR STANDARDS, 69- Concurrent Pos: Mem US nat comt, Int Sci Radio Union. Mem: Fel Inst Elec & Electronics Eng. Res: Impedance measurement techniques; equivalent network representations; strip transmission lines; electromagnetic and transmission line and modal theory; dielectric constant measurement techniques; microwave bridge measurement methods; nonreciprocal and active two-ports; electromagnetic standards; measurements, standards and laboratories. Mailing Add: Electromagnetic Div 272.10 Nat Bur of Standards Boulder CO 80302

ALTSCHULER, MARTIN DAVID, b Brooklyn, NY, Feb 25, 40; m 62; c 3. SOLAR PHYSICS, RADIOLOGY. Educ: Polytech Inst Brooklyn, BS, 60; Yale Univ, MS, 61, PhD(astron), 64. Prof Exp: Instr astron, Yale Univ, 64-65; SOLAR PHYSICIST, HIGH ALTITUDE OBSERV, NAT CTR ATMOSPHERIC RES, 66- Concurrent Pos: Adj assoc prof astrogeophys, Univ Colo, 74-; mem comts 10 & 12, Int Astron Union. Mem: Am Phys Soc; Am Astron Soc; Am Asn Physicists in Med. Res: Solar magnetic fields; structure and dynamics of solar corona; medical image reconstruction; computer roentgenography. Mailing Add: HAO-NCAR PO Box 3000 Boulder CO 80303

ALTSCHULER, MILTON, b New York, NY, Apr 9, 26; c 2. ANTHROPOLOGY. Educ: Univ Calif, Los Angeles, AB, 53, Am, 56; Univ Minn, Minneapolis, PhD(anthrop), 64. Prof Exp: Instr anthrop, Univ Minn, 61-64; asst prof, Univ Cincinnati, 64-66; asst prof, 66-69, ASSOC PROF ANTHROP, SOUTHERN ILL UNIV, CARBONDALE, 69- Mem: Am Anthrop Asn; Royal Anthrop Inst Gt Brit & Ireland; Soc Appl Anthrop; Latin Am Studies Asn; Am Ethnol Asn. Res: Problems of migrant labor within Ecuador and between Bolivia and Argentina; ethnic farming communities in Argentina; computer simulation of social systems. Mailing Add: Dept of Anthrop Southern Ill Univ Carbondale IL 62901

ALTSHULER, BERNARD, b Newark, NJ, June 22, 19; m 44; c 2. BIOMATHEMATICS. Educ: Lehigh Univ, BS, 40; NY Univ, PhD(math), 53. Prof Exp: Asst math, 49-51, from instr to assoc prof, 51-69, PROF ENVIRON MED, NY UNIV MED CTR, 69-. ASSOC DIR, INST ENVIRON MED, MED CTR, 75- Mem: Am Math Soc; Am Phys Soc; Biophys Soc; Am Indust Hyg Asn; Soc Indist & Appl Math. Res: Biophysics; environmental health; pulmonary handling of aerosols; quantitative aspects of carcinogenesis. Mailing Add: Dept of Environ Med NY Univ Med Ctr New York NY 10016

ALTSHULER, CHARLES HASKELL, b Detroit, Mich, Apr 14, 19; m 61; c 2. PATHOLOGY. Educ: Univ Mich, BS, 39, MD, 43. Prof Exp: Intern, Mt Sinai Hosp, New York, NY, 43-44; instr bact, Univ Mich, 46-47; resident path, Univ Wis, 47-50; res assoc, Univ Mich, 50-52; PATHOLOGIST, ST JOSEPH'S HOSP, MILWAUKEE, 53- Concurrent Pos: Clin asst prof, col med, Wayne State Univ, 50-52; assoc pathologist, Wayne County Gen Hosp, 50-52; lectr, col med, Marquette Univ. Mem: Fel Col Am Pathologists; AMA; Am Soc Clin Pathologists; Am Soc Exp Path; Am Fedn Clin Res. Mailing Add: St Joseph's Hosp 5000 W Chambers St Milwaukee WI 53210

ALTSHULER, HAROLD LEON, b New York, NY, June 8, 41; m 64; c 1. PHARMACOLOGY. Educ: Cornell Univ, BS, 64; Univ Calif, Davis, PhD(pharmacol), 72. Prof Exp: Res asst pharmacol, Sandoz Pharmaceut, 63-64; res asst pharmacol & toxicol, Parke, Davis & Co, 64-65; lab supvr clin biochem, Nat Ctr Primate Biol, 67-71; lab supvr psychopharmacol, Univ Calif, Davis, 71-72; ASST PROF PHARMACOL, BAYLOR COL MED, 72-; SECT HEAD NEUROPSYCHOPHARMACOL, TEX RES INST MENT SCI, 72- Concurrent Pos: Spec assoc, Univ Tex Grad Sch Biomed Sci Houston, 73-; Pharmaceut Mfrs Asn Found res starter grant, 74; CLIN ASSOC PROF, UNIV HOUSTON, 74- Mem: AAAS; Am Col Clin Pharmacol; Soc Neurosci. Res: Neuropharmacology; psychopharmacology; drug abuse; alcoholism; electroencephalography; drug self-administration; clinical psychopharmacology; epilepsy; laboratory primate husbandry and behavior. Mailing Add: Dept of Pharmacol Baylor Col of Med Houston TX 77025

ALTSHULLER, AUBREY PAUL, physical chemistry, see 12th edition

ALTSZULER, NORMAN, b Suwalki, Poland, Nov 20, 24; nat US; m 56; c 2. PHARMACOLOGY, ENDOCRINOLOGY. Educ: George Washington Univ, BS, 50, MS, 51, PhD(physiol), 54. Prof Exp: From instr to assoc prof, 55-69, PROF PHARMACOL, SCH MED, NY UNIV, 69- Mem: Am Physiol Soc; Am Soc Pharm & Exp Therapeut; Endocrine Soc; NY Acad Sci. Res: Endocrine regulation of metabolism. Mailing Add: Sch of Med NY Univ New York NY 10016

ALTURA, BELLA T, b Solingen, Ger, US citizen; m 61; c 1. PHYSIOLOGY, PHARMACOLOGY. Educ: Hunter Col, BA, 53, MA, 64; City Univ NY, PhD(physiol), 68. Prof Exp: Clin biochemist, Mem Hosp-Sloan Kettering Cancer Inst, 57-64; NASA fel, 64-67; from res assoc to instr anesthesiol, Albert Einstein Col Med, 69-74; ASST PROF PHYSIOL, STATE UNIV NY DOWNSTATE MED CTR, 74- Res: Physiology, pharmacology and biochemistry of blood vessels; excitation-contraction coupling mechanisms of vascular smooth muscles; drug abuse and the cardiovascular system. Mailing Add: Dept of Physiol State Univ of NY Downstate Med Ctr Brooklyn NY 11203

ALTURA, BURTON MYRON, b New York, NY, Apr 9, 36; m 61; c 1. PHYSIOLOGY, PHARMACOLOGY. Educ: Hofstra Univ, BA, 57; NY Univ, MS, 61, PhD(physiol), 64. Prof Exp: Res asst physiol, Sch Med, NY Univ, 61-62, instr exp anesthesiol, 64-65, asst prof anesthesiol, 65-66; from asst prof to assoc prof anesthesiol & physiol, Albert Einstein Col Med, 70-74; PROF PHYSIOL, STATE UNIV NY DOWNSTATE MED CTR, 74- Concurrent Pos: Res fel, Bronx Munic Hosp Ctr, 67-; NIH travel fel, Gothenburg, Sweden, 68; USPHS res career develop award, 68-72; mem coun on stroke, Am Heart Asn, 73-; vis prof anesthesiol & physiol, Albert Einstein Col Med, 74- Mem: Am Physiol Soc; Am Soc Pharmacol & Exp Therapeut; fel Am Col Nutrit; Soc Exp Biol & Med; Microcirc Soc. Res: Local regulation of blood flow; physiology and pharmacology of vascular smooth muscle; vascular smooth muscle pharmacology of antihistamines and neurohypophyseal hormones; physiology of reticuloendothelial system; physiology and therapeutics in experimental shock. Mailing Add: Dept of Physiol State Univ NY Downstate Med Ctr Brooklyn NY 11203

ALTWICKER, ELMAR ROBERT, b Wolfen-Bitterfeld, Ger, Apr 4, 30; US citizen; m 58; c 2. ORGANIC CHEMISTRY. Educ: Univ Dayton, BS, 52; Ohio State Univ, PhD(org chem), 57. Prof Exp: Asst, Ohio State Univ, 52-57; res chemist, Am Cyanamid Co, NJ, 57-62; res assoc, Univ Vt, 62-63; res chemist, Princeton Chem Res, Inc, 63-65, group leader, UOP Chem Div, 65-68; assoc prof chem, Bio-Environ Div, 68-75, PROF CHEM & ENVIRON ENG, RENSSELAER POLYTECH INST, 75- Concurrent Pos: Guest scientist, Dechema Inst, Frankfurt, 74-75. Mem: Am Chem Soc; Air Pollution Control Asn; Sigma Xi. Res: Air Pollution; atmospheric and control chemistry; ozone reactions; additives. Mailing Add: Dept Chem & Environ Eng Rensselaer Polytech Inst Troy NY 12181

ALUMBAUGH, ROBERT L, organic chemistry, see 12th edition

ALUOTTO, PATRICK F, organic chemistry, physical chemistry, see 12th edition

ALVAGER, TORSTEN KARL ERIK, b Halsberg, Sweden, Aug 22, 31; US citizen; m 58; c 2. BIOPHYSICS, NUCLEAR PHYSICS. Educ: Univ Stockholm, PhD(physics), 60. Prof Exp: Asst prof physics, Univ Stockholm, 60-66; vis prof, Princeton Univ, 66-68; PROF PHYSICS, IND STATE UNIV, TERRE HAUTE, 68- Mem: Am Phys Soc; AAAS; Sigma Xi. Res: Relativity. Mailing Add: Dept of Physics Ind State Univ Terre Haute IN 47802

ALVARADO, FRANCISCO, b Madrid, Spain, Nov 19, 32; m 61; c 5. BIOCHEMISTRY, PHYSIOLOGY. Educ: Univ Madrid, MD, 58. Prof Exp: Res fel biochem, NY Univ, 58-59; res fel med, Col Physicians & Surgeons, Columbia Univ, 59-60; res fel biochem, Sch Med, Washington Univ, 60-61; res assoc enzymol, Inst Maranon, Coun Sci Res, Spain, 62-64; asst prof biochem, Chicago Med Sch, 64-66; asst prof physiol, Med Sch, Rutgers Univ, 66-67; assoc prof pharmacol, Med Sch, Univ Louisville, 67-68; from assoc prof to prof pharmacol & physiol, Med Sch, Univ PR, San Juan, 69-75; VIS PROF, LAB COMP PHYSIOL, UNIV PARIS-SOUTH, CENTRE D'ORSAY, 75- Concurrent Pos: NIH grants, 62-77; Nat Inst Arthritis & Metab Dis res career develop award, 69-73. Honors & Awards: Leonardo Torres-Quevedo Prize, Span Res Coun, 58. Mem: Biophys Soc; Am Physiol Soc; NY Acad Sci; Span Soc Biochem; Royal Span Natural Hist Soc. Res: Glycolytic enzymes; solute transport across biological membranes. Mailing Add: Lab of Comp Physiol Univ of Paris-South Ctr d'Orsay Orsay France

ALVARADO, RONALD HERBERT, b San Bernardino, Calif, Dec 26, 33; m 55; c 3. ZOOLOGY. Educ: Univ Calif, Riverside, BA, 56; Wash State Univ, MS, MS, 59, PhD(zool physiol), 62. Prof Exp: Res fel physiol, Wash State Univ, 62; from asst prof to prof zool, Ore State Univ, 62-74; PROF ZOOL & CHMN DEPT, ARIZ STATE UNIV, 74- Mem: AAAS; Am Soc Zool. Res: Osmotic and ionic regulation in amphibians; mechanism of ion transport. Mailing Add: Dept of Zool Ariz State Univ Tempe AZ 85281

ALVARES, ALVITO PETER, b Bombay, India, Dec 25, 35; US citizen; m 69; c 1. PHARMACOLOGY, BIOCHEMISTRY. Educ: Univ Bombay, BSc, 55, BSc(tech), 57; Univ Detroit, MS, 61; Univ Chicago, PhD(pharmacol), 66. Prof Exp: USPHS grant, Univ Minn, 65-67; sr res biochemist, Burroughs Wellcome & Co, 67-70; res assoc biochem pharmacol, 70-71, asst prof, 71-74; ASSOC PROF BIOCHEM PHARMACOL, ROCKEFELLER UNIV, 74-; ADJ ASSOC PROF PHARMACOL, MED COL, CORNELL UNIV, 75- Concurrent Pos: Hirschl career scientist award, 74-78; NIH career develop award, 75-79. Mem: AAAS; NY Acad Sci; Am Soc Pharmacol & Exp Therapeut. Res: Effects of drugs, carcinogens and environmental chemicals on hepatic microsomal enzymes and heme biosynthetic pathway; cytochrome P-450 and mechanisms of enzyme induction. Mailing Add: Dept of Biochem Pharmacol Rockefeller Univ New York NY 10021

ALVAREZ, ANNE MAINO, b Rochester, Minn, Apr 14, 41; m. PLANT PATHOLOGY. Educ: Stanford Univ, BA, 63; Univ Calif, Berkeley, MS, 66, PhD(plant path), 72. Prof Exp: Instr plant path, Univ Neuquen, Arg, 69-70; jr specialist, 73-75, ASST PROF PLANT PATH, UNIV HAWAII, 75- Mem: Am Phytopath Soc. Res: Bacterial diseases of vegetable and fruit crops, epidemiology and disease control; survival of Xanthomonas in tropical soils; influence of nutrition on disease severity; orchard and post-harvest diseases of papaya. Mailing Add: Univ of Hawaii Dept Plant Path 3190 Maile Way Honolulu HI 96822

ALVAREZ, LAURENCE RICHARDS, b Jacksonville, Fla, Sept 27, 37; m 60; c 2. MATHEMATICS. Educ: Univ of the South, BS, 59; Yale Univ, MA, 62, PhD(graph theory), 64. Prof Exp: Instr math, Trinity Col, Conn, 63-64; from instr to asst prof, 64-71, ASSOC PROF MATH, UNIV OF THE SOUTH, 71- Concurrent Pos: Am Coun Educ acad admin intern, Pomona Col. Mem: Am Math Soc; Math Asn Am. Res: Graph theory. Mailing Add: Dept of Math Univ of the South Sewanee TN 37375

ALVAREZ, LUIS WALTER, b San Francisco, Calif, June 13, 11; m 36, 58; c 4. EXPERIMENTAL PHYSICS. Educ: Univ Chicago, BS, 32, PhD(physics), 36. Hon Degrees: DSc, Univ Chicago, 66, ScD, 68, Carnegie-Mellon Univ, 68 & Kenyon Col, 69. Prof Exp: Asst physics, 36-38, from instr to assoc prof, 38-45, assoc dir, Lawrence Berkeley Lab, 54-59, PROF PHYSICS, UNIV CALIF, BERKELEY, 45- Concurrent Pos: Staff mem, radiation lab, Mass Inst Technol, 40-43, metall lab, Univ Chicago, 43-44 & Los Alamos Lab, NMex, 44-45; mem, President's Sci Adv Comt, 71-72. Honors & Awards: Nobel Prize in Physics, 68; Collier Trophy, 46; Medal for Merit, 46; Scott Medal, 53; Calif Scientist of Year, 60; Einstein Medal, 61; Pioneer Award, Inst Elec & Electronics Engrs, 63; Nat Medal Sci, 64; A A Michelson Award, 65. Mem: Nat Acad Sci; Nat Acad Eng; Am Phys Soc (pres, 69); Am Acad Arts & Sci; Am Philos Soc. Res: Particle physics; astrophysics; ophthalmic optics. Mailing Add: Lawrence Berkeley Lab Univ of Calif Berkeley CA 94720

ALVAREZ, MARVIN RAY, cytology, cytogenetics, see 12th edition

ALVAREZ, RAYMOND ANGELO, JR, b Mobile, Ala, May 15, 34; m 61; c 1. PHYSICS. Educ: La State Univ, BS, 55; Stanford Univ, PhD(physics), 64. Prof Exp: Res assoc physics, Stanford Univ, 60-61; from instr to asst prof, Mass Inst Technol, 61-68; physicist, 68-74, HEAD LAB, LAWRENCE RADIATION LAB, 74- Mem: Am Phys Soc. Res: Nuclear physics. Mailing Add: Lawrence Radiation Lab L-221 PO Box 808 Livermore CA 94550

ALVAREZ, ROBERT, b New York, NY, Feb 6, 21; m 45; c 2. ANALYTICAL CHEMISTRY, PHYSICAL CHEMISTRY. Educ: City Col New York, BS, 42. Prof Exp: Chemist, Tenn Valley Authority, 42-43 & Manhattan Proj, 44-46; sr chemist, Carbide & Carbon Chem Corp & Union Carbide & Carbon Corp, Tenn, 46-49; phys chemist radiol div, Army Chem Corps, Md, 49-51; phys chemist, 51-55, ANAL CHEMIST, SPECTROCHEM ANAL SECT, NAT BUR STANDARDS, 55- Mem: Am Chem Soc; Soc Appl Spectros; Am Soc Testing & Mat. Res: Development of stable isotope dilution techniques for determinations by spark source mass spectrometry; chemical preconcentration techniques applied to spectrochemical analysis; developing optical emission and x-ray fluorescent methods of analysis. Mailing Add: 4113 Sampson Rd Wheaton MD 20906

ALVAREZ, VINCENT EDWARD, b Shanghai, China, Aug 15, 46; US citizen; m 72. INORGANIC CHEMISTRY. Educ: Calif State Univ, Hayward, BS, 69; Univ Calif, Santa Barbara, MA, 72, PhD(chem), 74. Prof Exp: RES CHEMIST, E I DU PONT DE NEMOURS & CO, INC, 74- Concurrent Pos: Asst prof chem, Del Tech & Community Col, 74- Mem: Am Chem Soc. Res: Reactions at metal oxide surfaces; reactions of singlet oxygen. Mailing Add: Pigments Dept E335-135 E I du Pont Exp Sta Wilmington DE 19898

ALVAREZ, WALTER, b Berkeley, Calif, Oct 3, 40; m 65. VOLCANOLOGY, TECTONICS. Educ: Carleton Col, BA, 62; Princeton Univ, PhD(geol), 67. Prof Exp: Geologist, Am Overseas Petrol, Ltd, Netherlands, 67-68, sr geologist, Libya, 68-70; NATO fel, Brit Sch Archaeol, Rome, 70-71; res scientist, 71-73, RES ASSOC, LAMONT-DOHERTY GEOL OBSERV, 73- Concurrent Pos: Lectr geol, Columbia Univ, 73-74. Mem: Ital Geol Soc; Geol Soc Am; Am Geophys Union. Res: Microplate and mountain belt tectonics; geological-archaeological studies; paleomagnetism; Alpine-Mediterranean region; northern South America. Mailing Add: Lamont-Doherty Geol Observ Palisades NY 10964

ALVAREZ-BUYLLA, RAMON, b Oviedo, Spain, June 22, 19; Mex citizen; m 56; c 4. NEUROENDOCRINOLOGY. Educ: Ashkhabad State Univ, USSR, MD, 43; Med Acad Sci, USSR, PhD(physiol), 46. Prof Exp: Prof & investr, Nat Polytech Inst, Mex, 47-54; investr, Nat Cardiol Inst Mex, 50-56 & Nat Inst Neumol Mex, 58-61; PROF PHYSIOL, CTR INVEST & ADVAN STUDY, NAT POLYTECH INST, 61- Concurrent Pos: Guggenheim award; vis prof human anat, Oxford Univ, 69-70. Mem: Arg Med Asn; Physiol Soc Mex; Latin Am Physiol Soc; Mex Acad Sci. Res: Neuroendocrinic integration; neuroendocrinic mechanism of the homeostasis; conditioned reflexes. Mailing Add: Ctr Invest & Advan Study IPN Apdo Postal 14-740 Mexico 14 DF Mexico

ALVAREZ-TOSTADO, CLAUDIO, inorganic chemistry, see 12th edition

ALVARINO DE LEIRA, ANGELES, b El Ferrol, Spain, Oct 3, 16; US citizen; m 40; c 1. BIOLOGICAL OCEANOGRAPHY, MARINE BIOLOGY. Educ: Univ Santiago, BSLett, 33; Univ Madrid, MS, 41, Cert Dr, 51, DSc, 67. Prof Exp: Prof biol, El Ferrol Col, 41-47; fishery biologist, Dept Sea Fisheries, Ministry Commerce, Madrid, 48-52; histologist, Super Coun Sci Res, 49-51; biol oceanogr, Span Inst Oceanog, Madrid, 50-57; biologist, Scripps Inst Oceanog, Univ Calif, La Jolla, 58-69; FISHERY RES BIOLOGIST, NAT MARINE FISHERIES SERV, FISHERY-OCEANOG CTR, LA JOLLA, 70- Concurrent Pos: Brit Coun grant, Marine Biol Lab, Plymouth, Eng, 53-54; Fulbright grant, Woods Hole Oceanog Inst, 56-57; NSF grants, 61-66; Bd Examr for theses, Univ Kerala, 72; dir doctoral thesis cand, Cent Univ Venezuela, 73-; vis prof, Nat Univ Mex, 75- Mem: Asn Trop Biol; Marine Biol Asn UK; fel Am Inst Fishery Res Biologists; Sigma Xi. Res: Zooplankton; Chaetognathae; Siphonophorae; Medusae; taxonomy; zoogeography; indicator organisms in water masses; fouling; sea fisheries; biology and ecology of Thunnidae. Mailing Add: 7535 Cabrillo Ave La Jolla CA 92037

ALVERSON, DAYTON L , b San Diego, Calif, Oct 7, 24; m 51; c 2. MARINE BIOLOGY. Educ: Univ Wash, BS, 50, PhD, 67. Prof Exp: Specialist method & equip, US Fish & Wildlife Serv, 50-53; biologist, Wash State Dept Fisheries, 53-58; chief, Explor Fishing & Gear Res Unit, US Fish & Wildlife Serv, 58-60, dir res base, Seattle, 60-67, Bur Com Fisheries, 67-69, assoc dir fisheries, Washington, DC, 69-70; assoc regional dir resource progs, 70, DIR NORTHWEST FISHERIES CTR, NAT OCEANIC & ATMOSPHERIC ADMIN, SEATTLE, 71- Concurrent Pos: Affil prof, Inst Marine Studies, Univ Wash; mem, Marine Bd, Nat Res Coun; chmn, Adv Comt Marine Resources Res, Food & Agr Orgn, UN. Honors & Awards: Distinguished Serv Award, Dept Interior, 66; Nat Oceanic & Atmospheric Admin Award. Mem: Am Inst Fisheries Res Biologists. Mailing Add: Nat Marine Fisheries Serv NOAA 2725 Montlake Blvd East Seattle WA 98112

ALVERSON, HOYT SUTLIFF, b Washington, DC, June 7, 42; m 64; c 2. ANTHROPOLOGY. Educ: George Washington Univ, AB, 64; Yale Univ, MPhil, 67, PhD(anthrop), 68. Prof Exp: Asst prof, 68-74, ASSOC PROF ANTHROP, DARTMOUTH COL, 74- Concurrent Pos: NSF grant develop comput based curric, Dartmouth Col, 70-71. Mem: AAAS. Res: Economic development and culture change in sub-Saharan Africa; anthropological linguistics, especially socio-linguistics; legal anthropology; effects of labor migration among Tswana of Botswana on values, on self-identity and on rural social organization. Mailing Add: Dept of Anthrop Dartmouth Col Hanover NH 03755

ALVERSON, ROY CARL, b Eureka, Calif, Apr 21, 22; m 52. APPLIED MATHEMATICS. Educ: Univ Wis, BS, 51; Brown Univ, PhD(appl math), 57. Prof Exp: Res engr, Boeing Airplane Co, Wash, 51-53; instr, Brown Univ, 53-57; res specialist, Boeing Airplane Co, 57-59; from res mathematician to sr res mathematician, Calif, 59-66, head appl math group, 66-74, HEAD APPL MATH GROUP, STANFORD RES INST, VA, 74- Mem: Am Math Soc; Soc Indust & Appl Math; Am Inst Aeronaut & Astronaut; Am Soc Eng Educ. Res: Elasticity; plasticity; partial differential equations; fluid dynamics; wave propagation; celestial mechanics. Mailing Add: Stanford Res Inst 1611 N Kent St Arlington VA 22209

ALVES, RONALD V, b Oakland, Calif, Apr 11, 35; m 67; c 2. SOLID STATE PHYSICS, QUANTUM ELECTRONICS. Educ: Univ San Francisco, BS, 58; Univ Calif, Berkeley, PhD(physics), 68. Prof Exp: Physicist, US Navy Radiol Defense Lab, 57-60; res scientist, Lockheed Missiles & Space Co, Lockheed Aircraft Corp, 68-74; MEM STAFF, COHERENT RADIATION, PALO ALTO, CALIF, 74- Mem: Am Phys Soc. Res: Solid state laser materials and rare earth activated luminescent materials. Mailing Add: Coherent Radiation 3210 Porter Dr Palo Alto CA 94304

ALVEY, DAVID DALE, b Harrisburg, Ill, Jan 2, 32; m 55; c 3. AGRONOMY. Educ: Univ Ill, BS, 57, MS, 58; Purdue Univ, PhD(genetics), 62. Prof Exp: Asst agron, Purdue Univ, 58-62, instr, 62; res agronomist, DeKalb Agr Asn, Inc, 62-70; CORN BREEDER, FARMERS FORAGE RES, 70- Mem: Am Soc Agron; Am Genetic Asn; Sigma Xi. Res: Genetics and plant breeding; coordinating corn breeding programs; developing commercial corn varieties. Mailing Add: 4112 E State Rd 225 West Lafayette IN 47906

ALVEY, FRANCIS BERTRAND, organic chemistry, physical chemistry, see 12th edition

ALVI, ZAHOOR MOHEM, b Sept 11, 32; US citizen; div; c 2. RADIOLOGICAL PHYSICS, BIOMEDICAL ENGINEERING. Educ: Carnegie-Mellon Univ, BS, 56; Univ Calif, Los Angeles, MS, 66, PhD(med physics, nuclear eng), 68; Am Bd Health Physics, dipl, 65; Am Bd Radiol, dipl & cert radiol physics, 71. Prof Exp: Radiological physicist, Radiation Ther & Nuclear Med, Cedars of Lebanon Hosp, Los Angeles, 58-64 & Dept Radiother, Kaiser-Permanente Med Ctr, Hollywood, 64-72; DIR MED PHYSICS, RADIOLOGICAL SYSTS, INC, 72- Concurrent Pos: Mem eng exten fac, Univ Calif, Los Angeles, 71-; attend med physicist, Martin Luther King, Jr Gen Hosp, 74-; res affil, Jet Propulsion Lab, Calif Inst Technol, 74- Mem: Am Asn Physicists in Med; Health Physics Soc; Int Radiation Protection Asn; Biomed Eng Soc. Res: Mathematical modeling of human physiological systems; ALVI-Henry Paul catheter probe system for gastrointestinal motility automatic monitoring; in-vivo neutron activation analysis using Californium-252 sealed ug sources. Mailing Add: RadioLogical Systs Inc 13224 G Admiral Ave Marina Del Rey CA 90291

ALVINO, WILLIAM MICHAEL, b New York, NY, Jan 12, 39; m 60; c 2. POLYMER CHEMISTRY. Educ: Iona Col, BS, 60; Seton Hall Univ, MS, 66. Prof Exp: Sr tech aide polymers, Bell Tel Labs, 60-66, assoc mem tech staff, 66; res chemist, 66-70, SR RES CHEMIST, WESTINGHOUSE ELEC CORP, 70- Mem: Am Chem Soc. Res: Thermal and ultraviolet stability of polymers; development and formulation of polymers as coating, films, adhesives and laminating resins for electrical insulation; synthesis and modification of polymers for high temperature applications. Mailing Add: Westinghouse Res Labs Beulah Rd Pittsburgh PA 15235

ALVIS, HARRY J, b East St Louis, Ill, Sept 15, 10; m 37; c 2. OCCUPATIONAL MEDICINE. Educ: Univ Ill, BS, 29; Univ Iowa, MD, 33; Harvard Univ, MPH, 49; Am Bd Prev Med, dipl occup med, 56. Prof Exp: Pvt med pract, 37-41; submarine med officer, Med Corps, US Navy, 41-64; assoc res prof, 64-66, assoc dean continuing med educ, 67-70, ASSOC PROF PREV MED, SCH MED, STATE UNIV NY BUFFALO, 66-, DIR HYPERBARIC MED PROG, 64-, DIR MED EDUC, MILLARD FILLMORE HOSP, 70- Concurrent Pos: Consult, Vet Admin Hosp, Buffalo, 64- Mem: Am Col Prev Med; Indust Med Asn. Res: Diving and submarine medicine; occupational health. Mailing Add: Dept of Med Educ Millard Fillmore Hosp Buffalo NY 14209

ALVORD, DONALD C, b Newton, Mass, Nov 12, 22; m 50; c 2. ECONOMIC GEOLOGY, CHEMISTRY. Educ: Univ Idaho, BS, 51. Prof Exp: GEOLOGIST, US GEOL SURV, 51- Mem: AAAS; fel Geol Soc Am; Am Chem Soc. Res: Geology and chemical characterization of coal; chemistry of organic mineral deposits; physical organic chemistry. Mailing Add: US Geol Surv 8422 Federal Bldg Salt Lake City UT 84138

ALVORD, ELLSWORTH CHAPMAN, JR, b Washington, DC, May 9, 23; m 43; c 4. NEUROPATHOLOGY. Educ: Haverford Col, BS, 44; Cornell Univ, MD, 46. Prof Exp: Asst path, Med Col, Cornell Univ, 47-48; instr neurol, Sch Med, Georgetown Univ, 50-55; assoc prof neurol & path, Col Med, Baylor Univ, 55-60; assoc prof, 60-62, PROF PATH, SCH MED, UNIV WASH, 62- Concurrent Pos: Prof lectr, Sch Med, George Washington Univ, 50-51; neurologist, Neurophysiol Sect, Army Med Serv Grad Sch, 51-53; instr, Wash Sch Psychiat, 52-55; chief clin neuropath sect, Nat Inst Neurol Dis & Blindness, NIH, 53-55; consult, Seattle hosps. Mem: AAAS; Am Asn Neuropath (pres, 64-); Am Soc Exp Biol & Med; Asn Res Nervous & Ment Dis; assoc Am Neurol Asn. Res: Spinal reflexes; allergic encephalomyelitis. Mailing Add: 5547 NE Windermere Rd Seattle WA 98105

ALWORTH, WILLIAM LEE, b Twin Falls, Idaho, Jan 3, 39; m 59; c 2. BIO-ORGANIC CHEMISTRY. Educ: Harvard Univ, AB, 60; Univ Calif, Berkeley, PhD(org chem), 64. Prof Exp: NIH fel, Harvard Univ, 64-65; asst prof org chem, 65-71, ASSOC PROF CHEM, TULANE UNIV, 71-, ADJ ASSOC PROF BIOCHEM, 75- Concurrent Pos: NIH career develop award, 72. Mem: Am Chem Soc; Sigma Xi. Res: Biosynthesis of vitamins; chemical carcinogenesis; enzyme mechanisms. Mailing Add: Dept of Chem Tulane Univ New Orleans LA 70118

ALYEA, ETHAN DAVIDSON, JR, b Orange, NJ, Mar 7, 31; m 57; c 4. PHYSICS. Educ: Princeton Univ, AB, 53; Calif Inst Technol, PhD(physics), 62. Prof Exp: From asst prof to assoc prof, 62-71, PROF PHYSICS, IND UNIV, BLOOMINGTON, 71- Mem: Am Asn Physics Teachers; Am Phys Soc. Res: Experimental high energy physics; hybrid bubble chamber and proportional wire chamber experiments. Mailing Add: Dept of Physics Ind Univ Bloomington IN 47401

ALYEA, FRED NELSON, b Harlingen, Tex, Oct 26, 38; m 62; c 2. DYNAMIC METEOROLOGY. Educ: Univ Wis-Madison, BS, 64; Colo State Univ, PhD(atmospheric sci), 72. Prof Exp: RES ASSOC METEOROL, MASS INST TECHNOL, 72- Concurrent Pos: Consult, Mfg Chemists Asn, 75- Mem: Am Meteorol Soc; Am Geophys Union; Sigma Xi. Res: Stratospheric modeling and climatology; large-scale dynamic processes in the atmosphere; numerical weather prediction methods; atmospheric predictability; paleoclimatology; climatological effects of anthropogenic pollution sources. Mailing Add: Dept of Meteorol 54-1517 Mass Inst of Technol Cambridge MA 02139

ALYEA, HUBERT NEWCOMBE, b Clifton, NJ, Oct 10, 03; m 29; c 1. PHYSICAL CHEMISTRY. Educ: Princeton Univ, AB, 25, PhD(phys chem), 28; Univ Stockholm, AM, 26. Hon Degrees: DSc, Beaver Col, 70. Prof Exp: Nat Res Coun fel, Univ Minn, 29; int res fel, Univ Berlin, 30; from instr to prof 30-72, EMER PROF CHEM, PRINCETON UNIV, 72- Concurrent Pos: Vis prof, Univ Hawaii, 48-49; lectr, Int Expos, Brussels, 58, Seattle, 62 & Montreal, 67; Fulbright, NSF, AEC, Dept Com, Dept State, US AID, Asia Found & UNESCO lectr, Europe, Africa, MidE, Orient, Mex & Cent Am, 62- Honors & Awards: Sci Apparatus Makers Asn Award in Chem Educ, 70; James Flack Norris Award, 70. Mem: Am Chem Soc. Res: Chain reactions; inhibition and catalysis; radiochemistry; polymerization; TOPS method of projecting chemical experiments and armchair chemistry. Mailing Add: 337 Harrison St Princeton NJ 08540

AMACHER, PETER, b Portland, Ore, Jan 29, 32; m 73; c 2. INFORMATION SCIENCE. Educ: Amherst Col, BA, 54; Univ Wash, PhD(hist sci), 62. Prof Exp: Fel hist neurol, Brain Res Inst, Univ Calif, Los Angeles Ctr Health Sci, 62-64, asst & assoc prof hist med, Ctr Health Sci, 64-71, asst to dir, Brain Info Serv, 64-69, dir, 69-71; prin, Amacher & Assocs Info Serv, 71-72; DIR, CONF PROG, KROC FOUND ADVAN MED SCI, 72- Mailing Add: Kroc Found PO Box 547 Santa Ynez CA 93060

AMADO, RALPH, b Los Angeles, Calif, Nov 23, 32; m. THEORETICAL PHYSICS. Educ: Stanford Univ, BS, 54; Oxford Univ, PhD(physics), 57. Prof Exp: Res assoc physics, 57-59, from asst prof to assoc prof, 59-65, PROF PHYSICS, UNIV PA, 65- Concurrent Pos: Consult, Arms Control & Disarmament Agency. Mem: Am Phys Soc. Res: Theoretical nuclear physics; many-body problem; particle physics; scattering theory; three-body problem. Mailing Add: Dept of Physics Univ of Pa Philadelphia PA 19104

AMADON, DEAN, b Milwaukee, Wis, June 5, 12; m 40; c 2. ORNITHOLOGY. Educ: Hobart Col, BS, 34; Cornell Univ, PhD, 49. Hon Degrees: DSc, Hobart Col, 60. Prof Exp: Wildlife technician, State Bd Fish & Game, Conn, 36-37; LAMONT CUR BIRDS, AM MUS NAT HIS, 37-, PRES BD DIRS & OFF MUS DIR, 74- Concurrent Pos: Lectr, Univ Wis, 54. Mem: AAAS; Am Soc Naturalists; Am Ornith Union (pres); Wilson Ornith Soc; Cooper Ornith Soc. Res: Avian taxonomy; evolution of birds. Mailing Add: Am Mus of Natural Hist Central Park W at 79th St New York NY 10024

AMADOR, ELIAS, b Mexico City, Mex, June 8, 32; m 61; c 3. MEDICINE, PATHOLOGY. Educ: Cent Univ Mex, BS, 59; Nat Univ Mex, MD, 56; Am Bd Path, dipl, 63. Prof Exp: Resident path, Peter Brent Brigham Hosp, Boston, 56-58; intern med, Pa Hosp, 58-59; resident, Univ Hosps, Boston Univ, 59-60; assoc path, Peter Bent Brigham Hosp, 64-66; assoc prof, Inst Path, Case Western Reserve Univ, 66-72; PROF PATH & CHMN DEPT, CHARLES R DREW POSTGRAD MED SCH, 72-; PROF UNIV SOUTHERN CALIF, 72-, CHIEF PATH, MARTIN LUTHER KING JR GEN HOSP, 72- Concurrent Pos: Teaching fel, Harvard Med Sch, 57-58 & Sch Med, Boston Univ, 59-60; Dazian Found fel, 60-61; fel med, Peter Bent Brigham Hosp & Biophysics Res Lab, 60-64; Med Found, Inc fel, 61-64. Mem: Am Chem Soc; Am Asn Path & Bact; Am Soc Exp Path; Am Asn Clin Chem. Res: Development of accurate and sensitive methods for diagnosis and detection of disease. Mailing Add: Dept of Path Charles R Drew Postgrad Med Sch Los Angeles CA 90059

AMADOR, JOSE MANUEL, b Calimete, Cuba, Mar 3, 38; US citizen; m 65; c 3. PLANT PATHOLOGY. Educ: La State Univ, BS, 60, MS, 62, PhD(plant path), 65. Prof Exp: AREA PLANT PATHOLOGIST, EXTEN SERV, TEX A&M UNIV, 65- Mem: Am Soc Plant Physiol; Am Phytopath Soc; Int Soc Sugar Cane Technol. Res: Physiology of diseased plants; sugar cane breeding; identification and control of diseases of field crops, citrus, vegetables and ornamentals; evaluation of fungicides and nematicides. Mailing Add: Tex Agr Exten Serv PO Drawer 1104 Weslaco TX 78596

AMAI, ROBERT LIN SUNG, b Hilo, Hawaii, Oct 19, 32; m 59; c 1. ORGANIC CHEMISTRY. Educ: Univ Hawaii, BA, 54, MS, 56; Iowa State Univ, PhD(org chem), 62. Prof Exp: Res assoc, 60-62, from asst prof to assoc prof, 62-74, PROF CHEM, NMEX HIGHLANDS UNIV, 74-, CHMN DEPT, 70- Mem: AAAS; Am Chem Soc. Res: Natural products; medicinal chemistry; correlation of structure with activity; psychopharmacology. Mailing Add: Dept of Chem NMex Highlands Univ Las Vegas NM 87701

AMANN, RUPERT PREYNOESSL, b Boston, Mass, Dec 27, 31; m 63; c 2. REPRODUCTIVE PHYSIOLOGY. Educ: Univ Maine, BS, 53; Pa State Univ, MS, 57, PhD(dairy sci), 61. Prof Exp: Asst reprod physiol, Pa State Univ, 55-61, res assoc, 61-62; NIH res fel, Royal Vet & Agr Col, Denmark, 62-63; from asst prof to assoc prof, 63-72, PROF DAIRY PHYSIOL, PA STATE UNIV, 72- Concurrent Pos: Vis prof, Dept Physiol Biophys, Colo State Univ, 75-76. Mem: AAAS; Am Dairy Sci Asn; Am Soc Animal Sci; Soc Study Reprod; Brit Soc Study Fertil. Res: Male reproductive physiology; testicular and epididymal physiology; spermatogenesis; spermatozoan maturation; semen physiology; reproductive capacity and sexual behavior of the male. Mailing Add: Dairy Breeding Res Ctr Pa State Univ University Park PA 16802

AMAR, HENRI, b Casablanca, French Morocco, July 14, 20; m 52. THEORETICAL PHYSICS. Educ: Univ Paris & Univ Algiers, Lic Math, 46; Ohio State Univ, AM, 48, PhD(physics), 52. Prof Exp: Instr & res assoc, Pa State Col, 50-51; asst prof physics, Lafayette Col, 52-56; sr res physicist, Franklin Inst Labs, 56-59; assoc prof, 59-64, PROF PHYSICS, TEMPLE UNIV, 64- Concurrent Pos: Vis prof, Univ Grenoble, 66-67. Honors & Awards: Lindback Found Award, 66. Mem: Fel Am Phys Soc; Am Asn Physics Teachers. Res: Relativity and field theory; magnetism; band theory of metallic alloys. Mailing Add: 330 Euclid Ave Ambler PA 19002

AMAROSE, ANTHONY PHILIP, b Oneonta, NY, Mar 17, 32. BIOLOGY, CYTOLOGY. Educ: Fordham Univ, BS, 53, MS, 57, PhD(cytol), 59. Prof Exp: Asst biol, Fordham Univ, 56-57; instr, Marymount Col, NY, 57-59; resident res assoc, Argonne Nat Lab, 59-61; res assoc radiobiol, Cancer Res Inst, New Eng Deaconess Hosp, Boston, 61-62; asst prof cytogenetics & lectr path, Albany Med Col, 63-66, res assoc prof, 66-67; asst prof, 67-71, ASSOC PROF OBSTET & GYNEC, CHICAGO LYING-IN HOSP, UNIV CHICAGO, 71- Concurrent Pos: Fel gen path, Harvard Med Sch, 61; personal consult to Shields Warren, MD, Cancer Res Inst, Mass, 62-67; abstractor, Excerpta Medica Found, 63-; consult, Albany Med Ctr Hosp, 63-67 & Inst Defense Anal, Arlington, Va, 65-67; hon trustee Charles A Berger Scholar Fund, Fordham Univ, 66-; mem int ed bd, Excerpta Medica Found, 73- Mem: AAAS; Am Soc Cell Biol; Am Inst Biol Sci; Environ Mutagen Soc; Am Soc Human Genetics. Res: Pharmacogenetics, radiation cytology; prediction of fetal sex; chromosomology; human bone marrow; cytogenetics of animal test model system in drug testing. Mailing Add: Dept of Obstet & Gynec Univ of Chicago Chicago IL 60637

AMASSIAN, VAHE EUGENE, b Paris, France, Nov 11, 24; nat US; m 56; c 1. NEUROPHYSIOLOGY. Educ: Cambridge Univ, BA, 46, MB, 48. Prof Exp: House physician, Middlesex Hosp, London, 48-49; from instr to assoc prof physiol, Sch Med, Univ Wash, 49-55; prof, Albert Einstein Col Med, 57-72; PROF PHYSIOL & CHMN DEPT, STATE UNIV NY DOWNSTATE MED CTR, 72- Concurrent Pos: Markle Found scholar, 52-59; mem postgrad training comt, USPHS, 59-63. Mem: Am Physiol Soc. Res: Neurophysiology of cerebral cortex; brainstem and sensory systems. Mailing Add: Dept of Physiol State Univ NY Downstate Med Ctr Brooklyn NY 11203

AMATA, CHARLES DAVID, b Agata, Italy, Feb 11, 41; US citizen; m 61; c 3. PHYSICAL CHEMISTRY. Educ: John Carroll Univ, BS, 64; Univ Notre Dame, PhD(chem), 68. Prof Exp: Res asst chem, Parma Tech Ctr, Union Carbide Corp, 60-64, res scientist, Carbon Prod Div, 67-71; sr res scientist, Addressograph-Multigraph Corp, 71-74; MGR CORP RES, CONWED CORP, 74- Mem: Am Chem Soc. Res: Molecular luminescence and energy transfer; thermal degradation of polymeric systems; organic photoconductors; high temperature inorganic fibers; composites. Mailing Add: 10535 33rd Ave N Plymouth MN 55441

AMATO, R STEPHEN S, b Brooklyn, NY, July 11, 36; m 59; c 3. HUMAN GENETICS, PEDIATRICS. Educ: Manhattan Col, BS, 58; Columbia Univ, MA, 59; NY Univ, PhD(genetics), 68; Univ Nebr, MD, 73. Prof Exp: Instr basic sci, St Johns Riverside Hosp, Yonkers, NY, 60-63; asst prof biol, State Univ NY Westchester, 63-65; fel genetics, NY Univ & Beth Israel Med Ctr, 65-67; cytogeneticist, Div Labs, Beth Israel Med Ctr, 68; asst prof human genetics, Univ Nebr Med Ctr, Omaha, 68-72, instr pediat & anat, 72-76, dir lab med & molecular genetics, 70-76, med geneticist, Muscular Dystrophy Clin, 74-76; ASSOC PROF PEDIAT, MED CTR, WVA UNIV, 76- Mem: Am Soc Human Genetics; Am Soc Cell Biol; Am Acad Pediat; Am Genetics Soc; AMA. Res: Chromosome structure and function; cell cycle and nucleic acid synthesis; slow virus infections and the central nervous system; genetic counseling and delivery of genetics and pediatrics services. Mailing Add: Dept of Pediat WVa Univ Med Ctr Morgantown WV 26500

AMATO, VINCENT ALFRED, b New York, NY, July 20, 15; m 43. PLANT PATHOLOGY, HORTICULTURE. Educ: Mich State Univ, BS, 54, MS, 55; Tex A&M Univ, PhD(plant path), 64. Prof Exp: Res instr hort, Mich State Univ, 54-56; asst exten horticulturist, WVa Univ, 56-57; instr floricult, Tex A&M Univ, 59-60 & plant path, 60-64, res asst plant physiol & microtech, 64; from asst prof to assoc prof, 64-73, PROF HORT, SAM HOUSTON STATE UNIV, 73- Mem: Am Phytopath Soc; Am Soc Hort Sci. Res: Horticultural crops; microtechniques; pathological and physiological plant parasitism. Mailing Add: Dept of Agr Sam Houston State Univ Huntsville TX 77340

AMAZIGO, JOHN C, b Onitsha, Nigeria, Aug 6, 39; m 65; c 3. APPLIED MATHEMATICS, APPLIED MECHANICS. Educ: Rensselaer Polytech Inst, BS, 64; Harvard Univ, SM, 65, PhD(appl math), 68. Prof Exp: Teacher math & physics, Govt Col, Ibadan, Nigeria, 61; res asst struct mech, Harvard Univ, 66-67, res fel, 67-68; asst prof, 68-72, ASSOC PROF MATH, RENSSELAER POLYTECH INST, 72- Concurrent Pos: Assoc scientist, Cambridge Acoust Assocs, Mass, 66-67; res fel, Harvard Univ, 74-75. Mem: Soc Indust & Appl Math; NY Acad Sci; Am Soc Mech Engrs; Am Geophys Union; AAAS. Res: Solid mechanics; elastic stability; asymptotic analysis; fracture mechanics; random processes; solid earth geophysics. Mailing Add: Dept of Math Sci Rensselaer Polytech Inst Troy NY 12181

AMBELANG, JOSEPH CARLYLE, b Bellevue, Ohio, Nov 7, 14; m 48; c 2. ORGANIC CHEMISTRY, RUBBER CHEMISTRY. Educ: Univ Akron, BS, 35; Yale Univ, PhD(chem), 38. Prof Exp: Asst chem, Yale Univ, 38-40; instr, D'Youville Col, 40-42; res chemist, Firestone Tire & Rubber Co, 42-56; sr compounder, 56-61, PRIN COMPOUNDER, GOODYEAR TIRE & RUBBER CO, 61- Mem: Am Chem Soc. Res: Pyrimidines; synthesis and catalytic hydrogenation; rubber chemistry; softeners; tackifiers; age-resistors; textile treating; antiozonants; curing agents; vulcanization; heat flow; analysis of test data. Mailing Add: 366 Dorchester Rd Akron OH 44320

AMBERG, CARL HELMUT, b Nuremberg, Ger, Dec 16, 23; Can citizen; m 50; c 3. CHEMISTRY. Educ: Queen's Univ, Ont, BA, 46, MA, 47; Univ Toronto, PhD(phys chem), 52. Prof Exp: Lectr chem, Univ NB, 47-49; res assoc surface chem, Univ St Andrews, 52-53; res assoc & instr, Amherst Col, 53-55; from asst res officer to sr res officer, div appl chem, Nat Res Coun Can, 55-64; assoc prof, 64-66, PROF CHEM, CARLETON UNIV, ONT, 66- Concurrent Pos: Vis prof, lab inorg chem & catalysis, Eindhoven Technol Univ, 71-72. Mem: Am Chem Soc; The Chem Soc; fel Chem Inst Can. Res: Adsorption; heterogeneous catalysis; gas chromatography; application of infrared spectroscopy to surfaces. Mailing Add: Dept of Chem Carleton Univ Ottawa ON Can

AMBLER, ERNEST, b Bradford, Eng, Nov 20, 23; nat US; m 55; c 2. CRYOGENICS, SCIENCE ADMINISTRATION. Educ: Oxford Univ, BA, 44, MA & PhD(physics), 53. Prof Exp: Physicist, Metall Lab, Armstrong Siddeley Motors, Ltd, Eng, 44-48; physicist, Cryogenic Physics Sect, 53-65, head div inorg mat, 65-68, dir, Inst Basic Stand, 68-73, DEP DIR, NAT BUR STANDARDS, 73- Concurrent Pos: Guggenheim Mem Found fel, 63. Honors & Awards: US Dept Com Gold Medal, 57; Wetherill Medal, Franklin Inst, 62; Stratton Award, Nat Bur Standards, 64. Mem: Am Phys Soc. Res: Low temperature research and nuclear orientation. Mailing Add: A-1134 Admin Bldg Nat Bur of Standards Washington DC 20234

AMBLER, JOHN EDWARD, b Bidwell, Ohio, Apr 5, 17; m 62; c 1. PLANT NUTRITION. Educ: Marshall Univ, BS, 52, MS, 53; Univ Md, PhD(plant physiol), 69. Prof Exp: PLANT PHYSIOLOGIST, STRESS LAB, USDA, 56- Mem: Soc Am Plant Physiologists; Am Soc Agron. Res: Effect of stress-nutrition, temperature, water, ultraviolet B light, on growth and development of food and fiber crops. Mailing Add: Stress Lab USDA-BARC-W Bldg 001 Beltsville MD 10705

AMBLER, JOHN RICHARD, b Denver, Colo, Jan 23, 34; m 57; c 3. ANTHROPOLOGY. Educ: Univ Colo, BA, 58, PhD(anthrop), 66; Univ Ariz, MA, 61. Prof Exp: Archaeologist, Mus NMex, 60-61, Mus Northern Ariz, 61-63 & Univ Utah, 64; exec dir, Tex archaeol salvage proj, Univ Tex, 65-67; ASSOC PROF ANTHROP, NORTHERN ARIZ UNIV, 67- Mem: Soc Am Archaeol. Res: Archaeological research in the American Southwest and coastal Texas; prehistoric technology. Mailing Add: Dept of Anthrop Box 15200 Northern Ariz Univ Flagstaff AZ 86001

AMBLER, MICHAEL RAY, b Wichita, Kans, Feb 20, 47; m 68; c 1. POLYMER SCIENCE. Educ: Wichita State Univ, BS, 68; Akron Univ, MS, 71, PhD(polymer sci), 75. Prof Exp: Develop engr, 68-75, PROJ LEADER, GOODYEAR TIRE & RUBBER CO, 75- Res: Polymer characterization; polymer rheology; physical chemistry of colloidal and macromolecular species; analytical chemistry; spectroscopy. Mailing Add: Chem Mat Develop Goodyear Tire & Rubber Co Akron OH 44316

AMBORSKI, LEONARD EDWARD, b Buffalo, NY, Aug 23, 21; m 44; c 2. PHYSICAL CHEMISTRY, POLYMER CHEMISTRY. Educ: Canisius Col, BS, 43; State Univ NY Buffalo, AM, 49, PhD(chem), 52. Prof Exp: Civilian instr physics, US Army Air Force, Canisius Col & civilian res physicist magnetism, Carnegie Inst, 44-45; res chemist polymer chem, 45-59, staff scientist, 59-62, group mgr, bldg mat div, 62-68, STAFF SCIENTIST, SPECIALTY MKT DIV, FILM DEPT, E I DU PONT DE NEMOURS & CO, INC, 68-, ENVIRON COORDR, 72- Mem: Am Chem Soc; Am Soc Testing & Mat; Air Pollution Control Asn; Water Pollution Control Fedn. Res: Chemistry of high polymers; physical and chemical properties of synthetic fibers; properties of polymeric films; electrical properties of polymers; structure-property relationship of polymers; rubber chemistry; mechanism of reinforcement plastic building materials. Mailing Add: 26 Cherrywood Dr Williamsville NY 14221.

AMBRE, JOHN JOSEPH, b Aurora, Ill, Sept 14, 37; m 72; c 4. CLINICAL PHARMACOLOGY. Educ: Notre Dame Univ, BS, 59; Loyola Univ, Chicago, MD, 63; Univ Iowa, PhD(pharmacol), 72. Prof Exp: Clin fel internal med, Mayo Clin, 66-68; fel clin pharmacol, 68-72, asst prof med & pharmacol, 72-75, ASSOC PROF MED & PHARMACOL, UNIV IOWA, 75- Concurrent Pos: Clin investigatorship, Vet Admin, 73. Mem: Am Soc Clin Pharmacol & Therapeut; Am Fedn Clin Res; Am Acad Clin Toxicol; AAAS. Res: Drug metabolism in man, particularly as it is influenced by disease states. Mailing Add: Dept of Med Univ of Iowa Hosps Iowa City IA 52242

AMBROMOVAGE, ANNE MARIE, b Gilberton, Pa, Aug 27, 36. PHYSIOLOGY. Educ: Susquehanna Univ, BA, 58; Jefferson Med Col, MS, 61, PhD(physiol), 68. Prof Exp: Res instr, 64-66, 68-69, res asst prof, 69-70, ASST PROF HUMAN PHYSIOL, HAHNEMANN MED COL, 70-, RES ASSOC, 63- Concurrent Pos: Merck Found fac develop grant, 70. Mem: AAAS; NY Acad Sci. Res: Gastrointestinal physiology; water of electrolyte absorption; pancreatic enzymes; mesenteric blood flow; physiology of shock and vasoactive substances. Mailing Add: Dept of Physiol & Biophys Hahnemann Med Col Philadelphia PA 19102

AMBROSE, CHARLES T, b Indianapolis, Ind, Nov 29, 29. IMMUNOLOGY. Educ: Ind Univ, AB, 51; Johns Hopkins Univ, MD, 55. Prof Exp: Intern med, New Eng Med Ctr, Boston, 55-56; asst resident infectious dis, Mass Mem Hosp, 56-57; resident, New Eng Med Ctr, 57-59; from instr to assoc prof bacteriol & immunol, Harvard Med Sch, 62-72; assoc prof, Col de France, Paris, 72-73; PROF CELL BIOL, SCH MED, UNIV KY, 73- Concurrent Pos: NSF fel, 59-60; res fel bacteriol & immunol, Harvard Med Sch, 59-62; dir res, NIH & Med Res, Paris, 72-73. Mem: Am Asn Immunol; Brit Soc Immunol; Fr Soc Immunol; Reticuloendothelial Soc; Soc Exp Biol & Med. Res: Regulation of antibody synthesis in vitro; organ cultures; antigenic competition; antimetabolites; salicylates; insecticides; corticosteroids. Mailing Add: Dept of Cell Biol Univ of Ky Lexington KY 40506

AMBROSE, ERNEST R, b Montreal, Que, May 12, 26; m 49; c 7. DENTISTRY. Educ: McGill Univ, DDS, 50. Prof Exp: From instr to asst prof, 50-66, ASSOC PROF OPER DENT, McGILL UNIV, 66-, CHMN DEPT, 58-, DEAN FAC DENT, 70- Concurrent Pos: Asst, Outdoor Dent Dept, Montreal Gen Hosp, 58- Mem: Fel Am Col Dent; Can Acad Restorative Dent. Res: Restorative dentistry; clinical use of dental materials; assessing and maintaining the pulp potential of human teeth. Mailing Add: Fac of Dent McGill Univ Montreal PQ Can

AMBROSE, HARRISON WILLIAM, III, b Winter Haven, Fla, Feb 26, 38; m; c 2. ECOLOGY, ANIMAL BEHAVIOR. Educ: Univ Fla, BS, 60; Univ Ky, MS, 62; Cornell Univ, PhD(ecol), 67. Prof Exp: From instr to asst prof biol, Cornell Univ, 66-73; ASST PROF BIOL, UNIV ILL, URBANA, 73- Mem: AAAS; Am Soc Mammal; Ecol Soc Am; Animal Behav Soc. Res: Behavioral animal ecology; social and orientation behavior; population regulation; predator-prey interactions. Mailing Add: Dept of Ecol Ethol & Evolution Univ of Ill Rm 515 Morrill Hall Urbana IL 61801

AMBROSE, JOHN AUGUSTINE, b Ft Dodge, Iowa, Feb 15, 23; m 64; c 1. NUTRITIONAL BIOCHEMISTRY, BIOCHEMICAL GENETICS. Educ: Johns Hopkins Univ, BA, 48; Marquette Univ, MS, 51; Univ Miami, PhD(microbiol, biochem), 65. Prof Exp: Res chemist, Ore State Univ, 51-52; res biochemist, Chicago Med Sch, 52-54 & Med Sch, Johns Hopkins Univ, 54-57; res biochemist, Commun Dis Ctr, 64-70, chief, Ment Retardation Lab, 67-70, Biochem Genetics & Metab Disorders Lab, Ctr for Dis Control, 70-72, Pediat & Genetic Chem Lab, 72-73 & Genetic Chem Lab, 73-74, res chemist, Metab Lab, 74-75, RES CHEMIST, NUTRIT BIOCHEM SECT, CTR DIS CONTROL, USPHS, 75- Concurrent Pos: Consult chem genetics, Ment Retardation Activ, State Health Dept; dir, Nat & State Health Dept Multi-State Fluorometric & Ment Retardation Workshops. Mem: Sigma Xi; fel Am Inst Chemists; fel Am Pub Health Asn; Am Chem Soc; NY Acad Sci. Res: Enzymology; radiation chemistry; protein structure; microbiology; protozoology; bacteriology; virology; clinical genetic metabolic disorders; cancer biochemical research; analytical biochemical methodology and automated instrumentation; microbiological and chemical methodology research on vitamins and amino acids. Mailing Add: Ctr for Dis Control 1600 Clifton Rd NE Atlanta GA 30333

AMBROSE, JOHN DANIEL, b Detroit, Mich, June 20, 43; m 71; c 1. BOTANY. Educ: Univ Mich, BS, 65, MS, 66; Cornell Univ, PhD(bot), 75. Prof Exp: Asst hydrographic officer, US Navy, 67-68; CUR ARBORETUM, UNIV GUELPH, 74- Mem: AAAS; Am Asn Bot Gardens & Arboreta. Res: Comparative anatomy and morphology of the Melanthioideae. Mailing Add: Univ of Guelph Arboretum Guelph ON Can

AMBROSE, JOHN RUSSELL, b Orange, NJ, Feb 25, 40; m 62; c 2. CHEMISTRY, CORROSION. Educ: Washington & Lee Univ, BSc, 61; Univ Md, College Park, PhD(phys chem), 72. Prof Exp: High sch teacher chem, Va Beach City Sch Bd, 62-64; res chemist, Newport News Shipbuilding & Dry Dock Co, 64-66; RES CHEMIST, NAT BUR STANDARDS, 66- Mem: Electrochem Soc; Nat Asn Corrosion Engr; Am Soc Testing & Mat. Res: Relationship between repassivation kinetics and susceptibilities of various materials to localized corrosion attack. Mailing Add: Nat Bur of Standards Rm B-254 Bldg 223 Washington DC 20234

AMBROSE, JOHN WILLIS, geology, deceased

AMBROSE, RICHARD JOSEPH, b Youngstown, Ohio, July 4, 42; m 65; c 2. POLYMER CHEMISTRY. Educ: Bowling Green State Univ, BS, 64; Univ Akron, PhD(polymer sci), 68. Prof Exp: Res scientist, 68-73, GROUP LEADER, PLASTICS DIV, FIRESTONE TIRE & RUBBER CO, 73- Mem: Am Chem Soc. Res: Chemical reactions of polymers; hydrolysis of acrylate and methacrylate polymers; structure physical property relationships of block copolymers; thermosetting resins; polar-nonpolar block polymers; cationic and anionic polymerization. Mailing Add: Plastics Div Firestone Cent Res 1200 Firestone Pkwy Akron OH 44301

AMBROSE, ROBERT T, b Palmdale, Calif, Sept 21, 33. ANALYTICAL CHEMISTRY, ENVIRONMENTAL CHEMISTRY. Educ: Univ Calif, Riverside, BA, 60; Northwestern Univ, PhD(anal chem), 65. Prof Exp: RES ASSOC, EASTMAN KODAK CO RES LABS, 64- Concurrent Pos: Instr, Rochester Inst Technol, 72-74. Mem: Am Chem Soc; Sigma Xi. Res: Analytical and environmental chemistry of photographic processing effluents, characterization and treatment of the chemical content of such solutions. Mailing Add: Eastman Kodak Co Res Labs B82 Kodak Park Rochester NY 14650

AMBROSIANI, VINCENT F, b Oneida, NY, Aug 4, 29; m 52; c 8. PHYSICAL CHEMISTRY. Educ: Le Moyne Col, NY, BS, 52; Syracuse Univ, MS, 57, PhD(phys chem), 64. Prof Exp: Res chemist, Exp Sta, E I du Pont de Nemours & Co, Inc, 59-63 & Textile Res Lab, 63-66; tech dir fiber & fabric develop, Blue Ridge-Winkler Textiles, Pa, 66-68; dir res & develop, Vanity Fair Mills, Inc, 68-70, DIR RES & DEVELOP, V F CORP, READING, 70- Mem: Am Chem Soc; Am Asn Textile Chem & Colorists. Res: Research and development to utilize synthetic yarns for new and improved fabrics for apparel and industrial applications; polymer and gaseous radiation chemistry. Mailing Add: 1326 Delaware Ave Wyomissing PA 19610

AMBROSIO, CESARE, organic chemistry, polymer chemistry, see 12th edition

AMBROSONE, JOSEPH PAUL, b Corning, NY, Nov 29, 19; m 45; c 3. CHEMISTRY. Educ: Niagara Univ, BS, 43. Prof Exp: Anal chemist, toxicol res lab, Edgewood Arsenal, Md, 44; phys chemist, 45-53, process engr, 53-56, sr qual control engr, 56-60, plant chemist, 60-64, SR PLANT CHEMIST, FALL BROOK PLANT, CORNING GLASS WORKS, 64- Mem: Am Chem Soc; Am Ceramic Soc; Am Crystallog Asn; fel Am Inst Chemists. Res: Analytical chemistry of war gases; glass chemistry; x-ray diffraction; vycor plant process engineering. Mailing Add: Corning Glass Works Corning NY 14830

AMBRUS, CLARA MARIA, b Rome, Italy, Dec 28, 24; nat US; m 45; c 7. HEMATOLOGY, PHARMACOLOGY. Educ: Univ Zurich, MD, 49; Jefferson Med Col, PhD, 55; Am Bd Clin Chem, dipl. Prof Exp: Asst histol, Budapest Med Sch, 43-46, demonstr pharmacol, 46-47; asst, Med Sch, Univ Zurich, 47-49; asst therapeut chem & virol, Pasteur Inst, Univ Paris, 49; asst pharmacol, Philadelphia Col Pharm, 50-51, from asst prof to assoc prof, 51-55; asst prof pharmacol & assoc med, Sch Med, State Univ NY Buffalo, 55-65, asst res prof pediat, 65-69; assoc cancer res scientist, 55-69, assoc res prof pharmacol, 66-69, PRIN CANCER RES SCIENTIST, ROSWELL PARK MEM INST, 69-, PROF PHARMACOL, 70-; ASSOC RES PROF PEDIAT, SCH MED, STATE UNIV NY BUFFALO, 69- Mem: Am Asn Cancer Res; Am Physiol Soc; Am Soc Pharmacol & Exp Therapeut; Am Soc Exp Biol; fel Am Col Physicians. Res: Pediatric hematology and oncology; leukemias; hyaline membrane disease of infants; biochemistry and pathology of the blood coagulation and fibrinolysin systems; transplantation and regeneration of hemic tissue; radiation sickness; experimental and clinical pharmacology; chemotherapy. Mailing Add: Roswell Park Mem Inst 666 Elm St Buffalo NY 14203

AMBRUS, JULIAN LAWRENCE, b Budapest, Hungary, Nov 29, 24; nat US; m 45; c 7. HEMATOLOGY, ONCOLOGY. Educ: Univ Zurich, MD, 49; Jefferson Med Col, PhD(med sci), 54; Am Bd Clin Chem, dipl. Prof Exp: Asst histol, Med Sch, Univ Budapest, 43-46, demonstr pharmacol, 46-47; asst, Med Sch, Univ Zurich, 47-49; asst therapeut chem & virol, Pasteur Inst, Paris, 49; from asst prof to assoc prof pharmacol, Philadelphia Col Pharm, 50-55; prin cancer res scientist, State Dept Health, NY, 55-65; assoc prof pharmacol, Grad Sch & asst prof med, Sch Med, 55-65, dir, Springville Labs, Roswell Park Mem Inst, 65-75, HEAD DEPT PATHOPHYSIOL, ROSWELL PARK MEM INST, 75-, PROF BIOCHEM PHARMACOL, EXP PATH & INTERNAL MED, SCH MED, STATE UNIV NY BUFFALO, 65- Concurrent Pos: Consult adv comt coagulation components, Comn

Plasma Fractionation & Related Processes; consult, Bur Drugs, Food & Admin; mem coun drugs, AMA; ed-in-chief, J Med & Hematol Rev; mem ed bd, Res Commun in Chem Path & Pharmacol Haemostasis & Folia Angiologica; mem comt thrombolytic agents, Nat Heart Inst; mem adv comt res grants & fels, United Health Found, NY; fel, AM Col Physicians & coun clin cardiol, Am Heart Asn. Mem: Fel AAAS; Am Soc Pharmacol & Exp Therapeut; Am Fedn Clin Res; Asn Am Med Cols; fel NY Acad Sci. Res: Hemorrhagic and thromboembolic diseases; blood coagulation, platelet and fibrinolysin systems; physiology of the leukocytes; leukemias; radiation sickness; biochemistry and chemotherapy of neoplastic diseases; experimental and clinical pharmacology; resistance to drugs. Mailing Add: Roswell Park Mem Inst Buffalo NY 14263

AMBRUS, LASZLO, b Tapolca, Hungary, Feb 11, 34. ORGANIC CHEMISTRY. Educ: Univ Chicago, MS, 59. Prof Exp: Starch chemist, Corn Prod Co, 57; assoc res chemist, 60-68, leader endocrine sect, 68-70, sr anal res chemist, 70-73, MAT ANAL RES LEADER, CUTTER LABS, CALIF, 73- Mem: Am Chem Soc; NY Acad Sci. Res: Polymers, plastics, rubbers analysis, including their extractables. Mailing Add: Dept of Chem Cutter Labs Fourth & Parker Sts Berkeley CA 94710

AMBS, WILLIAM JOSEPH, b Philadelphia, Pa, Dec 3, 29; m 57; c 1. SURFACE CHEMISTRY. Educ: Villanova Univ, BS, 52; Stevens Inst Technol, MS, 54; Cath Univ Am, PhD(phys chem), 61. Prof Exp: Asst chem, Stevens Inst Technol, 52-54 & Cath Univ Am, 54-56; phys chemist, Nat Bur Standards, 56-63; phys chemist, 63-71, SR RES CHEMIST, HOUDRY DIV, AIR PROD & CHEM, INC, 71- Mem: Am Chem Soc; Catalysis Soc; Sigma Xi. Res: Catalysis; chemical kinetics; surface chemistry of solids; field emission microscopy; thin film optics; x-ray diffraction. Mailing Add: Res Dept Houdry Labs Air Prod & Chem Inc PO Box 427 Marcus Hook PA 19061

AMBUEL, JOHN PHILIP, b Broadus, Mont, Mar 23, 18; m 46; c 2. PEDIATRICS. Educ: Luther Col, BA, 41; Univ Chicago, MD, 46. Prof Exp: Intern, Doctors Hosp, Seattle, 46-47; resident pediat, Children's Hosp, Detroit, 49-51; fel, Univ Chicago, 51-53; from asst prof to prof pediat, Ohio State Univ, 65-74; PROF PEDIAT, SCH MED, NORTHWESTERN UNIV, CHICAGO, 74- Mem: Am Acad Pediat; Sigma Xi; Am Pediat Soc. Res: Erythroblastosis; behavior development; medical care programs. Mailing Add: 2449 Marcy Ave Evanston IL 60201

AMBURGEY, TERRY L, b Trenton, NJ, Dec 11, 40; m 61; c 2. FOREST PATHOLOGY. Educ: State Univ Col Forestry, Syracuse Univ, BS, 63, MS, 65; NC State Univ, PhD(plant path), 69. Prof Exp: RES PLANT PATHOLOGIST, SOUTHERN FOREST EXP STA, 69- Mem: Am Phytopath Soc; Forest Prod Res Soc. Res: Prevention and control of wood decay; interactions between wood-inhabiting fungi and termites. Mailing Add: Box 2008 Evergreen Sta Gulfport MS 39501

AMDUR, MARY OCHSENHIRT, b Pittsburgh, Pa, Feb 18, 21; m 44. TOXICOLOGY. Educ: Univ Pittsburgh, BS, 43; Cornell Univ, PhD(biochem), 46. Prof Exp: Asst ophthal res, Howe Lab, Med Sch, Harvard, 47-48; res biochemist, Vet Admin Hosp, 48-49; res assoc, 49-57, asst prof physiol, 57-63, ASSOC PROF TOXICOL, SCH PUB HEALTH, HARVARD UNIV, 63- Mem: AAAS; Am Chem Soc; Am Indust Hyg Asn; NY Acad Sci; Soc Toxicol. Res: Bone formation in the rat; effect of manganese and choline on liver fat in the rat; effect of inhalation of sulfur dioxide and sulfuric acid mist on guinea pigs and humans; analyses for lead in air and urine; physiologic response to respiratory irritants. Mailing Add: Sch of Pub Health Harvard Univ Cambridge MA 02115

AMDUR, MILLARD JASON, b Pittsburgh, Pa, Aug 23, 37; m 63; c 2. PSYCHIATRY. Educ: Univ Pittsburgh, BA, 59; Yale Univ, MD, 64. Prof Exp: Fel psychiat, Sch Med, Yale Univ, 65-68, from instr to asst prof, 68-70; dir student ment health serv & asst prof psychiat, 70-72, ASST CLIN PROF PSYCHIAT, SCH MED, UNIV CONN, 72-; DIR MENT HEALTH CLIN, WINDHAM COMMUNITY MEM HOSP, 72- Concurrent Pos: Dir, Psychiat Outpatient Div & Dana Clin, Yale-New Haven Hosp, 68-70; consult, Undercliff Ment Health Ctr, 68-70, New Fairview Hall Convalescent Hosp, 68-70 & Student Ment Health Serv, Univ Conn, 68-70 & 72-; psychiat staff, Natchaug Hosp, 73-; pres, Eastern Conn Parent-Child Resource Syst, Inc, 74- Mem: Am Psychiat Asn; Am Group Psychother Asn; Am Col Health Asn. Res: Adolescent psychiatry; guilt and conscience. Mailing Add: Windham Community Mem Hosp Ment Health Clin Mansfield Ave Willimantic CT 06226

AMEEL, DONALD JULES, b Detroit, Mich, Apr 24, 07; m 37; c 3. HELMINTHOLOGY. Educ: Wayne State Univ, AB, 28; Univ Mich, AM, 30, ScD(zool), 33. Prof Exp: Assoc prof sci, Augustana Col, SDak, 33-35; res helminthol, Univ Mich, 35-37; from instr to prof, 37-72, head dept, 45-67, EMER PROF ZOOL, KANS STATE UNIV, 72- Concurrent Pos: Instr zool, Univ Mich, 36-37. Mem: AAAS; Am Soc Parasitol; Am Micros Soc. Res: Biology of trematodes. Mailing Add: Div of Biol Kans State Univ Manhattan KS 66502

AMEER, GEORGE ALBERT, b Norwalk, Conn, July 26, 31; m 72; c 6. OPTICAL PHYSICS. Educ: Univ Maine, BS, 53; Univ Pittsburgh, PhD(physics), 61. Prof Exp: Sr physicist, J W Fecker Div, Am Optical Co, Pa, 61-64; asst prof physics, Am Univ Beirut, 64-67; tech specialist optics, Autonetics Div, 67-73, SUPVR ADVAN SENSOR TECHNOL AUTONETICS GROUP, NAM ROCKWELL CORP, ANAHEIM, 73- Mem: Am Phys Soc; Optical Soc Am. Res: Molecular spectroscopy; optical instrumentation; physical optics. Mailing Add: 12872 Bubbling Well Rd Santa Ana CA 92705

AMELIN, CHARLES FRANCIS, b Worcester, Mass. MATHEMATICAL ANALYSIS. Educ: Col of the Holy Cross, AB, 64; Univ Calif, Berkeley, PhD(math), 72. Prof Exp: ASST PROF MATH, CALIF STATE POLYTECH UNIV, POMONA, 69- Concurrent Pos: Vis asst prof math, Univ Md, College Park, 72-73 & Ariz State Univ, 73-74. Mem: Am Math Soc. Res: Operator theory; functional analysis. Mailing Add: Dept of Math Calif State Polytech Univ Pomona CA 91768

AMELIO, GILBERT FRANK, b New York, NY, Mar 1, 43; m 63; c 3. SOLID STATE ELECTRONICS. Educ: Ga Inst Technol, BS, 65, MS, 67, PhD(physics), 69. Prof Exp: Mem tech staff device physics, Bell Tel Labs, Inc, 68-71; MGR CHARGE COUPLED DEVICES, FAIRCHILD CAMERA & INSTRUMENT CORP, 71- Mem: Am Phys Soc; Inst Elec & Electronics Engr; Sigma Xi. Res: Semiconductor surfaces with Auger spectroscopy; hot electron transport theory; silicon diode array camera tube; theory and experiment of charge coupled devices; management of design and manufacture of charge coupled devices. Mailing Add: Fairchild Camera & Instrument Corp 4001 Miranda Ave Palo Alto CA 94304

AMELL, ALEXANDER RENTON, b North Adams, Mass, Mar 3, 23; m 45; c 4. PHYSICAL CHEMISTRY. Educ: Univ Mass, BS, 47; Univ Wis, PhD(chem), 50. Prof Exp: Instr phys chem, Hunter Col, 50-52; asst prof chem, Lebanon Valley Col, 52-55; asst prof, 55-60, PROF CHEM & CHMN DEPT, UNIV NH, 60- Concurrent Pos: Fulbright lectr, San Marcos Univ, Lima, 63. Mem: Am Chem Soc. Res: Radiation chemistry; kinetics. Mailing Add: 4 Chesley Dr Durham NH 03824

AMELUNXEN, REMI EDWARD, b Kansas City, Mo, Jan 27, 28. MICROBIOLOGY. Educ: Rockhurst Col, BS, 49; Univ Kans, MA, 57, PhD(microbiol), 59. Prof Exp: USPHS fel biochem, 59-62, instr, 62-64, asst prof microbiol & biochem, 64-66, assoc prof microbiol, 66-71, PROF MICROBIOL, SCH MED, UNIV KANS, 71- Concurrent Pos: Res career develop award, USPHS; NIH res career develop award, 64-69. Mem: Am Soc Microbiol. Res: Physico-chemical characterization of thermophilic enzymes. Mailing Add: Dept of Microbiol Univ of Kans Med Ctr Kansas City KS 66103

AMEMIYA, FRANCES (LOUISE) CAMPBELL, b Riverside, Calif, June 16, 15; m 52; c 2. MATHEMATICS. Educ: Univ Calif, Los Angeles, AB, 35, MA, 36; Univ Mich, PhD(math), 45. Prof Exp: Asst math, Univ Calif, Los Angeles, 35-37; prof, George Pepperdine Col, 37-58; assoc prof, Calif Western Univ, 58-62 & Parsons Col, 62-64; assoc prof, 64-71, PROF MATH, CALIF STATE UNIV, HAYWARD, 71- Concurrent Pos: Fel, Univ Calif, Los Angeles & Ibaraki Christian Col, Japan, 49-50. Mem: Math Asn Am. Res: Mathematical statistics; truncated bivariate normal distributions; elementary geometry for liberal studies. Mailing Add: Dept of Math Calif State Univ Hayward CA 94542

AMEMIYA, MINORU, b San Francisco, Calif, Mar 17, 22; m 47; c 2. SOIL CONSERVATION, AGRONOMY. Educ: Univ Calif, Berkeley, BS, 42; Ohio State Univ, MS, 48, PhD(agron), 50. Prof Exp: Lab technician, Carpenter Bros, Inc, Wis, 43-44; asst agron, Ohio Agr Exp Sta, 48-50; soil scientist, Agr Res Serv, USDA, Colo, 50-58 & Tex, 58-60, res soil scientist, Iowa State Univ, 60-68, assoc prof agron, 68-71, PROF AGRON, IOWA STATE UNIV, 71-, EXTEN AGRONOMIST, AGR RES SERV, USDA, 68- Concurrent Pos: Assoc agronomist, Colo Agr Exp Sta, 50-58. Mem: Soil Sci Soc Am; Am Soc Agron; Soil Conserv Soc Am; AAAS. Res: Reclamation of saline and alkali soils; soil, water and crop management; soil and water conservation; soil-water-plant relationships; soil tilth and tillage. Mailing Add: Dept of Agron Iowa State Univ Ames IA 50011

AMEN, RALPH DUWAYNE, b Cheyenne, Wyo, Feb 26, 28; m 52; c 4. PHYSIOLOGICAL ECOLOGY. Educ: Univ Northern Colo, AB, 52, AM, 54; Univ Colo, MBS, 59, PhD(bot), 62. Prof Exp: Instr high sch, Colo, 55-58; instr biol, Univ Colo, 59-60, teaching assoc, 61-62; asst prof plant physiol, 62-67, chmn dept, 67-72, ASSOC PROF BIOL, WAKE FOREST UNIV, 67- Concurrent Pos: Sigma Xi res grant, 61-62; NC Bd Sci & Technol grant, 68-69; assoc ed, Ecology, 71-72. Mem: Ecol Soc Am. Res: Seed germination and dormancy; effects of thermal pollution on bud dormancy in trees. Mailing Add: Dept of Biol Wake Forest Univ Winston-Salem NC 27109

AMEN, RONALD JOSEPH, b Brooklyn, NY, Mar 3, 43; m 67; c 2. FOOD SCIENCE, NUTRITION. Educ: Bethany Col, BS, 63; Rutgers Univ, MS, 64, PhD(food sci), 71. Prof Exp: Food technologist, Dietetic Food Co Inc, Brooklyn, 64-67; res asst food sci, Rutgers Univ, 67-71; food scientist protein res, Thomas J Lipton Inc, NJ, 71-72; DEPT HEAD NUTRIT, SYNTEX RES INC, PALO ALTO, 72- Mem: AAAS; Am Pub Health Asn; Am Oil Chem Soc; Sigma Xi; Inst Food Technol. Res: Effect of special dietary foods in clinical situations where specific nutritional adjuncts are indicated; including the effects and mechanism of fiber and peptide absorption and transport in gastrointestinal tract. Mailing Add: 19717 Yuba Ct Saratoga CA 95070

AMEND, DONALD FORD, b Portland, Ore, Jan 8, 39; m 62; c 3. FISH PATHOLOGY. Educ: Ore State Univ, BS, 60, MS, 65; Univ Wash, PhD(fish path), 73. Prof Exp: Aquatic biologist, Ore Fish Comn, 60-63; res asst food toxicol, Ore State Univ, 63-65; proj leader virol, 66-73, RES MICROBIOLOGIST, US FISH & WILDLIFE SERV, 65-, SECT LEADER MICROBIOL, 74- Concurrent Pos: Affil asst prof, Univ Wash, 74-75. Mem: Am Fisheries Soc; Sigma Xi; Am Soc Microbiol; Am Inst Fishery Res Biologists. Res: Controlling infectious diseases of fish including chemotherapy, immunization, environmental and biological control. Mailing Add: Western Fish Dis Lab Sandpoint NSA Bldg 204 Seattle WA 98115

AMENDOLA, ALBERT, inorganic chemistry, see 12th edition

AMENT, MARVIN EARL, b Dec 5, 38. PEDIATRICS, GASTROENTEROLOGY. Educ: Ill Inst Technol, BS, 59; Univ Minn, Minneapolis, MD, 63. Prof Exp: Resident pediat, Hosps, Univ Wash, 64-65; sr resident, Sch Med, Univ Calif, Los Angeles, 65-66; instr gastroenterol, Dept Med, Univ Wash, 70-71, actg asst prof med, 71-73; ASST PROF PEDIAT & MED, CTR HEALTH SCI, UNIV CALIF, LOS ANGELES, 73- Concurrent Pos: NIH fel, Univ Wash, 68-70; mem attend staff, Children's Orthop Hosp, Seattle, 69-70, head gastroenterol clin, 71; mem attend staff gastroenterol, Olive View Med Ctr, 73- Mem: Am Fedn Clin Res; Am Gastroenterol Asn; Soc Pediat Res. Res: Intractable diarrhea in neonates; development of gastric secretion in infancy; peptic ulcer disease in childhood; inflammatory bowel disease in childhood; endoscopy in children. Mailing Add: Dept of Pediat Univ of Calif Sch of Med Los Angeles CA 90024

AMENTA, PETER SEBASTIAN, b Cromwell, Conn, Mar 26, 27; m 53; c 2. ANATOMY. Educ: Fairfield Univ, BS, 52; Marquette Univ, MS, 54; Univ Chicago, PhD(anat), 58. Prof Exp: Asst biol, Fairfield Univ, 49-52; asst zool, Marquette Univ, 52-54; asst anat, Univ Chicago, 55-58; from instr to assoc prof, 58-71, actg chmn dept, 73-75, PROF ANAT, HAHNEMANN MED COL, 71-, DIR ELECTRON MICROS LAB, 70-, CHMN DEPT, 75- Concurrent Pos: Instr, Marine Biol Lab, Woods Hole Oceanog Inst, 56; consult, Franklin Inst. Mem: Fel AAAS; Tissue Cult Asn; Am Soc Zool; Int Cong Photobiol; Am Asn Anat. Res: Culture; histology; cytology; hematology; nucleocytoplasmic relationships; effects of microbeams of ultraviolet and infrared light on parts of cells. Mailing Add: Hahnemann Med Col 235 N 15th St Philadelphia PA 19102

AMENTA, RODDY V, structural geology, petrology, see 12th edition

AMER, MOHAMED SAMIR, b Tanta, Egypt, Sept 2, 35; m 58; c 3. PHARMACOLOGY, BIOCHEMISTRY. Educ: Univ Alexandria, BS, 56, dipl hosp pharm, 57; Univ Ill, MS, 60, PhD(pharmacol), 62. Prof Exp: Assoc prof pharmacol, Univ Alexandria, 62-66; sr res pharmacologist & proj coordr, Wilson Labs Div, Wilson Pharmaceut & Chem Corp, Ling-Temco-Vought, Inc, 66-69; PRIN INVESTR PHARMACOL, MEAD JOHNSON RES CTR, BRISTOL-MYERS LABS, INC, 69- Mem: Am Fedn Clin Res; Acad Pharmaceut Sci; Soc Exp Biol & Med; Am Soc Pharmacol & Exp Therapeut; Am Soc Biol Chemists. Res: Biochemical pharmacology of the gastrointestinal tract; mechanisms of hormonal regulation; schistosomiasis. Mailing Add: Mead Johnson Res Ctr Evansville IN 47721

AMER, NABIL MAHMOUD, b Alexandria, Egypt, May 1, 42; m 67; c 2. SOLID STATE PHYSICS, CHEMICAL PHYSICS. Educ: Univ Alexandria, BS, 57; Univ Calif, Berkeley, PhD(biophys), 65. Prof Exp: Res assoc biophys, Lawrence Radiation Lab, 66-68, PHYSICIST, LAWRENCE BERKELEY LAB, UNIV CALIF, BERKELEY, 68- Concurrent Pos: Lectr physics, Univ Calif, Berkeley, 70- Mem: Am Phys Soc; Sigma Xi; AAAS. Res: Light scattering from liquid crystals, magnetic crystals and semiconductors; infrared laser spectroscopy and multiphoton processes in gases and solids. Mailing Add: Dept of Physics Univ of Calif Berkeley CA 94720

AMERAULT, THOMAS EUGENE, b Baltimore, Md, Dec 6, 24; m 48; c 3. MICROBIOLOGY, IMMUNOCHEMISTRY. Educ: Univ Md, BS, 50. Prof Exp: Res microbiologist, nat animal dis lab, 50-60, SR RES MICROBIOLOGIST, NAT ANIMAL PARASITE LAB, USDA, 60- Honors & Awards: Superior Serv Award, USDA, 74. Mem: Am Soc Microbiol; Am Asn Lab Animal Sci. Res: Microbiological research on causative agents associated with brucellosis of livestock; immunochemical studies, especially electrophoretic, chromatographic and ultracentrifugal analysis of antigens and antibodies prevailing in anaplasmosis of cattle. Mailing Add: Animal Parasitol Inst USDA Beltsville MD 20705

AMERINE, MAYNARD ANDREW, b San Jose, Calif, Oct 30, 11. AGRICULTURE. Educ: Univ Calif, BS, 32, PhD(plant physiol), 36. Prof Exp: Jr enologist, 36-40, from asst prof to prof enol, 40-74, chmn dept viticult & enol, 57-62, fac res lectr, 64, EMER PROF ENOL, COL AGR, UNIV CALIF, DAVIS, 74- Concurrent Pos: Guggenheim fel, 54; consult, Wine Adv Bd & Wine Inst, 74- Honors & Awards: Am Soc Enol Merit Award, 67. Mem: AAAS; Am Chem Soc; Am Soc Enol (pres, 58-59); Inst Food Technol. Res: Viticulture; enology; food science; psychophysics. Mailing Add: PO Box 208 St Helena CA 94574

AMES, ADELBERT, III, b Boston, Mass, Feb 25, 21; m 48; c 3. NEUROPHYSIOLOGY, NEUROCHEMISTRY. Educ: Harvard Univ, MD, 45. Prof Exp: Intern & sr resident internal med, Presby Hosp, 45-52; res assoc, 55-69, PROF PHYSIOL, DEPT SURG, HARVARD MED SCH, 69-; NEUROPHYSIOLOGIST IN NEUROSURG, MASS GEN HOSP, 69- Concurrent Pos: Res fel biochem, Harvard Med Sch, 52-55; NIMH res scientist award, 68- Mem: Am Physiol Soc; Am Soc Neurochem; Soc Neurosci; Int Soc Neurochem. Res: Brain function; correlation of electrophysiology with metabolism of electrolytes, proteins and neurotransmitters in an in vitro preparation of retina; formation of cerebrospinal fluid; cerebreal ischemia. Mailing Add: Mass Gen Hosp Fruit St Boston MA 02114

AMES, BRUCE NATHAN, b New York, NY, Dec 16, 28; m 60; c 2. BIOCHEMICAL GENETICS. Educ: Cornell Univ, BA, 50; Calif Inst Technol, PhD(biochem), 53. Prof Exp: USPHS fel, NIH, 53-54, biochemist, Nat Inst Arthritis & Metab Dis, 54-67; PROF BIOCHEM, UNIV CALIF, BERKELEY, 68- Honors & Awards: Eli Lilly Award, 64. Mem: Nat Acad Sci; Am Soc Biol Chemists; Genetics Soc Am; Am Acad Arts & Sci; Am Soc Microbiol. Res: Histidine biosynthesis; regulation of metabolism and protein synthesis; mutagens and mutations; bacterial biochemical genetics; environmental carcinogens and mutagens. Mailing Add: 1324 Spruce St Berkeley CA 94709

AMES, DAVID WASON, b Crawfordsville, Ind, May 30, 22; m 48; c 1. CULTURAL ANTHROPOLOGY. Educ: Wabash Col, AB, 47; Northwestern Univ, PhD(anthrop), 53. Prof Exp: Instr cultural anthrop, Ill Inst Technol, 49-50; asst prof, Univ Wis, 52-59; assoc prof, 61-66, PROF ANTHROP, CALIF STATE UNIV, SAN FRANCISCO, 66- Concurrent Pos: Am Coun Learned Socs-Soc Sci Res Coun Joint Comt African Studies res fel, Nigeria, 63-64 & 66-67. Mem: Fel Am Anthrop Asn; fel African Studies Asn; Soc Appl Anthrop. Res: Acculturation in West Africa and among New World Negroes; ethnography of the Wolof people in Senegal and the Gambia, Hausa of northern Nigeria, West Africa; economic anthropology; ethnomusicology. Mailing Add: Dept of Anthrop Calif State Univ 1600 Holoway Ave San Francisco CA 94132

AMES, DENNIS BURLEY, b Bristol, Eng, Sept 6, 06; nat US; m 38; c 2. MATHEMATICS. Educ: Bishop's Col, Can, BS, 27, MA, 28; Yale Univ, PhD(math), 31. Prof Exp: Instr math, Yale Univ, 31-32, hon fel, 33-34; from instr to prof, Rensselaer Polytech Inst, 34-49; prof & chmn dept, Univ NH, 49-56; res mathematician, Hughes Aircraft Co, 56-60; prof & chmn dept, 60-74, EMER PROF MATH, CALIF STATE COL, FULLERTON, 74- Mem: Am Math Soc; Soc Indust & Appl Math; Math Asn Am. Res: Celestial mechanics; theory of elasticity; internal ballistics; boundary value problems; analysis; abstract algebra, particularly group and ring theory. Mailing Add: Dept of Math Calif State Col Fullerton CA 92631

AMES, DONALD PAUL, b Brandon, Man, Sept 13, 22; m 49; c 2. PHYSICAL CHEMISTRY. Educ: Univ Wis, BS, 44, PhD, 49. Prof Exp: Staff chemist, Los Alamos Sci Lab, 50-52; asst prof, Univ Ky, 52-54; res chemist, E I du Pont de Nemours & Co, 54-56; res chemist, Monsanto Chem Co, 56-59, scientist, 59-61; scientist, McDonnell Aircraft Corp, 61-68, dep dir res, McDonnell Res Labs, 68-70, DIR RES, McDONNELL DOUGLAS RES LABS, 70- Concurrent Pos: Adv, Air Force Off Sci Res, 71- Mem: Am Chem Soc; Am Phys Soc. Res: Radiochemistry; exchange kinetics; diffusion; complex ions; microwave masers; electron and nuclear resonance spectroscopy; nuclear chemistry; actinide elements chemistry; beta and gamma ray spectroscopy; chemical and molecular lasers. Mailing Add: 914 Black Twig Ln St Louis MO 63122

AMES, EDWARD R, b Denver, Colo, Apr 10, 35; m 57; c 3. MEDICAL EDUCATION. Educ: Colo State Univ, BS, 57, DVM, 59, PhD(vet parasitol), 68. Prof Exp: Field sta mgr, eval vet therapeut, Merck Sharp & Dohme Res Labs, 60-68; from asst prof to assoc prof vet parasitol, Col Vet Med, Univ Mo-Columbia, 68-73; DIR CONTINUING EDUC, AM VET MED ASN, 73- Mem: Am Vet Med Asn; Am Soc Vet Parasitol; Am Soc Parasitol; Asn Am Vet Med Cols; Asn Teachers Vet Pub Health & Prev Med. Res: Human motivity and applied principles of adult learning; self-assessment evaluation; definition and measurement of competencies for veterinarians. Mailing Add: 930 N Meacham Rd Schaumburg IL 60172

AMES, IRA HAROLD, b Brooklyn, NY, Apr 27, 37; m 58; c 2. CELL BIOLOGY. Educ: Brooklyn Col, AB, 59; NY Univ, MS, 62, PhD(biol), 66. Prof Exp: Instr biol, Brooklyn Col, 60-63, lectr, 63-64; asst prof, 68-73, ASSOC PROF ANAT, STATE UNIV NY UPSTATE MED CTR, 73- Concurrent Pos: Fel, Brookhaven Nat Lab, 66-68; NSF grant, 72-76. Mem: AAAS; Am Soc Cell Biol; Bot Soc Am; Am Asn Anat; Am Soc Plant Physiol. Res: Control mechanisms in development; development of genetic tumors in plants. Mailing Add: Dept of Anat State Univ NY Upstate Med Ctr Syracuse NY 13210

AMES, IRVING, b New York, NY, June 15, 29; m 52; c 2. PHYSICS. Educ: Syracuse Univ, BS, 51; Cornell Univ, PhD(physics), 55. Prof Exp: Asst physics, Brown Univ, 51 & Cornell Univ, 51-55; STAFF MEM, IBM RES CTR, 55- Mem: Am Phys Soc; Am Vacuum Soc; Inst Elec & Electronics Engrs. Res: Thin films. Mailing Add: IBM Res Ctr Box 218 Yorktown Heights NY 10598

AMES, LLOYD LEROY, JR, b Norwich, Conn, Aug 23, 27; m 53; c 4. MINERALOGY. Educ: Univ NMex, BA, 52; Univ Utah, MS, 55, PhD(mineral), 56. Prof Exp: Res fel, Univ Utah, 56-57; sr scientist, Gen Elec Co, 57-65; RES ASSOC, WATER & WASTE MGT DEPT, PAC NORTHWEST LABS, BATTELLE MEM INST, 65- Mem: Mineral Soc Am; Mineral Asn Can. Res: Mineralogy and geochemistry applied to removal of radioisotopes from radioactive wastes; inorganic ion exchangers. Mailing Add: Pac Northwest Labs Battelle PO Box 999 Richland WA 99352

AMES, LYNFORD LENHART, b Fresno, Ohio, May 20, 38; m 63; c 2. PHYSICAL CHEMISTRY. Educ: Muskingum Col, BS, 60; Ohio State Univ, PhD(phys chem), 65. Prof Exp: NSF fel, 65-66; asst prof, 66-71, ASSOC PROF CHEM, NMEX STATE UNIV, 71-, ASST HEAD DEPT, 70- Mem: AAAS; Am Chem Soc. Res: High temperature chemistry; matrix isolation; infrared spectroscopy; mass spectroscopy; electronic band spectroscopy. Mailing Add: Dept of Chem NMex State Univ Las Cruces NM 88001

AMES, OAKES, physics, see 12th edition

AMES, PETER L, b St Paul, Minn, May 2, 31; m 58; c 2. ENVIRONMENTAL SCIENCES. Educ: Harvard Col, BA, 58; Yale Univ, MS, 62, PhD(biol), 65. Prof Exp: Pharmacologist, Smith, Kline & French Labs, 58-59; asst prof zool & asst cur birds, Mus Vert Zool, Univ Calif, Berkeley, 65-68; assoc ed life sci Encycl Britannica, 68-72; HEAD DEPT ENVIRON & RECREATION STUDIES, HARZA ENG CO, 73- Mem: AAAS; Am Ornith Union; Am Soc Zool; Cooper Ornith Soc; Wilson Ornith Soc. Res: Avian anatomy; ecology of raptorial birds; environmental conservation. Mailing Add: Harza Eng Co 150 S Wacker Dr Chicago IL 60606

AMES, RALPH WOLFLEY, b June 27, 20. PHYTOPATHOLOGY. Educ: Univ Wyo, BS, 40, MS, 41; Univ Ill, PhD, 50. Prof Exp: Asst prof bot, Univ Mass, 50-51; assoc plant pathologist, State Nat Hist Surv, Ill, 51-52; assoc prof bot, Utah State Agr Col, 52-54, prof & head dept, 54-58; plant pathologist, Los Angeles State Arboretum, 58-60; assoc prof, 60-73, PROF BIOL SCI & CHMN DEPT, CALIF STATE POLYTECH UNIV, POMONA, 73- Mem: AAAS; Am Phytopath Soc; Bot Soc Am; Mycol Soc Am. Res: Diseases of ornamentals. Mailing Add: Dept of Biol Sci Calif State Polytech Univ Pomona CA 91766

AMES, ROGER LYMAN, b Northampton, Mass, Nov 6, 33; m 57; c 4. GEOCHEMISTRY. Educ: Williams Col, BA, 55; Yale Univ, MS, 59, PhD(econ geol), 63. Prof Exp: SR RES SCIENTIST, AMOCO PROD CO RES CTR, 62- Mem: AAAS; Geol Soc Am; Am Geochem Soc. Res: Application of stable isotope studies to exploration for petroleum and metallic deposits. Mailing Add: Amoco Prod Co Res Ctr PO Box 591 Tulsa OK 74102

AMES, SMITH WHITTIER, b Clinton, Mass, June 5, 07; m 37; c 4. PHYSIOLOGY. Educ: Univ Maine, BA, 32, MA, 33; Univ Southern Calif, MS, 53, PhD(physiol), 55. Prof Exp: Teacher pub schs, Mass, 34-37, Ariz, 40-41, Calif, 46-49; supt schs, Westminster, Colo, 37-40; asst safety serv, Am Red Cross, Los Angeles, 41-42; civilian instr, US Army Cadet Training Ctr, Santa Ana, Calif, 42-43; res assoc, Sch Med, Univ Southern Calif, 49-51; proj engr, Impact Res Inst, Inglewood, Calif, 53-54; sr aviation physiologist, Off of Surg Gen, US Dept Air Force, 54-71; CONSULT, 71- Honors & Awards: Paul Bert Award, Aerospace Med Asn, 69. Mem: AAAS; Aerospace Med Asn; Asn Mil Surg US; Human Factors Soc (vpres, 59); hon mem Space & Flight Equip Asn. Res: Aviation physiology; acceleration on the human centrifuge; flight safety research; head impact studies. Mailing Add: 1532 Harle Pl Anaheim CA 92802

AMES, STANLEY RICHARD, b Madison, Wis, Dec 18, 18; m 43; c 4. NUTRITIONAL BIOCHEMISTRY, ANIMAL NUTRITION. Educ: Univ Mont, BA, 40; Columbia Univ, AM, 42, PhD(chem), 44. Prof Exp: Asst chem, Univ Mont, 37-40; asst, Columbia Univ, 40-41, statutory asst, 41-42, asst, 42-43; Rockefeller fel biochem, Univ Wis, 43-46; sr res chemist, Distillation Prod Industs, Eastman Kodak Co, 46-53, res assoc, 53-63, head biochem res dept, 63-65, dir biochem res labs & sr res assoc, 65-69, SR RES ASSOC & HEAD BIOCHEM RES LAB, HEALTH & NUTRIT RES DIV, TENN EASTMAN RES LABS, TENN EASTMAN CO, EASTMAN KODAK CO, 69- Concurrent Pos: Mem adv bd, Off Biochem Nomenclature, Nat Acad Sci-Nat Res Coun, 65-68; mem comt nomenclature, Int Union Nutrit Sci, 67-72, actg chmn, 69, 75, chmn, 76-79, mem comn I, nomenclature, procedures & stand, 76-79; actg chmn adv comt, NIH Guide to Nutrit Terminol, 68-69. Mem: Fel AAAS; Am Chem Soc; Am Soc Biol Chem; Asn Off Anal Chemists; Nutrit Soc. Res: Vitamins A and E; biochemistry of monoglycerides; acetylated monoglycerides; bioassay; nutrition; lipid metabolism; enzymes; organic and biological oxidations; feed preservatives; ruminant nutrition. Mailing Add: Tenn Eastman Res Labs Tenn Eastman Co PO Box 1911 Rochester NY 14603

AMES, SUSAN, b Biloxi, Miss, Aug 18, 45. ASTROPHYSICS. Educ: Bryn Mawr Col, BA, 67; Univ Calif, Berkeley, MA, 69, PhD(astron), 72. Prof Exp: Res asst radio astron, Univ Calif, Berkeley, 67-69; mem astrophys, Inst Advan Study, Princeton, NJ, 72-74; MEM STAFF, THEORET DIV, LOS ALAMOS SCI LAB, UNIV CALIF, 74- Concurrent Pos: NSF trainee astrophys, Univ Calif, Berkeley, 69-72; Am Acad Sci travel grant, Int Astron Union Gen Assembly, Poland, 73. Mem: Am Astron Soc; fel Royal Astron Soc. Res: Problems of interstellar clouds and star formation. Mailing Add: T-4 Mail Stop 212 Los Alamos Sci Lab Los Alamos NM 87545

AMES, WENDELL RUSSELL, b Lockport, NY, Aug 1, 12; m 36; c 2. EPIDEMIOLOGY. Educ: Univ Buffalo, MD, 35; Johns Hopkins Univ, MPH, 38. Prof Exp: Intern, Buffalo Gen Hosp, 35-36; intern, State Dept Health, NY, 36-37, asst & dist health officer, 38-39, epidemiologist, Commun Dis Div, 40-41; comnr health, Cattaraugus County Health Dept, 41-47; dir med care sect, Dept Health, Baltimore, 47-48; asst prof prev med & pub health, Sch Med, Univ Buffalo, 48-56; dir, Monroe County Health Dept, 58-73; RETIRED. Concurrent Pos: Instr, Sch Pharm, Univ Buffalo, 48-56; actg comnr health, Mt Vernon, NY, 39-40; dep comnr preventable dis serv, Erie County Health Dept, 48-56; assoc clin prof prev med & pub health, Sch Med & Dent, Univ Rochester, 56-73. Mem: Am Pub Health Asn. Res: Epidemiology of pneumococcus pneumonia, typhoid, tuberculosis, poliomyelitis, syphilis, lung malignancy and leukemias. Mailing Add: 225 Lanning Rd Honeoye Falls NY 14472

AMES, WILLIAM F, b Brandon, Man, Dec 8, 26; m 51; c 3. APPLIED MATHEMATICS. Educ: Univ Wis, BS, 49, MS, 50. Prof Exp: Instr math, Univ Wis, 54-56; sr serv engr appl math, E I du Pont de Nemours & Co, Inc, 56-59; from asst prof to prof mech eng, Univ Del, 59-67; prof mech & hydraul, Univ Iowa, 67-75; PROF MATH, GA INST TECHNOL, 75- Concurrent Pos: USPHS grant, 60-62; mem, NSF Panel, 61-, adv sci sem, 63, fac fel, 63-64, dir, 65; B F Goodrich res grant, 62-63; vis prof, Stanford Univ, 63-64; NSF grant, 65-67; dir, Proj Themis res contract, 68-; Nat Acad Sci exchange prof with Romania, 70; chmn dept appl math, Univ Iowa, 70-74; consult, Acad Press, 70-; vis prof, Univ Karlsruhe, WGer, 72-73; assoc ed, J Math Anal & Appln, 72-; Alexander von Humboldt Found res award, WGer, 74. Mem: Soc Indust & Appl Math. Res: Numerical analysis; application of mathematical techniques for the solution of engineering problems in the fields of continuum mechanics with particular emphasis on nonlinear partial differential equations. Mailing Add: Sch of Math Ga Inst of Technol Atlanta GA 30332

AMEY, RALPH LEONARD, b Huntington Park, Calif, June 5, 37; m 64. PHYSICAL CHEMISTRY. Educ: Pomona Col, AB, 59; Brown Univ, PhD(phys chem), 64. Prof

Exp: Instr chem, Barrington Col, 62-63; mat res & develop specialist missile & space systs div, Douglas Aircraft Co, Inc, 63-65; asst prof, 65-74, ASSOC PROF CHEM, OCCIDENTAL COL, 74- Mem: Am Phys Soc; Am Chem Soc. Res: Properties of the condensed state; intermolecular forces and local liquid structure; dielectric behavior of materials; thermal analysis of solids. Mailing Add: Dept of Chem Occidental Col 1600 Campus Rd Los Angeles CA 90041

AMICK, CHESTER ALBERT, b Scipio, Ind, Apr 19, 95; m 25; c 3. PHYSICAL CHEMISTRY, COLLOID CHEMISTRY. Educ: Ind Univ, AB, 20, AM, 21. Prof Exp: Instr tech sch, Ind, 21-25; res chemist, Pac Mills, 27-31; dir res textiles, US Finishing Co, RI, 31-32; supt resin finishing, Glenlyon Print Works, RI, 32-35; res chemist, Am Cyanamid Co, 35-49, patent chemist, 49-60; chem consult, 60-63; asst patent adv, Agr Res Serv Southern Res & Develop Div, USDA, 64-68; INDEPENDENT CHEM CONSULT, 68- Mem: Am Chem Soc; Am Inst Chemists; Am Asn Textile Chemists & Colorists; Sigma Xi. Res: Textile dyeing, finishing and printing, including cottons, synthetics, wool and resin treated fabrics. Mailing Add: 121 E Maple Ave Bound Brook NJ 08805

AMICK, JAMES ALBERT, b Lawrence, Mass, Feb 18, 28; m 61; c 1. SOLID STATE ELECTRONICS, SEMICONDUCTORS. Educ: Princeton Univ, AB, 49, AM, 51, PhD(phys chem), 52. Prof Exp: Res assoc electron micros, Princeton Univ, 52-53; res engr, Radio Corp Am Labs, 53-56, res engr, Zurich labs, 56-57, res phys chemist, 57-68, head process res, RCA Labs, 68-71, MGR MAT & PROCESSES, SOLID STATE DIV, RCA LABS, 71- Mem: AAAS; Electrochem Soc; Am Chem Soc (actg secy-treas, 53); fel Am Inst Chemists; Sigma Xi. Res: Infrared absorption intensities; electron microscopy of aerosols; mass spectrometry of hydrocarbons; x-ray diffraction; electrophotographic processes; surface chemistry of semiconductors; epitaxial growth of semiconductors; processes for electronic components. Mailing Add: 76 Leabrook Lane Princeton NJ 08540

AMICK, LAWRENCE DOUGLAS, b Millersburg, Iowa, Aug 18, 22; c 4. NEUROLOGY, PSYCHIATRY. Educ: Univ Iowa, BA, 42, MD, 45. Prof Exp: Head phys med & rehab, Lovelace Clin, 58-60; assoc prof phys med & rehab & head sect, Div Med, Univ Tenn, 60-66; prof neurol & dir stroke rehab proj, NMex Regional Med Prog, Sch Med, Univ NMex, 70-73; prof neurol, psychiat & rehab med & chmn dept rehab med, Med Col Va, Va Commonwealth Univ, 73-75; PROF NEUROL & PROF PSYCHIAT, UNIV OKLA HEALTH SCI CTR, 75- Concurrent Pos: Fel phys med & rehab, Mayo Clinic, 54-57; Nat Found scholar, Royal Free Hosp, Univ London, 57-58. 15 Mem: AM Acad Neurol; Am Acad Phys Med & Rehab; AMA; assoc mem Am Psychiat Asn. Mailing Add: Dept of Neurol Box 26901 Okla Health Sci Ctr Oklahoma City OK 73190

AMIDON, ELLSWORTH LYMAN, b West Barnet, Vt, Apr 3, 06; m 32; c 2. MEDICINE. Educ: Tufts Col, BS, 27; Univ Vt, MD, 32; Univ Pa, MS, 36; Am Bd Internal Med, cert, 39. Prof Exp: Instr path, Col Med, Univ Vt, 33-35, instr path & med, 36-37, from asst prof to prof med, 37-75, chmn dept, 42-64; RETIRED. Concurrent Pos: Med dir, Mary Fletcher Hosp, 34-67; coordr med affairs, Med Ctr Hosp Vt, 67- Mem: Fel AMA; master Am Col Physicians (vpres, 63-64); Am Heart Asn; Fedn Clin Res. Res: Leukocytic response to sulfonamides; electrocardiographic changes in patients with tricinosis; hematologic studies in acute infections. Mailing Add: Col of Med Univ of Vt Burlington VT 05401

AMIDON, ROGER WELTON, b Rockford, Ill, Nov 29, 14; m 38; c 3. CHEMISTRY. Educ: Univ Minn, BChem, 36, MS, 40, PhD(org chem), 49. Prof Exp: Petrol chemist, Midcontinent Petrol corp, 36-39; org res chemist, Dow Chem Co, 40-41; chemist, Gen Labs, US Rubber Co, 49-54 & Naugatuck Chem Div, 54-68; SR INFO SCIENTIST, CHEM DIV, UNIROYAL INC, 68- Mem: AAAS; Am Chem Soc; The Chem Soc; Sigma Xi. Res: New rubbers and elastomers; plastics and resins; rubber chemicals; polyester resins. Mailing Add: 127 Hurley Rd Oxford CT 06483

AMIN, OMAR M, b Minya El Kamh, Egypt, Jan 23, 39; nat US; m 64. MEDICAL ENTOMOLOGY, PARASITOLOGY. Educ: Cairo Univ, BSc, 59, MSc, 63; Ariz State Univ, PhD(zool), 68. Prof Exp: Res asst med zool, US Naval Med Res Unit, Cairo, 60-64; assoc agr & zool, Ariz State Univ, 66-67; res assoc & instr biol, Old Dominion Univ, 67-69; vis fel virol, Commun Dis Ctr, Ga, 69-70; ASST PROF BIOL, UNIV WIS-PARKSIDE, 71- Concurrent Pos: Sigma Xi res grant, 69. Mem: Entom Soc Am; Am Soc Parasitologists. Res: Helminth parasites of fresh water fishes; bio ecology of ectoparasites, of vertebrates, vectors of diseases, particularly ticks and fleas; host-parasite interrelationships. Mailing Add: Sci Div Univ of Wis-Parkside Kenosha WI 53140

AMINOFF, DAVID, b Kokand, Russia, Jan 16, 26; US citizen; m 56; c 3. BIOCHEMISTRY, IMMUNOGENETICS. Educ: Univ London, BSc, 45, PhD(biochem), 49, DSc(biochem), 74. Prof Exp: Res asst biochem, Lister Inst Prev Med, 49-50; res assoc, Israeli Inst Biol Res, 50-55, 56-57; exchange scientist, Col Physicians & Surgeons, Columbia Univ, 55-56; res assoc, Pub Health Res Inst, New York, 57-60; res assoc, Rackham Arthritis Res Unit, 60-62, res assoc, Simpson Mem Inst, 62-66, asst prof, 66-70, ASSOC PROF BIOL CHEM, DEPT INTERNAL MED, SIMPSON MEM INST, UNIV MICH, ANN ARBOR, 70- Concurrent Pos: Consult, Marcus Inst, Israeli Red Cross, 53-57. Mem: Soc Complex Carbohydrates; Am Asn Immunol; Am Chem Soc; Am Soc Biol Chem; Brit Biochem Soc. Res: Immunochemistry and biosynthesis of the blood group substances; glycoprotein chemistry and metabolism. Mailing Add: Dept of Internal Med Simpson Mem Inst Ann Arbor MI 48109

AMIRAIAN, KENNETH, b Richmond Hill, NY, Sept 29, 26; m 51; c 2. BIOCHEMISTRY. Educ: Brooklyn Col, BS, 49; Columbia Univ, MA, 50, PhD, 59. Prof Exp: Asst biochem, Sloan-Kettering Inst, 50-53; asst, Columbia Univ, 53-55; asst immunochem, Rutgers Univ, 55-57; from res scientist to sr res scientist biochem, NY State Dept Health, 57-68; lectr, 60-64, ASST PROF MICROBIOL, ALBANY MED COL, 64-; ASSOC RES SCIENTIST, IMMUNOCHEM, NY STATE DEPT HEALTH, 68- Mem: AAAS; Am Asn Immunol; Am Soc Microbiol. Res: Immunochemistry research on immune hemolytic reaction and gamma globulins. Mailing Add: Div of Labs & Res NY State Dept of Health Albany NY 12208

AMIR-MOEZ, ALI R, b Tehran, Iran, Apr 7, 19; US citizen. MATHEMATICS. Educ: Univ Tehran, BA, 43; Univ Calif, Los Angeles, MA, 51, PhD(math), 55. Prof Exp: Instr math, Tehran Technol Col, 43-47; asst prof, Univ Idaho, 55-56, Queens Col, NY, 56-60, Purdue Univ, 60-61 & Univ Fla, 61-63; res & writing, 63-64; prof math, Clarkson Col Technol, 64-65; PROF MATH, TEX TECH UNIV, 65- Mem: Am Math Soc; Math Asn Am; Soc Indust & Appl Math. Res: Singular values of linear transformations and matrices. Mailing Add: Dept of Math Tex Tech Univ Lubbock TX 79409

AMIS, EDWARD STEPHEN, b Himyar, Ky, Nov 9, 05; m 34; c 2. PHYSICAL CHEMISTRY. Educ: Univ Ky, BS, 30, MS, 33; Columbia Univ, PhD(chem), 39. Prof Exp: Prin & teacher pub schs, Ky, 23-27 & 30-31; asst & instr chem, Univ Ky, 31-36; asst & instr chem, Columbia Univ, 36-39; from instr to asst prof phys chem, La State Univ, 39-46; res chemist & sr res chemist, Carbide & Carbon Chem Corp, Tenn, 46-47; PROF CHEM, UNIV ARK, FAYETTEVILLE, 47- Concurrent Pos: Res partic, Oak Ridge Inst Nuclear Studies, 48- Honors & Awards: Southern Chemist Award, 59; Southwest Award, Am Chem Soc, 60, Tour Speakers Plaque, 75; Univ Ark Award for Distinguished Res, 67; Hon Scroll, Am Inst Chemists, 75. Mem: Am Chem Soc; fel NY Acad Sci; Brit Chem Soc. Res: Kinetics of reactions between ions and dipolar molecules; fluorocarbons; electrochemistry; polarography; solvent effects on chemical phenomena; solvation; solution chemistry. Mailing Add: 1655 Woolsey Ave Fayetteville AR 72701

AMITH, AVRAHAM, b Przemysl, Poland, Mar 16, 29; US citizen; m 52; c 2. SOLID STATE PHYSICS. Educ: Univ Mich, BS, 51; Harvard Univ, MA, 53, PhD(chem phys), 56. Prof Exp: Mem tech staff, RCA Labs, 56-70; prof physics, City Univ New York, 71-72; sr engr & leader, RCA Thermoelec Opers, 72-74; SR SCIENTIST & PROG MGR, INST ENERGY CONVERSION, UNIV DEL, 74- Concurrent Pos: Adj prof, City Univ New York, 72-74. Honors & Awards: RCA Outstand Achievement Award. Mem: Am Phys Soc. Res: Molecular quantum mechanics; photoeffects, lifetimes and transport in semiconductors; electrical properties of thin films and surface layers; thermoelectric properties of semiconductors; galvanomagnetic and thermomagnetic properties of semiconductors and semimetals; magnetism and thin films; thin film solar cells. Mailing Add: Four Whitemarsh Dr Lawrenceville NJ 08648

AMKRAUT, ALFRED A, b Saarbruecken, Ger, Sept 21, 26; US citizen; m 55; c 2. IMMUNOLOGY, IMMUNOCHEMISTRY. Educ: Univ Calif, Berkeley, BA, 49; NY Univ, MS, 55; Stanford Univ, PhD(med microbiol), 63. Prof Exp: Res asst immunohemat, Med Sch, Stanford Univ, 55-62; vis scientist, Ore Regional Primate Res Ctr, 64-67; asst prof immunol, Med Sch, Univ Ore, 65-67; asst prof med microbiol, Sch Med, Stanford Univ, 67-74; HEAD MICROBIOL, ALZA RES, 74- Concurrent Pos: Res fel immunochem, Calif Inst Technol, 62-64; Nat Acad Sci-Nat Res Coun fel, 63-64; NIH fel, 64- Mem: Fel AAAS; Am Asn Immunol; Am Chem Soc; Am Soc Microbiol. Res: Antigen-antibody interactions and their biological manifestations; mechanism of antibody formation; central nervous system and the immune response. Mailing Add: 3358 Kenneth Dr Palo Alto CA 94303

AMLING, HARRY JAMES, b Baltimore, Md, Jan 22, 31; m 53; c 1. HORTICULTURE. Educ: Rutgers Univ, BS, 52; Univ Del, MS, 54; Mich State Univ, PhD(hort), 58. Prof Exp: Asst, Univ Del, 52-54; res asst, Mich State Univ, 56-58; from asst prof to assoc prof, 58-68, PROF HORT, AUBURN UNIV, 68- Mem: Am Soc Hort Sci; Am Soc Plant Physiol; Weed Sci Soc Am. Res: Practical and fundamental aspects of use of herbicides and growth regulators on horticultural crops; growth and development; mineral nutrition of horticultural crops. Mailing Add: Dept of Hort Auburn Univ Auburn AL 36830

AMMA, ELMER LOUIS, b Cleveland, Ohio, Feb 13, 29. BIOINORGANIC CHEMISTRY, BIOPHYSICAL CHEMISTRY. Educ: Case Inst, BS, 52; Iowa State Univ, PhD(chem), 57. Prof Exp: Asst chem, Iowa State Univ, 52-54; asst, Ames Lab, AEC, 54-57; res assoc, Radiation Lab, Univ Pittsburgh, 57-58, lectr & res assoc, Radiation Lab & Physics Dept, 58-59, prof chem, 59-65; assoc prof, 65-70, PROF CHEM, UNIV SC, 70- Honors & Awards: Russell Award in Res, 70. Mem: AAAS; Am Inst Chem; Am Chem Soc; Am Crystallog Asn; Am Phys Soc. Res: Protein crystallography structure of model enzymes and proteins; structure of carriers of molecular oxygen and nitrogen; structure of molecules of importance to homogeneous catalysis. Mailing Add: Dept of Chem Univ of SC Columbia SC 29208

AMMAN, GENE DOYLE, b Greeley, Colo, Feb 26, 31; m 54; c 5. FOREST ENTOMOLOGY. Educ: Colo State Univ, BS, 56, MS, 58; Univ Mich, PhD(forest insect ecol), 66. Prof Exp: Assoc entomologist, Southeastern Forest Exp Sta, 58-66, PRIN ENTOMOLOGIST, INTERMOUNTAIN FOREST & RANGE EXP STA, US FOREST SERV, 66- Mem: AAAS; Int Orgn Biol Control; Int Union Forestry Res Orgn; Entom Soc Am; Entom Soc Can. Res: Population ecology; natural control of forest insects. Mailing Add: Intermt Forest & Range Exp Sta US Forest Serv 507 25th St Ogden UT 84401

AMMANN, GEORGE ANDREW, ornithology, see 12th edition

AMMAR, RAYMOND GEORGE, b Jamaica, WI, July 15, 32; US citizen; m 61; c 3. PHYSICS. Educ: Harvard Univ, AB, 53; Univ Chicago, SM, 55, PhD(physics), 59. Prof Exp: Res assoc physics, Enrico Fermi Inst, Chicago, 59-60; from asst prof to assoc prof, Northwestern Univ, 60-69; PROF PHYSICS, UNIV KANS, 69- Concurrent Pos: Consult, High Energy Physics Div, Argonne Nat Lab, 65-69, vis scientist, 71-72. Mem: Fel Am Phys Soc. Res: High energy physics. Mailing Add: Dept of Physics Univ of Kans Lawrence KS 66045

AMME, ROBERT CLYDE, b Ames, Iowa, May 15, 30; m 51; c 2. EXPERIMENTAL ATOMIC PHYSICS, ATMOSPHERIC PHYSICS. Educ: Iowa State Univ, BS, 53, MS, 55, PhD(physics), 58. Prof Exp: Res physicist, Humble Oil & Refining Co, 58-59; res physicist, Denver Res Inst, 59-68, from asst prof to assoc prof, 61-68, PROF PHYSICS, UNIV DENVER, 68-, SR RES PHYSICIST, DENVER RES INST, 68- Mem: Fel Am Phys Soc. Res: Atomic and molecular physics; collision phenomena; energy and charge transfers; ionization; gaseous transport properties; acoustics; stratospheric composition. Mailing Add: Dept of Physics Univ of Denver Denver CO 80210

AMMERMAN, CLARENCE BAILEY, b Cynthiana, Ky, May 21, 29; m 50; c 3. ANIMAL NUTRITION. Educ: Univ Ky, BS, 51, MS, 52; Univ Ill, PhD(animal sci), 56. Prof Exp: Asst animal nutrit, Univ Ky, 51-52; asst animal sci, Univ Ill, 54-55; from asst prof to assoc prof, 58-70, PROF ANIMAL NUTRIT, UNIV FLA, 70- Mem: AAAS; Am Soc Animal Sci; Am Inst Nutrit; Am Dairy Sci Asn; Soc Exp Biol & Med. Res: Animal nutrition mineral elements; forage utilization by ruminants. Mailing Add: 631 NW36th Dr Gainesville FL 32607

AMMIRATO, PHILIP VINCENT, b New York, NY, Nov 22, 43. PLANT PHYSIOLOGY. Educ: City Col New York, BS, 64; Cornell Univ, PhD(plant physiol), 69. Prof Exp: NY State Col teaching fel, Cornell Univ, 64-65; asst prof bot, Rutgers Univ, 69-72; PROF BOT, BARNARD COL, COLUMBIA UNIV, 72- Mem: Bot Soc Am; Am Soc Plant Physiol. Res: Plant growth and development; tissue culture; economic botany; morphogenesis; embryogenesis. Mailing Add: Dept of Biol Sci Barnard Col Columbia Univ New York NY 10027

AMMON, HERMAN L, b Passaic, NJ, Nov 24, 36; m 58; c 1. ORGANIC CHEMISTRY. Educ: Brown Univ, ScB, 58; Univ Wash, PhD(chem), 63. Prof Exp: NIH fel, 63-64; res instr x-ray crystallog, Univ Wash, 64-65; asst prof chem, Univ Calif, Santa Cruz, 65-69; asst prof, 69-72, ASSOC PROF CHEM, UNIV MD, COLLEGE PARK, 72- Mem: Am Chem Soc; Am Crystallog Asn. Res: Organic synthesis and structure; computer programming; x-ray crystallography. Mailing Add: Dept of Chem Univ of Md College Park MD 20742

AMMONDSON, CLAYTON JOHN, b Decorah, Iowa, Dec 7, 23; m 49; c 3. PHYSICAL CHEMISTRY. Educ: Luther Col, Iowa, BA, 44; Univ Minn, MS, 50; Rutgers Univ, PhD(chem), 56. Prof Exp: Res chemist, Am Cyanamid Co, 50-52; res chemist, Visking Co, 55-57; dir rheol lab, Cent Res & Eng Div, Continental Can Co, 57-60; PRES, ZARN, INC, 60- Mem: Am Chem Soc; Soc Plastics Engrs. Res: Flow properties of plastics; physical and mechanical properties of polymers; management. Mailing Add: 1828 Trentwood Circle Reidsville NC 27320

AMOLS, HOWARD IRA, b New York, NY, Feb 11, 49; m 70; c 2. RADIOLOGICAL PHYSICS, RADIOBIOLOGY. Educ: Cooper Union, BS, 70; Brown Univ, MS, 73, PhD(physics), 74. Prof Exp: MEM STAFF RADIOL PHYSICS & RADIOBIOL, LOS ALAMOS SCI LAB, UNIV CALIF, 74- Concurrent Pos: Nat Cancer Inst fel, Los Alamos Sci Lab, 74-76. Mem: Am Asn Physicists Med; Radiation Res Soc. Res: Physical and biological studies related to the use of heavy particles in radiotherapy; dose modeling, treatment planning, effects of dose fractionation and cell heterogeneity; modeling radiation effects in tissue. Mailing Add: Group MP-3 Los Alamos Sci Lab Los Alamos NM 87545

AMOORE, JOHN ERNEST, b Derbyshire, Eng, Apr 28, 30; US citizen; m 59; c 2. BIOCHEMISTRY. Educ: Oxford Univ, BA, 52, MA & PhD(biochem), 58. Prof Exp: Sci officer microanal, John Innes Hort Inst, Eng, 58-59; Nuffield Found res fel plant biochem, Univ Edinburgh, 59-62; Jane Coffin Childs Mem Fund res fel zool, Univ Calif, Berkeley, 62-63; RES CHEMIST, WESTERN REGIONAL RES LAB, USDA, 63- Res: Stereochemical theory of olfaction; measurement of of molecular shape; smell-blindness; primary odors. Mailing Add: Western Region Res Lab USDA Berkeley CA 94710

AMORE, SALVATORE THOMAS, organic chemistry, see 12th edition

AMORIM, DALMO DE SOUZA, b Rio de Janeiro, Brazil, Mar 31, 30; m 56; c 3. CARDIOVASCULAR DISEASES, CARDIOVASCULAR PHYSIOLOGY. Educ: Univ Sao Paulo, Dr(med), 66. Prof Exp: Instr cardiol, Univ Rio de Janeiro, 55-57; res asst physiol, Mayo Grad Sch Med, 58-61, Guggenheim fel & res assoc physiol, 68; livre-docente med, 69-73, mem postgrad progs comn, 70-74, ASSOC MED, UNIV SAO PAULO, 73-, COORDR POSTGRAD PROGS MED, 70- Concurrent Pos: Vis scientist, Univ Kampala & Univ Cape Town, 69, Univ London & Univ Birmingham, 74. Mem: Fel Royal Soc; AAAS; Sigma Xi; Brazilian Soc Advan Sci; Brazilian Soc Cardiol. Res: Physiological investigations concerning regulation of cardiac function in Chagas' disease; autonomic impairment in myocardiopathies. Mailing Add: Univ of Sao Paulo Fac of Med 14100 Ribeirao Preto Sao Paulo Brazil

AMORY, DAVID WILLIAM, b Newark, NJ, Jan 27, 28; m 60; c 2. ANESTHESIOLOGY, PHARMACOLOGY. Educ: St John's Univ, NY, BS, 52, MS, 55; Univ Wash, PhD(pharmacol), 61; Univ BC, MD, 67. Prof Exp: Pharmacologist toxicol, Ciba Pharmaceut Co, Inc, 55-57; res asst pharmacol, Univ Wash, 57-61; USPHS fel neuropharmacol, Univ Calif, San Francisco, 61-62; intern med, Virginia Mason Hosp, Seattle, 67-68; USPHS scholar, Cardiovasc Res Inst, Univ Calif, San Francisco, 68-69; resident anesthesiol, 69-71, ASSOC PROF ANESTHESIOL & PHARMACOL, SCH MED, UNIV WASH, 75- Concurrent Pos: Nat Heart & Lung Inst res grants, 72- Honors & Awards: Walter Stewart Baird Mem Prize, Sch Med, Univ BC, 67; Residents' Res Prize, Am Soc Anesthesiologists, 70; Res Prize in Anesthesiol, Mt Sinai Sch Med, 70. Mem: Am Soc Anesthesiologists; Sigma Xi. Res: Cardiovascular physiology and pharmacology, especially effects of drugs on total and regional blood flow with particular emphasis on coronary and cerebral circulation; physiology of deep hypothermia for cardiac surgery. Mailing Add: Dept of Anesthesiol Univ of Wash Sch of Med Seattle WA 98195

AMOS, DENNIS BERNARD, b Bromley, Eng, Apr 16, 23; m 49; c 5. IMMUNOGENETICS. Educ: Univ London, MB, BS, 51, MD, 63. Prof Exp: Intern, Guy's Hosp, London, 51; from assoc cancer res scientist to prin cancer res scientist, Roswell Park Mem Inst, 56-62; JAMES B DUKE PROF IMMUNOL & EXP SURG, MED CTR, DUKE UNIV, 62- Concurrent Pos: Res fel, Guy's Hosp, London, 52-55; sr res fel, Roswell Park Mem Inst, 55-56; chmn task force immunol & dis, NIH Nat Inst Allergy & Infectious Dis & nomenclature comt on Leukocyte Antigens, WHO, Int Union Immunol Socs. Mem: Genetics Soc Am; Am Asn Cancer Res; Transplantation Soc; Am Asn Clin Histocompatability Test; AAAS. Res: Tissue and tumor transplantation genetics; immunology; mouse genetics; antigen differentiation. Mailing Add: Duke Univ Med Ctr Box 3010 Durham NC 27710

AMOS, DEWEY HAROLD, b Harrisville, WVa, Feb 27, 25; m 48; c 3. PETROLOGY. Educ: Marietta Col, BS, 49; Univ Ill, MA, 50, PhD(geol), 58. Prof Exp: Asst geol, Univ Ill, 49-51; geologist, US Geol Surv, 52-55; asst prof geol, Southern Ill Univ, 55-65; PROF GEOL, EASTERN ILL UNIV, 65- Concurrent Pos: Consult, US Geol Surv & Defense Minerals Explor Admin, 55-64; consult geologist, US Geol Surv, 64- Mem: Geol Soc Am; Soc Econ Geologists. Res: Metamorphic petrology; geology of base metals deposits. Mailing Add: 2003 University Dr Charleston IL 61920

AMOS, DONALD E, b Lafayette, Ind, Feb 28, 29; m 52; c 2. MATHEMATICS, CHEMICAL ENGINEERING. Educ: Purdue Univ, BS, 51; Ore State Univ, MS, 56, PhD(math), 60. Prof Exp: Chem engr, Gen Elec Co, 51-53; staff mem appl math, Sandia Corp, 60-64; assoc prof math, Univ Mo-Columbia, 64-66; mathematician, 66, STAFF MEM, SANDIA LABS, 66- Concurrent Pos: Ed, Communications in Statist, 72- Mem: Am Math Soc; Soc Indust & Appl Math. Res: Special functions; numerical analysis and solution of ordinary and partial differential equations. Mailing Add: 7704 Pickard Ave NE Albuquerque NM 87110

AMOS, HAROLD, b Pennsauken, NJ, Sept 7, 19. BACTERIOLOGY. Educ: Springfield Col, BS, 41; Harvard Univ, PhD(bact), 52. Prof Exp: Instr bact, Springfield Col, 48-49; assoc, 54-59, from asst prof to assoc prof, 59-70, PROF BACT & IMMUNOL, HARVARD MED SCH, 70- Concurrent Pos: Fulbright res fel, Pasteur Inst, France, 51-52; res fel, Harvard Med Sch, 52-54; USPHS fel, 52-54, sr res fel, 58; mem, Nat Cancer Adv Bd, 72-; trustee, Josiah Macy Found, 73- Mem: Am Soc Microbiol; Am Soc Biol Chemists; Tissue Cult Asn. Res: Hexose metabolism in mammalian cells; surface changes and hormonal influences. Mailing Add: Dept Microbiol & Molec Genetics Harvard Med Sch Boston MA 02115

AMOS, MALCOLM FREDERICK, organic chemistry, see 12th edition

AMOSS, MAX ST CLAIR, b Baltimore, Md, May 9, 37. ENDOCRINOLOGY. Educ: Pa State Univ, BS, 62; Tex A&M Univ, MS, 65; Baylor Col Med, PhD(physiol), 69. Prof Exp: Fel, Baylor Col Med, 69, asst prof endocrinol, 69-70; asst res prof neuroendocrinol, Salk Inst, 70-75; ASST PROF VET PHYSIOL, TEX A&M UNIV, 75- Concurrent Pos: Instr, San Diego Eve Col, 72-75. Mem: Endocrine Soc; Am Physiol Soc; Soc Study Reproduction; Int Soc Neuroendocrinol. Res: Hypothalamic control of hypophyseal function with special emphasis on reproduction; development of radioimmunoassay procedures for use in veterinary medicine. Mailing Add: Dept of Vet Physiol & Pharmacol Tex A&M Univ Col of Vet Med College Station TX 77843

AMPLATZ, KURT, b Weistrach, Austria, Feb 25, 24; US citizen; m 56; c 3. RADIOLOGY. Educ: Univ Innsbruck, MD, 50. Prof Exp: From instr to assoc prof, 58-70, PROF RADIOL, SCH MED, UNIV MINN, MINNEAPOLIS, 70- Res: Detection of cardiac shunts with radioactive and nonradioactive gases; blood flow measurements by radiographic technics; radiographic evaluation of renovascular hypertension; pneumotomography; development of see-through film changesrs; magnification coronary angiography; development of angiographic injection; nonthrombogenic catheter. Mailing Add: Dept of Radiol Univ of Minn Sch of Med Minneapolis MN 55455

AMPULSKI, ROBERT STANLEY, b Grand Rapids, Mich, Sept 22, 42; m 65; c 3. MASS SPECTROMETRY. Educ: Aquinas Col, BS, 64; Marquette Univ, MS, 69; Fla State Univ, PhD(phys chem), 72. Prof Exp: Res chemist, Milwaukee Blood Ctr Inc, 67-68; GROUP LEADER MASS SPECTROMETRY, MIAMI VALLEY LABS, PROCTER & GAMBLE CO, 72- Mem: Am Soc Mass Spectrometry; Sigma Xi. Res: Stable isotopes; gas phase kinetics; ion molecule reactions. Mailing Add: Miami Valley Labs Procter & Gamble Co PO Box 39175 Cincinnati OH 45239

AMREIN, YOST URSUS LUCIUS, b Arosa, Switz, Jan 3, 18; nat US; m 48; c 1. ANIMAL PARASITOLOGY. Educ: Univ Calif, Los Angeles, BA, 47, MA, 48, PhD(parasitol), 51. Prof Exp: Teaching asst zool, Univ Calif, Los Angeles, 47-51; from instr to assoc prof, 51-59, chmn dept, 59-68, PROF ZOOL, POMONA COL, 59- Concurrent Pos: USPHS fel, Nat Inst Med Res, London, 57-58; NIH res grants, 59-64 & 66-74; res fel, Swiss Trop Inst, 64-65. Mem: AAAS; Am Soc Parasitol; fel Royal Soc Trop Med & Hyg. Res: Physiology of hemoflagellates; ecology of parasites in arid areas. Mailing Add: Dept of Zool Seaver Lab Pomona Col Claremont CA 91711

AMROMIN, GEORGE DAVID, b Gomel, Russia, Feb 27, 19; US citizen; m 42; c 5. PATHOLOGY. Educ: Northwestern Univ, BS, 40, MD, 43. Prof Exp: Asst pathologist, Michael Reese Hosp, Chicago, 49; pathologist, Mem Hosp, Exeter, Calif, 50-53; asst dir & assoc pathologist, Michael Reese Hosp, Chicago, 54-56; chmn div path, 56-71, CHMN DEPT NEUROPATH & RES PATH, CITY OF HOPE MED CTR, 71- Concurrent Pos: Nat Inst Neurol Dis & Blindness fel, Bellevue Med Ctr, NY Univ, 68-69; attend pathologist, Tulare Kings Joint Hosp, Springville, Calif, 51-54; pathologist, Tulare County Hosp & Visalia Munic Hosp, 51-54; consult staff path, Tulare Dist Hosp, 51-54; clin prof, Loma Linda Univ Med Sch, 70. Mem: AMA; fel Am Soc Clin Path; fel Am Col Path; fel Am Col Physicians; World Med Asn. Res: Neuropathology; leukemia; experimental atherosclerosis and diabetes; complications of therapy of neoplasms. Mailing Add: Dept of Neuropath & Res Path City of Hope Med Ctr Duarte CA 91010

AMSBURY, DAVID LEONARD, b Topeka, Kans, Dec 30, 32; m 54. GEOLOGY. Educ: Sul Ross State Col, BS, 52; Univ Tex, Austin, PhD(geol), 57. Prof Exp: Geologist, Shell Develop Co, 56-64, res geologist, 64-66; GEOLOGIST, NASA-MANNED SPACECRAFT CTR, 66- Concurrent Pos: Lectr, Univ Houston, 70- Mem: Soc Econ Paleont & Mineral; Geol Soc Am; Sigma Xi. Res: Physical stratigraphy; remote sensing in environmental geology. Mailing Add: 1422 Brookwood Ct Seabrook TX 77586

AMSDEN, ANTHONY AVERY, b Santa Fe, NMex, May 11, 39; m 60; c 2. FLUID DYNAMICS. Prof Exp: EDP operator, 59-63, systs programmer, 60-63, sci programmer, 63-66, STAFF MEM, LOS ALAMOS SCI LAB, 66-, ASST GROUP LEADER NUMERICAL FLUID DYNAMICS, 70- Res: Development of new methods for the numerical solution of problems in fluid dynamics and related phenomena. Mailing Add: Los Alamos Sci Lab Los Alamos NM 87545

AMSDEN, THOMAS WILLIAM, b Wichita, Kans, Jan 31, 15; m 40; c 2. INVERTEBRATE PALEONTOLOGY. Educ: Univ Wichita, AB, 39; Univ Iowa, MA, 41; Yale Univ, PhD(geol), 47. Prof Exp: Asst geol, Yale Univ, 42-43; geologist, US Geol Surv, 43-45; asst geol, Yale Univ, 45-46, Nat Res fel, 46-47; geologist, Md State Geol Surv, 47; from instr to assoc prof geol, Johns Hopkins Univ, 47-55; GEOLOGIST, OKLA GEOL SURV, 55- Concurrent Pos: NSF grants, Wales, Sweden & Czech, 64-66, Russia, 68, Poland, 71. Honors & Awards: Levorsen Award, Am Asn Petrol Geologists, 73. Mem: Paleont Soc; Geol Soc Am; Am Asn Petrol Geologists; Soc Econ Paleontologists & Mineralogists; Brit Palaeont Asn. Res: Late Ordovician, Silurian and early Devonian stratigraphy and paleontology; brachiopods; middle Paleozoic analysis of petroliferous deep sedimentary basins. Mailing Add: Okla Geol Surv Univ of Okla Norman OK 73069

AMSEL, LEWIS PAUL, b New York, NY, Mar 23, 42; m 64; c 2. PHARMACEUTICAL CHEMISTRY. Educ: Columbia Univ, BSc, 63; State Univ NY Buffalo, PhD(pharm), 69. Prof Exp: Assoc res pharmacist, Sterling-Winthrop Res Inst, 69-74; MGR PHARMACEUT DEVELOP, PHAMRACEUT DIV, PENNWALT CORP, 74- Mem: AAAS; Am Pharmaceut Asn; Acad Phamraceut Sci. Res: Biopharmaceutics and pharmacokinetics; drug dosage design and evaluation. Mailing Add: Pennwalt Corp Rochester NY 14623

AMSTER, ADOLPH BERNARD, b New York, NY, Nov 22, 24; m 53; c 1. PHYSICAL CHEMISTRY. Educ: City Col New York, BS, 43; Columbia Univ, AM, 47; Ohio State Univ, PhD(phys chem), 51. Prof Exp: Asst chem, Columbia Univ, 47 & Ohio State Univ, 49-51; chemist, US Bur Standards, 51-54; res assoc, Naval Ord Lab, 54-57, supv chemist, 57-60; mgr phys chem, Stanford Res Inst, 61-65, chmn phys & inorg chem div, 65-68; CHIEF EXPLOSIVES & PYROTECH BR, NAVAL SEA SYSTS COMMAND, 68- Mem: AAAS; Sigma Xi; Am Phys Soc; Combustion Inst. Res: Experimental physical chemistry; thermochemistry; explosives and propellant sensitivity; detonations; pollution abatement; hazard analysis; chemiluminescence; accident investigation. Mailing Add: Explosives & Pyrotechnics Br Naval Sea Systs Command Washington DC 20362

AMSTER, HARVEY JEROME, b Cleveland, Ohio, Sept 30, 28. THEORETICAL PHYSICS. Educ: Calif Inst Technol, BS, 50; Mass Inst Technol, PhD, 54. Prof Exp: Adv scientist, Westinghouse Bettis Plant, 54-61; assoc prof, 61-65, PROF NUCLEAR ENG, UNIV CALIF, BERKELEY, 65- Mem: AAAS; Am Nuclear Soc. Res: Nuclear physics; reactor theory; environmental effects, especially of nuclear technology. Mailing Add: Dept of Nuclear Eng Univ of Calif Berkeley CA 94720

AMSTER, ROBERT L, physical chemistry, spectroscopy, see 12th edition

AMSTUTZ, EDWARD DELBERT, b New Athens, Ohio, May 1, 09; m 35; c 3. ORGANIC CHEMISTRY. Educ: Col Wooster, BS, 30; Lawrence Col, MS, 31; Cornell Univ, PhD(org chem), 36. Prof Exp: Asst chem, Cornell Univ, 33-36; instr, Union Univ, NY, 36-38; from instr to prof, 38-70, Howard S Bunn Distinguished Prof, 70-74, DISTINGUISHED EMER PROF CHEM, LEHIGH UNIV, 74- Concurrent Pos: Chmn dept chem, Lehigh Univ, 60; Fulbright lectr, Spain, 63. Res: Synthetic organic chemistry; reaction mechanisms; heterocyclic compounds. Mailing Add: Dept of Chem Lehigh Univ Bethlehem PA 18015

AMSTUTZ, HARLAN CABOT, b Santa Monica, Calif, July 17, 31; m; c 3.

ORTHOPEDICS, BIOENGINEERING. Educ: Univ Calif, Los Angeles, BA, 53, MD, 56. Prof Exp: Intern, Los Angeles County Gen Hosp, 56-57; resident surg, Med Ctr Hosp, Univ Calif, Los Angeles, 57-58; resident orthop, Hosp Spec Surg, 58-61; res asst, Inst Orthop, London, 63-64; PROF ORTHOP & CHIEF DIV ORTHOP, SCH MED & ASSOC DIR, INST CHRONIC DIS & REHAB, UNIV CALIF, LOS ANGELES, 70- Concurrent Pos: Am-Brit-Can traveling fel, 70; lectr, Polytech Inst Brooklyn; chief prosthetics & orthotics, dir bioeng, assoc scientist & asst attend, Hosp Spec Surg; asst prof, Cornell Med Col; hon registr, Royal Nat Orthop Hosp, London, 63-64; mem comt skeletal systs & mem subcomt bioeng, Nat Acad Sci-Nat Res Coun; mem subcomt prosthetic & orthotics Educ, Nat Acad Sci; mem comt interplay eng biol & med, Nat Mat Adv Bd; physician & consult orthop surg, Vet Admin Ctr, Los Angeles. Mem: Am Acad Orthop Surgeons; Orthop Res Soc; NY Acad Sci; Am Soc Testing & Mat; Am Col Surgeons. Res: Total joint replacement; congenital anomalies. Mailing Add: Div of Orthop Surg Univ of Calif Sch of Med Los Angeles CA 90024

AMSTUTZ, HAROLD EMERSON, b Barrs Mill, Ohio, June 21, 19; m 49; c 4. VETERINARY MEDICINE. Educ: Ohio State Univ, BS, 42, DVM, 45. Prof Exp: From instr to prof vet med, Ohio State Univ, 47-61, actg chmn dept, 55-56, chmn, 56-61; PROF VET SCI & HEAD DEPT LARGE ANIMAL CLINS, PURDUE UNIV, WEST LAFAYETTE, 61- Concurrent Pos: Organizer, VI Int Conf Cattle Dis, US, 70. Mem: Am Asn Vet Clinicians; World Asn Buiatrics (pres, 72-76); Asn Bovine Practitioners; Am Col Internal Med (pres, 73). Res: Calf diseases; cattle lameness. Mailing Add: Lynn Hall Purdue Univ West Lafayette IN 47907

AMSTUTZ, LARRY IHRIG, b Wooster, Ohio, July 11, 41. SOLID STATE PHYSICS. Educ: Col Wooster, BA, 63; Duke Univ, PhD(physics), 69. Prof Exp: Physicist, Electrotech Lab, 69-76, PHYSICIST, ELEC EQUIP DIV, US ARMY, MOBILITY EQUIP RES & DEVELOP COMMAND, 76- Mem: Am Phys Soc. Res: Magnetic resonance; solid hydrogen; thermal and electrical conductivity in alloys; magnetic behavior of rare-earth alloys; applied superconductivity; numerical methods; high energy laser power supplies. Mailing Add: Fort Belvoir VA

AMUNDSEN, CLIFFORD C, b Norwalk, Conn, Dec 21, 33; m 56; c 2. PHYSIOLOGICAL ECOLOGY, MICROCLIMATOLOGY. Educ: Idaho State Col, BS, 61; Univ Colo, Boulder, PhD(bot, physiol ecol), 67. Prof Exp: Res ecologist, Inst Arctic & Alpine Res, Univ Colo, Boulder, 65-67; res assoc, 67-69, asst prof, 69-74, ASSOC PROF BOT, UNIV TENN, KNOXVILLE, 74- Concurrent Pos: Consult, Battelle Mem Inst, US Forest Serv; prin investr Aleutian res, US Energy Res & Develop Admin, 68-; consult, Nat Park Serv, 75- Mem: Ecol Soc Am; Am Polar Soc; Am Quaternary Asn. Res: Arctic-alpine and subarctic-subalpine physiological plant ecology and remote sensing. Mailing Add: Ecol Prog Univ of Tenn 408 Tenth St Knoxville TN 37916

AMUNDSEN, LAWRENCE HARDIN, b Pine River, Wis, Sept 23, 09; m 34; c 4. ORGANIC CHEMISTRY. Educ: Col Ozarks, BS, 31; Univ Fla, PhD(org chem), 35. Prof Exp: Asst org chem, Univ Fla, 31-35; asst, Inst Chem, Adelphi Col, 35-36; from asst to prof, 36-73, actg head dept, 45-46, EMER PROF CHEM, UNIV CONN, 73- Concurrent Pos: Nat Defense Res Comt fel, Purdue Univ, 42-43; chemist, Venereal Dis Res Lab, USPHS, Staten Island, NY, 43-44. Mem: AAAS; Am Chem Soc. Res: Preparation and properties of aliphatic diamines; preparation of sulfonamides; allylic rearrangements in animation reactions. Mailing Add: Dept of Chem Univ of Conn Storrs CT 06268

AMUNDSON, CLYDE HOWARD, b Nekoosa, Wis, Aug 15, 27; m 52; c 2. FOOD SCIENCE. Educ: Univ Wis-Madison, BS, 55, MS, 56, PhD(food sci, biochem), 60. Prof Exp: Chemist, Sherwin-Williams Co, 50-52; from instr to assoc prof food sci, 56-70, PROF FOOD SCI, UNIV WIS-MADISON, 70-, ASSOC CHMN DEPT, 64- Concurrent Pos: Consult, Bur Com Fisheries, Wash. Honors & Awards: Am Dairy Sci Asn Res Award, 71. Mem: Fel AAAS; fel NY Acad Sci; fel Am Inst Chem; Inst Food Technol; Am Chem Soc. Res: Flavor chemistry of food and food products; dehydration of foods and biological materials; fermentation chemistry of foods and food products; fat and oil chemistry and food enzymes. Mailing Add: 109 Babcock Hall Dept Food Sci Univ of Wis Madison WI 53706

AMUNDSON, MARY JANE, b Keokuk, Iowa, Jan 23, 36; m 58; c 1. PSYCHIATRIC NURSING. Educ: Univ Iowa, BS, 57; Univ Calif, Los Angeles, MS, 66, PhD(spec educ), 74. Prof Exp: Head nurse, Mendocino State Hosp, Talmage, Calif, 58-60; staff nursing, Langley Porter Inst, San Francisco, 60-61; clinical specialist & lectr nursing, Univ Calif, Los Angeles, Neuropsychiat Inst, 61-74; asst prof, Univ Ore, 74-75; ASST PROF NURSING, UNIV HAWAII, 75- Mem: Asn Women Sci; Am Asn Univ Prof. Res: Comparative study of three types of small group experience in teaching baccalaureate students nurse-patient relationship skills. Mailing Add: 2528 The Mall Webster Hall Univ Hawaii Sch of Nursing Honolulu HI 96822

AMUNDSON, MERLE E, b Sioux Falls, SDak, Aug 21, 36; m 58; c 3. ANALYTICAL CHEMISTRY, PHARMACEUTICAL CHEMISTRY. Educ: SDak State Univ, BS, 58; Mass Col Pharm, MS, 59, PhD(pharm), 61. Prof Exp: Sr anal chemist, Eli Lilly & Co, 61-67, res scientist, 67-68, head agr prod develop dept, 68-71, head agr biochem, Lilly Res Labs, 71-75, DIR AGR ANAL CHEM & AGR BIOCHEM, LILLY RES LABS, DIV ELI LILLY & CO, 75- Mem: Am Chem Soc. Res: Degradation of pesticides in soil and plants; analytical methodology for determination of residues in soil, plant tissue, and animal tissue. Mailing Add: Lilly Res Labs Greenfield Labs PO Box 708 Greenfield IN 46140

AMY, JONATHAN WEEKES, b Delaware, Ohio, Mar 3, 23; m 47; c 3. CHEMISTRY. Educ: Ohio Wesleyan Univ, BA, 48; Purdue Univ, MS, 50, PhD(molecular spectros), 55. Prof Exp: Res assoc chem, 54-60, assoc prof, 63-70, ASSOC DIR LABS, PURDUE UNIV, 60-, PROF CHEM, 70- Mem: Am Chem Soc; Sigma Xi; AAAS. Res: Molecular and mass spectroscopy; gas chromatography; chemical instrumentation; surface chemistry. Mailing Add: Dept of Chem Purdue Univ West Lafayette IN 47907

AMY, ROBERT LEWIS, b Jamestown, Pa, July 18, 19; m 44; c 3. ZOOLOGY. Educ: Thiel Col, BS, 41; Univ Pittsburgh, MS, 49; Univ Va, PhD(biol), 55. Prof Exp: From asst prof to assoc prof, Susquehanna Univ, 49-58; assoc prof, 58-61, PROF BIOL, SOUTHWESTERN AT MEMPHIS, 61-, CHMN DEPT, 68- Concurrent Pos: Res partic, Oak Ridge Nat Lab, 55 & 56; USPHS fel, Nat Blood Transfusion Ctr, France, 64-65; coinvestr, Habrobracon proj, NASA Biosatellite Prog, 66-69; USPHS fel, Inst Cellular Path, France, 71-72. Mem: AAAS; Radiation Res Soc; Am Soc Zool; Soc Develop Biol. Res: Effects of radiation on development in insects. Mailing Add: Dept of Biol Southwestern at Memphis Memphis TN 38112

ANACKER, ROBERT LEROY, microbiology, see 12th edition

ANAGNOSTOPOULOS, CONSTANTINE E, b Greece, Nov 1, 22; nat US; m 49; c 1. ORGANIC CHEMISTRY, POLYMER CHEMISTRY. Educ: Brown Univ, ScB, 49; Harvard Univ, MA, 50, PhD(chem), 52. Prof Exp: Res chemist, Monsanto Chem Co, 52-57, scientist, 57-61, from asst dir to dir res, Monsanto Co, 61-68, dir res, Org Chem Div, 68-69, Rubber Chem, 69-70 & Tire Textiles Div, 70-71, gen mgr, New Enterprise Div, 71-75, GEN MGR, RUBBER CHEM DIV, MONSANTO CO, 75- Concurrent Pos: Mem nat inventors coun, Brit Intel Serv, 41-43. Mem: Am Chem Soc; Sigma Xi (vpres, 59); Soc Chem Indust. Res: Synthesis of steroids, derivatives, amino acids and analogs; plasticization of vinyl polymers; polymer-diluent interactions; photocatalyzed oxidation of polyalkylenes. Mailing Add: 12101 Point Oak Rd St Louis MO 63131

ANAGNOSTOPOULOS, CONSTANTINE N, b Patras, Greece, July 4, 44; US citizen; m 69; c 2. SOLID STATE ELECTRONICS. Educ: Merrimack Col, BS, 67; Brown Univ, ScM, 71; Univ RI, PhD(elec eng), 75. Prof Exp: RES PHYSICIST, EASTMAN KODAK CO RES LABS, 75- Mem: Inst Elec & Electronics Engrs. Res: Solid state devices with application to imaging and signal processing such as charge coupled and charge injection devices. Mailing Add: Eastman Kodak Co Res Labs Physics Div 1669 Lake Ave Rochester NY 14650

ANAND, AMARJIT S, b New Delhi, India, June 1, 40; US citizen; m 68. REPRODUCTIVE PHYSIOLOGY, BIOCHEMISTRY. Educ: Univ Panjab, India, DVM, 62; Univ Wis-Madison, MS, 65, PhD(physiol), 67. Prof Exp: Vet, Vet Med Sch, Univ Panjab, India, 58-62; res asst physiol & biochem, Univ Wis-Madison, 62-67; from asst prof to assoc prof physiol, 67-74, ASSOC PROF BIOL, UNIV WIS-OSHKOSH, 74- Concurrent Pos: Co-recipient, Wis State Univ Bd Regents res grant, 68-69; NSF instnl res grant, 70-71. Mem: Am Soc Animal Sci; Soc Study Reproduction. Res: Veterinary medicine. Mailing Add: Dept of Biol Univ of Wis 936 Whittier Dr Appleton WI 54911

ANAND, SATISH CHANDRA, b India, Aug 9, 30; m 50; c 3. PLANT BREEDING. Educ: Univ Delhi, BS, 51; NC State Univ, MAgr, 59; Univ Wis, PhD(agron), 62. Prof Exp: Res assoc, Univ Ill, Urbana, 62-63; wheat breeder, Punjab Agr Univ, India, 63-71, sr geneticist, 72-73; SOYBEAN BREEDER, McNAIR SEED CO, 73- Concurrent Pos: Mem cereals & pulses subcomt, Indian Stand Inst, Indian Stand Inst, 64-72; zonal coordr wheat, Indian Coun Agr Res, 68-72. Mem: Am Soc Agron; fel Indian Soc Genetics & Plant Breeding. Res: Breeding and development of high yielding soybean varieties resistant to nematodes and phytophthora root-rot, especially improving the plant type in soybeans. Mailing Add: McNair Seed Co PO Box 706 Laurinburg NC 28352

ANANTHA NARAYANAN, VENKATARAMAN, b Madras City, India, Oct 22, 36. PHYSICS, SPECTROSCOPY. Educ: Annamalai Univ, Madras, BSc, 57, MSc, 58; Indian Inst Sci, Bangalore, PhD(physics, spectros), 62. Prof Exp: Vis fel, Mellon Inst, 63-64; Robert A Welch Found fel chem, Tex A&M Univ, 64-65; PROF PHYSICS & THEORET VIBRATIONAL SPECTRAL STUDIES, SAVANNAH STATE COL, 65- Concurrent Pos: Coop res partic, Oak Ridge Nat Lab, 70. Honors & Awards: Hon Chem Abstractor, Am Chem Soc, 69- Mem: Optical Soc Am; Am Asn Physics Teachers; fel Brit Inst Physics; Am Asn Univ Prof. Res: Vibrational spectra of crystals and hydrogen bonded systems; far infrared and Raman techniques; use of spectral data for evaluation of force constants and other chemical bond properties; vibrational spectra; molecular constants; crystal fields; computer oriented physics teaching. Mailing Add: Dept of Math & Physics PO Box 20473 Savannah State Col Savannah GA 31404

ANAST, CONSTANTINE SPIRO, b Chicago, Ill, Mar 23, 24; m 46; c 4. PEDIATRICS. Educ: Univ Chicago, MD, 47; Am Bd Pediat, dipl. Prof Exp: Intern, State Univ NY, 48-49, asst resident pediat, 49-50, resident, 50-51, instr, 51-52 & 54-56; from instr to assoc prof, 56-67, PROF PEDIAT, SCH MED, UNIV MO-COLUMBIA, 67- Mem: AMA. Res: Calcium and magnesium metabolism; parathyroid hormone and calcitonin. Mailing Add: Dept of Pediat Univ of Mo Sch of Med Columbia MO 65201

ANASTASIO, ANGELO, b Waterbury, Conn, Oct 29, 14; m 48; c 3. ANTHROPOLOGY. Educ: Julliard Sch, cert music, 41; Boston Univ, AA, 49; Univ Chicago, MA, 52, PhD(anthrop), 55. Prof Exp: Instr anthrop, Reed Col, 54-55; from asst prof to assoc prof anthrop & sociol, 55-68, actg chmn dept, 59-60, actg chmn, 65-66, PROF ANTHROP, WESTERN WASH STATE COL, 68- Concurrent Pos: Res assoc anthrop, Wash State Univ, 58-59; Western Wash State Col grant, 68-; Sigma Xi grant, 70- Mem: Fel Am Anthrop Asn; Am Ethnol Soc. Res: Area studies; science of wind instrumental music production; history of anthropology. Mailing Add: Dept of Sociol & Anthrop Western Wash State Col Bellingham WA 98225

ANASTASIO, SALVATORE, b Brooklyn, NY, Mar 12, 32; m 57; c 2. MATHEMATICS. Educ: Cathedral Col, AB, 54; NY Univ, MS, 61, PhD(math), 64. Prof Exp: Instr math, Iona Col, 59-61; asst prof, Fordham Univ, 64-70; assoc prof, 70-72, PROF MATH, STATE UNIV NY COL NEW PALTZ, 72- Mem: Am Math Soc; Math Asn Am. Res: Hilbert space; von Neumann algebras. Mailing Add: Dept of Math State Univ of NY Col New Paltz NY 12561

ANASTASIOU, CLIFFORD J, b Vancouver, BC, Feb 24, 29; m 52; c 3. MYCOLOGY. Educ: Univ BC, BA, 52, MEd, 57; Claremont Grad Sch, PhD(mycol), 63. Prof Exp: Res biologist, BC Res Coun, 53-54; teacher, High Sch, BC, 55-57 & Calif, 57-61; pres res comt grants, 62-71, assoc prof educ & bot, 69-73, RES ASSOC BOT, UNIV BC, 72-, PROF SCI EDUC, 73- Concurrent Pos: Nat Res Coun Can oper grants, 64-70; res fel mycol, Harvard Univ, 64-65; staff mycologist, Educ Serv Inc, 64-65; dir, Western Educ Develop Group, 72- Mem: Mycol Soc Am; Am Bot Soc; Pac Sci Asn. Res: Study of taxonomy, ecology and physiology of fungi occurring in marine and salt lake locations. Mailing Add: Fac of Educ Univ of BC Vancouver BC Can

ANASTASSAKIS, EVANGELOS M, b Iraklion, Greece, Feb 12, 38. SOLID STATE PHYSICS, SPECTROSCOPY. Educ: Nat Univ Athens, BS, 61; Univ Pa, MS, 65, PhD(physics), 68. Prof Exp: ASST PROF PHYSICS, NORTHEASTERN UNIV, 69- Concurrent Pos: Res asst, Greek Atomic Energy Comn, 61-63; res assoc, Univ Pa, 69. Mem: Am Phys Soc. Res: Morphic effects; effects of electric fields, stresses on infrared absorption and Raman scattering by optical phonons in crystals; group theoretical treatment of phonon Raman activity in magnetic crystals; resonance Raman scattering in II-IV semiconductors. Mailing Add: Dept of Physics Northeastern Univ Boston MA 02115

ANASTASSIADIS, PHOEBUS A, b Athens, Greece, Jan 1, 11; Can citizen; m 35; c 1. AGRICULTURAL CHEMISTRY. Educ: Col of Agr, Greece, BAgr, 31; Rutgers Univ, MEd, 47; McGill Univ, PhD(agr chem), 54. Prof Exp: From asst prof to assoc prof, 53-73, PROF AGR CHEM, MacDONALD COL, MCGILL UNIV, 73- Mem: Can Biochem Soc. Res: Mucopolysaccharides and glycoproteins in animal tissues. Mailing Add: Dept of Agr Chem Box 131 MacDonald Col of McGill Univ Ste Anne de Bellevue PQ Can

ANASTOS, GEORGE, b Akron, Ohio, Jan 9, 20; m 46; c 2. ACAROLOGY. Educ: Univ Akron, BS, 42; Harvard Univ, AM, 47, PhD(biol), 49. Prof Exp: Asst prof zool, Miami Univ, 49-51; assoc prof, 51-58, asst dir inst acarol, 58-61, head dept zool, 61-

68, PROF ZOOL, UNIV MD, 58- Concurrent Pos: Consult, Dept Defense, 57-, NSF, 61- & WHO, 67-; Guggenheim fel, 58; lectr, Ohio State Univ, 62; ed, J Econ Entom, 69-; dir regional ref ctr, WHO, 70-; secy-gen, XV Int Cong Entom, 75-76. Mem: Entom Soc Am. Mailing Add: Dept of Zool Univ of Md College Park MD 20742

ANBAR, MICHAEL, b Danzig, June 29, 27; m 53; c 2. PHYSICAL INORGANIC CHEMISTRY, MASS SPECTROMETRY. Educ: Hebrew Univ, Jerusalem, MSc, 50, PhD(phys org chem), 53. Prof Exp: Instr chem, Univ Chicago, 53-55; sr scientist, Weizmann Inst Sci, 55-60, prof, Frienberg Grad Sch, 60-67; sr res assoc exobiol, Ames Res Ctr, 67-68; dir phys sci, 68-72, DIR MASS SPECTROMETRY RES CTR, STANFORD RES INST, 72- Concurrent Pos: Dir, Radioisotope Training Ctr, Weizmann Inst Sci, 56-59 & Radiation Res Dept, AEC Soreq Nuclear Res Ctr, Israel, 59-66; head chem div, Israel AEC Res Lab, 62-66; sr res assoc, Argonne Nat Lab, 63-64; prof inorg chem, Univ Tel Aviv, 66-67. Honors & Awards: Zondek Award, Israel Med Asn, 62. Mem: Am Chem Soc; The Chem Soc; Am Soc Mass Spectrom; NY Acad Sci. Res: Mechanisms of organic and inorganic reactions; isotope methodology in biology and medicine; radiation chemistry and molecular radiobiology; hydrated electron chemistry; mass spectrometry, field ionization mass spectrometry in particular; molten salt electrochemistry. Mailing Add: Phys Sci Stanford Res Inst 333 Ravenswood Ave Menlo Park CA 94025

ANCHEL, MARJORIE WOLFF (MRS HERBERT RACKOW), b New York, NY, May 6, 10; m 42. BIOCHEMISTRY. Educ: Columbia Univ, BA, 31, MA, 33, PhD(biol chem), 39. Prof Exp: Fel, Col Physicians & Surg, Columbia Univ, 37; asst chem, Queens Col, NY, 39-41 & Columbia Univ, 41-43; res assoc, Squibb Inst Med Res, 43-46; from res assoc to sr res assoc, 46-74, SR CHEMIST & ADMINR LAB, NY BOT GARDEN, 74- Mem: Fel AAAS; Am Chem Soc; Mycol Soc Am; fel NY Acad Sci; Am Soc Biol Chemists. Res: Antibiotics; natural polyacetylenes; chemistry and biogenesis of fungal products. Mailing Add: NY Bot Garden Bronx NY 10458

ANCKER-JOHNSON, BETSY, b St Louis, Mo, Apr 27, 29; m 58; c 4. PHYSICS. Educ: Wellesley Col, BA, 49; Tübingen Univ, PhD(physics), 53. Prof Exp: Jr res physicist & lectr physics, Univ Calif, Berkeley, 53-54; mem staff, Inter-Varsity Christian Fel, 54-56; sr res physicist, Microwave Physics Lab, Sylvania Elec Prods, Inc, 56-58; mem tech staff, David Sarnoff Res Ctr, Radio Corp Am, 58-61; res specialist, Electronic Sci Lab, Boeing Sci Res Labs, 61-70, supvr solid state & plasma electronics, 70-71, mgr advan energy syst, Boeing Aerospace Co, 71-73; ASST SECY COM FOR SCI & TECHNOL, US DEPT COM, 73- Concurrent Pos: From assoc prof to prof elec eng, Univ Wash, 64-73. Mem: Nat Acad Eng; fel AAAS; fel Am Phys Soc; fel Inst Elec & Electonic Engr. Res: Solid state physics; plasma in solids; microwaves and molecular electronics; ferrimagnetism and nonreciprocal effects; x-ray studies of imperfections in nearly perfect crystals. Mailing Add: US Dept of Com Washington DC 20230

ANCONA, UMBERTO, b Milano, Italy, July 27, 21; US citizen; m 49; c 4. INDUSTRIAL CHEMISTRY, CHEMICAL ENGINEERING. Educ: Univ Milan, DSc(indust chem), 46; Columbia Univ, MS, 49. Prof Exp: Asst to tech dir, 50-53, asst chief chemist, 53-65, chief chemist, 65-73, tech dir, 73, VPRES, McCLOSKEY VARNISH CO, 74- Honors & Awards: Liberty Bell Award, 70. Mem: Am Chem Soc; Am Inst Chem Eng; Fedn Socs Paint Technol. Res: Polymers for coatings. Mailing Add: McCloskey Varnish Co 7600 State Rd Philadelphia PA 19136

ANDALAFTE, EDWARD ZIEGLER, b Springfield, Mo, Aug 7, 35. MATHEMATICS. Educ: Southwest Mo State Col, BS, 56; Univ Mo-Columbia, AM, 59, PhD(math), 61. Prof Exp: Assoc prof math, Southwest Mo State Col, 61-64; chmn dept, 65-67 & 68-69, ASSOC PROF MATH, UNIV MO-ST LOUIS, 64- Concurrent Pos: NSF fac fel, 70-71; vis assoc prof, Math Res Ctr, Univ Wis-Madison, 70-71. Mem: Am Math Soc; Math Asn Am; Asn Symbolic Logic. Res: Geometry of metric spaces and their generalizations; metric characterizations of Euclidean spaces. Mailing Add: Dept of Math Univ of Mo St Louis MO 63121

ANDEEN, CARL GUSTAV, b East Cleveland, Ohio, Oct 22, 37. SOLID STATE PHYSICS. Educ: Case Inst Technol, BS, 58, Western Reserve Univ, PhD(physics), 71. Prof Exp: Res asst chem physics, Union Carbide Corp, 58-64; fel physics, 64-71, RES ASSOC, CASE WESTERN RESERVE UNIV, 71- Res: Dielectric properties of solids. Mailing Add: Dept of Physics Case Western Reserve Univ Cleveland OH 44106

ANDELIN, ROBERT L, physical chemistry, radiochemistry, see 12th edition

ANDELMAN, JULIAN BARRY, b Boston, Mass, Sept 23, 31; m 53; c 3. WATER CHEMISTRY, ENVIRONMENTAL CHEMISTRY. Educ: Harvard Univ, AB, 52; Polytech Inst Brooklyn, PhD(chem), 60. Prof Exp: Res fel chem, NY Univ, 59-61; mem tech staff, Bell Tel Labs, Inc, 61-63; from asst prof to assoc prof water chem, 63-73, PROF WATER CHEM, GRAD SCH PUB HEALTH, UNIV PITTSBURGH, 73- Concurrent Pos: Consult, WHO, 69, 71 & 74 & Nat Acad Sci-Nat Res Coun, 71-75; vis lectr, Univ Col, London, 70-71. Mem: AAAS; Water Pollution Control Fedn; Am Water Works Asn; Am Chem Soc; Int Asn Water Pollution Res. Mailing Add: Grad Sch of Pub Health Univ of Pittsburgh Pittsburgh PA 15261

ANDELSON, JONATHAN GARY, b Chicago, Ill, Feb 10, 49. ETHNOLOGY, PHYSICAL ANTHROPOLOGY. Educ: Grinnell Col, BA, 70; Univ Mich, Ann Arbor, MA, 73, PhD(anthrop), 74. Prof Exp: ASST PROF ANTHROP, GRINNELL COL, 74- Mem: Am Anthrop Asn; Soc Med Anthrop. Res: Small scale communal societies of the nineteenth and twentieth centuries. Mailing Add: Dept of Anthrop Grinnell Col Grinnell IA 50112

ANDER, PAUL, b Brooklyn, NY, Apr 19, 31; m 56; c 2. PHYSICAL CHEMISTRY. Educ: City Col New York, BS, 53; Polytech Inst Brooklyn, MS, 54; Rutgers Univ, PhD(chem), 61. Prof Exp: Instr chem, Rutgers Univ, 60-61; from asst prof to assoc prof, 61-70, PROF CHEM, SETON HALL UNIV, 70- Mem: Am Chem Soc. Res: Physical chemistry of macromolecules; solution properties of synthetic and biological polyelectrolytes. Mailing Add: Dept of Chem Seton Hall Univ South Orange NJ 07079

ANDEREGG, ARNOLD HENRY, organic chemistry, see 12th edition

ANDEREGG, DOYLE EDWARD, b Uhrichsville, Ohio, Jan 19, 30; m 57; c 2. LICHENOLOGY. Educ: Ohio State Univ, BSc, 52, MSc, 57, PhD(bot, plant path), 59. Prof Exp: Asst bot, Ohio State Univ, 55-59; from instr bot & microbiol to assoc prof, Univ Okla, 59-67, chmn dept, 63-67; head dept biol sci, 67-75, asst grad dean, 69-70, PROF BIOL SCI, UNIV IDAHO, 67- Mem: AAAS; Mycol Soc Am; Am Inst Biol Sci; Bot Soc Am; Am Bryol & Lichenological Soc. Res: Biogeography, taxonomy and ecology of lichens. Mailing Add: Dept of Biol Sci Univ of Idaho Moscow ID 83843

ANDEREGG, JOHN WILLIAM, b White Lake, Wis, Nov 12, 23. BIOPHYSICS. Educ: Univ Wis, BS, 47, MS, 49, PhD(physics), 52. Prof Exp: Asst physics, 46-52, fel & instr, 53-56, proj assoc oncol, 56-57, asst prof zool & physics, 57-62, assoc prof physics, 62-69, PROF PHYSICS & BIOPHYS, UNIV WIS-MADISON, 69- Concurrent Pos: Fulbright scholar, Royal Inst, London, 52-53. Mem: Am Phys Soc; Biophys Soc; Am Crystallog Asn. Res: X-ray scattering; biomolecular structure. Mailing Add: Biophysics Lab Univ of Wis Madison WI 53706

ANDERLE, RICHARD, b New York, NY, Oct 8, 26; m 60. MATHEMATICS. Educ: Brooklyn Col, BA, 48. Prof Exp: Mathematician, Exterior Ballistics Br, 48-60, HEAD ASTRONAUT DIV, NAVAL WEAPONS LAB, 60- Honors & Awards: Superior Civilian Award, US Dept Navy, 60. Mem: AAAS; Am Geophys Union; Am Inst Aeronaut & Astronaut. Res: Geodesy; celestial mechanics. Mailing Add: 1320 Parcell St Fredericksburg VA 22401

ANDERMANN, GEORGE, b Szarvas, Hungary, Oct 14, 24; US citizen; m 50; c 1. SPECTROSCOPY. Educ: Univ Calif, Los Angeles, BS, 49; Univ Southern Calif, MS, 61, PhD(chem), 65. Prof Exp: Chemist, Southern Pac Co, 49-51; spectroscopist, Appl Res Labs, 51-53, lab supvr, 53-60, sr physicist, 60-61; sr chemist, Austin Robinson Lab, 61-62; instrument consult mat testing div, Magnaflux Corp, 62-63; res asst chem, Univ Southern Calif, 63-65, res assoc mat sci, 65; from asst prof to assoc prof, 65-74, PROF CHEM, UNIV HAWAII, MANOA, 74- Concurrent Pos: Assoc chemist, Hawaii Inst Geophys; NSF grant, 66-71. Mem: Am Chem Soc; Soc Appl Spectros. Res: Optical properties of solids; x-ray emission spectroscopy, fundamental and analytical; infrared spectroscopy; lattice dynamics, particularly electrical anharmonicity. Mailing Add: Dept of Chem Univ of Hawaii Honolulu HI 96822

ANDERS, EDWARD, b Latvia, June 21, 26; nat US; m 55; c 2. COSMOCHEMISTRY, RADIOCHEMISTRY. Educ: Columbia Univ, AM, 51, PhD(chem), 54. Prof Exp: Instr chem, Univ Ill, 54-55; from asst prof to prof, 55-73, HORACE B HORTON PROF CHEM, UNIV CHICAGO, 73- Concurrent Pos: Vis prof, Calif Inst Technol, 60; Nat Acad Sci resident res assoc, Goddard Space Flight Ctr, 61; consult, NASA, 61-69 & 73; NSF sr fel & vis prof, Univ Berne, 63-64, vis prof, 70; res assoc, Field Mus Natural Hist, 69; chmn comt meteorites, Int Astron Union, 70-73; Guggenheim fel, 73-74. Honors & Awards: Newcomb Cleveland Prize, AAAS, 59; Univ Medal for Excellence, Columbia Univ, 66; J Lawrence Smith Medal, Nat Acad Sci, 71; Medal Exceptional Sci Achievement, NASA, 73; Leonard Medal, Meteoritical Soc, 74. Mem: Nat Acad Sci; fel Am Acad Arts & Sci; assoc Royal Astron Soc; Am Chem Soc; fel Meteoritical Soc (vpres, 70-72). Res: Origin and age of meteorites; composition and origin of moon and planets. Mailing Add: Enrico Fermi Inst Univ of Chicago 5640 SEllis Ave Chicago IL 60637

ANDERS, EDWARD B, b Chatham, La, Jan 7, 30; m 63; c 2. MATHEMATICAL ANALYSIS. Educ: La Polytech Univ, BS, 50; Pa State Univ, BS, 54; Northwestern State Univ, MS & ME, 58; Auburn Univ, PhD(math), 65. Prof Exp: High sch teacher, La, 50; instr, Northeastern La State Col, 58-60, from asst prof to assoc prof, 63-66; assoc prof & chmn div math & physics, Ark State Univ, 66-67; assoc prof, 67-74, PROF MATH, NORTHWESTERN STATE UNIV, 74- Mem: Am Math Soc; Math Asn Am; Asn Comput Mach. Res: Numerical and error analysis; numerical filter theory; least squares curve fitting and numerical integration theory. Mailing Add: Dept of Math Northwestern State Univ Natchitoches LA 71457

ANDERS, MARION WALTER, b Alden, Iowa, May 10, 36; m 58; c 3. TOXICOLOGY. Educ: Iowa State Univ, DVM, 60; Univ Minn, PhD(pharmacol), 64. Prof Exp: Instr pharmacol, Univ Minn, 64-65; from asst prof to assoc prof, Cornell Univ, 65-69; assoc prof, 69-75, PROF PHARMACOL, UNIV MINN, MINNEAPOLIS, 75- Concurrent Pos: Mem ad hoc sci adv comt, Food & Drug Admin, 70-71; mem toxicol study sect, NIH; USP comt of revision, 75- Mem: AAAS; Soc Toxicol; Am Soc Biol Chem; Am Soc Pharmacol & Exp Therapeut. Res: Relationship of metabolism to toxicity; analytical chemistry. Mailing Add: Dept of Pharmacol Univ of Minn Minneapolis MN 55455

ANDERS, OSWALD ULRICH, b Karlsruhe, Ger, Nov 10, 28; nat US; m 53; c 3. RADIOCHEMISTRY. Educ: Georgetown Univ, BS, 52; Univ Mich, MS, 54, PhD(chem), 57. Prof Exp: Asst, Univ Mich, 54-57; radiochemist, 57-63, assoc scientist, Radiochem Res Lab, 63-67, REACTOR SUPVR, TRIGA REACTOR, RADIOCHEM RES LAB, DOW CHEM CO, 67- Mem: Am Inst Chem; Sigma Xi; Am Nuclear Soc (treas, 75). Res: Activation analysis; particle accelerators; nuclear reactor; nuclear instrumentation; nuclear reactions; ion exchange separations; tracer chemistry; decontamination of nuclear power plants. Mailing Add: 801 Linwood Dr Midland MI 48640

ANDERSEN, ALLEN CLARENCE, b Oakland, Calif, Apr 12, 17; m 47; c 2. VETERINARY MEDICINE. Educ: Univ Calif, BS, 41, MS, 43, PhD(comp path), 54; Univ Pa, VMD, 46. Prof Exp: Pvt vet pract, 47-51; prin investr & radiopathologist, AEC Proj 4 & 6, 51-64, specialist, 65-75, SPECIALIST, SCH VET MED, UNIV CALIF, DAVIS, 75- Res: Histology; reproduction; effects of irradiation in the dog. Mailing Add: Radiobiol Lab Univ of Calif Davis CA 95616

ANDERSEN, AXEL LANGVAD, b Askov, Minn, Sept 24, 14; m 42; c 2. PLANT PATHOLOGY. Educ: Univ Minn, BS, 37; Mich State Univ, MS, 41, PhD(plant path), 47. Prof Exp: Forest guard, Stanislaus Nat Forest, Calif, 37; asst plant pathologist, Exp Sta, Mich State Univ, 39-42; plant pathologist, Camp Detrick, Md, 46-48; res plant pathologist, Crops Res Div, Agr Res Serv, USDA, 49-65, res coordr res prog develop & eval staff, Off of Dir Sci & Educ, 65-68; assoc prof bot & plant path, 51-64, head exten plant path, 68-74, PROF BOT & PLANT PATH, MICH STATE UNIV, 64-, ASST TO DIR, AGR EXP STA, 71- Concurrent Pos: Mem, Nat Steering Comt Plant Dis Mgt, 74-75. Honors & Awards: Recognition Award, Mich Bean Prod, 64; Mich Bean Shippers, 71. Mem: AAAS; Am Phytopath Soc; Am Inst Biol Sci. Res: Plant diseases; bean disease and breeding investigations; turfgrass diseases; disease management, detection, development, loss appraisal and control strategies; epidemiology, air pollution injury, remote sensing. Mailing Add: Dept of Bot & Plant Path Mich State Univ East Lansing MI 48823

ANDERSEN, CARL MARIUS, b Detroit, Mich, Mar 10, 36; m 59; c 3. THEORETICAL PHYSICS. Educ: Univ Mich, BS, 58, MS, 59; Univ Pa, PhD(physics), 64. Prof Exp: Res assoc theoret physics, Purdue Univ, 64-65, asst prof, 65-67; asst prof theoret physics, 67-72, SR RES ASSOC MATH & COMPUT SCI, COL WILLIAM & MARY, 72- Mem: Am Phys Soc; Asn Comput Mach. Res: High energy physics; theory of group representations; classical relativistic dynamics; algebraic manipulation by computer; applied mathematics. Mailing Add: Dept of Math Col of William & Mary Williamsburg VA 23185

ANDERSEN, DEAN MARTIN, b Monroe, Utah, Dec 14, 31; m 61; c 4. MEDICAL ENTOMOLOGY, ENVIRONMENTAL BIOLOGY. Educ: Univ Utah, BS, 60, MS, 62, PhD(entom), 66. Prof Exp: Inspector entomologist, Salt Lake City Mosquito Abatement, 58-62; dir field res, Inst Environ Res, Univ Utah, 63-66; assoc prof, 66-74, PROF BIOL, BRIGHAM YOUNG UNIV, HAWAII CAMPUS, 74- Concurrent Pos: Vis prof, Brigham Young Univ, 73-74. Mem: Am Mosquito Control Asn. Res: Ecological control of mosquitoes through water management; host preferences of mosquitoes; audio-tutorial systems of teaching biological sciences and individualized approaches to learning for both science and nonscience majors. Mailing Add: Div of Math & Natural Sci Brigham Young Univ Laie HI 96762

ANDERSEN, DONALD EDWARD, b New Haven, Conn, June 3, 23; m 47; c 1. PHYSICAL CHEMISTRY, RESEARCH ADMINISTRATION. Educ: Brown Univ, ScB, 48, PhD(phys chem), 52; Stanford Univ, MS, 49. Prof Exp: Res chemist, 52-56, res supvr, 56-60, supt develop, 60-61, gen supt res & develop, 61-63, asst plant mgr, 63-65, plant mgr, 65-70, LAB DIR & MGR ELASTOMERS RES & DEVELOP, E I DU PONT DE NEMOURS & CO, INC, 70- Mem: AAAS; Am Chem Soc. Mailing Add: 353-204 DuPont Exp Sta Wilmington DE 19898

ANDERSEN, EMIL THORVALD, b Alta, Can, Jan 25, 17; m 43; c 3. HORTICULTURE, POMOLOGY. Educ: Univ Alta, BSc, 41, MSc, 43; Univ Minn, PhD(hort), 62. Prof Exp: Res off hort, Can Exp Farm, Alta, 43-44; from asst prof to assoc prof hort, Univ Man, 44-56; from instr to assoc prof pomol, Univ Minn, St Paul, 57-67; CHIEF PROD RES, HORT RES INST ONT, 67- Mem: Am Soc Hort Sci; Am Pomol Soc; Can Soc Hort Sci; Int Soc Hort Sci. Res: Orchard management systems studies with apples; rootstock problems of apples; iron uptake problems of fruits and plants. Mailing Add: Hort Res Inst of Ont Vineland Station ON Can

ANDERSEN, FERRON LEE, b Howell, Utah, July 10, 31; m 58; c 6. PARASITOLOGY. Educ: Utah State Univ, BS, 57, MS, 60, PhD(zool), 63; Univ Ill, MS, 62. Prof Exp: Asst prof vet med sci & assoc staff mem, Zoonoses Res Ctr, Univ Ill, 63-67; assoc prof zool, 67-72, PROF ZOOL, BRIGHAM YOUNG UNIV, 72- Mem: Am Soc Parasitol. Res: Coccidiosis; micrometeorology and ecology of parasites; hydatid disease. Mailing Add: Dept of Zool Brigham Young Univ Provo UT 84602

ANDERSEN, FRANK ALAN, b Lewiston, Maine, May 20, 44; m 75. RADIATION BIOPHYSICS. Educ: Muhlenberg Col, BS, 66; Pa State Univ, MS, 69, PhD(biophys), 72. Prof Exp: SUPVRY RES PHYSICIST, GENETICS STUDIES SECT, DIV BIOL EFFECTS, BUR RADIOL HEALTH, FOOD & DRUG ADMIN, 71- Concurrent Pos: Mem interagency panel environ mutagenesis, Dept Health, Educ & Welfare, 74-; liaison mem comt safe use of lasers, Am Nat Stand Inst, 74- Mem: AAAS; Radiation Res Soc; Am Soc Photobiol. Res: Biological effects of radiation, including acoustic energy with special reference to emissions from electronic products; development of criteria, based on such effects, in support of performance standards to control radiation emissions. Mailing Add: Food & Drug Admin BRH HFX-120 5600 Fishers Lane Rockville MD 20852

ANDERSEN, HANS CHRISTIAN, b Brooklyn, NY, Sept 25, 41; m 67. PHYSICAL CHEMISTRY. Educ: Mass Inst Technol, BS, 62, PhD(phys chem), 66. Prof Exp: Jr fel, soc fels, Harvard Univ, 65-68; asst prof, 68-74, ASSOC PROF CHEM, STANFORD UNIV, 74- Mem: Am Phys Soc; AAAS. Res: Statistical mechanics; transport processes; structure of fluids; scattering of light; physical properties of membranes. Mailing Add: Dept of Chem Stanford Univ Stanford CA 94305

ANDERSEN, HAROLD VERAL, b Cumberland, Iowa, Nov 12, 07; m 38. GEOLOGY. Educ: Univ Nebr, BA, 40; La State Univ, PhD(geol), 50. Prof Exp: Dir statewide archaeol invest, State Geol Surv, Ala, 40-42; from instr to asst prof, 47-53, assoc prof & cur dept mus, 56-61, PROF GEOL, LA STATE UNIV & PROF GEOL DEPT MUS, 61- Mem: Am Asn Petrol Geol; Am Paleont Soc; Soc Econ Paleont & Mineral. Res: Small foraminifera, recent and Tertiary; stratigraphy; general geology; micropaleontology. Mailing Add: Dept of Geol La State Univ Baton Rouge LA 70803

ANDERSEN, JON, physiology, biochemistry, see 12th edition

ANDERSEN, KENNETH J, b Brooklyn, NY, Dec 24, 36; m 62; c 2. RESEARCH ADMINISTRATION, MICROBIOLOGY. Educ: Davis & Elkins Col, BS, 59; Syracuse Univ, MS, 62, PhD(microbiol), 65. Prof Exp: Bacteriologist, Syracuse Univ Res Corp, 60-62, asst microbiol, Syracuse Univ, 62-65; res microbiologist, Battelle Mem Inst, 65-67, sr microbiologist, 67-70, assoc chief microbiol & environ biol div, 70-72; assoc dir diag serv, Johnson & Johnson Int, 72-73; dir chemother, 73-75, DIR BIOL RES, NORWICH PHARMACAL CO, 75- Mem: Fel AAAS; Am Soc Microbiol; Am Inst Biol Sci. Res: Virology; bacteriology; biochemical engineering; fermentation design and technology; microbial genetics and physiology; development of new drugs for human diseases. Mailing Add: Norwich Pharmacal Co Res & Develop PO Box 191 Norwich NY 13815

ANDERSEN, KENNETH K, b Perth Amboy, NJ, May 13, 34; m 57; c 3. ORGANIC CHEMISTRY. Educ: Rutgers Univ, BS, 55; Univ Minn, PhD(chem), 59. Prof Exp: USPHS fel, 59-60; from asst prof to assoc prof chem, 60-70, PROF CHEM, UNIV NH, 70- Concurrent Pos: NSF sci fac fel, Univ EAnglia, 66-67; Fulbright lectr, Tech Univ Denmark, 71; vis lectr, Polish Acad Sci, 73. Mem: Am Chem Soc. Res: Chemistry of organosulfur compounds with emphasis on stereochemistry. Mailing Add: Dept of Chemistry Univ of NH Durham NH 03824

ANDERSEN, NEIL RICHARD, b Lynn, Mass, Sept 22, 35; m 56; c 2. CHEMICAL OCEANOGRAPHY. Educ: Clark Univ, AB, 60; Mass Inst Technol, PhD(anal chem), 65. Prof Exp: Res asst oceanog, Woods Hole Oceanog Inst, 60-65; prin investr, US Naval Oceanog Off, 65-67, br head, 67-70; chief appl chem oceanog br, Off Res & Develop, US Coast Guard, 70-72; chem oceanogr, Off Naval Res, 72-73, prog dir, Chem Oceanog Prog, 73-75; PROG DIR, MARINE CHEM PROG, NSF, 75- Concurrent Pos: Fel, Woods Hole Oceanog Inst, 62-65; assoc prof, George Washington Univ, 68-; dep US deleg, GIPME Comt of Intergovt Oceanog Comn, UNESCO, 74- Mem: AAAS; Am Chem Soc; Marine Technol Soc; Am Geophys Union; NY Acad Sci. Res: Trace metal and radioisotope distributions in the oceans; chemical, geological and biological mechanisms causing variations such that measured distributions can aid in solving oceanographic problems and in identifying and regulating marine pollution. Mailing Add: Washington DC

ANDERSEN, NIELS HJORTH, b Copenhagen, Denmark, Oct 9, 43; US citizen; m 63; c 2. ORGANIC CHEMISTRY, BIOCHEMISTRY. Educ: Univ Minn, Minneapolis, BA, 63; Northwestern Univ, PhD(org chem), 67. Prof Exp: NIH res fel chem, Harvard Univ, 67-68; asst prof, Univ Wash, 68-70; prin scientist, Alza Corp, Calif, 70; asst prof chem, 70-72, ASSOC PROF CHEM, UNIV WASH, 70- Concurrent Pos: Consult, Worchester Found Exp Biol, 68-69 & Alza Corp, 70-75; Alfred P Sloan res fel, 72-74; Dreyfus Found scholar, 74-; NIH career develop award, 75- Mem: Am Chem Soc; The Chem Soc; Phytochem Soc; Intra-Sci Res Found; NY Acad Sci. Res: New synthetic methods; synthesis, structure elucidation and biogenesis of sesquiterpenes; circular dichroism of olefins; chemical models for physiological interactions; prostaglandin analogs, particularly synthesis and structure-activity correlations; molecular biochemistry. Mailing Add: Dept of Chem Univ of Wash Seattle WA 98105

ANDERSEN, PAUL GEORGE, b Brooklyn, NY, July 4, 49; m 72. POLYMER SCIENCE. Educ: Cornell Univ, BS, 71; Northwestern Univ, PhD(mat sci), 76. Prof Exp: RES CHEMIST POLYMER SCI, UNIROYAL CORP RES, UNIROYAL INC, 75- Honors & Awards: George L Kehl Award, Int Microstruct Anal Soc & Am Soc Metals, 73. Res: Effect of flow on crystallization in polyolefins and effect of composition on the relationship between flow properties and microstructure and the mechanical properties of thermoplastic elastomers. Mailing Add: Uniroyal Inc Corp Res R-1-2 Oxford Mgt & Res Ctr Middlebury CT 06749

ANDERSEN, RICHARD NICOLAJ, b Oakland, Calif, Mar 11, 30; m 58; c 3. REPRODUCTIVE ENDOCRINOLOGY. Educ: Abilene Christian Col, BS, 56; Baylor Univ, PhD(biochem), 60. Prof Exp: Asst res prof surg, Med Col Va, 63; res assoc biochem, Mayo Found, 63-66; asst prof pharmacol, Univ NC, Chapel Hill, 66, asst prof obstet & gynec, 67-69; ASSOC PROF BIOCHEM & OBSTET & GYNEC, UNIV TENN, MEMPHIS, 70- Concurrent Pos: USPHS fels, Univ Calif, Berkeley, 60-61 & Med Col Va, 61-63. Mem: AAAS; Soc Study Reproduction; NY Acad Sci; Endocrine Soc. Res: Control of gonadal function; steroid radioimmunoassay; hyperandrogenism; estrogen binding and mechanism of action. Mailing Add: Div of Reproduction Med Univ of Tenn Memphis TN 38163

ANDERSEN, ROBERT NEILS, b Steele City, Nebr, June 8, 28; m 52; c 2. PLANT PHYSIOLOGY. Educ: Univ Nebr, BS, 51, MS, 53; Univ Minn, PhD, 60. Prof Exp: Asst agronomist, Weed Invest Sect, Field Crops Res Br, Agr Res Serv, USDA, Agr Exp Sta, NDak, 53-57; res agronomist, 57-61, PLANT PHYSIOLOGIST, AGR RES SERV, USDA & UNIV MINN, ST PAUL, 61- Mem: Weed Sci Soc Am. Res: Weed control. Mailing Add: Dept of Agron & Plant Genetics Inst of Agr Univ of Minn St Paul MN 55101

ANDERSEN, ROGER ALLEN, b Milwaukee, Wis, June 24, 30; m 63; c 3. BIOCHEMISTRY, ONCOLOGY. Educ: Univ Wis, BS, 53, PhD(oncol), 63; Marquette Univ, MS, 58. Prof Exp: Anal chemist, Allis-Chalmers Mfg Co, Wis, 53-58; res asst oncol, McArdle Mem Labs, Univ Wis, 58-63; USPHS fel biochem, Roswell Park Mem Inst, NY, 63-65; RES CHEMIST, AGR RES SERV, USDA, 65-; ASSOC PROF AGRON, UNIV KY, 71- Concurrent Pos: Asst prof agron, Univ Ky, 65-71. Mem: Am Chem Soc. Res: Biochemistry of plant phenolics, chemical carcinogens, carbohydrates and related enzymes. Mailing Add: Tobacco & Health Res Inst Dept of Agron Univ of KY Lexington KY 40506

ANDERSEN, TERRELL NEILS, b Central, Idaho, Apr 22, 37; m 57; c 3. PHYSICAL CHEMISTRY. Educ: Univ Utah, BS, 58, PhD(phys chem), 62. Prof Exp: Fel & res assoc electrochem, Univ Utah, 62 & Univ Pa, 62-63; res assoc, Univ Utah, 63-64, asst res prof chem & metall, 64-69; SR SCIENTIST, KENNECOTT RES CTR, 69- Mem: Am Chem Soc; Sigma Xi; Am Inst Mining, Metall & Petrol Eng. Res: Electrochemistry; electrometallurgy; general analytical chemistry. Mailing Add: 1515 Mineral Sq Salt Lake City UT 84111

ANDERSEN, THORKILD WAINØ, b Copenhagen, Denmark, Nov 13, 20; m 40; c 4. ANESTHESIOLOGY. Educ: Univ Copenhagen, MD, 47. Prof Exp: Intern, Bispebjerg Hosp, Copenhagen, 48, resident, 49; PROF ANESTHESIOL, COL MED, UNIV FLA, 69- Concurrent Pos: Clin fel, Mass Gen Hosp, 51-52, sr fel, 52-56; NIH res grant, 68-72. Mem: AMA; Am Soc Anesthesiologists. Res: Drug interactions in anesthesia in man. Mailing Add: 777 NW 15th St Gainesville FL 32601

ANDERSEN, WILFORD HOYT, b Central, Idaho, Dec 15, 24; m 56; c 2. PHYSICAL CHEMISTRY. Educ: Univ Utah, BS, 49, PhD(chem), 52. Prof Exp: Res assoc chem, Univ Utah, 52-53 & Brown Univ, 53-55; tech specialist ordnance, Aerojet-Gen Corp, 55-64; prin res scientist, Heliodyne Corp, 64-65; MEM SR TECH STAFF, SHOCK HYDRODYNAMICS DIV, WHITTAKER CORP, 65- Mem: NY Acad Sci; AAAS; Combustion Inst. Res: Fundamental and applied research in shock, detonation and related ordnance phenomena; molecular theory of biological processes. Mailing Add: 1165 East Comstock Ave Glendora CA 91740

ANDERSEN, WILLIAM RALPH, b Rexburg, Idaho, Dec 11, 30; m 54; c 4. BIOCHEMICAL GENETICS. Educ: Utah State Univ, BS, 56, MS, 58; Univ Calif, Davis, PhD(genetics), 63. Prof Exp: Asst prof hort & genetics, Univ Minn, 63-66; PROF BOT, BRIGHAM YOUNG UNIV, 66- Concurrent Pos: Fel & grants, Univ Calif, 67-69. Mem: Am Soc Hort Sci; Genetics Soc Am. Res: Small fruit breeding and physiological genetics in garden tomato and in Solanum. Mailing Add: Dept of Bot Brigham Young Univ Provo UT 84601

ANDERSON, A KEITH, b Monticello, Calif, Mar 14, 24; m 52; c 1. MATHEMATICS. Educ: Pac Union Col, BA, 49; Loma Linda Univ, MD, 53; Colo State Univ, MA, 64, PhD(statist, math), 67. Prof Exp: Pvt pract, 55-60, 63-64; asst prof, 65-67, ASSOC PROF MATH, PAC UNION COL, 67- Mem: Math Asn Am. Res: Non-parametic statistics. Mailing Add: Dept of Math Pac Union Col Angwin CA 94508

ANDERSON, ADOLPH (GUSTOF), b Salem, SDak, Sept 7, 13; m 37; c 2. PHYSICAL CHEMISTRY, ACADEMIC ADMINISTRATION. Educ: Univ Pittsburgh, BS, 34, PhD(phys chem), 40. Prof Exp: Asst chem, Univ Pittsburgh, 34-37; instr, Geneva Col, 36-39, Univ Pittsburgh, 39-42 & City Col New York, 42-60; dean, New Col, Hofstra Univ, 60-69; PRES, HARTWICK COL, 69- Honors & Awards: Royal Order of the North Star, King of Sweden, 72. Mem: AAAS. Res: Heat capacities and entrophies of disaccharides; blood pH work. Mailing Add: Hartwick Col Oneonta NY 13820

ANDERSON, ALBERT DOUGLAS, b New York, NY, Jan 11, 28; m 73; c 2. PHYSICAL MEDICINE, REHABILITATION MEDICINE. Educ: Columbia Univ, BA, 48; Harvard Med Sch, MD, 52. Prof Exp: Asst attend, Med Div, Montefiore Hosp, 57-59, adj attend rehab med, 60-65, coordr amputee serv, 63-65; PROF CLIN REHAB MED & DIR REHAB MED, HARLEM HOSP AFFIL-COLUMBIA UNIV, 66- Concurrent Pos: Nat Med fel, Med Div, Montefiore Hosp, 55-56, Off Vocab Rehab fel phys med & rehab, 57-59; United Cerebral Palsy consult, Suffolk Rehab Ctr, 62-65; consult phys med & rehab, Hebrew Home Aged, 62-70; consult, Loeb Ctr Nursing & Rehab, 63-65 & Health Ins Plan Greater New York, 63-; clin consult prog, NY State Dept Health, 64-; asst attend, Presby Hosp-Columbia Univ, 67- Mem: Am Col Physicians; Am Heart Asn; Am Rheumatism Asn; Am Acad Phys Med & Rehab. Res: Work physiology in the elderly, the disabled and the chronically ill; disability in the poor, the black, and the ghetto dweller. Mailing Add: Dept of Rehab Med Harlem Hosp Ctr New York NY 10037

ANDERSON, ALBERT EDWARD, b Jamestown, NY, July 4, 28; m 50; c 1. ORGANIC CHEMISTRY. Educ: Allegheny Col, BS, 50; Univ Rochester, PhD(org chem), 53. Prof Exp: Res chemist, 53-59, sr res chemist, 59-65, RES ASSOC, EASTMAN KODAK CO, 65- Mem: Am Chem Soc; Soc Photog Sci & Eng. Res: Synthetic organic chemistry; photographic chemistry; film emulsion making. Mailing Add: Kodak Res Labs 1669 Lake Ave Rochester NY 14650

ANDERSON, ALBERT SYDNEY, computer science, nuclear physics, see 12th edition

ANDERSON, ALFRED TITUS, JR, b Port Jefferson, NY, Aug 3, 37. PETROLOGY, VOLCANOLOGY. Educ: Northwestern Univ, BA, 59; Princeton Univ, PhD(geol),

63. Prof Exp: NSF res fel, Univ Chicago, 63-66; geologist, US Geol Surv, 66-68; asst prof, 68-73, ASSOC PROF GEOPHYS SCI, UNIV CHICAGO, 73- Mem: AAAS; Am Geophys Union. Res: Physical-chemical conditions of crystallization of igneous rocks as revealed by field and mineralogical studies; role of gases in volcanism; igneous sources of excess volatiles. Mailing Add: Dept of Geophys Sci Univ of Chicago Chicago IL 60637

ANDERSON, ALLAN GEORGE, b New York, NY, Oct 30, 23; m 49, 55; c 4. MATHEMATICS. Educ: Univ Fla, BS, 46, MA, 47; Univ Mich, PhD(math), 51. Prof Exp: From instr to asst prof math, Oberlin Col, 50-53; from asst prof to assoc prof & chmn dept, Duquesne Univ, 53-55; mathematician, Jones & Laughlin Steel Corp, 55-56 & Pittsburgh Plate Glass Co, 56-57; chief statistician, Gen Tire & Rubber Co, 57-58; prof math & head dept, Western Ky State Col, 58-65; prof & chmn dept, Upsala Col, 65-66; prof, Parsons Col, 66-68; PROF MATH, QUEENSBOROUGH COMMUNITY COL, 68- Concurrent Pos: Vis lectr, Am Math Asn, 61-63 & Ky Acad Sci, 64-65. Mem: Am Math Soc; Math Asn Am; Soc Indust & Appl Math. Res: Mathematical genetics and statistics. Mailing Add: Dept of Math Queensborough Community Col Bayside NY 11364

ANDERSON, AMOS ROBERT, b Delavan, Wis, Feb 11, 20; m 45; c 1. ORGANIC CHEMISTRY. Educ: Adrian Col, BS, 42. Hon Degrees: DSc, Adrian Col, 65. Prof Exp: Anal chemist, Parker Rust-Proof Co, 41, res chemist, 46; res chemist gas process div, Girdler Corp, 43; chief chemist, Houdaille Hershey Corp, Manhattan Proj, 44-45; PRES, TEX ALKYLS, INC, 59-; VPRES, STAUFFER CHEM CO, 59-, GEN MGR, ANDERSON CHEM DIV, 59-, SILICONE DIV, 63-, PRES, ANDERSON DEVELOP CO, 66- Concurrent Pos: Pres, Anderson Chem Div, Stauffer Chem Co, 46-58. Mem: Am Chem Soc; Armed Forces Chem Asn. Res: Organometallics; silicones; organic chemicals; pharmaceutical intermediates. Mailing Add: Anderson Develop Co 1415 E Michigan St Adrian MI 49221

ANDERSON, ANSEL COCHRAN, b Warren, Pa, Sept 17, 33; m 55; c 2. LOW TEMPERATURE PHYSICS. Educ: Allegheny Col, BS, 55; Wesleyan Univ, MA, 57; Univ Ill, PhD(physics), 61. Prof Exp: Res assoc, 61-62, from asst prof to assoc prof, 62-69, PROF PHYSICS, UNIV ILL, URBANA, 69- Concurrent Pos: Guggenheim fel, 66; Fulbright-Hays res grant, 66; assoc, Ctr Advan Studies, 71-72. Mem: Fel Am Phys Soc. Res: Acoustics of quartz plates; properties of materials at temperatures below 1 Kelvin degree; ultralow temperature refrigeration and instrumentation. Mailing Add: Dept of Physics Univ of Ill Urbana IL 61801

ANDERSON, ANTHONY, b London, Eng, June 8, 35; m 61; c 2. MOLECULAR SPECTROSCOPY, SOLID STATE PHYSICS. Educ: Oxford Univ, BA, 56, MA & DPhil(low temperature physics), 60. Prof Exp: Sci officer far infrared spectros, Nat Phys Lab, Eng, 60-63; res assoc infrared & Raman spectros, Princeton Univ, 63-65; vis asst prof, Kans State Univ, 65-66; asst prof, 66-67, ASSOC PROF INFRARED & RAMAN SPECTROS, UNIV WATERLOO, 67- Concurrent Pos: Consult, Parsons & Co, Ltd, Eng, 63-65; mem ed bd, Spectros Lett, 71-; vis prof, Univ Grenoble, 72-73. Res: Infrared and Raman spectroscopic studies of molecular crystals, especially lattice vibrations and crystal field splittings; Brillouin scattering; lattice dynamics. Mailing Add: Dept of Physics Univ of Waterloo Waterloo ON Can

ANDERSON, ARCHIE DUANE, b Boulder, Colo, Dec 29, 40; m 64; c 2. PHARMACOLOGY. Educ: Univ Wyo, BS, 63; Univ Colo, Boulder, PhD(pharmacol), 70. Prof Exp: ASST PROF PHARMACOL, UNIV WYO, 69- Res: Fatty acid oxidation control mechanisms in the heart; cancer research. Mailing Add: Dept of Pharmacol Univ of Wyo Laramie WY 82070

ANDERSON, ARTHUR BERNHARDT, b North Fond du Lac, Wis, Feb 17, 07; m 30; c 2. ORGANIC CHEMISTRY, FOREST PRODUCTS. Educ: Univ Wis, AB, 29, PhD(wood chem), 33. Prof Exp: Fel biochem, Univ Wis, 33-34; chemist, Quaker Oats Co, Cedar Rapids, 34-37, res chemist, Chicago, 37-39 & tech sales, 39-41; res chemist, West Pine Asn, 41-48; dir res & develop, Ore Lumber Co, 48-50; LECTR WOOD CHEM & BIOCHEMIST, FOREST PRODS LAB, UNIV CALIF, 50- Concurrent Pos: Fulbright res scholar, Norwegian Pulp & Paper Res Inst, Oslo, 56-57. Mem: Am Chem Soc; Soc Am Foresters; Forest Prods Res Soc; Coun Agr & Chemurgic Res. Res: Wood chemistry utilization of logging and plant residues through chemistry. Mailing Add: Forest Prod Lab Univ of Calif 1301 S46th St Richmond CA 94804

ANDERSON, ARTHUR G, JR, b Sioux City, Iowa, July 1, 18; m 44; c 3. ORGANIC CHEMISTRY. Educ: Univ Ill, AB, 40; Univ Mich, MS, 42, PhD(chem), 44. Prof Exp: From jr to sr chemist, Tenn Eastman Corp, Oak Ridge, 44-45; spec asst, Univ Ill, 46; from instr to assoc prof chem, 46-57; PROF CHEM, UNIV WASH, 57- Concurrent Pos: NSF sr fel, 60-61; vis prof, Australian Nat Univ, 66. Mem: Am Chem Soc; The Chem Soc; fel NY Acad Sci; fel Am Inst Chem. Res: Nonclassical aromatic compounds; novel heterocyclic systems; polynuclear alicyclic compounds; new synthetic reactions. Mailing Add: Dept of Chem Univ of Wash Seattle WA 98195

ANDERSON, ARTHUR GEORGE, b Evanston, Ill, Nov 22, 26; m 47; c 3. PHYSICS. Educ: Univ San Francisco, BS, 49; Northwestern Univ, MS, 51; NY Univ, PhD(physics), 58. Prof Exp: Tech engr, IBM Corp, 51-53, comput & physics res, Watson Lab, 53-58, physicist, 58-59, res physicist, 59-61, dir, San Jose Res Lab, 61-63, asst dir res, 63-65, staff dir corp tech comt, 65-67, dir res, 67-70, vpres, 69-72, IBM VPRES & PRES, GEN PROD DIV, IBM CORP, 72- Concurrent Pos: Vis fel, Ctr Study Democratic Insts, 70-71. Mem: Nat Acad Eng; fel Am Phys Soc; Inst Elec & Electronics Eng; NY Acad Sci. Res: Nuclear magnetic resonance in solids; computer engineering. Mailing Add: IBM Corp 5600 Monterey Rd San Jose CA 95193

ANDERSON, ARTHUR JAMES, b Calgary, Alta, Feb 4, 16; m 56. PHARMACEUTICS. Educ: Univ Alta, PhC, 40, BSc, 41; Univ Wash, MS, 46. Prof Exp: Lectr, 49-53, from asst prof to assoc prof, 53-65, PROF PHARM, UNIV ALTA, 65- Res: Technology of clays; pharmaceutical formulations. Mailing Add: Fac of Pharm Univ of Alta Edmonton AB Can

ANDERSON, ARTHUR JAMES OUTRAM, b Phoenix, Ariz, Nov 26, 07; m 37. ANTHROPOLOGY. Educ: San Diego State Col, AB, 30; Claremont Col, AM, 31; Univ Southern Calif, PhD(anthrop), 40. Prof Exp: Asst instr eng, San Diego State Col, 34-37; instr, Riverside Jr Col, 38; from assoc prof to prof anthrop, Eastern NMex Col, 39-45; cur hist, Mus NMex & Sch Am Res, 45-57; instr anthrop, El Camino Col, 57-61; from asst prof to assoc prof, 61-71, PROF ANTHROP, CALIF STATE UNIV, SAN DIEGO, 71- Concurrent Pos: Dir, Roosevelt County Mus Art, Hist & Archaeol, NMex, 39-45; chmn div soc sci, Eastern NMex & Sch Am Res, 42-45; ed, El Palacio, Sch Am Res & Archaeol Soc NMex, 48-57; vis lectr anthrop, Calif State Univ, San Diego, 53-54, actg chmn dept, 64; Guggenheim Mem Found fels, 55 & 57; spec lectr, Occidental Col, 57-58; assoc, Sch Am Res, 57-; vis prof, Univ Madrid, 68-69; hon cur, Latin Am Ethnog, San Diego Mus Man. Mem: Mex Anthrop Soc; Int Nahuatl Asn. Res: Primary sources of information on ancient Mexico, particularly immediately prior to the conquest and the sixteenth to eighteenth centuries. Mailing Add: Dept of Anthrop Calif State Univ San Diego CA 92115

ANDERSON, ARTHUR W, b Lisbon, NDak, Dec 2, 14; m 48; c 3. BACTERIOLOGY. Educ: NDak State Col, BS, 43; Univ Wis, MS, 47; Ore State Col, PhD, 52. Prof Exp: Asst prof bact, Univ Calif, Berkeley, 52-53; ASSOC PROF FOOD MICROBIOL, ORE STATE UNIV, 53- Mem: Inst Food Technol; Am Soc Microbiol. Res: Food borne pathogens and irradiation microbiology. Mailing Add: Dept of Microbiol Sch of Sci Ore State Univ Corvallis OR 97331

ANDERSON, ARTHUR WILLIAM, b Chicago, Ill, Apr 14, 16; m 43; c 2. POLYMER CHEMISTRY. Educ: Northwestern Univ, BS, 37; Univ Ill, PhD(chem), 41. Prof Exp: RES CHEMIST & MGR PLASTICS DEPT, E I DU PONT DE NEMOURS & CO, 41- Mem: Am Chem Soc. Res: Plastics; chemistry of polymers and their intermediates. Mailing Add: Plastics Dept E I du Pont de Nemours & Co Wilmington DE 19898

ANDERSON, BARBARA GALLATIN, b San Francisco, Calif, May 13, 26; div; c 3. CULTURAL ANTHROPOLOGY. Educ: San Francisco Col Women, BA, 47, cert gen sec teaching, 55; Univ Sorbonne, DrUniv(ethnol), 59. Prof Exp: NSF fel field study village, France, 58-59; assoc anthropologist, Geriatrics Res Proj, Langley Porter Neuropsychiat Inst, San Francisco, 61-64; PROF ANTHROP, CALIF STATE UNIV, HAYWARD, 64- Concurrent Pos: NSF res grant French & Danish ethnog, 61; Nat Inst Ment Health res grant psychosocial anal aging, US, 67-68; Calif State Univ grant res urban aging adaptation, US, 69-70; Inst Int Studies-US Off Educ fac res abroad award, Morocco, 71; res assoc psychiat, Sch Med, Univ Calif, San Francisco, 61-64; ad hoc mem, Cult Anthrop Rev Comt, Nat Inst Ment Health, 68-69. Mem: Fel Am Anthrop Asn; Soc Appl Anthrop; Soc Woman Geog. Res: Europe, Morocco; culture change, childhood cross-culturally; social gerontology. Mailing Add: Dept of Anthrop Calif State Univ 25800 Hillary St Hayward CA 94542

ANDERSON, BERNARD A, b Thornton, Wash, Aug 23, 09; m 42; c 8. SPEECH PATHOLOGY, AUDIOLOGY. Educ: Univ Wash, BA, 33, dipl, 37; Univ Wis, PhD(speech path), 49. Prof Exp: Actg dir speech & hearing, Univ Okla, 39-40; instr speech path, Ind Univ, 46-48; assoc prof, Univ Utah, 49-57; PROF SPEECH PATH & DIR SPEECH & HEARING CLIN, UNIV NEV, RENO, 58- Concurrent Pos: Consult, Shriner's Hosp Crippled Children, 55-57. Mem: Fel Am Speech & Hearing Asn. Res: Photography of the human larynx; disorders of speech and hearing. Mailing Add: Speech & Hearing Clin Univ of Nev Reno NV 89507

ANDERSON, BERNARD JEFFREY, b Waco, Tex, Feb 23, 44; c 1. CLOUD PHYSICS. Educ: Univ Nev, Reno, BS, 66, MS, 68, PhD(physics), 74. Prof Exp: Res assoc cloud physics, Desert Res Inst, Lab Atmospheric Physics, Univ Nev, 74-75; AEROSPACE ENGR, GEORGE C MARSHALL SPACE FLIGHT CTR, NASA, 75- Res: Cloud physics experimentation at low gravity levels in space; ice crystal growth and nucleation. Mailing Add: ES-44 G C Marshall Space Flight Ctr Huntsville AL 35812

ANDERSON, BERTIL GOTTFRID, b Escanaba, Mich, Jan 5, 04; m 31; c 3. Educ: Augustana Col, Ill, AB, 26; Univ Iowa, MS, 29, PhD(zool), 30. Prof Exp: Asst biol, Augustana Col, Ill, 24-26; high sch instr, Ill, 26-27; asst zool, Univ Iowa, 27-30; from instr to asst prof biol, Western Reserve Univ, 30-41; res assoc, Stone Lab, Ohio State Univ, 41-42; from instr to prof zool, Univ WVa, 42-53; prof zool, 53-69, head dept zool & entom, 53-63, head dept zool, 63-67, EMER PROF ZOOL, PA STATE UNIV, 69- Concurrent Pos: Mem aquatic life adv comt, Ohio River Valley Water Sanit Comn; mem nat tech adv comt to Secy Interior on Water Qual Criteria, 67-68; adv comt standard methods exam water & waste water, Water Pollution Control Fedn, 67- Mem: AAAS; Am Soc Zool; Am Fisheries Soc; Am Soc Limnol & Oceanog; Am Inst Biol Sci. Res: Growth and regeneration in Daphnia; toxicity thresholds for Daphnia and for zooplankton; bioassay methods using Daphnia. Mailing Add: 315 Hillcrest Ave State College PA 16801

ANDERSON, BERTIN W, b Luverne, Minn, Dec 15, 39; m 69; c 1. ZOOLOGY. Educ: Mankato State Col, BS, 62; Univ Minn, MS, 66; Univ SDak, PhD(zool), 70. Prof Exp: Asst prof biol, Northwestern State Col, Okla, 69-73; ASSOC PROF ZOOL, ARIZ STATE UNIV, 73- Mem: Am Ornith Union; Cooper Ornith Soc; Wilson Ornith Soc. Res: Ornithology; evolutionary and systematic biology. Mailing Add: Dept of Zool Ariz State Univ Tempe AZ 85281

ANDERSON, BROR ERNEST, b Mazeppa, Minn, Nov 24, 14; m 41; c 3. PAPER CHEMISTRY. Educ: Gustavus Adolphus Col, BA, 35; Univ Minn, PhD(agr biochem), 40. Prof Exp: Instr high sch, Minn, 35-36; chemist, Northwest Paper Co, 36-37; assoc chemist, Exp Sta, Univ Minn, 40; chemist, Johnson & Johnson, New Brunswick, 40-42, in charge lab, Filter Prod Div, Chicago, 42-46; res mgr paper & fiber, A B Dick Co, Ill, 46-55, res mgr mimeograph, 55-58; tech dir, Wyomissing Corp, Pa, 58-65; VPRES & TECH DIR, WEBER MARKING SYSTS, INC, 65- Mem: Am Chem Soc; Tech Asn Pulp & Paper Indust. Res: Formulation of mimeograph stencils, both thermal and conventional; industrial marking systems; label papers. Mailing Add: 7 WCedar St Arlington Heights IL 60005

ANDERSON, BRUCE MURRAY, b Detroit, Mich, July 14, 29; m 50; c 3. BIOCHEMISTRY. Educ: Ursinus Col, BS, 53; Purdue Univ, MS, 54; Johns Hopkins Univ, PhD(biochem), 58. Prof Exp: NIH fel biochem, Brandeis Univ, 58-60; asst prof, Univ Louisville, 60-63; from assoc prof to prof, Univ Tenn, 63-70; PROF BIOCHEM & HEAD DEPT, VA POLYTECH INST & STATE UNIV, 70- Concurrent Pos: NSF res grants, 60-; NIH res grants, 64-69. Mem: AAAS; Am Chem Soc; Am Soc Biol Chem. Res: Mechanism of enzyme action; function of pyridine nucleotide coenzymes. Mailing Add: Dept of Biochem & Nutrit Va Polytech Inst & State Univ Blacksburg VA 24061

ANDERSON, BRYON DON, b Ft Riley, Kans, Dec 23, 44; m 69; c 2. NUCLEAR PHYSICS. Educ: Univ Idaho, BS, 66; Case Western Reserve Univ, PhD(physics), 72. Prof Exp: Res asst nuclear physics, Idaho Nuclear Corp, 66-67; res assoc, Calif Inst Technol, 71-73; sr res assoc, Case Western Reserve Univ, 73-75; ASST PROF NUCLEAR PHYSICS, KENT STATE UNIV, 75- Mem: Am Phys Soc. Res: Neutron and charged-particle spectrometry at intermediate energies with emphasis on experimental study of reaction mechanisms. Mailing Add: Dept of Physics Kent State Univ Kent OH 44242

ANDERSON, BURTON, b Chicago, Ill, Aug 27, 32; m 60; c 3. INTERNAL MEDICINE, IMMUNOLOGY. Educ: Univ Ill, BS, 55, MD & MS, 57. Prof Exp: Rotating intern, Minneapolis Gen Hosp, 57-58; resident internal med, Univ Hosps, Univ Ill, 58-59, assoc prof med & microbiol, Col Med, Northwestern Univ, 67-70; assoc prof, 70-72, PROF MED & MICROBIOL, COL MED, UNIV ILL, 72-; CHIEF SECT INFECTIOUS DIS, WEST SIDE VET ADMIN HOSP, 70- Concurrent Pos: Fel infectious dis, Univ Ill, 59-61; Nat Inst Allergy & Infectious Dis fel, 61-64; fel, Med Ctr, Univ Rochester, 64-67. Mem: Am Asn Immunol; Am Fedn Clin Res; Am

Soc Microbiol; Infectious Dis Soc Am. Res: Infectious diseases. Mailing Add: Sect of Infectious Dis West Side Vet Admin Hosp Chicago IL 60612

ANDERSON, BURTON CARL, b Kewanee, Ill, Oct 8, 30; m 52; c 3. ORGANIC POLYMER CHEMISTRY. Educ: Univ Ill, AB, 52; Mass Inst Technol, PhD(org chem), 55. Prof Exp: Res chemist, Cent Res Dept, Exp Sta, 55-66, res chemist, Elastomers Dept, 66-69, div head res, 69-72, develop supt, Chambers Works, 72-74, ASSOC DIR, CENT RES & DEVELOP DEPT, E I DU PONT DE NEMOURS & CO, INC, 74- Mem: Am Chem Soc. Res: Organic fluorine chemistry; polymer physical chemistry; polymer synthesis. Mailing Add: Cent Res & Develop Dept Exp Sta E I du Pont de Nemours & Co Inc Wilmington DE 19898

ANDERSON, BYRON, b Hammond, Ind, Dec 30, 41; m 64. BIOCHEMISTRY, IMMUNOLOGY. Educ: Kalamazoo Col, BA, 63; Univ Mich, PhD(biochem), 68. Prof Exp: ASST PROF BIOCHEM, MED & DENT SCHS, NORTHWESTERN UNIV, 71- Concurrent Pos: Helen Hay Whitney & Nat Cystic Fibrosis Res Found fel, Col Physicians & Surgeons, Columbia Univ, 68-71; NIH Career Develop Award, 74-; sr investr, Arthritis Found, 73-74; consult, Abbott Labs, 74- Mem: AAAS; Am Rheumatism Asn; Am Chem Soc; NY Acad Sci. Res: Autoimmune diseases, especially rheumatoid arthritis; cancer research and membrane glycoprotein antigens. Mailing Add: Dept of Biochem Northwestern Univ Med & Dent Schs Chicago IL 60611

ANDERSON, CARL ANDREW, soil chemistry, soil fertility, see 12th edition

ANDERSON, CARL DAVID, b New York, NY, Sept 3, 05; m 46; c 2. PHYSICS. Educ: Calif Inst Technol, BS, 27, PhD(physics), 30; Hon Degrees: ScD, Colgate Univ, 37 & Gustavus Adolphus Col, 63; LLD, Temple Univ, 49. Prof Exp: Res fel physics, 30-33, from asst prof to assoc prof, 33-39, chmn div physics, math & astron, 62-70, PROF PHYSICS, CALIF INST TECHNOL, 39- Concurrent Pos: With Nat Defense Res Comt & Off Sci Res & Develop, 41-45. Honors & Awards: Nobel Prize, 36; Gold Medal, Am Inst, 35; Cresson Medal, Franklin Inst, 37; Presidential Cert Merit, 45; John Ericsson Medal, Am Soc Swed Engr, 60. Mem: Nat Acad Sci; fel Am Phys Soc; Am Philos Soc. Res: X-rays; gamma rays; radioactivity; cosmic rays. Mailing Add: Dept of Physics Calif Inst of Technol Pasadena CA 91109

ANDERSON, CARL EDGAR, b Templeton, Calif, Apr 11, 14; m 36; c 3. ORTHOPEDIC SURGERY. Educ: Univ Calif, AB, 34, MD, 39; Am Bd Orthop Surg, dipl. Prof Exp: From clin instr to clin assoc prof, 42-65, CLIN PROF ORTHOP SURG, SCH MED, UNIV CALIF, SAN FRANCISCO, 65- Mem: AAAS; Orthop Res Soc; Am Acad Orthop Surgeons; Am Col Surgeons; AMA. Res: Connective tissue pathology and metabolism. Mailing Add: 1405 Montgomery Dr Santa Rosa CA 94505

ANDERSON, CARL EINAR, b Brooklyn, NY, Sept 15, 23; m 57. PHYSICS. Educ: Rensselaer Polytech Inst, BS, 49; Yale Univ, MS, 50, PhD(physics), 53. Prof Exp: Res assoc electron physics res lab, Gen Elec Co, 52-54; res assoc nuclear physics, Yale Univ, 54-61; sr scientist appl physics lab, Johns Hopkins Univ, 61; mgr radiation & space physics re-entry systs dept, 61-65, MGR APPL PHYSICS, SPACE SCI LAB, GEN ELEC CO, 65- Concurrent Pos: Consult, Wallingford Steel Co, 56-59. Mem: Am Phys Soc; Am Inst Aeronaut & Astronaut. Res: Nuclear, reentry and space physics. Mailing Add: Rm M9161 Missile & Space Div Gen Elec Co PO Box 8555 Philadelphia PA 19101

ANDERSON, CARL ELMORE, b South Manchester, Conn, July 14, 06; m 40; c 3. BIOCHEMISTRY. Educ: Univ Conn, BS, 35; Univ NC, PhD(biochem), 43. Prof Exp: Res assoc, Wayne Univ, 38-40; staff officer, US War Dept, Washington, DC, 44-46; asst prof biochem, Sch Med, Vanderbilt Univ, 46-50; assoc prof biochem, 50-59, assoc prof nutrit, 50-56, asst dean, Sch Med, 56-62, PROF BIOCHEM, SCH MED, SCH PUB HEALTH, DENT SCH & GRAD SCH, UNIV NC, CHAPEL HILL, 59-, PROF NUTRIT, 62- Concurrent Pos: Consult, Oak Ridge Inst Nuclear Sci, NC Res Triangle Inst & Health Serv & Ment Health Admin, US Dept Health, Educ & Welfare; co-dir, Univ NC Intestinal Malabsorption Proj, Guatemala, 71- Mem: Fel AAAS; Am Chem Soc; Am Soc Biol Chem; Am Inst Nutrit. Res: Metabolism of quinine; chemistry and metabolism of plasmalogens; bile, fat and phospholipid metabolism; intestinal absorption; phosphorus intoxication. Mailing Add: Dept of Biochem Univ of NC Sch of Med Chapel Hill NC 27514

ANDERSON, CARL LEONARD, b Ironwood, Mich, Feb 28, 01; m 30; c 3. HYGIENE, PUBLIC HEALTH. Educ: Univ Mich, BS, 28, MS, 32, DrPH, 34. Prof Exp: Dir health educ, Mich Bd Educ, 28-31; pub health adminr, Mich State Dept Health, 31-32; rural health adminr, Mich State Dept Health & USPHS, 33-35; from asst prof to prof physiol & pub health, Utah State Univ, 35-45; prof biol sci, Mich State Univ, 45-49; PROF HYG & HEALTH EDUC, ORE STATE UNIV, 49- Mem: AAAS; fel Am Pub Health Asn; fel Soc Res Child Develop; Am Genetic Asn; Am Asn Health, Phys Educ & Recreation. Res: Experimental production of convulsive seizures; inheritance of cataract; metabolism of the human skin; physiological indices of the human; neurone impulse; physiology. Mailing Add: Ore State Univ Sch of Sci Corvallis OR 97331

ANDERSON, CARL MARTIN, b Portland, Ore, Nov 11, 15; m 39; c 3. CHEMISTRY. Educ: Linfield Col, BS, 38; Univ Nebr, MS, 40; Ore State Col, PhD(phys chem), 42. Prof Exp: Res chemist, Hercules Powder Co, Del, 42-44; sr chemist, Phillips Petrol Co, Okla, 46-47; PROF CHEM, LINFIELD COL, 47-, HEAD DEPT, 58- Concurrent Pos: Asst prog dir spec projs prog, NSF, 67-68. Mem: Fel Am Inst Chem; Am Chem Soc. Res: Celulose derivatives, Fischer-Tropsch catalyst; calorimetry; heats of combustion of some organic nitrogen compounds; analytical instrumentation. Mailing Add: Dept of Chem Linfield Col McMinnville OR 97128

ANDERSON, CARL WILLIAM, inorganic chemistry, see 12th edition

ANDERSON, CARL WILLIAM, b Washington, DC, May 19, 44; m 68; c 2. MOLECULAR BIOLOGY, ANIMAL VIROLOGY. Educ: Harvard Col, BA, 66; Wash Univ, PhD(microbiol), 70. Prof Exp: Fel cancer viruses, Cold Spring Harbor Lab, 70-73, mem staff, 73-75; ASST GENETICIST, BROOKHAVEN NAT LAB, 75- Concurrent Pos: NIH fel, 70-72; consult, NSF, 75- Mem: Am Soc Microbiol. Res: Control of gene expression in mammalian cells. Mailing Add: 23 Shelbourne Lane Stony Brook NY 11790

ANDERSON, CHARLES ALFRED, b Bloomington, Calif, June 6, 02; m 27; c 1. GEOLOGY. Educ: Pomona Col, AB, 24; Univ Calif, PhD(geol), 28. Hon Degrees: DSc, Pomona Col, 60. Prof Exp: Assoc geol, Univ Calif, 26-28, from instr to assoc prof, 28-42; geologist, US Geol Surv, 42-53, chief mineral deposits br, 53-58, chief geologist, 59-64, res geologist, 64-72; RES ASSOC, UNIV CALIF, SANTA CRUZ, 72- Honors & Awards: Penrose Medal, Soc Econ Geologists, 74. Mem: Nat Acad Sci; fel Geol Soc Am; fel Mineral Soc Am; Soc Econ Geol; Am Acad Arts & Sci. Res: Aerial mapping of Precambrian rocks; structural control of ore deposits; petrology of metamorphic and igneous rocks. Mailing Add: Earth Sci Bd Appl Sci Bldg Univ of Calif Santa Cruz CA 95064

ANDERSON, CHARLES DEAN, b Redwood Falls, Minn, Mar 8, 30; m 53; c 3. BIO-ORGANIC CHEMISTRY. Educ: St Olaf Col, BA, 52; Harvard Univ, MA, 54, PhD(org chem), 59. Prof Exp: Org chemist, Stanford Res Inst, Calif, 56-59; assoc prof org chem, 59-62, chmn dept chem, 61-66, dean col arts & sci, 66-70, regency professorship, 74-75, PROF CHEM, PAC LUTHERAN UNIV, 62- Concurrent Pos: NSF fac fel, Univ Minn, 64-65; Northwest Col & Univ Asn Sci vis fac appointment, Pac Northwest Labs, Battelle Mem Inst. Mem: Am Chem Soc; NY Acad Sci; The Chem Soc. Res: Synthetic methods; natural products and analogs. Mailing Add: Dept of Chem Pac Lutheran Univ Tacoma WA 98447

ANDERSON, CHARLES EDWARD, b St Louis, Mo, Aug 13, 19; m 43; c 3. PHYSICAL METEOROLOGY. Educ: Lincoln Univ, Mo, BS, 41; Univ Chicago, cert meteorol, 43; Polytech Inst Brooklyn, MS, 48; Mass Inst Technol, PhD(meteorol), 60. Prof Exp: Chief, Cloud Physics Br, Air Force Cambridge Res Ctr, Mass, 48-61 & Atmospheric Sci Br, Douglas Aircraft Co, Calif, 61-65; dir, Off Fed Coord Meteorol, Environ Sic Serv Admin, US Dept Com, Washington, DC, 65-66; prof space sci & eng, 67-69, PROF METEOROL & CHMN CONTEMP TRENDS COURSE, UNIV WIS-MADISON, 66-, PROF AFRO-AM STUDIES & CHMN DEPT, 70- Concurrent Pos: Prin investr, NSF grant, 69-; consult, Sci Ctr, NAm Rockwell Corp, Calif, 67- & Rand Corp, 68-; chmn, Aviation Adv Panel, Nat Ctr Atmospheric Res, Colo, 63- Mem: AAAS; Sigma Xi. Res: Cloud and aerosol physics; high polymer chemistry; space and planetary science; meteorology of other planets; science and society; science, technology and race. Mailing Add: Dept of Meteorol Univ of Wis Madison WI 53706

ANDERSON, CHARLES EUGENE, b Winfield, Kans, Dec 3, 34; m 54; c 3. MORPHOLOGY. Educ: Purdue Univ, BSA, 56, MS, 61, PhD(biol sci), 63. Prof Exp: Asst prof bot, Univ Okla, 63-66; asst prof, 66-69, ASSOC PROF BOT, NC STATE UNIV, 69- Mem: AAAS; Bot Soc Am. Res: Developmental morphology of salt marsh and dune plants; environmental factors influencing cell and tissue differentiation. Mailing Add: Dept of Bot NC State Univ Raleigh NC 27607

ANDERSON, CHARLES H, JR, physical chemistry, see 12th edition

ANDERSON, CHARLES HAMMOND, b Mineola, NY, May 31, 35; m 60; c 3. SOLID STATE PHYSICS. Educ: Calif Inst Technol, BS, 57; Harvard Univ, PhD(molecular physics), 62. Prof Exp: Res assoc, Harvard Univ, 61-62, instr physics, 62-63; MEM TECH STAFF, RCA LABS, 63- Concurrent Pos: NSF fel, 62-63; RCA fel, Clarendon Lab, Oxford, Eng, 71-72. Mem: Fel Am Phys Soc. Res: Ion physics in solids; liquid helium; acoustics; low temperature physics; electron optics. Mailing Add: RCA Labs Princeton NJ 08540

ANDERSON, CHARLES THOMAS, b Fairmont, WVa, Feb 26, 21; m 50; c 2. PHYSICAL INORGANIC CHEMISTRY. Educ: Fairmont State Col, AB, 42; Ohio State Univ, PhD(chem), 55. Prof Exp: Teaching asst chem, Ohio State Univ, 43-46; instr, Ohio Univ, 46-51; from asst prof to assoc prof, 55-62, PROF CHEM, EASTERN MICH UNIV, 62- Mem: Am Chem Soc. Res: Coordination compounds; equilibrium constants. Mailing Add: 720 Kewanee Ave Ypsilanti MI 48197

ANDERSON, CHARLES V, b Little Sioux, Iowa, Aug 18, 33; m 66; c 1. AUDIOLOGY. Educ: Univ Nebr, BS, 55, MA, 57; Univ Pittsburgh, PhD(psychoacoust), 62. Prof Exp: Instr hearing ther, Univ Nebr, 56-58; asst prof audiol, Purdue Univ, 62-66; asst prof, 66-68, ASSOC PROF AUDIOL, UNIV IOWA HOSPS, 68- Mem: AAAS; fel Am Speech & Hearing Asn; Acoust Soc Am. Res: Psychoacoustics; hearing loss; auditory phenomena. Mailing Add: Dept of Otolaryngol Univ of Iowa Hosps Iowa City IA 52242

ANDERSON, CHRISTIAN DONALD, b Ft Dodge, Iowa, Apr 10, 31; m 57; c 6. EXPLORATION GEOPHYSICS. Educ: Univ Minn, Duluth, BA, 57; Univ NMex, MS, 61; Univ Utah, PhD(geophys), 66. Prof Exp: Physicist, Los Alamos Sci Lab, 58-61; geophysicist, Newmont Mining Corp, 66-67; ASSOC PROF GEOPHYS, UNIV MAN, 67- Concurrent Pos: Consult explor geophys, 67- Mem: Can Explor Geophysicists Soc; Sigma Xi; Soc Explor Geophysicists; Am Inst Mining Eng. Res: Development of geophysical methods and exploration strategies to improve discovery probabilities in mineral exploration. Mailing Add: Dept of Earth Sci Univ of Manitoba Winnipeg MB Can

ANDERSON, CHRISTOPHER MARLOWE, b Las Cruces, NMex, Feb 21, 41; m 70. ASTRONOMY. Educ: Univ Ariz, BS, 63; Calif Inst Technol, PhD(astron), 68. Prof Exp: Asst prof, 68-74, ASSOC PROF ASTRON, UNIV WIS-MADISON, 74- Mem: Am Astron Soc; Int Astron Union. Res: Photoelectric spectrophotometry; interstellar extinction; stellar rotation and chromospheric activity; spectrum variability. Mailing Add: Washburn Observ Univ of Wis Madison WI 53706

ANDERSON, CLAUDE M, b Denver, Colo, Apr 17, 07. ASTRONOMY. Educ: Univ Calif, Berkeley, AB, 29, MA, 32, PhD(astron), 33. Prof Exp: Res assoc astron, Univ Calif, 33-37; instr, Univ Va, 37-42; educ officer, US Merchant Marine Cadet Sch, 42-47; teacher astron, Col San Mateo, 47-72; RETIRED. Mem: Am Astron Soc. Res: Perturbations of minor planets; stellar parallaxes. Mailing Add: 247 W40th Ave San Mateo CA 94403

ANDERSON, CLIFFORD HAROLD, b Peoria, Ill, Jan 11, 39; m 62; c 2. MATHEMATICAL ANALYSIS. Educ: Ill Inst Technol, BS, 61; Purdue Univ, MS, 63; Univ Mo, PhD(math), 68. Prof Exp: Asst prof math, Ohio Univ, 67-74; ASST PROF MATH, CALIF STATE UNIV, LOS ANGELES, 74- Mem: Am Math Soc; Soc Indust & Appl Math. Res: Functional differential equations. Mailing Add: Dept of Math Calif State Univ Los Angeles CA 90032

ANDERSON, CLYDE LEE, b El Paso, Tex, Sept 17, 26; m 50; c 3. CHEMISTRY. Educ: Univ Utah, BS, 52, MS, 53; PhD(org chem), 67. Prof Exp: Head chem dept, Dixie Jr Col, 53-57; PROF CHEM, SAN BERNARDINO VALLEY COL, 57-, CHMN DEPT, 68- Concurrent Pos: Lectr, Westminster Col, 65-66; NSF sci fac fel, 65-66. Mem: Soc Appl Spectros; Am Chem Soc; Sigma Xi. Res: Organic thesis; natural products. Mailing Add: Dept of Chem San Bernardino Valley Col San Bernardino CA 92403

ANDERSON, CURTIS BENJAMIN, b Fargo, NDak, Feb 19, 32. ORGANIC CHEMISTRY. Educ: Univ Minn, BS, 56; Univ Calif, Los Angeles, PhD(org chem), 63. Prof Exp: Lectr, 62, assoc, 62-63, lectr, 63-64, actg asst prof, 64-65, asst prof, 65-69, ASSOC PROF CHEM, UNIV CALIF, SANTA BARBARA, 69- Mem: Am Chem Soc. Res: Physical organic chemistry; conformational analysis of heterocyclic compounds; reactions of olefins complexes of metal salts; neighboring group participation. Mailing Add: Dept of Chem Univ of Calif Santa Barbara CA 93106

ANDERSON, CYRUS VINCENT, b Sioux Falls, SDak, Nov 22, 16; m 47. ANIMAL

NUTRITION, PARASITOLOGY. Educ: Lenoir-Rhyne Col, AB, 34; Univ Kans, MA, 41. Prof Exp: Assoc prof biol, Salem Col, 38-39; entomologist, 46-50, nutritionist, 50-63, DIR RES & DEVELOP, WATKINS PROD INC, 63- Concurrent Pos: Mayor, Minnesota City, Minn, 58-64. Mem: Am Soc Parasitol. Res: Immunology of parasitic infections; beef, swine, dairy nutrition research; feed additives. Mailing Add: Res & Develop Watkins Prod Inc 150 Liberty St Winona MN 55987

ANDERSON, DANIEL CRAIG, b Orofino, Idaho, Mar 18, 44; m 72. ANIMAL SCIENCE. Educ: Univ Idaho, BS, 66; Ore State Univ, MS, 69, PhD(animal nutrit), 72. Prof Exp: ASST PROF ANIMAL SCI & ASST ANIMAL SCIENTIST, WASH STATE UNIV, 72- Mem: Am Soc Animal Sci; Sigma Xi. Res: Beef cow-calf nutrition and management; low quality roughage utilization; alternate systems of beef production. Mailing Add: Dept of Animal Sci Wash State Univ Pullman WA 99163

ANDERSON, DANIEL WILLIAM, b Underwood, NDak, Feb 5, 39; m 71; c 2. WILDLIFE ECOLOGY, POLLUTION BIOLOGY. Educ: NDak State Univ, BS, 61; Univ Wis-Madison, MS, 67, PhD(wildlife ecol, zool), 70. Prof Exp: Res biologist, US Fish & Wildlife Serv, 70-75; AVIAN ECOLOGIST, UNIV CALIF, DAVIS, 76- Concurrent Pos: Regional rep exec coun, Pac Seabird Group, 73-75; mem bd dirs, Bodega Bay Inst Pollution Ecol, 71- Mem: Wildlife Soc; AAAS; Am Ornithologists Union; Brit Ornithologists Union. Res: Effects of environmental pollution on wildlife populations and individuals; ecological and physiological effects of pollutants; study of avian ecology, with emphasis on marine birds. Mailing Add: Div of Wildlife & Fisheries Biol Univ of Calif Davis CA 95616

ANDERSON, DAVID EUGENE, b Ashby, Mass, May 16, 26; m 46; c 2. CANCER GENETICS. Educ: Univ Mass, BS, 50; Univ Conn, MS, 52; Iowa State Col, PhD(animal breeding & genetics), 54. Prof Exp: Assoc biologist & assoc prof biol, 58-65, BIOLOGIST & PROF BIOL, UNIV TEX M D ANDERSON HOSP & TUMOR INST, HOUSTON, 65-, ACTG HEAD DEPT BIOL, 75- Concurrent Pos: Nat Cancer Inst fel, Okla State Univ, 54-57 & spec fel human genetics, Univ Mich, 63-65. Mem: AAAS; Genetics Soc Am; Am Soc Human Genetics. Res: Human genetics as it relates to cancer and allied diseases. Mailing Add: Dept of Biol M D Anderson Hosp & Tumor Inst Houston TX 77030

ANDERSON, DAVID G, b Iron Mountain, Mich, Sept 30, 28; m 50; c 4. OBSTETRICS & GYNECOLOGY. Educ: Univ Mich, AB, 50, MD, 53. Prof Exp: From instr to asst prof, 60-66, ASSOC PROF OBSTET & GYNEC, MED CTR, UNIV MICH, ANN ARBOR, 66- Mem: AMA; Am Col Obstet & Gynec; Am Col Surg; Am Soc Study Fertility. Res: Gynecologic cancer chemotherapy; obstetric radiology; sterility; cancer therapy; operative care; obstetric complications; perinatal studies. Mailing Add: 2208 Women's Hosp Univ of Mich Med Ctr Ann Arbor MI 48104

ANDERSON, DAVID GENE, b Lethbridge, Alta, Dec 7, 31; m 58; c 2. PHYSICAL BIOCHEMISTRY, ENDOCRINOLOGY. Educ: Univ Alta, BSc, 55, MSc, 57; Univ Minn, PhD(biochem), 66. Prof Exp: Part-time instr, Univ Minn, 60 & 63; res assoc chem, Cornell Univ, 66-67; asst prof obstet & gynec, Col Med, Ohio State Univ, 67-71; ADJ ASST PROF OBSTET & GYNEC, SCH MED, UNIV CALIF, DAVIS, 71- Mem: AAAS; Am Chem Soc. Res: Physical and chemical properties of proteins and gonadotropic hormones; proteins found in reproductive system as related to diagnostic and contraceptive problems; tumor antigens and cancer immunology. Mailing Add: Dept of Obstet & Gynec Univ of Calif Sch of Med Davis CA 95616

ANDERSON, DAVID GORDON, b Huron, SDak, Jan 2, 23. BIOCHEMISTRY. Educ: Univ Chicago, BS, 48, PhD(biochem), 53. Prof Exp: Res assoc pharmacol, Western Reserve Univ, 53-54; res assoc biochem, Univ Wis, 54-56; biochemist, Vet Admin Hosp, 57-58; res assoc, Univ Wis, 58-61; asst prof, 63-67, ASSOC PROF BIOCHEM, SCH MED, UNIV MIAMI, 67- Concurrent Pos: Fel, Marine Lab, Univ Miami, 61-62. Mem: AAAS; Am Chem Soc; Am Soc Plant Physiol. Res: Plant biochemistry; terpene metabolism; enzymology; marine symbiosis. Mailing Add: PO Box 520875 Biscayne Annex Univ Miami Sch Med Miami FL 33152

ANDERSON, DAVID H, b Louisville, Ky, Dec 15, 39; m 70,. MATHEMATICS. Educ: Univ Miss, BS, 61; Univ Calif, Berkeley, MA, 66; Duke Univ, PhD(math), 69. Prof Exp: Asst prof math, Millsaps Col, 66-67; ASST PROF MATH, SOUTHERN METHODIST UNIV, 69- Mem: Math Asn Am; Am Math Soc. Res: Fourier analysis; measure theory. Mailing Add: Dept of Math Southern Methodist Univ Dallas TX 75222

ANDERSON, DAVID HAMEL, b New York, NY, May 21, 31; m 56; c 3. PHYSICAL CHEMISTRY, CHEMICAL PHYSICS. Educ: Northwestern Univ, BS, 53; Harvard Univ, MA, 55; Univ Ill, PhD(phys chem), 59. Prof Exp: Staff mem phys res, Sandia Corp, 59-62, div supvr, 62-71, DEPT MGR, SANDIA LABS, 71- Mem: Am Phys Soc; fel Am Inst Chemists. Res: Chemistry and physics of solids, using radio-frequency spectroscopic techniques, especially magnetic, ferroelectric, explosive and pyrotechnic materials. Mailing Add: Sandia Labs PO Box 5800 Albuquerque NM 87115

ANDERSON, DAVID J, b Rochester, NY, Apr 16, 39; m 67; c 1. NEUROPHYSIOLOGY, BIOENGINEERING. Educ: Rensselaer Polytech Inst, BSEE, 61; Univ Wis, MS, 63, PhD(elec eng), 67. Prof Exp: Trainee, Lab Neurophysiol, Univ Wis-Madison, 67-69; asst prof elec eng, Univ & asst prof hearing res, Inst, 70-75, ASSOC PROF ELEC & COMPUT ENG, UNIV MICH, ANN ARBOR & ASSOC PROF OTORHINOLARNGOLOGY, KRESGE HEARING RES INST, 75- Concurrent Pos: Fel life sci, NASA-Johnson Space Ctr, 75-76. Mem: Soc Neurosci; Acoust Soc Am; Inst Elec & Electronic Engr. Res: Neurophysiology of auditory and vestibular systems; application of digital computers to medicine and biomedical research. Mailing Add: Kresge Hearing Res Inst Univ of Mich Ann Arbor MI 48104

ANDERSON, DAVID KENT, b Waukegan, Ill, June 4, 37; m 59. ATOMIC PHYSICS, MOLECULAR PHYSICS. Educ: Univ Chicago, BS, 59, MS, 60, PhD(physics), 64. Prof Exp: From asst prof to assoc prof physics, Mont State Univ, 64-69; with Va Assoc Res Ctr, Newport News, 69-73; MEM FAC, DEPT PHYSICS, MONT STATE UNIV, 73- Concurrent Pos: Consult, Telemation Inc, 66- Res: Excited state atomic lifetimes; atomic and molecular collisions; plasma physics. Mailing Add: Dept of Physics Mont State Univ Bozeman MT 59715

ANDERSON, DAVID LEONARD, b Portland, Ore, Dec 19, 19; m 47; c 4. PHYSICS. Educ: Harvard Univ, SB, 41, AM, 47, PhD, 50. Prof Exp: Scientist, Manhattan Proj, Los Alamos Lab, NMex, 43-46; from asst prof to assoc prof physics, 48-62, chmn dept, 62-68, PROF PHYSICS, OBERLIN COL, 62- Concurrent Pos: Res fels, Univ Birmingham, 54-55, Harvard Univ, 61-62 & 75-76 & Univ Edinburgh, 68-69. Mem: AAAS; Am Phys Soc; Am Asn Physics Teachers; Hist Sci Soc. Res: Nuclear physics; history of modern physics. Mailing Add: Dept of Physics Oberlin Col Oberlin OH 44074

ANDERSON, DAVID MARTIN, b Boston, Mass, July 19, 30; m 58; c 4. ENVIRONMENTAL HEALTH. Educ: Northeastern Univ, BS, 53; Harvard Univ, SM, 55, PhD(indust hyg), 58. Prof Exp: Pub health engr, USPHS, 58-60; indust health engr, 60-67, asst mgr, 67-71, MGR ENVIRON QUAL CONTROL, BETHLEHEM STEEL CORP, 71- Concurrent Pos: Chmn coun tech adv, Pa Dept Health, 64- Mem: AAAS; Am Chem Soc; Am Inst Chem Eng; Indust Hyg Asn; Air Pollution Control Asn. Res: Air cleaning methods and devices; air pollution studies; control of industrial airborne contaminants. Mailing Add: Environ Qual Control Div Bethlehem Steel Corp Bethlehem PA 18016

ANDERSON, DAVID PREWITT, b Twin Falls, Idaho, Sept 14, 34; m 62; c 2. VETERINARY MICROBIOLOGY. Educ: Wash State Univ, BS, 59, DVM, 61; Univ Wis, MS, 64, PhD(vet sci, med microbiol), 65. Prof Exp: Asst prof vet sci & asst dir biotron, Univ Wis, 65-69; prof med microbiol & dir poultry dis res ctr, 69-70, chmn dept avian med, Col Vet Med, 70-71, assoc dean res & grad affairs, 71-73, prof avian med, 73-75, DEAN COL VET MED, UNIV GA, 75- Concurrent Pos: Ed, J Avian Dis, 73- Mem: Vet Med Asn; Am Asn Avian Path; Am Col Vet Microbiol. Res: Avian diseases, particularly those caused by mycoplasmas. Mailing Add: Col of Vet Med Univ of Ga Athens GA 30602

ANDERSON, DAVID ROBERT, b Cuba, NY, Feb 10, 40; m 63. ZOOLOGY, PARASITOLOGY. Educ: Lycoming Col, BA, 62; Pa State Univ, MS, 64; Colo State Univ, PhD(parasitol & zool), 67. Prof Exp: Asst prof, 67-74, ASSOC PROF ZOOL, PA STATE UNIV, FAYETTE, 74- Concurrent Pos: NSF inst grant, 68-69. Mem: Am Soc Parasitol; Soc Protozool; Wildlife Dis Asn; Wildlife Soc; Am Inst Biol Sci. Res: Ecology of helminths in relation to host's environment; all phases of research on coccidia, especially ecology, immunity and life cycle. Mailing Add: Dept of Biol Pa State Univ Fayette Campus Box 519 Uniontown PA 15401

ANDERSON, DAVID VICTOR, physics, see 12th edition

ANDERSON, DAVID W, JR, b Worcester, Mass, May 16, 21; m 48; c 2. FOOD CHEMISTRY. Educ: Univ Mass, BS, 47, MS, 48, PhD, 50. Prof Exp: Head, Nutrit & Bact Res Labs, H J Heinz Co, 50-53; dir tech serv, Spec Prod Div, Borden Co, 52-54, dir res, Pharmaceut Div, 54-67, Cosmetics & Toiletries Div, 67-69; dir, Res Labs, 69-73, V PRES RES & DEVELOP, MAX FACTOR & CO, INC, 73- Mem: Am Chem Soc; Am Soc Microbiol; Soc Cosmetic Chemists. Res: Nutrition; bacteriology. Mailing Add: Max Factor & Co Inc Res Labs 1655 N McCadden Pl Hollywood CA 90028

ANDERSON, DAVID WALTER, b Heron Lake, Minn, June 18, 37; m 60; c 2. MEDICAL PHYSICS, NUCLEAR PHYSICS. Educ: Hamline Univ, BS, 59; Iowa State Univ, PhD(phyiscs), 65. Prof Exp: Asst prof physics & radiation physics, 66-69, assoc prof radiol sci, 69-75, PROF RADIOL SCI, HEALTH SCI CTR, UNIV OKLA, 75- Concurrent Pos: AEC fel physics, Iowa State Univ, 65-66. Mem: AAAS; Soc Nuclear Med; Am Phys Soc; Am Asn Physicists in Med; Am Col Radiol. Res: Photonuclear reaction cross sections; thermoluminescent dosimetry; radiation spectroscopy. Mailing Add: Dept of Radiol Sci Univ of Okla Health Sci Ctr Oklahoma City OK 73104

ANDERSON, DEAN ALBERT, bacterial physiology, see 12th edition

ANDERSON, DENNIS ELMO, b Dunnell, Minn, Sept 2, 34; m 54; c 3. PLANT TAXONOMY. Educ: Univ Northern Iowa, BA, 55; Iowa State Univ, MS, 58, PhD(plant taxon), 60. Prof Exp: Instr bot, Fla Presby Col, 60-61; from asst prof to assoc prof, 61-70, PROF BOT, HUMBOLDT STATE UNIV, 70- Mem: AAAS; Am Soc Plant Taxon; Int Asn Plant Taxon. Res: Plant anatomy and morphology; cytology; biosystematics. Mailing Add: Dept of Bot Humboldt State Univ Arcata CA 95521

ANDERSON, DON LYNN, b Frederick, Md, Mar 5, 33; m 56; c 2. GEOPHYSICS, SEISMOLOGY. Educ: Rensselaer Polytech Inst, SB, 55; Calif Inst Technol, MS, 59, PhD(geophys, math), 62. Prof Exp: Geophysicist, Air Force Cambridge Res Ctr, Bedford, Mass, 56-58 & Arctic Inst NAm, Boston, Mass, 58-62; res fel, 62-63, asst prof, 63-64, ASSOC PROF GEOPHYS, CALIF INST TECHNOL, 64-, DIR SEISMOL LAB, 67- Concurrent Pos: Consult, var NASA, Nat Acad Sci & NSF comts; consult, Chevron Oil Co, 71-; Fulbright-Hayes Found sr scholar, 75. Honors & Awards: Macelwane Award, Am Geophys Union, 66. Mem: Am Geophys Union; Am Acad Arts & Sci; Sigma Xi; AAAS; Seismol Soc Am. Res: Structure and composition of the earth and planets; earthquake prediction. Mailing Add: Seismol Lab 252-21 Calif Inst of Technol Pasadena CA 91125

ANDERSON, DONALD ARTHUR, b Can, Feb 25, 18; m 49; c 2. ELECTRONIC PHYSICS. Educ: Mt Allison Univ, BA, 42; McGill Univ, PhD(nuclear physics), 49. Prof Exp: Lab instr physics, Mt Allison Univ, 41-43; jr physicist radar electronics, Nat Res Coun, Can, 43-45; res physicist electronics, Can Marconi Co, 49-54, chief physics, 54-58; PRES SEMICONDUCTORS & PHOTOCONDUCTORS, NAT SEMICONDUCTORS LTD, 58- Mem: Optical Soc Am; sr mem Inst Elec & Electronics Eng. Res: Photoconductor physics; semiconductor and solid state physics. Mailing Add: Nat Semiconductors Ltd 2150 Ward St St Laurent PQ Can

ANDERSON, DONALD FREDERICK, organic chemistry, see 12th edition

ANDERSON, DONALD GORDON MARCUS, b Sarnia, Ont, Jan 4, 37. APPLIED MATHEMATICS. Educ: Univ Western Ont, BSc, 59; Harvard Univ, AM, 60, PhD(appl math), 63. Prof Exp: Lectr & res fel, 63-65, asst prof, 65-69, PROF APPL MATH, HARVARD UNIV, 69- Mem: AAAS; Am Math Soc; Math Asn Am; Soc Indust & Appl Math; Am Phys Soc. Res: Numerical mathematics and scientific computation; applications of mathematics and computing in the physical sciences; chemical physics; mathematical economics. Mailing Add: Aiken Comput Lab Harvard Univ Cambridge MA 02138

ANDERSON, DONALD GRIGG, b New York, NY, Aug 2, 13; m 38; c 3. MEDICINE. Educ: Harvard Univ, AB, 35; Columbia Univ, MD, 39. Hon Degrees: DSC, NY Med Col, 61. Prof Exp: Intern, Boston City Hosp, 39-41; resident path, St Luke's Hosp, New York, 41; asst med, Col Physicians & Surgeons, Columbia Univ, 41-42; instr, Sch Med, Boston Univ, 42-45, asst prof & dean, 45-47; secy, Coun Med Educ & Hosps, AMA, 47-53; dean, 53-66, PROF MED, SCH MED & DENT, UNIV ROCHESTER, 53-, PROF PSYCHIAT, 73- Concurrent Pos: Res fel, Evans Mem Hosp, Boston, 43-45; asst resident, Presby Hosp, New York, 41-42; resident, Evans Mem Hosp, Boston, 42-43; pres, Nat Found Med Educ, 66-69. Mem: Fel AMA; Asn Am Med Cols (pres, 61-62); Am Fedn Clin Res. Res: Clinical application of penicillin and other chemotherapeutic agents. Mailing Add: Sch of Med & Dent Univ of Rochester Rochester NY 14642

ANDERSON, DONALD HERVIN, b Seattle, Wash, July 5, 16; m 40; c 2. ANALYTICAL CHEMISTRY. Educ: Univ Wash, BS, 38, PhD(chem), 43. Prof Exp: Instr physics, Univ Wash, 43; from instr to asst prof chem, Univ Idaho, 44-47; mem staff, 47-56, DIR INDUST LAB, EASTMAN KODAK CO, 56- Honors & Awards: Distinguished Serv Award Outstanding Achievement Anal Chem, Am Chem Soc, 71

& Soc Appl Spectros, 75. Mem: Soc Appl Spectros; Am Chem Soc. Res: Absorption and emission spectroscopy; instrumental methods of analysis; analytical chemical research; corrosion and surface treatment of metals; oceanographic chemistry; absorption phenomena. Mailing Add: Indust Lab Kodak Park Div Eastman Kodak Co Rochester NY 14650

ANDERSON, DONALD LEE, veterinary medicine, radiation biology, see 12th edition

ANDERSON, DONALD LINDSAY, b Cambridge, Mass, July 2, 25; m 48; c 3. ANIMAL NUTRITION. Educ: Univ Mass, BS, 50; Univ Conn, MS, 52; Cornell Univ, PhD, 55. Prof Exp: Asst nutrit, Univ Conn, 50-52; asst, Cornell Univ, 52-54, res assoc, 54-55; from asst prof to assoc prof, 55-71, PROF NUTRIT, UNIV MASS, 71- Mem: AAAS; Am Inst Nutrit; Poultry cci Asn. Res: Poultry nutrition; physiology; animal biochemistry; relationships between environmental stress and nutritional physiology. Mailing Add: Dept of Vet & Animal Sci Univ of Mass Amherst MA 01002

ANDERSON, DONALD MORGAN, b Washington, DC, Dec 27, 30. SYSTEMATIC ENTOMOLOGY. Educ: Miami Univ, BA, 53; Cornell Univ, PhD(entom), 58. Prof Exp: Res asst entom, Ohio Agr Exp Sta, 53; asst prof biol, State Univ NY Col Buffalo, 59-60; RES ENTOMOLOGIST, AGR RES SERV, USDA, 60- Mem: AAAS; Entom Soc Am; Am Inst Biol Sci; Soc Syst Zool. Res: Taxonomy of immature Coleoptera in general, and of adult and immature Curculionidae. Mailing Add: Syst Entom Lab USDA c/o US Nat Mus Washington DC 20560

ANDERSON, DONALD OLIVER, b Vancouver, BC, May 22, 30; m 56; c 4. EPIDEMIOLOGY, MEDICAL CARE ADMINISTRATION. Educ: Univ BC, BA, 50, MD, 54; FRCP(C), 59; Harvard Univ, SM, 61. Prof Exp: Asst prof prev med & med, 61-64, assoc prof, 64-68, head dept, 68-70, PROF HEALTH CARE & EPIDEMIOL, FAC MED, UNIV BC, 68-, DIR DIV HEALTH SERV RES & DEVELOP, HEALTH SCI CTR, 72- Concurrent Pos: Mead Johnson residency scholar med, 57-58; McLaughlin traveling fel, 60-61; Can Coun fel, 64; mem task force on cost of health serv, Dept Nat Health & Welfare, Can, 69; mem health serv res study sect, Nat Ctr Health Serv Res & Develop, Dept Health, Educ & Welfare, 67-71; chmn specialty comt pub health, Royal Col Physicians & Surgeons Can, 70-; fac mem, London Sch Hyg & Trop Med, 70-71; chmn health serv demonstration comt, Nat Health Grant, Can, 71-74; BC rep, Fed-Provincial Health Manpower Comt, 72-; consult epidemiol & statist, BC Med Ctr, 73-; mem exec comt, Int Epidemiol Asn; consult & secy health manpower planning, Province of BC; mem, Expert Adv Panel on Air Pollution, WHO, 68-, temp adv, Div Strengthening Health Servs, Columbia, 71-, Geneva, 75; mem study of int migration of health workers, Geneva, 75; mem steering comt children's dent health res study for minister of health, BC, 74-76. Mem: Am Col Physicians; Am Pub Health Asn; Asn Teachers Prev Med; Can Med Asn. Res: Epidemiology of non-infectious disease, especially non-tuberculous respiratory diseases; health manpower studies; health sciences education; medical school selection practices; hospital utilization; health resources planning; international studies of medical care utilization; internal medicine. Mailing Add: Div Health Serv Res & Develop Univ BC Health Sci Ctr Vancouver BC Can

ANDERSON, DONALD REX, b Ephraim, Utah, June 10, 16; m 47; c 3. RADIOBIOLOGY. Educ: Univ Utah, BA, 49, MA, 52, PhD(exp biol), 56. Prof Exp: Head exp biol group, Radiobiol Lab, Air Univ & Univ Tex, 57-60; chief exp radiobiol br, 60-68, DEP CHIEF, RADIOBIOL DIV, US AIR FORCE SCH AEROSPACE MED, 68- Mem: AAAS; Radiation Res Soc; Transplantation Soc. Res: Biological effects of ionizing radiations; protection and therapy. Mailing Add: 6010 Winterhaven San Antonio TX 78239

ANDERSON, DONALD T, b Lampman, Sask, Oct 2, 25; m 59; c 5. ECONOMIC GEOLOGY. Educ: Queen's Univ, Ont, BSc, 51; Univ Man, MSc, 59, PhD(geol), 63. Prof Exp: Field geologist, Falconbridge Nickel Mines, Ltd, 50-62; photogeologist, Geol Surv Can, 62-65; ASSOC PROF PHOTOGEOL, UNIV MAN, 65- Concurrent Pos: Asst head dept photogeol, Univ Man, 68. Mem: Fel Geol Asn Can; Can Inst Min & Metall. Res: Application of photo interpretation to solution of regional geological structure in areas of economic interest. Mailing Add: Dept of Earth Sci Univ of Man Winnipeg MB Can

ANDERSON, DONALD WERNER, b Atlantic, Iowa, Apr 16, 38; m 66; c 2. MATHEMATICS. Educ: Calif Inst Technol, BS, 60; Univ Calif, Berkeley, PhD(math), 64. Prof Exp: Nat Acad Sci-Nat Res Coun fel, Oxford Univ, 64-65; from asst prof to assoc prof math, Mass Inst Technol, 65-71; PROF MATH, UNIV CALIF, SAN DIEGO, 71-, CHMN DEPT, 73- Concurrent Pos: Sloan fel, 67-68. Mem: Am Math Soc. Res: K-theory; generalized cohomology theories. Mailing Add: Dept of Math Univ of Calif at San Diego La Jolla CA 92037

ANDERSON, DOUGLAS DORLAND, b Olympia, Wash, June 1, 36; m 63. ANTHROPOLOGY, ARCHAEOLOGY. Educ: Univ Wash, Seattle, BA, 60; Brown Univ, MA, 62; Univ Pa, PhD(anthrop), 67. Prof Exp: From instr to asst prof, 65-70, ASSOC PROF ANTHROP & CUR RES, HAFFENREFFER MUS, BROWN UNIV, 70- Mem: AAAS; Soc Am Archaeol; Am Anthrop Asn. Res: American archaeology; prehistoric archaeology and ecology in Alaska. Mailing Add: Dept of Anthrop Brown Univ Providence RI 02912

ANDERSON, DOUGLAS RICHARD, b Memphis, Tenn, Apr 7, 38; 64; c 3. OPTHALMOLOGY. Educ: Univ Miami, AB, 58; Wash Univ, MD, 62. Prof Exp: Rotating intern, Univ Hosps, Western Reserve Univ, 62-63; staff assoc, Nat Cancer Inst, 63-65; resident opthal, Med Ctr, Univ Calif, San Francisco, 65-68; asst prof, 69-74, ASSOC PROF OPTHAL, SCH MED, UNIV MIAMI, 74- Concurrent Pos: Res fel, Howe Lab Opthal, Mass Eye & Ear Infirmary, Boston, 68-69; Nat Eye Inst res grant, 69-77; Res to Prevent Blindness, Inc prof, 69-74. Mem: Asn Res Vision & Opthal. Res: Electron microscopy of cartilage, viruses, mycoplasma, leukemia, eye and optic nerve; eye diseases; especially papilledema,, optic atrophy and glaucoma, using electron microscopy and other techniques. Mailing Add: Bascom Palmer Eye Inst Univ Miami Sch Med PO Box 875 Miami FL 33152

ANDERSON, DUWAYNE MARLO, b Lehi, Utah, Sept 9, 27; m 53; c 3. EARTH SCIENCES, SOIL PHYSICS. Educ: Brigham Young Univ, BS, 54; Purdue Univ, PhD(soil chem), 58. Prof Exp: From asst prof to prof soil physics, Univ Ariz, 58-63; soil geologist, 63-67, CHIEF EARTH SCI BR, US ARMY COLD REGIONS RES & ENG LAB, 67- Concurrent Pos: Soil physicist, Ariz Agr Exp Sta, 61-63; vis prof, Dartmouth Col, 65, adj prof earth sci, 75-76; guest researcher, Royal Inst Technol, Sweden & vis prof, Univ Stockholm, 67; vis scientist, Soil Sci Soc Am, 68-71; mem molecular anal team, Viking Sci Team, 69-; distinguished vis prof, Univ Wash, 70, vis prof, 71, Sloan Found distinguished vis lectr, 75; mem, US Deleg 2nd Int Conf on Permafrost, Yakutsk, USSR, 73; mem panel on permafrost, Nat Acad Sci, 73. Honors & Awards: Dept of Army Commendation for Res Achievement, US Army Cold Regions Res & Eng Lab, 64, 65 & 68, Sci Achievement Award, 68. Mem: AAAS; Soil Sci Soc Am; Am Geophys Union; Int Soc Soil Sci; Int Asn Study Clays. Res: Physical chemistry of soils; thermodynamics of soil processes; surface phenomena in soils; phase relationships in permafrost and frozen ground; physicochemical processes in cold environments; physics and chemistry of planetary surface materials. Mailing Add: Earth Sci Br PO Box 282 US Army Cold Reg Res & Eng Lab Hanover NH 03755

ANDERSON, DWANE ELMER, mathematics, see 12th edition

ANDERSON, DWIGHT LYMAN, b Cokato, Minn, Oct 28, 35; m 60; c 2. MICROBIOLOGY. Educ: Univ Minn, BA, 57, MS, 59, PhD(microbiol), 61. Prof Exp: Asst bact, 57-60, from instr microbiol to asst prof microbiol & dent, 61-66, ASSOC PROF MICROBIOL & DENT, UNIV MINN, MINNEAPOLIS, 66- Concurrent Pos: Com Solvents Corp fel, Univ Minn, Minneapolis, 61-62. Mem: AAAS; Am Soc Microbiol; Electron Micros Soc Am; Brit Soc Gen Microbiol; Int Asn Dent Res. Res: Structure and genetics of microorganisms. Mailing Add: Dept of Microbiol & Dent Univ of Minn Owre Hall Minneapolis MN 55455

ANDERSON, EDMUND GEORGE, b Seattle, Wash, Sept 1, 28; m 60; c 4. PHARMACOLOGY. Educ: Univ Wash, BS, 50, MS, 55, PhD(pharmacol), 57. Prof Exp: Asst pharmacol, Univ Wash, 53-54; instr, Seton Hall Col Med, 57-60; from asst prof to assoc prof, 60-70, PROF PHARMACOL, UNIV ILL COL MED, 70-, HEAD DEPT, 76- Mem: AAAS; Am Soc Pharmacol & Exp Therapeut; Am Soc Neurosci. Res: Neuropharmacology; central synaptic transmission and biogenic amines. Mailing Add: 700 S Lombard Oak Park IL 60304

ANDERSON, EDMUND HUGHES, b Camden, Ark, Jan 8, 24; div; c 3. MATHEMATICS. Educ: Tex A&M Univ, BS, 48; La State Univ, MS, 62, PhD(math), 67. Prof Exp: Geophysicist, Shell Oil Co, 48-62; asst prof math, Univ NDak, 67-68; asst prof, 68-69, ASSOC PROF MATH, MISS STATE UNIV, 69- Mem: Am Math Soc. Res: Upper semicontinuous decompositions of three-space; Poincare's conjecture. Mailing Add: Dept of Math Miss State Univ Mississippi State MS 39762

ANDERSON, EDWARD EVERETT, b Peabody, Mass, Dec 20, 19; m 48; c 5. FOOD SCIENCE. Educ: Mass State Col, BS, 41, MS, 42; Univ Mass, PhD, 49. Prof Exp: Fed food inspector, War Food Admin, USDA, 42-44; from asst res prof to assoc res prof food technol, Univ Mass, 48-56; food technologist, A D Little, Inc, 56-63; chief plant prod br, Food Div, US Army Natick Labs, 63-69, SPEC ASST, DEPT OF DEFENSE FOOD PROG, US ARMY NATICK RES & DEVELOP COMMAND, 69- Mem: Fel AAAS; Inst Food Technol; fel Pub Health Asn; Soc Nutrit Educ; Sigma Xi. Res: Administration and evaluation of food research and development programs; design and development of new food and ration components and processing methods; sensory and objective methods of quality evaluation and improvement of existing canned or frozen foods; technological factors associated with flavor and packaging problems in foods. Mailing Add: Off of Tech Dir US Army Natick Res & Develop Com Natick MA 01760

ANDERSON, EDWARD FREDERICK, b Covina, Calif, June 17, 32; m 56; c 3. BOTANY, PLANT TAXONOMY. Educ: Pomona Col, BA, 54; Claremont Grad Sch, MA, 59, PhD(bot), 61. Prof Exp: Instr bot, Pomona Col, 61-62; asst prof biol, 62-67, ASSOC PROF BIOL, WHITMAN COL, 67- Concurrent Pos: Fulbright lectr, Ecuador, 65-67 & Malaysia, 69-70. Mem: Am Soc Plant Taxon; Am Inst Biol Sci; Bot Soc Am; Int Asn Plant Taxon; Orgn Trop Biol. Res: Taxonomy of several genera of Cactaceae, such as Lophophora, Obregonia, Thelocactus, Turbinicarpus, Stombocactus, Aztekium and Pelecyphora; Cactaceae of the Galapagos Islands. Mailing Add: Dept of Biol Whitman Col Walla Walla WA 99362

ANDERSON, EDWIN J, b Concord, Mass, Mar 16, 39; m 60; c 1. PALEOECOLOGY. Educ: Cornell Univ, AB, 61; Brown Univ, ScM, 64, PhD(geol), 67. Prof Exp: Instr geol, Univ NH, 65-66; ASST PROF GEOL, TEMPLE UNIV, 67- Concurrent Pos: Res Corp grant, 67; asst trip leader, NAm Paleont Conv, 69; partic, Penrose Conf, 70. Mem: AAAS; Soc Study Evolution; Soc Econ Paleontologists & Mineralogists; Int Paleont Union; Brit Palaeont Asn. Res: Paleoecology and paleoenvironments of Paleozoic skeletal limestones for the purpose of providing a base for community analysis and evaluation of speciation within selected invertebrate groups. Mailing Add: Dept of Geol Temple Univ Philadelphia PA 19122

ANDERSON, ELMER AGER, physical chemistry, see 12th edition

ANDERSON, ELMER E, b Ottawa, Ill, June 28, 22; m 43; c 5. SOLID STATE PHYSICS. Educ: Occidental Col, AB, 50; Univ Ill, MS, 56; Univ Md, PhD(physics), 64. Prof Exp: Instr physics, Deep Springs Col, 52-55; asst, Univ Ill, 55-57; res physicist & chief electromagnetics div, Naval Ord Lab, Silver Spring, Md, 57-65; from assoc prof to prof physics & chmn dept, 65-74, DEAN SCH ARTS & SCI, CLARKSON COL TECHNOL, 74- Mem: Fel Am Phys Soc; Am Asn Physics Teachers. Res: Magnetism in solids; critical phenomena; transport properties of solids. Mailing Add: RD 4 Potsdam NY 13676

ANDERSON, ERNEST CARL, b Rock Island, Ill, Aug 23, 20; m 42; c 3. BIOPHYSICS. Educ: Augustana Col & Theol Sem, AB, 42; Univ Chicago, PhD(chem), 49. Prof Exp: Asst anal chem, metall lab, Univ Chicago, 42-44; asst anal chem, 44-46, MEM STAFF BIOPHYS, LOS ALAMOS SCI LAB, 49- Concurrent Pos: Rask-Orsted fel, Copenhagen Univ, 51-52. Honors & Awards: E O Lawrence Award, AEC, 66. Mem: Fel AAAS; Biophys Soc. Res: Natural radiocarbon; liquid scintillation counters; low-level radio activity measurements; cellular biochemistry. Mailing Add: 1610 S Sage Los Alamos NM 87544

ANDERSON, ERNEST CLIFFORD, b Ashton, Idaho, Mar 16, 13; m; c 6. BIOLOGY. Educ: Univ Idaho, BS, 40; Ore State Col, MS, 42, PhD(entom), 47. Prof Exp: Field asst agr exp sta, Ore State Col, 41; sr clerk, USDA, 42; asst entomologist, USPHS, 42-44; asst entom, Ore State Col, 45-46; instr, 46, from asst prof to assoc prof, 48-57, PROF BIOL, EASTERN ORE COL, 57-, CHMN DEPT, 69- Res: Sod webworms and anopheles mosquitoes in Oregon; spider mites in Eastern Oregon. Mailing Add: Dept of Biol Eastern Ore Col La Grande OR 97850

ANDERSON, ETHEL IRENE, b Brooklyn, NY, June 20, 24. BIOCHEMISTRY. Educ: Ursinus Col, BS, 45; Univ Del, MS, 47, PhD(bio-org chem), 50. Prof Exp: Res fel biochem, Inst Cancer Res, Philadelphia, 50-52, res assoc, 52-58; sr res biochemist, Colgate Palmolive Co, 58-62; RES ASSOC BIOCHEM, DEPT COLUMBIA UNIV, 62- Concurrent Pos: NIH res grant, Nat Eye Inst, 71-76. Mem: AAAS; Am Chem Soc; Asn Res Vision & Ophthal. Res: Effects of the limiting cell layers on corneal function; the metabolic basis of transendothelial fluid transport and the nature and influence of epithelial proteolytic enzymes in pathophysiologies. Mailing Add: Dept of Ophthal Columbia Univ New York NY 10032

ANDERSON, EUGENE N, JR, b Washington, DC, Feb 2, 41; m 62; c 2. ANTHROPOLOGY. Educ: Harvard Univ, BA, 62; Univ Calif, Berkeley, PhD(anthrop), 67. Prof Exp: From actg asst prof to asst prof, 66-71, ASSOC PROF ANTHROP, UNIV CALIF, RIVERSIDE, 71- Mem: Fel Am Anthrop Asn; Am Folklore Soc; Am Ornith Union; Brit Ornith Union. Res: Fishing societies; Asia;

cognitive studies. Mailing Add: Dept of Anthrop 1224 Watkins Hall Univ of Calif Riverside CA 92502

ANDERSON, EVERETT, cytology, see 12th edition

ANDERSON, FLOYD EDMOND, b Racine, Wis, Oct 16, 15; m 47; c 1. ORGANIC CHEMISTRY. Educ: Univ Wis, BS, 39; Univ Mich, MS, 46, PhD(pharmaceut chem), 49. Prof Exp: Analyst, Ditzler Color Co, Mich, 39-40; analyst, Lakeside Labs, Inc, Wis, 40-41, org res chemist, 41-44; res engr, Off Sci Res & Develop, Battelle Mem Inst, 44-45; asst prof chem, WVa Univ, 48-49; sr org res chemist, Nepera Chem Co, 49-50, dir org chem res, 51-57; sr res assoc, Warner Lambert Res Inst, 57-60; head org res & develop, Pharmaceut Div, Wallace & Tiernan, Inc, 60-63; assoc prof med chem, Northeastern Univ, 64-66; chemist, Bur Med, Div New Drugs, Food & Drug Admin, Washington, DC, 66-70, drug control specialist, Bur Narcotics & Dangerous Drugs, 70-73; CHIEF CHEM-BIOL COORD, SPEC PROGS DIV, DRUG ENFORCEMENT ADMIN, DEPT JUSTICE, WASHINGTON, DC, 73- Concurrent Pos: Prof chem & chmn dept, Luther Rice Col, 67- Mem: Am Chem Soc. Res: Development of analytical methods; synthesis of plastics and polymers; structure activity relationships in biological activities; antituberculous agents; local anesthetics; insecticides; hypnotics; diuretics; antimetabolites; narcotics; enzyme inhibition; synthesis of organic heterocyclics. Mailing Add: 2500 N Van Dorn St Apt 1327 Alexandria VA 22302

ANDERSON, FRANCIS DAVID, b Cabano, Que, May 12, 25; m 48; c 3. GEOLOGY. Educ: Univ NB, BSc, 48; McGill Univ, MSc, 51, PhD(geol), 56. GEOLOGIST, GEOL SURV CAN, 52- Mem: AAAS; fel Geol Soc Am; fel Geol Asn Can. Res: General problems of field geology, particularly structural and economic geology. Mailing Add: Geol Surv of Can 601 Booth St Ottawa ON Can

ANDERSON, FRANK DAVID, b Duncan, Okla, May 24, 27; m 54; c 3. ANATOMY. Educ: Westminster Col, Mo, BA, 48; Cornell Univ, MS, 50, PhD, 52. Prof Exp: Instr anat, Med Col, Cornell Univ, 52-56; from asst prof to assoc prof, Seton Hall Col Med & Dent, 56-67; PROF ANAT & CHMN DEPT, MED COL WIS, 67- Concurrent Pos: Consult, Med Ctr, Univ Ala, 59; consult, Inst Crippled & Disabled, Med Ctr, NY Univ, 63-64. Mem: Am Acad Neurol; Am Asn Anat. Res: Neuroanatomy; anatomy of ductus arteriosus in newborn; central pathways of vagus nerve and of trigeminal nerve; experimental degeneration studies of spinal cord and brain stem; segmental anatomy of the spinal cord; ascending fiber systems in the spinal cord and brain stem. Mailing Add: Dept of Anat Med Col of Wis Milwaukee WI 53233

ANDERSON, FRANK WALLACE, b Milwaukee, Wis, Feb 12, 21; m 49; c 2. PHYSICAL CHEMISTRY, ELECTRON MICROSCOPY. Educ: Univ Ariz, BS, 51, MS, 52. Prof Exp: Res chemist, Martinez Res Lab, Shell Oil Co, 52-61, Shell Develop Co, Calif, 61-62; sr res chemist, Sprague Elec Co, Mass, 62-68; ADV CHEMIST, IBM CORP, E FISHKILL, 68- Mem: Am Chem Soc; Electron Micros Soc Am. Res: Transmission and scanning electron microscopy; analytical methods development; organic separations; electrophoresis in non-aqueous solvents, colloids and surfaces. Mailing Add: Carol Dr Hopewell Junction NY 12533

ANDERSON, FRANK WYLIE, b Omaha, Nebr, Feb 5, 28. MATHEMATICS. Educ: Univ Iowa, BA, 51, MS, 52, PhD, 54. Prof Exp: From instr to asst prof math, Univ Nebr, 54-57; from asst prof to assoc prof, 57-69, PROF MATH, UNIV ORE, 69- Mem: Am Math Soc; Math Asn Am. Res: Algebra; topological algebra. Mailing Add: Dept of Math Univ of Ore Eugene OR 97403

ANDERSON, FRANZ ELMER, b Cleveland, Ohio, July 23, 38; m 63; c 2. OCEANOGRAPHY, MARINE GEOLOGY. Educ: Ohio Wesleyan Univ, BA, 60; Northwestern Univ, MS, 62; Univ Wash, PhD(oceanog), 67. Prof Exp: Res asst geol, Northwestern Univ, 60-62; res asst oceanog, Univ Wash, 62-67, lectr, 66; ASSOC PROF OCEANOG, UNIV NH, 67- Concurrent Pos: NSF res grant, 66-68; Cottrell res grant, 70-72; Off Naval Res grant, 72-74; Fulbright grant, Turkey, 73-74. Mem: AAAS; Geol Soc Am; Soc Econ Paleontologists & Mineralogists. Res: Distribution of particulate matter in estuarine and marine environments; effects of small amplitude waves on estuarine erosion. Mailing Add: Dept of Earth Sci Univ of NH Durham NH 03824

ANDERSON, GARY, b Norristown, Pa, Oct 5, 48; m 70. PHYSIOLOGICAL ECOLOGY, PARASITOLOGY. Educ: Univ RI, BS, 70; Univ SC, MS, 72, PhD(marine sci), 74. Prof Exp: Asst prof zool, Univ Maine, Orono, 74-75; ASST PROF BIOL, UNIV SOUTHERN MISS, 75- Mem: AAAS; Am Soc Limnol & Oceanog; Am Soc Zoologists; Sigma Xi. Res: Ecology of host-symbiont interrelationships, host finding behavior, energy flow in host-parasite systems, physiological ecology of parasites; physiological ecology of aquatic invertebrates living in stressed systems. Mailing Add: Dept of Biol Univ Southern Miss Box 353 Southern Sta Hattiesburg MS 39401

ANDERSON, GARY CHESTER, polymer chemistry, surface chemistry, see 12th edition

ANDERSON, GARY L, applied mathematics, applied mechanics, see 12th edition

ANDERSON, GAYLORD WEST, b Minneapolis, Minn, Dec 31, 01; m 29; c 1. PUBLIC HEALTH. Educ: Dartmouth Col, AB, 22; Harvard Univ, MD, 28, DrPH, 42. Prof Exp: Epidemiologist, State Dept Pub Health, Mass, 29-30, asst dir div commun dis, 30-31, dir div commun dis & dep comnr pub health, 31-37; prof prev med & pub health, 37-69, head dept, 37-45, dir sch pub health, 45-69, dean, 69-70, EMER DEAN SCH PUB HEALTH, UNIV MINN, 70- Concurrent Pos: Asst Sch Pub Health, Harvard Univ, 31-37; dir div med intel, Off Surgeon Gen, War Dept, 43-45. Honors & Awards: Sedgwick Mem Medal, Am Pub Health Asn, 63; Order of Hipolito Unanue, Govt Peru, 67. Mem: Fel Am Pub Health Asn (pres, 52); Am Epidemiol Soc (pres, 51). Res: Epidemiology; geomedicine; communicable disease control; public health administration. Mailing Add: 2261 Folwell St Paul MN 55108

ANDERSON, GEORGE ALBERT, b New Britain, Conn, May 3, 37. MATHEMATICS. Educ: Trinity Col, BS, 59; Yale Univ, MA, 61, PhD(math), 64. Prof Exp: Actg instr, Yale Univ, 63-64; from instr to asst prof math, Trinity Col, Conn, 64-71; ASSOC PROF MATH, RI COL, 71- Concurrent Pos: Consult res div, United Aircraft Corp, 61-64. Mem: Am Math Soc; Math Asn Am; Inst Math Statist. Res: Mathematics of sound theory in an inhomogeneous media; mathematical statistics; multivariate analysis; asymptotic distributions for latent roots of Wishart matrices. Mailing Add: Dept of Math RI Col Providence RI 02908

ANDERSON, GEORGE BOINE, b Buffalo, NY, June 29, 34; m 57; c 4. MARINE GEOPHYSICS, MILITARY SYSTEMS. Educ: State Univ NY Maritime Col, BS, 56; Ohio State Univ, MSc, 62. Prof Exp: Mem staff, Oceanogr of Navy, Washington, DC, 62, head hydrographic-oceanog dept, USS Tanner, New York, 62-63; oceanog prog officer, US Navy Underwater Sound Lab, 64-66; asst coordr ocean sci, US Navy Electronics Lab, 66-67; dir undersea surveillance, Ocean Sci Dept, Naval Undersea Ctr, 67-75; VPRES EARTH SCI, WHITEHALL CORP, 75- Concurrent Pos: Mem ocean res planning & adv comt, Naval Ship Systs Command, 68-69; mem, Navy Deep Ocean Technol Planning Comt, 68-69; chmn, Navy Symp Underwater Acoust, 69. Honors & Awards: Super Achievement Award, US Naval Undersea Ctr, 74. Mem: Acoust Soc Am; Am Oceanic Orgn. Res: Hydrography; geophysics; ocean engineering; development of undersea surveillance; ASW systems plus the necessary support research in underwater acoustics, oceanography and signal processing; offshore geophysical exploration and continuing research and development with undersea surveillance systems. Mailing Add: VPres Earth Sci Group Whitehall Corp PO Box 30128 Dallas TX 75230

ANDERSON, GEORGE CAMERON, b Vancouver, BC, Oct 8, 26; m 57; c 1. BIOLOGY. Educ: Univ BC, BA, 47, MA, 49; Univ Wash, PhD(zool, bot), 54. Prof Exp: Teaching asst zool, Univ BC, 47-49; teaching fel, res asst & actg instr, Univ Wash, 50-54, res assoc, 54-58, from res asst prof to res prof, 58-71, actg asst chmn res, 70-71; marine biologist, AEC, 71-72; RES PROF ZOOL, UNIV WASH, 72- Concurrent Pos: Consult, US Bur Reclamation, 55, Seattle, Wash, 59 & State Fisheries Dept, Wash, 63-; chief scientist, Deep Ocean Mining Environ Study, Nat Oceanic & Atmospheric Admin, 65-; mem ship opers panel, NSF, 70-; mem atomic safety & licensing bd panel, US Nuclear Regulating Comn, 73- Mem: AAAS; Am Soc Limnol & Oceanog; Plankton Soc Japan. Res: Biological oceanography and limnology; phytoplankton ecology; marine primary productivity. Mailing Add: Dept of Oceanog Univ of Wash Seattle WA 98105

ANDERSON, GEORGE CARL, physics, see 12th edition

ANDERSON, GEORGE R, b Columbus, Ohio, May 24, 25; m 49; c 2. VETERINARY MEDICINE, VIROLOGY. Educ: Ohio State Univ, DVM, 50; Univ Mich, MPH, 51; Am Bd Microbiol, dipl. Prof Exp: From virologist to chief virologist, State Dept Health, Ohio, 51-58, lab supvr, 58-60; chief viral vaccines unit, 60-64, CHIEF BIOLOGICS SECT, STATE DEPT HEALTH LABS, MICH, 64- Concurrent Pos: Consult, Nat Cancer Inst, 59-61. Res: Efficacy of viral vaccines in animal and in man. Mailing Add: 915 N Hagadorn East Lansing MI 48823

ANDERSON, GEORGE ROBERT, b Burlington, Iowa, Jan 10, 34; m 58; c 1. PHYSICAL CHEMISTRY. Educ: Augustana Col, Ill, BA, 56; Univ Iowa, PhD(phys chem), 61. Prof Exp: Res chemist, E I du Pont de Nemours & Co, until 68; hoofdmedewerker, State Univ Groningen, 68-69; res fel chem, Wesleyan Univ, 69-70; instr, 70-71, ASST PROF CHEM, BOWDOIN COL, 71- Concurrent Pos: Vis lectr, Swarthmore Col, 65-66; vis prof, Univ Minn, 74-75. Mem: Am Chem Soc. Res: Molecular structure and spectroscopy; ion-radical salts; electronic, infrared and Raman spectroscopy; inflection spectroscopy; hydrogen bonding, studies of vibrational states; donor-acceptor complexes, symmetry and pi molecular complexes. Mailing Add: Dept of Chem Bowdoin Col Brunswick ME 04011

ANDERSON, GEORGE WASHINGTON, b Gainesville, Fla, Oct 7, 13; m 47; c 5. ORGANIC CHEMISTRY. Educ: Univ Fla, BS, 35, PhD(org chem), 39. Prof Exp: Res chemist, 39-44, group leader, 44-56, res assoc, 56-58, RES FEL, LEDERLE LABS, AM CYANAMID CO, 58- Mem: Am Chem Soc; Brit Chem Soc. Res: Synthesis of organic medicinal compounds; sulfa drugs; synthesis of certain alpha, omega, amino alcohols from cyclic amines; antithyroid and antihistamine compounds; bronchodilators; carbonic anhydrase inhibitors; peptides. Mailing Add: Lederle Labs Am Cyanamid Co Pearl River NY 10965

ANDERSON, GEORGE WATKINS, JR, physics, see 12th edition

ANDERSON, GEORGE WILSON, microbiology, see 12th edition

ANDERSON, GERALD CLIFTON, b Barre, Vt, Dec 13, 20; m 47; c 4. ANIMAL NUTRITION. Educ: Univ Mass, BS, 43; Univ Mo, MS, 48, PhD(animal nutrit), 50. Prof Exp: Asst instr, Univ Mo, 47-50; assoc prof, 50-58, PROF ANIMAL SCI, WVA UNIV, 58-, ANIMAL HUSBANDMAN, AGR EXP STA, 58- Concurrent Pos: Assoc animal husbandman, Agr Exp Sta, WVa Univ, 50-58, head dept animal husb, 58-63. Res: Nutritional problems pertinent to swine, poultry and beef cattle. Mailing Add: Dept of Animal Indust & Vet Sci WVa Univ Morgantown WV 26506

ANDERSON, GERALD S, b Montevideo, Minn, July 28, 30; m 57; c 2. SOLID STATE PHYSICS. Educ: Luther Col, Iowa, BA, 52; Iowa State Univ, PhD(physics), 57. Prof Exp: Res assoc, Iowa State Univ, 57-58; prin scientist, Electronics Div, Gen Mills, Inc, 58-63; tech specialist, Appl Sci Div, Litton Industs, Minn, 63-69; RES MGR, MINN MINING & MFG CO, 69- Mem: Am Phys Soc; Am Vacuum Soc. Res: Thin film physics; sputtering. Mailing Add: 2453 Cohansey St St Paul MN 55113

ANDERSON, GERALD WILLIAM, plant pathology, forestry, see 12th edition

ANDERSON, GLEN DOUGLAS, b Madison, SDak, Oct 18, 30; m 53; c 4. MATHEMATICS. Educ: Drury Col, AB, 58; Univ Mich, AM, 59, PhD(math), 65. Prof Exp: From asst prof to assoc prof math, Eastern Mich Univ, 60-65; asst prof, 65-70, ASSOC PROF MATH, MICH STATE UNIV, 70- Honors & Awards: Fulbright grant, Univ Halsinki, 72-73. Mem: Am Math Soc; Math Asn Am. Res: Complex variables; quasiconformal mappings. Mailing Add: Dept of Math Mich State Univ East Lansing MI 48824

ANDERSON, GLENN ARTHUR, b Mt Vernon, Wash, Nov 30, 24; m 49. ZOOLOGY. Educ: Wash State Univ, BS, 53, MA, 58; Ore State Univ, PhD(parasitol), 64. Prof Exp: Instr zool, Ore State Univ, 63-64; asst prof biol, Ariz State Col, 64-67; assoc prof, 67-71, PROF BIOL, NORTHERN ARIZ UNIV, 71- Mem: Am Soc Parasitol. Res: Life cycles of parasitic helminths. Mailing Add: Dept of Biol Sci Northern Ariz Univ Flagstaff AZ 86001

ANDERSON, GLORIA LONG, b Altheimer, Ark, Nov 5, 38; m 60; c 1. ORGANIC CHEMISTRY. Educ: Agr, Mech & Norm Col, Ark, BS, 58; Atlanta Univ, MS, 61; Univ Chicago, PhD(org chem), 68. Prof Exp: Instr chem, SC State Col, 61-62 & Morehouse Col, 62-64; assoc prof, 68-73, CALLOWAY PROF CHEM, MORRIS BROWN COL, 73-, CHMN DEPT, 68- Mem: AAAS; Am Chem Soc; Nat Sci Teachers Asn; Nat Inst Sci. Res: Synthetic organic fluorine chemistry; fluorine-19 nuclear magnetic resonance spectroscopy; mechanism of transmission of substituent effects. Mailing Add: 560 Lynn Valley Rd SW Atlanta GA 30311

ANDERSON, GREGOR MUNRO, b Montreal, Que, Aug 18, 32; m 57; c 2. GEOCHEMISTRY. Educ: McGill Univ, BEng, 54; Univ Toronto, MASc, 56, PhD(geol), 61. Prof Exp: Geologist, Ventures, Ltd, 56-57; res assoc, Pa State Univ, 61-64, asst prof geol, 64-65; from asst prof to assoc prof, 65-72, PROF GEOL, UNIV TORONTO, 72- Concurrent Pos: Nuffield fel, 73. Mem: Geochem Soc; Soc Econ Geologists; Geol Asn Can. Res: Chemistry of solutions and melts at high temperatures and pressures. Mailing Add: Dept of Geol Univ of Toronto Toronto ON Can

ANDERSON, GREGORY JOSEPH, b Chicago, Ill, Nov 26, 44; m 66; c 1. BIOSYSTEMATICS, POLLINATION BIOLOGY. Educ: St Cloud State Col, BS, 66; Ind Univ, Bloomington, MA, 68, PhD(bot), 71. Prof Exp: Asst prof bot, Univ Nebr, Lincoln, 71-73; ASST PROF BIOL SCI, UNIV CONN, 73- Concurrent Pos: Res grants, Sigma Xi, 72, Univ Nebr Res Coun, 72 & Univ Conn Res Found, 73-75. Mem: AAAS; Am Soc Plant Taxonomists; Bot Soc Am; Int Asn Plant Taxon. Res: Systematics and evolution of vascular plants. Mailing Add: Biol Sci Group U43 Univ of Conn Storrs CT 06268

ANDERSON, GUY RICHARD, b Idaho, Nov 19, 18; m 47; c 4. SOIL MICROBIOLOGY. Educ: Univ Idaho, BS & MS, 46; Wash State Univ, PhD, 56. Prof Exp: Asst prof & asst bacteriologist, 46-57, assoc prof & assoc bacteriologist, 57-68, PROF BACT & BACTERIOLOGIST, AGR EXP STA, UNIV IDAHO, 68-, DIR WAMI PROG MED EDUC, 73- Mem: AAAS; Am Soc Microbiol. Res: Nitrogen fixation; minor element studies; cryophilic fungus on cereals. Mailing Add: Dept of Bacteriol Univ of Idaho Moscow ID 83843

ANDERSON, HARLAN DWIGHT, b Plankinton, SDak, Mar 21, 14; m 44; c 3. BIOCHEMISTRY. Educ: SDak State Univ, BS, 37; Univ Wis, MS, 39, PhD(biochem), 41. Prof Exp: Student analyst, SDak State Univ, 36-37, assoc agr chemist, 40-42; asst, Univ Wis, 37-39; biochemist, State Pub Health Labs, Mich, 42-44, asst dir, 44-50, assoc dir & chief biol prods sect, 50-64; supt, Parkedale Biol Labs, Parke, Davis & Co, 64-66, dir biol div, 66-69; vpres opers, Metab Inc, 69-71; TECH DIR, SOUTHEASTERN MICH BLOOD CTR, AM NAT RED CROSS, 71- Concurrent Pos: Mem comt plasma & plasma substitutes, Nat Res Coun, 62-64. Mem: Am Soc Microbiol. Res: Immunization, boosting and recall of immunity using multiple antigens; blood and plasma processing. Mailing Add: 100 Mack Ave Blood Ctr PO Box 351 Detroit MI 48232

ANDERSON, HAROLD J b Green Bay, Wis, Sept 4, 28; m 50; c 6. PHYSICAL CHEMISTRY. Educ: St Norbert Col, BS, 52. Prof Exp: Scientist, Marathon Div, 52-58, sr scientist, 58-62, group leader, 62-66, supvr, 66-72, mgr prod develop & improv, 72-75, MGR MAJ PROJ, CONSUMER PROD, AM CAN CO, 75- Mem: Tech Asn Pulp & Paper. Res: Decorative and functional coatings for paper and paperboard; air laid, nonwoven technology. Mailing Add: Am Can Co Day St Green Bay WI 54305

ANDERSON, HARRISON CLARKE, b Louisville, Ky, Sept 2, 32; m 61; c 4. PATHOLOGY. Educ: Univ Louisville, BA, 54, MD, 58. Prof Exp: Intern path, Mass Gen Hosp, Boston, 58-59; res fel, Sloan-Kettering Inst, 62-63; from asst prof to assoc prof, 63-71, PROF PATH, STATE UNIV NY DOWNSTATE MED CTR, 71- Concurrent Pos: Fel, Mem Hosp, New York, 60-62. Mem: Am Soc Exp Path; Am Soc Cell Biol; Am Asn Pathologists & Bacteriologists. Res: Tissue culture; experimental bone induction; calcium metabolism; electron microscopy of ossification; virus cultivation; biological calcification; fine structure of cartilage and bone; experimentally induced ossification; differentiation of skeletal tissues in culture. Mailing Add: Dept of Path State Univ NY Downstate Med Ctr Brooklyn NY 11203

ANDERSON, HARRY CHOVIN, chemistry, see 12th edition

ANDERSON, HARRY GEORGE, ornithology, see 12th edition

ANDERSON, HARVEY GENE, organic chemistry, see 12th edition

ANDERSON, HELEN LESTER, b Lexington, Ky, Oct 18, 36. NUTRITION. Educ: Univ Ky, BS, 58; Univ Wis, MS, 65, PhD(nutrit sci), 69. Prof Exp: Dietetic internship, Univ Mich Hosps, 58-59; dietitian, Roanoke Mem Hosp, Va, 59-60; Santa Barbara Cottage Hosp, 60-61; unit mgr, De La Guerra Commons, Univ Calif, Santa Barbara, 61-63; asst prof nutrit, 69-71, ASSOC PROF NUTRIT, UNIV MO-COLUMBIA, 71- Mem: NY Acad Sci; Am Pub Health Asn; Am Dietetic Asn; Am Home Econ Asn; Soc Nutrit Educ. Res: Amino acid nutrition in humans and small animals. Mailing Add: 10 Gwynn Hall Univ of Mo Columbia MO 65201

ANDERSON, HENRY LEONARD, II, inorganic-physical chemistry, see 12th edition

ANDERSON, HENRY WALTER, b Des Moines, Iowa, Dec 27, 11; m 40; c 3. HYDROLOGY, WATERSHED MANAGEMENT. Educ: Univ Calif, BS, 43, MS, 47. Prof Exp: Res forester, Univ Calif, 43-46; hydrologist, Soil Conserv, 46-51, soil scientist, 51-55, snow res leader, 55-62, water source hydrol proj leader, 63-71, CHIEF RES HYDROLOGIST, US FOREST SERV, 71- Concurrent Pos: Mem nat comt, Int Asn Soil Hydrol, 72- Honors & Awards: Jr Hydrol Award, Am Geophys Union, 46; US Forest Serv Award, 58. Mem: Fel Am Geophys Union (vpres sect hydrol, pres); Soc Am Foresters; Int Water Resources Asn. Res: Forest effects on floods; sedimentation and water yield; snow accumulation; snow melt. Mailing Add: 995 Sunnyhill Rd Lafayette CA 94549

ANDERSON, HENRY WAYNE, physical chemistry, see 12th edition

ANDERSON, HERBERT GODWIN, JR, b Roanoke, Ala, Dec 29, 31. BIOLOGICAL SCIENCES. Educ: Auburn Univ, BS, 58, MS, 60; Univ Miami, PhD(marine sci), 64. Prof Exp: Fishery biologist, Sandy Hook Marine Lab, US Bur Sport Fisheries, 62-64; assoc prof, 64-74, PROF BIOL SCI, CENT CONN STATE COL, 74- Mem: Am Soc Parasitol. Res: Animal parasites; invertebrate taxonomy. Mailing Add: Dept of Biol Cent Conn State Col New Britain CT 06050

ANDERSON, HERBERT HALE, b Dayton, Ohio, Nov 6, 13; m 71; c 1. INORGANIC CHEMISTRY. Educ: Harvard Univ, AB, 34; Mass Inst Technol, PhD(inorg chem), 37. Prof Exp: Res assoc, Harvard Univ, 38-42, Mass Inst Technol, 42-43, Manhattan Dist, Univ Chicago, 43-46 & Harvard Univ, 46-47, 50-51; from asst prof to assoc prof, 52-63, PROF CHEM, DREXEL UNIV, 63- Mem: Am Chem Soc. Res: Organometallic derivatives of silicon, germanium, and tin; halides of silicon, phosphorus and plutonium. Mailing Add: Dept of Chem Drexel Univ Philadelphia PA 19104

ANDERSON, HERBERT LAWRENCE, b New York, NY, May 24, 14; m 47; c 4. PHYSICS. Educ: Columbia Univ, AB, 35, BSEE, 36, PhD(physics), 40. Prof Exp: Asst nuclear chain reaction, Columbia Univ, 40-42; physicist, Univ Chicago, 42-44 & Los Alamos Lab, Univ Calif, 44-46; from asst prof to assoc prof, 46-50, PROF PHYSICS, UNIV CHICAGO, 50- Concurrent Pos: Dir construct 170 inch synchrocyclotron, 47-51; consult physicist, US Naval Ord Test Sta, 53-54; Guggenheim fel, 55; Fulbright scholar, Italy, 56; dir Enrico Fermi Inst Nuclear Studies, 58-63; fel, Los Alamos Sci Lab, 72- Mem: Nat Acad Sci; fel Am Phys Soc. Res: High energy physics; neutron and meson physics; nuclear energy; nuclear magnetic moments; beta ray spectroscopy; moonic atomics; high energy moon-proton scattering. Mailing Add: 4923 Kimbark Ave Chicago IL 60615

ANDERSON, HERBERT RUDOLPH, JR, b Athol, Mass, May 24, 20; m 45; c 5. PHYSICAL CHEMISTRY. Educ: Univ NH, BS, 48; Cornell Univ, PhD(phys chem), 52. Prof Exp: Asst chem, Los Alamos Sci Lab, 44-46; asst, Cornell Univ, 48-52; res assoc, Knolls Atomic Power Lab, Gen Elec Co, 52-53; res chemist, Phillips Petrol Co, 54-61; res chemist, 61-69, mgr new mat progs, Systs Develop Div, 69-70, mgr mat sci dept, 70-73, MGR PRINTING & POLYMER CHEM, GENERAL PRODUCTS DIV, IBM CORP, 73-, ON LEAVE, ADHESION SCI RES COLLAB WITH LEHIGH UNIV, 74- Concurrent Pos: Fel, Cornell Univ, 53-54; cong sci counr, Am Chem Soc, 73- Mem: Am Chem Soc; Sigma Xi. Res: Photochemistry; nuclear fuel elements; low-angle x-ray scattering; rubber and polymer chemistry; radiation chemistry; colloids; adhesion science. Mailing Add: Dept K-15 Bldg 028-2 IBM Corp Cottle Rd San Jose CA 95114

ANDERSON, HOWARD ARNE, b Minneapolis, Minn, June 10, 16; m 42; c 3. MEDICINE. Educ: Univ Minn, BS, 40, BM, 42, MD, 43, MSc, 50. Prof Exp: Assoc prof, 57-69, PROF MED, MAYO GRAD SCH MED, UNIV MINN, 69-, PROF MED, MAYO MED SCH, 73-; CONSULT INTERNAL MED, THORACIC DIS & BRONCHO-ESOPHAGOLOGY, MAYO CLIN, 50- Mem: Am Thoracic Soc; AMA; fel Am Col Physicians; Am Col Chest Physicians. Res: Pulmonary and esophageal diseases. Mailing Add: Mayo Clin Rochester MN 55901

ANDERSON, HOWARD BENJAMIN, b Escanaba, Mich, Dec 13, 14; m 46; c 2. APPLIED MATHEMATICS. Educ: Northern Mich Univ, AB, 38; Univ Mich, MA, 39; Mich Technol Univ, BS, 54. Prof Exp: Teacher pub sch, Mich, 39-42 & 45-46; assoc prof, 46-67, PROF MATH, MICH TECHNOL UNIV, 67- Concurrent Pos: Chmn group paced modular math, Mich Technol Univ, 73- Mem: Math Asn Am; Am Astron Soc. Res: Education; astronomy. Mailing Add: Dept of Math Mich Technol Univ Houghton MI 49931

ANDERSON, HOWARD T, b La Center, Wash, Mar 15, 09; m 35; c 2. ENVIRONMENTAL GEOLOGY, EXPLORATION GEOLOGY. Educ: Univ Calif, Berkeley, AB, 31, MA, 35; Univ Calif, Los Angeles, 41-42. Prof Exp: Researcher for exhibits, Gold Gate Expos Proj, Univ Calif, 38; instr earth sci, Los Angeles City Col, 38-41; geol, Univ Ariz, 42-43; geologist, Stand Oil Co Calif, 43-51, area geologist, Western Opers, Inc, 51-54, dist geologist, 55-56; chief geologist, Iranian Oil Explor & Producing Co, Masjid-i Sulaimen, 57-58, asst head explor, Tehran, 58-64; sr geologist, Western Opers, Inc, Stand Oil Co Calif, 65-71; CONSULT GEOLOGIST, 72- Mem: Fel Geol Soc Am; Am Inst Prof Geol; Am Asn Petrol Geol; Paleont Soc; Soc Econ Paleontologists & Mineralogists. Res: Petroleum exploration and relative geological problems, particularly regional relations of marine sedimentary stratigraphy; restoration of musculature of extinct fossil reptiles and mammals as interpreted from study of bony skeleton; geological and environmental history of Santa Ana River watershed. Mailing Add: 4465 Ninth St Riverside CA 92501

ANDERSON, HUGH JOHN, b Winnipeg, Man, Mar 17, 26. SYNTHETIC ORGANIC CHEMISTRY. Educ: Univ Man, BSc, 47, MSc, 49; Northwestern Univ, PhD(chem), 52. Prof Exp: Lectr chem, Univ Man, 47-49; Nat Res Coun Can fel & Corday-Morgan Commonwealth fel, Oxford Univ, 52-53; assoc prof chem, 53-64, PROF CHEM, MEM UNIV NFLD, 64- Mem: Chem Inst Can; Chem Soc; Am Chem Soc. Res: Heterocyclic compounds. Mailing Add: Dept of Chem Mem Univ of Nfld St John's NF Can

ANDERSON, HUGH RIDDELL, b Iowa City, Iowa, June 16, 32; m 56. SPACE PHYSICS. Educ: Univ Iowa, BA, 54, MS, 58; Calif Inst Technol, PhD(physics), 61. Prof Exp: Sr scientist, Jet Propulsion Lab, Calif Inst Technol, 62-65; asst prof, 65-70, ASSOC PROF SPACE SCI, RICE UNIV, 71- Mem: Am Phys Soc; Am Geophys Union. Res: Primary cosmic radiation variations in space and time caused by solar modulation; ionizing radiation in space; auroral and ionospheric physics. Mailing Add: Dept of Space Physics & Astron Rice Univ Houston TX 77001

ANDERSON, HUGH VERITY, b Chicago, Ill, Aug 25, 21; m 46; c 1. ORGANIC CHEMISTRY. Educ: Kalamazoo Col, AB, 46; Univ Ill, PhD(org chem), 49. Prof Exp: Fel, Ohio State Univ, 49-50; res chemist, Mich, 50-58, sect head, 58-65, mgr chem process res & develop, 65-69, gen mgr, Lab Procedures Div, King of Prussia, 69-73, DIR CORP PURCHASING, THE UPJOHN CO, 73- Mem: Am Chem Soc; Am Asn Bioanalysts; Am Asn Clin Chem. Res: Steroids; pharmaceutical chemistry; clinical chemistry. Mailing Add: 9820 West H Ave Kalamazoo MI 49009

ANDERSON, INGRID, b Sioux Falls, SDak, Oct 7, 30. NUTRITION. Educ: Col St Benedict, Minn, BS, 53; Univ Minn, MS, 55, PhD(nutrit), 67. Prof Exp: Instr, St Benedicts High Sch, Minn, 53-54; instr home econ, Col St Benedict, Minn, 53-54, instr home econ & biol, 55-60; res asst nutrit, Univ Minn, 63-65; asst prof, 65-70, ASSOC PROF NUTRIT & PHYSIOL, COL ST BENEDICT, MINN, 70- Concurrent Pos: Vis assoc prof, Univ Minn, 74-75. Mem: AAAS; Am Pub Health Asn; Am Home Econ Asn; Am Dietetic Asn; Soc Nutrit Educ. Res: Congenital abnormalities in folic acid deficiency; growth factors; nutrition education; competency-based dietetic education. Mailing Add: Dept of Nutrit & Physiol Col of St Benedict St Joseph MN 56374

ANDERSON, IRVIN CHARLES, b Iowa, Apr 4, 28; m 56; c 2. PLANT PHYSIOLOGY. Educ: Iowa State Univ, BS, 51; NC State Col Agr & Eng, MS, 54, PhD, 56. Prof Exp: Fel biol, Brookhaven Nat Lab, 56-58; from asst prof to assoc prof agron, 58-72, PROF AGRON & BOT, IOWA STATE UNIV, 72- Mem: Am Soc Plant Physiol; Am Soc Agron. Res: Biochemical aspects of genetic mutants; plant growth regulatory chemicals and nodulation. Mailing Add: Dept of Agron Iowa State Univ Ames IA 50011

ANDERSON, J ROBERT, b Ames, Iowa, Sept 29, 34. PHYSICS. Educ: Iowa State Univ, BS, 55, PhD(physics), 62. Prof Exp: Asst prof, 64-72, ASSOC PROF PHYSICS, UNIV MD, COLLEGE PARK, 72- Concurrent Pos: NSF fel, Cambridge Univ, 63-64; AEC fel, Iowa State Univ, 64. Mem: Am Phys Soc. Res: Fermi surface studies; de Haas-van Alphen effect. Mailing Add: Dept of Physics Col of Arts & Sci Univ of Md College Park MD 20740

ANDERSON, JACK ROGER, physical chemistry, see 12th edition

ANDERSON, JAMES ARTHUR, b Aurelia, Iowa, Mar 25, 35; m 56. ECONOMIC GEOLOGY, RESOURCE MANAGEMENT. Educ: Univ Utah, BS, 57; Harvard Univ, MS, 60, PhD(econ geol), 65. Prof Exp: Explor & res geologist, Bear Creek Mining Co, 60-65, sr explor geologist, 65-66, staff geologist, 66-68; explor mgr, Occidental Minerals Corp, 68-72, vpres, US Explor & Develop, 72-75; VPRES EXPLOR DIV, HOMESTAKE MINING CO, 75- Mem: Am Inst Mining Eng; Geol Soc Am. Res: Mineral exploration and development. Mailing Add: Homestake Mining Co 650 California St 9th Fl San Francisco CA 94108

ANDERSON, JAMES DAVID, b Fargo, NDak, Sept 21, 40; m 63; c 1. PLANT PHYSIOLOGY. Educ: NDak State Univ, BS, 63, MS, 64; Ore State Univ, PhD(plant physiol), 67. Prof Exp: PLANT PHYSIOLOGIST, US DEPT AGR, 67- Mem: Am Soc Plant Physiol; Am Soc Agron. Res: Hormonal control of plant growth and development; metabolic changes occurring in seeds during storage and germination. Mailing Add: Post-Harvest Plant Physiol Lab BARC Beltsville MD 20705

ANDERSON, JAMES DONALD, b Newark, NJ, Aug 16, 30; m 54; c 2. HERPETOLOGY. Educ: Rutgers Univ, AB, 54; Univ Calif, PhD, 60. Prof Exp: Asst, Newark Mus, 47-52; asst, Univ Calif, 54-57, assoc zool, 57-58, technician, Mus Vert Zool, 58, actg instr zool, Univ & actg cur herpet, Mus, 58-59; from asst prof to assoc prof zool, 60-68, PROF ZOOL, RUTGERS UNIV, 68- Concurrent Pos: Res assoc herpet, Am Mus Nat Hist, 63- Mem: Am Soc Ichthyologists & Herpetologists; Ecol Soc Am; Soc Study Evolution; Soc Study Amphibians & Reptiles; Soc Syst Zool. Res: Systematics and ecology of amphibians and reptiles. Mailing Add: Dept of Zool Rutgers Univ Newark NJ 07102

ANDERSON, JAMES E, b Hartford, Conn, July 31, 38; m 62; c 4. PHYSICAL CHEMISTRY. Educ: Union Col, NY, BS, 60; Princeton Univ, PhD(chem), 63. Prof Exp: Appointee, Bell Tel Labs, Inc, 63-65; STAFF SCIENTIST, SCI LAB, FORD MOTOR CO, 65- Concurrent Pos: Vis scientist biophys, Max-Planck Inst, 75-76; vis scientist phys chem, Univ Mainz, Ger, 75-76. Mem: Am Chem Soc; Am Phys Soc. Res: Dielectric and nuclear relaxation in solids and liquids; high polymer physics; stochastic processes; reverse osmosis; membrane transport. Mailing Add: Sci Lab Ford Motor Co Dearborn MI 48121

ANDERSON, JAMES EDWARD, b Perth, Ont, Feb 23, 26; m 57. PHYSICAL ANTHROPOLOGY, ANATOMY. Educ: Univ Toronto, MD, 53. Prof Exp: Lectr anat, Univ Toronto, 56-58, asst prof, 58-61, assoc prof anat & anthrop, 61-63; assoc prof phys anthrop, State Univ NY, Buffalo, 63-65, prof dept anthrop, 65-67; PROF ANAT & CHMN DEPT, McMASTER UNIV, 67- Concurrent Pos: Consult, Nat Mus Can, 60-63, Burlington Orthod Res Centre, 62-; mem coun res, Can Dent Asn, 60-63. Honors & Awards: Starr Medal Res Basic Sci, 61. Mem: Am Asn Phys Anthrop; Can Asn Anat. Res: Use of hereditary variations in the human skeleton to illustrate microevolution; paleopathology of New World Indians; growth at adolescence. Mailing Add: Dept of Anat McMaster Univ Hamilton ON Can

ANDERSON, JAMES G, b Portland, Ore, Oct 22, 14. CHEMISTRY. Educ: Univ Notre Dame, AB, 37, PhD(chem), 46. Prof Exp: Instr, Univ Notre Dame, 46; dean col sci, 56-68, INSTR ORG CHEM, UNIV PORTLAND, 46-, HEAD DEPT CHEM, 48-, DEAN COL ARTS & SCI, 68- Mem: Am Chem Soc. Res: Synthetic organic chemistry; bromination of some alkylbenzenes. Mailing Add: Col of Arts & Sci Univ of Portland Portland OR 97203

ANDERSON, JAMES GERARD, b Ceylon, Minn, July 27, 24; m 70; c 1. SOLID STATE PHYSICS, ELECTRONICS. Educ: Univ Minn, BS, 45; St Mary's Col, Minn, BA, 51; Univ Colo, MS, 59, PhD(physics), 63. Prof Exp: Res elec engr, Univ Minn, 45-46; aircraft electronic engr, Northwest Orient Airlines, Minn, 46-48; instr physics, St John's Univ, Minn, 53-58; asst prof, Wis State Univ, Oshkosh, 63-64; asst prof, Colo State Univ, 64-66; assoc prof, 66-74, PROF PHYSICS, UNIV WIS-EAU CLAIRE, 74- Mem: Am Phys Soc; Am Asn Physics Teachers. Res: De-Haas-Van Alphen effect in metals measured in high pulsed magnetic fields. Mailing Add: Dept of Physics Univ of Wis Eau Claire WI 54701

ANDERSON, JAMES HOWARD, b Joliet, Ill, Nov 17, 44; m 70. PHYSIOLOGY, RADIOLOGY. Educ: Ill Wesleyan Univ, BA, 66; Univ Ill, MS, 68, PhD(physiol), 71. Prof Exp: ASST PROF RADIOL, M D ANDERSON HOSP & TUMOR INST, UNIV TEX SYST CANCER CTR, 71-, HEAD SECT EXP DIAG RADIOL, BIOL SCI DIV, 74-, ASST PROF PHYSIOL, UNIV TEX MED SCH HOUSTON, 72- ASST PROF RADIOL, 74- Mem: Am Physiol Soc. Res: Experimental diagnostic radiology; gastrointestinal physiology; scanning electron microscopy. Mailing Add: Dept of Diag Radiol Cancer Ctr Tex Med Ctr Houston TX 77025

ANDERSON, JAMES JAY, b Seattle, Wash, July 29, 46. CHEMICAL OCEANOGRAPHY. Educ: Univ Wash, BS, 69, cert, 75, PhD(oceanog), 76. Prof Exp: OCEANOGR, UNIV WASH, 69- Mem: Sigma Xi. Res: Interactions between metabolic rates, mixing and distributions of nonconservative chemical properties in estuaries and the oceans. Mailing Add: Dept of Oceanog WB-10 Univ of Wash Seattle WA 98195

ANDERSON, JAMES LEROY, b Chicago, Ill, Apr 16, 26; m 50; c 2. THEORETICAL PHYSICS. Educ: Univ Chicago, BS, 46, MS, 49; Syracuse Univ, PhD(physics), 52. Prof Exp: Asst instr physics, Ill Inst Technol, 47-49; instr, Rutgers Univ, 52-53; res assoc, Univ Md, 53-56; assoc prof, 56-66, PROF PHYSICS, STEVENS INST TECHNOL, 66- Concurrent Pos: Guggenheim fel, 60-61; res fel, Woods Hole Oceanog Inst, 66 & 69; vis res scientist, Cambridge Univ, 68-69; consult, Ramo-Wooldridge, 57, Combustion Engr, 57-58 & Sci Teaching Ctr, Mass Inst Technol, 66 & 67. Mem: Am Astron Soc. Res: General relativity; relativistic astrophysics. Mailing Add: Dept of Physics Stevens Inst of Technol Hoboken NJ 07030

ANDERSON, JAMES NELSON, b Burbank, Calif, Mar 20, 30; m 57; c 3. ANTHROPOLOGY. Educ: Occidental Col, AB, 52; Univ Calif, MA, 59, PhD(anthrop), 64. Prof Exp: Asst prof anthrop, Southern Ill Univ, 63-64; asst prof, 64-69, ASSOC PROF ANTHROP, UNIV CALIF, BERKELEY, 69- Mem: Am Anthrop Asn; Asn Asian Studies. Res: Social and economic anthropology; ecology; Southeast Asia, especially the Philippines; relationship of land tenure, labor organization, exchange and social structure and social and economic change. Mailing Add: Dept of Anthrop Univ of Calif Berkeley CA 94720

ANDERSON, JAMES RICHARD, b Whitaker, Ind, Nov 20, 19; m 45; c 4. RESOURCE GEOGRAPHY, AGRICULTURAL ECONOMICS. Educ: Ind Univ, BS, 41, BA & MA, 47; Univ Md, PhD(geog), 50. Prof Exp: Asst prof geog, Univ Md, 50-52; agr geogr, USDA, 52-60; prof geog & chmn dept, Univ Fla, 60-72; CHIEF GEOGR, US GEOL SURV, 72- Concurrent Pos: Mem tech comt, Nat Atlas US, Nat Acad Sci, 55-61; mem int comt, World Atlas Agr, 61-; consult, Nat Atlas & Remote Sensing, US Geol Surv, 62-; mem comn educ agr & natural resources, Nat Acad Sci, 69-71; mem comt land use planning, NSF, 71- Mem: Asn Am Geog; Nat Coun Geog Educ; Agr Hist Soc. Res: Land and water resource analysis, especially land use and land use inventory; agricultural geography. Mailing Add: Geog Prog US Geol Surv Reston VA 22092

ANDERSON, JAMES UBBE, soil mineralogy, soil classification, see 12th edition

ANDERSON, JAY EARL, JR, geology, see 12th edition

ANDERSON, JAY LAMAR, b Madison, Wis, Apr 22, 31; m 55; c 4. POMOLOGY, WEED SCIENCE. Educ: Utah State Univ, BS, 55; Univ Wis, PhD(plant path), 61. Prof Exp: From asst prof to assoc prof, 61-75, PROF HORT, UTAH STATE UNIV, 75- Mem: Am Soc Hort Sci; Weed Sci Soc Am. Res: Pomology; fruit growth and development; physiology of herbicides and plant growth regulators; weed control in fruit crops; diseases of fruit crops; control of fruit blossoming by evaporative cooling. Mailing Add: Dept of Plant Sci Utah State Univ Logan UT 84322

ANDERSON, JAY MARTIN, b Paterson, NJ, Oct 16, 39; m 63; c 1. PHYSICAL CHEMISTRY, ENVIRONMENTAL SCIENCES. Educ: Swarthmore Col, BA, 60; Harvard Univ, AM, 61, PhD(chem), 64. Prof Exp: From asst prof to assoc prof, 63-75, PROF CHEM, BRYN MAWR COL, 75- Concurrent Pos: Secy-treas, Exp Nuclear Magnetic Resonance Conf, 67-71; mem ed bd, J Magnetic Resonance; Alfred P Sloan res fel, 67-70; Kettering vis lectr, Univ Ill, 68-69; vis assoc prof, Sloan Sch Mgt, Mass Inst Technol, 71-72. Mem: Am Chem Soc; Inst Elec & Electronic Engrs; Cybernet Soc. Res: Nuclear magnetic resonance; chemical kinetics; environmental systems simulation. Mailing Add: Dept of Chem Bryn Mawr Col Bryn Mawr PA 19010

ANDERSON, JAY OSCAR, b Brigham, Utah, Dec 5, 21; m 44; c 6. POULTRY NUTRITION. Educ: Utah State Univ, BS, 43; Univ Md, MS, 48, PhD(poultry nutrit), 50. Prof Exp: Chemist, Merck & Co, 50-51; PROF ANIMAL SCI, UTAH STATE UNIV, 51- Concurrent Pos: Secy-treas, Utah Feed Mfrs & Dealers Asn, 66- Mem: AAAS; World Poultry Sci Asn; NY Acad Sci; Poultry Sci Asn; Am Inst Nutrit. Res: Amino acid and protein requirements of chickens and turkeys; amino acid content of feedstuffs. Mailing Add: Dept of Animal Sci Utah State Univ Logan UT 84322

ANDERSON, JEAN HACKETT, b Minneapolis Minn, Aug 12, 28; m 56; c 2. ASTRONOMY. Educ: Univ Minn, BA, 52. Prof Exp: Lab technologist astron, Univ Minn, 55-59; scientist, Anderson Assocs, 68-73; DIR ASTRON, ROSE HILL PLANETARIUM, 73- Mem: AAAS; Am Astron Soc; Math Asn Am. Res: Stellar statistics; optimization. Mailing Add: Rose Hill Planetarium 2400 Ione St Lauderdale MN 55113

ANDERSON, JEREMY, b Seattle, Wash, Jan 3, 35; m 58, 68; c 3. CULTURAL GEOGRAPHY, ENVIRONMENTAL PSYCHOLOGY. Educ: Yale Univ, BA, 56; Univ Wash, PhD(geog), 64. Prof Exp: Asst prof geog, Univ Md, 60-66; from asst prof to assoc prof, Clark Univ, 66-71; PROF GEOG, EASTERN WASH STATE COL, 71- Mem: Asn Am Geog; Am Geog Soc; Nat Coun Geog Educ. Res: Agricultural change in developing countries; environmental learning and behavior; landscape modification and other behavioral adaptations to urban environments. Mailing Add: Dept of Geog Eastern Wash State Col Cheney WA 99004

ANDERSON, JOHN ARTHUR, b Chicago, Ill, Mar 6, 32; m 56; c 2. BIOCHEMISTRY. Educ: Colo State Univ, BS, 52, MS, 54; Ore State Univ, PhD(biochem), 62. Prof Exp: Asst prof chem, Univ Tex, Arlington, 54-55; asst prof, 61-66, ASSOC PROF CHEM, TEX TECH UNIV, 66- Mem: AAAS; Am Chem Soc; Am Soc Biol Chemists. Res: Pyridoxal phosphate-containing enzymes; alkaloid biosynthesis. Mailing Add: Dept of Chem Tex Tech Univ Lubbock TX 79409

ANDERSON, JOHN B, b Mobile, Ala, Sept 5, 44; m 65; c 2. MARINE GEOLOGY. Educ: Univ SAla, BS, 68; Univ NMex, MS, 70; Fla State Univ, PhD(geol), 72. Prof Exp: Asst prof geol, Hope Col, 72-75; ASST PROF GEOL, RICE UNIV, 75- Mem: Sigma Xi; Geol Soc Am. Res: Glacio-marine geology of antarctic regions; environmental geology of coastal waters. Mailing Add: Dept of Geol Rice Univ Houston TX 77001

ANDERSON, JOHN D, b Jamestown, NY, Dec 10, 30; m 55; c 3. NUCLEAR PHYSICS. Educ: San Diego State Col, BA, 51; Univ Calif, Berkeley, MA, 53, PhD(physics), 56. Prof Exp: SR STAFF PHYSICIST, LAWRENCE RADIATION LAB, UNIV CALIF, 56- Mem: Fel Am Phys Soc. Res: Fast neutron physics; polarization phenomena; nuclear reactions. Mailing Add: 572 Tyler Ave Livermore CA 94550

ANDERSON, JOHN DENTON, b Didsbury, Alta, July 15, 12; m 43; c 4. PHYSIOLOGY. Educ: Univ Colo, BA, 38, MA, 40; Stanford Univ, PhD(biol), 49. Prof Exp: Assoc prof, 49-55, PROF PHYSIOL & ASSOC DEAN, SCH BASIC MED SCI, UNIV ILL, URBANA-CHAMPAIGN, 55- Mem: AAAS; Soc Gen Physiol; Am Soc Zool; Am Soc Cell Biol. Res: Amoeboid movement; galvanotaxis. Mailing Add: Dept of Physiol & Biophys Univ of Ill Urbana IL 61801

ANDERSON, JOHN EDWARD, organic chemistry, see 12th edition

ANDERSON, JOHN FRANCIS, b Hartford, Conn, July 25, 36; m 63. PHYSIOLOGY. Educ: Cent Conn State Col, BS, 62; Univ Fla, MS, 65, PhD(zool), 68. Prof Exp: Asst prof, 69-74, ASSOC PROF ZOOL, UNIV FLA, 74- Mem: Am Soc Zool. Res: Arthropod physiology; energetics in arachnids. Mailing Add: Dept of Zool Univ of Fla Gainesville FL 32601

ANDERSON, JOHN FREDRIC, b Fargo, NDak, Feb 25, 36; m 58; c 3. ENTOMOLOGY. Educ: NDak State Univ, BS, 57, MS, 59; Univ Ill, PhD(entom), 63. Prof Exp: NSF fel, 63-64; from asst entomologist to assoc entomologist, 64-69, HEAD DEPT ENTOM, CONN AGR EXP STA, 69- Mem: Entom Soc Am; Soc Invert Path; Am Mosquito Control Asn. Res: Bionomics of mosquitoes; developmental biology; medical entomology; insect pathology; forest entomology. Mailing Add: Dept of Entom Conn Agr Exp Sta Box 1106 New Haven CT 06504

ANDERSON, JOHN HOWARD, b Northfork, WVa, Jan 26, 24; m 46; c 3. CHEMICAL PHYSICS. Educ: Univ NC, BS, 44, MS, 47; Univ Chicago, PhD(chem), 55. Prof Exp: Asst prof physics & chem, Talladega Col, 46-49; fel glass sci, Mellon Inst, 54-58; assoc prof, 58-66, PROF PHYSICS, UNIV PITTSBURGH, 66- Concurrent Pos: Fulbright-Hays fel, Coun Int Exchange Scholars, 70. Mem: Am Phys Soc; Am Asn Physics Teachers; AAAS. Res: Electron spin resonance; color centers; energy conversion; energy utilization. Mailing Add: Dept of Physics Univ of Pittsburgh Pittsburgh PA 15260

ANDERSON, JOHN JEROME, b Port Arthur, Tex, Oct 10, 30; m 64; c 2. GEOLOGY. Educ: Carleton Col, BA, 52; Univ Minn, MS, 62; Univ Tex, Austin, PhD(geol), 65. Prof Exp: ASSOC PROF GEOL, KENT STATE UNIV, 65- Concurrent Pos: Prin investr, NSF grants to Kent State Univ, 67-71; vis scientist, Am Geol Inst, 70. Mem: Geol Soc Am. Res: Geology of Antartica and Ellsworth Mountains, Antarctica; Cenozoic geologic evolution of southern high plateaus, Utah. Mailing Add: Dept of Geol Kent State Univ Kent OH 44242

ANDERSON, JOHN JOSEPH BAXTER, b Cleveland, Ohio, June 12, 34; m 57; c 3. PHYSIOLOGY, NUTRITION. Educ: Williams Col, BA, 56; Harvard Univ, MAT, 58; Boston Univ, MA, 62; Cornell Univ, PhD(phys biol), 66. Prof Exp: Teacher high sch, Maine, 58-59; instr biol, Bradford Jr Col, 59-62; asst prof vet physiol & pharmacol, Col Vet Med, Univ Ill, Urbana, 66-71; ASSOC PROF NUTRIT, SCH PUB HEALTH, UNIV NC, CHAPEL HILL, 72- Concurrent Pos: NIH fel phys biol, Cornell Univ, 66. Mem: AAAS; Am Physiol Soc; Radiation Res Soc. Res: Mineral metabolism of calcium, strontium and phosphorus, especially bone, intestine, kidneys and endocrines. Mailing Add: Dept of Nutrit Univ of NC SCh of Pub Health Chapel Hill NC 27514

ANDERSON, JOHN LYNDE, organic chemistry, see 12th edition

ANDERSON, JOHN MAXWELL, b Minden, Nebr, July 14, 17; m 44; c 3.

ZOOLOGY. Educ: Southern Methodist Univ, BS, 38; NY Univ, MS, 41, PhD(invert zool), 48. Prof Exp: Teaching fel, NY Univ, 38-42, 46-48; from instr to asst prof biol, Brown Univ, 48-52; assoc prof zool, 52-59, PROF ZOOL, CORNELL UNIV, 59- Concurrent Pos: Instr, Marine Biol Lab, Woods Hole, 53-57 & 62-63; Guggenheim fel, 58. Mem: Am Soc Zool; Sigma Xi; Am Micros Soc. Res: Invertebrate zoology; functional histology and cytology of invertebrate organ-systems. Mailing Add: Div of Biol Sci Cornell Univ Ithaca NY 14850

ANDERSON, JOHN MURRAY, b Toronto, Ont, Sept 3, 26; m 51; c 4. BIOLOGY. Educ: Univ Toronto, BScF, 51, PhD(biol), 58. Prof Exp: Res assoc, Univ Toronto, 58; asst prof biol, Univ NB, 58-63; assoc prof animal physiol, Carleton Univ, 63-67; dir biol sta, Fisheries Res Bd Can, 67-72; dir gen, Can Dept Environ, 72-73; PRES FISHERIES RES & DEVELOP, UNIV N B, FREDERICTON, 73- Concurrent Pos: Treas & mem bd dirs, Huntsman Marine Lab, NB, 69-72, pres & chmn bd dir, 74-; mem comt on Canada's Energy Opportunities, Sci Coun Can, 73-74; mem bd gov, Inst Can Bankers, 74-; mem exec comt, St John Mus, 74- Mem: AAAS; Am Fisheries Soc; Can Soc Zool (vpres, 71, pres, 73-74. Res: Marine biology; physiological and behavioural effects of sub-lethal pollutants on aquatic organisms. Mailing Add: Univ of N B Fredericton NB Can

ANDERSON, JOHN NORTON, b Mannington, WVa, Aug 28, 37; m 62; c 2. POLYMER CHEMISTRY. Educ: Fairmont State Col, BS, 59; Univ Pittsburgh, PhD(acid-base equilibria), 64. Prof Exp: Res chemist, 65-69, sr res chemist, 69-71, GROUP LEADER, FIRESTONE TIRE & RUBBER CO, 71- Mem: Am Chem Soc. Res: Polymerization of alpha olefins; structural characterization of elastomers; thermal analysis; anionic polymerication of dienes; solution properties of polymers; gel permeation chromatography. Mailing Add: 113 Tallwood Dr Tallmadge OH 44278

ANDERSON, JOHN RAY, chemistry, see 12th edition

ANDERSON, JOHN RICHARD, b Fargo, NDak, May 5, 31; c 3. MEDICAL ENTOMOLOGY, PARASITOLOGY. Educ: Utah State Univ, BS, 57; Univ Wis, MS, 58, PhD(med entom, vet sci), 60. Prof Exp: Biol aide, USPHS, 55; proj assoc arbo viruses, Dept Vet Sci, Univ Wis, 60-61; asst prof, 61-67, assoc prof, 67-71, chmn div, 70-71, PROF PARASITOL, UNIV CALIF, BERKELEY, 71- Mem: AAAS; Entom Soc Am; Wildlife Dis Asn; Am Mosquito Control Asn; Biol Res Inst Am. Res: Ecology and behavior of Diptera of medical and veterinary importance; host-vector-parasite interrelationships of Hematozoa; arbo viruses. Mailing Add: Div of Entom & Parasitol Univ of Calif Berkeley CA 94720

ANDERSON, JOHN SEYMOUR, b Kearney, Nebr, Oct 27, 36. BIOCHEMISTRY. Educ: Nebr State Teachers Col, AB, 58; Univ Nebr, Lincoln, MS, 60, PhD(chem), 63. Prof Exp: NSF res fels, Sch Med, Washington Univ, 63-64 & Sch Med, Univ Wis-Madison, 64-65; USPHS fel, Med Res Coun Lab Molecular Biol, Cambridge, Eng, 65-67; asst prof biochem, 67-73, ASSOC PROF BIOCHEM, UNIV MINN-ST PAUL, 73- Mem: Am Chem Soc; Am Soc Microbiol; Am Soc Biol Chemists. Res: Biosynthesis and structure of bacterial cell walls and membranes; biosynthesis of polysaccharides; intermediary metabolism of monosaccharides. Mailing Add: Dept of Biochem Univ of Minn St Paul MN 55108

ANDERSON, JOHN THOMAS, b Paterson, NJ, Apr 7, 45; m 68. LOW TEMPERATURE PHYSICS. Educ: Case Inst Technol, BS, 67; Univ Minn, PhD(physics), 71. Prof Exp: From teaching asst physics to res asst, Dept Physics, Univ Minn, 67-71; res assoc, 71-74, RES PHYSICIST, HANSEN LAB PHYSICS, STANFORD UNIV, 74- Mem: Am Phys Soc. Res: Application of superconducting quantum inferference device magnetometry to measurement of very small angles, specifically in a gyroscope with a London moment readout, or to precise angle measurement; noise characteristics of superconducting quantum interference devices. Mailing Add: Hansen Lab of Physics Stanford Univ Stanford CA 94305

ANDERSON, JOHN WALBERG, b New York, NY, Dec 3, 27; m 64; c 3. ANATOMY, HISTOLOGY. Educ: Swarthmore Col, BA, 50; Univ NH, MS, 52; Cornell Univ, PhD(histol), 56. Prof Exp: From instr to assoc prof, 56-69, PROF ANAT, UNIV WIS-MADISON, 69- Mem: Am Soc Cell Biol; Am Asn Anat; Soc Study Reproduction. Res: Histophysiology, especially of reproduction; materno-fetal transfer of immunity. Mailing Add: Dept of Anat Univ of Wis Madison WI 53706

ANDERSON, JON C, b Worcester, Mass, Feb 5, 42; m 65; c 3. COSMETIC CHEMISTRY, PHARMACEUTICAL CHEMISTRY. Educ: Mass Col Pharm, BS, 63, MS, 65, PhD(pharmaceut chem), 67. Prof Exp: From teaching asst to instr chem, Mass Col Pharm, 64-67; sect head phys pharm, Astra Pharmaceut Prod, Inc, 67-69, dir pharmaceut develop, 70-72; mgr sci affairs, 72-74, MGR PROD RES & DEVELOP, AVON PRODS, INC, 74- Mem: Am Pharmaceut Asn; Am Chem Soc; Acad Pharmaceut Sci; Soc Cosmetic Chem. Res: Medicinal synthetic chemistry; analytical development; thermal analysis; aerosol technology; pl.ysical chemistry of pharmaceuticals; cosmetic chemistry; emulsion technology. Mailing Add: Avon Prods Inc Division St Suffern NY 10901

ANDERSON, JOSEPH TOMLINSON, b Hilton, NY, Dec 10, 09; m 35; c 2. NUTRITION. Educ: Univ Rochester, BS, 30, MS, 32, PhD(vital econ), 47. Prof Exp: Chemist, Bausch & Lomb Optical Co, 34-43; instr physiol & nutrit, Univ Rochester, 47-50; from asst prof to assoc prof, 50-59, PROF PHYSIOL HYG, UNIV MINN, MINNEAPOLIS, 59-, CHMN NUTRIT DEPT, 72- Concurrent Pos: Res scientist, Ment Health Res Prog, Minn, 50-54; mem coun arteriosclerosis, Am Heart Asn. Mem: Am Chem Soc; Am Inst Nutrit; Am Oil Chem Soc. Res: Energy and amino acid metabolism; body composition; diet, blood lipids and atherosclerosis. Mailing Add: Dept of Physiol Hyg Univ of Minn Minneapolis MN 55455

ANDERSON, JULIUS HORNE, JR, b Harrisburg, Pa, June 20, 39; m 63; c 3. PHARMACOLOGY, BIOCHEMISTRY. Educ: Princeton Univ, AB, 61; Yale Univ, PhD(pharmacol), 67, MD, 68. Prof Exp: ASST PROF PHARMACOL, SCH MED, UNIV PITTSBURGH, 70- Concurrent Pos: Johnson Res Found fel, Univ Pa, 68-70; internship in med, Univ Health Ctr, Univ Pittsburgh, 73-74. Mem: Am Soc Pharmacol & Exp Therapeut; AMA; Soc Comput Simulation. Res: Biochemical pharmacology. Mailing Add: Dept of Pharmacol Univ of Pittsburgh Pittsburgh PA 15261

ANDERSON, JUNE S, b Covington, Tenn, Feb 11, 26. INORGANIC CHEMISTRY, SCIENCE EDUCATION. Educ: Peabody Col, BS, 47, MA, 48; Fla State Univ, PhD(chem & sci ed), 64. Prof Exp: Teacher sci, Davidson County Schs, Tenn, 47-48; head dept sci, Nashville City Schs, 48-58; seed analyst, Davidson County Coop, 55-56; assoc prof chem, 58-59, PROF CHEM & PHYSICS, MID TENN STATE UNIV, 69- Concurrent Pos: Consult sci, State of Tenn, 58-59. Mem: Am Chem Soc. Res: Coordination chemistry; radiochemistry; testing and evaluation of chemistry curriculum. Mailing Add: Dept of Chem Mid Tenn State Univ Murfreesboro TN 37130

ANDERSON, KEITH PHILLIPS, b Morgan, Utah, Oct 6, 19; m 42; c 3. PHYSICAL CHEMISTRY. Educ: Brigham Young Univ, AB, 46; Cornell Univ, PhD, 50. Prof Exp: Mem staff, Univ Calif at Los Alamos, NMex, 50-51; instr, exten div, Univ NMex, 51; chief radiol warfare div, Dugway Proving Ground, Utah, 51-52; assoc res prof physics & develop dir, Univ Utah, 52-53; chmn dept chem, 58-60, PROF CHEM, BRIGHAM YOUNG UNIV, 53- Concurrent Pos: Fulbright lectr, Valencia & Barcelona, Spain, 60-61. Mem: AAAS; Am Chem Soc. Res: Tracer chemistry; calorimetry; amino acid chelates; equilibria in slightly soluble salt solutions in aqueous-nonaqueous media. Mailing Add: 315 E 3140 North Provo UT 84601

ANDERSON, KENNETH ELLSWORTH, b Ithaca, NY, Dec 21, 14; m 42; c 11. BACTERIOLOGY. Educ: Cornell Univ, BS, 37, PhD(bact), 43; Univ NH, MS, 40. Prof Exp: Asst chemist & bacteriologist, NY Water Serv Corp, 37-38; asst bot, Univ NH, 38-40; asst bact, Cornell Univ, 41-42; bacteriologist, Genesee Brewing Co, 46; asst prof biol, 46-48, prof & chmn dept, 48-67, dean col arts & sci, 66-69, DEAN GRAD SCH, ST BONAVENTURE UNIV, 69-, PROF BIOL, 74- Concurrent Pos: Ed, Sci Studies. Mem: AAAS; Am Soc Microbiol; Am Pub Health Asn; NY Acad Sci. Res: Physiology of bacteria; amino acid metabolism by the genus Proteus; bacterial genus Desulphovibrio; cysteine transaminase; bacteriophage for Clostridia. Mailing Add: Grad Sch St Bonaventure Univ St Bonaventure NY 14778

ANDERSON, KENNETH VERLE, b La Crosse, Wis, Sept 14, 38; m 67. NEUROPSYCHOLOGY, NEUROANATOMY. Educ: Carleton Col, AB, 60; Brown Univ, MSc, 62, PhD(exp psychol), 64. Prof Exp: Res psychologist, US Naval Electronics Lab, Calif, 62; lab instr neuroanat, Yale Univ, 65-66; from instr to asst prof, 66-70, ASSOC PROF ANAT, EMORY UNIV, 70- Concurrent Pos: US Pub Health Serv fel neurophysiol & neuroanat, Yale Univ, 64-66; NIMH res scientist develop award, 68-72, Nat Inst Neurol Dis & Blindness res grant, 69-71. Mem: AAAS; Am Psychol Asn; Am Asn Anat; Soc Neurosci; Int Asn Study Pain. Res: Analysis of neuroanatomical correlates of perception and learning, including microelectrode assessment of sensory codes and rigorous analysis of behavior concomitant with changes in sensory transmission. Mailing Add: Dept of Anat Emory Univ Atlanta GA 30322

ANDERSON, KENNING M, biochemistry, see 12th edition

ANDERSON, KINSEY AMOR, b Preston, Minn, Sept 18, 26; m 54; c 2. SOLAR PHYSICS; SPACE PHYSICS. Educ: Carleton Col, BA, 49; Univ Minn, Minneapolis, PhD(physics, 55. Prof Exp: Res assoc, Univ Minn, Minneapolis, 55; res assoc, Univ Iowa, 55-58, asst prof physics, 58-60; from asst prof to assoc prof, 60-66, assoc dir, 68-70, PROF PHYSICS, UNIV CALIF, BERKELEY, 66-, DIR SPACE SCI LAB, 70- Concurrent Pos: Guggenheim fel, 59-60. Honors & Awards: Space Sci Award, Am Inst Aeronaut & Astronaut, 68. Mem: Am Phys Soc; Am Geophys Union; Am Astron Soc; Int Astron Union. Res: Solar cosmic rays; auroral zone phenomena; radiation zone studies using high altitude ballons and satellites. Mailing Add: 8321 Buckingham Dr El Cerrito CA 94530

ANDERSON, LARRY BERNARD, b Cottonwood, Idaho, July 18, 37; m 62; c 3. ANALYTICAL CHEMISTRY. Educ: Univ Wash, Seattle, BSc, 59; Syracuse Univ, PhD(chem), 64. Prof Exp: Res assoc chem, Univ NC, 64-66; asst prof, 66-72, acad vchmn dept, 68-72, ASSOC PROF CHEM, OHIO STATE UNIV, 72- Concurrent Pos: Consult, Indust Nucleonics, 68- Mem: Am Chem Soc; Electrochem Soc (secy-treas, Cols Sect, 75-76). Res: Electroanalytical chemistry; electrode kinetics; electrochemistry of organic and inorganic materials in non-aqueous system; photovoltaic solar energy conversion; semiconductor electrolyte interfaces. Mailing Add: Dept of Chem Ohio State Univ 140 W 18th Columbus OH 43210

ANDERSON, LARRY ERNEST, b Corvallis, Ore, Jan 30, 43; m 67; c 4. BIOCHEMISTRY, NEUROCHEMISTRY. Educ: Brigham Young Univ, BSc, 68; Univ Ill, PhD(biochem), 73. Prof Exp: NIH SR FEL BIOCHEM, UNIV WASH, 73- Res: Axoplasmic transport; proteases-isolation and mechanism of action. Mailing Add: Dept of Biochem Univ of Wash Seattle WA 98105

ANDERSON, LARS WILLIAM JAMES, b Pasadena, Calif, Apr 22, 45. PHYCOLOGY, BIOLOGICAL OCEANOGRAPHY. Educ: Univ Calif, Irvine, BA, 67; Calif State Univ, San Diego, MA, 70; Univ Calif, Santa Barbara, PhD(biol), 74. Prof Exp: Consult-marine biologist, Oceanog Serv Inc, 72-73; AQUATIC PLANT PHYSIOLOGIST HERBICIDES, ENVIRON PROTECTION AGENCY, 74- Mem: AAAS; Phycol Soc Am; Am Soc Limnol & Oceanog; Sigma Xi. Res: Phytoplankton physiology pertaining to vertical distribution; intracellular ion regulation in relation to cell buoyancy; methods for assessing toxicity of pesticides to aquatic microorganisms. Mailing Add: Off Pesticide Progs Criteria & Eval Div Environ Protection Agency Washington DC 20460

ANDERSON, LAUREL ETHAN, b Upsala, Minn, May 17, 16; m 49; c 3. FIELD CROPS, WEED SCIENCE. Educ: Univ Minn, BS, 47, MS, 53, PhD(agron, bot), 55. Prof Exp: Instr sch agr, Univ Minn, Morris, 47-50; from instr to assoc prof agron, Kans State Col, 53-65; prof field crops, 65-68, PROF AGRON, UNIV MO-COLUMBIA, 68- Res: Crop production. Mailing Add: Dept of Agron Univ of Mo Columbia MO 65201

ANDERSON, LAUREN DAVIS, b Morganville, Kans, Apr 10, 09; m 35; c 3. ENTOMOLOGY. Educ: Univ Kans, AB, 30, AM, 31, Ohio State Univ, PhD(entom), 41. Prof Exp: Asst entom, Mus Technician, Univ Kans, 27-31; from asst entomologist to assoc entomologist, Veg Insect Control, Va Truck Exp Sta, 32-44; entomologist, Bur Entom & Plant Quarantine, USDA, Washington, 44-45; head entomologist, Va Truck Exp Sta, 45-47; from assoc entomologist to entomologist, 48-61, prof entom, 61-76, EMER PROF ENTOM, UNIV CALIF, RIVERSIDE, 76- Concurrent Pos: With Off Sci Res & Develop & Nat Defense Res Comt, 44-45; mem, Smithsonian Inst. Mem: AAAS; Entom Soc Am; Am Entom Soc; Lepidop Soc; Am Inst Biol Sci. Res: Vegetable insect pest control; effect of pesticides on honey bees; aquatic midges; teaching of entomology, especially taxonomy of immature stages of insects. Mailing Add: Dept of Entom Univ of Calif Riverside CA 92502

ANDERSON, LAURENS, b Belle Fourche, SDak, May 19, 20; m 45; c 3. BIOCHEMISTRY, BIO-ORGANIC CHEMISTRY. Educ: Univ Wyo, BS, 42; Univ Wis, MS, 48, PhD(biochem), 50. Prof Exp: Merck fel natural sci, Swiss Fed Inst Technol, 50-51; from asst prof to assoc prof, 51-61, PROF BIOCHEM, UNIV WIS, MADISON, 61- Concurrent Pos: NIH sr fel, Ind Univ, 71-72. Mem: Am Chem Soc; Am Soc Biol Chem. Res: Organic chemistry of carbohydrates; chemical synthesis of oligosaccharides; mutarotation of sugars. Mailing Add: Dept of Biochem Univ of Wis Madison WI 53706

ANDERSON, LAWRENCE B, chemical physics, see 12th edition

ANDERSON, LAWRENCE CONRAD, b Brainerd, Minn, 37; m 61; c 3. GEOGRAPHY OF THE POLAR LANDS. Educ: St Cloud State Col, BS, 60, MS, 62; Univ Idaho, EdD, 68. Prof Exp: Asst prof geog, Univ Wis-Oshkosh, 67-69; ASSOC PROF GEOG, MANKATO STATE COL, 69- Concurrent Pos: NSF grant, Mankato State Col, 69-70. Mem: Asn Am Geog; Nat Coun Geog Educ. Res: Cultural geography, especially spatial analysis of the North American Hutterian Brethren. Mailing Add: Dept of Geog Mankato State Col Mankato MN 56001

ANDERSON, LEE ROY, b Pueblo, Colo, July 20, 36; m 64; c 2. APPLIED PHYSICS, SCIENCE EDUCATION. Educ: Stanford Univ, BS, 61; Univ Ore, MA, 65; Ore State Univ, MS, 67, PhD(phys sci), 69. Prof Exp: Res asst radiol, Sch Med, Stanford Univ, 61-62; asst physicist solid state res & develop, Hewlett Packard Assocs, 62-63; res asst appl math & programming, Stanford Linear Accelerator Ctr, 67-68; asst prof phys sci, San Diego State Univ, 68-71; FAC MEM PHYS SCI, EVERGREEN STATE COL, 71- Res: Radiation physics, electroptics; computer science. Mailing Add: Arts & Sci Bldg Evergreen State Col Olympia WA 98505

ANDERSON, LEE WILLIAM, mathematics, deceased

ANDERSON, LERAY J, b Raymond, Alta, Can, Apr 5, 22; US citizen; m 46; c 10. MEDICINAL CHEMISTRY, ORGANIC CHEMISTRY. Educ: Idaho State Col, BS, 49; Wash State Univ, PhD(pharmaceut chem), 61. Prof Exp: Asst prof pharmaceut chem, Univ Utah, 52-54; pharmacist, Berntsen Pharm, Salt Lake Clin, 54-56; asst prof pharm, Univ Wyo, 58-60 & Univ of the Pac, 61-62; res chemist agr chem, Univ Calif, Davis, 62-63; chmn sci div, 72-73, MEM FAC CHEM, SANTA ANA COL, 63- Mem: Am Chem Soc; Sigma Xi. Res: Synthesis and study of compounds related to morphine with respect to chemical structure versus biological activity; unusual chemical compounds found in plants. Mailing Add: Fac of Chem Santa Ana Col 17th at Bristol Santa Ana CA 92706

ANDERSON, LEWIS DANIEL, b Greensboro, Ala, Oct 13, 30; m 51; c 4. ORTHOPEDIC SURGERY. Educ: Univ Pa, MD, 53; Univ Tenn, MS, 60; Am Bd Orthop Surg, 63. Prof Exp: From asst prof to assoc prof, 62-72, chief staff, William F Bowld Hosp, 64-73, PROF ORTHOP SURG, UNIV TENN CTR FOR HEALTH SCI, MEMPHIS, 72- Concurrent Pos: Med dir, City Memphis Hosp, 72-75, asst to chancellor for clin opers, 75- Mem: Fel Am Col Surg; fel Am Asn Surg Trauma; fel Am Acad Orthop Surg; Orthop Res Soc; fel Int Soc Orthop Surg & Traumatol. Mailing Add: Dept of Orthop Surg Univ of Tenn Col of Med Memphis TN 38103

ANDERSON, LEWIS EDWARD, b Batesville, Miss, June 16, 12; m 41; c 5. BOTANY. Educ: Miss State Col, BS, 31; Duke Univ, AM, 33; Univ Pa, PhD(bot), 36. Prof Exp: Asst bot, Duke Univ, 31-33; asst instr, Univ Pa, 36; from instr to assoc prof, 36-53, PROF BOT, DUKE UNIV, 54- Concurrent Pos: Mem, Systs Panel, NSF, 59-62, Inland Biol Sta Cmt, 63-64. Mem: AAAS; Am Soc Plant Taxon; Am Bryol & Lichenological Soc; Am Fern Soc; Int Bryol Asn (pres, 75-). Res: Cytology, taxonomy and ecology of bryophytes. Mailing Add: Dept of Bot Duke Univ Durham NC 27706

ANDERSON, LEWIS L, b Deadwood, SDak, Jan 22, 35; m 56; c 2. PHYSICAL BIOCHEMISTRY. Educ: SDak Sch Mines & Technol, BS, 56; Univ NMex, PhD(chem), 61. Prof Exp: RES CHEMIST, E I DU PONT DE NEMOURS & CO, 61- Res: Isotopic exchange; absorption spectroscopy; ultracentrifugation applied to biochemical systems; isolation of proteins; physical characterization and study of biological systems. Mailing Add: Cent Res Dept E I du Pont de Nemours & Co Wilmington DE 19898

ANDERSON, LLOYD JAMES, b Salt Lake City, Utah, Dec 18, 17; m 41; c 3. RADIOPHYSICS. Educ: Univ Calif, Los Angeles, AB, 39; Univ Calif, MA, 42. Prof Exp: Asst physicist, US Navy Electronics Lab, 42-44, head radio-meteorol sect, 50-53, head environ studies br, 53-55; supvry physicist, Smyth Res Assocs, 55-61; PRIN PHYSICIST, CALSPAN CORP, 61- Mem: Am Geophys Union. Res: Effects of weather and terrain on tropospheric radio propagation; atmospheric effects on refraction of tracking radar beams; radar/optical analysis of satellites. Mailing Add: Calspan Corp PO Box 235 Buffalo NY 14221

ANDERSON, LLOYD L, b Nevada, Iowa, Nov 18, 33; m 58; c 2. ANIMAL SCIENCE, PHYSIOLOGY. Educ: Iowa State Univ, BS, 57, PhD(animal sci), 61. Prof Exp: Asst animal reprod, 57-58, res assoc, 58-61, from asst prof to assoc prof animal sci, 61-71, PROF ANIMAL SCI, IOWA STATE UNIV, 71- Concurrent Pos: NIH res fel, 61-62; Lalor Found res fel, Nat Inst Agron Res, France, 63-64. Mem: Am Soc Animal Sci; Am Asn Anat; Am Physiol Soc; Endocrine Soc; Soc Exp Biol & Med. Res: Neuro-utero-ovarian relationships primarily in farm animals, especially endocrine and neural influences in rhythmic regulation of reproduction cycle. Mailing Add: 11 Kildee Hall Dept of Animal Sci Iowa State Univ Ames IA 50011

ANDERSON, LORAN C, b Idaho Falls, Idaho, Feb 7, 36; m 57; c 2. BOTANY. Educ: Utah State Univ, BS, 58, MS, 59; Claremont Grad Sch, PhD(bot), 62. Prof Exp: Asst prof bot, Mich State Univ, 62-63 & Kans State Univ, 63-74; ASSOC PROF & CUR HERBARIUM, FLA STATE UNIV, 74- Concurrent Pos: Assoc prog dir, NSF, 70-71. Mem: Am Soc Plant Taxonomists (treas, 74-); Bot Soc Am; Am Inst Biol Sci. Res: Comparative anatomy of Chrysothamnus and related genera of tribe Astereae of Compositae. Mailing Add: Dept of Biol Sci Fla State Univ Tallahassee FL 32306

ANDERSON, LOUIS WESTON, b Attleboro, Mass, May 12, 23; m 46; c 6. ANALYTICAL CHEMISTRY. Educ: Tufts Univ, BS, 49. Prof Exp: From control chemist to chief chemist, Federated Metals Div, Am Smelting & Ref Co, 49-57, chief anal chemist, Cent Res Labs, 57-74, SUPT, ANAL SERVS, ASARCO INC, 74- Mem: Am Chem Soc; Sigma Xi. Res: Analytical chemistry of non-ferrous metals and associated materials. Mailing Add: Cent Res Labs ASARCO Inc South Plainfield NJ 07080

ANDERSON, LOUIS WILMER, b Houston, Tex, Dec 24, 33; m 62; c 3. ATOMIC PHYSICS. Educ: Rice Inst Technol, BA, 56; Harvard Univ, AM, 57, PhD(physics), 60. Prof Exp: From asst prof to assoc prof, 60-68, PROF PHYSICS, UNIV WIS, MADISON, 68- Mem: Fel Am Phys Soc. Res: Radio frequency spectroscopy; atomic hyperfine structure; atomic structure; nuclear magnetic resonance in solids; atomic collisions; atom ion charge changing collisions; production of negative ions; magnetism; lasers. Mailing Add: Dept of Physics Univ of Wis Madison WI 53706.

ANDERSON, LOUISE ELEANOR, b Cleveland, Ohio, May 18, 34. BIOCHEMISTRY. Educ: Augustana Col, AB, 56; Cornell Univ, PhD(biochem), 61. Prof Exp: Res assoc bot, Wash Univ, 60-62; res assoc microbiol, Dartmouth Med Sch, 62-64, 66-67; Kettering Int fel, Food Preservation Div, Plant Physiol Unit, Commonwealth Sci Indust & Res Orgn, Australia, 64-65; res asst prof, Oak Ridge Biomed Grad Sch, Univ Tenn, 67-68; from asst prof to assoc prof, 68-75, PROF BIOL SCI, UNIV ILL, CHICAGO CIRCLE, 75- Concurrent Pos: Katzia-Kotchalsky fel, Weizmann Inst Sci, Israel, 74-75. Mem: AAAS; Am Chem Soc; Am Soc Plant Physiol; Am Soc Microbiol. Res: Carbohydrate metabolism; metabolic control in plants and photosynthetic bacteria; enzymology; chlorplast biogenesis. Mailing Add: Dept of Biol Sci Univ of Ill at Chicago Circle Chicago IL 60680

ANDERSON, LOWELL LEONARD, b Spokane, Wash, Sept 3, 30; m 55; c 2. BIOPHYSICS. Educ: Whitworth Col, Spokane, Wash, BS, 53; Univ Rochester, PhD(biophys), 58. Prof Exp: From asst biophysicist to assoc biophysicist, Argonne Nat Lab, 58-69; ASSOC ATTEND PHYSICIST, MEM SLOAN-KETTERING CANCER CTR, MED COL, CORNELL UNIV, 69-, ASST PROF PHYSICS IN RADIOL, 70- Mem: AAAS; Health Physics Soc; Am Phys Soc; Sigma Xi; Am Asn Physicists in Med. Res: Neutron measurement methods for health physics; nuclear accident dosimetry; intermediate-energy neutron calibration source; medical radiation dosimetry. Mailing Add: 429 E 51st New York NY 10022

ANDERSON, LOWELL RAY, b Burnt Prairie, Ill, Nov 4, 34; m 60. INORGANIC CHEMISTRY. Educ: Southern Ill Univ, BA, 56; Ohio State Univ, PhD(chem), 64. Prof Exp: Res chemist, 64-66, sr res chemist, 66-68, tech supvr, 68-69, RES GROUP LEADER, ALLIED CHEM CORP, 69- Mem: Am Chem Soc. Res: Synthesis of fluorinated compounds, particularly fluorinated hypohalites, peroxides and ethers; high energy oxidizers; synthesis of nitrogen-fluorine and chlorine-fluorine compounds; intercalation compounds; catalysis. Mailing Add: 27 Sunderland Dr Morristown NJ 07960

ANDERSON, LUCIA LEWIS, b Pittsburgh, Pa, Aug 9, 22; m 43, 55; c 3. MICROBIOLOGY. Educ: Univ Pittsburgh, BS, 43, MS, 44, PhD(microbiol), 46. Prof Exp: Asst prof, Univ Pittsburgh, 49-53 & Duquesne Univ, 54-55; res assoc, Univ Pittsburgh, 55-56; from assoc prof to prof biol, Western Ky State Col, 58-65; sr info res scientist, Squibb Inst Med Res, NJ, 65-66; assoc prof biol, Parsons Col, 66-68; assoc prof, 68-70, PROF BIOL, QUEENSBORO COL CITY UNIV NEW YORK, 70- Concurrent Pos: NSF guest lectr for high schs. Mem: Am Soc Microbiol; Am Acad Microbiol. Res: Antibiotics. Mailing Add: 41 Bourndale Rd North Manhasset NY 11030

ANDERSON, LUCY MACDONALD, b Huntington, WVa, Oct 10, 42; m 64; c 2. CANCER. Educ: Bryn Mawr Col, AB, 64; Univ Pa, PhD(zool), 68. Prof Exp: Res assoc biol, Univ PA, 68-70; asst prof biochem, Carleton Col, 71; asst prof biol, Macalester Col, 73; ASSOC CARCINOGENESIS, SLOAN-KETTERING INST CANCER RES, 73- Concurrent Pos: NIH training grant biochem, Univ Minn, Minneapolis, 70-72; lectr, Bryn Mawr Col, 68-69. Mem: Am Soc Zool. Res: Insect egg follicle development; metabolic and immunological factors in transplacental carcinogenesis. Mailing Add: Sloan-Kettering Inst Walker Lab 145 Boston Post Rd Rye NY 10580

ANDERSON, MARGARET, b Omaha, Nebr, June 17, 41. NEUROBIOLOGY. Educ: Augustana Col, SDak, BA, 63; Stanford Univ, PhD(biol), 67. Prof Exp: Fel neurobiol, Harvard Univ, 68-70; res assoc, Univ PR, 70-71; asst prof biol, Bennington Col, 73; ASST PROF BIOL, SMITH COL, 73- Mem: Am Soc Zool; Soc Neurosci; Soc Gen Physiol. Res: Processes occurring at nerve-muscle junctions, particularly facilitation and depression. Mailing Add: Clark Sci Ctr Smith Col Northampton MA 01060

ANDERSON, MARION C, b Concordia, Kans, Oct 9, 26; m 49; c 4. SURGERY. Educ: Northwestern Univ, BM, 50, MD, 53, MS, 62. Prof Exp: Resident surg, Passavant Mem Hosp, Chicago, 54-55; resident, Vet Admin Res Hosp, 55-56; resident, Passavant Mem Hosp, 56-57; assoc, Cook County Hosp, 57-58; clin asst, Sch Med, Northwestern Univ, 58-59, instr, 59-60, assoc, 60-61, from asst prof to assoc prof, 61-69; chmn dept, 69-72, PROF SURG, MED COL OHIO, 69-, PRES COL, 72- Concurrent Pos: Kemper res scholar, Am Col Surgeons, 61-64. Mem: AMA: Am Asn Surg of Trauma; Soc Univ Surg; Am Surg Asn. Res: Biliary and pancreatic diseases. Mailing Add: Dept of Surg Med Col of Ohio Toledo OH 43614

ANDERSON, MARLIN DEAN, b Stanton, Iowa, Nov 30, 34; m 68. ANIMAL NUTRITION, ANIMAL PHYSIOLOGY. Educ: Iowa State Univ, BS, 59, MS, 62, PhD(animal nutrit), 64. Prof Exp: ANIMAL NUTRITIONIST, HESS & CLARK DIV, ROHDIA, INC, 65- Mem: AAAS; Am Soc Animal Sci; NY Acad Sci. Res: Growth; reproduction; deficiencies; toxicities; influence of drugs. Mailing Add: RD 1 Box 326-B Ashland OH 44805

ANDERSON, MARLOWE GEORGE, b Oakland, Nebr, July 8, 08; m 33; c 2. ZOOLOGY. Educ: Nebr Wesleyan Univ, AB, 28; Northwestern Univ, AM, 30, PhD(parasitol), 34. Prof Exp: Field agent barberry eradication, USDA, Univ Nebr, 27-28; asst, Northwestern Univ, 28-31, instr zool & parasitol, 31-36; asst prof biol, Northern Mont Col, 36-37; from instr to prof, 37-74, EMER PROF BIOL, NMEX STATE UNIV, 74- Concurrent Pos: Instr, Loyola Univ, Ill, 34-36; investr, Shedd Aquarium, Chicago, 33; head dept biol, NMex StState Univ, 56-67. Mem: AAAS; Am Soc Zool; Am Soc Parasitol; Soc Syst Zool; Am Micros Soc. Res: Taxonomy and morphology of Helminths; germ cell cycles and physiology of Trematoda. Mailing Add: Dept of Biol NMex State Univ Las Cruces NM 88001

ANDERSON, MARVIN A, b Stanhope, Iowa, Jan 31, 13; m 40; c 4. SOILS. Educ: Iowa State Univ, BS, 39, MS, 49, PhD, 55. Prof Exp: Dist soils agent, Agr Exten Serv, 40-42, from assoc to assoc prof agron, 42-52, prof & assoc dir agr exten serv, 52-66, dean univ exten, 66-74, PROF AGRON & DIR COOP EXTEN SERV, IOWA STATE UNIV, 66-, EMER DEAN UNIV EXTEN, 74- Concurrent Pos: Consult, Ford Found, 59- Mem: Adult Educ Asn US; Soil Conserv Soc Am. Res: Economics of soil management and soil conservation. Mailing Add: Dept of Agron Iowa State Univ Ames IA 50010

ANDERSON, MARY LOUCILE, b Kansas City, Kans, Feb 16, 40; c 1. REPRODUCTIVE ENDOCRINOLOGY. Educ: Univ Kans, AB, 62; Univ Calif, San Francisco, MA, 68, PhD(endocrinol), 73. Prof Exp: Res fel antifertil agents, Pop Res Inst, Oak Ridge Assoc Univs, 72-75; RES ASSOC REPRODUCTION, VANDERBILT UNIV, 75- Mem: Soc Study Reproduction. Res: Enrichment of human y-sperm fractions; visualization of a possible fusogenic agent in human sperm by electron microscopy; fusion of ova with cancer cells and investigation of resulting karyotypes. Mailing Add: Dept of Obstet & Gynec Vanderbilt Univ Nashville TN 37203

ANDERSON, MARY PIKUL, b Buffalo, NY, Sept 30, 48; m 73. GROUNDWATER HYDROLOGY. Educ: State Univ NY Buffalo, BA, 70; Stanford Univ, MS, 71, PhD(hydrogeol), 73. Prof Exp: Adj asst prof stratig & hydrogeol, Southampton Col, Long Island Univ, 73-75; ASST PROF HYDROGEOL, UNIV WIS-MADISON, 75- Concurrent Pos: Lectr hist geol, State Univ NY Stony Brook, 74. Mem: Am Geophys Union. Res: Development and analysis of mathematical models of subsurface flow systems; nutrient transport to lake basins. Mailing Add: Weeks Hall Univ of Wis 1215 W Dayton St Madison WI 53706

ANDERSON, MARY RUTH, b Bellingham, Wash, Feb 3, 39; m 66; c 2. INDUSTRIAL ENGINEERING, STATISTICS. Educ: Hope Col, BA, 61; Univ Iowa, MS, 63, PhD(math & statist), 66. Prof Exp: Teaching asst math, Univ Iowa, 63-66; lectr math, 66-72, asst to dean, Col Lib Arts, 72-74, ASST PROF ENG, FAC INDUST ENG, ARIZ STATE UNIV, 74- Concurrent Pos: Consult, US Indian Health Serv, 73-74. Mem: Inst Math Statist; Am Inst Indust Engrs. Res: Exponential class characterizations; probability application to combinatorics; applied statistics in engineering and nursing. Mailing Add: 1235 E Pebble Beach Dr Tempe AZ 85282

ANDERSON, MAURITZ GUNNAR, b Chicago, Ill, Aug 11, 18; m 57; c 2. MYCOLOGY. Educ: Univ Mich, AB, 42; Ind Univ, MS, 62. Prof Exp: Supvr histol & micros div, Swift & Co, 49-57; chief histologist, Norwich Pharmacal Co, 57; histologist & microscopist, Swift & Co, 58-63; asst prof, 63-69, ASSOC PROF BIOL, TOWSON STATE COL, 69- Mem: Bot Soc Am; Mycol Soc Am; Am Asn Feed Micros. Res: Application of microscopy to biology; histopathology. Mailing Add: Dept of Biol Towson State Col Baltimore MD 21204

ANDERSON, MELVERN K, b Temple, Tex, Dec 14, 42; m 63; c 2. AGRONOMY. Educ: NDak State Univ, BS, 65, PhD(agron), 69. Prof Exp: Asst prof agron, Univ Ky, 69-74; ASSOC PROG AGRON, NDAK STATE UNIV, 74- Mem: AAAS; Am Soc Agron; Crop Sci Soc Am; Sigma Xi. Res: Barley breeding, genetics, cytogenetics. Mailing Add: Dept of Agron NDak State Univ Fargo ND 58102

ANDERSON, MELVIN JOSEPH, b Heber City, Utah, Feb 14, 27; m 55; c 4. ANIMAL NUTRITION. Educ: Utah State Univ, BS, 50; Cornell Univ, MS, 57, PhD(animal nutrit), 59. Prof Exp: Asst animal nutrit, Cornell Univ, 55-58; asst prof animal sci, Univ Maine, 58-61; DAIRY HUSBANDMAN, AGR RES SERV, UTAH STATE UNIV, 61- Concurrent Pos: Assoc prof, Utah State Univ, 71. Mem: Am Dairy Sci Asn; Am Soc Animal Sci. Res: Dairy nutrition; nutritional evaluation of forages, hay cubes and level of concentrates for livestock; utilization of liquid whey for feeding dairy animals; effect of exercising pregnant dairy cows. Mailing Add: Dept of Dairy Sci UMC 46 Agr Res Serv Utah State Univ Logan UT 84322

ANDERSON, MELVIN LEE, b Perkins, Mich, Jan 19, 28; m 55; c 2. INORGANIC CHEMISTRY. Educ: Mich Technol Univ, BS, 53, MS, 55; Mich State Univ, PhD(inorg chem), 65. Prof Exp: Resident assoc actinide ion solution chem, Argonne Nat Lab, 53-54, jr chemist, 54-55; chemist, Dow Chem Co, Midland, 56-61, res chemist, 63-65, Rocky Flats Div, Colo, 65-69; asst prof chem, 69-75, ASSOC PROF CHEM, LAKE SUPERIOR STATE COL, 75- Mem: Am Chem Soc; Sci Res Soc Am. Res: Actinides and lanthanides; coordination complexes; separation and purification; nonaqueous solvents; chemistry of fluorine, chlorine, nitrogen and sulfur; organometallics. Mailing Add: Dept of Chem Lake Superior State Col Sault Ste Marie MI 49783

ANDERSON, MICHAEL PETER, b San Francisco, Calif, Dec 12, 46; m 69. ALGEBRA. Educ: Univ Notre Dame, BA, 69; Princeton Univ, MA, 72, PhD(math), 74. Prof Exp: Vis mem, Sch Math, Inst Advan Study, 73-74; Gibbs instr math, Yale Univ, 74-76. Mem: Am Math Soc. Res: Theory of profinite groups and its application to arithmetical algebraic geometry. Mailing Add: Dept of Math Yale Univ New Haven CT 06520

ANDERSON, MILES EDWARD, b Ft Worth, Tex, Dec 13, 26; m 48; c 3. MAGNETIC RESONANCE. Educ: NTex State Univ, BS, 49, MS, 50; Stanford Univ, PhD(physics), 63. Prof Exp: Instr physics, NTex State Univ, 50-52; asst, Willow Run Res Ctr, Univ Mich, 52-54; asst prof, 54-60, assoc vpres acad affairs, 69-75, PROF PHYSICS, NTEX STATE UNIV, 60-, VPRES ACAD AFFAIRS, 75- Concurrent Pos: Lectr, Washington Univ, 62-63. Mem: Am Phys Soc; Am Asn Physics Teachers; Acoust Soc Am. Mailing Add: NTex State Univ Denton TX 76203

ANDERSON, MILO VERNETTE, b Wolsey, SDak, Dec 24, 24; m 43; c 4. MICROWAVE PHYSICS. Educ: Union Col, Nebr, BA, 49; Univ Nebr, MA, 55; Univ Colo, Boulder, PhD(elec eng), 71. Prof Exp: From instr to asst prof physics, Union Col, Nebr, 49-57, chmn dept, 57-59; physicist, Nat Bur Stand, 59-64; assoc prof, 64-71, PROF PHYSICS, PAC UNION COL, 71-, CHMN DEPT, 73- Concurrent Pos: High sch teacher sci & math, Platte Valley Acad, Shelton, Nebr, 49-52. Mem: Am Asn Physics Teachers; Am Phys Soc; Inst Elec & Electronics Engrs; Sigma Xi. Res: Applications of microwave resonant structures. Mailing Add: Dept of Physics Pac Union Col Angwin CA 94508

ANDERSON, MILTON WINFIELD, b Minn, Dec 30, 15; m 41; c 6. INTERNAL MEDICINE, CARDIOLOGY. Educ: Gustavus Adolphus Col, BA, 36; Univ Minn, MB & MD, 44, MS, 47. Prof Exp: From instr to assoc prof, 48-70, PROF CLIN MED, MAYO GRAD SCH MED, UNIV MINN, 70- Concurrent Pos: Coun Clin Cardiol fel, Am Heart Asn; consult, Mayo Clin, 48- Mem: Fel Am Col Physicians; fel Am Col Chest Physicians. Res: Clinical investigation in cardiology. Mailing Add: Mayo Grad Sch of Med Univ of Minn Rochester MN 55901

ANDERSON, NEAL SAMPLE, b San Diego, Calif, Jan 17, 21; m 51. PHYSICS. Educ: Univ Calif, Los Angeles, AB, 46, MA, 49, PhD(physics), 51. Prof Exp: Res physicist, Naval Res Lab, 51; assoc prof physics, US Naval Postgrad Sch, 51-54; assoc prof, US Dept Defense, 54-56; mem sr staff, Ramo-Wooldridge Corp, 56-60; SR STAFF SCIENTIST, AEROSPACE CORP, 60- Mem: Am Phys Soc; Acoust Soc Am. Res: Propagation of waves in fluids. Mailing Add: Aerospace Corp 2350 E El Segundo Blvd El Segundo CA 90245

ANDERSON, NEIL ALBERT, b Minneapolis, Minn, Oct 21, 28; m 60; c 2. PLANT PATHOLOGY, MYCOLOGY. Educ: Univ Minn, BS, 51, MS, 57, PhD(plant path), 60. Prof Exp: Instr mycol, 59-60, from asst prof to assoc prof mycol & plant pathogens genetics, 60-70, PROF MYCOL & PLANT PATHOGENS GENETICS, UNIV MINN, ST PAUL, 70- Concurrent Pos: Vis prof, Waite Inst, Univ Adelaide, 70-71. Mem: Am Phytopath Soc; Mycol Soc Am; Sigma Xi; Potato Asn Am; AAAS. Res: Genetic studies on the plant pathogen, Rhizoctonia solani. Mailing Add: Dept of Plant Path Univ of Minn St Paul MN 55101

ANDERSON, NEIL OWEN, b Chicago, Ill, Oct 10, 35; m 57; c 2. CELL PHYSIOLOGY. Educ: Ripon Col, BA, 57; Drake Univ, MS, 64; Univ Ill, PhD(physiol), 70. Prof Exp: PROF PHYSIOL, EAST STROUDSBURG STATE COL, 70- Mem: AAAS; Sigma Xi (secy). Res: An examination of the nuclear proteins and their role in the cell. Mailing Add: Dept of Physiol Sci Bldg East Stroudsburg State Col East Stroudsburg PA 18301

ANDERSON, NEIL VINCENT, b Monterey, Minn, Apr 14, 33; m 53; c 6. VETERINARY MEDICINE. Educ: Mankato State Col, BS, 53; Univ Minn, St Paul, BS, 59, DVM, 61, PhD(vet path), 68. Prof Exp: Res asst vet med, Univ Minn, St Paul, 61; NIH res fels, 62-63, res asst vet path, 63-65; asst prof vet med, Auburn Univ, 65-67; from asst prof to assoc prof, 67-75, PROF VET MED, KANS STATE UNIV, 75- Mem: Am Vet Med Asn; Com Gastroenterol Soc (pres, 70-71); Am Col Vet Internal Med. Res: Clinical investigation and experimental study of canine pancreas and small intestine. Mailing Add: 1636 Leavenworth St Manhattan KS 66502

ANDERSON, NELS CARL, JR, b Detroit Lakes, Minn, June 15, 36; m 63; c 2. ENDOCRINOLOGY, BIOPHYSICS. Educ: Concordia Col, Moorhead, Minn, BA, 58; Kans State Univ, MS, 60; Purdue Univ, PhD(endocrinol), 64. Prof Exp: NIH fel, 64-66; ASST PROF OBSTET, GYNEC, PHYSIOL & PHARMACOL, MED CTR, DUKE UNIV, 66- Mem: Soc Gen Physiol; Am Physiol Soc. Res: Hormonal control of myometrial structure and function, particularly the hormonal and ionic basis of excitation in uterine smooth muscle. Mailing Add: Dept of Physiol Duke Univ Med Ctr Durham NC 27706

ANDERSON, NORMAN GULACK, b Davenport, Wash, Apr 21, 19; m 43; c 2. PHYSIOLOGY. Educ: Duke Univ, BA, 47, MA, 49, PhD(physiol), 52. Prof Exp: Biologist, Oak Ridge Nat Lab, 52-66, coordr joint NIH-AEC Zonal Centrifuge Develop Prog, Biol Div, 66-69, dir joint NIH-AEC Molecular Anat Prog, 69-74; MEM FAC, DEPT BASIC & CLIN IMMUNOL & MICROBIOL, MED UNIV SC, 74- Mem: Am Physiol Soc; Am Soc Zool; Biophys Soc. Res: Cellular physiology; biochemistry; isolation of cell components; effects of ionizing radiations; embryonic antigens in cancer; development of zonal centrifuges; rotating fast chemical analyzers; isopyenometric serology; early cancer detection. Mailing Add: Dept of Basic & Clin Immunol & Microbiol Med Univ of SC Charleston SC 29401

ANDERSON, NORMAN HERBERT, b Edam, Sask, Mar 17, 33; m 56; c 4. AQUATIC ECOLOGY, ENTOMOLOGY. Educ: Univ BC, BSA, 55; Ore State Col, MS, 58; Imp Col, Univ London, dipl, 61; Univ London, PhD(entom), 61. Prof Exp: Res officer entom, Can Dept Agr, 55-62; from asst prof to assoc prof, 62-75, PROF ENTOM, ORE STATE UNIV, 75- Concurrent Pos: NSF, US Dept Interior & Agr Res Serv, US Dept Agr grants. Mem: Ecol Soc Am; Entom Soc Am; Entom Soc Can; Brit Freshwater Biol Asn. Res: Aquatic entomology; insect and animal ecology; biological control; ecology and taxonomy of aquatic insects. Mailing Add: Dept of Entom Ore State Univ Corvallis OR 97331

ANDERSON, NORMAN LAVERNE, chemistry, deceased

ANDERSON, NORMAN LEWIS, JR, b Tacoma, Wash, Jan 16, 24; m 51; c 1. ECONOMIC ENTOMOLOGY. Educ: Mont State Col, BS, 49, MS, 51, PhD(entom), 62. Prof Exp: Asst entom, Mont State Col, 50, 51; asst state entomologist, Rangeland Grasshopper Res, Mont, 51-57; from asst prof to assoc prof, 57-68, PROF ENTOM, MONT STATE UNIV, 68- Concurrent Pos: Fulbright award, Eng & Africa, 55-56. Mem: Am Inst Biol Sci; AAAS; Entom Soc Am; Ecol Soc Am. Res: Insect ecology. Mailing Add: Dept of Biol Mont State Univ Bozeman MT 59715

ANDERSON, NORMAN RODERICK, b Tacoma, Wash, Oct 12, 21; m 49; c 2. GEOLOGY, GEOMORPHOLOGY. Educ: Univ Puget Sound, BS, 46; Univ Wash, MS, 54; Univ Utah, PhD(geol), 65. Prof Exp: Geologist, US Army Engrs, 47-48; from instr to assoc prof, 49-66, chmn dept, 57-66, PROF GEOL, UNIV PUGET SOUND, 66- Mem: Geol Soc Am; Am Asn Petrol Geologists; Soc Econ Paleont & Mineral; Nat Asn Geol Teachers; Sigma Xi. Res: Pleistocene geology; glacial geology of Western Washington; geology of the Snake River Plain, Southwestern Idaho. Mailing Add: Dept of Geol Univ of Puget Sound Tacoma WA 98416

ANDERSON, ORLIN, b Stockholm, Wis, Aug 16, 16. BOTANY, MICROBIOLOGY. Educ: Wis State Col, River Falls, BS, 38; Univ Wis, MS, 48, PhD(plant ecol), 54. Prof Exp: High sch instr, Wis, 39-42; instr bot & bact, Hamline Univ, 49-51; instr, Beloit Col, 52-54; prof, Wis State Univ, La Crosse, 54-68; PROF BIOL, UNIV WIS-STOUT, 68- Res: Vegetational composition studies of Western Wisconsin; thin-soil prairie relics of Southwestern Wisconsin. Mailing Add: Dept of Biol Univ of Wis-Stout Menomonie WI 54751

ANDERSON, ORSON LAMAR, b Price, Utah, Dec 23, 24; m 46; c 3. GEOPHYSICS, MINERALOGY. Educ: Univ Utah, BS, 48, MS, 49, PhD(physics), 51. Prof Exp: Instr mech eng, Univ Utah, 50-51, asst res prof physics, 51; mem tech staff, Bell Tel Labs, 51-60; mgr mat dept, Res Div, Am Standard Co, 60-63; res assoc geophys, Lamont Geol Observ, Columbia Univ, 63-64, adj prof, 64-65, prof mineral, Dept Geol & Lamont Geol Observ, 66-71; PROF GEOPHYS, INST GEOPHYS & PLANETARY PHYSICS & DEPT GEOPHYS & SPACE PHYSICS, UNIV CALIF, LOS ANGELES, 71- Concurrent Pos: Ed-in-chief, J Geophys Res, 67-73. Honors & Awards: Meyer Award, Am Ceramic Soc, 54. Mem: Am Ceramic Soc; Am Phys Soc; Am Geophys Union. Res: Mechanical properties of solids; equations of state; high pressure physics; acoustics; physics of the earth's interior. Mailing Add: Inst Geophys & Planetary Physics Univ of Calif Los Angeles CA 90024

ANDERSON, ORVIL ROGER, b East St Louis, Ill, Aug 4, 37. CELL BIOLOGY. Educ: Washington Univ, AB, 59, MAEd, 61, EdD(bot), 64. Prof Exp: From asst prof to assoc prof, 64-70, PROF NATURAL SCI, TEACHERS COL, COLUMBIA UNIV, 70- Concurrent Pos: From res scientist to sr res scientist, Lamont-Doherty Geol Observ, 66-; biol consult, Bd Higher Educ, State NJ, 68; ed, J Res in Sci Teaching. Mem: AAAS; Sigma Xi; NY Acad Sci; Am Inst Biol Sci; Nat Asn Res Sci Teaching. Res: Ultrastructure and cytochemistry of marine microplankton. Mailing Add: 525 W 120th St New York NY 10027

ANDERSON, OSCAR EMMETT, b Englewood, Fla, Nov 4, 16; m 46; c 2. AGRONOMY, SOIL FERTILITY. Educ: Univ Fla, BS, 41; Rutgers Univ, PhD(soils), 55. Educ: Jr soil surveyor, Soil Conserv Serv, USDA, 41-46; asst soil fertility, Rutgers Univ, 52-55; from asst soil scientist to assoc soil scientist, 55-65, PROF AGRON & SOIL SCIENTIST, GA EXP STA, UNIV GA, 66-, HEAD DEPT AGRON, 71- Concurrent Pos: Mem, Int Soil Sci Coun Soil Testing & Plant Anal. Mem: Am Soc Agron; Soil Sci Soc Am. Res: Soil-plant relationships, particularly trace elements in soils and their availability to plants; microelements. Mailing Add: Ga Exp Sta Experiment GA 30212

ANDERSON, OWEN THOMAS, b Adams, Minn, May 11, 31; m 56; c 5. PHYSICS. Educ: St Olaf Col, BA, 53; Univ Wis, MS, 55, PhD(physics), 60. Prof Exp: Assoc engr, Adv Systs Develop Div, Int Bus Mach Corp, 58-60, adv engr, 60-61; from asst prof to assoc prof physics, 61-75, PROF PHYSICS, BUCKNELL UNIV, 75- Mem: Am Phys Soc. Am Asn Physics Teachers. Res: Low temperature physics; properties of the liquid helium II film; ellipsometry; magnetic susceptibility. Mailing Add: Dept of Physics Bucknell Univ Lewisburg PA 17837

ANDERSON, PAUL DEAN, b Grand Junction, Colo, Oct 28, 40; m 62; c 2. ANALYTICAL CHEMISTRY. Educ: Univ Calif, Berkeley, BS, 62; Mass Inst Technol, PhD(anal chem), 66. Prof Exp: RES CHEMIST, PHILLIPS PETROL CO, 66- Mem: Am Chem Soc. Res: Instrumental analysis; molecular spectroscopy; photochemistry; computer applications to analytical chemistry; atomic absorption spectrophotometry. Mailing Add: 2301 Avalon Rd Bartlesville OK 74003

ANDERSON, PAUL HAMILTON, mathematics, see 12th edition

ANDERSON, PAUL J, b Akron, Ohio, Oct 11, 25. NEUROPATHOLOGY, NEUROLOGY. Educ: Ohio Univ, BS, 49; Univ Chicago, MD, 53; Am Bd Psychiat & Neurol, dipl, 59; Am Bd Path, dipl, 60. Prof Exp: Clin instr neurol, NY Univ, 60-61; asst prof neuropath, Col Physicians & Surgeons, Columbia Univ, 61-67; assoc prof path, 65-66, assoc prof neurol, 66-68, PROF NEUROPATH, MT SINAI SCH MED, 67- Concurrent Pos: Nat Inst Neurol Dis & Blindness spec fel, 57-59; asst attend pathologist, Mt Sinai Hosp, New York, 58-61, from asst attend neurologist to assoc attend neurologist, 59-68, assoc attend pathologist in charge, Div Neuropath, 61-74, attend pathologist in charge, Div Neuropath, 74-; ed, J Neuropath & Exp Neurol, 61-; consult neuropathologist, City Hosp Elmhurst, 64-; ed-in-chief, J Histochem & Cytochem, 73- Mem: Fel Am Acad Neurol; Am Asn Neuropath; Am Asn Path & Bact; Asn Res Nerv & Ment Dis; Am Neurol Asn. Res: Electron microscopy and electron histochemistry of neuromuscular diseases. Mailing Add: Mt Sinai Sch of Med Fifth Ave & 100th St New York NY 10029

ANDERSON, PAUL KNIGHT, b Ludlow, Mass, Aug 6, 27; m 57, 68; c 2. VERTEBRATE ECOLOGY. Educ: Cornell Univ, AB, 49; Tulane Univ, MS, 51; Univ Calif, PhD(zool), 58. Prof Exp: Assoc zool, Univ Calif, 58, instr, 58-59; res assoc, Columbia Univ, 59-61; asst prof, Univ Alta, 61-65, assoc prof zool & asst dean arts & sci, 65-67; sr res fel, Nat Res Coun Can, 67-68; vis assoc prof zool, Stanford Univ, 67-68; assoc prof, 68-71, PROF ZOOL, UNIV CALGARY, 71- Concurrent Pos: Exchange scientist, USSR-Can, 67. Mem: AAAS; Am Soc Ichthyol & Herpet; Ecol Soc Am; Soc Study Evolution; Am Soc Naturalists; Am Inst Biol Sci. Res: Population structure, dynamics and evolution in vertebrate populations, particularly rodents; evolution in island populations of vertebrates; human populations and the planetary ecosystem; community ecology. Mailing Add: Dept of Biol Univ of Calgary Calgary AB Can

ANDERSON, PAUL LEROY, b Magnolia, Ill, Sept 20, 35; m 66. ORGANIC CHEMISTRY. Educ: Univ Ill, BS, 62; Univ Mich, PhD(org chem), 66. Prof Exp: Fel, Syntex Steroid Inst, Mex, 66-67; RES CHEMIST, SANDOZ PHARMACEUT CORP, HANOVER, 67- Mem: Am Chem Soc; NY Acad Sci. Res: Studies of carbene reactions, dipolar cycloaddition reactions; synthesis of small ring compounds, steroids, terpenes and heterocyclic compounds of medicinal interest. Mailing Add: Sandoz Pharmaceut Rte 10 Hanover NJ 07936

ANDERSON, PAUL M, b Jackson, Minn, May 1, 38; m 59; c 3. BIOCHEMISTRY. Educ: Univ Minn, BS, 59, PhD(biochem), 64. Prof Exp: NSF res fel biochem, Sch Med, Tufts Univ, 64-65, NIH res fel, 65-66; from asst prof to assoc prof, Southern Ill Univ, 66-70; res biochemist, Molecular Biol Dept, Miles Labs, Inc, 70-71; ASSOC PROF BIOCHEM, UNIV MINN-DULUTH, 71- Mem: AAAS; Am Chem Soc; Am Soc Biol Chemists. Res: Purification, mechanisms of catalytic action and regulation of enzymes related to nucleotide and urea biosynthesis. Mailing Add: Sch of Med Univ of Minn Duluth MN 55812

ANDERSON, PAUL NATHANIEL, b Omaha, Nebr, May 30, 37; m 65; c 2. INTERNAL MEDICINE, ONCOLOGY. Educ: Univ Colo, BA, 59, MD, 63; Am Bd Internal Med, dipl, 73. Prof Exp: Intern internal med, Johns Hopkins Hosp, 63-64, resident, 64; res assoc embryol, Sect Exp Embryol, Lab Neuroanat Sci, Nat Inst Neurol Dis & Blindness, 65-67, staff assoc carcinogenesis sect, Lab Biol, Nat Cancer Inst, 67-70; STAFF ASSOC, DIV ONCOL, DEPT MED, JOHNS HOPKINS HOSP, 70-; ASST PROF MED & ONCOL, JOHNS HOPKINS UNIV, 72- Concurrent Pos: Fel internal med, Johns Hopkins Hosp, 63-64; asst physician, Johns Hopkins Hosp; attend physician, Baltimore City Hosps. Mem: AAAS; AMA; Am Soc Clin Oncol; Am Asn Cancer Res; NY Acad Sci. Res: Cancer chemotherapy; bone marrow transplantation; tumor immunology; anti-leukemia cellular immunity; efficacy of granulocyte transfusions during therapy-induced aplasia. Mailing Add: Oncol Ctr Johns Hopkins Hosp Baltimore MD 21205

ANDERSON, PAUL SIGFRIED, JR, b New Haven, Conn, June 10, 27; m 50; c 3. BIOMETRY, BACTERIOLOGY. Educ: Wesleyan Univ, BA, 50; Yale Univ, MS, 51, PhD(pub health, bact), 57. Prof Exp: Instr pub health & sanit, Sch Med, Yale Univ, 51-54, from instr to assoc prof biometry, 57-68; PROF BIOSTATIST & EPIDEMIOL, DEPT PREV MED, MED SCH, & PROF BIOSTATIST & EPIDEMIOL & CHMN DEPT, SCH HEALTH, MED CTR, UNIV OKLA, 68- Concurrent Pos: Instr math, New Haven State Teachers Col, 56-59; mem, Atomic Bomb Casualty Comn, Japan, 60-61. Mem: AAAS; Am Pub Health Asn; Am Statist Asn; Biomet Soc; NY Acad Sci. Res: Virus-host relationships in Corynebacterium diphtheriae; mechanisms responsible for conversion from toxigenicity to nontoxigenicity; biometry, design and analysis of experiments; cancer epidemiology. Mailing Add: Dept of Biostatist & Pub Health Med Ctr Univ Okla Oklahoma City OK 73104

ANDERSON, PEARL ROSALIN, biochemistry, see 12th edition

ANDERSON, PETER JOHN, biochemistry, see 12th edition

ANDERSON, PETER OLE, b Chewelah, Wash, July 5, 41. MATHEMATICAL STATISTICS. Educ: Wash State Univ, BA, 63; Stanford Univ, PhD(statist), 69. Prof Exp: Asst prof statist, Ohio State Univ, 68-74; RES STATISTICIAN, HEALTH SCI COMPUT FACIL, UNIV CALIF, LOS ANGELES, 74- Mem: Inst Math Statist; Am Statist Asn. Res: Multivariate statistical analysis; least squares theory; regression methods; computer data analysis. Mailing Add: Health Sci Comput Facil Univ of Calif Los Angeles CA 90024

ANDERSON, PHILIP CARR, b Grand Rapids, Mich, Dec 25, 30; m 62; c 3. DERMATOLOGY. Educ: Univ Mich, AB, 51, MD, 55; Am Bd Dermat, dipl, 62. Prof Exp: From jr clin instr to sr clin instr dermat, Med Ctr, Univ Mich, 60-63; asst prof, 63-68, dir res training, 65-68, ASSOC PROF DERMAT & ASST DEAN FOR PLANNING, SCH MED, UNIV MO-COLUMBIA, 68-, CHMN DEPT, 70-, MEM BD DIRS, 75- Concurrent Pos: NIH res fel, 61-63; Markle scholar, 66-; spec asst to dir, NIH, 67-68; mem bd dirs, Nat Prog Dermat, 70- Mem: Fel Am Acad Dermat; Am Dermat Asn; Soc Invest Dermat; Tissue Cult Asn; Am Asn Prof Dermat. Res: Tissue culture; education; management; cholesterol in cell walls. Mailing Add: Dept of Dermat Univ of Mo Sch of Med Columbia MO 65201

ANDERSON, PHILIP WARREN, Indianapolis, Ind, Dec 13, 23; m 47; c 1. THEORETICAL PHYSICS. Educ: Harvard Univ, BS, 43, MS, 47, PhD(physics), 49. Prof Exp: Mem staff, Naval Res Lab, 43-45; vis prof theoret physics, Cambridge Univ, 67-75; mem tech staff, 49-74, ASST DIR PHYS RES, BELL LABS, 74-; PROF PHYSICS, PRINCETON UNIV, 75- Concurrent Pos: Fulbright lectr, Univ Tokyo, 53-54; overseas fel, Churchill Col, Cambridge Univ, 61-62. Honors & Awards: Buckley Prize, Am Phys Soc, 64; Dannie Heineman Prize, Am Phys Soc & Am Inst Physics, 75. Mem: Nat Acad Sci; fel Am Phys Soc; fel Am Acad Arts & Sci; fel Brit Inst Physics & Phys Soc; Europ Phys Soc. Res: Solid state physics; magnetism; breadths of spectral lines; relaxation; superconductivity; dielectrics. Mailing Add: Bell Labs Murray Hill NJ 07974

ANDERSON, R F V, b Montreal, Que, Dec 23, 43. MATHEMATICS. Educ: McGill Univ, BSc, 64; Princeton Univ, MA, 66, PhD(math), 67. Prof Exp: Instr math, Mass Inst Technol, 67-69; asst prof, 69-74, ASSOC PROF MATH, UNIV BC, 74- Mem: Am Math Soc; Can Math Cong. Res: Spectral theory of non-commuting operators, using methods of harmonic analysis, operator semigroups and lie groups. Mailing Add: Dept of Math Univ of BC Vancouver BC Can

ANDERSON, RALPH F, b Glenbeulah, Wis, Apr 13, 24; m 48; c 2. PESTICIDE CHEMISTRY, INDUSTRIAL CHEMISTRY. Educ: Univ Wis, BS, 48, MS, 51, PhD(biochem), 52. Prof Exp: Control chemist, Shell Oil Co, 48; instr, Univ Wis, 52-53; biochemist, Rohm & Haas Co, 53-54; supvr chemist, Northern Regional Res Lab, 54-63; dir res & develop, Bioferm Div, Int Minerals & Chem Corp, 63-67, div vpres res & develop, 67-71, vpres technol develop, 71-73; V PRES RES, VELSICOL CHEM CORP, 73- Mem: Am Chem Soc; Indust Res Inst; Agr Res Inst; Europ Weed Res Soc. Res: Development of pesticides, benzoic acid derivatives, hydrocarbon resins and polymers. Mailing Add: Velsicol Chem Corp 341 E Ohio Chicago IL 60611

ANDERSON, RALPH ROBERT, b Fords, NJ, Nov 1, 32; m 61; c 2. DAIRY HUSBANDRY. Educ: Rutgers Univ, BS, 53, MS, 58; Univ Mo, PhD(endocrinol), 61. Prof Exp: Instr endocrinol, Univ Mo, 61-62; asst prof reproductive physiol, Iowa State Univ, 62-64; trainee endocrinol, Univ Wis, Madison, 64-65; asst prof, 65-66, ASSOC PROF DAIRY HUSB, UNIV MO-COLUMBIA, 67- Concurrent Pos: Sr Fulbright scholar, Ruakura, NZ, 73-74. Mem: Am Dairy Sci Asn; Am Soc Animal Sci; Soc Exp Biol & Med; Endocrine Soc; Soc Study Reproduction. Res: Endocrine control of mammary gland growth and lactation; thyroid physiology in mammals with emphasis on secretion rates and relation to reproduction. Mailing Add: Dept of Dairy Husb Univ of Mo Columbia MO 65201

ANDERSON, RAY CARL, b Duluth, Minn, Sept 24, 17; m 41; c 2. MEDICINE. Educ: Gustavus Adolphus Col, BA, 39; Univ Minn, MA, 41, PhD(zool, genetics), 43, MD, 46. Prof Exp: PROF PEDIAT, UNIV MINN, MINNEAPOLIS & MEM ATTEND STAFF, UNIV HOSPS, 51- Concurrent Pos: Mem attend staff, Minneapolis Gen Hosp. Mem: Am Soc Human Genetics; Soc Pediat Res; Am Acad Pediat; Am Col Cardiol. Res: Congenital cardiac defects. Mailing Add: Univ of Minn Hosp Minneapolis MN 55455

ANDERSON, RAY HAROLD, b Minneapolis, Minn, Mar 23, 15; m 46. FOOD SCIENCE. Educ: St Thomas Col, BS, 36; Univ Minn, PhD(org chem), 50. Prof Exp: Chemist, Gen Mills Inc, 37-46; asst instr, Univ Minn, 46-48; PRIN SCIENTIST & HEAD CHEM SERV & NUTRIT RES, GEN MILLS, INC, 50- Mem: Am Chem Soc; Am Asn Cereal Chemists; Am Oil Chem Soc; Inst Food Technologists. Res: Nutritional enrichment of food products; food preservation; shelf-stable product development; food additives; chemistry of food ingredients and processes. Mailing Add: 10930 W River Rd Champlin MN 55316

ANDERSON, RAYMOND EDWARD, b Chicago, Ill, Nov 22, 19. HORTICULTURE. Educ: Univ Ill, BS, 51, MS, 52, PhD(hort), 57. Prof Exp: Asst hort, Univ Ill, 51-57; HORTICULTURIST, USDA, 57- Mem: Am Soc Hort Sci; Am Chem Soc. Res: Post harvest handling and physiology of horticultural crops, mainly fruits and vegetables. Mailing Add: Agr Mkt Res Inst Hort Crops Mkt Lab Beltsville MD 20705

ANDERSON, RAYMOND KENNETH, b White Lake, Wis, May 2, 28; m 53; c 4. WILDLIFE ECOLOGY. Educ: Wis State Univ, Stevens Point, BS, 54; Univ Mich, MA, 58; Univ Wis, PhD(wildlife ecol), 69. Prof Exp: High sch instr, Wis, 54-58; conserv, 58-61, PROF WILDLIFE, UNIV WIS-STEVENS POINT, 66- Mem: Am Inst Biol Sci; Wilson Ornith Soc; Wildlife Soc. Res: Prairie chicken behavior and management. Mailing Add: Col of Natural Resources Univ of Wis Stevens Point WI 54481

ANDERSON, RAYMOND PAUL, organic chemistry, see 12th edition

ANDERSON, RICHARD ALAN, b Rock Rapids, Iowa, Aug 9, 30; m 57; c 1. ATOMIC PHYSICS, MOLECULAR PHYSICS. Educ: Augustana Col, BA, 52; Kans State Univ, MS, 54, PhD(physics), 59. Prof Exp: From asst prof to assoc prof, 58-69, PROF PHYSICS, UNIV MO-ROLLA, 69- Mem: Am Phys Soc; Am Asn Physics Teachers. Res: Atomic spectra and structure; inelastic atomic collision phenomena; atomic and molecular lifetime transition probability and oscillator strength measurements. Mailing Add: Dept of Physics Univ of Mo Rolla MO 65401

ANDERSON, RICHARD C, b Venice, Utah, Sept 24, 33; m 56; c 5. ORGANIC CHEMISTRY. Educ: Brigham Young Univ, BS, 54, PhD(org chem), 61. Prof Exp: Chemist, Dugway Proving Ground, 55-57; chemist, Texaco Res Ctr, NY, 61-62; USPHS fel & res assoc org chem, Univ Ore, 62-63; asst prof, 63-68, ASSOC PROF ORG CHEM, UTAH STATE UNIV, 68-, ASST HEAD DEPT CHEM, 69- Mem: AAAS; Am Chem Soc. Res: Chemistry of heterocyclic systems; determination of structure of complex organic substances. Mailing Add: Dept of Chem Utah State Univ Logan UT 84321

ANDERSON, RICHARD CHARLES, b Moline, Ill, Apr 22, 30; m 53; c 3. GEOLOGY. Educ: Augustana Col, AB, 52; Univ Chicago, MS, 53, PhD(geol), 55. Prof Exp: Geologist, Geophoto Servs, Colo, 55-57; from asst prof to assoc prof geol & geog, 57-65; PROF GEOL, AUGUSTANA COL, ILL, 65-, CHMN DEPT, 69- Concurrent Pos: Res affil, Ill Geol Surv, 59-; asst dir earth sci summer inst, Iowa State Univ, 66- Mem: Geol Soc Am; Am Geophys Union; Int Union Quaternary Res; Am Asn Quaternary Environ; Am Meteorol Soc. Res: Glacial geology; geomorphology; Tertiary and Quaternary geomorphic history. Mailing Add: Dept of Geol Augustana Col Rock Island IL 61201

ANDERSON, RICHARD DAVIS, b Hamden, Conn, Feb 17, 22; m 43; c 5. MATHEMATICS. Educ: Univ Minn, BA, 41; Univ Tex, PhD(math), 48. Prof Exp: Instr math, Univ Tex, 41-42 & 45-48; from instr to assoc prof, Univ Pa, 48-56; prof, 56-59, BOYD PROF MATH, LA STATE UNIV BATON ROUGE, 59- Concurrent Pos: Inst Advan Study, 51-52 & 55-56; Alfred P Sloan res fel, 60-63; colleague, Mathematisch Centrum, Amsterdam, 62-63; panel mem, Sch Math Study Group & mem comt undergrad prog in math, chmn, 66-68. Mem: AAAS; Am Math Soc; Math Asn Am. Res: Topology; point set theory; collections of continua; dimension theory; transformation groups. Mailing Add: Dept of Math La State Univ Baton Rouge LA 70803

ANDERSON, RICHARD GILPIN WOOD, b Bryn Mawr, Pa, Mar 25, 40; m 64; c 2. CELL BIOLOGY, REPRODUCTIVE BIOLOGY. Educ: Ore State Univ, BS, 65; Univ Ore, PhD(anat), 70. Prof Exp: Res assoc reprod physiol, Ore Regional Primate Res Ctr, 70-73, asst scientist, 73; ASST PROF CELL BIOL, UNIV TEX HEALTH SCI CTR DALLAS, 73- Mem: Soc Develop Biol; Am Asn Cell Biol; Soc Reprod Biol. Res: Ciliogenesis in the oviduct; structure and function of oviduct basal bodies and cilia; biogenesis of cell membranes in ciliated epithelium of oviduct. Mailing Add: Dept of Cell Biol Univ of Tex Health Sci Ctr Dallas TX 75235

ANDERSON, RICHARD HAYDEN, b Hayden, Ariz, Sept 17, 21; m 44; c 3. PSYCHIATRY. Educ: Univ Utah, BA, 43, MD, 45. Prof Exp: Resident psychiat, Vet Admin Hosp, Palo Alto, Calif, 48-51, chief, Ft Douglas, Utah, 53-54, chief prof servs, 55-57; asst prof psychiat, Col Med, Univ Utah, 51-57; asst supt med servs, Mendocino State Hosp, 57-59; assoc supt med servs, Fairview State Hosp, 59-60; CONSULT, SANTA CRUZ MENT HEALTH SERVS, 60- Concurrent Pos: Dir psychiat servs, Dominican Santa Cruz Hosp, 73- Mem: AMA; fel Am Psychiat Asn. Res: Biological psychiatry. Mailing Add: 526 Soquel Ave Santa Cruz CA 95062

ANDERSON, RICHARD JASPER, b St Petersburg, Russia, Oct 25, 13; m 39; c 2. GEOLOGY. Educ: Columbia Univ, AB, 35, AM, 38. Prof Exp: Asst geol, Columbia Univ, 35-37; instr, Univ Minn, 38-41; geologist, Ark State Geol Surv, 42-43 & Alcoa Mining Co, NY, 43-47; managing engr, Raw Mat Surv, Inc, Ore, 47; res engr, 48-49, asst to dir, 49-69, dir res prom, 70-72, ASSOC DIR ENERGY PROG, BATTELLE MEM INST, 73- Concurrent Pos: Adj prof, Ohio State Univ, 65- Mem: AAAS; Geol Soc Am; Am Inst Mining, Metall & Petrol Eng; Am Asn Petrol Geol; Am Inst Prof Geol. Res: Industrial raw materials supplies; economics of transportation of bulk materials; fossil fuel resources; unconventional sources of energy. Mailing Add: 3820 Woodbridge Rd Upper Arlington OH 43220

ANDERSON, RICHARD JOHN, b Chicago, Ill, June 11, 38; m 60; c 2. ATOMIC PHYSICS. Educ: Coe Col, BA, 59; DePaul Univ, MS, 62; Univ Okla, PhD(physics), 66. Prof Exp: Appl physicist, Eastman Kodak Co, 61-62; asst prof physics, 66-71, ASSOC PROF PHYSICS, UNIV ARK, FAYETTEVILLE, 71- Mem: Am Phys Soc; Am Asn Physics Teachers. Res: Spectroscopic investigation of electronic and atomic collision processes. Mailing Add: Dept of Physics Univ of Ark Fayetteville AR 72701

ANDERSON, RICHARD L, b North Liberty, Ind, Apr 20, 15; m 46; c 2. EXPERIMENTAL STATISTICS. Educ: DePauw Univ, AB, 36; Iowa State Col, MS, 38, PhD(math statist), 41. Prof Exp: Asst math, Iowa State Col, 36-41; instr, NC State Univ, 41-42, asst prof statist, 42-44, from assoc prof to prof, Inst Statist, 45-66; res mathematician, Princeton Univ, 44-45; vis res prof statist, Univ Ga, 66-67; PROF STATIST & CHMN DEPT, UNIV KY, 67- Concurrent Pos: Consult, Off Sci Res & Develop, US Army & US Navy, 44-45; prof, Purdue Univ, 50-51 & London Sch Econ, 58; consult, Pan-Am World Airways Guided Missile Range Div, 63-67; partic, NSF-Japan Soc Promotion Sci Seminar Sampling of Bulk Mat, Tokyo, 65; Inaugural Conf for Sci Comput Ctr, Univ Cairo, 69; mem adv bd, Int Math & Statist Libr, 71-; mem census adv comt, Am Statist Asn, 72-77; consult, Merrill Labs, 73- Mem: Fel AAAS; fel Inst Math Statist; fel Am Statist Asn; Economet Soc; Biomet Soc. Res: Experimental designs and estimating procedures for variance components; time series analysis, including serial correlation; econometrics; regression analysis, selection of predictors, prior information, intersecting straight lines. Mailing Add: Dept of Statist Univ of Ky Lexington KY 40506

ANDERSON, RICHARD LEE, b Grinnell, Iowa, Feb 24, 45; m 65; c 2. OPHTHALMOLOGICAL SURGERY. Educ: Grinnell Col, BA, 67; Univ Iowa, MD, 71, cert ophthal specialist, 75. Prof Exp: Resident ophthal, Univ Iowa, 72-75; fel, Albany Med Ctr, 75 & Univ Calif Med Ctr, San Francisco, 76; Heed fel, Heed Ophthalmic Found, 76; ASST PROF OPHTHAL PLASTIC SURG, DEPT OPHTHAL, UNIV IOWA, 76- Concurrent Pos: NSF grants, 65 & 66. Mem: AMA. Res: Acquired Ptosis, levator aponeurosis defects and their repair; nasolacrimal duct obstructions; anatomy of the levator muscle. Mailing Add: Dept of Ophthal Univ of Iowa Hosp & Clin Iowa City IA 52242

ANDERSON, RICHARD LEE, b Astoria, Ore, June 14, 33; m 75. BIOCHEMISTRY, MICROBIOLOGY. Educ: Univ Wash, BS, 54, PhD(microbiol), 59. Prof Exp: Asst, Univ Wash, 54-59; NIH fel, 59-61, from asst prof to assoc prof, 61-70, PROF BIOCHEM, MICH STATE UNIV, 70- Concurrent Pos: Ed, J Bact, 70-75. Mem: AAAS; Am Chem Soc; Am Soc Microbiol; Am Soc Biol Chem. Res: Microbial physiology; enzymology and metabolism of microorganisms; pathways, mechanisms and control of carbohydrate metabolism. Mailing Add: Dept of Biochem Mich State Univ East Lansing MI 48824

ANDERSON, RICHARD LOUIS, b Hale Center, Tex, Nov 27, 35; m 60; c 2. PHYSICAL CHEMISTRY. Educ: Mass Inst Technol, BS, 58; Rice Univ, PhD(chem), 66. Prof Exp: Res chemist, Heat Div, Nat Bur Standards, 63-71; scientist, Nat Phys-Tech Inst, Bur Standards, Brunswick, Ger, 71-74; HEAD METEOROL RES & DEVELOP LAB, OAK RIDGE NAT LAB, 74- Mem: Am Vacuum Soc; Sigma Xi. Res: Low temperature calorimetry and thermometry; gas thermometry; high temperature resistance thermometry; vacuum devices; pressure measurement; thermocouple thermometry. Mailing Add: Bldg 3500 Oak Ridge Nat Lab Oak Ridge TN 37830

ANDERSON, RICHARD ORR, b Evanston, Ill, Oct 23, 29; m 51; c 2. AQUATIC ECOLOGY. Educ: Univ Wis, BS, 51; Univ Mich, MS, 53, PhD(fisheries), 59. Prof Exp: Fishery biologist, Mich Inst Fisheries Res, 52-58; instr dept fisheries, Univ Mich, 58-59; biologist in charge, Wolf Lake Fish Hatchery, 59-63; LEADER, MO COOP FISHERY UNIT, 63- Mem: Am Fisheries Soc; Am Soc Limnol & Oceanog; Int Asn Theoret & Appl Limnol; Wildlife Soc. Res: Aquatic and fishery biology and ecology; fish production and growth. Mailing Add: Stephens Hall Univ of Mo Columbia MO 65201

ANDERSON, RICHARD R, aquatic ecology, see 12th edition

ANDERSON, RICHMOND KARL, b Bangalore, India, Dec 6, 07; US citizen; m 35; c 4. PHYSIOLOGICAL CHEMISTRY, MEDICINE. Educ: Cornell Col, AB, 29; Northwestern Univ, MS, 31, PhD(physiol chem), 34, MD, 37; Johns Hopkins Univ, MPH, 48. Hon Degrees: DSc, Cornell Col, 58. Prof Exp: Asst chem, Med Sch, Northwestern Univ, 29-34, univ fel, 34-35; instr, Tulane Univ, 35-38; intern, Alameda County Hosp, 38-39; instr med, Sch Med, Univ Buffalo, 39-42; Rockefeller Found spec fel, Sch Pub Health, Univ NC, 42-43; mem field staff, Rockefeller Found, Mex, 43-45, field dir, India, 48-55, asst dir biol & med res, 56-58, assoc dir med & nat sci, 59-64; dir tech assistance div, Population Coun, 64-70; prog dir, Josiah Macy Jr Found, 70-71; dir int progs off, Carolina Population Ctr, Univ NC, Chapel Hill, 71-74, sr consult, 74-75; HEAD FIELD STUDIES DIV, INT FERTILITY RES PROG, 75- Concurrent Pos: Assoc dir labs & resident med, Buffalo City Hosp, 39-42. Mem: AAAS; Pop Asn Am; Am Inst Nutrit; fel Am Pub Health Asn; Am Soc Trop Med & Hyg. Res: Nutrition; public health; medical education; population. Mailing Add: Int Fertility Res Prog Research Triangle Park NC 27709

ANDERSON, ROBBIN COLYER, b DeRidder, La, June 8, 14; m 46; c 3. PHYSICAL CHEMISTRY. Educ: La State Univ, BS, 34, MS, 36; Univ Wis, PhD(phys chem), 39. Prof Exp: From instr to prof chem, Univ Tex, Austin, 39-67, assoc dean grad sch, 66-67; PROF CHEM & DEAN COL ARTS & SCI, UNIV ARK, FAYETTEVILLE, 67- Concurrent Pos: Mem adv coun col chem, AAAS, 63-66. Mem: AAAS; Am Chem Soc; Combustion Inst; The Chem Soc. Res: Complex ions; mechanism of flame propagation and carbon formation. Mailing Add: Col of Arts & Sci Univ of Ark Fayetteville AR 72701

ANDERSON, ROBERT, b Bessemer, Mich, Feb 17, 14; m 47; c 2. CULTURAL ANTHROPOLOGY. Educ: Northern Mich Col, AB, 37; Univ Mich, AM, 47, PhD(anthrop), 51. Prof Exp: Instr, Univ Mich, 50-51; asst prof anthrop, Fla State Univ, 51-54; from asst prof to assoc prof anthrop, 54-63, lectr psychiat, 60-70, PROF ANTHROP, UNIV UTAH, 63-, ADJ PROF PSYCHIAT, 70- Mem: Fel Am Anthrop Asn; Am Ethnol Soc. Res: Social Structure of Indians of the Northern Plains and Great Basin; kinship; primitive religions. Mailing Add: Dept of Anthrop Univ of Utah Salt Lake City UT 84112

ANDERSON, ROBERT ALAN, b Princeton, Ind, Sept 14, 42; m 65; c 2. EXPERIMENTAL PHYSICS. Educ: Univ Ill, Urbana, BS, 65, MS, 66, PhD(physics), 71. Prof Exp: Res asst & fel low-temperature physics, Univ Ill, Urbana, 67-71; MEM TECH STAFF PHYSICS, SANDIA LABS, 71- Mem: Am Phys Soc. Res: Electrical breakdown phenomena, including surface flashover of insulators in vacuum, breakdown of solid and liquid dielectrics, and high-field leakage currents. Mailing Add: Div 5814 Sandia Labs Albuquerque NM 87115

ANDERSON, ROBERT BERNARD, b Moline, Ill, Aug 31, 15; m 42; c 2. PHYSICAL CHEMISTRY, CATALYSIS. Educ: Augustana Col, BA, 38; Univ Iowa, MS, 40, PhD(phys chem), 42. Prof Exp: Godfrey L Cabot fel, Johns Hopkins Univ, 42-43; mem staff, US Bur Mines, 44-55, asst chief synthetic fuels res br, 52-55, chief br coal to oil res, 55-60, proj coordr catalysis, 60-65; Petrol Res Fund fel, Queen's Univ, Ireland, 65; PROF CHEM ENG, McMASTER UNIV, 65- Concurrent Pos: With Off Sci Res & Develop, 44. Honors & Awards: Ipatieff Award, Am Chem Soc, 53, Pittsburgh Award, 60; Distinguished Serv Award, US Dept of Interior, 66. Mem: Fel Chem Inst Can; Can Soc Chem Eng; Catalysis Soc; Am Chem Soc. Res: Catalytic, adsorption, and gas kinetics; structure of carbon black, charcoal, coal, and catalysts; Fischer-Tropsch synthesis and methanation; coal hydrogenation; hydrogenolysis; Raney nickel. Mailing Add: Dept of Chem Eng McMaster Univ Hamilton ON Can

ANDERSON, ROBERT CHRISTIAN, b Perth Amboy, NJ, Sept 26, 18; m 44; c 2. ORGANIC CHEMISTRY. Educ: Middlebury Col, AB, 40; Princeton Univ, PhD(chem), 48. Prof Exp: Asst chemist, Merck & Co, Inc, NJ, 42-45; assoc chemist, 48-57, ASST DIR, BROOKHAVEN NAT LAB, 57- Concurrent Pos: Fel, Woodrow Wilson Int Ctr Scholars, 70; chmn coun, State Univ NY Stony Brook, 75-; mem sci adv coun, NY Legis; mem Bd Atomic Energy Comn Nuclear Sci & Eng Fels. Mem: AAAS; Am Chem Soc; Am Soc Eng Educ; Am Nuclear Soc. Res: Synthesis and structure determination of physiologically active organic compounds; stereo-chemistry of organic compounds; nuclear transformations in organic systems. Mailing Add: Brookhaven Nat Lab Upton NY 11973

ANDERSON, ROBERT CLARKE, b Salem, Ind, Feb 19, 11; m 34; c 3. TOXICOLOGY. Educ: Purdue Univ, BS, 31, DSc, 53. Prof Exp: Lab asst, Eli Lilly & Co, 31-34, pharmacologist, 34-36, head dept toxicol, 46-61, dir toxicol div, 61-66, dir res biol, 66-70, dir res toxicol, 70-72; RETIRED. Mem: AAAS; Am Pharmaceut Asn (secy-treas, 58-64); Am Soc Toxicol; Am Soc Pharmacol & Therapeut. Res: Toxicology of antibiotics, analgesics, steroids and pesticides. Mailing Add: 318 N Franklin Rd Indianapolis IN 46219

ANDERSON, ROBERT CURTIS, b Columbus, Ohio, May 10, 41; m 67; c 1. ENTOMOLOGY. Educ: Calif State Col, Long Beach, BS, 63; Purdue Univ, MS, 66, PhD(entom), 68. Prof Exp: Instr entom, Purdue Univ, 68-69; asst prof, 69-74, ASSOC PROF BIOL, IDAHO STATE UNIV, 74-, CHMN DEPT, 74- Mem: AAAS; Entom Soc Am. Res: Biology and ecology of insect parasites; biological control; medical entomology, Trypanosoma cruzi, Tyrpanosoma rangeli and acarine parasites; scorpions. Mailing Add: Dept of Biol Idaho State Univ Pocatello ID 83201

ANDERSON, ROBERT E, b Carrizo Springs, Tex, Aug 26, 26; m 47; c 3. SOLID STATE PHYSICS. Educ: Tex Col Arts & Indust, BS, 47, MS, 49; Univ Tex, PhD(physics), 55. Prof Exp: Teacher math & physics, San Antonio Jr Col, 49-52; asst physics, Univ Tex, 52-55; res engr, Tex Instruments, Inc, 55-57; co res fel low temperature physics, Univ Tex, 57-58; prof physics, Tex Col Arts & Indust, 58-66; PROF PHYSICS & CHMN DEPT, TEX STATE UNIV, 66- Concurrent Pos: Res asst, Defense Res Lab, Tex, 54-55; consult, Tex Instruments, Inc, 60- Mem: Am Asn Physics Teachers. Res: Semiconductor physics; electronics; vacuum techniques; acoustics; transistor design. Mailing Add: Dept of Physics Southwest Tex State Univ San Marcos TX 78666

ANDERSON, ROBERT E, b Livingston, Tex, Oct 5, 40; m 62; c 3. BIOCHEMISTRY, OPTHALMOLOGY. Educ: Tex A&M Univ, BA, 63, MS, 65, PhD(biochem), 68; Baylor Col Med, MD, 75. Prof Exp: ASST PROF BIOCHEM, BAYLOR COL MED, 69- Concurrent Pos: Fel biochem, Oak Ridge Assoc Univs, 68-69. Honors & Awards: MacGee Award, Am Oil Chem Soc, 66. Mem: AAAS; Am Chem Soc; Am Soc Biol Chemists; Biophys Soc; Int Soc Neurochem. Res: Biochemistry of the visual process; chemistry of photoreceptor membranes; relationships between lipid structure and membrane function. Mailing Add: Dept of Opthal Baylor Col Med Houston TX 77025

ANDERSON, ROBERT EDWIN, b Los Angeles, Calif, Aug 20, 31; m 53; c 3. PATHOLOGY. Educ: Col Wooster, BA, 53; Western Reserve Univ, MD, 57. Prof Exp: Intern med, Strong-Mem-Rochester Munic Hosps, NY, 57-58; resident path, Univ Hosps, Cleveland, Ohio, 58-59; resident, Med Ctr, Univ Calif, Los Angeles, 59-62, from instr to asst prof, 60-64, consult, Clin Labs, 60-62; asst prof, 64-69, PROF PATH, SCH MED, UNIV NMEX, 69-, CHMN DEPT, 68- Concurrent Pos: Consult, Community Blood & Plasma Inc, 61-62 & Bernalillo County Indian Hosp, NMex, 64- Mem: AAAS; Am Soc Exp Path; Am Asn Path & Bact (secy-treas, 74-); Asn Path Chmn (pres, 75). Res: Hematologic pathology; radiation in the immune response. Mailing Add: Dept of Path Univ of NMex Sch of Med Albuquerque NM 87106

ANDERSON, ROBERT EMRA, b Mentone, Ind, Jan 6, 24; m 45; c 2. WATER CHEMISTRY. Educ: Ind Univ, BS, 49, MA, 51. Prof Exp: Chemist, Dow Chem Co, 50-54, proj leader ion exchange, 54-56, group leader, 56-68; GROUP LEADER ION EXCHANGE, DIAMOND SHAMROCK CHEM CO, 68- Concurrent Pos: Chmn, Gordon Res Conf Ion Exchange, AAAS, 71. Mem: Am Chem Soc; Sigma Xi. Res: Study of ion exchange and ion-exchange resins, particularly as applied to water treatment and industrial processing. Mailing Add: Diamond Shamrock Chem Co 800 Chestnut St Redwood City CA 94064

ANDERSON, ROBERT GLENN, b Stratton, Ont, Jan 2, 24; m 46; c 4. PLANT GENETICS, PLANT BREEDING. Educ: Univ Man, BSA, 50, MSc, 52; Univ Sask, PhD, 55. Prof Exp: Asst prof plant breeding & statist, Univ Sask, 54-56; agr res officer, Res Sta, Can Dept Agr, Winnipeg, 56-64; wheat breeder, 64, joint coordr, All India Wheat Improv Prog, 64-71, ASSOC DIR, INT MAIZE & WHEAT IMPROV CTR, ROCKEFELLER FOUND, 71- Concurrent Pos: Mem genetic stock cmt, Int Wheat Genetics Symp, 58-; sci ed, Agr Inst Can, 63-64. Mem: Agr Inst Can; Can Soc Agron; Genetics Soc Can; fel Indian Soc Genetics & Plant Breeding; AAAS. Res: Wheat breeding; international agriculture in developing countries. Mailing Add: Londres 40 Mexico 6 DF Mexico

ANDERSON, ROBERT GORDON, b Elwood, Ind, Sept 5, 26; m 56; c 5. BOTANY. Educ: Youngstown State Univ, BA, 53; Univ Nebr, MS, 55, PhD(bot, bact), 58. Prof Exp: Instr bot, biol & bact, Univ Nebr, Onamha, 58-59; assoc prof biol, Davis & Elkins Col, 59-60; PROF BOT, UNIV MO-KANSAS CITY, 60- Mem: AAAS; Phycol Soc Am; Bot Soc Am; Am Inst Biol Sci; Int Phycol Soc. Res: Physiology of algae; Chara. Mailing Add: Dept of Biol Univ of Mo Kansas City MO 64110

ANDERSON, ROBERT GRIFFIN, b Memphis, Tenn, Aug 4, 32; m 54; c 1.

ORGANIC CHEMISTRY. Educ: Harvard Univ, AB, 57; Univ Wash, PhD(org chem), 61. Prof Exp: Res chemist, Calif Res Corp, 61-65, res chemist, Chevron Res Co, 65-70, SR RES CHEMIST, CHEVRON RES CO, 70- Mem: Am Chem Soc. Res: Chemistry of azulene and related compounds; surfactant chemistry. Mailing Add: PO Box 1627 Richmond CA 94802

ANDERSON, ROBERT HUNT, b Kansas City, Kans, Oct 20, 24; m 50; c 6. PHYSICAL CHEMISTRY. Educ: Baker Univ, AB, 46; Columbia Univ, MA, 48, PhD(chem), 54. Prof Exp: Instr chem, Bloomfield Col, 51-56; asst res specialist, Rutgers Univ, 56-57; asst prof chem, 57-61, ASSOC PROF CHEM, WESTERN MICH UNIV, 61- Mem: AAAS; Am Chem Soc; Soc Social Responsibility in Sci. Res: Neutron spectroscopy in chemical analysis; phase rule; corrosion inhibitors; computers in chemistry. Mailing Add: Dept of Chem Western Mich Univ Kalamazoo MI 49008

ANDERSON, ROBERT I, b Crown Point, NY, Jan 31, 20; c 4. SEROLOGY, PARASITOLOGY. Educ: Cortland State Teachers Col, BS, 43; Cath Univ Am, MS, 55, PhD(parasitol), 61. Prof Exp: Asst chief lab, Sta Hosp, US Army, Ft Belvoir, Va, 46-47, chief bact, Fourth Army Med Lab, 47-48, chief lab, Sta Hosp, Clark AFB, Philippines, 48-50, chief bact, Fourth Army Med Lab, 50-51, asst chief serol, Walter Reed Army Inst Res, 51-55, chief dept serol, Trop Res Med Lab, San Juan, PR, 55-58, asst chief parasitol dept med zool, Walter Reed Army Inst Res, 60-62, chief dept serol, 62-65; ASSOC PROF, DEPT EPIDEMIOL, SCH HYG & PUB HEALTH, JOHNS HOPKINS UNIV, 65- Concurrent Pos: US Army rep bact & mycol study sect, NIH, 62-; chmn comt lab animal facil, Johns Hopkins Univ. Mem: AAAS; Am Soc Parasitol; Am Soc Trop Med & Hyg; Am Asn Immunol. Res: Immunodiagnosis and immunology of microbial infections. Mailing Add: Dept of Epidemiol Johns Hopkins Univ 615 N Wolfe St Baltimore MD 21205

ANDERSON, ROBERT J, veterinary medicine, see 12th edition

ANDERSON, ROBERT JAMES, b Chicago, Ill, Sept 26, 39; m 62. CHEMICAL PHYSICS, ELECTRONIC ENGINEERING. Educ: Univ Ill, Urbana, BSc, 62; Mich State Univ, PhD(phys chem), 67. Prof Exp: Staff electronics engr, Interstate Electronics Corp, Calif, 66-67; dir res & develop, Optics Technol Ctr, Singer Co, Md, 67-70; SR SCIENTIST, BECKMAN INSTRUMENTS, INC, 70- Concurrent Pos: Consult, Nat Acad Sci, 73- Mem: AAAS; Inst Elec & Electronics Engrs; Optical Soc Am. Res: Measurement of the depolarization of Rayleigh scattered laser light; electro-optical display systems, including laser deflection and the interaction of lasers with photochromic materials; optical methods for kinetic measurement and assay of antibody-antigen reactions; photochemical determination of biological materials. Mailing Add: Corp Res Beckman Instruments 2500 Harbor Blvd Fullerton CA 92634

ANDERSON, ROBERT JAMES, dairy industry, biochemistry, see 12th edition

ANDERSON, ROBERT JOHN, physical chemistry, see 12th edition

ANDERSON, ROBERT L, b Chicago, Ill, Feb 22, 32; m 58; c 2. BIOCHEMISTRY, NUTRITION. Educ: Colo State Univ, BS, 57; NC State Col, MS, 59, PhD(nutrit), 62. Prof Exp: RES BIOCHEMIST, PROCTER & GAMBLE CO, 62- Mem: Am Chem Soc; Am Inst Nutrit; Am Oil Chemists Soc; AAAS. Res: Lipid metabolism, especially fatty acid dynamics. Mailing Add: Procter & Gamble Co PO Box 39175 Cincinnati OH 45239

ANDERSON, ROBERT LEONARD, b Detroit, Mich, Aug 18, 33; m 55; c 2. ELEMENTARY PARTICLE PHYSICS. Educ: Wayne State Univ, BA, 55, MS, 60, PhD(physics), 63. Prof Exp: Physicist, Gen Elec Co, 55; instr physics, Wayne State Univ, 59; res physicist, Gen Motors Tech Ctr, 62-63, sr res physicist, 63-66; NSF fel, Int Centre Theoret Physics, Trieste, Italy, 66-67, Int Atomic Energy Agency vis scientist, 67; Nat Acad Sci-Polish Acad Sci exchange scientist, Inst Nuclear Res, Warsaw, Poland, 67-69; Carl-Bertel Nathhorsts Found fel, Inst Theoret Physics, Göteborg, Sweden, 69-70; ASSOC PROF PHYSICS, UNIV OF THE PAC, 70- Mem: AAAS; Am Phys Soc. Res: Theory of vehicular traffic flow; noncompact group representation theory; relativistic Hamiltonian particle theory. Mailing Add: Dept of Physics Univ of the Pac Stockton CA 95204

ANDERSON, ROBERT LESTER, b Parkersburg, WVa, Dec 30, 33; m 56; c 4. PHYSICS. Educ: Marietta Col, AB, 55; Pa State Univ, MS, 59, PhD(physics), 61. Prof Exp: From asst prof to assoc prof, 61-71, PROF PHYSICS, MARIETTA COL, 71-, HEAD DEPT, 66- Mem: AAAS; Am Phys Soc; Am Asn Physics Teachers. Res: Viscosity and density of pure hydrocarbon liquids at moderately elevated pressures; electrical and thermal transport properties of metals and semiconductors. Mailing Add: Dept of Physics Marietta Col Marietta OH 45750

ANDERSON, ROBERT LEWIS, b Springfield, Ill, Sept 30, 33; m 55; c 3. BIOCHEMISTRY, PROTEIN CHEMISTRY. Educ: Bradley Univ, BS, 55, MS, 67; Univ Nebr, PhD(biochem), 71. Prof Exp: BIOCHEMIST, NORTHERN UTILIZATION RES & DEVELOP DIV, US DEPT AGR, 57- Concurrent Pos: Res asst, Univ Nebr, 67-70. Mem: Am Chem Soc; Am Asn Cereal Chem; Inst Food Technol. Res: Enzymology; trypsin inhibitors. Mailing Add: Northern Regional Res Lab 1815 N University Peoria IL 61604

ANDERSON, ROBERT SIMPERS, b Bryn Mawr, Pa, Jan 4, 39; m 64; c 2. IMMUNOLOGY. Educ: Drexel Univ, BS, 61; Hahnemann Med Col, MS, 68; Univ Del, PhD(immunol, comp physiol), 71. Prof Exp: ASSOC IMMUNOL, SLOAN-KETTERING INST CANCER RES, 73-; ASST PROF, GRAD SCH MED SCI, CORNELL UNIV, 74- Concurrent Pos: USPHS training grant, Univ Minn, 70-73. Mem: Am Soc Zool; Soc Invertebrate Path; Am Entom Soc; Am Asn Immunol. Res: Phylogeny of immunity and cancer. Mailing Add: Sloan-Kettering Inst 145 Boston Post Rd Rye NY 10580

ANDERSON, ROBERT SPENCER, b Wilmington, Del, June 7, 22; m 45; c 2. INTERNAL MEDICINE. Educ: Lincoln Univ, BA, 46; Meharry Med Col, MD, 46; Am Bd Internal Med, dipl, 54. Prof Exp: Intern, Harlem Hosp, New York, 46-47; resident, George W Hubbard Hosp, 47-50; from instr to assoc prof internal med, 50-60, chmn dept, 60-69, chmn admin comt, 66-68, dir comprehensive health serv prog, 69-72, vpres health serv, 72-75, PROF INTERNAL MED, MEHARRY MED COL, 60- Concurrent Pos: Fel cardiol, George W Hubbard Hosp, Nashville, Tenn, 50-51; res fel internal med, Col Physicians & Surgeons, Columbia Univ, 52-53; instr internal med, George W Hubbard Hosp, 50-51, jr attend physician, 50-52, attend physician, 53-, actg med dir, 58-60; dir, Fisk-Meharry Student Health Ctr, 65-; mem bd dirs, Crestview Convalescent Ctr; mem adv comts, Regional Med Prog & Josiah Macy Fac fel; mem, Nat Adv Health Manpower Coun & Mid-Cumberland Health Planning Coun. Mem: AMA; fel Am Col Physicians; Am Diabetes Asn; Am Heart Asn; Nat Med Asn. Res: Heart disease; drug therapy; diabetes mellitus. Mailing Add: Meharry Med Col Nashville TN 37208

ANDERSON, ROBERT SVEN, b Delhi, NY, Mar 19, 45; m 67; c 2. Educ: Cornell Univ, BS, 67, MS, 70, PhD(elec eng), 72. Prof Exp: LASER PHYSICST, GEN ELEC CO, 72- Res: Study of operating characteristics of copper vapor laser leading toward development of a practical, reliable laser for uranium isotope separation. Mailing Add: 719 Raynham Rd Collegeville PA 19426

ANDERSON, ROBERT THOMAS, b Oakland, Calif, Dec 27, 26; c 3. CULTURAL ANTHROPOLOGY. Educ: Univ Calif, Berkeley, BA, 49, MA, 53, PhD(anthrop), 56. Prof Exp: PROF ANTHROP, MILLS COL, 60- Res: European ethnology. Mailing Add: Dept of Anthrop Mills Col Oakland CA 94613

ANDERSON, RODNEY EBON, b DeKalb, Ill, Nov 20, 20; m 43; c 6. MATHEMATICS. Educ: Northern Ill State Teachers Col, BE, 42; Univ Chicago, SM, 50; Univ Ind, EdD, 56. Prof Exp: From instr to asst prof math, Northern Ill State Teachers Col, 47-57; assoc prof, Northern Ill Univ, 57-58; from asst prof to assoc prof, 58-67, PROF MATH, SAN JOSE STATE COL, 67- Mem: Math Asn Am. Res: Teaching of collegiate mathematics; preparation of secondary and junior college teachers of mathematics. Mailing Add: Dept of Math San Jose State Col San Jose CA 95114

ANDERSON, ROGER ARTHUR, b Madison, SDak, Nov 7, 35; m 57; c 4. BOTANY, BIOLOGY. Educ: Augustana Col, SDak, AB, 58; Univ Colo, MA, 61, PhD(bot), 64. Prof Exp: Asst prof bot, Univ Mont, 64-65; asst prof bot & biol, 65-74, ASSOC PROF BIOL SCI, UNIV DENVER, 74- Mem: AAAS; Am Bryol Soc; Am Soc Plant Taxon; Bot Soc Am. Res: Taxonomy and world distribution of the lichenized fungi of North America, particularly western North America. Mailing Add: Dept of Biol Sci Univ of Denver Denver CO 80210

ANDERSON, ROGER CLARK, b Wausau, Wis, Oct 30, 41; m 67; c 1. BOTANY, ECOLOGY. Educ: Wis State Univ, La Crosse, BS, 63; Univ Wis-Madison, MS, 65, PhD(bot), 68. Prof Exp: Asst prof bot, Southern Ill Univ, Carbondale, 68-70; asst prof bot & managing dir arboretum, Univ Wis-Madison, 70-73, assoc prof, 73; ASSOC PROF BIOL, CENT STATE UNIV, OKLA, 73- Mem: Ecol Soc Am; AAAS; Sigma Xi. Res: Physiological plant ecology and phytosociology. Mailing Add: Dept of Biol Cent State Univ Edmond OK 73034

ANDERSON, ROGER E, b Donovan, Ill, May 14, 30; m 59; c 2. PHYSICAL CHEMISTRY, COMPUTER SCIENCES. Educ: Univ Ill, BS, 51; Wash State Univ, PhD(phys chem), 61. Prof Exp: CHEMIST, LAWRENCE RADIATION LAB, UNIV CALIF, LIVERMORE, 59- Mem: Inst Elec & Electronics Eng; Asn Comput Mach. Res: Instrumentation; laboratory automation; on-line computer utilization; data processing techniques; minicomputer systems analysis. Mailing Add: PO Box 808 Lawrence Livermore Lab Livermore CA 94550

ANDERSON, ROGER FABIAN, b St Paul, Minn, Sept 14, 14; m 50; c 2. FOREST ENTOMOLOGY. Educ: Univ Minn, PhD(entom), 45. Prof Exp: Entomologist, USDA, 40-48; asst prof entom, Univ Ga, 49-50; assoc prof, 50-59, PROF FOREST ENTOM, DUKE UNIV, 59- Mem: Am Entom Soc; Soc Am Foresters. Res: Ecology of forest insects. Mailing Add: Sch of Forestry Duke Univ Durham NC 27706

ANDERSON, ROGER HARRIS, b Seattle, Wash, Feb 3, 30; m 58; c 2. LOW TEMPERATURE PHYSICS. Educ: Univ Wash, BSc, 51, PhD(physics), 61. Prof Exp: Asst, Univ Wash, 52-58; res scientist, Boeing Co, 58-61; PROF PHYSICS, SEATTLE PAC COL, 61- Concurrent Pos: NSF sci fac fel, Univ Ill, 68-69. Mem: Am Phys Soc; Am Asn Physics Teachers. Res: Cloud chamber study of scattering of cosmic-ray muons; calculation of thermal properties of monomolecular layers of liquid helium-3; transport phenomena in Landau theory of Fermi liquids. Mailing Add: Dept of Physics Seattle Pac Col 3307 Third Ave Seattle WA 98119

ANDERSON, ROGER VINCENT, nematology, see 12th edition

ANDERSON, ROGER W, b Fargo, NDak, Jan 9, 43; m 64; c 1. PHYSICAL CHEMISTRY. Educ: Carleton Col, BA, 64; Harvard Univ, MA, 65, PhD(chem), 68. Prof Exp: Asst prof, 69-75, ASSOC PROF CHEM, UNIV CALIF, SANTA CRUZ, 75- Mem: AAAS; Am Phys Soc. Res: Molecular beam studies of nonadiabatic atomic and molecular collisions; collisional excitation; electronic quenching, reaction of excited atoms; theory of nonreactive scattering in reactive systems and scattering of excited atoms. Mailing Add: Natural Sci Bldg II Univ of Calif Santa Cruz CA 95060

ANDERSON, ROGER YATES, b Oak Park, Ill, Oct 4, 27; m 48; c 4. GEOLOGY. Educ: Univ Ariz, BS, 54, MS, 55; Stanford Univ, PhD, 60. Prof Exp: Asst biochem, Univ Ariz, 55-56; from instr to assoc prof, 56-74, PROF GEOL, UNIV NMEX, 74- Concurrent Pos: NSF prin investr varve res. Res: Paleoecology; palynology; varves. Mailing Add: Dept of Geol Univ of NMex Albuquerque NM 87106

ANDERSON, ROLAND CARL, b Trenton, NJ, Nov 27, 34; m 53; c 2. ASTROPHYSICS, FLUID MECHANICS. Educ: Univ Fla, BS, 60, MS, 61, PhD(turbulence), 65. Prof Exp: Assoc engr, Martin Co, Fla, 60; instr aerodyn, Univ Fla, 61-62, teaching assoc, 63-65; instr, WVa Univ, 62-63; Ford Found fel astrophys, Johns Hopkins Univ, 65-66; from asst prof to assoc prof, 66-73, asst dean res, Col Eng, 74-75, PROF ASTROPHYS, UNIV FLA, 73- Mem: AAAS; Optical Soc Am. Res: Lidar; remote sensing; optical instruments. Mailing Add: Dept of Eng Sci Univ of Fla Gainesville FL 32603

ANDERSON, RONALD EUGENE, b Sioux City, Iowa, Sept 15, 20; m 48; c 4. GENETICS, PLANT BREEDING. Educ: Univ Nebr, BS, 48; Univ Wis, MS, 49, PhD(genetics), 52. Prof Exp: Asst prof agron, Univ Ky, 52-54; asst prof plant breeding, 54-69, ASSOC PROF PLANT BREEDING & BIOMET, CORNELL UNIV, 69- Res: Cytology; breeding of forage crops species; effects of radiation on forage species. Mailing Add: Dept of Plant Breeding Cornell Univ Ithaca NY 14850

ANDERSON, ROWLAND C, b Cyrus, Minn, July 29, 12; m 36; c 3. MATHEMATICS. Educ: St Cloud State Col, BE, 31; Univ Minn, MA, 39; Columbia Univ, EdD, 50; Purdue Univ, MS, 60. Prof Exp: Prof, 39-74, EMER PROF MATH, ST CLOUD STATE UNIV, 74- Concurrent Pos: NSF fac fel, 59-60; vis prof from Columbia Univ to Inst Educ, Makerere Univ Uganda, 66-68; asst dir, Colegia Bautista, San Andres, Isla, 74-75. Mem: Math Asn Am. Res: Mathematics education. Mailing Add: 2108 7 Ave S St Cloud MN 56301

ANDERSON, ROY CLAYTON, b Camrose, Alta, Apr 26, 26; m 48; c 1. PARASITOLOGY. Educ: Univ Alta, BSc, 50; Univ Toronto, MA, 52, PhD(parasitol), 56; Sch Advan Studies, Paris, dipl helminth, 58. Prof Exp: Sr res scientist parasitol, Ont Res Found, 54-65; PROF ZOOL, UNIV GUELPH, 65- Concurrent Pos: Fel Rothmansted Exp Sta, London Sch Hyg & Trop Med, 56-57; med fac, Univ Paris, 57-58. Honors & Awards: Henry Baldwin Ward Medal, Am Soc Parasitol, 68. Mem: Am Soc Parasitol. Mailing Add: Dept of Zool Univ of Guelph Guelph ON Can

ANDERSON, ROY SCOTT, b New York, NY, Jan 30, 31; m 58; c 2. ORGANIC CHEMISTRY, POLYMER CHEMISTRY. Educ: Haverford Col, BS, 53; Tufts Univ, MS, 58; Purdue Univ, PhD(org chem), 61. Prof Exp: Biochem technician, Charles

Pfizer, 53-57; res scientist, Res Ctr, Uniroyal Corp, 61-63; res & develop mgr, Textile Div, 63-65; asst dir develop, Deering-Milliken Res Corp, 65-67; explor res & develop supvr, Tanatex Chem Corp, 67-68; mgr appl res & develop sect, Develop Div, Chem & Plastics Group, Borg-Warner Corp, 68-71; mgr chem res, Acushnet Corp, 71-73; TECH MGR NEW PRODUCT DEVELOP DEPT, PQ CHEMS, 73- Mem: Am Chem Soc; Brit Chem Soc; fel Am Inst Chem. Res: Polymers; inorganic polymers; silicas and silicates; aluminum compounds; flame retardants; chemical structure-performance correlations. Mailing Add: 514 Concord Dr Broomall PA 19008

ANDERSON, ROY STUART, b Springfield, Mass, Oct 16, 21; m 44; c 3. PHYSICS. Educ: Clark Univ, AB, 43; Dartmouth Col, AM, 48; Duke Univ, PhD(physics), 51. Prof Exp: Res engr, Stanford Res Inst, Calif, 51-52; from asst prof to assoc prof physics, Univ Md, 52-60; chmn dept, 60-70, PROF PHYSICS, CLARK UNIV, 60-, DEAN GRAD SCH, 70- Concurrent Pos: Res assoc, Duke Univ, 51, 53 & 54, Univ Calif, 58-59 & Woods Hole Oceanog Inst, 60; consult, United Aircraft Corp. Mem: Am Phys Soc; Am Asn Physics Teachers. Res: Microwave and radio spectroscopy; electron spin resonance; radiation damage in single crystals. Mailing Add: Dept of Physics Clark Univ Worcester MA 01610

ANDERSON, RUPERT SIGFRED, chemistry, physiology, deceased

ANDERSON, RUSSELL D, b Salt Lake City, Utah, Dec 27, 27; m 58; c 2. ENTOMOLOGY, INVERTEBRATE ZOOLOGY. Educ: Univ Utah, BS, 54, MS, 56, PhD(entom), 61. Prof Exp: Entomologist, South Salt Lake County Mosquito Abatement Dist, 55-58; from instr to assoc prof zool, Church Col, Hawaii, 58-64, head dept biol sci, 59-64; from asst prof to assoc prof zool, 64-71, PROF ZOOL, SOUTHERN UTAH STATE COL, 71- Mem: AAAS; Entom Soc Am. Res: Taxonomy of the Dytiscidae; revisionary study of the genus Hygrotus. Mailing Add: Dept of Life Sci Southern Utah State Col Cedar City UT 84720

ANDERSON, RUSSELL K, b LeRoy, Minn, Mar 13, 24; m 48; c 2. ANIMAL NUTRITION, PHYSIOLOGY. Educ: Univ Minn, BS, 48; Iowa State Univ, MS, 50, PhD(animal nutrit), 56. Prof Exp: Instr animal sci, Iowa State Univ, 48-55; from asst prof to assoc prof, 55-64, PROF ANIMAL SCI, CALIF POLYTECH STATE UNIV, SAN LUIS OBISPO, 64- Mem: Am Soc Animal Sci. Res: Ruminant nutrition; mineral requirement for rumen microorganisms and sources of phosphorus. Mailing Add: Dept of Animal Sci Calif Polytech State Univ San Luis Obispo CA 93401

ANDERSON, RUTH SUZANNE, biology, see 12th edition

ANDERSON, SAMUEL, b Clarksdale, Miss, Aug 26, 34; m 65; c 2. ANALYTICAL CHEMISTRY, INORGANIC CHEMISTRY. Educ: Univ Mo-Kansas City, BS, 57; Iowa State Univ, MS, 60, PhD(anal chem), 62. Prof Exp: Asst prof chem, Tenn State Univ, 62-63; PROF CHEM, NORFOLK STATE COL, 63- Concurrent Pos: Res partic, NSF, 64-65. Mem: Am Chem Soc; Sigma Xi. Res: Preparation, physical and chemical properties of vanadium B-diketone complexes; evaluation of an immobilized chelate for extraction of metal ions from solutions. Mailing Add: Dept of Chem Norfolk State Col Norfolk VA 23504

ANDERSON, SCOTT, physics, see 12th edition

ANDERSON, STANLEY H, b San Francisco, Calif, Aug 6, 39; m 65; c 2. ECOLOGY, ENVIRONMENTAL SCIENCES. Educ: Univ Redlands, BSc, 61; Ore State Univ, MA, 68, DrPhil(ecol), 70. Prof Exp: Asst prof biol, Kenyon Col, 70-75; ECOLOGIST, OAK RIDGE ASSOC UNIVS, 75- Concurrent Pos: Consult, Oak Ridge Nat Lab, 71- & Tenn Valley Authority, 75. Mem: AAAS; Am Inst Biol Sci; Am Ornith Union; Cooper Ornith Soc; Ecol Soc Am. Res: Community interaction of avifauna; determination of effects of habitat alteration on vertebrate populations. Mailing Add: ORAU Training Div Box 117 Oak Ridge TN 37830

ANDERSON, STANLEY ROBERT, b Rudyard, Mich, Mar 11, 20; m 46; c 3. AGRONOMY. Educ: Mich State Col, BS, 46, MS, 49; Iowa State Col, PhD, 54. Prof Exp: Field rep, Mich Crop Improv Asn, Mich State Col, 46-49; instr pub schs, Mich, 49-50; instr agron, Iowa State Col, 50-54; from asst prof to prof, Ohio State Univ, 54-67; DEAN AGR, TEX A&I UNIV, 67- Concurrent Pos: Chmn, Am Asn Univ Agr Adminr, 75-75. Mem: Am Soc Agron; Crop Sci Soc Am. Res: Crop production; management and utilization of pasture and forage establishment. Mailing Add: 1115 Elizabeth Ave Kingsville TX 78363

ANDERSON, STEVEN CLEMENT, b Grand Canyon Nat Park, Ariz, Sept 7, 36; m 60; c 1. SYSTEMATICS, ECOLOGY. Educ: Univ Calif, Riverside, BA, 57; San Francisco State Col, MA, 62; Stanford Univ, PhD(biol), 66. Prof Exp: Lab technician entom, Univ Calif, Riverside, 60; res asst herpet, Calif Acad Sci, 63-64, from asst cur to assoc cur, 64-70; asst prof, 70-72, ASSOC PROF ENVIRON SCI, CALLISON COL, UNIV OF THE PAC, 72- Concurrent Pos: Vis res assoc, Iran Dept Environ, 75- Mem: Am Soc Ichthyol & Herpetol; Soc Study Amphibians & Reptiles; Soc Vert Paleont. Res: Systematics, ecology, zoogeography of amphibians and reptiles of Southwest Asia. Mailing Add: Callison Col Univ of the Pac Stockton CA 95204

ANDERSON, SYDNEY, b Topeka, Kans, Jan 11, 27; m 51; c 3. VERTEBRATE ZOOLOGY. Educ: Univ Kans, AB, 50, MA, 52, PhD, 59. Prof Exp: Asst instr, Univ Kans, 50-52; NSF fel, 52-54; asst cur mammals, Mus Natural Hist, Kans, 54-59; from asst cur to assoc cur, 60-69, CUR, AM MUS NATURAL HIST, 69- Concurrent Pos: Instr, Univ Kans, 54-59; consult, Bk Div, Time, Inc & C S Hammond Co. Mem: Fel AAAS; Am Soc Mammal; Ecol Soc Am; Soc Study Evolution; Soc Syst Zool. Res: Distribution, variation and relationships of mammals; instrumentation and methodology; bibliography and information retrieval. Mailing Add: Am Mus of Natural Hist Central Park W at 79th St New York NY 10024

ANDERSON, TERRY GRANT, physical chemistry, see 12th edition

ANDERSON, THEODORE EDMUND, b Dallas, Tex, Nov 27, 29; m 56; c 2. MICROBIOLOGY, BIOCHEMISTRY. Educ: Univ Mich, BS, 53; Mich State Univ, MS, 58, PhD(microbiol), 63. Prof Exp: Chemist, Kalamazoo Veg Parchment, 56, Union Carbide Chem Co, 58-59 & Dow Chem Co, 63-66; MGR QUAL CONTROL DEVELOP, MILES LABS, INC, 66- Mem: Am Soc Microbiol. Res: Management of analytical chemistry, microbiology, statistics and computerization. Mailing Add: Qual Control Dept Miles Labs Inc 1127 Myrtle St Elkhart IN 46514

ANDERSON, THEODORE GUSTAVE, b New Haven, Conn, Nov 6, 02; m 36. BACTERIOLOGY, Educ: Brown Univ, PhB, 31; Yale Univ, PhD(bact), 35. Prof Exp: Assoc dairy husb, Univ Calif, 36-37; from instr to asst prof bact, Pa State Univ, 37-43, 46-47; bacteriologist, Vet Admin Hosp, New York, 47; from assoc prof bact & immunol to prof microbiol & bacteriologist, Hosp, 47-70, EMER PROF MICROBIOL, SCH MED, TEMPLE UNIV, 70-; DIR MICROBIOL LAB, EPISCOPAL HOSP, 70- Concurrent Pos: Consult microbiol, US Naval Hosp, Philadelphia. Mem: Am Soc Microbiol; fel Am Acad Microbiol. Res: Medical bacteriology; antibiotic susceptibility testing; urinary tract infection; dental caries; dairy microbiology. Mailing Add: Dept of Microbiol Episcopal Hosp Front & Lehigh Ave Philadelphia PA 19125

ANDERSON, THEODORE WILBUR, b Minneapolis, Minn, June 5, 18; m 50; c 3. MATHEMATICAL STATISTICS, ECONOMETRICS. Educ: Northwestern Univ, BS, 39; Princeton Univ, MA, 42, PhD(math), 45. Prof Exp: Asst math, Northwestern Univ, 39-40; instr, Princeton Univ, 41-43, res assoc, Appl Math Panel Contract, Nat Defense Res Comt, 43-45 & Cowles Comn, Chicago, Ill, 45-46; from instr to prof math statist, Columbia Univ, 46-67, actg exec officer, 50-51, chmn dept math statist, 56-60 & 64-65; PROF STATIST & ECON, STANFORD UNIV, 67- Concurrent Pos: Consult, Cowles Found, 46-60, Bur Appl Social Res, New York, 47 & Rand Corp, 49-66; Guggenheim fel, Univ Stockholm & Cambridge Univ, 47-48; ed, Ann Math Statist, 50-52; assoc prof, Stanford Univ, 54; mem, Nat Acad Sci-Nat Res Coun Comt Basic Res, adv to Off Ord Res, 55-58; fel, Ctr Advan Study Behav Sci, 57-58; mem, Nat Res Coun Comt Statist, 60-63, chmn, 61-63; mem, Nat Acad Sci-Nat Res Coun Panel Appl Math, adv to Nat Bur Standards, 64-65; mem, Nat Acad Sci Comt Support of Res in Math Sci, 65-68; acad vis, Imp Col & vis prof, Univ London, Univ Moscow & Univ Paris, 67-68; vis scholar, Ctr Advan Study Behav Sci, 72-73; acad vis, London Sch Econ & Polit Sci, 74-75. Mem: Nat Acad Sci; Inst Math Statist (pres, 63); Am Math Soc; fel Am Statist Asn (vpres, 71-73); fel Economet Soc. Res: Multivariate statistical analysis; time series analysis; theory of statistical inference; econometric methodology. Mailing Add: Dept of Statist Stanford Univ Stanford CA 94305

ANDERSON, THOMAS ALEXANDER, b Elkhart, Ind, Feb 23, 28; m 52; c 2. BIOCHEMISTRY, NUTRITION. Educ: Univ Wash, BA, 50; Calif State Polytech Col, BS, 56; Univ Ariz, MS, 61, PhD(agr biochem & nutrit), 62. Prof Exp: Sr scientist, Sci Info Dept, Smith, Kline & French Labs, Pa, 62-63; asst prof biochem, Univ SDak, 63-65; chief nutrit res lab, H J Heinz Co, 65-70; assoc prof pediat, 70-74, PROF PEDIAT, COL MED, UNIV IOWA, 74- Concurrent Pos: Partic, White House Conf Food, Nutrit & Health, 69. Mem: AAAS; Am Chem Soc; Am Soc Clin Nutrit; Brit Nutrit Soc; Am Soc Animal Sci. Res: Pediatric nutrition,especially influence of the pattern of food intake on body composition and digestibility of complex carbohydrates by the human infant. Mailing Add: Dept of Pediat Univ of Iowa Col of Med Iowa City IA 52240

ANDERSON, THOMAS BROWN, b Lancaster, Ky, May 30, 16; m 42; c 2. SPEECH PATHOLOGY, AUDIOLOGY. Educ: Wash Univ, BS, 39; Butler Univ, MS, 47; Ohio State Univ, PhD(speech), 52. Prof Exp: From instr to asst prof speech & hearing, Ohio State Univ, 47-55; assoc prof, 55-62, PROF SPEECH & HEARING, MIAMI UNIV, 62- Mem: Am Speech & Hearing Asn. Mailing Add: Dept of Speech Miami Univ Oxford OH 45056

ANDERSON, THOMAS DALE, b Cleveland, Ohio, Sept 23, 29; m 56; c 3. GEOGRAPHY. Educ: Kent State Univ, AB, 53, MA, 56; Univ Nebr, PhD(geog), 66. Prof Exp: Instr geog, State Univ NY Col Geneseo, 59-64; asst prof, 64-68, ASSOC PROF GEOG, BOWLING GREEN STATE UNIV, 68- Concurrent Pos: Corps Engrs grant, Cent Ohio Environ Anal, 72-73; Fulbright-Hays lectr geog, Inst Geog & Regional Develop, Cent Univ, Venezuela, 74. Mem: Asn Am Geog; Am Geog Soc; Nat Coun Geog Educ; Conf Latin Am Geographers; World Pop Soc. Res: Agricultural land use stability on muck soils; remapping of world coastlines in relation to supertanker navigation; world population distribution and dynamics; mapping of liberalism in United States political geography. Mailing Add: Dept of Geog Bowling Green State Univ Bowling Green OH 43403

ANDERSON, THOMAS FOXEN, b Manitowoc, Wis, Feb 7, 11; m 37; c 2. BIOPHYSICS. Educ: Calif Inst Technol, BS, 32, PhD(chem), 36. Prof Exp: Asst phys chem, Calif Inst Technol, 34-36; instr, Univ Chicago, 36-37; investr plant physiol, Univ Wis, 37-39, instr phys chem, 39-40; RCA Corp fel, Nat Res Coun, 40-42; assoc, Johnson Found, 42-46, from asst prof to assoc prof res viruses, 46-58, PROF ZOOL, UNIV PA, 58-; SR MEM, INST CANCER RES, 58- Concurrent Pos: Secy comt appln electron microscope, Nat Res Coun, 44; Fulbright & Guggenheim fels, Inst Pasteur, Paris, 55-57; pres, Int Fedn Electron Microscope Socs, 60-64; consult, NIH, 61-64 & 65-68; chmn, US Nat Comt Pure & Appl Biophys, 64-67; mem, Int Comt Nomenclature of Viruses, 68- Mem: Nat Acad Sci; AAAS; Am Soc Gen Physiol; Biophys Soc (pres, 65-66); Electron Micros Soc (pres, 55). Res: Raman spectra; surface chemistry; genetics of bacteria and bacterial viruses; biological applications of electron microscopy; molecular structure. Mailing Add: Inst for Cancer Res Philadelphia PA 19111

ANDERSON, THOMAS FRANK, b Chicago, Ill, Dec 7, 39; m 63. GEOCHEMISTRY. Educ: DePauw Univ, BA, 61; Columbia Univ, PhD(geochem), 67. Prof Exp: Res assoc geochem, Enrico Fermi Inst, Univ Chicago, 67-68; asst prof, 67-73, ASSOC PROF GEOL, UNIV ILL, URBANA, 73- Concurrent Pos: Exchange scientist, NSF and Nat Ctr Sci Res, France, 75-76. Mem: AAAS; Am Geophys Union; Geochem Soc; Geol Soc Am. Res: Kinetics and mechanisms of oxygen isotope exchange in mineral-fluid systems; stable isotope geochemistry of deep-sea sediments and applications to paleo-oceanography. Mailing Add: Dept of Geol Univ of Ill Urbana IL 61801

ANDERSON, THOMAS PAGE, b Anadarko, Okla, Aug 3, 18. PHYSICAL MEDICINE & REHABILITATION. Educ: Univ Okla, BS, 40, MD, 43; Univ Minn, MS, 51; Am Bd Phys Med & Rehab, dipl, 53. Prof Exp: Intern, St Paul's Hosp, Dallas, Tex, 44; clin instr phys med & rehab, Dartmouth Med Sch, 51-63, asst clin prof, 63-64; assoc clin prof, 64-68, clin prof, 68-70, ASSOC PROF PHYS MED & REHAB, MED SCH, UNIV MINN, MINNEAPOLIS, 70- Concurrent Pos: Consult, Hitchcock Clin, Hanover, NH, 51-64 & Vet Admin Hosp, 52-64; staff Physiatrist, Kenny Rehab Inst, 64-70; dir dept med educ, Am Rehab Found, 65-70. Mem: Fel Am Acad Phys Med & Rehab; fel Am Col Physicians; Am Cong Rehab Med; Am Geriat Soc; Am Rheumatism Asn. Res: Stroke; degenerative joint disease; group communications and leadership skills; disorders of the back. Mailing Add: Dept of Phys Med & Rehab Univ of Minn Hosp Minneapolis MN 55455

ANDERSON, TRUMAN OLIVER, microbiology, see 12th edition

ANDERSON, VICTOR CHARLES, b Shanghai, China, Mar 31, 22; m 43; c 3. PHYSICS. Educ: Univ Redlands, AB, 43; Univ Calif, Los Angeles, MA, 50, PhD(physics), 53. Prof Exp: Tech asst, Radiation Lab, Univ Calif, 43-45; asst, Los Alamos, NMex, 45-46; asst, Univ Calif, Los Angeles, 46-47; from asst res physicist to res physicist, 47-68, actg dir, Marine Physical Lab, 64, 68 & 74, PROF APPL PHYSICS & ASSOC DIR MARINE PHYSICAL LAB, SCRIPPS INST, UNIV CALIF, SAN DIEGO, 68- Concurrent Pos: Res fel, Harvard Univ, 54-55. Mem: Fel Acoust Soc Am; Inst Elec & Electronics Engrs. Res: Acoustics; experimental and theoretical work in field of underwater sound and electronics, methods and instrumentation for digital signal processing; remotely controlled sea floor work systems. Mailing Add: Marine Physical Lab Scripps Inst of Oceanog San Diego CA 92132

ANDERSON, VICTOR ELVING, b Stromsburg, Nebr, Sept 6, 21; m 46; c 4. HUMAN

GENETICS, BEHAVIORAL GENETICS. Educ: Univ Minn, BA, 45, MS, 49, PhD(zool), 53. Prof Exp: From instr to prof biol, Bethel Col, Minn, 46-60, chmn dept, 52-60; vis scientist, NIH, 60-61; asst scientist, 49-54, assoc prof zool, 61-65, assoc prof genetics, 65-66, PROF GENETICS, UNIV MINN, MINNEAPOLIS, 66-, ASST DIR, DIGHT INST HUMAN GENETICS, 54- Concurrent Pos: Regent, Bethel Col & Seminary, St Paul, 69-74; mem, Develop Behav Sci Study Sect, NIH, 72-75, chmn, 74-75. Mem: Fel AAAS; Am Sci Affil (pres, 63-65); Am Soc Human Genetics; Genetics Soc Am; Behav Genetics Asn (secy, 72-74). Res: Genetic factors in mental retardation, psychotic disorders, and other human behavioral problems; behavioral variation in phenylketonuria, porphyria, and other metabolic diseases. Mailing Add: 1775 N Fairview Ave St Paul MN 55113

ANDERSON, VIRGIL LEE, b North Liberty, Ind, May 2, 22; m 43; c 3. MATHEMATICS, STATISTICS. Educ: Iowa State Univ, BS, 47, PhD(statist), 53. Prof Exp: From asst prof to assoc prof statist, 51-60, dir statist lab, 56-66, PROF STATIST, PURDUE UNIV, WEST LAFAYETTE, 60- Concurrent Pos: Statist consult to var govt, mil & pvt orgn, 58- Mem: Fel Am Statist Asn; Biomet Soc. Res: Quantitative genetics; design of experiments; transportation; environmental effects of cadmium. Mailing Add: Dept of Statist Purdue Univ West Lafayette IN 47906

ANDERSON, W FRENCH, b Tulsa, Okla, Dec 31, 36; m 61. BIOCHEMISTRY, GENETICS. Educ: Harvard Univ, AB, 58, MD, 63; Cambridge Univ, MA, 60. Prof Exp: Intern pediat med, Children's Hosp Med Ctr, Boston, 63-64; res assoc, Lab Biochem Genetics, 65-67, res med officer, 67-68, head sect human biochem, 68-71, CHIEF MOLECULAR HEMAT BR, NAT HEART & LUNG INST, 71- Concurrent Pos: Am Cancer Soc res fel bact & immunol, Harvard Med Sch, 64-65; vol asst, Children's Hosp Med Ctr, 64-65; prof lectr, Sch Med, George Washington Univ, 67-75; mem heart fel bd, Nat Heart & Lung Inst, 68-70; mem fac, Dept Genetics, NIH Grad Prog, 69-; mem task force hemoglobinopathies, 72; chmn, Inter-Inst Coord Comt on Cooley's Anemia, NIH, 72-; mem hemoglobinopathies subcomt, Am Soc Hematol, 73-74; mem Med Resources Coun, Cooley's Anemia Blood & Res Found for Children, 74-; chmn, Inter-Agency Coord Comt on Cooley's Anemia, 75- Honors & Awards: Sci Achievement Biol Sci Award, Wash Acad Sci, 71; Superior Serv Award, Dept Health Educ & Welfare, 75. Mem: Am Chem Soc; Am Soc Clin Invest; Am Soc Hemat; Am Soc Human Genetics; Am Soc Biol Chem. Res: Human biochemical genetics; hematology; regulation of RNA and protein synthesis; hemoglobin biosynthesis; thalassemia and hemoglobinopathies. Mailing Add: Bldg 10 7D-20 Nat Heart & Lung Inst Bethesda MD 20014

ANDERSON, WALLACE ERVIN, b Florence County, SC, Oct 28, 13; m 39; c 2. PHYSICS. Educ: The Citadel, BS, 34; Univ Ky, MS, 36; Univ Mich, PhD(physics), 49. Prof Exp: From asst prof to prof physics, 36-66, head dept, 53-66, acad dean, 66-70, VPRES ACAD AFFAIRS, THE CITADEL, 70- Honors & Awards: Legion of Merit, 46. Mem: Am Phys Soc; Am Asn Physics Teachers; Am Soc Eng Educ. Res: Infrared and microwave spectroscopy; determination of molecular structure; optics. Mailing Add: The Citadel Charleston SC 29409

ANDERSON, WALTER CLINTON, b Erie, Pa, May 14, 23; m 48; c 3. FOREST ECONOMICS. Educ: Pa State Univ, BS, 48, MS, 50, PhD(agr econ), 66. Prof Exp: Jr forester, Southeastern Forest & Range Exp Sta, 53-54, forest economist, 54-66, PRIN ECONOMIST, SOUTHERN FOREST EXP STA, US FOREST SERV, USDA, 66- Mem: Am Econ Asn; Am Agr Econ Asn; Soc Am Foresters. Res: Economics of growing, harvesting and utilizing southern pine timber and related problems such as forest land ownership. Mailing Add: PO Box 53153 New Orleans LA 70153

ANDERSON, WARREN BOYD, b Oak City, Utah, Nov 1, 29; m 53; c 6. SOIL CHEMISTRY. Educ: Brigham Young Univ, BS, 58; Colo State Univ, MS, 62, PhD(soil sci), 64. Prof Exp: Phys sci technician, Agr Res Serv, USDA, 60; res assoc soil fertil, Colo Agr Exp Sta, 63-64; asst prof, 64-71, ASSOC PROF SOIL & CROP SCI, TEXAS A&M UNIV, 71- Mem: Am Soc Agron; Soil Sci Soc Am. Res: Fertility and plant nutrition; micronutrient nutrition of crops. Mailing Add: Dept of Soil & Crop Sci Tex A&M Univ College Station TX 77843

ANDERSON, WAYNE I, b Montrose, Iowa, Sept 14, 35; m 58; c 4. GEOLOGY. Educ: Univ Iowa, BA, 58, MS, 61, PhD(geol), 64. Prof Exp: Assoc prof, 63-73, PROF EARTH SCI, UNIV NORTHERN IOWA, 73-, HEAD DEPT, 63- Mem: Geol Soc Am; Soc Econ Paleontologists & Mineralogists. Res: Paleozoic stratigraphy and paleontology; geology of Iowa. Mailing Add: Dept of Earth Sci Univ of Northern Iowa Cedar Falls IA 50613

ANDERSON, WAYNE JAY, b Ft Meade, Md, July 14, 45; m 67; c 4. OPTICAL PHYSICS. Educ: Utah State Univ, BS, 69, MS, 70, PhD(physics), 73. Prof Exp: Res physicist, Aerospace Res Lab, 73-75, RES PHYSICIST, AVIONICS LAB, US AIR FORCE, 75- Mem: Optical Soc Am; Am Phys Soc; Electrochem Soc; Sigma Xi. Res: Ion implantation and radiation damage in GaAs and GaP compounds. Mailing Add: Air Force Avionics Lab-DHR Wright-Patterson AFB OH 45433

ANDERSON, WAYNE KEITH, b Pine Falls, Man, Apr 4, 41; m 62; c 2. MEDICINAL CHEMISTRY, ORGANIC CHEMISTRY. Educ: Univ Man, BSc, 62, MSc, 65; Univ Wis-Madison, PhD(med chem), 68. Prof Exp: Instr med chem, Univ Wis, 67 & 68; ASST PROF MED CHEM, STATE UNIV NY BUFFALO, 68- Mem: Am Chem Soc. Res: Synthetic approaches to and structural modifications of naturally occurring antitumor compounds; new types of olefinic cyclization reactions; synthesis and structure of mesoionic compounds; isolation and structure elucidation of naturally occurring antitumor compounds. Mailing Add: 505 Sprucewood Terr Williamsville NY 14221

ANDERSON, WAYNE PHILPOTT, b Jamestown, NY, Apr 1, 42. INORGANIC CHEMISTRY. Educ: Harpur Col, BA, 64; Univ Ill, MS, 66, PhD(chem), 68. Prof Exp: Asst prof chem, Univ Del, 68-75; ASST PROF CHEM, BLOOMSBURG STATE COL, 75- Mem: Am Chem Soc; AAAS; The Chem Soc; Nat Sci Teachers Asn. Res: Structure and bonding in organometallic compounds; metal carbonyls. Mailing Add: Dept of Chem Bloomsburg State Col Bloomsburg PA 17815

ANDERSON, WESTON ARTHUR, b Kingsburg, Calif, Mar 28, 28; m 52; c 2. PHYSICS. Educ: Stanford Univ, BS, 50, MS, 53, PhD(physics), 55. Prof Exp: Res physicist, Europ Coun Nuclear Res, 54-55; res physicist, 53-63, dir res, Anal Instrument Div, 63-72, DIR SYSTS & TECH LAB, VARIAN ASSOCS, 72- Mem: Inst Elec & Electronics Engrs; Am Asn Physics Teachers; Am Phys Soc; Am Soc Psychical Res; Am Inst Ultrasound in Med. Res: Magnetic resonance; biomedical engineering. Mailing Add: 763 La Para Ave Palo Alto CA 94306

ANDERSON, WILLARD EUGENE, b Vilas, SDak, Aug 27, 33; m 59. APPLIED PHYSICS. Educ: Huron Col, SDak, BS, 55. Prof Exp: Res asst & tech asst, Corp Res Ctr, Honeywell, Inc, 55-56; teaching asst math, SDak State Univ, 56-59; res scientist, Aeronaut Div, Res Dept, 59-64, sr res scientist, Mil Prod Group, 64-65, prin res scientist, 65-74, PRIN SYSTS ENGR, AEROSPACE & DEFENSE GROUP, SYSTS & RES CTR, HONEYWELL, INC, 74- Concurrent Pos: Consult gen physics, 59-; consult, Manned Orbiting Lab, Controls Subcontract, Aerospace Div, Honeywell Inc, 66-69. Mem: AAAS. Res: Gyroscopes, precision spheres, rotor spin-up, pickoffs, voltage breakdown; density, atmosphere, ocean; trace gas sensing, chemistry, ionic mobility; optics, laser Doppler, opdar, correlators, spatial filtering, scattering; space physics; atmospheric physics. Mailing Add: 2418 Ione St St Paul MN 55113

ANDERSON, WILLIAM ALAN, b St Peter's Bay, PEI, Can, Jan 5, 41; m 63; c 3. MOLECULAR BIOLOGY. Educ: Queen's Univ, Ont, BSc, 62; Mass Inst Technol, PhD(microbiol), 69. Prof Exp: Fel molecular biol, Univ Geneva, 69-70; ASST PROF BIOL, UNIV LAVAL, 71- Mem: French-Can Asn Advan Sci; Can Biochem Soc; Genetics Soc Can; Sigma Xi; Am Soc Microbiol. Res: Metabolism, protein synthesis and regulation in Escherichia coli. Mailing Add: Dept of Biol Univ Laval Quebec PQ Can

ANDERSON, WILLIAM ARNOLD DOUGLAS, b Ont, Aug 27, 10; nat US; m 34; c 4. PATHOLOGY. Educ: Univ Toronto, BA, 31, MD, 34, MA, 36. Prof Exp: Rotating intern, St Michael's Hosp, Toronto, 34-35; asst resident, Univ Hosp, Duke Univ, 36-37; instr, Univ Tenn, 37-40; from asst prof to assoc prof, St Louis Univ, 40-45; prof path & bact, Marquette Univ, 45-53; PROF PATH, UNIV MIAMI, 53- Concurrent Pos: Teaching fel path, Banting Inst, Univ Toronto, 35-36; dir labs, St Joseph's Hosp, Milwaukee, Wis, 45-53 & Jackson Mem Hosp, Miami, Fla, 53-72; vis prof, Univ Cape Town, 61. Honors & Awards: Sci Prod Found Award, Col Am Path, 59, Distinguished Serv Award, 68; Ward Burdick Award, Am Soc Clin Path, 65, Distinguished Serv Award, 68. Mem: Am Asn Path & Bact; Am Soc Exp Path; Am Soc Clin Path; fel Col Am Path (pres, 56-57); fel Royal Col Path Australia. Res: Histopathology; renal disease; oncology; endocrine glands. Mailing Add: Dept Path Univ Miami Sch Med PO Box 875 Biscayne Annex Miami FL 33152

ANDERSON, WILLIAM B, b Montevideo, Minn, Nov 8, 23; m; c 3. PSYCHIATRY. Educ: Univ Minn, BS, 45, MD, 48. Prof Exp: Instr pediat & asst dir pediat clin, Univ Pa, 54-56; asst prof pediat, asst dir state serv crippled children & dir pediat clin, State Univ Iowa, 56-61; ASST PROF PSYCHIAT, MED CTR, DUKE UNIV & DIR CHILD PSYCHIAT, DURHAM CHILD GUID CLIN, 65- Concurrent Pos: Fel pediat, Tulane Univ, 49-50; fel child psychiat, Med Ctr, Duke Univ, 63-65. Mem: Am Acad Pediat; AMA. Res: Adult and child psychiatry. Mailing Add: Dept of Psychiat Duke Univ Durham NC 27706

ANDERSON, WILLIAM DEWEY, JR, b Columbia, SC, June 4, 33; m 63; c 2. ICHTHYOLOGY. Educ: Univ SC, BS, 53, MS, 55, PhD(biol), 60. Prof Exp: Asst prof biol, Susquehanna Univ, 60-61; fishery biologist, Biol Lab, Bur Commercial Fisheries, US Fish & Wildlife Serv, Ga, 61-65; interim asst prof biol, Univ Fla, 65-66; assoc prof, Univ Chattanooga, 66-69; ASSOC PROF BIOL LAB, COL CHARLESTON, 69- Mem: Am Soc Ichthyol & Herpet; Am Fisheries Soc; Soc Syst Zool; Am Soc Zool; Soc Study Evolution. Res: Systematics of the fishes of the families Lutjanidae and Serranidae; ecology of marine and estuarine environments. Mailing Add: Grice Marine Biol Lab 205 Ft Johnson Charleston SC 29412

ANDERSON, WILLIAM EVAN, b Mankato, Minn, Sept 7, 27; m 51; c 1. INTERNAL MEDICINE. Educ: Gustavus Adolphus Col, AB, 50; Univ Minn, MD, 54. Prof Exp: Resident med, Vet Admin Hosp, Univ Minn, 55-58, instr, 58-60; from instr to assoc prof, 60-74, PROF MED, SCH MED, W VA UNIV, 74- Res: Gastroenterology. Mailing Add: Dept of Med WVa Univ Med Ctr Morgantown WV 26506

ANDERSON, WILLIAM JOHN, b Boston, Mass, Dec 9, 38; m 68. NEUROANATOMY. Educ: Univ Miami, BA, 62; Purdue Univ, MS, 73, PhD(anat), 75. Prof Exp: Res assoc psychol, Mass Inst Technol, 62-68; fac instr neurobiol, Purdue Univ, 68-71, grad instr anat, Sch Vet Med, 71-75; ASST PROF NEUROANAT, IND UNIV DIV, TERRE HAUTE CTR MED EDUC, IND STATE UNIV, TERRE HAUTE, 75- Mem: Int Soc Develop Psychobiol; Soc Neurosci; Am Asn Vet Anatomists; Sigma Xi. Res: Morphological analysis of brain development and brain abnormalities due to toxic elements, with special reference to postnatal development. Mailing Add: RR 31 Box 470 Terre Haute IN 47803

ANDERSON, WILLIAM KERMIT, physical chemistry, chemical engineering, see 12th edition

ANDERSON, WILLIAM LENO, b Carney, Okla, Mar 13, 35; m 62; c 2. WILDLIFE BIOLOGY. Educ: Okla State Univ, BS, 58; Southern Ill Univ, MA, 64. Prof Exp: Field asst pop ecol, 58-62, res asst, 62-63, res assoc ecol & physiol, 63-68, ASSOC WILDLIFE SPECIALIST, ILL NATURAL HIST SURV, 68- Mem: Wildlife Soc; Wilson Ornith Soc; Cooper Ornith Soc. Res: Population dynamics; field ecology; behavioral and physiological responses of birds and mammals to pesticides, trace elements and heated water. Mailing Add: Ill Natural Hist Surv 279 Natural Resources Bldg Urbana IL 61801

ANDERSON, WILLIAM NILES, JR, b Pittsburgh, Pa, Nov 27, 39; m 69; c 2. APPLIED MATHEMATICS. Educ: Carnegie-Mellon Univ, BS, 60, MS, 67, PhD(math), 68. Prof Exp: Res assoc math, Rockefeller Univ, 68-70; asst prof, Univ Md, 70-75; ASSOC PROF MATH & COMPUT SCI, WVA UNIV, 75- Mem: Soc Indust & Appl Math; Asn Comput Mach. Res: Combinatorial mathematics; numerical analysis; electrical networks. Mailing Add: Dept of Math WVa Univ Morgantown WV 26506

ANDERSON, WILLIAM RAYMOND, b St Charles, Ill, Mar 12, 11; wid. PHYSICS. Educ: Univ Ill, BS, 33; DePaul Univ, MS, 47. Prof Exp: Pvt & pub acct, 34-42; pub sch teacher, Ill, 43-47; from instr to asst prof physics, 47-59, ASSOC PROF PHYSICS, UNIV ILL, CHICAGO CIRCLE, 59- Mem: Am Asn Physics Teachers. Res: Spectroscopy; molecular physics. Mailing Add: Dept of Physics Univ of Ill Chicago IL 60680

ANDERSON, WILLIAM SUNLEY, polymer chemistry, organic chemistry, see 12th edition

ANDERSON, WILLIAM WESTERLIN, b Rockford, Ill, July 3, 24; m 52; c 2. NEUROLOGY. Educ: Augustana Col, AB, 46; Chicago Med Sch, MB, 50, MD, 51. Prof Exp: Intern, Charity Hosp, New Orleans, 50-51; res physician, 53-54; from jr clin instr to sr clin instr, Univ Mich, 54-56; ASST CLIN PROF NEUROL & PEDIAT, MED CTR, UNIV CALIF, SAN FRANCISCO, 56- Concurrent Pos: Chief neurol serv, Peninsula Hosp, Burlingame, Calif. Mem: Am Acad Neurol; AMA; Am Heart Asn; Soc Clin Neurol. Res: Neuro-otology; temporal lobe epilepsy; cerebrovascular disease. Mailing Add: Suite 201 101 San Mateo Dr San Mateo CA 94401

ANDERSON, WILLIAM WYATT, b Elton, La, Feb 8, 09; m 32; c 3. ICHTHYOLOGY. Educ: La State Univ, BS, 29, MS, 30. Prof Exp: Aquatic biologist, Bur Fisheries, US Fish & Wildlife Serv, 31-40, actg in charge shrimp invests, 40-46, chief, Gulf fishery invests, 46-52, chief, SAtlantic fishery invests, 52-59, dir biol lab, Bur Commercial Fisheries, 59-68, sr scientist, 68-70; CHIEF COASTAL FISHERIES,

RES & DEVELOP PROG, GA GAME & FISH COMN, 70- Concurrent Pos: Area coordr, Off Fishery Coord, Western Gulf of Mex, 43-46; mem Gulf & Caribbean Fisheries Inst. Mem: Am Fisheries Soc; Am Soc Ichthyol & Herpet; Am Inst Fishery Res Biol. Res: Life history and systematics of fishes and penaeid shrimp. Mailing Add: PO Box 1138 St Simons Island GA 31522

ANDERSON, WILMER CLAYTON, b Waco, Tex, Nov 24, 09; m 33; c 3. PHYSICS. Educ: Baylor Univ, AB, 29, AM, 30; Harvard Univ, PhD(physics), 36. Prof Exp: Instr physics, chem & math, Lamar Col, 31-33; instr physics, Harvard Univ, 35-38, fel, 38-40; physicist, Tex Co, 41-42 & Div Res, Columbia Univ, 42-44; sr res & develop engr, Arma Corp, NY, 44-45; chief engr, Aireon Mfg Corp, Conn, 45-46; head phys res group, Deering Milliken Res Trust, 46-49; head physics group & mem staff, New Prod Res Lab, Remington Rand, Inc, 49-52; dir electronic res, Talon, Inc, 52-54; dir res, Liquidometer Corp, 54-55; dir res & develop, Gen Time Corp, 55-69; vpres & tech dir, Quipu Corp, Conn, 69-71; CONSULT, 71- Concurrent Pos: Bayard Cutting fel, Harvard Univ, 37; with Off Sci Res & Develop, 44. Mem: AAAS; Am Phys Soc; Optical Soc Am; Inst Elec & Electronics Eng. Res: Industrial applications of physics and electronics; magnetometers for magnetic surveys; sonic compensators for land-mine detection; voltage regulators for A-C; various textile applications; new high frequency method for measuring velocity of light. Mailing Add: 56 Sound View Dr Greenwich CT 06830

ANDERSON, WYATT W, b New Orleans, La, Mar 27, 39; m 62; c 3. POPULATION GENETICS. Educ: Univ Ga, BS, 60, MS, 62; Rockefeller Univ, PhD(life sci), 67. Prof Exp: From lectr to assoc prof biol, Yale Univ, 66-72; assoc prof, 72-75, PROF ZOOL, UNIV GA, 75- Mem: Soc Study Evolution; Genetics Soc Am; Am Soc Nat; Ecol Soc Am; Am Soc Human Genetics. Res: Genetic mechanisms in evolution; selection in natural and experimental populations; relationships between genetical and ecological aspects of populations. Mailing Add: Dept of Zool Univ of Ga Athens GA 30602

ANDERSON, ZOE ESTELLE, b Glenview, Ill, Jan 20, 13. NUTRITION. Educ: Ill Inst Technol, BSAS, 39; Univ Ill, MS, 47, PhD(animal nutrit), 50. Prof Exp: Mgr plant cafeteria, Hydrox Ice Cream Co, 35-39; fac mem cafeteria mgt & home econ, Frankfort Community High Sch, 39-42; dietician sch food serv, J Sterling Morton High Sch, 42-43; asst dept nutrit res, Nat Dairy Coun, 49-51, dir, 51-60; asst prof home econ & chmn dept, Wayne State Univ, 60-62; assoc prof nutrit, Dept Internal Med, Col Med, Iowa State Univ, 62-64; ASSOC PROF FOODS & NUTRIT, SAN DIEGO STATE UNIV, 65- Mem: Am Dietetic Asn; Am Inst Nutrit; Am Home Econ Asn; Soc Nutrit Educ. Res: Nutrition in health and disease; effects of diet on voluntary activity; bone ash; growth; tissues; blood and urinary values. Mailing Add: 5065 54th St San Diego CA 92115

ANDERTON, LAURA GADDES, b Providence, RI, Sept 6, 18. EXPERIMENTAL EMBRYOLOGY, HUMAN CYTOGENETICS. Educ: Wellesley Col, BA, 40; Brown Univ, MS, 48; Univ NC, PhD, 59. Prof Exp: Instr biol & chem, Howard Sem, 41-43; instr biol & counsr, 48-55, instr biol, 56-58, from asst prof biol & embryol to assoc prof biol, 58-68, PROF BIOL, UNIV NC, GREENSBORO, 68-, DIR CYTOGENETICS LAB, 65- Concurrent Pos: NIH grant, 68-71; consult, Moses H Cone Mem Hosp, 69- Mem: AAAS; Am Inst Biol Sci; Am Soc Human Genetics; Am Soc Zool; Int Union Against Cancer. Res: Cell renewal system in the colon mucosa; familial polyposis; tissue culture of adenomas of the colon. Mailing Add: Dept of Biol Cytogenetics Lab Univ of NC Greensboro NC 27412

ANDERZHON, MAMIE LOUISE, b July 28, 06; US citizen. GEOGRAPHY. Educ: Univ Chicago, PhB, 38; Univ Chicago, MS, 48. Prof Exp: Pub sch teacher, Nebr, Iowa & Ill, 23-42; cartog researcher, Army Map Serv, Washington, DC, 42-45; teacher geog, Oak Park Pub Schs, Ill, 45-46; consult, All India Coun Sec Educ, 56-57; teacher, Oak Park Pub Schs, 57-61; prof, 61-72, EMER PROF GEOG, INDIANA UNIV PA, 72- Concurrent Pos: Mem exec bd, Nat Coun Geog Educ, 49-67, dir coord & vpres, 52-56, assoc ed, J Geog, 53-65, pres, 61-62; lectr, Adelaide Teachers Col, 68; mem, Int Geog Cong. Honors & Awards: Distinguished Serv Award, Nat Coun Geog Educ, 71. Mem: Asn Am Geographers; Nat Coun Geog Educ (pres, 63). Res: Geography education. Mailing Add: Rte 1 Box 176B Shenandoah IA 51601

ANDES, RALPH VERNE, physical chemistry, see 12th edition

ANDOSE, JOSEPH DAVID, b Philadelphia, Pa, July 26, 44. PHYSICAL ORGANIC CHEMISTRY, COMPUTER SCIENCE. Educ: Temple Univ, BA, 66; Princeton Univ, MA, 68, PhD(org chem), 71. Prof Exp: Res assoc phys org chem, Princeton Univ, 71-74; systs analyst comput-assisted molecular modeling, 74-75, SYSTS ASSOC COMPUT-ASSISTED ORG SYNTHESIS, MERCK & CO, INC, 75- Mem: Am Chem Soc; The Chem Soc; Sigma Xi. Res: Stereochemistry and conformational analysis; structure-property relationships; computer-assisted molecular modeling; computer-assisted organic synthesis. Mailing Add: Merck & Co Inc 126 E Lincoln Ave Rahway NJ 07065

ANDRADE, JOSEPH D, b Hayward, Calif, July 13, 41; m 66; c 2. BIOENGINEERING, MATERIALS SCIENCE. Educ: San Jose State Col, BS, 65; Univ Denver, PhD(metall, mat), 69. Prof Exp: ASSOC PROF MAT SCI, ENG & PHARM, UNIV UTAH, 69- Mem: AAAS; Am Chem Soc; Am Soc Artificial Internal Organs; Biomed Eng Soc. Res: Interface between non-living materials and living systems; bloodmaterials interface; interface conversion processes; polymer surface chemistry; adsorption from solution; biomaterials. Mailing Add: Dept of Mat Sci Univ of Utah Col of Eng Salt Lake City UT 84112

ANDRAKO, JOHN, b Perth Amboy, NJ, Jan 19, 24; wid; c 3. PHARMACEUTICAL CHEMISTRY. Educ: Rutgers Univ, BS, 47, MS, 49; Univ NC, PhD(pharmaceut chem), 53. Prof Exp: Instr pharmaceut chem, Univ NC, 49-50, instr pharm, 50-53, from asst prof to assoc prof pharmaceut chem, 53-56; assoc prof, 56-62, asst dean sch pharm, 65-73, asst vpres health sci, 73-75, PROF PHARMACEUT CHEM, MED COL VA, VA COMMONWEALTH UNIV, 62-, ASST PROVOST, 75- Mem: Am Chem Soc; Am Pharmaceut Asn. Res: Synthesis of derivatives of various heterocyclic systems as antihypertensives and psychotropic agents. Mailing Add: 1911 Barribee Lane Richmond VA 23229

ANDRAWES, NATHAN R, b UAR, Sept 16, 39. ENTOMOLOGY. Educ: Univ Alexandria, BS, 60; Auburn Univ, MS, 63; Tex A&M Univ, PhD(toxicol), 67. Prof Exp: Entomologist, Tex State Dept Health, 67; PROJ SCIENTIST, UNION CARBIDE CORP, 67- Mem: Am Chem Soc; Entom Soc Am. Res: Metabolism and toxicology of pesticides. Mailing Add: Union Carbide Corp PO Box 8361 South Charleston WV 25303

ANDRE, FLOYD, entomology, deceased

ANDRE, HERMAN WILLIAM, b Watertown, SDak, Apr 9, 38; m 60. NUCLEAR CHEMISTRY. Educ: Calvin Col, AB, 59; Purdue Univ, PhD(phys & nuclear chem), 64. Prof Exp: Instr chem, Purdue Univ, 63-64, asst prof, 64-69; asst to pres, 69-74, CORP SECY, GREAT LAKES CHEM CORP, 74- Concurrent Pos: Exec off dept, Purdue Univ, 68-69. Mem: Am Chem Soc; Am Inst Chemists. Res: Low-energy nuclear fission; low-level radiochemistry; nuclear instrumentation. Mailing Add: Great Lakes Chem Corp PO Box 2200 West Lafayette IN 47906

ANDREA, STEPHEN ALFRED, b Cuba, NY, July 10, 38. MATHEMATICS. Educ: Oberlin Col, BA, 60; Calif Inst Technol, PhD(math), 64. Prof Exp: Instr math, Harvard Univ, 64-67; asst prof, Univ Calif, San Diego, 67-75; MEM FAC, SIMON BOLIVAR UNIV, VENEZUELA, 75- Mem: Am Math Soc. Res: Differential topology; topological dynamics. Mailing Add: Dept of Physics & Math Simon Bolivar Univ Apt Postal 5354 Caracas Venezuela

ANDREADES, SAM, organic chemistry, see 12th edition

ANDREASEN, ARTHUR ALBINUS, b Springfield, Ill, Dec 18, 17; m 44; c 2. INDUSTRIAL MICROBIOLOGY. Educ: Univ Ill, BS, 40, MS, 42; Ind Univ, PhD(cytophysiol), 53. Prof Exp: From bacteriologist to chief control chemist & head fermentation res, 41-61, DIR RES, JOSEPH E SEAGRAM & SONS, INC, 61- Concurrent Pos: In chg fermentation, Exp Wood Hydrolysis Distillery, Vulcan Copper & Supply Co, Ore, 46-47. Mem: Am Chem Soc; Am Soc Microbiol; Am Soc Brewing Chem. Res: Anaerobic growth and nutrition of yeast; yeast and mold fermentations; enzyme production. Mailing Add: 508 Hill Ridge Rd Louisville KY 40214

ANDREE, RICHARD VERNON, b Minneapolis, Minn, Dec 16, 19; m 44; c 4. MATHEMATICS. Educ: Univ Chicago, BS, 42; Univ Wis, PhM, 45, PhD(math), 49. Prof Exp: From asst prof to prof math & chmn dept, 49-69, PROF INFO & COMPUT SCI & MATH, UNIV OKLA, 69-, RES ASSOC COMPUT SCI, 57- Concurrent Pos: Actg dir lab, Univ Okla, 58, fel, res inst; vis assoc prof, Carnegie Found grant, Haverford Col, 55-56; vis prof var univs, 55, 57-59; lectr, NSF Insts, 55-72; fel numerical anal, Nat Bur Standards, 59; dir, Can Math Inst, Alta, 59; Univ Okla Math Comput Consult, 62-; mem adv panel inst comput serv, NSF; mem bd gov, Ctr Res Col Instr Sci & Math; math ed, Dryden Press, 56-; assoc ed, Am Math Monthly, 57-62; ed-in-chief, Math Log, 60-73; founder & former ed, Univ Okla Math Letter. Honors & Awards: Demolay Legion of Honor, 67. Mem: Fel AAAS; Math Asn Am; Am Math Soc; Asn Symbolic Logic; Asn Comput Mach. Res: P-adic numbers; matrix theory; modern abstract algebra; computer mathematics; divergent sequences; group theory; expository articles and lectures. Mailing Add: Dept of Math Univ of Okla Norman OK 73069

ANDREEN, BRIAN HERBERT, b Superior, Wis, Aug 15, 34; m 56; c 3. ANALYTICAL CHEMISTRY. Educ: Wis State Univ, Superior, BS, 56; Fla State Univ, MS, 59. Prof Exp: Assoc chemist, Inst Gas Technol, 59-60, res chemist, 60-62, supvr chem res, 62-64; midwest rep, 64-69, REGIONAL DIR, RES CORP, 69- Mem: AAAS; Am Chem Soc. Res: Gas chromatographic analysis of fuel gases and determination of sulphur compounds; analysis of air pollutants; chemistry of sulphur compound alteration in pipelines; microbiological oxidation of hydrocarbons. Mailing Add: Res Corp 4570 W 77th St Minneapolis MN 55435

ANDREIS, HENRY JEROME, b Milwaukee, Wis, Sept 1, 31; m 54; c 5. SOIL FERTILITY. Educ: Univ Wis, BS, 54, MS, 55. Prof Exp: Soil technologist res, 57-69, FIRST ASST TO DIR RES DEPT & FIRST ASST TO VPRES RES, US SUGAR CORP, 69- Mem: Am Soc Agron; Int Soc Sugarcane Technologists; Am Soc Sugarcane Technologists. Res: Sugarcane and pasture fertility studies, silage research, seed cane germination studies. Mailing Add: Res Dept US Sugar Corp PO Drawer 1207 Clewiston FL 33440

ANDREOLI, ANTHONY JOSEPH, b New York, NY, Sept 13, 26; m 52. BACTERIOLOGY, BIOCHEMISTRY. Educ: Univ Southern Calif, AB, 50, PhD(bact), 55. Prof Exp: Asst bact, Univ Southern Calif, 50-52, clin lab technician, 52-53, lectr, 54-55; from asst prof to assoc prof chem & microbiol, 55-64, CHEM, CALIF STATE COL, LOS ANGELES, 64- Mem: Am Soc Microbiol. Res: Bacterial metabolism and physiology; oxidation of organic acids and hydrocarbons. Mailing Add: 5151 State University Dr Los Angeles CA 90032

ANDREOLI, KATHLEEN GAINOR, b Albany, NY, Sept 22, 35; m 60; c 3. MEDICAL EDUCATION. Educ: Georgetown Univ, BSN, 57; Vanderbilt Univ, MSN, 59. Prof Exp: Instr nursing, Sch Nursing, St Thomas Hosp, Nashville, Tenn, 58-59, Sch Nursing, Georgetown Univ, 59-60, Sch Nursing, Duke Univ, 60-61 & Sch Nursing, Bon Secours Hosp, Baltimore, 62-64; educ coord physician asst prog, Med Ctr, Duke Univ, 65-70; educ dir physician asst prog, Med Ctr, 70-75, ASSOC PROF NURSING, DEPT MED, UNIV ALA, BIRMINGHAM, 75- Concurrent Pos: Consult, NC Regional Med Prog, 66-70, Sch Nursing, Univ NC, 67-69, Ala Regional Med Prog, 70-73 & Med Ctr, Univ Ala, 73; mem adv bd physician asst exam, Nat Bd Med Examr, 72-; adv consult ambulatory care proj, Asn Am Med Cols, 75- Res: Identification of demand for and accessibility of primary health care services; improvement of educational programs for health professionals; patterns of health care delivery according to public need. Mailing Add: Dept of Med Univ of Ala Birmingham AL 35294

ANDREOLI, THOMAS EUGENE, b New York, NY, Jan 9, 35; m 60; c 3. INTERNAL MEDICINE, PHYSIOLOGY. Educ: St Vincent Col, BA, 56; Georgetown Univ, MD, 60; Am Bd Internal Med, dipl, 67. Prof Exp: Intern med, Duke Hosp, 60-61, jr asst, 61, jr asst resident, 64; chief resident, Durham Vet Admin Hosp, 65; instr med, Duke Univ, 65, assoc med & physiol, 65-67, asst prof physiol, 66-69, assoc prof med, 69-70, head div clin physiol, 67-70; assoc prof, 70-73, PROF PHYSIOL & BIOPHYS, MED CTR, UNIV ALA, BIRMINGHAM, 73-, PROF MED & DIR DIV NEPHROLOGY, 70- Concurrent Pos: Nat Inst Gen Med Sci res career develop award, 67. Mem: Am Soc Nephrology; Biophys Soc; Am Physiol Soc; fel Am Col Physicians; Am Heart Asn. Res: Membrane transport processes; effects of polyene antibiotics on lipid membranes; clinical nephrology. Mailing Add: Dept of Med Univ of Ala Med Ctr Birmingham AL 35233

ANDRES, CAL L, b Valmeyer, Ill, Aug 18, 21; m 46; c 2. FOOD SCIENCE. Educ: Univ Ill, BS, 47; Ill Inst Technol, MS, 56. Prof Exp: Fruit & veg buyer, Libby McNeill & Libby, 47-57, dir agr res, 57-65, asst to vpres, 65-67, assoc dir food technol res, 67-71; ASSOC ED, FOOD PROCESSING MAG, PUTMAN PUBL CO, 72- Concurrent Pos: Chmn seed qual comt & agrimeteorol comt, Nat Canners Asn, 62-65. Mem: Inst Food Tech; Am Soc Hort Sci; Am Asn Cereal Chem. Res: Business administration; vegetable genetics; vegetable and fruit production and preservation; food product formulation including ingredients, equipment and methods. Mailing Add: 36 Oxford Ave Clarendon Hills IL 60514

ANDRES, JOHN MILTON, b Santa Ana, Calif, Feb 4, 27; m 56; c 4. MICROWAVE ELECTRONICS. Educ: Calif Inst Technol, BS, 49, MS, 50; Mass Inst Technol, PhD(physics), 53. Prof Exp: Mem tech staff, Calif Res Corp, 53-54; mem tech staff, Space Technol Labs, Thompson-Ramo-Wooldridge, Inc, 54-60, sect head, 60-63, mgr quantum electronics dept, TRW Systs Group, 62-70, MEM SR STAFF, ELECTRONIC SYSTS DIV, TRW SYSTS GROUP, 70- Mem: Am Phys Soc. Res:

Field Effect Transistor amplifiers; oscillators; laser communications. Mailing Add: 1340 Via Margarita Palos Verdes Estates CA 90274

ANDRES, KLAUS, b Zurich, Switz, Mar 1, 34. PHYSICS. Educ: Swiss Fed Inst Technol, PhD(physics), 63. Prof Exp: MEM TECH STAFF PHYSICS, BELL TEL LABS, 63- Concurrent Pos: Acad guest, Swiss Fed Inst Technol, 68-69. Mem: Am Phys Soc. Res: Superconductivity; magnetism; nuclear magnetism at very low temperatures. Mailing Add: Bell Tel Labs I D 237 PO Box 261 Murray Hill NJ 07974

ANDRES, LLOYD A, b Santa Ana, Calif, May 17, 28; m 56; c 3. ENTOMOLOGY, WEED SCIENCE. Educ: Univ Calif, Berkeley, PhD(entom), 57. Prof Exp: Res entomologist, Univ Calif, Riverside, 56-58; res entomologist, 58-64, LEADER BIOL CONTROL WEEDS INVESTS, USDA, 64- Mem: AAAS; Entom Soc Am; Am Inst Biol Sci; Weed Sci Soc Am. Res: Biological control of weeds. Mailing Add: 1324 Arch St Berkeley CA 94708

ANDRES, MARTIN YORK, veterinary anatomy, see 12th edition

ANDRES, REUBIN, b Dallas, Tex, June 13, 23; m 48; c 4. MEDICINE, GERONOTLOGY. Educ: Southwestern Med Col, MD, 44; Am Bd Internal Med, dipl. Prof Exp: Intern, Gallinger Munic Hosp, Washington, DC, 45; resident med, Vet Admin Hosp, McKinney, Tex, 47-50; from instr to asst prof, 55-63, ASSOC PROF MED, JOHNS HOPKINS UNIV, 63-; CHIEF CLIN PHYSIOL BR & ASST CHIEF GERONT RES CTR, NAT INST ON AGING, 62-, ACTG CLIN DIR, 75- Concurrent Pos: Fel med, Johns Hopkins Univ, 50-55; asst physician, Outpatient Dept, Johns Hopkins Hosp, 50-58; instr, Sch Hyg & Pub Health, Johns Hopkins Univ, 53-55, asst prof, 55-60, lectr, 60-61; vis physician, Baltimore City Hosps, 55-57, 62-, asst chief med, 58-62; asst prof, Univ Md, 58-62. Mem: Am Physiol Soc; Am Soc Clin Invest; Geront Soc; Am Diabetes Asn; Endocrine Soc. Res: Carbohydrate and lipid metabolism; physiology of aging. Mailing Add: Geront Res Ctr Nat Inst Aging Baltimore City Hosps Baltimore MD 21224

ANDRES, WILLIAM WOLCOTT, b Trenton, NJ, Feb 9, 28; m 49; c 6. MICROBIAL BIOCHEMISTRY. Educ: Davis & Elkins Col, BS, 51; Univ Wis, MS, 56. Prof Exp: Biochemist, Chas Pfizer & Co, 51-54; biologist, Lederle Div, Am Cyanamid Co, 56-58, chemist, 58-62, res scientist, 62-69; MICROBIAL CHEMIST, ABBOTT LABS, 69- Res: Microbial transformations; natural products produced by microorganisms, isolation and characterization; antibiotics. Mailing Add: Dept 481 Sci Div Abbott Labs North Chicago IL 60064

ANDRESEN, BRIAN DEAN, b Reed City, Mich, Jan 20, 47. MASS SPECTROMETRY, ORGANIC CHEMISTRY. Educ: Fla State Univ, BS, 69; Mass Inst Technol, SM, 71, PhD(chem), 74; Woods Hole Oceanog Inst, SM, 72. Prof Exp: ASST PROF PHARMACEUT CHEM, COL PHARM, UNIV FLA, 74- Mem: Am Chem Soc; Int Oceanog Found. Res: Application of gas chromatography, mass spectrometry and computer analysis for the identification of biologically active compounds; application of synthetic organic chemistry for the preparation of useful pharmaceuticals; clinical chemistry and toxicology. Mailing Add: Univ of Fla Col of Pharm Box J-4 Gainesville FL 32610

ANDRESEN, JOHN WILLIAM, plant taxonomy, see 12th edition

ANDRESEN, MARVIN JOHN, b Chicago, Ill, July 14, 33; m 55; c 4. GEOLOGY. Educ: Univ Ill, BS, 55, MS, 56; Univ Mo, PhD(geol), 60. Prof Exp: Asst geol, Ill State Geol Surv, 54-56; from instr to asst prof, Univ Mo, 59-60; asst prof, Univ Alaska, 60-65; consult geologist, Geonomics, 65-69; PRES, ALASKA ENERGY CORP, 69- Concurrent Pos: Res Corp grant, 63-64. Mem: Am Asn Petrol Geol; Am Inst Prof Geol. Res: Paleodrainage patterns; regional facies analysis; petroleum and mineral exploration and development in Alaska. Mailing Add: Alaska Energy Corp Box 1912 Fairbanks AK 99707

ANDRESON, CLARK ALFRED, pharmacy, deceased

ANDRESON, PAUL S, b Sylvan Grove, Kans, May 17, 12; m 40; c 1. OBSTETRICS & GYNECOLOGY. Educ: Univ Kans, AB, 35, MD, 39; Am Bd Obstet & Gynec, dipl, 56. Prof Exp: Chief serv obstet & gynec, Murphy Gen Hosp, Waltham, Mass, 47-49, Wm Beaumont Gen Hosp, El Paso, Tex, 51-54 & Tripler Gen Hosp, Honolulu, 54-58; assoc prof obstet & gynec, Col Med NJ, 61-; chief dept obstet & gynec, St Elizabeth Hosp, 64-74; RETIRED. Concurrent Pos: Consult, Martland Hosp, Newark, NJ. Mem: AMA; fel Am Col Surg; fel Am Col Obstet & Gynec. Res: Culdoscopy. Mailing Add: 5D 2700 G Rd Grand Junction CO 81501

ANDRESS, HARRY JOHN, JR, chemistry, see 12th edition

ANDREW, BARBARA JEAN, b Hollywood, Calif, Aug 9, 43. RESEARCH ADMINISTRATION. Educ: Univ Calif, Los Angeles, AB, 65, MA, 66; Univ Southern Calif, PhD(educ psychol), 70. Prof Exp: Instr foreign lang, Ventura Col, 66-68; asst prof med educ, Med Sch, Univ Southern Calif, 70-72; assoc dir, 72-73, dir allied med eval, 73-74, DIR RES & DEVELOP, NAT BD MED EXAMR, 74- Concurrent Pos: Consult, Nat Cancer Inst, 75, Bur Health Serv Res, NIH, 75-, Health Resources Admin, US Dept Health, Educ & Welfare, 75-, Asn Am Med Cols, 75- & Am Acad Physicians' Assts, 76. Mem: Nat Coun Measurement in Educ; Am Educ Res Asn. Res: Development of evaluation methodologies for the assessment of competence in the health professions. Mailing Add: Nat Bd of Med Examr 3930 Chestnut St Philadelphia PA 19104

ANDREW, BRYAN HAYDN, b Glasgow, Scotland, Feb 26, 39; m 62; c 2. RADIO ASTRONOMY. Educ: Glasgow Univ, BSc, 61; Cambridge Univ, PhD(radio astron), 66. Prof Exp: Asst res officer, 65-72, ASSOC RES OFFICER, NAT RES COUN CAN, 72- Concurrent Pos: Vis lectr, Univ Toronto, 74-76. Mem: Fel Royal Astron Soc; Can Astron Soc. Res: Molecules; extragalactic variables; planets; comets. Mailing Add: Herzberg Inst of Astrophys Nat Res Coun of Can Ottawa ON Can

ANDREW, DAVID ROBERT, b Wink, Tex, Nov 10, 35; m 58; c 3. TOPOLOGY. Educ: Univ Southwestern La, BS, 58; Iowa State Univ, MS, 59; Univ Pittsburgh, PhD(topology), 61. Prof Exp: From asst prof to assoc prof, 61-66, head dept, 69-75, PROF MATH, UNIV SOUTHWESTERN LA, 66-, DEAN COL SCI, 75- Concurrent Pos: Consult, Minn Sch Math & Sci Teaching Proj, 64-66. Mem: Math Asn Am. Res: Point set topology. Mailing Add: Col of Sci Univ of Southwestern La Lafayette LA 70501

ANDREW, GEORGE MCCOUBREY, b New Glasgow, PEI, Sept 8, 29; m 53; c 3. PHYSIOLOGY, PHYSICAL EDUCATION. Educ: McGill Univ, BSc, 52, MSc, 63, PhD(physiol), 67. Prof Exp: Athletic dir, YMCA, Charlottetown, PEI, 52-53 & Prince of Wales Col, 53-57; lectr educ, McGill Univ, 58-67; from asst prof to assoc prof, 67-74, PROF EDUC, QUEEN'S UNIV, ONT, 74-, ASST PROF, DEPT PHYSIOL, 67- Mem: Can Asn Health, Phys Educ & Recreation; Can Asn Sports Sci; fel Am Col Sports Med; Can Physiol Soc. Res: Cardiorespiratory functions at rest and exercise and the effect of physical training through growth and aging on their adaptation to exercise. Mailing Add: Sch of Phys & Health Educ Queen's Univ Kingston ON Can

ANDREW, JAMES F, b Mt Airy, NC, Mar 5, 25; m 51; c 2. EXPERIMENTAL SOLID STATE PHYSICS. Educ: Guilford Col, BS, 48; NC State Col, MS, 52; State Univ NY Buffalo, PhD(physics), 62. Prof Exp: Weather observer, US Weather Bur, 44-46; instrument specialist, Carter Fabrics Inc, NC, 48-49; asst physics, NC State Col, 49-52; jr engr, Sprague Elec Co, Mass, 52-53; res assoc, State Univ NY Buffalo, 53-62; sr physicist, Thiokol Chem Co, Utah, 62-63; STAFF MEM, LOS ALAMOS SCI LAB, 63- Mem: Am Phys Soc; Am Vacuum Soc. Res: Electronic and mechanical properties of carbon base material; physics of solids at high pressures; electronic and thermal properties of plutonium; compatibility of reactor fuels with steel cladding. Mailing Add: Los Alamos Sci Lab CMB-11 MS-328 Los Alamos NM 87545

ANDREW, KENNETH L, b Wichita, Kans, June 14, 19; m 40; c 3. ATOMIC SPECTROSCOPY. Educ: Friends Univ, AB, 40; Johns Hopkins Univ, MA, 42; Purdue Univ, PhD(physics), 51. Prof Exp: Head dept physics, Friends Univ, 42-56; chmn dept, Dickinson Univ, 56-57; assoc prof, 57-68, PROF PHYSICS, PURDUE UNIV, WEST LAFAYETTE, 68- Concurrent Pos: Mem comt line spectra of elements, Nat Res Coun, 60-68; consult, Int Astron Union, 64- Mem: Am Phys Soc; fel Optical Soc Am; Am Asn Physics Teachers. Res: Atomic emission spectroscopy; interferometric measurements; analysis of atomic spectra; atomic energy levels; standard wavelengths; comparison of atomic theory with experiments. Mailing Add: 240 E Sunset Lane West Lafayette IN 47906

ANDREW, MERLE M, b St Joseph, Mo, Aug 27, 20; m 50; c 2. APPLIED MATHEMATICS. Educ: Univ Nebr, BSEE, 42; Mass Inst Technol, PhD(math), 48. Prof Exp: Staff mem, Radiation Lab, Mass Inst Technol, 43-45 & Opers Eval Group, Navy Dept, 48-49; mathematician, Nat Bur Stand, 49-51 & Air Res & Develop Command, US Air Force, 51-54; chief math div, 54-59, dir math sci, 59-70, DIR MATH & INFO SCI, AIR FORCE OFF SCI RES, 70- Concurrent Pos: Lectr, Cath Univ Am, 48-50; guest, Vienna Tech, Austria, 65-66; prof lectr, Am Univ, 69. Mem: Am Math Soc. Res: Electromagnetic theory; differential equations; numerical analysis; high speed computers; scientific research administration. Mailing Add: 7203 Radnor Rd Bethesda MD 20034

ANDREW, ROBERT HARRY, b Platteville, Wis, Aug 2, 16; m 44; c 4. AGRONOMY. Educ: Univ Wis, BA, 38, PhD(agron), 42. Prof Exp: From instr to assoc prof, 43-58, PROF AGRON, UNIV WIS-MADISON, 58- Concurrent Pos: Vis lectr, Wageningen, Neth, 53-54. Mem: Am Soc Agron; Ecol Soc Am; Am Genetics Asn; Crop Sci Soc Am. Res: Corn production; sweet corn improvement, genetics; plant ecology. Mailing Add: Dept of Agron Univ of Wis Madison WI 53706

ANDREW, WARREN, b Portland, Ore, July 19, 10; m 36; c 1. ZOOLOGY, ANATOMY. Educ: Carleton Col, AB, 32; Brown Univ, MS, 33;Univ Ill, PhD(zool), 36; Baylor Univ, MD, 43; Butler Univ & Christian Theol Sem, 74. Prof Exp: Asst instr zool, Yale Univ, 33-34 & Univ Ill, 34-36; instr anat, Univ Ga, 37-39; from instr to asst prof histol & embryol, Col Med, Baylor Univ, 39-43; from assoc prof to prof, Med Col, Southwestern Univ, 43-47; prof anat & chmn dept, Sch Med, George Washington Univ, 47-52; prof & dir dept, Bowman Gray Sch Med, Wake Forest Univ, 52-58; PROF ANAT & CHMN DEPT, SCH MED, IND UNIV, INDIANAPOLIS, 58- Concurrent Pos: Vis prof, Univ Montevideo, Uruguay, 45-46, Wash Univ, 48-49 & Univ Hawaii, 68; chmn biol, Int Res Comt Geront, 54-; vis scientist, Dept Anat, Univ Col, Univ London & Univ Capetown, 70; res fel, Divinity Sch, Yale Univ, 74-75. Honors & Awards: Nat Award, Geront Res Found, 59. Mem: AAAS; Am Soc Zoologists; Am Asn Anatomists; Soc Exp Biol & Med; Geront Soc. Res: Age changes in nervous system; neuronophagia; role of oligodendroglia; structure of nerve cell nucleus; senile involution of thyroid gland; age changes in liver, pancreas, spleen, lymph nodes and salivary glands; intracellular position of intestinal lymphocytes. Mailing Add: Dept of Anat Univ of Ind Sch of Med Indianapolis IN 46202

ANDREW, WILLIAM TRELEAVEN, b Lucknow, Ont, Sept 1, 21; m 47; c 5. VEGETABLE CROPS. Educ: Univ Alta, BSc, 44; Utah State Col, MS, 49; Mich State Col, PhD(hort), 53. Prof Exp: Student asst, Alta Hort Res Ctr, 39-42; lab asst plant path, Univ Alta, 43; tech asst veg crops, Res Br, Can Dept Agr, 44-45; trial ground & receiving supvr veg & flower seeds, BC Seeds Ltd, 45-47; from instr to assoc prof veg crops, Southern Ill Univ, 50-59; PROF PLANT SCI, UNIV ALTA, 59- Concurrent Pos: Head div hort, Univ Alta, 59-70. Mem: Am Soc Hort Sci; Can Soc Hort Sci; Agr Inst Can; Int Soc Hort Sci. Res: Flowering and reproduction; growth regulators; moisture temperature and nutritional relationships. Mailing Add: Dept of Plant Sci Univ of Alta Edmonton AB Can

ANDREWS, ALBERT H, JR, b Chicago, Ill, June 20, 07; m 38. MEDICINE. Educ: Northwestern Univ, MS, 32, MD, 33. Prof Exp: Clin asst ophthal, Med Sch, Northwestern Univ, 33-35, res asst physiol, 33-40, instr otolaryngol, 35-38; PROF BRONCHOESOPHAGOLOGY, UNIV ILL COL MED, 40-, HEAD DEPT OTOLARYNGOL, ABRAHAM LINCOLN SCH MED, 68- Concurrent Pos: Attend bronchoesophagologist, Presby-St Luke's Hosp & Skokie Valley Community Hosp, 66-; actg head dept otolaryngol, Abraham Lincoln Sch Med, 67-68; assoc attend bronchoesophagologist, Children's Mem Hosp; consult bronchoesophagologist, West Side Vet Admin Hosp, Chicago. Mem: Am Acad Ophthal & Otolaryngol; Am Col Chest Physicians (treas, 61-69, pres elect, 71, pres, 72); Am Laryngol, Rhinol & Otol Soc; fel AMA; Am Laryngol Asn. Res: Bronchoesophagology and laryngeal surgery; applied bronchopulmonary physiology; inhalation therapy. Mailing Add: 122 S Michigan Ave Chicago IL 60603

ANDREWS, ALICE E (MRS THEODORE B HUNT), b Estevan, Sask, Jan 30, 05; US citizen; m 46; wid; c 1. MATHEMATICS. Educ: NY Univ, BS, 38, MA, 41, PhD, 47. Prof Exp: Statistician, Am Tel & Tel Co, 30-42; instr math, Lafayette Col, 43-44; from instr to asst prof, Wilson Col, 44-59; from asst prof to assoc prof, 59-75, EMER ASSOC PROF MATH, UNION COL, CRANFORD, 75- Mem: Am Math Soc. Res: Modern mathematics. Mailing Add: Box 452c Rahway NJ 07065

ANDREWS, ARTHUR CLINTON, b Bloomer, Wis, Oct 16, 00; m 28; c 1. PHYSICAL CHEMISTRY. Educ: Univ Wis, BS, 24, PhD(phys chem), 38; Kans State Col, MS, 29. Prof Exp: Teacher schs, Nev, 24-25; from instr to prof, 26-74, EMER PROF CHEM, KANS STATE UNIV, 74- Mem: Am Chem Soc; NY Acad Sci; The Chem Soc. Res: Physical chemistry in fields of surface chemistry; allergy, colloids, proteins, bituminous materials and electrophoresis; physical chemistry of secalin; molecular structure of heterocyclic amines and metal chelates. Mailing Add: Dept of Chem Kans State Univ Manhattan KS 66504

ANDREWS, BILLY FRANKLIN, b Alamance Co, NC, Sept 22, 32; m; c 3. PEDIATRICS. Educ: Wake Forest Col, BS, 53; Duke Univ, MD, 57; Am Bd Pediat, cert, 63. Prof Exp: Intern, US Army Hosp, Ft Benning, GA, 57-58; resident pediat, Walter Reed Gen Hosp, Washington, DC, 58-60 & Walter Reed Army Inst Res, 60-61; chief pediat serv, Rodriguez US Army Hosp, Ft Brooke, PR, 61-63; from asst prof to assoc prof pediat, 64-69, co-dir genetic coun unit, 65-68, PROF PEDIAT, CHMN DEPT & DIR COMPREHENSIVE HEALTH CARE CTR FOR HIGH RISK INFANTS & CHILDREN, SCH MED, UNIV LOUISVILLE, 69-; DIR NEWBORN SERV, LOUISVILLE GEN HOSP, 64- Concurrent Pos: Consult div maternal & child health, Ky State Dept Health, 66-; civilian consult, US Army, 69-; chief of staff, Norton-Children's Hosp, 69- Mem: AAAS; AMA; Am Pediat Soc; Soc Pediat Res; Royal Soc Med. Res: Low birth weight infants; respiratory distress in infants; amniotic fluid studies; infant nutrition. Mailing Add: Dept of Pediat Health Sci Ctr Univ of Louisville Sch of Med Louisville KY 40202

ANDREWS, CATER WILSON, b Wesleyville, Nfld, Dec 6, 16; m 46; c 1. ZOOLOGY, ECOLOGY. Educ: Mt Allison Univ, BSc, 42; Univ Western Ont, 44; Univ Toronto, PhD(marine biol), 47. Prof Exp: Prof biol, Mem Univ Nfld, 47-48; from lectr to asst prof, NY Univ, 48-51; head dept, 51-63, PROF BIOL, MEM UNIV NFLD, 51- Res: Biology and ecological relationships; behavioral studies of Atlantic salmon and other salmonidae fishes; marine biology. Mailing Add: Dept of Biol Mem Univ of Nfld St John's NF Can

ANDREWS, CECIL HUNTER, b Starkville, Miss, July 26, 32; m 60; c 3. AGRONOMY. Educ: Miss State Univ, BS, 54, MS, 58, PhD(agron), 66. Prof Exp: Asst prof agron & asst agronomist, 58-70, ASSOC PROF AGRON, MISS STATE UNIV & ASSOC AGRONOMIST, MISS AGR & FORESTRY EXP STA, 70- Concurrent Pos: Asst agronomist, Chile & Taiwan, 59 & 60; chief of party, Miss State Univ-AID Brazil Contract, 67-68 & Seed Develop Progs in Ecuador, Colombia, Nicaragua, Honduras, Costa Rica & Thailand, 69-74. Mem: Am Soc Agron. Res: Biochemical and physiological research in seeds of field crops. Mailing Add: Box 5267 Miss State Univ Mississippi State MS 39762

ANDREWS, CHARLES EDWARD, b Stratford, Okla, Jan 22, 25; m 46; c 2. INTERNAL MEDICINE. Educ: Boston Univ, MD, 49. Prof Exp: Instr med, Univ Minn, 55-56; assoc, Med Sch, Univ Kans, 56-57, from asst prof to assoc prof, 57-60; assoc prof, 61-63, PROF MED, W VA SCH MED, 63-, PROVOST HEALTH SCI, 68- Concurrent Pos: Chief med serv, Vet Admin Hosp, Kansas City, Mo, 56-61; lectr, Sch Dent, Univ Kans, 58- Mem: Am Thoracic Soc; AMA; Fedn Clin Res; Am Col Physicians. Res: Pulmonary disease and function. Mailing Add: Dept of Med WVA Univ Med Ctr Morgantown WV 26505

ANDREWS, CHARLES LAWRENCE, b Atlanta, Ga, Mar 6, 38; m 65; c 2. WILDLIFE ECOLOGY, PARASITOLOGY. Educ: Ga State Col, BS, 63; Univ Ga, MS, 66, PhD(wildlife ecol), 69. Prof Exp: Res assoc wildlife dis, Sch Vet Med, Univ Ga, 66-69; PROF BIOL, BRENAU COL, 69- Mem: Wildlife Soc; Am Soc Mammal; Wildlife Dis Asn. Res: Cottontail rabbit, gray squirrel and white-tailed deer parasitism; wildlife leptospirosis. Mailing Add: Dept of Biol Brenau Col Gainesville GA 30501

ANDREWS, CHARLES LUTHER, b Berkshire, NY, May 6, 08; m 34; c 2. PHYSICS. Educ: Cornell Univ, AB, 30, PhD(physics), 38. Prof Exp: Instr, 31-44, chmn dept, 44-69, PROF PHYSICS, STATE UNIV NY ALBANY, 44- Concurrent Pos: Res physicist, Gen Elec Co, Schenectady, 44- Mem: Optical Soc Am; Am Phys Soc; Am Asn Physics Teachers. Res: Absorption of x-rays; diffraction of microwaves; design of microwave laboratory equipment for teaching microwave optics. Mailing Add: Dept of Physics State Univ of NY Albany NY 12222

ANDREWS, DANIEL KELLER, b Rockland, Maine, Sept 5, 24; m 50; c 6. POULTRY SCIENCE. Educ: Univ Maine, BS, 49; Kans State Univ, MS, 51; Univ Wis, PhD(poultry sci), 63. Prof Exp: Supvr, Swift & Co, Pa, 51-52; inspector, US Dept Agr, Iowa, 52-53; asst prof poultry teaching & admin, State Univ NY Agr & Tech Inst Delhi, 53-56; exten poultry specialist, Univ Conn, 56-60; res asst poultry nutrit, Univ Wis, 60-63; EXTEN POULTRY SCIENTIST, WESTERN WASH RES & EXTEN CTR, WASH STATE UNIV, 63- Mem: Poultry Sci Asn. Res: Effects of arsanilic acid on egg production; extension education for poultrymen and youth. Mailing Add: Western Wash Res & Exten Ctr Wash State Univ Puyallup WA 98371

ANDREWS, DAVID F, b Indianapolis, Ind, Apr 3, 43; Can citizen; m 65; c 2. STATISTICS. Educ: Univ Toronto, BSc, 65, MSc, 66, PhD(statist), 68. Prof Exp: Lectr statist, Imp Col, Univ London, 68-69; lectr, Princeton Univ, 69-71; ASST PROF STATIST, UNIV TORONTO, 71- Concurrent Pos: Consult, Bell Lab, 69-; vis asst prof, Univ Chicago, 73; assoc ed, Am Statist Asn, 73- Mem: Am Statist Asn. Res: Robust statistical procedures; graphical methods for data display. Mailing Add: Dept of Math Univ of Toronto Toronto ON Can

ANDREWS, DAVID HENRY, b Washington, DC, Jan 23, 33; m 56; c 2. ANTHROPOLOGY, ETHNOGRAPHY. Educ: Ohio Wesleyan Univ, BA, 54; Cornell Univ, MA, 59, PhD(cult anthrop), 63. Prof Exp: Res assoc anthrop, Cornell Univ, 63-64; asst prof, Univ Iowa, 64-68; ASSOC PROF ANTHROP, MIDDLEBURY COL, 68- Concurrent Pos: NSF travel grant, 70; Middlebury Col fac res grant field res in Costa Rica, 70-71. Mem: AAAS; Am Anthrop Asn; Am Ethnol Soc; Soc Appl Anthrop; Latin Am Studies Asn. Res: Change in social structure and values, especially ethnosemantics, kinesics and cultural ecology. Mailing Add: Dept of Sociol-Anthrop Middlebury Col Middlebury VT 05753

ANDREWS, DONALD HATCH, chemistry, deceased

ANDREWS, EDNA (BERNICE) (CAMPBELL), biochemistry, clinical chemistry, see 12th edition

ANDREWS, EDWIN RUFFIN, organic chemistry, see 12th edition

ANDREWS, EUGENE RAYMOND, b Rockford, Ill, Apr 29, 18; m 48; c 2. ORGANIC CHEMISTRY. Educ: Cent YMCA Col, BS, 41. Prof Exp: Chief chemist, US Indust Chem Co, 42-57; res chemist, 58-66, MGR RES LAB, ALLIED MILLS, INC, 66- Mem: AAAS; Am Chem Soc. Res: Scientific feeding of farm animals, effect of drugs and hormones on such animals. Mailing Add: 5810 Nicolet Ave Chicago IL 60631

ANDREWS, FRANK CLINTON, b Manhattan, Kans, May 29, 32; m 64; c 1. CHEMICAL PHYSICS. Educ: Kans State Univ, BS, 54; Harvard Univ, AM, 59, PhD(chem physics), 60. Prof Exp: NIH fel chem physics, Univ Calif, Berkeley, 60-61; asst prof phys chem, Univ Wis, 61-67, on leave, Theoret Physics Dept, Oxford Univ, 66-67; assoc prof phys chem, 67-74, PROF CHEM, CROWN COL, UNIV CALIF, SANTA CRUZ, 74- Concurrent Pos: Sloan fel, 63-67. Mem: AAAS; Am Phys Soc. Res: Statistical mechanics and thermodynamics, especially of irreversible processes; statistical theory of traffic flow on highways. Mailing Add: Dept of Physical Chem Crown Col Univ of Calif Santa Cruz CA 95060

ANDREWS, FRED ALBERT, b Cedar Rapids, Iowa, July 25, 24; m 51; c 2. BIOCHEMISTRY. Educ: Coe Col, BA, 47; Univ Iowa, MS, 49. Prof Exp: Histochemist, Am Meat Inst Found, 49-51; biochemist, Chicago Div, Kendall Co, 51-56; chief biochemist, Bjorksten Res Labs, 56-66; res biochemist, 66-72, RES CHEMIST, BACT RES LAB, GEN MED RES, MINNEAPOLIS VET ADMIN HOSP, 72- Mem: Am Chem Soc. Res: Isolation, characterization and alteration of proteins and enzymes from muscle, connective tissues, blood and cereals; muscle dehydration; lipid-protein interactions; protein tannage; biochemistry of aging; antimicrobial drugs. Mailing Add: Gen Med Res Minneapolis Vet Admin Hosp Minneapolis MN 55417

ANDREWS, FRED CHARLES, b Aylesbury, Sask, July 13, 24; nat US; m 44; c 3. MATHEMATICAL STATISTICS. Educ: Univ Wash, BS, 46, MS, 48; Univ Calif, PhD(statist), 53. Prof Exp: Lectr math, Univ Calif, 51-52; res assoc statist, Stanford Univ, 52-54; asst prof math & assoc statistician, Univ Nebr, 54-57; dir statist lab & comput ctr, 60-69, assoc prof, 57-66, PROF MATH, UNIV ORE, 66- Concurrent Pos: Fulbright-Hays sr lectr, Univ Tampere, Finland, 69-70. Mem: AAAS; Biomet Soc; Am Statist Asn; Inst Math Statist. Res: Statistical theory, including non-parametric inference; statistical computations. Mailing Add: Dept of Math Univ of Ore Eugene OR 97403

ANDREWS, FREDERICK NEWCOMB, b Boston, Mass, Feb 5, 14; m 38; c 2. ANIMAL PHYSIOLOGY. Educ: Univ Mass, BS, 35, MS, 36; Univ Mo, PhD(physiol reprod), 39. Hon Degrees: ScD, Univ Mass, 62. Prof Exp: From asst prof to assoc prof, 40-49, head dairy dept, 60, head animal sci dept, 62, PROF ANIMAL SCI, PURDUE UNIV, 49-, DEAN GRAD SCH & VPRES FOR RES, 63-, VPRES & GEN MGR, PURDUE RES FOUND, 64- Concurrent Pos: Agr consult, Rockefeller Found, 61; mem bd dirs, Indianapolis Ctr for Advan Res, 70-, vpres, 73- Honors & Awards: Morrison Award, 61. Mem: Fel AAAS; Am Asn Anat; fel Am Soc Animal Sci; Am Soc Zool; Am Dairy Sci Asn. Res: Physiology of reproduction; endocrinology; nutrition; environmental physiology; growth and development. Mailing Add: Grad Sch Purdue Univ West Lafayette IN 47906

ANDREWS, GEORGE EYRE, b Salem, Ore, Dec 4, 38; m 60; c 1. MATHEMATICS. Educ: Ore State Univ, BS & MA, 60; Univ Pa, PhD(math), 64. Prof Exp: From asst prof to assoc prof, 64-70, PROF MATH, PA STATE UNIV, 70- Concurrent Pos: Vis prof, Mass Inst Technol, 70-71; vis prof math, Univ Wis, 75-76. Mem: Am Math Soc; Edinburgh Math Soc. Res: Basic hypergeometric series; partitions; number theory. Mailing Add: Dept of Math McAllister Bldg Pa State Univ University Park PA 16802

ANDREWS, GEORGE HAROLD, b Syracuse, NY, July 31, 32; m 55; c 4. MATHEMATICS, NUMERICAL ANALYSIS. Educ: Oberlin Col, AB, 54; Univ Mich, AM, 55, PhD(math), 63. Prof Exp: From asst prof to assoc prof, 62-73, PROF MATH, OBERLIN COL, 73- Concurrent Pos: NSF fac fel, 68-69; consult numerical anal, Am Soc Actuaries, 74-; vis prof, Univ Mich, 75-76. Mem: Math Asn Am; Am Math Soc; Soc Indust & Appl Math; Am Soc Actuaries; Sigma Xi. Res: Mathematical statistics; foundations of mathematics; actuarial mathematics; numerical solution of differential equations. Mailing Add: Dept of Math Oberlin Col Oberlin OH 44074

ANDREWS, GEORGE WILLIAM, b Eau Claire, Wis, Oct 15, 29; m 56; c 3. GEOLOGY. Educ: Univ Wis, BA, 51, MA, 53, PhD(geol), 55. Prof Exp: Geologist, Tech Serv Div, Shell Oil Co, 55-58 & E&P Res Div, Shell Develop Co, 58-59; GEOLOGIST, US GEOL SURV, 59- Mem: Geol Soc Am; Soc Econ Paleontologists & Mineralogists. Res: Diatoms; paleontology, stratigraphy and paleoecology of Mesozoic and Cenozoic strata; petroleum geology. Mailing Add: US Geol Surv E-501 US Nat Mus Washington DC 20244

ANDREWS, GORDON LOUIS, b New Orleans, La, Sept 22, 45; m 74. ENTOMOLOGY. Educ: Southeastern La Col, BS, 67; La State Univ, MS, 69; Miss State Univ, PhD(entom), 72. Prof Exp: RES ASSOC ENTOM, MISS STATE UNIV, 72- Mem: Sigma Xi; Entom Soc Am. Res: Cotton insects; field evaluation and modeling of Heliothis species. Mailing Add: Dept of Entom Miss State Univ Drawer EM Mississippi State MS 39762

ANDREWS, GOULD ARTHUR, b Grand Rapids, Mich, May 8, 18; m 55; c 1. INTERNAL MEDICINE. Educ: Univ Mich, AB, 40, MD, 43; Am Bd Internal Med, cert, 50; Am Bd Nuclear Med, cert, 72. Prof Exp: Instr internal med, Univ Mich & res asst, Simpson Mem Inst, 43-48; dir cancer teaching, Stritch Med Sch, 48-49; chief hemat, Med Div, 49-50, clin serv, 50-59, assoc chmn, 59-62, chmn med div, 62-75, DIR CLIN APPLN, OAK RIDGE ASSOC UNIVS, 75- Mem: AAAS; Soc Nuclear Med; AMA; Soc Exp Biol & Med; fel Am Col Physicians. Res: Clinical hematology; nuclear medicine especially as it involves diagnosis and treatment of cancer; radiation effects in man and management of radiation accidents. Mailing Add: Med Div Oak Ridge Assoc Univ Oak Ridge TN 37830

ANDREWS, HENRY NATHANIEL, JR, b Melrose, Mass, June 15, 10; m 39; c 3. PALEOBOTANY. Educ: Mass Inst Technol, BS, 34; Wash Univ, MS, 37, PhD, 39. Prof Exp: From instr to prof bot, Wash Univ, 38-64, dean sch, 47-64; prof, 64-75, chmn dept, 64-70, EMER PROF BIOL, UNIV CONN, 75- Concurrent Pos: Paleobotanist, Mo Bot Garden, St Louis, 41-64; botanist, US Geol Surv, 50-54, 59-; Guggenheim Mem Found fel, 50-51, 58-59; Fulbright teaching fel, Univ Poona, 60-61; NSF sr fel, Sweden, 64. Honors & Awards: Cert of Merit, Bot Soc Am, 66. Mem: Nat Acad Eng; Bot Soc Am; Am Geol Soc; Torrey Bot Club. Res: Carboniferous and Devonian plants; Arctic paleobotany; history of paleobotany. Mailing Add: RFD 1 Laconia NH 03246

ANDREWS, HORACE PORTER, b NC, Oct 23, 21. EXPERIMENTAL STATISTICS. Educ: NC State Univ, BS, 41, MS, 48; Pa State Univ, PhD(agr & biol chem), 51. Prof Exp: Chemist, US Ord Dept, Pa, 42-44; head statist div, Res Labs, Swift & Co, Ill, 51-64; ASSOC PROF STATIST, RUTGERS UNIV, NEW BRUNSWICK, 64- Mem: AAAS; Biomet Soc; Am Oil Chemists Soc; Am Soc Qual Control; Am Statist Asn. Res: Application of statistics in design; analysis of experiments in biological, physical and engineering sciences. Mailing Add: Statist Ctr Rutgers Univ New Brunswick NJ 08901

ANDREWS, HOWARD LUCIUS, b Davisville, RI, Oct 27, 06; m 31; c 3. BIOPHYSICS. Educ: Brown Univ, BS, 27, MS, 28, PhD(physics), 31. Prof Exp: Instr physics, Brown Univ, 29-34, res assoc psychol, 34-37; assoc physicist, US Pub Health Serv, 37-41, physicist, 41-48, chief nuclear radiation biol sect, Nat Insts Health, 48-61, dept radiation safety, 62-65; asst dir health & safety, PR Nuclear Ctr, 65-67; prof radiation biol & biophysics, Univ Rochester, 67-71; CONSULT, RADIATION SAFETY, 71- Concurrent Pos: Exec secy biol effects of atomic radiation comt, Nat Acad Sci-Nat Res Coun, 59-64. Mem: Fel Am Phys Soc; Radiation Res Soc; Health Physics Soc (pres, 64-65). Res: Biological effects of high energy radiations; radiation health protection. Mailing Add: 143a Share Rd Jamestown RI 02835

ANDREWS, HUGH ROBERT, b Fredericton, NB, Apr 29, 40; m 71; c 1. EXPERIMENTAL NUCLEAR PHYSICS. Educ: Univ NB, BSc, 62; Harvard Univ, AM, 63, PhD(physics), 71. Prof Exp: RES OFFICER NUCLEAR PHYSICS, ATOMIC ENERGY CAN LTD, 71- Mem: Sigma Xi; Can Asn Physicists; AAAS. Res: Hyperfine interactions and perturbed angular correlations; in-beam gamma ray spectroscopy applied to measurement of short lifetimes and the nature of high spin states in nuclei; atomic physics-heavy ion stopping powers. Mailing Add: Chalk River Nuclear Labs Sta 49 Chalk River ON Can

ANDREWS, JAMES EINAR, b Minneapolis, Minn, Oct 31, 42; m 65. GEOLOGICAL OCEANOGRAPHY. Educ: Amherst Col, BA, 63; Univ Miami, PhD(oceanog), 67. Prof Exp: Asst prof, 67-74, ASSOC PROF OCEANOG, UNIV HAWAII, 74- Mem: Geol Soc Am. Res: Deep sea sediments; sea floor structure and topography. Mailing Add: Dept Oceanog Inst Geophys 342 Univ of Hawaii at Manoa Honolulu HI 96822

ANDREWS, JAMES TUCKER, b Memphis, Tenn, July 24, 21; m 42; c 4. DENTISTRY. Educ: Southwestern at Memphis, BA, 43; Univ Tenn, DDS, 52. Prof Exp: Asst, 52-53, from instr to assoc prof, 53-59, PROF OPER DENT & PEDODONT, COL DENT, UNIV TENN, MEMPHIS, 59-, HEAD OPER DEPT, 55- Mem: Fel Am Col Dent; Acad Oper Dent; Am Asn Dent Schs. Res: Pulp therapy and evaluation of restorative materials. Mailing Add: 847 Monroe Memphis TN 38163

ANDREWS, JAY DONALD, b Bloom, Kans, Sept 9, 16; m 48; c 2. ZOOLOGY. Educ: Kans State Col, BS, 38; Univ Wis, MS, 40, PhD(zool), 47. Prof Exp: From asst biologist to assoc biologist, 46-55, SR BIOLOGIST, VA INST MARINE SCI, 55-; ASSOC PROF MARINE SCI, UNIV VA, 60-; PROF, COL WILLIAM & MARY, 60- Concurrent Pos: Ed, Proc, Nat Shellfisheries Asn, 57-58. Mem: Atlantic Estuarine Res Soc; Am Soc Limnol & Oceanog; Soc Syst Zool; Nat Shellfisheries Asn (secy-treas, 60-62, vpres, 62-64, pres, 64-66); Am Inst Fishery Res Biologists. Res: Marine biology; biological oceanography; ecology of marine invertebrates; shellfish ecology, epizootiology and diseases. Mailing Add: Va Inst of Marine Sci Gloucester Point VA 23062

ANDREWS, JOHN EDWIN, b Selkirk, Man, Mar 15, 22; m 48; c 4. GENETICS, PLANT BREEDING. Educ: Univ Man, BSA, 49; Univ Minn, MS, 50, PhD(genetics), 53. Prof Exp: Cerealist, Rust Res Lab, Univ Man, 47-51; sr cerealist, Res Sta, Alta, 51-60, res dir, Res Br, Man, 60-65; res dir, Sask, 65-69, DIR RES STA, DEPT AGR, 69- Concurrent Pos: Dir Indo-Canadian Res Proj on Dryland Agr, India, 70-79. Mem: Fel Agr Inst Can; Can Soc Agron. Res: Physiology of cold hardiness; winter wheat breeding. Mailing Add: Res Sta Can Dept of Agr Lethbridge AB Can

ANDREWS, JOHN JACOB, b Ann Arbor, Mich, Jan 10, 06; m 29. MATHEMATICS. Educ: Westminster Col, AB, 29; St Louis Univ, MA, 40, PhD(math), 50. Prof Exp: Teacher math, Bd Educ, St Louis, Mo, 39-42; from instr to prof, 46-74, EMER PROF MATH, ST LOUIS UNIV, 74- Mem: Am Math Soc; Math Asn Am. Res: Mathematical probability. Mailing Add: Dept of Math St Louis Univ 221 N Grand Blvd St Louis MO 63103

ANDREWS, JOHN M, JR, solid state physics, semiconductor physics, see 12th edition

ANDREWS, JOHN ROBERT, b Kent, Ohio, June 10, 06; m 35, 51; c 3. RADIOLOGY. Educ: Brown Univ, PhB, 28; Western Reserve Univ, MD, 32; Univ Pa, DSc(med), 46. Prof Exp: Intern, Cleveland City Hosp, 32-33; instr radiol, Univ Pa, 33-35; asst roentgenologist, Univ Hosps, Cleveland, 35-38; attend radiologist & dir clin radiol, St Vincent Charity Hosp, 38-47; practicing radiologist, attend radiologist & dir clin radiol, Good Samaritan Hosp, Fla, 48-50; dir & prof radiol, Bowman Gray Sch Med, Wake Forest Col, 50-55; chief radiation br, Nat Cancer Inst, NIH, 55-64; prof radiol, Sch Med & dir radiother, Univ Hosp, Georgetown Univ, 64-72; CHIEF RADIOTHER, WASHINGTON VET ADMIN HOSP, 65- Concurrent Pos: Asst, Western Reserve Univ, 35-38; resident, Mem Hosp, NY, 36; consult, Admin Hosp, DC, 64- Mem: Am Radium Soc; Am Roentgen Ray Soc; fel AMA; Am Col Radiol; Radiol Soc NAm. Res: Clinical radiology; planigraphy; therapeutic radiation; oncology; radiological physics and diagnosis; cancer therapy and research. Mailing Add: Washington Vet Admin Hosp 50 Irving St NW Washington DC 20422

ANDREWS, JOHN SCOTT, b New Palestine, Ind, Oct 1, 05; m 41; c 3. VETERINARY PARASITOLOGY. Educ: Purdue Univ, BS, 27, MS, 29; Johns Hopkins Univ, DSc(helminth), 38. Prof Exp: Asst gen biol, Purdue Univ, 27-29; from jr zoologist to asst zoologist, Zool Div, Bur Animal Indust, 30-38, parasitologist, Exp Sta, PR, 38-41, assoc parasitologist, Zool Div, Ga, 41-53 & Zool Div, Md, 53-56, leader helminth invests, 56-70, leader trichinosis res, Animal Parasitol Inst, Agr Res Serv, USDA, 70-75; RETIRED. Mem: Am Soc Parasitol. Res: Internal parasites of ruminants, horses and swine; helminthology. Mailing Add: 10314 Naglee Rd Silver Spring MD 20903

ANDREWS, JOHN STEVENS, JR, b Lawrence, Mass, July 12, 27; m 55; c 4. BIOCHEMISTRY. Educ: Univ NH, BS, 51, MS, 52; NC State Col, PhD(animal nutrit), 55. Prof Exp: Asst res physiol chemist, Univ Calif, Los Angeles, 55-58; instr ophthal res, Harvard Univ, 59-66, assoc, 66-68; asst prof, 68-70, ASSOC PROF OPHTHAL, VANDERBILT UNIV, 70- Mem: Am Chem Soc; Am Soc Biol Chemists. Res: Abnormal tissue lipid deposition and metabolism; lipid-protein interrelationships. Mailing Add: Dept of Ophthal Vanderbilt Med Sch Nashville TN 37232

ANDREWS, JOHN THOMAS, b Millom, Eng, Nov 8, 37; m 61; c 1. GEOMORPHOLOGY. Educ: Univ Nottingham, BSc, 59, PhD(geomorphol), 65; McGill Univ, MSc, 61. Prof Exp: Res scientist, Can Govt, 61-67; asst prof geol, 68-74, PROF GEOL SCI, UNIV COLO, 74- Mem: Geol Soc; Asn Am Geog. Res: Glacial chronology and glacioisostatic uplift. Mailing Add: Inst Arctic & Alpine Res Univ of Colo Boulder CO 80302

ANDREWS, JOHN TIMOTHY SAWFORD, b Oxford, Eng, Mar 20, 38; m 69. PHYSICAL CHEMISTRY, THERMODYNAMICS. Educ: Oxford Univ, BA, 62; Univ Mich, MS, 65, PhD(phys chem), 69. Prof Exp: Fel, Liquid Crystal Inst, Kent State Univ, 69-74; ASST PROF CHEM, HIRAM COL, 74- Mem: AAAS; Am Chem Soc; The Chem Soc. Res: Chemical thermodynamics; calorimetry; relationship of structure to thermodynamic properties; mesomorphic state. Mailing Add: Dept of Chem Hiram Col Hiram OH 44234

ANDREWS, LAWRENCE JAMES, b San Diego, Calif, Sept 27, 20; m 44; c 2. CHEMISTRY. Educ: Univ Calif, Berkeley, BS, 40; Univ Calif, Los Angeles, AM, 41, PhD(chem), 43. Prof Exp: Asst chem, Univ Calif, 40-43, Sharp & Dohme fel & lectr, 43-44; chemist, Tenn Eastman Corp, Oak Ridge, 44-45; from instr to assoc prof chem, 45-57, chmn dept, 59-62, actg dean col lett & sci, 62-63, fac res lectr, 65, PROF CHEM, UNIV CALIF, DAVIS, 57-, DEAN COL LETT & SCI, 64- Concurrent Pos: Advan Educ Fund fel, 53-54; consult, Kasetsart Univ, Bangkok, 66-67; Fulbright res scholar, Univ Hull, 67-68; chmn comn arts & sci, Nat Asn State Univs & Land Grant Cols, 73-74. Mem: Am Chem Soc. Res: Mechanisms organic reactions in solution; coordination compounds; molecular complex formation; electrophilic aromatic substitution processes. Mailing Add: Dept of Chem Univ of Calif Davis CA 95616

ANDREWS, LUCY GORDON, b Washington, DC, Mar 27, 41; m 65; c 2. MEDICAL GENETICS. Educ: Agnes Scott Col, BA, 63; Univ Ga, PhD(bot), 67. Prof Exp: Prof biol, Brenau Col, 67-75, chmn dept, 67-70; GENETIC COUNR, GA DEPT HUMAN RESOURCES, 73- Mem: AAAS; Am Inst Biol Sci; Bot Soc Am; Am Soc Human Genetics. Res: Nucleic acid relationships during phloem differentiation in Populus deltoides; evaluation of effectiveness of genetic counseling. Mailing Add: PO Box 2395 Gainesville GA 30501

ANDREWS, MERRILL LEROY, b Albany, NY, Apr 5, 39; m 65; c 1. PLASMA PHYSICS. Educ: Cornell Univ, BA, 60; Mass Inst Technol, PhD(physics), 67. Prof Exp: Res assoc plasma physics, Lab Plasma Studies, Cornell Univ, 67-69, asst prof appl physics, 69-70; asst prof, 70-74, PROF PHYSICS, WRIGHT STATE UNIV, 74- Mem: Am Phys Asn; Am Asn Phys Teachers. Res: Far infrared interferometry; electron beam physics. Mailing Add: 425 Fawcett Hall Dept Physics Wright State Univ Dayton OH 45431

ANDREWS, MYRON FLOYD, b Huron, SDak, Dec 8, 24; m 48; c 2. VETERINARY PARASITOLOGY. Educ: Stanford Univ, AB, 50; Univ Calif, Davis, DVM, 58. Prof Exp: Fisheries res biologist, Chesapeake Shellfish Invests, US Fish & Wildlife Serv, MD, 51-52; sr lab technician, Dept Surg, Sch Med, Univ Calif, Los Angeles, 52-54; asst prof vet sci, NDak State Univ, 58-60; dir large animal clin, Norwich Pharmacal Co, 60-61; asst prof, 61-65, actg chmn dept, 64-65, PROF & CHMN DEPT VET SCI, N DAK STATE UNIV, 65- Concurrent Pos: Res consult, Norwich Pharmacal Co, 61-63 & Diamond Labs, 63-64; consult vet, NDak Livestock Sanitary Bd, 64- Mem: Am Vet Med Asn; Am Asn Vet Parasitol; Wildlife Dis Asn; US Animal Health Asn. Res: Veterinary parasitology and zoonoses. Mailing Add: 2914 Edgemont St N Fargo ND 58102

ANDREWS, NEIL CORBLY, b Spokane, Wash, Mar 31, 16; m 43, 70; c 2. THORACIC SURGERY. Educ: Univ Ore, BA, 40, MD, 43; Ohio State Univ, MSc, 50. Prof Exp: Intern surg, Union Mem Hosp, Baltimore, Md, 43, asst resident, 44; resident, Jefferson Hosp, Roanoke, Va, 44-45; from asst resident to resident, Col Med, Ohio State Univ, 47-50, from instr to prof thoracic surg, 47-70; PROF SURG, SCH MED, UNIV CALIF, DAVIS, 70-, CHMN DIV COMMUNITY & POSTGRAD MED, 71-, CHMN DEPT POSTGRAD MED, 73- Concurrent Pos: Chief surg, Ohio Tuberc Hosp, 50-68; consult, Vet Admin Hosp, Chillicothe, 53-70 & Dayton, 54-70; area consult, Vet Admin, 60-70; coordr, Ohio State Regional Med Prog, 66-70 & Area II, Calif Regional Med Prog, 70-73; mem attend staff, Sacramento Med Ctr, Calif, 70- Mem: AAAS; fel Am Col Surg; fel Am Col Chest Physicians; Am Heart Asn; AMA. Res: Cancer chemotherapy. Mailing Add: Sch of Med Univ of Calif Davis CA 95616

ANDREWS, OLIVER AUGUSTUS, b Plymouth, Wis, Dec 15, 31; m 51; c 5. CHEMISTRY, SCIENCE EDUCATION. Educ: Wis State Univ, Stevens Point, BS, 53; Univ Wis, MS, 57. Prof Exp: Teacher high schs, Wis, 53-59; traveling sci teacher, Mich State Univ & Nat Sci Found, 59-60; from instr to asst prof, 60-65, ASSOC PROF CHEM, UNIV WIS-STEVENS POINT, 65- Mem: Am Chem Soc; Nat Sci Teachers Asn. Res: Investigation of learner initiated cues. Mailing Add: Dept of Chem Univ of Wis Stevens Point WI 54481

ANDREWS, PETER BRUCE, b New York, NY, Nov 1, 37; m 64. MATHEMATICAL LOGIC. Educ: Dartmouth Col, AB, 59; Princeton Univ, PhD(math), 64. Prof Exp: Asst prof, 63-68, ASSOC PROF MATH, CARNEGIE-MELLON UNIV, 68- Mem: AAAS; Am Math Soc; Asn Symbolic Logic; Asn Comput Mach. Res: Symbolic logic, especially type theory and the formalization of mathematics; theorem proving by computer; artificial learning and intelligence. Mailing Add: Dept of Math Carnegie-Mellon Univ Pittsburgh PA 15213

ANDREWS, RICHARD D, b Mitchell Nebr, Jan 6, 33; m 55; c 3. ZOOLOGY. Educ: Nebr State Teachers Col, Kearney, BA, 58; Iowa State Univ, MS, 60; Univ Ill, Urbana, PhD(vet med sci), 66. Prof Exp: Ecologist, Univ Ill, Urbana, 60-63; ASSOC PROF ZOOL, EASTERN ILL UNIV, 66- Mem: Wildlife Soc; Am Soc Mammal; Wildlife Dis Asn. Mailing Add: Dept of Zool Eastern Ill Univ Charleston IL 61920

ANDREWS, RICHARD VINCENT, b Arapahoe, Nebr, Jan 9, 32; m 54; c 6. PHYSIOLOGY. Educ: Creighton Univ, BS, 58, MS, 59; Univ Iowa, PhD(physiol), 63. Prof Exp: Instr biol, Creighton Univ, 58-60; instr physiol, Univ Iowa, 60-63; from instr to prof biol, 63-70, prof physiol & asst dean med, 70-75, GRAD DEAN, CREIGHTON UNIV, 75- Concurrent Pos: NSF sci fac fel, Univ Iowa, 61-63; Arctic Inst NAm vis investr, Naval Arctic Res Lab, 63-72. Mem: Endocrine Soc; Am Physiol Soc; Tissue Cult Asn; AAAS. Res: Regulatory and environmental physiology. Mailing Add: Creighton Univ 2500 California Omaha NE 68178

ANDREWS, ROBERT SANBORN, b Minneapolis, Minn, Sept 20, 35; m 60; c 2. GEOLOGICAL OCEANOGRAPHY, MARINE GEOPHYSICS. Educ: Univ Minn, Minneapolis, BGE, 58; Univ Wash, MS, 65; Tex A&M Univ, PhD(oceanog), 71. Prof Exp: ASSOC PROF OCEANOG, NAVAL POSTGRAD SCH, 68- Mem: AAAS; Am Geophys Union; Geol Soc Am; Soc Explor Geophys; Nat Asn Geol Teachers. Res: Marine seismic exploration; theoretical seismograms; sedimentary petrology; acoustic and physical properties of marine sediments; hydrographic surveying; crustal tectonics; optical oceanography; antisubmarine warfare. Mailing Add: Dept of Oceanog Naval Postgrad Sch Code 58 Ad Monterey CA 93940

ANDREWS, RODNEY DENLINGER, JR, b Chicago, Ill, May 13, 22. PHYSICAL CHEMISTRY. Educ: Princeton Univ, AB, 43, AM, 47, PhD(chem), 48. Prof Exp: Asst, Princeton Univ, 43-48, 49-50; phys chemist, Phys Res Lab, Dow Chem Co, 50-58; res assoc, Plastics Res Lab, Mass Inst Technol, 58-61, asst prof mat res lab, Dept Civil Eng, 61-64, asst prof textile div, Dept Mech Eng, 64-69; PROF CHEM, DEPT CHEM & CHEM ENG, STEVENS INST TECHNOL, 69- Concurrent Pos: Inst Int Educ fel, Univ Basel, 48-49; ed, Soc Rheol, 57-58. Mem: Am Chem Soc; Am Phys Soc; Soc Rheol; Soc Plastics Eng; Fiber Soc. Res: Mechanical and optical properties of high polymers in the solid state; viscoelastic properties of materials. Mailing Add: Dept of Chem & Chem Eng Stevens Inst of Technol Hoboken NJ 07030

ANDREWS, RONALD ALLEN, b Pontiac, Mich, Mar 1, 40; m 67; c 1. LASERS, OPTICS. Educ: Wayne State Univ, BS, 62, PhD(physics), 66; Mass Inst Technol, MS, 76. Prof Exp: Fel solid state physics, Wayne State Univ, 66-67; res physicist quantum optics, 67-71, head optical physics br, 71-72, HEAD INTERACTION PHYSICS BR, NAVAL RES LAB, 72- Concurrent Pos: Sloan fel, Mass Inst Technol, 75-76. Mem: Am Phys Soc; Am Optical Soc; Inst Elec & Electronic Engr. Res: Nonlinear optics and parametric phenomenon; infrared up-conversion; optical waveguides and devices; lasers physics; x-ray lasers; laser-matter interactions. Mailing Add: Code 5520 Naval Res Lab Washington DC 20375

ANDREWS, RUSSELL S, JR, b Mobile Ala, July 22, 42; m 68. PULP CHEMISTRY. Educ: Univ Ala, BSCh, 64, PhD(org chem), 69. Prof Exp: Teaching asst, Univ Ala, 64-68; RES ASSOC, ERLING RIIS RES LAB, INT PAPER CO, 69- Mem: Am Chem Soc. Res: Organic photochemistry; cellulose and lignin chemistry; high yield pulping; chemical recovery. Mailing Add: 1005 Highpoint Dr E Mobile AL 36609

ANDREWS, STEPHEN BRIAN, b McKeesport, Pa, Apr 13, 44; m 69; c 1. CELL

BIOLOGY, BIOPHYSICAL CHEMISTRY. Educ: Providence Col, BS, 66; Mass Inst Technol, PhD(inorganic chem), 71. Prof Exp: ASST PROF CYTOLOGY, SCH MED, YALE UNIV, 73- Concurrent Pos: USPHS fel, Sch Med, Yale Univ, 71-73. Mem: Am Chem Soc; Am Soc Cell Biol. Res: Biomembrane structure and function; electron microscopy; physiology and physical chemistry of model biological membranes. Mailing Add: Dept of Cytology Yale Univ Sch of Med New Haven CT 06510

ANDREWS, THEODORE FRANCIS, b Atchison, Kans, Aug 17, 17; m 36, 69; c 4. ECOLOGY. Educ: Kans State Teachers Col, BS, 40; Univ Iowa, MS, 42; Ohio State Univ, PhD(zool), 48. Prof Exp: From asst prof to assoc prof biol, Kans State Teachers Col, 48-56, prof & head dept, 57-66; dir sci, Educ Res Coun Am, Ohio, 66-69; DEAN COL ENVIRON & APPL SCI, GOVERNORS STATE UNIV, 69- Concurrent Pos: Consult, Biol Sci Curriculum Study, 64-65 & Harvard Univ Nigerian Proj, 65; assoc dir comn undergrad educ biol sci, George Washington Univ, 65-66. Mem: AAAS; Nat Asn Biol Teachers (pres, 64); Am Inst Biol Sci; Nat Sci Teachers Asn; Ecol Soc Am. Res: Electrophysiology; biological assay of industrial wastes; freshwater biology; limnological trends in artificially impounded waters in Kansas; comparative limnology of Kansas streams; ecology of bluehorn Ardea herodias in Kansas; science curriculum development; interdisciplinary science program; environmental science curriculum. Mailing Add: Col of Environ & Appl Sci Governors State Univ Park Forest South IL 60466

ANDREWS, WALLACE HENRY, b Biloxi, Miss, Oct 6, 43. MICROBIOLOGY. Educ: Univ Miss, BA, 65, MS, 67, PhD(microbiol), 69. Prof Exp: Res microbiologist, Shellfish Sanit Br, Div Food Technol, 69-71, RES MICROBIOLOGIST, FOOD MICROBIOL BR, DIV MICROBIOL, FOOD & DRUG ADMIN, 71- Concurrent Pos: Consult, Nat Shellfish Sanit Prog, 73-75. Mem: Am Soc Microbiol; Asn Off Anal Chemists; Nat Shellfisheries Asn; Inst Food Technologists. Res: Development of sensitive, rapid methods for detecting Salmonella and other foodborne enterics; modes of action of antibiotics, antibiotic-resistant Salmonella; physiology of shellfish; metabolism of Klebsiella; bacteriological standards for foods. Mailing Add: FDA Div of Microbiol HFF-124 200 C St SW Washington DC 20204

ANDREWS, WALTER GLENN, b Graham, NC, Aug 31, 16; m 39; c 3. POULTRY SCIENCE. Educ: NC State Univ, BS, 39; Cornell Univ, MS, 59, PhD, 62. Prof Exp: Asst county agent, 40-49, exten poultry specialist, 49-64, DIST AGR AGENT & PROF POULTRY SCI, NC AGR EXTEN SERV, 64- Mem: Poultry Sci Asn. Res: Supervision and administration. Mailing Add: 1520 Trailwood Dr Raleigh NC 27606

ANDREWS, WARREN MCCORMICK, nuclear science, see 12th edition

ANDREWS, WILLIAM LESTER SELF, b Lincolnton, NC, Jan 31, 42; m 65. PHYSICAL CHEMISTRY, SPECTROSCOPY. Educ: Miss State Univ, BS, 63; Univ Calif, Berkeley, PhD(phys chem), 66. Prof Exp: Asst prof, 66-70, ASSOC PROF PHYS CHEM, UNIV VA, 70- Concurrent Pos: Univ Va Ctr Advan Studies sub-grant, 66-68; NSF grant, 68-76; A P Sloan fel, 73-75; Petrol Res Fund grant, 75-78. Mem: Am Chem Soc; Am Phys Soc. Res: Infrared, optical and laser-Raman spectroscopic studies of chemical intermediates and ions produced by matrix reactions and proton radiolysis. Mailing Add: Dept of Chem Univ of Va Charlottesville VA 22901

ANDRIA, GEORGE D, b Hibbing, Minn, Aug 4, 41; m 64; c 4. MATHEMATICAL ANALYSIS, ENERGY CONVERSION. Educ: St John's Univ, Minn, BA, 63; St Louis Univ, MS, 65, PhD(math), 68. Prof Exp: Asst prof math, Fontbonne Col, 67-68 & Univ Pittsburgh, 68-73; RES MATHEMATICIAN, BITUMINOUS COAL RES, INC, 74- Concurrent Pos: Math consult, Bituminous Coal Res, Inc, 74- Mem: Soc Indust & Appl Math; Inst Math Statist. Res: Numerical analysis; numerical solution of ordinary and partial differential equations; approximation theory; mathematical modeling and statistical analysis of coal conversion to synthetic gas and coal mining systems. Mailing Add: Bituminous Coal Res Inc 207 Hochberg Rd Monroeville PA 15146

ANDRICHUK, JOHN MICHAEL, b Downing, Alta, July 4, 26; m 55; c 3. PETROLEUM GEOLOGY, SEDIMENTOLOGY. Educ: Univ Alta, BSc, 46, MSc, 49; Northwestern Univ, PhD(geol), 51. Prof Exp: Res geologist, Gulf Oil Corp, 51-54; consult geologist, Alex McCoy Assocs, 54-56; CONSULT GEOLOGIST, ANDRICHUK & EDIE, 56- Honors & Awards: Pres Award, Am Asn Petrol Geologists, 58. Mem: Geol Soc Am; Soc Econ Paleontologists & Mineralogists; Am Asn Petrol Geologists; Can Geol Asn; Int Asn Sedimentol. Mailing Add: Andrichuk & Edie Third Floor 205 Ninth Ave SE Calgary AB Can

ANDRINGA, KEIMPE, b Salatiga, Indonesia, Oct 25, 35; US citizen. LASERS. Educ: Delft Univ Technol, MSc, 61. Prof Exp: Engr lasers, Laser Advan Develop Ctr, Raytheon Co, 63-66, sr scientist electro-optics, Res Div, 66-74; SR SCIENTIST LASERS, EXXON NUCLEAR CORP, 74- Mem: Inst Elec & Electronics Engrs. Res: High-power laser development; electro-optics; non-linear optics. Mailing Add: Exxon Nuclear Co 2955 George Washington Way Richland WA 99352

ANDRIOLE, VINCENT T, b Scranton, Pa, Aug 3, 31; m 56; c 3. INTERNAL MEDICINE, INFECTIOUS DISEASES. Educ: Col of Holy Cross, BS, 53; Yale Univ, MD, 57. Prof Exp: Intern med, NC Mem Hosp, 57-58, asst resident, 58-59; clin assoc infectious dis, NIH, 59-61; from asst prof to assoc prof, 63-74, PROF INTERNAL MED, SCH MED, YALE UNIV, 74- Concurrent Pos: USPHS res fel, Sch Med, Yale Univ, 61-63; clin investr, Vet Admin Hosp, West Haven, Conn, 63-66; estab investr, Am Heart Asn, 66-71. Mem: Am Soc Clin Invest; Am Soc Nephrology; Infectious Dis Soc Am; Am Col Physicians; Am Soc Microbiol. Res: Host defense mechanisms in the pathogenesis of infectious diseases. Mailing Add: Dept of Internal Med Yale Univ Sch of Med New Haven CT 06510

ANDRIST, ANSON HARRY, b Omaha, Nebr, Sept 28, 43; m 68; c 2. PHYSICAL ORGANIC CHEMISTRY. Educ: Calif State Univ, San Diego, BS, 66; Univ Ill, Urbana, PhD(chem), 70. Prof Exp: Assoc chem, Univ Ore, 70-71; res worker, Univ Sheffield, 71; vis lectr, Univ Colo, Boulder, 72-73; ASST PROF CHEM, CLEVELAND STATE UNIV, 73- Concurrent Pos: Cottrell res grant, 74; Petrol Res Fund res grant, 75. Mem: Am Chem Soc; The Chem Soc. Res: Mechanistic complexities, including reaction stereochemistry of thermal as well as photochemical molecular rearrangements are being probed through structural, kinetic and isotopic labeling studies. Mailing Add: Dept of Chem Cleveland State Univ Cleveland OH 44115

ANDRLE, ROBERT FRANCIS, b Buffalo, NY, Oct 28, 27; m 53; c 4. BIOGEOGRAPHY, ORNITHOLOGY. Educ: Canisius Col, BA, 48; Univ Buffalo, MA, 60; La State Univ, PhD(geog), 64. Prof Exp: Asst preparator exhibs, 56-59, cur div biogeog, 59-65, asst dir, 65-72, ASSOC DIR, BUFFALO MUS SCI, 72- Concurrent Pos: Res grants, Frank M Chapman Mem Fund, Am Mus Natural Hist, 65, Buffalo Soc Natural Sci & Am Philos Soc, 67 & Int Coun Bird Preserv, 73. Mem: Am Inst Biol Sci; Am Ornithologists Union; Asn Am Geogr. Res: Niagara Frontier region ornithology; rare and endangered bird species in Mexico and Guatemala. Mailing Add: Buffalo Mus of Sci Humboldt Park Buffalo NY 14211

ANDROS, GEORGE JAMES, b Lansing, Mich, Apr 30, 16; m 41; c 2. OBSTETRICS & GYNECOLOGY. Educ: Univ Mich, MD, 41; Am Bd Obstet & Gynec, dipl, 49. Prof Exp: From instr to asst prof obstet & gynec, Univ Chicago, 46-49; asst prof, Univ Mich, 49-51 & Univ Chicago, 51-52; from asst prof to assoc prof, Temple Univ, 56-67; assoc prof, 67-71, PROF OBSTET & GYNEC, JEFFERSON MED COL, 71- Mem: Am Col Obstetricians & Gynecologists; Am Fertil Soc; Am Soc Cytol; Asn Profs Gynec & Obstet. Res: Analgesia in labor and delivery; uterine cancer and its detection; urban community obstetric care problems; intrauterine contraceptive devices. Mailing Add: Dept of Obstet & Gynec Jefferson Med Col Philadelphia PA 19107

ANDRULIS, MARILYN ANN, b Ft Monmouth, NJ, Sept 6, 40; c 2. ACOUSTICS. Educ: Univ Mich, BS, 62; Univ Ill, MS, 64; Univ Tex, PhD(mech eng), 68. Prof Exp: Consult, Underwater Syst, Inc, 67-68, B-K Dynamics, 69-71 & Tracor Inc, 71-72; res assoc & dir Ocean Acoust Course, Cath Univ Am, 72-73; V PRES, MID-ATLANTIC RES INST, 73-; PRES, ANDRULIS RES CORP, 72- Concurrent Pos: Mem bd dirs, Mil Opers Res, 75-79. Mem: Acoust Soc Am. Res: Application of underwater acoustics to detection and classification of underwater systems; mathematical modelling. Mailing Add: 600E 7315 Wisconsin Ave Bethesda MD 20014

ANDRULIS, PETER JOSEPH, JR, b New York, NY, Apr 16, 40; m 64; c 2. PHYSICAL ORGANIC CHEMISTRY, BIO-ORGANIC CHEMISTRY. Educ: Canisius Col, BS, 62, MS, 64; Univ Tex, Austin, PhD(phys org chem), 67. Prof Exp: Proj scientist, Goddard Space Flight Ctr, NASA, Md, 62; res asst chem, Canisius Col, 62-64 & Univ Tex, Austin, 64-67; res chemist, US Army Mobility Equip Res & Develop Ctr, 67-69; asst prof chem, Am Univ, 69-71 & Trinity Col, DC, 71-75; MEM STAFF, MID-ATLANTIC RES INST, 75- Concurrent Pos: Lectr, NSF High Sch Teachers Inst, Am Univ, 68-69. Mem: Am Chem Soc; The Chem Soc. Res: Structure-reactivity correlations; correlation of semi-empirical SCF-MO calculations with empirical parameters; synthesis of antiaromatic biologically important compounds; synthesis of chemotherapeutic agents. Mailing Add: Mid-Atlantic Res Inst 7315 Wisconsin Ave Bethesda MD 20014

ANDRUS, CHARLES FREDERICK, b Ill, Jan 21, 06; m 32; c 2. FORESTRY, HORTICULTURE. Educ: George Washington Univ, AB, 30, AM, 31. Hon Degrees: DSc, Clemson Univ, 67. Prof Exp: Asst sci aide, Bur Plant Indust, 28-31, from jr pathologist to pathologist, 31-43, pathologist, Bur Plant Indust, Soils & Agr, 43-48, sr horticulturist, 48-59, res horticulturist in charge veg breeding lab, 59-70, RES COLLABR, VEG BREEDING LAB, AGR RES SERV, USDA, 70- Mem: AAAS; fel Am Soc Hort Sci; Am Forestry Asn; Am Hort Soc. Res: Plant breeding; population genetics. Mailing Add: US Veg Breeding Lab PO Box 3348 Charleston SC 29407

ANDRUS, GRANT MERRILL, organic chemistry, see 12th edition

ANDRUS, JAN FREDERICK, b Washington, DC, Sept 17, 32; m 61; c 2. APPLIED MATHEMATICS. Educ: Col Charleston, BS, 54; Emory Univ, MA, 55; Univ Fla, PhD(math), 58. Prof Exp: Sr mathematician, Ga Div, Lockheed Aircraft Corp, 58-61, math specialist, 61-62; consult mathematician, Huntsville Oper, Comput Dept, Gen Elec Co, 62-63, sub-oper mgr, 63-64, consult mathematician, 64-66; mem tech staff, Northrop-Huntsville, Northrop Corp, 66-68, gen supvr, 68-69, mem sr tech staff, 69-73; ASSOC PROF MATH, UNIV NEW ORLEANS, 73- Mem: Am Math Soc; Soc Indust & Appl Math; Am Inst Aeronaut & Astronaut. Res: Space flight optimization; numerical analysis; image processing. Mailing Add: 237 Dorrington Blvd Metairie LA 70005

ANDRUS, MILTON HENRY, JR, b Omaha, Nebr, Sept 2, 38; m 64; c 2. ORGANIC POLYMER CHEMISTRY, ORGANIC CHEMISTRY. Educ: Augustana Col, SDak, BS, 61; Univ Wash, PhD(chem), 67. Prof Exp: SR CHEMIST, 3M CO, 67- Mem: Am Chem Soc. Res: Reaction rates and mechanisms of solvolysis of allylic compounds; light sensitive and polymer materials used in printing; synthesis of acrylic monomers and polymers and other organic compounds and polymers. Mailing Add: 3M Co 3M Ctr 236-3B St Paul MN 55101

ANDRUS, PAUL GRIER, b Due West, SC, Dec 23, 25; m 48. ELECTROPHOTOGRAPHY. Educ: Univ Wis, PhB, 46. Prof Exp: Asst physics, Univ Wis, 46-47; res engr, 47-51, prin physicist, 51-57, asst chief appl physics div, 57-73, SR RESEARCHER, BATTELLE MEM INST, 73- Mem: Inst Elec & Electronics Eng. Res: Xerography; electrostatics; graphic arts. Mailing Add: Battelle Mem Inst Columbus OH 43201

ANDRUS, WILLIAM DEWITT, JR, b Cincinnati, Ohio, Sept 24, 28; m 56; c 3. ZOOLOGY, CELL PHYSIOLOGY. Educ: Oberlin Col, AB, 52; Stanford Univ, PhD(biol), 62. Prof Exp: From instr to assoc prof, 60-74, chmn dept, 70-73, PROF ZOOL, POMONA COL, 74- Mem: AAAS; Am Soc Zool; Tissue Cult Asn. Res: Brine algae and protozoa; mammalian hibernation; tissue culture applications. Mailing Add: Dept of Zool Pomona Col Claremont CA 91711

ANDRUS, WINFIELD SCOTT, physics, astronomy, see 12th edition

ANDRUSHKIW, JOSEPH WASYL, b Horodok, Ukraine, Mar 21, 06; US citizen; m 32; c 2. MATHEMATICS. Educ: J Casimirus Univ, Lviv, MS, 30, MEd, 32; Ukrainian Free Univ, PhD(math), 46, Dr Habil, 47. Hon Degrees: DPhil, Ukrainian Free Univ, 62. Prof Exp: Teacher math & physics, State Teachers Col, Kenty, Poland, 30-35 & State Gym & Lyceum, 36-39; instr math, Lviv Univ, 39-40; from prof to dir, Ukrainian Teachers Col, 45-48, asst prof, Ukrainian Free Univ, 47-49; from asst prof to prof, 49-72, chmn dept, 62-72, EMER PROF MATH, SETON HALL UNIV, 72- Mem: AAAS; fel NY Acad Sci; Am Math Soc; Math Asn Am; Shevchenko Sci Soc (vpres, 74-). Res: Algebra; mathematical analysis. Mailing Add: 149 Milton Pl South Orange NJ 07079

ANDRUSHKIW, ROMAN IHOR, b Lviv, Ukraine, May 3, 37; US citizen. MATHEMATICAL ANALYSIS. Educ: Stevens Inst Technol, BE, 59, PhD(math), 73; Newark Col Eng, MSEE, 64; Univ Chicago, MS, 67. Prof Exp: From elec engr to sr elec engr, Weston Instruments & Electronics Co, Schlumberger Inc, 59-64; instr math, Newark Col Eng, 64-66; res asst, Univ Chicago, 66-68; instr, Newark Col Eng, 68-70; ASST PROF MATH, NJ INST TECHNOL, 70- Mem: Am Math Soc; Soc Indust & Appl Math; Math Asn Am. Res: Spectral theory of non-selfadjoint unbounded operators; variational characterization of characteristic functions and values of operators; iterative methods for the solution of linear and nonlinear eigenvalue problems. Mailing Add: NJ Inst of Technol Dept of Math 323 High St Newark NJ 07102

ANDRYCHUK, DMETRO, b Fisher Branch, Man, Dec 10, 18; nat US; m 44; c 3. SPECTROSCOPY. Educ: Univ Man, BS, 41, MA, 42; Univ Toronto, PhD(spectros), 49. Prof Exp: Jr res officer, Spectros, Nat Res Coun, Ottawa, Ont, 49-51; spectroscopist, Diamond Alkali Co, Ohio, 51-57; res scientist, Leeds & Northrup Co, Pa, 57-58; MEM TECH STAFF RES, TEX INSTRUMENTS INC, 58- Mem: Am Chem Soc; Can Physics Soc. Res: Emission, raman, infrared and x-ray spectroscopy;

infrared studies of semiconductors; analytical instruments; application of laser and infrared technology to reconnaisance. Mailing Add: Cent Res Labs Tex Instruments 6000 Lemmon Ave Dallas TX 75222

ANDRYKOVITCH, GEORGE, b St Michael, Pa, Jan 1, 41. MICROBIAL PHYSIOLOGY. Educ: Univ Pittsburgh, BS, 62, MS, 65; Univ Md, PhD(microbiol), 68. Prof Exp: ASST PROF BIOL, GEORGE MASON COL, UNIV VA, 68- Mem: AAAS; Am Inst Biol Sci; Sigma Xi; Am Soc Microbiol. Res: Regulation and control of the biosynthesis of extracellular polysaccharides of yeasts and bacteria; mechanisms of action of bacteriolytic enzymes elaborated by soil microorganisms. Mailing Add: Dept Biol George Mason Col 4400 University Dr Fairfax VA 22030

ANDY, ORLANDO JOSEPH, b New Britain, Conn, Jan 21, 20. NEUROSURGERY. Educ: Ohio Univ, BS, 43; Univ Rochester, MD, 45. Prof Exp: Instr neurol surg, Univ & neurol surgeon, Hosp, Johns Hopkins Univ, 52-55; PROF NEUROSURG, SCH MED, UNIV MISS, 55-, CHMN DEPT, 60- Concurrent Pos: USPHS fel, Johns Hopkins Univ, 52-55. Mem: Neurosurg Soc Am; Am Asn Neurol Surgeons; AMA; Am Fedn Clin Res; Cong Neurol Surgeons. Res: Epilepsy. Mailing Add: Univ of Miss Med Ctr Jackson MS 39216

ANELLIS, ABE, b Russia, Feb 15, 14; nat US; m 61; c 2. BACTERIOLOGY. Educ: Univ Ill, BS, 40, MS, 41. Prof Exp: Jr bacteriologist, Northern Regional Res Lab, Ill, 41-45; bacteriologist, Qm Food & Container Inst, 45-63; RES MICROBIOLOGIST, US ARMY NATICK DEVELOP CTR, 63- Mem: Am Soc Microbiol. Res: Thermal and radiation resistance of microorganisms; thermal and radiation processes of foods; fermentation of farm products; food microbiology. Mailing Add: US Army Natick Develop Ctr Natick MA 01760

ANELLO, CHARLES, b Philadelphia, Pa, Dec 20, 35; m 58; c 2. BIOSTATISTICS, EPIDEMIOLOGY. Educ: Towson State Col, BS, 58; Johns Hopkins Univ, ScM, 63, ScD(biostatist), 64. Prof Exp: Consult statist, Booz-Allen Appl Res, Md, 64-65; prin investr, Res Anal Corp, Va, 65-67; asst prof biostatist & epidemiol, Sch Hyg & Pub Health, Johns Hopkins Univ, 67-69; DIR DIV BIOMET, BUR DRUGS, FOOD & DRUG ADMIN, 69- Res: Competing risk models; application of stochastic processes to medical problems; statistical aspects of epidemiology. Mailing Add: 515 Manakee Rockville MD 20850

ANET, FRANK ADRIEN LOUIS, b Doulcon/Meuse, France, Oct 24, 26; m 55. ORGANIC CHEMISTRY. Educ: Univ Sydney, BSc, 49, MSc, 50; Oxford Univ, DPhil(chem), 52. Prof Exp: Fel, Nat Res Coun Can, 53-54; from asst prof to prof chem, Univ Ottawa, Can, 54-64; PROF CHEM, UNIV CALIF, LOS ANGELES, 64- Mem: Am Chem Soc; fel Brit Chem Soc. Res: Natural products; reaction mechanisms in organic chemistry; nuclear magnetic resonance. Mailing Add: Dept of Chem Univ of Calif Los Angeles CA 90024

ANEX, BASIL GIDEON, b Seattle, Wash, May 4, 31; m 59; c 3. PHYSICAL CHEMISTRY. Educ: Wesleyan Univ, BA, 53; Univ Wash, PhD(chem), 59. Prof Exp: Res assoc quantum chem, Ind Univ, 59-60; from instr to asst prof chem, Yale Univ, 60-67; prof & head dept, NMex State Univ, 67-71; PROF CHEM, UNIV NEW ORLEANS, 71- Mem: Am Chem Soc; fel Am Inst Chem. Res: Visible, ultraviolet and reflection spectroscopy; electronic properties of highly absorbing crystals; electronic structure of organic molecules, inorganic complexes and charge-transfer complexes; quantum chemistry. Mailing Add: Dept of Chem Univ of New Orleans New Orleans LA 70122

ANFINSEN, CHRISTIAN BOEHMER, b Monessen, Pa, Mar 26, 16; m 41; c 3. BIOCHEMISTRY. Educ: Swarthmore Col, BA, 37; Univ Pa, MS, 39; Harvard Univ, PhD(biochem), 43. Hon Degrees: DSc, Swarthmore Col, 65, Georgetown Univ, 67, NY Med Col, 69, Univ of Pa, 73 & Gustavus Adolphus Col, 75. Prof Exp: Asst instr org chem, Univ Pa, 37-39; instr biochem, Harvard Med Sch, 43-45, assoc, 45-47, asst prof, 48-50; chief lab cellular physiol, Nat Heart Inst, 50-52, chief lab cellular physiol & metab, 52-62; prof biochem, Harvard Med Sch, 62-63; CHIEF LAB CHEM BIOL, NAT INST ARTHRITIS, METAB & DIGESTIVE DIS, 63- Concurrent Pos: Am Cancer Soc sr fel, Biochem Div, Med Nobel Inst, Sweden, 47-48; Guggenheim Found fel, 59; vis fel, All Souls Col, 70; hon fel & mem bd gov, Weizman Inst Sci. Honors & Awards: Nobel Prize Chem, 72; Pub Serv Award, Rockefeller Found, 54. Mem: Nat Acad Sci; Am Soc Biol Chem (pres, 71-72); Am Acad Arts & Sci. Res: Structure-function relationship in proteins; genetic basis of protein structure; protein isolation, synthesis and proteolysis. Mailing Add: Nat Inst Arthritis Metab Dig Dis 9000 Rockville Pike Bethesda MD 20014

ANFINSEN, JON ROBERT, b Chippewa Falls, Wis, Mar 28, 40; m 61; c 2. ANALYTICAL CHEMISTRY. Educ: Univ Ill, BS, 62, PhD(food chem), 67. Prof Exp: Sect head, Anal Dept, Armour & Co, 66-68; mgr anal systs develop, Miles Labs, Inc, 68-72, mgr qual control, Dome Div, 72-73; DIR QUAL CONTROL, MERRELL-NAT LABS, DIV OF RICHARDSON-MERRELL INC, CINCINNATI, 73- Mem: Am Chem Soc; Inst Food Technologists. Res: Development of quality control monitoring systems for pharmaceuticals, diagnostic products and foods; automation; thin-layer, gas and liquid chromatography; mass, nuclear magnetic resonance and infrared spectroscopy; statistics; computer technology. Mailing Add: Merrell-Nat Labs Div Richardson-Merrell Inc Cincinnati OH 45215

ANFINSON, OLAF P, b La Crosse, Wis, Aug 4, 14; m 46. PHYSICS. Educ: Winona State Col, BS, 39; Colo State Univ, MA, 52, EdD, 54. Prof Exp: Instr physics, Winona State Col, 46-47 & 54-55; asst prof, Fla Southern Col, 47-48; instr, Moorhead State Col, 49-52 & San Bernardino Valley Union Jr Col, 55-56; from asst prof to prof, 56-74, EMER PROF PHYS SCI, CALIF STATE COL, LONG BEACH, 74- Mem: AAAS; Nat Sci Teachers Asn. Res: Science education: testing procedures and testing instruments; evaluation of students achievement of objectives of science courses. Mailing Add: Dept of Phys Sci Calif State Col Long Beach CA 90840

ANG, CATHARINA YUNG-KANG WANG, b Mongolia, China, Aug 22, 40; US citizen. FOOD SCIENCE. Educ: Nat Taiwan Univ, BS, 63; NC State Univ, MS, 66; Mich State Univ, PhD(food sci), 70. Prof Exp: Res assoc food sci, Inst Human Nutrit, Columbia Univ, 70-72; lab mgr, 72-74, V PRES LAB DIR FOOD SCI, FOOD SCI ASSOCS, INC, DOBBS FERRY, NY, 74- Concurrent Pos: Expert consult, US Army Natick Develop Ctr, 75- Mem: Am Chem Soc; Inst Food Technologists; Am Oil Chemists Soc. Res: Effects of new heating methods on nutritional retention in foods; development of computer assisted nutritional auditing program; nutritional quality of processed, diet or convenience foods; food service systems studies. Mailing Add: Food Sci Assocs Inc Dobbs Ferry NY 10522

ANG, FRANK SOAN, organic chemistry, polymer chemistry, see 12th edition

ANG, JAN KEE, b Los Banos, Philippines, June 20, 22; US citizen; m 55. HORTICULTURE, PLANT PHYSIOLOGY. Educ: Nanking Univ, BSc, 46; Cornell Univ, MSc, 58, PhD, 63. Prof Exp: Teacher high sch, Philippines, 51-55; from res asst to res assoc, Cornell Univ, 55-64; PROF BIOL, STATE UNIV NY COL ONEONTA, 64- Mem: Am Soc Hort Sci. Res: Post-harvesting physiology of vegetables. Mailing Add: Dept of Biol State Univ NY Col Oneonta NY 13820

ANG, TJOAN-LIEM, b Bogor, Indonesia, Jan 7, 33; m 62; c 2. POLYMER SCIENCE. Educ: Inst Technol Bandung, Indonesia, Drs, 62; Univ Akron, MSc, 65, PhD(polymer sci), 75. Prof Exp: Lectr org chem, Inst Technol Bandung, 63-71; sr chemist polymer res, Swedlow, Inc, 74-75; sr res chemist, Swedcast Corp, 75-76; SR RES CHEMIST POLYMER APPLN & PROD DEVELOP, AVERY INT CORP, 76- Res: Impact modification of glassy polymers; activatable pressure sensitive adhesives; release and anti blocking agents. Mailing Add: Avery Int Corp 325 N Altadena Dr Pasadena CA 91107

ANGALET, GEORGE WILLIAM, b Cleveland, Ohio, Dec 15, 17; m 54; c 1. ENTOMOLOGY. Educ: Ohio State Univ, BSc, 49. Prof Exp: RES ENTOMOLOGIST, AGR RES SERV, USDA, 50- Mem: AAAS; Entom Soc Am; Entom Soc India. Res: Foreign exploration for parasites and predators of insect pests in Asia and Europe; beneficial insects and insect pathogens. Mailing Add: Beneficial Insects Lab USDA 501 S Chapel St Newark DE 19711

ANGEL, CHARLES, b Lockney, Tex, June 10, 23; m 45; c 3. BIOCHEMISTRY. Educ: Tex Tech Col, BS, 48, MS, 49; Univ Tex, PhD(biochem), 54. Prof Exp: Lab asst org chem, Tex Tech Col, 47-48, fel chem, 48-49; teaching & asst, Med Br, Univ Tex, 49-53; from asst prof to assoc prof physiol chem, Sch Med, Univ Miss, 53-57; chief res lab, Vet Admin Ctr, Gulfport Div, 57-72; PROF BIOCHEM, UNIV ARK MED CTR, 72-; RES CHEMIST, VET ADMIN HOSP, LITTLE ROCK, 72- Concurrent Pos: Lectr biochem & psychiat, Sch Med, Tulane Univ, 57-; clin assoc prof anat, La State Univ Med Ctr, 70- Honors & Awards: Sustained Superior Performance Award, US Civil Serv, 66; Cert, Nat Registry Clin Chem, 69. Mem: Fel AAAS; Am Chem Soc; fel Am Inst Chem; Soc Biol Psychiat; Pavlovian Soc Am. Res: Biochemical studies in mental diseases; studies relating behavior to physiological and biochemical variables; effects of stress and adrenal hormones on brain protein synthesis and permeability function in the central nervous system. Mailing Add: Neuropsychiat Res Vet Admin Hosp North Little Rock AR 72114

ANGEL, CHARLES ROBERT, radiation biology, see 12th edition

ANGEL, HENRY SEYMOUR, b London, Eng, Jan 20, 19; nat US; m 41; c 3. ORGANIC CHEMISTRY. Educ: Yale Univ, BS, 40; Univ Pa, MS, 47, PhD(org chem), 48. Prof Exp: Res chemist, Socony-Vacuum Oil Co, NJ, 40-45; asst instr org chem, Univ Pa, 45-47; sr chemist, Sherwin-Williams Co, 48-52; res chemist, Quaker Oats Co, 53 & Am Cyanamid Co, 53-60; RES CHEMIST, ENGELHARD MINERALS & CHEM CORP, 60- Mem: AAAS; Am Chem Soc. Res: Synthesis of organic compounds; synthesis of precious metal-organic compounds. Mailing Add: 79 Willow Ave North Plainfield NJ 07060

ANGEL, JOHN LAWRENCE, b London, Eng, Mar 21, 15; nat US; m 37; c 3. PHYSICAL ANTHROPOLOGY. Educ: Harvard Univ, AB, 36, PhD(phys anthrop), 42. Prof Exp: Traveling fel, Exped, Greece, 37-39; asst anthrop, Harvard Univ & Radcliffe Col, 39-41; instr, Univ Calif, 41-42 & Univ Minn, 42-43; from assoc to prof phys anthrop & anat, Jefferson Med Col, 43-62, Guggenheim fel, 49; CUR PHYS ANTHROP, US NAT MUS, SMITHSONIAN INST, 62- Concurrent Pos: Prof lectr, George Washington Univ, 64-; vis prof, Med Sch, Howard Univ, 65-70; lectr, Sch Pub Health, Johns Hopkins Univ, 65-; consult, US Naval Hosp, Philadelphia, 55-62. Mem: AAAS; Am Anthrop Asn; Am Soc Human Genetics; Am Asn Anat; Am Asn Phys Anthrop (secy-treas, 52-56, vpres, 58-60). Res: Social biology of the eastern Mediterranean region; forensic anthropology; skeletal form relative to function. Mailing Add: Div of Phys Anthrop Smithsonian Inst Washington DC 20560

ANGEL, JOSEPH FRANCIS, b Haifa, Palestine, Aug 31, 40; Can citizen; m 66; c 2. NUTRITION, BIOCHEMISTRY. Educ: Ain Shams Univ, Cairo, BSc, 63; Am Univ Beirut, MSc, 65; Univ Toronto, PhD(nutrit), 71. Prof Exp: Res asst biochem, Dept Pediat, Augusta Victoria Hosp, Jerusalem, 65-66; asst prof human nutrit, Univ BC, 69-75; ASSOC PROF BIOCHEM, UNIV SASK, 75- Mem: Nutrit Soc Can; Brit Nutrit Soc. Res: Energy utilization in pyridoxine deficiency; lipogenesis in meal-fed animals; evolutionary aspects of nutrition; nutrition and development in mammalia. Mailing Add: Dept of Biochem Col of Med Univ of Sask Saskatoon SK Can

ANGELAKOS, EVANGELOS THEODOROU, b Tripolis, Greece, July 15, 29; nat US; m 54; c 1. PHYSIOLOGY. Educ: Nat Univ Athens, BS, 48; Boston Univ, MA, 53, PhD(physiol), 55; Harvard Univ, MD, 59. Prof Exp: Asst physiol, Sch Med, Boston Univ, 52-53, from instr to prof, 55-68; PROF PHYSIOL & BIOPHYS & CHMN DEPT, HAHNEMANN MED COL, 68- Concurrent Pos: Med Found fel, 59-60; USPHS res career develop award, 60-68; res assoc biomath, Mass Inst Technol, 59-61 & physiol, Maine Med Ctr, Portland, 59-61; vis scientist, Karolinska Inst, Sweden, 62-63; consult, US Army Natick Labs Environ Med, 63-; dir, Biomed Res Inst, Univ of Maine, Portland, 73- Mem: AAAS; fel Am Col Cardiol; fel Am Col Angiol; Microcirc Soc; Am Physiol Soc. Res: Physiology and pharmacology of the heart and circulation and autonomic nervous system; electrophysiology of the heart; hypothermia; catecholamines; physiology of emotion and stress. Mailing Add: Dept of Physiol & Biophys Hahnemann Med Col Philadelphia PA 19102

ANGELICI, ROBERT JOE, b Rochester, Minn, July 29, 37; m 60; c 2. INORGANIC CHEMISTRY. Educ: St Olaf Col, BA, 59; Northwestern Univ, PhD(inorg chem), 62. Prof Exp: NSF fel, 62-63; from instr to assoc prof, 63-71, PROF CHEM, IOWA STATE UNIV, 71- Concurrent Pos: Alfred P Sloan Found fel, 70-72. Mem: Am Chem Soc. Res: Kinetic studies of substitution reactions of metal carbonyl complexes and of metal complex catalysis of biologically related reactions. Mailing Add: Dept of Chem Iowa State Univ Ames IA 50010

ANGELINE, JOHN FREDERICK, b Somerville, Mass, Sept 29, 29; m 57; c 3. ORGANIC CHEMISTRY. Educ: Northeastern Univ, BS, 52, MBA, 63. Prof Exp: From chemist to sr chemist, 52-61, PROJ DIR, ARTHUR D LITTLE, INC, 61- Res: Organization of research and development; evaluation of technical change on economic structure; utilization of technology in less developed economies; sensory properties of food and beverages in market performance. Mailing Add: Arthur D Little Inc 15 Acorn Park Cambridge MA 02140

ANGELINI, PIO, b Ipswich, Mass, May 2, 32; m 62; c 1. FOOD CHEMISTRY. Educ: Univ Mass, BS, 54; Mich State Univ, MS, 56, PhD(food sci), 63. Prof Exp: ANAL CHEMIST, US ARMY NATICK LABS, 62- Mem: AAAS; Inst Food Tech; Am Chem Soc; Sigma Xi. Res: Analysis of food flavors using gas chromatographic and mass spectrometric techniques. Mailing Add: FSL US Army NDC Natick MA 01760

ANGELL, C A, b Canberra, Australia, Dec 14, 33; m 58; c 3. PHYSICAL CHEMISTRY. Educ: Univ Melbourne, BSc, 54, MSc, 56; Univ London, PhD(chem), 61. Prof Exp: Lectr chem metall, Univ Melbourne, 62-64; res assoc & fel chem, Argonne Nat Lab, 64-66; from asst prof to assoc prof, 66-70, PROF CHEM, PURDUE UNIV, 71- Concurrent Pos: Mem ad hoc comt infra-red transmitting mat,

Nat Res Coun, 67-68; mem ed adv bd, J Phys Chem, 74-78. Mem: Am Chem Soc. Res: Mass Transport properties of ionic liquids; nature of supercooled liquids and glass transition; spectroscopic studies of coordination states in, and vibrational dynamics of, ionic liquids. Mailing Add: 424 Littleton St West Lafayette IN 47906

ANGELL, CHARLES LESLIE, b July 4, 26; US citizen; m 58; c 2. PHYSICAL CHEMISTRY. Educ: Univ Sydney, BSc, 49, MSc, 50; Cambridge Univ, PhD(chem), 55. Prof Exp: Res officer, New South Wales Univ Technol, 51-53; fel & res assoc spectros, Univ Mich, 56 & Purdue Univ, 57-58; res scientist, 58-69, GROUP LEADER SPECTROS, TARRYTOWN TECH CTR, UNION CARBIDE CORP, 70- Mem: Am Chem Soc; Optical Soc Am; Soc Appl Spectros; Coblentz Soc (pres, 69-71); Brit Chem Soc. Res: Structural investigations of molecules by infrared and Raman spectroscopy; mass spectroscopy; adsorbed molecules and catalysis; molecular and bond moments; tautomeric systems; infrared absorption band intensities; solvent effects. Mailing Add: Tarrytown Tech Ctr Union Carbide Corp PO Box 65 Tarrytown NY 10591

ANGELL, FREDERICK FRANKLYN, b Alto Pass, Ill, July 25, 37; m 64; c 4. PLANT BREEDING, GENETICS. Educ: Southern Ill Univ, BS, 60, MS, 61; Univ Wis, PhD(hort, plant path), 65. Prof Exp: Asst prof, 65-69, ASSOC PROF HORT, UNIV MD, COLLEGE PARK, 70- Concurrent Pos: Mem, Tomato Genetics Coop Am. Mem: Soc Hort Sci; Am Genetic Asn. Res: Vegetable breeding; development of disease resistant varieties; genetics of tomatoes, especially mutants, gene location and inheritance of disease resistance; cultural procedures for mechanized tomato production and harvest. Mailing Add: Dept of Hort Univ of Md College Park MD 20742

ANGELL, JAMES KENNEDY, meteorology, see 12th edition

ANGELL, ROBERT WALKER, b Milwaukee, Wis, Apr 24, 29; m 58. PROTOZOOLOGY, CYTOLOGY. Educ: Beloit Col, BS, 58; Univ Chicago, PhD(paleozool), 65. Prof Exp: USPHS traineeship, 66; asst prof zool, Univ NC, 66-67; ASST PROF ZOOL, UNIV DENVER, 67- Mem: Soc Protozool. Res: Biological mineralization, especially calcification in invertebrates; biology of Foraminifera. Mailing Add: 702 S Corona St Denver CO 80209

ANGELL, THOMAS STRONG, b Bakersfield, Calif, Jan 8, 42; m 67. MATHEMATICS. Educ: Harvard Univ, AB, 63; Univ Mich, MA, 62, PhD(math), 69. Prof Exp: ASST PROF MATH, UNIV DEL, 69- Mem: Am Math Soc; Soc Indust & Appl Math; Math Asn Am. Res: Mathematical theory of optimal control; calculus of variations; ordinary and functional differential equations. Mailing Add: Dept of Math Univ of Del Newark DE 19711

ANGELLO, STEPHEN JAMES, b Haddonfield, NJ, Mar 2, 18; m 44; c 3. PHYSICS, SEMICONDUCTORS. Educ: Univ Pa, BS, 39, MS, 40, PhD(physics), 42. Prof Exp: Res engr, Westinghouse Elec Co, 42-46; asst prof elec eng, Univ Pa, 46-47, head spec proj, Res Inst, 46-47; res engr, Westinghouse Elec Corp, 47-52, sect mgr, 52-54, eng mgr, Semiconductor Dept, 54-57, consult engr, 57-59, mgr thermoelec proj, 59-60, mgr molecular electronics, 60-62, mgr solid state res & develop, 62-64; prof elec eng, Univ Calif, Santa Barbara, 64-66; CONSULT, WESTINGHOUSE RES LABS, 66- Concurrent Pos: Spec lectr, Duquesne Univ, 47-48. Mem: Inst Elec & Electronics Eng; Am Phys Soc. Res: Electrical properties of metal-semiconductor contacts; oxide coated cathodes; Hall effect in semiconductors; low-loss, low-noise crystal rectifier for 3 cm microwave mixers; Hall effect and conductivity of cuprous oxide; silicon rectifier development and thermoelectricity; microelectronics. Mailing Add: Westinghouse Res Labs 1310 Beulah Rd Pittsburgh PA 15235

ANGELO, RUDOLPH J, b Ellwood City, Pa, June 16, 30; m 57; c 1. ORGANIC CHEMISTRY. Educ: Geneva Col, BS, 51; Univ Fla, MS, 53, PhD(chem), 55. Prof Exp: Technician anal, Callery Chem Co, 51; instr, Univ Fla, 51-52, asst, AEC, 52-55; res chemist, Exp Sta, 55-58, Iowa, 58-59, res chemist, Exp Sta, 59-73, STAFF SCIENTIST, EXP STA, E I DU PONT DE NEMOURS & CO, INC, 73- Mem: Am Chem Soc; Sigma Xi. Res: Syntheses of vinyl and condensation monomers; cyclic polymerization mechanism of non-conjugated diolefins; syntheses of polyelectrolytes, heterocyclics and various condensation polymers; emulsions and coatings; high pressure polymerizations. Mailing Add: Film Dept Res Lab E I du Pont de Nemours & Co Inc Wilmington DE 19898

ANGELONE, LUIS, b Alliance, Ohio, Oct 3, 19; m 46; c 3. PHYSIOLOGY. Educ: Ohio State Univ, BA, 47, MA, 49, PhD(physiol), 52. Prof Exp: Asst physiol, Ohio State Univ, 49-52, asst instr, 52; res physiologist, Army Chem Ctr, Md, 52-53; from instr to assoc prof physiol, Wash Univ, 53-65; exec secy hemat study sect, NIH, 65-67, dep chief pac off, 67-68, asst chief spec progs, Res Grants Review Br, Div Res Grants, NIH, 68-71, asst chief referral, 71-74, CHIEF REFERRAL BR, DIV RES GRANTS, NIH, 74-; PROF PHYSIOL, WASH UNIV, 65- Concurrent Pos: Expert adv, Aberdeen Proving Grounds, Md, 53-54; consult, Army Chem Ctr, 53-59; Nat Bd Dent Exam, 62-67. Mem: AAAS; Am Physiol Soc. Res: Hematology; cardiovascular physiology; neurophysiology. Mailing Add: Div Res Grants NIH Bethesda MD 20014

ANGELONI, FRANCIS M, b Butler, Pa, Apr 26, 28; m 52; c 2. ANALYTICAL CHEMISTRY. Educ: Pa State Univ, BS, 52, MS, 59, PhD(anal chem), 65. Prof Exp: Res asst petrol ref lab, Pa State Univ, 52-61; chief chemist, Appl Sci Lab, Inc, Pa, 61-65; sr scientist, 65-66, group leader physics & phys chem group, 66-69, MGR ANAL & RES SERV SECT, RES DEPT, KOPPERS CO, INC, 69- Mem: AAAS; Am Chem Soc; Am Soc Test & Mat. Res: Thermal methods of analysis; x-ray analysis; petroleum chemistry; instrumental analysis; polymer characterization. Mailing Add: Koppers Co Inc Res Dept 440 College Park Dr Monroeville PA 15146

ANGELOPOULOS, ANGELOS PANAYOTIS, b Pyrgos-Ilias, Greece, Nov 12, 36; m 67. DENTISTRY. Educ: Nat Univ Athens, DDS, 60; Univ Okla, MS & PhD(path), 66. Prof Exp: Clin asst oral surg, Sch Med, Univ Okla, 64-65, instr, 65; asst prof oral path, Sch Dent, Univ Minn, 65-66; asst prof, 67-69, ASSOC PROF ORAL PATH & HEAD DIV, DALHOUSIE UNIV, 69- Mem: AAAS; Am Acad Oral Path; Am Acad Oral Med; NY Acad Sci; Fel Int Asn Oral Surgeons. Res: Oral pathology; oral surgery; histopathological, histochemical and quantitative studies of mast cells in the human gingiva, as well as other oral structures in health and disease. Mailing Add: Div of Oral Path Dalhousie Univ Fac of Dent Halifax NS Can

ANGELOPOULOS, EDITH W, b Vienna, Austria, Jan 31, 36; US citizen; m 67. PARASITOLOGY, CYTOLOGY. Educ: Univ Minn, Minneapolis, BA & BS, 61, MS, 64, PhD(zool), 67. Prof Exp: ASST PROF PARASITOL, CELLULAR BIOL & ENTOM, DALHOUSIE UNIV, 67- Mem: AAAS; Soc Protozoologists. Res: Structural and functional aspects of parasitic flagellates; microtubules in Trypanosomatidae; scanning and transmission microscopy; freeze etching and histochemical studies of Trichomonas tenax. Mailing Add: Dept of Biol Dalhousie Univ Halifax NS Can

ANGELOTTI, ROBERT, b New York, NY, Oct 30, 27; m 48; c 3. MICROBIOLOGY. Educ: Transylvania Col, BA, 50; Ohio State Univ, MSc, 52, PhD(microbiol), 55. Prof Exp: Asst bact, Ohio State Univ, 52-54, asst instr, 54-55; res microbiologist, Taft Sanit Eng Ctr, USPHS, 55-60, chief food microbiol, 60-65, dep chief milk & food res, 65-67, dir off res & develop, Nat Ctr Urban & Indust Health, 67-68; dep dir div microbiol, Food & Drug Admin, 69-71; dep asst comnr res & develop, Environ Control Admin, Rockville, Md, 68-69; ASSOC DIR COMPLIANCE, BUR FOODS, FOOD & DRUG ADMIN, WASHINGTON, DC & DEP DIR, OFF FOOD SANIT, BUR FOODS, 71- Mem: Am Soc Microbiol; Am Pub Health Asn; Am Acad Microbiol; Inst Food Technol. Res: Food microbiology and food-borne disease; investigation and evaluation of microbiological hazards associated with the production, processing and service of foods in relation to environmental health and hygiene. Mailing Add: Food & Drug Admin 200 C St SW Washington DC 20204

ANGELOVIC, JOSEPH WILLIAM, ecology, fishery biology, see 12th edition

ANGER, CLIFFORD D, b Long Beach, Calif, Nov 15, 34; m 67; c 2. ATMOSPHERIC PHYSICS. Educ: Univ Calif, Berkeley, BA, 56, MA, 59, PhD(physics), 63. Prof Exp: From asst to assoc prof, 62-74, PROF PHYSICS, UNIV CALGARY, 74- Mem: Am Geophys Union; Can Asn Physicists; Sigma Xi. Res: Upper atmosphere physics; satellite, balloon and ground level studies of auroral phenomena. Mailing Add: Dept of Physics Univ of Calgary Calgary AB Can

ANGER, HAL OSCAR, b Denver, Colo, May 24, 20. INSTRUMENTATION, NUCLEAR MEDICINE. Educ: Univ Calif, Berkeley, BS, 43. Hon Degrees: DSc, Ohio State Univ, 72. Prof Exp: Res assoc, Harvard Univ, 43-46; BIOPHYSICIST, LAWRENCE RADIATION LAB & DONNER LAB, UNIV CALIF, BERKELEY, 46- Concurrent Pos: Guggenheim fel, 66-67. Honors & Awards: John Scott Award, 64; Nuclear Med Pioneer Citation, Soc Nuclear Med, 74. Mem: Inst Elec & Electronics Eng; Soc Nuclear Med; Soc Photo-Optical Instrument Eng. Res: Radiological use of high energy nuclear particle beams; instruments for detecting and measuring radioactive isotopes and for mapping their distribution; invention of the first clinically successful radioisotope camera. Mailing Add: 1771 Highland Pl Berkeley CA 94709

ANGERER, CLIFFORD ACKERMAN, b Philadelphia, Pa, Aug 9, 05; m 32; c 2. PHYSIOLOGY. Educ: Columbia Univ, AB, 29; Univ Pa, PhD(zool, physiol), 37. Prof Exp: Instr zool, Univ Pa, 35-39; from instr to assoc prof, 39-54, PROF PHYSIOL, COL MED, OHIO STATE UNIV, 54- Mem: Fel AAAS; fel Am Heart Asn; Am Inst Biol Sci; Am Physiol Soc; Soc Exp Biol & Med. Res: Physical properties of protoplasm; steroid hormones on water shift, tissue metabolism membrane potentials and permeability; hormone imbalance on metabolism of endocrine tissue; lipid diet on metabolism of tissues; atherogenesis. Mailing Add: Ohio State Univ Col of Med Columbus OH 43210

ANGERER, JOHN DAVID, b Columbus, Ohio, Mar 9, 42; m 63; c 3. ORGANIC CHEMISTRY, POLYMER CHEMISTRY. Educ: Ohio State Univ, BSc, 63; Univ Ill, Urbana, MSc, 65, PhD(org chem), 68. Prof Exp: Res chemist, 68-74, RES SUPVR, HERCULES INC, 74- Mem: Sigma Xi; Am Chem Soc; The Chem Soc. Res: Monomer and polymer synthesis; organic reaction mechanisms; chemistry and rheology of polysaccharides. Mailing Add: Res Ctr Hercules Inc B 8100-R 204 Wilmington DE 19899

ANGERER, LYNNE MUSGRAVE, b Ft Sill, Okla, Dec 7, 44; m 66; c 1. MOLECULAR GENETICS. Educ: Ohio State Univ, BSc, 66, MSc, 67; Johns Hopkins Univ, PhD(cell & develop biol), 73. Prof Exp: FEL MOLECULAR GENETICS, CALIF INST TECHNOL, 73- Concurrent Pos: Damon Runyan-Walter Winchell fel cancer res, 73-75; NIH fel, 76. Mem: Sigma Xi. Res: Study of eucaryotic chromosome organization by examining the physical relationships of specific DNA sequences in the genome and its RNA transcripts by electron microscopy. Mailing Add: Div of Chem Calif Inst of Technol Pasadena CA 91125

ANGERER, ROBERT CLIFFORD, b Columbus, Ohio, Nov 4, 44; m 66; c 1. MOLECULAR BIOLOGY. Educ: Ohio State Univ, BSc, 66; Johns Hopkins Univ, PhD(cell & develop biol), 73. Prof Exp: RES FEL MOLECULAR BIOL, CALIF INST TECHNOL, 73- Concurrent Pos: Am Cancer Soc fel, 73- Res: Regulation of gene activity in eukaryotes. Mailing Add: Dept of Biol Calif Inst of Technol Pasadena CA 91125

ANGEVINE, DANIEL MURRAY, b St John, NB, Oct 8, 03; nat US; m 33; c 3. PATHOLOGY. Educ: Mt Allison Univ, BA, 24; McGill Univ, MD, 29. Prof Exp: Demonstr biol, Mt Allison Univ, 24; from instr to asst prof path, Med Col, Cornell Univ, 32-41, instr med, 380438-40; pathologist & bacteriologist, Alfred I du Pont Inst, 41-45; prof, Med Sch & pathologist, Univ Hosp, Univ Wis, 45-68; ASSOC DIR RES, ARMED FORCES INST PATH, 68- Concurrent Pos: Vis asst prof path, Sch Med, Univ Pa, 41-45; mem staff, US Vet Admin, 44 & USPHS, 47-52; consult, US Dept Army, 48- & Nat Res Coun, 49-52 & 59-; chief ed, Arch Path, AMA, 64- Mem: Am Asn Pathologists & Bacteriologists (pres, 61-62); Am Soc Exp Path (pres, 53); Am Soc Exp Biol & Med; Am Fedn Biol Socs; Am Soc Clin Invest. Res: Streptococcus infections; hypersensitivity; immunity; tuberculosis; localization of infections; rheumatic diseases; pathology of bones and joints; connective tissue; muscle; amebiasis. Mailing Add: Armed Forces Inst of Path Washington DC 20305

ANGEVINE, JAY BERNARD, JR, b Boston, Mass, June 29, 28; m; c 3. NEUROANATOMY. Educ: Williams Col, BA, 49; Cornell Univ, MA, 52, PhD, 56. Prof Exp: Asst histol & embryol, Cornell Univ, 49-51, neuroanat, 52-54, embryol, 54, instr zool, 55-56; asst neuropath, Harvard Med Sch, 56-57, instr, 57-60, assoc anat, 59-64, asst prof, 64-67; assoc prof, 67-70, PROF ANAT, COL MED, UNIV ARIZ, 70- Concurrent Pos: USPHS career develop award, 64; mem, Neurol B Study Sect, Nat Inst Neurol Dis & Stroke, 72-74, chmn, 74-76; mem clin teaching fac, Dept Neurol Surg, St Joseph's Hosp & Med Ctr, Phoenix, 74- Mem: Am Asn Anat; Am Neurol Asn; Soc Neurosci. Res: Human and comparative neuroanatomy and neuroembryology. Mailing Add: Dept of Anat Univ of Ariz Col of Med Tucson AZ 85724

ANGHILERI, LEOPOLDO JOSE, b Lujan-Buenos Aires, Argentina, Aug 22, 28; m 54; c 1. RADIOCHEMISTRY, RADIOPHARMACOLOGY. Educ: Univ Buenos Aires, Dr chem, 57. Prof Exp: Researcher, Inst Radium, Univ Paris, 64-66; Ger Cancer Res Ctr, Inst Nuclear Med, 66-67; dept radiol sci, Med Inst, Johns Hopkins Univ, 67-69; asst prof radiol, Med Ctr, Univ Colo & res chemist, Vet Admin Hosp, Denver, 69-70; RESEARCHER, INST TUMOR RES, RUHR UNIV, 70- Mem: Am Chem Soc; Soc Nuclear Med. Res: Development of new radiopharmaceuticals; cancer research. Mailing Add: Tumor Res Clin Hufelandstrasse 55 43 Essen West Germany

ANGIER, DEREK JOHN, b London, Eng, May 17, 29; US citizen; m 50; c 2. POLYMER CHEMISTRY. Educ: Univ London, BSc, 53, PhD(chem), 56. Prof Exp: Res chemist, Gen Tire & Rubber Co, 56-58; res dir, Am Foam Rubber Corp, 58-60; sr scientist, Ethicon Inc, Johnson & Johnson Co, 60-63; res dir, Radiation Appln, Inc, 63-64; SR STAFF ADV RES PLANNING, EXXON RES & ENG CO DIV,

STANDARD OIL CO, NJ, 64- Concurrent Pos: Lectr, Northern Polytech Inst, Univ London, 54-56. Mem: Am Chem Soc. Res: Production and properties of block and graft polymers, especially by mechanochemical and irradiation techniques; research planning and coordination; plasma research. Mailing Add: Exxon Res & Eng Co PO Box 45 Linden NJ 07036

ANGIER, ROBERT BRUCE, b Litchfield, Minn, Mar 24, 17; m 45; c 2. ORGANIC CHEMISTRY. Educ: Hamline Univ, BS, 40; Univ Nebr, MS, 42, PhD(org chem), 44. Prof Exp: Res chemist, 44-56, res assoc, 56-62, GROUP LEADER, LEDERLE LABS DIV, AM CYANAMID CO, 62- Mem: Am Chem Soc. Res: Synthetic organic chemistry; folic acid; synthesis of pteroylglutamic acid; pteridine, pyrimidine and purine chemistry; antineoplastic, antifungal and antiviral compounds. Mailing Add: Lederle Labs Am Cyanamid Co Pearl River NY 10965

ANGINO, ERNEST EDWARD, b Winsted, Conn, Feb 16, 32; m 54; c 2. GEOCHEMISTRY. Educ: Lehigh Univ, BS, 54; Univ Kans, MS, 58, PhD(geochem), 61. Prof Exp: Instr, Univ Kans, 61-62; asst prof oceanog, Tex A&M Univ, 62-65; res assoc & head div geochem, Kans State Geol Surv, 65-70, assoc dir, 70-72; PROF GEOL & CHMN DEPT, UNIV KANS, 72- Concurrent Pos: Assoc prof civil eng, Univ Kans 67-71, prof, 71-; mem, US Nat Comt Geochem, Nat Acad Sci-Nat Res Coun, 71-76, co-chmn, Comt Geochem Rel Health & Dis, 74-76; assoc ed, Soc Econ Paleontologists & Mineralogists, 73-; vchmn tech adv comt, Comt Res & Develop, Fed Power Comn, 75- Honors & Awards: Antarctic Serv Medal, US Dept Defense. Mem: Geochem Soc (secy, 70-76); Am Chem Soc; Geol Soc Am; Soc Econ Paleontologists & Mineralogists; Soc Environ Geochem & Health. Res: Trace element complexing in natural waters, sediment-water interactions. Mailing Add: Dept of Geol Univ of Kans Lawrence KS 66045

ANGLE, WILLIAM DODGE, b Lincoln, Nebr, Jan 28, 26; m 52; c 1. INTERNAL MEDICINE. Educ: Univ Nebr, BS, 46, MS, 65; Harvard Univ, MD, 48. Prof Exp: Asst prof, 53-59, ASSOC PROF INTERNAL MED, COL MED, UNIV NEBR, OMAHA, 59-, ASSOC PROF PHYSIOL, COL & ASSOC PROF ELEC ENG, UNIV, 70- Res: Electrocardiography. Mailing Add: Clarkson Hosp Omaha NE 68105

ANGLEMIER, ALLEN FRANCIS, b Tiffin, Ohio, Sept 10, 26; m 52; c 4. FOOD SCIENCE. Educ: Fresno State Col, BS, 53; Ore State Univ, MS, 55, PhD(animal nutrit, physiol), 57. Prof Exp: Instr food technol, 56-58, from asst prof to assoc prof food sci, 58-73, PROF FOOD SCI, ORE STATE UNIV, 73- Concurrent Pos: Res grants, NIH, 62-69 & Bur Com Fisheries, 69-71. Mem: AAAS; Inst Food Technol; Am Meat Sci Asn. Res: Chemistry and physiology of muscle; food protein systems. Mailing Add: Dept of Food Sci Ore State Univ Corvallis OR 97331

ANGLETON, GEORGE M, b Pontiac, Mich, May 25, 27; m 55; c 5. RADIATION BIOLOGY, BIOMETRICS. Educ: Mich State Univ, BS, 49; Univ Rochester, MS, 52; Med Col Va, PhD(biophys, biomet), 61. Prof Exp: Asst health physics, Los Alamos Sci Lab, 50-55, staff mem, 55-57; instr biophys, Med Col Va, 61; asst prof, 62-65, ASSOC PROF BIOSTATIST & RADIATION BIOL, COLO STATE UNIV, 65-, HEAD BIOMETRY & RADIATION BIOPHYS, COLLAB RADIOL HEALTH LAB, 62- Concurrent Pos: Fel biomath, NC State Col, 61-62; prin investr radiation recovery studies & thermoluminescent dosimetry studies, US Air Force Contracts, 65-69 & 74; consult, Martin Co, 61-62, Los Alamos Sci Lab, 64-65, Datametrics, 67- & US Air Force Acad, 70-74. Mem: Am Statist Asn; Biomet Soc; Health Physics Soc; Radiation Res Soc. Res: Theoretical radiation biology; mathematical biology; computer applications. Mailing Add: Dept of Radiation Biol Colo State Univ Ft Collins CO 80523

ANGLIN, J HILL, JR, b Iowa Park, Tex, Oct 16, 22; m 48; c 2. ORGANIC BIOCHEMISTRY. Educ: York Col, BS, 47; Univ Nebr, MS, 50; Univ Okla, PhD(biochem), 63. Prof Exp: Teacher pub schs, NMex, 50-59; asst biochem, 59-62, instr biochem & dermat, 62-63, asst prof, 64-69, ASSOC PROF BIOCHEM & DERMAT, MED SCH, UNIV OKLA, 69- Mem: Biophys Soc; Soc Invest Dermat; fel Am Inst Chem. Res: Effect of ultraviolet light on biochemistry of skin; photochemistry; metabolism, especially of photo products, steriochemistry and chemistry of tungsten; affinity chromatography of glyco-proteins. Mailing Add: Dept Biochem & Molecular Biol OUHSC Box 26901 Oklahoma City OK 73190

ANGONA, FRANK ANTHONY, b Los Angeles, Calif, Apr 15, 20; m 44; c 4. GEOPHYSICS, ACOUSTICS. Educ: Univ Calif, Los Angeles, BA, 42, MA, 50, PhD(physics), 52. Prof Exp: Res asst, Univ Calif, Los Angeles, 50-52; technologist seismic explor, Socony Mobile Oil Co, 52-53, sr res technologist, 53-70, RES ASSOC, MOBIL RES & DEVELOP CORP, 70- Mem: Acoustical Soc Am; Soc Explor Geophys; Sigma Xi (pres, Sci Res Soc Am, 58). Res: Transmission of sound in rarified gases; propagation of seismic waves in the earth with seismic and Loggina applications. Mailing Add: Mobil Res & Develop Corp PO Box 900 Dallas TX 75221

ANGOTTI, RODNEY, b Ellsworth, Pa, Apr 14, 37; m 59; c 2. MATHEMATICS. Educ: Univ Pittsburgh, BS, 59, PhD(math), 63. Prof Exp: Asst prof math, Univ Akron, 63-64 & State Univ NY Buffalo, 64-67; asst head dept, 69-71, ASSOC PROF MATH, NORTHERN ILL UNIV, 67-, ASST DEAN, COL LIBERAL ARTS & SCI, 71- Mem: Am Math Soc; Math Asn Am; Ital Math Union. Res: Geometric invariant theory; algebraic and topological foundations of geometry. Mailing Add: Dept of Math Northern Ill Univ DeKalb IL 60115

ANGSTADT, CAROL NEWBORG, b Gladwyne, Pa, Oct 23, 35; m 58; c 2. BIOCHEMISTRY, BIONUCLEONICS. Educ: Juniata Col, BS, 57; Purdue Univ, West Lafayette, PhD(biochem), 62. Prof Exp: ASST PROF BIOCHEM, HAHNEMANN MED COL, PA, 62- Concurrent Pos: Consult, Thomas Jefferson Univ, 76. Mailing Add: Hahnemann Med Col & Hosp Philadelphia PA 19102

ANGSTADT, HOWARD PERCIVAL, organic chemistry, see 12th edition

ANGSTADT, ROBERT B, b Kutztown, Pa, Mar 13, 37; m 62; c 2. ANIMAL BEHAVIOR, ANIMAL PHYSIOLOGY. Educ: Ursinus Col, BS, 59; Cornell Univ, MS, 61, PhD(ethology), 69. Prof Exp: Asst prof, 67-74, ASSOC PROF BIOL, LYCOMING COL, 74- Mem: Animal Behav Soc. Res: Behavior of vertebrates; neural control of behavior. Mailing Add: Dept of Biol Lycoming Col Williamsport PA 17701

ANGUS, THOMAS ANDERSON, b Toronto, Ont, Sept 19, 15; m 50; c 2. INSECT PATHOLOGY. Educ: Univ Toronto, BSA, 49, MSA, 50; McGill Univ, PhD, 55. Prof Exp: Res scientist, 49-70, assoc dir, 70-75, DIR INSECT PATH RES INST, CAN DEPT ENVIRON, CAN FORESTRY SERV, 75- Mem: Soc Invert Path; Can Soc Microbiol; Entom Soc Can. Res: Bacterial diseases of insects. Mailing Add: Insect Path Res Inst PO Box 490 Sault Ste Marie ON Can

ANHORN, VICTOR JOHN, b Winner, SDak, Aug 12, 10; m 41; c 1. PHYSICAL CHEMISTRY. Educ: Huron Col, AB, 35; Purdue Univ, MS, 37, PhD(phys chem), 40. Prof Exp: Res chemist, Gulf Res & Develop Co, 40-53; phys chemist, Southwest Res Inst, 53-54; MGR ORG CHEM PROCESS RES, GOODYEAR TIRE & RUBBER CO, 54- Concurrent Pos: Teacher war training courses, Pa State Col, 42-43; mem staff, Nat Defense Res Comt. Mem: Am Chem Soc; Am Inst Chem Eng. Res: Catalytic treatment of petroleum crudes and fractions; plastic horizons; scientific glass blowing; process development. Mailing Add: Goodyear Tire & Rubber Co 142 Goodyear Blvd Akron OH 44316

ANIKOUCHINE, WILLIAM A, b Cleveland, Ohio, Dec 23, 29; m 54; c 2. OCEANOGRAPHY, GEOLOGY. Educ: Ohio State Univ, BS, 52; Univ Wash, MS, 61, PhD(oceanog), 66. Prof Exp: Res oceanogr, Environ Sci Serv Admin, Wash, 66-67; res oceanogr, 67-72, VPRES, OCEANOG SERV, INC, 72- Concurrent Pos: Marine Geol Expert, UN Develop Prog, 72-73. Mem: AAAS; Geochem Soc; NY Acad Sci; Am Soc Limnol & Oceanog; Geol Soc Am. Res: Physical and chemical processes in compacting marine sediments; geochemistry of economic minerals on land and in the ocean. Mailing Add: 1636 Hillcrest Rd Santa Barbara CA 93103

ANILINE, ORM, physical organic chemistry, see 12th edition

ANIMALU, ALEXANDER OBIEFOKA ENUKORA, b Oba, Biafra, Aug 28, 38; m 65; c 2. PHYSICS, MATHEMATICS. Educ: Univ Ibadan, BS, 62, PhD(physics), 65; Cambridge Univ, BA, 63, MA, 70. Prof Exp: Res assoc physics, Stanford Univ, 66-67; vis prof, Univ NC, Chapel Hill, 68; asst prof, Univ Mo-Rolla, 68-69; assoc prof, Drexel Univ, 69-74; MEM STAFF, LINCOLN LAB, MED INST TECHNOL, 74- Mem: Am Phys Soc. Res: Quantum theory of solids, particularly metals using the pseudopotential method. Mailing Add: Lincoln Lab Med Inst of Technol Lexington MA 02173

ANKEL-SIMONS, FRIDERUN ANNURSEL, b Krofdorf, WGer, Oct 23, 33; m 72; c 1. PRIMATOLOGY, COMPARATIVE ANATOMY. Educ: Univ Giessen, Dr rer nat(marine biol), 60; Univ Zurich, Habil(phys anthrop), 68. Prof Exp: Res grant primatol, Ger Res Team, Max Planck Inst Brain Res, Univ Giessen, 60-63; res asst & lectr, Inst Anthrop, Univ Zurich, 63-69; instr human anat, Univ Kiel Med Sch, 70-71; ASSOC CUR, PEABODY MUS, YALE UNIV, 71- Concurrent Pos: Res asst paleont, Univ Giessen, 60; Wenner-Gren Found res award, 70; Boise Fund res travel grant, Oxford Univ, 72; lectr human anat, Brown Univ, 72 & Yale Univ, 73-74; Leakey Found res grant, 73-75. Mem: Int Primatol Soc; Soc Vert Paleont. Res: Locomotor behavior and comparative anatomy of living and fossil primates and man; investigations of the anatomy of vertebrae, of hands and feet and of the limbs in various extinct primates. Mailing Add: 52 Morris St Hamden CT 06517

ANKENBRANDT, CHARLES MARTIN, b Cleveland, Ohio, Aug 20, 39; m 61; c 3. EXPERIMENTAL HIGH ENERGY PHYSICS. Educ: St Louis Univ, BS, 61; Univ Calif, Berkeley, PhD(physics), 67. Prof Exp: Fel high energy physics, Lawrence Radiation Lab, Univ Calif, Berkeley, 67-68; fel, Brookhaven Nat Lab, 68-70; asst prof physics, Ind Univ, 70-73; PHYSICIST HIGH ENERGY PHYSICS, FERMI NAT ACCELERATOR LAB, 73- Mem: Am Phys Soc. Res: Hadron scattering; hyperon beam experiments; meson form factors; accelerator physics. Mailing Add: Fermi Nat Accelerator Lab PO Box 500 Batavia IL 60510

ANKENEY, JAY LLOYD, b Cleveland, Ohio, June 7, 21; m 46; c 3. MEDICINE. Educ: Ohio Wesleyan Univ, BA, 43; Western Reserve Univ, MD, 45. Prof Exp: From instr to assoc prof thoracic surg, 55-69, PROF SURG, SCH MED, CASE WESTERN RESERVE UNIV, 69-, CHMN DIV CARDIOTHORACIC SURG, 74-; DIR DIV CARDIOTHORACIC SURG, UNIV HOSPS CLEVELAND, 74- Mem: Am Asn Thoracic Surg; fel Am Col Surgeons; Am Surg Asn; Soc Thoracic Surgeons; Soc Vascular Surg. Res: Physiological aspects of cardiovascular and thoracic surgery. Mailing Add: Univ Hosps 2065 Adelbert Rd Cleveland OH 44106

ANKER, HERBERT S, b Danzig, Ger, Sept 16, 12; nat US; m 63; c 3. BIOCHEMISTRY. Educ: Univ Basel, MD, 39; Columbia Univ, PhD(biochem), 43. Prof Exp: From instr to assoc prof, 45-59, PROF BIOCHEM, UNIV CHICAGO, 59- Mem: Biophys Soc; Am Chem Soc; Am Soc Biol Chem. Res: Fat metabolism; protein synthesis. Mailing Add: Dept of Biochem Univ of Chicago Chicago IL 60637

ANLYAN, WILLIAM GEORGE, b Alexandria, Egypt, Oct 14, 25; nat US; m 48; c 3. SURGERY. Educ: Yale Univ, BS, 45, MD, 49; Am Bd Surg, dipl, 55. Prof Exp: Intern & resident surg, 49-55, assoc, 52-53, from asst prof to assoc prof surg, 54-61, dean, Sch Med, 64-69, assoc provost, Med Ctr, 69, PROF SURG, MED CTR, DUKE UNIV, 61-, VPRES HEALTH AFFAIRS, 69- Concurrent Pos: Markle scholar med sci, Duke Univ, 53-58. Mem: Soc Univ Surg; Soc Vascular Surg; Am Col Surg; AMA; Soc Clin Surg. Res: Pancreatic physiology; abnormalities in blood clotting; serotonin metabolism and thromboembolic disease. Mailing Add: Box 3701 Duke Univ Med Ctr Durham NC 27710

ANNAN, MURVEL EUGENE, b Coin, Iowa, July 11, 20; m 45. GENETICS. Educ: Univ Nebr, PhD(zool), 54. Prof Exp: Instr zool, Univ Nebr, 50-51; from asst prof to assoc prof biol, 54-59, PROF BIOL, WAGNER COL, 59- Mem: Soc Study Evoluation; Genetics Soc Am. Res: Genetics of Drosophilas. Mailing Add: Dept of Biol Wagner Col Staten Island NY 10301

ANNAND, ROBERT RODNEY, electrochemistry, see 12th edition

ANNAU, ZOLTAN, b Szeged, Hungary, May 23, 26; Can citizen; m 60. PSYCHOPHYSIOLOGY. Educ: Carleton Univ, Ont, BA, 58; McMaster Univ, MA, 60, PhD(psychol), 64. Prof Exp: Res assoc & lectr psychol, Univ Mich, 64-65; fel environ med, 65-66, lectr med psychol, 66-68, asst prof, 68-70, assoc prof environ med, 70-73, ASST PROF MED PSYCHOL, DEPT MED PSYCHOL, JOHNS HOPKINS UNIV, 73- Mem: Am Soc Neurosci; NY Acad Sci. Res: Effect of motivational variables on learning and performance and the physiological concomitants of these interactions; effect of changes in the gaseous environment on behavior. Mailing Add: Dept of Med Psychol Univ Circular Johns Hopkins Univ 34th & Charles Sts Baltimore MD 21218

ANNEAR, PAUL RICHARD, b Cedar Rapids, Iowa, Jan 19, 15; m 39; c 3. ASTRONOMY. Educ: Drake Univ, BA, 36; Case Western Reserve Univ, MS, 38; Univ Mich, PhD, 49. Prof Exp: Tutor physics & astron, Hunter Col, 40-41; from instr to assoc prof, 41-51, PROF MATH & ASTRON, BALDWIN-WALLACE COL, 51-, DIR BURRELL MEM OBSERV. Mem: Am Astron Soc; Math Asn Am. Res: Astronomical photometry; galactic structure in the constellation of Cygnus; artificial satellite tracking and orbit computation. Mailing Add: Dept of Math & Astron Baldwin-Wallace Col Berea OH 44017

ANNEGERS, JOHN HERMAN, b Sask, June 24, 17; m 42; c 2. PHYSIOLOGY. Educ: Knox Col, AB, 39; Northwestern Univ, MD, 44, PhD(physiol), 47. Prof Exp: From asst prof to assoc prof, 46-68, PROF PHYSIOL, MED SCH, NORTHWESTERN UNIV, 68- Mem: AAAS; Am Soc Exp Biol; Am Physiol Soc. Res: Physiology of intestinal absorption. Mailing Add: Northwestern Univ Med Sch 303 E Chicago Ave Chicago IL 60611

ANNETT, ROBERT GORDON, b Windsor, Ont, Feb 24, 41; m 64; c 1. BIOCHEMISTRY. Educ: Univ Windsor, BSc, 64, PhD(biochem), 68. Prof Exp: Asst prof, 68-75, ASSOC PROF BIOCHEM, TRENT UNIV, 75- Mem: Am Chem Soc; Chem Inst Can. Res: Enzymology of oxaloacetic acid. Mailing Add: Dept of Chem Trent Univ Peterborough ON Can

ANNINO, RAYMOND, b NY, Sept 5, 27; m 50; c 4. ANALYTICAL CHEMISTRY, PHYSICAL CHEMISTRY. Educ: Columbia Univ, BA, 50; Okla State Univ, PhD, 56. Prof Exp: Asst scientist, Nat Dairy Res Labs, Inc, 50-53; supvr anal res, Westvaco Chlor-Alkali Div, 55-57; assoc prof chem, Northeastern La State Col, 57-60 & Canisius Col, 60-67; sr res scientist, Foxboro Co, Mass, 67-72; PROF CHEM, CANISIUS COL, 72- Concurrent Pos: Sci adv, Food & Drug Admin, Buffalo Dist, 73- Mem: Am Chem Soc; Soc Appl Spectros; The Chem Soc. Res: Application of signal enhancement techniques to various instrumental methods; design of new chromatographic techniques and instruments; electrochemical methods of analysis. Mailing Add: Dept of Chem Canisius Col Buffalo NY 14208

ANNIS, MARTIN, b New York, NY, Apr 27, 22; m 51; c 3. COSMIC RAY PHYSICS. Educ: Mass Inst Technol, PhD(physics), 51. Prof Exp: Asst prof cosmic rays, Wash Univ, 51-53; vis prof, Univ Padua, 53-54; chief proj scientist, Allied Res Assocs, Inc, 54-58; PRES, AM SCI & ENG, INC, 58- Concurrent Pos: Consult, AEC & US Air Force, 58. Mem: Am Phys Soc; Ital Physics Soc. Res: Cloud chamber investigations of new unstable particles; application of mathematical statistics to physics; high temperature effects on materials. Mailing Add: Am Sci & Eng Inc 955 Massachusetts Ave Cambridge MA 02142

ANSARI, ALI, b Hyderabad, India, Dec 29, 34; US citizen; m 67; c 2. CLINICAL CHEMISTRY, LIPID CHEMISTRY. Educ: Osmania Univ, India, BSc, 57; Kans State Univ, MS, 63; Tex A&M Univ, PhD(biochem), 66. Prof Exp: Res assoc lipid biochem dermat, Sch Med, Univ Southern Calif, 66-71, asst clin biochemist, Med Ctr, 70-71, ASST PROF PATH, SCH MED, UNIV SOUTHERN CALIF & CHAS R DREW POSTGRAD MED SCH, 71-; CLIN CHEMIST, LOS ANGELES COUNTY-MARTIN LUTHER KING, JR GEN HOSP, 71- Mem: Am Chem Soc; Am Asn Clin Chem; AAAS; Am Oil Chem Soc. Res: Biochemistry of plasma lipids and lipoproteins; development of ultramicro methodologies involving chromatographies, radioimmunoassays and electrophoresis. Mailing Add: Martin Luther King Jr Gen Hosp 12021 S Wilmington Ave Los Angeles CA 90059

ANSBACHER, RUDI, b Sidney, NY, Oct 11, 34; m 65; c 2. OBSTETRICS & GYNECOLOGY, REPRODUCTIVE BIOLOGY. Educ: Va Mil Inst, BA, 55; Univ Va, MD, 59; Univ Mich, MS, 70. Prof Exp: Intern, Univ Va Hosp, 59-60; Med Serv Corps, US Army, 60-, physician, Richmond Mem Hosp, 60-62, surg resident, Womack Army Hosp, Ft Bragg, NC, 62-63, resident obstet-gynec, Letterman Gen Hosp, San Francisco, Calif, 63-66, mem staff, US Army Hosp, Ryukyu Islands, Okinawa, 66-68, researcher mil med & allied sci, Walter Reed Army Inst Res, Washington, DC, 68-69, teaching assoc, Dept Obstet & Gynec, Med Ctr, Univ Mich, 69-71, chief family planning & consult serv, Dept Obstet & Gynec & chief clin invest serv, 71-74, ASST CHIEF DEPT OBSTET & GYNEC & CHIEF OBSTET SERV, BROOKE ARMY MED CTR, FT SAM HOUSTON, US ARMY, 74- Concurrent Pos: Fel reprod biol, Ctr Res Reprod Biol, Univ Mich, 69-71; clin instr, Dept Obstet & Gynec, Univ Hawaii, 67-68. Honors & Awards: Chmn Award for Clin Res, Armed Forces Dist, Am Col Obstet & Gynec, 70. Mem: AMA; Asn Mil Surg US; fel Am Col Obstet & Gynec; Int Fedn Gynec & Obstet; Am Fertil Soc. Res: Reproductive biology, keying on the clinical aspects, using immunologic techniques, especially in the fields of infertility and conception control. Mailing Add: Dept of Obstet & Gynec PO Box 103 Brooke Army Med Ctr Ft Sam Houston TX 78234

ANSBACHER, STEFAN, b Frankfurt-am-Main, Ger, Jan 27, 05; nat US; m 30, 51; c 4. BIOCHEMISTRY. Educ: Univ Frankfurt, BSc, 23; Univ Geneva, MSc, 29, ScD(med chem), 33. Prof Exp: Asst, Path Inst, Univ Geneva, 20-30; res chemist, State Food Res Comn, SC, 30-31, Res Lab, Borden Co, 31-36 & Squibb Inst, 37-41; dir res & med rels, Int Vitamin Corp, NY, 41-46; dir nutrit res, Schenley Industs, Cincinnati, 46-47; sci & med consult, 47-70; PROF LIFE SCI & FUTURISM & DEAN OF MEN, IND NORTHERN GRAD SCH PROF MGT, 70- Honors & Awards: Medal, AMA, 36. Mem: AAAS; AMA; Am Soc Biol Chemists; Soc Exp Biol & Med. Res: Enzymology; hormonology; vitaminology; pharmaceuticals; precipitation of proteinaceous matter; vitamin K and analogues; para-amino-benzoic acid; animal protein factor; vitamin B12 chemotherapy of tuberculosis; animal diseases; nutrition. Mailing Add: Ansie Acres PO Box 867 Marion IN 46952

ANSBACHER, THEODORE HENRY, b New York, NY, Feb 19, 39. SOLID STATE PHYSICS. Educ: Mass Inst Technol, BS, 60; Univ Vt, PhD(physics), 68. Prof Exp: Res metallurgist, Wright Aero Div, Curtiss-Wright Corp, 61-62; asst prof physics, Worcester Polytech Inst, 66-68 & New Col, Fla, 69-73; ASST PROF PHYSICS, LAFAYETTE COL, 73- Mem: Am Asn Physics Teachers. Res: Surface physics; electrical properties of metals. Mailing Add: Dept of Physics Lafayette Col Easton PA 18042

ANSBERRY, MERLE, b Guthrie Center, Iowa, Sept 20, 07; m 52; c 4. SPEECH PATHOLOGY, AUDIOLOGY. Educ: Univ Calif, AB, 29, MA, 31; Univ Wis, PhD(speech path, audiol), 37. Prof Exp: Assoc prof speech, Ariz State Col, 37-40; asst prof, Univ Va, 40-42; chief spec rehab procedures, US Vet Admin, 46-47; assoc prof, dir speech & hearing clin & dir grad studies speech, Univ Md, 48-54; prof speech & dir speech & hearing clin, 54-65, chmn dept speech, 65, PROF SPEECH PATH & AUDIOL & DIR SPEECH & HEARING CLIN, SCH MED, UNIV HAWAII, MANOA, 66- Concurrent Pos: Chief audiol & speech corrections, US Vet Admin, 47-54; consult, Hawaii Dept Educ & Dept Health & US Army Tripler Gen Hosp, Hawaii. Mem: Fel Am Speech & Hearing Asn. Res: Auditory deprivation; genetic factors in hearing sensitivity. Mailing Add: Div of Speech Path & Audiol Univ of Hawaii at Manoa Honolulu HI 96822

ANSCHEL, JOACHIM, b Ger, Oct 15, 12; nat US; m 39; c 2. PHARMACEUTICS. Educ: Pharmaceut Inst, Leipzig, MS, 35; Columbia Univ, PhG, 36. Prof Exp: Pharmacist, J L Lascoff & Son, 35-41; mem staff, Res Labs, Warner-Lambert Res Inst, 48-66; sr res pharmacist, 66-67, SR STAFF SCIENTIST STERILE PROD RES, CIBA-GEIGY PHARMACEUT CO, 67- Concurrent Pos: Lectr, Col Pharm, Rutgers Univ, 63- Mem: Am Pharmaceut Asn. Res: Pharmaceutical research. Mailing Add: Ciba-Geigy Pharmaceut Co 556 Morris Ave Summit NJ 07901

ANSCHEL, MORRIS, b Petach Tikvah, Israel, May 27, 40; US citizen; m 70. ORGANIC CHEMISTRY, COMPUTER SCIENCE. Educ: City Col New York, BS, 62; Lehigh Univ, PhD(org chem), 67. Prof Exp: TECH DIR CHEM & CHEM PROCESS DEVELOP LAB & STAFF CHEMIST, ADVAN PROCESS DEVELOP LABS, IBM CORP, 66-, ADV CHEMIST, 69- Honors & Awards: Outstanding Contribution Award, 69 & Invention Achievement Awards, IBM Corp, 72 & 75. Mem: Am Chem Soc. Res: Electrochemical and photochemical hardcopy systems; synthesis of photosensitive and organoconductive materials; synthesis of EMC, anti-glare, hostile environment coatings; development of novel printed circuitry and electrochemical display systems; dielectric coolant synthesis and applications; gas panel and ink jet technology. Mailing Add: IBM Corp Thomas J Watson Res Ctr Yorktown Heights NY 10598

ANSCHEL, STEVEN WILLIAM, b New York, NY, July 18, 42; m 63; c 1. NEUROPHYSIOLOGY. Educ: Ind Univ, BA, 64; Univ Md, MS, 73, PhD(zool), 75. Prof Exp: RES ASSOC NEUROPHYSIOL, YERKES REGIONAL PRIMATE RES CTR, EMORY UNIV, 75- Mem: Soc Neurosci. Res: Neurophysiological aspects of sexual and agonistic behavior in the male rhesus monkey. Mailing Add: Yerkes Regional Primate Res Ctr Emory Univ Atlanta GA 30322

ANSCOMBE, FRANCIS JOHN, b Elstree, Eng, May 13, 18; m 54; c 4. STATISTICS. Educ: Cambridge Univ, BA, 39, MA, 43. Prof Exp: With Brit Ministry Supply, 40-45 & Rothamsted Exp Sta, Eng, 45-47; lectr math, Cambridge Univ, 48-56; res assoc, Princeton Univ, 53-54, from assoc prof to prof, 56-63; PROF STATIST, YALE UNIV, 63- Concurrent Pos: Corresp consult, Higher Coun Sci Invests, Madrid, 53; vis assoc prof, Univ Chicago, 59-60. Mem: Economet Soc; fel Inst Math Statist; fel Am Statist Asn; Soc Indust & Appl Math; Royal Statist Soc. Res: Theory and practice of statistical method; statistical computing. Mailing Add: Dept of Statist Yale Sta Box 2179 New Haven CT 06520

ANSEL, HOWARD CARL, b Cleveland, Ohio, Oct 18, 33; m 60; c 3. PHARMACY. Educ: Univ Toledo, BS, 55; Univ Fla, MS, 57, PhD(pharmaceut sci & chem), 59. Prof Exp: Asst prof pharm, Univ Toledo, 59-62; from asst prof to assoc prof, 62-70, PROF PHARM, UNIV GA, 70-, HEAD DEPT, 68- Res: Pharmaceutical sciences; hemolytic effect of various chemical agents; binding of antibacterial preservatives by macromolecules; relationship between bacteriolysis, hemolysis and chemical structure. Mailing Add: Sch of Pharm Univ of Ga Athens GA 30601

ANSELL, JULIAN SAMUEL, b Portland, Maine, June 30, 22; m 51; c 5. UROLOGY. Educ: Bowdoin Col, BA, 47; Tufts Univ, MD, 51; Univ Minn, Minneapolis, PhD(urol), 59. Prof Exp: Instr urol, Med Col & mem staff, Univ Hosp, Univ Minn, Minneapolis, 56-59; from asst prof to assoc prof, 59-65, HEAD DIV UROL, SCH MED, UNIV WASH & AFFIL TEACHING HOSPS, 59-. PROF & CHMN DEPT, 65- Concurrent Pos: Chief urol, Vet Admin Hosp, Minneapolis, 56-59; urologist in chief, King County Hosp & chief consult, Vet Admin Hosp, Seattle, 59- Mem: AAAS; AMA; Am Col Surg; Am Urol Asn. Res: Physiology of urinary apparatus; potassium metabolism; pediatric urology. Mailing Add: Dept of Urol Univ of Wash Sch of Med Seattle WA 98195

ANSELME, JEAN-PIERRE L M, b Port-au-Prince, Haiti, Sept 22, 36; US citizen; m 60; c 3. ORGANIC CHEMISTRY. Educ: St Martial Col, Haiti, BA, 55; Fordham Univ, BS, 59; Polytech Inst Brooklyn, PhD(chem), 63. Prof Exp: Res fel org chem, Polytech Inst Brooklyn, 60-63, res assoc, 63, sr instr, 65; NSF fel, Inst Org Chem, Univ Munich, 64-65; from asst prof to assoc prof, 65-69, PROF ORG CHEM, UNIV MASS, BOSTON, 70- Concurrent Pos: AP Sloan Found fel, 69-71; founder & ed, Org Prep & Procedures, 67-70; founder & ed, Org Prep & Procedures Int, 71-; vis prof, Kyushu Univ, Japan, 72. Mem: Am Chem Soc; Brit Chem Soc; Ger Chem Soc; Chem Soc Japan; fel Japan Soc Promotion Sci. Res: Synthetic organic chemistry; reaction mechanisms; meso-ionic systems; N-nitrenes; diazo-alkanes; azides; thionylhydrazines; 3-pyrazolidinones; sulfinyl and sulfenylhydratines. Mailing Add: Dept of Chem Univ of Mass Harbor Campus Boston MA 02125

ANSELMI, ROBERT THEODORE, organic chemistry, see 12th edition

ANSELMO, VINCENT C, b New York, NY, July 29, 30; m 54; c 5. RADIO CHEMISTRY, PHYSICAL CHEMISTRY. Educ: Fordham Univ, BS, 51, MS, 59; Univ Kans, PhD(chem), 61. Prof Exp: Instr pharm, Fordham Univ, 53-57; asst prof chem, John Carroll Univ, 61-64; from asst prof to assoc prof, 64-74, PROF CHEM & CHMN DEPT, NMEX HIGHLANDS UNIV, 74- Mem: AAAS; Am Chem Soc. Res: Hot-atom chemistry of phosphorus and its oxy-anions; general inorganic chemistry of phosphorus oxy-anions. Mailing Add: Dept of Chem NMex Highlands Univ Las Vegas NM 87701

ANSELONE, PHILIP MARSHALL, b Tacoma, Wash, Feb 8, 26; m 51; c 1. MATHEMATICAL ANALYSIS. Educ: Col Puget Sound, BS, 49, MA, 50; Ore State Univ, PhD(math), 57. Prof Exp: Head math anal group, Hanford Atomic Prod Oper, 51-54; res assoc, Johns Hopkins Radiation Lab, 54-58; assoc prof math, Math Res Ctr, US Army, Wis, 58-63; PROF MATH, ORE STATE UNIV, 64- Concurrent Pos: Vis prof, Math Res Ctr, US Army, Wis, 66-67 & Mich State Univ, 70-71. Mem: Am Math Soc; Math Asn Am; Soc Indust & Appl Math. Res: Integral equations; approximation theory. Mailing Add: Dept of Math Ore State Univ Corvallis OR 97331

ANSEVIN, ALLEN THORNBURG, b Springfield, Ohio, Nov 22, 28; m 61; c 1. BIOPHYSICS. Educ: Earlham Col, AB, 51; Univ Pittsburgh, MS, 58, PhD(biophys), 61. Prof Exp: Res asst pharmacol, Christ Hosp Int Med Res, 53-56; res assoc biochem, Rockefeller Univ, 61-64; asst physics, Univ Tex, 64-70; asst prof, Grad Sch Biomed Sci, 65-70; ASSOC PHYSICIST, UNIV TEX, MD ANDERSON HOSP & TUMOR INST, 70-, ASSOC PROF BIOPHYS, GRAD SCH BIOMED SCI, 70- Mem: AAAS; Biophys Soc; Am Chem Soc. Res: Virus structure; protein-protein associations; ultracentrifuge methods; nucleoprotein structure; radiation biology; DNA thermal denaturation. Mailing Add: Dept Physics Univ Tex MD Anderson Hosp & Tumor Inst Tex Med Ctr Houston TX 77025

ANSEVIN, KRYSTYNA D, b Warsaw, Poland, Apr 1, 25; m 61; c 1. BIOLOGY. Educ: Jagiellonian Univ, BS & MS, 50; Univ Pittsburgh, PhD(biol), 61. Prof Exp: Asst biol of tumor cells, Inst Oncol, Poland, 55-57; asst cell biol, Univ Pittsburgh, 58-61; fel develop biol, Med Col, Cornell Univ, 62-63 & Columbia Univ, 63-65; lectr, 65-66, asst prof, 66-71, ASSOC PROF BIOL, RICE UNIV, 71- Concurrent Pos: Nat Cancer Inst fel, 64-65. Mem: AAAS; Tissue Cult Asn. Res: Cell differentiation; induction in development and regeneration; histogenesis. Mailing Add: Dept of Biol Rice Univ Houston TX 77001

ANSFIELD, FRED JOSEPH, b Milwaukee, Wis, Aug 30, 10; m 40; c 2. ONCOLOGY. Educ: Univ Wis, BS, 31, MD, 33. Prof Exp: From instr to assoc prof, 57-64, PROF CLIN ONCOL, SCH MED, UNIV WIS-MADISON, 64- Concurrent Pos: Consult, Vet Admin Hosp, 60- Mem: AAAS; AMA; NY Acad Sci; Asn Cancer Res. Res: Cancer chemotherapy in man employing new drugs alone, in combination with other drugs, and in combination with x-ray therapy. Mailing Add: Div of Clin Oncol Univ of Wis Hosps Madison WI 53706

ANSHEL, MICHAEL, b New York, NY, Nov 2, 41; m 65; c 1. MATHEMATICS, COMPUTER SCIENCE. Educ: Adelphi Univ, BA, 63, MS, 65, PhD(math), 67. Prof Exp: Asst prof math, Polytech Inst Brooklyn, 66-67 & Univ Ariz, 67-68; systs analyst, Lambda Corp, 68; asst prof, 68-73, ASSOC PROF COMPUT SCI, CITY COL NEW YORK, 73- Concurrent Pos: Consult, Princeton-Bethesda, 68 & Mt Sinai Sch Med, 75-; reviewer, Math Rev, Am Math Soc, 75- Mem: Am Math Soc; Math Asn Am;

Ann Symbolic Logic; Sigma Xi; Asn Comput Mach. Res: Group-theoretic decisions and problems and their relation to the theory of computation; applications of algebra and logic in the computer sciences. Mailing Add: 1140 Fifth Ave New York NY 10028

ANSON, BARRY JOSEPH, anatomy, deceased

ANSON, FRED (COLVIG), b Los Angeles, Calif, Feb 17, 33; m 59; c 2. ELECTROANALYTICAL CHEMISTRY. Educ: Calif Inst Technol, BS, 54; Harvard Univ, AM, 55, PhD(chem), 57. Prof Exp: From instr to assoc prof, 57-68, PROF CHEM, CALIF INST TECHNOL, 68- Concurrent Pos: Guggenheim fel, 64; Alfred P Sloan res fel, 65-69; Fulbright scholar, 72. Mem: Am Chem Soc. Res: Kinetics of electrode reactions; mechanisms of electrode processes; chemical education. Mailing Add: Div of Chem & Chem Eng Calif Inst of Technol Pasadena CA 91125

ANSPAUGH, BRUCE EDWARD, b Thermopolis, Wyo, Sept 26, 33; m 54. PHYSICS. Educ: Nebr Wesleyan Univ, BA, 55; Univ Nebr, MA, 58, PhD(physics), 65. Prof Exp: Asst physics, Univ Nebr, 55-64; RES ENGR, JET PROPULSION LAB, CALIF INST TECHNOL, 64- Mem: Am Phys Soc. Res: Radiation damage; solid state physics. Mailing Add: Jet Propulsion Lab Calif Inst of Technol Pasadena CA 91103

ANSPAUGH, LYNN RICHARD, b Rawlins, Wyo, May 25, 37; m 65; c 2. ENVIRONMENTAL SCIENCES. Educ: Nebr Wesleyan Univ, BA, 59; Univ Calif, Berkeley, MBioradiol, 61, PhD(biophys), 63. Prof Exp: From biophysicist to group leader environ sci, 63-75, PROJ LEADER ENVIRON EFFECTS GEOTHERMAL ENERGY, LAWRENCE LIVERMORE LAB, UNIV CALIF, 75- Mem: AAAS. Res: Study of the environmental effects of utilizing geothermal energy resources; experimental study of the resuspension of pollutant aerosols with emphasis on plutonium. Mailing Add: Lawrence Livermore Lab PO Box 808 L-523 Livermore CA 94550

ANSPON, HARRY DAVIS, b Washington, DC, Sept 25, 17; m 56; c 2. ORGANIC CHEMISTRY. Educ: Univ Md, BS, 39, PhD(org chem), 42. Prof Exp: Asst, Univ Md, 42-46; res chemist, US Rubber Co, NJ, 46-48; res chemist, Cent Res Labs, Gen Aniline & Film Corp, 48-56, res fel, 56-58; plant chemist, Acetylene Chem Plant, Ky, 58-59; sect leader, Res Ctr, Spencer Chem Co, 59-68; mgr prod-process develop, 68-69, TECH DIR, PLASTICS DEVELOP DEPT, USS CHEM DIV, UNITED STATES STEEL CORP, 69- Concurrent Pos: With Nat Defense Res Coun. Mem: Am Chem Soc; Soc Plastics Eng. Res: Hydrogenation and dehydrogenation of pyrethrosin; organic polymers; organic synthesis; plastics. Mailing Add: Plastic Develop Dept USS Chem Div US Steel Corp 1820 Grant Bldg Pittsburgh PA 15230

ANSTEY, ROBERT L, b Creston, Iowa, July 15, 21; m; c 2. GEOGRAPHY, CLIMATOLOGY. Educ: Univ Nebr, AB, 47, MA, 48; Univ Md, PhD, 57. Prof Exp: Geogr, US Bur Census, DC, 48-49, Off Qm Gen, 49-52 & US Army Natick Labs, 54-70; assoc prof, 70-74, PROF GEOG, FRAMINGHAM STATE COL, 74- Concurrent Pos: Vis lectr, Clark Univ, 57-58 & Mass State Col Framingham, 58; consult, Sylvania Elec Prod, Inc, 60-61. Mem: Asn Am Geog; Nat Coun Geog Educ. Res: Applications of critical data in climatology and military geography to problems of item design, test, issue and storage. Mailing Add: Dept of Geog Framingham State Col Framingham MA 01701

ANSTEY, THOMAS HERBERT, b Victoria, BC, Dec 27, 17; m 45; c 3. GENETICS, PLANT BREEDING. Educ: Univ BC, BSA, 41, MSA, 43; Univ Minn, PhD(plant breeding), 51. Prof Exp: Horticulturist, Exp Farm, Agassiz, BC, 46-49, sr horticulturist, 49-53, supt exp sta, Summerland, 53-59, dir res sta, Lethbridge, 59-69, ASST DIR GEN WESTERN RES BR, CENT EXP FARM, CAN DEPT AGR, OTTAWA, 69- Mem: Can Soc Hort Sci; Agr Inst Can (pres, 70-71); Genetics Soc Can. Res: Genetics and breeding for strawberries and broccoli. Mailing Add: Res Br Cent Exp Farm Ottawa ON Can

ANSUL, GERALD R, b Philadelphia, Pa, July 17, 25; m 58; c 3. ORGANIC CHEMISTRY. Educ: Temple Univ, AB, 50; Pa State Univ, PhD(org chem), 54. Prof Exp: Res chemist, Marshall Lab, Fabrics & Finishes Dept, 54-58, chem assoc, 58-60, mkt specialist, 60-62, proj supvr, Indust Sales Div, 62-64, tech coord, Wynnewood Automotive & Indust Sales Div, 64-65, Lincolnwood Automotive & Indust Prod, 65-68, mgr insulation sales, Fabrics & Finishes Dept, 68-69, MGR NEW PROD DEVELOP & ELEC SALES, INDUST PROD DIV, FABRICS & FINISHES DEPT, E I DU PONT DE NEMOURS & CO, INC, 70- Mem: AAAS; Am Chem Soc. Res: Organosilicon chemistry; urethane chemistry; chemistry of the vinyl dioxolanes; automotive finishes; diallyl phthalate pre-pregs and laminates; wood furniture finishes; adhesives; sealants and paper coatings; polyimide resins; wire enamels; epoxy pre-pregs; rigid and flexible printed circuitry. Mailing Add: 17 Stonecrop Rd Northminster Wilmington DE 19810

ANTAL, JOHN JOSEPH, b Taylor, Pa, Apr 23, 26; m 55; c 2. EXPERIMENTAL PHYSICS. Educ: Univ Scranton, BS, 48; St Louis Univ, MS, 49, PhD(physics), 52. Prof Exp: Res physicist, Watertown Arsenal Lab, US Army, 52-53, ord mat res off, 53-62, mat res agency, 62-67, RES PHYSICIST, MAT & MECH RES CTR, US ARMY, 67- Concurrent Pos: Guest assoc physicist, Brookhaven Nat Lab, 53-65; Secy of Army res & study fel, 61-62. Mem: Am Phys Soc; Am Crystallog Asn; Am Soc Nondestructive Test. Res: Thermal neutron spectroscopy; digital instrumentation. Mailing Add: Mat Sci Div Army Mat & Mech Res Ctr Watertown MA 02172

ANTCLIFFE, GAULT ANDERSON, b Adelaide, SAustralia, Jan 12, 37; m 61; c 2. SOLID STATE PHYSICS. Educ: Univ Adelaide, BSc, 58, Hons, 59, PhD(solid state physics), 64. Prof Exp: George Murray vis scholar, Clarendon Lab, Oxford Univ, 64-65; MEM TECH STAFF, PHYSICS RES LAB, TEX INSTRUMENTS INC, 65- Res: Band structure of semiconductors using Shubnikov de Haas effect; infrared properties of p-n junction diode lasers; imaging with charge coupled devices. Mailing Add: Advan Technol Lab Tex Instruments Box 5936 MS134 Dallas TX 75222

ANTE, ROBERT, b Covington, Ky, June 17, 37; m 73. ECONOMIC GEOGRAPHY, GEOGRAPHY OF CENTRAL ASIA. Educ: Univ Cincinnati, BA, 59; Columbia Univ, MA, 62, PhD(econ geog), 69. Prof Exp: Asst prof geog, Newark State Col, 63-64; instr, 64-67, ASST PROF GEOG, QUEENS COL, NY, 70- Mem: Mongolia Soc; Asn Am Geogrs; Am Geog Soc; Asn Asian Studies. Res: Economic development of the Mongolian People's Republic, China and Central Asia with special emphasis upon their population geography, regional development and social development. Mailing Add: Queens Col City Univ of New York 33 W 42nd New York NY 10036

ANTELMAN, GORDON RANDOLPH, statistics, see 12th edition

ANTHES, HARRISON INMAN, organic chemistry, see 12th edition

ANTHES, JOHN ALLEN, b Janesville, Wis, Aug 18, 13; m 38; c 3. ORGANIC CHEMISTRY. Educ: Univ Minn, BChE, 34, PhD(org chem), 39. Prof Exp: Asst, Univ Minn, 35-38; res chemist, Union Oil Co, Calif, 39-40; contracting off rep, Ord Dept, US Army, 41-42, asst chief spec mat sect, Manhattan Dist, 43-45; res chemist, Am Cyanamid Co, Conn, 46-48, asst chief chemist, Pa, 48-50, chief chemist, 50-53; proj engr, 53-58, asst dir res & develop, 58-65, MGR RES, DRAVO CORP, 65- Mem: AAAS; Am Chem Soc; Am Inst Chem Eng; Am Inst Min, Metall & Petrol Eng; NY Acad Sci. Res: Ore agglomeration processes; compounds related to perinaphthenone; synthesis of lubricating oil additives and of ion exchange resins; esterification of high boiling alcohols; dehydration of maleic acid; process metallurgy. Mailing Add: Dravo Corp Pittsburgh PA 15222

ANTHES, RICHARD ALLEN, b St Louis, Mo, Mar 9, 44; m 66; c 2. METEOROLOGY. Educ: Univ Wis, BS, 66, MS, 67, PhD(meteorol), 70. Prof Exp: Res meteorologist, Nat Hurricane Res Lab, Nat Oceanic & Atmospheric Admin, 68-71; asst prof, 71-75, ASSOC PROF METEOROL, PA STATE UNIV, 75- Concurrent Pos: Consult, Atmospheric Sci Lab, White Sands Missile Range, 73- & Nat Weather Serv, 73-; mem stormfury panel, Comt Atmospheric Sci, Nat Res Coun, 75- Mem: Am Meteorol Soc; Sigma Xi. Res: Numerical modeling of atmospheric phenomena, especially on the mesoscale; hurricane modeling and parameterization of physical processes in models. Mailing Add: Dept of Meteorol 543 Deike Bldg Pa State Univ University Park PA 16802

ANTHOLINE, WILLIAM E, b Milwaukee, Wis, July 1, 43; m 65; c 3. PHYSICAL BIOCHEMISTRY. Educ: Univ Wis-Madison, BS, 65; Iowa State Univ, PhD(phys chem), 71. Prof Exp: Fel phys biochem, Radiation Biol & Biophys Div, Med Col Wis, 72-74; NIH FEL & INSTR PHYS BIOCHEM, RADIATION BIOL & BIOPHYS DIV, NAT CANCER INST, MED COL WIS, 75- Concurrent Pos: Fac res specialist, Biochem Sect, Univ Wis-Milwaukee, 75-; lectr, Gordon Res Conf Magnetic Resonance in Biol & Med, 74. Res: Applying physical chemical principles in order to better understandthe origin and interaction of paramagnetic compounds found in tissues, especially tumor tissue. Mailing Add: Med Col Wis Co Med Complex 8700 W Wisconsin Ave Milwaukee WI 53226

ANTHON, EDWARD W, b Springville, Utah, Apr 20, 12; m 39. ENTOMOLOGY. Educ: Utah State Agr Col, BS, 35, MS, 39. Prof Exp: Entomologist & supvr mosquito control, Bur Entom & Plant Quarantine, Utah, 33-34, rodent control foreman, Biol Surv, 36, state dir Mormon Cricket Control, SDak, 38, rodent control foreman, Biol Surv, Utah, 39-41, res entomologist, Bur Entom & Plant Quarantine, Calif, 44-47, RES ENTOMOLOGIST, TREE FRUIT RES CTR, USDA & WASH STATE UNIV, WENATCHEE, 47- Mem: Fel Entom Soc Am. Res: Control of stone fruit insects; vector virus research of stone fruits; integrated control on prunes; pheromone studies on peach twig borer. Mailing Add: Tree Fruit Res Ctr Wash State Univ Rte 4 Wenatchee WA 98801

ANTHONISEN, NICHOLAS R, b Boston, Mass, Oct 12, 33; m 57; c 3. RESPIRATORY PHYSIOLOGY. Educ: Dartmouth Col, AB, 55; Harvard Univ, MD, 58; McGill Univ, PhD(exp med), 69. Prof Exp: Intern med, NC Mem Hosp, 58-59, jr asst resident, 59-60; sr asst resident, Respiratory Dept, Royal Victoria Hosp, 63-64; demonstr med, 64-66, asst prof exp med, 69-70, ASSOC PROF EXP MED, McGILL UNIV, 70- Concurrent Pos: Med Res Coun Can scholar, 69-71. Mem: Can Soc Clin Invest; Am Physiol Soc; Can Thoracic Soc. Res: Chest disease; pulmonary physiology; physiologic aspects of respiratory disease. Mailing Add: Respiratory Div Royal Victoria Hosp Montreal PQ Can

ANTHONY, ADAM, b Buffalo, NY, Oct 19, 23; m 49. PHYSIOLOGY. Educ: Univ Buffalo, BA, 43; Marquette Univ, MS, 48; Univ Chicago, PhD(zool), 52. Prof Exp: Lab asst comp anat & bact, Univ Buffalo, 43; asst gen zool, parasitol & histol, Marquette Univ, 46-48 & med parasitol, biol sci, endocrinol & vert embryol, Univ Chicago, 49-52; from asst prof to assoc prof zool, 52-61, chmn comt physiol, Col Sci, 69-73, PROF ZOOL, PA STATE UNIV, UNIVERSITY PARK, 61- Mem: Am Physiol Soc; Soc Exp Biol & Med; AAAS; NY Acad Sci; Am Soc Zool. Res: Biological effects of sound in animals; altitude physiology; histochemistry; microspectrophotometry. Mailing Add: 418 Life Sci I Pa State Univ University Park PA 16802

ANTHONY, DAVID HENRY, b Collingswood, NJ, June 29, 20; m 44. DENTAL MATERIALS. Educ: Univ Louisville, DMD, 43; Univ Mich, MS, 61. Prof Exp: Pvt pract, 47-56; res asst dent mat, Univ Mich, 56-61; mgr dent res dent mat, J M Ney Co, 61-73; CONSULT SELECTED LAB PROCEDURES IN GROUP PRACT, 73- Mem: Am Dent Asn; Int Asn Dent Res. Res: Evaluation of denture cleaners; evaluation of fit and repair processing of dentures; dental castings and soldering; plastics, ceramics and other dental materials. Mailing Add: PO Box 496 South Orleans MA 02662

ANTHONY, DONALD JOSEPH, b Troy, NY, May 30, 22; m 45; c 5. PHYSICS. Educ: Siena Col, BS, 47; Univ Notre Dame, PhD(physics), 53. Prof Exp: Instr chem & math, Siena Col, 46-47 & physics, Univ Notre Dame, 47-52; res assoc, Knolls Atomic Power Lab, 52-56, mgr exp physics, 56-59, digital reactor physics, 59-68 & operating nuclear plants, 68-74; MEM STAFF, ENERGY SYSTS & TECHNOL DIV, GEN ELEC CO, 74- Mem: Am Nuclear Soc. Res: Reactor and nuclear physics. Mailing Add: Gen Elec Co Bldg 300 Schenectady NY 12345

ANTHONY, E JAMES, b Jan 21, 16; US citizen; m 40; c 4. CHILD PSYCHIATRY. Educ: Univ London, BSc, 38, DPM, 48, MD, 49. Prof Exp: Sr lectr child psychiat, Univ London, 51-58; PROF CHILD PSYCHIAT, WASH UNIV, 58- Concurrent Pos: Fel child develop, Univ Geneva, 50-51; Ctr Advan Studies Behav Sci fel, 70-71; consult, NIMH, 62-65 & Nat Bd Med Examr, 63-64; prof lectr, Univ Chicago, 63- Mem: Am Acad Child Psychiat; Am Psychiat Asn; Group Advan Psychiat; Int Asn Child Psychiat (vpres, 62-70, pres, 70-). Res: Clinical research in child psychiatry; child development. Mailing Add: 369 N Taylor Ave St Louis MO 63108

ANTHONY, ELMER HAROLD, biology, see 12th edition

ANTHONY, HARRY D, b Fredonia, Kans, Apr 17, 21; m 55; c 1. VETERINARY MEDICINE. Educ: Kans State Univ, DVM, 52, MS, 57. Prof Exp: Pvt pract, Ill, 52-55; from instr to assoc prof vet med, 55-67, DIR KANS VET DIAG LAB, KANS STATE UNIV, 68- Honors & Awards: Vet of Year, Kans Vet Med Asn, 74. Res: Blood parasites of domestic animals; baby pig diseases; respiratory diseases of ruminants; infectious keratitis of the bovine. Mailing Add: 2802 Oregon Lane Manhattan KS 66502

ANTHONY, JAMES DOUGLAS, b Grayling, Mich, June 27, 18; m 43; c 3. PARASITOLOGY. Educ: Antioch Col, BA, 41; Univ Mich, MS, 48, PhD, 52. Prof Exp: Instr zool, Univ Mich, 51-52; head microscope slide dept, Wards Nat Sci Estab, 52-54; asst prof biol & chem dept, 54-64, assoc prof zool, 64-74, PROF ZOOL, UNIV WIS-MILWAUKEE, 74- Mem: Am Soc Parasitol; Am Micros Soc; Am Fisheries Soc. Res: Caryophyllid cestodes. Mailing Add: Dept of Zool Univ of Wis Milwaukee WI 53211

ANTHONY, JOHN WILLIAMS, b Brockton, Mass, Nov 25, 20; m 44; c 3.

GEOLOGY, MINERALOGY. Educ: Univ Ariz, BS, 46, MS, 51; Harvard Univ, PhD(geol), 65. Prof Exp: Mineralogist, Ariz Bur Mines, 46-51; from asst prof to assoc prof, 51-64, actg head dept, 66-67, PROF GEOL, UNIV ARIZ, 64- Mem: Fel Geol Soc Am; fel Mineral Soc Am; Soc Econ Geol; Am Crystallog Asn. Res: Mineralogy and economic geology of the Southwest; experimental and descriptive mineralogy, crystallography, economic geology. Mailing Add: 4462 E Seventh St Tucson AZ 85711

ANTHONY, LEE SAUNDERS, b Roanoke, Va, Sept 11, 32; m 53; c 2. PHYSICS. Educ: Roanoke Col, BS, 53; Va Polytech Inst, MS, 58, PhD(physics), 62. Prof Exp: Asst prof, 62-65, actg dean, 63-65, PROF PHYSICS, ROANOKE COL, 65-, CHMN DEPT, 62- Concurrent Pos: Consult physicist, Roanoke Mem Hosp, 64- Mem: Am Phys Soc; Am Nuclear Soc. Res: Nuclear, reactor and radiological physics. Mailing Add: Dept of Physics Roanoke Col Salem VA 24153

ANTHONY, LUEAN EVANGELINE, b Denver, Colo, Dec 16, 41; m 74. BIOCHEMISTRY, NUTRITION. Educ: Tufts Univ, BS, 63; Mass Inst Technol, SM, 67, PhD(nutrit biochem), 70. Prof Exp: Instr biochem, Univ Tex Health Sci Ctr Dallas, 71-74; RES BIOCHEMIST, CELL BIOL SECT, VET ADMIN HOSP, 74- Concurrent Pos: Robert Welch Found fel, Univ Tex Southwestern Med Sch, 71; lectr human biol, Stanford Univ, 74- Mem: AAAS; Soc Nutrit Educ. Res: Protein-calorie malnutrition; liver drug metabolism; hypolipidemic drugs. Mailing Add: Vet Admin Hosp Cell Biol Sect 4150 Clement St San Francisco CA 94121

ANTHONY, MARGERY STUART, b New York, NY, Feb 23- 24. BOTANY, RADIATION ECOLOGY. Educ: Univ Mich, BS, 45, MS, 46, PhD(bot), 50. Prof Exp: Asst bot, Univ Mich, 44-49; from asst prof to assoc prof biol & bot, 49-71, PROF BIOL, CALIF STATE UNIV, CHICO, 71- Concurrent Pos: Lectr, NSF Inst, Univ Wyo, 58-62. Mem: Ecol Soc Am; Health Physics Soc. Res: Plant ecology; radiation ecology. Mailing Add: Dept of Biol Sci Calif State Univ Chico CA 95929

ANTHONY, MAURICE LEE, applied mechanics, see 12th edition

ANTHONY, PAUL, physical chemistry, see 12th edition

ANTHONY, ROBERT GENE, b Smith Center, Kans, Jan 6, 44; m 65; c 2. WILDLIFE ECOLOGY. Educ: Ft Hays Kans State Col, BS, 66; Wash State Univ, MS, 68; Univ Ariz, PhD(zool), 72. Prof Exp: ASST PROF WILDLIFE MGT, PA STATE UNIV, UNIVERSITY PARK, 72- Mem: Wildlife Soc; Am Soc Mammalogists; Am Ecol Soc. Res: Ecology and population dynamics of mammals; effect of pollution and environmental alteration on wildlife; biometrics. Mailing Add: 313 Forest Resources Lab Pa State Univ University Park PA 16802

ANTHONY, ROBERT LOUIS, physics, see 12th edition

ANTHONY, ROMUALD, b Chicago, Ill, Dec 17, 17; div; c 2. Univ Chicago, BS, 44. Prof Exp: Supvr optical shop, Univ Chicago, 42-45; physicist photogram, US Naval Ord Test Sta, 45-47, optics, Calif, 47-52; asst res physicist, Visibility Lab, Scripps Inst Oceanog, Univ Calif, San Diego, 52-54; tech adv optics, Bill Jack Sci Instrument Co, 54-57; gen mgr photometric instruments, Photo Res Corp, 57-58; head optics & infrared group, Convair Div, Gen Dynamics Corp, 58-63; mgr, Optics-Photo Dept & staff scientist, FMA, Inc, 63-69, staff scientist, AIL Info Systs, 69-75; PVT CONSULT, 75- Mem: Optical Soc Am; Sigma Xi; Soc Photog Scientists & Engrs; Soc Photo-Optical Instrument Engrs. Res: High intensity fluorescent illumination, high resolution fluorescent display screen; grating ruling engines; photometry; instrumentation; optical shop techniques; solar infrared transmission and scatter through atmosphere; infrared and electro-optical instruments; information storage and retrieval equipment. Mailing Add: 428 24th St Manhattan Beach CA 90266

ANTHONY, RONALD LEWIS, b Ft Edward, NY, Sept 7, 38; m 66; c 1. IMMUNOLOGY, SEROLOGY. Educ: Susquehanna Univ, BA, 61; Univ Kans, PhD(zool), 65. Prof Exp: Res assoc serol, 65-69, asst prof internal med, Inst Internal Med, 69-70, ASST PROF PATH & DIR IMMUNOL, SCH MED, UNIV MD, BALTIMORE CITY, 70- Mem: AAAS. Res: Clinical immunology; tumor immunology. Mailing Add: Div of Clin Path Univ of Md Hosp 22 S Greene St Baltimore MD 21201

ANTHONY, WILSON BRADY, b Waco, Tex, Nov 2, 16; m 45; c 1. ANIMAL HUSBANDRY, NUTRITION. Educ: Univ Ill, BS, 48; Agr & Mech Col Tex, MS, 49; Cornell Univ, PhD, 53. Prof Exp: Asst prof dairy husb, Agr & Mech Col Tex, 52-53; assoc prof, 53-55, PROF ANIMAL SCI, AUBURN UNIV, 55- Mem: AAAS; Am Soc Animal Sci; Am Dairy Sci Asn; Am Inst Nutrit. Res: Ruminant nutrition; evaluation of chemical composition, nutritive value of forages and feed; developing procedures for fermenting feeds and by-products to upgrade nutritive value; other factors affecting nutritional states in livestock. Mailing Add: PO Box 447 Auburn AL 36830

ANTHROP, DONALD F, b Lafayette, Ind, June 29, 35; m 65; c 1. THERMODYNAMICS, GEOGRAPHY. Educ: Purdue Univ, BS, 57; Univ Calif, Berkeley, PhD(mat sci), 63. Prof Exp: Staff res scientist, Res & Develop Div, Avco Corp, 62-65; staff res chemist, Refractory Mat Div, Lawrence Radiation Lab, 65-69; asst prof geog, Calif State Col, Dominguez Hills, 69-74; ASSOC PROF ENVIRON STUDIES, SAN JOSE STATE UNIV, 74- Mem: Am Chem Soc. Res: Natural resources and environmental pollution, particularly environmental noise and environmental effects of energy production. Mailing Add: Dept of Environ Studies San Jose State Univ San Jose CA 95114

ANTIA, NAVAL JAMSHEDJI, b Bombay, India, Feb 12, 21; Can citizen; m 57; c 1. ORGANIC CHEMISTRY, MICROBIOLOGY. Educ: Univ Bombay, BSc, 41 & 43; Univ Zurich, DPhil(org chem), 51. Prof Exp: Res chemist, Fison's Pest Control, Ltd, Eng, 55 & Textile Fibers Div, Can Industs Ltd, 57-59; res scientist, Oceanog & Marine Microbiol, Nanaimo Biol Sta, Fisheries Res Bd Can, 60-63 & Marine Planktonology, Tech Sta, Fisheries Res Bd Can, 63-75, RES SCIENTIST, PAC ENVIRON INST, ENVIRON CAN, 75- Concurrent Pos: French Nat Ctr Sci Res fel, Univ Montpellier, 51-52; fel, Royal Col Sci & Technol, Scotland, 52-53; Nat Res Coun Can fel, 55-57; fel, Queen's Univ, Ont, 59-60; hon prof bot, Univ BC, Vancouver, 75- Mem: Am Soc Limnol & Oceanog; Phycol Soc Am; Am Inst Biol Sci; Soc Protozool; Int Phycol Soc. Res: Structural and synthetic organic chemistry of sulphoxides, alkaloids, steroids, plant-growth substances, sugars and their phosphate esters, algal carotenoids; seawater chemistry; marine microbiology; biochemistry, physiology and ecology of marine phytoplankters. Mailing Add: Pac Environ Inst Environ Can 4160 Marine Dr West Vancouver BC Can

ANTIPA, GREGORY ALEXIS, b San Francisco, Calif, Aug 9, 41; m 66; c 2. PROTOZOOLOGY, CELL BIOLOGY. Educ: Univ Calif, Berkeley, AB, 63; Calif State Univ, San Francisco, MA, 66; Univ Ill, Urbana-Champaign, PhD(zool), 71. Prof Exp: NIH fel cell biol, Univ Chicago, 70-71; AEC fel, Argonne Nat Lab, 71-74; ASST PROF BIOL, WAYNE STATE UNIV, 74- Mem: Am Micros Soc; Am Soc Cell Biol; Am Soc Zoologists; Electron Micros Soc Am; Soc Protozoologists. Res: Temporal events and control of eukaryotic and circadian cell cycles; fine structure; development and phylogeny of ciliates; ciliated protozoa as indicators of water quality and role in activated sludge process. Mailing Add: Dept of Biol Wayne State Univ Detroit MI 48202

ANTLE, CHARLES EDWARD, b East View, Ky, Nov 11, 30; m 53; c 3. MATHEMATICAL STATISTICS. Educ: Eastern Ky State Col, BS, 54, MA, 55; Okla State Univ, PhD(math), 62. Prof Exp: Aerophysics engr, Gen Dynamics/Convair, 55-57; from instr to asst prof math, Univ Mo, Rolla, 57-60; assoc prof, 62-64, PROF STATIST, PA STATE UNIV, 64- Mem: Am Statist Asn; Inst Math Statist; Royal Statist Soc; Biomet Soc. Res: Biostatistics; reliability; Bayesian decision rules; the Weibull Model. Mailing Add: 320 Pond Lab Pa State Univ University Park PA 16802

ANTLER, MORTON, b New York, NY, Apr 27, 28; m 50; c 3. APPLIED CHEMISTRY. Educ: NY Univ, BA, 48; Cornell Univ, PhD(inorg chem), 53. Prof Exp: Res chemist, Ethyl Corp, 53-58; supvr phys chem, Borg-Warner Res Ctr, 58-59; adv chemist, Eng Lab, Int Bus Mach Corp, 59-63; dep dir res, Burndy Corp, 63-70; MEM TECH STAFF, BELL LABS, 70- Concurrent Pos: Vchmn, Holm Sem Elec Contacts, Ill Inst Technol, 72- Honors & Awards: Precious Metal Plating Awards, Am Electroplaters Soc, 68 & 71; Capt Alfred E Hunt Mem Award, Am Soc Lubrication Eng, 71. Mem: Am Chem Soc; Electrochem Soc; Am Electroplaters Soc; Am Soc Lubrication Eng; Sigma Xi. Res: Inorganic and surface chemistry; friction, wear and lubrication; electric contacts. Mailing Add: 821 Strawberry Hill Rd E Columbus OH 43213

ANTLIFF, HAROLD ROY, medical science, see 12th edition

ANTMAN, STUART S, b Brooklyn, NY, June 2, 39; m 68; c 2. APPLIED MATHEMATICS, MECHANICS. Educ: Rensselaer Polytech Inst, BS, 61; Univ Minn, MS, 63, PhD(mech), 65. Prof Exp: Vis mem, Courant Inst Math Sci, NY Univ, 65-67, from asst prof to assoc prof math, 67-72; PROF MATH, UNIV MD, COLLEGE PARK, 72- Concurrent Pos: Sci Res Coun Brit sr vis fel, Oxford Univ; co-ed, Springer Tracts in Natural Philos, 72-; Soc Indust & Appl Math lectr, 73-75. Mem: Am Math Soc; Soc Natural Philos; Soc Indust & Appl Math. Res: Nonlinear equations of mechanics; bifurcation and stability theory; theories of rods and shells. Mailing Add: Dept of Math Univ of Md College Park MD 20742

ANTOGNINI, JOE, b San Rafael, Calif, Sept 15, 23; m 44; c 3. PLANT SCIENCE. Educ: Univ Calif, BS, 48; Cornell Univ, PhD(veg crops), 51. Prof Exp: Asst veg crops, Cornell Univ, 48-51; agronomist, Geigy Chem Corp, 51-54; agronomist, Biol Res Ctr, Stauffer Chem Co, 54-58, field res rep, 59-64, mgr field res, 64-71; from dir to vpres prod develop, 71-73, vpres prod planning, Zoecon Corp, 73-75; MGR MKT DEVELOP, AGR CHEM DIV, BASF WYANDOTTE CORP, 75- Res: Chelates in plant nutrition; herbicide, defoliant, insecticide and related agricultural chemicals. Mailing Add: BASF Wyandotte 100 Cherry Hill Rd Parsipanny NJ 07054

ANTOINE, ALAN DEXTER, biochemistry, microbiology, see 12th edition

ANTON, AARON HAROLD, b Hartford, Conn, Sept 4, 21; m 47; c 2. PHARMACOLOGY. Educ: Univ Conn, BS, 44; Trinity Col, Conn, BS, 52; Yale Univ, PhD(pharmacol), 56. Prof Exp: Instr pharmacol, Col Med, Univ Fla, 56-59, asst prof psychiat, 59-65, assoc prof cprof anesthesiol & pharmacol, Div Anesthesiol, 65-69; PROF ANESTHESIOL, SCH MED, CASE WESTERN RESERVE UNIV, 69- Mem: Am Soc Pharmacol. Res: Biochemical changes in hypertension and mental disease; pharmacological significance of protein binding. Mailing Add: Dept of Anesthesiol Sch of Med Case Western Reserve Univ Cleveland OH 44106

ANTON, HOWARD, b Philadelphia, Pa, July 27, 39; m 65; c 1. MATHEMATICS. Educ: Lehigh Univ, BA, 60; Univ Ill, MA, 63; Polytech Inst Brooklyn, PhD(math), 68. Prof Exp: Mathematician, Burroughs Corp, 60-61; lectr math, Hunter Col, City Univ New York, 64-66; ASST PROF MATH, DREXEL UNIV, 68- Mem: Am Math Soc; Math Asn Am; Soc Indust & Appl Math. Res: Functional analysis; Banach algebras; distribution theory; approximation theory. Mailing Add: Dept of Math Drexel Univ Philadelphia PA 19104

ANTON, JOHN RALPH, b Vienna, Austria, Aug 3, 17; US citizen; m 64; c 2. URBAN GEOGRAPHY. Educ: Univ Vienna, MA, 47, PhD(geog), 49; Univ Toronto, dipl town & regional planning, 58. Prof Exp: Prof geog & hist, Austrian Sec Schs, 48-56; res planner, City Toronto Planning Bd, 56-57, planning assoc, 58-64; dir master planning sect, City Peoria Planning Bd, 64-65; planning analyst, Dept Planning & Urban Renewal & Akron Metrop Area Transp Study, City Akron, 65-69; ASST PROF GEOG, YOUNGSTOWN STATE UNIV, 69- Concurrent Pos: Cent Mortgage & Housing Corp fel, Ottawa, 57-58; lectr & instr geog, Univ Akron, 66-69. Honors & Awards: Hon prof, Pres Austria, 54. Mem: Asn Am Geogrs; Regional Sci Asn; Am Geog Soc; Asn Soc Planning Offs. Res: Urban and regional planning; mapping of urban problems; Western and Eastern Europe, Union of Soviet Socialist Republics, China and South and East Asia; economics of urban problems, especially transportation. Mailing Add: Dept of Geog Youngstown State Univ Youngstown OH 44503

ANTONELLI, PETER LOUIS, b Syracuse, NY, Mar 5, 41; div; c 2. TOPOLOGY, GEOMETRY. Educ: Syracuse Univ, BS, 63, MA, 65, PhD(math), 67. Prof Exp: Asst prof math, Univ Tenn, Knoxville, 67-68; fel, Inst Advan Study, Princeton, 68-70; ASSOC PROF MATH, UNIV ALTA, 70- Concurrent Pos: Fel biol, Univ Sussex, Eng, 72-73. Mem: Am Math Soc; Can Math Cong. Res: Mathematical theory of evolution; nonequilibrium statistical mechanics; path integrals in quantum mechanics and relativity. Mailing Add: Dept of Math Univ of Alta Edmonton AB Can

ANTONIADES, HARRY NICHOLAS, b Thessaloniki, Greece, Mar 12, 23; US citizen; m 53; c 2. BIOCHEMISTRY. Educ: Nat Univ Athens, BS, 48, PhD(chem), 52. Prof Exp: Res assoc biochem, Evangelismos Med Ctr, Athens, Greece, 48-53; vis investr, Univ Lab Phys Chem, Harvard Univ, 53-54; from res asst to res assoc, Ctr Blood Res, 54-56, assoc investr, 56-61; res assoc biol chem, Dept Gynec, Med Sch, 56-63 & Dept Med, 63-66, asst prof biochem, Sch Pub Health, 65-70, ASSOC PROF BIOCHEM, SCH PUB HEALTH, HARVARD UNIV, 70-; SR INVESTR, CTR BLOOD RES, 61- Concurrent Pos: Vis assoc prof, Sch Med, Univ Southern Calif, 60; assoc staff med, Peter Bent Brigham Hosp, 61-65; lectr, NATO Advan Study Inst, Stratford-upon-Avon, Eng, 62; vis prof, Med Ctr, Univ Ala, 63; lectr, Cong Int Diabetes Fedn, Toronto, 64, Stockholm, 67 & Buenos Aires, 70; lectr, Cong Int Endocrine Soc, London, 64; vis prof, Inst Physiol, Med Sch, Univ Buenos Aires, 66; spec lectr, 41st Japan Endocrinol Soc, Kyoto, 68; vis prof, Ain Shams Univ Med Sch, Cairo & Alexandria Univ Med Sch, Alexandria, Egypt, 71; tutor biol chem, Harvard Univ, 73- Honors & Awards: Eli Lilly Award, Am Diabetes Asn, 62. Mem: AAAS; Am Diabetes Asn; Am Soc Biol Chem; NY Acad Sci; Int Soc Thrombosis & Haemostasis. Res: Mechanisms of hormone transport and regulation; plasma protein fractionation and characterization; interaction of proteins. Mailing Add: Ctr Blood Res 800 Huntington Ave Boston MA 02115

ANTONIAK, CHARLES EDWARD, b Norfolk, Va, Apr 11, 40; m 64; c 2. MATHEMATICAL STATISTICS. Educ: Calif Inst Technol, BS, 60; Calif State Univ, San Diego, MS, 63; Univ Calif, Los Angeles, PhD(math), 69. Prof Exp: Res physicist, US Navy Electronics Lab, San Diego, 60-61, opers analyst, 63-65, res math, 65-70; ASST PROF STATIST, UNIV CALIF, BERKELEY, 70- Concurrent Pos: Off Naval Res grant, Univ Calif, Berkeley, 71-; consult, Lawrence-Berkeley Lab, 74-75. Mem: Am Math Soc; Math Asn Am; Inst Elec & Electronics Eng; Inst Math Statist. Res: Bayesian, distribution free and sequential applied statistics especially signal detection and data compression. Mailing Add: Dept of Statist Univ of Calif Berkeley CA 94720

ANTONIEWICZ, PETER R, b Tarnow, Poland, Feb 5, 36; US citizen; m 61; c 3. THEORETICAL SOLID STATE PHYSICS, SURFACE PHYSICS. Educ: NC State Univ, BS, 59; Purdue Univ, MS, 64, PhD, 65. Prof Exp: Asst prof, 65-70, ASSOC PROF PHYSICS, UNIV TEX, AUSTIN, 70- Mem: AAAS; Am Phys Soc; Am Asn Physics Teachers. Mailing Add: Dept of Physics Univ of Tex Austin TX 78712

ANTONINI, GUSTAVO ARTHUR, b New York, NY, Feb 7, 38; m 64; c 1. RESOURCE GEOGRAPHY, RESOURCE MANAGEMENT. Educ: Columbia Univ, BS, 61, MA, 62, PhD(geog), 68. Prof Exp: Res geogr, HRB-Singer, Inc, Pa, 63-65; geogr, Secy State Agr & Inst Hydraul Resources, Govt of Dominican Repub, 66-67; asst prof geog, Univ Windsor, 68-69 & Univ PR, Rio Piedras, 69-70; geogr, Environ Consult, St Thomas, VI, 70; ASSOC PROF LATIN AM STUDIES IN GEOG, UNIV FLA & DIR RES, CTR LATIN AM STUDIES, 70- Concurrent Pos: Geogr, Presidency & State Secretariat of Agr of Dominican Repub, 71-; consult, Regional Agr Develop, Land Mgt Progs, State Secretariat of Agr of Dominican Repub, 74- & Asn Caribbean Univs & Res Insts, 75- Honors & Awards: Off Commendations, Press Dominican Repub, 71 & 73. Mem: Asn Am Geog; Latin Am Studies Asn; Am Geog Soc. Res: Resource utilization; human tropical ecology. Mailing Add: Ctr for Latin Am Studies Univ of Fla Gainesville FL 32601

ANTONIUS (KENNELLY), MARY, b St Thomas, NDak, May 8, 01. ORGANIC CHEMISTRY. Educ: St Catherine Col, AB, 26; Univ Munich, PhD(chem), 33. Prof Exp: Asst chem, St Catherine Col, 26-29, assoc prof, 33-38, head dept, 36-43, pres, 43-49; adminstr, St Joseph's Hosp, 50-56; chmn div natural sci, 56-65 & dept chem, 64-69, PROF CHEM, COL ST CATHERINE, 56- Mem: AAAS; Am Soc Med Technol; Am Chem Soc. Res: Bile acids; physiological chemistry; general chemistry. Mailing Add: Col of St Catherine 2004 Randolph Ave St Paul MN 55105

ANTONOFF, MARVIN M, b New York, NY, Nov 29, 30; m 53; c 3. THEORETICAL PHYSICS, SOLID STATE PHYSICS. Educ: NY Univ, BS, 52, MA, 53; Cornell Univ, PhD(solid state physics), 62. Prof Exp: Staff mem, Gen Elec Co, NY, 55-59; asst physics, Cornell Univ, 59-61; staff mem, Lincoln Lab, Mass Inst Technol, 61-62 & Sperry Rand Res Ctr, 62-65; asst prof, 65-67, ASSOC PROF PHYSICS, UNIV MASS, BOSTON, 67-, CHMN DEPT, 68- Concurrent Pos: Guest scientist, Nat Magnet Lab, Mass Inst Technol, 70-; consult, Kennecott Copper Corp, 74-75. Mem: Am Phys Soc; Inst Elec & Electronics Eng. Res: Theory of cooperative phenomena in solid state physics, especially ferromagnetism, plasma physics and superconductivity. Mailing Add: 11 Paul Revere Rd Lexington MA 02173

ANTONSEN, DONALD HANS, b Weehawken, NJ, Aug 11, 30; m 53; c 2. INDUSTRIAL CHEMISTRY. Educ: Davis & Elkins Col, BS, 52; Univ Del, PhD(org chem), 60. Prof Exp: Anal chemist, Am Cyanamid Co, 52-54; org chemist, Esso Res & Eng Co, 59 & Sun Oil Co, 60-64; res supvr indust chem, 64-69, appln mgr nickel chem, 69-72, NICKEL CHEM INDUST SALES MGR, INT NICKEL CO, INC, 72- Mem: Am Chem Soc; Commercial Develop Asn; Am Inst Chem Eng; NY Acad Sci; The Chem Soc. Res: Nickel chemicals, salts and catalysts; nickel oxide properties; organometallic chemistry; polymerization; synthetic fibers and films; catalysis; rheology of hydrocarbons; clay mineralogy; precious metals; electrochemistry; pesticides, crude oil production; petroleum products. Mailing Add: Int Nickel Co Inc One New York Plaza New York NY 10004

ANTONUCCI, FRANK RALPH, b Auburn, NY, Sept 8, 46; m 68; c 2. ORGANIC POLYMER CHEMISTRY. Educ: St Michaels Col, BA, 68; Col of Holy Cross, MS, 69; Rensselaer Polytech Inst, PhD(org chem), 73. Prof Exp: SR RES CHEMIST, ADDRESSOGRAPH-MULTIGRAPH CORP, 75- Mem: Am Chem Soc. Res: Synthesis and evaluation of organic photoconductors and resins for electrophotographic applications. Mailing Add: Multigraphics Develop Ctr 19701 S Miles Cleveland OH 44128

ANTONUCCI, JOSEPH MATTHEW, organic chemistry, see 12th edition

ANTONY, ARTHUR, physical inorganic chemistry, see 12th edition

ANTOPOL, WILLIAM (ARNOLD), pathology, experimental medicine, deceased

ANTOS, ROBERT JOHN, b Cleveland, Ohio, Mar 28, 20; m 45; c 1. PHARMACOLOGY. Educ: Western Reserve Univ, BS, 42, MS, 43, MD, 45. Prof Exp: Instr physiol, Western Reserve Univ, 43-44; mem adv coun & res assoc, Poisonous Animals Res Lab, Ariz State Univ, 54-64; DIR DEPT CLIN PHARMACOL, MEM HOSP & SOUTHWEST CANCER INST, 63- Concurrent Pos: Instr, Sch Nursing, Good Samaritan Hosp, Phoenix, 46-56, dir dept res, 48-60, lectr, Intern Training Sch, 48-; lectr, Sch Nursing, Mem Hosp, 48-52; head poison control, State Ariz, 60-72; head poison control ctr, Maricopa County, Ariz, 61-72. Mem: AAAS; Nat Soc Med Res; AMA; fel Am Inst Chem; Am Heart Asn. Res: Clinical toxicology and pharmacology. Mailing Add: 4034 N 15th Ave Phoenix AZ 85015

ANTOSHKIW, THOMAS, b New York, NY, Sept 10, 18; m 45; c 2. PHARMACEUTICAL CHEMISTRY. Educ: Brooklyn Col Pharm, PhG, 38, BS, 40. Prof Exp: Control chemist, R H Macy & Co, NY, 40-42; res chemist, Ives-Cameron Co, Inc, 46-47, asst tech dir in charge of res & develop labs, 48-49, chief chemist, 49-50; res chemist, 50-52, asst to dir pharmaceut res & develop labs, 52-53, mgr pharmaceut, Pilot Plant, 53-55, asst dir pharmaceut prod div, 55-63, dir div, 64-71, GROUP CHIEF PROD DEVELOP, HOFFMANN-LA ROCHE, INC, 71- Mem: Am Chem Soc; Am Pharmaceut Asn; Sigma Xi; Acad Pharmaceut Sci. Res: Water solubilization of fats and oils; high vacuum techniques; development of pharmaceutical dosage forms; pharmaceutical techniques and processes; vitamin enrichment of foods. Mailing Add: Hoffmann-La Roche Inc Nutley NJ 07110

ANTOSIEWICZ, HENRY ALBERT, b Wollersdorf, Austria, May 14, 25; US citizen; m 58, 70; c 1. MATHEMATICAL ANALYSIS. Educ: Univ Vienna, PhD(math, theoret physics), 47. Prof Exp: Assoc prof math, Mont State Col, 48-52; assoc prof, Am Univ, 52-55, adj prof, 55-57; assoc ed, Math Rev, 57-58; vis assoc prof, 58-59, assoc prof, 59-61, PROF MATH, UNIV SOUTHERN CALIF, 61-, CHMN DEPT, 68- Concurrent Pos: Mathematician, Nat Bur Stand, 52-57; consult, Space Technol Labs, Thompson-Ramo-Wooldridge, Inc, 59-70. Mem: Am Math Soc; Math Asn Am. Res: Qualitative theory of ordinary differential equations. Mailing Add: Dept of Math Univ of Southern Calif Los Angeles CA 90007

ANTOSZ, FREDERICK JOHN, organic chemistry, see 12th edition

ANVER, MIRIAM R, b Chicago, Ill, Mar 1, 43. COMPARATIVE PATHOLOGY. Educ: Univ Ill, BS, 54, DVM, 66; Kans State Univ, PhD(vet path), 70. Prof Exp: Instr vet path, Kans State Univ, 66-68; instr, 72-73, ASST PROF COMP PATH, UNIV MICH, ANN ARBOR, 73- Concurrent Pos: NIH spec fel, Harvard Univ, 71-72. Mem: Am Col Vet Path; Am Vet Med Asn; Int Acad Path; Am Asn Lab Animal Sci. Res: Comparative and viral oncology; diseases of laboratory animals. Mailing Add: Unit Lab Animal Med Univ Mich Med Sch Ann Arbor MI 48109

ANWAR, MOHAMMAD HAIDER, colloid chemistry, see 12th edition

ANWAR, RASHID AHMAD, b Nakodar, India, Oct 15, 30; m 63. BIOCHEMISTRY. Educ: Univ Panjab, WPakistan, BSc, 51, MSc, 52; Mich State Univ, PhD(chem), 57. Prof Exp: Lectr pharmaceut chem, Univ Panjab, WPakistan, 52-54; asst biochem, Mich State Univ, 55-57, res instr, 57-60; from res assoc to assoc prof, 63-74, PROF BIOCHEM, UNIV TORONTO, 74- Concurrent Pos: Nat Res Coun Can fel, 60-62. Mem: AAAS; Am Chem Soc; Am Soc Biol Chem; Can Biochem Soc. Res: Proteolytic enzymes; structure of proteins; bacterial cell wall biosynthesis. Mailing Add: Dept of Biochem Univ of Toronto Toronto ON Can

ANWAY, ALLEN R, b Cloquet, Minn, Mar 19, 41; m 64; c 3. PHYSICS, ELECTRONICS. Educ: Univ Minn, Duluth, BA, 63; Univ Chicago, MS, 65, PhD(physics), 68. Prof Exp: Asst prof physics, Chicago State Col, 68-69; asst prof, 69-72, LECTR PHYSICS, UNIV WIS-SUPERIOR, 72- Concurrent Pos: Chmn new prod div, Duluth Sci, 73-74; consult, Earl Ruble & Assocs, 74- Mem: Am Phys Soc; Am Asn Physics Teachers. Res: Electronic detector of leaks in water mains. Mailing Add: Dept of Physics Univ of Wis Superior WI 54880

ANWAY, JERRY C, botany, biosystematics, see 12th edition

ANYOS, TOM, organic chemistry, polymer chemistry, see 12th edition

ANYSAS, JURGIS ARVYDAS, b Hamburg, Ger, Sept 5, 34; Can citizen. CHEMICAL PHYSICS. Educ: Univ Toronto, BASc, 56; Ill Inst Technol, PhD(chem), 66. Prof Exp: Geophysicist, Imp Oil Ltd, Can, 56-60; ASST PROF CHEM, DePAUL UNIV, 66- Concurrent Pos: Fel, Ill Inst Technol, 66-67. Mem: AAAS; Am Chem Soc. Res: Quantum chemistry; molecular spectroscopy. Mailing Add: Dept of Chem DePaul Univ Chicago IL 60614

ANZALONE, LOUIS, JR, b Independence, La, Oct 7, 31; m 53; c 4. PHYTOPATHOLOGY. Educ: Southeastern La Col, BS, 54; La State Univ, MS, 56, PhD(plant path), 58. Prof Exp: Asst bot, bact & plant path, 54-58, asst prof, 58-62, assoc prof bot & plant path, 62-68, PROF PLANT PATH & SUPT BURDEN RES CTR BR, LA AGR EXP STA, LA STATE UNIV, BATON ROUGE, 68- Mem: Am Phytopath Soc; Sigma Xi. Res: Fungus diseases of ornamental plants; sugar cane breeding and pathology; virus diseases of sugar cane. Mailing Add: Dept of Plant Path La State Univ Baton Rouge LA 70803

ANZENBERGER, JOSEPH F, SR, b Altoona, Pa, Oct 27, 22; m 50; c 7. INDUSTRIAL ORGANIC CHEMISTRY. Educ: Duquesne Univ, BS, 48, MS, 50. Prof Exp: From res asst to res assoc, Mellon Inst, 50-54, from jr fel to fel, 54-62; from res chemist to sr res chemist, Monsanto Co, 62-70; SR RES CHEMIST, STAUFFER CHEM CO, 71- Mem: Am Chem Soc. Res: Nickel organic compounds; lubricating oil additives; organic flame retardants; development of industrial fire resistant hydraulic fluids and lubricants. Mailing Add: Eastern Res Ctr Stauffer Chem Co Dobbs Ferry NY 10522

APEL, JOHN RALPH, b Absecon, NJ, June 14, 30; m 56; c 2. FLUID PHYSICS, PHYSICAL OCEANOGRAPHY. Educ: Univ Md, BS, 57, MS, 61; Johns Hopkins Univ, PhD(solid state plasma physics), 70. Prof Exp: Phys sci aide, Nat Bur Stand, 55-57; assoc mathematician, Appl Physics Lab, Johns Hopkins Univ, 57-61, sr physicist, 61-70, asst supvr plasma dynamics, 66-70; SUPVRY PHYSICIST & DIR, OCEAN REMOTE SENSING LAB, ATLANTIC OCEANOG & METEOROL LABS, NAT OCEANOG & ATMOSPHERIC ADMIN, 70- Concurrent Pos: Sec Greater Wash Plasma Physics Colloquium , 68-70; adj prof physics, Univ Miami, 71-; consult, NASA, 72-, Dept Defense, 73- & UNESCO/Intergovt Oceanog Comn, 75-; mem Int Union Radio Sci, Comn II, 73- & Int Union Radio Meterol, 75-; chmn Seasat Interagency Steering Comt, 73-; mem NASA appln adv comt, 75-; expert witness, US House Rep & US Senate, 74-; chmn ocean dynamics subcomt, Space Appl Comn, NASA, 73- Honors & Awards: Gold Medal, US Dept Com, 74. Mem: Am Phys Soc; Am Geophys Union; Sigma Xi. Res: Fluid and plasma instabilities; waves in liquids, gases and solids; physical oceanography; remote sensing. Mailing Add: NOAA/AOML/ORSL 15 Rickenbacker Causeway Miami FL 33149

APELLANIZ, JOSEPH E P, b Caguas, PR, Mar 31, 26; m 50; c 5. PHOTOGRAPHIC CHEMISTRY. Educ: Columbia Univ, AB, 48; Rutgers Univ, PhD(chem), 53. Prof Exp: Asst, Rutgers Univ, 49-51, res asst, 51; chemist, Stand Oil Co, Ind, 52-55; res specialist, Ansco Div, Gen Aniline & Film Corp, 55-63; RES DIR, CHEMCO PHOTOPRODS, INC, 64- Mem: Am Chem Soc; Am Soc Quality Control; Soc Photog Sci & Eng; Brit Chem Soc; Royal Photog Soc Gt Brit. Res: Photography; graphic arts. Mailing Add: Chemco Photoprods Inc Glen Cove NY 11542

APFEL, ROBERT EDMUND, b New York, NY, Mar 16, 43; m 68; c 1. ACOUSTICS, FLUID PHYSICS. Educ: Tufts Univ, BA, 64; Harvard Univ, MA, 67, PhD(appl physics), 70. Prof Exp: Res fel acoust, Harvard Univ, 70-71; asst prof, 71-74, ASSOC PROF ENG & APPL SCI, YALE UNIV, 74- Concurrent Pos: Independent consult acoust, 74- Honors & Awards: A B Wood Medal & Prize, Inst Physics, Gt Brit, 71; Biennial Award, Acoust Soc Am, 76. Mem: Fel Acoust Soc Am; Am Phys Soc; Am Asn Physics Teachers. Res: Study of superheated, supercooled, supersaturated and/or tensilely stressed liquids, possibly irradiated, with emphasis on acoustical techniques for producing phase changes or probing liquid properties. Mailing Add: Mason Lab Yale Univ New Haven CT 06520

APFELBERG, BENJAMIN, b Vienna, Austria, Feb 28, 97; m 39. PSYCHIATRY. Educ: Long Island Col Med, MD, 20. Prof Exp: Asst physician, Kings Park State Hosp, NY, 22-24; med officer, US Vet Hosp, NY, 24-30; sr psychiatrist, Bellevue Hosp, 30-49, assoc dir psychiat div, 49-61; MED DIR, LAW-PSYCHIAT PROJ, LAW SCH & MED SCH, NY UNIV, 63- Concurrent Pos: Clin prof, NY Univ, 30- Mem: Fel Am Psychiat Asn. Res: Clinical neurology; criminology; experiences with a new criminal code; psychiatric study of sex offenders. Mailing Add: Med Sch NY Univ 100 Washington Square E New York NY 10003

APGAR, BARBARA JEAN, b Tyler, Tex, Mar 4, 36. NUTRITION. Educ: Tex

Woman's Univ, BA, 57; Cornell Univ, MS, 59, PhD(biochem), 64. Prof Exp: RES CHEMIST, SOIL & NUTRIT LAB, US PLANT, 59- Mem: Am Soc Animal Sci; Am Inst Nutrit. Res: Effect of zinc deficiency on reproduction in the female. Mailing Add: US Plant Plant Soil & Nutrit Lab Ithaca NY 14853

APGAR, EDWARD G, b Toronto, Ont, Nov 30, 25; US citizen; m 50; c 2. PHYSICS. Educ: Rutgers Univ, BS, 50, PhD(physics), 57. Prof Exp: Mem staff, David Sarnoff Res Ctr, RCA Corp, 50-51 & Plasma Physics Lab, Princeton Univ, 57-65; prof physics, Bennington Col, 65-67; physicist, NASA Electronics Res Ctr, 67-70; Dot Transportation Systs Ctr, 70-75; vis scientist, 73-75, STAFF MEM, MASS INST TECHNOL, 75- Mem: Am Phys Soc; Am Vacuum Soc. Res: Electron tube devices; nuclear magnetic resonance; hydrogen diffusion in tantalum; mass spectroscopy; thin films; sputtering; surface physics of plasma devices; cosmology and particle physics; plasma-wall interactions and impurity effects in thermonuclear fusion; acoustics and particle physics. Mailing Add: Magnet Lab MIT 170 Albany St Cambridge MA 02139

APGAR, VIRGINIA, teratology, deceased

APGAR, WILLIAM P, b Newark, NJ, Feb 24, 31; m 55; c 6. ANIMAL NUTRITION. Educ: Rutgers Univ, BS, 54, MS, 61, PhD(dairy cattle nutrit), 63. Prof Exp: Lab technician dairy cattle nutrit, Rutgers Univ, 54-55; sales rep, NJ Power & Light Co, 57-58; res asst dairy cattle nutrit, Rutgers Univ, 58-63; asst prof animal sci, 63-68, ASSOC PROF ANIMAL & VET SCI, UNIV MAINE, 68- Mem: Am Dairy Sci Asn; Am Soc Animal Sci. Res: Ruminant nutrition; comparisons of nitrogen-fertilized grasses and legumes for dairy cattle; factors related to relative intake of various forages by dairy cattle. Mailing Add: Dept of Animal Sci Univ of Maine Orono ME 04473

APLEY, MARTYN LINN, b Fairbury, Nebr, May 27, 38. INVERTEBRATE ZOOLOGY, PHYSIOLOGICAL ECOLOGY. Educ: Kans State Univ, BS, 60; Syracuse Univ, MS, 63, PhD(zool), 67. Prof Exp: Fel, Woods Hole Oceanog Inst, 68; fel, Freshwater Biol Lab, Copenhagen Univ, 69; asst prof biol, Brooklyn Col, 69-72, fac res award, 70-71; ASST PROF BIOL, MERRIMACK COL, 75- Mem: AAAS. Res: Physiological ecology of mollusks, biorhythms and reproductive periodicites neurosecretion and hormonal control mechanisms in invertebrates. Mailing Add: Dept of Biol Merrimack Col North Andover MA 01845

APLINGTON, HENRY WEBSTER, JR, b New York, NY, Oct 24, 09; m 36; c 3. ANATOMY. Educ: Amherst Col, BA, 30; Columbia Univ, AM, 37; Cornell Univ, PhD(zool), 39. Prof Exp: From instr to assoc prof biol, Colby Col, 39-47; from asst prof to assoc prof, Muhlenberg Col, 47-53; from asst prof to assoc prof, Univ Pa, 50-51; from assoc prof anat to prof, Ohio State Univ, 53-68; PROF BIOL, AM INT COL, 68- Mem: Am Asn Anat. Res: Endocrines of amphibia. Mailing Add: Dept of Biol Am Int Col Springfield MA 01109

APONTE, GONZALO ENRIQUE, b Santurce, PR, July 15, 29; m 63. MEDICINE. Educ: Georgetown Univ, BS, 48; Jefferson Med Col, MD, 52. Prof Exp: Chief labs, US Naval Hosp & dep med exam, Govt of Guam, 57-59; res collab, Med Dept, Brookhaven Nat Lab, NY, 60; from asst prof to assoc prof path, 61-67, coord oncol teaching, 61, PROF PATH & HEAD DEPT, JEFFERSON MED COL, 67- Concurrent Pos: Nat Cancer Inst fel, 55-57, grant, 61; Markle scholar, 60; res collabr, Brookhaven Nat Lab, 61-; mem educ coun for med grads, Nat Bd Med Examr, 61-; dir clin labs, Jefferson Hosp, 67- Honors & Awards: Lindback Award, Jefferson Med Col, 62. Mem: AAAS; Col Am Path; Am Soc Clin Path; Asn Clin Sci. Res: Radiation biology and carcinogenesis; diseases of the kidney; studies of the thymus and pineal glands. Mailing Add: Jefferson Med Col 1020 Locust St Philadelphia PA 19107

APOSHIAN, HURAIR VASKEN, b Providence, RI, Jan 28, 26; m 48; c 3. CELL BIOLOGY, PHARMACOLOGY. Educ: Brown Univ, BS, 48; Univ Rochester, MS, 51, PhD(physiol), 54. Prof Exp: Asst, Univ Rochester, 48-49; from instr to asst prof pharmacol, Sch Med, Vanderbilt Univ, 54-59; USPHS fel biochem, Stanford Univ, 59-62; assoc prof microbiol, Sch Med, Tufts Univ, 62-67; head dept, 67-72, prof cell biol & pharmacol, Sch Med, Univ Md, Baltimore City, 67-75; PROF CELL & DEVELOP BIOL & CHMN DEPT, COL LIB ARTS, UNIV ARIZ, 75- Mem: AAAS; Am Soc Microbiol; Am Chem Soc; Am Soc Biol Chem; Am Soc Pharmacol & Exp Therapeut. Res: Development of gene therapy; enzymology of DNA and its precursors; biochemistry of bacterial and animal virus infection. Mailing Add: Biol Sci West Bldg 88 Univ of Ariz Tucson AZ 85721

APOSTOL, TOM M, b Helper, Utah, Aug 20, 23; m 59. MATHEMATICS, STATISTICS. Educ: Univ Wash, BS, 44, MS, 46; Univ Calif, Berkeley, PhD(math), 48. Prof Exp: Lectr math, Univ Calif, Berkeley, 48-49; C L E Moore instr, Mass Inst Technol, 49-50; from asst prof to assoc prof, 50-62, PROF MATH, CALIF INST TECHNOL, 62- Mem: Am Math Soc; Math Asn Am. Res: Theory of numbers. Mailing Add: Dept of Math Calif Inst Technol Pasadena CA 91125

APOTHEKER, DAVID, b New York, Aug 25, 21; m 51; c 5. POLYMER CHEMISTRY. Educ: Brooklyn Col, BA, 42; Polytech Inst Brooklyn, MS, 51; Georgetown Univ, PhD, 53. Prof Exp: Chemist, Schering Corp, 47-51; RES CHEMIST, E I DU PONT DE NEMOURS & CO, INC, 54- Concurrent Pos: Lectr, Ursuline Col, 56-58 & Univ Louisville, 58-60. Mem: Am Chem Soc; Sigma Xi. Res: Emulsion polymerization; latex applications; coordination polymerization of olefins; coordination catalyst studies; synthesis of fluoroelastomers; vulcanization accelerators. Mailing Add: 2105 Dunhill Dr Wilmington DE 19810

APP, ALVA AGEE, b Bridgeton, NJ, Feb 19, 32; m 55; c 3. PLANT SCIENCE. Educ: Cornell Univ, BS, 53; Rutgers Univ, MS, 55, PhD(soils), 56. Prof Exp: Tech adv, Valley Green Mushroom Farms, Pa, 59-61; res assoc, McCollum Pratt Inst, Johns Hopkins Univ, 61-64; assoc seed physiologist, 64-69, PROG DIR CELL PHYSIOL & VIROL, BOYCE THOMPSON INST, YONKERS, 69- Concurrent Pos: Adj prof biochem, City Univ NY, 73-; consult, Biochem Prog, US Army Med Res & Develop Command, 75- Mem: Fel AAAS; Am Soc Plant Physiol; sr mem Am Chem Soc; Tissue Cult Asn. Res: Soil and plant chemistry. Mailing Add: 22 Stornowaye Chappaqua NY 10514

APPEL, JEFFREY ALAN, b Cleveland, Ohio, Aug 11, 42; m 65; c 2. ELEMENTARY PARTICLE PHYSICS. Educ: Williams Col, AB, 64; Harvard Univ, MA, 65, PhD(physics), 69. Prof Exp: Res assoc physics, Nevis Labs, Columbia Univ, 68-70, asst prof, Columbia Univ, 70-74, sr res assoc, Nevis Labs, 74-75; PHYSICIST, FERMI NAT ACCELERATOR LAB, 75- Mem: Am Phys Soc. Res: Discrete symmetries, resonance phenomena and weak and electromagnetic phenomena. Mailing Add: Fermi Lab PO Box 500 Batavia IL 60510

APPEL, KENNETH I, b Brooklyn, NY, Oct 8, 32; m 59; c 3. MATHEMATICS. Educ: Queens Col, NY, BS, 53; Univ Mich, MA, 56, PhD(math), 59. Prof Exp: Mathematician, Inst Defense Anal, 59-61; asst prof, 61-67, ASSOC PROF MATH, UNIV ILL, URBANA-CHAMPAIGN, 67- Mem: Am Math Soc; Math Asn Am; Asn Symbolic Logic. Res: Chromatic graph theory and combinations. Mailing Add: Dept of Math Univ of Ill Urbana IL 61801

APPEL, MAX J, b Giessen, Ger, Dec 13, 29; m 61; c 4. VETERINARY VIROLOGY. Educ: Vet Col, Hannover, DVM, 56; Cornell Univ, PhD(vet virol), 67. Prof Exp: Res asst vet med, Univ Munich, 57-59; fel, Univ Sask, 59-60; res off, Animal Dis Res Inst, Hull, Que, 61-64; res asst vet virol, Vet Virus Res Inst, 64-67, asst prof, 67-70, ASSOC PROF VET VIROL, NY STATE COL VET MED, CORNELL UNIV, 70- Mem: Electron Micros Soc Am; Am Vet Med Asn; Am Soc Microbiol. Res: Viral diseases of domestic animals. Mailing Add: NY State Col Vet Med Cornell Univ Ithaca NY 14850

APPEL, STANLEY HERSH, b Boston, Mass, May 8, 33; m 56; c 2. NEUROLOGY. Educ: Harvard Univ, AB, 54; Columbia Univ, MD, 60; Am Bd Psychiat & Neurol, dipl, 68. Prof Exp: Intern med, Mass Gen Hosp, 60-61; resident neurol, Mt Sinai Hosp, 61-62; res assoc, Lab Molecular Biol, NIH, 62-64; assoc neurol, Med Ctr, Duke Univ, 64-65; chief res assoc, Sch Med, Univ Pa, 65-66, asst prof, 66-67; assoc prof neurol, 67-71, ASSOC PROF BIOCHEM, MED CTR, DUKE UNIV, 68-, PROF NEUROL, 71-, CHIEF DIV, 69- Concurrent Pos: USPHS res career develop award, 65- Mem: Am Acad Neurol; Am Soc Biol Chem; Am Soc Clin Invest; Am Neurol Asn. Res: Molecular neurobiology; neurochemistry; demyelination; synapse function; muscle membranes and disease. Mailing Add: Dept of Med Duke Univ Med Ctr Durham NC 27706

APPEL, WARREN CURTIS, b Cheyenne, Wyo, Oct 19, 44; m 69; c 1. PHARMACOLOGY. Educ: Univ Wyo, BSc, 67; Univ Wash, MSc, 70, PhD(pharmacol), 71. Prof Exp: Lectr pharmacol, Univ Man, 72-73; SCI ADV, BUR DRUGS, HEALTH & WELFARE CAN, 73- Mem: AAAS. Res: Mechanisms of drug dependence. Mailing Add: Bur of Drugs 355 River Rd Vanier Tower Ottawa ON Can

APPELBAUM, EMANUEL, b Poland, Apr 14, 94; nat US; m 27; c 2. INTERNAL MEDICINE. Educ: Columbia Univ, AB, 16, MD, 18. Prof Exp: Instr med, Col Med, 33-43, assoc clin prof, 43-50, assoc clin prof, Postgrad Med Sch, 50-54, assoc prof, 54-57, PROF CLIN MED, POSTGRAD MED SCH, NY UNIV, 57- Concurrent Pos: Pathologist, NY Health Dept, 21-28, bacteriologist, 28-41, chief div acute infections of cent nervous syst, 41-64; asst vis physician, Bellevue Hosp, 26-38, vis physician, 38-; vis physician, Willard Parker Hosp, 43-55; dir med, Sydenham Hosp, 48-66; consult physician, Long Beach Mem Hosp, 48-, NY Infirmary, 49-, Knickerbocker Hosp, 57-, Trafalgar Hosp, 58-, Beth Israel Hosp, Newark, 60- & Sydenham Hosp, 66-; attend physician, Beth David Hosp, 50-61 & Univ Hosp, 54- Mem: Fel AMA; fel Am Pub Health Asn; fel Am Col Physicians; fel NY Acad Sci; fel NY Acad Med. Res: Diagnosis and treatment of acute infections of central nervous system; meningitis, encephalitis, poliomyelitis, rabies and tetanus; chemotherapy and antibiotic therapy of bacterial infections. Mailing Add: 910 Park Ave New York NY 10021

APPELBAUM, JOEL A, b Brooklyn, NY, Dec 30, 41; m 63; c 1. SOLID STATE PHYSICS. Educ: City Col New York, BS, 63; Univ Chicago, MS, 64, PhD(physics), 66. Prof Exp: Teaching asst physics, Univ Chicago, 63-64; mem tech staff, Bell Labs, NJ, 67; asst prof, Univ Calif, Berkeley, 67-68; MEM TECH STAFF, BELL LABS, 68- Mem: Am Phys Soc. Res: Solid state physics, particularly electron tunneling, surface physics and magnetism. Mailing Add: Lab 111 Bell Labs 600 Mountain Ave Murray Hill NJ 07974

APPELGREN, WALTER PHON, b Macomb, Ill, Nov 10, 28; m 61; c 3. MICROBIOLOGY, HEALTH SCIENCES. Educ: Northwestern Univ, BS, 50; Univ Mich, MS, 53, PhD(microbiol), 67. Prof Exp: Res chemist, Continental Can Corp, 54-55; asst plant mgr, Nat Can Corp, 55-57; instr org chem, Sacramento City Col, 57-61; asst prof microbiol, Wis State Univ, Oshkosh, 66-67; asst prof, 67-71, ASSOC PROF MICROBIOL, NORTHERN ARIZ UNIV, 71-, DIR CTR HEALTH SCI, 73- Mem: Am Soc Microbiol; AAAS; Am Soc Allied Health Professions; Am Soc Med Technol. Res: Host-parasite relationships in glomerulonephritis; oral microbiology; septic shock. Mailing Add: Box 15045 Northern Ariz Univ Flagstaff AZ 86001

APPELL, RAYNOR NORBERT, b Chicago, Ill, Oct 3, 23; m 51. CLINICAL PHARMACOLOGY. Educ: Augustana Col, Ill, BA, 48; Univ Wis, MS, 57, PhD(bact), 61. Prof Exp: Res bacteriologist, Baxter Labs, Ill, 50-55; sr bacteriologist, Animal Dis Diag Lab, Wis, 58-60; sr res microbiologist, 60-65, group leader, 65-68, SECT HEAD, ABBOTT LABS, 68- Mem: AAAS; Am Soc Microbiol; NY Acad Sci. Res: Antibacterial, antifungal and antiviral agents; bacterial vaccines. Mailing Add: Dept 421 Abbott Labs North Chicago IL 60064

APPELLA, ETTORE, b Castronuovo, Italy, Aug 5, 33; US citizen. IMMUNOLOGY. Educ: Univ Rome, MD, 59. Prof Exp: Res asst biol, Johns Hopkins Univ, 60-63; vis scientist, Lab Molecular Biol, Nat Inst Arthritis & Metab Dis, 63-64, MED OFFICER IMMUNOL, LAB BIOL, NAT CANCER INST, 67- Concurrent Pos: Am Cancer Soc fel, Nat Inst Arthritis & Metab Dis, 64-66. Res: Biochemistry; protein chemistry. Mailing Add: Lab of Biol Nat Cancer Inst Bethesda MD 20014

APPELMAN, EVAN HUGH, b Chicago, Ill, June 6, 35; m 60; c 2. INORGANIC CHEMISTRY. Educ: Univ Chicago, AB, 53, MS, 55; Univ Calif, Berkeley, PhD(chem), 60. Asst chemist, 60-63, ASSOC CHEMIST, ARGONNE NAT LAB, 63- Honors & Awards: Distinguished Performance Award at Argonne Nat Lab, Univ Chicago, 75. Mem: Fedn Am Scientists; AAAS; Am Chem Soc. Res: Inorganic reaction kinetics and equilibria in aqueous solution, particularly kinetics of oxidation-reduction reactions; chemistry of the less familiar elements, particularly astatine, xenon and radon; halogen chemistry. Mailing Add: Argonne Nat Lab 9700 S Cass Ave Argonne IL 60439

APPELQUIST, THOMAS, b Emmetsburg, Iowa, Nov 1, 41; m 65; c 1. THEORETICAL PHYSICS. Educ: Ill Benedictine Col, BS, 63; Cornell Univ, PhD(theoret physics), 68. Prof Exp: Res assoc theoret physics, Stanford Linear Accelerator Ctr, 68-70; from asst prof to assoc prof, Harvard Univ, 70-75; PROF THEORET PHYSICS, YALE UNIV, 75- Concurrent Pos: AP Sloan Found fel, 74-76. Mem: Fedn Am Sci; Am Phys Soc. Res: Applications of quantum field theory to the interactions of elementary particles. Mailing Add: Dept of Physics Yale Univ New Haven CT 06520

APPELT, GLENN DAVID, b Yoakum, Tex, Aug 24, 35; m 58. PHARMACOLOGY. Educ: Univ Tex, BS, 57, MS, 59; Univ Colo, PhD(pharm), 63. Prof Exp: Asst prof pharmacol, Univ Tex, Austin, 63-67; ASSOC PROF PHARMACOL, UNIV COLO, BOULDER, 67- Mem: Am Pharmaceut Asn. Res: Biochemical pharmacology; effects of drugs on brain coenzyme levels; cardiac oxidative processes. Mailing Add: Sch of Pharm Univ of Colo Boulder CO 80302

APPENZELLER, OTTO, b Czernowitz, Romania, Dec 11, 27; m 56; c 3. NEUROLOGY, MEDICINE. Educ: Univ Sydney, MB & BS, 57, MD, 66; Univ London, PhD, 63. Prof Exp: Jr resident, Royal Prince Alfred Hosp, Sydney, Australia, 57-58, sr resident, 58-59, med registr, 59-60; clin asst, Nat Hosp Nerv Dis, London,

Eng, 61-62; asst prof neurol, Col Med, Univ Cincinnati, 65-67; assoc prof med, 67-70, PROF NEUROL & MED, SCH MED, UNIV N MEX, 70- Concurrent Pos: Mass Gen Hosp & Harvard Med Sch fel, 63-64; Harvard Med Sch fel, 64-65; Boston Med Found fel, 65; chief neurol sect, Vet Admin Hosp, Cincinnati, Ohio, 65-67; consult neurologist, Vet Admin Hosp, Albuquerque, NMex, 68. Mem: AAAS; AMA; fel Am Acad Neurol; Brit Med Asn; Royal Australasian Col Physicians. Res: Clinical neurology; experimental neuropathology; physiology and pathology of autonomic and peripheral nervous systems; neuropathology; biochemistry of peripheral nerves; headache and its mechanism; immunology as applied to the nervous system. Mailing Add: Univ of NMex Sch of Med Albuquerque NM 87106

APPERSON, LESTER DONALD, organic chemistry, see 12th edition

APPINO, JAMES B, b Benton, Ill, Apr 27, 31; m 55; c 4. INDUSTRIAL PHARMACY. Educ: Purdue Univ, BS, 54, MS, 58, PhD(indust pharm), 60. Prof Exp: Sr res scientist, Armour Pharmaceut Co, 60-62; mgr pharm, Baxter Labs, Inc, 62-69; DIR PHARM RES, McNEIL LABS, INC, 69- Mem: Am Pharm Asn; Acad Pharmaceut Sci; AAAS. Res: Product development and dosage form design; analytical development in areas of purity, methods development and stability; biopharmaceutics, including dissolution methodology and physical characterization of drugs. Mailing Add: McNeil Labs Inc Camp Hill Rd Ft Washington PA 19034

APPLE, EUGENE FRED, inorganic chemistry, see 12th edition

APPLE, JAMES WILBUR, b Ann Arbor, Mich, Sept 3, 15; m 38; c 2. ENTOMOLOGY. Educ: Ohio State Univ, BS, 37; Iowa State Col, MS, 40; Univ Ill, PhD(entom), 49. Prof Exp: Asst entom, Iowa State Col, 39-42; assoc entomologist, State Nat Hist Surv, Ill, 42-49; assoc prof, 49-54, PROF ENTOM, UNIV WIS-MADISON, 54- Mem: Entom Soc Am. Res: Biology and control of vegetable and field crop insects. Mailing Add: Dept of Entom Univ of Wis Madison WI 53706

APPLE, JAY LAWRENCE, b Guilford Co, NC, Jan 8, 26; m 45. PHYTOPATHOLOGY. Educ: NC State Col, BS, 49, MS, 53, PhD(plant path), 55. Prof Exp: From res instr to assoc prof, 49-63, dir inst biol sci, 67-71, PROF PLANT PATH & GENETICS, 63-, ASST DIR ACAD AFFAIRS & RES FOR BIOL SCI, 71- Concurrent Pos: Plant path adv, NC State Univ-US AID Mission, Peru, 63-65, chief-of-party, 65-67; consult int develop, Univ Calif, Berkeley, 72-; consult pest mgt, Environ Protection Agency, 73-74. Mem: AAAS; Am Inst Biol Sci; Am Phytopath Soc. Res: Genetics of disease resistance in plants; genetics of microorganisms; international agricultural development. Mailing Add: Sch Agr & Life Sci NC State Univ Raleigh NC 27607

APPLE, MARTIN ALLEN, b Duluth, Minn, Sept 17, 39; m; c 4. MOLECULAR PHARMACOLOGY, ONCOLOGY. Educ: Univ Minn, Minneapolis, AB, 59, MSc, 62; Univ Calif, PhD(biochem), 67. Prof Exp: Asst microbiol, Med Sch, Univ Minn, 60-62; asst res biochemist, Sch Med, 66-68 & Cancer Res Inst, 68-75, lectr biochem, Sch Med, 68-70, asst prof pharmacol & exp therapeut in residence, 70-75, asst prof pharmaceut chem in residence, Sch Pharm, 71-75, ASSOC PROF PHARMACOL, & EXP THERAPEUT IN RESIDENCE, SCH MED, ASSOC PROF PHARMACEUT CHEM IN RESIDENCE, SCH PHARM & ASSOC RES BIOCHEMIST, CANCER RES INST, UNIV CALIF, SAN FRANCISCO, 75- Concurrent Pos: Consult enzyme, Vet Admin Hosps, 63-69 & Nat Cancer Inst, 73-74; mem bd of regents, Am Col Clin Pharmacol. Honors & Awards: Am Cancer Soc Award, 69. Mem: AAAS; Am Soc Microbiol; Am Fedn Clin Res; fel Am Inst Chem; fel Am Col Clin Pharmacol. Res: Biochemical and clinical pharmacology; enzyme rate modulation; biochemical oncology; new drug designs. Mailing Add: Dept of Pharm & Exp Therapeut Univ of Calif Sch of Med San Francisco CA 94143

APPLE, RICHARD SEYMOUR, chemistry, see 12th edition

APPLE, SPENCER BUTLER, JR, b Kansas City, Mo, June 11, 12; m 40; c 3. HORTICULTURE. Educ: Agr & Mech Col Tex, BS, 33, MS, 36; Wash State Univ, PhD(hort), 53. Prof Exp: From instr to asst prof hort, Agr & Mech Col Tex, 35-41; res asst & exten specialist, Mich State Col, 41-46; exten specialist, Agr & Mech Col Tex, 47-48, assoc prof, 49-50; assoc prof, 50-55, horticulturist & head dept hort, 55-73, prof, 73-75, EMER PROF HORT, ORE STATE UNIV, 75- Concurrent Pos: Gen Educ Bd fel, 48-49. Mem: Fel AAAS; Am Soc Plant Physiol; Am Soc Hort Sci. Res: Plant nutrition; physiological problems in vegetable crop production. Mailing Add: Dept of Hort Ore State Univ Corvallis OR 97331

APPLE, WILLIAM S, b Spokane, Wash, July 28, 18; m 42; c 1. PHARMACY. Educ: Univ Wis, BS, 49, MBA, 51, PhD(pharm admin), 54. Hon Degrees: DSc, Long Island Univ & Union Univ. Prof Exp: Instr pharm & coord pharm exten div, Univ Wis, 51-53, asst prof pharm admin & chmn dept, 53-58; vpres, pres & chmn bd dirs, Wis Pharmaceut Asn, 55-58; asst secy, 58-59, EXEC DIR, AM PHARMACEUT ASN, 59- Concurrent Pos: Past pres, Nat Drug Trade Conf; charter bd mem, Community Health, Inc; mem comt for nat health ins, NY Acad Sci; bd mem, Am Asn World Health, Inc; US comt, WHO; US rep, Coun Int Pharmaceut Fedn, 59-74, vpres, 74- Honors & Awards: Lascoff Mem Award, 61; Col Pharmaceut Chem Award, Chile, 61; Remington Honor Medal, 61; Distinguished Serv Award, Wayne State Univ, 62; Hugo H Schaefer Medal, 66. Res: Public health; economics of pharmaceutical service and products, including voluntary pre-payment plans and government welfare programs; executive management of scientific, practical and educational endeavors of professional associations. Mailing Add: Am Pharmaceut Asn 2215 Constitution Ave NW Washington DC 20037

APPLEBAUM, CHARLES H, b Newark, NJ, Nov 26, 42; m 64; c 2. MATHEMATICAL LOGIC. Educ: Case Inst Technol, BS, 64; Rutgers Univ, MS, 66, PhD(math), 69. Prof Exp: Asst math, Rutgers Univ, 64-69; asst prof, 69-74, fac res grant, summer 70, ASSOC PROF MATH, BOWLING GREEN STATE UNIV, 74- Mem: Math Asn Am; Am Math Soc; Asn Symbolic Logic. Res: Theory of recursive equivalence types and recursive functions and their application to various areas of algebra. Mailing Add: Dept of Math Bowling Green State Univ Bowling Green OH 43403

APPLEBAUM, EDMUND, b New York, NY, Feb 1, 99; m 38; c 1. DENTISTRY. Educ: NY Col Dent, DDS, 22. Prof Exp: From instr to asst prof, 28-40, ASSOC PROF DENT HISTOL, SCH DENT & ORAL SURG, COLUMBIA UNIV, 40- Mem: AAAS; Am Dent Asn; Int Asn Dent Res. Res: Dental pathology; dental caries; polarized light and hard dental tissues and embryology. Mailing Add: Sch of Dent & Oral Surg Columbia Univ New York NY 10032

APPLEBY, ALAN, b Newcastle, Eng, Apr 25, 37; m 60; c 2. RADIATION CHEMISTRY. Educ: Univ Durham, BSc, 58, PhD(radiation chem), 63. Prof Exp: Res assoc radiation chem, Univ Durham, 61-63; Brookhaven Nat Lab, 63-65; sr sci officer, Radiochem Ctr, UK Atomic Energy Authority, Eng, 65-67; asst prof, 67-71, ASSOC PROF RADIATION SCI, RUTGERS UNIV, 71- Mem: Am Chem Soc; Brit Chem Soc. Res: Radiation chemistry of water and aqueous systems; radiation induced isotope exchange reactions. Mailing Add: Radiation Sci Doolittle Bldg Busch Campus Rutgers Univ New Brunswick NJ 08903

APPLEBY, ARNOLD PIERCE, b Formoso, Kans, Oct 24, 35; m 56; c 2. WEED SCIENCE. Educ: Kans State Univ, BS, 57, MS, 58; Ore State Univ, PhD(herbicide physiol), 62. Prof Exp: Instr agron, 59-62, asst prof, Pendleton br exp sta, 62-63 & Univ, 63-67, assoc prof, 67-72, PROF AGRON, ORE STATE UNIV, 72- Mem: Am Soc Agron; Crop Sci Soc Am; Weed Sci Soc Am. Res: Agronomic and physiological aspects of herbicides and plant-growth regulators. Mailing Add: Dept of Agron Crop Sci Ore State Univ Corvallis OR 97331

APPLEBY, JAMES E, b Canton, Ohio, Aug 6, 36. ENTOMOLOGY. Educ: Ohio State Univ, BS, 59, MS, 60, PhD(entom), 64. Prof Exp: From asst prof to assoc prof, 64-75, PROF ENTOM, ILL NATURAL HIST SURV, 75- Res: Life histories and control of insects and mites injurious to trees, shrubs and garden flowers. Mailing Add: Ill Natural Hist Surv 165 Natural Resources Bldg Urbana IL 61801

APPLEBY, RALPH CARSON, b Eldon, Iowa, Nov 16, 19; m 50; c 2. DENTISTRY. Educ: Univ Iowa, BS, 41, DDS, 51, MS, 53. Prof Exp: From instr to asst instr, 51-56, assoc prof & head dept, 56-70, PROF PROSTHETICS & COORDR DENT STUDENT AFFAIRS & DIR CONTINUING EDUC, UNIV IOWA, 70- Concurrent Pos: Consult, Vet Admin Hosp, Iowa City. Mem: AAAS; fel Am Col Dent; Am Prosthodontic Soc; Int Asn Dent Res. Res: Prosthetic dentistry of both complete and partial dentures; vertical dimensions in denture construction. Mailing Add: Dept of Prosthetics Univ of Iowa Col of Dent Iowa City IA 52240

APPLEDORF, HOWARD, b Stoneham, Mass, Jan 16, 41. NUTRITION, FOOD SCIENCE. Educ: Tufts Univ, BS, 62; Mass Inst Technol, SM, 64, PhD(nutrit), 68. Prof Exp: Prof trainee nutrit, Gen Foods Res Ctr, summer 64; fel, Univ Calif, Berkeley, 68-69; asst prof, 69-74, ASSOC PROF NUTRIT, UNIV FLA, 74- Mem: AAAS; Inst Food Technol; Am Inst Nutrit; NY Acad Sci. Res: Appetite control; endocrine and vitamin interrelationships; thiamine metabolism; neural control of metabolism; novel protein sources. Mailing Add: Dept of Food Sci Univ of Fla Gainesville FL 32601

APPLEGARTH, DEREK A, b London, Eng, July 5, 37; Can citizen; m 61; c 2. CLINICAL BIOCHEMISTRY. Educ: Univ Durham, BSc, 58, PhD, 62. Prof Exp: Fel, Univ BC, 62-63; asst clin chemist, Vancouver Gen Hosp, 63-66; asst prof, 66-71, ASSOC PROF PEDIAT, UNIV BC, 71-; DIR BIOCHEM DIS LAB, CHILDREN'S HOSP, VANCOUVER, 69- Concurrent Pos: Clin instr, Univ BC, 63-66. Mem: Can Soc Clin Chem. Res: Biochemical diseases of children. Mailing Add: Dept of Pediat Univ of BC Vancouver BC Can

APPLEGATE, ARTHUR L, embryology, cell physiology, see 12th edition

APPLEGATE, HOWARD GEORGE, b Philadelphia, Pa, Mar 9, 22. PLANT PHYSIOLOGY. Educ: Colo State Univ, BS, 50, MS, 52; Mich State Univ, PhD(physiol), 56. Prof Exp: Asst prof floricult, Univ Conn, 56-58; asst biochemist, State Col Wash, 58-60; asst prof plant physiol, Southern Ill Univ, 60-61; assoc prof, Ariz State Univ, 61-62 & Tex A&M Univ, 62-69; ASSOC PROF PLANT PHYSIOL, UNIV TEX, EL PASO, 69- Mem: Am Soc Plant Physiol; Am Genetic Asn; Bot Soc Am. Res: Growth regulators; air pollution. Mailing Add: Dept of Civil Eng Univ of Tex El Paso TX 79902

APPLEGATE, JAMES EDWARD, b South Amboy, NJ, Aug 28, 42; m 65; c 2. WILDLIFE BIOLOGY. Educ: Rutgers Univ, BS, 64; Pa State Univ, MS, 66, PhD(zool), 68. Prof Exp: NIH trainee arbovirus ecol, Pa State Univ, 64-66, teaching asst zool, 66-67, USPHS fel parasitol, 67-68; res parasitologist, Naval Med Res Inst, 68-71; ASST PROF WILDLIFE BIOL, RUTGERS UNIV, 71- Mem: Wildlife Dis Asn; Wildlife Soc; Am Inst Biol Sci. Res: Ecology of malaria; social aspects of wildlife management. Mailing Add: Dept of Hort & Forestry Col Agr & Environ Sci Rutgers Univ New Brunswick NJ 08903

APPLEGATE, JAMES KEITH, b Macomb, Ill, Mar 26, 44; m 67; c 2. EXPLORATION GEOPHYSICS, GEOTECHNICAL ENGINEERING. Educ: Colo Sch Mines, BS, 66, MS, 69, PhD(geophys eng), 74. Prof Exp: Geophysicist, Marathon Oil Co, 68-70; ASST PROF GEOPHYS, BOISE STATE UNIV, 73- Concurrent Pos: Spec consult geophys, Woodward-Clyde Consult, 73- Mem: Soc Explor Geophysicists; Seismol Soc Am; Geol Soc Am; Asn Eng Geologists. Res: High resolution seismic reflection techniques; geothermal exploration; rock and soil properties from geophysical techniques. Mailing Add: 2887 Snowflake Dr Boise ID 83706

APPLEGATE, LYNN E, b Dayton, Ohio, Apr 3, 41; m 61; c 3. WATER CHEMISTRY. Educ: Univ Dayton, BS, 63; Ind Univ, PhD(chem), 67. Prof Exp: SR RES CHEMIST, E I DU PONT DE NEMOURS & CO, INC, 67- Mem: Am Chem Soc; Am Inst Chemists. Res: Desalination of water via reverse osmosis; water treatments for potable and industrial systems; water pollution control systems; polymer chemistry. Mailing Add: Permasep Prod E I du Pont de Nemours & Co Wilmington DE 19898

APPLEGATE, RICHARD LEE, b Mt Carmel, Ill, Jan 15, 36; m 57; c 2. ENTOMOLOGY, ZOOLOGY. Educ: Univ Southern Ill, BA, 59, MA, 61. Prof Exp: Fishery biologist, Va Comn Game & Inland Fisheries, 61-63; fishery res biologist, Bur Sport Fisheries & Wildlife, SCent Reservoir Invests, 63-67; fishery res biol & asst leader aquatic biol & fisheries, Bur Sport Fisheries & Wildlife, SDak Coop Fishery Unit, 67-72, INSTR WILDLIFE & FISHERIES SCI, SDAK STATE UNIV, 72- Concurrent Pos: Mem task force team 4 conserv aquatic ecosyst, Int Biol Prog, Nat Acad Sci-Nat Res Coun, 67- Mem: Am Fisheries Soc; Am Soc Limnol & Oceanog. Res: Aquatic biology and ecology; limnology; fisheries. Mailing Add: Dept Wildlife & Fisheries Sci SDak State Univ Brookings SD 57006

APPLEGATE, SHELTON P, b Richmond, Va, Nov 24, 28; m 53; c 3. VERTEBRATE PALEONTOLOGY, ICHTHYOLOGY. Educ: Univ Richmond, BA, 56; Univ Va, MS, 57; Univ Chicago, PhD(paleozool), 60. Prof Exp: Instr biol, Univ Chicago, 58-61; assoc prof, Ark State Col, 61-62; instr, Duke Univ, 62-63; ASSOC CUR VERT PALEONT, LOS ANGELES COUNTY MUS NATURAL HIST, 63- Concurrent Pos: Consult shark res panel, Am Inst Biol Sci. Mem: Soc Vert Paleont; Am Soc Ichthyologists & Herpetologists; Am Soc Zoologists. Res: Ecology and evolution of sharks and rays; origin and early evolution of the bony fishes. Mailing Add: LA County Mus Natural Hist Exposition Pk 900 Exposition Blvd Los Angeles CA 90007

APPLEGATE, VERNON CALVERT, b New York, NY, May 13, 19; m 44; c 3. FISHERIES. Educ: Univ Mich, BSc, 43, MSc, 47, PhD(zool), 50. Prof Exp: Jr fisheries biologist, Inst Fisheries Res, Mich State Conserv Dept, 45-50; supvry fishery res biologist, US Bur Com Fisheries, 50-71; CONSULT AQUATIC ECOL & MGT, 71- Honors & Awards: Award, US Dept Interior, 61. Mem: Am Soc Zool; Am Fisheries Soc; Am Soc Ichthyologists & Herpetologists; fel Am Inst Fishery Res Biologists; Wildlife Soc. Res: Biology and economics of predator fish populations;

fishery engineering, especially fish weirs, electric fish screens and electrofishing techniques; bioassay techniques and development of selective toxins for control of aquatic species; ecology and management of large freshwater lakes. Mailing Add: Box 29 Rogers City MI 49779

APPLEMAN, DANIEL EVERETT, b Berkeley, Calif, Apr 11, 31; m 67. GEOLOGY, CRYSTALLOGRAPHY. Educ: Calif Inst Technol, BS, 53; Johns Hopkins Univ, MA, 54, PhD(geol), 56. Prof Exp: Geologist, US Geol Surv, 54-74; CRYSTALLOGR, NAT MUS NATURAL HIST, SMITHSONIAN INST, 74- Concurrent Pos: Prof lectr, George Washington Univ, 64- Mem: Geol Soc Am; Mineral Soc Am; Am Crystallog Asn; Mineral Soc Gt Brit & Ireland; Am Geophys Union. Res: X-ray crystallography; crystal structures of uranium minerals; silicates; geochemistry. Mailing Add: NHB 119 Smithsonian Inst Washington DC 20560

APPLEMAN, HERBERT STANLEY, atmospheric physics, see 12th edition

APPLEMAN, M MICHAEL, b Los Angeles, Calif, June 13, 33; m 58; c 3. BIOCHEMISTRY. Educ: Univ Calif, Berkeley, BA, 57; Univ Wash, Seattle, PhD(biochem), 62. Prof Exp: Technician, Berkeley Lab, Nat Canners Asn, 57-58; NSF res fel metab, Inst Biochem Res, Argentina, 63-65; actg asst prof biochem, Univ Wash, Seattle, 65-66; asst prof, 66-70, ASSOC PROF BIOCHEM, UNIV SOUTHERN CALIF, 70- Mem: AAAS; Am Chem Soc; Am Soc Biol Chemists. Res: Mechanism of hormone action; regulation of metabolism; enzymology. Mailing Add: Dept of Biol Sci Univ of Southern Calif Los Angeles CA 90007

APPLEMAN, MARIA DUARTE, b Guatemala City, Guatemala, Jan 1, 46; US citizen. BACTERIOLOGY. Educ: Calif State Univ, Long Beach, BS, 68; Univ Southern Calif, PhD(bact), 72. Prof Exp: Res assoc bact, Univ Southern Calif, 73-74; dir anaerobe lab, Martin Luther King Jr Gen Hosp, 74-75; RES ASSOC MICROBIAL PHYSIOL, VA POLYTECH INST & STATE UNIV, 75- Mem: Am Soc Microbiol. Res: Physiology of intestinal anaerobes and their possible role in the development of colon cancer. Mailing Add: Anaerobe Lab Va Polytech Inst PO Box 49 Blacksburg VA 24060

APPLEMAN, MILO DON, b Wellston, Mo, Dec 3, 09; m 36; c 1. FOOD BACTERIOLOGY. Educ: Univ Ill, AB, 31, MS, 35, PhD(food bact), 40; Am Bd Med Microbiol, dipl microbiol & pub health; Am Intersoc Acad Cert Sanit, dipl. Prof Exp: Asst soil bact, Univ Ill, 35-37, first asst, 37-39, assoc, 39-42, asst prof, 42-47; from assoc prof to prof food bact, 47-75, chmn dept, 48-63, EMER PROF BACT, UNIV SOUTHERN CALIF, 75- Concurrent Pos: Fund Advan Educ fel, Aberdeen Univ, 54-55; NIH spec fel, Torry Res Sta, 62-63, mem study sect bact & mycol, 63-67; adv comn, Los Angeles County Pub Health, 68-; mem subcomt microbial stand foods, AFDOUS & US Food & Drug Admin, 63-; vpres, Res & Develop, Daylin Labs, Inc, 71-; consult, Pac Kenyon Inc, Long Beach, Calif. Mem: Am Soc Microbiol; Inst Food Tech; fel Am Acad Microbiol; Can Inst Food Tech; Brit Soc Appl Bact. Res: Bacterial physiology; public health; food spoilage control. Mailing Add: Dept of Biol Sci Univ Southern Calif Univ Park Los Angeles CA 90007

APPLEQUIST, DOUGLAS EINAR, b Salt Lake City, Utah, Oct 29, 30; m 66; c 2. ORGANIC CHEMISTRY. Educ: Univ Calif, Berkeley, BS, 52; Calif Inst Technol, PhD(chem), 55. Prof Exp: From instr to assoc prof, 55-64, PROF ORG CHEM, UNIV ILL, URBANA-CHAMPAIGN, 64- Mem: Am Chem Soc. Res: Small-ring compounds; photochemical reactions; bridgehead displacements and eliminations; free-radical reactions. Mailing Add: 345 Roger Adams Lab Univ of Ill Urbana IL 61801

APPLEQUIST, JON BARR, b Salt Lake City, Utah, Mar 19, 32; m 60; c 5. BIOPHYSICAL CHEMISTRY. Educ: Univ Calif, BS, 54; Harvard Univ, PhD(chem), 59. Prof Exp: Asst prof chem, Univ Calif, Berkeley, 58-60 & Columbia Univ, 61-65; assoc prof, 65-68, PROF BIOPHYS, IOWA STATE UNIV, 68- Mem: Am Chem Soc; Biophys Soc. Res: Molecular conformations; molecular optical properties. Mailing Add: Dept of Biochem & Biophys Iowa State Univ Ames IA 50011

APPLETON, B R, b Pampa, Tex, Nov 24, 37; m 59; c 3. SOLID STATE PHYSICS. Educ: Univ Mo, BS, 60; Rutgers Univ, MS, 64, PhD(physics), 66. Prof Exp: Mem res staff, Radio Corp Am, 61-63 & Bell Tel Labs Inc, 66-67; MEM RES STAFF, OAK RIDGE NAT LAB, 67- Mem: Am Phys Soc; AAAS; Böhmische Phys Soc. Res: Basic ion-solid interactions; alteration of materials properties by ion bombardment and ion implantation techniques. Mailing Add: Solid State Div Oak Ridge Nat Lab Oak Ridge TN 37830

APPLETON, GEORGE LUDWIG, b Pittsburgh, Pa, Mar 26, 26; m 47; c 2. SOLID STATE PHYSICS. Educ: Carnegie Inst Technol, BS, 49; Univ Southern Calif, PhD(physics), 54. Prof Exp: From instr to assoc prof, 53-61, PROF PHYSICS, CALIF STATE UNIV, LONG BEACH, 61- Mem: Am Phys Soc. Res: Low temperature transport properties of solids. Mailing Add: Dept of Physics Calif State Univ Long Beach CA 90804

APPLETON, GEORGE SANDERS, b Bowling Green, Ky, July 2, 27; m 49; c 3. AGRICULTURAL MICROBIOLOGY. Educ: Univ Tenn, BA, 51, MS, 55. Prof Exp: Jr bacteriologist, Cent Lab, State Health Dept, Nashville, Tenn, 49-50; bacteriologist, E Tenn Baptist Hosp, Knoxville, 50-51; res bacteriologist, Univ Tenn, 51-52; control bacteriologist, Am Sci Labs, Inc, 52-53, lab dir, 55-59, mgr animal health prod, 59-60; dir labs & vpres, L & M Labs, Inc, 60-64; opers mgr, Amdal Co Agr Div, 64-65, asst to dir prod planning & develop, 65-69, dir mkt, 69, gen mgr div, 69-72, VPRES & GEN MGR, AGR & VET PROD DIV, ABBOTT LABS, 72- Concurrent Pos: Mem subcomt avian dis, Agr Bd, Nat Acad Sci; secy vet biol licensees comt, Animal Health Inst, 68-, second vpres, 74-75, first vpres, 75-76. Mem: AAAS; Am Soc Microbiol; Am Inst Biol Sci; NY Acad Sci. Res: Immunology; viral and bacterial vaccines. Mailing Add: Abbott Labs Dept 385 Midway Exec Manor North Chicago IL 60064

APPLETON, MARTIN DAVID, b Bangor, Pa, Feb 11, 17; m 41; c 2. BIOCHEMISTRY. Educ: Univ Scranton, BS, 39; Pa State Univ, MS, 41, PhD, 61. Prof Exp: Group leader & sr res chemist, Nuodex Prod Co, NJ, 41-42; tech dir & partner, Com & Indust Prod Co, 45-55; from instr to assoc prof, 55-63, PROF CHEM, UNIV SCRANTON, 63-, CHMN DEPT, 75- Concurrent Pos: Consult indust chem & biochem. Honors & Awards: Rosmary Dybwad Int Award, Nat Asn Retarded Children, 70; Spec Citation, Pa Asn Retarded Children, 70. Mem: Fel AAAS; Am Chem Soc; fel Am Inst Chem. Res: Organic synthesis; clinical biochemistry; biochemistry of mental retardation. Mailing Add: 209 Claremont Ave Clarks Summit PA 18411

APPLEWHITE, PHILIP BOATMAN, b Los Angeles, Calif, Sept 14, 38; m 63; c 3. BEHAVIORAL PHYSIOLOGY. Educ: Pomona Col, BA, 60; Yale Univ, MIA, 62; Stanford Univ, PhD(bioeng), 65. Prof Exp: Instr indust eng, Stanford Univ, 63-64; fel molecular biophys & biochem, Yale Univ, 64-67, asst prof biol, 67-71, asst dean grad sch, 71-73, vis fel, 73-; ASSOC PROF BIOL, STATE UNIV NY, PURCHASE, NY, 73- Concurrent Pos: Consult logistics, Rand Corp, 62-63. Mem: AAAS; Inst Elec & Electronics Eng; Biophys Soc; Sigma Xi. Res: Molecular basis of learning and memory; biological rhythms. Mailing Add: Dept of Biol State Univ NY Purchase NY 10577

APPLEWHITE, THOMAS HOOD, b Imperial, Calif, Dec 30, 24; m 45; c 2. ORGANIC CHEMISTRY. Educ: Calif Inst Technol, BS, 53, PhD(chem, plant physiol), 57. Prof Exp: Res chemist, Dow Chem Co, 56-59; org chemist, Western Utilization Res & Develop Div, Agr Res Serv, USDA, 59-63, head oilseed invest, 63-67; res dir, Pac Veg Oil Corp, 67-69; MGR EDIBLE OIL PROD LAB, RES & DEVELOP, KRAFTCO CORP, 69- Mem: Am Chem Soc; Am Oil Chemists Soc. Res: Enzyme-catalyzed reactions and the chemistry of amino acids; physical chemistry of water-soluble high polymers; chemistry of fats and fatty acids; mechanisms of organic reactions; analytical chemistry. Mailing Add: Kraftco Corp 801 Waukegan Rd Glenview IL 60025

APPLEYARD, EDWARD CLAIR, b Strathroy, Ont, June 22, 34; m 62; c 2. GEOLOGY. Educ: Univ Western Ont, BSc, 56; Queen's Univ, Ont, MSc, 60; Cambridge Univ, PhD(petrol), 62. Prof Exp: Asst lectr geol, Bedford Col, Univ London, 62-65; asst prof, 65-68, actg chmn dept, 69-70, ASSOC PROF EARTH SCI, UNIV WATERLOO, 68- Concurrent Pos: Rec secy, Can Geosci Coun, 72- Mem: Fel Geol Asn Can; Mineral Asn Can; Can Inst Mining & Metall; Norweg Geol Soc; Mineral Soc Gt Brit & Ireland. Res: Petrology, structure, geochemistry and origin of syn-orogenic nepheline gneisses; Precambrian geology; geology of metasomatic rocks. Mailing Add: Dept of Earth Sci Univ of Waterloo Waterloo ON Can

APPLIN, ESTHER ENGLISH RICHARDS, micropaleontology, deceased

APPLIN, PAUL LIVINGSTON, b Keene, NH, Sept 17, 91; m 23; c 2. GEOLOGY. Educ: Dartmouth Col, AB, 14. Prof Exp: Instr, Dartmouth Col, 16; geologist, 17-31; consult geologist, 31-43; geologist, US Geol Surv, 43-61; RETIRED. Mem: Fel Geol Soc Am; Am Asn Petrol Geologists. Res: Mesozoic stratigraphy and structure of the Gulf Coast; oil geology. Mailing Add: 2305 Ben Ali Ct Owensboro KY 42301

APPLING, JOHN WILLIAM, microbiology, see 12th edition

APPLING, WILLIAM DAVID LOVE, b Chicago, Ill, Sept 13, 34; m 64. MATHEMATICS. Educ: Univ Tex, BA, 55, PhD(math), 58. Prof Exp: Instr math, Duke Univ, 60-63; from asst prof to assoc prof, 63-69, PROF MATH, NTEX STATE UNIV, 69- Mem: Am Math Soc. Res: Real analysis; integral theory; absolute continuity. Mailing Add: Dept of Math NTex State Univ Denton TX 76203

APRIL, ERNEST W, b Salem, Mass, Nov 6, 39. ANATOMY, MUSCULAR PHYSIOLOGY. Educ: Tufts Univ, BS, 61; Columbia Univ, PhD(anat), 69. Prof Exp: ASST PROF ANAT, COL PHYSICIANS & SURGEONS, COLUMBIA UNIV, 69- Concurrent Pos: USPHS grant, 71-; vis asst prof, Sch Med, Univ Miami, 70-71. Mem: Am Soc Zool; Am Asn Anat; Am Soc Cell Biol; Biophys Soc Am; Soc Gen Physiol. Res: Investigation, using electron microscopic x-ray diffraction and physiological methods, of the biophysical aspects of muscle contraction in single fibers by correlation of morphological ultrastructure, physiological processes and physical-chemical phenomena. Mailing Add: Col of Physicians & Surgeons Columbia Univ New York NY 10032

APRIL, ROBERT WAYNE, b Chicago, Ill, Nov 15, 46; m 72. ANALYTICAL CHEMISTRY, PHYSICAL CHEMISTRY. Educ: Univ Chicago, SB, 68; Mass Inst Technol, PhD(anal chem), 72. Prof Exp: Asst prof chem, Univ Colo, Denver, 72-74; VIS ASST PROF CHEM, COLO SCH MINES, 74- Res: Plasma emission spectroscopy; electrochemistry; computer software and hardware. Mailing Add: Dept of Chem Colo Sch of Mines Golden CO 80401

APRISON, MORRIS HERMAN, b Milwaukee, Wis, Oct 6, 23; m 49; c 2. NEUROCHEMISTRY, BIOPHYSICS. Educ: Univ Wis, BS, 45, cert, 47, MS, 49, PhD(biochem), 52. Prof Exp: Teaching asst physics, Univ Wis, 47-49; tech asst, Inst Paper Chem, Lawrence Col, 49-50; asst path, Univ Wis, 50-51, asst biochem & bact, 51-52; head biophys sect, Galesburg State Res Hosp, Ill, 52-56; from asst prof to assoc prof, Dept Biochem & Psychiat, 56-64, chief sect neurobiol, Inst Psychiat Res, 69-74, DIR, INST PSYCHIAT RES, IND UNIV, INDIANAPOLIS, 74-, PROF BIOCHEM, DEPT BIOCHEM & PSYCHIAT, MED CTR, 64- Concurrent Pos: Mem study sect neuropsychol, NIMH, 70-74; co-ed, Advances Neurochem, 73-; mem bd overseers, St Meinrad. Honors & Awards: Gold Medal Award Distinguished Res, Soc Biol Psychiat, 75. Mem: Am Physiol Soc; Sigma Xi; Biophys Soc; Soc Biol Psychiat; Int Soc Neurochem (secy, 75-77). Res: Transmitters and their enzyme systems; biochemical and biophysical correlates of behavior; synaptic mechanisms; biochemical mapping of the brain. Mailing Add: Inst of Psychiat Res Sect Neurol Ind Univ Med Ctr Indianapolis IN 46202

APSIMON, JOHN W, b Liverpool, Eng, Aug 5, 35; m 61; c 3. ORGANIC CHEMISTRY. Educ: Univ Liverpool, BSc, 56, PhD(org chem), 59. Prof Exp: Demonstr org chem, Univ Liverpool, 59-60; fel, Nat Res Coun Can, 60-62; from asst prof to assoc prof, 62-70, PROF ORG CHEM, CARLETON UNIV, ONT, 70- Mem: Am Chem Soc; Chem Inst Can; The Chem Soc. Res: Natural product chemistry; organic synthesis. Mailing Add: Dept of Chem Carleton Univ Ottawa ON Can

APT, CHARLES MAURICE, b New York, NY, June 15, 23; m 53; c 2. CHEMISTRY. Educ: Oxford Univ, BA, 49; Mass Inst Technol, PhD(phys chem), 52. Prof Exp: Instr chem, Amherst Col, 52-54; proj leader phys chem, Arthur D Little, Inc, 54-63; assoc dir explor develop lab, United-Carr Corp, 63-69; group mgr, Gillette Toiletries Co, Mass, 69; SR STAFF MEM, ARTHUR D LITTLE, INC, 69- Mem: Am Phys Soc; Am Chem Soc. Res: Thermodynamic properties of gaseous mixtures; electrical discharges; properties of the liquid state; physical-chemical properties of adsorbed enzyme systems; study of chemical processes employed in the manufacture of integrated circuits; investigation of kinetics of nucleation and crystal growth. Mailing Add: 80 Washington St Belmont MA 02178

APT, KENNETH E, b Bellingham, Wash, Apr 27, 45; m; c 2. NUCLEAR CHEMISTRY. Educ: Western Wash State Col, BA, 67; Mass Inst Technol, PhD(nuclear chem), 71. Prof Exp: Instr chem, Western Wash State Col, 63-67; asst chem, Mass Inst Technol, 67-71; staff mem, Chem & Nuclear Chem Group, 71-73 & Environ Studies Group, 73-75, SECT LEADER SURVEILLANCE & ANAL SECT, ENVIRON STUDIES GROUP, LOS ALAMOS SCI LAB, 75- Res: Research in nuclear chemistry, analytical chemistry and physics as applied to interdisciplinary areas of science including environmental studies, geosciences, materials analysis and epidemiology. Mailing Add: Los Alamos Sci Lab PO Box 1663 Los Alamos NM 87545

APT, WALTER JAMES, b Belfield, NDak, Jan 30, 22; m 48; c 3. PLANT PATHOLOGY, NEMATOLOGY. Educ: Wash State Univ, BS, 50, PhD(plant path), 58. Prof Exp: Agent nematologist, USDA, 55-56, nematologist, 56-63; nematologist, Pineapple Res Inst, 63-73; PLANT PATHOLOGIST, UNIV HAWAII, MANOA, 73- Concurrent Pos: Consult nematologist, Hawaiian Sugar Planters Asn, 58-59. Mem:

Am Phytopath Soc; Soc Nematol; Orgn Trop Am Nematologists. Res: Plant parasitic nematodes; root rot complex with nematodes as a factor; soil fumigation; systematic nematicides. Mailing Add: Dept of Plant Path Univ of Hawaii at Manoa Honolulu HI 96822

APTER, JULIA TUTELMAN, b Philadelphia, Pa, 1918; m 41; c 2. OPHTHALMOLOGY, MATHEMATICAL BIOLOGY. Educ: Univ Pa, BA, 39; Johns Hopkins Univ, MD, 43; Northwestern Univ, MS, 59; Univ Chicago, PhD(math biol), 64; Am Bd Ophthalmol, dipl, 57. Prof Exp: Instr ophthal, Wilmer Inst, 43-46; mem fac, Med Sch, Northwestern Univ, 51-59; res assoc neurophysiol, Univ Ill, 59-61; res assoc math biol, Univ Chicago, 64-66; from assoc prof to prof surg & dir math biol, Univ Ill Med Ctr, 66-71; prof surg, Rush Univ, 71-75, dir lab biomat & biomech, Rush-Presby-St Luke's Med Ctr, 71-75; DIR DEPT OPHTHAL, WHITING CLIN, INC, 75- Concurrent Pos: Dir neurophysiol res, Manteno State Hosp, 50-57; Ill Dept of Health grants, 51-57; USPHS grants, 56-60 & 66-76; Nat Heart Inst spec fel, 64-65; AEC grant, 69-70; attend surgeon, Presby-St Luke's Hosp, 67-75. Mem: AAAS; fel Soc Advan Med Systs; sr mem Inst Elec & Electronics Engrs; Am Physiol Soc; fel Am Acad Ophthal & Otolaryngol. Res: Medical engineering; computers in patient care; mechanical properties of tissues and organs. Mailing Add: Whiting Clin Inc 2450 169th St Hammond IN 46323

APTER, NATHANIEL STANLEY, b New York, NY, May 10, 13; m 41; c 2. PSYCHIATRY. Educ: Cornell Univ, AB, 33; Univ Buffalo, MD, 38; Chicago Inst Psychoanal, grad, 58. Prof Exp: Rotating intern, Beth Israel Hosp, NY, 38-39; psychiat intern, Bellevue Hosp, 39-40; resident neurol, Kings County Hosp, 40-42; asst psychiat, Johns Hopkins Univ, 42-44, house officer, Johns Hopkins Hosp, 42-44; from asst prof to assoc prof & head dept, 46-55, PROF LECTR PSYCHIAT, UNIV CHICAGO, 55- Concurrent Pos: Chief investr res proj schizophrenia, Manteno State Hosp, 51-69, chief psychiat consult, 69-; attend psychiatrist, Michael Reese Hosp, 54-73, sr attend psychiatrist, 73-; psychiat consult, State Dept Ment Health, 54-; chief consult, McLean County Ment Health Clin, Bloomington, Ill, 57-68; sr psychiat consult, Ill State Psychiat Inst, 59-71, distinguished consult, 71-; chief psychiat consult, Tinley Park Ment Health Clin, 70-74, 75- Mem: Fel AAAS; fel Am Psychiat Asn. Res: Schizophrenia. Mailing Add: 111 N Wabash Ave Chicago IL 60602

AQUILINA, JOSEPH THOMAS, b Farnham, NY, Nov 19, 16; m 46; c 4. MEDICINE. Educ: Univ Buffalo, MD, 41; Am Bd Internal Med, dipl. Prof Exp: Asst chief med serv, 46-63, exec secy res comt, 50-58, asst dir prof servs for res, 58-64, CHIEF MED SERV, VET ADMIN HOSP, 63-, ASSOC CHIEF OF STAFF FOR RES, 64- Concurrent Pos: Asst prof med, Sch Med, State Univ NY Buffalo, 58-67, assoc clin prof, 67-69, clin prof, 69- Mem: Fel AMA; fel Am Col Physicians; Am Heart Asn; NY Acad Med. Res: Cardiovascular disease. Mailing Add: Vet Admin Hosp 3495 Bailey Ave Buffalo NY 14215

AR, ERGUN, biomathematics, see 12th edition

ARABIAN, KAREKIN GASPAR, b Lawrence, Mass, June 20, 16; m 42; c 2. PETROLEUM, PHYSICAL CHEMISTRY. Educ: Mass Inst Technol, BS, 37. Prof Exp: Res chemist asphalt res, Shell Develop Co, 37-41, sr res chemist wax res, 45-48 & Shell Oil Co, 48-64, CHEMIST WAX RES, SHELL DEVELOP CO, 64- Mem: Am Chem Soc. Res: Asphalt; wax. Mailing Add: 376 Dover Dr Walnut Creek CA 94598

ARADINE, PAUL WILLIAM, b Bergen, NY, Jan 25, 10; m 37; c 2. INORGANIC CHEMISTRY, ANALYTICAL CHEMISTRY. Educ: Univ Rochester, BS, 32, MS, 34, PhD(anal chem), 36. Prof Exp: Asst chem, Univ Rochester, 32-36, instr, 36-38, exten div, 38, 40-48; chemist, Taylor Instrument Process Control Div, Sybron Corp, 38-47, asst chief chemist, 47-50, chief chemist, 50-75; RETIRED. Mem: Am Chem Soc. Res: Determination of quartz in the presence of silicates; composition of fluoboric acid. Mailing Add: 15 N Main Churchville NY 14428

ARAGAKI, MINORU, b Honolulu, Hawaii, June 26, 26; m 52; c 3. PLANT PATHOLOGY. Educ: Univ Hawaii, BS, 50, MS, 54, PhD(bot), 63. Prof Exp: Lab technician, 46-54, jr plant pathologist, 54-61, from asst prof to assoc prof, 61-72, PROF PLANT PATH, UNIV HAWAII, 72- Mem: Am Phytopath Soc; Mycol Soc Am. Res: Light induced fungal sporulation including inhibitory effects of light; biology of Phytophthora; anatomical changes following invasion by fungal pathogen, particularly host barrier formation. Mailing Add: Dept of Plant Path Univ Hawaii at Manoa Honolulu HI 96822

ARAGON, PEDRO J, chemistry, electrochemistry, see 12th edition

ARAI, HISAO PHILIP, b Los Angeles, Calif, Oct 8, 26; m 58. PARASITOLOGY. Educ: Univ Calif, Los Angeles, AB, 54, PhD(zool), 60. Prof Exp: USPHS fel zool, Univ BC, 60-61; assoc prof biol, Ill State Univ, 61-63; asst prof, Univ Alta, 63-65; ASSOC PROF ZOOL, UNIV CALGARY, 65- Concurrent Pos: Assoc mem ctr for zoonoses res, Univ Ill, 62. Mem: Am Soc Parasitol; Soc Syst Zool; Can Soc Zool. Res: Physiological ecology; taxonomy; ecology and physiology of animal parasites. Mailing Add: Dept of Biol Univ of Calgary Calgary AB Can

ARAJS, SIGURDS, b Taurkalne, Latvia, Sept 2, 27; US citizen; m 52; c 1. SOLID STATE PHYSICS. Educ: Iowa State Univ, BS, 53, PhD, 57. Prof Exp: Res technician ultrasonics, Eng Exp Sta, Iowa State Univ, 52-53, asst physics, 53-54, Inst Atomic Res, 54-57; scientist, Bain Lab Fundamental Res, US Steel Corp, 58-60, from supv scientist to sr scientist, 60-67; PROF PHYSICS, CLARKSON COL TECHNOL, 67- Mem: Fel Am Phys Soc. Res: Magnetism; transport properties of solids; electronic structures of metals; low temperature phenomena; fluid physics; biophysics. Mailing Add: Dept of Physics Clarkson Col of Technol Potsdam NY 13676

ARAKAWA, EDWARD TAKASHI, b Honolulu, Hawaii, Apr 8, 29; m 55; c 4. SOLID STATE PHYSICS. Educ: Howard Col, BS, 51; La State Univ, MS, 53; Univ Tenn, PhD(physics), 57. Prof Exp: Res assoc med physics, Oak Ridge Inst Nuclear Studies, 53-56, HEALTH PHYSICIST, OAK RIDGE NAT LAB, 57- Mem: Fel Optical Soc Am; Am Phys Soc. Res: Dosimetry of nuclear radiation; electron physics; vacuum ultraviolet spectroscopy; molecular spectroscopy; thin film physics. Mailing Add: 4500 S Oak Ridge Nat Lab Oak Ridge TN 37830

ARAKI, GEORGE SHOICHI, b Oakland, Calif, Jan 11, 32; m 58; c 3. BIOLOGY. Educ: San Francisco State Col, AB, 57; Stanford Univ, PhD(biol), 64. Prof Exp: From asst prof to assoc prof, 62-74, PROF BIOL, SAN FRANCISCO STATE UNIV, 74-; DIR CTR INTERDISCIPLINARY & INNOVATIVE SCI, 74- Mem: AAAS. Res: Comparative invertebrate physiology; cellular physiology; physiology of feeding and digestion in asteroids; intercellular digestive processes in asteroids and slime molds. Mailing Add: Dept of Biol San Francisco State Univ San Francisco CA 94132

ARALA-CHAVES, MARIO PASSALAQUA, b Fafe, Portugal, Dec 2, 39; m 65; c 2. IMMUNOBIOLOGY. Educ: Univ Lisbon, MD, 65; Univ Louvain, MS, 67, PhD(transfer factor), 75. Prof Exp: Fel res immunobiol, Dept Exp Med, Univ Louvain, 65-67; head dept blood bank, Univ Hosp Luand, 68-70; fel immunobiol, Transplantation Dept, Cent Blood Transfusions, Amsterdam, 70-71; head dept blood bank, Portugese Cancer Inst, Lisbon, 72-75; ASSOC IMMUNOBIOL, DEPT BASIC & CLIN IMMUNOL, MED UNIV SC, 75- Mem: Brit Soc Immunol; Europ Soc Clin Invest. Res: Study of leukocyte and erythrocyte antigens; immunodeficient syndromes; biological aspects of linfoquines; biological aspects and biological characterization of Transfer Factor-d. Mailing Add: Basic & Clin Immunol & Microbiol Med Univ of SC Charleston SC 29401

ARANDA, JUAN MANUEL, b San Juan, PR, Feb 3, 42; m 65; c 4. CARDIOLOGY. Educ: Univ PR, BS, 63, MD, 69. Prof Exp: From intern to resident internal med, Sch Med, Univ PR, San Juan, 67-70, fel cardiol, 70-71; chief dept med, Beach Army Hosp, Ft Walters, Tex, 71-73; fel cardiol, 73-74, ASST PROF MED, SCH MED, UNIV MIAMI, 74- Concurrent Pos: Staff physician, Vet Admin Hosp, Miami, 74; attend physician, Jackson Mem Hosp, Miami & Nat Cardiac Children Hosp, Miami, 74; asst prof med, Sch Med, Univ Miami, 74. Honors & Awards: AMA Physician Recognition Award, 73 & 76; US Army Nat Defense Serv Medal, 71; US Army Commendation Medal, 73; Harry Botick Mem Award, 74. Mem: AMA; assoc mem Am Col Cardiol; assoc mem Am Col Chest Physicians; Am Heart Asn. Res: Clinical electrophysiology, especially the pattern of retrograde conduction in patients with pre-excitation in order to assess the frequency of concealed atrio-ventricular nodal bypass tracts in patients with ventricular or supraventricular tachycardias. Mailing Add: Cardiol Sect Univ of Miami Sch of Med Miami FL 33152

ARANOFF, SANFORD, theoretical physics, see 12th edition

ARANOW, HENRY, b New York, NY, May 5, 13; m 41; c 4. INTERNAL MEDICINE. Educ: Harvard Univ, AB, 35; Columbia Univ, MD, 38, MedSciD(internal med), 44. Prof Exp: Asst physiol, Sch Med, Johns Hopkins Univ, 41-42; asst med, 42-44, instr, 44-47, assoc, 47-58, actg chmn dept, 70-71, from asst prof to assoc prof, 58-67, prof clin med, 67-76, SAMUEL LAMBERT PROF MED, COL PHYSICIANS & SURGEONS, COLUMBIA UNIV, 76-, ACTG CHMN DEPT, 75- Concurrent Pos: NY Acad Med Harlow Brooks scholar; consult, St John's Hosp, Yonkers, NY, 56-; actg dir med serv, Presby Hosp, Columbia Univ, 70-71 & 75- Mem: Fel AMA; Endocrine Soc; Am Thyroid Asn. Res: Pheochromocytoma; thyroid disease; common poisonings. Mailing Add: Dept of Med Columbia-Presby Med Ctr New York NY 10032

ARANOW, RUTH LEE HORWITZ, b Brooklyn, NY, Aug 25, 29; m 50; c 3. PHYSICAL CHEMISTRY. Educ: Brooklyn Col, BS, 51; Johns Hopkins Univ, MA, 52, PhD(chem), 57. Prof Exp: Jr instr chem, Johns Hopkins Univ, 51-53, asst, 53-54, instr, McCoy Col, 54-57; staff scientist, RIAS, Martin-Marietta Co, 57-69; NIH spec fel, 69-71, res scientist, 69-74, FEL CHEM, JOHNS HOPKINS UNIV, 74- Mem: AAAS; Am Phys Soc. Res: Statistical mechanics; physical chemistry of membranes; interaction of dispersion and chemical reaction in stack plumes; effect of noise on neural membranes. Mailing Add: Dept of Chem Johns Hopkins Univ Baltimore MD 21218

ARANT, FRANK SELMAN, b McKenzie, Ala, July 9, 04; m 28. ZOOLOGY. Educ: Auburn Univ, BS, 26, MS, 29; Iowa State Univ, PhD(entom), 37. Prof Exp: Instr zool & entom, Auburn Univ, 26-30; teaching fel, Iowa State Univ, 30-31; from asst prof to assoc prof, 31-42, entomologist, 46-49, head dept zool & entom, 49-75, prof zool & entom, 69-75, EMER PROF ZOOL & ENTOM, AUBURN UNIV, 75- Concurrent Pos: Mem, Nat Defense Educ Act Nat Tel Panel, Off Educ, 65; mem, Nat Res Coun & Nat Acad Sci comt imported fire ant, 67; mem, Nat Tech Guid Comt Pilot Boll Weevil Eradication Exp, 70-73; chmn subcomt pesticides, Nat Adv Comt Occup Safety & Health Agr, 72-73. Mem: Entom Soc Am (pres, 61); Am Soc Zool; Am Soc Parasitol. Res: Cotton, corn, peanut and vegetable insects; life history methods; rearing and mounting hookworm larvae; game animals in Alabama; rotenone and fluorine insecticides; use and limitations of new insecticides against cotton insects; spray residues; systemic insecticides for arthropods of animals. Mailing Add: Dept of Zool & Entom Auburn Univ Auburn AL 36830

ARASE, ELIZABETH MARTHA, b Bonn, Ger, Nov 29, 27; US citizen; m 51; c 2. PHYSICS. Educ: Univ Bonn, Univ Heidelberg & Univ Zurich, BS, 47; Columbia Univ, MA, 53; NY Univ, PhD(physics), 60. Prof Exp: Res asst high energy physics, Columbia Univ, 49-54; high sch teacher, Morningside Sch, NY, 54-55 & Barnard Sch Girls, 55-56; scientist, Walter Kidde Nuclear Lab, 56-58; instr physics, Hofstra Col, 58-60; instr, City Col New York, 61-63; res assoc underwater sound physics, Hudson Labs, Columbia Univ, 63-67, sr res assoc, 67-69; ASSOC PROF OCEAN ENG, STEVENS INST TECHNOL, 69- Mem: Am Phys Soc; Acoustical Soc Am. Res: Reactor, solid state and underwater sound physics. Mailing Add: Dept of Ocean Eng Stevens Inst Technol Castle Pt Hoboken NJ 07030

ARASE, TETSUO, b Seattle, Wash, Jan 2, 27; m 51; c 2. PHYSICS. Educ: Univ Wash, BS, 49; NY Univ, MS, 53, PhD(physics), 59. Prof Exp: Electronic scientist, Mat Lab, NY Naval Shipyard, Brooklyn, 51-53; res scientist underwater acoustics, Hudson Labs, Columbia Univ, 53-69; ASSOC PROF OCEAN ENG, STEVENS INST TECHNOL, 69- Concurrent Pos: Asst prof, NY Univ, 60-62. Mem: AAAS; Am Phys Soc. Res: Solid state physics; neutron physics; physical oceanography and underwater acoustics. Mailing Add: Dept of Ocean Eng Stevens Inst Technol Castle Pt Hoboken NJ 07030

ARATA, ANDREW ANTHONY, vertebrate zoology, mammalian ecology, see 12th edition

ARATA, DOROTHY, b New York, NY, Mar 8, 28. NUTRITION. Educ: Pratt Inst, BS, 48; Cornell Univ, MS, 50; Univ Wis, PhD(biochem), 56. Prof Exp: Asst dept foods & nutrit, Cornell Univ, 48-50, technician dept biochem, Med Sch, 50-52, fel, 56-57; asst biochem, Univ Wis, 53-56; from asst prof to prof foods & nutrit, 57-69, assoc dir, Honors Col, 69-71, PROF HUMAN DEVELOP, COL HUMAN MED, MICH STATE UNIV, 69-, ACTG DIR, HONORS COL, 75-, ASST PROVOST, 71- Concurrent Pos: Acad admin intern, Am Coun Educ, 68, mem comn educ credit, 74-77. Mem: AAAS; Am Chem Soc; Am Inst Nutrit; Am Inst Chem; Am Asn Higher Educ. Res: Biochemical disorders associated with nutritional deficiencies; cellular disruptions in fatty livers induced by threonine deficiency. Mailing Add: 443 Admin Bldg Mich State Univ East Lansing MI 48823

ARAUJO, JOSE EMILIO GONCALVES, b Rio de Janeiro, Brazil, Sept 8, 22; m 46; c 3. SOIL SCIENCE. Educ: Nat Sch Agron, Brazil, Agron Engr, 45; Eliseu Maciel Sch Agron, DSc(agron), 48. Prof Exp: Chief soils sect, Agron Inst, Brazil, 47-50; dir, Eliseu Maciel Sch Agron, 50-51; prof geog physics, hist of geol & petrol, Cath Fac Philos, Cath Univ, Brazil, 52-60; dir, Cetreisul, Brazil, 60-61; DIR, GEN INTERAM INST AGR SCI, 70- Concurrent Pos: Cathedratic prof, Eliseu Maciel Sch Agron, 46-; dir, S Agr Res Inst, Brazil, 52-55; prof regional course on soils, Interam Inst Geog & Hist, 58; mem sci & tech coun exp plan prom wheat cultivation, Ministry Agr, 62-64, adv wheat matters, 63-64; Brazilian rep, Interam Inst Agr Sci, 64; natural resources specialist & head, Interam Ctr Agrarian Reform, 64; mem var sci & tech cong &

meetings, Brazil, Latin Am & US. Mem: Soil Sci Soc Am; Int Soc Soil Sci; Latin Am Soc Soil Sci; Brazilian Soc Soil Sci (pres, 63); Brazilian Soc Geol. Res: Physical soil, particularly water studies on soil; soil fertility emphasizing study of aluminum and its toxicity effects; land use for planning and execution of agrarian reform. Mailing Add: Interam Inst Agr Sci PO Box 10281 San Jose Costa Rica

ARAUJO, OSCAR EDUARDO, b Brazil, June 2, 27; nat US; m 50; c 2. PHARMACY. Educ: Purdue Univ, BS, 54, MS, 55, PhD(pharm), 57. Prof Exp: Asst pharm, Purdue Univ, 54-55; asst prof, Ohio Northern Univ, 57-62; from asst prof to assoc prof, 62-74, PROF PHARM, UNIV FLA, 74- Mem: Am Pharmaceut Asn; Am Asn Cols Pharm; Acad Pharmaceut Sci. Res: Physical and practical pharmacy. Mailing Add: 406 SW 40th Terr Gainesville FL 32601

ARAUJO, ROGER JEROME, physical chemistry, see 12th edition

ARAVE, CLIVE W, b Idaho Falls, Idaho, May 12, 31; m 50; c 5. ANIMAL BREEDING. Educ: Utah State Univ, BS, 56, MS, 57; Univ Calif, Davis, PhD(genetics), 63. Prof Exp: Asst mgr, Lovacre Farms, Calif, 57-59; lab technician genetics, Univ Calif, 59-60; asst prof dairy husb, Chico State Col, 63-65; ASST PROF DAIRY HUSB, UTAH STATE UNIV, 65- Concurrent Pos: Mem health comt, Am Dairy Sci Asn, 73-76. Mem: Am Dairy Sci Asn; Am Soc Animal Sci; Am Behav Soc. Res: Dairy cattle production genetics and management; animal breeding; animal behavior of domisticated species. Mailing Add: Dept of Dairy Sci Utah State Univ Logan UT 84321

ARBENZ, JOHANN KASPAR, b Berne, Switz, Aug 27, 18; nat US; m 45; c 3. GEOLOGY. Educ: Univ Berne, PhD(geol), 45. Prof Exp: Asst geologist, R Helbling Surv Off, 46-47 & Swiss Geol Surv, 47; from asst prof to assoc prof geol, Univ Okla, 48-55; geologist, 55-75, SR STAFF GEOLOGIST, SHELL OIL CO, 75- Mem: Geol Soc Am; Am Asn Petrol Geol; Swiss Geol Soc. Res: Structural and petroleum geology. Mailing Add: Shell Oil Co Box 527 Houston TX 77001

ARBIB, MICHAEL A, b Eastbourne, Eng, May 28, 40; m 65; c 2. COMPUTER SCIENCES. Educ: Univ Sydney, BSc, 61; Mass Inst Technol, PhD(math), 63. Prof Exp: Assoc prof elec eng, Stanford Univ, 69-70; chmn dept comput & info sci, 70-75, PROF COMPUT & INFO SCI & ADJ PROF PSYCHOL, UNIV MASS, AMHERST, 70-, DIR CTR SYSTS NEUROSCI, 74- Res: Brain modelling; information processing in complex systems, in automata and system theory and in neurophysiology and psychology; social implications of computer science. Mailing Add: Dept of Comput & Info Sci Univ of Mass Amherst MA 01002

ARBOGAST, JOHN LYNN, b Forreston, Ill, Mar 27, 04; m 36; c 2. CLINICAL PATHOLOGY. Educ: Ind Cent Col, AB, 25; Ill Wesleyan Univ, BMus, 31; Ind Univ, MD, 36. Prof Exp: From intern to resident, Ind Univ, 36-40, instr path, 38-40; clin pathologist, Home Hosp, Ind, 40-42; dir clin labs, 46-65, assoc prof, 46-54, PROF PATH, SCH MED, IND UNIV, INDIANAPOLIS, 54-, DIR DIV ALLIED HEALTH SCI, 65- Concurrent Pos: Instr, Purdue Univ, 41-42. Mem: Fel AMA; Am Soc Clin Pathologists; Col Am Pathologists; Asn Schs Allied Health Professions. Res: Organization, administration and curricula of allied health programs. Mailing Add: Ind Univ Sch of Med Indianapolis IN 46207

ARBOGAST, RICHARD TERRANCE, b Freeport, Ill, Aug 7, 37; m 58; c 4. ENTOMOLOGY. Educ: Univ Ill, BS, 59; Univ Fla, PhD(entom), 65. Prof Exp: RES ENTOMOLOGIST, STORED-PROD INSECTS RES & DEVELOP LAB, SOUTHERN REGION, AGR RES SERV, USDA, 65- Mem: Entom Soc Am; Lepidop Soc. Res: Ecology and behavior of stored-product insects; taxonomy and distribution of butterflies. Mailing Add: 114 Monica Blvd Savannah GA 31406

ARBONA, GUILLERMO, b Maricao, PR, May 25, 10; m 38; c 4. PUBLIC HEALTH. Educ: St Louis Univ, MD, 34; Johns Hopkins Univ, MPH, 37. Prof Exp: Health officer, Dept Health, PR, 35-36, asst dir local health work, 37-39, health officer, Rio Piedras Pub Health Unit & Training Ctr, 39-41; assoc pub health pract, Columbia Sch Trop Med, 41-42, from asst prof to assoc prof, 42-49, head dept hyg, 42-49; prof pub health & prev med & head dept, Sch Med, Univ PR, 49-57; secy health, Commonwealth PR, 57-66; PROF PREV MED & PUB HEALTH, SCH PUB HEALTH, UNIV PR, SAN JUAN, 66- Concurrent Pos: Consult, Inst Inter-Am Affairs, 51-52 & Dept of Health, Commonwealth of PR, 66-; secy panel on aging, US Dept Health, Educ & Welfare; US deleg, Health Assembly, Geneva, 59 & 60; mem comt pub health admin, WHO, 60. Honors & Awards: Bronfman Award, Am Pub Health Asn, 65. Mem: AAAS; fel Am Pub Health Asn (vpres, 49); hon mem Am Hosp Asn; hon mem Asn State & Territorial Health Off; fel PR Pub Health Asn (secy, 40-46, pres, 47). Res: Typhus fever; Shigellosis; public health administration; health care administration; health planning. Mailing Add: Sch of Pub Health Univ of PR Med Sci Campus San Juan PR 00936

ARBUCKLE, WENDELL SHERWOOD, b Scottsburg, Ind, Mar 16, 11; m 41; c 2. DAIRY SCIENCE. Educ: Purdue Univ, BSA, 33; Univ Mo, AM, 37, PhD, 40. Prof Exp: Dairyman, Earlham Col, 33-36; from asst instr to instr dairy husb, Univ Mo, 38-41; asst prof, Agr & Mech Col Tex, 41-43, assoc dairy mfg exp sta, 43-46; assoc prof dairy husb, NC State Col, 46-49; prof dairy mfg & chief of dairy inspect serv, 49-72, EMER PROF DAIRY SCI, UNIV MD, COLLEGE PARK, 73- Concurrent Pos: With US Dept Agr, 44, dairy prod utilization specialist, 56-57; adv prof info serv, dairy & food indust, 72-; pres, Agr Ice Cream Inc, Mo, 72- Honors & Awards: Distinguished Serv Award, Dairy Technol Soc, Md & DC, 72. Mem: Am Dairy Sci Asn. Res: Milk analysis; optical analysis of crystalline structure of ice cream; optical properties of lactose; physical and chemical properties of ice cream; new products. Mailing Add: 4602 Harvard Rd College Park MD 20740

ARBULU, AGUSTIN, b Lima, Peru, Sept 15, 28; US citizen; c 3. SURGERY, THORACIC SURGERY. Educ: San Marcos Univ, Lima, MB, 54, MD, 55. Prof Exp: Asst chief surg, Wichita Vet Admin Hosp, Kans, 61-62; from instr to assoc prof, 62-72, PROF SURG, SCH MED, WAYNE STATE UNIV, 72- Concurrent Pos: Asst chief surg, Vet Admin Hosp, Allen Park, Mich, 62-66, actg chief surg, 66-67, consult, 67-; assoc, Hutzel & Harper Hosps, Detroit; attend surg, Detroit Gen Hosp; consult chest surg, Herman Kiefer & Grace Hosps. Honors & Awards: Frederick A Coller Award, Am Col Surgeons, 68 & 72; Cecile Lehman Mayer Res A Award, Am Col Chest Physicians, 68, 69 & 70, Regents Award, 72. Mem: Am Thoracic Soc; Soc Thoracic Surg; Peruvian Am Med S Soc (pres, 73-74). Mailing Add: 540 E Canfield Detroit MI 48201

ARCAND, GEORGE MYRON, b Ocean Falls, BC, Feb 9, 24; nat US; m 52. ANALYTICAL CHEMISTRY, INORGANIC CHEMISTRY. Educ: Calif Inst Technol, BS, 50, PhD(chem), 55. Prof Exp: Asst prof chem, Univ Mo, 55-58 & Mont State Col, 58-62; sr scientist, Jet Propulsion Lab, Calif Inst Technol, 62-65, scientist specialist, 65; assoc prof, 65-67, PROF CHEM, IDAHO STATE UNIV, 67- Mem: Am Chem Soc. Res: Electro-analytical methods; mechanisms of liquid-liquid extraction processes and their use in the study of aqueous systems; inorganic complex ions in aqueous solutions; electrochemistry of batteries; environmental trace analysis. Mailing Add: Dept of Chem Idaho State Univ Pocatello ID 83209

ARCE, GINA, b Cuba, Aug 28, 29; nat US. BIOLOGY. Educ: George Peabody Col, BA, 48, MA, 49; Vanderbilt Univ, PhD(biol), 56. Prof Exp: Teacher high sch, 49-51; instr bot, Vanderbilt Univ, 56-57; assoc prof, 58-69, PROF BOT, CALIF STATE UNIV FRESNO, 69- Mem: Bot Soc Am; Phycol Soc Am. Res: Taxonomy of green algae. Mailing Add: Dept of Bot Calif State Univ Fresno CA 93710

ARCE, JOSE EDGAR, b Lima, Peru, Apr 21, 37; m 61; c 4. GEOPHYSICS, GEOLOGY. Educ: San Marcos Univ, Lima, Geol Engr, 60, BSc, 66, PhD(geol), 72. Prof Exp: Geologist, Marcona Mining Co, Peru, 60-63, assoc consult geophys, 63-64; assoc prof, San Marcos Univ, Lima, 64-72; INDEPENDENT CONSULT GEOPHYSICIST & EXPLORATION GEOPHYSICIST, 64-; PRIN PROF GEOPHYS, UNIV ENG, LIMA, 69- Mem: AAAS; Soc Explor Geophys; Am Inst Min, Metall & Petrol Eng; Am Geophys Union; European Asn Explor Geophys. Res: Applied geophysics in mining and groundwater exploration; civil engineering studies with geophysical methods; regular use of electrical, magnetic, seismic, electromagnetic and well logging techniques. Mailing Add: Petit Thouars 4380 Lima Peru

ARCENEAUX, GEORGE, b Lafayette, La, Mar 9, 95; m 23; c 2. PLANT BREEDING, AGRONOMY. Educ: George Peabody Col, BS, 20; Cornell Univ, PhD(genetics), 39. Prof Exp: Teacher & prin high sch, La, 20-22; county agr agent, Houma, La, 22-25; agronomist & assoc mgr, Scully Reclamation & Farming Proj, La, 25-27; from agronomist to prin agronomist, Div Sugar Plants, Bur Plant Indust, USDA, 28-51; CONSULT SUGARCANE RES & DIR INT RES SERV, 51- Mem: AAAS; Am Genetic Asn; Int Soc Sugar Cane Technol. Res: Sugar cane variety and milling tests; field practices, sampling and maturity of sugar cane; theoretical sugar yield calculations; plot technique; weed control; sugar cane breeding. Mailing Add: 1581 E Second St Pass Christian MS 39571

ARCENEAUX, JOSEPH LINCOLN, b Lafayette, La, Aug 13, 41; m 70. MICROBIOLOGY, BACTERIAL PHYSIOLOGY. Educ: Univ Southwestern La, BS, 63; Univ Tex, Austin, PhD(microbiol), 68. Prof Exp: NIH fel, Princeton Univ, 67-70; asst prof, 70-75, ASSOC PROF MICROBIOL, MED CTR, UNIV MISS, 75- Mem: AAAS; Am Soc Microbiol. Res: Chromosome replication and genetic studies in Bacillus subtitis. Mailing Add: Dept of Microbiol Univ of Miss Med Ctr Jackson MS 39216

ARCESE, PAUL SALVATORE, b West New York, NJ, Mar 14, 30; m 53; c 2. MEDICAL RESEARCH. Educ: NY Univ, BA, 51, MS, 58, PhD(endocrinol), 64. Prof Exp: Lab technician biochem & endocrinol, Merck Inst Therapeut Res, 54-60, from res asst to res assoc, 60-65; from res assoc to asst dir clin res, 65-74, SR ASSOC DIR CLIN RES, SANDOZ PHARMACEUT, 74- Mem: Acad Psychosom Med; Am Asn Study Headache. Res: Adrenal steroid metabolism; hormonal effects on erythropoiesis in the hypophysectomized rat. Mailing Add: Dept of Clin Res Sandoz Pharmaceut Hanover NJ 07936

ARCESI, JOSEPH A, b Sayre, Pa, Sept 1, 38; m 61; c 3. ORGANIC CHEMISTRY, POLYMER CHEMISTRY. Educ: Gettysburg Col, BA, 60; Univ Del, PhD(org polymer chem), 64. Prof Exp: SR POLYMER CHEMIST, EASTMAN KODAK CO, 67- Res: Polymer preparation and evaluation; photo-polymers. Mailing Add: Eastman Kodak Co Res Labs Rochester NY 14650

ARCH, STEPHEN WILLIAM, b Los Angeles, Calif, May 15, 42; m 64. NEUROBIOLOGY. Educ: Stanford Univ, AB, 64; Univ Chicago, PhD(biol), 70. Prof Exp: Res fel neurobiol, Calif Inst Technol, 70-72; ASST PROF ANIMAL PHYSIOL, REED COL, 72- Mem: AAAS; Soc Gen Physiol; Am Soc Zool; Sigma Xi. Res: Analysis of biochemical and electrophysiological regulation at the cellular level in identified neuroendocrine cells. Mailing Add: Dept of Biol Reed Col Portland OR 97202

ARCHAMBAULT, GEORGE FRANCIS, pharmacy, see 12th edition

ARCHAMBAULT, JACQUES OLIVER, inorganic chemistry, see 12th edition

ARCHAMBEAU, JOHN ORIN, b Maine, Aug 5, 25; c 8. RADIOLOGY. Educ: Stanford Univ, AB, 50, MD, 55. Prof Exp: Resident path, Univ Chicago, 54-58; fel oncol, Swedish Hosp, 58-60; Fulbright scholar & fel oncol, Curie Found, 60-61; assoc scientist med, Brookhaven Nat Lab, 61-66; DIR RADIATION ONCOL, NASSAU COUNTY MED CTR, 66-; PROF RADIOL, STATE UNIV NY STONY BROOK, 70- Concurrent Pos: Res collabr, Brookhaven Nat Lab, 66-; mem cent comt radiation ther, Eastern Coop Oncol Group, 74-; mem adv comt, Harvard Cyclotron Biomed Group, 75- Mem: Radiation Res Soc; Am Col Radiol; Am Soc Therapeut Radiol. Res: Development of proton radiation therapy; cell kinetics of irradiated tissues. Mailing Add: Div Radiation Oncol Nassau County Med Ctr East Meadow NY 11554

ARCHANGELSKY, SERGIO, b Casablanca, Morocco, Mar 27, 31; m 59; c 2. PALEOBOTANY, PALYNOLOGY. Educ: Univ Buenos Aires, Lic en cie, 55, Dr(geol), 57. Prof Exp: Prof paleont, Inst Miguel Lillo, Nat Univ Tucuman, 56-61; PROF PALEOBOT, LA PLATA UNIV, 61- Concurrent Pos: Arg Nat Res Coun res fel, 61-; consult, Yacimientos Petroliferos Fiscales, 75- Mem: Arg Paleont Asn (vpres, 61-63 & 74-75, pres, 63-65 & 67-69); Latin Am Paleobot & Palynology Asn (pres, 74-); corresp mem Bot Soc Am. Res: Paleobotany and palynology of the Cretaceous in Argentina; palynology of the Lower Tertiary in Argentina; paleobotany and palynology of Carboniferous and Permian periods in Argentina. Mailing Add: Av Santa Fe 3344 p 12 Buenos Aires Argentina

ARCHARD, HOWELL OSBORNE, JR, b Yonkers, NY, Mar 25, 29; m 61; c 1. ORAL PATHOLOGY. Educ: Rutgers Univ, BSc, 51; Columbia Univ, DDS, 55; Am Bd Oral Path, dipl, 64. Prof Exp: Sr asst dent surgeon, Clin Dent, Alaska Native Health Serv, 55-57; intern dent, Columbia-Presby Med Ctr, 57-58; clin assoc stomatol, Sch Dent & Oral Surg, Columbia Univ, 58-60; resident oral path, Nat Inst Dent Res, 60-62; asst pathologist, Armed Forces Inst Path, 62-64; oral pathologist, 64-69, CHIEF SECT DIAG PATH, LAB EXP PATH, NAT INST DENT RES, 69- Mem: Fel AAAS; fel Am Acad Oral Path; Am Dent Asn; Int Asn Dent Res; affil AMA. Res: Clinical and light microscopic pathology of human oral mucosa; kinetics of oral epithelium; effects of metabolic disorders on oral structures; oral ecologic factors relative to pathogenesis of human oral disease. Mailing Add: Bldg 10 Rm 2B-11 Nat Inst of Dent Res Bethesda MD 20014

ARCHBOLD, NORBERT L, b Ft Wayne, Ind, Apr 7, 30; m 51; c 2. GEOLOGY. Educ: Western Reserve Univ, BS, 52; Univ Mich, MS, 56, PhD(geol), 62. Prof Exp: Geologist, US Geol Surv, 53-58, Homestake Mining Co, 61-64 & Nev Bur Mines, 64-68; PROF GEOL, WESTERN ILL UNIV, 68- Mem: Geol Soc Am; Am Inst Mining, Metall & Petrol Engrs; Soc Econ Geologists. Res: Diabase dikes in Canadian shield; mineral deposit exploration; ore deposits of Nevada. Mailing Add: Dept of Geol Western Ill Univ Macomb IL 61455

ARCHDEACON, JAMES WILLIAM, b Carlisle, Ky, Oct 29, 11. PHYSIOLOGY. Educ: Univ Ky, AB, 33, MS, 40; Univ Rochester, PhD(physiol), 43. Prof Exp: From asst prof to assoc prof, 46-55, PROF PHYSIOL, COL MED, UNIV KY, 55- Concurrent Pos: Fulbright-Hays grant & lectr, Univ Malaya, 64-65, China Med Bd of NY grant, 73-74; vis prof, Univ Benghazi, 75-76. Mem: Am Physiol Soc. Res: Iron transport in reticulocyte and placenta. Mailing Add: Dept of Physiol & Biophys Univ of Ky Col of Med Lexington KY 40506

ARCHER, ALFORD, b Garrettsville, Ohio, Apr 11, 08; m 38; c 2. POPULATION GEOGRAPHY, CARTOGRAPHY. Educ: Columbia Univ, BS, 35, MS, 36; Ohio State Univ, PhD(geog), 62. Prof Exp: Asst geog, Ohio State Univ, 36-41; instr geol & geog, Ind State Teacher's Col, 41-42; asst prof com & geog, Univ Toledo, 42-46; geogr, US Bur Census, 46-49, census geog adv int statist progs, Panama, 49-50, Bolivia, 50, Honduras, 51-55 & 60-72, geogr, 55-59, census geog adv, Costa Rica, 52-54, El Salvador, 53-55 & 61, chief cartog methods br, 59-61, census geog adv, Thailand, 61-63, chief for census res br, 63-66, census geog adv, Iran, 66-68, Paraguay, 71-72, GEOGR CENSUS & SURV METHODS, INT STATIST PROGS CTR, US BUR CENSUS, 72- Concurrent Pos: Mem adv comt geog, Pan-Am Inst Geog & Hist, US Dept State & Nat Acad Sci, 59-62; lectr, George Washington Univ & Am Univ, 50- Honors & Awards: Meritorious Serv Award, US Dept Com, 56. Mem: Am Statist Asn; Pop Asn Am; Asn Am Geog; Am Cong Surv & Mapping. Res: Compilation and use of maps for census and statistical purposes; statistical, census and population geography. Mailing Add: Bur of the Census Washington DC 20233

ARCHER, ALLAN FROST, b Roxbury, Mass, Jan 22, 08; m 42; c 2. ARACHNOLOGY, ANTHROPOLOGY. Educ: Harvard Univ, AB, 31; Univ Mich, AM, 33, PhD(zool), 36. Prof Exp: Asst, Ala Geol Surv, 35; asst, Dept Mollusca, Univ Mich, 36-37; Rackham fel, 37-38; cur, Mollusca & Arachnida, Ala Mus Natural Hist, 38-52; dir res, State Dept Conserv, Ala, 39-42, state ecologist, 46-47, geol surv, 47-52; head dept biol, Union Univ, Tenn, 52-60; prof biol, Tift Col, 62-75; RETIRED. Concurrent Pos: Coun Fund grant, Am Mus Natural Hist, 49; NSF grant, 57-58; Fulbright grant & lectr, Chile, 61-62 & Peru, 66-67; field expeds, southern States, Tex, WIndies, Cent Am & SAm. Mem: Assoc Entom Soc Am; Soc Syst Zool; Conchol Soc Gt Brit & Ireland; Nat Geog Soc; assoc Am Mus Natural Hist. Mailing Add: 87 Hillsdale Rd Forsyth GA 31029

ARCHER, CASS L, b Spearman, Tex, June 1, 24; m 53; c 3. MATHEMATICS EDUCATION. Educ: Univ Tex, Austin, BS, 50, MEd, 54, MA, 59, PhD(math educ), 67. Prof Exp: Teacher pub schs, Tex, 50-58; PROF MATH & HEAD DEPT, ANGELO STATE UNIV, 59- Concurrent Pos: Math consult, Tex Educ Agency, 62-64. Mem: Am Math Soc; Math Asn Am. Res: Mathematics programs of institutions changing from a two-year to a four-year college. Mailing Add: Dept of Math Angelo State Univ San Angelo TX 76901

ARCHER, DAVID ANDERSON, b Mercedes, Tex, Sept 7, 46; m 72. NUMERICAL ANALYSIS. Educ: Tex Christian Univ, BS, 69; Rice Univ, MA, 72, PhD(math sci), 73. Prof Exp: Asst prof math, Naval Postgrad Sch, 72-75; ASST PROF MATH, UNIV NC CHARLOTTE, 75- Mem: Soc Indust & Appl Math; Math Asn Am. Res: Study of applications of finite-element/galerkin and collocation methods to the numerical solution of time-dependent partial differential equations. Mailing Add: Dept of Math Univ of NC Charlotte NC 28223

ARCHER, DOUGLAS HARLEY, b Saskatoon, Sask, May 20, 25; m 59; c 2. PHYSICS. Educ: Univ BC, BA, 47, MS, 48; Harvard Univ, PhD(physics), 53. Prof Exp: Staff mem radiowave propagation, Lincoln Lab, Mass Inst Technol, 53-58, radiowave & auroral physics, Hughes Aircraft Co, 58 & atmospheric physics, Gen Elec Co, 58-71; STAFF MEM ATMOSPHERIC PHYSICS, MISSION RES CORP, 71- Mem: Am Geophys Union. Res: Atomic and molecular physics; radiowave propagation; atmospheric physics, especially ionization and deionization of perturbed regions of the atmosphere. Mailing Add: 735 State St Santa Barbara CA 93101

ARCHER, ELLEN GLEASON, b Dunns Station, Pa, Mar 24, 29. NEUROCHEMISTRY. Educ: Chatham Col, BS, 50; Univ Ark, MS, 56, PhD(biochem), 63. Prof Exp: Instr radiol, Sch Med, Univ Ark, 62-65, instr biochem, 64-66, asst prof radiol, 65-66, asst prof biochem, 66; res asst prof radiol, Sch Med, Wash Univ, 66-68; asst prof biol, Univ Mo-St Louis, 68-72; DIR LAB BIOCHEM PSYCHIAT, PSYCHIAT INST FOUND, 72- Concurrent Pos: Vis scientist, Dept Biol Chem, Sch Med, Univ Mich, 66; vis res asst prof radiol, Sch Med, Wash Univ, 69-71. Mem: AAAS; Am Chem Soc; Am Soc Neurochem; Biophys Soc. Res: Biochemistry of schizophrenia and affective illnesses; membrane transport of amino acids and peptides; blood-brain barrier; peptide synthesis. Mailing Add: Psychiat Inst Found 4460 MacArthur Blvd NW Washington DC 20007

ARCHER, FRANCIS L, b New York, NY, Feb 6, 19; m 48; c 2. PATHOLOGY. Educ: Univ Pa, AB, 48; Hahnemann Med Col, MD, 52. Prof Exp: From instr to asst prof path, Univ Chicago, 60-68; sr lectr, Aberdeen Univ, 68-69; ASSOC PROF PATH & ASSOC DIR AUTOPSY SERV, UNIV CHICAGO, 70- Res: In vitro characteristics of mammalian neoplasms; immunopathology of muscle proteins. Mailing Add: Dept of Path Univ of Chicago Chicago IL 60637

ARCHER, GILES ALLAN, organic chemistry, see 12th edition

ARCHER, JOHN DALE, b Brady, Tex, Mar 10, 23; m 52; c 2. MEDICINE, PHARMACOLOGY. Educ: Univ Tex, Austin, BA, 50; Univ Tex, Galveston, MD, 52. Prof Exp: Res asst pharmacol, Med Br, Univ Tex, 50-51, res assoc, 51-52, from instr to asst prof, 52-55; staff physician, Student Health Serv, Univ Tex, Austin, 55-57; asst med dir, State Dept Pub Welfare, Tex, 57-58; med officer, Div New Drugs, Bur Med, Food & Drug Admin, Dept Health, Educ & Welfare, Washington, DC, 58-61, dep dir, 61-62, actg dir, 62; staff physician, Med Serv, Vet Admin Ctr, Tex, 62-64; from asst dir to dir drug eval sect, Dept Drugs, 64-72, SR ED JOUR, AMA, 72- Concurrent Pos: Intern med, Brackenridge Hosp, Austin, Tex, 52-53; team physician, Dept Intercollegiate Athletics, Univ Tex, Austin, 55-57. Mem: AAAS; AMA; Drug Info Asn; Sigma Xi; Am Med Writers Asn. Res: Evaluation and publication of research and clinical experience with drugs; animal and clinical research in pharmacology and toxicology. Mailing Add: Div of Sci Publ AMA 535 N Dearborn St Chicago IL 60610

ARCHER, JUANITA ALMELTA, b Washington, DC, Nov 3, 34; m 58; c 1. INTERNAL MEDICINE. Educ: Howard Univ, BS, 56, MS, 58, MD, 65. Prof Exp: Intern, Freedmens Hosp, 65-66, resident internal med, 66-69, fel endocrinol, 69-70; with NIH, 70-73; instr, Dept Med, 73-75, ASST PROF MED, HOWARD UNIV HOSP, 75- Concurrent Pos: Josiah Macy fac fel, Dept Internal Med, Howard Univ, 74- Mem: Am Fedn Clin Res. Res: Physiological significance of insulin receptors in man with studies to determine whether insulin receptors can be altered so as to effect better diabetic therapy. Mailing Add: Howard Univ Hosp 2041 Georgia Ave NW Washington DC 20060

ARCHER, MILTON CALVERT, JR, plant virology, plant physiology, see 12th edition

ARCHER, NORMAN PHILLIP, nuclear physics, operations research, see 12th edition

ARCHER, ROBERT ALLEN, b Reading, Pa, Apr 3, 36; m 63; c 3. ORGANIC CHEMISTRY. Educ: Harvard Univ, AB, 58; Univ Del, MS, 60; Stanford Univ, PhD(chem), 63. Prof Exp: Sr org chemist, Chem Res Div, Lilly Res Labs, 64-73, RES SCIENTIST, ELI LILLY & CO, 73- Mem: AAAS; Am Chem Soc; Sigma Xi. Res: Synthetic organic chemistry; antibiotics; photochemistry; medicinal chemistry. Mailing Add: Lilly Res Labs Eli Lilly & Co Indianapolis IN 46206

ARCHER, ROBERT JAMES, b San Francisco, Calif, Jan 17, 26; m 61; c 1. SOLID STATE PHYSICS. Educ: Columbia Univ, BA, 51, PhD(phys chem), 54. Prof Exp: Mem tech staff, Bell Tel Labs, 54-63; dept head device physics, Hewlett-Packard Assocs, 63-66, DEPT HEAD DEVICE PHYSICS, HEWLETT PACKARD LABS, HEWLETT-PACKARD CO, 66- Res: Solid state device research and development in the fields of optoelectronics and microwaves, especially involving III-V compound semiconductors. Mailing Add: Hewlett-Packard Co 1501 Page Mill Rd Palo Alto CA 94304

ARCHER, ROBERT RAYMOND, b Omaha, Nebr, Sept 8, 28; m 50; c 5. APPLIED MATHEMATICS, ENGINEERING MECHANICS. Educ: Mass Inst Technol, SB, 52, PhD(math), 56. Prof Exp: Asst prof math, Mass Inst Technol, 56-59; from asst prof to assoc prof, Univ Mass, 59-61; assoc prof eng, Case Western Reserve Univ, 61-66; PROF CIVIL ENG, UNIV MASS, AMHERST, 66- Concurrent Pos: Fulbright lectr, Indian Inst Technol, Kanpur, India, 74-75. Mem: Soc Indust & Appl Math; Am Soc Mech Eng. Res: Mathematical elasticity; applied mechanics; bioengineering; anisotropic stress analysis; continuum mechanical models for study of tree growth stresses and reaction; wood mechanics. Mailing Add: Dept of Civil Eng Univ of Mass Amherst MA 01002

ARCHER, RONALD DEAN, b Rochelle, Ill, July 22, 32; m 54; c 4. INORGANIC CHEMISTRY. Educ: Ill State Univ, BS, 53, MS, 54; Univ Ill, PhD(chem), 59. Prof Exp: Asst prof chem, Univ Calif, 59-63; from asst prof to assoc prof, Tulane Univ, 63-66; assoc prof, 66-70, PROF CHEM, UNIV MASS, 70- Concurrent Pos: Vis prof, Tech Univ Denmark, 72; consult, Alden Res Found, 73- Mem: AAAS; Am Chem Soc; The Chem Soc; Am Crystallog Asn. Res: Coordination compounds; synthesis; properties; kinetics; biometallic studies; photochemistry; spectroscopy; inorganic polymers. Mailing Add: Dept of Chem Univ of Mass Amherst MA 01002

ARCHER, STANLEY J, b Stephenville, Tex, Aug 13, 44; m 66; c 2. IMMUNOLOGY. Educ: Abilene Christian Col, BS, 66, MS, 68; Univ Tenn, Knoxville, PhD(microbiol), 72. Prof Exp: Instr microbiol, Univ Tex, Austin, 72-74; ASST PROF MICROBIOL, ARIZ STATE UNIV, 74- Mem: Am Soc Microbiol; AAAS; NY Acad Sci. Res: Cellular and molecular events involved in the initiation of an immune response. Mailing Add: Dept of Bot & Microbiol Ariz State Univ Tempe AZ 85281

ARCHER, SYDNEY, b New York, NY, Jan 23, 17; m 46; c 3. MEDICINAL CHEMISTRY. Educ: Univ Wis, AB, 37; Pa State Col, MS, 38, PhD(org chem), 40. Prof Exp: Procter & Gamble fel, Northwestern Univ, 40-41; Lilly fel, Univ Chicago, 41-42; res chemist, Sun Oil Co, 42-43; res chemist, Sterling-Winthrop Res Inst, 43-60, asst dir chem res, 60-64, dir chem res & develop, 64-68, assoc dir res, 68-73; RES PROF MED CHEM, RENSSELAER POLYTECH INST, 73- Concurrent Pos: Consult, Spec Action Off Drug Abuse Prev, 73 & Nat Inst Drug Abuse, 74-; chmn med chem study sect, NIH, 76- Honors & Awards: Award, Am Chem Soc, 68. Mem: AAAS; Am Chem Soc; Brit Chem Soc; Am Soc Pharmacol & Exp Therapeut. Mailing Add: 52 Wisconsin Ave Delmar NY 12054

ARCHER, VERNON SHELBY, b Ada, Okla, Dec 5, 39; m 67. ANALYTICAL CHEMISTRY. Educ: Univ Okla, BS, 61, PhD(anal chem), 64. Prof Exp: Asst prof, 64-74, ASSOC PROF CHEM, UNIV WYO, 74- Mem: Am Chem Soc. Res: Electrochemistry; fluoride chemistry. Mailing Add: Dept of Chem Univ of Wyo Laramie WY 82070

ARCHER, VICTOR EUGENE, b Teigen, Mont, May 21, 22; m 49; c 2. MEDICINE, EPIDEMIOLOGY. Educ: Northwestern Univ, BS, 45, MB, 48, MD, 49; Am Bd Prev Med, dipl. Prof Exp: Instr physics, Mont State Univ, 43-44; intern clin med, USPHS Hosp, 48-49, instr radiol health, Environ Health Ctr, USPHS, 50-51, radiation safety officer, NIH, 51-55, field investr, Nat Cancer Inst, 56-59, chief, Coffman Res Lab, 59-61, chief epidemiol serv, Occup Health Field Sta, USPHS, 61-72, MED DIR & EPIDEMIOLOGIST, WESTERN AREA LAB FOR OCCUP SAFETY & HEALTH, NAT INST OCCUP SAFETY & HEALTH, 72- Concurrent Pos: Clin lectr, Med Sch, Univ Utah, 58-70. Honors & Awards: Commendation Medal, USPHS. Mem: Health Physics Soc; Am Indust Hyg Asn; Am Pub Health Asn; AAAS; Am Col Prev Med. Res: Radiobiology; environmental causes of lung disease and cancer. Mailing Add: 4370 Spruce Circle Salt Lake City UT 84117

ARCHER, WESLEY LEA, b Marietta, Ohio, Feb 20, 27; m 71. INDUSTRIAL ORGANIC CHEMISTRY. Educ: Kalamazoo Col, AB, 50; Ind Univ, PhD(chem), 53. Prof Exp: Pharmaceut chemist, Irwin Neisler, Ill, 53-56; chemist, Dow Chem Co, 56-73, RES SPECIALIST, DOW CHEM USA, 73- Mem: Am Chem Soc; Am Soc Metals. Res: Chemistry of diphenyl oxide derivatives; development of antioxidants for high temperature lubricants; reactions of polyhalohydrocarbons with aluminum and inhibitors for this reaction; use of chlorinated solvents in metal cleaning operations. Mailing Add: 4101 Swede Rd Midland MI 48640

ARCHER, WILLIAM HARRY, b Ambridge, Pa, Mar 6, 05; m 33; c 1. ORAL SURGERY. Educ: Univ Pittsburgh, DDS, 27, BS, 37, MA, 47; Am Bd Oral Surg, dipl. Prof Exp: Demonstr exodont, oral surg & anesthesia, 27-29, instr, 29-33, instr & lectr, 33-38, from asst prof to assoc prof oral surg & anesthesia & chmn dept, 38-46, prof oral surg & anesthesia & chmn dept oral surg, 46-70, univ prof, 70-75, EMER UNIV PROF, SCH DENT, UNIV PITTSBURGH, 75- Concurrent Pos: Spec lectr, Sch Pharm, Univ Pittsburgh, 41-44, Sch Nursing, 45- & Sch Pub Health, 55-, mem fac grad sch, 54-; asst dent staff, Elizabeth Steel Magee Hosp, 31-36, assoc, 36-41, sr mem, 41-46, chief, 46-65, emer staff, 65-; sr staff, Eye & Ear Hosp, 40-50, sr mem med staff, Oral Surg Sect, 50-65, emer staff, 65-; consult dent, Munic Hosp, 41-50; consult oral surg, Western Psychiat Inst & Clin, 42-65, Army Hosp, Tilton Gen, Ft Dix, NJ, 48-52 & Henry Clay Frick Mem Hosp, Mt Pleasant, Pa, 59-65; sr consult, Vet Admin Hosp, Oakland, Pittsburgh, Pa, 47-; emer sr mem staff dept dent & oral surg, South Side Hosp, 53-; emer sr mem staff & chief dept oral surg, St Clair Mem Hosp, 53-; emer sr mem staff dent, Children's Hosp Pittsburgh, 61- Honors & Awards: President's Distinguished Serv Award, Selective Serv Syst, 63; Rene Le Fort Medal, Brasilia, Brazil, 70; Prof Dr Halit Chazi Medal & Rosette, Istanbul, Turkey, 70; Heidbrink Award, Chicago, 71; Wm J Gies Award, San Francisco, 72. Mem: Am Dent Asn; Am Col Dent; Am Soc Oral Surg; Am Acad Oral Path; Int Asn Oral Surg. Res: General and local anesthetics for dental surgery; causes of Alveolalgia; dentifrices and mouth antiseptics; general oral medicine. Mailing Add: Dept of Oral Surgery Univ of Pittsburgh Sch of Dent Pittsburgh PA 15213

ARCHIBALD, FRANCIS MAGOUN, b Milton, NS, Oct 7, 99; US citizen; m 32; c 2.

LIPID CHEMISTRY. Educ: Acadia Univ, BSc, 19; McGill Univ, BSc, 23; Univ Toronto, MSc, 27, PhD(phys chem), 28. Hon Degrees: LLD, Acadia Univ, 60. Prof Exp: Res chemist, Standard Oil Develop Co, 28-29, lab head, Standard Alcohol Co, 29-50 & Bayway Refinery, 50-53, tech adv, Vacuum Oil Co SAfrica, 53-55 & Standard Vacuum Oil Co, NY, 55-60; res vol, Sloan-Kettering Inst, 60-62; RETIRED. Mem: Am Chem Soc. Res: Petrochemical manufacture; analytical biochemistry; lipid complex analysis. Mailing Add: Walker Lab Sloan-Kettering Inst Rye NY 10580

ARCHIBALD, JAMES, b Bellshill, Scotland, Jan 4, 19; Can citizen; m 48; c 3. VETERINARY MEDICINE. Educ: Ont Vet Col, DVM, 49; Univ Toronto, MVSc, 51; Univ Giessen, Dr med vet, 58; FRCVS. Prof Exp: From asst prof to assoc prof, 49-54, PROF MED VET, DIV SMALL ANIMAL MED & SURG, ONT VET COL, UNIV GUELPH, 54-, CHMN DEPT CLIN STUDIES, 63- Concurrent Pos: Consult, Defense Res Med Labs, Can, 58. Mem: Am Vet Med Asn; Can Vet Med Asn (pres, 63). Res: Experimental surgery; liver and tissue transplantation. Mailing Add: Dept Clin Studies Ont Vet Col Univ of Guelph Guelph ON Can

ARCHIBALD, JOHN ARTHUR, plant nutrition, see 12th edition

ARCHIBALD, KALMAN DALE, b Brockton, Mass, Jan 19, 10; m 35; c 3. ICHTHYOLOGY, EMBRYOLOGY. Educ: Denison Univ, BA, 33; Ohio State Univ, MA, 34, PhD(entom), 54; Colgate-Rochester Divinity Sch, BD, 46. Prof Exp: From asst zool to instr, Ohio State Univ, 35-36 & 40-41; assoc prof biol & head dept, Ouachita Col, 36-43, assoc prof physics, 42-43; assoc prof zool, Keuka Col, 46-47; assoc prof biol, Acadia Univ, 47-48; asst prof zool, 48-54, dir gen educ course life sci, 49-60 & 64-66, chmn dept biol, 60-63, from assoc prof to prof biol sci, 54-74, EMER PROF BIOL SCI, DENISON UNIV, 74- Concurrent Pos: Pastor, NY, 43-47, interim supply pastor, 48-; dir forest insect res, NS Res Found, 48-59. Mem: AAAS; Genetics Soc Am; Entom Soc Am. Res: Age determination of Ictalurus punctatus and Noturus flavus; histogenesis of blood cells in fresh water fishes; forest Aphidae of Nova Scotia. Mailing Add: RR 2 Granville OH 43023

ARCHIBALD, KENNETH C, b White Plains, NY, Aug 1, 27. PHYSICAL MEDICINE. Educ: Cornell Univ, MD, 53; St Lawrence Univ, BS, 53. Prof Exp: Sr instr phys med, Sch Med, Western Reserve Univ & assoc dir dept phys med & rehab, Univ Hosps Cleveland, 57-58; asst prof phys med, Med Col, Cornell Univ & dir dir phys med & rehab & asst attend physician, New York Hosp-Cornell Med Ctr, 58-66; from assoc prof to prof phys med & rehab, Sch Med, Temple Univ & dir dept, Episcopal Hosp, 66-69; DIR PHYS MED & REHAB, JERD SULLIVAN REHAB CTR, GARDEN HOSP & PRESBY HOSP OF PAC MED CTR, SAN FRANCISCO, 69-; DIR PHYS MED & REHAB, CHILDREN'S HOSP, 74- Concurrent Pos: Consult, Muscle Dis Clin, Univ Hosps Cleveland, 58-69; assoc clin prof orthop surg, Med Ctr, Univ Calif, San Francisco, 70- Mem: Am Cong Rehab Med; Am Acad Phys Med & Rehab; Am Asn Electromyography & Electrodiagnosis. Res: Physical medicine and rehabilitation; neuromuscular disorders. Mailing Add: 2360 Clay St San Francisco CA 94115

ARCHIBALD, PATRICIA ANN, b Olney, Ill, July 18, 34. ALGOLOGY, PHYCOLOGY. Educ: Ball State Univ, BS & MA, 61; Univ Tex, Austin, PhD(bot), 69. Prof Exp: Teacher high sch, Ind, 59-64; instr biol, Palm Beach Jr Col, 64-66 & Cuyahoga Community Col, 66-67; asst prof, 69-71, ASSOC PROF BIOL, SLIPPERY ROCK STATE COL, 71- Concurrent Pos: Fulbright lectr, Harrow Technol Col, Eng, 62-63. Mem: Phycol Soc Am; Brit Phycol Soc; Int Phycol Soc; Bot Soc Am; AAAS. Res: Phycology, morphology and physiology of zoosporic chlorococcales and chlorosarcinales; edaphic algae, distribution, morphology, physiology; algal flora of acid mine polluted streams. Mailing Add: Dept of Biol Slippery Rock State Col Slippery Rock PA 16057

ARCHIBALD, RALPH GEORGE, b Sackville, NB, May 23, 01; nat US; m 41; c 1. MATHEMATICS. Educ: Univ Man, BA, 22; Univ Toronto, MA, 24; Univ Chicago, PhD(math), 27. Prof Exp: Demonstr physics, Univ Man, 21-22; lectr math, Wesley Col, Can, 22-23; from assoc to asst prof, Columbia Univ, 27-38; from asst prof to prof, 38-71, EMER PROF MATH, QUEENS COL, CITY UNIV NY, 71- Mem: Am Math Soc; Math Asn Am; Am Sci Affil. Res: Theory of numbers. Mailing Add: 43-19 192nd St Flushing NY 11358

ARCHIBALD, REGINALD MACGREGOR, b Syracuse, NY, Mar 2, 10; m 48; c 2. BIOCHEMISTRY, PEDIATRIC ENDOCRINOLOGY. Educ: Univ BC, BA, 30, MA, 32; Univ Toronto, PhD(path chem), 34, MD, 39. Prof Exp: Asst, Univ BC, 30-32; asst, Univ Toronto, 32-33; intern, Hosp for Sick Children, Toronto, 37-38; intern, Toronto Gen Hosp, 39-40; asst resident, Rockefeller Inst Hosp, 41-46; prof biochem & head dept, Johns Hopkins Univ, 46-48; MEM STAFF, ROCKEFELLER UNIV, 48-, PROF MED, 59- Concurrent Pos: Nat Res Coun fel, 40-42; vis investr, Rockefeller Univ, 40-43, spec investr, 43-45, assoc, 46, physician, Univ Hosp, 48-59, sr physician, 59-; vpres, Am Bd Clin Chem, 57-63. Mem: AAAS; Am Chem Soc; Am Soc Biol Chemists; Soc Res Child Develop; Soc Adolescent Med. Res: Non-fermentable carbohydrates of urines; fractionation of urinary steroids; physiological roles of glutamine and other amino acids; measurement of enzymes; use of enzymes in analysis; influences of hormones on enzymes; problems of growth and development of children; skeletal anomalies in children. Mailing Add: Rockefeller Univ Hosp 1230 York Ave New York NY 10021

ARCHIBALD, WILLIAM JAMES, b Sydney, NS, Oct 30, 12; m 36; c 3. PHYSICS. Educ: Dalhousie Univ, BA, 33, MA, 36; Univ Va, PhD, 38. Hon Degrees: DSc, Univ NB, 61. Prof Exp: Sterling fel, Yale Univ, 38-39; res scientist, Nat Res Coun, 39-42; assoc prof, 42-47, dean arts & sci, 55-60, prof, 47-73, DR A C FALES PROF THEORET PHYSICS, DALHOUSIE UNIV, 73-, DEAN FRESHMEN, 73- Concurrent Pos: Fel, Univ Va, 51-52; mem, Defense Res Bd, Can, 56-59. Mem: Fel Royal Soc Can. Res: Ultra centrifuge; theory of molecular weight determination; theoretical physics; nuclear magnetism; field theory. Mailing Add: Dept of Theoret Physics Dalhousie Univ Halifax NS Can

ARCHIE, WILLIAM C, JR, b Brownsville, Tex, Mar 1, 44; m 70; c 2. PHYSICAL ORGANIC CHEMISTRY. Educ: Duke Univ, BS, 66; Stanford Univ, PhD(chem), 70. Prof Exp: Fel chem, Harvard Univ, 70-72; SR RES CHEMIST, EASTMAN KODAK CO, 72- Mem: Am Chem Soc. Res: Photographic science; use of physical organic chemical methods to elucidate reaction mechanisms in film; stability of photographic color images. Mailing Add: Res Labs Eastman Kodak Co Rochester NY 14560

ARCHIMOVICH, ALEXANDER S, b Novosybkov, Ukraine, Apr 23, 92; US citizen; m 35; c 3. BIOLOGY, AGRONOMY. Educ: St Vladimir Univ, Kiev, dipl natural sci, 17; Kiev Polytech Inst, dipl agr sci, 22; T H Shevchenko State Univ, Kiev, MS, 38, PhD(biol), 40. Prof Exp: Res botanist, Ukrainian Acad Sci, 19-23; sr specialist, Plant Breeding Sta, Bila Zerkva, Ukraine, 23-33; prof agron, Agr Cols, 24-43; prof plant breeding, Ukrainian Inst, Ger, 45-48; head dept selection sugar beet, Valladolid Selection Sta, Spain, 48-52; prof plant breeding, Ukrainian Tech Inst, NY, 53-64, pres, 55-62; pres, 62-70, HEAD DEPT NATURAL SCI & MED, UKRAINIAN ACAD ARTS & SCI US, 59- Concurrent Pos: Hon mem, Inst Study USSR, Munich & NY, 59; head chem, biol & med sect, Shevchenko Sci Soc, 74- Mem: AAAS; Am Soc Sugar Beet Technol; Royal Span Natural Hist Soc; Bavarian Bot Soc; hon mem Am-Ukrainian Vet Med Asn. Res: Botany, biology and selection of sugar beet; control of pollination in sugar beet; methods of producing sugar beet seed in north of Spain; analysis of agriculture in Soviet Union; geography of the field crops in the Ukraine. Mailing Add: 143 Mountain Rd Rosendale NY 12472

ARCOS, JOSEPH (CHARLES), b Hungary, Aug 22, 21; US citizen; m 46; c 1. BIOCHEMISTRY, ONCOLOGY. Educ: Univ Cluj, LChem, 47; Conserv Nat Arts et Metiers, France, ChE, 50; Univ Paris, DSc, 51. Prof Exp: Res engr surface chem, Tech Ctr Graphic Indust, 52; asst tech dir develop & prod, Jouan Co, 52-53; res assoc cancer res, McArdle Lab, Univ Wis, 53-57; asst res prof, Cancer Res Lab, Univ Fla, 57-60; assoc prof, 60-68, PROF MED & BIOCHEM, SCH MED, TULANE UNIV, 68- Mem: Am Soc Biol Chemists; Biophys Soc; Am Asn Cancer Res; Soc Exp Biol & Med; Am Soc Pharmacol & Exp Therapeut. Res: Structure-activity relationships and molecular geometry of chemical carcinogens; biochemistry of mitochondrial pathways of energy production; conformational changes in biological macromolecules; metabolism of carcinogens. Mailing Add: 210 State St New Orleans LA 70118

ARCOS, MARTHA, biochemistry, endocrinology, see 12th edition

ARCULUS, RICHARD J, b Calcutta, India, Jan 20, 49; Brit citizen; m 73. PETROLOGY. Educ: Univ Durham, BSc, 70, PhD(petrol), 73. Prof Exp: Fel petrol, Geophys Lab, Carnegie Inst Washington, 73-75; ASST PROF TETROL, RICE UNIV, 75- Mem: Mineral Soc Am; Am Geophys Union; Geol Soc Am. Res: Lanthanide geochemistry; high-pressure experimental phase equilibria; island-arc and continental volcanicity; ore genesis and association with volcanism. Mailing Add: Dept of Geol Rice Univ Houston TX 77001

ARD, WILLIAM BRYANT, JR, b Jackson, Tenn, Oct 8, 27; m 50; c 8. PHYSICS. Educ: Ala Polytech Inst, BS, 50, MS, 51; Duke Univ, PhD(physics), 55. Prof Exp: From asst prof to assoc prof, Univ Ala, 55-59; assoc prof, Univ Fla, 59-62; physicist, Oak Ridge Nat Lab, 62-73; SR RES SCIENTIST, UNITED TECHNOL RES CTR, 73- Mem: Am Phys Soc. Res: Plasma physics and controlled thermonuclear fusion research. Mailing Add: United Technol Res Ctr Silver Ln East Hartford CT 06108

ARDEN, DANIEL DOUGLAS, b Bainbridge, Ga, Sept 24, 22; m 43; c 4. EXPLORATION GEOLOGY. Educ: Emory Univ, AB, 48, MS, 49; Univ Calif, Berkeley, PhD(paleont), 61. Prof Exp: Asst prof geol, Birmingham-Southern Col, 49-51; geologist petrol geol, Standard Oil Co Calif, 54-56; staff geologist, Standard Oil Co (Ohio), 56-65; explor adv, Signal Oil & Gas Co, 65-70; PROF GEOL, GA SOUTHWESTERN COL, 70- Concurrent Pos: Chief geol consult, Geophys Serv Inc, 71-; mem, Int Comn Hist Geol, 75- Mem: Geol Soc Am; Am Geophys Union; Am Inst Prof Geol. Res: Geobotanical relationships; stratigraphy of Ocala and Tivola limestones in Georgia. Mailing Add: Dept of Geol Ga Southwestern Col Americus GA 31709

ARDEN, SHELDON BRUCE, b New York, NY, Sept 14, 42. MEDICAL MICROBIOLOGY. Educ: Rutgers Univ, BA, 64; NY Univ, PhD(microbiol), 70. Prof Exp: Fel microbiol, Sch Med, NY Univ, 70-72, asst res scientist, 72-75; CHIEF CLIN MICROBIOL, CABRINI HEALTH CARE CTR, COLUMBUS HOSP, 75- Concurrent Pos: Adj asst prof microbiol, Sch Med, NY Univ, 75- Mem: Am Soc Microbiol; Soc Gen Microbiol. Res: Genetics, physiology and taxonomy of corynebacteria; phage-host interrelationships in corynebacterial systems. Mailing Add: Dept of Path Clin Microbiol Cabrini Health Care Ctr 227 E 19th St New York NY 10003

ARDITTI, JOSEPH, b Sofia, Bulgaria, May 1, 32; US citizen. PLANT PHYSIOLOGY. Educ: Univ Calif, Los Angeles, BS, 59; Univ Southern Calif, PhD(biol, plant physiol), 65. Prof Exp: Teaching asst, Univ Southern Calif, 60-65, from lectr to asst prof, 65-72, ASSOC PROF BIOL, UNIV CALIF, IRVINE, 72- Mem: AAAS; Am Inst Biol Sci; Am Soc Plant Physiol; Bot Soc Am. Res: Phytochemistry; plant development; orchid biology. Mailing Add: Dept of Develop & Cell Biol Univ of Calif Irvine CA 92664

ARDLEY, HARRY MOUNTCASTLE, b Oakland, Calif, Jan 22, 26; m 48; c 3. MATHEMATICAL STATISTICS, APPLIED STATISTICS. Educ: Univ Calif, Berkeley, AB, 50. Prof Exp: Economist, US Dept Com, 51-53; statistician, Pac Tel Co, 53-59, gen statistician, 59-63, SUPV STATISTICIAN & HEAD DEPT MATH STATIST, PAC TEL CO, 63- Mem: Am Statist Asn; Opers Res Soc Am. Res: Demand modeling and cross-elasticities of demand; productivity measurement and production functions; probability sampling. Mailing Add: Pac Tel Co Mkt Surv & Statist 140 New Montgomery St San Francisco CA 94105

ARDREY, WILLIAM BOYLE, b Denver, Colo, Feb 2, 12; m 40; c 2. BACTERIOLOGY. Educ: Monmouth Col, BS, 34; Mich State Col, MS, 36, PhD(bact), 39. Prof Exp: From asst bacteriologist to assoc bacteriologist, 39-49, pathologist, 49-70, from instr to prof bact, 39-70, prof vet sci & vet microbiologist, 70-74, EMER PROF VET SCI & VET MICROBIOLOGIST, UNIV IDAHO, 74- Mem: AAAS; Am Soc Microbiol. Res: Poultry diseases; vibriosis of sheep; brucellosis of swine and cattle; avian leukosis; measurement of germicidal chlorine. Mailing Add: Dept of Vet Sci Univ of Idaho Moscow ID 83843

ARENA, JAY M, b Clarksburg, WVa, Mar 3, 09; m 31; c 7. PEDIATRICS. Educ: WVa Univ, BS, 30; Duke Univ, MD, 32. Prof Exp: Intern, Strong Mem Hosp, Rochester, NY, 32 & Johns Hopkins Hosp, 32-33; asst resident & resident, Univ Hosp, Duke Univ, 33-35; instr pediat, Vanderbilt Univ, 36; from asst prof to assoc prof, 36-50, PROF PEDIAT, SCH MED, DUKE UNIV, 50-, PROF COMMUNITY HEALTH SCI, 66- Concurrent Pos: Mem ed adv bd, Nutrit Today, Clin Pediat, Highlight, Emergency Med, Pediat News & Reportes Medicos; chmn Z66 standards comt, Am Nat Standards Inst. Mem: Am Acad Pediat (pres-elect, 70); Am Pediat Soc; Am Asn Poison Control Ctrs (pres, 68-70). Res: Poisonings and accidents in children. Mailing Add: Box 3024 Duke Hosp Durham NC 27706

ARENBERG, DAVID LEWIS, physics, see 12th edition

ARENDALE, WILLIAM FRANK, b South Pittsburg, Tenn, Nov 11, 21; m 45; c 3. CHEMICAL PHYSICS. Educ: Mid Tenn State Univ, BS, 42; Univ Tenn, MS, 48, PhD(chem & phys), 53. Prof Exp: Chemist, Ala Ord Works, 42-44, Tenn Eastman Co, 44-46 & Carbide & Carbon Co, 48-50; chief res, Alpha Div, Thiokol Chem Corp, 51-56, dir res, 56-59, tech dir & asst to gen mgr, 59-64; prof chem & asst dir res inst, 64-66, dir div natural sci & math, 66-70, PROF CHEM, UNIV ALA, HUNTSVILLE, 70- Mem: Fel AAAS; Am Chem Soc; Am Phys Soc; Am Inst Aeronaut & Astronaut; Optical Soc Am. Res: Infrared spectroscopy; organometallic compounds; materials properties; organo- selenium chemistry; enviromental chemistry; interaction of electromagnetic energy with matter. Mailing Add: Dept of Chem Univ of Ala Huntsville AL 35807

ARENDS, CHARLES BRADFORD, b Chicago, Ill, June 10, 31; m 55; c 4. PLASTICS CHEMISTRY. Educ: Ill Inst Technol, BS, 52; Univ Wash, PhD(phys chem), 55. Prof Exp: Chemist, 55-68, sr res chemist, 68-74, RES SPECIALIST, DOW CHEM CO, 74- Res: Micromechanics of composite plastics. Mailing Add: Dow Chem Co 672 Bldg Midland MI 48640

ARENDS, ROBERT LEANDER, b Aplington, Iowa, Oct 10, 17; m 46; c 3. CLINICAL CHEMISTRY. Educ: Iowa State Col, BS, 40; Univ Colo, MS, 49, PhD(biochem), 51. Prof Exp: Asst chemist, Iowa State Dept Agr, 40-42 & 46; asst physiol & pharmacol, Univ Colo, 47-51; instr biochem, State Univ NY, 51-54; biochemist, Milwaukee Children's Hosp, 54-67; CLIN CHEMIST, COLUMBIA HOSP, 67- Mem: Am Asn Clin Chemists; Asn Clin Scientists; Am Inst Chemists; AAAS. Res: Metabolic diseases. Mailing Add: 3321 N Maryland Ave Milwaukee WI 53211

ARENDT, BILLY DEAN, b Gibson, Iowa, May 1, 38; m 58; c 3. ALGEBRA. Educ: Iowa State Univ, BS, 60, MS, 62; Univ Iowa, PhD(math), 67. Prof Exp: Instr math, Kans State Univ, 62-64; vis lectr, Univ Wis-Madison, 67-68; asst prof, 68-73, ASSOC PROF MATH, UNIV MO-COLUMBIA, 73- Mem: Am Math Soc; Math Asn Am. Res: Algebraic semigroups. Mailing Add: Dept of Math Univ of Mo Columbia MO 65201

ARENDT, KENNETH ALBERT, b New Britain, Conn, July 24, 25; m 49; c 2. PHYSIOLOGY. Educ: Union Col, AB, 49; Boston Univ, AM, 52, PhD(biol), 55. Prof Exp: Res assoc, Biol Res Labs, Boston Univ, 53-57; from instr to assoc prof, 57-70, PROF PHYSIOL, SCH MED, LOMA LINDA UNIV, 70- Mem: Am Physiol Soc; Microcirculatory Soc; Sigma Xi. Res: Microcirculatory physiology. Mailing Add: Dept of Physiol Loma Linda Univ Loma Linda CA 92354

ARENDT, RONALD H, b Chicago, Ill, Apr 20, 41; m 64; c 2. PHYSICAL CHEMISTRY. Educ: Univ Chicago, BS, 62, MS, 65, PhD(phys chem), 68. Prof Exp: MEM RES STAFF, GEN ELEC RES & DEVELOP CTR, 68- Mem: Am Chem Soc; Am Phys Soc. Res: Molten salts; inorganic synthesis; thermodynamics. Mailing Add: Gen Elec Res & Develop Ctr One River Rd Schenectady NY 12301

ARENDT, VOLKER DIETRICH, b Berlin, Ger, Sept 18, 34; m 66; c 3. POLYMER CHEMISTRY, RUBBER CHEMISTRY. Educ: NC State Col, BS, 60; Princeton Univ, MA, 62, PhD(org chem), 64. Prof Exp: Textile tech supvr, Duncan Fox Co, Peru, 55-57 & W R Grace Co, Peru, 57; res chemist, 64-73, SR RES SCIENTIST, AM CYANAMID CO, 73- Mem: Am Chem Soc. Res: Cellulose and carbohydrate chemistry; chemical structure determinations; new analytical methods; synthetic organic and polymer chemistry; structure property relationships; research in elastomers; specialty polyacrylates; polyurethanes. Mailing Add: 8 Monroe Ct RD 4 Princeton NJ 08540

ARENS, JOHN FREDERIC, b New York, NY, July 30, 39. COSMIC RAY PHYSICS. Educ: Mass Inst Technol, BS, 61; Univ Calif, Berkeley, PhD(physics), 66. Prof Exp: ASTROPHYSICIST, GODDARD SPACE FLIGHT CTR, NASA, 66- Concurrent Pos: Assoc, Nat Acad Sci, 66-69. Mem: Am Phys Soc. Res: Experimental high energy physics; measurements of cross-sections and polarizations of hadrons; satellite measurements of electrons and protons in the radiation belts and interplanetary space; cosmic ray element and isotope measurements. Mailing Add: Goddard Space Flight Ctr NASA Code 612 Greenbelt MD 20771

ARENS, MAX QUIRIN, b South Bend, Ind, July 22, 45; m 67; c 2. VIROLOGY. Educ: Purdue Univ, BSA, 67; Va Polytech Inst & State Univ, PhD(microbiol), 71. Prof Exp: Res assoc, 71-75, INSTR MOLECULAR VIROL, INST MOLECULAR VIROL, SCH MED, ST LOUIS UNIV, 75- Mem: Sigma Xi. Res: Mechanism and enzymology of adenovirus DNA replication; specific functions of various enzymes involved in DNA replication. Mailing Add: Inst for Molecular Virol St Louis Univ Sch of Med St Louis MO 63110

ARENS, RICHARD FRIEDERICH, b Iserlohn, Ger, Apr 24, 19; nat US; m 43; c 1. MATHEMATICS. Educ: Univ Calif, Los Angeles, AB, 41; Harvard Univ, AM, 42, PhD(math), 45. Prof Exp: Tutor math, Harvard Univ, 42-45, staff asst, Inst Advan Study, 45-47; from asst prof to assoc prof, 47-57, PROF MATH, UNIV CALIF, LOS ANGELES, 57- Concurrent Pos: Mem, Inst Advan Study, 53-54; ed, Pac J Math, 64- Mem: Am Math Soc. Res: Analysis of function spaces; topological algebra; dynamics; theoretical physics. Mailing Add: Dept of Math Univ of Calif Los Angeles CA 90024

ARENSON, DONALD LEWIS, b Chicago, Ill, June 15, 26; m 48; c 3. MATHEMATICS. Educ: Ill Inst Technol, BS, 47, MS, 49. Prof Exp: Assoc res engr, Armour Res Found, 48-50; consult engr self-employed, 50-52; staff engr, Cook Res Lab Div, Cook Elec Co, 52-54, dir, Ex-cel Develop Lab, 54-57; asst gen mgr, Mech Res Div, Am Mach & Foundry Co, 57-62; asst dir mech res div, Armour Res Found, 63; res dir, Booz-Allen Appl Res, Inc, 63-69, vpres, 65-69; pres & mem bd dirs, 69-72, VPRES, CHASE MANHATTAN CONSULT INC, 72- Concurrent Pos: Mem bd dirs, Booz-Allen Pub Admin Serv Inc & Yuasa Ionics Co, 73-; Managing Dir, Mitsubshi Chase Manhattan Consult Co, 72- Mem: Am Math Soc; Am Inst Aeronaut & Astronaut; Opers Res Soc Am. Res: Systems analysis; operations research; research and development management. Mailing Add: 7544 N Karlov Ave Skokie IL 60076

ARENSTEIN, MARVIN, physics, see 12th edition

ARENSTORF, RICHARD F, b Hamburg, Ger, Nov 7, 29; US citizen; m 56; c 3. MATHEMATICS. Educ: Univ Göttingen, BS, 52, MS, 54; Univ Mainz, D rer nat(math), 56. Prof Exp: Develop engr, Telefonaktiebelaget L M Ericsson, Ger, 55-56; physicist, Army Ballistic Missile Agency, Redstone Arsenal, Ala, 57-60; staff scientist, Marshall Space Flight Ctr, NASA, 60-69; PROF MATH, VANDERBILT UNIV, 69- Concurrent Pos: Asst prof, Univ Ala, 58-59, from assoc prof to prof, 60-69; consult, Math Res Ctr, US Naval Res Lab, Washington, DC, 70- Honors & Awards: Medal for Except Sci Achievement, NASA, 66. Mem: Am Math Soc; Soc Indust & Appl Math. Res: Analytic algebraic number theory; nonlinear differential equations; celestial mechanics. Mailing Add: Dept of Math Vanderbilt Univ Nashville TN 37235

ARENTS, JOHN (STEPHEN), b Brooklyn, NY, Dec 11, 26; m 65. PHYSICAL CHEMISTRY. Educ: Columbia Univ, AB, 50, AM, 51, PhD(chem), 57. Prof Exp: Asst chem, Columbia Univ, 50-54, 55-56; lectr, 56-57, from instr to asst prof, 57-66, ASSOC PROF CHEM, CITY COL NEW YORK, 66- Mem: AAAS; Am Chem Soc; Am Phys Soc. Res: Quantum chemistry. Mailing Add: Dept of Chem City Col of New York New York NY 10031

AREWA, E OJO, b Okeagbe, Nigeria, July 22, 35; m 62; c 4. ANTHROPOLOGY. Educ: Univ Nebr, Lincoln, BA, 61; Univ Calif, Berkeley, PhD(anthrop), 66. Prof Exp: From lectr to asst prof sociol, Univ Ife, Nigeria, 66-68; asst prof anthrop, Calif State Univ, San Jose, 68-69; asst prof, 69-71, ASSOC PROF ANTHROP, OHIO STATE UNIV, 71- Concurrent Pos: Consult, Taba Prog Soc Sci, Addison-Wesley Publ Co, 68-71. Mem: AAAS; foreign fel Am Anthrop Asn; Int Soc Folk Narrative Res. Res: The communicative aspects of traditional African forms; morphology of African narratives; the continuity of African patterns in the new world; the ethnography of suburban life in America. Mailing Add: Dept of Anthrop Ohio State Univ Columbus OH 43210

ARFIN, STUART MICHAEL, b New York, NY, Sept 4, 36; m 64; c 2. BIOCHEMISTRY. Educ: City Col New York, BS, 58; Albert Einstein Col Med, PhD(biochem), 66. Prof Exp: Res assoc biol, Purdue Univ, 66-69; asst prof biochem, Sch Med, Univ Pittsburgh, 69-71; ASSOC PROF BIOL CHEM, SCH MED, UNIV CALIF, IRVINE, 71- Mem: Am Soc Microbiol; Am Soc Biol Chemists; AAAS. Res: Biochemical control mechanisms; enzymology. Mailing Add: Dept of Biol Chem Univ of Calif Sch of Med Irvine CA 92664

ARFKEN, GEORGE BROWN, JR, b Jersey City, NJ, Nov 20, 22; m 49; c 3. PHYSICS. Educ: Yale Univ, BE, 43, MS, 48, PhD(physics), 50. Prof Exp: Physicist, Oak Ridge Nat Lab, 50-52; from asst prof to assoc prof, 52-56, chmn dept, 56-72, PROF PHYSICS, MIAMI UNIV, 56- Concurrent Pos: Consult, Los Alamos Sci Lab, 62-66. Mem: Am Phys Soc; Am Asn Physics Teachers. Res: Theoretical physics; nuclear forces; nuclear spectroscopy. Mailing Add: 5301 Coulter Lane Oxford OH 45056

ARGABRIGHT, LOREN N, b Nemaha, Nebr, Jan 29, 33; m 61. MATHEMATICS. Educ: Nebr State Teachers Col, Peru, BS, 54; Univ Kans, MA, 58; Univ Wash, PhD(math), 63. Prof Exp: Instr math, Univ Calif, Berkeley, 63-65; asst prof, Univ Minn, 65-70; assoc prof & chmn dept, Univ Nebr, Lincoln, 70-74; PROF MATH & HEAD DEPT, DREXEL UNIV, 74- Mem: Am Math Soc; Math Asn Am. Res: Functional analysis; harmonic analysis; group representations. Mailing Add: Dept of Math Drexel Univ Philadelphia PA 19104

ARGABRIGHT, PERRY A, b Pueblo, Colo, June 6, 29; m 51; c 3. CHEMISTRY. Educ: Univ Denver, BS, 51; Univ Colo, PhD(org chem), 57. Prof Exp: Res chemist, Shell Develop Co, 51-52; res chemist, Esso Res & Eng Co, NJ, 56-58, group leader, 58-59, sr chemist, 59-62; sr res scientist org synthesis, 62-68, RES ASSOC, MARATHON OIL CO, 68- Honors & Awards: Gold Medallion, Am Chem Soc Colo Sect, 72. Mem: AAAS; Sigma Xi (vpres, Sci Res Soc Am, 63); Am Chem Soc; NY Acad Sci. Res: Synthesis and characterization of organic compounds containing N-F bonds; influence of polar-aprotic solvents on rate and course of organic reactions; synthesis of nitrogen heterocycles; preparation and study of polyelectrolytes; chemicals for oil recovery. Mailing Add: Marathon Oil Co 7400 S Broadway Littleton CO 80120

ARGALL, CLIFFORD IRVING, b Mich, Dec 4, 17; m 42; c 4. IMMUNOLOGY. Educ: Chapman Col, AB, 40; Stanford Univ, MA, 53; Univ Utah, PhD(immunol), 55. Prof Exp: Instr biol, Chapman Col, 40-42; chief res div, Bur Labs, Utah State Dept Health, 46-56; DIR BLOOD BANK, BAPTIST MEM HOSP, 56-; ASST IMMUNO-HEMATOL, UNIV TENN, MEMPHIS, 57- Mem: AAAS. Res: Immunohematology and transfusion therapy. Mailing Add: Baptist Mem Hosp 899 Madison Ave Memphis TN 38117

ARGANBRIGHT, ROBERT PHILIP, b Ft Leavenworth, Kans, Nov 7, 23; m 57; c 2. ORGANIC CHEMISTRY. Educ: Univ Kans, BS, 50; Univ Colo, PhD(chem), 56. Prof Exp: Chemist, Continental Oil Co, 50-52; chemist, Monsanto Chem Co, 56-60, res specialist, 60-67; res assoc, 67-74, RES MGR, PETRO-TEX CHEM CORP, 74- Mem: Catalysis Soc; Am Chem Soc. Res: Mechanisms of organic reactions; heterogeneous catalysis. Mailing Add: Petro-Tex Chem Corp 8600 Park Place Blvd Houston TX 77017

ARGAUER, ROBERT JOHN, b Buffalo, NY, Feb 23, 37; m 64; c 5. ANALYTICAL CHEMISTRY. Educ: Canisius Col, BS, 58; Univ Md, PhD(inorg chem), 63. Prof Exp: Res chemist, Mobil Oil Co, 63-65; RES CHEMIST, USDA, 65- Mem: Am Chem Soc; Entom Soc Am. Res: Application of light and flame absorption and emission spectrofluorometry and spectrophotometry, gas chromatography, infrared, mass spectroscopy and radiochemical methods to chemical problems, especially pesticide analysis. Mailing Add: Agr Res Ctr USDA Beltsville MD 20705

ARGERSINGER, WILLIAM JOHN, JR, b Chittenango, NY, Apr 14, 18; m 42; c 3. PHYSICAL CHEMISTRY. Educ: Cornell Univ, AB, 38, PhD(phys chem), 42. Prof Exp: Asst chem, Cornell Univ, 38-41, instr, 42-44; assoc chemist, Monsanto Chem Co, Ohio, 44-45, group leader, 45-46; from asst prof to assoc prof chem, 46-56, assoc dean grad sch, 56-63 & faculties, 63-70, dean res admin, 70-72, PROF CHEM, UNIV KANS, 56-, VCHANCELLOR RES & GRAD STUDIES & DEAN GRAD SCH, 72- Mem: Fel AAAS; fel Am Inst Chemists; Soc Eng Sci; Am Chem Soc. Res: Surface equilibrium in solutions; coordination compounds; chemical kinetics, cation exchange; thermodynamics of solutions. Mailing Add: 214 Strong Hall Univ of Kans Lawrence KS 66045

ARGO, HAROLD VIRGIL, b Walla Walla, Wash, Jan 20, 18; m 42; c 4. ASTRONOMY. Educ: Whitman Col, AB, 39; George Washington Univ, MA, 41; Univ Chicago, PhD, 48. Prof Exp: Jr physicist, Naval Ord Lab, Washington, DC, 42-44; physicist, 44-46, PHYSICIST, LOS ALAMOS SCI LAB, 48-, STAFF MEM, 48- Mem: AAAS; Am Geophys Union; Am Phys Soc; Am Astron Soc. Res: Experimental nuclear physics; space physics; satellites and probes; solar physics. Mailing Add: PO Box 1663 Los Alamos NM 87544

ARGOS, PATRICK, b Carbondale, Ill, Nov 24, 42. CRYSTALLOGRAPHY, MOLECULAR BIOPHYSICS. Educ: St Louis Univ, BS, 64, MS, 66, PhD(physics), 68. Prof Exp: USPHS fel protein crystallog, Sch Med, Wash Univ, 69-71; asst prof physics, Pomona Col, 71-73; res assoc, Purdue Univ, 73-74; ASST PROF BIOPHYS, SOUTHERN ILL UNIV, 74- Mem: Am Phys Soc; Am Crystallog Asn; Biophys Soc; Am Asn Physics Teachers; AAAS. Res: X-ray crystallographic studies of the structure, function, and evolution of proteins and viruses. Mailing Add: Dept of Physics Southern Ill Univ Edwardsville IL 62026

ARGOT, JEANNE, b Pocono Lake, Pa, Mar 12, 36. MICROBIOLOGY. Educ: Moravian Col, BS, 65; Lehigh Univ, MS, 67, PhD(biol), 69. Prof Exp: Biologist, Pharmachem Corp, Pa, 62-66; ASST PROF BIOL, LEBANON VALLEY COL, 69- Mem: AAAS; Am Soc Microbiol; Sigma Xi. Mailing Add: Dept of Biol Lebanon Valley Col Annville PA 17003

ARGOUDELIS, CHRIS J, b Piraeus, Greece, July 30, 29; nat US. ORGANIC CHEMISTRY, BIOCHEMISTRY. Educ: Nat Univ Athens, BS, 56; Univ Ill, PhD(food technol), 61. Prof Exp: Res asst food technol, 60-61, ASST PROF FOOD SCI, UNIV ILL, URBANA-CHAMPAIGN, 61- Concurrent Pos: NIH res grant, 64-71. Mem: AAAS; Am Chem Soc. Res: Chemistry and metabolism of vitamin B6. Mailing Add: Burnsides Res Lab Univ of Ill Urbana IL 61801

ARGUE, CHARLES WILLIAM, botany, deceased

ARGUE, GARY R, b Hanna, Alta, Nov 19, 31; nat US; m 58; c 2. INORGANIC

CHEMISTRY, ELECTROCHEMISTRY. Educ: Univ Alta, BSc, 53, MSc, 56; Purdue Univ, PhD(inorg chem), 60. Prof Exp: Jr chemist, Atomic Energy Can, Ltd, 53-54; chemist, Tex Instruments Inc, 60-61; res chemist, Tex Res & Electronic Corp, 61-63; res specialist electrochem, Atomic Int Div, NAm Aviation Inc, 63-66; proj engr advan batteries, Atomic Int, 66-68; dept mgr energy conversion, Gen Tel & Electronics Corp, NY, 68-70, dept mgr semiconductor device res & develop, 70-73, vpres new prod develop, GTE New Ventures Corp, 73-74, DIR ELECTRONIC COMPONENTS, FACTORY PROJS ORGN, GTE INT, 74- Res: Practical electrochemistry with emphasis on electrochemical devices of prime interest; study of battery systems and how they are affected by radiation. Mailing Add: GTE Int 32 Third Ave Burlington MA 01803

ARGUELLES, AMILCAR EMILIO, b Bahia Blanca, Arg, May 20, 17; m 40; c 3. ENDOCRINOLOGY. Educ: Univ Buenos Aires, BM, 40, MD, 42. Prof Exp: Dir, Nat Inst Aviation Med, 57-59; dir gen, Med Serv & Res, Arg Air Force, 60-66; CHIEF ENDOCRINOLOGIST, AERONAUT HOSP, BUENOS AIRES, 66- Concurrent Pos: Res fel, Lab Biol Chem, Med Sch, Univ Paris, 61-62; lectr, Buenos Aires Nat Med Sch, 62-; vis sr lectr, St Thomas' Hosp Med Sch, London, 64-68; endocrinologist, Arg Pub Health Ministry, 65-; mem sci res bd, Arg Ministry of Defense, 67- Honors & Awards: Award, Arg Soc Endocrinol & Metab, 61; Award, Arg Nat Acad Med, 62. Mem: NY Acad Sci; Arg Soc Clin Invest; Arg Soc Neurocrinol; Arg Soc Endocrinol & Metab; Arg Endocrine Soc (pres, 70-71). Res: Steroid hormones and stress. Mailing Add: Hosp Aeronautico V de la Vega y Einstein Buenos Aires Argentina

ARGUS, GEORGE WILLIAM, b Brooklyn, NY, Apr 14, 29; m 55; c 5. SYSTEMATIC BOTANY. Educ: Univ Alaska, BS, 52; Univ Wyo, MS; Harvard Univ, PhD(biol), 61. Prof Exp: Nat Res Coun Can fel plant taxon, Univ Sask, 61-63, asst prof plant ecol & lectr biol, 63-67, assoc prof, 67-69, cur, W P Fraser Herbarium, 63-69, actg dir, Inst North Studies, 65-66; cur herbarium, Mus Natural Hist, Univ Ore, 69-70; res scientist, Can Forest Serv, 70-72; ASSOC CUR, NAT MUS NATURAL SCI, 72- Concurrent Pos: Mem, Nat Res Coun Can, 73-, chmn plant biol grant selection comt, 75-76; sci authority plants, Int Conv Trade in Endangered Species, 75- Mem: Am Soc Plant Taxon; Int Asn Plant Taxon; Can Bot Asn. Res: Taxonomy of Salix including experimental hybridization and flavonoid chemistry; endemism in northwestern Saskatchewan; rare plants of Canada. Mailing Add: Bot Div Nat Mus Natural Sci Ottawa ON Can

ARGUS, MARY FRANCES, b Ironton, Ohio, July 16, 24. BIOCHEMISTRY. Educ: Col Mt St Joseph, BS, 45; Univ Cincinnati, MS, 47; Univ Fla, PhD(biochem), 52. Prof Exp: Instr chem, Marygrove Col, 47-49; from asst prof to assoc prof cancer res, Univ Fla, 52-60; assoc prof biochem & med, 60-68, PROF MED BIOCHEM, SCH MED, TULANE UNIV, 68- Mem: Am Soc Biol Chem; Am Asn Cancer Res; Am Soc Pharmacol & Exp Therapeut; Soc Exp Biol & Med. Res: Chemical carcinogenesis; drug metabolism; conformational changes in macromolecules during carcinogenesis; alterations in oxidative metabolism in cardiac hypertrophy. Mailing Add: USPHS Hosp Res Lab 210 State St New Orleans LA 70118

ARGYLE, BERNELL EDWIN, solid state physics, see 12th edition

ARGYRES, PETROS, b Lefkas, Greece, Mar 9, 27; nat US; m 58; c 3. THEORETICAL PHYSICS. Educ: Univ Calif, AB, 50, MA, 52, PhD(physics), 54. Prof Exp: Res physicist, Westinghouse Res Lab, 54-58 & Lincoln Lab, Mass Inst Technol, 58-67; PROF PHYSICS, NORTHEASTERN UNIV, 67- Concurrent Pos: Lectr, Univ Pittsburgh, 55-57; vis assoc prof, Mass Inst Technol, 65-66; vis prof, Ecole Normale Superieure, Univ Paris, 69-70. Mem: Fel Am Phys Soc; AAAS. Res: Solid state physics; irreversible phenomena; transport processes; statistical mechanics; spin resonance; many-body theories. Mailing Add: Dept of Physics Northeastern Univ Boston MA 02115

ARGYRIS, BERTIE, b Holland, June 27, 30; nat US; m 55. ZOOLOGY. Educ: Columbia Univ, BA, 51; Brown Univ, MS, 53; Syracuse Univ, PhD, 58. Prof Exp: Res assoc exp morphol, Syracuse Univ, 58-62 & immunobiol, 62-68; assoc prof urol, 68-71, ASSOC PROF MICROBIOL, STATE UNIV NY UPSTATE MED CTR, 71- Mem: Am Soc Zool; Soc Study Develop Biol; Am Inst Biol Sci; Am Asn Immunol; Transplantation Soc. Res: Endocrinology-insulin resistance; experimental morphology-tissue interaction; effects of tumors on mammary glands; transplantation immunity; acquired homograft tolerance; immunosuppression; cellular interactions during immune response. Mailing Add: Dept of Microbiol State Univ of NY Upstate Med Ctr Syracuse NY 13210

ARGYRIS, THOMAS STEPHEN, b Newark, NJ, July 16, 23; m 55. EXPERIMENTAL PATHOLOGY. Educ: Rutgers Univ, BS, 48; Brown Univ, PhD, 53. Prof Exp: Instr & fel, Harvard Univ, 53-55; from asst prof to prof zool, Syracuse Univ, 55-72; PROF PATH, STATE UNIV NY UPSTATE MED CTR, 72- Concurrent Pos: NSF sr fel embryol, Carnegie Inst, 61-62; NIH spec res fel zool, Univ Col, Univ London, 65-66. Mem: Fel AAAS; Am Asn Cancer Res; Soc Develop Biol; Am Asn Pathologists & Bacteriologists; Soc Invest Dermat. Res: Wound healing; carcinogenesis; drug induced growth. Mailing Add: 1250 Comstock Ave Syracuse NY 13210

ARHART, RICHARD JAMES, b Joliet, Ill, July 27, 44; m 70. ORGANIC POLYMER CHEMISTRY. Educ: Univ NC, Chapel Hill, BS, 66; Univ Ill, Urbana, PhD(org chem), 71. Prof Exp: Fel, Univ Alta, 71-72; RES CHEMIST ORG POLYMER CHEM, E I DU PONT DE NEMOURS & CO, INC, 72- Mem: Am Chem Soc. Res: Condensation polymerizations; initiation of addition polymerization of vinyl monomers; crosslinking reactions of elastomeric polymers. Mailing Add: Elastomer Chem Dept Exp Sta E I du Pont de Nemours & Co Wilmington DE 19898

ARHELGER, ROGER BOYD, b Green Bay, Wis, May 23, 32; m 53; c 3. MEDICINE, PATHOLOGY. Educ: Hamline Univ, BS, 54; Univ Minn, MD, 58. Prof Exp: Intern, USPHS Hosp, Boston, Mass, 58-59; resident, Med Ctr, Univ Miss, 59-62, from instr to asst prof path, 62-66; assoc prof path, Univ Tex Med Sch San Antonio, 66-68; assoc prof, Med Ctr, Univ Miss, 68-72; chief surg path & cytol, Fresno Community Hosp, 72-75; CHMN DEPT PATH, KERN MED CTR, 75- Concurrent Pos: USPHS res career develop award, 63- Mem: Electron Micros Soc Am; Am Soc Exp Path; Am Asn Path & Bact; Int Acad Path; AMA. Res: Experimental pathology; electron microscopy. Mailing Add: Kern Med Ctr 1830 Flower St Bakersfield CA 93305

ARIAS, IRWIN MONROE, b New York, NY, Sept 4, 26; m 53; c 4. MEDICINE, GASTROENTEROLOGY. Educ: Harvard Univ, SB, 46; Long Island Col Med, MD, 52; Am Bd Internal Med, dipl. Prof Exp: From intern to resident med, Boston City Hosp, 52-54; resident, Boston Vet Admin Hosp, 54-55; assoc, 56-64, assoc prof, 64-67, PROF MED, ALBERT EINSTEIN COL MED, 67- Concurrent Pos: Res fel, Boston City Hosp, 55-56; NY Heart Asn fel, 56-; instr, NY Heart Asn, 56- assoc vis prof, Bronx Munic Hosp Ctr; Suiter Mem lectr, 59; consult, New Rochelle Hosp, Mt Vernon Hosp, USPHS, Pan Am Union & Kellogg Found; attend physician, Albert Einstein Col Med Hosp, 67- Honors & Awards: Hon Achievement Award, Am Gastroenterol Asn, 69; Citation in Sci, Univ Recife, 70. Mem: AAAS; Am Soc Gastrointestinal Endoscopy; Am Gastroenterol Asn; Am Physiol Soc; Am Asn Study Liver Dis. Res: Academic gastroenterology; mechanisms of hepatic excretory function. Mailing Add: Dept of Med Albert Einstein Col of Med New York NY 10461

ARIEFF, ALEX J, b Ind, May 8, 08; m 52; c 3. NEUROLOGY, PSYCHIATRY. Educ: Northwestern Univ, BS, 29, MD, 33. Prof Exp: Asst prof nerv & ment dis, 48-58, PROF NEUROL & PSYCHIAT, NORTHWESTERN UNIV, CHICAGO, 58- Concurrent Pos: Asst psychiatrist, Munic Psychiat Inst, Chicago, 40; attend neurologist, Cook County Hosp, 45 & Passavant Mem Hosp, 46-; consult neurol, Vet Admin Hosp, Hines, Ill, 49- & Chicago Bd Educ, 69- Mem: AAAS; Am Psychiat Asn; Am Neurol Asn; Am EEG Soc; Am Acad Neurol. Res: Epilepsy. Mailing Add: 707 N Fairbanks Ct Chicago IL 60611

ARIEMMA, SIDNEY, b New York, NY, Apr 9, 22; m 51; c 4. POLYMER CHEMISTRY, ORGANIC CHEMISTRY. Educ: Wagner Col, BS, 47; Stevens Inst Technol, MS, 54. Prof Exp: Chemist, Armour & Co, 47-50 & Wallerstein Co, 50-52; group leader polymers, Air Reduction Co, Inc, 52-60; sr res scientist, Texaco, Inc, 60-62; mgr polymers, Tex Butadiene & Chem Co, 62-63; mgr natural & synthetic resins, Morningstar Paisley, Inc, 63-64; mgr polymers & coatings, Reichhold Chem Co, 64-66; group leader coatings & finishes, Huyck Corp, 66-69; ASSOC DIR POLYMERS, WAXES, COATINGS & RUBBER DEPT, FOSTER D SNELL, INC, FLORHAM PARK, 69- Concurrent Pos: Mem, Int Coun Fabric Flammability. Mem: Am Chem Soc; Soc Plastics Engrs; NY Acad Sci. Res: Development of petrochemicals, polymers and specialties applied to paper, textiles, paint, polish, adhesives, automotive, leather, moldings, extrusion and casting resins. Mailing Add: 29 Skytop Dr Ramsey NJ 07446

ARIES, LEON JUDAH, b Chicago, Ill, Feb 25, 09; m 45; c 3. SURGERY. Educ: Univ Ill, AB, 29, MS, 31, MD, 32; Northwestern Univ, PhD(surg), 40; Am Bd Surg, dipl, 40. Prof Exp: Asst anat, Univ Ill, 31-33, instr surg, 33-35; instr, Northwestern Univ, 35-39, assoc prof, Col Med, 39; prof surg, Chicago Med Sch, 45-75, SURGEON, CHICAGO MED SCH HOSP, 45- Concurrent Pos: Prof, Cook County Grad Sch Med, 35-49, attend surgeon, Hosps, 35-; mem surg staff, St Joseph's Hosp, 66- Mem: AMA; fel Am Col Surg; fel Int Col Surg. Res: General surgery. Mailing Add: St Joseph's Hosp 30 N Michigan Ave Chicago IL 60602

ARIETI, SILVANO, b Pisa, Italy, June 28, 14; US citizen; m 65; c 2. PSYCHIATRY. Educ: Lycee Galileo, Italy, BA, 32; Med Sch Pisa, MD, 38; William Alanson White Inst, dipl psychoanal, 52. Prof Exp: Dazian Found fel neuropath, NY State Psychiat Inst, 39-41; assoc prof clin psychiat, Col Med, State Univ NY, 53-61; PROF CLIN PSYCHIAT, NEW YORK MED COL, 61- Concurrent Pos: Mem fac, Training & Supv Analyst, William Alanson White Inst, 62- Honors & Awards: Gold Medal Award, Milan Group Advan Psychoanal, 64; Frieda Fromm-Reichman Award, Am Acad Psychoanal, 68; Nat Book Award Sci, 75. Mem: Am Acad Psychoanal; Asn Advan Psychother (vpres, 69-70); Am Asn Neuropath; Am Psychiat Asn. Res: Office treatment of schizophrenic patients; creative process; cultivation of creativity. Mailing Add: 125 E 84th St New York NY 10028

ARIMOTO, FRED SHUNJI, b Armona, Calif, Feb 15, 20; m 42; c 3. ORGANIC CHEMISTRY. Educ: Univ Calif, BS, 42; Univ Chicago, PhD(org chem), 51. Prof Exp: Res assoc, Michael Reese Res Found, 43-49; RES CHEMIST, JACKSON LAB, E I DU PONT DE NEMOURS & CO, 51- Mem: Am Chem Soc. Res: Organic chemistry; free radicals in solutions. Mailing Add: Jackson Lab E I du Pont de Nemours & Co Wilmington DE 19898

ARIMURA, AKIRA, b Kobe, Japan, Dec 26, 23; m 57; c 3. ENDOCRINOLOGY, PHYSIOLOGY. Educ: Nagoya Univ, BS, 43, MD, 51, PhD(physiol), 57. Prof Exp: From asst to instr med, Sch Med, Tulane Univ, 58-61; instr & res assoc physiol, Sch Med, Hokkaido Univ, 61-65; from asst prof to assoc prof, 65-73, PROF MED, SCH MED, TULANE UNIV, 73- Concurrent Pos: J H Brown fel, Sch Med, Yale Univ, 56-58; USPHS res grants, Sch Med, Hokkaido Univ, 63-64 & Sch Med, Tulane Univ, 65-; res consult, Vet Admin Hosp, New Orleans, 65- Mem: AAAS; Endocrine Soc; Am Physiol Soc; Soc Exp Biol & Med; Japan Endocrinol Soc. Res: Experimental and clinical neuroendocrinology. Mailing Add: Dept of Med Tulane Univ Sch of Med New Orleans LA 70112

ARING, CHARLES DAIR, b Dent, Ohio, June 21, 04; m 31; c 2. NEUROLOGY. Educ: Univ Cincinnati, BS & MD, 29. Prof Exp: Rotating intern, Cincinnati Gen Hosp, 29-30, resident physician, 30-31; resident, Longview State Hosp, Cincinnati, 31; house officer neurol, Boston City Hosp, 32, resident, 33-34; house officer, Children's Hosp, Boston, 33; from instr to assoc prof, Col Med, Univ Cincinnati, 36-46; prof, Col Med, Univ Calif, 46-47, dir neurol serv, 46-47; PROF NEUROL & DIR NEUROL SERV, COL MED, UNIV CINCINNATI, 47- Concurrent Pos: Fel neuro-physiol, Yale Univ, 34-35; Rockefeller fel neurol, Madrid, Breslau & Nat Hosp, London, 35-36; attend neurologist, Cincinnati Gen Hosp, 36-45; mem comn neurotropic virus dis, Bd Invest Control Influenza & Other Epidemic Dis. Mem: AMA; Am Neurol Asn (pres, 63); Am Psychiat Asn; Asn Res Nerv & Ment Dis (vpres, 47-52). Res: Psychiatry; medical administration; medical philosophy. Mailing Add: Dept of Neurol Cincinnati Gen Hosp Cincinnati OH 45267

ARION, WILLIAM JOSEPH, b Cando, NDak, May 31, 40; m 62; c 3. BIOCHEMISTRY. Educ: Jamestown Col, BS, 62; Univ ND, 62-65, MS, 64, PhD(biochem), 66. Prof Exp: NIH fel biochem, Cornell Univ, 66-68; asst prof, 68-73, ASSOC PROF BIOCHEM, NY STATE COL VET MED, 73- Mem: AAAS; Am Chem Soc; Am Soc Biol Chemists; Sigma Xi. Res: Membrane Structure-Function Relationships; microsomal glucose 6-phosphatase and the regulation of liver and kidney glucose production. Mailing Add: Dept Physiol Biochem Pharmacol NY State Col Vet Med Cornell U Ithaca NY 14853

ARISUMI, TORU, b Hawaii, July 28, 19; m 54; c 4. PLANT GENETICS. Educ: Univ Hawaii, BS, 47; Univ Ill, MS, 49, PhD, 51. Prof Exp: Asst storage physiologist, Univ Hawaii, 51-54; plant geneticist, 55-66, RES PLANT GENETICIST, AGR RES SERV, USDA, 66- Mem: Genetics Soc Am; Am Genetic Asn; Am Soc Hort Sci. Res: Genetics and cytogenetics of herbaceous and woody ornamentals. Mailing Add: 5217 Palco Pl Col Park MD 20740

ARIYAN, ZAVEN S, b Cairo, Egypt, Oct 10, 33; m 66. ORGANIC CHEMISTRY. Educ: Univ London, BSc, 57; PhD(org chem), 64. Prof Exp: Lectr chem, Medway Col Technol, Eng, 57-59; demonstr org chem, Royal Mil Col Sci, Eng, 60-64; fel, Univ Southern Calif, 64-65; sr res chemist, 65-69, coord pharmaceut prog, 69-71, sr group leader pharmaceut chem res, Royal, Inc, 71-75; MGR, PHARMACEUT DIAMOND SHAMROCK CORP, T R EVANS RES CTR, 75- Concurrent Pos: Grants, Gulbenkian Found, 56-57 & Stauffer Chem Co, 64-65. Mem: Am Chem Soc; Brit Chem Soc. Res: Chemistry of sulfur compounds. Mailing Add: Res & Develop Chem Div Uniroyal Inc Naugatuck CT 06770

ARKELL, ALFRED, b Chicago, Ill, Feb 16, 24; m 43; c 3. ORGANIC CHEMISTRY.

Educ: Kalamazoo Col, AB, 54; Ohio State Univ, PhD(chem), 58. Prof Exp: Asst gen & org chem, Ohio State Univ, 54-55; RES CHEMIST, TEXACO RES CTR, TEXACO, INC, BEACON, NY, 58- Honors & Awards: Upjohn Prize, 54; Clark MacKenzie Prize, 54; Texaco Res Award, Am Chem Soc, 68. Mem: Am Chem Soc; fel Am Inst Chem. Res: Synthesis of highly-branched aliphatic compounds; derivatives of acetylene; low temperature infrared spectroscopy; matrix isolation technique; isolation and identification of radicals and unstable compounds. Mailing Add: McFarland Rd Wappingers Falls NY 12590

ARKIN, ARTHUR MALCOLM, b New York, NY, Jan 25, 21; m 43; c 2. PSYCHIATRY, PHYSIOLOGICAL PSYCHOLOGY. Educ: Univ Chicago, BS, 42; New York Med Col, MD, 45. Prof Exp: Gen intern, Montefiore Hosp, Bronx, 45-46; resident psychiatrist, Bellevue Psychiatric Hosp, New York, 46-47, Menninger Sch Psychiat, Topeka, 47, Brooklyn State Hosp, 48 & Postgrad Ctr Psychother, New York, 48-51; mem headache clin, Montefiore Hosp, Bronx, 50-51; psychiatric consult, Bur Appl Social Res, Columbia Univ, 53-55; asst clin prof psychiat, Albert Einstein Col Med, 55-75, PROF PSYCHOL, GRAD SCH, CITY COL NEW YORK, 75- Concurrent Pos: Pvt pract, 49-51; adj attend physician, Montefiore Hosp & Med Ctr, 53-; psychiat consult, Dept Home Care, Montefiore Hosp, 54-55 & Barrett House, Florence Crittenton League, 54-56; assoc vis psychiatrist, Bronx Munic Hosp Ctr, 56-; adj assoc prof psychol, Grad Sch Arts & Sci, NY Univ, 61- & City Col New York, 66-, ctr staff psychologist, 66-; chmn, Task Force on Behav Ther, NY Dist Br, Am Psychiat Asn, 74; assoc clin prof, Dept Psychiat, Albert Einstein Col, 75- Mem: Am Med Asn; fel Am Psychiat Asn; Soc Psychophysiol Res; Asn Psychophysiol Study Sleep; Asn Advan Psychotherapy. Res: Psychophysiology of sleep and dreams; psychotherapy. Mailing Add: 12 E 97th St New York NY 10029

ARKIN, GERALD FRANKLIN, b Washington, DC, Sept 16, 42; m 66; c 2. AGRICULTURAL ENGINEERING. Educ: Cornell Univ, BS, 66; Univ GA, MS, 68; Univ Ill, PhD(agr eng), 72. Prof Exp: Res asst agr eng, Univ Ga, 67-68 & Univ Ill, 68-72; ASST PROF AGR ENG, TEX AGR EXP STA, TEX A&M UNIV, 72- Concurrent Pos: Consult comt climate & weather fluctuations & agr prod, Nat Acad Sci, 75-76. Mem: Am Soc Agr Engrs; Am Soc Agron; Soil Sci Soc Am; Sigma Xi. Res: Crop growth simulation modeling to optimize yield and water efficiency. Mailing Add: Blackland Res Ctr PO Box 748 Temple TX 76501

ARKIN, HERBERT, b New York, NY, Apr 30, 06; m 47; c 2. APPLIED STATISTICS, ANALYTICAL STATISTICS. Educ: City Col New York, BSS, 27, MBA, 28; Columbia Univ, PhD(statist), 40. Prof Exp: Prof Statist, 30-75, EMANUEL LAXE DISTINGUISHED PROFESSORSHIP, CITY UNIV NEW YORK, 75- Mem: Fel AAAS; Am Statist Asn; Am Soc Qual Control; Inst Math Statist. Mailing Add: Dept of Statist Baruch Col 17 Lexington Ave New York NY 10010

ARKING, ALBERT, b Brooklyn, NY, Nov 5, 32. PHYSICS. Educ: Columbia Univ, AB, 53; Cornell Univ, PhD(physics), 59. Prof Exp: Physicist, Artificial Intel, Hughes Aircraft Co, 59; physicist astrophys & atmospheric physics, Goddard Inst Space Studies, New York, 59-74, PHYSICIST, GODDARD SPACE FLIGHT CTR, NASA, GREENBELT, MD, 74- Concurrent Pos: Instr, City Col New York, 61-62; from adj asst prof to adj assoc prof, NY Univ, 62-70; vis res assoc, Univ Calif, Los Angeles, 70-71. Mem: AAAS; Am Phys Soc; Am Astron Soc; Am Geophys Union; Am Meteorol Soc. Res: Atmospheric and nuclear physics; radiative transfer; astrophysics. Mailing Add: Goddard Space Flight Ctr NASA Code 911 Greenbelt MD 20771

ARKING, ROBERT, b New York, NY, July 1, 36; m 59; c 2. DEVELOPMENTAL GENETICS. Educ: Dickinson Col, BS, 58; Temple Univ, PhD(develop biol), 67. Prof Exp: Fel NIH, 67-68; asst prof zool, T H Morgan Sch Biol, Univ Ky, 68-70; res assoc, Develop Biol Labs, Univ Calif, Irvine, 70-75; ASST PROF BIOL, WAYNE STATE UNIV, 75- Mem: AAAS; NY Acad Sci; Genetics Soc Am. Res: Developmental and physiological genetics of Drosophila; genetic analysis of pattern formation; gene-hormone interactions during insect metamorphosis. Mailing Add: Dept Biol Wayne State Univ Detroit MI 48202

ARKLE, THOMAS, JR, b Colerain, Ohio, Sept 10, 18; m 45; c 2. GEOLOGY. Educ: Marietta Col, AB, 40; Ohio State Univ, MS, 50. Prof Exp: Teacher high schs, WVa, 40-42 & Ohio, 46-47; asst geol, Ohio State Univ, 48-50; asst geologist, 50-51, asst to dir, 52-58, econ geologist, State Geol Surv, WVa, 58-73; SUPVR COAL RESOURCES STUDY, WVA GEOL ECON SURV, 73- Mem: Am Geol Soc; Geol Soc Am; Am Inst Mining Metall Engrs; Am Inst Prof Geol. Res: Stratigraphy, especially coal bearing strata; economic geology, especially of coal and sandstone; to collect, interpret and collate data for use in the determination of the physical and chemical characteristics of the remainder of coal seams and the associated rocks in West Virginia. Mailing Add: 899 Fairfax Dr Morgantown WV 26505

ARKLEY, RODNEY JOHN, b Santa Ana, Calif, Nov 15, 15; m 45; c 4. SOIL MORPHOLOGY. Educ: Univ Calif, AB, 40, PhD(soil sci), 61. Prof Exp: Asst soil scientist, USDA, 41-45; LECTR SOILS & PLANT NUTRIT, AGR EXP STA, UNIV CALIF, BERKELEY, 45-, SOIL MORPHOLOGIST, 69- Concurrent Pos: Assoc soil morphologist, 45-69. Mem: Am Soc Agron; Soil Sci Soc Am; Int Soc Soil Sci; Am Quaternary Soc; Friends of the Pleistocene. Res: Soil genesis, morphology and classification; climate and plant growth. Mailing Add: 554 Santa Barbara Rd Berkeley CA 94707

ARKO, ALOYSIUS JOHN, b Loski Potok, Slovenia, Yugoslavia, Feb 6, 40; US citizen; m 64; c 3. SOLID STATE PHYSICS. Educ: Ill Inst Technol, BS, 62; Northwestern Univ, PhD(physics), 67. Prof Exp: Res assoc solid state physics, Northwestern Univ, 67-68; asst physicist, 68-73, PHYSICIST, ARGONNE NAT LAB, 73- Mem: Am Phys Soc. Res: Band structure of plutonium and other actinide materials. Mailing Add: Argonne Nat Lab Argonne IL 60439

ARKOWITZ, MARTIN ARTHUR, b Brooklyn, NY, Apr 17, 35; m 57; c 3. MATHEMATICS. Educ: Columbia Univ, BA, 56, MA, 57; Cornell Univ, PhD(math), 60. Prof Exp: Teaching asst, Cornell Univ, 57-60; Off Naval Res assoc, Johns Hopkins Univ, 60-61; instr math, Princeton Univ, 61-63, lectr, 63-64; from asst prof to assoc prof, 64-71, PROF MATH, DARTMOUTH COL, 71- Concurrent Pos: Dartmouth fac fel, Math Inst, Oxford Univ, 67-68. Mem: Am Math Soc. Res: Topology. Mailing Add: Dept of Math Dartmouth Col Hanover NH 03755

ARLEDTER, HANNS FERDINAND, b Hamburg, Ger, Dec 22, 10; nat US; m 53; c 3. PHYSICAL CHEMISTRY. Educ: Col Technol, Vienna, BS, 32; Col Technol Aachen, Dipl Ing, 35; Col Technol, Darmstadt, DrSc(cellulose & phys chem), 44. Prof Exp: Consult engr & chemist, Tech Bur, Univ Cologne, 36-37; res chemist, Escanaba Paper Mill, Mich, 38; chief chemist & dir lab, Hoffmann & Engelmann A G, Ger, 39-41; chief chemist & group leader, Chem Lab, Col Technol Inst Theoret Physics, Dresden, 44-45; res chemist, Farbenfabriken Bayer, Ger, 45-50; dir res, Hurlbut Paper Co, 50-60 & Mead Cent Res Lab Indust Papers, 60-65, dir pioneering res, 65-68; DIR EXP & RES INST PULP & PAPER, GRAZ TECH UNIV, 68- Mem: Am Chem Soc; Electrochem Soc; Tech Asn Pulp & Paper Indust; Ger Chem Soc; Asn Pulp & Paper Chem Ger. Res: Adsorbent papers; inorganic and organic synthetic fiber paper; magnetic recording materials and ferrits; electrical and dielectrical behavior of organic materials; paper sizing; water clarification; beater theory; testing methods; air and liquid filters; photographic paper; environmental systems and technology; wood science and technology. Mailing Add: Reichsstr 122 Graz A 8046 Austria

ARLINGHAUS, FRANCIS JOSEPH, solid state physics, see 12th edition

ARLINGHAUS, RALPH B, b Newport, Ky, Aug 16, 35; m 57, 68; c 4. BIOCHEMISTRY. Educ: Univ Cincinnati, BS, 57, MS, 59, PhD(biochem), 61. Prof Exp: NIH fel protein synthesis, Univ Ky, 61-63, Am Cancer Soc fel, 63-65; res chemist, Plum Island Animal Dis Lab, USDA, NY, 65-69; asst prof, 69-73, ASSOC PROF BIOL, UNIV TEX M D ANDERSON HOSP & TUMOR INST, 73-, CHIEF SECT ENVIRON BIOL, 69- Mem: AAAS; Am Chem Soc; Am Soc Microbiol; Am Asn Biol Chem. Res: Protein biosynthesis and nucleic acid synthesis; animal virus replication at the molecular level; RNA-containing viruses such as mengovirus and Rauscher leukemia virus. Mailing Add: Sect of Environ Biol M D Anderson Hosp & Tumor Inst Houston TX 77025

ARLON, ROBERT, drug information, see 12th edition

ARLOW, JACOB, b New York, NY, Sept 3, 12; m 36; c 4. PSYCHIATRY. Educ: NY Univ, BS, 32, MD, 36. Prof Exp: Intern, Harlem Hosp, New York, 36-38; resident neuropsychiatrist, USPHS, 38-39; resident psychiatrist, NY Psychiat Inst, 40-41; assoc psychiatrist, Presby Hosp, New York, 44-52; clin assoc prof, 50-62, CLIN PROF PSYCHIAT, STATE UNIV NY DOWNSTATE MED CTR, 62- Concurrent Pos: Vis prof, Albert Einstein Col Med, Mt Sinai Col Med, Columbia Univ & La State Univ; practicing psychiatrist, 42-; ed, Psychoanal Quart. Honors & Awards: Int Clin Essay Prize, Brit Psychoanal Soc, 56. Mem: Am Psychoanal Asn (pres-elect, 59-60, pres, 60-61); Int Psychoanal Asn (treas, vpres). Res: Psychosomatic aspects of arthritis and angina pectoris. Mailing Add: 120 Central Park S New York NY 10019

ARLT, HERBERT GEORGE, JR, b Brooklyn, NY, Nov 28, 28; m 51; c 6. ORGANIC CHEMISTRY, POLYMER CHEMISTRY. Educ: Lehigh Univ, BS, 50; State Univ NY, PhD(org & polymer chem), 57. Prof Exp: Chemist, Am Cyanamid Co, 50-53, res chemist, Paper Chem Res, 56-57 & org chem, 57-62; MGR, ARIZ CHEM CO, 62- Mem: Tech Asn Pulp & Paper Indust; Sigma Xi; Am Oil Chem Soc; Am Chem Soc. Res: Synthesis of organic molecules suitable for pharmaceutical uses; theoretical organic chemistry; organic polymers; tall oil and terpene products. Mailing Add: Ariz Chem Co 1937 W Main St Stamford CT 06904

ARLUK, DAVID JAY, b Brooklyn, NY, July 8, 42. DERMATOLOGY, ANATOMY. Educ: Columbia Univ, AB, 64; NY Univ, PhD(anat), 68; NY Med Col, MD, 75. Prof Exp: Instr, 68-71, ASST PROF ANAT, NY MED COL, 71- Concurrent Pos: Partic, NATO Conf Uptake of Informational Molecules, Mol, Belg, 70. Mem: AAAS. Res: Hormonal control of sebaceous glands, lipogenesis and skin; analysis of skin surface lipids; electron microscopy of cells and tissues, especially cardiovascular system. Mailing Add: Dept of Dermat NY Med Col New York NY 10029

ARM, HERBERT GUNTHER, medical bacteriology, see 12th edition

ARMACOST, DAVID LEE, b Santa Monica, Calif, Mar 9, 44. MATHEMATICS. Educ: Pomona Col, BA, 65; Stanford Univ, MS, 66, PhD(math), 69. Prof Exp: Asst prof, 69-74, ASSOC PROF MATH, AMHERST COL, 74- Mem: Am Math Soc; Sigma Xi. Res: Topological groups; infinite Abelian groups; locally compact Abelian groups. Mailing Add: Dept of Math Amherst Col Amherst MA 01002

ARMACOST, WILLIAM L, b Santa Monica, Calif, Oct 6, 41. MATHEMATICS. Educ: Pomona Col, BA, 63; Univ Calif, Los Angeles, MA, 65, PhD(math), 68. Prof Exp: ASST PROF MATH, CALIF STATE COL, DOMINGUEZ HILLS, 68- Mem: Am Math Soc. Res: Group representation theory and functional analysis; Frobenius reciprocity theorem. Mailing Add: Dept of Math Calif State Col Dominguez Hills CA 90247

ARMANINI, LOUIS ANTHONY, b New York, NY, Feb 11, 30; m 58; c 4. PHYSICAL CHEMISTRY. Educ: Manhattan Col, BS, 57. Prof Exp: Res chemist, Curtiss-Wright Corp, Pa, 57-59; res chemist, Francis Earle Labs, 59-67, group leader interference films, 67-75, ASST DIR RES, MEARL CORP, 76- Mem: Am Chem Soc. Res: Optical interference pigments made by thin film deposition or controlled crystal growth and the instrumental measurement and optical properties of such films. Mailing Add: Mearl Corp 217 N Highland Ave Ossining NY 10562

ARMBRECHT, BERNARD HENRY, b Parkston, SDak, July 21, 16; m 45. BIOCHEMISTRY. Educ: Dakota Wesleyan Univ, BA, 39; Univ Md, MS, 50; Georgetown Univ, PhD(biochem), 51. Prof Exp: Teacher high sch, SDak, 39-41, asst, Univ Md, 46-49; chemist, Bur Entomol & Plant Quarantine, USDA, 49-50; res asst, Georgetown Univ, 50-51; res chemist, Ansco, Binghamton, NY, 51-52; dir chem lab, Georgetown Hosp, 52-57; asst prof, Georgetown Univ, 56-58; biochemist, 58-61, res biochemist, div pharmacol, 61-64, res biochemist, Div Vet Med, Food & Drug Admin, 64- Concurrent Pos: Mem comn quantities and units, Int Union Pure & Appl Chem; expert panel quantities & units, Int Fedn Clin Chemists. Mem: Asn Lab Animal Sci; Am Chem Soc; Am Asn Clin Chem; Am Asn Toxicol. Res: Biophysical instrumentation; clinical chemistry; evaluation of mycotoxins and their toxicology; assessment of analytical method performance. Mailing Add: Food & Drug Admin Agr Res Ctr Beltsville MD 20705

ARMBRECHT, FRANK MAURICE, JR, b Norfolk, Va, Jan 29, 42; m 65; c 2. ORGANOMETALLIC CHEMISTRY. Educ: Duke Univ, BS, 63; Mass Inst Technol, PhD(chem), 68; Prof Exp: RES CHEMIST, ELASTOMERS CHEM RES, E I DU PONT DE NEMOURS & CO, 68- Mem: Am Chem Soc. Res: Organometallic chemistry; homogeneous catalysis; ioniomeric elastomers. Mailing Add: Du Pont Exp Sta 353 Wilmington DE 19898

ARMBRUSTER, CHARLES WILLIAM, b St Louis, Mo, Mar 24, 37. ORGANIC CHEMISTRY. Educ: Univ Notre Dame, BS, 58; Wash Univ, PhD(org chem), 66. Prof Exp: From instr to asst prof, 62-66, div sci, 63-67, ASSOC PROF CHEM, UNIV MO-ST LOUIS, 66-, CHMN DEPT, 67- Mem: Am Chem Soc; AAAS. Res: Physical organic chemistry; biochemical mechanisms. Mailing Add: Dept of Chem Univ of Mo St Louis MO 63121

ARMBRUSTER, DAVID CHARLES, b Cincinnati, Ohio, Feb 21, 39; m 62; c 3. POLYMER CHEMISTRY, PLASTICS CHEMISTRY. Educ: Xavier Univ, Ohio, BS, 61; Univ Cincinnati, PhD(org chem), 65. Prof Exp: Sr res chemist, 65-66, GROUP LEADER CHEMIST, ROHM AND HAAS CO, 66- Mem: Am Chem Soc; Brit Chem Soc. Res: Heterocyclic chemistry, nitrogen chemistry, acrylate monomer synthesis; emulsion polymerization; crosslinking-network polymers, reactions of polymers; plastics intermediates and plastic sheet. Mailing Add: Rohm and Haas Co Philadelphia PA 19105

ARMBRUSTER, FREDERICK CARL, b Aurora, Ill, Apr 14, 31; m 54; c 3. MICROBIOLOGY, BIOCHEMISTRY. Educ: Purdue Univ, BS, 53, MS, 55. Prof Exp: Asst microbiol, Purdue Univ, 54-55; res chemist, 57-61, SECT HEAD FERMENTATION RES, CPC INT, INC, 61- Mem: Am Soc Microbiol; Am Chem Soc. Res: Fermentation processes; industrial production, purification, and use of enzymes. Mailing Add: Moffett Tech Ctr CPC Int Inc Argo IL 60501

ARMBRUSTER, GERTRUDE D, b Alta, Nov 29, 25; US citizen. FOOD SCIENCE. Educ: Univ Alta, BS, 47; Wash State Univ, MS, 50, PhD(food sci), 65. Prof Exp: County agent, exten home econ, Wash State Univ, 50-52; asst prof, 52-58, ASSOC PROF FOOD & NUTRIT, CORNELL UNIV, 59-61, 65- Mem: Am Home Econ Asn; Inst Food Technologists; Am Soc Hort Sci; Am Dietetic Asn. Res: Interrelationship of plant tissue properties and product quality; plant texture. Mailing Add: Dept of Human Nutrit Cornell Univ Ithaca NY 14850

ARMELAGOS, GEORGE JOHN, b Lincoln Park, Mich, May 22, 36; m 67. BIOLOGICAL ANTHROPOLOGY. Educ: Univ Mich, BA, 58; Univ Colo, MA, 64, PhD(anthrop), 68. Prof Exp: Res assoc, Univ Colo Nubian Exped, 63; instr, Univ Utah, 65-66, asst prof, 67-69; asst prof anthrop, Univ Mass, Amherst, 69-73, ASSOC PROF ANTHROP, UNIV MASS, 73-, DIR UNIV HONORS, 74- Concurrent Pos: NIH co-prin investr, 67-69; Nat Sci Found Co-prin investr. Mem: Am Asn Phys Anthrop; NY Acad Sci; Brit Soc Study Human Biol; Human Biol Coun; Am Anthrop Asn. Res: Bone growth and development in archeological populations, especially involution; paleopathology; hominid evolution; nutrition; skeletal biology and demography of archeological populations; paleopathology human evolution; race; history of physical anthropology. Mailing Add: Dept of Anthrop Univ of Mass Amherst MA 01002

ARMEN, HARRY A, JR, b New York, NY, Feb 4, 40; m 63; c 2. APPLIED MECHANICS. Educ: Cooper Union, BCE, 61; New York Univ, MCE, 62, ScD(struct mech), 64. Prof Exp: Head, Appl Mech Br, Grumman Aircraft Eng Corp, 64-70, HEAD, APPL MECH BR, GRUMMAN AEROSPACE CORP, 70- Concurrent Pos: Adj prof civil eng & appl mech, Cooper Union, 65- Mem: Am Inst Aeronaut & Astronaut. Res: Development of methods for the static and dynamic nonlinear analysis of structures; fracture mechanics; crashworthiness evaluation techniques; large displacement and stability analysis of structures. Mailing Add: Grumman Aerospace Corp Dept 584 Plant 35 Bethpage NY 11714

ARMENDAREZ, PETER X, b San Pedro, Calif, Sept 7, 30; m 54; c 7. PHYSICAL CHEMISTRY. Educ: Loyola Univ Los Angeles, BS, 52; Wash Univ, MS, 54; Univ Ariz, PhD(phys chem), 64. Prof Exp: Instr chem, Odessa Col, 58-59; asst prof, Univ Tenn, Martin, 63-65; asst prof physics & phys chem, 65-68, PROF PHYSICS & CHMN DEPT, BRESCIA COL, KY, 68-, CHMN DIV NATURAL SCI & MATH, 74- Mem: Am Crystallog Asn; Am Chem Soc; fel Am Inst Chemists. Res: Molecular spectra and structure of inorganic complexes; low temperature emission and absorption spectra of chromium complexes; infrared studies and normal coordinate analyses of metal chelate compounds. Mailing Add: Dept of Physics Brescia Col Owensboro KY 42301

ARMENDARIZ, EFRAIM PACILLAS, b Brownsville, Tex, July 9, 38; m 61; c 2. ALGEBRA. Educ: Agr & Mech Col, Tex, BA, 60, MS, 62; Univ Nebr, PhD(math), 66. Prof Exp: Instr math, Agr & Mech Col, Tex, 62-63 & Univ Nebr, 64-66; asst prof, Univ Tex, 66-67, Univ Southern Calif, 67-68 & Univ Tex, Austin, 68-71; assoc prof, Univ Southwestern La, 71-72; ASSOC PROF MATH, UNIV TEX, AUSTIN, 72- Mem: Am Math Soc. Res: Ring theory; mathematics. Mailing Add: Dept of Math Univ of Tex Austin TX 78712

ARMENTI, ANGELO, JR, b Bridgeport, Pa, Feb 13, 40. THEORETICAL PHYSICS. Educ: Villanova Univ, BS, 63; Temple Univ, MA, 65, PhD(physics), 70. Prof Exp: Asst prof physics, Temple Univ, 70-72; asst prof, 72-75, ASSOC PROF PHYSICS, VILLANOVA UNIV, 75- Mem: Am Phys Soc; Int Soc Gen Relativity & Gravitation. Res: Exact solutions of Einstein's field equations, in particular geodesic motion in the field of compact masses and black holes. Mailing Add: Dept of Physics Villanova Univ Villanova PA 19085

ARMENTO, WILLIAM HARRIS, organic chemistry, see 12th edition

ARMENTROUT, DAVID NOEL, b San Francisco, Calif, Dec 7, 38; m 63; c 3. ANALYTICAL CHEMISTRY. Educ: Univ Kans, BS, 61; Cornell Univ, PhD(anal chem), 65. Prof Exp: RES CHEMIST ANAL CHEM, DOW CHEM CO, 66- Concurrent Pos: Fel anal chem, NIH, 65-66. Mem: Am Chem Soc; Sigmz Xi. Res: Separation and identification of trace organic compounds in waste waters, brine streams and air samples. Mailing Add: Dow Chem Co 574 Bldg Midland MI 48640

ARMENTROUT, STEVE, b Eldorado, Tex, June 19, 30; m 62; c 3. TOPOLOGY. Educ: Univ Tex, BA, 51, PhD(math), 56. Prof Exp: Teaching asst math, Univ Tex, 52-55; from instr to prof, Univ Iowa, 56-70; PROF MATH, PA STATE UNIV, 70- Concurrent Pos: Res fel, Univ Iowa, 62; vis prof, Univ Wis, 66-67. Mem: Am Math Soc; Math Asn Am. Res: Topology of manifolds; upper semicontinuous decompositions of manifolds. Mailing Add: Dept of Math Pa State Univ 230 McAllister Bldg University Park PA 16802

ARMIJO, LARRY, b El Paso, Tex, Jan 6, 38. MATHEMATICS. Educ: St Edward's Univ, BS, 58; Rice Univ, MA, 60, PhD(math), 62. Prof Exp: Asst, Rice Univ, 58-62; sr res engr math, Gen Dynamics/Astronaut, 62; fel, Rice Univ, 62-63; asst res scientist, Denver Div, Martin Co, 63-64; res mathematician, Nat Eng Sci Co, 64-66; staff analyst, TRW Systs, 66-68; chief scientist, Tetra Tech, Inc, 68-74; SR STAFF ENGR, LOCKHEED AIRCRAFT CORP, HOUSTON, TEX, 74- Mem: Am Math Soc; Math Asn Am; Soc Indust & Appl Math; NY Acad Sci. Res: Classical and numerical analysis; differential equations and optimization techniques; operations research; biomedical engineering; radar meteorology. Mailing Add: 18100 Nassau Bay Dr Apt 143 Houston TX 77058

ARMINGTON, ALTON, b Everett, Mass, Sept 20, 27; m 52; c 2. PHYSICAL CHEMISTRY, FUEL TECHNOLOGY. Educ: Boston Univ, BA, 50; Tufts Univ, MS, 51; Pa State Univ, PhD(fuel technol), 61. Prof Exp: Chemist develop, Congoleum-Nairn, 52; phys res, Naval Res Labs, 52-54; asst fuel technol, Pa State Univ, 54-57; chief chemist, Pure Mat Univ, SR CHEMIST, SOLID STATE CHEM LAB, 57- Mem: Am Chem Soc; Am Asn Crystal Growth. Res: Preparation and properties of ultra-pure chemical elements and intermetallic compounds; surface properties; graphite surfaces; crystal preparation. Mailing Add: Rome Air Dev Cmd Electr Tech Dep Laurence G Hanscom Field Bedford MA 01730

ARMISTEAD, WILLIAM HOUSTON, JR, b Nashville, Tenn, July 28, 16; m 36; c 6. CHEMISTRY, GLASS TECHNOLOGY. Educ: Vanderbilt Univ, BE, 37, MS, 38, PhD(inorg & anal chem), 41. Prof Exp: Res chemist, 41-51, sr res assoc chem, 51-54, mgr melting dept & dir res & develop, 54-56, vpres & dir res & develop, 56-61, vpres, 61-71, DIR TECH STAFFS, CORNING GLASS WORKS, 61-, VCHMN BD, 71- Mem: Am Chem Soc; fel Am Ceramic Soc; Soc Glass Tech. Mailing Add: Corning Glass Works Houghton Park Corning NY 14830

ARMISTEAD, WILLIS WILLIAM, b Detroit, Mich, Oct 28, 16; m 38, 67; c 3. VETERINARY MEDICINE, MEDICAL EDUCATION. Educ: Agr & Mech Col, Tex, DVM, 38; Ohio State Univ, MSc, 50; Univ Minn, PhD(vet med), 55; Am Col Vet Surgeons, dipl, 75. Prof Exp: Gen vet practitioner, 38-40; from instr to prof vet med & surg, Agr & Mech Col, Tex, 40-53, dean col vet med, 53-57; dean col vet med, Mich State Univ, 57-74; DEAN COL VET MED, UNIV TENN, KNOXVILLE, 74- Concurrent Pos: Ed, NAm Vet, 50-56; consult, Southern Regional Educ Bd, 53-56, Surgeon Gen, US Air Force, 60-62 & Tenn Higher Educ Comn, 73-74; collabr, Animal Dis & Parasite Res Div, Agr Res Serv, USDA, 54-65; mem adv bd, Inst Lab Animal Resources, Nat Res Coun, 62-; mem vet med resident investr selection comt, US Vet Admin, 67-70; mem vet med rev comt, Bur Health Professions Educ & Manpower Training, US Dept Health, Educ & Welfare, 67-71. Mem: Hon mem Am Animal Hosp Asn; Fedn Asns Schs Health Professions (pres, 74-75); Am Vet Med Asn (pres, 57-58); Asn Am Vet Med Col (pres, 64-65, 73-74); NY Acad Sci. Res: General surgery; tissue transplantation; wound healing. Mailing Add: Box 1071 Univ of Tenn Col of Vet Med Knoxville TN 37901

ARMITAGE, JOHN BRIAN, b Ripon, Eng, Oct 28, 27; m 54; c 2. ORGANIC POLYMER CHEMISTRY. Educ: Univ Manchester, BSc, 48, PhD(chem), 51. Prof Exp: Fel, Ohio State Univ, 51-52; sr res fel, Cambridge Univ, 52-55; res chemist, Exp Sta, 55-60, res supvr, 60-61, sr res supvr, Univ Tex, 61-67, sr res supvr, Plastics Dept, 67-71, RES ASSOC, PLASTIC PRODUCTS & RESINS DEPT, E I DUPONT DE NEMOURS & CO, 71- Concurrent Pos: Sr studentship of 1851 Royal Exhib, 52-54; Consult, Chester Water Co, Eng, 54-55. Mem: Am Chem Soc; Soc Plastics Eng; fel Royal Inst Chem; Brit Chem Soc. Res: Synthesis of polyacetylenes; structure of vitamin B12; potato root eel worm chemistry; polymer chemistry; polyolefins; polyamides; fire retardance and toxicity of thermoplastics. Mailing Add: Plastic Prod & Resins Dept Exp Sta E I Du Pont de Nemours & Co Wilmington DE 19898

ARMITAGE, JOHN DENTON, JR, b San Antonia, Tex, Jan 17, 35; m 58, 75; c 2. COMMUNICATIONS SCIENCE, OPTICAL PHYSICS. Educ: Mass Inst Technol, BS, 57; Univ London, PhD(optical physics), 68. Prof Exp: Engr, Jarrell-Ash Co, 57-60; physicist of the atmosphere, US Air Force Cambridge Res Lab, 60; sr physicist, Black Assocs, 60-61; sr assoc engr, 61-65, staff engr, 68-69, ADV PHYSICIST, IBM CORP, 69- Concurrent Pos: Consult, Mass Gen Hosp, 59-61 & US Navy Electronics Lab, Calif, 64. Mem: Inst Elec & Electronics Eng. Mailing Add: IBM Corp 510-023 Boulder CO 80302

ARMITAGE, KENNETH BARCLAY, b Steubenville, Ohio, Apr 18, 25; m 53; c 3. ECOLOGY. Educ: Bethany Col, WVa, BS, 49; Univ Wis, MS, 51, PhD, 54. Prof Exp: Asst zool, Univ Wis, 49-52, proj asst bot, 53, instr bot & zool, 54-56; from asst prof to assoc prof zool, 56-66, chmn dept biol, 68-75, PROF ZOOL, UNIV KANS, 66- Concurrent Pos: Trustee, Rocky Mountain Biol Lab, 70- Mem: AAAS; Am Soc Limnol & Oceanog; Ecol Soc Am; Soc Study Evolution; Am Soc Zool. Res: Behavioral and physiological ecology; social ecology of marmots. Mailing Add: Div of Biol Sci Univ of Kans Lawrence KS 66045

ARMOR, JOHN N, b Philadelphia, Pa, Sept 14, 44; m 66; c 3. PHYSICAL INORGANIC CHEMISTRY, ORGANOMETALLIC CHEMISTRY. Educ: Pa State Univ, BS, 66; Stanford Univ, PhD(chem), 70. Prof Exp: Asst prof inorg chem, Boston Univ, 70-74; SR RES CHEMIST, CENT RES CTR, ALLIED CHEM CORP, 74- Mem: Am Chem Soc. Res: Homo and heterogeneous catalysis using inorganic and organometallic compounds. Mailing Add: Allied Chem Corp Cent Res Ctr PO Box 1021R Morristown NJ 07960

ARMOUDIAN, GARABED, b Houch-Hala, Lebanon, Aug 1, 38; m 62; c 2. ELECTRODYNAMICS, THEORETICAL PHYSICS. Educ: Am Univ Beirut, BS, 62; La State Univ, PhD(physics), 66. Prof Exp: Instr physics, La State Univ, 65-66; AEC res fel, Columbia Univ, 66-68; ASST PROF PHYSICS, SOUTHWESTERN STATE COL, OKLA, 68- Mem: Am Phys Soc; Am Asn Physics Teachers. Res: Quantum electrodynamics; theoretical studies on the structure of proton and other elementary particles. Mailing Add: Dept of Physics Southwestern State Col Weatherford OK 73096

ARMOUR, ALBERT GEORGE, organic chemistry, see 12th edition

ARMOUR, EUGENE ARTHUR, b Mercer, Pa, Sept 21, 46; m 68; c 2. ORGANIC CHEMISTRY, PHOTOGRAPHIC CHEMISTRY. Educ: Purdue Univ, BS, 68; Ohio State Univ, PhD(org chem), 73. Prof Exp: SR RES CHEMIST ORG CHEM, EASTMAN KODAK CO, 73- Mem: Am Chem Soc. Res: Design and synthesis of organic compounds for utilization in various photographic products and processes. Mailing Add: Eastman Kodak Co Res Labs Rochester NY 14650

ARMSON, KENNETH AVERY, b Newton Brook, Ont, Feb 19, 27; m 52; c 1. SILVICULTURE, SOIL SCIENCE. Educ: Univ Toronto, BScF, 51; Oxford Univ, dipl forestry, 55. Prof Exp: Forester, Soils Surv, Res Div, Ont Dept Lands & Forests, 51-52; lectr, 52-57, from asst prof to assoc prof, 57-68, PROF FORESTRY, UNIV TORONTO, 68- Concurrent Pos: Spec consult, Ont Ministry Natural Resources, 75-76. Mem: Soc Am Foresters; Soil Sci Soc Am; Can Inst Forestry. Res: Forest soils; effect of fertility on tree growth. Mailing Add: Fac of Forestry Univ of Toronto Toronto ON Can

ARMSTEAD, ROBERT LOUIS, b Blair, Nebr, Nov 5, 36; m 61; c 2. THEORETICAL PHYSICS. Educ: Univ Rochester, BS, 58; Univ Calif, Berkeley, PhD(theoret physics), 65. Prof Exp: ASSOC PROF PHYSICS, NAVAL POSTGRAD SCH, 64- Mem: Am Phys Soc; Sigma Xi. Res: Theoretical atomic scattering; laser light propagation. Mailing Add: Dept of Physics Naval Postgrad Sch Monterey CA 93940

ARMSTRONG, ALFRED RINGGOLD, b Washington, DC, July 6, 11; m 34; c 2. ANALYTICAL CHEMISTRY. Educ: Col William & Mary, BS, 32, AM, 34; Univ Va, PhD(chem), 45. Prof Exp: Instr, Col William & Mary, 33-35, asst prof chem, 36-42; asst to Dr J. H. Yoe, Nat Defense Res Comt & investigator, Chem Warfare Serv, Univ Va, 44-45; assoc prof, 45-61, PROF CHEM, COL WILLIAM & MARY, 61- Mem: Am Chem Soc. Res: Chemistry of sea water; organic analytical reagents. Mailing Add: Dept of Chem Col of William & Mary Williamsburg VA 23185

ARMSTRONG, ANDREW THURMAN, b Haslet, Tex, May 26, 35; m 58; c 3. PHYSICAL CHEMISTRY, SPECTROSCOPY. Educ: North Tex State Univ, BS, 58, MS, 59; La State Univ, PhD(chem), 67. Prof Exp: Instr chem, West Tex State Univ, 59-61 & La State Univ, 63-66, vis asst prof, 67-68; ASST PROF CHEM, UNIV TEX, ARLINGTON, 68- Mem: Am Chem Soc. Res: Molecular spectroscopy of inorganic and organic systems. Mailing Add: Dept of Chem Univ of Tex Arlington TX 76010

ARMSTRONG, AUGUSTUS K, b Charleston, WVa, June 6, 30; m 53; c 3. PALEOBIOLOGY, STRATIGRAPHY. Educ: Univ NMex, BS, 53; Univ Cincinnati,

MS, 57, PhD(geol), 60. Prof Exp: Asst prof geol, Portland State Col, 60-62; explor geologist, Northwest Div, Shell Oil Co, Arctic Alaska, 62-66; GEOLOGIST, PALEONT & STRATIG BR, US GEOL SURV, 66- Mem: Am Asn Petrol Geol; Am Paleont Soc; Soc Econ Paleont & Mineral; Brit Paleont Asn. Res: Systematics of upper Paleozoic corals, brachiopods and foraminifera; petrography, sedimentation, diagenesis and stratigraphy of carbonates sediments. Mailing Add: 753 Silver Tip Ct Sunnyvale CA 94086

ARMSTRONG, BAXTER HARDIN, b Portland, Ore, Jan 27, 29; m 54; c 4. ATOMIC PHYSICS. Educ: Univ Calif, Berkeley, AB, 51, MA, 53, PhD(physics), 56. Prof Exp: From assoc res scientist to res scientist, Lockheed Missiles & Space Co, 56-62, staff scientist, 62-63, sr staff scientist & sr mem res lab, 63-64; mem staff, Sci Ctr, 64-68, sci staff mgr, 69-73, DIR CENTRO CIENTIFICO DE AM LATINA, IBM DE MEX, IBM CORP, 73- Concurrent Pos: Asst prof, San Jose State Col, 58-60; guest staff mem, Queen's Univ, Belfast, 60-61. Mem: Am Phys Soc; Asn Am Physics Teachers; Am Meteorol Soc. Res: Theoretical atomic physics; interaction of atoms with radiation fields; atmospheric radiative transfer; plasma physics; air pollution; research administration. Mailing Add: IBM Sci Ctr PO Box 10500 Palo Alto CA 94304

ARMSTRONG, CLAY M, b Chicago, Ill, Sept 26, 34; m 63; c 3. PHYSIOLOGY, BIOPHYSICS. Educ: Rice Univ, BA, 56; Wash Univ, MD, 60. Prof Exp: Res assoc biophys, Lab Biophys, NIH, 61-64; hon res asst physiol, Univ Col, Univ London, 64-66; asst prof, Duke Univ, 66-69; ASSOC PROF PHYSIOL, UNIV ROCHESTER, 69- Mem: Biophys Soc; Soc Gen Physiol. Res: Permeability mechanisms in excitable membranes; excitation-contraction coupling. Mailing Add: 11 Sutherland Pittsford NY 14534

ARMSTRONG, DALE DEAN, b Salina, Kans, Aug 18, 27; m 52; c 3. NUCLEAR PHYSICS. Educ: Univ Colo, BS, 57; Univ NMex, MS, 62, PhD(physics), 65. Prof Exp: STAFF MEM PHYSICS, LOS ALAMOS SCI LAB, 57- Mem: Am Phys Soc. Res: Design and development of a high intensity neutron source for simulating the radiation damage effects that will be produced in fusion reactor materials. Mailing Add: Los Alamos Sci Lab PO Box 1663 MS-459 Los Alamos NM 87545

ARMSTRONG, DAVID ANTHONY, b Barbados, WI, Aug 27, 30; m 55; c 3. PHYSICAL CHEMISTRY. Educ: McGill Univ, BSc, 52, PhD, 55. Prof Exp: Brotherton res lectr phys chem, Univ Leeds, 55-57; Nat Res Coun Can fel chem, Univ Sask, 57-58; from asst prof to assoc prof, Univ Alta, 58-68; PROF CHEM, UNIV CALGARY, 68- Mem: Chem Inst Can. Res: Radiation chemistry; chemical kinetics. Mailing Add: Dept of Chem Univ of Calgary Calgary AB Can

ARMSTRONG, DAVID EDWIN, water chemistry, soil chemistry, see 12th edition

ARMSTRONG, DAVID THOMAS, b Kinburn, Ont, Nov 5, 29; m 56; c 3. PHYSIOLOGY. Educ: Ont Agr Col, BSA, 51; Cornell Univ, MS, 56, PhD(physiol), 59. Prof Exp: Asst agr rep, Ont Dept Agr, 51-54; asst dept animal husb, Cornell Univ, 54-57; res assoc, Dept Biol, Brookhaven Nat Lab, 58-60; from res assoc to assoc anat, Sch Dent Med, Harvard Univ, 60-68; assoc prof, 68-69, PROF PHYSIOL, OBSTET & GYNEC, UNIV WESTERN ONT, 69- Concurrent Pos: Res career develop award, NIH, 63-; assoc, Med Res Coun Can, 69- Res: Pituitary-gonad interrelationships; mechanism of hormone action. Mailing Add: Dept of Physiol Univ of Western Ont London ON Can

ARMSTRONG, DON LEIGH, b Alhambra, Calif, June 7, 16; m 40; c 2. Educ: Univ Calif, Los Angeles, AB, 37; Univ Southern Calif, MS, 38, PhD(chem), 42, MChE, 45. Prof Exp: Res assoc chem, 41-45; prin chemist, Aerojet-Gen Corp, Gen Tire & Rubber Co, 46-57, mgr chem div, 57-59, dir chem, 59-62, sr scientist, 62-56, mgr corp tech info ctr, 62-64; PROF CHEM & CHMN DEPT, WHITTIER COL, 65- Concurrent Pos: Pres, Tech Asst Group, Los Angeles, 64- Mem: Sigma Xi; Am Chem Soc; Marine Technol Soc. Res: Marine chemistry, desalinization by reverse osmosis; liquid and solid rocket propellants; technical information retrieval. Mailing Add: Dept of Chem Whittier Col Whittier CA 90608

ARMSTRONG, DONALD B, b Minneapolis, Minn, June 30, 37; m 64; c 2. ACOUSTICS, CRYSTALLOGRAPHY. Educ: Univ Minn, BS, 59, MSEE, 61, PhD(elec eng), 64. Prof Exp: Res fel solid state physics, Univ Minn, 61-64; sr engr, Electron Tube Div, Litton Indust, Calif, 64-67, sr scientist, 67-72, prog mgr electrooptic devices, 72-73; DIR NEW PROD DEVELOP, CRYSTAL TECHNOL INC, 73- Mem: Inst Elec & Electronics Engr. Res: Microwave properties of semiconductors; properties of acoustic surface waves on piezoelectric crystals at microwave frequencies. Mailing Add: Crystal Technol Inc 2510 Old Middlefield Way Mountain View CA 94043

ARMSTRONG, DONALD JAMES, b Marshall Co, WVa, Mar 14, 37; m 61; c 1. PLANT PHYSIOLOGY. Educ: Marshall Univ, AB, 59, MA, 61; Univ Wis, Madison, PhD(bot), 67. Prof Exp: Proj assoc bot, Univ Wis, Madison, 67-68; Nat Sci Found fel biochem, Univ BC, 68-70; proj assoc bot, Univ Wis, Madison, 70-74; ASST PROF, ORE STATE UNIV, 74- Mem: Am Soc Plant Physiol. Res: Mechanism of action of plant growth substances; nucleic acids and protein synthesis; regulation of plant growth and development. Mailing Add: Dept of Bot Plant Path Ore State Univ Corvallis OR 97331

ARMSTRONG, FRANK BRADLEY, JR, b Brownsville, Feb 15, 28; m 58; c 3. BIOCHEMISTRY. Educ: Univ Tex, BS, 50, MA, 53; Univ Calif, PhD(biochem), 59. Prof Exp: Fel biochem & genetics, Univ Tex, 59-62; from asst prof to assoc prof genetics & bact, 62-66, assoc prof biochem, 66-68, UNIV PROF BIOCHEM, NC STATE UNIV, 68- Concurrent Pos: Career develop award & Sigma Xi res award, 64; NSF panel, 70 & 71. Mem: AAAS; Genetics Soc Am; Am Soc Biol Chemists; Am Chem Soc; Am Inst Biol Sci. Res: Biochemical and microbial genetic studies on Salmonella typhimurium; enzyme studies of drosophila. Mailing Add: Dept of Biochem NC State Univ Raleigh NC 27607

ARMSTRONG, FRANK CLARKSON, b New York, NY, June 2, 13; m 47; c 2. GEOLOGY. Educ: Yale Univ, BA, 36; Univ Wash, MS, 48; Stanford Univ, PhD(geol), 63. Prof Exp: Assayer & engr, Tiblemont Island Mining Co, 36-37; engr, Hollinger Consol Gold Mines Ltd, 37-40; mine geologist, Dome Mines, Ltd, 40-43; geologist strategic minerals invests, 43-45, actg regional geologist, Wash, 45-46, geologist, Wyo, 46, western phosphate deposits, 47 & 48, Idaho, 49 & 50-51, Colo, 51-52, Pac Northwest, 52-54, Idaho, 54-61 & Wyo, 62-64, GEOLOGIST, US GEOL SURV, 64- Concurrent Pos: Part-time teacher, Ft Wright Col Holy Names, 66. Mem: AAAS; fel Geol Soc Am; Soc Econ Geologists. Res: Economic and structural geology; metals and nonmetals. Mailing Add: US Geol Surv W 920 Riverside Ave Spokane WA 99201

ARMSTRONG, GEORGE GLAUCUS, JR, b Houston, Miss, Feb 28, 24; m 48; c 2. AEROSPACE MEDICINE, PHYSIOLOGY. Educ: Univ Miss, BA, 48, MS, 50, BS, 52; Univ Ill, MD, 56. Prof Exp: Res assoc, Sch Med, Univ Miss, 48-49, instr physiol, 49-52, from asst prof to assoc prof physiol & biophys, 52-64; aerospace technologist, 64-68, chief biomed technol div, 68-70; dep chief, Med Opers Div, 70-72, CHIEF HEALTH SERV DIV, JOHNSON SPACE CTR, NASA, 72- Mem: Fel AAAS; Aerospace Med Asn; Soc Exp Biol & Med; Am Physiol Soc; Undersea Med Soc. Res: Cardiovascular reflexes; spaceborne medical systems. Mailing Add: Health Serv Div Johnson Space Ctr Mail Code DD Houston TX 77058

ARMSTRONG, GEORGE THOMSON, b Caster, Alta, Dec 8, 16; m 45; c 2. PHYSICAL CHEMISTRY, THERMODYNAMICS. Educ: Univ Fla, BS, 39, MS, 43; Johns Hopkins Univ, PhD(phys chem), 48. Prof Exp: Asst, Univ Fla, 40-42; staff mem, Radiation Lab, Mass Inst Technol, 42-45; jr instr, Johns Hopkins Univ, 45-47; instr, Yale Univ, 48-51; chief thermochem sect, 68-74, CHEMIST, US NAT BUR STAND, 51-, CHIEF THERMOCHEM MEASUREMENTS AND STANDARDS, 74- Concurrent Pos: Counr, Calorimetry Conf, 70-73; mem, Joint Int Union Pure & Appl Chem-Int Union Biochem-Int Union Pure & Appl Biophys Subcomn Biothermodyn, 73-; pvt pract, info serv, 74- Honors & Awards: Silver Medal, US Dept Commerce, 67. Mem: AAAS; Am Chem Soc; Am Phys Soc; Am Soc Testing & Mat; NY Acad Sci. Res: Vapor-liquid phase equilibria; calorimetry; thermochemistry; thermodynamics; reference materials for physical chemistry; biological materials. Mailing Add: 1402 Dale Dr Silver Springs MD 20910

ARMSTRONG, HAROLD LEWIS, b Picton, Ont, Apr 22, 21; m 52; c 3. PHYSICS. Educ: Queen's Univ, Ont, BSc, 50, MSc, 51. Prof Exp: Mem staff, elec eng, Nat Res Coun Can, 51-53, electronics, Clevite Corp, Ohio, 53-54, electronics, Pac Semiconductors Inc, Calif, 54-58 & Nat Res Coun Can, elec eng, 56-57; ASST PROF PHYSICS, QUEEN'S UNIV, ONT, 58- Concurrent Pos: Researcher, electronic studies, Can Postal Serv, 52-53; teacher math, Fenn Col, 54; semiconductor electronics, Univ Calif, Los Angeles, 58; consult, Electro Systs, Inc, 58-60; mem Can Comts Semiconductors & Mat, 60-; ed, Creative Res Soc, 64- Mem: Creative Res Soc. Res: Semiconductor devices; electroluminescence; electrical properties of materials; mechanics; philosophy of science; science and Christianity. Mailing Add: Dept of Physics Queen's Univ Kingston ON Can

ARMSTRONG, HERBERT STOKER, b Toronto, Ont, Nov 23, 15; m 41; c 2. GEOLOGY. Educ: Univ Toronto, BA, 38, MA, 39; Univ Chicago, PhD(econ geol), 42; McMaster Univ, DSc, 67. Prof Exp: Asst geol, Univ Toronto, 38-39 & Univ Chicago, 40-41; lectr, McMaster Univ, 41-44, from asst prof to assoc prof, 44-48, prof, 48-62, asst dean arts & sci, 46-48, from assoc dean to dean, 48-62; prof geol, Univ Alta, 62-64, dean sci, 62-63, acad vpres, 63-64; prof geol & pres, Univ Calgary, 64-68, vchancellor, 66-68; PROF GEOL & DEAN GRAD STUDIES, UNIV GUELPH, 68- Mem: Can Geol Found; Am Geochem Soc; fel Geol Asn Can; Can Inst Min & Metal; fel Royal Can Geog Soc. Res: Precambrian geology; igneous and metamorphic petrology. Mailing Add: Grad Studies Univ of Guelph Guelph ON Can

ARMSTRONG, HOWARD WAYNE, b Brownwood, Tex, July 6, 42. PARASITOLOGY, INVERTEBRATE ZOOLOGY. Educ: Lamar Univ, BS, 65; Tex A&M Univ, MS, 69, PhD(zool), 74. Prof Exp: Instr biol, Lamar Univ, 67-68, lectr, 75, FEL & RES ASSOC BIOL, TEX A&M UNIV, 74- Mem: Am Soc Parasitologists; Gulf Estuarine Res Soc; Estuarine Res Fedn. Res: Effects of oil and other pollutants on marine organisms and benthic communities; taxonomy and ecology of helminth parasites. Mailing Add: Dept of Biol Tex A&M Univ College Station TX 77840

ARMSTRONG, JAMES G, b London, Eng, Nov 14, 24; US citizen; m 51; c 4. MEDICAL ADMINISTRATION. Educ: Oxford Univ, MA, 49, BM, BCh, 50; Royal Col Physicians & Surgeons Can, cert pediat, 63; FRCP(C). Prof Exp: House physician, St Olave's Hosp, London, Eng, 50-51, house surgeon, 51-52; staff physician, Alta Ment Inst, 55-56; resident, St Joseph's Hosp, Toronto, Ont, 56-57; chief resident, Hosp Sick Children, 57-58; physician, Lilly Labs Clin Res, 58-66, sr physician, Lilly Res Labs, 66-68, asst dir med regulatory affairs, 68-69, dir med plans & regulatory affairs, 69-73, DIR MED LIAISON, LILLY RES LABS, 73- Concurrent Pos: Fel pediat, Hosp Sick Children, 57-58; fel pediat med, Univ Toronto, 57-58; instr, Ind Univ, 58-66, asst prof, 66-; staff physician, Dept Pediat, Marion County Gen Hosp, Indianapolis, Ind, 58-, physician, Tumor Bd, 60- Mem: Fel Am Col Physicians; Am Asn Cancer Res; Am Fedn Clin Res; fel Royal Soc Med; Brit Med Asn. Res: Pediatric medicine; clinical pediatrics; cancer chemotherapy; medical research administration. Mailing Add: Lilly Res Labs 307 E McCarty St Indianapolis IN 46206

ARMSTRONG, JAMES WALTER, JR, b Denver, Colo, Nov 11, 26; m 49; c 3. MATHEMATICS. Educ: Colo Col, BA, 48; Purdue Univ, PhD(math), 52. Prof Exp: Sr aerophys engr, Convair, Tex, 52-54; opers analyst, Hq Air Defense Command, 54-57; chief scientist, 57-60, sci adv, NAm Air Defense Command, 60-62; res scientist, Kaman Nuclear, 62-67; asst prof, 67-70, ASSOC PROF MATH, UNIV ILL, URBANA-CHAMPAIGN, 70-, GRAD SUPVR, GRAD ADV OFF, 73- Mem: Inst Elec & Electronics Engrs. Res: Nuclear weapons effects, including system analyses of air defense systems; ballistic missile defense systems; ballistic re-entry vehicle systems. Mailing Add: Grad Adv Off Dept of Math Univ of Ill 259 Altgeld Hall Urbana IL 61801

ARMSTRONG, JOHN ALEXANDER, b London, Ont, May 11, 29; m 63; c 2. INSECT TOXICOLOGY, FOREST METEOROLOGY. Educ: Univ Western Ont, BSc, 53; Queens Univ, Ont, MSc, 63, PhD(insect physiol), 67. Prof Exp: Sci off, Trop Pesticides Res Inst, Arusha, EAfrica, 54-61; consult, Vector Control, WHO, Geneva, Switz, 62, Can Int Develop Agency adv, Cayman Islands Govt, 66-68; RES SCIENTIST, CHEM CONTROL RES INST, CAN FORESTRY SERV, 69- Concurrent Pos: Mem, Int Agr Aviation Ctr, 69- Mem: Am Mosquito Control Asn; Can Agr Pesticide Tech Soc. Res: Susceptibility of insects of insecticides; studies associated with practical insect control schemes; spray cloud drift and deposition with particular interest in studies on air movement within and above the forest canopy. Mailing Add: Chem Control Res Inst Can Forestry Serv Ottawa ON Can

ARMSTRONG, JOHN BRIGGS, b Toronto, Ont. PEDIATRIC NEUROLOGY, BIOCHEMISTRY. Educ: McGill Univ, BSc, 63, MD & CM, 65, PhD(neurochem), 75. Prof Exp: Intern med, St Paul's Hosp, Vancouver, BC, 66-67; resident, Montreal Gen Hosp, 67-68; resident neurol, Montreal Neurol Inst, 68-69, res fel neurochem, 69-71; res fel neurochem, Hosp for Sick Children, Toronto, 71-74, chief resident pediat neurol, 74-75; ASST PROF PEDIAT, UNIV TORONTO, 75-; STAFF NEUROLOGIST & RES ASSOC, HOSP FOR SICK CHILDREN, TORONTO, 75- Concurrent Pos: Med Res Coun Can fel, 69-73; Muscular Dystrophy Asn Can fel, 73-74, clin fel, 74-75. Mem: Am Acad Neurol; Can Neurol Soc. Res: Energy metabolism in normal and pathological skeletal muscle; muscular dystrophy; organic mercury poisoning; anticonvulsant drugs. Mailing Add: Hosp for Sick Children 555 University Ave Toronto ON Can

ARMSTRONG, JOHN BUCHANAN, b Toronto, Ont, Oct 19, 18; m 44; c 4. INTERNAL MEDICINE. Educ: Univ Toronto, MD, 43; McGill Univ, dipl, 51. Prof Exp: Sr intern & asst resident med, Royal Victoria Hosp, 46-48; registr, Post-Grad Med Sch, Hammersmith Hosp, London, 49-50; asst prof physiol & med res & spec lectr med, Univ Man, 50-56; assoc prof pharmacol, Univ Toronto, 56-69; SR LECTR

PHARMACOL, UNIV OTTAWA, 69-; CHIEF MED ADV, CAN HEART FOUND, 72- Concurrent Pos: Fel, Duke Hosp, 48-49; Can Life Ins med fel, 50-S2; Markle scholar, 52-56; physician, Winnipeg Munic Hosps, 51-56; chmn & mem panel shock & plasma expanders, Defense Res Bd Can, 53-68; asst physician, Winnipeg Gen Hosp, 54-56; consult, Can Forces Med Coun, 54-60; vol asst, Toronto Gen Hosp, 57-68; exec dir, Can Heart Found, 57-72; pres, Biol Coun Can, 71. Mem: Pharmacol Soc Can (pres, 72); fel Am Col Cardiol; NY Acad Sci; Can Physiol Soc; Can Med Asn. Res: Clinical cardiovascular physiology and pharmacology. Mailing Add: Can Heart Found Suite 1200 1 Nicholas St Ottawa ON Can

ARMSTRONG, JOHN EDWARD, b Cloverdale, BC, Feb 18, 12; m 37; c 1. QUATERNARY GEOLOGY, GLACIAL GEOLOGY. Educ: Univ BC, BASc, 34, MASc, 35; Univ Toronto, PhD(econ geol), 39. Prof Exp: Asst geol, Univ Toronto, 35-37; asst geologist, Geol Surv Can, 36-42, from assoc geologist to Geologist, 43-47, grade 4 geologist, 48-56, geologist in charge BC Off, 56-65; chmn geol div, Royal Soc Can, 65-66 & Can Inst Mining & Metall, 67-68; res scientist, Geological Survey Can, 72-75; SECY GEN, INT GEOL CONG, 68- Concurrent Pos: Mem subcomt on Pleistocene, Am Comn Stratig Nomenclature, 54-59; secy-gen, 24th Int Geol Congr, 68-75. Honors & Awards: Merit Award, Public Serv Can, 73. Mem: Fel Geol Soc Am; fel Royal Soc Can; Geol Asn Can; Can Inst Min & Metall. Res: Cordilleran field geology; quaternary stratigraphy and geomorphology in the Pacific Northwest. Mailing Add: 206-2298 McBain Ave Vancouver BC Can

ARMSTRONG, JOHN RICHARD, organic chemistry, see 12th edition

ARMSTRONG, JOHN WILLIAM, b Roslyn, NY, Mar 15, 48; m 75. ASTRONOMY, SPACE PHYSICS. Educ: Harvey Mudd Col, BS, 69; Univ Calif, San Diego, MS, 71, PhD(appl physics), 75. Prof Exp: RES ASSOC ASTRON, NAT RADIO ASTRON OBSERV, 75- Mem: AAAS; Am Geophys Union. Res: Structure of compact radio sources; dynamics of interplanetary and interstellar media. Mailing Add: Nat Radio Astron Observ Edgemont Rd Charlottesville VA 22901

ARMSTRONG, JOSEPH CUNNINGHAM, parasitology, microbiology, see 12th edition

ARMSTRONG, KENNETH WILLIAM, b Brandon, Man, Sept 4, 35; m 61; c 2. MATHEMATICS. Educ: Univ Man, BSc, 57, MSc, 61; McGill Univ, PhD(math), 65. Prof Exp: Lectr, Univ Man, 57-58, 59-60, 63-64, from asst prof to assoc prof, 64-69, asst head dept, 66-69; ASSOC PROF MATH, UNIV WINNIPEG, 69-, DIR, BD EDUC PROG, 73- Mem: Math Asn Am; Can Math Cong. Mailing Add: Dept of Math Univ of Winnipeg Winnipeg 2 MB Can

ARMSTRONG, LLOYD, JR, b Austin, Tex, May 19, 40; m 65. ATOMIC PHYSICS. Educ: Mass Inst Technol, BS, 62; Univ Calif, Berkeley, PhD(physics), 66. Prof Exp: Physicist, Lawrence Radiation Lab, 65-67; sr physicist, Westinghouse Res Labs, Pa, 67-68; res assoc physics, 68, asst prof, 69-73, ASSOC PROF PHYSICS, JOHNS HOPKINS UNIV, 73- Concurrent Pos: Consult, Westinghouse Res Labs, 69-71. Mem: Am Phys Soc. Res: Theoretical atomic physics, investigation of atomic shell theory through use of group theory and second quantization techniques; many body theory applied to atoms; interaction of atoms with intense laser beams. Mailing Add: Dept of Physics Johns Hopkins Univ Baltimore MD 21218

ARMSTRONG, MARVIN DOUGLAS, b Wilmington, NC, Apr 15, 18; m 46; c 5. BIOLOGICAL CHEMISTRY. Educ: Univ SC, BS, 38; Univ Ill, MS, 39, PhD(org chem), 41. Prof Exp: Asst biochem, med col, Cornell Univ, 41-42 & 46; from asst res prof to assoc res prof biochem, col med, Univ Utah, 46-57; CHMN DEPT BIOCHEM, FELS RES INST, OHIO, 57- Mem: AAAS; Am Chem Soc; Am Soc Biol Chemists. Res: Chemistry and metabolism of the amino acids; metabolism of aromatic compounds; hereditary and metabolic disorders; biochemistry of mental disorders. Mailing Add: Fels Res Inst Yellow Springs OH 45387

ARMSTRONG, MAURICE R, textiles, see 12th edition

ARMSTRONG, PAUL DOUGLAS, b New Albany, Ind, Jan 4, 41; m 70. ORGANIC CHEMISTRY, MEDICINAL CHEMISTRY. Educ: Ind Univ, BS, 63; Univ Iowa, PhD(med chem), 68. Prof Exp: Res assoc chem under Prof John C Sheehan, Mass Inst Technol, 68-69; asst prof chem, Mass Col Pharm, 69-74; ASSOC PROF CHEM, CALIFORNIA BAPTIST COL, 74-, CHMN DIV NATURAL SCI, 75- Mem: Am Chem Soc. Res: Organic synthesis designed to make contributions to structure-activity relationships and/or to produce useful drugs. Mailing Add: Dept of Chem Calif Baptist Col 8432 Magnolia Ave Riverside CA 92504

ARMSTRONG, PETER BROWNELL, b Syracuse, NY, Apr 27, 39; m 62; c 3. CELL BIOLOGY, DEVELOPMENTAL BIOLOGY. Educ: Univ Rochester, BS, 61; Johns Hopkins Univ, PhD(biol), 66. Prof Exp: Asst prof zool, Univ Calif, Davis, 66-73; ASSOC PROF, UNIV CALIF DAVIS, 73- Concurrent Pos: Assoc, Clare Hall, Cambridge Univ, 73-74. Mem: Am Soc Cell Biol; Soc Develop Biol. Res: Physiological basis for cellular adhesion; mechanisms of morphogenetic movements; mechanisms of intercellular invasion; vitellogenesis; comparative hematology. Mailing Add: Dept of Zool Univ of Calif Davis CA 95616

ARMSTRONG, R WARWICK, b Auckland, NZ, May 22, 35; m 59; c 1. GEOGRAPHY. Educ: Univ Auckland, BA, 57, MA, 59; Univ Ill, PhD(geog), 63; Univ Mich, MPH, 64. Prof Exp: Jr lectr geog, Univ Auckland, 58-59; asst prof geog, Univ Ill, Urbana, 64-68; assoc prof, 68-72, PROF GEOG & PUB HEALTH, UNIV HAWAII, 72- Concurrent Pos: Res med geogr, Univ Calif, San Francisco, 72-74 & ICMR, Kuala Lumpur, Malaysia. Mem: Am Pub Health Asn; Asn Am Geogr; Royal Soc Health; Royal Soc NZ. Res: Gepgraphical epidemiology of cancer; mapping of disease; medical geography of local communities. Mailing Add: Dept of Geog Univ of Hawaii Honolulu HI 96822

ARMSTRONG, RICHARD LEE, b Seattle, Wash, Aug 4, 37; m 61; c 3. GEOCHRONOLOGY, TECTONICS. Educ: Yale Univ, BS, 59, PhD(geol), 64. Prof Exp: Actg instr, Yale Univ, 62-63; Nat Sci Found fel, Univ Berne, 63-64; from asst prof to assoc prof, Yale Univ, 70-73; ASSOC PROF GEOL, UNIV BC, 73- Concurrent Pos: Guggenheim fel, Canberra, Australia & Pasadena, Calif, 68-69; mem, Working Group 9, Int Geodynamics Comn, 72- Mem: Geol Soc Am; Geol Asn Can; Am Geophys Union. Res: Geochronometry and regional geology of the Western United States and Canada; lead and strontium isotope studies of igneous rocks. Mailing Add: Geol Sci Dept Univ of BC Vancouver BC Can

ARMSTRONG, ROBERT A, b Midland, Ont, July 3, 29; m 56; c 3. SURFACE PHYSICS. Educ: Univ Toronto, BA, 52; McGill Univ, MS, 54, PhD(physics), 57. Prof Exp: Asst res off, 57-62, assoc res off, 62-72, SR RES OFFICER, NAT RES COUN CAN, 72- Res: Surface physics; paramagnetic resonance and relaxation; high and low energy electron diffraction; physical adsorption; auger electron spectroscopy. Mailing Add: 1523 Chomley Cresc Ottawa ON Can

ARMSTRONG, ROBERT D, b Sligo, Pa, July 21, 28; m 51; c 6. PHARMACOLOGY. Educ: Col Wooster, BA, 52; Univ Rochester, MS, 58. Prof Exp: Technician atomic energy proj, Univ Rochester, 52-53, sr technician & res assoc aerosotoxicol, 53-58, res assoc psychotoxicol & psychopharmacol, 58-63; res pharmacologist, Pharmacol Sect, 63-74, RES PHARMACOLOGIST, TOXICOL DIV, BIOMED LAB, US ARMY, EDGEWOOD ARSENAL, 74- Mem: Am Indust Hyg Asn. Res: Toxicology; basic and applied research in psychopharmacology. Mailing Add: Physiotoxicol Br Toxicol Div Biomed Lab Edgewood Arsenal MD 21010

ARMSTRONG, ROBERT G, b Chicago, Ill, Mar 8, 28; m 52; c 2. POLYMER CHEMISTRY, ANALYTICAL CHEMISTRY. Educ: Univ Ill, BS, 49. Prof Exp: Lab asst, Loyola Univ, Ill, 53-54 & Northwestern Univ, 54-55; asst res chemist, Cent Res & Eng, Continental Can Co, Ill, 55-57, from res chemist to sr res chemist, Corp Res Develop, 57-67; RES SCIENTIST, PHILIP MORRIS INC, 67- Mem: AAAS; Am Chem Soc. Res: Organic materials; analysis of and determination of physical-chemical properties of materials; high pressure liquid chromatography. Mailing Add: Philip Morris Inc PO Box 3D Richmond VA 23206

ARMSTRONG, ROBERT JOHN, b Cropsey, Ill, Feb 12, 39. HORTICULTURE, GENETICS. Educ: Univ Ill, BS, 62, PhD(hort), 67; Univ Ariz, MS, 63. Prof Exp: GENETICIST, LONGWOOD GARDENS, 67- Concurrent Pos: Adj asst prof, Longwood Grad Prog, Univ Del, 68- Mem: Am Soc Hort Sci; Am Genetic Asn; Ecol Soc Am. Res: Breeding and genetics of ornamental and vegetable crops; air pollution damage to wild and cultivated plants; plant ecology as it is concerned with the cultivation of plants. Mailing Add: Longwood Gardens Kennett Square PA 19348

ARMSTRONG, ROBERT KRICK, b New Kensington, Pa, Nov 25, 30; m 55; c 3. ORGANIC CHEMISTRY. Educ: Grove City Col, BSc, 54; Ohio State Univ, PhD(chem), 58. Prof Exp: Res chemist, 58-67, col rels rep, 67-70, col rels supvr, 70-72, coordr, Placement Serv, 72, asst mgr, Col Rels Sect, 72-74, MGR, COL RELS SECT, PERSONNEL DIV, EMPLOYEE RELS DEPT, E I DU PONT DE NEMOURS & CO, 74- Mem: Am Chem Soc; Mfg Chemists Asn. Res: Organic synthesis. Mailing Add: Personnel Div Employee Rels Dept E I du Pont de Nemours & Co Wilmington DE 19898

ARMSTRONG, ROBERT LEE, b Xenia, Ohio, July 15, 39; m 63; c 2. BIOCHEMISTRY. Educ: Heidelberg Col, BS, 61; Ohio State Univ, MS, 63; Mich State Univ, PhD(biochem), 66. Prof Exp: NIH fel, Princeton Univ, 66-68; asst prof biochem, Univ Mich, 68-74; ASST PROF CHEM, ALBION COL, 74- Mem: Am Soc Microbiol; Sigma Xi; AAAS. Res: Spore germination; ribosome biosynthesis; antibiotic resistance. Mailing Add: Dept of Chem Albion Col Albion MI 49224

ARMSTRONG, ROBERT PLANT, b Wheeling, WVa, May 19, 19. ANTHROPOLOGY. Educ: Univ Ariz, BA, 44; Univ Iowa, MA, 45; Northwestern Univ, PhD(anthrop), 57. Prof Exp: Traveller, Houghton Mifflin & Co, 45-46; instr Eng, Mont State Univ, 46-50; instr, Inst Eng Studies, Jogjakarta, Indonesia, 55-56; lectr Eng, Eng Lang Inst, Univ Mich, 56-57; field ed, Harper & Bros, 56-58; ed, Alfred A Knopf, 58-59; founder & dir, Univ Ariz Press, 59-60; dir, Northwestern Univ Press, 60-73; Prof Arts & Sci, Northwestern Univ, 67-73; vis prof, Inst African Studies, Univ of Ibadan & vis dir, Univ Ibadan Press, 72-73; vis prof anthrop & African Studies, Ind Univ, 74; PROF ANTHROP, UNIV TEX DALLAS & MASTER, COL V SOCIAL SCI, 74- Concurrent Pos: Consult ed, Book Forum, 74; mem, Adv Bd, Exhib Contemp African Arts, The Field Mus Natural Hist, Chicago, 74; grant, The Rockefeller Found, Ibadan, Nigeria, 72-73; co-chmn, Conf African Lang Lit, Northwestern Univ, 66; Trustee, Dallas Mus Fine Arts, 75- Honors & Awards: Phoebe M Bogan Poetry Prize, Univ Ariz, 44. Mem: Fel Am Anthrop Asn; fel African Studies Asn; Mod Lang Asn; fel Am Folklore Soc. Res: Ethnoaesthetics; publishing; founding the nature of human culture in the nature of human consciousness. Mailing Add: 4240 Glenwood Ave Dallas TX 75205

ARMSTRONG, ROBERT THEXTON, b Chadron, Nebr, Dec 27, 09; m 42; c 1. CHEMISTRY. Educ: Mass Inst Technol, BS, 31, PhD(org chem), 35. Prof Exp: Asst chem, Mass Inst Technol, 28-35, instr org chem, 35-37; chemist, US Rubber Co, 37-41, group leader, 41-44; res chemist, NAm Rayon Corp, Tenn, 44-46; group leader, Celanese Corp, 46-49, tech supt, 49-50, dir tech control, 51-52, assoc dir res, 52-53, tech dir, textile div, 53-56, vpres, 56-66, sr vpres res, 66-75; RETIRED. Mem: AAAS; Am Chem Soc; Soc Chem Indust. Res: Spectroscopy; theory of rubber vulcanization; organic synthesis and reaction mechanism; polymerization; cellulose chemistry; synthetic fiber formation; rubber accelerators and antioxidants; copolymer plastics; textile technology. Mailing Add: Celanese Corp 522 Fifth Ave New York NY 10036

ARMSTRONG, ROBIN L, b Galt, Ont, May 14, 35; m 60; c 2. NUCLEAR MAGNETIC RESONANCE. Educ: Univ Toronto, BA, 58, MA, 59, PhD(physics), 61. Prof Exp: Nat Res Coun Can & Rutherford Mem fels, 61-62; from asst prof to assoc prof, 62-71, assoc chmn dept, 69-73, assoc prof, 69-71, PROF PHYSICS, UNIV TORONTO, 71-, PROF ENG SCI, 71- Concurrent Pos: Prof Physics, Univ Toronto, 71-, chmn, 74- Honors & Awards: Herzberg Medal, Can Asn Physicists, 74. Mem: Am Phys Soc; Can Asn Physicists; Int Soc Magnetic Resonance. Res: Nuclear spin relaxation studies of molecular motions in gases and liquids; nuclear quadrupole resonance and thermodynamic studies of lattice dynamics associated with structural phase transitions. Mailing Add: Dept of Physics Univ of Toronto Toronto ON Can

ARMSTRONG, ROSA MAE, b New Orleans, La, Apr 9, 37; m 58; c 2. HISTOLOGY, EMBRYOLOGY. Educ: Southern Univ, BS, 56; Univ Calif, Berkeley, MA, 58; Univ Calif, San Francisco, PhD(anat), 64. Prof Exp: Lectr, 67-73, ASST PROF ANAT, SCH MED, UNIV CALIF, SAN FRANCISCO, 73- Concurrent Pos: NIH fel, Stanford Univ, 65-67. Mem: Teratology Soc. Res: Organ culture; embryonic rudiments; organ culture of human breast tumors. Mailing Add: Dept of Anat Univ of Calif Sch of Med San Francisco CA 94143

ARMSTRONG, THOMAS PEYTON, b Atchison, Kans, Nov 24, 41; m 62; c 1. SPACE PHYSICS. Educ: Univ Kans, BS, 62; Univ Iowa, MS, 64, PhD(physics), 66. Prof Exp: Res assoc space physics, Univ Iowa, 66-67; res assoc plasma physics, Culham Lab, UK Atomic Energy Auth, Eng, 67-68; from asst prof to assoc prof, 68-74, PROF PHYSICS, UNIV KANS, 74- Mem: AAAS; Am Phys Soc; Am Geophys Union. Res: Geomagnetically trapped radiation; solar cosmic ray events; magnetospheric physics; nonlinear waves in plasmas; interplanetary energetic particles and plasmas. Mailing Add: Dept of Physics Univ of Kans Lawrence KS 66044

ARMSTRONG, WALLACE DAVID, b Hunt Co, Tex, July 8, 05; m 29; c 2. BIOCHEMISTRY. Educ: Univ Tex, BA, 26; NY Univ, MS, 28; Univ Minn, PhD(physiol chem), 33, MD, 37. Hon Degrees: DOdont, Univ Stockholm, 55. Prof Exp: Asst, Univ Tex, 25-26; asst, Nat Res Coun, 26-27; res chemist, Tex Co, 28-29; asst physiol chem, Med Sch, 29-32, from instr to prof, 32-66, dir dent res lab, 40-43, chmn dept physiol chem, 46-66, prof biochem & head dept, 66-74, REGENT'S PROF BIOCHEM, UNIV MINN, MINNEAPOLIS, 74-; DIR INTRAMURAL RES, NAT INST DENT RES, NIH, 74- Concurrent Pos: Commonwealth Fund fel, Denmark & Eng, 37-38, Sweden, 60; Rockefeller fel, Univ Stockholm, 50; mem dent study sect,

Nat Dent Res Inst, 47-52, mem bd sci adv, 57-61; mem nat adv dent coun, NIH, 52-56; mem dent res adv comt, Dept Army, Off Surgeon Gen, 52-; mem rev comt div biol & med, Argonne Nat Lab, 58-61; mem nat adv comt, AEC, 59-65. Honors & Awards: William John Gies Award, Am Col Dent, 66; Biol Mineralization Award, Int Asn Dent Res, 66, H Trendley Dean Award, 67. Mem: AAAS; Am Chem Soc; Am Soc Biol Chem; Soc Exp Biol & Med; Am Col Dent; Int Asn Dent Res (pres, 45-46). Res: Calcified tissues; fluoride analysis and metabolism; radioisotope techniques and studies. Mailing Add: Intramural Res Nat Inst of Dent Res NIH Bethesda MD 20014

ARMSTRONG, WILLIAM DAVID, b Clanton, Ala, Jan 15, 44. PHYSICAL CHEMISTRY. Educ: Univ Tenn, BSChem, 67; Univ Ga, MS, 70; Brown Univ, PhD(phys chem), 74. Prof Exp: Asst chem, Univ Ga, 67-70; res asst chem, Brown Univ, 70-74; RES SCIENTIST CATALYSIS, AM CYANAMID CO, 74- Mem: Am Chem Soc; Sigma Xi. Res: Development of improved hydrotreating catalysts and reforming catalysts. Mailing Add: Am Cyanamid Co 1937 W Main St Stamford CT 06904

ARMSTRONG, WILLIAM LAWRENCE, b Lorain, Ohio, Jan 14, 39. ORGANIC CHEMISTRY. Educ: Oberlin Col, BA, 60; Univ Rochester, PhD(chem), 66. Prof Exp: Asst prof chem, State Univ NY Col Geneseo, 64-65; asst prof, 65-68, ASSOC PROF CHEM, STATE UNIV NY COL ONEONTA, 68- Mem: Am Chem Soc; The Chem Soc; AAAS. Res: Synthetic organic chemistry. Mailing Add: Dept of Chem State Univ NY Col Oneonta Oneonta NY 13820

ARN, HEINRICH, chemistry, see 12th edition

ARNAL, ROBERT EMILE, b Orleans, France, Sept 10, 22; nat US; m 54; c 3. GEOLOGY. Educ: Univ Poitiers, France, BS, 46; Univ Nancy, France, MS, 48; Univ Southern Calif, PhD(geol), 57. Prof Exp: Geophysicist, Cie Gen de Geophysique, France, 48-51; paleontologist, Western Gulf Oil Co, 53-58; asst prof phys sci, San Jose State Col, 58-61, assoc prof geol, 61-69; PROF OCEANOG, MOSS LANDING MARINE LABS, 69- Mem: Fel Geol Soc Am; Am Asn Petrol Geol; Soc Econ Paleontologists & Mineralogists. Res: Shallow water oceanography; micropaleontology environmental studies of recent organisms and tertiary fossils. Mailing Add: Moss Landing Marine Labs PO Box 223 Moss Landing CA 95039

ARNASON, BARRY GILBERT WYATT, b Winnipeg, Man, Aug 30, 33; m 60; c 1. IMMUNOLOGY, NEUROLOGY. Educ: Univ Man, MD, 57. Prof Exp: Intern, Winnipeg Gen Hosp, Man, 56-57, asst resident internal med, 56-58; Nat Mult Sclerosis Soc prin investr, Lab of Dr Pierre Grabar, Inst Sci Res Cancer, Villejuif, France, 62-64; instr, 64-67, assoc, 67, asst prof, 68-71, ASSOC PROF NEUROL, HARVARD MED SCH, 71- Concurrent Pos: Teaching fel neurol, Harvard Med Sch, 58-59 & 61-62, res fel, 59-61; clin & res fel, Lab of Dr Byron H Waksman, 59-61, asst, 64-66, asst neurologist, 66-68, clin fel, Hosp, 61-62; asst resident neurol, Mass Gen Hosp, 58-59, assoc neurologist, 71- Honors & Awards: Dr Jon Stefansson Mem Prize, 55-56. Mem: Am Neurol Asn; Am Acad Neurol; Am Soc Clin Invest; Am Asn Neuropathologists; Am Asn Immunologists. Mailing Add: Dept of Immunol Mass Gen Hosp Boston MA 02114

ARNAUD, JACQUES, b Paris, France, Feb 24, 32; m 59; c 3. ELECTROMAGNETICS. Educ: Advan Sch Elec, Malakoff, dipl, 53; Univ Paris, Dr Ing, 63. Prof Exp: Asst elec eng, Advan Sch Elec, Malakoff, 53-55; engr, Compagnie Generale TSF, 55-65; engr, Warnecke Electron Tubes, Inc, 65-66; mem tech staff, 66-70, SUPVR, BELL TEL LABS, 70- Mem: Optical Soc Am; sr mem Inst Elec & Electronics Engrs. Res: Propagation of electromagnetic waves in multiperiodic structures; noise in crossed-field guns; traveling-wave tubes; heterodyne optical receivers; gaussian optical beams; optical resonators; millimeter wave antennae; fiber optics; transmission of optical pulses through glass fibers for communication purposes. Mailing Add: L123 Bell Tel Labs Crawford Hill Holmdel NJ 07733

ARNAUD, PAUL HENRI, JR, b San Francisco, Calif, Sept 15, 24; m 70. ENTOMOLOGY. Educ: San Jose State Col, AB, 49; Stanford Univ, MA, 50, PhD(biol), 61. Prof Exp: Syst entomologist, Bur Entom, Calif Dept Agr, 50-52, 55-56 & 57-59; prin investr med entom, Stanford Univ & Off Surgeon Gen, US Army, Japan, 54-55; entomologist, USDA at US Nat Mus, Washington, DC, 56-57; res entomologist, Calif Acad Sci, 59-61; res fel entom, Am Mus Natural Hist, NY, 61-62; res entomologist, 63-64, asst curator, 64-65, ASSOC CURATOR ENTOM, CALIF ACAD SCI, 65-, CHMN DEPT, 68- Concurrent Pos: Mem, Sefton Found Exped, Gulf Calif, 53 & entom expeds, Baja Calif & Mex, 53, 59 & 63; grants, NSF, 60, USDA, 66-68 & Am Philos Soc Penrose Fund, 72; assoc, Exp Sta, Univ Calif, Albany, 66- & Div Biol Control, Univ Calif, Berkeley, 66- Mem: Soc Syst Zool; Entom Soc Am; Am Entom Soc. Res: Classification; biology; systematics and distribution of Diptera; zoogeography; evolution. Mailing Add: Dept of Entom Calif Acad of Sci San Francisco CA 94118

ARNDT, RICHARD ALLEN, b Cleveland, Ohio, Jan 3, 33; m 53; c 4. PHYSICS. Educ: Case Inst Technol, BS, 57; Univ Calif, Berkeley, MA, 62, PhD(physics), 65. Prof Exp: Engr, Northrop Corp, Calif, 57-59; physicist, Lawrence Radiation Lab, Univ Calif, 59-67; asst prof, 67-70, ASSOC PROF PHYSICS, VA POLYTECH INST & STATE UNIV, 70- Res: Elementary particle physics; dispersion relations in nucleon-nucleon scattering; single boson exchange contribution to nucleon-nucleon scattering. Mailing Add: Dept of Physics Va Polytech Inst & State Univ Blacksburg VA 24061

ARNDT-JOVIN, DONNA JEANNE, biochemistry, animal virology, see 12th edition

ARNELL, J HAL, b Montpelier, Idaho, Oct 30, 41; m 68; c 1. SYSTEMATIC ENTOMOLOGY. Educ: Univ Utah, BS, 66, MS, 68, PhD(biol), 71. Prof Exp: Fel entom, 71-72, RES ZOOLOGIST, UNIV CALIF, LOS ANGELES, 72- Mem: Entom Soc Am; Am Mosquito Control Asn. Res: Systematics, biology and disease relations of mosquitoes. Mailing Add: Dept of Biol Univ of Calif Los Angeles CA 90024

ARNELL, JOHN CARSTAIRS, b Halifax, NS, Apr 4, 18; m 42; c 3. PHYSICAL CHEMISTRY. Educ: Dalhousie Univ, BSc, 39, MSc, 40; McGill Univ, PhD(phys chem), 42. Prof Exp: Supt, Defence Res Chem Labs, 49-54, dir sci intel, 55-58, dir plans, 58; sci adv, Royal Can Air Force, 58-63; sci adv, Royal Can Navy, 63-64; sci dep chief logistics, eng & develop, Can Forces Hq, Ottawa, 64, sci dep chief tech serv, 64-66; asst dep minister finance, Can Dept Nat Defence, Nat Defence Hq, 66-73; prog consult, Finance Ministry, 73-76, COORDR METRICATION, GOVT BERMUDA, 75- Mem: Fel Chem Inst Can; assoc fel Can Aeronaut & Space Inst; Sigma Xi. Res: Overall management of planning, programming budgeting and financial management systems for defence. Mailing Add: PO Box 1263 Hamilton Bermuda

ARNEMAN, HAROLD FREDERICK, b Mankato, Minn, May 31, 15; m; c 1. SOILS. Educ: Univ Minn, BS, 39, MS, 46, PhD, 50. Prof Exp: Soil surveyor, Univ Minn, 35-40; farm credit apprentice, Fed Land Bank, Minn, 40-42; from instr to assoc prof, 46-61, PROF SOILS, UNIV MINN, ST PAUL, 61- Res: Mineralogical and physical studies of soils. Mailing Add: Dept of Soil Sci Univ of Minn 130a S St Paul MN 55108

ARNER, DALE H, b Weissport, Pa, Feb 10, 20; m 45; c 1. WILDLIFE MANAGEMENT. Educ: Pa State Univ, BS, 49, MS, 54; Auburn Univ, PhD(wildlife mgt), 59. Prof Exp: Hatchery asst, Pa Fish Comn, 41; wildlife field supt, Md Game & Inland Fish Comn, 48-54; biologist, Soil Conserv Serv, USDA, Ala, 57-62; from asst prof to assoc prof wildlife mgt & ecol, 62-72, PROF WILDLIFE & FISHERIES & ZOOL, MISS STATE UNIV, 72-, HEAD DEPT WILDLIFE & FISHERIES, 69- Concurrent Pos: Consult, Weyerhaeuser Co, 63-64. Honors & Awards: USDA Cert of Merit, 62; Wildlife Conservationist of Year Award, Miss Wildlife Fedn & Sears-Roebuck Found, 70. Mem: Wildlife Soc; Ecol Soc Am; Wilson Ornith Soc. Res: Use of mechanical equipment, fire fertilizer and seed in utility line right-of-way development for wildlife and reduction of woody vegetation; beaver-pond and duck ecology; wild turkey nutrition. Mailing Add: Dept Wildlife & Fish Sch For Resrs Miss State Univ Drawer LW Mississippi State MS 39762

ARNESON, AXEL NORMAN, b Ft Worth, Tex, July 5, 05; m 31; c 1. OBSTETRICS & GYNECOLOGY. Educ: Tex Christian Univ, BS, 25; Wash Univ, MD, 28. Prof Exp: PROF CLIN OBSTET & GYNEC & ASSOC PROF CLIN RADIOL, WASH UNIV, 34- Concurrent Pos: Assoc attend gynecologist, Barnes Hosp & St Louis Maternity Hosp. Mem: Fel Am Gynec Soc; fel Am Asn Obstet & Gynec; Am Radium Soc (pres, 47); Am Col Obstet & Gynec (pres, 65-). Mailing Add: 4511 Forest Park Blvd St Louis MO 63108

ARNESON, DORA WILLIAMS, b Fayetteville, Ark, Aug 4, 47; m 73. ANALYTICAL CHEMISTRY, BIOCHEMISTRY. Educ: Univ Mo, BA, 67, PhD(anal chem), 72. Prof Exp: INSTR ANAL BIOCHEM, CTR HEALTH SCI, UNIV TENN, MEMPHIS, 72- Concurrent Pos: Memphis Heart Asn grant, Univ Tenn, Memphis, 74-75. Mem: Am Chem Soc. Res: Development of analytical methods suitable for clinical chemistry laboratories, with emphasis on competitive protein binding assays; diagnosis of coagulation disorders. Mailing Add: Dept of Biochem Univ of Tenn Ctr Health Sci Memphis TN 38163

ARNESON, PHIL ALAN, b Fergus Falls, Minn, May 17, 40; m 63; c 2. PLANT PATHOLOGY. Educ: Carleton Col, BA, 62; Univ Wis, PhD(plant path), 67. Prof Exp: Plant pathologist, Trop Res Div, United Fruit Co, 67-70; ASST PROF PLANT PATH, CORNELL UNIV, 70- Mem: Am Phytopath Soc. Res: Orchard replant problems; soil fumigation; nematicides. Mailing Add: Dept of Plant Path Cornell Univ Ithaca NY 14853

ARNESON, RICHARD MICHAEL, b Fergus Falls, Minn, Oct 18, 38; m 73. BIOCHEMISTRY. Educ: Univ Minn, BA, 60; Duke Univ, PhD(biochem), 71. Prof Exp: Instr, 71-72, ASST PROF BIOCHEM, CTR HEALTH SCI, UNIV TENN, MEMPHIS, 72- Res: Neurochemistry; free radical biochemistry. Mailing Add: Dept of Biochem Univ Tenn Ctr for Health Sci Memphis TN 38163

ARNETT, EDWARD MCCOLLIN, b Philadelphia, Pa, Sept 25, 22; m 51; c 2. PHYSICAL ORGANIC CHEMISTRY. Educ: Univ Pa, BA, 43, MS, 47, PhD(chem), 49. Prof Exp: Asst instr qual, inorg, anal & org chem, Univ Pa, 43-46, 47; res dir, Max Levy & Co, 49-53; asst prof, Western Md Col, 53-55; res fel, Harvard Univ, 55-57; from asst prof to assoc prof, 57-64, PROF GEN & ORG CHEM, UNIV PITTSBURGH, 64- Concurrent Pos: Vis lectr, Univ Ill, 63; adj sr fel, Mellon Inst, 64-; distinguished lectr, Howard Univ, 66 & Guggenheim fel, 68-69; dir, Pittsburgh Chem Info Ctr, 68-; mem adv bd, Petrol Res Fund, 68-; comt chem info, Nat Res Coun, 69-; vis prof, Univ Kent, 70. Mem: AAAS; Am Chem Soc; The Chem Soc. Res: Acid base behavior of organic compounds; solvent effects in organic chemistry. Mailing Add: Dept of Chem Univ of Pittsburgh Pittsburgh PA 15213

ARNETT, JAMES DELOS, JR, b Augusta, Ga, June 13, 42; m 65; c 3. PLANT PATHOLOGY. Educ: Univ SC, BS, 64, MS, 66; Clemson Univ, PhD(plant path), 71. Prof Exp: Instr biol, Clemson Univ, Sumter, 66-67; tech rep plant path, Buckman Labs, Inc, 67-68; exten specialist, Clemson Univ Coop Exten, 71-75; EXTEN SPECIALIST PLANT PATH, UNIV GA COOP EXTEN SERV, 75- Mem: Am Phytopath Soc; Am Peanut Res & Educ Asn. Res: Control of plant diseases of crop plants. Mailing Add: 410 Tennessee Dr Tifton GA 31794

ARNETT, RAYMOND LEE, chemical physics, see 12th edition

ARNETT, ROSS HAROLD, JR, b Medina, NY, Apr 13, 19; m 42; c 8. ENTOMOLOGY. Educ: Cornell Univ, BS, 42, MS, 46, PhD(entom), 48. Prof Exp: Asst entom, Cornell Univ, 46-48; entomologist, Div Insects, USDA, 48-54; head dept biol, St John Fisher Col, 54-58; from assoc prof to prof biol, Cath Univ, 58-66, head dept, 63-66; prof entom, Purdue Univ, 66-70; res biologist, Tall Timbers Res Sta, 71-73; PROF BIOL, SIENA COL, NY, 73- Concurrent Pos: Asst, State Conserv Dept, NY, 42; instr, US Army Sch Malariol, CZ, 44-45; founder & ed, Coleopterists Bull; exec dir, The Biol Res Inst Am, Inc, 73- Mem: AAAS; Am Soc Zool; Soc Syst Zool; Soc Study Evolution; Entom Soc Am. Res: Taxonomy of Coleoptera; biology of mosquitoes; Onychophora; revision of world Oedemeridae; experimental systematic biology; desert biology. Mailing Add: Dept of Biol Siena Col Loudonville NY 12211

ARNETT, WILLIAM HAROLD, b Louisville, Miss, June 13, 28; m 57; c 3. ECONOMIC ENTOMOLOGY. Educ: Miss State Univ, BS, 55, MS, 57; Kans State Univ, PhD(entom), 60. Prof Exp: Instr entom, Kans State Univ, 57-60; asst prof & asst entomologist, 60-65, assoc prof & assoc entomologist, 65-74, PROF & ENTOMOL, UNIV NEV, RENO, 74- Mem: AAAS; Entom Soc Am; Sigma Xi. Res: Biology, ecology and control of agricultural insect pests; forage and animal insects. Mailing Add: Col of Agr Univ of Nev Reno NV 89507

ARNEY, WILLIAM CHARLES, JR, organic chemistry, see 12th edition

ARNFIELD, ANTHONY JOHN, b Nuneaton, Eng, Jan 2, 45; m 69. PHYSICAL GEOGRAPHY, CLIMATOLOGY. Educ: Univ Wales, BA, 66; McMaster Univ, MA, 68, PhD(geog), 73. Prof Exp: ASST PROF GEOG, OHIO STATE UNIV, 72- Mem: Royal Meteorol Soc; Am Meteorol Soc; Asn Am Geographers. Res: Microclimatology; radiation climatology; investigation of the fluxes of energy, mass and momentum across the earth-atmosphere interface, especially as determined by surface properties; urban surface energy budgets. Mailing Add: Dept of Geog Ohio State Univ 1775 S College Rd Columbus OH 43210

ARNHEIM, NORMAN, b New York, NY, Dec 24, 38; m 68; c 2. MOLECULAR BIOLOGY, EVOLUTION. Educ: Univ Rochester, BA, 60, MS, 62; Univ Calif, Berkeley, PhD(genetics), 66. Prof Exp: Res biochemist, Dept Biochem, Univ Calif, Berkeley, 67-68; asst prof biochem, 68-74, ASSOC PROF BIOCHEM, STATE UNIV NY, STONY BROOK, 74- Mem: Genetics Soc Am; Am Soc Biol Chemists; Am Soc Naturalists. Res: Macromolecular evolution; the evolution of regulatory systems; multi-gene evolution. Mailing Add: Dept of Biochem State Univ of NY Stony Brook NY 11790

ARNHOLT, PHILIP JOHN, b Danville, Ill, May 16, 40; m 64; c 1. BOTANY. Educ: Eastern Ill Univ, BSEd, 63, MSEd, 67; Univ Nebr, PhD(bot), 73. Prof Exp: Teacher biol, Dixon High Sch, Ill, 63-67 & LaSalle-Peru Twp High Sch, Ill, 67-68; ASST PROF BIOL, CONCORDIA COL, WIS, 71- Mem: Bot Soc Am; Nat Asn Biol Teachers; Nat Audubon Soc. Res: Morphology and histochemistry of grasses. Mailing Add: Concordia Col 3201 W Highland Blvd Milwaukee WI 53208

ARNISON, PAUL GRENVILLE, b Clatterbridge, Eng, Aug 9, 49; Can citizen. PLANT PHYSIOLOGY. Educ: McGill Univ, BSc, 70, PhD(plant physiol), 75. Prof Exp: ASST PROF BIOL, NORTHEASTERN UNIV, 75- Mem: Am Bot Soc; Can Soc Plant Physiologists. Res: Isoenzymatic characterization of plant cell cultures in relation to growth and development; peroxidase asctivity and the growth of cell walls; flowering physiology in tissue cultures and Lemna. Mailing Add: Northeastern Univ Dept of Biol 360 Huntington Ave Boston MA 02115

ARNO, STEPHEN FRANCIS, b Seattle, Wash, Oct 1, 43. FOREST ECOLOGY. Educ: Wash State Univ, BS, 65; Univ Mont, MF, 66, PhD(forestry, plant sci), 70. Prof Exp: Park ranger & naturalist, US Nat Park Serv, 63-65; instr forestry, Univ Mont, 67; forester, 70-71, RES FORESTER, US FOREST SERV, 71- Mem: Ecol Soc Am; Nat Parks Asn. Res: Phytosociology; phytogeography; documentation of the historical role of wildfire in forests; dendrology. Mailing Add: Forestry Sci Lab US Forest Serv Missoula MT 59801

ARNOFF, E LEONARD, b Cleveland, Ohio, Oct 15, 22; m 48; c 2. MATHEMATICS. Educ: Western Reserve Univ, BS, 43; Case Inst Technol, MS, 48; Calif Inst Technol, PhD(math), 51. Prof Exp: Instr math, Case Inst Technol, 46-48; hydrodynamicist & mathematician, Naval Ord Test Sta, 50-51; mathematician & aeronaut res scientist, Nat Adv Comt Aeronaut, 51-52; assoc prof opers res & asst dir opers res group, Case Inst Technol, 52-61; PRIN & DIR MGT SCI, ERNST & ERNST, 60- Concurrent Pos: Ed, Mgt Sci, Inst Mgt Sci, 55-70. Mem: Fel AAAS; fel Opers Res Soc Am; Math Asn Am; Am Math Soc; Inst Mgt Sci(assoc secy, 62, secy-treas, 63, vchmn, 64, int pres, 68-69). Res: Mathematical programming; production and inventory control; development and application of operations research methods, techniques and tools to new problem areas. Mailing Add: Ernst & Ernst 1300 Union Com Bldg Cleveland OH 44115

ARNOLD, ALLEN PARKER, b Oshkosh, Wis, June 25, 23; m 48; c 2. ORGANIC CHEMISTRY. Educ: Oberlin Col, AB, 44; Case Inst Technol, MS, 57, PhD(chem), 58. Prof Exp: Jr anal chemist, Shell Chem Corp, Calif, 44-45 & 46-47; res assoc org chem, Case Inst Technol, 47-48; res chemist, Cleveland Indust Res, Inc, Euclid, 48-62, asst lab dir, 62-72; GROUP LEADER RES & DEVELOP, LUBRIZOL CORP, 72- Mem: Am Chem Soc. Res: Organic synthesis, particulary phosphorus chemistry. Mailing Add: 1975 Temblethurst Rd Cleveland OH 44121

ARNOLD, ARTHUR, b New York, NY, Sept 21, 21; m 43; c 2. NEUROSURGERY. Educ: Univ Ala, AB, 43; Harvard Med Sch, MD, 46. Prof Exp: From asst prof to assoc prof neurol surg, Univ Ill Col Med, 54-73; RES PROF NEUROL SURG, SCH MED, UNIV CHICAGO, 73- Concurrent Pos: USPHS res fel, 49-52; consult, Ill Psychiat Inst. Mem: Am Asn Neurol Surgeons; Cong Neurol Surgeons; Radiation Res Soc; NY Acad Sci; Int Cardiovasc Soc. Res: Effects of conventional and high energy radiations on central nervous system of man and monkey; radiations tumors of the brain; amyotrophic lateral sclerosis. Mailing Add: Chicago IL

ARNOLD, BRADFORD HENRY, b Chehalis, Wash, Oct 14, 16; m 41. PURE MATHEMATICS. Educ: Univ Wash, BS, 38, MS, 40; Princeton Univ, PhD(math), 42. Prof Exp: Instr math, Purdue Univ, 42-44, 46; major aerodynamicist, Boeing Airplane Co, 44-46; asst prof, Mont State Col, 46-47; from asst prof to assoc prof, 47-59, PROF MATH, ORE STATE UNIV, 59-, ASST CHMN MATH DEPT, 72- Concurrent Pos: Fulbright grants, Iraq, 57-58, China, 62-63 & 69-70. Mem: Am Math Soc; Math Asn Am. Res: Topology and abstract algebra. Mailing Add: Dept of Math Ore State Univ Corvallis OR 97331

ARNOLD, BRIGHAM ALICEN, b Boyne City, Mich, Nov 3, 17; m 49; c 1. GEOGRAPHY. Educ: Univ Ariz, BS, 42; Univ Calif, PhD(geog), 54. Prof Exp: Asst, Univ Calif, 47-50; instr, Univ Nev, 51; meteorologist & hydrologist, Eng Corps, US Army, 51-53; from instr to PROF GEOG, CALIF STATE UNIV, SACRAMENTO, 54- Res: Alluvial land forms; landscapes of arid regions; climatic change; early man in the New World. Mailing Add: Dept of Geog Calif State Univ Sacramento CA 95819

ARNOLD, CHARLES, JR, b Brooklyn, NY, May 29, 30; m 59; c 6. POLYMER CHEMISTRY. Educ: Yale Univ, BS, 52; Purdue Univ, MS, 54, PhD(org chem), 57. Prof Exp: From res chemist to sr res chemist, E I du Pont de Nemours & Co, 57-67; MEM STAFF POLYMER CHEM, SANDIA LABS, 67- Mem: Am Chem Soc; Soc Advan Mat & Process Engrs. Res: Thermal and radiative degradation of polymers; development of novel polyurethanes for encapsulation of electronic components. Mailing Add: Polymer Sci & Eng Div Sandia Labs Albuquerque NM 87115

ARNOLD, CHARLES YESBRA, b Anderson, Ind, Jan 1, 17; m 36; c 2. HORTICULTURE. Educ: Ohio State Univ, BSc, 39, MSc, 45; Univ Ill, PhD, 51. Prof Exp: Assoc horticulturist, Idaho, 40-42; supt, Cook County Exp Sta, 42-47, assoc prof, 47-60, PROF HORT, UNIV ILL, URBANA-CHAMPAIGN, 60- Mem: Am Soc Hort Sci. Res: Plant-soil relationships; climatic relationships with vegetable crop plants. Mailing Add: 401 George Juff Dr Urbana IL 61801

ARNOLD, CHESTER ARTHUR, b Leeton, Mo, June 25, 01; m 33; c 3. PALEOBOTANY. Educ: Cornell Univ, BS, 24, PhD(paleobot), 29. Prof Exp: Instr bot, Univ Mich, 28-35, from asst prof to prof bot, 35-71, cur paleobot, 29-71; RETIRED. Concurrent Pos: Prin investr, Arctic Res Lab, 51 & 53-; vis scientist, Birbal Sahni Inst Paleobot, Lucknow, India, 58-59. Honors & Awards: Silver Medal, Birbal Sahni Inst Paleobot, 72; Dist Achievement Award, Bot Soc Am, 74. Mem: AAAS; Bot Soc Am; Am Paleont Soc; Geol Soc Am; Ecuadorian Inst Natural Sci. Res: Flora of Devonian and Carboniferous periods; silicified plants of Tertiary age from western North America; fossil flora of northern Alaska; Carboniferous floras of Michigan and Alabama. Mailing Add: Apt 801A 2222 Fuller Rd Ann Arbor MI 48105

ARNOLD, DAVID BROWN, b Hackensack, NJ, Oct 24, 40; m 63, 71; c 1. PHYSICAL ORGANIC CHEMISTRY. Educ: Univ Pa, BS, 63; Bryn Mawr Col, PhD(org & phys chem), 70. Prof Exp: ASST PROF CHEM, WIDENER COL, 68- Mem: Am Chem Soc; The Chem Soc; AAAS. Res: Reactivity of aromatic compounds; polynuclear aromatic compounds; chemiluminescence; fungal metabolites. Mailing Add: Dept of Chem Widener Col Chester PA 19013

ARNOLD, DEAN EDWARD, b Elmira, NY, Apr 8, 39; m 64; c 2. AQUATIC ECOLOGY. Educ: Univ Rochester, AB, 61; Cornell Univ, MS, 65, DPhil(aquatic ecol), 69. Prof Exp: Res asst fisheries, Cornell Univ, 63-65, res assoc & asst proj leader warmwater fisheries invests, 65-66, teaching asst limnol, 66-68; asst res limnologist, Great Lakes Res Div, Univ Mich, 69-72; ASST PROF BIOL, PA STATE UNIV, 72-; ASST LEADER, PA COOP FISHERY RES UNIT, US FISH & WILDLIFE SERV, 72- Honors & Awards: W F Thompson Award, Am Inst Fishery Res Biologists, 74. Mem: Int Asn Great Lakes Res; Am Fisheries Soc; Int Asn Theoret & Appl Limnol; Am Soc Limnol & Oceanog; Am Inst Fishery Res Biologists. Res: Aquatic productivity; effects of manipulations of lakes and streams; lake restoration; pollution. Mailing Add: Dept of Biol 328 Life Sci Bldg University Park PA 16802

ARNOLD, DONALD ROBERT, b Buffalo, NY, Mar 1, 35; m 55; c 6. ORGANIC CHEMISTRY, PHOTOCHEMISTRY. Educ: Bethany Col, WVa, BS, 57; Univ Rochester, PhD(org chem), 61. Prof Exp: Res chemist, Union Carbide Res Inst, New York, 61-70; assoc prof, 70-71, PROF CHEM, UNIV WESTERN ONT, 71- Concurrent Pos: Alfred P Sloan Found fel, 72-74; consult, Energy Conversion Devices, Troy, Mich, 73-; Chem Inst Can fel, 74. Mem: Am Chem Soc; Chem Inst Can. Res: Synthesis of nonbenzinoid aromatic heterocycles and morphine analogs; heterocyclic compounds; small ring compounds; free radicals; synthetic applications and mechanisms of photochemistry. Mailing Add: Dept of Chem Univ Western Ont London ON Can

ARNOLD, EMIL, b Luck, Poland, May 20, 32; US citizen; m 65; c 2. SOLID STATE PHYSICS. Educ: Dalhousie Univ, BS, 54; Yale Univ, MS, 55, PhD(phys chem), 63. Prof Exp: Sr physicist, Gen Instrument Corp, 55-57; sr engr, Gen Tel & Electronics Labs, 57-60; sr physicist, 63-69, SR PROG LEADER, PHILIPS LABS INC DIV, NAM PHILIPS, 69- Concurrent Pos: Vis Scientist, Mullard Res Labs, Eng, 74-75. Mem: Am Phys Soc. Res: Solid state electronics; transport properties of semiconductors; electrical behavior of semiconductor surfaces; electrochemistry; optoelectronics. Mailing Add: Philips Labs Inc Briarcliff Manor NY 10510

ARNOLD, FREDERIC G, b Irvington, NJ, Jan 14, 23; m 47; c 2. BIOLOGY, HUMAN PHYSIOLOGY. Educ: Montclair State Col, BA, 46, MA, 47; Columbia Univ, EdD(preparation sci teachers), 56. Prof Exp: Teacher high schs, NJ, 47-55; prof biol, 56-73, PROF EDUC ARTS & SYSTS, KEAN COL NJ, 73-, CHMN DEPT BIOL, 69- Mem: AAAS; Nat Sci Teachers Asn. Res: Body fluids, blood and circulation; desert biology, especially the ecology and plant life of the desert with emphasis on the cacti and succulent plants. Mailing Add: Dept of Educ Arts & Systs Kean Col of NJ Union NJ 07083

ARNOLD, GAIL, cytology, see 12th edition

ARNOLD, GEORGE BENJAMIN, b North Platte, Nebr, June 25, 14; m 39; c 1. CHEMISTRY. Educ: Nebr State Teachers Col, BS, 35; Univ Nebr, MS, 38, PhD(org chem), 40. Prof Exp: Res & develop chemist, Tex Co, NY, 40-42, group leader, 43-44, proj leader, 45-50, asst supvr lubricants res, 51-53, supvr chem res, 54-58, asst to mgr, 58-60, commercial develop dir, 60-68, asst to vpres, 68-72, REGIONAL MGR PETROCHEM, TEXACO INC, TEX, 73- Mem: Am Chem Soc; Am Inst Chem Eng; Chem Mkt Res Asn; Commercial Develop Asn. Res: Petroleum chemistry; synthesis of hydrocarbons from CO; solvent extraction; azeotropic distillation; fuels; petrochemicals. Mailing Add: Texaco Inc 1111 Rusk Houston TX 77002

ARNOLD, GEORGE W, b Dayton, Ohio, Nov 24, 23; m 45; c 3. SOLID STATE PHYSICS. Educ: Univ Tex, BS, 48, MA, 49. Prof Exp: Physicist, Optical Res Lab, Univ Tex, 48-49, Nat Bur Standards, 49-50 & US Naval Res Lab, 50-61; PHYSICIST, SANDIA CORP, 61- Concurrent Pos: Fulbright prof, UAR, 60-61. Mem: Am Phys Soc. Res: Ion implantation and radiation effects studies in inorganic solids. Mailing Add: Sandia Labs Div 5112 Albuquerque NM 87115

ARNOLD, GORDON WILLIAM, b Gordon, Nebr, Nov 5, 17; m 46; c 1. PHYTOPATHOLOGY. Educ: Univ Wyo, BS, 52, MS, 54. Prof Exp: Supply instr agron, 53-55, res agronomist, Afghanistan contract, Kabul, 55, instr agron & coord, Afghanistan Prog & asst prof agron, 58-64, DIR INT PROGS, UNIV WYO, 64-, COORDR ADMIN SERV, UNIV EXTEN, 74- Concurrent Pos: Mem feasibility surv agr proj, AID Somalia, 64; mem review team, Univ Wyo-AID contracts, Somalia, 65, 66 & 69, Afghanistan, 60, 62, 65, 66, 68, 71, 72, 73. Res: Plant pathology; bacterial and virus diseases. Mailing Add: Rm 301 Hoyt Hall Univ of Wyo Box 3707 Univ Sta Laramie WY 82071

ARNOLD, HAROLD WILFRED, b Sterling, Nebr, Apr 18, 11; m 35; c 2. CHEMISTRY. Educ: Wittenberg Col, AB, 31; Ohio State Univ, PhD(org chem), 36. Prof Exp: Chemist, exp sta, E I du Pont de Nemours & Co, Del, 36-51, res supvr, textile fibers dept, Tenn, 51-53, tech supt, 53-54, res supvr, textile res lab, Wilmington, 54-75; RETIRED. Mem: Am Chem Soc. Res: Carbohydrates; synthetic resins; vinyl polymer chemistry; development of commercial applications for new organic compounds; textile fibers. Mailing Add: 401 Painters Crossing Chadds Ford PA 19317

ARNOLD, HARVEY JAMES, b Niagara Falls, Ont, Can, Mar 5, 33; US citizen; m 56; c 4. MATHEMATICAL STATISTICS. Educ: Queen's Univ, Ont, BA, 51, MA, 52; Princeton Univ, MA, 55, PhD(math), 58. Prof Exp: Asst, Princeton Univ, 57-58; asst prof, Univ Western Ont, 58-60 & Wesleyan Univ, 60-63; assoc prof, Bucknell Univ, 63-67; actg chmn dept, 70-71; assoc prof, 67-70, PROF MATH, OAKLAND UNIV, 70- Concurrent Pos: Vis res assoc, Princeton Univ, 62-63. Mem: Math Asn Am; Inst Math Statist; Am Statist Asn. Res: Nonparametric statistics; design of experiments and statistical applications to the physical and biological sciences. Mailing Add: Dept of Math Oakland Univ Rochester MI 48063

ARNOLD, HUBERT ANDREW, b Chicago, Ill, Nov 15, 12. MATHEMATICS. Educ: Univ Nebr, AB, 33; Calif Inst Technol, PhD(math), 39. Prof Exp: Asst instr math, Calif Inst Technol, 35-39; instr, Univ Minn, 39-40; vis res scholar, Univ Va, 40-41; instr, Princeton Univ, 41-42; from instr to asst prof, 48-56, ASSOC PROF MATH, UNIV CALIF, DAVIS, 56- Mem: AAAS; Math Asn Am; Am Math Soc. Res: Differentials in abstract spaces; topological structure of limit sets; numerical analysis and table making. Mailing Add: Dept of Math Univ of Calif Davis CA 95616

ARNOLD, JAMES RICHARD, b New Brunswick, NJ, May 5, 23; m 52; c 3. CHEMISTRY. Educ: Princeton Univ, AB, 43, MA, 45, PhD(chem), 46. Prof Exp: Asst, Princeton Univ, 43 & Manhattan porj, 43-46; fel, Inst Nuclear Studies, Univ Chicago, 46; Nat Res fel, Harvard Univ, 47; asst prof, Inst Nuclear Studies, Univ Chicago, 48-55; from asst prof to assoc prof, Princeton Univ, 55-58; assoc prof, 58-60, chmn dept, 60-63, PROF CHEM, UNIV CALIF, SAN DIEGO, 60- Honors & Awards: Ernest Orlando Lawrence Award, AEC, 68. Mem: Nat Acad Sci; Am Acad Arts & Sci; AAAS; Am Chem Soc. Res: Cosmic-ray produced nuclides; history of moon and meteorites; cosmochemistry. Mailing Add: Dept of Chem B-017 Univ of Calif San Diego La Jolla CA 92093

ARNOLD, JAMES TRACY, b Taiyuanfu, China, Oct 23, 20; US citizen; m 57; c 4. EXPERIMENTAL PHYSICS. Educ: Oberlin Col, AB, 42; Stanford Univ, PhD(physics), 54. Prof Exp: Lab instr physics, Oberlin Col, 47-48; spec asst, Europ Ctr Nuclear Res, 54-55; res assoc physics, Stanford Univ, 55-57; asst prof, Ore State Col, 57-58; physicist, 58-72, SR SCIENTIST, VARIAN ASSOCS, 72- Res: High resolution nuclear magnetic resonance spectroscopy, magnetometers, mass spectroscopy, ultrasonic imaging and computer application to instruments. Mailing Add: Varian Assocs 611 Hansen Way Palo Alto CA 94303

ARNOLD, JEFFREY, b Pittsburgh, Pa, Oct 19, 37; m 63; c 1. BIOCHEMISTRY, DATA PROCESSING. Educ: Yale Univ, BA, 58; Columbia Univ, PhD(biochem), 66; NY Univ, Cert comput prog & systs anal, 74. Prof Exp: Instr natural sci, Rockland Community Col, 66-68; res assoc biol, NY Univ, 68-69; lectr chem, Lehman Col, 69-73; INSTR CHEM, PINGRY SCH, 75- Concurrent Pos: Lectr phys sci, Pace Col, 69, instr molecular biol, 69-70; instr org chem, Fairleigh Dickinson Univ, 69, adj asst prof, 73-74; adj lectr & course coordr, Queens Col, NY, 73. Mem: AAAS; Am Chem Soc; fel Am Inst Chemists; NY Acad Sci. Res: Bio-organic chemistry of nucleotides and nucleotide analogs; biochemistry of electron transfer proteins from plants; synthesis and properties of metallorganic compounds. Mailing Add: 21 Marion St Nyack NY 10960

ARNOLD, JESSE CHARLES, b Bowie, Tex, Sept 28, 37; m 59; c 2. APPLIED STATISTICS, MATHEMATICAL STATISTICS. Educ: Southeastern State Col, BS, 60; Fla State Univ, MS, 65, PhD(statist), 67. Prof Exp: Asst chief exp design, Ctr Dis Control, USPHS, 61-63; asst prof educ res, Fla State Univ, 67-68; assoc prof, 68-74, PROF STATIST & HEAD DEPT, VA POLYTECH INST & STATE UNIV, 74- Concurrent Pos: Va State Water Control Bd grant water qual monitoring, Va Polytech Inst & State Univ, 69-71; consult, USPHS, 71, NASA-Langley Res Ctr, 72 & USAID, 72-; coordr Philippine Nat Nutrit Proj, Va Polytech Inst & State Univ & USAID, 74- Mem: Am Statist Asn; Biomet Soc (chmn regional adv bd, 75-). Res: Theory of estimation; application of statistics to environmental problems; sampling biometrics; international nutrition in the Philippines and Haiti. Mailing Add: Dept of Statist Va Polytech Inst & State Univ Blacksburg VA 24061

ARNOLD, JIMMY THOMAS, b Crowville, La, May 31, 41; m 67; c 1. MATHEMATICS. Educ: Northeastern La State Col, BS, 63; Fla State Univ, MS, 65, PhD(algebra), 67. Prof Exp: Asst prof, 69-74, ASSOC PROF MATH, VA POLYTECH INST & STATE UNIV, 74- Mem: Am Math Soc; Math Asn Am. Res: Dimension theory in power series rings and in polynomial rings; ideal theory in Kronecker function rings; extensions of valuations to power series rings; generalized transforms. Mailing Add: Dept of Math Va Polytech Inst & State Univ Blacksburg VA 24061

ARNOLD, JOHN D, b Bradford, Ohio, May 4, 22. CHEMOTHERAPY, CLINICAL PHARMACOLOGY. Educ: Univ Chicago, MD, 46. Prof Exp: Res asst med, Univ Chicago, 47-53, from asst prof to assoc prof, 53-63; PROF MED, SCH MED, UNIV MO-KANSAS CITY, 63-; DIR QUINCY RES CTR, 75- Mem: Am Soc Trop Med & Hyg; AMA; Am Soc Pharmacol & Exp Therapeut; Am Fedn Clin Res; NY Acad Sci. Res: Chemotherapy of malaria; cell growth and metabolism. Mailing Add: Quincy Res Ctr 5104 E 24th St Kansas City MO 64127

ARNOLD, JOHN MILLER, b St Paul, Minn, Oct 6, 36; m 58. ZOOLOGY, EMBRYOLOGY. Educ: Univ Minn, BA, 58, PhD(zool), 63. Prof Exp: Am Mus Natural Hist res fel, Lerner Marine Lab, 63-64; instr zool, Oberlin Col, 60-61; asst prof, Iowa State Univ, 64-66; from asst prof to assoc prof cytol, 66-75, PROF CYTOL, PAC BIOMED RES CTR, UNIV HAWAII, MANOA, 75- Concurrent Pos: Mem corp & summer staff mem, Marine Biol Lab, Woods Hole. Mem: AAAS; Am Soc Zool; Soc Develop Biol; Am Soc Cell Biol; Int Soc Develop Biologists. Res: Embryological development; biology of cephalopods. Mailing Add: Pac Biomed Res Ctr Univ of Hawaii at Manoa Honolulu HI 96822

ARNOLD, JOHN P, b Elmore, Minn, Feb 5, 11; m 44; c 2. VETERINARY MEDICINE. Educ: Iowa State Univ, DVM, 41, MS, 48; Univ Minn, St Paul, PhD(vet med), 56. Prof Exp: Pvt pract, Minn, 41-46 & 48-50; instr & asst prof vet surg, Iowa State Univ, 46-48; instr, 50-56, assoc prof, 56-57, prof vet surg & radiol & head dept, 57-71, actg head vet hosp, 71-72, PROF VET CLIN SCI, UNIV MINN, ST PAUL, 71- Concurrent Pos: Univ Minn adv, Col Vet Med, Seoul Nat Univ, 59-60; AID consult, Nat Univ Columbia & Univ Caldas, Columbia, 63. Mem: Am Vet Med Asn; Am Asn Equine Practitioners; Conf Res Workers Animal Dis; Educ Vet Radiol Sci; Am Asn Vet Clinicians. Res: Radiology of large animals; bovine mammary gland. Mailing Add: Col of Vet Med Univ of Minn St Paul MN 55101

ARNOLD, JOHN RONALD, b Hanford, Calif, June 29, 10; m 33; c 1. ZOOLOGY. Educ: Fresno State Col, BA, 32; Univ Calif, MA, 34; Cornell Univ, PhD(zool), 38. Prof Exp: Teacher, high sch & jr col, Calif, 33-36; teacher gen zool, Stockton Col, 38-41 & 46-61, chmn sci div, 46-58, dean, 58-61; PROF BIOL, CALIF STATE UNIV, SONOMA, 61- Concurrent Pos: Res assoc, Calif Acad Sci. Res: Ornithology; mammalogy; Mexican rodents. Mailing Add: Dept of Biol Calif State Univ Sonoma Rohnert Park CA 94928

ARNOLD, JOHN WALTER, b St Thomas, Ont, Jan 10, 21; m 57. ENTOMOLOGY. Educ: Univ Western Ont, BA, 44, MSc, 47, PhD(entomol), 51. Prof Exp: ENTOMOLOGIST, AGR CAN, 47- Concurrent Pos: Lectr, Carleton Col & Univ Ottawa. Mem: AAAS; Entom Soc Am; Entom Soc Can; Can Soc Zool; Agr Inst Can. Res: Insect haemocytes and circulation; insect physiology. Mailing Add: Biosyst Res Inst Agr Can Ottawa ON Can

ARNOLD, JOSEPH FREDERICK, b Colorado Springs, Colo, Oct 9, 11; m 46; c 2. ENVIRONMENTAL MANAGEMENT, PLANT ECOLOGY. Educ: Colo Col, BA, 34; Univ Ariz, MS, 36; Harvard Univ, MPA, 51. Prof Exp: Instr bot & assoc ecol, 35-41; asst range exam, Soil Conserv Serv, USDA, Nebr, 41-42, work unit conservationist, SDak, 45-46; range conservationist, US Forest Serv, 46-56; dir watershed mgr div, Ariz State Land Dept, 56-64; proj leader, Salt-Verde Watershed Proj, 64-70, FOREST SERV REP, RIVER BASIN STAFF, USDA, 70- Mem: AAAS; Ecol Soc Am; Am Soc Pub Admin; Soil Conserv Soc Am; Soc Range Mgt. Res: Range ecology; grazing and livestock management; watershed management. Mailing Add: 1220 W Highland Phoenix AZ 85013

ARNOLD, KEITH ALAN, b Jackson, Mich, Sept 23, 37; m 68; c 2. ORNITHOLOGY. Educ: Kalamazoo Col, AB, 59; Univ Mich, MS, 61; La State Univ, PhD(zool), 66. Prof Exp: Instr zool, La State Univ, 65-66; asst prof, 66-72, ASSOC PROF WILDLIFE SCI, TEX A&M UNIV, 72- Mem: Am Ornith Union; Soc Syst Zool; Cooper Ornith Soc; Wilson Ornith Soc. Res: Taxonomic, ecological and behavioral ornithology; avian ectoparasites; tropical vertebrate ecology. Mailing Add: Dept of Wildlife & Fisheries Sci Tex A&M Univ College Station TX 77843

ARNOLD, KENNETH JAMES, b Pawtucket, RI, Aug 20, 14; m 39; c 3. MATHEMATICAL STATISTICS. Educ: Mass Inst Technol, BS, 37, PhD(math), 41. Prof Exp: Asst, Mass Inst Technol, 38-39, 40-41, instr math, 41-43; asst prof, Univ NH, 43-44; sr math statistician, Statist Res Group, Columbia Univ, 44-45; asst prof math, Univ Wis, 45-52; assoc prof, 52-55, assoc prof statist, 55-62, chmn dept, 63-67, PROF STATIST, MICH STATE UNIV, 62- Concurrent Pos: Ed, J Soc Indust & Appl Math, 58-69. Mem: Am Math Soc; Math Asn Am; Inst Math Statist (secy-treas, 52-55); Am Statist Asn; Soc Indust & Appl Math. Res: Statistical theory; computational methods; spherical probability distributions. Mailing Add: Dept of Statist & Probability Mich State Univ East Lansing MI 48823

ARNOLD, KENNETH WILTSHIRE, b Aldershot, Eng, Dec 29, 33. PLASMA PHYSICS. Educ: Oxford Univ, BA, 57, MA & DPhil(physics), 61. Prof Exp: Sr physicist, Ion Physics Corp, Mass, 61-64; staff scientist, res & advan develop div, Avco Corp, 64-65; asst prof elec eng, Colo State Univ, 65-66; mem tech staff, res labs, 66-70 & LM cathode devices develop sect, 70-73, HEAD CONVERTER EQUIP SECT, RES LABS, HUGHES AIRCRAFT CO, 73- Mem: Brit Inst Physics. Res: Gas discharge physics; high voltage breakdown phenomena; vacuum techniques. Mailing Add: Res Labs Hughes Aircraft Co 3011 Malibu Canyon Rd Malibu CA 90265

ARNOLD, LESLIE K, b Larned, Kans, Oct 18, 38; m 63; c 2. MATHEMATICS. Educ: Rice Univ, BA, 61; Brown Univ, PhD(ergodic theory), 66. Prof Exp: Assoc, 66-72, SR ASSOC, DANIEL H WAGNER, ASSOCS, 72- Mem: Am Math Soc; Math Asn Am; Soc Indust Appl Math; London Math Soc. Res: Ergodic theory; invariant measures; naval operations analysis. Mailing Add: Daniel H Wagner, Assocs Sta Sq 1 Paoli PA 19301

ARNOLD, LLOYD L, II, philosophy of science, see 12th edition

ARNOLD, LUTHER BISHOP, JR, b Duluth, Minn, Aug 9, 07; m 47; c 1. TEXTILE CHEMISTRY. Educ: Carleton Col, AB, 29; Harvard Univ, AM, 30, PhD(org chem), 33. Prof Exp: Asst, Harvard Univ, 33; res chemist, E I du Pont de Nemours & Co, 33-41, res supvr, Rayon Dept, 41-43; asst dir metal lab, Univ Chicago, 43-45; asst to pres, Arthur D Little, Inc, 45-47; consult & proprietor, 47-65, PRES & TREAS, VIKON CHEM CO, 65- Concurrent Pos: Gen chmn, Southern Textile Res Conf, 61, treas, 62- Mem: Fel AAAS; Am Chem Soc; Am Asn; Textile Chem & Colorists; Am Soc Test & Mat. Res: Gas reaction kinetics; textile processing; surface active agents and detergents; emulsions; emulsion polymerization; radiochemistry. Mailing Add: 1614 Woodland Ave Burlington NC 27217

ARNOLD, MARY B, b Fitchburg, Mass, Sept 29, 24; wid; c 3. PEDIATRIC ENDOCRINOLOGY. Educ: Vassar Col, AB, 45; Univ Vt, MD, 50; Am Bd Pediat, dipl, 56. Hon Degrees: Brown Univ, MA, 74. Prof Exp: Intern, Hartford Hosp, 50-52; asst resident pediat, Babies Hosp, 52-54; asst pediat, Sch Med, Harvard Univ, 55-57; instr, Sch Med, Univ NC, 57-59, asst prof, 59-65; consult biochem, 66-68, DIR PEDIAT OUTPATIENT DEPT & ENDOCRINE CLIN, ROGER WILLIAMS GEN HOSP, 66-, CHMN DEPT PEDIAT, 74-; ASSOC PROF PEDIAT & DIR PEDIAT ENDOCRINOL, PROG MED, BROWN UNIV, 71- Concurrent Pos: Res fel pediat endocrinol, Mass Gen Hosp, Boston, 54-57; lectr med sci, Brown Univ, 66-71; attend pediatrician, Roger Williams Gen Hosp, 66-; consult pediat endocrinol, Women & Infants Hosp RI, 66-; asst physician pediat & attend pediatrician, RI Hosp, 70-, dir pediat endocrinol, 71- Honors & Awards: Carrbee Award Obstet, Univ Vt, 50. Mem: Sigma Xi; Endocrine Soc; Am Fedn Clin Res; AMA; Lawson Wilkins Pediat Endocrinol Soc. Res: Endocrine problems of childhood with particular reference to thyroid disorders and disorders of calcium and phosphorus metabolism; fetal endocrinology, and various mechanisms responsible for growth hormone release, including the role of somatostatin and somatomedin in normal growth and development. Mailing Add: Roger Williams Gen Hosp 825 Chalkstone Ave Providence RI 02908

ARNOLD, MARY TRYSON, b Philadelphia, Pa, Jan 12, 43; m 71. FORENSIC SCIENCE, MYCOLOGY. Educ: Rosemont Col, AB, 68; Bryn Mawr Col, PhD(org chem), 73. Prof Exp: Fel mycol, Skin & Cancer Hosp, Temple Univ Health Sci Ctr, 72-74; FORENSIC CHEMIST, DEL-CHESTER COUNTIES CRIME LAB, 74- Mem: Am Chem Soc; Am Soc Mass Spectrometry; Can Soc Forensic Sci; Brit Soc Forensic Sci; Sigma Xi. Res: Methods development in forensic science, specifically chemistry; isolation, purification and characterization of antigens from fungi and bacteria. Mailing Add: Del-Chester Counties Crime Lab Middletown Rd Lima PA 19037

ARNOLD, R KEITH, b Long Beach, Calif, Nov 17, 13; m 42; c 3. FORESTRY. Educ: Univ Calif, Berkeley, BS, 37; Yale Univ, MF, 38; Univ Mich, PhD(forestry), 50. Prof Exp: Teacher forestry, Univ Calif, Berkeley, 39-42; duty stas, US Navy, 42-46; teacher forestry, Univ Calif, Berkeley, 46-49 & 50-51; div chief fire res, Pac Southwest Exp Sta, US Forest Serv, 51-57, sta dir, 57-63, dir forest protection res, Washington, DC, 63-66; dean, sch natural resources, Univ Mich, 66-69; dep chief forestry res, US Forest Serv, 69-73; PROF & ASSOC DEAN, LYNDON B JOHNSON SCH PUB AFFAIRS & DIR DIV NATURAL RESOURCES & ENVIRON, UNIV TEX, AUSTIN, 73- Concurrent Pos: Mem conf fir res, Nat Acad Sci, 59-65, mem & panel chmn forestry-agr-geog summer study, 67-68, mem comt fire res, 67-, mem & subcomt chmn study conf on prog for outdoor recreation, 68; consult, Fed Civil Defense, 62 & 64; mem, NAm Forestry Comn, 62, chmn fire res, 64-66; mem, Int Union Forestry Res Orgn, 65-67, sect leader, 68- Mem: AAAS; Soc Am Foresters. Res: Forest products and engineering; forest economics and marketing; forest insects and disease; forest fire and atmospheric sciences; timber management and forest environmental research; international forestry. Mailing Add: Div Natural Resources & Environ Univ Tex Box Y Austin TX 78712

ARNOLD, RALPH GUNTHER, b Potsdam, Ger, Dec 8, 28; Can citizen; m 59; c 4. ENVIRONMENTAL GEOLOGY. Educ: Univ Toronto, BASc, 53, MASc, 54; Princeton Univ, PhD(geochem), 58. Prof Exp: Asst res off geol, 59-62, assoc res off, 62-67, HEAD GEOL DIV, SASK RES COUN, 67- Mem: Fel Carnegie Inst, 57-59; lectr, Univ Sask, 59, adj prof, 68. Mem: Mineral Soc Am; Soc Econ Geol; Geochem Soc; Mineral Asn Can. Res: Phase equilibrium; studies of mineral deposits. Mailing Add: 30 Campus Dr Saskatoon SK Can

ARNOLD, RICHARD C, b Owenton, Ky, May 4, 06; m 28; c 3. PUBLIC HEALTH, PREVENTIVE MEDICINE. Educ: Univ Louisville, BS, MS & MD, 30; Am Bd Prev Med, dipl, 49. Prof Exp: Intern, US Marine Hosp, 30-31; from ward surgeon to chief venereal dis serv, US Marine Hosp, La, USPHS, 31-34, chief venereal dis & genito-urinary serv, US Marine Hosp, Calif, 38-39, dir syphilis res, Venereal Dis Lab, NY, 39-51, chief tech serv br, Nat Heart Inst, 51-55, chief heart dis control prog, Div Spec Health Serv, Bur State Serv, 56-59, asst surgeon gen personnel & training, 59-62; dir community health & training, Div Health, Mo, 63-65; med dir, Mo Crippled Children's Serv, 65-74; RETIRED. Concurrent Pos: Fel, Mayo Clin, 37. Mem: Fel AMA; Am Pub Health Asn. Res: Penicillin treatment of syphilis and other venereal diseases; application of new research knowledge for the treatment and prevention of cardiovascular diseases. Mailing Add: 814 Yale St Columbia MO 65201

ARNOLD, RICHARD CHENEY, theoretical physics, see 12th edition

ARNOLD, RICHARD LANE, b Ortonville, Minn, Dec 13, 43; m 67; c 2. POULTRY NUTRITION. Educ: SDak State Univ, Brookings, BS, 65, MS, 67, PhD(animal sci), 71. Prof Exp: Nutritionist poultry & animal nutrit, Sci Nutrit Serv, Inc, Upland, Calif, 71-72; NUTRITIONIST POULTRY & ANIMAL NUTRIT, THOMPSON HAYWARD CHEM CO, KANSAS CITY, KANS, 72- Concurrent Pos: Mem poultry res comt of Nutrit Coun, Am Feed Mfr Asn, 75- Mem: Poultry Sci Asn; Sigma Xi. Res: Relationships of selenium and vitamin E in poultry; factors affecting nutrient requirements for various vitamins, minerals and amino acids for various animals. Mailing Add: 5200 Speaker Rd Kansas City KS 66110

ARNOLD, RICHARD THOMAS, b Indianapolis, Ind, June 18, 13; m 39; c 2. PHYSICAL ORGANIC CHEMISTRY. Educ: Southern Ill State Teachers Col, BEd, 34; Univ Ill, MS, 35, PhD, 37. Prof Exp: From asst prof to prof chem, Univ Minn, 37-55, adminr basic sci prog, Alfred P Sloan Found, 55-60; dir res, Mead Johnson & Co, Ind, 60-61, pres res ctr, 61-69; prof chem & chmn dept, 69-75, PROF CHEM, SOUTHERN ILL UNIV, CARBONDALE, 75- Concurrent Pos: Guggenheim fel, 48-49; sci attache, US High Comnr, Ger, 52-53; vis prof chem, Northwestern Univ, 75. Mem: Am Chem Soc; Brit Chem Soc. Res: Mechanism of organic reactions; stereochemistry and reaction mechanisms. Mailing Add: Dept of Chem Southern Ill Univ Carbondale IL 62901

ARNOLD, RICHARD WARREN, b Creston, Iowa, Aug 25, 29; m 53; c 3. SOIL MORPHOLOGY. Educ: Iowa State Univ, BS, 52, PhD(soil sci), 63; Cornell Univ, MS, 59. Prof Exp: Soil scientist, Soil Conserv Serv, USDA, 52-55; soil technologist, Cornell Univ, 55-59; res assoc, Iowa State Univ, 59-63; asst prof, Univ Guelph, 63-64, assoc prof, 64-66; assoc prof, 66-75. PROF SOIL SCI, CORNELL UNIV, 75- Mem: AAAS; Am Soc Agron; Soil Sci Soc Am; Can Soc Soil Sci; Int Soc Soil Sci. Res: Processes and changes involved in soil development as related to landscape evolution; classification of soils and application in soil survey program. Mailing Add: Dept of Agron Cornell Univ Ithaca NY 14850

ARNOLD, ROBERT FAIRBANKS, b Gilroy, Calif, July 3, 40; m 66. MATHEMATICS. Educ: Fresno State Col, BS, 63, MA, 64; Univ Calif, Berkeley, PhD(math), 69. Prof Exp: Asst prof math, 68-72, ASSOC PROF MATH, CALIF STATE UNIV, FRESNO, 72-, CHMN DEPT, 75- Mem: Math Asn Am. Mailing Add: Dept of Math Calif State Univ Fresno CA 93740

ARNOLD, ROY GARY, b Lyons, Nebr, Feb 20, 41; m 63; c 2. FOOD SCIENCE, BIOCHEMISTRY. Educ: Univ Nebr, BSc, 62; Ore State Univ, MS, 65, PhD(food sci), 67. Prof Exp: Proj leader food res, Res & Develop Lab, Fairmont Food Co, Nebr, 62-63; assoc prof, 67-74, PROF FOOD SCI & TECHNOL, UNIV NEBR, LINCOLN, 74-, CHMN DEPT, 67- Mem: Am Chem Soc; Inst Food Technologists; Am Dairy Sci Asn. Res: Flavor chemistry of foods. Mailing Add: Dept of Food Sci & Technol Univ of Nebr Lincoln NE 68503

ARNOLD, ROY TURNER, physics, see 12th edition

ARNOLD, STEPHEN REYNOLDS, solid state physics, see 12th edition

ARNOLD, VERNON W, physical chemistry, see 12th edition

ARNOLD, WILFRED NIELS, b Australia, May 28, 36; US citizen; m 61; c 1. BIOCHEMISTRY. Educ: Univ Queensland, BS, 56; Univ Calif, Los Angeles, MA, 58; Cornell Univ, PhD(plant biochem), 62. Prof Exp: Res specialist air pollution, Univ Calif, Riverside, 57-58; res fel, Waite Res Inst, Univ Adelaide, 62-63; res scientist, Commonwealth Sci & Indust Res Orgn, Adelaide, Australia, 63-66; sr fel, Med Sch, Univ Wis, 66-67; asst prof biochem, Med Sch, Wayne State Univ, 67-71; ASSOC PROF BIOCHEM, UNIV KANS MED CTR, KANSAS CITY, 72- Mem: AAAS; Am Chem Soc; Sigma Xi; Am Soc Biol Chemists. Res: Enzymology; structure and function of the cell wall in yeasts; carbohydrate and amino acid metabolism in higher plants and microorganisms. Mailing Add: Dept of Biochem Univ of Kans Med Ctr Kansas City KS 66103

ARNOLD, WILLIAM ARCHIBALD, b Douglas, Wyo, Dec 6, 04; m 29; c 2. PHYSIOLOGY. Educ: Calif Inst Technol, BS, 31; Harvard Univ, PhD(physiol), 35. Prof Exp: Sheldon fel biol, Univ Calif, 35-36; Gen Educ Bd fel, Hopkins Marine Sta, Stanford Univ, 36-37, asst, 37-38, res assoc biol, 39-40 & 41, asst prof, 41-46; PRIN BIOLOGIST, OAK RIDGE NAT LAB, 46- Concurrent Pos: Rockefeller fel, Inst Theoret Physics, Copenhagen Univ, 38-39; physicist, Princeton Univ, 42; sr physicist, Eastman Kodak Co, NY, 42-44 & Tenn, 44-46. Mem: Nat Acad Sci; Am Phys Soc; Soc Gen Physiol. Res: Photosynthesis. Mailing Add: 102 Balsam Rd Oak Ridge TN 37830

ARNOLD, WILLIAM H, JR, b St Louis, Mo, May 13, 31; m 52; c 4. EXPERIMENTAL PHYSICS. Educ: Cornell Univ, AB, 51; Princeton Univ, MA, 53, PhD(physics), 55. Prof Exp: Instr physics, Princeton Univ, 55; sr scientist reactor physics, atomic power div, Westinghouse Electric Corp, 55-57, mgr reactor physics design, 57-61; dir nuclear fuel mgt, Nuclear Utility Serv, Inc, 61-62; dep mgr reactor eng, astronuclear lab, 62-64, mgr test systs & opers, 64-66, dep prog mgr NERVA proj, 66-68, proj mgr, 68, mgr weapons dept astronuclear/underseas div, 68-70, mgr eng, 70-72, GEN MGR PRESSURIZED WATER REACTOR SYSTS, WESTINGHOUSE NUCLEAR ENERGY SYSTS, WESTINGHOUSE ELEC CORP, 72- Concurrent Pos: Lectr, Univ Pittsburgh, 58. Mem: Nat Acad Eng; Am Nuclear Soc; Am Phys Soc; Am Inst Aeronaut & Astronaut. Res: Cosmic ray and elementary particle physics; reactor physics; nuclear engineering. Mailing Add: Westinghouse Nuclear Energy Systs PO Box 355 Pittsburgh PA 15230

ARNOLD, WILLIAM JAMES, b Louisville, Ky, Sept 20, 28. CELL BIOLOGY. Educ: Univ Louisville, AB, 50; Univ Calif, Berkeley, MS, 57, PhD(entom), 60. Prof Exp: Asst prof biol, Wayne State Univ, 60-66; RES SCIENTIST, DETROIT INST CANCER RES, 66-, CHIEF ELECTRON MICROS, 68- Mem: Am Soc Cell Biol; Electron Micros Soc Am. Res: Histopathology of murine tumor virus infection; mouse hematopoiesis; morphology of cell-virus interaction and RNA tumor viruses. Mailing Add: Detroit Inst of Cancer Res 4811 John R St Detroit MI 48201

ARNOLD, ZACH MCLENDON, b Ft Gaines, Ga, Nov 14, 21; m 44; c 2. PROTOZOOLOGY, MICROPALEONTOLOGY. Educ: Emory Univ, AB, 42, MS, 47; Univ Calif, PhD(zool), 53. Prof Exp: Asst zool, 49-50, mus paleontologist, 50-53, res micropaleontologist, 54-57, from asst prof to assoc prof, 57-68, PROF PALEONT, UNIV CALIF, BERKELEY, 68-, CUR MUS, 57- Concurrent Pos: Guggenheim fel, 58-59. Mem: Am Soc Zool; Soc Protozool; Soc Syst Zool; Paleont Soc. Res: Biology and paleontology of the Foraminifera, life history, cytology and variation. Mailing Add: Dept of Paleont Univ of Calif 113 Earth Sci Bldg (B) 655 Santa Maria Rd El Sobrante CA 94803

ARNOLDI, LOUIS B, b St Louis, Mo, Apr 10, 12; m 47, 62; c 2. MEDICINE. Educ: Univ Mo, AB & BS, 36; NY Univ, MD, 38. Prof Exp: Med officer, US Air Force, Maxwell AFB, Ala, 40, flight surgeon, Sta Dispensary, CZ, 41 & Sta Hosp, Ecuador, 42, surgeon, Pre-Flight Div, Lackland AFB, Tex, 43-45, resident physician, Southern Baptist Hosp, New Orleans, La, 46-47, base surgeon, Carswell AFB, Tex, 48-49 & Chatham AFB, Ga, 49, chief control br & dir staffing & educ, Off Surgeon Gen Hq, 49-51, dep chief med inspection div, Kelly AFB, Tex, 51-52 & Norton AFB, Calif, 52-54, command surgeon, Hq Fifth Air Force, Japan, 54-58, chief med liaison & selection, Off Surgeon Gen Hq, 58-59, spec asst to surgeon gen, 59-62, command surgeon, Wright-Patterson AFB, Ohio, 62-67; MED DIR, DIV OCCUP MED & ENVIRON HEALTH, NASA, 67- Concurrent Pos: Mem coun, Fedn Med Dirs Occup Med; liaison mem nat adv heart & lung coun, NIH. Mem: Fel Aerospace Med Asn; fel Indust Med Asn; Am Acad Occup Med. Res: Aviation medicine; public administration and management. Mailing Add: NASA Hq BG/Div of Occup Med & Environ Health Washington DC 20546

ARNOLDY, ROGER L, b LaCrosse, Wis, May 30, 34; m 61; c 5. SPACE PHYSICS. Educ: St Mary's Col, Minn, BS, 56; Univ Minn, MS, 60, PhD(physics), 62. Prof Exp: Res assoc physics, Univ Minn, 62-64; sr res scientist, Honeywell Res Ctr, Minn, 64-67; ASSOC PROF PHYSICS, UNIV NH, 67-, DIR SPACE SCI CTR, 74- Mem: Am Geophys Union. Res: Experimentation to measure charged particle radiation in the earth's magnetosphere and in interplanetary space. Mailing Add: Dept of Physics DeMeritt Hall Univ of NH Durham NH 03824

ARNON, DANIEL ISRAEL, b Poland, Nov 14, 10; nat US, m 40; c 5. BIOCHEMISTRY. Educ: Univ Calif, BS, 32, PhD(plant physiol), 36. Prof Exp: Instr, 36-41, from asst prof to assoc prof, 41-50, prof plant physiol, 50-60, PROF CELL PHYSIOL, UNIV CALIF, BERKELEY, 60-, BIOCHEMIST, EXP STA, 58- Concurrent Pos: Guggenheim fel, Cambridge Univ, 47-48 & Hopkins Marine Sta, Stanford Univ, 62-63; ed, Annual Rev Plant Physiol, 48-55; Belg-Am Found lectr, Univ Liege, 48; Fulbright res scholar, Max Planck Inst, Ger, 55-56. Honors & Awards: Gold Medal, Univ Pisa, 58; Charles F Kettering award in photosynthesis, 63; Nat Medal Sci, President of the United States, 73; Newcomb Cleveland Prize, AAAS, 40; Stephen Hales Prize award, Am Soc Plant Physiol, 66. Mem: Nat Acad Sci; AAAS; Am Acad Arts & Sci; Am Chem Soc; Am Soc Biol Chem. Res: Photosynthesis; bioenergetics; nitrogen fixation; light reactions of photosynthesis; photophosphorylation; ferredoxins. Mailing Add: Dept of Cell Physiol Univ of Calif Berkeley CA 94720

ARNONE, ARTHUR RICHARD, b Syracuse, NY, Sept 25, 42; m 66; c 2. BIOCHEMISTRY. Educ: Mass Inst Technol, PhD(phys chem), 70. Prof Exp: Fel molecular biol, Med Res Coun Lab, Cambridge, Eng, 70-73; ASST PROF BIOCHEM, UNIV IOWA, 73- Concurrent Pos: NIH res career develop award, 75. Mem: Am Chem Soc; Am Crystallog Asn; Sigma Xi. Res: X-ray crystallographic studies of hemoglobin and other proteins. Mailing Add: Dept of Biochem Univ of Iowa Iowa City IA 52242

ARNOTT, HOWARD JOSEPH, b Los Angeles, Calif, Mar 9, 28; m 50; c 4. BOTANY, CELL BIOLOGY. Educ: Univ Southern Calif, AB, 52, MS, 53; Univ Calif, PhD(bot), 58. Prof Exp: Asst bot, Univ Calif, 55-58; asst prof biol, Northwestern Univ, 58-64; fel, cell res inst, Univ Tex, Austin, 64-65, vis assoc prof bot, 65-66, from assoc prof to prof bot, 66-72; prof biol & chmn dept, Univ South Fla, 72-74; PROF BIOL & DEAN COL SCI, UNIV TEX, ARLINGTON, 74- Concurrent Pos: Consult, Encycl Britannica & NASA; ed, Protoplasma, 74-; adv ed, J Calcified Tissue Res, 69- Mem: AAAS; Am Soc Cell Biol; Bot Soc Am; Am Micros Soc; Soc Invert Path. Res: Biological ultrastructure; development of crystals in biological systems; organic crystals in insect viruses and animal reflecting systems; inorganic crystals in plants and animals; ultrastructural ecology; thermophilic organisms. Mailing Add: Col Sci Univ Tex Arlington TX 76019

ARNOTT, ROBERT A, b Spencer, WVa, Dec 4, 41; m 63; c 2. AIR POLLUTION, ANALYTICAL CHEMISTRY. Educ: WVa Univ, BS, 63; Univ Wis, PhD(phys chem), 68. Prof Exp: Asst prof chem, Univ Wis, Oshkosh, 68-75; dir, NSF Environ Assessment Prog, 74-75; MGR, AMBIENT AIR MONITORING SECT, DIV AIR POLLUTION CONTROL, ILL EPA, 75- Mem: Am Chem Soc; Sigma Xi; Air Pollution Control Asn. Res: Environmental trace analysis; macro analysis of water pollutants; environmental air sampling and analysis. Mailing Add: Ill Environ Protection Agency 2200 Churchill Rd Springfield IL 62706

ARNOTT, RONALD JAMES, b Roblin, Man, Oct 31, 21; US citizen. GEOLOGY. Educ: Univ Man, BSc, 43 & 46, MSc, 49; Columbia Univ, PhD(x-ray mineral), 54. Hon Degrees: MA, Brown Univ, 71. Prof Exp: Geol asst, Geol Surv, Can, 46, jr mineralogist, 48; jr geologist, Int Nickel Co Can, 46-47; physicist, Res Lab, Am Cyanamid Co, Conn, 49-51; petrol geologist, La Div, Texaco, Inc, 52-53, WTex Div, 54, res geologist, Bellaire Lab, 54-57; x-ray crystallographer, Lincoln Lab, Mass Inst Technol, 57-70; ASSOC RES PROF ENG, BROWN UNIV, 70- Mem: Am Crystallog Asn; fel Mineral Soc Am. Res: X-ray crystallography of radioactive minerals; clay minerals; magnetic and thermoelectric materials; crystallographic studies at high and low temperatures. Mailing Add: Eng Div Brown Univ Providence RI 02912

ARNOTT, STRUTHER, b Larkhall, Scotland, Sept 25, 34; m; c 2. MOLECULAR BIOLOGY. Educ: Glasgow Univ, BSc, 56, PhD(chem), 60. Prof Exp: Scientist, Med Res Coun, biophys unit, King's Col, Univ London, 60-70, dir postgrad studies, dept biophys, col, 67-70; PROF BIOL, PURDUE UNIV, 70-, HEAD DEPT BIOL SCI, 75- Mem: Fel Royal Inst Chem; Am Crystallog Asn; Am Soc Biol Chemists. Res: X-ray diffraction analysis; fibrous polymers; nucleic acids; polysaccharides; proteins; viruses. Mailing Add: Dept Biol Sci Purdue Univ West Lafayette IN 47907

ARNOW, LESLIE EARLE, b Micanopy, Fla, June 22, 09; m 33; c 1. PHYSIOLOGICAL CHEMISTRY, RESEARCH ADMINISTRATION. Educ: Univ Fla, BS, 30; Univ Minn, PhD(biochem), 34, MB, MD, 40. Prof Exp: Asst physiol chem & biophysics, Univ Minn, 31-34, instr physiol chem, 34-40, asst prof, 40-42; dir biochem res, Res Labs, Sharp & Dohme Div, Merck & Co, 42-44, dir res, 44-53, vpres & dir res, 53-56, vpres, Merck Sharp & Dohme Res Labs & exec dir, Merck Inst Therapeut Res, 56-58; pres, Warner-Lambert Res Inst 58-65, sr sci consult, 65-74; RETIRED. Concurrent Pos: Vpres, Warner-Lambert Pharmaceut Co, 58-65. Mem: AAAS; Am Soc Biol Chem; Soc Exp Biol & Med; Am Chem Soc; Am Soc Clinical Pharmal Therapeut. Res: Protein and amino acid chemistry; research administration. Mailing Add: 14 Fairfield Dr Convent Sta NJ 07961

ARNOW, THEODORE, b New York, NY, July 27, 21; m 49; c 2. GEOLOGY. Educ: NY Univ, BA, 42; Columbia Univ, MA, 49. Prof Exp: Geologist, 46-68, DIST CHIEF, WATER RESOURCES DIV, US GEOL SURV, 68- Mem: Am Asn Petrol Geol; Geol Soc Am. Res: Ground-water geology. Mailing Add: Water Resources Div US Geol Surv 8002 Fed Bldg Salt Lake City UT 84117

ARNOWICH, BEATRICE, b New York, NY, June 14, 30. PHYSICAL ORGANIC CHEMISTRY. Educ: Vassar Col, AB, 51; Univ Mich, MS, 52; NY Univ, PhD(org chem), 57. Prof Exp: Asst, NY Univ, 56-57; chemist, Charles Bruning Co, 57-59; sr chemist, Interchem Corp, 59-66, group leader, 66-69; ASST PROF CHEM, QUEENSBOROUGH COMMUNITY COL, 69- Mem: AAAS; Am Chem Soc; NY Acad Sci. Res: Surface chemistry; photoconductivity. Mailing Add: Dept of Chem Queensborough Community Col Bayside NY 11364

ARNOWITT, RICHARD LEWIS, b New York, NY, May 3, 28. THEORETICAL PHYSICS. Educ: Rensselaer Polytech Inst, BS & MS, 48; Harvard Univ,

PhD(physics), 53. Prof Exp: Res assoc, Radiation Lab, Univ Calif, 52-54; mem, Inst Advan Study, 54-56; from asst prof to assoc prof, Syracuse Univ, 56-59; PROF PHYSICS, NORTHEASTERN UNIV, 59- Res: Elementary particle theory; quantum field theory. Mailing Add: Dept of Physics Northeastern Univ Boston MA 02115

ARNQUIST, CLIFFORD WARREN, mathematics, see 12th edition

ARNQUIST, WARREN NELSON, b New Richmond, Wis, Aug 12, 05; m 33; c 3. SOLAR PHYSICS. Educ: Whitman Col, BS, 27; Calif Inst Technol, PhD(physics), 30. Prof Exp: From asst prof to assoc prof physics, Ala Polytech Inst, 30-35; physicist, Gulf Res & Develop Co, 35-41; res fel, Nat Defense Res Comt, Calif Inst Technol, 41-42; sect head, Off Sci Res & Develop Contract, 42-45; tech asst, Naval Ord Test Sta, Calif, 45-46, asst head exp opers dept, 46-47; phys scientist, Off Naval Res, Calif, 47-53, physicist, London, 53-55, physicist, Ord Res, Western Regional Off, Pasadena, 55-58; physicist, Syst Develop Corp, 58-61; dir sci res, Douglas Aircraft Co Inc, 61-66, res scientist, Douglas Advan Res Labs, 66-70, staff asst, McDonnell Douglas Astronautics Co, 70-71; RES ASSOC ASTRON, UNIV CALIF, LOS ANGELES, 71- Concurrent Pos: Consult, NAm Aviation, 51, Pomona Div, Convair, 58, Army Ballistic Missile Agency, 58-60 & Duke Univ, 61; instr mech eng, Univ Southern Calif, 56-60; dir, Airborne Solar Eclipse Exped, 63. Honors & Awards: Tech Award, Bur Ord, US Navy, 45. Mem: Fel AAAS; fel Am Phys Soc; Am Astron Soc; Int Astron Union. Res: Scattering of electrons in gases; fluids in porous media; dielectric phenomena; rockets, internal and external ballistics; semiconductors; military infrared applications; jet propulsion; operations research; research administration. Mailing Add: Dept of Astron Univ of Calif Los Angeles CA 90024

ARNRICH, LOTTE, b Elberfeld, Ger, Sept 23, 11; US citizen. NUTRITION. Educ: Univ Calif, BS, 44, PhD(nutrit), 52. Prof Exp: Lectr nutrit, Univ Calif, 48-52, instr, 52-55; assoc prof, 55-60, PROF NUTRIT, IOWA STATE UNIV, 60- Mem: AAAS; Am Inst Nutrit; Am Inst Biol Sci; Inst Food Technologists. Res: Metabolic interactions of lipids, proteins and vitamins A and D. Mailing Add: 1121 Hyland Ave Ames IA 50010

ARNS, ROBERT GEORGE, b Buffalo, NY, July 24, 33; m 57; c 3. NUCLEAR PHYSICS. Educ: Canisius Col, BS, 55; Univ Mich, MS, 56, PhD(physics), 60. Prof Exp: Res assoc physics, Univ Mich, 59-60, instr, 60; from asst prof to assoc prof physics, State Univ NY Buffalo, 60-64; assoc prof, 64-69, PROF PHYSICS, OHIO STATE UNIV, 69- Mem: Am Phys Soc; Am Asn Physics Teachers. Res: Nuclear structure; nuclear reactions; gamma and particle-gamma directional correlations; nuclear lifetimes; particle accelerators. Mailing Add: Dept of Physics Ohio State Univ Columbus OH 43210

ARNTZEN, CLYDE EDWARD, b Crescent City, Ill, Dec 26, 15; m 48; c 2. ORGANIC CHEMISTRY. Educ: Univ Ill, BS, 38; Iowa State Col, PhD(org chem), 42. Prof Exp: Engr res labs, 42-45, engr & group leader, 45-46, mgr, Insulation Develop Sect, Mat Eng Dept, 46-48, Insulation Appln Sect, 48-52 & Chem Develop Sect, 52-53, sect mgr chem labs, 53-55, asst to mgr, Mat Eng Depts, 55-58, mgr, 58-62, mgr prog eval & div coord, Res Labs, 62-64, MGR TECH SERV, RES LABS, WESTINGHOUSE ELEC CORP, 64- Mem: Am Chem Soc. Res: Synthetic organic chemistry; organic resins and plastics; chemical engineering; synthetic resins and plastics; radio; evaluation of technical projects; administration. Mailing Add: Westinghouse Res & Dev Ctr 1310 Beulah Rd Pittsburgh PA 15235

ARNUSH, DONALD, b New York, NY, Feb 7, 36; m 66. THEORETICAL PHYSICS. Educ: Mass Inst Technol, SB, 57, PhD(physics), 61. Prof Exp: NSF fel, Max Planck Inst Physics & Astrophys, 61-62, mem tech staff, 63-72, MGR, PLASMA PHYSICS DEPT, TRW SYSTS, 72- Mem: Am Phys Soc; Am Inst Aeronaut & Astronaut; Am Geophys Union. Res: Theory of fundamental particles; electromagnetic wave propagation; plasma physics; meson scattering; ionospheric and fusion related plasma physics; electromagnetic wave propagation; plasma physics; meson scattering; ionospheric and fusion related plasma physics; electromagnetic wave propagation; theoretical physics. Mailing Add: R1/1070 TRW Systs Redondo Beach CA 90278

ARNY, DEANE CEDRIC, b St Paul, Minn, May 22, 17; m 47; c 5. PLANT PATHOLOGY. Educ: Univ Minn, BS, 39; Univ Wis, PhD(agron, plant path), 43. Prof Exp: From instr to assoc prof, 43-63, PROF PLANT PATH, UNIV WIS, MADISON, 63- Concurrent Pos: Vis prof, Univ Ife, Nigeria, 66-68. Mem: Am Phytopath Soc; Am Soc Agron. Res: Diseases of cereals. Mailing Add: Dept of Plant Path Univ of Wis Madison WI 53706

ARNY, THOMAS TRAVIS, b Montclair, NJ, June 2, 40; m 68. ASTROPHYSICS. Educ: Haverford Col, BA, 61; Univ Ariz, PhD(star formation), 65. Prof Exp: Res assoc astron, Amherst Col, 65-66; asst prof, 66-69, ASSOC PROF ASTRON, UNIV MASS, AMHERST, 69- Concurrent Pos: NSF grant, 67-71. Mem: Am Astron Soc; Int Astron Union. Res: Star formation; interstellar gas dynamics. Mailing Add: Dept of Astron Univ Mass Amherst MA 01002

AROIAN, LEO AVEDIS, b Holden, Mass, May 10, 07; m 41; c 2. MATHEMATICAL STATISTICS. Educ: Univ Mich, AB, 28, MA, 29, PhD, 40. Prof Exp: Asst prof math, Colo State Univ, 30-39; instr, Hunter Col, 39-47, asst prof, 48-50; head math sect, Hughes Aircraft Co, 50-59, sr math consult, 59-60; mem tech staff, space technol labs, Thompson-Ramo-Wooldridge, Inc, 60-67, sr staff mathematician, TRW Systs, 67-68; PROF INDUST ADMIN, 68-72, EMER PROF ADMIN & MGT, UNION COL, 72- Concurrent Pos: Mem Nat Defense Res Comt, 45. Honors & Awards: Res Award, Am Soc Qual Control, 70. Mem: Am Math Soc; Biomet Soc; Sigma Xi; fel Am Statist Asn; Inst Math Statist. Res: High speed computation; numerical methods; mathematical, applied and industrial statistics; reliability theory; sequential analysis; time series; security investments. Mailing Add: Bailey Hall Union Col Schenectady NY 12308

ARON, WALTER ARTHUR, b Milwaukee, Wis, Aug 30, 21; m 43; c 3. THEORETICAL PHYSICS. Educ: Univ Calif, PhD(physics), 51. Prof Exp: Physicist, Radiation Lab, Mass Inst Technol, 44-46; physicist, Radiation Lab, Univ Calif, 46-51; res assoc physics, Princeton Univ, 51-55; asst prof, Univ Va, 55-59; theoret physicist, Lawrence Radiation Lab, Univ Calif, Berkeley, 59-64; res physicist, Stanford Res Inst, 64-71; SCIENTIST, SCI APPLNS, INC, 71- Res: Penetration of heavy charged particles through matter. Mailing Add: Sci Applns Inc 2680 Hanover Palo Alto CA 94304

ARON, WILLIAM IRWIN, b Brooklyn, NY, June 26, 30; m 61; c 2. BIOLOGICAL OCEANOGRAPHY. Educ: Brooklyn Col, BS, 52; Univ Wash, MS, 57, PhD, 60. Prof Exp: Biologist, Wash State Dept Fisheries, 52-53, res instr oceanog, Univ Wash, 56-60, res asst prof, 60-61; res supvry biologist, defense res labs, Gen Motors Corp, 61-67; dep head, off oceanog & limnol, Smithsonian Inst, 67-69, dir oceanog & limnol prog, 69-71; DIR OFF ECOL & ENVIRON CONSERV, NAT OCEANIC & ATMOSPHERIC ADMIN, 71- Concurrent Pos: Consult, Vet Corps, US Army, 56-, President's Marine Sci Comn, 68-69 & Nat Water Comn, 69-; mem gen adv comt, Mediter Marine Sorting Ctr, Tunisia, 70-; US del, Coop Invests of the Mediter, 70-; mem comt oceanog biol methods panel, Nat Acad Sci; mem bd dirs, Nat Oceanog Instrumentation Ctr. Mem: AAAS; Am Soc Limnol & Oceanog; Am Soc Ichthyologists & Herpetologists; Soc Study Evolution. Res: Instrumentation for sampling; zoogeography of pelagic fish and plankton; research administration. Mailing Add: Off Ecol & Environ Conserv Nat Oceanic & Atmospheric Admin Washington DC 20230

ARONOFF, SAMUEL, b New York, NY, Feb 27, 15; m 36; c 2. BIOCHEMISTRY. Educ: Univ Calif, Los Angeles, AB, 36; Univ Calif, PhD(physico-chem biol), 42. Prof Exp: Asst agr, Univ Calif, 41-42, res fel, 42-43; instr agr, Boston Univ, 43-44; res instr, Univ Chicago, 44-46; mem staff radiation lab, Univ Calif, 46-48; from assoc prof to prof bot, Iowa State Univ, 48-60, prof biochem & biophys, 69-69; dean grad sch arts & sci & vpres res, Boston Col, 69-71; dean sci, 71-76, PROF CHEM, SIMON FRASER UNIV, 76- Concurrent Pos: Sr fel, NSF, 57-58, prog dir molecular biol, 63-64, consult, 64-65. Mem: Am Chem Soc; Am Soc Plant Physiol (secy, 62-63, vpres, 64, pres, 65); Am Soc Biol Chemists; Can Soc Plant Physiol. Res: Plant biochemistry; photosynthesis; chlorophyll; radiobiochemistry; boron biochemistry. Mailing Add: Dept of Chem Simon Fraser Univ Burnaby BC Can

ARONOFSKY, JULIUS S, b Dallas, Tex, July 29, 21; m 45; c 4. APPLIED MATHEMATICS. Educ: Southern Methodist Univ, BS, 44; Stevens Inst Technol, MS, 46; Univ Pittsburgh, PhD(math), 49. Prof Exp: Engr, Kellex Corp, 44-45; res engr, Westinghouse Elec Co, 47-49; sr res engr, Magnolia Petrol Co, 49-55, res assoc, 55-57; mgr electronic computer ctr, Socony Mobil Oil Co, Inc, 57-65, mgr opers res dept, 65-68; prof statist & opers res, Univ Pa, 68-70; lectr, 48-57, prof statist & opers res, 70-72, PROF MGT SCI & COMPUT, SOUTHERN METHODIST UNIV, 72- Concurrent Pos: Consult, Universal Comput Co, 70- Mem: Am Soc Mech Eng; Am Inst Mining, Metall & Petrol Eng; Am Phys Soc; Opers Res Soc Am; Inst Mgt Sci. Res: Flow of fluids through porous media; elasticity; plastic flow of metals; computer technology; operations research; management science. Mailing Add: Dept of Mgt Sci Southern Methodist Univ Dallas TX 75225

ARONOVIC, SANFORD MAXWELL, b New York, NY, June 10, 26; m 57; c 3. ANALYTICAL CHEMISTRY. Educ: Columbia Univ, BS, 45; Univ Wis, PhD(chem), 57. Prof Exp: Chemist microanal, Lederle Labs Div, Am Cyanamid Co, 47-52, res chemist anal develop, Stamford Res Labs, 57-62; res asst infrared spectros, Univ Wis, 52-57; res anal chemist, Maumee Chem Co, 62-65; res scientist, Union Bag-Camp Paper Corp, 65, res chemist, Union Camp Corp, 65-68; supvr qual control, 69-72, SUPVR ANAL SERV, THIOKOL CHEM CORP, 72- Mem: AAAS; Am Chem Soc; Coblentz Soc. Res: Microanalytical chemistry; infrared spectroscopy; instrumental methods of analysis; analytical separations, identification and separations, analytical methods development; gas chromatography. Mailing Add: Thiokol Corp 930 Lower Ferry Rd Trenton NJ 08607

ARONOW, SAUL, b Brooklyn, NY, Dec 10, 23; m 48; c 3. GEOLOGY. Educ: Brooklyn Col, BA, 45; Univ Iowa, MS, 46; Univ Wis, PhD, 55. Prof Exp: Draftsman, Army Map Serv, 45; geologist, US Geol Surv, 48-52; asst geol, Univ Wis, 53-54; PROF GEOL, LAMAR UNIV, 55- Concurrent Pos: Consult, Tex Bur Econ Geol, 65-70. Mem: AAAS; Am Geol Soc; Nat Asn Geol Teachers; Soc Econ Paleont & Mineral. Res: Geomorphology; glacial geology; ground water; Pleistocene; soils; Gulf Coast region. Mailing Add: Dept of Geol Box 10031 Lamar Univ Sta Beaumont TX 77710

ARONOW, SAUL, b Brooklyn, NY, Oct 4, 17; m 42; c 6. BIOMEDICAL ENGINEERING. Educ: Cooper Union, BEE, 39; Harvard Univ, MS, 46, PhD(appl sci), 53. Prof Exp: Electronics engr, Harvey Radio Labs, 46-49; from asst physicist to assoc appl physicist, 53-68, dir med eng group, 68-72, ASSOC PHYSICIST, DEPT RADIOL, MASS GEN HOSP, 72- Concurrent Pos: Prin assoc med, Harvard Med Sch, 54-; lectr, Northeastern Univ, 56- & Mass Inst Technol, 57-; vis prof, Denmark Tech Univ, 69; consult engr, Mass. Mem: Inst Elec & Electronics Engrs; Am Asn Physicists in Med; Soc Nuclear Med; Biomed Eng Soc; Int Fedn Med Electronics & Biol Eng. Res: Medical applications of electronic instrumentation; nuclear medicine. Mailing Add: Dept of Radiol Mass Gen Hosp Boston MA 02114

ARONOWITZ, FREDERICK, b New York, NY, July 3, 35. QUANTUM PHYSICS. Educ: Polytech Inst Brooklyn, BS, 56; NY Univ, PhD(physics), 69. Prof Exp: Physicist, David Taylor Model Basin, 56; consult, Eastern Res Group, 61-62; STAFF SCIENTIST, HONEYWELL, INC, 62- Mem: Am Phys Soc; Sigma Xi. Res: Quantum electronics with major emphasis of the theory and development of the laser gyroscope. Mailing Add: Res Dept Systs & Res Ctr Honeywell Inc 2600 Ridgway Pkwy Minneapolis MN 55413

ARONOWITZ, LEONARD, b New York, NY, Mar 9, 25; m 56; c 2. PHYSICS. Educ: Brooklyn Col, BA, 46; NY Univ, MS, 49, PhD(physics), 57. Prof Exp: Eng group leader, Sperry Gyroscope Co, 57-58; sr engr, W L Maxson Corp, 58-59; specialist sci res plasma physics, Repub Aviation Corp, 59-64; res scientist astrophys, Grumman Aircraft Eng Corp, 64-70, res scientist, Grumman Aerospace Corp, 70-74; CONSULT, 74- Mem: Am Phys Soc. Res: Plasma physics; electrical discharge; electrostatics. Mailing Add: Apt 7B 1500 Hornell Loop Brooklyn NY 11239

ARONS, ARNOLD BORIS, b Lincoln, Nebr, Nov 23, 16; m 42; c 4. PHYSICS. Educ: Stevens Inst Technol, ME, 37, MS, 40; Harvard Univ, PhD(phys chem), 43; Amherst Col, hon AM, 53. Prof Exp: Asst, Harvard Univ, 42-43; instr chem, Stevens Inst Technol, 37-40, from asst prof to assoc prof, 46-52; prof, Amherst Col, 52-68; PROF PHYSICS, UNIV WASH, 68- Prof Exp: Res assoc, Oceanog Inst, Woods Hole, 43-, mem corp, 63-, trustee, 64-; chief sect underwater blast measurements, Oper Crossroads, Navy Bur Ord, 46; Guggenheim fel, 57-58; NSF faculty fel, 62-63; consult, Naval Ord Lab, Waterways Exp Sta; mem, Comn Col Physics, 62-68; gov bd, Am Inst Physics, 66-72. Honors & Awards: Honor medal, Stevens Inst Technol, 68; Oersted Medal, Am Asn Physics Teachers, 72. Mem: AAAS; Am Phys Soc; Am Asn Physics Teachers (pres, 67); Am Geophys Union; Nat Sci Teachers Asn. Res: Explosion phenomena; development of tourmaline piezoelectric gauges for measurement of explosion pressure waves; physical oceanography; hydrodynamics of rotating systems; science education. Mailing Add: Dept of Physics Univ of Wash Seattle WA 98105

ARONS, MICHAEL GENE, b New York, NY, Mar 29, 39; m 61; c 1. PHYSICS. Educ: Cooper Union, BEE, 59; Univ Rochester, PhD(theoret physics), 64. Prof Exp: Asst res scientist theoret physics, NY Univ, 64-66; asst prof physics, 66-70; ASSOC PROF PHYSICS, CITY COL NEW YORK, 70- Mem: Am Phys Soc. Res: Field theory and quantum electrodynamics; theories of elementary particles, including symmetries and field-theoretic models. Mailing Add: Dept of Physics City Col of New York New York NY 10031

ARONSON, A L, b Minneapolis, Minn, Aug 24, 33; m 56; c 3. PHARMACOLOGY, TOXICOLOGY. Educ: Univ Minn, BS, 55, DVM, 57, PhD(pharmacol), 63; Cornell Univ, MS, 59. Prof Exp: Teaching asst pharmacol, Cornell Univ, 57-58; res fel, Univ Mo, 58-63, res assoc, 63-64; from asst prof to assoc prof, 64-71, PROF

PHARMACOL, STATE UNIV NY VET COL, CORNELL UNIV, 71- Mem: Am Vet Med Asn; Am Soc Pharmacol and Exp Ther; Am Soc Toxicol. Res: Mobilization of lead by chelating agents; toxicity of chelating agents; pharmacokinetics. Mailing Add: NY State Col Vet Med Cornell Univ Ithaca NY 14853

ARONSON, ARTHUR IAN, b Boston, Mass, July 2, 30; m 56; c 2. MICROBIOLOGY. Educ: Univ Chicago, BA, 50; Univ Mass, MS, 53; Univ Ill, PhD(microbiol), 58. Prof Exp: NSF fel, 58-59; fel, Carnegie Inst, 59-60; from asst prof to assoc prof, 60-67, PROF MICROBIOL, PURDUE UNIV, 67- Mem: AAAS; Am Soc Microbiol; Am Soc Biol Chemists; Soc Develop Biol. Res: Molecular biology; bacterial spore formations; macromolecule synthesis in sea urchins. Mailing Add: Dept of Biol Sci Purdue Univ West Lafayette IN 47907

ARONSON, CARL EDWARD, b Providence, RI, Mar 14, 36; m 60; c 2. PHARMACOLOGY, TOXICOLOGY. Educ: Brown Univ, AB, 58; Univ Vt, PhD(pharmacol), 66. Hon Degrees: MA, Univ Pa, 73. Prof Exp: Res technician, Worcester Found Exp Biol, Shrewsbury, Mass, 58-60; instr pharmacol, Sch Med, 67-70, assoc, 70-71, asst prof, Sch Vet Med, 71-73, ASSOC PROF PHARMACOL, DEPT ANIMAL BIOL, SCH VET MED, UNIV PA, 73-, HEAD LABS PHARMACOL, 72-, ASST PROF PHARMACOL, SCH MED, 71- Concurrent Pos: Fel pharmacol, Univ Pa, 65-67, Pa Plan scholar, 69-71; Heart Asn Southeastern Pa res fel, 69-72; lectr pharmacol, Div Grad Med Educ, Sch Med, Univ Pa, 65-, actg assoc dean student affairs, Sch Vet Med, 74-75. Mem: AAAS; Am Soc Pharmacol & Exp Therapeut. Res: Biochemical pharmacology and toxicology; effects of drugs and hormones on cardiac performance and metabolism. Mailing Add: Dept Animal Biol Lab of Pharmacol Univ of Pa Sch of Vet Med Philadelphia PA 19174

ARONSON, CASPER JACOB, b Canisteo, NY, Sept 1, 16; m 43; c 1. PHYSICS, EXPLOSIVES. Educ: Univ Rochester, BS, 38, MS, 39. Prof Exp: Asst, Univ Rochester, 38-40; physicist, Carnegie Inst, 40-41 & chief, Magnetic Model Sect, 44-46 & Magnetic Fields Sect, 46-49, sr res assoc, Explosives Res Dept, 49-52, dep chief, Explosion Effects Div, 52-53, asst chief atomic tests, 53-57, chief, Explosion Res Dept, 57-74, actg assoc head chem & explosions res, US Naval Ord Lab, 74-75; SCIENTIFIC ASST RES & TECHNOL, NAVAL SURFACE WEAPONS CTR, 75- Honors & Awards: Meritorious civilian awards, US Navy, 46, 51; Flemming award, 55. Mem: AAAS; Am Phys Soc; Optical Soc Am. Res: Ship's magnetic fields; shock waves; effects of explosions. Mailing Add: 3401 Oberon St Kensington MD 20795

ARONSON, DAVID L, b Bryn Mawr, Pa, July 6, 29; m 57; c 4. MEDICINE, PHYSIOLOGY. Educ: Yale Univ, BS, 51; Univ Pa, MD, 56. Prof Exp: Epidemiologist, USPHS, 57-58, INVESTR, NIH, 58- Mem: Int Soc Thrombosis & Haemostasis. Res: Radiation biology; epidemiology; blood coagulation. Mailing Add: Div of Biologics Stand NIH Bldg 29 Bethesda MD 20014

ARONSON, DONALD GARY, b Jersey City, NJ, Oct 2, 29; m 53; c 3. MATHEMATICS. Educ: Mass Inst Technol, SB, 51, SM, 52, PhD(math), 56. Prof Exp: Res assoc digital comput lab, Univ Ill, 56-57; from instr to assoc prof, 57-65, PROF MATH, UNIV MINN, MINNEAPOLIS, 65- Concurrent Pos: NSF fel, 61-62; Soc Indust & Appl Math vis lectr, 64-66. Mem: Am Math Soc; Math Asn Am. Res: Partial differential equations. Mailing Add: Sch of Math Univ of Minn Minneapolis MN 55455

ARONSON, IRVING, b New Britain, Conn, June 22, 33. THEORETICAL PHYSICS. Educ: Yeshiva Univ, BA, 54; Columbia Univ, MA, 57; NY Univ, PhD(physics), 65. Prof Exp: Engr, Int Tel & Tel Labs, Inc, 57-59; sr engr, Repub Aviation Corp, 59-61; from asst res scientist to assoc res scientist, Courant Inst Math Sci, NY Univ, 61-67; ASST PROF PHYSICS, HOFSTRA UNIV, 67- Mem: Am Phys Soc; Am Asn Physics Teachers. Res: Development and application of variational principles to atomic scattering problems and use of computers in solving these problems. Mailing Add: Dept of Physics Hofstra Univ Hempstead NY 11550

ARONSON, JAMES RIES, b Chicago, Ill, Sept 1, 32; m 63; c 5. SPECTROSCOPY. Educ: Northwestern Univ, BS, 54; Mass Inst Technol, PhD(chem), 58. Prof Exp: Res assoc, Mass Inst Technol, 58-59; RES CHEMIST, ARTHUR D LITTLE, INC, 59- Mem: Am Geophys Union; Optical Soc Am. Res: Infrared and far infrared spectroscopy; remote sensing; theory of emittance and reflectance of particulate media; optical properties of solids; applications of spectroscopy to earth sciences and pollution; tunable laser spectroscopy. Mailing Add: 8 Blossom Hill Rd Winchester MA 01890

ARONSON, JEROME MELVILLE, b Oakland, Calif, May 20, 30; m 55; c 2. PLANT PHYSIOLOGY. Educ: Univ Calif, BA, 52, PhD(bot), 58. Prof Exp: Lectr bot, Univ Calif, 57; fel NIH, 58-59; asst prof biol, Wayne State Univ, 59-63; lectr, Univ Calif, Berkeley, 64-66; assoc prof, 66-71, PROF BOT, ARIZ STATE UNIV, 71- Mem: Am Soc Plant Physiol; Bot Soc Am; Am Inst Biol Sci; Mycol Soc Am. Res: Chemistry of fungal cell walls and polysaccharide storage products; biochemical systematics of fungi. Mailing Add: Dept of Bot & Microbiol Ariz State Univ Tempe AZ 85281

ARONSON, JOAB KLAPP, physiological zoology, see 12th edition

ARONSON, JOHN FERGUSON, b Philadelphia, Pa. PHYSIOLOGY. Educ: Amherst Col, BA, 50; Univ Rochester, PhD(biol), 60. Prof Exp: USPHS fel, Dartmouth Med Sch, 60-63; ASST PROF MED, UNIV PA, 73- Mem: Soc Gen Physiologists. Res: Cell physiology. Mailing Add: 601 Childs Ave Drexel Hill PA 19104

ARONSON, JOHN NOEL, b Dallas, Tex, Mar 15, 34; m 58; c 3. BIOCHEMISTRY. Educ: Rice Inst, BA; Univ Wis, MS, 55, PhD(biochem), 59. Prof Exp: Asst prof chem, Ariz State Univ, 59-65; ASSOC PROF CHEM, STATE UNIV NY ALBANY, 65- Concurrent Pos: NIH res fel oncol, Univ Wis, 59; NSF fac fel, Ind Univ, 63-64; Am Soc Microbiol Pres fel, Univ Ill, 64. Mem: AAAS; Am Chem Soc; Am Soc Microbiol; Am Soc Biol Chemists; NY Acad Sci. Res: Biochemistry of bacterial sporulation; structure and function of proteins. Mailing Add: Dept Chem State Univ NY 1400 Washington Ave Albany NY 12222

ARONSON, LESTER RALPH, b New York, NY, Apr 9, 11; m 36; c 2. ANIMAL BEHAVIOR, NEUROSCIENCES. Educ: Cornell Univ, AB, 32, MA, 33; NY Univ, PhD(vert morphol), 45. Prof Exp: Asst, Dept Animal Behavior, 38-42, staff asst, 42-43, from asst curator to assoc curator, 43-56, actg chmn, 46-49, dean, Coun Sci Staff, 58-60 & 69-70, CUR, AM MUS NATURAL HIST, 56-, CHMN DEPT ANIMAL BEHAVIOR, 49- Concurrent Pos: Lectr, Hunter Col, 46-53; Fulbright res fel, 51-52, 53-54; adj assoc prof, NY Univ, 51-58, adj prof, 58-; assoc ed, Animal Behavior, 63-66, ed, 66-; adj prof, City Univ New York, 69- Mem: Fel AAAS; Am Soc Ichthyol & Herpet; Am Soc Zool; Ecol Soc Am; fel NY Acad Sci. Res: Animal behavior; ecology; endocrinology; neurology; physiological and psychological mechanism mediating reproductive behavior in cichild fishes; amphibians and mammals; brain function in mammals and lower vertebrates. Mailing Add: Dept of Animal Behav Am Mus Natural Hist New York NY 10024

ARONSON, NATHAN NED, JR, b Dallas, Tex, Dec 8, 40; m 64; c 3. BIOCHEMISTRY. Educ: Rice Univ, 62; Duke Univ, PhD(biochem), 67. Prof Exp: Guest investr cell biol, Rockefeller Univ, 66-68; fel molecular biol, Vanderbilt Univ, 68-69; asst prof biochem, 69-75, ASSOC PROF BIOCHEM, PA STATE UNIV, UNIVERSITY PARK, 75- Concurrent Pos: Helen Hay Whitney fel, 66-69. Mem: AAAS; Am Soc Biol Chemists. Res: Lysosomes; membranes; glycoproteins. Mailing Add: 308 Althouse Lab Pa State Univ University Park PA 16802

ARONSON, ROBERT BERNARD, b Seneca Falls, NY, Aug 29, 30; m 55; c 2. BIOCHEMISTRY. Educ: Harvard Univ, AB, 52; Boston Univ, AM, 58, PhD(biochem), 62. Prof Exp: Res assoc, 60-62, asst chmn med dept, 70-72, from asst scientist to assoc scientist, 62-71, scientist biochem, 71-72, ASSOC CHMN MED DEPT, BROOKHAVEN NAT LAB, 72- Mem: Am Chem Soc. Res: Collagen: structure, function and metabolism. Mailing Add: Med Dept Brookhaven Nat Lab Upton NY 11973

ARONSON, SEYMOUR, b New Britain, Conn, Jan 23, 29; m 57; c 3. PHYSICAL CHEMISTRY. Educ: Yeshiva Col, BA, 50; Polytech Inst Brooklyn, PhD(chem), 56; Univ Pittsburgh, MS, 60. Prof Exp: Res chemist, Westinghouse Atomic Power Div, 54-61 & Brookhaven Nat Lab, 61-68; assoc prof, 68-74, PROF CHEM, BROOKLYN COL, 74-Mem: Am Chem Soc. Res: Thermodynamic and electrical properties of solids; kinetics of gas-solid reactions. Mailing Add: Dept of Chem Brooklyn Col Brooklyn NY 11210

ARONSON, STANLEY MAYNARD, b New York, NY, May 28, 22; m 47; c 3. MEDICINE. Educ: City Col New York, BS, 43; NY Univ, MD, 47; Brown Univ, MA, 71. Prof Exp: Asst instr biol, City Col New York, 43; assoc neuropath, Col Physicians & Surgeons, Columbia Univ, 52-54; assoc prof neuropath, State Univ NY Downstate Med Ctr, 54-60, prof path & asst dean, 60-70; chmn sect path, 70-72, PROF MED SCI, BROWN UNIV, 70-, DEAN MED AFFAIRS, 72- Concurrent Pos: Res assoc, Mt Sinai Hosp, NY, S1-54; attend neuropathologist, Kingsbrook Jewish Med Ctr, NY, 52-; dir labs & Neuropathologist, Kings County Hosp, 54-70; consult neuropath, NIH, 62-, US Vet Admin, 63- & Brooklyn Hosp, Long Island Col Hosp, RI Hosp, Butler Hosp, Roger Williams Hosp, Mem Hosp, State Univ NY Hosp & Lutheran Med Ctr; lectr, Yale Sch Med, 64-65; pathologist-in-chief, Miriam Hosp, Providence, 70-; prof lectr, State Univ NY Downstate Med Ctr, 70- Mem: Am Soc Exp Biol & Med; Am Asn Neuropath; Am Acad Neurol; Am Neurol Asn; Am Asn Path & Bact. Res: Experimental neuropathology; population studies of vascular disorders of nervous system; sphingolipidoses. Mailing Add: Off of Med Affairs Brown Univ Providence RI 02912

ARONSZAJN, NACHMAN, b Warsaw, Poland, July, 1907; m 51; c 1. MATHEMATICS. Educ: Univ Warsaw, PhD, 30; Univ Paris, D es Sc, 35. Prof Exp: Nat Res Ctr, Univ Paris, 36-49; res prof math, Okla Agr & Mech Col, 49-52; prof, 51-63, SUMMERFIELD DISTINGUISHED PROF MATH, UNIV KANS, 63- Mem: Am Math Soc; Math Asn Am; Math Soc France; Polish Math Soc. Res: General analysis; topology; approximation methods. Mailing Add: 1015 Tennessee St Lawrence KS 66044

ARORA, HARBANS LALL, b Bharthanwala, India, Apr 14, 21; m 51; c 2. ZOOLOGY. Educ: Univ Panjab, India, BSc, 44, MSc, 45; Stanford Univ, PhD(fisheries biol), 49. Prof Exp: Asst zool, Univ Panjab, India, 44-45; from asst res officer to res officer, Marine Fisheries Inst, Govt India, 49-57; res fel biol, Calif Inst Technol, 57-65; res assoc, Rockefeller Univ, 65-68; assoc prof, 68-72, PROF BIOL SCI, CALIF STATE COL, DOMINGUEZ HILLS, 72- Concurrent Pos: Vis lectr, Univ Panjab, India, 51-53. Honors & Awards: First Prize, Am Soc Ichthyologists & Herpetologists, 48. Mem: AAAS; Am Soc Ichthyologists & Herpetologists; Am Soc Zoologists; Am Asn Anat; Int Inst Embryol. Res: Physiology of learning and behavior in marine organisms and fishes; neuroembryology and experimental biology. Mailing Add: Dept of Biol Sci Calif State Col Dominguez Hills CA 90247

ARORA, KASTURI LAL, b Warburton, Pakistan, Oct 1, 24; Indian citizen; m 50; c 2. BIOCHEMISTRY. Educ: Univ Panjab, WPakistan, BSc, 45, MSc, 46; Panjab Univ, India, PhD(biochem), 57; UNESCO, dipl microbiol, 65. Prof Exp: Res scholar chem, Panjab Univ, India, 49-50, lectr, Panjab Univ Cols, 50-51; sci asst biochem, Cent Drug Res Inst, Lucknow, India, 51-55, sci officer, 55-61; sr sci planning officer, Indian Drugs & Pharmaceut Ltd, Govt India, New Delhi, 61-64, dep chief res & develop antibiotics, 64-68; instr, 69-71, ASSOC BIOCHEM, MED SCH, NORTHWESTERN UNIV, CHICAGO, 71- Concurrent Pos: UN Tech Assistance fel, Czech, Hungary & USSR, 61; Govt India fel, USSR, 64; UNESCO fel, Czech, 65. Mem: Am Soc Microbiol. Res: Antibiotics. Mailing Add: Dept of Biochem Med Sch Northwestern Univ Chicago IL 60611

ARP, ADRIAN L, plant physiology, weed science, see 12th edition

ARP, GERALD KENCH, b Denver, Colo, Dec 6, 47; m 74. BIOSYSTEMATICS, REMOTE SENSING. Educ: Univ Colo, BA, 70, PhD(plant biosyst), 72. Prof Exp: Instr biol, Kans State Univ, 72-73; SR SCIENTIST REMOTE SENSING, LOCKHEED ELECTRONICS CO, 73- Mem: AAAS; Int Asn Plant Taxon; Am Soc Plant Taxonomists. Res: Remote sensing the natural vegetation of the United States and Mexico, using thermal and visible data from the Landsat and National Oceanic and Atmospheric Administration meteorological satellites. Mailing Add: Lockheed Electronics Co 16811 El Camino Real Houston TX 77062

ARP, HALTON CHRISTIAN, b New York, NY, Mar 21, 27; m 49, 62; c 3. ASTRONOMY. Educ: Harvard Univ, AB, 49; Calif Inst Technol, PhD(astron), 53. Prof Exp: Carnegie fel, Mt Wilson & Palomar Observs, 53-55; res assoc, Ind Univ, 55-57; asst astronr, Mt Wilson & Palomar Observs, 57-65, astronr, 65-69, ASTRONR, HALE OBSERVS, 69- Honors & Awards: Warner Prize, Am Astron Soc, 60; Newcomb Cleveland Prize, AAAS, 61. Mem: AAAS; Am Astron Soc; Int Astron Union. Res: Globular clusters and globular cluster variable stars; novae; cepheids and extragalactic nebula; peculiar galaxies; formation and evolution of galaxies; quasars, compact galaxies, companion galaxies and the nature of red shifts. Mailing Add: Hale Observs 813 Santa Barbara St Pasadena CA 91106

ARP, VINCENT D, b Grass Valley, Calif, May 9, 30; m 56; c 3. SOLID STATE PHYSICS. Educ: Univ Calif, Berkeley, BS, 53, PhD(physics), 53. Prof Exp: PHYSICIST, US NAT BUR STAND, 53- Mem: Am Phys Soc. Res: Superconductivity and magnetic properties of solids. Mailing Add: US Nat Bur of Stand 325 Broadway Boulder CO 80301

ARPER, WILLIAM BURNSIDE, b Duluth, Minn, Dec 10, 15; m 42. GEOLOGY. Educ: Univ Okla, BS, 40, MS, 42; Univ Kans, PhD(geol), 53. Prof Exp: Geologist, Phillips Petrol Co, 46-48; asst instr, Univ Kans, 48-50; instr, Univ Mo, 50-53; from asst prof to assoc prof, 53-60, PROF GEOL, TEX TECH UNIV, 60- Mem: Geol Soc Am; Am Asn Petrol Geologists; Am Soc Econ Paleontologists & Mineralogists; Am Chem Soc; Geochem Soc. Res: Stratigraphy and sedimentology. Mailing Add: Dept of Geosci Tex Tech Univ Lubbock TX 79409

ARPKE, CHARLES KENNETH, b Beatrice, Nebr, Dec 8, 21; m 50; c 3. SCIENCE ADMINISTRATION. Educ: Univ Nebr, AB, 48, MA, 50; Air Force Inst Technol, PhD(chem), 62. Prof Exp: Asst prof air sci, Univ Nebr, 55-59, dep chief, Chem & Mat Br, US Air Force Rocket Propulsion Lab, 61-63; res assoc, F J Seiler Res Lab, US Air Force Acad, 63-64, assoc prof chem & dir chem res, 64-66, prof chem & actg head dept chem & physiol, 66-68; chief, Weapons Div, Air Force Armament Lab, 68-69, Technol Div, 69-72 & Weapons Effect Div, 72-74; chief, Anal Div, Armament Develop & Testing Ctr, 74-75; PVT CONSULT, 75- Concurrent Pos: Consult environ sci study, Okaloosa-Walton Jr Col. Mem: Am Chem Soc; Am Inst Aeronaut & Astronaut; fel Am Inst Chem. Res: Bacterial enzymes, thermal stabiltiy of enzymes; toxicity of rocket propellants; high explosive thermal stability; high energy compounds; aeroballistic dispersion phenomena; exterior and terminal ballistics phenomena; computer simulation and modeling studies. Mailing Add: PO Box 1566 Eglin AFB FL 32542

ARQUEMBOURG, PIERRE CHARLES, b Lille, France, Nov 15, 19; US citizen; m 44; c 1. IMMUNOLOGY, IMMUNOCHEMISTRY. Educ: Univ Lille, Cert biol, physics & chem, 40, MD, 51; Pasteur Inst, Lille, Cert serol, 48. Prof Exp: Asst serol, Pasteur Inst, Lille, 40-44, first class asst, 46-51; med off lab method, WHA, Southern Tunisia, Iran & Switz, 52-60; chief lab cancer res, Hodgkin's Disease Res Ctr, Houston, 60-62; res asst cancer res, Dept Surg, Tulane Univ, 62-63; res asst, 63-69, INSTR IMMUNOL, MED CTR, LA STATE UNIV, NEW ORLEANS, 70- Honors & Awards: Cert, Nat Bd Clin Path, France, 51; honorable mention for immunoelectrophoresis of human serum, Am Med Asn, 70. Mem: AAAS; Int Leprosy Asn; Am Med Writers Asn. Res: Kinetics of antigen-antibody reactions; neoplastic tissue antigens; normal animal and bacterial antigens; enzyme versus anti-enzyme reactions; application of immunoelectrophoresis to human pathology and to research. Mailing Add: Res Dept A Ochsner Med Found 1520 Jefferson Hwy New Orleans LA 70121

ARQUETTE, GORDON JAMES, b North Lawrence, NY, Mar 3, 25; m 48; c 3. PHYSICAL CHEMISTRY. Educ: St Lawrence Univ, BS, 47, MS, 49; Cornell Univ, PhD(phys chem), 53. Prof Exp: Res assoc, Cornell Univ, 52-56; sr chemist, Cent Res Lab, Air Reduction Co, Inc, 56-57, sect head, 57-58, supvr, Chem Res Div, 58-61, assoc dir, 61-63, dir, 63-67, dir res, 67-68, dir sci & eng & mem operating comt, 68-70; DIR CORP PLANNING DEVELOP & TECHNOL, AIRCO, INC, 70- Mem: Am Chem Soc; NY Acad Sci; Am Inst Chemists. Res: Chemistry and physics of high polymers; phase equilibria; chemical kinetics. Mailing Add: AIRCO Inc 85 Chestnut Ridge Rd Montvale NJ 07645

ARQUILLA, EDWARD R, b Chicago, Ill, Sept 10, 22; m 49; c 4. PATHOLOGY, IMMUNOLOGY. Educ: Northern Ill Univ, BS, 47; Univ Ill, MS, 49; Western Reserve Univ, PhD(anat), 57. Prof Exp: Asst Physiol, Univ Ill, 48-49; asst prof path, Univ Southern Calif, 59-61; from asst prof to assoc prof, Univ Calif, Los Angeles, 61-68; PROF PATH & CHMN DEPT, UNIV CALIF, IRVINE-CALIF COL MED, 68- Concurrent Pos: USPHS grants, 57-65; Pop Coun grant, 64-65; mem coun, Midwinter Conf Immunologists, 62; mem path training comt, Nat Inst Gen Med Sci, 70- Mem: AAAS; Am Asn Immunol; Soc Exp Path; Am Diabetes Asn; Am Asn Path & Bact. Res: Immunological aspects of insulin as related to etiology and pathogenesis of diabetes; immunological properties of thyroid stimulating hormones and human chorionic gonadotropin. Mailing Add: Dept of Path Univ Calif-Calif Col Med Irvine CA 92664

ARRET, BERNARD, b New York, NY, Nov 11, 19; m 41; c 3. MICROBIOLOGY. Educ: City Col New York, BS, 41; George Washington Univ, MS, 55. Prof Exp: Chemist & sci aid, US Naval Powder Factory, Md, 41-42; chemist, cosmetic div, US Food & Drug Admin, Washington, DC, 45-46; bacteriologist & supvr control lab, E R Squibb & Sons, NJ, 46-49; from bacteriologist to DEP DIR NAT CTR ANTIBIOTICS ANAL, US FOOD & DRUG ADMIN, 49- Honors & Awards: Award of Merit, US Food & Drug Admin; Super Serv Award, US Dept Health, Educ & Welfare. Mem: Am Soc Microbiol. Res: Review of antibiotic new drug applications; certification of all antibiotic drugs manufactured for human use and those veterinary drugs subject to certification according to law. Mailing Add: Nat Ctr for Antibiotics Anal 200 C St SW Washington DC 20204

ARRHENIUS, GUSTAF OLOF SVANTE, b Stockholm, Sweden, Sept 5, 22; m 48; c 3. OCEANOGRAPHY. Educ: Univ Stockholm, DSc, 53. Prof Exp: Geologist, Swedish Skagerak Exped, 46 & Swedish Deep Sea Exped, 47-48; vis asst res oceanogr, 52, asst res oceanogr, 53-55, assoc prof biogeochem 55-59, asst dir, 67-71, PROF OCEANOG, SCRIPPS INST OCEANOG, UNIV CALIF, 59-, ASSOC DIR, INST PURE & APPL PHYS SCI, 71- Concurrent Pos: Vis res fel, Calif Inst Technol, 53-56; vis res fel, Univ Brussels, 55; vis assoc prof, Int Inst Meterol, Sweden, 57-58; Guggenheim fel, Brussels & Berne, 57-58; chmn dept earth sci, Univ Calif, La Jolla, 59-61, dir, Space Res Lab, San Diego, 61-63; mem, NASA Lunar Sample Analysis Team, 70-71 & NASA Asteroid-Comet Mission Study Panel, 71-72; vis comt mem, Max Planck Soc, Ger Fed Repub, 75. Honors & Awards: Petrol Res Fund award, Am Chem Soc, 60-; Group Achievement Award, Lunar Sci Team, NASA, 73. Mem: Am Chem Soc; Geophys Soc; Swedish Geol Soc; Swedish Acad Sci; Int Acad Astronaut. Res: Processes of sedimentation; geochemistry; cosmochemistry; solid state chemistry; oceanography. Mailing Add: Geol Res Div A-020 Scripps Inst of Oceanog La Jolla CA 92093

ARRICK, MYRON S, b New Haven, Conn, Jan 14, 17; m 48; c 3. MEDICINE. Educ: Wesleyan Univ, BA, 39, MA, 40; Long Island Univ, MD, 43. Prof Exp: ASST CLIN PROF MED, SCH MED, UNIV CALIF, SAN FRANCISCO, 49- Res: Clinical diabetes mellitus. Mailing Add: 595 Buckingham Way San Francisco CA 94132

ARRIGHI, FRANCES ELLEN, b Topeka, Kans, July 24, 24. CELL BIOLOGY. Educ: Marymount Col, BS, 47; La State Univ, MS, 61; Univ Tex, PhD(zool), 65. Prof Exp: Med technologist, Charity Hosp, New Orleans, La, 47-54, Self Mem Hosp, Greenwood, SC, 54 & Sch Med, La State Univ, 54-58; ASSOC BIOLOGIST, UNIV TEX M D ANDERSON HOSP & TUMOR INST HOUSTON, 65- Mem: Genetics Soc Am; Am Soc Mammal; Tissue Cult Asn. Res: Cell biology, particularly nucleic acid, protein metabolism and chromosome structure. Mailing Add: Dept of Biol M D Anderson Hosp & Tumor Inst Houston TX 77025

ARRINGTON, CHARLES HAMMOND, JR, b Rocky Mount, NC, Dec 23, 20; m 41; c 2. PHYSICAL CHEMISTRY. Educ: Duke Univ, BS, 41; Calif Inst Technol, PhD(phys chem), 49. Prof Exp: Res chemist, 49-52, res supvr, 52-57, lab dir, 57-67, dir res, 67-74, ASST GEN DIR RES, PHOTO PROD DEPT E I DU PONT DE NEMOURS & CO, 74- Mem: AAAS; Am Chem Soc; Am Phys Soc; Sigma Xi. Res: Physical chemistry of solutions; properties of macromolecules; solid state chemistry. Mailing Add: 711 Greenwood Rd Wilmington DE 19807

ARRINGTON, JACK PHILLIP, b Waterloo, Ohio, Dec 28, 42; m 67; c 2. ORGANIC CHEMISTRY. Educ: Ohio Univ, BS, 65; Ind Univ, PhD(org chem), 69. Prof Exp: Sr res chemist, Spec Assignments Group, 70-71 & Org Chem Res, 71-73, RES SPECIALIST, AGR SYNTHESIS LAB, DOW CHEM CO, 73- Mem: Am Chem Soc; Sigma Xi. Res: Synthesis and structure-activity relationships of agricultural chemicals. Mailing Add: Agr Chem Dept Dow Chem Co Walnut Creek CA 94598

ARRINGTON, LEWIS ROBERTS, b Kirksey, SC, May 1, 19; m 45; c 2. ANIMAL NUTRITION. Educ: Clemson Col, BS, 40; Va Polytech Inst, MS, 41; Univ Fla, PhD, 52. Prof Exp: Asst prof dairy sci, 46-50, assoc prof, 52-67, PROF ANIMAL NUTRIT & ANIMAL NUTRITIONIST, UNIV FLA, 67- Mem: Am Soc Animal Sci; Am Inst Nutrit. Res: Mineral nutrition and metabolism. Mailing Add: Nutrit Lab Univ of Fla Gainesville FL 32601

ARRINGTON, LOUIS CARROLL, b Baltimore, Md, July 5, 36; m 60; c 2. POULTRY SCIENCE. Educ: Univ Md, BS, 58; Univ Calif, Davis, MS, 61; Mich State Univ, PhD(poultry sci), 66. Prof Exp: Res asst, Univ Calif, 58-61 & Mich State Univ, 61-66; asst prof, 66-71, ASSOC PROF POULTRY SCI, UNIV WIS, MADISON, 71- Mem: Poultry Sci Asn; World's Poultry Sci Asn. Res: Physiology of the domestic fowl; reproductive and environmental physiology of the domestic fowl; artificial insemination of poultry; poultry products technology. Mailing Add: Dept of Poultry Sci Univ of Wis Madison WI 53706

ARRINGTON, RICHARD, JR, b Livingston, Ala, Oct 19, 34; m 53; c 4. INVERTEBRATE ZOOLOGY. Educ: Miles Col, AB, 55; Univ Detroit, MS, 57; Univ Okla, PhD(zool), 66. Prof Exp: Asst prof, 57-61, PROF BIOL, MILES COL, 66-, DEAN, 65- Concurrent Pos: Spec instr, Univ Okla, 65-66. Res: Insect morphology, particularly taxonomic characters; protozoology involving nucleocytoplasmic studies and their influences on cell division. Mailing Add: Dept of Biol Miles Col Birmingham AL 35208

ARRINGTON, WENDELL S, b Brooklyn, NY, Oct 3, 36. SCIENCE ADMINISTRATION, BIOSTATISTICS. Educ: Rensselaer Polytech Inst, BS, 58, PhD(physics), 64; Temple Univ, MBA, 73. Prof Exp: NIH trainee biophys, Rensselaer Polytech Inst, 64-65; asst prof physics, State Univ NY Albany, 65-66; head res comput appln, Merck & Co Inc, 66-67; mgr biostatist, 68-72, assoc dir sci comput appln, 72-74, DIR SCI COMPUT APPLN, WYETH LABS, DIV AM HOME PROD, 74- Mem: Drug Info Asn; Asn Comput Mach; Sigma Xi. Res: Development of new user-oriented computer programming languages and data base technology to maximize cost-effectiveness in the analysis of clinical and related research data. Mailing Add: Wyeth Labs Div of Am Home Prod PO Box 8299 Philadelphia PA 19101

ARROE, HACK, b Denmark, Feb 3, 20; US citizen; m 52; c 3. ATOMIC PHYSICS. Educ: Holte Col, Denmark, Cand Art, 38; Copenhagen Univ, DrPhil, 51. Prof Exp: Instr cloud chamber work, Inst Theoret Physics, Denmark, 43; instr atomic spectra, Agr Col, Denmark, 43-47; proj assoc, Univ Wis, 47-50; instr atomic spectra, Agr Col, Denmark, 50-52; proj assoc, Univ Wis, 52-54; prof, Mont State Col, 55-58; head physics div, Denver Res Inst, 58-61; prof, Mont State Col, 61-63; CHMN DEPT PHYSICS, STATE UNIV NY COL FREDONIA, 63- Mem: Fel Am Phys Soc; Math Asn Am. Res: Spectroscopic determination of hyperfine structure and isotope shift. Mailing Add: Dept of Physics State Univ NY Col at Fredonia Fredonia NY 14063

ARROTT, ANTHONY, b Pittsburgh, Pa, Apr 1, 28; m 53; c 4. MAGNETISM. Educ: Carnegie Inst Technol, BS, 48, PhD(physics), 54; Univ Pa, MS, 50. Prof Exp: Asst prof physics, Carnegie Inst Technol, 55-56; sr scientist, Sci Lab, Ford Motor Co, 56-69; PROF PHYSICS, SIMON FRASER UNIV, 68- Concurrent Pos: Guggenheim fel, 63-64; adj prof nuclear eng, Univ Mich, 67-69. Mem: Am Phys Soc; Can Asn Physicists; Am Asn Physics Teachers; Inst Electrical & Electronic Engrs. Res: Magnetism; neutron scattering; electronic structure of metals; liquid crystal elasticity; nuclear engineering. Mailing Add: 6054 Gleneagles Dr West Vancouver BC Can

ARROWSMITH, WILLIAM RANKIN, b Scio, Ohio, Aug 19, 10; m 37; c 4. MEDICINE. Educ: Muskingum Col, BS, 30; Ohio State Univ, MD, 38. Prof Exp: High school teacher, Ohio, 30-34; asst med, Med Sch, Ohio State Univ, 35-36, asst anat, 36-37; intern, Strong Mem & Rochester Munic Hosps, NY, 38-39; asst med, Sch Med, Wash Univ, 39-46; instr med, 46-48, from asst prof to assoc prof, 48-60, PROF CLIN MED, MED SCH, TULANE UNIV, 60- Concurrent Pos: Am Col Physicians res fel, Wash Univ, 41-42, Nat Res Coun fel, 42; asst resident, Barnes Hosp, St Louis, Mo, 39-40, resident, 40-41, asst physician, 41-42; mem staff dept med, Ochsner Clin, 46-, head sect hemat, 47-71, head dept internal med, 51-68; sr vis physician, Charity Hosp; consult dept med, Touro Infirmary, New Orleans; sr vis physician, Oak Ridge Inst Nuclear Studies, 50-62, mem bd dirs, 62-64. Mem: AMA; fel Am Col Physicians; Am Soc Hemat; Am Fedn Clin Res; Am Soc Internal Med. Mailing Add: Ochsner Clin 1514 Jefferson Hwy New Orleans LA 70121

ARSCOTT, GEORGE HENRY, b Hilo, Hawaii, July 20, 23; m 53; c 2. POULTRY NUTRITION. Educ: Ore State Col, BS, 49; Univ Md, MS, 50, PhD(poultry husb), 53. Prof Exp: Asst, Univ Md, 49-50 & 51-53; from asst prof to assoc prof, 53-65, actg head dept, 69-70, PROF POULTRY HUSB, ORE STATE UNIV, 65-, HEAD DEPT POULTRY SCI, 70- Mem: AAAS; Poultry Sci Asn; Am Inst Nutrit; World Poultry Sci Asn. Mailing Add: Dept of Poultry Sci Ore State Univ Corvallis OR 97331

ARSENAULT, GUY PIERRE, b Montreal, Que, Sept 25, 30; m 55; c 2. BIO-ORGANIC CHEMISTRY. Educ: Univ Toronto, BASc, 53; Ohio State Univ, PhD(org chem), 58. Prof Exp: From asst res officer to assoc res officer org chem, Atlantic Regional Lab, Nat Res Coun Can, 60-67; guest, dept chem, 65-66, RES CHEMIST, MASS INST TECHNOL, 67- Mem: Am Chem Soc; Am Soc Mass Spectrometry; Chem Inst Can. Res: Structure of fungal and bacterial metabolites; application of mass spectrometry to organic and biological chemistry; biomedicine and environmental research. Mailing Add: Dept of Chem Mass Inst Technol 77 Massachusetts Ave Cambridge MA 02139

ARSENAULT, HENRI H, b Montreal, Que, Sept 29, 37; m 65; c 2. OPTICS. Educ: Laval Univ, BSc, 63, MSc, 66, PhD(physics), 69. Prof Exp: Scientist x-ray spectros, Aluminium Labs Ltd, 63-64; teacher physics, Col Jonquiere, 65-68; Nat Res Coun Can fel, Inst Optics, Univ Paris, Orsay, 68-70; asst prof, 70-74, ASSOC PROF PHYSICS, LAVAL UNIV, 74- Concurrent Pos: Mem sensors working group, Can Adv Comt on Remote Sensing, 71-74. Mem: Optical Soc Am; Can Asn Physicists; Fr-Can Asn Advan Sci. Res: Optical data processing; digital image processing, holography, speckle phenomena. Mailing Add: Laser & Optics Res Lab Laval Univ Quebec PQ Can

ARSENEAU, DONALD FRANCIS, b St John, NB, Mar 18, 28; m 54; c 9. CELLULOSE CHEMISTRY. Educ: St Francis Xavier Univ, BSc, 50; Laval Univ, DSc(chem), 55. Prof Exp: Assoc prof, 55-70, prof chem, St Francis Xavier Univ, 70-74; prof chem, 74-75, DIR, BRAS D'OR INST, COL CAPE BRETON, 75- Mem: Am Chem Soc; Forest Prod Res Soc; Chem Inst Can; AAAS. Res: Thermal breakdown of wood; differential thermal analysis; thermal degradation of naturally occurring polymers; water quality; utilization of coal for coking. Mailing Add: 10 Xavier Dr Sydney NS Can

ARSHAD, MUHAMMAD AHMAD, b Jallunder, India, Jan 15, 36; m 67; c 2. SOIL

CONSERVATION. Educ: Panjab Univ, BSc, WPakistan, 58, MSc, 60; Univ Alta, PhD(soil sci), 64. Prof Exp: Res fel, Soils Div, Res Coun Alta, 64-66; sr soil scientist, Water & Power Develop Auth, Pakistan, 66-69; res fel soil sci, Univ Sask, 69-72; READER & HEAD DEPT OF AGR ENG, UNIV OF NIGERIA, 72- Mem: Am Soc Agron; Soil Sci Soc Am; Int Soc Soil Sci. Res: Mineral weathering in relation to soil genesis; isolation and characterization of amormous inorganic materials in soils; ion exchange and structural properties of clay minerals; salt movement in soils. Mailing Add: Dept of Agr Eng Univ of Nigeria Nsukka Nigeria

ARSOVE, MAYNARD GOODWIN, b Lincoln, Nebr, Mar 11, 22; m 44; c 4. MATHEMATICS. Educ: Lehigh Univ, BS, 43; Brown Univ, ScM, 48, PhD(math), 50. Hon Degrees: DrEd, Saipan Univ, 46. Prof Exp: Fulbright scholar, Grenoble Univ, 50-51; from instr to assoc prof, 51-60, PROF MATH, UNIV WASH, 60- Concurrent Pos: Guggenheim fel, 57-58; Fulbright fel, Univ Paris, 57-58; NATO fel, Univ Hamburg, 64-65. Mem: Am Math Soc. Res: Subharmonic function theory; potential theory; complex analysis; theory of bases. Mailing Add: Dept of Math Univ of Wash Seattle WA 98195

ART, HENRY WARREN, b Norfolk, Va, Sept 11, 44; m 66; c 2. FOREST ECOLOGY, ENVIRONMENTAL BIOLOGY. Educ: Dartmouth Col, BA, 66; Yale Univ, MPhil, 69, PhD(forest ecol), 71. Prof Exp: Res assoc, 70-71, asst prof & asst dir res, ctr environ studies, 71-76, ASSOC PROF BIOL, WILLIAMS COL, 76- Concurrent Pos: Vis prof environ studies, Univ Vt, 75. Mem: Ecol Soc Am; Brit Ecol Soc; AAAS. Res: Biomass and productivity of ecosystems; nutrient cycling; ecosystem dynamics; land use and successional relationships; coastal ecology. Mailing Add: Dept of Biol Williams Col Williamstown MA 01267

ARTEAGA, LUCIO, b Calatayud, Spain, May 22, 24; Can citizen; m 52; c 2. MATHEMATICS, STATISTICS. Educ: Univ Madrid, cert, 56; Dalhousie Univ, MSc, 59; Univ Sask, PhD(math), 64. Prof Exp: Statistician, NS Tumor Clin, 57-60; consult, Sask Res Coun, 60-62; asst prof math, Univ Windsor, 62-65; asst prof, Dalhousie Univ, 65-67; asst prof, Univ NC, Charlotte, 67-68; assoc prof, 68-69, PROF MATH, WICHITA STATE UNIV, 69- Concurrent Pos: Nat Res Coun Can grants, 64-66. Mem: Am Math Soc; Math Asn Am; Can Math Cong. Res: Integral equations; Fourier transforms. Mailing Add: Dept of Math Wichita State Univ Wichita KS 67208

ARTEMAN, ROBERT LLOYD, b Bloomington, Ill, Nov 2, 34; m 63; c 1. PLANT PATHOLOGY. Educ: Ill State Univ, BA, 61; Univ Ill, MS, 64, PhD(bot), 68. Prof Exp: Asst prof biol, Wis State Univ-Superior, 68-69; ASST PROF BIOL, ILL WESLEYAN UNIV, 69- Mem: AAAS; Bot Soc Am; Mycol Soc Am; Am Inst Biol Sci. Res: Developmental mycology. Mailing Add: Dept of Biol Ill Wesleyan Univ Bloomington IL 61701

ARTEMIADIS, NICHOLAS, b Constantinople, Turkey, May 17, 17; US citizen; m 63. MATHEMATICAL ANALYSIS. Educ: Univ Thessolaniki, MS, 39; French Inst, Athens, dipl, 51; Univ Paris, certs, 54, 55, DSc(math), 57. Prof Exp: Asst prof, Univ Wis, Milwaukee, 58-60, assoc prof, 61-67; assoc prof, Univ Thessolonki, 60-61; PROF, SOUTHERN ILL UNIV, CARBONDALE, 67- Concurrent Pos: Vis lectr, Univ Wis, 59 & Univ Chicago, 60; prof math, Univ Patrous, Greece, 75- Honors & Awards: Stravropoulos award, Greek Embassy, Paris, 63. Mem: Am Math Soc; Math Soc France; Greek Math Soc. Res: Mathematical and hormonic analysis. Mailing Add: Dept of Math Southern Ill Univ Carbondale IL 62901

ARTENSTEIN, MALCOLM S, b Haverhill, Mass, Feb 24, 30; m 56; c 2. MEDICINE, INFECTIOUS DISEASES. Educ: Brown Univ, AB, 51; Tufts Univ, MD, 55; Am Bd Internal Med, dipl, 63. Prof Exp: Intern, Mt Sinai Hosp, New York, 55-56; resident med, Boston Vet Admin Hosp, Mass, 56-58; med officer virol, 59-61, 62-66, CHIEF DEPT BACT DIS, WALTER REED ARMY INST RES, 66- Concurrent Pos: Concurrent Pos: USPHS fels, NEng Ctr Hosp, Boston, 58-59 & 61-62. Mem: Am Fedn Clin Res: Asn Am Physicians; Am Asn Immunol; Infectious Dis Soc Am; Am Epidemiol Soc. Res: Acute respiratory infections; meningococcal infections; virus diseases of skin; virus diagnostic investigations; immunology of infection. Mailing Add: Dept of Bact Dis Walter Reed Army Inst of Res Washington DC 20012

ARTERBURN, DAVID ROE, b Norfolk, Nebr, Dec 29, 39; m 63; c 2. MATHEMATICAL ANALYSIS. Educ: Southern Methodist Univ, BS, 61; NMex State Univ, MS, 63, PhD(functional anal), 64. Prof Exp: Asst prof math, Univ Mont, 64-67; ASSOC PROF MATH, NMEX INST MINING & TECHNOL, 67- Mem: Am Math Soc; Math Asn Am. Res: Functional analysis; rings of functions; rings of operators. Mailing Add: Dept of Math NMex Inst Mining & Technol Socorro NM 87801

ARTH, JOSEPH GEORGE, b Rockville Centre, NY, July 2, 45; m 71; c 1. GEOCHEMISTRY. Educ: State Univ NY, Stony Brook, BS, 67, MS, 70, PhD(petrol, geochem), 73. Prof Exp: Nat Res Coun res assoc isotope geol, Denver, 73-74, GEOLOGIST, US GEOL SURV, 74- Mem: Geol Soc Am; Geochem Soc; Am Geophys Union; Mineral Soc Am. Res: Geochronology, petrology, chemistry and isotopic character of igneous and meta-igneous rock suites to determine their age, origin and role in the earth's crustal history. Mailing Add: US Geol Surv 929 National Ctr Reston VA 22092

ARTHAUD, RAYMOND LOUIS, b Cambridge, Nebr, Apr 21, 21; m 47; c 5. ANIMAL SCIENCE. Educ: Univ Nebr, BSc, 47, MSc, 49; Univ Mo, PhD(animal breeding), 53. Prof Exp: Instr animal husb, Univ Nebr, 49; asst prof beef cattle res sta, Va Polytech Inst, 52-54; animal husbandman beef cattle breeding, Animal & Poultry Husb Res Br, Agr Res Serv, USDA, Univ Nebr, 54-59; PROF ANIMAL SCI, EXTEN, UNIV MINN, 59- Mem: Am Soc Animal Sci. Res: Swine breeding and nutrition; beef cattle breeding and nutrition. Mailing Add: Inst of Agr Univ of Minn St Paul MN 55108

ARTHAUD, VINCENT HENRY, b Cambridge, Nebr, Sept 7, 11; m 45; c 2. ANIMAL SCIENCE. Educ: Univ Nebr, MS, 53. Prof Exp: Asst prof, 45-68, ASSOC PROF ANIMAL SCI, UNIV NEBR, LINCOLN, 68- Mem: Am Soc Animal Sci; Sigma Xi. Res: Beef cattle breeding and management; meats; cattle nutrition; beef carcass evaluation. Mailing Add: Dept of Animal Sci Univ of Nebr Lincoln NE 68508

ARTHUR, ALAN THORNE, b Evanston, Ill, Feb 26, 42; m 65; c 2. REPRODUCTIVE BIOLOGY. Educ: Hanover Col, BA, 64; Drake Univ, MA, 66; Univ Colo, Boulder, PhD(biol), 71. Prof Exp: NIH fel, Worcester Found Exp Biol, 71-73; INSTR OBSTET & GYNEC, M S HERSHEY MED CTR, PA STATE UNIV, 73- Mem: Am Fertility Soc; AAAS. Res: Role of proteins in pre- and postimplantation embryonic development. Mailing Add: Reprod Biol Dept Obstet & Gynec Milton S Hershey Med Ctr Hershey PA 17033

ARTHUR, B WAYNE, b Phil Campbell, Ala, Jan 1, 25; m 55; c 2. ENTOMOLOGY, TOXICOLOGY. Educ: Auburn Univ, BS, 51, MS, 53; Univ Wis, PhD(entom, toxicol), 57. Prof Exp: From asst prof to assoc prof entom, Auburn Univ, 57-63; dir agr chem res, Ciba Corp, 63-67, pres, Ciba Agrochem Co, 67-72, VPRES AGR DIV, CIBA-GEIGY CORP, 72- Concurrent Pos: Consult toxicol study sect, USPHS, 61-65. Mem: Entom Soc Am; Am Chem Soc. Res: Chemistry and biochemistry of insecticides; relationship of structure to activity with emphasis on insect and mammalian toxicology. Mailing Add: Agr Div Ciba-Geigy Corp PO Box 11422 Swing Rd Greensboro NC 27409

ARTHUR, ELIZABETH JEAN, physiology, neurophysioloy, see 12th edition

ARTHUR, JAMES ALAN, b Stockton, Calif, June 19, 36; m 58; c S. POULTRY BREEDING. Educ: Univ Calif, Davis, BS, 61, PhD, 65. Prof Exp: Pop geneticist, 64-70, mgr appl res, 70-71, RES MGR, HY-LINE POULTRY FARMS, 71- Mem: Genetics Soc Am; Poultry Sci Asn. Res: Improvement of hybrid performance in the fowl. Mailing Add: Res Hq Hy-line Int Johnston IA 50131

ARTHUR, JETT CLINTON, b Hemphill, Tex, May 31, 18; m 41; c 3. PHYSICAL CHEMISTRY. Educ: Stephen F Austin State Col, BA, 39; Univ Tex, MA, 46; Environ Eng Intersoc Bd, dipl. Prof Exp: Chemist, 41-49, chemist in charge protein prod, 49-52, chemist in charge biochem, 52-56, prin chemist, head radiochem, 56-66, CHIEF CHEMIST, SOUTHERN REGIONAL RES CTR, AGR RES SERV, USDA, 66- Concurrent Pos: Abstractor, Chem Abstr, 46-, ed phys org chem, 63-, ed adv bd, 74-; lectr, Brit Asn Radiol Res, 65; lectr, NATO Advan Study Inst, Wales, 67. Del, Int Atomic Energy Comn, Austria, 62, Int Union Pure & Appl Chemists, Japan, 66, Can, 68, Australia, 69, Fourth UN Conf Peaceful Uses Atomic Energy, Switz, 71 & First Pac Chem Eng Cong, Japan, 72. Mem: Am Chem Soc; Fiber Soc; fel Am Inst Chemists; Am Inst Chem Engrs; Am Acad Environ Engrs. Res: Radiation chemistry of natural products; physical organic chemistry of natural polymers including cellulose, protein and starch; industrial chemistry; applied mathematics; radio; radar. Mailing Add: Southern Regional Res Ctr USDA 1100 Robert E Lee Blvd New Orleans LA 70179

ARTHUR, JOHN READ, JR, b Omaha, Nebr, Dec 17, 31; m 54; c 4. SURFACE PHYSICS. Educ: Iowa State Univ, BS, 54, PhD(phys chem), 61. Prof Exp: Res asst surface chem, Inst Atomic Res, Iowa State Univ, 56-61; MEM TECH STAFF, BELL TEL LABS, 61- Mem: Am Phys Soc; Am Vacuum Soc. Res: Adsorption and surface reaction kinetics on semi-conductors; thin film structures and growth mechanisms; electron diffraction and auger spectroscopy; molecular beam scattering from surfaces. Mailing Add: Semiconductor Electronics Res Lab Bell Labs Murray Hill NJ 07974

ARTHUR, PAUL, JR, b Detroit, Mich, June 20, 15. CHEMISTRY. Educ: Loyola Univ, Ill, BS, 35; Mass Inst Technol, PhD(chem), 38. Prof Exp: RES CHEMIST, E I DU PONT DE NEMOURS & CO, 38- Concurrent Pos: Res fel, Harvard Univ, 57-58. Mem: Am Chem Soc; Mineral Soc Am; Asn Comput Mach. Res: Organic and hydrothermal reactions; crystallography. Mailing Add: 11 Brandywine Blvd Wilmington DE 19809

ARTHUR, RANSOM JAMES, b New York, NY, Dec 5, 25; m 54; c 2. PSYCHIATRY, EPIDEMIOLOGY. Educ: Univ Calif, Berkeley, AB, 47; Harvard Univ, MD, 51. Prof Exp: Intern, Mass Gen Hosp, Boston, 51-52; asst & chief resident pediat, Children's Med Ctr, 52-54; resident, Queen's Hosp, Honolulu, 54-55; teaching consult pediat, Kauikeolani Children's Hosp, 55-56; resident psychiat, Hosp, Nat Naval Med Ctr, Bethesda, Md, US Navy, 57-60, officer-in-chg training in psychiat, Naval Hosp, Oakland, Calif, 60-63, cmndg officer, officer-in-chg, Med Neuropsychiat Res Unit, 63-74; PROF PSYCHIAT & EXEC VCHMN DEPT, UNIV CALIF, LOS ANGELES, 74-, ASST DEAN EDUC POLICY & CURRICULUM, SCH MED, 74-, EXEC ASSOC DIR, NEUROPSYCHIAT INST, 74- Concurrent Pos: Teaching fel, Harvard Med Sch, 52-54; clin instr, Sch Med, Univ Calif, 61-63, asst clin prof, Univ Calif, Los Angeles, 65-74; adj prof, Univ Calif, San Diego. Mem: AAAS; Am Psychiat Asn; Asn Advan Psychother; NY Acad Sci. Res: Clinical and social psychiatry; epidemiology of mental disease; medical education. Mailing Add: Dept of Psychiat Univ of Calif Ctr for Health Sci Los Angeles CA 90024

ARTHUR, ROBERT DAVID, b Union City, Ind, Oct 3, 42; m 65; c 2. BIOCHEMISTRY. Educ: Purdue Univ, BS, 64, MS, 66; Univ Mo, PhD(agr chem), 70. Prof Exp: Fel, Univ Mo, 70-72; ASST PROF BIOCHEM, MISS STATE UNIV, 72- Mem: Am Soc Animal Sci. Res: Study of the health effects of pesticides in humans, metabolism of pesticides, effects of pesticides on microorganisms and pesticide residues in the environment and their effect on human health. Mailing Add: Dept of Biochem Miss State Univ PO Drawer BB Mississippi State MS 39762

ARTHUR, ROBERT SIPLE, b Redlands, Calif, Mar 16, 16; m 41, 75; c 1. OCEANOGRAPHY. Educ: Univ Redlands, AB, 38; Univ Calif, Los Angeles, AM, 49, PhD(oceanog), 50. Prof Exp: Teacher, high sch, 39-40; asst, Univ Calif, Los Angeles, 40-42, asst prof naval sci, 43-44; assoc oceanog, 44-46, asst oceanog, 46-48, oceanogr, 48-50, from asst prof to assoc prof, 51-63, PROF OCEANOG, SCRIPPS INST OCEANOG, UNIV CALIF, 63- Mem: Am Geophys Union; Am Meteorol Soc. Res: Physical oceanography. Mailing Add: ORD A-030 Scripps Inst Oceanog Univ Calif La Jolla CA 92093

ARTHUR, WALLACE, b New York, NY, Nov 22, 32; m 60; c 3. NUCLEAR PHYSICS, SPACE PHYSICS. Educ: NY Univ, BEngSc & BEE, 57, PhD(physics), 62. Prof Exp: Asst physics, NY Univ, 57-61; lectr physics, Rutgers Univ, 61-62; prof physics & chmn dept, 62-73, DEAN COL SCI & ENG, FAIRLEIGH DICKINSON UNIV, 73- Mem: Am Asn Physics Teachers; Am Phys Soc. Res: Cosmic ray analyses; atmospheric physics and gas kinetics; low energy experimental nuclear physics; gamma ray lasers. Mailing Add: Col of Sci & Eng Fairleigh Dickinson Univ Teaneck NJ 07666

ARTIST, RUSSELL (CHARLES), b Francisville, Ind, Jan 5, 11; m 39, 65; c 2. BOTANY. Educ: Butler Univ, BS, 32; Northwestern Univ, MS, 34; Univ Minn, PhD(bot, geol), 38. Prof Exp: Head dept sci, Amarillo Col, 38-45; head dept natural sci, Westminster Col, 45-47; head dept biol, Abilene Christian Col, 47-48; prof apologetics, Kolleg Der Gemeinde Christi, Ger, 48-51; PROF BIOL, DAVID LIPSCOMB COL, 53- Mem: Creation Res Soc. Res: Paleoecology and pollen analysis; seedling survival studies. Mailing Add: Dept of Biol David Lipscomb Col Nashville TN 37203

ARTIZZU, MARIA, b Cagliari, Italy. HISTOPATHOLOGY. Educ: Univ Cagliari, Sardinia, MD, 52, PhD(pediat), 56; Univ Rome, PhD(path), 65; Univ Montreal, PhD(anat), 70. Prof Exp: Asst prof pediat, Univ Cagliari, Sardinia, 53-56 & path, 58-67; vis prof exp path, 67-68, LECTR HISTOL, UNIV MONTREAL, 69- Concurrent Pos: Sci res awards, Ital Univ Coun, 61 & 62. Mem: Can Asn Anat. Res: Biochemical and morphological studies on rat hepatic steatosis and necrosis; morphologic and radioautographic studies on stomach and liver of magnesium deficient rats. Mailing Add: Dept of Anat Sch Optom Univ Montreal Montreal PQ Can

ARTMAN, JOSEPH OSCAR, b New York, NY, Apr 22, 26. CHEMICAL PHYSICS, SOLID STATE PHYSICS. Educ: City Col New York, BS, 44; Columbia Univ, MA, 48, PhD(physics), 53. Prof Exp: Asst physics, Columbia Univ, 49-50, asst radiation lab, 50-52; staff mem, Lincoln Labs, Mass Inst Technol, 52-55; res fel, div eng & appl physics, Harvard Univ, 55-58; staff mem, appl physics lab, Johns Hopkins Univ, 58-64; assoc prof, 64-67, PROF PHYSICS & ELEC ENG, CARNEGIE-MELLON UNIV, 67- Concurrent Pos: Sr fel, Mellon Inst, 68- Mem: Am Phys Soc; Optical Soc Am; AAAS; Sigma Xi. Res: Microwave spectroscopy; ferromagnetic and paramagnetic resonance; masers; crystal physics; magnetism; optics; optical processing; visible and infrared spectroscopy; environmental monitoring; energy conversion; chemical physics. Mailing Add: Dept of Physics Carnegie-Mellon Univ Pittsburgh PA 15213

ARTMAN, NEIL ROSS, b Amarillo, Tex, Jan 11, 28; m 50; c 4. ORGANIC CHEMISTRY. Educ: Tex Tech Col, BS, 48; Univ Colo, MS, 50; Univ Tex, PhD(org chem), 55. Prof Exp: Chemist, Bancroft & Sons Co, 51-53; chemist, 55-71, MGR TECH PUBL, PROCTER & GAMBLE CO, 71- Concurrent Pos: Assoc ed, J Am Oil Chemists' Soc, 72- Mem: Am Chem Soc; Am Oil Chemists Soc. Res: Chemistry of fats and oils; flavor chemistry; technical writing. Mailing Add: Miami Valley Lab Procter & Gamble PO Box 39175 Cincinnati OH 45247

ARTNA-COHEN, AGDA, b Tartu, Estonia, Oct 19, 30; US citizen; m 67; c 1. INFORMATION SCIENCE, NUCLEAR PHYSICS. Educ: McMaster Univ, BSc, 57, PhD(nuclear physics), 61. Prof Exp: Nuclear physicist nuclear data proj, Nat Acad Sci-Nat Res Coun, 61-63, nuclear physicist nuclear data proj, Oak Ridge Nat Lab, 64-67, consult, 67-75; RES PHYSICIST, NAVAL RES LAB, WASHINGTON, DC, 75- Concurrent Pos: Lectr, USDA Grad Sch, 62-63; assoc ed, Nuclear Data, 65-67. Mem: Can Asn Physicists; Am Phys Soc. Res: Nuclear structure physics; information retrieval and evaluation; science writings. Mailing Add: 8801 Mansion Farm Pl Alexandria VA 22309

ARTUSIO, JOSEPH F, JR, b Jersey City, NJ, Nov 26, 17; m 45; c 6. ANESTHESIOLOGY. Educ: St Peter's Col, BS, 39; Cornell Univ, MD, 43; Am Bd Anesthesiol, dipl. Prof Exp: Intern, Bellevue Hosp, New York, 43-44; resident anesthesiol, New York Hosp, 46-47; from instr to assoc prof surg, 47-57, prof anesthesiol in surg & obstet & gynec, 57-67, PROF ANESTHESIOL & CHMN DEPT, MED COL, CORNELL UNIV, 67- Concurrent Pos: Asst attend anesthesiologist in chg, 48-57, anesthesiologist-in-chief, 57-; ed-in-chief, Clin Anesthesis; mem, Unitarian Serv Comt to Japan, 56. Mem: Fel Am Col Anesthesiol; AMA; Am Soc Anesthesiol; Asn Univ Anesthetists; fel NY Acad Med. Res: Effect of anesthetic and muscular relaxant agents on the physiology and pharmacology of surgical patients. Mailing Add: Ithaca NY

ARTUSY, RAYMOND L, economics, resource geography, see 12th edition

ARTZ, CURTIS PRICE, b Jerome, Ohio, Sept 29, 15; m 39, 73; c 3. SURGERY. Educ: Ohio State Univ, AB, 36, MD, 39; Baylor Univ, MS, 51; Am Bd Surg, dipl. Prof Exp: Mem surg team to Japan, 50, Korea, 51, dir team, Korea, 53; dir surg res unit, Brooke Army Med Ctr & prof surg res, Army Med Serv Sch, Univ Tex, 53-56; assoc prof surg, Sch Med, Univ Miss, 56-62, prof, 62-63; Shrine prof, Med Br, Univ Tex, 63-65; PROF SURG & CHMN DEPT, MED UNIV SC, 65- Concurrent Pos: Vis lectr, Walter Reed Res Inst, 53-56; consult surgeon, Vet Admin Ctr, 56-; consult, Surgeon Gen, US Army, 57-; mem, Surg Gen Adv Comt Nutrit; dir, First Int Cong Burns, 60. Honors & Awards: Surgeons Award Distinguished Serv to Safety, Am Col Surg, Am Asn Surg of Trauma & Nat Safety Joint Action Prog, 74. Mem: Am Surg Asn; Am Trauma Soc (pres, 73); AMA; Am Col Surg; Asn Mil Surg US. Res: Surgical metabolism and infection; nutrition; burns; physiology in surgery; shock. Mailing Add: Dept of Surg Med Univ of SC Charleston SC 29401

ARTZT, KAREN, b New York, NY, Sept 4, 42. DEVELOPMENTAL GENETICS. Educ: Cornell Univ, AB, 64, PhD(genetics), 72. Prof Exp: Fel immunogenetics, Pasteur Inst, Paris, 72-73; res assoc develop biol, Col Med, Cornell Univ, 73-75, asst prof, 75-76; ASSOC DEVELOP BIOL, SLOAN KETTERING INST CANCER RES, 76- Res: Developmental biology of embryonal tumors with particular reference to their cell surface antigens. Mailing Add: Sloan Kettering Inst Cancer Res 1275 York Ave New York NY 10021

ARTZY, RAFAEL, b Königsberg, Ger, July 23, 12; US citizen; m 34; c 3. GEOMETRY, ALGEBRA. Educ: Hebrew Univ, Jerusalem, MA, 34, PhD(math), 45. Prof Exp: Teacher & prin, high schs, Israel, 35-51; from instr to assoc prof math, Israel Inst Technol, 51-60; assoc prof math, Univ NC, 60-61; prof math, Rutgers Univ, 61-65; prof math, State Univ NY Buffalo, 65-67; PROF MATH, TEMPLE UNIV, 67- Concurrent Pos: Res assoc & vis lectr, Univ Wis, 56-58; NSF res grants, 62-67; visitor, Inst Advan Study, 64; vis prof, Israel Inst Technol, 70; ed, J Geom, 71-; prof, Univ Haifa, 73-75. Mem: Am Math Soc; Israel Math Union. Res: Algebraic properties and geometric applications of quasi-groups and loops; foundations of linear and nonlinear geometry. Mailing Add: Dept of Math Temple Univ Philadelphia PA 19122

ARUMI, FRANCISCO NOE, b Valparaiso, Chile, Feb 4, 40; m 63; c 2. THERMAL PHYSICS. Educ: Univ NC, BS, 62, MS, 64; Univ Tex, PhD(physics), 70. Prof Exp: Asst prof physics, Calif State Polytech Col, 64-65; vis prof, Univ Costa Rica, 65-66; regional specialist, Asn Am Univ, 66-67; res assoc physics, 70-72, asst prof archit, 71-75, ASSOC PROF ARCHIT, UNIV TEX, 75- Concurrent Pos: Fulbright fel, Comn Int Exchange of Persons, 71 & 72. Mem: Am Phys Soc; Am Asn Physics Teachers. Res: Thermal physics applied to energy analysis in architectural design; thermodynamic modelling of societal systems. Mailing Add: 2604 Parkview Dr Austin TX 78757

ARUNASALAM, VICKRAMASINGAM, b Jaffna, Ceylon, Aug 26, 35; m 68. PHYSICS. Educ: Univ Ceylon, BS, 57; Univ Mass, Amherst, MS, 60; Mass Inst Technol, PhD(physics), 64. Prof Exp: Asst lectr physics, Univ Ceylon, 57-58; instr physics, Univ Mass, summer 60; res assoc plasma physics, 64-67, MEM RES STAFF, PLASMA PHYSICS LAB, PRINCETON UNIV, 67- Mem: Am Phys Soc. Res: Plasma physics; quantum theory. Mailing Add: Plasma Physics Lab Princeton Univ PO Box 451 Princeton NJ 08540

ARVAN, DEAN ANDREW, b Arta, Greece, Sept 27, 33; US citizen; m 58; c 3. CLINICAL PATHOLOGY, BIOCHEMISTRY. Educ: Wilkes Col, BS, 55; Hahnemann Med Col, MD, 59. Prof Exp: Intern, St Agnes Hosp, Philadelphia, Pa, 59-60; resident path, Hosp, 60-64, asst dir chem sect, Pepper Lab, Hosp, 67-69, asst instr, Univ, 60-64, assoc, 64-69, ASSOC PROF PATH, SCH MED, UNIV PA & DIR CHEM SECT, PEPPER LAB, HOSP, 69- Concurrent Pos: Attend physician, Dept Path, Vet Admin Hosp, Philadelphia, Pa, 67-; consult biochemist, Coatesville Hosp, Pa, 68- Mem: AAAS; Am Asn Clin Chemists; Am Soc Clin Pathologists. Res: Clinical enzymology, use of serum and tissue enzymes in diagnosis; protein chemistry, application of newer fractionation techniques to the diagnosis of disease. Mailing Add: Hosp of the Univ of Pa 3400 Spruce St Philadelphia PA 19104

ARVAN, PETER GEORGE, inorganic chemistry, see 12th edition

ARVESEN, JAMES NORMAN, b Portland, Ore, Oct 22, 42; m 65; c 2. MATHEMATICAL STATISTICS, APPLIED STATISTICS. Educ: Univ Calif, AB, 64; Stanford Univ, MS, 66, PhD(statist), 68. Prof Exp: Mathematician, Daniel H Wagner & Assocs, 66; asst prof statist, Purdue Univ, Lafayette, 68-73; assoc dir sci affairs, 73-74, GROUP ASSOC DIR SCI AFFAIRS, PFIZER PHARMACEUT, 74- Concurrent Pos: Consult, Ill Geol Surv, 68-; adj asst prof math statist, Columbia Univ, 73-74; vis assoc prof math, Hunter Col, 74-; sr consult, JJJ Statist Appln, Inc, 75- Mem: Inst Math Statist; Biomet Soc; Am Statist Asn; Inst Mgt Sci; Opers Res Soc Am. Res: Robust procedures; regression analysis; biostatistics. Mailing Add: Sci Affairs Pfizer Pharmaceut 235 E 42nd St New York NY 10017

ARVESON, MAURICE HENSHAW, chemistry, deceased

ARVESON, WILLIAM BARNES, b Oakland, Calif, Nov 22, 34; m 61; c 3. MATHEMATICS. Educ: Calif Inst Technol, BS, 60; Univ Calif, Los Angeles, AM, 63, PhD(math), 64. Prof Exp: Mathematician, US Naval Ord Test Sta, 60-64; actg asst prof math, Univ Calif, Los Angeles, 65; Benjamin Peirce instr, Harvard Univ, 65-68; lectr, 68-69, assoc prof, 69-73, PROF MATH, UNIV CALIF, BERKELEY, 73- Mem: Am Math Soc. Res: Algebras of operators on Hilbert space; representations of Banach algebras; ergodic and prediction theory. Mailing Add: Dept of Math Univ of Calif Berkeley CA 94720

ARVEY, MARTIN DALE, b Los Angeles, Calif, Dec 6, 15; m 39; c 3. ZOOLOGY, ORNITHOLOGY. Educ: Univ Calif, BA, 37; Univ Idaho, MS, 39; Univ Kans, PhD(vert zool), 39. Prof Exp: Instr & chmn natural sci, Boise Jr Col, 39-44; instr zool, Univ Kans, 47-49; asst prof biol sci, Univ Okla, 49-50; prof biol & chmn dept, Long Beach State Col, 53-63; mem biomed div, NSF, 63-65, mem int sci activ, 65-67, asst prog dir grad educ sci, 66-67; chmn dept biol sci, Univ Pac, 67-72; CHMN DEPT BIOL, UNIV NC, CHARLOTTE, 72- Concurrent Pos: Fulbright Award, State Dept, Senegal, Africa, 75-76. Mem: Cooper Ornith Soc; Am Ornith Union. Res: Distribution of Western birds; avian anatomy, especially passerine birds; distribution, behavioral adaptations of birds. Mailing Add: Dept of Biol Univ of NC Charlotte NC 28223

ARYA, ATAM PARKASH, b Panjab, India, June 2, 34; m 61; c 2. NUCLEAR PHYSICS. Educ: Univ Rajasthan, India, BS, 53, MS, 55; Univ Panjab, India, MA, 56; Pa State Univ, PhD(physics), 60. Prof Exp: Vis res assoc physics, Pa State Univ, 60-62; asst prof, Univ Toledo, 62-64; from asst prof to assoc prof, 64-74, PROF PHYSICS, WVA UNIV, 74- Mem: Am Phys Soc; Am Asn Physics Teachers. Res: Beta and gamma spectroscopy; neutron inelastic scattering; energy and angular distribution of protons from neutron induced reactions. Mailing Add: Dept of Physics WVa Univ Morgantown WV 26506

ARYA, SATYA PAL SINGH, b Mavi Kalan, India, Aug 24, 39. METEOROLOGY, FLUID DYNAMICS. Educ: Univ Roorkee, India, BE, 61, ME, 64; Colo State Univ, Ft Collins, PhD(fluid mech), 68. Prof Exp: Asst engr civil eng, Irrigation Dept, United Prov, India, 61-62; lectr, Univ Roorkee, United Prov, India, 63-65; from res asst to res assoc fluid dynamics, Colo State Univ, Ft Collins, 65-69; res asst prof atmospher sci, 69-74, RES ASSOC PROF ATMOSPHER SCI, UNIV WASH, SEATTLE, 74- Mem: Am Geophys Union; AAAS; Sigma Xi; Am Soc Mech Engrs. Res: Research on planetary boundary layers; atmospheric turbulence and diffusion; air-sea and air-sea ice interactions; micrometeorology. Mailing Add: Dept of Atmospher Sci Univ of Wash Seattle WA 98195

ARZT, SHOLOM, b New York, NY, May 3, 29; m 60; c 1. MATHEMATICS. Educ: NY Univ, AB, 46, MS, 48, PhD(math), 51. Prof Exp: Rockefeller asst, Courant Inst Math Sci, NY Univ, 46-51; sr mathematician, Appl Physics Lab, Johns Hopkins Univ, 51-54; proj engr, Specialty Electronics & Eng Co, 55-57 & Universal Transistor Prod Corp, 57-58; asst prof, 58-63, assoc prof & head dept, 63-69, PROF MATH, COOPER UNION, 69- Mem: Am Math Soc; Math Asn Am; Soc Indust & Appl Math. Res: Number theory; analysis. Mailing Add: Dept of Math Cooper Union New York NY 10003

ASADULLA, SYED, b Channapatna, India, June 3, 33; m 59. MATHEMATICS. Educ: Cent Col, Bangalore, India, BSc, 55; Univ Karachi, MSc, 56; Univ Fla, PhD(math), 66. Prof Exp: Lectr math, Fed Col, Univ Karachi, Pakistan, 55-56, head dept, 56-57, chmn fac sci, 57-61; interim instr math, Univ Fla, 61-66; asst prof, Miami Univ, 66-68; asst prof, 68-71, ASSOC PROF MATH, ST FRANCIS XAVIER UNIV, 71- Concurrent Pos: Coop lectr, Univ Karachi, 57-61. Mem: Math Asn Am; Am Math Soc. Res: Number theory; quadratic forms. Mailing Add: PO Box 868 Antigonish NS Can

ASAI, GEORGE NAPOLEON, b Houston, Tex, Dec 25, 16; m 50; c 4. PLANT PATHOLOGY. Educ: Cornell Univ, BS, 38, PhD(plant physiol), 43. Prof Exp: Res instr floricult, Cornell Univ, 42-44; plant pathologist, Ft Detrick, US Army, 57-62, chief spec proj div, Rocky Mt Arsenal, Colo, 62-64; plant pathologist, US Army For Sci & Technol Ctr, Washington, DC, 65-71, SUPVRY RES SCIENTIST, US ARMY SCI & TECHNOL CTR, FAR EAST OFF, 71- Mem: Am Phytopath Soc. Res: Cereal rusts; rice blast; anti-crop warfare. Mailing Add: PSC Box 4488 APO San Francisco CA 96323

ASAKAWA, GEORGE, b San Diego, Calif, Jan 18, 18; m 45; c 2. PHYSICAL CHEMISTRY. Educ: Calif Inst Technol, BS, 39. Prof Exp: Res chemist, 46-47, from asst chief chemist to chief chemist, 48-59, vpres oper, 59-61, exec vpres, 62-74, DIR, VERNAY LABS, INC, 53-, PRES, 74- Res: Heat sensitive materials. Mailing Add: Vernay Labs Inc 116 East S College St Yellow Springs OH 45387

ASAKURA, TOSHIO, b Osaka, Japan, Aug 21, 35; m 67; c 1. BIOCHEMISTRY, HEMATOLOGY. Educ: Kyoto Med Col, BS, 56, MD, 60; Univ Tokyo, PhD(biochem), 65; Univ Pa, MA, 74. Prof Exp: Intern hosp, Univ Tokyo, 60-61; asst prof biochem, Univ, 65-67; assoc biophysics, Johnson Res Found, 67-69, asst prof, 69-74, ASSOC PROF PEDIAT, UNIV PA, 74- Concurrent Pos: Career develop award, NIH, 70-75. Mem: Japanese Biochem Soc; Am Soc Biol Chemists; Am Chem Soc. Res: Metabolism of red blood cells; heme modification; spin-labeling of hemes and porphyrins; hemoglobinopathies. Mailing Add: Children's Hosp of Philadelphia 34th St & Civic Ctr Blvd Philadelphia PA 19104

ASAL, NABIH RAFIA, b Haifa, Israel, Dec 21, 38; m 66. EPIDEMIOLOGY, BIOSTATISTICS. Educ: William Jewell Col, AB, 63; Univ Mo, MS, 65; Univ Okla, PhD(epidemiol), 68. Prof Exp: Res asst epidemiol, Univ Mo, 65-66; from instr to asst prof biostatist & epidemiol, 68-72, epidemiologist, 69-72, ASSOC PROF BIOSTATIST & EPIDEMIOL, SCH HEALTH, UNIV OKLA, 73- Concurrent Pos: Consult, Okla Regional Med Prog, 68-, Epidemiol Div, Okla State Health Dept, Fed Aviation Agency & Civil Aeromed Inst. Honors & Awards: Meritorious Award, Am Cancer Soc, 72-73. Mem: Soc Epidemiol Res; Am Pub Health Asn; Asn Teachers Prev Med. Res: Infectious and chronic disease epidemiology; cancer and cerebrovascular disease. Mailing Add: Dept of Biostatist & Epidemiol Univ of Okla Sch of Health Oklahoma City OK 73104

ASANO, AKIRA, b Stockton, Calif, Jan 20, 23; m 51; c 2. PHARMACEUTICS. Educ: Drake Univ, BS, 44; Univ Minn, MS, 45, PhD(pharmaceut chem), 48. Prof Exp: Res assoc, Merck Sharp & Dohme, 48-57; group leader, 57-59, asst dir pharmaceut res, 59-65, dir prod develop, Johnson's Prof Prod Co, 65-67, ASST MGR PHARMACEUT RES, JOHNSON & JOHNSON, 67- Mem: Am Pharmaceut Asn. Res: Drug formulation. Mailing Add: Res Ctr Johnson & Johnson New Brunswick NJ 08903

ASANO, TOMOAKI, b Tokyo, Japan, Nov 13, 29; m 58. PHYSIOLOGY. Educ: Keio Univ, MD, 51, DMS(physiol), 59; Univ Rochester, MS, 55. Prof Exp: Asst physiol, Rockefeller Inst Med Res, 55-56 & Sch Med, Keio Univ, 56-57; asst prof physiol, Sch Med, Kanazawa Univ, 57-64; asst prof, 64-68, ASSOC PROF MICROBIOL, LOBUND LAB, NOTRE DAME UNIV, 68- Mem: Am Physiol Soc; Soc Exp Biol & Med; Sigma Xi. Res: Physiological study of germfree life; carcinogenesis in germfree animals. Mailing Add: Dept of Microbiol Lobund Lab Univ of Notre Dame Notre Dame IN 46556

ASANUMA, HIROSHI, b Kobe, Japan, Aug 17, 26; m 53; c 2. NEUROPHYSIOLOGY. Educ: Keio Univ, Japan, MD, 52; Kobe Univ, DMedSci, 59. Prof Exp: Instr physiol, Kobe Med Col, 53-59; asst prof, Med Sch, Osaka City Univ, 59-61, 63-65; guest investr, Rockefeller Inst, 61-63; from assoc prof to prof physiol, NY Med Col, 65-72; PROF, ROCKEFELLER UNIV, 72- Mem: Am Physiol Soc; Physiol Soc Japan. Res: Physiology of mammalian motor system with reference to the function of the pyramidal tract. Mailing Add: Rockefeller Univ New York NY 10021

ASATO, GORO, b Mt View, Hawaii, May 29, 31; m 62. ORGANIC CHEMISTRY. Educ: Univ Hawaii, BA, 53, MS, 58; Purdue Univ, PhD(org chem), 61. Prof Exp: Res chemist, 61-72, GROUP LEADER, AGR DIV, AM CYANAMID CO, 72- Mem: Am Chem Soc. Res: Stereochemical and optical rotatory dispersion studies of cyclohexane derivatives; synthesis of pesticides, antifungals and antibacterials; synthesis and infrared spectroscopic studies of transition metal carbonyls; animal health products. Mailing Add: Agr Ctr Am Cyanamid Co Box 400 Princeton NJ 08540

ASATO, YUKIO, b Waipahu, Hawaii, Jan 19, 34; m 69; c 1. MICROBIOLOGY. Educ: Univ Hawaii, BA, 57, MS, 66, PhD(microbiol), 69. Prof Exp: Microbiologist, State Dept Health, Hawaii, 61-63; asst microbiol, Univ Hawaii, 63-65; res assoc, biol adaptation br, Ames Res Ctr, NASA, 69-71; ASST PROF MICROBIOL, SOUTHEASTERN MASS UNIV, 71- Mem: AAAS; Am Soc Microbiol; Genetics Soc Am. Res: Microbial genetics; biochemical genetics of cyanobacteria. Mailing Add: Southeastern Mass Univ North Dartmouth MA 02747

ASAY, KAY HARRIS, b Lovell, Wyo, Nov 20, 33; m 53; c 3. CROP BREEDING, AGRONOMY. Educ: Univ Wyo, BS, 57, MS, 59; Iowa State Univ, PhD(crop breeding), 65. Prof Exp: Instr high sch, Wyo, 59-61; asst crop breeding, Iowa State Univ, 61-65; asst prof, 65-70, ASSOC PROF GRASS BREEDING, UNIV MO-COLUMBIA, 70- Res: Development of improved forage grass varieties and related basic studies. Mailing Add: Dept of Agron 208 Waters Hall Univ of Mo Columbia MO 65201

ASBRIDGE, JOHN ROBERT, b Lakeside, Mont, Aug 26, 28; m 54; c 4. SPACE PHYSICS. Educ: Mont State Col, BS, 53; Lehigh Univ, MS, 55, PhD(physics), 59. STAFF MEM, LOS ALAMOS SCI LAB, UNIV CALIF, 59- Mem: Am Phys Soc; Am Geophys Union. Res: Satellite based study of space environment, the constitution of the solar wind, and the earth's magnetosphere and cosmic rays. Mailing Add: 15 Encino Los Alamos NM 87544

ASCAH, RALPH GORDON, b Montreal, Que, July 7, 18; m 44; c 3. PHYSICAL CHEMISTRY. Educ: McGill Univ, BSc, 39; NY Univ, PhD(chem), 44. Prof Exp: Res chemist, Can Indust, Ltd, 43-47; asst prof chem, 47-57, ASSOC PROF CHEM, PA STATE UNIV, 57- Mem: Am Chem Soc. Res: Photochemistry; vapor phase kinetics and low temperature calorimetry; infrared spectroscopy. Mailing Add: 108 Whitmore Lab Pa State Univ University Park PA 16802

ASCHBACHER, PETER WILLIAM, b Ashland, Wis, Apr 20, 28; m 55; c 3. ANIMAL PHYSIOLOGY. Educ: Univ Wis, BS, 51, MS, 53, PhD(dairy husb), 57. Prof Exp: From asst prof to assoc prof dairy husb, NDak State Univ, 56-64, asst dairy husbandman, 56-61; ANIMAL PHYSIOLOGIST, METAB & RADIATION RES LAB, AGR RES SERV, USDA, 64- Concurrent Pos: Assoc scientist agr res lab, AEC, 61. Mem: AAAS; Am Dairy Sci Asn; Am Soc Animal Sci. Res: Physiology of farm animals; metabolic fate of agricultural chemicals in farm animals. Mailing Add: Metab & Radiation Res Lab Agr Res Serv USDA Fargo ND 58102

ASCHENBRENNER, JOYCE CATHRYN, b Salem, Ore, Mar 24, 31; m 52. ANTHROPOLOGY. Educ: Tulane Univ, BA, 54, MA, 56; Univ Minn, PhD(anthrop), 67. Prof Exp: Instr sociol, Wis State Univ, River Falls, 65-67; asst prof family soc sci, Univ Minn, Minneapolis, 67-70; asst prof, 70-74, ASSOC PROF ANTHROP, SOUTHERN ILL UNIV, EDWARDSVILLE, 74- Concurrent Pos: Am Philos Soc res grant, 67; assoc prof, Augsburg Col, 68-70. Mem: Fel Am Anthrop Asn. Res: Kinship and marriage among Muslim societies in the subcontinent; transition to urban life among Chippewa Indians; black families in Chicago. Mailing Add: Dept of Anthrop Southern Ill Univ Edwardsville IL 62025

ASCHER, EDUARD, b Vienna, Austria, Nov 23, 15; nat US; m 54; c 4. PSYCHIATRY. Educ: Washington Univ, BS & MD, 42. Prof Exp: Asst prof, 49-73, ASSOC PROF PSYCHIAT, JOHNS HOPKINS UNIV, 73-, PSYCHIATRIST, OUTPATIENT DEPT, JOHNS HOPKINS HOSP, 49- Concurrent Pos: Asst clin prof, Sch Med, Univ Md, 64-69, assoc clin prof, 69-; instr, Washington Sch Psychiat, 71-; consult, US Vet Admin. Mem: Am Group Psychother Asn; Am Psychiat Asn. Res: Psychotherapy; depressive disorders; expressive behavior; group psychotherapy. Mailing Add: 2 E Read St Baltimore MD 21202

ASCHER, MARCIA, b New York, NY, Apr 23, 35; m 56. NUMERICAL ANALYSIS, APPLIED MATHEMATICS. Educ: Queens Col, NY, BS, 56; Univ Calif, Los Angeles, MA, 60. Prof Exp: Comput analyst math, Douglas Aircraft Co, 57-60; specialist tech discipline, Gen Elec Co, 60-61; from asst prof to assoc prof, 61-72, PROF MATH, ITHACA COL, 72- Mem: Math Asn Am; Asn Comput Mach. Res: Applications of mathematics to anthropology. Mailing Add: Dept of Math Ithaca Col Ithaca NY 14850

ASCHER, ROBERT, b New York, NY, Apr 28, 31; m 50. ANTHROPOLOGY, ARCHAEOLOGY. Educ: Queens Col, NY, BA, 54; Univ Calif, Los Angeles, MA, 59, PhD, 60. Prof Exp: Asst gen anthrop, Univ Calif, Los Angeles, 57-59, instr, Exten Div, 58-60; from asst prof to assoc prof, 60-66, PROF ANTHROP & ARCHAEOL, CORNELL UNIV, 66- Mem: AAAS; Am Archaeol Inst; Soc Am Archaeol; Am Asn Phys Anthrop; Am Anthrop Asn. Res: Theory and method in anthropology and archaeology; biocultural evolution. Mailing Add: Dept of Anthrop Cornell Univ Ithaca NY 14850

ASCHKENASY, HERBERT, polymer chemistry, organic chemistry, see 12th edition

ASCHMANN, H HOMER, b San Francisco, Calif, May 5, 20; m 43; c 7. GEOGRAPHY. Educ: Univ Calif, Los Angeles, AB, 40, MA, 42; Univ Calif, Berkeley, PhD, 54. Prof Exp: From instr to asst prof geog, San Diego State Col, 46-48; instr, Univ Nebr, 50-51; asst prof, Los Angeles State Col, 51-54; from asst prof to assoc prof, 54-64, PROF GEOG, UNIV CALIF, RIVERSIDE, 64- Mem: Am Geog Soc; Asn Am Geog. Res: Demography of primitive societies; ecology of low latitude dry lands; historical geography of Baja California, Mexico and Guajira Peninsula, Colombia. Mailing Add: Dept of Earth Sci Univ of Calif Riverside CA 92502

ASCHNER, JOSEPH FELIX, b Vienna, Austria, Jan 23, 22; US citizen; m 45, 62; c 1. SOLID STATE PHYSICS, SEMICONDUCTORS. Educ: Univ Chicago, BS, 43; Univ Ill, PhD(physics), 54. Prof Exp: Asst physics, Univ Ill, 50-54; mem tech staff solid state device tech, Bell Tel Labs, 54-62; asst prof physics, 62-71, ASSOC PROF PHYSICS, CITY COL NEW YORK, 71- Concurrent Pos: Fac assoc, Gen Atomic Co, 67-68 & Zenith Radio Co, 69. Mem: Am Phys Soc; Sigma Xi. Res: Alkali halide crystals; solid state diffusion; semiconductor device technology and surfaces; radiation effects; infrared phenomena in semiconductors. Mailing Add: 340 E 64th St New York NY 10021

ASCHNER, THOMAS CARL, organic chemistry, see 12th edition

ASCIONE, RICHARD, b Brooklyn, NY, June 3, 36; m 58; c 4. ONCOLOGY, BIOCHEMISTRY. Educ: Brooklyn Col, BA, 58; Princeton Univ, MA, 63, PhD(biochem), 66. Prof Exp: Res technician biochem, Inst Muscle Dis, NY, 58-59; res asst, Dept Biochem, NY Univ-Bellevue Med Ctr, 60-61; Nat Acad Sci-USDA fel, 66-67; res virologist, US Animal Dis Lab, 67-72; VIRAL ONCOLOGIST, NAT CANCER INST, 72- Concurrent Pos: Consult-adv, Med Sch, Georgetown Univ, 74-; XIth Int Cancer Cong fel award, Nat Acad Sci-Int Union Cong Cancer, 74. Mem: AAAS; Am Soc Microbiol; Am Chem Soc. Res: Nucleic acid biosynthesis; protein biosynthesis; animal virology; molecular biology of oncornaviruses. Mailing Add: Human Tumor Stud Sect Nat Cancer Inst NIH Bldg 41 Suite 100 Bethesda MD 20014

ASCOLI, GIULIO, b Milano, Italy, Oct 26, 22; nat US; m 50; c 3. EXPERIMENTAL PHYSICS. Educ: Mass Inst Technol, PhD(physics), 51. Prof Exp: Chem engr nuclear reactor develop, Oak Ridge Nat Lab, 46-47; res asst, Cosmic Rays, Mass Inst Technol, 49-50; from instr to assoc prof, 50-72, PROF PHYSICS, UNIV ILL, URBANA, 72- Mem: Am Phys Soc. Res: Investigation on cosmic radiation. Mailing Add: Dept of Physics Univ of Ill Urbana IL 61801

ASEN, SAM, b Philadelphia, Pa, Aug 24, 20; m 44; c 2. PLANT PHYSIOLOGY. Educ: Pa State Univ, BS, 42, MS, 47; Rutgers Univ, PhD(hort), 50. Prof Exp: Asst prof hort, Mich State Univ, 50-55; SR PLANT PHYSIOLOGIST, PLANT INDUST STA, USDA, 55- Honors & Awards: Vaughn Award, 51, 59 & 60. Mem: Am Soc Plant Physiol; Am Soc Hort Sci; Phytochem Soc NAm. Res: Nutrition of floriculture plants; plant growth regulators for controlling flowering; effect of light on floriculture plants; anthocyanins; post harvest physiology of cut flowers. Mailing Add: Plant Indust Sta USDA Beltsville MD 20705

ASENDORF, ROBERT HARRY, b Philadelphia, Pa, Mar 5, 27; m 53; c 2. THEORETICAL PHYSICS, INFORMATION SCIENCE. Educ: Univ Pa, BA, 47, PhD(physics), 56. Prof Exp: Physicist, Selas Corp Am, 52-53; asst physics, Bryn Mawr Col, 53-56; physicist res labs, Westinghouse Elec Co, 56-58; sr mem tech staff, Hughes Res Labs, Hughes Aircraft Co, 58-68; with US Govt, 68-73; sr corp scientist, Systs Control, Inc, 73-75; PRIN SCIENTIST, AERONUTRONIC FORD WESTERN DEVELOP LAB, 75- Mem: Inst Elec & Electronics Eng; Am Phys Soc; Pattern Recognition Soc; Am Asn Computing Mach. Res: Solid state theory; group theory; crystal physics; semiconductors; ferrites; relaxation mechanisms; localized energy; pattern recognition; artificial intelligence; information processing systems. Mailing Add: 14510 Manuela Rd Los Altos Hills CA 94022

ASENJO, CONRADO FEDERICO, b San Juan, PR, Dec 6, 08; m 45; c 6. CHEMISTRY. Educ: Rensselaer Polytech Inst, ChE, 33; Univ Wis, MS, 38, PhD(phytochem, biochem), 40; Am Bd Nutrit, dipl, 51; Am Bd Clin Chem, dipl, 52. Prof Exp: Asst chem, Sch Trop Med, 33-35, instr, Sch Med, 35-39, assoc, 39-42, from asst prof to assoc prof, 42-52, chmn dept chem, 45-50, head dept biochem & nutrit, 52-73, asst dean sch med, 60-66, PROF BIOCHEM & NUTRIT, SCH MED, UNIV PR, SAN JUAN, 52-, ASSOC DEAN SCH MED, 66- Concurrent Pos: Vis prof biochem, Tulane Univ, 56; mem, Tulane-Columbia Med Educ Prog, 56; mem, Int Union Nutrit Sci, 67-; consult, Pan-Am Sanit Bur, 65, 68, 70 & Orgn Am States, 68, 71. Honors & Awards: Grand Prize Physico-Chem Sci, PR Acad Arts & Sci, 67; Honor Plate Award, Dietetic Asn PR, 67. Mem: Fel AAAS; Am Chem Soc; Am Soc Biol Chem; Soc Exp Biol & Med; Am Inst Nutrit. Res: Chemical composition of economic and medicinal plants; vitamin survey of tropical foods; food yeast nutritional value; folic acid deficiency; fecal fat; biological evaluation of proteins; public health nutrition; nutritional surveys. Mailing Add: Dept of Biochem & Nutrit Univ PR Sch Med GPO Box 5067 San Juan PR 00905

ASENJO, FLORENCIO GONZALEZ, b Buenos Aires, Arg, Sept 28, 26; US citizen; m 60; c 2. MATHEMATICAL LOGIC. Educ: La Plata Nat Univ, Lic math, 54, PhD(math), 56. Prof Exp: Res mathematician, Lab Testing Mat & Tech Invests, Arg, 48-57, head calculus & statist sect, 57-58; asst prof math, Georgetown Univ, 58-61; assoc prof math, Univ Southern Ill, 61-63; assoc prof, 63-66, PROF MATH, UNIV PITTSBURGH, 66- Concurrent Pos: Asst, La Plata Nat Univ, 53-55, instr, 55-57, titular prof, 57-58; Fulbright fel, Univ Lisbon, 70. Mem: Am Math Soc; Asn Symbolic Logic. Res: Formalization of internal relations; arithmetic of term-relation numbers; calculus of antinomies; theory of multiplicities; sampling processes; model theory. Mailing Add: Dept of Math Univ of Pittsburgh Pittsburgh PA 15260

ASERINSKY, EUGENE, b New York, NY, May 6, 21; m 42; c 2. PHYSIOLOGY. Educ: Univ Chicago, PhD(physiol), 53. Prof Exp: Asst physiol, Univ Chicago, 49-52; res assoc, Univ Wash, 53-54; from instr to assoc prof, 54-67, PROF PHYSIOL, JEFFERSON MED COL, 67- Concurrent Pos: Vis res scientist, Eastern Pa Psychiat Inst, 61-70; vis prof, NY Col Psychiat, 66- & Cajal Inst, Madrid, 71. Mem: Am Physiol Soc; Am Med Writers Asn; sr mem Am Astronaut Soc; NY Acad Sci. Res: Physiology of sleep, eye movements and blinking; electro-oculography; neurophysiology; circulatory and respiratory reflexes; chronaxie; long term effects on synaptic resistance. Mailing Add: Jefferson Med Col 1020 Locust St Philadelphia PA 19107

ASGAR, KAMAL, b Tabriz, Iran, Aug 28, 22; c 2. BIOENGINEERING. Educ: Tech Col Tehran, BA, 45; Univ Mich, Ann Arbor, MS, 48, BS, 50, PhD(mat & metall), 59. Prof Exp: From res asst to res assoc, 52-59, from asst prof to assoc prof, 59-66, PROF DENT MAT, SCH DENT, UNIV MICH, ANN ARBOR, 66- Concurrent Pos: Mem, Base Metal Alloys Comt, Am Nat Stand Comt & chmn, Casting Investment Comt & Gypsum Subcomt, 73; mem consult team, Nat Inst Dent Res, 74- Honors & Awards:

Paul Gibbons Award, Sch Dent, Univ Mich & Souder Award, Int Asn Dent Res, 70. Mem: Int Asn Dent Res; Am Soc Metallurgists; Microbeam Anal Soc. Res: Cast alloys used in dentistry; both noble and base metal alloys; porcelain-metal restorations; dental amalgam; dental investments. Mailing Add: Dept of Dent Mat Univ of Mich Sch of Dent Ann Arbor MI 48109

ASH, ARTHUR BURR, b Dry Branch, Ga, July 21, 13; m 61; c 3. ORGANIC CHEMISTRY. Educ: Wayne Univ, BS, ChE, 40, MS, 45, PhD(org chem), 48. Prof Exp: Res chemist, Wyandotte Chem Corp, 39-40; develop chemist, US Rubber Co, Ind, 41-42; sect head org div, Wyandotte Chem Corp, 42-46; dir res & eng, Am Mach & Foundry Co, 47; sect head org div, Wyandotte Chem Corp, 48-50, mgr, High Pressure Lab, 50-52, proj supvr, Contract Res Dept, 53-55; proj supvr, Turbo Div, Am Mach & Foundry Co, 55-57; tech dir, Sundstrand Turbo Div, 58-59; dir res, Tenn Prod & Chem Corp, 59-61; res assoc chem, Wayne State Univ, 61-63; PRES, ASH STEVENS INC, 63- Mem: Am Chem Soc. Res: Organic synthesis and reaction mechanisms; chemical engineering; chemical process development. Mailing Add: Ash Stevens Inc 5861 John C Lodge Freeway Detroit MI 48202

ASH, CECIL GRANT, bacteriology, see 12th edition

ASH, CLIFFORD L, b Edmonton, Alta, July 9, 09; m 36; c 4. RADIOLOGY. Educ: Univ Alta, BSc, 30; Univ Toronto, MD, 34, DMR, 40; FRCPS(C), 64. Prof Exp: DIR, ONT CANCER INST, 52-; PROF THERAPEUT RADIOL, UNIV TORONTO, 52-, PROF BIOPHYS, 58- Concurrent Pos: Radiotherapist-in-chief, Toronto Gen Hosp, 52- Mem: Fel Am Col Radiol; Am Roentgen Ray Soc; Radiol Soc NAm; Am Radium Soc; Can Asn Radiologists. Res: Therapeutic radiology. Mailing Add: Ont Cancer Inst 500 Sherbourne St Toronto ON Can

ASH, J MARSHALL, b New York, NY, Feb 18, 40; m 62; c 1. PURE MATHEMATICS. Educ: Univ Chicago, SB, 61, SM, 63, PhD(math), 66. Prof Exp: Joseph Fels Ritt instr math, Columbia Univ, 66-69; from asst prof to assoc prof, 69-74, PROF MATH, DePAUL UNIV, 74- Concurrent Pos: Partic harmonic anal conf, Warwick Univ, 68; Am Math Soc partic, Int Math Cong, 70 & 74. Mem: Am Math Soc; Math Asn Am. Res: Generalized derivatives of functions of a real variable; multiple trigonometric series; real variable; harmonic analysis; measure theory; singular integrals; Fourier series. Mailing Add: Dept of Math DePaul Univ 2323 N Seminary Chicago IL 60614

ASH, JACOB R, chemistry, see 12th edition

ASH, KENNETH OWEN, b Provo, Utah, Aug 21, 36; m 56; c 4. BIOCHEMISTRY, CLINICAL CHEMISTRY. Educ: Brigham Young Univ, BS, 58, PhD(biochem), 61. Prof Exp: Proj leader protein res, Gen Mills, Inc, 61-64; sr res scientist, Honywell Regulator Co, 64-66; prin scientist, honeywell, Inc, 66-70, tech dir clin instrumentations dept, 70-71; dir lab, A&M Labs, 71-74; dir, Br Lab, Bio-Sci Labs, 75; DIR CLIN CHEM, MED CTR, UNIV UTAH, 75- Concurrent Pos: Lab consult, Vet Admin Hosp, Salt Lake City, 75- Mem: Am Asn Clin Chemists; Acad Clin Lab Physicians & Scientists. Res: Protein isolation, modification and function evaluation; enzymes isolation, identification and olfaction; clinical chemistry. Mailing Add: Dept of Path Univ of Utah Med Ctr Salt Lake City UT 84132

ASH, LAWRENCE ROBERT, b Holyoke, Mass, Mar 5, 33; m 60; c 1. PARASITOLOGY, TROPICAL MEDICINE. Educ: Univ Mass, BS, 54, MA, 56; Tulane Univ, PhD(parasitol), 60. Prof Exp: Asst parasitologist, Univ Hawaii, 60-61; instr parasitol, Tulane Univ, 62-65; med parasitologist, NIH, 65-67; from asst prof to assoc prof infectious & trop dis, 67-75, PROF EPIDEMIOL & INFECTIOUS & TROP DIS, SCH PUB HEALTH, UNIV CALIF, LOS ANGELES, 75- Concurrent Pos: Consult, US Naval Med Res Unit 2, 70-; NIH res grant, Univ Calif, Los Angeles, 71-; mem US panel parasitic dis, US-Japan Coop Med Sci Prog, NIH, 72-; mem adv sci bd, Gorgas Mem Inst, 74-; mem ad hoc study group on parasitic dis, US Army Med Res & Develop Command, 75- Mem: Am Soc Parasitol; Am Soc Trop Med & Hyg; Royal Soc Trop Med & Hyg; Int Filariasis Asn. Res: Parasitic diseases of man and animals; biology, pathology and systematics of filariae, metastrongyles, ascarids, spirurids and other helminths. Mailing Add: Div of Epidemiol Univ of Calif Sch of Pub Health Los Angeles CA 90024

ASH, MAJOR MCKINLEY, JR, b Bellaire, Mich, Apr 7, 21; m 47; c 4. PERIODONTICS. Educ: Mich State Col, BS, 47; Emory Univ, DDS, 51; Univ Mich, MS, 54. Hon Degrees: DrMed, Univ Bern, 75. Prof Exp: Instr oral path, Emory Univ, 52-53; teaching fel, Sch Dent, Univ Mich, 53-54; from instr to assoc prof periodont & oral path, 54-63, PROF PERIODONT & ORAL PATH, SCH DENT, UNIV MICH, ANN ARBOR, 63-, CHMN DEPT OCCLUSION, 69- Concurrent Pos: Attend physician, Vet Admin Hosp, Atlanta, Ga, 52-53; consult, Vet Admin Hosp, Ann Arbor, 61-; pres, Mich Basic Sci Bd, 66-70. Mem: AAAS; Int Asn Dent Res. Res: Evaluation of electric and manual toothbrushes; telemetry of intra-oral occlusal forces, pH, muscle forces and jaw movements; evaluation of dental pain thresholds and hypersensitivity; occlusion and temporomandibular joint pathology. Mailing Add: Dept of Occlusion Univ of Mich Sch of Dent Ann Arbor MI 48103

ASH, MICHAEL EDWARD, b Detroit, Mich, June 26, 37; m 63; c 2. APPLIED MATHEMATICS. Educ: Mass Inst Technol, BS, 59; Princeton Univ, MA, 60, PhD(math), 63. Prof Exp: Instr math, Princeton Univ, 62-63; res assoc, Brandeis Univ, 63-64; MEM STAFF CELESTIAL MECH & GEN RELATIVITY, LINCOLN LAB, MASS INST TECHNOL, 64- Concurrent Pos: Lectr, Northeastern Univ Grad Eve Sch, 68-71 & Boston Univ Eve Sch, 73-75. Mem: Am Math Soc. Res: Celestial mechanics; scientific computer programming; general relativity. Mailing Add: 16 Baskin Rd Lexington MA 02173

ASH, MILTON, physics, see 12th edition

ASH, ROBERT B, b New York, NY, May 20, 35; m 56. MATHEMATICS. Educ: Columbia Univ, BA, 55, BS, 56, MS, 57, PhD(elec eng), 60. Prof Exp: From instr to asst prof elec eng, Columbia Univ, 58-62; vis asst prof, Univ Calif, Berkeley, 62-63; ASSOC PROF MATH, UNIV ILL, URBANA-CHAMPAIGN, 63- Mem: Am Math Soc. Res: Information theory; probability theory. Mailing Add: Dept of Math Univ of Ill Urbana IL 61801

ASH, RONALD JOSEPH, b Covington, Ky, Dec 24, 41; m 70; c 2. MICROBIOLOGY, VIROLOGY. Educ: Thomas More Col, Ky, AB, 61; Univ Cincinnati, MS, 64, PhD(microbiol), 68. Prof Exp: From instr to asst prof, 67-74, ASSOC PROF MICROBIOL, EMORY UNIV, 74- Concurrent Pos: Vis asst prof, Inst Microbiol, Univ Basel, 73-74. Mem: Am Soc Microbiol; Soc Cryobiol. Res: Animal viruses containing envelopes; viruses in studies on membrane phenomena. Mailing Add: Dept of Microbiol Emory Univ Atlanta GA 30322

ASH, SIDNEY ROY, b Albuquerque, NMex, Nov 25, 28; m 62. GEOLOGY, PALEOBOTANY. Educ: Midland Lutheran Col, AB, 51; Univ NMex, AB, 57, MS, 61; Univ Reading, PhD, 66. Prof Exp: Phys sci aid ground water br, US Geol Surv, NMex, 56-58, geologist, 58-61, geologist paleont & stratig br, 61-64; instr natural sci, Midland Lutheran Col, 66-67, asst prof earth sci, 67-69; asst prof geol, Ft Hays Kans State Col, 69-70; assoc prof, 70-75, PROF GEOL, WEBER STATE COL, 75- Mem: Bot Soc Am; Paleont Soc; Paleont Asn. Res: Stratigraphy and paleobotany of the Mesozoic. Mailing Add: Dept of Geol Weber State Col Ogden UT 84408

ASH, WILLARD OSBORNE, b Cumberland, Md, Nov 14, 14; m 39; c 2. STATISTICS. Educ: St John's Col, Md, AB, 37; Univ Md, AM, 41; Va Polytech Inst, PhD(statist), 57. Prof Exp: Statistician, War Prod Bd, 42-45; statistician, War Assets Admin, 45-47; consult, Bur Aeronaut Navy Dept, 47-48; asst prof statist, Univ Md, 47-55; asst prof, Va Polytech Inst, 55-57; prof, Univ Fla, 57-67; DEAN COL ARTS, SCI & TECHNOL, UNIV N FLA, 70- Concurrent Pos: Chmn fac math & statist, Univ W Fla, 67-70; consult, J I Thompson Co, 55-56 & statist lab, Univ Fla, 57- Mem: Am Statist Asn; Biomet Soc. Res: Experimental design; sample survey design; stochastic processes; econometrics. Mailing Add: Col Arts Sci Technol Univ N Fla PO Box 17074 Pottsburg Sta Jacksonville FL 32216

ASH, WILLIAM JAMES, b New York, NY, Nov 3, 31; m 53; c 5. ANIMAL GENETICS, DEVELOPMENTAL GENETICS. Educ: Cornell Univ, BS, 53, MS, 58, PhD(animal genetics), 60. Prof Exp: Res geneticist, Cornell Univ, 59-64; dir res duck breeding, path & mgt, Crescent Duck Inc, NY, 64-65; mem fac dept biol, WVa Univ, 65-66; assoc prof, 66-75, PROF BIOL, ST LAWRENCE UNIV, 75- Concurrent Pos: Vis lectr, Univ RI, 62; Cornell Univ res grant, Europe, 63. Mem: Am Genetic Asn; Genetics Soc Am; World Poultry Sci Asn. Res: Basic genetics and biology of avian and marine animals. Mailing Add: Dept of Biol St Lawrence Univ Canton NY 13617

ASHBAUGH, JAMES G, b Lawrenceville, Ill, June 24, 28; m 50; c 2. GEOGRAPHY. Educ: Cent Wash State Col, BA, 50; Univ Colo, MA, 53; Univ Calif, Los Angeles, PhD(geog), 65. Prof Exp: Asst prof geog, Western Ill Univ, 56-57; instr, 57-59, from asst prof to assoc prof, 59-70, head dept, 70-75, PROF GEOG, PORTLAND STATE UNIV, 70- Mem: Asn Am Geog; Am Geog Soc; Am Inst Urban & Regional Affairs. Res: Changes in the form and function of Western European cities; geography of disaster. Mailing Add: Dept of Geog Portland State Univ Box 751 Portland OR 97207

ASHBROOK, JOSEPH, b Philadelphia, Pa, Apr 4, 18; m 42; c 4. ASTRONOMY. Educ: Johns Hopkins Univ, BA, 39; Case Inst Technol, MSc, 41; Harvard Univ, PhD(astron), 47. Prof Exp: Physicist, US Dept Navy, 42-45; fron instr to asst prof astron, Yale Univ, 46-53; RES ASSOC ASTRON, HARVARD OBSERV, 53-; ED, SKY & TELESCOPE, 64- Concurrent Pos: Asst ed, Sky & Telescope, 53-64. Mem: Am Astron Soc; Am Asn Variable Star Observers; Int Astron Union. Res: Variable stars; Mars; history of astronomy; period changes in variable stars; selenography. Mailing Add: 16 Summer St Weston MA 02193

ASHBURN, ALLEN DAVID, b Clarkrange, Tenn, Mar 6, 33; m 55; c 3. ANATOMY, PHYSIOLOGY. Educ: Tenn Technol Univ, BS, 58, MA, 60; Univ Miss, PhD, 64. Prof Exp: From instr to assoc prof, 62-75, PROF ANAT, SCH MED, UNIV MISS, 75- Mem: Am Asn Anat. Res: Experimental pathology of the cardiovascular system. Mailing Add: Dept of Anat Univ of Miss Med Sch Jackson MS 39216

ASHBURN, EDWARD VICTOR, b Pittsburgh, Pa, May 13, 10; m 37; c 2. PHYSICS. Educ: Univ Calif, Berkeley, AB, 39; Mass Inst Technol, SM, 40. Prof Exp: Meteorologist, US Weather Bur, 32-46; physicist, atmospheric optics, US Naval Ord Test Sta, 46-62; scientist, Lockheed Calif Co, 62-73; RETIRED. Concurrent Pos: Assoc prof, San Fernando Valley State Col, 63-71. Mem: AAAS; fel Am Phys Soc; Sigma Xi. Res: Atmospheric optics; aurora; night airglow; planetary physics. Mailing Add: 12306 Woodley Ave Granada Hills CA 91344

ASHBURN, GILBERT, b Winston-Salem, NC, Feb 25, 11; m 44; c 4. ORGANIC CHEMISTRY. Educ: Univ Ill, BS, 45; Univ Va, PhD(chem), 49. Prof Exp: Res chemist, R J Reynolds Tobacco Co, 48-76; RETIRED. Mem: Am Chem Soc; Am Inst Chemists. Res: Tobacco chemistry; antimalarials; microanalysis; isolation of natural products. Mailing Add: 714 Finsbury Rd Winston-Salem NC 27104

ASHBURN, HOWARD GILMER, organic chemistry, see 12th edition

ASHBURN, WILLIAM LEE, b New York, NY, Jan 18, 33; m 60; c 3. NUCLEAR MEDICINE. Educ: Western Md Col, AB, 55; Univ Md, MD, 59; Am Bd Radiol, dipl, 66. Prof Exp: Intern, Ohio State Univ Hosp, 59-60; med officer, USPHS Hosp, Savannah, Ga, 60-62, resident radiol, Clin Ctr, NIH, 62-66, chief diag radioisotopes sect, Dept Nuclear Med, 66-68; PROF RADIOL & CHIEF DIV NUCLEAR MED, SCH MED, UNIV CALIF, SAN DIEGO, 68- Mem: AMA; Soc Nuclear Med. Res: Use of radioisotopes in medical diagnosis; development of improved techniques of radioisotopic organ visualization using computers and other electronic aids with particular emphasis on cardiopulmonary diagnosis and diseases of the central nervous system. Mailing Add: Univ Hosp 225 W Dickenson St San Diego CA 92103

ASHBY, BRUCE ALLAN, b Elmhurst, NY, Mar 28, 22; m 48; c 2. ORGANIC CHEMISTRY. Educ: Rutgers Univ, BS, 48, PhD(org chem), 54; Univ Mich, MS, 49. Prof Exp: Chemist, Dow Chem Co, Mich, 49-50; instr chem, Rutgers Univ, 50-53; chemist silicones, 54-60, specialist, 60-66, mgr spec projs, Silicone Prod Dept, 66-70, SPECIALIST SILICONE PROD DEPT, GEN ELEC CO, 70- Mem: AAAS; Am Chem Soc; Am Inst Chem. Res: Reaction mechanisms; synthetic routes to morphine; agriculturally important sulfur-nitrogen compounds; organo-silicon chemistry; fluorine compounds; organic and organometallic polymers. Mailing Add: Silicone Prod Dept Gen Elec Co Waterford NY 12188

ASHBY, CARL TOLIVER, b Vale, Ark, June 26, 05; m 29; c 2. PHYSICAL CHEMISTRY. Educ: Univ Tex, AB, 29, AM, 31, PhD(phys chem), 34. Prof Exp: Tutor chem, Univ Tex, 27-32, instr chem, 33-34; res engr, Servel, Inc, 34-43, dir develop, 43-53, chief engr, 53-56; dir eng & pres, Conrad, Inc, Mich, 56-60; dir eng, Norge Div, Borg-Warner Corp, 60-63; adv eng, Arkla Air Conditioning Co, 63-65; RES ASSOC AGR-BIOCHEM, UNIV ARIZ, 65- Mem: Am Chem Soc; Am Soc Heating, Refrig & Air-Conditioning Engrs. Res: Absorption refrigeration; low temperature refrigeration; environmental engineering; corrosion; crystallography; hydrothermal growth of crystals. Mailing Add: 6611 N Swan Rd Tucson AZ 85718

ASHBY, DAVID L, soil fertility, plant nutrition, see 12th edition

ASHBY, EBERT ALTON, comparative physiology, zoology, see 12th edition

ASHBY, EUGENE CHRISTOPHER, b New Orleans, La, Oct 25, 30; m 52; c 7. ORGANIC CHEMISTRY, INORGANIC CHEMISTRY. Educ: Loyola Univ, La, BS, 51; Auburn Univ, MS, 52; Univ Notre Dame, PhD(chem), 56. Prof Exp: Res assoc, Ethyl Corp, 56-63; asst prof chem, 63-65, from assoc prof to prof, 65-73, REGENTS' PROF CHEM, GA INST TECHNOL, 73- Concurrent Pos: Sloan fel, 65-67. Honors & Awards: Lavoisier Medal, French Chem Soc, 71. Mem: Am Chem Soc. Res: Organometallic chemistry; organoaluminum, magnesium, beryllium and boron;

complex metal hydrides; organometallic reaction mechanisms; stereochemistry of alkylation and reduction reactions. Mailing Add: Dept of Chem Ga Inst of Technol Atlanta GA 30332

ASHBY, GEORGE ELLIOTT, physics, see 12th edition

ASHBY, NEIL, b Dalhart, Tex, Mar 5, 34; m 61; c 2. THEORETICAL PHYSICS. Educ: Univ Colo, BA, 55; Harvard Univ, MA, 56, PhD(physics), 61. Prof Exp: Asst prof, 61, Sheldon fel, 61-62, from asst prof to assoc prof, 62-70, PROF PHYSICS, UNIV COLO, BOULDER, 70- Mem: Am Phys Soc. Res: Transport phenomena; quantum statistical mechanics. Mailing Add: Dept of Physics & Astrophys Univ of Colo Boulder CO 80302

ASHBY, ROBERT MORRELL, b American Fork, Utah, May 9, 12; m 46; c 3. SOLID STATE ELECTRONICS. Educ: Brigham Young Univ, AB, 34, MA, 39; Univ Wis, PhD, 42. Prof Exp: Mem staff adv develop group, Radiation Lab, Mass Inst Technol, 42-46; consult & assoc technol dir naval res lab, Boston Field Sta, 46-49; asst prof physics, Univ Utah, 49; group leader, Electronics Group, NAm Aviation, Inc, 49-51, sect chief fire control electro-mech dept, 51-54, asst chief engr, 54-55, chief engr autonetics div, 55-59, vpres res & develop, 59-63, vpres technol, 63-69, chief scientist microelectronics div, NAm Rockwell Corp, 69-73; CHIEF SCIENTIST & MEM BD, AM TELECOMMUN CORP, 73- Mem: Am Phys Soc; fel Inst Elec & Electronics Engrs. Res: Inertial navigation; digital computers; radar equipment; microelectronics; communications. Mailing Add: Am Telecommun Corp 9620 Flair Dr El Monte CA 91731

ASHBY, VAL JEAN, b Kansas City, Mo, June 24, 23; m 48; c 3. APPLIED PHYSICS, ELECTROOPTICS. Educ: Univ Kans, BS, 44; Univ Calif, PhD(nuclear physics), 53. Prof Exp: Physicist, Lawrence Radiation Lab, Univ Calif, 46-59; physicist, Ramo-Woodridge, 59-61; physicist, Space Technol Labs, 61-62; PHYSICIST, LITTON INDUST, INC, 62- Mem: Am Phys Soc; Optical Soc Am. Res: Signal processing for noise-limited measurement systems; random noise, correlation, filtering and Fourier analyses; multidimensional spatial filtering; geometrical and physical optics; research and development of electrooptical systems. Mailing Add: Guid & Control Systs Div Litton Indust Inc 5500 Canoga Woodland Hills CA 91364

ASHBY, WILLIAM, b Hammond, Ind, May 2, 30; m 49; c 4. PHYSICS. Educ: Ind Univ, BA, 51. Prof Exp: Metallurgist, Inland Steel Co, 52-56; head dept physics, Steel City Testing & Eng Labs, 56; res metallurgist, Continental Can Co, 56-61; mgr appln lab, Picker X-Ray Corp, 61-66, anal sales mgr, Picker Nuclear, Inc, 66-67, prod adminstr, Picker Corp, 67-69, prod safety dir, 69-74, DIR INDUST RELS, PICKER CORP, 74- Concurrent Pos: Indust rep radiol device classification panel, Food & Drug Admin, 74- Mem: Am Soc Metals; Soc Appl Spectros; Am Crystallog Asn; Am Asn Physicists Med. Res: Physical metallurgy, crystallographic and analytical. Mailing Add: 595 Miner Rd Cleveland OH 44143

ASHBY, WILLIAM CLARK, b Duluth, Minn, July 6, 22; m 51; c 3. PHYSIOLOGICAL ECOLOGY. Educ: Univ Chicago, SB, 47, PhD(bot), 50. Prof Exp: Plant physiologist, US Forest Serv, 50-53; res fel, Calif Inst Technol, 53-54; Fulbright res scholar, Sydney, Australia, 54; asst prof bot, Univ Chicago, 55-60; assoc prof, 60-69, PROF BOT, SOUTHERN ILL UNIV, CARBONDALE, 69- Concurrent Pos: Fac res partic, Argonne Nat Lab, 66 & 70-71. Mem: AAAS; Ecol Soc Am; Bot Soc Am; Am Inst Biol Sci; Brit Ecol Soc. Res: Forest ecology; land reclamation; ecology of Tilia americana. Mailing Add: Dept of Bot Southern Ill Univ Carbondale IL 62901

ASHCRAFT, THOMAS LEE, b Sheridan, Ark, Aug 19, 26; m 48; c 3. ANALYTICAL CHEMISTRY. Educ: Ouachita Baptist Univ, BS, 49. Prof Exp: Chemist lab br, 49-52, chief, 52-63, chief lab div, 63-66, chief tech plans div, 67-69, chief qual surety off, 70-73, CHIEF QUAL ASSURANCE OFF, PINE BLUFF ARSENAL, 73- Mem: Am Chem Soc. Res: Analytical chemistry method development; nondestructive testing; differential thermal analysis applied to solid fuel mixtures in burning type munitions; quality assurance for chemicals and munition components. Mailing Add: 401 Parkway Dr Pine Bluff AR 71601

ASHCROFT, GAYLEN LAMB, climatology, see 12th edition

ASHDOWN, DONALD, b Bountiful, Utah, Apr 11, 18; m 41; c 7. ENTOMOLOGY. Educ: Utah State Univ, BS, 42; Cornell Univ, PhD(entom), 48. Prof Exp: Res asst entom, Cornell Univ, 42-48; assoc prof hort & entom, Okla State Univ, 48-52; PROF ENTOM, TEX TECH UNIV, 52- Concurrent Pos: Consult, Cosden Oil Co, 62- Mem: AAAS; Entom Soc Am. Res: Applied entomology; development and use of agricultural chemicals; green bug research; environmental pollution by pesticides. Mailing Add: Dept of Entom Tex Tech Univ Lubbock TX 79409

ASHE, ARTHUR JAMES, III, b New York, NY, Aug 5, 40; m 62; c 2. ORGANIC CHEMISTRY. Educ: Yale Univ, BA, 62, MS, 65, PhD(org chem), 66. Prof Exp: Asst prof, 66-71, ASSOC PROF CHEM, UNIV MICH, 71- Concurrent Pos: A P Sloan fel, 73; vis scholar, Phys Chem Inst, Univ Basel, 74. Mem: Am Chem Soc; The Chem Soc. Res: Physical organic chemistry; organometallic compounds. Mailing Add: Dept of Chem Univ of Mich Ann Arbor MI 48104

ASHE, JOHN BERRY, nuclear physics, see 12th edition

ASHE, WARREN (KELLY), b Halifax, NC, Aug 20, 29; m 51; c 5. MICROBIOLOGY, BIOCHEMISTRY. Educ: Howard Univ, BS, 51, MS, 62. Prof Exp: Med biol technician, Nat Inst Dent Res, 54-56, supvr med biol technol, 56-57, from microbiologist to res microbiologist, 57-70, health scientist adminr, 70-71; ASST DEAN RES & INSTR, COL MED, HOWARD UNIV, 71- Mem: AAAS; NY Acad Sci; Am Soc Microbiol; Soc Exp Biol & Med; Int Asn Dent Res. Res: Virology; immunology; pathology; etiology, epidemiology, pathophysiology, therapy and prevention of oral-facial ulcerative disorders, oral neoplasms and salivary gland disorders; research administration. Mailing Add: Howard Univ Col of Med Washington DC 20059

ASHENHURST, ROBERT LOVETT, b Paris, France, Aug 9, 29; div; c 4. APPLIED MATHEMATICS, COMPUTER SCIENCES. Educ: Harvard Univ, AB, 50, SM, 54, PhD(appl math), 56. Prof Exp: Res assoc, Harvard Comput Lab, 50-56, instr appl math, Harvard Univ, 56-57; from asst prof to assoc prof, 57-65, chmn comt info sci, 69-74, PROF APPL MATH, UNIV CHICAGO, 65-, DIR INST COMPUT RES, 69- Concurrent Pos: Ed-in-chief, Commun of Asn Comput Mach, 73- Mem: Soc Indust & Appl Math; Inst Elec & Electronics Engrs; Asn Comput Mach; Inst Mgt Sci; AAAS. Res: Computer and information systems and their application to scientific and management problems. Mailing Add: Inst for Comput Res Univ of Chicago Chicago IL 60637

ASHER, DAVID MICHAEL, b Chicago, Ill, Nov 10, 37; m 71; c 1. INFECTIOUS DISEASES, PEDIATRICS. Educ: Harvard Univ, AB, 59; Harvard Med Sch, MD, 63; Am Bd Pediat, dipl, 75. Prof Exp: Intern med, King County Hosp, Seattle, 63-64; jr asst resident pediat, Boston City Hosp, 64-65; jr resident, Mass Gen Hosp, Boston, 65-66; res assoc virol, Nat Inst Child Health & Human Develop, Bethesda, Md, 66-69; Nat Inst Neurol Dis & Stroke spec res fel virol, USA-USSR Health Scientist Exchange, 69-70; RES MED OFFICER VIROL, NAT INST NEUROL & COMMUN DIS & STROKE, NIH, 70- Concurrent Pos: Vis scientist, Div Pediat Oncol, Sch of Med, Johns Hopkins Univ, 76. Mem: Am Soc Microbiol; AAAS; Infectious Dis Soc Am. Res: Persistent viral infections, especially pathogenesis of infections of the nervous system and the urinary tract; viral infections of children with impaired immune defenses. Mailing Add: Nat Insts Health Bldg 8 Rm 100 Bethesda MD 20014

ASHER, IRVIN MARK, b Savannah, Ga, Nov 11, 44. BIOPHYSICS, LASERS. Educ: Mass Inst Technol, BS, 66, PhD(laser physics), 71. Prof Exp: NIH fel commun biophys group, Mass Inst Technol, 71-72; asst prof physics, Northeastern Univ, 72-74; res assoc biophys, Harvard-Mass Inst Technol Health Sci, 73-75; SCI ADV BIOPHYS, OFF SCI, FOOD & DRUG ADMIN, DEPT HEALTH EDUC & WELFARE, 75- Concurrent Pos: Guest scientist, NIH, 75- Honors & Awards: Cert of Recognition, NASA, 74. Mem: AAAS; Sigma Xi; Biophys Soc; NY Acad Sci. Res: Laser Raman spectroscopy of biological materials. Mailing Add: HFS-50 Off of Sci FDA Dept of Health Educ & Welfare Rockville MD 20852

ASHFORD, ROSS, b Stalwart, Sask, Nov 1, 26; m 53; c 3. AGRONOMY. Educ: Univ BC, BSA, 53; Univ Sask, MSc, 59, PhD, 67. Prof Exp: Asst economist, Hops Mkt Bd, London, Eng, 47-48; res off, Res Br, Can Dept Agr, 53-60 & 61-66; Nat Res Coun Can scholar, 60-61; ASSOC PROF CROP SCI, UNIV SASK, 66- Concurrent Pos: Mem, Can Weed Comt. Mem: Agr Inst Can; Can Soc Agron; Weed Sci Soc Am. Res: Chemical control of weeds in field crops; selective action of herbicides. Mailing Add: Dept of Crop Sci Univ of Sask Saskatoon SK Can

ASHFORD, THEODORE ASKOUNES, b Greece, Feb 27, 08; nat US; m 33; c 3. ORGANIC CHEMISTRY. Educ: Univ Chicago, BS, 32, MS, 34, PhD(chem), 36. Prof Exp: Chemist, Rapid Roller Co, 30-32; teacher schs, Ill, 32-36; from instr to asst prof, Univ Chicago, 36-50; from assoc prof to prof, St Louis Univ, 50-60; dir div natural sci & math, 60-71, assoc dean col lib arts, 66-71, PROF CHEM, UNIV S FLA, 60-, DEAN COL NATURAL SCI, 71- Concurrent Pos: Examr, Univ Chicago, 38-45; mem exam staff, US Armed Forces Inst, 42-44; dir inst teaching chem, St Louis Univ, 50-60; mem, New Era Educ Study Task Force, State of Fla, 62-63; mem bd & secy, Fla State Bd Exam Basic Sci, 63-; consult, Sci Teaching Div, UNESCO, Paris, 66, 68 & 70; consult, Royal Inst Chem, 70. Honors & Awards: Sci Apparatus Makers Award Chem Educ, Am Chem Soc, 64; Citation for Distinguished Serv to Sci Educ, Nat Sci Teachers Asn, 71. Mem: Fel AAAS; Am Chem Soc; Nat Sci Teachers Asn. Res: Organometallic compounds; chemical education; examinations and tests in science. Mailing Add: 1832 Bearss Ave Tampa FL 33612

ASHFORD, VICTOR AARON, b Bremerton, Wash, Sept 12, 42; m 73. EXPERIMENTAL HIGH ENERGY PHYSICS. Educ: Long Beach State Col, BS, 65; Univ Calif, San Diego, MS, 68, PhD(physics), 73. Prof Exp: RES ASSOC PHYSICS, FERMI NAT ACCELERATOR LAB, 73- Mem: Am Phys Soc. Res: Study of internal structure of elementary particles by means of large momentum transfer, high energy hadron; hadron scattering using a large aperture multiparticle spectrometer. Mailing Add: Dept of Physics Fermilab PO Box 500 Batavia IL 60510

ASHFORD, WALTER RUTLEDGE, b Winnipeg, Man, July 4, 14; m 43; c 2. ORGANIC CHEMISTRY. Educ: Univ BC, BA, 39, MA, 41; McGill Univ, PhD(org chem), 43. Prof Exp: Res chemist, Nat Res Coun Can, 43-44; res & develop chemist, Dom Rubber Res Lab, 44-50; mgr chem develop, Merck & Co, Ltd, 50-58; res assoc, Connaught Med Res Labs, Univ Toronto, 58-64, res mem, 64-70, asst dir, 70-72, head qual control dept, 63-72, DIR QUAL CONTROL, CONNAUGHT LABS, LTD, 72- Mem: Fel Chem Inst Can; Parenteral Drug Asn. Res: Synthesis of organic compounds; development of pharmaceutical and biological products; biological testing procedures; quality control methods in parenteral drug industry. Mailing Add: Connaught Labs Ltd 1755 Steeles Ave W Willowdale ON Can

ASHIKAWA, JAMES KATASHI, biophysics, see 12th edition

ASHKIN, ARTHUR, b Brooklyn, NY, Sept 2, 22; m 54; c 3. EXPERIMENTAL PHYSICS. Educ: Columbia Col, AB, 47; Cornell Univ, PhD(physics), 52. Prof Exp: Staff mem, Radiation Lab, Columbia Univ, 42-46; asst, Nuclear Studies Lab, Cornell Univ, 48-52; mem tech staff, Res Dept, 52-63, HEAD DEPT, RES LAB, BELL LABS, 63- Mem: Fel Am Phys Soc; Inst Elec & Electronics Eng. Res: Quantum electronics. Mailing Add: Dept 4E422 Bell Labs Holmdel NJ 07733

ASHKIN, JULIUS, b Brooklyn, NY, Aug 23, 20; m 46; c 2. MAGNETIC RESONANCE. Educ: Columbia Univ, AB, 40, AM, 41, PhD(physics), 43. Prof Exp: Res assoc, Metal Lab, Chicago, Ill, 42-43; mem staff, Los Alamos Lab, NMex, 43-46; asst prof physics, Univ Rochester, 46-50; from asst prof to assoc prof, 50-58, chmn dept physics, 61-73, PROF PHYSICS, CARNEGIE-MELLON UNIV, 58- Concurrent Pos: Guggenheim fel, 68-69; Pulitzer scholar. Mem: AAAS; Am Phys Soc. Res: Scattering of high energy nuclear particles; electron paramagnetic resonance of transition metal ions. Mailing Add: Dept of Physics Carnegie-Mellon Univ Pittsburgh PA 15213

ASHLEY, CHARLES ALLEN, b Bronxville, NY, Oct 10, 23; m 48; c 5. PATHOLOGY. Educ: Cornell Univ, AB, 44, MD, 47; Univ Ill, MS, 51. Prof Exp: Intern, Mary Imogene Bassett Hosp, 47-49; resident path, Presby Hosp, Chicago, 49-52, asst attend pathologist, 52-53; asst chief path, Walter Reed Army Hosp, 53-55; assoc pathologist, 55-67, DIR, MARY IMOGENE BASSETT HOSP, 67- Concurrent Pos: Instr, Univ Ill, 52-53; instr path, Columbia Univ, 55-62, assoc clin prof, 69- Mem: Am Asn Path & Bact; Am Soc Exp Path. Res: Structural and chemical studies of muscular contractions; chemical carcinogenesis; pathology of radiation; electron microscopy. Mailing Add: Mary Imogene Bassett Hosp Cooperstown NY 13326

ASHLEY, DOYLE ALLEN, b Collinsville, Ala, May 18, 32; m 58; c 2. AGRONOMY, APPLIED PHYSIOLOGY. Educ: Auburn Univ, BS, 54, MS, 57; NC State Univ, PhD(agron), 67. Prof Exp: Asst agron, Auburn Univ, 56-58; soil scientist, agr res serv, USDA, 58-67, res soil scientist, Coastal Plains Soil & Water Conserv Res Ctr, 67-69; asst prof, 69-73, ASSOC PROF AGRON, UNIV GA, 73- Mem: Am Soc Agron; Soil Sci Soc Am; Crop Sci Soc Am. Res: Nutrient and water utilization by field crops; photosynthate translocation and utilization by crop plants; plant-soil relationships. Mailing Add: Dept Agron Miller Plant Sci Bldg Univ of Ga Athens GA 30602

ASHLEY, FRANKLIN LONGLEY, b St Petersburg, Fla, May 27, 15; m 52; c 2. PLASTIC SURGERY. Educ: Northwestern Univ, BA, 37, MS, 40, MD, 41; Am Bd Otolaryngol, dipl, 51; Am Bd Plastic Surg, dipl, 56. Prof Exp: Intern, Passavant Mem Hosp, 41-42; instr surg, Med Col, Cornell Univ, 46-47; resident otolaryngol & gen surg, US Naval Hosp, 47-50; resident gen surg, US Vet Admin Hosp, 50-51, surgeon, 51-53; asst prof, 52-55, asst clin prof, 55-59, assoc prof, 59-70, PROF SURG &

PLASTIC SURG, CTR HEALTH SCI, UNIV CALIF, LOS ANGELES, 70-, CHIEF PLASTIC SURG, 59- Concurrent Pos: Am Cancer Soc fel, Francis Delafield Hosp, 53-54; resident otolaryngol, Cornell Hosp, NY, 46-47; attend plastic surgeon & mem plastic clin, Hollywood Presby Hosp, 47-; attend plastic surgeon, St John's Hosp & Santa Monica Hosp, 54-, Univ Calif Hosp, 56- & Hosp Good Samaritan, 58-; consult plastic surgeon, US Vet Admin Hosp, City of Hope Nat Med Ctr, Harbor Gen Hosp & Orthop Hosp, 54- Mem: Am Asn Plastic Surgeons; fel Am Col Surgeons; Am Soc Plastic & Reconstruct Surgeons; AMA; Am Soc Surg of Hand. Res: Mammalian homotransplants of skin; studies of tendon and nerve healing and replacement. Mailing Add: Dept of Surg Univ of Calif Ctr for Health Sci Los Angeles CA 90024

ASHLEY, JOHNSON WELLS, JR, analytical chemistry. see 12th edition

ASHLEY, KENNETH R, b Louisville, Ky, Nov 7, 41; m 63; c 2. INORGANIC CHEMISTRY. Educ: Southern Ill Univ, BA, 63; Wash State Univ, PhD(chem), 66. Prof Exp: Fel inorg chem, Univ Southern Calif, 66-68; asst prof, 68-71, ASSOC PROF CHEM, E TEX STATE UNIV, 71- Mem: Am Chem Soc; Sigma Xi. Res: Inorganic reaction mechanisms of transition metals. Mailing Add: Dept of Chem E Tex State Univ Commerce TX 75428

ASHLEY, MARSHALL DOUGLAS, b Portland, Maine, Apr 18, 42; m 65; c 1. FOREST BIOMETRY, REMOTE SENSING. Educ: Univ Maine, BS, 65; Purdue Univ, MS, 68, PhD, 69. Prof Exp: Forester, Int Paper Co, 65-66; asst prof, 69-75, ASSOC PROF FOREST RESOURCES, UNIV MAINE, ORONO, 75- Mem: Am Soc Photogram; Soc Am Foresters. Res: Forest sampling design; application of remote sensors to forest measurement; crop yield from satellites; insect damage from aerial photography. Mailing Add: Sch of Forest Resources Univ of Maine Orono ME 04473

ASHLEY, RICHARD ALLAN, b Wilmington, Del, Mar 20, 41; m 63; c 2. WEED SCIENCE. Educ: Univ Del, BS, 63, MS, 65, PhD(biol sci), 68. Prof Exp: Asst prof, 68-74, ASSOC PROF HORT, UNIV CONN, 74- Mem: Weed Sci Soc Am. Res: Effect of herbicides on plant-soil water relations and the effect of varying environmental conditions on herbicide selectivity. Mailing Add: Dept of Plant Sci U67 Univ of Conn Storrs CT 06268

ASHLEY, ROGER PARKMAND, b Portland, Ore, Oct 1, 40; m 64; c 2. ECONOMIC GEOLOGY. Educ: Carleton Col, BA, 62; Stanford Univ, PhD(geol), 67. Prof Exp: GEOLOGIST, US GEOL SURV, 66- Mem: AAAS; Geol Soc Am; Soc Econ Geologists. Res: Geology of gold deposits in western United States; fission-track dating applied to ore deposits; detection and mapping of hydrothermal alteration using multispectral imagery; ore deposits. Mailing Add: US Geol Surv 345 Middlefield Rd Menlo Park CA 94025

ASHLEY, SAMUEL EDWARD QUALTROUGH, b Brooklyn, NY, Dec 15, 08. ANALYTICAL CHEMISTRY. Educ: NY Univ, BS, 30; Princeton Univ, MA, 33. Prof Exp: Asst photochem, NY Univ, 30-31; asst electrochem, Princeton Univ, 31-34; head anal sect, Gen Elec Co, 34-53, mgr major appliance lab, 53-68, consult nat & int standards, 68-71; RETIRED. Mem: Am Chem Soc; Optical Soc Am; fel NY Acad Sci; fel Am Inst Chemists; The Chem Soc. Res: Spectrophotometry; low pressure techniques in analytical chemistry; photochemistry; electrochemistry of the Weston cell; metal analysis. Mailing Add: 132 Mercer St Princeton NJ 08540

ASHLEY, WARREN COTTON, b Yorkville, Ill, Dec 25, 04; m 29; c 2. ORGANIC CHEMISTRY. Educ: Univ Ill, BS, 29, MS, 30, PhD, 35. Prof Exp: Org chemist, Eastman Kodak Co, 34; res org chemist, Battelle Mem Inst, 35-38; res org chemist, Pyroxylin Prod, Inc, 38-62; res org chemist, G J Aigner Co, 62-69; INDEPENDENT FORMULATOR, 69- Honors & Awards: Outstanding Serv Award, Chicago Soc Paint Technol, 62. Mem: Am Chem Soc; Fedn Socs Paint Technol. Res: Emulsion and lacquer type and hot melt coatings and adhesives. Mailing Add: 565 N Washington Hinsdale IL 60521

ASHLOCK, PETER DUNNING, b San Francisco, Calif, Aug 22, 29; m 56; c 2. TAXONOMIC ENTOMOLOGY. Educ: Univ Calif, BS, 52, PhD, 66; Univ Conn, MS, 56. Prof Exp: Entomologist insect identification and parasite introd lab, Agr Res Serv, USDA, 58-60; entomologist, Bishop Mus, 60-68; asst prof entom, 68-69, ASSOC PROF ENTOM, SYSTS & ECOL, UNIV KANS, 69- Mem: Entom Soc Am; Soc Syst Zool. Res: Systematics of Hemiptera-Heteroptera especially Lygaeidae; systematic theory. Mailing Add: Dept of Entom Univ of Kans Lawrence KS 66045

ASHMAN, MICHAEL NATHAN, b Baltimore, Md, Oct 15, 40; m 66; c 2. ANESTHESIOLOGY, COMPUTER SCIENCES. Educ: Johns Hopkins Univ, BA, 60; Univ Md, MD, 64; Polytech Inst Brooklyn, MS, 69; Am Bd Anesthesiologists, dipl, 70. Prof Exp: Instr, 69-71, ASST PROF ANESTHESIOL, SCH MED, UNIV MD, BALTIMORE CITY, 71- Concurrent Pos: Fel, Dept Anesthesiol, Columbia-Presby Med Ctr, NY, 67-69; consult, Ctr Educ Comput Develop, Univ Md, Baltimore City, 69- Mem: Fel Am Col Anesthesiologists; Asn Advan Med Instrumentation; Asn Comput Mach; Biomed Eng Soc. Res: Uptake and distribution of inhalation anesthetics; continuous system simulation; continuous system simulation languages. Mailing Add: 7206 Denberg Rd Baltimore MD 21209

ASHMORE, CHARLES ROBERT, b Camden, NJ, Dec 1, 34; m 57; c 4. BIOCHEMISTRY, DEVELOPMENTAL BIOLOGY. Educ: Univ Conn, BS, 56, PhD(develop biol), 68; Univ Bridgeport, MS, 61. Prof Exp: ASSOC PROF ANIMAL SCI, UNIV CALIF, DAVIS, 68- Mem: Soc Exp Biol & Med; Histochem Soc; Am Asn Animal Sci. Res: Hereditary muscular dystrophy; muscle metabolism; muscle growth and development. Mailing Add: Dept of Animal Sci Univ of Calif Davis CA 95616

ASHMORE, JAMES, b Russelville, Ark, Aug 6, 26; m 48; c 4. PHARMACOLOGY. Educ: NTex State Col, BS, 47, MS, 48; St Louis Univ, PhD(biochem), 53. Prof Exp: Res assoc biochem, Southwestern Med Sch, 48-50; instr, Sch Nursing, Baylor Univ, 49-50; asst prof, Harvard Univ, 54-58; assoc prof, Ind Univ, 58-60, prof pharmacol & chmn dept, 60-68; prof biochem, Med Sch, Univ Mass, 68-69; PROF PHARMACOL & CHMN DEPT, SCH MED, IND UNIV, 69- Mem: AAAS; Soc Pharmacol & Exp Therapeut; Am Diabetes Asn; Soc Exp Biol & Med; Endocrine Soc. Res: Endocrine regulation of metabolism; carbohydrate; protein; lipids. Mailing Add: Dept of Pharmacol Ind Univ Sch of Med Indianapolis IN 46202

ASHORN, THEODORE H, organic chemistry, wood chemistry, see 12th edition

ASHTON, DAVID HUGH, b Rivers, Man, Nov 25, 39; m 66; c 2. FOOD MICROBIOLOGY. Educ: Univ Man, BS, 62, MS, 64; NC State Univ, PhD(food sci, microbiol), 67. Prof Exp: Nat Acad Sci-Nat Res Coun fel, USDA, 67-69; sr res microbiologist, Campbell Inst Food Res, Campbell Soup Co, 69-72, div head microbiol res, 72-75; LAB MGR MICROBIOL RES & DEVELOP DEPT, HUNT-WESSON FOODS, INC, 75- Mem: Am Soc Microbiol; Inst Food Technol. Res: Heat processed microbiology; rapid techniques of Salmonella detection; toxin detection including Clostridium botulinum, Salmonella and Staphylococcus aureus. Mailing Add: Hunt-Wesson Foods Inc 1645 W Valencia Dr Fullerton CA 92634

ASHTON, FLOYD MILTON, b Indianapolis, Ind, Jan 27, 22; m 42; c 1. PLANT PHYSIOLOGY, WEED SCIENCE. Educ: Univ Ill, BS, 47; Ohio State Univ, PhD, 55. Prof Exp: Anal chemist, Presto-Lite Battery Co, Inc, 40-42; lab asst, Eli Lilly & Co, 45-46; asst biochemist exp sta, Hawaiian Sugar Planters' Asn, 47-55; instr bot, 56-58, from asst prof to assoc prof, 58-68, PROF BOT, UNIV CALIF, DAVIS, 68- Concurrent Pos: NIH fel, 62-63; consult, Nat Acad Sci, 67-68. Mem: Am Soc Plant Physiol; Weed Sci Soc Am; Am Soc Biol Chem; Scand Soc Plant Physiol. Res: Mode of action of herbicides; weed control; storage protein metabolism in germinating seeds. Mailing Add: Dept of Bot Univ of Calif Davis CA 95616

ASHTON, GEOFFREY C, b Croydon, Eng, July 5, 25; nat; m 51; c 4. GENETICS. Educ: Univ Liverpool, BSc, 45, PhD(genetics), 58, DSc, 67. Prof Exp: Asst, Univ Toronto, 48-50; sect leader, fermentation res div, Glaxo Labs, Ltd, 51-56; sr sci officer, Farm Livestock Res Ctr, Eng, 56-58; prin sci officer, div animal genetics, Commonwealth Sci & Indust Res Orgn, Australia, 58-64; chmn dept genetics, 65-72, asst vchancellor, 72-74, PROF GENETICS, UNIV HAWAII, 64-, VCHANCELLOR ACAD AFFAIRS, 74- Concurrent Pos: Mem panel blood group scientists, Food & Drug Orgn, UN, 63-68. Mem: Genetics Soc Am; Am Soc Human Genetics; Am Soc Naturalists; Int Soc Animal Blood Group Res. Res: Factors maintaining protein and enzyme polymorphisms; genetic aspects of fertility and disease susceptibility; human behavorial genetics. Mailing Add: Dept Genetics Biomed Sci Bldg Univ of Hawaii Honolulu HI 96822

ASHTON, GORDON CLEMENCE, b Ont, June 4, 06; m 44. STATISTICS. Educ: Univ Toronto, BSA, 35; McGill Univ, MSc, 39; NC State Col, PhD(animal indust, exp statist), 55. Prof Exp: Asst, Macdonald Col, McGill Univ, 35-46, asst prof animal nutrit, 46-51; asst prof animal husb, Iowa State Col, 51-56; assoc prof statist, Univ Guelph, 56-59, prof statistics, 59-74; RETIRED. Mem: Biomet Soc; Agr Inst Can. Res: Animal nutrition; protein and mineral requirements; least cost rations; characteristic variability in animal husbandry experiments. Mailing Add: 24 Moore Ave Guelph ON Can

ASHTON, JOSEPH BENJAMIN, b Brownwood, Tex, Sept 17, 30; m 60; c 2. ORGANIC CHEMISTRY, MATHEMATICS. Educ: Tex Technol Col, BS, 52; Univ Tex, Austin, PhD(chem), 59. Prof Exp: Chemist res & develop lab, Shell Chem Co, Tex, 59-60, chemist, Shell Develop Co, Calif, 60-61, group leader indust chem, Shell Co, Tex, 61-65, proj eval mgr indust chem div, NY, 65-68, asst mgr transportation & supplies, Shell Oil Co, 68-69, mgr chem plant indust chem div, Shell Chem Co, Calif, 69-70, mgr supply forecasts, transp & supplies, Shell Oil Co, Houston, 70, mgr, Dominguez Chem Plant, Shell Chem Co, Calif, 70-71, mgr forecasting, Shell Oil Co, Houston, 71-72, MGR ELASTOMERS, SHELL CHEM CO, HOUSTON, 72- Mem: Am Chem Soc. Res: Mechanism of attack of ozone on organic compounds; oxidation of organic compounds; forecasting of energy requirements and development of energy model for interfuel competition. Mailing Add: Shell Chem Co One Shell Plaza Houston TX 77002

ASHWELL, G GILBERT, b Jersey City, NJ, July 16, 16; m 42; c 2. BIOCHEMISTRY. Educ: Univ Ill, BA, 38, MS, 41; Columbia Univ, MD, 48. Prof Exp: Chemist, Merck & Co, Inc, NJ, 41-44; res fel biochem, Columbia Univ, 48-50; MED DIR, NIH, 50- Mem: Am Chem Soc; Am Soc Biol Chem. Res: Pentose metabolism; uronic acid metabolism in bacteria; sugar nucleotide biosynthesis and metabolism; mechanism of polysaccharide biosynthesis; structure and function of carbohydrate moiety of glycoproteins. Mailing Add: 9522 Forest Rd Bethesda MD 20014

ASHWIN, JAMES GUY, b Prince Albert, Sask, Nov 18, 26; m 60; c 1. PHYSIOLOGY, PHARMACOLOGY. Educ: Univ Sask, BSc, 48, MSc, 50; McGill Univ, PhD(physiol), 53. Prof Exp: Lectr physiol, Christian Med Col, India, 53-55; spec lectr, Univ Sask, 56-60; res assoc, Can Heart Found, 60-63; assoc med writer, Sterling-Winthrop Res Inst, 63-66; BIOLOGIST, BUR DRUGS, DEPT HEALTH & WELFARE, PUB SERV CAN, 66- Concurrent Pos: Fel, Univ Miami, 63; mem, Intervarsity Christian Fel. Mem: Am Sci Affil; Pharmacol Soc Can. Res: Thrombosis; hemorrhage; histamines; medical writing. Mailing Add: Bur of Drugs Dept of Health & Welfare Pub Serv of Can Ottawa ON Can

ASHWORTH, JOHN NEWTON, physical chemistry, see 12th edition

ASHWORTH, LEE JACKSON, JR, b Oroville, Calif, Apr 1, 26; m 47; c 6. PLANT PATHOLOGY. Educ: Univ Calif, BS, 51, MS, 54, PhD(plant path), 59. Prof Exp: Asst, Univ Calif, 51-58; asst prof plant path, Tex A&M Univ, 58-65; res plant pathologist, Cotton Res Sta, USDA, Calif, 65-69; LECTR & PLANT PATHOLOGIST, UNIV CALIF, BERKELEY, 69- Concurrent Pos: Assoc ed, Phytopath, 73-75. Mem: Am Phytopath Soc; AAAS. Res: Soil borne diseases; epidemiology and control of soil borne plant pathogens. Mailing Add: Univ of Calif Exp Sta 147 Hilgard Berkeley CA 94720

ASHWORTH, MURRAY ALEXANDER, b Toronto, Ont, Sept 20, 15. HUMAN PHYSIOLOGY. Educ: Univ Toronto, MD, 41, BScMed, 45, PhD(physiol), 58. Prof Exp: Res asst, 45-47, sr demonstr, 46-48, lectr, 48-54, asst prof, 54-69, ASSOC PROF PHYSIOL, UNIV TORONTO, 69- Mem: Am Diabetes Asn; Can Med Asn. Res: Pancreas and carbohydrate metabolism. Mailing Add: Dept of Physiol Univ of Toronto Toronto ON Can

ASHWORTH, RALPH P, b Asheville, NC, June 1, 18; m 41; c 1. BOTANY. Educ: Wake Forest Col, BS, 39; Univ NC, MA, 45, PhD(bot), 60. Prof Exp: Teacher biol & chmn dept, Mars Hill Col, 41-56; assoc prof bot, 60-68, PROF BOT & BACT, CLEMSON UNIV, 68- Res: Invertebrate morphology; morphology of Dero Limosa; foliar ontogeny; comparative ontogeny of foliage of southeastern species of Ilex; morphological and physiological investigations of Xanthium. Mailing Add: Dept of Bot & Bact Clemson Univ Clemson SC 29631

ASHWORTH, RAYMOND BERNARD, biochemistry, see 12th edition

ASHWORTH, ROBERT DAVID, b Dallas, Tex, Apr 22, 40; m 62, 71; c 3. ANATOMY, NEUROPHYSIOLOGY. Educ: NTex State Univ, BS, 63; Univ Tex, Dallas, PhD(anat, neurophysiol), 68. Prof Exp: Asst prof anat, Queen's Univ, Ont, 68-69 & Emory Univ, 69-71; ASST PROF ANAT, UNIV TEX HEALTH SCI CTR DALLAS, 71- Concurrent Pos: Med Res Coun Can res grant, 69-71; NIH grant, 70-72. Mem: Am Asn Anatomists. Res: Reflex control of limb musculature; electrical events associated with neurosecretion and hormonal release. Mailing Add: Dept of Anat Univ of Tex Health Sci Ctr Dallas TX 75230

ASHWORTH, T, b Colne, Eng, Sept 18, 40; m 64; c 2. SOLID STATE PHYSICS, THERMODYNAMICS. Educ: Univ Manchester, BSc, 61, PhD(physics), 67. Prof Exp: Turner & Newall res fel physics, Univ Manchester, 66-68; asst prof, 68-71, ASSOC PROF PHYSICS, S DAK SCH MINES & TECHNOL, 71- Concurrent Pos:

Sr res fel, Nat Ctr Atmospheric Res, 74-75. Mem: AAAS; Am Phys Soc; Brit Inst Physics. Res: Thermal conductivities; specific heats; condensation and coalescence. Mailing Add: Dept of Physics SDak Sch Mines & Technol Rapid City SD 57701

ASHWORTH, URAL STEPHEN, b Walla Walla, Wash, Sept 11, 05; m 35; c 5. FOOD SCIENCES. Educ: Univ Mo, AB, 29, PhD(agr chem), 33. Prof Exp: Asst agr chem, Univ Mo, 29-34, res instr, 34-35; Coxe fel, Yale Univ, 35-37; asst prof med sch, Univ Ark, 37-39; res dairy chemist, 39-50, from asst prof to prof dairy sci, 50-71, from assoc dairy scientist to dairy scientist, 50-69, food scientist, 69-71, EMER PROF DAIRY SCI, WASH STATE UNIV, 71- Mem: Am Chem Soc; Am Dairy Sci Asn. Res: Energy and protein metabolism during growth; nutrition; protein and enzyme chemistry; methods of analysis for food products; powdered milk research. Mailing Add: Dept of Food Sci Wash State Univ Pullman WA 99163

ASHY, PETER JAWAD, b Latakia, Syria, Aug 5, 40; US citizen; m 62; c 3. CHEMICAL PHYSICS, MATHEMATICS. Educ: Univ Southwestern La, BS, 62; Clemson Univ, MS, 65, PhD(chem physics), 67. Prof Exp: Chem engr, textile fiber div, E I du Pont de Nemours & Co, 63-64, res physicist, 67-68; asst prof math, 68-72, ASSOC PROF COMPUT SCI-MATH, FURMAN UNIV, 72- Concurrent Pos: Consult, Philips Fibers Res Orgn, 68-; statist consult, Union Carbide Corp, Greenville, 73-75. Mem: Am Inst Chem Engrs; Am Chem Soc; Math Asn Am. Res: Fundamental and applied work in thermal reactions; kinetics and sorption; phenomena and intermolecular forces; structure and physical behavior of fibers; thermal diffusivity of gases; quantum theory of surface tension; application of Markov processes; chemical engineering. Mailing Add: Regent Dr Rte 7 Greenville SC 29609

ASIK, JOSEPH R, b Lorain, Ohio, Aug 8, 37; m 59; c 2. APPLIED PHYSICS. Educ: Case Inst Technol, BS, 59; Univ Ill, Urbana, MS, 61, PhD(physics), 66. Prof Exp: Res staff mem phys sci, T J Watson Res Ctr, IBM Corp, 66-69; prin res investr, 69-72, SCIENTIST ASSOC, ENG & RES STAFF, ELEC SYSTS DEPT, FORD MOTOR CO, 72- Mem: Am Phys Soc; Soc Automotive Engrs; Inst Elec & Electronics Engrs. Res: Combustion initiation; instrumentation for combustion initiation studies; electromagnetic devices. Mailing Add: 465 Weybridge Dr Bloomfield Hills MI 48013

ASIMOV, ISAAC, b Petrovichi, Russia, Jan 2, 20; nat US; m 42, 73; c 2. SCIENCE WRITING, BIOCHEMISTRY. Educ: Columbia Univ, BS, 39, MA, 41, PhD(enzyme chem), 48. Prof Exp: Chemist, Naval Air Exp Sta, Pa, 42-45; from instr to asst prof, 49-55, ASSOC PROF BIOCHEM, SCH MED, BOSTON UNIV, 55-; SCI WRITER, 54- Mem: Am Chem Soc. Res: Kinetics of enzyme inactivation; photochemistry of antimalarials; enzymology of malignant tissues; irradiation of nucleic acids. Mailing Add: 10 W 66th St New York NY 10023

ASKARI, AMIR, b Ahwaz, Iran, Dec 24, 30; US citizen; m 57; c 2. PHARMACOLOGY, BIOCHEMISTRY. Educ: Univ Dubuque, BS, 53; NY Univ, MS, 57; Cornell Univ, PhD(biochem), 60. Prof Exp: From instr to assoc prof, 63-73, PROF PHARMACOL, MED COL, CORNELL UNIV, 73- Concurrent Pos: Res fel pharmacol, Med Col, Cornell Univ, 60-63; USPHS grant, 63. Mem: AAAS; Am Chem Soc; Am Soc Pharmacol & Exp Therapeut; Harvey Soc; Biophys Soc. Res: Mechanism of ion transport through cell membranes; effects of drugs on membranes. Mailing Add: Dept of Pharmacol Cornell Univ Med Col New York NY 10021

ASKEVOLD, ROBERT JAMES, b Chicago, Ill, Nov 3, 12; m 46; c 2. PETROLEUM CHEMISTRY. Educ: Univ Chicago, SB, 34. Prof Exp: Res chemist & div dir res ctr, Pure Oil Co, 34-57, asst res dir, 58-66, MGR PROD RES, RES DEPT, UNION OIL CO CALIF, 66- Mem: Am Chem Soc; Am Petrol Inst; Soc Automotive Engrs. Res: Marketing and refining research and development. Mailing Add: Res Dept Union Oil Co of Calif PO Box 76 Brea CA 92621

ASKEW, CORNELIUS, JR, b Jackson, Tenn, July 17, 40; c 1. CARDIOVASCULAR DISEASES, EPIDEMIOLOGY. Educ: Tenn State Univ, BS, 61, MSc, 63; Univ Okla, MSPH, 66, DrPh(epidemiol), 68. Prof Exp: Res assoc biochem, Meharry Med Col, 62-63; instr biol, Jarvis Christian Col, 63-67, assoc prof community health, 67-69; asst prof epidemiol, Sch Pub Health, Univ Tex Health Sci Ctr, 69-74; ASSOC PROF MED CARE ORGN, UNIV ILL SCH PUB HEALTH & ASSOC PROF PREV MED, ABRAHAM LINCOLN SCH MED, UNIV ILL MED CTR, 75- Concurrent Pos: Adj asst prof, Baylor Col Med, 69-74; dir epidemiol anal br, Nat Ctr Health Serv Res & Develop, 70-72; mem coun epidemiol, Am Heart Asn, 70-; mem epidemiol & biomet training study sect, Nat Inst Gen Med Sci, 72-74; mem, Gov's Adv Coun on Aging, 72-; mem cerebrovascular dis adv comt, Dept Epidemiol, Johns Hopkins Univ, 74- Mem: Soc Epidemiol Res; Asn Teachers Prev Med; Am Pub Health Asn. Res: Application of epidemiologic techniques in the qualitative and quantitative evaluation of health care systems. Mailing Add: Univ of Ill Med Ctr Chicago IL 60612

ASKEW, HAROLD COCHRAN, b Hattiesburg, Miss, Dec 10, 23; m 52; c 1. ANATOMY, DENTISTRY. Educ: Univ Chicago, BS, 47; Wash Univ, DDS, 51; Univ Ala, MS, 58. Prof Exp: Asst prof dent & instr anat, 58-62, assoc prof dent, 62-65, ASST PROF ANAT, MED CTR, UNIV ALA, BIRMINGHAM, 62-, PROF DENT, 65-, CHMN DEPT ODONTOL, 73- Mem: Am Dent Asn. Res: Physiology of mastication; dental caries; periodontal disease. Mailing Add: Dept of Anat Univ of Ala Med Ctr Birmingham AL 35233

ASKEW, RAYMOND FIKE, b Birmingham, Ala, Oct 29, 35; m 56; c 3. PLASMA PHYSICS. Educ: Birmingham-Southern Col, BS, 56; Univ Va, MS, 58, PhD(physics), 60. Prof Exp: Sr physicist res lab eng sci, Univ Va, 60; from asst prof to assoc prof physics, 60-71, PROF PHYSICS, AUBURN UNIV, 71-, ACTG DIR NUCLEAR SCI CTR, 69- Concurrent Pos: Consult, US Air Force Systs Command, Eglin AFB, Fla, 72-74. Mem: Am Asn Physics Teachers; Am Phys Soc. Res: Nuclear and atomic physics; gaseous discharges. Mailing Add: Dept of Physics Auburn Univ Auburn AL 36830

ASKEW, WILLIAM CREWS, b Athens, Ala, Jan 5, 40; m 63; c 3. RADIATION CHEMISTRY, ANALYTICAL CHEMISTRY. Educ: Auburn Univ, BS, 62, MS, 63; Univ Fla, PhD(radiolysis of fluorocarbons), 66. Prof Exp: Res asst chem eng, Univ Fla, 63-66; ASST PROF CHEM ENG, AUBURN UNIV, 67- Concurrent Pos: Proj leader, US AEC res grant, 69- Mem: Am Inst Chem Eng. Res: Synthesis by radiolysis, identification and purification of perfluoroalkanes. Mailing Add: Dept of Chem Eng Auburn Univ Auburn AL 36830

ASKEY, RICHARD ALLEN, b St Louis, Mo, June 4, 33; m 58; c 2. MATHEMATICAL ANALYSIS. Educ: Wash Univ, AB, 55; Harvard Univ, MA, 56; Princeton Univ, PhD(math), 61. Prof Exp: Instr math, Wash Univ, 58-61; instr, Univ Chicago, 61-63; from asst prof to assoc prof, 63-68, PROF MATH, UNIV WIS, MADISON, 68- Concurrent Pos: Guggenheim fel, 69-70. Mem: Am Math Soc; Math Asn Am; Soc Indust & Appl Math. Res: Special functions; Fourier analysis. Mailing Add: Dept of Math Univ of Wis Madison WI 53706

ASKILL, JOHN, b Kent, Eng, Apr 27, 39; m; c 1. SOLID STATE PHYSICS, METALLURGY. Educ: Univ Reading, BSc, 60 & 61, PhD(physics), 64. Prof Exp: Fel, Univ Reading, 64-65; res physicist, Oak Ridge Nat Lab, 65-66; ASSOC PROF PHYSICS & CHMN DEPT, MILLIKIN UNIV, 66- Concurrent Pos: Consult metals & ceramics div, Oak Ridge Nat Lab, 66-67. Mem: Am Asn Physics Teachers; Brit Inst Physics & Phys Soc. Res: Diffusion of metals in metals using radioactive tracers. Mailing Add: Dept of Physics Millikin Univ Decatur IL 62522

ASLESON, JOHAN ARNOLD, b Stoughton, Wis, Sept 13, 18; m 43; c 3. SOILS. Educ: Univ Wis, BS, 42, MS, 47, PhD, 57. Prof Exp: From asst prof to assoc prof soils, 47-54, asst to dir, Agr Exp Sta, 54-57, from asst dir to assoc dir, 57-61, DIR AGR EXP STA, MONT STATE UNIV, 61-, PROF SOILS, 54-, DEAN COL AGR, 65- Mem: Am Soc Agron; Soil Sci Soc Am. Res: Applied fertilizer research; agronomy and soil science; plant physiology. Mailing Add: Col of Agr Mont State Univ Bozeman MT 59715

ASLING, CLARENCE WILLET, b Duluth, Minn, June 17, 13; m 36; c 2. ANATOMY. Educ: Univ Kans, AB, 34, AM, 37, MD, 39; Univ Calif, PhD(anat), 47. Prof Exp: Asst instr micros anat, Univ Kans, 34-37; intern, Huntington Mem Hosp, Pasadena, Calif, 39-40; instr anat, Sch Med, Vanderbilt Univ, 40-41; instr, Univ Kans, 41-42, asst prof anat & lectr surg anat, 42-46; res asst, Univ Calif, Berkeley, 44-47, lectr anat, 45-47, from asst prof to prof, 47-66, res assoc, Inst Exp Biol, 51-58, co-chmn dept anat & physiol, 58-62; PROF ANAT & VCHMN DEPT, MED CTR, UNIV CALIF, SAN FRANCISCO, 66- Concurrent Pos: Fulbright award, Eng, 53-54; Guggenheim fel, 62-63; guest prof, Univ Geneva, 62-63. Mem: Am Asn Anatomists; Am Asn Phys Anthropologists; Soc Exp Biol & Med; Anat Soc Gt Brit & Ireland. Res: Physical anthropology of the head form; endocrine and nutritional control of skeletal development; localization of radioactive minerals; teratology and congenital abnormalities. Mailing Add: Dept of Anat Univ of Calif Med Ctr San Francisco CA 94122

ASMUS, JOHN FREDRICH, b Chicago, Ill, Jan 20, 37; m 63; c 2. LASERS. Educ: Calif Inst Technol, BS, 58, MS, 59, PhD(elec eng, physics), 65. Prof Exp: Staff assoc high energy fluid dynamics, Gen Atomic Div, Gen Dynamics Corp, 64-67; mem staff spec nuclear effects, Gulf Gen Atomic Inc, 67-69; res staff mem laser technol, Inst for Defense Anal, 69-71; vpres, Sci Applns, Inc, 71-73; RES PHYSICIST, UNIV CALIF, SAN DIEGO, 73- Concurrent Pos: Assoc mem spec group on optical masers, Dir Defense Res & Eng Adv Group Electron Devices, 69-70, mem, 70- Mem: Int Inst Conserv; Am Inst Conserv. Res: Interaction of high-power laser beams with solid surfaces; laser-induced stress wave propagation and micrometeorite generation; laser excited chemistry. Mailing Add: Inst of Geophys & Planetary Physics Univ of Calif San Diego PO Box 1529 La Jolla CA 92037

ASNER, BERNARAD A, JR, b Oklahoma City, Okla, Dec 28, 33; m 60; c 2. APPLIED MATHEMATICS. Educ: Univ Okla, BS, 57; Univ Ala, MA, 64; Northwestern Univ, PhD(eng sci), 68. Prof Exp: Mathematician control, NASA, 60-65; assoc prof math, Bogazici Univ, Turkey, 70-72; exchange scientist, Nat Acad Sci, 72-73; ASSOC PROF MATH & BUS MGT, UNIV DALLAS, 73- Mem: Math Asn Am; Am Math Soc; Soc Indust & Appl Math. Res: Degeneracy in functional differential equations and computer science. Mailing Add: Dept of Math Univ of Dallas Irving TX 75061

ASOFSKY, RICHARD MARCY, b New York, NY, Sept 25, 33; m 60; c 2. IMMUNOLOGY, PATHOLOGY. Educ: State Univ NY, MD, 58. Prof Exp: Intern internal med, Kings County Hosp, NY, 58-59; resident, Med Ctr, NY Univ, 59-63; staff assoc immunol, 63-66, head exp path sect, Lab Germfree Animal Res, 65-70, asst chief, 70-72, CHIEF, LAB MICROBIAL IMMUNITY, NAT INST ALLERGY & INFECTIOUS DIS, 72- Concurrent Pos: NIH trainee path, Med Ctr, NY Univ, 59-63. Res: Sites and control of immunoglobulin synthesis; interaction among lymphocytes in immune responses. Mailing Add: 8501 Pelham Rd Bethesda MD 20014

ASP, CARL W, b Cleveland, Ohio, Aug 14, 31; m 65; c 2. SPEECH & HEARING SCIENCES, STATISTICS. Educ: Ohio State Univ, BS, 63, MA, 64, PhD(speech & hearing sci), 67. Prof Exp: PROF SPEECH & HEARING SCI, UNIV TENN, KNOXVILLE, 67- Concurrent Pos: Mem field read & site visit team, US Off Educ, 69- Mem: AAAS; Acoust Soc Am; Am Speech & Hearing Asn; Audio Eng Soc. Res: Psychoacoustics; speech communication; psychophysical experiments; perception of hearing impaired. Mailing Add: Dept of Audiol & Speech Path Univ of Tenn Knoxville TN 37916

ASPELIN, GARY BERTIL, b Bristol, Conn, Dec 31, 39; m 67; c 2. ORGANIC CHEMISTRY. Educ: Brown Univ, AB, 61; Univ Wis, PhD(org chem), 66. Prof Exp: Res chemist, Am Cyanamid Co, Bound Brook, 66-71; SR RES SCIENTIST, SURGIKOS, NEW BRUNSWICK, 71- Mem: Am Asn Textile Chemists & Colorists; Tech Asn Pulp & Paper Indust. Res: Organic reaction mechanisms; cellulosic and synthetic organic chemistry; nonwoven fabrics; film and fabric laminates; bacterial filtration. Mailing Add: 268 Farmer Rd Bridgewater NJ 08807

ASPER, SAMUEL PHILLIPS, JR, b Oak Park, Ill, July 14, 16; m 42; c 2. MEDICAL ADMINISTRATION. Educ: Baylor Univ, AB, 36; Johns Hopkins Univ, MD, 40. Prof Exp: House officer med, Johns Hopkins Hosp, 40-41; res fel, Thorndike Mem Lab, Harvard Univ, 41-42 & 46-47; from instr to prof, Sch Med, Johns Hopkins Univ, 47-73, vpres med affairs, Johns Hopkins Hosp, 70-73; PROF MED & DEAN MED SCH, UNIV & CHIEF-OF-STAFF, HOSP, AM UNIV BEIRUT, 73- Mem: Endocrine Soc; Asn Am Physicians; Am Col Physicians; Am Soc Clin Invest. Res: Clinical endocrinology; metabolism. Mailing Add: Am Univ of Beirut Med Sch Beirut Lebanon

ASPERGER, ROBERT GEORGE, b Detroit, Mich, Oct 1, 37; m 61; c 1. CHEMISTRY. Educ: Eastern Mich Univ, AB, 60; Univ Mich, MS, 63, PhD(chem), 65. Prof Exp: Res chemist, 65-75, SR RES SPECIALIST, DOW CHEM CO, 75- Mem: AAAS; Am Chem Soc; NY Acad Sci; Sigma Xi; Nat Asn Corrosion Engrs. Res: Coordination chemistry; general phenomenon of catalysis, heterogeneous and homogeneous, especially stereochemistry of reaction products; fundamentals of corrosion and its inhibition. Mailing Add: 12 Dennis Ct Midland MI 48640

ASPEY, WAYNE PETER, b Pittsburgh, Pa, Mar 26, 46; m 70. ANIMAL BEHAVIOR, ETHOLOGY. Educ: Ohio Univ, AB, 68, PhD(zool), 74; Dartmouth Col, MA, 70. Prof Exp: Teaching fel psychol, Dartmouth Col, 68-70; teaching asst zool, Ohio Univ, 70-74; FEL ANIMAL BEHAV, MARINE BIOMED INST, UNIV TEX MED BR GALVESTON, 74- Concurrent Pos: Mem sci adv panel, NASA, 75-; zool/marine sci book reviewer, Sci Mag, 75- Mem: Animal Behav Soc; Int Soc Develop Psychobiol; AAAS; Am Soc Zoologists; Brit Arachnol Soc. Res: Field, laboratory and multivariate statistical analysis of communication systems, including the ethology and behavioral ecology of aggression and reproduction in marine gastropod molluscs and arthropods, especially sea hares, fiddler crabs and lycosid spiders. Mailing Add: Univ of Tex Marine Biomed Inst 200 University Blvd Galveston TX 77550

ASPIN, NORMAN, biophysics, see 12th edition

ASPINALL, GERALD OLIVER, b Chesham Bois, Eng, Dec 30, 24; m 53; c 2. ORGANIC CHEMISTRY. Educ: Bristol Univ, BSc, 44, PhD(chem), 48; Univ Edinburgh, DSc(chem), 58. Prof Exp: From lectr to sr lectr chem, Univ Edinburgh, 48-63, reader, 63-67; prof & chmn dept, Trent Univ, 67-72; PROF CHEM & CHMN DEPT, YORK UNIV, 72- Mem: Am Chem Soc; Chem Inst Can; The Chem Soc; Royal Inst Chem; Brit Biochem Soc. Res: Chemistry of carbohydrates, especially plant polysaccharides. Mailing Add: Dept of Chem York Univ 4700 Keele St Downsview ON Can

ASPINALL, SAMUEL RUSMISELL, b Philippi, WVa, June 27, 15; m 41; c 2. ORGANIC CHEMISTRY, RESEARCH ADMINISTRATION. Educ: Univ WVa, AB, 35; Yale Univ, PhD(org chem), 38. Prof Exp: Instr chem, Swarthmore Col, 38-41; tech aide div B, Off Sci Res & Develop Off Emergency Mgt, 41-46; tech investr, E I du Pont de Nemours & Co, 46-47; asst prof chem, Williams Col, 47-50; asst prof, Off Naval Res Br Off, London, 50-56, sci dir, 53-56; asst dir, 56-64, MGR, UNION CARBIDE RES INST, 64- Mem: Am Chem Soc. Res: Aliphatic diamines; preparation and cyclization of monoacylethylenediamines. Mailing Add: Union Carbide Res Inst Tarrytown Tech Ctr Tarrytown NY 10591

ASPITARTE, THOMAS (ROBERT), b Gooding, Idaho, Nov 13, 20; m 48; c 4. MICROBIOLOGY. Educ: Univ Idaho, BS, 52; Ore State Univ, MS, 53, PhD, 59. Prof Exp: Microbiologist, Com Solvents Corp, 53-56; plant pathologist, Res Lab, United Fruit Co, Honduras, 58-60; from sr res chemist to supvr, Environ Res, Cent Res Labs, 60-71, MGR ENVIRON SERV DIV, CROWN ZELLERBACH CORP, 71- Mem: Am Soc Microbiol; Am Chem Soc. Res: Biologics of paper; paper machines and related systems; water pollution. Mailing Add: Cent Res Labs Crown Zellerbach Corp Camas WA 98607

ASPLAND, JOHN RICHARD, b Leeds, Eng, Oct 22, 36; m 61; c 2. TEXTILE CHEMISTRY. Educ: Univ Leeds, BSc, 58, MSc, 60; Univ Manchester, PhD(textile chem), 64. Prof Exp: Prof textile chem, Univ Manchester Inst Sci & Technol, 61-66; sr res chemist dye applns res, Southern Dyestuff Co, Martin-Marietta Corp, 66-68, group leader dye applns & eval, 68-74; TEXTILE RES MGR, REEVES BROS, INC, 74- Concurrent Pos: Perkin Travel Scholar, The Chem Soc, 65. Mem: Am Chem Soc; Am Asn Textile Chem & Colorists; fel Brit Soc Dyers & Colourists. Res: Interactions between dyestuffs and polymeric substances, particularly disperse dyes and hydrophobic fibers and reactive dyes and cellulose; instrumental color measurement; sulfur dyes; flame-resistant fabrics; energy conservation. Mailing Add: 4501 Town & Country Dr Charlotte NC 28211

ASPLUND, KENNETH KARL, biology, zoology, see 12th edition

ASPLUND, RUSSELL OWEN, b Lethbridge, Can, May 5, 28; m 51; c 6. BIOCHEMISTRY. Educ: Univ Alta, BSc, 49; Utah State Univ, MS, 55; WVa Univ, PhD(biol chem), 58. Prof Exp: Asst, Utah State Univ, 49-50 & Univ Utah, 51-52; chemist, exp sta, Utah State Univ, 53-56; asst, WVa Univ, 56-58; from asst prof to assoc prof, 58-65 & 67-72, PROF CHEM, UNIV WYO, 72-73 & 75- Concurrent Pos: Clayton Found fel, 65-66; vis prof, Univ Calif, Los Angeles, 74. Mem: AAAS; Am Chem Soc. Res: Chemistry of biologically active indole compounds; natural products; protein chemistry. Mailing Add: Dept of Chem Univ of Wyo Laramie WY 82070

ASPNES, DAVID E, b Madison, Wis, May 1, 39; m 64; c 2. PHYSICS. Educ: Univ Wis, Madison, BS, 60, MS, 61; Univ Ill, Urbana, PhD(physics), 65. Prof Exp: Res assoc physics, Univ Ill, Urbana, 65-66 & Brown Univ, 66-67; MEM TECH STAFF, BELL TEL LABS, 67- Mem: Am Phys Soc; Optical Soc Am. Res: Experimental solid state physics; semiconductors; electric field effects on crystals; optical properties of solids. Mailing Add: Bell Tel Labs Murray Hill NJ 07974

ASPREY, LARNED BROWN, b Sioux City, Iowa, Mar 19, 19; m 44; c 7. PHYSICAL CHEMISTRY, INORGANIC CHEMISTRY. Educ: Iowa State Univ, BS, 40; Univ Calif, PhD(chem), 49. Prof Exp: Chemist, Campbell Soup Co, 40-42, metall lab, Univ Chicago, 44-46 & radiation lab, Univ Calif, 46-49; MEM STAFF, LOS ALAMOS SCI LAB, 49- Mem: AAAS; Am Chem Soc. Res: Physical and inorganic chemistry of actinide and lanthanide elements; fluorine chemistry. Mailing Add: 720 46th St Los Alamos NM 87544

ASPREY, WINIFRED ALICE, b Sioux City, Iowa, Apr 8, 17. MATHEMATICS, COMPUTER SCIENCES. Educ: Vassar Col, AB, 38; Univ Iowa, MS, 43, PhD(math), 45. Prof Exp: Teacher, Brearley Sch, New York, 38-40 & Girls Latin Sch, Chicago, 40-42; asst math, Univ Iowa, 42-45; from instr to prof, 45-61, chmn dept math, 58-62, ELIZABETH STILLMAN WILLIAMS CHAIR PROF MATH, VASSAR COL, 61- Concurrent Pos: IBM Corp indust res fel, 57-58; NSF grant, Univ Calif, Los Angeles, 62, NSF fac fel & vis scholar, 64-65; dir, Acad Year Inst Math, State Dept Educ, NY, 62-64 & Acad Year Inst Comput Math, Vassar Col, spring, 64, 66 & 67; IBM Corp exchange grant, 69-70; vis staff mem & consult, Los Alamos Sci Lab, 72- Mem: Fel AAAS; Math Asn Am; Am Math Soc; Soc Indust & Appl Math; Asn Comput Mach. Res: Analysis and topology; families of total oscillators and their derived functions. Mailing Add: Dept of Math Vassar Col Poughkeepsie NY 12601

ASQUITH, DEAN, b Lowell, Mass, Mar 3, 12; m 32; c 2. ECONOMIC ENTOMOLOGY. Educ: Univ Mass, BS, 33, MS, 39. Prof Exp: Entomologist tech develop, Rohm and Haas Co, 39-46; consult entomologist, 46-48; assoc prof econ entom fruit insect invest, 48-57, prof econ entom, 57-67, PROF ENTOM, PA STATE UNIV, 67- Mem: AAAS; Entom Soc Am; Am Inst Biol Sci. Res: Integrated biological and chemical control of fruit insects and mites. Mailing Add: Pa State Univ Fruit Res Lab Biglerville PA 17307

ASQUITH, GEORGE BENJAMIN, b Chicago, Ill, June 23, 36; m 63; c 1. PETROLOGY, MINERALOGY. Educ: Tex Tech Univ, BS, 61; Univ Wis, Madison, MS, 63, PhD(geol), 66. Prof Exp: Instr geol, Univ Wis, Madison, 66; res geologist, Atlantic-Richfield Res Lab, 66-70; asst prof, 70-74, ASSOC PROF GEOL, W TEX STATE UNIV, 74- Mem: Soc Econ Paleontologists & Mineralogists; Clay Minerals Soc. Res: Igneous petrology and volcanology; sandstone and carbonate petrology and sedimentation and carbonate diagenesis. Mailing Add: Kilgore Res Ctr WTex State Univ Canyon TX 79015

ASSADOURIAN, FRED, b Panderma, Turkey, Apr 13, 15; nat US; m 54; c 2. MATHEMATICS. Educ: NY Univ, BS, 35, MS, 36, PhD(math), 40. Prof Exp: Instr math, Col Eng, NY Univ, 37-42; assoc prof, Tex Tech Col, 42-44; res engr, Westinghouse Res Labs, Pa, 44-46; develop engr, Fed Telecommun Labs, NJ, 46-51, proj engr, 51-56; sr engr, Radio Corp Am, NY, 56-59, leader tech staff, 59-62, sr staff scientist & head adv transmission, 62-68; ASSOC PROF, DEPT ELEC ENG, PRATT INST, 68- Mem: Am Math Soc; Inst Elec & Electronics Eng. Res: Almost-periodic functions; pulse transformers; integral equations in gas discharge; microwave transmission; reflex klystron; distortion of amplitude modulation and frequency modulation signals; special microwave components; special problems in communications systems; communications satellite systems. Mailing Add: Dept of Elec Eng Pratt Inst Brooklyn NY 11205

ASSAYKEEN, TATIANA ANNA, b South Orange, NJ, Sept 27, 39. ENDOCRINOLOGY, PHARMACOLOGY. Educ: Trinity Col, DC, BA, 61; Univ Va, PhD(pharmacol), 65. Prof Exp: Asst prof surg & pharmacol, 67-73, adj prof surg, Sch Med, Stanford Univ, 73-76; MED DIR, ASTRA CHEM, SYDNEY, AUSTRALIA, 76- Concurrent Pos: USPHS fel, Med Ctr, Univ Calif, San Francisco, 65-67; NIH res grant, 69-76. Mem: AAAS; Endocrine Soc; Am Soc Pharmacol & Exp Therapeut; Am Soc Nephrology. Res: Renin-angiotensin-aldosterone system; hypertension; catecholamines. Mailing Add: Astra Chem PO Box 131 North Ryde NSW Australia

ASSEFF, PETER ANTHONY, organic chemistry, see 12th edition

ASSELBERGS, EDWARD ANTON MARIA, b Bergen-op-Zoom, Neth, Jan 17, 27; nat US; m 50; c 5. PLANT PHYSIOLOGY. Educ: State Agr Univ Wageningen, BS, 50; Univ Toronto, MSA, 52; Cornell Univ, PhD, 54. Prof Exp: Sr food technologist, Can Dept Agr, 54-62; food technologist nutrit div, 62-69, CHIEF FOOD & AGR INDUST SERV, FOOD & AGR ORG, UN, 69- Res: Food processing. Mailing Add: Food & Agr Orgn of the UN Viale delle Terme di Caracalla Rome Italy

ASSENZO, JOSEPH ROBERT, b Boston, Mass, Jan 1, 32; m 60; c 2. BIOSTATISTICS. Educ: Northeastern Univ, BS, 54; Harvard Univ, SM, 55; Univ Okla, PhD(biostatist), 63. Prof Exp: Asst engr, Camp Dresser & McKee, Mass, 51-54, engr, 54-56, proj engr, 58; instr sanit engr, Univ Okla, 58-61, instr biostatist & sanit eng, 61-63, assoc prof, 63-66; RES MGR BIOSTATIST & RES DATA & INFO SYSTS, UPJOHN CO, 66- Concurrent Pos: Consult, Bur Water Resources Res, 58-, Civil Aeromed Res Inst, Fed Aviation Agency, 61-66, med ctr, Univ Okla, 61-66 & pub sch syst, Oklahoma City, 62-66; mem water resources develop prog, Okla, 63-66; statist analyst, hydrol of small watersheds, US Bur Pub Rds, 64-66. Mem: Am Statist Asn; Water Pollution Control Fedn; Biomet Soc; Drug Info Asn; Am Soc Pharmacol & Therapeut. Res: Applications of statistics, biomathematics, epidemiology and computer science to the discovery of and laboratory and medical research and development of drugs for human use. Mailing Add: Biostatist & Res Data & Info Upjohn Co Kalamazoo MI 49001

ASSINK, ROGER ALYN, b Holland, Mich, Sept 28, 45; m 67; c 2. POLYMER CHEMISTRY. Educ: Mich State Univ, BS, 67; Univ Ill, MS, 69, PhD(chem), 72. Prof Exp: MEM STAFF MAT RES, SANDIA LABS, 72- Mem: Am Phys Soc; Am Chem Soc. Res: Study of the molecular dynamics and structure of polymeric materials by nuclear magnetic resonance spectroscopy. Mailing Add: Sandia Labs Div 5811 Albuquerque NM 87115

ASSMUS, EDWARD FERDINAND, JR, b Nutley, NJ, Apr 19, 31; m 61, 72; c 1. ALGEBRA. Educ: Oberlin Col, AB, 53; Harvard Univ, AM, 55, PhD(math), 58. Prof Exp: Off Naval Res res assoc, Columbia Univ, 58-59, Ritt instr math, 59-62; lectr, Wesleyan Univ, 62-66; assoc prof, 66-70, PROF MATH, LEHIGH UNIV, 70- Concurrent Pos: Consult appl res lab, Sylvania Electronic Systs, 61-71; vis prof, Queen Mary Col, London Univ, 68-69. Mem: Am Math Soc; Math Asn Am; London Math Soc. Res: Homological algebra; algebraic coding theory; combinatorial mathematics. Mailing Add: Dept of Math Bldg 14 Lehigh Univ Bethlehem PA 18015

ASSONY, STEVEN JAMES, b Czech, Apr 13, 20; nat; m 48; c 2. ORGANIC CHEMISTRY. Educ: Rutgers Univ, BS, 50; Univ Southern Calif, MS, 53, PhD(chem), 57. Prof Exp: Analyst, Geigy Co, Inc, 49-50; assoc, Univ Southern Calif, 51-56, res asst, off ord res, 52-55; asst tech dir, Am Latex Prod Corp, 56-61; dir res & develop, 61-70, HEAD TECH SERV, CPR DIV, UPJOHN CO, 70- Mem: Fel Am Inst Chemists; Am Chem Soc; Soc Plastics Engrs; The Chem Soc; NY Acad Sci. Res: Organic and polymer chemistry; polyurethanes and polymers of boron, phosphorus and nitrogen. Mailing Add: 5511 N Fidler Ave Lakewood CA 90712

ASSOUSA, GEORGE ELIAS, b Jerusalem, Palestine, Mar 15, 36; US citizen; m 60; c 2. ASTROPHYSICS, SCIENCE POLICY. Educ: Earlham Col, BA, 57; Columbia Univ, MA, 62; Fla State Univ, PhD(exp nuclear physics), 68. Prof Exp: Fac mem physics, Earlham Col, 60-63; Carnegie fel atomic physics, 68-70, SR FAC MEM ATOMIC PHYSICS & RADIO ASTRON, CARNEGIE INST WASH, 70- Concurrent Pos: Consult, Princeton Univ Observ, 71-72; sr consult & vpres sci & educ affairs, Inst Develop & Econ Affairs Serv Inc, 74-; US rep, Arab Phys Soc, 74-; consult, Comn on Middle East, Nat Acad Sci, 75- & Iraqi Atomic Energy Comn, 75- Mem: Am Phys Soc; Am Astron Soc; Am Asn Univ Prof; Lebanese Asn Adv Sci; Arab Phys Soc. Res: Supernova remnant studies at radio frequencies; supernovae and the interstellar medium; beam-foil spectroscopy and heavy ion produced x-ray spectroscopy; planning and development of accelerator based research in universities and research institutions in the Middle East. Mailing Add: Carnegie Inst of Wash DTM 5241 Broad Branch Rd NW Washington DC 20015

ASSUR, ANDREW, b Pjatigorsk, Caucasus, June 16, 18; nat US; m 48; c 4. GEOPHYSICS. Educ: Univ Hamburg, Dr rer nat (phys sci), 51. Prof Exp: Mem sci staff ice serv, Hydrol Inst, Riga Univ, 41-44; mem staff, Ger Marine Observ, Univ Hamburg, 44-46; staff mem abstracts & bibliog, Am Meterol Soc, 52-58, mem snow, ice & permafrost res estab, 54-61; sci adv, 61-66, chief scientist, 66-73, CHMN, COLD REGIONS RES & ENG LAB, US ARMY, 73- Concurrent Pos: Mem, Nat Comt Int Hydrol Decade, Nat Acad Sci, 67- Honors & Awards: Distinguished Serv Award, US Navy; Sci Award, US Army, 57, Except Serv Award, 66. Mem: Am Geophys Union; Inst Soc Hydraul Eng. Res: Ice engineering; sea ice physics; water balance; applied mathematics. Mailing Add: Cold Regions Res & Eng Lab US Army 48 Rip Rd Hanover NH 03755

AST, DAVID BERNARD, b New York, NY, Sept 30, 02; m 28; c 1. DENTISTRY, PUBLIC HEALTH ADMINISTRATION. Educ: NY Univ, DDS, 24; Univ Mich, MPH, 42. Prof Exp: Pvt pract, 24-38; asst dir oral hyg, NY State Dept Health, 38-45, chief dent health sect, 45-48, dir bur dent health, 48-66, assoc dir div med serv, 66-70, asst comnr div med care serv & eval, 70-72; RETIRED. Concurrent Pos: Dept Health, Educ & Welfare grants, 45-55, 63-64 & 67-69; lectr pub health pract, Columbia Univ, 44-; spec lectr pub health, Albany Med Col, 50-; consult, Prof Exam Serv, Am Pub Health Asn, 50-; mem community health proj rev comt, Dept Health, Educ & Welfare, 63-65. Honors & Awards: Gov Alfred E Smith Award, Am Soc Pub Admin, 55; Award, NY State Soc Dent Children, 63; Award, USPHS & Am Dent Asn, 66; Award, NY Acad Prev Med, 66; H Trendley Dean Award, Int Asn Dent Res, 68; Hermann M Biggs Mem Award, NY State Pub Health Asn, 71. Mem: Am Dent Asn; fel Am Pub Health Asn; fel Am Col Dent; Asn State & Territorial Dent Dirs (pres, 50-51). Res: Effectiveness of supplementing a fluoride deficient community water supply with a fluoride compound to bring the fluoride ion concentration up to 1.0 part per million of water to prevent dental caries; public health aspects of malocclusion; administration and medical assistance programs. Mailing Add: 954-A Calle Aragon Laguna Hills CA 92653

ASTHEIMER, ROBERT W, b Jersey City, NJ, Oct 16, 22; m 48; c 3. ELECTROOPTICS. Educ: Stevens Inst Technol, ME, 44, MS, 49. Prof Exp: Staff instr physics & res asst, Stevens Inst Technol, 46-51; res physicist, Naval Ord Lab, 51-53; proj engr, 53-57, dept mgr, 57-60, chief engr, 60-63, tech dir, 63-67, VPRES, BARNES ENG CO, 67- Mem: Fel Optical Soc Am; Am Phys Soc. Res: Electrooptical systems and components; explosion hydrodynamics; infrared systems. Mailing Add: Barnes Eng Co 30 Commerce Rd Stamford CT 06902

ASTILL, BERNARD DOUGLAS, b Nottingham, Eng, Feb 11, 25; US citizen; m 55; c 2. BIOCHEMISTRY. Educ: Univ Nottingham, BSc, 50, PhD(org chem), 53. Prof Exp: Asst anal chemist, Boots Pure Drug Co, Nottingham, Eng, 43-44; chemist, Lab Indust Med, 55-60, sr biol chemist, 60-66, res assoc, 67-72, SUPVR BIOCHEM, HEALTH & SAFETY LAB, EASTMAN KODAK CO, 72- Concurrent Pos: Cerebral Palsy Found grant, Univ Rochester, 52-54, NIH fel, 54-55; clin instr, Dept Community Health & Prev Med, Univ Rochester Med Ctr, 62- Mem: Am Chem Soc; NY Acad Sci; Soc Toxicol; Am Indust Hyg Asn; Permanent Comn & Int Asn Occup Health. Res: Fate of organic chemicals in mammalian and aquatic species; biological monitoring for environmental exposures; clinical chemistry; biochemical mechanism of toxicity. Mailing Add: Health & Safety Lab Eastman Kodak Co Rochester NY 14650

ASTIN, ALLEN VARLEY, b Salt Lake City, Utah, June 12, 04; m 27; c 2. PHYSICS. Educ: Univ Utah, AB, 25; NY Univ, MS, 26, PhD(physics), 28. Hon Degrees: DSc, Lehigh Univ, 53, George Washington Univ, 58, NY Univ, 60. Prof Exp: Teaching fel physics, NY Univ, 25-28; Nat Res Coun fel, Johns Hopkins Univ, 28-30; res assoc, 30-32, from assoc physicist to physicist, 32-42, sr physicist, 42-44, asst chief ord & electronic div, 45-48, chief electronic div, 48-50, from assoc dir to dir, 51-69, EMER DIR ELECTRONIC DIV, NAT BUR STANDARDS, 69- Concurrent Pos: Consult & ed div 4, Nat Defense Res Comt; mem, US Air Force Sci Adv Bd, 49-53; mem, Nat Adv Comt Aeronaut, 52-58; mem, Naval Res Adv Comt, 53-59; chmn, Interdept Comt Sci Res & Develop, 54 & 55; mem, Int Comt Weights & Measures, 54-69; mem, Nat Motor Vehicle Safety Adv Coun, 68-72; US coordr, US-French Sci Coop Prog, 69-75; trustee, Sci Serv, 67- Honors & Awards: King's Medal, Eng; President's Cert of Merit; Award to Executives, Am Soc Testing & Mat, 25; Nat Civil League Career Serv Award, 60; Rockefeller Pub Serv Award, 63; Standards Medal, Am Nat Standards Inst, 69; Harry Diamond Award, Inst Elec & Electronics Eng, 70; Astin-Polk Int Standards Award, Am Nat Standards Inst, 74. Mem: Nat Acad Sci (home secy, 71-75); fel Am Phys Soc; fel Inst Elec & Electronics Eng; hon mem Am Inst Aeronaut & Astronaut; hon mem Am Soc Heat, Refrig & Air Conditioning Eng. Res: Dielectric constants and power factors of liquids; electronic instruments and ordnance; electrical measurements; proximity fuses; guided missiles. Mailing Add: 5008 Battery Lane Bethesda MD 20014

ASTLE, MELVIN JENSEN, b Afton, Wyo, May 8, 10; m 40; c 4. ORGANIC CHEMISTRY. Educ: Univ Utah, BS, 32, MS, 35; Purdue Univ, PhD(org chem), 42. Prof Exp: Asst chem, Univ Utah, 32-34; instr high sch, Utah, 35-38; asst chem, Purdue Univ, 38-41; instr org chem, Univ Ky, 41-44; chief chemist, Shell Chem Corp, Calif, 44-46; asst prof, Case Western Reserve Univ, 46-49, from assoc prof to prof, 49-62, actg dean grad sch, 58-60; asst mgr corp res, Glidden Co, 62-64, mgr corp org res, 64-68, mgr res, SCM Corp, Glidden-Durkee Div, 68-75; RETIRED. Mem: AAAS; Am Chem Soc. Res: Petrochemicals; ion-exchange surfaces; catalysis with ion-exchange resins. Mailing Add: SMC Corp Glidden-Durkee Div 900 Union Commerce Bldg Cleveland OH 44115

ASTLING, ELFORD GEORGE, b Sycamore, Ill, Aug 13, 37. METEOROLOGY. Educ: Univ Utah, BS, 60; Univ Wis, Madison, MS, 64, PhD(meteorol), 70. Prof Exp: Res meteorologist, Nat Environ Satellite Ctr, Environ Sci Serv Admin, 64-66; asst prof meteorol, Fla State Univ, 69-72; asst prof geophys sci, Old Dom Univ, 72-75; ASST PROF METEOROL, UNIV UTAH, 75- Mem: AAAS; Am Meteorol Soc; Am Geophys Union. Res: Applications of satellite technology in synoptic meteorology; energy transfer and energy budgets. Mailing Add: Dept of Meteorol Univ of Utah Salt Lake City UT 84112

ASTON, DUANE RALPH, b Strabane, Pa, Aug 3, 32; m 55; c 5. SOLID STATE PHYSICS. Educ: Brigham Young Univ, BS, 55, MS, 57; Temple Univ, PhD(physics), 69. Prof Exp: Engr, Gen Dynamics/Convair, 55-56; engr, Boeing Airplane Co, 57-58; assoc scientist, Bettis Atomic, Westinghouse Elec Corp, 58-59; instr physics, Drexel Inst Technol, 59-67; ASST PROF PHYSICS, CALIF STATE UNIV, SACRAMENTO, 68- Mem: Am Asn Physics Teachers. Res: Nuclear reactor calculations; nuclear shielding studies; effects of demagnetization coefficients on type II superconductors; magnetization studies on the strong-coupling Pb-In alloy systems; intermediate and mixed states in type II superconductors. Mailing Add: 2500 Waterton Way Sacramento CA 95826

ASTON, ROY, b Windsor, Ont, Dec 31, 29; m 73; c 1. PHARMACOLOGY. Educ: Univ Windsor, BA, 50; Wayne State Univ, MSc, 54; Univ Toronto, PhD(pharmacol), 58. Prof Exp: Res assoc pharmacol, Med Sch, Univ Mich, 58-59; res assoc, 59-62, from instr to asst prof, 62-70, ASSOC PROF PHARMACOL, SCH MED, WAYNE STATE UNIV, 70- Mem: Am Soc Pharmacol; Pharmacol Soc Can. Res: Pharmacodynamics and psychopharmacology of drugs acting on the central nervous system; drug protection against pathological effects of accerleation.acceleration. Mailing Add: Dept of Anesthesiol Wayne State Univ Sch of Med Detroit MI 48207

ASTRACHAN, BORIS MORTON, b New York, NY, Dec 1, 31; m 56; c 4. PSYCHIATRY. Educ: Alfred Univ, BA, 52; Albany Med Col, MD, 56. Prof Exp: Asst dir inpatient psychiat serv, Yale-New Haven Hosp, 63-66; from asst prof to assoc prof psychiat, Sch Med, Yale Univ, 65-71; chief day hosp, 66-68, dir gen clin div, 68-70, actg dir, Ctr, 70, DIR, CONN MENT HEALTH CTR, 71-; PROF CLIN PSYCHIAT, SCH MED, YALE UNIV, 71- Concurrent Pos: NIMH fel, Yale Univ, 61-63; mem ed bd, J Appl Behav Sci, 71; Bertrum Roberts Mem lectr psychiat, Yale Univ, 72; Paul Mason de la Vergne Mem lectr, Undercliff Ment Health Ctr, 74. Mem: AAAS; fel Am Psychiat Asn; Am Pub Health Asn; NY Acad Sci. Res: Organizational group dynamics; schizophrenia, clinical epidemiology and outcome; evaluation research. Mailing Add: Dept of Psychiat Yale Univ Sch of Med New Haven CT 06508

ASTRACHAN, LAZARUS, b New York, NY, Aug 7, 25; m 45; c 2. BIOCHEMISTRY. Educ: City Col NY, BS, 44; Yale Univ, PhD(physiol chem), 51. Prof Exp: Biochemist, Biol Div, Oak Ridge Nat Lab, 54-62; asst prof, 62-64, ASSOC PROF MICROBIOL, SCH MED, CASE WESTERN RESERVE UNIV, 64- Concurrent Pos: Fel, McCollum-Pratt Inst, Johns Hopkins Univ, 51-53, Polio Found fel, Sch Hyg & Pub Health, 54. Mem: Am Chem Soc; Am Soc Microbiol; Am Soc Biol Chem; Genetics Soc Am. Res: Biochemistry of nucleic acids; virology; bacterial lipids. Mailing Add: Dept of Microbiol Case Western Reserve Sch Med Cleveland OH 44106

ASTRACHAN, MAX, b Rochester, NY, Mar 30, 09; m 31; c 3. MATHEMATICS, STATISTICS. Educ: Univ Rochester, AB, 29; Brown Univ, AM, 30, PhD(math), 35. Prof Exp: Instr math, Brown Univ, 29-35; from instr to prof math, Antioch Col, 35-50, chmn dept, 42-50; prof statist, Air Force Inst Technol, Wright-Patterson AFB, 49-60, head dept acct & statist, sch bus, 55-60; mathematician, logistics dept, Rand Corp, 60-65; dir educ & training inst, Am Soc Qual Control, 65-67; prof sta- tist, 67-70, PROF MGT SCI, SCH BUS, CALIF STATE UNIV, NORTHRIDGE, 70-, CHMN DEPT, 75- Concurrent Pos: Instr, exten div, Pa State Col, 41; statistician, Wright Field, Ohio, 44; lectr, Wright-Patterson Grad Ctr, Ohio State Univ, 46-60, Air Inst Technol, 48-49 & Univ Calif, Los Angeles, 61-; opers analyst, Korea & Japan, 53, Eng, 54; consult, Rand Corp, 56-60. Mem: Am Statist Asn; Inst Mgt Sci; Am Soc Qual Control. Res: Summability of Fourier series by Norlund means. Mailing Add: Calif State Univ Sch Bus 18111 Nordhoff St Northridge CA 91324

ASTROMOFF, ANDREW, b Paris, France, Nov 30, 32; US citizen; m 57; c 3. MATHEMATICS. Educ: Univ Calif, Berkeley, AB, 54, MA, 58, PhD(math), 63. Prof Exp: From asst prof to assoc prof, 62-70, PROF MATH, SAN FRANCISCO STATE UNIV, 70- Mem: Math Asn Am. Res: Foundations of mathematics; theory of models; general algebraic systems. Mailing Add: Dept of Math San Francisco State Univ San Francisco CA 94132

ASTRUC, JUAN A, b Utrera, Spain, June 5, 33; m 61; c 7. ANATOMY, NEUROANATOMY. Educ: Univ Granada, MD, 57, DrMed & Surg, 59. Prof Exp: From instr to asst prof anat, Sch Med, Univ Granada, 57-61; asst prof, Sch Med, Lit Univ Salamanca, 58-60; from asst prof to assoc prof, Sch Med, Univ Navarra, 62-67; ASSOC PROF ANAT, MED COL VA, 67- Concurrent Pos: Span Govt res fel, 59-60; Ger Govt res fel, 60; March Found res grant, 62-63; USPHS res fel, 63-64 & res grants, 65-67 & 69- Res: Influence of the central nervous system upon the endocrine glands, control of the movement of the eyes; structure and connections in the central nervous system. Mailing Add: Dept of Anat Med Col of Va Richmond VA 23298

ASTRUE, ROBERT WILLIAM, b San Francisco, Calif, Feb 2, 28; m 56; c 5. SOLID STATE PHYSICS. Educ: Univ Calif, Berkeley, AB, 51; Univ Wash, MS, 59; Wayne State Univ, PhD(physics), 66. Prof Exp: Res physicist, Ford Motor Co Sci Res Lab, 63-66; asst prof, 66-70, ASSOC PROF PHYSICS, HUMBOLDT STATE UNIV, 70- Mem: Am Phys Soc; Am Asn Physics Teachers. Res: Magnetic resonance; ferromagnetic thin films. Mailing Add: Dept of Physics & Phys Sci Humboldt State Univ Arcata CA 95521

ASTWOOD, EDWIN BENNETT, b Bermuda, Dec 29, 09; nat US; m 37; c 2. ENDOCRINOLOGY, INTERNAL MEDICINE. Educ: Columbia Union Col, BS, 29; McGill Univ, MD & CM, 34; Harvard Univ, PhD(biol), 39. Hon Degrees: ScD, Univ Chicago, 67. Prof Exp: Med house officer, Royal Victoria Hosp, Montreal, 34-35; assoc obstet, Johns Hopkins Univ & asst obstetrician, Johns Hopkins Hosp, 39-40; asst prof pharmacother, Harvard Med Sch, 40-45; res prof, 45-52, prof, 52-72, EMER PROF MED, MED SCH, TUFTS UNIV, 73- Concurrent Pos: Nat Inst Arthritis & Metab Dis res career award, 62-72; assoc, Peter Bent Brigham Hosp, 40-45; ed jour, Endocrine Soc, 43-45; endocrinologist, New Eng Ctr Hosp, 45-48, sr physician, 48-73, consult physician, 73-; physician, Boston Dispensary, 45-73; mem adv coun, Nat Inst Arthritis & Metab Dis, 65-68; consult med, King Edward VII Mem Hosp, Bermuda, 72-; pvt pract internal med, Bermuda, 72- Honors & Awards: Ciba Award, Endocrine Soc, 44, Fred Konrad Koch Award, 67; Cameron Prize, Edinburgh, 48; Philips Medal, Am Col Physicians, 49; Borden Award, 52; Lasker Award, 54. Mem: Nat Acad Sci; AAAS; Endocrine Soc (pres, 61-62); Am Physiol Soc; Am Chem Soc. Res: Mammary gland development; sex hormones; corpus luteum function; thyroid gland and antithyroid compounds; radioactive iodine; pituitary hormones; metabolism of fat. Mailing Add: PO Box 1146 Hamilton Bermuda

ASUNMAA, SAARA K, b Mikkeli, Finland. PHYSICAL CHEMISTRY. Educ: Univ Helsinki, PhD(phys chem), 43. Prof Exp: Instr phys chem, Inst Technol, Helsinki, 44-46; res assoc, Swed Forest Prod Res Lab, 47-57; fel physics, Stanford Univ, 57-58, res assoc x-ray micros, W W Hansen Labs Physics, 58-60; sr res scientist, Owens-Ill Glass Co, 60-65 & Astropower Lab, Douglas Aircraft Co, Inc, Calif, 65-74; MEM STAFF, GEOL RES DIV, UNIV CALIF AT SAN DIEGO, LA JOLLA, 74- Concurrent Pos: Mem staff, Stanford Res Inst, 59. Mem: Am Phys Soc. Res: Diffusion in solutions; adsorption-desorption; macromolecular structures; electron microscopy; x-ray microscopy with high resolution; surface adhesion and interface reactions. Mailing Add: Geol Res Div Univ of Calif at San Diego La Jolla CA 92037

ASZALOS, ADORJAN, b Sweged, Hungary, May 9, 29; US citizen; m 60; c 2. BIO-ORGANIC CHEMISTRY, CANCER. Educ: Tech Univ Budapest, MS, 51; Tech Univ Vienna, PhD(natural prod chem), 61. Prof Exp: Plant engr, Hungary, 51-53; supvr, Biochem Res Inst, 53-56; res chemist, Nat Starch & Chem Co, 57-59; sr res scientist, Squibb Inst Med Res, 62-74; HEAD BIOCHEM, FREDERICK CANCER RES CTR, 74- Concurrent Pos: Fel, Rutgers Univ, 61-62, coadj prof, 71-73; res prof, Princeton Univ, 73-75. Mem: Am Chem Soc; NY Acad Sci; Intra Sci Found. Res: Isolation and structural studies of antibiotics; synthetic organic chemistry in field of carbohydrates; biochemistry; studies of enzymes; membrane studies of mammalian cells. Mailing Add: Frederick Cancer Res Ctr Frederick MD 21701

ATAC, MUZAFFER, b Turkey, Aug 24, 33; m 58; c 1. PHYSICS. Educ: Univ Ankara, BS, 57, PhD(physics), 67; Univ Ill, MS, 64. Prof Exp: Physicist, Maden Teknik Arastirma Enstitüsü, 57-61; PHYSICIST, NAT ACCELERATOR LAB, 68- Res: Nuclear physics; high energy and solid state physics. Mailing Add: Nat Accelerator Lab PO Box 500 Batavia IL 60510

ATACK, DOUGLAS, b Wakefield, Eng, Aug 6, 23; m 50; c 1. PHYSICAL CHEMISTRY. Educ: Univ Leeds, BSc, 44, PhD(chem), 46. Prof Exp: Res assoc high pressure, PTX Data, Univ Leeds, 46-48; fel critical phenomena, Nat Res Coun Can, 48-50; fel PVI data, Univ Manchester, 50-51; res assoc, Off Naval Res Contract, Univ NC, 51-53; asst prof phys chem, Syracuse Univ, 53-54; head wood & fiber physics div, 54-67, chmn wood & fiber res dept, 67-69, DIR APPL PHYSICS DIV, PULP & PAPER RES INST CAN, 69- Concurrent Pos: Hon lectr mech eng, McGill Univ, 64-; fel, Asn Pulp & Paper Indust, 75. Honors & Awards: Weldon Medal, Can Pulp & Paper Asn, 63. Mem: Am Phys Soc; The Chem Soc. Res: High speed friction phenomena; mechanical properties of polymeric materials. Mailing Add: Pulp & Paper Res Inst of Can 570 St John's Rd Pointe Claire PQ Can

ATALLA, MARTIN MOHAMED, physics, engineering, see 12th edition

ATALLA, RAJAI HANNA, b Jerusalem, Palestine, Feb 21, 35; nat US; m 63; c 2. CHEMICAL PHYSICS. Educ: Rensselaer Polytech Inst, BChE, 55; Univ Del, MChE, 58, PhD(chem eng), 61. Prof Exp: Res engr res ctr, Hercules Inc, 60-62, res chemist, 62-68; ASSOC PROF CHEM PHYSICS, INST PAPER CHEM, 68- Mem: AAAS; Am Asn Physics Teachers; Am Chem Soc; Tech Asn Pulp & Paper Indust. Res: Vibrational spectroscopy; thermodynamics; water chemistry; molecular biophysics. Mailing Add: Dept of Chem Inst of Paper Chem Appleton WI 54911

ATALLA, ROBERT E, b Brooklyn, NY, Oct 31, 29. MATHEMATICS. Educ: Univ Calif, Los Angeles, AB, 54; Univ Idaho, MA, 62; Univ Rochester, PhD(math), 66. Prof Exp: ASST PROF MATH, OHIO UNIV, 66- Mem: Am Math Soc. Res: Mathematical analysis and point set topology; applications of topology and functional analysis to matrix summability. Mailing Add: Dept of Math Ohio Univ Athens OH 45701

ATASSI, ZOUHAIR, b Homs, Syria, Dec 20, 34; m 63. ORGANIC CHEMISTRY, BIOCHEMISTRY. Educ: Bristol Univ, BSc, 57; Univ Birmingham, MSc, 58, PhD(chem), 60, DSc(chem), 73. Prof Exp: Res fel chem, Univ Birmingham, 60-61; res assoc biochem, Albany Med Col & res fels lab protein chem, Div Labs & Res, NY State Dept Health, 62-63; asst prof biochem sch med, State Univ NY Buffalo, 63-68; prof chem, Wayne State Univ, 68-75; PROF IMMUNOL, MAYO MED SCH, 75-; PROF BIOCHEM, UNIV MINN, MINNEAPOLIS, 75- Concurrent Pos: Grants, Off Naval Res, 64-70, Nat Inst Arthritis & Metab Dis, 64-74 & Am Heart Asn, 71-74; estab investr, Am Heart Asn, 66-71; consult, Mayo Clin, 75- Mem: Am Soc Biol Chem; Am Asn Immunol; Brit Biochem Soc; Am Chem Soc; AAAS. Res: Protein structure; chemistry and immunochemistry; protein chemistry, immunochemistry and conformation; chemical modification and cleavage of proteins; organic synthesis of peptides. Mailing Add: Dept of Immunol Mayo Med Sch Rochester MN 55901

ATCHESON, J D, b London, Ont, July 27, 18; m 42; c 3. PSYCHIATRY. Educ: Univ Western Ont, MD, 41; Univ Toronto, dipl psychiat, 47. Prof Exp: Dir psychiat clin, Juv & Family Court, Toronto, 47-57; treatment serv, Dept Reform Inst, Prov Ont, 57-58; supt, Thistletown Childrens Psychiat Hosp, Toronto, 58-69; med dir, Thistletown Regional Ctr Children & Adolescents, 69-71; DIR OUTPATIENT DEPT FORENSIC SERV, CLARKE INST PSYCHIAT, UNIV TORONTO, 71- Concurrent Pos: Assoc psychiat, Univ Toronto, 47-50, asst prof, 50-68, assoc prof, 68-, lectr, Sch Nursing, 48- & Inst Child Study, 69-; consult, Northern Med Br, Dept Health & Welfare, 67- Honors & Awards: Centennial Medal, Dom Can, 67. Mem: Fel Am Psychiat Asn; Can Psychiat Asn; Can Med Asn. Res: Juvenile delinquency; residential care of children; child psychiatry; trans-cultural psychiatry, especially problems of emotional disorder in the Canadian Arctic. Mailing Add: Clarke Inst of Psychiat Univ of Toronto Toronto ON Can

ATCHISON, JOSEPH EDWARD, b Barnum, WVa, Dec 25, 14; m 51; c 4. PULP CHEMISTRY, PAPER CHEMISTRY. Educ: La State Univ, BS, 38; Lawrence Col, MS, 40, Inst Paper Chem, PhD(pulp & paper technol), 42. Prof Exp: Mem staff, Various Paper Mills, 40-42; tech dir, John Strange Paper Co, Wis, 46-48; chief pulp & paper br, Econ Coop Admin, Washington, DC, 48-52; resident mgr paperboard mill & dir, Bagasse Pulp Pilot Plant, Portorican Paper Prod, PR, 52-53; sr proj engr, Parsons & Whittemore, Inc, NY, 53-54, vpres pulp & paper proj div & dir various co, 54-65, sr vpres, 65-66; PRES, JOSEPH E ATCHISON CONSULTS, INC, 67- Concurrent Pos: Guest lectr & mem exchange group, Moscow State Univ & Leningrad State Univ, 59; consult, Hawaiian Sugar Planters Asn, 51; mem symp pulp & paper res & technol, Mid East & NAfrica, UN Food & Agr Orgn, Lebanon, 62; coordr, Bagasse Pulping Proj; dir, Am-Arab Asn Indust & Com; partic & chmn various int meetings on utilization of agr fibers, UN, chmn int conf chem utilization agr residues, Food & Agr Orgn, Wis, 51, chmn int conf pulp & paper develop, Asia & Far East, Japan, 60; chmn, Int Conf Pulp & Paper Experts, Argentina, 54; mem task group, President's Comn Increased Use of Agr Prod, 57; mem pulp & paper comt, US Del, Int Mat Conf, 50-51. Mem: Am Tech Asn Pulp & Paper Indust; Am Mgt Asn; Am Chem Soc; Int Soc Sugar Cane Technol; Coun Agr & Chemurgic Res. Res: Cellulose chemistry; pulp and paper engineering; distribution of chain lengths; production of paper from sugar cane; influence of cooking and bleaching upon the chain length distribution of the carbohydrate fraction in pulpwood. Mailing Add: 30 E 42nd St New York NY 10017

ATCHISON, ROBERT WAYNE, b Pratt, Kans, Sept 19, 30; m 52; c 2. VIROLOGY, IMMUNOLOGY. Educ: Univ Kans, AB, 52, AM, 55, PhD(virol), 60. Prof Exp: Res assoc, 60-62, from asst res prof to assoc prof virol & immunol, 63-69, ASSOC PROF MICROBIOL, GRAD SCH PUB HEALTH, UNIV PITTSBURGH, 69- Mem: Am Soc Microbiol. Res: Electron microscopy; biochemistry; histochemistry; anatomy; bacteriology. Mailing Add: 1001 Delyla Dr Bethel Park PA 15102

ATCHISON, THOMAS ANDREW, b Brady, Tex, July 3, 37; m 59; c 2. NUMERICAL ANALYSIS. Educ: Univ Tex, BA, 59, MS, 60, PhD(math), 63. Prof Exp: Instr math, Howard Payne Col, 59; spec instr, Univ Tex, 61-63; from asst prof to assoc prof, Tex Tech Univ, 63-66; mem staff, LTV Electrosysts, Inc, 66-67; assoc prof math, Tex Tech Univ, 67-72; PROF MATH & HEAD DEPT, MISS STATE UNIV, 72- Mem: Am Math Soc; Math Asn Am. Res: Complex variables, specifically Riemann surface theory; functional analysis and topology; the determination of approximating functions for probability distributions and the approximations of other integrals. Mailing Add: Dept of Math Miss State Univ Mississippi State MS 39762

ATCHISON, THOMAS CALVIN, JR, b Fremont, Ohio, June 11, 22; m 50; c 3. PHYSICS. Educ: Princeton Univ, AB, 43. Prof Exp: Electronic technician, Palmer Lab, Princeton Univ, 43; physicist, Bur Ships, US Navy Dept, 46-49; supvr physicist, 49-70, RES DIR US BUR MINES, 70- Mem: Acoust Soc Am; Seismol Soc Am; Int Soc Rock Mech. Res: Mining research; rock mechanics; blasting; lunar surface research. Mailing Add: Mining Res Ctr US Bur of Mines PO Box 1660 Twin Cities MN 55111

ATCHISON, WILLIAM FRANKLIN, b Smithfield, Ky, Apr 7, 18; m 47; c 4. MATHEMATICS, COMPUTER SCIENCE. Educ: Georgetown Col, AB, 38; Univ Ky, MA, 40; Univ Ill, PhD(math), 43. Prof Exp: Asst, Physics Lab, Georgetown Col, 36-38; asst math, Univ Ky, 39-40; asst, Univ Ill, 40-42, instr, 43-44 & 46-48, asst prof, 49-50 & 51-55; asst prof, Harvard Univ, 50-51; res assoc prof, Ga Inst Technol, 55-63, res prof, 63-66; PROF COMPUT SCI & DIR COMPUT SCI CTR, UNIV MD, COLLEGE PARK, 66- Concurrent Pos: Head prog & coding group, Rich Electronic Comput Ctr, 55-57, dir ctr, 57-66, actg dir sch info sci, 63-64. Mem: Asn Comput Mach; Am Math Soc; Math Asn Am; Soc Indust & Appl Math. Res: Digital computation; information science; numerical analysis; algebraic geometry; numerical solution of ordinary differential equations. Mailing Add: Comput Sci Ctr Univ of Md College Park MD 20742

ATCHLEY, FLOYD OWEN, b Alliance, Ohio, Oct 11, 08; m 40; c 2. PUBLIC HEALTH. Educ: Mt Union Col, BS, 30; Johns Hopkins Univ, ScD, 34. Prof Exp: Teacher, 34-42 & 47-48; microbiologist, USPHS, 49-65, SCIENTIST-ADMINR, NIH, 65- Mailing Add: Nat Inst of Health Bethesda MD 20014

ATCHLEY, RALPH WARREN, b Cincinnati, Ohio, July 31, 21; m 58; c 1. ORGANIC CHEMISTRY. Educ: Univ Cincinnati, BS, 52, PhD(org chem), 64. Prof Exp: Res assoc toxicol, Kettering lab, col med, Univ Cincinnati, 52-57, asst org chem, 57-60; org chemist, Vet Admin Hosp, Cincinnati, 62-64; USPHS fel pharmacol, sch med, Yale Univ, 64-65; instr, 65-67, ASSOC PROF CHEM, SCHOOLCRAFT COL, 67- Concurrent Pos: Res assoc, Ment Health Res Inst, 69. Mem: Am Chem Soc; Sigma Xi. Res: Toxicology of lubricants; hydraulic fluids and insecticides; synthesis of steroids, purines, pyrimidines, pteridines, nitrogen mustards and folic acid analogs; cancer chemotherapy; ascorbic acid; brain lipids. Mailing Add: 480 Auburn St Plymouth MI 48170

ATEN, CARL FAUST, JR, b Lorain, Ohio, Aug 15, 32; m 55; c 4. PHYSICAL CHEMISTRY. Educ: Col Wooster, BA, 54; Brown Univ, PhD(phys chem), 59. Prof Exp: Chemist, Texaco, Inc, NY, 58-60; asst chem kinetics in shock tubes, Cornell Univ, 60-62; from asst prof to assoc prof phys chem, 62-72, PROF PHYS CHEM, HOBART & WILLIAM SMITH COLS, 72- Mem: Am Chem Soc. Res: Chemical kinetics at high temperatures in shock tubes; pyrolysis of hydrocarbons. Mailing Add: Dept of Chem Hobart & William Smith Cols Geneva NY 14456

ATENCIO, ALONZO C, b Ortiz, Colo, June 24, 29; m 53; c 5. BIOCHEMISTRY. Educ: Univ Colo, Boulder, BA, 58; Univ Colo, Denver, MS, 64, PhD(med), 67. Prof Exp: Res technician, Sch Med, Univ Colo, Denver, 58-67; ASST PROF BIOCHEM, SCH MED, UNIV N MEX, 70- Concurrent Pos: NIH fel chem, Northwestern Univ, 67-70. Mem: AAAS; Am Chem Soc; Am Heart Asn; Am Physiol Soc; Int Soc Thrombosis & Haemostasis. Res: Hormonal control of protein biosynthesis in mammals using the fibrinogen system as a model. Mailing Add: Dept of Biochem Univ of NMex Albuquerque NM 87106

ATERMAN, KURT, b Bielitz, Sept 9, 13; m 55; c 2. PATHOLOGY. Educ: Charles Univ, Prague, MD, 38; Queen's Univ, Belfast, MB & BCh(hons), 42, DSc, 65; Conjoint Bd, London, DCH, 43; Univ Birmingham, PhD, 59. Prof Exp: Sr lectr, Univ Birmingham, 50-58; assoc prof path, Dalhousie Univ, 58-61; prof, Woman's Med Col Pa, 61-63; prof, State Univ NY Buffalo & Children's Hosp, 63-67; PROF PATH, DALHOUSIE UNIV & I W KILLIAM HOSP CHILDREN, 67- Mem: Am Asn Path & Bact; Am Soc Exp Path; Path Soc Gt Brit & Ireland;-Ger Asn Path; Can Asn Path. Res: Experimental pathology of liver. Mailing Add: I W Killam Hosp for Children Halifax NS Can

ATHAR, MOHAMMED AQEEL, b Lucknow, India, Mar 9, 39; Can citizen; m 65; c 2. MEDICAL MICROBIOLOGY, PUBLIC HEALTH. Educ: Univ Karachi, BS, 57, MS, 59; Univ Mich, MPH, 63; Univ London, PhD(med microbiol), 69. Prof Exp: Lectr microbiol, D J Sci Col, Univ Karachi, 60-62; res asst clin microbiol, Med Ctr, Univ MIch, 63-65; bacteriologist, 65-66, HEAD, MICROBIOL DIV, DEPT LABS, HOLY CROSS HOSP, CALGARY, ALTA, 70- Concurrent Pos: Study fel, Inst Int Educ, NY, 62-64; asst prof path, Fac Med, Univ Calgary, 70- Mem: AAAS; Int Soc Human & Animal Mycol; Am Soc Microbiol; Brit Inst Biol. Res: Effects of polyene antibiotics on Candida species; experimental Candida infection of the chick embryo; development of resistance in microorganisms to antibiotics. Mailing Add: Holy Cross Hosp Dept of Labs 2210 Second St SW Calgary AB Can

ATHAY, RUSSELL GRANT, astrophysics, see 12th edition

ATHELSTAN, GARY THOMAS, b Minneapolis, Minn, Dec 26, 36; m 60; c 2. PHYSICAL MEDICINE & REHABILITATION, PSCYHOLOGY. Educ: Univ Minn, BA, 60, PhD(psychol), 66. Prof Exp: Voc rehab counr, Minn Div Voc Rehab, 60-61; res psychologist, Am Rehab Found, 65-68; dir res, E& Occup Res Div, Inst Interdisciplinary Studies, 68-70; assoc prof, 70-75, PROF PHYS MED & REHAB & PSYCHOL, SCH MED, UNIV MINN, MINNEAPOLIS, 75- Concurrent Pos: Consult continuing educ, Am Acad Phys Med & Rehab, 69-70; sect ed, Arch Phys Med & Rehab, 74-75; consult psychol, Vet Admin Hosp, Minneapolis, 73- Mem: Am Psychol Asn; Am Cong Rehab Med. Res: Development of psychosocial outcome criteria in rehabilitation; study of psychosocial factors related to rehabilitation success; development of improved methods to select, train and utilize health professionals. Mailing Add: 5527 Zenith Ave S Minneapolis MN 55410

ATHENS, JOHN WILLIAM, b Buhl, Minn, Oct 2, 23; m 49; c 3. INTERNAL MEDICINE, HEMATOLOGY. Educ: Univ Mich, BA, 45; Johns Hopkins Univ, MD, 48; Am Bd Internal Med, dipl. Prof Exp: Intern med, Peter Bent Brigham Hosp, Boston, Mass, 48-49; asst res, 52-54, from res instr to asst res prof med, 55-62, assoc prof, 62-68, PROF MED & HEAD DIV HEMAT & ONCOL, COL MED, UNIV UTAH, 68- Concurrent Pos: AEC fel, Johns Hopkins Hosp, 49-50; Nat Cancer Inst res fel hemat, Col Med, Univ Utah, 54-56; dir outpatient dept, Salt Lake County Gen Hosp, 58-68. Mem: Am Fedn Clin Res; Am Col Physicians; Am Soc Clin Invest; Asn Am Physicians. Res: Leukocyte physiology; oncology. Mailing Add: Div of Hemat & Oncol Univ of Utah Sch of Med Salt Lake City UT 84112

ATHERLY, ALAN G, b Kalamazoo, Mich, Nov 19, 36; m 59; c 1. GENETICS, MOLECULAR BIOLOGY. Educ: Western Mich Univ, BS, 60; Univ NC, PhD(biochem), 65. Prof Exp: Fel, Dept Biol, Western Reserve Univ, 65-66 & Inst Molecular Biol, Univ Ore, 66-68; asst prof, 68-74, ASSOC PROF GENETICS, IOWA STATE UNIV, 74- Mem: AAAS; Am Soc Microbiol. Res: Protein biosynthesis in bacteria; RNA synthesis and regulation. Mailing Add: Dept of Genetics Iowa State Univ Ames IA 50010

ATHERTON, ELWOOD, b Chicago, Ill, Dec 5, 09; m 39. GEOLOGY. Educ: Univ Chicago, BS, 30, PhD(geol), 37. Prof Exp: From asst geologist to assoc geologist, 37-52, GEOLOGIST, ILL STATE GEOL SURV, 52- Mem: AAAS; Geol Soc Am; Am Asn Petrol Geologists. Res: Sub-surface geology; stratigraphy of Illinois. Mailing Add: Natural Resources Bldg State Geol Surv Urbana IL 61801

ATHERTON, HENRY VERNON, b Eden, Vt, Nov 12, 23; m 47; c 3. DAIRY MANUFACTURING. Educ: Univ Vt, BS, 48, MS, 50; Pa State Univ, PhD(dairy mfg & agr biochem), 53. Prof Exp: Instr agr & dairy husb, 50-51, from asst prof to assoc prof animal & dairy husb, 53-69, PROF ANIMAL SCI, UNIV VT, 69- Mem: Am Dairy Sci Asn; Int Asn Milk, Food & Environ Sanitarians (pres-elect, 75). Res: Milk quality; farm sanitation; dairy bacteriology; dairy sanitation; farm water supplies. Mailing Add: Dept of Animal Sci Univ of Vt Burlington VT 05401

ATHERTON, JOHN HARVEY, b San Francisco, Calif, Sept 12, 40. CULTURAL ANTHROPOLOGY, ARCHAEOLOGY. Educ: Univ Ore, BS, 64, MA, 66, PhD(anthrop), 69. Prof Exp: Asst anthrop, Univ Ore, 65-66; res asst, African Studies Comt, Inst Int Studies, 66-67; instr anthrop, 67, asst prof, 69-73, ASSOC PROF ANTHROP, PORTLAND STATE UNIV, 73- Concurrent Pos: Vis res fel, Inst African Studies, Fourah Bay Col, Sierra Leone, 67-68; vis prof dept anthrop, Colo Col, 75; dir Champoeg archaeol proj, Ore State Hwy Dept, 73-; dir Clarno Basin archaeol proj, Ore Mus Sci & Indust, 75. Honors & Awards: Award Stud Training Prog, NSF, 75. Mem: Fel AAAS; fel Am Anthrop Asn; fel Royal Anthrop Inst Gt Brit & Ireland; Int African Inst; Soc Am Archaeol. Res: African prehistory and ethnohistory; African ethnology; archaeological and ethnohistorical method and theory; Oregon culture history. Mailing Add: Dept of Anthrop Portland State Univ Portland OR 97207

ATHERTON, ROBERT W, b Wichita, Kans, Jan 16, 38; m 64; c 1. DEVELOPMENTAL BIOLOGY, PHYSIOLOGY. Educ: Univ Okla, BS, 61; Wichita State Univ, MS, 64; Univ Md, PhD(zool), 67. Prof Exp: Asst zool, Wichita State Univ & Univ Md, 62-67; fel physiol, Univ Calif, Berkeley, 67-69; asst prof zool & physiol, 69-75, ASSOC PROF ZOOL & PHYSIOL, UNIV WYO, 75- Mem: AAAS; Am Soc Zool; Develop Biol Soc; Am Soc Cell Biol; Gen Embryol Serv. Res: Changes in multiple molecular compounds; fetal to adult hemoglobin; isozymes during development; neurochemicals, sperm motility, and reproductive physiology of oviduct. Mailing Add: Dept of Zool & Physiol Univ of Wyo Laramie WY 82070

ATHEY, ROBERT JACKSON, b Washington, DC, Mar 7, 25; m 49; c 3. ORGANIC CHEMISTRY. Educ: Ga Inst Technol, BS, 48, MS, 50; Univ Wis, PhD(org chem), 54. Prof Exp: RES CHEMIST, E I DU PONT DE NEMOURS & CO, 53- Mem: Am Chem Soc. Res: Spectral investigation of nitryl chloride; preparation of morphine analogs; rubber chemistry; isocyanate chemistry; urethane polymers. Mailing Add: Elastomers Lab E I du Pont de Nemours & Co Wilmington DE 19898

ATHOW, KIRK LELAND, b Tacoma, Wash, Jan 22, 20; m 51. PLANT PATHOLOGY. Educ: State Col Wash, BSA, 46; Purdue Univ, MS, 48, PhD(plant path), 51. Prof Exp: Instr plant path, 49-51, asst plant pathologist, 51-56, assoc prof bot & plant path, 56-65, PROF BOT & PLANT PATH, PURDUE UNIV, 65- Concurrent Pos: Mem team, Purdue-Brazilian Proj, 66-68; USAID consult, Brazil, 70, 71, 72 & 73. Mem: Am Phytopath Soc. Res: Forage crop and soybean diseases; seed treatment; disease resistance. Mailing Add: Dept of Bot & Plant Path Lilly Hall Life Sci Purdue Univ West Lafayette IN 47906

ATIK, MOHAMMAD, b Kabul, Afghanistan, Mar 10, 20; US citizen; m 53; c 3. SURGERY. Educ: Univ Calif, Berkeley, BA, 47; Harvard Med Sch, MD, 51. Prof Exp: From instr to assoc prof surg, La State Univ, 57-64; assoc prof, 64-69, PROF SURG, SCH MED, UNIV LOUISVILLE, 69- Concurrent Pos: Res fel surg, La State Univ, 56-57; res grants, US Army, 56-64, NIH, 58-64 & Schlieder Found, 62-65. Mem: AAAS; Am Asn Surg Trauma; Am Heart Asn; Soc Surg Alimentary Tract; fel Am Col Surgeons. Res: Shock, especially related to pulmonary hepatic and renal hemodynamics, acute renal failure, intestinal obstruction, hypothermia, hypertension, peripheral vascular disease and thromboembolism. Mailing Add: Univ of Louisville Sch of Med Louisville KY 40201

ATKIN, JOHN DWAIN, genetics, plant breeding, see 12th edition

ATKINS, CHARLES GILMORE, b Stambaugh, Mich, July 4, 39; m 58; c 3. MICROBIAL GENETICS. Educ: Albion Col, BA, 61; Eastern Mich Univ, MS, 63; NC State Univ, PhD(genetics), 69. Prof Exp: Teaching asst biol, Eastern Mich Univ, 62-63; instr biol, Coe Col, 63-66; asst prof, 69-74, ASSOC PROF GENETICS & MICROBIOL, OHIO UNIV, 74- Concurrent Pos: Lectr, Cornell Col, 64-65; dir training prog, Appalachian Life Sci Col, 72-74. Mem: AAAS; Am Soc Microbiol; Genetics Soc Am. Res: Transductional studies and biochemical analysis of Salmonella typhimurium, Salmonella montevideo and hybrids of these species, with emphasis on the isoleucine-valine biosynthetic pathway; recombination in T-even bacteriophage. Mailing Add: 12 Zool Bldg Ohio Univ Athens OH 45701

ATKINS, DAVID LYNN, b Wichita Falls, Tex, July 12, 35; m 59. VERTEBRATE ANATOMY, NEUROANATOMY. Educ: Univ Tex, Austin, BA, 57; ETex State Univ, MA, 63; Tex A&M Univ, PhD(comp neuroanat), 70. Prof Exp: Teacher high schs, Tex, 57-59 & 61-62; asst prof zool, Tarleton State Col, 65-67; ASST PROF ZOOL, GEORGE WASHINGTON UNIV, 70- Mem: Am Soc Mammal; Am Soc Zool; Soc Study Amphibians & Reptiles; Sigma Xi; Soc Ichthyol & Herpetol. Res: Comparative neuroanatomy; systematics; comparative ontogeny and evolution of the mammalian cerebellum, particularly related to the phylogeny of the major taxa. Mailing Add: Dept of Biol Sci George Washington Univ Washington DC 20052

ATKINS, DON CARLOS, JR, b Denver, Colo, Apr 11, 21; m 47; c 2. INDUSTRIAL CHEMISTRY. Educ: Univ Calif, Los Angeles, BA, 43, MS, 48. Prof Exp: Tech dir, US Chem Milling Corp, 57-62; vpres, Chem & Aerospace Prod, Inc, 62-64; vpres, Wilco Prod, Inc, 64-68; exec vpres, Sanitek Prod, Inc, 68-73; lab dir non-foods chem specialties, Hunt-Wesson Foods, 73-75; PRES, D C ATKINS & SON, 75- Concurrent Pos: Staff consult, Solder Removal Co, 70-, Inst Tricology, Inc, 72-, Chemrite Corp, 75- & Clorox Co, 75- Honors & Awards: Honor Scroll Award, Am Inst Chemists, 74. Mem: Am Inst Chemists; Am Chem Soc; Am Soc Metals; Am Electroplater Soc. Res: Interaction of chemicals to surfaces with particular emphasis on chemical specialty compounds for consumer and industrial applications; development of cleaning materials, coatings and cosmetics; special expertise in chemical milling and plating of metals onto plastics. Mailing Add: 11282 Foster Rd Los Alamitos CA 90720

ATKINS, ELISHA, b Belmont, Mass, Nov 16, 20; m 44; c 5. INTERNAL MEDICINE. Educ: Harvard Univ, AB, 42; Univ Rochester, MD, 50. Prof Exp: From intern to asst resident, Barnes Hosp, 50-52; instr med, Wash Univ, 54-55; from asst prof to assoc prof, 55-67, PROF MED, YALE UNIV, 67- Concurrent Pos: Res fel infectious dis, Wash Univ, 52-54; USPHS res grant, 57-; asst attend physician, Yale-New Haven Hosp, Conn, 55-; consult, Vet Admin Hosp, West Haven, Conn, 55-; sabbatical, Radcliffe Infirmary, Eng, 62-63 & Scripps Clin & Res Found, Calif, 69-70. Mem: Am Soc Clin Invest; Asn Am Physicians; Infectious Dis Soc Am. Res: Pathogenesis of fever, especially experimental fevers induced by microbial agents and antigens in specifically sensitized animals. Mailing Add: Dept of Med Yale Univ Med Sch New Haven CT 06510

ATKINS, FERREL, b West York, Ill, Feb 15, 24. MATHEMATICS. Educ: Eastern Ill State Col, BS, 45; Univ Ill, MS, 46; Univ Ky, PhD(math), 50. Prof Exp: Asst math, Univ Ill, 45-46; instr math, Univ Ky, 46-50; asst prof, Bowling Green State Univ, 50-52; from asst prof to assoc prof math, 52-58, chmn dept, 57-58; from asst prof to assoc prof, 58-63, PROF MATH, EASTERN ILL UNIV, 63- Concurrent Pos: NSF fac fel, Stanford Univ, 64-65. Mem: Math Asn Am; Asn Comput Mach. Res: Computer sciences; analysis. Mailing Add: Dept of Math Eastern Ill Univ Charleston IL 61920

ATKINS, HAROLD LEWIS, b Newark, NJ, Sept 24, 26; m 63; c 4. NUCLEAR MEDICINE, RADIOLOGY. Educ: Yale Univ, BS, 48; Harvard Med Sch, MD, 52. Prof Exp: Intern med, Grace-New Haven Hosp, 52-53; resident radiol, Hosp Univ Pa, 53-56; instr radiol, Sch Med, Yale Univ, 56-59; from instr to asst prof, Col Physicians & Surgeons, Columbia Univ, 59-63; from assoc scientist to scientist, 63-72, SR SCIENTIST RADIOL NUCLEAR MED, BROOKHAVEN NAT LAB, 72- Concurrent Pos: Am Cancer Soc fel, Hosp Univ Pa, 55-56 & Yale-New Haven Med Ctr, 56-57; from adj assoc prof to prof radiol & nuclear med, State Univ NY Stony Brook, 68-; consult, USPHS Hosp, Staten Island, NY, 61-63, Vet Admin Hosp, Northport, 71-, St Charles Hosp, Port Jefferson, 71-, Southside Hosp, Bayshore, 72- & Nassau County Med Ctr, East Meadow, 72- Mem: AAAS; Soc Nuclear Med (trustee, 73); Radiol Soc NAm; fel Am Col Radiol. Mailing Add: Med Dept Brookhaven Nat Lab Upton NY 11973

ATKINS, HENRY PEARCE, b Birmingham, Ala, Jan 12, 15; m 41; c 2. MATHEMATICS. Educ: Cornell Univ, AB, 36; Brown Univ, MSc, 37; Univ Rochester, PhD(math), 47. Prof Exp: Instr, Brown Univ, 38-39; asst, Univ Rochester, 39-42, from instr to assoc prof, 42-58, dean of men, 54-58; PROF MATH, UNIV RICHMOND, 58- Res: Fractional derivatives of univalent functions and bounded functions. Mailing Add: Dept of Math Univ of Richmond Richmond VA 23173

ATKINS, IRVIN MILBURN, b Corning, Kans, July 24, 04; m 32; c 2. PLANT BREEDING, AGRONOMY. Educ: Kans State Col, BS, 28, MS, 36; Univ Minn, PhD(plant genetics), 43. Prof Exp: Asst supt & jr agronomist field sta, USDA, San Antonio, Tex, 28-30, agronomist & asst supt div cereal crops & dis, Tex Substa No 6, 30-54; agronomist dept agron, 54-69, EMER PROF SOIL & CROP SCI, TEX A&M UNIV, 69-; RES DIR, GRAIN RES ASSOCS, 69- Mem: Am Soc Agron. Res: Plant breeding; improved varieties of wheat, oats and barley for Texas conditions; effects of awns on yield; lodging of small grains; variental resistance to rust and loose rust smut. Mailing Add: 1215 Marsteller St College Station TX 77840

ATKINS, JASPARD HARVEY, b Boston, Mass, May 9, 26; m 54; c 4. PHYSICAL CHEMISTRY, ANALYTICAL CHEMISTRY. Educ: Univ Mass, BS, 50; Rensselaer Polytech Inst, PhD(chem), 55. Prof Exp: Anal chemist, US Food & Drug Admin, 50-51; AEC res assoc, Rensselaer Polytech Inst, 53-56; sr scientist, Westinghouse Atomic Power Div, Bettis Atomic Power Lab, 56-57; Portland Cement Asn res fel, Nat Bur Standards, 57-59; proj mgr, Nat Res Corp, 59-61; group leader, 61-70, SR RES ASSOC, CABOT CORP, 70- Mem: Am Chem Soc; Sigma Xi. Res: Physical chemistry measurement; gas-solid interfaces; vacuum techniques; gaseous kinetics; instrumentation. Mailing Add: Cabot Corp Concord Rd Billerica MA 01821

ATKINS, KENNETH EARL, b Kimberly, WVa, Feb 20, 39; m 60; c 2. ORGANIC CHEMISTRY. Educ: WVa Inst Technol, BS, 60; WVa Univ, MS, 65. Prof Exp: Res chemist, Union Carbide Olefins Div, 60-67, PROJ SCIENTIST, UNION CARBIDE CHEM & PLASTICS DIV, 67- Res: Transition metal catalysis; base catalyzed isomerization; alkylation reactions; electrodeposition of coatings; unsaturated polyesters. Mailing Add: 1311 Martha Rd South Charleston WV 25303

ATKINS, RICHARD ELTON, b Corning, Kans, Feb 10, 19; m 52; c 3. AGRONOMY. Educ: Kans State Univ, BS, 41; Iowa State Univ, MS, 42, PhD(crop breeding), 48. Prof Exp: Res asst prof, 48-50, assoc prof, 50-60, PROF AGRON, IOWA STATE UNIV, 60- Mem: Am Soc Agron; Crop Sci Soc Am. Res: Breeding of improved varieties of small grains and grain sorghum for agronomic and disease characters and allied genetic and production studies; crop breeding. Mailing Add: Dept of Agron Iowa State Univ Ames IA 50011

ATKINS, ROBERT CHARLES, b Norwood, Mass, Aug 19, 44; m 67; c 2. ORGANIC CHEMISTRY. Educ: Mass Inst Technol, SB, 66; Univ Wis, PhD(org chem), 70. Prof Exp: Assoc org chem, Columbia Univ, 70-71; asst prof chem, 71-75, ASSOC PROF CHEM, MADISON COL, 75- Concurrent Pos: NIH fel, 70-71. Mem: Am Chem Soc. Res: pyrolytic reactions; reactions involving ylides; new synthetic methods. Mailing Add: Dept of Chem Madison Col Harrisonburg VA 22801

ATKINS, ROBERT W, b Belmont, Mass, Nov 30, 17; m 40; c 5. PSYCHIATRY, PSYCHOANALYSIS. Educ: Yale Univ, BA, 40; Harvard Univ, MD, 43. Prof Exp: DIR COMMUNITY MENT HEALTH CTR DIV, DEPT PSYCHIAT, SCH MED & DENT, UNIV ROCHESTER & STRONG MEM HOSP, 68-, PROF PSYCHIAT, UNIV & HOSP, 70- Mem: Am Psychiat Asn. Res: Mental illness and unemployment; psychological effects of induced abortion and surgical sterilization in men and women; cerebral metabolic disturbances in hypothyroidism; conceptual thinking in psychiatric patients. Mailing Add: Community Health Ctr Div Univ of Rochester Rochester NY 14620

ATKINS, RONALD LEROY, b Martinez, Calif, Mar 27, 39; m 64; c 2. ORGANIC CHEMISTRY. Educ: Univ Wyo, BS, 66, MS, 68; Univ NH, PhD(org chem), 71. Prof Exp: Nat Res Coun res assoc, 71-73, RES CHEMIST, NAVAL WEAPONS CTR, DEPT NAVY, 73- Mem: Am Chem Soc; Sigma Xi. Res: Synthesis and photochemistry of small ring heterocyclic compounds; synthesis and study of the photostability of dye molecules. Mailing Add: Code 6056 Naval Weapons Ctr China Lake CA 93555

ATKINSON, BURR GERVAIS, b Elizabeth, NJ, Sept 17, 37; m 68; c 2. BIOCHEMISTRY, DEVELOPMENTAL BIOLOGY. Educ: Col Steubenville, BA, 61; Univ Conn, PhD(biochem, biophys), 68. Prof Exp: NIH fel, Fla State Univ, 68-71, res assoc develop biochem, 71-72; asst prof zool, 72-76, ASSOC PROF ZOOL, UNIV WESTERN ONT, 76- Concurrent Pos: Lab instr, Marine Biol Labs, Woods Hole, 66-68, investr, 69. Mem: Sigma Xi; NY Acad Sci; Am Soc Biol Chemists; Am Soc Cell Biol. Res: Nucleic acid biochemistry; hormone-induced synthesis of nucleic acids; tissue-specific proteins in development; regeneration; vertebrate muscle biochemistry; tissue culture; proteins and nucleic acids of genetic machinery; muscular dystrophy. Mailing Add: Dept of Zool Univ of Western Ont London ON Can

ATKINSON, DANIEL EDWARD, b Pawnee City, Nebr, Apr 8, 21; m 48; c 5. BIOCHEMISTRY. Educ: Univ Nebr, BS, 42; Iowa State Col, PhD(chem), 49. Hon Degrees: Dsc, Univ Nebr, 75. Prof Exp: Res fel, Calif Inst Technol, 49-50; assoc plant physiologist, Argonne Nat Lab, 50-52; from asst prof to assoc prof, 52-62, PROF CHEM, UNIV CALIF, LOS ANGELES, 62- Mem: Am Chem Soc; Am Soc Plant Physiologists; Am Soc Microbiol; Am Soc Biol Chemists. Res: Metabolic regulation and correlation; properties of regulatory enzymes; energy metabolism. Mailing Add: Dept of Chem Univ of Calif Los Angeles CA 90024

ATKINSON, EDWARD REDMOND, b Boston, Mass, Feb 15, 12; m 44; c 2. ORGANIC CHEMISTRY. Educ: Mass Inst Technol, BS, 33, PhD(org chem), 36. Prof Exp: Teaching fel chem, Mass Inst Technol, 33-36; instr org chem, Trinity Col, Conn, 36-38; from asst prof to assoc prof chem, Univ NH, 38-51; group leader, Dewey & Almy Chem Co, 51-57; SR CHEMIST, ARTHUR D LITTLE, INC, 57- Concurrent Pos: Lectr, Northeastern Univ, 51-63 & Boston Univ, 64-65; indust consult. Mem: Am Chem Soc; AAAS. Res: Diazonium compounds; history of chemistry; condensation polymers; antiradiation drugs; pharmaceuticals; biphenylene; chemical hazard evaluation. Mailing Add: Arthur D Little Inc Cambridge MA 02140

ATKINSON, EUGENE RONALD, electrooptics, see 12th edition

ATKINSON, GENE, b Houston, Tex, June 20, 25; m 47; c 1. ACADEMIC ADMINISTRATION, PHYSICS EDUCATION. Educ: Rice Inst, BA, 47; Univ Houston, EdD, 57. Prof Exp: Teacher pub schs, Tex, 49-51 & 53-55; from asst prof to assoc prof phys sci, Calif State Col, Long Beach, 57-64, prof phys sci & assoc dean instr, 64-66; assoc prof educ admin, 66-70, assoc dean col educ, 67-70, PROF EDUC ADMIN & ASST DEAN FAC, UNIV HOUSTON, 70- Mem: AAAS; Am Asn Physics Teachers; Nat Asn Res Sci Teaching. Res: Studies of the organization and administration of higher education. Mailing Add: 8015 Glen Valley Dr Houston TX 77107

ATKINSON, GEORGE FRANCIS, b Toronto, Ont, Feb 25, 32; m 61; c 4. ANALYTICAL CHEMISTRY, INSTRUMENTATION. Educ: Univ Toronto, BA, 53, MA, 54; PhD(chem), 60. Prof Exp: Lectr chem, Univ Western Ont, 60-61; from lectr to asst prof, 61-66, ASSOC PROF CHEM, UNIV WATERLOO, 66- Concurrent Pos: Mem sci study comt, Ont Curric Inst, 64-68; vis prof, Univ Southampton, 69-70. Mem: Am Chem Soc; Chem Inst Can; The Chem Soc. Res: Stability of metal-organic coordination complexes; applications of computers in analysis; instrumentation for electroanalytical and spectrophotometric methods; continuous analysis and control of process streams; elementary and secondary science education. Mailing Add: Dept of Chem Univ of Waterloo Waterloo ON Can

ATKINSON, GLENN FRANCIS, biometry, see 12th edition

ATKINSON, GORDON, b Brooklyn, NY, Aug 29, 30; m 54; c 3. PHYSICAL CHEMISTRY, INORGANIC CHEMISTRY. Educ: Lehigh Univ, BS, 52; Iowa State Col, PhD(chem), 56. Prof Exp: Res assoc, Ames Lab, US AEC, 56-57; instr chem, Univ Mich, 57-61; from asst prof to prof, Univ Md, College Park, 61-71; PROF CHEM & CHMN DEPT, UNIV OKLA, NORMAN, 71- Concurrent Pos: Consult, US Naval Ord Lab, Md, 62- Mem: AAAS; Am Chem Soc; Electrochem Soc. Res: Electrolyte solutions; conductance; ultrasonic absorption in solution; kinetics and thermodynamics of ion association; structure of aqueous solutions; chemical oceanography. Mailing Add: Dept of Chem Univ of Okla 620 Parrington Oval Norman OK 73069

ATKINSON, HAROLD RUSSELL, b Hamilton, Ont, Sept 25, 37; m 62; c 4. MATHEMATICAL ANALYSIS. Educ: Univ Western Ont, BA, 60; Assumption Univ, MSc, 61; Queen's Univ, Ont, PhD(math), 64. Prof Exp: Asst prof, 64-69, ASSOC PROF MATH, UNIV WINDSOR, 69- Mem: Can Math Cong. Res: Function spaces in which all linear functionals are represented by integrals; ordered Banach spaces; spaces of affine continuous functions on compact convex sets. Mailing Add: Dept of Math Univ of Windsor Windsor ON Can

ATKINSON, JOE WILLIAM, b Eldon, Mo, Sept 13, 17; m 46; c 2. VETERINARY MEDICINE. Educ: Kans State Univ, DVM, 50; Am Bd Vet Pub Health, dipl, 61. Prof Exp: Poultry inspector, USDA, Iowa & Kans, 50-52, vet consult milk & food prog, Mo, 52-53 & DC, 53-57, asst chief food sanit sect, 58-59; asst lab aids br, 60, EXEC SECY SURG STUDY SECT B, NIH, 61- Concurrent Pos: Tech consult, Poultry Indust Liaison Comt, 52-57. Mem: Conf Pub Health Vets (secy-treas, 58-60, pres, 61-62); fel Am Pub Health Asn; Am Vet Med Asn; US Animal Health Asn; NY Acad Sci. Res: Science administration; comparative medicine; epidemiology and zoonoses; public health practice and administration; environmental health and hygiene; laboratory animal production, medicine and care. Mailing Add: Surg Study Sect B Nat Insts of Health Bethesda MD 20014

ATKINSON, JOHN BRIAN, b Louth, Eng, Aug 15, 42; m 70; c 1. ATOMIC PHYSICS. Educ: Oxford Univ, BA, 63, DPhil(physics), 67. Prof Exp: Fel physics, Univ Windsor, 67-69, vis asst prof, 69-72; res assoc, Wayne State Univ, 69-72; ASST PROF PHYSICS, UNIV WINDSOR, 72- Mem: Can Asn Physicists; Am Phys Soc. Res: Study of atomic collision cross-sections and atomic lifetimes using tunable dye lasers; applications of atomic properties to development of tunable lasers. Mailing Add: Dept of Physics Univ of Windsor Windsor ON Can

ATKINSON, JOHN CHARLES, biostatistics, see 12th edition

ATKINSON, JOSEPH GEORGE, b St Paul, Alta, Mar 9, 35. ORGANIC CHEMISTRY. Educ: Univ Alta, BSc, 57; Mass Inst Technol, PhD(photochem), 62. Prof Exp: Group leader, Merck, Sharp & Dohme Can, Ltd, 62-70; GROUP LEADER MED CHEM, MERCK FROSST LABS, 70- Mem: Am Chem Soc; fel Chem Inst Can. Res: Organic photochemistry; synthesis of isotopically labeled organic compounds; synthesis of organometallic compounds and metal hybrides and deuterides; medicinal chemistry. Mailing Add: 5643 Plantagenet St Montreal PQ Can

ATKINSON, KENDALL E, b Centerville, Iowa, Mar 23, 40; m 61; c 2. MATHEMATICS. Educ: Iowa State Univ, BS, 61; Univ Wis, Madison, MS, 63, PhD(math), 66. Prof Exp: Assoc prof math, Ind Univ, Bloomington, 66-72; PROF MATH, UNIV IOWA, 72- Concurrent Pos: Vis res fel, Australian Nat Univ, 70-71. Mem: Soc Indust & Appl Math; Asn Comput Mach. Res: Numerical analysis; numerical solution of integral equations; mathematical computer software. Mailing Add: Dept of Math Univ of Iowa Iowa City IA 52242

ATKINSON, LARRY P, b Ames, Iowa, Aug 6, 41; m 66. CHEMICAL OCEANOGRAPHY. Educ: Univ Wash, BS, 63, MS, 66; Dalhousie Univ, PhD(oceanog), 73. Prof Exp: Res assoc oceanog, Marine Lab, Duke Univ, 66-68; res assoc, 72-74, ASST PROF OCEANOG, SKIDAWAY INST OCEANOG, 74- Mem: AAAS; Am Geophys Union; Am Soc Limnol & Oceanog. Res: Air sea gas exchange; continental shelf oceanography. Mailing Add: Skidaway Inst of Oceanog PO Box 13687 Savannah GA 31406

ATKINSON, LENETTE ROGERS, b South Carver, Mass, Mar 30, 99; m 28; c 2. PLANT CYTOLOGY, PLANT MORPHOLOGY. Educ: Mt Holyoke Col, BA, 21; Univ Wis, MA, 22, PhD(bot), 25. Prof Exp: Asst bot, Univ Wis, 21-25; Comn Relief Belg fel, Cath Univ Louvain, 25-27; actg asst prof bot, Mt Holyoke Col, 27-28; RES ASSOC BIOL DEPT, AMHERST COL, 46- Mem: AAAS; Am Fern Soc (secy, 63-69); Am Soc Plant Taxon; Bot Soc Am; Int Soc Plant Morphol. Res: Fern gametophyte, cytology in pteridophytes. Mailing Add: 415 S Pleasant St Amherst MA 01002

ATKINSON, MARSHALL B, b Willets, Calif, June 23, 17; m 46; c 3. OPHTHALMOLOGY. Educ: Univ Calif, AB, 40, MD, 46. Prof Exp: ASST CLIN PROF OCULAR PATH, SCH OPTOM, UNIV CALIF, BERKELEY, 56- Mem: Am Acad Ophthal & Otolaryngol; Asn Res Ophthal. Res: Etiology of refractive errors; headache mechanisms, interpretation of vascular changes in ocular fundi. Mailing Add: 450 Sutter Bldg San Francisco CA 94108

ATKINSON, REILLY, theoretical physics, see 12th edition

ATKINSON, RICHARD FOSTER, organic chemistry, see 12th edition

ATKINSON, ROBERT D'ESCOURT, b Rhayader, Wales, Apr 11, 98; m 31. ASTRONOMY. Educ: Oxford Univ, BA, 22; Univ Göttingen, DPhil, 28. Prof Exp: Demonstr physics, Clarendon Lab, Oxford Univ, 22-26; asst, Tech Univ, Berlin, 28-29; asst prof, Rutgers Univ, 29-34, assoc prof, 34-37; chief asst astron, Royal Observ Greenwich & Herstmonceux, 37-64; vis prof, 64-73, ADJ PROF ASTRON, IND UNIV, BLOOMINGTON, 73- Concurrent Pos: Mem Harvard Univ-Mass Inst Technol Eclipse Exped, Russia, 36; with mine design dept, Brit Admiralty, 40-46, liaison scientist degaussing, Del & DC, 42-46, with Ballistic Res Lab, Aberdeen, Md, 44-46; designer, Astron Clock, York Minster, Eng, 52-55; designer, Twilight-Setting Star Globe, Kelvin & Hughes, 54-55; mem, Brit Nat Comt Astron, 60-62. Honors & Awards: Award, Royal Comn Awards to Inventors, 48; Eddington Medal, Royal Astron Soc, 60. Mem: Am Phys Soc; Am Astron Soc; Royal Astron Soc (secy, 39-40); Brit Astron Asn (pres, 60-62); Brit Inst Navig. Res: Atomic synthesis and stellar energy; positional astronomy; instrument design. Mailing Add: Dept of Astron Ind Univ Bloomington IN 47401

ATKINSON, ROBERT GEORGE, b Vancouver, BC, June 27, 18; nat US; m 47; c 1. PLANT PATHOLOGY. Educ: Univ BC, BSA, 40; Univ Toronto, PhD(plant path), 49. Prof Exp: PLANT PATHOLOGIST RES STA, RES BR, CAN DEPT AGR, 47- Mem: Mycol Soc Am; Can Phytopath Soc. Res: Root rot and wilt of greenhouse tomatoes and cucumbers. Mailing Add: Res Sta Can Agr 8801 E Saanich Rd Sidney BC Can

ATKINSON, ROBERT LEON, b Iowa Park, Tex, Feb 14, 14; m 41. POULTRY NUTRITION. Educ: Agr & Mech Col Tex, BS, 49, MS, 50; Univ Calif, PhD(nutrit), 58. Prof Exp: From res asst to res assoc, Agr & Mech Col Tex, 48-53, instr, 53-54; res asst, Univ Calif, 54-55, sr lab technician, 56-57; asst prof poultry sci, 55-68, ASSOC PROF POULTRY SCI, TEX A&M UNIV, 68- Honors & Awards: Res Award, Nat Turkey Fedn, 74. Mem: AAAS; Soc Exp Biol & Med; Am Inst Nutrit; Poultry Sci Asn; Am Chem Soc. Res: Poultry, particularly turkey nutrition, breeding and management. Mailing Add: Dept of Poultry Sci Tex A&M Univ College Station TX 77843

ATKINSON, RUSSELL H, b Brooklyn, NY, Aug 22, 22; m 44; c 2. PHYSICAL CHEMISTRY, ENGINEERING MANAGEMENT. Educ: Polytech Inst Brooklyn, BS, 48, PhD(chem), 52. Prof Exp: Jr chemist, Manhattan Proj, Tenn Eastman Corp, 44-46; sr engr, 51-56, sect mgr metals res group, 56-59, asst res dir, 59-61, mgr parts eng, 61-63, asst mgr div eng, 63-66, mgr parts mfg, 67-69, mgr div eng, 69-76, PROD ENG MGR, LAMP DIV, WESTINGHOUSE ELEC CORP, 76- Concurrent Pos: Instr, Polytech Inst Brooklyn, 52-54. Mem: AAAS; Am Illum Eng Soc; Electrochem Soc; Sigma Xi. Res: Chemistry of refractory metals; gas-metal reactions; analysis of refractory metals; vapor deposition reactions; lamp development and materials. Mailing Add: 62 Oakwood Court Fanwood NJ 07023

ATKINSON, THOMAS GRISEDALE, b Vancouver, BC, Apr 20, 29; m 56; c 3. PLANT PATHOLOGY. Educ: Univ BC, BSA, 52; Univ Sask, MSc, 53, PhD(plant physiol), 56. Prof Exp: Proj assoc plant physiol, Univ Wis, 56-58; PLANT PATHOLOGIST, CAN DEPT AGR RES STA, 58- Mem: Am Phytopath Soc; Can Phytopath Soc; Sigma Xi. Res: Cereal root rots; genetic and microbiological analysis of root rot reaction of cereals; resistance to wheat streak mosaic; disease loss assessment; non-parasitic disorders of cereals. Mailing Add: Plant Path Sect Can Dept of Agr Res Sta Lethbridge AB Can

ATKINSON, WILLIAM ALLEN, b Berkeley, Calif, Mar 14, 33; m 68; c 3. FOREST MANAGEMENT. Educ: Univ Calif, Berkeley, BS, 55, MS, 56, PhD(forestry), 74. Prof Exp: Forester, Soper-Wheeler Co, 56-67; asst prof forestry, Univ Calif, Berkeley, 70-71; ASSOC PROF FORESTRY, UNIV WASH, 71- Mem: Soc Am Foresters. Res: Forest fertilization; control of stocking. Mailing Add: Col of Forest Resources Univ of Wash Seattle WA 98195

ATKISSON, ARTHUR ALBERT, b Omaha, Nebr, Oct 5, 30; m 48; c 5. ENVIRONMENTAL MANAGEMENT. Educ: Lewis & Clark Col, BS, 50; Univ Southern Calif, MPA, 68, DPA, 73. Prof Exp: Res asst polit sci, Univ Ore, 50-51; admin analyst, Bonneville Power Admin, 51-53; admin analyst, Chief Admin Off, County of Los Angeles, 53-55; dep dir, Los Angeles County Air Pollution Dist, 55-65; exec dir, Inst Urban Ecol, Univ Southern Calif, 66-69; prof urban health & dir urban health unit, Univ Tex Sch Pub Health Houston, 69-73; pres, Inst Urban Ecol & Pub Affairs, Inc, 72-75; PROF ENVIRON & CHMN FAC ENVIRON ADMIN, UNIV WIS-GREEN BAY, 75- Concurrent Pos: Mem, Nat Air Pollution Manpower Develop & Training Adv Comt, 66-71; consult, US Dept Labor, 68 & Dept Environ Qual, City of Austin, Tex, 72-74; mem, Nat Air Conserv Comn, 68-; adj fac mem, Sch Archit, Rice Univ, 72-73. Mem: AAAS; Air Pollution Control Asn; Am Soc Pub Admin; Am Pub Health Asn. Res: Relationship between community variables and rates of morbidity, mortality and net migration; roles and functions of public health and environmental management executives; metropolitan decision chains and decisional processes. Mailing Add: Social Ecol Bldg Rm 436 Univ of Wis Green Bay WI 54302

ATLAS, DAVID, b Brooklyn, NY, May 25, 24; m 48; c 2. METEOROLOGY. Educ: NY Univ, BSc, 46; Mass Inst Technol, MSc, 51, DSc(meteorol), 55. Prof Exp: Res meteorologist, Materiel Command, US Air Force, 45-48, atmospheric physicist, Air Force Cambridge Res Labs, 48-66; prof meteorol, Univ Chicago, 66-72; dir atmospheric technol div, 72-74, dir nat hail res exp, 74-75, SR SCIENTIST, ADVAN STUDY PROG, NAT CTR ATMOSPHERIC RES, 75- Concurrent Pos: Assoc ed jour, Am Meteorol Soc, 57-75; NSF sr fel, Imp Col, Univ London, 59; pres inter-union comn radio meteorol, Int Sci Radio Union & Int Union Geod & Geophys, 69-72, mem, US Nat Comn, Int Sci Radio Union, 76-; chmn panel remote atmospheric probing, Nat Acad Sci, 68-69 & mem comt atmospheric sci, 75-; mem meteorol adv comt, Environ Protection Agency, 69-72; mem meteorol adv comt, Chicago Dept Environ Control, 70-72; mem US deleg to VII Cong, World Meteorol Orgn, 75. Honors & Awards: Meisinger Award, Am Meteorol Soc, 57; Loeser Award, Air Force Cambridge Res Labs, 57, O'Day Award, 66; Losey Award, Am Inst Aeronaut & Astronuat, 66. Mem: Fel Am Meteorol Soc (pres-elect, 74, pres, 75); fel Am Geophys Union; Royal Meteorol Soc. Res: Weather modification; cloud physics; radar meteorology; atmospheric wave motions and clear air turbulence. Mailing Add: 265 Fox Dr Boulder CO 80303

ATLAS, LEON MAURICE, physical chemistry, ceramics, see 12th edition

ATNEOSEN, RICHARD ALLEN, b St James, Minn, Sept 11, 34; m 61. PHYSICS. Educ: Univ Minn, BS, 56, MS, 58; Ind Univ, PhD(physics), 63. Prof Exp: Res assoc nuclear chem, Princeton Univ, 63-65; asst res prof physics, Mich State Univ, 65-68; ASSOC PROF PHYSICS, WESTERN WASH STATE COL, 68- Mem: Am Phys Soc; Am Asn Physics Teachers. Res: Reaction mechanisms in alpha particle scattering; particle-induced fission. Mailing Add: Dept of Physics Western Wash State Col Bellingham WA 98225

ATNIP, ROBERT LEE, b Bridgeport, Ala, Aug 10, 28; m 51; c 5. ANATOMY. Educ: David Lipscomb Col, BA, 51; George Peabody Col, MA, 52; Univ Tenn, PhD(anat), 64. Prof Exp: Teacher high schs, Ga & Ala, 52-54; instr biol, Freed-Hardeman Col, 54-59; instr anat & pediat, 66-68, asst prof anat & child develop & res assoc pediat, 68-74, ASSOC PROF ANAT & CHILD DEVELOP, UNIV TENN MED UNITS, 74- Concurrent Pos: Nat Inst Child Health & Human Develop fel anat, Col Med, Univ Tenn, 65-66. Mem: Am Soc Human Genetics. Res: Mammalian embryology and experimental teratology; experimentally-produced cleft palate in mice; chromosomal aberrations of malformed mentally retarded children. Mailing Add: Dept of Anat Univ of Tenn Memphis TN 38103

ATOJI, MASAO, b Osaka, Japan, Dec 21, 25; US citizen; m 57; c 2. PHYSICAL CHEMISTRY. Educ: Shizuoka Univ, BS, 46; Osaka Univ, MS, 48, DSc(chem), 56. Prof Exp: Res fel chem, Univ Minn, Minneapolis, 51-56; assoc res, Iowa State Univ, 56-58, asst prof, 58-60; chemist, 60-69, SR CHEMIST & GROUP LEADER NEUTRON GROUP, DIV CHEM, ARGONNE NAT LAB, 69- Mem: AAAS; Am Crystallog Asn; Am Chem Soc; fel Am Phys Soc; Phys Soc Japan. Res: Crystal and magnetic structure and analyses by means of neutron and x-ray diffraction;

intermolecular force and molecular structure theories. Mailing Add: Div of Chem Argonne Nat Lab 9700 S Cass Ave Argonne IL 60439

ATREYA, SUSHIL KUMAR, b Ajmer, India, Apr 15, 46; m 70; c 1. ATMOSPHERIC SCIENCES, PLANETARY ATMOSPHERES. Educ: Univ Rajasthan, India, BSc, 63, MSc, 65; Yale Univ, MS, 68; Univ Mich, PhD(atmospheric sci), 73. Prof Exp: Res fel physics, Univ Delhi, India, 65-66; res assoc physics, Univ Pittsburgh, 73-74; RES SCIENTIST ATMOSPHERIC SCI, UNIV MICH, ANN ARBOR, 74- Mem: AAAS; Am Geophys Union; Am Astron Soc; Sigma Xi. Res: Atmospheres and ionospheres of earth and planets. Mailing Add: Dept Atmospheric & Oceanic Sci Space Physics Res Lab Univ Mich Ann Arbor MI 48109

ATTALLA, ALBERT, b Cuyahoga Falls, Ohio, Sept 29, 31; m 56; c 3. PHYSICAL CHEMISTRY, ANALYTICAL CHEMISTRY. Educ: Kent State Univ, BA, 55, MA, 59; Univ Cincinnati, PhD(phys chem), 62. Prof Exp: Sr chemist, Corning Glass Works, NY, 62-63; RES SPECIALIST INSTRUMENTAL ANAL, MONSANTO RES CORP, 63- Res: Nuclear magnetic resonance and x-ray fluorescence spectroscopy; gas chromatography. Mailing Add: Monsanto Res Corp Miamisburg OH 45342

ATTAWAY, DAVID HENRY, b Sterling, Okla, June 9, 38. BIOCHEMISTRY. Educ: Univ Okla, BS, 60, PhD(biochem), 68. Prof Exp: Phys oceanogr, US Naval Oceanog Off, 62-65; res assoc org geochem, Marine Sci Inst, Tex, 68-69; sr res assoc, State Geol Surv, Univ Kans, 69-71; res chemist, US Coast Guard Hq, 71-72; ASST DIR GRANTS MGT, OFF SEA GRANT, NAT OCEAN & ATMOSPHER ADMIN, 72- Concurrent Pos: Univ Tex grad sch fel, 68-69. Mem: AAAS; Am Geochem Soc; Am Chem Soc; Brit Chem Soc. Res: Comparative chemistry of natural products from marine plants and animals; toxic substances of marine origin; organic geochemistry; petroleum in marine environment. Mailing Add: Nat Sea Grant Proj 3300 Whitehaven Bldg Washington DC 20235

ATTAWAY, JOHN ALLEN, b Atlanta, Ga, July 19, 30; m 57; c 4. ORGANIC CHEMISTRY, PLANT BIOCHEMISTRY. Educ: Fla Southern Col, BS, 51; Univ Fla, MS, 53; Duke Univ, PhD(chem), 57. Prof Exp: Asst, Off Ord Res, Univ Fla, 53; asst, Duke Univ, 53-54, Off Ord Res & Off Naval Res, 54-55, fel plant biochem, 57; res chemist, Monsanto Chem Co, 57-58; res chemist org chem & biochem, Resources Res Inc, 58-59; res chemist, 59-68, DIR RES, FLA DEPT OF CITRUS, 68- Mem: Am Chem Soc; Am Soc Plant Physiologists; Inst Food Technologists. Res: Chromatographic analysis; instrumentation; natural products and flavors; organic fluorine compounds; plant growth regulators. Mailing Add: PO Box 148 Lakeland FL 33802

ATTEBERY, BILLY JOE, b Bearden, Ark, Dec 26, 27; m 58; c 2. MATHEMATICS. Educ: Ark State Teachers Col, BSE, 50; Univ Ark, MA, 54; Univ Mo, PhD, 58. Prof Exp: Instr math, Univ Mo, 54-58; from asst prof to assoc prof, Univ Ark, 58-67; assoc prof, 67-68, PROF MATH, LA TECH UNIV, 68- Mem: Math Asn Am. Res: Probability theory. Mailing Add: Dept of Math La Tech Univ Ruston LA 71270

ATTERBOM, HEMMING A, b Ystad, Sweden, Aug 7, 38; m 65; c 5. PHYSIOLOGY. Educ: Royal Cent Inst, Sweden, BS, 63; Univ NMex, MS, 66; Univ Ore, PhD(phys educ), 71. Prof Exp: Teacher, Stockholm's Sch Bd, 63-64; part-time instr phys educ, Univ NMex, 64-65; asst, Univ Ore, 66-67; res assoc physiol, Lovelace Found Med Educ & Res, 67-69; ASST PROF EXERCISE PHYSIOL & DIR HUMAN PERFORMANCE LAB, UNIV NMEX, 69- Mem: AAAS; Aerospace Med Asn; Am Asn Health, Phys Educ & Recreation; Int Coun Health, Phys Educ & Recreation. Res: Physiology of work under environmental stress; cardiopulmonary rehabilitation; acid-base physiology; maximal work; ergogenic aids; exercise metabolism; physical rehabilitation of cardiac patients. Mailing Add: Human Performance Lab Carlisle Gym Univ of NMex Albuquerque NM 87106

ATTIG, THOMAS GEORGE, b Pontiac, Ill, Oct 2, 46; m 68; c 1. INORGANIC CHEMISTRY, ORGANOMETALLIC CHEMISTRY. Educ: DePauw Univ, BA, 68; Ohio State Univ, PhD(chem), 73. Prof Exp: Teaching asst, Ohio State Univ, 69-70, NSF trainee, 70-73; fel chem, Univ Western Ont, 73-75; ASST PROF CHEM, UNIV KY, 75- Mem: Am Chem Soc. Res: Preparation, structure and reactions of organometallic compounds of the nickel triad; chemistry of transition metal carbonyl complexes; synthesis and stereochemistry of chiral organometallic compounds. Mailing Add: Dept of Chem Univ of Ky Lexington KY 40506

ATTINGER, ERNST OTTO, b Zurich, Switz, Dec 27, 22. BIOENGINEERING. Educ: Winterthur Cantonal Sch, Switz, BA, 41; Univ Zurich, MD, 48; Drexel Inst Technol, MS, 61; Univ Pa, PhD(biomed eng), 65. Prof Exp: Chief res, Dis of Chest, Heilstaette Du Midi, Davos, Switz, 50-52; asst res, Internal Med, Lincoln Hosp, NY, 52-53; asst prof med, Med Sch, Tufts Univ, 56-59; from asst prof to prof physiol, Sch Vet Med, Univ Pa, 61-67; res dir, Res Inst, Presby Hosp, 62-67; PROF PHYSIOL & CHMN BIOMED ENG, UNIV VA, 67- Concurrent Pos: Res fels, Cardiopulmonary Lab, Nat Jewish Hosp, Denver, Colo, 43-54, Lung Sta, Boston City Hosp, Tufts Univ, 54-59 & Res Inst, Presby Hosp, Philadelphia, 59-62. Mem: AAAS; Am Physiol Soc; Fedn Clin Res; Inst Elec & Electronics Engrs; Biophys Soc. Res: Analysis of biological systems; systems analysis with particular emphasis on control hierarchies in biological and social systems. Mailing Add: Div of Biomed Eng Univ of Va Med Ctr Charlottesville VA 22901

ATTIX, FRANK HERBERT, b Portland, Ore, Apr 2, 25; m 59; c 2. RADIOLOGICAL PHYSICS. Educ: Univ Calif, Berkeley, AB, 49; Univ Md, MS, 53. Prof Exp: Res physicist, NIH, 49-50; res physicist, Nat Bur Standards, 50-57; reactor shielding physicist, ACF Industs, DC, 57-58; head dosimetry br, 63-68, consult nuclear sci div, 69-75, RES PHYSICIST, NAVAL RES LAB, 58-, CONSULT RADIATION TECHNOL DIV, 75- Concurrent Pos: Vis scientist, UK Atomic Energy Res Estab, Harwell, 68-69. Mem: Am Phys Soc; Health Phys Soc; Sigma Xi; Am Asn Physicists Med. Res: Radiation physics and dosimetry of ionizing radiations; dosimetry of fast neutrons for radiotherapy applications. Mailing Add: Code 6603A Radiation Technol Div Naval Res Lab Washington DC 20375

ATTREP, MOSES, JR, b Alexandria, La, Jan 2, 39; m 65. RADIOCHEMISTRY, GEOCHEMISTRY. Educ: La Col, BS, 60; Univ Ark, MS, 62, PhD(chem), 65. Prof Exp: Res assoc chem, Clark Univ, 65-66; asst prof, 66-74, PROF CHEM, E TEX STATE UNIV, 74- Mem: Geochem Soc; Am Chem Soc. Res: Analytical chemistry. Mailing Add: Dept of Chem ETex State Univ Commerce TX 75428

ATWATER, EDWARD CONGDON, b Rochester, NY, Feb 6, 26; m 50; c 2. INTERNAL MEDICINE, HISTORY OF MEDICINE. Educ: Univ Rochester, BA, 50; Harvard Univ, MD, 55; Johns Hopkins Univ, MA, 74. Prof Exp: From intern to asst resident, 55-57, chief resident, 59-60, from instr to asst prof, 59-69, ASSOC PROF MED & ASST PROF HIST MED, SCH MED, UNIV ROCHESTER, 69- Concurrent Pos: USPHS trainee arthritis & metab dis, Univ Rochester, 57-59; Macy fel, Sch Med, Johns Hopkins Univ, 70-71; from asst physician to assoc physician, Strong Mem Hosp, Rochester, 60-69, sr assoc physician, 69- Mem: Am Rheumatism Asn; Am Asn Hist Med. Res: Arthritis and related rheumatologic problems; history of medicine education. Mailing Add: Dept of Med Univ of Rochester Sch of Med Rochester NY 14642

ATWATER, GORDON INGHAM, geology, deceased

ATWATER, HARRY ALBERT, b Boston, Mass, Jan 10, 21; m 58. SOLID STATE PHYSICS. Educ: Tufts Col, BS, 40; Harvard Univ, MS, 41, PhD(physics), 57; Boston Univ, MA, 49. Prof Exp: Instr physics, Univ Ore, 56-59; assoc prof elec eng, 59-65, ASSOC PROF PHYSICS, PA STATE UNIV, 65- Mem: Am Phys Soc; Am Asn Physics Teachers. Res: Electron paramagnetic resonance in solids. Mailing Add: Dept of Physics Osmond Lab Pa State Univ University Park PA 16802

ATWATER, NORMAN WILLIS, b Paterson, NJ, Mar 16, 26; m 47; c 3. ORGANIC CHEMISTRY. Educ: Rensselaer Polytech Inst, BS, 49; Johns Hopkins Univ, PhD(chem), 53. Prof Exp: DIR QUAL CONTROL, G D SEARLE & CO, 53- Mem: AAAS; Am Chem Soc. Res: Steroids; terpenes. Mailing Add: 5 W South St Arlington Heights IL 60005

ATWATER, TANYA MARIA, b Los Angeles, Calif, Aug 27, 42; m 72; c 1. MARINE GEOPHYSICS, PALEOMAGNETISM. Educ: Univ Calif, Berkeley, AB, 65, Univ Calif, San Diego, PhD(earth sci), 72. Prof Exp: Vis res assoc seismol, Univ Chile, 66-67; res assoc paleomagnetism, Stanford Univ, 70-71; asst prof marine geophys, Univ Calif, San Diego, 72-73; Nat Acad Sci exchange scientist, USSR, 73-74; ASST PROF MARINE GEOPHYS, MASS INST TECHNOL, 74- Concurrent Pos: Sloan fel, 75-77; nat lectr, Sigma Xi, 75-76. Mem: Fel Geol Soc Am; fel Am Geophys Union; AAAS; Am Geol Inst; Asn Women Sci. Res: Detailed tectonic nature of mid ocean ridges and the mechanisms by which ocean floor is created and modified; plate tectonic reconstructions and their implications for continental geologic structure and history. Mailing Add: Mass Inst of Technol Dept of Earth & Planetary Sci Cambridge MA 02139

ATWELL, ROBERT JAMES, b Gary, Ind, Sept 1, 19; m 45; c 3. MEDICINE. Educ: Duke Univ, BA, 41, MD, 45. Prof Exp: Instr med, Duke Univ, 48-50; from asst prof to assoc prof, 51-66, actg dean, Col Med, 72-73, PROF MED, COL MED & DIR SCH ALLIED MED PROFESSIONS, OHIO STATE UNIV, 66- Concurrent Pos: Consult, Vet Admin, 49-; chief med serv, Ohio Tuberc Hosp, 50-65. Mem: Am Fedn Clin Res; Am Col Physicians; Am Soc Allied Health Professions (secy-treas, 68-69, pres, 70-71). Res: Clinical medicine; pulmonary disease. Mailing Add: Sch of Allied Med Professions Ohio State Univ Columbus OH 43210

ATWELL, WILLIAM HENRY, b Milwaukee, Wis, Dec 13, 36; m 60; c 4. ORGANIC CHEMISTRY. Educ: Marquette Univ, BS, 59, MS, 60; Iowa State Univ, PhD(org chem), 64. Prof Exp: From res chemist to sr res chemist, 65-70, GROUP MGR, DOW CORNING CORP, 70- Mem: Am Chem Soc; Sigma Xi. Res: Organometallic chemistry, particularly organosilicon chemistry. Mailing Add: Dow Corning Corp PO Box 1592 Midland MI 48640

ATWOOD, BRUCE, b Newark, NJ, Apr 7, 47; m 67. ASTROPHYSICS. Educ: Bard Col, BA, 70; Wesleyan Univ, PhD(physics), 75. Prof Exp: Tutor physics, Wesleyan Univ, 70-73; LECTR PHYSICS, CALIF POLYTECH STATE UNIV, SAN LUIS OBISPO, 74- Mem: Am Astron Soc; Am Phys Soc; Am Asn Physics Teachers; Sigma Xi. Res: Astrometry of double stars by area scanning and image processing for speckle interferometry. Mailing Add: Calif Polytech State Univ Dept of Physics San Luis Obispo CA 93407

ATWOOD, CARL EDMUND, b Clyde River, NS, June 19, 06; m 35; c 3. ENTOMOLOGY. Educ: McGill Univ, BScA, 31; Univ Toronto, MScA, 33, PhD(entom), 37. Prof Exp: Insect pest investr, Dept Agr, Ottawa, 29, from jr entomologist to asst entomologist, 29-45; from asst prof to assoc prof, 45-62, PROF ZOOL, UNIV TORONTO, 62- Concurrent Pos: Asst & sr demonstr, 31-36; consult forest entomologist, Dept Lands & Forests, Ont, 45-56. Mem: AAAS; Entom Soc Am; Entom Soc Can; Royal Can Inst. Res: Behavior, ecology and taxonomy of primitive bees and sawflies. Mailing Add: Dept of Zool Univ of Toronto Toronto ON Can

ATWOOD, DONALD KEITH, b Burlington, Vt, June 5, 33; m 54; c 4. CHEMICAL OCEANOGRAPHY. Educ: St Michael's Col (Vt), BS, 55; Purdue Univ, PhD(chem), 60. Prof Exp: Res engr, Humble Oil & Refining Co, 60-63, sr res chemist, 63-67, minerals prod, 67-69; ASSOC PROF CHEM OCEANOG, UNIV PR, MAYAGUEZ, 69- Concurrent Pos: Subj leader chem oceanog, Coop Invest Caribbean & Adj Regions, 72-; consult, NSF, 75- Mem: Am Chem Soc; Sigma Xi; Soc Limnol & Oceanog; Am Geophys Union. Res: Seasonal variations in ocean chemistry; descriptive chemical oceanography; ocean thermal energy conversion. Mailing Add: Dept of Marine Sci Univ of PR Mayaguez PR 00708

ATWOOD, FRANCIS CLARKE, b Salem, Mass, May 7, 93; m 18; c 2. CHEMISTRY. Educ: Mass Inst Technol, BS, 14. Prof Exp: Instr, Mass Inst Technol, 14-16; res engr, Exolon Co, 16-17; jr partner, Kalmus, Comstock & Westcott, Inc, 17-25; VPRES, BONNAR-ATWOOD, INC, 25-; PRES, ATLANTIC RES ASSOCS, 31- Concurrent Pos: Chief chemist, Craftex Co, 27-34; with Off Prod Res & Develop, 44. Honors & Awards: Am Paint J Award, 31. Mem: AAAS; Am Chem Soc; fel Am Inst Chem Eng; fel Acoust Soc Am; Fedn Socs Coating Technol. Res: Metallurgy; electric arc lamps; photography; proteins; textiles; rubber; resins; emulsion and resin emulsion paints; synthetic protein fibers; mica specialties; dairy by-products; acrylates; beef blood albumen. Mailing Add: Box 1098 Edgartown MA 02539

ATWOOD, GILBERT RICHARD, b Taunton, Mass, Apr 9, 28; m 55; c 2. PHYSICAL CHEMISTRY. Educ: Carnegie Inst Technol, BS, 49; Purdue Univ, MS, 51; Univ Pittsburgh, PhD(chem), 58. Prof Exp: From asst to assoc, Mellon Inst, 51-54, from jr fel to fel, 54-58; sr chemist, Koppers Co, Inc, Pa, 58-60, mgr physics & phys chem, 60-63; proj leader, 63-68, from res scientist to sr res scientist, 68-74, RES ASSOC, UNION CARBIDE CORP, 74- Mem: AAAS; Am Chem Soc. Res: Thermodynamics; phase equilibria; thermometry; cryoscopy; ebulliometry; calorimetry; separation and purification; gas absorption; liquid-liquid extraction. Mailing Add: Union Carbide Corp PO Box 65 Tarrytown NY 10591

ATWOOD, HAROLD LESLIE, b Montreal, Que, Feb 15, 37; m 59; c 3. PHYSIOLOGY. Educ: Univ Toronto, BA, 59; Univ Calif, Berkeley, MA, 60; Glasgow Univ, PhD(zool), 63. Prof Exp: Res assoc biol, Univ Ore, 62-64; res fel, Calif Inst Technol, 64-65; from asst prof to assoc prof, 65-71, PROF ZOOL, UNIV TORONTO, 71- Mem: Am Physiol Soc; Brit Soc Exp Biol. Res: Comparative and neuromuscular physiology. Mailing Add: Dept of Zool Univ of Toronto Toronto ON Can

ATWOOD, JERRY LEE, b Springfield, Mo, July 27, 42; m 64; c 2. INORGANIC CHEMISTRY, PHYSICAL CHEMISTRY. Educ: Southwest Mo State Col, BS, 64; Univ Ill, MS, 66, PhD(inorg chem), 68. Prof Exp: ASST PROF CHEM, UNIV ALA,

TUSCALOOSA, 68- Mem: AAAS; Am Chem Soc; Am Crystallog Asn. Res: Organometallic chemistry; x-ray crystallography. Mailing Add: Dept of Chem Univ of Ala University AL 35486

ATWOOD, JOHN WILLIAM, b Sherbrooke, Que, Feb 23, 41; m 70; c 2. COMPUTER SCIENCE. Educ: McGill Univ, BEng, 63; Univ Toronto, MASc, 65; Univ Ill, Urbana, PhD(elec eng), 70. Prof Exp: Res assoc elec eng, Univ Ill, Urbana, 65-70; asst prof, Univ Toronto, 70-72; asst prof comput sci, 73-75, ASSOC PROF COMPUT SCI, CONCORDIA UNIV, 75- Concurrent Pos: Consult, Metrop Toronto Police Comn, 71-72, Amdahl Corp, 76- Mem: Inst Elec & Electronics Engrs; Can Info Processing Soc. Res: Operating systems; computer architecture; computer performance measurement; interactive program editing. Mailing Add: Dept Comput Sci Concordia Univ Sir George Williams Campus Montreal PQ Can

ATWOOD, KENTON, b Cincinnati, Ohio, June 11, 16; m 43; c 2. PHYSICAL CHEMISTRY. Educ: Wilmington Col, BS, 36; Haverford Col, AM, 37; Univ Calif, PhD(chem), 40. Prof Exp: Asst chem, Univ Calif, 37-40; res chemist, Joseph E Seagram & Sons, Inc, 41-44; res chemist, Girdler Corp, 44-51; asst to dir res, 51-57; tech consult, Catalysts & Chems Inc, Louisville, Ind, 57-71; CONSULT, TECH COMPUT SERV, INC, 71- Mem: Am Chem Soc. Res: Photochemical reactions in solution; production of butadiene from 2-3 butanediol; development of catalysts for gas reactions; catalytic-environmental studies; thermodynamics of hydrogen and sy-x synthesis gas production. Mailing Add: 1742 Millerwood Dr New Albany IN 47150

ATWOOD, LINDA, b Rochester, NY, Nov 13, 46. BIO-ORGANIC CHEMISTRY. Educ: Bard Col, BA, 68; Wesleyan Univ, MA, 72, PhD(chem), 74. Prof Exp: Teaching & res asst chem, Wesleyan Univ, 70-74; ASST PROF CHEM, CALIF POLYTECH STATE UNIV, SAN LUIS OBISPO, 74- Mem: Am Chem Soc; AAAS. Res: Bio-organic chemistry of phosphorus and metal ion catalysis in biological systems. Mailing Add: Dept of Chem Calif Polytech State Univ San Luis Obispo CA 93407

ATWOOD, MARK TREVOR, b Okmulgee, Okla, Nov 21, 27; m 50; c 1. INDUSTRIAL ORGANIC CHEMISTRY. Educ: Park Col, BA, 50; Purdue Univ, MS, 52, PhD(org chem), 54. Prof Exp: Res chemist, Continental Oil Co, 54-63, petrochem develop coordr, 63-66; chief chemist, 66-68, MGR LABS, OIL SHALE CORP, 68- Mem: Am Chem Soc. Res: Vapor phase nitration; chemistry of oil shale and coal; aliphatic amines; organometallics; petrochemicals. Mailing Add: Oil Shale Corp 18200 W Hwy 72 Golden CO 80401

ATWOOD, MARK WYLLIE, b Fairfield, Iowa, Mar 23, 42; m 62; c 2. ENTOMOLOGY, PHYSIOLOGY. Educ: Parsons Col, BS, 62; Univ Wis, Madison, MS, 65, PhD(zool), 68. Prof Exp: Asst prof biol, Purdue Univ, Ft Wayne, 67-70; ENTOMOLOGIST, IOWA STATE DEPT AGR, 70- Mem: AAAS; Am Soc Zool; Cent Plant Bd (pres, 75). Res: Excitation in insect visual system and use of visual information in behavior; mechanism of action of pesticides and relation of sub-lethal doses of pesticides to insect behavior. Mailing Add: Iowa State Dept of Agr E Seventh & Court Des Moines IA 50319

ATWOOD, SANFORD SOVERHILL, b Janesville, Wis, Dec 3, 12; m 36; c 4. CYTOLOGY, GENETICS. Educ: Univ Wis, BA & MA, 34, PhD(plant cytol), 37. Hon Degrees: LHD, Gettysburg Col, 66. Prof Exp: From asst geneticist to assoc geneticist US regional pasture res lab, Pa State Col, 37-44; asst prof plant breeding, Cornell Univ, 44-45, from assoc prof to prof, 45-63, head dept, 49-53, dean grad sch, 53-55, provost, 55-63; PRES, EMORY UNIV, 63- Concurrent Pos: Mem agr adv comt, W K Kellogg Found, 54-60; chmn comt agr land use & wildlife resources, Nat Acad Sci-Nat Res Coun, 65-70; trustee, Assoc Univs, Inc, 61-63; dir, Oak Ridge Assoc Univs, 65-71; trustee, Comt Econ Develop, 65-; dir, Metrop Atlanta Rapid Transit Auth, 65-; mem adv panel sea grant inst support, Nat Oceanic & Atmospher Admin, 66-; dir, Nat Med Fels, 69- Mem: Fel AAAS; Am Bot Soc; Genetics Soc Am; Am Genetics Asn; fel Am Soc Agron. Res: Cytogenetics fundamental to breeding forage plants. Mailing Add: Emory Univ Atlanta GA 30322

ATWOOD, WALLACE WALTER, JR, b Chicago, Ill, June 7, 06; m 32; c 2. GEOMORPHOLOGY. Educ: Univ Chicago, BS, 26; Clark Univ, MS, 27, PhD(physiog), 30. Prof Exp: Asst chief br res & educ, Nat Park Serv, Washington, 30-32; prof physiog & regional geog, Clark Univ, 32-43; chief topog model sect, US Off Strategic Servs, 43-45; chief relief map div, US Army Map Serv, 45-46; dep exec dir comt geog explor res & develop bd, Nat Mil Estab, 47-48, dep exec dir comt geophys & geog, 48-50; dir off int rels, Nat Acad Sci-Nat Res Coun, 50-63, spec asst to pres, 64-71; RETIRED. Concurrent Pos: Geogr, Babson Inst, 38-41; pres, Rappahannock Marine Lodge Inc, 61-74; ed, News Report, Nat Acad Sci, 51-63; chmn US comn, Int Geog Union, 49-59, mem US del, 49, chmn, 52-56; mem US del, Int Coun Sci Unions, 52-55, 58 & 61; mem, UNESCO Int Adv Comt Sci Res, 53-59, chmn, 57-59, mem US comt, Int Geophys Year, 53-63. Honors & Awards: Meritorious Award, Asn Am Geog, 52; Distinguished Serv Award, Am Hungarian Found, 74. Mem: AAAS (vpres, 55); Am Geog Soc; Asn Am Geog (treas, 57-59). Res: International relations in science; cartography; relief model construction. Mailing Add: Windmill Point VA 22578

ATYEO, WARREN THOMAS, b Highland Park, Mich, Feb 15, 27; m 55; c 3. ENTOMOLOGY. Educ: Western Ill Univ, MS, 53; Univ Kans, PhD(entom), 59. Prof Exp: From asst prof to assoc prof entom, Col Agr, Univ Nebr, 58-67, cur mus, 58-67; participant, Grants Assocs Prog, NIH, 67-68; PROF ENTOM & CUR, UNIV GA, 68- Mem: Soc Study Evolution; Entom Soc Am; Acarological Soc Am; Soc Systematic Zool. Res: Acarology; systematics of Acarina, Bdelloidea, Analgoidea. Mailing Add: Dept of Entom Univ of Ga Athens GA 30602

AU, ANDREW TAICHIU, b Hong Kong, Dec 12, 42; m 69; c 1. ORGANIC CHEMISTRY. Educ: Univ Mo, BS, 65; Mich State Univ, MS, 68; State Univ NY Buffalo, PhD(org chem), 74. Prof Exp: Res chemist, Upjohn Co, 69-71; assoc org chem, Mass Inst Technol, 74-75; SR RES CHEMIST, DOW CHEMICAL CO, 75- Mem: Am Chem Soc. Res: Various types of synthesis using phase transfer catalysis. Mailing Add: New Eng Res Lab Dow Chem Co PO Box 400 Wayland MA 01778

AU, CHI-KWAN, b Macau, China, Jan 21, 46; m 70; c 1. THEORETICAL PHYSICS. Educ: Hong Kong Univ, BSc, 68; Columbia Univ, MA, 70, PhD(physics), 72. Prof Exp: Res assoc physics, Univ Ill, Urbana, 72-74; res assoc, Yale Univ, 74-75, lectr, 75; ASST PROF PHYSICS, UNIV S C, COLUMBIA, 75- Mem: Am Phys Soc. Res: Theoretical astrophysics, neutron star matter; condensed matter; high density matter and quantum electrodynamics of simple atomic systems. Mailing Add: Dept of Physics & Astron Univ S C Columbia SC 29208

AUB, JOSEPH CHARLES, medicine, deceased

AUBEL, JOSEPH LEE, b Lansing, Mich, Sept 7, 36; m 67; c 2. PHYSICS. Educ: Mich State Univ, BS, 54, PhD(physics), 64. Prof Exp: Asst prof, 64-74, ASSOC PROF PHYSICS, UNIV S FLA, 74- Mem: Optical Soc Am; Am Asn Physics Teachers; Am Sci Affil. Res: Infrared spectroscopy; instrumentation and molecular structure; plasma spectroscopy; computer-based and individualized competency-based education. Mailing Add: Dept of Physics Univ of SFla Tampa FL 33620

AUBERTIN, GERALD MARTIN, b Kankakee, Ill, Sept 23, 31; m 57; c 4. WATERSHED MANAGEMENT. Educ: Univ Ill, Urbana-Champaign, BS, 58, MS, 60; Pa State Univ, PhD(agron), 65. Prof Exp: Asst agron, Univ Ill, 58; res asst, Pa State Univ, 64; res soil physicist & fel, Univ Calif, Riverside, 64-65; RES SOIL SCIENTIST, US FOREST SERV, 65- Concurrent Pos: Vis prof dept earth resources, Colo State Univ, 75. Mem: Am Soc Agron; Soil Sci Soc Am; Sigma Xi. Res: Water quality in forested watersheds; subsurface water movement and soil macropores in forested watersheds; oxygen-salt-moisture and root relations; physical edaphology; soil structure problems; geologic/soil/water quality investigations. Mailing Add: Timber & Watershed Lab Northeastern Forest Exp Sta Parsons WV 26287

AUBORN, JAMES JOHN, b Portland, Ore, Feb 21, 40; m 62; c 2. PHYSICAL CHEMISTRY. Educ: Ore State Univ, BS, 61, MS, 62; Univ Utah, PhD(chem), 71. Prof Exp: Mem tech staff electrochem, Bayside, NY, 71-72 & Waltham, Mass, 72-73, mem tech staff mat sci, 74-75, TECH PROJ MGR ELECTROCHEM, GTE LAB, INC, GEN TEL & ELECTRONICS CORP, 75- Concurrent Pos: Instr, Northeastern Univ, 74- Mem: Electrochem Soc; Am Chem Soc; AAAS; Am Inst Physics; Sigma Xi. Res: Chemistry and physics of inorganic non-aqueous liquids relating to electrochemical power sources; dielectric and breakdown properties; photochemistry and non-ohmic conduction; electrical and optical properties of refractory materials. Mailing Add: GTE Labs Inc 40 Sylvan Rd Waltham MA 02154

AUBRECHT, GORDON JAMES, II, b Bedford, Ohio, May 2, 43; m 65; c 2. HIGH ENERGY PHSYICS. Educ: Rutgers Univ, BA, 65; Princeton Univ, PhD(theoret physics), 70. Prof Exp: Res assoc high energy theory, Ohio State Univ, 70-72; vis asst prof, Inst Theoret Sci, Univ Ore, 72-75; ASST PROF HIGH ENERGY, OHIO STATE UNIV, 75- Mem: Am Phys Soc; Am Asn Physics Teachers; Sigma Xi; AAAS. Res: Theoretical structure of the weak and electromagnetic interaction, especially as applied to neutrino scattering, SV 4 symmetry and its effect on decay rates and strong interaction cross sections. Mailing Add: Dept of Physics Ohio State Univ 1465 Mt Vernon Ave Marion OH 43302

AUBREY, NORMAN EDWARD, organic chemistry, see 12th edition

AUCHAMPAUGH, GEORGE FREDRICK, b Chicago, Ill, Jan 19, 39. EXPERIMENTAL NUCLEAR PHYSICS. Educ: Univ Ill, BS, 61; Univ Calif, MS, 66, PhD(appl sci), 68. Prof Exp: Staff mem neutron physics, Lawrence Livermore Lab, 61-68; STAFF MEM NEUTRON PHYSICS, LOS ALAMOS SCI LAB, 68- Mem: Am Phys Soc. Res: Neutron physics research with monoenergetic and continuous neutron sources; fission barrier physics; from study of the resonance region of sub threshold fission isotopes; nuclear spectroscopy of light elements from R-matrix analysis of neutron cross section data. Mailing Add: Los Alamos Sci Lab P3 Los Alamos NM 87545

AUCHINCLOSS, JOSEPH HOWLAND, JR, b Lawrence, NY, June 28, 21; m 46; c 4. INTERNAL MEDICINE. Educ: Yale Univ, BS, 43; Columbia Univ, MD, 45. Prof Exp: From instr to assoc prof, 51-71, PROF MED, COL MED, STATE UNIV NY UPSTATE MED CTR, 71- Mem: Am Fedn Clin Res; Am Thoracic Soc; Am Physiol Soc; Am Col Physicians. Res: Cardiopulmonary physiology. Mailing Add: Dept of Med State Univ Hosp Syracuse NY 13210

AUCHMUTY, JAMES FRANCIS GILES, b Dublin, Eire, June 1, 45. APPLIED MATHEMATICS. Educ: Australian Nat Univ, BSc, 65; Univ Chicago, MS, 68, PhD(math), 70. Prof Exp: Tutor appl math, Australian Nat Univ, 66; lectr math, State Univ NY Stony Brook, 70-72; vis asst prof, 72-73, ASST PROF MATH, IND UNIV, 73- Concurrent Pos: NSF res grants, 71, 72, 74 & 75; vis prof math, Int Inst Physics & Chem, Belg, 74; fel, Fluid Mech Res Inst, Univ Essex, 75-76. Res: Nonlinear functional and numerical analysis and applications to astronomy and biology. Mailing Add: Dept of Math Univ of Ind Bloomington IN 47401

AUCHTER, HARRY A, b St Louis, Mo, Aug 31, 20; m 54; c 2. PHYSICS. Educ: Southeast Mo State Col, AB, 42; Univ Iowa, MS, 46. Prof Exp: Instr physics, Kearney State Col, 46-47, Bangkok Christian Col, Thailand, 47-50 & Augustana Col (SDak), 51-53; from instr to asst prof physics & math, 53-58, ASSOC PROF PHYSICS, CARROLL COL (WIS), 58- Mem: Am Asn Physics Teachers. Res: Adiabatic compressibilities of aqueous solutions using ultrasonics. Mailing Add: Dept of Physics Carroll Col Waukesha WI 53186

AUCLAIR, JACQUES LUCIEN, b Montreal, Que, Apr 2, 23; m 51; c 2. ENTOMOLOGY. Educ: Univ Montreal, BSc, 42; McGill Univ, MSc, 45; Cornell Univ, PhD(insect physiol), 49. Prof Exp: Asst prof physiol, Univ Montreal, 49-53; res entomologist, Res Lab, Can Dept Agr, 53-64; prof, Dept Bot & Entom, NMex State Univ, 64-67; dir, 67-73, PROF, DEPT BIOL SCI, UNIV MONTREAL, 67- Mem: AAAS; Entom Soc Am; Entom Soc Can. Res: General entomology; insect physiology and biochemistry; factors of plant resistance to insects. Mailing Add: Dept of Biol Sci Univ of Montreal Montreal PQ Can

AUCLAIR, WALTER, b Manchester, NH, Sept 13, 33; m 54; c 3. CELL BIOLOGY, DEVELOPMENTAL BIOLOGY. Educ: Univ Conn, BA, 55; NY Univ, MS, 58, PhD(biol), 60. Prof Exp: NIH res fel, George Washington Univ, 60-61; from asst prof to assoc prof zool, Univ Cincinnati, 61-66; ASSOC PROF BIOL, RENSSELAER POLYTECH INST, 66- Concurrent Pos: Investr, Zool Sta, Naples, Italy, 65-66; NSF sci consult, India, 68-69; trustee & secy, Bermuda Biol Sta Res, Inc. Mem: Am Inst Biol Sci; Am Soc Cell Biol; Soc Develop Biol; Am Soc Animal Sci; Am Soc Zoologists. Res: Differentiation of cilla and other organelles in marine invertebrate embryos; renal hemopoiesis in amphibians; bovine reproductive physiology. Mailing Add: Dept of Biol Rensselaer Polytech Inst Troy NY 12181

AUCOIN, ANTHONY ANDREW, mathematics, see 12th edition

AUE, DONALD HENRY, b Columbus, Ohio, June 19, 42; m 64; c 2. ORGANIC CHEMISTRY. Educ: Ohio State Univ, BSc, 63; Cornell Univ, PhD(org chem), 67. Prof Exp: NSF fel org chem, Columbia Univ, 67-68; ASSOC PROF CHEM, UNIV CALIF, SANTA BARBARA, 68- Mem: Am Chem Soc; Brit Chem Soc. Res: Synthesis and properties of strained carbocyclic and heterocyclic systems; additions to strained multiple bonds; carbonium ion reactions in solution and gas phase; organic photochemistry; organic synthesis. Mailing Add: Dept of Chemistry Univ of Calif Santa Barbara CA 93106

AUE, WALTER ALOIS, b Vienna, Austria, Jan 20, 35. ORGANIC CHEMISTRY, ANALYTICAL CHEMISTRY. Educ: Univ Vienna, PhD(org chem), 63. Prof Exp: Res investr org chem, Western Reserve Univ, 63-65; asst prof anal chem, Univ Mo-Columbia, 65-69, assoc prof chem & res assoc Space Sci Ctr, 69-73; PROF CHEM, DALHOUSIE UNIV, 73- Concurrent Pos: USDA grant with Dr Billy G Tweedy, 67-

70; USPHS grant, 67-70; NSF grant, 70-72; US Environ Protection Agency grant, 72-75; Nat Res Coun Can grant, 73-; Can Defense Res Bd grant, 74-76; Agr Can grant, 74- Res: Pesticide residue analysis; specific detectors for gas chromatography; support-bonded chromatographic phases. Mailing Add: Dept of Chem Dalhousie Univ Halifax NS Can

AUER, EDWARD EVERETT, b East Orange, NJ, Mar 20, 11; m 64. POLYMER CHEMISTRY. Educ: Columbia Univ, AB, 33, ChE, 35. Prof Exp: Anal res chemist, Calco Chem Co, 35-39; chemist, Res Ctr, US Rubber Co, 39-41 & Ord Plant, Iowa, 41-43, res & develop chemist, NJ, 45-63; info specialist, Res & Develop Div, Mobil Chem Co, Metuchen, 63-65, supvr tech info, Cent Res Lab, 65-71; RETIRED. Concurrent Pos: With Div War Res, Columbia Univ, 43-45. Mem: AAAS; Am Chem Soc. Res: Evaluation of synthetic rubbers and plastics; information science; polymerization kinetics; analytical chemistry; electrochemical preparations. Mailing Add: 36 Riggs Pl South Orange NJ 07079

AUER, JAN WILLEM, b Utrecht, Neth, Apr 10, 42; Can citizen; c 2. MATHEMATICS. Educ: McGill Univ, BEng, 64, MSc, 66; Univ Toronto, PhD(math), 71. Prof Exp: Lectr, 69-70, ASST PROF MATH, BROCK UNIV, 70- Mem: Am Math Soc; Math Asn Am; Tensor Soc. Res: Algebraic topology and differential geometry, specifically in smooth bundles. Mailing Add: Dept of Math Brock Univ St Catharines ON Can

AUER, LAWRENCE H, b Englewood, NJ, Dec 26, 41. ASTROPHYSICS. Educ: Princeton Univ, PhD(astron), 67. Prof Exp: Res assoc & fel astron, Joint Inst Lab Astrophys, Univ Colo, 66-68; asst prof, 68-71, ASSOC PROF ASTRON, YALE UNIV, 71-, DIR GRAD STUDIES, WATSON ASTRON CTR, 74- Mem: Am Astron Soc; Asn Comput Mach; Royal Astron Soc. Res: Stellar atmospheres; radiative transfer; numerical analysis. Mailing Add: Dept of Astron Yale Univ Watson Astron Ctr 60 Sachem St New Haven CT 06053

AUER, PETER LOUIS, b Budapest, Hungary, Jan 12, 28; nat US; m 52; c 4. PLASMA PHYSICS, THEORETICAL PHYSICS. Educ: Cornell Univ, BA, 47; Calif Inst Technol, PhD(chem & physics), 51. Prof Exp: Fel & res asst, Calif Inst Technol, 48-50; res chemist & physicist, Calif Res & Develop Co, 50-54; physicist, Gen Elec Res Lab, 54-61; head plasma physics dept, Sperry Rand Res Ctr, 62-64; dep dir ballistic missile defense, Advan Res Proj Agency, 64-67; PROF, GRAD SCH AEROSPACE ENG, CORNELL UNIV, 67-, DIR LAB PLASMA STUDIES, 69- Concurrent Pos: Guggenheim fel, 60-61. Mem: Fel Am Phys Soc. Res: Theoretical physics and chemistry; gaseous electronics. Mailing Add: Lab of Plasma Studies Cornell Univ Ithaca NY 14850

AUERBACH, ANDREW BERNARD, b New York, NY, Jan 23, 48; m 72. POLYMER CHEMISTRY. Educ: Brooklyn Col, BS, 68; City Univ New York, PhD(org chem), 75. Prof Exp: RES CHEMIST, ITT RAYONIER INC, 74- Mem: Am Chem Soc. Res: Improving the end use performance characteristics of cellulose acetate fibers, films and plastics and elucidating the mechanism of acetylation. Mailing Add: 30 Merle Pl Staten Island NY 10305

AUERBACH, ARTHUR HENRY, b Philadelphia, Pa, Mar 12, 28. PSYCHIATRY Educ: Jefferson Med Col, MD, 51. Prof Exp: Resident psychiat, St Elizabeth's Hosp, Washington, DC, 52-53; resident, West Haven Vet Admin Hosp, Conn, 53-54; resident, Hosp, Temple Univ, 56-57; NIMH career develop award, Dept Psychiat, 64-69, ASST PROF PSYCHIAT, SCH MED, UNIV PA, 64- Concurrent Pos: Consult, Magee Mem Hosp, Philadelphia, 58-68. Mem: AAAS; Am Psychiat Asn; Am Acad Psychother; Soc Psychother Res. Res: Psychotherapy research. Mailing Add: Dept of Psychiat Sch of Med 239 Med Lab Bldg Univ of Pa Philadelphia PA 19174

AUERBACH, CLEMENS, b Berlin, Ger, Nov 30, 23; US citizen; m 64; c 1. CHEMISTRY. Educ: Robert Col, Istanbul, BS, 43; Harvard Univ, AM, 48, PhD(chem), 51. Prof Exp: Res fel, Univ Minn, 50-53; asst prof chem, Univ Buffalo, 53-56; from asst chemist to assoc chemist, 56-62, CHEMIST, DEPT APPL SCI, BROOKHAVEN NAT LAB, 62- Concurrent Pos: Ed, Chem Instrumentation. Mem: AAAS; Am Chem Soc. Res: Analytical chemistry; analysis of uranium and other reactor materials; nuclear materials safeguards. Mailing Add: Dept of Appl Sci Brookhaven Nat Lab Upton NY 11973

AUERBACH, EARL, b Lodz, Poland, Feb 22; nat US; m 50; c 3. PARASITOLOGY, FOOD SCIENCE. Educ: Univ Ill, BS, 46, MS, 47; Northwestern Univ, PhD(biol), 51. Prof Exp: Histochemist, Am Meat Inst Found, 51-54, asst to dir, 54-59, chief div histol, 59-64; oper mgr, Polo Food Prod Co, 64-65, dir res & develop, 65-70; dir res & develop, Pronto Food Corp, 70-73; DIR RES & DEVELOP, DAVID BERG & CO, 73- Mem: Am Soc Parasitol; Soc Protozool; Inst Food Technol. Res: Parasitic protozoa; cytochemistry and histochemistry; freeze dehydration. Mailing Add: David Berg & Co 165 S Watermarket Chicago IL 60608

AUERBACH, ELLIOT H, b New York, NY, July 21, 32; m 71. NUCLEAR PHYSICS. Educ: Columbia Univ, AB, 53, AM, 57. Prof Exp: Mathematician & physicist, Knolls Atomic Power Lab, Gen Elec Co, 58-61; assoc physics, Brookhaven Nat Lab, 61-64; vis res scientist, Lab Nuclear Sci, Mass Inst Technol, 64-66; STAFF MEM, PHYSICS DEPT, BROOKHAVEN NAT LAB, 66- Concurrent Pos: Vis staff mem, Los Alamos Sci Lab, 65 & 66. Mem: Am Phys Soc; AAAS. Res: Nuclear structure and scattering with emphasis on heavy-ion induced reactions; applications of large scale computers to analysis of nuclear models. Mailing Add: Dept of Physics Brookhaven Nat Lab Upton NY 11973

AUERBACH, HARRY, b New York, NY, Mar 28, 14; m 41; c 2. BIOSTATISTICS. Educ: Brooklyn Col, BS, 34; St John's Univ, NY, LLB, 37, JSD, 38; Yale Univ, MPH, 49. Prof Exp: Mem lab staff dept hosps, New York, 35-42; mem lab staff, Greenville Hosp, Pa, 46-48; asst prof biostatist sch med, Yale Univ, 49-52; statistician, Comn on Financing Hosp Care, 52-54; assoc statistician, Argonne Nat Lab, 54-70; asst dir res & eval, 70-73, assoc exec dir, 73-74, EXEC DIR, ILL REGIONAL MED PROG, 74- Mem: Am Statist Asn; fel Am Pub Health Asn; Radiation Res Soc; fel Royal Soc Health; Sigma Xi. Res: Geographic pathology; epidemiology of cancer; aging; health care delivery systems, evaluation. Mailing Add: Ill Regional Med Prog 122 S Michigan Ave Chicago IL 60603

AUERBACH, IRVING, b Cleveland, Ohio, May 24, 19; m 69. PHYSICAL CHEMISTRY, ORGANIC CHEMISTRY. Educ: Ohio State Univ, BSc, 42, PhD(chem), 48. Prof Exp: Asst org chem, Res Found, Ohio State Univ, 44-46; res assoc, Cleveland Indust Res, 48-50; res assoc, Goodyear Tire & Rubber Co, 50-57; STAFF MEM, SANDIA CORP, 57- Concurrent Pos: Res fel, Case Inst Technol, 48. Mem: Fel AAAS; Am Chem Soc; fel Am Inst Chem. Res: Physical properties of polymers; radiation chemistry and effects of radiation on polymers; graphite and high temperature chemistry. Mailing Add: Sandia Labs Kirtland AFB Albuquerque NM 87115

AUERBACH, LEONARD B, b New York, NY, Aug 11, 29; m 51; c 2. PHYSICS. Educ: City Col New York, BS, 51; Univ Ill, MS, 52; Univ Calif, Berkeley, PhD(physics), 62. Prof Exp: Res assoc particle physics, Segre Group, Lawrence Radiation Lab, 62-63; res assoc, Univ Pa, 63-65, asst prof, 65-67; ASSOC PROF, TEMPLE UNIV, 67- Mem: Am Phys Soc. Res: Elementary and experimental high energy particle physics. Mailing Add: Dept of Physics Temple Univ Philadelphia PA 19122

AUERBACH, LEWIS EDWARD, b New York, NY, June 2, 41; Can citizen; m 65; c 2. SCIENCE POLICY, SCIENCE WRITING. Educ: Harvard Univ, AB, 63, AM, 64. Prof Exp: Res assoc hist physics, Am Inst Physics, 66-67; producer sci broadcasting, Can Broadcasting Corp & Ont Educ TV, 67-74; SCI ADV, SCI POLICY, SCI COUN CAN, 74- Concurrent Pos: Proj mgr study northern develop, Sci Coun Can, 74- Mem: Can Gerontol Soc; Can Futures Studies Asn; Can Sci Writers Asn; Can Asn Hist Sci. Res: Technology assessment and northern development; implications of changing age structure of population; critical analysis of future studies and science films and broadcasting. Mailing Add: Sci Coun of Can 150 Kent St Ottawa ON Can

AUERBACH, MELVIN, b Perth Amboy, NJ, July 29, 35; m 58; c 2. ORGANIC CHEMISTRY, POLYMER CHEMISTRY. Educ: Monmouth Col, NJ, BS, 60; Purdue Univ, MS, 67, PhD(org chem), 70. Prof Exp: Technician, Catalin Corp Am, 56-60, chemist, 60-61; chemist, Carter-Wallace Inc, 61-64; sr res chemist, Goodyear Tire & Rubber Co, 69-73; TECH DIR, SWISS LAB INC, AKRON, OHIO, 73- Mem: Am Chem Soc; Soc Plastics Engrs. Res: Synthesis and development of antioxidants; synthesis of nitrogen compounds for pharmaceutical evaluation; mass spectroscopy of nitrogen heterocycles; alkyl nitrate nitration of sulfonate esters; research and development of new synthetic elastomers; reinforced plastics; polyester application; thermosetting polyesters. Mailing Add: 553 Garnette Rd Akron OH 44313

AUERBACH, MICHAEL HOWARD, b Akron, Ohio, June 24, 43; m 66; c 3. ANALYTICAL CHEMISTRY. Educ: Mass Inst Technol, BS, 64; Cornell Univ, PhD(anal chem), 69. Prof Exp: Technician, Res Div, Goodyear Tire & Rubber Co, Ohio, summer 65; sr res chemist, Hydrocarbons & Polymers Div, Monsanto Co, 69-74; RES SPECIALIST, MONSANTO POLYMERS & PETROCHEM CO, 74- Concurrent Pos: Ed, Valchemist, 74- Mem: Am Chem Soc; Coblentz Soc. Res: Infrared spectrometry of polymer systems; laboratory automation; reactions of aromatic hydrocarbons in a radiofrequency plasma discharge; determination of copolymer composition distributions by thin layer chromatography. Mailing Add: Monsanto Polymers & Petrochem Co 730 Worcester St Indian Orchard MA 01151

AUERBACH, MORRIS BALINE, b New York, NY, June 8, 08; m 72. ENDODONTICS. Educ: NY Univ, DDS, 32; Am Bd Endodontics, dipl, 67. Prof Exp: From instr to prof, 35-67, HOLZMAN PROF ENDODONTICS, COL DENT, NY UNIV, 67-, CHMN DEPT, 56- Concurrent Pos: Consult, US Vet Admin, 51-56, Misericordia Hosp & NY Mem Inst Clin Oral Path; dir, Am Bd Endodontics, 68-71, pres, 70-71. Honors & Awards: Achievement Award, NY Univ, 73. Mem: Am Acad Oral Med; fel Am Col Dent; Am Asn Endodontics; Sigma Xi; Am Dent Asn. Res: Methods of filling root canals; bacteriological control; antibiotics in endodontics; diagnosis of pulp abnormalities. Mailing Add: 9 W 64th St New York NY 10023

AUERBACH, OSCAR, b New York, NY, Jan 1, 05; m 32; c 2. MEDICINE. Educ: New York Med Col, MD, 29; Am Bd Path, dipl, 44. Prof Exp: Pathologist, Sea View Hosp, NY, 32-47; chief lab serv, Halloran Vet Admin Hosp, 47-51; chief lab serv, 52-59, SR MED INVESTR, VET ADMIN HOSP, EAST ORANGE, NJ, 60- Concurrent Pos: Res fel, Univ Vienna, 32; consult, Richmond Mem Hosp, 38-47, US Vet Admin Hosp, 46-47 & US Naval Hosp, 47-49; vis instr, Sch Med, Wash Univ, 44; assoc prof, New York Med Col, Flower & Fifth Ave Hosps, 49-61, prof, 61-71; prof path, Col Med & Dent, NJ, 71- Mem: Am Thoracic Soc; Am Asn Thoracic Surg; fel Am Col Physicians; fel NY Acad Med; hon mem Mex Tuberc Soc. Res: Relationship of smoking to lung cancer and other diseases both human and experimental. Mailing Add: Vet Admin Hosp East Orange NJ 07019

AUERBACH, ROBERT, bBerlin, Ger, Apr 12, 29; nat; m 50; c 2. DEVELOPMENTAL BIOLOGY, IMMUNOLOGY. Educ: Berea Col, AB, 49; Columbia Univ, AM, 50, PhD(zool), 54. Prof Exp: Assoc, Biol Div, Oak Ridge Nat Lab, 54-55; fel, Nat Cancer Inst, 55-57; from asst prof to assoc prof zool, 57-65, PROF ZOOL, UNIV WIS-MADISON, 65- Concurrent Pos: Guggenheim fel, 68; vis prof surg, Harvard Med Sch, 73-74; vis prof cell biol, Southwestern Med Sch, Dallas, 74-75; Rockefeller fel, 74-75. Mem: Am Soc Zool; Soc Develop Biol; Soc Exp Biol & Med; Transplantation Soc; Am Asn Immunol. Mailing Add: Dept of Zool Univ of Wis Madison WI 53706

AUERBACH, STANLEY IRVING, b Chicago, Ill, May 21, 21; m 54; c 4. RADIATION ECOLOGY. Educ: Univ Ill, BS, 46, MS, 47; Northwestern Univ, PhD(zool), 49. Prof Exp: Asst zool & animal ecol, Northwestern Univ, 47-48; lectr biol, Roosevelt Univ, 50-51, instr, 51-54, asst prof, 54-70; assoc scientist, 54-55, health physicist ecol, 54-59, scientist, 55-59, sect chief radiation ecol, 59-70, dir ecol sci div, 70-72, DIR ENVIRON SCI DIV, OAK RIDGE NAT LAB, 72- Concurrent Pos: Lectr, Univ Tenn, 60; adj res prof, Univ Ga, 64-; dir eastern deciduous forest biome, Int Biol Prog, 68-, vpres US exec comt, 71-; mem bd energy studies & various subcomt, Nat Acad Sci-Nat Res Coun, 74- Mem: Am Inst Biol Sci; Am Soc Zoologists; fel AAAS; Econ Soc Am (secy, 64-69, pres-elect, 70-71, pres, 71-72); Brit Ecol Soc. Res: Ecosystem analysis; radioactive waste cycling in terrestrial ecosystems. Mailing Add: Environ Sci Div Bldg 2001 Oak Ridge Nat Lab PO Box X Oak Ridge TN 37830

AUERBACH, STEWART HART, b Montgomery, Ala, Jan 12, 11; m 38; c 2. PATHOLOGY. Educ: Univ Ala, AB, 31; Univ Ga, MD, 34; Am Bd Path, dipl, 39. Prof Exp: From asst instr to instr path, Vanderbilt Univ, 35-38; pathologist, Nan Travis Mem Hosp, 38-40; from asst to assoc prof path, Univ Ga, 40-46; assoc prof clin path, Vanderbilt Univ, 46; chief lab serv, Thayer Vet Admin Hosp, 46-56; PATHOLOGIST, BARONESS ERLANGER HOSP, 56- Mem: AAAS; Am Soc Clin Path; Am Col Path; Am Soc Exp Path; AMA. Res: Clinical pathology. Mailing Add: Dept of Path Baroness Erlanger Hosp Chattanooga TN 37403

AUERBACH, VICTOR, b Philadelphia, Pa, July 4, 17; m 46; c 3. POLYMER CHEMISTRY. Educ: Brooklyn Col, BA, 40; Polytech Inst Brooklyn, PhD(chem), 45. Prof Exp: Chemist urea resins, Bakelite Co, 42-43; res assoc thin films, Polytech Inst Brooklyn, 43-45; group leader reactive resins, Bakelite Co, 45-54; assoc coordr patents, Union Carbide Plastics Co, 54-62, coordr patents plastics div, 62-67, PATENT MGR CHEM & PLASTICS, UNION CARBIDE CORP, 67- Res: Thermosetting resins; thermoplastic resins. Mailing Add: 118 Tappan Ave North Plainfield NJ 07060

AUERBACH, VICTOR HUGO, b New York, NY, Oct 2, 28; m 56. BIOCHEMISTRY. Educ: Columbia Col, AB, 51; Harvard Univ, AM, 55, PhD(biochem), 57. Prof Exp: Asst, Col Physicians & Surgeons, Columbia Univ, 45-47; biochemist, Rheumatic Fever Res Inst, 48; asst, Columbia Univ, 48-52; res biochemist, Fleishmann Labs, 52; res assoc pediat & instr physiol chem, Univ Wis, 57-58; asst prof physiol chem & res pediat, 58-64, assoc prof biochem & res pediat, 64-68, RES PROF PEDIAT, SCH MED, TEMPLE UNIV, 68-; DIR ENZYME LAB & DIR RES CHEM, ST CHRISTOPHER'S HOSP FOR CHILDREN, 58- Mem: Am Asn Clin Chem; Am Asn Cancer Res; Am Soc Biol Chem; Am Soc Human Genetics; Soc Pediat Res; Am Inst Nutrit. Res: Mammalian amino acid metabolism; control mechanisms such as adaptive enzyme formation and negative feedback regulation of enzyme synthesis; inborn errors of metabolism and genetic disorder in man. Mailing Add: St Christopher's Hosp Children 2600 N Lawrence St Philadelphia PA 19133

AUERBACK, ALFRED, b Toronto, Ont, Sept 20, 15; nat US; m 42; c 3. PSYCHIATRY. Educ: Univ Toronto, MD, 38. Prof Exp: Intern, Toronto Gen Hosp, 38-39; resident med, French Hosp, Calif, 39-40; resident neurol & psychiat, Univ Calif Hosp, 40-42; psychiatrist, Sheppard Pratt Hosp, Md, 42-43; clin asst, Med Sch, 43-46, clin instr, 46-53, asst clin prof psychiat, 53-62, assoc clin prof, 62-70, CLIN PROF PSYCHIAT, UNIV CALIF, SAN FRANCISCO, 70- Concurrent Pos: Consult, Vet Admin, 46-52; chmn, Ment Health Adv Bd, San Francisco, 66-70; pvt practr psychiat; mem, Adv Comt, Div Rehab, Calif Dept Pub Health. Honors & Awards: Royer Award, Univ Calif Bd Regents, 66. Mem: Fel Am Psychiat Asn (vpres, 66-67); fel Am Col Physicians; hon fel Royal Col Psychiatrists; hon fel Australian & NZ Col Psychiatrists. Res: Psychosomatic relationships; community and social aspects of psychiatry; alcoholic rehabilitation. Mailing Add: 450 Sutter St San Francisco CA 94108

AUERSPERG, NELLY, b Vienna, Austria, Dec 13, 28; Can citizen; m 55; c 2. CANCER, CELL BIOLOGY. Educ: Univ Wash, MD, 55; Univ BC, PhD(cell biol), 68. Prof Exp: Intern med, Vancouver Gen Hosp, BC, 55-56, res asst tissue cult, 59-60; cytologist, Cytol Lab, Vancouver, 60-65; assoc prof, 68-72, PROF ZOOL, UNIV BC, 72-, MEM STAFF, CANCER RES CTR, 68- Mem: Am Asn Cancer Res; Am Soc Cell Biol; Tissue Cult Asn; Can Soc Cell Biol. Res: Regulation of phenotypic expression in malignant tumors. Mailing Add: Cancer Res Ctr Univ of BC Vancouver BC Can

AUFDEMBERGE, THEODORE PAUL, b Winnipeg, Man, Feb 1, 34; US citizen; m 60; c 4. PHYSICAL GEOGRAPHY. Educ: Concordia Teachers Col, BSc, 56; Wayne State Univ, MA, 64; Univ Mich, Ann Arbor, PhD(geog), 71. Prof Exp: Prin & teacher, Immanual Lutheran Sch, Brownton, Minn, 56-58; teacher sci & english, Trinity Lutheran Sch, Mt Clemens, Mich, 58-64; instr geog, 64-66, asst prof, 66-70, ASSOC PROF GEOG, CONCORDIA LUTHERAN JR COL, ANN ARBOR, 70- Mem: Am Asn Geogrs; Nat Coun Geog Educrs. Res: Spatial distributions of economic activities particularly agricultural activities in Anglo-America; physical geography of Middle-America and glacial surging applied to Pleistocene chronology. Mailing Add: Concordia Lutheran Jr Col 4090 Geddes Rd Ann Arbor MI 48105

AUFDERHEIDE, KARL JOHN, b Minneapolis, Minn, Aug 17, 48; m 69. CELL BIOLOGY, DEVELOPMENTAL BIOLOGY. Educ: Univ Minn, BS, 70, MS, 72, PhD(cell biol), 74. Prof Exp: RES ASSOC CELL BIOL, IND UNIV, BLOOMINGTON, 74- Concurrent Pos: NIH fel, 75- Mem: Sigma Xi; Soc Protozoologists. Res: Intracellular motility to differentiation and morphogenesis; motility and cortical placement of trichocysts in Paramecium; relation of this to other cellular processes. Mailing Add: Dept of Zool Jordan Hall Ind Univ Bloomington IN 47401

AUFDERMARSH, CARL ALBERT, JR, b Cincinnati, Ohio, Nov 9, 32; m 55; c 4. ORGANIC CHEMISTRY. Educ: Univ Cincinnati, BS, 54; Yale Univ, MS, 55, PhD(chem), 58. RES CHEMIST, E I DU PONT DE NEMOURS & CO, 58- Res: Polymer chemistry; organic synthesis; isocyanates and polyurethanes; elastomers; free radical reactions. Mailing Add: Exp Sta E I du Pont de Nemours & Co Wilmington DE 19899

AUFENKAMP, DON, b Brock, Nebr, Mar 31, 27; m 55; c 2. MATHEMATICS, COMPUTER SCIENCE. Educ: Southwestern Univ Memphis, BA, 49; Univ Paris, Dr(math physics), 52. Prof Exp: Instr math, Princeton Univ, 52-53; instr, Reed Col, 53-56; vis asst prof elec eng, Univ Ill, 56-57; res scientist missiles & space div, Lockheed Aircraft Corp, 57-60; sr mathematician, Gen Elec Co, 60-62; mem staff systs planning, Philco Corp, 62-65; prof math & elec & electronic eng & dir comput ctr, Ore State Univ, 65-68; head comput appln res sect, 68-75, staff assoc, 68-71, SR STAFF ASSOC US-USSR ACTIV, OFF COMPUT ACTIV, NSF, 75- Concurrent Pos: Mem tech staff, Bell Tel Labs, 54 & 55; appl mathematician, Syst LabsCorp, 56-57. Mem: Math Asn Am; Asn Comput Mach. Res: Applied mathematics; digital data processing; systems analysis; computer applications in research. Mailing Add: Div of Math & Comput Sci NSF Washington DC 20550

AUFFENBERG, WALTER, b Dearborn, Mich, Feb 6, 28; m 49; c 4. HERPETOLOGY. Educ: Stetson Univ, BS, 51; Univ Fla, MS, 53, PhD(biol), 56. Prof Exp: Cur vert paleont, Charleston Mus, 53; asst prof biol, Univ Fla & assoc cur vert paleont, Fla State Mus, 56-59; assoc dir, Biol Sci Curric Study, Univ Colo, 59-63; chmn dept natural sci, 63-73, CUR HERPET, FLA STATE MUS, 73- Concurrent Pos: Res assoc, Univ Colo Mus, 59-63 & NY Zool Soc, 69-; consult, AEC, India, 62 & -64 & NSF, India, 67. Honors & Awards: 1st Prize Honorarium, Am Soc Ichthyol & Herpet, 53. Mem: Fel AAAS; Soc Vert Paleont; Am Soc Ichthyol & Herpet; Am Soc Naturalists; Am Soc Zool. Res: Behavior; field biology of large reptiles; evolutionary history of reptiles of the United States and Caribbean area; biosystematics of fossil and living land tortoises of the world. Mailing Add: Fla State Mus Univ of Fla Gainesville FL 32611

AUFRANC, WILL H, b Mo, Jan 1, 06; m 38; c 2. MEDICINE. Educ: Univ Mo, AB, 30, BS, 31; Med Col Va, MD, 33; Vanderbilt Univ, cert pub health, 37; Johns Hopkins Univ, MPH, 41; Am Bd Prev Med, dipl. Prof Exp: Med officer, US Army Reserve, 33-37; dist health officer, Mo State Bd Health, 37-39, state epidemiologist & asst dir local health, 39-42; venereal dis control officer, USPHS, Columbus, Ga, 42 & Portland, Ore, 43, venereal dis control officer, Ore State Bd Health, 43-45, dir local health serv, 45-46, consult gen health & venereal dis control, San Francisco, Calif, 46-48, asst chief div venereal dis control, Washington, DC, 48-50, dir health resources, Exec Off of Pres, 50-53, asst prof officer, 53-54, regional med dir, Ga, 55-63; DIR WESTERN REGIONAL OFF, AM PUB HEALTH ASN, INC, 63- Concurrent Pos: Lectr, Sch Med, Wash Univ, 39-42, Univ Ore, 43-46 & Univ Calif, Berkeley, 63- Mem: AMA; Am Pub Health Asn; Am Col Prev Med. Res: Public health administration. Mailing Add: Am Pub Health Asn Inc 655 Sutter St San Francisco CA 94102

AUGELLI, JOHN PAT, US citizen. CULTURAL GEOGRAPHY, GEOGRAPHY OF LATIN AMERICA. Educ: Clark Univ, BA, 43; Harvard Univ, MA, 49, PhD(geog), 51. Prof Exp: Assoc prof geog, Univ PR, 48-52; prof, Univ Md, 52-61; prof geog & dir, Ctr Latin Am Studies, Univ Kans, 61-70; prof geog & dir Latin Am Studies Ctr, Univ Ill, 70-71; PROF GEOG & DEAN INT PROGS, UNIV KANS, 71- Concurrent Pos: Mem, Nat Acad Sci-Nat Res Coun Adv Comt to US State Dept, Pan-Am Inst Geog & Hist, 61-69 & Comn Geog, 65-69; mem, Soc Sci Res Coun-Am Coun Learned Socs Joint Comt Latin Am Studies, 64-70, Screening Comt, Student Fulbright-Hays Prog to Latin Am, 65-67, Bd For Scholars, 67-70 & Govt Comt Partners Alliance Prog, Kans, 68-69 & 70-72; Orgn Am States fel, Dominican Repub; Univ Md grant, Brazil; Univ Kans grant, Guatemala; Am Coun Learned Socs grant, Sweden, India & UK. Mem: AAAS; Asn Am Geog (secy, 66-69); Am Geog Soc; Latin Am Studies Asn (vpres, 68, pres, 69); Int Studies Asn. Res: Cultural, political and regional geography of Latin America and the Caribbean. Mailing Add: Dept of Geog Univ of Kans Lawrence KS 66044

AUGENFELD, JOHN MARTIN, zoology, physiology, see 12th edition

AUGENLICHT, LEONARD HAROLD, b New York, NY, Aug 16, 46; m 70. CELL BIOLOGY, BIOCHEMISTRY. Educ: State Univ NY Binghamton, BA, 67; Syracuse Univ, PhD(biol), 71. Prof Exp: Trainee path, Med Sch, Temple Univ, 71-74; ASSOC, MEM SLOAN-KETTERING CANCER CTR, 74- Res: Chromatin structure and function; control of transcription and of cell proliferation. Mailing Add: Mem Sloan-Kettering Cancer Ctr 1275 York Ave New York NY 10021

AUGL, JOSEPH MICHAEL, b Pasching, Austria, Jan 8, 32; m 62; c 1. ORGANIC CHEMISTRY. Educ: Univ Vienna, PhD(org chem), 59. Prof Exp: Asst prof org chem, Univ Vienna, 59-61; sr res organometallic chemist, Sohio Res Ctr, Ohio, 61-63, proj leader, 63; org chem, Melpar Inc, 63-64, res leader, 64-68; RES CHEMIST, NAVAL ORD LABS, 68- Honors & Awards: Meritorious Civilian Serv Award, US Navy, 70. Mem: Am Chem Soc. Res: Synthesis of heterocyclic compounds; transition metal coordination compounds; synthesis of high temperature stable polymers; composite materials. Mailing Add: 606 W Nettle Tree Rd Sterling VA 22170

AUGOOD, DEREK RAYMOND, b London, Eng, Mar 7, 28; nat US; m 55; c 2. PHYSICAL ORGANIC CHEMISTRY, CHEMICAL ENGINEERING. Educ: Univ London, BSc, 49, PhD(phys org), 52; Cambridge Univ, PhD(chem eng), 55. Prof Exp: Harwell fel, UK Atomic Energy Res Estab, 54-56; engr, E I du Pont de Nemours & Co, 56-62; head sect prod develop & mgr, Res & Develop Lab, 63-67, mgr res & develop planning, 67-72, PROG MGR, RES & DEVELOP LAB, KAISER CHEM, 72- Honors & Awards: Moulton Medal, Brit Inst Chem Engrs, London, 55. Mem: Inst Elec & Electronics Engrs; Am Chem Soc; Am Inst Chem Engrs; Brit Inst Chem Engrs. Res: Homolytic aromatic substitution; separation of stable isotopes hydrogen-deuterium; chemistry of nylon intermediates; fluorocarbon and vinyl butyral polymers; isocyanates. Mailing Add: Kaiser Aluminum & Chem Corp Ctr for Technol Pleasanton CA 94566

AUGURT, THOMAS A, organic chemistry, polymer chemistry, see 12th edition

AUGUST, LEON STANLEY, b New Orleans, La, Sept 30, 26; m 52. NUCLEAR SCIENCE, RADIOLOGICAL PHYSICS. Educ: La State Univ, BS, 50, PhD(physics), 57; Tulane Univ, MS, 52. Prof Exp: RES PHYSICIST, US NAVAL RES LAB, 57- Concurrent Pos: Mem cyclotron group, Univ Calif, Los Angeles, 65-66. Mem: Am Phys Soc; Sigma Xi; AAAS. Res: Applied nuclear science utilizing a sector-focused cyclotron; radioisotope applications and production; fast-neutron physics and dosimetry with application to cancer therapy. Mailing Add: 6920 Baylor Dr Alexandria VA 22307

AUGUSTIN, JORG A L, b Heerbrugg, Switz, Jan 5, 30; US citizen; m 62. FOOD SCIENCE, PLANT BIOCHEMISTRY. Educ: Swiss Fed Inst Technol, dipl agrotech, 55; Univ Ill, MS, 57; Mich State Univ, PhD(food sci), 64. Prof Exp: Res assoc, Hero Konserven, Switz, 57-60; asst plant supt food processing, Libby, McNeil & Libby, 60; res asst food sci, Mich State Univ, 61-64; food technologist food res div, Armour & Co, 64-66, from asst sect head to sect head meat prod develop, 66-68; assoc res prof agr biochem, Br Exp Sta, 68-75, HEAD FOOD RES CTR, UNIV IDAHO, 75- Mem: AAAS; Inst Food Technol; Am Chem Soc; Am Soc Microbiol. Res: Process and product development and improvement of foods, especially canned foods; studies on thermal destruction of bacterial spores; sugar beet storage; nutrient composition of fresh, stored and processed potatoes. Mailing Add: Food Res Ctr Univ of Idaho Moscow ID 83843

AUGUSTINE, JAMES ROBERT, b Belleville, Ill, Jan 27, 46; m 70; c 1. NEUROANATOMY. Educ: Millikin Univ, BA, 68; St Louis Univ, Mo, MS, 70; Univ Ala, Birmingham, PhD(anat), 73. Prof Exp: From instr to asst prof anat, Sch Med & Sch Dent, Univ Ala, Birmingham, 73-76; ASST PROF ANAT, SCH MED, UNIV SC, COLUM BIA, 76- Mem: Am Asn Anatomists. Res: Functional neuroanatomy of the primate nervous system particularly as it relates to our understanding of human nervous system. Mailing Add: Dept of Anat Univ of SC Sch Med Columbia SC 29208

AUGUSTINE, ROBERT LEO, b Omaha, Nebr, Nov 15, 32; m 57; c 2. ORGANIC CHEMISTRY. Educ: Creighton Univ, 54; Columbia Univ, MA, 55, PhD(chem), 57. Prof Exp: From instr to asst prof chem, Univ Tex, 57-61; from instr to assoc prof, 61-69, PROF CHEM, SETON HALL UNIV, 69- Mem: Am Chem Soc; NY Acad Sci; Brit Chem Soc; Org Reactions Catalysis Soc (dir, 75-77). Res: Synthetic and mechanistic applications of catalytic hydrogenation, oxidation and related reactions. Mailing Add: Dept of Chem Seton Hall Univ South Orange NJ 07079

AUGUSTINE, ROBERTSON J, b Grand Rapids, Mich, Jan 28, 33; m 56; c 3. HEALTH PHYSICS. Educ: Univ Mich, BS, 55, MS, 57, MPH, 59, PhD(environ health), 62. Prof Exp: Health physicist, Univ Mich, 56-57, res assoc radiol health, 57-62; asst officer-in-charge, SE Radiol Health Lab, USPHS, 62-65; tech dir, Appl Health Physics, Inc, 65-66; chief environ radioactivity sect, Nat Ctr Radiol Health, USPHS, 66-68, dep dir div environ radiation, Bur Radiol Health, 68-70; SCIENTIST DIR, OFF RADIATION PROG, ENVIRON PROTECTION AGENCY, 70- Mem: Health Physics Soc; Am Pub Health Asn. Res: Radiological health; radiation protection and measurement. Mailing Add: EPA Off of Radiation Progs (AW-458) 401 M St SW Washington DC 20460

AUGUSTINE (SMALT), MARIE, b Rochester, NY, July 7, 08. CHEMISTRY. Educ: Nazareth Col, BA, 37; Cath Univ Am, MS, 42, Inst Divi Thomae, PhD, 53. Prof Exp: PROF CHEM, NAZARETH COL, 41-, CHMN DEPT, 74- Mem: Am Chem Soc. Res: Electrochemistry; enzyme chemistry; complex ions; accelerator of catalase. Mailing Add: Nazareth Col of Rochester 4245 East Ave Rochester NY 14610

AUGUSTYN, JOAN MARY, b Buffalo, NY. BIOCHEMISTRY. Educ: D'Youville Col, BA, 62; State Univ NY Buffalo, PhD(biochem), 69. Prof Exp: Trainee steroid biochem, Worcester Found Exp Biol, Shrewsbury, Mass, 69-71; RES BIOCHEMIST, RES SERV, VET ADMIN HOSP, ALBANY, 71-, RES ASST PROF ATHEROSCLEROSIS, ALBANY MED COL, 76- Mem: Tissue Cult Asn. Res: Biochemical mechanisms involved in atherosclerotic heart disease. Mailing Add: Res Serv Vet Admin Hosp Albany NY 12208

AUKERMAN, LEE WILLIAM, b North Grove, Ind, June 7, 23; m 62; c 4. SOLID STATE PHYSICS. Educ: Purdue Univ, BS, 49, MS, 53, PhD(physics), 58. Prof Exp: Asst, Purdue Univ, 49-58; prin physicist, Battelle Mem Inst, 58-63; MEM TECH STAFF, AEROSPACE CORP, LOS ANGELES, 63- Mem: AAAS; Am Phys Soc.

Res: Semiconductors and insulators; radiation damage; electrical and optical properties; infrared detectors. Mailing Add: Aerospace Corp PO Box 92957 Los Angeles CA 90045

AULD, DAVID STUART, b Newton, NJ, Jan 8, 37; m 61; c 2. ORGANIC CHEMISTRY, BIOLOGICAL CHEMISTRY. Educ: Lehigh Univ, BA, 60, MS, 62; Cornell Univ, PhD(chem), 67. Prof Exp: Am Cancer Soc res fel biol chem, 67-69, assoc, 69-70, ASST PROF BIOL CHEM, HARVARD MED SCH, 70-; ASSOC MED, PETER BENT BRIGHAM HOSP, 67- Concurrent Pos: Res fel, Cancer Div, NIH, 69-70, spec fel, 70-71; consult, Monsanto Res Corp, 75- Mem: Am Chem Soc; Am Soc Exp Biol Chemists; Sigma Xi; NY Acad Sci. Res: Mechanism of enzyme action; relationship of diseased states and abnormal enzyme activities. Mailing Add: Biophys Res Lab Peter Bent Brigham Hosp Boston MA 02115

AULD, EDWARD GEORGE, b Chilliwack, BC, Apr 27, 36; m 60; c 2. NUCLEAR PHYSICS. Educ: Univ BC, BASc, 59, MASc, 61; Univ Southampton, PhD(nuclear physics), 64. Prof Exp: Sci officer, Rutherford High Energy Lab, Eng, 63-64, sr sci officer, 64-66; ASSOC PROF NUCLEAR PHYSICS, UNIV BC, 66- Mem: Am Asn Physics Teachers; Can Asn Physicists. Res: Particle physics; nuclear physics, especially deuteron interactions; high energy physics; meson resonance and pion inelastic interactions; cyclotron magnetic design for the TRIUMF negative ion cyclotron. Mailing Add: Dept of Physics Univ of BC Vancouver BC Can

AULD, PETER A MCF, b Toronto, Ont, Feb 5, 28; m 51; c 2. MEDICINE, PEDIATRICS. Educ: Univ Toronto, BA, 48; McGill Univ, MD, 52; FRCP(C), 57. Prof Exp: Instr pediat, Harvard Med Sch, 59-60; res assoc, McGill Univ, 60-62; from asst prof to prof pediat, 62-74, PROF PERINATAL MED IN OBSTET & GYNEC, MED COL, CORNELL UNIV, 74- Concurrent Pos: Res fel pediat, Harvard Med Sch, 57-59; career investr, NY Health Res Coun, 62- Mem: Fel Am Acad Pediat; Perinatal Res Soc; Soc Pediat Res; Am Pediat Soc. Res: Cardiology; cardio-pulmonary physiology; neonatal pulmonary disease. Mailing Add: Dept of Pediat Cornell Univ Med Ctr New York NY 10021

AULERICH, RICHARD J, b Detroit, Mich, Mar 12, 36; m 63; c 2. ANIMAL SCIENCE. Educ: Mich State Univ, BS, 58, MS, 64, PhD(poultry sci), 67. Prof Exp: Technician fur animal proj, 62-67, asst prof poultry sci, 67-73, ASSOC PROF POULTRY SCI, MICH STATE UNIV, 73- Concurrent Pos: Mem subcomt furbearer nutrit, Nat Res Coun, 67. Mem: Am Soc Mammal; Wildlife Soc; Am Soc Animal Sci. Res: Nutrition, physiology and management of furbearing animals, particularly mink. Mailing Add: 3609 E Arbutus Dr Okemos MI 48864

AULETTA, ANGELA ELAINE, b Torrington, Conn, Nov 8, 38. ONCOLOGY, CELL BIOLOGY. Educ: St Joseph Col, Conn, BA, 61; Cath Univ Am, MS, 63, PhD(biol), 66. Prof Exp: Virologist, 65-68, sr scientist biochem-pharmacol, Cancer Chemother Dept, 68-75, PROJ DIR VIROL & CHEM CARCINOGENESIS & HEAD MUTAGENESIS/CARCINOGENESIS SCREENING SERV, MICROBIOL ASSOCS, 75- Mem: Am Soc Microbiol; AAAS; Sigma Xi; Tissue Cult Asn. Res: Use of microsomal enzyme preparations in established in vitro assays for chemical carcinogens. Mailing Add: Microbiol Assocs 4733 Bethesda Ave Bethesda MD 20014

AULL, CHARLES EDWARD, b US, Sept 1, 27. MATHEMATICS. Educ: Columbia Univ, BS, 49; Univ Ore, MS, 53; Univ Colo, PhD(math), 62. Prof Exp: Asst math, Univ Ore, 52-53; instr, Univ Ariz, 53-55; instr, Univ Colo, 55-62; asst prof, Kent State Univ, 62-65; assoc prof, 65-68, PROF MATH, VA POLYTECH INST & STATE UNIV, 68- Mem: Am Math Soc; Math Asn Am. Res: Topological spaces; separation and base axioms; covering properties and sequences; algebra, especially ideal theory; analysis, especially generalized variations. Mailing Add: Dept of Math Va Polytech Inst & State Univ Blacksburg VA 24061

AULL, FELICE, b Vienna, Austria, Aug 12, 38; US citizen; m 62; c 1. PHYSIOLOGY. Educ: Columbia Univ, AB, 60; Cornell Univ, PhD(physiol), 64. Prof Exp: Physiologist, Radiobiol Lab, Bur Com Fisheries, NC, 64-65; from instr to asst prof, 66-72, ASSOC PROF PHYSIOL, SCH MED, NY UNIV, 72- Concurrent Pos: USPHS fel physiol, Med Col, Cornell Univ, 65-66. Mem: Harvey Soc; Soc Gen Physiol; Am Physiol Soc; Biophys Soc. Res: Membrane permeability and transport processes of malignant cells; role of membrane glycoproteins in transport and malignancy. Mailing Add: Dept of Physiol NY Univ Sch of Med New York NY 10016

AULL, LUTHER BACHMAN, III, b Greenwood, SC, Mar 1, 29; m 50; c 4. SCIENCE ADMINISTRATION, NUCLEAR PHYSICS. Educ: US Mil Acad, BS, 50; Univ Va, PhD(physics), 59. Prof Exp: Physicist nuclear weapon effects, Defense Atomic Support Agency, 59-63; physicist nuclear physics, US Army Res Off, 63-67; tech asst nuclear weapon effects, Defense Res & Eng, Off Secy Defense, 67-70; planning consult nuclear weapon progs, Union Carbide Combined Opers Planning Group, 70-73, MGR PLANNING & SPEC TASKS GAS CENTRIFUGE DEVELOP, UNION CARBIDE NUCLEAR DIV-SEPARATION SYSTS DIV, 73- Mem: Sigma Xi; Am Phys Soc. Res: Planning continuing development program for the gas centrifuge for use in uranium isotope separation. Mailing Add: Union Carbide Nuclear Div Oak Ridge Gaseous Diffusion Plant Separation Systs Div PO Box P Oak Ridge TN 37830

AULSEBROOK, LUCILLE HAGAN, b Houston, Tex, Dec 31, 25; m 52; c 4. ANATOMY. Educ: Univ Tex, Austin, BA, 46, MA, 47; Univ Ark, Little Rock, PhD(anat), 66. Prof Exp: Instr zool, Univ Tex, Austin, 48; res assoc cancer res, M D Anderson Hosp Cancer Res, Houston, Tex, 48-50; teaching asst zool, Univ Wis, Madison, 50-53; instr biol & chem, Univ Ark, Little Rock, 57-60, teaching asst anat, Sch Med, 60-64; instr, 69-71, ASST PROF ANAT, SCH MED, VANDERBILT UNIV, 71-, ASSOC PROF, SCH NURSING, 73- Concurrent Pos: USPHS fel, Sch Med, Vanderbilt Univ, 66-69. Mem: Am Asn Anat; Am Soc Neurosci. Res: Hypothalamic-endocrine interrelations, particularly neural regulation of releasing factors and factors inhibiting release involved in reproduction, parturition and lactation, lactogenesis and milk-ejection. Mailing Add: Dept of Anat Vanderbilt Univ Sch of Med Nashville TN 37232

AULT, ADDISON, b Boston, Mass, July 3, 33; m 58; c 5. ORGANIC CHEMISTRY. Educ: Amherst Col, AB, 55; Harvard Univ, PhD(chem), 60. Prof Exp: Asst prof chem, Grinnell Col, 59-61; resident res assoc, Argonne Nat Lab, 61-62; from asst prof to assoc prof, 62-70, PROF CHEM & CHMN DEPT, CORNELL COL, 70- Concurrent Pos: NSF sci fac fel, Pa State Univ, 68-69. Mem: Am Chem Soc; AAAS; Brit Chem Soc. Res: Structure determination of natural products; reaction mechanisms; physical organic chemistry; nuclear magnetic resonance; chemical education. Mailing Add: Dept of Chem Cornell Col Mt Vernon IA 52314

AULT, CHARLES ROBERT, zoology, see 12th edition

AULT, JANET E (MILLS), b Bremerton, Wash, Jan 20, 43. ALGEBRA. Educ: Western Wash State Col, BA, 65; Pa State Univ, PhD(math), 70. Prof Exp: Asst prof math, Univ Fla, 70-71; ASST PROF MATH, MADISON COL, 74- Mem: Am Math Soc; Math Asn Am. Res: Algebraic semigroups, particularly in constructions and classifications of inverse and regular semigroups. Mailing Add: Dept of Math Madison Col Harrisonburg VA 22801

AULT, JOHN WILLARD, b Findlay, Ohio, Jan 14, 10; m 34; c 1. MATHEMATICS. Educ: Bowling Green State Univ, BS, 32; Ohio State Univ, MA, 35. Prof Exp: Prof math, Cedarville Col, 32-38; instr, Ohio State Univ, 38-41; assoc prof, US Air Force Acad, 54-56, prof & head dept, 56-65; assoc prof, 65-75, EMER ASSOC PROF MATH, TEX TECH UNIV, 75- Mem: Math Asn Am. Mailing Add: Dept of Math Tex Tech Univ Lubbock TX 79413

AULT, WAYNE URBAN, b Monroe Co, Mich, Jan 20, 23; m 47; c 4. GEOCHEMISTRY. Educ: Wheaton Col, BA, 50; Columbia Univ, MA, 56, PhD(geochem), 57. Prof Exp: Asst geochem, Lamont Geol Observ, 50-57; res geochemist, US Geol Surv, 57-62; sr res scientist, Isotopes, Inc, 62-64, dir tech representation, 64-68; assoc prof, Wheaton Col, 68-69; ASSOC PROF CHEM & ACTG CHMN DEPT & CHMN DEPT GEOL, KINGS COL, NY, 69- Concurrent Pos: Asst prof, Shelton Col, 52-54; asst prof, Nyack Missionary Col, 64-72, assoc prof geol, 72- Mem: Geochem Soc; Geol Soc Am; Geophys Union; Am Nuclear Soc; Marine Technol Soc. Res: Isotope geology; mass spectrometry; vacuum fusion; age determination; volcanic fluids; subsurface tracing; radioisotope applications; environmental geology. Mailing Add: Dept of Geol Kings Col Briarcliff Manor NY 10510

AUMAN, JASON REID, b High Point, NC, Feb 5, 37; m 58; c 2. ASTROPHYSICS. Educ: Duke Univ, BS, 59; Northwestern Univ, PhD(astron), 65. Prof Exp: Res assoc astron, Princeton Univ, 64-68; asst prof, 68-71, ASSOC PROF ASTRON, UNIV BC, 71- Mem: Am Astron Soc; Int Astron Union. Res: Structure and evolution of stars; atmospheres of late-type stars; infrared opacity of water vapor at elevated temperatures; nucleii of galaxies. Mailing Add: Dept of Geophys & Astron Univ of BC Vancouver BC Can

AUMANN, GLENN D, b Elbowoods, NDak, June 17, 30; m 50; c 3. ECOLOGY, ETHOLOGY. Educ: Wis State Univ, Eau Claire, BS, 57; Univ Wis, MS, 58, PhD(ecol), 64. Prof Exp: Asst prof biol, Gogebic Community Col, 57-61; NIH fel electron micros, Sch Med, Univ Wis, 64-65; assoc prof biol, 65-70, PROF BIOL, UNIV HOUSTON, 70-, CHMN DEPT, 67- Mem: AAAS; Animal Behav Soc; Ecol Soc Am. Res: Vertebrate population control mechanisms. Mailing Add: Dept of Biol Univ of Houston Houston TX 77004

AUNDERSON, AUBREY LEE, b Ft Worth, Tex, Apr 24, 40; m 60; c 2. ACOUSTICS, OCEANOGRAPHY. Educ: Baylor Univ, BS, 62; Univ Tex, Austin, PhD(eng acoust), 74. Prof Exp: Physicist, Naval Ord Lab, Corona, Calif, 62-63 & Navy Electronics Lab, San Diego, 63-64; RES SCIENTIST UNDEREATER SOUND, APPL RES LABS, UNIV TEX, AUSTIN, 65- Mem: Acoust Soc Am; Am Inst Physics; Marine Technol Soc. Res: Underwater sound propagation; mathematical modeling of sound propagation; use of acoustics as a tool in geophysics and oceanography; marine sediment studies; physical and geological oceanography. Mailing Add: Appl Res Labs Univ of Tex Austin TX 78712

AUNE, JANET L, b St Louis, Mo. CELL BIOLOGY. Educ: St Louis Univ, BS, MS, PhD(biol), 68. Prof Exp: Assoc prof biol, Harris Teachers Col, 64-66; ASSOC PROF BIOL, TEX WOMAN'S UNIV, 68-, EXEC DEAN, 75- Mem: Am Soc Cell Biol; Tissue Cult Asn; Sigma Xi; AAAS. Res: Ageing of normal diploid cells in culture. Mailing Add: Tex Woman's Univ Houston Ctr 1130 M D Anderson Blvd Houston TX 77025

AUNE, KIRK CARL, b Winona, Minn, Nov 5, 42; m 64; c 2. PHYSICAL BIOCHEMISTRY. Educ: Univ Minn, Minneapolis, BCh, 64; Duke Univ, PhD(biochem), 68. Prof Exp: Mem staff biochem, Brandeis Univ, 68-70; asst prof, Ohio State Univ, 70-73; ASST PROF BIOCHEM, BAYLOR COL MED, 73- Concurrent Pos: Am Cancer Soc res fel, 68-70; NIH career develop award, 73 & 75. Mem: AAAS; Am Chem Soc; Am Soc Biol Chemists; Sigma Xi. Res: Physical chemistry of macromolecules; energetics of interaction in complex biological structures such as the ribosomal subunits from ultracentrifugal techniques. Mailing Add: Dept of Biochem Baylor Col of Med Houston TX 77025

AURAND, HENRY SPIESE, JR, b Columbus, Ohio, Feb 24, 24; m 46; c 2. UNDERWATER ACOUSTICS. Educ: US Mil Acad, BS, 44; State Univ Iowa, MS, 47. Prof Exp: Design engr ballistic missiles, Douglas Aircraft Co, 54-56; mem tech staff weapon syst, Tech Mil Planning Oper, Gen Elec, 57-63; mgr eng develop dept, Electronics Div, Gen Mill Inc, 63; sr res engr ocean acoustics, Lockheed Missiles & Space Co, 63-69; phys sci adminr, Off Naval Res, Washington, 69-71; SUPVR PHYSICIST OCEAN ACOUSTICS, NAVAL UNDERSEA CTR, SAN DIEGO, 71- Mem: Acoustical Soc Am; Am Geophys Union; Am Soc Civil Engr. Res: Large scale computers and sensors in ocean acoustics. Mailing Add: Naval Undersea Ctr Code 3002 San Diego CA 92132

AURAND, LEONARD WILLIAM, b Shamokin Dam, Pa, Feb 5, 20; m 43; c 3. BIOCHEMISTRY. Educ: Pa State Univ, BS, 41, PhD(biochem), 49; Univ NH, MS, 47. Prof Exp: From res asst prof to res assoc prof animal indust, 49-60, PROF FOOD CHEM, NC STATE UNIV, 60- Mem: Am Chem Soc; Am Dairy Sci Asn; Inst Food Technol; Am Inst Nutrit. Res: Off-flavors in milk; enzyme chemistry; oxidation of lipids. Mailing Add: Dept of Food Sci NC State Univ Raleigh NC 27607

AURBACH, GERALD DONALD, b Cleveland, Ohio, Mar 24, 27; m 60; c 2. MEDICINE. Educ: Univ Va, BA, 50, MD, 54. Prof Exp: Intern med, New Eng Ctr Hosp, 54-55, assoc, 58-59; resident, Boston City Hosp, 55-56; res assoc, 59-61, mem sect endocrinol, 61-65, mem endocrinol study sect, 67-69, chief mineral metab sect, 65-73, CHIEF METAB DIS BR, NAT INST ARTHRITIS, METAB & DIGESTIVE DIS, 73- Concurrent Pos: Res fel med, New Eng Ctr Hosp, 56-58; USPHS fel, Nat Inst Arthritis & Metab Dis, 56-59; asst, Sch Med, Tufts Univ, 56-58, instr, 58-59; secy-treas, Int Cong Endocrinol, 72; Am Clin & Climatol Asn Gordon Wilson lectr, 73; vol ed sect endocrinol, Handbook Physiol, Am Physiol Soc; mem ed bd, Year in Metab; assoc ed, Endocrine Soc, 68-72; mem ed comt, Am Soc Clin Invest, 71- Honors & Awards: John Horsley Mem Prize, Univ Va, 60; Andre Lichtwitz Prize, 68; Meritorious Serv Medal, USPHS, 69. Mem: AAAS; Am Fedn Clin Res; AMA; Am Soc Biol Chem; Asn Am Physicians. Res: Endocrinology; biochemistry. Mailing Add: Metab Dis Br Nat Inst Arthritis Metab & Digestive Dis Bethesda MD 20014

AURELIAN, LAURE, b Bucharest, Rumania, June 17, 39; m 64. VIROLOGY. Educ: Tel Aviv Univ, MSc, 62; Johns Hopkins Univ, PhD(microbiol), 66. Prof Exp: Res asst microbiol, Tel Aviv Univ, 60-62; from instr to asst prof, 66-74, ASSOC PROF MICROBIOL & LAB ANIMAL MED, SCH MED, JOHNS HOPKINS UNIV, 74- Concurrent Pos: Res fel, Sch Med, Johns Hopkins Univ, 65-66. Mem: AAAS; Soc Exp Biol & Med; Am Soc Microbiol; Am Asn Immunol. Res: Biological, biochemical and biophysical properties of viruses; basis of host-range restrictions; role of viruses in cancer. Mailing Add: Dept of Animal Med Johns Hopkins Univ Sch of Med Baltimore MD 21205

AUSBAND, JOHN R, b Winston-Salem, NC, Oct 14, 20; m 49; c 2. OTOLARYNGOLOGY. Educ: Asbury Col, BA, 40; Bowman Gray Sch Med, MD, 43. Prof Exp: From instr to assoc prof, 52-67, PROF OTORHINOLARYNGOL, BOWMAN GRAY SCH MED, 67- Mem: Am Acad Ophthal & Otolaryngol; Am Laryngol, Rhinol & Otol Soc; Am Broncho-Esophagol Asn (secy, 65-70, pres-elect, 70, pres, 71); Int Broncho-Esophagol Soc; Am Laryngol Soc. Res: Effects of sectioning of efferent bundle of eighth nerve on cochlear microphonics. Mailing Add: Bowman Gray Sch of Med Winston-Salem NC 27103

AUSLANDER, BERNICE LIBERMAN, b Brooklyn, NY, Nov 21, 30; m 50; c 2. MATHEMATICS. Educ: Columbia Univ, AB, 51; Univ Chicago, MS, 54; Univ Mich, PhD(math), 63. Prof Exp: Asst prof, Wellesley Col, 65-71; ASSOC PROF MATH, UNIV MASS, BOSTON, 71- Concurrent Pos: Fel, Radcliffe Inst Independent Study, 63-65. Mem: Am Math Soc; Math Asn Am. Res: Algebra, particularly the theory of commutative rings. Mailing Add: 16 Everett St Newton Centre MA 02159

AUSLANDER, DAVID E, b Brooklyn, NY, Apr 27, 40; m 72. PHARMACEUTICS. Educ: Brooklyn Col Pharm, BS, 61; Columbia Univ, MS, 65; Rutgers Univ, PhD(pharmaceut), 73. Prof Exp: Staff pharmacist, Vet Admin Hosp, New York, 61-62; scientist, Schering Corp, 64-70; RES INVESTR PHARMACEUT, SQUIBB CORP, 72- Concurrent Pos: Schering Corp res fel, Rutgers Univ, 70-72. Mem: Sigma Xi. Res: Interaction of drugs with biological interfaces; surface chemistry of antibiotics and tranquilizers; theory of compression of solids. Mailing Add: The Squibb Inst for Med Res New Brunswick NJ 08903

AUSLANDER, JOSEPH, b New York, NY, Sept 10, 30; m 57; c 2. MATHEMATICS. Educ: Mass Inst Technol, SB, 52; Univ Pa, PhD(math), 57. Prof Exp: Asst prof math, Carnegie Inst Technol, 57-60; mem staff, Res Inst Advan Study, 60-62; assoc prof, 62-66, PROF MATH, UNIV MD, COLLEGE PARK, 66- Concurrent Pos: Res assoc, Yale Univ, 64-65; vis sr lectr, Imp Col, Univ London, 67-68. Mem: AAAS; Am Math Soc; Math Asn Am. Res: Topological dynamics; ergodic theory. Mailing Add: Dept of Math Univ of Md College Park MD 20742

AUSLANDER, LOUIS, b Brooklyn, NY, July 12, 28; m 51; c 2. MATHEMATICS. Educ: Columbia Univ, BA, 49, MA, 50; Univ Chicago, PhD, 54. Prof Exp: Instr, Yale Univ, 53-55; asst prof math, Univ Pa, 56-57; from asst prof to assoc prof, Ind Univ, 57-62; prof, Purdue Univ, 62-64 & Belfer Grad Sch Sci, Yeshiva Univ, 64-65; PROF MATH, GRAD SCH & UNIV CTR, CITY UNIV NY, 65- Concurrent Pos: NSF fel, Inst Advan Study, 55-57; vis assoc prof, Yale Univ, 60-62; vis prof, Univ Calif, Berkeley, 63-64; Guggenheim fel, 71-72. Mem: Am Math Soc. Res: Structure of solvmanifolds and harmonic analysis on solvable Lie groups and solvmanifolds. Mailing Add: Dept of Math Grad Sch City Univ NY New York NY 10036

AUSLANDER, MAURICE, b Brooklyn, NY, Aug 3, 26; m 50; c 2. MATHEMATICS. Educ: Columbia Univ, BA, 49, PhD(math), 54. Prof Exp: Instr math, Univ Chicago, 53-54 & Univ Mich, 54-56; NSF fel, Inst Advan Study, 56-57; from asst prof to assoc prof, 57-63, PROF MATH, BRANDEIS UNIV, 63- Res: Homological, associative and communitative algebra. Mailing Add: Dept of Math Brandeis Univ Waltham MA 02154

AUSMAN, ROBERT K, b Milwaukee, Wis, Jan 31, 33. MEDICAL ADMINISTRATION, RESEARCH ADMINISTRATION. Educ: Marquette Univ, MD, 57. Prof Exp: Intern, Univ Minn, 57-58; dir, Health Res, Inc, 61-66; asst comnr hosp affairs, NY State Dept Health, 66-69; dep dir, Fla Regional Med Prog, 69-70; dir clin res, 70-73, VPRES CLIN RES, BAXTER LABS, 73- Concurrent Pos: Res fel, Univ Minn, 58-61; Damon Runyon res fel, 59-61; sci consult, Northern Eng Co, 61-; assoc clin prof surg, Med Col Wis. Mem: AAAS; Am Asn Cancer Res; Am Geriat Soc; Int Soc Hemat. Res: Medical systems and data processing; cancer chemotherapy; public health statistics. Mailing Add: Baxter Labs Morton Grove IL 60053

AUSPOS, LAWRENCE ARTHUR, b Anoka, Minn, Aug 27, 17; m 45; c 2. CHEMISTRY. Educ: Univ Portland, BS, 40; Univ Notre Dame, MS, 41, PhD(org chem), 43. Prof Exp: Sr lab asst chem, Univ Portland, 39-40; lab asst, Univ Notre Dame, 40-41; instr, Holy Cross Sem, 41-42; mem staff, Pioneering Res Div, Rayon Tech Div, 43-50, MEM STAFF, TEXTILE FIBRE DEPT, E I DU PONT DE NEMOURS & CO, 50- Mem: Am Chem Soc. Res: Synthetic plastics and fibers; preparation of fiberforming polymers; preparation of dibutyl benezenes. Mailing Add: Textile Fibre Dept Christina Lab E I du Pont de Nemours & Co Wilmington DE 19899

AUST, CATHERINE COWAN, b Atlanta, Ga, Apr 26, 46; m 67. ALGEBRA. Educ: Univ Ga, BS, 68; Emory Univ, PhD(math), 73. Prof Exp: From instr to asst prof math, Ga Inst Technol, 72-75; ASST PROF MATH, CLAYTON JR COL, 75- Mem: Am Math Soc; Math Asn Am; Sigma Xi. Mailing Add: 589 Forrest Ave Fayetteville GA 30214

AUST, J BRADLEY, b Buffalo, NY, Sept 8, 26; m 49; c 6. SURGERY. Educ: Univ Buffalo, MD, 49; Univ Minn, MS, 57, PhD(surg), 58; Am Bd Surg, dipl, 58; Am Bd Thoracic Surg, dipl, 65. Prof Exp: Res asst, Univ Minn, 53-57, from instr to prof surg, 57-66, coordr cancer res, 60-66; PROF SURG & CHMN DEPT, UNIV TEX MED SCH SAN ANTONIO, 66- Concurrent Pos: Am Cancer Soc scholar, 57-62; surg consult, Minn State Prison, 57-63 & Anoka State Hosp, 63-66; lectr, Univ Buffalo, 59. Mem: Soc Univ Surg; Am Asn Cancer Res; fel Am Col Surg; Transplantation Soc; Am Surg Asn. Res: Cancer immunity; homotransplantation; shock. Mailing Add: Dept of Surg Univ of Tex Health Sci Ctr San Antonio TX 78284

AUST, STEVEN DOUGLAS, b South Bend, Wash, Mar 11, 38; m 72; c 2. BIOCHEMISTRY. Educ: Wash State Univ, BS, 60, MS, 62; Univ Ill, PhD(dairy sci), 65. Prof Exp: ASSOC PROF BIOCHEM, MICH STATE UNIV, 67- Concurrent Pos: USPHS fel, Karolinska Inst, Sweden, 66; Swed Med Res Coun fel, 74; Ministry Agr & Fisheries NZ fel, Ruakura Agr Res Ctr, Hamilton, NZ, 75-76. Mem: AAAS; Am Soc Biol Chem; Am Soc Pharmacol & Exp Therapeut. Res: Mixed function oxidation of drugs and the peroxidation of lipids; mechanism of tissue necropsy caused by the fungal toxin sporidesinin. Mailing Add: Dept of Biochem Mich State Univ East Lansing MI 48824

AUSTEN, K FRANK, b Akron, Ohio, Mar 14, 28; m 59; c 4. INTERNAL MEDICINE, IMMUNOLOGY. Educ: Amherst Col, BA, 50; Harvard Univ, MD, 54; Am Bd Internal Med, dipl; Am Bd Allergy, dipl, 65. Prof Exp: Chief rheumatology, Walter Reed Army Med Ctr, 56-58; resident med, Mass Gen Hosp, 58-59, asst physician, 59-64, chief pulmonary unit, 64-65; assoc, 64-65, from asst prof to prof, 65-72, THEODORE BEVIER BAYLES PROF MED, HARVARD MED SCH, 72-; PHYSICIAN-IN-CHIEF, ROBERT B BRIGHAM HOSP, 66-; PHYSICIAN, PETER BENT BRIGHAM HOSP, 66- Concurrent Pos: USPHS fel, Nat Inst Med Res, Eng, 59-61; USPHS res career develop award, NIH, 61-66; mem, Nat Comn Arthritis & Related Musculoskeletal Dis; mem bd dirs, Arthritis Found & chmn res comt; mem ed bd, J Immunol, Arthritis & Rheumatism, Proc Transplantation Soc, J Infectious Dis, Immunol Commun & Clin Immunol, Immunopath, J Clin Invest, J Cyclic Nucleotide Res & Am Rev Respiratory Dis; adv ed, J Exp Med; Am Asn Immunol rep, Nat Res Coun; mem, Am Bd Allergy & Immunol & chmn res & develop comt. Mem: Nat Acad Sci; fel Am Col Physicians; Asn Am Physicians; Am Asn Immunol; fel Am Acad Arts & Sci. Res: Basic and clinical immunology. Mailing Add: Robert B Brigham Hosp Boston MA 02120

AUSTEN, WILLIAM GERALD, b Akron, Ohio, Jan 20, 30; m 61; c 4. SURGERY. Educ: Mass Inst Technol, BS, 51; Harvard Med Sch, MD, 55. Prof Exp: Intern, Mass Gen Hosp, Boston, 55-56, asst resident surg, 56-59; sr registr surg, Kings Col Hosp, London, Eng, 59; hon sr registr, Thoracic Unit, Gen Infirmary, Univ Leeds, 59-60; resident, E Surg Serv, Mass Gen Hosp, 60-61; clin surgeon, NIH, 61-62; assoc, 63-65, assoc prof, 65-66, PROF SURG, HARVARD MED SCH, 66-; CHIEF SURG CARDIOVASC RES UNIT, MASS GEN HOSP, 63-; VIS SURGEON, 66-, CHIEF SURG, 69- Concurrent Pos: Teaching fel surg, Harvard Med Sch, 60-61. Mem: Am Asn Thoracic Surg; Soc Univ Surg; Soc Vascular Surg; Am Col Surg; Am Heart Asn. Res: Cardiovascular physiology. Mailing Add: Dept of Surg Mass Gen Hosp Boston MA 02114

AUSTENSON, HERMAN MILTON, b Viscount, Sask, Nov 28, 24; m 52; c 3. AGRONOMY. Educ: Univ Sask, BSA, 46, MSc, 48; State Col Wash, PhD, 51. Prof Exp: Asst agronomist, Wash State Univ, 51-53; asst prof, State Univ NY Col Agr, Cornell Univ, 53-54; asst agronomist, Wash State Univ, 54-59, assoc agronomist, 59-66; assoc prof crop sci, 66-69, PROF CROP SCI, UNIV SASK, 69-, HEAD DEPT CROP SCI & DIR CROP DEVELOP CTR, 75- Mem: Am Soc Agron; Agr Inst Can. Res: Ecology and production of grain and forage crops; seed quality. Mailing Add: Dept of Crop Sci Univ of Sask Saskatoon SK Can

AUSTERN, BARRY M, b New York, NY, Nov 2, 42; m 68; c 1. ANALYTICAL CHEMISTRY, BIOCHEMISTRY. Educ: Columbia Univ, AB, 63; NY Univ, MS, 65; Univ Mass, PhD(biochem), 69. Prof Exp: Res chemist, US Dept Interior, 69-70; RES CHEMIST, US ENVIRON PROTECTION AGENCY, 70- Concurrent Pos: Abstractor, Chem Abstracts Serv, 66- Mem: Am Chem Soc; Sigma Xi. Res: Analytical chemistry and biochemistry of wastewaters and effluents; chromatography; mass spectroscopy. Mailing Add: Munic Environ Res Lab Environ Protection Agency Cincinnati OH 45268

AUSTERN, NORMAN, b New York, NY, Feb 23, 26; m 64; c 4. THEORETICAL NUCLEAR PHYSICS. Educ: Cooper Union, BSEE, 46; Univ Wis, PhD(physics), 51. Prof Exp: Asst, Univ Wis, 46-48; AEC fel physics, Cornell Univ, 51-52, res assoc, 52-54; staff scientist, Comput Facility, AEC & math inst, NY Univ, 54-55; from asst prof to assoc prof physics, 56-70, PROF PHYSICS, UNIV PITTSBURGH, 70- Concurrent Pos: Fulbright res scholar, Univ Sydney, 57-58, Fulbright lectr, 68-69; NSF sr fel, Inst Theoret Physics, Copenhagen Univ, 61-62. Mem: Am Phys Soc. Res: Nuclear reactions and structure. Mailing Add: Dept of Physics Univ of Pittsburgh Pittsburgh PA 15260

AUSTIC, RICHARD EDWARD, b Ithaca, NY, Apr 10, 41; m 63; c 2. NUTRITION. Educ: Cornell Univ, BS, 63; Univ Calif, Davis, PhD(nutrit), 68. Prof Exp: Res assoc nutrit, 68-70, asst prof animal nutrit, 70-75, ASSOC PROF ANIMAL NUTRIT, CORNELL UNIV, 75- Mem: AAAS; Poultry Sci Asn; Am Inst Nutrit; World Poultry Sci Asn; NY Acad Sci. Res: Poultry nutrition; amino acid requirements and interactions; hyperuricemia and gout; nutrition and embryonic development. Mailing Add: Dept of Poultry Sci Cornell Univ Ithaca NY 14853

AUSTIN, ALFRED ELLS, b Waltham, Mass, Oct 26, 20; m 43; c 2. PHYSICAL INORGANIC CHEMISTRY, SOLID STATE SCIENCE. Educ: Alfred Univ, BA, 42; Yale Univ, PhD(chem), 44. Prof Exp: Res chemist, Corning Glass Works, 44-47; res engr, 47-56, div consult, 56-60, FEL, BATTELLE MEM INST, 60- Concurrent Pos: With AEC, 44. Mem: Am Chem Soc; Am Crystallog Asn; Am Phys Soc; fel Am Inst Chemists. Res: Solid state chemistry and physics; crystallography. Mailing Add: Dept of Physics Columbus Lab Battelle Mem Inst Columbus OH 43201

AUSTIN, CARL FULTON, b Oakland, Calif, July 18, 32; m 53; c 3. GEOLOGY. Educ: Univ Utah, BS, 54, MS, 55, PhD(geol eng), 58. Prof Exp: Geologist, NMex Bur Mines & Mineral Resources, 58-61; GEOLOGIST, US NAVAL WEAPONS CTR, 61- Mem: Am Inst Mining, Metall & Petrol Eng; Sigma Xi. Res: Exploration geochemistry, geothermal deposits, lined-cavity and other impulsive loading phenomena in brittle solids; fracture and penetration of rock and concrete; sub-sea floor construction. Mailing Add: Star Rte 1 Box 240 Inyokern CA 93527

AUSTIN, CHARLES WARD, b Seattle, Seattle, Wash, Nov 5, 32; m 55; c 4. MATHEMATICS. Educ: Univ Wash, BS, 54, MS, 60, PhD(math), 62. Prof Exp: Actg instr math, Univ Wash, 61-62; asst prof, Univ Colo, Boulder, 62-66; assoc prof, 66-71, chmn dept, 67-73, PROF MATH, CALIF STATE UNIV, LONG BEACH, 71-, ASSOC DEAN SCH LETT, 73- Mem: Am Math Soc; Math Asn Am. Res: Topological algebra, especially semigroups and continuous functions on them. Mailing Add: Dept of Math Calif State Col Long Beach CA 90804

AUSTIN, DONALD GUY, b Chicago, Ill, Sept 24, 26. MATHEMATICS. Educ: Univ Ill, BS, 47; Northwestern Univ, MA, 50, PhD(math), 51. Prof Exp: Asst, Northwestern Univ, 47-49; sr mathematician, US Air Force Proj, Univ Chicago, 51; instr math, Syracuse Univ, 51-55; asst prof, Ohio State Univ, 55-56; NSF fel, Yale Univ, 57-58; assoc prof, Univ Miami, 59-61; assoc prof, 61-66, PROF MATH, NORTHWESTERN UNIV, 66- Mem: Am Math Soc. Res: Mathematical analysis; measure theory; theory of probability. Mailing Add: Dept of Math Northwestern Univ Evanston IL 60201

AUSTIN, DONALD MAC, b Erie, Pa, July 5, 44. BIOLOGICAL ANTHROPOLOGY. Educ: Pa State Univ, BA, 67, MA, 70, PhD(anthrop), 74. Prof Exp: Res asst anat, Med Col Ohio, 69-70; ASST PROF ANTHROP, SOUTHERN METHODIST UNIV, 74- Mem: Am Asn Phys Anthrop. Res: Culture change and its effects on nutrition and health; realationships between energy flow and cultural variables in peasant communities; human adaptability. Mailing Add: Dept of Anthrop Southern Methodist Univ Dallas TX 75275

AUSTIN, GEORGE M, b Philadelphia, Pa, May 10, 16; m 42; c 4. NEUROPHYSIOLOGY. Educ: Lafayette Col, AB, 38; Univ Pa, MD, 42; McGill Univ, MSc, 51; Am Bd Neurol Surg, dipl. Prof Exp: Assoc neurosurg, Univ Pa, 51-54, asst prof, 54-57; prof & head div, Med Sch, Univ Ore, 57-68; PROF NEUROSURG & CHIEF SECT, LOMA LINDA UNIV, 68- Concurrent Pos: USPHS spec fel, Dept Theoret Chem, Cambridge Univ; chief neurosurg, Philadelphia Gen Hosp & Lankenau Hosp. Mem: Am Physiol Soc; Am Col Surgeons; AMA; Soc Univ Surgeons; Am EEG Soc. Res: Brain edema and water flux in nerve cells; mechanisms of spasticity in spinal cord and brain lesions; new techniques and measurement of cerebral blood flow in conditions of cerebrovascular disease and brain edema. Mailing Add: Sect of Neurol Surg Loma Linda Univ Sch of Med Loma Linda CA 92354

AUSTIN, GEORGE STEPHEN, b Roanoke, Va, Mar 24, 36; m 70; c 2. GEOLOGY. Educ: Carleton Col, BA, 58; Univ Minn, MS, 62; Univ Iowa, PhD(geol), 71. Prof Exp: Teacher geol, Col of St Thomas, 62-67; stratigrapher, Minn Geol Surv, 67-69; clay mineralogist, Ind Geol Surv, 71-74; INDUST MINERALS GEOLOGIST, N MEX BUR MINES, 74- Mem: Soc Econ Paleontologists & Mineralogists; Clay Mineral Soc; Am Inst Mining Engrs. Res: The study of industrial minerals, especially clay and shales, potash, perlite, fluorite and coal. Mailing Add: NMex Bur of Mines Campus Station Socorro NM 87801

AUSTIN, JAMES HENRY, b Cleveland, Ohio, Jan 4, 25; m 48; c 3. NEUROLOGY. Educ: Harvard Univ, MD, 48; Brown Univ, AB, 61. Prof Exp: Intern med, Boston City Hosp, 48-49, asst resident neurol, 49-50; asst resident, Neurol Inst NY, 53-55; assoc neurol, Med Sch, Univ Ore, 55-57, from asst prof to assoc prof, 57-67; PROF NEUROL & CHMN DIV, SCH MED, UNIV COLO, 67-, CHMN DEPT NEUROL, 75- Concurrent Pos: Univ fel neuropath, Col Physicians & Surgeons, Columbia Univ, 53; Kenny Found fel, 58-63; Commonwealth Fund fel, 62-63. Mem: Am Acad Neurol; Asn Res Nervous & Ment Dis; Am Asn Neuropath; Am Neurol Asn. Res: Hypertrophic neuritis; recurrent polyneuropathy; metachromatic leukodystrophy; globoid leukodystrophy; genetically-determined neurological diseases; sulfatase enzymes; cerebrovascular diseases; aging of the brain; biogenic amines. Mailing Add: Div of Neurol Univ of Colo Sch of Med Denver CO 80220

AUSTIN, JAMES MURDOCH, b Dunedin, NZ, May 25, 15; US citizen; m 41; c 2. METEOROLOGY. Educ: Univ NZ, BA, 35, MA, 36; Mass Inst Technol, ScD, 41. Prof Exp: Asst meteorologist, Apia Observ, West Samoa, 37-39; asst, 40-41, from asst prof to assoc prof meteorol, 41-69, PROF METEOROL, MASS INST TECHNOL, 69-, DIR SUMMER SESSION, 56- Concurrent Pos: Consult, Army Air Forces, Washington, DC, 43-45; expert consult, Joint Res & Develop Bd, 47-48; consult, AEC, 47-53 & US Navy, 55-62; vis lectr, Harvard Univ, 55-; mem adv coun air pollution emergencies, Dept Pub Health, Mass, 69- Honors & Awards: Medal of Freedom, 46. Mem: Fel Am Acad Arts & Sci; fel Am Meteorol Soc. Res: Development of quantitative forecasting methods; growth of cumulus clouds; climatology; meteorological aspects of air pollution. Mailing Add: 36 Wood St Concord MA 01742

AUSTIN, JANET EVANS, b St Louis, Mo, May 30, 03; m 30; c 2. CHEMISTRY. Educ: Mt Holyoke Col, AB, 25; AM, 27; Yale Univ, PhD(chem), 29. Prof Exp: Asst chem, Mt Holyoke Col, 25-27, instr, 29-30; asst, Yale Univ, 27-29; abstractor, 30-71, EMER ABSTRACTOR, CHEM ABSTR, 71- Concurrent Pos: Chemist, Magnus Chem Co, NJ, 34-35 & Merck & Co, NJ, 35-36. Mem: Am Chem Soc. Res: Organic chemistry; ultraviolet absorption spectra and synthesis of hydantoins and pyrimidines. Mailing Add: 114 Buckingham Rd Pittsburgh PA 15215

AUSTIN, JOSEPH WELLS, b Snowville, Utah, Mar 10, 30; m 54; c 5. ANIMAL SCIENCE. Educ: Utah State Univ, BS, 58; Univ Tenn, MS, 61; Tex A&M Univ, PhD(physiol reprod), 67. Prof Exp: Res asst radiation, Tex A&M Univ, 62-65, from asst radiobiologist to assoc radiobiologist, 65-67; asst prof biol, Ill Wesleyan Univ, 67-74; AREA LIVESTOCK SPECIALIST, COOP EXTEN SERV, 74- Mem: Am Soc Animal Sci; Sigma Xi. Res: Mammalian reproduction. Mailing Add: Coop Exten Serv Logan UT 84321

AUSTIN, MARY LELLAH, b Austinburg, Ohio, 1896. GENETICS. Educ: Wellesley Col, BA, 20, MA, 22; Columbia Univ, PhD, 27. Prof Exp: Asst & custodian, 20-22, from instr to prof, 28-61, EMER PROF ZOOL, WELLESLEY COL, 61-; RES SCHOLAR, IND UNIV, BLOOMINGTON, 61- Concurrent Pos: Exchange prof, Isabella Thoburn Col, India, 32-33 & 35-36. Mem: AAAS; Am Genetic Asn; Am Soc Naturalists; Am Soc Zool; Soc Protozool. Res: Protozoology and physiological genetics; study of nitroso-guanidine-induced mutations; building up of multigene stocks of Paramecium aurelia; two cases of linkage discovered. Mailing Add: 506 1/2 N Indiana Ave Bloomington IN 47401

AUSTIN, MAX E, b Pine Grove, Pa, July 17, 33; m 53; c 4. HORTICULTURE. Educ: Univ RI, BS, 55, MS, 60; Mich State Univ, PhD(hort), 64. Prof Exp: Jr res asst hort, Univ RI, 57-60; res assoc, Mich State Univ, 60-64, dist agt, Coop exten serv, 64-65; asst prof, Va Polytech Inst & State Univ, 65-70, assoc prof, 70-72; HEAD DEPT HORT, COASTAL PLAIN EXP STA, 72- Concurrent Pos: Mem, Int Hort Cong. Mem: Am Soc Hort Sci. Res: Harvesting efficiency of vegetable crops; machine harvest; small fruit culture. Mailing Add: Dept of Hort Coastal Plain Exp Sta Tifton GA 31794

AUSTIN, MORRIS EDWIN, b Galesburg, Mich, May 18, 10; m 42. SOILS. Educ: Mich State Col, BS, 37. Prof Exp: Soil scientist, Univ Tenn, 37-45, US Geol Surv, 45-46 & Exp Sta, Cornell Univ, 46-50; soil surv reports, Soil Conserv Serv, USDA, 50-55, sr soil correlator, 55-62, soil scientist, Soil Classification & Correlation, 62-75; RETIRED. Concurrent Pos: Staff mem nat resources sect, Supreme Comdr Allied Powers, 45-46. Mem: AAAS; Soil Sci Soc Am; Soil Conserv Soc; Int Soil Sci Soc. Res: Soil classification, geography, morphology and genesis. Mailing Add: 8524 63rd Ave College Park MD 20740

AUSTIN, OLIVER LUTHER, JR, b Tuckahoe, NY, May 24, 03; m 30; c 2. ORNITHOLOGY. Educ: Wesleyan Univ, BS, 26; Harvard Univ, PhD(zool), 31. Prof Exp: Asst biologist, Bur Biol Surv, USDA, 30-35; dir, Austin Ornith Res Sta, Mass, 32-58; cur birds, 57-73, EMER CUR BIRDS, FLA STATE MUS, 73- Concurrent Pos: Tech consult, Dept of Army, Tokyo, 46-50; Guggenheim fel, 51-52; res & ed specialist, Arctic, Desert, Tropic Info Ctr, Res Studies Inst, Air Univ, Maxwell AFB, Ala, 53-57; ed, The Auk, 68- Honors & Awards: A A Allen Medal, Cornell Univ, 73. Mem: Fel Am Ornith Union; Cooper Ornith Soc; Wilson Ornith Soc; Ornith Soc Japan. Res: Zoogeography; distribution, ecology and taxonomy of birds. Mailing Add: Fla State Mus Gainesville FL 32611

AUSTIN, PAUL ROLLAND, b Monroe, Wis, Dec 9, 06; m 34; c 4. ORGANIC CHEMISTRY. Educ: Univ Wis, AB, 27; Northwestern Univ, MS, 29; Cornell Univ, PhD(org chem), 30. Prof Exp: Asst chem, Cornell Univ, 28-30; Nat Res fel, Univ Ill, 30-32; res chemist, Rockefeller Inst Technol, 32-33; res chemist, Chem Dept, E I du Pont de Nemours & Co, 33-36, group leader, 36-39, asst lab dir, 39-49, dir res div, Electrochem Dept, 49-59, mgr patent & licensing div, Int Dept, 59-66; counsel, Tech Serv Div, 67-70, ADJ PROF MARINE STUDIES, COL MARINE STUDIES, UNIV DEL, 70- Mem: AAAS; Am Chem Soc. Res: Organometallic compounds; hormones of the anterior pituitary; polymer chemistry; protein nutrition. Mailing Add: Col of Marine Studies Univ of Del Newark DE 19711

AUSTIN, PAULINE MORROW, b Kingsville, Tex, Dec 18, 16; m 41; c 2. METEOROLOGY. Educ: Wilson Col, BA, 38; Smith Col, MA, 39; Mass Inst Technol, PhD(physics), 42. Hon Degrees: ScD, Wilson Col, 64. Prof Exp: Computer, Radiation Lab, Mass Inst Technol, 41-42, mem staff, 42-45, res staff, 46-53; lectr, Wellesley Col, 53-55; SR RES ASSOC, MASS INST TECHNOL, 56- Concurrent Pos: Mem Comn II, Int Sci Radio Union. Mem: Am Meteorol Soc (counr, 74-). Res: Radar scattering cross sections; weather radar; propagation of electromagnetic waves in the atmosphere; precipitation physics. Mailing Add: 36 Wood St Concord MA 01742

AUSTIN, ROBERT ANDRAE, b Wilmington, Del, Sept 26, 38; m 68; c 2. PAPER CHEMISTRY. Educ: Bucknell Univ, BS, 61; Univ Mass, MS, 67, PhD(chem), 68. Prof Exp: Teacher, High Sch, Pa, 61-63; develop chemist, E I du Pont de Nemours & Co, 68-74; RES CHEMIST, MEAD CORP, 74- Mem: Am Chem Soc. Res: Conformation of substituted indans by nuclear magnetic resonance; building products from polymers; carbonless copy papers. Mailing Add: Mead Cent Res Chillicothe OH 45601

AUSTIN, ROGER SETH, b Wilmington, Del, Oct 30, 40. GEOLOGY, CLAY MINERALOGY. Educ: Lafayette Col, BA, 62; Univ Ga, MS, 65, PhD(geol), 73. Prof Exp: From asst prof to assoc prof geol, West Ga Col, 68-74; CHIEF GEOL, FREEPORT KAOLIN CO, 74- Mem: Clay Minerals Soc; AAAS. Res: Origin of Kaolin and Bauxite. Mailing Add: Freeport Kaolin Co Gordon GA 31031

AUSTIN, ROLAND W, medical physiology, see 12th edition

AUSTIN, SAMUEL MANLY, b Columbus, Wis, June 6, 33; m 59; c 2. NUCLEAR PHYSICS. Educ: Univ Wis, BS, 55, MS, 57, PhD(physics), 60. Concurrent Pos: Res assoc physics, Univ Wis, 60; NSF fel, Oxford Univ, 60-61; asst prof, Stanford Univ, 61-65; assoc prof, 65-69, PROF PHYSICS, MICH STATE UNIV, 69- Concurrent Pos: Sloan res fel, 63-66; vis scientist, Neils Bohr Inst, 70; guest prof, Univ Munich, 72-73. Mem: Fel Am Phys Soc; AAAS; Am Asn Physics Teachers; Fedn Am Scientists. Res: Experimental study of the structure of nuclei of nuclear reaction mechanisms and of the two body force; nuclear astrophysics; nitrogen fixation. Mailing Add: Dept of Physics Mich State Univ East Lansing MI 48824

AUSTIN, THOMAS HOWARD, b Mt Pleasant, Tex, Aug 12, 37; m 57; c 2. ORGANIC CHEMISTRY. Educ: Cent State Col, Okla, BS, 61; Okla State Univ, PhD(org chem), 65. Prof Exp: Asst chem, Okla State Univ, 61-65; RES CHEMIST, JEFFERSON CHEM CO, INC, 65- Mem: Am Chem Soc. Res: Nucleophilic reactions of organophosphorus esters; catalytic reactions of petrochemical derivatives; chemistry of isocyanates and polyurethanes. Mailing Add: Jefferson Chem Co Inc PO Box 4128 N Austin Sta Austin TX 78751

AUSTIN, THOMAS LEROY, JR, b Memphis, Tenn, Sept 8, 29; m 55; c 1. MATHEMATICS. Educ: Univ Ga, MA, 55. Prof Exp: Mathematician, US Dept Defense, 55-60; sr analyst, Tech Opers, Inc, 60; mgr opers res, Am Systs, Inc, 61-62; design specialist, Gen Dynamics Corp, 62-64; sr tech specialist, Autonetics Div, NAm Aviation, Inc, 64-67 & NAm Rockwell Corp, 67-70; MEM STAFF, DEFENSE COMMUN AGENCY, 70- Mem: Inst Math Statist; Am Math Soc. Res: Combinatorial analysis; graph theory; probability and statistics and applications; stochastic processes; operations research. Mailing Add: 1768 Ivy Oak Sq Reston VA 22090

AUSTIN, WILLIAM CAREY, marine biology, invertebrate zoology. see 12th edition

AUSTON, DAVID H, b Toronto, Ont, Nov 14, 40; m 62; c 2. LASERS. Educ: Univ Toronto, BA, 62, MA, 63; Univ Calif, Berkeley, PhD(elec eng), 69. Prof Exp: MEM TECH STAFF PHYSICS RES, BELL LABS, INC, 69- Mem: Am Phys Soc: Inst Elec & Electronics Engrs. Res: Physical properties of materials, especially semiconductors, utilizing picosecond optical pulses. Mailing Add: 1043 Sunny Slope Dr Mountainside NJ 07092

AUSTRIAN, ROBERT, b Baltimore, Md, Apr 12, 16; m 63. INTERNAL MEDICINE, BACTERIAL GENETICS. Educ: Johns Hopkins Univ, AB, 37, MD, 41; Am Bd Internal Med, dipl. Prof Exp: Asst med, Sch Med, Johns Hopkins Univ, 42-43, instr, 43-47; res assoc microbiol, Col Med, NY Univ, 47-48; instr med, Sch Med, Johns Hopkins Univ, 49-52; from assoc prof to prof med, Col Med, State Univ NY Downstate Med Ctr, 52-62; JOHN HERR MUSSER PROF RES MED & CHMN DEPT, SCH MED, UNIV PA, 62- Concurrent Pos: Vis physician, Univ Div Med Serv, Kings County Hosp, 52-62 & Univ Pa Hosp, 62-; consult, Brooklyn & Maimonides Hosps, 53-62 & Univ Pa Hosp, 62-; vis scientist, Dept Microbial Genetics, Pasteur Inst, 60-61; trustee, Johns Hopkins Univ, 63-69; Tyndale vis lectr & prof, Col Med, Univ Utah, 64; mem study sect A, Allergy & Immunol, NIH, 65-69, mem bd sci counr, Nat Inst Allergy & Infectious Dis, 67-70, chmn, 69-70; mem subcomt streptococci & pneumococci, Int Comn Bact Nomenclature; mem comt meningococcal infections, Comn Acute Respiratory Dis, Armed Forced Epidemiol Bd, 66-72. Mem: Asn Am Physicians; Am Soc Clin Invest; Am Soc Microbiol; fel Am Col Physicians; Infectious Dis Soc Am (pres, 71). Res: Infectious disease; pneumococcal transformation reactions; pneumococcal vaccines. Mailing Add: Dept of Res Med Univ of Pa Sch of Med Philadelphia PA 19104

AUSUBEL, FREDERICK MICHAEL, b New York, NY, Sept 2, 45. MOLECULAR GENETICS. Educ: Univ Ill, Urbana, BS, 66; Mass Inst Technol, PhD(biol), 72. Prof Exp: ASST PROF BIOL, HARVARD UNIV, 75- Res: Molecular genetic analysis of nitrogen-fixation; somatic cell genetics of plant cell tissue cultures. Mailing Add: Dept of Biol Harvard Univ Cambridge MA 02138

AUTEN, JOHN THOMPSON, b White Hall, Ill, Nov 6, 87; m 20; c 1. SOIL SCIENCE. Educ: Univ Ill, BS, 16; Iowa State Col, MS, 23, PhD, 29. Prof Exp: Soil analyst, Iowa State Col, 20-23; prof soils & chem, Forest Sch, Pa State Univ, 23-29; silviculturist, Cent States Forest Exp Sta, US Forest Serv, 29-51; FOREST SOIL CONSULT, 51- Mem: Soc Am Foresters; Soil Sci Soc Am. Res: Organic phosphorus of soils; soils of Mont Alto State Forest; litter in forest practice; base exchange calcium and magnesium of forest soil profiles; porosity and water absorption of forest soils; soil profile-tree species relations; forest tree nursery soil; soil-site relations of Virginia Pine in the Virginia Piedmont and of the Loblolly Pine in the coastal plain of South Carolina; soilsite reconnaissance of Dare County, North Carolina; relation of soils to trafficability of forest machinery in Carolina Coastal Plains. Mailing Add: College Park Rte 1 Box 187 Lewisburg PA 17837

AUTENRIETH, JOHN STORK, b Weehawken, NJ, Dec 28, 16; m 46. ORGANIC CHEMISTRY. Educ: City Col New York, BS, 37; Mass Inst Technol, PhD(org chem), 41. Prof Exp: Res chemist, Exp Sta, Hercules Powder Co, 41-44, tech serv rep, 44-48, tech sales rep, 48-53, dist mgr, Synthetics Dept, 53-56, indust sales mgr, 56-59, mgr mkt serv, 59-61, mgr hard resin develop, Pine & Paper Chem Dept, 61-63, sales mgr resins, 63-65, mgr sales develop, 65, mgr mkt develop, 65-67, MGR TECH SERV PINE CHEM, PINE & PAPER CHEM DEPT, HERCULES INC, 67- Concurrent Pos: Dir, Adhesive & Sealant Coun. Mem: Am Chem Soc; Am Soc Testing & Mat; Tech Asn Pulp & Paper Indust. Res: Identification of aromatic hydrocarbons; research on development of synthetic resins; synthetic rosin ester resins; paints; varnishes; enamels; lacquers; adhesives; printing inks; floor tiles. Mailing Add: Pine & Paper Chem Dept Hercules Inc 910 Market St Wilmington DE 19899

AUTHEMENT, RAY PAUL, b Chauvin, La, Nov 19, 29; m 51; c 2. MATHEMATICS. Educ: Southwestern La Inst, BS, 50; La State Univ, MS, 52, PhD, 56. Prof Exp: Instr math, La State Univ, 52-56; assoc prof, McNeese State Col, 56-57; assoc prof, 57-60, vpres, 66-73, PRES, UNIV SOUTHWESTERN LA, 73-, PROF MATH, 60- Concurrent Pos: Vis prof, Univ NC, 62-63. Mem: Math Asn Am. Res: Algebraic number theory; matrix theory; Galois theory. Mailing Add: PO Drawer 1008 Univ of Southwestern La Sta Lafayette LA 70501

AUTIAN, JOHN, b Philadelphia, Pa, Aug 20, 24; m 62; c 1. PHARMACY. Educ: Temple Univ, BS, 50; Univ Md, MS, 52, PhD(pharm chem), 54. Prof Exp: Instr pharm, Univ Md, 52-53, Franklin Sq Hosp Sch Nursing, Md, 53 & Sinai Hosp Sch Nursing, Md, 54; asst prof, Sch Pharm, Temple Univ, 54-56, Sch Pharm, Univ Md, 56-57 & Col Pharm & Rackham Sch Grad Study, Univ Mich, 57-60; assoc prof pharm, Col Pharm, Univ Tex, 60-67, dir drug-plastic res lab, 61-67; PROF PHARMACEUT & DENT & DIR MAT SCI TOXICOL LAB, COL DENT & COL PHARM, UNIV TENN, MEMPHIS, 67-, DEAN COL PHARM, 75- Concurrent Pos: Mem, Nat Formulary Adv Panel, 57; fel, Am Found Pharmaceut Educ; consult, Clin Ctr, NIH, 60- Honors & Awards: Citation, Sch Med, Univ Md, 64. Mem: AAAS (vpres, 62, secy Np-pharm & pharmaceut sci sect, 71-); Am Pharmaceut Asn; Am Chem Soc. Res: Toxicology; plastic materials and their use in pharmacy and medicine. Mailing Add: Col Dent & Col Pharm Univ Tenn Ctr Health Sci Memphis TN 38103

AUTOR, ANNE POMEROY, b Prince George, BC, Jan 26, 35; m 57; c 2. BIOCHEMISTRY. Educ: Univ BC, BA, 56, MSc, 57; Duke Univ, PhD(biochem), 70. Prof Exp: Res asst biochem, Duke Univ, 61-65; asst res biochemist, Univ Mich, 70-72; res assoc, 72-73, ASST PROF PHARMACOL, UNIV IOWA, 73- Concurrent Pos: Damon Runyon fel, 70; Nat Found Basil O'Connor starter grant, 74; Nat Inst Child Health & Human Develop res career develop award, 75. Mem: Am Soc Pharmacol & Exp Therapeut; Soc Toxicol; Biophys Soc. Res: Biochemistry and toxicology of the lung; oxygen and paraquat toxicity; enzymology of the developing lung; pulmonary superoxide dismutase; collagen synthesis; infant and adult respiratory distress syndrome. Mailing Add: Dept of Pharmacol Univ of Iowa Iowa City IA 52242

AUTREY, KENNETH MAXWELL, b Mansfield, La, Nov 11, 17; m 42; c 2. DAIRY SCIENCE, NUTRITION. Educ: La State Univ, BS, 38; Iowa State Univ, MS, 39, PhD(dairy husb, animal nutrit), 41. Prof Exp: Asst prof dairying, Univ Ga, 41-44; assoc prof dairy husb, Pa State Col, 46-47; PROF DAIRY & ANIMAL SCI, AUBURN UNIV, 47- Concurrent Pos: Adv grad prog, Nat Univ Colombia-Colombian Land & Cattle Inst, SAm, 69-71. Mem: AAAS; Am Dairy Sci Asn. Res: Dairy cattle nutrition; physiologic and economic efficiency of rations containing different amounts of grain for dairy cattle; a double changeover design for dairy cattle feeding experiments; physiology of ruminant nutrition; technique for quantitative collection of paratid saliva of the bovine. Mailing Add: Dept of Animal & Dairy Sci Auburn Univ Auburn AL 36830

AUTREY, ROBERT LUIS, b Indio, Calif, Feb 24, 32; m 55; c 3. ORGANIC CHEMISTRY, SCIENCE EDUCATION. Educ: Reed Col, AB, 53; Harvard Univ, AM, 55, PhD(org chem), 58. Prof Exp: Instr chem, Reed Col, 57-58; from instr to asst prof org chem, Univ Rochester, 59-65; lectr, Harvard Univ, 65-67; assoc prof chem, Ore Grad Ctr, 67-72, chmn fac, 70-72; SECY & TREAS, CHIRON PRESS, INC, 71- Concurrent Pos: NSF fel, Imp Col, Univ London, 58-59; asst ed, Am Chem Soc Jour, 62-65. Mem: AAAS; Brit Chem Soc. Res: Organic chemical reactions of synthetic utility; structure and synthesis of natural products and related substances. Mailing Add: Chiron Press Inc 2177 SW Main St Portland OR 97205

AUVIL, PAUL R, JR, b Charleston, WVa, Aug 4, 37; m 60; c 2. ELEMENTARY PARTICLE PHYSICS, THEORETICAL HIGH ENERGY PHYSICS. Educ: Dartmouth Col, BA, 59; Stanford Univ, PhD(physics), 62. Prof Exp: NSF fel physics, Imperial Col, London, 62-64; res assoc, 64-65, asst prof, 65-68, ASSOC PROF PHYSICS, NORTHWESTERN UNIV, 68- Concurrent Pos: Vis scientist, CERN, Switz, 69-70. Mem: Am Phys Soc; Sigma Xi. Res: Theory of elementary particles, especially pion-nucleon scattering, Coulomb corrections, current algebra and symmetry breaking. Mailing Add: Dept of Physics Northwestern Univ Evanston IL 60201

AUXIER, JOHN ALDEN, b Paintsville, Ky, Oct 7, 25; m 48; c 2. HEALTH PHYSICS, NUCLEAR ENGINEERING. Educ: Berea Col, BS, 51; Vanderbilt Univ, MS, 52; Ga Inst Technol, PhD(nuclear eng), 72; Am Bd Health Physics, cert. Prof Exp: Head dept physics & eng, Radiobiol Lab, Univ Tex & US Air Force, 52-55; head radiation dosimetry, 55-72, DIR, HEALTH PHYSICS DIV, OAK RIDGE NAT LAB, 72- Concurrent Pos: Mem subcomt on shielding, Adv Comt Civil Defense, Nat Acad Sci-Nat Res Coun; managing ed, Health Physics J. Mem: Health Physics Soc; AAAS; Am Phys Soc. Res: Nuclear radiation measurements and effects in relation to health physics and radiobiology. Mailing Add: PO Box X Bldg 45005 Oak Ridge TN 37830

AUYANG, KING, b Canton, China. ORGANIC CHEMISTRY. Educ: Sun-Yat-Sen Univ, BS, 50; Temple Univ, MA, 62, PhD(chem), 67. Prof Exp: Chemist electroplating, Sunbeam Mfg Co, Ltd, Hong Kong, 56-59; instr chem, Temple Univ, 66-69; Air Force Off Sci Res fel, Univ Pa, 66-67, NIH fel, Med Sch, 67-68; sr org chemist, 68-70, GROUP LEADER MED CHEM, RES DIV, WILLIAM H RORER, INC, 70- Mem: Am Chem Soc. Res: Design and synthesis of biologically active organic compounds; structure-activity correlations of drugs; organic reaction mechanism. Mailing Add: Res Div William H Rorer Inc Ft Washington PA 19034

AUYONG, THEODORE KOON-HOOK, b Honolulu, Hawaii, Jan 18, 25; m 70. PHARMACOLOGY. Educ: Univ Mo, Kansas City, BSc, 54, MSc, 55; Univ Mo, Columbia, PhD(pharmacol), 62. Prof Exp: Res assoc pharmacol, Univ Mo, Kansas City, 55-56; instr, Univ Mo, Columbia, 62-63; asst prof, 63-72, ASSOC PROF PHARMACOL, SCH MED, UNIV N DAK, 72- Mem: AAAS; Sigma Xi; NY Acad Sci. Res: Isolation of pharmacologically active extracts of plants and determination of its mechanism of action; structure-function relationships of drugs; mechanism of renal toxicity by lithium. Mailing Add: Sch of Med Univ of NDak Grand Forks ND 58202

AVADHANI, NARAYAN G, b Honavar, India, Jan 28, 43; m 72; c 1. BIOCHEMISTRY, MOLECULAR GENETICS. Educ: Karnatak Univ, India, BSc, 61; Univ Bombay, PhD(biochem), 69. Prof Exp: Res assoc physiol, Univ Ill, Urbana, 69-70, res asst prof, 70-72; res investr, 72-73, ASST PROF BIOCHEM, UNIV PA, 73- Concurrent Pos: Consult, Vet Admin Hosp, Coatesville, Pa, 73- Mem: Am Chem Soc; Brit Biochem Soc. Res: Biogenesis of mitochondria with special reference to the information contents and the expression of the mitochondrial genome in normal and tumor cells. Mailing Add: Sch of Vet Med Univ of Pa Philadelphia PA 19174

AVAKIAN, PETER, b Tabriz, Iran, May 15, 33; US citizen; m 57; c 2. POLYMER PHYSICS. Educ: Univ Rochester, BS, 55; Mass Inst Technol, PhD(physics), 60. Prof Exp: Res physicist, Lab Insulation Res, Mass Inst Technol, 60; RES PHYSICIST, CENT RES & DEVELOP DEPT, E I DU PONT DE NEMOURS & CO, 61- Concurrent Pos: Fulbright fel, Univ Stuttgart, 60-61. Mem: AAAS; Am Phys Soc; Optical Soc Am; Sigma Xi. Res: Optical properties; luminescence; energy transfer in molecular crystals; structure-property relationships. Mailing Add: Cent Res & Develop Dept E I du Pont de Nemours & Co Wilmington DE 19898

AVAKIAN, SOUREN, organic chemistry, see 12th edition

AVANN, SHERWIN PARKER, b Seattle, Wash, Mar 11, 13; m 55. MATHEMATICS. Educ: Univ Wash, BS, 38; Calif Inst Technol, MS, 40, PhD(math), 42. Prof Exp: Instr math, Wash Univ, 42-43, Yale Univ, 43-44 & Ore State Col, 44-46; asst prof, 46-62, ASSOC PROF MATH, UNIV WASH, 62- Concurrent Pos: NSF fel, 62-63. Mem: Am Math Soc; Math Asn Am. Res: Lattice theory; graph theory; sigma and tau functions in semi-modular lattices. Mailing Add: C339 Padelford Hall GN 50 Univ of Wash Seattle WA 98195

AVAULT, JAMES W, JR, b East St Louis, Ill, May 20, 35; m 66; c 1. BIOLOGY. Educ: Univ Mo, BS, 61; Auburn Univ, MS, 63, PhD(fisheries), 66. Prof Exp: Res asst fish parasitol, Auburn Univ, 61-63 & biol weed control, 63-66; from asst prof to assoc prof fisheries, 66-75, PROF FISHERIES, LA STATE UNIV, BATON ROUGE, 75- Mem: Am Fisheries Soc; World Maricult Soc (secy-treas, 70-71, vpres, 74, pres, 75); Int Asn Astacology (pres, 74-). Res: Fish culture; marine fisheries; ichthyology; fishery biology. Mailing Add: 1845 Stafford Dr Baton Rouge LA 70808

AVCIN, MATTHEW JOHN, JR, b Pittsburgh, Pa, June 8, 43; m 75. ECONOMIC GEOLOGY, STRATIGRAPHY. Educ: Lafayette Col, BA, 65; Univ Ill, Urbana, MS, 69, PhD(geol), 74. Prof Exp: Res asst geol, Ill State Geol Surv, 69-73; CHIEF COAL SECT GEOL, IOWA STATE GEOL SURV, 73- Concurrent Pos: Staff rep, Nat Gov Conf Coal Mine Reclamation Task Force, 75- Mem: Soc Econ Paleontologists & Mineralogists; Paleont Res Inst; Sigma Xi; AAAS. Res: Coal research and the practical applications of stratigraphy, biostratigraphy and paleobotany. Mailing Add: Iowa Geol Surv 123 N Capitol Iowa City IA 52240

AVE LALLEMANT, HANS GERHARD, b Benkulen, Indonesia, May 2, 38; m 66. STRUCTURAL GEOLOGY. Educ: State Univ Leiden, BSc, 60, MSc, 64, PhD(struct petrol), 67. Prof Exp: Res staff geologist, Yale Univ, 67-70; asst prof geol, 70-74, ASSOC PROF GEOL, RICE UNIV, 74- Mem: AAAS; Am Geophys Union; Geol Soc Am. Res: Structural and experimental structural geology; tectonics; tectonophysics; petrology. Mailing Add: Dept of Geol Rice Univ Houston TX 77001

AVELSGAARD, ROGER A, b Minneapolis, Minn, Jan 5, 32; m 63. ALGEBRA. Educ: Univ Minn, Minneapolis, BChE, 54, MA, 63; Univ Iowa, PhD(math), 69. Prof Exp: Teaching assoc math, Univ Minn, 59-63; asst prof, 64-66, ASSOC PROF MATH, BEMIDJI STATE COL, 69- Mem: Am Math Soc; Math Asn Am. Res: Cohomology of non-associative algebra. Mailing Add: Dept of Math Bemidji State Col Bemidji MN 56601

AVEN, MANUEL, b Tallinn, Estonia, Dec 25, 24; nat US; m 47; c 2. SOLID STATE PHYSICS. Educ: Univ Pittsburgh, BS, 51, PhD(phys chem), 55. Prof Exp: Chemist, Mercy Hosp, 50-51; res asst, Univ Pittsburgh, 51-55, lectr, 54; res phys chemist, Lamp Develop Lab, Cleveland, Ohio, 55-59, phys chemist, Res Lab, Schenectady, NY, 59-68, mgr luminescence br, Gen Elec Res & Develop Ctr, 68-72, MGR, PHYS CHEM LAB, GEN ELEC RES & DEVELOP CTR, GEN ELEC CO, SCHENECTADY, 72- Mem: Am Chem Soc; Am Inst Chemists; Am Phys Soc. Mailing Add: 38 Forest Rd Burnt Hills NY 12027

AVENA, REMEDIOS M, b Aringay, Philippines, June 18, 28; US citizen; m 64. BIOCHEMISTRY, LIPID CHEMISTRY. Educ: Univ Philippines, BS, 53; Georgetown Univ, PhD(biochem), 62. Prof Exp: Asst res instr agr chem, Col Agr, Univ Philippines, 54-58; res asst, Georgetown Univ, 59-62; res assoc biochem, Pomona Col, 63-64 & Food Res Inst, Chicago, 64-65; res biochemist, NIH, 66-69, sr res fel, 69-71; LECTR BIOCHEM, UNIV MD, 74- Mem: AAAS; Sigma Xi. Res: Fatty acid synthesis; contractile adenosine triphosphatases. Mailing Add: 8717 Sleepy Hollow Ln Potomac MD 20854

AVENI, ANTHONY, b New Haven, Conn, Mar 5, 38; m 59; c 2. ASTRONOMY. Educ: Boston Univ, AB, 60; Univ Ariz, PhD(astron), 65. Prof Exp: From instr to asst prof, 63-68, ASSOC PROF ASTRON, COLGATE UNIV, 68- Concurrent Pos: NSF grants, 66-71; vis assoc prof, Univ SFla, 71; consult, Ferson Optical Div, Bausch & Lomb. Mem: AAAS; Am Astron Soc. Res: Galactic structure; star formation; works on observational evidence relating to formation of stars in low mass primary condensations; astroarcheology of Mesoamerica; studies on orientation of buildings in Maya and Zapotec zone of Mexico. Mailing Add: Dept of Physics & Astron Colgate Univ Hamilton NY 13346

AVENS, JOHN STEWART, b Geneva, NY, Mar 23, 40; m 63; c 2. FOOD MICROBIOLOGY. Educ: Syracuse Univ, BS, 62; Colo State Univ, MS, 69, PhD(avian sci), 72. Prof Exp: Res technician hemat, NY Univ Med Ctr, 62-63; res scientist food microbiol, Syracuse Univ Res Corp, 63-65; sr microbiologist, Food & Drug Admin, Buffalo, 65-67; from instr to asst prof, 67-74, ASSOC PROF AVIAN SCI & FOOD MICROBIOL, COLO STATE UNIV, 74- Mem: Am Soc Microbiol; Inst Food Technologists; Int Asn Milk Food & Environ Sanitarians; Poultry Sci Asn; World Poultry Sci Asn. Res: Improvement of microbiological analytical methodology for food; microbiology related to food safety in food processing and food service establishments; food egg production from ducks. Mailing Add: Dept of Animal Sci Colo State Univ Ft Collins CO 80523

AVENT, JON C, b Billings, Mont, July 4, 34. GEOLOGY. Educ: Univ Colo, BA, 56; Univ Wash, Seattle, MS, 62, PhD(phys stratig), 65. Prof Exp: Asst prof, 65-74, ASSOC PROF GEOL, CALIF STATE UNIV, FRESNO, 74- Mem: Geol Soc Am; Am Asn Petrol Geol. Res: Cenozoic stratigraphy and structure; trace element distribution in basalt flows. Mailing Add: Dept of Geol Calif State Univ Fresno CA 93726

AVERA, FITZHUGH LEE, b Pocahontas, Ark, July 4, 06; m 50; c 1. FOOD SCIENCE, CHEMISTRY. Prof Exp: Sales mgr, 26-38, chief chemist, 38-43, DIR RES, BEST FOODS DIV, CORN PROD CO, SOUTHERN CALIF, 46- Concurrent Pos: Consult, CPC Int Inc. Mem: Nat Soc Prof Engrs; Am Oil Chem Soc; NY Acad Sci. Res: Edible fats and oils; quantal chemical analysis; food products. Mailing Add: 1809 Yale Dr Alameda CA 94501

AVERELL, JOHN P, applied physics, see 12th edition

AVERETT, JOHN E, b Coleman, Tex, Apr 19, 43; m 63; c 2. SYSTEMATIC BOTANY, PHYTOCHEMISTRY. Educ: Sul Ross State Univ, BS, 66, MA, 67; Univ Tex, Austin, PhD(bot), 70. Prof Exp: ASST PROF BIOL, UNIV MO-ST LOUIS, 70- Prof Exp: Res assoc, Mo Bot Garden, 70- Mem: Int Asn Plant Taxon; Phytochem Soc NAm; Am Soc Plant Taxon; Bot Soc Am. Res: Biosystematic problems in which chemical and experimental techniques may be effectively employed; research in Chamaesaracha and in Gaillardia. Mailing Add: Dept of Biol Univ of Mo 8001 Natural Bridge Rd St Louis MO 63121

AVERILL, FRANK WALLACE, b Anniston, Ala. QUANTUM MECHANICS, APPLIED MATHEMATICS. Educ: Univ Fla, BS, 67, PhD(physics), 71. Prof Exp: Res assoc quantum mech, Univ Fla, 72; res assoc, Northwestern Univ, 72-73; ASST PROF MATH, JUDSON COL, 73- Mem: Math Asn Am; Am Phys Soc. Res: Developing computer programs which produce approximate solutions to Schroedingers equation for large electronic systems. Mailing Add: Dept of Math Judson Col Elgin IL 60120

AVERILL, SEWARD JUNIOR, b Punxsutawney, Pa, Jan 5, 21; m 43; c 3. RUBBER CHEMISTRY. Educ: Kent State Univ, BS, 42. Prof Exp: Instr chem, Williams Col, Mass, 43-45; res chemist, 46-49, mgr raw mat develop, 59-70, SR DEVELOP ENGR, B F GOODRICH CO, 70- Concurrent Pos: Consult, Sprague Elec Co, 43-45. Mem: Am Chem Soc; Am Soc Testing & Mat. Res: Rubber reinforcement and natural rubber technical characterization; rubber compounding; organo metallics; synthesis and reactions of vinylidene cyanide; plasticizers for vinyl polymers. Mailing Add: B F Goodrich Co D/2055 B/25-A Akron OH 44318

AVERITT, PAUL, b Lexington, Ky, Sept 1, 08; m 45; c 2. ECONOMIC GEOLOGY. Educ: Univ Ky, BS, 30, MS, 31. Prof Exp: Asst, Northwestern Univ, 31-33; instr geol, Univ Ky, 34-38; jr geologist, 37-40, asst geologist, 40-42, assoc geologist, 42-44, geologist, 44-47, sr geologist, 47-49, chief coal resources sect, 49-54, geologist, Fuels Br, 54-72, GEOLOGIST, BR COAL RESOURCES, US GEOL SURV, 72- Honors & Awards: Distinguished Serv Award, US Dept Interior, 72. Mem: Fel Geol Soc Am; Soc Econ Geol; Am Asn Petrol Geol. Res: Geology of fuels; coal resources. Mailing Add: Br Coal Resources US Geol Surv Denver CO 80225

AVERRE, CHARLES WILSON, III, b Puerto Castilla, Honduras, June 3, 32; US citizen; m 55; c 2. PLANT PATHOLOGY. Educ: NC State Univ, BS, 55, MS, 60; Purdue Univ, PhD(plant path), 63. Prof Exp: Asst plant pathologist, Subtrop Exp Sta, Univ Fla, 63-67; asst prof plant path, Ga Exp Sta, 67-68; EXTEN ASSOC PROF PLANT PATH, NC STATE UNIV, 68- Concurrent Pos: Consult, PPG Indust, Inc, 73; ed, Fungicide & Nematicide Tests, Am Phytopath Soc, 74- Mem: Am Phytopath Soc. Res: Extension; diseases of vegetable crops; disease control systems; evaluation of fungicides and nematicides. Mailing Add: Dept of Plant Path NC State Univ PO Box 5397 Raleigh NC 27607

AVERS, CHARLOTTE JO, b Brooklyn, NY, Apr 4, 26. CYTOGENETICS. Educ: Cornell Univ, BS, 48; Ind Univ, PhD(bot), 53. Prof Exp: Instr & res assoc bot, Conn Col, 53-56; asst prof, Univ Miami, 56-59; assoc prof, 59-69, PROF BOT, DOUGLASS COL, RUTGERS UNIV, 69- Mem: AAAS; Soc Develop Biol; Bot Soc Am; Genetics Soc Am; Am Soc Cell Biol. Res: Organelle genetic systems; peroxisome functions; mitochondria. Mailing Add: Dept of Biol Sci Douglass Col Rutgers The State Univ New Brunswick NJ 08903

AVERY, GORDON B, b Beirut, Lebanon, Dec 10, 31; US citizen; m 54; c 3. MEDICINE, EMBRYOLOGY. Educ: Harvard Univ, AB, 53; Univ Pa, MD, 58, PhD(embryol), 59; Am Bd Pediat, dipl. Prof Exp: Dir clin res ctr, Children's Hosp DC, 66-71, DIR NEWBORN NURSERY, CHILDREN'S HOSP DC, 63-, PROF CHILD HEALTH, SCH MED, GEORGE WASHINGTON UNIV, 71- Concurrent Pos: Grants, Gen Res Support, 63-65, Children's Bur, 65-67, NSF, 65-67 & Clin Res Ctr, 66-73; consult, Wash Hosp Ctr, Bethesda Naval Hosp & Nat Inst Child Health & Human Develop, 64-; assoc prof pediat, Sch Med, George Washington Univ, 67-71; mem comt ment retardation res & training, NIH, 69- Honors & Awards: Leavonson Prize, 57; Pepper Prize, 58. Mem: AMA; Am Acad Pediat; Transplantation Soc; Soc Pediat Res; Am Pediat Soc. Res: Pediatrics; neonatology; immunology; genetics. Mailing Add: Children's Hosp of DC 2125 13th St NW Washington DC 20009

AVERY, JAMES KNUCKEY, b Holly, Colo, Aug 6, 21; m 50; c 3. EMBRYOLOGY, ANATOMY. Educ: Univ Kansas City, DDS, 45; Univ Rochester, BA, 48, PhD(anat), 52. Prof Exp: Res assoc & instr, Univ Rochester, 52-54; from asst prof to assoc prof dent, Sch Dent, 54-63, assoc prof anat, Sch Med, 61-71, PROF ANAT, SCH MED, UNIV MICH, ANN ARBOR, 71-, PROF DENT, SCH DENT, 63- Concurrent Pos: Dir, Dent Res Inst, Univ Mich & res & educ consult, Ann Arbor Vet Admin Hosp, 63-; consult dent training comt, NIH. Mem: Fel AAAS (chmn sect dent, 75-76); fel Am Col Dent; Int Asn Dent Res (pres, 74-75); Am Asn Anat; Electron Micros Soc Am. Res: Embryology of teeth, jaws, and face, comparative and human; histology and histochemistry of oral and related structures; in vivo and in vitro study of developing organs and structures; cytology of formative cells by means of light, phase, interference and electron microscopy. Mailing Add: Dent Res Inst Univ of Mich Sch of Dent Ann Arbor MI 48104

AVERY, JOHN SCALES, b Beirut, Lebanon, May 26, 33; US citizen; m 59; c 3. QUANTUM CHEMISTRY. Educ: Mass Inst Technol, BSc, 54; Univ Chicago, MSc, 55; Univ London, PhD(chem), 65. Prof Exp: Lectr theoret physics, Tait Inst, Univ Edinburgh, 61-62; lectr theoret chem, Imp Col, Univ London, 63-73; LECTR, UNIV COPENHAGEN, 73- Concurrent Pos: Res assoc, Inst Muscle Res, Woods Hole, 61-62; Royal Soc Europ Prog fel, Univ Copenhagen, 68-69, amanuensis vikar, 69-70; managing ed, J Bioenergetics, 69- Mem: Royal Inst Gt Brit; Danish Chem Soc. Res: Bioenergetics; exciton theory; resonance energy transfer theory; applications of Fourier transforms; second quantization. Mailing Add: H C Ørsted Inst Universitetsparken 5 2100 Copenhagen Ø Denmark

AVERY, MARY ELLEN, b Camden, NJ, May 6, 27. MEDICINE, PEDIATRICS. Educ: Wheaton Col, Mass, BA, 48; Johns Hopkins Univ, MD, 52. Hon Degrees: MA, Harvard Univ; ScD, Wheaton Col, Mass & Univ Mich. Prof Exp: From asst prof to assoc prof pediat, Johns Hopkins Univ, 61-69; prof, Montreal Children's Hosp, McGill Univ, 69-74; THOMAS MORGAN ROTCH PROF PEDIAT, HARVARD MED SCH, 74-; PHYSICIAN-IN-CHIEF, CHILDREN'S HOSP, BOSTON, 74- Concurrent Pos: Res fel pediat, Harvard Med Sch, 57-59; Nat Inst Neurol Dis & Blindness spec trainee, 57-60; fel, Sch Med, Johns Hopkins Univ, 59-60; USPHS res grants, 61-; Markel scholar acad med, 61-; pediatrician in chg newborn nurseries, Johns Hopkins Hosp, 61-69; consult, US Dept Health, Educ & Welfare, 64; mem, Med Res Coun Can, 70-74. Mem: Fel Am Acad Pediat; Soc Pediat Res; Am Pediat Soc; Am Physiol Soc; Am Acad Arts & Sci. Res: Respiratory physiology of newborn infants; surface tension in lungs; lung metabolism. Mailing Add: Children's Hosp 300 Longwood Ave Boston MA 02115

AVERY, ROBERT, b Ft Worth, Tex, Sept 7, 21. THEORETICAL PHYSICS. Educ: Univ Ill, BS, 43; Univ Wis, MS, 47, PhD(physics), 50. Prof Exp: Assoc physicist, 50-56, sr physicist & head theoret physics sect, Reactor Eng Div, 56-63, DIR, REACTOR PHYSICS DIV, ARGONNE NAT LAB, 63- Mem: Am Phys Soc; Am Nuclear Soc. Res: Nuclear reactor theory. Mailing Add: Reactor Physics Bldg Argonne Nat Lab 9700 S Cass Ave Argonne IL 60440

AVERY, THOMAS EUGENE, forestry, see 12th edition

AVERY, WILLIAM HINCKLEY, b Ft Collins, Colo, July 25, 12; m 38; c 2. PHYSICAL CHEMISTRY, PHYSICS. Educ: Pomona Col, AB, 33; Harvard Univ, AM, 35, PhD(photochem), 37. Prof Exp: Asst chem, Harvard Univ, 34-39; res chemist, Shell Oil Co, 39-43; res assoc & chief propellants div, Nat Defense Res Comt, Allegany Ballistics Lab, 43-46; res assoc, A D Little, Inc, 46-47; res assoc, Appl Physics Lab, 47-50, supvr, Launching & Propulsion, 50-54, Res & Develop, 54-60 & Appl Res, 60-73, ASST DIR EXPLOR DEVELOP, APPL PHYSICS LAB, JOHNS HOPKINS UNIV, 73-, SUPVR, AERONAUT DIV, 61- Concurrent Pos: Mem res adv comt air breathing propulsion systs, NASA; mem advan systs panel, Hwy Res Bd, 70- Honors & Awards: President's Cert Merit, 47; Hickman Medal, 50; Sir Alfred C Egerton Gold Medal, Combustion Inst, 71. Mem: Am Chem Soc; Am Inst Aeronaut & Astronaut; Combustion Inst. Res: Photochemistry; spectroscopy; molecular structure; chemical kinetical rocket propellants and internal ballistics; ramjet propulsion; combustion; urban transportation; ocean thermal energy conversion. Mailing Add: 724 Guilford Ct Silver Spring MD 20901

AVGEROPOULOS, GEORGE N, b Greece, Jan 25, 34; m 64; c 1. PHYSICAL CHEMISTRY, QUANTUM CHEMISTRY. Educ: Nat Univ Athens, MA, 61; Wayne State Univ, MSc, 67, PhD(quantum chem), 68. Prof Exp: Chief res chemist, Michaelides Food Indust, Greece, 61-62; asst chem, Wayne State Univ, 62-68; eng scientist, 68-72, POLYMER PHYSICIST-RHEOLOGIST, CENT RES LABS, FIRESTONE TIRE & RUBBER CO, 72- Mem: Am Chem Soc; Am Phys Soc; Soc Rheol; AAAS; Asn Greek Chemists. Res: Applied mathematics; rheology of polymers and polymeric blends; mechanical properties and characterization of polymers; morphology of heterogeneous blends; mathematical models for the study of synthesis and characterization of polymers. Mailing Add: 1556 Kingsley Ave Akron OH 44313

AVIADO, DOMINGO MARIANO, b Manila, Philippines, Aug 28, 24; m 53; c 4. PHARMACOLOGY, PHYSIOLOGY. Educ: Univ Pa, MD, 48. Prof Exp: From asst instr to instr, 48-50, assoc, 50-52, from asst prof to assoc prof, 52-65, PROF PHARMACOL, MED SCH, UNIV PA, 65- Concurrent Pos: NIH res fel, 48-50; Guggenheim Found fel, 62-63; assoc ed, Circulation Res, 58-62; treas sect pharmacol, Int Union Physiol Sci, 59- Mem: AAAS; AMA; Am Soc Clin Pharmacol & Therapeut; Am Soc Pharmacol & Exp Therapeut; Am Physiol Soc. Res: Respiratory and circulatory reflexes; physiology and pharmacology of pulmonary circulation. Mailing Add: Dept of Pharmacol Univ of Pa Sch of Med Philadelphia PA 19174

AVIGAD, GAD, b Jerusalem, Israel, Apr 30, 30; nat US; m 65; c 1. BIOLOGICAL CHEMISTRY. Educ: Hebrew Univ Jerusalem, MSc, 55, PhD(biochem), 58. Prof Exp: Asst biochem, Hadassah Med Sch, Hebrew Univ Jerusalem, 54-57, instr, 57-59; res fel, Med Col, NY Univ, 59-61; sr lectr, Hebrew Univ Jerusalem, 61-65, assoc prof, 65-69; ASSOC PROF BIOCHEM, RUTGERS MED SCH, 70- Concurrent Pos: Vis scientist, Carlsberg Labs, Copenhagen, Denmark, 58; Jane Coffin Childs Mem Fund fel, 60-61; vis asst prof, Albert Einstein Col Med, 64, vis assoc prof, 67-70. Mem: Soc Gen Microbiol; NY Acad Sci; Am Chem Soc; Am Soc Microbiol; Am Soc Biol Chem. Res: Carbohydrate chemistry; biosynthesis and degradation of polysaccharides; mechanisms of enzyme action; structure and function in microorganisms. Mailing Add: Dept of Biochem Rutgers Med Sch PO Box 101 Piscataway NJ 08854

AVIGAN, JOEL, b Warsaw, Poland, Jan 8, 20; m 48; c 3. BIOCHEMISTRY. Educ: Hebrew Univ, Israel, MSc, 43; McGill Univ, PhD(biochem), 54. Prof Exp: Res chemist, Israel Res Coun, 50-52; vis scientist, Bio-Med Res, Lab Metab, 55-64, RES CHEMIST, MOLECULAR DIS BR, BIO-MED RES, NAT HEART & LUNG INST, 64- Concurrent Pos: Vis fel, Hadassah Med Sch, Hebrew Univ, 63; vis scientist, Lab Chem Enzymol, Shell Res Ltd, Eng, 63; mem coun arteriosclerosis, Am Heart Asn. Mem: Am Soc Biol Chem. Res: Biochemistry of sterols and fatty acids in relation to atherosclerosis and to other medical problems. Mailing Add: Molecular Dis Br Nat Heart Inst Bethesda MD 20014

AVIGNONE, FRANK TITUS, III, b New York, NY, May 9, 32; m 54; c 2. NUCLEAR PHYSICS. Educ: Ga Inst Technol, BS, 60, MS, 62, PhD(physics), 65. Prof Exp: From asst prof to assoc prof physics, 65-73, PROF PHYSICS & ASTRON, UNIV SC, 73- Res: Low energy nuclear physics. Mailing Add: Dept of Physics Univ of SC Columbia SC 29208

AVILA, VERNON LEE, b Segundo, Colo, Apr 5, 41; m 62; c 2. COMPARATIVE ENDOCRINOLOGY, ANIMAL BEHAVIOR. Educ: Univ NMex, BS, 62; Northern Ariz Univ, MA, 66; Univ Colo, PhD(biol), 73. Prof Exp: Teacher, Los Lunas High Sch, 62-65 & Valley High Sch, 65-71; lectr biol, Univ Colo, 72-73; ASST PROF ZOOL, SAN DIEGO STATE UNIV, 73- Concurrent Pos: Sci adv, Goodyear Publ Co, 74-; sci consult, Scott Foresman & Co Publ, 75- Mem: AAAS; Am Soc Zoologists; Animal Behav Soc; Am Inst Biol Sci. Res: Endocrinological aspects of behavior in teleost fishes; endocrine relationships in marsupial frog brood pouch formation; hormonal aspects of behavior in vertebrates. Mailing Add: Dept of Zool San Diego State Univ San Diego CA 92182

AVILES, JOSEPH B, b Atlantic City, NJ, Aug 21, 27; m 59; c 2. THEORETICAL PHYSICS. Educ: Rutgers Univ, BS, 50; Johns Hopkins Univ, PhD, 58. Prof Exp: THEORET PHYSICIST, US NAVAL RES LAB, 56- Res: Many-body systems with strong interactions; nuclear physics. Mailing Add: US Naval Res Lab Washington DC 20390

AVIOLI, LOUIS, b Coatesville, Pa, Apr 13, 31; m 55; c 5. NUCLEAR MEDICINE, ENDOCRINOLOGY. Educ: Princeton Univ, AB, 53; Yale Univ, MD, 47. Prof Exp: From intern to resident med, NC Mem Hosp, 57-59; clin investr, NIH, 59-61; from instr to asst prof med, Seton Hall Col Med, 61-66, dir isotope lab & asst sci dir clin res ctr, 61-66; asst prof, 66-70, PROF MED, MED SCH, WASH UNIV, 70-; CHIEF ENDOCRINOL & METAB, JEWISH HOSP ST LOUIS, 66- Concurrent Pos: Grants, Nat Inst Arthritis & metab Dis, 60-67, AEC, 62-67 & NIH, 65-70; attend physician, Jersey City Med Ctr, 62-66. Mem: AAAS; Soc Nuclear Med; Radiation Res Soc; Am Nuclear Soc; Am Fedn Clin Res. Res: Metabolism; radioactive isotopes. Mailing Add: Jewish Hosp of St Louis 216 South Kingshighway St Louis MO 63110

AVIRAM, ARI, b Braila, Romania, Aug 18, 37; m 59; c 3. ORGANIC CHEMISTRY, CHEMICAL PHYSICS. Educ: Israel Inst Technol, BSc, 65; NY Univ, MSc, 71, PhD(theoret chem), 75. Prof Exp: Synthetic chemist pharmaceut, Ayerst Lab, Can, 67-68; RES STAFF MEM ORG & THEORET CHEM, T J WATSON RES LAB, IBM CORP, 68- Mem: Am Chem Soc. Res: The ability of molecules to perform device functions; molecular rectifiers and molecular memory elements; theoretical and synthetic aspects. Mailing Add: T J Watson Res Ctr IBM Corp Yorktown Heights NY 10598

AVIS, KENNETH EDWARD, b Elmer, NJ, June 3, 18; m 43; c 3. PHARMACEUTICS. Educ: Philadelphia Col Pharm, BSc, 42, MSc, 47, DSc(pharm), 56. Prof Exp: From instr to assoc prof pharm, Philadelphia Col Pharm, 46-61; assoc prof pharmaceut, 61-68, vchmn dept, 68-72, PROF PHARMACEUT, COL PHARM,

UNIV TENN CTR HEALTH SCI, 68-, DIR DIV PARENTERAL MEDICATIONS, 72- Concurrent Pos: Consult, Jefferson Med Col Hosp, 56-60, Carron Prod Co, 59-67, Nat Cancer Inst, 66-67, Food & Drug Admin, 68- & Vet Admin Hosp, Memphis; mem, Nat Formulary Comt on Specifications, 70-74; mem nat joint coord comt on large vol parenterals, US Pharmacopeia & Food & Drug Admin, 72- Honors & Awards: Parenteral Drug Asn Res Award, 71 & 74. Mem: Am Soc Hosp Pharmacists; Am Pharmaceut Asn; Am Asn Cols Pharm; fel Acad Pharmaceut Sci; Parenteral Drug Asn (dir, 60-, pres, 68-69). Res: Sterile pharmaceutical products; formulation, evaluation and effects of environmental and processing factors; radiopharmaceuticals. Mailing Add: Room 107 Col of Pharm Univ of Tenn Health Sci Ctr Memphis TN 38103

AVIZONIS, PETRAS V, b Lithuania, Aug 17, 35; US citizen; m 59; c 2. LASERS, PHYSICAL OPTICS. Educ: Duke Univ, BS, 57; Univ Del, MS, 59, PhD(phys chem), 62. Prof Exp: Chemist, Sun Oil Co, Pa, 61-62; sr proj scientist, 62-68, tech dir laser div, 68-71, TECH DIR, ADVAN RADIATION TECHNOL OFF, US AIR FORCE WEAPONS LAB, 71- Honors & Awards: Arthur S Flemming Award, Civil Serv Comn, Washington, DC, 71. Mem: Am Phys Soc; Am Optical Soc. Res: High energy lasers; optical systems; exited state chemistry and physics. Mailing Add: US Air Force Weapons Lab Kirtland AFB NM 87117

AVOLIZI, ROBERT JOSEPH, b Worcester, Mass, Apr 5, 38; m 66; c 2. ZOOLOGY. Educ: Mass State Col, Worcester, BS, 62, MEd, 65; Syracuse Univ, MS, 66, PhD(zool), 70. Prof Exp: ASST PROF BIOL, AM UNIV BEIRUT, 70- Mem: AAAS; Ecol Soc Am; Am Soc Zool; Malacol Soc London. Res: Marine and fresh water invertebrate ecological physiology; animal population dynamics and energy flow; molluscan biology. Mailing Add: Dept of Biol Am Univ of Beirut Beirut Lebanon

AVONDA, FRANK PETER, b New York, NY, Nov 26, 24; m 56; c 3. ORGANIC CHEMISTRY. Educ: City Col New York, BS, 48; Columbia Univ, AM, 49; Ohio State Univ, PhD(chem), 53. Prof Exp: Res assoc org fluorides, Duke Univ, 53-54; res chemist, Hooker Electrochem Co, 54-56; instr org chem, Univ Tampa, 56-58; asst prof, Univ Md, 58-59 & Seton Hall Univ, 59-61; proj leader res & develop, Maxwell House Div, Gen Foods Corp, 61-62; prof Am indust technol, Ger Employees Acad, 62-63; sr res chemist, Allied Chem Corp, NJ, 63-64; assoc prof chem, Northern State Col, 64-65; prof, Nicholls State Col, 65-67; chmn div sci & math, Southern State Col, SDak, 67-68; prof chem & asst dean arts & sci, Delgado Col, 68-69; PROF CHEM, NICHOLLS STATE UNIV, 69- Concurrent Pos: NSF grant, 70-72. Mem: AAAS; Am Chem Soc; Am Inst Chem; Brit Chem Soc. Res: Synthesis and chemistry of organic fluorine compounds. Mailing Add: Dept of Chem Nicholls State Univ Thibodaux LA 70301

AVRAMI, LOUIS, b Atlantic City, NJ, May 7, 22; m 61; c 3. PHYSICS. Educ: Rutgers Univ, BS, 49; Stevens Inst Technol, MS, 52. Prof Exp: Physicist, Atomic Ammunition Br, 50-55, asst chief, 55-60, supvry physicist, Reactor Requirements Off, 60-62, actg chief, 62-63, chief physics br, Explosives Res Lab, Feltman Res Labs, 63-68, CHIEF RADIATION EFFECTS & RES SUPPORT BR, EXPLOSIVE RES LAB, FELTMAN RES LABS, PICATINNY ARSENAL, 68- Honors & Awards: Res & Develop Award, US Army Picatinny Arsenal, 73. Mem: Sigma Xi; Am Phys Soc; Am Nuclear Soc. Res: Radiation effects on explosives; explosives research; detonation physics; reactor physics, site survey, hazards analysis for steady-state research reactor; high speed photography. Mailing Add: Explosives Res Lab Feltman Res Labs Picatinny Arsenal Dover NJ 07801

AVRETT, EUGENE HINTON, b Atlanta, Ga, Oct 28, 33; m 61; c 2. ASTROPHYSICS. Educ: Ga Inst Technol, BS, 57; Harvard Univ, PhD(physics), 62. Prof Exp: PHYSICIST, SMITHSONIAN ASTROPHYS OBSERV, 62- Concurrent Pos: Lectr, Harvard Univ, 64- Res: Radiative transfer in stellar atmospheres; spectral line formation; applied mathematics. Mailing Add: Smithsonian Astrophys Observ 60 Garden St Cambridge MA 02138

AWAD, ALBERT T, b Cairo, Egypt, Nov 25, 28; m 58; c 3. NATURAL PRODUCTS CHEMISTRY, PHARMACOGNOSY. Educ: Cairo Univ, BSc, 52, MSc, 60; Ohio State Univ, PhD(pharmacog), 66. Prof Exp: Pharmacist, Drug Stores, Egypt, 52-54; analyst, Oil & Soap Factory, 54-55; instr & researcher pharmacog, Col Pharm, Cairo Univ, 56-61; teacher & res asst, Col Pharm, Ohio State Univ, 61-66; assoc prof, 66-71, PROF PHARMACOG, COL PHARM, OHIO NORTHERN UNIV, 71- Concurrent Pos: Lederle pharm fac award, 67; Gilford instrument apparatus grant, 68; NSF col sci improvement grant, 69-73. Mem: Am Asn Col Pharm; Am Soc Pharmacog; Acad Pharmaceut Sci; Marine Technol Soc. Res: Medicinal plant alkaloids and glycosides; study of active principles in medicinal plants, including the screening, isolation and charaacterization of these constituents which might possess certain pharmacological and clinical activities in treatment. Mailing Add: Raabe Col Pharm Ohio Northern Univ Ada OH 45810

AWAD, ESSAM A, b Cairo, Egypt, Mar 5, 32; m 65; c 5. PHYSICAL MEDICINE & REHABILITATION. Educ: Cairo Univ, MD, 57; Univ Minn, PhD(phys med rehab), 64. Prof Exp: Instr phys med & rehab, Univ Minn, Minneapolis, 64; from asst prof to assoc prof, Temple Univ, 65-67; lectr, Med Sch, Al-Azhar Univ, 67-68; PROF PHYS MED & REHAB, UNIV MINN, MINNEAPOLIS, 69- Honors & Awards: Gold Medal Award, Am Acad Phys Med & Rehab, 66. Mem: Am Cong Rehab Med; NY Acad Sci. Res: Neuromuscular diseases; electromyography and nerve conduction studies. Mailing Add: Univ of Minn Hosps Box 297 Minneapolis MN 55455

AWAD, MOHAMED ZEINELABIDEEN, b Kafr-Sakr, Egypt, Nov 3, 25; US citizen; m; c 3. PHYSIOLOGY. Educ: Univ Alexandria, BSc, 50; Univ London, PhD(physiol), 54. Prof Exp: Instr physiol, Sch Med, Univ Alexandria, 50-51; res fel, King's Col, Univ London, 51-54; lectr, Sch Med, Univ Cairo, 54-59; vis lectr, Univ Khartoum, 59-61; prof, Sch Med, Univ Assiut, Egypt, 61-68; res prof, Psychiat Inst, Univ Md, 68-71; PROF PHYSIOL, NAT COL CHIROPRACTIC, 71- Concurrent Pos: Fulbright exchange fel, USA, 61-62; consult, Medi-Ctr Lab Inc, Md, 70- & Inter-African Soc on Traditional Pharmcopoeia & African Med Plants, 71- Honors & Awards: Order of Merit, Egyptian Govt, 59. Mem: Islamic Med Asn US & Can; AAAS; Egyptian Physiol Soc. Res: Studies on mechanisms of memory; function of the spine; physiological effects of the active principals of medicinal plants. Mailing Add: 827 E School St Lombard IL 60148

AWAD, WILLIAM MICHEL, JR, b Shanghai, China, Nov 5, 27; US citizen; m 57; c 2. INTERNAL MEDICINE, ONCOLOGY. Educ: Manhattan Col, BS, 50; State Univ NY, MD, 54; Univ Wash, PhD(biochem), 65. Prof Exp: Intern med, Kings County Hosp, Brooklyn, NY, 54-55; asst resident path, St Vincent's Hosp, Manhattan, 55-56; asst resident med, Bronx Munic Hosp Ctr, 56-57; physician, NY Tel Co, 59-60; asst prof, 65-73, ASSOC PROF MED, UNIV MIAMI, 73-, ASSOC PROF BIOCHEM, 74- Mem: Am Soc Biol Chem; Am Asn Oncol Soc; Am Chem Soc; Biochem Soc. Res: Protein and peptide chemistry; biosynthesis of methyl containing metabolites; metabolism of ethanol. Mailing Add: Univ of Miami Sch of Med PO Box 520875 Biscayne Annex Miami FL 33152

AWADA, MINORU, b Hilo, Hawaii, Nov 23, 14; m 49; c 2. PLANT PHYSIOLOGY. Educ: Univ Hawaii, BS, 38, MS, 49. Prof Exp: Asst pineapple prod, Hawaiian Pineapple Co, 39-45; asst plant physiologist, 59-70, ASSOC PLANT PHYSIOLOGIST, AGR EXP STA, UNIV HAWAII, 70- Mem: Am Soc Hort Sci; Am Soc Plant Physiol; Am Soc Agron. Res: Sex expression of papaya; soil moisture in relation to growth, yield, fruit development and quality of horticultural plants; mineral nutrition of papaya plants. Mailing Add: Agr Exp Sta Plant Sci Bldg Univ Hawaii 3190 Maile Way Honolulu HI 96822

AWAPARA, JORGE, b Arequipa, Peru, Dec 15, 18; nat US; m 61; c 1. BIOCHEMISTRY. Educ: Mich State Univ, BS, 41, MS, 42; Univ Southern Calif, PhD(biochem), 47. Prof Exp: Biochemist, Univ Tex M D Anderson Hosp & Tumor Inst, 48-57, assoc biochemist, 57-74; assoc prof biol, 57-62, PROF BIOCHEM, RICE UNIV, 62- Mem: AAAS; Am Soc Biol Chem. Res: Intermediary metabolism of amino acids; metabolism of amino acids in the brain; mechanism of enzymatic decarboxylation of amino acids in the brain; sulfur metabolism. Mailing Add: Dept of Biochem Rice Univ Houston TX 77001

AWASTHI, YOGESH C, b July 13, 39; Indian citizen; m 61; c 4. BIOLOGICAL CHEMISTRY. Educ: Univ Lucknow, India, BS, 57, MS, 59, PhD(chem), 67. Prof Exp: Jr res fel plant chem, 59-60, from sr sci asst to jr sci officer, 60-70, SR SCI OFFICER PLANT CHEM, NAT BOT GARDENS, LUCKNOW, INDIA, 70-; SR RES ASSOC, UNIV TEX MED BR, GALVESTON, 74- Concurrent Pos: Fel biol sci, Purdue Univ, 68-70; res assoc hemat, City of Hope Nat Med Ctr, Duarte, Calif. Res: Biochemical genetics of Tay-Sach's disease and other lysosomal storage diseases; red cell metabolism with particular interest in the metabolism of glutathione; protein chemistry and enzymology. Mailing Add: Dept Human Biol Chem & Genetics Hendrix Bldg Univ Tex Med Br Galveston TX 77550

AWBREY, FRANK THOMAS, b Carlsbad, NMex, Oct 16, 32; m 56; c 3. BIOLOGY. Educ: Univ Calif, Riverside, AB, 60; Univ Tex, MA, 63, PhD(bioacoust), 65. Prof Exp: Asst prof, 64-68, ASSOC PROF BIOL, SAN DIEGO STATE UNIV, 68- Concurrent Pos: NSF res grant, 65-67. Mem: AAAS; Soc Study Evolution; Am Soc Ichthyologists & Herpetologists; Sigma Xi. Res: Biological significance of vocalization in amphibians. Mailing Add: Dept of Biol San Diego State Univ San Diego CA 92182

AWSCHALOM, MIGUEL, b Buenos Aires, Arg, Dec 20, 27; m 53; c 2. NUCLEAR PHYSICS. Educ: Rutgers Univ, BS, 50; Univ Rochester, PhD(physics), 55. Prof Exp: Res assoc physics, La State Univ, 55-57; res staff physicist & head health physics group, Princeton, 57-68; head radiation physics sect, Nat Accelerator Lab, 68-73, head res serv, Fermi Nat Accelerator Lab, 73-75, DEP HEAD CANCER THERAPY DEPT, FERMI NAT ACCELERATOR LAB, 75- Concurrent Pos: Mem adv panel accelerator radiation safety, US AEC, 61-68; consult, Vet Admin Hosp, Hines, Ill, 75- Mem: Am Phys Soc; Health Physics Soc; Am Asn Physicists Med. Res: Use of fast neutron beams in cancer therapy; high energy physics; radiation safety; nuclear instrumentation; medical physics. Mailing Add: Fermi Nat Accelerator Lab PO Box 500 Batavia IL 60510

AXE, JOHN DONALD, b Denver, Colo, Sept 6, 33; m 63. CHEMICAL PHYSICS. Educ: Univ Denver, BS, 55; Univ Calif, Berkeley, PhD(chem), 60. Prof Exp: NSF res fel, Dept Physics, Johns Hopkins Univ, 61-62; mem res staff, Thomas J Watson Res Ctr, Int Bus Mach Corp, 62-69; MEM RES STAFF, BROOKHAVEN NAT LAB, 69- Res: Neutron spectroscopy; lattice dynamics; structural phase transformations in solids. Mailing Add: Dept of Physics Brookhaven Nat Lab Upton NY 11973

AXE, WILLIAM NELSON, b Pulaski, Pa, Dec 19, 09; m 38; c 3. ORGANIC CHEMISTRY. Educ: Westminster Col (Pa), BS, 31; Univ Tex, MA, 36, PhD(org chem), 38. Prof Exp: Asst, Western Reserve Univ, 31-32; lit compilator, Carleton Ellis, NJ, 34; tutor org chem, Univ Tex, 34-36, instr, 36-38; sr chemist, Phillips Petrol Co, 38-45, assoc dir res, 45-47, sr sect chief, 47-57, mgr hydrocarbon conversion br, 57-64, res adv, 64-75; RETIRED. Mem: Am Chem Soc; Sigma Xi. Res: Isolation of nitrogen bases from petroleum; distillation of fatty acids; germicidal properties of nitrogen bases; gas sweetening; desulfurization of natural gasoline; alkylation research; gasoline stability studies; olefin and diolefin polymerization; hydrochlorination; lubricants research; refining process applications of gasoline; lubricants and asphalt. Mailing Add: 2003 S Johnstone St Bartlesville OK 74003

AXEL, PETER, b Brooklyn, NY, May 12, 23; m 54; c 1. PHYSICS. Educ: Brooklyn Col, AB, 43; Univ Ill, MS, 47, PhD(physics), 49. Prof Exp: Mem staff elec develop, Radiation Lab, Mass Inst Technol, 43-46; from asst prof to assoc prof physics, 49-59, PROF PHYSICS, UNIV ILL, URBANA, 59- Concurrent Pos: Guggenheim fel, 55-56; NSF sr fel, 62-63. Mem: Am Phys Soc. Res: Nuclear physics; radioactivity; photo induced reactions; electronics; nuclear instrumentation. Mailing Add: Dept of Physics Univ of Ill Urbana IL 61801

AXEL, RICHARD, b Brooklyn, NY, July 2, 46; m 75. MOLECULAR BIOLOGY. Educ: Columbia Univ, AB, 67; Johns Hopkins Sch Med, MD, 70. Prof Exp: Intern path, Columbia Univ, Col Physicians & Surgeons, 70-71; fel path & oncol, Inst Cancer Res, Joint appointment vis fel, Dept Path, Columbia Univ, 71-72; res assoc, NIH, 72-74; ASST PROF DEPT PATH & INST CANCER RES, COLUMBIA UNIV, 74-, ASST ATTEND PHYSICIAN, PRESBY HOSP, NY, 74- Mem: AAAS. Res: Control of gene expression in normal and transformed cells. Mailing Add: Columbia Univ 99 Ft Washington Ave New York NY 10032

AXELRAD, ARTHUR AARON, b Montreal, Que, Dec 30, 23; m 60; c 3. HEMATOLOGY, HISTOLOGY. Educ: McGill Univ, BSc, 45, MD & CM, 49, PhD(anat), 54. Prof Exp: Demonstr zool, McGill Univ, 43-44, comp anat, 44-45, med histol, 46 & exp morphol, 52-54; from asst prof to assoc prof, 56-66, PROF ANAT & HEAD DIV HISTOL, UNIV TORONTO, 66- Concurrent Pos: Nat Cancer Inst Can fel, Inst Cell Res & Genetics, Karolinska Inst, Sweden, 54-56; scientist & head subdiv immunogenetics & cytol, Ont Cancer Inst, Toronto, 57-68. Mem: Asn Cancer Res; Can Soc Cell Biol; Int Soc Exp Hemat; Am Soc Hemat. Res: Leukemia, genetics and virology; hemopoietic cell differentiation. Mailing Add: Dept of Anat Univ of Toronto Toronto ON Can

AXELRAD, GEORGE, b Trutnov, Czech, Oct 3, 29; US citizen; m 62; c 2. ORGANIC CHEMISTRY. Educ: City Col New York, BS, 54; Univ Kans, PhD(org chem), 60. Prof Exp: Fel chem, Ohio State Univ, 60-61; asst prof, 61-70, ASSOC PROF CHEM, QUEENS COL, 70- Mem: Am Chem Soc; The Chem Soc. Res: Synthetic organic chemistry; reaction mechanisms. Mailing Add: Dept of Chem Queens Col Flushing NY 11367

AXELROD, ABRAHAM EDWARD, b Cleveland, Ohio, June 10, 12; m 39; c 2. BIOCHEMISTRY. Educ: Western Reserve Univ, BA, 33, MA, 36; Univ Wis, PhD(biochem), 39. Prof Exp: Commercial Solvents fel, Univ Wis, 39-40, Rockefeller Found fel, 40-42; res chemist, Western Pa Hosp, Pittsburgh, 42-51; assoc prof chem, Univ Pittsburgh, 46-51, lectr nutrit, Sch Pub Health, 50-51; assoc prof biochem, Western Reserve Univ, 51-54; actg chmn dept, 64-66, assoc dean sch med, 65-70;

PROF BIOCHEM, UNIV PITTSBURGH, 54- Mem: AAAS; Am Chem Soc; Am Soc Biol Chem; Am Inst Nutrit; Am Asn Immunol. Res: Relationships of vitamins and enzymes; physiological studies in nutritional deficiencies; environmental factors and bacterial virulence; mode of action of vitamins; role of vitamins in antibody formation; chemistry of skin; skin transplantation. Mailing Add: Apt 610 5 Bayard Rd Pittsburgh PA 15213

AXELROD, ARNOLD RAYMOND, b Cleveland, Ohio, Jan 1, 21; m 43; c 3. HEMATOLOGY. Educ: Ohio Univ, AB, 41; Wayne Univ, MD, 44; Am Bd Internal Med, dipl, 51. Prof Exp: Intern, Billings Hosp, Chicago, 44-45; first asst resident med, Detroit Receiving Hosp, 48-49; instr clin med, 49-51, asst prof med, 51-53, from asst prof to assoc prof clin med, 53-70, PROF MED, COL MED, WAYNE STATE UNIV, 70-; CHMN DEPT MED, SINAI HOSP, 74- Concurrent Pos: Teaching fel hemat, Col Med, Wayne Univ, 45-48. Mem: Am Fedn Clin Res; Cent Soc Clin Res; fel Am Col Physicians; Int Soc Hemat; Am Soc Hemat. Mailing Add: 14800 McNichols Rd Detroit MI 48235

AXELROD, BERNARD, b New York, NY, Oct 16, 14; m 34; c 2. BIOCHEMISTRY. Educ: Wayne Univ, BS, 35; George Washington Univ, MS, 39; George Washington Univ, PhD(biochem), 43. Prof Exp: Chemist, USDA, Washington, DC, 38-43 & Calif, 43-50, chief enzyme sect, Western Regional Res Lab, Bur Agr & Indust Chem, 52-54; assoc prof biochem, 54-58, head dept, 65-75, PROF BIOCHEM, PURDUE UNIV, WEST LAFAYETTE, 58- Concurrent Pos: Sr res fel, Calif Inst Technol, 50-52; NSF sr fel, Carlsberg Lab, Copenhagen, 60-61; NSF sr res fel, Univ Calif, Santa Cruz, 70-71. Mem: Am Chem Soc; Am Soc Biol Chem; Am Soc Plant Physiol. Res: Enzymology; carbohydrate metabolism; plant biochemistry. Mailing Add: Dept of Biochem Purdue Univ West Lafayette IN 47906

AXELROD, DANIEL ISAAC, b Brooklyn, NY, July 16, 10; m 39; c 1. PALEOBOTANY. Educ: Univ Calif, AB, 33, MA, 36, PhD(paleobot), 38. Prof Exp: Technician, Calif Forest & Range Exp Sta, 34, jr forester, 36; Nat Res fel, US Nat Mus, 39-41; asst cur, Carnegie Inst, 41-42 & 46; from asst prof to prof geol, Univ Calif, Los Angeles, 46-68; PROF BOT, UNIV CALIF, DAVIS, 68- Concurrent Pos: John Simon Guggenheim fel, 52-53. Mem: Fel Geol Soc Am; Paleont Soc; Soc Study Evolution. Res: Tertiary paleobotany; evolution of vegetation; paleoecology; paleoclimate; plate tectonics and angiosperm evolution; evolution of Madrean-Tethyan Sclerophyll vegetation; biogeography. Mailing Add: Dept of Botany Univ of Calif Davis CA 95616

AXELROD, DAVID E, b Chicago, Ill, Aug 25, 40. MICROBIOLOGY, GENETICS. Educ: Univ Chicago, BS, 62; Univ Tenn, PhD(radiation biol), 67. Prof Exp: Nat Cancer Inst fel develop biol & cancer, Albert Einstein Col Med, 68-70; ASST PROF BIOL, DOUGLASS COL, RUTGERS UNIV, 70- Concurrent Pos: Prin investr, USPHS Res Grant, Douglass Col, Rutgers Univ; prin investr, NSF Res Grant, 75-77. Mem: Soc Develop Biol; Am Soc Microbiol; Genetics Soc Am. Res: Microbial and developmental genetics. Mailing Add: Dept of Biol Sci Douglass Col Rutgers Univ New Brunswick NJ 08903

AXELROD, HERMAN DAVID, analytical chemistry, see 12th edition

AXELROD, JOSEPH MEYER, chemistry, see 12th edition

AXELROD, JULIUS, b New York, NY, May 30, 12; m 38; c 2. BIOCHEMICAL PHARMACOLOGY. Educ: City Col New York, BS, 33; NY Univ, MS, 41; George Washington Univ, PhD(chem pharmacol), 55. Hon Degrees: DSc, Univ Chicago, 66, NY Univ & Med Col Wis, 71; LLD, George Washington Univ, 71. Prof Exp: Chemist, Lab Indust Hyg, Inc, NY, 35-45; res assoc, Res Div, Goldwater Mem Hosp, Welfare Island, NY Univ, 45-49; biochemist, Nat Heart Inst, 49-55, CHIEF SECT PHARMACOL, LAB CLIN SCI, NIMH, 55- Concurrent Pos: Otto Loewi Mem lectr, NY Univ Med Sch, 64; NIH Lectr, 67; Claude Bernard prof, Univ Montreal, 69; Parkinson lectr, Columbia Univ, 71; Hodge lectr, Rochester Univ, 71; panel mem, Bd US Civil Serv Exam; mem, Int Brain Res Orgn. Honors & Awards: Nobel Prize in Med Physiol, 70; Distinguished Serv Awards, Asn Res Nerv & Ment Dis, 65, Dept Health, Educ & Welfare, 69 & Mod Med Mag, 70; Gardner Award, 67; Distinguished Alumni Award, George Washington Univ, 68; Albert Einstein Award, Yeshiva Univ, 71. Mem: Nat Acad Sci; AAAS; fel Am Acad Arts & Sci; hon mem Am Psychophysiol Asn; Am Soc Pharmacol & Exp Therapeut. Res: Biochemical mechanisms of drug and hormone action; drug and hormone metabolism; enzymology. Mailing Add: 10401 Grosvenor Pl Rockville MD 20852

AXELROD, LEONARD RICHARDSON, biochemistry, deceased

AXELROD, NORMAN NATHAN, b New York, NY, Aug 26, 34. OPTICS, OPTICAL ENGINEERING. Educ: Cornell Univ, AB, 54; Univ Rochester, PhD(physics, optics), 59. Prof Exp: Aerospace scientist, Goddard Space Flight Ctr, Nat Aeronaut & Space Admin, 59-60; res fel, Univ London, 60-61; asst prof physics, Univ Del, 61-65; mem tech staff, Bell Tel Labs, 65-72; PRES, NORMAN N AXELROD ASSOC, 72- Concurrent Pos: Univ res found grant, Univ Del, 63-64, Off Naval Res grant, 64-65; dir, World Resources Develop Co, 71-; consult, Gen Elec, 73-, Recognition Equipment Inc, 74- & Timex, 74- Mem: Inst Elec & Electronic Engr; fel AAAS; NY Acad Sci; Am Phys Soc; Optical Soc Am. Res: Optical-electronic measurement, control, and display systems and techniques; dynamic materials characterization; optical pattern recognition; non-contact defect detection, metrology, and alignment; applications to commercial, industrial, consumer and professional products. Mailing Add: Norman N Axelrod Assocs 445 E 86th St New York NY 10028

AXELROD, ROBERT JAY, organic chemistry, see 12th edition

AXELROD, SOLOMON JACOB, b Gloversville, NY, Sept 25, 12; m 35; c 2. PUBLIC HEALTH. Educ: Dartmouth Col, BA, 34; Jefferson Med Col, MD, 38; Univ Mich, MPH, 49. Prof Exp: Venereal dis control officer, State Dept Health, Tenn, 40-43; sr surgeon, USPHS, 43-49; res lectr, 49-50, assoc prof pub health econ, 50-55, chmn dept med care orgn, 65-70, PROF MED CARE ORGN, SCH PUB HEALTH, UNIV MICH, ANN ARBOR, 55- Concurrent Pos: Assoc dir, Bur Pub Health Econ, 50-59, dir, 59-70. Mem: Fel Am Pub Health Asn. Res: Medical care. Mailing Add: Dept of Med Care Orgn Univ of Mich Sch Pub Health Ann Arbor MI 48104

AXEN, DAVID, b Brackendale, BC, Aug 6, 38; m 65. NUCLEAR PHYSICS. Educ: Univ BC, BASc, 60, PhD(physics), 65. Prof Exp: Nat Res Coun Can-NATO res fel physics, 65-68, asst prof, 68-74, ASSOC PROF PHYSICS, UNIV BC, 74- Res: A polarized helium-3 ion source. Mailing Add: Dept of Physics Univ of BC Vancouver BC Can

AXEN, UDO FRIEDRICH, b Siegburg, Ger, Sept 10, 35; m 62; c 2. ORGANIC CHEMISTRY. Educ: Univ Bonn, Dipl Chem, 61, Dr rer nat(org chem), 63. Prof Exp: Res assoc org chem, Mass Inst Technol, 64-66; SCIENTIST, UPJOHN CO, 66-, HEAD RES, 74- Mem: Am Chem Soc. Res: Structure elucidation of natural products; saponines; macrolides; prostaglandin synthesis. Mailing Add: Upjohn Co Kalamazoo MI 49001

AXENROD, THEODORE, b New York, NY, Aug 27, 35; m 61; c 1. ORGANIC CHEMISTRY. Educ: NY Univ, BA, 56, PhD(chem), 61. Prof Exp: Res fel, Univ Wis, 60-61; from asst prof to assoc prof, 61-75, PROF CHEM & CHMN DEPT, CITY UNIV NEW YORK, 75- Mem: Am Chem Soc; Brit Chem Soc. Mailing Add: Dept of Chem City Univ of New York New York NY 10021

AXFORD, WILLIAM IAN, b Dannevirke, NZ, Jan 2, 33; m 55; c 4. ASTROPHYSICS, SPACE SCIENCE. Educ: Univ Caterbury, BE & BS, 55, 55, ME & MS, 57; Univ Manchester, PhD(math), 60. Prof Exp: Mem staff, Defense Res Bd, Can, 60-62; from assoc prof to prof astron, Cornell Univ, 63-67; PROF PHYSICS & APPL PHYSICS, UNIV CALIF, SAN DIEGO, 67- Concurrent Pos: Appleton mem lectr, Int Sci Radio Union, 69; ed, J Geophys Res, 70- Honors & Awards: Space Sci Award, Am Inst Aeronaut & Astronaut, 71. Mem: Fel Am Geophys Union; Am Astron Soc; Royal Astron Soc. Res: Theoretical work in astrophysics and space science, including magnetic storms, the solar wind and the magnetosphere, galactic and solar cosmic rays; magnetohydrodynamics; dynamics of the interstellar medium. Mailing Add: Dept of Physics 3426 Mayer Hall Univ of Calif at San Diego La Jolla CA 92037

AXLER, DAVID ALLAN, b Philadelphia, Pa, Feb 6, 42; m 65; c 1. VIROLOGY. Educ: Pa State Univ, BS, 63; Hahnemann Med Col, MS, 66, PhD(virol), 69. Prof Exp: Asst prof vet path, Ohio State Univ, 70-73; prin virologist, Battelle Mem Inst, 73-75; ASSOC PROF MICROBIOL, PA COL PODIATRIC MED, 75- Mem: AAAS; Am Soc Microbiol; Tissue Cult Asn. Res: Mechanism of carcinogenesis induced by viruses and development of methods for extraction and in vitro assay in tumor specific transplantation antigens. Mailing Add: Dept of Microbiol Pa Col of Podiatric Med Philadelphia PA 19107

AXLEY, JOHN HAROLD, b Butternut, Wis, June 2, 15; m; c 3. SOILS. Educ: Univ Wis, BA, 37, PhD(soil chem), 45. Prof Exp: Mem staff, Riverside Exp Sta, Univ Calif, 46; agronomist, Zeloski Potato Farm, 47; assoc prof soils, 48-68, PROF AGRON, UNIV MD, COLLEGE PARK, 68- Concurrent Pos: Chmn phosphorus & lime work group, Regional Soil Nitrogen Proj & mem soil test work group, Northeast Soil Res Comt. Mem: Am Soc Agron; Soil Sci Soc Am; Potato Asn Am. Res: Response of Maryland tobacco to liming; clay separation and identification by a density gradient method; chemistry and phytotoxicity of arsenic in soils—arsenic soil test correlation; technique for determining phototoxicity of water extracts of plants; nitrogen removal from sewage waters by soils and plants. Mailing Add: Dept of Agron Univ of Md College Park MD 20740

AXMAN, MARY CLAUDINE, b Olmitz, Kans. BIOLOGY. Educ: Friends Univ, BA, 37; Univ Wichita, MS, 40; Cath Univ, PhD(biol), 47. Prof Exp: Head sci dept, 37-44 & 47-71, dean col, 53-57, PROF BIOL, TEACHER & DIR STUDENT RES, KANS NEWMAN COL, 71- Concurrent Pos: Exchange teacher, St Mary of the Plains Col, Kans, 59-60. Mem: AAAS; Nat Asn Biol Teachers; Nat Sci Teachers Asn. Res: Morphological studies on glycogen deposition in schistosomes and other flukes. Mailing Add: Dept of Sci Kans Newman Col Wichita KS 67213

AXTELL, DARRELL DEAN, b Grants Pass, Ore, July 27, 44; m 69. INORGANIC CHEMISTRY. Educ: Linfield Col, BA, 67; Ore State Univ, PhD(chem), 73. Prof Exp: Fel chem, Univ Ariz, 73-74; FEL & LECTR CHEM, TEX A&M UNIV, 75- Mem: Am Chem Soc; Sigma Xi. Res: Synthesis of metalloporphyrins; coordination chemistry of main group and transition elements with nitrogen and phosphorus ligands. Mailing Add: Dept of Chem Tex A&M Univ College Station TX 77840

AXTELL, JAMES CLINTON, physics, see 12th edition

AXTELL, JOHN DAVID, b Minneapolis, Minn, Feb 5, 34; m 57; c 3. GENETICS, PLANT BREEDING. Educ: Univ Minn, BS, 57, MS, 65; Univ Wis, PhD(genetics), 67. Prof Exp: Res asst genetics, Univ Wis, 59-65, res assoc, 67; ASST PROF GENETICS, PURDUE UNIV, 67- Mem: Genetics Soc Am; Am Soc Agron; Am Genetic Asn; Crop Sci Soc Am. Res: Controlling elements in maize; paramutation in maize; chemical paramutagenesis; mutagenesis in higher plants; regulation of gene action in higher plants; plant genetics and breeding; genetic improvement of protein quality in cereals. Mailing Add: Dept of Agron Life Sci Bldg Purdue Univ Lafayette IN 47907

AXTELL, RALPH WILLIAM, b Norfolk, Nebr, Apr 20, 28; m 67; c 1. VERTEBRATE ZOOLOGY. Educ: Univ Tex, BA, 53, MA, 54, PhD, 58. Prof Exp: Res sci zool, Univ Tex, 55-57; instr biol, ETex State Col, 57-58; asst prof, Sul Ross State Col, 58-60; asst prof zool, 60-65, assoc prof biol, 65-70, chmn fac biol sci, 65-67, PROF BIOL, SOUTHERN ILL UNIV, EDWARDSVILLE, 70- Concurrent Pos: Herpet ed, Copeia, Am Soc Ichthyol & Herpet, 68- Honors & Awards: Student Paper Award, Am Soc Ichthyol & Herpet, 57. Mem: Fel AAAS; Am Soc Ichthyol & Herpet; Soc Study Evolution; Soc Syst Zool; Am Inst Biol Sci. Res: North American herpetology and zoogeography; lizard genus Holbrookia; lizard family Iguanidae; reptilian evolution. Mailing Add: Fac of Biol Southern Ill Univ Edwardsville IL 62025

AXTELL, RICHARD CHARLES, b Medina, NY, Aug 4, 32; m 61; c 2. MEDICAL ENTOMOLOGY. Educ: State Univ NY Albany, BS, 54, MS, 55; Cornell Univ, PhD(entom), 62. Prof Exp: Vis instr zool, Col Agr, Univ Philippines, 55-56; from asst prof to assoc prof entom, 62-68, PROF ENTOM, NC STATE UNIV, 68- Concurrent Pos: Consult, Nat Develop Co, Philippines, 55-56; prin investr, USPHS Grant, 63-70; dir, NIH med entom training grant, 66-73; Sigma Xi res award, NC State Univ, 69; Off Naval Res contract, 70-75; dir dept com, Nat Oceanic & Atmospheric Admin sea grant proj, 70- Mem: AAAS; Entom Soc Am; Am Mosquito Control Asn; Acarological Soc Am. Res: Medical and veterinary entomology; biology and control of Diptera; acarology; ultrastructure of arthropod sensory structures. Mailing Add: Dept of Entom NC State Univ Raleigh NC 27607

AXTMANN, ROBERT CLARK, b Youngstown, Ohio, Feb 25, 25; m 49; c 3. APPLIED PHYSICS, CHEMICAL ENGINEERING. Educ: Oberlin Col, AB, 47; Johns Hopkins Univ, MA, 49, PhD(phys chem), 50. Prof Exp: Instr org chem, McCoy Col, 48-49; res physicist, Ballistic Res Lab, Aberdeen Proving Ground, Md, 50; res physicist, E I du Pont de Nemours & Co, 50-53, res supvr reactor physics, 53-54, area tech supvr reactor physics & eng, 55-56, sr res supvr chem physics, 57-59; assoc prof chem eng, 59-65, chmn prog nuclear studies, 60-68, chmn coun environ studies, 70-73, PROF CHEM ENG, PRINCETON UNIV, 65- Concurrent Pos: Vis fel, Israel AEC, 64 & Nat Comn Nuclear Energy, Mex, 69; mem, NJ Comn Radiation Protection, 66-70; mem vis comt, Brookhaven Nat Lab, 68-71; consult, Del River Basin Comn, 70 & UN Off Sci & Technol, 70; vis scientist, Dept Sci & Indust Res, Govt NZ, 74. Mem: AAAS; Am Phys Soc; Am Chem Soc; Am Inst Chem Eng. Res: Nuclear fusion technology; geothermal power; environmental science; chemical physics; radiation chemistry; nuclear reactor physics; Mössbauer effect; nuclear magnetic resonance. Mailing Add: Sch Eng & Appl Sci Princeton Univ Princeton NJ 08540

AXWORTHY, ARTHUR EDWARD, JR, b Whittier, Calif, Mar 26, 29; m 49; c 2. PHYSICAL CHEMISTRY. Educ: Whittier Col, AB, 51; Univ Southern Calif, PhD(chem), 59. Prof Exp: Res chemist, Shell Oil Co, 56-64, prin scientist, Chem Kinetics Unit, 64-70, MEM TECH STAFF, ROCKETDYNE DIV, ROCKWELL INT, CANOGA PARK, 70- Mem: Am Chem Soc; Combustion Inst. Res: Chemical kinetics; photochemistry; fuel and propellant pyrolysis; combustion kinetics; formation of nitrogen oxides; interhalogen compounds. Mailing Add: 1808 Garvin Simi Valley CA 93065

AYALA, FRANCISCO JOSE, b Madrid, Spain, Mar 12, 34; m 68; c 2. GENETICS. Educ: Univ Madrid, BS, 55; Lit Univ Salamanca, STL, 60; Columbia Univ, MA, 63, PhD(genetics), 64. Prof Exp: Res assoc genetics, Rockefeller Inst, 64-65; prof biol, Providence Col, 65-67; asst prof, Rockefeller Univ, 67-71; assoc prof genetics, 71-74, PROF GENETICS, UNIV CALIF, DAVIS, 74- Mem: AAAS; Genetics Soc Am; Soc Study Evolution (vpres, 72); Am Soc Naturalists (secy, 74-77). Res: Population genetics, fitness of natural and experimental populations; reproductive isolation and evolutionary processes; biochemical variation in natural populations; philosophy of science. Mailing Add: Dept of Genetics Univ of Calif Davis CA 95616

AYALA, REYNALDO, b Saltillo, Mex, Sept 28, 34; m 58; c 3. GEOGRAPHY. Educ: Univ Minn, Minneapolis, BA, 60; Southern Ill Univ, Carbondale, MA, 64, PhD(geog), 71. Prof Exp: Instr geog & dir, Latin Am Inst, Southern Ill Univ, 64-66; instr, Univ NMex, 66-69; asst prof, 69-72, ASSOC PROF GEOG, SAN DIEGO STATE UNIV, CALEXICO CAMPUS, 72- Concurrent Pos: Prof sociol, Sch Social & Polit Sci, Univ Baja Calif, 71- Mem: AAAS; Asn Am Geogrs. Res: Human geography of Baja California; historical geography of settlements in Northern Mexico; recreational environmental impact studies of Imperial and Mexicali Valleys. Mailing Add: San Diego State Univ Dept Geog 720 Heber Ave Calexico CA 92231

AYALA-CASTANARES, AGUSTIN, micropaleontology, marine ecology, see 12th edition

AYCOCK, BENJAMIN FRANKLIN, b Washington, DC, May 18, 22; m 48; c 5. TEXTILE CHEMISTRY. Educ: Univ NC, BS, 42; Univ Ill, MS, 45, PhD(org chem), 47. Prof Exp: Asst, Nat Defense Res Comt, Univ Ind, 44-45; res assoc, Explosives Res Lab, Carnegie Inst Technol, 45; from instr to asst prof chem, Univ Wis, 47-50; chemist, Rohm and Haas Co, Pa, 50-59, sect head, Redstone Arsenal Res Div, 59-69; head chem dept, Burlington Industs Res Ctr, 69-72, SR SCIENTIST, BURLINGTON INDUST CORP RES & DEVELOP, 72- Mem: Am Chem Soc; Swiss Chem Soc; Am Asn Textile Chemists & Colorists. Res: Chemistry of natural products; nitrogeneous resins; plasticizers; propellant chemistry; textile chemistry. Mailing Add: Burlington Indust Corp Res & Dev PO Box 21327 Greensboro NC 27420

AYCOCK, MARVIN KENNETH, JR, b Warrenton, NC, Nov 12, 35; m 60; c 2. PLANT BREEDING, PLANT GENETICS. Educ: NC State Univ, BS, 59, MS, 63; Iowa State Univ, PhD(plant breeding), 66. Prof Exp: Teacher, Bear Grass High Sch, 59-60; ASSOC PROF PLANT BREEDING, UNIV MD, COLLEGE PARK, 66- Mem: Am Soc Agron; Crop Sci Soc Am; Am Genetic Asn. Res: Effects of a male-sterile cytoplasm in Nicotiana tabacum and inbreeding depression in Medicago sativa; developing new and improved varieties of Maryland tobacco. Mailing Add: Dept of Agron Univ of Md College Park MD 20740

AYCOCK, NANCY RAE, b Fremont, NC, May 29, 45. HUMAN ANATOMY. Educ: Wake Forest Univ, BS, 67; Med Col Va, PhD(anat), 75. Prof Exp: ASST PROF ANAT, HAHNEMANN MED COL & HOSP, 74- Res: Ultrastructure of adenocarcinoma of the endometrium and ultrastructural changes induced in normal post-menopausal endometrium by estrogen replacement therapy. Mailing Add: Dept of Anat Hahnemann Med Col & Hosp Philadelphia PA 19102

AYCOCK, ROBERT, b Lisbon, La, Dec 23, 19; m 41; c 2. PHYTOPATHOLOGY. Educ: La State Univ, BS, 40, NC State Col, MS, 42, PhD(plant path), 49. Prof Exp: Assoc plant pathologist, 53-55; res assoc prof, 55-61, PROF PLANT PATH, NC STATE UNIV, 61-, HEAD DEPT, 73- Concurrent Pos: Ed-in-chief, Phytopathology, Am Phytopath Soc, 70-72. Mem: Am Phytopath Soc (pres, 75-76). Res: Plant disease diagnosis and diseases of ornamental crops. Mailing Add: Dept of Plant Path Box 5397 NC State Univ Raleigh NC 27607

AYENGAR, PADMASINI (MRS FREDERICK ALADJEM), b Bangalore, India, July 31, 24; m 57; c 1. BIOCHEMISTRY, MICROBIOLOGY. Educ: Travancore Univ, India, BSc, 44; Univ Wash, St Louis, PhD, 53. Prof Exp: Fel enzymechem, Enzyme Inst, Univ Wis, 53-54; res assoc biochem, City of Hope Med Ctr, Calif, 54-56; vis scientist, Nat Inst Arthritis & Metab Dis, NIH, Md, 57; RES ASSOC IMMUNOCHEM, DEPT MED MICROBIOL, SCH MED, UNIV SOUTHERN CALIF, 58- Mem: Am Chem Soc; Am Soc Biol Chem. Res: Microbial fermentation; growth factors and inhibitors; biosynthesis and metabolism of phosphodiester of serine and ethanolamine; enzymes; purification; properties and mechanism of action. Mailing Add: 2025 Zonal Ave Los Angeles CA 90033

AYER, DARRELL, (JR), b Atlanta, Ga, Oct 15, 08; m 38, 56, 64; c 8. PATHOLOGY. Educ: Va Mil Inst, BS, 29; Emory Univ, MD, 33. Prof Exp: Intern, Grady Hosp, Atlanta, 33-34 & Brigham Hosp, Boston, 34-36; res pathologist, Children's Hosp, 36 & Boston Lying-in Hosp, 36-37; instr path, Harvard Med Sch, 37-40; pathologist, Bassett Hosp, NY, 40-46; pathologist, Long Mem Hosp, 47-65; med dir, Linden Labs, Inc, Upjohn Co, 65-74; MED DIR, AYER LAB, 74- Concurrent Pos: Asst prof path, Sch Med, Emory Univ; res pathologist, Brigham Hosp, 37-39; dir, Otsego County Pub Health Lab, 40. Mem: Am Soc Clin Path; Am Asn Path & Bact; fel Col Am Path. Res: Pathology of kidney diseases. Mailing Add: Ayer Lab 401 Peachtree St NE Atlanta GA 30308

AYER, MIRIAM CLOUGH, mathematics, see 12th edition

AYER, RANDALL PLAISTED, organic chemistry, see 12th edition

AYER, WILLIAM ALFRED, b Sackville, NB, July 4, 32; m 54; c 6. ORGANIC CHEMISTRY. Educ: Univ NB, BSc, 53, PhD(chem), 56. Prof Exp: Res assoc, Harvard Univ, 57-58; from asst prof to assoc prof, 58-67, PROF CHEM, UNIV ALTA, 67- Concurrent Pos: Fel, Alfred P Sloan Found, 65-67. Honors & Awards: Merck Sharp & Dohme Award, Chem Inst Can, 70. Mem: Sr mem Am Chem Soc; fel Chem Inst Can; The Chem Soc. Res: Structural and synthetic studies of naturally occurring substances. Mailing Add: Dept of Chem Univ of Alta Edmonton AB Can

AYERS, ALVIN DEARING, b Alvin, Tex, Apr 27, 09; m 39; c 2. SOIL SCIENCE, PLANT PHYSIOLOGY. Educ: Univ Ariz, BS, 32, MS, 34; Univ Calif, PhD(soil sci), 39. Prof Exp: Asst agr chemist, Exp Sta, Univ Ariz, 32; chemist, Salt River Valley Water Users Asn, Ariz, 34 & Asn Lab, Calif, 35; asst plant nutrit, Univ Calif, 36-37; agent, Salinity Lab, USDA, 38-41, asst chemist, 41-42; chemist, Calif Inst Technol, 42-43; supvr safety & personnel, 43-45; assoc chemist, Salinity Lab, USDA, 45-48, chemist, 48-50, agriculturist field comt, Ark White & Red Basins Inter-Agency Comt, Okla, 51-53, soil scientist, US Salinity Lab, 53-59, asst dir, Europ Res Off, Agr Res Serv, Italy, 59-62, dir, Far Eastern Regional Res Off, 62-66; soil & water mgt specialist, Agr & Rural Develop Serv, War on Hunger, AID, 66-70 & Off Agr & Fisheries Tech Assistance Bur, 70; CONSULT, 70- Mem: Soil Sci Soc Am; Am Soc Agron. Res: Agricultural research administration; soil salinity and alkali; salt tolerance of plants; soil moisture-plant relationships; soil chemistry. Mailing Add: 5621 Del Rio Ct Cape Coral FL 33904

AYERS, AUGUSTUS SIDNEY, chemistry, see 12th edition

AYERS, CARLOS R, b Oakvale, WVa, Apr 2, 32; m 58; c 3. INTERNAL MEDICINE. Educ: Lincoln Mem Univ, BS, 53; Univ Va, MD, 58. Prof Exp: Intern med, Univ Va, 58-S9; resident univ affil hosp, Univ Utah, 61-62; resident, 63-64, from instr to assoc prof internal med, 64-75, PROF INTERNAL MED, SCH MED, UNIV VA, 75-; DIR, VA HEART RES LAB, 69- Concurrent Pos: Fel cardiol, Univ Hosp, Univ Va, 62-63; mem coun high blood pressure res, Am Heart Asn. Mem: Am Fedn Clin Res; fel Am Col Physicians. Res: Peripheral vascular disease; hypertension; aldosterone metabolism; renin; angiotensin metabolism. Mailing Add: Va Heart Res Lab Univ of Va Hosp Charlottesville VA 22901

AYERS, CAROLINE LEROY, b Augusta, Ga, Oct 30, 41; m 63. PHYSICAL CHEMISTRY. Educ: Univ Ga, BS, 62, PhD(phys chem), 66. Asst prof chem, Portland State Col, 66-67; ASST PROF CHEM, ECAROLINA UNIV, 67- Mem: Am Chem Soc; Sigma Xi. Res: Electron paramagnetic resonance studies of gamma irradiated solids, particularly triphenylmethyl derivatives and organo-metallic compounds. Mailing Add: Dept of Chem ECarolina Univ Greenville NC 27834

AYERS, JERRY BART, b Atlanta, Ga, Apr 17, 39; m 71. SCIENCE EDUCATION. Educ: Oglethorpe Univ, BS, 60; Univ Ga, MEd, 66, EdD(sci educ), 67. Prof Exp: Assoc physicist, Union Carbide Nuclear-Oak Ridge Nat Labs, 60-61; instr chem, Lenoir Rhyne Col, 64-65; asst prof sci educ, Univ Ga, 67-70; PROF SCI EDUC & ASST DEAN EDUC, TENN TECHNOL UNIV, 70- Mem: Am Chem Soc; Nat Sci Teachers Asn. Res: Preparation of improved teaching materials for use at the college level in training teachers; evaluation of science teaching; scientific information retrieval. Mailing Add: 580 Pleasant Hill Dr Cookeville TN 38501

AYERS, JOHN CARR, b Marcellus, Mich, Oct 4, 12; m 38; c 4. BIOLOGICAL OCEANOGRAPHY. Educ: Kalamazoo Col, AB, 34; Kans State Col, MS, 36; Duke Univ, PhD(zool), 39. Prof Exp: Instr biol, Univ SC, 39-41, adj prof, 41-43; instr physics & theory of flight, US Naval Flight Prep Sch, 43-44; res assoc, Woods Hole Oceanog Inst, 44-49; from asst prof to assoc prof oceanog, Cornell Univ, 49-56; assoc prof, 56-58, PROF OCEANOG, UNIV MICH, ANN ARBOR, 58-, RES OCEANOGR, GREAT LAKES RES DIV, 56- Mem: Am Soc Limnol & Oceanog (pres, 63-64). Res: Cytology and histology of hypophysis; cytology of pituitary; endocrinology of growth hormones; respiration of aquatic crabs; development and testing of antifouling paint; hydrography of estuaries; shellfish biology; limnology of Great Lakes. Mailing Add: Great Lakes Res Div Univ of Mich Ann Arbor MI 48105

AYERS, JOHN E, b Indianapolis, Ind, Dec 20, 41; m 63; c 2. PLANT PATHOLOGY, PLANT GENETICS. Educ: Purdue Univ, BS, 63, MS, 65; Pa State Univ, PhD(genetics), 69. Prof Exp: Asst prof, 69-74, ASSOC PROF PLANT PATH, PA STATE UNIV, UNIVERSITY PARK, 74- Mem: Am Phytopath Soc. Res: Nature and inheritance of disease resistance in plants. Mailing Add: Buckhout Lab Rm 211 Pa State Univ Dept Plant Path University Park PA 16802

AYERS, ORVAL EDWIN, b Grant, Ala, Mar 29, 32; m 52; c 3. ORGANIC CHEMISTRY. Educ: Berea Col, Ky, BA, 55; Auburn Univ, MS, 58; Univ Ala, Tuscaloosa, PhD(org chem), 73. Prof Exp: Chemist qual control, Nat Cash Regist Co, Dayton, Ohio, 55-56; RES CHEMIST PROPELLANT RES, US ARMY MISSILE COMMAND, REDSTONE ARSENAL, ALA, 59- Honors & Awards: Army Res & Develop Achievement Awards, Dept Army, 75. Mem: Am Chem Soc (treas, 75-76). Res: Development and evaluation of solid propellants for rocket motors; development of storable solid propellants for hydrogen and fluorine gas generators for use in chemical lasers. Mailing Add: 7805 Martha Dr SE Huntsville AL 35802

AYERS, PAUL WAYNE, b Winter Garden, Fla, Oct 18, 36; m 63. PHYSICAL ORGANIC CHEMISTRY. Educ: David Lipscomb Col, BA, 60; Univ Ga, MS, 62, PhD(chem), 66. Prof Exp: Asst prof chem, Portland State Col, 66-67; asst prof, 67-73, ASSOC PROF CHEM, ECAROLINA UNIV, 73- Mem: Am Chem Soc. Res: Free radical rearrangements and oxidation processes. Mailing Add: Dept of Chem ECarolina Univ Greenville NC 27834

AYERS, RAYMOND DEAN, b Ossining, NY, July 11, 40; m 69. EXPERIMENTAL SOLID STATE PHYSICS. Educ: Calif Inst Technol, BS, 63, MS, 64; PhD(mat sci), 71. Prof Exp: Asst prof, 67-74, ASSOC PROF PHYSICS & ASTRON, CALIF STATE UNIV, LONG BEACH, 74- Mem: Am Asn Physics Teachers. Res: Electronic properties and structure of amorphous metallic phases. Mailing Add: Dept of Physics & Astron Calif State Univ Long Beach CA 90840

AYERS, ROBERT CLAYTON, JR, physical chemistry, see 12th edition

AYERS, WILLIAM ARTHUR, b Highgrove, Calif, Dec 22, 24; m 51; c 2. MICROBIOLOGY. Educ: Univ Calif, Los Angeles, AB, 49; Rutgers Univ, MS, 51; Univ Wis, PhD(bact), 54. Prof Exp: Asst bact, Rutgers Univ, 49-51 & Univ Wis, 51-54; asst prof, Univ NH, 54-59; microbiologist, Brooklyn Bot Garden, 59-62; MICROBIOLOGIST, USDA, 62- Mem: Am Soc Microbiol. Res: Soil microbiology; ecology and physiology of soilborne plan pathogenic fungi. Mailing Add: 262 Biosci Bldg Agr Res Serv-W Beltsville MD 20705

AYERS, WILLIAM ROSCOE, chemistry, see 12th edition

AYKAN, KAMRAN, b Istanbul, Turkey, May 19, 30; m 57. INORGANIC CHEMISTRY. Educ: Univ Istanbul, MSc, 54. Prof Exp: Vis chemist, Bergbau A G Koenig Ludwig, Ger, 56-57; chemist, Wm T Burnett & Co, Inc, 57-58; from chemist to staff scientist, E I du Pont de Nemours & Co, 58-71; mgr chem res, 71-72, tech dir auto exhaust catalyst group, 72-74, DIR RES & DEVELOP, ENGELHARD INDUSTS DIV, ENGELHARD MINERALS & CHEMICALS CORP, 74- Mem: Am Chem Soc; Am Crystallog Asn; Am Ceramic Soc; Indust Res Inst; Catalysis Soc. Res: Heterogeneous catalysts and catalytic processes; precious metals chemistry; auto exhaust catalysis; materials technology; solid state chemistry. Mailing Add: Engelhard Industs Div Menlo Park Edison NJ 08817

AYLESWORTH, THOMAS GIBBONS, b Valparaiso, Ind, Nov 5, 27; m 49; c 2. ZOOLOGY. Educ: Univ Ind, AB, 50, MS, 53; Ohio State Univ, PhD(sci educ), 59. Prof Exp: Asst prof sci educ, Mich State Univ, 57-62; lectr sci & sr ed, Current Sci, Wesleyan Univ, 62-65; SR ED SCI BKS, DOUBLEDAY & CO, 65- Concurrent Pos: Vis fac mem, Ohio State Univ, 62 & Wis State Univ, Whitewater, 64; New Eng ed, Am Biol Teacher, 62-64. Mem: Nat Asn Biol Teachers; Nat Sci Teachers Asn; Nat

Asn Res Sci Teaching; NY Acad Sci. Res: Critical thinking and its applications in the field of teaching; uses of printed materials in the teaching of science. Mailing Add: Doubleday & Co 245 Park Ave New York NY 10017

AYO, DONALD JOSEPH, b Bourg, La, Apr 1, 34; m 58; c 2. AGRONOMY, HORTICULTURE. Educ: La State Univ, BS, 56, MS, 58, PhD, 64. Prof Exp: Asst prof hort, La State Univ, 58; from instr to prof plant sci, 58-67, head dept agr, 66-68, dean div sci, 68-69, FORTIER DISTINGUISHED HONOR PROF PLANT SCI, NICHOLLS STATE UNIV, 67-, DEAN DIV LIFE SCI & TECHNOL, 69-, VPRES, 71-, PROVOST, 75- Concurrent Pos: Asst ed, Nat Asn Col Teachers Am J, 63-65. Mem: Am Soc Hort Sci; Am Soc Agron; Soil Sci Soc Am; NY Acad Sci; Am Inst Biol Sci. Res: Dormancy and rest periods in decidious plants. Mailing Add: Box 2013 Nicholls State Univ Thibodaux LA 70301

AYOUB, CHRISTINE WILLIAMS, b Cincinnati, Ohio, Feb 7, 22; m 50; c 2. MATHEMATICS. Educ: Bryn Mawr Col, AB, 42; Radcliffe Col, AM, 43; McGill Univ, MA, 44; Yale Univ, PhD(math), 47. Prof Exp: Fel, Inst Advan Study, Off Naval Res, 47-48; instr math, Cornell Univ, 48-51; from instr to assoc prof, 53-69, PROF MATH, PA STATE UNIV, 69- Concurrent Pos: Res asst, Radcliffe Col, 51-52. Mem: Am Math Soc. Res: Theory of groups and normal chains; modern algebra. Mailing Add: Dept of Math McAlister Bldg Pa State Univ University Park PA 16802

AYOUB, ELIA MOUSSA, b Haifa, Palestine, Apr 12, 28; m 54; c 5. PEDIATRICS. Educ: Am Univ Beirut, BS, 49, MD, 53; Am Bd Pediat, dipl. Prof Exp: Intern, Am Univ Beirut, 49; physician in-chg, Qaisumah Hosp, Beirut, 53-56; resident pediat, Univ Wis, 56-57; from instr to prof, Univ Minn, 59-69; PROF PEDIAT, COL MED, UNIV FLA, 69- Concurrent Pos: Fel, Col Med, Univ Minn, 57-58, res fel, 58-59; Helen Hay Whitney res fel, 60-63; guest investr, Rockefeller Univ, 63-65; estab investr, Helen Hay Whitney Res Found, 63-68. Mem: AAAS; Am Soc Microbiol; Soc Pediat Res; Am Asn Immunol; Am Acad Pediat. Mailing Add: Dept of Pediat Univ of Fla Col of Med Gainesville FL 32601

AYOUB, RAYMOND G DIMITRI, b Sherbrooke, Can, Jan 2, 23; m 50; c 2. MATHEMATICS. Educ: McGill Univ, BSc, 43, MSc, 46; Univ Ill, PhD(math), 50. Prof Exp: Lectr math, McGill Univ, 47-49; Pierce instr, Harvard Univ, 50-52; from asst prof to assoc prof, 52-66, PROF MATH, PA STATE UNIV, 66- Mem: Am Math Soc; Can Math Cong. Res: Analytic theory of numbers; theory of algebraic numbers. Mailing Add: 203 McAlister Bldg Pa State Univ University Park PA 16802

AYOUB, SADEK M, b Cairo, Egypt, Nov 2, 22; US citizen; m 56; c 1. PLANT NEMATOLOGY, PLANT PATHOLOGY. Educ: Okla State Univ, BS, 55, MS, 56. Prof Exp: Jr plant pathologist, 58-59, asst plant pathologist, 59, assoc plant nematologist, 59-60, PLANT NEMATOLOGIST, CALIF DEPT FOOD & AGR, 60- Mem: Soc Nematol; Am Phytopath Soc. Res: Nematology regulatory control. Mailing Add: Calif Dept Food & Agr 1220 N St Sacramento CA 95814

AYRE, CHARLES A, b Winnipeg, Man, May 9, 09; m 36; c 4. BIOCHEMISTRY. Educ: Univ Man, BSc, 32, MSc, 33; Univ London, PhD(biochem), DIC, 35. Prof Exp: Researcher malt & barley, Nat Res Coun, Can, 37-38; Dominion analyst, Dept Nat Health & Welfare, 38-43; gen mgr, BDH Chem Div, Glaxo Can Ltd, 43-73, pres, BDH Chem Can Ltd, 73-74; RETIRED. Concurrent Pos: Grant, Milk Mkt Bd, Nat Inst Res Dairying, Eng, 36-37. Mem: Brit Soc Hist Sci. Res: Plant biochemistry; history of science. Mailing Add: 4 Loyalist Ct RR 3 Garrison Village Niagara-on-the-Lake ON Can

AYRES, BARBARA CHARTIER, b Clyde, Kans, July 21, 25; m 52. CULTURAL ANTHROPOLOGY. Educ: Coe Col, BA, 47; Univ NC, MA, 49; Harvard Univ, PhD(anthrop), 56. Prof Exp: Res assoc ment health, Sch Pub Health, Harvard Univ, 54-59, res assoc obstet & gynec, Harvard Med Sch, 59-61; from lectr to asst prof social sci, Boston Univ, 62-64; res assoc anthrop, Columbia Univ, 64-65; lectr anthrop, Harvard Univ, 65-67; from lectr to asst prof anthrop, 68-71, chmn dept, 72-74, ASSOC PROF ANTHROP, HARBOR CAMPUS, UNIV MASS, BOSTON, 71- Concurrent Pos: Res assoc & lectr res, Sch Social Work, Boston Univ, 56-64; instr human rels, Peter Bent Brigham Hosp Sch Nursing, 60-61; lectr, NY Univ, 64-65; res fel, Lab Social Rels, Harvard Univ, 65-67; lectr, Radcliffe Inst, 71-72; fac res grant, Univ Mass, Boston, 72. Mem: AAAS; Am Anthrop Asn; Soc Cross-Cult Res (pres elect, 75); Am Ethnol Soc. Res: Psychological anthropology; comparative socialization; cross-cultural method; cross-cultural analysis of sex roles; bio-social anthropology. Mailing Add: Dept of Anthrop Col I Univ of Mass Harbor Campus Boston MA 02125

AYRES, DAVID SMITH, b Boston, Mass, June 14, 39; m 65. EXPERIMENTAL HIGH ENERGY PHYSICS. Educ: Williams Col, BA, 61; Univ Calif, Berkeley, MA, 63, PhD(physics), 68. Prof Exp: US Peace Corps lectr physics, Univ Nigeria, 63-65; res asst physics, Univ Calif, Berkeley, 65-68, res assoc, Lawrence Berkeley Lab, 68-69; res assoc, 69-70, asst physicist, 71-74, PHYSICIST, ARGONNE NAT LAB, 74- Mem: Am Phys Soc; AAAS. Res: Planning, execution, and analysis of spectrometer experiments in strong interaction elementary particle physics, with particular emphasis on meson spectroscopy and polarization phenomena. Mailing Add: Argonne Nat Lab Bldg 362 Argonne IL 60439

AYRES, GILBERT HAVEN, b Upland, Ind, Aug 29, 04; m 26; c 2. ANALYTICAL CHEMISTRY. Educ: Taylor Univ, AB, 25; Univ Wis, PhD(chem), 30. Prof Exp: Instr chem, Taylor Univ, 25-27; asst, Univ Wis, 27-30, instr, 30-31; from asst prof to assoc prof, Smith Col, 31-47; from assoc prof to prof, 47-74, chmn dept chem, 50-52, grad adv, 57-61 & 65-69, EMER PROF CHEM, UNIV TEX, AUSTIN, 74- Mem: AAAS; Am Chem Soc. Res: Spectrophotometric methods of analysis; analytical chemistry of platinum elements. Mailing Add: Dept of Chem Univ of Tex Austin TX 78712

AYRES, JAMES T, analytical chemistry, see 12th edition

AYRES, JAMES WALTER, b Boise, Idaho, Apr 14, 42; m 64; c 2. BIOPHARMACEUTICS. Educ: Idaho State Univ, BS, 65; Univ Kans, PhD(org med chem), 70. Prof Exp: ASSOC PROF BIOPHARMACEUT, ORE STATE UNIV, 70- Mem: Acad Pharmaceut Sci; Am Pharmaceut Asn; Am Asn Cols Pharm. Res: Drug product formulation and evaluation. Mailing Add: Dept of Pharmaceut Sci Ore State Univ Sch Pharm Corvallis OR 97331

AYRES, JOHN AUGUSTUS, physical chemistry, deceased

AYRES, JOHN CLIFTON, b Beckemeyer, Ill, Apr 17, 13; m 35; c 2. FOOD SCIENCE. Educ: Ill State Normal Univ, BEd, 36; Univ Ill, MS, 38, PhD(microbiol), 42. Prof Exp: Teacher, Pub Schs, Ill, 31-34 & 36-41; res microbiologist, W S Merrell Co, Ohio, 42-43; food microbiologist, Res Div, Gen Mills, Inc, Minn, 43-46; from asst prof to prof bact, Iowa State Univ, 46-54, prof food technol, 54-67; head dept & chmn div, 67-73, PROF FOOD SCI, UNIV GA, 67-, D W BROOKS DISTINGUISHED PROF AGR, 75- Concurrent Pos: Mem adv comt, Qm Food & Container Inst Armed Forces, 51-62 & Iowa Agr Adjust Ctr, 56-57; assoc ed, Food Technol, 54-57 & Appl Microbiol, 58-63; mem, NIH Toxicol Study Sect, 60-64; mem food technol comt, Nat Acad Sci-Nat Res Coun, 62-70, mem food protection comt & chmn subcomt food microbiol, 65-, mem microbiol subcomt mil personnel supplies & comt Salmonella, Div Biol & Agr; US mem comt food microbiol & Hyg Sect, Int Asn Microbiol Socs, 64-66; mem environ health training study sect, USPHS, 66-70; Inst Food Technol rep, Int Comt Food Sci & Technol; mem, Citizens' Comn Sci, Law & Food Supply, 73- Honors & Awards: Achievement Award, Inst Am Poultry Indust, 66; Nicholas Appert Medalist, Inst Food Technol, 72. Mem: Am Soc Microbiol; fel Inst Food Technol (pres, 76-77); Soc Appl Bact; Am Meat Sci Asn; Inst Am Poultry Indust. Res: Microbiology of meats, poultry and eggs; ecology of food spoilage microorganisms; psychrophilic flora of flesh foods; aflatoxin and other mycotoxins in foods. Mailing Add: Dept of Food Sci Univ of Ga Athens GA 30602

AYRES, PAUL E, b Detroit, Mich, Apr 13, 21; m 44; c 3. INDUSTRIAL HYGIENE. Educ: Mich State Col, DVM, 43. Prof Exp: Mgr polio vet dept, 44-71, MGR TISSUE CULT DEPT, BIOL DIV, PARKE, DAVIS & CO, 71-, TRAINING COORDR, 73- Mem: Am Vet Med Asn. Res: Viruses. Mailing Add: 755 Parkdale Rd Rochester MI 48063

AYRES, WESLEY P, b Los Angeles, Calif, Sept 26, 24; m 48; c 2. PHYSICS Educ: Fresno State Col, BS, 51; Stanford Univ, MS, 53, PhD(physics), 54. Prof Exp: Sr engr, Electronic Defense Lab, Sylvania Elec Prod, Gen Tel & Electronics Corp, 54-56; vpres eng, Melabs, Inc, 56-69; DIR SYSTS ENG, BUS EQUIP DIV, SCM CORP, 69- Mem: Am Phys Soc; Inst Elec & Electronics Eng; Sigma Xi. Res: Application of solid state physics at microwave frequencies to achieve new devices; study and design of microwave systems; study and development of systems engineering techniques to office equipment research and development. Mailing Add: Bus Equip Div SCM Corp 3210 Porter Dr Palo Alto CA 94304

AYRES, WILLIAM LEAKE, b Gatesville, Tex, June 26, 05; m 26; c 3. MATHEMATICS. Educ: Southwestern Univ (Tex), AB, 23; Univ Pa, PhD(math), 27. Hon Degrees: ScD, Southwestern Univ (Tex), 57. Prof Exp: Nat Res fel, 27-28; Int Educ Bd fel, 28-29; from asst prof to assoc prof math, Univ Mich, 29-41; prof, Purdue Univ, 41-62, head dept, 41-48, dean sch sci, educ & humanities, 47-62; vpres & provost, 62-66, prof math, 62-70, Frensley prof, 70-72, EMER PROF MATH, SOUTHERN METHODIST UNIV, 72- Honors & Awards: Henry Russel Prize, Univ Mich, 32. Mem: Am Math Soc (assoc secy, 38-46); Math Asn Am (gov, 41-44 & 48-50, vpres, 46-47). Res: Topology; theory of sets and functions. Mailing Add: 5440 Del Roy Dr Dallas TX 75229

AYRES, WILLIAM STANLEY, b Eau Claire, Wis, Jan 16, 44. ANTHROPOLOGY. Educ: Univ Wyo, BA, 66; Tulane Univ, PhD(anthrop), 73. Prof Exp: Archeologist, Int Fund for Monuments, Inc, 68; archeologist, Bernice P Bishop Mus, 70-71; instr, 71-73, ASST PROF ANTHROP, UNIV SC, 73- Concurrent Pos: Collabr, Inst Archeol & Anthrop, Univ SC, 71-; NSF res grant, 73. Mem: Fel Am Anthrop Asn; Soc Am Archeol; Ling Soc Am. Res: Cultural adaptation in isolated environments, particularly prehistoric cultures of the Pacific islands and East Asia; megalithic architecture; marine subsistence and fishing cultures, current and prehistoric. Mailing Add: Dept of Anthrop Univ of SC Columbia SC 29208

AYROUD, ABDUL-MEJID, b Aleppo, Syria, Nov 24, 26; nat Can; m 54; c 3. CHEMISTRY. Educ: Univ Lille, ChE, 49; French Sch Paper Making, Papermaking E, 50; Univ Grenoble, PhD(chem), 53. Prof Exp: Head res dept, French Sch Paper Making, 53-57; chem engr, Pulp & Paper Res Inst Can, 57-58; sr res chemist, Res Ctr, Consol Paper Corp, Ltd, 58-60, leader chem pulping, 60-66, sect chief, Res & Develop Ctr, 66-67, mgr chem pulping & bleaching, Res & Develop Ctr, Consol-Bathurst Ltd, 67-71, ASST DIR RES & TECH SERV, RES & DEVELOP CTR, CONSOL-BATHURST LTD, 71- Mem: Tech Asn Pulp & Paper Indust; Can Pulp & Paper Asn; French Pulp & Paper Asn. Res: Applied chemistry; wood chemistry; pulp and paper technology; pollution abatement; administration. Mailing Add: Res & Develop Ctr Consolidated-Bathurst Ltd Grand'Mere PQ Can

AYUSO, KATHARINE, b New York, NY, June 25, 18. FOOD SCIENCE. Educ: St Lawrence Univ, BS, 40. Prof Exp: Asst chief chemist, Great Atlantic & Pac Tea Co, 42-65; assoc res dir, Foster D Snell, Inc, Booz Allen & Hamilton, 65-73; DIR FOOD SCI, ROSNER-HIXSON LABS, 73- Mem: Am Inst Chemists; Am Asn Cereal Chemists; Inst Food Technologists; Am Chem Soc; Am Asn Candy Technologists. Res: New product development, product reformulation, utilization of by-products; development of foods for the future; consumer evaluation studies. Mailing Add: Rosner-Hixson Labs 3570 N Avondale Ave Chicago IL 60618

AYVAZIAN, L FRED, b Ordu, Turkey, Oct 3, 19; US citizen; m 47; c 3. INTERNAL MEDICINE. Educ: Columbia Col, BA, 39; NY Univ, MD, 43. Prof Exp: Fel med, Sch Med, NY Univ, 44-45; res fel, Thorndike Med Lab, Boston City Hosp, Mass, 47-48; fel, Harvard Med Sch, 47-48; investr sterilization blood plasma, Goldwater Mem Hosp, Welfare Island, NY, 53-54; asst chief med serv, Vet Admin Hosp, NY, 54-61; med dir, Will Rogers Hosp & O'Donnell Mem Res Labs, 61-71; CHIEF CHEST SERV, VET ADMIN HOSP, 71-; PROF MED, NJ MED SCH, NEWARK, 71- Concurrent Pos: Assoc clin prof, Sch Med, NY Univ, 61-71. Mem: Harvey Soc; NY Acad Sci; fel Am Col Physicians; Am Thoracic Soc; Fedn Clin Res. Res: Tuberculosis and pulmonary diseases; sterilization of blood plasma; uric acid metabolism; hematologic changes in arterial oxygen desaturation states secondary to pulmonary disease; big ACTH in lung cancer. Mailing Add: Vet Admin Hosp East Orange NJ 07019

AZAM, FAROOQ, b Lahore, Pakistan. MICROBIAL ECOLOGY. Educ: Panjab Univ, BSc, 61, MSc, 63; Czech Acad Sci, Prague, CSc(microbiol), 68. Prof Exp: Res assoc anal biochem, West Regional Labs, Coun Sci & Indust Res, Pakistan, 63-65; fel biochem, State Univ NY Stony Brook, 68-69; res biochemist, Scripps Inst Oceanog, 69-73, ASST RES BIOLOGIST MARINE MICROBIAL ECOL, SCRIPPS INST OCEANOG, INST MARINE RESOURCES, UNIV CALIF, SAN DIEGO, 73- Mem: AAAS; Am Soc Microbiol. Res: Biochemistry; role of bacteria in marine food webs; transport of sugars and amino acids in marine bacteria; metabolic regulation in marine bacteria. Mailing Add: Inst of Marine Resources Scripps Inst Univ of Calif La Jolla CA 92093

AZAR, HENRY A, b Heliopolis, Egypt, Dec 21, 27; US citizen; m 60; c 2. PATHOLOGY. Educ: Am Univ Beirut, BA, 48, MD, 52; Am Bd Path, dipl, cert anat path, 58, cert clin path, 74. Prof Exp: Asst prof path, Am Univ Beirut, 58-60; from asst prof to assoc prof, Col Physicians & Surgeons, Columbia Univ, 60-70, assoc dir path, Francis Delafield Hosp, 68-70, assoc attend pathologist, Presby Hosp, 68-70; prof path & dir surg path, Univ Kans Med Ctr, Kansas City, 70-72; PROF PATH, COL MED, UNIV S FLA, 72-; CHIEF LAB SERV, VET ADMIN HOSP, TAMPA, 72- Mem: Am Asn Path & Bact; Am Soc Exp Path; Col Am Path; Harvey Soc; Int Acad Path. Res: Lymphomas; myeloma and amyloid. Mailing Add: Dept of Path Univ of SFla Col of Med Tampa FL 33620

AZAR, JOSEPH E, b Bshamoun, Lebanon, Jan 1, 19; m 51; c 4. INFECTIOUS

DISEASES, EPIDEMIOLOGY. Educ: Am Univ Beirut, BS, 42, MD, 46; London Sch Hyg & Trop Meu, dipl, 50. Prof Exp: Univ physician, 49, 51, asst prof commun dis, 52-57, from asst prof to assoc prof infectious dis & epidemiol, 58-65, PROF INFECTIOUS DIS & EPIDEMIOL & CHMN DEPT EPIDEMIOL & BIO STATIST, AM UNIV BEIRUT, 65- Concurrent Pos: Vis prof, Howard Univ, 63-64. Honors & Awards: Hon Lebanese Gold Medal, 59. Mem: Fel Am Pub Health Asn; fel Royal Soc Trop Med & Hyg; Lebanese Pub Health Asn (pres). Res: Tropical health; schistosomiasis. Mailing Add: Schs of Pub Health & Med Am Univ of Beirut Beirut Lebanon

AZAR, MIGUEL M, b Cordoba, Arg, Oct 21, 36; m 60; c 6. IMMUNOLOGY, PATHOLOGY. Educ: Nat Col Dean Funes, Arg, BA, 53; Cordoba Nat Univ, MD, 58; Univ Tenn, PhD(path), 65. Prof Exp: Assoc physician, Children's Hosp, Cordoba, 58-60; intern, Mercy Hosp, Des Moines, Iowa, 60-61; resident pediatrician, Med Units, Univ Tenn, 61-63, res assoc, 66-67; asst prof lab med & dir blood bank, 69-70, assoc prof lab med & dir clin lab, 70-75, PROF LAB MED, DIR LABS & DIR GRAD STUDIES LAB MED & PATH, VET ADMIN HOSPS, SCH MED, UNIV MINN, MINNEAPOLIS, 75- Concurrent Pos: Nat Inst Allergy & Infectious Dis spec fel immunol, Sch Med, Univ Minn, Minneapolis, 67-69. Mem: Am Asn Immunol; Soc Exp Biol & Med; Am Soc Exp Path. Res: Newborn pediatrics; pediatric pathology; immunopathology. Mailing Add: Vet Admin Hosp Univ of Minn Minneapolis MN 55417

AZARI, PARVIZ, b Baku, Russia, Feb 3, 30; US citizen; m 55; c 5. BIOCHEMISTRY. Educ: Univ Calif, AB, 55; Univ Nebr, MS, 58, PhD(chem), 61. Prof Exp: Asst prof chem & physiol, 63-70, assoc prof biochem, 70-75, PROF BIOCHEM, COLO STATE UNIV, 75- Concurrent Pos: USPH fel, 61-63, grant, 64- Mem: Am Chem Soc; Am Soc Biol Chem. Res: Structure and biological function of iron-transferrins; biochemistry of cataract. Mailing Add: Dept of Biochem Colo State Univ Ft Collins CO 80523

AZARNIA, ROOBIK, b Tabriz, Iran, Jan 12, 41. CELL PHYSIOLOGY, BIOPHYSICS. Educ: San Francisco State Univ, BA, 65; Univ Miami, PhD(physiol), 70. Prof Exp: Res assoc physiol, Col Physicians & Surgeons, Columbia Univ, 70-71; res assoc, 71-72, res scientist, 72-73, RES ASST PROF PHYSIOL & BIOPHYS, SCH MED, UNIV MIAMI, 74- Mem: AAAS; Am Physiol Soc; Tissue Cult Asn. Res: Electrophysiology; tissue culture; intercellular communication in normal and cancer cells; fertilization and ion permeability in echinoderm eggs. Mailing Add: Dept Physiol & Biophys Sch Med Univ Miami PO Box 520875 Miami FL 33152

AZARNOFF, DANIEL LESTER, b Brooklyn, NY, Aug 4, 26; m 51; c 3. CLINICAL PHARMACOLOGY. Educ: Rutgers Univ, BS, 47, MS, 48; Univ Kans, MD, 55. Prof Exp: Res assoc, Univ Kans, 49-52, intern, 55-56; asst prof med, Sch Med, St Louis Univ, 60-62; from asst prof med to assoc prof med & dir clin pharmacol study unit, 64-68, prof med & pharmacol, 68-73, DISTINGUISHED PROF MED & PHARMACOL, MED CTR, UNIV KANS, 73- Concurrent Pos: USPHS res fel med, Univ Kans, 56-58; Nat Inst Neurol Dis & Blindness spec trainee, Sch Med, Wash Univ, 58-60; Markle scholar, 62-; Burroughs-Wellcome scholar clin pharmacol, 64-69; Fulbright scholar, Karolinska Inst, Sweden, 68; mem drug res bd, Nat Res Coun, mem adv coun & pharmacol-toxicol prog adv comt, Nat Inst Gen Med Sci & mem comt on revision, US Pharmacopoeia; ed, Rev Drug Interactions & Year Bk Drug Ther. Mem: Fel Am Col Physicians; Brit Pharmacol Soc; Am Soc Clin Nutrit; Am Soc Pharmacol & Exp Therapeut; Am Soc Clin Pharmacol & Exp Therapeut. Res: Lipid metabolism; atherosclerosis. Mailing Add: Univ of Kans Med Ctr 39th & Rainbow Blvd Kansas City KS 66103

AZAROFF, LEONID VLADIMIROVICH, b Moscow, Russia, June 19, 26; nat US; m 46. X-RAY CRYSTALLOGRAPHY. Educ: Tufts Univ, BS, 48; Mass Inst Technol, PhD(crystallog), 54. Prof Exp: Asst res engr, Raytheon Mfg Co, 41-44; asst, Mass Inst Technol, 50-52; assoc, 52-53; assoc physicist, Armour Res Found, 53-55, res physicist, 55-56, sr scientist, 56-57; from assoc prof to prof metall eng, Ill Inst Technol, 57-66; PROF PHYSICS & DIR INST MAT SCI, UNIV CONN, 66- Mem: Am Phys Soc; Am Crystallog Asn; Mineral Soc Am; Am Inst Min, Metall & Petrol Eng; Am Soc Eng Educ. Res: Physics of metals; x-ray diffraction; soft x-ray spectroscopy. Mailing Add: PO Box 103 Storrs CT 06268

AZBELL, WILLIAM, b Manito, Ill, Jan 1, 06; m; c 3. PHYSICS. Educ: Ill State Norm Univ, BS, 33; Univ Ill, AM, 34. Prof Exp: Pub sch instr, Ill, 34-41; instr physics, Exten Div, Univ Ill, 41-43, Army Specialized Training Prog, Ball State Teachers Col, 43-44, Navy V-12, DePauw Univ, 44-45 & Univ Ill, 45-46; asst prof, Bradley Univ, 46-52; ASSOC PROF PHYSICS & HEAD DEPT, WARTBURG COL, 52- Mem: Am Asn Physics Teachers. Res: Atomic physics. Mailing Add: 502 Fourth St SW Waverly IA 50677

AZIZ, PHILIP MICHAEL, b Toronto, Ont, Mar 12, 24. PHYSICAL CHEMISTRY. Educ: Univ Toronto, BASc, 46, MA, 47, PhD(phys chem), 50. Prof Exp: Head theoret corrosion sect, Aluminum Labs, Ltd, Can, 49-51; res assoc, Inst Study Metals, Univ Chicago, 52; head theoret corrosion sect, Aluminum Labs, Ltd, Can, 53-59; mgr opers res & tech comput div, Systs Dept, Aluminum Co Can, Ltd, 59-70; CORP SYSTS MGR, ALCAN ALUMINUM CORP, 70- Concurrent Pos: Jr Author Award, Nat Asn Corrosion Engrs. Mem: Fel Chem Inst Can. Res: Mechanisms of the corrosion and oxidation of aluminum and its alloys; operations research; technical computing. Mailing Add: Alcan Aluminum Corp Box 6977 Cleveland OH 44101

AZIZ, RONALD A, b Toronto, Ont, Sept 27, 28; m 59; c 3. CHEMICAL PHYSICS. Educ: Univ Toronto, BA, 50, MA, 51, PhD(physics), 55. Prof Exp: Lectr physics, Royal Mil Col, Can, 55-57; asst prof, Univ Windsor, 57-58; from asst prof to assoc prof, 58-67, PROF PHYSICS, UNIV WATERLOO, 67- Concurrent Pos: Heinemann Found scholar, 65-66. Mem: Am Phys Soc; Am Asn Physics Teachers; Can Asn Physicists. Res: Low temperature physics; solidified inert gas solids; intermolecular potentials; ultrasonics; thermodynamic and transport properties. Mailing Add: Dept of Physics Univ of Waterloo Waterloo ON Can

AZMON, EMANUEL, marine geology, see 12th edition

AZPEITIA, ALFONSO GIL, b Madrid, Spain, Feb 22, 22; nat US. MATHEMATICS. Educ: Univ Madrid, BA, 39, MA, 46, PhD, 52. Prof Exp: Res assoc, Nat Res Coun, Spain, 52-55; res assoc, Brown Univ, 55-57; from instr to prof math, Univ Mass, Amherst, 57-65; PROF MATH, UNIV MASS, BOSTON, 65- Concurrent Pos: Fulbright lectr, 69-70. Mem: Am Math Soc; Math Asn Am; Soc Indust & Appl Math; Opers Res Soc Am. Res: Theory of functions; linear and mathematical programming. Mailing Add: Dept of Math Univ of Mass Boston MA 02125

AZUMAYA, GORO, b Tsurumi, Japan, Feb 26, 20; m 50; c 1. ALGEBRA. Educ: Tokyo Univ, BS, 42; Nagoya Univ, PhD(math), 49. Prof Exp: From lectr to asst prof math, Nagoya Univ, 45-53; prof, Hokkaido Univ, 53-68; vis prof, 65-66, PROF MATH, IND UNIV, BLOOMINGTON, 68- Concurrent Pos: Fel math, Off Naval Res, Yale Univ, 56-58; NSF fel, Northwestern Univ, 58-59; vis prof, Univ Mass, 64-65. Mem: Math Soc Japan; Am Math Soc. Res: Ring theory; number theory; group theory; homological algebra. Mailing Add: Dept of Math Ind Univ Bloomington IN 47401

B

BAAD, MICHAEL FRANCIS, b Hillsdale, Mich, Jan 5, 41; m 65; c 1. PLANT TAXONOMY, ECOLOGY. Educ: Univ Mich, BS, 63, MS, 66; Univ Wash, PhD(bot), 69. Prof Exp: ASST PROF BIOL SCI, CALIF STATE UNIV, SACRAMENTO, 69- Mem: Ecol Soc Am; Am Soc Plant Taxon; Int Soc Plant Taxon; Soc Study Evolution; Am Inst Biol Sci. Res: Taxonomy of the Caryophyllaceae; pollination ecology; plant succession; genecology. Mailing Add: Dept of Biol Sci Calif State Univ Sacramento CA 95819

BAADSGAARD, HALFDAN, b Minneapolis, Minn, Apr 16, 29; m 57; c 5. GEOCHEMISTRY. Educ: Univ Minn, BS, 51; Swiss Fed Inst Technol, PhD, 55. Prof Exp: Sr res assoc, Univ Minn, 55-57; from asst prof to assoc prof, 57-66, PROF GEOCHEM, UNIV ALTA, 66- Concurrent Pos: NATO exchange prof, Copenhagen Univ, 64. Mem: Geochem Soc; Am Geophys Union; fel Geol Asn Can. Res: Methods of geologic dating; geochronology; trace elements in geochemical research. Mailing Add: Dept of Geol Univ of Alta Edmonton AB Can

BAAK, TRYGGVE, b Stockholm, Sweden, Nov 6, 17; m 47; c 3. PHYSICAL CHEMISTRY, INORGANIC CHEMISTRY. Educ: Univ Stockholm, Fil Kand, 50, Fil Lic, 55. Prof Exp: Fel metall, Mass Inst Technol, 55-56; res assoc glass, Pa State Univ, 56-57; assoc prof silicate sci, Silicate Inst, Univ Toledo, 57-59; sr res scientist phys & inorg chem, Owens-Ill Glass Co, 59-65; sr staff engr & sr glass physicist in chg optical mat lab, Perkin-Elmer Corp, Conn, 65-69; vpres res, United Sierra Div, Cyprus Mines Corp, 69-75; RETIRED. Concurrent Pos: Swedish Iron Masters Asn grant, 50-55; Nat Tech Coun Sweden res grant, 53-55. Mem: AAAS; Am Chem Soc; Am Ceramic Soc; Brit Soc Glass Technol; Swedish Soc Tech Sci. Res: Molten salts and silicates; high-temperature electrochemistry; phase diagrams and cyrstal structure; metallurgical processes; glass technology; clays, talcs, diatomaceous earth and their industrial applications in paper, plastics, rubber and paints. Mailing Add: 15 Catamaran St Apt 5 Marina Del Ray CA 90291

BAALMAN, ROBERT J, b Grinnell, Kans, Oct 28, 39; m 64; c 2. BOTANY. Educ: Ft Hays Kans State Col, BS, 60, MS, 61; Univ Okla, PhD(bot), 65. Prof Exp: Teaching asst bot, Ft Hays Kans State Col, 60-61; teacher, High Sch, Kans, 61-62; teaching asst bot, Univ Okla, 62-64; from asst prof to assoc prof biol, 65-74, PROF BIOL, CALIF STATE UNIV, HAYWARD, 74- Mem: AAAS; Ecol Soc Am; Am Forestry Asn; Am Inst Biol Sci. Res: Plant succession; fire as a successional factor; seed dissemination and germination; saline plant as communities in inland and coastal areas; plant competition. Mailing Add: Dept of Biol Calif State Univ Hayward CA 94542

BAARDA, DAVID GENE, b Newton, Iowa, Apr 23, 37; m 62; c 3. ORGANIC CHEMISTRY, ENVIRONMENTAL CHEMISTRY. Educ: Cent Col, Iowa, BA, 59; Univ Fla, MS, 60, PhD(org chem), 62. Prof Exp: Res asst chem, Univ Fla, 62-63; chemist, Texaco, Inc, 63-64, sr chemist, 64-65; from asst prof to assoc prof, 65-74, PROF CHEM, GA COL, 74- Concurrent Pos: Danforth assoc, 70. Mem: Am Chem Soc. Res: Use of laboratory investigation and computerized data handling to teach chemistry at the undergraduate level. Mailing Add: Dept of Chem Ga Col Milledgeville GA 31061

BAARDA, ISAAC F, microbiology, biochemistry, see 12th edition

BAARS, DONALD LEE, b Oregon City, Ore, May 27, 28; m 48; c 3. STRATIGRAPHY, SEDIMENTARY PETROLOGY. Educ: Univ Utah, BS, 52; Univ Colo, Boulder, PhD(geol), 65. Prof Exp: Explor geologist, Shell Oil Co, 52-61; res geologist, Continental Oil Co, 61-62; assoc prof geol, Wash State Univ, 65-68; ASSOC PROF GEOL, FT LEWIS COL, 68- Concurrent Pos: Proj dir, NSF Res Grant, 68-; dir short course for col teachers proj, NSF, 70-71. Mem: Am Asn Petrol Geol; fel Geol Soc Am; Soc Econ Paleont & Mineral. Res: Paleozoic stratigraphy of Colorado Plateau; carbonate sedimentation and sedimentary petrology; early Paleozoic paleotectonic history of the ancestral Rockies and related tectonic system. Mailing Add: Dept of Geol Ft Lewis Col Durango CO 81301

BABA, ANTHONY JOHN, b Bethleham, Pa, May 2, 36; m 59; c 6. RADIATION PHYSICS. Educ: Georgetown Univ, BS, 57. Prof Exp: Physicist, Nat Bur Standards, 58-60; RES PHYSICIST, HARRY DIAMOND LABS, 60- Concurrent Pos: Lectr, Trinity Col, DC, 63-69; assoc mem adv group electron devices, Dept Defense, 72- Mem: Inst Elec & Electronics Eng. Res: Chemical vapor depositions for silicon integrated circuit applications; transient radiation effects on integrated circuits. Mailing Add: Harry Diamond Labs 2800 Powder Mill Rd Adelphi MD 20783

BABA, NOBUHISA, b Tokyo, Japan, Oct 15, 32; US citizen; m 63; c 2. PATHOLOGY, ELECTRON MICROSCOPY. Educ: Univ Tokyo, MD, 57; Ohio State Univ, MSc, 61, PhD(path), 65. Prof Exp: Intern, Univ Tokyo Hosps, 57-58; resident, 58-63, from instr to assoc prof, 63-74, PROF PATH, OHIO STATE UNIV, 74-, CHIEF DIV PATH ANAT, 71- Mem: AMA; Am Asn Path & Bact; Am Acad Forensic Sci; Am Soc Exp Path; Am Soc Cell Biol. Res: Electron microscopy of rat hearts; investigation of ischemia; anoxia; effects of anions; electron histochemistry of lactic dehydrogenase and creatine kinase; study of cardiac conduction system; immunoelectron microscopy of actomyosin; forensic medicine. Mailing Add: Dept of Path Ohio State Univ Columbus OH 43210

BABAD, HARRY, b Vienna, Austria, Mar 12, 36; US citizen; m 65; c 2. ORGANIC CHEMISTRY, NUCLEAR CHEMISTRY. Educ: Polytech Inst Brooklyn, BS, 56; Univ Ill, MS & PhD(org chem), 61. Prof Exp: Fel org chem, Mass Inst Technol, 61-62; res assoc, Univ Chicago, 62-63; asst prof, Univ Denver, 63-67; group leader new prod res, Ott Chem Co, 67-69, head sect res & develop, 69-74; MGR, CHEM TECHNOL LAB, ATLANTIC RICHFIELD HANFORD CO, 74- Concurrent Pos: Consult, Colo Int, 74- Mem: AAAS; Am Chem Soc; NY Acad Sci; fel Am Inst Chemists. Res: Industrial synthesis of fine organics; physical organic and synthetic chemistry; industrial chemical research on chemical intermediates, including specialty monomers, phosgene derivatives, amino acids, aromatics for pharmaceuticals, new products development, research management and project selection criteria. Mailing Add: 200W 202-S 112-D Atlantic Richfield Hanford Co Richland WA 99352

BABAYAN, VIGEN KHACHIG, b Armenia, Jan 1, 13; nat US; m 42; c 3. FOOD SCIENCE, LIPID CHEMISTRY. Educ: NY Univ, BA, 38. Prof Exp: Chemist, E F Drew & Co, 38-40; res chemist, Warwick Chem Co, 40-41; dir res & develop, Ridbo Labs, 41-47; dir labs, Theobald Industs, 47-48; dir res & develop, E F Drew & Co, 48-61, vpres & dir labs, Drew Chem Corp, 61-65; VPRES & DIR RES & DEVELOP & QUAL CONTROL, STOKELY-VAN CAMP, INC, 65-; ASST PROF MED, SCH MED, IND UNIV, 69- Concurrent Pos: Mem res & develop tech comts, Nat Canners

Asn & Res & Develop Assocs. Honors & Awards: Glycerin Asn Award, 64. Mem: Am Chem Soc; Am Oil Chem Soc; Inst Food Technol; Am Asn Cereal Chemists; Am Asn Candy Technologists. Res: Foods, canned and frozen; edible oils; fatty derivatives; lipids. Mailing Add: 3910 Cranbrook Dr Indianapolis IN 46240

BABB, DANIEL PAUL, b Red Wing, Minn, Aug 1, 39. INORGANIC CHEMISTRY. Educ: Mankato State Col, BA, 63; Univ Idaho, PhD(inorg chem), 68. Prof Exp: Asst prof chem, Kearney State Col, 67-69 & Va Polytech Inst, 69-74; ASST PROF CHEM, MARSHALL UNIV, 74- Concurrent Pos: Asst scientist, Inorg Chem Inst, Univ Göttingen, 68-69. Mem: Am Chem Soc. Res: Fluorine and inorganic synthesis chemistry. Mailing Add: Dept of Chem Marshall Univ Huntington WV 25701

BABB, DAVID DANIEL, b Saugus, Mass, Sept 20, 28; m 49. Educ: Mass Inst Technol, BS & MS, 50, MS, 58. Prof Exp: Student engr, Philco Corp, 47-49; proj engr, Radio Div, Bendix Corp, 49-52; res asst atomic physics, Mass Inst Technol, 55-58; sr nuclear engr, Gen Dynamics/Ft Worth, 58-61; sr res physicist, Dikewood Corp, NMex, 61-66, head physics div, 66-68, physics prog mgr, 68-69, prog mgr theoret res, 69-70; pres, Albuquerque Res Assocs, Inc, 70-73; SR RES PHYSICIST, DIKEWOOD CORP, N MEX, 73- Mem: Am Phys Soc; Inst Elec & Electronics Engrs. Res: Nuclear weapons effects, including electromagnetic pulse, neutron and gamma transport and x-ray vulnerability; air and space borne reactor hazards, including reactor shielding and space radiation; atomic frequency standards; radio design; lasers. Mailing Add: 524 Chamiso Lane NW Albuquerque NM 87107

BABB, ROBERT MASSEY, b Durham, NC, Apr 24, 38; m 66. ORGANIC CHEMISTRY. Educ: Univ NC, Chapel Hill, BS, 60; Univ Utah, PhD(org chem), 69. Prof Exp: Jr chemist, Merck Sharp & Dohme Res Labs, NJ, 60-63; res chemist, Res Ctr, Burlington Industs, Inc, NC, 68-72; group leader, Dyestuffs & Chem Div, 72-75, STAFF CHEMIST, AGR DIV, CIBA-GEIGY CORP, 75- Mem: Am Chem Soc. Res: Reactions of cyclopropenes and methylenecyclopropanes; organophosphorus chemistry; textile chemistry; pesticides and process development. Mailing Add: PO Box 11 St Gabriel LA 70776

BABB, STANLEY ERNEST, JR, b Galveston, Tex, Sept 22, 34; m 57; c 5. PHYSICS. Educ: Univ Tex, BS, 54, MA, 55, PhD(physics), 57. Prof Exp: Res assoc chem eng, Univ Ill, 57-58; from asst prof to assoc prof physics, 58-68, chmn dept physics & astron, 74-75, PROF PHYSICS, UNIV OKLA, 68- Res: High pressure phenomena; physical optics; thermodynamics; history of astronomy. Mailing Add: Dept of Physics Univ of Okla Norman OK 73069

BABBITT, DONALD GEORGE, b Detroit, Mich, Feb 24, 36; m 64; c 2. MATHEMATICAL ANALYSIS. Educ: Univ Detroit, BS, 57, Univ Mich, MA, 58, PhD(math), 62. Prof Exp: From asst prof to assoc prof, 62-73, PROF MATH, UNIV CALIF, LOS ANGELES, 73- Concurrent Pos: Mem, Inst Advan Study, 70-71. Mem: AAAS; Am Math Soc. Res: Study of mathematical structures which arise in theoretical physics. Mailing Add: Dept of Math Univ of Calif Los Angeles CA 90024

BABBITT, JERRY, b Moscow, Idaho, Jan 16, 43; m 62; c 2. FOOD SCIENCE, BIOCHEMISTRY. Educ: Wash State Univ, BS, 65, MS, 67, PhD(food sci & technol), 70. Prof Exp: ASST PROF FOOD SCI & TECHNOL, ORE STATE UNIV, 69- Mem: AAAS; Inst Food Technol. Res: Effect of growth-regulating substances on enzymatic softening of tomato fruit; effects of chemical, physiological and biochemical changes in fish and shellfish on quality and consumer acceptance. Mailing Add: Seafoods Labs Ore State Univ 250 36th St Astoria OR 97103

BABBITT, JOHN DAVID, b Fredericton, NB, Feb 2, 08; m 48; c 5. PHYSICS. Educ: Univ NB, BA, 29, Oxford Univ, BA, 32, PhD(physics), 35. Prof Exp: Res physicist, Nat Res Coun Can, 35-50, Can sci liaison officer, Washington, DC, 50-57, secy int rels, 57-70, spec adv, Nat Sci Libr, 70-73; RETIRED. Mem: Fel Royal Soc Can; Can Asn Physicists. Res: Thermal conductivity; diffusion of water vapor; international relations. Mailing Add: 436 Meadow Dr Ottawa ON Can

BABBOTT, FRANK LUSK, JR, b New York, NY, Feb 6, 19; m 50; c 3. EPIDEMIOLOGY. Educ: Amherst Col, BA, 47; State Univ NY, MD, 51; Harvard Univ, MPH, 53, MS, 54; Am Bd Prev Med, dipl. Prof Exp: Instr epidemiol, Sch Pub Health, Harvard Univ, 54-55, assoc, 55-58; asst prof prev med, Sch Med, Univ Pa, 58-62; vis fel epidemiol, Pub Health Lab Serv, London, Eng, 62-63; ASSOC PROF EPIDEMIOL, COL MED, UNIV VT, 63- Concurrent Pos: Vchmn, Am Bd Prev Med, 74- Mem: Fel Am Pub Health Asn; fel Am Col Prev Med; Am Epidemiol Soc; Asn Teachers Prev Med; Arctic Inst NAm. Mailing Add: Dept of Epidemiol Univ of Vt Col of Med Burlington VT 05401

BABCOCK, CHARLOTTE GERTRUDE, b Jackson Center, Ohio, May 12, 07; m 53. PSYCHIATRY. Educ: Milton Col, AB, 29; Univ Chicago, MD, 38; Am Bd Psychiat & Neurol, dipl, 44. Hon Degrees: DSc, Milton Col, 54. Prof Exp: Instr psychiat, Univ Chicago, 41-44; pvt pract, 44-53; assoc dir, Psychoanal Inst, 61-74, PROF PSYCHIAT, UNIV PITTSBURGH, 53-; MEM FAC, PITTSBURGH PSYCHOANAL INST, 53- Concurrent Pos: Consult, Cook County Psychopath Hosp, 43-44, United Charities of Chicago, 47-52 & Chicago Child Care Soc, 49-53; staff mem eve clin, Vet Rehab Ctr, 44-46; lectr, Sch Nursing Educ, Univ Chicago, 44-53, lectr, Sch Social Serv Admin, 45-49; res assoc, Inst Psychoanal, Chicago, 46-49, staff mem, 49-53; fel, Ctr Advan Study Behav Sci, 61-62. Mem: AAAS; fel Am Psychiat Asn; Am Psychoanal Asn; AMA. Res: Psychoanalytic research, education and problems of minority groups. Mailing Add: 3811 O'Hara St Pittsburgh PA 15261

BABCOCK, CLARENCE LLOYD, b Paoli, Ind, Aug 30, 04; m 29; c 3. PHYSICS. Educ: Ball State Teachers Col, BS, 28; Purdue Univ, MS, 31, PhD(physics), 36. Prof Exp: Instr high sch, Ind, 28-30; asst & instr physics, Purdue Univ, 30-35; chief glass technologist, Owens-Ill Glass Co, 35-66, assoc dir res, 66-68; PROF OPTICAL SCI, UNIV ARIZ, 68- Mem: Nat Inst Ceramic Engrs; fel Am Ceramic Soc; Am Soc Testing & Mat; Optical Soc Am; fel Brit Soc Glass Technologists. Res: Measurement of all glass properties and their engineering applications in the manufacture and commercial use of glass. Mailing Add: Optical Sci Ctr Univ of Ariz Tucson AZ 85721

BABCOCK, ELKANAH ANDREW, b Elizabeth, NJ, Nov 23, 41; m 64; c 1. STRUCTURAL GEOLOGY, REMOTE SENSING. Educ: Union Col, NY, BS, 63; Syracuse Univ, MS, 65; Univ Calif, Riverside, PhD(geol), 69. Prof Exp: ASST PROF GEOL, UNIV ALTA, 69- Mem: AAAS; Geol Soc Am. Res: Structural geology of the Salton Trough, Southern California; terrain analysis and geologic mapping applications of remote sensing. Mailing Add: Dept of Geol Univ of Alta Edmonton AB Can

BABCOCK, GEORGE, JR, b NJ, Apr 26, 16; m 52; c 3. PHARMACOLOGY. Educ: Rutgers Univ, BA, 42, MSc, 43; Western Reserve Univ, MD, 47. Prof Exp: Intern, East Orange Gen Hosp, 47-48; mem res staff, Schering Corp, 48-52, asst dir clin res, 52-54, assoc dir, 55-60, dir med res div, 60-63, DIR SCI AFFAIRS, IVES LABS, INC, 64-, VPRES, 65- Mem: Fel Am Col Allergists; fel Am Geriat Soc; Am Acad Allergy; Sigma Xi. Res: Allergy; endocrinology; clinical pharmacology. Mailing Add: 55 Hilltop Terr Kinnelon NJ 07405

BABCOCK, HORACE MAXSON, b Los Angeles, Calif, Aug 28, 16; m 40; c 5. HYDROLOGY. Educ: Univ Ariz, BS in CE, 39, CE, 55. Prof Exp: Proj chief water resources studies, Water Resources Div, US Geol Surv, 39-41 & Ground Water Br, Duncan, Ariz, 41-42, engr-in-chg, Phoenix, 46-49, Wyo & Nebr, 49-51, dist engr, Wyo, 51-58, chief opers sect, Washington, DC, 58-62, asst chief entire US, 62-63, dist engr, Ariz, 63-64, DIST CHIEF WATER RESOURCES INVESTS, GROUND WATER BR, WATER RESOURCES DIV, US GEOL SURV, ARIZ, 64- Mem: Am Soc Civil Eng. Res: Ground water hydrology, especially arid zone hydrology. Mailing Add: 5417 E Rosewood Ave Tucson AZ 85711

BABCOCK, HORACE WELCOME, b Pasadena, Calif, Sept 13, 12; m 40, 58; c 3. ASTRONOMY. Educ: Calif Inst Technol, BS, 34; Univ Calif, PhD(astron), 38. Hon Degrees: DSc, Univ Newcastle-upon-Tyne, 65. Prof Exp: Asst, Lick Observ, Univ Calif, 38-39; instr, McDonald Observ, Univ Chicago & Univ Tex, 39-41; res assoc, Radiation Lab, Mass Inst Technol, 41-42; res assoc, Rocket Proj, Calif Inst Technol, 42-46; staff mem, Mt Wilson & Palomar Observ, 46-57; asst dir, 57-64, DIR, HALE OBSERV, 64- Concurrent Pos: Counr, Nat Acad Sci, 73-76. Honors & Awards: Draper Medal, Nat Acad Sci, 57; Eddington Medal, Royal Astron Soc, 58, Gold Medal, 70; Bruce Gold Medal, Astron Soc of the Pac, 69. Mem: Nat Acad Sci; Am Astron Soc; Am Acad Arts & Sci; Int Astron Union; Am Philos Soc. Res: Magnetic fields of sun and stars; stellar spectroscopy; diffraction gratings; astronomical instrumentation; observatory development; telescope design and instrumentation. Mailing Add: Hale Observ 813 Santa Barbara St Pasadena CA 91101

BABCOCK, JOHN CLAUDE, b Phoenixville, Pa, Oct 20, 25; m 49; c 3. ORGANIC CHEMISTRY. Educ: Harvard Univ, BS, 48, MS, 49, PhD(chem), 52. Prof Exp: Chemist, Merck & Co, Inc, 49-50; CHEMIST, UPJOHN CO, 52- Mem: Am Chem Soc; Endocrine Soc. Res: Steroids; prostaglandins. Mailing Add: 3210 Magnolia Circle Kalamazoo MI 49001

BABCOCK, KENNETH LESLIE, b Riverside, Calif, Oct 10, 26; m 60; c 4. SOIL CHEMISTRY. Educ: Univ Calif, BS, 50, PhD(soil sci), 55. Prof Exp: Chmn dept soils & plant nutrit, 69-75, PROF SOIL CHEM, UNIV CALIF, BERKELEY, 55-, ASSOC DEAN ACAD AFFAIRS, COL NATURAL RESOURCES, 75- Concurrent Pos: Guggenheim fel, 64. Mem: Soil Sci Soc Am. Res: Thermodynamics and soil chemistry; behavior of radioisotopes in soils and plants; alkali soils. Mailing Add: 2428 Russell St Berkeley CA 94705

BABCOCK, MACLEAN JACK, b State College, Pa, Feb 20, 18; m 40; c 2. NUTRITION. Educ: Pa State Univ, BS, 38; Cornell Univ, MS, 39, PhD(nutrit), 46. Prof Exp: Instr biochem, Northwestern Univ, 46-48; assoc prof, 48-74, EMER ASSOC PROF BIOCHEM, RUTGERS UNIV, 74- Concurrent Pos: Consult, Radiol Defense, NJ, 58. Mem: AAAS; Am Chem Soc; Am Inst Nutrit; Inst Food Technologists. Res: Nutrition surveys; methods for measuring nutritional status; vitamin B-6; protein nutrition; nutrition labeling. Mailing Add: Dept of Food Sci Cook Col Rutgers Univ New Brunswick NJ 08903

BABCOCK, PHILIP ARNOLD, b Clinton, Mass, May 11, 32; m 59; c 2. PHARMACOGNOSY, BOTANY. Educ: Mass Col Pharm, BS, 54, MS, 59; Univ Iowa, PhD(pharm), 62. Prof Exp: Asst prof, 62-66, ASSOC PROF PHARMACOG, COL PHARM, BUSCH CAMPUS, RUTGERS UNIV, 66- Mem: AAAS; Am Pharmaceut Asn; Am Soc Pharmacog; Bot Soc Am. Res: Biosynthesis of drugs in plants; tissue culture of medicinal plants; chemotaxonomy. Mailing Add: Col of Pharm Busch Campus Rutgers Univ New Brunswick NJ 08903

BABCOCK, ROBERT FREDERICK, b Ft Wayne, Ind, Feb 26, 30; m 53; c 2. ANALYTICAL CHEMISTRY. Educ: DePauw, BA, 51; Ind Univ, PhD(anal chem), 58. Prof Exp: From proj chemist to sr proj chemist, Am Oil Co, Whiting, Ind, 57-74; RES CHEMIST, STAND OIL CO (IND), 74- Mem: Am Chem Soc. Res: Development of analytical methods; spectrophotometry; environmental analytical research. Mailing Add: 305 Gettysburg Park Forest IL 60466

BABCOCK, SIDNEY HENRY, JR, organic chemistry, see 12th edition

BABCOCK, WILLIAM EDWARD, b Buhl, Idaho, Oct 15, 22; m 45; c 4. VETERINARY MEDICINE. Educ: State Col Wash, BS, 44, DVM, 45; Ore State Col, MS, 51. Prof Exp: Assoc vet, Dept Vet Med, Ore State Col, 49-63; res vet, Agr Res Dept, 63, MGR AGR RES & DEVELOP DEPT, CHAS PFIZER & CO, 63- Mem: Am Vet Med Asn; Poultry Sci Asn; Am Asn Avian Path. Res: Poultry diseases. Mailing Add: Animal Health Res Dept Chas Pfizer & Co Terre Haute IN 47808

BABCOCK, WILLIAM JAMES VERNER, b Sydenham, Ont, July 5, 12; US citizen; m 32; c 4. ECOLOGY, SCIENCE EDUCATION. Educ: Eastern Nazarene Col, AB, 36; Boston Univ, AM, 39. Prof Exp: Instr gen biol & health sci, Eastern Nazarene Acad, 37-39, instr comp anat & physiol, 39-40, asst prof biol sci, 40-42, assoc prof biol & chmn dept, 42-57, PROF BIOL, EASTERN NAZARENE COL, 57- Concurrent Pos: Dir, NSF In-Serv Inst Environ Studies, Eastern Nazarene Col, 70, head prof develop BS prog in environ sci & AS prog in hort; res grant, W K Kellogg Found, 71-73. Mem: AAAS; Ecol Soc Am; Nat Sci Teachers Asn; Nat Asn Biol Teachers; Am Forestry Asn. Res: Ecology of northeast United States, especially New England and New York; analysis of Blue Hills Reservation in Massachusetts; bog and fresh water studies; marine life. Mailing Add: 114 Willet St Wollaston MA 02170

BABEL, FREDERICK JOHN, b Traverse City, Mich, Oct 22, 11; m 39. FOOD MICROBIOLOGY. Educ: Mich State Univ, BS, 35; Purdue Univ, MS, 36; Iowa State Univ, PhD(dairy bact), 39. Prof Exp: Res chemist & bacteriologist, NAm Creameries, Inc, Minn, 39-40; res assoc dairy bact, Iowa State Univ, 40-43, asst prof, 43-44, res assoc prof, 44-47; res assoc prof, 47-50, PROF DAIRY & FOOD MICROBIOL, PURDUE UNIV, WEST LAFAYETTE, 50- Concurrent Pos: Consult, C Hansen's Lab, Milwaukee & Sealtest Foods, Pa. Honors & Awards: Paul-Lewis Pfizer Award, 61. Mem: Am Dairy Sci Asn; Int Asn Milk & Food Sanitarians. Res: Microbiology and enzymology of dairy products; bacteriophage; flavor of cultured dairy products; milk-coagulating enzymes; dairy plant sanitation. Mailing Add: Food Sci Inst Smith Hall Purdue Univ West Lafayette IN 47906

BABERO, BERT BELL, b St Louis, Mo, Oct 9, 18; m 50; c 2. PARASITOLOGY. Educ: Univ Ill, BS, 49, MS, 50, PhD, 57. Prof Exp: Med parasitologist, USPHS, Alaska, 50-53; asst parasitol, Univ Ill, 54-57; prof zool & head dept, Ft Valley State Col, 57-59; prof, Southern Univ, 59-60; lectr, Fed Emergency Sci Scheme, Lagos, Nigeria, 60-62; parasitologist, Col Med, Univ Baghdad, 62-65; assoc prof biol sci, 65-67, PROF BIOL SCI, UNIV NEV, LAS VEGAS, 67- Concurrent Pos: Fulbright-Hayes fel, US Dept State, Iraq, 63-65; La State Univ fel, 68. Mem: Soc Syst Zool; Soc Protozool; Am Soc Parasitol; Wildlife Dis Asn; Am Micros Soc. Res: Animal and human parasitology; ascariasis; diphyllobothriasis; echinococcosis; zoonotic diseases. Mailing Add: Dept of Biol Sci Univ of Nev Las Vegas NV 89109

BABIANIAN, ROSTOM, b Addis-Ababa, Ethiopia, Nov 24, 29; US citizen; m 55; c 1. VIROLOGY, BIOLOGY. Educ: Am Univ Cairo, BS, 53, BA, 55; NY Univ, MS, 59,

PhD(cytochem), 64. Prof Exp: Res assoc virol, Rockefeller Univ, 64-65; asst prof, 66-72, ASSOC PROF MICROBIOL & IMMUNOL, STATE UNIV NY DOWNSTATE MED CTR, 72- Mem: Am Soc Microbiol. Res: Mechanism of animal virus-induced cell damage. Mailing Add: Dept of Microbiol & Immunol State Univ NY Downstate Med Ctr Brooklyn NY 11203

BABICH, GEORGE LEON, b New York, NY, Aug 4, 44; m 66; c 2. DEVELOPMENTAL BIOLOGY. Educ: Atlantic Union Col, BA, 66; Univ NH, MS, 69, PhD(zool), 72. Prof Exp: Res assoc med physiol, Loma Linda Univ, 66-67; asst prof biol, Atlantic Union Col, 67-68; assoc prof biol, 71-73, DIR ALLIED HEALTH-HEALTH SCI, QUINSIGAMOND COMMUNITY COL, 74- Mem: AAAS; Am Zool Soc; Soc Develop Biol; Am Asn Allied Health Professions; Sigma Xi. Res: Effects of cyclic nucleotides on amputated amphibian limbs regenerating in vitro. Mailing Add: Quinsigamond Community Col W Boylston St Worcester MA 01606

BABIEC, JOHN STANLEY, JR, b Providence, RI, Mar 6, 40; m 63; c 3. INDUSTRIAL CHEMISTRY, RESEARCH ADMINISTRATION. Educ: Providence Col, BS, 61, MS, 63; Univ Mass, Amherst, PhD(chem), 67. Prof Exp: Res assoc chem, Olin Mathieson Chem Corp, 66-68; asst prof, New Haven Col, 68-70; sr res assoc & group supvr, 70-75, SECT LEADER, OLIN CORP, 75- Mem: Sigma Xi; Am Chem Soc. Res: Fire retardance of plastics; new product development; urethane research and application. Mailing Add: Olin Corp 275 Winchester Ave New Haven CT 06511

BABINEAU, G RAYMOND, b New York, NY, Apr 8, 37; m 60; c 3. PSYCHIATRY. Educ: Bowdoin Col, BA, 59; Harvard Med Sch, MD, 63. Prof Exp: Chief ment hyg psychiat, US Army Hosp, Berlin, Ger, 67-70; ASST PROF PSYCHIAT, UNIV ROCHESTER, 70-, CHIEF MENT HEALTH SECT, UNIV HEALTH SERV, 73- Mem: Am Psychiat Asn; Am Col Health Asn. Res: Psychological problems of university students. Mailing Add: Strong Mem Hosp Box 217 260 Crittenden Blvd Rochester NY 14642

BABIONE, ROBERT WILLIAM, b Luckey, Ohio, Sept 9, 04; m 30; c 3. EPIDEMIOLOGY. Educ: Oberlin Col, AB, 24; Western Reserve Univ, MD, 30; Johns Hopkins Univ, MPH, 48. Prof Exp: US Navy, 30-59, epidemiologist, 41-45, dir prev med div, Bur Med, 48-52, officer-in-chg naval prev med, Far East, 53-54; exec secy, Armed Forces Epidemiol Bd, 55-59; mem, Pan Am Sanit Bur, 59-68; consult, AID, Washington, DC, 70-73; RETIRED. Concurrent Pos: Liaison mem exec bd, Am Bd Prev Med, 50-52. Mem: AMA; Am Soc Trop Med & Hyg; Am Mosquito Control Asn. Res: Epidemiology of infectious diseases; malaria. Mailing Add: 418 N Jackson St Arlington VA 22201

BABIOR, BERNARD M, b Los Angeles, Calif, Nov 10, 35; m 61; c 2. BIOCHEMISTRY, HEMATOLOGY. Educ: Univ Calif, San Francisco, MD, 59; Harvard Univ, PhD(biochem), 65. Prof Exp: Asst prof med, Harvard Med Sch, 69-71, assoc prof, 71-72; ASSOC PROF MED, MED SCH, TUFTS UNIV, 72- Concurrent Pos: Tutor biochem, Harvard Univ, 68- Mem: AAAS; Am Soc Biol Chem; Am Soc Clin Invest. Res: Mechanism of action of coenzyme B12; granulocyte physiology. Mailing Add: Dept of Med Tufts-New Eng Med Ctr Hosp Boston MA 02111

BABITCH, JOSEPH AARON, b Detroit, Mich, July 14, 42; m 64; c 1. NEUROCHEMISTRY. Educ: Univ Mich, BS, 65; Univ Calif, Los Angeles, PhD(biol chem), 71. Prof Exp: ASST PROF CHEM, TEX CHRISTIAN UNIV, 73- Concurrent Pos: Fel, US Nat Mult Sclerosis Soc, 71-73; dir chem of behav prog, Tex Christian Univ, 75- Mem: Am Soc Neurochem; Sigma Xi; Soc Neurosci. Res: Structure and function of synaptic membranes. Mailing Add: Dept of Chem Tex Christian Univ Ft Worth TX 76129

BABLER, BERNARD JOSEPH, chemistry, see 12th edition

BABLER, JAMES HAROLD, b Evanston, Ill, June 14, 44. SYNTHETIC ORGANIC CHEMISTRY. Educ: Loyola Univ, Ill, BS, 66; Northwestern Univ, PhD(chem), 71. Prof Exp: ASST PROF CHEM, LOYOLA UNIV CHICAGO, 70- Mem: AAAS; Am Chem Soc. Res: Natural products; heterolytic fragmentation reactions; chemicals controlling insect behavior; allylic rearrangements. Mailing Add: Loyola Univ Dept Chem 6525 N Sheridan Rd Chicago IL 60626

BABOIAN, ROBERT, b Watertown, Mass, Nov 17, 34; m 59; c 3. ELECTROCHEMISTRY. Educ: Suffolk Univ, BS, 59; Rensselaer Polytech Inst, PhD(chem), 64. Prof Exp: Ford Found fel, Univ Toronto, 64-65, sr res assoc, 65-66; mem tech staff, 66-73, MGR LAB, TEX INSTRUMENTS INC, 73- Honors & Awards: Accredited Corrosion Specialist, Nat Asn Corrosion Engr, 74. Mem: AAAS; Am Chem Soc; Electroche Soc; Am Soc Testing & Mat; Nat Asn Corrosion Engr. Res: Electrochemical behavior of metals in electrolyte solutions; corrosion properties of materials; processes for surface treatment of metals, electrocleaning, electropolishing, electromachining; ionic interactions in molten salts. Mailing Add: Tex Instruments Inc Attleboro MA 02703

BABROV, HAROLD J, b New York, NY, Apr 21, 26; m 52; c 4. MOLECULAR SPECTROSCOPY. Educ: Univ Calif, Los Angeles, AB, 50, MA, 51; Univ Pittsburgh, PhD(physics), 59. Prof Exp: Physicist, US Naval Ord Test Sta, 50; asst, Inst Geophys, Univ Calif, 51; physicist, Knolls Atomic Power Lab, Gen Elec Co, 51-54; asst, Univ Pittsburgh, 54-60; sr physicist, Control Instrument Div, Warner & Swasey Co, NY, 60-63, chief physicist, 63-68; assoc prof physics, Ind Univ, South Bend, 68-72; MEM TECH STAFF, SPACE DIV, ROCKWELL INT CO, 72- Concurrent Pos: Cottrell grant, Res Corp, 69. Mem: Fel Optical Soc Am; Am Phys Soc; Combustion Inst; Sigma Xi. Res: Infrared physics; intensity spectra of small molecules and atoms, especially the broadening of spectral lines and determination of transition probabilities; applications of intenisty spectroscopy of problems of radiant energy transfer. Mailing Add: 9881 Cheshire Ave Westminster CA 92683

BABSON, ARTHUR LAWRENCE, b Orange, NJ, Mar 3, 27; m 50; c 2. CLINICAL CHEMISTRY. Educ: Cornell Univ, BA, 50; Rutgers Univ, MS, 51, PhD(physiol, biochem), 53. Prof Exp: Res assoc, Radiation Res Lab, Univ Iowa, 53-54; sr res assoc biochem, 54-66, DIR, DEPT DIAG RES, WARNER-LAMBERT RES INST, 66- Honors & Awards: Bernard F Gerault Mem Award, Am Asn Clin Chemists, 75. Mem: AAAS; Am Asn Clin Chemists; Am Chem Soc; NY Acad Sci; Can Soc Clin Chemists. Res: Blood coagulation; diagnostic enzymology. Mailing Add: Warner-Lambert Res Inst Morris Plains NJ 07950

BABSON, ROBERT DANIEL, b Lincoln, Nebr, July 11, 18; m 41; c 2. ORGANIC CHEMISTRY. Educ: Univ Nebr, BS, 40, MS, 41. Prof Exp: Chemist, 41-58, sect leader org res, 58-60, mgr process develop, 60-64 & cancer prep lab, 64-75, MGR REGULATORY AFFAIRS, MERCK & CO, INC, 75- Mem: Am Chem Soc. Res: Organic medicines, vitamins and antibiotics; organic synthesis, isolation and fermentation technology. Mailing Add: 52 Cambridge Dr Berkeley Heights NJ 07922

BABUSKA, IVO MILAN, b Prague, Czech, Mar 22, 26. NUMERICAL ANALYSIS, APPLIED MATHEMATICS. Educ: Tech Univ Prague, PhD(civil eng), 51; Czech Acad Sci, PhD(math), 55, DSc(math), 60. Prof Exp: Res scientist math, Math Inst, Czech Acad Sci, 50-69; RES PROF MATH, UNIV MD, 69- Mem: Am Math Soc. Res: Numerical analysis of partial differential equations; applied mathematics related to continuum theory. Mailing Add: Inst Fluid Dynamics & Appl Math Univ Md College Park MD 20742

BACA, GLENN, b Socorro, NMex, Nov 22, 43. PHYSICAL INORGANIC CHEMISTRY, ELECTROCHEMISTRY. Educ: NMex State Univ, BS, 65, MS, 67, PhD(inorg chem), 69. Prof Exp: Teaching asst chem, NMex State Univ & Rose-Hulman Inst Technol, 65-69; asst prof chem, Mercer County Community Col, 69-70; ASSOC PROF CHEM, ROSE-HULMAN INST TECHNOL, 70- Mem: Sigma Xi; Am Chem Soc. Res: The determination of transference numbers in non-aqueous solvents in conjunction with the development and construction of batteries in water-like solvents. Mailing Add: Rose-Hulman Inst of Technol Box 357 5500 E Wabash Terre Haute IN 47803

BACANER, MARVIN BERNARD, b Chicago, Ill, Mar 18, 23; m 48; c 4. PHYSIOLOGY, MEDICINE. Educ: Boston Univ, MD, 53. Prof Exp: Intern path, Boston City Hosp, Mass, 53-54; resident internal med, Cambridge City Hosp, 54-55 & Mt Sinai Hosp, New York, 55-56; res physician biophys & physiol, Donner Lab, Lawrence Radiation Lab, Univ Calif, Berkeley, 59-61; PROF PHYSIOL, UNIV MINN, MINNEAPOLIS, 61- Concurrent Pos: Percy Klingenstein fel cardiol, Mt Sinai Hosp, New York, 56-57; USPHS res fel, Med Sch, Stanford Univ, 57-59; Burroughs-Wellcome fel, Rambam Hosp, Haifa, Israel, 68-69. Honors & Awards: William & Dorothy Fish Kerr Award, San Francisco Heart Asn, 57. Mem: Am Physiol Soc. Res: Metabolic determinants of heart performance; antiarrhythmic drugs, bretylium tosylate; gut circulation; x-ray microanalysis of deep frozen tissue with the scanning electron microscope; electron optical imaging of frozen muscle. Mailing Add: Dept of Physiol Univ of Minn Minneapolis MN 55455

BACCANARI, DAVID PATRICK, b Wilkes-Barre, Pa, Apr 15, 47; m 68; c 2. ENZYMOLOGY. Educ: Wilkes Col, BS, 68; Brown Univ, PhD(biochem), 72. Prof Exp: Fel, Brown Univ, 72-73; SR RES MICROBIOLOGIST, WELLCOME RES LAB, 73- Mem: AAAS. Res: Comparative enzymology and its relationship to chemotherapy. Mailing Add: Dept of Microbiol Wellcome Res Lab Reasearch Triangle Park NC 27709

BACCEI, LOUIS J, organic chemistry, polymer chemistry, see 12th edition

BACCHI, CYRUS JOSEPH, b New York, NY, Nov 29, 42; m 66; c 1. PARASITOLOGY, BIOLOGICAL CHEMISTRY. Educ: Manhattan Col, BS, 64; Fordham Univ, MS, 66, PhD(biol), 68. Prof Exp: Res asst protozoan biochem, Haskins Labs, 64-68, instr biol, Univ, 69-74, ASST PROF BIOL, PACE UNIV, 74-, RES ASSOC PROTOZOAN BIOCHEM, HASKINS LABS, 68- Honors & Awards: Merck Found Award, 70. Mem: AAAS; Am Soc Microbiol; Soc Protozool; Am Soc Trop Med & Hyg. Res: Biochemistry and physiology of protozoa, especially Trypanosomatidae; drug mode-of-action studies, including respirometric and spectrophotometric analyses of subcellular fractions; nutriton of protozoa; enzyme systems. Mailing Add: Haskins Labs Pace Univ 41 Park Row New York NY 10038

BACCHUS, HABEEB, b Triumph, Brit Guiana, Oct 15, 28; US citizen; m 56; c 6. INTERNAL MEDICINE, ENDOCRINOLOGY. Educ: Howard Univ, BS, 47; George Washington Univ, MS, 48, PhD(physiol), 50, MD, 54; Am Bd Internal Med, dipl, 63 & 74, cert endocrinol & metab, 75. Prof Exp: Res assoc physiol, George Washington Univ, 51-54, asst res prof, 54-57, asst clin prof, 57-59; prin investr, Res Labs Biochem, Providence Hosp, Washington, DC, 60-65, dir metab lab & sr attend physician internal med, 65-69, res mem dept path, 65-67, chief res div, 67-69; ASSOC PROF MED, SCH MED, LOMA LINDA UNIV, 69-; ASSOC CHIEF INTERNAL MED, RIVERSIDE GEN HOSP, 69-, ASSOC CLIN PATHOLOGIST, 70- Concurrent Pos: Clin assoc, Nat Cancer Inst, 57-59; coordr health serv, Am Univ, 68-69; med consult, Vet Admin, 61- Mem: AAAS; AMA; Am Fedn Clin Res; Am Physiol Soc; NY Acad Sci. Res: Pituitary-adrenal axis; ascorbic acid adrenal hormone formation and breakdown; ascorbic acid and intermediary metabolism; adrenogenital syndromes; metabolic diseases; Marfan's syndrome; connective tissue diseases; hypertensive vascular disease; experimental hypertension. Mailing Add: Riverside Gen Hosp 9851 Magnolia Ave Riverside CA 92503

BACDAYAN, ALBERT SOMEBANG, anthropology, see 12th edition

BACH, DAVID RUDOLPH, b Shimonoseki, Japan, Apr 24, 24; US citizen; m 48; c 4. NUCLEAR PHYSICS, REACTOR PHYSICS. Educ: Univ Mich, BA, 48, MS, 50, PhD(physics), 55. Prof Exp: Res assoc physics, Knolls Atomic Power Lab, Gen Elec Co, NY, 55-58, mgr flexible plastic reactor, 58-64; assoc prof nuclear eng, 64-68, PROF NUCLEAR ENG, UNIV MICH, ANN ARBOR, 68- Concurrent Pos: Consult, Knolls Atomic Power Lab, 65-69. Mem: Am Phys Soc. Res: Experimental nuclear and reactor physics; neutron physics, especially pulsed neutron research; plasma physics, especially laser interaction with dense plasmas. Mailing Add: Dept of Nuclear Eng Univ of Mich Ann Arbor MI 48104

BACH, FREDERICK LOUIS, b Corning, NY, Feb 5, 21; m 47; c 2. ORGANIC CHEMISTRY. Educ: Univ Md, BS, 43; NY Univ, MS, 57, PhD(chem), 63. Prof Exp: Sr res chemist, Lederle Labs Div, 49-71, DIR OFF REGULATORY AFFAIRS, CYANAMID INT, AM CYANAMID CO, 71- Concurrent Pos: Lectr, NY Univ, 63- Mem: AAAS; Am Chem Soc; Brit Chem Soc. Res: Tuberculostatic agents; tropical diseases; cardiovascular drugs; hypocholesteremic agents; free radical chemistry; computer solutions of problems in organic chemistry and photochemistry. Mailing Add: 44 Sunrise Dr Montvale NJ 07645

BACH, HARTWIG C, b Kiel, Ger, May 24, 30; m 57. POLYMER CHEMISTRY. Educ: Kiel Univ, BS, 54, MS, 57, PhD(org chem), 59. Prof Exp: Res assoc fels, org chem, Univ SC, 59-60 & Fla State Univ, 60-61; res chemist polymer chem, Chemstrand Res Ctr, Inc, 61-69, res chemist nylon res, 69-71, SCI FEL NYLON RES, MONSANTO CO, 71- Mem: Am Chem Soc; Soc Ger Chem. Res: Organic chemistry; catalysis. Mailing Add: Nylon Res Monsanto Co PO Box 1507 Pensacola FL 32502

BACH, JOHN ALFRED, b Detroit, Mich, July 26, 35; m 57; c 3. MICROBIOLOGY. Educ: Mich State Univ, BS, 58, MS, 61, PhD(microbiol), 63. Prof Exp: Sr microbiologist, Bristol Labs, NY, 63-67; ASSOC SCIENTIST, UPJOHN CO, 67- Mem: Am Soc Microbiol. Res: Physiology of bacterial endospores; antibiotics; chemotherapy; fluorescent antibody techniques; fermentation microbiology. Mailing Add: Unit 1400-89-1 The Upjohn Co Kalamazoo MI 49001

BACH, L MATTHEW N, b San Francisco, Calif, Dec 30, 19; m 46; c 2. NEUROPHYSIOLOGY. Educ: Univ Calif, AB, 40, MA, 43, PhD(physiol), 45. Prof Exp: Asst physiol, Univ Calif, 41-44; from instr to prof, Tulane Univ, 44-70, actg chmn dept, 68-69, assoc prof psychiat & neurol, 51-59, exec dir surv physiol sci, 53-

56, lectr, Sch Eng, 53-58, Sch Social Work, 56-68; dir div biomed sci, 70-75, PROF PHYSIOL, SCH MED SCI, UNIV NEV, RENO, 70-, ASST DEAN RES & GRAD STUDIES, 75- Concurrent Pos: Lectr, Sch Law, Univ Tex, 52-68; lectr, New Orleans Psychoanal Training Inst, 58-60; mem postdoctoral res associateship panel, Nat Res Coun-Nat Acad Sci, 69-73; mem adv comt, Health Sci Advan Awards, NIH, 70-71. Mem: AAAS; Am Physiol Soc; Soc Exp Biol & Med; Soc Neurosci; Animal Behav Soc. Res: Nervous regulation of sleep wake activity; feeding; growth; swallowing; coma; conditioning; biotelemetry; comparative neurophysiology; telemetry of limbic-hypothalamic activity during species specific behavior. Mailing Add: Div of Biomed Sci Univ of Nev Sch of Med Sci Reno NV 89507

BACH, MARY JEAN, b Pittsburgh, Pa, Feb 12, 25; m 48; c 5. ANESTHESIOLOGY, EXPERIMENTAL PSYCHOLOGY. Educ: Northwestern Univ, BA, 69, MA, 71, PhD(psychol), 72. Prof Exp: Res assoc, 72-73, instr, 73, lectr, 73-74, ASST PROF ANESTHESIA, MED SCH, NORTHWESTERN UNIV, 74- Mem: Affil Am Soc Anesthesiol; Am Psychol Asn. Res: Effects of anesthetics on perceptual, motor and cognitive function; effects of maternal anesthetics on neonatal development; learning and memory; language development; developmental psychology. Mailing Add: Dept of Anesthesia Northwestern Univ Med Sch Chicago IL 60611

BACH, MICHAEL KLAUS, b Stuttgart, Ger, Oct 2, 31; US citizen; m 54; c 2. BIOCHEMISTRY, IMMUNOLOGY. Educ: Queens Col (NY), BS, 53; Univ Wis, MS, 55, PhD, 57. Prof Exp: Res plant biochemist, Union Carbide Chem Co, 57-61; biochem res assoc, 61-70, SR RES SCIENTIST, UPJOHN CO, 70- Concurrent Pos: Mem, Int Cong Biochem, 64 & 67; vis lectr med, Harvard Med Sch, 69-70. Mem: Am Asn Immunol; Am Chem Soc; Am Soc Biol Chemists. Res: Interactions of small molecules with nucleic acids; molecular biology of antimutagens; separation of subpopulations of lymphoid cells; antilymphocyte serum; cellular aspects of immunoglobulin E-mediated tissue damage. Mailing Add: 2115 Frederick Ave Kalamazoo MI 49001

BACH, RICARDO O, b Ulm, Ger, Dec 25, 17; US citizen; m 45; c 3. INORGANIC CHEMISTRY, PHYSICAL CHEMISTRY. Educ: Univ Zurich, PhD(phys chem), 42. Prof Exp: Dir res electrometall & inorg chem, Meteor Est Met, Buenos Aires, Arg, 42-54; res investr, Cent Res Labs, Am Smelting & Ref Co, NJ, 54-59; dir inorg res, 59-67, MGR RES DEPT, LITHIUM CORP AM, INC, 67- Mem: Am Chem Soc; Am Inst Mining, Metall & Petrol Eng; fel Am Inst Chem; Electrochem Soc; Sigma Xi. Res: Chemical atmosphere regeneration; extractive metallurgy of lithium beryllium, cadmium, zinc and copper; complex chemistry of antimony and arsenic; physical chemistry of alkyl boron-amine compounds; inorganic chromatography. Mailing Add: Ellestad Res Labs Lithium Corp PO Box 795 Bessemer City NC 28016

BACH, ROBERT DREYER, organic chemistry, see 12th edition

BACH, SHIRLEY, b Williston Park, NY, Nov 22, 31; m 54; c 2. ORGANIC CHEMISTRY. Educ: Queens Col, NY, BS, 53; Univ Wis, PhD(org chem), 57. Prof Exp: Instr chem, 61-64, res assoc, 64-74, ASSOC PROF NATURAL SCI, WESTERN MICH UNIV, 74- Mem: Am Chem Soc. Res: Chemical reactions in monomolecular layers; natural product chemistry; human serum complement system in health and in disease. Mailing Add: Dept of Natural Sci Western Mich Univ Kalamazoo MI 49001

BACH, SVEN AAGE, b Moji, Japan, May 20, 18; US citizen; m 65; c 3. MEDICINE, BIOPHYSICS. Educ: Univ Nebr, BS, 41, MD, 43. Prof Exp: Jr resident physician, Montreal Gen Hosp, Que, 43-44; instr nuclear med, Med Field Serv Sch, US Army, Ft Sam Houston, Tex, 48-50, action officer, Res & Develop Div, G-4, Dept Army, 51-53, surgeon, Hq Armed Forces Spec Weapons Proj, DC, 56-59, chief microwave br & biophys div, Med Res Lab, Ft Knox, Ky, 59-61, commanding officer, 61-64, res coordr life sci div, Off Chief Res & Develop, 65-67, res assoc, Inst Marine Sci, Miami, Fla, 67-70, med mem, Phys Eval Bd, Letterman Gen Hosp, 70-75; CONSULT, FED AM SOC EXP BIOL, 75- Concurrent Pos: Guest worker, NIH, 56-59; adj prof, Univ Louisville, 62-64. Res: Nuclear weapon effects; drug, radio frequency and ablation effects on primate cerebral activity; radio frequency and microwave effects on proteins; structure of water; mechanisms of memory. Mailing Add: 13094 Portofino Dr Del Mar CA 92014

BACH, WALTER DEBELE, JR, meteorology, oceanography, see 12th edition

BACHA, JOHN D, b Minneapolis, Minn, Aug 21, 41; m 66; c 3. ORGANIC CHEMISTRY. Educ: St Mary's Col (Minn), BA, 63; Case Inst Technol, PhD(chem), 67. Prof Exp: Res chemist, 67-72, SR RES CHEMIST, GULF RES & DEVELOP CO, 72- Mem: Am Chem Soc. Res: Free radical chemistry; carbonium ion chemistry and aromatic substitution; metal catalysis of organic reactions; solvent effects upon organic reactions. Mailing Add: Gulf Res & Develop Co PO Drawer 2038 Pittsburgh PA 15230

BACHA, WILLIAM JOSEPH, JR, b US, Mar 1, 30; m 53; c 2. PARASITOLOGY. Educ: Long Island Univ, BS, 51; NY Univ, PhD(biol), 59. Prof Exp: Asst biol, NY Univ, 51-54 & 56-58; lectr, City Col New York, 58-59; from instr to assoc prof, 59-70, actg dean, 69-70, chmn dept biol & dir grad prog, 71-74, PROF BIOL, RUTGERS UNIV, CAMDEN, 70- Mem: AAAS; Am Soc Parasitol; Am Micros Soc. Res: Biology; life cycles, behavior and development of trematodes. Mailing Add: Dept of Biol Rutgers Univ Camden NJ 08102

BACHARACH, MARTIN MAX, b Munich, Ger, Dec 18, 25; nat US. BIOLOGY, GENETICS. Educ: Rutgers Univ, BS, 50; Univ Wis, MS, 52, PhD(poultry genetics), 55. Prof Exp: Asst, Univ Wis, 50-55, assoc, 55-57; cytologist, Tumor Res, Vet Admin Hosp, Hines, Ill, 57-58; instr biol, NCent Col, Ill, 59; geneticist, Honegger Farms Co, Inc, Ill, 59-60; res nutritionist & librn, Dawe's Labs, Inc, Chicago, 60-62; info res scientist, Squibb Inst Med Res, NJ, 62-64; mem staff, Med Serv Dept, Sandoz, Inc, NJ, 64; dir sci info, Cortez F Enloe, Inc, NY, 65-66; SUPVR INFO SERV, WILLIAM D McADAMS, INC, NEW YORK, 66- Mem: AAAS; NY Acad Sci. Res: Immuno-genetics of species and species hybrids; developmental genetics; physiological reactions of normal and malignant cells; science information. Mailing Add: 390 Prospect Ave Hackensack NJ 07601

BACHELIS, GREGORY FRANK, b Los Angeles, Calif, Mar 9, 41. MATHEMATICS. Educ: Reed Col, BA, 62; Univ Ore, MA, 63, PhD(math), 66. Prof Exp: Asst, Univ Ore, 62-63 & 64-66; asst prof math, State Univ NY Stony Brook, 66-69; vis assoc prof, Kans State Univ, 70-74; ASSOC PROF MATH, WAYNE STATE UNIV, 74- Mem: Am Math Soc; Math Asn Am. Res: Functional analysis; non-commutative Banach algebras. Mailing Add: Dept of Math Wayne State Univ Detroit MI 48202

BACHELOR, FRANK WILLIAM, b New York, NY, Nov 2, 28; m 55; c 3. ORGANIC CHEMISTRY. Educ: Univ Calif, Berkeley, BS, 51; Mass Inst Technol, PhD(org chem), 60. Prof Exp: Asst chemist, Merck & Co, Inc, NJ, 51-54; NSF fel, Inst Org Chem, Munich Tech Univ, 60-61; res assoc org chem, Johns Hopkins Univ, 61-63; asst prof, 63-68, ASSOC PROF ORG CHEM, UNIV CALGARY, 68- Mem: Am Chem Soc; Brit Chem Soc; fel Chem Inst Can. Res: Structure and synthesis of natural products; organic reaction mechanisms. Mailing Add: Dept of Chem Univ of Calgary Calgary AB Can

BACHENHEIMER, STEVEN LARRY, b Chicago, Ill, Sept 19, 45; m 67; c 1. ANIMAL VIROLOGY. Educ: Univ Ill, Urbana, BS, 67; Univ Chicago, PhD(microbiol), 72. Prof Exp: From res assoc to lectr animal virol, Columbia Univ, 72-74; fel, Rockefeller Univ, 74-75; ASST PROF ANIMAL VIROL, DEPT BACT & IMMUNOL, UNIV NC, CHAPEL HILL, 75- Concurrent Pos: Damon Runyon fel, 72-74; NIH res fel. Mem: Am Soc Microbiol; AAAS. Res: Synthesis and post-transcriptional modification of DNA-tumor virus transcripts; insertion of viral genetic information into DNA of productively infected and transformed cells. Mailing Add: Dept of Bact & Immunol Univ of NC Chapel Hill NC 27514

BACHER, ANDREW DOW, b Ithaca, NY, Sept 5, 38; m 62. NUCLEAR PHYSICS. Educ: Harvard Univ, AB, 60; Calif Inst Technol, PhD(nuclear physics), 67. Prof Exp: Res fel physics, Calif Inst Technol, 66-67; NSF fel, Lawrence Radiation Lab, Univ Calif, 67-68, res physicist, 68-70; asst prof, 70-75, ASSOC PROF PHYSICS, IND UNIV, BLOOMINGTON, 75- Mem: Am Phys Soc. Res: Nuclear astrophysics. Mailing Add: Dept of Physics Ind Univ Bloomington IN 47401

BACHER, FREDERICK ADDISON, b Northampton, Mass, July 18, 15; m 44, 54; c 3. PHYSICAL CHEMISTRY. Educ: Harvard Univ, SB, 36; Rutgers Univ, MSc, 45. Prof Exp: Asst chemist, Plastics Dept, E I du Pont de Nemours & Co, Mass, 36-38; chemist, 39-57, mgr new prod control, 57-70, DIR PHARMACEUT ANAL, MERCK SHARP & DOHME RES LABS, 70- Mem: AAAS; Am Chem Soc; Acad Pharmaceut Sci. Res: Stability-indicating assays for drug formulations; purity of organic compounds; analytical methods; bio-availability of drug dosage forms. Mailing Add: Bean & Whitehall Rd RD 3 Norristown PA 19401

BACHER, ROBERT FOX, b Loudonville, Ohio, Aug 31, 05; m 30; c 2. PHYSICS. Educ: Univ Mich, BS, 26, PhD(physics), 30. Hon Degrees: ScD, Univ Mich, 48. Prof Exp: Nat res fel physics, Calif Inst Technol, 30-31 & Mass Inst Technol, 31-32; Lloyd fel, Univ Mich, 32-33; instr, Columbia Univ, 34-35; from instr to prof, Cornell Univ, 35-49, dir nuclear studies lab, 46; res assoc radiation lab, Mass Inst Technol, 41-45; head, Exp Physics Div, Atomic Bomb Proj, Los Alamos Lab, 43-44, head bomb physics div, 44-45; mem, US AEC, 46-49; chmn div physics, math & astron, 49-62, provost, 62-70, PROF PHYSICS, DOWNS LAB, CALIF INST TECHNOL, 49- Concurrent Pos: Mem, President's Sci Adv Comt, 57-60 & Naval Res Adv Comt, 57-62; trustee, Assoc Univs, 46, Rand Corp, 50-60, Carnegie Corp, 59-, US AEC, & Univ Res Assoc, 66-, chmn, 70- Mem: Nat Acad Sci; AAAS; Am Phys Soc (pres, 64); Am Philos Soc; Int Union Pure & Appl Sci (vpres, 66-69, pres, 69-72). Res: High energy physics; photoproduction of mesons; nuclear physics; atomic energy. Mailing Add: Downs Lab 405-47 Calif Inst Technol Pasadena CA 91125

BACHHUBER, EDWARD A, b Mayville, Wis, Mar 13, 12; m 41; c 7. SURGERY, ANATOMY. Educ: Univ Wis, BS, 35; Harvard Univ, MD, 37; Am Bd Surg, dipl, 47. Prof Exp: From instr anat to clin instr surg, 47-52, assoc anat, 48-70, from asst prof to assoc prof surg, 52-70, assoc dean admin, 56-65, coordr med educ for nat defense, 57-69, PROF SURG & ANAT, MED COL WIS, 70- Concurrent Pos: Attend surgeon, Vet Admin Hosp, Wood, Wis, 53-, consult, 57-; attend surgeon, St Joseph's Hosp, Deaconess Hosp & Milwaukee County Gen Hosp; trustee, Alverno Col, 69-71. Mem: AAAS; AMA; Asn Am Med Cols; Am Col Surg; Am Heart Asn. Res: Medical education and administration; mechanisms of acute injury with emphasis on abdominal trauma; color television in the teaching of anatomy. Mailing Add: Div of Surg Med Col of Wis Milwaukee WI 53226

BACHMAN, CHARLES HERBERT, b Cedar Falls, Iowa, Dec 8, 08; m 44; c 4. PHYSICS. Educ: Iowa State Col, BS, 32, MS, 33, PhD(physics), 35. Prof Exp: Physicist, Hygrade Sylvania Corp, 35-39 & Gen Elec Co, 39-46; from assoc prof to prof, 46-74, EMER PROF PHYSICS, SYRACUSE UNIV, 74-; DIR, SHADOW MOUNTAIN BIOPHYS LAB, 74- Concurrent Pos: Fulbright prof, Univ Calcutta, 59-60; consult, Ramo Wooldridge Corp, 56-57. Mem: Fel AAAS; fel Inst Elec & Electronic Engr; fel Am Phys Soc; Am Vacuum Soc (pres, 63-64); Am Asn Physics Teachers. Res: Electron physics; atmospheric electricity; biophysics. Mailing Add: Shadow Mountain Biophys Lab Box 161 Moose WY 83012

BACHMAN, GEORGE, b New York, NY, Jan 17, 29. MATHEMATICS. Educ: NY Univ, BEE, 50, MS, 52, PhD(math), 56. Prof Exp: From instr to asst prof math, Rutgers Univ, 57-60; from asst prof to assoc prof, 60-65, PROF MATH, POLYTECH INST BROOKLYN, 65- Honors & Awards: Distinguished Teacher Award, Polytech Inst NY, 74. Mem: Am Math Soc; Math Asn Am; Math Soc France; Indian Math Soc; Can Math Cong. Res: Functional analysis; algebraic number theory; topological measure theory. Mailing Add: Dept of Math Polytech Inst of Brooklyn Brooklyn NY 11201

BACHMAN, GERALD LEE, b Alton, Ill, Oct 6, 32; m 57; c 2. ORGANIC CHEMISTRY. Educ: Univ Ill, BS, 54; Wash Univ, PhD(org chem), 63. Prof Exp: Res chemist biochem, 56-58, RES CHEMIST, EXPLOR SYNTHESIS, MONSANTO CO, 63- Mem: Am Chem Soc. Res: Chemistry of carbenes, particularly phenylcarbene; synthesis of heterocyclic compounds; amino acid chemistry; asymmetric hydrogenation. Mailing Add: Org Res Dept Monsanto Co 800 N Lindbergh Blvd St Louis MO 63166

BACHMAN, GUSTAVE BRYANT, b Kansas City, Mo, Aug 8, 05; m; c 3. CHEMISTRY. Educ: Univ Colo, AB, 26; Yale Univ, PhD(org chem), 30. Prof Exp: From instr to asst prof chem, Ohio State Univ, 30-36; res chemist, Eastman Kodak Co, 36-39; from assoc prof to prof, 39-71, EMER PROF CHEM, PURDUE UNIV, WEST LAFAYETTE, 71- Mem: Am Chem Soc; fel Am Inst Chem. Res: Plastics; rubbers; antimalarials and other drugs; nitration; synthetic organic chemistry. Mailing Add: Dept of Chem Purdue Univ West Lafayette IN 47907

BACHMAN, KENNETH CHARLES, b West New York, NJ, Mar 1, 22; m 51; c 3. PHYSICAL CHEMISTRY, FUEL SCIENCE. Educ: Rutgers Univ, BS, 44, MS & PhD(phys chem), 50. Prof Exp: Chemist, Benzol Prod Co, 44 & 46; asst, Rutgers Univ, 46-49, res asst, 49-50; res chemist, M W Kellogg Co, 51-55; res chemist, Esso Res & Eng Co, Standard Oil Co NJ, 55-66, sr chemist, 66-70, RES ASSOC, EXXON RES & ENG CO, EXXON CORP, LINDEN, 70- Honors & Awards: Arch T Colwell Merit Award, Soc Automotive Eng, 66. Mem: Am Chem Soc; Sigma Xi. Res: Cause of and control methods for ignitions induced by static discharges during fuel transfer, such as truck loading and aircraft fueling; methods for control of automotive exhaust emissions, particularly sulfates. Mailing Add: 404 Wells St Westfield NJ 07090

BACHMAN, MARVIN CHARLES, b Decorah, Iowa, Feb 8, 21; m 44; c 3. MICROBIOLOGY. Educ: Luther Col, BS, 42; Okla State Univ, MS, 47. Prof Exp: Chemist, E I du Pont de Nemours & Co, 42-44; teaching fel, Okla State Univ, 46-47; microbiologist, Com Solvents Corp, 47-51, dir microbiol res, 51-56, lab res, 56-62, dir res, 62-75; DIR RES & DEVELOP, IMC CHEM GROUP, 75- Mem: Am Soc Microbiol. Mailing Add: 108 S 25th St Terre Haute IN 47803

BACHMAN, PAUL LAUREN, b Cleveland, Ohio, July 2, 39; m 65; c 2. ORGANIC CHEMISTRY. Educ: Case Inst Technol, BS, 61; Ohio State Univ, PhD(org chem), 65. Prof Exp: Sr chemist, Merck Sharp & Dohme Res Labs, 65-68; group leader, Horizon's Res Inc, 68-72, mgr, Photo Systs Dept, 72-74, DIR, RES & DEVELOP BR, PHOTO HORIZONS DIV, HORIZON'S RES INC, 74- Concurrent Pos: NIH fel, Stanford Univ, 65-66. Mem: Am Chem Soc. Res: Organic synthesis; isolation and structural elucidation of natural occuring phenols; process development in organic chemistry; research and development of non-silver photographic systems. Mailing Add: 1188 Pennfield Rd Cleveland Heights OH 44121

BACHMANN, BARBARA JOYCE, b Ft Scott, Kans, May 16, 24; div. MICROBIAL GENETICS. Educ: Baker Univ, Univ Ky, AB, 45; Univ Ky, MS, 47; Stanford Univ, PhD(microbiol), 54. Prof Exp: Assoc bact, Univ Calif, 53-56; res assoc biochem, Columbia Univ, 57-58; res assoc microbiol, Yale Univ, 58-64; asst prof, Sch Med, NY Univ, 64-68; LECTR MICROBIOL, SCH MED, YALE UNIV, 68-, CUR, COLI GENETIC STOCK CTR, 70- Concurrent Pos: Res asst, Oak Ridge Nat Lab; ed, Neurospora Newslett, 61- Mem: AAAS; Am Soc Microbiol; Genetics Soc Am. Mailing Add: Dept of Human Genetics Yale Univ Sch of Med New Haven CT 06510

BACHMANN, JOHN HENRY, b Minneapolis, Minn, Nov 19, 10; m 37; c 2. PHYSICAL CHEMISTRY. Educ: Univ Minn, BChE, 31, PhD(phys chem), 39. Prof Exp: Asst, Univ Minn, 35-36; res chemist, B F Goodrich Co, 39-42; res supvr, 42-59; mgr appl inorg res, Columbia-Southern Chem Corp, 59-61; PROF CHEM & HEAD DEPT, UNIV AKRON, 61- Res: Inorganic synthesis; silica and silicates; applied research. Mailing Add: Dept of Chem Univ of Akron Akron OH 44325

BACHMANN, KENNETH ALLEN, b Columbus, Ohio, Mar 21, 46; m 69. PHARMACOLOGY. Educ: Ohio State Univ, BS, 69, PhD(pharmacol), 73. Prof Exp: ASST PROF PHARMACOL, COL PHARM, UNIV TOLEDO, 73- Mem: AAAS; NY Acad Sci. Res: Pharmacokinetic aspects of oral anticoagulant-drug interactions; influence of environmental chemical exposure upon oral anticoagulant disposition. Mailing Add: Dept of Pharmacol Univ of Toledo Col of Pharm Toledo OH 43606

BACHMANN, RALPHAEL OTTO, pharmaceutical chemistry, see 12th edition

BACHMANN, ROGER WERNER, b Ann Arbor, Mich, Dec 11, 34; m 60; c 2. LIMNOLOGY. Educ: Univ Mich, 56, PhD(zool), 62; Univ Idaho, MS, 58. Prof Exp: Res zoologist, Univ Calif, Davis, 62-63; from asst prof to assoc prof limnol, 63-71, PROF LIMNOL, IOWA STATE UNIV, 71- Mem: AAAS; Am Soc Limnol & Oceanog; Ecol Soc Am; Am Fisheries Soc. Res: Biological productivity of lakes; aquatic mineral cycles. Mailing Add: Dept of Animal Ecol Iowa State Univ Ames IA 50011

BACHOP, WILLIAM EARL, b Youngstown, Ohio, Aug 31, 26; m 58; c 2. DEVELOPMENTAL BIOLOGY, GROSS ANATOMY. Educ: Western Reserve Univ, AB, 50; Ohio State Univ, MSc, 58, PhD(zool), 63. Prof Exp: Asst prof biol, Univ Omaha, 63-65; training prog fel develop biol res, Sch Med, Univ Wash, 65-69; asst prof zool, Clemson Univ, 69-73; ACTG CHMN DEPT ANAT, NAT COL, LOMBARD, 74- Mem: Am Micros Soc; Am Soc Zool; Am Soc Cell Biol; Soc Develop Biol; Am Anatomists Asn. Res: Transmission electron microscopy of demyelinating diseases; scanning electron microscopy of vertebrate embryos; yolk sac syncytium of bony fishes. Mailing Add: Dept Anat Nat Col 200 E Roosevelt Rd Lombard IL 60148

BACHRACH, HOWARD LLOYD, b Faribault, Minn, May 21, 20; m 43; c 2. BIOCHEMISTRY. Educ: Univ Minn, BA, 42, PhD(biochem), 49. Prof Exp: Res asst, Explosives Res Lab, Nat Defense Res Comt Proj, Carnegie Inst Technol, 42-45; res asst, Univ Minn, 45-49; biochemist, Foot & Mouth Dis Res Mission, USDA, Denmark, 49-50; res biochemist, Biochem & Virus Lab, Univ Calif, 50-53; CHIEF SCIENTIST & HEAD BIOCHEM & PHYS INVESTS, PLUM ISLAND ANIMAL DIS CTR, USDA, 53- Honors & Awards: Naval Ord Develop Award, 45; USDA Cert Merit, 60; Presidential Citation, 65. Mem: Am Chem Soc; NY Acad Sci; Electron Micros Soc Am; Soc Exp Biol & Med; Am Col Vet Microbiologists. Res: Starch; explosives; virology; tissue culture; identification of foot and mouth disease and poliomyelitis virus particles; immunization with a capsid protein of foot and mouth disease virus; structure and molecular biology of viruses. Mailing Add: Box 848 Greenport NY 11944

BACHRACH, JOSEPH, b Ger, Feb 9, 18; US citizen; m 58. CARBOHYDRATE CHEMISTRY. Educ: Queen's Univ (Ont), BA, 44, MA, 45; Purdue Univ, PhD(bio-org-chem), 50. Prof Exp: Asst prof chem, Concord Col, 50-53; from asst prof to assoc prof, Univ Ill, Chicago Circle, 53-66; PROF CHEM, NORTHEASTERN ILL UNIV, 66- Mem: Am Chem Soc; Sigma Xi. Res: Carbohydrates; hemicelluloses. Mailing Add: Dept of Chem Northeastern Ill Univ Chicago IL 60625

BACHUR, NICHOLAS R, SR, b Baltimore, Md, July 21, 33; m 52; c 3. BIOCHEMISTRY, PHARMACOLOGY. Educ: Johns Hopkins Univ, AB, 54; Univ Md, MD & PhD(biochem), 61. Prof Exp: Res asst biochem, McCollum-Pratt Inst, Johns Hopkins Univ, 51-55; instr biochem, Univ Md, 61, intern med, Univ Hosp, 61-62; res biochemist, Lab Clin Biochem, Nat Heart Inst, 62-65; HEAD SECT BIOCHEM, LAB PHARMACOL, BALTIMORE CANCER RES CTR, NAT CANCER INST, 66- Mem: AAAS; AMA; Am Fedn Clin Res: Am Soc Pharmacol & Exp Therapeut; Am Asn Cancer Res. Res: Biochemical mechanisms of pathological processes; enzymology. Mailing Add: Cancer Res Ctr Nat Cancer Inst Baltimore MD 21211

BACHYNSKI, MORREL PAUL, b Bienfait, Sask, July 19, 30; m 59. PLASMA PHYSICS. Educ: Univ Sask, BEng, 52, MSc, 53; McGill Univ, PhD(physics), 55. Prof Exp: Res assoc, Eaton Electronics Lab, McGill Univ, 55; mem sci staff, Res Labs, RCA Corp, 55-57, assoc labs dir, 58-59, dir microwave & plasma physics labs, 59-65, dir res, RCA Ltd, 65-72, dir res & develop, 72-75, VPRES RES & DEVELOP, RCA LTD, 75- Concurrent Pos: Pres, Asn Sci, Eng & Tech Community of Can, 74-75 Honors & Awards: David Sarnoff Gold Medal, 63; Prix Scientifique du Quebec, Que. Mem: Inst Elec & Electronics Engrs; fel Am Phys Soc; fel Can Aeronaut & Space Inst; fel Royal Soc Can; Can Asn Physicists (pres, 68-69). Res: Electromagnetic wave propagation; microwave optics; plasma physics; geophysics and space physics; plasma and laser technology, including controlled fusion and communications. Mailing Add: RCA Ltd Res & Develop Labs 21001 N Service Rd Ste Anne de Bellevue PQ Can

BACH-Y-RITA, PAUL, b New York, NY, Apr 24, 34; m 60; c 2. NEUROPHYSIOLOGY, REHABILITATION. Educ: Nat Univ Mex, MD, 59. Prof Exp: Res physiologist dept physiol, Sch Med, Univ Calif, Los Angeles, 54-56, jr res pharmacologist, Dept Biophys & Nuclear Med, 60; sci transl & writer, Ctr Sci & Tech Doc Mex, UNESCO, 55-57, rural pub health physician, 58-59; intern, Presby Med Ctr, San Francisco, 60-61; SR RES MEM, PAC MED CTR, SMITH-KETTLEWELL INST VISUAL SCI, 63-, ASSOC DIR, 68-; PROF VISUAL SCI, UNIV OF THE PAC, 69- Concurrent Pos: USPHS ment health trainee, Univ Calif, Los Angeles, 59-60, fel, Dept Clin Neurophysiol, Univ Freiburg, 62-63 & res career award, 63-73; Bank of Am-Giannini Found fel, Ctr Study Nerv Physiol & Electrophysiol, Nat Ctr Sci Res, Paris, 61-62. Honors & Awards: Silver Hektoen Medal, AMA, 72; Franceschetti-Liebrecht Prize, Ger Ophthal Soc, 74. Mem: AAAS; Am Physiol Soc; NY Acad Sci; Asn Res Vision & Ophthal. Res: Brain control of eye movement; tactile vision substitution based on brain plasticity; helping blind persons learn to see by means of images from a television camera delivered to the skin. Mailing Add: Pac Med Ctr Smith-Kettlewell Inst Visual Sci San Francisco CA 94115

BACK, KENNETH CHARLES, b Wharton, NJ, Nov 17, 25; m 50; c 7. PHARMACOLOGY. Educ: Muhlenberg Col, BS, 51; Univ Okla, MS, 54, PhD(med sci), 57. Prof Exp: Sr scientist, Warner-Lambert Res Inst, 57-58; pharmacologist, Pitman-Moore Co, 58-60; pharmacologist-toxicologist, 60-66, SUPVRY PHARMACOLOGIST-TOXICOLOGIST & CHIEF TOXICOL BR, 6570th AEROSPACE MED RES LABS, WRIGHT-PATTERSON AFB, 66- Concurrent Pos: Adj prof biol sci, Wright State Univ, 73-; mem, Am Conf Govt Indust Hygienists. Mem: Sigma Xi; Soc Toxicol; Am Soc Pharmacol & Exp Therapeut; NY Acad Sci. Res: General pharmacology; intermediary metabolism; aerospace toxicology; propellant toxicology. Mailing Add: 5229 Bayside Dr Dayton OH 45431

BACK, MARGARET HELEN, b Ottawa, Ont, Dec 7, 29; m 54; c 2. PHYSICAL CHEMISTRY. Educ: Acadia Univ, BSc, 50; McGill Univ, MSc, 54, PhD(chem), 59. Prof Exp: Fel chem, Nat Res Coun Can, 59-61; res assoc, 61-65, asst prof, 65-70, ASSOC PROF CHEM, UNIV OTTAWA, 70- Concurrent Pos: Nat Res Coun Can res grants, 63-; Ont Res Found grant, 64-67. Mem: Fel Chem Inst Can. Res: Kinetics of gas phase reactions; unimolecular theory. Mailing Add: Dept of Chemistry Univ of Ottawa Ottawa ON Can

BACK, NATHAN, b Philadelphia, Pa, Nov 30, 25; m 51; c 3. PHARMACOLOGY. Educ: Pa State Univ, BS, 48; Philadelphia Col Pharm, MS, 52, DSc(pharmacol), 55. Prof Exp: Asst admin & tech orgn med labs, Israel, 48-49; biochemist, Wyeth Inst, 50-51, pharmacologist, 51-52; res asst pharmacol, Philadelphia Col Pharm, 52-54, instr, 54-55; cancer res scientist, Roswell Park Mem Inst, 55-58, sr cancer res scientist, 58-66; PROF BIOCHEM PHARMACOL, STATE UNIV NY BUFFALO, 66- Concurrent Pos: Res fel, State Univ NY Buffalo, 56-58, from asst prof to prof pharmacol, Sch Pharm, 56-66, prof, Grad Sch Arts & Sci, 58-66, from actg chmn to chmn dept biochem pharmacol, 67-71; vis prof & actg dir sch pharm, Fac Med, Hebrew Univ, Jerusalem, 69-71, vis prof, 71-; exec ed, Pharmacol Res Commun; chmn, State Univ Task Force on Israel Progs; UN expert pharmacol, 75- Honors & Awards: E K Frey Medal, 70. Mem: Fel AAAS; Am Soc Hemat; Int Soc Hemat; Fedn Am Soc Exp Biol; fel Am Soc Clin Pharmacol & Therapeut. Res: Chemotherapy of cancer; alkylating agents; physiology and pathology of blood and blood-forming organs; pharmacology of fibronolytic agents and anticoagulants; extracorporeal circulation; thrombo-embolic phenomena; mechanisms of drug resistance; shock mechanisms; vasoactive polypeptides. Mailing Add: Dept of Biochem Pharmacol State Univ of NY Buffalo NY 14216

BACK, ROBERT ARTHUR, b Delhi, Ont, Aug 13, 29; m 54; c 2. CHEMICAL KINETICS. Educ: Univ Western Ont, BSc, 50, MSc, 51; McGill Univ, PhD(chem), 53. Prof Exp: Dewar res fel natural philos, Univ Edinburgh, 53-56; Nat Res Coun Can fel chem, McGill Univ, 56-59; from asst res officer to sr res officer, 59-71, PRIN RES OFFICER PHOTO CHEM & KINETICS, NAT RES COUN CAN, 71- Mem: Fel Chem Inst Can. Res: Kinetics of gas-phase free-radical and atomic reactions; photochemistry; mercury photosensitization; active nitrogen; radiation chemistry of gases. Mailing Add: Div of Chem Nat Res Coun of Can Ottawa ON Can

BACK, WILLIAM, b East St Louis, Ill, Aug 9, 25; m 50; c 4. HYDROGEOLOGY. Educ: Univ Ill, AB, 48; Univ Calif, MS, 55; Harvard Univ, MPA, 56; Univ Nev, Reno, PhD(chem hydrogeol), 69. Prof Exp: Asst, Ill State Geol Surv, 45-48; asst geol, Univ Calif, 49-50; geologist, Calif, 50-54 & DC, 54-75, GEOLOGIST, WATER RESOURCES DIV, US GEOL SURV, VA, 75- Concurrent Pos: Instr hist & Eng, Fla Mil Acad, 47; asst geol, Harvard Univ, 56-57; lectr, Am Univ, 57-59; vis scholar, Desert Res Inst, Nev, 67-68. Honors & Awards: O E Meinzer Award, Geol Soc Am, 73. Mem: Int Asn Hydrogeologists; Int Asn Sci Hydro; Geol Soc Am; Am Geophys Union; Am Geochem Soc. Res: Isotopes and geochemistry of ground water; regional ground water systems. Mailing Add: US Geol Surv MS 432 Reston VA 22092

BACKDERF, RICHARD HAROLD, organic chemistry, see 12th edition

BACKER, RONALD CHARLES, b Newark, NJ, May 3, 43; m 64; c 2. TOXICOLOGY. Educ: Univ Ariz, BA, 64, PhD(pharmaceut chem), 70. Prof Exp: Asst chief forensic chem, Milwaukee Health Dept, 69-74; asst toxicol, Off Chief Med Examr, State Md, 74-76; CHIEF TOXICOL, OFF CHIEF MED EXAMR, STATE W VA, 76- Concurrent Pos: Instr toxicol, Dept Path, Div Forensic Path, Univ Md, Baltimore City. 74-76; supvr toxicol, Cent Labs Assoc Md Pathologists, Ltd, 74-76; lectr toxicol, Sch Pub Health & Hyg, Johns Hopkins Univ, 74-76. Mem: Am Chem Soc; Am Acad Forensic Sci. Res: Improvement and development of analytical procedures for drugs from biological specimens. Mailing Add: Off of Chief Med Examr State Govt WVa 701 Jefferson Rd South Charleston WV 25309

BACKMAN, PAUL ANTHONY, plant pathology, plant physiology, see 12th edition

BACKUS, EDWARD JAMES, b Madison, Wis, Mar 10, 16; m 40; c 2. MYCOLOGY. Educ: Univ Wis, AB, 37, AM, 39, PhD(bot), 41. Prof Exp: Asst bot, Univ Wis, 37-39, asst plant path, 39-40; mycologist, Lederle Labs Div, Am Cyanamid Co, 41-56, head, Microbiol Res Dept, 56-73; RETIRED. Res: Strain development in antibiotic producing molds through use of induced mutation; fungus survey for antibiotics; speciation in the genus Streptomyces. Mailing Add: 801 E Pine Knoll Dr Prescott AZ 86301

BACKUS, GEORGE EDWARD, b Chicago, Ill, May 24, 30; m 61; c 3. GEOPHYSICS. Educ: Univ Chicago, BS, 48, MS, 50 & 54; PhD(physics), 56. Prof Exp: Physicist, Proj Matterhorn, Princeton Univ, 57-58; from asst prof to assoc prof math, Mass Inst Technol, 58-60; assoc prof geophys, 60-62, PROF GEOPHYS, UNIV CALIF, SAN DIEGO, 62- Concurrent Pos: Guggenheim Mem fel, 63-64 & 70-71; mem comt on sci & pub policy, Nat Acad Sci, 70-73, mem report rev comt, 73-76, chmn day fund selection comt, 75- Mem: Nat Acad Sci; fel Am Geophys Union; Seismol Soc Am; Soc Indust & Appl Math; Am Acad Arts & Sci. Res: Origin of geomagnetic field; plasma oscillations; seismology of multilayered and anisotropic media; normal modes of elastic-gravitational oscillation of the earth; general theory of geophysical inverse problems; mantle viscosity; seismic source theory. Mailing Add: Inst of Geophys & Planetary Phys A-025 Univ of Calif at San Diego La Jolla CA 92093

BACKUS, JOHN (GRAHAM), b Portland, Ore, Apr 29, 11; m 45; c 2. PHYSICS. Educ: Reed Col, BA, 32; Univ Calif, MA, 36, PhD(physcis), 40; Univ Southern Calif, MM, 59. Prof Exp: Asst, Radiation Lab, Univ Calif, 36-40, res physicist, 40-45; assoc prof, 45-65, PROF PHYSICS, UNIV SOUTHERN CALIF, 65- Mem: Fel Acoustical Soc Am. Res: Musical acoustics. Mailing Add: Dept of Physics Univ of Southern Calif Los Angeles CA 90007

BACKUS, JOHN KING, b Buffalo, NY, May 22, 25; m 50; c 4. PHYSICAL CHEMISTRY. Educ: Hamilton Col, AB, 47; Cornell Univ, MS, 50, PhD(chem), 52. Prof Exp: Res chemist, Procter & Gamble Co, 52-53; res chemist, O-celo Dept, Gen Mills, Inc, 53-62, res supvr, 56-62; res specialist, 62-63, group leader, 63-67, sr group leader, 67, MGR, MOBAY CHEM CORP, 67- Concurrent Pos: Chmn, Gordon Conf Chem & Physics of Cellular Mat, 70. Mem: AAAS; NY Acad Sci; Am Chem Soc. Res: Physical polymer chemistry; flammability; degradation; molecular structure. Mailing Add: Mobay Chem Corp Penn Lincoln Pkwy W Pittsburgh PA 15205

BACKUS, MILO M, b Ill, May 3, 32; m 52; c 5. GEOPHYSICS. Educ: Mass Inst Technol, BS, 52, PhD(geophys), 56. Prof Exp: From res physicist to vpres, Geophys Serv Inc, 56-74; consult, 74-75; PROF GEOPHYS, UNIV TEX, AUSTIN, 75- Mem: Soc Explor Geophysicists; Acoustical Soc Am; Sigma Xi; Europ Asn Geophysicists. Res: Petroleum exploration. Mailing Add: Dept of Geol Sci Univ of Tex PO Box 7909 Austin TX 78712

BACKUS, MYRON PORT, b Madison, Wis, Apr 14, 08; m 31; c 1. MYCOLOGY. Educ: Univ Wis, AB, 28, AM, 29, PhD(bot), 31. Prof Exp: Nat Res fel bot, NY Bot Garden & Columbia Univ, 31-33; assoc mycol, Boyce Thompson Inst, 33-34; from instr to asst prof bot, 34-44, from assoc prof to prof bot & plant path, 44-71, EMER PROF BOT & PLANT PATH, UNIV WIS-MADISON, 71- Mem: Mycol Soc Am; Am Phytopath Soc. Res: Cytology of fungi; sexuality in Ascomycetes; parasitic fungi; cytology of host-parasite relations; soil fungi. Mailing Add: 1811 Regent St Madison WI 53705

BACKUS, RICHARD HAVEN, b Rochester, NY, Dec 5, 22; m 49; c 3. MARINE BIOLOGY. Educ: Dartmouth Col, BA, 47; Cornell Univ, MS, 48, PhD(ichthyol), 53. Prof Exp: Supvry res assoc oceanog, Cornell Univ, 51-52; res assoc marine biol, 52-59, marine biologist, 59-63, chmn dept biol, 70-74, SR SCIENTIST, WOODS HOLE OCEANOG INST, 63- Mem: AAAS; Ecol Soc Am. Res: Biology of marine vertebrates, especially mesopelogic fishes; marine biogeography. Mailing Add: Woods Hole Oceanog Inst Woods Hole MA 02543

BACKUS, ROBERT COBURN, molecular biology, see 12th edition

BACLAWSKI, LEONA MARIE, b Akron, Ohio, Dec 26, 45. ORGANIC CHEMISTRY. Educ: Univ Akron, BS, 67, PhD(chem), 74. Prof Exp: Fel org chem, State Univ NY Col Environ Sci & Forestry, Syracuse Univ, 74; fel chem, 75, LAB DIR ORG CHEM, BUCKNELL UNIV, 75- Mem: Sigma Xi; Am Chem Soc. Res: Diaziridine chemistry; 1,3-dipolar addition; heterocyclic chemistry; carbanion synthesis. Mailing Add: Dept of Chem Bucknell Univ Lewisburg PA 17837

BACON, CHARLES VINCENT, b New York, NY, Mar 24, 85. CHEMISTRY. Educ: Cooper Union, BS, 19, ChemE, 32. Hon Degrees: ChemE, St Mary's Col, 19. Prof Exp: CONSULT CHEMIST & CHEM ENGR, 32-; DIR, CHAS V BACON LABS, INC, 70-, PRES, 74- Mem: Am Chem Soc; fel Am Inst Chem; Asn Consult Chemists & Chem Engrs; Am Soc Testing & Mat; Am Oil Chem Soc. Res: Experimental, analytical and industrial research; surveying and inspection of organic products; chemicals and fatty oils; process development and plant supervision; preparation of technical work for court testimony and presentation; marine cargo surveying transportation and handling of oils and chemicals in bulk. Mailing Add: Chas V Bacon Labs Inc 34 Exchange Pl Jersey City NJ 07302

BACON, DAVID W, b Peterborough, Ont, Sept 12, 35; m 63; c 1. STATISTICS, CHEMICAL ENGINEERING. Educ: Univ Toronto, BASc, 57; Univ Wis, MS, 62, PhD(statist), 65. Prof Exp: Comput analyst, Can Gen Elec Co, Ont, 57-61; comput programmer, Math Res Ctr, Univ Wis-Madison, 61-63; statist & math group leader, Du Pont Can, Ont, 65-67; assoc prof chem eng, 68-73, PROF CHEM ENG, QUEEN'S UNIV, ONT, 73- Concurrent Pos: Vis assoc prof, Univ Wis-Madison, 69-70. Mem: Am Statist Asn; Can Soc Chem Eng; Royal Statist Soc. Res: Adaptive control of industrial processes; mathematical modeling of processes; design of experimental programs; short term forecasting. Mailing Add: Dept of Chem Eng Queen's Univ Kingston ON Can

BACON, EDMOND JAMES, b Chidester, Ark, Jan 10, 44; m 73. AQUATIC ECOLOGY. Educ: Southern State Col, Ark, BS, 66; Univ Ark, MS, 68; Univ Louisville, PhD(biol), 73. Prof Exp: From teaching asst to res asst zool, Univ Ark, 66-68; technician radiation, Univ Louisville, 68-70, res asst water resources, 72-74; ASST PROF BIOL, UNIV ARK, MONTICELLO, 74- Concurrent Pos: Environ consult, Aquatic Control, Inc & Dames & Moore, 72-74; ichthyologist, Water Resources Lab, Univ Louisville, 73-74; proj engr, E D'Appolonia Consult Engrs, 74. Mem: Am Soc Limnol & Oceanog; Int Soc Limnol; Am Fisheries Soc; Brit Freshwater Biol Asn; assoc Sigma Xi. Res: Environmental deterioration appraisal and the impact of costly restoration efforts; natural history of lower vertebrates of Arkansas and primary productivity, water quality and limiting factors in oxbow lakes. Mailing Add: Dept of Biol Univ of Ark Monticello AR 71655

BACON, EGBERT KING, b Sault Ste Marie, Mich, Aug 20, 00; m 26; c 1. CHEMISTRY. Educ: Univ Mich, BS, 22, MS, 23, PhD(chem), 26. Prof Exp: Asst quantitative anal, Univ Mich, 20-22, asst gen chem, 22-26; instr chem, Brown Univ, 26-30; from instr to prof, 30-66, chmn dept, 63-64, chmn div sci, 56-57, EMER PROF CHEM, UNION UNIV, NY, 66-; PROF, ALBANY COL PHARM, 68- Concurrent Pos: City chemist, Schenectady, NY, 41-51; lectr, Albany Col Pharm, 45-46. Mem: AAAS; Am Chem Soc. Res: Diffusion potentials; equivalent weight of gelatin; viscosity of gelatin; analytical chemistry; chemical history. Mailing Add: 1964 Eastern Pkwy Schenectady NY 12309

BACON, FRANK RIDER, b Wilton Junction, Iowa, Apr 28, 14; m 38; c 3. PHYSICAL INORGANIC CHEMISTRY. Educ: Iowa State Univ, BS, 35; Washington Univ, MS, 37. Prof Exp: Asst chem, Washington Univ, 35-37; res chemist, Owens-Ill Glass Co, 37-58; CHIEF SURFACE CHEM SECT, GLASS CONTAINER DIV, OWENS-ILL, INC, 58- Mem: Am Chem Soc; fel Am Ceramic Soc. Res: Chemical durability of glass; light transmission of glass; strength of glass. Mailing Add: Owens-Ill Inc PO Box 1035 Toledo OH 43666

BACON, GEORGE EDGAR, b New York, NY, Apr 13, 32; m 56; c 3. PEDIATRIC ENDOCRINOLOGY. Educ: Wesleyan Univ, BA, 53; Duke Univ, MD, 57; Univ Mich, MS, 67. Prof Exp: Intern pediat, Duke Hosp, 57-58; resident, Columbia Presby Med Ctr, 61-63; from instr to assoc prof, 63-74, PROF PEDIAT, UNIV MICH, ANN ARBOR, 74- Concurrent Pos: Fel physiol, Univ Mich, Ann Arbor, 65, fel pharmacol, 65-67; pediat endocrinologist, Univ Pittsburgh, 70; consult pediat endocrinol, Wayne County Gen Hosp, Mich, 75- Mem: Soc Pediat Res; Endocrine Soc; Am Fedn Clin Res. Res: Adrenal disorders, particularly modes of therapy and metabolic effects of treatment in congenital adrenal hyperplasia. Mailing Add: F 2438 Mott Hosp Univ of Mich Med Ctr Ann Arbor MI 48109

BACON, HAROLD MAILE, b Los Angeles, Calif, Jan 13, 07; m 46; c 1. MATHEMATICS. Educ: Stanford Univ, AB, 28, AM, 29, PhD(math), 33. Prof Exp: Actg instr, 31-32, from instr to prof, 32-72, EMER PROF MATH, STANFORD UNIV, 72- Concurrent Pos: Instr, Calif State Col, San Jose, 35. Mem: AAAS; Am Math Soc; Math Asn Am. Res: Analysis. Mailing Add: Box 4144 Stanford CA 94305

BACON, HARRY ELLIOTT, b Philadelphia, Pa, Aug 25, 00; m 35; c 2. GASTROENTEROLOGY. Educ: Villanova Col, BS, 21, Temple Univ, MD, 25; Am Bd Surg, dipl, 41; Am Bd Colon & Rectal Surg, dipl, 49. Hon Degrees: Dr, Univ Rome, 49, Univ Bordeaux, 52, Univ Padua, 52, Univ Minas Gerais, 56, Univ Chiba, Japan, 56, Istanbul Univ, 58, Univ Guayaquil, 59 & Univ Bologna, 60; LLD, Univ Buenos Aires, 50 & Univ of the Repub, 63; ScD, Ursinus Col, 53. Prof Exp: Lectr anat, Sch Med, Temple Univ, 32-34; asst chief surgeon, Dept Radiol, Philadelphia Gen Hosp, 34-38; assoc prof, 38-42, PROF PROCTOL SURG & HEAD DEPT, SCH MED, TEMPLE UNIV, 42- Concurrent Pos: Fels, Univ Pa, 28 & Univ Vienna, 32; consult, St Marks Hosp, London, Eng & St Antoine Hosp, Paris, France. Mem: Gastroenterol Res Group; AMA; fel Am Proctol Soc (secy, 44-46, pres, 49); fel Am Col Surg; hon fel Royal Australian Col Surg. Res: Diseases of the anus, rectum and colon; editing. Mailing Add: St Luke's & Children's Hosp 8th & Girard Ave Philadelphia PA 19122

BACON, HILARY EDWIN, b Evansville, Ind, Feb 8, 01; m 26; c 2. WATER CHEMISTRY, POLLUTION CHEMISTRY. Educ: Johns Hopkins Univ, PhD(chem), 34. Prof Exp: Comptroller, H E Bacon Co, 23-28; from prin asst to sr engr, 34-58, PARTNER, SHEPPARD T POWELL ASSOCS, 58- Mem: Am Chem Soc; Am Water Works Asn. Res: Water treatment for steam generation, cooling, process and potable use; power station chemistry. Mailing Add: Sheppard T Powell Assocs 501 St Paul Pl Baltimore MD 21202

BACON, JOHN ALVIN, b Holton, Kans, July 28, 15; m 46; c 2. ENTOMOLOGY. Educ: Univ Kans, PhD(entom), 50. Prof Exp: PROF BIOL, OTTAWA UNIV, 49- Mailing Add: Dept of Biol Ottawa Univ Ottawa KS 66067

BACON, LARRY DEAN, b Topeka, Kans, Jan 24, 38; m 69. GENETICS, IMMUNOGENETICS. Educ: Kans State Univ, BS, 61, MS, 67, PhD(genetics), 69. Prof Exp: Instr path, NY Med Col, 69-72; ASST PROF MICROBIOL, SCH MED, WAYNE STATE UNIV, 72- Concurrent Pos: NIH fel, NY Med Col, 69-71. Res: Immunological responses to sex-linked histocompatibility antigens in chickens; utilizing the Jerne plaque technique to study isoimmune responses in chickens; genetics of autoimmune disease and the relationship of alloantigens to autoimmunity. Mailing Add: Dept of Microbiol Wayne State Univ Sch Med Detroit MI 48201

BACON, MARION, b Hemingford, Nebr, May 26, 14; m 50. MEDICAL MICROBIOLOGY. Educ: Iowa State Col, BSc, 50; Wash State Univ, MS, 52, PhD(bact), 56. Prof Exp: Asst entomol, Wash State Univ, 50-52; med entomologist, Commun Dis Ctr, USPHS, 52-53; asst bact, Wash State Univ, 53-56; from instr to asst prof microbiol, Univ Nebr, 56-59; assoc prof biol, 59-68, PROF BIOL, EASTERN WASH STATE COL, 68- Mem: Am Soc Microbiol. Res: Diseases of nature communicable to man; medical entomology; bacterial indicators of fecal pollution. Mailing Add: Dept of Biol Eastern Wash State Col Cheney WA 99004

BACON, OSCAR GRAY, b Sanger, Calif, Nov 8, 19; m 45; c 2. ENTOMOLOGY. Educ: Fresno State Col, AB, 41; Univ Calif, MS, 44, PhD(entom), 48. Prof Exp: Field aide, Div Fruit Insects, Bur Entom & Plant Quarantine, USDA, 41-43; assoc div entom, 46-47, sr lab technician, 47-48, instr econ entom & jr entomologist, 48-50, from asst prof to assoc prof econ entom, 50-63, chmn dept entom, 67-74, PROF ECON ENTOM & ENTOMOLOGIST, EXP STA, UNIV CALIF, DAVIS, 63- Mem: Entom Soc Am. Res: Economic entomology; biology and control of insect pests of field crops; small seeded legumes for seed; field corn and grain sorghums; insect pests of potato. Mailing Add: Dept of Entom Univ of Calif Davis CA 95616

BACON, PHILLIP, b Cleveland, Ohio, July 10, 22; m 51; c 2. GEOGRAPHY. Educ: Univ Miami, AB, 46; George Peabody Col, MA, 51, EdD in Soc Sci(geog), 55. Prof Exp: Teacher social studies, Castle Heights Mil Acad, 46-47; teacher, Army & Navy Acad, 48-53; asst geog, George Peabody Col, 53-55; asst prof, Univ Pittsburgh, 55-57; from assoc prof to prof, Columbia Univ, 57-63; dean grad sch, George Peabody Col, 63-64; prof geog, Columbia Univ, 64-66; prof, Univ Wash, 66-71; PROF GEOG & CHMN DEPT, UNIV HOUSTON, 71- Concurrent Pos: Field reader, Bur Res, US Off Educ, 64-70; consult, Educ Personnel Div, 64-71; mem bd consults, World Bk Atlas, 65-69; mem ed adv bd, World Bk Encycl, 65-; pres, Nat Coun Geog Educ, 66; co-dir, Tri-Univ Proj Elem Educ, 67-69; mem, Wash Social Studies Adv Comn, 68-71; co-dir, TTT Proj, Univ Wash, 69-71; vis geog scholar, NSF-Asn Am Geog Vis Sci Prog, 69-71. Mem: Asn Am Geog. Res: Geographic education. Mailing Add: Dept of Geog Univ of Houston Houston TX 77004

BACON, RALPH HOYT, b Kenosha, Wis, July 16, 07. ENGINEERING PHYSICS. Educ: Fordham Univ, AB, 29; Columbia Univ, AM, 32; NY Univ, PhD(physics), 40. Prof Exp: Fel & tutor physics, City Col, NY, 29-32; asst Drew Univ, 35-36; instr, St Joseph's Col, NY, 36-37 & Wash Sq Col, NY Univ, 37-39; ballistician & mathematician, US War Dept, Frankford Arsenal, 38-44; sr physicist & proj engr, Fairchild Camera & Instrument Corp, NY, 45-47; asst prof physics, Newark Col Eng, 46-47; assoc prof, Bradley Univ, 47-48; head anal sect, Elmhurst Lab, Perkin-Elmer Corp, 48-50; sr staff mem, Gen Precision Lab, 50-57, sr scientist, Singer-Gen Precision, Inc, 57-72; RETIRED. Concurrent Pos: Instr, Adelphi Col, 36; instr eng sci & mgt war training, St Joseph's Col, Pa, 43-45. Res: Statistical methods for research, engineering and quality control; analogue and digital computers; servomechanisms. Mailing Add: PO Box 190 Brewster NY 10509

BACON, ROBERT ELWIN, b Lansdowne, Pa, Apr 24, 34; m 57; c 2. PHOTOGRAPHIC SCIENCE. Educ: Univ Mich, BSChem, 56, BSChE, 56; Mass Inst Technol, PhD(phys chem), 60. Prof Exp: Sr chemist, 60-64, res assoc, 65-73, LAB HEAD RES, EASTMAN KODAK CO, 73- Concurrent Pos: Fel, Dept Mat Sci, Stanford Univ, 70-71. Mem: Soc Photog Scientists & Engr. Res: Physical and chemical aspects of photographic response in silver halides. Mailing Add: Res Labs Eastman Kodak Co Rochester NY 14650

BACON, ROBERT LEWIS, b Olean, NY, Feb 4, 18; m 46; c 2. ANATOMY. Educ: Hamilton Col, NY, BS, 40; Yale Univ, MS, 42, PhD(zool), 44. Prof Exp: From instr to asst prof anat, Sch Med, Stanford Univ, 44-51; assoc prof, Univ Tenn, 51-53 & Johns Hopkins Univ, 53-55; assoc prof, 55-59, PROF ANAT, MED SCH, UNIV ORE, 59- Concurrent Pos: Vis fac, Ore Inst Marine Biol, 57-; assoc ed, J Morphol, 58-60. Mem: AAAS; Am Soc Zool; Am Asn Anat; Soc Develop Biol; Microcirc Soc. Res: Mammalian and amphibian experimental embryology; endocrines and cancer; aging, self-differentiation and induction in the heart; electron microscopy of eggs and embryos. Mailing Add: Dept of Anat Univ of Ore Med Sch Portland OR 97201

BACON, ROGER, b Cleveland, Ohio, Apr 16, 26; m 51; c 3. MATERIALS SCIENCE. Educ: Haverford Col, AB, 51; Case Inst Technol, MS, 53, PhD, 56. Prof Exp: Res physicist, 55-62, GROUP LEADER, RES LABS, CARBON PROD DIV, UNION CARBIDE CORP, 62- Mem: Am Phys Soc; Am Ceramic Soc. Res: Mechanical properties of solids. Mailing Add: Union Carbide Corp Parma Tech Ctr PO Box 6116 Cleveland OH 44101

BACON, WILLIAM EDWARD, inorganic chemistry, organic chemistry, see 12th edition

BACSKAI, ROBERT, b Ujpest, Hungary, Feb 17, 30; US citizen; m 56; c 2. ORGANIC CHEMISTRY, POLYMER CHEMISTRY. Educ: Eötvös Lorand Univ, Budapest, dipl, 52. Prof Exp: Res chemist, Plastics Res Inst, Budapest, Hungary, 52-56; res assoc polymer chem, Princeton Univ, 57-60; from res chemist to sr res chemist, 60-68, SR RES ASSOC POLYMER CHEM, CHEVRON RES CO, 68- Mem: Am Chem Soc. Res: Ferrocene chemistry; hydrogen bonding; water soluble, stereoregular, cationic and free radical polymers; nuclear magnetic resonance of polymers. Mailing Add: 254 Stanford Ave Kensington CA 94708

BACUS, JAMES WILLIAM b Alton, Ill, Apr 21, 41; m 62; c 2. BIOENGINEERING, PHYSIOLOGY. Educ: Mich State Univ, BS, 64; Ill Inst Technol, 64-66; Univ Ill, PhD(physiol), 71. Prof Exp: Sci asst radiation biol, Argonne Nat Lab, 64-66; ASST BIOMED ENGR, BIOENG DEPT & CHIEF AUTOMATION, SECT CLIN HEMAT, RUSHPRESBYST LUKE'S MED CTR, 70- Concurrent Pos: Consult, Electronics Res Div, Corning Glass Works, 71- Mem: AAAS; Pattern Recognition Soc; Inst Elec & Electronics Eng; Asn Advan Med Instrumentation. Res: Automatic pictorial pattern recognition; medical image processing and classification. Mailing Add: Bioeng Dept Rush-Presby-St Luke's Med Ctr Chicago IL 60612

BACZEWSKI, ALEXANDER, b Vienna, Austria, Oct 9, 10; US citizen; m 41; c 2. COLLOIDAL CHEMISTRY. Educ: Univ Vienna, PhD, 37. Prof Exp: Chemist, France Emblems, 39-41, RSA, 41-42, Paisley, 42-43, Arvey Corp, 43-46 & 54-56, Carlisle Paint, Chem, 46-47, Gordon Lacey, 47-51 & Cilco, 51-54; RES ENGR, GEN TEL & ELECTRONICS LABS, INC, 56- Mem: Am Chem Soc; Tech Asn Pulp & Paper Indust; Soc Plastics Eng; Am Soc Photog Sci & Eng. Res: Application of polymer chemistry to electronics. Mailing Add: 8319 116th St Kew Gardens NY 11418

BADA, JEFFREY L, b San Diego, Calif, Sept 10, 42; m 65; c 2. GEOCHEMISTRY, GEOCHRONOLOGY. Educ: San Diego State Col, BS, 65; Univ Calif, San Diego, PhD(chem), 68. Prof Exp: Instr chem, Univ Calif, San Diego, 68-69; res fel geophys & environ chem, Harvard Univ, 69-70; asst prof oceanog, 70-74, ASSOC PROF MARINE CHEM, SCRIPPS INST OCEANOG, UNIV CALIF, SAN DIEGO, 74- Concurrent Pos: Res grants, NSF, 70-71 & 73-75 & Petrol Res Fund, 70-73; Alfred P Sloan res fel, 75-77. Mem: AAAS; Am Chem Soc. Res: Organic chemistry of natural waters and sediments; stability of organic compounds on the primitive earth; geochemical implications of the kinetics of reactions involving biologically produced organic compounds; biochemistry of fishes. Mailing Add: Scripps Inst Oceanog Univ Calif San Diego PO Box 109 La Jolla CA 92037

BADCOCK, CHARLES CORYDON, physical chemistry, organic chemistry, see 12th edition

BADDING, VICTOR GEORGE, b Buffalo, NY, Dec 15, 35; m 61; c 2. ORGANIC CHEMISTRY. Educ: Canisius Col, BS, 57; Univ Notre Dame, PhD(org chem), 61. Prof Exp: Fel, Univ Wis, 61-62; res chemist, Silicones Div, Union Carbide Corp, 62-65; asst prof, 65-70, ASSOC PROF CHEM, MANHATTAN COL, 70- Concurrent Pos: Fel, State Univ NY Buffalo, 65. Mem: Am Chem Soc; Brit Chem Soc. Res: Chemistry of acetals; reaction mechanisms; complex metal hydride reductions. Mailing Add: Dept of Chem Manhattan Col Bronx NY 10471

BADDLEY, WILLIAM H, inorganic chemistry, see 12th edition

BADE, MARIA LEIPELT, b Hamburg, Ger, Dec 13, 25; US citizen; m 49; c 1. BIOCHEMISTRY. Educ: Univ Nebr, BS, 51, MS, 54; Yale Univ, PhD(biochem), 60. Prof Exp: Fel biochem, Harvard Univ, 60-62, fel, Sch Med, 63-64; USPHS fel, Mass Inst Technol, 64-65, res assoc, 65-67; asst prof biol, 67-72, ASSOC PROF BIOL, BOSTON COL, 67- Mem: AAAS; Am Chem Soc; Am Soc Zool; Am Physiol Soc; Am Soc Biol Chem. Res: Biochemistry and structure of arthropod cuticle; water quality; human mycoses. Mailing Add: Dept of Biol Boston Col Chestnut Hill MA 02167

BADE, WILLIAM GEORGE, b Oakland, Calif, May 29, 24; m 52; c 5. MATHEMATICS. Educ: Calif Inst Technol, BS, 45; Univ Calif, Los Angeles, MA, 48, PhD(math), 51. Prof Exp: Instr math, Univ Calif, 51-52 & Yale Univ, 52-55; from asst prof to assoc prof, 55-64, PROF MATH, UNIV CALIF, BERKELEY, 64- Concurrent Pos: NSF sr fel, 58-59. Mem: Am Math Soc. Res: Functional analysis. Mailing Add: Dept of Math Univ of Calif Berkeley CA 94720

BADE, WILLIAM LEMOINE, b Lincoln, Nebr, Apr 5, 28; m 49; c 1. THEORETICAL PHYSICS. Educ: Univ Nebr, BS, 49, MA, 51, PhD(physics), 54. Prof Exp: Fel, Univ Utah, 54; NSF fel, Yale Univ, 56-57; sr consult scientist, Res & Advan Develop Div, 57-69, PHYSICIST, AVCO CORP, 57-, PRIN STAFF SCIENTIST, 69- Mem: Am Phys Soc. Res: Atomic and molecular physics; gas dynamics. Mailing Add: Avco Syst Div Wilmington MA 01887

BADEER, HENRY SARKIS, b Mersine, Turkey, Jan 31, 15; nat US; m 48; c 2. PHYSIOLOGY. Educ: Am Univ Beirut, MD, 38. Prof Exp: Asst physiol, Am Univ Beirut, 38-41, instr, 41-45, from adj prof to asst clin prof, 45-51, from assoc prof to prof, 51-65, actg head dept, 51-56, chmn dept, 56-65; vis prof physiol, State Univ NY Downstate Med Ctr, 65-67; PROF PHYSIOL & PHARMACOL, SCH MED, CREIGHTON UNIV, 67- Concurrent Pos: Rockefeller res fel physiol, Harvard Med Sch, 48-49; vis prof, Univ Iowa, 57-58; actg chmn dept physiol & pharmacol, Creighton Univ, 71-72; mem, Int Study Group Res in Cardiac Metab. Mem: AAAS; NY Acad Sci; Am Physiol Soc. Res: Hypothermia on heart; ventricular fibrillation after coronary occlusion; myocardial oxygen uptake; cardiac hypertrophy. Mailing Add: Dept of Physiol-Pharmacol Creighton Univ Sch of Med Omaha NE 68178

BADEN, ERNEST, b Berlin, Ger, Feb 2, 24; nat US. ANATOMIC PATHOLOGY, ORAL PATHOLOGY. Educ: Univ Algiers, PhB, 42; Univ Paris, Sorbonne, Lic es Let, 46, Chirugien-Dentiste, 47; Odont Sch France, Laureat, 47; NY Univ, DDS, 50, MS, 64; Univ Geneva, MD, 63. Prof Exp: Asst stomatol, Univ Hosp Cochin, Univ Paris, 45-46 & St Louis Univ Hosps, 46-47; intern oral surg, Mt Sinai Hosp, New York, 50-51, asst dentist, 51-59; from asst prof to assoc prof path, 58-68, chmn dept, 56-72, PROF PATH, SCH DENT, FAIRLEIGH DICKINSON UNIV, 68- Concurrent Pos: Clin asst oral surg, City Hosp New York, 52-58; resident, Queens Gen Hosp, 56-57; vis asst prof, Albert Einstein Col Med, 59-69; consult, Hackensack Gen Hosp, 62-72; resident, Francis Delafield Hosp Div, Columbia Med Ctr, 64, fels path, 65 & 67-68, fels path, Presby Hosp Div, 65 & 66; consult oral path, Holy Name Hosp, Teaneck, NJ, 67-, Out-Patient Dent Clin, USPHS, New York, 68-, Englewood Hosp, 70- & St Joseph Hosp, Paterson, 72-; asst clin prof path, Col Physicians & Surgeons, Columbia Univ, 68-72, adj assoc prof, 73; consult surg, Harlem Hosp Ctr, 68-72, consult path, 68-; consult path, Cath Med Ctr, Brooklyn & Queens, 74- Mem: Fel Am Acad Oral Path; Am Dent Asn; fel Col Am Pathologists. Res: Experimental oral pathology and surgery; pathobiology of neoplasia; pathology of salivary glands; oral manifestations of systemic diseases. Mailing Add: Sch of Dent Fairleigh Dickinson Univ Hackensack NJ 07601

BADEN, HARRY CHRISTIAN, b Brooklyn, NY, Oct 22, 23; m 47; c 2. ANALYTICAL CHEMISTRY. Educ: Col Educ Albany, BA, 48, MA, 49; Rensselaer Polytech Inst, PhD(chem), 55. Prof Exp: Teacher, High Sch, NY, 48-50; asst chem, Rensselaer Polytech Inst, 50-52, asst chem physics, 52-54; anal chemist, 54-58, sr anal chemist, 58-64, TECH ASSOC, PHOTOG TECHNOL DIV, EASTMAN KODAK CO, ROCHESTER, 65- Mem: Am Chem Soc. Res: Explosive oxidation of boron hydrides; instrumental analysis; infrared spectra; gas chromatography; photographic chemistry. Mailing Add: 205 Curtice Park Webster NY 14580

BADEN, HOWARD PHILIP, b Boston, Mass, Feb, 23, 31; m 54; c 3. DERMATOLOGY. Educ: Harvard Univ, AB, 52; Harvard Med Sch, MD, 56. Prof Exp: Intern med, Peter Bent Brigham Hosp, Boston, 56-57; asst resident dermat, Mass Gen Hosp, Boston, 59-60, clin fel, 60-62; res assoc, 62-63, instr, 63-64, assoc, 64-67, from asst prof to assoc prof, 67-75, PROF DERMAT, HARVARD MED SCH, 75-, MEM FAC, CTR HUMAN GENETICS, 71-; ASSOC DERMATOLOGIST, MASS GEN HOSP, 76- Concurrent Pos: NIH fel, Res Lab, Harvard Med Sch, Dept Dermat, Mass Gen Hosp & Grad Dept Biochem, Brandeis Univ, 61-62; clin assoc, Mass Gen Hosp, Boston, 62-64, asst, 64-67, from asst dermatologist to dermatologist, 67-75; mem, Gen Med A Study Sect, NIH, 69-73; Am Cancer Soc-Eleanor Roosevelt int cancer fel, Galton Lab, Univ Col, Univ London, 70-71; ed, Progress in Dermat, 71-; assoc ed, Jour Soc Invest Dermat, 72- Mem: Am Fedn Clin Res; Soc Invest Dermat; Am Soc Human Genetics; AAAS; Am Soc Clin Invest. Res: Differentiation of epidermal tissues with emphasis on the structural proteins; diagnosis and management of genetic disorders of keratinization. Mailing Add: Dept of Dermat Harvard Med Sch Mass Gen Hosp Boston MA 02114

BADENHOP, ARTHUR FREDRICK, b Napoleon, Ohio, Oct 8, 41; m 63; c 1. FOOD SCIENCE, BIOCHEMISTRY. Educ: Ohio State Univ, BSc, 63, PhD(hort), 66. Prof Exp: NIH fel, 66-68; res assoc biochem & nutrit, NY State Agr Exp Sta, 68-70; asst prof food sci, Coop Exten Serv, Col Agr, Univ Ga, 70-75; ASSOC PROF HORT, SCH AGR, PURDUE UNIV, WEST LAFAYETTE, 75- Mem: Inst Food Technologists; Am Chem Soc. Res: Food processing methods and effects on product quality; energy efficiency; waste utilization. Mailing Add: Dept of Hort Purdue Univ West Lafayette IN 47907

BADENHUIZEN, NICOLAAS PIETER, b Zaandam, Netherlands, June 14, 10; Can citizen; m; c 6. BOTANY. Educ: Univ Amsterdam, DSc, 38. Prof Exp: Asst bot, Univ Amsterdam, 31-39; geneticist & biochemist, Tobacco Exp Sta, Java, 39-42; biochemist, Royal Netherlands Yeast Factory, 46-50; prof bot & head dept, Univ Witwatersrand, 50-60; head dept, 61-71, PROF BOT, UNIV TORONTO, 61- Concurrent Pos: Teacher sec sch, Netherlands, 35-39. Honors & Awards: Thomas Burr Osborne Gold Medal, Am Asn Cereal Chemists, 69. Mem: Am Asn Cereal Chemists; Can Bot Asn; Royal Netherlands Bot Soc; fel Royal Netherlands Acad Sci. Res: Fine structure and physiology of cells in relation to starch granule production and composition. Mailing Add: Dept of Bot Univ of Toronto Toronto ON Can

BADER, ALFRED ROBERT, b Vienna, Austria, Apr 28, 24; US citizen; m 52; c 2. ORGANIC CHEMISTRY. Educ: Queen's Univ (Ont), BSc, 45, BA, 46, MSc, 47; Harvard Univ, MA, 49, PhD(chem), 50. Prof Exp: Res chemist, Pittsburgh Plate Glass Co, 50-53, group leader org res, 53-54; chief chemist, Aldrich Chem Co, Inc, 54-55, pres, 55-75, PRES, SIGMA-ALDRICH CORP, 75- Mem: Am Chem Soc; Brit Chem Soc. Res: Fatty acids; quinones; reaction mechanisms; alkenylphenols; indoles. Mailing Add: Sigma-Aldrich Corp 940 W St Paul Ave Milwaukee WI 53233

BADER, FRANK, b Hampton, Va, Dec 9, 18; m 66; c 2. PHYSICS. Educ: Col William & Mary, BS, 40; Univ Va, MS & PhD(physics), 44. Prof Exp: Lab asst, Friez Instrument Div, Bendix Aviation Corp, 40-41; res assoc, US Navy Proj, Univ Va, 44-46; physicist, Kellex Corp, Md, 46-48; PHYSICIST, APPL PHYSICS LAB, JOHNS HOPKINS UNIV, 48- Mem: Am Phys Soc. Res: Combustion research for ram jet supersonic propulsion; ultracentrifuge techniques; instrumentation and telemetering. Mailing Add: Johns Hopkins Univ Appl Phys Lab 11100 Johns Hopkins Rd Laurel MD 20810

BADER, HENRI, b Brugg, Switz, Jan 15, 07; nat US; m 38. GLACIOLOGY, PARTICLE PHYSICS. Educ: Univ Zurich, PhD(mineral), 34. Prof Exp: Avalanche specialist, Swiss Govt, 35-38; mineralogist, Nat Chem Lab, Bogota, 40-41; quarry supt, Curacao Mining Co, Neth WI, 41-45; asst dir, Bur Mineral Res, Rutgers Univ, 45-49; res assoc, Univ Minn, 49-51; chief scientist, Snow, Ice & Permafrost Res Estab, US Army, Wilmette, Ill, 52-60; res prof, Sch Eng, Univ Miami, 60-63; sci attache, US Dept State, Am Embassy, Bonn, Ger, 64-65 & Bern, Switz, 65-66; SR RES SCIENTIST, INST MARINE & ATMOSPHERIC SCI, UNIV MIAMI, 67-, ADJ PROF CHEM OCEANOG, 74- Concurrent Pos: Lectr, Nat Univ Colombia, 40-41; consult, Arctic Inst NAm. Honors & Awards: Seligman Crystal Award, Glaciol Soc, 67. Mem: AAAS; Am Geophys Union; Arctic Inst NAm; Glaciol Soc. Res: Crystallography, particularly geometrical symmetry; arctic research, particularly structure of snow and ice and snow mechanics; oceanography, particularly small particles in sea water. Mailing Add: Dept of Chem Oceanog Univ of Miami Coral Gables FL 33124

BADER, HENRY, b Warsaw, Poland, Apr 26, 20; nat US; m 50; c 2. ORGANIC CHEMISTRY. Educ: Univ Strasbourg, Ing Chem, 40; Univ Paris & Univ Montpellier, LSc, 42; Univ London, PhD(chem), 48, Imperial Col, dipl, 48. Prof Exp: Res chemist, Beecham Res Labs, Eng, 48-50 & May & Baker, Ltd, 50-51; group leader & sr chemist, Dominion Tar & Chem Co, Can, 51-53; sr res assoc, Ortho Res Found, Raritan, NJ, 53-61 & Am Cyanamid Co, Bound Brook, 61-63; dir labs, Aldrich Chem Co, 63-66; sr scientist, 66, group leader, 67-69, SR GROUP LEADER, POLAROID CORP, 69- Mem: Am Chem Soc; The Chem Soc. Res: Synthetic organic chemistry; acetylenes and heterocyclics; chemotherapy; process research. Mailing Add: Polaroid Corp 600 Main St Cambridge MA 02139

BADER, JOHN PAUL, microbiology, see 12th edition

BADER, KENNETH L, b Carroll, Ohio, May 4, 34; m 55; c 2. AGRONOMY, ACADEMIC ADMINISTRATION. Educ: Ohio State Univ, BSc, 56, MSc, 57, PhD(agron), 60. Prof Exp: Asst agron, Ohio State Univ, 56-57, from instr to prof agron, 57-72, actg asst dean col agr & home econ, 63-64, asst dean col agr, 64-68, from assoc dean students to dean students, 68-72; PROF AGRON & VCHANCELLOR STUDENT AFFAIRS, UNIV NEBR-LINCOLN, 72- Concurrent

Pos: Am Coun Educ fel acad admin, 67-68. Mem: Am Soc Agron; Sigma Xi; Nat Asn Student Personnel Adminr; AAAS. Res: Grass physiology and turf management. Mailing Add: Univ of Nebr Lincoln NE 68508

BADER, MICHEL, b Paris, France, Dec 24, 30; US citizen. RESEARCH ADMINISTRATION. Educ: Calif Inst Technol, BS, 52, PhD(physics, math), 55. Prof Exp: Res scientist physics, 55-61, chief physics br, 61-65, chief airborne sci off, 65-70, chief space sci div, 70-73, DEP DIR ASTRONAUTICS, NASA AMES RES CTR, 74- Honors & Awards: Except Sci Achievement Medal, NASA, 67, Except Serv Medal, 73. Res: Management policy. Mailing Add: NASA Ames Res Ctr Moffett Field CA 94035

BADER, RICHARD FREDERICK W, b Kitchener, Ont, Oct 15, 31; m 58; c 2. THEORETICAL CHEMISTRY. Educ: McMaster Univ, BSc, 53, MSc, 55; Mass Inst Technol, PhD(chem), 57. Prof Exp: Fel, Cambridge Univ, 58-59; from asst prof to assoc prof phys chem, Univ Ottawa, 59-63; assoc prof, 63-66, PROF PHYS CHEM, McMASTER UNIV, 66- Mailing Add: Dept of Chem McMaster Univ Hamilton ON Can

BADER, RICHARD GEORGE, oceanographic geochemistry, deceased

BADER, ROBERT SMITH, b Falls City, Nebr, June 18, 25; m 48; c 4. EVOLUTIONARY BIOLOGY. Educ: Kans State Col, BS, 49; Univ Chicago, PhD(paleozool), 54. Prof Exp: From instr to asst prof biol sci, Univ Fla, 52-56; from asst prof to prof zool, Univ Ill, 56-68; DEAN COL ARTS & SCI, UNIV MO-ST LOUIS, 68- Mem: Soc Vert Paleont; Soc Study Evolution; Am Soc Zool; Am Soc Mammal. Res: Vertebrate evolution; variation in rodent dentition. Mailing Add: Col of Arts & Sci Univ of Mo St Louis MO 63121

BADERTSCHER, DARWIN EARL, b Beaverdam, Ohio, Sept 10, 08; m 30; c 3. ORGANIC CHEMISTRY, INFORMATION SCIENCE. Educ: Ind Univ, AB, 29; Pa State Col, MS, 31, PhD(org chem), 34. Prof Exp: Res chemist, Exp Sta, Hercules Powder Co, 29-30; asst, Pa State Col, 33-34; res org chemist, Paulsboro Lab, Mobil Res & Develop Corp, 34-36, proj leader, 36-43, from sect leader to res assoc, 43-53, res assoc, Brooklyn Lab, 53-60, supvr tech info group, Paulsboro Lab, 60-70, mgr tech info serv, 70-73; RETIRED. Mem: AAAS; Am Chem Soc; Am Soc Info Sci; fel Am Inst Chem. Res: Petroleum additives; petrochemical technical and market development; patent coordination; technical information processing and dissemination. Mailing Add: 118 S Broad St Woodbury NJ 08096

BADGER, ALISON MARY, b Croyden, Eng, Nov 25, 35; US citizen; m 61; c 2. IMMUNOBIOLOGY. Educ: Univ London, BSc, 58; Boston Univ, PhD(microbiol), 72. Prof Exp: Res asst microbiol, United Fruit Co, Boston, 58-64; res asst biochem, Blood Res Inst, Boston, 66-68; res assoc microbiol, 68-73, instr, 73-74, ASST PROF MICROBIOL, SCH MED, BOSTON UNIV, 74-, ASST PROF SURG, 75- Mem: Am Asn Immunologists. Mailing Add: Boston Univ Sch of Med 80 E Concord St Boston MA 02118

BADGER, BLANCHE CRISP, b Mountville, SC, Dec 4, 10; m 38. MATHEMATICS. Educ: Winthrop Col, AB, 31; Univ Tenn, MA, 32; George Peabody Col, PhD(math), 56. Prof Exp: From instr to asst prof math, Ball State Teachers Col, 36-38; from asst prof to assoc prof, Winthrop Col, 42-46; asst prof, McMurray Col, 47-48; teacher, Memphis State Col, 48-50; instr, Fla State Univ, 50-53; asst prof, Tenn Polytech Inst, 55-56; from assoc prof to prof, Longwood Col, 56-73, chmn dept, 57-73; RETIRED. Concurrent Pos: Mem panel for eval fac sci fels, NSF, 70. Mem: Math Asn Am. Res: Mathematics programs used in secondary schools and for undergraduates in colleges. Mailing Add: Nine Hearths Mountville SC 29370

BADGER, DONALD W, b Portland, Ore, Oct 1, 28; m 54; c 3. PHYSIOLOGY. Educ: Ore State Univ, BS, 50; Univ Calif, PhD(physiol), 64. Prof Exp: Res physiologist, Univ Calif, 60-63; scientist, Cardiovasc Dis, WHO, 63-66; asst prof physiol, Univ Ill, Urbana-Champaign, 66-74; MEM STAFF, NAT INST OCCUP SAFETY & HEALTH, 74- Res: Adaption to hypoxic environments; relationships between hypoxia and incidence of essential hypertension; development of methodology for epidemiological studies. Mailing Add: Nat Inst for Occup Safety & Health 550 Main St Cincinnati OH 45202

BADGER, GEORGE FRANKLIN, b Everett, Mass, May 14, 07; m 34; c 3. BIOSTATISTICS. Educ: Mass Inst Technol, BS, 29; Johns Hopkins Univ, MPH, 32; Univ Mich, MD, 38. Prof Exp: Assoc epidemiologist, Dept Health, Mich, 29-34; assoc biostatist, Johns Hopkins Univ, 38-45, asst prof biostatist, 45-46; from assoc prof to prof, 46-72, dir dept biomet, 63-69, EMER PROF BIOSTATIST, SCH MED, CASE WESTERN RESERVE UNIV, 72- Concurrent Pos: Consult to Secy War, 42-46; mem comn acute respiratory dis, Armed Forces Epidemiol Bd, 42-44 & 53-60; mem adv comt epidemiol & biomet, NIH, 58-62, human ecol study sect, 63-65; med adv comt clin invest, Nat Found, 59-61; mem ed bd, Am Rev Respiratory Dis, 66-70. Mem: Am Epidemiol Soc (pres, 61). Res: Epidemiology of communicable diseases; the family as an epidemiological unit; statistical evaluation of clinical and laboratory procedures. Mailing Add: 750 Spafford Oval Northfield OH 44067

BADGER, RICHARD MCLEAN, physical chemistry, physics, deceased

BADGER, RODNEY ALLAN, b Fullerton, Calif, Feb 7, 43; m 66. ORGANIC CHEMISTRY. Educ: Ore State Univ, BA & BS, 64; Univ Calif, Berkeley, MS, 66, PhD(chem), 68. Prof Exp: Asst chemist, Univ Hawaii, 68-69; ASSOC PROF CHEM, SOUTHERN ORE STATE COL, 69- Mem: Am Chem Soc; Sigma Xi. Res: Isolation and synthesis of natural products; total synthesis of sequiterpenes; isolation and structure determination of pharmacologically-active compounds. Mailing Add: Dept of Chem Southern Ore State Col Ashland OR 97520

BADGETT, JAMES THOMAS, physical chemistry, see 12th edition

BADGLEY, FRANKLIN ILSLEY, b Mansfield, Ohio, Dec 20, 14; m 43; c 2. METEOROLOGY. Educ: Univ Chicago, BS, 35; NY Univ, MS, 49, PhD(meteorol), 51. Prof Exp: Chemist, Swift & Co, 36-42; meteorologist, Trans World Airlines, Inc, 47; from instr to assoc prof, 50-67, PROF METEOROL, UNIV WASH, 67-, PROF ATMOSPHERIC SCI, 73-, ASSOC DIR, QUARTERNARY RES CTR, 73- Mem: AAAS; Am Meteorol Soc; Am Geophys Union. Res: Atmospheric turbulence. Mailing Add: Dept of Atmospheric Sci Univ of Wash Seattle WA 98105

BADGLEY, PETER COLES, b Montreal, Que, May 15, 25; US citizen; m 48; c 4. EARTH SCIENCES, REMOTE SENSING. Educ: McGill Univ, BSc, 48; Princeton Univ, AM, 50, PhD(geol), 52. Prof Exp: Party chief, Can Geol Surv, 48-51; staff geologist, Standard Oil Co Calif, 51-52; dist geologist, Husky Oil & Refining Co, 52-54; explor mgr, Ranger Oil Ltd, 54-56; from asst prof to assoc prof struct geol, Colo Sch Mines, 56-63; chief adv missions, NASA, 63-65, prog chief, Earth Resources Surv Prog, 65-67; exec dir, Gulf Univ Res Corp, 67-69; DIR EARTH SCI DIV, OFF NAVAL RES, 69- Concurrent Pos: Consult, Tidewater Oil Co, 57-61; US Air Force Off Sci Res grant, 61-63. Mem: AAAS; fel Geol Soc Am; Am Geophys Union; Am Asn Petrol Geol. Res: Application of remotely sensed data for mineral exploration purposes, for analysis of the terrestrial environment and for global study of the earth's natural and cultural resources; tectonic control of mineral deposits; global tectonic mechanisms. Mailing Add: Earth Sci Div Off Naval Res Code 460 Arlington VA 22217

BADGLEY, WILFRID JOHN, b New York, NY, Mar 13, 12. CHEMISTRY. Educ: St Francis Col, NY, BS, 36; Columbia Univ, MA, 39; Polytech Inst Brooklyn, PhD(polymer chem), 45. Prof Exp: Instr org chem, St Francis Col, NY, 36-42; res assoc, Polytech Inst Brooklyn, 44-45; prof chem & chmn dept, St Francis Col, NY, 46-52; asst prof, 52-59, ASSOC PROF CHEM, UNIV MIAMI, 59- Concurrent Pos: Instr, Hunter Col, 40-48; grad lectr, Polytech Inst Brooklyn, 46-47; adminr QMC Proj, Polytech Inst Brooklyn, 45-47, consult, 47-48. Mem: Am Chem Soc; Soc Rheol. Res: Polymer chemistry; theory of polymers in solution; plastics. Mailing Add: Dept of Chem Univ of Miami Coral Gables FL 33146

BADHWAR, GAUTAM DEV, cosmic ray physics, space science, see 12th edition

BADIN, ELMER JOHN, b Canonsburg, Pa, Aug 21, 22. PHYSICAL ORGANIC CHEMISTRY. Educ: Cooper Union, BChE, 42; Princeton Univ, MA, 44, PhD(org chem), 45. Prof Exp: Res assoc combustion chem, Princeton Univ, 45-46, instr chem, 46-50; sr res chemist, Celanese Corp Am, 56-59; prof chem & chmn dept, Union Col, NJ, 59-63; sr res chemist, Cities Serv Oil Co, 65-72; STAFF MEM, MITRE CORP, 74- Concurrent Pos: Consult, Wesco Paint Co, 45-47; assoc, US Navy Proj Squid, Princeton Univ, 45-47; researcher, Univ Calif, Berkeley, 49-50. Honors & Awards: Citation, US Comt Sci Res Personnel, 45; US Navy Bur Ord Develop Award, 46. Mem: Am Chem Soc; Combustion Inst; Sigma Xi. Res: Inorganic-organic-physical organic chemistry; combustion-oxidation; petroleum and polymer chemistry; coal sciences; synthesis and chemical mechanisms. Mailing Add: 2202 Mohegan Dr 102 Falls Church VA 22043

BADLER, NORMAN IRA, b Los Angeles, Calif, May 3, 48; m 68; c 1. INFORMATION SCIENCE. Educ: Univ Calif, Santa Barbara, BA, 70; Univ Toronto, MS, 71, PhD(comput sci), 75. Prof Exp: ASST PROF COMPUT SCI, UNIV PA, 74- Mem: Asn Comput Mach. Res: The computer representation, analysis and description of movement in digitized pictures; the understanding and processing of movement concepts in natural languages; graphic notational systems for movement. Mailing Add: Dept of Comput & Info Sci Moore Sch Elec Eng D 2 Univ Pa Philadelphia PA 19174

BADMAN, DAVID GEORGE, comparative physiology, physiological ecology, see 12th edition

BADMAN, WINONA SUE, zoology, see 12th edition

BADRE, ALBERT NASIB, b Beirut, Lebanon, Aug 15, 45; US citizen; m 69. INFORMATION SCIENCE. Educ: Univ Iowa, BA, 68; Univ Mich, Ann Arbor, MA, 71, PhD(behav sci), 73. Prof Exp: Soc worker, New York City Dept Soc Serv, 68-69; asst res, Ment Health Res Inst, Univ Mich, 70-71, res asst comput applns, 71-73; ASST PROF, SCH INFO & COMPUT SCI, GA INST TECHNOL, 73- Mem: AAAS; Am Asn Univ Prof; Am Soc Info Sci; Am Educ Res Asn; Am Psychol Asn. Res: The study, design, and analysis of man-computer problem-solving, information-processing, learning, pattern recognition, memory and decision processing systems. Mailing Add: Sch of Info & Comput Sci Ga Inst of Technol Atlanta GA 30332

BAECHLER, CHARLES ALBERT, b Lima, Ohio, Dec 15, 34; m 57; c 3. PHYSIOLOGY. Educ: Ohio State Univ, BScEduc, 57; Univ Toledo, MS, 64; Wayne State Univ, PhD(med microbiol), 71. Prof Exp: Teacher biol & chem, Libby High Sch, Toledo, Ohio, 60-63; teacher chem, Sylvania High Sch, 63-64; microbiol res specialist, Parke-Davis & Co, 64-71; res assoc, 71-74, ASST PROF PHYSIOL, SCH MED, WAYNE STATE UNIV, 74- Concurrent Pos: Lectr, Univ Toledo Community Col, 64; consult electron micros, Parke-Davis & Co, 71- Mem: AAAS; Soc Exp Biol & Med; Electron Micros Soc Am; Am Soc Microbiol; NY Acad Sci. Res: Function, structure and developmental aspects of blood and cardiovasculature interactions and microbial interactions at the ultrastructural level. Mailing Add: Dept of Physiol Wayne State Univ Sch of Med Detroit MI 48201

BAECHLER, RAYMOND DALLAS, b Brooklyn, NY, Mar 11, 45; m 67; c 2. ORGANIC CHEMISTRY. Educ: Fordham Univ, BS, 66; Princeton Univ, MA, 68, PhD(chem), 71. Prof Exp: Instr, Princeton Univ, 71-73; teaching fel, Miami Univ, 73-74; ASST PROF CHEM, RUSSELL SAGE COL, 74- Mem: Am Chem Soc. Res: Elucidation of reaction mechanisms; correlations between structure and reactivity; conformational processes of bond torsion and pyramidal inversion; silicon, phosphorus and sulfur chemistry; thermal rearrangements; unstable intermediates. Mailing Add: Dept of Chem Russell Sage Col Troy NY 12180

BAECHLER, ROY HERMAN, b Fountain City, Wis, Mar 9, 97; m 27; c 1. CHEMISTRY, FOREST PRODUCTS. Educ: Univ Wis, BS, 21, MS, 24, PhD(chem of forest prod), 27. Prof Exp: Asst chemist, Wilhoit Labs, Minneapolis, 21-22; chemist, Forest Prod Lab, US Forest Serv, 22-65; CONSULT CHEM, 66-, CONSULT WOOD PRESERVATION, 70- Honors & Awards: Award of Merit, Am Wood Preservers Asn, 70. Mem: AAAS; Am Chem Soc; Forest Prod Res Soc. Res: Pharmacology; relation of chemical constitution to toxicity; wood preservatives; fire retardants; corrosion. Mailing Add: 1105 Seminole Hwy Madison WI 53711

BAEDECKER, PHILIP A, b East Orange, NJ, Dec 19, 39; m 66. NUCLEAR CHEMISTRY, GEOCHEMISTRY. Educ: Ohio Univ, BS, 61; Univ Ky, MS, 64, PhD(chem), 67. Prof Exp: Res assoc nuclear chem, Mass Inst Technol, 67-68; asst res chemist, Univ Calif, Los Angeles, 68-73, asst prof in residence, 70-71; CHEMIST, US GEOL SURV, 74- Mem: AAAS; Am Chem Soc; Meteoritical Soc. Res: Radiochemistry; trace element geochemistry; activation analysis. Mailing Add: Mail Stop 924 US Geol Surv Reston VA 22092

BAEKELAND, FREDERICK, b Bronxville, NY, Sept 3, 28; m 50; c 2. PSYCHIATRY. Educ: Columbia Univ, BS, 51; Yale Univ, MS, 54, MD, 58; State Univ NY Downstate Med Ctr, DMSc(psychiat), 67. Prof Exp: From instr to asst prof, 64-71, ASSOC PROF PSYCHIAT, STATE UNIV NY DOWNSTATE MED CTR, 71- Mem: Asn Psychophysiol Study Sleep; fel Am Psychiat Asn. Res: Sleep: effects of drugs, physiology, dream content, requirements; alcoholism; correlates of treatment outcome; psychology of art. Mailing Add: Dept of Psychiat State Univ NY Downstate Med Ctr 450 Clarkson Ave Brooklyn NY 11203

BAENSCH, WILLY EDWARD, medicine, radiology, deceased

BAENZIGER, H, b Rebstein, Switz, June 6, 27; Can citizen; m 56; c 4. PLANT BREEDING, GENETICS. Educ: Swiss Fed Inst Technol, Ing agr, 51; Univ Sask, MSc, 57, PhD(genetics, plant breeding), 61. Prof Exp: Res scientist, Res Sta, Alta, 58-68, RES SCIENTIST, OTTAWA RES STA, CAN DEPT AGR, 68- Mem: Genetics

Soc Can; Can Soc Agron. Res: Cytology of supernumerary chromosomes in Agropyron cristatum and A desertorum; effects of supernumerary chromosomes on agronomic traits; breeding red clover for resistance to Kabatiella caulivora; alfalfa breeding, resistance to bacterial wilt, studies of breeding methods, particularly hybrid breeding using male-sterile lines. Mailing Add: Ottawa Res Sta Can Dept of Agr Cent Exp Farm Ottawa ON Can

BAENZIGER, NORMAN CHARLES, b New Ulm, Minn, Sept 23, 22; m 44; c 4. PHYSICAL CHEMISTRY. Educ: Hamline Univ, BS, 43; Iowa State Col, PhD(phys chem), 48. Prof Exp: Assoc chemist, Manhattan Proj & AEC, Iowa State Col, 44-48; instr phys chem, 46-48; fel, Mellon Inst, 48-49; from asst prof to assoc prof, 49-57, PROF PHYS CHEM, UNIV IOWA, 57- Mem: Am Chem Soc; Am Crystallog Asn. Res: X-ray crystallography; intermetallic compounds; metal-olefin complexes. Mailing Add: Dept of Chem Univ of Iowa Iowa City IA 52240

BAEPLER, DONALD H, b Edmonton, Alta, July 15, 32; US citizen; m 55. ORNITHOLOGY. Educ: Carleton Col, BA, 54; Univ Okla, MS, 56, PhD(zool), 60. Prof Exp: Assoc prof biol, Cent Wash State Col, 60-68, asst to pres, 66-68; vpres acad affairs, 68-74, PROF BIOL, UNIV NEV, 68-, PRES UNIV, 74- Mem: Am Ornith Union; Wilson Ornith Soc; Cooper Ornith Soc; Am Soc Mammal; Am Soc Zool. Res: Neotropical birds; avian ecology and distribution. Mailing Add: Dept of Biol Univ of Nev Las Vegas NV 89109

BAER, ADELA DEE, b Blue Island, Ill, Apr 4, 31; c 2. GENETICS. Educ: Univ Ill, BS, 53; Purdue Univ, 54-56; Univ Calif, Berkeley, PhD(genetics), 63. Prof Exp: From asst prof to assoc prof, 62-68, PROF BIOL, SAN DIEGO STATE UNIV, 68-, CHMN DEPT, 75- Concurrent Pos: Fulbright prof genetics, Univ Malaya, 67-68; res scientist, Int Ctr Med Res, Inst Med Res, Malaysia, 71-72. Mem: Genetics Soc Am; Am Soc Human Genetics; Am Eugenics Soc. Res: Human ecological and biochemical genetics. Mailing Add: Dept of Biol San Diego State Univ San Diego CA 92182

BAER, ADRIAN DONALD, b St Louis, Mo, Nov 3, 42; m 64; c 2. SOLID STATE PHYSICS. Educ: Univ Denver, BS, 64; Stanford Univ, MS, 65, PhD(elec eng), 71. Prof Exp: Res assoc physics, Mont State Univ, 71-73 & Stanford Univ, 73; RES PHYSICIST, NAVAL WEAPONS CTR, 74- Mem: Am Vacuum Soc. Res: Measuring various electron emission and optical spectra and utilizing these spectra to study electronic states at surfaces and interfaces, and in thin films. Mailing Add: 59 B Burroughs China Lake CA 93555

BAER, ALEC JEAN, b Geneva, Switz, m 55; c 2. STRUCTURAL GEOLOGY. Educ: Univ Neuchatel, MSc, 53, PhD(struct geol), 58. Prof Exp: Geologist, Gewerkschaft Elwerath, 58-60; fel geol, Polytech Sch, Montreal, 60-62; res scientist, Geol Surv Can, 62-71; lectr geol, Carleton Univ, 67-71; PROF GEOL & CHMN DEPT, UNIV OTTAWA, 72- Mem: Geol Soc Am; Geol Asn Can; Swiss Geol Soc; Swiss Acad Sci. Res: Alpine Precambrian and structural geology; geology of the Precambrian Grenville Province; metamorphic petrology. Mailing Add: Dept of Geol Univ of Ottawa Ottawa ON Can

BAER, CHARLES HENRY, b Columbus, Ohio, Sept 1, 19; m 50. PLANT PHYSIOLOGY, ECOLOGY. Educ: Ohio State Univ, BS, 47, MSc, 48; Univ Md, PhD, 61. Prof Exp: From instr to assoc prof plant physiol & ecol, 48-69, assoc prof biol, 69-74, PROF BIOL, WVA UNIV, 74- Res: Plant water relations; physiological instrumentation; microclimatology; indicator species; forest ecology. Mailing Add: Dept of Biol WVa Univ Morgantown WV 26506

BAER, DONALD RAY, b Warren, Ohio, Nov 12, 47; m 71. SOLID STATE PHYSICS. Educ: Carnegie-Mellon Univ, BS, 69; Cornell Univ, PhD(physics), 74. Prof Exp: RES ASSOC PHYSICS, MAT RES LAB & DEPT PHYSICS, UNIV ILL, URBANA, 74- Mem: Am Phys Soc; Am Asn Physics Teachers. Res: Rare-gas solids; defect properties; x-ray diffraction; phase transitions; thermodynamics; metals physics; electrical and thermal transport properties; radio frequency size effect; electron scattering; low temperature physics. Mailing Add: Dept of Physics Univ of Ill Urbana IL 61801

BAER, DONALD ROBERT, b Paterson, NJ, Apr 20, 28; m 52; c 3. RESEARCH ADMINISTRATION. Educ: Cornell Univ, BA, 48; Univ Mich, MS, 50, PhD(org chem), 52. Prof Exp: Res chemist dyes, Flourine Chem, 52-57, res supvr dyes, 57-62, div head dyes-textile chem, 62-75, CHIEF SUPVR MFG TECH, E I DU PONT DE NEMOURS & CO INC, 75- Mem: Am Chem Soc; AAAS; Am Asn Textile Chemists & Colorists. Res: Fluorochemicals; dyes; textile chemicals. Mailing Add: Chambers Works E I Du Pont de Nemours & Co Inc Wilmington DE 19898

BAER, EDWARD F, bacteriology, deceased

BAER, ERIC, b Nieder-Weisel, Ger, July 18, 32; US citizen; m 56; c 2. POLYMER CHEMISTRY, PLASTICS ENGINEERING. Educ: Johns Hopkins Univ, MA, 53, PhD(chem eng), 57. Prof Exp: Res engr, Polychem Dept, E I du Pont de Nemours & Co, 57-60; asst prof chem eng, Univ Ill, 60-62; assoc prof, 62-66, prof-in-chg polymer sci & eng, 62-66, PROF ENG, CASE WESTERN RESERVE UNIV, 66-, CHMN DEPT MACROMOLECULAR SCI, 67- Concurrent Pos: Indust consult ed, Polymer Eng & Sci. Honors & Awards: McGraw Award, Am Soc Eng Educ, 68. Mem: Am Inst Chem Engrs; Am Soc Mech Engrs; Am Chem Soc; fel Am Phys Soc; fel Am Inst Chemists. Res: Physical behavior and structure of polymers; engineering design for plastics; reactions on polymer crystal surfaces; properties of polymeric materials under high pressure and at cryogenic temperatures; nucleation phenomena in phase changes; heat transfer during condensation. Mailing Add: Dept of Macromolec Sci Case Western Reserve Univ Cleveland OH 44106

BAER, ERICH, b Berlin, Ger, March 8, 01; nat US; m 44. ORGANIC CHEMISTRY. Educ: Univ Berlin, PhD(org chem), 27. Prof Exp: Asst, Univ Berlin, 27-32; asst, Univ Basel, 32-36, privat-docent chem, 36-37; asst prof org chem, 37-47, from assoc prof to prof, Banting & Best Dept Med Res, 47-69, EMER PROF ORG CHEM & HON PROF, BANTING & BEST DEPT MED RES, UNIV TORONTO, 69- Concurrent Pos: Asst, Kaiser Wilhelm Inst Arbeitphysiol, 27-28. Honors & Awards: Glycerine Res Award, 53; Neuberg Medal, Soc Prom Int Sci Rels, 61; Chem Inst Can Medal, 62; Am Oil Chem Soc First Award, 64; Flavelle Medal, Royal Soc Can, 66; Norman Medal, Ger Soc Lipid Sci, 75. Mem: AAAS; Am Chem Soc; Am Soc Biol Chemists; fel Chem Inst Can; fel Royal Soc Can. Res: Carbohydrates; optically active glycerol derivatives; lipids; phospholipids; oxidative cleavage; terpenes; thrombine; intermediaries of carbohydrate metabolism. Mailing Add: 121 St Joseph St Toronto ON Can

BAER, FERDINAND, b Dinkelsbuhl, Ger, Aug 30, 29; US citizen; div; c 4. METEOROLOGY. Educ: Univ Chicago, AB, 50, MS, 54, PhD(geophys sci), 61. Prof Exp: Asst atmospheric physics, Univ Ariz, 55-56; asst meteorol, Univ Chicago, 56-61; from asst prof to assoc prof atmospheric sci, Colo State Univ, 61-71; PROF ATMOSPHERIC SCI, UNIV MICH, 72- Concurrent Pos: Vis res fel, Geophys Fluid Dynamics Lab, Princeton Univ, 68-69; consult, McGraw Hill Info Systs, 71-75. Mem: Am Meteorol Soc; Am Geophys Union; Royal Meteorol Soc; Meteorol Soc Japan. Res: Dynamics of the atmosphere, especially the method of expansion of atmospheric space variables in series of orthogonal polynominals; numerical analysis of atmospheric space fields. Mailing Add: Dept of Atmospheric Sci Univ of Mich Ann Arbor MI 48108

BAER, GEORGE M, b London, Eng, Jan 12, 36; US citizen; m 60; c 2. EPIDEMIOLOGY, VIROLOGY. Educ: Cornell Univ, DVM, 59; Univ Mich, MPH, 61. Prof Exp: Practicing vet, 59-60; vet epidemiologist, NY State Health Dept, 61-63; actg chief rabies invests lab, Commun Dis Ctr, Ga, 63-64, actg chief, Southwest Rabies Invests Lab, NMex, 64-66, chief lab, 66-69, CHIEF LAB INVESTS UNIT, VIRAL ZOONOSES SECT, CTR DIS CONTROL, USPHS, 69- Concurrent Pos: Rabies consult, Pan Am Health Orgn, Mex, 66-69. Mem: Am Vet Med Asn; Am Pub Health Asn. Res: Veterinary epidemiology, especially zoonotic diseases such as rabies, brucellosis and tuberculosis. Mailing Add: Ctr for Dis Control PO Box 363 Lawrenceville GA 30245

BAER, HANS HELMUT, b Karlsruhe, Ger, July 3, 26; Can citizen; m 56; c 2. ORGANIC CHEMISTRY. Educ: Karlsruhe Tech Univ, cand chem, 47; Univ Heidelberg, dipl chem, 50, Dr rer nat(chem), 52. Prof Exp: Res assoc chem, Max Planck Inst Med Res, Heidelberg, 52-57; vis scientist biochem, Univ Calif, Berkeley, 57-59; vis scientist chem, Nat Inst Arthritis & Metab Dis, 59-61; assoc prof, 61-65, chmn dept, 69-75, PROF CHEM, UNIV OTTAWA, 65- Honors & Awards: C S Hudson Award, Am Chem Soc, 75. Mem: Am Chem Soc; Chem Inst Can; Soc Ger Chemists. Res: Chemistry of natural products, especially carbohydrates; general synthetic organic chemistry. Mailing Add: Dept of Chem Univ of Ottawa Ottawa ON Can

BAER, HAROLD, b New York, NY, Oct 3, 18; m 46; c 2. ALLERGY, IMMUNOLOGY. Educ: Brooklyn Col, BA, 38; Columbia Univ, MA, 40; Harvard Univ, MA, 42, PhD(chem), 43. Prof Exp: From asst to res assoc, Col Physicians & Surgeons, Columbia Univ, 43-50; from asst prof to assoc prof, Med Sch, Tulane Univ, 50-60; chief sect allergenic prod, Div Biologics Stand, NIH, 60-71, chief lab bact prod, 71-74, ASSOC DIR DIV BACT PROD, BUR BIOLOGICS, FOOD & DRUG ADMIN, 74- Concurrent Pos: Instr, City Col New York, 48-50. Mem: AAAS; Sigma Xi; Can Soc Allergy & Clin Immunol; Int Asn Biol Stand; Am Soc Microbiol. Res: Allergen chemistry and standardization of allergenic extracts; delayed contact sensitivity. Mailing Add: Bur Biologics Food & Drug Admin 8800 Rockville Pike Bethesda MD 20014

BAER, HELMUT W, b Hunan, China, Aug 16, 39; US citizen. NUCLEAR PHYSICS. Educ: Franklin & Marshall Col, BA, 61; Univ Mich, Ann Arbor, PhD(physics), 67. Prof Exp: Res asst physics, Univ Mich, Ann Arbor, 67-69; res asst, Univ Colo, Boulder, 69-71; physicist, Lawrence Berkeley Lab, Univ Calif, 71-74; ASST PROF PHYSICS, CASE WESTERN RESERVE UNIV, 74- Mem: Am Phys Soc. Res: Experimental nuclear research at low and medium energies. Mailing Add: Dept of Physics Rockefeller Bldg Case Western Reserve Univ Cleveland OH 44106

BAER, HERMAN, b Uetikon, Switz, Sept 11, 33; US citizen; m; c 3. MEDICAL MICROBIOLOGY. Educ: Univ Basel, Switz, MD, 60; Am Acad Med Microbiol, dipl med bact, 69. Prof Exp: Fel, Inst Med Microbiol, Sch Med, Univ Zurich, 60-62, head diag unit, 62-63; fel, Div Infectious Dis, Dept Med, Med Sch, Univ Pittsburgh, 63-64, fel, Dept Bact, Div Natural Sci, 65-66; asst prof, 66-70, ASSOC PROF IMMUNOL & MED MICROBIOL, COL MED, UNIV FLA, 70-, ASSOC PROF PATH, 74-, DIR CLIN MICROBIOL LAB, 66- Mem: Am Soc Microbiol. Res: Role of hand contamination of personnel in the epidemiology of gram negative nosocomal infections; role of airborne contamination in orthopedic surgery. Mailing Add: Clin Microbiol Lab Shands Teaching Hosp Univ Fla Gainesville FL 32611

BAER, JAMES L, b Van Wert, Ohio, Oct 6, 35; m 61; c 4. PALEOECOLOGY, STRUCTURAL GEOLOGY. Educ: Ohio State Univ, BS, 57; Brigham Young Univ, MS, 62, PhD(paleoecol), 68. Prof Exp: Asst prof geol, Northeast La State Col, 68-69; asst prof, 69-74, ASSOC PROF GEOL, BRIGHAM YOUNG UNIV, 74- Mem: Geol Soc Am; Paleont Soc. Res: Paleoecology of ancient lake sediments; structural evolution of western United States; continental drift; geochemistry of sediments. Mailing Add: Dept of Geol Brigham Young Univ Provo UT 84601

BAER, JOHN ELSON, b Cleveland, Ohio, Apr 25, 17; m 47; c 4. PHARMACOLOGY, DRUG METABOLISM. Educ: Swarthmore Col, AB, 38; Univ Pa, MS, 40, PhD(org chem), 48. Prof Exp: Chemist, Med Sch, NY Univ, 43-46; instr chem, Haverford Col, 47-48; asst prof, Carleton Col, 48-51; res assoc, Pharmacol Sect, Sharp & Dohme Div, Merck & Co, 51-58, dir pharmacol chem, Merck Inst, 58-69, assoc dir, Merck Inst Therapeut Res, 69-71, SR DIR DRUG METAB, MERCK SHARP & DOHME RES LABS, 71- Honors & Awards: Albert Lasker Spec Award, Lasker Found, 75. Mem: AAAS; Am Soc Pharmacol & Exp Therapeut; Am Soc Nephrology; Am Soc Clin Chemists. Res: Chemical methods for analysis of compounds in biological systems; pharmacology of diuretics and saluretics. Mailing Add: Merck Sharp & Dohme Res Labs West Point PA 19486

BAER, LEDOLPH, b Monroe, La, Nov 21, 29; m 57; c 3. OCEANOGRAPHY, METEOROLOGY. Educ: La Polytech Inst, BS, 50; Tex Agr & Mech Col, MS, 55; NY Univ, PhD(ocean waves), 62. Prof Exp: Meteorologist, Gulf Consult, Tex, 55-56; res specialist, Lockheed Missiles & Space Co, 59-62, res & develop scientist, Lockheed-Calif Co, 62-66, mgr ocean sci dept, Lockheed Ocean Lab, 66-74; DIR OCEANOG SERV, NAT OCEANIC & ATMOSPHERIC ADMIN, 74- Mem: AAAS; Am Meteorol Soc; Am Geophys Union. Res: Physical oceanography; surface wave forecasting; internal waves; oceanic variability; buoy systems; synoptic oceanography; storm surges; missile meteorology; wind statistics and kinematics; rainfall distributions; marine environmental pollution. Mailing Add: Nat Oceanic & Atmospheric Admin 6010 Executive Blvd EM3 Rockville MD 20852

BAER, MASSIMO, b Milano, Italy, Dec 23, 16; nat US; m 46; c 2. ORGANIC CHEMISTRY. Educ: Mass Inst Technol, BS, 40. Prof Exp: Res chemist dielectrics, Cornell-Dubillier Elec Corp, 40-43; res chemist, 43-52, scientist, 52-71, SR SCI FEL, POLYMER RES, MONTSANTO CO, 71- Mem: Am Chem Soc. Res: Technology of polymerization of vinyl monomers; graft polymerization of vinyl monomers; anionic polymerization; block copolymers; polyblend systems; correlation of polymer structure and mechanical properties; glass fiber composites. Mailing Add: 730 Worcester St Indian Orchard MA 01151

BAER, PAUL NATHAN, b New York, NY, Oct 13, 21; m 50; c 1. PERIODONTOLOGY. Educ: Brooklyn Col, BA, 42; Columbia Univ, DDS, 45; Am Bd Periodont, dipl, 56. Prof Exp: Asst prof periodont, Univ Southern Calif, 55-56; peridontist, Nat Inst Dent Res, 56-70, chief adolescent periodont unit, 70-73; PROF PERIODONT & CHMN DEPT, SCH DENT MED, STATE UNIV NY STONY BROOK, 73- Concurrent Pos: Vis assoc prof, Grad Sch Dent, Boston Univ; assoc ed, J Dent Res; consult, Long Island Jewish-Hillside Med Ctr Hosp, Vet Admin Northport Hosp, Nassau County Med Ctr & Sagamore Childrens Ctr. Mem: Am Dent

Asn; Am Acad Periodont; Am Acad Oral Path; Int Asn Dent Res; hon mem Colombia Soc Periodont. Res: Periodontal disease. Mailing Add: Dept of Periodont State Univ NY Sch of Dent Med Stony Brook NY 11790

BAER, ROBERT M, b Chicago, Ill, May 15, 25; m 51; c 3. MATHEMATICS. Educ: Univ Chicago, MS, 49; Univ Ill, PhD(math), 53. Prof Exp: Physicist, Oak Ridge Inst Nuclear Studies, 45-46 & Argonne Nat Lab, 46-47; instr physics, Roosevelt Univ, 47-48; res assoc, Control Systs Lab, Univ Ill, 52; asst prof math, Purdue Univ, 54-57; res mathematician, Calif Res Corp, 58-61; consult phys sci comput ctr, Univ Calif, Berkeley, 61-70, lectr comput sci, 70-73; DIR METADATA CORP, 73- Mem: AAAS; Am Math Soc; Am Phys Soc. Res: Medical programming applications. Mailing Add: Metadata Corp PO Box 746 Mill Valley CA 94941

BAER, RUDOLF L, b Strasbourg, Alsace-Lorraine, July 22, 10; nat US; m 41; c 2. DERMATOLOGY, IMMUNOLOGY. Educ: Univ Frankfurt, MD, 34; Am Bd Dermat, dipl, 40. Prof Exp: Teacher dermat allergy, Med Sch, Columbia Univ, 39-46, instr dermat & syphil, 46-48, asst clin prof, Postgrad Med Sch & Hosp, 48; asst prof clin dermat & syphil, Postgrad Med Sch, 49-50, from assoc prof to prof, 50-61, PROF DERMAT & CHMN DEPT, SCH MED, NY UNIV, 61- Concurrent Pos: Morrow-Miller-Taussig mem lectr, San Francisco, 61; Dohi lectr, Japanese Dermat Soc, 65; Von Zumbusch Mem lectr, Univ Munich, 67; mem comt cutaneous dis, Armed Forces Epidemiol Bd, 67; consult, Surgeon Gen, US Army, 67; mem, Int Comt Dermat, 67-, pres, 72-77; consult, Food & Drug Admin, 69; Hellerstrom lectr, Karolinska Inst, Sweden, 70; mem comt on revision, US Pharmacopeia, 70; pres, Am Bd Dermat, 70; O'Leary lectr, Mayo Clin, Rochester, Minn, 71; Robinson lectr, Univ Md, 72; Barrett Kennedy Mem lectr, SCent Dermat Soc, 73; Duhring lectr, Pa Acad Dermat, 74; Bluefarb lectr, Chicago Dermat Soc, 75. Honors & Awards: Dohi Mem Medal, Japanese Dermat Soc, 65; Hellerstrom Medal-Karolinska Inst, Sweden, 70; Stephen Rothman Medal, Soc Invest Dermat, 73. Mem: AAAS; fel Am Acad Allergy; fel Am Acad Dermat (pres, 74-75); fel Am Col Allergists (vpres, 54); Soc Invest Dermat (vpres, 56, pres, 63). Res: Various forms of allergic, cutaneous, sensitization; tolerance to simple chemicals; cross-sensitization; photobiology of the skin; biology of fungous infections of the skin. Mailing Add: Dept of Dermat NY Univ Sch of Med New York NY 10016

BAER, TOMAS, b Zurich, Switz, Aug 27, 39; US citizen; m 62; c 3. CHEMICAL PHYSICS. Educ: Lawrence Univ, BA, 62; Wesleyan Univ, MA, 64; Cornell Univ, PhD(chem), 69. Prof Exp: Fel chem, 69-70; asst prof, 70-75, ASSOC PROF CHEM, UNIV NC, CHAPEL HILL, 75- Mem: Am Chem Soc; Am Phys Soc; Am Soc Mass Spectrometry. Res: Photoion-photoelectron coincidence spectroscopy used to study the dissociation rates of energy selected ions and the role of ion internal energy in ion-molecule reactions. Mailing Add: Dept of Chem Univ NC Chapel Hill NC 27514

BAER, WALTER S, b Chicago, Ill, July 27, 37; m 59; c 2. SOLID STATE PHYSICS, TELECOMMUNICATIONS. Educ: Calif Inst Technol, BS, 59; Univ Wis, MS, 61, PhD(physics), 64. Prof Exp: Mem tech staff, Bell Tel Labs, 64-67; White House fel, Off of Vice President, 66-67; staff mem, Off Sci & Technol, Exec Off of President, 67-69; sr staff mem, Laird Systs Inc, 69-70; CONSULT, RAND CORP, 70- Concurrent Pos: Mem comput sci & eng bd, Nat Acad Sci, 69-72; consult to UN & maj US corps, 70-; mem cable TV adv comt, Fed Commun Comn, 72-74; dir, Aspen Calbe Workshop, 72-73; mem adv coun, Aspen Prog Commun & Soc, 74- Mem: AAAS; Asn Comput Mach; Am Phys Soc; NY Acad Sci. Res: Communications systems; science and public policy; telecommunications policy; technical management. Mailing Add: 1700 Main St Santa Monica CA 90402

BAER, WILLIAM, b New York, NY, Dec 24, 24; m 49; c 1. NUCLEAR PHYSICS. Educ: Amherst Col, AB, 44; Dartmouth Col, AM, 47; Yale Univ, MS, 49. Prof Exp: Jr scientist, 49-50, from assoc scientist to sr scientist, 50-58, fel scientist, 58-70, mgr nuclear mat mgt, 71-74, MGR LAB OPERS SAFEGUARDS, BETTIS ATOMIC POWER LAB, WESTINGHOUSE ELEC CORP, 74- Mem: Am Phys Soc; Am Nuclear Soc. Res: Radiation detectors; experimental reactor physics; neutron cross sections; data storage and retrieval. Mailing Add: Bettis Atomic Power Lab Westinghouse Box 79 West Mifflin PA 15122

BAER, WILLIAM KERN, physical chemistry, analytical chemistry, see 12th edition

BAERG, DAVID CARL, b Dinuba, Calif, June 23, 38; m 62; c 2. MEDICAL ENTOMOLOGY. Educ: Univ Calif, Davis, BS, 64, MS, 65, PhD(entom), 67. Prof Exp: Res asst mosquito res, Univ Calif, Davis, 64-67; RES ENTOMOLOGIST, GORGAS MEM LAB, 67-; CLIN ASST PROF & ASSOC MEM GRAD FAC TROP MED & MED PARASITOL, LA STATE UNIV MED CTR, NEW ORLEANS, 73- Mem: AAAS; Am Soc Trop Med & Hyg; NY Acad Sci; Entom Soc Am; Mosquito Control Asn. Res: Mosquito biology and ecology; transmission, development and chemotherapy of malarias. Mailing Add: Gorgas Mem Lab Box 2016 Balboa Heights CZ

BAERNSTEIN, ALBERT, II, b Birmingham, Ala, Apr 25, 41; m 62; c 2. MATHEMATICAL ANALYSIS. Educ: Cornell Univ, AB, 62; Univ Wis, MA, 64, PhD(math), 68. Prof Exp: Cost analyst, Prudential Ins Co, 62-63; instr math, Wis State Univ, Whitewater, 66-68; asst prof, Syracuse Univ, 68-72; assoc prof, 72-74, PROF MATH, WASH UNIV, 74- Mem: Am Math Soc. Res: Complex analysis and related areas. Mailing Add: Dept of Math Wash Univ St Louis MO 63130

BAERREIS, DAVID ALBERT, b New York, NY, Nov 2, 16; m 43. ANTHROPOLOGY. Educ: Univ Okla, BA, 41, MA, 43; Columbia Univ, PhD(anthrop), 49. Prof Exp: From instr to assoc prof anthrop, 47-55, prof integrated liberal studies, 55-73, PROF ANTHROP, UNIV WIS-MADISON, 55- Concurrent Pos: Mem, Div Anthrop & Psychol, Nat Res Coun, 59-61. Mem: AAAS (vpres sect H, 63); Am Anthrop Asn; Soc Am Archaeol (secy, 56-59, pres, 62). Res: Archaeology and ethnology of North and South America. Mailing Add: Dept of Anthrop Univ of Wis Madison WI 53706

BAES, CHARLES FREDERICK, JR, b Cleveland, Ohio, Dec 2, 24; m 48; c 3. PHYSICAL INORGANIC CHEMISTRY. Educ: Rutgers Univ, BSc, 46; Univ Southern Calif, MSc, 48, PhD(phys inorg chem), 50. Prof Exp: Fel metal, Columbia Univ, 50-51; RES CHEMIST, OAK RIDGE NAT LAB, 51- Concurrent Pos: Vis prof chem, Col William & Mary, 72-73. Mem: AAAS; Am Chem Soc; Sigma Xi. Res: Solution chemistry; thermodynamics; molten salt chemistry; environmental chemistry. Mailing Add: X-10 Area Oak Ridge Nat Lab Oak Ridge TN 37830

BAETJER, ANNA MEDORA, b Baltimore, Md, July 7, 99. PHYSIOLOGY. Educ: Wellesley Col, AB, 20; Johns Hopkins Univ, DSc(physiol, physiol hyg & occup health), 24; Am Bd Indust Hyg, dipl, 62. Hon Degrees: DPH, Woman's Med Col Pa, 53; DSc, Wheaton Col, 66. Prof Exp: Asst physiol, 23-24, from instr to asst prof, 24-50, from asst prof to prof environ med, 50-70, EMER PROF ENVIRON MED, SCH HYG & PUB HEALTH, JOHNS HOPKINS UNIV, 70- Concurrent Pos: Consult, Prev Med Div, Off Surgeon Gen Army, 47-; mem, Comt Sanit Eng & Environ, Nat Res Coun, 53-65, chmn, Subcomt, Atmospheric & Indust Hyg, 60-65, mem, Comt Environ Physiol, 65-67, Comt Biol Effects Atmospheric Pollutants, 70-73; chem, Toxicol Comt, Indust Hyg Found, Mellon Inst, 53-61, mem, Bd Trustees, 58-, vchmn, 64-70; indust hyg fel, Bd, Atomic Energy Comt, 53-62; mem, Comn Environ Hyg, Armed Forces Epdiemiol Bd, 54-73; mem, Permanent Comn & Int Asn Occup Health; mem, Adv Comt Safety Pesticide Residues in Foods, Food & Drug Admin, 66-70, foods, USPHS-Nat Inst Indust Occup Safety & Health Study Sect, 68-70 & 72-74, Nat Air Qual Criteria Adv Comn, Environ Protection Agency, 72-, Baltimore City Noise Control Adv Bd, 73- & Standards Adv Comn Heat Stree, US Dept Labor, 73-74. Honors & Awards: Cummings Mem Award, Am Indust Hyg Asn, 64. Mem: Am Physiol Soc; Am Pub Health Asn; Am Indust Hyg Asn (pres, 51); Conf Govt Indust Hygienists; hon mem Am Acad Occup Med. Res: Sympathetic nervous systems; physiological hygiene; effects of sodium, potassium and calcium on myocardium and myoskeleton, industrial dusts on pneumonia and cancer, high temperatures and dehydration on toxic reactions and respiratory clearance, and food and water deprivation on liver metabolism; occupational cancer and arsenic. Mailing Add: Dept of Environ Med Johns Hopkins Sch Hyg & Pub Health Baltimore MD 21205

BAETZ, ALBERT L, b Cleveland, Ohio, Dec 25, 38; m 65. CLINICAL CHEMISTRY. Educ: Purdue Univ, BS, 61; Iowa State Univ, MS, 63, PhD(anal chem), 66. Prof Exp: RES CHEMIST, NAT ANIMAL DIS LAB, AGR RES SERV, USDA, 66- Mem: AAAS; Am Asn Clin Chem; Am Chem Soc; Am Asn Vet Physiologists & Pharmacologists. Res: Study of body fluids from normal and diseased farm animals to discover diagnostic significance and to outline pathogenesis of various diseases. Mailing Add: Nat Animal Dis Lab PO Box 70 Ames IA 50010

BAETZOLD, ROGER C, b Warsaw, NY, Feb 26, 42; m 64; c 4. PHYSICAL CHEMISTRY, CHEMICAL PHYSICS. Educ: Univ Buffalo, BA, 63; Univ Rochester, PhD(chem), 67. Prof Exp: Sr res chemist, 66-71, RES ASSOC, EASTMAN KODAK CO, NY, 71- Mem: Am Chem Soc. Res: Experimental and theoretical kinetics of chemical reactions; molecular orbital calculations; solid state chemistry. Mailing Add: Res Labs Bldg 81 Eastman Kodak Co Rochester NY 14650

BAEUMLER, HOWARD WILLIAM, b Buffalo, NY, Sept 29, 21. MATHEMATICS, ELECTRICAL ENGINEERING. Educ: State Univ NY teachers Col Buffalo, BS, 43; Univ Buffalo, MA, 50; Ohio State Univ, PhD(elec eng), 64. Prof Exp: Instr math & sci, Park Sch, Buffalo, 43-44; instr math, Univ Buffalo, 46-53; asst prof, Marshall Univ, 54-55; asst instr, Ohio State Univ, 55-56, res assoc elec eng, Res Found, 56-61, instr, univ, 59-64; ASSOC PROF MATH, OLD DOM UNIV, 65- Concurrent Pos: Lectr, Va High Sch Vis Scientist Prog, 66-67. Mem: AAAS; Am Soc Eng Educ; Asn Comput Mach; Inst Elec & Electronics Engrs; Math Asn Am. Res: Electromagnetic field theory; numerical methods, computation and analysis; Boolean algebra and Boolean matrices; electronic and mechanical switching circuits; symbolic binary and ternary logic; computer science, programming languages and non-numerical applications. Mailing Add: Dept of Math Old Dom Univ 5215 Hampton Blvd Norfolk VA 23508

BAEVSKY, MELVIN M, physical chemistry, organic chemistry, see 12th edition

BAEZ, ALBERT VINICIO, b Puebla, Mex, Nov 15, 12; nat US; m 35; c 3. PHYSICS. Educ: Drew Univ, BA, 33; Syracuse Univ, MA, 36; Stanford Univ, PhD, 50. Hon Degrees: Dr, Open Univ, Gt Brit, 74. Prof Exp: Instr physics & math, Morris Jr Col, 36-38; instr math, Drew Univ, 38-40; prof physics, Wagner Col, 40-44; instr, Stanford Univ, 44-45, actg instr math, 45-46, res asst physics, 46-49; physicist, Aeronaut Lab, Cornell Univ, 49-50; prof physics, Univ Redlands, 51-56, physicist, Film Group, Phys Sci Study Comt, 58-60; assoc prof physics, Harvey Mudd Col, 60-61; dir div sci teaching, UNESCO, 61-67; consult sci educ to dir sci & technol, UN, NY & UNESCO, Paris, 67-74; ASSOC, LAWRENCE HALL OF SCI, UNIV OF CALIF, BERKELEY, 74- Concurrent Pos: Prof, Col Arts & Sci, Univ Bagdad & head, UNESCO Tech Asst Mission, 51-52; vis prof, Stanford Univ, 56-58; hon res assoc, Dept Physics, Harvard Univ, 70-; sci dir, Physics Loop Film Proj, Encycl Britannica, 67-75; consult int sci educ, NSF, 70; vchmn, Comt Sci Teaching, Int Comn Sci Unions, 70-74, chmn, 74-; vis prof, Open Univ, Gt Brit, 71-; mem, Bd Int Orgns & Progs, Nat Acad Sci, 74- Mem: Fel AAAS; Am Phys Soc; Optical Soc Am; Am Asn Physics Teachers; Nat Sci Teachers Asn. Res: Absolute intensity of x-ray radiation; x-ray optics and microscopy; x-ray optical images; holography; science education. Mailing Add: Lawrence Hall of Sci Univ of Calif Berkeley CA 94720

BAEZ, SILVIO, b July 6, 15; US citizen; m 50; c 3. PHYSIOLOGY. Educ: Nat Univ Paraguay, BS, 35, MD, 42. Prof Exp: Res assoc, Cornell Univ, 48-52, asst prof, 52-58; asst prof anesthesiol, NY Univ, 58-61; assoc prof, 61-69, PROF ANESTHESIOL, ALBERT EINSTEIN COL MED, 69- Concurrent Pos: Fel physiol, Int Training Admin, Cornell Univ, 44, res fel med, Med Col, 45-48; hon prof, Univ Montevideo, 68. Mem: AAAS; Am Heart Asn; Am Physiol Soc; Harvey Soc; NY Acad Sci. Res: Peripheral vascular physiology; shock; hypertension; adaptation; anesthesiology. Mailing Add: Dept of Anesthesiol & Physiol Albert Einstein Col of Med New York NY 10461

BAGATELL, FILLMORE KENNETH, b Brooklyn, NY, Jan 24, 29; m 51; c 5. DERMATOLOGY, BIOCHEMISTRY. Educ: Cornell Univ, BS, 50; Iowa State Col, MS, 51; Western Reserve Univ, PhD(biochem), 58, MD, 61; Am Bd Dermat, dipl. Prof Exp: Instr dermat, Case Western Reserve Univ, 65-67; assoc clin res dir, 67-71, CLIN RES DIR, SQUIBB INST, 71- Concurrent Pos: Fel dermat, Sch Med, Case Western Reserve Univ, 62-65. Mem: AAAS; Am Chem Soc; Am Fedn Clin Res; Am Inst Chem; AMA. Res: Drug therapy for benign proliferative skin diseases; percutaneous absorption of corticosteroids; pharmacology and biochemistry of the skin. Mailing Add: Squibb Inst Med Res Georges Rd New Brunswick NJ 08903

BAGBY, JOHN R, JR, b Aurora, Mo, Mar 3, 19; m 43; c 3. PUBLIC HEALTH, BIOLOGY. Educ: Univ Ark, BS, 54, MS, 55; Emory Univ, PhD(parasitol), 62. Prof Exp: Malaria control aid, USPHS, 46-51, info officer, Econ Stabilization Agency, 51-53, vector control specialist, 55-59, asst to chief tech br, 59-62, asst chief tech br, 62-63 & 64-66, dep chief tech br, 63-64, dept dir, Nat Commun Dis Ctr, 66-69; DIR INST ENVIRON HEALTH, COLO STATE UNIV, 69- Concurrent Pos: Consult, NASA, Jet Propulsion Lab & Am Inst Biol Sci. Mem: AAAS; Entom Soc Am; Am Pub Health Asn; Am Soc Trop Med & Hyg. Res: Parasitology; entomology; bacteriology; serology; control of vectors; reservoirs; etiologic agents of communicable diseases. Mailing Add: Inst of Environ Health Colo State Univ Ft Collins CO 80523

BAGBY, MARVIN ORVILLE, b Macomb, Ill, Sept 27, 32; m 57; c 2. ORGANIC CHEMISTRY. Educ: Western Ill Univ, BS & MS, 57. Prof Exp: Org chemist, 57-75, RES LEADER, NORTHERN REGIONAL RES CTR, AGR RES SERV, USDA, 57- Mem: AAAS; Am Chem Soc; Am Oil Chem Soc; Tech Asn Pulp & Paper Indust. Res: Natural products; lipids, especially isolation, characterization, and reactions of fatty acids and sugars; annual plants for pulp and paper raw materials. Mailing Add: US Dept of Agr 1815 N University St Peoria IL 61604

BAGCHI, AMITABHA, b Calcutta, India, June 12, 45; m 74. SOLID STATE PHYSICS, SURFACE PHYSICS. Educ: Calcutta Univ, BSc, 64; Univ Calif, San

Diego, MS, 67, PhD(physics), 70. Prof Exp: Res assoc physics, Univ Ill, Urbana, 70-71; res assoc surface physics, James Franck Inst, Univ Chicago, 71-73; res fel physics, Battelle Mem Inst, Columbus, Ohio, 73-74; ASST PROF PHYSICS, UNIV MD, 74- Mem: Am Phys Soc; Am Vacuum Soc. Res: Electronic properties of metal surfaces including chemisorption and photoemission and optical reflectance from clean and adsorbate-covered surfaces; excitation spectrum of He3-He4 mixtures. Mailing Add: Dept of Physics & Astron Univ of Md College Park MD 20742

BAGCHI, MIHIR, b Ranchi, India, Nov 28, 38. CELL BIOLOGY, OPHTHAMLMOLOGY. Educ: Univ Bihar, BS, 59; Univ Ranchi, MS, 62; Univ Vt, PhD(zool), 69. Prof Exp: Res asst biol, Ciba Res Ctr, India, 63-65; res assoc, Oakland Univ, 63-75; ASST PROF ANAT, SCH MED, WAYNE STATE UNIV, 75- Concurrent Pos: NIH spec fel, 73-75. Mem: Am Soc Cell Biol; Sigma Xi. Res: Biochemistry and morphology of mammalian lens in culture. Mailing Add: Dept of Anat Wayne State U Sch of Med Detroit MI 48201

BAGCHI, PRANAB, b Calcutta, India, Jan 14, 46. COLLOID CHEMISTRY, SURFACE CHEMISTRY. Educ: Jadavpur Univ, India, BSc, 65; Univ Southern Calif, PhD(phys chem), 70. Prof Exp: SR RES CHEMIST, EASTMAN KODAK CO, 70- Mem: Am Chem Soc; AAAS; NY Acad Sci; The Chem Soc; Fedn Am Scientists. Res: Investigation of the mechanism of the stability of lyophobic dispersions in nonaqueous media and in the presence of nonionic surfactants; long chain nonionic molecules and polymeric nonionic molecules, both experiment and theory considered; surface colloidal properties and stability of silver salts in nonaqueous media. Mailing Add: Eastman Kodak Res Labs Kodak Park Rochester NY 14650

BAGCHI, SAKTI PRASAD, b Tatanagar, India, Dec 26, 31; Can citizen; m 59; c 2. BIOCHEMISTRY. Educ: Univ Calcutta, BS, 51, MS, 54, PhD(biochem), 59. Prof Exp: Res assoc biochem, Okla State Univ, 59-60; jr res biochemist, Sch Med, Univ Calif, San Francisco, 60-62; res assoc, Kinsmen Lab of Neurol Dis, Univ BC, 63-66; res biochemist, Psychiat Res Unit, Dept Pub Health, Univ Hosp, 66-73; SR RES SCIENTIST, NY STATE DEPT MENT HYG, ROCKLAND RES INST, 73- Mem: AAAS; Int Soc Neurochem; Can Inst Chem; Am Soc Neurochem; Soc Biol Psychiat. Res: Biochemistry of the central nervous system; psychotropic drug action; biochemistry of mental disorders. Mailing Add: Rockland Res Inst Bldg 37 Orangeburg NY 10962

BAGDASARIAN, ANDRANIK, b Tehran, Iran, Dec 5, 35; m 69; c 3. BIOCHEMISTRY. Educ: Univ Tehran, BS, 60, DrPharm, 61; Univ Louisville, PhD(biochem), 67. Prof Exp: Fel protein chem, Calif Inst Technol, 67-68; fel enzymol, Syntex Res, Palo Alto, 68-70; res fel plasma proteases, Mass Gen Hosp, Harvard Med Sch, 70-73; res assoc plasma proteases, 73-75, RES ASST PROF BIOCHEM, DEPT MED, UNIV PA, 75- Concurrent Pos: NIH award studies prekallikrein activation, 71; Nat Cancer Inst & Am Cancer Soc grant kinin pathway normal & malignant tissue, 75. Mem: Sigma Xi; Am Chem Soc; Am Soc Biol Chemists. Res: Purification and characterization of proteases and protease inhibitors involved in the kallikrein-kinin, coagulation and fibrinolysis systems of human plasma, platelets and normal and malignant tissue. Mailing Add: Hosp of Univ of Pa 578 Maloney 3400 Spruce St Philadelphia PA 19104

BAGDON, ROBERT EDWARD, b Newark, NJ, Sept 18, 27; m 56; c 2. PHARMACOLOGY. Educ: Upsala Col, BS, 49; Univ Chicago, PhD, 55. Prof Exp: Chemist, Ciba Pharmaceut Prod Inc, 49-52; pharmacologist, Am Cyanamid Co, 56-57; sr pharmacologist, Hoffmann-La Roche, Inc, 57-67; HEAD TOXICOL, TOXICOL & PATH SECT SANDOZ PHARMACEUT DIV, SANDOZWANDER, INC, 67- Mem: Am Chem Soc; NY Acad Sci; Am Soc Pharmacol & Exp Therapeut; Soc Toxicol; Soc Exp Biol & Med. Res: Toxicology; mechanism; drug action. Mailing Add: Toxicol & Path Sect Sandoz Pharmaceut Hanover NJ 07936

BAGDON, WALTER JOSEPH, b Camden, NJ, Apr 22, 38; m 62; c 3. TOXICOLOGY, PHARMACOLOGY. Educ: Philadelphia Col Pharm, BS, 59; Temple Univ, MS, 61, PhD(pharmacol), 64. Prof Exp: Sr res toxicologist, 64-68, res fel, 68-74, SR RES FEL, MERCK INST THERAPEUT RES, 74- Mem: AAAS; Am Pharmaceut Asn. Res: Safety evaluation of current and potential therpeutic agents; evaluation of clinical tests concerning normal body function of animals. Mailing Add: 1360 Slayton Dr Maple Glen PA 19002

BAGGA, HARMAHINDER SINGH, b Amritsar, India, Sept 24, 36; m 61; c 2. PLANT PATHOLOGY. Educ: Panjab Univ, India, BSc, 57; Univ Wis-Madion, PhD(plant path), 66. Prof Exp: Agr inspector, Dept Agr, Panjab, India, 60-61; assoc plant pathologist, 66-68, PLANT PATHOLOGIST, DELTA BR EXP STA, 68- Mem: Am Phytopath Soc; Nat Cotton Dis Coun; Int Phytopath Soc. Res: Genetics of microorganisms; host-parasite interaction; inheritance of pathogenicity and disease resistance; cotton boll rot; fungicides; epidemiology; survey. Mailing Add: Delta Br Exp Sta Stoneville MS 38776

BAGGENSTOSS, ARCHIE HERBERT, b Richardton, NDak, Apr 13, 08; m 34; c 3. PATHOLOGY. Educ: Univ NDak, AB, 30, BS, 31; Univ Cincinnati, BM, 33, MD, 34; Univ Minn, MS, 38. Prof Exp: Consult path anat, 38-52, head sect exp & anat path, 55-68, SR CONSULT, MAYO CLIN, 68-, PROF PATH, MAYO GRAD SCH MED, UNIV MINN, 52- Mem: Fel AMA; fel Am Soc Clin Path; Am Asn Path & Bact; Am Gastroenterol Asn; Am Asn Study Liver Dis (pres, 68-69). Res: Pathologic anatomy; liver; pancreas; gastrointestinal tract. Mailing Add: Mayo Clin 200 SW First St Rochester MN 55901

BAGGERLY, LEO L, b Wichita, Kans, Mar 13, 28. PHYSICS. Educ: Calif Inst Technol, BS, 51, MS, 52, PhD(physics), 56. Prof Exp: Sr res engr, Jet Propulsion Lab, Calif, 55-56; Fulbright lectr physics, Univ Ceylon, 56-59; from asst prof to prof, Tex Christian Univ, 59-69; assoc prof dir, NSF, 69, prog dir, 70-71; vis fel, Cornell Univ, 71-72; prof physics, Calif State Col, Bakersfield, 72-75; VIS PROF PHYSICS, POMONA COL, 75- Concurrent Pos: Sr scientist, LTV Res Ctr Div, Ling-Temco-Vought Corp, 63-64. Mem: AAAS; Am Phys Soc; Am Asn Physics Teachers; Nat Asn Res Sci Teaching. Res: Radiation physics; science education. Mailing Add: Dept of Physics Pomona Col Claremont CA 91711

BAGGEROER, ARTHUR BERNARD, electrical engineering, oceanography, see 12th edition

BAGGETT, BILLY, b Oxford, Miss, Oct 23, 28; m 49; c 5. BIOCHEMISTRY, PHARMACOLOGY. Educ: Univ Miss, BA, 47; St Louis Univ, PhD(biochem), 52. Prof Exp: Instr biochem, Harvard Med Sch, 52-57; asst, Mass Gen Hosp, 52-57; from asst prof to prof pharmacol & biochem, Univ NC, Chapel Hill, 57-69; PROF BIOCHEM & CHMN DEPT, MED UNIV SC, 69- Concurrent Pos: USPHS sr res fel, Univ NC, Chapel Hill, 57-59; consult, Res Triangle Inst, 66-71. Mem: AAAS; Endocrine Soc; Am Soc Biol Chem; Am Chem Soc; Am Asn Cancer Res. Res: Metabolism and actions of steroid hormones. Mailing Add: Dept of Biochem Med Univ of SC Charleston SC 29401

BAGGETT, JAMES RONALD, b Boise, Idaho, Apr 24, 28; m 51; c 2. HORTICULTURE. Educ: Univ Idaho, BS, 52; Ore State Col, PhD(hort), 56. Prof Exp: From asst prof to assoc prof, 56-71, PROF HORT, ORE STATE UNIV, 71- Mem: Am Soc Hort Sci. Res: Genetics and breeding of disease resistance in vegetables. Mailing Add: Dept of Hort Ore State Univ Corvallis OR 97331

BAGGETT, LARRY W, b Morehead, Miss, Mar 3, 39; m 64; c 1. MATHEMATICS. Educ: Davidson Col, BS, 60; Univ Wash, Seattle, PhD(math), 66. Prof Exp: Asst prof, 66-69, ASSOC PROF MATH, UNIV COLO, BOULDER, 69- Concurrent Pos: NSF grant, 67-69. Mem: Am Math Soc. Res: Topological groups and their representations. Mailing Add: Dept of Math Univ of Colo Boulder CO 80302

BAGGETT, LESTER MARCHANT, b Oakland, Calif, Nov 28, 20; m 51; c 2. NUCLEAR PHYSICS, HYDRODYNAMICS. Educ: Southwestern at Memphis, BA, 43; Ga Inst Technol, MS, 48; Rice Univ, PhD(physics), 51. Prof Exp: GROUP LEADER, LOS ALAMOS SCI LAB, 51- Mem: Am Phys Soc. Res: Energy levels of light nuclei. Mailing Add: 996 Nambe Pl Los Alamos NM 87544

BAGGOT, JOHN DESMOND, b Dublin, Ireland, Oct 11, 39; m 65; c 2. VETERINARY PHARMACOLOGY. Educ: Nat Univ Ireland, BSc, 62, MVB & MRCVS, 66, MVM, 68; Ohio State Univ, PhD(vet pharmacol), 71. Prof Exp: Lectr vet pharmacol & toxicol, Col Vet Med, Dublin, 66-69; teaching assoc vet physiol & pharmacol, Ohio State Univ, 69-72; sr lectr vet pharmacol, Massey Univ, NZ, 72-74; asst prof, 74-75, ASSOC PROF VET PHARMACOL, OHIO STATE UNIV, 74-, DIR EQUINE PHARMACOL RES, 75- Mem: Am Soc Pharmacol & Exp Therapeut; Australian Physiol & Pharmacol Soc; Am Asn Vet Physiologists & Pharmacologists. Res: Absorption, distribution, biotransformation and excretion of drugs; bioavailability, disposition kinetics and dosage of drugs in domestic animals. Mailing Add: 3731 Kennybrook Lane Columbus OH 43220

BAGGOTT, JAMES PATRICK, b Milwaukee, Wis, Jan 19, 41; m 66; c 5. BIOCHEMISTRY. Educ: Marquette Univ, BS, 62; Johns Hopkins Univ, PhD(biochem), 66. Prof Exp: Sr instr, 68-71, ASST PROF BIOCHEM, HAHNEMANN MED COL, 71- Concurrent Pos: Fel biol chem, Univ Utah, 66-68. Mem: Am Chem Soc. Res: Enzymology and physical biochemistry as applied to enzymological problems; medical education. Mailing Add: Dept of Biol Chem Hahnemann Med Col Philadelphia PA 19102

BAGINSKI, EUGENE S, b Warsaw, Poland, Oct 11, 28; US citizen; m 55; c 3. BIOCHEMISTRY, PHYSIOLOGY. Educ: Wayne State Univ, BS, 56, MS, 64, PhD(physiol), 68. Prof Exp: Res assoc chem, Wayne State Univ, 56-57; chief chemist, Crittenton Gen Hosp, 57-58, Sinai Hosp, 58-68 & Oakwood Hosp, 68-69; CHIEF BIOCHEMIST, ST JOSEPH MERCY HOSP, 69-; ASST PROF PATH, SCH MED, WAYNE STATE UNIV, 69- Mem: Am Chem Soc; Am Asn Clin Chemists. Res: Methodology; enzymology. Mailing Add: St Joseph Mercy Hosp Pontiac MI 48053

BAGLEY, BRIAN G, b Racine, Wis, Nov 20, 34; m 59; c 3. SOLID STATE PHYSICS. Educ: Univ Wis-Madison, BS, 58, MS, 59; Harvard Univ, AM, 64, PhD(appl physics), 68. Prof Exp: Mem res staff, Univ Wis, 59-60; metallurgist, Ladish Co, Wis, 61; MEM TECH STAFF, BELL LABS, INC, MURRAY HILL, NJ, 67- Mem: AAAS; Am Phys Soc. Res: Physics of amorphous materials. Mailing Add: Bell Labs Inc Murray Hill NJ 07974

BAGLEY, GEORGE EVERETT, b Corry, Pa, Feb 14, 33; m 56; c 4. POLYMER CHEMISTRY. Educ: Houghton Col, BS, 54; Univ Pa, PhD(phys chem), 59. Prof Exp: Res chemist, 58-69, res supvr, 69-70, SR RES SCIENTIST, ARMSTRONG CORK CO, 70- Mem: Am Chem Soc; Soc Plastics Engrs. Res: Inorganic synthesis and inorganic polymers; polyvinyl chloride technology; radiation curable polymers. Mailing Add: 2425 Mayfair Dr Lancaster PA 17603

BAGLEY, ROBERT WALLER, b Wesson, Miss, Apr 14, 21; m 44; c 3. MATHEMATICS. Educ: Univ Mich, BS, 47; Tulane Univ, MS, 49; Univ Fla, PhD(math), 54. Prof Exp: Asst prof math, Univ Ky, 54-55; assoc res scientist, Lockheed Missile Systs, 55-58; prof math, Miss Southern Univ, 58-60 & Univ Ala, 60-63; assoc prof, 63-64, PROF MATH, UNIV MIAMI, 64- Concurrent Pos: Consult, Grad Sch Bus, Stanford Univ, 57. Mem: Am Math Soc; Math Asn Am. Res: General topology. Mailing Add: Dept of Math Univ of Miami Coral Gables FL 33146

BAGLEY, WALTER THAINE, b Eckley, Colo, Jan 3, 17; m 42; c 1. FORESTRY, PLANT ECOLOGY. Educ: Colo State Col, BS, 38; Iowa State Col, MS, 40. Prof Exp: With forest serv, USDA, 40-42 & soil conserv serv, 46-54; asst prof, 55-65, ASSOC PROF HORT & FORESTRY, UNIV NEBR, 65- Mem: Soil Conserv Soc Am; Soc Am Foresters; Ecol Soc Am. Res: Ecology of native woodlands and plantations in the Great Plains; influence of windbreaks on micro-climate and crops; tree improvement by selection from provenance plantings. Mailing Add: Dept of Hort Univ of Nebr Lincoln NE 68508

BAGLI, JENANBUX FRAMROZ, b Bombay, India, Sept 25, 28; m 56; c 2. ORGANIC CHEMISTRY. Educ: Univ Bombay, BSc, 49 & 51; Univ London, PhD(pharmaceut chem), 55. Prof Exp: Fel, Johns Hopkins Univ, 55-59; Nat Res Coun Can fel, Laval Univ, 59-60; sr res scientist, 60-69, SR RES ADV, AYERST, McKENNA & HARRISON RES LABS, 69- Mem: Am Chem Soc; The Chem Soc. Res: Elucidation of structure and stereochemistry of natural products; mechanism of the organic reactions and the biogenesis of the steroids and terpenoids. Mailing Add: Ayerst McKenna & Harrison Res Labs PO Box 6115 Montreal PQ Can

BAGLIO, JOSEPH ANTHONY, b New York, NY, May 16, 37; m 60; c 1. PHYSICAL CHEMISTRY. Educ: St John's Univ, NY, BS, 58; Rutgers Univ, PhD(phys chem), 65. Prof Exp: Chemist, AEC, 58-60; chemist phys chem, David Sarnoff Res Ctr, Radio Corp Am, Inc, 60-61; MEM TECH STAFF PHYS CHEM, GEN TEL & ELECTRONICS LABS, INC, 65- Mem: Am Crystallog Asn; NY Acad Sci. Res: Determination of crystal and molecular structures by the methods of x-ray diffraction; correlation of electrical and optical properties of compounds with their structure. Mailing Add: Gen Tel & Electronics Labs Inc 40 Sylvan Rd Waltham MA 02154

BAGLIONI, CORRADO, b Rome, Italy, Aug 1, 33; m 57; c 3. BIOCHEMICAL GENETICS, IMMUNOLOGY. Educ: Univ Rome, MD, 57. Prof Exp: Vis lectr biochem, Mass Inst Technol, 59-61, asst prof genetics, 61-62; Ital Nat Res Coun intern genetics, Biophys Lab, 62-66; from assoc prof to prof biol & chmn dept biol sci, Mass Inst Technol, 73-75; PROF BIOL, STATE UNIV NY ALBANY, 75- Concurrent Pos: Mem dir, Ital Asn Biophys & Molecular Biol, 64-66; mem, Immunol & Allergy Study Sect, NIH, 67- Mem: Int Inst Embryol; Europ Molecular Biol Orgn; Am Soc Immunologists. Res: Human genetics; genetic control of protein structure; action of suppressor genes in Drosophila; human abnormal hemoglobins; mechanism of assembly of hemoglobin and immunoglobulin peptide chains; genetic basis of antibody variability; regulation of histone synthesis. Mailing Add: Bi-126 State Univ of NY 1400 Washington Ave Albany NY 12222

BAGNALL, RICHARD HERBERT, b Hazel Grove, PEI, Feb 23, 23; m 52; c 4. PLANT PATHOLOGY. Educ: Prince of Wales Col, jr col dipl, 44; McGill Univ, BSc, 46, MSc, 49; Univ Wis, PhD, 56. Prof Exp: Agr scientist, 46, res scientist, Plant Path Lab, 46-73, RES SCIENTIST, POTATO PROG, DEPT VIRUS RESISTANCE, CAN AGR RES STA, PEI, 73- Mem: Am Phytopath Soc; Can Phytopath Soc. Res: Potato viruses; virus serology; virus nomenclature; general virology. Mailing Add: Can Agr Res Sta PO Box 280 Fredericton NB Can

BAGNARA, JOSEPH THOMAS, b Rochester, NY, July 26, 29; m 55. EMBRYOLOGY. Educ: Univ Rochester, BA, 52; Univ Iowa, PhD, 56. Prof Exp: Lab instr, Univ Iowa, 52-56, res asst, 55; from instr to prof zool, 56-74, PROF BIOL SCI, UNIV ARIZ, 74- Concurrent Pos: Fulbright res scholar, Univ Paris, 64 & Zool Sta, Naples, 69-70; managing ed, Am Zoologist, Am Soc Zool, 71- Mem: Am Soc Zool; Am Asn Anat; Soc Develop Biol; Am Soc Cell Biol. Res: Experimental embryology of amphibia; endocrinology; physiology of pigmentation. Mailing Add: Dept of Biol Sci Univ of Ariz Tucson AZ 85721

BAGNE, FARIDEH, b Persia, Sept 24, 45; m 68; c 2. MEDICAL PHYSICS. Educ: Mich State Univ, BS, 66; Univ Pa, MS, 67, PhD(physics), 70. Prof Exp: ASST PROF MED PHYSICS, DARTMOUTHHITCHCOCK MED CTR, 72- Concurrent Pos: NIH fel, Thomas Jefferson Univ, 70-72. Honors & Awards: Sci Exhibit Award, Am Asn Physicists in Med, 74. Mem: Am Col Radiol; Am Asn Physicists in Med; Soc Nuclear Med; Health Physics Soc; Radiation Res Soc. Res: Radiological medical physics. Mailing Add: Dept of Radiation Ther Dartmouth-Hitchcock Med Ctr Hanover NH 03755

BAGSHAW, JOSEPH CHARLES, b Niagara Falls, NY, Sept 2, 43; m 71. MOLECULAR BIOLOGY, BIOCHEMISTRY. Educ: Johns Hopkins Univ, BA, 65; Univ Tenn, PhD(biochem), 69. Prof Exp: ASST PROF BIOCHEM, SCH MED, WAYNE STATE UNIV, 71- Concurrent Pos: Res fel, Harvard Med Sch-Mass Gen Hosp, Boston, 70-71. Mem: AAAS; Am Soc Cell Biol; Biophys Soc. Res: Transcriptional regulation in eukaryotes; eukaryotic RNA polymerases; biochemical aspects of development and differentiation. Mailing Add: Dept of Biochem Wayne State Univ Sch of Med Detroit MI 48201

BAGSHAW, MALCOLM A, b Adrian, Mich, June 24, 25; m 48; c 3. RADIOTHERAPY, RADIOBIOLOGY. Educ: Wesleyan Univ, BA, 46; Yale Univ, MD, 50. Prof Exp: From instr to assoc prof, 55-69, PROF RADIOL, SCH MED, STANFORD UNIV, 69-, CHMN DEPT, 72-, DIR DIV RADIOTHER, 60- Concurrent Pos: Consult, Vet Admin Hosp, Palo Alto, 60- Mem: Am Col Radiol; Asn Univ Radiol; Radiol Soc NAm; Radiation Res Soc; Am Soc Therapeut Radiol. Mailing Add: Stanford Univ Sch of Med Stanford CA 94305

BAGUS, PAUL SAUL, b New York, NY, Nov 19, 37; m 65; c 1. ATOMIC PHYSICS, QUANTUM CHEMISTRY. Educ: Univ Chicago, BS & MS, 58, PhD(physics), 65. Prof Exp: Fel, Argonne Nat Lab, 62-64, resident res assoc physics, Solid State Sci Div, 64-66; assoc prof, Fac Sci, Univ Paris, 66-67; res assoc physics, Univ Chicago, 67-68; PROF STAFF MEM, INT BUS MACH RES LAB, 68- Concurrent Pos: Consult, Lockheed Missiles & Space Corp, 62-63. Mem: Am Phys Soc. Res: Ab initio calculation of the electronic structure of atoms and molecules to determine properties of atoms, molecules and solids; application to the theory of x-ray photo-electron spectroscopy and related matters. Mailing Add: IBM Res Lab Monterey & Cottle Rds San Jose CA 95114

BAGWELL, ERVIN EUGENE, b Honea Path, SC, Jan 24, 36; m 58; c 2. PHARMACOLOGY. Educ: Furman Univ, BS, 58; Med Univ SC, MS, 60, PhD, 63. Prof Exp: From asst prof to assoc prof pharmacol, 65-75, PROF PHARMACOL, MED UNIV SC, 75- Concurrent Pos: SC Heart Asn advan res fel pharmacol, Med Univ SC, 63-65; NIH res grant, 66-72. Mem: Am Soc Pharmacol & Exp Therapeut. Res: Autonomic and cardiovascular drugs; catechol amines; coronary flow and myocardial metabolism. Mailing Add: Dept of Pharmacol Med Univ of SC Charleston SC 29401

BAHADUR, RAGHU RAJ, b Delhi, India, Apr 30, 24; m 50; c 2. MATHEMATICAL STATISTICS. Educ: Univ Delhi, BA, 43, MA, 45; Univ NC, Chapel Hill, PhD(math statist), 50. Prof Exp: Instr statist, Univ Chicago, 50-51, asst prof, 54-56; prof, Indian Coun Agr Res, 51-52; asst prof math statist, Columbia Univ, 52-53; prof statist, Indian Statist Inst, 56-61; assoc prof, 62-64, PROF STATIST, UNIV CHICAGO, 65- Mem: Inst Math Statist; Int Statist Inst; Nat Inst Sci India; Indian Soc Agr Statist. Res: Theory of statistical tests and decisions; approximations to classical distributions; large sample theory. Mailing Add: Dept of Statist Eckhart Hall Univ of Chicago Chicago IL 60637

BAHAL, SURENDRA MOHAN, b India, July 1, 35; m 61; c 2. PHYSICAL PHARMACY. Educ: Univ Bombay, BS, 54, BS, 56, MS, 59; Temple Univ, PhD(phys pharm), 65. Prof Exp: Instr pharm, Dept Chem Technol, Univ Bombay, 58-61; res pharmacist, 64-72, SUPVR ORAL LIQUIDS & TOPICAL PROD UNIT, WYETH LABS, INC, 73- Honors & Awards: Lunsford-Richardson Pharm Award, Eastern Region, Richardson Merrell Co, Inc, 65; Pharmaceut Res Discussion Group Award, 70. Mem: Am Pharmaceut Asn; Acad Pharmaceut Sci. Res: Excretion kinetics of drugs as modified by protein binding and renal tubular secretion; antiinfective agents; powder caking; antacid technology; effect of sugars on penicillin degradation kinetics; stabilization of pharmaceutical formulations; drug sterilization techniques and patents. Mailing Add: 1021 Longspur Rd Audubon PA 19407

BAHAM, ARNOLD, b Folsom, La, Oct 7, 43; m 68; c 2. DAIRY NUTRITION. Educ: La State Univ, Baton Rouge, BS, 66, MS, 68; Auburn Univ, PhD(dairy nutrit), 71. Prof Exp: Asst dairy sci, La State Univ, 66-68; asst dairy nutrit, Auburn Univ, 68-71; asst prof agr, Southeastern La Univ, 71-74; ASST PROF DAIRY SCI, LA STATE UNIV, 74- Mem: Am Dairy Sci Asn; Nat Asn Animal Breeders. Res: Nutrition and reproduction of dairy animals. Mailing Add: Box B D La State Univ Baton Rouge LA 70803

BAHAR, LEON Y, b Turkey, Apr 4, 28; US citizen; m 65; c 3. APPLIED MECHANICS. Educ: Robert Col, Istanbul, BS, 50; Lehigh Univ, MS, 59, PhD(mech), 63. Prof Exp: Instr mech, Lehigh Univ, 57-63; asst prof eng, City Univ New York, 63-66; ASSOC PROF APPL MECH, DREXEL UNIV, 66- Concurrent Pos: Consult, Data Processing Div, IBM Corp, NY, 65, United Engrs & Constructors, Inc, 74-75, Gulf Res & Develop Corp & Western Res & Develop, 74. Mem: Am Soc Mech Engrs; Am Acad Mech. Res: State space approach to boundary value problems; dynamic response of nuclear power plants; application of integral transforms to engineering. Mailing Add: Dept of Mech Eng & Mech Drexel Univ Philadelphia PA 19104

BAHARY, WILLIAM S, b Kermanshah, Iran, Jan 20, 36; US citizen. BIOPHYSICAL CHEMISTRY. Educ: Harvard Univ, AB, 57; Columbia Univ, AM, 58, PhD(chem), 61. Prof Exp: Sr res chemist, Tex-US Chem Co, 61-68; vis asst prof chem, Farleigh Dickinson Univ, 68-73; res scientist, 73-75, ADJ ASST PROF CHEM, STEVENS INST TECHNOL, 75-; CONSULT CHEMIST, WALLACE CLARK CO, 75- Mem: AAAS; Am Chem Soc; NY Acad Sci; Am Soc Microbiol. Res: Structure and properties of synthetic and biological macromolecules; viscoelasticity of concentrated polymer solutions; structure and function of polysaccharides of dental plaque; chemistry of biomimetic blood anticoagulation. Mailing Add: 291 N Middletown Rd Pearl River NY 10965

BAHCALL, JOHN NORRIS, b Shreveport, La, Dec 30, 34; m 66; c 2. ASTROPHYSICS. Educ: Univ Calif, Berkeley, BA, 56; Univ Chicago, MA, 57; Harvard Univ, PhD(physics), 61. Prof Exp: Res fel physics, Ind Univ, 60-62; res fel physics, Calif Inst Technol, 62-63, sr res fel, 63-65, from asst prof to assoc prof theoret physics, 65-68; mem, 68-70, PROF THEORET PHYSICS, INST ADVAN STUDY, 71- Honors & Awards: Warner Prize, Am Astron Soc, 70. Mem: Nat Acad Sci; fel Am Phys Soc. Mailing Add: Inst for Advan Study Princeton NJ 08540

BAHCALL, NETA ASSAF, b Tel-Aviv, Israel, Dec 16, 42; m 66; c 2. ASTROPHYSICS. Educ: Hebrew Univ, Jerusalem, BSc, 63; Weizmann Inst Sci, Israel, MSc, 65; Tel-Aviv Univ, PhD(astrophys), 70. Prof Exp: Res asst astrophys, Calif Inst Technol, 65-66, res fel, 70-71; res assoc, 71-74, res staff mem, 74-75, RES ASTRONR, PRINCETON UNIV OBSERV, 75- Mem: Am Astron Soc. Res: Galactic and extra-galactic astronomy; optical properties of galaxies and clusters of galaxies; optical identification and properties of x-ray sources. Mailing Add: 127 Peyton Hall Princeton Univ Princeton NJ 08540

BAHE, LOWELL W, b Sycamore, Ill, Jan 30, 27; m 54; c 3. PHYSICAL CHEMISTRY. Educ: Purdue Univ, BS, 49; Princeton Univ, AM, 51, PhD(chem), 52. Prof Exp: Res chemist, Allis-Chalmers Mfg Co, 53-57; from asst prof to assoc prof phys chem, 57-65, PROF PHYS CHEM, UNIV WIS-MILWAUKEE, 65- Mem: AAAS; Am Chem Soc. Res: Electrolytic solutions in aqueous and nonaqueous solvents; thermodynamics of solutions; dielectric constants. Mailing Add: Dept of Chem Univ of Wis Milwaukee WI 53201

BAHILL, ANDREW TERRY, b Washington, Pa, Jan 31, 46; m 71. BIOENGINEERING, NEUROSCIENCE. Educ: Univ Ariz, BSEE, 67; San Jose State Univ, MSEE, 70; Univ Calif, Berkeley, PhD(elec eng & comput sci), 75. Prof Exp: ASST PROF BIOENG, CARNEGIE-MELLON UNIV, 76- Concurrent Pos: Consult, Solid State Commun, Fargo Co & Eastergraves Co, 72-74. Mem: Inst Elec & Electronics Engrs; Sigma Xi; Soc Neurosci. Res: Neurological control of human movement studied primarily through the window of eye movements, utilizing measurements on normal subjects and clinical patients, instrumentation, modeling and data reduction. Mailing Add: Biotechnol Prog Carnegie-Mellon Univ Pittsburgh PA 15213

BAHL, OM PARKASH, b Lyallpur, Punjab, Pakistan, Jan 10, 27; m 52; c 3. CHEMISTRY. Educ: Punjab Univ, India, MSc, 50; Univ Minn, PhD, 62. Prof Exp: Lectr chem, Arya Col, Ludhiana, India, 50-52 & Govt Col, 52-57; res assoc biochem, Univ Minn, 62-63 & Univ Calif, 63-64; Am Cancer Soc career investr & Dernham fel, Univ Southern Calif, 64-65, asst prof biochem, 65-66; from asst prof to assoc prof, 66-71, PROF BIOCHEM, STATE UNIV NY BUFFALO, 71- Concurrent Pos: Nat Found res grant, 65-66; Am Cancer Soc career develop award, 65-66; USPHS res grant, 66 & 69-72. Mem: Am Chem Soc; Am Soc Biol Chem; Brit Chem Soc; Brit Biochem Soc. Res: Proteins and enzymes; polysaccharides; glycoproteins. Mailing Add: Dept of Biochem State Univ NY 3435 Main St Buffalo NY 14214

BAHLER, THOMAS LEE, b Walnut Creek, Ohio, Feb 1, 20; m 48; c 2. ZOOLOGY. Educ: Col Wooster, BA, 43; Univ Wis, PhD(zool), 49. Prof Exp: From res asst to teaching asst, Univ Wis, 44-49; from asst prof to assoc prof, 49-59, prof zool & physiol, 59-73, PROF BIOL, UTAH STATE UNIV, 73- Concurrent Pos: NSF fac fel, 58-59. Res: Physiology of parasites; physiological effects of insecticides; physiology of eosinophils. Mailing Add: Dept of Zool Utah State Univ Logan UT 84321

BAHN, ARTHUR NATHANIEL, b Boston, Mass, Jan 5, 26; m 52; c 2. MICROBIOLOGY, DENTAL RESEARCH. Educ: Boston Univ, AB, 49; Univ Kans, MA, 52; Univ Wis, PhD(bact), 56. Prof Exp: Asst bacteriologist, Kans State Bd Health, 51-52; asst bact, Univ Wis, 55-66; instr, Col Med, Univ Ill, 56-69; asst prof, Sch Dent, Northwestern Univ, 59-63; assoc prof bact, Sch Dent & assoc prof microbiol, Sch Med, 63-71; PROF MICROBIOL & CHMN DEPT, SCH DENT MED, SOUTHERN ILL UNIV, ALTON, 71- Concurrent Pos: Proj asst, Univ Wis, 52-56; mem, Am Bd Med Microbiol, 70-; consult, Food & Drug Admin Oral Cavities Prod Panel, 74- Mem: AAAS; Am Soc Microbiol; Int Asn Dent Res. Res: Resistance to infection; immunochemistry, particularly of streptococcal enzymes; oral streptococci; experimental endocarditis; viruses. Mailing Add: Dept of Microbiol So Ill Univ Sch of Dent Med Alton IL 62002

BAHN, EMIL LAWRENCE, JR, b Cape Girardeau, Mo, Sept 6, 24; m 49; c 2. INORGANIC CHEMISTRY, NUCLEAR CHEMISTRY. Educ: Mo Sch Mines, BSChE, 46; Washington Univ, MSChE, 49, MA, 61, PhD(chem), 62. Prof Exp: Head dept, 62-73, PROF CHEM, SOUTHEAST MO STATE COL, 55- Mem: AAAS; Am Chem Soc; Am Nuclear Soc; Health Physics Soc; Soc Appl Spectros. Res: X-ray fluorescence; trace elements emitted from coal burning; utilization of drinking water and waste water sludges. Mailing Add: Rte 1 Box 141 Cape Girardeau MO 63701

BAHN, ROBERT CARLTON, b Newark, NY, July 24, 25; m 49; c 4. PATHOLOGY. Educ: Univ Buffalo, MD, 47; Univ Minn, PhD(path), 53; Am Bd Path, dipl path anat, 53, cert clin path, 54 & cert neuropath, 59. Prof Exp: Asst to staff, Mayo Clin, 53; sr asst surgeon, NIH, 54-55; CONSULT SECT PATH ANAT, MAYO CLIN, 56-, PROF PATH, 69- Concurrent Pos: Instr grad sch, Univ Minn & Mayo Med Sch, 57-; from asst prof to assoc prof path, Mayo Med Sch, 59-69; mem spec interest group biomed comput. Honors & Awards: Heinrich Leonard Award, Univ Buffalo, 47; Mayo Alumni Award Meritorious Res, 54. Mem: Am Soc Exp Biol & Med; Am Physiol Soc; Endocrine Soc; Histochem Soc; Biol Stain Comn. Res: Cytophysiology of hypothalmus and pituitary; human and experimental pituitary neoplasms; medical applications of digital computers; biomedical digital image reconstruction and analysis. Mailing Add: Dept of Path Mayo Clin Rochester MN 55901

BAHNER, CARL TABB, b Conway, Ark, July 14, 08; m 31; c 3. ORGANIC CHEMISTRY, BIOMEDICAL ENGINEERING. Educ: Hendrix Col, AB, 27; Univ Chicago, MS, 28; Southern Baptist Theol Sem, ThM, 31; Columbia Univ, PhD(chem eng), 36. Prof Exp: Asst prof chem & head dept physics, Union Univ, Tenn, 36-37; prof chem, Carson-Newman Col, 37-73, res coordr, 67-73; ASSOC PROF CHEM, WALTERS STATE COMMUNITY COL, 73- Concurrent Pos: Consult, Tenn Valley Authority, 42-45, Carbide & Carbon Chem Corp, 48- & Oak Ridge Inst, 50-55. Mem: AAAS; Am Chem Soc; Am Asn Cancer Res; fel Am Inst Chem. Res: Reactions of aliphatic nitro compounds and of organic halides; amines; heterocyclic nitrogen compounds; protein extraction; immersion freezing solutions; hydrotropic solutions; organic complexes of radioisotopes; cancer chemotherapy. Mailing Add: PO Box 549 Jefferson City TN 37760

BAHNG, JOHN DEUCK RYONG, b Bookchung, Korea, Mar 21, 27; US citizen; m 61; c 3. ASTRONOMY. Educ: St Norbert Col, BS, 50; Univ Wis, MS, 54, PhD(astron), 57. Prof Exp: Res assoc astron, Princeton Univ, 57-62; asst prof, 62-66, ASSOC PROF ASTRON, NORTHWESTERN UNIV, 66-, CHMN DEPT, 75- Mem: Am Astron Soc; Royal Astron Soc; Int Astron Union. Res: Photoelectric spectrophotometry of stars; stellar structure and evolution. Mailing Add: Dearborn Observ Northwestern Univ Evanston IL 60201

BAHR, GUNTER F, b Hamburg, Ger, Oct 25, 22; US citizen; m 60; c 2. BIOPHYSICS, PATHOLOGY. Educ: Univ Würzburg, MD, 52; Karolinska Inst, MD, Sweden, 57. Prof Exp: Asst prof cell res, Nobel Inst Cell Res, Karolinska Inst, Sweden, 50-57, asst prof path, Inst Path, 57-60; chief biophys br, 60-74, CHMN DEPT CYTOL & CELL PATH, ARMED FORCES INST PATH, 74- Concurrent Pos: Consult, Int Acad Cytol, 58-; vis prof, Northwestern Univ, 58; clin prof, Georgetown Univ, 63-; mem, NIH Comput & Biomath Sci Study Sect, 74-; registr, Cytol Registry, Am Registry Path, 74- Honors & Awards: Maurice Goldblatt Award, Int Acad Cytol, 66; Army Res & Develop Award, 67; Exceptional Civilian Serv Award, US Army, 69. Mem: Electron Micros Soc Am; Am Soc Cell Biol; Am Soc Exp Path; Histochem Soc; Int Acad Path. Res: Development of quantitative electron microscopy for the determination of dry-masses of biological objects; chromosome structure; pattern recognition of cell images. Mailing Add: 3206 Chestnut St NW Washington DC 20015

BAHR, GUSTAVE KARL, b Chicago, Ill, Mar 7, 29; m 61; c 2. RADIOLOGICAL PHYSICS, NUCLEAR MEDICINE. Educ: Xavier Univ, Ohio, BS, 51; Univ Cincinnati, MS, 59, PhD(physics), 64; Am Bd Radiol, dipl, 66. Prof Exp: Instrumentation engr, Keleket Instrument Co, Ohio, 51-53; exp physicist, Ohmart Corp, 53-54; res asst appl mech, Univ Cincinnati, 54-56; tech engr, Gas Turbine Div, Gen Elec Co, 56-58; radiol physicist, 58-59, from instr to asst prof, 59-70, ASSOC PROF RADIOL, COL MED, UNIV CINCINNATI, 70- Concurrent Pos: Adj prof physics, Xavier Univ Ohio, 65-; lectr eng physics, Eve Col, Univ Cincinnati, 66-; consult, Bur Dis Prev & Environ Control, Nat Ctr Radiol Health, 66-, Christian Holmes Hosp & Good Samaritan Hosp, Cincinnati, 68- Mem: AAAS; Am Asn Physicists in Med; Am Col Radiol. Res: Application of digital computer techniques to radiation therapy planning and nuclear medicine diagnostic techniques; application of linear programming to radiation therapy; application of stochastic simulation of radiation to radiologic physics problems. Mailing Add: Radioisotope Lab Cincinnati Gen Hosp Cincinnati OH 45229

BAHR, JAMES THEODORE, b East Orange, NJ, Nov 19, 42; m 66; c 1. BIOPHYSICAL CHEMISTRY. Educ: Mass Inst Technol, BS, 64; Univ Pa, PhD(biophys), 71. Prof Exp: Res assoc, 71-75, RES ASSOC BIOCHEM, UNIV ARIZ, 75- Mem: Am Soc Plant Physiologists; Japanese Soc Plant Physiologists. Res: Regulation of photosynthetic carbon metabolism; regulation and mechanism of ribulose diphosphate carboxylase-oxygenase; cyanide-insensitive respiration in plant mitochondria; electron transport in plant microsomes. Mailing Add: Dept of Chem Univ Ariz Tucson AZ 85721

BAHR, THOMAS GORDON, b La Crosse, Wis, Apr 17, 40; m 60; c 1. LIMNOLOGY. Educ: Univ Idaho, BS, 63; Mich State Univ, MS, 66, PhD(limnol, physiol), 68. Prof Exp: Asst prof zool, Colo State Univ, 68-70; asst dir, 70-73, DIR INST WATER RES, MICH STATE UNIV, 73- Mem: AAAS; Am Fisheries Soc; Am Soc Limnol & Oceanog; Int Soc Theoret & Appl Limnol. Res: Applied limnology; fish physiology; nutrient cycling; waste water management; dynamics of toxic materials in natural systems. Mailing Add: Inst of Water Res Mich State Univ East Lansing MI 48823

BAIAMONTE, VERNON D, b Greeley, Colo, Apr 29, 34; m 58; c 4. PHYSICAL CHEMISTRY. Educ: Colo State Col, AB, 59, MA, 60; Univ Ind, PhD(phys chem), 66. Prof Exp: Instr chem, Ball State Univ, 60-62; fel phys chem, Los Alamos Sci Lab, 66-67; assoc prof, 67-69, PROF CHEM, MO SOUTHERN STATE COL, 70-, HEAD DEPT, 69- Mem: Am Chem Soc. Res: Kinetic spectroscopy, especially the fast reactions initiated by flash photolysis and detected by spectroscopic methods. Mailing Add: Dept of Chem Mo Southern State Col Joplin MO 64801

BAIARDI, JOHN CHARLES, b Brooklyn, NY, Feb 9, 18; m 43; c 2. HEMATOLOGY, PHYSIOLOGY. Educ: St Francis Col, NY, BS, 40; Brooklyn Col, MA, 43; NY Univ, PhD(biol), 53. Prof Exp: Instr biol, St Francis Col, NY, 42-43, asst prof & chmn dept, 46-48; assoc prof, St John's Univ, NY, 48-54; prof biol, Long Island Univ, 54-70, chmn dept, 54-62, assoc dean sci, 58-62, vpres & provost, 62-67, vchancellor, 67-70; PRES, AFFIL COLS & UNIVS, INC, 70-; DIR, NY OCEAN SCI LAB, MONTAUK, 70- Concurrent Pos: Mem nat adv bd, Cooley's Anemia Found, NY, vchmn, 63-; mem, Int Oceanog Found. Mem: Fel NY Acad Sci; Harvey Soc. Res: Research administration; blood and hemopoietic organs of mammals, fish and invertebrates. Mailing Add: 16 Theodore Dr Plainview NY 11848

BAIC, DUSAN, b Zagreb, Yugoslavia, Feb 11, 18; US citizen; m 54. EXPERIMENTAL MORPHOLOGY. Educ: Univ Zagreb, MD, 52. Prof Exp: Asst prof histochem, Pathophysiol Inst, Sch Med, Univ Zagreb, 53-61; lectr exp path, Inst Exp Med & Surg, Univ Montreal, 64-67; LECTR ELECTRON MICROS, DIV BIOL SCI, UNIV MICH, 67- Concurrent Pos: Head histochem lab, Rudjer Boskovic Inst, Zagreb, 55-56; trainee anat, Univ Chicago, 61-63; vis scientist, Dept Zool, Univ Mich, 63-64. Mem: Electron Micros Soc Am. Res: Ultrastructure of liver under experimental conditions; ultrastructure of cadmium-induced lesions in the nervous system. Mailing Add: Div of Biol Sci Univ of Mich Ann Arbor MI 48109

BAICH, ANNETTE, b Chicago, Ill, Mar 2, 30; m 50. BIOCHEMISTRY. Educ: Roosevelt Univ, BS, 51; Univ Ore, MS, 54, PhD(chem), 60. Prof Exp: Asst chem, Univ Ore, 51-54, instr, 55-56, asst, 56-57; lab technician, Agr Exp Sta, Univ Calif, 54-55; fel biochem, Ore State Univ, 60-61; fel microbiol, Rutgers Univ, 61-62; asst prof biochem, Ore State Univ, 63-69; assoc prof, 69-74, PROF BIOCHEM, SOUTHERN ILL UNIV, EDWARDSVILLE, 74- Concurrent Pos: NIH career develop award, 65-69. Mem: AAAS; Am Chem Soc; Am Soc Microbiol. Res: The relation and response of an organism to its environment and how this relationship may be explained in chemical terms. Mailing Add: Dept of Biol Sci Southern Illinois Univ Edwardsville IL 62025

BAIDINS, ANDREJS, b Dauguli, Latvia, Dec 16, 30; US citizen; m 60; c 2. PHYSICAL CHEMISTRY. Educ: Franklin & Marshall Col, BS, 54; Rutgers Univ, PhD(phys chem), 58. Prof Exp: CHEMIST, RES DIV, PIGMENTS DEPT, E I DU PONT DE NEMOURS & CO, INC, 58- Concurrent Pos: Abstractor, Chem Abstr Serv, 59- Mem: AAAS; Am Chem Soc. Res: Kinetics of decomposition of light and heat sensitive compounds; physical and chemical properties of inorganic crystal surfaces and or liquid-liquid interfaces. Mailing Add: Exp Sta 335/136 Pigments Dept E I du Pont de Nemours & Co Inc Wilmington DE 19898

BAIER, EDWARD JOHN, b Pittsburgh, Pa, Apr 1, 25; m 47; c 2. INDUSTRIAL HYGIENE. Educ: Univ Pittsburgh, BS, 46, MPH, 55; Am Bd Indust Hyg, dipl, 61, cert comprehensive pract indust hyg; Bd Cert Safety Professionals, cert. Prof Exp: Res chemist, Gulf Res & Develop Co, 46; asst indust hygienist, Pa Dept Health & Dept Environ Resources, 46-48, dist indust hygienist, 48-54, regional indust hygienist, 55-56, chief indust hyg sect, 56-68, dir div occup health, 68-72; DEP DIR, NAT INST OCCUP SAFETY & HEALTH, DEPT HEALTH, EDUC & WELFARE, 72- Concurrent Pos: Lectr, Pa State Univ, 48- & Sch Pub Health, Univ Pittsburgh, 58-; dir bd, Am Bd Indust Hyg, 71-, dir, 70-76; adj prof, Drexel Inst, Capitol Campus Grad Ctr & Temple Univ; mem conf, State Sanit Eng; mem secy's coal mine health res adv coun, adv comt nat surveillance network & adv group on comprehensive environ health planning, US Dept Health, Educ & Welfare. Mem: Am Conf Govt Indust Hygienists; Am Indust Hyg Asn (secy, 69-72, vpres, 73-74, pres, 75-76); Am Acad Indust Hyg. Res: Administration of occupational health and safety program. Mailing Add: Nat Inst Occup Safety & Health 5600 Fishers Lane Rockville MD 20852

BAIER, HAROLD LEWIS, immunology, see 12th edition

BAIER, JOSEPH GEORGE, b New Brunswick, NJ, Feb 12, 08; m 29; c 3. IMMUNOBIOLOGY. Educ: Rutgers Univ, BS, 28, MS, 29; Univ Wis, PhD(zool), 32. Prof Exp: Asst zool, Rutgers Univ, 28-29; asst, Univ Wis, 29-32, from instr to prof, 32-68, dean col lett & sci, 56-66, Michael F Guyer Prof, 69-75, EMER MICHAEL F GUYER PROF ZOOL, UNIV WIS-MILWAUKEE, 75- Mem: Fel AAAS; Am Soc Zoologists; fel Linnaean Soc London. Res: Immunotaxonomy; quantitative studies on precipitins; electronic instrumentation; photoelectric densitometry; horological research. Mailing Add: Dept of Zool Univ of Wis-Milwaukee Milwaukee WI 53201

BAIER, ROBERT EDWARD, b Buffalo, NY, Oct 31, 39; m 61; c 2. SURFACE CHEMISTRY, BIOPHYSICS. Educ: Cleveland State Univ, BES, 62; State Univ NY Buffalo, PhD(biophys), 66; Environ Eng Intersoc Bd, dipl. Prof Exp: Res asst, Buffalo Gen Hosp, 59-60; nucleonics engr, Electromech Res Lab, Repub Steel Corp, 61-62; res chemist & Nat Res Coun-Nat Acad Sci fel surface chem, Naval Res Lab, 66-68; prin physicist, Cornell Aeronaut Lab, 68-71, head chem sci sect, 71-74; MEM TECH STAFF, CALSPAN CORP, 74- Concurrent Pos: Guest worker, Nat Heart & Lung Inst, 68; consult, Roswell Park Mem Inst, NY, 70-; mem, Adv Bd, Prog Biomed Eng, Clemson Univ, 70-; chmn, Columbia Univ Biomat Sem, 71-72; adj assoc prof chem eng, Cornell Univ, 71-; res asst prof biophys, State Univ NY Buffalo, 71-, Dent Mat, 75- Honors & Awards: Union Carbide Chem Award, Am Chem Soc, 71. Mem: Soc Biomat; Am Acad Environ Engrs; Biophys Soc; Am Chem Soc; Am Soc Cell Biol. Res: Interdisciplinary studies in surface chemistry and physics of materials at liquid-gas, solid-gas and solid-liquid interfaces, including biomedical and environmental systems. Mailing Add: Environ & Energy Systs Calspan Corp PO Box 235 Buffalo NY 14221

BAIER, WOLFGANG, b Liegnitz, Ger, Dec 10, 23; nat US; m 50; c 2. AGRICULTURAL METEOROLOGY. Educ: Hohenheim Agr Univ, Dipl, 49, Dr Agr, 52; Univ Pretoria, MSc, 64. Prof Exp: Officer in charge agrometeorol sect, Agr Res Inst Highveld Region, SAfrican Dept Agr, 55-64; res scientist agrometeorol sect, Plant Res Inst, 64-69, chief sect, Res Br, 69-73, HEAD AGROMETEOROL RES & SERV, AGROMETEOROL SECT, CHEM & BIOL RES INST, CAN DEPT AGR, 73- Concurrent Pos: Secy, Can Comt Agr Meteorol; pres, Comn Agr Meteorol, World Meteorol Orgn, 71- Mem: Am Meteorol Soc; Can Meteorol Soc; Int Soc Biometeorol. Res: Problems related to the impact of weather and climate on agricultural crops and production; development and application of computer models to evaluate plant-weather relationships. Mailing Add: Chem & Biol Res Inst Res Br Can Dept of Agr Ottawa ON Can

BAIERLEIN, RALPH FREDERICK, b Springfield, Mass, Dec 19, 36; m 65; c 2. COSMOLOGY. Educ: Harvard Univ, AB, 58; Princeton Univ, PhD(gen relativity), 62. Prof Exp: Instr physics, Harvard Univ, 62-65; from asst prof to assoc prof, 66-74, PROF PHYSICS, WESLEYAN UNIV, 74-, PHYSICIST, 69- Concurrent Pos: Res assoc, Smithsonian Astrophys Observ, 65-66. Mem: Am Phys Soc. Res: Cosmology and gravitational radiation. Mailing Add: Dept of Physics Wesleyan Univ Middletown CT 06457

BAIG, MIRZA MANSOOR, b Jabulpar, India, June 16, 42; m 72; c 1. BIOCHEMISTRY. Educ: Univ Karachi, BS, 60, MS, 62; State Univ NY Buffalo, PhD(biol), 69. Prof Exp: Lectr biol, Jamia Col, Karachi, 62-64; fel chem, Fla State Univ, 69-70; fel internal med & biol chem, Univ Mich, Ann Arbor, 70-71; fel, Dept Biochem & Pediat, 72-75, ASST PROF PEDIAT, COL MED, UNIV FLA, 75- Mem: Am Chem Soc; Biochem Soc; AAAS. Res: Chemistry and biology of complex carbohydrates; glycoproteins and glycosamynoglycans. Mailing Add: Dept of Pediat Col of Med Univ of Fla Gainesville FL 32601

BAILAR, BARBARA ANN, b Monroe, Mich, Nov 24, 35; m 66; c 2. STATISTICS. Educ: State Univ NY Albany, AB, 56; Va Polytech Inst, MS, 65; Am Univ, PhD(statist), 72. Prof Exp: Math statistician, 58-72, CHIEF, RES CTR MEASUREMENT METHODS, BUR CENSUS, 72- Concurrent Pos: Chmn rev comt for grant to Am Statist Asn from NSF, 75-76; chmn dept math & statist, USDA Grad Sch, 75- Mem: Fel Am Statist Asn; Pop Asn Am; Int Asn Surv Statisticians. Res: Effect of different methods of measurement on final published census statistics. Mailing Add: Rm 3555 Bldg 3 Bur of the Census Washington DC 20233

BAILAR, JOHN CHRISTIAN, JR, b Golden, Colo, May 27, 04; m 31; c 2. INORGANIC CHEMISTRY. Educ: Univ Colo, AB, 24; Univ Mich, PhD(org chem), 28. Hon Degrees: DSc, Univ Colo, 59 & Univ Buffalo, 59. Prof Exp: Asst chem, Univ Mich, 26-28; from instr to prof, 28-74, secy dept, 37-54, EMER PROF CHEM, UNIV ILL, URBANA, 74- Concurrent Pos: With Nat Defense Res Comt, 44; treas, Int Union Pure & Appl Chem, 63-71. Honors & Awards: Chem Educ Award, Am Chem Soc, 61, Priestley Medal, 64, Distinguished Serv Award, 72; John R Kuebler Award, Alpha Chi Sigma, 62; Frank P Dwyer Medal, Chem Soc NSW, 65; Alfred Werner Gold Medal, Swiss Chem Soc, 66. Mem: Am Chem Soc (pres, 59); fel Am Inst Chemists. Res: Complex inorganic ions, including stereochemistry; polymerization through coordination; valence stabilization through coordination; selective catalyses by complex compounds. Mailing Add: Dept of Chem Univ of Ill Urbana IL 61801

BAILAR, JOHN CHRISTIAN, III, b Urbana, Ill, Oct 9, 32; m 66; c 4. RESEARCH ADMINISTRATION, BIOSTATISTICS. Educ: Univ Colo, BA, 53; Yale Univ, MD, 55; Am Univ, PhD, 73. Prof Exp: Field investr cancer res, Nat Cancer Inst, 56-62, head demog sect, 62-70, dir, Third Nat Cancer Surv, 67-70; dir, res serv, US Vet Admin, 70-72; dep assoc dir cancer control, 72-74, SR CONSULT COOP STUDIES & ED-IN-CHIEF, JOUR, NAT CANCER INST, 74- Concurrent Pos: Lectr, Sch Med, Yale Univ, 59-; assoc ed, J Nat Cancer Inst, 64-66; instr, Grad Sch, USDA, 66-; assoc ed, Cancer Res, 68-72; mem, Task Force Cancer of Prostate & Bladder, 68-70 & Clin Cancer Training Grants Comt, 69-73; vis prof, State Univ NY Buffalo, 74-; instr, George Washington Univ, 75- Mem: Am Asn Cancer Res; fel Am Statist Asn; Biomet Soc; Inst Math Statist; fel Am Pub Health Asn. Res: Administration of medical research; cancer epidemiology; randomized clinical trials; development of related statistical methods and theory. Mailing Add: Nat Cancer Inst Bethesda MD 20014

BAILDON, JOHN DAVID, b Johnstown, Pa, Nov 14, 43; m 68; c 2. GEOMETRY, TOPOLOGY. Educ: Lafayette Col, BS, 65; Rutgers Univ, MS, 67; State Univ NY Binghamton, PhD(math), 71. Prof Exp: Instr, 70-71, ASST PROF MATH, PA STATE UNIV, WORTHINGTON SCRANTON CAMPUS, 71- Mem: Sigma Xi; Am Math Soc; Math Asn Am. Res: Studies in convexity on visibility problems concerning conditions under which sets will be star shaped or unions of star shaped sets; light open maps on 2-manifolds. Mailing Add: Worthington Scranton Campus Pa State Univ Dunmore PA 18512

BAILE, CLIFTON AUGUSTUS, III, b Warrensburg, Mo, Feb 8, 40; m 60. NUTRITION, PHYSIOLOGY. Educ: Cent Mo State Col, BS, 62; Univ Mo, PhD(nutrit), 64. Prof Exp: NIH fel, Sch Pub Health, Harvard Univ, 64-66, from instr to asst prof nutrit, 66-71; sr investr res & develop, Animal Health Prod Div, Smith Kline & French Labs, 71-73, mgr neurobiol res, 73-75; ASSOC PROF NUTRIT, DEPT CLIN STUDIES, SCH VET MED, UNIV PA, 75- Concurrent Pos: Lectr, Univ Pa & res assoc, Monell Chem Senses Ctr, 71-75; adj assoc prof, Pa State Univ, 71-75. Mem: Am Physiol Soc; Am Inst Nutrit; Am Dairy Sci Asn; Am Pub Health Asn. Res: Regulatory mechanisms for control of food intake and energy balance. Mailing Add: Sch Vet Med Univ of Pa New Bolton Ctr RD 1 Kennett Square PA 19348

BAILES, RICHARD HAZEL, b Hinton, WVa, Apr 10, 15; m 39; c 4. PHYSICAL CHEMISTRY, ORGANIC CHEMISTRY. Educ: Antioch Col, BS, 38; Univ Calif, PhD(chem), 42. Prof Exp: Instr chem, Antioch Col, 38-39; res assoc, Nat Defense Res Comt proj, Univ Calif, 41-44; instr chem, Boston Univ, 44-47; asst prof, St Lawrence Univ, 47-48; res supvr, Dow Chem Co, 48-74; RETIRED. Mem: Am Chem Soc. Res: Synthesis, structure and stability of metal chelates; organic synthesis; reduction of metal chelate compounds. Mailing Add: 320 Muller Rd Walnut Creek CA 94598

BAILEY, ALAN JAMES, b Tacoma, Wash, Feb 16, 09. WOOD CHEMISTRY. Educ: Univ Wash, BS, 33, MS, 34, PhD(wood chem), 36. Prof Exp: Supvry technician, Bur Plant Indust, USDA, 34; in charge wood technol, Inst Paper Chem, Lawrence Col, 36-37; asst prof forestry, Univ Minn, 37-39; assoc prof lignin & cellulose res, Univ Wash, 39-51; CONSULT, 51- Res: Wood chemistry and technology; chemical technology. Mailing Add: Univ Sta PO Box 5122 Seattle WA 98105

BAILEY, ALFRED MARSHALL, b Iowa City, Iowa, Feb 18, 94; m 17; c 2. ORNITHOLOGY, MAMMALOGY. Educ: Univ Iowa, AB, 16; Univ Denver, DPS, 54. Hon Degrees: DSc, Norwich Univ, 44. Prof Exp: Cur birds & mammals, La State Mus, 16-19; rep bur biol surv, USDA, Alaska, 19-21; cur birds & mammals, Denver Mus Natural Hist, 21-26; cur zool, Field Mus Natural Hist, 26-27; dir, Chicago Acad Sci, 27-36; dir, 36-70, EMER DIR, DENVER MUS NATURAL HIST, 70- Concurrent Pos: Leader Arctic exped, Colo Mus Natural Hist, 21-22; mem Abyssinian exped, Field Mus Natural Hist-Chicago Daily News, 26-27; leader Denver Mus exped, Mex, 41, Alaska, 45, Labrador, 46, mid-Pac, Australia, N & Fiji, 49, 52 & 54, Campbell Island, subantarctic of NZ, 58-59 & Ecuador & Galapagos Islands, 60; mem SAfrica exped, Rhodesia & Botswana, 69. Mem: AAAS; fel Am Ornith Union; Cooper Ornith Soc; Wilson Ornith Soc; Am Soc Mammal; Am Forestry Asn. Res: Museum exhibits; ecology; birds of arctic Alaska and Colorado. Mailing Add: 4340 Mount View Blvd Denver CO 80207

BAILEY, ALSON HUNNICUTT, mathematics, see 12th edition

BAILEY, ARTHUR W, b Chilliwack, BC, July 22, 38; m 62; c 2. RANGE SCIENCE. Educ: Univ BC, BSA, 60; Ore State Univ, MS, 63, PhD(range mgt), 66. Prof Exp: Asst prof, 66-71, ASSOC PROF PLANT SCI, UNIV ALTA, 71- Concurrent Pos: Grants, Can Dept Agr, 66-, Alta Agr Res Trust, 67- & Nat Res Coun Can, 69- Mem: Soc Range Mgt; Agr Inst Can; Wildlife Soc; Am Soc Agron. Res: Range ecology; synecology of native and cultivated grazing lands; brush control; game habitat manipulation; rangeland fire; grazing effects on vegetation. Mailing Add: Dept of Plant Sci Univ of Alta Edmonton AB Can

BAILEY, BYRON JAMES, b Oklahoma City, Okla, Apr 5, 34; m 57; c 5. OTOLARYNGOLOGY. Educ: Univ Okla, BA, 55, MD, 59. Prof Exp: Surg intern, Univ Calif, Los Angeles, 59-60, asst resident gen surg, 60-61, resident head & neck surg, 61-62; ASSOC PROF OTOLARYNGOL, WEISS PROF & CHMN DEPT, UNIV TEX MED BR GALVESTON, 64- Concurrent Pos: USPHS fel, Univ Calif, Los Angeles, 62-64; NIH res grant, 67- Mem: AAAS; Soc Univ Otolaryngol; fel Am Acad Ophthal & Otolaryngol; fel Am Col Surg; fel Am Soc Head & Neck Surg. Res: Laryngeal reconstruction; tracheal transplantation. Mailing Add: Dept of Otolaryngol Univ of Tex Med Br Galveston TX 77550

BAILEY, CARL A, microbiology, see 12th edition

BAILEY, CARL LEONARD, b Grafton, NDak, Aug 2, 18; m 42; c 2. PHYSICS. Educ: Concordia Col, BA, 40; Univ Minn, MA, 42, PhD(physics), 47. Prof Exp: Assoc scientist, Off Sci Res & Develop proj, Univ Minn, 42-43; scientist, Los Alamos Lab, Univ Calif, 43-46; with AEC, 46; PROF PHYSICS, CONCORDIA COL, 47- Mem: Am Phys Soc. Res: Development of electrostatic generators and ion sources; efficiency of nuclear reactions; neutron scattering; charged particle scattering. Mailing Add: Dept of Physics Concordia Col Moorhead MN 56560

BAILEY, CARL WILLIAMS, III, b New Haven, Conn, Feb 7, 41; m 63; c 2. WOOD CHEMISTRY, PAPER CHEMISTRY. Educ: State Univ NY Col Environ Sci & Forestry & Syracuse Univ, BS, 63, MS, 65, PhD(pulp & paper), 68. Prof Exp: From res chemist to sr res chemist, 68-73, group leader res lignin chem, 73-74, GROUP LEADER RES TALL OIL CHEM, CHARLESTON RES CTR, WESTVACO CORP, 74- Mem: Am Oil Chem Soc. Res: Preparation of derivatives of rosin and fatty acids; characterization of these products and finding commercial areas where they can be utilized. Mailing Add: Westvaco Corp PO Box 5207 North Charleston SC 29406

BAILEY, CARROLL EDWARD, b Lewiston, Maine, Jan 14, 40. SOLID STATE PHYSICS. Educ: Bates Col, BS, 62; Dartmouth Col, MA, 64, PhD(physics), 68. Prof Exp: Nat Res Coun-Naval Res Lab res assoc physics, US Naval Res Lab, 68-70; ASST PROF PHYSICS, UNIV PR, MAYAGUEZ, 70- Mem: Am Phys Soc; Am Asn Physics Teachers; Sigma Xi. Res: Electron spin resonance and electron-nuclear double resonance in the study of defect centers in solids and phase transitions. Mailing Add: Dept of Physics Univ of PR Mayaguez PR 00708

BAILEY, CATHERINE HAYES, b New Brunswick, NJ, May 9, 21. POMOLOGY. Educ: Douglass Col, BA, 42; Rutgers Univ, PhD(fruit breeding), 57. Prof Exp: Tech asst, Agr Exp Sta, 48-54, res assoc, 54-57, asst prof, 57-66, assoc res prof, 66-72, PROF POMOL, RUTGERS UNIV, 72- Concurrent Pos: Exchange scientist, Res Sta, Can Dept Agr, BC, 65-66. Honors & Awards: Recognition Award, Nat Peach Coun, 69. Mem: Am Inst Biol Sci; Am Soc Hort Sci; Am Pomol Soc; Torrey Bot Club; Int Soc Hort Sci. Res: Fruit breeding; breeding for disease resistance; pears, peaches, apricots, apples; inheritance of season of ripening in progenies from certain early ripening peach varieties and selections. Mailing Add: Dept of Hort & Forestry Rutgers Univ New Brunswick NJ 08903

BAILEY, CECIL DEWITT, b Ayers, Miss, Oct 25, 21; m 42; c 2. DYNAMICS. Educ: Miss State Univ, BS, 51; Purdue Univ, MS, 54, PhD(trapezoidal plates), 62. Prof Exp: Instr B-47 bomber sch, US Air Force, 51-52, eng mech & aircraft structure, Air Force Inst Technol, 54-56, asst prof, 56-58, chief aerospace res, Air Force Systs Command Off, Langley Res Ctr, NASA, 59-63, aeronaut engr, Hq Air Force Systs Command, 64-65, assoc prof mech, Air Force Inst Technol, 66-67; assoc prof, 67-70, PROF AERONAUT & ASTRONAUT ENG, OHIO STATE UNIV, 70- Mem: Soc Exp Stress Anal; Am Inst Aeronaut & Astronaut; Am Soc Eng Educ. Res: Alternatives to the theory of differential equations; systems dynamics; particles, rigid and deformable body mechanics; dynamics of thermally stressed structures. Mailing Add: Dept of Aeronaut Eng Ohio State Univ Columbus OH 43210

BAILEY, CHARLES BASIL MANSFIELD, b Vancouver, BC, Feb 10, 30; m 57; c 3. ANIMAL PHYSIOLOGY, NUTRITION. Educ: Univ BC, BSA, 54, MSA, 56; Univ Reading, PhD(animal physiol), 59. Prof Exp: RES SCIENTIST ANIMAL PHYSIOL, CAN DEPT AGR, 59- Concurrent Pos: Nat Res Coun Can fel, 59-60. Mem: Can Soc Animal Prod; Brit Nutrit Soc. Res: Renal physiology, particularly urinary calculi formation in cattle; effect of diet on muscle protein synthesis in cattle. Mailing Add: Can Dept of Agr Res Sta Lethbridge AB Can

BAILEY, CHARLES EDWARD, b Johnson City, Tenn, May 25, 32; m 58; c 3. PHYSICAL CHEMISTRY. Educ: Univ Tenn, BS, 53, MS, 55, PhD(phys chem), 58. Prof Exp: Sr physicist, Polymer Intermediates Dept, Savannah River Lab, 58-69, sr res supvr comput appln div, 69-74, STAFF CHEMIST, ENV ANAL & PLANNING DIV, SAVANNAH RIVER LAB, E I DU PONT DE NEMOURS & CO, INC, 74- Mem: Am Chem Soc; Am Nuclear Soc. Res: Environmental management; nuclear reactor safety; environmental sciences. Mailing Add: 563 Coker Spring Rd Aiken SC 29801

BAILEY, CURTISS MERKEL, b Cleveland, Ohio, Aug 31, 27; m 54; c 4. ANIMAL BREEDING. Educ: Univ Wis, BS, 52, PhD(genetics, animal husb), 60; Tex A&M Univ, MS, 54. Prof Exp: From asst prof to assoc prof, 60-71, PROF ANIMAL SCI, UNIV NEV, RENO, 71-, GENETICIST IN ANIMAL SCI, 74- Mem: Am Genetic Asn; Am Soc Animal Sci. Res: Effect of environment on selection responses. Mailing Add: Animal Sci Div Univ of Nev Reno NV 89507

BAILEY, DANA KAVANAGH, b Clarendon Hills, Ill, Nov 22, 16. GEOPHYSICS, SYSTEMATIC BOTANY. Educ: Univ Ariz, BS, 37; Oxford Univ, BA, 40, MA, 43, DSc, 67. Prof Exp: Observ asst, Steward Observ, Univ Ariz, 33-37; astronr, Hayden Planetarium, Am Mus Natural Hist, New York, 37; physicist, Anarctic Serv, US Dept Interior, 40-41; proj engr, Rand, Douglas Aircraft Co, Calif, 46-48; consult to chief, Cent Radio Propagation Lab, Nat Bur Standards, 48-55; sci dir, Page Commun Engrs, Inc, DC, 55-59; CONSULT, SPACE ENVIRON LAB, ENVIRON RES LABS, NAT OCEANIC & ATMOSPHERIC ADMIN, 59-; RES ASSOC, UNIV COLO MUS, 72- Concurrent Pos: Mem, Hayden Planetarium-Grace Eclipse exped, Peru, 37 & US Antarctic exped, 40-41; chmn, Int Study Group Ionospheric Radio Progagation, Int Radio Consult Comt, Int Telecommun Union, 56-; hon res assoc, Rhodes Univ, 70-71. Honors & Awards: US Antarctic Serv Medal, 47; Flemming Award, US Govt; Meritorious Serv Award, US Dept Com, 51, Except Serv Award, 55. Mem: Am Astron Soc; fel Am Phys Soc; Am Geophys Union; fel Royal Geog Soc; fel Royal Astron Soc. Res: Solar terrestial relationships; cosmic ray physics; ionosphere; radio wave propagation; phytogeography; dendrochronology; systematic and evolutionary aspects of gymnosperms. Mailing Add: 1441 Bluebell Ave Boulder CO 80302

BAILEY, DAVID GEORGE, b New York, NY, Feb 21, 40; m 61; c 3. BIOCHEMISTRY, ENZYMOLOGY. Educ: Univ Del, BS, 63, MS, 64; Univ Vt, PhD(biochem), 67. Prof Exp: RES CHEMIST, AGR RES SERV, USDA, PA, 67- Mem: Am Chem Soc; Am Leather Chemists Asn. Res: Investigation of non-salt processes to preserve fresh animal hides and skins; qualitative and quantitative changes in proteins during processing into leather. Mailing Add: 1213 Emma Lane Warminster PA 18974

BAILEY, DAVID NEWTON, b Pittsburgh, Pa, Dec 1, 41; m 61; c 3. ANALYTICAL CHEMISTRY. Educ: Juniata Col, BS, 63; Mass Inst Technol, PhD(anal chem), 68. Prof Exp: Asst prof anal chem, Gustavus Adolphus Col, 68-71; ASST PROF ANAL CHEM, LEBANON VALLEY COL, 71- Mem: AAAS; Am Chem Soc. Res: Organic photochemistry; charge transfer complexes. Mailing Add: Dept of Chem Lebanon Valley Col Annville PA 17003

BAILEY, DAVID SCOTT, organic chemistry, see 12th edition

BAILEY, DAVID TIFFANY, b Olney, Tex, Aug 26, 42; m 66. ORGANIC CHEMISTRY, BIOCHEMISTRY. Educ: Univ Colo, Boulder, BA, 64; Iowa State Univ, PhD(org chem), 68. Prof Exp: NIH res fel & assoc chem, Yale Univ, 68-69; asst prof, 69-74, ASSOC PROF CHEM, CALIF STATE COL, FULLERTON, 74- Mem: AAAS; Am Chem Soc. Res: Organic chemistry and biosynthesis of natural products. Mailing Add: Dept of Chem Calif State Col Fullerton CA 92631

BAILEY, DENNIS MAHLON, b St Louis, Mo, Aug 14, 35; m 61; c 4. ORGANIC CHEMISTRY. Educ: Washington Univ, BA, 57, PhD(org chem), 61. Prof Exp: Qual control chemist, Uranium Div, Mallinckrodt Chem Works, 57-58; NIH fel, Stanford Univ, 62-63; res assoc, 63-65, group leader, 65-68, sr res chemist, 68-74, SECT HEAD, STERLING WINTHROP RES INST, 68-, SR RES ASSOC, 74- Mem: Am Chem Soc; Brit Chem Soc. Res: Medicinal chemistry; cardiovascular and metabolic diseases; central nervous system, analgesic and antiinflammatory agents. Mailing Add: Sterling Winthrop Res Inst Columbia Turnpike Rensselaer NY 12144

BAILEY, DONALD ETHERIDGE, b Moore Co, NC, May 4, 31; m 54; c 2. SCIENCE EDUCATION, BIOLOGY. Educ: Univ NC, BS, 53, MEd, 58, EdD(sci educ), 62. Prof Exp: Teacher pub schs, NC, 53-54 & 56-60; assoc prof sci educ & biol, 61-69, DEAN GEN COL, E CAROLINA UNIV, 69- Mem: Nat Sci Teachers Asn. Res: Incidence and prevalence of misconceptions of matters scientific in secondary schools of North Carolina. Mailing Add: Box 2761 East Carolina Univ Greenville NC 27834

BAILEY, DONALD FOREST, b Cliffside, NC, Apr 10, 39; m 61; c 2. MATHEMATICS. Educ: Wake Forest Univ, BS, 61; Vanderbilt Univ, MA, 63, PhD(math), 65. Prof Exp: From asst prof to assoc prof math, ECarolina Univ, 65-69; ASSOC PROF MATH, CORNELL COL, 69- Mem: Math Asn Am; Am Math Soc. Res: Iteration techniques for locating fixed points of continuous mappings. Mailing Add: Dept of Math Cornell Col Mt Vernon IA 52314

BAILEY, DONALD LEROY, b Sterling, Okla, Feb 22, 39; m 61; c 2. ENTOMOLOGY. Educ: Okla State Univ, BS, 61, MS, 62, PhD(entom), 64. Prof Exp: Entomologist, Walter Reed Army Inst Res, Washington, DC, 64-66;

ENTOMOLOGIST, USDA, 66- Mem: Entom Soc Am. Res: Mass rearing of insects; insects; insect nutrition; mosquito and house fly pathology; biological control of snails. Mailing Add: USDA Insect Res Lab PO Box 14565 Gainesville FL 32601

BAILEY, DONALD LEROY, b Benzonia, Mich, July 29, 22; m 43. ORGANIC CHEMISTRY. Educ: Mich State Col, BS, 43, MS, 44; Pa State Univ, PhD(org chem), 49. Prof Exp: Res chemist, Linde Co, 48-55, res supvr, 55-56, asst mgr res, 56-57; mgr res, NY, 57-64, res & develop, Calif, 64-69, ASST DIR CHEMS & PLASTICS, SILICONES DIV, UNION CARBIDE CORP, 69- Honors & Awards: Schoelkopf Medal, Am Chem Soc, 68. Mem: AAAS; Am Chem Soc. Res: Organic and silicon chemistry; organic polymers. Mailing Add: Chems & Plastics Div Union Carbide Corp Sisterville WV 26175

BAILEY, DONALD WAYNE, b Hutchinson, Kans, Feb 28, 26; m 48; c 2. IMMUNOGENETICS. Educ: Univ Calif, AB, 49, PhD(genetics), 53. Prof Exp: Fel, USPHS, Jackson Lab, 53-55, lab fel, 55-56, staff scientist, 56-57; asst prof, Univ Kans, 57-59; geneticist, Lab Aids Br, NIH, 59-61; assoc res geneticist, Univ Calif, San Francisco, 61-67; staff scientist, 67-70, SR STAFF SCIENTIST, JACKSON LAB, 70- Mem: Genetics Soc Am; Transplantation Soc. Res: Linkage analysis of complex traits of the mouse; nutation, linkage, expression and function of histocompatability genes in the mouse. Mailing Add: Jackson Lab Bar Harbor ME 04609

BAILEY, DONALD WYCOFF, b Emory, Va, Dec 17, 33; m 58; c 3. PHYSIOLOGY. Educ: Vanderbilt Univ, BA, 55; Emory Univ, MS, 56, PhD(biol), 58. Prof Exp: Prof biol & head dept sci, Tift Col, 58-62; PROF BIOL, WESTERN KY UNIV, 62- Mem: Am Soc Zool. Res: Chromatography of iodinated compounds; histochemical localizations of alkaline phosphatase; effects of testicular function on alloxan diabetes; physiology of exercise; lipid heat increment; metabolic cycles. Mailing Add: Dept of Biol Western Ky Univ Bowling Green KY 42101

BAILEY, DUANE W, b Moscow, Idaho, Sept 22, 36; m 59; c 3. MATHEMATICS. Educ: Wash State Univ, BA, 57; Univ Ore, MA, 59, PhD(math), 61. Prof Exp: Instr math, Yale Univ, 61-63; from asst prof to assoc prof, 63-71, PROF MATH, AMHERST COL, 71-, CHMN DEPT, 67- Mem: Am Math Soc; Math Asn Am; London Math Soc. Res: Banach and topological algebras. Mailing Add: Dept of Math Amherst Col Amherst MA 01002

BAILEY, EDGAR HERBERT, b Washington, DC, Aug 3, 14; m 46; c 5. GEOLOGY. Educ: Univ Redlands, AB, 34; Stanford Univ, PhD(mineral), 41. Prof Exp: From jr geologist to geologist, 38-53, admin geologist, 53-56, GEOLOGIST, US GEOL SURV, 56- Concurrent Pos: Asst, Pomona Col, 39; instr, Stanford Univ, 46, lectr, 47; dir Cent Treaty Orgn geol field training prog, Turkey, Iran & Pakistan, 66-70; vis scientist, Acad Sci, Moscow, USSR, 70. Honors & Awards: Distinguished Serv Award, US Dept Interior, 73. Mem: Fel Geol Soc Am; fel Mineral Soc Am; Soc Econ Geol. Res: Geology of mercury ore deposits; field geology; geologic cartography; petrology of ophiolites and low-grade metamorphic rocks, especially glaucophane schists; mineralogy, petrology and structure of Franciscan rocks of California coast range. Mailing Add: US Geol Surv 345 Middlefield Rd Menlo Park CA 94025

BAILEY, EDWARD D, b Port Clinton, Pa, Sept 22, 31; m 56. ANIMAL BEHAVIOR. Educ: Mont State Univ, BS, 58, MS, 60; Pa State Univ, PhD(zool), 63. Prof Exp: Instr zool, Mont State Univ, 63-64; asst prof, Ont Agr Col, 64-69; ASSOC PROF ZOOL, UNIV GUELPH, 69- Mem: AAAS; Animal Behav Soc. Res: Social behavior in wild mammals and birds; laboratory investigation of learning and retention of learned behavior in wild mammals; development of communication and early learning. Mailing Add: Dept of Zool Univ of Guelph Guelph ON Can

BAILEY, EDWARD THOMAS, b Quincy, Ill, Aug 7, 37; m 61; c 2. ORGANIC CHEMISTRY, POLYMER CHEMISTRY. Educ: Ill Wesleyan Univ, BS, 59; Univ Ill, PhD(chem), 65. Prof Exp: Mgr res dept, Freeman Chem Corp, 64-66; res chemist, Celanese Res Co, 66-68 & Celanese Plastics Co, 68-69; mgr polymer res, 69-73, ADV SCIENTIST, CONTINENTAL CAN CO, INC, 73- Mem: Am Chem Soc. Res: Formulation and preparation of thermoplastic and thermosetting polymers; relationship of structure with physical performance. Mailing Add: Continental Can Co Inc 7622 S Racine Ave Chicago IL 60620

BAILEY, ERVIN GEORGE, chemistry, mechanical engineering, deceased

BAILEY, EVERETT MURL, JR, b Big Spring, Tex, Mar 24, 40; m 61; c 2. VETERINARY TOXICOLOGY. Educ: Tex A&M Univ, DVM, 64; Iowa State Univ, MS, 66, PhD(physiol), 68; Am Bd Vet Toxicol, dipl. Prof Exp: Staff pathologist, US Army Inst Environ Med, 68-70; asst prof, 70-75, ASSOC PROF PHYSIOL & PHARMACOL, COL VET MED, TEX A&M UNIV, 75- Mem: Am Vet Med Asn; Am Col Vet Toxicol; Am Asn Vet Physiol & Pharmacol. Res: Chemical restraining agents in horses; experimental surgery; toxicology; anesthesiology; toxic plants; treatment of intoxications. Mailing Add: Dept of Physiol & Pharmacol Tex A&M Univ Col of Vet Med College Station TX 77843

BAILEY, F WALLACE, b Britt, Iowa, Nov 6, 29; m 50; c 3. PHYSICAL CHEMISTRY. Educ: Univ Okla, BS, 53, MS, 54. Prof Exp: Chemist, Silicones Div, Union Carbide Corp, 54-61; CHEMIST, RES & DEVELOP DEPT, PHILLIPS PETROL CO, 61- Mem: Soc Plastics Engrs. Res: Product development and application research of polymers. Mailing Add: 87F PRC Phillips Petrol Co Bartlesville OK 74004

BAILEY, FREDERICK EUGENE, JR, b Brooklyn, NY, Oct 8, 27. PHYSICAL CHEMISTRY. Educ: Amherst Col, AB, 48; Yale Univ, MS, 50, PhD(chem), 52. Prof Exp: Asst, Yale Univ, 50-52; res chemist, 52-59, group leader, 59-61, asst dir res & develop dept, Chem Div, 61-69, tech mgr calendering, flooring & rec prod dept, Plastics Div, 64-65, tech mgr vinyl resins, 65-68, mgr mkt res, 69-71, SR RES SCIENTIST POLYMER SCI, UNION CARBIDE CORP, 71- Concurrent Pos: Lectr, Kanawha Valley Grad Ctr, WVa Univ, 59-60 & Morris Harvey Col, 61-62, 65-66; adj prof chem, Marshall Univ, 75- Mem: Fel AAAS; Am Chem Soc; Am Phys Soc; NY Acad Sci; fel Am Inst Chem. Res: Polymer chemistry and physics; polymer morphology-property relationships and polymer synthesis. Mailing Add: 848 Beaumont Rd Charleston WV 25314

BAILEY, FREDERICK GEORGE, b Liverpool, Eng, Feb 24, 24. ANTHROPOLOGY. Educ: Oxford Univ, BA, 47; Univ Manchester, PhD(anthrop), 52. Prof Exp: Lectr anthrop, Sch Orient & African Studies, Univ London, 56-59, reader, 59-63; vis prof, Univ Rochester & Univ Chicago, 63-64; prof, Univ Sussex, 64-71; PROF ANTHROP, UNIV CALIF, SAN DIEGO, 71- Res: Political anthropology; areal interest in India and Western Europe. Mailing Add: Dept of Anthrop Univ of Calif at San Diego La Jolla CA 92093

BAILEY, GARLAND HOWARD, b Giatto, WVa, Nov 6, 90; m 20; c 1. IMMUNOLOGY. Educ: WVa Univ, BS, 15; Johns Hopkins Univ, MD, 20, DrPH, 21; Harvard Univ, MA, 65. Prof Exp: Instr path, Univ Wis, 21-22; instr epidemiol, Harvard Univ, 22-23, prev med & hyg, 23-25; assoc immunol, Johns Hopkins Univ, 25-26, assoc prof, 26-46, instr serol, Army Med Off, 43-45; MED & RES CONSULT BACT & IMMUNOL, 46- Mem: Am Soc Microbiol; Am Pub Health Asn; fel Royal Soc Health. Res: Fatigue and susceptibility to infection; cold, influenza, diphtheria and pneumococcus immunity; respiratory immunity in rabbits; accessory etiological factors of infection; measles; heterophile antigen and antibody; infectious mononucleosis; anaphylaxis; bacterial polysaccharides; organ specificity; experimental cancer. Mailing Add: 378 Toll Gate Rd Warwick RI 02886

BAILEY, GEORGE WILLIAM, b Monmouth, Ill, May 14, 30; m 53; c 4. ANALYTICAL CHEMISTRY. Educ: Monmouth Col, BS, 52; Univ Ill, PhD(chem), 55. Prof Exp: Anal chemist, Esso Res & Eng Co, 55-56 & Borg-Warner Corp, 56-58; microscopist, Esso Res Labs, Humble Oil & Refining Co, 58-72, MICROSCOPIST, EXXON RES & DEVELOP LABS, EXXON CO, USA, 72- Mem: Am Chem Soc; Electron Micros Soc Am. Res: Optical and electron microscopy; electron diffraction; electron probe analysis; thermal analysis; x-ray diffraction and spectrography. Mailing Add: Exxon Res & Develop Labs PO Box 2226 Baton Rouge LA 70821

BAILEY, GEORGE WILLIAM, b Des Moines, Iowa, Oct 30, 33; m 58; c 2. WATER POLLUTION. Educ: Iowa State Univ, BS, 55; Purdue Univ, MS, 58, PhD(soil chem, mineral), 61. Prof Exp: Soil surveyor, Soil Conserv Serv, 55-56; res fel, Purdue Univ, 61-64; soil-pesticide chemist, Div Water Supply & Pollution Control, USPHS, 64-66; soil phys chemist, Fed Water Pollution Control Admin, US Dept Interior, 66-70, Fed Water Qual Admin, 70, Southeastern Environ Lab, 70-74, SUPVRY RES SOIL PHYS CHEMIST, ENVIRON RES LAB, US ENVIRON PROTECTION AGENCY, ATHENS, GA, 74- Mem: AAAS; Am Soc Agron; Clay Minerals Soc; Soil Sci Soc Am; Am Chem Soc. Res: Soil pesticide chemistry; water quality; clay and soil mineralogy; mathematical modeling and simulation of the runoff and movement of pesticides and plant nutrients from agricultural land; trends in agriculture and environmental quality. Mailing Add: Environ Res Lab US EPA Athens GA 30601

BAILEY, GLENN CHARLES, b Pekin, Ill, Aug 21, 30; m 58; c 5. SOLID STATE PHYSICS. Educ: Univ Ill, BS, 52; Univ Wis, MS, 54; Cath Univ Am, PhD(physics), 65. Prof Exp: Physicist, Caterpillar Tractor Co, Ill, 55-57; RES PHYSICIST MAT SCI DIV, US NAVAL RES LAB, 57- Mem: Am Phys Soc; Sigma Xi. Res: Non-linear electron conductivity effects in metals; ferromagnetic resonance of thin films and single crystals; magnetostriction and magnetic anisotropy; radiation damage in semiconductors, ferrites and thin films; x-ray diffraction and fluorescent analysis; residual stresses in metals. Mailing Add: Code 6411 Mat Sci Div US Naval Res Lab Washington DC 20375

BAILEY, GORDON BURGESS, b Worcester, Mass, Feb 13, 34; m 58; c 3. BIOCHEMISTRY. Educ: Brown Univ, BA, 56; Univ Mass, MA, 61; Univ Fla, PhD(biochem), 66. Prof Exp: Vis prof biochem, Field Staff, Rockefeller Found, 66-73; dep dir, Anemia & Malnutrition Res Ctr, Fac Med, Chiang Mai Univ, Thailand, 73-76; V CHMN DEPT BIOCHEM, MED EDUC PROG, MOREHOUSE COL, 76- Concurrent Pos: Vis prof, Mahidol Univ, Bangkok, 66-; consult human reproduction unit, WHO, 75-76. Honors & Awards: Future Leader Award, Nutrit Found, 74. Mem: AAAS; Am Chem Soc; Am Soc Microbiol; Am Soc Parasitol. Res: Biochemistry and factors controlling encystation and exystation of Entamoeba; growth and differentiation in Entamoeba; etiology of malnutrition in humans. Mailing Add: Med Educ Prog Morehouse Col Atlanta GA 30314

BAILEY, GRANT CARTER, b Cedar Falls, Iowa, Nov 27, 10; m 36; c 2. PETROLEUM CHEMISTRY. Educ: Univ Wis, BS, 32; State Col Wash, MS, 33; Univ Iowa, PhD(chem), 38. Prof Exp: Fel, State Col Wash, 33; chemist, Zapon-Brevolite Lacquer Co, 34-35; asst, Univ Iowa, 35-38; res chemist, Phillips Petrol Co, 38-47; res mgr, Parker Rust Proof Co, Mich, 47-49; res chemist, Phillips Petrol Co, 49-56, mgr fundamental catalysis sect, 56-59, catalysis br, 56-69, electrochem br, 69-72, chem processes br, 72-73, plastics br, 73-75; RETIRED. Mem: Fel Am Inst Chemists; Am Chem Soc. Res: Polymerization of olefins; hydrocarbon reactions; catalysis; electrochemistry. Mailing Add: 1501 Hillcrest Dr Bartlesville OK 74003

BAILEY, HAROLD EDWARDS, b Salt Lake City, Utah, July 26, 06; m 33. PHARMACOGNOSY. Educ: Univ Calif, Los Angeles, AB, 30, PhD(microbiol), 35. Prof Exp: Teaching fel bot, Univ Calif, Los Angeles, 31-34; asst forester, Nat Park Serv, 35-38, jr park naturalist, 39-42; instr bot, Univ Tenn, 38; res asst, Parke Davis & Co, 42-47; asst prof biol sci & pharmacog, 47-53, assoc prof biol sci, 53-58, prof pharmacog, 59-74, EMER PROF PHARMACOG, WAYNE STATE UNIV, 74- Mem: AAAS; Am Pharmaceut Asn; Am Soc Pharmacog; Soc Econ Bot; Acad Pharmaceut Sci. Res: Antibiotics; physiology of fungi; chemotaxonomy. Mailing Add: Dept of Pharmacog Wayne State Univ Col Pharm Detroit MI 48202

BAILEY, HAROLD STEVENS, b Springfield, Mass, Apr 18, 22; m 46; c 5. PHARMACEUTICAL CHEMISTRY. Educ: Mass Col Pharm, BS, 44, MS, 48; Purdue Univ, PhD(pharmaceut chem), 51. Prof Exp: Asst chem, Mass Col Pharm, 46-48; instr pharm, Purdue Univ, 50-51; asst prof pharmacol, 51-52, from assoc prof to prof pharmaceut chem, 52-73, dean acad affairs, 61-73, DEAN GRAD SCH, S DAK STATE UNIV, 65-, VPRES ACAD AFFAIRS, 73- Mem: Fel AAAS; Sigma Xi. Res: Application of biochemical research to pharmacology and radiation biology; dental research. Mailing Add: SDak State Univ Brookings SD 57007

BAILEY, HARRY HUDSON, b Burkeville, Va, Jan 22, 21; m 48; c 3. SOIL SCIENCE. Educ: Va Polytech Inst, BS, 42; Mich State Univ, MS, 49, PhD(soil sci), 56. Prof Exp: Soil scientist, USDA, 46-55; asst agronomist, 55-57, from asst prof to assoc prof soils, 56-61, assoc prof agron, 61-72, assoc agronomist, 57-61, PROF AGRON, UNIV KY, 72- Mem: Am Soc Agron; Soil Conserv Soc Am; Soil Sci Soc Am; Clay Minerals Soc; Int Soc Soil Sci. Res: Soil origin, classification, mapping. Mailing Add: Dept of Agron Univ of Ky Lexington KY 40506

BAILEY, HARRY PAUL, b Allentown, Pa, Feb 12, 13; m 42; c 3. CLIMATOLOGY. Educ: Univ Calif, Los Angeles, BA, 39, MA, 42, PhD, 50. Prof Exp: Assoc prof geog, Los Angeles State Col, 50-51; asst res geogr, Univ Calif, Los Angeles, 51-52, from asst prof to assoc prof, 52-64, PROF GEOG, UNIV CALIF, RIVERSIDE, 64- Concurrent Pos: Guggenheim fel, 58-59; guest prof, Univ Aarhus, 66-67. Mem: Asn Am Geogrs; Am Meteorol Soc. Res: Climatology. Mailing Add: Dept of Geog Univ of Calif Riverside CA 92502

BAILEY, HERBERT R, b Denver, Colo, Nov 2, 25; m 51; c 5. MATHEMATICS. Educ: Rose Polytech Inst, BS, 45, 46; Univ Ill, MS, 47; Purdue Univ, PhD(math), 55. Prof Exp: Instr math, Gen Motors Inst, 47-49; mathematician, US Naval Ord Plant, Ind, 49-55 & Denver Res Ctr, Ohio Oil Co, 56-62; assoc prof math, Colo State Univ, 62-66; prof & chmn dept, 66-75, VPRES ACAD AFFAIRS & DEAN FAC, ROSE-HULMAN INST TECHNOL, 75- Mem: Am Math Soc; Math Asn Am. Res: Differential difference equations; fluid flow and diffusion. Mailing Add: 328 Potomac Ave Terre Haute IN 47803

BAILEY, JACK CLINTON, b Austin, Tex, Aug 10, 36; m 61; c 2. ENTOMOLOGY.

Educ: Tex A&M Univ, BS, 59; Miss State Univ, MS, 65, PhD(entom), 67; Am Registry Prof Entomologists, cert. Prof Exp: Jr entomologist, Tex Agr Exp Sta, Westlaco, 59-63; res asst entom, Miss State Univ, 63-67; res entomologist, Soybean Insects, 67-71, RES ENTOMOLOGIST & RES LEADER QUARANTINE INSECTS RES, AGR RES SERV, USDA, 71- Concurrent Pos: Adj asst prof entom & assoc mem grad coun, Miss State Univ, 68- Mem: Sigma Xi; Entom Soc Am. Res: Biological control both of plants and insects; pest management systems. Mailing Add: Insect Quarantine Agr Res Serv USDA PO Box 225 Stoneville MS 38776

BAILEY, JAMES ALLEN, b Chicago, Ill, May 14, 34; m 59; c 1. WILDLIFE ECOLOGY. Educ: Mich Technol Univ, BS, 56; State Univ NY Col Forestry, MS, 58, PhD(zool), 66. Prof Exp: Res assoc ecol, Ill Natural Hist Surv, 64-68; asst prof wildlife biol, Univ Mont, 68-69; ASST PROF WILDLIFE BIOL, COLO S7ATE UNIV, 69- Mem: Wildlife Soc. Res: Population dynamics; nutrition of wild vertebrates. Mailing Add: Dept of Fishery & Wildlife Biol Colo State Univ Ft Collins CO 80521

BAILEY, JAMES L, b Tiffin, Ohio, Mar 21, 30; m 52; c 3. APPLIED MATHEMATICS. Educ: Heidelberg Col, BS, 52; Mich State Univ, MS, 54, PhD(math), 58. Prof Exp: Instr math, Mich State Univ, 58; asst prof, Case Inst Technol, 58-63; ASSOC PROF MATH, UNIV TOLEDO, 63- Mem: Math Asn Am; Soc Indust & Appl Math; Am Acad Math. Res: Mechanics. Mailing Add: 2423 Middlesex Toledo OH 43606

BAILEY, JAMES MICHAEL, b Williamsburg, Va, Jan 19, 47; m 69. PHYSICAL BIOCHEMISTRY. Educ: Davidson Col, BS, 69; Univ NC, Chapel Hill, PhD(chem), 73. Prof Exp: Helen Hay Whitney res fel chem, Calif Inst Technol, 73-75; ASST PROF CHEM, SOUTHERN ILL UNIV, CARBONDALE, 75- Mem: Am Soc Microbiol. Res: Structure and other physical properties of RNA; organization and function of RNA in RNA tumor viruses. Mailing Add: Dept of Chem & Biochem Southern Ill Univ Carbondale IL 62901

BAILEY, JAMES STUART, b Chicago, Ill, Dec 23, 21; m 45. GEOLOGICAL OCEANOGRAPHY. Educ: Univ Wash, BS, 50; Northwestern Univ, MS, 53; Univ Ariz, PhD(geol), 55. Prof Exp: Explor geologist, Calif Explor Co, Stand Oil Co Calif, 55-60; indepndent consult geol, 60-64; geol oceanogr, US Naval Oceanog Off, 64-67; res oceanogr fisheries, Bur Com Fisheries, Dept Interior, 67-68; sr eng scientist electro-optics, TRW Systs, 68-70; PROF DIR GEOG, OFF NAVAL RES, DEPT NAVY, 70- Concurrent Pos: Geol oceanogr, Spacecraft Oceanog Proj, US Naval Oceanog Off, 65-67; mem, Am Cong Surv & Mapping Marine Surv & Mapping Comt, 75- Honors & Awards: Sustained Superior Achievement Award, US Naval Oceanog Off, 67. Mem: Am Inst Physics; Am Soc Photogram. Res: Research on marine geology and coastal processes including sea floor geology and structure; paleostratigraphic and structural reconstructions. Mailing Add: Chief of Naval Res Code 462 Off of Naval Res Arlington VA 22217

BAILEY, JOHN H, b Elliott Co, Ky, Sept 5, 07; m 39; c 1. ZOOLOGY, SCIENCE EDUCATION. Educ: Morehead State Col, AB, 39; Univ Ky, MA, 40; Cornell Univ, EdD(sci educ), 58. Prof Exp: Teacher high schs, Ky, 30-44; mem fac biol jr col, Tenn, 46-47; prof sci surv & sci educ, 47-73, EMER CHMN GEN SCI & SCI EDUC, E TENN STATE UNIV, 73- Concurrent Pos: Mem, Tenn Sci Acad, 60-64; Tenn affil rep, Nat Wildlife Fedn, 60-70. Res: Conservation education through community organizations. Mailing Add: Dept of Sci Educ ETenn State Univ Johnson City TN 37601

BAILEY, JOHN MARTIN, b Lakewood, Ohio, Feb 23, 28; m 60; c 2. SOLID STATE PHYSICS. Educ: Hiram Col, AB, 49; Mass Inst Technol, MS, 51; Univ Va, PhD(physics), 59. Prof Exp: Aeronaut res scientist, Nat Adv Comt Aeronaut, Ohio, 51-55; from asst prof to assoc prof, 59-70, PROF PHYSICS, BELOIT COL, 70- Concurrent Pos: NSF res partic grant, 61; res, Cavendish Lab, Cambridge, 65-66; consult, Gen Motors & Warner Elec Brake & Clutch Co. Mem: Am Asn Physics Teachers; Am Soc Planning Off. Res: High temperature friction and wear; solid lubricants; friction and wear of single crystals of copper; imperfections in solids; philosophy of science; urban transportation. Mailing Add: 1215 Chapin St Beloit WI 53511

BAILEY, JOHN MARTYN, b Hawarden, Eng, May 13, 29; US citizen; m 63; c 3. BIOCHEMISTRY. Educ: Univ Wales, BSc, 49, PhD(biochem), 52; Univ Wales, DSc, 70. Prof Exp: Res assoc, Iowa State Univ, 54-55; res assoc physiol chem, Johns Hopkins Univ, 55-56, res assoc biochem, Dept Surg, Med Sch, 56-59; from asst prof to assoc prof, 59-69, PROF BIOCHEM, SCH MED, GEORGE WASHINGTON UNIV, 69- Concurrent Pos: Fel chem, Prairie Regional Lab, Nat Res Coun Can, 52-54; USPHS career develop award, 63-73; consult, Div Res Grants, USPHS, 69- & NSF, 74-; fel coun arteriosclerosis, Am Heart Asn, 69-; vis prof biochem, Univ Miami Sch Med, 71. Mem: Am Heart Asn; AAAS; Am Soc Biol Chem; Soc Exp Biol & Med; Tissue Cult Asn. Res: Lipid and lipoprotein metabolism in cultured mammalian cells, genetic disorders, essential fatty acids, prostaglandins and thromboxanes; experimental atherosclerosis; tumor viruses and cell transformations. Mailing Add: Dept of Biochem George Washington Univ Sch Med Washington DC 20037

BAILEY, JOSEPH RANDLE, b Fairmont, WVa, Sept 17, 13; m 46; c 2. ZOOLOGY. Educ: Univ Mich, AB, 35, PhD(zool), 40; Haverford Col, MA, 37. Prof Exp: Int exchange fel, Brazil, 40-41 & fel, 41-42; from instr to assoc prof, 46-65, PROF ZOOL, DUKE UNIV, 65- Concurrent Pos: Fulbright lectr, Sao Paulo Univ, 61-62; Guggenheim Mem Found fel, 53-54; Fulbright sr scholar, James Cook Univ NQueensland, Townsville, 71. Mem: Ecol Soc Am; Soc Study Amphibians & Reptiles; Herpetologists League; AAAS; Am Soc Ichthyol & Herpet (pres, 72). Res: Systematics; life history and distribution of herpetology and ichthyology; zoogeography. Mailing Add: Dept of Zool Duke Univ Durham NC 27706

BAILEY, KEITH, b Finedon, Eng, Dec 15, 40; m 65; c 3. SYNTHETIC ORGANIC CHEMISTRY, ANALYTICAL CHEMISTRY. Educ: Oxford Univ, BA, 63, MA, 66, PhD(chem), 66. Prof Exp: Res assoc org synthesis, Oxford Univ, 65-67; res fel, Trent Univ, 67-69; res scientist drug synthesis, 69-73, HEAD DRUG ANAL SECT, HEALTH PROTECTION BR, GOVT CAN, 73- Mem: Chem Inst Can; fel The Chem Soc. Res: Investigations of structure activity relationships in drugs; development of analytical methods for drugs of abuse; mechanisms and stereochemistry in organic synthesis and reactions. Mailing Add: Health Protection Br Tunney's Pasture Ottawa ON Can

BAILEY, LEO L, b Blossom, Tex, Jan 17, 22; m 45; c 4. HORTICULTURE, ENTOMOLOGY. Educ: Tex A&M Univ, BS, 43; Sam Houston State Teachers Col, MS, 49; La State Univ, PhD(hort), 54. Prof Exp: Instr voc agr, Red River County Voc Sch, Tex, 46-48; instr hort, Sam Houston State Teachers Col, 48-49; from asst prof to prof, 49-74, for student adv, 65-73, TEACHER AGR, TEX A&I UNIV, 75- Honors & Awards: Hon Agr Educ Award, Am Soc Hort Sci. Mem: Am Soc Hort Sci. Res: Maturity test for vegetable crops; propagation of ornamental plants; teaching techniques. Mailing Add: Col of Agr Tex A&I Univ Kingsville TX 78363

BAILEY, LEONARD CHARLES, b Jersey City, NJ, Feb 5, 36; m 64; c 2. PHARMACEUTICAL CHEMISTRY. Educ: Fordham Univ, BS, 57; Rutgers Univ, MS, 65, PhD(pharmaceut sci), 69. Prof Exp: Sci drug analyst, Ortho Pharmaceut Corp, 69-73; ASST PROF PHARMACEUT CHEM, RUTGERS UNIV, NEW BRUNSWICK, 73- Mem: Am Pharmaceut Asn; Am Chem Soc; Sigma Xi. Res: Application of modern chromatographic techniques to the analysis of drugs in formulations and in biological fluids. Mailing Add: Col of Pharm Busch Campus Rutgers Univ New Brunswick NJ 08903

BAILEY, LLOYD EVAN, JR, physics, see 12th edition

BAILEY, LOUIS GLEN, physical chemistry, see 12th edition

BAILEY, LOWELL FREDERICK, b Holton, Kans, June 15, 11; m 37; c 1. PLANT CHEMISTRY, PLANT PHYSIOLOGY. Educ: Southern Ill State Teachers Col, BEd, 32; Univ Mich, PhD(plant physiol), 39. Prof Exp: Instr bot, Grand Rapids Jr Col, Mich, 38-42; res plant chemist, Exp Sta, Univ Ky, 42-44; res forester, Forest Rels Dept, Tenn Valley Authority, 44-47; from asst prof to assoc prof, Univ Tenn, 47-50; assoc prof, 50-55, PROF BOT, UNIV ARK, FAYETTEVILLE, 55-, CHMN DEPT, 63- Mem: Am Soc Plant Physiologists; Am Chem Soc. Res: Water relations of plants; dormancy in buds; growth inhibitors; gibberellins. Mailing Add: Dept of Bot & Bacteriol Univ of Ark Fayetteville AR 72703

BAILEY, MAURICE EUGENE, b Portsmouth, Ohio, Mar 14, 16; m 37; c 4. SCIENCE EDUCATION, MINING. Educ: Ohio Wesleyan Univ, AB, 37; Purdue Univ, MS, 39, PhD, 41. Prof Exp: Res chemist, Purdue Univ Res Found, 40-41; res chemist, Allied Chem Corp, 41-55, mgr appln res, 55-59, tech liaison, 59-63, chem planning, 63-68, dir res, 68-70; PROF CHEM, PIKEVILLE COL, 70-, DIR MINING TECHNOL & CHMN DIV SCI & TECHNOL, 71- Concurrent Pos: Consult, Ky Coun Pub Higher Educ, 75. Mem: AAAS; Am Chem Soc; Soc Mining Engrs; Am Asn Univ Profs. Res: Coal liquefaction, mine water quality; urethane polymers; surface coatings; catalysis; polyesters; diene synthesis. Mailing Add: Pikeville College Pikeville KY 41501

BAILEY, MILTON, b New York, NY, May 20, 17; m 54; c 1. INDUSTRIAL CHEMISTRY, LEATHER CHEMISTRY. Educ: City Col NY, BBA, 40, MS, 49; Pratt Inst, cert, 51. Prof Exp: Ed newspaper & educr, Adj Gen Off, US Dept Army, 41-43; chem supt, Ruderman, Inc, NY, 46-52; leather chemist, US Naval Supply Res & Develop Facil, NJ, 52-66, footwear specialist, 66-67; leather chemist & opers res analyst, US Naval Clothing Res & Develop Unit, Mass, 67; LEATHER CHEMIST & FOOTWEAR SPECIALIST, US NAVY CLOTHING & TEXTILE RES UNIT, 67- Concurrent Pos: Lectr, City Col NY, 51-52 & NY Community Col, 53-61; arbitrator, Am Arbit Asn, 73-; chmn elec hazard & conductive safety shoe comt, Am Nat Standards Inst, 73- Mem: Am Leather Chem Asn; Am Soc Testing & Mat. Res: Impregnation of leather and protein fibers; coloring of hair and protein fibers; instrumentation; leather and footwear technology; chemical specialties. Mailing Add: 18 Bayfield Rd Wayland MA 01778

BAILEY, MILTON (EDWARD), b Shreveport, La, June 7, 24; m 58; c 3. FOOD CHEMISTRY. Educ: Tulane Univ, BS, 49; La State Univ, MS, 53, PhD(biochem), 58. Prof Exp: Res assoc, La State Univ, 53-58; asst prof animal husb, 58-60, assoc prof food sci & nutrit, 60-69, PROF FOOD SCI & NUTRIT, UNIV MO-COLUMBIA, 69- Mem: Am Chem Soc; Inst Food Technologists. Res: Muscle physiology and biochemistry; flavoring constituents of foods; enzymology. Mailing Add: Dept of Food Sci & Nutrit Univ of Mo Columbia MO 65201

BAILEY, NORMAN SPRAGUE, b Malden, Mass, June 27, 12; m 52; c 4. BIOLOGY. Educ: Boston Univ, SB, 40, AM, 41; Harvard Univ, PhD(biol, entom), 51. Prof Exp: Instr biol, Boston Univ, 42-46, asst prof, 48-52; res biologist, State Dept Sea & Shore Fisheries, Maine, 55-56; chmn dept sci, Bradford Jr Col, 56-75; RETIRED. Concurrent Pos: Co-founder & trustee, Swans Island Marine Sta, 65-; independent researcher, 75- Mem: AAAS. Res: Bioecology of insects; the Tingidae and Tabanidae. Mailing Add: Box 86 Swans Island ME 04685

BAILEY, ORVILLE TAYLOR, b Jewett, NY, May 28, 09. PATHOLOGY. Educ: Syracuse Univ, AB, 28; Albany Med Col, MD, 32. Prof Exp: Rotating intern, Albany Hosp, NY, 32-33; house officer path, Peter Bent Brigham Hosp, Boston, 33-34; resident, Children's Hosp, 34-35; instr path, Harvard Med Sch, 35-40, assoc, 40-46, asst prof, 46-51; prof neuropath, Sch Med, Ind Univ & chief neuropath, Larue Carter Mem Hosp, 51-59; prof neurol, 59-70, prof neurol surg, Univ Ill Col Med, 70-72, PROF NEUROPATH, ABRAHAM LINCOLN SCH MED, UNIV ILL COL MED, 72- Concurrent Pos: Guggenheim fel, Univ Cambridge, 46-47; resident, Peter Bent Brigham Hosp, 35-37, assoc pathologist, 40-43, assoc pathologist neuropath, 46-51; jr fel, Soc Fels, Harvard Univ, 37-40; lectr, Lowell Inst, 40; mem, Off Sci Res & Develop, 41-46; neuropathologist & vchmn, Neurol Inst, Children's Hosp, 47-51; chmn deleg & vpres, US Int Cong Neuropath, 59-67. Mem: AAAS; AMA; Am Asn Path & Bact; Am Soc Exp Path; Am Asn Neuropath (pres, 60). Res: Dural sinus thrombosis; alloxan diabetes; myelin degeneration; brain tumors; radioactive materials in brain; pediatric neuropathology; hydrocephalus. Mailing Add: Abraham Lincoln Sch of Med Univ of Ill Col of Med Chicago IL 60612

BAILEY, PAUL BERNARD, b Goshen, Ind, June 1, 25; m 57. MATHEMATICS. Educ: Univ London, BSc, 51; Univ Wash, PhD(math), 61. Prof Exp: Teacher math & physics, St Mary's Col, Eng, 52-54; instr math, Milwaukee Sch Eng, 54-56; res engr, Boeing Co, Wash, 56-61; STAFF MEM, SANDIA CORP, ALBUQUERQUE, 61- Mem: Am Math Soc. Res: Differential equation eigenvalues; invarient imbedding. Mailing Add: S Star Rte Box 161 Corrales NM 87048

BAILEY, PAUL C, b Baileyton, Ala, Oct 26, 21; m 44; c 2. CYTOLOGY. Educ: Ala State Teachers Col, Jacksonville, BS, 42; Vanderbilt Univ, MA, 46, PhD, 49. Prof Exp: Prof biol, Ala Col, 47-48 & 49-63; prof, 63-71, DEAN, BIRMINGHAM-SOUTHERN COL, 71- Concurrent Pos: Res grants, Nat Cancer Inst & US AEC. Mem: AAAS; Am Genetic Asn. Res: Natural factors influencing division of and radiation damage to cells. Mailing Add: 1220 Greensboro Rd Birmingham AL 35208

BAILEY, PAUL TOWNSEND, b Sydney, Australia, Nov 14, 39; m 65; c 2. PHYSICS. Educ: Univ New Eng, Australia, BSc, 62; Mass Inst Technol, ScD(physics), 66. Prof Exp: Sr res physicist, 66-69, group leader, Electronic Prod Div, 69-72, MGR COM DEVELOP, MONSANTO CO, 72- Mem: Am Phys Soc. Mailing Add: Monsanto Co 800 N Lindbergh Blvd St Louis MO 63166

BAILEY, PEARCE, clinical neurology, see 12th edition

BAILEY, PERCIVAL, neurology, neurosurgery, deceased

BAILEY, PHILIP SIGMON, b Chickasha, Okla, June 9, 16; m 41, 73; c 5. ORGANIC CHEMISTRY. Educ: Okla Baptist Univ, BS, 37; Univ Okla, MS, 40; Univ Va, PhD(org chem), 44. Prof Exp: Asst chem, Univ Okla, 37-40; chemist,

Halliburton Oil Well Cementing Co, Okla & Kans, 40-41; asst under contract, Off Sci Res & Develop, 42-45; from asst prof to assoc prof, 45-57, PROF CHEM, UNIV TEX, AUSTIN, 57- Concurrent Pos: Fulbright res grant, Tech Inst, Ger, 53-54; consult. Mem: Am Chem Soc. Res: Reactions of ozone with organic compounds; peroxides; radicals; complexes. Mailing Add: Dept of Chem Univ of Tex Austin TX 78712

BAILEY, R L, b Trenton, Tenn, Sept 14, 16; m 44; c 1. AVIAN PHYSIOLOGY, VETERINARY ANATOMY. Educ: Tenn State Univ, BS, 42; Iowa State Univ, MS, 46, PhD(avian physiol, vet anat), 50. Prof Exp: Instr animal husb, Tenn State Univ, 42-45; instr poultry husb, WVa State Col, 46-48, asst prof, 50-53; prof, Agr & Tech Col, NC, 53-58; PROF POULTRY HUSB, GRAMBLING COL, 58-, HEAD DEPT AGR, 63- Mem: Am Inst Biol Sci; Poultry Sci Asn; World Poultry Sci Asn. Res: Reproduction in the male animal. Mailing Add: Dept of Agr Grambling State Univ Grambling LA 71245

BAILEY, REEVE MACLAREN, b Fairmont, WVa, May 2, 11; m 39; c 4. ZOOLOGY. Educ: Univ Mich, AB, 33, PhD(zool), 38. Prof Exp: Asst biol surv, NY Conserv Dept, 35-37; biologist, Biol Surv, NH Fish & Game Dept, 38; from instr to asst prof zool, Iowa State Col, 38-44; from asst prof to assoc prof, 44-59, assoc cur, 44-48, PROF ZOOL, UNIV MICH, 59-, CUR FISHES, MUS ZOOL, 48- Concurrent Pos: Leader, Iowa Fisheries Res, Univ, 41-44; with Tenn Valley Authority, 44; res assoc, Am Mus Natural Hist, 64-, mem, Bolivian Exped, 64; mem, US Expeds, 30-, Bermuda, 51, Guatemala, 66, 68 & 71 & Zambia, 70; mem coun, AAAS, 69-72. Mem: AAAS; Am Soc Ichthyol & Herpet (vpres, 54, pres, 59); Am Fisheries Soc; Ecol Soc Am; Am Soc Limnol & Oceanog. Res: Ichthyology; taxonomy, variation, distribution, ecology and hybridization of American fresh-water fishes; herpetology of Iowa; revision of Centrarchidae; Percidae. Mailing Add: Mus of Zool Univ of Mich Ann Arbor MI 48104

BAILEY, RICHARD ELMORE, b Cleveland, Ohio, Nov 4, 29; m 53; c 4. MEDICINE. Educ: Stanford Univ, BA, 51, MD, 55. Prof Exp: Trainee diabetes & metab dis, Sch Med, Stanford Univ, 57-59; trainee steroid biochem training prog, Clark Univ & Worcester Found Exp Biol, 60-61; dir res training prog diabetes & metab, Med Sch, Univ Ore, 61-71, asst prof, 61-66, assoc prof med, 66-73; ASSOC PROF MED, COL MED, UNIV UTAH, 73- Concurrent Pos: Fel diabetes, Sch Med, Univ Southern Calif, 56-57; NASA fac fel award, 69; USPHS spec res fel, Case Western Reserve Univ, 69-70; asst prof biochem, Case Western Reserve Univ, 69-70. Honors & Awards: Ayerst-Squibb Res Award, Endocrine Soc, 69; Estab Investigatorship Award, Am Heart Asn, 70. Mem: Endocrine Soc; Am Diabetes Asn; Am Fedn Clin Res; AMA; Am Physiol Soc. Res: Metabolic diseases; hypertension; endocrinology and diabetes. Mailing Add: Dept Internal Med Div Kidney Dis Univ of Utah Col of Med Salt Lake City UT 84132

BAILEY, RICHARD HENDRICKS, b Rocky Mt, NC, Oct 30, 46; m 72; c 1. INVERTEBRATE PALEONTOLOGY. Educ: Old Dominion Univ, BS, 68; Univ NC, MS, 71, PhD(geol), 73. Prof Exp: ASST PROF GEOL, NORTHEASTERN UNIV, 72- Mem: Am Asn Petrol Geologists; Soc Econ Paleontologists & Mineralogists; Sigma Xi. Res: Cenozoic molluscan communities from Atlantic Coastal Plain; systematic and evolutionary studies of Cenozoic Mollusca; ecology of modern benthic molluscan communities. Mailing Add: Dept of Earth Sci Northeastern Univ Boston MA 02115

BAILEY, ROBERT VERNON, b Hanna, Wyo, Sept 16, 32. ECONOMIC GEOLOGY. Educ: Univ Wyo, BS, 56. Prof Exp: Geologist uranium geol, Utah Int, 59-64; geologist coal geol, Page T Jenkins, Geologist, 64-65; dist geologist uranium, Union Carbide Corp, 65-68; independent consult geol, 68-71; pres, Aquarius Resources Corp, 71-73; PRES, POWER RESOURCES CORP, 73- Mem: Soc Econ Geol; Soc Mining Engrs. Mailing Add: Power Resources Corp 1660 S Albion Suite 827 Denver CO 80222

BAILEY, RODNEY ALBERT, b Hartford, Conn, Apr 8, 42; m 64; c 4. ENVIRONMENTAL CHEMISTRY. Educ: Univ Conn, BA, 64; Wash State Univ, PhD(org chem), 68. Prof Exp: Res asst chem, Case Western Reserve Univ, 67-68; asst prof, Cleveland State Univ, 68-72; ASST DEAN & MEM FAC, WILLIAM JAMES COL, 72- Mem: Am Chem Soc. Res: Solar energy, alternative energy systems; environmental chemistry. Mailing Add: William James Col Grand Valley State Cols Allendale MI 49401

BAILEY, RONALD ALBERT, b Winnipeg, Man, June 2, 33; m 61. INORGANIC CHEMISTRY. Educ: Univ Man, BSc, 56, MSc, 57; McGill Univ, PhD(radiochem), 60. Prof Exp: NATO sci fel, Univ Col, Univ London, 60-61; from asst prof to assoc prof, 61-71, PROF INORG CHEM, RENSSELAER POLYTECH INST, 71- Mem: Am Chem Soc; Chem Inst Can; Brit Chem Soc. Res: Preparation and characterization of coordination complexes; study of complex species in molten salt solutions. Mailing Add: Dept of Chem Rensselaer Polytech Inst Troy NY 12180

BAILEY, ROY ALDEN, b Providence, RI, July 28, 29; m 58; c 3. GEOLOGY. Educ: Brown Univ, AB, 51; Cornell Univ, MSc, 54. Prof Exp: GEOLOGIST, US GEOL SURV, 53-55, 57- Mem: Geol Soc Am; Mineral Soc Am; Am Geophys Union. Res: Volcanology, petrology of igneous rocks, volcano tectonics, geothermal energy. Mailing Add: US Geol Surv Nat Ctr 951 Reston VA 22092

BAILEY, ROY HORTON, JR, b Maxton, NC, Nov 7, 21; m 47; c 3. PHYSICAL ORGANIC CHEMISTRY. Educ: Univ NC, BS, 48, PhD(chem), 58. Prof Exp: Asst anal chem, Univ NC, 48-53; asst prof chem, King Col, 53-57, prof & chmn dept, 57-63; asst prof, 63-71, ASSOC PROF CHEM, CLEMSON UNIV, 63- Mem: Am Chem Soc. Res: Heterocyclic reaction kinetics; analytical instrumentation. Mailing Add: Dept of Chem Clemson Univ Clemson SC 29631

BAILEY, SAMUEL DAVID, b Cedar Falls, Iowa, July 14, 15; m 42; c 1. PHYSICAL CHEMISTRY. Educ: Univ Northern Iowa, BA, 37; Univ Iowa, MS, 40, PhD(phys chem), 42. Prof Exp: From instr to asst prof chem, Univ Northern Iowa, 41-48; res chemist, Rohm and Haas Co, 42-46; head phys chem sect, Smith Kline & French Lab, 48-51; dir pioneering res lab, 51-74, DIR FOOD SCI LAB, NATICK DEVELOP CTR, 74- Mem: AAAS; Am Chem Soc; Am Inst Chemists. Res: Physical and optical properties of matter; research administration; radiation and flavor chemistry; photochemistry; food chemistry; pollution abatement; food nutrition and microbiology; behavioral sciences. Mailing Add: Food Sci Lab Natick Develop Ctr Natick MA 01760

BAILEY, STANLEY FULLER, b Middleboro, Mass, Aug 1, 06; m 31; c 1. ENTOMOLOGY. Educ: Univ Mass, BS, 29; Univ Calif, PhD(entom), 32. Prof Exp: Sci aid, Bur Entom, USDA, Univ Mass, 28-29; asst entom, Univ Calif, Davis, 29-31, jr entomologist, 32-37, instr entom, 34-37, from asst entomologist & asst prof to entomologist & prof, 38-68, res biologist, 68-73, vchmn dept, Berkeley & Davis, 46-57, EMER PROF ENTOM, UNIV CALIF, SANTA CRUZ, 68- Mem: Am Entom Soc. Res: Crop insect control; mosquito ecology; taxonomy of Thysanoptera. Mailing Add: 225-183 Mt Hermon Rd Scotts Valley CA 95066

BAILEY, STURGES WILLIAMS, b Waupaca, Wis, Feb 11, 19; m 49; c 2. CLAY MINERALOGY, CRYSTALLOGRAPHY. Educ: Univ Wis, BA, 41, MA, 48; Cambridge Univ, PhD(physics), 54. Prof Exp: From instr to assoc prof x-ray crystallog, Dept Geol, 51-61, chmn dept geol & geophys, 68-71, PROF X-RAY CRYSTALLOG, DEPT GEOL, UNIV WIS-MADISON, 61- Concurrent Pos: Ed, Clays & Clay Minerals, 65-69; chmn joint comt on nomenclature, Int Union Crystallog & Int Mineral Asn, 70- Mem: Clay Minerals Soc (vpres, 70-71, pres, 71-72); Mineral Soc Am (vpres, 72-73, pres, 73-74); Int Asn Study Clays (pres, 75-78); Geol Soc Am; Am Crystallog Asn. Res: Potassium feldspars. Mailing Add: Dept of Geol & Geophys Univ of Wis Madison WI 53706

BAILEY, SYLVIA M, physical chemistry, see 12th edition

BAILEY, THOMAS DANIEL, b Birmingham, Ala, July 15, 45; m 66. ORGANIC CHEMISTRY. Educ: Univ Montevallo, BS, 69; Univ Ala, PhD(chem), 74. Prof Exp: RES CHEMIST, REILLY TAR & CHEM CORP, 75- Mem: Am Chem Soc. Res: Synthesis of heterocyclic compounds; synthesis and chemistry of heterocyclic-N-oxides; process development. Mailing Add: Reilly Labs 1500 S Tibbs Ave PO Box 41076 Indianapolis IN 46241

BAILEY, THOMAS L, III, b Newnan, Ga, Dec 24, 23; m 55; c 2. CHEMICAL PHYSICS. Educ: Carnegie Inst Technol, BS, 49; Univ Chicago, SM, 50, PhD(chem), 53. Prof Exp: Asst res elec eng, 53-54, asst res prof, 54-57, assoc prof, 57-63, PROF PHYSICS & ELEC ENG, UNIV FLA, 63- Concurrent Pos: Guggenheim fel, 59-60. Mem: Fel Am Phys Soc. Res: Molecular and atomic structure; experimental mass spectroscopy; collisions of positive ions, negative ions and electrons with atoms and molecules. Mailing Add: Dept of Physics Univ of Fla Gainesville FL 32601

BAILEY, THOMAS LAVAL, b Berkeley, Calif, Oct 5, 97; m 24. GEOLOGY. Educ: Univ SC, AB, 17; Univ Calif, AM, 21, PhD(geol), 26. Prof Exp: Instr geol & biol, Univ of the South, 17-18; instr geol, Pomona Col, 21-22; assoc geologist, Bur Econ Geol, Tex, 22-24; instr geol, Univ Calif, 24-25; dist geologist, Shell Oil Co, 25-34; geologist, Bataafsche Petrol MIJ, The Hague, Netherlands, 34-35; regional geologist, Shell Oil Co, 35-40, chief res geologist, 40-45; CONSULT GEOLOGIST, 46- Concurrent Pos: Dir, Ventura River Munic Water Dist, 52-56; mem, Ventura City Planning Comn, 56-58 & Ventura City Coun, 58-72; mem adv bd, County Flood Control Dist, 62-72. Honors & Awards: Petit Award; City of San Buena Ventura, 72. Mem: AAAS; fel Geol Soc Am; Am Asn Petrol Geologists. Res: Structural, petroleum, engineering and underwater geology; stratigraphy; sedimentation; petrology; petrography; core barrel and other equipment for taking shallow oriented cores for purposes of working out structure of areas covered with alluvium and other shallow coverings. Mailing Add: 3714 Foothill Rd Ventura CA 93003

BAILEY, VIRGINIA LONG, b Mariposa, Calif, July 28, 08; m 33. BOTANY. Educ: Univ Calif, AB, 30, MA, 32; Univ Mich, PhD, 59. Prof Exp: Asst bot, Univ Calif, 30-33, asst & bot illustrator, 33-38; instr biol, Wayne State Univ, 46-59; from asst prof to prof, 59-74, chmn dept biol sci, 62-74, EMER PROF BIOL, DETROIT INST TECHNOL, 74- Mem: AAAS; Am Inst Biol Sci; Am Soc Plant Taxon; Bot Soc Am; Nat Asn Biol Teachers. Res: Taxonomic botany; plants of the western national parks. Mailing Add: 4727 Second Ave Detroit MI 48201

BAILEY, WILFORD SHERRILL, b Somerville, Ala, Mar 2, 21; m 42; c 4. VETERINARY PARASITOLOGY. Educ: Auburn Univ, DVM, 42, MS, 46; Johns Hopkins Univ, ScD(hyg), 50. Prof Exp: From instr to assoc prof, Sch Vet Med, Auburn Univ, 42-48, head prof, 50-62, res prof parasitol & assoc dean grad sch & coordr res, 62-66, vpres acad affairs, 66-69, vpres acad & admin affairs, 69-72; health scientist adminr, Nat Inst Allergy & Infectious Dis, 72-74; PROF PATH & PARASITOL, AUBURN UNIV, 74- Concurrent Pos: NSF sci fac fel, 59; mem, Nat Res Coun, 63-69; mem comt vet med educ & res, Nat Acad Sci, 68-70; mem, Nat Adv Allergy & Infectious Dis Coun, NIH, 71-72 & 75-78. Mem: Am Soc Parasitol (pres, 71); Am Vet Med Asn; Am Soc Trop Med & Hyg; World Asn Advan Vet Parasitol. Res: Life cycle and immunity to Hymenolepis nana; immunology and pathology of Cooperia punctata and ruminant stomach worms; toxic hepatitis in dogs; life cycle and pathology of Spirocerca lupi and its relationship to neoplasia; prevalence of Pneumocystis carinii in dogs and cats. Mailing Add: Dept of Path Auburn Univ Auburn AL 36830

BAILEY, WILFRID CHARLES, b Cicero, Ill, May 3, 18; m 41; c 4. ANTHROPOLOGY. Educ: Univ Ariz, BS, 40, MA, 42; Univ Chicago, PhD(anthrop), 55. Prof Exp: Asst, Ariz State Mus & Univ Ariz, 37-41; asst anthrop, Univ Chicago, 42 & 46-47, mem comt human develop, 42-43 & 46; instr, Univ Tex, 47-51, asst prof, 51-55; from asst prof to prof, Miss State Univ, 55-62; PROF ANTHROP & HEAD DEPT, UNIV GA, 62- Mem: Fel Am Anthrop Asn; fel Am Sociol Asn; Rural Sociol Soc; Am Ethnol Soc; Royal Anthrop Inst Gt Brit & Ireland. Res: Social and educational anthropology; regional and rural development. Mailing Add: Dept of Anthrop Univ of Ga Athens GA 30602

BAILEY, WILLIAM AQUILA, JR, organic chemistry, see 12th edition

BAILEY, WILLIAM BEST, b Yarmouth, NS, Aug 30, 24; m 45; c 4. PHYSICAL OCEANOGRAPHY. Educ: Acadia Univ, BSc, 48. Prof Exp: Demonstr physics, Acadia Univ, 47-48 & Dalhousie Univ, 48-50; asst oceanogr, Atlantic Oceanog Group, Fisheries Res Bd, Can, 50-57, assoc oceanogr, 57-62; staff officer oceanog, 62-72, SR STAFF OFFICER OCEANOG, HQ, CAN MARITIME COMMAND, 72- Concurrent Pos: Sr oceanogr, Circumnavig NAm, HMCS Labrador, 54; sr sci officer, Bedford Inst Oceanog, 62-72. Res: Physical oceanography of Atlantic and Arctic waters; synoptic oceanography; climatic changes and their effect on the oceans and man. Mailing Add: 26 Richards Dr Dartmouth NS Can

BAILEY, WILLIAM CHARLES, b Albany, NY, Aug 21, 39. NUCLEAR MAGNETIC RESONANCE. Educ: Siena Col, BS, 63; State Univ NY Albany, MS, 66, PhD(physics), 71. Prof Exp: RES ASSOC PHYSICS, STATE UNIV NY ALBANY, 71- Concurrent Pos: J Corbett fel physics, AEC, State Univ NY Albany, 71-73, fel Inst Study of Defects in Solids, 75- & instr phys sci, Educ Opportunities Prog, 75- Mem: Am Phys Soc. Res: Ion diffusion in superionic conducting solids such as sodium beta alumina. Mailing Add: Dept of Physics State Univ NY Albany NY 12222

BAILEY, WILLIAM FRANCIS, b Jersey City, NJ, Dec 8, 46. ORGANIC CHEMISTRY. Educ: St Peter's Col, NJ, BS, 68; Univ of Notre Dame, PhD(chem), 73. Prof Exp: Assoc chem, Univ NC, 73 & Yale Univ, 74-75; ASST PROF CHEM, UNIV CONN, 75- Mem: Am Chem Soc; Sigma Xi. Res: Conformational analysis; molecular structure and energetics; Carbon-13 magnetic resonance; synthesis and reactions of strained-ring systems; preparative electrochemistry. Mailing Add: Dept of Chem Univ of Conn Storrs CT 06268

BAILEY, WILLIAM JOHN, b East Grand Forks, Minn, Aug 11, 21; m 49; c 3. ORGANIC CHEMISTRY. Educ: Univ Minn, BChem, 43; Univ Ill, PhD(org chem), 46. Prof Exp: Asst org chem, Univ Ill, 43, asst War Prod Bd prog, 43-46; Little fel,

Mass Inst Technol, 46-47; from asst prof to assoc prof org chem Wayne State Univ, 47-51; PROF ORG CHEM, UNIV MD, 51- Concurrent Pos: Welch Found lectr, 70-71; consult, Am Cyanamid Co, Goodyear Tire & Rubber Co, Phillips Fibers, BASF Wyandotte Corp, Naval Surface Weapons Lab & Hydron Labs; chmn, Gordon Res Conf Org Reactions, 60; mem, Fel Selection Comt, NSF, 65-68; mem, Nat Res Coun Elastomers Adv Comt, US Army Natick Lab; chmn, Comt Macromolecules, Nat Res Coun; nat rep to macromolecular div, Int Union Pure & Appl Chem, 68-75. Honors & Awards: Res Award, Fatty Acid Producers, 55; Chem Soc Wash Serv Award, 69; Hon Scroll Award, DC Inst Chemists, 75. Mem: AAAS; Am Chem Soc (pres, 75); Am Oil Chemists Soc. Res: Cyclic dienes; pyrolysis of esters and unsaturated compounds; polypeptides; phosphorus compounds; spiro and ladder polymers; biodegradable polyamides; monomers that expand on polymerization; mechanism of thermal decompostion of vinyl polymers. Mailing Add: Dept of Chem Univ of Md College Park MD 20742

BAILEY, WILLIAM ROBERT ARTHUR, microbiology, deceased

BAILEY, WILLIAM T, b Buffalo, NY, Apr 4, 36; m 62; c 2. MATHEMATICS. Educ: State Univ NY Buffalo, BA, 59, MA, 62, EdD(math educ), 71. Prof Exp: Instr statist, State Univ NY Buffalo, 60-64; ASSOC PROF MATH, STATE UNIV NY COL BUFFALO, 64- Mem: Math Asn Am; Sch Sci & Math Asn; Nat Coun Teachers Math. Res: Group reaction to and productivity on a mathematical task involving productive thinking. Mailing Add: 19 Woodcrest Blvd Kenmore NY 14223

BAILEY, ZENO EARL, b Frisco City, Ala, Aug 9, 21; m 56; c 2. BOTANY, GENETICS. Educ: Auburn Univ, BS, 50, MS, 53; Ohio State Univ, PhD(agr educ), 55; ETex State Univ, MS, 61. Prof Exp: Teacher high sch, Ala, 50-53; instr biol, Snead Jr Col, 55-56; assoc prof agr educ, ETex State Univ, 56-60, assoc prof biol, 61-67; prof, Livingston Univ, 67-69; PROF BOT, EASTERN ILL UNIV, 69- Concurrent Pos: Tex State Univ study grant, 65-67. Mem: AAAS. Res: Phases of the life history of the boat-tailed grackle. Mailing Add: Dept of Bot Eastern Ill Univ Charleston IL 61920

BAILIE, JAMES CLYDE, organic chemistry, see 12th edition

BAILIE, MICHAEL DAVID, b South Bend, Ind, Sept 27, 36; m 58; c 2. PEDIATRICS, PHYSIOLOGY. Educ: Ind Univ, Bloomington, AB, 59, MA, 60, PhD(physiol), 66; Ind Univ, Indianapolis, MD, 64. Prof Exp: Asst prof, 70-73, ASSOC PROF HUMAN DEVELOP & PHYSIOL, COL HUMAN MED, MICH STATE UNIV, 73- Concurrent Pos: Fel, Univ Tex Southwestern Med Sch Dallas, 68-70. Mem: Am Fedn Clin Res; Am Soc Nephrology; Int Soc Nephrology; Soc Pediat Res; Am Physiol Soc. Res: Renal physiology, specifically, control of renal hemodynamics; relation of renin-angiotensin system to renal hemodynamics; intra renal actions and formation of angiotensin II; development of renal function; renal metabolism. Mailing Add: Dept of Human Develop Mich State Univ East Lansing MI 48824

BAILIE, WAYNE E, b Bondurant, Iowa, Aug 28, 32; m 55; c 3. VETERINARY MICROBIOLOGY. Educ: Kans State Univ, BS & DVM, 57, PhD(path), 69. Prof Exp: Vet, self-employed, Nebr, 57-64; instr path, Kans State Univ, 64-69; assoc prof, SDak State Univ, 69-70 & Ahmadu Bello Univ, Nigeria, 70-72; ASSOC PROF PATH, KANS STATE UNIV, 72- Mem: Am Vet Med Asn; Am Soc Microbiol. Res: Infectious diseases of domestic animals; taxonomy of unnamed species of bacteria associated with domestic animals. Mailing Add: Col of Vet Med Kans State Univ Manhattan KS 66506

BAILIN, GARY, b New York, NY, Apr 2, 36. BIOCHEMISTRY. Educ: City Col New York, BS, 58; Adelphi Univ, PhD(biochem), 65. Prof Exp: Res asst, Adelphi Univ, 59-61; res assoc protein chem, Inst Muscle Dis, 64-74, ASST PROF BIOCHEM, MT SINAI SCH MED, 74- Concurrent Pos: Adj asst prof, Brooklyn Col, 68-75. Mem: Biophys Soc. Res: Determination of structure of enzymes and proteins; correlation of structure and function of enzymes and proteins. Mailing Add: 25 Lucille Ct Edison NJ 08817

BAILIN, LIONEL J, b New York, NY, Oct 28, 28; m 58; c 4. PHYSICAL INORGANIC CHEMISTRY. Educ: NY Univ, BA, 49; Polytech Inst Brooklyn, MS, 52; Tulane Univ, PhD(inorg chem), 58. Prof Exp: Develop chemist plaster of Paris, Davis & Geck Co, Am Cyanamid Co, 51-52; phys chemist chem warfare, Army Chem Ctr, Md, 52-54; chemist pigment technol, E I du Pont de Nemours & Co, 57-62; res scientist, 62-73, STAFF SCIENTIST, CHEM DEPT, PALO ALTO RES LAB, LOCKHEED MISSILES & SPACE CO, 74- Mem: Fel Am Inst Chemists. Res: Microwave and radiation field plasma chemistry; inorganic and modified inorganic materials stable to electromagnetic radiation; protective coatings technology. Mailing Add: Palo Alto CA

BAILLIE, ANDREW DOLLAR, b Dollar, Scotland, Nov 20, 12; m 46. STRATIGRAPHY. Educ: Univ Man, BSc, 49, MSc, 50; Northwestern Univ, PhD(geol), 53. Prof Exp: Geologist, Mines Br, Man, 50-53; Can Gulf Oil Co, 53-57; from regional geologist to res geologist, Brit Am Oil Co, 57-70; dir res geol, 70-72, MGR GEOL SERV, GULF OIL CAN, 72- Honors & Awards: Spencer Gold Medal, Univ Man, 53; Medal Merit, Alta Soc Petrol Geol, 55. Mem: Am Asn Petrol Geologists; fel Geol Soc Am; Soc Econ Paleont & Mineral; fel Geol Asn Can; Can Soc Petrol Geol. Res: Sedimentation and allied fields. Mailing Add: 917 Rideau Rd Calgary AB Can

BAILLIF, RALPH NORMAN, b Minneapolis, Minn, May 5, 09; m 39; c 3. CYTOLOGY. Educ: St Olaf Col, AB, 30; Univ Minn, AM, 33, PhD(zool), 35. Prof Exp: Teaching asst zool, Univ Minn, 31-35, teaching fel anat, 35-37; instr, Sch Med, La State Univ, 37-42, asst prof, 43-47; from asst prof to prof, 47-74, EMER PROF ANAT, SCH MED, TULANE UNIV, 74- Mem: Am Asn Cancer Res; Reticuloendothelial Soc; Endocrine Soc; Am Asn Anat. Res: Cytophysiology of secretion; reticuloendothelial activity; hematology; relation of RE system to cancer growth; Ehrlich ascites tumor and mouse host aging; aging and immunity. Mailing Add: 221 Ridgeway Dr Metairie LA 70001

BAILY, NORMAN ARTHUR, b New York, NY, July 2, 15; m 40; c 2. MEDICAL PHYSICS, RADIOBIOLOGY. Educ: St John's Univ, NY, BS, 41; NY Univ, MA, 43; Columbia Univ, PhD(physics), 52. Prof Exp: Res scientist, Columbia Univ, 46-52; sci adv, US Air Force, 52-54; prin cancer res scientist & chief physicist, Dept Radiation Ther, Roswell Park Mem Inst, 54-59; from assoc clin prof to prof in residence radiol, Univ Calif, Los Angeles, 59-68; prof radiol & physics & dir div radiol sci, Emory Univ, 67-68; PROF RADIOL, UNIV CALIF, SAN DIEGO, 68- Concurrent Pos: NASA grants, Univ Calif, San Diego, 68-73 & 73-75, AEC grant, 74-75 & NIH grant, 75-76; radiation physicist, Marine Biol Lab, Woods Hole, Mass, 46-52; assoc radiol, Sch Med, Univ Buffalo, 54-59; asst res prof biophys, Roswell Park Div, 57-59; lectr chem, Canisius Col, 57-59; mem sci adv bd, US Air Force, 63-74; consult, Rand Corp, 67-71, US Naval Hosp, San Diego, 68- & Vet Admin Hosp, San Diego, 71-; vis scientist, Europ Orgn Nuclear Res, 70; vis prof, Hebrew Univ, Jerusalem, 72; mem NIH radiation study sect, 75- Mem: AAAS; Am Asn Physicists in Med; fel Am Col Radiol; Am Phys Soc; Radiation Res Soc. Res: Quantitation of diagnostic radiological procedures; dose reduction in diagnostic radiology; microdosimetry. Mailing Add: 5046 Basic Sci Bldg Univ of Calif San Diego La Jolla CA 92093

BAILY, WALTER LEWIS, JR, b Waynesburg, Pa, July 5, 30; c 1. MATHEMATICS. Educ: Mass Inst Technol, SB, 52; Princeton Univ, MA, 53, PhD(math), 55. Prof Exp: Instr math, Princeton Univ, 55-56 & Mass Inst Technol, 56-57; from asst prof to assoc prof, 57-63, PROF MATH, UNIV CHICAGO, 63- Concurrent Pos: Mathematician, Bell Tel Labs, 57-58; NSF sr fel, 65-66. Mem: Am Math Soc; Math Soc Japan. Res: Algebraic groups; modular forms; arithmetic properties of automorphic forms. Mailing Add: Dept of Math Univ of Chicago Chicago IL 60637

BAILYN, MARTIN H, b Conn, June 24, 28. PHYSICS. Educ: Williams Col, BA, 48; Harvard Univ, PhD, 56. Prof Exp: From asst prof to assoc prof, 58-66, PROF PHYSICS, NORTHWESTERN UNIV, 66- Res: Magnetism in metals. Mailing Add: Dept of Physics Northwestern Univ Evanston IL 60201

BAIN, BARBARA, b Montreal, Que, May 8, 32. EXPERIMENTAL MEDICINE. Educ: McGill Univ, BSc, 53, MSc, 57, PhD(exp med), 65. Prof Exp: Res asst hemat, Royal Victoria Hosp, 65-68; demonstr, 65-67, lectr, 67-68, ASST PROF EXP MED, McGILL UNIV, 68-; RES ASSOC HEMAT, ROYAL VICTORIA HOSP, 68- Concurrent Pos: Med Res Coun Can scholar, 66-71, grant, 70- Mem: Tissue Cult Asn; Transplantation Soc; Can Soc Immunol; Can Soc Cell Biol; Am Soc Cell Biol. Res: Lymphocyte transformation to blast cells in vitro, using allogenic leukocytes and other stimuli; application of mixed leukocyte reaction to histocompatibility testing; mechanisms of blast cell transformation. Mailing Add: Hemat Div Royal Victoria Hosp Montreal PQ Can

BAIN, DOUGLAS COGBURN, plant pathology, deceased

BAIN, GEORGE HOWARD, inorganic chemistry, see 12th edition

BAIN, GORDON ORVILLE, b Lethbridge, Alta, Nov 1, 26; m 51; c 5. PATHOLOGY. Educ: Univ Alta, BSc, 49, MD, 51. Prof Exp: From asst prof to assoc prof, 56-67, PROF PATH, UNIV ALTA, 67-, CHMN DEPT, 70- Mem: Am Soc Exp Path; Reticuloendothelial Soc; Transplantation Soc; Int Acad Path. Res: Hemotological pathology; immunopathology; tissue transplantation. Mailing Add: Dept of Path Univ of Alta Edmonton AB Can

BAIN, JAMES ARTHUR, b Langdon, NDak, May 22, 18; m 47; c 2. BIOCHEMICAL PHARMACOLOGY. Educ: Univ Wis, BS, 40, PhD(physiol), 44. Prof Exp: Asst, Univ Wis, 40-44; res assoc psychiat & pharmacol, Med Sch, Univ Ill, 47-50, from asst prof to assoc prof, 50-54; chmn dept pharmacol, 57-62, PROF PHARMACOL, EMORY UNIV, 54-, DIR DIV BASIC HEALTH SCI, 60- Concurrent Pos: Rockefeller fel, Univ Wis, 46-47. Mem: AAAS; Asn Cancer Res; Am Chem Soc; Am Soc Pharmacol & Exp Therapeut; Soc Exp Biol & Med. Res: Biochemical basis of drug action; epilepsy; carcinogenesis; neurochemistry. Mailing Add: Div of Basic Health Sci Emory Univ Atlanta GA 30322

BAIN, LEE J, b Newkirk, Okla, Jan 11, 39; m 62; c 3. MATHEMATICAL STATISTICS. Educ: Okla State Univ, BS, 60, MS, 62, PhD(math), 63. Prof Exp: From asst prof to assoc prof, 63-73, PROF MATH, UNIV MO-ROLLA, 73- Mem: Am Statist Asn; Math Asn Am. Res: Statistical inference concerning life-testing distribution and reliability problems. Mailing Add: Dept of Math Univ of Mo-Rolla Rolla MO 65401

BAIN, RALPH LEE, b Los Angeles, Calif, May 8, 33; m 64; c 1. CHEMISTRY. Educ: Univ Ill, BSCh, 56; Ore State Univ, PhD(chem), 64. Prof Exp: Asst prof chem, Univ Sask, 64-66; asst prof, 66-70, assoc prof & chmn dept, 70-73, PROF CHEM, SOUTHERN ILL UNIV, 75- Concurrent Pos: Vis scientist, Argonne Nat Lab, 69-; NSF fel, 69; hon res fel, Univ Col, Univ London, 73. Mem: Am Chem Soc; Can Inst Chem. Res: Equilibrium studies with heavy metal amine complexes. Mailing Add: Southern Ill Univ Dept of Chem Edwardsville IL 62026

BAIN, ROGER J, b Kenosha, Wis, Mar 29, 40; m 66; c 2. GEOLOGY. Educ: Univ Wis, BS, 62, MS, 64; Brigham Young Univ, PhD(geol), 68. Prof Exp: Asst prof geol, Univ RI, 67-68 & Univ Va, 68-70; asst prof, 70-74, ASSOC PROF GEOL, UNIV AKRON, 74- Concurrent Pos: Res grants, 68-70, 71, 73 & 75; consult, NAm Explor, 70. Mem: Soc Econ Paleont & Mineral; Am Asn Petrol Geologists; Geol Soc Am. Res: Sedimentary and environmental geology; sedimentology; provenance; paleoecology; paleoenvironments. Mailing Add: Dept of Geol Univ of Akron Akron OH 44325

BAIN, WILLIAM MURRAY, b Indianapolis, Ind, Dec 28, 28; m 64; c 2. MICROBIAL PHYSIOLOGY. Educ: Ind Univ, AB, 51, MA, 53, PhD(bact), 59. Prof Exp: From asst prof to assoc prof, 59-75, PROF BACT, UNIV MAINE, 75- Mem: AAAS; Am Soc Microbiol. Res: Endogenous metabolism of bacteria; survival of bacteria, terrestial and aquaric bacteria. Mailing Add: Dept of Microbiol Univ of Maine Orono ME 04473

BAINBOROUGH, ARTHUR RAYMOND, b Can, Mar 8, 18; m 45; c 2. PATHOLOGY. Educ: Univ Western Ont, BA, 43, MD, 46; McGill Univ, MSc, 51. Prof Exp: Pathologist & dir labs, Med Hat Gen & Galt Hosps, 51-55; PATHOLOGIST & DIR LABS, ST MICHAEL'S GEN & LETHBRIDGE MUNIC HOSPS & LETHBRIDGE CANCER CLIN, ALTA, 51- Mem: Fel Am Soc Clin Pathologists; Can Asn Pathologists (pres, 65-66); Int Acad Path. Res: Atherosclerosis; histopathology of prostate, bladder and lungs. Mailing Add: Lethbridge Munic Hosp Lethbridge AB Can

BAINBRIDGE, ARNOLD ERNEST, b Invercarghill, NZ, Dec 16, 30; m 58; c 1. GEOPHYSICS, OCEANOGRAPHY. Educ: Univ NZ, BSc, 52. Prof Exp: Physicist isotope geophys, Inst Nuclear Sci, NZ, 52-60, Univ Calif, San Diego, 60-62 & Nat Ctr Atmospheric Res, 60-66; dir geochem, 66-70, PROJ DIR, NSF OPERS GROUP, OCEAN SECT STUDY, SCRIPPS INST OCEANOG, UNIV CALIF, SAN DIEGO, 70-, HEAD, DATA PROCESSING GROUP, MARINE LIFE RES GROUP, 75- Mem: AAAS; Am Geophys Union; Sigma Xi; Am Vacuum Soc. Res: Geophysical applications of naturally and artificially produced isotopes, particularly the isotope tritium and its occurrence in the oceans and atmospheric gases. Mailing Add: Scripps Inst of Oceanog S-001 La Jolla CA 92037

BAINBRIDGE, KENNETH TOMPKINS, b Cooperstown, NY, July 27, 04; m 31, 69; c 3. NUCLEAR PHYSICS. Educ: Mass Inst Technol, BS, 25, MS, 26; Princeton Univ, AM, 27, PhD(physics), 29. Hon Degrees: AM, Harvard Univ, 42. Prof Exp: Nat res fel physics, 29-31; fel, Bartol Res Found, 31-33; Guggenheim fel, 33-34; from asst prof to prof, 34-61, chmn dept, 53-56, Leverett Prof, 61-75, EMER LEVERETT PROF PHYSICS, HARVARD UNIV, 75- Concurrent Pos: Wagner Inst lectr, 40; div leader & mem steering comt, Radiation Lab, Mass Inst Technol, 40-43; div leader, mem steering comt & dir first atom bomb test, Los Alamos, NMex, 43-45; mem, Solvay Cong Chem, 47; chmn, Task Force Acad Planning Sci, Iran/Harvard Univ-Reza Shah Kabir Univ, 75. Honors & Awards: Levy Medal, Franklin Inst, 33; Presidential Cert Merit Radar, 48. Mem: Nat Acad Sci; Am Acad Arts & Sci; fel Am Phys Soc. Res: Nuclear physics; mass-spectroscopy, mass-spectrometry; photoelectric effect; radar development; high sensitivity caesium-oxygen-silver photocell; amplification of photocell currents by secondary emission; electromagnetic mercury pump; linear direct current motor. Mailing Add: Lyman Lab Harvard Univ Cambridge MA 02138

BAINES, A D, b Toronto, Ont, July 17, 34. PHYSIOLOGY, CLINICAL BIOCHEMISTRY. Educ: Univ Toronto, MD, 59, PhD(path chem), 65. Prof Exp: Assoc prof path chem, 68-72, PROF CLIN BIOCHEM, UNIV TORONTO, 72- Concurrent Pos: Res fel, Dept Med, Univ NC, Chapel Hill, 65-67; res fel, Dept Physiol, AEC, France, 67-68; mem, Inst Biomed Eng & lectr, Dept Med, Univ Toronto, 72-; assoc physician, Toronto Gen Hosp, 72-; prin, New Col, Univ Toronto, 74- Res: Normal and abnormal renal physiology; cellular transport mechanisms; histochemistry. Mailing Add: Dept of Clin Biochem Univ of Toronto Toronto ON Can

BAINS, MALKIAT SINGH, b Montgomery, India, Jan 29, 32; m 66; c 3. PHYSICAL INORGANIC CHEMISTRY. Educ: Panjab Univ, India, BSc, 55, MSc, 57; Univ London, PhD(inorg chem), 59; Atomic Energy Estab, Trombay, Bombay, cert radiation nuclear chem, 63. Prof Exp: Nat Res Coun Can fel, Univ Western Ont, 59-61; lectr inorg chem, Panjab Univ, India, 61-65; reader, Marathwada Univ, India, 65-66; Nat Acad Sci-Nat Res Coun resident res assoc inorg free radicals & electron spin resonance, Agr Res Serv, USDA, 66-68; assoc prof, 68-73, PROF GEN & INORG CHEM, SOUTHERN UNIV, 73- Concurrent Pos: New Delhi Univ Grants Comn grant, 64-66. Mem: Am Chem Soc; Sigma Xi; The Chem Soc; life mem Indian Chem Soc. Res: Coordination chemistry of metal alkoxides and metal carboxylates; flame retardation; inorganic free radicals in solution; application of nuclear magnetic resonance and electron spin resonance to biomedical problems; synthesis of transition metal compounds. Mailing Add: Dept of Chem Southern Univ New Orleans LA 70126

BAINTER, MONICA EVELYN, b Lanesboro, Minn, Feb 28, 10. ATOMIC PHYSICS. Educ: Col St Teresa, Minn, BA, 30; Univ Minn, MA, 38; Univ Wis, PhD(physics, educ), 55. Prof Exp: Teacher physics & math, Chatfield High Sch, Minn, 30-35, Spring Valley High Sch, Minn, 35-38; sr prin, New Richmond High Sch, Wis, 38-43; vis prof math, Univ Wis-Madison, 43, vis prof physics, 43-47, asst prof, Univ Wis-Stevens Point, 47-55, chmn dept, 59-73, PROF PHYSICS, UNIV WIS-STEVENS POINT, 73- Concurrent Pos: Consult nuclear physics & 2nd vpres, Wis Energy Coalition, 75- Mem: Optical Soc Am; Am Asn Physics Teachers. Res: Innovative methods of teaching laboratory using problem solving; techniques for offering service courses to other departments in the university such as communicative disorders, medical technology and others. Mailing Add: Dept of Physics Univ of Wis Stevens Points WI 54481

BAINTON, DOROTHY FORD, b Magnolia, Miss, June 18, 33; m 59; c 3. PATHOLOGY, INTERNAL MEDICINE. Educ: Millsaps Col, BS, 55; Tulane Univ, MD, 58; Univ Calif, San Francisco, MS, 66. Prof Exp: Intern med, Strong Mem Hosp, Univ Rochester, 58-59, resident internal med, 59-60; resident hemat, Sch Med, Univ Wash, 60-62; ASST PROF PATH, SCH MED, UNIV CALIF, SAN FRANCISCO, 66- Mem: Am Soc Cell Biol. Res: Hematology and cell biology; leukocyte maturation and function using the light and electron microscopes, histochemistry and cell fractionation procedures. Mailing Add: Dept of Path Univ of Calif San Francisco CA 94143

BAIR, EDWARD JAY, b Ft Collins, Colo, June 30, 22; m 58. PHYSICAL CHEMISTRY, ANALYTICAL CHEMISTRY. Educ: Colo Agr & Mech Col, BS, 43; Brown Univ, PhD(chem), 49. Prof Exp: Chemist, Manhattan Proj, Tenn Eastman Corp, 44-46; res assoc chem, Univ Wash, 49-54; from instr to assoc prof, 54-65, PROF CHEM, IND UNIV, BLOOMINGTON, 65- Mem: Am Chem Soc; Am Phys Soc; Faraday Soc. Res: Photochemistry; spectroscopy; kinetics of atmospheric and combustion reactions. Mailing Add: Dept of Chem Ind Univ Bloomington IN 47401

BAIR, JOE KEAGY, b Massillon, Ohio, Mar 10, 18; m 41; c 1. NUCLEAR PHYSICS. Educ: Rice Inst, BA, 40. Prof Exp: Asst physics, Columbia Univ, 40-41; physicist, Naval Ord Lab, 41-47 & Fairchild Eng & Airplane Co, 47-51; PHYSICIST, OAK RIDGE NAT LAB, 51- Concurrent Pos: Civilian technician, Naval Mine Modification Unit, 44-45. Mem: Am Phys Soc. Res: Low energy nuclear physics. Mailing Add: 200 W Fairview Rd Oak Ridge TN 37830

BAIR, KENNETH WALTER, b Detroit, Mich, Mar 20, 48; m 69; c 1. SYNTHETIC ORGANIC CHEMISTRY. Educ: Wayne State Univ, BS, 70, MS, 73; Brandeis Univ, PhD(org chem), 76. Prof Exp: Lab technician clin chem, Clin Chem Lab, Harper Hosp, 69-72; res assoc enzyme chem, Dept Internal Med, Hutzel Hosp, 72-73; res assoc org chem, Wayne State Univ, 73; RES ASSOC ORG CHEM, BRANDEIS UNIV, 73- Mem: Am Chem Soc; The Chem Soc. Res: Development of new synthetic methods; application to synthesis of important complex organic molecules. Mailing Add: Dept of Chem Brandeis Univ Waltham MA 02154

BAIR, THOMAS DE PINNA, b New York, NY, Mar 19, 22; m 69; c 2. ZOOLOGY, PHYSIOLOGY. Educ: DePauw Univ, AB, 46; Ind Univ, MA, 47; Univ Ill, PhD(zool, physiol), 51. Prof Exp: Asst zool, Ind Univ, 47; asst zool & physiol, Univ Ill, 47-49, vet parasitol, 49-50; instr biol, Utica Col, Syracuse Univ, 50-53; instr physiol, Albany Med Col, 53-56; from asst prof to assoc prof, 59-72, PROF ZOOL, CALIF STATE UNIV, LOS ANGELES, 72- Concurrent Pos: Vis asst prof, Calif Col Med, 58-64; lectr & consult, Gormac Polygraph Inst, 59-; sr lectr, Univ Southern Calif, 64-; Fulbright lectr, Nangrahar Med Fac, Jalalabad, Afghanistan, 70-71. Mem: AAAS; Am Soc Parasitol. Res: Cardiovascular physiology; comparative physiology; parasitology; marine zoology; history and philosophy of science; physiology of parasites; history and philosophy of science. Mailing Add: Dept of Biol Calif State Univ Los Angeles CA 90032

BAIR, THOMAS IRVIN, b Grier City, Pa, Feb 12, 38; m 66. POLYMER CHEMISTRY, PHYSICAL ORGANIC CHEMISTRY. Educ: Pa State Univ, BS, 60; Univ Wis, PhD(electrophilic substitution), 66. Prof Exp: Res chemist, 65-69, sr res chemist, 69-72, SUPVR RES & DEVELOP, E I DU PONT DE NEMOURS & CO, 72- Mem: Am Chem Soc; Sigma Xi; NY Acad Sci. Res: Cyclopropanol chemistry; nuclear magnetic resonance spectroscopy of cyclopropane compounds; electrophilic substitution at saturated carbon; synthesis and characterization of condensation polymers; fiber technology; rubber reinforcing materials; adhesives and finishes. Mailing Add: 2303 Mousley Pl Beacon Hill Wilmington DE 19810

BAIR, WILLIAM J, b Jackson, Mich, July 14, 24; m 52; c 3. RADIOBIOLOGY, RADIOLOGICAL HEALTH. Educ: Ohio Wesleyan Univ, BA, 49; Univ Rochester, PhD(radiation biol), 54. Prof Exp: Res assoc, Univ Rochester, 51-54; biol scientist, Hanford Labs, Gen Elec Co, 54-56, mgr pharmacol, 56-65; mgr inhalation toxicol sect, Biol Dept, 64-68, mgr biol dept, 68-74, dir life sci prog, 73-75, MGR BIOMED & ENVIRON RES PROG, PAC NORTHWEST LABS, BATTELLE MEM INST, 75- Concurrent Pos: Lectr, Gen Elec Sch Nuclear Eng & Joint Ctr Grad Study, Univ Wash, 55-; mem subcomt on inhalation hazards, Comt on Path Effects of Atomic Radiation, Nat Acad Sci, 57-64; Japan AEC lectr, 69; mem task group biol effects of inhaled particles, Int Comn Radiol Protection, 69-70, chmn, 70-, mem comt 2 on permissible dose for internal radiation, 73-; mem comt 34, Nat Coun Radiation Protection & Measurement, 70-, mem coun, 74-, mem comt 1 basic radiation protection criteria, 75-, chmn ad hoc comt on hot particles, 74-; consult, Adv Comt on Reactor Safeguards, AEC, 71-75; chmn tech group, Am Inst Biol Sci, 72-; mem ad hoc comt on hot particles, subcomt biol effects of ionizing radiation, Nat Acad Sci-Nat Res Coun, 74- Honors & Awards: E O Lawrence Mem Award, US AEC, 70. Mem: AAAS; Am Soc Exp Biol & Med; Radiation Res Soc; Health Physics Soc; Reticuloendothelial Soc. Res: Radiation biology of inhaled radionuclides; health and environmental effects of transuranium elements; establishment of radiation protection criteria and standards for radionuclides. Mailing Add: 102 Somerset Richland WA 99352

BAIRD, ALEXANDER KENNEDY, b Pasadena, Calif, Nov 22, 32. GEOCHEMISTRY, PETROLOGY. Educ: Pomona Col, BA, 54; Claremont Col, MA, 57; Univ Calif, Berkeley, PhD(geol), 60. Prof Exp: Instr geol, Pomona Col, 55-56; from instr to assoc prof, 58-71, PROF GEOL, POMONA COL, 71- Concurrent Pos: Scientist, Viking 1975 Mission to Mars, NASA, 72-77. Mem: AAAS; Microbeam Anal Soc; Geol Soc Am; Nat Asn Geol Teachers. Res: Geochemistry of Mars; geochemistry and petrology of granitic rocks. Mailing Add: Dept of Geol Pomona Col Claremont CA 91711

BAIRD, BRUCE LLOYD, b Ioka, Utah, June 25, 20; m 45; c 5. SOILS, STATISTICS. Educ: Utah State Univ, BS, 47, MS, 49; NC State Univ, PhD(soils, statist), 59. Prof Exp: Asst agr res, Utah & Idaho Sugar Co, 48-50; soil scientist, SDak Agr Exp Sta, USDA, 50-54; res instr soils, NC State Univ, 54-58; head math serv, Wasatch Div, 58-64, SCIENTIST, THIOKOL CHEM CORP, 64- Mem: Am Statist Asn; Am Soc Qual Control. Res: Soil fertility, water relations and crop management; statistical analysis of data, interpretation and experimental design for engineering in rocket industry. Mailing Add: Thiokol Chem Corp Brigham City UT 84302

BAIRD, CRAIG RISKA, b Woodland, Utah, May 6, 39; m 63; c 5. ENTOMOLOGY. Educ: Utah State Univ, BS, 67, MS, 70; Wash State Univ, PhD(entom), 73. Prof Exp: Sr sci aide entom, Wash State Univ, 72-73, pest mgt specialist, Dept Entom, 73-74; ASST PROF ENTOM & EXTEN ENTOMOLOGIST, UNIV IDAHO, 74- Mem: Entom Soc Am; Entom Soc Can. Res: Life history and biology of rodent and rabbit botflies; livestock parasites and control; alfalfa seed production and pest management. Mailing Add: Univ of Idaho Exten Serv PO Box 1058 Caldwell ID 83605

BAIRD, D C, b Edinburgh, Scotland, May 6, 28; m 54; c 4. PHYSICS. Educ: Univ Edinburgh, BSc, 49; Univ St Andrews, PhD(physics), 52. Prof Exp: Lectr, 52-54, from asst prof to assoc prof, 54-65, PROF PHYSICS, ROYAL MIL COL CAN, 65- Concurrent Pos: Defence Res Bd Can res grant, 58- Res: Low temperature physics; magnetic properties of superconductors. Mailing Add: Dept of Physics Royal Mil Col Kingston ON Can

BAIRD, DAVID MCCURDY, b Fredericton, NB, July 28, 20; m 51; c 3. GEOLOGY. Educ: Univ NB, BSc, 41; Univ Rochester, MS, 43; McGill Univ, PhD(geol), 47. Hon Degrees: DSc, Mem Univ Nfld, 72 & Univ NB, 73. Prof Exp: Asst prof geol, Mt Allison Univ, 46-47; from asst prof to assoc prof, Univ NB, 47-52; prof, Mem Univ Nfld, 53-58 & Univ Ottawa, 58-66; DIR, NAT MUS SCI & TECHNOL, 68- Concurrent Pos: Field geologist, Geol Surv Nfld, 42-51 & Geol Surv Can, 58-64; prof geologist, Nfld, 52-58. Honors & Awards: Bancroft Award, Royal Soc Can, 70. Mem: Fel Royal Soc Can; fel Geol Asn Can; Can Inst Mining & Metall. Res: Geology of Newfoundland, evaporites and Appalachian Carboniferous; national parks of Canada. Mailing Add: Nat Mus of Sci & Technol 1867 St Laurent Blvd Ottawa ON Can

BAIRD, DAVID WILLIAM ECCLES, internal medicine, deceased

BAIRD, DERWOOD MCVEY, b Moorefield, Ky, July 13, 22; m 49; c 4. ANIMAL SCIENCE, NUTRITION. Educ: Univ Ky, BS, 47, MS, 48; Univ Ill, PhD(animal sci), 51. Prof Exp: Asst res, Univ Ky, 47-48 & Univ Ill, 48-51; asst, 51-52, ASSOC ANIMAL SCI, GA EXP STA, UNIV GA, 52- Mem: AAAS; Am Soc Animal Sci; Am Soc Range Mgt. Res: Swine nutrition, management and carcass research with special emphasis on energy utilization of feeds in the nutrition of swine. Mailing Add: Ga Exp Sta Dept Animal Sci Univ of Ga Experiment GA 30212

BAIRD, DONALD, b Pittsburgh, Pa, May 12, 26; m 48; c 2. VERTEBRATE PALEONTOLOGY. Educ: Univ Pittsburgh, BS, 48; Univ Colo, MS, 49; Harvard Univ, PhD(biol), 55. Prof Exp: Asst geologist, Pa Topog & Geol Surv, 47-48; cur, Univ Mus, Univ Cincinnati, 49-51; asst cur vert paleont, Mus Comp Zool, Harvard Univ, 54-57; from asst cur to assoc cur, Mus Comp Zool, 57-67, cur, Mus Natural Hist, 67-73, DIR, MUS NATURAL HIST, PRINCETON UNIV, 73- Concurrent Pos: Res assoc, Am Mus Natural Hist, 64-; pres, Princeton Jr Mus, 68-70. Mem: Soc Vert Paleont; Paleont Soc; Soc Study Evolution; Soc Syst Zool; Brit Palaeont Asn. Res: Carboniferous and Triassic reptiles, Amphibians and footprints. Mailing Add: Dept of Geol Princeton Univ Princeton NJ 08540

BAIRD, DONALD HESTON, b Milwaukee, Wis, July 3, 21; m 51; c 2. PHYSICAL CHEMISTRY. Educ: Haverford Col, BS, 43; Harvard Univ, MA, 50, PhD(phys chem), 50. Prof Exp: Asst, Manhattan Proj, 43-46; MEM TECH STAFF, GEN TEL & ELECTRONICS LABS, INC, 50- Mem: Am Chem Soc; Am Phys Soc; Electrochem Soc; Inst Elec & Electronic Engr; AAAS. Res: Solid state physics; inorganic and physical chemistry; magnetic materials; semiconductors. Mailing Add: GTE Labs Inc 40 Sylvan Rd Waltham MA 02168

BAIRD, HENRY W, III, b Leavenworth, Kans, Oct 10, 22; m 50; c 4. PEDIATRICS, NEUROLOGY. Educ: Yale Univ, BS, 45, MD, 49; Am Bd Pediat, dipl, 56. Prof Exp: Intern med, San Francisco Hosp, Calif, 49-50; resident, 51-53, from instr to assoc prof, 53-68, PROF PEDIAT, SCH MED, TEMPLE UNIV, 68- Concurrent Pos: Fel pediat neurol, Sch Med, Temple Univ, 50-51; attend pediatrician, St Christopher's Hosp Children, 66- Mem: Soc Pediat Res; Am Acad Cerebral Palsy; Am Acad Pediat. Res: Convulsive disorders; experiment production and treatment of movement disorders; subcortical electroencephalography; data processing; medical records. Mailing Add: Dept of Pediat St Christopher's Hosp Children Philadelphia PA 19133

BAIRD, HERBERT WALLACE, b Morganton, NC, Sept 2, 36; m 58; c 2. STRUCTURAL CHEMISTRY. Educ: Berea Col, BA, 58; Univ Wis, PhD(chem), 63. Prof Exp: Asst chem, Univ Wis, 59-60; from asst prof to assoc prof, 63-75, PROF CHEM, WAKE FOREST UNIV, 75- Mem: Am Chem Soc; Am Crystallog Asn. Res: X-ray diffraction structure analysis. Mailing Add: Dept of Chem Wake Forest Univ Winston-Salem NC 27109

BAIRD, IRWIN LEWIS, b St Joseph, Mo, Mar 11, 25; m 49; c 3. ANATOMY. Educ: Univ Kans, BA, 48, MA, 49; Harvard Univ, PhD, 60. Prof Exp: Instr zool, Univ Mass, 49-50; from instr to asst prof anat, Univ Kans, 52-62, assoc prof, Univ Tenn, 62-67; assoc prof, 67-70, PROF ANAT, MILTON S HERSHEY MED CTR, PA STATE UNIV, 70- Concurrent Pos: Consult, Fed Bur & Kans Bur Invest, 55-62; mem communicative sci study sect, Nat Inst Neurol Dis & Stroke, 69-73. Mem: Am Asn Anat; Am Soc Zool. Res: Comparative morphology, histology, fine structure, and evolution of the vertebrate inner ear, particularly in reptiles. Mailing Add: Milton S Hershey Med Ctr Pa State Univ Hershey PA 17033

BAIRD, JACK VERNON, b Grand Island, Nebr, July 7, 28; m 49; c 3. SOIL SCIENCE. Educ: Univ Nebr, BS, 49, MS, 51; Wash State Univ, PhD(soil sci), 55. Prof Exp: Exten agronomist soil mgt, Univ Ill, 58-60; gen agron, Kans State Univ, 61-64; exten agronomist soil sci, 64-71, actg head dept, 70-71, EXTEN PROF SOIL SCI, NC STATE UNIV, 71- Mem: Am Soc Agron; Soil Sci Soc Am. Res: Soil fertility problems in North Carolina; proper fertilizer usage and soil management. Mailing Add: Dept of Soil Sci NC State Univ Raleigh NC 27607

BAIRD, JAMES, b Glasgow, Scotland, Feb 10, 25; nat US; m 52; c 4. ORNITHOLOGY. Educ: Univ Mass, BS, 51. Prof Exp: Dir, Norman Bird Sanctuary, 55-60; asst to exec vpres, 61-68, DIR NATURAL HIST SERV, MASS AUDUBON SOC, 68- Concurrent Pos: Pres, Defenders of Wildlife, 71- Mem: Am Ornith Union; Wilson Ornith Union; Ecol Soc Am; Cooper Ornith Soc; Brit Ornith Union. Res: Bird migration, weights, molts and life histories. Mailing Add: 69 Hartwell Ave Littleton MA 01460

BAIRD, JAMES CLYDE, b Montgomery, Ala, Oct 26, 31; m 59; c 1. PHYSICAL CHEMISTRY. Educ: Stanford Univ, BS, 53; Rice Inst, PhD(chem), 59. Prof Exp: Fel chem, Harvard Univ, 58-60; res chemist, Calif Res Corp, 60-62; from asst prof to assoc prof chem, 62-71, PROF CHEM & PHYSICS, BROWN UNIV, 71- Concurrent Pos: Res assoc, Univ Calif, Berkeley, 60-62; Alfred Sloan fel, 65-67; vis fel, Joint Inst Lab Astrophys, Univ Colo, 70-71. Mem: Am Phys Soc. Res: Atomic and molecular physics and chemistry; radio frequency spectroscopy. Mailing Add: Dept of Chem Brown Univ Providence RI 02912

BAIRD, JAMES HAYTHORN, b Newark, NJ, June 26, 22; m 44. ANALYTICAL CHEMISTRY. Educ: Rutgers Univ, BA, 43; Polytech Inst Brooklyn, MS, 52. Prof Exp: Anal res chemist, Merck & Co, NJ, 47-49, chief micros sect, 49-53; sr res chemist, Celanese Corp, 53-56; group leader anal chem, 56-63, mgr anal res dept, Wash Res Ctr, 63-67, dir, 67-70, DIR ANAL RES, PHYS TESTING & INFO CTR, WASH RES CTR, W R GRACE & CO, 70- Mem: AAAS; Am Chem Soc; Am Crystallog Asn; NY Acad Sci; Brit Soc Anal Chem. Res: Microscopy; functional group analysis; radiochemistry; x-ray diffraction. Mailing Add: Wash Res Ctr W R Grace & Co 7379 Rte 32 Columbia MD 21044

BAIRD, JAMES LEROY, JR, b Bridgeport, Conn, Aug 5, 34; m 56; c 3. COMPARATIVE PHYSIOLOGY. Educ: Tufts Univ, BS, 56; Univ Minn, MS, 59; Univ Conn, PhD(zool), 64. Prof Exp: Res entomologist, US Forest Serv, 59; instr biol, Conn Col, 62-63 & zool, Univ Conn, 63-64; asst prof, Lafayette Col, 64-71, actg dean studies, 71; assoc prof biol & head dept, Rochester Inst Technol, 71-74, fac assoc student affairs, 74-75; DIR, SOUTHEASTERN BR, UNIV CONN, 75- Mem: Entom Soc Am; Am Soc Zoologists. Res: Comparative physiology of insect flight; insect neuromuscular biology; biological control systems; arthropod behavior. Mailing Add: Southeastern Br Univ of Conn Groton CT 06340

BAIRD, JOHN JEFFERS, b North English, Iowa, Jan 1, 21; m 45; c 1. NEUROEMBRYOLOGY. Educ: Iowa State Teachers Col, AB, 48; Univ Iowa, MS, 53, PhD, 57. Prof Exp: Teacher high sch, Iowa, 48-54; instr, Univ Iowa, 54-56; from asst prof to prof biol embryol, Calif State Col Long Beach, 56-73, chmn dept biol, 61-67, assoc dean acad planning, 67-73, DEP DEAN INSTRNL PROGS, DIV EDUC PROGS & RESOURCES, CALIF STATE UNIV & COLS, 73- Mem: AAAS; Am Soc Zool; Am Inst Biol Sci. Res: Peripheral control of development of motor and sensory elements of central nervous system. Mailing Add: Div of Educ Progs & Resources Calif State Univ & Cols 5670 Wilshire Blvd Los Angeles CA 90036

BAIRD, KENNETH MACCLURE, b Hawaiking fu, China, Jan 23, 23; Can citizen; m 51; c 4. PHYSICS. Educ: Univ New Brunswick, BSc, 43; Univ Bristol, PhD(physics), 53. Prof Exp: PRIN RES SCIENTIST, NAT RES LABS, 43-48 & 50- Res: Optics; interferometry. Mailing Add: Appl Physics Dept Nat Res Labs Ottawa ON Can

BAIRD, LEEMON CLAUDE, b Meridian, Miss, Sept 6, 40; m 59; c 2. MATHEMATICS. Educ: Park Col, BA, 61; Univ Nebr, Lincoln, MS, 63; Ind Univ, Bloomington, PhD(math, physics), 70. Prof Exp: Instr math & physics, Brescia Col, Ky, 64-66; ASST PROF MATH, VA POLYTECH INST & STATE UNIV, 70- Mem: AAAS; Am Phys Soc; Am Asn Physics Teachers; Am Math Soc; Math Asn Am. Res: Mathematical physics; scattering theory; functional analysis; approximation methods for differential equations; integral equations; foundations of quantum mechanics. Mailing Add: Dept of Math Va Polytech Inst & State Univ Blacksburg VA 24061

BAIRD, MALCOLM BARRY, b Wilkes Barre, Pa, Feb 4, 43; m 64; c 4. PHYSIOLOGY, BIOCHEMICAL PHARMACOLOGY. Educ: Wilkes Col, AB, 64; Univ Del, MA, 68, PhD(genetics), 70. Prof Exp: Olsen Mem fel, 70-72, RES SCIENTIST EXP GERONT, MASONIC MED RES LAB, 72- Concurrent Pos: Vis prof, Syracuse Univ, 73-75. Mem: Genetics Soc Am; Am Asn Cancer Res; AAAS; Am Aging Asn. Res: Genetics of development and mechanisms of aging; chemical carcinogenesis; metabolism of xenobiotics via mixed-function oxidase activity. Mailing Add: Masonic Med Res Lab 2150 Bleecker St Utica NY 13503

BAIRD, MERTON DENISON, b Ft Wayne, Ind, Sept 9, 40; m 67; c 2. ORGANIC CHEMISTRY. Educ: Purdue Univ, BS, 62; Univ Wis, PhD(org chem), 69. Prof Exp: Asst prof, 68-72, chmn dept, 74-76, ASSOC PROF CHEM, SHIPPENSBURG STATE COL, 68- Mem: Am Chem Soc. Res: Synthesis of organic compounds related to natural products; thermal and photochemical isomerization reactions; applications of nuclear magnetic resonance to conformational analysis of molecules and correlations of structure and reactivity. Mailing Add: Dept of Chem Shippensburg State Col Shippensburg PA 17257

BAIRD, NORMAN COLIN, b Montreal, Que, May 22, 42; m 65; c 1. PHYSICAL CHEMISTRY. Educ: McGill Univ, BSc, 63, PhD(phys chem), 67. Prof Exp: Robert A Welch fel, Univ Tex, 66-68; asst prof, 68-71, ASSOC PROF PHYS CHEM, UNIV WESTERN ONT, 71- Mem: Am Chem Soc. Res: Quantum-mechanical investigations of the structure, energy and chemical bonding in molecules. Mailing Add: Dept of Chem Univ of Western Ont London ON Can

BAIRD, QUINCEY LAMAR, b Hoschton, Ga, Mar 22, 32; m 56; c 4. REACTOR PHYSICS. Educ: Berry Col, AB, 52; Emory Univ, MS, 54; Vanderbilt Univ, PhD(physics), 58. Prof Exp: Teacher pvt sch, Tenn, 52-53; physicist, Argonne Nat Lab, 58-63; engr Nuclear Technol Div, Idaho Opers Off, US AEC, 63-66; sr res scientist, Pac Northwest Lab, Battelle Mem Inst, 66-70; sr res scientist, Wadco Corp, 70-74, SR ENGR & TECH ASSOC, WESTINGHOUSE-HANFORD CO, 74- Mem: Am Phys Soc; Am Nuclear Soc. Res: Experimental reactor physics related to power reactor design and operation; nuclear engineering aspects of system design; liquid metal and fast breeder reactor safety, design requirements and operations analysis. Mailing Add: 2404 Swift Blvd Richland WA 99352

BAIRD, RICHARD LEROY, b Sioux City, Iowa, Aug 15, 31; m 59; c 2. POLYMER CHEMISTRY. Educ: Reed Col, BA, 53; Univ Calif, Los Angeles, PhD, 58. Prof Exp: Instr chem, Yale Univ, 58-60, asst prof, 60-63, lectr, 63-64; res chemist, Cent Res Dept, 64-69, Elastomers Dept, 69-73, develop supvr, 73-74, DIV HEAD, ELASTOMERS DEPT, E I DU PONT DE NEMOURS & CO, 74- Res: Physical organic chemistry; coordination chemistry. Mailing Add: Elastomers Dept Exp Sta E I du Pont de Nemours & Co Wilmington DE 19898

BAIRD, RONALD C, b New Albany, Ind, Apr 1, 36; m 62; c 1. ICHTHYOLOGY, MARINE ECOLOGY. Educ: Yale Univ, BS, 58; Univ Tex, Austin, MA, 65; Harvard Univ, PhD(biol), 69. Prof Exp: Asst ecol & physiol, Univ Tex, Austin, 63-65; asst to cur fishes, Mus Comp Zool, Harvard Univ, 68-69; asst prof, 69-73, ASSOC PROF MARINE SCI, MARINE SCI INST, UNIV SOUTH FLA, 69- Concurrent Pos: Res asst, Inst Marine Sci, Tex, 64. Mem: Am Soc Fishery Res Biologists. Res: Ecology and behavior of fishes; application of field and laboratory results to general and theoretical problems; biology of the mesopelagic and midwater fish fauna; computer techniques in ecological problems. Mailing Add: Marine Sci Inst Univ of South Fla St Petersburg FL 33701

BAIRD, RONALD JAMES, b Toronto, Ont, May 3, 30; m 55; c 3. CARDIOVASCULAR SURGERY. Educ: Univ Toronto, MD, 54, BSc, 56, MSurg, 64; FRCS(C), 59, cert cardiovasc & thoracic surg, 64. Prof Exp: PROF SURG, UNIV TORONTO, 72-, DIR SURG RES, 72-; CHIEF CARDIOVASC SURG, TORONTO WESTERN HOSP, 72- Concurrent Pos: Res assoc, Ont Heart Found, 64-; dir, Can Heart Found, 67- & Ont Heart Found, 67-; assoc ed, Can J Surg, 67-; mem, Med Res Coun Can, 67- Honors & Awards: Royal Col Medal Surg, 69. Mem: Can Cardiovasc Soc (secy-treas, 70-73); fel Am Col Surg; Int Cardiovasc Soc (vpres, 73-75); Am Surg Asn; Soc Univ Surg. Res: Cardiovascular surgery and physiology; surgery of the ischemic heart and the ischemic limb. Mailing Add: Suite 304 Toronto Western Hosp 399 Bathurst St Toronto ON Can

BAIRD, STEPHEN SYDNEY, b Laredo, Tex, Nov 20, 29; m 50; c 2. ANALYTICAL CHEMISTRY, ELECTRONICS. Educ: Sul Ross State Col, BS, 49; Ore State Col, MA, 52; Univ Tex, PhD, 58. Prof Exp: Lab asst, Sul Ross State Col, 47-49; asst, Ore State Col, 47-51, analyst, Agr Exp Sta, 50-51; anal chemist, US Naval Powder Factory, 51-53, actg head prod lab, 53; anal chemist, Res Dept, Columbia-Southern Chem Corp, 54-58; anal chemist, Math Res Dept, Tex Instruments, Inc, 58-59, Device Res Dept, 59-64, mgr surface chem div tech progs, 64-65, mgr eng germanium small signal transistor dept, 65-67, mgr environ test & test equip control labs, 67-68, plastic reliability & technol consult, 68-71, tech consult, 72-75; CONSULT, BAIRD TECH SERV, 75- Mem: Am Chem Soc; Electrochem Soc; Soc Plastics Engrs; Am Mgt Asn. Res: Analytical chemistry of pyridine thiocyanates; catalysis in analytical chemistry; trace analysis; wet analysis; spectrophotometry; surface chemistry; semiconductor chemistry. Mailing Add: 8906 Park Lane No 160 Dallas TX 75231

BAIRD, WALTER SCOTT, b Long Green, Md, Oct 2, 08; m 37; c 4. PHYSICS. Educ: St John's Col, Md, BS, 30; Johns Hopkins Univ, PhD(elec eng), 36. Prof Exp: Instr elec eng, Harvard Univ, 34-35; physicist, Watertown Arsenal, 35-36; pres, 36-56 & 64-74, CHMN BD, BAIRD-ATOMIC, INC, 57- Mem: AAAS; fel Am Acad Arts & Sci; Optical Soc Am; Am Inst Mining, Metall & Petrol Eng. Res: Ultraviolet, visible and infrared spectroscopy; effect of hydrogen on tantalum and columbium; automatic recording infrared spectrophotometer; direct reading spectrometer; instrumentation. Mailing Add: 125 Middlesex Turnpike Bedford MA 01730

BAIRD, WILLIAM CHALMERS, organic chemistry, see 12th edition

BAISDEN, CHARLES ROBERT, b Logan, WVa, Apr 23, 39; m 62; c 3. MEDICINE, PATHOLOGY. Educ: WVa Univ, AB, 61, MD, 65. Prof Exp: Intern med, WVa Univ Hosp, 65-66, resident path, 66-68; resident, Baptist Mem Hosp, Memphis, Tenn, 68-70; ASST PROF LAB MED, SCH MED, JOHNS HOPKINS UNIV, 70-, PATHOLOGIST, JOHNS HOPKINS HOSP, 70- Concurrent Pos: Path consult, Vet Admin Hosp, Perry Point, Md, 70-; dir clin labs, Franklin Sq Hosp, Baltimore, 72- Mem: Am Soc Clin Path. Res: Types of bisalbuminemia and methods of detection. Mailing Add: Sch of Med Johns Hopkins Univ Baltimore MD 21205

BAISTED, DEREK JOHN, b London, Eng, Oct 5, 34; m 60; c 1. ORGANIC CHEMISTRY, BIOCHEMISTRY. Educ: Exeter Univ, BSc, 57, PhD(org chem), 60. Prof Exp: USPHS trainee steroid biochem, Clark & Worcester Found Exp Biol, 60-62; NSF fel, Ind Univ, 62-63; NIH fel, 63-64, USPHS career develop award, 65-70, asst prof biochem, 65-69, ASSOC PROF BIOCHEM, ORE STATE UNIV, 69- Mem: The Chem Soc; Am Soc Plant Physiologists; Brit Biochem Soc. Res: Steroid and terpene synthesis; biosynthesis and metabolism of isoprenoids; membrane-associated enzymes. Mailing Add: Dept of Biochem & Biophys Ore State Univ Corvallis OR 07330

BAITINGER, WILLIAM F, JR, b Bridgeton, NJ, Nov 24, 35; m 58; c 3. TEXTILE CHEMISTRY. Educ: Albright Col, BS, 58; Princeton Univ, MA, 62, PhD(chem), 64. Prof Exp: Chemist, Am Cyanamid Co, 58-60, res chemist, 63-66, group leader textile chem, 66-71, sr res scientist, 71-74; MGR FIRE RETARDANT RES, COTTON, INC, 74- Mem: Am Chem Soc; The Chem Soc; Am Asn Textile Chemists & Colorists. Res: Synthesis or aromatic and aliphatic light stabilizers; effects of ultra-violet light on polymers and organic compounds; chemistry of textile wet processing. Mailing Add: Cotton Inc 4505 Creedmoor Rd Raliegh NC 08520

BAIZER, MANUEL M, b Philadelphia, Pa, May 20, 14; m 39; c 3. ORGANIC CHEMISTRY. Educ: Univ Pa, BS, 34, MS, 37, PhD(chem), 40. Prof Exp: Asst res chemist, Philadelphia Inst Med Res, 38-39; tutor chem, Brooklyn Col, 41-44, instr, 45-48; res chemist, NY Quinine & Chem Works, 46-47, head res dept, 47-58; scientist, 58-66, adv scientist, 66-68, sr scientist, 68-71, DISTINGUISHED SR FEL, MONSANTO CO, 71- Concurrent Pos: Res assoc, Nat Defense Res Comt, Univ Pa, 42-44; res chemist, Gen Chem Co LI, 42-44; vis prof, Univ Southampton, Eng, 73-74. Honors & Awards: Creative Invention Award, Am Chem Soc, 76. Mem: AAAS; Am Chem Soc; fel Am Inst Chem; NY Acad Sci; Electrochem Soc. Res: Opium alkaloids; explosives; mercuration; surface active agents; organic synthesis; quaternary ammonium compounds; activating influence of quaternary ammonium group; electro-organic syntheses. Mailing Add: Monsanto Co 800 N Lindbergh Blvd St Louis MO 63166

BAJAJ, PREM NATH, b Ferozepur City, India, Oct 6, 32; m 56; c 5. MATHEMATICAL ANALYSIS, TOPOLOGY. Educ: Punjab Univ, India, BA, 51, MA, 54; Case Western Reserve Univ, MS, 67, PhD(math), 68. Prof Exp: Teacher, 51-52; lectr, RKSD Col, Kaithal, 54-57 & DAV Col, Jullundur, 57-65; vis asst prof, Case

Western Reserve Univ, 68; asst prof, 68-72, ASSOC PROF MATH, WICHITA STATE UNIV, 72- Concurrent Pos: Mem, Vishveshvaranand Vedic Res Inst. Mem: Am Math Soc; Indian Math Soc. Res: Ordinary differential equations; dynamical systems; topology. Mailing Add: Dept of Math Wichita State Univ Wichita KS 67208

BAJCSY, RUZENA K, b Czech, May 28, 33; c 2. COMPUTER SCIENCE. Educ: Slovak Tech Univ, Bratislava, PhD(elec eng), 67; Stanford Univ, PhD(comput sci), 72. Prof Exp: Comput engr, Comput Ctr, Slovak Tech Univ, Bratislava, 62-64, asst prof comput sci, 64-67; res assoc, Stanford Univ, 67-72; ASST PROF COMPUT SCI, UNIV PA, 72- Concurrent Pos: NSF res grant, 74. Mem: Asn Comput Mach. Res: Artificial intelligence; scene analysis; pattern recognition of outdoor scenes; analysis of biological images. Mailing Add: Dept of Comput & Info Sci Moore Sch of Elec Eng Univ of Pa Philadelphia PA 19174

BAJEMA, CARL J, b Plainwell, Mich, May 25, 37; m 59; c 2. HUMAN ECOLOGY, BIOLOGICAL ANTHROPOLOGY. Educ: Western Mich Univ, BS, 59, MA, 61; Mich State Univ, PhD(zool), 63. Prof Exp: Instr sci, Grand Rapids Pub Sch Syst, Mich, 59-60; asst zool, Mich State Univ, 62-63; asst prof biol, Mankato State Col, 63-64; from asst prof to assoc prof, 64-72, PROF BIOL, GRAND VALLEY STATE COLS, 72- Concurrent Pos: Pop Coun sr fel demog & pop genetics, Univ Chicago, 66-67; res assoc pop studies, Ctr Pop Studies, Harvard Univ, 67-73, vis prof anthrop, 74-75. Mem: Soc Study Social Biol; Eugenics Soc Gt Brit; Ecol Soc Am; Am Inst Biol Sci; Nat Asn Biol Teachers. Res: Measurement of natural selection in human populations; interactions between genetic and cultural systems for adapting to the environment; history of evolutionary thought; teaching of human ecology and evolution. Mailing Add: Dept of Biol Col of Arts & Sci Grand Valley State Cols Allendale MI 49401

BAJER, ANDREW, b Czestochowa, Poland, Jan 3, 28; US citizen; m 51; c 2. CELL BIOLOGY. Educ: Jagiellonian Univ, MA, 49, PhD(cytol), 50, DSc, 56. Prof Exp: Asst cytol, Jagiellonian Univ, 48, assoc prof, 56-63; assoc prof, Polish Acad Sci, 63-64; assoc prof, 64-69, PROF BIOL, UNIV ORE, 69- Concurrent Pos: NIH career develop award, 67-72. Honors & Awards: Sci Awatd, Alfred Jurzykowski Found, 74. Mem: AAAS; Am Soc Cell Biol; Bot Soc Fund; Int Soc Cell Biol; Am Sci Film Asn. Res: Physiology of cell division and mechanism of chromosome movements; mitosis and meiosis, especially microcinematrographic technique on plant endosperm; developed technique permitting study of the same cell with light microscopy and electron microscopy. Mailing Add: Dept of Biol Univ of Ore Eugene OR 97403

BAJPAI, PRAPHULLA K, b Charkhari, India, Sept 24, 36; m 66. PHYSIOLOGY, IMMUNOLOGY. Educ: Agra Univ, BVSc, 58, MVSc, 60; Ohio State Univ, MSc, 63, PhD(avian physiol), 65. Prof Exp: Vet surgeon, Govt India, 58; instr physiol, 64-66, asst prof physiol & immunol, 66-70, ASSOC PROF PHYSIOL & IMMUNOL, UNIV DAYTON, 70- Concurrent Pos: Res assoc & fel, Reprod Endocrinol Prog, Dept Path, Univ Mich, Ann Arbor, 72-73; adj assoc prof, Sch Med & Sch Sci & Eng, Wright State Univ, 75- Mem: Am Fedn Clin Res; Am Physiol Soc; Soc Study Reprod; Fedn Am Socs Exp Biologists; Sigma Xi. Res: Physiology and immunology of heart, reproductive system and bone. Mailing Add: Dept of Biol Univ of Dayton Dayton OH 45469

BAJUSZ, EÖRS, experimental pathology, see 12th edition

BAJZA, CHARLES C, b Bacs, Hungary, May 9, 14; US citizen; m 44; c 5. GEOGRAPHY. Educ: Univ Ind, BA, 41, MA, 42, PhD(geog), 53. Prof Exp: Instr geog, Univ Ind, 41-43 & 47-49; geologist, Stanolind Oil & Gas Co, 43-47; assoc prof geog, 49-56, PROF GEOG, TEX A&I UNIV, 56-, CHMN DEPT GEOG & GEOL, 62- Concurrent Pos: Vis prof, Orange State Col, 62; curriculum consult, Tex Pub Sch Systs. Mem: Asn Am Geog; Nat Coun Geog Educ. Res: The geography of Texas, North America and Central and East Europe. Mailing Add: Dept of Geog & Geol Tex A&I Univ Kingsville TX 78363

BAJZER, WILLIAM XAVIER, b Cleveland, Ohio, June 17, 40; m 63; c 4. ORGANIC CHEMISTRY, MEDICINAL CHEMISTRY. Educ: Case Inst Technol, BS, 63; Ohio State Univ, MS, 66, PhD(org chem), 68. Prof Exp: Chemist, 63-65, res chemist, 69-70, supvr chem, 70-73, develop mgr, 73-75, RES SECT MGR, DOW CORNING CORP, 75- Mem: NY Acad Sci; Am Pharmaceut Asn; Sigma Xi. Res: Chemistry of fluorinated compounds; methods of fluorine introduction in organic molecules; synthesis and chemistry of fluoroalklsilicon compounds and polymers derived therefrom; fluoroalkyl triazine chemistry; biocompatible polymers and elastomers; flexible silicone contact lens; biologically active compounds containing silicon; polymer bound drugs. Mailing Add: 4201 McKeith Rd Midland MI 48640

BAK, DAVID ARTHUR, b Yankton, SDak, Feb 6, 39; m 64; c 2. PHYSICAL CHEMISTRY, ORGANIC CHEMISTRY. Educ: Augustana Col, BA, 61; Kans State Univ, PhD(org chem), 66. Prof Exp: Res assoc org chem, Mich State Univ, 65-66; asst prof, 66-71, ASSOC PROF CHEM, HARTWICK COL, 71-, CHMN DEPT, 74- Concurrent Pos: Am Chem Soc-Petrol Res Fund res grant, 67-69. Mem: Am Chem Soc. Res: Reactions of cyclooctatetraenyl dianion; electron transfer reactions; small ring ketone photochemistry; preparation of new 4n plus 2 pi electron righ systems. Mailing Add: Dept of Chem Hartwick Col Oneonta NY 13820

BAKALE, GEORGE, radiation chemistry, radiation physics, see 12th edition

BAKAN, JOSEPH A, b Dayton, Ohio, Feb 13, 35; m 59; c 3. CHEMISTRY. Educ: Univ Dayton, BS, 57. Prof Exp: Sr res chemist, Fundamental Res Dept, 57-63, sect head, Capsular Res & Prod Develop Dept, 63-68, MGR RES & DEVELOP, CAPSULAR PROD DIV, NCR CORP, 68- Mem: Am Chem Soc. Res: Particle size analysis; gas chromatography; general physical and colloid chemistry; general encapsulation procedures; capsular programs in the areas of pharmaceutical products, foods, flavors, perfumes and detection systems for military applications. Mailing Add: Capsular Prod Div NCR Corp Dayton OH 45409

BAKAY, LOUIS, b Pozsony, Hungary, June 18, 17; US citizen; m 54; c 2. NEUROSURGERY, NEUROPHYSIOLOGY. Educ: Univ Budapest, MD, 41. Prof Exp: Asst prof surg, Univ Budapest, 45-47; clin asst neurosurg, Serafimer Hosp, Stockholm, Sweden, 47-48; instr neurosurg, Harvard Med Sch, 53-61; PROF SURG, SCH MED, STATE UNIV NY BUFFALO, 61- Concurrent Pos: Res fel, Harvard Univ, 48-50; clin assoc, Mass Gen Hosp, Boston, 52-61; chief neurosurg serv, E J Meyer Mem Hosp, Buffalo Gen Hosp & Children's Hosp, Buffalo, 61- Mem: Am Col Surg; Harvey Cushing Soc; Am Acad Neurol. Res: Surgery of pituitary tumors and cervical spine; cerebral circulation and metabolism; hydrodynamics of cerebrospinal fluid; blood-brain barrier; radioactive isotopes in nervous system; cerebral edema. Mailing Add: 462 Grider St Buffalo NY 14215

BAKER, AARON SIDNEY, b St Thomas, Ont, Jan 27, 24; nat US; m 54; c 4. SOIL FERTILITY. Educ: Cornell Univ, BS, 50, MS, 51; Mich State Univ, PhD(soil sci), 55. Prof Exp: Asst soils chemist, N Fla Exp Sta, Univ Fla, 55-58; from asst soil scientist to assoc soil scientist, 58-73, SOIL SCIENTIST, WESTERN WASH RES & EXTEN CTR, WASH STATE UNIV, 73- Mem: Soil Sci Soc Am; Sigma Xi. Res: Toxic elements in sewerage sludge and particulate matter from smelter. Mailing Add: Western Wash Res & Exten Ctr Wash State Univ Puyallup WA 98371

BAKER, ABE BERT, b Minneapolis, Minn, Mar 27, 08; m 33; c 4. NEUROLOGY. Educ: Univ Minn, Minneapolis, BA, 28, BS, 29, MB, 30, MD, 31, MS, 32, PhD, 39; Am Bd Psychiat & Neurol, dipl, 40. Prof Exp: Asst neuropath, 31-34, asst neuropsychiat, 34-37, from instr to assoc prof neuropsychiat & neuropath, 37-46, prof neurol & head dept, 46-73, REGENTS' PROF, UNIV MINN, MINNEAPOLIS, 73- Concurrent Pos: Fulbright scholar, Oslo, Norway, 59; consult, Minneapolis Vet Admin Hosp, Vet Admin Hosp & Vet Admin Ctr, SDak & Vet Admin, DC. Honors & Awards: Presidential Award, 57; Merit Citation, President's Comt Employ Physically Handicapped, 58; Citation, Norweg Acad Sci, 59. Mem: Am Soc Exp Path; Am Asn Neuropath; Asn Res Nerv & Ment Dis; Norweg Acad Sci. Res: Cerebrovascular disease; causes of cerebral arteriosclerosis in aging population; infections of nervous system. Mailing Add: Dept of Neurol Univ of Minn Minneapolis MN 55455

BAKER, ADOLPH, b Russia, Nov 15, 17; US citizen; m 42; c 3. PHYSICS. Educ: City Col New York, BA, 38, MS, 39; Polytech Inst Brooklyn, BME, 46; NY Univ, MS, 49; Brandeis Univ, PhD(physics), 64. Prof Exp: Stress analyst, Repub Aviation Corp, NY, 46-47; sr engr, Ranger Aircraft Engines, 47-48; mem staff, Int Bus Mach Corp, 48-49; sr engr, Raytheon Co, Mass, 49-55; mgr airborne digital comput develop, Radio Corp Am, 55-59; PROF PHYSICS & APPL PHYSICS, UNIV LOWELL, 63- Concurrent Pos: Consult, Radio Corp Am, Mass, 59-64; sr Fulbright-Hays scholar, USSR, 74-75. Mem: Am Phys Soc. Res: Scattering theory; high energy electron scattering from nuclei; potential scattering theory; digital computers. Mailing Add: Dept of Physics Univ of Lowell Lowell MA 08154

BAKER, ALAN PAUL, b Saltsburg, Pa, Aug 6, 38; m 61; c 2. BIOCHEMISTRY Educ: Philadelphia Col Pharm, BSc, 60; Hahnemann Med Col, PhD(biochem), 64. Prof Exp: Damon Runyon Mem Fund fel cancer res, Royal Univ Umea, Sweden, 64-66; fel enzymol, Brandeis Univ, 66-67; sr chemist, Wyeth Labs, 67-68; SR CHEMIST, SMITH KLINE & FRENCH LABS, 68- Mem: AAAS; Am Chem Soc; NY Acad Sci; Am Soc Biol Chemists. Res: Structure and function of myeloperoxidase; uterine peroxidase; eosinophils; mast cells; electron spin resonance of flavin enzymes; microbial cell wall synthesis; mucopolysaccharides. Mailing Add: Smith Kline & French Labs 1500 Spring Garden St Philadelphia PA 19101

BAKER, ALBERT LEROY, biochemistry, enzymology, see 12th edition

BAKER, ANDREW NEWTON, JR, b Wellington, Kans, Oct 21, 28; m 51; c 2. PHYSICS. Educ: Univ Rochester, BS, 49; Univ Calif, Los Angeles, MS, 51, PhD(physics), 54. Prof Exp: Assoc physics, Univ Calif, Los Angeles, 52-54; mem tech staff, Bell Tel Labs, 54-60; head solid state lab, Lockheed-Calif Co, 60-63, mgr phys sci, 64-66; Sloan exec fel, Stanford Univ, 66-67; dep chief engr, 68-70, dir res, 70-72, dir develop planning, 72-73, DIR RES, LOCKHEED-CALIF CO, 73- Mem: Am Phys Soc; Am Inst Aeronaut & Astronaut. Res: Infrared spectroscopy; hydrogen bonds; semiconductors and semiconductor devices; electro-optics and ferroelectrics; titanium metallurgy and corrosion. Mailing Add: 12324 Woodley Ave Granada Hills CA 91344

BAKER, ARTHUR ALAN, b New Britain, Conn, Oct 31, 97; m 25; c 1. GEOLOGY. Educ: Yale Univ, PhD(geol), 31. Prof Exp: Geol aid, US Geol Surv, 21-24, from asst geologist to admin geologist, 24-56, assoc dir, 56-69, spec asst to dir, 69-73; RETIRED. Concurrent Pos: Dept of Interior rep on US Bd on Geog Names, 53-73. Honors & Awards: Distinguished Serv Award, Dept of Interior. Mem: Fel Geol Soc Am; Am Asn Petrol Geol; Am Geophys Union. Res: Geology of fuels; stratigraphy and structural geology. Mailing Add: 5201 Westwood Dr Washington DC 20016

BAKER, BARTON SCOFIELD, b Paint Bank, Va, Oct 4, 41; m 64. AGRONOMY. Educ: Berea Col, BS, 64; WVa Univ, MS, 66, PhD(agron), 69. Prof Exp: Asst prof, 69-73, ASSOC PROF AGRON, ALLEGHENY HIGHLANDS PROJ, W VA UNIV, 73- Mem: Am Soc Agron; Crop Sci Soc Am. Res: Influence of environmental conditions on the growth and composition of forage crops. Mailing Add: Allegheny Highlands Proj WVa Univ Box 149 Elkins WV 26241

BAKER, BERNARD RANDALL, bio-organic chemistry, deceased

BAKER, BERNARD RAY, b Wheatland, Wyo, Dec 9, 32; m 60; c 2. SURFACE CHEMISTRY. Educ: Univ Denver, BS, 54; Northwestern Univ, PhD(chem), 58. Prof Exp: Instr chem, Northwestern Univ, 57-58 & Univ Nev, 58-60; from instr to asst prof, Univ Rochester, 60-65; res assoc, Brookhaven Nat Lab, 65-66; RES CHEMIST, CTR TECHNOL, KAISER ALUMINUM & CHEM CORP, 66- Concurrent Pos: Sect ed, Chem Abstr, 62-63. Mem: Am Soc Testing & Mat. Res: Kinetics and mechanisms of inorganic reactions; electrochemistry; surface chemistry; chemistry of thin films; metal finishing. Mailing Add: Ctr for Technol Kaiser Aluminum & Chem Corp Pleasanton CA 94566

BAKER, BERTSIL BURGESS, b Fairfield, Ala, Oct 25, 24; m 48. ANALYTICAL CHEMISTRY. Educ: Ohio State Univ, BS, 45, MS, 50, PhD(chem), 51. Prof Exp: Fel, Ohio State Univ, 51-52; sr scientist, Southern Res Inst, 52-58; SR RES CHEMIST, PLASTICS DEPT, E I DU PONT DE NEMOURS & CO, INC, 58- Mem: Am Chem Soc. Res: Electroanalytical chemistry; organometallics; polyolefins; air pollution; trace analysis; gas chromatography; infrared spectroscopy. Mailing Add: Plastics Dept Du Pont Exp Sta Wilmington DE 19898

BAKER, BRENDA SUE, b Oakland, Calif, Dec 19, 48; m 72. COMPUTER SCIENCE. Educ: Radcliffe Col, BA, 69; Harvard Univ, MS, 70, PhD(appl math), 73. Prof Exp: Res fel comput sci, Harvard Univ, 73-74; MEM TECH STAFF COMPUT SCI, BELL LABS, 74- Mem: Asn Comput Mach; Soc Indust & Appl Math; AAAS. Res: Theoretical computer science. Mailing Add: Bell Labs 600 Mountain Ave Murray Hill NJ 07974

BAKER, BRUCE EARLE, physical chemistry, see 12th edition

BAKER, BRYAN, JR, b Grenada, Miss, Feb 24, 23; m 46; c 1. ANIMAL HUSBANDRY, ANIMAL PHYSIOLOGY. Educ: Miss State Univ, BS, 47, BS, 48, MS, 52; Univ Ill, PhD(animal sci), 55. Prof Exp: Teacher voc agr, Biggersville High Sch, 48-51; asst animal husb, Miss State Univ, 51-52; asst animal sci, Univ Ill, 52-55; assoc prof animal husb, Va Polytech Inst, 55-56; assoc prof, 56-60, PROF ANIMAL HUSB, MISS STATE UNIV, 60- Mem: AAAS; Am Soc Animal Sci. Res: Effect of environmental factors on physiology of reproduction of sheep, cattle and swine. Mailing Add: Dept of Animal Sci Miss State Univ Mississippi State MS 39762

BAKER, BURTON LOWELL, b Fife Lake, Mich, Apr 2, 12; m 40; c 2. ANATOMY. Educ: Kalamazoo Col, BA, 33; Kans State Col, MS, 35; Columbia Univ, PhD(anat), 41. Hon Degrees: ScD, Kalamazoo Col, 58. Prof Exp: Asst zool, Kans State Col, 33-35; instr anat, Columbia Univ, 36-38 & 40-41; from instr to assoc prof, 41-52, PROF

ANAT, UNIV MICH, ANN ARBOR, 52- Concurrent Pos: Assoc ed, Am J Anat, 57-68, managing ed, 68-74. Honors & Awards: Russel Award, Univ Mich, 47; Hon Prof, Fed Univ Pernambuco, 69. Mem: AAAS; Am Asn Anat; Endocrine Soc; Soc Exp Biol & Med; Am Physiol Soc. Res: Histochemistry; endocrinology; microanatomy; immunochemistry. Mailing Add: Dept of Anat Univ of Mich Ann Arbor MI 48104

BAKER, CARL GWIN, b Louisville, Ky, Nov 27, 20; m 49; c 2. RESEARCH ADMINISTRATION. Educ: Univ Louisville, AB, 41, MD, 44; Univ Calif, MA, 48. Prof Exp: Intern, Milwaukee County Gen Hosp, 44-45; sr asst surgeon, Nat Cancer Inst, 49-53, surgeon, 53-55, med dir, 55-70; asst to dir intramural res, NIH, 56-58; asst dir, Nat Cancer Inst, 58-61, actg sci dir, 60-61, assoc dir for prog, 61-67, sci dir etiology, 67-69, dir, Inst, 69-72; pres & sci dir, Hazleton Labs, 72-73; consult res admin, 73-75; SR SCI ADV PROG COORD, OFF ADMINR, HEALTH RESOURCES ADMIN, DEPT HEALTH, EDUC & WELFARE, 75- Concurrent Pos: Spec fel, Nat Cancer Inst, 49; assoc ed, Jour Nat Cancer Inst, 54-55; spec lectr, Sch Med, Georgetown Univ, 54-64; mem subcomt amino acids, Comt Biochem, Nat Res Coun, 56-58; mem planning comt, Nat Cancer Conf, 62-64; mem ed adv bd, Cancer, 65-73; vpres, Tenth Int Cancer Cong, 70; mem sci adv comt, Ludwig Inst Cancer Res, 71- Mem: Am Chem Soc; Soc Exp Biol & Med; Am Soc Biol Chem; Am Asn Cancer Res; Soc Toxicol. Res: Enzymatic resolution of amino acids; cancer research; health services administration. Mailing Add: Off of Adminr Health Resources Admin Parklawn Bldg Rockville MD 20852

BAKER, CARLETON HAROLD, b Utica, NY, Aug 2, 30; m 63; c 2. PHYSIOLOGY. Educ: Utica Col, BA, 52; Princeton Univ, MA, 54, PhD(biol), 55. Prof Exp: Asst instr biol, Princeton Univ, 52-54, asst res biol, 54-55; from asst prof to prof physiol, Med Col Ga, 55-67; prof physiol & biophys, Sch Med, Univ Louisville, 67-71; PROF PHYSIOL & CHMN DEPT, COL MED, UNIV S FLA, TAMPA, 71- Mem: AAAS; Am Heart Asn; Am Physiol Soc; Microcirc Soc; Soc Exp Biol & Med. Res: Adrenal cortex; blood volume regulation; cardiovascular system. Mailing Add: Dept of Physiol Univ of SFla Col of Med Tampa FL 33620

BAKER, CHARLES ALBERT, b Catonsville, Md, Jan 5, 24; m 53; c 4. RESEARCH ADMINISTRATIONx ADMINISTRATION, BIOLOGICAL CHEMISTRY. Educ: Loyola Col, Md, BS, 47; Georgetown Univ, MS, 48, PhD(biochem), 52; George Washington Univ, MEA, 59. Prof Exp: Res chemist, Fermentation Res Lab, US Indust Chem Co, 49-52; dir biol, Penniman & Browne, Inc, 53-59; sr biochemist, Eli Lilly & Co, 59-61 & Joseph E Seagrams & Sons, Inc, 61-63; opers analyst, Res Anal Corp, 63-65 & A Epstein & Sons, Inc, 65-66; chief opers res unit, Res Anal Br, 67-70, SCI EVAL OFFICER, DIV RES GRANTS, NIH, 70- Mem: Am Chem Soc; Inst Mgt Sci. Res: Mathematical programming applied to research resources allocation; interrelationships among health related activities; research analysis; pharmaceutical research; production and control methods for biochemical products; economics of chemical processes. Mailing Add: 4500 Sunflower Dr Rockville MD 20853

BAKER, CHARLES EDWARD, b Manchester, Ky, Nov 8, 31; m 52; c 4. PHYSICAL CHEMISTRY. Educ: Berea Col, BA, 53; Univ Fla, PhD(phys chem), 60. Prof Exp: Chemist, Res Div, Monsanto Chem Co, Ohio, 53-54; asst, Univ Fla, 56-60; sr res engr, Gen Dynamics/Convair, Calif, 60-62; AERONAUT RES SCIENTIST, LEWIS RES CTR, NASA, 62- Mem: AAAS; Am Chem Soc. Res: Atmospheric chemistry; low-energy scattering of negative-ion beams; thermal conductivities of ordinary and isotopically substituted polar gases; diffusion of polar gases; recombination of atoms. Mailing Add: 510 Wyleswood Dr Berea OH 44017

BAKER, CHARLES PARKER, b Leominster, Mass, Aug 2, 10; m 53. PHYSICS. Educ: Denison Univ, AB, 33; Cornell Univ, AM, 35, PhD(physics), 41. Prof Exp: Res assoc physics, Cornell Univ, 40-42; res fel, Purdue Univ, 42-43; scientist, Los Alamos Sci Lab, Univ Calif, 43-46; res assoc physics, Cornell Univ, 46-47, asst prof, 47-49; PHYSICIST, BROOKHAVEN NAT LAB, 49- Mem: Fel Am Phys Soc. Res: Nuclear physics; accelerators. Mailing Add: 18 Livingston Rd Bellport NY 11713

BAKER, CHARLES RAY, b Pine Bluff, Ark, May 22, 32; m 52; c 1. APPLIED MATHEMATICS, ENGINEERING. Educ: Univ Southwestern La, BS, 57; Univ Calif, Los Angeles, MS, 63, PhD(eng), 67. Prof Exp: Staff engr, Land-Air, Inc, Holloman Air Develop Ctr, NMex, 57-59; engr aerospace systs div, Bendix Corp, Mich, 59-60, sr engr electrodynamics div, Calif, 59-67, eng res specialist, 67-68; assoc prof statist, 68-73, PROF STATIST, UNIV NC, CHAPEL HILL, 73- Mem: Inst Elec & Electronics Eng; Inst Math Statist; Am Math Soc. Res: Statistical communication theory; probability theory; stochastic processes and applications to communication systems. Mailing Add: Dept of Statist Univ of NC Chapel Hill NC 27514

BAKER, CLINTON LYLE, b Hamburg, Ark, Aug 5, 04; m 28; c 3. ZOOLOGY, LIMNOLOGY. Educ: Emory Univ, BS, 25, MS, 26; Columbia Univ, PhD(protozool), 32. Prof Exp: Asst prof biol, Milsaps Col, 26-28; asst, NY Univ, 29; asst zool, Columbia Univ, 29-31; assoc prof biol, Univ Detroit, 31-32; from assoc prof to prof & head dept, 32-71, EMER PROF BIOL, SOUTHWESTERN UNIV, MEMPHIS, 71- Concurrent Pos: Dir, Reelfoot Biol Sta, Tenn, 36-68. Mem: AAAS; Am Soc Zoologists; Ecol Soc Am; Am Soc Ichthyol & Herpet; Am Soc Limnol & Oceanog. Res: Life history of Amphiuma and distribution in southern states; cytology of Euglena gracilis; uro-genital system and spermateleosis in Urodeles. Mailing Add: 1620 Galloway Ave Memphis TN 38112

BAKER, DALE E, b Marble Hill, Mo, May 5, 30; m 49; c 3. AGRONOMY, SOIL FERTILITY. Educ: Univ Mo, BS, 57, MS, 58, PhD(soils), 60. Prof Exp: Agronomist, NE Exp Sta, Duluth, Minn, 60-61; from asst prof to assoc prof, 61-70, PROF SOIL CHEM, PA STATE UNIV, 70- Concurrent Pos: Consult, Amax, Inc, 72-75 & Food & Drug Admin, Dept Health, Educ & Welfare, 74. Honors & Awards: Res Award, Gamma Sigma Delta, Pa State Univ, 71 & Northeast Am Soc Agron, 72. Mem: Fel Am Soc Agron; AAAS. Res: Physical and colloidal chemistry of soils and genetic control of physiological processes in plants in relation to ion uptake and improved quality and production of field crops. Mailing Add: 221 Tyson Bldg Pa State Univ University Park PA 16802

BAKER, DAVID BRUCE, b Akron, Ohio, May 29, 36; m 60; c 1. PLANT PHYSIOLOGY. Educ: Heidelberg Col, BS, 58; Univ Mich, MS, 60, PhD(bot), 63. Prof Exp: Asst prof bot, Rutgers Univ, 63-66; asst prof biol, 66-72, ASSOC PROF BIOL, HEIDELBERG COL, 72- Concurrent Pos: NSF fel, 63-64. Mem: Am Soc Plant Physiol. Res: Relationship between cell wall synthesis and cell elongation and the effect of auxins on these processes. Mailing Add: Dept of Biol Heidelberg Col Tiffin OH 44883

BAKER, DAVID H, b DeKalb, Ill, Feb 26, 39; m 57; c 3. ANIMAL NUTRITION. Educ: Univ Ill, BS, 61, MS, 63, PhD(animal nutrit), 65. Prof Exp: Res asst nutrit, Univ Ill, 61-65; sr scientist, Greenfield Labs, Eli Lilly & Co, Ind, 65-67; from asst prof to assoc prof, 67-74, PROF NUTRIT, UNIV ILL, URBANA-CHAMPAIGN, 74- Honors & Awards: Res Award, Am Soc Animal Sci, 71; Nutrit Res Award, Am Feed Mfrs Asn, 73. Mem: Am Soc Animal Sci; Poultry Sci Asn; Am Inst Nutrit. Res: Amino acid nutrition of swine and poultry; nutrition and pregnancy. Mailing Add: Dept of Animal Sci Univ of Ill Urbana IL 61801

BAKER, DAVID H, b Concord, NH, Aug 25, 25; m 52; c 3. PEDIATRICS, RADIOLOGY. Educ: Boston Univ, MD, 51. Prof Exp: From instr to asst prof pediat & radiol, NY Hosp-Cornell Med Ctr, 54-59; asst prof pediat & radiol, 59-64, assoc prof radiol, 64-68, PROF RADIOL, COL PHYSICIANS & SURGEONS, COLUMBIA UNIV, 68- Concurrent Pos: Consult, New York Foundling Hosp, 56-, Roosevelt Hosp, 57-, Montefiore Hosp & USPHS Hosp, Staten Island, 62- Mem: Soc Pediat Radiol; fel Am Acad Pediat; Am Col Radiol; Asn Univ Radiol; Am Roentgen Ray Soc. Res: Pathophysiologic states in children involving the gastrointestinal, cardiovascular, brochovascular and urologic systems. Mailing Add: Col of Physicians & Surgeons Columbia Univ New York NY 10032

BAKER, DAVID KENNETH, b Glasgow, Scotland, Oct 2, 23; nat US; m 47; c 2. PHYSICS. Educ: McMaster Univ, BSc, 46; Univ Pa, PhD, 53. Prof Exp: Instr physics, Univ Pa, 52-53; prof, Union Col, 53-65; prof personnel & univ rels, Res & Develop Ctr, Gen Elec Co, 65-67; VPRES & DEAN COL LETT & SCI, ST LAWRENCE UNIV, 67- Mem: Am Phys Soc; Am Asn Physics Teachers. Res: Semiconductors and solid state physics. Mailing Add: St Lawrence Univ Canton NY 13617

BAKER, DAVID THOMAS, b Granite Falls, Minn, June 27, 25; m 64; c 3. NUCLEAR PHYSICS, SCIENCE EDUCATION. Educ: US Mil Acad, BS, 46; Purdue Univ, MS, 51; Am Univ, MS, 71. Prof Exp: US Army, 46-70, instr nuclear physics & elec eng, US Mil Acad, 51-54, asst prof nuclear physics, 54-55, res & develop officer, nuclear weapons, Army Artil & Guided Missile Off, Ft Sill, Okla, 56-58, nuclear weapons effects, Atomics off, Off Chief Res & Develop, Hq Dept Army, 60-64, dir Nuclear Weapons Assembly Dept, Army Sch Europe, Ger, 65-66, develop officer, Nuclear Weapons Req Div, US Army Mobility Equip Res & Develop Ctr, 69-71; INSTR PHYSICS, CULVER MIL ACAD, 71- Mem: AAAS. Res: Nuclear weapons, their development, testing, effects and employment. Mailing Add: Culver Mil Acad Culver IN 46511

BAKER, DAVID WARREN, b Great Falls, Mont, Nov 9, 39; m 62; c 3. STRUCTURAL GEOLOGY. Educ: Mass Inst Technol, BS, 61; Swiss Fed Inst Technol, Diplom Nat Sci, 64; Univ Calif, Los Angeles, PhD(geol), 69. Prof Exp: ASST PROF GEOL SCI, UNIV ILL, 70- Concurrent Pos: Res geophysicist, Inst Geophys & Plantetary Physics, Univ Calif, Los Angeles, 69-71; res geologist, Dept Geol & Geophys, Yale Univ, 70. Mem: Am Geophys Union; Swiss Mineral & Petrog Soc; AAAS. Res: X-ray analysis of preferred orientation in experimentally and naturally deformed rocks; deep-seated thrust zones; seismic velocity anisotropy in the upper mantle; plate and regional tectonics. Mailing Add: Dept of Geol Sci Univ of Ill Chicago IL 60680

BAKER, DON ROBERT, b Salt Lake City, Utah, Apr 6, 33; m 54. ORGANIC CHEMISTRY. Educ: Sacramento State Col, AB, 55; Univ Calif, PhD(chem), 59. Prof Exp: CHEMIST, STAUFFER CHEM CO, 58- Mem: Am Chem Soc; Soc Indust Microbiol; Plant Growth Regulator Working Group. Res: Organic synthesis; agricultural chemistry; microbiology; computer applications; mineralogy. Mailing Add: 15 Muth Dr Orinda CA 94563

BAKER, DONALD GARDNER, b St Paul, Minn, July 20, 23; m 53; c 1. MICROCLIMATOLOGY. Educ: Univ Chicago, prof cert, 44; Univ Minn, BS, 49, MS, 51, PhD(soils), 58. Prof Exp: Asst soils, 49-51, 53-58, from instr to assoc prof, 58-69, PROF SOILS, UNIV MINN, ST PAUL, 69- Mem: AAAS; Am Meteorol Soc; Am Geophys Union; Soil Sci Soc Am; Am Soc Agron. Res: Agricultural meteorology and climatology. Mailing Add: Dept of Soil Sci Inst of Agr Univ of Minn St Paul MN 55108

BAKER, DONALD GRANVILLE, b Toronto, Ont, Oct 19, 24; m 52; c 2. PHYSIOLOGY. Educ: Univ Toronto, BA, 51, MA, 52, PhD(physiol), 55. Prof Exp: Res assoc, Banting & Best Dept Med Res, Univ Toronto, 55-62, lectr, Sch Hyg, 58-64, from asst prof to assoc prof physiol & spec lectr zool & radiobiol, 59-64; radiobiologist, Biol Div, Brookhaven Nat Lab, 64-68; RADIOBIOLOGIST, CLAIRE ZELLERBACH SARONI TUMOR INST, MT ZION HOSP & MED CTR, 68- Mem: Radiation Res Soc; NY Acad Sci; Royal Soc Med; Am Asn Cancer Res. Res: Radiation injury; radiation biology; physiological responses to low environmental temperature. Mailing Add: Zellerbach Saroni Tumor Inst Mt Zion Hosp & Med Ctr 1600 Divisadero St San Francisco CA 94115

BAKER, DONALD JAMES, JR, b Long Beach, Calif, Mar 23, 37; m 68. PHYSICAL OCEANOGRAPHY. Educ: Stanford Univ, BS, 58; Cornell Univ, PhD(physics), 62. Prof Exp: Res assoc oceanog, Univ RI, 62-63; NIH fel biophys, Univ Calif, Berkeley, 63-64; res fel geophys fluid dynamics, Harvard Univ, 64-66, asst prof oceanog, 66-70, assoc prof phys oceanog, 70-73; res assoc prof oceanog, 73-75, RES PROF OCEANOG & SR OCEANOGRAPHER, APPL PHYSICS LAB, UNIV WASH, 75- Concurrent Pos: Mem, Polar Res Bd, Nat Res Coun, 72-, chmn, Joint Polar Exp Panel, 73-, co-chmn, Int Southern Ocean Studies Prog, 74- & mem, Climate Dynamics Panel, US Global Atmospheric Res Prog, 76-; consult, Polar Exp, World Meteorol Orgn, 75- Mem: Am Geophys Union; Am Asn Univ Profs; AAAS; Fedn Am Sci. Res: Physical oceanography, physics of large-scale ocean circulation and climate; antarcticAntarctic oceanography; ocean instrumentation. Mailing Add: Dept of Oceanog Univ of Wash Seattle WA 98195

BAKER, DONALD ROY, b Norfolk, Va, May 8, 27; m 48; c 2. ORGANIC GEOCHEMISTRY, GEOLOGY. Educ: Calif Inst Technol, BS, 50; Princeton Univ, PhD(geol), 55. Prof Exp: Instr petrol & chem geol, Northwestern Univ, 54-56; sr res geologist, Denver Res Ctr, Marathon Oil Co, 56-66; assoc prof, 66-72, PROF GEOL, RICE UNIV, 72- Mem: Fel Geol Soc Am; Am Asn Petrol Geol; Soc Econ Paleontologists & Mineralogists; Geochem Soc; Mineral Soc NAm. Res: Organic geochemical and stable isotope geochemical research with focus on problems related to petroleum evolution and fundamental petrology. Mailing Add: Dept of Geol Rice Univ Houston TX 77001

BAKER, DORIS, b Pt Marion, Pa, Nov 16, 21. ANALYTICAL CHEMISTRY, CEREAL CHEMISTRY. Educ: Univ Md, BS, 48. Prof Exp: Chemist grain & oil seed technol, 48-52, res chemist, 52-74, RES CHEMIST & PROJ LEADER FOOD COMPOSITION METHODOLOGY, USDA, 74- Concurrent Pos: Am Asn Cereal Chemists rep, Agr Res Inst, Nat Acad Sci, 69-70. Mem: Am Asn Cereal Chemists; Am Oil Chemists Soc; Am Chem Soc. Res: Analytical chemistry as applied to nutrients in foods; methodology for determining composition of food fiber and its physical and chemical relationship to other food nutrients. Mailing Add: Nutrit Inst Agr Res Serv USDA Agr Res Ctr E Beltsville MD 20705

BAKER, DUDLEY DUGGAN, III, b Seguin, Tex, Feb 1, 36; m 59; c 3. UNDERWATER ACOUSTICS. Educ: Univ Tex, BS, 57, MA, 58. Prof Exp: Res scientist assoc, Defense Res Lab, 59-67, HEAD ELECTROACOUST DIV, APPL

RES LABS, UNIV TEX, AUSTIN, 67- Mem: Acoust Soc Am; Sigma Xi. Res: Underwater acoustical measurements, especially related to sonar transducer testing. Mailing Add: PO Box 8029 Austin TX 78712

BAKER, DURWOOD L, b Algona, Iowa, June 16, 19; m 45; c 4. VETERINARY MEDICINE. Educ: Iowa State Univ, DVM, 43. Prof Exp: From instr to assoc prof vet med, 47-59, asst dean, Col Vet Med, 64-68, PROF VET MED, IOWA STATE UNIV, 59-, ASSOC DEAN COL VET MED, 68- Mem: Am Vet Med Asn. Res: Small animal medicine and surgery. Mailing Add: Col of Vet Med Iowa State Univ Ames IA 50011

BAKER, DWIGHT LYNDS, b Amherst, Mass, Sept 1, 10; m 37; c 2. CHEMISTRY. Educ: Amherst Col, AB, 33; Columbia Univ, MA, 34, PhD(chem), 40. Prof Exp: Res chemist, E I du Pont de Nemours & Co, Mich, 37-39; Upjohn fel, Columbia Univ, 40-43; tech dir, Vita Zyme Labs, 43-51; res scientist, J E Siebel Sons, Inc, 51-52; dir res, Froedtert Malt Corp, 52-62; sr proj engr, Linde Div, Union Carbide Corp, 62-65; PROF CHEM, SOUTHEASTERN MASS UNIV, 65- Mem: AAAS. Res: Chemistry and physics of surfaces; protective coatings; enzyme production and application; malting and brewing; food technology; biochemistry of fish. Mailing Add: 3 School St South Dartmouth MA 02748

BAKER, EARL WAYNE, b Lewistown, Mont, Sept 26, 28; m 49; c 3. ORGANIC CHEMISTRY. Educ: Mont State Univ, BS, 52; Johns Hopkins Univ, MA, 62, PhD(org chem), 64. Prof Exp: Jr chemist, Lago Oil & Transport Co, Ltd, Aruba, 52-53; chemist, Stand Oil Co, NJ, 53-60; instr chem, Johns Hopkins Univ, 60-63; res fel petrol chem, Mellon Inst, 64-70; assoc prof chem, Univ Pittsburgh, 69-70; PROF CHEM & HEAD DEPT, NORTHEAST LA UNIV, 70- Concurrent Pos: Am Chem Soc grant & lectr, Carnegie-Mellon Univ, 67-69; NSF grant, 69-71; mem adv panel, Joint Oceanog Inst for Deep Earth Sampling, 68- Mem: AAAS; Am Chem Soc. Res: Trace metal analysis; emission spectroscopy; chromatography; synthesis and structure of tetrapyrrole pigments; chemistry of transition metal complexes; organic geochemistry. Mailing Add: Dept of Chem Northeast La Univ Monroe LA 71201

BAKER, EDGAR EUGENE, JR, b Visalia, Calif, Oct 12, 13; m 38; c 1. MEDICAL MICROBIOLOGY. Educ: Univ Calif, Los Angeles, AB, 35, MA, 37, PhD(microbiol), 41. Prof Exp: Res assoc, Hooper Found, Univ Calif, 42-46; assoc path & bact, Rockefeller Inst, 46-49; assoc prof, 49-52, PROF MICROBIOL & HEAD DEPT, SCH MED, BOSTON UNIV, 52- Mem: Am Soc Microbiol; Am Asn Immunol; Soc Exp Biol & Med. Res: Coccidioidomycosis; immunology of plague; antigenic structure of enterobacteriaceae; antibiotics. Mailing Add: Dept of Microbiol Boston Univ Sch of Med Boston MA 02118

BAKER, EDGAR GATES STANLEY, b Peotone, Ill, June 7, 09; m 35; c 3. BIOLOGY. Educ: DePauw Univ, AB, 31; Stanford Univ, PhD(biol), 43. Prof Exp: From instr to asst prof zool, Wabash Col, 32-38, actg head dept, 38-39; asst biol, Stanford Univ, 39-42; head biol sect, Del Mar Col, 46; asst prof biol, Cath Univ Am, 46-50; instr, Cath Sisters Col, 47-50; from assoc prof to prof, 50-74, head dept, 51-70, EMER PROF ZOOL, DREW UNIV, 74- Concurrent Pos: Consult surv physiol sci, Am Physiol Soc, 52-54; NSF sci fac fel, 57-58; chief reader biol, Adv Placement Prog, Col Entrance Exam Bd, 64-67; staff biologist, Comn on Undergrad Educ in Biol Sci, 67-68. Mem: Fel AAAS; Am Soc Zool; Soc Protozool. Res: Physiology of protozoan populations; bacteria-free cultures; effects of nutrition on rate of growth. Mailing Add: Box Syc-2 Drew Univ Madison NJ 07940

BAKER, EDWARD GEORGE, b New York, NY, Oct 20, 08; m 31; c 5. MATHEMATICS. Educ: Columbia Univ, AB, 30, AM, 31, DEd, 39. Educ: From instr to assoc prof math, Newark Col Eng, 31-42; eng mem mach tech staff, 42-66, MGR ELECTRONIC DATA PROCESSING STAFF, AM BUR SHIPPING, 66-73; RETIRED. Mem: AAAS; Am Math Soc; Am Soc Mech Eng; Soc Naval Archit & Marine Eng. Res: Ship propulsion systems. Mailing Add: Carob Ctr Rte 1 Pine Knoll Shores Morehead City NC 28557

BAKER, EDWARD WILLIAM, b Porterville, Calif, Dec 29, 14; m 35; c 1. ENTOMOLOGY. Educ: Univ Calif, BS, 36, PhD(entom), 38. Prof Exp: Mem staff, Mex Fruit-fly Lab, Bur Entom & Plant Quarantine, USDA, 39-44, entomologist, Div Insect Identification, 44-53, Div Insects, Entom Res Br, Agr Res Serv, 53-58, ENTOMOLOGIST, INSECT IDENTIFICATION & PARASITE INTROD LABS, ENTOM RES DIV, AGR RES SERV, USDA, 58- Concurrent Pos: Vis lectr, Univ Md, 53 & Ohio State Univ, 62. Mem: Entom Soc Am. Res: Acarina; biology and taxonomy of mites; plant feeding mites of importance to agriculture. Mailing Add: Syst Entom Lab ARS-USDA Agr Res Ctr-W Beltsville MD 20705

BAKER, ELIZABETH MCINTOSH, b Washington, DC, Sept 30, 45; m 68. INVERTEBRATE ZOOLOGY. Educ: George Washington Univ, BA, 67; Univ Mich, MS, 68; Univ Va, PhD(develop biol), 73. Prof Exp: Vis asst prof, 75, ELECTRON MICROS TECHNICIAN BIOL, UNIV NC, CHARLOTTE, 75- Res: Insectan prothoracic gland development and function. Mailing Add: Dept of Biol Univ of NC Charlotte NC 28213

BAKER, EUGENE (MANIGUALT), III, biochemistry, see 12th edition

BAKER, FLOYD B, b Bishop, Calif, Sept 21, 28; m 52; c 3. PHYSICAL CHEMISTRY. Educ: Univ Southern Calif, BA, 54; Univ NMex, PhD(chem), 60. Prof Exp: ASST GROUP LEADER, LOS ALAMOS SCI LAB, 60- Mem: Am Inst Chem; Am Chem Soc. Res: Inorganic kinetics of oxidation reductions in aqueous solution. Mailing Add: 161 El Carto Dr Los Alamos NM 87544

BAKER, FRANCIS TODD, b Chicago, Ill, Feb 22, 42; m 64; c 1. NUCLEAR PHYSICS. Educ: Miami Univ, AB, 63, MA, 64; Univ Mich, PhD(physics), 70. Prof Exp: Asst prof, Carroll Col, Wis, 66-68 & St Lawrence Univ, 70-71; res assoc, Univ Mich, 71 & Rutgers Univ, 71-74; ASST PROF PHYSICS, DEPT PHYSICS & ASTRON, UNIV GA, 74- Mem: Am Phys Soc. Res: Experimental nuclear physics, particularly nuclear structure and reactions. Mailing Add: Dept of Physics & Astron Univ of Ga Athens GA 30683

BAKER, FRANK HAMON, b Stroud, Okla, May 2, 23; m 46; c 4. ANIMAL SCIENCE. Educ: Okla Agr & Mech Col, BS, 47, MS, 51, PhD(animal nutrit), 54. Prof Exp: County agr agent, Del County, Okla, 47-48; instr high sch, Okla, 49-50; asst animal husb, Okla Agr & Mech Col, 51-53; instr, Kans State Col, 53-54, asst prof, 54-55; assoc prof, Univ Ky, 55-58; exten livestock specialist, Okla State Univ, 58-62; exten animal scientist, USDA, Washington, DC, 62-66; prof animal sci & chmn dept, Univ Nebr, Lincoln, 66-74; PROF & DEAN AGR, OKLA STATE UNIV, 74- Honors & Awards: Distillers Feed Res Coun Award, 64. Mem: Am Soc Animal Sci (pres, 73-74); Am Meat Sci Asn; AAAS; Am Inst Biol Sci; Coun Agr Sci & Technol. Res: Animal production; ruminant Mailing Add: Div of Agr Okla State Univ Stillwater OK 74074

BAKER, FRANK SLOAN, JR, b Brownwood, Tex, May 20, 21; m 42; c 3. ANIMAL HUSBANDRY, ANIMAL NUTRITION. Educ: Agr & Mech Col Tex, BS, 42; Univ Fla, MSA, 57. Prof Exp: From asst animal husbandman to assoc animal husbandman, N Fla Exp Sta, 45-63, ANIMAL HUSBANDMAN & PROF ANIMAL HUSBANDRY, AGR RES & EDUC CTR, INST FOOD & AGR SCI, UNIV FLA, QUINCY, 63- Mem: Am Soc Animal Sci; Am Dairy Sci Asn. Res: Production of beef cattle, swine and sheep. Mailing Add: Agr Res & Educ Ctr Univ of Fla Quincy FL 32351

BAKER, FRANK WEIR, b Anderson, Ind, Nov 22, 38; m 64. ORGANIC CHEMISTRY. Educ: Col Wooster, BA, 60; Univ Chicago, MS, 62, PhD(org chem), 66. Prof Exp: Res chemist, Miami Valley Labs, 66-71, Ivorydale Tech Ctr, 71-75, ASSOC DIR PROF & REGULATORY SERV, WINTON HILL TECH CTR, PROCTER & GAMBLE CO, 75- Mem: AAAS; Am Chem Soc; Am Soc Photobiol. Res: Photochemical stability of biologically active compounds; toxicology; biological effects of environmental contaminants. Mailing Add: Winton Hill Tech Ctr Procter & Gamble Co Cincinnati OH 45224

BAKER, FREDERICK D, microbiology, see 12th edition

BAKER, GEORGE ALLEN, b Robinson, Ill, Oct 31, 03; m 30; c 2. MATHEMATICAL STATISTICS. Educ: Univ Ill, BS, 26, PhD(math statist), 29. Prof Exp: Assoc statistician, USPHS, 29; Milbank Mem Fund fel, Columbia Univ, 29-30; statistician, State Dept Health, NY, 30-31; head dept math, Shurtleff Col, 31-34; prof & head dept, Miss Woman's Col, 34-36; with consumers purchase study, Bur Home Econ, USDA, 36-37; from instr math & jr statistician to prof math & statistician, 37-74, EMER PROF MATH, UNIV CALIF, DAVIS, 74- Mem: Am Math Soc; Math Asn Am; fel Inst Math Statist; Am Statist Asn; Biomet Soc. Res: Application of field trials and growth curves; selection, prediction and transformation of data; random sampling from nonhomogeneous populations; taste-testing; factor analysis; applications to astronomy, physics and education. Mailing Add: Dept of Math Univ of Calif Davis CA 95616

BAKER, GEORGE ALLEN, JR, b Alton, Ill, Nov 25, 32; m 56; c 3. STATISTICAL MECHANICS, APPLIED MATHEMATICS. Educ: Calif Inst Technol, BS, 54; Univ Calif, Berkeley, PhD(physics), 56. Prof Exp: NSF fel, Columbia Univ, 56-57; mem staff, Los Alamos Sci Lab, 57-66; physicist, Brookhaven Nat Lab, 66-71, sr physicist, 71-75; MEM STAFF, LOS ALAMOS SCI LAB, 75- Concurrent Pos: Assoc res physicist, Univ Calif, San Diego, 61-62; vis prof, Kings Col, Univ London, 64-65, Univ Nice, 70 & Cornell Univ, 71-72. Mem: Fel Am Phys Soc. Res: Statistical mechanics; mathematical methods of theoretical physics; quantum theory; field theory; nuclear physics. Mailing Add: Theoret Div Los Alamos Sci Lab Los Alamos NM 87545

BAKER, GEORGE LEROY, pharmacy, see 12th edition

BAKER, GEORGE SEVERT, b Chicago, Ill, Aug 2, 27. SOLID STATE PHYSICS. Educ: Purdue Univ, BS, 50; Univ Ill, PhD(physics), 57. Prof Exp: Res assoc metall, Univ Ill, 56-57; asst prof physics, Univ Utah, 57-62; tech specialist, Aerojet Gen Corp, Calif, 62-67; asst dean, Col Eng & Appl Sci, 72-76, PROF, DEPT MAT, UNIV WIS-MILWAUKEE, 67- Mem: Soc Mfg Eng; Am Soc Metals. Res: Defect properties of crystals; mechanical properties; refractory metals; forging; manufacturing processes. Mailing Add: 8065 N Mohawk Fox Point WI 53217

BAKER, GLADYS ELIZABETH, b Iowa City, Iowa, July 22, 08. BOTANY. Educ: Univ Iowa, AB, 30, MS, 32; Washington Univ, PhD(mycol), 35. Prof Exp: Asst bot, Washington Univ, 35-36; instr biol, Hunter Col, 36-40; from instr to prof plant sci, Vassar Col, 40-63, chmn dept, 48-60; vis prof, 61-62, prof, 63-73, EMER PROF BOT, UNIV HAWAII, 73- Concurrent Pos: Vassar Col fac fel, Stanford Univ, 45 & Univ Calif, 54. Mem: AAAS; Mycol Soc Am; Bot Soc Am; Am Soc Microbiol; Brit Soc Gen Microbiol. Res: Cytology and morphology of myxomycetes, lichens and basidiomycetes; cytogenetics of imperfect fungi; distribution of fungi. Mailing Add: 11091 Burntwood Dr Sun City AZ 85351

BAKER, GLENN JACKSON, b Clearfield, Iowa, Dec 21, 04; m 35; c 2. GEOLOGY. Educ: Univ Wis, BA, 27, MA, 32. Prof Exp: Field geologist, Rhokana Corp, NRhodesia, 28-31 & Grand Falls Lab Mining, Labrador, 33; develop geologist, Gold Coast, 34-35; party chief, Schlumberger Co, Tex, 35-38; GEOLOGIST & VPRES, W M BARRET, INC, 38- Concurrent Pos: Instr, Centenary Col, 46-49. Mem: Am Inst Mining, Metall & Petrol Engrs; Soc Explor Geophysicists; Am Geophys Union. Res: Base exchange in zeolites; electromagnetic wave propagation. Mailing Add: 215 E Wyandotte Shreveport LA 71101

BAKER, GRAEME LEVO, b Kalispell, Mont, Mar 7, 25; m 49; c 3. BIOCHEMISTRY. Educ: Mont State Univ, BSc, 47, MS, 53, PhD(chem), 59. Prof Exp: Instr chem, Mont State Col, 49-51; asst state feed control chemist, State Dept Agr, Mont, 51-52; asst chem, Mont State Col, 52-53, from instr to prof, 53-68; PROF CHEM & CHMN DEPT, FLA TECH UNIV, 68- Mem: Am Chem Soc. Res: Lipids of insect origin. Mailing Add: Dept of Chem Fla Tech Univ Orlando FL 32816

BAKER, GRIFFIN JONATHAN, b Marion, Ill, July 19, 17; m 47; c 2. ENTOMOLOGY. Educ: Univ Ill, BSc, 40. Prof Exp: Asst, United Fruit Co, Honduras, 40-42 & Univ Ill, 42-43; asst proj chemist, Standard Oil Co (Ind), 46-50, proj chemist, 50-59; with McLaughlin, Gormley, King & Co, 59-70, mgr tech serv, 70-71, MGR RES & DEVELOP, MCLAUGHLIN, GORMLEY, KING & CO, 71- Mem: Entom Soc Am. Res: Control of the banana thrips; product development of insecticidal aerosols; space and residual sprays; stock sprays and general garden sprays. Mailing Add: McLaughlin Gormley King & Co 8810 Tenth Ave N Minneapolis MN 55427

BAKER, HAROLD LAWRENCE, b Ogden, Utah, Jan 14, 18; m 46; c 2. FOREST ECONOMICS. Educ: Utah State Agr Col, BS, 39; Univ Calif, Berkeley, MS, 42, PhD(forest econ), 65. Prof Exp: Asst forestry res, Sch Forestry, Univ Calif, Berkeley, 40-42, asst forestry instr & res, 46-49; forest economist, Pac Southwest Forest & Range Exp Sta, US Forest Serv, 49-58; land economist, Hawaii State Land Study Bur, 58-64, prof forest econ & mem grad fac, 61-74, land economist & dir, Hawaii State Land Study Bur, 64-74, PROF RESOURCE ECON, DEPT AGR & RESOURCE ECON, UNIV HAWAII, 74- Concurrent Pos: Forest econ consult, US Forest Serv, 61-74; resource economist, Water Resources Res Ctr, Univ Hawaii, 64-74, land econ consult, Am Factors, 64; UN Food & Agr Orgn forest & land use econ consult, Philippine Govt, 66 & Malaysian Govt, 69-; land econ consult, Govt Am Samoa, 72. Mem: Soc Am Foresters. Res: Forest mensuration, management and policy; agricultural policy; production and land economics; land tenure; regional economic analyses; economics of resource allocation and development; land use planning; land use controls. Mailing Add: Univ of Hawaii 2444 Dole St Honolulu HI 96822

BAKER, HAROLD NORDEAN, b Iowa City, Iowa, May 18, 43; m 66; c 4. BIOCHEMISTRY, CLINICAL CHEMISTRY. Educ: Lamar Univ, BS, 65; Tulane Univ, PhD(org chem), 70. Prof Exp: Robert A Welch Found fel, Univ Tex M D

Anderson Hosp, Houston, 69-71, res assoc protein struct, 71; res assoc, 71-72, instr, 72-74, ASST PROF PROTEIN STRUCT IN ATHEROSCLEROSIS, BAYLOR COL MED, 74- Concurrent Pos: Dir, Lipid Res Clin, Core Lab, 72- Mem: Am Chem Soc. Res: Protein structure; metabolic basis of atherosclerosis; lipid transport. Mailing Add: Dept of Med Baylor Col Med Methodist Hosp Houston TX 77025

BAKER, HAROLD THEODORE, b Renville, Minn, Oct 13, 13; m; c 4. PHYSICAL CHEMISTRY. Educ: Univ Minn, BS, 39; Univ Iowa, PhD(phys chem), 43. Prof Exp: Teacher pub schs, Minn, 32-36, prin, 39-41; prin pub sch, Iowa, 41-43; res chemist, Esso Standard Oil Co, 43-54; assoc prof chem & physics, Tex Woman's Univ, 54-62; PROF CHEM & HEAD DEPT, LAMAR UNIV, 62- Concurrent Pos: Consult, Oak Ridge Inst Nuclear Studies. Res: Raman spectroscopy; petroleum catalyst developments; solvent extraction; radio chemistry. Mailing Add: 5170 Cambridge Beaumont TX 77707

BAKER, HAROLD WELDON, b Lincoln, Nebr, Jan 23, 31; m 57; c 2. ANALYTICAL CHEMISTRY. Educ: Nebr Wesleyan Univ, BA, 54; Purdue Univ, MS, 56; Univ Iowa, PhD, 61. Prof Exp: Instr chem, Butler Univ, 56-57; from assoc prof to prof, Parsons Col, 59-62; from asst prof to assoc prof, 62-74, PROF CHEM, SCH PHARM, TEMPLE UNIV, 74- Mem: Am Chem Soc. Res: Atomic absorption and fluorescence; trace analysis of metals; air and water pollution; heavy metal toxicity of life systems; pharmaceutical analysis. Mailing Add: Dept of Chem Temple Univ Sch of Pharm Philadelphia PA 19140

BAKER, HARRIS MITCHELL JR, b Jacksonville, Fla, Sept 21, 22; m 46; c 3. ANALYTICAL CHEMISTRY. Educ: Emory Univ, AB, 48, MS, 50. Prof Exp: Chemist, Commun Dis Ctr, USPHS, 49-50; CHEMIST, BIOCHEM DEPT, E I DU PONT DE NEMOURS & CO, INC, 50- Mem: Am Chem Soc; Sigma Xi. Res: Agrichemicals technical development. Mailing Add: Biochem Dept E I du Pont de Nemours & Co Wilmington DE 19898

BAKER, HERBERT GEORGE, b Brighton, Eng, Feb 23, 20; m 45; c 1. EVOLUTION, ECOLOGY. Educ: Univ London, BSc, 41, PhD, 45. Prof Exp: Res chemist & plant physiologist, Hosa Res Labs, 40-45; lectr bot, Univ Leeds, 45-54; sr lectr, Univ Col of Gold Coast, 54-55, prof & head dept, 55-57; dir bot garden, 57-69, assoc dir, 69-74, assoc prof bot, 57-60, PROF BOT, UNIV CALIF, BERKELEY, 60- Concurrent Pos: Res fel, Carnegie Inst, 48-49; assoc ed, Evolution, 56-59, 62-65 & Ecology, 63-66; res prof, Miller Inst, 66-67; mem bd gov, Orgn Trop Studies, 64- Mem: Am Bot Soc; Soc Study Evolution (secy, 67-69, pres, 69); Ecol Soc Am; Int Asn Bot Gardens (vpres, 64-69); Am Soc Naturalists. Res: Ecology and evolution of higher plants, especially on reproductive biology; palynology; cytogenetics; history of biology; general tropical botany; nectar chemistry. Mailing Add: Dept of Bot Univ of Calif Berkeley CA 94720

BAKER, HERMAN, b New York, NY, Jan 22, 26; m 52; c 2. METABOLISM, NUTRITION. Educ: City Col New York, BS, 46; Emory Univ, MS, 48; NY Univ, PhD(metab), 56; Am Bd Nutrit, dipl, 68. Prof Exp: Assoc prof med, 60-70, PROF MED & PREV MED & DIR DIV NUTRIT-PREV MED, COL MED & DENT NJ, 70- Mem: Soc Protozool; Soc Exp Biol & Med; Am Soc Clin Nutrit. Res: Vitamin metabolism; nutrition and analysis. Mailing Add: 27 Wilk Rd Edison NJ 08817

BAKER, HINTON JOSEPH, microbiology, medicine, see 12th edition

BAKER, HOWARD CRITTENDON, b Lexington Ky, Sept 4, 43; m 67. THEORETICAL PHYSICS. Educ: Berea Col, BA, 65; Washington Univ, St Louis, MA, 67, PhD(physics), 72. Prof Exp: ASST PROF PHYSICS, BEREA COL, 72- Concurrent Pos: Vis asst prof physics, Univ Conn, 74-75. Mem: Am Phys Soc. Res: Quantum field theory; elementary particle theory. Mailing Add: Dept of Physics Berea Col Berea KY 40403

BAKER, JAMES ADDISON, b Eugene, Ore, Aug 20, 22; m 46; c 1. COMPUTER SCIENCE. Educ: Pomona Col, AB, 44. Prof Exp: Head math & comput, Lawrence Radiation Lab, Univ Calif, Berkeley, 52-58; assoc head data systs div, Broadview Res Corp, 58-61; HEAD MATH & COMPUT GROUP, LAWRENCE RADIATION LAB, UNIV CALIF, BERKELEY, 61- Concurrent Pos: Consult, Comt Uses of Comput, Nat Acad Sci, 63; lectr elec eng, comput sci, indust eng & opers res, Univ Calif, Berkeley, 63-; vis scholar, Europ Orgn Nuclear Res, Geneva, 74-75. Mem: Asn Comput Mach. Res: Simulation of communications systems; programming systems; network analysis. Mailing Add: Math & Comput Group Lawrence Radiation Lab Berkeley CA 94720

BAKER, JAMES (ANDREW), animal pathology, deceased

BAKER, JAMES BERT, b Bernice, La, Feb 7, 39; m 60; c 2. FOREST SOILS, SILVICULTURE. Educ: Univ Ark, Monticello, BSF, 61; Duke Univ, MF, 62; Miss State Univ, PhD(forest soils), 70. Prof Exp: Res forester silvicult, 62-71, RES SOIL SCIENTIST FOREST SOILS, SOUTHERN FOREST EXP STA, USDA FOREST SERV, 71- Mem: Am Soc Agron; Soil Sci Am; Soc Am Foresters; Sigma Xi. Res: Intensive culture of southern hardwoods; forest fertilization; forest irrigation; timber and water management in greentree reservoirs. Mailing Add: Southern Hardwoods Lab USDA Forest Serv PO Box 227 Stoneville MS 38776

BAKER, JAMES DENNARD, b San Angelo, Tex, May 2, 39; m 59; c 2. MATHEMATICAL ANALYSIS. Educ: Univ Tex, BA, 61, MA, 62; Iowa State Univ, PhD(math), 69. Prof Exp: Opers analyst, Gen Dynamics Corp, 62-64; assoc prof math, Appl Physics Lab, Johns Hopkins Univ, 64-69; SR PRIN RES SCIENTIST, HONEYWELL CORP RES CTR, 70- Mem: Am Math Soc. Res: Mathematical theory and analysis relating to systems analysis problems. Mailing Add: 13001 Upton Ave S Burnsville MN 55337

BAKER, JAMES EARL, b Cowen, WVa, Dec 1, 31; m 61. PLANT PHYSIOLOGY. Educ: Univ Md, BS, 53, MS, 55; NC State Univ, PhD(plant physiol), 58. Prof Exp: Plant physiologist, Plant, Soil & Nutrit Lab, USDA, NY, 58-60, PLANT PHYSIOLOGIST, POSTHARVEST PLANT PHYSIOL LAB, USDA, 60- Honors & Awards: Japanese Govt Res Award for For Specialists, 71. Mem: AAAS; Am Inst Biol Sci; Am Chem Soc; Am Soc Plant Physiologists; Brit Biochem Soc. Res: Nitrogen metabolism of plants; structure-funtion interrelationships of plant organelles; biochemical and ultrastructural aspects of senescence in plants; respiratory mechanisms in plants. Mailing Add: Postharvest Plant Physiol Lab Agr Res Ctr-West Beltsville MD 20705

BAKER, JAMES GILBERT, b Louisville, Ky, Nov 11, 14; m 38; c 4. OPTICAL PHYSICS, ASTRONOMY. Educ: Univ Louisville, AB, 35, ScD, 48; Harvard Univ, MA, 36, PhD(astron, astrophys), 42. Prof Exp: Res fel, 42-45, dir, Optical Res Lab, 43-45, assoc prof, Harvard Observ, 46-48, res assoc, 49-62, ASSOC, HARVARD OBSERV, HARVARD UNIV, 62-; CONSULT OPTICAL PHYSICS & OPTICAL ASTRON, AEROSPACE CORP, 66- & PHOTOG OPTICS, POLAROID CORP, 66- Concurrent Pos: Lowell lectr, 40; res assoc, Lick Observ, 48-60; consult optical physics & aerial photog, US Air Force, 49-57, mem sci adv bd, 52-57; chmn, US Nat Comn, Int Comn Optics, 56-59, vpres, 59-62; assoc, Ctr Astrophys, Harvard & Smithsonian Observs, 66-; trustee, The Perkin Fund, 70- Honors & Awards: Exceptional Civilian Serv Award, USAF, 57; Adolph Lomb Medal, 42; Magellanic Medal, Am Philos Soc, 53; Elliott Cresson Medal, Franklin Inst, 62; Frederick Ives Medal, Optical Soc Am, 65. Mem: Nat Acad Sci; Am Astron Soc; fel Optical Soc Am (pres, 60); Am Acad Arts & Sci; Sigma Xi. Res: Instrumentation and optical design; astrophysics. Mailing Add: 7 Grove St Winchester MA 01890

BAKER, JAMES HASKELL, b Ft Worth, Tex, Sept 8, 40. MARINE ZOOLOGY. Educ: Tex Christian Univ, BA, 62, MS, 65; Univ Houston, PhD(biol), 75. Prof Exp: Mus technician, Smithsonian Oceanog Sorting Ctr, 67; res assoc aquatic biol, TCU Res Found, Tex Christian Univ, 68-70; SR RES SCIENTIST AQUATIC BIOL, SOUTHWEST RES INST, 73- Mem: AAAS; Am Soc Limnol & Oceanog; Asn Meiobenthologists; Sigma Xi; Soc Syst Zool. Res: Ecology and systematics of marine invertebrates. Mailing Add: Southwest Res Inst 3600 Yoakum Blvd Houston TX 77006

BAKER, JEFFREY JOHN WHEELER, b Montclair, NJ, Feb 2, 31; m 55; c 4. DEVELOPMENTAL BIOLOGY, SCIENCE WRITING. Educ: Univ Va, BA, 53, MS, 59. Prof Exp: Supvr qual control, Libby, McNeill & Libby Food Co, Inc, 53-54; teacher biol, Mt Hermon, Mass, 54-62; lectr sci, Wesleyan Univ, 62-66; vis assoc prof biol, George Washington Univ, 66-68; vis prof, Univ PR, 68-69; LECTR SCI, WESLEYAN UNIV, 69-, SR FEL, COL SCI IN SOCIETY, 75- Concurrent Pos: Staff biologist & educ dir, Comn Undergrad Educ Biol Sci, 66-68. Mem: AAAS; Am Inst Biol Sci; Nat Asn Sci Writers. Res: Amphibian development. Mailing Add: Col of Sci in Society Wesleyan Univ Middletown CT 06457

BAKER, JOHN BEE, b Clarksdale, Miss, Mar 30, 27; m 56; c 2. PLANT PHYSIOLOGY. Educ: Miss State Col, BS, 50; Univ Wis, MS, 51, PhD(bot), 55. Prof Exp: Asst prof plant path & asst pathologist weed control, 53-59, assoc prof bot & plant path, 59-66, PROF BOT & PLANT PATH, LA STATE UNIV, BATON ROUGE, 66- Mem: Am Soc Plant Physiol: Weed Sci Soc Am. Res: Mechanism of action of herbicides and growth regulators; autecology of weeds. Mailing Add: Dept of Plant Path La State Univ Baton Rouge LA 70803

BAKER, JOHN CUMMINS, b Sacramento, Calif, Oct 24, 40. PLASMA PHYSICS. Educ: Univ Calif, AB, 61, PhD(physics), 67. Prof Exp: Nat Res Coun fel, NASA, 68-69; mem staff, Philco-Ford Corp, 69-72; SR PHYSICIST, CALSPAN CORP, 72- Mem: Am Phys Soc; Am Geophys Union; NY Acad Sci. Res: Non-linear mechanisms in inhomogeneous plasmas; plasma synchrotron radiation; plasma-magnetic field interfaces; plasma confinement; effects of nuclear detonations on radar and communication systems. Mailing Add: Calspan Corp PO Box 235 Buffalo NY 14221

BAKER, JOHN KEITH, b San Antonio, Tex, Dec 1, 42; m 65; c 1. MEDICINAL CHEMISTRY, PHYSICAL CHEMISTRY. Educ: Univ Tex, Austin, BS, 66; Univ Calif, San Francisco, PhD(pharmaceut chem), 70. Prof Exp: ASST PROF MED CHEM, SCH PHARM, UNIV MISS, 70- Mem: Am Pharmaceut Asn. Res: Application of spectroscopic methods in molecular pharmacological studies; kinetic and thermodynamic studies of the interaction of adenosine triphosphate with divalent metal ions and phospholipids; quantitative drug analysis. Mailing Add: Dept of Med Chem Sch of Pharm Univ of Miss University MS 38677

BAKER, JOHN RICHARD, b Eng, Nov 23, 13; nat US; m 51; c 3. PATHOLOGY, HISTOCHEMISTRY. Educ: AMT, Inst Med Lab Technol, Eng, 45. Prof Exp: Lab chief histol, Post-grad Med Sch, Univ London, 35-48; lab chief & lab supt path, Univ Col West Indies, 48-51; asst cancer, Med Col Ga, 51-52; asst histochem cancer res unit, Tufts Univ, 52-57; lab dir & histochemist, St Margaret's Hosp, Mass, 57-59; res assoc histochem, Bio-Res Inst Inc & chief tissue res exp path, Bio-Res Consult Inc, 59-67; HEAD HISTOL & HISTOCHEM & PROJ MGR CARCINOGENESIS BIOASSAY, MASON RES INST, INC, 67- Concurrent Pos: Univ fel, Univ Col West Indies, 55. Mem: Histochem Soc; fel Royal Micros Soc; Royal Soc Health. Res: Cancer; experimental pathology; histochemistry; cytochemistry; fluorescence microscopy; electron microscopy; microtomy; cytology. Mailing Add: Mason Res Inst 25 Harvard St Worcester MA 01608

BAKER, JOHN WARREN, b El Paso, Tex, Aug 24, 36; m 70. MATHEMATICAL ANALYSIS, TOPOLOGY. Educ: Hardin-Simmons Univ, BS, 58; Univ Tex, Austin, MA, 65, PhD(math), 68. Prof Exp: Asst prof math, Fla State Univ, 68-73; asst prof, 73-75, ASSOC PROF MATH, KENT STATE UNIV, 75- Concurrent Pos: Undergrad coordr, Dept Math, Kent State Univ, 74-76. Mem: Am Math Soc; Math Asn Am. Res: Spaces of continuous functions; projections in Banach spaces; zero-dimension topological spaces.

BAKER, JOSEPH, soil chemistry, see 12th edition

BAKER, JOSEPH WILLARD, b Luray, Va, Sept 15, 24; m 52; c 2. ORGANIC CHEMISTRY. Educ: Bridgewater Col, BA, 45; Univ Va, MS, 49, PhD(org chem), 52. Prof Exp: Instr, Univ Va, 51-52; res chemist, 52-56, RES GROUP LEADER, MONSANTO CO, 56- Mem: Am Chem Soc; Sigma Xi. Res: Organophosphorus chemistry; agricultural chemicals; food and fine chemicals; bacteriostats; plasticizers; functional fluids; amino ketones and alcohols; organic heterocycles; catalytic reductions and chemical process development. Mailing Add: Res Ctr Monsanto Co 800 N Lindbergh Blvd St Louis MO 63166

BAKER, JUNE MARSHALL, b Napton, Mo, Nov 11, 22; m 47; c 2. ANALYTICAL CHEMISTRY, INSTRUMENTAL CHEMISTRY. Educ: Mo Valley Col, BA, 44; Ohio State Univ, MSc, 50; Univ Mo, PhD(chem), 55. Prof Exp: Instr chem, Mo Valley Col, 47-49; analyst, Univ Mo, 52-55; asst prof, Tenn Polytech Univ, 55-56; assoc prof, Eastern Ill State Col, 56-57; assoc prof, 57-65, PROF CHEM, AUBURN UNIV, 65- Concurrent Pos: Consult, Orradio Industs & Auburn Res Found. Mem: Am Chem Soc. Res: Ion-exchange method for study of chemical reactions; determinations for chromium. Mailing Add: Dept of Chem Auburn Univ Auburn AL 36830

BAKER, KAY DAYNE, b Escalante, Utah, Jan 31, 34; m 55; c 3. METEOROLOGY, ELECTRICAL ENGINEERING. Educ: Univ Utah, BS, 56, MS, 57, PhD(elec eng), 66. Prof Exp: Res asst, Upper Air Res Lab, Univ Utah, 54-57, res engr, 57-62, asst dir lab, 62-69, dir lab, 69-70; DIR SPACE SCI LAB, UTAH STATE UNIV, 70- Concurrent Pos: Mem comn 3, Inst Sci Radio Union, 67- Mem: Am Geophys Union. Res: Aerology; development of measuring techniques and investigations of the upper atmosphere of the earth, with special emphasis on ionospheric auroral measurements. Mailing Add: Space Sci Lab Utah State Univ Logan UT 84321

BAKER, KENNETH FRANK, b Ashton, SDak, June 3, 08; m 44. PHYTOPATHOLOGY, SOIL MICROBIOLOGY. Educ: State Col Wash, BS, 30, PhD(plant path), 34. Prof Exp: Asst plant path, State Col Wash, 30-34; Nat Res fel bot, Univ Wis, 34-35; jr pathologist, Div Forest Path, USDA, 35-36; assoc pathologist, Exp Sta, Pineapple Producers Coop Asn, 36-39; from asst plant pathologist & asst prof to plant pathologist, Exp Sta & prof plant path, Univ Calif, Los Angeles, 39-60; prof & plant pathologist, 61-74, EMER PROF PLANT PATH & EMER PLANT PATHOLOGIST, UNIV CALIF, BERKELEY, 74- Concurrent Pos: Supvry technician, Div Forest Path, USDA, 34; Mem, Comt Biol Control Agr Bd, Nat Res Coun, 58-65 & Bot Exped, SAm, 38-39 & Cent Am, 57; ed, J Am Phytopath Soc, 58-60; Fulbright res scholar, Univ Australia, 61-62. Mem: Fel Am Phytpath Soc; Mycol Soc Am; Brit Mycol Soc; Brit Asn Appl Biol; Netherlands Soc Plant Path. Res: Storage decays of apples; diseases of ornamental plants; seedborne pathogens; heat therapy; soil steaming; biological control of root pathogens. Mailing Add: Dept of Plant Path Univ of Calif Berkeley CA 94720

BAKER, KIRBY ALAN, b Boston, Mass, June 17, 40; m 66. MATHEMATICS. Educ: Harvard Univ, AB, 61, PhD(math), 66. Prof Exp: Ford Found res fel math, Calif Inst Technol, 66-68; asst prof, 68-71, ASSOC PROF MATH, UNIV CALIF, LOS ANGELES, 71- Mem: Am Math Soc; Math Asn Am; Asn Comput Mach. Res: Lattice theory and partial order; identities in abstract algebras. Mailing Add: Dept of Math Univ of Calif Los Angeles CA 90024

BAKER, LEE EDWARD, b Springfield, Mo, Aug 31, 24; m 48; c 2. PHYSIOLOGY, ELECTRICAL ENGINEERING. Educ: Univ Kans, BS, 45; Rice Univ, MS, 60; Baylor Univ, PhD(physiol), 65. Prof Exp: Design engr, Radio Corp Am, 46-47; consult engr, 47-51; chief engr & part owner, Radio Sta KDKD, Mo, 53-54; owner & mgr, Radio Sta KOKO, 54-55; from instr to asst prof elec eng, Rice Univ, 55-65; asst prof, 65-69, ASSOC PROF PHYSIOL, BAYLOR COL MED, 69- Concurrent Pos: Consult, Sci Res Ctr, Int Bus Mach Corp, 66-67. Mem: Am Inst Elec & Electronics Eng. Res: Use of electrical impedance for the measurement of physiological events. Mailing Add: Dept of Physiol Baylor Col of Med Houston TX 77025

BAKER, LENOX DIAL, b DeKalb, Tex, Nov 10, 02; m 33, 67; c 2. ORTHOPAEDIC SURGERY. Educ: Duke Univ, MD, 34. Prof Exp: Athletic trainer, Univ Tenn, 25-29, lab asst zool, 27-29; athletic trainer, Duke Univ, 29-33; intern orthop, Johns Hopkins Hosp, 33-34, intern surg, 34-35, asst resident orthop surg, 35-36, resident, 36-37; from instr to prof orthop, 37-72, dir, Sch Phys Ther, 45-63, EMER PROF ORTHOP, SCH MED, DUKE UNIV, 72- Concurrent Pos: Med dir, NC Cerebral Palsy Hosp; consult, US Vet Admin, 46; med consult, US War Dept, 48-49; secy, Dept Human Resources, State of NC, 71-73. Mem: AMA; fel Am Acad Orthop Surg; Am Orthop Asn. Res: Osteomyelitis; arthritis; fractures; cerebral palsy. Mailing Add: Dept of Orthop Duke Univ Sch of Med Durham NC 27710

BAKER, LEONARD MORTON, b Medford, Mass, Oct 2, 34; m 58; c 3. ORGANIC CHEMISTRY, POLYMER CHEMISTRY. Educ: Harvard Univ, AB, 56; Mass Inst Technol, PhD(org chem), 59. Prof Exp: Chemist, Plastics Div, 59-62, proj scientist, 62-64, group leader, 64-69, asst dir res & develop, Chem & Plastics, 69-74, DIR RES & DEVELOP, CHEM & PLASTICS, UNION CARBIDE CORP, 74- Mem: Am Chem Soc; Sigma Xi. Res: Catalysis; process research and development; polymer applications development. Mailing Add: Chem & Plastics Union Carbide Corp Bound Brook NJ 08805

BAKER, LEONARD SAMUEL, b Clarksdale, Miss, Aug 26, 22; m 55; c 2. GEODESY. Educ: Miss State Univ, BS, 44. Prof Exp: Officer-in-chg surv, US Coast & Geod Surv, Cape Canaveral, 60-63, field works officer hydrog, Ship Explorer, 64-65, exec officer, Ship Pioneer, 65-66, officer-in-chg, Div Geod, 67-73; DIR, NAT GEOD SURV, 73- Mem: Am Cong Surv & Mapping (pres, 76); Am Soc Photogram; Am Geophys Union; Am Soc Mil Engrs. Res: Surveying instrument automation. Mailing Add: 8612 Fox Run Potomac MD 20854

BAKER, LOIS VAN METER, b Pittsburgh, Pa, May 13, 41; m 63; c 2. INFORMATIONIENCE. Educ: Western Reserve Univ, BA, 63; Drexel Univ, MS, 66. Prof Exp: Chemist biomed res, Med Dir, US Army Edgewood Arsenal, Md, 63-64; chem indexer chem pharmacol, Inst Sci Info, 65-66 & 69-70; lit searcher & tech writer aerospace sci & technol, Jet Propulsion Labs, Calif Inst Technol, 67-68; SR INFO ANALYST CHEMOTHER & BIOCHEM, RES LABS, FRANKLIN INST, PHILADELPHIA, 71- Mem: Am Chem Soc. Res: Creation of abstracts and indices to provide current awareness of major worldwide literature in the fields of pesticide effects on humans and the environment, cancer therapy, gastroenterology and carcinogenesis. Mailing Add: 657 N Bishop Ave Springfield PA 19064

BAKER, LOUIS, JR, b Chicago, Ill, Dec 31, 27; m 51; c 4. PHYSICAL CHEMISTRY. Educ: Ill Inst Technol, BS, 49, MS, 51, PhD(chem), 55. Prof Exp: Aeronaut res scientist, Lewis Lab, nat Adv Comt Aeronaut, 54-58; assoc chemist, 58-70, SR CHEMIST, ARGONNE NAT LAB, 70- Honors & Awards: E O Lawrence Award, US AEC, 73. Mem: Am Chem Soc; Am Nuclear Soc. Res: Reactor safety; chemical kinetics and heat transfer. Mailing Add: 9700 Cass Ave Argonne IL 60439

BAKER, LOUIS COOMBS WELLER, b New York, NY, Nov 24, 21; m 64; c 2. INORGANIC CHEMISTRY, ACADEMIC ADMINISTRATION. Educ: Columbia Univ, AB, 43; Univ Pa, MS, 47, PhD(chem), 50. Prof Exp: Asst instr chem, Univ Pa, 43-50, instr, 45-50, assoc, Johnson Found, 50-51; from asst prof to assoc prof, Boston Univ, 51-62; PROF CHEM & CHMN DEPT, GEORGETOWN UNIV, 62- Concurrent Pos: Co-proj dir, Off Prod Res & Develop, 43-46; instr, Pa Area Cols, 45-49; external examr for doctorate, Univ Calcutta, 56-; Guggenheim fel, 60-; fel, Wash Acad Sci, 64-; consult, John Wiley & Sons, 68- & Nat Res Coun, 69-; lectr, Tour Univs, Romanian Ministry Educ, 73; chmn, Comt Recommendation US Army Basic Sci Res, Nat Acad Sci, 74-; consult sci educ, Ferdowsi Univ, Iran, 75- Honors & Awards: Tchugaev Medal, USSR Acad Sci, 73. Mem: Am Chem Soc; Sigma Xi. Res: Structures, properties and syntheses of heteropoly molybdates, tungstates and vanadates; coordination complexes; ion exchange, magnetochemistry; high thermal efficiency engines. Mailing Add: Dept of Chem Georgetown Univ Washington DC 20057

BAKER, LOUIS REED, b Philadelphia, Pa, May 23, 32; m 61. ANESTHESIOLOGY. Educ: Univ Pa, AB, 53; Jefferson Med Col, MD, 57; Am Bd Anesthesiol, dipl. Prof Exp: Instr anesthesiol, Sch Med, WVa Univ, 60-62; from instr to asst prof sch med, Univ Md, 62-66; asst prof anesthesia med sch, Northeastern Univ, 66-70; asst prof anesthesiol, Pritzker Sch Med, Univ Chicago, 70-71; assoc prof, Stritch Sch Med, Loyola Univ, 71; ASSOC PROF ANESTHESIOL COL MED, UNIV OKLA, 71- Mem: Am Soc Anesthesiologists; AMA; Int Anesthesia Res Soc; fel Am Col Anesthesiologists; fel Am Acad Pediat. Res: Short- and long-term respiratory care; intensive care; newborn and small child anesthesia. Mailing Add: Dept of Anesthesiol PO Box 26901 Univ of Okla Health Sci Ctr Oklahoma City OK 73190

BAKER, M MICHELLE, b Colorado Springs, Colo, July 25, 27; c 2. FRESH WATER BIOLOGY, ECOLOGY. Educ: Univ Colo, BA, 59, MA, 61, PhD(zool), 64. Prof Exp: Asst prof biol, Adams State Col, 64-68; asst prof biol sci, Univ Colo, 68-70; ASSOC PROF BIOL, COLO MOUNTAIN COL, 70- Mem: AAAS; Am Inst Biol Sci; Nat Asn Biol Teachers. Res: Animal behavior; wildlife ecology; consciousness studies. Mailing Add: PO Box 1797 Aspen CO 81611

BAKER, MARY ANN, b Los Angeles, Calif, Oct 11, 40; m 70. MAMMALIAN PHYSIOLOGY. Educ: Univ Redlands, BA, 61; Univ Calif, Santa Barbara, MA, 64; Univ Calif, Los Angeles, PhD(anat), 68. Prof Exp: Bank of Am-Giannini Found res fel, 68-69; NIH trainee physiol & biophys, Univ Wash, 69-70; asst prof, 70-73, ASSOC PROF PHYSIOL, SCH MED, UNIV SOUTHERN CALIF, 73- Mem: AAAS; Am Physiol Soc; Am Asn Anat; Soc Neurosci. Res: Mammalian thermoregulation; neurophysiology of temperature sensation and control in mammals. Mailing Add: Dept of Physiol Univ Southern Calif Sch of Med Los Angeles CA 90033

BAKER, MARY REBECCA, b Washington, DC, Aug 20, 38; m 63. ANESTHESIOLOGY. Educ: Madison Col, Va, BS, 59; WVa Univ, MD, 63. Prof Exp: Instr anesthesiol, Sch Med, Northwestern Univ, 67-69; asst prof, Stritch Sch Med, Loyola Univ Chicago, 69-71; asst prof, 71-73, ASSOC PROF ANESTHESIOL, COL MED, UNIV OKLA, 73- Mailing Add: 119 Lake Aluma Dr Oklahoma City OK 73121

BAKER, MAURICE FRANK, b Britt, Iowa, Feb 25, 14; m 38; c 4. ECONOMIC ZOOLOGY. Educ: Iowa State Col, BS, 37, MS, 39; Univ Kans, PhD(zool), 52. Prof Exp: Biologist, US Soil Conserv Serv, 39-41; wildlife biologist, Mo Conserv Comn, 41-44, 46-49; assoc prof biol, Southwestern Col, 53-56; leader, Ala Coop Wildlife Res Unit, US Fish & Wildlife Serv, Auburn Univ, 58-67; res biologist, Intermountain Forest & Range Exp Sta, Bur Sport Fisheries & Wildlife, US Forest Serv, 67-75; RETIRED. Mem: Wildlife Soc (treas, 50); Am Ornith Union; Am Inst Biol Sci. Res: Ecology of birds and mammals. Mailing Add: Rte 1 Clarion IA 50525

BAKER, MAX LESLIE, b Batesville, Ark, Aug 4, 43; m 69. RADIOBIOLOGY, BIOPHYSICS. Educ: Ark Col, BA, 65; Univ Ark, Little Rock, MS, 67, PhD(physiol, biophys), 70. Prof Exp: Instr radiol, 69, instr radiobiol, 70-71, ASST PROF RADIOBIOL, MED CTR, UNIV ARK, LITTLE ROCK, 71- Concurrent Pos: Fel biol, Univ Tex M D Anderson Hosp & Tumor Inst Houston, 69-70. Mem: Radiation Res Soc; Biophys Soc; Am Soc Cell Biol. Res: Repair of radiation injury in bacterial and mammalian systems. Mailing Add: Div Nuclear Med & Radiation Biol Univ of Ark Med Ctr Little Rock AR 72201

BAKER, MELVIN C, physical chemistry, see 12th edition

BAKER, MICHAEL ALLEN, b Toronto, Ont, Jan 24, 43; m 67; c 2. HEMATOLOGY, CANCER. Educ: Univ Toronto, MD, 66; FRCP(C), cert clin haematol, 71 & internal med, 74; Am Bd Internal Med, dipl & cert hemat, 72. Prof Exp: Intern, Toronto Gen Hosp, 66-67; asst resident internal med, Mt Sinai Hosp, New York, 67-69, chief resident med & clin resident hemat, 69-70, chief resident hemat, 70-71, res fel, 70-72; assoc med, 72-73, asst prof, 73-76, ASSOC PROF MED, UNIV TORONTO, 76-, ASSOC, INST MED SCI, SCH GRAD STUDIES, 74-; CLIN HAEMATOLOGIST & DIR ONCOL CLIN, TORONTO WESTERN HOSP, 72- Concurrent Pos: Asst med, Mt Sinai Sch Med, 67-69, instr, 69-72; Med Res Coun Can fel, 71-72; assoc, Ont Cancer Treatment & Res Found, 73- Mem: Am Fedn Clin Res; Am Soc Hemat; Can Soc Hemat; NY Acad Sci; Can Soc Clin Invest. Res: Studies of tumor immunology in acute and chronic leukemia; detection and isolation of leukemia associated antigens. Mailing Add: Toronto Western Hosp 399 Bathurst St Toronto ON Can

BAKER, MICHAEL HARRY, b Roanoke, Va, Oct 25, 16; m 40; c 3. APPLIED CHEMISTRY. Educ: Pratt Inst, ChE, 38; Va Polytech Inst, BS, 39, MS, 40; Univ Md, cert, 42. Prof Exp: Owner, Chem Prod Distrib Co, 37-40; plant-prod chem engr, Joseph Seagram Sons Co, 40-42; field develop engr, Davison Chem Co, 42-47; head chem prod eval, Gen Mills Inc, 47-51; PRES, CHEM/SERV INC, 51- Concurrent Pos: Ed, Minn Chemist, 60-62; consult, Vols for Int Tech Asst. Honors & Awards: Honor Medal, Minn Fedn Eng Socs, 63; Honor Award, Minn Indust Chemist's Forum, 68. Mem: AAAS; Am Inst Chemists; Am Chem Soc; Inst Food Technol; Am Inst Chem Eng. Res: Application of raw materials of a chemical nature in the production of food products, paints, paper, cosmetics and related industries; chemical economics. Mailing Add: 606 Washington Ave N Minneapolis MN 55401

BAKER, NEAL KENTON, b Boston, Mass, Mar 20, 45; m 68. SOLAR PHYSICS. Educ: Harvard Univ, AB, 67; Pa State Univ, PhD(astron), 75. Prof Exp: Mem tech staff solar physics, 68-71, MEM TECH STAFF OPTICS, AEROSPACE CORP, 75- Mem: Am Astron Soc; Am Soc Photogram; Sigma Xi. Res: Solar radio astronomy—microwave bursts, solar flares. Mailing Add: Aerospace Corp PO Box 92957 Los Angeles CA 90009

BAKER, NOME, b Los Angeles, Calif, July 19, 27; m 50; c 4. BIOCHEMISTRY. Educ: Univ Calif, Los Angeles, AB, 49; Univ Calif, Berkeley, PhD(physiol), 52. Prof Exp: Sr instr biochem, Sch Med, Western Reserve Univ, 52-56; CHIEF BIOCHEMIST RADIOISOTOPE RES, VET ADMIN CTR, LOS ANGELES, 56- Concurrent Pos: Prin scientist, Radioisotope Unit, Vet Admin Hosp, Ohio, 52-56; asst clin prof biochem, Sch Med, Univ Calif, Los Angeles, 56-; Multiple Sclerosis Found fel, 61-62. Mem: AAAS; Am Soc Biol Chem. Res: Carbohydrates; fat metabolism; diabetes; radioisotope kinetics; muscular dystrophy; lipid autoxidation. Mailing Add: Radioisotope Res Vet Admin Hosp Los Angeles CA 90073

BAKER, NORMAN FLETCHER, b Santa Barbara, Calif, June 18, 26; m 46; c 2. VETERINARY PARASITOLOGY. Educ: Univ Calif, BS, 50, DVM, 52, PhD(comp path), 54. Prof Exp: Lectr vet med & asst specialist, 52-53, lectr & jr parasitologist, 53-54, from asst prof parasitol & asst parasitologist to assoc prof & assoc parasitologist, 54-66, PROF PARASITOL & PARASITOLOGIST, SCH VET MED & AGR EXP STA, UNIV CALIF, DAVIS, 66- Mem: Am Vet Med Asn; Soc Exp Biol & Med; Am Asn Vet Parasitol; World Asn Adv Vet Parasitol. Res: Efficiency of newer anthelmintics against parasites of animals; mechanism of disease production by parasites; host response to parasitic infections; binomics and control of parasites affecting animals and man. Mailing Add: Sch of Vet Med Univ of Calif Davis CA 95616

BAKER, PATRICIA COOPER, b Greenville, Tex, Dec 9, 36; m 61; c 1. DEVELOPMENTAL BIOLOGY. Educ: Univ Hawaii, BA, 60; Univ Calif, PhD(zool), 64. Prof Exp: USPHS fel develop biol, Med Sch, Univ Ore, 64-67, asst prof path, 67-69; mem fac, 69-71, RES ZOOLOGIST, DEPT ZOOL, UNIV CALIF, BERKELEY, 71- Mem: Am Soc Zoologists; Electron Micros Soc Am; Soc Develop Biol; Soc Cell Biol. Res: Ultrastructure of early amphibian development; mechanics of morphogenetic movement. Mailing Add: Dept of Zool Univ of Calif Berkeley CA 94720

BAKER, PAUL, JR, b Ashland, Ky, Feb 10, 21; m 48; c 3. NUCLEAR PHYSICS. Educ: Washington & Lee Univ, BA, 42; US Mil Acad, BS, 45; NC State Univ, MS, 52; Univ Denver, PhD(physics), 66. Prof Exp: Proj officer nuclear propulsion, Directorate Requirements, Hq, US Air Force, 52-56, chief tech br, Hartford Area, US AEC, 56-61, from instr to prof physics, US Air Force Acad, 61-67, head dept, 64-65, res assoc, 65-66, dir fac res, 66-67, chief tech div, Directorate Space, Pentagon, 67-71, prog mgr, Defense Advan Res Proj Agency, Md, 71-75; CONSULT, NUCLEAR

REGULATORY COMN, 75- Mem: AAAS; Am Asn Physics Teachers. Res: Cosmic rays. Mailing Add: 4404 Random Ct Annandale VA 22003

BAKER, PAUL EUGENE, b Ark, Nov 20, 16; c 2. PHYSICS Educ: Univ Ark, AB, 38; Univ Calif, BS, 48; Rutgers Univ, PhD(physics), 51. Prof Exp: Teacher high sch, 38-41; registration officer, US Vet Admin, 45-46; SR RES ASSOC, CHEVRON OIL FIELD RES CO, 51- Concurrent Pos: Lectr petrol eng, Univ Southern Calif. Mem: Am Phys Soc; Soc Petrol Eng. Res: Heat transfer; fluid flow in porous media; numerical analysis; ice mechanics. Mailing Add: Chevron Oil Fluid Res Co PO Box 446 La Habra CA 90631

BAKER, PAUL THORNELL, b Burlington, Iowa, Feb 28, 27; m 49; c 4. BIOLOGICAL ANTHROPOLOGY, ENVIRONMENTAL HEALTH. Educ: Univ NMex, BA, 51; Harvard Univ, PhD(anthrop), 56. Prof Exp: Res phys anthropologist, Environ Protection & Res Div, US Army Qm Res & Develop Ctr, 52-57; res assoc biophys lab, 57-58, NIH grant, 59-62, actg head dept sociol & anthrop, 64-65, exec officer anthrop, 65-68, actg head dept anthrop, 68-69, PROF ANTHROP, PA STATE UNIV, UNIVERSITY PARK, 64- Concurrent Pos: Fulbright lectr, Brazil, & res scholar, Peru, 62; US Army grant, 62-68; mem, NIH Behav Fels Rev Comt, 66-70; NATO sr sci fel, Oxford Univ, 68; prog dir, US Int Biol Prog Biol Human Pop at High Altitude, 69-74, mem, US Exec Comt Int Biol Prog, 70-74; Nat Inst Gen Med Sci grant, 69-; NSF grant, 71-74; mem, Int Coun Sci Unions Comn Predictive World Ecosysts, 72-; ed, Am Anthropologist, 73-; Guggenheim fel, 74-75; chmn, US Man & Biosphere Prog 12 Directorate, 74-; consult, UNESCO, 75; Nat Inst Child Health & Human Develop grant, 75- Mem: AAAS; fel Am Anthrop Asn; Am Asn Phys Anthropologists (pres, 69-71); Human Biol Coun (pres, 74-); Soc Study Human Biol. Res: Adaptation ot high altitude; biological effects of migration in human populations. Mailing Add: Dept of Anthrop Pa State Univ University Park PA 16802

BAKER, PERCY HAYES, b Williamsburg, Va, April 17, 06; m 31; c 2. GENETICS. Educ: Univ Pittsburgh, BS, 29, MS, 30; Univ Mich, PhD(zool, genetics), 44. Prof Exp: Instr biol, NC Col for Negroes, 30-34; from instr to prof, Va State Col, 34-49; prof, Morgan State Col, 49-50; prog dir, Am Friends Serv Comt, 50-58; prof biol, 58-69, chmn div natural sci, 63-66, dept biol, 66-69, dean col, 69-72, PROF BIOL, MORGAN STATE UNIV, 69- Mem: AAAS; Soc Exp Biol & Med; Genetics Soc Am. Res: Genetics of Mormoniella. Mailing Add: Dept of Biol Morgan State Univ Baltimore MD 21234

BAKER, PETER C, b San Francisco, Calif, Feb 14, 33; m 59; c 4. DEVELOPMENTAL BIOLOGY, NEUROBIOLOGY. Educ: Univ Calif, Berkeley, AB, 56, MA, 61, PhD(zool), 66. Prof Exp: Instr biol, City Col San Francisco, 66-67; from asst prof to assoc prof, 67-74, PROF BIOL, CLEVELAND STATE UNIV, 74- Concurrent Pos: Lectr genetics, Dept Biol, San Francisco State Col, 66-67. Mem: AAAS; Am Soc Zool; Soc Develop Biol. Res: Developmental and maturational biochemistry of indoleamines in the brain and eye of amphibians and mammals; mammals; drug effects upon the developing and maturing brain and eye. Mailing Add: Dept of Biol Cleveland State Univ Cleveland OH 44115

BAKER, PHILIP SCHAFFNER, b Minneapolis, Minn, Nov 30, 16; m 44; c 3. CHEMISTRY. Educ: DePauw Univ, AB, 38; Univ Ark, MA, 39; Univ Ill, PhD(inorg chem), 43. Prof Exp: Analyst, Dow Chem Co, Mich, 35-36; lab asst, DePauw Univ, 36-38; grad asst chem, Univ Ill, 40-43, Pan Am Refining Corp, Tex, 43-45 & Inst Paper Chem, Lawrence Col, 45-46; asst prof chem, Univ Vt, 46-48; assoc prof, Bradley Univ, 48-52; head chemist, Stable Isotopes Div, 52-58, supt isotopes sales dept, 58-61, adv isotopes develop dept, 61-62, asst supt, Isotopes Develop Ctr & head isotopes info ctr, 62-72, mgr stable isotopes separation facil, 72-75, MGR RECORDS MGT & SPEC PUBL, OAK RIDGE NAT LAB, 75- Concurrent Pos: Ed, Isotopes & Radiation Technol, 63-72. Mem: Fel AAAS; Am Chem Soc; NY Acad Sci; fel Am Inst Chemists. Res: Inorganic chemistry; analytical chemistry, metal reduction; isotope production and applications; records management. Mailing Add: 104 Euclid Pl Oak Ridge TN 37830

BAKER, PHILLIP JOHN, b East Chicago, Ind, Aug 21, 35; m 67; c 2. MICROBIOLOGY, IMMUNOLOGY. Educ: Ind State Univ, BA, 60; Univ Wis, MS, 62, PhD(bact), 65. Prof Exp: Res asst bact, Univ Wis, 60-65; MICROBIOLOGIST, LAB MICROBIAL IMMUNITY, NIH, 70- Concurrent Pos: USPHS fel immunol, Lab Immunol, NIH, 65-67, staff fel microbiol & immunol, Lab Germfree Animal Res, 67-69, sr staff fel, Lab Microbial Immunity, 69-70. Mem: AAAS; Am Asn Immunol; Reticuloendothelial Soc; Am Soc Microbiol. Res: Antibody formation at cellular level, particularly microbial antigens or microbial products; initiation of antibody synthesis and tolerance to microbial polysaccharide antigens; factors related to virulence of bacteria. Mailing Add: Lab of Microbial Immunity Nat Inst Allergy & Infect Dis Bethesda MD 20014

BAKER, R RALPH, b Houston, Tex, Aug 31, 24; m 65; c 5. PLANT PATHOLOGY, MYCOLOGY. Educ: Colo Agr & Mech Col, BS, 48, MS, 50; Univ Calif, Berkeley, PhD(plant path), 54. Prof Exp: PROF BOT & PLANT PATH, COLO STATE UNIV, 54- Concurrent Pos: Assoc ed, Phytopath, 61-63; mem comt control soil borne pathogens, Nat Acad Sci, 62-; asst dir biol res, Res Found, 62-64; vis prof, Univ Calif, Berkeley, 63-64; chmn comt planning manned orbital res & lab exp in space biol, Am Inst Biol Sci, 65-68, chmn Study Group II, 69-70, comt study role of the lunar receiving lab in the post Apollo biol & biomed activities; mem plant sci comt, Comn Undergrad Educ Biol Sci, 66, partic workshop for teaching biol to nonsci majors, Bar Harbor, 68; NSF sr fel & vis prof, Cambridge Univ, 68-69; Lunar Sample Review Bd, 70-73. Honors & Awards: Florists Mutual Award, 59; Pennock Distinguished Serv Award, 65. Mem: Fel AAAS; Am Phytopath Soc. Res: Ecology of soil microorganisms; physiology of sexual reproduction in fungi; mathematical biology; ornamental pathology; space biology. Mailing Add: Dept of Bot & Plant Path Colo State Univ Ft Collins CO 80521

BAKER, RALPH ROBINSON, b Baltimore, Md, Dec 30, 28; m 53; c 4. SURGERY. Educ: Johns Hopkins Univ, AB, 50, MD, 54. Prof Exp: From instr to prof, 62-75, PROF SURG, SCH MED, JOHNS HOPKINS UNIV, 75- Concurrent Pos: Am Cancer Soc adv clin fel, 65-68; consult, US Vet Admin Hosp, Perry Point & Loch Raven, Md, 64-68. Mem: Soc Head & Neck Surg; Soc Univ Surg; Asn Acad Surg. Res: Academic surgery; cancer and transplantation research. Mailing Add: Dept of Surg Johns Hopkins Hosp Baltimore MD 21205

BAKER, RALPH STANLEY, b LeRaysville, Pa, Aug 1, 27; m 54; c 2. WEED SCIENCE. Educ: Univ Del, BS, 57; Purdue Univ, MS, 59, PhD(plant physiol), 61. Prof Exp: RESEARCHER WEED SCI, DELTA BR, MISS AGR & FORESTRY EXP STA, 61- Mem: Weed Sci Soc Am; Sigma Xi; Coun Agr & Sci Technol. Res: Weeds, herbicides and cultural methods for controlling weeds in cotton. Mailing Add: Delta Br Exp Sta Box 197 Stoneville MS 38776

BAKER, RAYMOND MILTON, b Kansas City, Mo, Nov 7, 40; m 63; c 2. GENETICS, CELL BIOLOGY. Educ: Yale Univ, BS, 62; Univ Calif, Berkeley, PhD(biophys), 69. Prof Exp: Lectr med biophys, Univ Toronto, 71-73, asst prof, 73-74; ASST PROF BIOL, CTR CANCER RES, MASS INST TECHNOL, 74- Concurrent Pos: USPHS fel med biophys, Univ Toronto-Ont Cancer Inst, 69-70; staff scientist, Ont Cancer Inst, 71-74 & Hosp Sick Children, Toronto, 73-74. Mem: AAAS; Biophys Soc; Genetics Soc Am; Am Soc Cell Biol. Res: Somatic cell genetics, particularly concerning membrane structure and function; molecular genetics, particularly recombination and radiobiology. Mailing Add: E17-220 Ctr for Cancer Res Mass Inst Technol Cambridge MA 02139

BAKER, RICHARD ALAN, physical chemistry, see 12th edition

BAKER, RICHARD DEAN, b Hot Springs, SDak, June 9, 13; m 46; c 1. CHEMISTRY. Educ: SDak Sch Mines & Technol, BS, 36; Iowa State Col, PhD(phys chem), 41. Prof Exp: Asst chem, Iowa State Col, 37-41; res assoc, 42-43; res chemist, US Gypsum Co, Ill, 41-42; sr scientist, 43-45, group leader, 45-56, DIV LEADER, LOS ALAMOS SCI LAB, 56- Mem: Am Chem Soc; Am Soc Metals. Res: Atomic energy; lime and building materials; low pressure carburization. Mailing Add: Los Alamos Sci Lab Los Alamos NM 87544

BAKER, RICHARD FRELIGH, b Westfield, Pa, Feb 14, 10; m 39; c 1. BIOPHYSICS. Educ: Pa State Col, BS, 32, MS, 33; Univ Rochester, PhD(physics), 38. Prof Exp: Asst physics, Univ Minn, 37-38; fel physiol, Col Physicians & Surgeons, Columbia Univ, 39-40; assoc anat, Johns Hopkins Univ, 40-41; physicist, Radio Corp Am Labs, 41-47; from asst prof to assoc prof exp med, 47-59, PROF MICROBIOL, UNIV SOUTHERN CALIF, 59- Concurrent Pos: Sr res fel, Calif Inst Technol, 53-56, Commonwealth Fund fel, 57-, res assoc, 68- Mem: Electron Micros Soc Am. Res: Mass spectrometry; photoelectricity for ultraviolet; biological electron microscopy; photoelectric properties of cadmium in the Schumann region; membrane structure; human red cells. Mailing Add: 526 Mudd Bldg Univ Southern Calif Sch of Med Los Angeles CA 90033

BAKER, RICHARD GRAVES, b Merrill, Wis, June 12, 38; m 61; c 2. PALYNOLOGY, QUATERNARY GEOLOGY. Educ: Univ Wis-Madison, BA, 60; Univ Minn, Minneapolis, MS, 64; Univ Colo, Boulder, PhD(geol), 69. Prof Exp: Res assoc, Ctr Climatic Res, Univ Wis-Madison, 69-70; ASST PROF GEOL, UNIV IOWA, 70- Mem: AAAS; Geol Soc Am; Am Asn Quaternary Environ; Am Asn Stratig Palynologists. Res: Quaternary palynology in Yellowstone Park and Disterhaft Farm Bog, Wisconsin; quaternary pollen and plant macrofossils in Northern Minnesota; pollen analysis from Lake Wingra, Madison, Wisconsin and in Iowa. Mailing Add: Dept of Geol Univ of Iowa Iowa City IA 52240

BAKER, RICHARD H, b Hayfield, Minn, Sept 14, 26; m 61; c 1. GENETICS. Educ: Univ Ill, BS, 59, MS, 62, PhD(zool, genetics), 65. Prof Exp: Res assoc mosquito genetics, Univ Ill, 65; res assoc, 65-66, asst prof mosquito genetics, 66-69, ASSOC PROF INT MED, SCH MED, UNIV MD BALTIMORE CITY, 69- Concurrent Pos: Consult, WHO, 65. Mem: Genetics Soc Am; Entom Soc Am; Am Soc Trop Med & Hyg; Am Soc Zool; Am Mosquito Control Asn. Res: Mosquito genetics and cytogenetics. Mailing Add: Inst of Int Med Univ of Md Sch of Med Baltimore MD 21201

BAKER, RICHARD WILLIAM, inorganic chemistry, see 12th edition

BAKER, RICHARD WILLIAM, b Boreham Wood, Eng, Aug 18, 41; m 68. PHYSICAL CHEMISTRY, POLYMER CHEMISTRY. Educ: London Univ, BS, 63, DIC, 65, PhD(phys chem), 66. Prof Exp: Sr chemist, Amicon Corp, 66-68; res fel, Polymer Res Inst, Brooklyn Polytech Inst, 68-70; sr scientist, Alza Corp, 70-74; DIR RES, BEND RES INC, 74- Concurrent Pos: Adj prof, Ore State Univ, 76- Mem: The Chem Soc; AAAS; Am Chem Soc. Res: Physical chemistry of synthetic polymer membranes as applied to separation processes such as ultrafiltration and reverse osmosis, and the application of membranes to sustained release drug-delivery systems. Mailing Add: Bend Res Inc 64550 Research Rd Bend OR 97701

BAKER, ROBERT ANDREW, b Lakewood, Ohio, Sept 8, 25; m 47. ENVIRONMENTAL SCIENCES. Educ: NC State Univ, BS, 49; Villanova Univ, MS, 55 & 58; Univ Pittsburgh, DSc(environ health eng), 69. Prof Exp: Chem engr, Atlantic Refining Co, 49-57; sr staff eng environ sci, Franklin Inst, 57-64; sr fel, Mellon Inst Sci, 64-70; dir environ sci, Teledyne Brown Eng, 71-73; REGIONAL RES HYDROLOGIST & CHIEF, GULF COAST HYDRO-SCI CTR, US GEOL SURV, 73- Honors & Awards: Award of Merit, Am Soc Testing & Mat, 68, Max Hecht Award, 70 & Award Sensory Res, 73; Buswell-Porges Award, Inst Advan Sanitation Res, 72. Mem: Am Chem Soc; Am Water Works Asn; Am Soc Testing & Mat; AAAS; Am Inst Chem Engr. Res: Direction of interdisciplinary research involving various aspects of the hydrological cycle with personal emphasis on trace organic contaminants in the environment. Mailing Add: US Geol Surv Gulf Coast Hydrosci Ctr Bay St Louis MS 39520

BAKER, ROBERT CARL, b Newark, NY, Dec 29, 21; m 44; c 6. FOOD SCIENCE. Educ: Cornell Univ, BS, 43; Pa State Univ, MS, 49; Purdue Univ, PhD(food sci), 58. Prof Exp: Asst prof poultry exten, Pa State Univ, 46-49; PROF FOOD SCI, CORNELL UNIV, 49-, DIR FOOD MKT, 69- Concurrent Pos: Inst Am Poultry Industs food res award, 59; mem, Comt Animal Prod, Nat Acad Sci, 65-68. Mem: AAAS; Poultry Sci Asn; Inst Food Technologists. Res: Bacteriology; biochemistry. Mailing Add: Dept of Food Sci Rice Hall Cornell Univ Ithaca NY 14850

BAKER, ROBERT CHARLES, b Blackduck, Minn, June 17, 30; m 54; c 6. ECOLOGY, SCIENCE EDUCATION. Educ: Bemidji State Col, BS, 54; Univ NDak, MSEd, 55; Ore State Univ, MS, 63. Prof Exp: Teacher high sch, 55-56; instr sci & math, Lab Sch, 56-68, actg prin, 59-62, asst prof sci educ, 63-65, Minn State Col Bd fel, 65-66; asst prof sci educ, 66-67, dir continuing educ, 67-68, ASSOC PROF BIOL & SCI EDUC, BEMIDJI STATE UNIV, 67- Concurrent Pos: Adv, State Curric Environ Educ Kindergarten to 12th grade, 70- Mem: AAAS; Am Inst Biol Sci; Nat Soc Stud Educ; Nat Asn Biol Teachers; Nat Sci Teachers Asn. Res: Environmental pollution, especially solutions to land use problems and waste disposal; statistical evaluation of general biology program, traditional versus experimental; biological and attitudinal analysis of Upper Mississippi, especially Iowa-Itasca; environmental impact, Minnesota Highway Department. Mailing Add: Dept of Biol Bemidji State Univ Bemidji MN 56601

BAKER, ROBERT DALE, b Cottage Hills, Ill, Jan 11, 38; m 58; c 4. ANIMAL SCIENCE. Educ: Southern Ill Univ, BSc, 60; Univ Ill, Urbana, PhD(animal sci), 65. Prof Exp: Assoc prof animal physiol, 65-73, ASSOC PROF ANIMAL SCI, MACDONALD COL, McGILL UNIV, 73- Honors & Awards: Lalor Found Award. Mem: Soc Study Reprod; Am Soc Animal Sci; Can Soc Animal Prod; Can Soc Study Fertil; Brit Soc Study Fertil. Res: Reproductive physiology. Mailing Add: Dept of Animal Sci Macdonald Col McGill Univ PO Box 228 Ste Anne de Bellevue PQ Can

BAKER, ROBERT DAVID, b Chicago, Ill, July 7, 29; m 54, 75; c 3. PHYSIOLOGY. Educ: Iowa Wesleyan Col, BS, 54; Univ Iowa, MS, 57, PhD(physiol), 59. Prof Exp: From instr to asst prof, 59-64, ASSOC PROF PHYSIOL, UNIV TEX MED BR GALVESTON, 64- Concurrent Pos: USPHS grants, 60-64 & 68-; Liberty Muscular

Dystrophy Res Asn grant, 61-64. Mem: AAAS; Am Physiol Soc; Biophys Soc. Res: Intestinal absorption; transport of materials across biological membranes. Mailing Add: Dept of Physiol Univ of Tex Med Br Galveston TX 77550

BAKER, ROBERT DONALD, b Chico, Calif, Dec 7, 27; m 58; c 1. FOREST MANAGMENT. Educ: Univ Calif, BS, 51, MF, 52; State Univ NY, PhD, 57. Prof Exp: From asst prof to prof forest mgt, Stephen F Austin State Univ, 56-74; PROF FOREST MGT, TEX A&M UNIV, 75- Mem: Soc Am Foresters; Am Soc Photogram. Res: Forest mensuration; photogrammetry; land information systems using photographic interpretation. Mailing Add: Dept of Forest Sci Tex A&M Univ College Station TX 77843

BAKER, ROBERT FRANK, b Weiser, Idaho, Apr 9, 36; m 65; c 2. MOLECULAR BIOLOGY, GENETICS. Educ: Stanford Univ, BS, 59; Brown Univ, PhD(biol), 66. Prof Exp: Fel microbial genetics, Stanford Univ, 66-68; asst prof molecular biol & develop genetics, 68-72, ASSOC PROF MOLECULAR BIOL & DEVELOP GENETICS, UNIV SOUTHERN CALIF, 72- Concurrent Pos: Vis prof, Harvard Med Sch, 75-76. Mem: Am Soc Microbiol; Am Soc Zoologists; Sigma Xi. Res: Microbial and developmental genetics, particularly the nature of the eukaryotic genome and the regulation of biosynthesis of DNA, RNA and protein in prokaryotic and eukaryotic cells. Mailing Add: Dept of Biol Sci Univ of Southern Calif Los Angeles CA 90007

BAKER, ROBERT G, b Kindersley, Sask, Dec 17, 18; m 42; c 4. BIOPHYSICS, NUCLEAR PHYSICS. Educ: Univ Sask, BA, 50, MA, 52; Univ Toronto, PhD(physics), 60. Prof Exp: Res officer radiol, Nat Res Coun Can, 52-54; physicist radiation ther, Ont Cancer Found, 54-58; physicist isotopes, Ont Cancer Inst, 60-68; asst prof med biophys, Univ Toronto, 62-68; CHIEF PHYSICIST, ONT CANCER FOUND, OTTAWA CIVIC HOSP, 68- Mem: Can Asn Physicists; Soc Nuclear Med. Res: Medical biophysics; instrumentation and application of isotopes to diagnosis and treatment of disease, especially cancer. Mailing Add: Ont Cancer Found Ottawa Civic Hosp Ottawa ON Can

BAKER, ROBERT HENRY, b Central City, Ky, June 14, 08; m 32; c 2. ORGANIC CHEMISTRY. Educ: Univ Ky, BS, 29, MS, 31; Univ Wis, PhD(org chem), 40. Hon Degrees: ScD, Univ Ky, 68. Prof Exp: From instr to asst prof chem, Univ Ky, 31-41; from asst prof to assoc prof, 41-49, from asst dean to assoc dean, 49-64, PROF CHEM, NORTHWESTERN UNIV, 49-, DEAN GRAD SCH, 64- Concurrent Pos: Mem exec comt grad deans, African-Am Inst, 64-; vis comt grad sch, Vanderbilt Univ, 68- Mem: AAAS; Am Chem Soc; Asn Grad Schs (pres elect, 70). Res: Synthesis, stereochemistry and the mechanism of organic reactions; hydrogenation. Mailing Add: Grad Sch Northwestern Univ Evanston IL 60201

BAKER, ROBERT HENRY, JR, b Lexington, Ky, Apr 5, 34; m 57; c 3. COMPUTER SCIENCES. Educ: Trinity Col, Conn, BS, 56; Ind Univ, PhD(biochem), 60. Prof Exp: NSF fels, Univ Wis, 60-62; res assoc phys & anal chem, 62-66, sect head comput systs, Res Comput Ctr, 66-69, mgr comput systs res & develop, 69-72, MGR ADMIN COMPUT CTR, UPJOHN CO, 72- Mem: Am Chem Soc. Res: Computer utilization monitoring and resource loading; computer scheduling and turnaround monitoring; computer charging. Mailing Add: Upjohn Co 7000 Portage Rd Kalamazoo MI 49001

BAKER, ROBERT JOHN, b Cold Lake, Alta, June 10, 38; m 60; c 2. PLANT BREEDING, QUANTITATIVE GENETICS. Educ: Univ Sask, BSA, 61, MSc, 62; Univ Minn, PhD(genetics), 66. Prof Exp: RES SCIENTIST PLANT BREEDING, CAN DEPT AGR, 66- Mem: Crop Sci Soc Am; Agr Inst Can. Res: Methods of improving quantitative characteristics in self-pollinated species. Mailing Add: Agr Can Res Sta 25 Dafoe Rd Winnipeg MB Can

BAKER, ROBERT LEWIS, b Baltimore, Md, Mar 24, 37. HORTICULTURE, BOTANY. Educ: Swarthmore Col, BA, 59; Univ Md, MS, 62, PhD(bot), 65. Prof Exp: From instr to asst prof, 63-73, ASSOC PROF HORT, UNIV MD, COLLEGE PARK, 73- Mem: AAAS; Am Soc Hort Sci; Am Hort Soc; Royal Hort Soc; Int Plant Propagators Soc. Res: Propagation of woody ornamental plants; cytogenetics of maize; landscape use of woody plant materials. Mailing Add: Dept of Hort Univ of Md College Park MD 20742

BAKER, ROBERT M L, JR, celestial mechanics, mathematical analysis, see 12th edition

BAKER, ROBERT NORTON, b Inglewood, Calif, Mar 25, 23; m 54. MEDICINE. Educ: Park Col, BA, 44; Univ Southern Calif, MD, 50. Prof Exp: Lab asst comp anat, Univ Mo, 42-43; asst biochem, Univ Southern Calif, 46-48, asst anat, 49-50, instr neuroanat, 50-53; from instr to assoc prof med neurol, Med Ctr, Univ Calif, Los Angeles, 53-70, prof neurol, Univ Nebr Med Ctr, Omaha, 70-72, PROF NEUROL & PATH, UNIV NEBR & CREIGHTON COL MED, 72-, CHIEF NEUROL SERV, CREIGHTON-ST JOSEPH HOSP, 74- Concurrent Pos: Chief neurol sect, Vet Admin Ctr, Gen Med & Surg Hosp, 53-54 & 56-70; Vet Admin liaison, Neurol A Study Sect, NIH, 71-75. Mem: AMA; Am Psychiat Asn; Am Acad Neurol; Am Asn Neuropath; Am Neurol Asn. Res: Experimental neuropathology; electron microscopic studies of cerebral vascular disease especially cerebral edema, hypertensive encephalopathy and cerebral embolism; clinical and pathological studies of cerebral vascular diseases and their metabolic aspects. Mailing Add: Creighton-Nebr Neurol Prog Rm 205 Creighton Health Ctr Tenth & Dorcas Omaha NE 68108

BAKER, ROBERT P, synthetic organic chemistry, philosophy of science, see 12th edition

BAKER, ROBERT SWAIN, biochemistry, see 12th edition

BAKER, ROGER (CARROLL), (JR), b New York, NY, June 18, 19; m 46; c 5. SURGERY, UROLOGY. Educ: Boston Col, BS, 41; Tufts Col, MD, 44; Univ Chicago, PhD(surg), 52. Prof Exp: Resident surg, St Vincent's Hosp, 44-46; resident urol, Mass Mem Hosp, 48-49; asst prof, Univ Chicago, 50-52, assoc prof & head dept, 52-53; PROF UROL, SCH MED, GEORGETOWN UNIV, 53- Concurrent Pos: Consult, NIH & Vet Admin; consult kidney dis, Surgeon Gen, USPHS; pres & chmn bd, Baker Res Corp. Honors & Awards: Nat Res Prize, Am Urol Asn, 56. Mem: AAAS; Am Asn Cancer Res; Soc Exp Biol & Med; fel Am Col Surgeons; fel Am Acad Pediat. Res: Renal calculus disease; bladder cancer; regeneration smooth muscle; pediatric urologic anomalies. Mailing Add: Georgetown Univ Hosp 3800 Reservoir Rd Washington DC 20007

BAKER, ROGER DENIO, b East Lansing, Mich, Apr 10, 02; m 29; c 3. PATHOLOGY. Educ: Univ Wis, AB, 24; Harvard Univ, MD, 28. Prof Exp: Asst resident house officer, Johns Hopkins Hosp, 28-30, asst path, Johns Hopkins Univ, 28-29, instr, 29-30; instr anat, Duke Univ, 30-32, instr path, 32-34, from asst prof to assoc prof, Sch Med, 34-44; prof & chmn dept, Med Col, Univ Ala, 44-52; prof, Sch Med, Duke Univ, 52-61; prof, Sch Med, La State Univ, 61-70; prof, 70-74, EMER PROF PATH, RUTGERS MED SCH, COL MED & DENT NJ, 74- Concurrent Pos: Trustee, Am Bd Path, 48-61, pres, 60-61; chief lab serv, Vet Admin Hosp, Durham, NC, 52-61; consult path, Lyons Vet Admin Hosp & Muhlenberg Hosp, NJ; NIH grants, 68-71; vis scientist, Armed Forces Inst Path, Washington, DC, 74-75; prof lectr, George Washington Univ Sch Med, 75-76. Honors & Awards: Bronze Medal, Am Heart Asn, 49. Mem: Am Asn Path & Bact; Am Soc Clin Path; Int Acad Path; Int Soc Human & Animal Mycol; Am Soc Microbiol. Res: Infectious, mycologic and tropical diseases of man, especially opportunistic mycotic infections; experimental fungus infections; technique of human postmortem examination; histochemistry of arteriolosclerosis, cardiac disease, burns; cellular content of chyle; surgical pathology; neuropathology; cancer. Mailing Add: 411 Pershing Dr Silver Spring MD 20910

BAKER, ROLLIN HAROLD, b Cordova, Ill, Nov 11, 16; m 39; c 3. ZOOLOGY Educ: Univ Tex, BA, 37; Tex A&M Univ, MS, 38; Univ Kans, PhD(zool), 48. Prof Exp: Asst, Tex A&M Univ, 38; field biologist, State Coop Wildlife Res Unit, Tex, 38-39; wildlife biologist, State Game, Fish & Oyster Comn, 39-43; vis investr, Rockefeller Inst, 44; asst instr zool, Univ Kans, 46-48, from instr to assoc prof & asst cur mammals, Mus Natural Hist, 48-55; PROF ZOOL, FISH & WILDLIFE & DIR MUS, MICH STATE UNIV, 55- Mem: AAAS; Am Soc Mammal; Wildlife Soc; Soc Study Evolution; Soc Syst Zool. Res: Mammalian taxonomy, distribution and ecology. Mailing Add: The Museum Mich State Univ East Lansing MI 48824

BAKER, ROSCOE JUNIOR, dairy bacteriology, food bacteriology, see 12th edition

BAKER, SAMUEL I, b Cannelton, Ind, Nov 5, 34; m 62; c 1. RADIATION PHYSICS. Educ: Ind Univ, BS, 56; Univ Ill, MS, 58; Ill Inst Technol, PhD(physics), 67. Prof Exp: Appointee reactor training, Int Sch Nuclear Sci & Eng, 58-60; physicist, Argonne Nat Lab, 60-62; res assoc nuclear physics, Univ Iowa, 67 & Cyclotron Br, US Naval Res Lab, 67-69; res physicist, IIT Res Inst, 69-72; PHYSICIST, FERMI NAT ACCELERATOR LAB, 72- Mem: AAAS; Am Phys Soc; Inst Elec & Electronics Eng; Sigma Xi; Am Nuclear Soc. Res: Non-linear response of thallium-activated sodium iodide to gamma rays and x-rays measurement of nuclear lifetimes by the Doppler-Shift attenuation method; fast neutron spectrometry; radioactivation by 400 GeV protons. Mailing Add: CL-7E Fermilab PO Box 500 Batavia IL 60510

BAKER, SAUL PHILLIP, b Cleveland, Ohio, Dec 7, 24. GERIATRICS, CARDIOLOGY. Educ: Case Inst Technol, BS, 45; Ohio State Univ, MSc, 49, MD, 53, PhD(physiol), 57. Prof Exp: Asst physiol, Ohio State Univ, 48, 50 & 52; health educ & field coordr, Greater Cleveland Tuberc X-ray Surv, 49; intern, Cleveland Metrop Gen Hosp, 53-54; asst vis staff physician, Dept Med, Baltimore City & Johns Hopkins Hosps, 54-56; sr asst resident physician, Dept Med, Billings Hosp & Univ Chicago Hosps, 56-57; asst prof med, Chicago Med Sch, 57-62; pvt pract, 62-70; Medicare consult, Gen Am Life Ins Co, 70-71; pvt pract geriat, cardiol & internal med, 72- Concurrent Pos: Med officer geront br, Nat Heart Inst, 54-56; Nat Heart Inst res grants, 58-62; Chicago Heart Asn res grants, 58-62; assoc attend physician & physician-in-chg, Chicago Med Sch Med Serv, Cook County Hosp, 57-62; res assoc, Hektoen Inst Med Res, 58-62; assoc prof med, Grad Sch Med, Cook County Hosp, 58-62; consult internal med, Div Disability Determination, Ohio Bur Voc Rehab, Social Security Admin, 63-; consult, Ohio Bur Workmen's Compensation, 63-; mem comt older people, Welfare Fedn Cleveland, 63-70; head dept geriatrics, St Vincent Charity Hosp, 64-67; mem adv comt, Sr Adult Div, Jewish Community Ctr, 64-; consult internal med, City of Cleveland, 64-; attend physician, Hillcrest Hosp, 68-; cardiovasc consult, Fed Aviation Admin, 73-; mem proj comt aging, Fedn Community Planning, 74-; partic, Fifth & Eighth Int Cong Geront; partic, 19th Int Physiol Cong; fel coun arteriosclerosis, Am Heart Asn. Mem: Fel AAAS; fel Geront Soc; fel Am Geriat Soc; Am Heart Asn; AMA. Res: Effects of heparin and the clearing factor or lipoprotein lipase; blood lipid enzymes; hepato-biliary disease; relation of aging to cardiovascular-endocrine interrelationships, thyroid function, basal metabolism, functional body mass and body fluid compartments; digitalis, heart failure and potassium in the geriatric patient. Mailing Add: 6803 Mayfield Rd Cleveland OH 44124

BAKER, SIMON, b Revere, Mass, Aug 3, 24; m 53; c 3. GEOGRAPHY. Educ: Univ Ariz, BS, 51, MS, 52; Clark Univ, PhD(geog), 65. Prof Exp: Specialist land use, Photog Surv Corp, Ceylon, 57-58; asst prof geog, Univ Ariz, 60-66; geogr, USDA, 66-68; assoc prof geog, Fla Atlantic Univ, 68-74; ADV SERV AGENT, UNIV NC SEA GRANT PROG, NC STATE UNIV, 74- Concurrent Pos: Prof lectr geog, George Washington Univ, 66-68; sr fel, Food Inst, East-West Ctr, Univ Hawaii, 71; vis assoc prof, Sch Design, NC State Univ, 74- Mem: Asn Am Geogr; Am Soc Photogram; Asn Asian Studies. Res: Land use; aerial photographic interpretation; remote sensing; radio communications; coastal area management, resources and development planning; Asia. Mailing Add: Univ NC Sea Grant Prog NC State Univ Raleigh NC 27607

BAKER, SOL RONALD, b London, Eng, Nov 7, 10; nat US; m 35; c 3. THERAPEUTICS, RADIOLOGY. Educ: Wayne State Univ, AB, 31; Univ Mich, MA & MD, 35. Prof Exp: Assoc dir radiol ther, Cedars of Lebanon Hosp, 37-43; dir radiol ther, Tumor Clin & Pan-Am Fel Prog, Marine Hosp, Baltimore, 43-46; dir radiol ther, Midway Hosp, Los Angeles, 47-55; from asst clin prof to assoc clin prof, 55-69, CLIN PROF RADIOL, UNIV CALIF, LOS ANGELES, 69- Concurrent Pos: Consult, Mt Sinai Hosp, 56- & Governors State Cancer Adv Coun & State Radiol Defense Adv Coun; chmn comt new & unproved methods & mem, Int Union Against Cancer. Mem: AAAS; Am Radium Soc; Int Col Surg; Radiol Soc NAm; Soc Nuclear Med. Res: Etiology and radiation therapy in relation to malignant tumors. Mailing Add: 300 S Beverly Dr Beverly Hills CA 90212

BAKER, STEPHEN DENIO, b Durham, NC, Nov 30, 36; m 62; c 2. NUCLEAR PHYSICS. Educ: Duke Univ, BS, 57; Yale Univ, MS, 59, PhD(physics), 63. Prof Exp: Res assoc & lectr, 63-66, from asst prof to assoc prof, 66-73, PROF PHYSICS, RICE UNIV, 73- Concurrent Pos: A P Sloan res fel, 67-71. Mem: Am Phys Soc; Am Asn Physics Teachers. Res: Elastic and inelastic scattering of heavy nuclei; nuclear reactions with polarized target; production of beams of polarized nuclei. Mailing Add: Dept of Physics Rice Univ Houston TX 77001

BAKER, SUSAN PARDEE, b Atlanta, Ga, May 31, 30; m 51; c 3. EPIDEMIOLOGY. Educ: Cornell Univ, BA, 51; Johns Hopkins Univ, MPH, 68. Prof Exp: Comput analyst, Dept Chronic Dis, Sch Hyg & Pub Health, 62-67, ASSOC PROF PUB HEALTH ADMIN, DIV FORENSIC PATH, SCH HYG & PUB HEALTH, JOHNS HOPKINS UNIV, 68- Concurrent Pos: Res assoc, Off Chief Med Exam Md, 68-; mem bd dirs, Baltimore Safety Coun, 69-74; mem nat hwy safety adv comt, 75- Honors & Awards: Prince Bernhard Medal for Dissertation in Traffic Med, 74. Mem: Am Asn Automotive Med (pres, 74-75); Am Trauma Soc; Am Pub Health Asn. Res: Injury control; driver and pedestrian fatalities; alcohol; drowning; carbon monoxide poisoning; emergency medical services; evaluation. Mailing Add: Sch Hyg & Pub Health Johns Hopkins Univ Baltimore MD 21205

BAKER, THEODORE PAUL, b Abington, Pa, Dec 15, 49; m 73; c 1. COMPUTER SCIENCE. Educ: Cornell Univ, AB, 70, PhD(comput sci), 74. Prof Exp: ASST PROF COMPUT SCI, FLA STATE UNIV, 73- Mem: Asn Comput Mach; Soc Indust & Appl Math. Res: Computational complexity; relative computability; theory of computation; formal languages. Mailing Add: 225 Love Bldg Fla State Univ Tallahassee FL 32306

BAKER, THOMAS IRVING, b La Rue, Ohio, Sept 28, 31; m 53; c 4. MICROBIOLOGY, BIOCHEMISTRY. Educ: Kent State Univ, BS, 54; Ohio State Univ, MS, 56; Western Reserve Univ, PhD(microbiol), 65. Prof Exp: Assoc chemist fermentation lab, Northern Utilization Res & Develop Div, USDA, 57-60; asst prof, 67-74, ASSOC PROF MICROBIOL, SCH MED, UNIV NMEX, 74- Concurrent Pos: Res fel microbiol, Western Reserve Univ, 65-67. Mem: Genetics Soc Am; Am Soc Microbiol. Res: Microbiol genetics, specifically gene-enzyme interrelationships. Mailing Add: Dept of Microbiol Univ of NMex Sch of Med Albuquerque NM 87106

BAKER, THOMAS NELSON, III, organic chemistry, see 12th edition

BAKER, TIMOTHY D, b Baltimore, Md, July 4, 25; m 51; c 3. PREVENTIVE MEDICINE. Educ: Johns Hopkins Univ, BA, 48, MPH, 54; Univ Md, MD, 52; Am Bd Prev Med, dipl, 59. Prof Exp: Clinician, Med Care Clin, Johns Hopkins Univ, 53-54; NY State actg dist health officer, Geneva, 54-56; from asst chief to actg chief health div, Tech Coop Mission, India, 56-58; asst prof prev med, State Univ NY Upstate Med Ctr, 58-59; PROF INT HEALTH & PUB HEALTH ADMIN, ASST DEAN CURRIC & ASSOC DIR DEPT INT HEALTH, JOHNS HOPKINS UNIV, 59- Concurrent Pos: Intern, Univ Baltimore Hosp & pub health resident, NY State, 52-54; NY State Syracuse Dist Health Officer, 58-59; consult health manpower, Taiwan, Thailand, Indonesia & US Dept Health, Educ & Welfare; consult health planning, Brazil, Saudi Arabia & El Salvador; consult med educ, Peru & Saudi Arabia; vpres & dir, Univ Assocs Inc, dir, Intermed Inc, 72-74; asst secy-treas, Am Bd Prev Med. Mem: AAAS; fel Am Pub Health Asn; Am Soc Trop Med & Hyg; Am Col Prev Med. Res: Health manpower and planning; medical education; international health and population dynamics; public health administration; medical care; epidemiology. Mailing Add: 615 N Wolfe St Baltimore MD 21205

BAKER, VICTOR RICHARD, b Waterbury, Conn, Feb 19, 45; m 67; c 1. GEOMORPHOLOGY, ENVIRONMENTAL GEOLOGY. Educ: Rensselaer Polytech Inst, BS, 67; Univ Colo, PhD(geol), 71. Prof Exp: Geophysicist hydrol, US Geol Surv, 67-69; city geologist urban geol, Boulder, Colo, 69-71; ASST PROF GEOL SCI, UNIV TEX, AUSTIN, 71- Concurrent Pos: Res scientist, Bur Econ Geol, Univ Tex, 73. Mem: Geol Soc Am; Am Geophys Union; AAAS; Am Quaternary Asn. Res: Geomorphic processes; paleohydrology; quaternary geology; environmental geomorphology. Mailing Add: Dept of Geol Sci Univ of Tex Austin TX 78712

BAKER, VIOLET EVA SIMMONS, inorganic chemistry, see 12th edition

BAKER, WALTER WOLF, b Philadelphia, Pa, Oct 29, 24; m 52; c 3. NEUROPHARMACOLOGY. Educ: Franklin & Marshall Col, BS, 48; Univ Iowa, MS, 50; Jefferson Med Col, PhD(pharmacol), 53. Prof Exp: Res asst, Radiation Res Lab, Univ Iowa, 48-50; admin asst & pharmacologist, Wyeth Inst Appl Biochem, 50-52; from instr to assoc prof pharmacol, 53-60, assoc prof neuropharmacol, 60-70, PROF PSYCHIAT, JEFFERSON MED COL, 70-, PROF PHARMACOL, 73-; DIR NEUROPHARMACOL, EASTERN PA PSYCHIAT INST, 63- Mem: AAAS; Soc Exp Biol & Med; Am Soc Pharmacol & Exp Therapeut; Am Col Clin Pharmacol; Int Col Neuropsychopharmacol. Res: Electron microscopy of muscle; cerebral evoked potentials; central autonomic regulation; chemically-induced tremors; psychomotor epilepsy; local action of drugs on brain electrical activity; neurotransmitter regulatory mechanisms in caudate nucleus and hippocampus. Mailing Add: Dept of Neuropharmacol Eastern Pa Psychiat Inst Philadelphia PA 19129

BAKER, WELDON NICHOLAS, b Mt Vernon, Iowa, July 15, 09; m 36; c 5. ANALYTICAL CHEMISTRY. Educ: Morningside Col, AB, 30; Univ Iowa, MS, 31; Columbia Univ, PhD(phys chem), 35. Prof Exp: Instr chem, Allegheny Col, 35-36; jr sea food inspector, Food & Drug Admin, USDA, 36-37; instr chem, Morningside Col, 37; instr chem eng, Rose Polytech Inst, 37-40, asst prof, 40-44; jr chemist, US Rubber Co, Conn, 44-47; from assoc prof to prof phys sci, Kans State Teachers Col, 47-58; assoc prof chem, 58-64, PROF CHEM, EASTERN ILL UNIV, 64- Mem: Am Chem Soc. Res: Science education. Mailing Add: Dept of Chem Eastern Ill Univ Charleston IL 61920

BAKER, WILBER WINSTON, b Ponca City, Okla, Nov 2, 34; m 57; c 3. BIOCHEMISTRY, CELL BIOLOGY. Educ: Grinnell Col, AB, 56; Iowa State Univ, MS, 56; Ore State Univ, PhD(biochem), 64. Prof Exp: Fel biol chem, Harvard Med Sch, 64-66; res assoc biol, Cancer Res Inst, Philadelphia, 66-68; ASST PROF BIOL, VILLANOVA UNIV, 68- Mem: Am Chem Soc; Am Inst Biol Sci. Res: Chemical enzymology; biochemical genetics; isozymes; quaternary structure of enzymes. Mailing Add: Dept of Biol Villanova Univ Villanova PA 19085

BAKER, WILLIAM ALBERT, physical organic chemistry, see 12th edition

BAKER, WILLIAM BRYAN, b Springfield, Mo, Apr 14, 20; m 42. GEOGRAPHY. Educ: Southwest Mo State Col, BS, 50; Univ Nebr, MA, 53, PhD, 58. Prof Exp: Instr geog, Univ Nebr, 56-57; asst prof, Univ Tulsa, 57-63; PROF EARTH SCI, SOUTHERN ILL UNIV, EDWARDSVILLE, 63- Mem: Am Geog Soc; Asn Am Geog. Res: Urban geography. Mailing Add: Social Sci Div Southern Ill Univ Edwardsville IL 62025

BAKER, WILLIAM HUDSON, b Portland, Ore, Dec 16, 11; m 34; c 1. BOTANY. Educ: Ore State Col, BS, 35, MS, 42, PhD(bot), 49. Prof Exp: Adminr pub schs, Ore, 35-43; asst bot, Ore State Col, 46-48; from asst prof to prof, 48-72, from actg chmn dept to chmn dept, 53-67, from actg head dept biol sci to head dept biol sci, 55-68, EMER PROF BOT, UNIV IDAHO, 72- Concurrent Pos: Ranger-naturalist, Crater Lake Nat Park, Nat Park Serv, 49-50 & 70. Mem: AAAS; Bot Soc Am; Am Soc Plant Taxon; Am Fern Soc; Sigma Xi. Res: Flowering plants of the northwest; floristics; plant distribution. Mailing Add: Dept of Bot Univ of Idaho Moscow ID 83843

BAKER, WILLIAM J, b East Florenceville, NB, Dec 25, 32; m 53; c 3. IMMUNOHEMATOLOGY. Educ: Univ NB, BSc, 57, Wayne State Univ, MS, 60, PhD, 67. Prof Exp: Sr scientist, Ortho Res Found, 60-68, dir diag serv int, Ortho Pharmaceut Corp, 68-74, DIR MFG, ORTHO DIAGNOSTICS INC, 74- Mem: AAAS; Am Asn Clin Chemists; NY Acad Sci; fel Am Inst Chemists; Can Soc Lab Technol. Res: Coagulation; hematology. Mailing Add: Ortho Diagnostics Inc Rte 202 Raritan NJ 08869

BAKER, WILLIAM KAUFMAN, b Portland, Ind, Dec 2, 19; m 44; c 3. GENETICS. Educ: Col of Wooster, BA, 41; Univ Tex, MS, 43, PhD(genetics), 48. Prof Exp: Nat Res Coun fel, Univ Tex, 46-48; asst prof zool, Univ Tenn, 48-52; biologist, Oak Ridge Nat Lab, 51-53, sr biologist, 53-55; assoc prof zool, 55-59, chmn dept biol, 68-72, PROF ZOOL, UNIV CHICAGO, 59- Concurrent Pos: NSF sr fel, 63-64; co-ed, Am Naturalist, 65-70; NIH spec fel, 72-73. Mem: Am Soc Naturalists (secy, 62-64, vpres, 76); Genetics Soc Am; Soc Study Evolution. Res: General genetics and development of Drosophila. Mailing Add: Dept of Biol Univ of Chicago Chicago IL 60637

BAKER, WILLIAM LAIRD, b Amarillo, Tex, June 17, 30; m 52; c 2. GEOPHYSICS. Educ: Yale Univ, BA, 52; Univ Calif, Berkeley, MA, 57, PhD(geophys), 60. Prof Exp: Jr geophysicist, Pan Am Petrol Co, 52-53; res physicist, Calif Res Corp, 60-63, sr res geophysicist, 63-68, SR RES GEOPHYSICIST, CHEVRON OIL CO, 68- Concurrent Pos: Sr staff geophysicist, West Australian Petrol Pty Ltd, 71-76. Mem: Seismol Soc Am; Soc Explor Geophys. Res: Theoretical studies of the propagation of waves in layered elastic media; propagation of surface waves; model seismology; design of geophysical exploration programs. Mailing Add: Stand Oil Co Calif 575 Market St San Francisco CA 94105

BAKER, WILLIAM OLIVER, b Chestertown, Md, July 15, 15; m 41; c 1. PHYSICAL CHEMISTRY, POLYMER SCIENCE. Educ: Washington Col, BS, 35; Princeton Univ, PhD(phys chem), 38. Hon Degrees: ScD, Washington Col, 57; DEng, Stevens Inst Technol, 62; DSc, Georgetown Univ, 62, Univ Pittsburgh, 63, Seton Hall Univ, 65, Univ Akron, 68, Univ Mich, 70, St Peter's Col, 72, Polytech Inst New York, 73 & Trinity Col, Dublin, 75; LLD, Glasgow Univ, 65 & Univ Pa, 74; LHD, Monmouth Col, 73 & Clarkson Col Technol, 74. Prof Exp: Mem tech staff, 39-51, asst dir chem & metall res, 51-54, dir res phys sci, 54-55, vpres res, 55-73, PRES, BELL LABS, 73- Concurrent Pos: With Off Rubber Reserve, 41-44, Off Sci Res & Develop, 42-44; mem panel phys chem, Off Naval Res, 48-51; mem, President's Sci Adv Comt, 57-60; mem sci info coun, NSF, 58-61, chmn, 59-61; consult, US Dept Defense, 58-71, spec asst sci & technol, 63-73; mem adv bd Mil Personnel Supplies, Nat Res Coun, 58-, chmn, 64-, mem Comn Sociotech Systs, 63-, mem Comt Phys Chem, Div Chem & Chem Technol, 63-70; mem, President's Foreign Intel Adv Bd, 59-; mem sci adv bd, Nat Security Agency, 60- Consult, Panel Opers Eval Group, US Navy, 60-62 & Nat Sci Bd, 60-66; mem bd visitors, US Air Force Systs Command, 62-73; mem Liaison Comt Sci & Technol, Libr Cong, 63-73, Comt Sci & Technol, US Chamber of Com, 63-66, Coun Trends & Perspective, 66-74; mem sci adv bd, Robert A Welch Found, 68-; mem bd regents, Nat Libr Med, 69-73; chmn adv tech panel, Nat Bur Stand & Nat Acad Sci-Nat Res Coun, 69-; mem coun, Nat Acad Sci, 69-72 & Inst Med, 73-; mem mgt adv coun, Oak Ridge Nat Lab, 70-; mem Nat Comn Libr & Info Sci, 71-, UNA Nat Policy Panel on Future UN Role in Sci & Technol, 72-, Nat Coun Educ Res, Nat Inst Educ, 73-75, Energy Res & Develop Adv Coun, Energy Policy Off, 73-75, Coun on Critical Choices for Americans, 73-, Proj Independence Adv Comt, Fed Energy Admin, 74-, Nat Cancer Adv Bd, 74-, Governor's Comn Evaluate Capital Needs NJ, 74-75, Steering Comt, President's Food & Nutrit Study, Comn Int Rels, Nat Acad Sci, 75-; mem vis comt, Tulane Univ, 63-, Sci & Math, Drew Univ, 69- & Metall & Mat Sci, Mass Inst Technol, 73- Honors & Awards: Honor Scroll, Am Inst Chemists, 62, Gold Medal, 75; Perkin Medal, 63; Priestley Medal, 66; Edgar Marburg Award, 67; Exec Award, Am Soc Testing & Mat, 67; Indust Res Inst Medal, 70; Frederik Philips Award, Inst Elec & Electronics Engrs, 72; Indust Res Man of Year Award, 73; Procter Prize, Sigma Xi, 73; James Madison Medal, Princeton Univ, 75; Mellon Inst Award, 75. Mem: Nat Acad Eng; Am Philos Soc; Am Acad Arts & Sci; Am Inst Chemists; Am Phys Soc. Res: Study of solid state materials and macromolecules, the basic elements of plastics, fibers, natural and synthetic rubbers, and the tissues of growing plants and animals. Mailing Add: Bell Labs 600 Mountain Ave Murray Hill NJ 07974

BAKER, WILLIE ARTHUR, JR, b San Antonio, Tex, Nov 7, 33; m 54; c 2. INORGANIC CHEMISTRY. Educ: Tex Col Arts & Indust, BS, 55; Univ Tex, PhD(chem), 59. Prof Exp: From asst prof to prof chem, 59-70, chmn dept, 66-70; PROF CHEM & DEAN GRAD SCH, UNIV TEX, ARLINGTON, 70-, VPRES ACAD AFFAIRS, 72- Mem: AAAS; Am Chem Soc; The Chem Soc. Res: Coordination compounds; magnetic properties; spectra structure. Mailing Add: Davis Hall Univ of Tex Arlington TX 76010

BAKER, WINSLOW FURBER, b Brockton, Mass, m 67; c 2. HIGH ENERGY PHYSICS, ELEMENTARY PARTICLE PHYSICS. Educ: Bowdoin Col, AB, 50; Columbia Univ, MA, 53, PhD(physics), 57. Hon Degrees: DSc, Allegheny Col, 75. Prof Exp: Res assoc physics, Columbia Univ, 57-59; physicist, Brookhaven Nat Lab, 59-64, Europ Orgn Nuclear Res, 64-69; PHYSICIST, FERMI LAB, UNIV RES ASN, 69- Mem: Am Phys Soc. Res: Strong interactions at high energies; backward scattering, baryon exchange, elastic scattering, total cross sections. Mailing Add: Fermi Lab PO Box 500 Batavia IL 60510

BAKER-BLOCKER, ANITA LINDA, b St Paul, Minn, July 20, 48; m 73. GEOENVIRONMENTAL SCIENCE, ATMOSPHERIC CHEMISTRY. Educ: Univ Minn, BS, 70, MPH, 72; Univ Mich, PhD(atmospheric & oceanic environ health sci), 76. Prof Exp: CONSULT GEOPHYSICS, DeWITT & ASSOCS, 74- Mem: Geol Soc Am; Am Meteorol Soc; Am Geophys Union; AAAS. Res: Flux of trace gases from terrestrial and aquatic ecosystems to the atmosphere; human evolution as a result of atmospheric changes; environmental geochemistry of trace elements; ice forecasting. Mailing Add: 2316 Brockman Blvd Ann Arbor MI 48104

BAKER-COHEN, KATHERINE FRANCE, b Winnipeg, Man, Mar 1, 28; US citizen; m 57. ZOOLOGY. Educ: Antioch Col, BA, 47; Columbia Univ, MA, 52, PhD(zool), 57. Prof Exp: USPHS fel biochem, Columbia Univ, 57-59; fel physiol, 60-61, interdisciplinary prog fel, 61-62, instr anat, 62-67, ASST PROF ANAT, ALBERT EINSTEIN COL MED, 67- Mem: AAAS; Am Asn Anatomists; Am Soc Zoologists. Res: Neurohistochemistry; biochemical embryology; thyroid development and function in fish. Mailing Add: Dept of Anat Albert Einstein Col of Med Bronx NY 10461

BAKERMAN, SEYMOUR, b Brooklyn, NY, May 26, 24; m 54; c 3. BIOPHYSICS, PATHOLOGY. Educ: NY Univ, BA, 49; Purdue Univ, MS, 52, PhD(chem), 57; Western Reserve Univ, MD, 59. Prof Exp: From isntr to assoc prof path, Sch Med, Univ Kans, 60-68; PROF PATH, MED COL VA, 68- Mem: Soc Exp Path; Am Chem Soc; Geront Soc. Res: Molecular biophysics; molecular structure of collagen and membranes and alterations with age and disease; computer diagnosis of disease. Mailing Add: Dept of Path Med Col of Va Richmond VA 23219

BAKHRU, HASSARAM, b Rohri, India, June 2, 37. NUCLEAR PHYSICS. Educ: Banaras Hindu Univ, India, BS, 58, MS, 60; Calcutta Univ, PhD(nuclear physics), 64. Prof Exp: Fel, Heavy Ion Accelerator, Yale Univ, 65-69, res staff fel nuclear physics, 65-66, res assoc & assoc dir, 66-69; asst prof, 70-73, ASSOC PROF NUCLEAR PHYSICS, STATE UNIV NY ALBANY, 73-, DIR ACCELERATOR LAB, 72- Concurrent Pos: USAEC res grant, 71- Honors & Awards: Frederick Gardner Cottrell Res Award, Res Corp, 71. Mem: Am Phys Soc. Res: Nuclear reactions and spectroscopy; new isotopes; heavy ion and neutron induced reactions; x-ray spectroscopy. Mailing Add: Nuclear Accelerator Lab Dept of Physics State Univ of NY Albany NY 12222

BAKHSHI, VIDYA SAGAR, b India, Sept 23, 39. APPLIED MATHEMATICS. Educ: Univ Delhi, BA, 59, MA, 63; Ore State Univ, PhD(math), 68. Prof Exp: Lectr math, Kurukshetra Univ, India, 63-65; teaching asst, Ore State Univ, 65-68; assoc prof, 68-75, PROF MATH, VA STATE COL, 75- Mem: Math Asn Am. Res: Fluid dynamics and mathematical seismology; two and three dimensional boundary layer problems. Mailing Add: Dept of Math Va State Col Petersburg VA 23803

BAKIS, RAIMO, b Tartu, Estonia, May 7, 33; US citizen; m 65; c 2. PHYSICS. Educ: Sterling Col, BA, 54; Kans State Univ, MS, 56, PhD(physics), 59. Prof Exp: Assoc physicist res ctr, 59, staff physicist, 59-60, res staff mem speech recognition, 60-61, res staff mem pattern recognition, 61-66, res staff mem speech processing res lab, Ruschlikon ZH, Switz, 66-68, RES STAFF MEM, IMAGE PROCESSING, RES CTR, INT BUS MACH CORP, 68- Mem: AAAS. Res: Speech processing—automatic recognition of continuous speech. Mailing Add: Int Bus Mach Corp Res Ctr Yorktown Heights NY 10598

BAKKE, JEROME E, b Mapes, NDak, July 14, 31; m 54; c 2. BIOCHEMISTRY. Educ: NDak State Univ, BS & MS, 56; Univ NDak, PhD(biochem), 60. Prof Exp: Chemist, Lignite Res Lab, US Bur Mines, NDak, 55-58; res biochemist, Lipid Sect, Fundamental Food Res Dept, Gen Mills, Inc, Minn, 60-61, Natural Prod Sect, 61-63, Explor Food Res Dept, 63-64; asst prof biochem, NDak State Univ, 68-73; RES BIOCHEMIST RADIATION & METAB LAB, USDA, 64- Concurrent Pos: Adj prof biochem, NDak State Univ, 73- Res: Natural product isolation, identification and synthesis; lipid metabolism; food research; insecticide metabolism in the animal. Mailing Add: Radiation & Metab Lab State Univ Sta Fargo ND 58102

BAKKE, JOHN LANGUM, b Spokane, Wash, Jan 5, 22; m 43; c 3. MEDICINE. Educ: State Col Wash, BS, 43; Harvard Univ, MD, 45; Am Bd Internal Med, dipl, 52. Prof Exp: Asst med, Med Col, Cornell Univ, 48-49; sr resident, King County Hosp, 49-50; from instr to asst prof, 51-61, CLIN ASSOC PROF MED, UNIV WASH, 61-; DIR LABS, PAC NORTHWEST RES FOUND, 61- Concurrent Pos: NIH fel, Univ Wash, 50-51; attend staff physician, King County Hosp, 51-; asst chief med, US Vet Admin Hosp, 51-61, asst dir prof serv for res, 58-61; consult, Madigan Gen Hosp, US Army & Vet Admin. Mem: AMA; Endocrine Soc; Am Fedn Clin Res; fel Am Col Physicians. Res: Pituitary-thyroid physiology. Mailing Add: Pac Northwest Res Found 1102 Columbia St Seattle WA 98104

BAKKEN, AIMEE HAYES, b Camden, NJ, Oct 18, 41; m 63; c 1. CELL BIOLOGY, DEVELOPMENTAL BIOLOGY. Educ: Univ Chicago, BA, 63; Univ Iowa, PhD(zool), 70. Prof Exp: Res asst reprod biol, Dept Obstet & Gynec, Univ Chicago, 63-66; investr, Biol Div, Oak Ridge Nat Lab, 70-72; res assoc biol, Yale Univ, 72-73; ASST PROF ZOOL, UNIV WASH, 73- Concurrent Pos: NSF res grant, 74- Mem: AAAS; Genetics Soc Am; Am Soc Cell Biol. Res: Correlation of chromosome structure and function in eukaryotic germ cells, embryonic cells and somatic cells, using molecular cytogenetic techniques. Mailing Add: Dept of Zool Univ of Wash Seattle WA 98195

BAKKEN, ARNOLD, b Antelope, Mont, Feb 19, 21; m 47; c 2. ZOOLOGY. Educ: Univ Mont, BA, 43; Univ Wis, MS, 49, PhD(zool, mammal, ecol), 53. Prof Exp: Teaching & res zool & anat, Reed Col, 52-53; teaching, 53-55, assoc prof biol, 55-61, PROF BIOL, UNIV WIS, EAU CLAIRE, 61- Concurrent Pos: Fulbright grant, Norway, 51-52; NSF award, 58. Mem: AAAS; Am Inst Biol Sci; Animal Behav Soc; Am Soc Mammal; Ecol Soc Am. Res: Behavior and interrelationships between fox squirrels and grey squirrels in mixed populations; population problems of Norwegian lemmings; microclimate and behavior. Mailing Add: Dept of Biol Univ of Wis Eau Claire WI 54701

BAKKEN, GEORGE STEWART, b Denver, Colo, Feb 17, 43; m 73. PHYSIOLOGICAL ECOLOGY, BIOPHYSICS. Educ: NDak State Univ, Bs, 65; Rice Univ, MA, 67, PhD(physics), 70. Prof Exp: Res assoc physics, Rice Univ, 69-70; fel physiol ecol biophys, Mo Bot Garden, 70-71, Univ Mich, 71-73; res assoc, Div Biol Sci, 73-75, ASST PROF LIFE SCI, IND STATE UNIV, 75- Mem: AAAS; Am Inst Biol Sci; Brit Ecol Soc; Ecol Soc Am; Optical Soc Am. Res: The study of energy and mass flow between organisms and their physical environment and the consequences for physiology, ecology and evolution; biological, physical and mathematical methods are employed. Mailing Add: Dept of Life Sci Ind State Univ Terre Haute IN 47809

BAKKER, CORNELIS BERNARDUS, b Rotterdam, Holland, Jan 6, 29; US citizen; m 55; c 3. PSYCHIATRY. Educ: State Univ Utrecht, MD, 52. Prof Exp: Intern, Clin of Rotterdam, Holland, 52-53; intern, Sacred Heart Hosp, Spokane, Wash, 53-54; resident psychiat, Eastern State Hosp, Medical Lake, 54-56; resident, Psychiat Clin, State Univ Utrecht, 56-57; resident, Med Sch, Univ Mich, Ann Arbor, 57-59, instr, 58-60, res assoc ment health res inst, 59-60; from instr to assoc prof, Univ Wash, 60-72, dir adult psychiat inpatient serv, Univ Hosp, 61-68, actg chmn psychiat, Sch Med, 69-70, DIR ADULT DEVELOP PROG, UNIV HOSP, UNIV WASH, 68-, PROF PSYCHIAT, SCH MED, 72- Concurrent Pos: Mem, World Cong Psychiat; Fulbright & Comt Int Exchange Persons grants, Sacred Heart Hosp, Spokane, Wash & Eastern State Hosp, Medical Lake, 53-56. Honors & Awards: Significant Achievement Award, Am Psychiat Asn, 75. Mem: Am Psychiat Asn; Am Col Psychiat. Res: Behavior change techniques; human territoriality. Mailing Add: Dept of Psychiat & Behav Sci Univ of Wash Sch of Med Seattle WA 98195

BAKKER, GERALD ROBERT, b Chicago, Ill, May 16, 33; m 57; c 2. ORGANIC CHEMISTRY. Educ: Calvin Col, AB, 55; Univ Ill, PhD(chem), 59. Prof Exp: Res assoc chem, Univ Kans, 58-59; from asst prof to assoc prof, 59-71, consult teaching & learning, 75-77, PROF CHEM, EARLHAM COL, 71- Mem: AAAS; Am Chem Soc; Sigma Xi; Am Asn Univ Prof. Res: Physical organic chemistry; acid-catalyzed decarboxylation reactions. Mailing Add: Dept of Chem Earlham Col Richmond IN 47374

BAKOS, GUSTAV ALFONS, b Trnava, Czech, Aug 2, 18; Can citizen; m 62; c 1. ASTRONOMY. Educ: Comenius Univ, Bratislava, BA, 42; Univ Toronto, MA, 53, PhD(astron), 60. Prof Exp: Astronr, Smithsonian Astrophys Observ, 59-61; asst prof astron, Northwestern Univ, 61-65; PROF PHYSICS, UNIV WATERLOO, 65- Mem: Am Astron Soc; Royal Astron Soc Can; Int Astron Union. Res: Photometry and spectroscopy; image orthicaon techniques. Mailing Add: Dept of Physics Univ of Waterloo Waterloo ON Can

BAKSHI, PRADIP M, b Baroda, India, Aug 21, 36; m 67; c 2. THEORETICAL PHYSICS. Educ: Univ Bombay, BSc, 55; Harvard Univ, AM, 57, PhD(physics), 52. Prof Exp: Res fel physics, Harvard Univ, 63; res physicist, Air Force Cambridge Res Labs, 63-66; sr res assoc physics, Brandeis Univ, 66-68, assoc prof, 68-70; assoc prof, 70-75, PROF PHYSICS, BOSTON COL, 75- Mem: Am Phys Soc. Res: New techniques in mathematical physics; quantum field theory; plasma physics. Mailing Add: Dept of Physics Boston Col Chestnut Hill MA 02167

BAKSHI, TRILOCHAN SINGH, b Wardha, India, Sept 13, 25; Can citizen; m 48; c 3. ECOLOGY. Educ: Univ Bombay, BSc, 47; Univ Saugar, MSc, 49; Wash State Univ, PhD(bot), 58. Prof Exp: Asst prof bot, Birla Col, India, 50-52; lectr, Univ Delhi, 52-53; head dept biol, SNat Col, India, 54; asst bot, Wash State Univ, 54-58; fel ecol & anat, Univ Sask, 58-59; syst botanist & ecologist, Ministry Nat Resources, Sierra Leone, WAfrica, 60-63; lectr bot, Univ Ghana, 63-64; from assoc prof to prof biol, Notre Dame Univ, BC, 64-73; HEAD ENVIRON SCI, ATHABASCA UNIV, 73- Concurrent Pos: Fulbright travel grantee, 54-57; Nat Res Coun Can study grants plant ecol southeast BC. Mem: Am Inst Biol Sci; Can Bot Asn; Int Soc Plant Morphol; Indian Bot Soc; Int Asn Ecol. Res: Ecology of vascular plants; evolutionary morphology of the flower; photography of wild flowers and natural vegetation; producing university courses for home-based study. Mailing Add: Athabasca Univ 14515-122 Ave Edmonton AB Can

BAKSHY, STANLEY, b Brooklyn, NY, Oct 2, 32; m 63; c 1. BIOCHEMISTRY. Educ: NY Univ, BS, 53; Univ Ark, MS, 56; Ill Inst Technol, PhD(biochem), 60. Prof Exp: Asst dir biochem dept, Michael Reese Hosp, Chicago, Ill, 59-63; dir biochem, Louis A Weiss Mem Hosp, 65-69; DIR BIOCHEM, MERCY HOSP, CHICAGO, 69- Concurrent Pos: Instr, Chicago Med Sch, 61-; consult, St Joseph's Hosp, Joliet, 59-63 & St Mary of Nazareth Hosp, Chicago, 63- Mem: Am Asn Clin Chemists. Res: Clinical biochemistry; metabolic aspects of liver disease. Mailing Add: Mercy Hosp & Med Ctr Biochem Dept Stevenson Expressway at King Dr Chicago IL 60616

BAKST, HENRY JACOB, medicine, deceased

BAKULE, RONALD DAVID, b Chicago, Ill, Nov 19, 36; m 68; c 2. PHYSICAL ORGANIC CHEMISTRY. Educ: Ill Inst Technol, BS, 57; Cornell Univ, PhD(phys chem), 62. Prof Exp: Chemist, 62-73, PROJ LEADER, ROHM AND HAAS CO, 73- Mem: Am Chem Soc. Res: Polymer chemistry. Mailing Add: Res Labs Rohm and Haas Co Springhouse PA 19477

BAKUS, GERALD JOSEPH, b Thorpe, Wis, Dec 5, 34; m 53; c 2. MARINE ECOLOGY. Calif State Univ Los Angeles, BA, 55; Univ Mont, MA, 57; Univ Wash, PhD(zool), 62. Prof Exp: Asst prof biol, Calif State Univ Northridge, 61-62; asst prof & asst cur, Allan Hancock Collections, 62-67, ASSOC PROF BIOL & CUR SPONGES, ALLAN HANCOCK FOUND, UNIV SOUTHERN CALIF, 67- Concurrent Pos: NSF & Cocos Found grants, Fanning Island Exped, 63; NSF grant, 64-66; NIH biomed sci grant, 68-70; staff officer, Div Biol & Agr, Nat Acad Sci, 69-70; various environ contracts, 74-; mem, Mirex Comt, Environmental Protection Agency, 71-72; chmn comt marine biol, Orgn Trop Studies, 72-; prin biologist, Tetra Tech Inc, Calif, 74-75, consult, 75-; hon mem, Great Barrier Reef Comt, Univ Queensland, Australia, 76. Mem: AAAS; Ecol Soc Am. Res: Biology of sponges; environmental impact assessments. Mailing Add: Dept of Biol Sci Col LAS Univ of Southern Calif Los Angeles CA 90007

BAKUZIS, EGOLFS VOLDEMARS, b Kuldiga, Latvia, May 27, 12; US citizen; m 40; c 6. FOREST ECOLOGY. Educ: Univ Latvia, ForestE, 35; Univ Minn, PhD(forestry), 59. Prof Exp: Instr forest mgt, Sch Forestry, Acad Agr Jelgava, 40-41; asst forest growth & yield, Eberswalde Forest Acad, Ger, 44-45; lectr forest mensuration, valuation & mgt, Baltic Univ, Ger, 47-50; res fel forest ecol, Sch Forestry, 56-59, res assoc, 59-61, asst prof forest synecol, 61-63, assoc prof, 63-69, PROF FOREST SYNECOL, SCH FORESTRY, UNIV MINN, ST PAUL, 69- Concurrent Pos: Assoc mem, Forest Res Inst Latvia, 36-44; res grant, Univ Minn, 60-63; NSF travel grant, Int Union Forest Res Orgn Cong, Vienna, Austria, 61, Munich, Ger, 67 & Gainesville, Fla, 71; Int Bot Cong, Seattle, Wash, 69. Mem: Fel AAAS; Ecol Soc Am; Soc Am Foresters; Soil Sci Soc Am; Am Inst Biol Scientist. Res: Forest measurements and silviculture; forest ecology, particularly theoretical aspects of forest synecology; growth and yield studies; systems ecology. Mailing Add: Univ of Minn Col Forestry St Paul MN 55101

BAKWIN, HARRY, pediatrics, deceased

BAL, ARYA KUMAR, b Jorhat, India, Aug 24, 34. CELL BIOLOGY. Educ: Univ Calcutta, BSc, 51, MSc, 53, PhD(bot), 60. Prof Exp: Lectr bot, Bangabasi Col, India, 54-58; res asst cytol, Carnegie Inst Genetics Res Unit, 58-60; lectr bot, Bangabasi Col, India, 60-62; fel biol, Brown Univ, 62-65; res assoc anat, McGill Univ, 65-66; res assoc biol, Univ Montreal, 66-69; asst prof, 69-73, ASSOC PROF BIOL, MEM UNIV NFLD, 73- Concurrent Pos: Nat Res Coun Can grants, 68- Mem: Am Soc Cell Biol; Can Soc Cell Biol; Electron Micros Soc Am; Bot Soc Am. Res: Ultrastructural analysis of differentiating cells of plants and animals; syntheses of macromolecules during development and differentiation. Mailing Add: Dept of Biol Mem Univ of Nfld St John's NF Can

BALABAN, MARTIN, b New York, NY, Oct 5, 30; m 50; c 3. ZOOLOGY. Educ: Univ Chicago, BA, 50, PhD(biopsychol), 59. Prof Exp: USPHS fel neurochem, Univ Ill, 59-61; USPHS fel neuroembryol, Univ Wash, St Louis, 61-62, res assoc, 61-64; from asst prof to assoc prof, 64-71, PROF ZOOL, MICH STATE UNIV, 71- Mem: AAAS; Am Soc Zool. Res: Behavioral biology; neuroembryology; ontogeny of behavior; ontogeny of central nervous system development. Mailing Add: Dept of Zool Mich State Univ East Lansing MI 48823

BALACHANDRAN, KASHI RAMAMURTHI, b Madras, India, June 16, 41; m 69; c 1. OPERATIONS RESEARCH, APPLIED STATISTICS. Educ: Univ Madras, India, BE, 62; Univ Calif, MS, 64; Univ Calif, Berkeley, PhD(opers res), 68. Prof Exp: Field engr, Neyveli Lignite Corp, India, 62; designer mech eng, Qual Casting Systs, 63; mem opers res staff planning, Gen Mills, Inc, 63; opers res analyst planning & designing, Nat Cash Regist Co, 67-68; asst prof mgt sci, Univ Wis-Milwaukee, 68-72; ASSOC PROF MGT SCI, GA INST TECHNOL, 72- Concurrent Pos: Res fel, Univ Wis, 69; mem, Govr's Comn Educ, Wis, 69-70; assoc ed, J Oper Res Soc India, 75. Mem: Inst Mgt Sci; Opers Res Soc Am; Oper Res Soc India. Res: Individual and collective optima in queues; operations research applications. Mailing Add: Col of Indust Mgt Ga Inst of Technol Atlanta GA 30332

BALAGNA, JOHN PAUL, b Florence, Colo, Aug 19, 20; m 49; c 2. ANALYTICAL CHEMISTRY, RADIOCHEMISTRY. Educ: Holy Cross Col, Colo, AB, 41. Prof Exp: Chemist, Chem Warfare Serv, US Army, 43-44; staff mem anal chem, Manhattan Dist, Corps Engrs, 44-47; staff mem radiochem, 47-74, ASSOC GROUP LEADER, LOS ALAMOS SCI LAB, UNIV CALIF, 74- Concurrent Pos: Consult, 66-69. Mem: AAAS; Inst Elec & Electronics Eng; Am Chem Soc. Res: Design of ultra low level gas filled proportional counters; fission fragment energy measurements on fermium isotopes. Mailing Add: 223 Rio Bravo Dr Los Alamos NM 87544

BALAGOT, REUBEN CASTILLO, b Philippines, July 28, 20; nat US; m 46; c 4. ANESTHESIOLOGY. Educ: Univ Philippines, BS, 41, MD, 44; Am Bd Anesthesiol, dipl, 55. Prof Exp: Resident, Res & Educ Hosps, Ill, 49-50; clin instr, 52-54, from asst prof to prof anesthesiol, Univ Ill Med Ctr, 54-75, chmn dept, 69-75; PROF ANESTHESIOL & CHMN DEPT, CHICAGO MED SCH, 75-, CHMN DEPT ANESTHESIOL, DOWNEY VET ADMIN HOSP, 75- Concurrent Pos: Res fel anesthesiol, Res & Educ Hosps, Ill, 51; asst head div anesthesiol, Res & Educ Hosps, 54-; consult, Coun Pharmacol, AMA, 55 & 66-, Coun Drugs, 63 & 67-; chmn dept anesthesiol, Grant Hosp, 57-, Ill Masonic Hosp, 66-67, Presby-St Luke's Hosp, 67- & Vet Admin Hosp, Hines, 71- Honors & Awards: Distinguished Physician, Philippine Med Asn, Chicago, 68. Mem: AAAS; AMA; Am Soc Anesthesiol; Am Fedn Clin

Res; NY Acad Sci. Res: Physiology and pharmacology in anesthesiology. Mailing Add: 4264 N Hazel Chicago IL 60613

BALAGURA, SAUL, b Cali, Colombia, Jan 11, 43; US citizen; m 64. NEUROSURGERY, PHYSIOLOGICAL PSYCHOLOGY. Educ: Univ Valle, Colombia, MD, 64; Princeton Univ, MA, 66, PhD(psychol), 67. Prof Exp: Asst prof psychol, Univ Chicago, 67-71; assoc prof, Univ Mass, Amherst, 71-73; resident neurosurg, Downstate Med Ctr, Brooklyn, 73-76; RESIDENT NEUROSURG, ALBERT EINSTEIN SCH MED, YESHIVA UNIV, 76- Concurrent Pos: NIMH res grant, 67-71. Honors & Awards: Physicians Recognition Award, AMA, 69. Mem: Am Psychol Asn; fel Am Physiol Soc; Soc Neurosci. Res: Neurophysiological and endocrine regulation of feeding behavior; motivation to eat; feedback systems involved in the control of food intake. Mailing Add: 500 E 77th St Apt 1609 New York NY 10021

BALAM, BAXISH SINGH, b Kukar Pind, India, Aug 16, 30; US citizen; m 65; c 1. SOIL CHEMISTRY. Educ: Punjab Univ, India, BSc, 51, MSc, 53; Ohio State Univ, PhD(soil chem), 65. Prof Exp: Agr asst soil chem, Dept Agr, Punjab, 53-55; asst prof chem, Col Agr, Univ Nagpur, 55-56 & Univ Jabalpur, 56-58; from asst prof to assoc prof, Punjab Univ, India, 58-66; res assoc agron, Ohio State Univ, 66; adv chem, Int Atomic Energy Agency, UN, 66-68; PROF CHEM, MISS VALLEY STATE UNIV, 69- Mem: Am Soc Agron; Soil Sci Soc Am. Res: Availability, plant uptake and interactions of soil and fertilizer phosphates; fertilizer technology; assessment of fertilizer requirements; radioisotopes in agricultural research; partial acidulation of rock phosphate; soil salinity and alkalinity. Mailing Add: PO Box 256 Itta Bena MS 38941

BALAMUTH, DAVID, b New York, NY, Nov 4, 42; m 69. NUCLEAR PHYSICS. Educ: Harvard Col, AB, 64; Columbia Univ, PhD(physics), 68. Prof Exp: Res asst physics, Univ Pa, 68-70, asst prof, 70-74; prog officer for nuclear physics, NSF, 74-75; ASSOC PROF PHYSICS, UNIV PA, 75- Mem: Am Phys Soc. Res: Study of nuclear structure using angular correlation techniques, both particle-gamma and particle-particle; study of nuclear reaction mechanisms using similar techniques. Mailing Add: Dept of Physics 2E5 Rittenhouse Lab Univ of Pa Philadelphia PA 19174

BALAMUTH, LEWIS, b New York, NY, Dec 31, 05; m 49; c 2. PHYSICS. Educ: City Col New York, BS; Columbia Univ, PhD(physics), 34. Prof Exp: Instr physics, City Col New York, 27-41; res dir, Tech Res Labs, NJ, 42-44, Gussack Machined Prod Co, Long Island City, 44-45 & Balco Res Labs, 46-50; dir res & develop, Cavitron Ultrasonics, Inc, 52-55, vpres res & develop, 55-69, res dir, 69-74, CONSULT, ULTRASONIC SYSTS, INC, 74- Mem: AAAS; Am Phys Soc; Inst Elec & Electronics Engrs; Acoust Soc Am; Am Ord Asn. Res: Theory of liquids, solids and fusion; ultrasonics both analytical and power generation; high and low reflectance optical coatings; development of miniature stable inductors, capacitors and resistors capable of high temperature operation; industrial and dental ultrasonic devices. Mailing Add: PO Box 413 Southampton NY 11968

BALAMUTH, WILLIAM, b New York, NY, Jan 16, 14; m 38; c 2. PROTOZOOLOGY, PARASITOLOGY. Educ: City Col New York, BS, 35; Univ Calif, PhD(zool), 39. Prof Exp: Teaching asst zool, Univ Calif, 35-38; instr, Univ Mo, 39-40; from instr to assoc prof, Northwestern Univ, 40-53; assoc prof, 53-55, PROF ZOOL, UNIV CALIF, BERKELEY, 55-, ASSOC DEAN COL LETT & SCI, 68- Concurrent Pos: LeClair fel, Univ Freiburg, 33-34; biol club fel, City Col New York, Woods Hole, 35. Mem: AAAS; Soc Cell Biol; Am Soc Trop Med & Hyg; Am Soc Parasitol; Soc Protozool (pres, 69-70). Res: Morphogenesis and nutrition of amoeboid and amoebo-flagellate protozoa; reorganization of ciliate protozoa. Mailing Add: Dept Zool 4079 Life Sci Bldg Univ of Calif Berkeley CA 94720

BALANT, CHARLES PAUL, b Mons, Belg, Sept 14, 19; nat US; m 49; c 5. MEDICINAL CHEMISTRY, RESEARCH ADMINISTRATION. Educ: Swiss Fed Inst Technol, dipl nat sci, 47; Univ Pa, PhD(physiol chem), 52. Prof Exp: Res chemist, Nat Drug Co, 52-54; asst dir res med chem, 54-57; res assoc, Smith Kline & French Labs, 57-66; asst dir res, 66-70, assoc dir res, 70-72, DIR REGULATORY AFFAIRS, MERCK SHARP & DOHME RES LABS, 72- Mem: Am Chem Soc; NY Acad Sci; Swiss Chem Soc. Res: Natural products; steroids; cardiovascular drugs; psychopharmacological drugs. Mailing Add: Merck Sharp & Dohme Res Labs West Point PA 19486

BALASKO, JOHN ALLAN, b Morgantown, WVa, July 16, 41; m 63; c 3. AGRONOMY, CROP PHYSIOLOGY. Educ: WVa Univ, BS, 63, MS, 67; Univ Wis, Madison, PhD(agron), 71. Prof Exp: Asst prof agron, 70-74, ASSOC PROF AGRON, WVA UNIV, 74- Mem: Crop Sci Soc Am; Am Soc Agron. Res: Carbohydrate metabolism in perennial forage plants; forage management and forage quality. Mailing Add: Div of Plant Sci WVa Univ Morgantown WV 26506

BALASSA, LESLIE LADISLAUS, b Bacsfoldvar, Yugoslavia, Sept 6, 03; nat US; m 33; c 1. PHARMACEUTICAL CHEMISTRY. Educ: Univ Vienna, PhD(chem), 26. Prof Exp: Chemist, Photophot, Vienna, Austria, 26 & United Piece Dye Works, NJ, 27; res fel, State Col Iowa, 27; res chemist, E I du Pont de Nemours & Co, Pa & Mich, 28 & 30-44; chemist & metallurgist, Am Smelting & Refining Co, Mex, 29-30; res dir, US Finishing Co, 44-49; dir, Balassa Res Labs, 49-66; pres, Balchem Corp, 66-72; PRES, LESCARDEN LTD, 63- Concurrent Pos: Vpres, Aula Chem, Inc, 49-57; pres, Barrington Industs, Inc, 53-68; mgr textile pigment develop labs, Geigy Dyestuffs Div, Geigy Chem Corp, 57-59; pres, Barinco Corp, 70- Mem: AAAS; Am Chem Soc; fel Am Inst Chemists. Res: Paints; textile printing and dyeing; pharmaceuticals; wound healing; encapsulation processes; soluble coffee; tissue regeneration accelerators; microencapsulated food additives; preparations for the treatment of osteoarthritis, psoriosis, rheumatoid arthritis and other autoallergic diseases. Mailing Add: Shore Dr Blooming Grove NY 10914

BALASUBRAHMANYAN, VRIDDHACHALAM K, b Tiruchendoor, India, Nov 11, 26; m 54; c 2. COSMIC RAY PHYSICS. Educ: Univ Calcutta, MSc, 49; Univ Bombay, PhD(cosmic radiation, physics), 61. Prof Exp: Lectr physics, G S Col Sci, Khamgaon, India, 49-50; fel cosmic rays, Tata Inst Fundamental Res, India, 50-62; Nat Acad Sci-Nat Res Coun resident res assoc, 62-65, PHYSICIST, GODDARD SPACE FLIGHT CTR, NASA, 65- Mem: Am Phys Soc; Am Geophys Union. Res: Primary cosmic rays; composition; energy spectrum interstellar medium; solar activity; interplanetary medium; high energy cosmic ray detection techniques. Mailing Add: Goddard Space Flight Ctr Code 661 Greenbelt MD 20771

BALAZS, ENDRE ALEXANDER, b Budapest, Hungary, Jan 10, 20; nat US; m 45; c 2. BIOCHEMISTRY. Educ: Univ Budapest, MD, 43. Hon Degrees: MD, Royal Univ Uppsala, 67. Prof Exp: Res asst histol, biol & embryol, Univ Budapest, 40-43, asst prof, 43-44, asst prof pharmacol, 45; dir, Lab Biol Anthrop, Mus Natural Hist, Budapest, 45-47; assoc exp histol, Karolinska Inst, Sweden, 48-51; assoc dir, Retina Found-Int Biol & Med Sci, 51-62, pres, 62-63 & 65-68, vpres, 63-65, dir dept connective tissue res, 62-69; res dir, Dept Connective Tissue Res, Boston Biomed Res Inst, 69-76, exec dir, 74-75; MALCOLM A ALDRICH PROF OPHTHAL, COL PHYSICIANS & SURGEONS, COLUMBIA UNIV, 76-, DIR RES, 76- Concurrent Pos: Vis scientist, Swed Inst, 47; instr, Harvard Med Sch, 51-56, lectr, 56-, mem bd freshman adv, 57-61; Mass Eye & Ear Infirmary fel ophthal, 51-56, biochemist, 56-6S; co-ed-in-chief, Exp Eye Res, 62-; Guggenheim fel, 68-69; vis scientist, Dept Tumor Biol, Karolinska Inst, Sweden, 68-69; vis prof chem, Univ Salford, Eng, 69-76; mem, Coun Biol Educ. Honors & Awards: Friedenwald Award, Asn Res Vision & Ophthal, 63. Mem: Reticuloendothelial Soc; fel Geront Soc; Asn Res Vision & Ophthal; fel NY Acad Sci; Int Soc Eye Res (pres protem, 74-). Res: Biochemistry and physiology of connective tissues, especially of the eye and joint; molecular biology of the intercellular matrix; physical biochemistry of glycosaminoglycans. Mailing Add: Dept of Ophthal Col Phys & Surg Columbia Univ 630 W 168th St New York NY 10032

BALAZS, LOUIS A P, b Tsing-Tao, China, July 31, 37; US citizen. THEORETICAL HIGH ENERGY PHYSICS. Educ: Univ Calif, Berkeley, BA, 59, PhD(physics), 62. Prof Exp: Mem, Sch Math, Inst Advan Study, 62-63; res fel, Calif Inst Technol, 63-64; vis prof, Tata Inst Fundamental Res, India, 64-65; asst prof, Univ Calif, Los Angeles, 65-70; ASSOC PROF PHYSICS, PURDUE UNIV, 70- Concurrent Pos: Alfred P Sloan Found fel, 67-69; visitor, Imp Col, London, 68-69. Mem: Am Phys Soc. Res: Bootstrap calculations in strong-interaction physics; studies of duality and multiperipheral integral equations for calculating Regge trajectories and for obtaining some information on amplitudes at low energies. Mailing Add: Dept of Physics Purdue Univ Sch of Sci West Lafayette IN 47906

BALAZS, NANDOR LASZLO, b Budapest, Hungary, July 7, 26; nat US. THEORETICAL PHYSICS. Educ: Univ Budapest, MSc, 48; Univ Amsterdam, PhD(physics), 51. Prof Exp: Fel, Dublin Inst Advan Study, 51-52; Nat Res Coun Can fel, 52; mem, Inst Advan Studies, Princeton Univ, 53; vis assoc prof physics, Univ Ala, 53, assoc prof, 53-54; sr theoret physicist, Am Optical Co, 55-56; res assoc, Enrico Fermi Inst Nuclear Studies, Chicago, 56-59; mem res staff, Plasma Physics Lab, Princeton Univ, 59-61; PROF PHYSICS, STATE UNIV NY STONY BROOK, 61- Concurrent Pos: Consult, Gen Atomics, La Jolla, Calif, 59, Am Optical Co, Mass, 60 & dept chem, Univ Calif, San Diego, 62; NSF grants, 62-64 & 64-; consult comt col physics, 63 & 65, Gen Tel & Electronics, Bayside, NY, 63-65 & Laser Inc, 66-67; vis prof, Univ Wash, 65, Dublin Inst Advan Study, 67, 70 & 71, Univ Cambridge, 68 & Oxford Univ, 72; State of NY distinguished fac res fel, 68. Mem: Am Phys Soc. Res: Statistical mechanics; relativity theory; quantum mechanics. Mailing Add: Dept of Physics State Univ of NY Stony Brook NY 11790

BALAZS, TIBOR, b Sarbogard, Hungary, Mar 1, 22; US citizen; m 49; c 1. TOXICOLOGY. Educ: Univ Agr Sci, Hungary, Vet Surg, 45, DrMedVet, 49; Pazmany Peter Univ, MPharm, 48. Prof Exp: Asst prof pharmacol, Univ Agr Sci, Hungary, 46-50; sr biologist, Nat Inst Pub Health, Hungary, 50-56; res toxicologist, Ayerst Lab Div, Am Home Prod Co, Quebec, 57-59; res toxicologist, Food & Drug Directorate, Can Dept Nat Health & Welfare, 59-63; res pharmacologist, Lederle Labs Div, Am Cyanamid Co, 63-67, group leader toxicol, 67-69; asst dir toxicol, Smith Kline & French Labs, Pa, 69-71; CHIEF DRUG TOXICOL BR, DIV DRUG BIOL, PHARMACEUT RES & TESTING, BUR DRUGS, FOOD & DRUG ADMIN, 71- Mem: Vet Med Asn; Soc Toxicol; Am Soc Pharmacol & Exp Therapeut. Res: Drug toxicology; pharmacology; experimental pathology. Mailing Add: Drug Toxicol Br Div Drug Biol Bur of Drugs Food & Drug Admin Washington DC 20204

BALBACH, HAROLD EDWARD, b Chicago, Ill, Sept 16, 36; m 61. PHYTOGEOGRAPHY, ECOLOGY. Educ: Chicago Teachers Col, BEd, 59; Univ Ill, MS, 61, PhD(bot), 65. Prof Exp: Asst prof biol sci, Chico State Col, 65-66; ASST PROF BOT, EASTERN ILL UNIV, 66- Mem: Bot Soc Am; Am Soc Plant Taxon; Asn Trop Biol; Ecol Soc Am; Int Asn Plant Taxon. Res: Variations in old-field successional patterns as studied in both temperate and tropical regions of the New World. Mailing Add: Dept of Bot Eastern Ill Univ Charleston IL 61920

BALBACH, MARGARET KAIN, paleobotany, see 12th edition

BALBES, RAYMOND, b Los Angeles, Calif, Dec 14, 40; m 62; c 2. MATHEMATICS. Educ: Univ Calif, Los Angeles, BS, 62, MA, 64, PhD(math), 66. Prof Exp: Asst math, Univ Calif, Los Angeles, 62-66; asst prof, 66-70, ASSOC PROF MATH, UNIV MO-ST LOUIS, 70- Mem: Math Asn Am; Am Math Soc. Res: Lattice theory, particularly distributive lattices and Boolean algebras; projective and injective distributive lattices; order sum of distributive lattices; representation theory for prime and implicative semi-lattices. Mailing Add: Dept of Math Univ of Mo St Louis MO 63121

BALBINDER, ELIAS, b Warsaw, Poland, Jan 22, 26; US citizen; m 55; c 2. GENETICS. Educ: Univ Mich, BS, 49; Ind Univ, PhD(cytogenetics), 57. Prof Exp: Res assoc & instr zool, Univ Pa, 56-57; res assoc genetics, Carnegie Inst, 57-60; Am Cancer Soc fel biol, Univ Calif, La Jolla, 60-63; vis investr Fulbright travel grant, Univ Buenos Aires, 63; from asst prof to assoc prof genetics, 63-71, PROF GENETICS, SYRACUSE UNIV, 71- Mem: Fel AAAS; Am Soc Microbiol; Genetics Soc Am; Brit Soc Gen Microbiol. Res: Evolution of tryptophan synthetase; mutation and recombination in bacteria; fine structure genetics; regulation of gene expression. Mailing Add: Dept of Biol Biol Res Labs Syracuse Univ Syracuse NY 13210

BALBONI, EDWARD RAYMOND, b Springfield, Mass, Dec 7, 30; m 54. INSECT PHYSIOLOGY. Educ: Boston Univ, AB, 57, AM, 58; Univ Mass, PhD(entom), 62. Prof Exp: From instr to asst prof, 62-69, ASSOC PROF BIOL SCI, HUNTER COL, 69- Concurrent Pos: NSF grant, 65-70. Mem: AAAS; Entom Soc Am; Am Soc Zool; NY Acad Sci. Res: Insect flight muscle metabolism. Mailing Add: Dept of Biol Sci Hunter Col City Univ New York 695 Park Ave New York NY 10021

BALCERZAK, STANLEY PAUL, b Pittsburgh, Pa, Apr 27, 30; m 53; c 7. INTERNAL MEDICINE, HEMATOLOGY. Educ: Univ Pittsburgh, BS, 53; Univ Md, Baltimore, MD, 55. Prof Exp: Instr med, Univ Chicago, 59-60; consult hemat & med, Walter Reed Army Inst, 60-62; from instr to asst prof med, Univ Pittsburgh, 62-67; assoc prof, 67-71, PROF MED, OHIO STATE UNIV, 71-, DIR DIV HEMATONCOL, 69- Concurrent Pos: Consult hemat, Dayton Vet Admin Hosp, 67- & Wright Patterson Air Force Hosp, 67- Mem: Am Col Physicians; Cent Soc Clin Res; Am Fedn Clin Res; Am Soc Hemat; AMA. Res: Iron metabolism; oxygen transport; abnormal hemoglobins; tumor immunology. Mailing Add: Ohio State Univ Hosp 410 West Tenth Ave Columbus OH 43210

BALCH, ALAN LEE, b Pottstown, Pa, Aug 21, 40; m 64; c 1. INORGANIC CHEMISTRY. Educ: Cornell Univ, BA, 62; Harvard Univ, PhD(chem), 67. Prof Exp: Asst prof chem, Univ Calif, Los Angeles, 66-70, asst prof, 70-73, ASSOC PROF CHEM, UNIV CALIF, DAVIS, 73- Mem: Am Chem Soc. Res: Synthesis, structure and reactions of transition metal complexes; electrochemistry. Mailing Add: Dept of Chem Univ of Calif Davis CA 95616

BALCH, ALFRED HUDSON, b Manhattan, Kans, May 22, 28; m 53; c 3. GEOPHYSICS. Educ: Stanford Univ, BS, 50; Colo Sch Mines, DSc(geophys), 64. Prof Exp: Asst petrol eng, Stanford Univ, 50-52; seismic computist, Phillips Petrol Co, 55-57 & Chevron Oil Co div, Stand Oil Co Calif, 57-58; asst, Colo Sch Mines, 58-63;

adv res geophysicist, Marathon Oil Co, 63-74; MEM STAFF, US GEOL SURV, DENVER, COLO, 74- Mem: AAAS; Soc Explor Geophys; Am Geophys Union; Sigma Xi; Europ Asn Explor Geophys. Res: Wave propagation; geophysical data analysis techniques from random noise theory. Mailing Add: US Geol Surv Denver Fed Ctr Denver CO 80225

BALCH, DONALD JAMES, b Hanover, NH, July 15, 22; m 46; c 2. ANIMAL BREEDING. Educ: Univ NH, BS, 48, MS, 52; Va Polytech Inst, PhD(animal breeding), 62. Prof Exp: Instr animal husb, Univ NH, 50-51; from instr to assoc prof, 52-71, PROF ANIMAL SCI, UNIV VT, 71-, DIR MORGAN HORSE FARM, 72- Mem: Am Soc Animal Sci. Mailing Add: Dept of Animal Sci Univ of Vt Burlington VT 05401

BALCHUM, OSCAR JOSEPH, b Detroit, Mich, Nov 9, 17; m 49; c 2. INTERNAL MEDICINE, PHYSIOLOGY. Educ: Wayne Univ, BS, 38, MS, 42, MD, 43; Univ Colo, PhD, 53. Prof Exp: Res asst, Cardio-Pulmonary Lab, Nat Jewish Hosp, Denver, Colo, 50-51; intern, Med Ctr, Univ Colo, 53-54, resident internal med, 56-57; instr med & asst dir, Clin Radioisotope Ctr, Vanderbilt Univ Hosp, 57-59; asst prof, 59-61, Hastings assoc prof, 61-67, HASTINGS PROF MED, SCH MED, UNIV SOUTHERN CALIF, 68- Concurrent Pos: Am Heart Asn res fel cardiol, Univ Colo, 54-56; dir pulmonary dis sect, Los Angeles County-Univ Southern Calif Med Ctr, 69-, dir pulmonary med sect, 70- Mem: AAAS; Am Thoracic Soc; fel Am Col Cardiol; fel Am Col Physicians; fel Am Col Chest Physicians. Res: Diseases of the chest and cardiovascular system; pulmonary physiology. Mailing Add: Med Sch Univ of Southern Calif Los Angeles CA 90033

BALCZIAK, LOUIS WILLIAM, b Duluth, Minn, Feb 11, 18; m 46; c 1. EDUCATIONAL STATISTICS. Educ: Duluth State Teachers Col, BS, 40; Univ Minn, AM, 49, PhD(ed statist), 53. Prof Exp: Instr chem, Duluth State Teachers Col, 41-44 & 46-47 & Duluth Br, Univ Minn, 47-48; from asst prof to assoc prof, 50-63, PROF SCI, MANKATO STATE COL, 63- Concurrent Pos: Mem, NSF Summer Insts, 59-63. Mem: Am Chem Soc. Res: Inorganic chemistry; radiochemistry. Mailing Add: Dept of Chem Mankato State Col Mankato MN 56001

BALD, JOHN GRIEVE, b Camberwell, Australia, Sept 1, 05; m 36; c 1. PLANT PATHOLOGY. Educ: Univ Melbourne, BAgrSci, 29, MAgrSci, 33; Cambridge Univ, PhD(plant path), 36. Prof Exp: Asst res officer, Coun Sci & Indust Res, Australia, 28-35, from res officer to sr res officer, 36-48; assoc prof, Univ Calif, Los Angeles, 48-53, prof plant path & plant pathologist, Exp Sta, 53-67; prof plant path & plant pathologist, 67-70, EMER PROF PLANT PATH, UNIV CALIF, RIVERSIDE, 70- Mem: Fel Am Phytopath Soc. Res: Virus diseases of plants and other phases of plant pathology; diseases of bulb crops. Mailing Add: Dept of Plant Path Univ of Calif Riverside CA 92502

BALD, KENNETH CHARLES, b Rock Island, Ill, Oct 27, 30; m 65. PHYSICAL CHEMISTRY. Educ: St Ambrose Col, AB, 52; St Louis Univ, PhD(phys chem), 62. Prof Exp: Chemist, Callery Chem Co, 52-57; teaching fel, St Louis Univ, 57-62; res chemist, M W Kellogg Co, 62-69; RES CHEMIST, RIEGEL PROD CORP, 69- Mem: Am Chem Soc; Tech Asn Pulp & Paper Indust. Res: Development of electrochemical processes for industrial applications; development of electrophotographic products. Mailing Add: Riegel Prod Corp Milford NJ 08848

BALDA, RUSSELL PAUL, b Oshkosh, Wis, June 14, 39; m 61; c 3. ANIMAL ECOLOGY. Educ: Wis State Univ, Oshkosh, BS, 61; Univ Ill, MS, 63, PhD(zool), 67. Prof Exp: Asst prof biol, 66-70, ASSOC PROF BIOL, NORTHERN ARIZ UNIV, 70- Mem: Ecol Soc Am; Ornith Union; Cooper Ornith Soc (secy, 75-76); Wilson Ornith Soc; Asn Study Animal Behav. Res: Effects of habitat alteration on birds and mammals; habitat selection and plant utilization by birds of various communities; social behavior of birds. Mailing Add: Dept of Biol Sci Northern Ariz Univ Flagstaff AZ 86001

BALDAUF, MILTON PAUL, food technology, see 12th edition

BALDAUF, RICHARD JOHN, b Reading, Pa, May 14, 20; c 3. ENVIRONMENTAL EDUCATION. Educ: Albright Col, BS, 49; Tex A&M Univ, MS, 51, PhD, 56. Prof Exp: Instr zool & biologist, Res Found, Tex A&M Univ, 52-56; from asst prof to prof wildlife mgt, 56-72; PROF BIOL, UNIV MO-KANSAS CITY, 72-; DIR EDUC, KANSAS CITY MUS, 72- Mem: AAAS; Am Soc Ichthyologists & Herpetologists; Am Soc Zoologists; Soc Syst Zoologists; fel Acad Zool. Res: Vertebrate morphology and phylogeny; conservation education; environmental quality. Mailing Add: Kansas City Mus 3218 Gladstone Blvd Kansas City MO 64123

BALDERRAMA, FRANCISCO E, b Mexico City, Mex, Aug 12, 28; m 57; c 2. INTERNAL MEDICINE. Educ: Nat Univ Mex, MD, 55. Prof Exp: Intern & resident, Nat Inst Nutrit, Mexico City, 55-59; asst physician, Hosp Secretaria de Comunicaciones, 59-61, instr, Sch Nursing, 62; regional med dir, 61-67, ASSOC DIR CLIN RES, INT DIV, SYNTEX SA, 67- Mem: Am Fertil Soc; Family Planning Asn Americas. Res: Gastroenterology. Mailing Add: Syntex Int ATSA Cerrada de Bezares No 9 Lomas Mexico DF Mexico

BALDES, EDWARD JAMES, biophysics, deceased

BALDESCHWIELER, JOHN DICKSON, b Elizabeth, NJ, Nov 14, 33 33; m. BIOCHEMISTRY, CHEMICAL PHYSICS. Educ: Cornell Univ, BChemE, 56; Univ Calif, Berkeley, PhD(phys chem), 59. Prof Exp: Lectr chem, Harvard Univ, 60, instr, 60-62, asst prof, 62-65; assoc prof, Stanford Univ, 65-67, prof, 67-71; dep dir, Off Sci & Technol, Washington, DC, 71-73; PROF CHEM & CHMN DIV CHEM & CHEM ENG, CALIF INST TECHNOL, 73- Concurrent Pos: Consult, Ballistic Res Labs, Aberdeen Proving Grounds, Md, 60-, Monsanto Chem Co, Mo & Res Systs, Inc, Mass; Alfred P Sloan Found fel, 62-66; mem, Army Sci Adv Panel, 66-69; mem, President's Sci Adv Comt, 69-; vis scientist, Nat Cancer Inst, 73; consult, NSF, Nat Security Coun, Nat Cancer Inst & Merck Sharp & Dohme, 73- Honors & Awards: Am Chem Soc Award, 67; Fresenius Award, Phi Lambda Upsilon, 68; Am Acad Arts & Sci Award, 72. Mem: Nat Acad Sci; Am Inst Physics; Am Chem Soc. Res: Molecular structure and spectroscopy, including nuclear magnetic resonance, nuclear spectroscopy, mass spectroscopy, x-ray diffraction and applications of these methods to biological systems. Mailing Add: Div of Chem & Chem Eng Calif Inst Technol Pasadena CA 91125

BALDINI, JAMES THOMAS, b Paterson, NJ, Jan 21, 27; m 51; c 4. NUTRITION, DERMATOLOGY. Educ: Rutgers Univ, BS, 47; Purdue Univ, MS, 49, PhD(nutrit), 51; Univ Del, MBA, 69. Prof Exp: Jr asst nutrit, Purdue Univ, 47-51; from res biologist to sr res biologist, Stine Lab, E I du Pont de Nemours & Co, 51-69; med serv mgr, White Labs Div, 69-70, dir sci affairs, Schering Labs, 70-71, DIR PROF SERV, SCHERING LABS, SCHERING CORP, 71- Mem: Am Inst Nutrit; Am Acad Dermat; Am Acad Allergy; Am Col Allergists. Res: Toxicology; pharmacology; virology; biochemistry and low protein diets in nutrition; amino acid nutrition; endocrinology; adrenal ascorbic acid; estrogen and calcium utilization; energy metabolism; appetite control; dermatology; allergy; antibiotics. Mailing Add: Schering Labs Kenilworth NJ 07033

BALDINI, MARIO G, b Santarcangelo, Italy, Jan 15, 17; US citizen; m 58; c 3. HEMATOLOGY. Educ: Univ Rome, MD, 42; Brown Univ, MA, 71. Prof Exp: From asst prof to assoc prof, Sch Med, Univ Siena, 50-53; asst, Sch Med, Tufts Univ, 54-55, from asst prof to assoc prof, 56-65; DIR DIV HEMAT RES & DIV BIOL & MED SCI, MEM HOSP, BROWN UNIV, 65-, PROF MED SCI, 68-, CHIEF MED, MEM HOSP, 73- Concurrent Pos: Assoc prof, Sch Med, Univ Genoa, 53-57; lectr, US Naval Hosp, Newport, RI, 65- Mem: Am Soc Hemat; Am Physiol Soc; Soc Exp Hemat; Am Fedn Clin Res; Soc Exp Biol & Med. Res: Hematology, especially bone marrow culture in vitro; DiGuglielmo syndrome; platelet antigens, disorders, preservation, circulation and survival. Mailing Add: Div Hemat Res Mem Hosp Pawtucket RI 02860

BALDONI, ANDREW ATELEO, b Pekin, Ill, May 16, 16; m 42; c 2. ORGANIC CHEMISTRY. Educ: Bradley Univ, BS, 47; Univ Notre Dame, PhD(chem), 51. Prof Exp: Chemist, Control Lab, Am Distilling Co, 37-42; res chemist & group leader, Ringwood Chem Corp, 50-56, dir res, 57; asst dir res, Morton Chem Co, 58-64, dir tech serv, 64-67; dir res, Simoniz Co, 67-70; dir corp res, Morton Int, Inc, 70-72, dir res & develop dyes, Morton Chem Co, 72-74, DIR NEW PROD RES, MORTON CHEM CO, 74- Mem: Am Chem Soc. Res: Organic research. Mailing Add: Morton Chem Co 1275 Lake Ave Woodstock IL 60098

BALDRIDGE, JOE DANIEL, agronomy, soil science, see 12th edition

BALDRIDGE, ROBERT CRARY, b Herington, Kans, Jan 9, 21; m 43; c 5. BIOCHEMISTRY. Educ: Kans State Univ, BS, 43; Univ Mich, MS, 48, PhD(biol chem), 51. Prof Exp: Instr biol chem, Univ Mich, 51-53; from asst prof to prof biochem, Sch Med, Temple Univ, 53-70; assoc dean grad sch, 65-70; PROF BIOCHEM, JEFFERSON MED COL, 70-, DEAN, COL GRAD STUDIES, THOMAS JEFFERSON UNIV, 70- Mem: Am Soc Biol Chem. Res: Amino acid metabolism; biochemical genetics; histidinemia. Mailing Add: Col of Grad Studies Thomas Jefferson Univ Philadelphia PA 19107

BALDUS, WILLIAM PHILLIP, b Milwaukee, Wis, Mar 11, 32; m 56; c 5. INTERNAL MEDICINE, GASTROENTEROLOGY. Educ: Marquette Univ, BS, 54, MD, 57; Univ Minn, MS, 64. Prof Exp: Intern, Boston City Hosp, 57-58; resident, Univ Minn Hosps, 58-61; instr, 65-68, ASST PROF GASTROENTEROL, MAYO GRAD SCH MED, UNIV MINN, 68- Mem: Am Gastroenterol Asn; fel Am Col Physicians; Am Fedn Clin Res. Res: Renal failure and disturbances of water and electrolyte metabolism in liver disease. Mailing Add: Mayo Clin Rochester MN 55901

BALDUZZI, PIERO, b Voghera, Italy, Mar 22, 28; m 56; c 2. MICROBIOLOGY, VIROLOGY. Educ: Univ Pavia, MD, 52. Prof Exp: Asst genetics, Univ Pavia, 52-53; res in microbiol, Pharmaceut Co, Italy, 56-58; asst, Univ Florence, 58-62; asst prof, 62-67, ASSOC PROF MICROBIOL, SCH MED, UNIV ROCHESTER, 67- Concurrent Pos: Fel, Univ Rochester, 59-60. Res: Tumor viruses; inhibitors of viral replication; genetics of animal viruses. Mailing Add: Sch of Med Univ of Rochester Rochester NY 14642

BALDWIN, ARTHUR DWIGHT, JR, b Boston, Mass, Oct 28, 38; m 62; c 2. GEOLOGY. Educ: Bowdoin Col, BA, 61; Univ Kans, MA, 63; Stanford Univ, PhD(geol), 67. Prof Exp: ASSOC PROF GEOL, MIAMI UNIV, 66- Mem: Geol Soc Am; Nat Water Well Asn; Sigma Xi. Res: Problems related to finding sufficient water, both in quantity and in quality, to satisfy our nation's increasing needs. Mailing Add: Dept of Geol Miami Univ Oxford OH 45056

BALDWIN, ARTHUR RICHARD, b Palmyra, Mich, Feb 20, 18; m 42; c 3. CHEMISTRY. Educ: Stetson Univ, BS, 40; Univ Pittsburgh, PhD(biochem), 43. Prof Exp: Lab asst chem, Univ Pittsburgh, 40-42, Swift & Co fel, 43, Nutrit Found fel, 44; sect leader food & lipid chem, Corn Prod Refining Co, 44-54; dir res, 54-64, VPRES & EXEC DIR RES, CARGILL, INC, 64- Concurrent Pos: Ed, J Am Oil Chem Soc; mem space applns bd, Nat Acad Eng, 73. Honors & Awards: Normann Medal, Ger Oil Sci Soc, 51; Alton E Bailey Award, Am Oil Chem Soc, 63; Chevreul Medal, Groupement Technique des Corps Gras, 65. Mem: AAAS; Am Chem Soc; Am Oil Chem Soc; Inst Food Technologists. Res: Purification of inositol and phytates; fat and oil analyses recoveries and processing; development of food products; fat digestibilities and metabolism; compounding hydrogenated shortening; preparation of synthetic lipids; carbohydrate oxidations; new packaged food products; plant breeding; feed and animal nutrition research; drying oils and protection coatings. Mailing Add: Cargill Bldg Cargill Inc Minneapolis MN 55402

BALDWIN, BERNARD ARTHUR, b Chicago, Ill, Sept 7, 40; m 62; c 2. PHYSICAL CHEMISTRY, MOLECULAR SPECTROSCOPY. Educ: La Verne Col, BA, 62; San Diego State Col, MS, 65; Univ Calif, Santa Barbara, PhD(phys chem), 68. Prof Exp: Asst chem, San Diego State Col, 62-64, asst phys chem, 63-65; asst chem, Univ Calif, 65-66, asst phys chem, 66-67; SR RES CHEMIST, RADIATION LAB, PHILLIPS PETROL CO, 68- Mem: AAAS; Am Chem Soc. Res: Perturbation effects on the population and depopulation processes of electronically excited molecules; chemical character of natural adsorbates and mechanism for adsorption of molecular species on solid surfaces; chemical characterization of surfaces, adsorbed molecular species and their reaction products; mechanism(s) responsible for boundary lubrication of metals. Mailing Add: Phillips Petrol Co RB-5 Bartlesville OK 74004

BALDWIN, BERNELL ELWYN, b Angwin, Calif, Jan 21, 24; m 60. NEUROPHYSIOLOGY. Educ: Columbia Union Col, BA, 56; George Washington Univ, MA, 58, PhD, 63. Prof Exp: Instr physiol, 63-67, asst prof physiol & biophys, 67-69, ASST PROF PHYSIOL, SCH MED, LOMA LINDA UNIV, 69- Concurrent Pos: Consult, Patton State Hosp, 64- Res: On line computer analysis of autonomic nervous system activity secondary to electrical and chemical stimulation of the brain. Mailing Add: Dept of Physiol & Biophys Loma Linda Univ Sch of Med Loma Linda CA 92354

BALDWIN, BREWSTER, b Brooklyn, NY, Nov 10, 19; m 46; c 4. GEOLOGY. Educ: Williams Col, Mass, AB, 41; Columbia Univ, AM, 44, PhD(geol), 51. Prof Exp: Jr geologist, US Geol Surv, 42-44; instr geol, Univ SDak, 47-51; econ geologist, Bur Mines, NMex, 51-58; from asst prof to assoc prof geol, 58-71, PROF GEOL, MIDDLEBURY COL, 71- Concurrent Pos: Mem, Coun Educ in Geol Sci, 67-70. Mem: Geol Soc Am; Soc Econ Paleontologists & Mineralogists; Nat Asn Geol Teachers. Res: Sedimentology of taconics; environmental geology; curriculum philosophy; granite exploration. Mailing Add: RD 1 Middlebury VT 05753

BALDWIN, DAVID ELLIS, b Bradenton, Fla, June 12, 36; m 61; c 2. PLASMA PHYSICS. Educ: Mass Inst Technol, BSc, 58, PhD(physics), 62. Prof Exp: Res assoc physics, Stanford Univ, 62-64 & Theory Div, Culham Lab, Eng, 64-66; from asst prof to assoc prof appl sci, Yale Univ, 66-70; RES PHYSICIST, LAWRENCE LIVERMORE LAB, 70- Mem: Fel Am Phys Soc. Res: Theoretical plasma physics;

controlled thermonuclear fusion. Mailing Add: L-388 Lawrence Livermore Lab Livermore CA 94550

BALDWIN, DONALD E, veterinary medicine, microbiology, see 12th edition

BALDWIN, ELDON DEAN, b Ashley, Ohio, Sept 23, 39; m 64; c 2. AGRICULTURAL ECONOMICS. Educ: Ohio State Univ, BS, 63; Univ Ill, MS, 67, PhD(agr econ), 70. Prof Exp: Agr statistician, Bur Census, US Dept Com, 63-65; res asst agr econ, Univ Ill, 65-69; asst prof econ, Eastern Ill Univ, 69-70 & Miami Univ, 70-74; ASST PROF AGR ECON, OHIO STATE UNIV, 74- Concurrent Pos: Mem, Subcomt SM-42, Southern Regional Grain Mkt Res Comt, 71- Mem: Am Agr Econ Asn. Res: Impact of technology, policy and economic changes on the structure, conduct and performance of feed grain and livestock markets; alternative sets of market strategies. Mailing Add: Dept Agr Econ Ohio State Univ 2120 Fyffe Rd Columbus OH 43210

BALDWIN, EWART MERLIN, b Pomeroy, Wash, May 17, 15; m 42; c 2. GEOLOGY. Educ: State Col Wash, BS, 38, MS, 39; Cornell Univ, PhD(geol), 43. Prof Exp: Asst, State Col Wash, 38-39 & Cornell Univ, 40-43; from asst geologist to geologist, Ore Dept Geol & Mining Industs, Portland, 43-47; from asst prof to assoc prof, 47-58, PROF GEOL, UNIV ORE, 58- Concurrent Pos: Fulbright award, Dacca, EPakistan, 59-60. Mem: Fel Geol Soc Am; Am Asn Petrol Geologists; Paleont Soc; Sigma Xi; Soc Econ Paleontologists & Mineralogists. Res: Oregon coast range stratigraphy; Coos Bay coal fields; Hambaek coal field of Korea; geology of Oregon and parts of Washington and Idaho; regional geology of Oregon, in particular the Oregon Coast Range and Klamath Mountains. Mailing Add: Dept of Geol Univ of Ore Eugene OR 97403

BALDWIN, FRANCIS PAUL, physical chemistry, see 12th edition

BALDWIN, GEORGE C, b Denver, Colo, May 5, 17; m 44, 52; c 3. PHYSICS. Educ: Kalamazoo Col, BA, 39; Univ Ill, MA, 41, PhD(physics), 43. Prof Exp: Instr physics, Univ Ill, 43-44; res physicist, Gen Elec Co, 44-55, physicist, Aircraft Nuclear Propulsion Dept, 55-57; mgr, Argonaut Reactor, Argonne Nat Lab, 57-58; appl physicist, Adv Tech Labs, Gen Elec Co, 58-65, sr physicist, 65-67; adj prof, 61-67, PROF NUCLEAR ENG & SCI, RENSSELAER POLYTECH INST, 67- Mem: AAAS; Am Vacuum Soc; fel Am Phys Soc. Res: Nuclear photo-effect; high energy radiation and neutron physics; nuclear instrumentation; reactor kinetics and physics; nuclear and ion propulsion; particle accelerators; stimulated emission devices; electron collision spectrometry of molecules. Mailing Add: Dept of Nuclear Eng & Sci Rensselaer Polytech Inst Troy NY 12181

BALDWIN, HEBER ROSS, b New Castle, Pa, July 10, 16; m 56; c 3. BIOCHEMISTRY, FOOD TECHNOLOGY. Educ: Westminster Col, Pa, BS, 38; Syracuse Univ, MS, 40; Univ Iowa, PhD(biochem), 49. Prof Exp: Asst animal nutrit, Mich Agr Exp Sta, 41-43; chemist labs, Gen Foods Corp, 43-45; food technologist, Kraft Foods Labs, 49-52, group leader, Food Prod Develop, 52-56; lab mgr res ctr, Nat Dairy Prod Corp, 57-69, dir prod develop, 69-75, DIR RES & DEVELOP SERV, KRAFTCO CORP, 75- Mem: Inst Food Technologists. Res: Animal and human nutrition; vitamin and amino acid metabolism; food product development; enzymology; dairy protein research. Mailing Add: Kraftco Corp 801 Waukegan Rd Glenview IL 60025

BALDWIN, HENRY IVES, b Saranac Lake, NY, Aug 23, 96; m 24; c 4. FORESTRY. Educ: Yale Univ, AB, 19, MF, 22, PhD(silvicult), 31. Prof Exp: High sch instr, NY, 19; asst to dean sch forestry, Yale Univ, 22-23; res forester, Brown Co, NH, 24-32; vis prof forestry, Pa State Col, 32-33; res forester, State Forestry & Recreation Comn, 33-64; forestry consult, 64-66; from lectr ecol to prof bot, 66-71, EMER PROF BOT, FRANKLIN PIERCE COL, 71- Concurrent Pos: Consult, Nat Resources Planning Bd, 42, Fed Reserve Bank, Boston, 48 & Food & Agr Orgn, UN, 52; mem adv bd, Mt Wash Observ, NH; mem subcomt forestry, Nat Res Coun, 34, mem comt preservation natural conditions; Int Union Forest Res Orgns, 29- Mem: AAAS; fel Soc Am Foresters; Ecol Soc Am; fel Am Acad Arts & Sci; corresp mem Swedish Forestry Asn. Res: Silviculture; seed germination; forest genetics, management and ecology. Mailing Add: RFD 2 Center Rd Hillsboro NH 03244

BALDWIN, HOWARD WESLEY, b Winnipeg, Man, June 11, 28; m 58; c 3. INORGANIC CHEMISTRY. Educ: Univ Sask, BA, 50, MA, 51; Univ Chicago, PhD(chem), 59. Prof Exp: Asst prof, 59-67, ASSOC PROF CHEM & ASST DEAN FAC GRAD STUDIES, UNIV WESTERN ONT, 69- Mem: Chem Inst Can. Res: Kinetic and isotopic studies of reactions of complex ions in solution. Mailing Add: Dept of Chem Univ of Western Ont London ON Can

BALDWIN, JACK EDWARD, b London, Eng, Aug 8, 38. ORGANIC CHEMISTRY. Educ: Univ London, BSc, 60, PhD(chem), 64. Prof Exp: Res asst chem, Imp Col, Univ London, 61-63, from asst lectr to lectr, 63-67; asst prof, Pa State Univ, 67-70; assoc prof, 70-74, PROF CHEM, MASS INST TECHNOL, 74- Concurrent Pos: Consult, Eli Lilly & Co, 67-; A P Sloan fel, 69-71. Mem: Fel Am Chem Soc; The Chem Soc. Res: Organic chemistry and its relationship to biological systems. Mailing Add: Dept of Chem Mass Inst of Technol Cambridge MA 02139

BALDWIN, JACK NORMAN, b Nephi, Utah, Dec 6, 19; m 46; c 3. MICROBIOLOGY. Educ: Univ Utah, BA, 42, MA, 47; Purdue Univ, PhD, 50. Prof Exp: From asst prof to prof bact, Ohio State Univ, 50-63; prof microbiol, Univ Ky, 63-67; RES PROF MICROBIOL, UNIV GA, 67- Mem: Am Soc Microbiol; Am Acad Microbiol; Soc Exp Biol & Med; Brit Soc Gen Microbiol. Res: Microbial genetics; genetics of production and regulation of staphylococcal toxins. Mailing Add: Dept of Microbiol Univ of Ga Athens GA 30601

BALDWIN, JAMES GORDON, b May 24, 45; m 68. PLANT NEMATOLOGY, PLANT PATHOLOGY. Educ: Bob Jones Univ, BS, 67; NC State Univ, MS, 70, PhD(plant path, bot), 73. Prof Exp: Asst prof biol & bot, Bryan Col, 73-74; RES ASSOC PLANT NEMATOL, NC STATE UNIV, 74- Concurrent Pos: USDA fel, NC State Univ, 74-75, AID grant, 75- Mem: Soc Nematol; Sigma Xi. Res: The fine structure of plant parasitic nematodes with emphasis on the Heteroderidae; taxonomy of the Heteroderidae and in host-nematode relationships. Mailing Add: Dept of Plant Path Box 5397 NC State Univ Raleigh NC 27607

BALDWIN, JOE G, JR, b Houston, Tex, Dec 11, 30; m 57; c 1. MATHEMATICS. Educ: Univ Göttingen, dipl, 61, Dr rer nat, 63. Prof Exp: Mem staff, Statist Res Div, Res Triangle Inst, 63-68; ASSOC PROF MATH, UNIV HOUSTON, 68- Concurrent Pos: Adj prof, NC State Univ, 64-68. Mem: Inst Math Statist; Am Math Soc. Res: Information theory and the theory of stochastic processes. Mailing Add: Dept of Math Col Arts & Sci Univ of Houston Houston TX 77004

BALDWIN, JOHN ARNOLD, JR, b Seattle, Wash, Aug 9, 29. PHYSICS. Educ: Univ Calif, Berkeley, AB, 51, PhD(physics), 56. Prof Exp: Mem tech staff, Bell Tel Labs, 56-61 & Sandia Corp, 61-64; sr tech specialist, Autonetics Div, NAm Aviation Inc, 64-67; assoc prof elec eng, 66-69, PROF ELEC ENG, UNIV CALIF, SANTA BARBARA, 69- Mem: Am Phys Soc; Inst Elec & Electronics Engrs. Res: Ferromagnetic memory and logic devices; radiation effects; ferromagnetism. Mailing Add: Dept of Elec Eng Univ of Calif Santa Barbara CA 93106

BALDWIN, JOHN E, b Berwyn, Ill, Sept 10, 37; m 61; c 3. PHYSICAL ORGANIC CHEMISTRY. Educ: Dartmouth Col, AB, 59; Calif Inst Technol, PhD, 63. Prof Exp: From instr to assoc prof org chem, Univ Ill, 62-68; PROF CHEM, UNIV ORE, 68-, DEAN COL ART, 75- Concurrent Pos: Alfred P Sloan res fel, 66-68; Guggenheim Mem Found fel, 67; consult, Stauffer Chem Co & Off Sci & Technol; mem bd ed, Org Reactions & Chem Rev; Alexander von Humboldt Found sr US scientist award, 74-75. Mem: Am Chem Soc; The Chem Soc. Res: Stereochemistry; reaction mechanisms; molecular rearrangements; cycloadditions; stereochemical definition of reaction paths for hydrocarbon molecular rearrangements. Mailing Add: Dept of Chem Univ of Ore Eugene OR 97403

BALDWIN, JOHN THEODORE, b Danbury, Conn, Aug 14, 44; m 65; c 1. MATHEMATICAL LOGIC. Educ: Mich State Univ, BSc, 66; Univ Calif, Berkeley, MS, 67; Simon Fraser Univ, PhD(math), 71. Prof Exp: Res assoc math, Mich State Univ, 71-73; ASST PROF MATH, UNIV ILL, CHICAGO CIRCLE, 73- Mem: Am Math Soc; Asn Symbolic Logic; Math Asn Am. Res: Model theory; universal algebra. Mailing Add: Dept of Math Box 4398 Univ Ill Chicago Circle Chicago IL 60680

BALDWIN, JOHN THOMAS, JR, cytogenetics, deceased

BALDWIN, JON MICHAEL, b Dayton, Ky, Apr 22, 42; m 64; c 2. ANALYTICAL CHEMISTRY, SPECTROSCOPY. Educ: Thomas More Col, AB, 62; Univ Ill, Urbana-Champaign, PhD(anal chem), 68. Prof Exp: Sr chemist, Anal Chem Br, Idaho Nuclear Corp, 67-68, assoc scientist, 68-71, ASSOC CHEMIST, ALLIED CHEM CORP, 71- Concurrent Pos: Affil prof chem, Univ Idaho, 68- Mem: Am Chem Soc; Soc Appl Spectros. Res: Interactions of high-power lasers with solids and their applications to sampling for analytical chemistry; automation and computer applications in analytical chemistry; kinetics of flame reactions; atomic absorption, fluorescence and emmision spectroscopy. Mailing Add: Allied Chem Corp 550 Second St Idaho Falls ID 83401

BALDWIN, JOSEPH, b Houston, Tex, Dec 11, 30; m 64; c 3. INFORMATION SCIENCE, MATHEMATICS. Educ: Univ Göttingen, Hauptdiplom, 61, Dr rer nat, 63. Prof Exp: Mathematician, Statist Res Div, Res Triangle Inst, 63-65; ASSOC PROF MATH, UNIV HOUSTON, 65- Concurrent Pos: Adj prof, NC State Univ, 63-65. Mem: Am Math Soc; Math Asn Am. Res: Mathematical models. Mailing Add: Dept of Math Univ of Houston Houston TX 77004

BALDWIN, KEITH MALCOLM, b Buffalo, NY, May 25, 28; m 51; c 4. PHYSICS, ELECTRONICS. Educ: Mich State Univ, BS, 50; Univ Maine, MS, 55, Prof Exp: Assoc scientist, Raytheon Mfg Co, 55-57; assoc scientist, Res & Adv Develop Div, Avco Corp, 57-59, from sr scientist to sr proj scientist, 59-63; ASSOC PROF PHYSICS, MICH TECHNOL UNIV, 63- Mem: AAAS; Am Asn Physics Teachers; Sigma Xi. Res: Analysis of properties of ionized gas by microwave, optical and ultrasonic diagnostic techniques; microwave transmission through ionized air during missile re-entry; electronic instrumentation. Mailing Add: Point Mills Dollar Bay MI 49922

BALDWIN, LYNNE JUEDEMAN, b Great Falls, Mont. COMPUTER SCIENCES. Educ: Univ Tex, Austin, BA, 67, PhD(comput sci), 74. Prof Exp: Asst prof comput sci, Univ Tex, Austin, 74-75; ASST PROF COMPUT SCI, UNIV NEBR, OMAHA, 75- Mem: Asn Comput Mach. Res: Programming language design; program structure design; program design methodology. Mailing Add: Dept of Math & Comput Sci Univ Nebr Box 688 Omaha NE 68101

BALDWIN, MAYNARD MARTIN, b Walla Walla, Wash, Apr 25, 04; m 32; c 3. ORGANIC CHEMISTRY. Educ: Whitman Col, BS, 26; Univ Wash, PhD(chem), 35. Prof Exp: High sch teacher, Wash & Idaho, 26-30; asst mineral industs, Exp Sta, Pa State Col, 35-38; res chemist, Battelle Mem Inst, 38-44, asst supvr, Chem Res Div, 44-47, supvr, 47-54, from asst tech dir to tech dir, 54-66, actg mgr chem & chem eng dept, 66-67, asst dir technol, Columbus Labs, 67-69; CONSULT, 69- Mem: Am Chem Soc; Soc Indust Microbiol. Res: Condensation of aliphatic aldehydes with ketones; coal hydrogenation; plasticizers; fungicides; lubricant additives; industrial chemistry. Mailing Add: 1990 W Third Ave Columbus OH 43212

BALDWIN, PAUL CLAY, b Tully, NY, May 19, 14; wid; c 3. CHEMISTRY. Educ: Syracuse Univ, BSChE, 36; Inst Paper Chem, MS, 38, PhD(chem), 40. Hon Degrees: DSc, Lawrence Univ, 72. Prof Exp: Res chemist, 40-42, tech dir, 42-46, gen plant mgr, 46-51, asst vpres, 51-53, vpres, 53, dir, 55, vpres mfg, 57-60, exec vpres mfg, res eng & develop, 60-62, exec vpres, 62-68, VCHMN, SCOTT PAPER CO, 68- Concurrent Pos: Dir, Bowater-Scott Australia Ltd, Bowater-Scott Corp Ltd, BC Forest Prod Ltd; dir & vchmn, Burgo Scott, SpA, dir & vpres, Celulosa Jujuy, SA & Compania Indust de San Cristobal, SA; dir, Gureola-Scott, SA, Papeles Scott de Colombia, SA, Bouton-Brochard Scott; dir & exec vpres, Sanyo Scott Co Ltd; dir, Scott Continental, SA, Scott Paper Co de Costa Rica, SA; dir & exec comt mem, Scott Paper Ltd; dir & vpres, Scott Paper Philippines, Inc; dir, Taiwan Scott Paper Corp & Thai-Scott Paper Co Ltd; chmn bd, Brunswick Pulp & paper Co; mem exec comt, Brunswick-S Coast Realty Co; former chmn bd trustees, Inst Paper Chem; dir-at-lg, Syracuse Pulp & Paper Found; trustee, Syracuse Univ; mem, Syracuse Corp Adv Coun; chmn bd, Syracuse Univ Res Corp. Mem: Am Paper Inst; fel Tech Asn Pulp & Paper Indust. Res: Cellulose and lignin; development of cine-photo-micrographic technique for the study of the action of pulping liquors on the microstructure of wood. Mailing Add: Scott Paper Co Scott Plaza I Philadelphia PA 19113

BALDWIN, PAUL HERBERT, b Berkeley, Calif, Feb 26, 13; m 40; c 3. ORNITHOLOGY, VERTEBRATE ECOLOGY. Educ: Univ Calif, AB, 36, PhD(zool), 50; Univ Hawaii, MS, 39. Prof Exp: Asst to supt, Hawaii Nat Park, 43-46; cur collections, Bernice P Bishop Mus, Honolulu, 46-47; assoc prof zool, Univ Calif, 49-50; assoc prof, 50-64, PROF ZOOL, COLO STATE UNIV, 64- Mem: Cooper Ornith Soc; Wilson Ornith Soc; Ecol Soc Am; Am Ornith Union; Entom Soc Can. Res: Feeding ecology of grassland birds; predator-prey relations of woodpeckers and bark-beetles; ecology of Hawaiian land birds; life history of swifts. Mailing Add: Dept of Zool Colo State Univ Ft Collins CO 80521

BALDWIN, RALPH BELKNAP, b Grand Rapids, Mich, June 6, 12; m 40; c 3. ASTROPHYSICS. Educ: Univ Mich, BS, 34, MS, 35, PhD(astron), 37. Hon Degrees: LLD, Univ Mich, 75. Prof Exp: Asst astron, Univ Mich, 35-36; asst, Flower Observ, Pa, 37-38; instr, Dearborn Observ, Northwestern Univ, 38-42; sr physicist, Off Sci Res & Develop, Appl Physics Lab, Johns Hopkins Univ, 42-46, consult, Grand Rapids, Mich, 46-47; prod mgr, 47-58, vpres, 56-70, PRES, OLIVER MACH CO, 70- Concurrent Pos: Lectr, Adler Planetarium, Chicago, Ill, 41-42. Honors & Awards: Presidential Cert of Merit, 47. Mem: Am Astron Soc; fel AAAS; fel Am Geophys Union. Res: Stars; stellar atmospheres; spectral classification; novae; eclipsing binaries;

proximity fuze design characteristics; the moon. Mailing Add: 3110 Manhattan Lane SE East Grand Rapids MI 49506

BALDWIN, RANSOM LELAND, JR, b Meriden, Conn, Sept 21, 35; m 57; c 3. NUTRITION, PHYSIOLOGY. Educ: Univ Conn, BS, 57; Mich State Univ, MS, 58, PhD(biochem), 62. Prof Exp: Asst prof animal husb, 62-70, PROF ANIMAL SCI, UNIV CALIF, DAVIS, 70- Concurrent Pos: Guggenheim Mem Found fel, 68-69. Honors & Awards: Am Feed Mfrs Award, 70. Mem: AAAS; Am Dairy Sci Asn; Am Inst Nutrit. Res: Ruminant digestion; physiology of lactation; nutritional energetics; mechanisms and quantitative aspects of regulation of animal and tissue metabolism; computer simulation modeling of animal systems. Mailing Add: Dept of Animal Sci Univ of Calif Davis CA 95616

BALDWIN, RICHARD WILLIAM, plant physiology, see 12th edition

BALDWIN, ROBERT CHARLES, b Oakland, Calif, Sept 13, 42; m 65; c 2. TOXICOLOGY, ANALYTICAL CHEMISTRY. Educ: Univ Calif, Davis, BS, 64; Stanford Univ, PhD(prg chem), 72. Prof Exp: Chemist anal chem, Vacuum Anal Lab, Lawrence Livermore Lab, 65-66; trainee toxicol, Univ Calif, San Francisco, 72-74; TOXICOLOGIST INHALATION TOXICOL, FLAMMABILITY RES CTR, UNIV UTAH, 74- Mem: NY Acad Sci; Environ Mutagen Soc; AAAS. Res: Inhalation toxicity of products from thermal degradation of natural and synthetic polymers. Mailing Add: Flammability Res Ctr Univ of Utah PO Box 8089 Salt Lake City UT 84108

BALDWIN, ROBERT EDMUND, b Emsworth, Pa, Feb 11, 31; m 57; c 2. PLANT PATHOLOGY. Educ: Waynesburg Col, BS, 52; WVa Univ, MS, 59, PhD(plant path), 64. Prof Exp: PLANT PATHOLOGIST, VA TRUCK & ORNAMENTALS RES STA, 63- Mem: Am Phytopath Soc; Mycol Soc Am; Soc Nematol. Res: Vegetable and ornamental disease control; nematology; mycology. Mailing Add: Va Truck & Ornamentals Res Sta Painter VA 23420

BALDWIN, ROBERT LESH, b Madison, Wis, Sept 30, 27. PHYSICAL BIOCHEMISTRY. Educ: Univ Wis, BA, 50; Oxford Univ, PhD, 54. Prof Exp: From asst prof to assoc prof biochem, Univ Wis, 55-59; assoc prof, 59-64, PROF BIOCHEM, STANFORD UNIV, 64- Concurrent Pos: Guggenheim fel, 58-59; mem ed bd, J Molecular Biol, 64-68; mem, USPHS Fel Comt Biochem & Molecular Biol, 69-71; mem, NSF Adv Panel Biochem & Biophys, 74- Mem: Am Chem Soc; Am Soc Biol Chemists; Biophys Soc. Res: Physical chemistry of nucleic acids and proteins. Mailing Add: Dept of Biochem Stanford Univ Stanford CA 94305

BALDWIN, ROBERT RUSSEL, b Chicago, Ill, Nov 15, 16; m 44; c 2. BIOCHEMISTRY. Educ: DePauw Univ, AB, 38; Iowa State Col, PhD(plant chem), 43. Prof Exp: Instr chem, Iowa State Col, 39-43, assoc chemist, Manhattan Dist Proj, 43-46; biophys sect head, Gen Foods Corp, NY, 46-48, dir biochem lab, 48-56 & chem res lab, 56-61; res supvr, Continental Baking Co, 61-66, ASSOC DIR RES, ITT CONTINENTAL BAKING CO, 66- Concurrent Pos: Mem, Nat Res Coun Comt Civil Defense, 56. Mem: AAAS; fel Am Inst Chem; Inst Food Technol; Am Asn Cereal Chem; NY Acad Sci. Res: Physico-chemical studies; carbohydrates, fats and proteins; toxicology; quality control; product development. Mailing Add: Res Labs ITT Continental Baking Co PO Box 731 Rye NY 10580

BALDWIN, ROBERT SAMUEL, microbiology, see 12th edition

BALDWIN, ROGER ALLAN, b Decatur, Ill, June 2, 31; m 54; c 2. ORGANIC CHEMISTRY, FUEL SCIENCE. Educ: Millikin Univ, BS, 53; Mich State Univ, MS, 56, PhD(chem), 59. Prof Exp: Asst, Mich State Univ, 53-59; sr res chemist, Am Potash & Chem Corp, 59-60, res proj chemist, 60-63, group leader, 63-64, head org sect, 64-69; GROUP LEADER, KERR McGEE CORP, 69- Mem: Am Chem Soc; The Chem Soc. Res: Phosphorus and boron chemistry; azide chemistry; thiophene and sulfur chemistry; solvent extraction; coal conversion. Mailing Add: Tech Ctr Kerr McGee Corp PO Box 25861 Oklahoma City OK 73125

BALDWIN, SIDNEY, organic chemistry, see 12th edition

BALDWIN, THOMAS O, b Ironton, Mo, Sept 18, 38; m 61; c 2. SOLID STATE PHYSICS. Educ: Univ Mo, BS, 59, MS, 61, PhD(physics), 64. Prof Exp: Res physicist, Solid State Div, Oak Ridge Nat Lab, 64-69; ASSOC PROF PHYSICS, SOUTHERN ILL UNIV, EDWARDSVILLE, 69-, CHMN FAC PHYSICS, 71- Concurrent Pos: Consult, Oak Ridge Nat Lab, 69- Mem: Am Phys Soc; Am Asn Physics Teachers; Am Crystallog Asn. Res: Debye-Waller factors; expansion coefficients in ionic crystals; anomalous transmission in perfect metal crystals; x-ray studies of defects in crystals; x-ray diffraction. Mailing Add: Fac of Physics Southern Ill Univ Edwardsville IL 62025

BALDWIN, THOMAS OAKLEY, b Jackson, Miss, June 3, 47; m 67; c 1. BIOCHEMISTRY, EVOLUTIONARY BIOLOGY. Educ: Univ Tex, Austin, BS, 69, PhD(zool), 71. Prof Exp: Fel biochem, Harvard Univ, 72-75; ASST PROF BIOCHEM, UNIV ILL, URBANA, 75- Mem: AAAS; Biophys Soc; Am Chem Soc. Res: Structure and action of bioluminescent proteins; structure of bacterial luciferase; analysis of the structures of mutant forms of bacterial luciferase, particularly those exhibiting a bioluminescence color change. Mailing Add: Dept of Biochem Univ of Ill Urbana IL 61801

BALDWIN, W GEORGE, b Winnipeg, Man, May 3, 38; m 65; c 2. INORGANIC CHEMISTRY. Educ: Univ Man, BSc, 59, MSc, 61; Univ Melbourne, PhD(chem), 66. Prof Exp: Fel, Univ Adelaide, 66 & Royal Inst Technol, Sweden, 66-68; asst prof, 68-74, ASSOC PROF INORG CHEM, UNIV MAN, 74- Mem: Chem Inst Can. Res: Thermodynamic and kinetic studies of complex ions. Mailing Add: Dept of Chem Univ of Man Winnipeg MB Can

BALDWIN, WILLIAM F, b St Thomas, Ont, Dec 29, 13; m 43; c 4. RADIOBIOLOGY. Educ: Univ Guelph, BSA, 40; Univ Western Ont, MA, 42; Univ Toronto, PhD(biol), 52. Prof Exp: Asst entom, Biol Control Inst, Ont, 39-40, res officer, 40-55; assoc res officer, 55-60, sr res officer, 60-72, HEAD BIOL BR, CHALK RIVER NUCLEAR LABS, ATOMIC ENERGY CAN, LTD, 72- Concurrent Pos: Mem entom adv bd, Defence Res Bd Can, 59-64; consult, Malariologia, Dept Health, Maracay, Venezuela, 62-; WHO/Int Atomic Energy Agency expert, Greece, 66-67. Mem: Radiation Res Soc; Entom Soc Can (pres, 70-71); Can Soc Zool. Res: Radiation induced eye color mutations; latent radiation damage in dividing tissue cells; electrophysiology of irradiated insects; ecology of biting flies. Mailing Add: Biol Br Chalk River Nuclear Labs Atomic Energy of Can Ltd Chalk River ON Can

BALDWIN, WILLIAM JOHN, b Gillingham, Eng, Dec 12, 04; US citizen; m; c 2. ANALYTICAL CHEMISTRY. Educ: Univ Buffalo, BS, 26. Prof Exp: Chemist, Am Radiator & Stand Sanit Corp, NY, 26-34; fel, Mellon Inst, 34-40; mgr, Enamel Div, Humphreys Mfg Co, Ohio, 40-41; res engr, Crane Co, Ill & Tenn, 41-44; chief ceramic res, Titanium Alloy Mfg Div, Nat Lead Co, NY, 44-68; consult, 68-75; RETIRED. Mem: Fel Am Ceramic Soc. Res: Zirconia and compounds; refractories; glazes; enamels; dielectric ceramics glass polishing. Mailing Add: 2127 Greenbriar Blvd Clearwater FL 33515

BALDWIN, WILLIAM RUSSELL, b Danville, Ind, July 29, 26; m 47; c 2. OPTOMETRY, PHYSIOLOGY. Educ: Pac Univ, DOptom, 51; Ind Univ, MS, 56, PhD(physiol optics), 64. Prof Exp: Instr gen & ocular path, Ind Univ, 54-58, asst prof, 59-63, lectr, 56-59; prof optom & dean col optom, Pac Univ, 63-69; PRES, MASS COL OPTOM, 69- Concurrent Pos: Mem health facil rev comt, NIH, 67-71; mem bd dirs, Am Optom Found, 69-, Optom Res Inst, Inc, 74- & Am Optical Corp, 74-; mem rev panel, Vet Admin, 74-; mem adv coun, Asn Acad Health Ctrs, 75- Mem: AAAS; Asn Schs & Cols Optom (pres, 73-75); Am Optom Asn; Am Acad Optom; Optical Soc Am. Res: Nature and causes of myopia; evaluation of changes in the optical elements of the eye in vivo. Mailing Add: Mass Col of Optom 424 Beacon St Boston MA 02115

BALDWIN, WILLIAM WALTER, b Ft Wayne, Ind, July 27, 40. MICROBIOLOGY, BIOCHEMISTRY. Educ: Ind Cent Col, AB, 62; Ind Univ, MSEd, 67, PhD(microbiol), 73. Prof Exp: ASST PROF MICROBIOL, MED SCH, IND UNIV NORTHWEST, 73- Mem: AAAS; Am Soc Microbiol. Res: Lipid metabolism and its role in the cell cycle in bacteria. Mailing Add: Ind Univ Northwest Med Sch 3400 Broadway Gary IN 46408

BALDWIN, WILLIS HARFORD, b Havre de Grace, Md, Nov 13, 14; div; c 2. ORGANIC CHEMISTRY. Educ: Univ Md, BS, 35, MS, 37, PhD(org & physiol chem), 42. Prof Exp: Res fel, Nat Cottonseed Prods Asn, Tenn, 38-39; res chemist, Fish & Wildlife Serv, US Dept Interior, 39-42; res chemist, E I du Pont de Nemours & Co, 42-43; RES CHEMIST, OAK RIDGE NAT LABS, 43- Mem: Am Chem Soc. Res: Radioactive isotopes; organo-phosphorus compounds; solvent extraction. Mailing Add: Oak Ridge Nat Lab PO Box X Oak Ridge TN 37830

BALDWIN, WINFIELD MORGAN, JR, b Wilmington, NC, May 4, 26; m 50; c 2. ORGANIC CHEMISTRY. Educ: Univ NC, BS, 50, PhD(org chem), 63. Prof Exp: Chemist, Tenn Eastman Co, 55-61; br mgr, Perkin-Elmer Corp, 61-67; ASSOC PROF CHEM, UNIV GA, 67- Mem: Am Chem Soc. Res: Cellulose chemistry; organic and analytical chemistry; scientific instrumentation. Mailing Add: 160 Mockingbird Circle Athens GA 30601

BALDY, MARIAN WENDORF, b Cleveland, Ohio, June 18, 44; m 65. GENETICS, ENOLOGY. Educ: Univ Calif, Davis, AB, 65, PhD(genetics), 68. Prof Exp: Res assoc biochem, Med Sch, Univ Ore, Portland, 69-70, NIH fel, 70; asst prof plant sci, Calif State Univ, Chico, 71-73; inst agr, Butte Community Col, 73; ASST PROF PLANT SCI, CALIF STATE UNIV, CHICO, 73- Concurrent Pos: Winemaker, Butte Creek Vineyards, Inc, 72- Mem: Genetics Soc Am; Am Genetics Asn; AAAS; Am Soc Enologists. Mailing Add: Dept of Plant & Soil Sci Calif State Univ Chico CA 95926

BALE, HAROLD DAVID, b Fargo, NDak, Oct 3, 27; m 54; c 4. X-RAY CRYSTALLOGRAPHY. Educ: Concordia Col, BA, 50; Univ NDak, MS, 53; Univ Mo, PhD, 59. Prof Exp: Phys sci aide, US Bur Mines, 50-53; instr physics, Univ NDak, 53-55 & Univ Mo, 55-59; from asst prof to assoc prof, 59-70, actg chmn dept physics, 63-67, PROF PHYSICS, UNIV NDAK, 70- Concurrent Pos: Fel, Univ Mo, 69-70. Res: Small angle x-ray scattering. Mailing Add: Dept of Physics Univ of NDak Grand Forks ND 58201

BALE, WILLIAM FREER, b Augusta, NJ, Jan 2, 11; m 39; c 3. BIOPHYSICS. Educ: Cornell Univ, AB, 32; Univ Rochester, PhD(biophys), 36. Prof Exp: Res asst, Med Sch, 36-39, assoc radiol, 39-46, ASSOC PROF RADIOL, MED SCH, UNIV ROCHESTER, 46-, PROF RADIATION BIOL & BIOPHYS, 47- Concurrent Pos: Mem bd sci consults, Sloan-Kettering Inst Cancer Res, 62-72; adv comt biol & med, US AEC, 63-70. Mem: AAAS; Am Physiol Soc; Soc Exp Biol & Med; Am Phys Soc; Radiation Res Soc; NY Acad Sci. Res: Physiological effects and medical uses of radiation; utilization of labeled isotopes as tracers in biological processes; protein metabolism; cancer immunology. Mailing Add: Dept of Radiation Biol & Biophys Univ of Rochester Sch of Med Rochester NY 14642

BALEK, RICHARD WILLIAM, b Chicago, Ill, Nov 15, 31; m 61. PHARMACOLOGY. Educ: St Mary's Col, Minn, BS, 53; Univ Chicago, PhD(pharmacol), 56. Prof Exp: Asst & instr, Univ Chicago, 57-58; instr, Natural Sci Dept, Loyola Univ Chicago, 58-60; asst prof chem, 60-69, ASSOC PROF BIOL, UNIV DETROIT, 69- Mem: AAAS; Am Chem Soc; NY Acad Sci. Res: Enzymology of regeneration tissue; biochemical taxonomy; subcellular pharmacology. Mailing Add: Dept of Biol Univ of Detroit 4001 W McNichols Detroit MI 48221

BALES, BARNEY LEROY, b Amarillo, Tex, Jan 4, 39. BIOPHYSICS. Educ: Univ Colo, BS, 62, PhD(physics), 68. Prof Exp: Staff scientist biophys, Nat Inst Sci Res, Mex; ASSOC PROF PHYSICS, CALIF STATE UNIV, NORTHRIDGE, 75- Mem: Am Phys Soc. Res: Structure and dynamics of the biological membrane and of liquid crystals. Mailing Add: Dept of Physics Calif State Univ Northridge CA 91324

BALES, HAROLD W, JR, plastic surgery, deceased

BALES, HOWARD E, b Ohio, Dec 22, 12; m 37; c 2. SPECTROSCOPY, RESEARCH ADMINISTRATION. Educ: Wilmington Col, BS, 34. Prof Exp: Teacher, pub schs, Ohio, 34-43; chemist, Nat Cash Register Co, 43-46, x-ray diffractionist, 46-49, spectroscopist, 49-53; spectroscopist, C F Kettering Res Lab, 53-57, res physicist, 57-62, staff scientist, 62-64; instr biol & asst dir Col Sci & Eng, Wright State Campus, Miami Univ-Ohio State Univ, 64-68, from assoc dir to dir, Ohio State Tech Serv, 68-70, ASSOC DIR RES & DEVELOP & ED, WSU RES NEWS, WRIGHT STATE UNIV, 70- Concurrent Pos: Fel & vis scientist, Ohio Acad Sci. Mem: AAAS; Am Chem Soc; Am Phys Soc; Soc Appl Spectros; Am Crystallog Asn. Res: Application of instrumental methods to research allied with the study of photosynthesis and nitrogen fixation. Mailing Add: Res Develop Wright State Univ 7751 Colonel Glenn Hwy Dayton OH 45431

BALES, PAUL DOBSON, b Greenville, Tenn, Feb 25, 04; m 40. BIOMEDICAL ENGINEERING. Educ: Univ Chattanooga, BS, 25; Ind Univ, AM, 29. Prof Exp: Asst physics, Univ Chattanooga, 22-25, instr, 25-27; asst, Ind Univ, 27-28; from instr to assoc prof, Howard Col, 30-41, actg head dept, 35-41; res assoc radiation lab, Mass Inst Technol, 41-45, tech asst to ed off publ, 45-46, res physicist lab electronics, 46-47; physicist & chief climatic facil sect, Med Res Labs, 48-56, TECH ASST TO DIR, MED RES LABS, EDGEWOOD ARSENAL, 56- Honors & Awards: Spec Serv Award, Dept Army, 57, Qual Performance Award, Med Res Labs, 67. Mem: AAAS; Am Phys Soc; Am Asn Physics Teachers; Sigma Xi; Inst Environ Sci. Res: Design of climatic chambers for producing physiological stress over a wide range of simulated environmental conditions with bio-medical instrumentation for measuring and recording physiological parameters. Mailing Add: PO Box 86 Bel Air MD 21014

BALFOUR, MARSHALL COULTER, b Marlboro, Mass, Oct 14, 96; m 20; c 2. PUBLIC HEALTH. Educ: Mass Inst Technol, BS, 18; Harvard Univ, MD, 26; Johns Hopkins Univ, MPH, 48. Prof Exp: Asst dir div popular health instr, League Red Cross Socs, Geneva, 20-22; field dir, Int Health Div, Rockefeller Found, 26-38, regional dir, Far East, China, 39-40, regional dir, Philippines, 40-41, regional dir, India, 42-49 & 53-58; rep Asia & Far East, Population Coun, New York, 58-65; vis prof pop studies, Carolina Pop Ctr, Univ NC, Chapel Hill, 70-73; RETIRED. Concurrent Pos: Tech adv, Ministry of Health & dir div malaria, Sch Hyg, Athens, Greece, 30-39; lectr, Sch Hyg & Pub Health, Johns Hopkins Univ, 47-48; consult, Pop Coun, New York, 65-70. Mem: Am Soc Trop Med & Hyg; Pop Asn Am. Res: Medical education; public health demography. Mailing Add: 411 Morgan Creek Rd Chapel Hill NC 27514

BALFOUR, WILLIAM MAYO, b Pasadena, Calif, Nov 26, 14; m 39; c 4. PHYSIOLOGY. Educ: Univ Minn, MD, 40, MS, 48. Prof Exp: Intern internal med, Univ Rochester, 39-40, resident path, 40-42; consult, Mayo Clin & Mayo Found, 48-56; asst prof physiol, 57, res assoc, 59-64, assoc prof, 62-66, dean student affairs, 68-70, PROF PHYSIOL, SCH MED, UNIV KANS, 66-, VCHANCELLOR STUDENT AFFAIRS, 70- Concurrent Pos: Fel internal med, Mayo Found, 42-48, instr, Univ Minn, 50-56; USPHS fel, Uni Kans, 57-59. Mem: AAAS; Am Diabetes Asn; AMA; Am Physiol Soc; NY Acad Sci. Res: Cerebral energy metabolism; cellular structure and function of brain; endocrine physiology. Mailing Add: Univ of Kans Lawrence KS 66044

BALGLEY, ELY, b New York, NY, June 27, 16; m 40; c 4. CHEMISTRY. Educ: Brooklyn Col, AB, 37; NY Univ, MS, 39. Prof Exp: Chemist, Gen Foods Corp, NJ, 40-45; develop engr, Gen Elec Co, 45-50; asst mgr mkt develop, Heyden Newport Corp, 50-56; asst dir mgt res, Wyandotte Chem Corp, 56-60; DIR MKT RES, A E STALEY MFG CO, 61- Mem: Am Chem Soc; Am Inst Chemists; Chem Mkt Res Asn; Com Develop Asn. Res: Commercial chemical development; market development and research; new product development. Mailing Add: 147 Hightide Dr Decatur IL 62521

BALGOOYEN, THOMAS GERRIT, b Grand Haven, Mich, July 5, 43; c 1. ORNITHOLOGY. Educ: San Jose State Univ, BA, 66; MA, 67; Univ Calif, Berkeley, PhD(zool), 72. Prof Exp: Asst prof biol, Univ Calif, Riverside, 73, fel ornith, 72-74; ASST PROF BIOL, SAN JOSE STATE UNIV, 75- Mem: Am Ornithologists Union; Cooper Ornith Soc; Wilson Ornith Soc; Sigma Xi. Res: Adaptive strategies of raptorial birds in terms of time and energy; interactions of avian behavior and ecology. Mailing Add: Dept of Biol ScedSan Sci San Jose State Univ San Jose CA 95192

BALIN, HOWARD, b Philadelphia, Pa, Jan 22, 27; m; c 4. OBSTETRICS & GYNECOLOGY. Educ: Univ Pa, BA, 47, MD, 51, MSc, 57; Am Bd Obstet & Gynec, dipl, 60. Prof Exp: Intern, Grad Hosp Univ Pa, 51-52, resident gen surg, 52-53 & gynec, 53-55; resident obstet, Philadelphia Naval Hosp, 55-56; asst prof obstet & gynec, Sch Med, Univ Pa, 64-70; dir div reproductive biol, 70-74, PROF OBSTET & GYNEC, HAHNEMANN MED COL & HOSP, 70-, CHMN DEPT, 74- Concurrent Pos: Researcher, Wistar Inst, 52-57; Benjamin Rosenthal res grant & Schwartz Fund, Pa Hosp, 60-61. Mem: Fel Am Col Obstet & Gynec; fel Am Col Surg; AMA; fel Am Col Surgeons; Asn Prof Gynec & Obstet. Mailing Add: Suite 1323 Med Arts Bldg 1601 Walnut St Philadelphia PA 19103

BALINSKI, MICHEL LOUIS, b Geneva, Switz, Oct 6, 33; US citizen; m 57; c 2. APPLIED MATHEMATICS. Educ: Williams Col, BA, 54; Mass Inst Technol, MS, 56; Princeton Univ, PhD(math), 59. Prof Exp: Instr math, Princeton Univ, 59-61, lectr & res assoc, 61-63; assoc prof econ, Univ Pa, 63-65; assoc prof, 65-69, PROF MATH, GRAD CTR, CITY UNIV NY, 69- Concurrent Pos: Consult, Mathematica, 59-, Rand Corp, 60-66 & Mobil Oil Res Labs, 66-69; lectr, Univ Mich Summer Eng Conf, 62-69; IBM Corp World Trade Corp prof fel, Univ Paris, 69-70; mem opers res coun adv body, Mayor, City of New York, 67-70; ed-in-chief, Math Programming and Math Programming Studies, 70-; consult, Off Radio-diffusion Television, France, 71; vis prof, Swiss Fed Inst Technol, 72-73 & Univ Grenoble, 74-75; chmn syst & div sci area, Int Inst Appl Systs Anal, Luxemburg, 75- Honors & Awards: Lanchester Prize, 65, Opers Res Soc Am, 65. Res: Mathematical programming; graph theory; combinatorial applied mathematics; mathematical economics, including activity analysis, operations research and game theory; mathematical models for political science. Mailing Add: Spring Grove Lawrenceville NJ 08648

BALINT, FRANCIS JOSEPH, b Johnstown, Pa, Mar 2, 32; m 56; c 6. MATHEMATICS. Educ: Indiana State Col Pa, BS, 54; Univ Pittsburgh, MS, 59. Prof Exp: Customs engr, Int Bus Mach Corp, 54-55; programmer, Gulf Res & Develop Co, 55-60, opers leader, 60-62, sr proj engr, 63-65; chief mgt & prog br, US Weather Bur, 65-66, chief mgt & planning br, 66-70, chief mgt systs div, 70-72, DEP DIR OFF MGT & COMPUT SYSTS, NAT OCEANIC & ATMOSPHERIC ADMIN, 72- Concurrent Pos: Instr, Indiana State Col Pa, 62; mem exec bd, Share, Inc, 68-69, mgr basic systs div, 70. Honors & Awards: US Dept Com Silver Medal, 71. Mem: NY Acad Sci; Asn Comput Mach. Res: Computing operations and management. Mailing Add: 2400 Jackson Pkwy Vienna VA 22180

BALINT, JOHN ALEXANDER, b Budapest, Hungary, Feb 11, 25; m 49; c 2. MEDICINE. Educ: Univ Cambridge, BA, 45, MB, BCh, 48. Prof Exp: Fel gastroenterol, Col Med, Univ Cincinnati, 58-59; fel med, Sch Med, Johns Hopkins Univ, 59-60; asst prof med & gastroenterol, Med Col Ala, 60-63; assoc prof, 63-68, PROF MED & GASTROENTEROL, ALBANY MED COL, 68- Concurrent Pos: NIH trainee lipid biochem, Sinai Hosp, Baltimore, 59-60. Mem: Am Fedn Clin Res; Am Physiol Soc; Am Soc Clin Invest; Brit Med Asn; fel Royal Soc Med. Res: Lipid metabolism in relation to fat absorption and intrahepatic phospholipid metabolism; lipid and protein disorders in hereditary lipid storage diseases. Mailing Add: Dept of Med Albany Med Col Albany NY 12208

BALIS, EARL WILLIAM, b Ft Collins, Colo, July 9, 11; m 40; c 2. CHEMISTRY. Educ: Colo State Univ, BS, 32; Ohio State Univ, PhD(chem), 41. Prof Exp: Asst chemist, Exp Sta, Colo State Univ, 32-37; asst chem, Ohio State Univ, 37-40; res assoc, Res Lab, 41-52, mgr anal chem unit, 52-65, mgr anal chem oper, 65-70, anal chemist, 70-73, MGR ANAL CHEM OPER, RES & DEVELOP CTR, GEN ELEC CO, 73- Mem: Fel AAAS; fel Am Inst Chem; Am Chem Soc; Am Microchem Soc; Nat Asn Corrosion Eng. Res: Microanalysis; corrosion in aqueous systems; flammability limits of organosilicon materials; inorganic chemistry; separations; measurement of physical properties and molecular weight distributions. Mailing Add: Res & Develop Ctr Gen Elec Co PO Box 8 Schenectady NY 12301

BALIS, MOSES EARL, b Philadelphia, Pa, June 19, 21; m 45; c 2. BIOCHEMISTRY. Educ: Temple Univ, BA, 43; Univ Pa, MA, 47, PhD(org chem), 49. Prof Exp: Assoc, 49-54, assoc mem, 54-65, MEM, SLOAN-KETTERING INST CANCER RES, 65-, CHIEF DIV CELL METAB, 71-; PROF BIOCHEM, SLOAN-KETTERING DIV, CORNELL UNIV, 66- Concurrent Pos: USPHS res career award, 63; assoc prof biochem, Sloan-Kettering Div, Cornell Univ, 54-56, chmn dept, 69-72; mem ed adv bd, Cancer Res, 70-73, assoc ed, 74-; partic, Nat Cancer Planning Prog, 72; mem adv comt pathogenesis, Am Cancer Soc. Mem: AAAS; Am Chem Soc; Am Soc Biol Chem; Harvey Soc; Am Asn Cancer Res. Res: Biochemistry of differentiation and malignancy; regulation of enzyme function; purine metabolism; action of antimetabolites; genetic defects. Mailing Add: Sloan-Kettering Inst Sect 5801 410 E 68th St New York NY 10021

BALISH, EDWARD, b Scranton, Pa, Apr 23, 35; m 61; c 1. MICROBIOLOGY, BIOCHEMISTRY. Educ: Univ Scranton, BS, 57; Syracuse Univ, MS, 59, PhD(microbiol), 63. Prof Exp: Asst, Syracuse Univ, 57-60, res assoc, 62-63; resident res assoc, Argonne Nat Lab, 63-65; res scientist-microbiologist, Oak Ridge Assoc Univs, 66-70; ASSOC PROF MED MICROBIOL & SURG, MED SCH, UNIV WISMADISON, 70- Mem: AAAS; Am Soc Microbiol; Radiation Res Soc; Can Soc Microbiol; Int Soc Human & Animal Mycol. Res: Microbial biochemistry; enzymology; gnotobiology; mechanism of pathogenicity of fungi, particularly Candida albicans. Mailing Add: 185 Med Sci Bldg Univ of Wis Med Sch Madison WI 53706

BALKA, DON STEPHEN, b South Bend, Ind, Aug 14, 46; m 69; c 2. MATHEMATICS. Educ: Mo Valley Col, BS, 68; Ind Univ, MS, 70; St Francis Col, MS, 71; Univ Mo, Columbia, PhD(math educ), 74. Prof Exp: Jr high sch teacher math, Plymouth Community Sch Corp, 68-69; sr high sch teacher math, Union-North United Sch Corp, 69-71; lectr math educ, Univ Mo, Columbia, 73-74; ASST PROF MATH, ST MARY'S COL, IND, 74- Concurrent Pos: Math consult, Creative Publ, Inc. 74-, Ind Dept Pub Instr, 74-75; metric consult, South Bend Community Sch Corp, 74-75; assoc fac, Ind Univ, South Bend, 75- Mem: Nat Coun Teachers Math; Am Educ Res Asn; Nat Educ Asn. Res: Ability of young children to identify three-dimensional objects from orthographic drawings; creativity in mathematics; improving basic mathematics skills of students. Mailing Add: Dept of Math St Mary's Col Notre Dame IN 46556

BALKE, BRUNO, b Braunschweig, Ger, Sept 6, 07; US citizen; m 35; c 4. PHYSIOLOGY. Educ: Acad Phys Educ, Ger, dipl, 31; Univ Berlin, MD, 36; Univ Leipzig, Dr med habil, 45. Prof Exp: Consult sports med, Univ Berlin, 36-42; head dept performance physiol, Med Sch Ger Army Mt Corps, 42-45; physiologist, Sch Aviation Med, US Air Force, 50-60; chief biodynamics br, Civil Aeromed Res Inst, Fed Aviation Agency, 60-64; prof, 64-73, EMER PROF PHYS EDUC & PHYSIOL, UNIV WIS-MADISON, 73-; DIR ASPEN CARDIO-PULMONARY REHAB UNIT, 73- Concurrent Pos: Assoc prof, Air Univ, 58-60; prof, Col Med, Univ Okla, 60-; attend physician, Vet Admin Hosp, 60- Mem: Soc Exp Biol & Med; Aerospace Med Asn; Am Physiol Soc; Am Col Sports Med (pres, 65). Res: Medical and physiological research applied to human stress tolerance under condition of exercise, climate, altitude and high speed flying. Mailing Add: PO Box 630 Aspen CO 81611

BALKE, CLAIRE CODDINGTON, b Urbana, Ill, June 17, 08; m 43; c 1. CHEMISTRY. Educ: Northwestern Univ, BS, 32. Prof Exp: Instr chem, Univ Ill, 32-33; asst, Fansteel Metall Corp, Chicago, 33-39, group leader res, 40-43, head iron powder div, 45-48; asst prof powder metall, Stevens Inst Technol, 39-40; group leader metall dept, Chem Div, Manhattan Proj, Los Alamos, 43-45; dir res, Roberts & Mander Corp, 48-50; PRES, BALKE RES ASSOCS, 50- Res: Powder metallurgy of refractory metals, ferrous metals and hard carbides; ceramics of refractory oxides; processes for making and recovering tantalum and columbium; extrusion of metal powders; commercial testing. Mailing Add: 20 Livezey St Doylestown PA 18901

BALL, BILLY JOE, b Crowell, Tex, Nov 29, 25; m 47; c 1. MATHEMATICS. Educ: Univ Tex, BA, 48, PhD(math), 52. Prof Exp: Instr pure math, Univ Tex, 49-52; actg asst prof math, Univ Va, 52-53, asst prof, 53-59; assoc prof, 59-63, head dept math, 63-69, PROF MATH, UNIV GA, 63- Mem: Am Math Soc; Math Asn Am. Res: Point set topology. Mailing Add: Dept of Math Univ of Ga Athens GA 30602

BALL, CARROLL RAYBOURNE, b Leakesville, Miss, Oct 11, 25; m 47, 63; c 3. ANATOMY. Educ: Univ Miss, BA, 47, MS, 48, PhD(anat), 63. Prof Exp: Asst zool, Univ Miss, 46-48; asst, Duke Univ, 48-50, instr, 50-51; instr micros & gross anat, Med Sch, WVa Univ, 51-57; asst prof biol, Univ Southern Miss, 57-60, instr anat, 62-63; from asst prof to assoc prof, 63-71, PROF ANAT, SCH MED, UNIV MISS, 71- Mem: Am Asn Anat; Am Soc Exp Path; Soc Exp Biol & Med. Res: Experimental pathology; nutrition; cardiovascular disease; hepatic liposis. Mailing Add: Dept of Anat Univ of Miss Sch of Med Jackson MS 39216

BALL, CLAYTON GARRETT, b Wauconda, Ill, Dec 16, 06; m 35; c 2. GEOLOGY. Educ: Northwestern Univ, BS, 28; Harvard Univ, AM, 32, PhD(geol), 35. Prof Exp: Geologist, Ziegler Coal Co, Chicago, 28-29; asst geologist, Ill Geol Surv, 29-35; geologist, Bell & Zoller Coal Co, Chicago, 35-37; geologist, 37-54, pres, 54-70, bd chmn, 70-74, CONSULT, PAUL WEIR CO, 74- Mem: Fel Geol Soc Am; Am Inst Mining, Metall & Petrol Eng. Res: Mining and preparation of coal and minerals of similar occurrence in North America and abroad. Mailing Add: Paul Weir Co 20 N Wacker Dr Chicago IL 60606

BALL, DAVID RALPH, b Swansea, Wales, July 20, 40; m 64; c 3. INDUSTRIAL ORGANIC CHEMISTRY, FUEL SCIENCE. Educ: Univ Leeds, Eng, BSc, 64, PhD(coal chem), 69. Prof Exp: Chem engr & group leader, Brit Acheson Electrodes Ltd, 68-73; SCIENTIST INDUST GRAPHITE RES & DEVELOP, PARMA TECH CTR, UNION CARBIDE CORP, 73- Mem: Brit Inst Fuel; Soc Chem Indust; Brit Nuclear Eng Soc. Res: Preparation, composition, structure, chemical reactivity, carbonisation and pyrolysis of fossil fuel derivatives. Mailing Add: Parma Tech Ctr Box 6116 Union Carbide Corp Cleveland OH 44101

BALL, DEREK HARRY, b London, Eng, Sept 1, 31; m 65; c 1. CARBOHYDRATE CHEMISTRY. Educ: Bristol Univ, BSc, 53; Queen's Univ, Ont, PhD(chem), 56. Prof Exp: Res fel carbohydrate chem, Div Appl Biol, Nat Res Coun Can, 56-58; fel catalysis, State Univ NY Col Ceramics, Alfred Univ, 59-60; res chemist, Org Chem, 60-72, HEAD ORG CHEM GROUP, PIONEERING RES LAB, US ARMY NATICK LABS, 72- Mem: Am Chem Soc; fel Brit Chem Soc. Res: Selective substitution of carbohydrates; branched chain sugars; sucrochemistry; H-1 and C-13 nmr of carbohydrates; liquid chromatographic methods for the analysis of free sugars in foods. Mailing Add: Pioneering Res Lab US Army Natick Labs Natick MA 01760

BALL, DONALD LEE, b Kalamazoo, Mich, June 9, 31; m 69. CHEMISTRY. Educ: Kalamazoo Col, AB, 53; Brown Univ, PhD(chem), 56. Prof Exp: Proj assoc chem, Univ Wis, 56-57; sr res chemist res labs, Gen Motors Corp, Mich, 57-67; chief phys chem prog, Chem Sci Directorate, 67-70, prog mgr chem energetics, 70-73, DIR CHEM SCIS, AIR FORCE OFF SCI RES, 73- Mem: Am Chem Soc. Res: Physical and inorganic chemistry. Mailing Add: Chem Sci Directorate Air Force Off of Sci Res Bolling AFB Bldg 410 Washington DC 20332

BALL, EDWIN LAWRENCE, b Collins, Iowa, Jan 31, 13; m 48; c 2. STATISTICAL ANALYSIS, PHARMACEUTICS. Educ: State Col Iowa, BA, 38; Univ Wis, PhM, 39, PhD(physiol), 47. Prof Exp: Lab asst, State Col Iowa, 35-38; Nat Res Coun asst, Univ Wis, 38-39, asst biol, 39-42; res mycologist, Lederle Labs Div, 43-56, DEVELOP CHEMIST, LEDERLE LABS DIV, AM CYANAMID CO, 56- Mem: AAAS; Am

Soc Microbiol; Am Soc Plant Physiol; Soc Indust Microbiol; NY Acad Sci. Res: Stability of pharmaceutical products. Mailing Add: Pearl River NY

BALL, ELLEN MOORHEAD, plant pathology, see 12th edition

BALL, ERIC GLENDINNING, b Coventry, Eng, July 12, 04; US citizen; m 27. BIOCHEMISTRY. Educ: Haverford Col, BS, 25, AM, 26; Univ Pa, PhD(physiol chem), 30. Hon Degrees: AM, Harvard Univ, 42; DSc, Haverford Col, 49. Prof Exp: Asst, Sch Med, Univ Pa, 26-28; instr med, Johns Hopkins Univ, 30-33, assoc, 33-40; from asst prof to assoc prof, 40-46, actg head dept, 43-46 & 58-59, chmn div med sci, 52-68, prof biochem, 46-62, Edward S Wood prof, 62-71, EDWARD S WOOD EMER PROF BIOCHEM, HARVARD MED SCH, 71- Concurrent Pos: Vis prof, Univ Brazil, 45; Int Physiol Cong fel, Rome, 32; Guggenheim fels, 37-38 & 62-63; trustee, Marine Biol Lab, Woods Hole Oceanog Inst, 53-57; Commonwealth Fund fels, Gt Brit, 59 & Brazil, 64; vis investr, Scripps Clin & Res Found, 63. Honors & Awards: Lilly Award, Am Chem Soc, 40; Cruzeiro do Sol, Order of Southern Cross, Brazil; Cert Merit, 48. Mem: Nat Acad Sci; AAAS; Am Soc Biol Chemists; Am Chem Soc; Soc Gen Physiol. Res: Composition of pancreatic juice; oxidation-reduction potentials of biological systems; biological oxidations; xanthine oxidase; reaction of mustard gas with biological systems; in vitro growth of malarial parasites; pigments and enzymes of marine organisms; properties of the electron transmitter system; y-glutamyl transpeptidase; action of hormones on metabolic processes; adipose tissue metabolism. Mailing Add: Marine Biol Lab Woods Hole MA 02543

BALL, ERNEST, b Mangum, Okla, Dec 22, 09; m 41; c 3. BOTANY. Educ: Univ Okla, BS, 37, MS, 38; Univ Calif, PhD(bot), 41. Prof Exp: Asst bot, Univ Calif, 38-41; Nat Res fel, Yale Univ, 41-42; res asst, Carnegie Inst, 42-43; res asst, Harvard Forest, Harvard Univ, 43, lectr, 43-46; from asst prof to prof, Univ NC, Raleigh, 46-69; PROF DEVELOP & CELL BIOL, UNIV CALIF, IRVINE, 69- Concurrent Pos: Fulbright prof, Univ Delhi, 60-61; Gen Educ Bd fel travel in bot insts of Europe, 47; Guggenheim fel, 60; res partic, Oak Ridge Nat Lab, 51; vchmn morphol & anat, Int Bot Cong, Can, 59; vpres, Eighth Int Bot Cong, Paris. Mem: AAAS; Biol Stain Comn; Bot Soc Am; Soc Develop Biol; Am Soc Nature. Res: Morphogenesis of floral apex; mixed callus cultures; virus in callus cultures; culture of isolated mesophyll; cytokinesis in isolated callus cell; structure and function of the shoot apex. Mailing Add: Sch of Biol Sci Univ of Calif Irvine CA 92664

BALL, FRANCES LOUISE, b Murfreesboro, Tenn, Oct 6, 24. ORGANIC CHEMISTRY, ELECTRON MICROSCOPY. Educ: Mid Tenn State Col, BS, 45; Vanderbilt Univ, MS, 48. Prof Exp: Jr chemist, E I du Pont de Nemours & Co, 45; instr high sch, Tenn, 45-46; instr chem, Limestone Col, 48-49; chemist, Oak Ridge Gaseous Diffusion Plant, 49-68, PHYSICIST, MOLECULAR ANAT PROG, OAK RIDGE NAT LAB, UNION CARBIDE CORP, 68- Mem: Electron Micros Soc Am; Sigma Xi. Res: Beckmann rearrangement of oximes; oxidation of single crystals of copper; structure and reactions of thin metal films. Mailing Add: 502 Riverside Dr Clinton TN 37716

BALL, FRANK JERVERY, b Charleston, SC, Jan 29, 19; m 50; c 4. ORGANIC CHEMISTRY, WOOD CHEMISTRY. Educ: Univ of the South, BS, 41; Univ Rochester, PhD(org chem), 44. Prof Exp: Res chemist, 44-46; group leader, 46-53; dir, Charleston Res Ctr, 53-75, dir, Tyrone Res Lab, 58-60, ASSOC CORP RES DIR, WESTVACO CORP, 75- Concurrent Pos: Pres, Empire State Pulp Res Assocs, 74-75. Mem: Am Chem Soc; Tech Asn Pulp & Paper Industs; Can Pulp & Paper Asn. Res: Sulfate lignin; fatty and rosin acids from tall oil; activated carbon; pulping and bleaching chemical recovery; air pollution control. Mailing Add: Charleston Res Ctr Westvaco Corp PO Box 5207 North Charleston SC 29406

BALL, FRED, b Manly, Australia, Aug 29, 18; US citizen; m 44; c 3. ORGANIC CHEMISTRY, INDUSTRIAL CHEMISTRY. Educ: Cornell Univ, BCh, 41. Prof Exp: Chemist, Thompson & Co, 41-46; asst to dir res, Pierce & Stevens Chem Co, 46-51; sr chemist, Tenn Eastman Co, 51-53; chief chemist, 53-70, MGR COATINGS CHEM LAB, EASTMAN CHEM PROD, INC, 70- Concurrent Pos: Vpres, Buffalo Sect, Fedn Soc Paint Technol, 51- Mem: Am Chem Soc; Am Soc Testing & Mat; Am Tech Asn Pulp & Paper Indust. Res: Cellulose acetate extrusion; synthesis of polyesters and alkyds; plasticizers; chemicals for coatings, paper and inks. Mailing Add: Eastman Chem Prod Inc PO Box 431 Kingsport TN 37663

BALL, GENE V, b Rivesville, WVa, June 28, 31. INTERNAL MEDICINE. Educ: WVa Univ, BS, 57; Vanderbilt Univ, MD, 59; Am Bd Internal Med, dipl; Am Bd Rheumatol, dipl. Prof Exp: Intern, Cincinnati Gen Hosp, 59-60; resident, Hosp Univ Pa, 60-61; resident, Jackson Mem Hosp, 61-64; pract internal med, Univ Miami, 64-65; from instr to assoc prof, 65-71, PROF INTERNAL MED, MED CTR, UNIV ALA, BIRMINGHAM, 71- Concurrent Pos: Fel, Jackson Mem Hosp, 61-64; assoc ed, Arthritis & Rheumatism. Mem: Fel Am Col Physicians. Res: Gout; Saturnine gout; systemic lupus erythemactoues. Mailing Add: Dept of Internal Med Univ of Ala Med Ctr Birmingham AL 35233

BALL, GEORGE EUGENE, b Detroit, Mich, Sept 25, 26; m 49; c 2. SYSTEMATICS. Educ: Cornell Univ, AB, 49, PhD, 54; Univ Ala, MS, 50. Prof Exp: From asst prof to assoc prof entomol, 54-65, PROF ENTOM, UNIV ALTA, 65- Mem: Soc Syst Zoologists; Soc Study Evolution; Royal Entom Soc London. Res: Systematic entomology, especially the taxonomy, ecology and zoogeography of beetles of the family Carabidae. Mailing Add: Dept of Entom Fac of Agr Univ of Alta Edmonton AB Can

BALL, GORDON HAROLD, b Warren, Pa, Sept 20, 99; m 40. ZOOLOGY, PARASITOLOGY. Educ: Univ Pittsburgh, BS, 20, MS, 21; Univ Calif, PhD(zool), 24. Prof Exp: Asst biol, Univ Pittsburgh, 20-21; teaching fel zool, 21-23, from instr to prof, 24-67, EMER PROF ZOOL, UNIV CALIF, LOS ANGELES, 67- Mem: AAAS; hon mem Soc Protozool (vpres, 51, pres, 58); Soc Exp Biol & Med; Am Soc Trop Med & Hyg; Am Soc Parasitol (vpres, 61, pres, 63). Res: Parasites of marine invertebrates; in vitro culture of malaria; nutrition of protozoa; Paramecium life history; hydrogen ion concentration of living tissues; variation in mussels; life histories of protozoan parasites; physiology of mosquitos. Mailing Add: Dept of Biol Univ of Calif Los Angeles CA 90024

BALL, HAROLD JAMES, b Beaver Dam, Wis, Mar 13, 19; m 60; c 1. INSECT PHYSIOLOGY. Educ: Univ Wis, BA, 43, MA, 46, PhD(entom), 51. Prof Exp: Asst prof naval sci & tactics, Univ SC, 46; from asst prof to assoc prof entom, 51-60, PROF ENTOM, UNIV NEBR, LINCOLN, 60- Mem: Am Entom Soc; Entom Soc Am. Res: Insect photoreceptors. Mailing Add: Insectary Bldg Univ of Nebr Lincoln NE 68503

BALL, JAMES BRYAN, b Portland, Ore, Oct 2, 32; m 60. NUCLEAR CHEMISTRY. Educ: Ore State Col, BS, 54; Univ Wash, PhD, 58. Prof Exp: Resident res assoc, 58-59, PHYSICIST, OAK RIDGE NAT LAB, 59- Mem: AAAS; Am Phys Soc. Res: Experimental studies of reaction mechanisms and validity of nuclear models in medium energy nuclear reactions. Mailing Add: Oak Ridge Nat Lab PO Box X Oak Ridge TN 37831

BALL, JAMES STUTSMAN, b Reno, Nev, Sept 13, 34; m 57. PHYSICS. Educ: Calif Inst Technol, BS, 56; Univ Calif, Berkeley, PhD(physics), 60. Prof Exp: Res assoc physics, Lawrence Radiation Lab, Univ Calif, Berkeley, 60, asst res physicist, La Jolla, 60-63, asst prof physics, Los Angeles, 63-68; assoc prof, 68-72, PROF PHYSICS, UNIV UTAH, 72- Mem: Am Phys Soc. Res: Elementary particle physics; structure of elementary particles based on the S-matrix approach to strong interactions. Mailing Add: Dept of Physics Univ of Utah Salt Lake City UT 84112

BALL, JOHN MILLER, b Highland Park, Mich, Jan 7, 23; m 46; c 3. GEOGRAPHY. Educ: Cent Mich Univ, AB, 48; Univ Mich, MA, 50; Univ Chicago, MS, 52; Mich State Univ, PhD(geog), 61. Prof Exp: Teacher geog & math, Gladwin High Sch, Mich, 48-50; teacher, Lakeview High Sch, Decatur, Ill, 52-54; teacher geog, Excelsior Union High Sch Dist, Artesia, Calif, 54-57; instr, Cent Mich Univ, 57-61; prof, Slippery Rock State Col, 61-65; assoc prof geog & educ, Univ Ga, 65-70; prog geog, 70-74, PROF GEOG & EDUC, GA STATE UNIV, 74- Concurrent Pos: Pres, Nat Coun Geog Educ, 72-73. Mem: Asn Am Geog; Am Geog Soc; Latin Am Studies Asn; Asn Social & Behav Sci. Res: Teaching and learning geography; Latin America, especially Mexico. Mailing Add: Dept of Geog Ga State Univ Atlanta GA 30303

BALL, JOHN SIGLER, b Hamlin, Tex, Aug 2, 14; m 38; c 6. PETROLEUM CHEMISTRY. Educ: Tex Tech Col, BS, 34, MS, 36. Prof Exp: Jr chem engr petrol & oil-shale exp sta, US Bur Mines, 38-42, asst chemist, 42-43, assoc petrol chemist, 43-44, chemist, 44-45, head petrol chem & refining sect & shale oil res & anal sect, 45-56, chief br petrol, Region III, 56-58, projs coordr petrol res, 58-63, dir, Am Petrol Inst Projs, 48-63, RES DIR, BARTLESVILLE ENERGY RES CTR, US ENERGY RES & DEVELOP ADMIN, 63- Honors & Awards: Dept of Interior Distinguished Serv Medal, 54. Mem: AAAS; fel Am Inst Chemists; Am Chem Soc; Soc Petrol Engrs; Am Geochem Soc. Res: Sulfur and nitrogen compounds of petroleum; composition of shale oil; analytical methods for petroleum. Mailing Add: Bartlesville Energy Res Ctr Energy Res & Develop Admin Box 1398 Bartlesville OK 74003

BALL, JOSEPH ANTHONY, b Washington, DC, June 4, 47. PURE MATHEMATICS. Educ: Georgetown Univ, BS, 69; Univ Va, MS, 70, PhD(math), 73. Prof Exp: ASST PROF MATH, VA POLYTECH INST & STATE UNIV, 73- Mem: Am Math Soc; Math Asn Am. Res: Operator theory; function theory; transport theory. Mailing Add: Dept of Math Va Polytech Inst & State Univ Blacksburg VA 24601

BALL, LAWRENCE ERNEST, b Akron, Ohio, Nov 11, 35. POLYMER CHEMISTRY, ORGANIC CHEMISTRY. Educ: Univ Akron, BS, 57, MS, 58, PhD(polymer org chem), 61. Prof Exp: Proj assoc, 61-64, proj leader, 64-68, RES ASSOC POLYMER SYNTHEISS, RES LAB, STANDARD OIL CO OHIO, 68- Mem: Am Chem Soc. Mailing Add: Res Dept Standard Oil Co of Ohio 4440 Warrensville Center Rd Cleveland OH 44128

BALL, M ISABEL, b Elmendorf, Tex, June 1, 29. BIO-ORGANIC CHEMISTRY. Educ: Our Lady of the Lake Col, BA, 50; Univ Tex, MA, 63, PhD, 69. Prof Exp: Teacher parochial sch, Okla, 52-54; from instr to asst prof chem, Our Lady of the Lake Col, 54-65; res assoc, Clayton Found Biochem Inst, Univ Tex, 65-68; CHMN DEPT CHEM, OUR LADY OF THE LAKE COL, 68-, DIR DIV SCI & MATH, 73- & PROF CHEM, 75- Concurrent Pos: AEC res partic, Oak Ridge Nat Lab, 69. Mem: AAAS; Am Chem Soc. Res: Synthesis and assay of analogs of metabolites; studies of Aspergillus niger using purine analogs and derivatives; metabolic changes caused by phthalic acid esters in mice and bacteria. Mailing Add: Dept of Chem Our Lady of the Lake Col 411 SW 24th St San Antonio TX 78285

BALL, MAHLON M, b Lawrence, Kans, Apr 7, 31; m 53; c 2. GEOLOGY, GEOPHYSICS. Educ: Univ Kans, BSc, 53, MSc, 58, PhD(geol), 60; Univ Birmingham, MSc, 59. Prof Exp: Geologist, Shell Develop Co, 59-62, res geologist, 62-63, sr geologist, Shell Oil Co, 63-65, div geologist, Shell Develop Co, 65-66, res assoc, 66-67; assoc prof geophys, 67-70, PROF GEOPHYS, ROSENSTIEL SCH MARINE & ATMOSPHERIC SCI, UNIV MIAMI, 70-, RES GEOPHYSICIST, INST MARINE SCI, 67- Concurrent Pos: Mem geol adv comt, US Air Force Weapons Res, 73; mem site surv comt, Int Prog Ocean Drilling, 74. Mem: Geol Soc Am; Soc Econ Paleontologists & Mineralogists; Am Geophys Union; Soc Explor Geophys. Res: Recent carbonate sediments; marine geophysical investigations of the Bahamas, Caribbean and equatorial Atlantic Ocean. Mailing Add: Rosenstiel Sch Marine & Atmos Sci Univ of Miami 4600 Rickenbacker Causeway Miami FL 33149

BALL, MARY UHRICH, b Pittsburg, Kans, May 26, 44; m 62; c 2. QUANTITATIVE GENETICS. Educ: Trinity Univ, BS, 66; Tex A&M Univ, MS, 72, PhD(genetics), 74. Prof Exp: ASST PROF ZOOL-ENTOM, AUBURN UNIV, 74- Mem: Am Genetic Asn; Genetics Soc Am; AAAS. Res: Genetics of disc disease in Dachshunds; genetics of Spodoptera exigua; relationship between bioelimination and fecundity in insects. Mailing Add: Dept of Zool Entom Auburn Univ Auburn AL 36830

BALL, RALPH HENRY, b Kelowna, BC, May 24, 05; US citizen; m 31, 56, 74; c 3. PLASTICS CHEMISTRY, POLYMER CHEMISTRY. Educ: Univ BC, BA, 26, MA, 28; McGill Univ, PhD(chem), 31. Prof Exp: Chemist, Celluloid Corp, 31-40, asst tech dir, 40-41; head plastics & resins sect, War Prod Bd, 42-43; asst tech dir plastics, Celanese Corp Am, 43-46, asst mgr div cent res dept, 47-48, mgr cellulose div, 48-50; gen mgr, Columbia Cellulose Co, 51-52; tech dir plastics div, Celanese Corp Am, 52-60, asst tech dir, Celanese Polymer Co, 60-64, dir qual control & tech progs, Resin Prods, Celanese Plastics Co, 64-70; consult polymers & plastics, 70-72; RETIRED. Mem: Am Chem Soc. Res: Cellulose research; research management. Mailing Add: 135 Canoe Brook Pkwy Summit NJ 07901

BALL, RALPH WAYNE, b Los Angeles, Calif, Oct 14, 25; m 49; c 2. MATHEMATICS. Educ: Pomona Col, BA, 47; Univ Southern Calif, MA, 52; NMex State Univ, MS, 63, PhD(math), 64. Prof Exp: Teacher, Los Angeles City Schs, 49-59; secy curric coord, Inyo County Schs, Calif, 59-61; instr math, NMex State Univ, 61-64; assoc prof, Appalachian State Teachers Col, 64-66; PROF MATH, NMEX INST MINING & TECHNOL, 66- Mem: Am Math Soc; Math Asn Am. Res: Group theory, particularly infinite symmetric groups. Mailing Add: Dept of Math NMex Inst of Mining & Technol Socorro NM 87801

BALL, RICHARD WILLIAM, b Streator, Ill, Aug 16, 23. MATHEMATICS. Educ: Univ Ill, BA, 44, MA, 45, PhD(math), 48. Prof Exp: Instr math, Univ Wash, 48-54; from asst prof to assoc prof, 54-60, PROF MATH, AUBURN UNIV, 60- Mem: Am Math Soc; Math Asn Am. Res: Finite projective geometry; theory of groups. Mailing Add: Dept of Math Auburn Univ Auburn AL 36830

BALL, ROBERT CRAGIN, b Elyria, Ohio, Dec 10, 12; m 41; c 2. ZOOLOGY. Educ: Ohio State Univ, AB, 36, MSc, 37; Univ Mich, PhD(zool), 43. Prof Exp: Fisheries biologist, Ohio Div Conserv, 36-38; aquatic biologist, State Dept Conserv, Mich, 38-

43; entomologist, USPHS, 43-45; limnologist, Inst Fisheries Res, Mich, 45-46; from asst prof to assoc prof zool & res, 46-51, PROF ZOOL & RES, MICH STATE UNIV, 51-, DIR INST WATER RES, 66- Concurrent Pos: Vis prof, Oak Ridge Inst Nuclear Studies, 57. Mem: Am Fisheries Soc; Ecol Soc Am; Am Soc Limnol & Oceanog (vpres, 56). Res: Limnology; stream ecology; pollution and pesticide biology; radioecology. Mailing Add: Inst of Water Res Mich State Univ East Lansing MI 48823

BALL, ROBERT F, poultry pathology, see 12th edition

BALL, ROBERT P, b Harlan, Ky, July 14, 02; m 27; c 2. RADIOLOGY. Educ: Univ Louisville, MD, 24. Hon Degrees: Hon DSc, Centre Col, 48. Prof Exp: Instr path, Sch Med, Univ Louisville, 24-25, sr intern surg, 25-26, actg head dept path, 28-29; surg pathologist, Cleveland Clin, 26-28; practicing surgeon, Harlan, Ky, 30-32; dir labs, Baroness Erlanger Hosp, Chattanooga, Tenn, 32-36; from asst prof to prof radiol, Col Physicians & Surgeons, Columbia Univ, 36-49; prof, Med Col, Cornell Univ, 49-51; RADIOLOGIST, OAK RIDGE HOSP, 52- Concurrent Pos: Surg pathologist, Louisville City Hosp, Ky, 28-30; practicing surgeon, Louisville, 28-30; radiologist, Presby Hosp & Vanderbilt Clin, New York, 36-49; radiologist-in-chief, New York Hosp, 49-51. Mem: Fel AMA; Am Roentgen Ray Soc; fel Am Col Radiol (vpres, 52-); fel Radiol Soc NAm; Brit Inst Radiol. Res: Clinical medicine and surgery. Mailing Add: Oak Ridge Hosp Oak Ridge IN 37830

BALL, RUSSELL MARTIN, b Chicago, Ill, Oct 28, 27; m 55; c 3. REACTOR PHYSICS, NUCLEAR PHYSICS. Educ: Univ Ill, BS, 49, MS, 50; Univ Va, PhD, 70. Prof Exp: Process engr, Elec Storage Battery Co, 46-47; asst, Univ Ill, 47-50; chief engr, Nuclear Instrument & Chem Corp, 50-55; with Oak Ridge Sch Reactor Technol, 55; SUPVR, BABCOCK & WILCOX, 55- Concurrent Pos: Lectr, Univ Wis. Mem: Am Phys Soc; Am Asn Physics Teachers; Inst Elec & Electronics Engrs; Am Nuclear Soc. Res: Isometric transitions; radiation detectors and computers; industrial application of nucleonics and isotopes; nuclear instrumentation; process control; computers. Mailing Add: 1620 Belfield Pl Lynchburg VA 24503

BALL, STANTON MOCK, b Lawrence, Kans, Aug 20, 33; m 55; c 2. GEOLOGY. Educ: Univ Kans, BS, 56, MS, 58, PhD(geol), 64. Prof Exp: Geologist, Kans State Geol Surv, 57-64; geologist, Pan Am Petrol Corp, 64-70, res scientist, Okla, 70-74, RES SCIENTIST, AMOCO PRODS CO, TEX, 74- Mem: Geol Soc Am; Am Asn Petrol Geol; Soc Econ Paleont & Mineral. Res: Stratigraphy and carbonate petrology. Mailing Add: Amoco Prods Co 2007 Running Springs Humble TX 77338

BALL, WILBUR PERRY, b Briggsdale, Colo, Jan 2, 23; m 48; c 3. AGRICULTURE. Educ: Colo State Univ, BS, 48, MEd, 53; Iowa State Univ, PhD(agr), 56. Prof Exp: Teacher pub sch, Colo, 48-53; instr agr mech, Iowa State Univ, 53-56; consult agr educ, Stanford Univ Contract, Int Coop Admin, Philippines, 56-58; from asst prof to assoc prof agr educ, 58-68, PROF AGR & EDUC, CALIF STATE UNIV, FRESNO, 68-, GRAD COORDR, SCH AGR & SCH SUPVR TEACHER EDUC, 65- Concurrent Pos: Consult rural educ, Calif State Univ, Fresno, contract, US AID, Sudan, 61-63; vis prof agr educ, Ohio State Univ, 67-68; res assoc, Nat Ctr Voc & Tech Educ, 67-68. Mem: Assoc Am Soc Agr Eng; Soil Sci Soc Am. Res: Agricultural engineering; international agriculture. Mailing Add: Sch of Agr Sci Calif State Univ Fresno CA 93740

BALL, WILLIAM, b Philadelphia, Pa, Mar 2, 21; m 54; c 2. CLINICAL MICROBIOLOGY, PARASITOLOGY. Educ: Philadelphia Col Pharm, BSc, 43, MSc, 48, DSc, 50. Prof Exp: Instr zool & entom, Philadelphia Col Pharm, 46-50; instr biol sci, Drexel Inst Technol, 50-52; parasitologist, Albert Einstein Med Ctr, 52-54; microbiologist & co-dir, Park Labs, Pa, 54-70; CO-DIR, PHILADELPHIA MED LAB, 70- Concurrent Pos: Instr, Philadelphia Col Pharm, 65-67. Mem: Am Soc Parasitol; Am Soc Microbiol; Am Pub Health Asn; Conf Pub Health Lab Dirs. Mailing Add: Philadelphia Med Lab 3190 Tremont Ave Trevose PA 19047

BALL, WILLIAM DAVID, b Newark, NJ, July 22, 37; m 65. DEVELOPMENTAL BIOLOGY. Educ: Univ Wis, BS, 58, MS, 61; Univ Chicago, PhD(zool), 65. Prof Exp: Proj asst zool, Univ Wis, 58-61; res assoc biochem, Univ Ill, 65; res assoc, Univ Wash, 65-67, asst prof zool, 67-73, res assoc zool, 73-75; ASSOC PROF ORAL BIOL, SCH DENT, UNIV COLO, 75- Concurrent Pos: NIH res grant, 69-72. Mem: Am Soc Cell Biol; Am Soc Zoologists; Soc Develop Biol. Res: Epithelio-mesenchymal interactions in development of rat salivary glands; formation of specific enzymes in embryonic tissues. Mailing Add: Oral Biol Sch of Dent C-285 Univ of Colo Med Ctr Denver CO 80220

BALL, WILLIAM LEE, b North Hatley, Que, July 18, 08; m 42; c 5. ORGANIC CHEMISTRY. Educ: McGill Univ, BSc, 31, PhD, 35. Prof Exp: Res chemist, Ayerst, McKenna & Harrison, 35-41; tech supvr nylon div, Can Industs, 41-50; indust toxicologist, Occup Health Div, Dept Nat Health & Welfare, 50-55, sr scientist, 55-62, sci adv, 69-75, sci writer, Drug Adv Bur, Food & Drug Directorate, 70-75; SCI WRITING, 75- Concurrent Pos: Consult, Can Standards Asn, 55-64; mem, Conf Govt Indust Hyg; lectr, Univ Ottawa, 57-66. Mem: Chem Inst Can. Res: Foods, drugs and hazardous chemicals. Mailing Add: 1374 Cavendish Rd Ottawa ON Can

BALL, WILLIAM PAUL, b San Diego, Calif, Nov 16, 13; m 41; c 2. NUCLEAR PHYSICS, RADIATION SCIENCE. Educ: Univ Calif, Los Angeles, AB, 40; Univ Calif, Berkeley, PhD(physics), 53. Prof Exp: High sch teacher math & phys educ, Montebello Unified Sch Dist, 41-42; instr math & physics, Santa Ana Army Air Base, 42-43; physicist, Radiation Lab, Univ Calif, 43-58 & Ramo Wooldridge Corp, 58-59; sr scientist & mgr nuclear technol staff, Hughes Aircraft Co, 59-64; SR STAFF ENGR SPACE & NUCLEAR PHYSICS, TRW SYSTEMS, 64- Mem: AAAS; Am Phys Soc; Am Nuclear Soc; Am Inst Aeronaut & Astronaut; NY Acad Sci. Res: Space and nuclear radiation environments and their effects on properties and responses of materials, sensors, electronics and systems. Mailing Add: 209 Via El Toro Redondo Beach CA 90277

BALLAL, SRIKRISHNA, b Trichur, Kerala, India, Dec 31, 38. BOTANY. Educ: Univ Madras, BS, 59, MS, 61; Univ Tenn, PhD(bot), 64. Prof Exp: Asst bot, Univ Tenn, 61-64; from asst prof to assoc prof, 65-74, PROF BIOL, TENN TECHNOL UNIV, 74- Concurrent Pos: Shell Merit fel, Stanford Univ, 70. Mem: AAAS; Bot Soc Am. Res: Nutritional physiology of fungi. Mailing Add: Dept of Biol Tenn Tech Univ Cookeville TN 38501

BALLAM, JOSEPH, b Boston, Mass, Jan 2, 17; m 38; c 2. PHYSICS. Educ: Univ Mich, BS, 39; Univ Calif, PhD, 51. Prof Exp: Physicist, US Navy Dept, 41-45; asst physics, Univ Calif, 45-46, physicist, Radiation Lab, 46-48, asst physics, 48-51; asst, Princeton Univ, 51-52, instr, 52-53, res assoc, 53-56; from assoc prof to prof, Mich State Univ, 56-61; assoc prof, 61-64, PROF PHYSICS, STANFORD UNIV, 64-, ASSOC DIR, RES DIV, LINEAR ACCELERATOR CTR, 61- Concurrent Pos: Res collabr, Brookhaven Nat Lab, 56-; Ford Found fel, Europ Orgn Nuclear Res, 60-61; Guggenheim Found fel, 71-72. Mem: Fel Am Phys Soc. Res: Elementary particles; cosmic rays; high energy physics; infrared signaling and detection. Mailing Add: Stanford Linear Accelerator Ctr Stanford Univ PO Box 4349 Stanford CA 94305

BALLANTINE, C S, b Seattle, Wash, Aug 27, 29. MATHEMATICS. Educ: Univ Wash, BS, 53; Stanford Univ, PhD(math), 59. Prof Exp: Asst math, Stanford Univ, 56-58; actg instr, Univ Calif, 58-59, instr, 59-60; from instr to assoc prof, 60-72, PROF MATH, ORE STATE UNIV, 72- Mem: Am Math Soc. Res: Matrix theory. Mailing Add: Dept of Math Ore State Univ Corvallis OR 97331

BALLANTINE, DAVID STEPHEN, b New York, NY, Dec 26, 22; m 49; c 7. ENVIRONMENTAL CHEMISTRY. Educ: St Francis Col, BS, 43; Univ Notre Dame, PhD(chem), 52. Prof Exp: Jr chemist, Clinton Lab, 44-46; res chemist, Kellex Corp, 49-51; chemist, Brookhaven Nat Lab, 51-67; sr scientist, 67-75, ATMOSPHERIC SCIENTIST, US ENERGY RES & DEVELOP ADMIN, 75- Mem: Am Chem Soc; Am Soc Test & Mat; Sigma Xi. Res: Atmospheric science; environmental pollution; applied radiation chemistry. Mailing Add: Box 203 Rte 8 Frederick MD 21701

BALLANTINE, LARRY GENE, b Webster City, Iowa, July 6, 44; m 73; c 1. AGRICULTURAL CHEMISTRY. Educ: Iowa State Univ, BS, 66; Univ Mass, PhD(soil chem), 71. Prof Exp: Teacher chem, Essex Agr & Tech Inst, 71-73; asst prof soil chem, State Univ NY Agr & Tech Col, Cobleskill, 73-74; ENVIRONMENTALIST, CIBA-GEIGY CORP, 74- Mem: Am Chem Soc; Am Soc Agron; Sigma Xi. Res: Environmental research; emphasis on pesticide usage. Mailing Add: 3 Cranwood Ct Greensboro NC 27405

BALLANTYNE, DAVID JOHN, b Victoria, BC, May 1, 31; m 62; c 3. HORTICULTURE, PLANT PHYSIOLOGY. Educ: Univ BC, BCom, 54; Wash State Univ, MS, 57; Univ Md, PhD(hort), 60. Prof Exp: Nat Res Coun Can fel, 60-61; res officer plant physiol, Exp Farm, Can Dept Agr, 61-63; asst prof, 62-68, ASSOC PROF PLANT PHYSIOL, UNIV VICTORIA, BC, 68- Mem: Am Soc Plant Physiol; Am Soc Hort Sci; Can Soc Plant Physiol. Res: Effect of air pollutants on plant metabolism; pollen physiology. Mailing Add: Dept of Biol Univ of Victoria Victoria BC Can

BALLANTYNE, DONALD LINDSAY, JR, b Peking, China, Nov 8, 22; m 52; c 3. TRANSPLANTATION BIOLOGY. Educ: Princeton Univ, AB, 45; Cath Univ Am, MS, 48, PhD(biol), 52. Prof Exp: Parasitologist, E R Squibb & Sons, 51-53 & Ill State Psychopath Inst, 53-54; from asst prof to assoc prof, 62-75, RES ASSOC TISSUE TRANSPLANTATION, BELLEVUE HOSP, 54-; PROF EXP SURG, INST RECONSTRUCT PLASTIC SURG, NY UNIV MED CTR, 75- Mem: AAAS; Transplantation Soc; Am Soc Plastic & Reconstruct Surg; Reticuloendothelial Soc. Mailing Add: Dept of Exp Surg NY Univ Med Ctr 550 First Ave New York NY 10016

BALLANTYNE, JOHN HUBER, anatomy, deceased

BALLARD, BERTON ETIENNE, b Oakland, Calif, Aug 28, 28; m 53; c 4. PHARMACEUTICAL CHEMISTRY. Educ: Univ Calif, Berkeley, BA, 51; Univ Calif, San Francisco, BS, 55, PharmD, 56, PhD(pharmaceut chem), 61. Prof Exp: USPHS trainee neuropharmacol, Univ Calif, San Francisco, 61-62, from asst prof to assoc prof pharm & pharmaceut chem, 62-74; MEM FAC, DEPT PHARM, UNIV RI, 74- Concurrent Pos: Mem, Gen Comt Rev, US Pharmacopeia, 70- Mem: AAAS; Am Pharmaceut Asn; Am Asn Cols Pharm; NY Acad Sci. Res: Biopharmaceutics; kinetics of drug absorption, distribution, metabolism and excretion. Mailing Add: Dept of Pharm Univ of RI Kingston RI 02881

BALLARD, HAROLD NOBLE, b Little Rock, Ark, Feb 11, 19; m 45, 60; c 4. ATMOSPHERIC PHYSICS. Educ: Univ Tex, El Paso, BS, 48; Tex A&M Univ, MS, 50. Prof Exp: Res physicist, NMex Inst Mining & Technol, 50; asst prof physics, Univ Tex, El Paso, 50-54; physicist, Los Alamos Sci Lab, 54-57; res physicist, Univ Tex, El Paso, 57-64; RES PHYSICIST, ATMOSPHERIC SCI LAB, US ARMY ELECTRONICS COMMAND, WHITE SANDS MISSILE RANGE, 64- Concurrent Pos: Mem, Joint Comt Space Environ Forecasting, 68-; assoc prof & lectr, Univ Tex, El Paso, 70- Honors & Awards: 20 Commendations, Dept Army, 64-75; Commendation, NSF, 66; Cert of Appreciation, US AEC, 75. Mem: Am Phys Soc; Am Meteorol Soc; Am Geophys Union; AAAS. Res: Composition and thermal structure of stratosphere; multiple instrument balloon-borne experiments; development of rocket and balloon-borne atmospheric sensory instruments. Mailing Add: Atmospheric Sci Lab US Army Electronics Command White Sands Missile Range NM 88002

BALLARD, IAN MATHESON, b Norristown, Pa, Jan 13, 37; m 71; c 3. INTERNAL MEDICINE, CLINICAL PHARMACOLOGY. Educ: Lafayette Col, AB, 58; Temple Univ, MD, 62. Prof Exp: From intern to resident internal med, St Lukes Hosp, Bethlehem, Pa, 62-66; pvt pract, 66-71; assoc dir clin res, Merck Sharp & Dohme Res Labs, 71-73; ASSOC DIR CLIN RES, WYETH LABS, 73- Concurrent Pos: Mem, Consult Staff Clin Pharm, St Lukes Hosp, 65 & Attend Staff Internal Med, Bryn Mawr Hosp, 75; mem consult staff clin pharm, Montgomery Hosp, Norristown, Pa, 72 & US Naval Hosp, Philadelphia, 74. Mem: AMA; fel Am Acad Family Physicians; Am Rheumatism Asn. Res: Nonsteroidal anti-inflammatory agents; prostaglandins. Mailing Add: 308 Devon State Rd Devon PA 19333

BALLARD, KATHRYN WISE, b Waverly Hills, Ky, June 10, 30. CARDIOVASCULAR PHYSIOLOGY. Educ: Howard Univ, BS, 51; Univ Mich, MS, 53; Western Reserve Univ, MS, 59; Univ Southern Calif, PhD(physiol), 67. Prof Exp: Res asst physiol, Western Reserve Univ, 59 & Univ Southern Calif, 59-67; NIH fel, Sch Med, Univ Southern Calif, 67-68; trainee & vis scientist, Dept Pharmacol, Karolinska Inst, Sweden, 68-70, fel, 70; RES ASSOC, CARDIOVASC RES LAB, UNIV SOUTHERN CALIF, 70-, ASST PROF PHYSIOL, SCH MED, 71- Mem: AAAS; Microcirc Soc; Am Asn Univ Prof; Am Physiol Soc; Asn Women Sci. Res: Effects of various stimuli on nutritive and total blood flow in skeletal muscle and skin; vascular and metabolic responses in canine adipose tissue; effects of sympathetic nerve stimulation and of catecholamines. Mailing Add: Cardiovasc Res Lab Univ of Southern Calif Los Angeles CA 90033

BALLARD, KENNETH J, b Highland Park, Ill, Jan 22, 30; m 56; c 2. PHARMACY. Educ: Univ Calif, Berkeley, BA, 50, BSc, 54, PharmD, 55. Prof Exp: Staff pharmacist, USPHS Hosp, San Francisco, 55-57; sr asst pharmacist, Div Biologics Standards, Lab Blood & Blood Prod, NIH, 57-58; asst prof pharm, Sch Pharm & Grad Div Arts & Sci, Univ Buffalo, 58-60; assoc prof, Col Pharm, Northeastern Univ, 60-64; res assoc, Allergan Pharmaceut, 64-66; dir pharm serv, Student Health Serv, Univ Calif, Los Angeles, 66-70; ASST CLIN PROF PHARM & DIR CLIN PHARM SERV, LOS ANGELES COUNTY/UNIV SOUTHERN CALIF MED CTR, 70- Mem: Am Pharmaceut Asn; Am Soc Hosp Pharmacists. Res: Developing innovations in delivery of clinical pharmacy services; education of clinical pharmacists. Mailing Add: Dept of Pharm Los Angeles County Univ of Southern Calif Med Ctr 1200 N State St Los Angeles CA 90033

BALLARD, LARRY EUGENE, physical chemistry, see 12th edition

BALLARD, LEWIS FRANKLIN, b Mooresville, NC, Dec 17, 34; m 61; c 3. SOLID STATE PHYSICS, ELECTRICAL ENGINEERING. Educ: NC State Univ, BS, 58; Duke Univ, MS, 67, PhD(elec eng), 69. Prof Exp: Rocket engr, Thiokol Chem Corp, 58-59; nuclear engr, Res Triangle Inst, 59-65, staff scientist, 65-69, supvr anal chem & physics sect, 69-73; PRES, NUTECH CORP, 73- Mem: Inst Elec & Electronics Engrs; Am Phys Soc. Res: Electrical and optical properties of insulators, semiconductors, and organometallics; design of radiation detectors; applications of nuclear techniques in industry and highway engineering; ambient air monitoring instrumentation. Mailing Add: Nutech Corp 2806 Cheek Rd Durham NC 27704

BALLARD, MARGUERITE CANDLER, b Atlanta, Ga, June 14, 20. INTERNAL MEDICINE, HEMATOLOGY. Educ: Vassar Col, BA, 42; Emory Univ, MS, 43, MD, 48. Prof Exp: Bacteriologist, US War Dept, 43-44; intern & med resident, Md Gen Hosp, 48-51; exam physician, Lockheed Aircraft Corp, 51-52; SR HEMATOLOGIST, LAB TRAINING & CONSULT DIV, BUR LABS, CTR DIS CONTROL, USPHS, 54- Mem: AAAS; Sigma Xi; AMA; Am Pub Health Asn; Am Soc Hemat. Res: Development and presentation of lectures and benchtraining courses in clinical laboratory; hematology consultant in laboratory improvement program. Mailing Add: Lab Training & Consult Div Bur of Labs CDCPHS-US DHEW Atlanta GA 30333

BALLARD, NEIL BRIAN, b Mankato, Minn, Jan 26, 38. ZOOLOGY, PARASITOLOGY. Educ: Mankato State Col, BS, 59; Colo State Univ, MS, 65, PhD(zool), 68. Prof Exp: ASSOC PROF BIOL, MANKATO STATE COL, 68- Concurrent Pos: Brown-Hazen Fund grant, 69-70. Mem: Soc Protozool; Am Soc Parasitol; Wildlife Disease Asn. Res: Host-specificity and immunity phenomena with regard to coccidian infections in rodents. Mailing Add: Dept of Biol Mankato State Col Mankato MN 56001

BALLARD, RALPH CAMPBELL, b Washington, DC, Jan 4, 26; m 49; c 2. ENTOMOLOGY. Educ: George Washington Univ, BS, 51; Ohio State Univ, MS, 53; Rutgers Univ, PhD(entom), 56. Prof Exp: Assoc prof biol, Pac Union Col, 56-58; assoc prof, 58-69, dir res, 65-66, PROF BIOL, SAN JOSE STATE UNIV, 69- Concurrent Pos: Environ toxicol fel, Univ Calif, Davis, 75-76. Mem: AAAS; NY Acad Sci; Entom Soc Am; Am Soc Zoologists. Res: General physiology; toxicology. Mailing Add: 1015 Roy Ave San Jose CA 95125

BALLARD, ROBERT WILSON, b Trenton, NJ, Mar 4, 22; m 58; c 6. MEDICINE. Educ: Cornell Univ, 40-43; New York Med Col, MD, 47. Prof Exp: Gen pract, Mt St Mary's Hosp, Nelsonville, Ohio, 49-50, 55-59; asst med dir, White Labs, NJ, 59-60; asst dir clin res, Warner-Lambert Res Inst, NJ, 60-61; med dir, Pitman-Moore Co, Ind, 61-63; vpres & dir med res, Winthrop Labs Div, Sterling Drug, Inc, 63-66; exec dir med div, McNeil Labs, Inc, 66-69; MED DIR, VASSAR BROS HOSP, 69- Concurrent Pos: Fel, Duke Univ & Oak Ridge Inst Nuclear Studies, 51-52; gen pract, Mem Hosp, Syracuse, NY, 55-59; pvt pract, Wappingers Med Group, 69- Mem: AMA; Am Col Clin Pharmacol & Therapeut. Res: Radiation biological research; clinical pharmacology; medico-legal aspects of experimental therapy. Mailing Add: Vassar Bros Hosp Poughkeepsie NY 12601

BALLARD, STANLEY SUMNER, b Los Angeles, Calif, Oct 1, 08; m 35, 74; c 2. PHYSICS, OPTICS. Educ: Pomona Col, AB, 28; Univ Calif, MA, 32, PhD(physics), 34. Hon Degrees: DSc, Pomona Col, 74. Prof Exp: Asst physics, Dartmouth Col, 28-30; asst, Univ Calif, 30-34, res fel, 34-35; from instr to asst prof, Univ Hawaii, 35-41; prof physics & chmn dept, Tufts Univ, 46-54; res physicist, Scripps Inst, Univ Calif, 54-58; chmn dept physics & astron, 58-71, dir div phys & math sci, 68-71, PROF PHYSICS, UNIV FLA, 58- Concurrent Pos: Consult exp sta, Hawaiian Sugar Planters Asn, 37-40; collabr, Hawaii Agr Exp Sta & res assoc, Hawaii Nat Park, 39-41; consult physicist, Polaroid Corp, Mass, 46-54 & Baird Assocs, Inc, 46-51; US deleg int comn optics, Int Union Pure & Appl Physics, Delft, 48, London, 50, Madrid, 53, Stockholm, 59, Munich, 62, Paris, 66, Reading, Eng, 69, Calif, 72 & Czech, 75, vpres, 48-56, pres, 56-59, chmn US Nat Comt, 48-56; consult, Rand Corp, 52-53, physicist, 53-54, consult, 54-71; mem vision comt, Armed Forces-Nat Res Coun, exec secy, 56-59; consult, Inst Sci & Technol, Univ Mich, 57-65, Bendix Systs Div, 57-59, Perkin-Elmer Corp, 60-62 & Atlantic Missile Range, 60-63; consult astrionics div, Aerojet-Gen Corp, 61-66; mem bd dirs, Am Inst Physics, 62-65 & 67-73; mem phys sci div, Nat Res Coun, 63-65; consult, Douglas Advan Res Labs, 66-70, Aerospace Corp, 67-70 & Honeywell Radiation Ctr, 69. Honors & Awards: Royal Order of the North Star, King of Sweden, 72; Distinguished Serv Citation, Am Asn Physics Teachers, 75. Mem: AAAA (secy physics sect, 60-67, vpres, 68); fel Am Phys Soc; fel Optical Soc Am (pres, 63); Am Asn Physics Teachers (pres, 68-69); fel Brit Phys Soc. Res: Spectroscopy; optical and infrared instrumentation; properties of optical materials. Mailing Add: Dept of Physics & Astron Univ of Fla Gainesville FL 32611

BALLARD, WILLIAM WHITNEY, b Griswoldville, Mass, Apr 4, 06; m 38; c 4. VERTEBRATE EMBRYOLOGY, EXPERIMENTAL MORPHOLOGY. Educ: Dartmouth Col, BS, 28; Yale Univ, PhD(zool), 33. Prof Exp: Asst biol, Yale Univ, 28-30; instr zool, 30-35, instr embryol, Med Sch, 34-35, from asst prof to prof zool & anat, 35-71, EMER PROF & ADJ RES PROF ZOOL & ANAT, MED SCH, DARTMOUTH COL, 71- Concurrent Pos: Instr, Marine Biol Lab, Woods Hole, 38-41, 45 & 49. Mem: Am Soc Zool; Soc Develop Biol; Soc Study Evolution. Res: Amphibian anatomy; embryology of amblystoma; comparative vertebrate embryology; morphogenetic movements in fish embryos. Mailing Add: Dept of Biol Dartmouth Col Hanover NH 03755

BALLENTINE, ALVA RAY, b Irmo, SC, Feb 19, 31. PHYSICAL ORGANIC CHEMISTRY, SYNTHETIC ORGANIC CHEMISTRY. Educ: Univ SC, BS, 50, MS, 52, PhD(org chem), 56. Prof Exp: Asst prof chem, Presby Col (SC), 55-56; from asst prof to assoc prof, 56-68, PROF CHEM, THE CITADEL, 68- Mem: Am Chem Soc. Res: Reaction mechanisms; organic synthesis. Mailing Add: Dept of Chem The Citadel Charleston SC 29409

BALLENTINE, LESLIE EDWARD, b Wainwright, Alta, Apr 19, 40. PHYSICS. Educ: Univ Alta, BSc, 61, MSc, 62; Cambridge Univ, PhD(physics), 66. Prof Exp: Fel, Theoret Phys Inst, Univ Alta, 65-67; from asst prof to assoc prof physics, 67-73, PROF PHYSICS, SIMON FRASER UNIV, 73- Mem: Can Asn Physicists. Res: Electron properties of liquid metals; quantum mechanics. Mailing Add: Dept of Physics Simon Fraser Univ Burnaby BC Can

BALLENTINE, ROBERT, b Orange, NJ, Mar 6, 14; m 68. BIOCHEMISTRY, MICROBIAL ECOLOGY. Educ: Princeton Univ, AB, 37, PhD(biol), 40. Prof Exp: Proctor fel, Princeton Univ, 40-41; Nat Res fel zool, Rockefeller Inst, 41-42; lectr, Columbia Univ, 42-43, instr, 43-48; scientist, Brookhaven Nat Lab, 48-49; ASSOC PROF BIOL, JOHNS HOPKINS UNIV, 49- Concurrent Pos: Guggenheim fel, Calif Inst Technol, 47; assoc res specialist, Rutgers Univ, 48-49. Mem: AAAS; Am Chem Soc; Soc Gen Physiol. Res: Radioisotope assay; inorganic ion accumulation; phylogeny by DNA hybridization; surface colonization by bacteria in estuarine environments. Mailing Add: Dept of Biol Johns Hopkins Univ Baltimore MD 21218

BALLERT, ALBERT GEORGE, b Toledo, Ohio, Dec 26, 14; c 3. GEOGRAPHY OF THE GREAT LAKES, ECONOMIC GEOGRAPHY. Educ: Univ Toledo, BEd, 38; Syracuse Univ, MS, 40; Univ Chicago, PhD(geog), 47. Prof Exp: Instr geog, Univ Chicago, 42-45; mgr transp & world trade, Toledo Chamber Com, 45-47; asst prof geog, Univ Calif, Los Angeles, 47-48; chief res econ, Mich Dept Econ Develop, 48-50; head res, Chicago Plan Comm, 50-53; metrop area analyst, Ford Div, Ford Motor Co, 53-55; area studies supvr, Chrysler Div, Chrysler Corp, 55-56; RES DIR, GREAT LAKES COMN, 56- Concurrent Pos: Instr, Univ Chicago Home-Study, 53-64; prof, Easterm Mich Univ, 62-; lectr, Univ Ctr Adult Educ, Univ Mich, 72. Mem: Asn Am Geog; Am Soc Planning Off; Sigma Xi; Int Asn Gt Lakes Res. Res: Great Lakes interlakes and overseas trade and shipping; Great Lakes resources and water quality control; traffic of major world waterways; geography of Canada. Mailing Add: Gt Lakes Comn 2200 Bonisteel Blvd Ann Arbor MI 48105

BALLEW, DAVID WAYNE, b Mangum, Okla, Aug 20, 40; m 60; c 1. NUMBER THEORY. Educ: Univ Okla, BA, 62, MA, 64; Univ Ill, PhD(math), 69. Prof Exp: Asst math, Univ Okla, 62-64 & Univ Ill, 64-67; asst prof, 67-75, PROF MATH, SDAK SCH MINES & TECHNOL, 75- Mem: Am Math Soc; Math Asn Am. Res: Extensions of results in commutative ring theory to results in noncommutative ring theory, especially orders and separable algebras; application of computers to number theory. Mailing Add: Dept of Math SDak Sch of Mines & Technol Rapid City SD 57701

BALLIF, JAE R, b Provo, Utah, July 5, 31; m 53; c 7. SOLAR PHYSICS, TERRESTRIAL PHYSICS. Educ: Brigham Young Univ, BS, 53; Univ Calif, Los Angeles, MA, 61, PhD(atmospheric physics), 62. Prof Exp: From asst prof to assoc prof, 62-70, PROF PHYSICS, BRIGHAM YOUNG UNIV, 70-, DEAN PHYS & MATH SCI, 74- Mem: Am Geophys Union. Res: Atmospheric airglow, particularly nocturnal emission of sodium and hydroxyl emissions; interplanetary magnetic field structure, its solar origin and terrestrial effects. Mailing Add: Off of the Dean of Phys Sci Brigham Young Univ Provo UT 84602

BALLIN, JOHN CHRISTIAN, b Chicago, Ill, Dec 3, 24; m 51; c 3. CLINICAL PHARMACOLOGY. Educ: Univ Chicago, PhB, 50, MS, 53, PhD(pharmacol), 55. Prof Exp: Res asst pharmacol, Univ Chicago, 52-55; asst secy, Coun Drugs, 55-59, ASST DIR DIV SCI ACTIVITIES, AMA, 59-, DIR DEPT DRUGS, 71- Mem: AAAS; Am Soc Clin Pharmacol & Therapeut; Am Fedn Clin Res; Asn Am Med Cols; AMA. Res: Administration; medical writing. Mailing Add: Am Med Asn 535 N Dearborn Chicago IL 60610

BALLINA, RUDOLPH AUGUST, b New York, NY, Nov 30, 34; m 67; c 1. ORGANIC CHEMISTRY. Educ: St John's Univ, NY, BS, 56, MS, 61, PhD(org chem), 65. Prof Exp: Anal chemist, US Treas Dept, 59-60; PROD DEVELOP CHEMIST, TOMS RIVER CHEM CORP, 65- Mem: Am Chem Soc. Res: Natural products, specifically peptides and polypeptides; organic synthesis; dyestuffs. Mailing Add: 30 Smith Rd Toms River NJ 08753

BALLING, JAN WALTER, b Louisville, Ky, Apr 8, 34. ANIMAL BEHAVIOR. Educ: Univ Louisville, BA, 56, PhD(radiation biol), 64; Purdue Univ, MS, 57. Prof Exp: Instr biol, Univ Louisville, 60-64; dir labs, Cent State Hosp, Louisville, 64-65; asst prof biol, George Peabody Col, 65-68; PROF BIOL, CALIFORNIA STATE COL, PA, 68- Mem: Am Soc Parasitol; Am Nuclear Soc; Am Soc Ichthyologists & Herpetologists; Animal Behav Soc. Res: Turtle ethology; operant behavior of rats; trematode systematics; chemical protection of cold-blooded vertebrates against radiation damage; operant behavior of chronic schizophrenics. Mailing Add: Dept of Biol California State Col California PA 15419

BALLINGER, CARTER M, b Lewistown, Mont, Mar 9, 22; m 47; c 5. ANESTHESIOLOGY. Educ: Univ Iowa, BA, 43, MD, 45; Univ Pa, MMedSci, 61; Am Bd Anesthesiol, dipl, 55. Prof Exp: Intern, Fresno County Gen Hosp, Calif, 45-46; asst resident surg, Crile Vet Admin Hosp, Cleveland, Ohio, 48-S0; resident anesthesiol, Columbia Univ, 51-53, instr, 53-54; chmn div anesthesiol, Univ Utah, 54-73, anesthesiologist-in-chief, 65-73, assoc prof anesthesiol, 54-76; PROF ANESTHESIOL & CHMN DEPT, NJ MED SCH, COL MED & DENT, 76- Concurrent Pos: Examr, Vet Admin Regional Off, Ohio, 48; anesthetist, Assoc Anesthetists Cleveland, 52; asst attend, Presby Hosp, NY & adj attend, Valley Hosp, Ridgewood, NJ, 53-54; chmn div anesthesiol, Salt Lake County Hosp, Utah, 54-65; consult, Salt Lake Vet Admin Hosp, 54-75, Shriners Hosp, 57-72, Primary Childrens Hosp, 67 & McKay Hosp, Ogden, Utah, 70-75; secy-treas, Biennial Western Conf Anesthesiol, 61-65, pres, 65-67, prog chmn, 67-69; NIH spec fel neurophysiol, Univ Wis, 62-63; vis lectr, Cook County Hosp & var med schs, Japan & Korea, 63; pres, Guedel Mem Anesthesia Libr, San Francisco, Calif, 68-70. Mem: AAAS; AMA; Am Soc Anesthesiol; Asn Univ Anesthetists; Int Asn Study Pain. Res: Antiemetic and anesthetic agents in dogs and man; blood transfusion and intravenous therapy; mechanism of action anesthetic and convulsive agents; treatment of chronic pain; training of general practitioners in anesthesiology; development of a disposable anesthesia bacterial filter system. Mailing Add: Dept of Anesthesiol NJ Med Sch Col Med & Dent 65 Bergen St Newark NJ 07107

BALLINGER, PETER RICHARD, b London, Eng, Oct 15, 32; US citizen; m 59, 70; c 3. ORGANIC CHEMISTRY. Educ: Univ London, BSc, 54, PhD(org chem), 57. Prof Exp: Fel phys org chem, Cornell Univ, 57-58, instr chem, 58-59; res chemist petrol prod, Calif Res Corp, 59-64, sr res chemist, 64-67, SR RES ASSOC PETROL PROD, CHEVRON RES CO, 67- Concurrent Pos: AEC fel, 57-59. Mem: Am Chem Soc. Res: Mechanism of combustion of hydrocarbons; chemistry of organometallic compounds and detergents; composition of automobile exhaust; mechanism of detergency; detailed analysis of hydrocarbon fuels. Mailing Add: Chevron Res Co Standard Ave Richmond CA 94802

BALLINGER, WALTER ELMER, b Burlington, NJ, Jan 22, 26; m 62; c 3. HORTICULTURE. Educ: Rutgers Univ, BS, 52; Mich State Univ, Univ, MS, 55, PhD(hort), 57. Prof Exp: Res asst, Mich State Univ, 54-57; from asst prof to assoc prof hort, 57-65, PROF HORT, NC STATE UNIV, 65-, COORDR ACAD AFFAIRS, DEPT HORT SCI, 73- Concurrent Pos: Guest lectr, Peach Cong, Verona, Italy, 65 & Univ Berlin, 65. Mem: Am Soc Hort Sci; Int Soc Hort Sci. Res: Pre- and post-harvest physiology of blueberries, especially anthocyanin characterization, physical and chemical measurement of quality and maintenance of quality. Mailing Add: Dept of Hort Sci NC State Univ Raleigh NC 27607

BALLINGER, WALTER F, II, b Philadelphia, Pa, May 16, 25; m 53; c 3. SURGERY. Educ: Univ Pa, MD, 48. Prof Exp: From asst to assoc prof surg, Jefferson Med Col, 56-64; assoc prof, Sch Med, Johns Hopkins Univ, 64-67; BIXBY PROF SURG & CHMN DEPT, SCH MED, WASH UNIV, 67- Concurrent Pos: Markle scholar acad med, 61-66; surgeon-in-chief, Barnes & Allied Hosps, 67- Mem: Am Col Surg; Soc Univ Surg; Soc Clin Surg; Soc Vascular Surg. Res: Gastrointestinal, vascular and metabolic research. Mailing Add: Dept of Surg Barnes & Allied Hosps St Louis MO 63110

BALLMAN, RICHARD LEA, b St Louis, Mo, Nov 1, 25; m 52; c 4. RHEOLOGY. Educ: Mass Inst Technol, BS, 46. Prof Exp: Chemist plastics res, Firestone Tire & Rubber Co, 46-48; chemist plastics res, 48-60, res specialist, 60-66, sci fel, 66-73, SR SCI FEL PLASTICS RES, MONSANTO CO, 73- Mem: Am Chem Soc; Soc Rheology; Brit Soc Rheology; Ger Rheology Soc. Res: Mechanical properties of polymers; rheology of polymer melts and solutions; polymer processing; spinning of synthetic fibers. Mailing Add: Tech Ctr Monsanto Co PO Box 12830 Pensacola FL 32575

BALLMANN, DONALD LAWRENCE, b Dayton, Ohio, Apr 25, 27. GEOLOGY. Educ: St Joseph's Col, Ind, AB, 54; Univ Ill, BS, 55, MS, 56, PhD(geol), 59. Prof Exp: Asst prof geol, St Joseph's Col, Ind, 56-63, dir geol field camp, 56-60, dir undergrad res, 58-60, asst dir develop, 61-62, exec asst to pres, 62-63, assoc prof geol, 63-72, acad dean, 63-68, dir develop found & govt rels, 68-72, SR GEOLOGIST, DAMES & MOORE, 73- Mem: AAAS; Am Asn Petrol Geol; Geol Soc Am; Asn Prof Geol Scientists; Soc Econ Paleontologists & Mineralogists. Res: Areal geology; geomorphology; historical geology; engineering geology; soil mechanics. Mailing Add: 1780 Sunnyview Rd Libertyville IL 60048

BALLOU, CLINTON EDWARD, b King Hill, Idaho, June 18, 23; m 49; c. BIOCHEMISTRY. Educ: Ore State Col, BS, 44; Univ Wis, MS, 48, PhD(biochem), 50. Prof Exp: USPHS res fels, Univ Edinburgh, 50-51 & Univ Calif, Berkeley, 51-52; asst res biochemist, 52-55, from asst prof to assoc prof biochem, 55-61, chmn dept, 64-68, PROF BIOCHEM, UNIV CALIF, BERKELEY, 61- Concurrent Pos: NSF sr fel, 61-62; Guggenheim fel, Kyoto & Paris, 68-69; vis scientist, Basel Inst Immunol, 75-76. Mem: Nat Acad Sci; AAAS; Am Chem Soc; Am Soc Biol Chemists. Res: Structure and biosynthesis of complex polysaccharides, lipids and lipopolysaccharides and the role these substances play in the biochemistry of the cell envelope. Mailing Add: Dept of Biochem Univ of Calif Berkeley CA 94720

BALLOU, DAVID PENFIELD, b New Britain, Conn, Aug 25, 42; m 72; c 2. BIOCHEMISTRY, ENZYMOLOGY. Educ: Antioch Col, BS, 65; Univ Mich, MS, 67, PhD(chem), 71. Prof Exp: Asst res biochemist, 71-72, instr, 72-74, ASST PROF BIOCHEM, UNIV MICH, ANN ARBOR, 74- Concurrent Pos: Consult, Henry Ford Hosp, Detroit, 72- Res: Study of oxygen activation in biological systems; reactions of oxygen with flavins and cytochromes employing computerized rapid kinetics, absorption, fluorescence and magnetic resonance spectroscopy; development of instrumentation for the above. Mailing Add: Dept of Biol Chem Univ of Mich Ann Arbor MI 48109

BALLOU, DONALD HENRY, b Chester, Vt, Mar 28, 08; m 33; c 1. MATHEMATICS. Educ: Yale Univ, AB, 28; Harvard Univ, AM, 31, PhD(math), 34. Prof Exp: Instr, Harvard Univ, 33-34; instr math, Ga Inst Technol, 34-35, asst prof, 35-42; instr math, 42-47, from assoc prof to prof, 47-64, Beman prof, 64-71, prof, 71-75, head dept, 54-68, CHARLES A DANA EMER PROF MATH, 75- Concurrent Pos: Vis assoc prof, Yale Univ, 51-52. Mem: Am Math Soc; Math Asn Am. Res: Laplace transforms; roots of polynomial equations. Mailing Add: 27 Weybridge Middlebury VT 05753

BALLOU, JACK WAYNE, b Buffalo, NY, Oct 31, 13; m 43. POLYMER PHYSICS. Educ: Univ Buffalo, BA, 39; Johns Hopkins Univ, PhD(physics), 43. Prof Exp: Res physicist, E I du Pont Co, Buffalo, 42-43; res assoc, Underwater Sound Lab, Harvard Univ, 43-44, sect leader, 44-45; assoc prof eng res, Pa State Col, 45; res supvr, 45-58, RES FEL, E I DU PONT DE NEMOURS & CO, 58- Concurrent Pos: Instr, Univ Buffalo, 47. Mem: Fiber Soc. Res: Design of acoustic devices; underwater noise measurements; torpedo control; mechanical properties of high polymers and textiles; thermal properties of fabrics and body comfort. Mailing Add: Pioneering Res Lab Exp Sta E I du Pont de Nemours & Co Wilmington DE 19898

BALLOU, JOHN EDGERTON, b King Hill, Idaho, Oct 25, 25; m 52; c 3. INHALATION TOXICOLOGY. Educ: Ore State Col, BS, 50; Univ Rochester, PhD(toxicol), 69. Prof Exp: Biol scientist, Hanford Atomic Prod Oper, Gen Elec Co, 51-64; biol scientist, 65-69, SR SCIENTIST, PAC NORTHWEST LAB, BATTELLE MEM INST, 69- Mem: Am Chem Soc; Health Physics Soc. Res: Absorption, distribution and retention of inhaled radioisotopes. Mailing Add: Biol Sect Pac NW Lab Battelle Mem Inst Richland WA 99532

BALLOU, KENNETH DOUGLAS, organic chemistry, see 12th edition

BALLOU, NATHAN ELMER, b Rochester, Minn, Sept 28, 19; m 45, 73; c 2. NUCLEAR CHEMISTRY. Educ: Minn State Teachers Col, BS, 41; Univ Ill, MS, 42; Univ Chicago, PhD(chem), 47. Prof Exp: Asst chem, Univ Ill, 41-42; asst, Metall Lab, Univ Chicago, 42-43; asst, Clinton Labs, Oak Ridge, 43-44, assoc chemist, 45-46; res chemist, Hanford Eng Works, Wash, 44-45; assoc, Radiation Lab, Univ Calif, 47-48; head nuclear chem br, Naval Radiol Defense Lab, 48-69; mem staff, 69-74, SR RES MGR, PAC NORTHWEST LABS, BATTELLE MEM INST, 74- Concurrent Pos: Head chem dept, Belgian Nuclear Ctr, 59-60; mem comt nuclear sci, Nat Acad Sci-Nat Res Coun, chmn subcomt radiochem; mem exec comt, Div Nuclear Chem & Technol, Am Chem Soc, 69- Mem: AAAS; Am Chem Soc. Res: Fission process; radiochemical techniques; ultrasensitive analytical techniques. Mailing Add: Battelle-Northwest PO Box 999 Richland WA 99352

BALMAT, JEAN L, analytical chemistry, see 12th edition

BALMER, CLIFFORD EARL, b Brownstown, Pa, Sept 17, 21; m 44; c 3. ORGANIC CHEMISTRY. Educ: Albright Col, BS, 47. Prof Exp: Org chemist, Pioneer Res Labs, US Army Qm Depot, 47-53; chemist, Appln Labs, 53-59, tech sales, 59-68, rep, 68-70, tech sales rep, Indust Chem Dept, 70-73, TECH SALES REP, PLASTICS INT DEPT, ROHM AND HAAS CO, 73- Mem: Soc Plastics Engrs; Am Chem Soc. Res: Development of plasticizers for polyvinyl chloride. Mailing Add: Rohm and Haas Co 40 Grove St Wellesley MA 02181

BALMER, LOUIS WHITESIDE, b Ardmore, Pa, Nov 25, 16; m 45; c 3. ORGANIC CHEMISTRY. Educ: Wagner Col, BS, 38; Pa State Univ, MS, 52. Prof Exp: Chemist, L A Dreyfus Co, 38-41; instr chem, Rutgers Univ, 45-49; asst prof, 48-58, ASSOC PROF CHEM, PA STATE UNIV, BEHREND CAMPUS, 58- Mem: AAAS; Am Chem Soc. Res: Colorimetric determination of iron; attitudes of entering college students toward chemistry. Mailing Add: Pa State Univ Behrend Campus RD 6 Erie PA 16510

BALMFORTH, DENNIS, b Pudsey, Eng, Apr 29, 30; m 56; c 3. CHEMISTRY. Educ: Univ Leeds, BSc, 54, PhD(color chem, dyeing), 61. Prof Exp: Res chemist, James Anderson & Co, Ltd, Scotland, 54-56; chemist, Courtaulds Ltd, Eng, 57-58 & 61-63; supvr dyeing res, Chemstrand Co, 63-67; HEAD CHEM DIV, INST TEXTILE TECHNOL, 67- Mem: Royal Inst Chem; Brit Soc Dyers & Colourists. Res: Color chemistry; fine structure of fibrous materials and its influence on dyeing and finishing behavior. Mailing Add: Inst of Textile Technol PO Box 391 Charlottesville VA 22902

BALOG, GEORGE, b Gunn, Wyo, Oct 12, 28; m 51; c 4. LASERS. Educ: Univ Wyo, BS, 50; Ill Inst Technol, MS, 54, PhD(chem), 58. MEM STAFF, LOS ALAMOS SCI LAB, 56- Mem: Am Chem Soc. Res: Polarographic studies in non-aqueous solvents; refractory materials, specifically the thermionic emission microscopy; high temperature x-ray studies; chemical lasers; dye lasers and laser dye development. Mailing Add: 4 Erie Lane Whiterock Sta Los Alamos NM 87545

BALOGH, CHARLES B, b Beled, Hungary, Sept 13, 29; US citizen; m 54; c 1. MATHEMATICS. Educ: Eötvös Lorand Univ, Budapest, MS, 54; Ore State Univ, PhD(math), 65. Prof Exp: Res fel, Astrophys Observ, Budapest, 54-56; instr math, Jr Col, Austria, 57-58 & Univ Portland, 59-61; asst, Ore State Univ, 61-64; asst prof, 64-68, ASSOC PROF MATH, PORTLAND STATE UNIV, 68- Mem: Am Math Soc; Soc Inudst & Appl Math. Res: Asymptotic solutions of differential equations; special functions; computer languages. Mailing Add: Dept of Math Portland State Univ PO Box 751 Portland OR 97207

BALOGH, JOHN CHARLES, physics, see 12th edition

BALOGH, KAROLY, b Budapest, Hungary, Sept 24, 30; m 55; c 2. PATHOLOGY, HISTOCHEMISTRY. Educ: Med Univ Budapest, MD, 54. Prof Exp: Resident path anat, Univ Budapest, 54-56; dir otolaryngol path, Mass Eye & Ear Infirmary, 62-68; asst prof path, Sch Med, Harvard Univ, 68; from assoc prof to prof path, Sch Med, Boston Univ, 68-75, chief path, Univ Hosp, 68-74; sr staff mem, Mallory Inst Path, 74-75; CHIEF PATH, WORCESTER CITY HOSP, 75-; PROF PATH, SCH MED, UNIV MASS, 75- Concurrent Pos: Rockefeller fel, Sch Med, Tulane Univ, 57-S8; USPHS trainee path, Mass Gen Hosp, Boston, 58-61. Mem: Am Asn Path & Bact; Int Acad Path; Histochem Soc; Ger Path Soc. Res: Pathological anatomy; pediatric pathology; bone pathology; enzyme histochemistry; histochemistry of calcified tissues; enzymes in the inner ear. Mailing Add: Worcester City Hosp 26 Queen St Worcester MA 01610

BALOWS, ALBERT B, b Denver, Colo, Jan 3, 21; m 56; c 2. MICROBIOLOGY. Educ: Colo Col, BA, 42; Syracuse Univ, MS, 48; Univ Ky, PhD(microbiol), 52; Am Bd Med Microbiol, cert, 65. Prof Exp: Instr bact & genetics, Syracuse Univ, 46-48; instr microbiol, Univ Ky, 50-51, from asst prof to assoc prof med, Col Med, 52-69, asst prof cell biol, 65-69; DIR BACT DIV, CTR DIS CONTROL, USPHS, 69- Concurrent Pos: Instr, Nazareth Sch Nursing, 51-62; microbiologist, Lexington Clin, 52-69; microbiologist & assoc dir blood bank, St Joseph Hosp, 52-59; mem nat comt Clin Lab Standards; prof biol, Ga State Univ, 70-; assoc prof, Med Sch, Emory Univ, 70-; ed-in-chief, J Clin Microbiol, 75- Mem: Am Soc Microbiol; Am Soc Clin Path; Brit Soc Gen Microbiol; Med Mycol Soc Am; Brit Soc Appl Bact. Res: Clinical microbiology; enteric bacteriology; automation clinical microbiology and systemic mycoses; antibiotics; infectious diseases. Mailing Add: Bact Div Ctr for Dis Control USPHS 1600 Clifton Rd NE Atlanta GA 30333

BALPH, DAVID FINLEY, b Kashmir, India, Aug 28, 31; US citizen; m 62. ANIMAL BEHAVIOR, ECOLOGY. Educ: Hiram Col, BA, 55; Utah State Univ, MS, 61, PhD(behav), 64. Prof Exp: Asst prof wildlife resources, 64-69, assoc prof, 69-74, PROF WILDLIFE SCI, UTAH STATE UNIV, 74- Concurrent Pos: Co-prin investr res grant, NIH, 64-66; prin investr res grant, NSF, 66- Mem: Ecol Soc Am; Am Soc Mammal; Animal Behav Soc. Res: Behavior and ecology of wild populations of mammals. Mailing Add: Dept of Wildlife Resources Utah State Univ Logan UT 84321

BALSANO, JOSEPH SILVIO, b Kenosha, Wis, Apr 19, 37; m 60; c 5. POPULATION BIOLOGY. Educ: Marquette Univ, BS, 59, MS, 62, PhD(biol), 68. Educ: From instr to assoc prof biol, Dominican Col (Wis), 61-68; asst prof, Marquette Univ, 68-69; ASSOC PROF LIFE SCI, UNIV WIS-PARKSIDE, 69- Mem: AAAS; Am Inst Biol Sci; Am Soc Ichthyologists & Herpetologists; Ecol Soc Am; Sigma Xi. Res: Ecological genetics of fish and toad populations; use of electrophoretic analyses to assay variation in the frequencies of specific plasma proteins of natural populations. Mailing Add: Div of Sci Univ of Wis-Parkside Kenosha WI 53140

BALSBAUGH, EDWARD ULMONT, JR, b Harrisburg, Pa, Jan 12, 33; m 68; c 2. ENTOMOLOGY. Educ: Lebanon Valley Col, BS, 55; Pa State Univ, MS, 61; Auburn Univ, PhD(entom), 66. Prof Exp: Entomologist, Bur Plant Indust, Pa Dept Agr, 58-62; ASSOC PROF ENTOM, SDAK STATE UNIV, 65- Concurrent Pos: Consult, Grassland Biome, US Int Biol Prog, 70-73 & Nat Prog Trop Ecol & Nat Coun Sci & Technol, Mex, 74-; assoc, NAm Beetle Fauna Proj, 73- Mem: Coleopterists Soc; Entom Soc Am; Am Inst Biol Sci. Res: Taxonomy, biology, ecology, evolution and zoogeography of Coleoptera, especially Chrysomelidae; biological control of weeds; livestock insect pests. Mailing Add: Entom-Zool Dept SDak State Univ Brookings SD 57006

BALSER, BENJAMIN HARRIS, psychiatry, deceased

BALSER, DONALD S, b Oak Park, Ill, June 19, 23; m 49; c 4. WILDLIFE MANAGEMENT. Educ: Univ Idaho, BS, 54. Prof Exp: Game mgr, Wis Conserv Dept, 54-56; res biologist, Minn Dept Game & Fish, 56-61; CHIEF SECT PREDATOR DAMAGE RES, WILDLIFE RES CTR, US BUR SPORT FISHERIES & WILDLIFE, 61- Mem: Wildlife Soc; Am Soc Mammalogists Res: Ecology; damage control. Mailing Add: Wildlife Res Ctr Bldg 16 Fed Ctr Denver CO 80225

BALSLEY, JAMES ROBINSON, JR, b Pittsburgh, Pa, Dec 27, 16; m 43; c 2. GEOLOGY, GEOPHYSICS. Educ: Calif Inst Technol, BS, 38; Harvard Univ, AM, 41, PhD, 60. Prof Exp: Geologist, US Geol Surv, 41-47, airborne geophysicist, 47-53, chief geophys br, 53-59, asst chief geologist, 59-62; prof geol, Wesleyan Univ, 62-70; asst dir res, Washington, DC, 70-74, ASST DIR RES, US GEOL SURV, RESTON, VA, 74- Concurrent Pos: Am Geol Inst lectr, 59-; Am Geophys Union lectr, 63; with Off Critical Tables, 62-; mem, Antarctic Exped, 46-47. Honors & Awards: Distinguished Serv Award, US Dept Interior, 62. Mem: Geol Soc Am; Am Geophys Union. Res: High pressure work on rocks; occurrence of vanadium in titaniferous magnetite; design and use of airborne magnetometers; remanent magnetization of rocks; use of earth science data in land-use management. Mailing Add: US Geol Surv Nat Ctr Stop 104 12201 Sunrise Valley Dr Reston VA 22092

BALSLEY, MERLE, genetics, physiology, see 12th edition

BALSLEY, RICHARD BENJAMIN, b Endicott, NY, May 8, 38; m 61; c 4. ORGANIC CHEMISTRY. Educ: Univ Rochester, BS, 60; Univ Pa, PhD(org chem), 64. Prof Exp: Asst inst chem, Univ Pa, 60-64; res chemist, Org Chem Div & Lederle Labs, 64-69, GROUP LEADER ORG CHEM DIV, AM CYANAMID CO, BOUND BROOK, 70- Mem: AAAS; Am Chem Soc; NY Acad Sci. Res: Dideoxypentoses; methanolysis of furfuryl alcohol; sulfa drugs; synthesis of dyes, brighteners and pharmaceuticals; reversible bromination; asolvent stoichiometric reactions. Mailing Add: Deer Hill Rd RD 3 Lebanon NJ 08833

BALSTER, CLIFFORD ARTHUR, b Monmouth, Iowa, Feb 27, 22; m 43; c 4. GEOLOGY. Educ: Iowa State Univ, BS, 48, MS, 50. Prof Exp: Geologist, Pure Oil Co, 51-53, div stratigrapher, 53-54; stratigrapher, Am Stratig Co, 54-56, mgr northern div, 56-60; group leader carbonate res, Continental Oil Co, 60-61; area geologist, Soil Conserv Serv, USDA, 61-67; asst prof geol, 67-70, ASSOC PROF GEOL, MONT COL MINERAL SCI & TECHNOL, 70-; RES PETROL GEOLOGIST, MONT BUR MINES & GEOL, 67- Concurrent Pos: Asst prof, Ore State Univ, 61-67; consult geologist, 74- Mem: Geol Soc Am; Am Asn Petrol Geol; Am Inst Prof Geologists; Soil Sci Soc Am. Res: Petrology of sedimentary rocks, especially carbonate rocks; geology of soils and soil genesis, particularly weathering and geomorphology; energy mineral resources and supplies. Mailing Add: 3011 Beech Ave Billings MT 59102

BALTAY, CHARLES, b Budapest, Hungary, Apr 15, 37; US citizen; m 61; c 4. NUCLEAR PHYSICS. Educ: Union Col, BS, 58; Yale Univ, MS, 59, PhD(physics), 63. Prof Exp: Res physicist, Yale Univ, 63-64; from instr to asst prof, 64-69, ASSOC PROF PHYSICS, COLUMBIA UNIV, 69- Concurrent Pos: Alfred P Sloane fel, 66-68. Mem: Am Phys Soc. Res: Elementary particle physics. Mailing Add: Dept of Physics Columbia Univ 936 Pupin Physics Lab New York NY 10027

BALTAZZI, EVAN S, b Izmir, Turkey, Apr 11, 19; US citizen; m 45; c 3. ORGANIC CHEMISTRY. Educ: Univ Paris, DSc(org biochem), 49; Oxford Univ, PhD(org reaction mechanisms), 54. Prof Exp: Head org chem res, French Nat Res Ctr, 46-59; group leader org synthesis, Nalco Chem Co, 59-61; mgr org res, IIT Res Inst, 61-63; dir org res, 63-66, dir phys & chem res, 66-70, DIR APPL SCI, MULTIGRAPHICS DEVELOP CTR, ADDRESSOGRAPH MULTIGRAPH CORP, 70- Concurrent Pos: Brit Coun Exchange scholar, Oxford Univ, 52-54; Can Nat Res Coun fel, 55-56; chmn, Gordon Res Conf Photoconductors, 76. Honors & Awards: Distinguished Serv Award in Sci, 65. Mem: Sr mem Am Chem Soc; The Chem Soc; fel Soc Chem Indust; sr mem French Chem Soc; Soc Photog Scientists & Engrs. Res: Heterocyclic chemistry; synthetic methods; biologically active compounds; photosensitive compounds; photocopying and duplicating materials and systems. Mailing Add: Multigraphics Develop Ctr 19701 S Miles Rd Warrensville Heights OH 44128

BALTENSPERGER, ARDEN ALBERT, b Kimball, Nebr, Dec 25, 22; c 4. PLANT BREEDING. Educ: Univ Nebr, BS, 47, MS, 49; Iowa State Col, PhD(agron), 58. Prof Exp: Jr agronomist field crop res, Agr Exp Sta, Agr & Mech Col, Tex, 52-54; instr agron, Iowa State Col, 54-56, res assoc, 56-58; asst plant breeder sorghum res, Univ Ariz, 58-59, assoc prof & assoc agronomist grass breeding, 59-62, prof & agronomist, 62-63; PROF AGRON & HEAD DEPT, NMEX STATE UNIV, 63- Mem: Am Soc Agron. Res: Grass, corn and sorghum breeding and production. Mailing Add: Dept of Agron NMex State Univ PO Box 3Q Las Cruces NM 88001

BALTHASER, LAWRENCE HAROLD, b Camden, NJ, Feb 19, 37; m 67; c 1. GEOLOGY. Educ: Rutgers Univ, BA, 60; Ind Univ, Bloomington, MA, 63, PhD(geol), 69. Prof Exp: Asst prof geol, Southampton Col, Long Island Univ, 67-69, ASSOC PROF PHYSICS, CALIF STATE POLYTECH COL, SAN LUIS OBISPO, 69- Concurrent Pos: E R Cumings award, Ind Univ, 70; map draftsman, Sun Oil Co, field asst, NJ Agr Exp Sta & teaching asst, Indiana Univ, 75- Mem: Soc Econ Paleont & Mineral. Res: Sedimentary petrology and paleoecology of carbonate rocks; paleocurrent analysis of Franciscan and Great Valley sequence rocks. Mailing Add: Dept of Physics Calif State Polytech Col San Luis Obispo CA 93407

BALTHAZOR, TERRELL MACK, b Hays, Kans, Aug 25, 49; m 68; c 1. ORGANIC CHEMISTRY. Educ: Ft Hays Kans State Col, BS, 71; Univ Ill, PhD(org chem), 75. Prof Exp: SR RES CHEMIST ORG CHEM, MONSANTO CO, 75- Res: Polyvalent iodine chemistry; synthesis of plant growth hormones and their derivatives. Mailing Add: Monsanto Co 800 N Lindbergh St Louis MO 63141

BALTHIS, JOSEPH HENDRICKSON, JR, b Laytonsville, Md, Jan 25, 07; m 32; c 2. INORGANIC CHEMISTRY. Educ: Randolph-Macon Col, BS, 29; WVa Univ, MS, 31; Univ Ill, PhD(chem), 34. Prof Exp: Asst chem, Univ Ill, 31-33; res chemist, Cent Res Dept, Exp Sta, E I du Pont de Nemours & Co, 34-42, Metall Lab, Univ Chicago, 43 & Clinton Lab, Oak Ridge, Tenn, 44; res chemist, Cent Res Dept, E I du Pont de Nemours & Co, 45-67, patent liaison, 67-72; RETIRED. Mem: Am Chem Soc; fel Am Inst Chemists. Res: Inorganic and organoinorganic chemistry, particularly of titanium, silicon, molybdenum and chromium; catalysis. Mailing Add: Box 1 Mendenhall PA 19357

BALTIMORE, DAVID, b New York, NY, Mar 7, 38; m 68. VIROLOGY, BIOCHEMISTRY. Educ: Swarthmore Col, BA, 60; Rockefeller Inst, PhD(biol), 64. Prof Exp: Fel microbiol, Mass Inst Technol, 63-64; fel molecular biol, Albert Einstein Col Med, 64-65; res assoc virol, Salk Inst Biol Studies, 65-68; assoc prof microbiol, 68-72, PROF BIOL, MASS INST TECHNOL, 72-, AM CANCER SOC PROF MICROBIOL, 73- Concurrent Pos: Mem adv panel genetics, NSF, 69-72; mem cancer res ctr res comt, NIH, 71-73. Honors & Awards: Nobel Prize, Physiology & Med, 75; Gustav Stern Award Virol, 70; Warren Trienniel Prize, Mass Gen Hosp, 71; Eli Lilly Award, 71; US Steel Found Award, Molecular Biol, Nat Acad Sci, 74. Mem: Nat Acad Sci; Am Acad Arts & Sci; AAAS; Am Soc Microbiol; Am Soc Biol Chem. Res: Protein and nucleic acid synthesis of RNA animal viruses, especially poliovirus and the RNA tumor viruses. Mailing Add: Dept of Biol & Ctr for Cancer Res Bldg E17-517 Mass Inst Technol Cambridge MA 02139

BALTON, ISIDORE ALFRED, b Philadelphia, Pa, Aug 4, 14; m 40; c 1. GEOPHYSICS. Educ: City Col New York, BS, 34, MSEd, 46. Prof Exp: Supvry tech publ officer, Inst Explor Res, US Army Electronics Command, Ft Monmouth, 58-62, supvry phys scientist, 62-66, phys sci adminr, 66-71, Electronics Technol & Devices Lab, 71-75; RETIRED. Concurrent Pos: Mem staff, Oper Hardtack, Joint Task Force 7, Defense Atomic Support Agency, 58, Oper Dominic, Joint Task Force 8, 62; exec secy, Tech Adv Comt, Joint Serv Electronics Prog, 70-75. Mem: AAAS; sr mem Inst Elec & Electronics Engrs. Res: Field experiments of high energy lasers in connection with study of propagation through turbulent media. Mailing Add: 35 Sternberger Ave West End NJ 07740

BALTZ, ALFRED, b Ger, Aug 5, 28; US citizen; m 55; c 2. PHYSICS, PHYSICAL CHEMISTRY. Educ: Univ Heidelberg, MA, 49. Prof Exp: Res physicist, Atlantic Ref Co, 54-60; sr physicist, Franklin Inst Pa, 60-65; prin physicist, 65-69, staff scientist, 69-72, MGR, MAT ANAL LAB, UNIVAC DIV, SPERRY RAND CORP, 72- Mem: Am Phys Soc; Electron Micros Soc Am; Sigma Xi. Res: Direct structural research on thin magnetic films used in computer memories; direct activities of electron microscope; electron probe micro-analyser; scanning electron microscopy and x-ray laboratories. Mailing Add: Sperry Rand Corp Univac Div PO Box 500 Blue Bell PA 19422

BALTZ, ANTHONY JOHN, b Indianapolis, Ind, Mar 10, 42; m 68; c 1. THEORETICAL NUCLEAR PHYSICS. Educ: Spring Hill Col, BS, 66; Case Western Reserve Univ, MS, 68, PhD(physics), 71. Prof Exp: From res assoc to assoc physicist, 71-76, PHYSICIST THEORET NUCLEAR PHYSICS, BROOKHAVEN NAT LAB, 76- Mem: Am Phys Soc; AAAS. Res: Theory of nuclear reactions and nuclear structure. Mailing Add: Dept of Physics Brookhaven Nat Lab Upton NY 11973

BALTZ, ELMER HAROLD, JR, structural geology, see 12th edition

BALTZ, HOWARD BURL, b St Louis, Mo, July 27, 30; m 68; c 2. APPLIED STATISTICS, MANAGEMENT SCIENCES. Educ: Baylor Univ, BBA, 55, MS, 56; Okla State Univ, PhD(econ statist), 64. Prof Exp: Instr statist, Baylor Univ, 57-59, asst dir Bur Bus & Econ Res, 58-59; teaching asst, Okla State Univ, 59-62; asst prof statist & dir Comput Ctr, Midwestern Univ, 62-64; asst prof, Univ Tex, Austin, 64-68; ASSOC PROF STATIST, UNIV MO-ST LOUIS, 68- Concurrent Pos: Consult, Tex Concrete Masonry Asn, 66-67; coordr, Quant Mgt Sci Group, Univ Mo, 68-73. Mem: Am Statist Asn; Am Inst Decision Sci. Res: Simulation study of the effect of non-normality on confidence interval estimates of variance. Mailing Add: Sch Bus Admin Univ Mo-St Louis 8001 Natural Bridge Rd St Louis MO 63121

BALTZER, OTTO JOHN, b St Louis, Mo, Dec 24, 16; m 41; c 4. PHYSICS. Educ: Washington Univ, AB, 38, MS, 39, PhD(physics), 41. Prof Exp: Res assoc radiation lab, Mass Inst Technol, 41-45; asst dir & supvr, Defense Res Lab, Tex, 45-56; pres & tech dir, Textran Corp, 56-62, PRIN SCIENTIST, TRACOR, INC, 62- Mem: Am Phys Soc. Res: Electromagnetic propagation; guided missile research and development; precision timing; navigation. Mailing Add: Tracor Inc 6500 Tracor Lane Austin TX 78721

BALTZER, PHILIP KEENE, b Quincy, Mass, Feb 11, 24; m 44; c 4. PHYSICS. Educ: Northeastern Univ, SB, 52; Mass Inst Technol, SM, 55; Rutgers Univ, PhD, 63. Prof Exp: Draftsman, Fore River Shipyard, Mass, 46-47; student engr, Charles T Main, Inc, 48-52; res staff magnetic mat, Digital Comput Lab, Mass Inst Technol, 52-55; mem res staff, RCA Corp Res Labs, 55-67, dir res, Tokyo, 67-72, STAFF SCIENTIST, RCA CORP RES LABS, 72- Res: Microelectronics; microprocessors; solid state physics; magnetic materials. Mailing Add: RCA Corp Res Labs Princeton NJ 08540

BALTZO, RALPH M, b Seattle, Wash, June 24, 29; m 51; c 4. HEALTH PHYSICS. Educ: Univ Wash, BA, 50, MS, 62; Am Bd Health Physics, dipl, 62. Prof Exp: Dir radiol safety, Univ Wash, 62-73; INDEPENDENT CONSULT MED & INDUST HEALTH PHYSICS, 73- Mem: Health Physics Soc; Am Nuclear Soc. Res: Operational health physics; personnel dosimetry. Mailing Add: 3841 NE 87th St Seattle WA 98115

BALUDA, MARCEL A, b France, June 30, 30; nat US; c 3. VIROLOGY, ONCOLOGY. Educ: Univ Pittsburgh, BS, 51, MS, 53; Calif Inst Technol, PhD(virol, chem), 56. Prof Exp: Res assoc pediat, City of Hope Med Ctr, 56-62, sr virologist, Biol Dept, 62-66; PROF VIRAL ONCOL, SCH MED, UNIV CALIF, LOS ANGELES, 66- Mem: Am Soc Microbiol. Res: Tumor viruses; animal viruses; oncogenesis and cellular differentiation. Mailing Add: Dept of Microbiol & Immunol Univ of Calif Sch of Med Los Angeles CA 90024

BAMBENEK, MARK A, b Watertown, SDak, Nov 22, 34; m 57; c 2. ANALYTICAL CHEMISTRY. Educ: Col St Thomas, BS, 56; Univ Iowa, MS, 59, PhD(chem), 61. Prof Exp: Asst org photochem, Res Dept, Minneapolis-Honeywell Regulator Co, 56-57; res asst, Univ Iowa, 57-59; res assoc, Cornell Univ, 61-62; res assoc, Univ Pittsburgh, 62-65; from asst prof to assoc prof, 65-75, PROF CHEM, ST MARY'S COL, INC, 75- Mem: Am Chem Soc. Res: Organic reagents for analysis; ion exchange; optical methods of analysis. Mailing Add: Dept of Chem St Mary's Col Notre Dame IN 46556

BAMBER, EDWARD WAYNE, geology, paleontology, see 12th edition

BAMBERGER, CARLOS ENRIQUE LEOPOLDO, b Offenbach, Ger, Mar 26, 33; m 59; c 3. PHYSICAL INORGANIC CHEMISTRY. Educ: Univ Buenos Aires, Licenciado & Dr en Quimica, 58. Prof Exp: Chemist, Nat Atomic Energy Comn, Arg, 57-60, head lab, 60-63, head div beryllium chem, 63-66; CHEMIST, CHEM DIV, OAK RIDGE NAT LAB, 66- Concurrent Pos: Int Atomic Energy Agency fel, 61-63. Mem: Am Chem Soc; AAAS; Am Inst Chemists; Sigma Xi. Res: Beryllium chemistry; preparation of high purity compounds, solvent extraction, stability of complexes; molten (fluorides) salts physical chemistry; actinide chemistry in molten salts; inorganic chemistry for the development of thermochemical cycles for hydrogen production. Mailing Add: Chem Div Oak Ridge Nat Lab PO Box X Oak Ridge TN 37830

BAMBERGER, CURT, b Stettin, Ger, Jan 8, 00; nat US; m 30; c 2. INDUSTRIAL ORGANIC CHEMISTRY. Educ: Univ Würzburg, PhD(chem), 23. Prof Exp: Asst, Univ Würzburg, 23-25; res chemist, I G Farbenindust Co, Works Elberfeld & Leverkusen, 25-39; res & develop chemist, Belg Chem Union, Brussels, 39-40, Nat Aniline Div, Allied Chem & Dye Corp, 41-42 & Patent Chem, Inc, 43-68; sr res chemist, Morton Chem Co Div, Morton-Norwich Prod, Inc, 68-71; CHEM CONSULT, 71- Honors & Awards: Golden Doctor Dipl, Univ Würzburg, 73. Mem: Am Chem Soc; fel Am Inst Chemists. Res: Inorganic salts and organic pesticides; intermediates and dyestuffs of the anthraquinone series; fast wool dyes; vat dyes; fast pigments; oil . soluble colors dyestuffs for plastics; fluorescent dyes of the anthraquinone series. Mailing Add: 1050 George St Apt 11-L New Brunswick NJ 08901

BAMBERGER, JOAN THATCHER, b New York, NY, Feb 27, 34. ANTHROPOLOGY. Educ: Smith Col, BA, 56; Radcliffe Col, MA, 59; Harvard Univ, PhD(anthrop), 67. Prof Exp: Chmn Latin Am studies prog, Brandeis Univ, 70-71, asst prof anthrop, 67-74; assoc prof, Yale Univ, 74-75. Concurrent Pos: Mem exec comt, New Eng Coun Latin Am Studies, 72-74; vis scholar, Mass Inst Technol, 73-74. Mem: AAAS; fel Am Anthrop Asn; Latin Am Studies Asn; Royal Anthrop Inst Gt Brit & Ireland. Res: Social structure; systems of classification; indigeneous and modern South America. Mailing Add: 1558 Massachusetts Ave Cambridge MA 02138

BAMBERGER, ROBERT LEE, electrochemistry, analytical chemistry, see 12th edition

BAMBURG, JAMES ROBERT, b Chicago, Ill, Aug 20, 43; m 70; c 1. BIOCHEMISTRY. Educ: Univ Ill, Urbana, BS, 65; Univ Wis-Madison, PhD(biochem), 69. Prof Exp: From res asst biochem to res assoc physiol chem, Univ Wis-Madison, 65-69; fel biochem, Med Sch, Stanford Univ, 69-71; ASST PROF BIOCHEM, COLO STATE UNIV, 71- Concurrent Pos: Acad coordr, Grad Prog Cellular & Molecular Biol, Colo State Univ, 75-76. Mem: Sigma Xi; Am Chem Soc; AAAS. Res: Structural and functional role of microtubules, microfilaments and other proteins on the processes of nerve growth, cytoplasmic transport, secretion and mitosis. Mailing Add: Dept of Biochem Colo State Univ Ft Collins CO 80523

BAMBURY, RONALD EDWARD, b Aberdeen, SDak, Dec 26, 32; m 54; c 3. MEDICINAL CHEMISTRY. Educ: Univ Minn, BS, 65; Univ Nebr, MS, 68, PhD(chem), 60. Prof Exp: Sr res chemist anthelmintics, Agr Div, Am Cyanamid Co,

60-64; sect head antibact antiprotozoals, Hess & Clark, Div Richardson-Merrell Inc, 64-70; SECT HEAD ANTIBACT ANTIFUNGALS, MERRELL NAT LAB, DIV RICHARDSON-MERRELL INC, 70- Mem: Am Chem Soc; Sigma Xi. Res: Synthesis of new antibiotics related to penicillins and cephalosporins. Mailing Add: 9138 Shadetree Dr Cincinnati OH 45242

BAME, SAMUEL JARVIS, JR, b Lexington, NC, Jan 12, 24; m 56; c 3. SPACE PHYSICS. Educ: Univ NC, BS, 47; Rice Inst, AM, 49, PhD(physics), 51. Prof Exp: MEM STAFF, SPACE PHYSICS GROUP, LOS ALAMOS SCI LAB, 51- Concurrent Pos: Mem, NASA Particles & Fields Subcomt, 67-70. Mem: AAAS; Am Phys Soc; Am Geophys Union; Am Astron Soc. Res: Light particle reactions; neutron physics; particles and radiations in space. Mailing Add: Space Physics Group Los Alamos Sci Lab Los Alamos NM 87545

BAMFORD, ROBERT WENDELL, b Glendale, Calif, July 13, 37. ECONOMIC GEOLOGY, GEOCHEMISTRY. Educ: Univ Wash, BSc, 59; Stanford Univ, PhD(geol geochem), 70. Prof Exp: Res chem engr by-prod chem res, Ga Pac Corp, 60-63; res geologist geol & geochem res, 70, chief site geologist ore deposit eval, 71, SR RES GEOLOGIST GEOL & GEOCHEM, KENNECOTT EXPLOR, INC, 72- Mem: Geol Soc Am; Mineral Soc Am; Soc Mining Engrs; Am Inst Mining & Metall Engrs. Res: Development of ore deposit empirical models, chemical and mineralogical, and of new or refined techniques in exploration geochemistry. Mailing Add: 1138 Gilmer Dr Salt Lake City UT 84105

BAMFORTH, BETTY JANE, b New Britain, Conn, Jan 20, 23. ANESTHESIOLOGY. Educ: Bates Col, BS, 44; Boston Univ, MD, 47. Prof Exp: Intern, Mt Auburn Hosp, 48; resident, Univ Wis Hosp, 51; asst prof anesthesiol, Sch Med, Univ Okla, 51-54; assoc prof, 54-64, actg chmn dept, 69-71, PROF ANESTHESIOL, SCH MED, UNIV WIS, MADISON, 64-, ASST DEAN EDUC ADMIN, 73- Concurrent Pos: Consult, Madison Vet Admin Hosp, 54-; mem anesthetic & respiratory adv comt, Food & Drug Admin, 73- Mem: Am Soc Anesthesiol; Int Anesthesia Res Soc; Asn Univ Anesthesiol. Res: Medical education. Mailing Add: Univ of Wis Hosps 1300 University Ave Madison WI 53706

BAMFORTH, STUART SHOOSMITH, b White Plains, NY, Oct 23, 26; m 52; c 2. PROTOZOOLOGY. Educ: Temple Univ, AB, 51; Univ Pa, AM, 54, PhD(zool), 57. Prof Exp: Asst instr biol, Univ Pa, 51-57; from instr to assoc prof, 57-72, PROF BIOL, NEWCOMB COL, TULANE UNIV, 72- Mem: Am Soc Zool; Soc Study Evolution; Soc Protozool; Am Soc Limnol & Oceanog. Res: Ecology of protozoa and plankton. Mailing Add: Dept of Biol Newcomb Col Tulane Univ New Orleans LA 70118

BAMRICK, JOHN FRANCIS, b Rockwell, Iowa, Dec 15, 26; m 56; c 3. BIOLOGY, GENETICS. Educ: Loras Col, BS, 53; Marquette Univ, MS, 56; Iowa State Univ, PhD(zool), 60. Prof Exp: From instr to assoc prof biol, 55-70, PROF BIOL, LORAS COL, 70- Mem: AAAS; Soc Invert Path; Bee Res Asn; Sigma Xi. Res: Disease resistance; insect pathology. Mailing Add: Dept of Biol Loras Col Dubuque IA 52001

BAN, THOMAS ARTHUR, b Budapest, Hungary, Nov 16, 29; Can citizen; m 63. PSYCHIATRY, PSYCHOPHARMACOLOGY. Educ: Med Univ Budapest, MD, 54; McGill Univ, dipl, 60. Prof Exp: Demonstr, 60-63, lectr, 64-65, asst prof, 65-70, ASSOC PROF PSYCHIAT, McGILL UNIV, 71-, DIR DIV PSYCHOPHARMACOL, DEPT PSYCHIAT, 70- Concurrent Pos: Sr psychiatrist, Douglas Hosp, 60-61, sr res psychiatrist, 61-65, asst dir res, 66-70, chief res serv, 70-72; psychiat res consult, Hopital des Laurentides, 63-72, Lakeshore Gen Hosp, 67- & St Marys Hosp, 71- Mem: AAAS; Am Psychiat Asn; Pavlovian Soc NAm; Can Psychiat Asn; Int Brain Res Orgn. Res: Application of classical conditioning in psychiatry; influence of psychotropic drugs on psychopathology. Mailing Add: Rm 137 Res & Training Bldg McGill Univ Montreal PQ Can

BAN, VLADIMIR SINISA, b Bjelovar, Yugoslavia, Sept 7, 41; US citizen. SOLID STATE SCIENCE. Educ: Univ Zagreb, dipl Ing, 64; Pa State Univ, PhD(solid state sci), 69. Prof Exp: Res asst solid state sci, Mat Res Lab, Pa State Univ, 64-69; MEM TECH STAFF SOLID STATE SCI, MAT RES LAB, RCA LAB, PRINCETON, NJ, 69- Concurrent Pos: Vis fel, Imp Col, Univ London, 72. Honors & Awards: Outstanding Achievement Award, RCA Lab, 73. Mem: Electrochem Soc; Am Crystal Growth Soc; Sigma Xi. Res: Chemical vapor deposition of semiconducting films; fundamental research on thermodynamics and kinetics of vapor deposition of semiconductors and the subsequent optimization of their synthesis; laser solid interactions; high temperature mass spectrometry. Mailing Add: RCA Labs Princeton NJ 08540

BANAS, EMIL MIKE, b East Chicago, Ind, Dec 5, 21; m 48; c 2. PHYSICS. Educ: St Procopius Col, BA, 43; Univ Notre Dame, PhD(physics), 55. Prof Exp: Instr math & physics, St Procopius Col & Acad, 47-49; asst proj physicist, Standard Oil Co, 55-58, proj physicist, 58-61; sr proj physicist, Am Oil Co, 61-71; MOLECULAR SPECTROSCOPIST, STANDARD OIL RES CTR, 71- Concurrent Pos: Instr, Calumet Ctr, Purdue Univ, 56-60. Mem: Am Phys Soc: Am Chem Soc. Res: Nuclear magnetic and electron paramagnetic resonance; molecular physics; fuel cells; instrument development; nuclear quadrupole resonance; x-rays; infrared, microwaves and cryogenics; chromatography. Mailing Add: Standard Oil Res Ctr PO Box 400 Naperville IL 60540

BANASIK, ORVILLE JAMES, b Wales, NDak, Nov 17, 19; m 47; c 4. BIOCHEMISTRY. Educ: NDak State Univ, BS, 43, MS, 46. Prof Exp: Lab asst & fel, 46-47, asst cereal technologist, 47-58, assoc cereal technologist, 58-63, assoc prof cereal technol, 63-70, PROF CEREAL CHEM & TECHNOL, NDAK STATE UNIV, 70-, CHMN DEPT, 71- Mem: Am Asn Cereal Chem; Am Soc Brewing Chem. Res: Cereal research, especially barley; development of testing methods and basic research. Mailing Add: Dept of Cereal Technol NDak State Univ Fargo ND 58102

BANASZAK, LEONARD JEROME, b Milwaukee, Wis, Feb 1, 33; m; c 4. CRYSTALLOGRAPHY, BIOCHEMISTRY. Educ: Univ Wis, Madison, BS, 55; Loyola Univ, Chicago, MS, 60, PhD(biochem), 61. Prof Exp: Asst prof physiol & biophys, 66-71, assoc prof biol chem, 71-75, PROF BIOL CHEM, SCH MED, WASH UNIV, 75- Concurrent Pos: USPHS fels biochem, Ind Univ, 61-63 & Med Res Coun Lab Molecular Biol, Eng, 63-65; USPHS career develop award, 69-74. Mem: AAAS; Am Chem Soc; Am Crystallog Asn. Res: Structure of enzymes and the relationship of these structures to biological function. Mailing Add: Dept of Biol Chem Wash Univ Sch of Med St Louis MO 63110

BANAUGH, ROBERT PETER, b Los Angeles, Calif, Oct 27, 22; m 46; c 8. APPLIED MATHEMATICS, GEOPHYSICS. Educ: Univ Calif, Berkeley, AB, 46, MA, 52, PhD(eng sci), 62. Prof Exp: High sch teacher, Calif, 47-50; mathematician, Calif Res & Develop Corp, 52-54; aeronaut engr, Pioneer Industs, Nev, 54-55; theoret physicist, Lawrence Radiation Lab, Univ Calif, 55-62; head appl math group, Ventura Div, Northrop Corp, 62-64; dir comput ctr, 64-72, PROF COMPUT SCI, UNIV MONT, 64-, CHMN DEPT, 72- Concurrent Pos: Consult, Air Force Shock Tube Facility, Univ NMex, 65-66 & Boeing Co, 65- Mem: AAAS; Am Geophys Union; Seismol Soc Am. Res: Computer based methods of solution of biological and physical problems. Mailing Add: Rte 5 Upper Miller Creek Rd Missoula MT 59801

BANAY-SCHWARTZ, MIRIAM, b Sibiu, Roumania, Oct 9, 29; US citizen; m 51; c 2. BIOCHEMISTRY, ORGANIC CHEMISTRY. Educ: Hebrew Univ, Jerusalem, MSc, 54, PhD(biochem), 61. Prof Exp: Instr biochem, 62-68, ASST PROF BIOCHEM, ALBERT EINSTEIN COL MED, 68-; SR RES SCIENTIST, NY STATE RES INST NEUROCHEM & DRUG ADDICTION, 68- Mem: Am Soc Neurochem; NY Acad Sci; Am Inst Chemists. Res: Electron transport; intermediate metabolism. Mailing Add: NY State Res Inst Neurochem & Drug Addiction Ward's Island New York NY 10035

BANCROFT, DENNISON, b Newton, Mass, Oct 3, 11; m 39; c 3. PHYSICS. Educ: Amherst Col, AB, 33; Harvard Univ, PhD(physics), 39. Prof Exp: Res assoc geophys, Harvard Univ, 36-41; physicist, David W Taylor Model Basin, US Navy Dept, 41-42; asst prof physics, Princeton Univ, 46-47; from asst prof to assoc prof, Swarthmore Col, 47-59; prof physics & chmn dept, 59-74, EMER PROF PHYSICS, COLBY COL, 74- Concurrent Pos: Consult, Los Alamos Sci Labs, 55-56. Mem: Am Phys Soc. Res: Velocity of sound in nitrogen at high pressures; elastic constants of solids at high pressures and high temperatures; stress-strain relations under explosive load; thermodynamic properties of gases. Mailing Add: RD Box 171 Brooksville ME 04617

BANCROFT, GEORGE HERBERT, b Halifax, NS, Mar 2, 10; nat US; m 39; c 2. PHYSICS. Educ: WVa Univ, AB, 30, MS, 31; Univ Pa, PhD(physics), 37. Prof Exp: Instr physics, Hobart Col, 36-40; physicist & group leader develop group, Vacuum Equip Div, Distillation Prod, Inc, 43-53; dir eng, Consol Electrodynamics Corp, 53-54, res, 54-57, mgr new prod develop, 57-62, consult, 62-63; exec asst to pres, Bendix-Blazers Vacuum, Inc, NY, 63-65, chief scientist, Instrument & Life Support Div, Bendix Corp, Iowa, 65-70, prin engr, Kansas City Div, 70-75; RETIRED. Mem: Fel AAAS; Am Phys Soc; Optical Soc Am; Am Vacuum Soc (pres, 65). Res: Excitation of soft x-rays; evaporation coating under high vacuum; high vacuum dehydration; high vacuum evaporation coating unit for the continuous coating of sheet stock in rolls; an evaporation source; vacuum equipment component development. Mailing Add: 2325 W 98th St Leawood KS 66206

BANCROFT, GEORGE MICHAEL, b Saskatoon, Sask, Apr 3, 42; m 67; c 2. PHYSICAL INORGANIC CHEMISTRY. Educ: Univ Man, BSc, 63, MSc, 64; Univ Cambridge, PhD(chem), 67. Hon Degrees: MA, Univ Cambridge, 70. Prof Exp: Demonstr & teaching fel chem, Univ Cambridge, 67-70; from asst prof to assoc prof, 70-74, PROF CHEM, UNIV WESTERN ONT, 74- Concurrent Pos: Prin investr, NASA, 70-72; E W Steacie mem fel, Nat Res Coun Can, 73-76; vis investr, Phys Sci Lab, Univ Wis, 75-76; mem, Int Comn Applns Mossbauer Effect, 75- Honors & Awards: Harrison Mem Prize, Brit Chem Soc, 72; Meldola Medal, Royal Inst Chem, 72. Mem: The Chem Soc; Chem Inst Can; Am Chem Soc. Res: Mossbauer spectroscopy of iron and tin organometallic compounds, and iron-containing silicate minerals; photoelectron spectroscopy of crystal field effects on core energy levels, and adsorption of ions on ionic crystals. Mailing Add: Dept of Chem Univ of Western Ont London ON Can

BANCROFT, HAROLD RAMSEY, b Meridian, Miss, May 3, 32; m 59; c 1. ENTOMOLOGY. Educ: Miss State Univ, BS, 58, MS, 59, PhD(entom), 62. Prof Exp: Asst prof, 62-66, ASSOC PROF BIOL, MEMPHIS STATE UNIV, 66- Mem: Entom Soc Am. Res: Physiology of diapause in Anthonomus grandis Boh; diapause phenomenon in Coleoptera in general. Mailing Add: Dept of Biol Memphis State Univ Memphis TN 38111

BANCROFT, JOHN BASIL, b Vancouver, BC, Can, Dec 31, 29; m 54; c 4. BIOLOGY. Educ: Univ BC, BA, 52; Univ Wis, PhD(plant path), 55. Prof Exp: Asst prof, 55-64, PROF BOT & PLANT PATH, PURDUE UNIV, 64- Mem: Am Phytopath Soc. Res: Characterization of plant viruses and related nucleoproteins. Mailing Add: Dept of Bot and Plant Path Sch of Agr Purdue Univ Lafayette IN 47907

BANCROFT, RICHARD WOLCOTT, b Denver, Colo, Mar 28, 16; m 50. PHYSIOLOGY. Educ: Stanford Univ, BA, 38; Univ Southern Calif, MS, 47, PhD(physiol), 51. Prof Exp: Res asst aviation physiol, US Air Force Sch Aerospace Med, 47-48, physiologist, 50-68, chief appl physiol br, 68-72; RETIRED. Mem: AAAS; fel Aerospace Med Asn; Am Physiol Soc; Soc Exp Biol & Med. Res: Aerospace physiology; respiration; circulation. Mailing Add: 7709 Prospect Pl La Jolla CA 92037

BANCROFT, THEODORE ALFONSO, b Columbus, Miss, Jan 2, 07; m; c 2. STATISTICS. Educ: Univ Fla, AB, 27; Univ Mich, AM, 34; Iowa State Univ, PhD(math statist), 43. Prof Exp: Asst math, Vanderbilt Univ, 37-38; head dept, Mercer Univ, 38-41; asst prof, Iowa State Univ, 41-46; assoc prof, Univ Ga, 46-47; res prof statist & dir lab, Ala Polytech Inst, 47-49; assoc prof, 49-50, head dept & dir lab, 50-72, PROF STATIST, IOWA STATE UNIV, 50- Concurrent Pos: Fel, Vanderbilt Univ; prof, Univ Training Command, US Army, Italy, 45. Mem: Fel AAAS; fel Inst Math Statist; fel Am Statist Asn (pres, 70); Biomet Soc; Economet Soc. Res: Mathematical statistics; application to agriculture, biology and social sciences; physical and engineering sciences; statistical inference. Mailing Add: 102 E Snedcor Hall Statist Lab Iowa State Univ Ames IA 50011

BAND, HANS EDUARD, b Vienna, Austria, Oct 14, 24; US citizen; m 48; c 3. PHYSICS, MATERIALS SCIENCE. Educ: Harvard Univ, AB, 47; Boston Univ, AM, 53. Prof Exp: Elec engr, Gen Elec Co, 48-52; staff mem, Lincoln Lab, Mass Inst Technol, 52-53; lit analyst, Gillette Safety Razor Co, 53-56; sr res engr, Melpar, Inc, 56-57; opers analyst, Tech Opers, Inc, 57-58; staff physicist, Pickard & Burns, Inc, 58-61; sr res scientist, Raytheon Co, 61-63; optical physicist, Concord Radiance Lab, Utah State Univ, 63-66; prin res scientist, Avco Everett Res Lab, Mass, 66-67; PHYSICIST, ARMY MAT & MECH RES CTR, 68- Mem: Am Phys Soc; Am Sci Affil. Res: Optical and radiation physics; atomic spectroscopy; radio communication; solid state physics; materials science and engineering. Mailing Add: Army Mat & Mech Res Ctr Arsenal St Watertown MA 02172

BAND, HENRETTA TRENT, b Danville, Va, June 28, 32; m 55; c 1. GENETICS, ZOOLOGY. Educ: Col William & Mary, BS, 54; Univ Calif, Berkeley, PhD(genetics), 59. Prof Exp: Nat Cancer Inst fel, Amherst Col, 58-59; lectr zool, Univ BC, 59-60; res assoc, 60-63, lectr, 61-62; res assoc, Mich State Univ, 63-65, asst prof, 65-72, prof natural sci, 67-68; res fel, Dept Genetics, Univ Calif, Davis, 72-73. Concurrent Pos: Nat Res Coun Can res grant, 59-63 & NSF res grant, 60-64; USPHS spec fel genetics, Univ Cambridge, 65-66. Mem: Genetics Soc Am; Soc Study Evolution; Am Soc Naturalists. Res: Genetics of Drosophila; population genetics; ecological genetics; history of science; Friedrich Albert Lange's impact on evolution, philosophy and communist ideology. Mailing Add: 5854 Smithfield Ave East Lansing MI 48823

BAND, RUDOLPH NEAL, b Oakland, Calif, Aug 2, 29; m 55. ZOOLOGY. Educ:

Univ San Francisco, BS, 52, MS, 54; Univ Calif, Berkeley, PhD(zool), 59. Prof Exp: NIH fel, Amherst Col, 58-59; from instr to asst prof, 59-63; from asst prof to assoc prof, 63-69, PROF ZOOL, MICH STATE UNIV, 69- Concurrent Pos: USPHS spec res fel, physiol lab, Cambridge Univ, 65-66. Mem: AAAS; Soc Protozool; Am Soc Parasitol; Am Soc Cell Biol; Brit Soc Gen Microbiol. Res: Biochemistry and physiology of protozoan organelles; locomotion; phagocytosis; physiology and biochemistry of amoebae. Mailing Add: Dept of Zool Mich State Univ East Lansing MI 48823

BAND, YEHUDA BENZION, b Munich, Ger, Dec 1, 46; US citizen; m 70. ATOMIC PHYSICS, QUANTUM PHYSICS. Educ: Cooper Union, BS, 68; Univ Chicago, MS, 70, PhD(physics), 73. Prof Exp: Fel chem physics, Univ Chicago, 73-75; ASST SCIENTIST PHYSICS, ARGONNE NAT LAB, 75- Mem: Am Inst Physics. Res: Collision theory; charge exchange processes; ionization, dissociation of molecules; beam foil collisions; surface phenomena; many body physics. Mailing Add: Physics Div Argonne Nat Lab Argonne IL 60439

BANDEEN, JOHN DRUMMOND, b London, Ont, Dec 19, 33; m 58; c 3. AGRONOMY. Educ: Ont Agr Col, BSA, 57, MSA, 60; Univ Wis, PhD(agron), 65. Prof Exp: Lectr agron, Ont Agr Col, 60-61, from asst prof to assoc prof, 64-73, PROF CROP SCI, UNIV GUELPH, 73- Concurrent Pos: Consult, Eastern Sect, Can Weed Comt, 65-, exec mem, 69-74, nat exec comt, 74- Mem: Weed Sci Soc Am; Agr Inst Can. Res: Weed physiology as it affects the control of weeds in crop production programs. Mailing Add: Dept of Crop Sci Univ of Guelph Guelph ON Can

BANDEL, DAVID, polymer chemistry, see 12th edition

BANDEL, HERMAN WILLIAM, b Great Falls, Mont, Sept 23, 16; m 45. PLASMA PHYSICS. Educ: Univ Calif, Berkeley, AB, 47, PhD(physics), 54. Prof Exp: Sr engr, Appl Physics Sect, Electron Defence Lab, Sylvania Elec Prod, Inc, 54-57, adv res physicist, Microwave Physics Lab, 57-63; RES SCIENTIST, LOCKHEED MISSILES & SPACE CO, 63- Mem: Am Phys Soc. Res: Discharge through gases. Mailing Add: 1753 Mayflower Ct Mountain View CA 94040

BANDEL, VERNON ALLAN, b Baltimore, Md, May 8, 37; m 59; c 2. AGRONOMY, PLANT PHYSIOLOGY. Educ: Univ Md, BS, 59, MS, 63, PhD(agron, soil fertil), 65. Prof Exp: Asst prof soils, 64-69, ASSOC PROF SOILS, UNIV MD, COLLEGE PARK, 69-, EXTEN SOILS SPECIALIST, 64- Mem: Am Soc Agron; Soil Sci Soc Am. Res: Soil fertility; nutrient balance in plants; micro-nutrient fertilization of corn, soybeans; nitrogen fertilization of corn and small grains; no-tillage corn fertilization. Mailing Add: Dept of Agron Univ of Md College Park MD 20742

BANDELIN, FRED JOHN, b Cincinnati, Ohio, Mar 24, 13; m 40; c 2. PHARMACEUTICAL CHEMISTRY. Educ: Univ Cincinnati, BS, 36, AB, 39. Prof Exp: Res chemist germicides & local anesthetics, William S Merrell Co, 36-39; instr pharmacol & pharmaceut chem, Col Pharm, Univ Cincinnati, 39-41; tech dir germicides, anesthetics & pharmaceut formulations, Fidelity Med Supply Co, 41-43; res chemist pharmaceut formulations & antibiotics, Vick Chem Co, 43-44; tech dir lipotropic compounds, anesthetics, germicides & pharmaceut formulations, Flint, Eaton & Co, 44-48, vpres & dir res, Strong, Cobb-Arner, Inc, Ohio, 58-63; dir labs control & develop, 63-67, dir res & control, 67-68, DIR LABS, RES & DEVELOP, PLOUGH, INC, 68- Concurrent Pos: Chmn sci sect, Am Pharmaceut Mfrs Asn, 49-50; mem, Nat Formulary Subcomt Alkaloids & Heterocyclic Compounds; asst prof pharmaceut, Univ Tenn Med Units, 65- Honors & Awards: Off Vitamin Develop Admin Award, 36. Mem: AAAS; Am Chem Soc; Am Pharmaceut Asn; fel Am Inst Chem; fel Walter Reed Soc; NY Acad Sci. Res: Germicides; ultra-violet absorption compounds; pharmaceutical formulation. Mailing Add: 253 Brierview Rd Memphis TN 38138

BANDER, MYRON, b Belzyce, Poland, Dec 11, 37; US citizen; m 67. THEORETICAL PHYSICS. Educ: Columbia Univ, BA, 58, MA, 59, PhD(physics), 62. Prof Exp: NSF fel, 62-63; res assoc theoret physics, Linear Accelerator Ctr, Stanford Univ, 63-66; assoc prof physics, 66-72, PROF PHYSICS, UNIV CALIF, IRVINE, 72- Concurrent Pos: Sloan Found fel, 67-69. Mem: Am Phys Soc. Res: Elementary particle physics. Mailing Add: Dept of Physics Univ of Calif Irvine CA 92664

BANDERMANN, LOTHAR W, b Waltrop, Ger, June 25, 36; US citizen; m 70. PHYSICS, ASTROPHYSICS. Educ: Univ Calif, AB, 63; Univ Md, PhD(physics), 68. Prof Exp: Res assoc physics, Univ Md, 67-68; asst astronr, 68-75, ASSOC ASTRONR, INST ASTRON, UNIV HAWAII, 75- Mem: Am Astron Soc; fel Royal Astron Soc. Res: Physics of the solar system; interplanetary and interstellar matter. Mailing Add: Inst for Astron Univ of Hawaii Honolulu HI 96822

BANDES, DEAN, b New York, NY, Jan 16, 1944; m 68; c 2. MATHEMATICS. Educ: Williams Col, BS, 65; Brandeis Univ, MA, 67, PhD(math), 74. Prof Exp: ASST PROF MATH, UNIV MASS, BOSTON, 74- Mem: Am Math Soc; Math Asn Am. Res: Topology-knot theory; computer combinatorics; partitions, transportation problem. Mailing Add: 30 Hillcrest Circle Watertown MA 02172

BANDES, HERBERT, b New York, NY, May 23, 14; m 72; c 6. ELECTROCHEMISTRY. Educ: Univ Mich, BS, 35, MS, 36, PhD(electrochem), 38. Prof Exp: Res chemist, Alrose Chem Co, RI, 39; plant chemist, Cohn & Rosenberger, Inc, 40; physicist, Brooklyn Naval Shipyard, Bur Ord, 40-43; electrochemist, Kellex Corp, NY, 43-44; sr engr, Sylvania Elec Prod, Inc, 44-50, mgr chem lab, Cent Res Labs, 50-55, chief engr, Semiconductor Div, 55-58; mem sr staff & dir res, Western Div, Arthur D Little, Inc, 58-63; dir res, Eitel-McCullough, Inc, Calif, 63-65; chief engr, Electronics Dept, Hamilton Stand Div, United Aircraft Corp, 65-69; DEAN ADMIN AFFAIRS, MANCHESTER COMMUNITY COL, 69- Mem: Am Chem Soc; Electrochem Soc. Res: Kinetics of electrode reactions; mechanism of electrodeposition; solid state chemistry. Mailing Add: Manchester Community Col Box 1046 Manchester CT 06040

BANDI, WILLIAM R, b Tarentum, Pa, July 26, 24; m 48; c 2. ANALYTICAL CHEMISTRY. Educ: Univ Pittsburgh, BS, 49. Prof Exp: Res chemist, Allegheny Ludlum Steel Corp, 50-54; from asst technologist to supv technologist, 54-66, ASSOC RES CONSULT, APPL RES LAB, US STEEL CORP, 66- Mem: Am Chem Soc; Am Soc Testing & Mat; NAm Thermal Anal Soc; Int Confedn Thermal Anal. Res: Inorganic analysis; spectrophotometric methods; ion exchange separations; thermal methods of analysis; isolation and analysis of second phases from alloy matrix. Mailing Add: MS 15 US Steel Res Ctr 125 Jamison Lane Monroeville PA 15146

BANDICK, NEAL RAYMOND, b Orange, Calif, Mar 29, 38; m 63; c 2. HUMAN PHYSIOLOGY, MEDICAL EDUCATION. Educ: Univ Calif, Davis, BS, 60; Trinity Univ, MS, 65; Univ Mich, EdD(physiol), 70. Prof Exp: Instr physiol, Univ Mich, 69-70; ASST PROF BIOL, ORE COL EDUC, 70- Mem: Sigma Xi; Asn Am Med Cols. Res: Cardiovascular physiology; visco-elastic and contractile properties of arterial walls during renal hypertension; selection of medical curriculum content. Mailing Add: Dept of Sci & Math Ore Col Educ Monmouth OR 97361

BANDMAN, EVERETT, b New York, NY, June 29, 47; c 1. MOLECULAR BIOLOGY. Educ: City Col New York, BS, 69; Univ Calif, Berkeley, PhD(molecular biol), 74. Prof Exp: Res fel surg, Mass Gen Hosp, 74-76; fel biochem, Harvard Med Sch & fel, Shriner's Burns Inst, 74-76; FEL ZOOL, UNIV CALIF, BERKELEY, 76- Concurrent Pos: Res technician biochem, Sloan Kettering Inst, 69; USPHS traineeship, 69-74. Honors & Awards: Nat Res Serv Award, NIH, 76. Mem: Am Soc Cell Biol; Tissue Cult Asn. Res: Biochemical control of the regulation of cell growth in tissue culture. Mailing Add: Dept of Zool Univ of Calif Berkeley CA 94710

BANDONI, ROBERT JOSEPH, b Weeks, Nev, Nov 9, 26; Can citizen; m 56; c 1. MYCOLOGY. Educ: Univ Nev, BSc, 53; Univ La, MSc, 56, PhD(mycol), 57. Prof Exp: Instr & asst prof bot, Univ Wichita, 57-58; PROF MYCOL, UNIV BC, 58- Concurrent Pos: Gertrude S Burlingam fel mycol, NY Bot Garden, 57. Mem: AAAS; Can Bot Soc; Mycol Soc Am; Brit Mycol Soc; Japanese Mycol Soc. Res: Studies of leaf-decay fungi, especially dispersal; taxonomic studies of lower Basidiomycetes. Mailing Add: Dept of Bot Univ of BC Vancouver BC Can

BANDTEL, KENNETH CHARLES, b San Francisco, Calif, Mar 12, 24; m 47; c 5. EXPERIMENTAL NUCLEAR PHYSICS, RESEARCH ADMINISTRATION. Educ: Univ Calif, BS, 49, PhD(physics), 54. Prof Exp: Engr microwave res lab, Univ Calif, Berkeley, 49-50; physicist, Lawrence Radiation Lab, Berkeley, 50-55, PHYSICIST, LAWRENCE LIVERMORE LAB, 55- Mem: Am Phys Soc; Inst Elec & Electronics Engrs. Res: High energy nuclear physics; reactor physics; nuclear explosive physics; military system applications of nuclear explosives; vulnerability and effects of nuclear explosives; foreign technology evaluation in the military and civilian areas. Mailing Add: Lawrence Livermore Lab Univ of Calif PO Box 808 Livermore CA 94550

BANDURSKI, ROBERT STANLEY, b Chicago, Ill, May 11, 24; m 44. PLANT BIOCHEMISTRY. Educ: Univ Chicago, BS, 46, PhD(plant physiol), 49. Prof Exp: Instr bot, Univ Chicago, 48-49; Nat Res Coun fel biol, Calif Inst Technol, 49-50, res fel, 50-52, sr res fel, 52-53; asst biochem, Mass Gen Hosp, 53-54; assoc prof bot, 54-58, PROF BOT, MICH STATE UNIV, 58- Mem: Fedn Biol Chem; Am Soc Plant Physiol. Res: Plant pigments; respiratory enzymes; auxins; chemistry of plant hormones and enzymes of sulfur metabolism. Mailing Add: Dept of Bot Mich State Univ East Lansing MI 48824

BANDY, ALAN RAY, b Indiahoma, Okla, Sept 13, 40; m 63; c 3. ATMOSPHERIC CHEMISTRY. Educ: Okla State Univ, BS & MS, 64; Univ Fla, PhD(chem), 68. Prof Exp: Res assoc chem, Univ Md, College Park, 68-70; from asst prof to assoc prof, Old Dom Univ, 70-75; ASSOC PROF CHEM, DREXEL UNIV, 75- Mem: Am Chem Soc; Air Pollution Control Asn. Res: Photochemistry of trace atmospheric constituents; atmospheric chemistry of atmospheric sulfur and and nitrogen compounds; molecular structure and spectra. Mailing Add: Dept of Chem Drexel Univ Philadelphia PA 19104

BANDY, ORVILLE LEE, geology, see 12th edition

BANDY, PERCY JOHN, b Mexico City, Mex, Aug 13, 27; Can citizen; m 53; c 3. ZOOLOGY. Educ: Univ BC, BA, 52, MA, 55, PhD(zool), 65. Prof Exp: Biologist, Wildlife Res Div, 58-62, res biologist, 62-73, ASST DIR PLANNING, ADMIN & RES, WILDLIFE RES DIV, BC FISH & WILDLIFE BR, 73- Concurrent Pos: Chmn, Nat Adv Comt Land Capability Classification for Wildlife, 66; chmn, BC Wildlife & Recreation Comt, Can Land Inventory, 66-71: hon lectr fac agr, Univ BC, 67-71, res assoc inst animal resource ecol, 69-71; ed, Fish & Wildlife Publ, 70- Res: Wildlife management research of various types including population dynamics, diseases, parasites and nutrition. Mailing Add: BC Fish & Wildlife Br Parliament Bldg Victoria BC Can

BANDY, WILLIAM ROBERT, b Cincinnati, Ohio, Mar 26, 43; m 65. THEORETICAL SOLID STATE PHYSICS. Educ: Univ Okla, BS, 66, MS, 67; Univ Md, PhD(physics), 74. Prof Exp: Scientist semiconductor device physics, 67-74, MGR INTEGRATED CIRCUIT PROCESSING FACIL, DEPT DEFENSE, 74- Mem: Inst Elec & Electronic Engrs; Am Phys Soc. Res: Tight binding Green's function calculations exploring the relationship between electronic structure of tunneling systems and the tunneling current; modeling the physics of metal-oxide-semiconductor device processing and operation. Mailing Add: 1788 Regents Park Rd Crofton MD 21114

BANE, GILBERT WINFIELD, b San Diego, Calif, Dec 11, 31; m 70; c 3. MARINE ECOLOGY, BIOLOGICAL OCEANOGRAPHY. Educ: San Jose State Col, BA, 54; Cornell Univ, MS, 61, PhD(vert zool), 63. Prof Exp: From jr scientist to sr scientist, Inter-Am Trop Tuna Comn, 55-58; scientist, Starkist Foods, Inc, 59-60; lab asst vert zool, Conserv Dept, Cornell Univ, 60-63; asst prof marine biol, Univ PR, Mayagüez, 63-65 & Univ Calif, Irvine, 65-69; assoc prof, Southampton Col, LI Univ, 69-73; chmn marine & natural sci, St Francis Col Maine, 73-75; DIR MARINE SCI & ENVIRON STUDIES, UNIV OF NC, WILMINGTON, 75- Concurrent Pos: Danforth assoc, Danforth Found, 68. Mem: Am Fisheries Soc; Am Soc Ichthyol & Herpet; Am Soc Mammal; Am Soc Limnol & Oceanog; Am Soc Zool. Res: Biology and population studies of marine fishes, especially sharks; environmental stresses on coastal fishes, especially offshore petroleum operations and coastal nuclear facilities; fishes of southern Mexico. Mailing Add: Dir Marine Sci & Environ Studies Univ of N C Wilmington NC 28401

BANE, JOHN MCGUIRE, JR, b Kalamazoo, Mich, May 3, 46; m 70; c 2. PHYSICAL OCEANOGRAPHY. Educ: Western Mich Univ, BS, 70; Fla Atlantic Univ, ME, 71; Fla State Univ, PhD(phys oceanog), 75. Prof Exp: Res assoc phys oceanog, La State Univ, 73-74; ASST PROF PHYS OCEANOG, UNIV NC, CHAPEL HILL, 75- Mem: Am Geophys Union; Am Meteorol Soc. Res: Theoretical and observational studies of circulation on the continental shelves; estuarine dynamics; numerical modeling of marine systems. Mailing Add: Curric Marine Sci Univ of NC Chapel Hill NC 27514

BANERJEE, BHOLA NATH, b Calcutta, India, Jan 1, 34; m 65. TOXICOLOGY, TERATOLOGY. Educ: Univ Calcutta, BS, 53, DVM, 57; Univ Ky, MS & PhD(physiol), 66. Prof Exp: Vet officer, Govt India, 57-59; instr animal husb, Birla Col Agr, 59-61; res asst physiol, Univ Ky, 61-65; res assoc, Creighton Univ, 65-66; res instr, Univ Nebr, 66-68; DIR DRUG SAFETY ASSESSMENT, WALLACE LAB, 68- Concurrent Pos: Damon Runyon Mem Fund cancer res grant, 68-; consult, Nat Libr Med; course dir teratology, Ctr Prof Advan, Sommerville, NJ. Mem: AAAS; Teratology Soc; Int Fertil Asn; Soc Study Reproduction; Am Soc Animal Sci. Res: Cancer metabolism; toxicity of drugs and chemicals; fertility and infertility; reproductive physiology. Mailing Add: Wallace Lab Cranbury NJ 08512

BANERJEE, CHANDRA MADHAB, b Calcutta, India, Aug 28, 32; m 66; c 3. RESPIRATORY PHYSIOLOGY, CARDIOVASCULAR PHYSIOLOGY. Educ:

Univ Calcutta, MB & BS, 55; Med Col Va, PhD(physiol), 67. Prof Exp: Clin asst med, Calcutta Nat Med Col, 55-56, house surgeon venereal dis, 56-57; med officer leprosy, Cent Leprosy Teaching & Res Inst, India, 57-58; rotating intern, St Vincent's Hosp, NY, 59; intern, St Mary's Hosp, 59-60; staff scientist respiratory physiol, Hazleton Labs Inc, Va, 67-68; from asst prof to assoc prof physiol & anesthesiol, Jefferson Med Col, Pa, 68-74; PROF PHYSIOL, SCH MED, SOUTHERN ILL UNIV, 74- Mem: AAAS; assoc Am Physiol Soc; NY Acad Sci; Am Heart Asn. Res: Air pollution; myocardial infarction; pulmonary edema; right duct lymph. Mailing Add: Dept Physiol Sch Med Southern Ill Univ Carbondale IL 62901

BANERJEE, DILIP KUMAR, b Edinburgh, Scotland, June 7, 23; nat US. ANALYTICAL CHEMISTRY. Educ: Univ Calcutta, BSc, 43; Cornell Univ, BS, 47, MS, 48; Univ Ill, PhD, 52. Prof Exp: Asst, Univ Ill, 49-52; res anal chemist, Chase Brass & Cooper Co, 53-56; proj leader, 57-60, res supvr, 60-73, SR RES ASSOC, US INDUST CHEMS CO, 73- Mem: Am Chem Soc; Am Soc Testing & Mat. Res: Analytical chemistry of metals, organic chemicals, polymers and elastomers. Mailing Add: 7866 Newbedford Ave Cincinnati OH 45237

BANERJEE, KALI SHANKAR, b Dacca, India, Sept 1, 14; m; c 2. MATHEMATICAL STATISTICS. Educ: Univ Calcutta, BA, 35, MA, 37, DPhil(statist), 50. Prof Exp: Sr asst statist, Cent Sugarcane Res & Develop, India, 42-44, statistician, 46-51; statist officer, Govt of Orissa, India, 44-46; dep dir, State Statist Bur, WBengal, 57-62, 63-64 & 66, additional dir, 66-67: vis assoc prof, Cornell Univ, 62-63; vis prof, Kans State Univ, 65-66, prof, 68-69; additional dir eval develop & planning dept, India, 67-68; prof statist, 69-74, H RODNEY SHARP PROF STATIST, UNIV DEL, 74- Concurrent Pos: Fulbright travel assistance, 62-63; NSF grants, 64-66, 68-69 & 70-71. Mem: Int Statist Inst; fel AAAS; fel Royal Statist Soc; fel Am Statist Asn; fel Inst Math Statist. Res: Weighing designs; index numbers and quadratic forms; design of experiments in general. Mailing Add: Dept of Statist & Comput Sci Univ of Del Newark DE 19711

BANERJEE, R L, b Bengal, India, Mar 1, 33; m 66; c 1. SOLID STATE PHYSICS. Educ: Univ Calcutta, BSc, 53, MSc, 56; Univ Paris, Sorbonne, DSc(physics), 66. Prof Exp: Jr lectr physics, Univ Calcutta, 57; res fel, Bengal Eng Col, Calcutta, 57-58; crystallog, Univ Madrid, 58-60; from jr researcher to full researcher physics, Univ Paris, Sorbonne, 60-67; sr lectr, Univ WI, 67-69; asst prof, 69-70, ASSOC PROF PHYSICS, UNIV MONCTON, 71- Concurrent Pos: Nat Res Coun Can Res grants, 69-70. Mem: Fr Soc Mineral & Crystallog. Res: Clay-mineral contents of Indian soils; polymorphic transitions in single crystals; thermal diffuse scattering of x-rays from molecular crystals; lattice dynamics; Compton scattering of x-rays; structures and properties of metals. Mailing Add: Dept of Physics Univ of Moncton Moncton NB Can

BANERJEE, SATYENDRA NATH, b Barala, India, Mar 1, 32; m 62; c 2. RADIOBIOLOGY, CANCER. Educ: Univ Calcutta, BSc, 54, MSc, 59; McMaster Univ, PhD(biol), 65. Prof Exp: Prin, Seva Bharati Farm Sch, Kapgari, India, 55-57; lectr bot, S S Col, Gauhati Univ, 59-60; M B B Col, Calcutta Univ, 60-61; RES FEL, ONT CANCER FOUND, HAMILTON CLIN, 65- Mem: Bot Soc Am; Radiation Res Soc; Can Soc Cell Biol; Bot Soc Japan; Japanese Soc Plant Physiol. Res: Cellular radiobiology; cancer research; plant physiology; cytogenetics; experimental embryology; algal physiology. Mailing Add: Ont Cancer Found Hamilton Clin Henderson Gen Hosp Hamilton ON Can

BANERJEE, SUBIR KUMAR, b Jamshedpur, India, Feb 19, 38; m 63; c 3. GEOPHYSICS, PALEOMAGNETISM. Educ: Univ Calcutta, BSc, 56; Indian Inst Technol, Kharagpur, MTech, 59; Univ Cambridge, PhD(geophys), 63. Prof Exp: Res physicist magnetism of ferrites, Mullard Res Labs, Redhill, Eng, 63-64; sr res assoc rock magnetism, Dept Geophys & Planetary Physics, Univ Newcastle, 64-66, univ lectr geophys, 66-69; sr staff scientist geophys & magnetic mat, Dept Mat Sci & Eng, Franklin Inst Res Labs, Philadelphia, 69-71; assoc prof geophys, 71-74, PROF GEOPHYS, UNIV MINN, MINNEAPOLIS, 74- Concurrent Pos: Consult, Atomic Energy Res Estab, Eng, 65-69; mem res staff, Ampex Corp & res assoc, Stanford Univ, 67-68; lectr, Univ Pa, 69-71; vis sr res assoc, Lamont-Doherty Geol Observ, Columbia Univ, 70-71. Mem: Am Geophys Union; Europ Phys Soc; Brit Inst Physics & Phys Soc. Res: Remanent magnetism of synthesized minerals relevant to paleomagnetism; geomagnetic field fluctuations; continental drift and sea floor spreading; magnetism of lunar samples, meteorites, recording tapes, ferrites and permanent magnets. Mailing Add: 210B Pillsbury Hall Univ Minn Minneapolis MN 55455

BANERJEE, SUBRATA, b Khulna, India, Oct 15, 40; m 75. ANALYTICAL CHEMISTRY. Educ: Univ Calcutta, BSc, 58, MSc, 61; Univ Sheffield, PhD(glass technol), 69. Prof Exp: Chief glass technologist, B K Shaw Industs,India, 61-66; sci officer ceramics, Col Ceramic Technol, Coun Sci & Indust Res, India, 69-70; res assoc geochem, Dept Geol, Univ Ga, 71-73; sr chemist, 73-74, CHIEF CHEMIST REFRACTORIES, RES CTR, GEN REFRACTORIES CO, 74- Concurrent Pos: Consult chem, C E Minerals Co, Ga, 72-73. Mem: Am Ceramic Soc; Inst Ceramic Engrs; Soc Appl Spectros; Am Chem Soc; fel Am Inst Chemists. Res: Refractory materials analysis by rapid instrument methods; use of organic and inorganic binders and additives for improved physical, chemical and thermal properties of refractory brick and associated products. Mailing Add: Gen Refractories Res Ctr PO Box 1673 Baltimore MD 21203

BANERJEE, SUSHANTA KUMAR, b Ranchi, India, Sept 25, 27; m 52; c 2. SOIL SCIENCE, BOTANY. Educ: Patna Univ, BS, 48; Tex A&M Univ, MS, 57, PhD(plant & soil sci), 64. Prof Exp: Owner & mgr, Liluah Dairy Farm, India, 48-50; asst storage & transp off, Govt WBengal, 50-55, subdiv agr off, 55-57; supt agr, Govt India, 57-60; instr biol, Mitchell Col, Conn, 64-67; from asst prof to assoc prof, 67-72, PROF & CHMN DEPT BIOL SCI, POINT PARK COL, 72- Mem: Am Inst Biol Sci; Am Soc Agron. Res: Soil fertility and plant nutrition; response of rice crop to fertilizers; protective effects of chemicals on radiation damage of plants. Mailing Add: Dept of Biol Sci Point Park Col Pittsburgh PA 15222

BANES, DANIEL, b Chicago, Ill, Apr 19, 18; m 41; c 3. PHARMACEUTICAL CHEMISTRY. Educ: Univ Chicago, BS, 38, MS, 40; Georgetown Univ, PhD(biochem), 50. Prof Exp: Chemist, US Food & Drug Admin, 39-48, res chemist, 48-63, dep dir bur sci res, 63-68, assoc comnr sci, 68-69, dir pharmaceut res & testing, 69-73; DIR DRUG STANDARDS DIV, UNITED STATES PHARMACOPEIA, 73- Concurrent Pos: Adj prof, Am Univ, 52-; adj prof grad sch, Howard Univ, 53-; sci ed, J Asn Off Anal Chemists, 60- Honors & Awards: Harvey W Wiley Award, Asn Off Anal Chemists, 68. Mem: AAAS; Am Chem Soc; Asn Off Anal Chemists. Res: Analysis of estrogenic substances, adrenocortical hormones, cardioactive glycosides and alkaloids. Mailing Add: United States Pharmacopeia 12601 Twinbrook Pkwy Rockville MD 20852

BANEWICZ, JOHN JOSEPH, physical chemistry, see 12th edition

BANEY, RONALD HOWARD, b Alma, Mich, Nov 26, 32; m 55; c 2. INDUSTRIAL CHEMISTRY. Educ: Alma Col, BS, 55; Univ Wis, PhD(inorg chem), 60. Prof Exp: Group leader, 60-74, ASSOC SCIENTIST, DOW CORNING CORP, 74- Mem: Am Chem Soc; Sigma Xi. Res: Physical organic chemical investigations, especially organosilicon chemistry; development of new silicone materials. Mailing Add: Dow Corning Corp Midland MI 48640

BANFIELD, ALEXANDER WILLIAM FRANCIS, b Toronto, Ont, Mar 12, 18; m 42; c 3. ZOOLOGY. Educ: Univ Toronto, BA, 42, MA, 46; Univ Mich, PhD, 52. Prof Exp: Mus asst Royal Ont Mus Zool & Paleont, 45-46; mammalogist, Nat Parks Serv, 46-47; chief mammalogist, Can Wildlife Serv, Dept Northern Affairs & Nat Resources, 48-57, chief zoologist, Nat Mus Can, 57-63, dir Mus Nat Sci, Ottawa, 64-69; PROF BIOL SCI & ENVIRON STUDIES, BROCK UNIV, 69-; DIR, INST URBAN & ENVIRON STUDIES, 75- Concurrent Pos: Mem, Can Comt, Int Biol Prog. Mem: Fel AAAS; fel Arctic Inst NAm; Am Soc Mammal; Can Soc Zool. Res: Status, ecology and utilization of the barren-ground caribou; ecology; mammalian systematics; zoogeography of mammals. Mailing Add: Dept of Biol Sci Brock Univ St Catharines ON Can

BANFIELD, ARMINE FREDERICK, b Winnipeg, Man, Jan 27, 09; m 41; c 4. ECONOMIC GEOLOGY. Educ: McGill Univ, BSc, 30; Northwestern Univ, MSc, 33, PhD(geol), 40. Prof Exp: Instr geol, Univ Man, 30-31 & Northwestern Univ, 31-33; geologist, Island Lake Gold Mines, Man, 33-35; chief geologist, Beattie Gold Mines, Que, 35-41; consult geologist, 45-46; mgr, Aruba Gold Mines, NWI, 47 & Equatorial Mining Corp, French Guiana, 48; CONSULT GEOLOGIST, BEHRE DOLBEAR & CO, INC, 48- Concurrent Pos: Geologist subsidiary co, Ventures, Ltd, Que, 35-41; leader, Exped Galapagos Islands, 53; mem bd, Eng Socs Libr, NY, 56-64; in-chg mineral reconnaissance, Southern & MidE Iran, 58 & 59. Mem: Am Inst Mining, Metall & Petrol Engrs; Mining & Metall Soc Am (treas, 67-); Soc Econ Geologists; Am Asn Petrol Geologists; fel Geol Soc Am. Res: Mine examination and valuation; exploration; geological mapping; ore estimation; mine development; metallic deposits in Canada, South America, United States, Africa and Middle East. Mailing Add: Behre Dolbear & Co Inc 299 Park Ave New York NY 10017

BANFIELD, WILLIAM GETHIN, b Hartford, Conn, Mar 2, 20; m 44; c 3. EXPERIMENTAL PATHOLOGY. Educ: RI State Col, BS, 41, MS, 43; Yale Univ, MD, 46; Am Univ, Washington, DC, JD, 74. Prof Exp: Am Cancer Soc fel, Yale Univ, 49-52, from instr to asst prof path, 52-54; PATHOLOGIST, NAT CANCER INST, NIH, 54- Concurrent Pos: Assoc pathologist, Grace-New Haven Community Hosp, 52-54. Mem: AMA; Electron Micros Soc Am; Am Soc Cell Biol; Microbeam Anal Soc. Res: Connective tissues; viruses. Mailing Add: Bethesda MD

BANG, FREDERIK BARRY, b Philadelphia, Pa, Nov 5, 16; m 40; c 3. MEDICINE, PATHOBIOLOGY. Educ: Johns Hopkins Univ, AB, 35, MD, 39. Prof Exp: Sci asst, USPHS, 38-39; intern, US Marine Hosp, Baltimore, 39-40; asst, Rockefeller Inst, Princeton Univ, 41-47; asst prof, 47-49, ASSOC PROF MED, JOHNS HOPKINS UNIV, 49-, PROF PATHOBIOL & CHMN DEPT, SCH HYG & PUB HEALTH, 53-, DIR INT CTR MED RES & TRAINING, INDIA, 61- Concurrent Pos: Consult, US War Dept, 42-43; Nat Res Coun fel med, Sch Med, Vanderbilt Univ, 40; Fulbright scholar, Nat Inst Med Res, Eng, 55-56; Guggenheim fel, 61-64. Mem: Am Soc Trop Med & Hyg; Soc Exp Biol & Med; Am Soc Exp Path; Am Soc Immunol; Am Soc Clin Invest. Res: Behavior of anopheles; treatment and suppression of malaria; experimental chemotherapy of schistosomiasis; growth of virus in embryos and tissue culture; electron microscopy of viruses and cells; comparative pathology of virus infections; pathogenesis of respiratory virus infections; mucus regulation in invertebrates. Mailing Add: Sch of Hyg & Pub Health Johns Hopkins Univ Baltimore MD 21205

BANG, NILS ULRIK, b Copenhagen, Denmark, Sept 24, 29; m 60; c 3. MEDICINE, HEMATOLOGY. Educ: Copenhagen Univ, MD, 55. Prof Exp: Intern, Munic Hosp, Copenhagen, 55-56; resident, Univ Hosp, Univ of Copenhagen, 57; resident cardiol, Mem Ctr Cancer & Allied Dis, Med Col, Cornell Univ, 57-58; NIH res trainee enzymol, Sch Med, Wash Univ, 59-61; asst prof med, NY Hosp-Cornell Med Ctr, 62-66; assoc prof, 66-72, PROF MED, SCH MED, IND UNIV, INDIANAPOLIS, 72-; CHIEF HEMAT SERV, LILLY LAB CLIN RES, MARION COUNTY GEN HOSP, 66- Concurrent Pos: Spec fel enzymol, Munic Hosp, Copenhagen, 56; res fel, Sloan-Kettering Inst Cancer Res, 58-S9; fel couns thrombosis, stroke & arteriosclerosis, Am Heart Asn. Mem: AAAS; Harvey Soc; Am Fedn Clin Res; fel Am Col Cardiol; Am Hemat Soc. Res: Isolation, purification and study of interreactions between coagulation proteins from blood and tissues with specific reference to comparative physical chemical and electron microscopic studies of fibrinogen fibrin conversion. Mailing Add: Hemat Serv Lilly Lab Clin Res Marion Co Gen Hosp Indianapolis IN 46202

BANGASSER, SUSAN ANDRETTA, b Memphis, Tenn, May 18, 49; m 71. IMMUNOCHEMISTRY. Educ: Northwestern Univ, BA, 71; Univ Ill Med Ctr, PhD(biochem), 75. Prof Exp: FEL & JR RES SCIENTIST IMMUNOL, NAT MED CTR, CITY OF HOPE, 75- Res: Isolation and characterization of tumor associated antigens; isolation method and assay for human cholycystokinin. Mailing Add: City of Hope Nat Med Ctr Dept of Immunol 1500 E Duarte Duarte CA 91010

BANGDIWALA, ISHVER SURCHAND, b Surat, India, Jan 9, 22; m 47; c 2. STATISTICS. Educ: Bombay, BSc, 43, MSc, LLB, 46; Univ NC, MS, 50, PhD(exp statist, math statist), 58. Prof Exp: Head statist sect & legal adv, K A Pandit, Actuary, India, 44-48; head statist sect, Agr Exp Sta, PR, 52-58, subdir res & consult statistician, Superior Ed Coun, 58-66, PROF STATIST & RES, UNIV PR, RIO PIEDRAS, 66- Concurrent Pos: Lectr, Inst Statist, Univ PR, 53-, Col Agr & Mech Arts, 59-61. Mem: AAAS; Am Statist Asn; fel Inst Math Statist; Am Biomet Soc; Am Soc Agr Sci. Res: Statistical methodology; sampling; experimental designs; educational and social science research. Mailing Add: Box 21648 Univ of PR Rio Piedras PR 00931

BANGERTER, BENEDICT W, b Huron, SDak, Sept 14, 41. PHYSICAL CHEMISTRY. Educ: Macalester Col, BA, 63; Calif Inst Technol, PhD(chem), 69. Prof Exp: Asst prof chem, Univ Ill, Chicago Circle, 68-75; MGR MAGNETIC RESONANCE FACIL, DEPT CHEM, HARVARD UNIV, 75- Res: Nuclear magnetic resonance spectroscopy; magnetic relaxation; biomolecular interactions. Mailing Add: Dept of Chem Harvard Univ 12 Oxford St Cambridge MA 02138

BANGHART, FRANK W, b Michael, Ill, Oct 30, 23; m 46; c 2. BIOSTATISTICS. Educ: Quincy Col, AB, 49; Drake Univ, MA, 50, EdD(statist), 57. Prof Exp: Instr psychol, Quincy Col, 49-52; asst instr ed, Univ Va, 55-56, instr, 56-57, from asst prof to assoc prof biostatist, 57-65; PROF EDUC MGT SYSTS, FLA STATE UNIV, 65- Concurrent Pos: Prin investr, Off Naval Res, 57-61; Air Force Syst Command Hqs, 57-; mem adv comt new educ media, US Health, Educ & Welfare Dept, 61- Mem: AAAS; Am Statist Asn; Biomet Soc; NY Acad Sci. Res: Statistical methodology, research designs. Mailing Add: Dept of Educ Admin Fla State Univ Tallahassee FL 32306

BANGLE, RAYMOND, JR, b San Francisco, Calif, Jan 16, 25; m 47; c 3. PATHOLOGY. Educ: Univ Southern Calif, MD, 49. Prof Exp: Clin intern, Los Angeles County Hosp, 48-49; resident path, Mallory Inst Path, 49-50; sr asst surgeon, NIH, 51-54; instr, 53-60, ASST CLIN PROF PATH, UNIV SOUTHERN CALIF, 60-; LAB DIR & PATHOLOGIST, LAB PROCEDURES (SM) DIV, UPJOHN CO, 67- Concurrent Pos: Fel, Mem Ctr, NY Univ, 50-51; dir labs & res, Cancer Detection Ctr, Cancer Prev Soc, Los Angeles, 60-67. Mem: AMA; Am Asn Path & Bact; Histochem Soc; Am Soc Cytol; Am Soc Clin Path. Res: Pathologic anatomy; histochemistry; analytical cytology. Mailing Add: Upjohn Co Lab Procedures Div 6330 Variel Ave Woodland Hills CA 91364

BANICK, WILLIAM MICHAEL, JR, b Scranton, Pa, Feb 6, 32; m 56; c 3. ANALYTICAL CHEMISTRY. Educ: Kings Col, BS, 53; Univ Ill, PhD(chem), 57. Prof Exp: Anal proj chemist, 57-65, sr res chemist, 66-68, GROUP LEADER, AM CYANAMID CO, 68- Mem: Am Chem Soc. Res: Analytical chemistry of basic intermediates; rubber chemicals; brighteners; ultraviolet absorbers; antioxidants, especially acid-base behavior, thin-layer and liquid chromatography. Mailing Add: Res & Develop Dept Am Cyanamid Co Bound Brook NJ 08805

BANIGAN, THOMAS FRANKLIN, JR, b Dover, NJ, Sept 22, 20; m 46; c 2. ORGANIC CHEMISTRY. Educ: Univ Notre Dame, BS, 42, MS, 43, PhD(org chem), 46. Prof Exp: Chemist, Sinclair Refining Co, Ind, 46-48; res chemist, rubber res sta, USDA, 48-53; sr res chemist, Tide Water Oil Co, 53-57; chemist, Arthur D Little, Inc, 57-61; vpres & dir res, Ben Holt Co, 62-65; tech dir, Pilot Chem Co, 65-68; mgr chem res, Avery Prod Corp, 68-70; tech dir, Pilot Chem Co, 70-73; PRES, COAST CHEM RES, 74- Res: Organic synthesis; pesticides; detergent chemicals; adhesive polymers. Mailing Add: 925 Paloma Dr Arcadia CA 91006

BANIK, NARENDRA LAL, b Ganganagar, India, Jan 2, 38; m 68; c 1. NEUROCHEMISTRY. Educ: Univ Calcutta, BSc, 59; Univ London, MSc, 66, PhD(biochem), 70. Prof Exp: Res asst biochem, Inst Psychiat, London, 61-65; res asst, Charing Cross Hosp Med Sch, London, 65-71; lectr neurochem, Inst Neurol, London, 71-74; RES ASSOC NEUROL, SCH MED, STANFORD UNIV, 74- Mem: Biochem Soc London; Int Soc Neurochem; Am Soc Neurochem. Res: Brain development process of myelination and dissolution of the myelin membrane in demyelinating diseases; studies on the synthesis of myelin components by isolated myelin-forming oligodendroglial cells. Mailing Add: Neurol Serv Vet Admin Hosp 3801 Miranda Ave Palo Alto CA 94304

BANIK, UPENDRA K, b Chittagong, India, Nov 1, 30; Can citizen; m 58; c 2. REPRODUCTIVE PHYSIOLOGY, ENDOCRINOLOGY. Educ: Univ Calcutta, MSc, 52, DSc(zool, endocrinol), 60; Univ Minn, MS, 54; Worcester Found Exp Biol, dipl, 62. Prof Exp: Pop Coun NY fel reprod physiol, Worcester Found Exp Biol, 61-62, fel, 62-65; sr sci officer, Twyford Lab, London, Eng, 65; SR BIOLOGIST, AYERST LABS, 65- Mem: Can Soc Study Fertil; Can Physiol Soc; Int Soc Res Reprod; Can Soc Endocrinol Metab; Am Asn Marriage & Family Coun. Res: Antifertility and fertility promoting compounds. Mailing Add: Ayerst Labs 1025 Laurentian Blvd Montreal PQ Can

BANISTER, JOHN ROBERT, b Wayne, Nebr, Feb 4, 23; m 50; c 3. PHYSICS. Educ: Iowa State Col, BS, 48, PhD(physics), 53. Prof Exp: Physicist staff mem, 53-60, div leader, 60-64, DEPT MGR, SANDIA LAB, 64- Mem: Fel Am Phys Soc. Res: Plasma physics; hydromagnetics; fluid mechanics; particulate physics; weapon effects. Mailing Add: 1205 Jefferson NE Albuquerque NM 87110

BANITT, ELDEN HARRIS, b Red Wing, Minn, Apr 7, 37; m 59; c 2. ORGANIC CHEMISTRY. Educ: St Olaf Col, BA, 59; Univ Wis, PhD(org chem), 64. Prof Exp: Fel, Univ Calif, Berkeley, 64-65; res specialist, 65-75, SR RES SPECIALIST, RIKER LABS, 3M CO, 75- Mem: Am Chem Soc. Res: Synthetic medicinal chemistry. Mailing Add: Riker Res Lab Bldg 218-1 3M Co Ctr St Paul MN 55101

BANK, NORMAN, b New York, NY, Oct 19, 25; m 54; c 2. NEPHROLOGY. Educ: NY Univ, AB, 49; Columbia Univ, MD, 53. Prof Exp: From instr to assoc prof med, Sch Med, NY Univ, 60-71; assoc prof, 71-73, PROF MED, ALBERT EINSTEIN SCH MED, 73- Concurrent Pos: Life Ins Med Res Found res fel, New Eng Med Ctr, Tufts Univ, S7-59; assoc attend physician, NY Univ Med Ctr, 68-71, attend physician, 71-; chief nephrology, Montefiore Hosp, 71- Mem: Am Soc Clin Invest; Am Physiol Soc; Am Fedn Clin Res; Am Soc Nephrology. Res: Regulation of acid secretion by kidney; renal concentrating defects in hypercalcemia; renal function in pyelonephritis and mercury poisoning; regulation of sodium transport. Mailing Add: Montefiore Hosp Dept Nephrol 111 E 210th St Bronx NY 10467

BANK, SHELTON, b New York, NY, Jan 28, 32; m 57; c 2. PHYSICAL ORGANIC CHEMISTRY. Educ: Brooklyn Col, BS, 54; Purdue Univ, PhD(phys chem), 60. Prof Exp: Res fel, Harvard Univ, 60-61; res chemist, Esso Res & Eng Co, 61-64, sr res chemist, 64-66; assoc prof chem, 66-72, PROF CHEM, STATE UNIV NY ALBANY, 72- Concurrent Pos: Consult, Col Chem Consults, Am Chem Soc, 72- Mem: Am Chem Soc. Res: Chemistry of aromatic radical anions; base catalysed olefin isomerization; thermodynamic and kinetic properties of cyclic and bicyclic compounds. Mailing Add: Dept of Chem State Univ of NY Albany NY 12222

BANK, STEVEN BARRY, b New York, NY, Mar 14, 39. MATHEMATICS. Educ: Columbia Univ, AB, 59, AM, 60, PhD(math), 64. Prof Exp: INSTR MATH, UNIV ILL, URBANA-CHAMPAIGN, 64- Mem: Am Math Soc. Res: Pure mathematics; asymptotic theory of ordinary differential equations in the complex domain. Mailing Add: Dept of Math Univ of Ill Urbana IL 61803

BANKER, GILBERT STEPHEN, b Tuxedo Park, NY, Sept 12, 31; m 56; c 2. PHARMACY. Educ: Albany Col Pharm, Union Univ (NY), BS, 53; Purdue Univ, MS, 55, PhD(pharm), 57. Prof Exp: From asst prof to assoc prof pharm, 57-64, PROF INDUST PHARM, PURDUE UNIV, WEST LAFAYETTE, 64-, HEAD DEPT INDUST & PHYS PHARM, 66- Concurrent Pos: Consult, Sandoz Inc, 69- & Rohm and Haas Inc, 72- Honors & Awards: Lederle Fac Res Award, 61; Am Pharm Asn Indust Pharm Award, 71. Mem: Fel AAAS; Am Pharmaceut Asn; Am Chem Soc; NY Acad Sci; fel Acad Pharmaceut Sci (vpres, 70-71). Res: Quantitative evaluations of pharmaceutical unit operations; applications of radioactive tracer techniques to the evaluation of pharmaceutical products; emulsion and suspension rheology and stability studies; applications of synthetic polymers to pharmaceutical product development. Mailing Add: Dept of Indust & Phys Pharm Purdue Univ West Lafayette IN 47907

BANKERT, RALPH ALLEN, b York, Pa, Jan 19, 18; m 46; c 2. ORGANIC CHEMISTRY. Educ: Gettysburg Col, BA, 40; Pa State Col, MS, 41, PhD(org chem), 44. Prof Exp: Res chemist, Res Ctr, P F Goodrich Co, 43-50; RES CHEMIST, RES CTR, HERCULES INC, 50- Mem: Am Chem Soc. Res: Antioxidants; preparation of aliphatic ketimines; reactions of N-monochloramines; terpenes; amine chemistry; peroxides; polymers. Mailing Add: Res Ctr Hercules Inc Wilmington DE 19899

BANKERT, RICHARD BURTON, b St Louis, Mo, Apr 22, 40; m 70; c 1. IMMUNOLOGY. Educ: Gettysburg Col, BA, 62; Univ Pa, VMD, 68, PhD(immunol), 73. Prof Exp: CANCER RES SCIENTIST IMMUNOL, ROSWELL PARK MEM INST, 73- Concurrent Pos: Res fel, Sch Vet Med, Univ Pa, 68-70 & 70-73. Mem: AAAS; Sigma Xi. Res: Mechanisms that control the proliferation of normal as well as neoplastic antibody-forming cells with emphasis upon cell associated receptors and auto-anti-receptor antibodies. Mailing Add: Dept of Immunol Res Roswell Park Mem Inst 666 Elm St Buffalo NY 14263

BANKIER, ROBERT G, b Scotland, June 1, 26; Can citizen; m 51; c 4. PSYCHIATRY, ELECTROENCEPHALOGRAPHY. Educ: Glasgow Univ, MB, ChB, 49; Univ Man, dipl psychiat, 65, MSc, 68; Royal Col Physicians & Surgeons Can, cert psychiat, 65. Prof Exp: Intern med & surg, Royal Hosp Sick Children & Western Infirmary, Glasgow, 49-50; hon physician, Outpatient Dept, Huddersfield Royal Infirmary, Eng, 53-56; pvt pract, 56-61; asst resident psychiat, Hosp Ment Dis, Selkirk, Man, 61-62; resident, Winnipeg Gen hosp, 62-63; Childrens Hosp & Child Guid Clin, Winnipeg, 63-64; sr resident, Psychiat Inst, Winnipeg, 64-65; asst res psychiatrist, 65-66, DIR PHYS THER, SCH MED REHAB & ASSOC PROF PSYCHIAT, UNIV MAN, 66- Concurrent Pos: Teaching fel, Univ Man, 62-63, 64-65; clin dir & dir res & residency traning prog, Hosp Ment Dis, Selkirk, Man; specialist in electroencephalog, Prov Man; psychiat consult, Nat Parole Bd Can. Mem: Fel Am Psychiat Asn; NY Acad Sci; Can Med Asn; Can Psychiat Asn; Can Soc Electroencephalog. Res: Clinical psychiatry; psychopharmacology. Mailing Add: Dept of Psychiat Univ of Man Winnipeg MB Can

BANKIEWICZ, M C, organic chemistry, see 12th edition

BANKO, WINSTON EDGAR, b Spokane, Wash, May 22, 20; m 45; c 4. WILDLIFE ECOLOGY. Educ: Ore State Col, BS, 43. Prof Exp: Wildlife biologist, SDak Dept Game, Fish & Parks, 46-48; refuge mgr, 48-58, chief sect wildlife mgt, Bur Sports Fisheries & Wildlife, 58-65, WILDLIFE BIOLOGIST, RARE & ENDANGERED WILDLIFE RES PROG, BUR SPORTS FISHERIES & WILDLIFE, US FISH & WILDLIFE SERV, 65- Concurrent Pos: Asst to Dr Philip S Humphrey on Pac Ocean Biol Surv Prog, Div Birds, US Nat Mus, Smithsonian Inst, 63-65. Mem: Wildlife Soc; Am Ornith Soc; Cooper Ornith Soc. Res: Endangered wildlife on oceanic islands; waterfowl, endangered species. Mailing Add: PO Box 54 Hawaii National Park HI 96718

BANKOWSKI, RAYMOND ADAM, b Chicago, Ill, Feb 4, 14; m 40; c 2. VETERINARY MEDICINE. Educ: Mich State Col, DVM, 38; Univ Calif, MS, 40, PhD(comp path), 46. Prof Exp: Asst, Univ Calif, Berkeley, 38-40; assoc, Agr Exp Sta, 40-42, asst prof vet sci & asst vet, 46-52, assoc prof & assoc vet, 52-57, PROF VET SCI & VET, EXP STA, UNIV CALIF, DAVIS, 57- Concurrent Pos: Collabr bur animal indust, USDA, Mex, 48-49; Fulbright award, 57-58; pres, Conf Vet Lab Diag, 59-60; mem bd dirs, Am Col Vet Microbiologists, 66-70, pres, 69-70; mem virus study sect, NIH, 68-70; consult animal & plant health inspection serv, USDA, 68-74. Honors & Awards: Mark Morris Found Award, 61; Am Vet Med Asn Res Award, 71. Mem: Fel Am Acad Microbiol; Am Vet Med Asn; NY Acad Sci; Am Asn Avian Path. Res: Virus diseases, especially vesicular and poultry; diagnosis; prevention and control of diseases of animals. Mailing Add: Sch of Vet Med Univ of Calif Davis CA 95616

BANKS, CLARENCE KENNETH, b Kansas City, Mo, June 1, 14; m 40; c 2. ORGANIC CHEMISTRY, GRAPHICS. Educ: Univ Kansas City, AB, 36; Univ Nebr, MS, 38, PhD(chem), 40. Prof Exp: Sr res chemist, Parke, Davis & Co, 40-42, div head, 42-49; dir res & develop, 49-51, vpres res & develop, 51-62, VPRES & SCI DIR, M&T CHEM INST, AM CAN CO, 62- Mem: AAAS; Am Chem Soc; Am Pharmaceut Asn; Am Electroplaters Soc; Soc Chem Indust. Res: Organometallic compounds; medicinals; reaction mechanisms; heterocyclic compounds; tropical diseases; welding electroplating. Mailing Add: M&T Chem Inc Am Can Co 100 Park Ave New York NY 10017

BANKS, DALLAS O, b Campbell, Nebr, Oct 21, 28; m 52; c 2. APPLIED MATHEMATICS. Educ: Ore State Univ, BS, 50, MS, 52; Carnegie Inst Technol, PhD(math), 59. Prof Exp: Math asst, Gen Elec Co, Wash, 52-54; from asst prof to assoc prof math, 59-70, PROF MATH, UNIV CALIF, DAVIS, 70- Concurrent Pos: Grant, Air Force Off Sci Res, 62-65. Mem: Math Asn Am; Am Math Soc; Soc Indust & Appl Math. Res: Differential equations of engineering and physics. Mailing Add: Dept of Math Univ of Calif Davis CA 95616

BANKS, DAVID FRANKLIN, organic chemistry, photographic chemistry, see 12th edition

BANKS, DAVID LEE, b Lynchburg, Va, Mar 8, 25; m 54. ZOOLOGY. Educ: Bluefield State Col, BS, 48; WVa Univ, MS, 50. Prof Exp: From instr to assoc prof biol, Bluefield State Col, 52-69; PERSONNEL STAFF, NAVAL NUCLEAR FUEL DIV, BABCOCK & WILCOX CO, 69- Concurrent Pos: Vis scientist high schs, WVa, 59-; NSF stipend, Bot Conf, Univ NC, 59; NSF grant, Inst Col Teachers Biol, Williams Col, Mass, 65; consult pub schs. Mem: AAAS; NY Acad Sci. Res: Comparative vertebrate anatomy; human anatomy; improvement of science teaching at elementary and junior high school levels. Mailing Add: Box 785 Naval Nuclear Fuel Div Babcock & Wilcox Co Lynchburg VA 24505

BANKS, DONALD JACK, b Sentinel, Okla, July 11, 30; m 52; c 2. BOTANY. Educ: Okla State Univ, BS, 53, MS, 58; Univ Ga, PhD(bot), 63. Prof Exp: Instr agron, Auburn Univ, 58-60; asst prof biol, Stephen F Austin State Col, 63-66; RES GENETICIST, PLANT SCI RES DIV, AGR RES SERV, USDA, 66-; ASSOC PROF AGRON, OKLA STATE UNIV, 68- Mem: Bot Soc Am; Am Soc Plant Taxon; Am Soc Agron; Int Asn Plant Taxon. Res: Plant taxonomy, agrostology, cytology, genetics and breeding; speciation in genus Paspalum; studies in the Paspalum setaceum Michaux compex; breeding, genetics, cytology, speciation and evolution of peanuts. Mailing Add: Dept of Agron Okla State Univ Stillwater OK 74074

BANKS, EDWIN MELVIN, b Chicago, Ill, Mar 21, 26; m 50; c 3. BIOLOGY, ZOOLOGY. Educ: Univ Chicago, BS, 49, MS, 50; Univ Fla, PhD(biol), 55. Prof Exp: Instr biol, Univ Ill, 55-57, asst prof, 57-62; assoc prof zool, Univ Toronto, 62-65; assoc prof, 65-68, PROF ZOOL, ANIMAL SCI & PSYCHOL, UNIV ILL, URBANA, 68-, HEAD DEPT ECOL, ETHOLOGY & EVOLUTION, 73- Concurrent Pos: NIH spec fel, Yerkes Regional Primate Ctr, 70. Mem: AAAS; Ecol Soc Am; Am Soc Zoologists; Animal Behav Soc (pres-elect, 70). Res: Social behavior of vertebrates; social organization, aggressive and reproductive behavior of birds and mammals. Mailing Add: 515 Morrill Hall Univ of Ill Urbana IL 61801

BANKS, EPHRAIM, b Norfolk, Va, Apr 21, 18; m 45; c 2. INORGANIC CHEMISTRY. Educ: City Col New York, BS, 37; Polytech Inst Brooklyn, PhD(chem), 49. Prof Exp: Jr metallurgist, NY Naval Shipyard, 41-46; res fel, Polytech Inst Brooklyn, 46-49, res assoc, 49-50, from instr to assoc prof inorg chem, 50-58, actg head, 68-71, PROF INORG CHEM, POLYTECH INST NY, 58-, HEAD DEPT CHEM, 71- Concurrent Pos: Consult, Westinghouse Elec Corp, 57-63, Gen

Motors Corp, 65-70 & Mallinckrodt Chem Works, 67-73; fel, Weizmann Inst, 63-64; NSF fac fel, 71-72. Mem: Fel AAAS; Am Chem Soc; Am Phys Soc; Electrochem Soc; fel NY Acad Sci. Res: Solid state chemistry and physics; crystal chemistry; luminescence; semiconductors; magnetic materials and structures; Mössbauer spectroscopy. Mailing Add: Dept of Chem Polytech Inst of NY Brooklyn NY 11201

BANKS, EUGENE PENDLETON, b Florence, SC, Nov 5, 23; m 45; c 4. ANTHROPOLOGY. Educ: Furman Univ, BA, 43; Harvard Univ, MA, 50, PhD(anthrop), 54. Prof Exp: Field dir, Psychiat Res Proj, Univ NC, 52-53; instr anthrop, Duke Univ, 53-54; from asst prof to assoc prof sociol & anthrop, 54-62, PROF SOCIOL & ANTHROP, WAKE FOREST UNIV, 62-, CHMN DEPT, 64- Concurrent Pos: Fulbright lectr, Rangoon, Burma, 60-61 & Zagreb, 66-67; NSF vis lectr, Am Anthrop Asn, 64-; Fulbright res scholar, Bucharest, 72-73. Mem: Fel Am Anthrop Asn; Am Ethnol Soc; Soc Appl Anthrop; Soc Am Archaeol. Res: Southeast Asia; West Indies; Island Carib Indians; themes and values in culture; Eastern Europe. Mailing Add: Dept of Sociol & Anthrop Wake Forest Univ Winston-Salem NC 27109

BANKS, HARLAN PARKER, b Cambridge, Mass, Sept 1, 13; m 39; c 2. PLANT MORPHOLOGY, ANATOMY. Educ: Dartmouth Col, AB, 34; Cornell Univ, PhD(bot), 40. Prof Exp: Instr bot, Dartmouth Col, 34-36; asst, Cornell Univ, 37-39, instr, 39-40; from instr to assoc prof biol, Acadia Univ, 40-47; asst prof bot, Univ Minn, 47-49; assoc prof, 49-50, head dept bot, 52-61, PROF BOT, CORNELL UNIV, 50- Concurrent Pos: Fulbright res scholar, Univ Liege, 57-58; Guggenheim fel, 63-64; fel, Clare Hall, Cambridge, 68-; mem ed bd, Rev Paleobot & Palynology, 68-; lectr, Am Inst Biol Sci Vis Biologists Prog, 69-70 & 71-73; mem, Paleont Res Inst. Honors & Awards: Cert of Merit, Bot Soc Am, 75. Mem: AAAS; Bot Soc Am (secy pro-tem, 52-53, treas, 64-67, vpres, 68, pres, 69); Torrey Bot Club; Int Orgn Paleobot (vpres, 64-69, pres, 69-75); foreign mem Geol Soc Belg. Res: Paleobotany, origin and early evolution of land plants (Devonian and other early Paleozoic Periods). Mailing Add: 214 Plant Sci Bldg Div of Biol Sci Cornell Univ Ithaca NY 14850

BANKS, HAROLD DOUGLAS, b Brooklyn, NY, Apr 28, 43; m 67. ORGANIC CHEMISTRY. Educ: Brooklyn Col, BS, 63, MA, 65; Cornell Univ, PhD(org chem), 70. Prof Exp: ASST PROF CHEM, UNIV BRIDGEPORT, 70- Concurrent Pos: Am Chem Soc Petrol Res Fund grant, Univ Bridgeport, 70, Mem: AAAS; Am Chem Soc. Res: Amination with N-chlorodialkylamines; synthesis and solvolysis of norbornyl derivatives; conformational analysis in substituted 1,3-dioxanes; infrared studies of hydrogen bonding; stereochemistry of base-catalyzed decarboxylation. Mailing Add: Dept of Chem Univ of Bridgeport Bridgeport CT 06602

BANKS, HARVEY WASHINGTON, b Atlantic City, NJ, Feb 7, 23; m 52; c 4. ASTROPHYSICS. Educ: Howard Univ, BS, 46, MS, 48; Georgetown Univ, PhD(astron), 61. Prof Exp: Res asst physics, Howard Univ, 48-50; engr, Nat Electronics Inc, 52-54; teacher pub schs, Washington, DC, 54-56; res asst astron, Georgetown Col Observ, 56-61, lectr & res assoc, 63-67; fel, Georgetown Univ, 61-62; prof astron & math & dir observ, Del State Col Observ, 67-69; assoc prof astron, 69-71, ASSOC PROF PHYSICS, HOWARD UNIV, 71-, COORDR ASTRON & ASSOC DIR COMPREHENSIVE SCI, 70- Mem: Am Astron Soc. Res: Determination of orbits; celestial mechanics; geodetic determinations from the observations of solar eclipse and satellites; high dispersion spectroscopy. Mailing Add: Dept of Physics and Astron Howard Univ Washington DC 20001

BANKS, HENRY H, b Boston, Mass, Mar 9, 21; m 45; c 3. ORTHOPEDIC SURGERY. Educ: Harvard Univ, BA, 42; Tufts Univ, MD, 45; Am Bd Orthop Surg, cert, 56. Prof Exp: Geog full time, Peter Bent Brigham Hosp, 53-70; PROF ORTHOP SURG & CHMN DEPT, SCH MED, TUS UNIV, 70-, ASSOC DEAN AFFIL HOSPS, 72- Concurrent Pos: Asst clin prof orthop surg, Harvard Med Sch, 65-70; head div orthop, Peter Bent Brigham Hosp, 68-70; surgeon-in-charge pediat orthop surg, New Eng Med Ctr Hosps, 70-71; dir orthop surg, Boston City Hosp, 70-75, chmn surg serv, 61-73; mem bd trustees, Kennedy Mem Hosp, 74-; secy-treas, Am Bd Orthop Surg, 74-; consult orthop surg, numerous hosps. Mem: Am Orthop Asn; Am Acad Orthop Surg; Int Soc Orthop Surg & Traumatology; Orthop Res Soc; Am Acad Cerebral Palsy (past pres). Mailing Add: New Eng Med Ctr Hosp 171 Harrison Ave Boston MA 02111

BANKS, JAMES E, analytical chemistry, see 12th edition

BANKS, JOHN HOUSTON, b Ripley, Tenn, Feb 9, 11; m 33; c 2. GEOMETRY. Educ: Tenn Polytech Inst, BS, 35; George Peabody Col, MA, 38, PhD(math), 49. Prof Exp: Prof math & dean, ECent Jr Col, 44-46; prof, Ala State Teachers Col, Florence, 46-49; assoc prof, 49-58, PROF MATH, GEORGE PEABODY COL, 58-, HEAD DEPT, 59- Concurrent Pos: Vis prof, Auburn Univ, 56 & Univ NB, 60; consult textbk proj, AID, 64-70. Mem: Math Asn Am. Res: Improvement of critical thinking; research in training college teachers and television instruction; teaching of arithmetic. Mailing Add: 3708 Lealand Lane Nashville TN 37204

BANKS, NORMAN GUY, b Philadelphia, Pa, Feb 19, 40. GEOLOGY. Educ: NMex Inst Mining & Technol, BS, 62; Univ Calif, San Diego, MS, 65, PhD(geol), 67. Prof Exp: GEOLOGIST, US GEOL SURV, 67- Mem: AAAS; Geol Soc Am. Res: Geologic mapping; economic geology; petrology; geochemistry. Mailing Add: US Geol Surv 345 Middlefield Rd Menlo Park CA 94025

BANKS, PETER MORGAN, b San Diego, Calif, May 21, 37; m 60; c 4. ATMOSPHERIC PHYSICS, AERONOMY. Educ: Stanford Univ, MS, 60; Pa State Univ, PhD(physics), 65. Prof Exp: Res asst physics, Pa State Univ, 63-65; res assoc aeronomy, Inst Space Aeronomy, Brussels, Belgium, 65-66; from asst prof to assoc prof electrophys, 66-74, PROF ELECTROPHYS, UNIV CALIF, SAN DIEGO, 74- Concurrent Pos: Mem, Int Sci Radio Union. Mem: AAAS; Am Geophys Union. Res: Radio propagation; structure of the upper atmosphere; aeronomy; magnetospheric physics. Mailing Add: Dept of Appl Phys & Info Sci Univ of Calif San Diego La Jolla CA 92037

BANKS, PHILIP OREN, b Palo Alto, Calif, Jan 31, 37. GEOLOGY, GEOCHEMISTRY. Educ: Mass Inst Technol, BS, 58; Calif Inst Technol, PhD(geochem), 63. Prof Exp: Asst prof, 63-68, ASSOC PROF GEOL, CASE WESTERN RESERVE UNIV, 68- Mem: AAAS; Geol Soc Am; Mineral Soc Am; Am Geophys Union. Res: Geochronology; isotope geochemistry. Mailing Add: Dept of Geol Case Western Reserve Univ Cleveland OH 44106

BANKS, RICHARD C, b Green Bay, Wis, Aug 19, 40; m 65; c 2. ORGANIC CHEMISTRY. Educ: Col Idaho, BS, 63; Ore State Univ, PhD(org chem), 68. Prof Exp: Asst prof org chem, 68-75, ASSOC PROF CHEM, BOISE STATE COL, 75- Mem: Am Chem Soc. Res: Isolations and identification problem on terpenes and terpene derivatives; some mechanism study on the oxycope rearrangement. Mailing Add: Dept of Chem Boise State Col Boise ID 83707

BANKS, RICHARD CHARLES, b Steubenville, Ohio, Apr 19, 31; m 67; c 2. ORNITHOLOGY, MAMMALOGY. Educ: Ohio State Univ, BS, 53; Univ Calif, Berkeley, MS, 58, PhD(zool), 61. Prof Exp: NSF grant & res biologist, Birds & Mammals, Calif Acad Sci, 61-62; cur & ed, San Diego Nat Hist Mus, 62-66; chief bird sect, Bird & Mammal Labs, 66-71, dir, 71-73, STAFF SPECIALIST, US FISH & WILDLIFE SERV, 73- Concurrent Pos: Res assoc, Dept Vert Zool, Smithsonian Inst, 67- Mem: Am Ornithologists Union (secy, 69-73); Am Soc Mammal; Cooper Ornith Soc; Wilson Ornith Soc; Soc Study Evolution. Res: Systematics, distribution, molt and hybridization of American birds; endangered species; biology of introduced species. Mailing Add: US Fish & Wildlife Serv Washington DC 20240

BANKS, WILLIAM ALDEN, b Preston, Ga, Jan 7, 33; m 54; c 4. ENTOMOLOGY. Educ: Univ Ga, BS, 54; Clemson Univ, MS, 55. Prof Exp: Inspector, Plant Pest Control Div, 55-61, entomologist, Agr Res Serv, 61-65, ENTOMOLOGIST, ENTOM RES DIV, AGR RES SERV, USDA, 65- Mem: Entom Soc Am. Res: Investigations on biology and control of imported fire ants, cockroaches, fleas and bedbugs. Mailing Add: PO Box 14565 Gainesville FL 32604

BANKS, WILLIAM JOSEPH, JR, b Port Chester, NY, Nov 3, 38; m 60. VETERINARY HISTOLOGY, CYTOLOGY. Educ: Calif State Polytech Col,eCol Kellogg Voorhis, BS, 64; Colo State Univ, MS, 66, PhD(anat), 68. Prof Exp: Res chemist, Aerojet Gen Corp, Calif, 63; asst prof zool, Calif State Col, Fullerton, 67-68; asst prof zool, Calif State Polytech Col Kellogg Voorhis, 68; vet anat, Vet Col, Univ Sask, 68-70, assoc prof, 70; assoc prof, 71-74, PROF ULTRASTRUCTURAL CYTOL & ELECTRON MICROS & ACTG CHMN DEPT ANAT, COL VET MED & BIOMED SCI, COLO STATE UNIV, 74-, DIR ELECTRON MICROS TRAINING LABS, 71- Concurrent Pos: Med Res Coun Can res grant, 69-71. Mem: AAAS; Am Inst Biol Sci; Electron Micros Soc Am; Am Asn Anat; Am Asn Vet Anat. Res: Elucidation of chondrogenic and osteogenic processes, using cervine antler as model system. Mailing Add: Dept of Anat Col of Vet Med Colo State Univ Ft Collins CO 80521

BANKS, WILLIAM LOUIS, b Paterson, NJ, Mar 25, 36; m 65. BIOCHEMISTRY, NUTRITION. Educ: Rutgers Univ, BS, 58, PhD(physiol, biochem), 63; Bucknell Univ, MS, 61. Prof Exp: Clin lab officer, US Air Force Sch Aerospace Med, 63-65; from asst prof to prof biochem, 65-74, PROF BIOCHEM & SURG, MED COL VA, 74-, CO-DIR CANCER CTR, 74- Concurrent Pos: Lectr, St Mary's Univ, Tex, 64-65. Mem: Am Asn Cancer Res; Am Chem Soc; Am Inst Nutrit; NY Acad Sci; Soc Exp Biol & Med. Res: Cancer cell biology; hydrazine drug toxicology; protein nutrition; protein and nucleic acid metabolism; breast cancer. Mailing Add: MCV VCU Cancer Ctr Med Col of Va Richmond VA 23298

BANKS, WILLIAM MICHAEL, b New York, NY, Jan 10, 14; m 43; c 1. ZOOLOGY. Educ: Univ Iowa, BA, 42; Ohio State Univ, MA, 49, PhD, 53. Prof Exp: From assoc prof to prof biol, Agr & Technol Col NC, 52-56; assoc prof, Grambling Col, 56-58; prof & chmn dept, Clark Col, 58-61; from asst prof to assoc prof, 61-67, PROF ZOOL, HOWARD UNIV, 70- Mem: AAAS; Am Micros Soc; Am Soc Zool; Entom Soc Am; Am Inst Biol Soc. Res: Taxonomy; life cycles and nutritional requirements of trematodes; invertebrate physiology; carbohydrases; composition of hemolymph; cardiology of cockroaches; characterization and biological effects of bacterial toxins. Mailing Add: 4207 Blagden Ave & Terr NW Washington DC 20011

BANKS, WILLIAM PATRICK, b Little Rock, Ark, Jan 12, 25; m 53; c 2. ELECTROCHEMISTRY. Educ: Univ Ark, BS, 49; Univ Mo, MS, 51; Univ Okla, PhD(chem), 53. Prof Exp: Res chemist, Carter Oil Co, 53-60; res scientist corrosion chem, Continental Oil Co, 60-62, sr res scientist, 62-73; DEVELOP CHEMIST, HALLIBURTON SERV, 73- Mem: Am Chem Soc; Nat Asn Corrosion Engrs. Res: High pressure phase behavior studies; anodic corrosion control; organic electrochemistry; lattice hydrate structure; phase behavior; corrosion inhibitors; electrochemical methods of pollution control; electrochemistry of corrosion; corrosion control during industrial cleaning of power plant systems. Mailing Add: 1106 Harville Rd Duncan OK 73533

BANN, ROBERT (FRANCIS), b New Baden, Ill, June 30, 21; m; c 4. ORGANIC CHEMISTRY. Educ: Univ Ill, BS, 43; Rutgers Univ, MS, 59. Prof Exp: Sect chief chemist, 56-65, chief chemist, 65-68, gen supt dyes mfg, 68-73, MGR DYES & CHEM INTERMEDIATES PROD, DYES DEPT, ORG CHEM DIV, AM CYANAMID CO, 73- Mem: Am Chem Soc; Am Asn Textile Chemists & Colorists. Res: Azo dyes and intermediates; triphenylmethane dyes; dyes for synthetic fibers. Mailing Add: 489 Knollwood Dr Bridgewater NJ 08807

BANNAN, ELMER ALEXANDER, b Wilmington, Del, Sept 5, 28; m 58; c 3. MICROBIOLOGY. Educ: Univ Mich, BS, 58, MS, 60. Prof Exp: MICROBIOLOGIST, PROCTER & GAMBLE CO, 60- Mem: Am Soc Microbiol; Soc Indust Microbiol. Res: Bacterial physiology, the ultra structure of bacterial spores; antimicrobial cleansing products; skin degerming. Mailing Add: Ivorydale Tech Ctr Procter & Gamble Co Cincinnati OH 45217

BANNAN, MARVIN WILLIAM, b Parry Sound, Ont, Mar 23, 08; m 58. BOTANY. Educ: Univ Toronto, BA, 29, PhD(plant anat), 33. Prof Exp: Asst, 29-39, sr demonstr & res asst, 39-41, lectr, 41-47, from asst prof to prof, 47-70, assoc chmn dept, 66-69, EMER PROF BOT, UNIV TORONTO, 70- Honors & Awards: George Lawson Medal, Can Bot Asn, 73. Mem: Bot Soc Am; Royal Soc Can. Res: Plant anatomy. Mailing Add: 68 Edenbrige Dr Islington ON Can

BANNARD, ROBERT ALEXANDER BROCK, b Edmonton, Alta, Apr 28, 22; m 46; c 2. ORGANIC CHEMISTRY. Educ: Queen's Univ, BSc, 45, MSc, 46; McGill Univ, PhD(chem), 49. Prof Exp: Jr res officer, Atomic Energy Proj, Nat Res Coun Can, 49-50, asst res officer, Div Appl Chem, 50-51; SCI SERV OFFICER, DEFENCE RES ESTAB OTTAWA, DEFENCE RES BD, 52- Mem: Am Chem Soc; fel Chem Inst Can; The Chem Soc. Res: Stereochemistry; organophosphorus chemistry; heterocyclics; alicyclic compounds; gas-liquid chromatographic separations and analysis. Mailing Add: 618 Westminster Ave Ottawa ON Can

BANNER, ALBERT HENRY, b Bellingham, Wash, Aug 23, 14; m 38; c 4. MARINE BIOLOGY. Educ: Univ Wash, BS, 35, PhD(zool), 43; Univ Hawaii, MS, 39. Prof Exp: Jr biologist, US Fish & Wildlife Serv, 41-42; biologist, Wash State Dept Fisheries, 42-43; from asst prof to assoc prof zool, 46-56, PROF ZOOL, UNIV HAWWII, 56- Concurrent Pos: Consult, US Nat Mus, 47-; mem Coral Atoll res team, Pac Sci Bd, 51; fel, Bishop Mus, 54, hon assoc, 62-; dir, Hawaii Marine Lab, 55-63; mem, Jaluit Typhoon Invest Team, 58; Fulbright scholar, Thailand, 60-61. Mem: Am Soc Limnol & Oceanog; Soc Syst Zool; Int Soc Toxicol. Res: Marine toxicology; taxonomy and zoogeography of shrimps; biology of coral reefs; taxonomy and distribution and Mysidacea and Euphausiacea; marine pollution. Mailing Add: Hawaii Inst of Marine Biol PO Box 1346 Kaneohe HI 96744

BANNER, EDWARD ARTHER, b Chicago, Ill, Mar 15, 12; m 41; c 4. OBSTETRICS & GYNECOLOGY. Educ: Loyola Univ, Ill, BS, 38, MD, 39; Univ Minn, MS, 47; Am Bd Obstet & Gynec, dipl, 47. Prof Exp: From instr to assoc prof, 48-69, PROF OBSTET & GYNEC, MAYO GRAD SCH MED, UNIV MINN, 69- Concurrent

Pos: Consult obstet & gynec, Mayo Clin, 47- Mem: AMA; Am Col Obstet & Gynec; fel Am Col Surg; Continental Gynec Soc; Cent Asn Obstet & Gynec. Mailing Add: Mayo Grad Sch of Med Univ of Minn Rochester MN 55901

BANNERMAN, DOUGLAS GEORGE, b Calgary, Alta, Can, Feb 22, 17; US citizen; m 42; c 4. TEXTILE CHEMISTRY. Educ: Beloit Col, BS, 39; Colo Col, MA, 41; Mass Inst Technol, PhD(chem), 45. Prof Exp: Instr chem, Colo Col, 39-41 & Mass Inst Technol, 42-45; asst prof, Colo Col, 45-46; res chemist, 45-56, merchandizing mgr, 56-60, MKT DEVELOP MGR, E I DU PONT DE NEMOURS & CO, 60- Mem: Am Chem Soc. Res: Organic research in heterocyclic chemistry; high polymer chemistry; synthetic fiber research; textile development; synthetic fiber paper. Mailing Add: Textile Fibers Dept E I du Pont de Nemours & Co Wilmington DE 19898

BANNERMAN, HAROLD MACCOLL, b Barney's River, NS, Apr 9, 97; nat US; m 29; c 1. GEOLOGY, MINERAOLOGY. Educ: Acadia Univ, BSc, 24; Princeton Univ, PhD(geol), 27. Hon Degrees: AM, Dartmouth Col, 34; LLD, St Francis Xavier Univ, 53; DSc, Acadia Univ, 55. Prof Exp: Asst geol, Princeton Univ, 24-25; from instr to prof, Dartmouth Col, 27-42; geologist, US Geol Surv, 42-44, chief sect nonmetalliferous geol, 43-44; chief div econ geol, 44-46, asst chief geologist, 48-60, assoc chief geologist, 60-61, staff geologist, 61-67; vis lectr, Wesleyan Univ, 67-68, adj prof geol, 68-71; RETIRED. Concurrent Pos: Asst geologist, Geol Surv Can, 27-30; consult geologist, US & Can, 31-42; mem, Mat Adv Bd, 57-62. Honors & Awards: US Dept Interior Distinguished Serv Award, 59. Mem: Fel Geol Soc Am; fel Am Mineral Soc; Soc Econ Geol (secy, 57-61, pres, 64); Am Inst Mining, Metall & Petrol Eng. Res: Precambrian geology; mineral deposits; igneous and metamorphic geology. Mailing Add: 509 Coleman Rd Middletown CT 06457

BANNERMAN, ROBIN MOWAT, b Alton, Eng, Feb 2, 28; m 53; c 3. MEDICAL GENETICS, HEMATOLOGY. Educ: Oxford Univ, BA, 49, BM & BCh, 52, MA & DM, 60; MRCP, 55; FRCP, 73. Prof Exp: Mem house staff, St Thomas Hosp, London, 52-57; lectr, Radcliffe Infirmary & Oxford Univ, 59-62, tutor, 62-63; assoc prof, 63-70, PROF MED, STATE UNIV NY BUFFALO, 70-; CHIEF MED GENETICS UNIT, DEPT MED, BUFFALO GEN HOSP, 63- Concurrent Pos: Fel med, Barnes Hosp, St Louis, Mo, 57-58 & Johns Hopkins Hosp, Baltimore, Md, 58-59; Jr Bd Prog vis prof, Buffalo Gen Hosp, 62; vis prof, Sch Med, Nat Univ Paraguay, 67; vis prof hemat, St Thomas Sch Med, London, 71. Mem: Am Soc Human Genetics; Am Soc Hemat; Soc Exp Biol & Med; fel Am Col Physicians; Cent Soc Clin Res. Res: Mechanisms and inheritance of anemias, particularly thalassemias; iron metabolism. Mailing Add: Dept of Med State Univ of NY Buffalo NY 14203

BANNISTER, BRIAN, b Gateshead, Eng, Mar 10, 26; m 57. ORGANIC CHEMISTRY. Educ: Oxford Univ, BA, 47, BSc, 47, MA & DPhil(org chem), 50. Prof Exp: Nuffield Found res fel, Dyson Perrins Lab, Univ Oxford, 50-52; Alumni Res Found fel, Univ Wis, 52-53, NSF fel, 53-54; proj leader dept chem, 54-73, SR SCIENTIST, UPJOHN CO, 73- Mem: Am Chem Soc; The Chem Soc. Res: Degradative and synthetic investigations in field of natural products, with major emphasis on antibiotics. Mailing Add: Dept of Infectious Dis Res Upjohn Co Kalamazoo MI 49008

BANNISTER, BRYANT, b Phoenix, Ariz, Dec 2, 26; m 51; c 2. DENDROCHRONOLOGY, ACADEMIC ADMINISTRATION. Educ: Yale Univ, BA, 48; Univ Ariz, MA, 53, PhD(anthrop), 60. Prof Exp: Lab asst dendrochronology, Lab Tree-Ring Res, 51-53, res asst, 53-54, cur archaeol collections, 54-59, from instr to assoc prof dendrochronology, 59-65, asst dean col earth sci, 71-72, actg dean, 72, ASSOC DEAN COL EARTH SCI, UNIV ARIZ, 72-, PROF DENDROCHRONOLOGY, LAB TREE-RING RES, 65-, DIR LAB, 64- Concurrent Pos: Ed, Tree-Ring Bull, 58-69; collabr, Nat Park Serv, 60-; res assoc, Mus Northern Ariz, 60- Mem: Fel AAAS; fel Am Anthrop Asn; Soc Am Archaeol; Tree-Ring Soc. Res: Derivation and application of archaeological tree-ring dates. Mailing Add: Lab of Tree-Ring Res Univ of Ariz Tucson AZ 85721

BANNISTER, LOREN WILLARD, organic chemistry, see 12th edition

BANNISTER, PEGGY JEAN, analytical chemistry, see 12th edition

BANNISTER, ROBERT GRIMSHAW, b Terre Haute, Ind, Oct 27, 25; m 56; c 4. TEXTILE CHEMISTRY. Educ: Rose Polytech Inst, BS, 46; Univ Ill, MS, 48, PhD(org chem), 51. Prof Exp: Chemist anal chem, Com Solvents Corp, 47-48; res chemist textile chem, 51-53, patent chemist, 53-63, SR PATENT CHEMIST TEXTILE CHEM, E I DU PONT DE NEMOURS & CO, 64- Mem: Am Chem Soc. Res: Fiber forming polymers; fiber structure; textile machinery; chemical and textile patents. Mailing Add: E I du Pont de Nemours & Co Wilmington DE 19898

BANNISTER, THOMAS TURPIN, b Orange, NJ, Apr 20, 30; m 53; c 2. BIOLOGY. Educ: Duke Univ, BS, 51; Univ Ill, MS, 53, PhD(biophys), 58. Prof Exp: Res asst, Duke Univ, 48-51 & Univ Ill, 51-53, 55-58; instr biol, 58-60, from asst prof to assoc prof, 61-69, chmn dept biol, 69-75, PROF RADIATION BIOL & BIOPHYSICS, UNIV ROCHESTER, 69- Concurrent Pos: NIH res fel, Nat Cent Sci Res, France, 62-63. Mem: Am Soc Plant Physiol; Biophys Soc. Res: Photosynthesis; photochemistry of chlorophyll and energy transfer; plant and cell physiology; biophysics; ecology of algae. Mailing Add: Dept of Biol Univ of Rochester Rochester NY 14627

BANNISTER, WILLIAM WARREN, b Terre Haute, Ind, Feb 24, 29; m 52; c 5. PHYSICAL ORGANIC CHEMISTRY, MARINE CHEMISTRY. Educ: Purdue Univ, BS, 53, PhD(org chem), 61. Prof Exp: Chemist, Miami Valley Labs, Procter & Gamble Co, Ohio, 60-65; asst prof org chem, Tenn Tech Univ, 65-66; asst prof, Univ Cincinnati, 66-67; ASSOC PROF CHEM, UNIV LOWELL, 67- Concurrent Pos: Res grants, Sigma Xi, 66, NSF, 67, 68 & US Navy, 75 & 76. Mem: Am Chem Soc; Sigma Xi. Res: Structure-property correlations of organic compounds; acidities of organic compounds; organic reaction mechanisms; environmental chemistry with regard to inland, coastal and ocean waterways. Mailing Add: Dept of Chem Univ of Lowell Lowell MA 01854

BANNON, ROBERT EDWARD, b Schenectady, NY, Feb 27, 10; m 39; c 1. OPHTHALMOLOGY, OPTOMETRY. Educ: Columbia Univ, BS, 34. Hon Degrees: DOS, Chicago Col Optom, 50 & Mass Col Optom, 57. Prof Exp: Fel med sch, Dartmouth Col, 35-36, clin fel, 36-40, from instr to asst prof physiol optics, 40-47; assoc optom, Columbia Univ, 47-49, asst prof, 49-50; res optometrist bur visual sci, 50-58, CONSULT OPHTHALMIC INSTRUMENT, AM OPTICAL CO, 58- Concurrent Pos: Res ed, Am J Optom, 47-51; vis lectr, Mass Col Optom, 57-; assoc res prof ophthal, Sch Med, State Univ NY Buffalo, 67-73. Mem: AAAS; assoc Optical Soc Am; fel Am Acad Optom (vpres, 50-52); Asn Res in Vision & Ophthal; Nat Soc Prev Blindness. Mem: Refraction; aniseikonia; physiologic optics. Mailing Add: Instrument Div Am Optical Co Buffalo NY 14215

BANOS, ALFREDO, JR, b Mexico City, Mex, Nov 14, 05; nat US; m 30, 52; c 2. THEORETICAL PHYSICS. Educ: Johns Hopkins Univ, BE, 28, Dr Eng, 36; Mass Inst Technol, PhD(theoret physics), 38. Prof Exp: Mex Am scholar, Interchange Found foreign scholar, Johns Hopkins Univ & asst, Brooklyn Edison Co, NY, 28; res assoc elec eng, Johns Hopkins Univ, 28-32; jr engr, Mex Dept Petrol, 33; lectr applied math, faculty philos, Nat Univ Mex, 33-35, elec eng, sch eng, 34-35, 38-43, dir inst physics & head dept fac sci, 38-43; mem staff, radiation lab, Mass Inst Technol, 43-45; from assoc prof to prof, 46-75, EMER PROF PHYSICS, UNIV CALIF, LOS ANGELES, 75- Concurrent Pos: In charge physics lab, Sch Advan Mech Eng & Electronics, Mexico City, 34-35, lectr, 38-43; with Off Sci Res & Develop, 44; Guggenheim fel & Fulbright scholar, 58; consult, AEC, 55-61; Aerospace Corp, 62; liaison scientist, London br, Off Naval Res, 67-68, 70-71. Mem: Fel Am Phys Soc; Am Math Soc. Res: Electromagnetic wave theory; mathematical methods of theoretical physics; magnetohydrodynamics and plasma dynamics. Mailing Add: Dept of Physics Univ of Calif Los Angeles CA 90024

BANOVITZ, JAY BERNARD, b Los Angeles, Calif, Nov 10, 25. IMMUNOCHEMISTRY. Educ: Univ Ill, BS, 46; Univ Wis, PhD(zool), 58. Prof Exp: Asst immunochem, Calif Inst Technol, 48-52; asst zool, Univ Wis, 52-57; res assoc biochem, Inst Cancer Res, Pa, 57-60; res immunologist, Nat Jewish Hosp, 60-66 & Childrens Asthma Res Inst & Hosp, 66-69; mgr immunol proj, Smith Kline Instruments, Inc, 69-73; mgr immunochem, Oxford Labs, 73-74; CONSULT, 75- Mem: Am Asn Clin Chemists; AAAS; Am Asn Immunologists; Am Chem Soc; NY Acad Sci. Res: Immunodiagnostic product development. Mailing Add: PO Box 11464 Sta A Palo Alto CA 94306

BANSAL, KRISHAN MURARI, b Ludhiana, India, June 29, 40; m 70. PHYSICAL CHEMISTRY. Educ: Panjab Univ, India, BSc, 59; Vikram Univ, MSc, 61; Univ Alta, PhD(phys chem), 68. Prof Exp: Lectr chem, Meerut Col, Agra Univ, 61; sr sci asst, Oil & Natural Gas Comn of India, 61-63; FEL RADIATION RES LABS, MELLON INST SCI, CARNEGIE-MELLON UNIV, 68- Concurrent Pos: Vis scientist, Hahn-Meitner Inst Kernforschüng, 72-74. Mem: Am Chem Soc. Res: Radiation chemistry; reaction kinetics; chain reactions in the gas phase; polarographic studies of free radicals in the microsecond time scale in aqueous and nonaqueous solutions. Mailing Add: Radiation Res Labs Carnegie-Mellon Univ Pittsburgh PA 15213

BANSE, KARL, b Koenigsberg, Ger, Feb 20, 29. BIOLOGICAL OCEANOGRAPHY. Educ: Univ Kiel, PhD(oceanog zool), 55. Prof Exp: From asst prof to assoc prof, 60-66, PROF OCEANOG, UNIV WASH, 66- Concurrent Pos: Mem comt oceanog, Nat Acad Sci-Nat Res Coun, 64-70; mem, Sci Comt Oceanic Res Working Groups on Oceanog Methods. Mem: AAAS; Am Soc Limnol & Oceanog. Res: Marine zooplankton and phytoplankton; polychaete taxonomy and ecology. Mailing Add: Dept of Oceanog WB-10 Univ of Wash Seattle WA 98195

BANTA, BENJAMIN HARRISON, b Reno, Nev, Jan 2, 27; m 57; c 3. EVOLUTIONARY BIOLOGY, HERPETOLOGY. Educ: Univ Nev, BS, 50; Stanford Univ, PhD(biol sci), 61. Prof Exp: Instr zool, Pomona Col, 58-60; res biologist, Calif Acad Sci, 60-63; asst prof zool, Colo Col, 63-65; asst prof natural sci, Univ Col, Mich State Univ, 65-69; assoc prof biol, US Int Univ, Elliott Campus, 69-75, actg chmn sci div, 70-75; LECTR ZOOL, SAN DIEGO STATE UNIV, 75- Concurrent Pos: Consult, Tex Meat Packers, Inc, Calif, 61; res grants, Am Philos Soc, 61, Sigma Xi, 62, 63 & NSF, 64-65; lectr, Univ San Francisco, 61-63 & Golden Gate Col, 62-63; res assoc dept amphibians & reptiles, Calif Acad Sci, 64-69, field assoc dept entom, 70- Mem: AAAS; Am Inst Biol Sci; Am Soc Ichthyol & Herpet; Soc Syst Zool. Res: Systematics of western North American amphibians and reptiles; biogeography of western North America; history of natural scientific explorations in western North America; extinction as a biological process. Mailing Add: 421 Santa Helena Solana Beach CA 92075

BANTA, JAMES E, b Tucumcari, NMex, July 1, 27; m 50; c 2. EPIDEMIOLOGY. Educ: Marquette Univ, MD, 50; Johns Hopkins Univ, MPH, 54; Am Bd Prev Med, dipl, 60. Prof Exp: Intern, Hosp Good Samaritan, US Navy, Los Angeles, Calif, 50-51, epidemiologist, US Naval Med Sch, 51-52 & Streptococcal Dis Res Unit, Naval Training Ctr, Md, 52-53, prev med officer, Third Marine Div, Far East, 54-55, sr investr virol sect, Naval Res Inst, Md, 55-57, chief virol sect, US Naval Med Res Unit, Egypt, 57-59, chief mil med, Naval Dispensary, Va, 59-60; chief coronary artery dis unit, USPHS, 60-62, dir ecol field sta, Univ Mo, Columbia, 62-63, dir & dep dir, Med Prog Div, Peace Corps, 63-65, head spec int progs, Off Int Res, NIH, 65-68; assoc prof, Sch Med, Georgetown Univ, 68-69, clin assoc prof, 69-70; prof pub health, Univ Hawaii, Manoa, 70-73; dep dir, Off Health, US AID, 73-75; PROF PUB HEALTH & DEAN, SCH PUB HEALTH & TROP MED, TULANE UNIV, 75- Concurrent Pos: Lectr, US Naval Med Sch, 56-57; mem subcomt criteria & methodology, Comt Epidemiol, Am Heart Asn, 61-63; clin asst prof, Univ Mo, 62-; chronic dis adv, WHO-Pan-Am Health Orgn, 69-70. Mem: Fel AAAS; fel Am Pub Health Asn; fel Am Col Prev Med; fel Am Heart Asn; NY Acad Sci. Res: Research epidemiology in infectious and chronic disease; virology; tissue culture studies; epidemiologic studies ischemic heart disease; hypertension; public health operations. Mailing Add: Sch Pub Health & Trop Med Tulane Univ New Orleans LA 70112

BANTA, MARION CALVIN, b Marshall, Tex, July 19, 34; m 61; c 2. PHYSICAL CHEMISTRY, ANALYTICAL CHEMISTRY. Educ: NTex State Univ, BS, 56, MS, 57; Univ Tex, Austin, PhD(chem), 63. Prof Exp: Res chemist, Tracor, Inc, Tex, 60-61; sr res scientist, Esso Prod Res Co, 62-66; asst prof chem, ETex Baptist Col, 66-68; ASST PROF CHEM, SAM HOUSTON STATE UNIV, 68- Mem: Am Chem Soc. Res: Electrochemical kinetics; kinetics of reaction in solution. Mailing Add: Dept of Chem Sam Houston State Univ Huntsville TX 77340

BANTA, WILLIAM CLAUDE, b Long Beach, Calif, Nov 13, 41; m 65. MARINE ZOOLOGY. Educ: Univ Calif, Berkeley, BA, 63; Univ Southern Calif, PhD(biol), 69. Prof Exp: Res fel, Div Invert Paleont, Smithsonian Inst, 69-70; ASSOC PROF BIOL, AM UNIV, 70- Concurrent Pos: Assoc ed, Chesapeake Sci, 75- Mem: NY Acad Sci; Sigma Xi; Am Micros Soc; Am Soc Zoologists. Res: Recent cheilostome and ctenostome Bryozoa. Mailing Add: Dept of Biol Am Univ Washington DC 20016

BANTER, JOHN C, b Marion, Ind, Dec 25, 31; m 55; c 4. PHYSICAL CHEMISTRY. Educ: DePauw Univ, BA, 54; Western Reserve Univ, MS, 56, PhD(chem), 57. Prof Exp: Sr res chemist, Oak Ridge Nat Lab, Union Carbide Nuclear Corp, 57-66; assoc prof chem, 66-69, PROF CHEM, FLA ATLANTIC UNIV, 69- Concurrent Pos: Consult, Metals & Ceramics Div, Oak Ridge Nat Lab, 67-69, Van Prod Corp, Chem Form Div, KMS Industs, Int Bus Mach Corp & City of Boca Raton, Fla, 68-69. Mem: Am Chem Soc; Electrochem Soc; Sigma Xi. Res: Kinetics and mechanisms of metallic corrosion; anodization of metals; solid state electrochemistry; thin oxide films; spectroscopy applied to the study of thin dielectric films. Mailing Add: Dept of Chem Fla Atlantic Univ Boca Raton FL 33432

BANTING, JAMES DANIEL, b Richlea, Sask, May 4, 16; m 43; c 3. PLANT PHYSIOLOGY. Educ: Univ Sask, BSA, 50, MSc, 52; Univ Alta, PhD(plant physiol), 56. Prof Exp: With res off weed control, Res Br Exp Farm, 57-67, RES SCIENTIST, REGINA RES STA & HEAD, WEED CONTROL SECT, CAN DEPT AGR, 67- Concurrent Pos: Mem, Nat Weed Comt, Can, 57- Mem: Agr Inst Can; Can Soc Plant

Physiol; Can Soc Agron; Can Agr Pesticide Tech Soc; Prof Inst Pub Serv Can. Res: Chemical control of wild oats in cereals; mode of action of wild oat herbicides; persistence and control of green foxtail. Mailing Add: Can Dept of Agr Res Sta Box 440 Regina SK Can

BANTTARI, ERNEST E, b Bismarck, NDak, Aug 2, 32. PLANT PATHOLOGY. Educ: Univ Minn, BS, 54, MS, 59, PhD(plant path), 62. Prof Exp: From asst prof to assoc prof, 63-73, PROF PLANT PATH, UNIV MINN, ST PAUL, 73- Mem: Am Phytopath Soc. Res: Plant virology. Mailing Add: Dept of Plant Path Univ of Minn St Paul MN 55108

BANUCCI, EUGENE GEORGE, b Racine, Wis, June 5, 43; m 66; c 3. ORGANIC POLYMER CHEMISTRY. Educ: Beloit Col, BA, 65; Wayne State Univ, PhD(org chem), 70. Prof Exp: Staff chemist polymer chem, 70-75, TECH COORDR PHYS CHEM LAB, GEN ELEC CORP RES & DEVELOP CTR, 75- Mem: Am Chem Soc. Res: Heterocyclic chemistry; cycloaddition reactions; condensation polymers; high temperature materials; cyanopolymers. Mailing Add: Gen Elec Corp Res & Develop Ctr One River Rd Schenectady NY 12345

BANUS, MARIO DOUGLAS, b Geneva, Switz, Nov 11, 21; US citizen; m 43; c 4. MARINE ECOLOGY, ENVIRONMENTAL CHEMISTRY. Educ: Mass Inst Technol, SB, 44, PhD(inorg chem), 49. Prof Exp: Sr res chemist, Metal Hydrides, Inc, 49, asst dir chem res lab, 49-51, assoc dir, 51-52, dir, 52-55, dir res & develop labs, 55-60; mem sr staff chem, Lincoln Lab, Mass Inst Technol, 60-71; mem sr staff chem, B U Marine Prog, Marine Biol Lab, Woods Hole, 71-73; SCIENTIST, MARINE ECOL DIV, PR NUCLEAR CTR, 73- Mem: NY Acad Sci; AAAS; Am Chem Soc. Mailing Add: Marine Ecol Div PR Nuclear Ctr College Sta Mayaguez PR 00708

BANVILLE, BERTRAND, b Bic, Rimouski Co, Can, Feb 19, 31; m 54; c 3. PHYSICS. Educ: Univ Montreal, BSc, 53, MSc, 55, PhD(physics), 60. Prof Exp: Lectr physics, Univ Montreal, 57-59; assoc res officer health physics, Atomic Energy Can, Ltd, 59-64; asst prof, 64-66, ASSOC PROF PHYSICS, UNIV MONTREAL, 66- Mem: AAAS; Pattern Recognition Soc; Can Asn Physicists; Can Standards Asn. Res: Nuclear physics instrumentation; health physics; biomedical engineering. Mailing Add: Dept Physics Univ of Montreal Case Postale 6128 Montreal PQ Can

BANVILLE, MARCEL, b St Octave, Que, Jan 29, 33; m 62; c 4. THEORETICAL PHYSICS. Educ: Univ Montreal, BSc, 56; Univ BC, MSc, 59, PhD(physics), 65. Prof Exp: Asst prof, 63-73, PROF PHYSICS, UNIV SHERBROOKE, 73- Concurrent Pos: Nat Res Coun Can res grant, 64-66. Mem: Am Asn Physics Teachers; Can Asn Physicists. Res: Theoretical study of phenomenological models for two-body nuclear potential; methods of nuclear structure calculations associated with a transcription of the particle operations in the ideal space; phase transition in layers. Mailing Add: Dept of Physics Univ of Sherbrooke Sherbrooke PQ Can

BANWART, GEORGE J, b Algona, Iowa, Sept 15, 26; m 55; c 2. FOOD MICROBIOLOGY. Educ: Iowa State Univ, BS, 50, PhD(bact), 55. Prof Exp: Asst prof food technol, Univ Ga, 55-57; head egg prod sect, USDA, 57-62; assoc prof bact, Purdue Univ, 62-65; res microbiologist, Plant Indust Sta, USDA, Md, 65-69; PROF FOOD MICROBIOL, OHIO STATE UNIV, 69- Mem: Am Soc Microbiol; Inst Food Technologists; Poultry Sci Asn; Brit Soc Appl Bact. Res: Salmonella in foods; methodology of Salmonella; methods to destroy Salmonella; egg products; food spoilage; public health organisms. Mailing Add: Dept of Microbiol Ohio State Univ Columbus OH 43210

BANWART, WAYNE LEE, b West Bend, Iowa, Jan 9, 48; m 70; c 2. SOILS. Educ: Iowa State Univ, BS, 69, MS, 72, PhD(soil fertil), 75. Prof Exp: Res asst soil fertil, Iowa State Univ, 69-72, res assoc, 72-75; ASST PROF SOILS, UNIV ILL, 75- Honors & Awards: George D Scarseth Award, Am Soc Agron, 74. Mem: Am Soc Agron; Soil Sci Soc Am. Res: Factors affecting ionic balance of plants as it related to plant growth and yield; organic transformations of sulfur in soils and other natural systems. Mailing Add: N-215 Turner Univ Ill Urbana IL 61801

BANZIGER, RALPH FREDERICK, b South River, NJ, June 1, 27; m 49; c 4. PHARMACOLOGY. Educ: Purdue Univ, BSc, 50, MSc, 52, PhD(pharmacol), 53. Prof Exp: Prof pharmacol & chmn dept, NDak Agr Col, 53-56; pharmacologist, Sci Assoc, 56-57 & Wallace Labs, 57-60; ASST DIR DRUG REGULATORY AFFAIRS, HOFFMAN-LaROCHE, INC, 60- Mem: Am Soc Pharmacol & Exp Therapeut; Soc Toxicol. Res: General pharmacological screening; anticonvulsants; biological evaluation of pharmaceutical formulations; acute and chronic toxicology. Mailing Add: Hoffman-LaRoche Inc Nutley NJ 07110

BAPATLA, KRISHNA M, b Guntur, India, Feb 18, 36; m 67. PHARMACEUTICS, PHARMACEUTICAL CHEMISTRY. Educ: Andhra Univ, India, BPharm, 55, MPharm, 59; Univ Southern Calif, PhD(pharmaceut chem), 66. Prof Exp: Lectr pharm, Andhra Univ, India, 59-60; asst, Univ Southern Calif, 60-66; Cancer Chemother Nat Serv Ctr res fel pharmaceut chem, Univ Ariz, 66; asst prof pharm, SDak State Univ, 66-69; res fel, Syntex Res Ctr, Calif, 69-70; res assoc, State Univ NY Buffalo, 70-71; SR SCIENTIST, WARNER-LAMBERT RES INST, 71- Mem: Am Pharmaceut Asn; Acad Pharmaceut Sci; fel Am Inst Chemists. Res: Synthesis of organic medicinal compounds; bio-pharmaceutics; pharmacokinetics; research and development of new dosage forms; development of new drug products. Mailing Add: Warner-Lambert Res Inst Morris Plains NJ 07950

BAPTIST, JAMES (NOEL), b Shelbyville, Ill, June 6, 30. MICROBIAL BIOCHEMISTRY. Educ: Case Inst Technol, BS, 52; Univ Ill, PhD(biochem), 57. Prof Exp: Chemist, Phillips Petrol Co, Okla, 52-54; Damon Runyon Fund fel enxymatic study omega oxidation, Univ Mich, 57-59; biochemist, W R Grace & Co, Md, 59-63; microbiologist, Int Minerals & Chem Co, Ill, 63-65; self-employed, Fla, 65-68; res assoc biol, 68-69, ASST BIOLOGIST, UNIV TEX M D ANDERSON HOSP & TUMOR INST, 69- Mem: Am Chem Soc; Am Soc Microbiol; Sigma Xi. Res: Protein synthesis and isozymes; metabolism of chemical carcinogens. Mailing Add: Dept of Biol Univ of Tex M D Anderson Hosp & Tumor Inst Houston TX 77025

BAPTIST, JEREMY EDUARD, b Chicago, Ill, Mar 22, 40; m 62; c 2. BIOPHYSICS. Educ: Univ Chicago, BS, 60, PhD(biophys), 66. Prof Exp: ASST PROF RADIATION BIOPHYS, UNIV KANS, 66- Mem: AAAS; Biophys Soc; Radiation Res Soc; Am Soc Microbiol. Res: Radiation repair mechanisms and their interactions in microorganisms; mechanism of mutation induction; mechanism of effect of modifiers of radiation effects; partial cell irradiation; heavy particle irradiation of cells. Mailing Add: Dept of Radiation Biophys Univ of Kans Lawrence KS 66044

BAPTIST, VICTOR HARRY, b Redlands, Calif, Mar 26, 23; m 46; c 2. PHYSICAL BIOCHEMISTRY. Educ: Northwestern Univ, BS, 48, PhD(phys biochem), 52. Prof Exp: Control chemist, 42-43 & 46-48; res assoc protein chem, Northwestern Univ, 51-52 & Univ Iowa, 52-53; res chemist in chg chem control, Don Baxter, Inc, 53-63, control adminr, 63-68; PRIN RES CHEMIST, MAX FACTOR & CO, 68- Mem: Am Chem Soc; Soc Cosmetic Chemists; Soc Appl Spectros; Am Soc Testing Mat; Coblentz Soc. Res: Methods of analysis; solar energy; making potable water from sea water; levitation. Mailing Add: Max Factor & Co 1655 N McCadden Pl Hollywood CA 90028

BAPTISTA, LUIS FELIPE, b Hong Kong, Aug 9, 41; US citizen; m 70; c 1. BIOACOUSTICS, BEHAVIORAL GENETICS. Educ: Univ San Francisco, BS, 65, MS, 68; Univ Calif, Berkeley, PhD(zool), 71. Prof Exp: Cur asst bot, Calif Acad Sci, 65-67; cur asst ornith, Mus Vert Zool, Univ Calif, Berkeley, 67-70, actg asst prof zool, 71-72, asst cur, 71; ASST PROF BIOL, OCCIDENTAL COL, 73-, CUR MOORE LAB ZOOL, 73- Concurrent Pos: Max Planck Soc fel, Max Planck Inst Physiol & Behav, 71-72 & NATO fel, 72-73; referee, NSF Grant Proposals, 75- Mem: Am Ornithologists Union. Res: Biosystematics of neotropical birds; avian song dialects; behavior genetics of tropical estrildid finches. Mailing Add: Moore Lab of Zool Occidental Col Los Angeles CA 90041

BAR, ASHER, b Jerusalem, Israel, Aug 7, 38. SPEECH PATHOLOGY, AUDIOLOGY. Educ: Hebrew Univ Jerusalem, BA(physiol) & BA(sociol), 62; Univ Pittsburgh, MA, 64, PhD(speech path, audiol), 66. Prof Exp: Asst prof speech path, Emerson Col, 66-69; ASSOC PROF OTOLARYNGOL, MT SINAI SCH MED, 69- Concurrent Pos: Adj prof, Hunter Col, 69- & City Univ New York, 70- Honors & Awards: Honors, Fac Med, Nat Univ Rosario, Arg, 72. Mem: Am Speech & Hearing Asn; NY Acad Sci; Int Asn Logopedics & Phoniatrics. Res: Communication disorders; small group interaction. Mailing Add: Dept of Otolaryngol Mt Sinai Sch of Med New York NY 10029

BAR, HANS-PETER, b Koenigsberg, Ger, Apr 16, 37; m 68. ORGANIC CHEMISTRY, PHARMACOLOGY. Educ: Dipl org chem, Tech Univ Darmstadt, 63, PhD(org chem), 64. Prof Exp: Lectr chem, Gettysburg Col, 60-61; sci asst org chem, Max Planck Soc, Ger, 62-65; res assoc biochem, Univ BC, 65-67; asst mem hormone res, Inst Biomed Res, AMA Educ & Res Found, 67-69; asst prof, 69-71, ASSOC PROF PHARMACOL, UNIV ALTA, 71- Concurrent Pos: Can Heart Found fel, 66-67. Mem: NY Acad Sci; Pharmacol Soc Can; Can Fedn Biol Sci; Ger Chem Soc. Res: Chemical stepwise synthesis of oligonucleotides; enzymology of cardiac nucleotide-catabolizing enzymes related to autoregulation of coronary flow; mechanism of adenosine deaminase action; adenyl cyclase and molecular mechanisms of hormone action. Mailing Add: Dept of Pharmacol Univ of Alta Edmonton AB Can

BARABAS, EUGENE S, b Pestszentlorinc, Hungary, Mar 30, 20; US citizen; m 45; c 2. POLYMER CHEMISTRY, ORGANIC CHEMISTRY. Educ: Miklos Horthy Univ Sci, PhD(chem, physics), 43. Prof Exp: Res chemist, Agr Chem Co, Hungary, 45-47; group leader cereal chem, Res Inst Grain & Flour, Hungary, 47-50; dept head acetylene chem, Res Inst Plastics, Hungary, 50-53; chief engr chem, Hungarian Bur Standards, 53-57; sr res chemist polymer chem, Cent Res Lab, Air Reduction Co, Inc, 58-62; group leader, Borden Chem Co, 62-63; sr res chemist, 63-66, tech assoc, 66-68, SR TECH ASSOC POLYMER CHEM, GAF CORP, 68- Concurrent Pos: Mem, Nat Tech Comt Utilization Furfural & Nat Tech Comt Utilization Natural Gas, 51-53; lectr, Inst Continuation Course Engrs, 52; asst prof chem, Tech Univ, Budapest, 53-57; consult, Nat Mus Fine Arts & Cent Indust Court Arbitration, 54-57. Res: Polymerization; mechanism and kinetics of polymerization reactions; synthesis of monomers and acetylenic chemicals; preparation of new polymers for various applications; emulsion polymerization. Mailing Add: 41 Stanie Brae Dr Watchung NJ 07060

BARABAS, SILVIO, b Sarayevo, Yugoslavia, June 10, 20; m 59; c 2. WATER POLLUTION, ENVIRONMENTAL HEALTH. Educ: Univ Padua, DrChS, 50. Prof Exp: Biochemist, Gen & Tuberc Hosps, 50-51; anal chemist, Noranda Copper & Brass Co, 51-52; res chemist, Aluminum Res Labs, 52-53; from sr chemist to chief chemist, Can Copper Refiners, 53-63; head anal chem dept, Noranda Res Centre, 63-66; indust res mgr, Technicon Controls, NY, 66-69, managing dir, Technicon Int Can, 69-72; head anal methods res, Can Centre Inland Waters, 72-75; COORDR, WHO COLLABORATING CENTRE ON SURFACE & GROUND WATER QUAL, 75- Concurrent Pos: Ed, WHO Water Qual Bulletin. Honors & Awards: Fisher Sci Lect Award, Chem Inst Can, 75. Mem: Chem Inst Can; Spectros Soc of Can; Am Chem Soc; Am Soc Test & Mat; Nat Geog Soc. Res: Coordinating on behalf of WHO development of uniform or compatible analytical methodologies, instrumentation and data storage and retrieval systems. Mailing Add: WHO/CC PO Box 5050 Burlington ON Can

BARACH, ALVAN LEROY, b Newcastle, Pa, Feb 22, 95; m 33; c 2. INTERNAL MEDICINE. Educ: Columbia Univ, MD, 19. Prof Exp: Intern, Presby Hosp, 19-20; asst med, Harvard Med Sch & Boston Gen Hosp, 20-21; asst, 22-25, instr, 25-28, assoc, 28-36, from asst prof to assoc prof clin med, 36-52, CLIN PROF MED, COL PHYSICIANS & SURGEONS, COLUMBIA UNIV, 52- Concurrent Pos: Asst, Presby Hosp, New York, 22-23, assoc attend physician, 46-, attend physician in med, 58, consult med, 62. Honors & Awards: Harris Medal, City Col New York, 40; Gold Medal, Am Col Chest Physicians, 62; Redway Medal, NY State Med Soc, 64; Edward Henderson Medal, 71. Mem: Fel Am Col Physicians. Res: Pneumonia; heart disease; tuberculosis; asthma; development of oxygen tents and oxygen chambers; physiology of anoxemia in clinical disease and aviation medicine; effects of oxygen treatment in cardio-respiratory disease; physiologic therapy; pulmonary emphysema. Mailing Add: 72 E 91st St New York NY 10028

BARACH, JOHN PAUL, b New York, NY, Dec 21, 35; m 57; c 2. PLASMA PHYSICS. Educ: Princeton Univ, BA, 57; Univ Md, PhD(physics), 61. Prof Exp: From asst prof to assoc prof, 61-73, PROF PHYSICS, VANDERBILT UNIV, 73- Mem: Am Phys Soc. Res: Electromagnetic shock tubes; magnetic interactions; spectroscopy; dense plasma focus. Mailing Add: 1911 18th Ave S Nashville TN 37212

BARACH, JOSEPH LEONARD, b Pittsburgh, Pa, Oct 19, 17; m 41; c 5. POLYMER SCIENCE. Educ: Cornell Univ, AB, 39; Univ Pa, MA, 40. Prof Exp: Engr, Curtiss Propellor Div, Caldwell, NJ, 40-46; mgr res sect, Alexander Smith & Sons Carpet Co, 46-52; sect head, Textile Eng Res, Celanese Fibers Co, 52-59, mgr sales develop, 59-60, polyester coordr, 60-62, mgr new fiber prod develop, 62-66, MGR PROG EVAL & ADMIN, CELANESE FIBERS MKT CO, 66- Mem: Fiber Soc (pres, 70); Am Soc Testing Mat; Int Standards Orgn; Nat Fire Protection Asn. Res: Fiber testing; experimental stress analysis; electronic instrumentation; apparel, industrial yarn and fabric; technological forecasting; advanced management techniques; flammability research. Mailing Add: 2108 Beverly Dr Charlotte NC 28207

BARACH, RICHARD L, b Pittsburgh, Pa, Sept 24, 22; m 49; c 2. MEDICINE. Educ: Univ Pittsburgh, BS, 47; Yale Univ, MD, 49. Prof Exp: From instr to asst prof radiol, Yale Univ, 56-59; RADIOLOGIST, PRINCETON HOSP, 59-, CHMN DEPT RADIOL, 70- Concurrent Pos: Am Cancer Soc fel, Yale Univ, 54-55, USPHS trainee, 55-56; clin asst prof radiol, Sch Med, Temple Univ, 61-65; consult, US Vet Admin Hosp, Princeton Univ, Carrier Clin & Lawrenceville Sch, 60- Mem: AMA; Radiol Soc NAm. Res: Cardiovascular radiology. Mailing Add: 86 Poe Rd Princeton NJ 08540

BARAD, MORTON L, meteorology, see 12th edition

BARAFF, GENE ALLEN, b Washington, DC, Dec 27, 30; m 58; c 3. SOLID STATE PHYSICS. Educ: Columbia Univ, AB, 52; NC State Col, MS, 56; NY Univ, PhD(physics), 61. Prof Exp: Reactor physicist, Astra Assocs, Inc, 56-58; PHYSICIST, SOLID STATE THEORY, BELL TEL LABS, 61- Mem: Am Phys Soc. Res: Electronic properties of metals, semimetals, semiconductors and surfaces. Mailing Add: Bell Tel Labs Murray Hill NJ 07974

BARAGER, WILLIAM ROBERT ARTHUR, b Brandon, Man, Mar 7, 26; m 55; c 2. GEOLOGY. Educ: Univ BC, BASc, 50; Queens Col, NY, MSc, 52; Columbia Univ, PhD, 59. Prof Exp: Geologist, Falconbridge Nickel Mines Ltd, 52-54; TECH OFF & GEOLOGIST, GEOL SURV CAN, 57- Mem: Geol Soc Am; Geol Asn Can; Mineral Asn Can. Res: Petrology; petrology of basic igneous rocks; volcanism and volcanic rocks of the Canadian shield. Mailing Add: Geol Survey of Can Geol Survey Bldg 601 Booth St Ottawa ON Can

BARAJAS, LUCIANO, b Santa Cruz de Tenerife, Canary Islands, Spain, Oct 10, 33; m 60; c 3. PATHOLOGY. Educ: Univ Madrid, MD, 56. Prof Exp: Rotating intern med, Alexian Bros Hosp, Elizabeth, NJ, 57-58; res path, Malmonides Hosp, Brooklyn, NY, 58-59; res path, Barnes Hosp, Wash Univ, 59-61, USPHS trainee exp path, 59-60 & Nat Cancer Inst trainee, 60-61; from instr to asst prof path, Sch Med, 61-69, assoc prof zool, Univ, 69-74, PROF PATH, UNIV CALIF, LOS ANGELES & HARBOR HOSP, TORRANCE, 74- Concurrent Pos: NIH spec fel, 66-69. Mem: Am Soc Cell Biol; Am Asn Anat; Soc Exp Biol & Med; Int Acad Path. Res: Normal and pathologic anatomy of the kidney. Mailing Add: Dept of Path UCLA Harbor Hosp Torrance CA 90509

BARAK, ANTHONY JOSEPH, b Petersburg, Nebr, July 18, 22; m 50; c 5. BIOCHEMISTRY. Educ: Creighton Univ, BS, 48, MS, 50; Univ Mo, PhD(biochem), 53. Prof Exp: From instr to asst prof, 53-68, ASSOC PROF BIOCHEM, UNIV NEBR MED CTR, OMAHA, 68-; BIOCHEMIST, VET ADMIN HOSP, 56- Res: Alcohol metabolism; liver disease. Mailing Add: Vet Admin Hosp 42nd & Woolworth Ave Omaha NE 68105

BARAM, PETER, b St Paul, Minn, Aug 2, 26. IMMUNOLOGY, MICROBIOLOGY. Educ: Univ Ill, BS, 49, MS, 52, PhD(microbiol), 57. Prof Exp: Sr sanit bacteriologist, Chicago Park Dist, Ill, 50-S6; instr microbiol, Bowman Gray Sch Med, 56-59; from asst dir to dir dept allergy res, Michael Reese Hosp & Med Ctr, 59-73; assoc prof microbiol, 68-73, PROF MICROBIOL & SR SCI OFFICER DIV IMMUNOL, UNIV ILL MED CTR & COOK COUNTY HOSP, CHICAGO, 73-; VIS PROF MICROBIOL, RUSH-PRESBY-ST LUKES MED CTR, 73- Concurrent Pos: From assoc adj prof to adj prof, Ill Inst Technol; consult, Armour Pharmaceut Co, 53- & Am Dent Asn, 64-, dir div immunol, 68-; consult dir, Rheumatoid Arthritis Res, 73 & 74. Res: Isolation of the factor responsible for delayed hypersensitivity from human cells; purification of ragweed allergens. Mailing Add: Hektoen Inst Med Res 625 S Wood St Chicago IL 60612

BARANGER, ELIZABETH UREY, b Baltimore, Md, Sept 18, 27; m 51; c 3. THEORETICAL PHYSICS. Educ: Swarthmore Col, BA, 49; Cornell Univ, MA, 51, PhD, 54. Prof Exp: Res fel, Calif Inst Technol, 53-55; res fel, Univ Pittsburgh, 55-58, from asst prof to prof, 58-69; res assoc, Mass Inst Technol, 69-73; PROF PHYSICS & ASSOC DEAN, UNIV PITTSBURGH, 73- Mem: Am Phys Soc. Res: Theoretical nuclear physics. Mailing Add: Dept of Physics Univ of Pittsburgh Pittsburgh PA 15260

BARANGER, MICHEL, b Le Mans, France, July 31, 27; nat US; m 51; c 3. THEORETICAL PHYSICS. Educ: Ecole Normale Superieure, Paris, 45-49; Cornell Univ, PhD(physics), 51. Prof Exp: Res assoc physics, Cornell Univ, 51-53; res fel, Calif Inst Technol, 53-55; res physicist, Carnegie Inst Technol, 55-56, from asst prof to assoc prof, 56-64, prof, 64-69; PROF PHYSICS, MASS INST TECHNOL, 69- Concurrent Pos: NSF sr fel, Univ Paris, 61-62. Mem: Fel Am Phys Soc. Res: Nuclear physics; plasma spectroscopy; quantum electrodynamics. Mailing Add: Dept of Physics Rm 6-308A Mass Inst of Technol Cambridge MA 02139

BARANKIN, EDWARD WILLIAM, mathematics, statistics, see 12th edition

BARANOWSKI, RICHARD MATTHEW, b Utica, NY, Mar 1, 28; m 51; c 3. ENTOMOLOGY. Educ: Syracuse Univ, BA, 51; Univ Conn, MS, 53, PhD, 59. Prof Exp: Asst entomologist, 58-63, assoc entomologist, 63-67, ENTOMOLOGIST, SUB-TROP EXP STA, UNIV FLA, 67- Mem: Entom Soc Am. Res: Economic entomology; biology of hemiptera. Mailing Add: Univ of Fla Agr Res & Educ Ctr Homestead FL 33030

BARANSKI, MICHAEL JOSEPH, b Wheeling, WVa, Dec 8, 46; m 69. PLANT TAXONOMY. Educ: West Liberty State Col, BS, 68; NC State Univ, PhD(bot), 74. Prof Exp: Instr bot, NC State Univ, 73-74; ASST PROF BIOL, CATAWBA COL, 74- Mem: Assoc Sigma Xi; Int Asn Plant Taxon; Bot Soc Am; Soc Study Evolution; Ecol Soc Am. Res: Taxonomy and ecology of woody plants. Mailing Add: Dept of Biol Catawba Col Salisbury NC 28144

BARANWAL, KRISHNA CHANDRA, b Bahadurpur, India, Jan 1, 36. POLYMER CHEMISTRY, PHYSICAL CHEMISTRY. Educ: Univ Allahabad, BSc, 57, MSc, 59; Indian Inst Technol, Kharagpur, MTech, 60; Univ Akron, PhD(polymer chem), 65. Prof Exp: SECT MGR, B F GOODRICH RES CTR, 65- Honors & Awards: Best Paper Award, Am Chem Soc, 74. Mem: Am Chem Soc; World Future Soc. Res: Basic understanding and physical, mechanical and dynamic properties of new polymers; mechanical and dynamic properties of fibers and fiber-rubber composites. Mailing Add: B F Goodrich Res Ctr Brecksville OH 44141

BARANY, KATE, b Bekescsaba, Hungary; US citizen; m 49; c 2. PHYSICAL CHEMISTRY, BIOPHYSICS. Educ: Eötvös Lorand Univ, Budapest, 52; Univ Frankfurt, PhD(phys chem), 59. Prof Exp: Res asst, Electron Microscope Lab, Hungarian Acad Sci, Budapest, 50-57; res assoc, Max Planck Inst Physiol, Ger, 58-60; ASST MEM, INST MUSCLE DISEASE, INC, 60- Res: Preparation and characterization of myosin and its subunits from different types of muscle and various sources. Mailing Add: Inst for Muscle Dis Inc 515 East 71st St New York NY 10021

BARANY, MICHAEL, b Budapest, Hungary, Oct 29, 21; nat US; m 49; c 2. BIOCHEMISTRY. Educ: Univ Budapest, MD, 51, PhD(biochem, physiol), 56. Prof Exp: Lectr gen biochem, Inst Biochem, Univ Budapest, 51-56; intermediate scientist protein chem, Weizman Inst, 56-57; assoc mem muscle biochem, Max Planck Inst, 58-60; mem contractile proteins, Inst Muscle Dis, Inc, 61-74; PROF BIOL CHEM, UNIV ILL COL MED, 74- Mem: Am Soc Biol Chem; Am Physiol Soc; Am Chem Soc; Biophys Soc. Res: Contractile proteins of muscle; actin and myosin, isolation, characterization and mechanism of their interaction; muscle contraction; hereditary muscular dystrophy. Mailing Add: Dept of Biol Chem Univ of Ill Med Ctr Chicago IL 60612

BARANY, RONALD, b Bronx, NY, Mar 11, 28; m 60; c 4. PHYSICAL CHEMISTRY. Educ: City Col New York, BS, 49; NY Univ, MS, 51; Yale Univ, PhD(chem), 55. Prof Exp: Phys chemist thermochem measurements, US Bur Mines, 55-67; sr math analyst, Kaiser Engrs Div, Kaiser Indust Corp, 67-72; SR ENGR, BECHTEL CORP, 72- Mem: Am Chem Soc; Asn Comput Mach. Res: Thermochemistry and thermodynamics; theoretical chemistry. Mailing Add: 6553 Chabot Rd Oakland CA 94618

BARASCH, GUY ERROL, b Baltimore, Md, Aug 28, 37; wid; c 2. ATMOSPHERIC PHYSICS. Educ: Johns Hopkins Univ, AB, 58, PhD(physics), 65. Prof Exp: Staff scientist atmospheric physics, Los Alamos Sci Lab, Univ Calif, 65-69; sci specialist weapon physics, Santa Barbara Opers, EG&G, Inc, 69-71; STAFF SCIENTIST ATMOSPHERIC PHYSICS, LOS ALAMOS SCI LAB, UNIV CALIF, 71- Mem: Sigma Xi; Optical Soc Am. Res: Physics of atmospheric nuclear detonations; light emission and propagation in highly dosed air; production and dispersal of pollutants due to nuclear, chemical and natural sources. Mailing Add: MS 664 Los Alamos Sci Lab PO Box 1663 Los Alamos NM 87545

BARASCH, MURRAY LEONARD, b Brooklyn, NY, Mar 29, 27; m 55. THEORETICAL PHYSICS. Educ: Univ Pa, BA, 48, PhD(physics), 54. Prof Exp: Asst, Univ Pa, 50-54; asst, Univ Mich, 54-55, res assoc, 55-57, assoc res physicist, Radiation Lab, 57-58, res physicist, 58-67; physicist, Dikewood Corp, 67-69; PHYSICIST, APPL PHYSICS LAB, JOHNS HOPKINS UNIV, 69- Mem: Am Phys Soc; Am Geophys Union. Res: Electromagnetic theory; ionospheric physics; irreversible thermodynamics; satellite and missile problems; military operations research. Mailing Add: Appl Physics Lab Johns Hopkins Univ 8621 Georgia Ave Silver Spring MD 20910

BARASCH, WERNER, b Breslau, Ger, May 27, 19; nat US. PLASTICS CHEMISTRY. Educ: Univ Calif, AB, 47; Mass Inst Technol, SM, 49; Univ Colo, PhD(org chem), 52. Prof Exp: Asst chem, Mass Inst Technol, 47-49 & Univ Colo, 49-52; res chemist, Olin Industs, Inc, 52-56; chief chemist resins, US Plywood Corp, 56-62; dir chem res, Plastomeric, Inc, 62-67: DIR CHEM RES, BECTON, DICKINSON OF CALIF, INC, 67- Concurrent Pos: Study supvr, Harvard Univ, 49. Mem: Am Chem Soc. Res: Carbohydrate chemistry; mechanisms of organic reactions; polymers. Mailing Add: 23049 Santa Cruz Hwy Los Gatos CA 95030

BARASH, PAUL G, b Brooklyn, NY, Feb 22, 42; m 67; c 2. ANESTHESIOLOGY. Educ: City Col New York, BA, 63; Univ Ky, MD, 67. Prof Exp: From instr to asst prof, Sch Med, Yale Univ, 72-73, attend physician, 73-74, ASST CLIN DIR ANESTHESIOL, YALE NEW HAVEN HOSP, 74-, DIR SURG INTENSIVE CARE UNIT, 74- Mem: Int Anesthesia Res Soc; fel Am Col Anesthesiol; Am Soc Anesthesiol; Soc Comput Med. Res: Critical care medicine; cardiovascular and respiratory physiology; pharmacology of cocaine; computer applications in medicine. Mailing Add: Dept of Anesthesiol Yale Univ Sch of Med New Haven CT 06510

BARATOFF, ALEXIS, b France, May 23, 37; US citizen; m 61; c 2. THEORETICAL SOLID STATE PHYSICS. Educ: Mass Inst Technol, BS, 59; Cornell Univ, PhD(theoret physics), 64. Prof Exp: Res fel appl physics, Harvard Univ, 64-66; asst prof physics, Brown Univ, 66-71; res physicist, Inst Festkörperforschung, WGer, 71-73; RES PHYSICIST, IBM RES LAB, ZURICH, 73- Mem: Am Phys Soc. Res: Electronic properties of metals; superconductivity; theoretical physics. Mailing Add: IBM Forschungslabor Saumerstr 4 Rueschlikon Switzerland

BARATTA, EDMOND JOHN, b Somerville, Mass, June 22, 28; m 53; c 1. ANALYTICAL CHEMISTRY, RADIOCHEMISTRY. Educ: Northeastern Univ, BSc, 53. Prof Exp: Chemist, Control Lab, Shell Oil Co, 53-56; chemist, Petrol Lab, Naval Supply Depot, US Navy, 57-59; sr prof biol & med, Nat Lead Co, Inc, 59-61, chief anal serv, 61-68; chief anal qual control serv, Bur Radiol Health, USPHS, 68-70; chief radiation off, Environ Protection Agency, 71-72; CHIEF RADIONUCLIDE SECT, WINCHESTER ENG & ANAL CTR, FOOD & DRUG ADMIN, 72- Concurrent Pos: Consult, Standard Methods, 12th ed, Am Pub Health Asn, 63-65, 13th ed, 67-70. Honors & Awards: US Dept Health, Educ & Welfare-USPHS Superior Performance Award, 63; Am Pub Health Asn Cert, 70. Mem: Am Chem Soc; Health Physics Soc; Asn Off Anal Chem. Res: Radionuclide concentrations in human tissues including human bone; development, testing and standardization of methods for various radionuclides in environmental media; quality control for radionuclides in environment from fallout, nuclear reactors and fuel processing plants; radiopharmaceutical methodology investigations. Mailing Add: Winchester Eng & Anal Ctr 109 Holton St Winchester MA 01890

BARATZ, ROBERT SEARS, b New London, Conn, July 15, 46; m 75. DEVELOPMENTAL BIOLOGY. Educ: Boston Univ, AB, 69; Northwestern Univ, DDS, 72, PhD(anat), 75. Prof Exp: Nat Inst Dent Res fel, 72-74; INSTR ANAT, MED SCH, NORTHWESTERN UNIV, 74- Mem: Am Soc Cell Biol; AAAS; Int Asn Dent Res; Am Dent Asn; Soc Develop Biol. Res: Keratinization, cell differentiation, organ culture, tooth development and epithelio-mesenchymal interactions. Mailing Add: Dept of Anat Northwestern Univ Med Sch Chicago IL 60611

BARBA, WILLIAM P, II, b Philadelphia, Pa, Sept 25, 22; m 47; c 4. PEDIATRICS. Educ: Princeton Univ, BS, 43; Univ Pa, MD, 46. Prof Exp: Instr pediat, Res & Educ Hosp, Univ Ill, 51-52; instr, Sch Med, Temple Univ, 52-54, assoc attend pediatrician, 57-61; dir med educ, Babies Unit, United Hosps, Newark, NJ, 61-64; actg asst dean, 65-67, assoc dean, 67-69, actg dean, 69-71, ASSOC PROF PEDIAT, SCH MED, TEMPLE UNIV, 64-, ASSOC V PRES HEALTH SCI CTR, 71- Concurrent Pos: Mem adv comt, Child Guid Clin, 62-64; assoc attend pediatrician, St Christopher's Hosp Children, 64- Mem: AMA; Asn Am Med Cols; Asn Hosp Dirs Med Educ. Res: Pediatrics and medical education; systems of providing health care. Mailing Add: Sch of Med Temple Univ Philadelphia PA 19140

BARBACK, JOSEPH, b Buffalo, NY, Oct 27, 37. MATHEMATICS. Educ: Univ Buffalo, AB, 59; Rutgers Univ, MS, 61, PhD(math), 64. Prof Exp: Asst prof math, Univ Pittsburgh, 63-65 & State Univ NY Buffalo, 65-67; vis asst prof, Ariz State Univ, 67-68; NIH fel biostatist, Stanford Univ, 68-69; assoc prof math, 69-72, PROF MATH, STATE UNIV NY COL BUFFALO, 72- Mem: Asn Symbolic Logic; Am Math Soc. Res: Recursive function theory; isols. Mailing Add: Dept of Math State Univ NY Col Buffalo Buffalo NY 14222

BARBAN, STANLEY, b New York, NY, Mar 16, 21; m 50; c 2. BIOCHEMISTRY. Educ: City Col New York, BS, 43; Univ Mich, MS, 49; Wash Univ, PhD(bact & biochem), 52. Prof Exp: Bacteriologist, Dept Health, Syracuse, NY, 50; instr bact, Med Col, State Univ NY, 52-53; BIOCHEMIST, NAT INST ALLERGY & INFECTIOUS DIS, 54- Concurrent Pos: USPHS fel, 53-54. Mem: Am Soc Microbiol; Am Soc Biol Chem. Res: Tissue cell culture metabolism; bacterial metabolism; enzymology; experimental chemotherapy; biochemistry of tumor viruses. Mailing Add: Nat Insts of Health Bethesda MD 20014

BARBARO, JOHN FERDINAND, immunochemistry, bacteriology, see 12th edition

BARBAT, WILLIAM FRANKLIN, b San Francisco, Calif, Sept 18, 05; m 26; c 2. GEOLOGY. Educ: Univ Calif, AB, 26. Prof Exp: From geologist to chief geologist, Stand Oil Co Calif, 27-70; RES ASSOC, CALIF ACAD SCI, 70- Mem: Fel Geol Soc Am; Soc Econ Paleont & Mineral; Paleont Soc; Am Asn Petrol Geol. Res: Geology of California. Mailing Add: Calif Acad of Sci Golden Gate Park San Francisco CA 94118

BARBE, G DOUGLAS, plant pathology, mycology, see 12th edition

BARBEAU, ANDRE, b Montreal, Que, May 27, 31; m 56; c 4. NEUROLOGY. Educ: Univ Paris, BA, 48; Univ Montreal, MD, 56; FRCP(C), 60. Prof Exp: Markle scholar med, 61-66; DIR DEPT NEUROBIOL, CLIN RES INST MONTREAL, 67-; PROF NEUROL, UNIV MONTREAL & PROF EXP MED, McGIL UNIV, 70- Concurrent Pos: Assoc ed, Can J Neurol Sci, 74; chmn, Ctrs Excellence Neurol Prog, Pan-Am Health Orgn, 75; chief neurol prog, Hotel-Dieu Hosp, Montreal, 76. Honors & Awards: Res Award, Can Ment Health Asn, 65 & Asn French Speaking Physicians Can, 74; ICN Award Res Med Sci, 70. Mem: Can Soc Clin Invest (pres, 73); Can Neurol Soc (pres, 73); Am Neurol Asn; fel Am Acad Neurol; fel Am Col Physicians. Res: Neuropharmacology and neurogenetics of extrapyramidal and ataxia disorders in man; introduction of new drugs. Mailing Add: Clin Res Inst of Montreal 110 W Pine Ave Montreal PQ Can

BARBEAU, EDWARD JOSEPH, JR, b Toronto, Ont, June 25, 38; m 61; c 2. PURE MATHEMATICS. Educ: Univ Toronto, BA, 60, MA, 61; Univ Newcastle, PhD(semialgebras), 64. Prof Exp: Lectr math, Univ Newcastle, 63-64; asst prof, Univ Western Ont, 64-67; asst prof, 67-69, ASSOC PROF MATH, UNIV TORONTO, 69- Mem: Am Math Soc; Math Asn Am; Can Math Cong; Can Soc Hist & Philos Math. Res: Theory of semialgebras; Leonhard Euler. Mailing Add: Dept of Math Univ of Toronto Toronto ON Can

BARBEHENN, ELIZABETH KERN, b Washington, DC, June 21, 33; m 57; c 3. BIOCHEMISTRY. Educ: Cornell Univ, BS, 55; Wash Univ, PhD(biochem pharmacol), 74. Prof Exp: STAFF FEL, NIH, 74- Mem: Am Soc Biol Chemists. Res: Purification of enzymes involved in nucleic acid metabolism; stoichiometry of binding and conformational changes occurring on binding of substitutes. Mailing Add: NIH Bethesda MD 20014

BARBEHENN, KYLE RAY, b Cumberland, Md, Dec 24, 28; m 56; c 3. VERTEBRATE ECOLOGY. Educ: Rutgers Univ, BS, 50; Cornell Univ, MS, 52, PhD(vert zool), 55. Prof Exp: Biologist, Nat Res Coun, 57-58; res fel, Nat Inst Ment Health, 59-62; prin investr, NSF grant, 62-64; lectr biol, Univ Pa, 64-65; dir field ctr, Smithsonian Inst, 65-68; res fel, Ctr for Biol Natural Systs, Wash Univ, 68-74; PROJ MGR, OFF SPEC PESTICIDE REV, US ENVIRON PROTECTION AGENCY, 74- Mem: AAAS; Am Soc Mammal; Ecol Soc Am; Am Soc Naturalists; Wildlife Soc. Res: Intraspecific and interspecific behavioral relationships among small mammals and their relevance to population dynamics; urban rat ecology; management of vertebrate pest species by ecological methods; pharmacodynamics of pesticides. Mailing Add: 8208 Thoreau Dr Bethesda MD 20034

BARBER, ALBERT ALCIDE, b Providence, RI, July 13, 29; m 56; c 2. CELL BIOLOGY. Educ: Univ RI, BS, 50, MS, 52; Duke Univ, PhD, 58. Prof Exp: Assoc, Duke Univ, 57-58; from instr to assoc prof zool, 58-68, PROF ZOOL, UNIV CALIF, LOS ANGELES, 68-, ASSOC VCHANCELLOR RES, 71- Concurrent Pos: Consult, NIH, 73-75. Mem: Am Soc Biol Chemists; AAAS; Am Soc Zoologists; Am Physiol Soc. Res: Cell membranes; aging and antioxidant activity. Mailing Add: Dept of Biol Univ of Calif Los Angeles CA 90024

BARBER, DONALD E, b Harrisburg, Pa, Apr 1, 31; m 56; c 3. HEALTH PHYSICS. Educ: Dickinson Col, BS, 53; Univ Mich, MPH, 59, PhD(radiol health), 61. Prof Exp: Health physicist, Elec Boat Div, Gen Dynamics Corp, 54-55; asst prof environ health, Univ Mich, 61-66; assoc prof, 66-74, PROF HEALTH PHYSICS, SCH PUB HEALTH, UNIV MINN, MINNEAPOLIS, 74- Mem: Health Physics Soc; Radiation Res Soc; Am Indust Hyg Asn; Am Asn Phys Med. Res: Radiation dosimetry; personal dosimeters; x-ray exposure evaluations; environmental radioactivity. Mailing Add: Sch of Pub Health 1112 Mayo Bldg Univ of Minn Minneapolis MN 55455

BARBER, EUGENE DOUGLAS, b Philadelphia, Pa, Jan 20, 43; m 65; c 1. BIOCHEMISTRY, CLINICAL CHEMISTRY. Educ: Drexel Univ, BS, 65; Univ Pa, PhD(biochem), 70. Prof Exp: NIH fel lipid chem, Univ Mich, 70-72; SR RES CHEMIST CLIN CHEM, EASTMAN KODAK CO, 72- Concurrent Pos: Alt deleg, Nat Comt Clin Lab Stand, 75- Mem: AAAS; Am Chem Soc; Am Asn Clin Chemists. Res: Clinical chemical analysis of body fluids; composition and function of biological membranes. Mailing Add: Eastman Kodak Co Kodak Park Bldg 320 Rochester NY 14650

BARBER, EUGENE JOHN, b Kit Carson, Colo, Jan 8, 18; m; c 4. PHYSICAL INORGANIC CHEMISTRY. Educ: Univ Nev, BS, 40; Univ Wash, PhD(chem), 48. Prof Exp: Jr analyst, US Bur Mines, 40-41; res chemist, Div War Res, Columbia Univ, 42-45; res chemist, 48, SR DEVELOP CONSULT CHEMIST, UNION CARBIDE NUCLEAR DIV, 48- Mem: AAAS; Am Chem Soc; Sigma Xi. Res: Thermo-dynamics of nonaqueous solutions; gas phase corrosion; chemistry of fluorine and uranium. Mailing Add: Union Carbide Corp Nuclear Div PO Box P Oak Ridge TN 37830

BARBER, FRANKLIN WESTON, b New York, NY, July 3, 12; m 37; c 2. FOOD SCIENCE. Educ: Aurora Col, BS, 34; Univ Wis, MS, 42, PhD(agr bact), 44. Prof Exp: Lab technician, H P Hood & Sons, Boston, 37-40; instr bact, Univ Wis, 40-44; asst, Golden State Co, Ltd, San Francisco, 44-45; new prod develop, Res & Develop Div, Kraftco Corp, 45-46, bacteriologist & head dept bact res, 46-53, sr scientist & chief div fundamental res, 53-57, assoc mgr fundamental res, Ill, 58-60, div res dir, 60-64, asst mgr patents & regulatory compliance, 64-66; from mgr to dir regulatory compliance, 66-75; CONSULT, FOOD REGULATIONS & QUAL CONTROL, FOOD & DRUG ADMIN, 75- Concurrent Pos: Chmn, Food & Agr Orgn-WHO Joint Expert Comt Milk Hyg, Geneva, 59; mem expert adv panel environ health, WHO, 59-78. Honors & Awards: Int Asn Milk, Food & Environ Sanitarians Citation Award, 62. Mem: Am Soc Microbiol; Am Dairy Sci Asn; Inst Food Technologists; Int Asn Milk, Food & Environ Sanitarians (pres, 58-59); Asn Food & Drug Off. Res: Dairy bacteriological research; factors affecting accuracy of the phosphatase test; dissociation in lactobacilli; coliform bacteria in ice cream; psychrophilic bacteria in dairy products; milk plate count media; millipore filter techniques; patent and regulatory problems in dairy and food products. Mailing Add: 41 Fairview Blvd Ft Myers Beach FL 33931

BARBER, GEORGE ALFRED, b New York, NY, May 7, 25. BIOCHEMISTRY. Educ: Rutgers Univ, AB, 51; Columbia Univ, PhD(plant physiol, chem), 55. Prof Exp: Jr asst res biochemist, Univ Calif, Berkeley, 55-57; asst biochemist, Conn Agr Exp Sta, 57-59; biochemist, Stanford Res Inst, 59-60; asst res biochemist, Univ Calif, Berkeley, 60-65; assoc prof biochem, Univ Hawaii, 65-68; PROF BIOCHEM, OHIO STATE UNIV, 68- Mem: Am Soc Biol Chem. Res: Intermediary metabolism of higher plants. Mailing Add: Dept of Biochem Ohio State Univ Columbus OH 43210

BARBER, GEORGE WINSTON, b Arlington, Mass, May 28, 21; m 45. BIOCHEMISTRY. Educ: Yale Univ, BS, 42, PhD(chem), 49. Prof Exp: Instr physiol chem, Univ Pa, 51-53, assoc, 53-59; CHIEF BIOCHEM, WILLS EYE RES INST, 60- Concurrent Pos: Res fel physiol chem, Univ Pa, 48-51; prof ophthal, Sch Med, Temple Univ; mem visual sci study sect, NIH, 70-74. Mem: Am Chem Soc; Asn Res Vision & Ophthal. Res: Opthalmic biochemistry; steroil chemistry; amino acid metabolism and analysis. Mailing Add: 531 Narberth Ave Haddonfield NJ 08033

BARBER, JOHN CLARK, b Liberty, NC, Jan 6, 25; m 51; c 2. FOREST GENETICS. Educ: NC State Col, BS, 50, MS, 51; Univ Minn, PhD, 61. Prof Exp: Res asst forest mgt, NC State Col, 50-51; from res forester to forester, Southeastern Forest Exp Sta, 51-57, proj leader, 57-65, proj leader, Southern Forest Exp Sta, 64-67, br chief forest genetics res, 67-71, asst to dep chief res, 71-72, DIR, SOUTHERN FOREST EXP STA, US FOREST SERV, 72- Mem: AAAS; Soc Am Foresters; Am Forestry Asn. Res: Forest tree improvement; silviculture; environmental forestry. Mailing Add: Southern Forest Exp Sta US Forest Serv New Orleans LA 70113

BARBER, JOHN THRELFALL, b Lancaster, Eng, Mar 13, 37; m 61; c 2. PLANT PHYSIOLOGY. Educ: Univ Liverpool, BSc, 58 & 59, PhD(plant physiol), 62. Prof Exp: Res assoc, Lab Cell Physiol, Growth & Develop, Cornell Univ, 62-69; asst prof biol, 69-71, ASSOC PROF BIOL, TULANE UNIV, 71- Mem: AAAS; Am Soc Plant Physiol; Am Mosquito Control Asn. Res: Plant proteins in relation to growth, development and morphogenesis; ecology of mucilaginous plant seeds. Mailing Add: Dept of Biol Tulane Univ New Orleans LA 70118

BARBER, LOUIS MELVIN, pharmacy, see 12th edition

BARBER, MARY LEE, b Ft Wayne, Ind, May 20, 34; div; c 2. EMBRYOLOGY. Educ: Univ Miami, BS, 55; Duke Univ, MA, 58; Univ Calif, Los Angeles, PhD(embryol, zool), 62. Prof Exp: Part-time lectr gen zool, Univ Calif, Los Angeles, 62-64; lectr embryol, 66-68, asst prof, 66-72, ASSOC PROF EMBRYOL, CALIF STATE UNIV, NORTHRIDGE, 72- Concurrent Pos: Consult, Oak Ridge Nat Lab, 58-59; NSF grant, 71-73. Mem: Sigma Xi; Am Soc Cell Biol; Soc Develop Biol; Am Soc Zoologists. Res: Protein and lipid changes in cell membranes during early development in normal embryos and in those treated with agents producing abnormalities in development. Mailing Add: Dept of Biol Calif State Univ Northridge CA 91324

BARBER, MELVIN CLYDE, III, b Mantachie, Miss, Aug 3, 36; m 62; c 2. ECONOMIC GEOGRAPHY, URBAN GEOGRAPHY. Educ: Memphis State Univ, BS, 58; George Peabody Col, MA, 59; Southern Ill Univ, PhD(geog), 71. Prof Exp: Asst geog, George Peabody Col, 58-59; instr, Memphis State Univ, 59-61; asst, Northwestern Univ, 61-63; instr, Memphis State Univ, 63-67; asst, Southern Ill Univ, 67-70; ASST PROF GEOG, MEMPHIS STATE UNIV, 70- Mem: Asn Am Geog. Res: Economic geography of Memphis and its surrounding region. Mailing Add: 3045 Domar Memphis TN 38118

BARBER, PATRICK GEORGE, b Santa Barbara, Calif, Dec 14, 42; m 67; c 1. PHYSICAL CHEMISTRY, X-RAY CRYSTALLOGRAPHY. Educ: Stanford Univ, BS, 64; Cornell Univ, PhD(phys chem), 69. Prof Exp: Res technician radiol chem, US Naval Radiol Defense Lab, 63; Nat Sci Found res assoc chem, Duke Univ, 69-71; asst prof chem & math, 71-75, chmn div arts & sci, 73-75, PROF CHEM, DANIEL CAMPUS, SOUTHSIDE VA COMMUNITY COL, 75- Mem: AAAS; Am Chem Soc; Am Crystallog Asn; Sigma Xi. Res: Crystallographic studies on derivatives of Gramicidin-S; structure and properties of liquid crystalline compounds; conformation of polymers in the solid state. Mailing Add: Rte 2 Box 29B Keysville VA 23947

BARBER, ROBERT CHARLES, b Sarnia, Ont, Apr 20, 36; m 62; c 3. PHYSICS. Educ: McMaster Univ, BSc, 58, PhD(physics), 62. Prof Exp: Fel physics, McMaster Univ, 62-65; from asst prof to assoc prof, 65-75, PROF PHYSICS, UNIV MAN, 75- Concurrent Pos: Sabbatical leave, Sch Physics, Univ Minn, Minneapolis, 71-72; mem, Comn Atomic Masses & Fundamental Constants, Int Union Pure & Apl Physics, 73- Mem: Am Phys Soc; Can Asn Physicists. Res: Determination of atomic mass differences by high resolution mass spectrometer. Mailing Add: Dept of Physics Univ of Manitoba Winnipeg MB Can

BARBER, SAUL BENJAMIN, b Somerville, Mass, Sept 3, 20; m 55; c 3. ZOOLOGY. Educ: RI State Col, BS, 41; Yale Univ, PhD(zool), 54. Prof Exp: Instr zool, RI State Col, 46-48; instr biol, Williams Col, 52-54; instr zool, Smith Col, 54-55; res assoc, Narragansett Marine Lab, Univ RI, 55-56; from asst prof to assoc prof biol, 56-65, PROF BIOL & CHMN DEPT, LEHIGH UNIV, 65- Concurrent Pos: NIH res contract; Nat Inst Neurol Dis & Blindness spec res fel zool, Oxford, 63-64. Mem: AAAS; Am Soc Zoologists. Res: Chemoreception and proprioception in the horseshoe crab; optomotor behavior in Crustacea; fish sound production; physiology of insect flight and muscle. Mailing Add: Dept of Biol Lehigh Univ Bethlehem PA 18015

BARBER, SHERBURNE FREDERICK, b Nunda, NY, Oct 27, 07; m 44; c 3. MATHEMATICS. Educ: Univ Rochester, AB, 29, AM, 30; Univ Ill, PhD(math), 33. Prof Exp: Nat res fel, Johns Hopkins Univ, 33-34 & Princeton Univ & Inst Advan Study, 34-35; assoc math, Univ Iowa, 35-37; tutor, 37-40; from instr to assoc prof, 40-56, from asst dean to dean, Col Lib Arts & Sci, 53-71, PROF MATH, CITY COL NEW YORK, 56- Mem: Am Math Soc. Res: Cremona and birational transformations; algebraic geometry. Mailing Add: Dept of Math City Col of New York New York NY 10031

BARBER, STANLEY ARTHUR, b Wolseley, Can; Mar 29, 21; nat; m 50; c 2. SOIL CHEMISTRY, SOIL FERTILITY. Educ: Univ Sask, BSA, 45, MSc, 47; Univ Mo, PhD(soils), 49. Prof Exp: Instr soils, Univ Sask, 45-47; from asst prof to assoc prof, 49-59, PROF SOILS, PURDUE UNIV, WEST LAFAYETTE, 59- Concurrent Pos: Vis prof, Univ Calif, Berkeley, 61 & Univ Adelaide, 68-69; NSF sr fel, 68-69. Honors & Awards: Soil Res Award, Am Soc Agron, 74. Mem: Fel Am Soc Agron; Soil Sci Soc Am; Int Soc Soil Sci; Am Soc Plant Physiol. Res: Mechanisms for the movement of nutrients through the soil to the plant root; influence of the plant root on the soil; soil nutrients availability and crop growth. Mailing Add: Dept of Agron Purdue Univ West Lafayette IN 47907

BARBER, THOMAS D, b Plainview, Tex, Jan 23, 19; m 43; c 2. GEOLOGY. Educ: Tex Christian Univ, BA, 40, MA, 42. Prof Exp: Chemist, Dow Magnesium Corp, 43-44; geologist, Stanolind Oil & Gas Co, 46-52, dist geologist, 52-53, asst div geologist, 53-54, dist explor supt, 54-57, div geologist, 57, asst div explor supt, Pan Am Petrol Corp, 57-59; explor mgr, 59-68, GEN MGR, MICHEL T HALBOUTY INTERESTS, 68-, VPRES, HALBOUTY ALASKA OIL CO, 69- Mem: Am Asn Petrol Geologists; Soc Explor Geophys; Am Inst Prof Geologists; Geol Soc Am. Res: Petroleum exploration and production. Mailing Add: Michel T Halbouty Interests 5100 Westheimer Houston TX 77027

BARBER, THOMAS KING, b Highland Park, Mich, Sept 26, 23; m 47; c 3. DENTISTRY. Educ: Mich State Col, BS, 45; Univ Ill, BS, 47, DDS & MS, 49. Prof Exp: Instr pedodontics, Marquette Univ, 50-51; from instr to prof, Univ Ill, 51-69; PROF PEDIAT, SCH MED & CHMN PEDIAT DENT, SCH DENT, UNIV CALIF, LOS ANGELES, 69- Concurrent Pos: Consult, Coun Dent Educ, Am Dent Asn, Am Fund Dent Health & Educ Testing Serv. Mem: Fel Am Col Dent; Am Dent Asn; Am Soc Dent for Children; Am Acad Pedodontics. Res: Pedodontics, especially growth and dental development of children and minor orthodontics in children; educational research in dentistry. Mailing Add: Div Prev Dent Sci Univ of Calif Sch of Dent Los Angeles CA 90024

BARBER, THOMAS LYNWOOD, b Dothan, Ala, Feb 4, 34; m 47; c 3. VETERINARY MEDICINE, MICROBIOLOGY. Educ: Auburn Univ, DVM, 58; Cornell Univ, MS, 61, PhD, 69. Prof Exp: Res vet, Plum Island Animal Dis Lab, 58-69; VET MED OFFICER, ARTHROPOD-BORNE ANIMAL DIS RES LAB, AGR RES SERV, USDA, 69- Concurrent Pos: Fac affil, Colo State Univ, 73- Mem: AAAS; Am Soc Microbiol; US Animal Health Asn; Am Vet Med Asn; NY Acad Sci. Res: Arthropod borne diseases of animals; tissue culture; serologic characterization of blue tonge virus strains. Mailing Add: USDA Agr Res Serv Bldg 45 Ent B Denver Fed Ctr Denver CO 80225

BARBER, WALTER CARLISLE, b Logan, Utah, June 10, 19; m 42; c 4. NUCLEAR PHYSICS. Educ: Utah State Col, BS, 40; Univ Calif, PhD(physics), 48. Prof Exp: Asst physics, Univ Calif, 40-41, physicist, Radiation Lab, 42-45; physicist, Naval Ord Plant, Ind, 45-46; cosmic ray proj, Univ Calif, 47-48; instr physics, Stanford Univ, 48-51, res assoc, 51-54, assoc prof, 54-62, prof physics & dir high energy physics lab, 62-68; PROF PHYSICS, MASS INST TECHNOL, 68- Mem: Assoc Am Phys Soc. Res: Experimental research in nuclear physics; cosmic rays; particle counters. Mailing Add: Dept of Physics Bldg 26 Rm 443 Mass Inst of Technol Cambridge MA 02139

BARBER, WILLIAM AUSTIN, b Brooklyn, NY, Oct 2, 24; m 52; c 7. PHYSICAL CHEMISTRY. Educ: Holy Cross Col, BS, 49; Cornell Univ, PhD(phys chem), 52. Prof Exp: Res assoc, Cornell Univ, 52-53; phys res chemist, 53-60, sr res chemist, 60-62, GROUP LEADER RES DIV, AM CYANAMID CO, 62- Mem: Am Chem Soc. Res: Clay-water systems; soap-hydrocarbon gelation; polymers; organometallic chemistry; fused salts; fuel cells; heterogeneous catalysis. Mailing Add: Res Div Am Cyanamid Co Stamford CT 06901

BARBER, WILLIAM D, agronomy, genetics, see 12th edition

BARBERO, GIULIO J, b Mt Vernon, NY, Oct 13, 23; m 47; c 6. PEDIATRICS, GASTROENTEROLOGY. Educ: Univ Maine, BS, 43; Univ Pa, MD, 47. Prof Exp: Sr physician, dir div gastroenterol & cystic fibrosis res group, Children's Hosp Philadelphia, 59-67; prof pediat & chmn dept, Hahnemann Med Col, 67-73; PROF PEDIAT & CHMN DEPT, SCH MED, UNIV MO-COLUMBIA, 73- Concurrent Pos: Chmn, gen med & sci adv coun, Nat Cystic Fibrosis Res Found, 71-74. Mem: Soc Pediat Res; Am Acad Pediat; Am Gastroenterol Asn; Am Psychosom Soc. Res: Cystic fibrosis; gastrointestinal disturbances of children; psychosomatic aspects of gastrointestinal disorders of childhood. Mailing Add: Dept of Pediat Univ of Mo Columbia MO 65201

BARBIERS, ARTHUR ROBERT, JR, b Osceola, Ark, Feb 10, 24; m 45; c 3. BACTERIOLOGY. Educ: Bowling Green State Univ, BS, 50, BS & MA, 51. Prof Exp: Asst biol, Bowling Green State Univ, 50-51; asst, Dept Infectious Dis, 51-58, RES ASSOC, VET RES DEPT, UPJOHN CO, 58- Concurrent Pos: Adj asst prof, Western Mich Univ. Mem: Am Soc Microbiol. Res: Antibiotic combinations and methods for determining in vitro synergism; effects of sulfur bacteria on concrete; veterinary bacteriology; assay of antibiotic in tissues. Mailing Add: Upjohn Co Kalamazoo MI 49001

BARBORAK, JAMES CARL, organic chemistry, see 12th edition

BARBORIAK, JOSEPH JAN, b Slovakia, Feb 19, 23; nat US; m 56; c 3. BIOCHEMICAL PHARMACOLOGY. Educ: Inst Technol, Bratislava, BS, 44; Swiss Fed Inst Technol, ScD(nutrit biochem), 53. Prof Exp: Res assoc, Swiss Fed Inst Technol, 47-53; res assoc, Yale Univ, 54-58; group leader, Mead Johnson & Co, 59-61; assoc prof, 62-71, PROF PHARMACOL, MED COL WIS, 71- Concurrent Pos: Chief biochem sect, Res Serv, Vet Admin Ctr, 61- Mem: AAAS; Am Soc Pharmacol & Exp Therapeut; Am Inst Nutrit; Am Acad Allergy; Soc Exp Biol & Med. Res: Lipid metabolism; pharmacology of alcholism; immunology. Mailing Add: Vet Admin Ctr Wood WI 53193

BARBOSA, OCTAVIO, b Ituverava, Brazil, Apr 29, 07; m 33, 55; c 6. GEOLOGY. Educ: Min & Civil Eng, Sch Mining & Metall, Ouro Preto, Brazil, 30; Univ Sao Paulo, PhD(geol), 42. Prof Exp: Asst geologist, Brazilian Geol Surv, 31-34; geologist, Brazilian Bur Mines, 34-38, dir, 38-40; prof, polytech sch eng, Univ Sao Paulo, 40-42, ful prof, 42-56; CHIEF GEOLOGIST, PROSPEC S A, 56- Mem: Fel Geol Soc Am; Am Geophys Union; Brazilian Geog Soc; Brazilian Geol Soc; Brazilian Acad Sci. Res: Petrology of metamorphic rocks; basic geological mapping; economic geology and geochemical prospecting. Mailing Add: Caixa Postal 294 Petropolis Rio de Janiero Brazil

BARBOUR, ALLEN BABCOCK, b Oakland, Calif, Aug 11, 18; m 49; c 4. INTERNAL MEDICINE. Educ: Univ Calif, AB, 40, MD, 43; Am Bd Internal Med, dipl, 54. Prof Exp: Clin asst prof med, Sch Med, Univ Calif, San Francisco, 54-63; asst prof, 63-70, assoc prof clin med, 70-74, PROF CLIN MED, SCH MED, STANFORD UNIV, 74- Mailing Add: Dept of Med Stanford Univ Sch of Med Stanford CA 94305

BARBOUR, CLYDE D, b New York, NY, Oct 30, 35; m 62; c 2. ICHTHYOLOGY. Educ: Stanford Univ, AB, 48; Tulane Univ, PhD(zool), 66. Prof Exp: Asst prof biol, Univ Utah, 66-72; vis scientist, Mus Zool, Univ Mich, 72-73; vis asst prof, Miss State Univ, 73-74; asst prof biol, Tuskegee Inst, 74-75; ASSOC PROF BIOL, WRIGHT STATE UNIV, 75- Mem: AAAS; Am Soc Ichthyologists & Herpetologists; Soc Study Evolution; Soc Syst Zool. Res: Ichthyology, including systematics, zoogeography and ecology. Mailing Add: Dept of Biol Sci Wright State Univ Dayton OH 45431

BARBOUR, DONALD CURTISS, b Oakland, Calif, Dec 15, 19; m 50; c 3. INTERNAL MEDICINE. Educ: Univ Calif, AB, 41, MD, 44; Am Bd Internal Med, dipl, 58. Prof Exp: Instr, Sch Med, Harvard Univ, 52-53; from instr to asst clin prof, 56-68, ASSOC CLIN PROF MED, SCH MED, UNIV CALIF, SAN FRANCISCO, 68- Mailing Add: 45 Vista Dr Kentfield CA 94904

BARBOUR, HELEN F, b Okla, May 8, 09. HUMAN NUTRITION. Educ: Univ Okla, MHome Econ Educ, 40; Iowa State Col, MS, 48, PhD(nutrit & home econ educ), 53. Prof Exp: High sch teacher, Okla, 35-37 & 38-39; dist supvr, Works Prog Admin Sch Lunch Proj, Okla, 39-40; asst prof foods & nutrit, Ohio Wesleyan Univ, 48-50; instr, Iowa State Col, 50-52, 53; head dept home econ, NMex State Univ, 53-60; head dept food, nutrit & inst admin, 60-74, EMER PROF FOOD & NUTRIT, OKLA STATE UNIV, 74- Mem: Am Dietetic Asn; Am Home Econ Asn; Soc Nutrit Educ. Res: Nutritional status of Iowa school children; number of erythrocytes; concentration of hemoglobin and relative red cell volume as indices of evaluation; relationships of values and process concepts of selected students to generalizations in nutrition. Mailing Add: 2215 W Third Ave Stillwater OK 74074

BARBOUR, MICHAEL G, b Jackson, Mich, Feb 24, 42; m 63; c 2. BOTANY, ECOLOGY. Educ: Mich State Univ, BSc, 63; Duke Univ, PhD(bot), 67. Prof Exp: Asst prof, 67-71, ASSOC PROF BOT, UNIV CALIF, DAVIS, 71- Concurrent Pos: NSF grants, Study Coastal Vegetation, 69-71 & Study Beach Vegetation, 73-75; NSF/Int Biol Prog grant desert vegetation in NAm & SAm, 70-73; Univ Calif Sea grant mgt dune vegetation, 74-76. Mem: Am Inst Biol Sci; Ecol Soc Am; Brit Ecol Soc. Res: Physiological ecology; ecological life histories of coastal and desert plant species. Mailing Add: Dept of Bot Univ of Calif Davis CA 95616

BARBOUR, RICHARD VON, organic chemistry, see 12th edition

BARBOUR, ROGER WILLIAM, b Morehead, Ky, Apr 5, 19; m 38; c 3. VERTEBRATE ZOOLOGY. Educ: Morehead State Univ, BS, 38; Cornell Univ, MS, 39, PhD(vert zool), 49. Prof Exp: Instr zool, Moorhead State Col, 39-40, Western Ky State Col, 42-43 & Moorhead State Col, 46-47; dir nature educ, Oglebay Inst, 49-50; from instr to assoc prof zool, 50-68, PROF ZOOL, UNIV KY, 68- Concurrent Pos: Vis prof, Bandung Technol Inst, 57-59. Honors & Awards: Award of Merit, Am Conserv Asn, 67; Cert of Award, Nat Wildlife Fedn, 69; Wildlife Publ Award, Wildlife Soc, 74. Mem: Am Soc Mammal; Am Soc Ichthyol & Herpet; Wildlife Soc; Am Ornith Union. Res: Natural history and taxonomy of vertebrates. Mailing Add: Rte 1 Tates Creek Pike Lexington KY 40503

BARBOUR, STEPHEN D, b St Louis, Mo, Oct 7, 42; m 67; c 2. GENETICS, BIOCHEMISTRY. Educ: Temple Univ, AB, 64; Princeton Univ, MA, 66, PhD(biol), 67. Prof Exp: Jane Coffin Childs Fund fel, 67-69; ASST PROF MICROBIOL, CASE WESTERN RESERVE UNIV, 69- Concurrent Pos: Sabbatical, Inst Molecular Biol, Paris, 75-76. Res: Mechanism of genetic recombination; DNA replication and repair; control of enzyme synthesis. Mailing Add: Dept of Microbiol Case Western Reserve Univ Cleveland OH 44106

BARBOUR, WILLIAM E, JR, b Evanston, Ill, Nov 1, 09; m 50; c 2. APPLIED PHYSICS, ELECTRICAL ENGINEERING. Educ: Mass Inst Technol, SB, 33. Prof Exp: Consult elec instrumentation, 33-36; engr, Raytheon Co, 36-39 & Boston Edison Co, 39-41; pres & chmn nuclear instrumentation & chem, Tracerlab, Inc, 45-56; pres nuclear controls, Controls for Radiation, Inc, 57-59; pres & chmn, Magnion, Inc, 60-64; dir, Alloyd Gen Corp, 66-66; CONSULT, BARBOUR & ASSOCS, INC, 66- Concurrent Pos: Mem, Engrs Joint Coun, Int Nuclear Cong, 54-56, New Eng Coun Comt Patent Revision, AEC, 55 & Southern Regional Educ Bd Nuclear Develop, 55-56; observer, Atoms for Peace Cong, Geneva, 55; exec mgr, Asn Nuclear Instrument Mfr, Inc, 66-72; dir, QSC Indust, Inc, 71-; dir, Gen Aircraft Corp, 75- Mem: Am Nuclear Soc; Inst Elec & Electronics Engrs; Asn Advan Med Instrumentation. Res: Instrumentation; applied nuclear physics; high magnetic fields; superconductivity; nuclear medicine; x-ray, beam electronics and related vacuum technology. Mailing Add: Box 460 Concord MA 01742

BARCELLONA, WAYNE J, b Chicago, Ill, Sept 22, 40; m 70. CELL BIOLOGY. Educ: Univ Southern Calif, AB, 62, MS, 65, PhD(biol), 70. Prof Exp: Fel biol, Sect Cell Biol, M D Anderson Hosp & Tumor Inst, Univ Tex, 70-72; res assoc, Univ Tex Med Br, 72-73; ASST PROF BIOL, TEX CHRISTIAN UNIV, 73- Mem: AAAS; Am Inst Biol Sci; Am Soc Cell Biol; Soc Develop Biol; Soc Study Reproduction. Res: Reproductive biology; differentiation; control and regulation of cell division and mammalian spermatogenesis. Mailing Add: Dept of Biol Tex Christian Univ Ft Worth TX 76129

BARCELO, RAYMOND, b Montreal, PQ, July 14, 31; m 54; c 4. PHYSIOLOGY, BIOCHEMISTRY. Educ: Univ Montreal, BSc, 50, MD, 57; Univ Pa, DSc(biochem), 60; FRCPC, 62. Prof Exp: Intern, Hosp Necker, Paris, France, 60-61; asst prof, 62-70, ASSOC PROF MED, UNIV MONTREAL, 71-, CHIEF SERV NEPHROLOGY, MAISONEUVE HOSP, 64- Concurrent Pos: Head dept, Clin & Res Lab, Maisoneuve Hosp, 67-70. Mem: Am Soc Clin Res; Can Soc Clin Res; Can Soc Immunol; Can Soc Nephrol; Int Soc Nephrology. Res: Protein biochemistry; immunochemical studies of serum and urinary proteins; plasma seromucoids in various inflammatory conditions; perfusion of isolated kidney; diuretics; antibiotics; antihypertensive drugs. Mailing Add: Maisonneuve Hosp 5415 L'Assomption Blvd Montreal PQ Can

BARCH, STEPHANIE, zoology, see 12th edition

BARCHET, WILLIAM RICHARD, cloud physics, surface chemistry, see 12th edition

BARCILON, ALBERT I, b Alexandria, Egypt, July 21, 37; m 61; c 2. DYNAMIC METEOROLOGY. Educ: McGill Univ, BEngPhysics, 60; Harvard Univ, MA, 61, PhD(fluid dynamics), 65. Prof Exp: Res assoc, Meteorol Dept, Mass Inst Technol & Harvard Univ, 65-66; scientist, Space Sci Lab, Gen Elec Co, Philadelphia, 66-68; asst prof meteorol & res assoc, 68-72, ASSOC PROF METEOROL, FLA STATE UNIV, 72- Concurrent Pos: Adj asst prof, Physics Dept, Drexel Inst & Villanova Univ, 66-68; consult, Naval Res Lab, Washington, DC, 70-; liaison scientist, Off Naval Res, London, 75- Mem: Am Meteorol Soc; Royal Meteorol Soc. Res: Dynamics of intense atmospheric vortices; geophysical fluid dynamics, internal gravity waves and problems dealing with nearshore circulations. Mailing Add: Inst of Geophys Fluid Dynam Fla State Univ Tallahassee FL 32306

BARCILON, VICTOR, b Alexandria, Egypt, Oct 10, 39; m 61; c 1. APPLIED MATHEMATICS, FLUID DYNAMICS. Educ: McGill Univ, BSc, 59; Harvard Univ, AM, 60, PhD(math), 63. Prof Exp: Res assoc meteorol & oceanog, Harvard Univ, 63-64; fel meteorol, Univ Oslo, 64-65; res assoc, Mass Inst Technol, 65-66, asst prof appl math, 66-69; assoc prof math, Univ Calif, Los Angeles, 69-72; PROF MATH, DEPT GEOPHYS SCI, UNIV CHICAGO, 72- Concurrent Pos: Ed, J Soc Indust & Appl Math, 75- Mem: Soc Indust & Appl Math; Royal Astron Soc; Am Geophys Union. Res: Inverse problems of geophysics; dynamics of rotating/stratified flows. Mailing Add: 5734 S Ellis Ave Chicago IL 60615

BARCLAY, ARTHUR S, b Minneapolis, Minn, Aug 5, 32. PLANT TAXONOMY. Educ: Univ Tulsa, BS, 54; Harvard Univ, MA, 58, PhD(biol), 59. Prof Exp: Res botanist, New Crops Res Br, 59-72, RES BOTANIST, MED PLANT RESOURCES LAB, USDA, 72- Mem: Am Soc Plant Taxon; Soc Study Evolution; Soc Econ Bot. Res: Evolution; economic botany Mailing Add: Plant Genetics & Germplasm Inst Med Plant Resources Lab Beltsville MD 20705

BARCLAY, EUGENE SAMUEL, b Norfolk, Va, Jan 5, 11; m 38; c 2. BIOLOGY. Educ: Col William & Mary, BS, 36; Philadelphia Col Pharm, ScD(bact), 61. Prof Exp: Res assoc virol, Merck Sharp & Dohme, 39-40, mgr prod depts, 40-43, asst dir biol & sterile pharmaceut labs, 43-53, dir biol labs, 53-70; RETIRED. Concurrent Pos:

Consult adv comt serum albumins, Comn Plasma Fractionation & Related Processes; mem, Nat Formulary Adv Panel, 65-69 & 70-74. Mem: AAAS; Am Soc Microbiol; Am Mgt Asn; Pharmaceut Mfrs Asn; NY Acad Sci. Res: Development of process improvements and inventions in the field of human and veterinary biological production. Mailing Add: Wee Loch 36 Lochwood Lane West Chester PA 19380

BARCLAY, FRANK HUNT, b Dalton, Ga, Nov 30, 09; m 31. BIOLOGY. Educ: King Col, AB, 30; Univ Va, MA, 42; Univ Tenn, PhD(bot), 57. Prof Exp: Teacher pub sch, NC, 30-35 & Tenn, 35-47; from asst prof to assoc prof, 47-58, PROF BIOL, E TENN STATE UNIV, 58- Mem: AAAS; Ecol Soc Am; Bot Soc Am. Res: Fossil pollen analysis; vegetation surveys in upper East Tennessee; plant geography; taxonomy of vascular plants. Mailing Add: Dept of Biol East Tenn State Univ Johnson TN 37601

BARCLAY, HAROLD BARTON, b Newton, Mass, Jan 3, 24; m 53; c 2. CULTURAL ANTHROPOLOGY, ETHNOLOGY. Educ: Boston Univ, BA, 52; Cornell Univ, MA, 54, PhD(anthrop), 61. Prof Exp: Instr anthrop, Am Univ Cairo, 56-59; asst prof, Knox Col (Ill), 60-63; asst prof, Univ Ore, 63-66; assoc prof, 66-71, actg chmn dept, 67-68, PROF ANTHROP, UNIV ALTA, 71- Mem: Am Anthrop Asn. Res: Middle Eastern, especially Arab, ethnography; folk Islam; agrarian social movements; utopian communities; religious sectarianism; cross cultural study of the role of the horse. Mailing Add: Dept of Anthropology Univ of Alta Edmonton AB Can

BARCLAY, HARRIETT G, b Minneapolis, Minn, Aug 31, 01; m 28; c 2. BOTANY. Educ: Univ Minn, BA, 23, MA, 24; Univ Chicago, PhD(ecol), 28; Univ Tulsa, BA, 45. Prof Exp: Asst bot, Univ Minn, 23-25; res fel, Univ Chicago, 25-28; assoc prof, Univ Okla, 38-39; from lectr to prof bot, 39-72, EMER PROF BOT, UNIV TULSA, 72- Concurrent Pos: Head dept bot, Univ Tulsa, 53-58; res staff mem, Natural Sci Inst, Nat Univ Colilbia, 59; vis prof bot, NC State Univ, 72-73. Honors & Awards: NSF grants, Suppl Grants, Creoly Found Venezuela, Green Leaf Award, Nature Conservancy, 72. Mem: Asn Trop Bot; Int Asn Plant Taxon. Res: Ecology, plant taxonomy, species diversity of tropical mountains, especially above timberline and comparison with temperate mountains; survey throughout paramos of northern, equatorial Andes and comparison with punas of central Andes. Mailing Add: Fac of Natural Sci Univ of Tulsa Tulsa OK 74104

BARCLAY, LAWRENCE ROSS COATES, b Wentworth, NS, Oct 24; 28; m 50; c 3. ORGANIC CHEMISTRY. Educ: Mt Allison Univ, BSc, 50, MSc, 51; McMaster Univ, PhD(org chem), 57. Prof Exp: Asst prof chem, Mt Allison Univ, 51-55; vis lectr, McMaster Univ, 55-56; from assoc prof to prof, 56-67, CARNEGIE PROF CHEM & HEAD DEPT, MT ALLISON UNIV, 67- Concurrent Pos: Grants, Am Chem Soc Petrol Res Fund, 59-61, Res Corp, 60 & Nat Res Coun Can. Mem: Chem Inst Can. Res: Orthodiquaternary aromatic compounds; solvolysis hinereed benzyl compounds; chemistry of crowded and sterically hindered aromatics; photochemistry of aryl-nitro, nitroso, and halo compounds. Mailing Add: Box 459 Sackville NB Can

BARCLAY, LEE ARMSTEAD, (JR), b Auburn, Ala, May 23, 44. FISHERIES, ICHTHYOLOGY. Educ: Univ Montevallo, BS, 68; Samford Univ, MS, 70; Auburn Univ, PhD(fisheries), 73. Prof Exp: Asst prof biol, Calif Polytech State Univ, San Luis Obispo, 73-75; FISHERY BIOLOGIST, US FISH & WILDLIFE SERV, 75- Concurrent Pos: Consult, Calif Dept Fish & Game, 75. Mem: Am Fisheries Soc; Am Soc Ichthyologists & Herpetologists. Res: Home ranges and movement patterns of fishes; warmwater aquaculture; parasites and diseases of fish; reactions of aquatic organisms to freshwater and marine developmental activities. Mailing Add: US Fish & Wildlife Serv Custom House Galveston TX 77550

BARCLAY, RALPH KINNEY, b New Lenox, Ill, Aug 12, 15; m 39; c 2. BIOCHEMISTRY. Educ: Univ Ill, BS, 38; Iowa State Col, PhD(chem), 49. Prof Exp: Res chemist, Eastman Kodak Co, 38-46; asst, Sloan-Kettering Inst Cancer Res, 49-51, assoc, 51-67; MOLECULAR BIOLOGIST, CIBA-GEIGY CORP, 67- Concurrent Pos: From asst prof to assoc prof biochem, Sloan-Kettering Div, Med Col, Cornell Univ, 53-67. Mem: Am Chem Soc; Am Soc Biol Chemists; Am Asn Cancer Res. Res: Drug-macromolecule interactions; nucleic acid metabolism. Mailing Add: Ciba-Geigy Corp Saw Mill River Rd Ardsley NY 10502

BARCLAY, ROBERT, JR, b Mt Vernon, NY, Apr 1, 28. POLYMER CHEMISTRY, INDUSTRIAL ORGANIC CHEMISTRY. Educ: Cornell Univ, AB, 48; Univ Md, PhD(chem), 57. Prof Exp: Patent chemist, Barrett Div, Allied Chem Corp, 48-51; res chemist, Am Cyanamid Co, 51-52; res chemist, Union Carbide Corp Plastics Div, 56-60, proj scientist, 60-69; sr res chemist, 69-71, res specialist, 71-73, RES SCIENTIST, THIOKOL CORP, 73- Mem: Am Chem Soc. Res: Elastomers; condensation polymers; organic coatings; plasticizers. Mailing Add: Thiokol Corp 930 Lower Ferry Rd Trenton NJ 08607

BARCLAY, WILLIAM R, b Golden, BC, May 25, 19; m 44; c 2. MEDICINE. Educ: Univ BC, BA, 41; Univ Alta, MD, 45. Prof Exp: Instr biochem, Univ Alta, 42; mem, Dept Nat Health & Welfare, Can, 46-47; asst, Dept Med, Univ Chicago, 48-49; mem, Dept Nat Health & Welfare, Can, 49-51; instr med, Univ Chicago, 51-52, asst prof chest dis, 52-56, assoc prof, 56-64, prof med, 64-70; dir sci activities & asst exec vpres, 70-75, SR V PRES, AMA & ED J AMA, 75- Concurrent Pos: Palmer sr fel, Univ Chicago, 54-56; instr surg, Univ Alta, 50-51; bd mem, Nat Tuberc & Respiratory Dis Asn, 70; chmn tuberc panel, US-Japan Med Sci Prog, 70; mem adv comt to dir, NIH, 70. Mem: Am Thoracic Soc (pres, 63-64); fel NY Acad Sci; Am Col Chest Physicians; Am Soc Clin Invest. Res: Metabolic studies of tubercle bacillus. Mailing Add: Am Med Asn 535 N Dearborn St Chicago IL 60610

BARCUS, WILLIAM DICKSON, JR, b Mineola, NY, Mar 19, 29; m 55; c 2. MATHEMATICS. Educ: Mass Inst Technol, SB, 50; Oxford Univ, DPhil(math), 55. Prof Exp: Instr, Princeton Univ, 55-56; from instr to asst prof, Brown Univ, 56-61; assoc prof, 61-66, PROF MATH, STATE UNIV NY, STONY BROOK, 66- Concurrent Pos: Vis sr res fel, Jesus Col, Oxford Univ, 65-66 & Math Inst, 68-69. Mem: Am Math Soc. Res: Algebraic topology. Mailing Add: Dept of Math State Univ of NY Stony Brook NY 11790

BARCZAK, VIRGIL J, b Toledo, Ohio, Nov 29, 31; m 59; c 4. MINERALOGY, GEOLOGY. Educ: Univ Mich, BS, 58, MS, 59; Oklahoma City Univ, MBA, 66. Prof Exp: Petrographer, Ceramic Div, Champion Spark Plug Co, 59-64; res petrographer, 64-68, RES SEPCIALIST, CENT RES LAB, KERR McGEE CORP, 68- Honors & Awards: Purdy Award, Am Ceramic Soc, 66. Mem: Am Ceramic Soc; fel Geol Soc Am; Mineral Soc Am; Mineral Asn Can; Nat Inst Ceramic Engrs. Res: Determinative mineralogy; instrumental techniques applicable to quantitative determination of minerals and chemical compositions; general geology; ceramic technology; crystal chemistry; geochemistry; management. Mailing Add: 2500 NW 109th St Oklahoma City OK 73120

BARD, ALLEN JOSEPH, b New York, NY, Dec 18, 33; m 57; c 2. ELECTROANALYTICAL CHEMISTRY, PHYSICAL CHEMISTRY. Educ: City Col New York, BS, 55; Harvard Univ, AM, 57, PhD, 58. Prof Exp: Anal chemist, Gen Chem Co Res Lab, 55; asst chem, Harvard, 55-56; from instr to assoc prof, 58-67, PROF CHEM, UNIV TEX, AUSTIN, 67- Concurrent Pos: Consult, Phillips Petrol Co, Radian Corp & E I du Pont de Nemours & Co; Fulbright prof, Univ Paris, 73-74. Honors & Awards: Ward Medal, 55. Mem: AAAS; Am Chem Soc; Electrochem Soc. Res: Electroorganic chemistry; kinetics and mechanisms of electrode reactions; electron spin resonance spectroscopy, electrogenerated chemiluminescence; electrochemical techniques and instrumentation. Mailing Add: Dept of Chem Univ of Tex Austin TX 78712

BARD, CHARLETON CORDERY, b Chicago, Ill, Feb 2, 24; m 46; c 4. PHOTOGRAPHIC CHEMISTRY. Educ: Univ Chicago, PhB & BS, 46, MS, 48, PhD, 50. Prof Exp: Instr, Wilson Jr Col, 47-49; org chemist, 49-68, SUPVR PROCESSING CHEM SECT, PHOTOG TECHNOL DIV, EASTMAN KODAK CO, 68- Mem: Am Chem Soc; Soc Photog Sci & Eng; Int Color Coun. Res: Infrared spectroscopy and molecular structure; physical organic chemistry. Mailing Add: 74 Cornwall Lane Rochester NY 14617

BARD, EUGENE DWIGHT, b Blackwell, Okla, Nov 8, 28; m 50; c 4. SCIENCE EDUCATION. Educ: Okla A&M Col, BS, 50, MS, 54; Univ Northern Colo, EdD(sci educ), 74. Prof Exp: Teacher phys sci, Planeview Pub Sch, 54-55; teacher phys sci, Puelbo Pub Sch, 55-65, dept head sci, 58-65, curric specialist math-sci, 63-65; instr chem, Southern Colo State Col, 65-66, from asst prof to assoc prof physics, 66-74; ASSOC PROF PHYSICS, UNIV SOUTHERN COLO, 74- Mem: Am Asn Physics Teachers. Res: Development and evaluation of non-traditional teaching materials and techniques. Mailing Add: Dept of Physics Univ Southern Colo Pueblo CO 81001

BARD, GILY EPSTEIN, b Berlin, Ger, Sept 14, 24; nat US; m 45; c 2. PLANT ECOLOGY. Educ: Hunter Col, BA, 45; Cornell Univ, MS, 47; Rutgers Univ, PhD(bot), 51. Prof Exp: Res asst forest soils, Cornell Univ, 45-46, res asst bot, 46-47; res fel plant physiol, Brooklyn Bot Garden, 47-48; asst biol, Rutgers Univ, 48-51; lectr bot, 51-53, from instr to asst prof biol, 53-70, ASSOC PROF BIOL, LEHMAN COL, 70- Concurrent Pos: Res assoc, Columbia Univ, 53-54; ed, Torrey Bot Club Bull, 70- Mem: Am Bryol Soc; Bot Soc Am; Ecol Soc Am; Torrey Bot Club (treas, 55-56). Res: Mineral nutrition of forest species; secondary succession. Mailing Add: Dept of Biol Sci Herbert H Lehman Col Bronx NY 10468

BARD, JOHN C, b Huron, SDak, Apr 7, 25; m 47; c 3. FOOD SCIENCE. Educ: Huron Col, SDak, BS, 49; Iowa State Univ, MS, 50. Prof Exp: Technologist, Meat Prod Control, 51-52, head dept, 52-58, gen dept head, 58-61, dir res, Meat Prod & Packaging, 61-75, VPRES RES, OSCAR MAYER & CO, 75- Concurrent Pos: Mem sci adv comt, Am Meat Inst, 66-; mem indust liaison panel to Food & Nutrit Bd, Nat Acad Sci, coun mem, 74, vchmn coun, 75- Mem: Indust Res Inst; Inst Food Technologists; Am Chem Soc; Am Soc Qual Control. Res: Product, package and processing developments of application to meat industry, particularly an improved control of quality, costs and sanitation. Mailing Add: Oscar Mayer & Co PO Box 1409 Madison WI 53701

BARD, JOHN WILLIAM, b Antigo, Wis, Aug 21, 14; m 42; c 3. CHEMISTRY. Educ: Iowa State Col, BS, 36; Lawrence Col, MS, 38, PhD(cellulose chem), 40. Prof Exp: Chem engr, Marathon Paper Mills Co, 40-45, tech adv, 45-50, supt paper mill, 51-52, plant supt, 52-53, asst to exec vpres, 53-54, plant mgr, Green Bay Plant, 54-55, gen prod mgr, Marathon Div, Am Can Co, 55-64, VPRES MFG PULP & PAPER MILLS, AM CAN CO, 64- Mem: Am Tech Asn Pulp & Paper Indust. Res: Cellulose chemistry. Mailing Add: 22 Dundee Rd Stamford CT 06903

BARD, PHILIP, b Hueneme, Calif, Oct 25, 98; m 22; c 2. PHYSIOLOGY. Educ: Princeton Univ, AB, 23; Harvard Univ, AM, 25, PhD(physiol), 27. Hon Degrees: DSc, Princeton Univ, 47; ScD, Washington & Lee Univ, 49; Dr, Cath Univ Chile, 51; LLD, Johns Hopkins Univ, 68. Prof Exp: Instr physiol, Harvard Med Sch, 26-28, asst prof, 31-33; asst prof, Princeton Univ, 28-31; prof & dir dept, 33-64, dean med fac, 53-57, EMER PROF PHYSIOL, SCH MED, JOHNS HOPKINS UNIV, 64- Concurrent Pos: Mem staff, Div Med Sci, Nat Res Coun, 35-46; trustee, Rockefeller Univ; Harvey lectr, New York, 38; Hughlings Jackson lectr, Montreal Neurol Inst, 43; Graves lectr, Ind Univ, 48; hon mem fac biol & med sci, Univ Chile; Mitchell lectr, Col Physicians, Philadelphia, 53. Honors & Awards: Jacobi Award, Am Neurol Asn, 59; Karl Spencer Lashley Award, Am Philos Soc, 62; Achievement Award Med Sci, Am Col Physicians, 68. Mem: Nat Acad Sci; AAAS; Asn Am Physicians; Am Physiol Soc (secy, 39-41, pres, 41-46); Soc Exp Biol & Med (pres, 59-61). Res: Neural bases of emotional reactions; temperature regulation; postural reactions; localization of function in cerebral cortex; motion sickness; neural control of pituitary gland; central nervous system control of fever. Mailing Add: Dept of Physiol Johns Hopkins Univ Sch of Med Baltimore MD 21205

BARD, RAYMOND CAMILLO, b New Britain, Conn, Aug 26, 18. MICROBIOLOGY. Educ: City Col New York, BS, 38; Ind Univ, MA, 47, PhD(bact), 49; Am Bd Microbiol, dipl. Prof Exp: Med technologist, St John's Hosp, New York, 38-40 & St Joseph's Hosp, 40-42; from instr to assoc prof bact, Ind Univ, 49-53; head microbiol sect, Smith Kline & French Labs, 53-56, assoc dir res & develop, 56-60; res dir, Nat Drug Co, 60-62; prof dent, Univ Ky, 62-65, prof cell biol, 65-67, dir res, Sch Dent, 62-63, asst dean, 63-64, asst vpres res & exec dir univ res found, 64-67; vpres, Col, 67-72, actg dean sch allied health sci, 68-72, PROF MICROBIOL, MED COL GA, 67-, DEAN SCH ALLIED HEALTH SCI, 72- Concurrent Pos: Assoc prof, Hahnemann Med Col, 54-62. Mem: Fel AAAS; fel Am Acad Microbiol; Am Soc Microbiol; Am Chem Soc; fel Am Inst Chemists. Res: Microbiological metabolism; chemotherapy; dental research. Mailing Add: Med Col of Ga Augusta GA 30902

BARD, RICHARD JAMES, b Mt Kisco, NY, July 19, 23; m 46; c 3. PHYSICAL CHEMISTRY. Educ: Univ Mich, BS, 44, MS, 48, PhD(chem), 51. Prof Exp: Instr colloid & surface chem, Univ Mich, 50-51; mem staff, Phys Chem Res, 51-56; group leader uranium chem group, 56-73, LEADER PHYS CHEM & METALL GROUP, LOS ALAMOS SCI LAB, 73- Mem: Am Chem Soc; Am Inst Chemists. Res: Physical and inorganic chemistry; gas-solid reaction kinetics; uranium processing; development of coated nuclear fuel particles; studies of pyrolytic carbons. Mailing Add: 975 Nambe Loop Los Alamos NM 87544

BARD, ROBERT CHARLES, b Neenah, Wis, Jan 11, 29; m 56; c 3. PHYSICAL GEOGRAPHY, CARTOGRAPHY. Educ: Univ Wis, BS, 51; Univ Wis, MS, 57; Univ Calif, Los Angeles, PhD(geog), 72. Prof Exp: Cartog aide, Army Mapping Serv, 51; map compiler & ed, Army Corps Engrs, 52-53; planning aide, County San Mateo Planning Comn, Calif, 58-59; planning technician, City San Mateo Planning Comn, 59-60; asst prof geog, Bakersfield Col, 61-66; ASST PROF GEOG, ORE STATE UNIV, 69- Mem: Asn Am Geogrs; Am Geog Soc. Res: Landforms and settlement patterns in arid lands; climatic aspects of slope evolution; thematic cartography. Mailing Add: Dept of Geog Ore State Univ Corvallis OR 97331

BARDACH, JOHN E, b Vienna, Austria, Mar 6, 15; US citizen; m 47. AQUATIC ECOLOGY, FISH BIOLOGY. Educ: Queen's Univ, Ont, BA, 46; Univ Wis, MSc & PhD(zool), 49. Prof Exp: From instr to asst prof biol, Iowa State Teachers Col, 49-53; from asst prof to prof fisheries zool, Univ Mich, 53-71; DIR HAWAII INST

MARINE BIOL, UNIV HAWAII, 71- Concurrent Pos: Dir fisheries res prog, Bermuda Govt, 55-58; adv, Royal Cambodian Govt, 58-59; ecol adv, UN Mekong Comt; sr vis fel sci, OEEC, 61; independent investr, Int Indian Ocean Exped, 64. Mem: Fel AAAS; Ecol Soc Am; Am Soc Limnol & Oceanog; Am Soc Zoologists; Am Fisheries Soc. Res: Aquaculture; physiology and behavior of fishes; coral reef fishes; the senses of fishes; fisheries development; natural resources ecology; resource management. Mailing Add: Hawaii Inst of Marine Biol PO Box 1346 Coconut Island HI 96744

BARDACK, DAVID, b New York, Apr 11, 32; m 60; c 2. VERTEBRATE PALEONTOLOGY, ICHTHYOLOGY. Educ: Columbia Univ, AB, 54, AM, 55; Univ Kans, PhD(zool), 63. Prof Exp: Asst zool, Columbia Univ, 55-56; vert paleont & ichthyol, Am Mus Natural Hist, 56-58; instr zool, Hunter Col, 57-58; asst, Univ Kans, 59-62, curatorial asst vert paleont, Mus Natural Hist, 63, res assoc, 63-64; from asst prof to assoc prof, 64-75, PROF VERT PALEONT, UNIV ILL, CHICAGO CIRCLE, 75- Concurrent Pos: Res assoc, Field Mus Natural Hist. Mem: Fel AAAS; Soc Vert Paleont; Am Soc Ichthyologists & Herpetologists; Am Soc Zoologists; Paleont Soc. Res: Anatomy and evolution of living and fossil teleostean fishes. Mailing Add: Dept of Biol Sci Univ of Ill at Chicago Circle Box 4348 Chicago IL 60680

BARDANA, EMIL JOHN, JR, b New York, NY, May 21, 35; m 60; c 2. ALLERGY, IMMUNOLOGY. Educ: Georgetown Univ, BS, 57; McGill Univ, MD, CM, 61. Prof Exp: Intern gen med, Univ Calif Med Ctr, San Francisco, 61-62; gen med officer, Bremerton Naval Hosp, Wash, 62-64; from resident med to instr med allergy immunol, Univ Ore Health Sci Ctr, 65-69; res trainee, Dept Allergy & Immunol Clin, Nat Jewish Hosp & Res Ctr, Denver, 69-71; asst prof, 71-75, ASSOC PROF MED, UNIV ORE HEALTH SCI CTR, 74-, ASST PROF MICROBIOL & IMMUNOL, 75- Concurrent Pos: Allergy Found Am training grant, 68-70; Am Acad Allergy travel grant, 69. Mem: Fel Am Col Physicians; fel Am Acad Allergy; fel Am Asn Cert Allergists; Am Fedn Clin Res; Am Thoracic Soc. Res: Development of better understanding of the immunopathogenesis of mycotic hypersensitivity disorders of the lung; establishment of sensitive radioassays which will permit their earlier diagnosis. Mailing Add: Univ of Ore Health Sci Ctr 3181 SW Sam Jackson Park Rd Portland OR 97201

BARDASIS, ANGELO, b New York, NY, Nov 26, 36. SOLID STATE PHYSICS. Educ: Cornell Univ, AB, 57; Univ Ill, MS, 59, PhD(physics), 62. Prof Exp: Asst prof, 63-69, ASSOC PROF PHYSICS, UNIV MD, COLLEGE PARK, 69- Concurrent Pos: NSF fel, 61-63. Mem: Am Phys Soc. Res: Superconductivity; helium three; helium four; semiconductors. Mailing Add: Dept of Physics Univ of Md College Park MD 20742

BARDAWIL, WADI ANTONIA, b Mexico, May 13, 21; nat US; m 47; c 6. PATHOLOGY. Educ: Nat Univ Mex, MD, 46; Am Bd Path, dipl. Prof Exp: Instr path, Univ Vt, 50-52; asst, Med Sch, Harvard Univ, 52-53, instr, 53-57; from asst prof to assoc prof path, 57-71, from asst prof to assoc prof obstet & gynec, 57-66, PROF OBSTET & GYNEC, CREIGHTON UNIV, 66-, PROF PATH, 71-, CHMN DEPT PATH, 74- Concurrent Pos: Pathologist, Robert Breck Brigham Hosp & asst, Peter Bent Brigham Hosp, 54-55; res assoc, Med Sch, Harvard Univ, 57-59; dir dept path & med res, St Margaret's Hosp, 57-74. Mem: Am Soc Exp Path; Tissue Cult Asn; Am Rheumatism Asn; AMA. Res: Gynecological and obstetrical research; regression of trophoblast and possible immune reactions involved; pathogenesis of collagen diseases and pathogenesis of vascular pathology. Mailing Add: Dept of Path Creighton Univ Omaha NE 68131

BARDEEN, JAMES MAXWELL, b Minneapolis, Minn, May 9, 39; m 68. THEORETICAL ASTROPHYSICS. Educ: Harvard Univ, AB, 60; Calif Inst Technol, PhD(physics), 65. Prof Exp: Res assoc physics, Calif Inst Technol, 65; res physicist, Univ Calif, Berkeley, 66; from asst prof to assoc prof astron, Univ Wash, 66-72; assoc prof, 72-74, PROF PHYSICS, YALE UNIV, 74- Concurrent Pos: NSF res grants, 67-75; Sloan fel, 68-72. Mem: AAAS; Am Phys Soc; Am Astron Soc; Int Astron Union. Res: Astrophysics of black holes and compact x-ray sources; stability of disks; dynamics of spiral structure in galaxies. Mailing Add: Dept of Physics Yale Univ New Haven CT 06520

BARDEEN, JOHN, b Madison, Wis, May 23, 08; m 38; c 3. PHYSICS. Educ: Univ Wis, BS, 28, MS, 29; Princeton Univ, PhD(math, physics), 36. Prof Exp: Geophysicist, Gulf Res & Develop Corp, 30-33; jr fel, Soc Fel, Harvard Univ, 35-38; asst prof physics, Univ Minn, 38-41; physicist, US Naval Ord Lab, Washington, DC, 41-45; res physicist, Bell Tel Labs, 45-51; PROF ELEC ENG & PHYSICS, UNIV ILL, URBANA, 51- Concurrent Pos: Consult & mem bd dirs, Xerox Corp. Honors & Awards: Nobel Prize, physics, 56; Fritz London award, 62; Nat Medal Sci, 65; Michelson-Morley award, 68; medal of honor, Inst Elec & Electronics Engrs, 71. Mem: Nat Acad Sci; Am Acad Arts & Sci; Am Philos Soc; fel Am Phys Soc. Res: Theory of solid state; semi-conductors; superconductivity. Mailing Add: Dept of Physics Univ of Ill Urbana IL 61801

BARDEEN, THOMAS, b Madison, Wis, Apr 10, 12; m 34 & 50; c 6. GEOPHYSICS. Educ: Univ Wis, BSc, 33, MSc, 34. Prof Exp: Appl mathematician, Gulf Res & Develop Co, 34-36, geophysicist, 34-41, res assoc, Nat Defense Res Comt Proj, 41-45, res assoc geophys, 46-69, sr scientist, 69-73, SR SCIENTIST, EXPLOR DEPT, GULF ENERGY & MINERALS CO, 73- Mem: Soc Explor Geophys. Res: Mines and mine counter measures; techniques and instrumentation in applied geophysics. Mailing Add: 3945 S Poplar St Casper WY 82601

BARDEEN, WILLIAM A, b Washington, Pa, Sept 15, 41; m 61; c 2. PARTICLE PHYSICS. Educ: Cornell Univ, AB, 62; Univ Minn, Minneapolis, PhD(physics), 68. Prof Exp: Res assoc physics, State Univ NY Stony Brook, 66-68; mem, Inst Advan Study, 68-69; from asst prof to assoc prof physics, Stanford Univ, 69-75; PHYSICIST, FERMI NAT ACCELERATOR LAB, 75- Concurrent Pos: Vis scientist, CERN, 71-72; Sloan Found fel, 71. Mem: Am Phys Soc. Res: Theoretical particle physics and quantum field theory. Mailing Add: Theory Group CL3E Fermi Lab PO Box 500 Batavia IL 60510

BARDELL, DAVID, b Braintree, Eng; US citizen. VIROLOGY. Educ: City Univ New York, BA, 66; Univ NH, MS, 68, PhD(microbiol), 72. Prof Exp: Fel virol, Harvard Univ, 72-73; res assoc, Univ Wis, 73-75; ASST PROF BIOL, KEAN COL NJ, 75- Mem: Am Soc Microbiol. Res: Effect of viruses on host cell metabolism; virus induced cytopathology; applied viral immunology. Mailing Add: Dept of Biol Kean Col of NJ Union NJ 07083

BARDELL, EUNICE BONOW, b Milwaukee, Wis, Feb 8, 15; m 72. PHARMACY. Educ: Univ Wis, BS, 38, MS, 49, PhD(pharm), 52. Prof Exp: From instr to prof pharm, 48-73, EMER PROF PHARM & HEALTH SCI, UNIV WIS-MILWAUKEE, 73- Mem: Am Pharmaceut Asn; Am Inst Hist Pharm. Res: Evaluation of action of antibacterial agents in pharmaceutical formulations; history of pharmacy. Mailing Add: 1539 N 51 St Milwaukee WI 53208

BARDEN, ALBERT ARNOLD, JR, b Providence, RI, Apr 11, 11; m 41; c 3. ZOOLOGY. Educ: Brown Univ, AB, 32, ScM, 34; Northwestern Univ, PhD(zool), 41. Prof Exp: Instr, 46-48, from asst prof to assoc prof, 46-66, PROF ZOOL, UNIV MAINE, 66- Mem: Ecol Soc Am. Res: Activity and food habits of animals; vertebrate ecology. Mailing Add: Dept of Zool Univ of Maine Orono ME 04473

BARDEN, HOWARD STAVERS, b Ft Jackson, SC, Sept 15, 45; m 70; c 1. BIOLOGICAL ANTHROPOLOGY. Educ: Brown Univ, BA, 67; Univ Wis-Madison, PhD(phys anthrop), 74. Prof Exp: ASST PROF ANTHROP, UNIV ILL, CHICAGO CIRCLE, 74- Mem: AAAS; Am Asn Phys Anthropologists. Res: Investigation of growth and maturation, dental morphological and structural abnormalities; bone mineral values and vitamin A utilization and health status of Down's syndrome and other mentally retarded subjects. Mailing Add: Dept of Anthrop Box 4348 Univ of Ill at Chicago Circle Chicago Il 60680

BARDEN, JOHN ALLAN, b Providence, RI, Oct 30, 36; m 58; c 2. HORTICULTURE. Educ: Univ RI, BS, 58; Univ Md, MS, 61, PhD(hort), 63. Prof Exp: Instr hort, Univ Md, 63; asst prof, 63-72, ASSOC PROF HORT, VA POLYTECH INST & STATE UNIV, 72- Mem: Am Soc Hort Sci. Res: Horticultural physiology; tree fruit physiology with emphasis on apple production intensification, including growth regulators, size controlling rootstocks, tree spacing and photosynthetic efficency as influenced by various factors. Mailing Add: Dept of Hort Va Polytech Inst & State Univ Blacksburg VA 24061

BARDEN, NED THORSON, b Fairfield, Iowa, July 11, 46. MICROBIOLOGY. Educ: Iowa State Univ, BS, 68; Univ Wis, MS, 72, PhD(bact), 73. Prof Exp: ASST PROF MICROBIOL, UNIV MISS, 73- Mem: Am Soc Microbiol. Res: Chemistry and serology of Salmonella flagella; chemotactic response of Bdellovibrio. Mailing Add: Dept of Biol Univ of Miss University MS 38677

BARDIN, RUSSELL KEITH, b Fresno, Calif, Mar 22, 32; m 65; c 1. NUCLEAR PHYSICS. Educ: Calif Inst Technol, BS, 53, PhD(physics), 61. Prof Exp: Res fel, Calif Inst Technol, 61-62; res assoc, Columbia Univ, 62-67; RES SCIENTIST PHYSICS, LOCKHEED PALO ALTO RES LAB, 67- Mem: Am Phys Soc. Res: Experimental nuclear physics and applications of nuclear techniques to instrumentation. Mailing Add: Lockheed Palo Alto Res Lab Bldg 203 Dp 5210 3251 Hanover St Palo Alto CA 94304

BARDIN, TSING TCHAO, b Tsintao, China, July 31, 38; US citizen; m 65; c 2. NUCLEAR PHYSICS. Educ: Univ London, BS & ARCS, 61; Columbia Univ, PhD(physics), 66. Prof Exp: Res asst physics, Columbia Univ, 62-66; asst prof, Haverford Col, 66-67; assoc, Univ Pa, 67; CONSULT, PALO ALTO RES LAB, LOCKHEED CORP, 68- Mem: Am Phys Soc. Res: Experimental physics in nuclear spectroscopy and nuclear structure studies. Mailing Add: 120 Bear Gulch Dr Portola Valley CA 94025

BARDO, RICHARD DALE, b Garner, Iowa. QUANTUM CHEMISTRY. Educ: Univ Wyo, BS, 66; Iowa State Univ, PhD(phys chem), 73. Prof Exp: RES ASSOC, DEPT CHEM, UNIV CALIF, IRVINE, 73- Mem: Am Phys Soc. Res: Electronic structure calculations on atoms and molecules using variational and perturbation techniques. Mailing Add: Dept of Chem Univ of Calif Irvine CA 92664

BARDOLPH, MARINUS PETER, b Chicago, Ill, May 5, 13; m 43; c 2. ORGANIC CHEMISTRY. Educ: Univ Ill, BS, 43; Univ Iowa, PhD(chem), 47. Prof Exp: Instr chem, Univ Iowa, 43-45; asst prof, WVa Univ, 47-48; assoc prof, Univ Omaha, 48-51; res chemist, Commercial Solvents Corp, 51-53 & Olin Mathieson Chem Corp, 53-57; asst prof chem, 57-59, ASSOC PROF CHEM, SOUTHERN ILL UNIV, EDWARDSVILLE, 59- Mem: AAAS; Am Chem Soc. Res: Carbohydrates; antibiotics; explosives. Mailing Add: Dept of Chem Southern Ill Univ Sci & Technol Div Edwardsville IL 62025

BARDON, MARCEL, b Paris, France, Sept 16, 27; US citizen; m 64. PHYSICS. Educ: Univ Paris, dipl, 52; Columbia Univ, MA, 56, PhD(physics), 61. Prof Exp: Instr physics, Columbia Univ, 59-63, res assoc & asst dir, Nevis Labs, 63-64, assoc dir, 64-66, sr res assoc, 64-70, dep dir, 66-70; prog dir, Phys Sect, 70-71, head physics sect, 71-75, DEP DIV DIR PHYSICS, NAT SCI FOUND, 75- Mem: Am Phys Soc. Res: Intermediate and high energy physics. Mailing Add: Physics Div Nat Sci Found Washington DC 20550

BARDOS, THOMAS JOSEPH, b Budapest, Hungary, July 20, 15; nat US; m 51. MEDICINAL CHEMISTRY. Educ: Tech Univ Budapest, ChE, 38; Univ Notre Dame, PhD(chem), 49. Prof Exp: Chem engr, Vacuum Oil Co, 38-46; res fel, Biochem Inst, Univ Tex, 48-51; res chemist, Armour & Co, 51-55, sect head, 55-59, lab head, 59-60; PROF MED CHEM, STATE UNIV NY BUFFALO, 60- Concurrent Pos: Consult, Roswell Park Mem Inst, 60- Honors & Awards: Ebert Prize, Am Pharmaceut Asn, 71; Schoellkopf Medal, Am Chem Soc, 74. Mem: Fel AAAS; Am Chem Soc; fel NY Acad Sci; fel Acad Pharmaceut Sci; Am Soc Biol Chemists. Res: Organic synthesis; isolation and identification of natural growth factors and inhibitors; biosynthetic pathways; design of antimetabolites; chemotherapy; synthesis of anticancer agents; modification of nucleic acids. Mailing Add: Dept of Medicinal Chem State Univ of NY Buffalo NY 14214

BARDSLEY, CHARLES EDWARD, agronomy, soil science, see 12th edition

BARDSLEY, JAMES NORMAN, b Ashton-U-Lyne, Eng, May 10, 41; m 64; c 2. ATOMIC PHYSICS, QUANTUM CHEMISTRY. Educ: Univ Cambridge, BA, 61, MA, 65; Manchester Univ, PhD(physics), 65. Prof Exp: Asst lectr physics, Manchester Univ, 65-67; res assoc, Univ Pittsburgh, 68-69; asst prof, Univ Tex, Austin, 69-70; ASSOC PROF PHYSICS, UNIV PITTSBURGH, 70- Concurrent Pos: Lectr, Manchester Univ, 68-70. Mem: Brit Inst Physics & Phys Soc; Am Phys Soc. Res: Electron scattering by atoms and molecules and atom-atom collisions; the theory of dissociative attachment and recombination, electron detachment and molecule formation; motion of ions in electric fields. Mailing Add: Dept of Physics Univ of Pittsburgh Pittsburgh PA 15260

BARDWELL, GEORGE, b Denver, Colo, Jan 6, 24; m 46; c 5. MATHEMATICS, STATISTICS. Educ: Univ Colo, BS, 44, MS, 49, PhD(math), 61, MAA, 65; Bowdoin Col & Mass Inst Technol, certs, 45. Prof Exp: Res technician, Philadelphia Navy Yard, 46-47; instr statist, Univ Colo, 48-49; res assoc, Bur Bus & Social Res, 49-54, asst prof statist & res, 54-61, assoc prof & chmn dept, 61-62, joint assoc prof math & statist, 63-71, PROF MATH, UNIV DENVER, 71- Concurrent Pos: Gulf Oil Co fac res grant, 61; consult, indust firms & govt agencies; impartial labor arbitrator, 62-; mem, Fed Mediation & Conciliation Serv. Mem: Am Math Asn; Am Statist Asn; Inst Math Statist; Am Arbit Asn. Res: Mathematical analysis of discrete probability distributions; sampling theory; social and economic research. Mailing Add: Col of Arts & Sci Univ of Denver Denver CO 80210

BARDWELL, JOHN ALEXANDER EDDIE, b Appin, Ont, Dec 25, 21; m 54; c 3.

PHYSICAL CHEMISTRY. Educ: Univ Western Ont, BA, 44, MSc, 46; McGill Univ, PhD(chem), 48; Oxford Univ, DPhil, 50. Prof Exp: Spec lectr chem, 50-51, from asst prof to assoc prof, 51-64, asst dean arts & sci, 61-67, univ secy, 68-74, PROF CHEM, UNIV SASK, 64-, ASST TO PRES, 67- Mem: Chem Inst Can. Res: Polymers; combustion of hydrocarbons; radiation chemistry. Mailing Add: 121 Albert Ave Saskatoon SK Can

BARDWICK, JOHN, III, b Harvey, Ill, Mar 10, 31; m 54; c 3. PHYSICS. Educ: Purdue Univ, BS, 53; Univ Mich, MS, 59, PhD(physics), 64. Prof Exp: Res assoc, 64, from instr to asst prof, 64-69, ASSOC PROF PHYSICS, UNIV MICH, 69- Mem: Am Phys Soc. Res: Accelerator design and development; experimental nuclear physics. Mailing Add: Dept of Physics Univ of Mich Ann Arbor MI 48104

BARE, BARRY BRUCE, b South Bend, Ind, Apr 24, 42. FOREST MANAGEMENT, OPERATIONS RESEARCH. Educ: Purdue Univ, BSF, 64, PhD(quant methods), 69; Univ Minn, MS, 65. Prof Exp: Res asst, Univ Minn, 64-65, comput programmer, 65-66; asst, Purdue Univ, 66 & 68, instr forest financial mgt, 68-69; ASST PROF FOREST RESOURCES, UNIV WASH, 69- Mem: Soc Am Foresters; Am Forestry Asn; Inst Mgt Sci; Opers Res Soc Am. Res: Application of mathematical programming to forest production processes; development of management games for teaching of resource management; application of financial management principles to forestry-oriented problems; forest taxation. Mailing Add: Col of Forest Resources Univ of Wash Seattle WA 98195

BARE, PAUL ORVILLE, b Chadron, Nebr, July 4, 09; m 42; c 2. ORGANIC CHEMISTRY. Educ: Nebr State Teachers Col, BS, 31; Univ Nebr, MS, 34, PhD(org chem), 37. Prof Exp: Lab asst, Nebr State Teachers Col, 31-32; lab asst, Univ Nebr, 32-33; lab asst, Nebr State Teachers Col, 33, instr chem & physics, 35-36; res chemist, E I du Pont de Nemours & Co, Inc, 37-41, res group leader, 41-43, supvr area lab, 43-51, supt tech sect, 51-55, tech asst, 55-57, mgr planning sect, Mfg Div, Orchem Dept, 57-61, mem staff, Patent & Licensing Div, Int Dept, 61-67, sr tech investr, Tech Div, 68-71; RETIRED. Mem: Am Chem Soc. Res: Organic arsenicals; synthetic rubber; camphor; synthetic tanning agents; emulsion polymerization of chloroprene; detergents; intermediates; dyes; tetraethyl lead; fluorinated hydrocarbons; patents and licensing activities. Mailing Add: 1204 Brook Dr Normandy Manor Wilmington DE 19803

BARE, THOMAS M, b Lancaster, Pa, Nov 24, 42; m 67; c 1. SYNTHETIC ORGANIC CHEMISTRY. Educ: Pa State Univ, BS, 64; Mass Inst Technol, MS, 66, PhD(org chem), 68. Prof Exp: Sr res chemist, Lakeside Labs div, Colgate-Palmolive Co, 69-75; SR CHEMIST, MORTON CHEM CO DIV, MORTON-NORWICH PROD, INC, 75- Mem: Am Chem Soc; Sigma Xi. Res: Synthetic organic chemistry; medicinal chemistry; pesticide chemistry. Mailing Add: 2010 Broadway Crystal Lake IL 60014

BAREFIELD, EDWARD KENT, b Hopkinsville, Ky, Feb 26, 43; m 66; c 2. INORGANIC CHEMISTRY, ORGANOMETALLIC CHEMISTRY. Educ: Western Ky Univ, BS, 65; Ohio State Univ, PhD(chem), 69. Prof Exp: Assoc, Cent Res Dept, E I du Pont de Nemours & Co, Inc, 69-70; ASST PROF CHEM, UNIV ILL, URBANA, 70- Mem: Am Chem Soc. Res: Inorganic and organometallic synthesis; homogeneous catalysis. Mailing Add: Dept of Chem Univ of Ill Urbana IL 61801

BAREFOOT, ALDOS CORTEZ, JR, b Angier, NC, Feb 25, 27; m 49; c 3. WOOD SCIENCE & TECHNOLOGY. Educ: NC State Univ, BS, 50; MWT, 51; Duke Univ, DF(wood anat), 58. Prof Exp: Qual control engr, Henry County Plywood Corp, Va, 51-52; asst exp statistician, NC State Univ, 52-54, technologist, 54, asst prof, 54-59; forestry adv, Int Coop Admin, Pakistan, 59-61; assoc prof wood technol, 62-68, PROF WOOD TECHNOL, NC STATE UNIV, 68-, LEADER WOOD PROD EXTEN, 71-, HEAD DIV UNIV STUDIES, 75- Concurrent Pos: Fulbright-Hays fel, Univ Oxford, 73-74. Mem: Tech Asn Pulp & Paper Indust; Forest Prod Res Soc; Soc Wood Sci & Technol; fel Brit Inst Wood Sci. Res: Wood anatomy; tropical woods; statistical quality control; dendrochronology. Mailing Add: 3401 Hampton Rd Raleigh NC 27607

BAREISS, ERWIN HANS, b Schaffhausen, Switz, May 10, 22; US citizen; m 60; c 3. APPLIED MATHEMATICS. Educ: Schaffhausen Teachers Col, BA, 43; Univ Zurich, PhD(math), 50; Lehigh Univ, MS, 52. Prof Exp: Mem staff, C H Knorr, Switz, 43-45; instr math, Schaffhausen Teachers Col & Winterthur Tech Col, 45-51; mathematician, David Taylor Model Basin, US Dept Navy, 52-54, specialist math physics, 54-56, consult numerical anal, 56-57; consult appl math, Argonne Nat Lab, 57-63; vis lectr, 69-70, PROF COMPUT SCI & ENG SCI, NORTHWESTERN UNIV, 70-; SR MATHEMATICIAN, ARGONNE NAT LAB, 63- Concurrent Pos: Janggen-Pohen fel, 49-50; Baldwin Res fel, 51-52; lectr, Univ Md, 55-56; vis lectr, Harvard Univ, 64. Mem: Am Math Soc; Asn Comput Mach; Sigma Xi; Soc Indust & Appl Math; Swiss Math Soc. Res: Function theory of a hypercomplex variable; theoretical and applied mechanics; applied mathematics and numerical analysis; mathematical structure and methods of transport theory. Mailing Add: Dept of Comput Sci Northwestern Univ Evanston IL 60201

BARELARE, BRUNO, b Scranton, Pa, July 9, 13; m 39; c 5. MEDICINE, UROLOGY. Educ: Johns Hopkins Univ, AB, 34, MD, 38. Prof Exp: Asst, Johns Hopkins Univ, 38-39, intern, Hosp, 39-40; asst resident urol, Univ Va, 41-42, chief resident, 42-43, asst prof, Sch Med, 43-46; CLIN PROF UROL, UNIV ALA, BIRMINGHAM, 46-; CHIEF UROLOGIST, ST VINCENT'S HOSP, 48- Concurrent Pos: Res fel urol, Univ Va, 40-41; mem nat resources comt, Johns Hopkins Univ, 63- Mem: AMA; Am Urol Asn. Mailing Add: 2660 S Tenth Ave Birmingham AL 35205

BARFIELD, MARY ASHTON, b Jacksonville, Fla, June 22, 42. REPRODUCTIVE PHYSIOLOGY. Educ: Sweet Briar Col, BA, 64; Mt Holyoke Col, MA, 67; Princeton Univ, PhD(biol), 72. Prof Exp: Fel reproductive physiol, Dept Obstet & Gynec, Med Sch, Univ Pa, 72-74; STAFF SCIENTIST, INT COMT CONTRACEPTIVE RES, BIOMED DIV, POP COUN, 74- Mem: Am Soc Andrology; Soc Study Reprod; AAAS; Animal Behav Soc; Sigma Xi. Res: Reproductive physiology and contraception, particularly in the male. Mailing Add: Pop Coun Rockefeller Univ York Ave & 66th St New York NY 10021

BARFIELD, MICHAEL, b Charleston, SC, Oct 14, 34; m 60; c 3. PHYSICAL CHEMISTRY. Educ: San Diego State Col, BA, 57; Univ Utah, PhD(chem), 62. Prof Exp: Eng chemist, Gen Dynamics/Astronaut, Calif, 57-59; res fel, Harvard Univ, 62-63; asst prof chem, Univ SFla, 63-65; from asst prof to assoc prof, 65-72, PROF CHEM, UNIV ARIZ, 72- Mem: Am Phys Soc; Am Chem Soc. Res: Application of molecular quantum mechanics to problems in nuclear magnetic resonance spectroscopy and molecular structures. Mailing Add: Dept of Chem Univ of Ariz Tucson AZ 85721

BARFIELD, WALTER DAVID, b Gainesville, Fla, Nov 25, 28. PHYSICS. Educ: Univ Fla, BS, 50, MS, 51; Rice Univ, PhD(physics), 61. Prof Exp: Asst, Theoret Physics Div, Los Alamos Sci Lab, 51-53, mem staff, 53-63; mem tech staff, Res Eng Support Div, Inst Defense Anal, 63-68; STAFF MEM, LOS ALAMOS SCI LAB, 68- Mem: Am Phys Soc. Res: Theoretical and nuclear physics; radiation transfer; opacities; fluid dynamics; numerical methods; gamma ray spectroscopy. Mailing Add: 4647 Ridgeway Dr Los Alamos NM 87544

BARG, WILLIAM FREDRIC, JR, biochemistry, see 12th edition

BARGER, ABRAHAM CLIFFORD, b Greenfield, Mass, Feb 1, 17; m 43; c 3. PHYSIOLOGY, MEDICINE. Educ: Harvard Univ, AB, 39, MD, 43. Prof Exp: Res asst, Fatigue Lab, Harvard Univ, 38-41, instr, 48-49, assoc, 50-52, from asst prof to prof, 53-63, ROBERT HENRY PFEIFFER PROF PHYSIOL, HARVARD MED SCH, 63-, CHMN DEPT, 74- Concurrent Pos: Res fel physiol, Harvard Med Sch, 46-47; med house officer, Peter Bent Brigham Hosp, 43, asst resident, 45, asst med, 46-53, assoc staff mem, 53-58, consult physiol, 58-; assoc ed, Circulation Res, 63-66; mem bd sci counr, Nat Cancer Inst, 69-74, chmn, 73-74; mem fedn bd, Fedn Am Socs Exp Biol, 69-73; Harvard Apparatus Found, 70- Mem: Nat Inst Med; Am Physiol Soc (pres, 70-71); Soc Exp Biol & Med; Am Acad Arts & Sci. Res: Cardiovascular-renal physiology; congestive failure; hypertension. Mailing Add: Dept of Physiol Harvard Med Sch Boston MA 02115

BARGER, JAMES DANIEL, b Bismarck, NDak, May 17, 17; m 52; c 4. PATHOLOGY. Educ: Univ NDak, AB & BS, 39; Univ Pa, MD, 41; Univ Minn, MS, 49. Prof Exp: Dir labs, Pima County Hosp, Tucson, Ariz, 49-50 & Maricopa County Hosp, Phoenix, 50-51; med dir, Southwest Blood Bank Ariz, 51-64; dir clin path, 64-69, CHMN DEPT PATH, SUNRISE HOSP, 69- Concurrent Pos: Dir labs, Good Samaritan Hosp, Ariz, 51-63; prof med technol, Ariz State Univ; asst clin prof path, Univ Nev, Reno, 74-; consult, US Vet Admin, USPHS Indian Hosp & St Luke's Hosp, Phoenix. Mem: AAAS; Am Soc Clin Path; Am Asn Pathologists & Bacteriologists; AMA; fel Col Am Pathologists (secy-treas, 71-). Res: Blood group serology and clinical biochemistry as related to hematologic diseases. Mailing Add: Dept of Path Sunrise Hosp PO Box 14157 Las Vegas NV 89114

BARGER, JAMES EDWIN, b Manhattan, Kans, Dec 28, 34; m 57; c 4. ACOUSTICS. Educ: Univ Mich, BS, 57; Univ Conn, MS, 60; Harvard Univ, MA, 62, PhD(appl physics), 64. Prof Exp: Sr consult appl physics, 64-68, VPRES, BOLT BERANEK & NEWMAN, INC, 68-, CHIEF SCIENTIST, 75- Mem: Acoust Soc Am. Res: Underwater sound; acoustic cavitation in water; sonar development. Mailing Add: Bolt Beranek & Newman Inc 50 Moulton St Cambridge MA 02138

BARGER, JOHN WALTER, b Huntsville, Mo, Feb 24, 18; m 39; c 3. PHYSICAL CHEMISTRY. Educ: Univ Mo, BS, 48, PhD(chem), 51. Prof Exp: Res chemist, Textile Chem, E I du Pont de Nemours & Co, 51-57; head inorg chem sect, Midwest Res Inst, 57-63, asst dir chem div, 63-68; head dept chem, Southwestern Baptist Col, 68-71; CHMN MATH & SCI, McDONALD COUNTY R1 SCHS, 71- Mem: Sigma Xi. Res: Reaction rates; enzyme systems; textile chemistry; fiber and film preparation; polymers and plastics; boron chemistry. Mailing Add: Box 236 Anderson MO 64831

BARGER, RICHARD LEE, physics, see 12th edition

BARGER, VERNON DUANE, b Curllsville, Pa, June 5, 38; m 67; c 3. THEORETICAL HIGH ENERGY PHYSICS, ELEMENTARY PARTICLE PHYSICS. Educ: Pa State Univ, BS, 60, PhD(physics), 63. Prof Exp: Res assoc theoret physics, 63-65, from asst prof to assoc prof physics, 65-68, PROF PHYSICS, UNIV WIS-MADISON, 68- Concurrent Pos: Vis prof, Univ Hawaii, 70; vis scientist, CERN, Switz & Rutherford Labs, Eng, 72 & SLAC, Stanford, Calif, 75; Guggenheim fel, 72. Mem: Am Phys Soc. Res: Fundamental particle theory. Mailing Add: Dept of Physics Univ of Wis Madison WI 53706

BARGERON, LIONEL MALCOLM, JR, b Savannah, Ga, Nov 4, 23; m 50; c 3. PEDIATRICS, CARDIOLOGY. Educ: Univ Ala, BS, 48; Med Col Ala, MD, 52; Am Bd Pediat, dipl & cert cardiol. Prof Exp: Intern, Jefferson-Hillman Hosp, Birmingham, Ala, 52-53, resident, 53; asst resident, Babies Hosp, Columbia-Presby Med Ctr, NY, 54-55, sr resident, 55; instr pediat, 56-58, asst prof, 58-59, assoc prof pediat cardiol, 60-66, PROF PEDIAT, MED COL ALA, 66- Concurrent Pos: USPHS res fel pediat cardiol, 55-57. Mem: Fel Am Acad Pediat. Res: Pediatric cardiology. Mailing Add: Dept of Pediat Med Col of Ala Birmingham AL 35223

BARGHOORN, ELSO STERRENBERG, b NY, June 30, 15; m 41; c 1. PALEOBIOLOGY. Educ: Miami Univ, AB, 37; Harvard Univ, AM, 38, PhD(bot), 41. Prof Exp: Asst biol, Harvard Univ, 38-40; Sheldon fel & instr, Amherst Col, 41-43, asst prof, 45-46; asst prof bot, 46-49, assoc prof, 49-55, PROF BOT, HARVARD UNIV, 55-, CUR PALEOBOT COLLECTIONS, 48- Honors & Awards: Navy Civilian Serv Medal; Hayden Mem Award Medal; C D Walcott Medal, Nat Acad Sci, 72. Mem: Nat Acad Sci; AAAS; Bot Soc Am; fel Geol Soc Am; Linnaean Soc London. Res: Evolution and dispersal of vascular plants; plant micropaleontology; Pre-Cambrian fossils and organic geochemistry. Mailing Add: Dept of Biol Harvard Univ Cambridge MA 02138

BARGHUSEN, HERBERT RICHARD, b Englewood, NJ, July 9, 33; m 57; c 2. PALEOZOOLOGY, ANATOMY. Educ: Lafayette Col, AB, 55; Univ Chicago, PhD(paleozool), 60. Prof Exp: Asst prof zool, Smith Col, 61-64; instr anat, Stritch Sch Med, Loyola Univ, Ill, 64-67; asst prof, 67-68, ASSOC PROF ANAT & ORAL ANAT, COL MED & COL DENT, UNIV ILL MED CTR, 68- Concurrent Pos: NSF fel, 60-61; res grants, Loyola Univ, Ill, 64-67 & Univ Ill, 67-70; NSF grant, 73-75. Mem: AAAS; Soc Vert Paleont. Res: Biomechanics of vertebrate feeding mechanism; evolutionary origin of mammals and the mammalian feeding mechanism. Mailing Add: Col of Med Univ of Ill at the Med Ctr Chicago IL 60680

BARGMANN, ROLF ERWIN, b Glueckstadt, Ger, May 13, 21; nat; m 49; c 4. MATHEMATICAL STATISTICS. Educ: Univ NC, PhD(math statist), 57. Prof Exp: Court interpreter, Nuremberg Trials, 47-48; textbook reviewer math & sci, US State Dept, Ger, 48-49, consult educ res, 49-51; head dept statist, Inst Educ Res, Frankfurt, 52-55; asst, Psychomet Lab, Univ NC, 55-57; from assoc prof to prof statist, Va Polytech Inst, 58-61; mgr appl math & info sci, Systs Develop div, Thomas J Watson Res Ctr, Int Bus Mach Corp, 61-65; PROF STATIST, UNIV GA, 65- Concurrent Pos: Vis lectr, Columbia Univ, 63-64, Univ Paris, 65 & Univ Ill, 66; assoc ed, Psychometrika & J Statist Comp. Mem: Fel AAAS; fel Am Statist Asn; Economet Soc; Psychomet Soc. Res: Multivariate analysis; statistical computation. Mailing Add: Dept of Statistics Univ of Ga Athens GA 30601

BARGMANN, VALENTINE, b Berlin, Ger, Apr 6, 08; nat US; m 41. MATHEMATICAL PHYSICS. Educ: Univ Zurich, PhD(physics), 36. Prof Exp: Mem staff, Inst Adv Study, 37-46; vis lectr, Princeton Univ, 46-47; assoc prof math, Pittsburgh Univ, 48; assoc prof, 48-57, PROF MATH PHYSICS, PRINCETON UNIV, 57- Mem: Am Math Soc; Am Phys Soc; Am Acad Arts & Sci. Res: Quantum theory; relativity; group theory. Mailing Add: Dept of Math & Physics Princeton Univ Princeton NJ 08540

BARGON, JOACHIM, b Wiesbaden, Ger, Apr 13, 39; m 63; c 3. PHYSICAL CHEMISTRY. Educ: Darmstadt Tech Univ, BS, 61, MS, 65, PhD(physics), 68. Prof Exp: Staff mem spectroscopy, Ger Plastics Inst, Darmstadt, 65-69; fac mem org chem, Calif Inst Technol, 69-70; staff mem gen chem, Res Div, Yorktown NY, IBM Corp, 70-71; staff mem, 71-74, MGR RADIATION CHEM, SAN JOSE RES LAB, IBM CORP, 74- Mem: Ger Phys Soc. Res: Reaction mechanisms in radiation and free radical chemistry using spectroscopic techniques combined with molecular orbital calculations. Mailing Add: San Jose Res Lab IBM Corp Monterey & Cottle Rd San Jose CA 95193

BARHAM, ERIC GEORGE, biology, see 12th edition

BARHAM, WARREN SANDUSKY, b Prescott, Ark, Feb 15, 19; m 40; c 4. PLANT BREEDING, HORTICULTURE. Educ: Univ Ark, BSA, 41; Cornell Univ, PhD, 50. Prof Exp: Asst, Fruit & Truck Exp Sta, Univ Ark, 41-42, supt, 45-46; asst, Cornell Univ, 46-49; from asst prof to assoc prof hort, NC State Col, 49-58; DIR RAW MAT RES & SEED PROD, BASIC VEG PROD, INC, 58- Mem: Fel Am Soc Hort Sci; Am Genetic Asn; Crop Sci Soc Am; Soil Sci Soc Am; Am Soc Agron. Res: Plant physiology, genetics and breeding; onions, watermelons, canteloupes, tomatoes and garlic; disease and insect resistance; quality characteristics; seed production problems. Mailing Add: Basic Veg Prod Inc PO Box 599 Vacaville CA 95688

BARIANA, DILBAGH SINGH, b Nangal Ishar, India, Dec 30, 30. ORGANIC CHEMISTRY, PHARMACEUTICAL CHEMISTRY. Educ: Panjab Univ, India, BSc, 53, MSc, 55; Univ Kans, PhD(med chem), 60. Prof Exp: Res asst med chem, Sch Pharm, Univ Kans, 56-60; res assoc, Univ Mich, 60-62; res assoc org chem, Ariz State Univ, 62-63; res assoc immunol chem, McGill Univ, 63-67; SR RES CHEMIST, ABBOTT LABS, 67- Mem: Am Chem Soc. Res: Synthetic organic chemistry; malaria and cancer chemotherapy. Mailing Add: Abbott Labs PO Box 6150 Montreal PQ Can

BARIBO, LESTER E, b Exeland, Wis, July 2, 21. AGRICULTURAL MICROBIOLOGY. Educ: Wis State Univ-River Falls, BS, 43; Univ Wis, MS, 49, PhD(bact & biochem), 51. Prof Exp: Microbiologist, Western Condensing Co, 51-55, Arctic Aeromed Lab, 55-57 & A E Staley Mfg Co, 57-64; assoc dir, Syracuse Univ Res Corp, 64-66; PROF SPECIALIST, WEYERHAEUSER CO, 66- Mem: Am Chem Soc; Soc Am Microbiologists; Soc Indust Microbiol. Res: Packaging sanitation and development; development of animal feeds from wood products. Mailing Add: Weyerhaeuser Co 3400 13th Ave SW Seattle WA 98134

BARIC, LEE WILMER, b Carlisle, Pa, Nov 19, 32; m 56; c 3. MATHEMATICS. Educ: Dickinson Col, BSc, 56; Lehigh Univ, MSc, 62, PhD(math), 66. Prof Exp: Instr math, Lafayette Col, 58-63; mathematician, McCoy Electronics, Pa, 63-64; from asst prof to assoc prof, 64-75, PROF MATH & CHMN DEPT, DICKINSON COL, 75- Concurrent Pos: Consult, McCoy Electronics, 64-70. Mem: Am Math Soc; Math Asn Am. Res: Schauder bases in Banach spaces; summability theory. Mailing Add: Dept of Math Dickinson Col Carlisle PA 17013

BARIE, WALTER PETER, JR, b Pittsburgh, Pa, Oct 28, 26; m 64; c 1. ORGANIC POLYMER CHEMISTRY, INDUSTRIAL ORGANIC CHEMISTRY. Educ: Duquesne Univ, BS, 50, MS, 51; Pa State Univ, PhD(chem), 54. Prof Exp: Asst, Duquesne Univ, 50-51; res chemist, 54-67, SR RES CHEMIST, GULF RES & DEVELOP CO, 67- Mem: Am Chem Soc. Res: Specialty organic chemicals, monomers; petrochemicals; resins. Mailing Add: Gulf Res & Develop Co PO Drawer 2038 Pittsburgh PA 15230

BARIEAU, ROBERT (EUGENE), b Lindsay, Calif, Mar 12, 15; m 46 & 64; c 1. PHYSICAL CHEMISTRY. Educ: Univ Calif, BS, 37, PhD(chem), 40. Prof Exp: Asst chem, Univ Calif, 37-40, instr, 40-41, investr, Nat Defense Res Comt, 41-44; res chemist, Sharples Corp, Philadelphia, 44-45 & Calif Res Corp, 45-62; res chemist, Helium Res Ctr, 62-75; RES PROF PHYSICS, WTEX STATE UNIV, 73- Concurrent Pos: Res fel, Calif Inst Technol, 49-50. Mem: AAAS; Am Chem Soc. Res: X-ray and electron diffraction; electron microscopy; analytical chemistry by x-ray fluorescence and x-ray absorption; thermodynamics of galvanic cells; visible light scatterings; crystal structures; atomic emission and absorption spectroscopy; thermodynamics of gases and gas mixtures; cryogenics; wind energy conversion systems. Mailing Add: Dept of Physics WTex State Univ Canyon TX 79016

BARIL, ALBERT, JR, b New Orleans, La, Aug 19, 26; m 49; c 6. RESEARCH ADMINISTRATION. Educ: Loyola of the South, BS, 49; Tulane Univ, MS, 52. Prof Exp: Dir mfg photog, Kalvar Corp, La, 62-66; dir spec prod res, Allied Paper Co, Mich, 66-67; res physicist textiles, Mach Develop, 67-73, res leader, New Systs Res, 73-74, CHIEF RES ADMIN, COTTON TEXTILE PROCESSING LAB, SOUTHERN REGIONAL RES CTR, AGR RES SERV, USDA, 74- Mem: Electrostatic Soc Am; Am Asn Textile Chemists & Colorists; Inst Elec & Electronics Engrs. Res: New systems for textile processing, particularly, integrated system, producing yarn continuously from a supply of fibers by one machine; electrostatic manipulation of fibers; electrostatic spinning of yarns. Mailing Add: Southern Regional Res Ctr PO Box 19687 New Orleans LA 70179

BARILE, CONRAD ALBERT, physical chemistry, see 12th edition

BARILE, MICHAEL FREDERICK, b Bound Brook, NJ, Jan 9, 24; m 46; c 4. MICROBIOLOGY. Educ: Univ Ga, BS, 49; Univ Mich, MS, 51, PhD(bact), 54. Prof Exp: Instr microbiol, Univ Mich, 54; bacteriologist, 406th Gen Med Lab, 54-56; supvry bacteriologist, Med Lab, Tokyo, Japan, 56-57; bacteriologist, 58-62, res bacteriologist, 62-68, CHIEF SECT MYCOPLASMA, NIH, 68- Mem: Am Soc Microbiol; Soc Exp Biol & Med; Tissue Cult Asn. Res: Immunization against neonatal tetanus; diphtheria toxin for Schick test; mycoplasmatales; primary atypical pneumonia; ultrastructure; contamination of cell cultures; arginine dihydrolase pathway; mycoplasma species of man, dogs, monkeys, goats and sheep; mycoplasma-virus interactions. Mailing Add: Bldg 29 Rm 424 BOB(NIH)FDA 5600 Fischers Lane Rockville MD 20852

BARILE, RAYMOND CONRAD, b New York, NY, June 30, 36; m 59; c 2. PHYSICAL CHEMISTRY, INORGANIC CHEMISTRY. Educ: Manhattan Col, BS, 58; Fordham Univ, MS, 60, PhD(phys chem), 65. Prof Exp: Asst chem, Fordham Univ, 58-59; asst, 59-60, from instr to asst prof, 60-68, ASSOC PROF CHEM, MANHATTAN COL, 68- Concurrent Pos: Res assoc, Brookhaven Nat Lab, 66-67. Mem: Am Chem Soc. Res: Kinetics and mechanism of reactions of transition metal complexes, electron transfer, formation reaction and effect of solvent. Mailing Add: Dept of Chem Manhattan Col Bronx NY 10471

BARIOLA, LOUIS ANTHONY, b Lake Village, Ark, July 7, 32; m 59; c 2. ECONOMIC ENTOMOLOGY. Educ: Univ Ark, Fayetteville, BSA, 58, MS, 62; Tex A&M Univ, PhD(entom), 69. Prof Exp: Res assoc entom, Univ Ark, Fayetteville, 59-62; entomologist, Entom Res Div, Tex, 62-69, RES ENTOMOLOGIST, WESTERN COTTON RES LAB, AGR RES SERV, USDA, ARIZ, 69- Mem: Entom Soc Am; Sigma Xi. Res: Ecology of cotton insects; control of cotton insects using alternative methods, such as non-insecticides—cultural or biological methods; biology of cotton insects in the arid West, with major emphasis on the pink bollworm, cotton leafperforator, cotton bollworm and lygus bugs. Mailing Add: Western Cotton Res Lab ARS-USDA 4135 E Broadway Rd Phoenix AZ 85040

BARISH, BARRY C, b Omaha, Nebr, Jan 27, 36; m 60; c 1. HIGH ENERGY PHYSICS. Educ: Univ Calif, Berkeley, AB, 57, PhD(physics), 62. Prof Exp: Physicist, Radiation Lab, Univ Calif, 62-63; res fel physics, 63-66, from asst prof to assoc prof, 66-72, PROF PHYSICS, CALIF INST TECHNOL, 72- Mem: Am Phys Soc. Res: Investigation of the interaction of pions and nucleons; time-like structure of proton by studying antiproton-proton annihilations into electron-positron pairs; neutrino scattering at very high energies. Mailing Add: Dept of Physics Calif Inst of Technol Pasadena CA 91109

BARISH, LEO, b New Bedford, Mass, Jan 19, 30; m 52; c 2. MICROSCOPY, TEXTILE CHEMISTRY. Educ: Southeastern Mass Univ, BS, 52; Univ Lowell, MS, 54. Prof Exp: Cost-finder, Dartmouth Finishing Co, Mass, 51-52; res fel, Am Asn Textile Chemists & Colorists, 53-54; sr res assoc, Fabric Res Labs, Inc, Dedham, 56-72, SR RES ASSOC, FRL, AN ALBANY INT CO, DEDHAM, 72- Mem: Fiber Soc. Res: Optical and scanning electron microscopy; fiber structure; polymer morphology; scientific photography; textile mechanics; literature searches; fire resistant fabrics evaluation and development; stereoscopy; optical analysis; composite structures. Mailing Add: 2 Pole Plain Rd Sharon MA 02067

BARISH, NATHANIEL H, research administration, see 12th edition

BARK, LAURENCE DEAN, b Chanute, Kans, Mar 18, 26; m 64; c 3. AGRICULTURAL METEOROLOGY. Educ: Univ Chicago, BSc, 48, MSc, 50; Rutgers Univ, PhD(agron), 54. Prof Exp: Asst, Rutgers Univ, 51-54; bioclimatologist, US Weather Bur, 55; assoc prof physics & assoc climatologist, 56-67, PROF CLIMAT & CLIMATOLOGIST, AGR EXP STA, KANS STATE UNIV, 67- Honors & Awards: Nat Weather Serv Centennial Medallion. Mem: AAAS; Am Meteorol Soc; Am Soc Agron; Crop Sci Soc Am; Am Asn Physics Teachers. Res: Relation of weather to the life processes of plants, animals and humans. Mailing Add: Dept of Physics Kans State Univ Manhattan KS 66506

BARKA, TIBOR, b Debrecen, Hungary, Mar 31, 26; US citizen; m. EXPERIMENTAL MEDICINE. Educ: Debrecen Univ, MD, 50. Prof Exp: Asst, Inst Histol & Embryol, Med Univ Budapest, 51-54; res assoc, Inst Exp Med, Hungarian Acad Sci, 54-56; res assoc, Inst Cell Res & Genetics, Karolinska Inst, Sweden, 56-58; res assoc, 58-62, asst attend pathologist, 62-64, ASSOC ATTEND PATHOLOGIST, MT SINAI HOSP, 64-, PROF ANAT & CHMN DEPT, MT SINAI SCH MED, 67- Concurrent Pos: Ed-in-chief, J Histochem & Cytochem, 65-73; prof path, Mt Sinai Sch Med, 66- Mem: Histochem Soc; Am Soc Exp Path; Am Soc Cell Biol; Am Asn Anat. Res: Histochemistry; experimental pathology. Mailing Add: Dept of Anat Mt Sinai Sch of Med New York NY 10029

BARKALOW, FREDERICK SCHENCK, JR, b Marietta, Ga, Feb 23, 14; m 37; c 1. ZOOLOGY. Educ: Ga Inst Technol, BS, 36; Univ Mich, MS, 39, PhD(zool), 48. Prof Exp: Instr zool & mammal, Ala Polytech Inst, 36-39; sr biologist, Ala State Dept Conserv, 39-41; assoc prof zool, 47-48, head dept, 50-63, PROF ZOOL, NC STATE UNIV, 48-, PROF FORESTRY, 67- Concurrent Pos: Asst, Univ Mich, 38-39; prin biologist, NC Wildlife Resources Comn, 47-50; Orgn European Econ Coop sr vis fel, Gt Brit, 60; mem, Secy Agr Adv Comt Multiple Use Nat Forests, 63-68. Mem: Fel AAAS; Am Soc Mammal; Am Ornithologists Union; Soc Syst Zool; Soc Am Foresters. Res: Game inventories on a state-wide basis; management of wildlife resources; game populations; biology of tree squirrels. Mailing Add: NC State Univ Dept of Zool Box 5577 Raleigh NC 27607

BARKATE, JOHN ALBERT, b Sulphur, La, Dec 4, 36; m 64; c 4. MICROBIOLOGY, FOOD SCIENCE. Educ: Northwestern State Univ, MS, 63; La State Univ, PhD(food microbiol), 67. Prof Exp: DIR MICROBIOL DEPT, RALSTON PURINA CO, 67- Mem: Am Soc Microbiol; Inst Food Technologists; Int Asn Milk, Food & Environ Sanitarians, Inc. Res: Microbiological research and analysis. Mailing Add: Ralston Purina Co Checkerboard Sq St Louis MO 63188

BARKE, HARVEY ELLIS, b Plymouth, Mass, Nov 20, 17; m 41; c 5. ECONOMIC ENTOMOLOGY, BOTANY. Educ: Univ Mass, BS, 39, MS, 43; Univ Ga, PhD(acarology, biol), 70. Prof Exp: Head res plant breeding, Arnold-Fisher Rose Co, Mass, 40-43; lab instr basic bot & crypt, Univ Mass, 46-47; instr bot & entom, Long Island Agr Tech Inst, 47-56; from asst prof to assoc prof, 56-62, PROF BIOL, STATE UNIV NY AGR & TECH COL FARMINGDALE, 64-, COORDR BIOL TECHNOL, 70-, CHMN DEPT BIOL SCI, 74- Concurrent Pos: Consult, Am Rose Soc, 60-70; consult rosarian, 65- Mem: Entom Soc Am; Weed Sci Soc Am; Am Inst Biol Sci. Res: Screening new pesticides for ornamental horticulture; turf insect and disease control, weed control; house and tropical plants. Mailing Add: Dept of Biol State Univ of NY Agr & Tech Col Farmingdale NY 11735

BARKER, A S, JR, b Vancouver, BC, Dec 18, 33; m 55; c 3. SOLID STATE PHYSICS. Educ: Univ BC, BA, 55, MS, 57; Univ Calif, Berkeley, PhD(physics), 62. Prof Exp: Br mgr, Yukon Tel Co, 53-55; lab instr math, Univ BC, 55-57, res asst physics, 57; res asst, Univ Calif, Berkeley, 60-62; MEM STAFF, BELL TEL LABS, INC, 62- Mem: Fel Am Phys Soc. Res: Infrared dielectric behavior of ferroelectric materials; lattice vibrations in ionic crystals and semiconductors; free carriers in semimetals. Mailing Add: Bell Tel Labs Murray Hill NJ 07974

BARKER, ALLEN VAUGHAN, b McLeansboro, Ill, Aug 15, 37; m 67. PLANT PHYSIOLOGY, SOIL SCIENCE. Educ: Univ Ill, BS, 58; Cornell Univ, MS, 59, PhD(agron), 62. Prof Exp: Fel soil sci, NC State Univ, 62-64; asst prof plant physiol, Univ Mass, 64-66 & Mich State Univ, 66-67; asst prof, 67-70, ASSOC PROF PLANT PHYSIOL, UNIV MASS, AMHERST, 70- Concurrent Pos: Plant physiologist, Agr Res Serv, USDA, 66-67. Honors & Awards: Dow Chem Co Environ Qual Res Award, 75. Mem: Crop Sci Soc Am; Am Soc Agron; Soil Sci Soc Am; Am Soc Plant Physiol; fel Am Inst Chemists. Res: Ammonium toxicity in plants; ammonium fixation in soils; nitrate and heavy metal accumulation in vegetables. Mailing Add: Dept of Plant & Soil Sci Univ of Mass Amherst MA 01002

BARKER, CLIFFORD ALBERT VICTOR, b Ingersoll, Ont, Jan 5, 19; m 42; c 3. VETERINARY MEDICINE. Educ: Univ Toronto, DVM, 41, DVSc, 48; McGill Univ, MSc, 45. Prof Exp: Veterinarian & lectr vet sci, Macdonald Col, McGill Univ, 41-45; from asst prof to assoc prof, 45-65, PROF ANIMAL REPROD & DEP CHMN DEPT CLIN STUDIES, ONT VET COL, UNIV GUELPH, 65- Concurrent Pos: Consult, Caprine Res Found. Mem: Can Fertility Soc; Can Vet Med Asn; Am Col Theriogenology. Res: Theriogenology, particularly in ruminants; male and female infertility in domestic animals, particularly in ruminants. Mailing Add: 105 Stuart St Guelph ON Can

BARKER, CLYDE FREDERICK, b Salt Lake City, Utah, Aug 16, 32; m 56; c 4.

SURGERY, IMMUNOLOGY. Educ: Cornell Univ, BA, 54, MD, 58; Am Bd Surg, dipl, 65. Prof Exp: Intern, Hosp Univ Pa, 58-59, from asst resident to chief resident surg, 59-64; assoc, 64-68, from asst prof to assoc prof, 68-73, PROF SURG, SCH MED, UNIV PA, 73-, ASSOC MED GENETICS, 66- Concurrent Pos: Fel, Harrison Dept Surg Res, Sch Med, Univ Pa, 59-64, Am Cancer Soc fel, 61-62, USPHS fel med genetics, 65-66; Hartford fel vascular surg, Hosp Univ Pa, 64-65; Markle Found scholar acad med, 68; attend surgeon & assoc chief sect peripheral vascular surg, Hosp Univ Pa, 66-, chief sect renal transplantation, 69- Mem: Transplantation Soc; Am Diabetes Asn; Am Col Surg; Am Fedn Clin Res; Asn Acad Surg. Res: Gastrointestinal and cardiovascular physiology; transplantation biology; immunologically privileged sites; microvascular transplantation surgery; transplantation and pancreatic islets; clinical renal transplantation. Mailing Add: Hosp Univ Pa 3400 Spruce St Philadelphia PA 19104

BARKER, COLIN G, b Plymouth, Eng, Aug 3, 39; m 65; c 2. ORGANIC GEOCHEMISTRY. Educ: Oxford Univ, BA, 62, DPhil(geol), 65. Prof Exp: Res fel geol, Univ Tex, Austin, 65-67; asst prof chem, 69-72, ASSOC PROF CHEM, UNIV TULSA, 72- Concurrent Pos: Sr res chemist, Esso Prod Res Co, Tex, 67-69; consult, Amoco Prod Res Co, 70-, Sun Oil Co, 74-75, Agrico, 74 & Rockwell Int, 75; Res Corp grant; NASA grants; NSF grant, 73- Mem: Geochem Soc; Am Asn Petrol Geologists. Res: Generation and migration of petroleum; analysis of inorganic gases in minerals from igneous rocks and ore deposits. Mailing Add: Dept of Chem Univ of Tulsa Tulsa OK 74104

BARKER, DANIEL STEPHEN, b Waltham, Mass, Feb 27, 34; m 64; c 2. GEOLOGY. Educ: Yale Univ, BS, 56; Calif Inst Technol, MS, 58; Princeton Univ, PhD(geol), 61. Prof Exp: Res assoc geol, Cornell Univ, 61-62; res asst, Yale Univ, 62-63; from asst prof to assoc prof, 63-72, PROF GEOL, UNIV TEX, AUSTIN, 72- Concurrent Pos: Fulbright-Hays res fel, 74. Mem: Geol Soc Am; Mineral Soc Am; Mineral Asn Can. Res: Igneous and metamorphic petrology; hydrothermal phase equilibria and mineral synthesis. Mailing Add: Dept of Geol Sci Univ of Tex Austin TX 78712

BARKER, DAVID LOWELL, b Price, Utah, May 3, 41; m 64; c 1. NEUROBIOLOGY. Educ: Calif Inst Technol, BS, 63; Brandeis Univ, PhD(biochem), 69. Prof Exp: Res fel neurobiol, Harvard Med Sch, 69-71; ASST PROF BIOL, UNIV ORE, 71- Mem: Soc Neurosci; AAAS. Res: Identification of neurotransmitters in individual invertebrate neurons; analysis of biochemical control mechanisms that regulate neurotransmitter metabolism and electrical activity in single cells and in interconnected neurons generating motor patterns. Mailing Add: Dept of Biol Univ of Ore Eugene OR 97403

BARKER, DONALD YOUNG, b Glenboro, Man, Apr 4, 25. PHARMACY. Educ: Univ Man, BS, 49; Purdue Univ, MS, 53, PhD(pharm), 55. Prof Exp: Instr pharm, Univ Man, 49-53; PROF PHARM, UNIV OF THE PAC, 60- Mem: Am Pharmaceut Asn; Soc Cosmetic Chem; Royal Soc Health. Res: Pharmaceutical education; product development; dermatological vehicles; emulsion technology. Mailing Add: Sch of Pharm Univ of the Pac Stockton CA 95207

BARKER, EARL STEPHENS, b Salt Lake City, Utah, Sept 21, 20. INTERNAL MEDICINE. Educ: Univ Utah, BA, 41; Univ Pa, MD, 45. Prof Exp: Asst instr pharmacol, 48-49, from asst instr to instr, 51-53, assoc, 53-55, asst prof, 55-61, ASSOC PROF MED, SCH MED, UNIV PA, 61-, DIR, DIAG CLIN, 70- Concurrent Pos: Godey-Seger fel med, Univ Pa, 49-51; NIH res fel, 51-53; asst attend physician, Univ Hosps, 51-55, attend physician, 55-; investr, Am Heart Asn, 55-60. Mem: Fel Am Col Physicians; Am Fedn Clin Res; Am Soc Clin Invest; Am Physiol Soc. Res: Renal physiology; tubular transport mechanisms; acid-base balance; electrolyte excretion; oxygen metabolism of kidney. Mailing Add: Hosp Univ of Pa 36th and Spruce St Philadelphia PA 19104

BARKER, FRANKLIN BRETT, b Shawnee, Okla, Nov 12, 23; m 51; c 1. RADIOCHEMISTRY. Educ: Univ Okla, BA, 44; Univ NMex, PhD(chem), 54. Prof Exp: Asst chemist, Cities Serv Oil Co, 44-45, control chemist, 45-46; mem staff, Los Alamos Sci Lab, 47-51; res asst, Univ NMex, 51-54; chemist, US Geol Surv, Colo, 54-57, res chemist & proj chief, 57-61; sr scientist, 61-67, FEL SCIENTIST, BETTIS ATOMIC POWER LAB, WESTINGHOUSE ELEC CORP, 67- Mem: AAAS; Am Chem Soc; Am Nuclear Soc; Am Soc Testing & Mat; Am Soc Mass Spectrometry. Res: Mass spectrometry of reactor materials; automatic control and data acquisition; computer programming; radiochemical analysis; gamma-ray spectroscopy; nuclear fuel burn-up measurement. Mailing Add: Bettis Atomic Power Lab PO Box 79 West Mifflin PA 15122

BARKER, FRED, b Seekonk, Mass, Nov 4, 28; m 61; c 3. GEOLOGY. Educ: Mass Inst Technol, BS, 50; Calif Inst Technol, MS, 52, PhD, 54. Prof Exp: GEOLOGIST, US GEOL SURV, 54- Concurrent Pos: Vis res fel, Univ Witwatersrand, 74. Mem: Fel Geol Soc Am; fel Mineral Soc Am; Sigma Xi. Res: Petrology and geochemistry; geology of Precambrian rocks; isotope geology. Mailing Add: US Geol Surv Box 25046 Denver Fed Ctr Denver CO 80225

BARKER, GEORGE ERNEST, b Cortland, NY, Aug 8, 07; m 41; c 2. ORGANIC CHEMISTRY. Educ: Mass Inst Technol, BS, 30, PhD(chem), 35. Prof Exp: Asst, Mass Inst Technol, 30-34; res chemist, Nat Aniline & Chem Co, 34-38; indust fel, Mellon Inst, 38-42, sr indust fel, 42-45; res chemist, Atlas Powder Co, 45-55, dir metal res, Quaker Chem Prod Corp, 55-58, dir metal labs, 58-60; vpres & dir res, Van Straaten Chem Co, 60-74; CHEM CONSULT, 74- Mem: Am Chem Soc; Am Oil Chem Soc; Am Soc Lubrication Eng; Soc Mfg Eng. Res: Ultraviolet spectroscopy; development of surface active agents; development of instrument lubricants; investigation of utilization of products of sugar industry; synthetic detergents and lubricants; hydraulic fluids; petroleum additives; metal working lubricants; metal-treating processes. Mailing Add: 4908 Oakwood Dr McHenry IL 60050

BARKER, HAL B, b Palmer, Tenn, Feb 5, 25; m 52; c 2. ANIMAL PHYSIOLOGY, ANIMAL NUTRITION. Educ: Tenn Polytech Inst, BS, 47; Iowa State Col, MS, 49; Ala Polytech Inst, PhD, 59. Prof Exp: Assoc prof dairying, 49-58, head dept animal indust, 53-64, dean sch agr & forestry, 64-71, PROF ANIMAL HUSB, LA TECH UNIV, 58-, DEAN COL LIFE SCI, 71- Mem: AAAS; Am Soc Animal Sci; Am Dairy Sci Asn. Res: Milkfat replacements in rations for dairy calves; physiology of reproduction. Mailing Add: Box 5277 Tech Sta Ruston LA 71270

BARKER, HAROLD CLINTON, b Akron, Ohio, Aug 6, 22; m 44; c 3. ORGANIC CHEMISTRY, POLYMER CHEMISTRY. Educ: Univ Akron, BS, 44; Ohio State Univ, MSc, 48, PhD(org chem), 51. Prof Exp: Res chemist, 51-54, sales technologist, 54-59, indust specialist, 59-71, sales develop plastic pipe, 71-74, MKT DEVELOP POLYIMIDE PLASTICS, E I DU PONT DU NEMOURS & CO, 74- Mem: Am Chem Soc; Sigma Xi. Res: Organic free radical chemistry; degradation of polymers present; sales development of polyolefin plastics; market development of glass reinforced thermoplastics. Mailing Add: 3204 Tanya Dr Delwynn Wilmington DE 19803

BARKER, HAROLD GRANT, b Salt Lake City, Utah, June 10, 17; m 49; c 2. SURGERY. Educ: Univ Utah, AB, 39; Univ Pa, MD, 43; Am Bd Surg, dipl. Prof Exp: Intern, Hosp Univ Pa, 43-44, asst instr pharmacol, Med Sch, 46-47, asst instr surg, 47-51, instr, 51-52, assoc surg, 52-53; from asst prof surg to assoc prof, 53-68, PROF CLIN SURG, COL PHYSICIANS & SURGEONS, COLUMBIA UNIV, 68- Concurrent Pos: Res fel pharmacol, Univ Pa, 46-47, Harrison res fel surg, 47-51 & Runyon res fel, 51-52; asst resident surg, Hosp, Univ Pa, 47-51, sr resident, 51-52, asst attend surgeon, 52-53; from asst attend surgeon to attend surgeon, Presby Hosp, 53-, dir med affairs, 74- Mem: Am Surg Asn; Soc Univ Surgeons; Soc Surg Alimentary Tract; Am Col Surgeons; Am Physiol Soc. Res: Surgical physiology of gastrointestinal tract. Mailing Add: 161 Ft Washington Ave New York NY 10032

BARKER, HORACE ALBERT, b Oakland, Calif, Nov 29, 07; m 33; c 3. BIOCHEMISTRY. Educ: Stanford Univ, AB, 29, PhD(chem), 33. Hon Degrees: ScD, Western Reserve Univ, 64. Prof Exp: Asst zool, Univ Chicago, 30-31; Nat Res Coun fel, Hopkins Marine Sta, Pacific Grove, 33-35; Gen Educ Bd fel, Technol Univ Delft, 35-36; instr soil microbiol, Univ & jr soil microbiologist, Agr Exp Sta, 36-41, asst prof & asst soil microbiologist, 41-45, assoc soil microbiologist, 45-46, prof, 46-51, prof plant biochem, 51-59, prof biochem, 59-75, microbiologist, 46-75, chmn dept biochem, 62-64, EMER PROF BIOCHEM, AGR EXP STA, UNIV CALIF, BERKELEY, 75- Concurrent Pos: Guggenheim fels, 41-42 & 62; assoc ed, Ann Rev Microbiol; dir, Ann Rev, Inc, 46-62; mem ed bd, Arch Biochem Biophys, 50-53 & 70-, J Bact, 55-60, J Biol Chem, 60-65, Biochim Biophys Acta, 68-72 & Anal Biochem, 74- Honors & Awards: Sugar Res Award, 45; Carl Neuberg Medal, 59; Borden Award, 62; Calif Scientist of Year, 65; F G Hopkins Medal, Brit Biochem Soc, 67; Nat Medal Sci, 68; Citation, Univ Calif, Berkeley, 75. Mem: Nat Acad Sci; Am Soc Microbiol; Am Chem Soc; Am Soc Biol Chem; Brit Biochem Soc. Res: Soil microbiology; physiology and biochemistry of microorganisms; vitamin B-12 coenzymes. Mailing Add: Dept of Biochem Univ of Calif Berkeley CA 94720

BARKER, JAMES EMORY, b Rutledge, Ga, July 13, 27; m 50; c 1. INORGANIC CHEMISTRY. Educ: Emory Univ, BA, 49. Prof Exp: Res analyst, Tenn Corp, 49-57, chief chemist, 57-60, lab supvr, 60-62, proj supvr, 62-65, sr group leader, 65-68, mgr process res, Tenn Corp Div, Cities Serv Co, 68-70, mgr process res, Chem & Metals Group, 70-73, RES ASSOC, CHEM & METALS GROUP, CITIES SERV CO, 73- Mem: Am Chem Soc; Am Inst Chem. Res: Process research in agricultural chemicals and industrial metallic salts, including copper, iron, zinc and aluminum; analytical chemistry. Mailing Add: Cities Serv Co PO Drawer 8 Cranbury NJ 08512

BARKER, JANE ELLEN, b Bangor, Maine, June 21, 35. DEVELOPMENTAL BIOLOGY, HEMATOLOGY. Educ: Univ Maine, BA, 57; Wellesley Col, MA, 59; Univ Wis, PhD(zool), 67. Prof Exp: Asst develop hemat, 67-68; res asst oncol, Wellesley Col, 59-61; asst develop genetics, Univ Wis, 66-67; ASST INVESTR DEVELOP HEMAT, INST MED RES, PUTNAM MEM HOSP, 68- Concurrent Pos: Fel develop genetics, Univ Wis, 66-67; fel develop hemat, Jackson Lab, 67-68. Mem: AAAS; Soc Develop Biol; Am Soc Zool; Genetics Soc Am. Res: Embryonic development of mouse and hamster hematopoietic system; hematopoietic alterations in pregnant hamsters; effect of steroids on immune response; induction-repression of embryonic hemoglobin synthesis. Mailing Add: Inst for Med Res Putnam Mem Hosp Bennington VT 05201

BARKER, JEFFERY LANGE, b New York, NY, Jan 29, 43; m 70; c 2. NEUROBIOLOGY, NEUROPHARMACOLOGY. Educ: Harvard Col, BA, 64; Boston Univ, MD, 68. Prof Exp: Intern surg, Boston Univ Hosp, 68-69; res assoc neurobiol, Nat Inst Neurol Dis & Stroke, 69-72, spec fel, 72-73, MED OFFICER NEUROBIOL, NAT INST CHILD HEALTH & HUMAN DEVELOP, NIH, 73- Concurrent Pos: NSF grants res consult neurobiol, 75- Mem: AAAS; Soc Gen Physiologists; Soc Neurosci; Am Physiol Soc; Am Soc Neurochem. Res: Physiological roles of peptides and neurotransmitters in neuronal function; cellular mechanisms of neuropharmacologic agents, such as anesthetics and convulsants; biophsyical mechanisms of neuronal pacemaker activity; function of rapidly transported proteins in axons. Mailing Add: Behav Biol Br Nat Inst Child Health & Human Dev Bethesda MD 20014

BARKER, JOHN ADAIR, b Corrigin, Australia, Mar 27, 25; m 50; c 3. CHEMICAL PHYSICS, STATISTICAL MECHANICS. Educ: Demonstr physics, Univ Melbourne, 47-48; res scientist, Commonwealth Sci & Res Org, Australia, 49-67; fel Australian Acad Sci, 67; prof appl math & physics, Univ Waterloo, 68-69; RES STAFF, SAN JOSE RES LAB, IBM CORP, 69- Concurrent Pos: Adj prof, Univ Waterloo, 69- Mem: Fel Australian Acad Sci. Res: Intermolecular forces and the properties of matter; experimental determination of intermolecular forces; statistical mechanics theories of fluids and solids; perturbation theories; computer simulations; surface and interfacial structure and properties. Mailing Add: San Jose Res Lab IBM Corp Monterey & Cottel Rd San Jose CA 95193

BARKER, JOHN GROVE, b Ocala, Fla, June 24, 26; m 48; c 2. ZOOLOGY. Educ: Concord Col, BS, 51; Univ Md, MS, 53; Va Polytech Inst, PhD(zool), 57. Hon Degrees: DP, Concord Col, 72; LittD, Shenandoah Col & Conserv Music, 73. Prof Exp: Asst zool, Univ Md, 52-53; from instr to prof biol, Radford Col, 53-66, chmn dept, 58-62, head div natural sci, 62-66, vpres, Col, 66-68; assoc exec secy comn on cols, Southern Asn Cols & Schs, 68-71; pres, Marshall Univ, 71-74; PRES, MIDWESTERN STATE UNIV, 74- Res: Embryology of anura; genetics of insecticide resistance in Blattella; taxonomy of culicidae. Mailing Add: Midwestern State Univ Wichita Falls TX 76308

BARKER, JOHN L, JR, b New York, NY, Sept 4, 37; m 59; c 2. EARTH SCIENCES, REMOTE SENSING. Educ: Johns Hopkins Univ, AB, 58; Univ Chicago, MS, 62, PhD(phys chem), 67. Prof Exp: Res assoc chem, Univ Chicago, 67-68; asst prof, Univ Md, College Park, 68-72; radiochemist, US Geol Surv, 72-73; PHYS SCIENTIST, GODDARD SPACE FLIGHT CTR, NASA, 74- Concurrent Pos: Nat Acad Sci sr fel, 73 & 74. Mem: AAAS; Am Chem Soc; Sigma Xi. Res: Water quality; land cover; earth resources; applications of remote sensing from satellites; evaluation and improvement of digital image processing of multispectral scanner imagery; identifying needs for future satellite sensors. Mailing Add: 9102 Tuckahoe Lane Adelphi MD 20783

BARKER, JOHN ROGER, b Sewanee, Tenn, Oct 17, 43; m 72. CHEMICAL KINETICS. Educ: Hampden-Sydney Col, BS, 65; Carnegie-Mellon Univ, MS, 69, PhD(chem), 70. Prof Exp: Res assoc chem kinetics, Univ Wash, 69-71; res assoc, Brookhaven Nat Lab, 71-73; sr res assoc, Yeshiva Univ, 73-74; CHEMIST CHEM KINETICS, STANFORD RES INST, 74- Mem: Am Chem Soc; AAAS. Res: Gas phase chemical kinetics; inter- and intra-molecular energy transfer processes. Mailing Add: Thermochem & Chem Kinetics Stanford Res Inst 333 Ravenswood Menlo Park CA 94025

BARKER, JUNE NORTHROP, b Milwaukee, Wis, June 29, 28; m 51. PHYSIOLOGY. Educ: Univ Rochester, BS, 52; Duke Univ, MA, 54, PhD(physiol), 56. Prof Exp: Instr physiol, Med Sch, Duke Univ, 57-58; from instr to asst prof, Jefferson Med Col, 58-

64; ASST PROF PHYSIOL & REHAB MED, SCH MED, NY UNIV, 64- Mem: AAAS; Am Physiol Soc; Biomed Eng Soc; Am Soc Neurochem. Res: Developmental and cerebrovascular physiology. Mailing Add: Dept of Rehab Med NY Univ Sch of Med New York NY 10016

BARKER, KENNETH LEROY, b Columbus, Ohio, July 15, 39; m 64; c 1. BIOCHEMISTRY, ENDOCRINOLOGY. Educ: Ohio State Univ, BS, 60, MS, 62, PhD(biochem), 64. Prof Exp: Res asst dairy sci, Ohio State Univ, 61-64; res assoc obstet & gynec, Univ Kans, 65-66; from res instr to res assoc prof obstet & gynec & from instr to assoc prof biochem, 66-74, RES PROF OBSTET & GYNEC & PROF BIOCHEM, UNIV NEBR MED CTR, OMAHA, 74- Concurrent Pos: Fel biochem, Univ Kans, 65-66; Pop Coun res grant, 67; Nat Inst Child Health & Human Develop res grant, 67-; NIH res career develop award, 68-73; Nat Acad Sci exchange fel to Inst Molecular Biol, Moscow, USSR, 73; mem, Reproductive Biol Study Sect, NIH, 75- Mem: Endocrine Soc; Soc Gynec Invest; Am Physiol Soc; Soc Exp Biol & Med; Am Soc Biol Chem. Res: Investigations of hormonal control mechanisms in the mammalian uterus, particularly nuclear and cytoplasmic control of uterine nucleic acid, protein synthesis and enzyme activities. Mailing Add: Dept of Biochem Univ of Nebr Med Ctr Omaha NE 68105

BARKER, KENNETH NEIL, b Spring Valley, Ohio, Mar 25, 37; m 57; c 3. PHARMACY ADMINISTRATION. Educ: Univ Fla, BS, 59, MS, 61; Univ Miss, PhD(pharm admin), 70. Prof Exp: Residency, Hosp Pharm, Univ Fla, 61; proj coordr, Hosp Systs Res, Sch Pharm, Univ Miss, 67-70; dir admin res, US Pharmacopeial Conv, Inc, 70-72; assoc dir res inst & assoc prof pharm admin, Sch Pharm, Northeast La Univ, 72-75; CHMN DIV PHARM ADMIN & ALUMNI ASSOC PROF PHARM ADMIN, AUBURN UNIV, 75- Concurrent Pos: Mem vis scientist prog, Am Asn Col Pharm, 70-72; consult & ed dir, Health Care Facil Serv, Health Resources Admin, Dept Health, Educ & Welfare, 70-73; mem adv panel hosp pharm res needs, Am Soc Hosp Pharmacist, 73; proj dir & chmn, Nat Coord Comt Large Volume Parenterals, 72-; consult, Bd Trustees, US Pharmacopeial Conv, Inc, 74-75. Honors & Awards: Res Award, Am Soc Hosp Pharmacists, 73. Mem: Acad Pharmaceut Sci; Am Asn Col Pharm; Am Soc Hosp Pharmacists; Am Pharmaceut Asn. Res: Socioeconomics of health care with emphasis on pharmacy services; hospital medication systems, errors; co-originator of unit-dose concept; organizational research; utilization of pharmacists: design of hospital pharmacy facilities. Mailing Add: Sch of Pharm Auburn Univ Auburn AL 36830

BARKER, KENNETH RAY, b Memphis, Tenn, Oct 30, 39; m 71. ZOOLOGY, CELL BIOLOGY. Educ: Southwestern at Memphis, BS, 61; Univ Miss, MS, 63; Univ Tex, PhD(zool), 66. Prof Exp: Instr zool, Univ Tex, 66, NIH fel, 66-67; res fel embryol, Univ Witwatersrand, 67-68; asst prof biol, 69-74, ASSOC PROF BIOL, CANSIUS COL, 74- Concurrent Pos: Mem fertilization & gamete physiol prog, Woods Hole, Mass, 63. Mem: Am Soc Cell Biologists. Res: Ultrastructure of sperm and sperm formation of decapods, insects and urodeles; molecular control of eucaryotic cells. Mailing Add: Dept of Biol Canisius Col Buffalo NY 14208

BARKER, KENNETH REECE, b Roaring River, NC, Feb 1, 32; m 58; c 2. PLANT PATHOLOGY. Educ: NC State Col, BS, 56, MS, 59; Univ Wis, PhD(plant path), 61. Prof Exp: Asst prof, Univ Wis, Madison, 61-66; assoc prof, 66-71, PROF PLANT PATH, NC STATE UNIV, 71- Mem: Am Phytopath Soc; Soc Nematol. Res: Ecology and physiology of plant parasitic nematodes and their interaction with other plant pathogens. Mailing Add: Dept of Plant Path NC State Univ Raleigh NC 27607

BARKER, LAREN DEE STACY, b Honolulu, Hawaii, Mar 1, 42; m 64; c 4. VERTEBRATE PHYSIOLOGY. Educ: Univ Minn, BS, 64, MS, 66; Pa State Univ, PhD(physiol), 69. Prof Exp: Res asst dairy husb, Univ Minn, 64-66; ASSOC PROF BIOL, SOUTHWEST STATE UNIV, 69- Mem: Soc Study Reproduction; Am Inst Biol Sci; Brit Soc Study Fertil. Res: Physiology of the male reproductive system; epididymal physiology; immunoreproduction. Mailing Add: Dept of Biol Southwest State Univ Marshall MN 56258

BARKER, LEROY N, b Brigham City, Utah, Oct 18, 28; m 56; c 5. AGRONOMY, PLANT BREEDING. Educ: Utah State Univ, BS, 53, MS, 57; Univ Wis, PhD(agron), 64. Prof Exp: Plant breeder & res sta mgr, Asgrow Seed Co, Wis, 60-65; from asst prof to assoc prof agron, 65-73, PROF AGRON, CALIF STATE UNIV, CHICO, 73- Concurrent Pos: Vis prof, US AID, Univ Wis Proj, Univ Ife, Nigeria, 68-70. Mem: Am Soc Agron; Crop Sci Soc Am. Res: Breeding of cowpeas, peppers and rice; agronomic research on cowpeas and tomatoes; basic genetic studies on cowpeas; field crop yield and fertilizer trials and genetic studies. Mailing Add: Dept of Plant & Soil Sci Calif State Univ Chico CA 95926

BARKER, LOUIS ALLEN, b Charleston, WVa, Nov 23, 41; m 66; c 1. PHARMACOLOGY, NEUROCHEMISTRY. Educ: WVa Univ, BSc, 64; Tulane Univ, PhD(pharmacol), 68. Prof Exp: Grant researcher biochem, Queens Col, NY, 70-72, adj asst prof psychol, 72; instr, 72-73, ASST PROF PHARMACOL, MT SINAI SCH MED, 73- Concurrent Pos: Nat Inst Neurol Dis & Stroke fel, NY State Inst Basic Res Ment Retardation, 68-70. Mem: Am Soc Pharmacol & Exp Therapeut; Soc Neurosci; NY Acad Sci. Res: Pharmacology and neurochemistry of synaptic transmission; pharmacology of natural products. Mailing Add: Dept Pharmacol Mt Sinai Sch Med 100th St & Fifth Ave New York NY 10029

BARKER, LYNN MARSHALL, b Florence, Ariz, Apr 4, 28; m 64; c 2. APPLIED PHYSICS. Educ: Univ Ariz, BS, 54, MS, 55. Prof Exp: Staff mem, Sandia Labs, 55-74; STAFF CONSULT, TERRA TEK, INC, 74- Mem: Am Phys Aoc. Res: Shock wave instrumentation; experimental fracture mechanics; creep; thermal property measurements; geophysical instrumentation. Mailing Add: Terra Tek Inc 420 Wakara Way Salt Lake City UT 84108

BARKER, MARVIN WINDEL, b Flora, Ill, SEpt 26, 36; m 57; c 2. ORGANIC CHEMISTRY. Educ: Southern Ill Univ, BA, 58; Duke Univ, MA, 60, PhD(org chem), 63. Prof Exp: Res assoc, Univ Ill, 62-63; from asst prof to assoc prof, 63-70, PROF CHEM, MISS STATE UNIV, 70- Mem: Am Chem Soc; Int Soc Heterocyclic Chem. Res: Heterocyclic synthetic chemistry and chemistry of heterocumulenes. Mailing Add: Dept of Chem Miss State Univ Mississippi State MS 39762

BARKER, NORVAL GLEN, b Lincoln, Nebr, Aug 6, 25; m 45; c 3. NUTRITIONAL BIOCHEMISTRY. Educ: Univ Nebr, BS, 47, MS, 49, PhD(chem), 50. Prof Exp: Sect leader chem res, Gen Mills, Inc, 50-58; VPRES RES, DELMARK CO, 58- Mem: Am Chem Soc; Inst Food Technologists. Res: Therapeutic nutritional products; institutional foods. Mailing Add: 1566 Sumter Ave N Minneapolis MN 55427

BARKER, PHILIP SHAW, b Queretaro, Mex, Aug 2, 33. ENTOMOLOGY, INSECT TOXICOLOGY. Educ: Univ Calif, Berkeley, MSc, 60; McGill Univ, PhD, 65. Prof Exp: RES OFFICER, CAN DEPT AGR, 65- Mem: Entom Soc Am; Entom Soc Can. Res: Control of insects and mites in stored cereals; effect of freezing temperatures on fumigants. Mailing Add: Can Dept Agr Res Dept 25 Dafoe Rd Winnipeg MB Can

BARKER, RICHARD GORDON, b Rochester, NY, Feb 8, 37; m 57; c 2. ORGANIC CHEMISTRY. Educ: Hamilton Col, BA, 58; Lawrence Univ, MS, 60, PhD(org chem), 63. Prof Exp: Res scientist, Union Bag-Camp Paper Corp, 62-69, group leader, Union Camp Corp, 69-72, sect leader, 72-74, DIR RES & DEVELOP PROJS, UNION CAMP CORP, 74- Concurrent Pos: Mem indust liaison comt, Forest Prod Lab, 74- Mem: Am Chem Soc; Tech Asn Pulp & Paper Indust. Res: Chemical wood pulping processes; bleaching processes; pulp and paper. Mailing Add: Union Camp Corp Box 412 Princeton NJ 08540

BARKER, ROBERT, b Bedlington, Eng, Sept 21, 28; m 55; c 2. BIOCHEMISTRY. Educ: Univ BC, BA, 52, MA, 53; Univ Calif, Berkeley, PhD(biochem), 58. Prof Exp: Atlas Powder Co fel chem, Washington Univ, 58-59; vis scientist, NIH, 59-60; from asst prof to assoc prof biochem, Univ Tenn, 60-64; from assoc prof to prof, Univ Iowa, 64-74, assoc dean med, 70-74; PROF BIOCHEM & CHMN DEPT, MICH STATE UNIV, 74- Concurrent Pos: Vis prof, Duke Univ, 70-71; chmn biochem test comt, Nat Bd Med Examr, 74-; mem physiol chem rev comt, Vet Admin, 75- Mem: AAAS; Am Soc Biol Chemists; Am Inst Biol Sci; Am Chem Soc; Brit Chem Soc. Res: Effects of configuration and substitution on reactions of carbohydrates; mechanism of action of glycolytic enzymes and glycosyl transferases. Mailing Add: Dept of Biochem Mich State Univ East Lansing MI 48824

BARKER, ROBERT EDWARD, JR, b Bonifay, Fla, Oct 8, 30; m 54; c 2. POLYMER PHYSICS. Educ: Univ Ala, BS, 52, MS, 54, PhD(physics), 60. Prof Exp: High sch teacher, Fla, 50-51; physicist, US Navy Mine Defense Lab, 54 & 55; engr, Hayes Aircraft Corp, Ala, 56 & 57, scientist, 58; from instr to asst prof physics, Univ Ala, 58-60; physicist, Gen Elec Res Lab, 60-67; chmn eng sci prog, 69-73, ASSOC PROF MAT SCI, UNIV VA, 67- Concurrent Pos: Consult, Hayes Aircraft Corp, 56-59; mem exec bd & mem tech adv comt, Conf Elec Insulation & Dielectric Phenomena, Nat Acad Sci. Mem: Am Phys Soc; Am Asn Physics Teachers; Am Chem Soc; Metric Asn. Res: Thermomechanical properties of solids; radiation damage of polymers; electron spin resonance; wave motion; thermodynamics; thermal conductivity; diffusion; ionic mobility; glass transition phenomena in polymers. Mailing Add: Dept of Mat Sci Univ of Va Thornton Hall Charlottesville VA 22903

BARKER, ROBERT HENRY, b Washington, DC, Aug 17, 37; m 58; c 3. TEXTILE CHEMISTRY. Educ: Clemson Col, BS, 59; Univ NC, PhD(org chem), 63. Prof Exp: From instr to asst prof org chem, Tulane Univ, 62-67; assoc prof textiles & chem, 67-74, J E SIRRINE PROF TEXTILE & POLYMER CHEM, CLEMSON UNIV, 74- Concurrent Pos: Consult, Am Enka Co, 70- & Hooker Chem & Plastics Co, 73- Mem: AAAS; Info Coun Fabric Flammability; Am Chem Soc; Am Asn Textile Chemists & Colorists. Res: Chemistry of coordinated organic ligands and natural polymers; polymer pyrolysis and combustion; textile flammability. Mailing Add: Col of Indust Mgt & Textile Sci Clemson Univ Clemson SC 29631

BARKER, ROBERT SAMUEL, organic chemistry, see 12th edition

BARKER, ROY JEAN, b Norborne, Mo, July 9, 24; m 48; c 2. ENTOMOLOGY. Educ: Univ Mo, BS, 48; Univ Ill, PhD(entom), 53. Prof Exp: Asst entom, Univ Ill, 48-50 & Ill State Natural Hist Surv, 50-53; entomologist, E I du Pont de Nemours & Co, 53-55; sr insect physiologist, Pioneering Lab, USDA, Md, 55-64; entomologist, Res Labs, Rohm and Haas Co, Pa, 64-67; RES ENTOMOLOGIST, USDA, 67- Mem: Entom Soc Am; Am Chem Soc. Res: Relation of chemical structure to insecticidal action; insect physiology and biochemistry. Mailing Add: USDA 2000 E Allen Rd Tucson AZ 85719

BARKER, SAMUEL BOOTH, b Montclair, NJ, Mar 3, 12; m 34. PHYSIOLOGY, PHARMACOLOGY. Educ: Univ Vt, BS, 32; Cornell Univ, PhD(physiol), 36. Prof Exp: Asst physiol, Med Col, Cornell Univ, 37-40; from instr to asst prof physiol, Col Med, Univ Tenn, 41-44; from asst prof to assoc prof, Col Med, Univ Iowa, 44-52; prof pharm, Med Col & Sch Dent, Med Ctr, Univ Ala, 52-62; prof, Col Med, Univ Vt, 62-65; dir grad studies, 65-70, PROF PHYSIOL & BIOPHYS, UNIV ALA, BIRMINGHAM, 65-, DEAN GRAD SCH, 70- Concurrent Pos: Fel med, Cornell Univ, 38-41; Krichesky fel, Univ Calif, Los Angeles, 51-52, USPHS spec fel, 61-62, career res award, 62-65. Mem: AAAS; Am Soc Exp Pharmacol & Therapeut; Am Physiol Soc; Soc Exp Biol & Med; Endocrine Soc. Res: Fat, carbohydrate and protein metabolism; tissue metabolism; effect of endocrines on metabolism; thyroid and iodine. Mailing Add: Grad Sch Univ of Ala Birmingham AL 35294

BARKER, SHIRLEY HUGH, b Beloit, Wis, Nov 27, 15; m 44; c 3. ZOOLOGY. Educ: Wis State Teachers Col, Whitewater, BEd, 38; Univ Wis, MPh, 40, PhD(zool), 42. Prof Exp: Asst zool & genetics, Univ Wis, 38-42; fel, Univ Minn, 46; PROF BIOL, ST CLOUD STATE UNIV, 46- Res: Physiology; chemistry. Mailing Add: Dept of Biol Sci St Cloud State Univ St Cloud MN 56301

BARKER, STEVEN JOSEPH, b San Diego, Calif, Feb 5, 45; m 65; c 2. FLUID DYNAMICS. Educ: Harvey Mudd Col, BS, 67; Calif Inst Technol, MS, 68, PhD(eng sci), 71. Prof Exp: Physicist hydrodyn, US Navy Undersea Res & Develop Ctr, 62-68; res fel, Calif Inst Technol, 71-73, sr res fel, 73-74; ASST PROF FLUID MECH, UNIV CALIF, LOS ANGELES, 74- Concurrent Pos: Res scientist, Poseidon Res, 74- Mem: Am Phys Soc; Am Inst Aeronaut & Astronaut; Am Soc Mech Engr; Sigma Xi. Res: Experimental imcompressible fluid dynamics, including trailing vortex wake behavior, radiated sound from turbulent boundary layers and boundary layer transition. Mailing Add: Dept of Mech & Struct 5731 Boelter Hall Univ of Calif Los Angeles CA 90024

BARKER, WALTER MARLIN, biochemistry, see 12th edition

BARKER, WILEY FRANKLIN, b Santa Fe, NMex, Oct 16, 19; m 43; c 3. MEDICINE. Educ: Harvard Univ, BS, 41, MD, 55; Am Bd Surg, dipl. Prof Exp: Intern & resident surg, Peter Bent Brigham Hosp, 44-46; asst, 49-51, clin instr, 51-54, from asst prof to assoc prof, 54-63, PROF SURG, SCH MED, UNIV CALIF, LOS ANGELES, 63-, CHIEF, DIV GEN SURG, 69- Concurrent Pos: Jr assoc, Peter Bent Brigham Hosp, 48-49; resident surg, Vet Admin Hosp, Los Angeles, 49-51, attend physician, 54-; chief gen surg sect & asst chief gen surg serv, Wadsworth Vet Admin Hosp, 51-54; mem bd trustees, Am Bd Surg, 64-70. Mem: Fel Am Col Surg; Soc Univ Surg; Soc Vascular Surg; Am Surg Asn; Soc Clin Surg. Res: Clinical surgical aspects of arteriosclerosis, gastroenterological disease, and aseptic technique. Mailing Add: Univ of Calif Sch of Med Los Angeles CA 90024

BARKER, WILLIAM ALFRED, b Los Angeles, Calif, May 9, 19; m 41; c 5. PHYSICS. Educ: Yale Univ, BA, 41; Calif Inst Technol, MS, 48; St Louis Univ, PhD(physics), 52. Prof Exp: Instr, St Louis Univ, 52-55, from asst prof to assoc prof, 55-62, prof, 62-64; PROF PHYSICS, UNIV SANTA CLARA, 64-, CHMN DEPT, 69- Concurrent Pos: Res theoret physics, Swiss Fed Inst Technol, 53-55; consult, Argonne Nat Lab, 58-63, McDonnell Aircraft Corp, 60-, Hewlett-Packard, NASA-Ames, 68- Mem: Am Phys Soc; Am Asn Physics Teachers. Res: Nuclear orientation; quantum electronics; solid state; photoelectron spectroscopy; atmospheric physics;

interaction physics; cosmology; acoustics. Mailing Add: Dept of Physics Univ of Santa Clara Santa Clara CA 95053

BARKER, WILLIAM GEORGE, b Stratford, Ont, Apr 2, 22; m 46; c 3. PLANT PHYSIOLOGY. Educ: Univ Western Ont, BSc, 48, MSc, 49; Univ Mich, PhD(bot), 53. Prof Exp: Lectr bot, Univ Western Ont, 51-52; asst prof, Ont Col, 52-54; plant physiologist, Coto Res Sta, United Fruit Co, Costa Rica, 54-55; plant Vining G Dunlap Labs, La Lima, Honduras, 55-59 & Boston, Mass, 59-60; res officer, Can Dept Agr, 60-64; from assoc prof to prof biol sci, Univ Man, 64-72, dir biol teaching unit, Div Biol Sci, 70-72; PROF BOT & GENETICS & CHMN DEPT, UNIV GUELPH, 72- Mem: Bot Soc Am. Res: Plant tissue culture; activity of the pith region of 50 year old basswood stem; in vitro potato tuberization; banana growth and propagation; low bush blueberry. Mailing Add: Dept of Bot & Genetics Univ of Guelph Guelph ON Can

BARKER, WILLIAM HAMBLIN, II, b Albuquerque, NMex, Aug 10, 48. OPERATIONS RESEARCH, PURE MATHEMATICS. Educ: Univ Calif, Santa Barbara, BA, 70; Stanford Univ, MS, 72, PhD(math), 75. Prof Exp: Res asst math, Ames Res Ctr, NASA, 69-72, consult, 73; mathematician & asst, Stanford Univ, 70-75; MATHEMATICIAN, DANIEL H WAGNER ASSOCS, 75- Mem: Am Math Soc; Acoust Soc Am; Soc Indust & Appl Math. Res: Theory of search; theory of surveillance; complex analysis, conformal mappings; calculus of variations, hyperelliptic Reimann surfaces. Mailing Add: Daniel H Wagner Assocs Sta Square One Paoli PA 19355

BARKER, WILLIAM T, b Larned, Kans, Aug 3, 41; m 62; c 2. SYSTEMATIC BOTANY. Educ: Kans State Teachers Col, BA, 63, MA, 67; Univ Kans, PhD(bot), 68. Prof Exp: ASST PROF BOT, NDAK STATE UNIV, 68- Mem: AAAS; Am Soc Plant Taxon; Am Inst Biol Sci; Int Asn Plant Taxon. Res: Floristic studies of the United States prairies and plains; cytotaxonomy of the Cyperaceae. Mailing Add: Dept of Bot NDak State Univ Fargo ND 58102

BARKER, WINONA CLINTON, b New York, NY, Sept 5, 38; m 59; c 2. HUMAN PHYSIOLOGY. Educ: Conn Col, BA, 59; Univ Chicago, PhD(physiol), 66. Prof Exp: Sr res technician physiol, Univ Chicago, 65-66, trainee comt math, 66-67, res assoc med, 67-68; SR RES SCIENTIST MOLECULAR EVOLUTION, NAT BIOMED RES FOUND, 68- Concurrent Pos: Staff scientist, Atlas Protein Sequence & Struct, 68-; lectr, Sch Med, Georgetown Univ, 72-, assoc mem grad fac, 74-; NIH res grant, 75- Mem: Biophys Soc; Int Soc Study Origin Life; NY Acad Sci. Res: Homologous physiological mechanisms: evolutionary history, genetic basis, role in human development and differentiation; computer analysis of related proteins. Mailing Add: 9102 Tuckahoe Lane Adelphi MD 20783

BARKEY, KENNETH THOMAS, b Auburn, Wash, Dec 29, 16; m 42; c 1. ORGANIC CHEMISTRY. Educ: Univ Wash, BS, 38, MS, 40; Mass Inst Technol, PhD(org chem), 43. Prof Exp: Oceanog chemist, Univ Wash, 36-38; supvr, 43-75, SR TECH ASSOC, EASTMAN KODAK CO, 75- Mem: Am Chem Soc. Res: Cellulose esters; plasticizers; polyesters. Mailing Add: Eastman Kodak Co Rochester NY 14650

BARKHUFF, RAYMOND ADDISON, JR, b Amsterdam, NY, July 26, 18; c 4. POLYMER CHEMISTRY. Educ: Colgate Univ, BS, 40; Rensselaer Polytech Inst, MS, 41, PhD(phys chem), 42. Prof Exp: RES CHEMIST & GROUP LEADER POLYMERS, MONSANTO CO, 43- Mem: Am Chem Soc. Res: Polystyrene impact reinforcement and flame retardancy. Mailing Add: 155 Mountain Rd Hampden MA 01036

BARKHURST, RODNEY CHARLES, b Oakland, Md, Oct 16, 42; m 64; c 4. ORGANIC CHEMISTRY. Educ: Washington Col, BS, 65; Univ Kans, PhD(org chem), 71. Prof Exp: INSTR CHEM, HASKELL INDIAN JR COL, 71- Mem: Am Chem Soc. Res: Steroids, optical rotary dispersion-capacitor diode studies of dienes systems. Mailing Add: 125 E 17th Lawrence KS 66044

BARKIN, STANLEY MELVIN, physical chemistry, see 12th edition

BARKLEY, DWIGHT G, b Indiana, Pa, Nov 13, 32; m 59; c 2. HORTICULTURE. Educ: Pa State Univ, BS, 55; Va Polytech Inst, MS, 63, PhD(seed germination), 69. Prof Exp: Instr hort, Va Polytech Inst, 63-68; from asst prof to assoc prof agr, 68-71, PROF AGR, EASTERN KY UNIV, 71- Mem: Am Soc Agron; Crop Sci Soc Am. Res: Microclimate and microenvironment as they affect germination of grass and woody plant seeds; nursery crop production; turf-grass management. Mailing Add: Dept of Agr Eastern Ky Univ Richmond KY 40475

BARKLEY, FRED ALEXANDER, botany, bacteriology, see 12th edition

BARKLEY, JOHN ELBERT, physical chemistry, see 12th edition

BARKLEY, LLOYD BLAIR, b Ellwood City, Pa, Mar 10, 25; m 50; c 1. ORGANIC CHEMISTRY. Educ: Pa State Univ, BS, 47; Univ Pittsburgh, MS, 50, PhD(org chem), 52. 6PXRes chemist, Monsanto Chem Co, 52-56, res group leader, 56-61; dir res & vpres, Pa Indust Chem Corp, 62-69; mgr res, Newport Div, Tenneco Inc, 69-74; MEM RES STAFF, SOUTHERN RESINS, INC, 74- Mem: Am Chem Soc. Res: Carbon to carbon condensations; steroids; polymers. Mailing Add: Southern Resins Inc Hwy 82 South Tuscaloosa AL 35401

BARKLEY, RICHARD ANDREW, b Philadelphia, Pa, Apr 4, 29; m 54. OCEANOGRAPHY. Educ: La State Univ, BSc, 50; Univ Wash, MSc, 58, PhD(phys oceanog), 60. Prof Exp: Asst marine chem, Scripps Inst Oceanog, 50-51; proj engr, Alleghany Ballistics Lab, Hercules Powder Co, 51-53 & 55-56; chief oceanog prog, Honolulu Lab, Bur Com Fisheries, US Fish & Wildlife Serv, 60-68, RES OCEANOGR, NAT MARINE FISHERIES SERV, 68- Concurrent Pos: Grad affil fac, Univ Hawaii, 62- Honors & Awards: US Dept Interior Silver Medal, 70. Mem: Am Chem Soc; Am Geophys Union. Res: Theoretical aspects of distributions of variables in the ocean and of properties subject to biological effects in particular; actual distributions of properties in the Pacific Ocean; island wakes. Mailing Add: Nat Marine Fisheries Serv PO Box 3830 Honolulu HI 96812

BARKLEY, THEODORE MITCHELL, b Modesto, Calif, May 14, 34; m 55; c 3. SYSTEMATIC BOTANY. Educ: Kans State Univ, BS, 55; Ore State Univ, MS, 57; Columbia Univ, PhD(bot), 60. Prof Exp: Instr biol, Occidental Col, 60-61; PROF BOT & CUR HERBARIUM, KANS STATE UNIV, 61- Mem: Am Soc Plant Taxon; Bot Soc Am. Res: Monographic studies of the genus Senecio; floristics of the Central Great Plains. Mailing Add: Div of Biol Kans State Univ Herbarium Manhattan KS 66506

BARKLIS, SAM STEVEN, biochemistry, see 12th edition

BARKMAN, ERIK FREDRIK, b Eskilstuna, Sweden, Apr 26, 25; nat US; m 47; c 1. INORGANIC CHEMISTRY. Educ: Allmanna Laroverk, Eskil Hogre(MS), 45. Prof Exp: Anal chemist, Metro Smelting Co, Pa, 47-51 & NY Testing Labs, 51; chief chemist, Proveedora de Minbrales, SA, Mexico City, 51-53; chief chemist, G Kahn Testing Labs, 53-54; res supvr, Chem Finishing Sect, Metall Res Labs, 54-67, asst gen dir, Co, 67-70, DIR APPL CHEM & MATH, REYNOLDS METALS CO, 70- Honors & Awards: Sam Tour Award, Am Soc Testing & Mat, 67; Am Electroplaters Soc Aluminum Finishing Award, 67. Mem: Am Soc Testing & Mat; Am Soc Metals; Am Chem Soc; Electrochem Soc; Am Ord Asn. Res: Surface finishing; corrosion technology; computer science; process control. Mailing Add: Reynolds Metals Co Fourth & Canal St Richmond VA 23219

BARKMAN, ROBERT CLOYCE, b Massillon, Ohio, Oct 10, 42; m 65; c 2. COMPARATIVE PHYSIOLOGY. Educ: Wittenburg Univ, BA, 64; Univ Cincinnati, MS, 66, PhD(zool), 69. Prof Exp: Teaching asst biol, Univ Cincinnati, 64-66, 68-69; ASST PROF BIOL, SPRINGFIELD COL, 69- Mem: Am Soc Zoologists; Am Inst Biol Sci; Sigma Xi. Res: Development of culture methods for all life stages of the Atlantic silverside, Menidia menidia for purposes of comparing mechanisms of temperature adaptation of larva, juvenile and adult fish. Mailing Add: Dept of Biol Springfield Col Springfield MA 01109

BARKO, JOHN WILLIAM, b Detroit, Mich, June 28, 47; m 68; c 1. AQUATIC ECOLOGY. Educ: Mich State Univ, BS, 72, PhD(bot), 75. Prof Exp: Teaching asst biol, Mich State Univ, 72-75, asst prof bot, 75; RES BIOLOGIST, WATERWAYS EXP STA, US ARMY CORPS ENGRS, 75- Mem: AAAS; Am Inst Biol Sci; Am Soc Limnol & Oceanog; Ecol Soc Am. Res: Field and laboratory emphasis on chemical, physical and biological factors of importance in the establishment of marshland flora, in fresh water and marine coastal areas affected by dredging. Mailing Add: US Army Corps of Engrs Waterways Exp Sta Box 631 Vicksburg MS 39180

BARKS, PAUL ALLAN, b Ft Morgan, Colo, May 28, 36; m 60; c 3. ORGANIC CHEMISTRY. Educ: Grinnell Col, BA, 58; Iowa State Univ, PhD(org chem), 63. Prof Exp: Instr chem, Hamline Univ, 63-64; asst prof, St Norbert Col, 64-68; asst prof, Monmouth Col, Ill, 68-71; HEAD DEPT CHEM, NORTH HENNEPIN COMMUNITY COL, 71- Concurrent Pos: NSF undergrad instr equip grant, 65-, res participation for chem teachers grant, 65-67. Mem: Am Chem Soc. Res: Photochemical reactions of organic chemicals and natural products, particularly amino acids; effectual teaching. Mailing Add: Dept of Chem North Hennepin Community Col Minneapolis MN 55445

BARKSDALE, ALMA WHIFFEN, b Hammonton, NJ, Oct 25, 16; m 59. MYCOLOGY. Educ: Maryville Col, BA, 37; Univ NC, MA, 39, PhD(bot), 41. Prof Exp: Mycologist, Upjohn Co, 43-52; RES ASSOC, NY BOT GARDEN, 53 & 55- Concurrent Pos: Guggenheim fel, 51-52. Mem: AAAS; Bot Soc Am; Mycol Soc Am; Am Soc Microbiol. Res: Aquatic phycomycetes; sexual hormones of fungi; characterization of the sexual hormones of Achlya. Mailing Add: 37 W 12th St New York NY 10011

BARKSDALE, HENRY COMPTON, b Carters Bridge, Va, July 26, 98; m 28; c 2. HYDROLOGY. Educ: Univ Va, CE, 22. Prof Exp: Instrumentman & comput, Albemarle Co, Va, 22; asst city engr, Charlottesville, 23; hydraul engr, US Geol Serv, 24-27, State Dept Conserv & Develop, NJ & State Water Policy Comn, 27-39; hydraul engr, US Geol Surv, 39-45, dist engr, 45-52, staff engr, 52-57, br area chief, 57-64, staff hydrologist, Atlantic Coast Region, 64-68; WATER RESOURCES SPECIALIST, GEOL SURV OF ALA, 68- Honors & Awards: US Dept Interior Distinguished Serv Award. Mem: Am Geophys Union; Am Water Works Asn; Soc Econ Geol; Soil Conserv Soc Am. Res: Ground-water; salt-water intrusion; water resources planning. Mailing Add: Geol Surv of Ala PO Drawer O University AL 35486

BARKSDALE, JAMES BRYAN, JR, b Blytheville, Ark, Dec 29, 40. MATHEMATICS. Educ: Univ Ark, BA, 64, MS, 66, PhD(math), 69. Prof Exp: Instr high sch, 63-64; ASST PROF MATH, WESTERN KY UNIV, 68- Concurrent Pos: Mu Alpha Theta vol lectr, 69- Mem: AAAS; Math Asn Am; Am Math Soc. Res: Analysis in normed vector spaces, especially research involving integrals and differentials of vector functions. Mailing Add: Dept of Math Western Ky Univ Bowling Green KY 42101

BARKSDALE, JELKS, inorganic chemistry, see 12th edition

BARKSDALE, LANE, b Emporia, Va, Nov 23, 14; m. MICROBIOLOGY. Educ: Univ NC, AB, 38, MA, 40; NY Univ, PhD(microbiol), 53. Prof Exp: Chief bact, 406 Med Gen Lab, Tokyo, 46-49; fel, Pasteur Inst, Paris, 53-54; from asst prof to assoc prof microbiol, 54-69, PROF MICROBIOL, SCH MED, NY UNIV, 69- Concurrent Pos: Vis prof, Osaka, 60-61; Guggenheim fel, 70-71; chmn subcomt Corynebacterium & Related Organisms, Int Comt Systematic Bact, 74- Mem: Am Soc Microbiol; Soc Gen Physiol; Harvey Soc; NY Acad Med; Am Acad Microbiol. Res: Biology of Corynebacterium, Mycobacterium, Nocardia group and its viruses. Mailing Add: Dept of Microbiol NY Univ Sch of Med New York NY 10016

BARKSDALE, THOMAS HENRY, b Trenton, NJ, Nov 7, 32; m 62; c 2. PLANT PATHOLOGY. Educ: Rutgers Univ, BS, 54; Cornell Univ, PhD(plant path), 59. Prof Exp: Asst, Cornell Univ, 54-58; res plant pathologist, US Army Biol Labs, 59-66; RES PLANT PATHOLOGIST, AGR RES SERV, USDA, 66- Mem: Am Phytopath Soc; Sigma Xi. Res: Vegetable diseases and breeding for disease resistance. Mailing Add: Plant Genetics & Germplasm Inst USDA Agr Res Ctr-West Beltsville MD 20705

BARLAZ, JOSHUA, b Newark, NJ, July 19, 21; m 55; c 2. MATHEMATICS. Educ: City Col NY, BS, 41; Univ Cincinnati, AM, 42, PhD(math), 45. Prof Exp: Instr math & mech, Univ Cincinnati, 42-44; statistician, Allis-Chalmers Mfg Co, 44-45; instr, Ohio State Univ, 45-46; from asst prof to assoc prof, 46-64, PROF MATH, RUTGERS UNIV, 64- Concurrent Pos: Ed, J Asn Symbolic Logic, 52. Mem: Am Math Soc; Math Asn Am; Asn Symbolic Logic (secy-treas, 52-). Res: Infinite series; theory of functions; interpolation theory; triangular summability methods. Mailing Add: Dept of Math Rutgers State Univ New Brunswick NJ 08903

BARLOW, ANTHONY, b Southport, Eng, June 25, 38; m 61; c 3. POLYMER CHEMISTRY. Educ: Univ Birmingham, BSc, 59, PhD(phys chem), 62. Prof Exp: Proj leader polymer struct, Res Div, 65-68, asst mgr res div, 68-75, MGR RES DIV, US INDUST CHEM CO, 75- Mem: Am Chem Soc. Res: Polymer degradation by pyrolysis; radiation effect on polyethylene and ethylene copolymers; polyethylene structure and its relation to physical properties and synthesis conditions; crosslinking and fire retardancy of polymer compounds. Mailing Add: US Indust Chem Co 1275 Section Rd Cincinnati OH 45237

BARLOW, CHARLES F, b Mason City, Iowa, Nov 10, 23; m 53; c 3. MEDICINE. Educ: Univ Chicago, SB, 45, MD, 47. Prof Exp: Intern, Johns Hopkins Hosp, 47-48; jr asst resident, Boston Children's Hosp, 48-49; resident neurol, Clins Univ Chicago, 51-53, instr, 53-55, asst prof, 55-63; BRONSON CROTHERS PROF NEUROL, HARVARD MED SCH, 63- Concurrent Pos: Chief neurologist, Children's Hosp,

Med Ctr, Boston; consult, Peter Bent Brigham Hosp & Boston Lying-In Hosp. Mem: Am Acad Neurol; Am Asn Neuropath; Am Neurol Asn. Res: Clinical neurology of children; blood-brain barrier with isotope labeled compounds; neuropathology. Mailing Add: Dept of Neurol Harvard Med Sch Boston MA 02115

BARLOW, GEORGE, b Springfield, Mass, Dec 13, 26; m 51; c 2. PHYSIOLOGY. Educ: Syracuse Univ, AB, 50; Princeton Univ, MS, 52, PhD, 53. Prof Exp: Instr physiol, Col Med, Univ Tenn, 53-57, asst prof, 57-63, instr clin physiol, 53-55, from asst prof to assoc prof, 55-63, asst dir clin physiol labs, 57-63; assoc prof, 64-66, PROF BIOL, HEIDELBERG COL, 66- Concurrent Pos: Nat Acad Sci-Nat Res Coun travel award, Int Cong Physiol Sci, Buenos Aires, Arg, 58; Rockefeller Found travel award, Latin Am, 59; vis prof, Univ Valle, Colombia, 60-61; consult, John Gaston Hosp, Tenn. Mem: AAAS; Am Physiol Soc. Res: Adrenal cortex, water and electrolyte metabolism; capillary permeability; circulation and renal physiology. Mailing Add: Dept of Biol Heidelberg Col Tiffin OH 44883

BARLOW, GEORGE WEBBER, b Long Beach, Calif, June 15, 29; m 55; c 3. ANIMAL BEHAVIOR: ICHTHYOLOGY. Educ: Univ Calif, Los Angeles, AB, 51, MA, 55, PhD(zool), 58. Prof Exp: From asst prof to assoc prof zool, Univ Ill, Champaign, 60-66; assoc prof, 66-70, Miller prof, 70-71, PROF ZOOL, UNIV CALIF, BERKELEY, 70- Concurrent Pos: NIMH fel, Max Planck Inst Physiol Behav, Ger, 58-60; mem, NSF Adv Panel Psychobiol, 65-68; Am Inst Biol Sci vis biologist, 69; US rep, Int Ethology Comt, 69-75. Mem: Fel AAAS; Am Soc Zoologists; Am Soc Ichthyologists & Herpetologists; Asn Trop Biol; fel Animal Behav Soc. Res: Lab and field studies on social behavior of New World cichlid fishes; sociobiology of coral-reef fishes; nature of patterned movements; biology of social systems. Mailing Add: Dept of Zool Univ of Calif Berkeley CA 94720

BARLOW, GRANT HAROLD, b Wilkes-Barre, Pa, Sept 20, 25; m 48; c 2. PHYSICAL BIOCHEMISTRY. Educ: Wilkes Col, BS, 50. Prof Exp: Phys biochemist, Nat Med Res Inst, 50-52; phys biochemist, 52-70, RES FEL, ABBOTT LABS, 70- Concurrent Pos: Lectr molecular biol, Northwestern Univ, 74- Mem: Int Soc Hemostasis & Thrombosis; Int Soc Toxinology; Am Chem Soc; Biophys Soc; Brit Biochem Soc. Res: Electrophoresis; protein chemistry; ultracentrifugation; fibrinolysis; thrombosis research. Mailing Add: Dept of Molecular Biol Abbott Labs North Chicago IL 60064

BARLOW, IRMELA CHRISTIANE, b Berlin, Ger, Aug 29, 36; US citizen; m 62. PHYSICAL CHEMISTRY, ELECTRON MICROSCOPY. Educ: Univ Calif, Los Angeles, BS, 58, MS, 59, PhD(phys chem), 62. Prof Exp: STAFF CHEMIST, IBM CORP, 62- Mem: Electron Micros Soc Am; Am Soc Metals; Am Crystallog Asn. Res: X-ray crystallography; crystal and molecular structures; electron microscopy of thin films; electron microprobe analysis. Mailing Add: IBM Corp Monterey & Cottle Rd San Jose CA 95050

BARLOW, JAMES A, JR, b Englewood, NJ, Sept 4, 23; m 46; c 2. PHYSICAL GEOLOGY. Educ: Middlebury Col, BA, 49; Univ Wyo, MA, 50, PhD(geol), 53. Prof Exp: Geologist, Ohio Oil Co, 52-53 & Forest Oil Corp, 53-57; GEOLOGIST, BARLOW & HAUN, INC, 57- Mem: Geol Soc Am; Am Asn Petrol Geol. Res: Oil and gas exploration, stratigraphy and structure. Mailing Add: Barlow & Haun Inc 311 S Center Casper WY 82601

BARLOW, JAMES LAWRENCE, b Detroit, Mich, July 2, 26; m 49; c 1. MEDICAL MICROBIOLOGY, INFECTIOUS DISEASES. Educ: Western Mich Univ, BS, 48; Johns Hopkins Univ, MS, 54; Albany Med Col, PhD, 60. Prof Exp: Asst biochem, Johns Hopkins Univ, 52-55; biochemist, NY State Dept Health, 55-58, res scientist biochem, 58-60, sr res scientist, 60-67; HEAD MICROBIOL, ST PETER'S HOSP, 67- Concurrent Pos: Assoc prof, Union Col, NY, 65-67; asst prof nursing, State Univ NY Albany, 68-71. Mem: Am Asn Immunol. Res: Immunology. Mailing Add: Microbiol Labs St Peter's Hosp Albany NY 12208

BARLOW, JOHN PELEG, b Kingston, RI, Nov 6, 18; m 52; c 2. OCEANOGRAPHY. Educ: Univ RI, BS, 41; Harvard Univ, MA, 48, PhD(biol), 53. Prof Exp: Res assoc, Dept Oceanog, Univ Wash, Seattle, 52-53; asst prof, Agr & Mech Col, Tex, 53-56; ASSOC PROF OCEANOG, SECT ECOL & SYSTEMATICS, CORNELL UNIV, 56- Mem: Am Soc Limnol & Oceanog; Ecol Soc Am. Res: Ecology of plankton. Mailing Add: Div of Biol Sci Cornell Univ Ithaca NY 14850

BARLOW, JOHN SLANEY, b Leamington, Ont, Jan 30, 21; m 47; c 4. BIOCHEMISTRY. Educ: Ont Agr Col, BSC, 42; Univ Toronto, MA, 47, PhD, 53. Prof Exp: Biochemist, Defence Res NLab, 50-53 & Res Inst, Ont, 53-67; PROF BIOL SCI, PESTOLOGY CENTRE, SIMON FRASER UNIV, 67- Res: Tissue electrolytes; cold physiology; insect metabolism. Mailing Add: Pestology Centre Dept of Biosci Simon Fraser Univ Burnaby BC Can

BARLOW, JOHN SUTTON, b Raleigh, NC, June 10, 25; m 50; c 3. NEUROPHYSIOLOGY, BIOPHYSICS. Educ: Univ NC, BS, 44, MS, 48; Harvard Univ, MD, 53. Prof Exp: Asst neurol, 57-61, NEUROPHYSIOLOGIST, NEUROL SERV, MASS GEN HOSP, 61- Concurrent Pos: Clin & res fel neurol, Mass Gen Hosp, 53-57; Nat Inst Neurol Dis & Stroke spec trainee, 58-61, res career develop award, 62-71, res grant, 62-; asst, Harvard Med Sch, 57-61, res assoc, 61-69, prin res assoc, 69-; res assoc, Mass Inst Technol, 54-64, res affil, 64-; res assoc, Sch Med, Univ Calif, Los Angeles, 66; mem neurol study sect, NIH, 66-70; consult ed, EEG Clin Neurophysiol, 70; mem ed bd, Am J Chinese Med, 73-; mem panel neurocommun & biophys, Int Brain Res Orgn; mem Fed Drug Admin panel rev neurol devices, 74-76. Mem: Am EEG Soc (pres, 75-76); Am Neurol Asn; Am Acad Neurol; Am Geophys Union; Soc Neurosci. Res: Electrical activity of the nervous system in man; computer design and development; medical geography and geophysics; animal navigation. Mailing Add: Neurol Serv Mass Gen Hosp Boston MA 02114

BARLOW, JON CHARLES, b Jacksonville, Ill, Oct 31, 35; m 57; c 3. ORNITHOLOGY. Educ: Knox Col, Ill, BA, 57; Univ Kans, MA, 60, PhD(zool), 65. Prof Exp: Res chemist, Corn Prod Refining Co, 57-58; teaching asst zool, Univ Kans, 58-62 & 63-64; field assoc mammal, Am Mus Natural Hist, NY, 62-63; asst prof biol, Rockhurst Col, 64-65; asst prof zool, 65-69, ASSOC PROF ZOOL, UNIV TORONTO, 69-; CUR & HEAD DEPT ORNITH, ROYAL ONT MUS, 65- Mem: AAAS; Am Ornithologists Union; Am Soc Mammal; Am Soc Zoologists; Animal Behav Soc. Res: Avian systematics and mammalian ecology and zoogeography; avian behavior, especially of vireos. Mailing Add: Dept of Ornith Royal Ont Mus Toronto ON Can

BARLOW, MALCOLM, crystallography, polymer chemistry, see 12th edition

BARLOW, RICHARD EUGENE, b Galesburg, Ill, Jan 12, 31; m 56; c 4. MATHEMATICAL STATISTICS. Educ: Knox Col, BA, 53; Univ Ore, MA, 55; Stanford Univ, PhD, 60. Prof Exp: Adv res engr math, Sylvania Electronic Defense Lab, Gen Tel & Electronics Corp, 57-60, mem tech staff, Res Labs, 61-62; mem tech staff, Inst Defense Anal, 60-61; assoc prof opers res & statist, 63-69, PROF OPERS RES & STATIST, UNIV CALIF, BERKELEY, 69- Concurrent Pos: Consult, Rand Corp, Calif, 63-69; vis prof statist, Fla State Univ, 75-76; assoc ed, J Inst Math Statist, 75- Mem: Fel Inst Math Statist; fel Am Statist Asn. Res: Statistics; probability theory and its applications, especially mathematical theory of reliability. Mailing Add: Opers Res Ctr Univ of Calif Berkeley CA 94720

BARLOW, ROBERT BROWN, JR, b Trenton, NJ, July 31, 39; m 61; c 3. NEUROPHYSIOLOGY. Educ: Bowdoin Col, AB, 61; Rockefeller Univ, PhD(life sci), 67. Prof Exp: Asst prof, 67-71, ASSOC PRO SENSORY SCI, SYRACUSE UNIV, 71- Concurrent Pos: Mem corp, Marine Biol Lab, Woods Hole, Mass, 72- Mem: Asn Res Vision & Ophthal; Soc Neurosci; Sigma Xi. Res: Processing of contrast and intensity information by the eye and the brain; neurophysiology of the visual systems of Limulus and Macaque monkeys; psychophysics of normal and visually deficient humans. Mailing Add: Inst Sensory Res Syracuse Univ Syracue NY 13210

BARMACK, NEAL HERBERT, b New York, NY, Aug 23, 42; m 64; c 2. NEUROBIOLOGY, NEUROPSYCHOLOGY. Educ: Univ Mich, BS, 63; Univ Rochester, PhD, 70. Prof Exp: Asst lectr psychol, Univ Rochester, 68-69; from res assoc to sr res assoc neurophysiol, 70-74, ASSOC SCIENTIST, DEPT OPHTHAL, NEUROL SCI INST, GOOD SAMARITAN HOSP & MED CTR, PORTLAND, 75- Mem: Soc Neurosci; Am Physiol Soc; Asn Res Vision; Int Brain Res Orgn. Res: Neural control of eye movements. Mailing Add: Neurol Sci Inst 1120 NW 20th Portland OR 97210

BARMATZ, MARTIN BRUCE, b Los Angeles, Calif, May 25, 38; m 61; c 2. ACOUSTICS. Educ: Univ Calif, Los Angeles, BA, 60, MA, 62, PhD(physics), 66. Prof Exp: Asst prof physics in residence, Univ Calif, Los Angeles, 66-67; MEM TECH STAFF, CONDENSED STATE PHYSICS RES DEPT, BELL TEL LABS, 67- Mem: Acoust Soc Am; Am Phys Soc. Res: Critical point phenomena; sound velocity, sound attenuation and sound dispersion measurement. Mailing Add: Rm 1D-150 Bell Tel Labs Mountain Ave Murray Hill NJ 07974

BARMBY, DAVID STANLEY, b Hull, Eng, Mar 4, 28; m 57; c 3. PHYSICS. Educ: Univ Leeds, BSc, 49, PhD(physics), 54. Prof Exp: Physicist, Wool Industs Res Asn, Eng, 54-57; physicist, 57-66, chief petrol processing sect, Basic Res Div, 66-70, chief explor petrol res, Corp Res Div, 70-75, MGR TECHNOL ASSESSMENT, SUN OIL CO, ST DAVIDS CORP HQ, 75- Mem: Am Phys Soc; Am Chem Soc. Res: Solid state physics and catalysis; paraffin wax structures and phase transitions; correlations between physical properties and structure of solids; infrared spectroscopy of polymers. Mailing Add: 1237 Hunt Club Lane Media PA 19063

BARNAAL, DENNIS E, b Sacred Heart, Minn, Jan 5, 36; m 62; c 2. SOLID STATE PHYSICS. Educ: Univ Minn, BS, 58, MS, 62, PhD(physics), 65. Prof Exp: From asst prof to assoc prof, 64-74, PROF PHYSICS, LUTHER COL, 74- Mem: Am Phys Soc; Am Asn Physics Teachers. Res: Pulsed nuclear magnetic resonance in solids; ice physics. Mailing Add: Dept of Physics Luther Col Decorah IA 52101

BARNABY, BRUCE E, b Milwaukee, Wis, Sept 24, 29; m 57; c 6. PHYSICS, ELECTRICAL ENGINEERING. Educ: DePaul Univ, BS, 51; Univ Notre Dame, PhD(physics), 60. Prof Exp: Res physicist, Eitel-McCullough, Inc, 60-63, proj engr, 63-65; sr proj engr, Eimac Div, Varian Assocs, 65-67; staff mem, Los Alamos Sci Lab, 67-70; DIV SUPVR, TUBE DEVELOP DIV, SANDIA LABS, 70- Mem: Am Phys Soc; Am Soc Testing & Mat; Am Nuclear Soc. Res: Electron physics technology, especially use of technology in industry. Mailing Add: 3917 La Hacienda Dr NE Albuquerque NM 87110

BARNARD, ADAM JOHANNES, b Murraysburg, SAfrica, June 9, 29; m 57; c 4. PLASMA PHYSICS. Educ: Univ SAfrica, BSc, 49, MSc, 51; Glasgow Univ, PhD(physics), 54. Prof Exp: Sr lectr, Univ Natal, 54-59; from asst prof to assoc prof, 59-68; PROF PHYSICS, UNIV BC, 68- Mem: Am Phys Soc; Can Asn Physicists. Res: Spectroscopic investigation of plasmas. Mailing Add: Dept of Physics Univ of BC Vancouver BC Can

BARNARD, ALFRED JAMES, JR, b Hamilton, Ont, Jan 5, 20; nat US; m 58; c 1. CHEMISTRY. Educ: Tufts Col, BS, 42; Harvard Univ, MA, 44; Lehigh Univ, PhD(org chem), 50. Prof Exp: Mem staff, Tech Info Serv, J T Baker Chem Co, 47-48, dir tech info serv, 51-68, mgr res anal & info serv, 68-69, DIR ANAL SERV, RICHARDSON-MERRELL, INC, 69- Concurrent Pos: Asst chemist, Godfrey L Cabot Co, 42, 43; assoc, W F Greenwald & MG Mulinos, MD, Consult, 50-51; tech adv, Chemiis Analyst, 51-52, ed, 52-67; gen secy, Int Symposia Microchem Tech, 61, 65; co-ed, Chelates in Anal Chem, 67-; prog dir, Ctr Prof Advan, 70. Mem: Am Microchem Soc; Am Chem Soc; Soc Appl Spectros; Am Soc Qual Control. Res: Analytical reagents; chelation; high-purity chemicals; retrieval chemical information. Mailing Add: Richardson-Merrell Inc Baker Chem Co 222 Red School Lane Phillipsburg NJ 08865

BARNARD, ANTHONY C L, b Birmingham, Eng, Apr 30, 32; m 64; c 2. COMPUTER SCIENCE. Educ: Univ Birmingham, Eng, BSc, 53, PhD(nuclear physics), 57, DSc(physics), 74. Prof Exp: Res physicist, Res Lab, Assoc Elec Industs, Ltd, 56-58; res assoc physics, Univ Iowa, 58-60; from instr to asst prof, Rice Univ, 60-66; mgr physics dept, Int Bus Mach Sci Ctr, 66-68; assoc prof biomath, 68-71, PROF BIOMATH & INFO SCI, UNIV ALA, BIRMINGHAM, 71-, CHMN DEPT INFO SCI, 72- Mem: Am Phys Soc; Asn Comput Mach; Data Processing Mgt Asn; Inst Elec & Electronic Engrs Comput Soc. Res: Medical applications of physics and computer science. Mailing Add: 119 Rust Bldg Univ Sta Birmingham AL 35294

BARNARD, ERIC A, b London, Eng, July 2, 27; c 4. BIOCHEMISTRY, BIOCHEMICAL PHARMACOLOGY. Educ: Univ London, BSc, 52, PhD, 57. Prof Exp: Asst lectr, King's Col, Univ London, 58-59, lectr, 59-63; vis prof, Univ Marburg, 63; from assoc prof to prof biochem pharmacol, 64-66, dir molecular enzym, 64-65, PROF BIOCHEM PHARMACOL & BIOCHEM, STATE UNIV NY BUFFALO, 66-, CHMN DEPT BIOCHEM, 69- Concurrent Pos: Rockefeller fel, Univ of Calif, Berkeley, 60-61; State Univ NY univ fac award, 69; Guggenheim fel, 69; Macy Found fac scholar award, 74. Mem: Am Soc Biol Chem; Am Chem Soc; Am Soc Cell Biol; Brit Biochem Soc; Histochem Soc. Res: Molecular basis of synaptic function; molecular basis of evolution of enzymes; hexokinases; ribonucleases. Mailing Add: Dept of Biochem G-56 Capen Hall State Univ of NY Buffalo NY 14214

BARNARD, GARLAND RAY, b Victoria, Tex, Apr 12, 32; m 54; c 2. ACOUSTICS. Educ: Univ Tex, BS, 57, MA, 60. Prof Exp: SPEC RES ASSOC ACOUSTICS, APPL RES LAB, UNIV TEX, AUSTIN, 56- Mem: Acoust Soc Am. Res: Underwater acoustics; basic and applied research. Mailing Add: 3307 Stardust Dr Austin TX 78757

BARNARD, JERRY LAURENS, b Pasadena, Calif, Feb 27, 28; m 49; c 3. MARINE ZOOLOGY. Educ: Univ Southern Calif, AB, 49, MS, 50, PhD(zool), 53. Prof Exp: Instr biol sci, Calif State Polytech Col, 52-53; investr, US Air Force Arctic res grant, Univ Southern Calif, 53-59; res assoc, Beaudette Found, 59-64; CUR CRUSTACEA, NAT MUS NATURAL HIST, SMITHSONIAN INST, 64- Mem: Marine Biol Asn

UK. Res: Taxonomy and ecology of marine Amphipoda; estuarine and coastal shelf biology. Mailing Add: Div of Crustacea Nat Mus of Natural Hist Washington DC 20560

BARNARD, WALTHER M, b Hartford, Conn, May 30, 37. MINERALOGY, GEOCHEMISTRY. Educ: Trinity Col, Conn, BS, 59; Dartmouth Col, MA, 61; Pa State Univ, PhD(mineral), 65. Prof Exp: Asst prof geol, 64-70, ASSOC PROF GEOL, STATE UNIV NY COL FREDONIA, 70- Concurrent Pos: Res Corp res grant, 65-66; State Univ NY Res Found res grants-in-aid, 65-66, 67-68, 69-70 & 72-73; res grant, Energy & Resources Develop Agency, 75-76. Mem: AAAS; Nat Asn Geol Teachers; Soc Appl Spectros; Am Soc Testing & Mat; NY Acad Sci. Res: Environmental geochemistry. Mailing Add: Dept of Geol State Univ of NY Fredonia NY 14063

BARNARD, WILLIAM SPRAGUE, b Hillsboro, Ill, May 20, 25; m 48; c 3. PHYSICAL CHEMISTRY. Educ: Harvard Univ, AB, 47; Princeton Univ, PhD(phys chem), 51. Prof Exp: With res dept, Nat Lead Co, 50-53; group leader res div, 53-57, dir woven prod, 57-64, dir res & develop, 64-69, MEM BD DIRS, CHICOPEE MFG CORP, 66-, VPRES RES, 69- Mem: Am Chem Soc; Am Asn Textile Chemists & Colorists. Res: Textile chemistry and resin technology; fibers and finishes. Mailing Add: Res Div Chicopee Mfg Corp Milltown NJ 08850

BARNARTT, SIDNEY, b Toronto, Ont, July 31, 19; nat US; m 43; c 2. ELECTROCHEMISTRY, CORROSION. Educ: Univ Toronto, BA, 41, MA, 42, PhD(chem), 44. Prof Exp: Asst, Univ Toronto, 41-43; res chemist, Westinghouse Res Labs, 46-56, adv chemist, 56-62, mgr electrochem sect, 62-64; res chemist, Fundamental Res Lab, US Steel Corp, 64-72; dir electrochem res, Technicon Instruments Corp, 72; ADV SCIENTIST, WESTINGHOUSE BETTIS ATOMIC POWER LAB, 72- Concurrent Pos: Instr, Carnegie Inst Technol, 54. Honors & Awards: Bronze Medal, Brit Asn Advan Sci, 41. Mem: Am Chem Soc; Electrochem Soc; fel Am Inst Chemists; Nat Asn Corrosion Eng. Res: Adsorption; batteries; electrodeposition; electrode kinetics. Mailing Add: Westinghouse Bettis Atomic Power Lab PO Box 79 West Mifflin PA 15122

BARNAWELL, EARL B, b Maryville, Tenn, Nov 26, 22; m 46; c 2. COMPARATIVE ENDOCRINOLOGY. Educ: Univ Calif, Berkeley, AB, 51, MA, 54, PhD(endocrinol, zool), 64. Prof Exp: Exp animal biologist, Univ Calif, Berkeley, 54-60; asst prof, 64-69, dir inst cellular res, 66-71, ASSOC PROF ZOOL & PHYSIOL, UNIV NEBR, LINCOLN, 69- Mem: Sigma Xi; Am Soc Zool. Res: Physiology and endocrinology; organ culture. Mailing Add: Sch of Life Sci Univ of Nebr Lincoln NE 68588

BARNEIS, ZACHARY J, organic chemistry, see 12th edition

BARNEKOW, RUSSELL GEORGE, JR, b Cleveland, Ohio, Jan 26, 32; m 57; c 4. BACTERIOLOGY, BACTERIAL PHYSIOLOGY. Educ: Miami Univ, AB, 55; Kans State Univ, MS, 61, PhD(bact), 67. Prof Exp: Asst prof, 62-68, asst dean, 68-69, ASSOC PROF BACT, UNIV MO-KANSAS CITY, 68-, ASSOC DEAN, SCH GRAD STUDIES, 69-, LECTR SCH MED, 75- Mem: AAAS; Am Soc Microbiol. Res: Bacterial ecology in soil and water; function of bacterial membranes; physiology of bacterial growth. Mailing Add: Sch of Grad Studies Univ of Mo Kansas City MO 64110

BARNER, HENDRICK BOYER, b Seattle, Wash, Feb 23, 33; m 61; c 3. SURGERY. Educ: Univ Wash, BS, 54, MD, 57; Am Bd Surg, dipl, 67; Am Bd Thoracic Surg, dipl, 68. Prof Exp: Instr surg, Univ Rochester, 64-65; from instr to assoc prof, 65-73, PROF SURG, ST LOUIS UNIV, 73- Concurrent Pos: USPHS res fel, 64-65, fel cardiovasc surg, 65-66. Mem: Am Col Surg; AMA; Cent Surg Asn; Am Fedn Clin Res; Sigma Xi. Res: Cardiovascular surgery and physiology; cryobiology; organ preservation by freezing; general surgery. Mailing Add: Dept of Surg St Louis Univ Sch of Med St Louis MO 63104

BARNES, ADELE, b San Antonio, Tex, May 6, 19. BIOLOGY. Educ: Tex Woman's Univ, 41; NTex State Univ, MA, 44. Prof Exp: Teacher trainer natural sci, WTex State Univ, 41-46; PROF BIOL SCI & CHMN DEPT, AMARILLO COL, 46- Mem: AAAS; Am Asn Biol Teachers; Nat Audubon Soc; Am Inst Biol Sci. Res: Ecology of Rocky Mountain region; botany; zoology; bacteriology. Mailing Add: 3625 Doris Dr Amarillo TX 79109

BARNES, ALLAN CAMPBELL, b Coldwater, Mich, Dec 18, 11; m 37; c 4. MEDICINE. Educ: Princeton Univ, AB, 33; Univ Pa, MD, 37; Univ Mich, MS, 41; Am Bd Obstet & Gynec, dipl. Prof Exp: Instr obstet & gynec, Med Sch, Univ Mich, 42-45; assoc prof, Ohio State Univ, 45-47, prof & chmn dept, 47-53; prof & chmn dept, Sch dept, Sch Med, Case Western Reserve Univ, 53-60; prof & dir dept, Sch Med, Johns Hopkins Univ, 60-70; V PRES, ROCKEFELLER FOUND, 70- Mem: Soc Gynec Invest; Am Asn Obstet & Gynec; Am Gynec Soc. Res: Obstetrics and gynecology. Mailing Add: Rockefeller Found 1133 Ave of the Ams New York NY 10036

BARNES, ALLAN MARION, b Stirling City, Calif, Dec 8, 24; m 62; c 2. MEDICAL ENTOMOLOGY. Educ: San Diego State Col, AB, 53; Univ Calif, PhD(med entom), 63. Prof Exp: Vector control specialist, Calif State Health Dept, 55-67; chief ecol & control unit, 67-70, from asst chief to actg chief, 70-75, CHIEF PLAGUE BR, VECTOR BORNE DIS DIV, BUR LAB, CTR DIS CONTROL, 75- Concurrent Pos: Consult, Indonesian Ministry Health, 68; affil prof zool & affil prof grad sch, Colo State Univ; vis prof & lectr, Simon Fraser Univ. Mem: AAAS; Wildlife Dis Asn; Ecol Soc Am; Am Soc Trop Med & Hyg. Res: Research on the ecology of bubonic plague, tularemia, Colorado tick fever and other vector-borne diseases with emphasis on the dynamics of interactions between host-ectoparasite-pathogen complexes. Mailing Add: Ctr Dis Control Plague Br PO Box 2087 Ft Collins CO 80521

BARNES, ANNIE SHAW, b Red Level, Ala, m 63; c 1. ANTHROPOLOGY, SOCIOLOGY. Educ: Shaw Univ, AB, 53; Atlanta Univ, MA, 55; Univ Va, PhD(anthrop), 71. Prof Exp: Teacher Am govt & hist, Huntington High Sch, Newport News, Va, 54-65; instr anthrop & sociol, Hampton Inst, 65-68; ASSOC PROF ANTHROP & SOCIOL, NORFOLK STATE COL, 71- Concurrent Pos: Ford Found consult, 72. Mem: Am Anthrop Asn; Asn Social & Behav Sci. Res: Family life of Blacks in the United States; formal and informal weekend behavior of Blacks in the United States; voluntary association participation of middle class Blacks. Mailing Add: Dept of Social Sci Noroflk State Col Norfolk VA 23504

BARNES, ARNOLD APPLETON, JR, b Charleston, WVa, June 10, 30; m 60; c 3. METEOROLOGY. Educ: Princeton Univ, AB, 52; Mass Inst Technol, MSc, 57, PhD(meteorol), 62. Prof Exp: Asst meteorol, Mass Inst Technol, 56-61; res scientist, Meteorol Develop Lab, Tech Br, 61-62, res scientist, Meteorol Res Lab, Upper Atmosphere Br, 62-63, res scientist, Meteorol Lab, 63-71, res scientist, Weather Radar Br, 71-74, CHIEF CONVECTIVE CLOUD PHYSICS BR, METEOROL LAB, AIR FORCE CAMBRIDGE RES LABS, HANSCOM AFB, 74- Concurrent Pos: Consult, Allied Res Corp, 59-60. Mem: AAAS; Am Meteorol Soc; Am Geophys Union. Res: Large scale dynamics of the atmosphere; transport of of water vapor; radar meteor trail winds and densities; radar acoustic low level temperature soundings; weather radar; cloud physics aircraft. Mailing Add: 425 North Ave E Weston MA 02193

BARNES, ASA, JR, b Cape Girardeau, Mo, Jan 30, 33; m 57; c 3. PATHOLOGY, HEMATOLOGY. Educ: Univ Ky, BA, 55; Yale Univ, MD, 59. Prof Exp: Intern, Univ Hosps, Cleveland, Ohio, 59-60; resident anat path, Grace-New Haven Community Hosp, New Haven, Conn, 63-65; resident clin path, Walter Reed Gen Hosp, Washington, DC, 65-67, asst chief, 67; blood prog officer, US Army, Vietnam, 67-68; pathologist, Armed Forces Inst Path, Washington, DC, 69-70; ASSOC PROF PATH, SCH MED, UNIV MO-COLUMBIA, 70-; CHIEF CLIN PATH, ELLIS FISCHEL CANCER HOSP, 74-; SR SCIENTIST, CANCER RES CTR, 74- Honors & Awards: US Maj Gary Wratten Award, Asn Mil Surg, 69. Mem: AMA; Am Soc Clin Path; Asn Mil Surg; Acad Clin Lab Physicians & Scientists. Res: Immunohematology; morphologic hematology; blood banking. Mailing Add: Dept of Path Univ of Mo Sch of Med Columbia MO 65201

BARNES, BARRY KEITH, b Odessa, Tex, Apr 23, 39; m 61; c 2. PHYSICS. Educ: Rice Univ, BA, 61, MA, 63, PhD(nuclear spectros), 65. Prof Exp: Systs analyst, Baylor Col Med, 65; fel & mem staff, Los Alamos Sci Lab, Univ Calif, 68-69; ASST PROF PHYSICS, LOWELL TECHNOL INST, 69- Mem: Am Phys Soc. Res: Charged particle low energy nuclear spectroscopy; utilization of computers in cardiac stress studies; charged particle activiation analysis and channeling studies. Mailing Add: Dept of Physics Lowell Technol Inst Lowell MA 01854

BARNES, BRUCE HERBERT, b Minneapolis, Minn, May 28, 31; m 53; c 3. COMPUTER SCIENCE, MATHEMATICS. Educ: Mich State Univ, BS, 53, MS, 58, PhD(math), 60. Prof Exp: Teaching asst math, Mich State Univ, 56-58, res instr comput sci, 58-60; sr res engr, Jet Propulsion Labs, 60-61; asst prof math, Pa State Univ, 61-65, assoc prof, 65-68; vis assoc prof, Univ Iowa, 68-69; ASSOC PROF COMPUT SCI, PA STATE UNIV, 69- Mem: Am Math Soc; Asn Comput Mach. Res: Automata theory, especially groups of automorphisms of finite automata; numerical solution of ordinary differential equations; computer software and computer science education. Mailing Add: 424 McAllister Bldg Pa State Univ University Park PA 16802

BARNES, BURTON B, b Lapeer, Mich, Sept 19, 30; m 75; c 4. EXPLORATION GEOPHYSICS, MARINE GEOLOGY. Educ: Univ Calif, Berkeley, BA, 57. Prof Exp: Sr geophysicist petrol explor, Amoco Prod Co, 57-63; res engr ocean mining, Lockheed Missiles & Space Co, 63-67; res mgr & res geophysicist, Marine Minerals Tech Ctr, Dept Commun, Nat Oceanic & Atmospheric Admin, 67-73; chief engr appl earth sci, Geotesting, Inc, 73; chief scientist & sr res geophysicist petrol explor, Gulf Res & Develop Co, 73-75; BR CHIEF GEOTHERAML ENERGY, FED ENERGY ADMIN, ENERGY RES & DEVELOP ADMIN, 75- Concurrent Pos: Mem, Seabottom Surv Panel, US Japan Coop Natural Resources, 70-71, mem, Marine Mining Panel, 70-73; chmn, Nat Oceanic & Atmospheric Admin Geophys Coord Comt, 71-73; consult ocean explor & surv methodology, Nat Acad Engrs, 72; prin investr, Rapid Excavation & Tunneling Res Prog, Dept Defense, 72-74. Mem: Soc Explor Geophysicists; Marine Technol Soc; Am Inst Mining Metall & Petrol Engrs; Int Soc Geothermal Eng. Res: Marine and terrestrial mineral exploration; geothermal resource development; acoustics and application to measuring parameters; navigation and positioning at sea; mapping and survey methodology; marine mineral deposit delineation; petroleum exploration. Mailing Add: Fed Energy Admin 1200 Pennsylvania Ave Washington DC 20461

BARNES, BURTON VERNE, b Bloomington, Ill, Nov 4, 30; m 57; c 3. FORESTRY, BOTANY. Educ: Univ Mich, BSF, 52, MF, 53, PhD(forestry), 59. Prof Exp: Forester, Region VI, US Forest Serv, 53-59, res forester, Intermountain Forest & Range Exp Sta, 59-63; from asst prof to assoc prof, 63-70, PROF FORESTRY, SCH NATURAL RESOURCES, UNIV MICH, ANN ARBOR, 70-, FOREST BOTANIST, BOT GARDEN, 67- Concurrent Pos: NSF fel, Ger, 63-64; Danforth assoc, 69-; mem staff, Syst-Geobot Inst, Univ Göttingen, 70. Honors & Awards: Distinguished Serv Award, Univ Mich, 67. Mem: Ecol Soc Am; Soc Am Foresters. Res: Forest ecology; dendrology; genecology; forest genetics; silviculture; biosystematics of Populus and Betula. Mailing Add: Sch of Natural Resources Univ of Mich Ann Arbor MI 48104

BARNES, BYRON ASHWOOD, b Peoria, Ill, Mar 14, 27; m 48; c 2. PHARMACOLOGY. Educ: St Louis Col Pharm, BS, 51; Univ Fla, MS, 53, PhD(pharmacol), 54. Prof Exp: Lab instr pharm, St Louis Col Pharm, 49-51; instr, Clin Lab, Sch Aviation Med, US Air Force, Gunter Air Force Base, 54-S7; head dept pharmacol & physiol, 57-70, DEAN, ST LOUIS COL PHARM, 70- Concurrent Pos: Ed, Pharmaceut Trends. Mem: Am Pharmaceut Asn; Am Asn Cols Pharm; Royal Soc Health; Sigma Xi. Res: Continuing pharmacy education; drug abuse. Mailing Add: 10232 Richview Dr St Louis MO 63127

BARNES, CARL EDMUND, b Lewiston, Maine, Feb 16, 08; m 38; c 4. ORGANIC CHEMISTRY. Educ: Bates Col, BS, 30; Harvard Univ, MS, 33, PhD(org chem), 35. Prof Exp: Asst chem, Harvard Univ, 30-35; res chemist, Norton Co, 36-40; group leader, Polaroid Corp, Mass, 40-41; res chemist, Simplex Wire & Cable Co, Mass, 41-42; group leader, Gen Aniline & Film Corp, 42-55, sect leader, 45-47, assoc dir, 47-50; dir res, Arnold, Hoffman & Co, 50-53; asst to vpres, Minn Mining & Mfg Co, 53-54, dir cent res dept, 54-58, vpres res, 58-62; vpres res, FMC Corp, 61-63; dir Barnes Res Assocs, 63-67; PRES, ALRAC CORP, 67-, CHMN BD, 58- Concurrent Pos: Mem div chem & chem technol, Nat Res Coun. Mem: Am Chem Soc; fel Am Inst Chem; Soc Motion Picture & TV Eng. Res: Organic chemical research; synthetic resins and fibers; vinyl polymerization; color photography; acetylene chemistry. Mailing Add: Alrac Corp 649 Hope St Stamford CT 06907

BARNES, CARL ELDON, b Clovis, NMex, June 9, 36; m 54; c 2. AGRONOMY, FIELD CROPS. Educ: NMex State Univ, BS, 59, MS, 60. Prof Exp: Asst miller, Western Wheat Qual Lab, 60-61; res asst crops, Kans State Univ, 61-63; asst agronomist, Colo State Univ, 63-65; asst agronomist, 65-68, asst prof agron, 68-75, ASSOC PROF AGRON, NMEX STATE UNIV, 75-, SUPT CROPS RES, SOUTHEASTERN BR EXP STA, 68- Mem: Am Soc Agron. Res: Crop production research; cotton breeding; water requirements. Mailing Add: Southeastern Br Exp Sta Rte 1 Box 121 Artesia NM 88210

BARNES, CHARLES ANDREW, b Toronto, Ont, Dec 12, 21; nat US; m 50; c 2. NUCLEAR PHYSICS, ASTROPHYSICS. Educ: McMaster Univ, BA, 43; Univ Toronto, MA, 44; Cambridge Univ, PhD(physics), 50. Prof Exp: Res physicist, Can-Brit Atomic Energy Proj, 44-46; asst prof physics, Univ BC, 50-53; res fel, Calif Inst Technol, 53-55; asst prof, Univ BC, 55-56; res fel, 56-57, assoc prof, 57-62, PROF PHYSICS, CALIF INST TECHNOL, 62- Concurrent Pos: NSF sr fel, 62-63; guest prof, Nordita, Copenhagen, Denmark, 73-74. Mem: Fel Am Phys Soc. Res: Astrophysical nuclear reactions; nuclear structures physics. Mailing Add: Div of Physics Math & Astron Calif Inst Technol Pasadena CA 91125

BARNES, CHARLES DEE, b Carroll, Iowa, Aug 17, 35; m 57; c 4. PHYSIOLOGY.

Educ: Mont State Col, BS, 58; Univ Wash, MS, 61; Univ Iowa, PhD(physiol), 62. Concurrent Pos: Fel, Univ Calif, 62-64; from asst prof to assoc prof physiol, Ind Univ, 64-71; prof life sci, Ind State Univ, Terre Haute, 71-75; PROF PHYSIOL & CHMN DEPT, SCH MED, TEX TECH UNIV, 75- Concurrent Pos: Vis scientist, Inst Physiol, Univ Pisa, 68-69; prof physiol, Sch Med, Ind Univ, 71-75. Honors & Awards: Career Develop Award, USPHS, 67. Mem: Int Brain Res Orgn; Am Physiol Soc; Am Soc Pharmacol & Exp Therapeut; Radiation Res Soc; Soc Neurosci. Res: Neurophysiology, neuropharmacology and radiation effects on the nervous system; interaction of different levels of the central nervous system, particularly brain stem-spinal cord. Mailing Add: Dept of Physiol Tex Tech Univ PO Box 4569 Lubbock TX 79409

BARNES, CHARLES M, b Rising Star, Tex, July 21, 22; m 43; c 4. VETERINARY MEDICINE, MEDICAL PHYSICS. Educ: Tex A&M Univ, DVM, 44; Univ Calif, PhD(comp path), 57. Prof Exp: Pvt pract, 44-47; livestock inspector, Comn Eradication Foot & Mouth Dis, USDA, 47-50; base vet, US Air Force, 50-52, res radiobiologist, Hanford Atomic Prod Oper, Gen Elec Co, 52-54, life sci proj officer, Aircraft Nuclear Propulsion Off, US AEC, 56-60, safety officer, Systs Nuclear Auxiliary Power Prog, 60-62, pathologist, Armed Forces Inst Path, 62-66, dir vet med serv, Air Proving Ground Ctr & Spec Air Warfare Ctr, Eglin AFB, Fla, 66-68; mgr radiol health & pub health ecol, Manned Spacecraft Ctr, 68-71, MGR HEALTH APPLNS OFF, L B JOHNSON SPACE CTR, NASA, 71- Concurrent Pos: Mem, Nat Comt Radiation Protection & Measurements, 60-66; mem subcomt livestock damage, Adv Comt Civil Defense, Nat Acad Sci-Nat Res Coun, 61-62; prof comp path, Grad Sch Biomed Sci & clin prof, Grad Sch Pub Health, Univ Tex; chmn bd dirs, Int Vet Med Found, 73- Mem: Am Vet Med Asn; Health Physics Soc. Res: Remote sensing; radiation toxicology; radiation biology; geographic pathology; use of aircraft and spacecraft remote sensor data in application to public health. Mailing Add: NASA L B Johnson Space Ctr Houston TX 77058

BARNES, CHARLES WINFRED, b Oklahoma City, Okla, Oct 21, 34; m 58; c 3. PETROLOGY, STRUCTURAL GEOLOGY. Educ: Univ Okla, BS, 57; Univ Idaho, MS, 62; Univ Wis, PhD(geol), 65. Prof Exp: Asst prof geol, Eastern NMex Univ, 65-68; from asst prof to assoc prof, 68-74, chmn dept, 73-74, chmn fac, 74-75, PROF GEOL, NORTHERN ARIZ UNIV, 75- Mem: Soc Econ Paleontologists & Mineralogists; Nat Asn Geol Teachers; Geol Soc Am. Res: Metamorphic petrology; structural analysis of tectonites; structural geology of monoclines. Mailing Add: Dept of Geol Box 6030 Northern Ariz Univ Flagstaff AZ 86001

BARNES, CHARLIE JAMES, b Greenville, NC, Aug 23, 30; m 57; c 1. PHYSICAL CHEMISTRY, ORGANIC CHEMISTRY. Educ: Va Union Univ, BS, 51; Va State Col, MS, 57; Howard Univ, PhD, 59. Prof Exp: Phys & org chemist, Res & Develop Dept, US Naval Propellant Plant, 59-63; ANAL CHEMIST, FOOD & DRUG ADMIN, 63- Res: Development of pesticide reference standards; stability of pesticide materials; analysis of food additives; drugs in edible animal tissue. Mailing Add: 2203 Piermont Dr Oxon Hill MD 20022

BARNES, CLIFFORD ADRIAN, b Goldendale, Wash, Oct 14, 05; m 38; c 3. OCEANOGRAPHY. Educ: Univ Wash, BS, 30, PhD(chem), 36. Prof Exp: Res chemist, Fuels Div, Battelle Mem Inst, 36-40; assoc phys oceanogr, US Coast Guard, 40-46; oceanogr, US Navy Hydrographic Off, Washington, DC, 46-47; from assoc prof to prof oceanog, 47-73, actg chmn dept, 64-65, EMER PROF OCEANOG, UNIV WASH, 73- Concurrent Pos: Oceanogr, Exped to Aleutian Islands USN, 33 & oceanog cruise of US Coast Guard cutter Chelan, 34; officer in chg, Oceanog Surv Subsect JTF-1, Oper Crossroads, 46 & Beauford Sea Exped, 50, 51, 52; consult, President's Sci Adv Comt, 60 & 61; consult, Off Sci & Technol, 62-64. Mem: Am Geophys Union; Am Chem Soc; Am Soc Limnol & Oceanog; Arctic Inst NAm. Res: Ocean currents; chemistry of sea water; sea ice and icebergs; oceanography of arctic and inshore waters; river water at sea. Mailing Add: 1806 NE 73rd St Seattle WA 98115

BARNES, DAVID EDWARD, organic chemistry, see 12th edition

BARNES, DAVID FITZ, b Boston, Mass, May 23, 21; m 66; c 2. GEOPHYSICS. Educ: Harvard Univ, BS, 43, MA, 55. Prof Exp: Asst marine geophys, Woods Hole Oceanog Inst, 43-49; geophysicist, US Geol Surv, 49-54 & Air Force Cambridge Res Ctr, 56-58; GEOPHYSICIST, US GEOL SURV, 58- Mem: Am Geophys Union; Soc Explor Geophys; fel Arctic Inst NAm; Geol Soc Am. Res: Application of geophysics to Arctic research; Alaskan regional gravity; precision gravimetry; Alaskan aeromagnetics; permafrost; perennially frozen lakes; infrared luminescence. Mailing Add: US Geol Surv Br Regional Geophys 345 Middlefield Rd Menlo Park CA 94025

BARNES, DAVID KENNEDY, b Concordia, Kans, Apr 23, 23; m 42; c 4. ORGANIC CHEMISTRY. Educ: Olivet Col, BS, 43; Ind Univ, AM, 44, PhD(chem), 47. Prof Exp: Off Sci Res & Develop contract, Ind, 43-44, asst, 46; sr chemist, Process Res Dept, Stanolind Oil & Gas Co, 47-53; with Dacron process develop, Textile Fibers Dept, 53-54, Dacron plant tech, 54-55, group supvr, 55-57, sr supvr, Nylon Plant Technol, 57-59, supt, 59-60, asst mgr, Seaford Nylon Plant, 61-62, prod mgr, Nylon Mfg Div, 63-66, dir mfg, Orlon-acetate-lycra Div, 66-67, with Indust & Biochem Dept, 67-69, ASST GEN MGR INDUST CHEM DEPT, E I DU PONT DE NEMOURS & CO, 69- Mem: Am Chem Soc; fel Am Inst Chem. Res: Anti-malarials synthesis; diphenylmethanes; azlactones; hydrocarbon synthesis process; petrochemicals; synthetic textile fibers. Mailing Add: Indust Chem Dept E I du Pont de Nemours & Co Wilmington DE 19898

BARNES, DEREK, b Manchester, Eng, Dec 24, 30; Can citizen; m 53; c 4. ANALYTICAL CHEMISTRY, RESEARCH ADMINISTRATION. Educ: Salford Univ, cert chem, 56; Simon Fraser Univ, MBA, 71. Prof Exp: Res chemist, Wood Prod Res Div, MacMillan Bloedel, Ltd, 57-66, supvr plywood sect, 66-70, supvr lumber & logging sect, 67-70, supvr panelboard sect, 70-73, ASST MGR BLDG MAT DIV, MacMILLAN BLOEDEL RES LTD, 73- Mem: Forest Prod Res Soc; Chem Inst Can. Res: Adhesive technology; high energy cutting methods; high frequency heating methods; wood drying and preservation; wood-glue composite products. Mailing Add: 1641 W 60th Ave Vancouver BC Can

BARNES, DEREK A, b Sussex, Eng, Sept 10, 33; US citizen; m 58; c 2. PLASMA PHYSICS, FLUID DYNAMICS. Educ: Oxford Univ, BA, 56 & 57, MA, 60, PhD(plasma physics), 63. Prof Exp: Theoret physicist, UK Atomic Energy Authority, 58-60; res scientist, Courant Inst Math Sci, NY Univ, 60-61; mem tech staff, Bell Tel Labs, Inc, 61-64; assoc prof, 64-66, chmn dept, 66-75, PROF PHYSICS, MONMOUTH COL, NJ, 66- Concurrent Pos: Consult, Areonaut Res Assocs Princeton, Inc, 64-68. Mem: AAAS; Am Phys Soc; Am Inst Aeronaut & Astronaut. Res: Laminar to turbulent transition and jet impingement. Mailing Add: Dept of Physics Monmouth Col West Long Branch NJ 07764

BARNES, DONALD GEORGE, b Akron, Ohio, Feb 11, 39; m 62; c 1. PHYSICAL CHEMISTRY, CHEMICAL PHYSICS. Educ: Col Wooster, BA, 61; Fla State Univ, PhD(chem), 67. Prof Exp: Asst prof, 67-70, ASSOC PROF CHEM & PHYSICS, ST ANDREWS PRESBY COL, 70- Mem: Am Chem Soc; Am Inst Physics; Am Asn Physics Teachers; Am Sci Affiliation. Res: Electronic spectra of molecules; molecular spectroscopy; chemical applications to pollution studies. Mailing Add: Dept of Chem St Andrews Presby Col Laurinburg NC 28352

BARNES, DONALD KAY, b Minneapolis, Minn, Jan 15, 35; m 54; c 3. GENETICS, PLANT BREEDING. Educ: Univ Minn, BS, 57, MS, 58; Pa State Univ, PhD(genetics), 62. Prof Exp: Res geneticist, Regional Pasture Res Lab, Agr Res Serv, USDA, Pa, 62-63, res geneticist, Fed Exp Sta, PR, 63-65 & Plant Indust Sta, Md, 65-68; PROF AGRON & PLANT GENETICS, UNIV MINN, ST PAUL, 68- Mem: Am Soc Agron; Crop Sci Soc Am; Am Genetic Asn. Res: Alfalfa genetics; breeding for disease and insect resistance; development of pollen control procedures for hybrid seed production. Mailing Add: Dept of Agron & Plant Genetics Univ of Minn St Paul MN 55108

BARNES, DONALD MCLEOD, b New Lisbon, Wis, Sept 25, 21; m 49; c 2. VETERINARY PATHOLOGY, MICROBIOLOGY. Educ: Univ Minn, BSc, 53, DVM, 55, PhD(vet med), 63; Am Col Vet Path, dipl. Prof Exp: Instr vet diag, 55-63, from asst prof to assoc prof, 63-72, PROF VET PATH, UNIV MINN, ST PAUL, 72- Mem: Wildlife Disease Asn; Am Vet Med Asn. Res: Wildlife disease. Mailing Add: Vet Basic Sci Bldg Col of Vet Med Univ of Minn St Paul MN 55101

BARNES, DOUGLAS, b Empress, Alta, May 23, 22; m 46; c 2. ECONOMIC ENTOMOLOGY, ECOLOGY. Educ: Univ Alta, BSc, 46; Univ Minn, MS, 50, PhD, 52. Prof Exp: Entomologist, Rockefeller Found Agr Prog, Mex, 54-59, asst dir, 59-62; mem staff agr res, Productos de Maiz, SA, Mex, 62-65; MGR PLANT PROD & DIR AGR TECHNOL IN LATIN AM, CYANAMID INT, 65- Mem: Entom Soc Am; NY Acad Sci. Res: Economic entomology in Mexican agriculture. Mailing Add: 393 Dunham Pl Glen Rock NJ 07452

BARNES, EARL RUSSELL, b Bennettsville, SC, May 2, 42; m 66. MATHEMATICS. Educ: Morgan State Col, BS, 64; Univ Md, PhD(math), 68. Prof Exp: RES MATHEMATICIAN, IBM WATSON RES CTR, 68- Concurrent Pos: Adj assoc prof, Columbia Univ. Mem: Am Math Soc; Soc Indust & Appl Math. Res: Variational analysis and differential equations. Mailing Add: IBM Watson Res Ctr Yorktown Heights NY 10598

BARNES, EDWARD B, b Covington, Ky, Nov 20, 22; m 53; c 9. BIOPHYSICS, PHYSICS. Educ: Villa Madonna Col, BS, 51; Inst Divi Thomae, MS, 53; Univ Houston, PhD(biophys), 67. Prof Exp: Sr res physicist, Cincinnati Milling Mach Co, 54-60; asst prof, 60-63, ASSOC PROF PHYSICS, WVA STATE COL, 66- Mem: AAAS; Am Chem Soc; Am Inst Physics. Res: Biomagnetics; physical properties of macromolecules and their sub-units; electroencephalography; spectroscopy; x-ray diffraction and fluorescence; emission spectroscopy; ultracentrifugation; electron microscopy. Mailing Add: Dept of Physics WVa State Col Institute WV 25112

BARNES, EDWIN ELLSWORTH, b Kinzua, Ore, Sept 1, 29; m 52; c 3. ANALYTICAL CHEMISTRY. Educ: Univ Puget Sound, BS, 51; Univ Wash, PhD(chem), 61. Prof Exp: Chemist, Hooker Chem Corp, 51-53; res chemist, Am Marietta Co, 60-62; prof specialist phys chem, 62-68, SCIENTIST, WEYERHAEUSER CO, 68- Mem: Am Chem Soc. Res: Properties of forest products and synthetic polymers. Mailing Add: Res Div Weyerhaeuser Co 3400 13th SW Seattle WA 98134

BARNES, EUGENE MILLER, JR, b Versailles, Ky, Dec 9, 43; m 69. BIOCHEMISTRY, BIOLOGY. Educ: Univ Ky, BS, 65; Duke Univ, PhD(biochem), 70. Prof Exp: ASST PROF BIOCHEM, BAYLOR COL MED, 71- Concurrent Pos: Am Cancer Soc fels, Nat Heart & Lung Inst, 69-70 & Roche Inst Molecular Biol, Nutley, NJ, 70-71. Mem: AAAS; Am Soc Biol Chem. Res: Membrane transport; mechanism of sugar transport in bacterial membrane vesicles; regulation of transport in mammalian cells. Mailing Add: Dept of Biochem Baylor Col of Med Tex Med Ctr Houston TX 77025

BARNES, FREDERICK WALTER, JR, b Cleveland, Ohio, Mar 3, 09; m 40; c 2. MEDICINE. Educ: Yale Univ, BA, 30; Johns Hopkins Univ, MD, 34; Columbia Univ, PhD(chem), 43. Prof Exp: Intern med, Johns Hopkins Hosp, 34-35, pediat, 35-36; resident, Children's Hosp, Boston, 36-37, asst physician, 37-38; asst biochem, Col Physicians & Surgeons, Columbia Univ, 40-41; asst prof pediat, Col Med, Univ Cincinnati, 42-46, biol chem, 42-46; assoc prof med & physiol chem, Sch Med, Johns Hopkins Univ, 46-62; PROF MED SCI, DIV MED SCI, BROWN UNIV, 62- Mem: Am Soc Biol Chem; Soc Pediat Res; Am Heart Asn. Res: Protein regeneration and interaction after harmful stimuli; studies in behavioral science. Mailing Add: Div of Med Sci Brown Univ Box G Providence RI 02912

BARNES, GARRETT HENRY, JR, b Akron, Ohio, Mar 7, 26; m 57; c 3. ORGANIC CHEMISTRY. Educ: Case Inst Technol, BS, 51; Pa State Univ, PhD(chem), 54. Prof Exp: Dow-Corning fel, Mellon Inst, 54-66; res chemist, 66-68, SR RES CHEMIST, DOW CORNING CORP, 68- Res: Organosilicon compounds. Mailing Add: Dow Corning Corp Midland MI 48640

BARNES, GENE A, b Topeka, Kans, Feb 9, 35; m 68. SOLID STATE PHYSICS. Educ: Calif Inst Technol, BS, 56; Univ Ill, Urbana-Champaign, MS, 57; Univ Ore, PhD(physics), 67. Prof Exp: Sr res engr, Autonetics Div, NAm-Rockwell Corp, 60-63; ASSOC PROF PHYSICS, CALIF STATE UNIV, SACRAMENTO, 67- Mem: AAAS; Am Asn Physics Teachers; Am Phys Soc. Res: Inertial navigation instruments; electrical conductivity in high magnetic fields; solar energy. Mailing Add: Dept of Physics-Phys Sci Calif State Univ Sacramento CA 95819

BARNES, GEORGE, b Denver, Colo, May 27, 21; m 44; c 3. ELECTRONICS. Educ: Pomona Col, AB, 42; Univ Colo, MS, 46; Ore State Col, PhD(physics), 55. Prof Exp: Instr physics, Pomona Col, 43; jr physicist, US Naval Res Lab, 43-44; asst physics, Univ Colo, 44-45, instr eng math, 45-47; teacher, Taft Jr Col, 47-48; from asst prof to assoc prof physics, Linfield Col, 48-55; asst prof, Humboldt State Col, 55-57; assoc prof, 57-61, PROF PHYSICS, UNIV NEV, RENO, 61- Mem: Am Asn Physics Teachers. Res: Allotropic tranformations in refractory metals by field-electron emission. Mailing Add: Dept of Physics Univ of Nev Reno NV 89507

BARNES, GEORGE LEWIS, b Detroit, Mich, Aug 21, 20; m 47; c 4. PLANT PATHOLOGY. Educ: Mich State Univ, BS, 48, MS, 50; Ore State Univ, PhD(plant path), 53. Prof Exp: Res asst, Ore State Univ, 50-53; res assoc, Ohio State Univ, 53-55; plant pathologist, Olin Mathieson Chem Corp, 55-58; from asst prof to assoc prof plant path, 58-72, PROF PLANT PATH, OKLA STATE UNIV, 72- Concurrent Pos: Plant pathologist, Agr Res Serv, USDA, 58-61; consult, Allergy Labs, Inc, 75- Mem: Am Phytopath Soc. Res: Fungicides; mycology; entomology; fungus physiology; peanut mold fungi; diseases of pecans and alfalfa; insect pathology. Mailing Add: Dept of Plant Path Okla State Univ Stillwater OK 74074

BARNES, GLOVER WILLIAM, b Birmingham, Ala, Sept 7, 23; m 52; c 2.

IMMUNOLOGY, BACTERIOLOGY. Educ: Univ Akron, BSc, 49; Univ Buffalo, MA, 56, PhD(bact, immunol), 62. Prof Exp: Asst cancer res scientist, Roswell Park Mem Inst, 55-58; res assoc bact & immunol, State Univ NY Buffalo, 61-63, from instr to asst prof path, 63-69; ASSOC PROF UROL & LECTR MICROBIOL, UNIV WASH, 69- Concurrent Pos: Res consult, Millard Fillmore Hosp, 60-64. Mem: Sigma Xi; AAAS; Soc Study Reproduction; Am Fertil Soc. Res: Immunologic studies on the organ-specific character and function of the male reproductive system. Mailing Add: Dept of Urol Univ of Wash Sch of Med Seattle WA 98195

BARNES, HARLEY, b Painesville, Ohio, Sept 23, 16; m 42, 58; c 4. GEOLOGY. Educ: Wheaton Col, BS, 37; Northwestern Univ, MS, 39; Johns Hopkins Univ, PhD(geol), 54. Prof Exp: Asst, Northwestern Univ, 37-39; miner & mining engr, Slide Mines, Inc, 39-40 & NJ Zinc Co, Colo, 40-41; GEOLOGIST, US GEOL SURV, 48- Concurrent Pos: Tech adv coal geol, Philippine Bur Mines, Int Coop Admin, Philippines, 54-56. Mem: Geol Soc Am; Am Asn Petrol Geol; Am Eng Geol; Am Sci Affiliation. Res: Stratigraphy; structural geology; geology of fuels. Mailing Add: US Geol Surv Bldg 25 Fed Ctr Denver CO 80225

BARNES, HARRY MILFORD, organic chemistry, see 12th edition

BARNES, HERBERT M, b Paulding, Ohio, Mar 27, 18; m 45; c 2. ANIMAL SCIENCE. Educ: Ohio State Univ, BSc, 40. Prof Exp: ASSOC PROF & EXTEN SPECIALIST ANIMAL SCI, OHIO STATE UNIV, 46- Mem: Am Soc Animal Sci. Res: Swine production; nutrition; physiology; swine husbandry. Mailing Add: Animal Sci Bldg Ohio State Univ 2029 Fyffe Rd Columbia OH 43210

BARNES, HUBERT LLOYD, b Chelsea, Mass, July 20, 28; m 50; c 2. GEOCHEMISTRY. Educ: Mass Inst Technol, BS, 50; Columbia Univ, PhD(geol), 58. Prof Exp: Res geologist, Peru Mining Co, 50-52; lectr geol, Columbia Univ, 52-54; fel, Geophys Lab, Carnegie Inst, 56-60; from asst prof to assoc prof, 60-67, PROF GEOCHEM, PA STATE UNIV, UNIVERSITY PARK, 67-, DIR ORE DEPOSITS RES SECT, 69- Concurrent Pos: Guggenheim fel, Geochem Inst, Univ Göttingen, 66-67; Davidson Mem lectr, St Andrews, 71; vis prof, Univ Heidelberg, 74; exchange scientist, USSR Acad Sci, 74. Mem: AAAS; Geochem Soc; Soc Econ Geol; Geol Soc Am; Am Geophys Union. Res: Geochemistry of hydrothermal processes, especially associated with the formation of ore deposits. Mailing Add: 208 Deike Bldg Pa State Univ University Park PA 16802

BARNES, ISABEL JANET, b Union City, NJ, Sept 22, 36. MICROBIOLOGY. Educ: Pa State Univ, BS, 58; Cornell Univ, MS, 60; Hahnemann Med Col, PhD(microbiol), 69. Prof Exp: Res assoc microbiol, Hahnemann Med Col, 60-63, instr, 63-65; from instr to asst prof, Col Med, Pa State Univ, 68-73, assoc mem, Grad Sch, 70-73; asst prof microbiol & med technol, 73-74, ASSOC PROF MICROBIOL & MED TECHNOL, SANGAMON STATE UNIV, 74- Concurrent Pos: Res Corp res grant, 71; trustee, Found Microbiol, 73-, treas, 74-; adj assoc prof, Sch Med, Southern Ill Univ, 74- Mem: AAAS; Am Soc Med Technol; Am Soc Microbiol; Sigma Xi. Res: Antibiotic resistance in staphylococcus aureus, including function, transfer and replication of plasmids. Mailing Add: Med Technol Prog Sangamon State Univ Springfield IL 62708

BARNES, IVAN, b Boston, Mass, Mar 7, 31; m 58; c 3. GEOCHEMISTRY. Educ: Tufts Univ, BS, 55; Harvard Univ, PhD(geol), 61. Prof Exp: GEOLOGIST, US GEOL SURV, 61- Mem: Geol Soc Am; Geochem Soc; Mineral Soc Am. Res: Geochemistry of water, specifically the states of reactions between natural waters and coexisting phases. Mailing Add: US Geol Surv 345 Middlefield Rd Menlo Park CA 94025

BARNES, JAMES ALLEN, b Denver, Colo, Dec 7, 33; m 59; c 3. PHYSICS. Educ: Univ Colo, BS, 56, PhD(physics, 66; Stanford Univ, MS, 58. Prof Exp: Physicist, Atomic Frequency & Time Interval Stand Sect, 56-65, from actg sect chief to sect chief, 65-67, CHIEF TIME & FREQUENCY DIV, NAT BUR STAND, 67- Concurrent Pos: Mem comn I, Int Sci Radio Union; mem study group VII, Int Radio Consultative Comt; consult, comn 31, Int Astron Union, 72- Honors & Awards: Silver Medal Award, US Dept Com, 65, Gold Medal Award, 75. Mem: Sr mem Inst Elec & Electronics Engrs. Res: Atomic and molecular spectroscopy; noise limitations of devices; statistics. Mailing Add: Time & Frequency Div Nat Bur of Stand Boulder CO 80302

BARNES, JAMES CLARKSON, b Bristol, Conn, Feb 25, 38; m 65; c 2. METEOROLOGY, NATURAL RESOURCES. Educ: Middlebury Col, BA, 60; NY Univ, MS, 65. Prof Exp: Meteorologist-in-chg, US Weather Bur, Ellsworth Sta, Antarctica, 62-63; meteorologist, Travelers Res Ctr, 64-65; res meteorologist, Allied Res Assocs, 65-71; RES SCIENTIST & MGR GEOPHYS STUDIES, ENVIRON RES & TECHNOL INC, 72- Mem: Am Meteorol Soc; Am Geophys Union; Sigma Xi. Res: Developing applications of satellite data to earth resources problems in the areas of meteorology, hydrology and marine resources. Mailing Add: Environ Res & Technol Inc 696 Virginia Rd Concord MA 01742

BARNES, JAMES CROWELL, b Geneseo, NY, June 1, 16; m 41, 69; c 3. TEXTILE PHYSICS. Educ: Colgate Univ, AB, 37; Univ Va, AM, 39, PhD(physics), 41. Prof Exp: Mem res staff, 41-45. head lab, 45-5S, asst dir res, 56-74, MGR TECH SERV, KENDALL CO, 74- Mem: Am Phys Soc; Fiber Soc (pres, 52). Res: Transient arc discharge to a moving medium; impulse ion accelerator; physical properties of cotton fibers; fiber testing methods; production control. Mailing Add: Kendall Co Box 1828 Charlotte NC 28201

BARNES, JAMES EDWARD, radiation biophysics, see 12th edition

BARNES, JAMES MILTON, b Ypsilanti, Mich, July 5, 23; m 49. PHYSICS. Educ: Eastern Mich Univ, BS, 48; Mich State Univ, MS, 50, PhD(physics), 55. Prof Exp: From asst prof to assoc prof, 55-62, head dept, 61-74, PROF PHYSICS, EASTERN MICH UNIV, 62- Mem: AAAS; Am Asn Physics Teachers; Acoust Soc Am; Nat Sci Teachers Asn. Res: Ultrasonics in glass by optical methods. Mailing Add: 4872 Whitman Circle Ann Arbor MI 48103

BARNES, JAMES RAY, b Woodland, Calif, Feb 16, 40; m 62; c 3. AQUATIC ECOLOGY. Educ: Brigham Young Univ, BS, 63; Ore State Univ, MS, 66, PhD(zool), 72. Prof Exp: Asst prof, 69-73, ASSOC PROF ZOOL, BRIGHAM YOUNG UNIV, 73- Concurrent Pos: Adj prof, Kellogg Biol Sta, Mich State Univ, 75. Mem: AAAS; Am Soc Limnol & Oceanog; NAm Benthol Soc; Orgn Inland Biol Field Sta. Res: Terrestrial litter decomposition in lakes and streams; annual reproductive cycles of marine and freshwater invertebrates. Mailing Add: Dept of Zool Brigham Young Univ Provo UT 84602

BARNES, JOHN DAVID, b Allentown, Pa, Aug 12, 39; m 65. POLYMER PHYSICS. Educ: DePauw Univ, BA, 60; Univ Md, MS, 65; Cath Univ Am, PhD(physics), 72. Prof Exp: Physicist ceramics, Eastern Res Ctr, US Bur Mines, College Park, Md, 61-64; PHYSICIST POLYMER PHYSICS, POLYMERS DIV, NAT BUR STAND, 64- Mem: Am Phys Soc; AAAS; Am Soc Testing & Mat. Res: Permeation, diffusion, and solubility of gases, vapors and liquids in polymers. Mailing Add: Polymers Bldg Rm A209 Nat Bur of Standards Washington DC 20234

BARNES, JOHN FAYETTE, b Santa Cruz, Calif, Jan 28, 30; m 55; c 2. ATOMIC PHYSICS. Educ: Univ Calif, Berkeley, BA, 51; Univ Denver, MS, 52; Univ NMex, PhD(physics), 63. Prof Exp: Staff mem, 53-68, asst group leader, 68-71, alt group leader, 71, GROUP LEADER, LOS ALAMOS SCI LAB, 71- Mem: Am Phys Soc. Res: Thomas-Fermi theory; atomic structure; solid state physics; hydrodynamics; equation of state and opacity research for applied problems. Mailing Add: 723 Meadow Lane White Rock Los Alamos NM 87544

BARNES, JOHN LANDES, b Haddonfield, NJ, Oct 16, 06; m 35; c 2. MATHEMATICS, ELECTRICAL ENGINEERING. Educ: Mass Inst Technol, SB, 28, SM, 29; Princeton Univ, AM, 30, PhD(math), 34. Prof Exp: Asst math, Princeton Univ, 32-33, asst elec eng, 33-34; instr, Mass Inst Technol, 34-35; from asst prof to assoc prof math, Tufts Univ, 35-41, prof, 41-42 & 45-47, chmn dept appl math, 39-42 & 45-47, actg chmn dept elec eng, 40-41; mem tech staff, Bell Tel Labs, NY, 42-43, NJ, 43-45; prof eng, 47-74, EMER PROF ENG, UNIV CALIF, LOS ANGELES, PRES, SYSTS CORP AM, 57- Concurrent Pos: Consult, Hazeltine Serv Corp, NY, 41-42 & Raytheon Mfg Co, Mass, 45-47; consult, NAm Aviation Inc, 47-49, chief guid, 49-51, assoc dir electro-mech eng dept, 51-53; consult, Ramo-Wooldridge Corp, 53-55; radio engr, Radio Corp Am Mfg Co, 39-40; dir, Control & Comput Lab, Lockheed Missiles & Space Co, 55; vpres, Systs Res Corp, 55-56; pres, Systs Lab Corp, 56-57. Mem: Fel AAAS; Am Math Soc; Am Phys Soc; Math Asn Am; fel Inst Elec & Electronics Engrs. Res: Application of the Laplace transformation in integral and differential equations of mathematical physics; biotechnology; information theory; interplanetary space systems; probabilistics; statistics; stochastics; uncertainty principles; macrosystems engineering. Mailing Add: Sch Eng & Appl Sci BH 4731 H Univ of Calif Los Angeles CA 90024

BARNES, JOHN MAURICE, b Washington, DC, Apr 22, 31; m 57; c 5. PLANT PATHOLOGY. Educ: Univ Md, BS, 54; Cornell Univ, MS, 57, PhD(plant path), 60. Prof Exp: Opers analyst, Opers Res Off, Johns Hopkins Univ, 60-63; plant scientist, Hazleton Labs, TRW Systs, 63-67; plant pathologist, Coop State Res Serv, 67-75, coordr, Corn Blight Info Ctr, 71-72, PESTICIDE COORDR, OFF SECY, USDA, 75- Concurrent Pos: Report writing & consult pesticides, NSF, 75. Honors & Awards: Cert of Merit, USDA, 72. Mem: Am Phytopath Soc; AAAS. Res: Coordination and policy planning relative to agricultural pest control. Mailing Add: Off of Environ Qual Activities Off of the Secy USDA Washington DC 20250

BARNES, KAREN K, analytical chemistry, see 12th edition

BARNES, LESTER E, b Kalamazoo, Mich, Sept 29, 24; m 46 & 69; c 3. ZOOLOGY, PHYSIOLOGY. Educ: Univ Wis, BS, 49, MS, 51, PhD(zool), 69. Prof Exp: Jr scientist, Upjohn Co, 51-58; asst dir, Endocrine Labs Madison, Inc, 63-70; DIR, ALTECH LABS, 70- Mem: Soc Study Reproduction; Endocrine Soc. Res: Anti-inflammatory steroids; anabolic-androgenic steroids; female reproduction, control of ovulation and mechanisms controlling implantation and early development; biological assays; toxicology and teratology studies. Mailing Add: Altech Labs Rte 4 Syone Rd Madison WI 53711

BARNES, LOWELL R, veterinary medicine, see 12th edition

BARNES, LUCIEN, JR, b New Rochelle, NY, Apr 10, 17; m 44; c 2. ANALYTICAL CHEMISTRY. Educ: Lafayette Col, AB, 39; Mich State Univ, MS, 49. Prof Exp: From chemist to sr chemist, Anal Sect, Cent Res Labs, 49-53, sect head org anal, 53-62, group leader anal sect, 62-70, SUPVR ANAL SECT, CENT RES LABS, AIRCO, INC, 70- Mem: Am Chem Soc. Res: Application in fields of organic functional group, gas chromatographic, atomic absorption and metallurgical analysis; trace analysis metallurgical and organic samples. Mailing Add: 6 Wells Lane Warren NJ 07060

BARNES, MARTIN MCRAE, b Calgary, Alta, Aug 3, 20; US citizen; m 46; c 4. ECONOMIC ENTOMOLOGY. Educ: Univ Calif, Berkeley, BS, 41; Cornell Univ, PhD(entom), 46. Prof Exp: Jr entomologist, 46-49, from asst entomologist to assoc entomologist, 49-61, ENTOMOLOGIST, UNIV CALIF, RIVERSIDE, 61-, PROF ENTOM, 63- Mem: Fel AAAS; Entom Soc Am; Can Entom Soc; Mex Soc Entom. Res: Biology, ecology and management of insects and mites of deciduous fruit and nut orchards and vineyards. Mailing Add: Dept of Entom Univ of Calif Riverside CA 92502

BARNES, PETER DAVID, b Garden City, NY, Oct 10, 37; m 61; c 2. NUCLEAR PHYSICS. Educ: Univ Notre Dame, BS, 59; Yale Univ, MS, 60, PhD(physics), 65. Prof Exp: Res assoc nuclear physics, Niels Bohr Inst, Copenhagen, Denmark, 64-66 & Los Alamos Sci Lab, 66-68; ASSOC PROF NUCLEAR PHYSICS, CARNEGIE-MELLON UNIV, 68- Concurrent Pos: NATO fel, 65-66. Mem: Am Phys Soc. Res: Investigations and experimental tests of nuclear structure theories and nuclear reaction calculations. Mailing Add: Dept of Physics Carnegie-Mellon Univ Pittsburgh PA 15213

BARNES, RALPH CRAIG, b Summerfield, Ohio, May 4, 14; m 37; c 1. PUBLIC ENTOMOLOGY, PUBLIC HEALTH. Educ: Ohio Univ, BS, 39; NC State Col, MS, 41. Prof Exp: Eng aide. Commun Dis Ctr, 42-43, asst sanitarian, 43-44, sr asst sanitarian, 44-47, sr asst scientist, 47-48, scientist, 48-52, sr scientist, 52-56, scientist dir, 56-70, regional prog dir, Ctr Dis Control, 70-74, DIR DIV PREV, PUBLIC HEALTH SERV, US DEPT HEALTH, EDUC & WELFARE, REG VIII, 74- Honors & Awards: Commendation Medal, USPHS, 72. Mem: Entom Soc Am; Am Pub Health Asn. Res: Ecology of mosquitoes; medical entomology; communicable disease control. Mailing Add: Div of Prev USPHS 11037 Fed Off Bldg 1961 Stout St Denver CO 80202

BARNES, RAMON M, b Pittsburgh, Pa, Apr 24, 40; m 69. ANALYTICAL CHEMISTRY. Educ: Ore State Univ, BS, 62; Columbia Univ, AM, 63; Univ Ill, Urbana-Champaign, PhD(anal chem), 66. Prof Exp: Spectrochemist, NASA Lewis Res Ctr, Ohio, 67-68; assoc anal chem, Iowa State Univ, 68-69; asst prof, 69-75, ASSOC PROF ANAL CHEM, UNIV MASS, AMHERST, 75- Concurrent Pos: Lectr chem, Baldwin-Wallace Col, 67-68. Mem: Am Chem Soc; Soc Appl Spectros; Spectros Soc Can; Am Soc Test & Mat; Optical Soc Am. Res: Chemical reactions in high energy spectroscopic discharges. including sparks and radio frequency, induction heated plasmas for spectrochemical analysis. Mailing Add: Dept of Chem GRC Tower I Univ of Mass Amherst MA 01002

BARNES, RAYMOND D, b Oakland, Calif, Mar 29, 34; m 53; c 3. REPRODUCTIVE BIOLOGY, LABORATORY ANIMAL SCIENCE. Educ: Univ Calif, Berkeley, AB, 58; Univ Calif, Davis, PhD(zool), 66. Prof Exp: Res anatomist, Sch Vet Med, Univ Calif, Davis, 65-67, asst prof anat, 67-74; ASST PROF VET MED, UNIV MINN, 74- Mem: Soc Study Reprod; Am Asn Vet Anat. Res: Functional morphology of the mammalian ovary. Mailing Add: Dept of Vet Biol Col of Vet Med Univ of Minn St Paul MN 55108

BARNES, RICHARD GEORGE, b Milwaukee, Wis, Dec 19, 22; m 50; c 4. SOLID STATE PHYSICS, NUCLEAR MAGNETIC RESONANCE. Educ: Univ Wis, BA, 48; Dartmouth Col, MA, 49; Harvard Univ, PhD(physics), 52. Prof Exp: From asst prof to assoc prof physics, Univ Del, 52-56; assoc prof, Iowa State Univ, 56-60; sr physicist, Ames Lao, US AEC, 60-71, chief physics div, 71-75; chmn dept, 71-75, PROF PHYSICS. IOWA STATE UNIV, 60- Concurrent Pos: Vis assoc, Calif Inst Technol, 62-63; US sr scientist award, Alexander von Humooldt Found, 75-76; guest prof, Sch of Technol, Darmstadt, Ger, 75-76. Mem: Fel Am Phys Soc; Am Asn Physics Teachers; AAAS. Res: Nuclear magnetic resonance and Mössbauer effect in metallic solids, especially metal-hydrogen systems; diffusion in solids. Mailing Add: Dept of Physics Iowa State Univ Ames IA 50010

BARNES, RICHARD HENRY, b San Diego, Calif, June 29, 11; m 42; c 3. NUTRITION. Educ: San Diego State Teachers Col, AB, 33; Univ Minn, PhD(physiol chem), 40. Prof Exp: Res chemist, Scripps Metab Clin, Univ Calif, 33-37; asst physiol chem, Univ Minn, 37-40, from instr to asst prof, 40-44; dir biochem res, Sharp & Dohme, Inc, 44-49, from asst dir res to assoc dir res, 49-56; dean grad sch nutrit, 56-73, JAMES JAMISON PROF NUTRIT, CORNELL UNIV, 73- Concurrent Pos: Hon prof, Rutgers Univ, 48-56; ed, J Nutrit, Am Inst Nutrit, 59-69. Honors & Awards: Borden Award, Am Inst Nutrit, 67, Conrad Elvehjem Award, 75. Mem: AAAS; Am Chem Soc; Am Soc Biol Chem; Soc Exp Biol & Med; fel NY Acad Sci. Res: General biochemistry; metabolism; intestinal microflora and nutrition; experimental protein-calorie malnutrition and its effect upon biochemical and behavioral factors. Mailing Add: Div of Nutrit Sci Cornell Univ Ithaca NY 14853

BARNES, RICHARD N, b Washington, DC. Dec 10, 28; m 57; c 2. ECOLOGY. Educ: Univ Calif, Berkeley, AB, 52; Univ Calif, Davis, MA, 57, PhD(zool), 62. Prof Exp: Instr biol, Sacramento State Col, 57-62; from asst prof to assoc prof, 62-72, PROF BIOL, BEREA COL, 72- Mem: Sigma Xi; Am Soc Limnol & Oceanog; Ecol Soc Am. Res: Limnology; aquatic ecology. Mailing Add: Dept of Biol Berea Col Berea KY 40403

BARNES, RICHARD T, mathematics, see 12th edition

BARNES, ROBERT BOWLING, b Montgomery, Ala, June 9, 06; m 33; c 2. PHYSICS. Educ: Birmingham-Southern Col, AB, 25; Johns Hopkins Univ, PhD(physics), 29. Hon Degrees: DSc, Birmingham-Southern Col, 57. Prof Exp: Instr physics, Birmingham-Southern Col, 24-25; teacher high sch, Ala, 25-26 & John Hopkins Univ, 28-30; Nat Res Coun fel, Berlin & Breslau, 30-32; instr physics, John Hopkins Univ, 32-33, Princeton Univ, 33-36; dir physics div, Stamford res labs, Am Cyanamid Co, 36-48; vpres in charge res & develop, Am Optical Co, 48-51; pres, Olympic Develop Co, 51-55; pres, 55-71, CHMN BD, BARNES ENG CO, 71- Concurrent Pos: Civilian with Manhattan Proj; mem infrared panel, Comt Electronics, Res & Develop Bd; sci mission to Europe, Tech Indust Intel Br, US Dept Commerce, 45. Honors & Awards: Beckman Award, 55. Mem: Fel Am Phys Soc; Am Chem Soc; Optical Soc Am; Electron Micros Soc Am (pres, 42-44); fel NY Acad Sci. Res: Electron microscopy; infrared spectroscopy; applied optics. Mailing Add: Barnes Eng Co 30 Commerce Rd Stamford CT 06902

BARNES, ROBERT DRANE, b New Orleans, La, Apr 3, 27; m 53; c 3. INVERTEBRATE ZOOLOGY. Educ: Davidson Col, BS, 49; Duke Univ, PhD, 53. Prof Exp: Instr zool, Duke Univ, 52-53 & Smith Col, 53-54; res assoc biol, Rice Inst, 54-55; from asst prof to assoc prof, 55-63, chmn dept, 65-70, PROF BIOL, GETTYSBURG COL, 63- Concurrent Pos: Mem, Foreign Currency Prog, Res Coun Biol Studies, Smithsonian Inst, 71-74. Mem: AAAS; Am Soc Zool; Soc Syst Zool; Marine Biol Asn UK; Paleont Soc. Res: Polychaete physiology and behavior; anomuran crustaceans; pterobranchs. Mailing Add: Dept of Biol Gettysourg Col Gettysburg PA 17325

BARNES, ROBERT F. b Estherville, Iowa, Feo 6, 33; m 55; c 4. AGRONOMY. Educ: Iowa State Univ, BS, 57; Rutgers Univ. MS, 59; Purdue Univ, PhD(agron), 63. Prof Exp: Res agronomist. Ind, 59-70, res agronomist, Pa, 70-75, dir US regional pasture res lab, 70-75, STAFF SCIENTIST FORAGE & RANGE, NAT PROG STAFF, PLANT & ENTOM SCI, AGR RES SERV, USDA, 75- Concurrent Pos: From instr to assoc prof, Purdue Univ, 59-70; adj prof, Pa State Univ, 70-76; partic, Int Grassland Cong, Helsinki, Finland, 66 & Surfers Paradise, Australia, 70. Honors & Awards: Merit Cert, Am Forage Grassland Coun, 73. Mem: AAAS; fel Am Soc Agron; Crop Sci Soc Am; Am Soc Animal Sci; Am Forage & Grassland Coun. Res: Forage, pasture, range and turfgrass breeding, production and utilization; crop physiology. development and application of forage evaluation and utilization techniques; biochemistry of forage and range plants. Mailing Add: USDA Agr Res Serv Nat Prog Staff Beltsville MD 20705

BARNES, ROBERT FREDERICK, solid state physics, see 12th edition

BARNES, ROBERT KEITH, b Connersville, Ind, Apr 25, 25; m 57; c 3. ORGANIC CHEMISTRY. Educ: Purdue Univ, BS, 49; Univ Idaho, MS, 51; Univ Wash, PhD(chem), 55. Prof Exp: ASSOC DIR RES & DEVELOP DEPT, CHEM & PLASTICS DIV, UNION CARBIDE CORP, 55- Mem: Am Chem Soc. Res: Agricultural chemicals and intermediates; management of pesticide research, screening, formulations and product and process development; alkylamines processes; alkylene oxide monomer and polymer research. Mailing Add: Res & Develop Dept Union Carbide Corp PO Box 8361 South Charleston WV 25303

BARNES, ROBERT LEE, b Niagara Falls, NY, Aug 1, 35. INORGANIC CHEMISTRY. Educ: Hamilton Col, BA. 57; Pa State Univ, MS, 60, PhD(chem', 63. Prof Exp: Res fel chem. Harvard Univ, 63-65; staff scientist, Aerospace Div, Res Ctr, Gen Precision, Inc, Little Falls, 65-67 & Gen Precision/Aerospace, 67-69, res mgr chem dept, Singer-Gen Precision Kearfott Res Ctr, 69-74; MGR INORG CHEM, US TESTING CO, HOBOKEN, NJ. 74- Mem: Am Chem Soc; Am Soc Testing & Mat; Am Inst Aeronaut & Astronaut; Soc Appl Spectros. Res: Electron deficient compounds, especially boron hydrides and metal alkyls; mechanism of inorganic reactions; synthesis of inorganic materials; organometallics; adhesives; chemiluminescence; analytical chemistry; environmental analysis; thermal analysis. Mailing Add: 48 High Ridge Rd RD 3 Dover NJ 07801

BARNES, ROBERT LLOYD, b Royersford, Pa, Nov 13, 26; m 51; c 2. FOREST PHYSIOLOGY. Educ: Duke Univ, MF, 51, PhD(bot), 58. Prof Exp: From instr to asst prof forestry, Univ Fla, 51-57; plant physiologist, Southeastern Forest Exp Sta, USDA, 58-65; PROF FOREST BIOCHEM, DUKE UNIV, 65- Mem: Am Soc Plant Physiol; Soc Am Foresters. Res: Physiology of forest trees, especially biochemistry of wood formation and effects of air pollutants on physiology and metabolism. Mailing Add: Sch of Forestry Duke Univ Durham NC 27706

BARNES, RODERICK ARTHUR, b Madison, Wis, June 21, 20; m 44, 66; c 2. ORGANIC CHEMISTRY. Educ: Univ Wis, BS, 39; Univ Minn, PhD(org chem), 43. Prof Exp: Asst chem, Univ Minn, 39-42, instr & fel, 44-45; instr, Columbia Univ, 45-47; asst prof, Rutgers Univ, 47-53, from assoc prof to prof, 53-63; prof, Florida Atlantic Univ, 63-67; coordr res & grad prog, Engenharia Mil Inst, 70-75; PROF, FED UNIV RIO DE JANEIRO, 75- Concurrent Pos: NSF sr fel, 57-58; Conselho Nacional de Pesquisas fel, Brazil, 57-58; Fulbright lectr, Sao Paulo, 62-63; vis prof, Brazil, 63-65; temp prof, Univ Rio de Janeiro, 67. Mem: Am Chem Soc. Res: Biogenesis of natural products; reaction mechanisms; molecular orbital calculations. Mailing Add: Estrada do Tindiba 421 Rio de Janeiro Brazil

BARNES, RONNIE C, b Union City, Tenn, Sept 10, 41; m 66; c 1. ASTROPHYSICS. Educ: Vanderbilt Univ, BA, 63; Ind Univ, AM, 66, PhD(astrophys), 68. Prof Exp: From instr to asst prof physics & astron, Univ Mo-Columbia, 68-75; DIR M D ANDERSON PLANETARIUM, LAMBUTH COL, 75- Mem: Am Astron Soc; fel Royal Astron Soc. Res: Close binary systems; cosmology. Mailing Add: M D Anderson Planetarium Lambuth Col Jackson TN 38301

BARNES, ROSS OWEN, b Wahroonga, Australia, May 18, 46; m 67; c 2. EARTH SCIENCES. Educ: Andrews Univ, BA, 67; Univ Calif, San Diego, PhD(earth sci), 73. Prof Exp: Res chemist, Univ Calif, San Diego, 73-74; RES ASSOC MARINE SCI, MARINE STA, WALLA WALLA COL, 74- Mem: AAAS; Geol Soc Am; Am Geophys Union; Geochem Soc; Am Soc Limnol & Oceanog. Res: Marine biogeochemical cycles of carbon, nitrogen and sulfur; helium flux from the earth's crust. Mailing Add: Walla Walla Col Marine Sta 174 Rosario Beach Anacortes WA 98221

BARNES, STEPHEN NOBLE, b Los Alamos, NMex, Mar 6, 45; m 72. VISION. Educ: Univ Rochester, BA, 66; Univ Colo, PhD(anat), 72. Prof Exp: Fel biol, Yale Univ, 72-74, res assoc, 74-75; ASST PROF BIOL & MEM INST NEUROSCI & BEHAV, UNIV DEL, 75- Mem: AAAS; Am Soc Zoologists; Sigma Xi; Soc Neurosci; Asn Res Vision & Ophthal. Res: Physiology, anatomy and photopigment biochemistry of invertebrate visual systems. Mailing Add: Dept of Biol Sci Univ of Del Newark DE 19711

BARNES, THOMAS CUNLIFFE, physiology, deceased

BARNES, THOMAS GROGARD, b Blanket, Tex, Aug 14, 11; m 38; c 4. PHYSICS. Educ: Hardin-Simmons Col, AB, 33; Brown Univ, ScM, 36. Hon Degrees: ScD, Hardin-Simmons Col, 60. Prof Exp: Asst physics, Brown Univ, 35-36; teacher high sch, Tex, 36-38; instr math & physics, Col Mines & Metall, Univ Tex, El Paso, 38-43; elec engr, Div Phys War Res, Duke Univ, 43-45; from asst prof to assoc prof physics, 45-48, dir, Schellenger Res Labs, 55-65, PROF PHYSICS, UNIV TEX, EL PASO, 48- Concurrent Pos: Vis fel, Brown Univ, 52-53; consult physicist, US Army Res Off, Durham, 63; consult, Res Adv Proj Div, Globe Universal Sci, 65- Mem: Am Asn Physics Teachers; Am Meteorol Soc; Soc Explor Geophysicists; Acoust Soc Am; Creation Res Soc (pres, 73-). Res: Electromagnetism; acoustics; atmospheric physics; electronic instrumentation. Mailing Add: Dept of Physics Univ of Tex El Paso TX 79968

BARNES, VIRGIL EVERETT, b Chehalis, Wash, June 11, 03; m 32; c 3. GEOLOGY. Educ: State Col Wash, BS, 25, MS, 27; Univ Wis, PhD(geol), 30. Prof Exp: Cur dept geol, Univ Wis, 27-29; res fel, Am Petrol Inst, 30-31 & US Geol Surv, 33-35; assoc dir tektite res, 61-68, GEOLOGIST, BUR ECON GEOL, UNIV TEX, AUSTIN, 35-, PROF GEOL, DEPT GEOL SCI, 61-, DIR TEKTITE RES, 60- Concurrent Pos: NSF grants study of origin & compos of tektites, 60-75; mem, Am Comn Stratig Nomenclature, 70- Mem: Fel Geol Soc Am; fel Mineral Soc Am; Asn Petrol Geologists; Geochem Soc; Soc Econ Paleontologists & Mineralogists. Res: Cambrian, Ordovician and Devonian stratigraphy; geophysics; mineralogy; world wide occurrence of tektites; Lunar glass; meteorites; economic geology of Texas. Mailing Add: Bur of Econ Geol Box X Univ Sta Austin TX 78712

BARNES, VIRGIL EVERETT, II, b Galveston, Tex, Nov 2, 35; m 57, 72; c 3. HIGH ENERGY PHYSICS. Educ: Harvard Univ, AB, 57; Cambridge Univ, PhD(physics), 62. Prof Exp: Res assoc, Brookhaven Nat Lab, 62-64, from asst physicist to assoc physicist, 64-69; ASSOC PROF PHYSICS, PURDUE UNIV, WEST LAFAYETTE, 69- Mem: Am Inst Physics. Res: Experimental elementary particle physics. Mailing Add: Dept of Physics Purdue Univ West Lafayette IN 47906

BARNES, WALLACE EDWARD, b Buffalo, NY, Oct 22, 20; c 2. MATHEMATICS. Educ: Univ Buffalo, BA, 42; Brown Univ, ScM, 43; Cornell Univ, PhD(math), 49. Prof Exp: Asst prof math, Col William & Mary, 49-51; mathematician, US Naval Proving Ground, Va, 51-53, chief ballistics & statist theory br, Comput & Ballistics Dept, 53-55; math consult, Opers Res & Synthesis Sect, Gen Elec Co, 55-60, mathematician, Internal Automation Oper, 60-69; dir, Westinghouse Learning Corp, 69-70; asst prof math, 71-72, ASSOC PROF MATH & CHMN DEPT NATURAL SCI, ROBERT MORRIS COL, 72- Mem: Am Math Soc; Opers Res Soc Am; Math Asn Am; Inst Mgt Sci. Res: Application of operations research to building models of business problem areas; simulation on large scale digital computers; application of critical path technology; mass transportation system scheduling; short range forecasting. Mailing Add: 321 Spahr St Pittsburgh PA 15232

BARNES, WILFRED E, b Oak Park, Ill, June 3, 24; m 46; c 2. MATHEMATICS. Educ: Univ Chicago, BS, 49, MS, 50; Univ BC, PhD(math), 54. Prof Exp: From instr to prof math, Wash State Univ, 54-66; PROF MATH & HEAD DEPT, IOWA STATE UNIV, 66- Mem: AAAS; Am Math Soc; Soc Indust & Appl Math; Math Asn Am. Res: Algebra; theory of rings and ideals. Mailing Add: Dept of Math Iowa State Univ Ames IA 50011

BARNES, WILLIAM ALEXANDER, b New York, NY, Jan 7, 12; m 45; c 4. MEDICINE. Educ: City Col New York, BA, 33; Cornell Univ, MD, 37. Prof Exp: Instr surg, 44-45, from asst prof to assoc prof clin surg, 45-67, CLIN PROF SURG, COL MED, CORNELL UNIV, 67-, ATTEND SURGEON, NY GEN HOSP AT CORNELL, 72- Concurrent Pos: Consult, US Army, 47-55 & Valley Hosp, Ridgewood, NJ, 51- Mem: Fel Soc Univ Surg; fel Am Asn Cancer Res; fel Am Col Surg. Res: Experimental cancer; surgery. Mailing Add: Dept of Surg Med Col Cornell Univ 1300 York Ave New York NY 10021

BARNES, WILLIAM CARROLL, horticulture, deceased

BARNES, WILLIAM CHARLES, b Chicago, Ill, Dec 12, 34; m 67; c 1. GEOLOGY. Educ: Colo Sch Mines, Geol E, 56; Univ Wyo, MS, 58; Princeton Univ, PhD(geol), 63. Prof Exp: Geologist, Humble Oil & Ref Co, 58-59; asst prof geol, Ore State Univ, 63-67; geologist, US Geol Surv, 67-68; ASST PROF GEOL, UNIV BC, 68- Mem: Geol Soc Am; Geol Asn Can. Res: Sedimentology; organic geochemistry of recent marine and lacustrine sediments; overthrust fault mechanics. Mailing Add: Dept of Geol Univ of BC Vancouver BC Can

BARNES, WILLIAM WAYNE, b Amsterdam, Ohio, May 26, 27; m 50; c 3.

MEDICAL ENTOMOLOGY, INSECT TOXICOLOGY. Educ: Ohio State Univ, BSc, 50, MSc, 58, PhD(insect toxicol), 64. Prof Exp: Entomologist, US Army, 52-70, chief med entom div, 66-70; CHIEF ENVIRON BIOL BR, TENN VALLEY AUTH, 70- Mem: Entom Soc Am; Am Mosquito Control Asn. Res: Medical acarology; insect pathology; insect ecology; epidemiology of arthropod-borne diseases. Mailing Add: Tenn Valley Auth Environ Biol Br E&D Bldg Muscle Shoals AL 35660

BARNESS, LEWIS ABRAHAM, b Atlantic City, NJ, July 31, 21; m 53; c 3. NUTRITION, PEDIATRICS. Educ: Harvard Univ, AB, 41, MD, 44. Prof Exp: From instr to prof pediat, Med Sch, Univ Pa, 51-72; PROF PEDIAT & CHMN DEPT, COL MED, UNIV S FLA, TAMPA, 72- Concurrent Pos: Res fel, Children's Med Ctr, Boston, 47. Mem: AAAS; Soc Pediat Res: Am Pediat Soc; NY Acad Sci. Res: Liver necrosis; nitrogen requirements of infants. Mailing Add: Dept of Pediat Univ of SFla Col of Med Tampa FL 33620

BARNET, ANN B, b Chicago, Ill; m 53; c 3. PEDIATRIC NEUROLOGY. Educ: Sarah Lawrence Col, AB, 51; Harvard Univ, MD, 56. Prof Exp: Trainee ment health, Seizure Unit, Children's Med Ctr, Boston, Mass, 52; asst pediat, US Army 130th Field Hosp, Heidelberg, Ger, 56-57; clin dir, Child Develop Clin, Cambridge, Mass, 58-61; res assoc exp psychophysiol, Walter Reed Army Inst Res, 61-73; DIR ELECTROENCEPHALOG RES LAB, DEPT NEUROL, CHILDREN'S HOSP, DC, 65-; ASSOC PROF NEUROL & CHILD HEALTH & DEVELOP, SCH MED, GEORGE WASHINGTON UNIV, 67- Concurrent Pos: Fel pediat, Mass Gen Hosp, Boston, 58-61; USPHS grant, 64-; NIMH res scientist develop award, 70-75; W T Grant Found res grant, 73-; guest physician, Nerve Clin, Heidelberg Univ, 57; asst physician, Wrentham State Sch, Mass, 58; prin investr, Army Med Res & Develop Command Contract, 61-64; res assoc, Washington Sch Psychiat, 61-65; consult, Dept Neurol, Walter Reed Army Hosp, 61-65 & Children's Bur Proj 237, DC, 64-66; deleg, Conf Young Deaf Child, Toronto, Ont, 64, Int Cong Physiol, Moscow, USSR, 66 & Int Electroencephalog Evoked Response Audiometry Symp, Freiburg, Ger, 70; mem, Int Elec Response Audiometry Study Group; mem, NIH Commun Sci Study Sect, 73-77. Mem: AAAS; Soc Neurosci. Res: Research on development of sensory and perceptual processes in normal and abnormal infants and young children using the method of computer analysis of electroencephalographic responses to sensory stimuli. Mailing Add: Childrens Hosp Nat Med Ctr 2125 13th St NW Washington DC 20009

BARNET, HARRY NATHAN, b San Diego, Calif, Apr 30, 23; m 46; c 2. BIOCHEMISTRY. Educ: San Diego State Col, AB, 47; Univ Southern Calif, MS, 50, PhD(biochem), 53. Prof Exp: Asst, Med Sch, Univ Southern Calif, 49-53; res assoc, Scripps Metab Clin, 53-57; from asst prof to prof chem, US Int Univ, 57-75; PROF CHEM, PALOMAR COL, 75- Mem: AAAS; Am Chem Soc; Sigma Xi. Res: tracer carbohydrate metabolism; serum protein chemistry; erythrocyte metabolism; blood storage. Mailing Add: Dept of Chem Palomar Col San Marcos CA 92069

BARNETT, ALBERT GERALD, b Weston, WVa, Feb 24, 39; m 63; c 2. ENVIRONMENTAL PHYSICS, HEALTH PHYSICS. Educ: WVa Univ, BS, 61, MS, 63, PhD(physics), 66. Prof Exp: Sr res physicist, 66-69, group leader physics, 69-72, process develop mgr, Component Develop Sect, 72-75, MGR PERSONNEL SAFETY, WASTE MGR & ENVIRONMENTAL CONTROL, MOUND LAB, MONSANTO RES CORP, 75- Mem: Am Phys Soc. Res: Nondestructive test engineering; nuclear spectroscopy and reaction studies; applied physics research and development. Mailing Add: Mound Lab Monsanto Res Corp Miamisburg OH 45342

BARNETT, ALLEN, b Newark, NJ, May 5, 37; m 65; c 2. NEUROPHARMACOLOGY. Educ: Rutgers Univ, BS, 59; State Univ NY Buffalo, PhD(pharmacol), 65. Prof Exp: Pharmacologist, Hoffman-La Roche Inc, 65-66; PHARMACOLOGIST, SCHERING CORP, 66- Mem: AAAS; Am Chem Soc; Am Soc Pharmacol & Exp Therapeut; sr mem Acad Pharmaceut Sci; Am Pharmaceut Asn. Res: Central effects of alcohol; parkinsonism and dopamine receptors. Mailing Add: Dept of Pharmacol Schering Corp Bloomfield NJ 07003

BARNETT, AUDREY, b Somerset, Pa, May, 1933. GENETICS, PROTOZOOLOGY. Educ: Wilson Col, AB, 55; Ind Univ, MA, 57, PhD(genetics), 62. Prof Exp: Asst, Ind Univ, 55-58; asst prof biol, Western Col, 61-63; NIH vis fel, Princeton Univ, 63-65; asst prof, Bryn Mawr Col, 65-70; ASSOC PROF ZOOL, UNIV MD, COLLEGE PARK, 71- Concurrent Pos: NIH spec res fel, 68-69; vis scientist, Argonne Nat Lab, 68-70. Mem: AAAS; Soc Protozool; Genetics Soc Am. Res: Genetics of mating types and circadian rhythm of mating type reversals in Paramecium multimicronucleatum. Mailing Add: Dept of Zool Univ of Md College Park MD 20742

BARNETT, BENJAMIN, chemistry, see 12th edition

BARNETT, BOBBY DALE, b Elm Springs, Ark, Aug 12, 27; m 48. POULTRY NUTRITION. Educ: Univ Ark, BSA, 50, MS, 54; Univ Wis, PhD(biochem, poultry huso). 57. Prof Exp: Asst poultryman, 56-58, assoc poultryman, 58-59, PROF POULTRY SCI & HEAD DEPT, CLEMSON UNIV. 59- Mem: Poultry Sci Asn; Am Inst Nutrit. Res: Dietary toxins; nutrition-disease relationships. Mailing Add: Dept of Poultry Sci Clemson Univ Clemson SC 29631

BARNETT, BUFORD F, chemistry, see 12th edition

BARNETT, CHARLES JACKSON, b Chattanooga, Tenn, Feb 15, 42; m 70. ORGANIC CHEMISTRY. Educ: Vanderbilt Univ, BA, 64; Rice Univ, PhD(org chem), 69. Prof Exp: Sr org chemist, 69-75, RES SCIENTIST, LILLY RES LABS, ELI LILLY & CO, 75- Mem: Am Chem Soc; The Chem Soc; NY Acad Sci. Res: Chemistry and synthesis of indole alkaloids; development of synthetic procedures for pharmaceutical and agricultural chemical production. Mailing Add: 7540 N Pennsylvania St Indianapolis IN 46240

BARNETT, CHARLES WILLIAM HENRY, b Elmhurst, NY, Jan 20, 17; m 46; c 2. PETROGRAPHY. Educ: Va Polytech Inst, BS, 41, MS, 47. Prof Exp: Petrographer & spectrographer, Res Lab, Gen Refractories Co, 47-51; physicist, Res Lab, Fort Belvoir, 51; chief petrographer, Norton Co, 52-59; proj mgr & physicist, Advan Res Sect, Am Mach & Foundry Co, 59-61; RES ENGR, MICROMINIATURE SECT, HARRY DIAMOND LAB, US ARMY MAT COMMAND, WASHINGYON, DC, 61- Mem: Am Ceramic Soc. Res: Piezoelectrics; spectrographic analysis of refractories; design of infrared optics; properties of abrasives. Mailing Add: 800 E Broad St Falls Church VA 22046

BARNETT, CLARENCE FRANKLIN, b Sweetwater, Tenn, Oct 7, 24; m 41; c 3. PHYSICS. Educ: Univ Tenn, BS, 49, MS, 51. Prof Exp: ENGR & PHYSICIST, OAK RIDGE NAT LAB, 43- Concurrent Pos: Lectr, Univ London, 61-62. Mem: AAAS; Sigma Xi; Am Phys Soc. Res: Ion sources; charge exchange; energy loss charged particles in matter; electron multiplier; thermonuclear and plasma physics. Mailing Add: Rt 4 Lenoir City TN 37771

BARNETT, CLAUDE C, b Woodinville, Wash, Nov 8, 28; m 51; c 2. THEORETICAL PHYSICS. Educ: Walla Walla Col, BS, 52; Wash State Univ, MS, 56, PhD(physics), 60. Prof Exp: Instr, 57-58, from asst prof to assoc prof, 58-64, PROF PHYSICS, WALLA WALLA COL, 64-, CHMN DEPT, 61- Mem: Am Phys Soc; Am Asn Physics Teachers. Res: Electroluminescence; quantum hydrodynamics; relativity; bio-electronics; radar; many-body problem. Mailing Add: Dept of Physics Walla Walla Col College Place WA 99324

BARNETT, DAVID, b Arlington, Wash, Sept 4, 40; m 65. GEOMETRY. Educ: Univ Wash, BS, 64, MS, 66, PhD(math), 67. Prof Exp: Asst prof, Western Wash State Col, 67; asst prof, 67-70, ASSOC PROF MATH, UNIV CALIF, DAVIS, 70- Concurrent Pos: NSF res grant, Univ Calif, Davis, 68- Mem: Math Asn Am. Res: Combinatorial structure of convex polytopes, particularly problems dealing with the relationships between the numbers of faces of different dimensions of convex polytopes. Mailing Add: Dept of Math Univ of Calif Davis CA 95616

BARNETT, DONALD RAY, b Morton, Tex, Oct 19, 35. HUMAN GENETICS, BIOCHEMICAL GENETICS. Educ: Univ Tex, Austin, BA, 67; Univ Tex Med Br Galveston, MA, 69, PhD(human genetics), 71. Prof Exp: Res scientist assoc biochem genetics, Dept Zool, Univ Tex, Austin, 62-63; res asst, Dept of Genetics, Rockefeller Inst, 63-65; res scientist assoc, Genetics Found, Dept Zool, Univ Tex, Austin, 65-68; res scientist, 68-71, instr, 71-73, ASST PROF HUMAN BIOCHEM GENETICS, UNIV TEX MED BR & GRAD SCH BIOMED SCI, GALVESTON, 73- Concurrent Pos: Nat Found March of Dimes & Cystic Fibrosis Found grants, Univ Tex Med Br Galveston, 74-76, USPHS grant, 74-77. Honors & Awards: Award for Excellence in Res, Sigma Xi, 72. Mem: AAAS; Am Soc Human Genetics; Am Chem Soc. Res: Biochemical basis for inherited disease; genetic control of protein structure and function. Mailing Add: Dept Human Biol Chem & Genetics Univ of Tex Med Br Galveston TX 77550

BARNETT, DOUGLAS ELDON, b Walnut Ridge, Ark, Feb 11, 44; m 73. ENTOMOLOGY. Educ: Ark State Univ, BSE, 67; Northwestern State Univ, MS, 70; Univ Ky, PhD(entom), 74. Prof Exp: Lab asst biol, Northwestern State Univ, 67-68; sci instr, Maynard Pub Schs, 68-69; instr biol, Paragould Pub Schs, Ark, 69-70; researcher systs, Dept Entom, Univ Ky, 70, surv entomologist, 71-76; AGRICULTURIST-ENTOMOLOGIST, ANIMAL & PLANT HEALTH INSPECTION SERV, USDA, 76- Mem: Entom Soc Am; Speleol Soc; Sigma Xi. Res: Revision of various genera in the Cicadellidae; taxonomic methods, cave ecology. Mailing Add: Nat Prog Planning Staff USDA/APHIS/PPQ New Pest Detect Hyattsville MD 20782

BARNETT, EUGENE VICTOR, b New York, NY, Mar 28, 32; m 58; c 4. BIOLOGY, MEDICINE. Educ: Univ Buffalo, BA, 55, MD, 56; Am Bd Internal Med, dipl, 63. Prof Exp: Asst resident med, Rochester Med Ctr, 59-60; sr instr med & microbiol, Sch Med & Dent, Univ Rochester, 62-63, asst prof med & sr instr microbiol, Med Ctr, 63-65; assoc prof, 65-70, PROF MED, CTR HEALTH SCI, UNIV CALIF, LOS ANGELES, 70-, DIR INST REHAB & CHRONIC DIS, 71- Concurrent Pos: USPHS trainee med & immunol, 60-61; fel rheumatic dis, Med Res Coun Rheumatism Ctr, Taplow, Eng, 61-62; fel med & Microbiol, Univ Rochester, 62-63; dir immunol sect, Rochester Gen Hosp, 63-65. Mem: Am Fedn Clin Res; Am Rheumatism Asn; Am Asn Immunol; Am Soc Clin Invest; Am Col Physicians. Res: Immunology; internal medicine. Mailing Add: Dept of Med Univ of Calif Los Angeles CA 90024

BARNETT, GENE PAUL, theoretical chemistry, see 12th edition

BARNETT, GUY OCTO, b Chula Vista, Calif, Sept 18, 30; m 58; c 3. MEDICINE. Educ: Vanderbilt Univ, BA, 52; Harvard Univ, MD, 56. Prof Exp: Intern med, Peter Bent Brigham Hosp, 56-57, resident, 57-58 & 60-61, jr assoc, 62-64; clin assoc biophys, Nat Heart Inst, 58-60; asst med, 64-68, ASSOC PHYSICIAN, MASS GEN HOSP, 68-; ASSOC PROF MED, HARVARD MED SCH, 68- Concurrent Pos: Res fel, Harvard Univ, 57-58; spec res fel, Univ Wash, 61-62; lectr elec eng, Mass Inst Technol, 72- Mem: Inst Elec & Electronics Engrs; Am Fedn Clin Res; Asn Comput Mach; Biomed Eng Soc; Am Col Physicians. Res: Computer applications in patient care; computer-aided instruction in medical education. Mailing Add: Lab of Comput Sci Mass Gen Hosp Boston MA 02114

BARNETT, H J M, b Newcastle-on-Tyne, Eng, Feb 10, 22; Can citizen; m 46; c 4. NEUROLOGY. Educ: Univ Toronto, MD, 44; FRCP(C), 52. Prof Exp: Assoc prof neurol, Univ Toronto, 67-69; PROF NEUROL & CHMN DIV, UNIV WESTERN ONT, 69- Mailing Add: Dept of Neurol Univ of Western Ont London ON Can

BARNETT, HENRY LEWIS, b Detroit, Mich, June 25, 14; m 40; c 2. PEDIATRICS. Educ: Wash Univ, BS & MD, 38; Am Bd Pediat, dipl. Prof Exp: Instr pediat, Sch Med, Wash Univ, 41-43; from asst prof to assoc prof, Med Col, Cornell Univ, 46-55; prof & head dept, 55-69, assoc dean clin affairs, 69-72, UNIV PROF PEDIAT, ALBERT EINSTEIN COL MED, 72- Concurrent Pos: Vis pediatrician, Bronx Munic Hosp, Ctr, 55- Honors & Awards: Mead-Johnson Award, 49. Mem: AAAS; Am Soc Clin Invest; Am Pediat Soc; Am Physiol Soc; Soc Pediat Res. Res: Physiology of infants and children with particular reference to kidney physiology and electrolyte metabolism as affected by growth; pediatric clinical epidemiology; pediatric and medical education. Mailing Add: Albert Einstein Col of Med 1300 Morris Park Ave Bronx NY 10461

BARNETT, HERALD ALVA, b Orlando, WVa, Oct 2, 23; m 47; c 2. PHYSICAL CHEMISTRY. Educ: Salem Col, BS, 44; WVa Univ, MS, 50. Prof Exp: SECT HEAD ABSORPTION & MASS SPECTROMETRY, APPL RES LAB, US STEEL CORP, 50- Mem: Am Chem Soc; Am Soc Testing & Mat; Coblentz Soc; Am Phys Soc. Res: Applications of mass spectrometry; absorption and fluorescence spectrophotometry; atomic absorption spectrometry; gas chromatography to coal-chemical recovery processes, steel-making processes and air and water pollution abatement studies. Mailing Add: 106 Arlene Dr North Versailles PA 15137

BARNETT, HERBERT CHESTER, b Brooklyn, NY, Nov 19, 17; m 43, 75; c 3. MEDICAL ENTOMOLOGY. Educ: Cornell Univ, BS, 39; Univ Minn, MS, 46; Univ Pittsburgh, MPH, 53, PhD(epidemiol), 54. Prof Exp: Assoc prof med, 62-65, dir inst int med, 65-66, PROF INT MED, SCH MED, UNIV MD, BALTIMORE, 65-, DIR DIV MED ENTOM & ECOL, 62-65, 66- Concurrent Pos: Adj prof, Am Univ, 57-60; dir, Pakistan Med Res Ctr, Lahore, 66-68 & Brazilian-Am Biomed Prog, Bahia, Brazil, 71-75; on leave, Inst Microbiol, Fed Univ Rio de Janeiro, 76- Mem: AAAS; Am Soc Trop Med & Hyg; Entom Soc Am; Am Inst Biol Sci. Res: Ecology of arthropod-borne viral agents; malariology; biology, ecology and control of medically important arthropods. Mailing Add: Div of Med Entom and Ecol Univ of Md Sch of Med Baltimore MD 21201

BARNETT, HOMER GARNER, b Bisbee, Ariz, Apr 25, 06; m 41; c 2. ANTHROPOLOGY, APPLIED ANTHROPOLOGY. Educ: Stanford Univ, AB, 27; Univ Calif, PhD(anthrop), 38. Prof Exp: Instr anthrop, Univ NMex, 38-39; from instr to prof, 39-71, EMER PROF ANTHROP, UNIV ORE, 71- Concurrent Pos: Ethnologist, Smithsonian Inst, 44-46; Nat Acad Sci coord investr anthrop, Micronesia, 47-48; staff anthropologist, Territory Pac Islands, 51-53; mem, Res Coun, S Pac

Comn, 52-53; vis prof, Harvard Univ, 54, Columbia Univ, 57, Univ Hawaii, 58 & 67, Univ BC, 60, Stanford Univ, 64 & Univ Colo, 71; consult, Neth New Guinea Govt, 55; NSF sr fel, Columbia Univ, 56-57; fel, Ctr Study Behav Sci, 64-65. Honors & Awards: Distinguished Serv Award, Ore Acad Sci, 71. Mem: AAAS; Am Anthrop Asn; Soc Appl Anthrop (pres, 61); Am Ethnol Soc; hon fel Asn Social Anthrop Oceania. Res: Processes of social and cultural change. Mailing Add: Dept of Anthrop Univ of Ore Eugene OR 97403

BARNETT, HORACE LESLIE, b Indianapolis, Ind, Feb 19, 09; m 38; c 3. MYCOLOGY. Educ: DePauw Univ, AB, 31; NDak Agr Col, MS, 33; Mich State Col, PhD(mycol, plant path), 37. Prof Exp: Instr bot, Mich State Col, 37-38; instr biol, NMex State Col, 38-39; asst prof bot, Mich State Col, 39-40; instr biol, NMex State Col, 40-41; asst prof biol, NMex State Col, 41-43; plant pathologist, Emergency Plant Dis Prev Proj, USDA, Calif, 43-44; vis assoc prof bot, Univ Iowa, 44-45; assoc prof mycol, 45-48, mycologist, Agr Exp Sta & prof, 48-74, chmn dept plant path, 60-69, EMER PROF MYCOL, W VA UNIV, 74- Mem: Mycol Soc Am (pres, 63); Am Phytopath Soc (pres, 73); Bot Soc Am. Res: Imperfect fungi; mycoparasitism. Mailing Add: Dept of Plant Path WVa Univ Morgantown WV 26506

BARNETT, JAMES P, b Mena, Ark, May 19, 35; m 65; c 2. PLANT PHYSIOLOGY, FORESTRY. Educ: La State Univ, BSF, 57, MF, 63; Duke Univ, DF, 68. Prof Exp: RES FORESTER, SOUTHERN FOREST EXP STA, US FOREST SERV, 57- Mem: Soc Am Foresters; Am Soc Plant Physiol. Res: Seed dormancy and physiology in the genus Pinus; artificial regeneration of forest stands. Mailing Add: 1236 Canterbury Dr Alexandria LA 71301

BARNETT, JOHN DEAN, b Payson, Utah, May 21, 30; m 54; c 6. PHYSICS. Educ: Univ Utah, BA. 54, PhD(physics). 59. Prof Exp: Assoc prof, 58-65, PROF PHYSICS, BRIGHAM YOUNG UNIV, 65- Res: Ultra high pressure; high temperature physics. Mailing Add: 621 E Sagewood Ave Provo UT 84601

BARNETT, KENNETH WAYNE, b Mobile, Ala, Apr 8, 40; m 67; c 2. INORGANIC CHEMISTRY, ORGANIC CHEMISTRY. Educ: Univ Tex, Arlington, BS, 63; Univ Wis, PhD(inorg chem), 67. Prof Exp: Res chemist, Shell Develop Co, 67-70; ASST PROF CHEM, UNIV MO-ST LOUIS, 70- Mem: Am Chem Soc; The Chem Soc. Res: Organometallic chemistry of the transition elements; organic reactions catalyzed by transition metals; inorganic photochemistry. Mailing Add: Dept of Chem Univ of Mo St Louis MO 63121

BARNETT, LELAND BRUCE, b Los Angeles, Calif, Aug 21, 35; m 61; c 2. PHYSIOLOGICAL ECOLOGY. Educ: Brigham Young Univ, BA, 62, MS, 64; Univ Ill, PhD(zool), 68. Prof Exp: Asst prof, 68-73, ASSOC PROF BIOL, WAYNESBURG COL, 68- Mem: Ecol Soc Am; Am Ornith Union; Cooper Ornith Soc; Am Inst Biol Sci. Res: Ecology; cold temperature adaptations. Mailing Add: Dept of Biol Waynesburg Col Waynesburg PA 15370

BARNETT, LEWIS BRINKLY, b Lexington, Ky, Jan 29, 34; m 56; c 4. PHYSICAL BIOCHEMISTRY. Educ: Univ Ky, BS, 55; Univ Iowa, MS, 57, PhD(biochem), 59. Prof Exp: NSF fel phys chem, Univ Wis, 59-60, univ res fel, 60-61; Am Heart Asn advan res fel, Van't Hoff Lab, Neth, 61-63; asst prof, 63-67, ASSOC PROF BIOCHEM & NUTRIT, VA POLYTECH INST & STATE UNIV, 67- Mem: Am Chem Soc; Am Soc Biol Chem; Sigma Xi. Res: Protein chemistry. Mailing Add: Dept of Biochem & Nutrit Va Polytech Inst & State Univ Blacksburg VA 24061

BARNETT, NEAL MASON, b Lafayette, Ind, May 31, 37; m 60; c 2. PLANT PHYSIOLOGY. Educ: Purdue Univ, BS, 59; Duke Univ, PhD(bot), 66. Prof Exp: Res assoc plant physiol, Purdue Univ, 65-67; asst prof, 68-75, ASSOC PROF PLANT PHYSIOL, UNIV MD. COLLEGE PARK, 75- Concurrent Pos: NIH fel. Mich State Univ & AEC Plant Res Lab, 67-68. Mem: AAAS; Am Soc Plant Physiol; Japanese Soc Plant Physiologists. Res: Biochemistry of water stress in plants; plant cell wall protein; nitrogen assimilation. Mailing Add: Dept of Bot Univ of Md College Park MD 20742

BARNETT, ORTUS WEBB, JR, b Pine Bluff, Ark. July 2, 39; m 64; c 1. PLANT VIROLOGY, PLANT PATHOLOGY. Educ: Univ Ark, BSA, 61, MS, 65; Univ Wis, PhD(plant path), 69. Prof Exp: Sr sci officer, Scottist Hort Res Inst. 68-69; asst prof plant virol & path, 69-74, ASSOC PROF PLANT PATH & PHYSIOL, CLEMSON UNIV, 74- Mem: Am Phytopath Soc; Int Soc Plant Path. Res: White clover viruses; virus purification and serology; plant virus inactivation in vitro; free amino acids in virus-infected plants. Mailing Add: Dept of Plant Path & Physiol Clemson Univ Clemson SC 29631

BARNETT, PAUL EDWARD, b Jerryville, WVa, Mar 19, 36; m 59; c 2. FOREST GENETICS. Educ: WVa Univ, BS, 64; Univ Tenn, MS, 67, PhD(genetics), 72. Prof Exp: Res asst forest genetics, Forestry Dept, Univ Tenn, 64-69; botanist forest physiol, 69-73, staff forester forest genetics, 73-74, STAFF FORESTER & PROJ LEADER FOREST GENETICS, DIV FORESTRY, FISHERIES & WILDLIFE DEVELOP, TENN VALLEY AUTHORITY, 74- Mem: Soc Am Foresters; Sigma Xi; AAAS. Res: Genetics and breeding of forest trees with superior growth and form; geographic and altitudinal variation; establishment of seed production areas and hardwood plantation establishment and growth. Mailing Add: Forestry Fish & Wildlife Dev Tenn Valley Authority Norris TN 37828

BARNETT, RONALD DAVID, b Texarkana, Ark, Nov 20, 43; m 61; c 3. AGRONOMY, PLANT GENETICS. Educ: Univ Ark, BS, 65. MS, 68; Purdue Univ, PhD(genetics, plant breeding), 70. Prof Exp: ASSOC PROF AGRON, UNIV FLA, 70- Mem: Am Soc Agron; Crop Sci Soc Am; Am Genetic Asn; AAAS. Res: Plant breeding and genetics with emphasis on varietal development in wheat, barley, rye and oats. Mailing Add: Agr Res & Educ Ctr Univ of Fla PO Box 470 Quincy FL 32351

BARNETT, RONALD E, b Pueblo, Colo, Dec 26, 42; m 63; c 1. PHYSICAL ORGANIC CHEMISTRY, ENZYMOLOGY. Educ: Univ Colo, BA, 65; Brandeis Univ, PhD(biochem), 69. Prof Exp: Res assoc biochem, Stanford Univ, 68-69; ASST PROF CHEM, UNIV MINN, MINNEAPOLIS, 69- Concurrent Pos: NSF fel, 68-69; Merck Co Found Faculty develop award, 69; NIH res grant, 70- Mem: AAAS; Am Chem Soc. Res: Carbonyl and acyl group reactions with diffusion limited rate determining steps; models for enzymatic reactions; conformational changes in enzymes; mechanisms of active transport of cations through cell membranes. Mailing Add: Dept of Chem Univ of Minn Minneapolis MN 55455

BARNETT, STOCKTON GORDON, III, b East Orange, NJ, July 18, 39; m 66. GEOLOGY. Educ: Dartmouth Col, BA, 61; State Univ Iowa, MS, 63; Ohio State Univ, PhD(geol), 66. Prof Exp: Asst prof geol, 66-73, ASSOC PROF EARTH SCI, STATE UNIV NY COL PLATTSBURGH, 66- Concurrent Pos: State Univ NY Res Found grants-in-aid, 67-71. Mem: Soc Econ Paleont & Mineral; Paleont Soc. Res: Application of biometrical techniques to the study of evolution of invertebrate fossils, particularly conodonts; paleoecology; carbonate petrology. Mailing Add: Dept of Physics & Earth Sci State Univ of NY Plattsburgh NY 12901

BARNETT, THOMAS BUCHANAN, b Lewisburg, Tenn, May 27, 19; m 44; c 3. MEDICINE. Educ: Univ Tenn, BA, 44; Univ Rochester, MD, 49. Prof Exp: Instr internal med, Univ Rochester, 51-52; from instr to assoc prof, 52-64, PROF INTERNAL MED, SCH MED, UNIV NC, CHAPEL HILL, 64- Concurrent Pos: Vis prof & researcher, Copenhagen Univ, 66-67. Mem: Am Thoracic Soc; Am Col Physicians; Am Fedn Clin Res; Am Physiol Soc. Res: Pulmonary diseases and clinical pulmonary physiology; disorders of control of respiration. Mailing Add: Sch of Med Univ of NC Chapel Hill NC 27514

BARNETT, WILLIAM ARNOLD, b Boston, Mass, Oct 30, 41; m 69. ECONOMIC STATISTICS, MATHEMATICAL STATISTICS. Educ: Mass Inst Technol, BS, 63; Univ Calif, Berkeley, MBA, 65; Carnegie-Mellon Univ, PhD(statist), 74. Prof Exp: Res engr, Rocketdyne Div, Rockwell Int Corp, 63-69; RES ECONOMIST, BD GOVRS FED RESERVE SYST, 73- Mem: Sigma Xi; Am Statist Asn; Econometric Soc; Am Econ Asn; Fedn Am Scientists. Res: Development of asymptotic theory of inference with nonlinear models; demand theory and empirical systems of consumer demand functions; construction of food price forecasting model. Mailing Add: Apt 402 5055 S Chesterfield Rd Arlington VA 22206

BARNETT, WILLIAM EDGAR, b Kempner, Tex, Aug 10, 34; m 54; c 3. MOLECULAR BIOLOGY. Educ: Southwestern Univ, Tex, BS, 56; Northwestern Univ, MS, 57; Fla State Univ, PhD(biol genetics), 61. Prof Exp: Instr genetics, Fla State Univ, 61; USPHS fel, 61-63, biologist, 63-69, SCI DIR GENETICS & DEVELOP BIOL, OAK RIDGE NAT LAB, 69- Concurrent Pos: Vis lectr, Univ Tenn, 62-67, prof, 67- Mem: AAAS; Genetics Soc Am; Am Soc Biol Chem. Res: Genetic translational apparatus of cellular organelles. Mailing Add: Biol Div Oak Ridge Nat Lab Box Y Oak Ridge TN 37830

BARNETT, WILLIAM OSCAR, b Tuscola, Miss, Sept 20, 22; m 47; c 3. SURGERY. Educ: Univ Miss, BS, 44; Univ Tenn, MD, 46; Univ Md, 51; Am Bd Surg, dipl, 56. Prof Exp: Intern, Tampa Gen Hosp, Fla, 46-47; resident surg, Lutheran Hosp, Baltimore, Md, 50-53; assoc prof, 63-67, PROF SURG, SCH MED, UNIV MISS, 67- Concurrent Pos: Mem, Miss Comn Hosp Care, 62-69, vchmn, 67-68, chmn, 68-69; vis lectr surg, Med Ctr, La State Univ, 67, Sch Med, Emory Univ, 67, Med Col SC, 68 & Lubbock Med Soc, 74. Mem: Am Surg Asn; Soc Univ Surg; Soc Surg Alimentary Tract; Int Soc Surg; Am Col Surg (chmn, 70-). Res: Shock in late gangrenous bowel obstruction; pre-operative management of strangulation obstruction. Mailing Add: Dept of Surg Univ of Miss Med Ctr Jackson MS 39216

BARNEY, ARCHIE FAY, b Kanosh, Utah, July 4, 92; m 17; c 4. AGRONOMY. Educ: Utah Col, BS, 20; Cornell Univ, PhD(plant breeding), 24. Prof Exp: Plant breeding, Exp Sta, Utah Col, 24; AGRONOMIST, DEPT AGR RES, AM SMELTING & REFINING CO, 25- Res: Inheritance of smut resistance in oats; effects of smelter fumes and gases on vegetation. Mailing Add: 1256 Crystal Ave Salt Lake City UT 84106

BARNEY, ARTHUR LIVINGSTON, b Homer, Minn, Apr 4, 18; m 42; c 4. POLYMER CHEMISTRY. Educ: Middlebury Col, BS, 38; Syracuse Univ, MS, 40; Purdue Univ, PhD(org chem), 43. Prof Exp: Asst chem, Syracuse Univ, 38-40 & Purdue Univ, 40-41; res chemist, Cent Res Dept, 42-61, res assoc, Elastomer Chem Dept, 61-62, RES DIV HEAD, ELASTOMER CHEM DEPT, E I DU PONT DE NEMOURS & CO, INC, 62- Mem: Am Chem Soc. Res: Organic synthesis; fluorine chemistry; chemistry of high polymers; catalysis. Mailing Add: Elastomer Chem Dept E I du Pont de Nemours & Co Wilmington DE 19898

BARNEY, CHARLES WESLEY, b Brewster, NY, Apr 17, 15; m 43; c 3. FOREST ECOLOGY, SILVICULTURE. Educ: Syracuse Univ, BS, 38; Univ Vt, MS, 39; Duke Univ, DF(root growth), 47. Prof Exp: Asst agr aide, Soil Conserv Serv, USDA, 41; from asst prof to assoc prof, 47-51, head dept forest mgt & utilization, 53-65, PROF SILVICULTURE, COLO STATE UNIV, 65- Mem: Soc Am Foresters; Soil Conserv Soc Am; Am Soc Photogram; Forest Hist Soc; Sigma Xi. Res: Arid zone forestry; seed physiology. Mailing Add: Dept of Forest & Wood Sci Colo State Univ Ft Collins CO 80521

BARNEY, DUANE LOWELL, b Topeka, Kans, Aug 3, 28; m 50; c 2. INORGANIC CHEMISTRY. Educ: Kans State Univ, BS, 50; Johns Hopkins Univ, MA, 51, PhD(chem), 53. Prof Exp: Instr, Kans State Univ, 50; jr instr chem, Johns Hopkins Univ, 50-52; res assoc, Knolls Atomic Power Lab, 53-57, res chemist, 57-66, mgr battery tech lab, Battery Bus Sect, 66-68, mgr eng, 68-72, mgr battery bus sect, 72-74, GEN MGR HOME LAUNDRY PROD, ENG DEPT, GEN ELEC CO, 74- Mem: Am Chem Soc. Res: Phosphate chemistry; fission product chemistry; corrosion; inorganic insulation; materials development; battery technology. Mailing Add: Gen Elec Co Appliance Park 1 Gainesville FL 32601

BARNEY, GARY SCOTT, b Monroe, Utah, Sept 17, 42; m 63; c 4. PHYSICAL INORGANIC CHEMISTRY. Educ: Brigham Young Univ, BS, 64, PhD(inorg chem), 70. Prof Exp: RES STAFF CHEMIST, ATLANTIC RICHFIELD HANFORD CO, 68- Mem: Am Chem Soc. Res: Plutonium process chemistry; development of chemical processes for conversion of radioactive wastes from nuclear fuel processing to forms which can be safely stored. Mailing Add: Atlantic Richfield Hanford Co 234-5 Bldg 200 W Area Richland WA 99352

BARNEY, GERALD O, operations research, physics, see 12th edition

BARNEY, JAMES EARL, II, b Rossville, Kans, Sept 1, 26; m 46; c 2. ANALYTICAL CHEMISTRY. Educ: Univ Kans, BS, 46, PhD(anal chem), 50. Prof Exp: Mem staff, AEC, Oak Ridge, 46-47; res chemist, Stand Oil Co, Ind, 50-54, group leader, 55-60; res scientist, Spencer Chem Co, Kans, 60-62; from sr chemist to prin chemist, Midwest Res Inst, 62-65, head anal chem sect, 65-69; sect mgr, Western Res Ctr, 69-74, SR SECT MGR, WESTERN RES CTR, STAUFFER CHEM CO, RICHMOND, 74- Concurrent Pos: Instr, Purdue Univ, 51-53; vis lectr, Univ Kans, 61. Mem: Am Chem Soc; Asn Off Anal Chemists; Am Soc Testing & Mat. Res: Pesticide residue analysis; gas chromatography; applied spectroscopy; spectrophotometry; determination of anions; physical testing of polymers; rheology of gluten; spot tests; complex ions; analysis of spices; analysis of medical gases; air pollution. Mailing Add: 1147 Sanders Dr Moraga CA 94556

BARNHARD, HOWARD JEROME, b New York, NY, July 18, 25; m 47; c 3. MEDICINE, RADIOLOGY. Educ: Univ Miami, BS, 44; Med Col SC, MD, 49; Am Bd Radiol, dipl, 55. Prof Exp: Intern, US Naval Hosp, Charleston, SC, 49-50; resident radiol, Roper Hosp, 50-51; radiologist, US Naval Hosp, Quantico, Va, 52-53; resident radiol, Roper Hosp, 53-54; from instr to asst prof radiol, Med Ctr, Univ Ark, 54-59; asst prof, Hahnemann Med Col, 59-60; chmn dept, 60-73, PROF RADIOL, MED CTR, UNIV ARK, LITTLE ROCK, 60-, DIR PLANNING, 73- Concurrent Pos: Teaching fel radiol, Med Col SC, 53-54; attend radiologist, Vet Admin Hosp, Little

Rock, Ark, 54-59, consult, 60- Mem: Radiol Soc NAm; fel Am Col Radiol; Asn Univ Radiol; Am Roentgen Ray Soc. Res: Osseus system; medical computer applications. Mailing Add: Dept of Radiol Univ of Ark Med Sci Little Rock AR 72201

BARNHART, BENJAMIN J, b Winchester, Ind, July 27, 35; m 57; c 2. GENETICS. Educ: Ind Univ, AB, 58, AM, 59; Johns Hopkins Univ, ScD(biochem), 62. Prof Exp: Teaching asst zool, Ind Univ, 58-59; NIH trainee biochem, Johns Hopkins Univ, 62-63; Nat Inst Allergy & Infectious Dis fel, Lab Genetics, Brussels, 63-64; STAFF MEM CELLULAR BIOL, LOS ALAMOS SCI LAB, UNIV CALIF, 64- Concurrent Pos: Nat Acad Sci-Nat Res Coun travel grant, Int Biophys Cong, Austria, 66. Mem: Am Soc Microbiol; fel Am Inst Chemists; Am Soc Cell Biol. Res: Bacterial and bacteriophage genetics; photobiology; mammalian cell genetics; cell biology. Mailing Add: Biomed Res H-9 Los Alamos Sci Lab PO Box 1663 Los Alamos NM 87544

BARNHART, CHARLES ELMER, b Windsor, Ill, Jan 25, 23; m 46; c 3. ANIMAL NUTRITION. Educ: Purdue Univ, BS, 45; Iowa State Univ, MS, 48, PhD(animal nutrit), 54. Prof Exp: Instr, Iowa State Univ, 47-48; instr, 48-50, from asst prof to assoc prof, 50-57, assoc dir agr exp sta, 62-67, assoc dean col agr, 66-69, PROF ANIMAL HUSB, UNIV KY, 57-, DIR AGR EXP STA, 67-, DEAN & DIR COL AGR, 69- Mem: Am Soc Animal Sci. Res: Swine management, production, breeding and nutrition research. Mailing Add: Col of Agr Univ of Ky Lexington KY 40506

BARNHART, CLYDE STERLING, SR, b Clark Co, Ohio, May 1, 16; m 44; c 2. ENTOMOLOGY. Educ: Ohio State Univ, BSc, 42, MSc, 50, PhD, 58; Am Registry Prof Entomologists, cert med entom. Prof Exp: Med Entomologist, US Army Mobility Equip Res & Develop Labs, Ft Belvoir, Va, 59-63; med entomologist, US Army Land Warfare Lab, Aberdeen Proving Ground, Md, 63-73; RETIRED. Mem: Entom Soc Am; Am Mosquito Control Asn. Res: Pest control equipment; mosquito control; Thysanura. Mailing Add: Rte 4 Box 207A Athens OH 45701

BARNHART, DAVID M, b Wenatchee, Wash, May 28, 33; m 54; c 3. PHYSICAL CHEMISTRY. Educ: Univ Ore, BS, 57, MS, 59; Ore State Univ, PhD(chem), 64. Prof Exp: Fel, Harvey Mudd Col, 64-65; ASSOC PROF PHYSICS, EASTERN MONT COL, 65- Mem: Am Phys Soc; Am Crystallog Asn; Am Chem Soc. Res: Molecular structure studies by electron and x-ray diffraction. Mailing Add: Dept of Sci Eastern Mont Col Billings MT 59101

BARNHART, DONALD DELBERT, b Laona, Wis, Apr 26, 36; m 60; c 3. MEDICAL MICROBIOLOGY. Educ: Univ Wis-Milwaukee, BS, 63, Univ Wis-Madison, MS, 68, PhD(med microbiol), 71. Prof Exp: Fel immunol, Oak Ridge Assoc Univs, 71-73; ASST PROF MICROBIOL, MIAMI UNIV, 73- Mem: Am Soc Microbiol; Sigma Xi. Res: Immunosuppression by microorganisms, the specific bacterial components and/or metabolic products involved and their site of action, for example, T or B cells; applied research in rapid measurement of primary antigen-antibody interaction. Mailing Add: Dept of Microbiol Miami Univ Oxford OH 45056

BARNHART, JAMES WILLIAM, b Wadena, Minn, Apr 23, 35; m 56; c 3. BIOCHEMISTRY. Educ: Columbia Union Col, BA, 57; Univ Md, PhD(biochem), 63. Prof Exp: CHEMIST, DOW CHEM CO, 62- Res: Biochemistry and methodology of lipids and their relationship to atherosclerosis; drug metabolism; trace drug analysis; bioavailability of drugs. Mailing Add: Dow Chem Co Box 68511 Health & Consumer Prod Dept Indianapolis IN 46268

BARNHART, JOHN LOVE, b Greensburg, Pa, Sept 20, 08; m 38; c 3. DAIRY SCIENCE. Educ: Pa State Univ, BS, 30, PhD(dairy tech), 40; Univ WVa, MS, 32. Prof Exp: Assoc prof dairy mfg. Okla Agr & Mech Col, 40-41 & Kans State Col, 45-47; prof food tech, Nat Agr Col, 47-48, dean sch agr, 48-49; dir sci & head dept dairy tech, Temple Univ, 49-53; tech dir. Dairy Industs Supply Asn, 53-57; assoc prof dairy mfg & assoc food scientist, 57-75, EMER PROF FOOD SCI, UNIV IDAHO, 75- Mem: AAAS; Am Chem Soc; Am Dairy Sci Asn. Res: Dairy chemistry; bacteriology; food technology; dairy engineering. Mailing Add: Dept of Food Sci Univ of Idaho Moscow ID 83843

BARNHART, MARION ISABEL, b Webb City, Mo, Sept 23, 21. PHYSIOLOGY. Educ: Univ Mo, AB, 44, PhD, 50; Northwestern Univ, MS, 46. Prof Exp: Instr zool, Univ Mo, 46-49; from instr to assoc prof, 50-67, PROF PHYSIOL, SCH MED, WAYNE STATE UNIV, 67- Mem: AAAS; Am Soc Hemat; Am Physiol Soc; Soc Exp Biol & Med; Am Acad Neurol. Res: Cell physiology of microvasculature, liver, bone marrow, spleen and blood in relation to hemostasis, thrombosis and fibrinolysis; inflammation; pathophysiology of platelets and hemolytic anemias; scanning electron microscopy applied to medicine. Mailing Add: Wayne State Univ Sch of Med Detroit MI 48201

BARNHART, WILLIAM SIDDALL, b Dayton, Ohio, Apr 14, 18; m 42; c 3. ORGANIC POLYMER CHEMISTRY. Educ: Allegheny Col, AB. 40; Cornell Univ, MS, 42; Purdue Univ, PhD(org chem), 48. Prof Exp: Res chemist, Imp Paper & Color Corp, 42-43 & Firestone Tire & Rubber Co, 47-50; res supvr org chem, M W Kellogg Co, 50-57; dir polymer res, 57-73, TECH ADV, RES & DEVELOP, PENNWALT CORP, 73- Mem: AAAS; Soc Plastic Eng; Am Chem Soc; Sigma Xi. Res: Fluorine-containing chemicals and polymers; oiomedical polymers. Mailing Add: 4117 Hain Dr Lafayette Hill PA 19444

BARNHILL, MAURICE VICTOR, III, b Rocky Mount, NC, Mar 16, 40. THEORETICAL PHYSICS. Educ: Univ NC, Chapel Hill, BS, 62; Stanford Univ, MS, 65, PhD(physics), 67. Prof Exp: Res assoc physics, Univ Va, 67-68; asst prof, 68-74, ASSOC PROF PHYSICS. UNIV DEL, 74- Mem: Am Phys Soc; Am Asn Physics Teachers. Res: Nuclear and high energy theory. Mailing Add: Dept of Physics Univ of Del Newark DE 19711

BARNHILL, ROBERT E, b Lawrence, Kans, Oct 31, 39; m 63. MATHEMATICAL ANALYSIS. Educ: Univ Kans, BA, 61; Univ Wis, MA, 62, PhD(math), 64. Prof Exp: From asst prof to assoc prof, 64-71, PROF MATH, UNIV UTAH, 71- Mem: Am Math Soc; Math Asn Am; Soc Indust & Appl Math; Asn Comput Mach. Res: Numerical analysis; computer aided geometric design. Mailing Add: Dept of Math Univ of Utah Salt Lake City UT 84112

BARNHISEL, RICHARD I, b Peru, Ind, Mar 1, 38; m 58; c 1. SOIL MINERALOGY, SOIL CHEMISTRY. Educ: Purdue Univ, BS, 60; Va Polytech Inst, MS. 62, PhD(soil mineral), 65. Prof Exp: Asst prof soil mineral, 64-69, ASSOC PROF SOIL MINERAL, UNIV KY, 69- Mem: Am Soc Agron; Soil Sci Soc Am; Clay Minerals Soc. Res: Formation and stability of hydroxy-aluminum interlayers in clay minerals; weathering of primary minerals in clay formation; reclamation methods of surface-mined coal spoils. Mailing Add: Dept of Agron Univ of Ky Lexington KY 40506

BARNOSKI, MICHAEL K, b Williamsport, Pa, Aug 19, 40; m 63; c 2. SOLID STATE PHYSICS. Educ: Univ Dayton, BSEE, 63; Cornell Univ, MS, 65, PhD(solid state physics), 68. Prof Exp: Asst res engr, Res Inst, Univ Dayton, 63; res asst microwave electronics, Cornell Univ, 63-65, infrared spectros, 65-68; prin sr res engr, Honeywell Radiation Ctr, 68-69; MEM TECH STAFF, HUGHES RES LABS, HUGHES AIRCRAFT CO, 69- Mem: Optical Soc Am. Res: Infrared spectra of solids; Fourier transform spectroscopy; coherence properties of radiation; microwave electronics; lasers. Mailing Add: Chem Physics Dept Hughes Res Labs Malibu CA 90265

BARNOTHY, JENO MICHAEL, b Kassa, Hungary, Oct 28, 04; nat US; m 38. NUCLEAR PHYSICS, ASTROPHYSICS. Educ: Royal Hungarian Univ, PhD(physics), 33. Prof Exp: From instr to prof physics, Royal Hungarian Univ, 35-48; prof, Barat Col, 48-53; res physicist, Nuclear Instrument & Chem Corp, 53-55; OWNER & TECH DIR, FORRO SCI CO, 55-; PRES, BIOMAGNETIC RES FOUND, 60- Honors & Awards: Medal, Royal Hungarian Acad Sci, 39; Eotvos Medal, 48. Mem: Am Astron Soc; Am Phys Soc; Biophys Soc; Ger Astron Soc; Int Astron Union. Res: Cosmic radiation; instrumentation; magnetobiology. Mailing Add: 833 Lincoln St Evanston IL 60201

BARNOTHY, MADELEINE FORRO, b Zsambok, Hungary, Aug 21, 04; nat US; m 38. PHYSICS. Educ: Royal Hungarian Univ, Budapest, PhD(physics), 27. Prof Exp: From asst prof to assoc prof physics, Royal Hungarian Univ, Budapest, 29-48; prof math, Barat Col Sacred Heart, 48-53; from asst prof to prof. 55-72, EMER PROF PHYSICS, COL PHARM, UNIV ILL, 72- Concurrent Pos: Fel, Univ Göttingen, 28-29; res assoc, Northwestern Univ, 53-59. Honors & Awards: Award, Royal Hungarian Acad Sci, 38; Eotvos Medal, Hungarian Acad, 48. Mem: AAAS; Am Phys Soc; Biophys Soc; Am Astron Soc. Res: Cosmic radiation; nuclear physics; biological effect of magnetic fields; cosmological theories; quasars. Mailing Add: 833 Lincoln St Evanston IL 60201

BARNOUW, VICTOR, b The Hague, Neth, May 25, 15; US citizen; m 64. ANTHROPOLOGY, ETHNOLOGY. Educ: Columbia Univ, AB, 40, PhD(anthrop), 48. Prof Exp: Vis asst prof anthrop, Univ Buffalo, 48-51 & Univ Ill, 56-57; asst prof, 57-61, assoc prof, 62-65, PROF ANTHROP, UNIV WIS-MILWAUKEE, 65- Concurrent Pos: Am Inst Indian Studies res grant to study Sindhi community, Poona, 63. Honors & Awards: Stirling Award, Am Anthrop Asn, 68. Mem: Fel Am Anthrop Asn; Am Ethnol Soc. Res: Culture and personality, especially among Chippewa Indians of Wisconsin. Mailing Add: Dept of Anthrop Univ of Wis Milwaukee WI 53201

BARNS, ROBERT L, b Wichita, Kans, Mar 11, 27; m 55; c 1. CRYSTALLOGRAPHY. Educ: Wichita State Univ, BS, 49, MS, 51. Prof Exp: Res chemist, Found Indust Res, Wichita State Univ, 50-55; mem tech staff, Phys Chem Res Dept, 55-60, MEM TECH STAFF, CRYSTAL CHEM RES DEPT, BELL LABS, 60- Mem: Am Crystallog Asn. Res: Combustion; electrical contacts; crystal growth; imperfections in crystals; crystal characterization; x-ray diffraction; petrography. Mailing Add: Crystal Chem Res Dept Bell Labs Murray Hill NJ 07974

BARNSLEY, ERIC ARTHUR, b Birmingham, Eng, July 31, 34; c 2. MICROBIAL PHYSIOLOGY. Educ: Univ London, BSc, 60, PhD(biochem), 63. Prof Exp: ASSOC PROF BIOCHEM, MEM UNIV, 71- Mem: Biochem Soc. Res: Regulation of the degradation of polycyclic armatic hydrocarbons by micro-organisms. Mailing Add: Dept of Biochem Mem Univ St John's NF Can

BARNSTEIN, CHARLES HANSEN, b Newton, Wis, June 25, 25; m 54; c 3. PHARMACEUTICAL CHEMISTRY. Educ: Univ Wis, BSc, 52, MSc, 55, PhD(pharm), 60. Prof Exp: From instr to assoc prof pharm, Idaho State Col, 56-63; assoc prof, Univ Wis-Milwaukee, 63-70; assoc dir, Nat Formulary, Am Pharmaceut Asn, 70-75; SR SCIENTIST. US PHARMACOPEIAL CONV, INC, 75- Mem: AAAS; Am Chem Soc; Am Pharmaceut Asn. Res: Organic analytical chemistry; physical pharmacy and chemistry; chemical kinetics. Mailing Add: 16900 Freedom Way Rockville MD 20853

BARNSTORFF, HENRY DRESES, b Cincinnati, Ohio, July 26, 25; m 50; c 3. INDUSTRIAL ORGANIC CHEMISTRY. Educ: Mo Valley Col, AB, 46; Univ Mo, BS, 48, MA, 49; Univ Colo, PhD(chem), 53. Prof Exp: Asst chem, Univ Mo, 47-49 & Univ Colo, 49-52; staff mem, Res Dept, Spencer Chem Co, 53-56; from res chemist to sr res chemist, Org Div, Monsanto Co, 56-63, res specialist, 63-65, mgr prof employ, 65-67, res staff mgr, 67-71, res specialist, Org Chem Div, 71-72, SR RES SPECIALIST, MONSANTO INDUST CHEM CO, 72- Mem: Am Chem Soc. Res: Organic chemicals process development. Mailing Add: Monsanto Co 800 N Lindbergh Blvd St Louis MO 63166

BARNUM, DENNIS W, b Portland, Ore, June 30, 31; m 54; c 2. CHEMISTRY. Educ: Univ Ore, BA, 53, MA, 55; Iowa State Univ, PhD(chem), 57. Prof Exp: Chemist, Shell Develop Co, 57-64; from asst prof to assoc prof, 64-75, PROF CHEM, PORTLAND STATE UNIV, 75- Mem: Am Chem Soc. Res: Coordination chemistry; geochemistry; soils. Mailing Add: Dept of Chem Portland State Univ Portland OR 97207

BARNUM, DONALD ALFRED, b Ont, June 20, 18; m 42; c 3. VETERINARY SCIENCE, MICROBIOLOGY. Educ: Ont Vet Col, DVM, 41; Univ Toronto, DVPH, 49, DVSc, 52. Prof Exp: Bacteriologist, Ont Dept Health Labs, 41-43; lectr bact, 45-47, from asst prof to assoc prof, 47-58, head dept vet bact, 64-69, PROF BACT, ONT VET COL, UNIV GUELPH, 58-, HEAD DEPT VET MICROBIOL & IMMUNOL, 69- Mem: Am Soc Microbiol; fel Am Acad Microbiol; NY Acad Sci; Am Vet Med Asn; Path Soc Gt Brit & Ireland. Res: Veterinary bacteriology and mycology; bacterial infections of animals, especially of the mammary glands of dairy cattle and gastrointestinal tract of the young. Mailing Add: Ont Vet Col Univ of Guelph Guelph ON Can

BARNUM, EMMETT RAYMOND, b Rushville, Nebr, May 4, 13; m 41; c 1. ORGANIC CHEMISTRY. Educ: Nebr State Teachers Col, BS, 37; Univ Nebr, MS, 39, PhD(org chem), 42. Prof Exp: Asst, Univ Nebr, 37-38; res supvr, Shell Develop Co, 52-75. Mem: Am Chem Soc. Res: Organic synthesis; lubrication and industrial oil; dicarboxylic acids to prevent rusting of steel; sulfur compounds to reduce wear in lubricating oils; synthetic lubricants; polymers; petrochemicals; chemical intermediates. Mailing Add: 6100 Estates Dr Oakland CA 94611

BARNUM, HORACE GARDINER, b New London, Conn, May 10, 34; m 59; c 3. HISTORICAL GEOGRAPHY, URBAN GEOGRAPHY. Educ: Middlebury Col, AB, 57; Univ Chicago, MS, 61, PhD(geog), 65. Prof Exp: Asst prof, 65-69, ASSOC PROF GEOG, UNIV VT, 69- Mem: AAAS; Asn Am Geogr; Am Geog Soc; Nat Coun Geog Educ. Res: Historical geography of Europe and New England; toponyms. Mailing Add: Dept of Geog Univ of Vt Burlington VT 05401

BARNWELL, FRANKLIN HERSHEL, b Chattanooga, Tenn, Oct 14, 37; m 59; c 1. COMPARATIVE PHYSIOLOGY. Educ: Northwestern Univ, BA, 59, PhD(biol sci), 65. Prof Exp: Instr biol sci, Northwestern Univ, 64, res assoc, 65-67; asst prof, Univ Chicago, 67-70; asst prof, 70-73, ASSOC PROF ZOOL, UNIV MINN, MINNEAPOLIS, 73- Concurrent Pos: Mem fac, Orgn Trop Studies, Costa Rica, 66 &

69-71. Res: Persistent daily and tidal rhythms; invertebrate behavior; biochemical systematics and zoogeography of marine crustacea. Mailing Add: Dept of Zool Univ of Minn Minneapolis MN 55455

BAROFF, JAMES HARLEY, theoretical physics, see 12th edition

BAROFSKY, DOUGLAS FRED, b Bremerton, Wash, July 8, 41; m 69; c 2. MASS SPECTROMETRY. Educ: Wash State Univ, BS, 63; Pa State Univ, MS, 65, PhD(physics), 67. Prof Exp: Res assoc physics, Pa State Univ, 67-68; Alexander von Humboldt stiftung, Inst Phys Chem, Univ Bonn, 68-69; asst prof, 69-75, ASSOC PROF PHYSICS & ENVIRON TECHNOL, ORE GRAD CTR, 75- Mem: AAAS; Am Vacuum Soc; Am Soc Mass Spectrometry. Res: Field ionization; field desorption mass spectrometry; surface physics and chemistry; physical electronics. Mailing Add: Ore Grad Ctr 19600 NW Walker Rd Beaverton OR 97005

BARON, ARTHUR L, organic chemistry, polymer chemistry, see 12th edition

BARON, FRANK A, b Brooklyn, NY, May 6, 33; m 56; c 2. ORGANIC CHEMISTRY. Educ: City Col New York, BS, 54; Univ Kans, PhD(org chem), 61. Prof Exp: Chemist, Esso Res & Eng Co, 60-62 & Millmaster-Onyx Corp, 63-65; chief chemist. Cosan Chem Corp, 65-67; sr chemist, Mallinckrodt Chem Works, 67-71, TECH MGR, MALLINCKRODT, INC, 71- Mem: AAAS; Am Chem Soc; Soc Indust Microbiol. Res: Elucidation of chemical reactions, air oxidations of organic compounds; synthesis, processing and purification of medicinals; Baeyer-Villiger and Bamberger rearrangements; reactions of aminophenolic compounds. Mailing Add: 98 Hobart Ave Short Hills NJ 07644

BARON, HARRY, biochemistry, see 12th edition

BARON, LOUIS SOL, b New York, NY, Jan 2, 24; m 61; c 1. BACTERIOLOGY. Educ: City Col New York, BS, 47; Univ Ill, MS, 48, PhD(bact), 51. Prof Exp: Asst bact, Univ Ill, 49-52; bacteriologist, 52-56, CHIEF DEPT BACT IMMUNOL, DIV IMMUNOL, WALTER REED ARMY INST RES, 56- Concurrent Pos: Squibb & Sons fel, 51; President's fel, Soc Bact, 57; Army rep, Genetics Study Sect, USPHS, 58-; spec lectr, Dept Microbiol, Sch Med, George Washington Univ, 61; mem subcomt maintenance of genetic stocks, NSF, 64-68, chmn, 68-; ed, J Bact, 65-70; mem genetics fac, Grad Prog, NIH, 67-69; mem adv comt, Am Type Cult Collection, 69-; prof lectr, Sch Med, Georgetown Univ, 69- Mem: Am Soc Microbiol; Am Asn Immunol; Genetic Soc Am; Biophys Soc; Am Soc Biol Chem. Res: Bacterial genetics, virulence and immunology; enteric diseases; molecular biology. Mailing Add: Dept of Bact Immunol Walter Reed Army Inst of Res Washington DC 20012

BARON, ROBERT, b Chicago, Ill, May 16, 32; m 61; c 2. SOLID STATE PHYSICS. Educ: Univ Chicago, BA, 52, MS, 55, PhD(physics). 62. Prof Exp: SR STAFF PHYSICIST, SOLID STATE PHYSICS, HUGHES RES LABS, 62- Mem: Am Inst Physics; Am Phys Soc; Inst Elec & Electronics Eng. Res: Behavior of insulators and semi-insulators under high injection current conditions; physics of ion-implanted layers in semiconductors; current transport and materials for extrinsic infrared detectors. Mailing Add: Hughes Res Lab 3011 S Maliou Canyon Rd Malibu CA 90265

BARON, ROBERT RICHARD, b Denhoff, NDak, May 15, 30; m 57; c 4. PARASITOLOGY. Educ: Concordia Col, Moorhead, Minn, BA, 52; Kans State Univ, MS, 56, PhD(parasitol), 59. Prof Exp: Asst endocrinol, Kans State Univ, 54-56 & parasitol, 56-59; asst microbiologist, Br Exp Sta, Vet Sci Dept, Univ Idaho, 59-61; parasitol dept mgr, 61-70, govt rels mgr, 71-72, BIOL DEVELOP & GOVT RELS MGR, SALSBURY LABS, 73- Mem: Am Soc Parasitol; Poultry Sci Asn. Res: Animal health. Mailing Add: Salsbury Labs 2000 Rockford Rd Charles City IA 50616

BARON, RONALD L, b New York, NY, Sept 7, 33; m 57; c 1. BIOCHEMISTRY, ENTOMOLOGY. Educ: State Univ NY Col Long Island, ABS, 53; NC State Col, BS, 58, MS, 59; Univ Wis, PhD(entom), 62. Prof Exp: Pharmacologist, Pesticide Res, USFDA, 62-64, biochemist, 64-71; biochemist pesticide progs, 71-73, PHYS SCI ADMINR, OFF RES & DEVELOP, ENVIRON PROTECTION AGENCY, 73- Res: Biochemical effects of pesticide chemicals and their metabolic products; isolation and identification of metabolic fate of pesticides and evaluation of toxicological hazard of the parent compound and the residue products. Mailing Add: Environ Protection Agency Health Effects Res Lab Research Triangle Park NC 27711

BARON, SHIRLEY HAROLD, b Salt Lake City, Utah, Oct 22, 04; m 42; c 3. SURGERY. Educ: Univ Ore, AB, 24; Cornell Univ, MD, 27; Am Bd Otolaryngol, dipl. Prof Exp: Asst clin prof surg, Med Sch, Stanford Univ, 46-56, assoc prof otolaryngol, 56-68; clin prof, 68-74, EMER CLIN PROF OTOLARYNGOL, MED CTR, UNIV CALIF, SAN FRANCISCO, 74- Concurrent Pos: Chief Dept Ear Nose & Throat, Mt Zion Hosp, 46-64; mem exec med staff, French Hosp, 46-, chief Dept Ear, Nose & Throat, 64-; consult otorhinolaryngologist, Letterman Army Hosp, 47- & US Naval Hosp, Oakland, Calif, 70-; consult, Children's Hosp, 59. Honors & Awards: Honor Award, Am Acad Ophthal & Otolaryngol, 64; Distinguished Civilian Serv Medal, US Dept Army, 67; Spec Award, Soc Mil Otolaryngologists, 70. Mem: Am Otol Soc; Am Laryngol, Rhinological & Otol Soc (pres, 66-67); fel Am Col Surgeons; fel Am Rhinological Soc; fel Am Acad Ophthal & Otolaryngol. Res: Conservation of hearing in mastoid surgery; otorhinolaryngology; ear, nose and throat surgery. Mailing Add: 516 Sutter St San Francisco CA 94102

BARONDES, SAMUEL HERBERT, b Brooklyn, NY, Dec 21, 33; m 63; c 2. PSYCHIATRY, NEUROBIOLOGY. Educ: Columbia Col, AB, 54; Columbia Univ, MD, 58. Prof Exp: From intern to asst resident, Peter Bent Brigham Hosp, 58-60; clin assoc, NIH, 60-63; resident, McLean & Mass Gen Hosps, 63-66; from asst to assoc prof psychiat & molecular biol, Albert Einstein Col Med, 66-68, dir training behav & neurol sci, 66-68; PROF PSYCHIAT, UNIV CALIF, SAN DIEGO, 69- Concurrent Pos: NIH fel, 60-63; teaching fel psychiat, Harvard Med Sch, 63-66; NIH career develop award, 67-69; mem alcoholism & alcohol probs rev comt, NIMH, 67-70; mem neurobiol rev comt, NSF, 70-73; mem neurobiol merit rev bd, Vet Admin Cent Off, 72-; mem bd dirs, Found Fund for Res in Psychiat, 74- Mem: AAAS; fel Am Col Neuropsychopharmacol; Am Soc Neurochem (mem coun, 71-); Soc Biol Psychiat; Am Soc Biol Chem. Res: Biochemical basis of synaptic plasticity and neuronal recognition; biochemical basis of memory storage. Mailing Add: Dept of Psychiat Univ of Calif San Diego La Jolla CA 92093

BARONE, JOHN A, b Dunkirk, NY, Aug 30, 24; m 47. ORGANIC CHEMISTRY. Educ: Univ Buffalo, BA, 44; Purdue Univ, MS, 48, PhD(chem), 50. Prof Exp: Asst chem, Purdue Univ, 47-48; from instr to assoc prof, 50-62, dir, NSF In-Serv Inst, 61-68, dir res & grad sci, 63-66, vpres planning, 66-70, PROF CHEM, FAIRFIELD UNIV, 62-, PROVOST, 70- Concurrent Pos: Dir, Jesuit Res Coun Am, 63-70, chmn, 68-70; consult, Conn Regional Med Prog, 70-; proj mgr, Housing & Urban Develop-New Rural Soc Contracts, 72- Mem: AAAS; Am Chem Soc. Res: Heterocyclic and medicinal chemistry; organohalogen compounds; academic and research administration; resource management. Mailing Add: 1283 Round Hill Rd Fairfield CT 06430

BARONE. MILO C, b Throop, Pa, Dec 4, 41. ANIMAL PHYSIOLOGY. Educ: Univ Scranton. BS. 63; John Carroll Univ, MS, 65; St Bonaventure Univ, PhD(physiol), 68. Prof Exp: ASST PROF BIOL, FAIRFIELD UNIV, 68- Mem: AAAS; NY Acad Sci. Res: Reptilian hematology; seasonal variations in serum proteins and other blood properties in Pseudemys and Chrysemys turtles; etiology of Atherosclerosis. Mailing Add: Dept of Biol Fairfield Univ Fairfield CT 06430

BARONOWSKY, PAUL E, b Evansville, Ind, Feb 6, 30; m 57; c 4. BIOCHEMISTRY. Educ: Purdue Univ. BS, 58; Harvard Univ, PhD(biochem), 63. Prof Exp: NIH fel, Harvard Univ, 63-64; biochemist, Arthur D Little, Inc, 64-68; res assoc & group leader, 68-70, PRIN INVESTR, DEPT BIOL RES, MEAD JOHNSON RES CTR, 70- Mem: Am Chem Soc; Can Soc Immunol; AAAS. Res: Immunology; inflammation; cancer chemotherapy. Mailing Add: Dept of Biol Res Mead Johnson Res Ctr Evansville IN 47721

BAROODY, EUGENE MICHAEL, b Richmond, Va, Oct 26, 14. PHYSICS. Educ: Univ Richmond, BS, 35; Cornell Univ, PhD(theoret physics), 40. Prof Exp: Asst, Cornell Univ, 35-40; instr physics, NDak State Col, 40-41 & Univ Mo, 41-43; res physicist, Battelle Mem Inst, 43-44; scientist, Los Alamos Sci Labs, 44-46; RES PHYSICIST, BATTELLE MEM INST, 46- Concurrent Pos: Consult, Los Alamos Sci Labs, 46-52; lectr, Ohio State Univ, 53-54. Mem: Am Phys Soc. Res: Theoretical physics; directional dependence of electrical conductivity in metals; electron emission from solids; radiation effects in solids; spectra of isoelectronic atomic ions. Mailing Add: Battelle Mem Inst 505 King Ave Columbus OH 43201

BAROSH, PATRICK JAMES, b Inglewood, Calif, Jan 27, 36; m 58; c 2. GEOLOGY. Educ: Univ Calif, Los Angeles, AB, 57, MA, 59; Univ Colo, PhD(geol), 64. Prof Exp: Explor geologist, Utah Construct & Mining Co, 59-60; geologist, Astrogeol Br, Ariz, 64-65, spec proj br, Colo, 65-68, geologist, Off Int Geol, Turkey, 68-71, GEOLOGIST, EASTERN ENVIRON BR, US GEOL SURV, 71- Mem: AAAS; Geol Soc Am; Am Asn Petrol Geol. Res: Petrol Geologists. Res: Regional tectonics and seismicity of New England integrating geologic, geophysical and remote sensing data; earthquake effects and associated faulting; Turkish iron deposits; geology of the Basin and Range province. Mailing Add: Eastern Environ Geol Br US Geol Surv 150 Causeway St Boston MA 02114

BAROSS, JOHN ALLEN, b San Francisco, Calif, Aug 27, 41; m 67. MARINE MICROBIOLOGY. Educ: San Francisco State Univ, BS, 63, MA, 65; Univ Wash, PhD(marine microbiol), 72. Prof Exp: Res asst marine microbiol, Col Fisheries, Univ Wash, 66-70; RES ASSOC MARINE MICROBIOL, ORE STATE UNIV, 70- Mem: Am Soc Microbiol; AAAS; Soc Indust Microbiol; Sigma Xi; Audubon Soc. Res: Effects of temperature, pressure and salinity on the physiology and molecular biology of inshore and deepsea marine bacteria; incidence, pathogenicity and possible significance of genetic exchange among in-shore marine vibrio populations, including the human pathogen, Vibrio parahaemolyticus. Mailing Add: Dept of Microbiol Ore State Univ Corvallis OR 97331

BARR, ALLAN RALPH, b Ft Worth, Tex, Aug 13, 26; m 52. MEDICAL ENTOMOLOGY. Educ: Southern Methodist Univ, BS, 48; Johns Hopkins Univ, ScD(parasitol), 52. Prof Exp: Instr med, econ entom & parasitol, Univ Minn, 52-55; asst prof gen & med entom, Univ Kans, 55-58; supvr vector res, Bur Vector Control, State Dept Health, Calif, 58-66; assoc entomologist, Mosquito Control Res-Fresno Proj, 66-67, assoc prof, 67-70, PROF INFECTIOUS & TROP DIS, UNIV CALIF, LOS ANGELES, 70- Concurrent Pos: Vis prof, Fac Med, Univ Singapore, 62-63. Mem: Entom Soc Am; Am Soc Trop Med & Hyg; Am Mosquito Control Asn. Res: Systematics and biology of mosquitoes. Mailing Add: Div of Infectious and Trop Dis Univ of Calif Sch of Pub Health Los Angeles CA 90024

BARR, ALVIN FRANCIS, b Vernon, La, Apr 22, 42; m 67; c 2. MATHEMATICS. Educ: La Tech Univ, BS, 64, MS, 66; Univ Miss, PhD(math), 71. Prof Exp: Instr math, La Tech Univ, 66-67; asst prof, 71-75, ASSOC PROF MATH, LIVINGSTON UNIV, 75- Mem: Am Math Soc. Res: Radius of univalence of certain classes of analytic functions; complex variables. Mailing Add: Dept of Math Livingston Univ Sta 23 Livingston AL 35470

BARR, CHARLES E, b Chicago, Ill, Sept 20, 29; m 56; c 2. CELL PHYSIOLOGY. Educ: Iowa State Col. BS. 54; Univ Calif, Berkeley, PhD(plant physiol), 61. Prof Exp: Trainee biophys, Sch Med, Univ Md, 61-63, instr, 63-64; asst prof biol, Duquesne Univ, 64-68; assoc prof, 68-74, PROF BIOL, STATE UNIV NY COL BROCKPORT. 74- Res: Electrophysiology and ion transport in plant cells; origin of the resting potential in Nitella. Mailing Add: Dept of Biol Sci State Univ NY Brockport NY 14420

BARR, CHARLES RICHARD, b Dakota, Ill, May 16, 32. BIOCHEMISTRY. Educ: NCent Col, BA, 54; Mich State Univ, MS, 57, PhD, 60. Prof Exp: Asst Mich State Univ, 54-58; asst prof, Dubuque Univ, 58-62; asst prof, 62-70, ASSOC PROF CHEM, AUSTIN COL, 70- Mem: Am Chem Soc. Res: Synthetic organic chemistry; isolation and synthesis of biologically active compounds; binding of metals in enzymes; Mossbauer spectroscopy. Mailing Add: Dept of Chem Austin Col Sherman TX 75090

BARR, CHARLES (ROBERT), b Altoona, Pa, Mar 21, 22; m 43; c 4. PHOTOGRAPHIC CHEMISTRY. Educ: St Vincent Col, BS, 46. Prof Exp: Res chemist, Remington-Rand, Inc, 46-48; res chemist, 48-57, sr res chemist, 57-62, res assoc, 62, asst to div head, Color Photog Div, 62-65, HEAD EXP COLOR PHOTOG LAB, EASTMAN KODAK CO, 65- Mem: Am Chem Soc; Soc Photog Sci & Eng. Res: Synthesis of organic compounds for use in color photography; color and constitution of indoaniline dyes; new color photographic products and processes. Mailing Add: Exp Color Photog Lab Eastman Kodak Co Kodak Park Works 1669 Lake Ave Rochester NY 14615

BARR, DAVID ROSS, b Madison, Wis, Aug 21, 32; m 57; c 2. MATHEMATICAL STATISTICS. Educ: Miami Univ, BA & MA, 54, MS, 57; Univ Iowa, PhD(math), 64. Prof Exp: Programmer, Appl Math Lab, Eglin Air Force Base, US Air Force, 55-57, instr math, US Air Force Acad, 59-62, asst prof, 62, res math statistician, Aerospace Res Labs, 64-68; ASST PROF MATH, US AIR FORCE INST TECHNOL, 68- Mem: Inst Math Statist; Am Statist Asn; Math Asn Am. Res: Power function of the likelihood ratio test and ranking and selection problems of the uniform distribution and related distribution. Mailing Add: 2278 Yorkshire Place Kettering OH 45419

BARR, DONALD EUGENE, b June 30, 34; US citizen; m 56; c 2. CHEMISTRY. Educ: Elizabethtown Col, BS, 56; Bucknell Univ, MS, 61; Univ Mass, PhD(chem), 65. Prof Exp: Mem staff, Bucknell Univ, 61; res chemist, Chem Res Div, Armstrong Cork Co, 65-69; res specialist new imaging processes res, 69-71, mgr photoimaging, 71-74, MGR RAW MAT QUAL ASSURANCE DEPT, GAF CORP, 74- Mem: Am Chem Soc; The Chem Soc; Soc Photog Sci & Eng. Res: Seven-azabenzonorbornadiene synthesis; polyvinyl chloride polymerization, characterization and stabilization; high resolution nuclear magnetic resonance, polyvinyl chloride and model compounds;

photosensitive compounds and polymeric coatings for microelectronics industry. Mailing Add: 1508 Drexel Dr Binghamton NY 13903

BARR, DONALD JOHN STODDART, b Surrey, Eng, Sept 18, 37; Can citizen; m 61; c 2. MYCOLOGY. Educ: McGill Univ, BSc, 60, MSc, 62; Univ Western Ont, PhD(mycol, phycol), 65. Prof Exp: Res scientist mycol, Plant Res Inst, 65-73, RES SCIENTIST MYCOL, BIOSYSTS RES INST, CAN DEPT AGR, 73- Mem: Mycol Soc Am; Can Phytopath Soc; Can Bot Asn. Res: Fungal parasites of freshwater algae; taxonomy, ecology and physiology of lower Phycomycetes, particularly the Chytridiales; virus transmission by zoosporic fungi. Mailing Add: Biosysts Res Inst Can Dept of Agr Ottawa ON Can

BARR, DONALD R, b Durango, Colo, Dec 10, 38; m 58; c 2. MATHEMATICAL STATISTICS. Educ: Whittier Col, BA, 60; Colo State Univ, MS, 62, PhD(statist), 65. Prof Exp: Instr math. Colo State Univ, 63-65; asst prof, Wis State Univ, Oshkosh, 65-66; asst prof opers res, 66-70, ASSOC PROF OPERS RES, NAVAL POSTGRAD SCH. 70- Mem: Inst Math Statist; Math Asn Am; Opers Res Soc Am. Res: Probability theory; sequential decision theory; reliability growth and classification models. Mailing Add: Dept of Opers Anal Naval Postgrad Sch Monterey CA 93940

BARR, ERNEST SCOTT, b Lincolnton, NC, Nov 27, 05; m 31. PHYSICS. Educ: Univ NC, AB, 26, AM, 33, PhD(physics), 36. Prof Exp: Instr physics, Univ NC, 35-36; from instr to assoc prof, Tulane Univ, 36-47; prof, 47-70, EMER PROF PHYSICS, UNIV ALA, TUSCALOOSA, 71- Concurrent Pos: Sr physicist, Appl Physics Lab, Johns Hopkins Univ, 45-46; consult, Redstone Arsenal, 52-65. Mem: Fel AAAS; fel Am Phys Soc; fel Optical Soc Am; Am Asn Physics Teachers; Sigma Xi. Res: Infrared; history of physics. Mailing Add: PO Box 3174 Tuscaloosa AL 35401

BARR, FRANK THEODORE, b Sacramento, Calif, Dec 23, 31; m 57; c 2. GEOLOGY. Educ: Univ Calif, Berkeley, BA, 53, MA, 57; Univ London, PhD(paleont), 62. Prof Exp: Geologist, Standard Oil Co Calif, 53-54, 56, 57-59; lectr geol, Univ Sheffield, 61-62; sr geologist, Oasis Oil Co, Libya, 62-66, supvr geol lab, 66-72; sr geologist, Independent Indonesian Am Petrol Co, 72-75; COORDR GEOL PROJS, HOUSTON OIL & MINERALS CORP, 75- Mem: Palaeont Asn; Am Asn Petrol Geol; Geol Soc Malaysia; Geol Soc Philippines. Res: Petroleum geology; tectonics, basin evaluation, and hydrocarbon exploration of Africa, Asia, and North America; Cretaceous and tertiary micropaleontology and biostratigraphy of Western Europe, North Africa and California. Mailing Add: Houston Oil & Minerals Corp 1212 Main St Houston TX 77002

BARR, FRED S, b Benhams, Va, Jan 22, 26; m 50; c 2. MEDICAL MICROBIOLOGY. Educ: Univ Tenn, BS, 52; Eastern Tenn State Univ, MA, 57. Prof Exp: Prod asst, S E Massengill Co, Tenn, 52-53, res bacteriologist, 53-60, res assoc microbiol & biochem, 60-70, sect head biopharmaceut res, 70-71; MGR MICROBIOL SECT, BEECHAM LABS, 71- Mem: AAAS; Am Soc Microbiol; Am Soc Trop Med & Hyg; Am Chem Soc. Res: Antibiotic isolation, identification and purification; antibiotic combinations in human and animal medications; microbiological diagnostic reagents for clinical use; drug levels in body tissues and fluids; drug metabolism and absorption-excretion kinetics. Mailing Add: 500 Carter St Bristol VA 24201

BARR, GEORGE E, b Milwaukee, Wis, Oct 16, 37; m 75. APPLIED MATHEMATICS. Educ: Univ Wis, Madison, BA, 58; Ore State Univ, PhD(theoret physics), 64. Prof Exp: Jr res assoc math physics, Brookhaven Nat Lab, 62-63; MEM TECH STAFF APPL MATH, SANDIA LAB, 65- Mem: Soc Indust & Appl Math. Res: Special functions; theory of diffraction; fluid dynamics of viscous media; nuclide transport in geologic media. Mailing Add: Div 1141 Sandia Lab Albuquerque NM 87115

BARR, HAROLD JAY, b Philadelphia, Pa, July 25, 36; m 61; c 1. BIOLOGY. Educ: Temple Univ, BA, 58; Columbia Univ, MA, 59, PhD(zool), 62. Prof Exp: Asst zool, Columbia Univ, 58-60; vis scholar embryol, Oxford Univ, 60-61; NIH fel develop biol, Western Reserve Univ, 62-64, sr instr anat, 64-65; lectr zool, Univ Wis-Madison, 65-67, asst prof anat, 67-74; MEM FAC, CTR FOR GENETICS, UNIV ILL MED CTR, 74- Concurrent Pos: NIH grant, 64-; res career develop award, Nat Inst Gen Med Sci, 68. Mem: AAAS; Am Soc Zool; Soc Develop Biol; Am Asn Anat; Am Soc Cell Biol. Res: Experimental embryology, especially regeneration; cytogenetics, especially chromosome function and the nucleolus; philosophy of science. Mailing Add: Ctr for Genetics Univ of Ill Med Ctr Chicago IL 60612

BARR, HARRY L, b Bethesda, Ohio, Feb 15, 22; m 44; c 4. REPRODUCTIVE PHYSIOLOGY, GENETICS. Educ: Ohio State Univ, BSc, 54, MS, 55, PhD(genetics), 60. Prof Exp: From instr to asst prof physiol, 55-67, ASSOC PROF DAIRY SCI & SPECIALIST, COOP EXTEN SERV, OHIO STATE UNIV, 67- Concurrent Pos: Vis prof, Cornell Univ, 70. Mem: Am Dairy Sci Asn; Am Soc Animal Sci. Res: Reproductive physiology and population genetics. Mailing Add: 231 Plumb Hall Ohio State Univ Columbus OH 43210

BARR, JAMES K, b Washington, Pa, Sept 26, 33; m 60; c 3. PHYSICAL CHEMISTRY, RESEARCH ADMINSTRATION. Educ: Univ Calif, Los Angeles, BS, 56; Univ Wis, MS, 59; Univ Calif, Riverside, PhD(phys chem), 64. Prof Exp: Res chemist spectroscopy, E I du Pont de Nemours & Co, Del, 64-67, res assoc mat res, 67-69, res supvr coatings, Mich, 69-71, res mgr coatings, Pa, 71-72; DIR RES INORG MAT, PFIZER, INC, 72- Concurrent Pos: Mem sci adv comt, Winterthur Mus, 66-, US Senate, 67-69; adj assoc prof, Univ Del, 75- Mem: Am Chem Soc; Sigma Xi. Res: Materials research into new synthetic pigments and other new inorganic powdered products for the coatings, aerospace, plastics, paper and concrete industries. Mailing Add: Pfizer Inc 640 N 13th St Easton PA 18042

BARR, JOHN BALDWIN, b Niagara Falls, NY, Nov 8, 32; m 54; c 4. PHYSICAL CHEMISTRY. Educ: Univ Buffalo, BA, 54; Univ Mich, MS, 56; Pa State Univ, PhD(phys chem), 61. Prof Exp: Sr chemist, Corhart Refractories Div, Corning Glass Works. NY, 60-62; sr res chemist, 62-71, SR RES SCIENTIST, CARBON PROD DIV, RES LAB, UNION CARBIDE CORP, PARMA, 71- Mem: Am Chem Soc; Am Ceramic Soc. Res: High temperature chemistry; refractory compounds; carbon and graphite chemistry; boron halide reactions. Mailing Add: 17179 Ridge Point Circle Strongville OH 44136

BARR, JOHN TILMAN, b Norman, Ark, Sept 16, 25; m 44; c 3. ORGANIC CHEMISTRY. Educ: Ark Col, AB, 46; Univ Ark, MS, 48. Prof Exp: Chemist, K-25 Plant, Carbide & Carbon Chem Co, 48-52; sr res chemist, White-Marsh Res Lab, Pa Salt Mfg Co, 52-55; proj mgr, Escambia Chem Corp, 55-62, group leader, 62-64; proj engr, Latin Am Div, W R Grace & Co, 64-66, tech dir div plastics, Paramonga Div, 66-67; tech mgr, Calvert City Works, Ky, 67-73, TECH MGR MFG GROUP CHEM, AIR PROD & CHEM, INC, 73- Mem: Am Chem Soc; Am Inst Chem Eng. Res: Aliphatic fluorine compounds; fluorine; radical reactions; polymerization; polyvinyl chloride; aliphatic chemistry; polyvinyl acetate; process development. Mailing Add: Air Prod & Chem Inc PO Box 538 Allentown PA 18105

BARR, LLOYD, b Chicago, Ill, Dec 27, 29; m 52; c 4. PHYSIOLOGY. Educ: Univ Chicago, BS, 54; Univ Ill, MS, 56, PhD(physiol), 58. Prof Exp: Instr physiol, Univ Mich, 58-61, from asst prof to assoc prof, 61-65; prof, Med Col Pa, 65-70; PROF PHYSIOL & BIOPHYS, UNIV ILL, URBANA-CHAMPAIGN, 70- Concurrent Pos: Spec fel, USPHS, 63-64. Mem: AAAS; Soc Gen Physiol; NY Acad Sci; Biophys Soc. Res: Electrophysiology; cellular transport processes; kinetics and thermodynamics. Mailing Add: Dept of Physiol & Biophys Univ of Ill Urbana IL 61801

BARR, MARTIN, b Philadelphia, Pa, Nov 11, 25; m 51; c 4. PHYSICAL PHARMACY. Educ: Temple Univ, BSc, 46; Philadelphia Col Pharm, MSc, 47; Ohio State Univ, PhD, 50. Prof Exp: Asst pharm, Ohio State Univ, 47-49, instr, 50; from asst prof to prof, Philadelphia Col Pharm, 50-61; chmn dept, 61-64, DEAN, COL PHARM, WAYNE STATE UNIV, 64-, VIS LECTR PHARM, COL MED, 64-, VPRES SPEC ASSIGNMENTS, 72- Concurrent Pos: Mem staff, Ford Hosp. Mem: Am Pharmaceut Asn; Am Chem Soc; Am Col Apothecaries. Res: Clays; ion exchange; emulsions; ointments; colloids; product development; aerosols. Mailing Add: Detroit MI

BARR, MICHAEL, b Philadelphia, Pa, Jan 22, 37; m 64; c 2. MATHEMATICS. Educ: Univ Pa, AB, 59, PhD(math), 62. Prof Exp: Instr, Columbia Univ, 62-64; from asst prof to assoc prof, Univ Ill, Urbana, 64-68; ASSOC PROF MATH, McGILL UNIV, 68- Concurrent Pos: Guest, Res Inst Math, Swiss Fed Inst Technol, 67; res assoc, Math Inst, Univ Fribourg, 70-71. Mem: Math Asn Am. Res: Categorical algebra, especially the study of those aspects of mathematical structures describable in terms of mapping properties. Mailing Add: Dept of Math McGill Univ Montreal PQ Can

BARR, MURRAY LLEWELLYN, b Belmont, Ont, June 20, 08; m 34; c 4. ANATOMY. Educ: Univ Western Ont, BA, 30, MD, 33, MSc, 38; FRCPS(C), 64; FRCOG. Hon Degrees: LLD, Queen's Univ, Ont, 63, Univ Toronto, 64, Univ Alta, 67, Dalhousie Univ, 68 & Univ Sask, 73; Dr Med, Univ Basel, 66; DSc, Univ Western Ont, 74. Prof Exp: From instr to prof anat, 36-52, prof micros anat & head dept, 52-64, head dept anat, 64-67, PROF ANAT, UNIV WESTERN ONT, 64- Honors & Awards: Award, Soc Obstet & Gynec Can, 56; Borden Award, 57; Flavelle Medal, 59; Charles Mickle Award, Univ Toronto, 60; Ortho Medal, Am Soc Study Sterility, 62; Medal, Am Col Physicians, 62; Joseph P Kennedy Jr Found Award, 62; Gairdner Found Award, 63; Papanicalaou Award, 64; F N G Starr Medal, Can Med Asn, 67; Maurice Goldblatt Cytol Award, Int Acad Cytol, 68. Mem: Am Asn Anat; Asn Res Nerv & Ment Dis; Am Soc Human Genetics; Can Neurol Soc; fel Royal Soc Can. Res: Neuroanatomy; cytogen etics; mental deficiency. Mailing Add: Health Sci Ctr Univ of Western Ont London ON Can

BARR, NATHANIEL FRANK, b Union City. NJ, Dec 28, 27; m 56; c 3. RADIATION CHEMISTRY. Educ: Queens Col, NY, BS. 50; Columbia Univ, PhD(chem). 54. Prof Exp: Asst, Columbia Univ, 51-54; res assoc, Brookhaven Nat Lao, 54-56; asst, Div Biophys, Sloan Kettering Inst, 56-61; biophysicist, US AEC, 61-62, chief radiol physics & instrumentation br, Div Biol & Med, 62-67, tech adv to asst gen mgr res & develop, 67-69, asst dir div biol & med, 69-75; ASST DIR, DIV BIOMED & ENVIRON RES, US ENERGY RES & DEVELOP ADMIN, 75- Concurrent Pos: Asst prof, Med Sch, Cornell Univ & lectr, Columbia Univ, 58-61. Mem: Am Chem Soc; Radiation Res Soc (pres, 71-72); Health Physics Soc; Pattern Recognition Soc. Res: Dosimetry and properties of ionizing radiation. Mailing Add: US Energy Res & Develop Admin Washington DC 20545

BARR, RICHARD ARTHUR, b Southport, NY, Mar 12, 25; m 61; c 2. PLANT PHYSIOLOGY, PHYTOCHEMISTRY. Educ: Univ Vt, BS, 50, MS. 55; Cornell Univ, PhD(plant physiol), 63. Prof Exp: Res assoc plant physiol, Cornell Univ, 61-63, asst prof, 64-66; asst prof biol, Univ Mo-St Louis, 66-68; assoc prof, 68-72, PROF BIOL, SHIPPENSBURG STATE COL, 72- Mem: AAAS; Am Inst Biol Sci; Bot Soc Am; NY Acad Sci. Res: Amino acid metabolism in plants; nitrogen metabolism of banana; biosynthesis of allergens in poison ivy; plant growth and development; tissue culture; membrane permeability. Mailing Add: 201 Franklin Sci Ctr Shippensburg State Col Shippensburg PA 17257

BARR, RITA, b Riga, Latvia, Sept 30, 29. PLANT PHYSIOLOGY. Educ: Northern Ill Univ, BS, 54, MS, 56; Purdue Univ, PhD(genetics), 60. Prof Exp: Fel genetics, 60-62, res asst plant physiol, 63-65, ASSOC BIOL SCI, PURDUE UNIV, 65- Mem: Genetics Soc Am; Am Soc Plant Physiol. Res: Electron transport reactions in spinach chloroplasts; chelator inhibition of electron transport and photophosphorylation. Mailing Add: Dept of Biol Sci Purdue Univ Lafayette IN 47907

BARR, ROGER COKE, b Jacksonville, Fla, Feb 21, 42; m 67. BIOMEDICAL ENGINEERING, COMPUTER SCIENCE. Educ: Duke Univ, BS, 64, PhD(elec eng), 68. Prof Exp: Res assoc, 68-69, asst prof, 69-72, ASSOC PROF BIOMED ENG & PEDIAT, DUKE UNIV, 72- Mem: Am Heart Asn; Inst Elec & Electronics Engrs. Res: Electrocardiology; electrophysiology of the heart; digital computing. Mailing Add: Box 3305 Duke Univ Med Ctr Durham NC 27710

BARR, RONALD EDWARD, b St Louis, Mo, Sept 9, 36; m 64; c 4. MEDICAL PHYSIOLOGY. Educ: St Louis Univ, BS, 58, MS, 63, PhD(physics), 66. Prof Exp: Res asst, St Louis Univ, 61-65; res fel biophys. Space Sci Res Ctr, 65-66, asst prof radiol, 66-74, ASSOC PROF OPHTHALMOL, UNIV MO-COLUMBIA, 74- Concurrent Pos: NIH spec fel, Cambridge Univ, 69-70. Mem: AAAS; Am Phys Soc; Asn Res Vision & Ophthalmol; Am Asn Physicists Med. Res: Oxygen transfer properties of the cornea and other ocular tissues; development of long term staoility for oxygen electrodes. Mailing Add: Dept of Ophthalmol Sch of Med Univ of Mo-Columbia Columbia MO 65201

BARR, TERY LYNN, b Renova, Pa, Feb 9, 39; m 59; c 4. CHEMICAL PHYSICS, SURFACE CHEMISTRY. Educ: Univ Va, BS, 60; Univ SC, MS, 62; Univ Ore, PhD(chem), 68. Prof Exp: Res chemist, Union Oil Co, 63-64 & Shell Develop Co, 69-71; vis asst prof chem, Univ Calif, Berkeley, 71; asst prof, Harvey Mudd Col, 71-73; asst prof, Claremont Men's Col, 73-74; SR RES CHEMIST, CORP RES CTR, UOP, INC, 74- Concurrent Pos: Res assoc, Univ Wash, 68-69; consult, R&D Assoc, 72-74; vis prof chem, Harvey Mudd Col, 74-75. Mem: Sigma Xi; Am Phys Soc; Am Vacuum Soc; Soc Electron Spectros (pres, 75-). Res: X-ray photoelectron spectroscopy; surface chemistry and physics; quantum mechanics; spectroscopic characterization of materials. Mailing Add: Mat Sci Dept Corp Res Ctr UOP Inc 10 UOP Plaza Des Plaines IL 60016

BARR, THOMAS ALBERT, JR, b Chattanooga, Tenn, Aug 18, 24; m 49; c 5. PLASMA PHYSICS. Educ: Univ Chattanooga, BS, 47; Vanderbilt Univ, MS, 50, PhD(physics), 53. Prof Exp: Mem staff, Vanderbilt Univ, 50-51; asst prof physics, Univ Ga, 51-56; SUPVY RES PHYSICIST, PHYS SCI LAB, US ARMY MISSILE COMMAND, REDSTONE ARSENAL, 56- Concurrent Pos: Govt liaison rep, Plasma Phenomena Comt, Nat Acad Sci, 60. Mem: Fel AAAS; Am Asn Physics Teachers; assoc fel Am Inst Aeronaut & Astronaut. Res: Investigation of the properties of metal membranes; x-ray and neutron dosimetry; basic and applied research on partially and fully ionized gases; chemical and high power gas lasers. Mailing Add: 7803 Martha Dr SE Huntsville AL 35802

BARR, WILLIAM FREDERICK, b Oakland, Calif, Oct 20, 20; m 46; c 3. ENTOMOLOGY. Educ: Univ Calif, BS, 45, MS, 47, PhD(syst entom), 50. Prof Exp: Teaching asst entom, Univ Calif, 45-47; asst entomologist, 47-53, assoc prof entom & assoc entomologist, 53-58, asst prof entom, 48-53, PROF ENTOM & ENTOMOLOGIST, EXP STA, UNIV IDAHO, 58- Concurrent Pos: NSF fac fel, 59-60. Mem: Entom Soc Am; Soc Study Evolution; Soc Syst Zool; Royal Entom Soc London. Res: Systematic entomology; distribution and biology of Coleoptera; insect pollination; insects affecting range plants. Mailing Add: Dept of Entom Univ of Idaho Moscow ID 83843

BARR, WILLIAM HENRY, pharmaceutics, pharmacology, see 12th edition

BARR, WILLIAM J, b Reading, Pa, Feb 5, 19. ENERGY CONVERSION. Educ: Princeton Univ, AB, 39; Columbia Univ, MA, 40. Prof Exp: Instr math, Polytech Inst Brooklyn, 41-42 & Brooklyn Col, 42; mathematician, Naval Res Lab, 42-46; head anal group, Exp, Inc, 46-50; tech asst to dir Proj Squid, Princeton Univ, 51; mathematician, Res Div, Detroit Controls Corp, 52, asst res dir, 53-55, res dir, 55-57, asst to pres, 57-58; mgr physics res, Am Stand Inc, 58-60 & opers res, 60-72; mgr mfg systs develop, Am Can Co, 72-73; CONSULT, MATHEMATICA INC, 73- Mem: Am Math Soc; Am Chem Soc; Opers Res Soc Am. Res: Thermochemistry; interior ballistics; thermodynamics; combustion; propulsion. Mailing Add: 127 Westerly Rd Princeton NJ 08540

BARR, WILLIAM LEE, spectroscopy, see 12th edition

BARRACLOUGH, CHARLES ARTHUR, b Vineland, NJ, July 13, 26; m 52; c 2. ENDOCRINOLOGY. Educ: St Joseph's Col, Pa, BS, 47; Rutgers Univ, MS, 51, PhD(zool), 52. Prof Exp: Asst, Rutgers Univ, 50-53; jr res anatomist, Sch Med, Univ Calif, Los Angeles, 53-54, from instr to asst prof, 54-61; assoc prof, 62-65, actg chmn dept, 71-72, PROF PHYSIOL, SCH MED, UNIV MD, BALTIMORE CITY, 65- Concurrent Pos: Spec res fel, Cambridge Univ, 61-62; spec res fel, Univ Milan, 69-70; consult res div, Vet Admin Hosp, Calif, 59-61; mem reproductive biol study sect, NIH, 66-68 & 70-74, chmn, 73-74. Mem: AAAS; Am Asn Anat; Endocrine Soc; Am Physiol Soc; Soc Study Reproduction (dir, 71-73). Res: Reproductive physiology; neuroendocrinology. Mailing Add: Dept of Physiol Univ of Md Sch of Med Baltimore MD 21201

BARRACO, ROBIN ANTHONY, b Detroit, Mich, Mar 24, 45. NEUROCHEMISTRY. Educ: Georgetown Univ, BA, 66; Wayne State Univ, PhD(physiol), 71. Prof Exp: Res assoc, 70-71, ASST PROF PHYSIOL, WAYNE STATE UNIV, 71- Concurrent Pos: Mem task force, Off Drug Abuse & Alcoholism, Gov State of Mich, 72-73. Res: Behavioral neurochemistry; psychopharmacology of memory and learning; developmental neurobiology. Mailing Add: Dept of Physiol Wayne State Univ Detroit MI 48201

BARRADAS, REMIGIO GERMANO, b Hong Kong, Oct 23, 28; Can citizen; m 53; c 3. PHYSICAL CHEMISTRY. Educ: Univ Liverpool, BSc, 52; Royal Col Sci & Technol, Scotland, dipl, 53; Univ Ottawa, PhD(electrochem), 60. Prof Exp: Govt chemist, Govt Lab, Hong Kong, 53-56; chemist inspection serv, Chem Div, Dept Nat Defence, Can, 56-57; demonstr phys chem, Univ Ottawa, 57-60; res fel chem, Royal Mil Col, Can, 60-62; from asst prof to assoc prof, Univ Toronto, 62-68; PROF CHEM, CARLETON UNIV, ONT, 68- Concurrent Pos: Consult, Hooker Chem Corp, NY, 63. Mem: Electrochem Soc; Chem Inst Can; fel Royal Inst Chem

BARRALL, EDWARD MARTIN, II, b Louisville, Ky, Dec 8, 34; m 59; c 2. ANALYTICAL CHEMISTRY. Educ: Univ Louisville, BS, 55, MS, 57; Mass Inst Technol, PhD(anal chem), 61. Prof Exp: Instr anal chem, Univ Ga, 57-58; asst, Mass Inst Technol, 58-60; mem part-time staff, Dewey & Almy Chem Co Div, W R Grace Corp, 60-61; anal chemist, Chevron Res Co, Va, 61-69; RES CHEMIST, IBM RES, IBM CORP, 69- Concurrent Pos: Mem part-time staff, E I du Pont de Nemours & Co, Inc, 57-59; vis prof, Univ Conn, 74-75. Honors & Awards: Mettler Award, NAm Thermal Anal Soc, 73. Mem: AAAS; Am Chem Soc; NAm Thermal Anal Soc; Int Confedn Thermal Anal. Res: Thermal methods of analysis; gas chromatography; polymer characterization; thermodynamics of mesophase transitions; physics of polymers; thermal properties of organic compounds. Mailing Add: IBM Res Monterey & Cottle Rd San Jose CA 95115

BARRAN, LESLIE ROHIT, b Berbice, Guyana, July 29, 39; Can citizen; m 61; c 1. AGRICULTURAL MICROBIOLOGY. Educ: McGill Univ, BSc, 63, MSc, 65; Mich State Univ, PhD(biochem), 69. Prof Exp: RES SCIENTIST AGR MICROBIOL, CHEM & BIOL RES INST, AGR CAN, 71- Mem: Can Soc Microbiol. Res: Fungal biochemistry; cell wall structure; membrane structure and function; enzymology. Mailing Add: Chem & Biol Res Inst Agr Can Ottawa ON Can

BARRANCO, SAM CHRISTOPHER, b Beaumont, Tex, Nov 17, 38; m; c 2. RADIOBIOLOGY, CANCER. Educ: Tex A&M Univ, BS, 60, MS, 62; Johns Hopkins Univ, PhD(cellular radiobiol), 69. Prof Exp: Asst prof surg, M D Anderson Hosp & Tumor Inst, 71-72; ASSOC PROF BIOL, UNIV TEX MED BR GALVESTON, 72- Concurrent Pos: NIH fel, M D Anderson Hosp & Tumor Inst, 69-71, Damon Runyon res grant, 69-71; NIH cancer grant, Univ Tex Med Br Galveston, 72- Mem: Am Soc Cell Biol; Am Asn Cancer Res. Res: Normal and tumor cell kinetics; cancer chemotherapy; radiobiology and cystic fibrosis. Mailing Add: Dept of Human Biol Chem & Genetics Univ Tex Grad Sch Biomed Sci Galveston TX 77550

BARRANTE, JAMES RICHARD, b Torrington, Conn, Apr 30, 38; m 65; c 1. PHYSICAL CHEMISTRY. Educ: Univ Conn, BA, 60; Harvard Univ, MA, 62, PhD(chem), 64. Prof Exp: Sr res chemist, Olin-Mathieson Chem Corp, Conn, 64-66; asst prof phys chem, 66-69, ASSOC PROF CHEM, SOUTHERN CONN STATE COL, 68- Concurrent Pos: Instr, Eve Sch, Southern Conn State Col, 65-66. Mem: Am Inst Chem; Am Chem Soc. Res: Nuclear magnetic resonance studies of inorganic polymers; dehydration and deamminization kinetics. Mailing Add: 32 Rockview Dr Cheshire CT 06410

BARRAR, RICHARD BLAINE, b Dayton, Ohio, Oct 12, 23; m 47; c 4. MATHEMATICS. Educ: Univ Mich, MS, 48, PhD(math), 53. Prof Exp: Res assoc, Harvard Univ, 51-52; head theoret dept, McMillan Lab, 52-54; res physicist, Hughes Aircraft Co, 54-57; head, Antenna Group, Hoffman Lab, Inc, 57-58; sr opers res scientist, Syst Develop Corp, 58-66, sr scientist, 66-68; PROF MATH, UNIV ORE, 68- Mem: Am Math Soc; Soc Indust & Appl Math; Inst Elec & Electronics Eng; Asn Comput Mach; Am Astron Soc. Res: Electromagnetic theory; existence theorems for partial differential equations; system analysis; celestial mechanics. Mailing Add: Dept of Math Univ of Ore Eugene OR 97403

BARRAS, DONALD J, b St Martinsville, La, Jan 11, 32; m 61; c 3. PHYSIOLOGY, BIOCHEMISTRY. Educ: Univ Southwestern La, BS, 54; Univ Miss, MS, 64; Miss State Univ, PhD(physiol), 70. Prof Exp: Teaching assoc physiol, Univ Miss, 62-64; res assoc biol warfare, Univ Okla, 64-65; asst prof physiol, Nicholls State Col, 65-67; res assoc, Miss State Univ, 67-70; prof biol, Norman Col, 70; COLLAB RES EXPERT BIOL CONTROL, AGR RES SERV, USDA, 70-; ASSOC PROF PHYSIOL, TROY STATE UNIV, 70- Mem: Sigma Xi. Res: Biochemical and physiological studies of parasitoids in the use of biological control of harmful insects. Mailing Add: Dept of Biol Troy State Univ Troy AL 36081

BARRAS, STANLEY J, b New Orleans, La, Jan 4, 36. FOREST ENTOMOLOGY. Educ: La State Univ, BS, 59, MS, 61; Univ Wis, PhD(entom), 65. Prof Exp: Res entomologist, 65-75, PROJ LEADER, SOUTHERN FOREST EXP STA, US FOREST SERV, 75- Mem: Entom Soc Am; Entom Soc Can; Am Inst Biol Sci. Res: Fungal repositories and associated microorganisms of bark beetles; interactions between bark beetle symbiotic fungi; ecology of symbiotic associations in pine phloem. Mailing Add: Southern Forest Exp Sta 2500 Shreveport Hwy Pineville LA 71360

BARRAT, JOSEPH GEORGE, b New Haven, Conn, May 30, 22; m 48; c 3. PLANT PATHOLOGY. Educ: RI State Col, BS, 48; Univ RI, MS, 51; Univ NH, PhD(bot, plant path), 58. Prof Exp: Exten specialist entom & plant path, State Agr Exten Serv, Univ RI, 48-50; plant pathologist, Nursery Improv Prog, Dept Hort, State Dept Agr, Washington, 50-55; state exten specialist plant path, Coop Exten Serv, 58-70, assoc prof plant path, 67-70, PROF PLANT PATH & SUPT EXP FARM, W VA UNIV, 70- Mem: Am Phytopath Soc. Res: Stone and pome fruit disease and insect control; virus diseases of stone and pome fruits. Mailing Add: WVa Univ Exp Farm Kearneysville WV 25430

BARRATT, MICHAEL GEORGE, b London, Eng, Jan 26, 27; m 52; c 5. PURE MATHEMATICS. Educ: Oxford Univ, BA, 48, MA, 52, DPhil, 52; Manchester Univ, MSc, 68. Prof Exp: Jr lectr math, Oxford Univ, 50-52; fel, Oxford Univ, 52-56; lectr, Brasenose Col, 55-59; sr lectr, Manchester Univ, 59-63, reader, 63-64, prof pure math, 64; PROF MATH, NORTHWESTERN UNIV, 69-70, 72- Prof Exp: Vis asst prof, Princeton Univ, 56-57; prof, Univ Chicago, 63-64. Mem: Am Math Soc; London Math Soc. Res: Algebraic topology, particularly homotopy theory; homological algebra; topology of manifolds. Mailing Add: Dept of Math Lunt Hall Northwestern Univ Evanston IL 60201

BARRATT, RAYMOND WILLIAM, b Holyoke, Mass, May 4, 20; div; c 2. MICROBIAL GENETICS. Educ: Rutgers Univ, BSc, 41; Univ NH, MSc, 43; Yale Univ, PhD(microbiol), 48. Prof Exp: Asst plant path & hort, Univ NH, 43-44; res assoc, Crop Protection Inst & asst plant pathologist, Conn Agr Exp Sta, 44-46; res assoc biol, Stanford Univ, 48-53, res biologist & actg asst prof, 53-54; asst prof bot, Dartmouth Col, 54-57, prof, 58-62, prof biol, 62-70, lectr microbiol, Med Sch, 61-70, chmn dept biol sci, 65-69; PROF BIOL & DEAN SCH SCI, HUMBOLDT STATE UNIV, 70-, DIR FUNGAL GENETICS STOCK, 65- Concurrent Pos: USPHS spec fel, 61-62. Mem: Am Inst Biol Sci; Genetics Soc Am. Res: Genetics and biochemistry of microorganisms. Mailing Add: 468-D Sci Bldg Humboldt State Univ Arcata CA 95521

BARREKETTE, EUVAL S, b New York, NY, Feb 18, 31; m 56; c 3. ELECTROOPTICS, APPLIED MECHANICS. Educ: Columbia Univ, AB, 52, BS, 53, MS, 56, PhD(appl mech), 59. Prof Exp: Asst prof civil eng, Columbia Univ, 59-60; mem tech staff, 60-63, tech asst to dir res, 63, mgr tech studies, 64-65, explor systs, 65-66, electrooptical technol, 67-71, prog dir advan eng, Corp Staff, 71-72, mgr advan technol, Res Div, 73-76, ASST DIR APPL RES, RES DIV, IBM CORP, 76- Concurrent Pos: Adj assoc prof, Columbia Univ, 60-70. Mem: Am Inst Aeronaut & Astronaut; Am Soc Civil Eng; Optical Soc Am; Soc Eng Sci; Inst Elec & Electronics Eng. Res: Optical information processing; laser applications; microwave acoustics; computer printer and display technologies; semiconductor lasers. Mailing Add: IBM Corp T J Watson Res Ctr PO Box 218 Yorktown Heights NY 10598

BARRERA, CECILIO RICHARD, b Rio Grande City, Tex, Nov 30, 42; m 70; c 1. MICROBIAL PHYSIOLOGY. Educ: Univ Tex, Austin, BA, 65, MA, 67, PhD, 70. Prof Exp: Fel biochem, Clayton Found Biochem Inst, Univ Tex, Austin, 70, res assoc, 70-75; ASST PROF BIOL, NMEX STATE UNIV, 75- Mem: Am Soc Microbiol. Res: Enzyme isolation and characterization; regulation of enzyme activity. Mailing Add: Dept of Biol NMex State Univ Las Cruces NM 88003

BARRERA, FRANK, b Havana, Cuba, Nov 7, 17; m 49; c 5. PHYSIOLOGY. Educ: Belen Col, BS, 34; Univ Havana, MD, 41. Prof Exp: Instr med, Sch Med, Univ Havana, 46-47; from instr to assoc prof physiol, Sch Med, Temple Univ, 55-73, from asst prof to assoc prof med, 63-73; CHIEF RESPIRATORY SERV, AM HOSP, 73- Concurrent Pos: Chief clin & cardio-respiratory physiol dept, Inst Cardiovasc & Thoracic Surg, Cuba, 51-60. Honors & Awards: Tamayo Award, 52-59; Farinas Award, 58. Mem: Am Physiol Soc. Res: Cardiopulmonary physiology. Mailing Add: Am Hosp 11750 Bird Rd Miami FL 33165

BARRERA, JOSEPH S, b Brandon, Vt, Mar 13, 41; m 63; c 2. SOLID STATE PHYSICS. Educ: Harvey Mudd Col, BS, 62; Carnegie Inst Technol, MS, 63, PhD(solid state elec eng), 66. Prof Exp: Res scientist, Hewlett-Packard Labs, 67-74, RES & DEVELOP SECT MGR, HEWLETT-PACKARD ASSOC, 74- Mem: Inst Elec & Electronics Engrs. Res: Fluctuation studies in space charge and recombination limited, single and double injection devices; solid state microwave bulk oscillator; III-V microwave devices; field effect transistors, Gunn and microwave diode devices. Mailing Add: 640 Page Mill Rd Palo Alto CA 94304

BARRERAS, RAYMOND JOSEPH, b Albuquerque, NMex, June 8, 40; m 59; c 5. ORGANIC CHEMISTRY. Educ: Univ NMex, BS, 61; Mich State Univ, PhD(chem), 66. Prof Exp: From asst prof to assoc prof chem, Tuskegee Inst, 66-74; PROF NATURAL SCI & HEAD DIV, NAVAJO COMMUNITY COL, TSAILE, ARIZ, 74- Concurrent Pos: Res Corp Cottrell grant, 67-68; consult, Los Alamos Sci Lab, Univ Calif. Mem: AAAS; Am Chem Soc; The Chem Soc. Res: Synthesis of important organic intermediates; stereochemistry; reaction mechanisms; photochemistry. Mailing Add: Natural Sci Div Navajo Community Col Tsaile AZ 86556

BARRES, HERSTER, forest genetics, plant physiology, see 12th edition

BARRETO, ERNESTO, b Colombia, SAm, Nov 9, 34; US citizen; m 60; c 2. ELECTRODYNAMICS, FLUID DYNAMICS. Educ: NY Univ, BA, 58, MS, 60. Prof Exp: Res assoc physics, Dept Indust Med, NY Univ-Bellevue Med Ctr, 59-60; sci dir physics, Marks Polarized Corp, 60-64; proj engr, Curtiss-Wright Corp, 64-69; SR RES ASSOC, ATMOSPHERIC SCI RES CTR, STATE UNIV NY ALBANY, 66- Concurrent Pos: Consult, Mobil Res Labs & US Coast Guard, 72-; mem comt hazardous mat, Nat Res Coun, 74-76; comt mem, Subcomn IV, Ions-Aerosols-Radioactivity, Int Comn Atmospheric Elec, 74- Mem: Am Phys Soc; Am Geophys Union; Am Inst Aeronaut & Astronaut. Res: Transformation of stored electrical energy into heat in electrical discharges; oil transportation hazards, lightning studies, charged aerosol physics. Mailing Add: Atmospheric Sci Res Ctr State Univ of NY Albany 130 Saratoga Rd Scotia NY 12302

BARRETT, ALAN H, b Springfield, Mass, June 7, 27; m 49; c 2. RADIO ASTRONOMY. Educ: Purdue Univ, BS, 50; Columbia Univ, MS, 53, PhD(physics), 56. Prof Exp: Asst microwave spectros, Columbia Univ, 53-56; physicist radio astron, US Naval Res Lab, Washington, 56-57; assoc, Univ Mich, 57-61, assoc prof, 61-65, prof elec eng, 65-67, PROF PHYSICS, MASS INST TECHNOL, 67- Concurrent Pos: Nat Res Coun-Naval Res Lab fel, 56-57. Mem: Am Astron Soc; Am Geophys Union; Int Astron Union; Int Sci Radio Union. Res: Interstellar radio spectroscopy; microwave spectroscopy of planetary atmospheres; satellite meteorology. Mailing Add: Dept of Physics Mass Inst of Technol Cambridge MA 02139

BARRETT, BRUCE RICHARD, b Kansas City, Kans, Aug 19, 39. THEORETICAL NUCLEAR PHYSICS. Educ: Univ Kans, BS, 61; Stanford Univ, MS, 64, PhD(physics), 67. Prof Exp: Res fel physics, Weizmann Inst Sci, Israel, 67-68; Andrew Mellon res fel, Univ Pittsburgh, 68-69, res assoc, 69-70; asst prof, 70-72, ASSOC PROF PHYSICS, UNIV ARIZ, 72- Concurrent Pos: Alfred P Sloan Found res fel, 72-76; Alexander von Humboldt fel, 76-77. Mem: Am Phys Soc. Res: Structure of finite nuclei; particle-hole and particle-particle states with realistic nuclear forces; effective interaction calculations with realistic nuclear forces; exact reaction matrix calculations; real nuclear three-body forces. Mailing Add: Dept of Physics Univ of Ariz Tucson AZ 85721

BARRETT, BURDETTE EUGENE, invertebrate zoology, fisheries, see 12th edition

BARRETT, CHARLES SANBORN, b Vermillion, SDak, 1902; m 28; c 1. PHYSICS. Educ: Univ SDak, BS, 25; Univ Chicago, PhD(physics), 28. Prof Exp: Asst physicist, Naval Res Lab, 28-32; metallurgist, Metall Res Lab, Carnegie Inst Technol, 32-46, lectr metall, 32-41, assoc prof, 41-44, prof, 45-46; res prof, Inst Study Metals, 46-71, EMER PROF METALL, UNIV CHICAGO, 71-; PROF, UNIV DENVER, 70- Concurrent Pos: With Nat Res Coun, 48-65; vis prof, Univ Birmingham, 52, Univ Denver, 62, Stanford Univ, 63, Univ Va, 68, 70 & Ga Inst Technol, 73; George Eastman prof, Oxford Univ, 65-66. Consult, Argonne Nat Lab, Univ Chicago, 47-, Gen Elec Co, NY, 53-58, US Steel Corp, Pa, 59-64, Bell Tel Labs, 64 & Marmara Inst, Turkey, 73. Honors & Awards: Mathewson Gold Medal, Am Inst Mining, Metall & Petrol Engrs, 34, 44, 50, Hume Rothery Award, 75; Clamer Medal, Franklin Inst, 50; Howe Medal & Sauveur Medal, Am Soc Metals; Heyn Medal, Ger Metall Soc. Mem: Nat Acad Sci; fel Am Phys Soc; fel Am Inst Mining, Metall & Petrol Engrs; hon mem Am Soc Metals; Int Union Crystallog. Res: Preferred orientations in metals and alloys; x-ray equipment and methods; structure, deformation and transformation of metals; crystallography at low temperatures; asbestos identification. Mailing Add: Dept of Metall Univ of Denver Denver CO 80210

BARRETT, DENNIS, b Philadelphia, Pa, Feb 13, 36; m 64. DEVELOPMENTAL BIOLOGY, BIOCHEMICAL GENETICS. Educ: Univ Pa, AB, 57; Calif Inst Technol, PhD(biochem), 63. Prof Exp: USPHS res fel biochem, Univ Calif, Berkeley, 63-65; asst prof zool, Univ Calif, Davis, 65-73; ASST PROF BIOL SCI, UNIV DENVER, 73- Concurrent Pos: Spec consult, US AID Mission, Tunisia, 69; instr physiol, Marine Biol Lab, Woods Hole, 72-76. Mem: AAAS; Am Soc Zool; Soc Develop Biol; NY Acad Sci. Res: Developmental biochemistry; control of gene expression; enzyme changes in embryogenesis; substrate specificity of proteases. Mailing Add: Dept of Biol Sci Univ of Denver Denver CO 80210

BARRETT, EAMON BOYD, mathematics, operations analysis, see 12th edition

BARRETT, EARL WALLACE, b San Francisco, Calif, June 1, 19; m 55; c 2. METEOROLOGY. Educ: Univ Calif, BS, 41; Univ Chicago, SM, 48, PhD, 58. Prof Exp: Res asst org chem, Iowa State Col, 41-42; from res asst to instr meteorol, Univ Chicago, 46-53, res investr, Oceanog Inst, Woods Hole, 53-55; res assoc meteorol, Univ Chicago, 55-61; vis assoc prof, Johns Hopkins Univ, 61-62; assoc prof eng sci, Northwestern Univ, 62-67; RES METEOROLOGIST, ATMOSPHERIC PHYSICS & CHEM LAB, ENVIRON RES LABS, NAT OCEANOG & ATMOSPHERIC ADMIN, 67- Mem: Am Meteorol Soc; Am Geophys Union. Res: Meteorological instruments; cloud physics; atmospheric chemistry; dynamic meteorology. Mailing Add: Atmospheric Physics & Chem Lab Environ Res Labs NOAA Boulder CO 80302

BARRETT, EDWARD JOSEPH, b New York, NY, July 4, 31; m 55; c 2. ORGANIC CHEMISTRY. Educ: Fordham Univ, BA, 53; Columbia Univ, MA, 60, PhD(chem), 62. Prof Exp: From asst prof to assoc prof chem, 64-70, chmn dept, 68-73, PROF CHEM, HUNTER COL, 71-, DEAN & ASSOC PROVOST, 74- Mem: AAAS; Am Chem Soc; fel Am Inst Chem; NY Acad Sci; The Chem Soc. Res: Design of atomic and molecular models; differential thermal analysis; natural products. Mailing Add: Hunter Col 695 Park Ave New York NY 10021

BARRETT, ELTON RAY, b Georgetown, La, Apr 15, 33; m 55; c 2. PLANT PATHOLOGY, BREEDING. Educ: La Polytech Inst, BS, 55; La State Univ, MS, 57, PhD(plant path), 59. Prof Exp: From asst prof to assoc prof, 59-71, PROF BIOL, NORTHEAST LA UNIV, 71- Mem: Am Phytopath Soc; Mycol Soc Am. Res: Pollen germination studies in sugarcane; comparisons of methods of sugar cane breeding; ecology of the fungi. Mailing Add: Dept of Biol Northeast La Univ Monroe LA 71201

BARRETT, FRED FUNSTON, food technology, see 12th edition

BARRETT, FRED OLIVER, b Montgomery, Ala, June 1, 16; m 46; c 2. CHEMISTRY. Educ: Ala Polytech Inst, BS, 37; Ohio State Univ, MSc, 40, PhD(chem eng), 42. Prof Exp: Chemist, State Hwy Dept, Ala, 37, Southern Cotton Oil Co, La, 37-39 & State Dept Agr, Ala, 39; chem engr, Procter & Gamble Co, Ohio, 42-48; from res chemist to mgr polymerization res ctr, 48-67, mgr derivatives res, 67-71, MGR NEW DIMER & OZONE PROD, ORG CHEM DIV, EMERY INDUSTS, 71- Mem: Am Inst Chem Eng; Am Chem Soc; Am Oil Chem Soc. Res: Soaps; synthetic detergents; fats and oils; waxes; fineness versus porosity in hydrogenation; liquid-liquid extraction two film concept. Mailing Add: Bldg 53 Emery Industs 4900 Este Ave Cincinnati OH 45232

BARRETT, GARY WAYNE, b Princeton, Ind, Jan 3, 40; m 69. ECOLOGY. Educ: Oakland City Col, Ind, BS, 61; Marquette Univ, MS, 63; Univ Ga, PhD(zool), 67. Prof Exp: Asst prof biol, Drake Univ, 67-68; asst prof, 68-75, actg dir, Inst Environ Sci, 70-75, ASSOC PROF ZOOL & DEP DIR INST ENVIRON SCI, MIAMI UNIV, 75- Concurrent Pos: Drake Univ Res Coun grant, 67-68; NSF grant, 70. Mem: AAAS; Ecol Soc Am; Am Soc Mammal; Wildlife Soc; Am Inst Biol Sci. Res: Pesticide stresses on total ecosystems; mammalian population regulation and dynamics; species diversity in nature; bioenergetics of small mammal populations. Mailing Add: Dept of Zool & Physiol Miami Univ Oxford OH 45046

BARRETT, HAROLD SPENCER, b Providence, RI, Nov 18, 15; m 41; c 3. PREVENTIVE MEDICINE, PUBLIC HEALTH. Educ: Brown Univ, AB, 37; Harvard Univ, MD, 41; Univ Mich, MPH, 47; Am Bd Prev Med, dipl, 49. Prof Exp: Intern & resident, RI Hosp & intern, Chapin Hosp, Providence, 41-43; indust physician, Walsh-Kaiser Ship Yard, RI, 44; asst out-patient vis physician, Med Staff, RI Hosp, 44-47; pub health internist, 50-51, chief med serv sect, 51-53, DEPT COMNR, STATE DEPT HEALTH, CONN, 53- Concurrent Pos: Spec consult, tuberc control, USPHS, 50-53; lectr, Dept Pub Health, Sch Med, Yale Univ, 57- Mem: AAAS; fel Am Col Chest Physicians; fel Am Pub Health Asn; Am Col Prev Med; Am Col Sports Med. Res: Statistical techniques in public health administration. Mailing Add: State Dept of Health 79 Elm St Hartford CT 06115

BARRETT, HAROLD WHILBERT, b Ligonier, Pa, Mar 19, 09; m 42. BIOCHEMISTRY. Educ: Univ Colo, BA, 39, MA, 43, PhD(chem), 47. Prof Exp: Asst prof chem, Colo A&M Col, 47-49; assoc prof biochem, Univ Mass, 49-50; from asst prof to assoc prof, Sch Med, Univ Kans, 50-59; res prof chem & dir Millar Wilson Lab, Jacksonville Univ, 60-75; RETIRED. Concurrent Pos: Chem Found Inc grant; NIH fel, Calif Inst Technol; NSF fel, Univ Colo, 68-69. Mem: AAAS; assoc Soc Exp Biol & Med; Am Chem Soc; Am Soc Biol Chem. Res: Antithyroid compounds; steroids; pyrimidine metabolism; protein analysis. Mailing Add: 704 Pleasant St Boulder CO 80302

BARRETT, HARRISON HOOKER, b Springfield, Mass, July 1, 39; m 59; c 2. MEDICAL INSTRUMENTATION, OPTICS. Educ: Va Polytech Inst, BS, 60; Mass Inst Technol, SM, 62; Harvard Univ, PhD(appl physics), 69. Prof Exp: From res scientist to prin res scientist, Res Div, Raytheon Co, 62-74; ASSOC PROF, RADIOL DEPT & OPTICAL SCI CTR, UNIV ARIZ, 74- Mem: Am Phys Soc; Optical Soc Am; Soc Photo-Optical Instrumentation Engrs. Res: Optical instrumentation for radiology and nuclear medicine; medical ultrasound; holography. Mailing Add: Optical Sci Ctr Univ of Ariz Tucson AZ 85721

BARRETT, IZADORE, b Vancouver, BC, Oct 4, 26; m 58; c 4. FISHERIES. Educ: Univ BC, BA, 47, MA, 49. Prof Exp: Biologist, Trout Hatcheries, BC Game Comn, 52-56; scientist fishery biol, Inter-Am Trop Tuna Comn, Scripps Inst Oceanog, 56-61, sr scientist tuna physiol, 61-67, ed, Bull, 64-65; chief fishery biologist, Inst Fishery Develop, Chile, 67-70; asst dir, Fishery-Oceanog Ctr, 70-72, DEP DIR SOUTHWEST FISHERIES CTR, NAT MARINE FISHERIES SERV, NAT OCEANIC & ATMOSPHERIC ADMIN, US DEPT COM, 72- Concurrent Pos: Marine fisheries adv, Govt Chile, Food & Agr Orgn of UN, 69-70. Mem: AAAS; Am Soc Ichthyologists & Herpetologists; Am Inst Fishery Res Biologists (vpres, 73-74); Am Soc Limnol & Oceanog; Am Fisheries Soc. Res: Fisheries-oceanography research and administration. Mailing Add: Southwest Fisheries Ctr PO Box 271 La Jolla CA 92037

BARRETT, JAMES HENRY, organic chemistry, polymer chemistry, see 12th edition

BARRETT, JAMES MARTIN, b Chippewa Falls, Wis, July 4, 20; m 47; c 5. PROTOZOOLOGY. Educ: Marquette Univ, BS, 47, MS, 49; Univ Ill, PhD(zool), 53. Prof Exp: Asst zool, Marquette Univ, 47-48; asst, Univ Ill, 48-50; from instr to asst prof, 51-60, ASSOC PROF ZOOL, MARQUETTE UNIV, 60- Mem: Soc Protozool; Am Soc Zool. Res: Biology of protozoa. Mailing Add: Dept of Biol Marquette Univ Milwaukee WI 53233

BARRETT, JAMES PASSMORE, b Atlanta, Ga, June 20, 31; m; c 2. FORESTRY. Educ: NC State Univ, BS, 54; Duke Univ, MF, 57, PhD(forest mensuration), 62. Prof Exp: Forest researcher, US Forest Serv, 57-62; from asst prof forestry to assoc prof, 62-75, PROF FOREST BIOMET, UNIV N H, 75- Mem: Soc Am Foresters. Res: Applications of mathematical and statistical techniques in forestry. Mailing Add: Inst of Environ & Nat Resources Univ of NH Durham NH 03824

BARRETT, JAMES THOMAS, b Iowa, May 20, 27; m 49, 67; c 4. IMMUNOLOGY, MICROBIOLOGY. Educ: Univ Iowa, BA, 50, MS, 51, PhD, 53. Prof Exp: Asst prof microbiol, Sch Med, Univ Ark, 53-57; from asst prof to assoc prof, 57-69, PROF MICROBIOL, SCH MED, UNIV MO-COLUMBIA, 69- Concurrent Pos: NIH spec fel, Sweden, 63-64; sabbatical, Dept Immunol, Med Microbiol Inst, Fac Med, Gothenberg Univ, 70-71; Nat Acad Sci exchange scientist, Romanian Acad Sci, 71. Mem: AAAS; Am Asn Immunol; Am Soc Microbiol. Mailing Add: Dept of Microbiol Univ of Mo Columbia MO 65201

BARRETT, JERRY WAYNE, b Marshall, Tex, Apr 29, 36; m 59; c 2. PHYSICAL CHEMISTRY. Educ: ETex Baptist Col, BS, 61; Baylor Univ, PhD(chem), 68. Prof Exp: From asst prof to assoc prof, 66-75, PROF CHEM, SAMFORD UNIV, 75- Mem: Am Chem Soc; Electrochem Soc. Res: Electrodeposition, corrosion, electrodissolution and chemical plating of metals; crystallography of metal deposits; electroanalytical methods; electrode kinetics and electrode reaction mechanism studies. Mailing Add: Dept of Chem Samford Univ Birmingham AL 35209

BARRETT, JOHN HAROLD, b Springfield, Mo, Oct 9, 26; m 52; c 1. THEORETICAL SOLID STATE PHYSICS. Educ: Rice Inst, BS, 48, MA, 50, PhD(physics), 52. Prof Exp: Eli Lilly fel physics, Mass Inst Technol, 52-53; asst prof, NC State Col, 53-56 & La State Univ, 56-59; PHYSICIST, OAK RIDGE NAT LAB, 59- Mem: Am Phys Soc. Res: Radiation damage and defects in solids. Mailing Add: Solid State Div Oak Ridge Nat Lab Oak Ridge TN 37830

BARRETT, JOHN WILLIAM, b Columbus, Ohio, July 27, 13; m 34; c 2. FORESTRY. Educ: Univ Mich, BSF, 37, MF, 48; State Univ NY Col Forestry, Syracuse Univ, PhD(forestry), 53. Prof Exp: Asst timber foreman & logging engr, Long-Bell Lumber Co, 37-40, timber foreman & logging engr, 45-46; asst prof forestry, Farragut Col & Technol Inst, 46; from instr to prof silvicult, State Univ NY Col Forestry, Syracuse Univ, 48-63; PROF FORESTRY & HEAD DEPT, UNIV TENN, KNOXVILLE, 64- Mem: Soc Am Foresters. Res: Northern hardwood silviculture; thinning; fire control; forestry education. Mailing Add: Dept of Forestry Univ of Tenn Knoxville TN 37901

BARRETT, JOSEPH JOHN, b Scranton, Pa, Mar 11, 36; m 67; c 2. QUANTUM OPTICS, MOLECULAR SPECTROSCOPY. Educ: Univ Scranton, BS, 58; Fordham Univ, MS, 60, PhD(physics), 64. Prof Exp: Res asst physics, Fordham Univ, 62-64; sr scientist physics, Perkin-Elmer Corp, 64-70; SR PHYSICIST, ALLIED CHEM CORP, 70- Mem: Optical Soc Am; Am Phys Soc; Sigma Xi. Res: Lasers for spectroscopic applications with emphasis on the interferometric analysis of rotational Raman spectra of gases and the generation of coherent anti-Stokes Raman radiation. Mailing Add: Allied Chem Corp PO Box 1021R Morristown NJ 07960

BARRETT, LIDA KITTRELL, b Houston, Tex, May 21, 27; wid; c 3. MATHEMATICS. Educ: Rice Inst, BA, 46; Univ Tex, MA, 49; Univ Pa, PhD(math), 54. Prof Exp: Mathematician, Schlumberger Well Surv Corp, 46-47; teacher math, Tex State Col Women, 47-48; res mathematician, Defense Res Lab, Univ Tex, 49-50; asst instr math, Univ Pa, 53-54; instr, Univ Conn, 55-56; lectr, Univ Utah, 56-61; assoc prof, 61-70, PROF MATH, UNIV TENN, KNOXVILLE, 70-, HEAD DEPT, 73- Concurrent Pos: Vis lectr, Univ Wis, 59-60; consult, Oak Ridge Nat Lab, 64- Mem: Am Math Soc; Math Asn Am. Res: Point set topology; applications of topology in metallurgy. Mailing Add: Dept of Math Univ of Tenn Knoxville TN 37916

BARRETT, LOUIS CARL, b Murray, Utah, Jan 23, 24; m 47; c 4. MATHEMATICS. Educ: Univ Utah, BS, 48, MS, 51, PhD(appl math), 56. Prof Exp: Instr math, Univ Utah, 53-56; assoc prof math & physics, Ariz State Univ, 56-57; from assoc prof to prof math, SDak Sch Mines & Technol, 57-65, head dept, 60-65; prof & chmn dept, Clarkson Col Technol, 65-67; prof & head dept, 67-72, PROF MATH, MONT STATE UNIV, 72- Concurrent Pos: Res mathematician, Holloman Air Develop Ctr, NMex, 55, math consult, 56; res mathematician, US Naval Ord Test Sta, Calif, 57, 58, 63-64, consult, 63-64; lectr, Ariz State Univ, 57 & NSF Inst, SDak, 59. Mem: Am Math Soc; Math Asn Am; Soc Indust & Appl Math. Res: Eigenvalue problems; differential equation theory; boundary value problems; optimization problems; Sturm-Liouville systems; linear algebra; statistics; probability, mathematical physics and analysis; calculus variations; logic and foundations. Mailing Add: Dept of Math Mont State Univ Bozeman MT 59715

BARRETT, MARY OLIVIA, b Chicago, Ill, Oct 3, 20. ORGANIC CHEMISTRY. Educ: St Xavier Col, BS, 42; Univ Notre Dame, MS, 53, PhD(chem), 57. Prof Exp: From instr to asst prof, 56-62, chmn div natural sci, 61-63, pres, 63-68, PROF CHEM, ST XAVIER COL, 62- Mem: Am Chem Soc. Res: Reaction rates in aqueous solutions. Mailing Add: St Xavier Col 103rd & Central Park Chicago IL 60655

BARRETT, O'NEILL, JR, b Baton Rouge, La, Mar 21, 29; m 52; c 3. HEMATOLOGY, ONCOLOGY. Educ: La State Univ, BS, 49, MD, 53; Baylor Univ, MSc, 58. Prof Exp: Med Corps, US Army, 53-, clin investr hemat & asst dir basic sci course, Walter Reed Army Inst Res, 59-60, chief gen med & hemat, Madigan Gen Hosp, 60-62, dep chief dept med, Letterman Gen Hosp San Francisco, Calif, 63-68, chief dept med, Tripler Gen Hosp, Honolulu, 68-71, CHIEF DEPT MED, WALTER REED GEN HOSP, 71-; PROF MED, COL MED, UNIV S FLA, 73-, PROF COMPREHENSIVE MED & CHMN DEPT, 74- Concurrent Pos: Clin asst prof, Sch Med, Univ Calif, San Francisco, 63-68; clin prof, Sch Med, Univ Hawaii, 69-71. Mem: Fel Am Col Physicians; Am Soc Hemat; Am Fedn Clin Res; Soc Nuclear Med; AMA. Mailing Add: Dept of Comprehensive Med Univ S Fla Med Ctr Tampa FL 33620

BARRETT, PAUL HENRY, b Petaluma, Calif, Dec 4, 22; m 48; c 3. PHYSICS. Educ: Mont State Col, BS, 44; Univ Calif, PhD(physics), 51. Prof Exp: Res assoc physics, Cornell Univ, 51-53; asst prof, Syracuse Univ, 53-55; assoc prof, 55-66, PROF PHYSICS, UNIV CALIF, SANTA BARBARA, 66- Mem: Am Phys Soc. Res: Solid state physics. Mailing Add: Dept of Physics Univ of Calif Santa Barbara CA 93106

BARRETT, PAUL HOWARD, b Youngstown, Ohio, Nov 7, 17; m 44; c 3. LIMNOLOGY. Educ: Mich State Univ, BS, 40, PhD(zool), 52; Univ Conn, MS, 43. Prof Exp: Fisheries technician, State Dept Conserv, Mich, 41-42; instr biol sci, 47-52, from asst prof to assoc prof, 52-68, PROF NATURAL SCI, MICH STATE UNIV, 68- Mem: Am Soc Limnol & Oceanog; Nat Asn Geol Teachers; Soc Study Evolution; Hist Sci Soc. Res: History of science. Mailing Add: Dept of Natural Sci Mich State Univ East Lansing MI 48823

BARRETT, PETER FOWLER, b Port Dover, Ont, May 27, 39; m 65; c 2. INORGANIC CHEMISTRY. Educ: Queen's Univ, Ont, BSc, 61, MSc, 62; Univ Toronto, PhD(inorg chem), 65. Prof Exp: Nat Res Coun overseas fel, 65-67; asst prof inorg chem, 67-72, actg chmn dept chem, 75-76, ASSOC PROF INORG CHEM, TRENT UNIV, 72- Mem: Chem Inst Can. Res: Kinetics and mechanisms of reactions of transition metal carbonyl complexes. Mailing Add: Dept of Chem Trent Univ Peterborough ON Can

BARRETT, PETER VAN DOREN, b Los Angeles, Calif, Oct 22, 34; m 59; c 3. INTERNAL MEDICINE, GASTROENTEROLOGY. Educ: Univ Calif, Los Angeles, BA, 56; Harvard Med Sch, MD, 60; Am Bd Internal Med, dipl, cert internal med & gastroenterol. Prof Exp: From intern to asst resident med, Mass Gen Hosp, Boston, 60-62; clin assoc, NIH, 62-65; resident, Mass Gen Hosp, Boston, 65-66; asst prof med, Sch Med, Univ Calif, Los Angeles, 67-71; assoc chief div gastroenterol, Harbor Gen Hosp, Torrance, 67-75; ASSOC PROF MED, SCH MED, UNIV CALIF, LOS ANGELES, 71-; DIR MED EDUC, ST MARY MED CTR, LONG BEACH, CALIF, 75- Concurrent Pos: Fel gastroenterol, Wadsworth Vet Admin Hosp, Los Angeles, 66-67; NIH res grants, 67-76. Mem: Am Gastroenterol Asn; fel Am Col Physicians; Am Fedn Clin Res. Res: Hepatic physiology and bile pigment metabolism. Mailing Add: 29377 Quailwood Dr Ranchos Palos Verdes CA 90274

BARRETT, RICHARD ALLAN, b San Fernando, Calif, Nov 21, 38; m 61; c 1. ANTHROPOLOGY, ETHNOLOGY. Educ: Univ Calif, Los Angeles, BA, 64; Univ Mich, MA, 65, PhD(anthrop), 70. Prof Exp: Instr anthrop, Temple Univ, 68-70; ASST PROF ANTHROP, UNIV N MEX, 70- Mem: Am Anthrop Asn. Res: Modernization of traditional cultures; peasant society; peoples of the Mediterranean and Latin America. Mailing Add: Dept of Anthrop Univ of NMex Albuquerque NM 87106

BARRETT, ROBERT, b Los Angeles, Calif, Jan 4, 14; m 38; c 1. NEUROPHYSIOLOGY. Educ: Univ Calif, Los Angeles, BA, 34, MA, 65, PhD(zool), 69. Prof Exp: Tutor & instr sociol, Harvard Univ, 34-36; mem, NSF Cetacean Exped, Tierra del Fuego, 68; NIMH fel, Brain Res Inst, 69-70, DIR DEPT BIOL & PHYS SCI, EXTEN & LECTR ZOOL, UNIV CALIF, LOS ANGELES, 70- Mem: Am Inst Biol Sci; Am Soc Zool. Res: Sensory physiology, particularly in areas of heat sensing, olfaction and taste; comparative neurophysiology. Mailing Add: 1520 Roscomare Rd Los Angeles CA 90024

BARRETT, ROBERT EARL, b Wooster, Ohio, Mar 8, 23; m 44; c 5. PHYSICAL CHEMISTRY, POLYMER CHEMISTRY. Educ: Mich State Univ, BS, 47; Ohio State Univ, MS, 49, PhD(chem), 51. Prof Exp: Res chemist, Visking Corp Div, Union Carbide Corp, 51-52; prin res chemist, Battelle Mem Inst, 52-57; res supvr, 57-67, res lab mgr & asst to vpres res & develop, 67-75, RES MGR, COPOLYMER RUBBER & CHEM CORP, 75- Mem: Am Chem Soc. Res: Analytical, polymer and colloidal chemistry; studies of enzymes of tobacco, proteins in wheat seedlings and synthetic rubber. Mailing Add: Copolymer Rubber & Chem Corp PO Box 2591 Baton Rouge LA 70821

BARRETT, TERENCE WILLIAM, b London, Eng, Apr 22, 39; m 68; c 3. BIOPHYSICS. Educ: Edinburgh Univ, Scotland, MA, 64; Stanford Univ, PhD(neurophysiol), 67. Prof Exp: Res fel neurophysiol, Sch Med, Stanford Univ, 67-68; res assoc, NY Med Col, 68-69; asst prof, Carnegie-Mellon Univ, 69-71; asst prof neurobiol, 71-74, ASSOC PROF NEUROBIOL, CTR HEALTH SCI, UNIV TENN, 74- Mem: Biophys Soc; Am Physiol Soc; Acoust Soc Am; Soc Math Biol; Soc Neurosci. Res: Quantum physics and its application to the study of the central nervous system; physical chemistry of biological macromolecules; energy coupling and energy transfer. Mailing Add: Dept of Physiol & Biophys Univ of Tenn Ctr for Health Sci Memphis TN 38163

BARRETT, THOMAS WILSON, b Orem, Utah, Sept 6, 17; m 43; c 4. AGRONOMY. Educ: Brigham Young Univ, BS, 40; Cornell Univ, MS, 48, PhD(agron), 50. Prof Exp: Chemist, US Steel Co, Utah, 46; PROF AGRON, AGR DIV, ARIZ STATE UNIV, 50- Concurrent Pos: Chmn, Sulphur Springs Valley Arbit Comt, 55- Res: Decomposition of cyanamid in soil; effects of tillage and fertilizers on crop yields; soil fertility; effect of air pollutants on agricultural crops. Mailing Add: Agr Div Ariz State Univ Tempe AZ 85281

BARRETT, WALTER EDWARD, b Stamford, Conn, May 26, 21; m; c 4. PHARMACOLOGY. Educ: Columbia Univ, AB, 44; Princeton Univ, MA, 57, PhD, 58. Prof Exp: From asst pharmacologist to sr pharmacologist, Ciba Pharm Co, 44-61, assoc dir gen pharmacol, 62-64, head, 64-66, from asst dir macrobiol to dir pharmacol, 66-69; dir pharmacol, Parke, Davis & Co, 69-70; clin res assoc, Sandoz Pharmaceut, 71-72, head short term res & develop, 72-73, DIR SHORT TERM RES & DEVELOP, PHARMACEUT RES & DEVELOP, SANDOZ, INC, 74- Concurrent Pos: Mem med adv bd, Coun High Blood Pressure Res, Am Heart Asn. Mem: Am Soc Pharmacol; NY Acad Sci; Int Soc Hypertension; Sigma Xi. Res: Mechanism of action of autonomic drugs; cardiovascular drugs; clinical pharmacology; biopharmaceutics. Mailing Add: Short Term Res & Develop Sandoz Inc Hanover NJ 07936

BARRETT, WAYNE THOMAS, b Youngstown, Ohio, Dec 21, 19; m 43; c 2. PHYSICAL CHEMISTRY. Educ: Mich State Univ, BS, 41, MS, 42; Univ Pittsburgh, PhD(phys chem), 51. Prof Exp: Chemist, Phillips Petrol Co, 42-46 & Mellon Inst, 46-50; from supvr catalyst res to dir inorg chem res dept, W R Grace & Co, 50-63; vpres res & develop, 63-68, vpres & gen mgr chem & minerals div, 68-70, pres & chief operating officer, 70-71, PRES & CHIEF EXEC OFFICER, FOOTE MINERAL CO, 71-, MEM BD DIRS, 70- Mem: Am Chem Soc. Res: Inorganic fluorides, silica gel and silicalumina cracking catalysts; synthesis of hydrazine and hydrazine derivatives; inorganic colloids; rare earth and thoria chemistry; nuclear fuel systems; semiconductors. Mailing Add: Foote Mineral Co Rte 100 Exton PA 19341

BARRETT, WILLIAM A, b Pittsburgh, Pa, May 10, 03. UROLOGY. Educ: Duquesne Univ, BS, 25; Univ Pittsburgh, MD, 27; Am Bd Urol, dipl. Prof Exp: Instr bact, Pa State Univ, 32; resident urol, Cornell Med Ctr, 33-36; instr anat, 36-40, asst prof urol, 40-75, EMER PROF UROL, MED SCH, UNIV PITTSBURGH, 75- Concurrent Pos: Fel urol, Mercy Hosp, Pittsburgh, 28-30. Mem: AMA; Am Urol Asn; Am Col Surg; Int Col Surg; Int Soc Urol. Mailing Add: 121 Hibiscus Dr Pittsburgh PA 15235

BARRETT, WILLIAM JORDAN, b Harrison, Ga, Nov 7, 16; m 46; c 4. ANALYTICAL CHEMISTRY. Educ: Mercer Univ, AB, 36, MA, 37; Univ Fla, PhD(chem), 50. Prof Exp: Instr chem, Univ Fla, 47-50; head anal chem sect, 50-67, HEAD ANAL & CHEM DIV, SOUTHERN RES INST, 67- Mem: Am Chem Soc; Am Defense Preparedness Asn. Res: Electroanalytical chemistry; polarography; spot reactions; air pollution; water pollution; military defense. Mailing Add: 2000 Ninth Ave S Birmingham AL 35205

BARRETT, WILLIAM LOUIS, b Phoenix, Ariz, June 3, 33; m 54; c 3. SOLID STATE PHYSICS, NUCLEAR ENGINEERING. Educ: Univ Idaho, BS, 56; Univ Wash, MS, 61, PhD(physics), 68. Prof Exp: Res engr, Boeing Co, Wash, 61-64; asst prof, 68-71, ASSOC PROF PHYSICS, WESTERN WASH STATE COL, 71-, CHMN DEPT PHYSICS & ASTRON, 75- Mem: AAAS; Am Inst Physics. Res: Ferromagnetic resonance in metals; crystal growth; nuclear rocket concepts. Mailing Add: Dept of Physics & Astron Western Wash State Col Bellingham WA 98225

BARRICK, ELLIOTT ROY, b Worthington, Minn, Feb 7, 15; m 45; c 3. ANIMAL SCIENCE. Educ: Okla State Univ, BS, 38; Purdue Univ, MS, 41, PhD(animal nutrit), 47. Prof Exp: Asst prof animal husb, Purdue Univ, 47-48; assoc prof, Univ Tenn, 48-49; assoc prof animal indust, 49-54, head animal husb sect, 54-72, PROF ANIMAL SCI, NC STATE UNIV, 62- Mem: Am Soc Animal Sci; Am Dairy Sci Asn. Res: Animal production and nutrition. Mailing Add: Dept of Animal Sci NC State Univ Raleigh NC 27607

BARRICK, LLOYD D, organic chemistry, see 12th edition

BARRIE, ROBERT, b Newmains, Scotland, Sept 19, 27; m 54. THEORETICAL PHYSICS. Educ: Univ Glasgow, BSc, 49, PhD(physics), 54. Prof Exp: From sci officer to sr sci officer, Servs Electronics Res Lab, Baldock, Herts, England, 53-57; instr, 57-67, PROF PHYSICS, UNIV BC, 67- Res: Electronic properties of solids. Mailing Add: Dept of Physics Univ of BC Vancouver BC Can

BARRIENTOS, CELSO SAQUITAN, b Bacoor, Philippines, Jan 9, 36; m 59; c 2. OCEANOGRAPHY. Educ: Feati Univ, Philippines, BSc, 60; NY Univ, MSc, 65, PhD(meteorol), 69. Prof Exp: Asst res scientist geophys sci lab, NY Univ, 62-66; res scientist, Isotopes, Inc, 66-68 & Geophys Sci Lab, NY Univ, 68-69; RES METEOROLOGIST, TECH DEVELOP LAB, NAT WEATHER SERV, NAT OCEANIC & ATMOSPHERIC ADMIN, 69- Mem: AAAS; Am Meteorol Soc; Am Geophys Union; foreign mem Royal Meteorol Soc; Meteorol Soc Japan. Res: Hurricane dynamics and structures; tropical meteorology; atmospheric diffusion; turbulence; micrometeorology; ocean waves; air-sea interactions; marine environmental predictions; Great Lakes wind-wave forecasting. Mailing Add: 4724 Iris St Rockville MD 20853

BARRIER, GEORGE EDGAR, b Crossnore, NC, Aug 14, 28; m 53; c 1. PLANT PHYSIOLOGY. Educ: Berea Col, BS, 51; NC State Col, MS, 54; Iowa State Col, PhD(plant physiol), 56. Prof Exp: Res plant physiologist, 56-65, RES SUPVR, E I DU PONT DE NEMOURS & CO, 65- Mem: Am Soc Plant Physiol; Weed Sci Soc Am. Res: Translocation in plants; soil-plant relationships; response of plants to chemical treatment; plant-water relationships; plant growth; weed control. Mailing Add: E I du Pont de Nemours & Co Plant Res Lab Wilson Rd Wilmington DE 19898

BARRIGA, OMAR OSCAR, b Santiago, Chile, Mar 1, 38; m 60; c 2. PARASITOLOGY, IMMUNOBIOLOGY. Educ: Univ Chile, BA, 58, DVM, 63; Univ Ill, PhD(parasitol), 72. Prof Exp: Teacher biol, Pvt High Schs, Santiago, 58-62; small animal vet, 63; asst prof parasitol, Sch Med, Univ Chile, 64-71, prof immunol, Grad Sch Med & assoc prof parasitol, Schs Med & Pub Health, 72; ASST PROF PARASITOL, SCH VET MED, UNIV PA, 73- Concurrent Pos: Ed, J Chilean Soc Vet Med, 66-67; consult parasitic zoonoses, J Am Vet Med Asn, 74- Mem: World Asn Advan Vet Parasitol; AAAS; Am Asn Univ Profs; Am Asn Parasitologists; Am Asn Immunologists. Res: Immunity to parasitic infections; modifications of the immune response by parasitic infections. Mailing Add: 3800 Spruce St Philadelphia PA 19174

BARRINGER, ANTHONY R, b Bognor Regis, Eng, Oct 20, 25; Can citizen; m 48; c 5. GEOLOGY, GEOPHYSICS. Educ: Univ London, BSc, 51, PhD(econ geol), 54. Prof Exp: Geologist, Selco Explor Co Ltd, Can, 54-55, mgr airborne & tech serv div, 55-61; PRES, BARRINGER RES LTD, CAN, 61-, BARRINGER RES INC, MASS, 65- Concurrent Pos: Mem exec comt, Res Conf Instrumentation Sci; mem nat adv comt res geol sci, Nat Res Coun Can, 62-, subcomt explor geophys, 64-; chmn bd consults remote sensing terrain properties, US Army Engrs, 65-; vis prof, Univ London, 68- Mem: Am Geophys Union; Soc Explor Geophys; Inst Elec & Electronics Eng; Brit

Inst Mining & Metall; Am Inst Mining, Metall & Petrol Eng. Res: Airborne mineral exploration techniques; rapid natural resource evaluation using aircraft and satellites. Mailing Add: Barringer Res Ltd 304 Carlingview Dr Rexdale ON Can

BARRINGER, DONALD F, JR, b Cleveland, Ohio, June 3, 32; m 58; c 4. METABOLISM. Educ: Denison Univ, BS, 54; Ohio State Univ, PhD(org chem), 60. Prof Exp: Res chemist, E I du Pont de Nemours & Co, 59-60; RES CHEMIST, AM CYANAMID CO, 60- Mem: AAAS; Am Chem Soc. Res: Metabolism of pesticides and veterinary drugs in plants, animals and the environment. Mailing Add: 7 Stonicker Dr Lawrenceville NJ 08638

BARRINGER, WILLIAM CHARLES, b Cleveland, Ohio, Feb 28, 34; c 4. PHARMACEUTICAL CHEMISTRY. Educ: Denison Univ, BS, 56; NY Univ, MS, 64, PhD(photochem), 68. Prof Exp: Develop chemist, 57-75, GROUP LEADER PREFORMULATION CHEM, LEDERLE LAB DIV, AM CYANAMID CO, 75- Concurrent Pos: Prof chem, The King's Col, 69-71. Mem: Am Chem Soc. Res: The relationship between the physical and chemical properties of chemical compounds and their performance in pharmaceutical formulations. Mailing Add: Am Cyanamid Co Lederle Lab N Middletown Rd Pearl River NY 10965

BARRINGTON, BURNESS AUSTIN, JR, b Hawthorne, Fla, June 12, 16; m 42; c 4. ANATOMY, MAMMALOGY. Educ: Univ Fla, BS, 39, MS, 40, PhD(biol), 49. Prof Exp: Prof biol, King Col, 49-62; res asst, 64-65, assoc, 65-68, asst prof, 68-71, ASSOC PROF ANAT, MED UNIV SC, 71- Concurrent Pos: Res fel anat, Med Univ SC, 62-64. Mem: Am Asn Anat; Microcirc Soc. Res: Mammalian anatomy and ecology; pathologic mammalian microcirculatory anatomy and physiology. Mailing Add: Dept of Anat Med Univ of SC Charleston SC 29401

BARRINGTON, DAVID STANLEY, b Boston, Mass, Sept 4, 48. BOTANY. Educ: Bates Col, BS, 70; Harvard Univ, PhD(biol), 75. Prof Exp: CUR PRINGLE HERBARIUM, UNIV VT, 74- Mem: Bot Soc Am; Am Fern Soc. Res: Systematics and evolution of tropical American tree ferns; anatomy of the fossil tree fern stems of the Cyatheaceae and Dicksoniaceae. Mailing Add: Dept of Bot Univ of Vt Burlington VT 05401

BARRINGTON, LEONARD FLOYD, biochemistry, see 12th edition

BARRINGTON, RONALD ERIC, b Toronto, Ont, Aug 1, 31; m 56; c 3. PHYSICS. Educ: Univ Toronto, BA, 54, MA, 55, PhD(physics), 57. Prof Exp: Sci officer, Defence Res Telecommun Estab, 57-60 & 62-65 & Norweg Defence Res Estab, 60-62; SECT HEAD PLASMA PHYSICS, COMMUN RES CTR, 65- Concurrent Pos: Exec mem, Assoc Comt Space Res, 67- Mem: Can Asn Physicists. Res: Ionospheric physics; electromagnetic wave propagation; plasma physics. Mailing Add: Communications Res Ctr Box 11490 Sta H Ottawa ON Can

BARRIOS, EARL P, b New Orleans, La, Dec 14, 17; m 46. HORTICULTURE. Educ: La State Univ, PhD(bot, hort), 60. Prof Exp: Chief, Nebr Potato Develop Div, State Dept Agr & Univ Nebr, 48-54; sales mgr agr chem, P V Fertilizer Co, 54-57; from instr to asst prof, 58-64, ASSOC PROF HORT, LA STATE UNIV, BATON ROUGE, 64- Mem: Am Soc Hort Sci. Res: Breeding and genetics of horticultural crops, especially vegetables. Mailing Add: Dept of Hort La State Univ Baton Rouge LA 70803

BARRNETT, RUSSELL JOFFREE, b Boston, Mass, July 27, 20; m 48; c 3. ANATOMY. Educ: Ind Univ, AB, 43; Yale Univ, MD, 48. Prof Exp: Asst zool, Ind Univ, 41-43; physiol, Med Sch, Yale Univ, 45-46, pharmacol, 46-48; from instr to asst prof anat, Harvard Med Sch, 51-59; assoc prof, 59-62, PROF ANAT, MED SCH, YALE UNIV, 62-, CHMN DEPT & DIR GRAD STUDIES ANAT, 68- Concurrent Pos: Harvard fel anat, Harvard Med Sch, 49-51; mem panel cytochem, Biochem Sect, Comt Growth, Div Biol & Agr, Nat Res Coun, 67- Mem: AAAS; Am Asn Anat; Histochem Soc (pres, 62-63); Am Soc Cell Biol; Int Soc Cell Biol. Res: Histochemistry; cytochemistry; electron microscopy; cytology. Mailing Add: Yale Univ Sch of Med New Haven CT 06510

BARRON, ALMEN LEO, b Toronto, Ont, Jan 19, 26; nat US; m 49; c 1. VIROLOGY. Educ: Univ Toronto, BSA, 48, MSA, 49; Queen's Univ (Can), PhD(bact), 53. Prof Exp: Prof microbiol, State Univ NY Buffalo, 68-74; PROF MICROBIOL & CHMN DEPT MICROBIOL & IMMUNOL, COL MED, UNIV ARK, LITTLE ROCK, 74- Concurrent Pos: Fulbright scholar, 64; Commonwealth Fund travel fel, 64; Fight for Sight travel award, 72; consult microbiol, Little Rock Vet Admin Hosp, 74- Mem: Am Acad Microbiol; Infectious Dis Soc Am; Am Asn Immunologists; Soc Exp Biol & Med; Am Soc Microbiol. Res: Animal viruses, biological characteristics with relation to immunology and pathogenesis; Chlamydia group, biology and immunology; venereal diseases. Mailing Add: Dept of Microbiol & Immunol Univ of Ark Col of Med Little Rock AR 72201

BARRON, BRUCE ALBRECHT, b New York, NY, Nov 3, 34; m 75; c 1. OBSTETRICS & GYNECOLOGY. Educ: Allegheny Col, AB, 55; Yale Univ, MPH, 60, PhD(biomet), 65; NY Univ, MD, 71. Prof Exp: Res asst, Dept of Surg, Bellevue Med Ctr, NY Univ, 56-57; biochemist, Walter Reed Army Inst Res, 57-59; res assoc biomet, Rockefeller Univ, 65-67, asst prof, 67-74; intern, Dept Med, Bellevue Med Ctr, 71-72, asst res obstet & gynec, 72-73, chief res, 73-74; ASSOC PROF OBSTET & GYNEC, COL PHYSICIANS & SURGEONS, COLUMBIA UNIV, 74-; SR ATTEND PHYSICIAN, SLOAN HOSP FOR WOMEN, PRESBY MED CTR, 74- Concurrent Pos: Consult, Biomed Div, Population Coun, 65-71. Mem: AAAS; Biomet Soc; Am Statist Asn; Harvey Soc; Sigma Xi. Res: Mathematical models in carcinogenesis. Mailing Add: Columbia-Presby Med Ctr 161 Ft Washington Ave New York NY 10032

BARRON, CHARLES IRWIN, b Chicago, Ill, July 21, 16; m 45; c 4. AEROSPACE MEDICINE. Educ: Univ Ill, BS, 40, MD, 42. Prof Exp: Surgeon, US Dept Army Qm Depot, Ill, 48-50; flight surgeon, 50-51, sr exam physician, 51-53, MED DIR, LOCKHEED-CALIF CO, 53- Concurrent Pos: Lectr, Univ Southern Calif, 54-, assoc clin prof, 63-; med examr, Fed Aviation Agency, 55-; mem comt hearing & bio-acoustics, Nat Res Coun, 55-57; lectr, Univ Calif, Los Angeles, 55, assoc clin prof, Sch Pub Health, 63-; chmn adv coun, Civil Air Surgeon, 60-; lectr, Ohio State Univ, 60, vis prof, 63; chmn res adv comt advan biotechnol & human res, NASA, 63- Honors & Awards: Arnold D Tuttle Award, 63. Mem: AMA; fel Indust Med Asn; Am Inst Aeronaut & Astronaut; fel Aerospace Med Asn (pres, 63-64); Civil Aviation Med Asn (vpres, 55-59). Res: Industrial and aerospace medicine; life sciences and human factors engineering; education in aerospace safety; basic and applied research in aviation physiology, pathology and toxicology; environmental medicine. Mailing Add: Lockheed-Calif Co PO Box551 Burbank CA 91503

BARRON, CHARLIE NELMS, b Texas, Jan 8, 22; m 43; c 3. VETERINARY PATHOLOGY. Educ: A&M Col, Tex, DVM, 43, MS, 50; Univ Mich, PhD(path), 61. Prof Exp: Instr vet anat, A&M Col, Tex, 46-49; assoc prof & head vet diag lab, Okla A&M Col, 49-51; asst chief vet path sect, Armed Forces Inst, 51-53 & 55-59; asst chief path & toxicol sect, Smith Kline & French Labs, 59-71; res prof toxicol, Albany Med Col, 71-73; dir path, Flow Labs, 73-75; PATHOLOGIST, TRACOR JITCO, INC, 75- Concurrent Pos: Co-ed, Pathologia Veterinaria, 64-66, 70-, ed, 66-70. Mem: Am Vet Med Asn; Am Col Vet Path; Am Asn Path & Bact; Int Acad Path. Res: Parasitic diseases; fungous diseases; neoplasia, eye orbital region and skin; cardiovascular toxicology. Mailing Add: 1776 E Jefferson St Rockville MD 20852

BARRON, DONALD HENRY, b Flandreau, SDak, Apr 9, 05; m 32; 2. ANATOMY. Educ: Carleton Col, AB, 28; Iowa State Col, MS, 29; Yale Univ, PhD(zool), 32; Cambridge Univ, MA, 36. Prof Exp: Asst physiol, Iowa State Col, 28-29; asst zool, Yale Univ, 29-31, asst anat, 31-32; instr, Albany Med Col, 32-33, asst prof, 34-35; asst & demonstr anat, Cambridge Univ, 35-36, lectr, 36-37, fel & supvr anat & physiol, St John's Col, 37-40; from asst prof to assoc prof zool, Univ Mo, 40-43; from assoc prof to prof physiol, Sch Med, Yale Univ, 43-69, asst dean, 45-48; J WAYNE REITZ PROF REPROD BIOL & MED, UNIV FLA, 69- Concurrent Pos: Nat Res Coun fel, 33-34; Rockefeller Found traveling fel, 37; Sterling fel, Yale Univ, 40. Mem: Am Asn Anat; Am Physiol Soc; Anat Soc Gt Brit & Ireland. Res: Physiology of the spinal cord; function development of nervous system; fetal circulation; physiology of blood and circulation; neurological anatomy. Mailing Add: Miller Health Ctr Univ of Fla Col of Med Gainesville FL 32601

BARRON, EDWARD J, b Sudan, Tex, June 2, 27; m 53; c 3. CLINICAL BIOCHEMISTRY. Educ: Univ Wash, BS, 50, PhD(biochem), 59. Prof Exp: Clin chemist, Group Health Co-op Puget Sound, 50-55; NIH trainee biochem, Univ Calif, Davis, 60-62; assoc dir, Virginia Mason Res Ctr, 62-68, ASSOC DIR LABS, THE MASON CLIN, 68- Mem: Am Chem Soc; Am Asn Clin Chemists; NY Acad Sci. Res: Lipid chemistry; metabolism of lipids; biosynthesis of fatty acids. Mailing Add: Mason Clin 1118 Ninth Ave Seattle WA 98109

BARRON, EUGENE ROY, b Somerset, Pa, Sept 10, 41; m 63; c 2. ORGANIC CHEMISTRY. Educ: Univ Md, BS, 63, PhD(chem), 67. Prof Exp: SR RES CHEMIST, E I DU PONT DE NEMOURS & CO, INC, 67- Mem: Am Chem Soc. Res: Synthesis and evaluation of polymeric materials for use in fibers. Mailing Add: 1111 Dardel Dr Wilmington DE 19803

BARRON, JOHN ROBERT, b Niagara Falls, Ont, Dec 23, 32; m 55; c 1. ENTOMOLOGY, TAXONOMY. Educ: McGill Univ, BSc, 61, MSc, 62; Univ Alta, PhD(entom, taxon), 69. Prof Exp: Tech asst taxon, McGill Univ, 58-61; res asst exten entom, Univ Alta, 62-68; RES SCIENTIST, ENTOM RES INST, CAN DEPT AGR, 69- Mem: Entom Soc Am; Soc Study Evolution; Soc Syst Zool; Can Soc Zool; Entom Soc Can. Res: Taxonomy of Gryllus of North America; taxonomy and zoogeography of Cleroidea of the world, particularly the family Trogositidae; taxonomy and zoogeography of Ichneumonidae, particularly of North America. Mailing Add: Entom Res Inst Can Dept of Agr Cent Exp Farm Ottawa ON Can

BARRON, KEVIN D, b St John's, Nfld, Apr 21, 29; US citizen; m 56; c 2. MEDICINE, NEUROLOGY. Educ: Dalhousie Univ, MD & CM, 52. Prof Exp: Instr neurol, Col Physicians & Surgeons, Columbia Univ, 57-59; assoc neurol & psychiat, Med Sch, Northwestern Univ, 59-61, from asst prof to prof, 61-69; PROF NEUROL & CHMN DEPT & PROF PATH, ALBANY MED COL, 69- Concurrent Pos: USPHS fel neuropath, Montefiore Hosp, NY, 56-59; USPHS grants, 60-; Nat Multiple Sclerosis Soc grant, 61-63. Mem: Am Neurol Asn; Am Soc Neurochem; Asn Univ Prof Neurol; fel Am Acad Neurol; Am Asn Neuropath. Res: Histochemistry and cytology; electron microscopy of the nervous system; neuroenzymology, particularly hydrolytic enzymes. Mailing Add: Dept of Neurol Albany Med Col Albany NY 12208

BARROS, FERNANDO MONCKEBERG, nutrition, see 12th edition

BARROW, EMILY MILDRED STACY, b Washington, DC, July 2, 27; div; c 3. PHYSIOLOGY. Educ: Meredith Col, AB, 50; Univ NC, Chapel Hill, AM, 52, PhD(physiol), 63. Prof Exp: Asst bot, 50-52, path, 56-59 & physiol, 60-61, res assoc path, 62-68, asst prof, 68-71, ASSOC PROF PATH, UNIV NC, CHAPEL HILL, 71- Mem: AAAS; NY Acad Sci; Am Physiol Soc. Res: Biochemical genetics of blood coagulation disorders. Mailing Add: Dept of Path Univ of NC Sch of Med Chapel Hill NC 27514

BARROW, GORDON M, b Vancouver, BC, Nov 13, 23; m 57. PHYSICAL CHEMISTRY. Educ: Univ BC, BASc, 46, MASc, 47; Univ Calif, PhD(chem), 50. Prof Exp: Instr chem, Northwestern Univ, 51-52, asst prof, 53-57, assoc prof, 58-59; prof & head dept, Case Western Reserve Univ, 59-69; adj prof, Dartmouth Col, 69-70. Mailing Add: Middle Canyon Rd Carmel Valley CA 93924

BARROW, JAMES HOWELL, JR, b West Point, Ga, June 28, 20; m 50. PROTOZOOLOGY, PARASITOLOGY. Educ: Emory Univ, BA, 43; Yale Univ, PhD(zool), 51. Prof Exp: Instr, Emory Univ, 43-44; asst, Yale Univ, 44-47; lectr biol, Albertus Magnus Col, 47-48; prof biol & chmn dept, Huntingdon Col, 48-57; chmn dept, 63-71, PROF BIOL, HIRAM COL, 57-, DIR BIOL STA, 70- Concurrent Pos: Ford Found fel, Cambridge Univ, 53-54; vis prof protozool, biol sta, Univ Mich, 56-64. Mem: Wildlife Disease Asn; Soc Protozool; Am Micros Soc; Am Soc Zool. Res: Trypanosomes of poikilothermic vertebrates; haemosporidians of waterfowl and invertebrates; avian behavior and parasitism. Mailing Add: Dept of Biol Hiram Col PO Box 1388 Hiram OH 44234

BARROW, THOMAS D, b San Antonio, Tex, Dec 27, 24; m 50; c 4. PETROLEUM GEOLOGY. Educ: Univ Tex, BS, 45, MA, 48; Stanford Univ, PhD(geol), 53. Prof Exp: Jr geologist, Humble Oil & Ref Co, Calif, 51-55, area explor geologist, 55-59, div explor geologist, La, 59-61, regional explor mgr, 62-64, exec vpres & dir, Esso Explor, Inc, NY, 64-65, dir, Hunble Oil & Ref Co, Houston, 65-75, pres, 70-75, sr vpres, 67-70; SR V PRES, EXXON CORP, 75- Concurrent Pos: Mem, Oceanog Adv Comt, 68; mem, Sea Grant Adv Panel, 69. Mem: Am Asn Petrol Geol; Geol Soc Am; Am Geophys Union; Am Geog Soc; Marine Technol Soc. Res: Petroleum exploration. Mailing Add: Exxon Corp 1251 Ave of the Americas New York NY 10020

BARROWMAN, JAMES ADAMS, b Edinburgh, Scotland, June 4, 36; m 64; c 3. GASTROENTEROLOGY. Educ: Edinburgh Univ, BSc, 58, MBChB, 61; Univ London, MRCP, 64, PhD(physiol), 66. Prof Exp: From lectr to sr lectr physiol, London Hosp Med Col, 63-71, from lectr to sr lectr med, 71-75; ASSOC PROF MED, MEM UNIV, NFLD, 75- Mem: Brit Soc Gastroenterol; Brit Physiol Soc. Res: Physiology of intestinal lymph in relation to digestion and absorption; trophic action of gastrointestinal hormones on the alimentary tract. Mailing Add: 37 Forest Rd St John's NF Can

BARROWS, AUSTIN WILLARD, JR, b Addison, Vt, July 3, 37; m 60; c 4. APPLIED PHYSICS. Educ: Univ Vt, BA, 60, MS, 61; Univ Ky, PhD(physics), 65. Prof Exp: Res assoc nuclear physics, Univ Ky, 65; res physicist, US Army Nuclear Defense Lab, 66-67, br chief, 67-70; asst to dir admin, US Army Ballistic Res Lab, 70-71; BR CHIEF APPL PHYSICS ADMIN, COMBUSTION & PROPULSION BR,

BALLISTIC RES LABS, 71- Mem: Am Phys Soc; AAAS; Combustion Inst; Am Defense Preparedness Asn. Res: The phenomena of combustion of propellants and interior ballistics of guns and rockets. Mailing Add: 3103 Whitefield Rd Churchville MD 21028

BARROWS, CHARLES HARRY, JR, b Stelton, NJ, Nov 29, 24; m 47; c 5. BIOCHEMISTRY. Educ: Rutgers Univ, BS, 48; Johns Hopkins Univ, ScD, 54. Prof Exp: Chief, Sect Nutrit Biochem, Nat Heart Inst, 54-68, CHIEF SECT ENVIRON & GENETICS, NAT INST CHILD HEALTH & HUMAN DEVELOP, NIH, 68- Concurrent Pos: Asst prof, Sch Hyg & Pub Health, Johns Hopkins Univ, 63- Mem: Am Inst Nutrit; Geront Soc. Res: Gerontology; nutrition; protein; unknown accessory factors; cellular metabolism. Mailing Add: Gerontol Res Ctr Baltimore City Hosp Baltimore MD 21224

BARROWS, HAROLD LINDSEY, b Lookout Mountain, Tenn, Dec 14, 26; m 46; c 2. SOIL CHEMISTRY. Educ: Univ Chattanooga, BS, 49; Univ Fla, MS, 55, PhD(soils), 59. Prof Exp: Jr chemist field lab, Tung Invests, 49-53, asst chemist, 53-55, assoc chemist, 55-56, assoc plant physiologist, 56-58, plant physiologist, 58-60, soil scientist, 60-61, res soil scientist, 61-68, chief northeast br, Soil & Water Conserv Res Div, 68-72, staff scientist, Nat Prog Staff, 72-75, DEP ASST ADMINR, SOIL, WATER & AIR SCI, AGR RES SERV, USDA, 75- Mem: AAAS; Am Chem Soc; Soil Sci Soc Am; Int Soc Soil Sci; Soil Conserv Soc Am. Res: Soil-plant-water relations; developing agricultural management practices for improved efficiency in use of land and water resources for production of high quality food, especially role of macro- and micronutrients. Mailing Add: Agr Res Serv USDA Washington DC 20250

BARROWS, ROBERT S, organic chemistry, see 12th edition

BARRUETO, RICHARD BENIGNO, b Guatemala City, Guatemala, Feb 16, 28; US citizen; m 54; c 3. BIOCHEMISTRY, ORGANIC CHEMISTRY. Educ: Norm Sch, Guatemala, AB, 46; Eastern Nazarene Col, BS, 52; Boston Univ, MA, 54. Prof Exp: Res chemist, Sch Med, Boston Univ, 52-54; asst biochemist, US Qm Res & Eng Command, US Army, Mass, 54-60, res biochemist, Army Inst Environ Med, 60-62; bioscientist, 62-65, SR BIOSCIENTIST, BIOTECH SECT, AEROSPACE DIV, BOEING CO, 65- Mem: Am Chem Soc; NY Acad Sci; Am Inst Chemists; Sigma Xi; AAAS. Res: Life support and protection problems associated with man in a space vehicle; high altitude chamber; space-suit-vehicle integration; closed environment toxicological problems; optimizing man-machine interface in missile systems from psycho-physiological viewpoint. Mailing Add: 5155 NE Laurelcrest Lane Seattle WA 98105

BARRY, ALEXANDER, b Ayer, Mass, Mar 5, 13; m 39; c 2. EMBRYOLOGY. Educ: Harvard Univ, AB, 34, MA, 37; PhD(biol, embryol), 38. Prof Exp: Asst comp anat & physiol, Harvard Univ, 34-38; from instr to prof anat, Med Sch, Univ Mich, 38-68, asst dean, Med Sch, 63-68; assoc dean, 68-74, PROF HUMAN ANAT, SCH MED, UNIV CALIF, DAVIS, 68-, ACTG CHMN DEPT, 74- Mem: AAAS; Am Asn Anat; Am Soc Zool. Res: Morphogenesis; developmental physiology; functional development of human cardiovascular system. Mailing Add: Sch of Med Univ of Calif Davis CA 95616

BARRY, ARTHUR JOHN, b Buffalo, NY, Mar 11, 09; m 36; c 3. ORGANIC CHEMISTRY, PHYSICAL CHEMISTRY. Educ: State Univ NY Col Forestry, BS, 32, PhD(org chem), 36. Prof Exp: Actg prof org chem & carbohydrate chem, State Univ NY Col Forestry, 36-37; res chemist, Dow Chem Co, 37-47; res group supvr, 47-50, assoc dir res, 50-65; dir chem res, 65-70, RES CONSULT, DOW CORNING CORP, 70- Mem: AAAS; Am Chem Soc. Res: Preparation of cellulose derivatives; plastics; reactions in liquid ammonia, metallo-organic chemistry; synthesis of organosilicon compounds; investigation of polysiloxane resins, fluids and elastomers. Mailing Add: Dept of Res & Develop Dow Corning Corp Midland MI 48640

BARRY, ARTHUR LELAND, b Spokane, Wash, Aug 2, 32; m 55; c 4. MICROBIOLOGY, BACTERIOLOGY. Educ: Gonzaga Univ, BS, 55; St Lukes Sch Med Technol, MT, 55; Wash State Univ, MS, 57; Ohio State Univ, PhD(microbiol), 62. Prof Exp: Chief microbiologist, Clin Lab Med Group, Los Angeles, 64-66; med microbiologist, Los Angeles County Gen Hosp, 66-68; asst clin prof microbiol, Univ Calif-Calif Col Med, 67-68; lectr internal med & path, 68-75, ASSOC PROF CLIN MICROBIOL, DEPT MED & PATH, SCH MED, UNIV CALIF, DAVIS, 75-, DIR MICROBIOL LABS, SACRAMENTO MED CTR, 68- Concurrent Pos: NIH trainee clin microbiol, Univ Wash Hosp, 62-64; sr ed, Am J Med Technol, 68-74; mem ed bd, Antimicrobial Agents & Chemotherapy; chmn subcomt antimicrobic susceptibility, Nat Comt Clin Lab Stand, 69- Mem: Am Soc Med Technol; Am Soc Microbiol; Infectious Dis Soc Am. Res: Methods for rapid identification of bacterial pathogens; antimicrobic susceptibility tests; quality control in microbiology. Mailing Add: Sacramento Med Ctr 2315 Stockton Blvd Sacramento CA 94817

BARRY, BILLY DEAN, b Kahoka, Mo, Oct 11, 34; m 55; c 2. ENTOMOLOGY. Educ: Univ Mo, BS, 60, MS, 61; Tex A&M Univ, PhD(entom), 65. Prof Exp: Res entomologist, Ohio Agr Res & Develop Ctr, 64-69, Serere Res Sta, Soroti, Uganda, 69-71, Ohio Agr Res & Develop Ctr, 71-72 & Ahmadu Bello Univ, Inst Agr Res, Zaria, Nigeria, 72-74; RES ENTOMOLOGIST, UNIV MO, VINCENNES. IND, 75- Mem: Entom Soc Am. Res: Host plant resistance to insects. Mailing Add: Dept of Entom Univ of Mo 1-87 Agr Columbia MO 65201

BARRY, CORNELIUS, b Canandaigua, NY, Aug 23, 34; m 59. ENTOMOLOGY. Educ: St John Fisher Col, BS, 56; Univ Md, MS, 62, PhD(entom), 64. Prof Exp: Asst zool & parasitol, Univ Md, 56-59; teacher, Md Bd Educ, 59-61; asst insect physiol, Univ Md, 61-62, NSF res asst, 62-64; res entomologist, US Army Biol Labs, 64-68; asst prof int med, Sch Med, Univ Md, 68-73; MEM STAFF, BRAZIL-US BIOMED PROG, 73- Mem: AAAS; Am Soc Trop Med & Hyg; Sigma Xi; Entom Soc Am. Res: Medical entomology; insect physiology and biochemistry. Mailing Add: Brazil-US Biomed Prog PO 1838 40000 Salvador Bahia Brazil

BARRY, DAVID GEORGE, biology, history of science, see 12th edition

BARRY, DON CARY, b Los Angeles, Calif, Jan 4, 41; m 67. ASTRONOMY. Educ: Univ Southern Calif, BS, 62; Univ Ariz, PhD(astron), 67. Prof Exp: Asst prof, 67-71, ASSOC PROF ASTRON, UNIV SOUTHERN CALIF, 71-, CHMN DEPT, 74- Mem: AAAS; Am Astron Soc; Int Astron Union; Astron Soc Pac. Res: Astronomical spectroscopy; photoelectric photometry; spectral quantification. Mailing Add: Dept of Astron Univ of Southern Calif Los Angeles CA 90007

BARRY, EDWARD GAIL, b Butte, Mont, May 4, 33; m 58; c 3. GENETICS. Educ: Dartmouth Col, AB, 55; Stanford Univ, PhD(biol), 61. Prof Exp: USPHS trainee genetics, Yale Univ, 60-62; from asst prof to assoc prof bot, 62-75, PROF BOT, UNIV, NC, CHAPEL HILL, 75- Mem: Genetics Soc Am; Genetical Soc Gt Brit. Res: Microbial genetics; cytogenetics of Neurospora. Mailing Add: Dept of Bot Univ of NC Chapel Hill NC 27514

BARRY, GUY THOMAS, b Montreal, Que, Apr 14, 20; nat US; m 48; c 2. ORGANIC CHEMISTRY. Educ: Sir George Williams Col, BSc, 42; McGill Univ, PhD(org chem), 46. Prof Exp: Demonstr physics, Sir George Williams Col, 40-42; demonstr chem, McGill Univ, 43-45; NIH fel, Rockefeller Inst, 46-48, Am Cancer Soc fel, 48-50, asst, 50-53, assoc, 53-58; res prof, Mem Res Ctr & Hosp, Univ Tenn, 58-64; dir biochem res, Squibb Inst Med Res, 64-68; DEP DIR DIV TECH SERV, US CUSTOMS SERV, 68- Concurrent Pos: Consult, E Tenn Baptist Hosp Clin Lab, 59-64. Mem: Am Chem Soc; Am Soc Biol Chem. Res: Synthesis of insecticides; explosive and antibiotic research; nature of bacitracin; bacteriophage; cytoplasmic microchemistry; bacteriocines; colominic acid discoverer; bacterial polysaccharides. Mailing Add: US Customs Serv 1301 Constitution Ave NW Washington DC 20229

BARRY, HERBERT, III, b New York, NY, June 2, 30. PSYCHOPHARMOCOLOGY. Educ: Harvard Col, BA, 52; Yale Univ, MS, 53, PhD(psychol), 57. Prof Exp: From instr to asst prof psychol, Yale Univ, 58-61; asst prof, Univ Conn, 61-63; res assoc prof pharmacol, 63-70, PROF PHARMACOL, SCH PHARM, UNIV PITTSBURGH, 70-, PROF ANTHROP, 70-, PROF PHARMACEUT, 71- Concurrent Pos: NIMH fel, Yale Univ, 57-59; consult, Nat Inst Alcohol Abuse & Alcoholism, 72- Honors & Awards: Res Sci Develop Award, NIMH, 67. Mem: Am Psychol Asn; Am Anthrop Asn; Acad Pharmaceut Sci; Soc Cross-Cult Res (pres, 73-74); Am Soc Pharmacol & Exp Therapeut. Res: Psychopharmacology, testing effects of drugs, especially alcohol and marihuana components; cross-cultural research, correlating child training practices with other customs; relationships of birth order to psychiatric illness and personality characteristics. Mailing Add: 1100 Salk Hall Univ of Pittsburgh Pittsburgh PA 15261

BARRY, JAMES DALE, b Washington, DC, Feb 8, 42; m 64; c 2. SPACE PHYSICS, LASERS. Educ: Univ Calif, Los Angeles, BS, 64, PhD(space physics), 69; Calif State Col, Los Angeles, MS, 66. Prof Exp: Asst high energy physics, Univ Calif, Los Angeles, 64-65, res asst space physics, 65-68, res geophysicist, 68-69; laser scientist, Air Force Atomics Lab, 69-73, tech dir, Space Laser Commun, 73-75, chief scientist, Air Force Space & Missiles Orgn, Tech Div, 75-76, SR SCIENTIST, SPACE LASER COMMUN, AIR FORCE SPACE & MISSILES SYSTS ORGN, 76- Concurrent Pos: Chmn, Dept of Defense/NASA Optical Commun Working Group, 74-76. Honors & Awards: Most Outstanding Sci Achievement, Air Force Asn, 73. Mem: Am Geophys Union; Inst Electronic & Elec Engrs. Res: Interplanetary particles and fields; ionosphere; magnetosphere; wave-particle interactions; stimulated emission; gaseous lasers; space laser communications; laser design and theory, medical applications of lasers. Mailing Add: Space & Missiles Systs Orgn SKX PO Box 92960 Worldway Postal Ctr Los Angeles CA 90009

BARRY, JOHN YOUNG, b Syracuse, NY, May 4, 29; m 51; c 4. MATHEMATICS. Educ: Cornell Univ, AB, 50; Yale Univ, MA, 52, PhD(math), 55. Prof Exp: Mem tech staff, Opers Eval Group, 55-58; sr staffman opers res, Arthur Andersen & Co, 59-61; mem sci staff, Inst Defense Anal, 61-67; asst dir opers res, Merck & Co, 67-68; independent consult, 68-70; OPERS RES OFF, MORGAN GUARANTY TRUST CO NY, 70- Concurrent Pos: Sr res fel opers res, Univ Lancaster, 65-66. Mem: Opers Res Soc Am; Economet Soc; Inst Mgt Sci; Am Math Soc; Brit Opers Res Soc. Res: Governmental operations research; systems analysis on military and national health problems; functional analysis; applications of operations research to financial and economic analysis. Mailing Add: Opers Res Group Morgan Guaranty Trust Co 23 Wall St New York NY 10015

BARRY, KEVIN GERARD, b Newton, Mass, May 12, 23; m 42; c 4. INTERNAL MEDICINE. Educ: Georgetown Univ, MD, 49; Am Bd Internal Med, dipl, 56. Prof Exp: Intern, Walter Reed Gen Hosp, Washington, DC, US Army, 49-50, mem staff internal med, Soldiers Home Hosp, 50-51 & 98th Gen Hosp, Munich, Ger, 51-52, resident internal med, Walter Reed Gen Hosp, 53-56, batallion surgeon, 43rd Combat Engrs, 52-53, chief med serv, Army Hosp, NY, 56-58; instr, 58-63, clin asst prof, 63-69, ASSOC PROF MED, SCH MED, GEORGETOWN UNIV, 69- Concurrent Pos: Res internist, Walter Reed Army Inst Res, 58-61, chief dept metab, Inst & chief med sect, Hosp, 61-65, dir div med, Inst, 65-66; dir med educ & chief renal & metab sect, Washington Hosp Ctr, 66-69; chief Georgetown Univ med div, DC Gen Hosp, 69-71; med dir, MOrris Cafritz Mem Hosp, 71-74. Mem: Fel Am Col Physicians; fel Am Col Clin Pharmacol & Chemother; AMA; Am Fedn Clin Res; Am Soc Artificial Internal Organs. Res: Renal, water and electrolyte metabolism; prevention and therapy of acute renal failure in man and animals; pathogenesis of acute renal failure in animals; peritoneal membranes used as absorptive surfaces. Mailing Add: 6200 Westchester Park Dr College Park MD 20740

BARRY, P J S, b London, Eng, Feb 1, 29; nat Can; m 57; c 3. HEALTH PHYSICS, MICROMETEOROLOGY. Educ: Univ London, BSc, 50, PhD(phys chem), 53. Prof Exp: Nat Res Coun Can fel, Prairie Regional Lab, 53-54; res assoc radiation chem, Univ Sask, 54-55; asst res officer, 55-61, assoc res officer, 61-66, SR RES OFFICER, HEALTH PHYSICS, ATOMIC ENERGY CAN, LTD, 66- Concurrent Pos: Sci secy, UN Sci Comt on the Effects of Atomic Radiation, 67-69; sr vis res fel, Monitoring & Assessment Res Ctr, Chelsea Col, 75-76. Mem: Can Meteorol Soc; The Chem Soc; Brit Geol Asn; Royal Hort Soc. Res: Dispersion of stack effluents in atmosphere and exchange of gases and vapors between atmosphere and natural surfaces. Mailing Add: Biol & Health Physics Div Atomic Energy of Can Ltd Chalk River ON Can

BARRY, RICHARD HENRY, pharmaceutical chemistry, see 12th edition

BARRY, ROBERT MERRITT, b Pensacola, Fla, Nov 22, 42; m 65; c 1. ENTOMOLOGY. Educ: Troy State Univ, BS, 65; Miss State Univ, MS, 67; Univ Mo, PhD(entom), 72. Prof Exp: VD epidemiologist, Ala Dept Pub Health, 65-66; regional entomologist, Fla State Bd Health, 68-69; EXTEN ENTOMOLOGIST, UNIV GA, 72- Concurrent Pos: Tech dir, Tech Serv Pest Control Co, 74- Mem: Entom Soc Am; Nat Pest Control Asn. Res: Lawn and ornamental insect protection; household pest control; structural pest control. Mailing Add: Coop Exten Serv Univ of Ga Tifton GA 31794

BARRY, ROGER DONALD, b Columbus, Ohio, Nov 26, 35; m 57; c 4. ORGANIC CHEMISTRY, BIOCHEMISTRY. Educ: Univ Cincinnati, BS, 57, PhD(org chem), 60. Prof Exp: Res fel, Ohio State Univ, 60-61, from instr to asst prof obstet & gynec, Col Med, 61-68; PROF CHEM, NORTHERN MICH UNIV, 68- Concurrent Pos: Comnr, Accrediting Bur Med Lab Schs, 69- Mem: The Chem Soc. Res: Organic synthesis; spectroscopy; steroid chemistry and biochemistry; gas chromatography; analytical organic chemistry; chemical education. Mailing Add: Dept of Chem Northern Mich Univ Marquette MI 49855

BARRY, ROGER GRAHAM, b Sheffield, Eng, Nov 13, 35. CLIMATOLOGY. Educ: Univ Liverpool, BA, 57; McGill Univ, MSc, 59; Univ Southampton, PhD(geog), 65. Prof Exp: Lectr geog, Univ Southampton, 60-66; res scientist climat, Geog Br, Dept Energy, Mines & Resources, Ont, 66-67; lectr geog, Univ Southampton, 67-68; assoc prof, 68-71, PROF GEOG, UNIV COLO, BOULDER, 71- Concurrent Pos: Mem glaciol panel, Comt Polar Res, Nat Acad Sci, 76-77. Mem: Am Meteorol Soc; Am Quaternary Asn; Asn Am Geogr; Am Geog Soc. Res: Climates of arctic and alpine regions; climatic change and paleoclimatology, interrelations between land-sea ice and climate. Mailing Add: Inst of Arctic & Alpine Res Univ of Colo Boulder CO 80302

BARRY, SUE-NING C, b Shanghai, China, Nov 5, 32; US citizen; m 59. CELL PHYSIOLOGY. Educ: Barat Col, BA, 55; Univ Md, PhD(zool), 61. Prof Exp: Asst instr physiol, 61, asst prof, 62-68, assoc prof histol, 68-75, PROF ANAT, UNIV MD, 75- Concurrent Pos: Res fel histol, Univ Md, 61-62. Mem: AAAS; Am Soc Cell Biol; Am Soc Zool; Am Soc Protozool. Res: Carbohydrate metabolism of fresh water protozoan astasia longa; effects of sugar metabolism in streptococcus salivarius on oral tissues; effects of trace elements on tooth caries formation. Mailing Add: Dept of Anat Univ of Md Sch of Dent Baltimore MD 21201

BARRY, WILLIAM EARL, b Sept 5, 37; US citizen; m 65; c 1. AEROSPACE MEDICINE, BIOENGINEERING. Educ: US Mil Acad, BS, 59; Med Col Ga, MD, 68; Harvard Univ, MPH, 70. Prof Exp: Eng test pilot, US Air Force Europe Cent Aeromed Lab, Weisbaden, Ger, 60-64, proj test pilot-engr, Bioastronaut Div, Air Force Flight Test Ctr, Edwards Air Force Base, Calif, 65, res life scientist, Spacecraft Environ Cent Syst, Directorate of Bioastronautics, Projs Gemini & Apollo, Kennedy Space Ctr, Cape Canaveral, Fla, 66 & 67, intern med & surg, Wilford Hall, US Air Force Med Ctr, Lackland Air Force Base, Tex, 68-69, resident aerospace med, US Air Force Sch Aerospace Med, 70-72, dir aeromed serv, Air Force Spec Weapons Ctr, Sandia Labs, Albuquerque, NMex, 72-73, hosp comdr, 73, SPEC ASST TO THE ADMINR, HQ, NASA, 73-, US AIR FORCE, 59- Concurrent Pos: White House fel, President's Comn on White House Fels, Washington, DC, 73-74; consult high altitude physiol, Indian Air Force, Hq Palam Airport, New Delhi, 61-64; consult occup med, Armament Develop & Test Ctr, Eglin Air Force Base, Fla, 65-67; lectr aerospace med, Univ Tex Med Sch San Antonio, 68-72 & Med Sch, Univ NMex, 72-73; consult environ med, Air Force Spec Weapons Ctr, Air Force Weapons Lab & Lovelace Clin, Sandia Labs, 72-73. Concurrent Pos: Comdrs Award, Order of Daedalians, 60; Julian E Ward Award, Aerospace Med Asn, 72; Flight Surgeon of the Year Award, US Air Force Soc Flight Surgeons, 73; John Shaw Billings Award, Asn Mil Surg US, 74. Mem: Fel Aerospace Med Asn; fel Am Col Prev Med; fel Am Col Physicians; Asn Mil Surg US; Soc Exp Test Pilots. Res: Aerospace physiology; aircrew performance; human factors in flight; aviation pathology. Mailing Add: 1443 Laurel Hill Rd Vienna VA 22180

BARRY, WILLIAM EUGENE, b Staten Island, NY, Aug 29, 28; m 54; c 5. MEDICINE. Educ: Villanova Univ, BS, 50; NY Med Col, MD, 54; Am Bd Internal Med, dipl, 61. Prof Exp: Asst resident internal med & resident, Health Ctr, Ohio State Univ, 57-59, sr asst resident hemat & chief resident, 59-61; from instr to assoc prof, 61-72, PROF MED, MED CTR, TEMPLE UNIV, 72- Concurrent Pos: Am Col Physicians Brower travelling scholar, 65; res fel, Div Hemat, Univ Wash, 69-70. Mem: Am Fedn Clin Res; fel Am Col Physicians; Am Soc Hemat. Res: Clincial hematology; iron metabolism. Mailing Add: Med Ctr Temple Univ Philadelphia PA 19140

BARRY, WILLIAM JAMES, b Hudson, NY, May 14, 48; m 72. ANIMAL BEHAVIOR. Educ: Cornell Univ, BS, 70; Mich State Univ, MS, 71, PhD(zool), 74. Prof Exp: ASST PROF ZOOL, CONN COL, 75- Mem: Animal Behav Soc; Am Soc Mammalogists; The Wildlife Soc; AAAS; Ecol Soc Am. Res: The effect of photoperiod, temperature and other environmental factors on food hoarding behavior in small mammals, especially deermice; the study of the ecological relationships of small mammals on tidal marshes. Mailing Add: Dept of Zool Conn Col New London CT 06320

BARRY, WILLIAM THOMAS, aquatic biology, taxonomy, see 12th edition

BARSA-NEWTON, MARY CLAIRE, b New York, NY, July 31, 28; m 57; c 4. INSECT PHYSIOLOGY. Educ: Col New Rochelle, AB, 49; Fordham Univ, MS, 52, PhD(gen biol), 54. Prof Exp: Asst biol, Fordham Univ, 50-53, res assoc, 53-58; instr, Marymount Col, 58-59; from instr to asst prof, 59-65, assoc prof & chmn dept, 65-71, PROF BIOL, MANHATTANVILLE COL, 71- Mem: AAAS; Am Soc Zool; Entom Soc Am; NY Acad Sci. Res: Effects of ions on insect hearts; respiratory studies on insects; effect of dichloro-diphenyl-trichloroethane on respiratory enzymes; plant tissue culture; biochemical changes in the retina of the eye. Mailing Add: 55 Hillcrest Dr Pelham Manor NY 10803

BARSCH, GERHARD RICHARD, b Berlin, Ger, June 22, 27; m 59; c 3. SOLID STATE PHYSICS. Educ: Tech Univ Berlin, MS, 51; Univ Göttingen, PhD(theoret physics), 55. Prof Exp: Mem sci staff theoret physics, Battelle Inst, Ger, 56-62; sr res assoc appl physics, 62-64, assoc prof solid state sci, 64-70, PROF PHYSICS, PA STATE UNIV, 70- Res: Lattice dynamics of anharmonic properties and crystal defects; nonlinear crystal elasticity; high pressure physics. Mailing Add: Dept of Physics Pa State Univ University Park PA 16802

BARSCHALL, HENRY HERMAN, b Berlin, Ger, Apr 29, 15; nat US; m; c 2. NUCLEAR SCIENCE. Educ: Princeton Univ, AM, 39, PhD(physics), 40. Prof Exp: Instr physics, Princeton Univ, 40-41 & Univ Kans, 41-43; mem staff, Los Alamos Sci Lab, NMex, 43-46; from asst prof to prof physics, 46-73, chmn dept, 51, 54, 56-57, 63-64, BASCOM PROF PHYSICS & NUCLEAR ENG, UNIV WIS-MADISON, 73- Concurrent Pos: Consult, Los Alamos Sci Lab, 51-52 & Lawrenee Livermore Lab, 71-73; assoc ed, Rev Mod Physics, 51-53 & Nuclear Physics, 59-72; vis prof, Univ Calif, Davis, 71-73; ed, Phys Rev C, 72- Honors & Awards: T W Bonner Prize, Am Phys Soc, 65. Mem: Nat Acad Sci; Am Nuclear Soc; fel Am Phys Soc. Res: Nuclear physics; fast neutrons; neutron physics; applications of neutron physics to radiotherapy and fusion technology. Mailing Add: Eng Res Bldg Univ of Wis Madison WI 53706

BARSDATE, ROBERT JOHN, b Richmond Hill, NY, Sept 4, 34; m 59; c 2. GEOCHEMISTRY. Educ: Allegheny Col, BS, 59; Univ Pittsburgh, PhD(geochem), 64. Prof Exp: Asst prof, 63-69, ASSOC PROF MARINE SCI, UNIV ALASKA, 69- Mem: AAAS; Am Soc Limnol & Oceanog; Geochem Soc. Res: Biologically mediated movement of trace metals in natural waters. Mailing Add: Inst of Marine Sci Univ of Alaska College AK 99701

BAR-SELA, MILDRED ELWERS, b Neenah, Wis, Mar 24, 25; m 61; c 1. ANATOMY, PHYSIOLOGY. Educ: Lawrence Col, BA, 47; Univ Wis, MS, 48; Baylor Univ, PhD(anat), 61. Prof Exp: Asst zool, Univ Wis, 47-49; from instr to asst prof biol & actg chmn dept, McMurry Col, 53-56; instr, Univ Houston, 56-58; asst anat, Col Med, Baylor Univ, 59-62, from instr to asst prof, 62-74; PRES, LAMALO, INC, 74- Concurrent Pos: Instr, Sch Nursing, Tex Woman's Univ, 62-64. Mem: Endocrine Soc; Am Physiol Soc; Am Cong Rehab Med; Am Asn Anat. Res: Neural control of pituitary function, particularly of gonadotropin secretion. Mailing Add: 1929 Dunstan Rd Houston TX 77005

BARSHAD, ISAAC, b Safad, Israel, Dec 14, 13; nat US; m 41; c 1. SOIL CHEMISTRY. Educ: Univ Calif, BS, 36, MS, 37, PhD(soil sci), 44. Prof Exp: Assoc soil chemist, 44-59, LECTR, COL AGR, UNIV CALIF, BERKELEY, 59-, SOIL CHEMIST, EXP STA SOILS & PLANT NUTRITION, 69- Concurrent Pos: Guggenheim fel, 56- Mem: AAAS; Soc Soil Sci. Res: Clay chemistry and crystallography; genesis of soils and clay minerals; physical chemistry of absorption and swelling of clay minerals. Mailing Add: Dept of Soils and Plant Nutrit Col of Agr Univ of Calif Berkeley CA 94720

BARSHAY, JACOB, mathematics, see 12th edition

BARSKE, PHILIP, b Fairfield, Conn, Jan 12, 17; m 48; c 1. WILDLIFE MANAGEMENT. Educ: Univ Conn, BS, 40; Univ Mich, MS, 43. Prof Exp: Game biologist, Conn State Bd Fish & Game, 42-43, 46; FIELD CONSERVATIONIST, WILDLIFE MGT INST, WASHINGTON, DC, 46- Concurrent Pos: Consult environ systs to several major indust concerns, 70-; mem, Gov Coun Environ Qual, Conn. Honors & Awards: Am Motors Award, 63; Cert of Recognition, Wildlife Soc, 67, John Pearce Mem Award, 75. Mem: Wildlife Soc; Soil Conserv Soc Am; Soc Am Foresters. Res: Applied habitat development; wildlife areas and natural areas development. Mailing Add: 200 Audubon Lane Fairfield CT 06430

BARSKY, CONSTANCE KAY, b Newark, NJ, Nov 3, 44; m 75. GEOCHEMISTRY. Educ: Denison Univ, BS, 66; Washington Univ, PhD(geochem), 75. Prof Exp: RES ASSOC GEOCHEM PETROL, UNIV MO-COLUMBIA, 71- Mem: Sigma Xi; Am Geophys Union. Res: Applications of electron microprobe analysis to problems in geochemistry and petrology; major and trace element geochemistry of igneous rocks applied to problems of magma origin and evolution. Mailing Add: Dept of Geol Univ Mo Columbia MO 65201

BARSKY, JAMES, b Regina, Sask, Nov 28, 25; US citizen; m 57; c 2. INFORMATION SCIENCE. Educ: Univ Sask, BA, 46; Univ Toronto, MA, 51; Northwestern Univ, PhD(biochem), 58. Prof Exp: Asst biochem, Univ Sask, 46-48; fel physiol chem, Med Sch, Ment Health Prog & Vet Admin Ctr, Los Angeles, 58-60; mgr dept biochem, Res Div, Ethicon, Inc, 60-63; biol sci ed, 63-67, exec ed, 67-68, vpres ed, 68-73, SR VPRES, ACAD PRESS INC, 73- Concurrent Pos: Fel, Univ Calif, Los Angeles; mem adv panel, NSF Off Sci Info Serv, 75 & adv comt, Nat Comn Libr & Info Sci, 75- Mem: AAAS; Am Chem Soc; Am Asn Clin Chem; Inst Food Technol; NY Acad Sci. Res: Enzymology; scientific technical, medical and scholarly publications. Mailing Add: Acad Press Inc 111 Fifth Ave New York NY 10003

BARSKY, MARVIN, mathematics, see 12th edition

BARSKY, MURRAY HAROLD, analytical chemistry, see 12th edition

BARSS, WALTER MALCOMSON, b Corvallis, Ore, Jan 29, 17; nat Canada; m 47; c 1. PHYSICS. Educ: Univ BC, BA, 37, MA, 39; Purdue Univ, PhD(physics), 42. Prof Exp: Asst, Univ BC, 38-39 & Purdue Univ, 40-42; res physicist, Nat Res Coun Can, 42-52 & Atomic Energy Can Ltd, 52-64; ASSOC PROF PHYSICS, UNIV VICTORIA, 64- Concurrent Pos: Can Asn Physicists; Acoust Soc Am. Res: Acoustics, especially reverberation problems in underwater sound using laboratory models. Mailing Add: Dept of Physics Univ of Victoria Victoria BC Can

BARSTON, EUGENE MYRON, b Chicago, Ill, Aug 7, 35; m 62; c 2. APPLIED MATHEMATICS. Educ: Calif Inst Technol, BS, 57, MS, 58; Stanford Univ, PhD(physics), 64. Prof Exp: Math physicist, Stanford Res Inst, 63-65; assoc res scientist, Courant Inst Math Sci, NY Univ, 65-67, assoc prof math, 67-72; ASSOC PROF MATH, UNIV ILL, CHICAGO CIRCLE, 72- Mem: Soc Indust & Appl Math. Res: Stability theory; plasma physics. Mailing Add: Dept of Math Box 4348 Univ of Ill at Chicago Circle Chicago IL 60680

BART, GEORGE JAMES, b Cleveland, Ohio, Aug 24, 19; m 54; c 2. PLANT PATHOLOGY. Educ: Ohio State Univ, BS, 48, PhD(plant path, bot), 56. Prof Exp: Asst bot, 48 & 49, res found, 49, ASST PROF PLANT PATH & PLANT PATHOLOGIST, AGR EXP STA, OHIO STATE UNIV, 52- Mem: Am Phytopath Soc. Res: Diseases of woody ornamentals, especially root rots; vascular wilts of forest and shade trees; teaching. Mailing Add: Dept of Bot Ohio State Univ Mansfield Camp 2375 Springmill Rd Mansfield OH 44903

BART, GEORGE RAYMOND, b Oak Park, Ill. MATHEMATICAL PHYSICS, ELEMENTARY PARTICLE PHYSICS. Educ: Loyola Univ, Chicago, BS, 61; Ill Inst Technol, PhD(theoret physics), 70. Prof Exp: Physicist & mgr crystal contracts, Electronics & Res Ctr, Victor-Comptometer Corp, 63-66; resident assoc, High Energy Physics Div, Argonne Nat Lab, 67-69; asst prof physics, Loyola Univ, Chicago, 69-70; asst prof physics & math, Mayfair Col, 70-74; ASSOC PROF PHYSICS & MATH, NORTHEAST COL, 74- Mem: Am Phys Soc; Am Asn Physics Teachers. Res: Theories of elementary particles; null-plane quantum field theories; S-matrix theory; dispersion relations; analysis of non-linear equations; soliton theory. Mailing Add: Dept of Physics Northeast Col 1145 W Wilson Ave Chicago IL 60640

BARTA, ALLAN LEE, b Iowa City, Iowa, June 3, 42; m 64; c 2. AGRONOMY, CROP PHYSIOLOGY. Educ: Iowa State Univ, BS, 64; Wash State Univ, MS, 66; Purdue Univ, PhD(crop physiol), 69. Prof Exp: ASST PROF FORAGE PHYSIOL, OHIO AGR RES & DEVELOP CTR, 69- Mem: Am Soc Agron. Res: Applied aspects of photosynthesis; plant metabolism, especially organic acid and metabolism. Mailing Add: Ohio Agr Res & Develop Ctr Wooster OH 44691

BARTA, CHARLES IRA, analytical chemistry, physical chemistry, see 12th edition

BARTA, FRANK R, SR, psychiatry, see 12th edition

BARTALOS, MIHALY, b Pozsony, Czech, May 27, 35; US citizen; m 57; c 3. MEDICAL GENETICS, INTERNAL MEDICINE. Educ: Univ Heidelberg, MD, 60. Prof Exp: Asst physician, Czerny Hosp, Univ Heidelberg, 60; res assoc surg, Sinai Hosp, Baltimore, Md, 60-61; med geneticist, chief med genetics unit & res scholar pediat, Col Med, Howard Univ, 65-68; resident internal med, asst chief med & chief primary care clin, USPHS Hosp, Baltimore, 68-70; ADJ ASST PROF HUMAN GENETICS & DEVELOP, COL PHYSICIANS & SURGEONS, COLUMBIA UNIV, 71-, ASST PROF CLIN PEDIAT, 74- Concurrent Pos: Fel path, Sch Med, Johns Hopkins Univ, 61-62, fel med, 62-64; dir heredity clin, Freedman's Hosp, Washington, DC, 66-68; scientist, Dept Med Genetics, NY State Psychiat Inst, 71-74. Mem: Am Soc Human Genetics; NY Acad Sci. Res: Clinical genetics; human cytogenetics. Mailing Add: 14 E 60th St New York NY 10022

BARTEK, METHODIUS JOSEPH, biochemistry, see 12th edition

BARTEL, ALLEN HAWLEY, b San Diego, Calif, July 26, 23; m 51; c 4. BIOPHYSICS. Educ: Univ Calif, AB, 47, MS, 49, PhD, 54. Prof Exp: Instr biol sci, Univ Calif, 55-56; res fel chem, Calif Inst Technol, 56-58; from asst prof to assoc prof biol, 58-67, PROF BIOPHYS SCI & CHMN DEPT, UNIV HOUSTON, 67- Mem: Biophys Soc; Am Chem Soc. Res: Immunochemistry and biophysics of macromolecules; comparative biochemistry. Mailing Add: Dept of Biophys Sci Univ of Houston Houston TX 77004

BARTEL, LEWIS CLARK, b Hillsboro, Kans, Dec 5, 34; m 56; c 2. THEORETICAL SOLID STATE PHYSICS, GEOPHYSICS. Educ: Univ Kans, BS, 58; Iowa State Univ, PhD(physics), 64. Prof Exp: Asst prof physics, Colo State Univ, 64-67; STAFF MEM, PHYSICS RES, SANDIA LABS, 67- Concurrent Pos: Consult, PEC Res Assocs, 65-67. Mem: Am Phys Soc. Res: Theoretical solid state physics, magnetism and ferroelectrics; electromagnetic modeling calculations. Mailing Add: Orgn 5732 Sandia Labs Albuquerque NM 87115

BARTEL, MONROE H, b Newton, Kans, Oct 3, 36; m 57; c 2. BIOLOGY. Educ: Tabor Col, AB, 58; Kansas State Univ, MS, 60, PhD(parasitol), 63. Prof Exp: Instr zool, Kans State Univ, 62-63; from asst prof to assoc prof, 63-71, PROF BIOL, MOORHEAD STATE COL, 71- Concurrent Pos: Fel trop med, La State Univ sch med, 66, adv med technol, 65- Mem: Am Soc Parasitol; Am Micros Soc; Wildlife Disease Asn. Mailing Add: Dept of Biol Moorhead State Col Moorhead MN 56560

BARTELINK, DIRK JAN, b Heumen, Neth, Oct 28, 33; Can citizen; m 57; c 2. SOLID STATE ELECTRONICS. Educ: Univ Western Ont, BSc, 56; Stanford Univ, MS, 59, PhD(elec eng), 62. Prof Exp: Mem tech staff solid state electronics res, Bell Labs, NJ, 61-66, supvr semiconductor devices, 66-73; MGR DEVICE PHYSICS, XEROX PALO ALTO RES CTR, 73- Mem: Inst Elec & Electronics Engrs; Am Phys Soc. Res: Fundamental and applied phenomena in semiconductor and dielectrics and interfaces between these. Mailing Add: Xerox Palo Alto Res Ctr 3333 Coyote Hill Rd Palo Alto CA 94304

BARTELL, CLELMER KAY, b Kingstree, SC, Nov 1, 34. CELL PHYSIOLOGY, INVERTEBRATE PHYSIOLOGY. Educ: Davidson Col, BS, 57; Univ Tenn, MS, 63; Duke Univ, PhD(zool), 66. Prof Exp: Instr zool, Duke Univ, 65-66; res assoc, Tulane Univ, 66-69; ASST PROF ZOOL, UNIV NEW ORLEANS, 69- Mem: Marine Biol Asn; AAAS; Am Soc Zoologists. Res: Neuro-secretion in Crustacea; endocrine control of chromatophores; physiological effects of toxic substances on aquatic organisms. Mailing Add: Dept of Biol Sci Univ of New Orleans Lake Front New Orleans LA 70122

BARTELL, GILBERT D, anthropology, see 12th edition

BARTELL, LAWRENCE SIMS, b Ann Arbor, Mich, Feb 23, 23; m 52; c 1. PHYSICAL CHEMISTRY. Educ: Univ Mich, BS, 44, MS, 47, PhD(chem), 51. Prof Exp: Res asst, Manhattan Proj, Univ Chicago, 44-45; Rackham fel, Univ Mich, 51-52, res assoc, 52-53; from asst prof to prof chem, Iowa State Univ, 53-65; PROF CHEM, UNIV MICH, ANN ARBOR, 65- Concurrent Pos: Consult, Gillette Co, Ill, 56-62 & Mobil Oil Corp, NJ, 60-; chemist, AEC, 56-65; assoc ed, J Chem Physics, 63-66; mem adv bd, Petrol Res Fund, 70-73; vis prof, Moscow State Univ, 72 & Univ Paris XI, 73. Mem: Am Chem Soc; Am Crystallog Asn; Am Phys Soc. Res: Atomic and molecular structure by electron diffraction; surface and quantum chemistry. Mailing Add: Dept of Chem Univ of Mich Ann Arbor MI 48104

BARTELL, MARVIN H, b Wanatah, Ind, Sept 16, 38; m 63; c 2. VERTEBRATE ZOOLOGY, ENDOCRINOLOGY. Educ: Concordia Teachers Col, Ill, BS, 61; Univ Mich, MS, 62, PhD(zool), 69. Prof Exp: Instr biol, Concordia Teachers Col, Ill, 62-64; lectr endocrinol, Univ Mich, 67; ASST PROF BIOL, CONCORDIA TEACHERS COL, ILL, 68- Mem: AAAS; Am Soc Zool. Res: Developmental endocrinology; control of carbohydrate metabolism in lower vertebrates. Mailing Add: Dept of Biol Concordia Teachers Col River Forest IL 60305

BARTELL, PASQUALE, b Philadelphia, Pa, Oct 7, 22; m 47; c 3. MEDICAL MICROBIOLOGY. Educ: Villanova Col, BS, 49; Jefferson Med Col, MS, 51; Temple Univ, PhD(med microbiol), 55. Prof Exp: Sr res scientist microbiol, Wyeth Inst Med Res, Pa, 54-61; assoc, Harrison Dept Surg Res, Schs Med, Univ Pa, 61-67; assoc prof, 67-71, PROF MICROBIOL, COL MED & DENT NJ, 71- Mem: Am Soc Microbiol; NY Acad Sci. Res: Virology; virus-host cell interaction. Mailing Add: Dept of Microbiol Col of Med & Dent of NJ Newark NJ 07103

BARTELLI, LINDO JOSEPH, b Gaastra, Mich, June 20, 1917; m 41; c 2. SOIL SCIENCE, SOIL CONSERVATION. Educ: Mich State Univ, BS, 40, MS, 52; Univ Ill, PhD(soils & econ), 58. Prof Exp: Employee, Soil Conserv Serv, 41-52, state soil scientist, Champaign, Ill, 52-61, asst prin soil correlator, Knoxville, Tenn, 61-64, head soil correlation staff southern region, Ft Worth Tex, 64-73, DIR SOILS SURV INTERPRETATIONS DIV, SOIL CONSERV SERV, USDA, 73- Mem: Fel Soil Conserv Soc Am; Soil Sci Soc Am; Am Soc Agron; Am Soc Planning Off. Res: Investigations in the behavior of soil under various land uses in both farm and nonfarm sectors to provide basic data to develop land management systems that maintain the resource base without deterrent impacts on the environment and still economically sound. Mailing Add: 10010 Branch View Ct Silver Spring MD 20903

BARTELS, GEORGE WILLIAM, JR, b Hershey, Pa, June 23, 28; m 51; c 2. ORGANIC CHEMISTRY. Educ: Lebanon Valley Col, BS, 50; Univ Del, MS, 51, PhD(org chem), 53. Prof Exp: Asst chem, Univ Del, 50-51, Armstrong res fel, 51-53; res chemist org chem, 53-58, from res supvr to sr res supvr, 58-65, tech supt, 65-67, process supt dacron, 67-69, indust prod supt, 69-72, ASST TO MFG DIR, DACRON DIV, TEXTILE FIBERS DEPT, E I DU PONT DE NEMOURS & CO, 72- Mem: AAAS; Am Chem Soc; Sigma Xi. Res: Organic research in linear high polymers; synthetic fibers; spun-bonded nonwoven fabrics. Mailing Add: 707 Severn Rd Wilmington DE 19803

BARTELS, PAUL GEORGE, b Yuma, Colo, Apr 9, 34; m 56; c 2. PLANT PHYSIOLOGY. Educ: Vanderbilt Univ, PhD(biol), 64. Prof Exp: NIH fel, 64-65; asst prof biol, 66-70, ASSOC PROF BIOL SCI & ASSOC BIOLOGIST, EXP STA, UNIV ARIZ, 70- Concurrent Pos: NSF grant, 67-69. Mem: Am Soc Plant Physiol. Res: Plant cell physiology; inhibition of chloroplast development by aminotriazole and herbicides. Mailing Add: Dept of Biol Sci Univ of Ariz Tucson AZ 85721

BARTELS, PETER H, b Danzig, Ger, Jan 25, 29; US citizen; m 54; c 3. OPTICS, COMPUTER SCIENCE. Educ: Univ Göttingen, PhD(biophys), 54; Univ Giessen, Dr Habil, 58. Prof Exp: Asst prof, Univ Giessen, 54-58; dir res, E Leitz, Optical Co, Inc, 58-66; assoc prof, 66-70, PROF MICROBIOL & MED TECHNOL, UNIV ARIZ, 70- , OPTICAL SCI CTR & COMT COMPUT SCI, 70- Concurrent Pos: Assoc prof obstet & gynec, Univ Chicago, 67-69, prof, 69- Mem: Hon fel Inst Acad Cytol. Res: Image processing; pattern recognition; objectivation of diagnostic procedures through self learning computer programs; machine recognition of tumor cells. Mailing Add: Dept of Optical Sci Univ of Ariz Tucson AZ 85721

BARTELS, RICHARD ALFRED, b Saginaw, Mich, May 10, 38; m 59; c 3. SOLID STATE PHYSICS. Educ: Case Inst Technol, BS, 60, MS, 63, PhD(physics), 65. Prof Exp: Sloan fel, Princeton Univ, 64-66; asst prof, 66-69, ASSOC PROF PHYSICS, TRINITY UNIV, TEX, 69- Mem: Am Asn Physics Teachers. Res: Effects of high pressures on the physical properties of solids; elastic and dielectric constants of solids. Mailing Add: Dept of Physics Trinity Univ San Antonio TX 78284

BARTELS, RICHARD HAROLD, b Ann Arbor, Mich, Jan 10, 39; m 68. COMPUTER SCIENCE. Educ: Univ Mich, BS, 61, MS, 63; Stanford Univ, PhD, 68. Prof Exp: Asst prof comput sci, Univ Tex, Austin, 68-74; ASST PROF MATH SCI, JOHNS HOPKINS UNIV, 74- Res: Computational stability and round off error properties of mathematical programming algorithms. Mailing Add: Dept of Math Sci Johns Hopkins Univ Baltimore MD 21218

BARTELS, ROBERT CHRISTIAN FRANK, b New York, NY, Oct 24, 11; m 38; c 1. MATHEMATICS. Educ: Univ Wis, PhB, 33, PhM, 36, PhD(math), 38. Prof Exp: Asst math, Univ Wis, 33-36, 37-38; instr, Univ Mich, 38-42; aeronaut engr, Bur Aeronaut, US Navy, 42-45; from asst prof to assoc prof, 45-57, PROF MATH, UNIV MICH, ANN ARBOR, 57-, DIR COMPUT CTR, 59- Concurrent Pos: Res partic, Oak Ridge Inst Nuclear Studies, 54; mem comput res study sect, NIH, 62- Mem: AAAS; Soc Indust & Appl Math; Asn Comput Mach; Am Math Soc; Math Asn Am. Res: Mathematical theory of elasticity; hydrodynamics; dynamics of compressible fluid; applied mathematics; Saint-Venant's flexure problem for the regular polygon; numerical analysis. Mailing Add: Dept of Math Univ of Mich Ann Arbor MI 48104

BARTELS, WILLIAM CHARLES JOSEPH, b New York, NY, Sept 21, 23; m 46; c 7. PHYSICS. Educ: Fordham Univ, BS, 44; Polytech Inst Brooklyn, MS, 47. Prof Exp: Develop engr, Knolls Atomic Power Lab, 48-51, Brookhaven Nat Lab, 51-52 & Walter Kidde Nuclear Labs, 52-58; reactor physicist, Div Reactor Develop, US AEC, 58-61, tech asst to mem of comn, 61-67, chief terrestrial low power reactors br, 67-69 & tech studies br, Off Safeguards & Mats Mgt, 69-75; ASST DIR POLICY & ANAL, DIV SAFEGUARDS & SECURITY, US ENERGY RES & DEVELOP ADMIN, 75- Mem: Am Nuclear Soc; Inst Nuclear Mat Mgt. Res: Engineering. Mailing Add: Div of Safeguards & Security US Energy Res & Develop Admin Washington DC 20545

BARTELS-KIETH, JAMES RICHARD, b London, Eng, Oct 15, 26; nat US; m 56; c 1. ORGANIC CHEMISTRY. Educ: Cambridge Univ, BA, 48, MA & PhD(org chem), 52. Prof Exp: Tech officer, Akers Res Labs, Imp Chem Industs, Ltd, Eng, 51-60; sr chemist, Smith Kline & French Res Inst, Eng, 60-63; sr scientist photog chem, 64-74, RES ASSOC, POLAROID CORP, 74- Mem: Am Chem Soc; The Chem Soc. Res: Alicyclic and heterocyclic chemistry; organic synthesis; photographic chemistry; application of infrared and nuclear magnetic resonance (helium and carbon 13) to structure determination. Mailing Add: Polaroid Corp 730 Main St Cambridge MA 02139

BARTELSTONE, HERBERT JEROME, pharmacology, deceased

BARTER, CYRIL, physical chemistry, see 12th edition

BARTER, JAMES T, b South Portland, Maine, May 31, 30; m 54; c 3. PSYCHIATRY. Educ: Antioch Col, BA, 52; Univ Ariz, MA, 55; Univ Rochester, MD, 61. Prof Exp: Asst prof psychiat, Sch Med, Univ Colo, 65-69, assoc chief, Colo Psychiat Hosp, 68-69; dep dir, 69-73, DIR, SACRAMENTO MENT HEALTH SERV, 73-; ASSOC CLIN PROF PSYCHIAT, UNIV CALIF, DAVIS, 69- Concurrent Pos: Consult, USPHS Indian Health Serv, 68-71. Mem: Fel Am Psychiat Asn; AMA. Res: Cross cultural psychiatry; drugs and drug abuse; adolescents; suicide. Mailing Add: 630 Laurel Dr Sacramento CA 95825

BARTFELD, HARRY, b New York, NY, May 8, 13. IMMUNOLOGY, CELL BIOLOGY. Educ: NY Univ, AB, 36; NY Med Col, MD, 44. Prof Exp: Trainee tissue cult, Crocker Inst Cancer Res, Columbia Univ, 38-39; res asst animal oncol, Cancer Study Group, Post-grad Med Sch, 40, asst prof clin med, 50-56, ASSOC PROF CLIN MED, MED CTR, NY UNIV, 57-; HEAD IMMUNOL, CONNECTIVE TISSUE SECT, ST VINCENT'S HOSP & MED CTR, 64- Concurrent Pos: Fel med, Med Ctr, NY Univ, 48-49; dir lab connective tissue dis, Sch Med, NY Univ, 57-64; co-chmn conf rheumatoid factors & their biol significance, NY Acad Sci, 69; mem adv comt fundamental res related to multiple sclerosis, Nat Multiple Sclerosis Soc, 69-72; mem conf orgn comt, NY Acad Sci. Mem: Am Soc Cell Biol; Soc Exp Biol & Med; Am Fedn Clin Res: Am Asn Immunologists. Res: Cell biology studies of immunopathological phenomena associated with experimental and natural autoimmune disease. Mailing Add: St Vincent's Hosp & Med Ctr 153 W 11th St New York NY 10011

BARTGES, REX J, b Coeur d'Alene, Idaho, Nov 24, 11; m 41; c 2. ECONOMIC ENTOMOLOGY. Educ: San Jose State Col, AB, 40; Univ Calif, MS, 49, PhD(econ entom), 52. Prof Exp: Plant quarantine inspector, Dept Agr, San Mateo County, Calif, 37-41; field asst, Bur Entom & Plant Quarantine, US Dept Agr, 41-42; plant quarantine & nursery inspector, State Dept Agr, Calif, 42; clerk ord dept, Ford Motor Co, Calif, 42-43; chem analyst, Aluminum Co Am, 43; teacher pub sch, Calif, 43-44; air mechanic, Moffett Field, 44-45; teacher sci, Long Beach State Col, 52-56; teacher zool & biol, Bakersfield Col, 56-62; lectr biol, San Francisco State Col, 62-65; PROF BIOL, COL SAN MATEO, 65- Concurrent Pos: NSFAward, Univ CAlif, 58; Fulbright scholar & prof, guest lectr & consult, Biol Dept, Nat Univ El Salvador, 63-65. Mem: AAAS. Res: Synonymy of two tortricid moths, Argyrotaenia citrana and Argyrotaenia Franciscana; biology of orange tortrix Argyrotaenia citrana on deciduous fruits. Mailing Add: Dept of Biol Col of San Mateo 1700 W Hillsdale San Mateo CA 94402

BARTH, CHARLES ADOLPH, b Philadelphia, Pa, July 12, 30; m 54; c 4. PHYSICS. Educ: Lehigh Univ, BS, 51; Univ Calif, Los Angeles, MA, 55, PhD(physics), 58. Prof Exp: Res geophysicist, Inst Geophys, Univ Calif, Los Angeles, 57-58; res physicist, Jet Propulsion Lab, Calif Inst Technol, 58-65; assoc prof, 65-67, PROF ASTRO-GEOPHYS, UNIV COLO, BOULDER, 67-, DIR LAB ATMOSPHERIC & SPACE PHYSICS, 65- Concurrent Pos: NSF fel, Bonn, Ger, 58-59. Mem: AAAS; Am Astron Soc; Am Geophys Union. Res: Aeronomy; planetary atmospheres. Mailing Add: Lab Atmos & Space Physics Univ of Colo Boulder CO 80302

BARTH, HOWARD GORDON, b Boston, Mass, Nov 21, 46; m 72. ANALYTICAL CHEMISTRY. Educ: Northeastern Univ, BA, 69, PhD(anal chem), 73. Prof Exp: Res assoc clin chem, Hahnemann Med Col & Hosp, 73-74; RES CHEMIST, HERCULES RES CTR, 74- Mem: Am Chem Soc; AAAS. Res: High pressure liquid chromatography; separation science; pharmaceutical and industrial analysis. Mailing Add: 2650 Longfellow Dr Wilmington DE 19808

BARTH, KARL FREDERICK, b Houston, Tex, Sept 25, 38; m 63; c 2. MATHEMATICAL ANALYSIS. Educ: Rice Univ, BA, 60, MA, 62, PhD(math), 64. Prof Exp: Res mathematician, Ballistic Res Labs, 64-66; asst prof math, 66-70, ASSOC PROF MATH, SYRACUSE UNIV, 70- Concurrent Pos: Sr vis fel, Brit Sci Res Coun, 75-76. Mem: Am Math Soc; Math Asn Am. Res: Functions of a complex variable; potential theory. Mailing Add: Dept of Math Syracuse Univ Syracuse NY 13210

BARTH, KARL M, b Elizabeth, NJ, Dec 19, 27. ANIMAL NUTRITION. Educ: Del Valley Col, BS, 56; Rutgers Univ, MS, 58, PhD(nutrit), 64; Univ WVa, MS, 62. Prof Exp: Instr, Del Valley Col, 62; res assoc, Purdue Univ, 64-65; asst prof, 65-68,

ASSOC PROF ANIMAL NUTRIT, UNIV TENN, KNOXVILLE, 68- Mem: Am Inst Nutrit; Am Soc Animal Sci. Res: Forage evaluation; non-protein nitrogen utilization. Mailing Add: Dept of Animal Sci Univ of Tenn Knoxville TN 37916

BARTH, LESTER GEORGE, zoology, see 12th edition

BARTH, LUCENA J, b Kokomo, Ind, Jan 29, 18; m 48. ZOOLOGY. Educ: Univ Mo, BA, 38, MA, 39; Columbia Univ, PhD(zool), 45. Prof Exp: Res asst, Univ Mo, 38-39; instr biol, Brooklyn Col, 44-45; res assoc zool, Columbia Univ, 45-61, asst prof zool & dir, NSF Undergrad Res Prog, Barnard Col, 61-64; RES BIOLOGIST, MARINE BIOL LAB, WOODS HOLE, 64- Concurrent Pos: Lectr biol, Barnard Col, 56-61. Honors & Awards: Newberry Prize, Columbia Univ, 45. Mem: Am Soc Zool; Soc Develop Biol. Res: Chemical and experimental embryology; glycogenolysis and phosphate metabolism; changes in proteins during development of frog's egg; control of cellular differentiation. Mailing Add: 26 Quissett Ave Woods Hole MA 02543

BARTH, MAX, b Paterson, NJ, Apr 25, 19; m 49; c 4. PHYSICAL CHEMISTRY. Educ: NY Univ, BA, 49, PhD(chem), 55. Prof Exp: Asst instr, NY Univ, 51-55; from asst instr to asst prof, 55-61, assoc prof, 61-68, PROF CHEM, LA SALLE COL, 68-, CHMN DEPT, 64- Mem: Am Chem Soc. Res: Solution kinetics. Mailing Add: Dept of Chem La Salle Col Philadelphia PA 19141

BARTH, ROBERT HOOD, JR, b Midland, Mich, Feb 10, 34; m 67. ZOOLOGY. Educ: Princeton Univ, AB, 56; Harvard Univ, AM, 59, PhD(biol), 62. Prof Exp: Instr biol, Harvard Univ, 61-63, lectr, 63-64; NSF fel zool, Univ Sheffield, 64-65; asst prof, 65-70, ASSOC PROF ZOOL, UNIV TEX, AUSTIN, 70-, DIR FAC & STAFF, 73- Concurrent Pos: NSF res grants develop & regulative biol, 61-64, 66- Mem: AAAS; Am Soc Zool; Soc Develop Biol; Am Soc Naturalists; Animal Behav Soc. Res: Insect physiology, especially the neuroendocrine control of behavior; reproduction and development of insects. Mailing Add: Dept of Zool Univ of Tex Austin TX 78712

BARTH, ROLF FREDERICK, b New York, NY, Apr 4, 37; m 65; c 4. PATHOLOGY, IMMUNOLOGY. Educ: Cornell Univ, AB, 59; Columbia Univ, MD, 64; Am Bd Path, dipl, 70. Prof Exp: Intern surg, Columbia-Presby Hosp, New York, 64-65; fel tumor immunol, Dept Tumor Biol, Karolinska Inst, Sweden, 65-66; res assoc immunol, Nat Inst Allergy & Infectious Dis, NIH, 66-68, resident path, Path Br, Nat Cancer Inst, 68-70; from asst prof to assoc prof, 70-76, PROF PATH, UNIV KANS MED CTR, KANSAS CITY, 76- Concurrent Pos: Consult, Vet Admin, Kansas City, Mo, 70- & Nat Cancer Inst, 72- Mem: Transplantation Soc; Am Asn Immunologists; Am Asn Pathologists & Bacteriologists; Am Asn Cancer Res; Soc Nuclear Med. Res: Cell mediated immunity; transplantation and tumor immunology; immunosuppression; cell surface antigens; lymphocyte migration; cellular labeling with radioisotopes. Mailing Add: Dept of Path & Oncol Univ of Kans Med Ctr Kansas City KS 66103

BARTHA, RICHARD, b Budapest, Hungary, Nov 14, 34; US citizen; m 67; c 2. AGRICULTURAL ECOLOGY, MICROBIAL ECOLOGY. Educ: Univ Budapest, Eötvös Lorand, 52-56; Univ Göttingen, PhD(microbiol), 61. Prof Exp: USPHS fel microbiol, Univ Wash, 62-64; res assoc, 64-66, from asst prof to assoc prof, 66-73, PROF MICROBIOL, RUTGERS UNIV, 73- Mem: Am Soc Microbiol; AAAS. Res: Chemoautotrophic bacteria; microbial degradation of pesticides and oil pollutants; microbial ecology. Mailing Add: Dept of Biochem & Microbiol Rutgers Univ New Brunswick NJ 08903

BARTHAKUR, NAYANA N, b Sibsagar Town, India, Mar 1, 36; m 66. PHYSICS. Educ: Cotton Col, India, BSc, 56; Univ Allahabad, MSc, 59; Univ Sask, PhD(cloud physics), 65. Prof Exp: Demonstr chem, Assam Med Col, India, 56-57; lectr physics, St Anthony's Col, 59-61; asst prof, 66-69, ASSOC PROF AGR PHYSICS, MACDONALD COL, McGILL UNIV, 69- Mem: Am Meteorol Soc; Can Meteorol Soc; Royal Meteorol Soc. Res: Cloud physics, particularly ice nucleation by organic crystals; agricultural physics, particularly protecting plants from frosts. Mailing Add: Dept of Agr Physics Macdonald Col Ste Anne de Bellevue PQ Can

BARTHAUER, GERALD LEE, b Degraff, Ohio, Oct 4, 17; m 43; c 1. ENVIRONMENTAL CHEMISTRY, RESOURCE MANAGEMENT. Educ: Hiram Col, AB, 39; Purdue Univ, MS, 41, PhD(inorg chem), 43. Prof Exp: Group leader develop anal instruments, tech & methods, Magnolia Petrol Co, Tex, 43-46 & Gen Elec Co, 46-49; mgr tech serv dept, Res & Develop Div, 49-67, VPRES ENVIRON AFFAIRS, CONSOL COAL CO, 67- Honors & Awards: S A Braley Award, Ohio River Valley Sanit Comn, 74. Mem: AAAS; Am Chem Soc; Optical Soc Am. Res: High vacuum techniques; spectroscopy; x-ray diffraction; precision of fractionation; industrial radiography; analytical chemistry of the rare earths; quality control; statistical analysis. Mailing Add: Environ Qual Control Dept Consol Coal Co McMurray PA 15317

BARTHEL, CHRISTOPHER ERNEST, JR, physics, see 12th edition

BARTHEL, ROMARD, b Evansville, Ind, Apr 8, 24. ULTRASOUND. Educ: Univ Notre Dame, BS, 47; Univ Tex, PhD(physics), 51. Prof Exp: From instr to assoc prof physics, 47-57, chmn Div Physics & Natural Sci, 54-68, PROF PHYSICS, ST EDWARD'S UNIV, 57-, ON LEAVE AS DIR SW PROV, BROS OF THE HOLY CROSS, 68- Mem: Am Asn Physics Teachers; Acoust Soc Am. Res: Ultrasonic interferometers; theory of solutions; philosophy of science; atomic and nuclear physics and electronics. Mailing Add: St Edward's Univ Austin TX 78704

BARTHEL, WILLIAM FREDERICK, b Arbutus, Md, Mar 12, 15; m 38; c 1. AGRICULTURAL CHEMISTRY. Prof Exp: Chemist, USDA, 40-45; chief chemist insecticide div, Victor Prods Corp, 46-48; chief chemist & vpres, Edco Corp, 48; chief chemist in-chg develop insecticide div, Innis Speiden & Co, 49-50; owner, W F Barthel Chem Co, 50-51; chemist pesticide chem res lab, USDA, 51-57, chemist-in-chg chem lab, Plant Pest Control, 58-62 & Methods Improv Labs, 62-67, supvry chemist, Pesticides Lab, Nat Commun Dis Ctr, 67-68, chief toxicol lab, 68; supvry res chemist, Pesticides Div, Atlanta Toxicol Br, US Food & Drug Admin, 68-71; toxicologist, Ctr Dis Control, 71-74; CONSULT, 74- Honors & Awards: Serv Awards, USDA, 58 & 60. Mem: AAAS; Am Chem Soc; Entom Soc Am; Asn Official Agr Chem; Am Inst Chem. Res: Organic insecticides; structural studies of pyrethrins; cinerins; Barthel rearrangement; Barthrin and related insecticides of low mammalian toxicity. Mailing Add: Box 105 R 2 Mt Vernon IA 52314

BARTHOLD, STEPHEN WILLIAM, b San Francisco, Calif, Nov 10, 45; m 71; c 1. VETERINARY PATHOLOGY. Educ: Univ Calif, Davis, BS, 67, DVM, 69; Univ Wis-Madison, MS, 73, PhD(vet sci), 74. Prof Exp: Chief animal care, US Army Res Inst Environ Med, 69-71; res asst vet path, Univ Wis-Madison, 71-74; ASST PROF COMP MED, SCH MED, YALE UNIV, 74- Mem: Am Vet Med Asn; Vet Cancer Soc; Wildlife Dis Asn. Res: Comparative pathology; laboratory animal pathology; neoplasia. Mailing Add: Sect Comp Med Sch of Med Yale Univ 375 Congress Ave New Haven CT 06520

BARTHOLOMAI, C W, b Terre Haute, Ind, Nov 2, 21; m 53, 66. ENTOMOLOGY. Educ: Purdue Univ, BSA, 42, MS, 48; Univ NC, MPH, 52. Prof Exp: Entomologist, Bur Entom & Plant Quarantine, USDA, 46-47; staff entomologist, Hq Tactical Air Command, 48, 51 & Hq Continental Air Command, 49-51; entomologist, Indust Test Lab, US Naval Shipyard, Philadelphia, 53-54, Admin Off Chief Engrs, 54-56 & Hq Third US Army, 56-73; CONSULT, 73- Mem: Entom Soc Am. Res: Control of arthropods of medical importance. Mailing Add: 3148 Lenox Rd NE Atlanta GA 30324

BARTHOLOMAY, ANTHONY FRANCIS, mathematical biology, see 12th edition

BARTHOLOMEW, DUANE P, b Fargo, NDak, Sept 25, 34; m 56; c 3. PLANT PHYSIOLOGY. Educ: Calif State Polytech Col, BS, 61; Iowa State Univ, PhD(plant physiol), 65. Prof Exp: ASSOC PLANT PHYSIOLOGIST, PINEAPPLE RES INST, 65-, ASST AGRONOMIST, UNIV HAWAII, 66- Mem: AAAS; Am Soc Plant Physiol; Am Soc Agron. Res: Crop physiology; photosynthesis; plant-water relations; plant-environment relations. Mailing Add: Dept of Agron & Soil Sci Univ of Hawaii Honolulu HI 96822

BARTHOLOMEW, ELEANOR RACHEL, b Chicago, Ill. CHEMISTRY. Educ: Univ Nebr, BSc, 28, MSc, 30; Univ Chicago, PhD(chem), 33. Prof Exp: Instr inorg chem & math, Greenbrier Jr Col, 34-35; instr org chem, Hollins Col, 35-39; librn, Nat Aniline & Dye Corp, Buffalo, 41-42; patent agent, Hercules Inc, 42-71; RETIRED. Mem: Am Chem Soc. Res: Organic chemistry; pyrimidine derivatives. Mailing Add: 216 Wooddale Ave New Castle DE 19720

BARTHOLOMEW, GEORGE ADELBERT, b Independence, Mo, June 1, 19; m 42; c 2. ZOOLOGY. Educ: Univ Calif, AB, 40, MA, 41; Harvard Univ, PhD(zool), 47. Prof Exp: Asst, Mus Vert Zool, Univ Calif, 40-41; asst zool & comp anat, Harvard Univ, 45-47; from instr to assoc prof zool, 47-59, PROF ZOOL, UNIV CALIF, LOS ANGELES, 59- Mem: Ecol Soc Am; Am Soc Mammal; Am Ornith Union; Cooper Ornith Soc; Am Soc Zool. Res: Ecology and physiology of vertebrates. Mailing Add: Dept of Biol Univ of Calif Los Angeles CA 90024

BARTHOLOMEW, GILBERT ALFRED, b Nelson, BC, Apr 8, 22; m 52. NUCLEAR PHYSICS. Educ: Univ BC, BA, 43; McGill Univ, PhD(physics), 48. Prof Exp: Res officer nuclear physics, 48-62, head neutron & solid state physics br, 62-71, DIR PHYSICS DIV, ATOMIC ENERGY CAN LTD, 71- Mem: Fel Am Phys Soc; Can Asn Physicists; fel Royal Soc Can. Res: Neutron capture gamma rays; low energy nuclear physics; nuclear structure. Mailing Add: Box 1258 Deep River ON Can

BARTHOLOMEW, JAMES COLLINS, b Dec 18, 42; m 65; c 2. CELL BIOLOGY. Educ: Hobart Col, BS, 65; Cornell Univ, PhD(biochem), 70. Prof Exp: Am Cancer Soc res fel, The Salk Inst, 70-72; SR STAFF CELL BIOLOGIST, LAWRENCE BERKELEY LAB, UNIV CALIF, 72- Mem: AAAS; Am Soc Cell Biol. Res: The regulation of growth of mammalian cells in cultures and the factors that affect this regulation such as serum, viruses and chemical carcinogens. Mailing Add: Lab of Chem Biodynamics Lawrence Berkeley Lab Univ Calif Berkeley CA 94720

BARTHOLOMEW, JAMES WILLIAM, b Ashtabula, Ohio, May 10, 16; m 43. BACTERIOLOGY. Educ: Univ Ohio, BS, 39, MS, 41; Univ Wis, PhD(bact), 44. Prof Exp: PROF BACT, UNIV SOUTHERN CALIF, 44- Concurrent Pos: Fulbright res scholar, Inst Pasteur, Paris, 50-51; lectr, Univ Queensland, 58-59; consult, Douglas Aircraft Co. Mem: Am Soc Microbiol; Biol Stain Comn; Brit Soc Gen Microbiol; AAAS. Res: Cytology and cytochemistry of microorganisms; chemistry of staining reactions; halophilic amylase. Mailing Add: Dept of Biol Sci Univ of Southern Calif Los Angeles CA 90007

BARTHOLOMEW, LLOYD GIBSON, b Whitehall, NY, Sept 15, 21; m 43; c 5. INTERNAL MEDICINE, GASTROENTEROLOGY. Educ: Union Col, BA, 41; Univ Vt, MD, 44; Univ Minn, MS, 52. Prof Exp: Asst, Div Med, Mayo Clin, 49-50, 52, staff, 52-53; from instr to assoc prof, 53-67, PROF MED, MAYO GRAD SCH MED, UNIV MINN & HEAD DEPT GASTROENTEROL, MAYO CLIN, 67- Concurrent Pos: Attend physician, St Mary's Methodist & assoc hosps, 53- Honors & Awards: Woodbury & Carbee Prizes, 44. Mem: AMA; Am Gastroenterol Asn. Res: Pancreatic and liver diseases. Mailing Add: Mayo Clin 200 First St SW Rochester MN 55901

BARTHOLOMEW, ROGER FRANK, b Workington, Eng, May 21, 37; m 61; c 3. PHYSICAL CHEMISTRY. Educ: Univ London, BSc, 58, PhD(phys chem), 61. Prof Exp: Res fel, Nat Res Coun Can, 61-63; from res chemist to sr res chemist, 64-73, RES ASSOC CHEM, TECH STAFFS DIV, CORNING GLASS WORKS, 73- Mem: Am Chem Soc; Am Ceramic Soc; The Chem Soc; fel Am Inst Chemists. Res: Ion-exchange and diffusion studies in glasses and naturally occurring aluminosilicates; flow of gases through microporous activated carbon plugs; physical chemistry of molten salts in particular molten nitrates; low temperature glass forming systems. Mailing Add: 8 Overbrook Rd Painted Post NY 14870

BARTHOLOMEW, WILLIAM HOLDEN, b Cuba, NY, Jan 26, 15; m 41; c 2. BIOCHEMISTRY. Educ: Cornell Univ, AB, 36; Pa State Univ, MS, 38, PhD, 41. Prof Exp: Chemist petrol testing, Penn Grade Crude Oil Asn, 40-41; chem engr, Merck & Co, Inc, 41-50; asst dir, Pabst Labs, 51-59; mgr microbial develop, IMC, 59-60, mgr glutamate opers, 60-65, plant mgr, 65-67, mgr res admin, 67-71; asst to pres, Accent Int, 71-73; MGR MICROBIAL DEVELOP, RES, STAUFFER CHEM CORP, 73- Mem: Am Chem Soc; Am Soc Microbiol; Soc Indust Microbiol; AAAS. Res: Research, development, process engineering, project direction, plant design, economic and business evaluations in fermentation; pharmaceutical, chemical and food industries; vitamins, sulfas, antibiotics, enzymes, amino acids and biological pesticides. Mailing Add: Stauffer Chem Co Westport CT 06880

BARTHOLOMEW, WILLIAM VICTOR, b Monroe, Utah, Dec 19, 08; m 40; c 6. SOILS. Educ: Brigham Young Univ, BS, 39; Iowa State Col, MS, 41, PhD(soil bact), 47. Prof Exp: Asst soil surveyor, Emergency Rubber Proj, Calif, 43-44; assoc soil chemist, Bur Plant Indust, Soils & Agr Eng, USDA, Md, 44-45, soil scientist, 45-47; res assoc prof soil microbiol, Iowa State Col, 47-56; PROF SOILS & MICROBIOL & ASST DIR INT SOIL TESTING PROG, NC STATE UNIV, 56- Concurrent Pos: Vis prof agron, Univ Ibadan, Nigeria, 74-76. Mem: Am Soc Agron; Soil Sci Soc Am. Res: Microbiological transformations of nitrogen in soil; soil microbiology; nitrogen availability; nitrogen loss processes; influence of soil water on nitrogen use by crop plants. Mailing Add: Dept of Soil Sci NC State Univ Raleigh NC 27606

BARTHOLOW, GEORGE WILLIAM, b Yale, Iowa, July 28, 30; c 3. PSYCHIATRY. Educ: Univ Iowa, BS, 51, MD, 55. Prof Exp: Intern, Wayne County Gen Hosp, Eloise, Mich, 55-56; resident psychiat, Univ Iowa, 56-59; assoc clin dir adult inpatient serv, Nebr Psychiat Inst, 61-74, from asst prof to assoc prof, 61-74, PROF PSYCHIAT, COL MED, UNIV NEBR, OMAHA, 74-, ACTG CLIN DIR ADULT INPATIENT SERV, NEBR PSYCHIAT INST, 74- Concurrent Pos: Chief psychiat serv, Vet Admin Hosp, Omaha, 62-; consult, Vet Admin Hosps, Hot Springs, SDak,

66- & Knoxville, Iowa, 69- Mem: Am Psychiat Asn. Res: Vocational rehabilitation of the mentally ill; suicidology; adult inpatient psychiatry. Mailing Add: Nebr Psychiat Inst 602 S 45th St Omaha NE 68106

BARTH-WEHRENALP, GERHARD, b Teplitz-Schoenau, Oct 19, 20; nat US; m 52; c 3. INDUSTRIAL CHEMISTRY. Educ: Innsbruck Univ, PhD(chem), 49. Prof Exp: Asst prof, Innsbruck Univ, 49-51; res assoc, Temple Univ, 51-52; res chemist, 53-55, group leader, 55-57, dir inorg res, 57-63, mgr res, 63-70, asst to chmn, 71, vpres & tech dir, 71-74, SR VPRES & TECH DIR, PENNWALT CORP, 74- Concurrent Pos: Asst prof, LaSalle Col, 52-54, lectr, 54-57. Mem: Am Chem Soc. Res: Inorganic and organic industrial chemicals; fluorine chemistry; polymers; research administration. Mailing Add: Johns Lane Ambler PA 19002

BARTILUCCI, ANDREW J, b New York, NY, Nov 29, 22; m 50; c 3. PHARMACY. Educ: St John's Univ, NY, BS, 44; Rutgers Univ, MS, 49; Univ Md, PhD(pharm), 53. Prof Exp: Anal chemist, US Armed Serv Med Procurement Off, 47-48; assoc res pharmacist, Merck & Co, 49-50; asst dean col pharm, 52-56, PROF PHARM, ST JOHN'S UNIV, NY, 52-, DEAN COL PHARM & ALLIED HEALTH PROFESSIONS, 56- Mem: AAAS; Am Col Apothecaries; Am Pharmaceut Asn; fel NY Acad Sci. Res: Pharmaceutical formulation and analysis. Mailing Add: Col Pharm & Allied Health Prof St John's Univ Jamaica NY 11432

BARTIS, JAMES THOMAS, b Pawtucket, RI, Oct 22, 45; m 72; c 1. STATISTICAL MECHANICS, SYSTEMS ANALYSIS. Educ: Brown Univ, ScB, 67; Mass Inst Technol, PhD(chem physics), 72. Prof Exp: Asst prof chem, Cornell Univ, 74-75; RES ASSOC SYST ANAL, INST DEFENSE ANAL, 75- Concurrent Pos: Fel, NATO, 72; vis scientist, Weizmann Inst, 72; fel, Cornell Univ, 72-74. Mem: Am Phys Soc; Am Chem Soc; AAAS. Res: Statistical mechanics of fluids, the stoichastic theory of chemical reactions and the equilibrium properties of multicomponent solutions; defense systems evaluation and analysis. Mailing Add: Inst for Defense Anal 400 Army-Navy Dr Arlington VA 22202

BARTKE, ANDRZEJ, b Krakow, Poland, May 23, 39; m 66. REPRODUCTIVE PHYSIOLOGY, ENDOCRINOLOGY. Educ: Jagiellonian Univ, MSc, 62; Univ Kans, PhD(zool), 65. Prof Exp: Vis scientist, Inst Cancer Res, Philadelphia, 65; asst prof dept animal genetics, Jagiellonian Univ, 65-67; fel, 67-69, staff scientist, 69-72, SR SCIENTIST, WORCESTER FOUND EXP BIOL, 72- Concurrent Pos: Res career develop award, NIH, 72-77. Mem: AAAS; Genetics Soc Am; Soc Study Reproduction; Endocrine Soc; Brit Soc Endocrinol. Res: Pituitary control of testicular steroidogenesis; role of prolactin in the male; mammalian genetics. Mailing Add: Worcester Found for Exp Biol Shrewsbury MA 01545

BARTKO, FRANK, astronomy, see 12th edition

BARTKO, JOHN, b New York, NY, Mar 11, 31; m 58. NUCLEAR PHYSICS. Educ: Columbia Univ, BA, 54; Fairleigh Dickinson Univ, BSEE, 60; Pa State Univ, PhD(nuclear physics), 66. Prof Exp: Res scientist, Nuclear Div, Martin Marietta Corp, 67-68; sr physicist, 68-73, FEL SCIENTIST, RADIATION & NUCLEONICS LAB, RES & DEVELOP CTR, WESTINGHOUSE ELEC CORP, 73- Mem: Am Phys Soc. Res: Proton induced reactions in potassium; radiation effects on materials and electronic devices; nondestructive testing via radiation techniques. Mailing Add: Res Labs Westinghouse Elec Corp Pittsburgh PA 15235

BARTKO, JOHN JAROSLAV, b Massillon, Ohio, Nov 17, 37. MATHEMATICAL STATISTICS. Educ: Univ Fla, BA, 59; Va Polytech Inst, MS, 61, PhD(math statist), 62. Prof Exp: RES MATH STATISTICIAN, NIMH, 62-, INSTR, FOUND ADVAN EDUC IN SCI, INC, NIH GRAD SCH, 63-, CHMN DEPT STATIST, 65- Concurrent Pos: Consult int pilot study schizophrenia, WHO, Geneva, Switz, 69- Mem: Fel Am Statist Asn; Biomet Soc. Res: Statistical theory and methodology; application of statistics in the life and behavioral sciences; research in cluster analysis. Mailing Add: Nat Inst of Ment Health NIH Campus Bldg 36 Rm 1D24 Bethesda MD 20014

BARTKUS, EDWARD ALFRED, physical organic chemistry, see 12th edition

BARTKY, IAN ROBERTSON, physical chemistry, see 12th edition

BARTL, PAUL, b Prague, Czech, Apr 8, 28; m 53; c 1. BIOPHYSICS, SCIENCE ADMINISTRATION. Educ: Prague Tech Univ, MA, 51; Czech Acad Sci, PhD(phys chem), 54. Prof Exp: Scientist, Inst Org Chem & Biochem, Prague, 54-64; vis scientist, Inst Cancer Res, Villejuif, France, 65; fel biophys, Johns Hopkins Univ, 66-67, vis assoc prof, 67-68; sr scientist, Hoffmann-La Roche, Inc, 68-71; ASST TO DIR SCI AFFAIRS, ROCHE INST MOLECULAR BIOL, 71- Mem: AAAS; NY Acad Sci; Soc Res Adminr; Nat Coun Univ Res Adminr. Res: High resolution electron microscopy of nucleic acids; water-miscible embedding media for electron microscopy; physical chemistry of nucleic acids and their constituents; research administration. Mailing Add: Roche Inst of Molecular Biol Nutley NJ 07110

BARTLE, GLENN GARDNER, b Borden, Ind, Feb 7, 99; m 21; c 2. GEOLOGY. Educ: Ind Univ, AB, 21, AM, 23, PhD(stratig, econ geol), 32. Prof Exp: Sch supt, Ill, 22-24; asst, Univ Chicago, 24-25; instr geol, Kansas City Jr Col, 25-33; chmn dept geol & geog, Univ Kansas City, 33-38, prof geol & dean lib arts, 38-42; prof geol & dean, Triple Cities Col, Syracuse Univ, 46-50; pres, 50-64, EMER PRES, HARPUR COL, STATE UNIV NY BINGHAMTON, 64-; GEN PARTNER, BARTLE ENTERPRISES, 66- Concurrent Pos: Consult geologist, Panhandle Eastern Pipe Line Co, 29-42, Australian Oil & Gas Co, 57- & Am sponsored schs abroad, AID, 64- Mem: AAAS; Am Asn Petrol Geol; Am Inst Mining, Metall & Petrol Eng. Res: Economic geology; stratigraphy; natural gas reserves. Mailing Add: Harpur Col State Univ of NY Binghamton NY 13901

BARTLE, ROBERT GARDNER, b Kansas City, Mo, Nov 20, 27; m 51; c 2. MATHEMATICS. Educ: Swarthmore Col, BA, 47; Univ Chicago, SM, 48, PhD(math), 51. Prof Exp: AEC fel, Yale Univ, 51-52, instr math, 52-55; from asst prof to assoc prof, 55-64, PROF MATH, UNIV ILL, URBANA-CHAMPAIGN, 64- Mem: Am Math Soc; Math Asn Am. Res: Functional analysis; spectral theory. Mailing Add: Dept of Math UniY of Ill at Urbana-Champaign Urbana IL 61801

BARTLESON, CHRISTIAN JAMES, b Buffalo, NY, May 11, 29; m 51; c 2. PSYCHOPHYSICS. Prof Exp: Mem res staff, Tech Dept, Pavelle Color, Inc, 51-52 & Res Labs, Eastman Kodak Co, 52-67; vpres & dir res, Macbeth Corp, 67-70; GROUP V PRES RES, KOLLMORGEN CORP, 70- Concurrent Pos: Deleg, Inter-Soc Color Coun, 54-; tech expert, US Standards Inst, 64-; US expert, Int Standards Orgn, 66-; US deleg, Int Comn Illum; mem, Colour Group Gt Brit. Mem: AAAS; fel Optical Soc Am; Soc Photog Sci & Eng; Soc Motion Picture & TV Eng; Int Asn Color (vpres). Res: Color, vision, photographic technology and illumination. Mailing Add: Kollmorgen Corp Box 950 Newburgh NY 12550

BARTLESON, JOHN DAVID, b Detroit, Mich, Mar 17, 17; m 40; c 2. CHEMISTRY. Educ: Mich State Univ, BS, 38, MS, 39; Western Reserve Univ, PhD(org chem), 45. Prof Exp: Asst chem, Mich State Univ, 38-39; sect leader, Chem Res Div, Standard Oil Co, 39-52; proj leader, 52-58, mgr antioxidant sales, 58-66, MGR NEW AREA DEVELOP, PETROL CHEM DIV, ETHYL CORP, 66- Mem: Am Chem Soc; Am Soc Testing & Mat; Chem Develop Asn; Chem Mkt Res Asn; Independent Oil Compounders Asn. Res: Lubricating oil additives; hydrocarbon synthesis and analysis; engine evaluation of fuels and lubricants; refining, compounding and testing of lubricants; combustion chamber deposit studies; technical sales management; commercial development; market research; corporate planning; acquisition studies. Mailing Add: Ethyl Corp 1600 W Eight Mile Rd Ferndale MI 48220

BARTLETT, ALAN C, b Price, Utah, June 17, 34; m 56; c 4. GENETICS. Educ: Univ Utah, BA, 56, MS, 57; Purdue Univ, PhD(pop genetics), 62. Prof Exp: Instr pop genetics, Purdue Univ, 61-62; geneticist, Boll Weevil Res Lab, Entom Res Div, 62-67 & Western Cotton Insects Invests, 69-70, GENETICIST, WESTERN COTTON RES LAB, AGR RES SERV, USDA, 70- Mem: AAAS; Entom Soc Am; Am Genetic Asn; Genetics Soc Am. Res: Radiation genetics of Drosophila, Tribolium castaneum and Anthonomus grandis; population genetics, selection and reproduction; genetic techniques for insect control and eradication; genetics and biology of pink bollworm and lygus bugs. Mailing Add: Western Cotton Res Lab 4135 E Broadway Phoenix AZ 85040

BARTLETT, ALBERT ALLEN, b Shanghai, China, Mar 21, 23; US citizen; m 46; c 4. PHYSICS. Educ: Colgate Univ, BA, 44; Harvard Univ, MA, 48, PhD(physics), 51. Prof Exp: Mem staff, Los Alamos Sci Lab, 44-46; fel, Harvard Univ, 48-49; from asst prof to assoc prof physics, 50-62, PROF PHYSICS, UNIV COLO, BOULDER, 62- Concurrent Pos: Res vis, Nobel Inst Physics, Sweden, 63-64. Honors & Awards: Distinguished Serv Citation, Am Asn Physics Teachers, 70. Mem: Am Phys Soc; Am Asn Physics Teachers (vpres, 76-77); Sigma Xi. Res: Nuclear physics. Mailing Add: Dept of Physics Univ of Colo Boulder CO 80309

BARTLETT, BLAIR RALPH, b Bend, Ore, Aug 26, 11; m 40; c 3. ENTOMOLOGY. Educ: Univ Calif, BS, 36, MS, 38, PhD, 48. Prof Exp: Assoc entom, Citrus Exp Sta, Univ Calif, Riverside, 39-42; asst econ entomologist, State Bur Chem, Calif, 42-44; from asst entomologist to entomologist, Citrus Res Ctr & Agr Exp Sta, Dept Biol Control, 44-74, EMER ENTOMOLOGIST, CITRUS RES CTR, DEPT BIOL CONTROL, UNIV CALIF, RIVERSIDE, 74-, LECTR BIOL CONTROL, DEPT ENTOM, 66- Mem: AAAS; Entom Soc Am. Res: Importation culture-colonization of beneficial insects; effects of insecticides upon beneficial insects. Mailing Add: Citrus Res Ctr Dept Biol Control Univ of Calif Riverside CA 92502

BARTLETT, CHARLES J, fluid mechanics, plasma physics, see 12th edition

BARTLETT, DAVID E, b Bloomfield, NJ, Mar 18, 17; m 45; c 2. VETERINARY PATHOLOGY. Educ: Colo Agr & Mech Col, DVM, 40; Univ Minn, PhD(vet med), 52. Prof Exp: Jr veterinarian, Tuberc Eradication Div, Bur Animal Indust, USDA, DC, 40-41, vet parasitologist, Bur Animal Indust, Zool Div, USDA, Md, 41-48; instr vet med, Sch Vet Med, Univ Minn, 48-52; veterinarian, 53-74, VPRES PROD, AM BREEDERS SERV, INC, 67- Concurrent Pos: Mem subcomt infertility livestock, Nat Res Coun; consult, Int Coop Admin, Brazil, 58, Dominican Republic, 63, Azores, 65 & Trinidad-Tobago, 69. Honors & Awards: Borden Award, Am Vet Med Asn, 70 & Dipl, 71. Mem: Am Vet Med Asn; US Animal Health Asn; Conf Pub Health Vet; Int Fertility Asn; Col Theriogenologists (pres, 71-73). Res: Bovine venereal trichomoniasis and other causes of infertility of livestock; instruction in artificial insemination techniques; cattle improvement through artificial insemination in developing countries. Mailing Add: 6240 S Highlands Ave Madison WI 53705

BARTLETT, DAVID FARNHAM, b New York, NY, Dec 13, 38; m 60; c 4. ELEMENTARY PARTICLE PHYSICS, ELECTROMAGNETISM. Educ: Harvard Univ, AB, 59; Columbia Univ, AM, 61, PhD(physics), 65. Prof Exp: From instr to asst prof physics, Princeton Univ, 64-71; ASSOC PROF PHYSICS, UNIV COLO, BOULDER, 71- Mem: Am Phys Soc. Res: Experimental investigations of laws of electricity and magnetism in particle physics and classical physics. Mailing Add: Dept of Physics & Astrophys Univ of Colo Boulder CO 80302

BARTLETT, DONALD, JR, b Hanover, NH, Aug 4, 37; m 65; c 3. PHYSIOLOGY. Educ: Dartmouth Col, AB, 59; Dartmouth Med Sch, BMS, 61; Harvard Med Sch, MD, 64. Prof Exp: Intern, Internal Med, Strong Mem Hosp, Rochester, NY, 64-65, asst res, 65-66; physician, Epidemiol Sect, Field Studies Br, Div Air Pollution, USPHS, 66-67; chief med res sect, Health Effects Prog, Nat Ctr Air Pollution Control, 67-68; asst prof, 71-75, ASSOC PROF PHYSIOL, DARTMOUTH MED SCH, 75- Concurrent Pos: USPHS res fel physiol, Dartmouth Med Sch, 68-71. Res: Respiratory and comparative physiology; health effects of environmental pollution. Mailing Add: Dept of Physiol Dartmouth Med Sch Hanover NH 03755

BARTLETT, FRANK DAVID, b Clarksburg, WVa, Nov 6, 28; m 50; c 4. SOILS. Educ: Univ WVa, BS, 50, MS, 51; Univ Fla, PhD(soils), 55. Prof Exp: Conserv aid, 47-51, lab asst res, Fla Agr Exp Sta, 51-55, soil scientist, 55-58, supvry soil scientist, 58-62, soil scientist, Ark River Basin, 62-72, RESOURCE CONSERVATIONIST, ARK RIVER BASIN, SOIL CONSERV SERV, USDA, 72- Res: Soil hydrology and soil-water-plant relationships; environmental affairs watershed planning. Mailing Add: Rte 1 Box 21 Alexander AR 72002

BARTLETT, GLENN WILFRED, b Curling, Nfld, May 12, 33; m 57; c 3. MICROBIOLOGY. Educ: Mt Allison Univ, BA, 53; Univ London, MSc(plant physiol) & MSc(microbiol), 55; Oxford Univ, DPhil(phys chem), 58. Prof Exp: From asst prof to prof microbiol, Mem Univ, 58-63; assoc prof bact, McGill Univ, 63-66; dir sci planning, 66-75, VPRES RES & TECHNOL, SCI SPECIALTIES GROUP, AM HOSP SUPPLY CORP, 75- Concurrent Pos: Consult, Nfld Prov Dept Health, 58-61, Dept Educ, 61-63; mem, Ninth Int Bot Cong & Eighth Int Cong Microbiol. Mem: Am Soc Microbiol; Asn Advan Med Instrumentation; NY Acad Sci; Brit Soc Gen Microbiol; Brit Biochem Soc. Res: Adaptation of micro-organisms to drugs and chemicals; genetics and metabolism of bacteria and molds. Mailing Add: Am Hosp Supply Corp 1740 Ridge Ave Evanston IL 60201

BARTLETT, GRANT AULDEN, b Riverside, NS, Can, Dec 19, 38; m 62; c 1. GEOENVIRONMENTAL SCIENCE. Educ: Mt Allison Univ, BSc, 59; Carleton Univ, MSc, 62; NY Univ, PhD(micropaleont), 64. Prof Exp: Head micropaleont, Bedford Inst Oceanog, 64-68; ASSOC PROF GEOL SCI, QUEEN'S UNIV, ONT, 68- Concurrent Pos: Lectr, Queen's Univ, Ont, 66; mem, Can Comt Oceanog & Sci Comt Oceanic Res, 66-, Nat Res Coun, Negotiation grants, 73 & Scholarship Comn, 74-77; prof, Mem Univ Nfld, 74 & Dalhousie Univ, Halifax, 75. Mem: Geol Asn Can; fel Brit Geol Soc; Soc Econ Paleont & Mineral; AAAS; Soc Environ Geol. Res: Laboratory and natural ecological investigations of Foraminifera and their significance in the interpretation of water pollution, paleocology, paleoclimatology and paleogeography; geological, biological, chemical and physical modeling of marine coastal zone environments. Mailing Add: Dept of Geol Sci Queen's Univ Kingston ON Can

BARTLETT, GRANT ROGERS, b Berkeley, Calif, June 16, 12; m 46; c 3. BIOCHEMISTRY. Educ: Stanford Univ, BS, 34; Univ Chicago, PhD(biochem), 42. Prof Exp: Res assoc, Dept Med, Univ Chicago, 42-45; Scripps Clin, 46-61; DIR LAB COMP BIOCHEM, 62- Concurrent Pos: With Off Sci Res & Develop; mem comt blood & related probs, Nat Acad Sci-Nat Res Coun. Mem: Am Chem Soc; Am Soc Biol Chem; Am Soc Hemat. Res: Carbohydrate metabolism; biochemistry of red blood cells. Mailing Add: Lab for Comp Biochem 4620 Santa Fe St San Diego CA 92109

BARTLETT, JAMES HOLLEY, JR, b Brooklyn, NY, Nov 2, 04. PHYSICS. Educ: Northeastern Univ, BCE, 24; Harvard Univ, AM, 26, PhD, 30. Prof Exp: From asst prof to prof theoret physics, Univ Ill, Urbana, 30-65; prof physics, Univ Ala, 65-75; EMER PROF THEORET PHYSICS, UNIV ILL, URBANA, 65-; EMER PROF PHYSICS, UNIV ALA, 75- Concurrent Pos: Fel, Rockefeller Found, 40-41; consult, Lockheed Aircraft Corp, 58-59; exchange prof to USSR, 61-63. Mem: Fel Am Phys Soc; Electrochem Soc; Biophys Soc; Am Astron Soc. Res: Ionization; quadropole radiation; chemical valency; nuclear structure; properties of fast electrons; biophysics; dissolution of metals; anodic oxidation; passivity; stability of orbits; artificial intelligence. Mailing Add: Dept of Physics Univ of Ala University AL 35486

BARTLETT, JAMES KENNETH, b Lynden, Wash, Feb 2, 25; m 48; c 1. CHEMISTRY. Educ: Willamette Univ, BS, 49; Stanford Univ, PhD(chem), 55. Prof Exp: Instr chem, Univ Santa Clara, 53-54; asst prof, Long Beach State Col, 54-56; asst prof, 56-63, PROF CHEM, SOUTHERN ORE STATE COL, 63- Mem: Am Chem Soc. Res: Colorimetric methods of analysis. Mailing Add: Dept of Chem Southern Ore State Col Ashland OR 97520

BARTLETT, JAMES WILLIAMS, JR, b Baltimore, Md, Feb 2, 26; m 54; c 3. PSYCHIATRY. Educ: Harvard Univ, AB, 48; Johns Hopkins Univ, MD, 52; State Univ NY, 59-66. Prof Exp: From instr to asst prof psychiat, 57-68, asst dean, 58-65, PROF PSYCHIAT, PROF HEALTH SERV & CHMN DEPT, MED SCH, UNIV ROCHESTER, 68-, ASSOC DEAN, 65- Concurrent Pos: Actg med dir, Strong Mem Hosp, 67-68, med dir, 68- Mem: AAAS; Am Psychiat Asn; Asn Am Med Cols. Res: Medical education; psychoanalysis. Mailing Add: Strong Mem Hosp 601 Elmwood Ave Rochester NY 14642

BARTLETT, JOHN RICHARD, b Ovid, Mich, Nov 2, 36; m 60; c 2. NEUROSCIENCE. Educ: Univ Mich, BS, 59; McMaster Univ, MA, 61, PhD(psychol), 65. Prof Exp: Res assoc, 65-74, ASST PROF NEUROSCI, CTR BRAIN RES, UNIV ROCHESTER, 74- Concurrent Pos: USPHS fel, Ctr Brain Res, Univ Rochester, 65-67. Mem: Soc Neurosci; AAAS; Sigma Xi. Res: Single unit analysis of central visual processes; electrical stimulation for neural prosthetic purposes; single unit activity underlying memory processes; extraretinal effects on vision. Mailing Add: Ctr for Brain Res Strong Mem Hosp Univ Rochester Rochester NY 14642

BARTLETT, LAWRENCE MATTHEWS, b Concord, NH, July 13, 16; m 50; c 1. VERTEBRATE ANATOMY, ORNITHOLOGY. Educ: Univ Mass, BS, 39, MS, 42; Cornell Univ, PhD, 49. Prof Exp: Investr entom, NY Exp Sta, Geneva, 42-44; from instr to assoc prof zool, 44-59, exec officer dept, 63-66, assoc head dept, 66-70, assoc dir, NSF Acad Year Inst & Col Teacher Prog, 64-65, consult undergrad educ sci, 65-68, PROF ZOOL, UNIV MASS, AMHERST, 59- Concurrent Pos: Am Inst Biol Sci high sch lectr, 59-63. Mem: AAAS; Am Ornith Union; Wilson Ornith Soc; Cooper Ornith Soc; Am Soc Zool. Res: Field ornithology; avian anatomy and behavior. Mailing Add: 83 Spring St Amherst MA 01002

BARTLETT, M FREDERICK, organic chemistry, see 12th edition

BARTLETT, MORTON COVELL, bacteriology, chemical engineering, see 12th edition

BARTLETT, NEIL, b Newcastle-upon-Tyne, Eng, Sept 15, 32; m 57; c 2. INORGANIC CHEMISTRY. Educ: Univ Durham, BSc, 54, PhD(inorg fluorine chem), 58. Hon Degrees: DSc, Univ Waterloo, 68, Colby Col, Maine, 72. Prof Exp: Sr chem master, Duke's Sch, Eng, 57-58; lectr chem, Univ BC, 58-59, from instr to prof chem, 59-66; prof chem, Princeton Univ, 66-69; PROF CHEM, UNIV CALIF, BERKELEY, 69- Concurrent Pos: Steacie Mem fel, 64-66; Sloan fel, 64-; Res Corp award, 65; William Lloyd Evans Mem lectr, 66; Miller vis prof, Univ Calif, Berkeley, 67-68. Honors & Awards: Corday-Morgan Medal, The Chem Soc, 62; Noranda Award, Chem Inst Can, 63; Steacie Prize, 65; Cresson Medal, Franklin Inst, Pa, 68; Kirkwood Medal & Award, 69; Inorg Chem Award, Am Chem Soc, 70; Dannie-Heineman Prize, Göttingen Acad, 71. Mem: Am Chem Soc; Am Crystallog Asn; fel Chem Inst Can; fel Royal Soc; The Chem Soc. Res: Fluorine inorganic chemistry; noble gas chemistry; high-energy oxidizers; x-ray crystallography; coordination chemistry; nonaqueous solvent chemistry; thermochemistry. Mailing Add: Dept of Chem Latimer Hall Univ of Calif Berkeley CA 94720

BARTLETT, NOEL SLOANE, b Brooklyn, NY, July 7, 38; m 62; c 2. APPLIED STATISTICS. Educ: Amherst Col, BA, 60; Mass Inst Technol, BS, 60; Case Inst Technol, MS, 62, PhD(math), 65. Prof Exp: Tech asst mgt methods unit, 60, staff tech specialist, 61-63, sr analyst, 63-65, proj leader mgt sci staff, 65-69, staff assoc, 69-70, mgr anal, 71-73, mgr chem, res & eng systs, 73-74, MGR CONSUMER RES & ANAL, STANDARD OIL CO (OHIO), 74- Concurrent Pos: Spec lectr, Case Inst Technol, 63-64, 65-66. Mem: Am Statist Asn; Sigma Xi; Asn Consumer Res. Res: Developments in trade-off analysis and market segmentation. Mailing Add: Standard Oil Co (Ohio) Midland Bldg Cleveland OH 44115

BARTLETT, PAUL DEVERE, b Pellston, Mich, May 26, 11; m 38; c 3. BIOCHEMISTRY. Educ: Mich State Normal Col, BS, 35; Wayne Univ, MS, 38, PhD(chem), 48; Am Bd Clin Chem, dipl. Prof Exp: Res chemist, Henry Ford Hosp, 38-39; instr chem, Edison Inst Technol, 39-43; sr assoc biochem, 48-56, chief bio-org div, biochem dept, 56-66, CHMN DEPT BIOCHEM & MOLECULAR BIOL, EDSEL B FORD INST MED RES, 66- Mem: AAAS; Am Chem Soc; Am Asn Clin Chemists; Can Biochem Soc; Am Soc Biol Chemists. Res: Biochemistry, biophysics, physiology and pharmacology of renal tissue; pathogenesis of renal disease; glomerular basement membrane permeability; kidney cortex sulphydryl and disulfide components; kidney enzymes. Mailing Add: Edsel B Ford Inst for Med Res 2799 W Grand Blvd Detroit MI 48202

BARTLETT, PAUL DOUGHTY, b Ann Arbor, Mich, Aug 14, 07; m 31; c 3. CHEMISTRY. Educ: Amherst Col, AB, 28; Harvard Univ, AM, 29, PhD(chem), 31. Hon Degrees: ScD, Amherst Col, 53; ScD, Univ Chicago, 54; Dr, Univ Montpellier, 67 & Univ Paris, 69. Prof Exp: Nat Res fel chem, Rockefeller Inst, 31-32; instr, Univ Minn, 32-34; from instr to prof, Harvard Univ, 34-48, Erving prof, 48-74; ROBERT A WELCH RES PROF, TEX CHRISTIAN UNIV, 74- Concurrent Pos: Off investr, Nat Defense Res Comt, 41-61; pres org div, Int Union Pure & Appl Chem, 67-69. Honors & Awards: Am Chem Soc Award, 38, Willard Gibbs & Roger Adams Medals, 63, T W Richards Award, 66 & J F Norris Award, 69; A W von Hofmann Medal, Soc Ger Chem, 62; Nat Medal Sci, 68; Wetherill Medal, Franklin Inst, 70. Mem: Nat Acad Sci; Am Chem Soc; Am Acad Arts & Sci; Franklin Inst; Soc Ger Chem. Res: Stereochemistry; kinetics and mechanism of organic reactions; polymerization; Walden inversion; molecular rearrangements; paraffin alkylation; highly branched compounds; free radicals; reactions of sulphur; cycloaddition; organdithium reactions; singlet oxygen. Mailing Add: Dept of Chem Tex Christian Univ Ft Worth TX 76129

BARTLETT, RICHMOND J, b Columbus, Ohio, Sept 23, 27; m 52; c 4. SOIL CHEMISTRY, PLANT NUTRITION. Educ: Ohio State Univ, BA, 49, PhD(agron), 58. Prof Exp: Newspaper reporter, 50-52; promotional writer, 52-55; asst soils, Ohio State Univ, 55-58; from asst prof to assoc prof, 58-70, PROF SOILS, UNIV VT, 70- Mem: Fel AAAS; Am Soc Agron; Soil Sci Soc Am; Int Soc Soil Sci; Can Soc Soil Sci. Res: Soil chemistry in relation to mineral nutrition of plants; chemistry of the rhizospere; characterization of northern soils. Mailing Add: Dept of Plant & Soil Sci Univ of Vt Burlington VT 05401

BARTLETT, RODNEY JOSEPH, b Memphis, Tenn, Mar 31, 44; m 66; c 2. QUANTUM CHEMISTRY. Educ: Millsaps Col, BS, 66; Univ Fla, PhD(theoret chem), 71. Prof Exp: NSF fel theoret chem, Aarhus Univ, Denmark, 71-72; res assoc chem, Johns Hopkins Univ, 72-74; RES SCIENTIST THEORET CHEM, PAC NORTHWEST LAB, BATTELLE MEM INST, 74- Concurrent Pos: Adj asst prof chem, Wash State Univ, 74-77. Mem: Am Chem Soc; Sigma Xi. Res: Development of the theory and applications of quantum mechanics in chemistry; the correlation problem, particularly diagrammatic perturbation theory, molecular conformations, and studies of large molecules including actinide complexes. Mailing Add: Pac Northwest Lab Battelle Mem Inst Richland WA 99352

BARTLETT, ROGER JAMES, b Ft Madison, Iowa, Mar 19, 42; m 69. SOLID STATE PHYSICS. Educ: Iowa State Univ, BS, 64, MS, 68, PhD(physics), 70. Prof Exp: Fel optical physics, Ga Inst Technol, 70-72; STAFF MEM LOW TEMPERATURE PHYSICS, LOS ALAMOS SCI LAB, 72- Mem: Am Phys Soc. Res: Superconductivity with emphasis on critical current and current and field profiles in superconductors. Mailing Add: Los Alamos Sci Lab Q-26 MS764 PO Box 1663 Los Alamos NM 87545

BARTLETT, THOMAS JEFFERSON, b Great Falls, Mont, June 29, 14; m 37; c 3. ASTRONOMY. Educ: Univ Denver, BS, 35; Univ Calif, MA, 39. Prof Exp: Instr, Long Beach Jr Col, 39-41; high sch instr, Calif, 41-42; interim instr astron, Northwestern Univ, 42-45; head dept math, Pomona Jr Col, 45-46; asst prof astron & math, Univ Denver, 46-57; assoc res engr, Boeing Airplane Co, Wash, 57-58, res engr, 58; opers analyst & sci adv to Commanding Gen, US Army Air Defense Command, 58-67; ASSOC PROF MATH, UNIV SOUTHERN COLO, 67- Concurrent Pos: Mem, Int Cong Math. Mem: Am Astron Soc; Am Math Soc; Math Asn Am; Sigma Xi. Res: Spectra and distribution of faint red stars; orbits of comets and asteroids; occultations of stars by the moon; photoelectric magnitudes and colors of stars; orbits of eclipsing binary stars; motions of gases in solar flare-prominence; operations research; nuclear effects; ballistic missile phenomenology; lasers; military weapon systems; infrared and other optical techniques. Mailing Add: Dept of Math Univ of Southern Colo Pueblo CO 81001

BARTLETT, WILLIAM WALKER, physics, see 12th edition

BARTLEY, DAVID LAUREN, b Detroit, Mich, Nov 10, 43; m 68; c 1. SOLID STATE PHYSICS, CLOUD PHYSICS. Educ: Univ Mich, BA, 65, PhD(physics), 73. Prof Exp: Res assoc physics, Brown Univ, 74-75; FEL PHYSICS, UNIV MO-ROLLA, 75- Mem: Am Phys Soc. Res: Investigations into the spectra, interactions and detection of bulk and surface excitations in pure and isotopic solutions of liquid helium and study of ice crystal growth. Mailing Add: Cloud Physics Res Ctr Univ of Mo Rolla MO 65401

BARTLEY, EDWARD FRANCIS, b Scranton, Pa, June 24 24, 16; m 40; c 2. MATHEMATICS. Educ: Univ Scranton, BA, 38; Columbia Univ, MA, 56. Prof Exp: Pub sch teacher, Pa, 39-43; assoc prof & chmn dept, 46-68, PROF MATH, UNIV SCRANTON, 68- Concurrent Pos: Consult, Scranton Schs. Mem: Math Asn Am. Res: Algebra; secondary school curriculum. Mailing Add: Dept of Math Univ of Scranton Scranton PA 18510

BARTLEY, ERLE EDWIN (ST CLAIR), b Bangalore, India, Oct 23, 22; nat US; m 47; c 4. ANIMAL NUTRITION. Educ: Allahabad Univ, BS, 44; Iowa State Col, MS, 46, PhD, 49. Prof Exp: Assoc prof dairy sci, 49-58, PROF DAIRY & POULTRY SCI, UNIV & DAIRY NUTRITIONIST, AGR EXP STA, KANS STATE UNIV, 58- Concurrent Pos: Comt mem dairy cattle nutrit, Nat Res Coun-Nat Acad Sci, 66-71. Honors & Awards: Am Feed Mfrs Award, 57; Borden Award, 75. Mem: Am Dairy Sci Asn; Am Soc Animal Sci; Am Inst Nutrit. Res: Antibiotics in dairy cattle nutrition; nutrition of calves; bloat; rumen metabolism; rumen protein synthesis; feed processing; urea utilization; ammonid toxicity. Mailing Add: Dept of Dairy & Poultry Sci Kans State Univ Manhattan KS 66506

BARTLEY, MURRAY HILL, JR, b Jamestown, NY, June 15, 33; m 56; c 3. ANATOMY, ORAL PATHOLOGY. Educ: Univ Ore, DMD, 58, cert, 65; Univ Utah, PhD(anat), 68. Prof Exp: Asst oral path, Univ Ore, 62-64; assoc dir res, D N Sharp Hosp, San Diego, 64-65; actg assoc prof oral biol, Sch Dent, Univ Calif, Los Angeles, 68-69; ASSOC PROF PATH, DENT SCH, UNIV ORE, 69-, CHMN DEPT PATH, 71- Concurrent Pos: Univ Ore & NIH teaching grants, 61-64; resident, Providence Hosp, 61-62; consult, Vet Admin, 70. Mem: AAAS; Am Dent Asn; Reticuloendothelial Soc; Geront Soc; NY Acad Sci. Res: Cortisol effects on bone; dental pulp pathology; aging in skeletal tissues determined by densitometric or histological methods; aging in skeletal tissues of anthropological material; heritable and metabolic diseases of the hard dental tissues. Mailing Add: Dept of Path Univ of Ore Dent Sch Portland OR 97201

BARTLEY, SAMUEL HOWARD, b Pittsburgh, Pa, June 19, 01; m; c 3. PSYCHOPHYSIOLOGY. Educ: Greenville Col, BS, 23; Univ Kans, AM, 28, PhD(psychol), 31. Prof Exp: From asst instr to instr psychol, Univ Kans, 27-31; Nat Res Coun fel biol, Dept Ophthal, Sch Med, Wash Univ, 31-33, res assoc psychol & biophys, 32-42; asst prof res physiol optics, Eye Inst, Med Sch, Dartmouth Col, 42-46; prof res visual sci, 46-47; prof psychol, Mich State Univ, 47-71, dir lab study vision & related sensory processes, 66-71; distinguished vis prof, 72-74, DISTINGUISHED RES PROF, MEMPHIS STATE UNIV, 74-; EMER PROF PSYCHOL, MICH STATE UNIV, 71- Honors & Awards: Meritorious Res Award, 62; Apollo Award, Am Optom Asn, 70; Prentice Medal, Am Acad Optom, 72. Mem: Fel Am Acad Optom; Optical Soc Am; fel Am Psychol Asn; Am Physiol Soc. Res: Vision; physiological psychology; theory, vision and electrophysiology of the brain; neurophysiology of visual pathway; binocular vision; fatigue. Mailing Add: 348 Cowley Ave East Lansing MI 48823

BARTLEY, WILLIAM CALL, b Mason, Mich, Dec 4, 32; m 56; c 3. SCIENCE POLICY, SCIENCE ADMINISTRATION. Educ: Mich State Univ, BS, 55, MS, 59. Prof Exp: Asst elec eng, Mich State Univ, 55, instr, 58-59; mgg engr, Apparatus Div, Tex Instruments, 59-60, mkt engr, Semiconductor Div, 60, design engr, 60-61, br mgr,

61-63; res scientist & mgr exp develop, Grad Res Ctr Southwest, 63-67; exec secy US comt solar terrestrial res & exec secy SST comt on ozone, 67-71, exec secy, Climatic Impact Comt, Nat Acad Sci, 71-74; EXEC DIR ADV GROUP ON SCI PROGS, OFF OF THE PRESIDENT'S SCI ADV, 74- Concurrent Pos: Sr investr NASA res contracts, 63-67; dir Space Sci Bd Woods Hole study earth & space, 67-68; secy steering panel, Educ Comt Joint Sponsorship of Acoust Lect, Nat Res Coun-Acoust Soc Am, 68-70; dir comt solar-terrestrial res, Aspen Study on Ground Based Res, NSF contract, 69-74; staff dir, Photovoltaic Study, Space Sci Bd-Solid State Sci Panel, 69 & study on space sci & earth observations priorities, Space Sci Bd, 70; sr policy analyst, Sci Technol & Int Affairs Directorate, NSF, 76-; exec secy & mem, Fed Coun for Sci & Technol, 76- Honors & Awards: Award for Development of Satellite Aspect Computers, NASA, 74. Mem: Am Astronaut Soc; Am Geophys Union; Inst Elec & Electronics Eng; NY Acad Sci. Res: Experimental cosmic ray research using detectors on satellites and deep space probes, especially anisotropies; analyzing public policy implications of advances in science and technology; identifying emerging national and international problems. Mailing Add: 10841 Stanmore Dr Potomac MD 20854

BARTLEY, WILLIAM J, b Richfield, Utah, Feb 27, 35; m 56; c 2. INDUSTRIAL CHEMISTRY. Educ: Brigham Young Univ, BS, 57, MS, 58; Univ Wash, PhD(org chem), 62. Prof Exp: RES SCIENTIST CHEM & PLASTICS, UNION CARBIDE CORP, 62- Mem: Am Chem Soc. Res: Agricultural chemicals; pesticide metabolism studies; organic synthesis; heterogeneous catalysis; new process development. Mailing Add: Chem & Plastics Union Carbide Corp PO Box 8361 South Charleston WV 25303

BARTLOW, THOMAS L, b Johnson City, NY, May 14, 42. MATHEMATICS. Educ: State Univ NY Albany, BS, 63; State Univ NY Buffalo, MA, 66, PhD(math), 69. Prof Exp: Instr, 68-69, ASST PROF MATH, VILLANOVA UNIV, 69- Mem: Math Asn Am; Hist Sci Soc; Can Soc Hist & Philos Math. Res: Loops and quasigroups and their generalizations; power-associative quasigroups and loops; multiplicative systems; history of mathematics. Mailing Add: Dept of Math Villanova Univ Villanova PA 19085

BARTMAN, FRED LESTER, atmospheric physics, see 12th edition

BARTNICKI-GARCIA, SALOMON, b Mexico City, Mex, May 18, 35; m 61. BIOCHEMISTRY. Educ: Rutgers Univ, PhD(microbiol), 61. Prof Exp: Res assoc microbiol, Rutgers Univ, 61-62; asst microbiologist & lectr, 62-68, assoc prof & microbiologist, 68-71, PROF PLANT PATH & MICROBIOLOGIST, UNIV CALIF, RIVERSIDE, 71- Concurrent Pos: Vis prof, org chem inst, Univ Stockholm, 69-70. Honors & Awards: New York Bot Garden Award, Bot Soc Am, 75. Mem: AAAS; Am Soc Microbiol; Am Phytopath Soc; Mycol Soc Am; Brit Soc Gen Microbiol. Res: Microbial biochemistry; biochemistry of morphogenesis; physiology of fungi; cell wall structure and biosynthesis; biochemistry of parasitism. Mailing Add: Dept of Plant Path Univ of Calif Riverside CA 92502

BARTNOFF, SHEPARD, b Kobryn, Poland, Nov 6, 19; nat US; m 44 & 75; c 3. NUCLEAR PHYSICS. Educ: Syracuse Univ, AB, 41, MA, 44; Mass Inst Technol, PhD(physics), 49. Prof Exp: Asst physics, Syracuse Univ, 41-43, instr, 43-45; from instr to asst prof, Tufts Univ, 48-53, assoc prof & exec secy, 68; nuclear fuels mgr, Gen Pub Utilities Corp, 68-69, mgr eng, 69-71, dir environ affairs, 71-72; PRES, JERSEY CENT POWER & LIGHT CO, 72- Mem: AAAS; Am Phys Soc; Am Nuclear Soc. Res: Compressible fluid flow; electromagnetic theory and properties of dielectrics; piezoelectric crystals; radar meteorology; nuclear power systems; ecology. Mailing Add: Jersey Cent Power & Light Co PO Box 1279-R Morristown NJ 07960

BARTOK, WILLIAM, b Budapest, Hungary, May 1, 30; US citizen; m 57; c 3. PHYSICAL CHEMISTRY, CHEMICAL ENGINEERING. Educ: McGill Univ, BEng, 54, PhD(phys chem), 57. Prof Exp: From res chemist to res assoc, Esso Res & Eng Co, 57-73, SR RES ASSOC, EXXON RES & ENG CO, 73- Mem: Am Chem Soc; Am Inst Chem Eng; Combustion Inst; Air Pollution Control Asn. Res: Combustion; air pollution control; kinetics and catalysis; free radical processes in oxidation and high temperature chemistry; rheology of disperse systems; nitrogen oxides formation and control. Mailing Add: 304 Woods End Rd Westfield NJ 07090 Linden NJ

BARTOLINI, ROBERT ALFRED, b Waterbury, Conn, Apr 4, 42; m 64; c 3. ELECTROOPTICS. Educ: Villanova Univ, BS, 64; Case Inst of Technol, MS, 66; Univ Pa, PhD(elec eng), 72. Prof Exp: MEM TECHSTAFF ELEC ENG, DAVID SARNOFF RES CTR, RCA LABS, 66- Concurrent Pos: Lectr, LaSalle Col, Philadelphia, 72- Mem: Inst Elec & Electronic Engr; Optical Soc Am; Am Inst Physics; Sigma Xi. Res: Development of materials suitable for optical application such as information storage and retrieval systems, holography and optical wave guides. Mailing Add: David Sarnoff Res Ctr RCA Labs Princeton NJ 08540

BARTON, ALEXANDER JAMES, b Mt Pleasant, Pa, May 9, 24; m 45; c 3. HERPETOLOGY, ECOLOGY. Educ: Franklin & Marshall Col, BS, 46; Univ Pittsburgh, MS, 57. Prof Exp: Park naturalist, Riverview Park, Pa, 45-46; herpetologist, Highland Park Zool Gardens, 46-52; instr biol, Stony Brook Sch, NY, 52-63, dir admis & financial aid, 57-63; dir, Youth Mus, Ga, 57; prof asst, Undergrad Educ Sci Div, 63-65, prof assoc, 65-70, prog dir undergrad instrnl progs, 70-73, PROG MGR STUDENT-ORIENTED PROGS, NSF, 73- Concurrent Pos: Pres, Asn Admis Officers Independent Sec Schs, 59-62; adj asst prof, C W Post Col, Long Island Univ, 61-63; consult, Doubleday & Co, 62-64; mem inst serv subcomt, Col Entrance Exam Bd, 62-65, mem col scholar serv comt, 65-68; chmn, Schs Scholar Serv, 62-63; mem comt endangered reptiles & amphibians, Int Union Conserv of Nature, 67-; judge, Am Rose Soc, 70-; mem environ educ subcomt, Fed Interagency Comt Educ, 74- Mem: AAAS; Am Inst Biol Sci; Ecol Soc Am; Brit Herpet Soc. Res: Ecology and classification of amphibians and reptiles, especially turtles and salamanders; husbandry of wild animals in captivity; design of programs for improving undergraduate education in science. Mailing Add: 3818 N Vernon St Arlington VA 22207

BARTON, AMBROSE DONALD, b Toronto, Ont, June 25, 17; nat US; m 50; c 1. BIOCHEMISTRY. Educ: Univ Toronto, BA, 40, MA, 44; Univ Tex, PhD(org chem), 49. Prof Exp: Elec engr, Res Enterprises, Ltd, Can, 42-45; assoc biochemist, Argonne Nat Lab, 54-70; res biochemist, Vet Admin West Side Hosp, 70-72; res asst prof biochem, 72-73, RES ASSOC PROF MED & BIOCHEM, COL MED, UNIV ILL, 73- Concurrent Pos: Fel, McArdle Mem Lab, Univ Wis, 49-54. Mem: Am Soc Biol Chem; Am Chem Soc; Am Soc Cell Biol; Am Thoracic Soc. Res: Biochemical and cytological studies of growth processes and secretions in pulmonary disease. Mailing Add: Univ of Ill Med Sch Box 6998 Chicago IL 60680

BARTON, BERNARD CHARLES, physical chemistry, see 12th edition

BARTON, BETTE LEE, physical chemistry, nuclear quadrupole resonance, see 12th edition

BARTON, BYRON KURTZ, b Peotone, Ill, June 4, 16; m 38; c 2. GEOGRAPHY. Educ: Ill State Univ, BEd, 38; Univ Nebr, MA, 39, PhD(geog), 49. Prof Exp: Instr geog, Warton Sch, Univ Pa, 41-43, Univ Nebr, 46-47 & Ill State Univ, 47-48; from asst prof to prof & head dept, Eastern Ill Univ, 48-55; dir conserv educ, State of Ill, 55-60; exec dir, Wabash Valley Interstate Comn, 60-69; PROF GEOG & DIR CTR URBAN-REGIONAL STUDIES, IND STATE UNIV, 69- Mem: AAAS; Asn Am Geog. Res: Conservation of soil and agricultural water; regional resource management; industrial geography and regional development. Mailing Add: Ind State Univ Dept of Geog Terre Haute IN 47809

BARTON, CHARLES JULIAN, SR, b Jellico, Tenn, Jan 16, 12; m 40; c 3. ENVIRONMENTAL HEALTH. Educ: Univ Tenn, BS, 33, MS, 34; Univ Va, PhD(anal chem), 39. Prof Exp: Jr phys sci aide, Tenn Valley Authority, 34-36; chemist, Chesapeake & Ohio Rwy Co, 39-40, Nat Aluminate Corp, 40 & Indust Rayon Corp, 40-47; res chemist & group leader, Int Minerals & Chem Corp, 47-48; RES STAFF MEM, OAK RIDGE NAT LAB, 48- Mem: Am Chem Soc; Am Nuclear Soc; Health Physics Soc. Res: Photoelectric spectrophotometry applications in analytical chemistry; phase studies; plutonium chemistry; nuclear safety studies; radiological considerations in nuclear gas well stimulation; health effects of non-radioactive pollutants. Mailing Add: 237 Outer Dr Oak Ridge TN 37830

BARTON, DONALD WILBER, b Fresno, Calif, June 12, 21; m 44; c 4. PLANT BREEDING. Educ: Univ Calif, BS, 46, PhD(genetics), 49. Prof Exp: Asst genetics, Univ Calif, 46-49; asst prof, Univ Mo, 50-51; assoc prof, 51-59, head dept, 59-60, PROF VEG CROPS, CORNELL UNIV, 59-, DIR NY STATE AGR EXP STA & ASSOC DIR RES, COL AGR, 60- Concurrent Pos: AEC fel, Univ Mo, 49-50, res grant genetics, 50-51; res grant hort, Northwest Canners & Freezers Asn, 59-; vis prof, Ore State Col, 59-60. Mem: Genetics Soc Am; Am Soc Hort Sci. Res: Breeding and genetics of horticultural crops; genetic disease resistance. Mailing Add: Dept of Veg Crops NY State Agr Exp Sta Geneva NY 14456

BARTON, EVAN MANSFIELD, b Chicago, Ill, Nov 7, 03; m 37; c 2. INTERNAL MEDICINE. Educ: Williams Col, AB, 24; Johns Hopkins Univ, MD, 29; Am Bd Internal Med, dipl, 42. Prof Exp: Intern, Presby Hosp, Chicago, 29-30; asst, Rush Med Col & Presby Hosp, 31-34; clin prof med, Med Col, Univ Ill, 51-71; PROF MED, RUSH MED COL, 71- Concurrent Pos: Fel path, Rush Med Col & Presby Hosp, 31-34, fel internal med, 36; vol asst, Hamburg, Ger, 31, vol asst med res lab, Royal Infirmary, Aberdeen, Scotland, 35; attend physician, Presby-St Luke's Hosp, 36-; consult, rheumatol, Vet Admin Hosp, Hines, Ill, 46- Mem: Fel AMA; fel Am Col Physicians; Am Rheumatism Asn. Res: Rheumatic diseases. Mailing Add: 5817 S Blackstone Ave Chicago IL 60637

BARTON, GEORGE WENDELL, JR, b Los Angeles, Calif, Oct 9, 25; m 47; c 3. ANALYTICAL CHEMISTRY, RADIOCHEMISTRY. Educ: Calif Inst Technol, BS, 46; Univ Calif, PhD, 50. Prof Exp: Chemist, Argonne Nat Lab, 50-52; group leader mass spectros, 52-66, dep div leader, Radiochem Div, 66-69, assoc div leader, Gamma Sect, 69-74, SPEC PROJ LEADER, GEN CHEM DIV, LAWRENCE LIVERMORE LAB, UNIV CALIF, 74- Mem: AAAS; Am Phys Soc; Am Chem Soc; Am Soc Test & Mat. Res: Chemical laboratory automation; computers and instrumentation; spectroscopy; inorganic and nuclear chemistry. Mailing Add: Gen Chem Div Univ of Calif Lawrence Livermore Lab Box 808 Livermore CA 94550

BARTON, GERALD BLACKETT, b Arco, Idaho, Oct 16, 17; m 43; c 6. APPLIED CHEMISTRY. Educ: Brigham Young Univ, BS, 39, MS, 41; Ohio State Univ, PhD, 50. Prof Exp: Chemist, Am Smelting & Refining Co, Utah, 41-43, res investr, NJ, 43-45; asst chem, Ohio State Univ, 45-48; chemist, Gen Elec Co, 48-65; sr res scientist, Chem Dept, Pac Northwest Lab, Battelle Mem Inst, 65-70; RES SCIENTIST, WESTINGHOUSE HANFORD CO, 70- Mem: Am Chem Soc. Res: Chemical processing of nuclear reactor fuels; disposal of radioactive wastes; analysis for trace impurities in gases; stress corrosion cracking of stainless steel. Mailing Add: 101 W 31st Ave Kennewick WA 99336

BARTON, HARVEY EUGENE, b Couch, Mo, Aug 26, 36; m 54; c 5. ECONOMIC ENTOMOLOGY. Educ: Ark State Univ, BSE, 62, MSE, 63; Iowa State Univ, PhD(entom), 69. Prof Exp: Asst prof, 67-70, ASSOC PROF ZOOL, ARK STATE UNIV, 70- Mem: Entom Soc Am. Res: Aquatic entomology; artificial diets and rearing methods for soil insects. Mailing Add: Dept of Biol Ark State Univ Box 501 State University AR 72467

BARTON, JAMES CLYDE, b Westminster, SC, Sept 28, 16; m 41; c 2. CHEMISTRY. Educ: Berea Col, AB, 37. Prof Exp: Res chemist, Wannamaker Chem Co, Inc, 42-45; chemist & group leader, Carbide & Carbon Chem Corp, 45-51; supt, Works Lab, 45-64, MGR, LAB DIV, OAK RIDGE GASEOUS DIFFUSION PLANT, NUCLEAR DIV, UNION CARBIDE CORP, 64- Mem: Am Chem Soc; fel Am Inst Chemists; Am Soc Testing & Mat; AAAS; Inst Nuclear Mat Mgt. Res: Textile desizing agents, synthesis of ascorbic acid, synthesis of gluconic acid and gluconates, inorganic analysis, uranium isotopic standards, ultrahigh purity uranium compounds, nuclear materials management and safeguards, uranium hexafluoride specifications and handling criteria. Mailing Add: Union Carbide Corp Nuclear Div PO Box P Oak Ridge TN 37830

BARTON, JAMES DON, JR, b Anna, Ill, Oct 25, 29. PLANT ECOLOGY. Educ: Northern Ill Univ, BS, 52, MS, 53; Purdue Univ, PhD(plant ecol), 56. Prof Exp: From instr to asst prof sci, Boston Univ Jr Col, 56-63; assoc prof ecol & dir div nat sci, Southampton Col, Long Island Univ, 63-64, assoc dean gen educ, 64-66, dean col, 66-68; prof bot, provost & vpres acad affairs, 68-75, PROF SPEC STUDIES, ALFRED UNIV, 75- Concurrent Pos: Consult, Appel & Assocs & Southampton Town Conserv Comn. Mem: Ecol Soc Am. Res: Forest ecology; preparation of computer data base of all trees for 20 acres of Donaldsons Woods, Mitchell, Indiana. Mailing Add: 15 Reynolds St Alfred NY 14802

BARTON, JANICE SWEENY, b Trenton, NJ, Mar 22, 39; m 67. BIOPHYSICAL CHEMISTRY. Educ: Butler Univ, BS, 62; Fla State Univ, PhD(chem), 70. Prof Exp: Assoc res chemist, Eli Lilly & Co, 62-65; NIH fel, Johns Hopkins Univ, 70-72; ASST PROF CHEM, E TEX STATE UNIV, 72- Concurrent Pos: NSF res grant, 75. Mem: Am Chem Soc; AAAS; Sigma Xi. Res: The role and properties of proteins that undergo self-assembly into functional cellular macrostructures; and the mechanism of assembly; microtubular protein. Mailing Add: 3013 Monroe Commerce TX 75428

BARTON, JAY, II, b Chicago, Ill, June 22, 22; m 46; c 7. CELL PHYSIOLOGY. Educ: Univ Mo, AB, 47, MA, 48, PhD(zool), 50. Prof Exp: From instr to asst prof zool, Columbia Univ, 50-55; from asst prof to prof biol, St Joseph's Col, Ind, 55-65; staff biologist, Comn Undergrad Educ Biol Sci, 65-67; prof biol & chmn dept, 67-69, PROVOST FOR INSTR, WVA UNIV, 68- Concurrent Pos: Lalor fel, Marine Biol Lab, Woods Hole, 51 & 52; NSF sci fac fel & Fulbright grant, Carlsberg Found, Copenhagen, 61-62; vis prof, George Washington Univ, 65-67; consult, NSF, 68-; bd dirs & exec comt, Biol Sci Curriculum Study, 71- Honors & Awards: Curtis Award,

49. Mem: AAAS; Nat Asn Biol Teachers; Am Inst Biol Sci. Res: Nuclear structure and function. Mailing Add: 105 Stewart Hall Morgantown WV 26506

BARTON, JOHN M, b North Platte, Nebr, Sept 22, 32; m 64; c 1. ANATOMY. Educ: St Norbert Col, BS, 54; Creighton Univ, MS, 57; Univ Nebr, PhD(anat), 62. Prof Exp: Instr anat, Univ Nebr, 59-62 & Med Col Va, 62-64; ASST PROF ANAT, CREIGHTON UNIV, 64-, ASSOC PROF ORAL BIOL, 71- Mem: Am Soc Zool; Int Asn Dent Res. Res: Dental research, especially external and internal factors affecting the growth, development and maintenance of the periodontal ligament and the dental arches. Mailing Add: Dept of Oral Biol Creighton Univ Omaha NE 68131

BARTON, JOHN SELBY, b Boise, Idaho, Mar 23, 18; m 42; c 5. PULP CHEMISTRY, PAPER CHEMISTRY. Educ: Univ Wash, BS, 40; Lawrence Col, MS, 42, PhD(chem), 48. Prof Exp: Sr res chemist, 48-53, chief paper prod develop sect, 53-55, mgr paper res, 55-56, dir packaging res & develop, Western Waxide Div, Calif, 56-60, dir res, Cent Res Div, 60-69, vpres res, 69-74, VPRES RES & DEVELOP, CROWN ZELLERBACH CORP, 74- Mem: Am Chem Soc; assoc Am Tech Asn Pulp & Paper Indust. Res: Lignin chemistry; plastics and resins; pulp and paper. Mailing Add: Crown Zellerbach Corp 1 Bush St San Francisco CA 94119

BARTON, KENNETH RAY, b Elberton, Ga, July 12, 36; m 63; c 2. POLYMER CHEMISTRY, ORGANIC CHEMISTRY. Educ: Wofford Col, BS, 58; Univ Tenn, MS, 60, PhD(phys org chem), 63. Prof Exp: Res chemist, 64-66, SR RES CHEMIST, TENN EASTMAN CO, EASTMAN KODAK CO, 66- Mem: Am Chem Soc; Am Asn Textile Chemists & Colorists. Res: Flame retardancy of textiles; textile chemicals; dyes applications. Mailing Add: Res Labs Tenn Eastman Co Kingsport TN 37660

BARTON, LAWRENCE, b Preston, Eng, Aug 5, 38; m 64; c 3. INORGANIC CHEMISTRY. Educ: Univ Liverpool, BSc, 61, PhD(chem), 64. Prof Exp: Res assoc chem, Cornell Univ, 64-66; asst prof, 66-71, ASSOC PROF CHEM, UNIV MO-ST LOUIS, 71- Concurrent Pos: Petrol Res Fund award, 67-69; NSF award, 69-72; sr res fel, Explosives Res & Develop Estab, Waltham, Abbey, Eng, 70-71. Mem: Am Chem Soc; The Chem Soc. Res: High temperature inorganic chemistry; lower boranes and related compounds; mass spectrometry. Mailing Add: Dept of Chem Univ of Mo-St Louis St Louis MO 63121

BARTON, LUCIAN ANTHONY, b Wilno, Poland, Mar 27, 21; US citizen; m 49; c 1. INORGANIC CHEMISTRY, ORGANIC CHEMISTRY. Educ: Rutgers Univ, BA, 57. Prof Exp: Lab asst chem, Radio Corp Am Labs, 55-57, tech staff assoc, 57-62, MEM TECH STAFF CHEM, RCA LABS, 62- Honors & Awards: David Sarnoff Gold Medal, 69. Mem: Sr mem Am Chem Soc; fel Am Inst Chemists. Res: Photoconductors; phosphors; cathode-ray-tube color screens; liquid crystal materials and systems; video systems. Mailing Add: RCA Labs Princeton NJ 08540

BARTON, M XAVERIA, b Detroit, Mich, May 25, 10. CULTURAL GEOGRAPHY. Educ: Univ Mich, AB, 31, MS, 37, PhD(chem), 38. Prof Exp: Teacher, Gesu Sch, Mich, 32-34; assoc prof chem & head dept, Marygrove Col, 38-44; teacher sci & social sci, St Mary High Sch, Mich, 44-46 & Ohio, 46-49; assoc prof geog & hist & registr, Marygrove Col Monroe Campus, 49-55; Ford Found field res ed, Sister Formation Conf, 55-56; teacher sci & social sci, St Michael High Sch, Mich, 56-62; Fulbright study grant, Univ Mysore, 63; from assoc prof to prof geog & hist, Marygrove Col, 69-75, dir student financial aid, 68-74; RETIRED. Concurrent Pos: Primary trainer for Mich, US Off Educ Training Proj, 73-; consult, Mich Student Financial Aid Asn, 74- Mem: Am Geog Soc; Asn Am Geogr; Asn Asian Studies; Asia Soc; Nat Coun Geog Educ. Res: Organic chemistry of ketoximes; pre-service and in-service education of sisters; geography of Europe and the Mediterranean region; man-land relationships in developing areas, particularly India and Africa. Mailing Add: 8500 Marygrove Dr Detroit MI 48221

BARTON, MARK Q, b Kansas City, Mo, June 5, 28; m 54; c 2. PHYSICS. Educ: Cent Methodist Col, AB, 50; Univ Ill, PhD(physics), 56. Prof Exp: Asst physics, Univ Ill, 50-55; physicist, 56-74, CHMN ACCELERATOR DEPT, BROOKHAVEN NAT LAB, 74- Mem: Fel Am Phys Soc. Res: Elementary particle physics; intensity limitations of particle accelerators; accelerator design and development. Mailing Add: Brookhaven Nat Lab Upton NY 11973

BARTON, PAUL BOOTH, JR, b New York, NY, Sept 30, 30; m 55; c 2. GEOLOGY. Educ: Pa State Univ, BS, 52; Columbia Univ, AM, 54, PhD(geol), 55. Prof Exp: GEOLOGIST, US GEOL SURV, 55- Mem: Fel Mineral Soc Am; fel Geol Soc Am; Geochem Soc; Soc Econ Geol; Mineral Asn Can. Res: Genesis of mineral deposits; chemical and physical nature of ore forming fluids; phase relations between minerals; thermodynamic properties of minerals. Mailing Add: US Geol Surv Washington DC 20242

BARTON, PRESTON NICHOLS, b Amherst, Mass, June 14, 13; m 39, 51; c 2. MEDICINE. Educ: Bowdoin Col, BS, 35; Harvard Univ, MD, 39; Am Bd Prev Med, dipl. Prof Exp: Asst med dir, Colt Mfg Co, 41-45; plant physician & med dir new departure div, 45-67, MED DIR, NEW DEPARTURE-HYATT BEARING DIV, GEN MOTORS CORP, 67- Concurrent Pos: Med dir, Allen Mfg Co, 43-45; lectr, Med Sch, Yale Univ, 48-67. Mem: Fel Am Col Prev Med; AMA; fel Indust Med Asn; fel Am Acad Occup Med. Res: Prevention and care on industrial back injuries; visual skills; rehabiliation in industry; occupational dermatoses. Mailing Add: New Departure-Hyatt Bearing Div Gen Motors Corp Clark NJ 07066

BARTON, RANDOLPH, JR, b Wilmington, Del, Aug 15, 41; m 63; c 1. PHYSICAL CHEMISTRY, CRYSTALLOGRAPHY. Educ: Princeton Univ, AB, 63; Johns Hopkins Univ, MA, 65, PhD(phys chem), 68. Prof Exp: RES CHEMIST, TEXTILE FIBERS DEPT, E I DU PONT DE NEMOURS & CO, INC, 67- Mem: Am Crystallog Asn; Am Phys Soc. Res: Polymer structure and morphology. Mailing Add: Carothers Res Lab Textile Fibers E I du Pont de Nemours & Co Inc Wilmington DE 19898

BARTON, RICHARD DONALD, b Marlboro, Mass, Nov 20, 36; Can citizen; m 58; c 5. NUCLEAR PHYSICS. Educ: McGill Univ, BSc, 59, MSc, 61, PhD(nuclear physics), 63. Prof Exp: Asst prof physics, Loyola Col, Que, 63-64 & McGill Univ, 64-65; asst res officer, Nat Res Coun Can, 65-67; asst prof, 67, ASSOC PROF PHYSICS, CARLETON UNIV, 67- Concurrent Pos: Nat Res Coun Can grants in aid, 63-65 & 67- Mem: Am Phys Soc; Am Asn Physics Teachers; Can Asn Physicists. Res: Delayed proton emission; muonic x-rays; positron and nuclear lifetimes; nuclear electronics; precision testing of speeds of photons under various conditions. Mailing Add: Dept of Physics Carleton Univ Ottawa ON Can

BARTON, RICHARD J, b Painesville, Ohio, Aug 2, 28; m 55; c 2. PHYSICAL CHEMISTRY, PHYSICAL METALLURGY. Educ: Ohio Univ, BS, 50; Iowa State Col, PhD(chem), 56. Prof Exp: Sr scientist, Mat Lab, Wroght Air Develop Ctr, Wright Patterson Air Force Base, Ohio, 57-61; asst prof metall, Colo Sch Mines, 61-65; ASSOC PROF PHYSICS, UNIV SASK, REGINA, 65- Mem: Am Chem Soc; Electrochem Soc. Res: Structure and thermodynamic properties of non-stoichiometric compounds and their effect upon oxidation; solid phase reactions and electrochemical phenomena. Mailing Add: Dept of Physics Univ of Sask Regina SK Can

BARTON, RICHARD RUSSELL, chemistry, see 12th edition

BARTON, STUART SAMUEL, b Toronto, Ont, Nov 16, 22; m 56. PHYSICAL CHEMISTRY. Educ: Univ Toronto, BA, 47, MA, 49; McGill Univ, PhD(phys chem), 56. Prof Exp: Res chemist, Courtaulds Can Ltd, Ont, 54-56; Defense Res Bd res assoc chem, 56-62, from lectr to asst prof, 62-70, ASSOC PROF CHEM, ROYAL MIL COL CAN, 70- Concurrent Pos: Defense Res Bd Can grant, 64- Mem: The Chem Soc. Res: Reactions in electrical discharges; application of calorimetry to surface and solution chemistry. Mailing Add: Dept of Chem & Chem Eng Royal Mil Col of Can Kingston ON Can

BARTON, SYDNEY CHARLES, physical chemistry, environmental science, see 12th edition

BARTON, THOMAS J, b Dallas, Tex, Nov 5, 40; m 66. ORGANIC CHEMISTRY. Educ: Lamar State Col, BS, 62; Univ Fla, PhD(chem), 67. Prof Exp: NIH fel chem, Ohio State Univ, 67; instr, 67-69, ASST PROF ORG CHEM, IOWA STATE UNIV, 69- Mem: Am Chem Soc; The Chem Soc. Res: Nonbenzenoid aromatic chemistry; heterocyclic and organometallic chemistry; synthetic photochemistry. Mailing Add: Dept of Chem Iowa State Univ Ames IA 50010

BARTON, WALTER E, b Oak Park, Ill, July 29, 06; m 32; c 3. PSYCHIATRY. Educ: Univ Ill, BS, 28, MD, 31; Am Bd Neurol & Psychiat, dipl. Prof Exp: Intern, West Suburban Hosp, Ill, 30-31; resident psychiat, Worcester State Hosp, 31-34, sr psychiatrist, 34-38, asst supt, 38-42; supt, Boston State Hosp, 45-63; med dir, Am Psychiat Asn, 63-74; EMER PROF PSYCHIAT, DARTMOUTH MED SCH & SR PHYSICIAN, VET ADMIN GEN HOSP, WHITE RIVER JUNCTION, 74- Concurrent Pos: Lectr, Simmons Col, 36-37, Smith Col, 37-42, Clark Univ, 40-42, Columbia Univ & Yale Univ, 56-59, Med Sch, Tufts Univ, 58 & Med Sch, Georgetown Univ, 63-74; asst supt neurol, Nat Hosp, Eng, 38; assoc prof, Dept Psychiat, Sch Med, Boston Univ, 52-63, clin prof, 63-74; dir, Am Bd Neurol & Psychiat, 63-70; consult, NIMH & NIH: mem bd trustees, Joint Comn Ment Illness & Health; mem spec res projs comt, NIMH & adv comt, Neurol-Psychiat Div, Vet Admin. Mem: Fel AMA; fel Am Psychiat Asn (pres, 61-62). Res: Clinical psychiatry; occupational therapy; hospital administration; geriatrics; ward training program for psychiatric aides. Mailing Add: RFD 1 Hartland VT 05048

BARTON, WILLIAM R, b Boston, Mass, May 13, 28; m 50; c 3. GEOLOGY. Educ: Boston Univ, BA, 51, MA, 55; Indust Col Armed Forces, cert, 60. Prof Exp: Geologist, US Geol Surv, 51-57; phys scientist, 57-61, geologist, 61-64, phys scientist, 64-68, supvr phys scientist, 68-70, STATE LIAISON OFFICER, US BUR MINES, 70- Concurrent Pos: Mem raw mat panel, mat adv bd, Nat Acad Sci, 58-59; mem, New Eng Coun Nat Res Comt, 73-; allocation officer, Fed Energy Off, 73-74; mem, NH Gov Energy & Mineral Resource Coun, 74-; alt chmn, Fed Regional Coun Coal Comt, 75- Mem: Am Inst Mining, Metall & Petrol Engrs; Am Inst Prof Geologists; Soc Econ Geologists; Int Asn Math Geol. Res: Geology and economics of energy resources, industrial minerals and rare earth metals; New England mineral deposits and geology; mining environmental problems and land use. Mailing Add: US Bur of Mines Newmarket NH 03857

BARTONEK, JAMES CLOYD, b Ogden, Utah, Dec 13, 32. ECOLOGY, WILDLIFE MANAGEMENT. Educ: Utah State Univ, BS, 55, MS, 62; Univ Wis, PhD(wildlife ecol, zool), 68. Prof Exp: WILDLIFE BIOLOGIST, US FISH & WILDLIFE SERV, 66- Mem: Wildlife Soc; Am Ornith Union; Ecol Soc Am; Arctic Inst NAm. Res: Ecology of waterfowl and wetland habitat; management of marine birds and upland game-bird populations; environmental influence of resource development upon Arctic and sub-Arctic wildlife populations. Mailing Add: 800 A St Suite 110 Anchorage AK 99501

BARTOO, HARRIETTE VALLETTA (KRICK), b Dayton, Ohio, July 13, 03; m 39, 48. ECONOMIC BOTANY, PALEOBOTANY. Educ: Hiram Col, AB, 25; Univ Chicago, PhD(bot), 30. Prof Exp: Lab asst, Hiram Col, 23-25, instr biol, 25-27; instr bot, Ind State Teachers Col, 30; assoc prof biol, Eastern Ky State Teachers Col, 30-39; asst prof, Austin Peay State Col, 43; prof, Williamsport Dickinson Jr Col, 44-47; instr bot, Oberlin Col, 47-48; from asst prof to prof, 48-72, EMER PROF BOT, WESTERN MICH UNIV, 72- Mem: AAAS; Bot Soc Am; Am Inst Biol Sci; Soc Econ Bot; Am Hort Soc. Res: Paleobotany of coal measures; science teaching in elementary grades and high school; plant geography of economic plants; structure of seedlike fructifications in coalballs from Harrisburg, Illinois; geographic distribution and uses of Annatto. Mailing Add: 1341 Hillcrest Ave Kalamazoo MI 49008

BARTOO, JAMES BREESE, b Swanton, Vt, July 2, 21; m 43; c 5. MATHEMATICS. Educ: Pa State Teachers Col, Edinboro, BS, 47; Univ Iowa, MS, 49, PhD(math), 52. Prof Exp: Instr math, Univ Iowa, 47-52; from asst prof to assoc prof, 52-60, prof math & head dept, 60-68, prof math statist, 68-69, DEAN GRAD SCH, PA STATE UNIV, 69- Mem: AAAS; Am Math Soc; Soc Indust & Appl Math; Am Statist Asn; Math Asn Am. Res: Mathematical statistics. Mailing Add: 203 Kern Grad Bldg Pa State Univ University Park PA 16802

BARTOS, DAGMAR, b Prague, Czech, Oct 18, 29; US citizen; m 44; c 2. CLINICAL BIOCHEMISTRY. Educ: Charles Univ, Prague, PhD(biochem), 53. Prof Exp: Dir biochem, Dept Nutrit Chem, Regional Off Hyg & Epidemiol, Hradec Kralove, 53-55; asst prof biochem, Charles Univ, Prague, 55-57, assoc prof, 57-67; res assoc med, 67-70, ASST PROF PEDIAT, SCH MED, UNIV ORE, PORTLAND, 73- Res: Polyamine and their importance in cancer; application of the polyamine radioimmunoassay for establishment of library of normal serum values and of serum of cancer patients. Mailing Add: Univ Ore Med Sch Hlth Sci Ctr 3181 SW Sam Jackson Pk Rd Portland OR 97201

BARTOS, FRANTISEK, b Prague, Czech, Dec 31, 26; m 51; c 2. CLINICAL BIOCHEMISTRY. Educ: Charles Univ, Prague, PhD(biochem), 53. Prof Exp: Dir clin biochem, Med Sch Hosp, Czech, 53-55; dir biochem lab, Dept Plastic Surg, Charles Univ, Prague, 55-60; head dept radiotoxicol, Res Inst Radiation Hyg, 61-66; res assoc med, Sch Med, Univ Ore, 66-70; res scientist clin chem, United Med Lab Inc, Ore, 71-72; RES ASSOC, HEALTH SCI CTR, UNIV ORE, 72- Res: Development of radioimmunoassay for determination of hormones, peptides, drugs and other substances; developing radioimmunoassay for determination of polyamines. Mailing Add: Univ of Ore Health Sci Ctr 3181 SW Sam Jackson Rd Portland OR 97201

BARTOS, HENRY R, b New York, NY, Oct 2, 36; m 64; c 2. MEDICINE. Educ: Univ Wis, BA, 56, MD, 59. Prof Exp: Intern, Mt Sinai Hosp, 59-60; resident, Jersey City Med Ctr, 60-62; instr med, NJ Col Med, 64-65; ASST PROF MED, STATE UNIV NY UPSTATE MED CTR, 67- Concurrent Pos: Res fel Hemat, Boston City Hosp, Tufts Univ, 65-67; asst vis physician, I & II Med, Boston City Hosp, Mass, 66-67; asst attend physician, State Univ Hosp, 67-; attend physician, Syracuse Vet Admin

Hosp, NY, 67-; consult hematologist, Crouse Irving Mem Hosp, 67- Mem: AMA; Am Soc Hemat; Am Fedn Clin Res; fel Am Col Physicians. Res: Erythrocyte metabolism and enzymes in hemolytic anemias; cancer chemotherapy. Mailing Add: 600 E Genesee St Syracuse NY 13202

BARTOVICS, ALBERT, b Roosevelt, NY, Dec 1, 16; m 44; c 2. POLYMER CHEMISTRY. Educ: Polytech Inst Brooklyn, BS, 37, MS, 39, PhD(chem), 43. Prof Exp: Asst chem, Polytech Inst Brooklyn, 37-39; analyst, Shellac Res Bur, 39-40; res chemist, Firestone Tire & Rubber Co, 43-50 & Armstrong Cork Co, 50-52; from develop chemist to prod develop mgr, Borg Fabric Div, Amphenol-Borg Electronics Corp, 52-64; SR RES CHEMIST, TEXTILE RES LAB, E I DU PONT DE NEMOURS & CO, 64- Mem: AAAS; Am Chem Soc; Am Inst Chemists. Res: Viscometric and osmotic molecular weight studies; chlorination of synthetic polymers; polyvinyl chloride and related copolymers for various applications; development of pile fabrics from natural and synthetic fibers; end-use research on synthetic fibers. Mailing Add: 707 Potter Dr Cedarcroft Kennett Square PA 19348

BARTOW, DENNIS STUART, chemical physics, see 12th edition

BARTRAM, JOHN BOWMAN, b Arlington, NJ, Apr 7, 10; m 40, 75; c 2. PEDIATRICS. Educ: Hamilton Col, BS, 32; Harvard Univ, MD, 36. Prof Exp: Resident pediat path, Children's Hosp, Boston, Mass, 38-39; resident pediat, Children's Hosp, Cincinnati, Ohio, 39-40; resident, Sch Med, Temple Univ, 40-41; pvt pract, Pa, 41-49; dir out-patient clins, 49-54, DIR SERV HANDICAPPED CHILDREN & SR ATTEND PEDIATRICIAN, ST CHRISTOPHER'S HOSP CHILDREN, 54-; PROF PEDIAT, SCH MED, TEMPLE UNIV, 59- Concurrent Pos: Consult, Bancroft Sch, 59-, Summit Sch, Philadelphia Sch Dist, Pa Dept Pub Welfare, Pa Develop Disability Coun, Ken-Crest Ctrs, United Cerebral Palsy Asn & Asn Retarded Citizens. Mem: Am Acad Pediat; Am Pediat Soc; Am Acad Neurol; Am Acad Cerebral Palsy; Am Asn Ment Deficiency. Res: Development of programs and services for children with handicaps; medical administration; investigation in prevention and treatment of mental retardation and developmental disabilities. Mailing Add: St Christopher's Hosp Children 2600 N Lawrence St Philadelphia PA 19133

BARTRAM, RALPH HERBERT, b New York, NY, Aug 16, 29; m 53; c 2. SOLID STATE PHYSICS. Educ: NY Univ, BA, 53, MS, 56, PhD(physics), 60. Prof Exp: Adv res physicist, Gen Tel & Electronics Labs, Inc, NY, 53-61; from asst prof to assoc prof, 61-71, PROF PHYSICS, UNIV CONN, 71- Concurrent Pos: Guest assoc physicist, Brookhaven Nat Lab, 66-71, consult, 71-; tech adv, US Army, 66- & Am Optical Co, 66-; res assoc, Theoret Physics Div, Atomic Energy Res Estab, Harwell, Eng, 67-68. Mem: Am Phys Soc. Res: Theory of color centers and radiation damage in solids; microwave electronics. Mailing Add: Dept of Physics Univ of Conn Storrs CT 06268

BARTRAM, STANLEY F, b Winooski, Vt, May 8, 18; m 42; c 3. X-RAY CRYSTALLOGRAPHY. Educ: Univ NH, BS, 49, MS, 50; Rutgers Univ, PhD(phys chem), 58. Prof Exp: Res chemist, Anal Dept Res Lab, Titanium Pigment Div, Nat Lead Co, NJ, 50-59; sr engr, Metallog & Crystallog Sect, Nuclear Mat & Propulsion Oper, 59-63, prin engr, 63-69, SUPVR X-RAY DIFFRACTION SECT, MAT CHARACTERIZATION OPER, CORP RES & DEVELOP, GEN ELEC CO, 69- Mem: AAAS; Am Crystallog Asn. Res: X-ray diffraction and crystallographic analysis of metals, semimetals, ceramics, polymers and glasses. Mailing Add: Mat Characterization Oper Gen Elec Co Res & Develop Ctr Schenectady NY 12301

BARTRON, LESTER RAY, b Easton, Pa, May 2, 22; m 44; c 5. ORGANIC CHEMISTRY. Educ: Lehigh Univ, BS, 47, MS, 49, PhD, 52. Prof Exp: Jr analyst spectrog anal steel, Bethlehem Steel Corp, 41-42; res asst, Manhattan Proj, Substituted Alloy Mat Labs, 44-45, process control supvr gaseous diffusion plant, 45-46; res asst confectionary, Lehigh Univ, 47-51; res chemist polymers, 51-55, res chemist mkt develop, 55-57, res chemist sales develop, 57-59, group mgr indust & prod develop, 59-65, group mgr prod planning, 65-68, group mgr mkt develop & customer serv, 68-70, TECH CONSULT PACKAGING SYSTS & MKT DEVELOP, E I DU PONT DE NEMOURS & CO, INC, 70- Mem: Am Chem Soc. Res: Polymers; polymeric film; confections; 3-peperidones; market development. Mailing Add: 7 Little Leaf Ct Foulk Woods Wilmington DE 19810

BARTSCH, ALFRED FRANK, b Kaukauna, Wis, Nov 30, 13; m 37; c 2. AQUATIC BIOLOGY. Educ: Univ Minn, AB, 36; Univ Wis, PhD(bot), 39. Prof Exp: Instr bot, Milwaukee ctr, exten div, Univ Wis, 39-42, asst prof bot, 44-45; biologist, Wis State Comt Water Pollution, 45-49; biologist, div water supply & pollution control, USPHS, 49-64; dir res, Pac Northwest Water Lab, 64-68, dir, 68-71, dir, Nat Environ Res Ctr, Corvallis, Ore, 71-75, DIR, ENVIRON RES LAB, US ENVIRON PROTECTION AGENCY, CORVALLIS, 75- Concurrent Pos: Consult, Int Coop Admin, Brazil Water Pollution Authority & Trust Territory of Pac Islands; co-chmn, Great Lakes Res Adv Bd of Int Joint Comn, 74- Mem: Fel Mycol Soc Am; fel Torrey Bot Club; fel Am Fisheries Soc; Ecol Soc Am. Res: Lower fungi; aquatic Chytridiales; effects of growth substances upon algae; biotic responses to stream pollution; water resources and water quality management. Mailing Add: Environ Res Lab 200 S 35th St Corvallis OR 97330

BARTSCH, GLENN EMIL, b Mankato, Minn, Sept 11, 28. BIOSTATISTICS. Educ: Univ Minn, BS, 50, MA, 51; Johns Hopkins Univ, ScD, 57. Prof Exp: Asst, Univ Minn, 50-51 & Johns Hopkins Univ, 51-54; biostatistician, Army Chem Ctr, US Dept Army, 54-56; asst, Johns Hopkins Univ, 56-57, res assoc, 57-58; NIH fel, 58-59; asst prof biostatist, Western Reserve Univ, 60-65; ASSOC PROF BIOMET, UNIV MINN, MINNEAPOLIS, 65- Mem: Am Statist Asn. Res: Effects of non-normality upon common statistical tests; statistical techniques as applied to biological research. Mailing Add: Dept of Biomet Univ of Minn Minneapolis MN 55455

BARTSCH, RICHARD ALLEN, b Portland, Ore, June 7, 40; m 66; c 2. ORGANIC CHEMISTRY. Educ: Ore State Univ, BA, 62, MS, 63; Brown Univ, PhD(org chem), 67. Prof Exp: Instr chem, Univ Calif, Santa Cruz, 66-67; NATO fel, Univ Würzburg, 67-68; asst prof chem, Wash State Univ, 68-73; asst prog adminr, Petrol Res Fund, 73-74; ASSOC PROF CHEM, TEX TECH UNIV, 74- Mem: Am Chem Soc; The Chem Soc; Sigma Xi. Res: Organic reaction mechanisms and synthesis; elimination reactions; crown ethers and other neutral cation carriers; diazonium ion chemistry; phase transfer catalysis. Mailing Add: Dept of Chem Tex Tech Univ Lubbock TX 79409

BARTSCHMID, BETTY RAINS, b Shreveport, La, Dec 27, 49. ANALYTICAL CHEMISTRY. Educ: Univ Tex, Austin, BS, 70; Univ Houston, PhD(anal chem), 74. Prof Exp: Fel anal chem, Colo State Univ, 74-75; ASST PROF ANAL CHEM, VA POLYTECH INST & STATE UNIV, 75- Mem: Am Chem Soc; Optical Soc Am; Soc Appl Spectros. Res: Atomic absorption, fluorescense and emission applied to practical problems of trace metal analysis; development of a versatile, automatic background correcting atomic fluorescence system; molecular emission techniques to determine non-metals. Mailing Add: Dept of Chem Va Polytech Inst & State Univ Blacksburg VA 24061

BARTTER, FREDERIC CROSBY, b Manila, Philippines, Sept 10, 14; m 46; c 3. MEDICINE. Educ: Harvard Univ, BA, 35, MD, 40. Prof Exp: Dir labs, Hosp, USPHS, NY, 42-44; mem trop dis lab, NIH, Washington, DC, 45-46; tutor biochem sci, Harvard Univ, 46-51; chief hypertension-endocrine br, 51-71, CLIN DIR, NAT HEART & LUNG INST, 71- Concurrent Pos: Res & clin fel med, Harvard Med Sch & Mass Gen Hosp, 46-48; med officer in charge onchocerciasis invests, Pan-Am Sanit Bur, 44-45; asst med, Harvard Med Sch, 46-48. Mem: Endocrine Soc; Am Soc Clin Invest; Asn Am Physicians; Am Physiol Soc. Res: Adrenal cortex; parathyroids. Mailing Add: Nat Heart & Lung Inst Bethesda MD 20014

BARTZ, JERRY A, b Fond du Lac, Wis, Mar 4, 42; m 68. PLANT PATHOLOGY. Educ: Univ Wis-Madison, BS, 64, MS, 66, PhD(plant path), 68. Prof Exp: Res fel, Environ Sci Prog, Univ Calif, Riverside, 68-69; asst prof, 68-75, ASSOC PROF PLANT PATH, UNIV FLA, 75- Mem: Am Phytopath Soc. Res: Mode of action of fungicides; treatment and prevention of post-harvest diseases of vegetables. Mailing Add: Dept of Plant Path Univ Fla Inst of Food & Agr Sci Gainesville FL 32601

BARUCH, SULAMITA B, b Medellin, Colombia, Jan 6, 36. PHYSIOLOGY, MEDICINE. Educ: Univ Valle, Colombia, MD, 59; Cornell Univ, PhD(physiol), 63. Prof Exp: Intern, Univ Hosp, Sch Med, Univ Valle, Colombia, 57-58, instr physiol sci, Sch Med, 59-60, asst prof, 63-65; asst prof, 65-70, ASSOC PROF PHYSIOL, MED COL, CORNELL UNIV, 70- Concurrent Pos: USPHS res grants, 64-; estab investr, Am Heart Asn, 67-72. Mem: AAAS; Harvey Soc; Am Soc Nephrology; NY Acad Sci; Am Physiol Soc. Res: Metabolic activity of the kidney as influenced by acid-base balance of the body; excretion of ammonia by the kidney; renal handling of organic acids. Mailing Add: Dept of Physiol Cornell Univ Med Col New York NY 10021

BARUSCH, MAURICE R, b Yokohama, Japan, Sept 21, 19; m 42; c 2. CHEMISTRY. Educ: Stanford Univ, AB, 40, MA, 41, PhD(org chem), 44. Prof Exp: Res chemist, Calif Res Corp, 43-52, sr res chemist, 52-57, group supvr, 57-59, res assoc, 59-61, sr res assoc, 61-63, supvr fuel additives sect, 63-67, mgr fuel additives div, 67-69, mgr grease & indust oils div, 69-73, MGR LUBRICATING OIL ADDITIVES DIV, CHEVRON RES CO, 73- Mem: Am Chem Soc; Combustion Inst. Res: Combustion of hydrocarbons; fuel and lubricating oil additives. Mailing Add: Chevron Res Co PO Box 1627 Richmond CA 94802

BARUT, ASIM ORHAN, b Turkey, June 23, 26; m 54; c 3. THEORETICAL PHYSICS. Educ: Swiss Fed Inst Technol, dipl, 49, PhD, 52. Prof Exp: Asst prof, Swiss Fed Inst Technol, 51-53; fel physics, Univ Chicago, 53-54; asst prof, Reed Col, 54-55; Nat Res Coun Can fel, 55-56; from asst prof to assoc prof, Syracuse Univ, 56-61; mem staff, Lawrence Radiation Lab & Univ Calif, Berkeley, 61-62; PROF THEORET PHYSICS, UNIV COLO, BOULDER, 62- Concurrent Pos: Sr staff mem, Int Ctr Theoret Physics, Trieste, Italy, 64-65, 68-69 & 72-73; Erskin fel, Univ Canterbury, 71; vis prof, Warsaw, 72, Dijon, 73, Stockholm, 73, Santiago, 74 & Munich, 74-75. Honors & Awards: Alexander von Humboldt Award, Alexander von Humboldt Found, Bonn, Ger, 74. Mem: AAAS; fel Am Phys Soc; Europ Phys Soc; Swiss Phys Soc. Res: Quantum theory of fields and particles; elementary particles; statistical mechanics; mathematical physics. Mailing Add: Dept of Physics Univ of Colo Boulder CO 80302

BARVENIK, FRANK W, b Bridgeport, Conn, July 16, 43; m 65; c 2. MARINE MICROBIOLOGY, MICROBIAL ECOLOGY. Educ: Univ Conn, BA, 65; Univ NH, PhD(microbiol), 70. Prof Exp: Fel bact, Univ Wis, 70-71; asst prof biol, Univ Bridgeport, 71-74; asst biologist, 74-75, ASSOC BIOLOGIST OCEANOG SCI, BROOKHAVEN NAT LAB, 75- Concurrent Pos: Int Biol Prog res grant, 70; res collabr & consult, Brookhaven Nat Lab, 71-74; Sigma Xi res grant, 72; Energy Res & Develop Admin res grant, 74; adj prof, Southampton Col, 75- Mem: Am Soc Limnol & Oceanog; Am Soc Microbiol; Sigma Xi. Res: Microbial mineralization of dissolved organic compounds; association of bacteria with detritus; microbial sulfur cycle; denitrification in aquatic and soil environments. Mailing Add: Oceanog Sci Div Dept of Appl Sci Brookhaven Nat Lab Upton NY 11973

BARWISE, KENNETH JON, b Independence, Mo, June 29, 42; m 64; c 2. MATHEMATICAL LOGIC. Educ: Yale Univ, BA, 63; Stanford Univ, MS, 65, PhD(math), 67. Prof Exp: NSF fel, Univ Calif, Los Angeles, 67-68; asst prof math, Yale Univ, 68-70; assoc prof, Univ Wis-Madison, 70-74; MEM FAC MATH, UNIV CALIF, LOS ANGELES, 74- Mem: Am Math Soc; Asn Symbolic Logic. Res: Mathematical logig, especially infinitary logic and generalizations of recursion theory. Mailing Add: Dept of Math Univ of Calif Los Angeles CA 90024

BAR-YAM, ZVI H, b Krakow, Poland, Apr 12, 28; US & Israeli citizen; m 51; c 3. ELEMENTARY PARTICLE PHYSICS, HIGH ENERGY PHYSICS. Educ: Mass Inst Technol, BS, 58, MS, 59, PhD(physics), 63. Prof Exp: Res asst high energy physics, Synchrotron Lab, Mass Inst Technol, 57-63, staff mem, Lab Nuclear Sci, 63-64; from assoc prof to prof physics, Southeastern Mass Univ, 64-70; assoc prof, Israel Inst Technol, 70-72; COMMONWEALTH PROF PHYSICS, SOUTHEASTERN MASS UNIV, 72- Concurrent Pos: Vis scientist, Div Sponsored Res, Mass Inst Technol, 64-65; prin investr, Cambridge Electron Accelerator, 65-71; vis scientist, Brookhaven Nat Lab, 73-, prin investr, 75- Mem: Am Phys Soc; Am Asn Physics Teachers. Res: Strong and electromagnetic interactions at high energies. Mailing Add: 19 Jordan Rd Brookline MA 02146

BAR-ZEV, ASHER, b Brooklyn, NY, Aug 18, 32; m 54; c 4. MOLECULAR BIOLOGY. Educ: City Col New York, BS, 53; Univ Mass, MS, 71, PhD(zool), 73. Prof Exp: Fel entom, Hebrew Univ, 73-74; RES ASSOC ZOOL, UNIV MASS, 74- Res: Mechanisms of tissue response to hormones leading to differentiation; ethical implications of bio-medical research. Mailing Add: 48 Massasoit St Northampton MA 01060

BASAN, PAUL BRADLEY, b Union City, Tenn, Jan 8, 43; m 75; c 1. SEDIMENTOLOGY, PALEOECOLOGY. Educ: Ind Univ, AB, 65; State Univ NY Binghamton, MA, 70; Univ Ga, PhD(geol), 75. Prof Exp: Explor geologist, Texaco, Inc, 70-71; RES SCIENTIST SEDIMENTOLOGY & PALEOECOL, AMOCO PROD CO RES CTR, 74- Mem: Sigma Xi; Soc Econ Paleontologists & Mineralogists; Brit Palaeont Soc; Am Asn Petrol Geologists; Paleont Soc. Res: Sedimentary processes in relation to stratification features; biological and physical energy in sedimentary processes. Mailing Add: Amoco Prod Co Res Ctr PO Box 591 Tulsa OK 74102

BASARABA, JOSEPH, b Ukraine, Mar 13, 21. MICROBIOLOGY. Educ: Ont Agr Col, Univ Guelph, BSA, 54, MSA, 56; Rutgers Univ, PhD, 60. Prof Exp: Asst bact & res fel, Ont Agr Col, Univ Guelph, 54-56; res fel agr microbiol, Rutgers Univ, 56-59; instr soil microbiol, Fac Agr, Univ BC, 59-61, asst prof soil sci, 61-63; ASSOC PROF BIOL, ACADIA UNIV, 63- Mem: AAAS; Am Soc Microbiol; Can Soc Microbiol. Res: Influence of plant polyphenols on the oxidation and utilization of the Krebs' cycle metabolites by pseudomonads; degradation of plant polyphenols by the soil- and water-borne microorganisms. Mailing Add: Dept of Biol Acadia Univ Wolfville NS Can

BASART, JOHN PHILIP, b Des Moines, Iowa, Feb 26, 38; m 60; c 2. RADIO ASTRONOMY. Educ: Iowa State Univ, BS, 62, MS, 63, PhD(elec eng), 67. Prof Exp: Instr electronics technol, Iowa State Univ, 64-67; res assoc radio astron, Nat Radio Astron Observ, Va, 67-69; asst prof, 69-73, ASSOC PROF RADIO ASTRON & ELEC ENG, IOWA STATE UNIV, 73- Mem: Am Astron Soc; Inst Elec & Electronics Engrs; Am Geophys Union; Royal Astron Soc. Res: Radio interferometer instrumentation; observations with long baseline interferometers; planetary radio astronomy. Mailing Add: Dept of Elec Eng Iowa State Univ Ames IA 50011

BASAVAPPA, PARANNARA, b Hodonahalli, Mysore, India, June 7, 27; m 55; c 3. MATHEMATICS. Educ: Univ Mysore, BSc, 49; Benares Hindu Univ, MSc, 50; Univ Sask, PhD(math), 60. Prof Exp: Lectr math, Govt Intermediate Col, Bangalore, 53-55 & Cent Col, 55-65; asst prof, Dalhousie Univ, 65-67; assoc prof, Algoma Col, 67-68; vis prof, 68-70, ASSOC PROF MATH, UNIV NC, CHARLOTTE, 70- Mem: Am Math Soc; Can Math Cong. Res: Matrix theory on solutions of the matrix equations in monomials. Mailing Add: Dept of Math PO Box 20428 Univ of NC Charlotte NC 28202

BASCH, JAY JUSTIN, b Philadelphia, Pa, May 23, 32; m 57; c 2. BIOLOGICAL CHEMISTRY. Educ: Univ Pa, BA, 56; Drexel Inst Technol, MS, 60; Temple Univ, PhD(chem), 68. Prof Exp: RES CHEMIST, EASTERN REGIONAL RES CTR, AGR RES SERV, USDA, 56- Mem: Am Chem Soc. Res: Preparation, isolation and characterization of the glycoproteins from milk fat globule membranes; interaction of milk proteins with radioisotopes. Mailing Add: Eastern Regional Res Ctr USDA 600 E Mermaid Lane Philadelphia PA 19118

BASCH, PAUL FREDERICK, b Vienna, Austria, Nov 10, 33; nat US; m 66; c 2. PARASITOLOGY, PUBLIC HEALTH. Educ: City Col New York, 54; Univ Mich, MS, 56, PhD(zool), 58; Univ Calif, Berkeley, MPH, 67. Prof Exp: Assoc US Army grant, Univ Mich, 58-59; asst prof biol, Kans State Teachers Col, 59-62; from asst res zoologist to assoc res zoologist, Hooper Found, Univ Calif, San Francisco, 62-70; ASSOC PROF INT HEALTH, SCH MED, STANFORD UNIV, 70- Concurrent Pos: Res zoologist, Inst Med Res, Malaya, 63-65 & 69-70; consult parasitol, Calif State Dept Pub Health; mem panel parasitic dis, US-Japan Coop Med Sci Prog, NIH, 74- Mem: Am Micros Soc; Am Soc Parasitol; Am Soc Trop Med & Hyg. Res: Mollusk-trematode interactions; epidemiology and control of parasites; international health. Mailing Add: Dept Family Community & Prev Med Stanford Univ Sch of Med Stanford CA 94305

BASCO, N, b London, Eng, July 13, 29; m 56. PHOTOCHEMISTRY. Educ: Univ Birmingham, BSc, 53, PhD(chem), 56. Prof Exp: Sr sci off, Civil Serv, US, 56-58; Imp Chem Indust fel chem, Univ Cambridge, 58-61; lectr, Univ Sheffield, 61-64; asst prof, 64-68, ASSOC PROF CHEM, UNIV BC, 68- Mem: The Chem Soc. Res: Chemical kinetics and the production and properties of vibrationally excited species studied by flash photolysis and kinetic spectroscopy; electronic absorption spectra of transient species. Mailing Add: Dept of Chem Univ of BC Vancouver BC Can

BASCOM, WILLARD, b New York, NY, Nov 7, 16; m 47; c 1. OCEANOGRAPHY. Prof Exp: Mining engr, Idaho, Ariz, Colo & NY, 40-45; res engr, waves & beaches, Univ Calif, 45-51 & oceanog instruments, Scripps Inst Oceanog, 51-54; tech dir, Comt Civil Defense, 54-55; exec secy, Meteorol Comt, 56; sabbatical yr study Polynesian hist, Tahiti, 57; exec secy maritime res comt, Nat Res Coun, 58; dir, Mohole Proj, AMSOC Comt, Nat Acad Sci, 59-62; pres, Ocean Sci & Eng, Inc, 62-70; pres, Seafinders, Inc, 71-73; DIR, SOUTHERN CALIF COASTAL WATER RES PROJ, 73- Concurrent Pos: Consult, Comt Amphibious Opers, 49, Rockefeller Bros spec study group, 56-57 & sci progs, Columbia Broadcasting Syst TV, 58-59; mem panel underwater swimmers, Nat Acad Sci-Nat Res Coun, 50-53, proj Nobska, 56; mem, Naval Res Adv Comt, 70-; mem Lab Bd for Undersea Warfare, 73- Honors & Awards: Compass Award for Outstanding Contrib to Oceanog, Marine Technol Soc, 70. Mem: AAAS; Marine Technol Soc; Int Asn Water Pollution Res. Res: Oceanographic engineering and instrumentation; drilling in the deep sea; waves; beaches; science writing; undersea diamond mining; deep water archaeology; marine ecology and environmental studies. Mailing Add: 1900 E Ocean Blvd Long Beach CA 90802

BASCOM, WILLARD D, b Watertown, Conn, Oct 27, 31; m 52; c 3. PHYSICAL CHEMISTRY. Educ: Worcester Polytech Inst, BS, 53; Georgetown Univ, MS, 60; Cath Univ Am, PhD, 71. Prof Exp: Res biochemist, Worcester Found Exp Biol, 53-56; HEAD ADHESION SECT, US NAVAL RES LAB, 56- Mem: Am Chem Soc; Sigma Xi; Soc Plastics Engrs. Res: Colloid and surface chemistry; nonaqueous systems; adhesion; polymer fracture. Mailing Add: US Naval Res Lab Code 6170 Washington DC 20390

BASCOM, WILLIAM RUSSEL, b Princeton, Ill, May 23, 12; m 48. ETHNOLOGY. Educ: Univ Wis, BA, 33, MA, 36; Northwestern Univ, PhD(anthrop), 39. Prof Exp: Asst anthrop, Univ Wis, 35-36; from asst to instr anthrop, Northwestern Univ, 38-42; sr analyst, Bd Econ Warfare, 42-43, asst spec rep, Off Econ Warfare, 43-44, actg spec rep, Foreign Econ Admin, 44-45, spec rep, 45-46; from asst prof to prof anthrop, Northwestern Univ, 46-57; PROF ANTHROP, UNIV CALIF, BERKELEY, 57-, DIR, LOWIE MUS ANTHROP, 57- Concurrent Pos: Fulbright fel, 50-51; NSF fel, Cambridge Univ, 58; mem, Int Coun Mus. Honors & Awards: Giuseppe Pietre Int Folklore Prize, Italy, 69. Mem: Fel Am Folklore Soc (pres, 52-54); Am Anthrop Asn; Am Asn Mus; Int African Inst; Int Soc Ethnol & Folklore. Res: African ethnology, with emphasis on art, folklore and religion, in particular the Yoruba of Nigeria; Afro-American religion. Mailing Add: Lowie Mus Anthrop 103 Kroeber Hall Univ of Calif Berkeley CA 94720

BASDEKIS, COSTAS H, b Manchester, NH, Feb 20, 21; m 45; c 4. POLYMER CHEMISTRY. Educ: Univ NH, BS, 42. Prof Exp: Polymer chemist, Plastics Div, 42-48, fibers chemist, Cent Res Dept, Dayton, Ohio, 48-50, group leader, 50-52, from group leader to sr group leader polymers, Plastics Div, 52-60, asst res dir polystyrene plastics, 60-64, MGR RES NEW POLYMERS, PLASTICS DIV, MONSANTO CO, 64- Mem: Am Chem Soc. Res: Polystyrene and acrylonitrile-butadiene-styrene copolymer plastics; high nitrile polymers for barrier applications; synthesis and application studies of new types of engineering thermoplastics; process and product development of reinforced thermoplastics. Mailing Add: 57 Warwick St Longmeadow MA 01106

BASDEN, EDWIN H, II, microbiology, see 12th edition

BASEHART, HARRY WETHERALD, b Zanesville, Ohio, Feb 15, 10; m 42. ANTHROPOLOGY, ETHNOLOGY. Educ: Harvard Univ, MA, 50, PhD(soc anthrop, 53. Prof Exp: Asst prof anthrop, Goucher Col, 50-53; lectr soc anthrop, Harvard Univ, 53-54; from assoc prof to prof anthrop, 54-75, chmn dept, 72-75, EMER PROF ANTHROP, UNIV N MEX, 75- Concurrent Pos: Co-ed, Southwestern J Anthrop, 62-70, ed, 70-75; mem adv panel anthrop, NSF, 64-65; mem cultural anthrop fel rev comt, NIMH, 67-71, chmn, 69-71. Mem: fel Am Anthrop Asn; fel African Studies Asn; Int African Inst; assoc Current Anthrop. Res: Problems of social structure and kinship; culture change; East Africa and American Indians; political anthropology. Mailing Add: Dept of Anthrop Univ of N Mex Albuquerque NM 87131

BASEK, MILOS, b Czech, June 3, 17; m 53; c 2. OTOLARYNGOLOGY, OTOLOGY. Educ: Charles Univ, MD, 45; Am Acad Otolaryngol, dipl, 56. Prof Exp: PROF CLIN OTOLARYNGOL, COLUMBIA UNIV, 67- Concurrent Pos: Mem, Otosclerosis Study Group, 62- Mem: Am Laryngol Rhinol & Otol Soc; Am Otol Soc. Res: Otosclerosis; Meniere's disease; ultrasonic treatment of inner ear disorders. Mailing Add: 161 Ft Washington Ave New York NY 10032

BASEMAN, JOEL BARRY, b Boston, Mass, Apr 28, 42; m 68; c 2. MICROBIOLOGY. Educ: Tufts Univ, BS, 63; Univ Mass, MS, 65, PhD(microbiol), 68. Prof Exp: ASST PROF MICROBIOL, SCH MED, UNIV NC, CHAPEL HILL, 71- Concurrent Pos: NIH fel, Harvard Univ & Harvard Med Sch, 68-71; USPHS grant, Univ NC, Chapel Hill, 75-76, US Army Res & Develop Command grant, 75-76. Mem: AAAS; Am Soc Microbiol; Sigma Xi. Res: Biochemistry of pathogenesis; regulation of animal cell growth. Mailing Add: Dept of Bact & Immunol Univ of NC Sch of Med Chapel Hill NC 27514

BASERGA, RENATO, b Milan, Italy, Apr 11, 25; nat US; m 53; c 2. PATHOLOGY. Educ: Univ Milan, MD, 49. Prof Exp: Asst path, Univ Milan, 49-51; assoc oncol, Med Sch, Univ Chicago, 53-54; from instr to assoc prof path, Med Sch, Northwestern Univ, 58-65; res prof, Fels Res Inst, 65-68, PROF DEPT PATH, SCH MED, TEMPLE UNIV, 68- Concurrent Pos: Consult, Argonne Nat Lab, 59; chmn path study sect, NIH, 71-73 & mem cancer spec prog adv comt, 75- Mem: AAAS; Am Asn Cancer Res; Am Asn Path & Bact; Am Soc Exp Path; Am Soc Cell Biol. Res: Experimental pathology; control of cell division in normal and pathologic tissues. Mailing Add: Dept of Path Temple Univ Philadelphia PA 19140

BASFORD, ADELPHIA MEYER, b St Louis, Mo, June 3, 07. ECOLOGY. Educ: George Peabody Col Teachers, BS & MA, 28, PhD(zool), 36. Prof Exp: Instr, Woman's Col Ga, 28-29 & Teachers Col, Murfreesboro, Tenn, 30-31, Johnston City, 32; asst prof biol, Ball State Teachers Col, 32-35; instr, Teachers Col, Washington, DC, 36-37; asst prof, Ball State Teachers Col, 37-38; prof biol & head dept, Harding Col, 42-45; prof biol sci & chmn dept, 45-73, EMER PROF BIOL, HENDERSON STATE COL, 73- Mem: Fel AAAS. Res: Invertebrate ecology. Mailing Add: Dept of Biol Sci Henderson State Col Arkadelphia AR 71923

BASFORD, ROBERT EUGENE, b Montpelier, NDak, Aug 21, 23. BIOCHEMISTRY. Educ: Univ Wash, BS, 51, PhD(biochem), 54. Prof Exp: From asst prof to assoc prof, 58-70, PROF BIOCHEM, SCH MED, UNIV PITTSBURGH, 70- Concurrent Pos: Res fel, Inst Enzyme Res, Univ Wis, 54-58. Mem: AAAS; Am Chem Soc; Am Soc Biol Chem. Res: Terminal electron transport in the mitochondria of heart and brain; energy metabolism of brain; biochemical basis of phagocytosis. Mailing Add: Dept of Biochem Univ of Pittsburgh Sch of Med Pittsburgh PA 15261

BASH, FRANK NESS, b Medford, Ore, May 3, 37; m 60; c 2. RADIO ASTRONOMY. Educ: Willamette Univ, AB, 59; Harvard Univ, MA, 62; Univ Va, PhD(astron), 67. Prof Exp: Assoc astronomer, Nat Radio Astron Observ, 62-64, res assoc, 65-67; univ fel & fac assoc, 67-69, asst prof, 69-73, ASSOC PROF ASTRON, UNIV TEX, AUSTIN, 73- Mem: Am Astron Soc; Int Sci Radio Union; Int Astron Union. Res: Observational radio astronomy; radio source positions and brightness distributions with radio interferometers; interpretation of neutral hydrogen and carbon monoxide observations in the galaxy. Mailing Add: Dept of Astron Univ of Tex Austin TX 78712

BASHAM, CHARLES W, b Ponca City, Okla, July 25, 34; m 63. HORTICULTURE, PLANT PHYSIOLOGY. Educ: Okla State Univ, BS, 56; Univ Md, PhD(hort), 64. Prof Exp: Instr hort, Okla State Univ, Contract Imp Ethiopian Agr Col, 59-61; asst prof, Kans State Univ, 64-65; ASST PROF HORT, COLO STATE UNIV, 65- Mem: Am Soc Hort Sci; Am Soc Plant Physiol. Res: Post-harvest physiology of fruits and vegetables; production of vegetable crops under irrigation. Mailing Add: Dept of Hort Colo State Univ Ft Collins CO 80521

BASHAM, JACK TUCKER, b Lachine, Que, Nov 2, 26; m 56; c 3. FORESTRY, BOTANY. Educ: Univ Toronto, BSc, 48, MA, 50; Queen's Univ, Ont, PhD(forest path), 58. Prof Exp: Res scientist, Can Dept Agr, 48-60; RES SCIENTIST, CAN FORESTRY SERV, 60- Mem: Ecol Soc Am; Can Inst Forestry; Can Bot Asn. Res: Ecology and succession patterns of fungi in their invasion and destruction of heartwood in living trees and of sapwood and heartwood in recently killed trees. Mailing Add: Forest Res Lab Box 490 Sault Ste Marie ON Can

BASHAW, ELEXIS COOK, b Mt Juliet, Tenn, July 21, 23; m 45; c 2. CYTOGENETICS, PLANT BREEDING. Educ: Purdue Univ, BS, 47, MS, 48; Tex Agr & Mech Col, PhD(genetics), 54. Prof Exp: Asst agronomist, La Agr Exp Sta, 48-50; asst prof, Agr Exp Sta, 52-55, GENETICIST, AGR RES SERV, USDA, TEX A&M UNIV, 55- Mem: AAAS; fel Am Soc Agron; Crop Sci Soc Am. Res: Cytogenetics of grasses, especially genetics of apomixis, reproductive systems; radiation breeding and interspecific hybridization. Mailing Add: Dept of Soil & Crop Sci Tex A&M Univ College Station TX 77843

BASHAW, JOHN DARRELL, physical chemistry, see 12th edition

BASHE, WINSLOW JEROME, JR, b Chicago, Ill, Mar 10, 20; m 53; c 3. PEDIATRICS, PREVENTIVE MEDICINE. Educ: Seton Hall Col, BS, 42; Loyola Univ, MD, 45; Columbia Univ, MPH, 59. Prof Exp: Intern gen med, Gorgas Hosp, CZ, 45-46, resident med, 46-47; resident pediat, Sea View Hosp, NY, 48-49; resident, Children's Hosp Philadelphia, 50, res assoc, S2-53; pvt pract, 53-57; epidemiologist commun dis, Ohio Dept Health, 57-58, chief div, 59-63; assoc clin prof pediat, Col Med, Univ Cincinnati, 63-70; ASSOC PROF PREV MED & PEDIAT, OHIO STATE UNIV, 71- Concurrent Pos: Instr, Univ Pa, 51-53; instr & asst prof, Ohio State Univ, 57-63. Mem: AAAS; Am Pub Health Asn. Res: Virology; mumps and hepatitis; staphylococcal disease; bone maturation of children; influenza; trichinosis; sudden cardiac death; general epidemiology. Mailing Add: Col of Med Ohio State Univ Columbus OH 43210

BASHEY, REZA ISMAIL, b Bombay, India, Aug 28, 32; m 61; c 3. BIOCHEMISTRY. Educ: Univ Bombay, BSc, 52, MSc, 54; Rutgers Univ, PhD(biochem), 58. Prof Exp: Mem fac, Sch Med, Univ Miami, 62-64; res assoc biochem, Albert Einstein Col Med, 64-66, asst prof, 66-70; res asst prof, 70, res assoc prof med, 71-72, assoc prof med & biochem, Hahnemann Med Col, 72-75, VIS ASSOC PROF BIOL & CHEM, HAHNEMANN MED COL & HOSP, 75-; SR INVESTR, PHILADELPHIA GEN HOSP & ASSOC, RHEUMATOLOGY RES LAB, DEPT MED, UNIV PA, 75- Concurrent Pos: Fel biochem, Univ Southern Calif, 58-60; partic fel, Howard Hughes Med Inst, Fla, 60-62; pool officer med sci, Indian Coun Med Res, Ministry Health, India, 64. Mem: AAAS; Am Chem Soc. Res: Connective tissue; collagen and glycosammoglycans, biosynthesis and involvement in fibrillogenesis; heart valve biochemistry in normal and diseased conditions; scleroderma and other skin diseases. Mailing Add: Philadelphia Gen Hosp 700 Civic Center Blvd Philadelphia PA 19104

BASHIR, NASIR AHMAD, b WPakistan, Jan 9, 35; div; c 2. PHYSIOLOGY, ZOOLOGY. Educ: Forman Christian Col, Lahore, Pakistan, BSc, 53; Univ Panjab, WPakistan, MSc, 56; Tulane Univ, PhD(physiol), 67. Prof Exp: Lectr zool, Talimul Islam Col, Pakistan, 56-58; instr, Otero Jr Col, Colo, 61-63; chmn dept physiol, Sch Dent, Loyola Univ, La, 66-68; ASST PROF PHYSIOL, MEHARRY MED COL, 68- Mem: AAAS; Am Soc Zool; assoc Am Physiol Soc. Res: Termites of the Punjab; blood volume and extracellular fluid circulation in the bullfrog Rana catesbeiana. Mailing Add: Dept of Physiol Meharry Med Col Nashville TN 37208

BASHKIN, STANLEY, b Brooklyn, NY, June 20, 23; m 57; c 3. ATOMIC PHYSICS, NUCLEAR PHYSICS. Educ: Brooklyn Col, BA, 44; Univ Wis, PhD(physics), 50. Prof Exp: Asst physics, Manhattan Proj, 44-46; asst prof physics, La State Univ, 50-53; from asst prof to assoc prof, 53-62, PROF PHYSICS, UNIV ARIZ, 62-, DIR VAN DE GRAAFF LAB, 70- Concurrent Pos: Res fel, Calif Inst Technol, 59; Fulbright res scholar, Australian Nat Univ, 60, res fel, 69; mem comt atomic & molecular physics, Nat Acad Sci-Nat Res Coun. Mem: AAAS; Am Phys Soc; Am Astron Soc; Royal Soc Arts; for mem Royal Soc Gothenburg. Res: Nuclear energy levels; atomic lifetimes; nuclear astrophysics; structures of highly ionized atoms. Mailing Add: Dept of Physics Univ of Ariz Tucson AZ 85721

BASHOUR, FOUAD A, b Tripoli, Lebanon, Jan 3, 24; US citizen; m 56. BIOLOGY, MEDICINE. Educ: Am Univ Beirut, BA, 44, MD, 49; Univ Minn, PhD(med), 57. Prof Exp: Intern, Am Univ Beirut, 49-50; med officer, UN Relief & Works Agency, 50-51; resident internal med, Hosps, Univ Minn, 51-54, instr med, 55-57; res assoc, Med Sch, Am Univ Beirut, 57, asst prof med & in chg cardiopulmonary lab sect, 57-59; from instr internal med to assoc prof med, 59-71, PROF MED, UNIV TEX HEALTH SCI CTR DALLAS, 71-; DIR CARDIOL SECT, CARDIOPULMONARY INST, METHODIST HOSP DALLAS, 67- Concurrent Pos: Mem sr med staff, Parkland Mem Hosp, Dallas; med & cardiac consult, St Paul Hosp, Dallas, John Peter Smith Hosp, Ft Worth & Methodist Hosp, Dallas; cardiac consult, Wilford Hall Hosp & Lackland AFB; mem ad hoc comt coronary care unit & coun basic sci, Am Heart Asn; ad hoc proj site visit, Nat Heart Inst, 65; trustee, Cardiol Fund. Honors & Awards: Officer, Order of the Cedar of Lebanon, 71. Mem: Am Heart Asn; fel Am Col Chest Physicians; Am Fedn Clin Res; Am Physiol Soc; AMA. Res: Cardiovascular physiology and diseases. Mailing Add: Cardiopulmonary Inst Methodist Hosp of Dallas Dallas TX 75222

BASHOUR, JOSEPH TAMIR, b Hartford, Conn, Jan 6, 06; m 33; c 1. ORGANIC CHEMISTRY. Educ: Trinity Col, Conn, BS, 27; NY Univ, PhD(org chem), 34. Prof Exp: Chemist, food & drug admin, USDA, 27-30; asst chem, NY Univ, 30-34; res chemist, dept surg, col physicians & surgeons, Columbia Univ, 34-39, asst chem, 39-40; res chemist, Endo Prod Inc, NY, 40-42 & E R Squibb & Sons, 42-44; dir Yonkers Lab, Stauffer Chem Co, NY, 44-52, dir eastern res div, 52-64; instr gen & anal chem, Hunter Col, 64-65; vis asst prof org chem, Rutgers Univ, 65-66; prof biochem & chmn dept, Lewi Col Podiatry, 66-74; RETIRED. Concurrent Pos: Instr eve eng div, NY Univ, 35-38. Mem: AAAS; fel Am Inst Chemists; Am Chem Soc. Res: Synthesis of organic compounds for testing as agricultural chemicals; extraction and purification of curare alkaloids; synthesis of sulfa drugs; extraction and oxidation of bile acids; extraction and concentration of sex hormone from urine; synthesis of naturally occurring bile acids. Mailing Add: 120 W 70th St New York NY 10023

BASILA, MICHAEL ROBERT, b Scranton, Pa, June 5, 30; m 54; c 6. PHYSICAL CHEMISTRY. Educ: Norwich Univ, BS, 52; Rensselaer Polytech Inst, PhD(phys chem), 58. Prof Exp: Res chemist, Gulf Res & Develop Co, Pa, 58-61, supvr chem physics sect, 61-64, supvr catalysis sect, 64-67, supvr catalytic mechanisms sect, 67-68; dir res & develop, Howe Baker Eng, Inc, Tex, 68-70; TECH DIR CATALYST, NALCO CHEM CO, CHICAGO, ILL, 70- Mem: Am Chem Soc; Catalysis Soc. Res: Infrared spectroscopy; molecular complexes; kinetics of catalytic reactions; process and catalyst development. Mailing Add: 1512 Fran-Lin Pkwy Munster IN 46321

BASILE, DAVID GIOVANNI, b Youngstown, Ohio, Sept 2, 14; m 41; c 2. GEOGRAPHY. Educ: Washington & Lee Univ, AB, 36; Columbia Univ, MA, 39, PhD, 64. Prof Exp: Instr geog, Columbia Univ, 37-41; Soc Sci Res Coun fel, 41-43; procurement specialist, Defense Supplies Corp, US Govt, 43-44; chmn, Coord Comt Ecuador, 45; actg pub affairs officer, US Info Serv, US Dept State, Ecuador, 46-48; ASSOC PROF GEOG, UNIV NC, CHAPEL HILL, 49-, CHMN DEPT, 67- Mem: Asn Am Geographers; Am Geog Soc. Res: Geography of Middle and South America and Latin America, particularly Ecuador; economic geography; problems affecting agrarian land use in Ecuador. Mailing Add: Dept of Geog Univ of NC Chapel Hill NC 27514

BASILE, DOMINICK V, b Yonkers, NY, Oct 15, 31; m 59; c 3. MORPHOLOGY, BRYOLOGY. Educ: Manhattan Col, BS, 58; Columbia Univ, MA, 62, PhD(bot), 64. Prof Exp: Teaching asst bot, Columbia Univ, 58-63, preceptor, 63-64; guest investr plant biol, Rockefeller Univ, 64-65, res assoc, 65-66; asst prof bot, Columbia Univ, 66-71; ASSOC PROF BOT, LEHMAN COL, 71- Concurrent Pos: NSF fel, 64-65; adj cur, NY Bot Garden, 71- Mem: Am Bryol & Lichenological Soc; Torrey Bot Club (vpres, 75); Am Inst Biol Sci; Sigma Xi. Res: Chemical regulation of morphogenesis and phylogeny of plants, especially the Bryophyta. Mailing Add: NY Bot Garden Bronx NY 10458

BASILE, LOUIS JOSEPH, b Chicago, Ill, Mar 20, 24; m 53; c 3. MOLECULAR SPECTROSCOPY, INORGANIC CHEMISTRY. Educ: Univ Chicago, BS, 48, MS, 49; St Louis Univ, PhD(chem), 54. Prof Exp: ASSOC CHEMIST, ARGONNE NAT LAB, 52- Concurrent Pos: Mem bd, sect comt N15, methods of nuclear mat control, Am Nat Standards Inst; assoc ed, Appl Spectros, 71- Honors & Awards: Meggers Award, Soc Appl Spectros, 75. Mem: Fel AAAS; Am Chem Soc; Sigma Xi; Soc Appl Spectros. Res: Energy transfer; plastics; hydrides of boron; rare earths; transplutonium chemistry; infrared spectroscopy; high pressure molecular spectroscopy. Mailing Add: Chem Bldg 200 Argonne Nat Lab 9700 S Cass Ave Argonne IL 60439

BASILE, ROBERT MANLIUS, b Youngstown, Ohio, Mar 12, 16; m 45; c 3. PHYSICAL GEOGRAPHY. Educ; Washington & Lee Univ, BS, 38; Mich State Univ, MS, 40; Ohio State Univ, PhD(geog), 53. Prof Exp: Instr soils & geog, Northwestern State Col, Okla, 40-42; from instr to prof geog, Ohio State Univ, 50-69; PROF GEOG, UNIV TOLEDO, 69-, CHMN DEPT, 72- Mem: AAAS; Asn Am Geographers; Wilderness Soc. Res: Climatology and the morphology of soils. Mailing Add: Dept of Geog Univ of Toledo Toledo OH 43606

BASINISKI, ZBIGNIEW STANISLAW, metal physics, see 12th edition

BASINSKI, DANIEL HENRY, b Buffalo, NY, Jan 9, 12; m 38. CLINICAL CHEMISTRY. Educ: Emory Univ, AB, 38, MS, 39; Univ Rochester, PhD(physiol chem), 43. Prof Exp: Asst, Emory Univ, 39-40 & Univ Rochester, 40-43; res assoc, Univ Iowa, 46-47 & Children's Fund, Mich, 47-53; RES ASSOC, HENRY FORD HOSP, 53- Concurrent Pos: Mem & dir, Am Bd Clin Chemists, 65-71. Mem: AAAS; Am Chem Soc; Am Soc Biol Chemists; Soc Exp Biol & Med; Am Asn Clin Chemists. Res: Enzyme assays and trace metal determination in biological fluids. Mailing Add: Dept of Path Henry Ford Hosp Detroit MI 48202

BASINSKI, JOHN EDWARD, b San Pedro, Calif, July 9, 29. ORGANIC CHEMISTRY. Educ: Johns Hopkins Univ, BA, 54; Yale Univ, MS, 58, PhD(org chem), 61. Prof Exp: Chemist, Gasparcolor, Inc, 61-62 & Nat Eng Sci Co, 62-66; sr chemist, Space Gen Corp, 66-68; sr polymer chemist, Avery Prod Corp, 68-69; LAB MGR, PILOT CHEM CO, 69- Mem: Am Chem Soc; The Chem Soc. Res: Organic mechanisms; photochemistry; photographic chemistry; small ring compounds; fire extinguishant mechanisms; nonbenzenoid aromatics. Mailing Add: Pilot Chem Co 11756 Burke St Santa Fe Springs CA 90670

BASKERVILL, MARGARET (MALONE), b Birmingham, Ala, Aug 17, 14. MATHEMATICS. Educ: Randolph-Macon Woman's Col, AB, 34; Univ Mich, MA, 42; Ala Polytech Inst, PhD(numerical anal), 58. Prof Exp: Actuarial clerk, Protective Life Ins Co, 36-41; actuarial asst, Provident Life Ins Co, 42-43; instr math, Ala Polytech Inst, 43-47; prof, Shorter Col, 47-59; asst prof, 59-65, ASSOC PROF MATH, AUBURN UNIV, 65- Mem: Soc Indust & Appl Math; Math Asn Am. Mailing Add: Dept of Math Auburn Univ Auburn AL 36830

BASKERVILLE, CHARLES ALEXANDER, b New York, NY, Aug 19, 28; m 53, c 2. ENGINEERING GEOLOGY. Educ: City Col New York, BS, 53; NY Univ, MS, 58, PhD(micropaleont, stratig), 65. Prof Exp: Asst civil engr, NY State Dept Transp, 53-66; from asst prof to assoc prof, 66-73, PROF ENG GEOL, CITY COL NEW YORK, 73-, DEAN SCH GEN STUDIES, 70- Concurrent Pos: Eng geol consult, Madigan-Hyland-Praeger Cavanaugh-Waterbury Engrs, 68-69, St Raymond's Cemetery, 70-73 & Consol Edison Co NY, 73; mem nat adv comt minority partic in geosci, Dept Interior, 72-75. Mem: Fel Geol Soc Am; Nat Asn Geol Teachers; Asn Eng Geologists. Res: Cretaceous spores and pollen and stratigraphy on Staten Island, New York; relationship between organic content of silts and the shear strength of these silts. Mailing Add: Dept of Earth & Planetary Sci City Col of New York New York NY 10031

BASKERVILLE, GORDON LAWSON, b Emerson, Man, Can, Feb 20, 33; m 58; c 4. FORESTRY, ECOLOGY. Educ: Univ New Brunswick, BScF, 55; Yale Univ, MF, 57, PhD(forest ecol), 64. Prof Exp: Res scientist silvicult, Can Dept Forestry, 55-74; PROF FORESTRY, UNIV NEW BRUNSWICK, 74- Concurrent Pos: Mem Can subcomt terrestial productivity, Int Biol Prog, 65- Mem: Ecol Soc Am; Soc Am Foresters; Can Inst Forestry; Can Pulp & Paper Asn. Res: Physiological and ecological basis for silviculture; total biomass and productivity in forests. Mailing Add: Dept of Forest Resources College Hill Univ New Brunswick Fredericton NB Can

BASKETT, THOMAS SEBREE, b Liberty, Mo, Jan 23, 16; m. WILDLIFE BIOLOGY. Educ: Cent Col, Mo, AB, 37; Univ Okla, MS, 39; Iowa State Col, PhD(zool), 42. Prof Exp: Res asst, Iowa State Col, 38-41, exten wildlife specialist, USDA, 41, asst prof zool, 46-47; asst prof wildlife mgt, Univ Conn, 47-48; biologist, US Fish & Wildlife Serv, Mo, 48-68; chief div wildlife res, US Bur Sport Fisheries & Wildlife, Washington, DC, 68-73; BIOLOGIST & LEADER, MO COOP WILDLIFE RES UNIT, US FISH & WILDLIFE SERV, 73- Concurrent Pos: Ed, J Wildlife Mgt, 66-68; prof wildlife & fisheries, Univ Mo-Columbia, 73- Mem: Wildlife Soc (pres, 71); Am Soc Ichthyologists & Herpetologists; assoc Am Ornithologists Union; Coun Biol Educ. Res: Wildlife management; ecology of pheasant, bobwhite, swamp rabbit and bullfrog; effects of timber harvest on deer foods; breeding biology and behavior of mourning dove. Mailing Add: Mo Coop Wildlife Res Unit Univ of Mo Columbia MO 65201

BASKIN, AARON DAVID, b New York, NY, Dec 7, 16; m 46; c 2. PLANT PATHOLOGY. Educ: NY Univ, AB, 37; Rutgers Univ, MS, 47; Univ Minn, PhD(plant path), 50. Prof Exp: Hon fel exp biol, Am Mus Natural Hist, 37-38; res asst fungus physiol, Univ Minn, 47-50; plant pathologist, USDA, 51-54; supvry microbiologist, US Qm Res & Develop Ctr, 54-60, from asst tech dir to tech dir indust microbiol, 60-64, res admin asst planning & develop, 64-67; mgr fungicide, nematocide & fumigant develop, Velsicol Chem Corp, 67-68, assoc dir biol res, 68-70; environ researcher, T S Leviton & Assoc, 70-72; ed, Handling Guide for Transport Hazardous Chem, Railway Systs & Mgt Asn, 72-74; SECT HEAD MICROBIOL, INDUST BIO-TEST LABS, INC, 74- Concurrent Pos: Lectr, Gordon Res Conf, 59. Mem: AAAS; Am Phytopath Soc; Am Soc Microbiol; Soc Nematol; NY Acad Sci. Res: Toxicology biologically active compounds; selective toxicity; metabolic pathways and antimetabolites; environmental contamination; microbial mutagenics; degradation of chemicals by soil microflora. Mailing Add: 1802 Winthrop Rd Highland Park IL 60035

BASKIN, DENIS GEORGE, b Minneapolis, Minn, Feb 7, 41; m 64; c 2. COMPARATIVE ENDOCRINOLOGY, BIOLOGICAL STRUCTURE. Educ: San Francisco State Col, BA, 62; Univ Calif, Berkeley, PhD(zool), 69. Prof Exp: NIH fel, Albert Einstein Col Med, 69-71; ASST PROF ZOOL, POMONA COL, 71- Mem: Am Soc Zoologists; Am Soc Cell Biol; Am Asn Anatomists; Electron Micros Soc Am; Sigma Xi. Res: Comparative neuroendocrinology; cell junctions; endocrinology and reproductive biology of polychaetes; biology of cephalochordates; electron microscopy. Mailing Add: Dept of Zool Pomona Col Claremont CA 91711

BASKIN, JERRY MACK, b Covington, Tenn, July 27, 40; m 68. PLANT ECOLOGY. Educ: Union Univ, Tenn, BS, 63; Vanderbilt Univ, PhD(biol), 67. Prof Exp: Res assoc plant physiol, Univ Fla, 67-68; asst prof, 68-73, ASSOC PROF BOT, UNIV KY, 73- Mem: Ecol Soc Am; Bot Soc Am; Torrey Bot Club. Res: Autecology of herbaceous, vascular plants. Mailing Add: Sch of Biol Sci Univ of Ky Lexington KY 40506

BASKIN, LEONARD S, biochemistry, see 12th edition

BASKIN, RONALD J, b Joliet, Ill, Nov 25, 35; m 58; c 2. BIOPHYSICS. Educ: Univ Calif, Los Angeles, AB, 57, MA, 59, PhD(biophys), 60. Prof Exp: Fel biophysics, Univ Calif, Los Angeles, 60-61; asst prof biol, Rensselaer Polytech Inst, 61-64; from asst prof to assoc prof, 64-71, assoc dean col lett & sci, 67-71, PROF ZOOL & CHMN DEPT, UNIV CALIF, DAVIS, 71- Mem: AAAS; Biophys Soc Am; Am Physiol Soc. Res: Muscle biophysics; thermodynamics of biological systems; membrane structure and function. Mailing Add: Dept of Zool Univ of Calif Davis CA 95616

BASKIN, YEHUDA, b Chicago, Ill, Jan 11, 29; m 64; c 3. GEOLOGY, MINERALOGY. Educ: Univ Chicago, BS, 51, MS, 52, PhD(geol), 55. Prof Exp: Res scientist, Armour Res Found, 55-62; assoc ceramist, Argonne Nat Lab, 62-69; GROUP LEADER ADVAN CERAMICS, FERRO-TECH CTR, FERRO CORP, 69- Mem: Am Ceramic Soc. Res: Research implementation and supervision in specialty industrial ceramics, refractories and inorganic materials. Mailing Add: Tech Ctr Ferro Corp 7500 E Pleasant Valley Rd Independence OH 44131

BASKIR, EMANUEL, b New York, NY, July 13, 29; m 52; c 3. PHYSICS. Educ: Columbia Univ, BA, 51; Rochester Univ, PhD(physics), 57. Prof Exp: PHYSICIST, SHELL DEVELOP CO, 56- Mem: Am Phys Soc; Am Geophys Union; Sigma Xi. Res: Nuclear physics and applications to geophysical problems; physics of rock magnetism. Mailing Add: Shell Develop Co PO Box 481 Houston TX 77001

BASLER, EDDIE, JR, b Blanchard, Okla, Mar 25, 24; m 56; c 4. PLANT PHYSIOLOGY. Educ: Univ Okla, BS, 50, MS, 52; washington Univ, PhD(bot), 54. Prof Exp: Res assoc, Washington Univ, 54-55, asst prof bot, 55-57; assoc prof, 57-67, PROF BOT, OKLA STATE UNIV, 67- Mem: Am Soc Plant Physiologists; Bot Soc Am; Weed Sci Soc Am. Res: Mechanisms of herbicidal action; translocation of herbicides and auxins; metabolism of herbicides. Mailing Add: Sch of Biol Sci Okla State Univ Stillwater OK 74074

BASLER, ROY PRENTICE, b Florence, Ala, Aug 12, 35; m 58; c 4. RADIOPHYSICS. Educ: Hamilton Col, AB, 56; Univ Alaska, MS, 61, PhD(physics), 64. Prof Exp: Res assoc, Geophys Inst, Univ Alaska, 58-64; staff physicist, ITT Electro-Physics Labs, 64-73; SR PHYSICIST, STANFORD RES INST, 73- Concurrent Pos: Mem comn III, Int Sci Radio Union. Mem: Am Geophys Union; AAAS. Res: Interstellar communications; solar-terrestrial relations; ionospheric radio wave absorption; scintillation of satellite radio signals. Mailing Add: Stanford Res Inst Menlo Park CA 94025

BASMAJIAN, JOHN V, b Constantinople, Turkey, June 21, 21; Can citizen; m 47; c 3. ANATOMY. Educ: Univ Toronto, MD, 45. Prof Exp: Demonstr anat, Univ Toronto, 46-47, lectr, 49-51, from asst prof to prof, 51-57; prof & head dept, Queen's Univ, Ont, 57-69; PROF ANAT & PHYS MED & DIR REGIONAL REHAB CTR, EMORY UNIV, 69- Concurrent Pos: Asst resident surg, Hosp for Sick Children, 47-48, res assoc, 55-57; clin asst phys med, St Thomas Hosp, London, 53; hon secy, Banting Res Found, 55-57; Nat Res Coun exchange scientist, Soviet Acad Sci, 63; chief neurophysiol lab, Ga Ment Health Inst, 69- Honors & Awards: Starr Medal, 56. Mem: AAAS; fel Am Col Angiol; Am Acad Neurol; Am Asn Anat; Can Asn Anat. Res: Electromyography of normal anatomic and physiologic functions and of diseased muscle; nerve-muscle electrophysiology; studies of normal vascular patterns; biomechanics; bioengineering; psychophysiology. Mailing Add: Ga Ment Health Inst 1256 Briarcliff Rd NE Atlanta GA 30306

BASOCO, MIGUEL ANTONIO, b Mex, July 10, 00; nat US; m 35; c 3. MATHEMATICS. Educ: Univ Calif, AB, 24; Univ Chicago, MS, 26; Calif Inst Technol, PhD(math), 29. Prof Exp: Res fel, Calif Inst Technol, 29-30; from asst prof to prof, 30-70, chmn dept, 47-54, univ res fel, 56 & 59, EMER PROF MATH, UNIV NEBR, LINCOLN, 70- Mem: Am Math Soc; Math Asn Am. Res: Doubly periodic and pseudo-periodic functions and their applications to theory of numbers. Mailing Add: 2626 Rathbone Rd Lincoln NE 68502

BASOLO, FRED, b Coello, Ill, Feb 11, 20; m 47; c 4. INORGANIC CHEMISTRY. Educ: Southern Ill Norm Univ, BEd, 40; Univ Ill, MS, 42, PhD(chem), 43. Prof Exp: Res chemist, Rohm and Haas Chem Co, Pa, 43-46; from instr to assoc prof, 46-59, chmn dept, 69-72, PROF CHEM, NORTHWESTERN UNIV, EVANSTON, 59- Concurrent Pos: Guggenheim fel, Tech Univ Denmark, 54-55; NSF sr fel, Inst Inorg Chem, Univ Rome, 61-62; NATO vis prof, WGer, 69; bd trustees, Gordon Res Conf & bd chmn, 76. Honors & Awards: Award in Org Chem, Am Chem Soc, 64, Award for Mechanisms of Inorg Reactions, 71 & Award for Distinguished Serv in Org Chem, 75; Bailar Medalist, 73. Mem: Am Chem Soc; Sigma Xi. Res: Coordination compounds; reaction mechanisms of inorganic complex compounds; metal nitrenes; synthetic oxygen-carriers. Mailing Add: Dept of Chem Northwestern Univ Evanston IL 60201

BASOM, CHARLES RAY, b Twin Falls, Idaho, 43; m 63; c 4. ANATOMY. Educ: Col Idaho, BS, 64; Univ NDak, MS, 66, PhD(anat), 68. Prof Exp: Instr, Univ Cincinnati, 68-70, asst prof anat, 70-74; CLIN COORDR, MERRELL-NAT LABS, 74- Mem: Electron Micros Soc Am; Am Asn Anat. Res: Electron microscopy of connective tissue elements. Mailing Add: Merrell-Nat Labs Div Richardson-Merrell Inc Cincinnati OH 45215

BASRI, SAUL ABRAHAM, b Baghdad, Iraq, Feb 15, 26; nat US; m 50; c 2. ELEMENTARY PARTICLE PHYSICS. Educ: Mass Inst Technol, BS, 48; Columbia Univ, PhD(physics), 53. Prof Exp: Res asst physics, Columbia Univ, 52-53, res scientist, 53; from asst prof to assoc prof, 53-67, PROF PHYSICS, COLO STATE UNIV, 67- Concurrent Pos: Fulbright lectr, Univ Rangoon, Burma, 56 & Univ Ceylon, 65-66; vis prof, Israel Inst Technol, 73-74. Mem: Am Phys Soc; Am Asn Physics Teachers. Res: Elementary particle theory; algebraic s-matrix theory. Mailing Add: Dept of Physics Colo State Univ Ft Collins CO 80521

BASS, ABRAHAM, b Baltimore, Md, Oct 28, 06; m 47. BIOCHEMISTRY. Educ: Univ Chicago, BS, 27, PhD(biochem), 33. Prof Exp: Asst plant control, Cudahy Packing Co, 27-29; chief chemist, Chicago Mercantile Exchange, Ill, 34-37; researcher, Univ Chicago, 38-40; chief chemist, Harold N Simpson Co, Ill, 42-43; chemist, Off Res & Develop contract, Univ Ill, 43-44; lab dir, Plough, Inc, 46-50, res dir, 51-57, dir spec proj, 57-75; RETIRED. Mem: Am Chem Soc. Res: Proteins; nutrition; absorption of drugs through skin; analgesics. Mailing Add: 94 St Albans Fairway Memphis TN 38111

BASS, ALLAN DELMAGE, b Marcus, Iowa, Feb 12, 10; m 44; c 2. PHARMACOLOGY. Educ: Simpson Col, BS, 31; Vanderbilt Univ, MS, 32, MD, 39. Prof Exp: Intern, Vanderbilt Hosp, 39-40, asst resident physician, 40-41, resident & instr med, 43-44; instr pharmacol, Yale Univ, 42-43; prof & chmn dept, Syracuse Univ, 45-52; prof & chmn dept, 52-72, dir neurosci prog, 71-73, assoc dean biomed sci, 73-75, actg dean med sch, 73-74, EMER PROF PHARMACOL, VANDERBILT UNIV, 75-; ASSOC CHIEF STAFF EDUC, VET ADMIN HOSP, NASHVILLE, TENN, 75- Concurrent Pos: Mem gen med res prog proj comt, NIH, 64-67; mem, US Food & Drug Admin, 67-70; consult, Div Res Grants, NIH Res Anal & Eval, 68; mem res adv comt, NIMH, 68-70; chmn bd, Tenn Neuropsychiat Inst, 70-75; Nat Bd Med examr & Ann Rev Pharmacol; mem ed bd, Int Quart Sci Rev J, 73-; mem drug res bd, Nat Acad Sci, 74- Mem: AAAS (vpres med sci sect, 69); Am Soc Pharmacol & Exp Therapeut (pres, 67-68); AMA; Am Col Physicians; Fedn Am Socs Exp Biol. Res: Autonomic and endocrine pharmacology; neuropsychopharmacology; steroid metabolism; cyanide poisoning; hypnotics; nucleoproteins. Mailing Add: Vanderbilt Univ Med Sch Nashville TN 37232

BASS, ARNOLD MARVIN, b New York, NY, Dec 22, 22; m 47; c 3. PHYSICS. Educ: City Col New York, BS, 42; Duke Univ, MA, 43, PhD(physics), 49. Prof Exp: Asst physics, Duke Univ, 42-43, instr, 43-44; staff mem lab insulation res, Mass Inst Technol, 49-50; physicist, Temperature Measurement Sect, 50-54, asst chief, 54-56, from asst chief to chief, Free Radicals Res Sect, 56-61, physicist, 61-66, physicist, Molecular Energy Levels Sect, Heat Div, 66-69, ASST CHIEF PHYS CHEM DIV, NAT BUR STANDARDS, 69- Honors & Awards: Gold Medal, US Dept Com, 60. Mem: Fel Optical Soc Am; Am Phys Soc. Res: Fluorescence spectroscopy; molecular spectra and structure; spectra of liquids, solids, flames, hot gases and trapped radicals; fluorescence and vacuum ultraviolet spectra; infrared emissivities of hot gases; temperature measurements by radiation methods; low temperature spectra of solids; flash photolysis and pyrolysis. Mailing Add: Nat Bur of Standards Rm B-164 Chem Bldg Washington DC 20234

BASS, ARTHUR, b New York, NY, Apr 7, 41; m 66; c 1. AIR POLLUTION. Educ: Columbia Univ, BA, 61; Yale Univ, MS, 62; Mass Inst Technol, PhD(meteorol), 74. Prof Exp: Staff scientist atmospheric physics, Mitre Corp, 63-68; sr scientist, Am Sci & Eng Inc, 68-69; res asst, Dept Meteorol, Mass Inst Technol, 69-74; res scientist, Flow Res Inc, 74-75; SR STAFF SCIENTIST AIR POLLUTION, ENVIRON RES & TECHNOL INC, 75- Mem: Am Meteorol Soc; Air Pollution Control Asn. Res: Air pollution meteorology; geophysical fluid dynamics; numerical simulation of turbulent flows in complex terrain. Mailing Add: Environ Res & Technol Inc 3 Militia Dr Lexington MA 02173

BASS, BERL G, b Newark, NJ, May 21, 29; c 2. PHYSIOLOGY. Educ: Hamilton Col, AB, 50; Columbia Univ, MD, 54; Am Bd Internal Med, dipl, 63. Prof Exp: Intern med, Bellevue Hosp, Columbia Med Div, 54-55, jr asst resident, 58-59; asst resident cardiol, Columbia-Presby Med Ctr, 59-60, sr asst resident med, 60-61; res assoc, 63-65, asst prof physiol, 65-73, ASSOC PROF PHYSIOL, ALBERT EINSTEIN COL MED, 73- Concurrent Pos: NIH res fel med, 61-63; sr investr, NY Heart Asn. Res: Internal medicine; cardiac muscle physiology. Mailing Add: Dept of Physiol Albert Einstein Col of Med New York NY 10461

BASS, DAVID ELI, b Lowell, Mass, Aug 15, 12; m 45; c 3. PHYSIOLOGY. Educ: Brown Univ, AB, 32; Boston Univ, MA, 51, PhD(med physiol), 53. Prof Exp: Biochemist, Qm Climatic Res Lab, Lawrence, Mass, 47-55; chief physiol br, Qm Res & Eng Ctr, Natick, 56-61; dir, US Army Res Inst Environ Med, 61-63, sci dir, 63-73; VIS PROF PHYSIOL, SIMON FRASER UNIV, 75- Concurrent Pos: Secy Army res & study fel, 58-59; lectr, Sch Med, Boston Univ, 50-53, assoc prof, 54-; consult, Inst Environ Psychophysiol, Univ Mass; adj prof zool, Univ RI, 67- Honors & Awards: Qm Res Dir Award, 53. Mem: AAAS; Am Physiol Soc; Am Asn Clin Chem; Am Fedn Clin Res. Res: Human physiology and biochemistry; physiology of heat, cold, exercise and human temperature. Mailing Add: Dept of Physiol Simon Fraser Univ Burnaby BC Can

BASS, EDMUND P, b Poland, Aug 10, 16; US citizen; m 50; c 4. VETERINARY VIROLOGY. Educ: Vet Acad, Lwow, Poland, DVM, 41; Univ Turin, DSc(virol), 48. Prof Exp: Asst lab dir res prod, Affiliated Labs Corp, 53-65; res dir, Armour-Baldwin Labs, 65-68; RES SCIENTIST, NORDEN LABS, 68- Mem: Am Vet Med Asn; Tissue Cult Asn. Res: Developments of immunological agents against infectious diseases of animals; isolation, propagation and attenuation of viruses and development of vaccines; identification of the infectious agents; clinical evaluation of the preparations for efficacy and safety. Mailing Add: Norden Labs PO Box 80809 Lincoln NE 68501

BASS, EUGENE LAWRENCE, comparative physiology, neurophysiology, see 12th edition

BASS, GARLAND BOOKER, b Reidsville, NC, Jan 25, 19; m 39; c 3. AGRONOMY. Educ: Agr & Tech Col, NC, BS, 37; Univ Mass, MS, 48, PhD(agron), 49. Prof Exp: Teacher pub sch, 37-43; prof agron, Southern Univ & Agr & Mech Col, 49-50; PROF CHEM & PHYS SCI, ALA A&M UNIV, 50- Mem: Am Chem Soc. Res: Soil chemistry. Mailing Add: Dept of Chem Ala A&M Univ Normal AL 35762

BASS, HYMAN, b Houston, Tex, Oct 5, 32; m 58; c 2. MATHEMATICS. Educ: Princeton Univ, BA, 55; Univ Chicago, MS, 56, PhD(math), 59. Prof Exp: Ritt instr math, Columbia Univ, 59-62; NSF fel, Col of France, 62-63; from asst prof to assoc prof, 63-65, chmn dept, 75-78, PROF MATH, COLUMBIA UNIV, 65- Concurrent Pos: Chmn dept, Barnard Col, Columbia Univ, 64-65. Honors & Awards: Cole Prize in Algebra, Am Math Soc, 75. Mem: Am Math Soc. Res: Homological algebra; algebraic number theory; algebraic geometry. Mailing Add: Dept of Math Columbia Univ New York NY 10027

BASS, J CARL, b Clemscott, Okla, Feb 14, 25; m 46; c 1. BIOLOGY, SCIENCE EDUCATION. Educ: Southeastern State Col, BS, 49; Univ Okla, MS, 54, EdD(zool), 59. Prof Exp: From instr to asst prof biol, Hardin-Simmons Univ, 50-55; teacher high sch, Okla, 55-61; from asst prof to assoc prof, 61-67, PROF BIOL, KANS STATE COL PITTSBURG, 67- Res: Freshwater ecology; fisheries; venomous arthropods; limnology. Mailing Add: Dept of Biol Kans State Col Pittsburg KS 66762

BASS, JACK, b New York, NY, Apr 1, 38; m 59; c 3. SOLID STATE PHYSICS. Educ: Calif Inst Technol, BS, 59; Univ Ill, MS, 61, PhD(physics), 64. Prof Exp: From asst prof to assoc prof, 64-73, PROF PHYSICS, MICH STATE UNIV, 73- Concurrent Pos: Guest docent, Swiss Fed Inst Technol, 70-71. Mem: Am Phys Soc; Am Asn Physics Teachers; AAAS. Res: Electron transport properties of metals; low temperature physics; point defects in metals. Mailing Add: Dept Physics Col of Natural Sci Mich State Univ East Lansing MI 48824

BASS, JAMES W, b Shreveport, La, May 25, 30; m; c 2. PEDIATRICS, INFECTIOUS DISEASES. Educ: Tulane Univ, BS, 52, MPH, 68; La State Univ, New Orleans, MD, 57. Prof Exp: Researcher infectious dis, US Walter Reed Army Inst Res, Md, 60-63; asst chief pediat, Madigan Gen Hosp, Tacoma, Wash, 63-66; instr pediat, Sch Med, Tulane Univ & vis pediatrician, Charity Hosp La, New Orleans, 66-68; from asst dir to dir intern training, 68-71, asst chief pediat serv, 68, chief dept pediat, Tripler Army Med Ctr, Honolulu, 68-75; from asst clin prof to assoc clin prof pediat, Sch Med, Univ Hawaii, 68-73, clin prof pediat, Sch Med, Univ Hawaii, Manoa, 73-75 & trop med & med microbiol, 74-75; CHIEF DEPT PEDIAT, WALTER REED ARMY MED CTR, 75-; CLIN PROF PEDIAT, GEORGETOWN UNIV SCH MED, WASHINGTON, DC, 75- Concurrent Pos: From asst clin prof to assoc clin prof pediat, Sch Med, Univ Hawaii, 68-73; consult pediat infectious dis, Kauikeolani Childrens Hosp, Honolulu, 70-75; assoc clin prof pediat, Univ Southern Calif, 72-75. Mem: AMA; Soc Pediat Res: Am Pediat Soc; Infectious Dis Soc Am; fel Am Col Physicians. Res: Microbiology. Mailing Add: Dept of Pediat Walter Reed Army Med Ctr Washington DC 20012

BASS, JON DOLF, b Saginaw, Mich, Oct 31, 33; m 59; c 4. PHOTOGRAPHIC CHEMISTRY. Educ: Univ Mich, BS, 56; Univ Wis, PhD(org chem), 60. Prof Exp: Lab asst, Mellon Inst, 51-52; asst, Univ Wis, 58-59; res chemist, Air Force Cambridge Res Labs, Mass, 60-62; res fel chem, Harvard Univ, 62-63; from res chemist to sr res chemist, 63-70, RES ASSOC, KODAK RES LABS, 70- Mem: Am Chem Soc; Soc Photog Sci & Eng. Res: Synthesis of steroid intermediates; non-benzenoid aromaticity; organic semiconductors and photoconductors; photochemical organic synthesis; photographic chemistry; physical development; lithographic chemistry; spectral sensitization and image amplification processes. Mailing Add: Kodak Res Labs Bldg 59 Rochester NY 14650

BASS, JONATHAN LANGER, b New York, NY, May 12, 38. PHYSICAL CHEMISTRY. Educ: Univ Rochester, BS, 59; Univ Minn, Minneapolis, PhD(phys chem), 65. Prof Exp: Chemist, maintenance lab, hq Ogden Air Mat Area, Hill AFB, Utah, 65-66 & weapons lab, Kirtland AFB, NMex, 66-68; CHEMIST, WASHINGTON RES CTR, W R GRACE & CO, 68- Mem: Am Chem Soc; Soc Appl Spectros; Catalysis Soc. Res: Characterization of catalysts by infrared spectroscopy, porosimetry, scanning electron microscopy and x-ray analysis; oxidation of metals and alloys; degradation of polymers. Mailing Add: Petrol Chem Dept W R Grace & Co Columbia MD 21044

BASS, JOSEPH ALONZO, b Eagle Pass, Tex, Aug 14, 18; m 52; c 2. BACTERIOLOGY, IMMUNOLOGY. Educ: Univ Tex, BA, 49; Ohio State Univ, MSc, 51, PhD(bact), 53. Prof Exp: Asst prof bact & immunol, Med Br, Univ Tex, 53-55, from assoc prof to prof microbiol, 55-70; prof biol sci, NTex State Univ, 70-73; PROF MICROBIOL, UNIV TEX HEALTH SCI CTR SAN ANTONIO, 73- Concurrent Pos: McLaughlin fel infection & immunity, 55; ed, Tex Reports Biol & Med, 60-64; regional biol specialist, NSF Biol Sci Curriculum Study, AID Prog, Cent Am, 64-65; chief div microbiol, Shriners Burns Inst, 67-70. Mem: AAAS; Am Soc Microbiol; Tissue Cult Asn; NY Acad Sci; fel Am Pub Health Asn. Res: Pathogenic bacteriology; mechanisms of infection and resistance; burn bacteriology; staphylococcal and streptococcal infections; nosocomial infections; immunity against P. aeruginosa. Mailing Add: Dept of Microbiol Univ of Tex Health Sci Ctr San Antonio TX 78284

BASS, LEONARD JOEL, b Kalamazoo, Mich, Oct 30, 43; m 67; c 3. COMPUTER SCIENCES. Educ: Univ Calif, Riverside, BA, 64, MA, 66; Purdue Univ, PhD(comput sci), 70. Prof Exp: ASSOC PROF COMPUT SCI, UNIV RI, 70- Mem: Asn Comput Mach. Res: Computer system performance analysis; operating systems; theory of computation. Mailing Add: Dept of Comput Sci Univ of RI Kingston RI 02881

BASS, LOUIS NELSON, b Iola, Kans, Mar 7, 19; m 44; c 2. PLANT PHYSIOLOGY. Educ: Upper Iowa Univ, BS, 40; Univ Iowa, MS, 43; Iowa State Col, PhD(bot), 49. Prof Exp: Asst bot, Univ Iowa, 42-44; exten assoc, Iowa State Col, 45-49, asst prof bot & plant path, 49-58; plant physiologist, 58-63, res plant physiologist, 63-70, PLANT PHYSIOLOGIST & LAB DIR, NAT SEED STORAGE LAB, AGR RES SERV, USDA, 70- Honors & Awards: Award of Merit, Asn Off Seed Analysts, 75. Mem: Am Soc Plant Physiologists; Am Soc Agron; hon mem Soc Commercial Seed Technologists; assoc Asn Off Seed Analysts (secy-treas, 67-70, vpres, 70-71, pres, 71-72); Am Soc Hort Sci. Res: Factors affecting seed viability; methods of making laboratory germination tests of various kinds of seed; seed longevity and storage. Mailing Add: Nat Seed Storage Lab Colo State Univ Ft Collins CO 80521

BASS, MANUEL N, b Houston, Tex, July 20, 27; m 62; c 2. GEOLOGY, GEOCHEMISTRY. Educ: Calif Inst Technol, BS, 49, MS, 51; Princeton Univ, PhD(geol), 56. Prof Exp: Geologist, US Geol Surv, 51; asst prof geol, Northwestern Univ, 56-58; fel, Carnegie Inst, 58-60; asst prof geol, Calif Inst Technol, 60-62; asst prof, Univ Calif, San Diego, 62-69; mem staff, Geochem Br, NASA, 69-74; MEM STAFF, CHEVRON OIL FIELD RES CO, 74- Mem: Geol Soc Am; Mineral Soc Am; Am Geophys Union; Am Asn Petrol Geol. Res: Tectonics of age of Canadian Shield and buried Precambrian rocks of Central United States. Mailing Add: Chevron Oil Field Res Co PO Box 446 La Habra CA 90631

BASS, MARY ANNA, b Clanton, Ala, June 1, 30; m 53; c 3. FOOD SCIENCE, NUTRITION. Educ: Ala Col Women, BS, 51; Walter Reed Army Hosp, dipl, 52; Univ Ky, MS, 56; Kans State Univ, PhD(foot & nutrit), 72. Prof Exp: Instr food & nutrit, Univ Nebr, 59-60 & Univ Kans, 60-68; ASST PROF FOODS & NUTRIT, COL HOME ECON, UNIV TENN, 71- Concurrent Pos: Consult, Standing Rock Indian Reservation, 70- & Human Resources Inst, 73- Mem: Am Dietetic Asn; Am Home Econ Asn; Soc Nutrit Educ; Nutrit Today Soc. Res: Determining food intake patterns and identifying those factors that influence them. Mailing Add: 8201 Bennington Dr Knoxville TN 37919

BASS, MAX HERMAN, b Troy, Ala, Dec 27, 34; m 56; c 4. ENTOMOLOGY. Educ: Troy State Col, BS, 57; Ala Polytech Inst, MS, 59; Auburn Univ, PhD(entom), 64. Prof Exp: Instr zool, 59-60, res entom, 60-63, from asst prof to assoc prof, 63-70, PROF ENTOM, AUBURN UNIV, 70- Mem: Entom Soc Am. Res: Bionomics and control of insects affecting soybeans and peanuts. Mailing Add: Dept of Zool-Entom Auburn Univ Auburn AL 36830

BASS, NORMAN HERBERT, b New York, NY, July 10, 36; m 61; c 3. NEUROCHEMISTRY, NEUROLOGY. Educ: Swarthmore Col, BA, 58; Yale Univ, MD, 62. Prof Exp: Intern med, Sch Med, Univ Wash, 62-63; resident neurol, Sch Med, Univ Va, 63-65; Nat Inst Neurol & Commun Disorders & Stroke fels neurochem, McLean Hosp Res Lab, Harvard Univ, 65-67; asst prof neurol, 67-70, assoc prof pharmacol, 70-73, PROF NEUROL & DIR CLIN NEUROSCI RES CTR, SCH MED, UNIV VA, 74- Concurrent Pos: Markle scholar award, 69-74; Nat Inst Neurol & Commun Disorders & Stroke res career develop award, 71-76; vis prof, Inst Pharmacol, Univ Gothenburg, Sweden, 72-73. Honors & Awards: S Weir Mitchell Res Award, Am Acad Neurol, 67. Mem: Am Soc Neurochem; Int Soc Neurochem; Am Acad Neurol; Am Neurol Asn; Am Asn Anatomists. Res: Neurochemistry of developing brain; microchemistry of cerebral cortex maturation; malnutrition and hormonal imbalance during critical periods of brain development; experimental epilepsy; blood-brain-barrier; inborn errors of human brain metabolism; neuropharmacology of morphine abuse. Mailing Add: Dept of Neurol Univ of Va Sch of Med Charlottesville VA 22901

BASS, PAUL, b Winnipeg, Man, Aug 12, 28; m 53; c 2. PHARMACOLOGY, PHYSIOLOGY. Educ: Univ BC, BSP, 53, MA, 55; McGill Univ, PhD(pharmacol), 57. Prof Exp: Asst, Ayerst, McKenna & Harrison, Can, 56; res pharmacologist, Parke Davis & Co, 60-70; PROF PHARMACOL, SCH PHARM & SCH MED, UNIV WIS-MADISON, 70- Concurrent Pos: Fel biochem, McGill Univ, 57-58; fel physiol, Mayo Found, 58-60. Mem: Am Soc Pharmacol; Soc Exp Biol & Med; Pharmacol Soc Can; Am Gastroenterol Asn. Res: Motility, secretion and absorption of the gastro-intestinal tract; cardiovascular and autonomic nervous system pharmacology. Mailing Add: Sch of Pharm Univ of Wis Madison WI 53706

BASS, ROBERT EUGENE, b Vienna, Austria, Dec 31, 03; nat US; m 33. PHYSICS. Educ: Univ Vienna, PhD(physics), 27. Prof Exp: Lectr physics, Viennese Insts Adult Educ, 30-38; master, King Edward VI Grammar Sch, Eng, 42-44; instr physics, Univ Pa, 45; from asst prof to assoc prof, 47-74, EMER PROF PHYSICS, UNIV TOLEDO, 74- Mem: Am Phys Soc; Am Asn Physics Teachers; Am Philos Asn. Res: General foundations of science; philosophy of science. Mailing Add: 3150 Middlesex Toledo OH 43606

BASS, ROBERT GERALD, b Durham, NC, May 2, 33; m 59; c 2. ORGANIC CHEMISTRY. Educ: Va Polytech Inst, BS, 54; Univ Va, PhD(chem), 61. Prof Exp: Res chemist, E I du Pont de Nemours & Co, Inc, 59-62; assoc prof, 62-72, PROF CHEM, VA COMMONWEALTH UNIV, 72- Concurrent Pos: Fel, Univ Va, 62. Mem: Am Chem Soc. Res: Grignard addition, oxidative rearrangement and vinylogous pinacol rearrangement of phenylated unsaturated 1,4-diketones and their derivatives; steroids. Mailing Add: Dept of Chem Va Commonwealth Univ Richmond VA 23220

BASS, SHAILER LINWOOD, b Paducah, Ky, Apr 5, 06; m 31; c 7. SILICON CHEMISTRY. Educ: Butler Univ, AB, 26; Yale Univ, PhD(org chem), 29. Prof Exp: Asst, Yale Univ, 28-29; res chemist, Dow Chem Co, 29-44, asst gen mgr, Dow Corning Corp, 44-54, from vpres to pres, 54-67, chmn bd, 67-71, MEM BD DIRS, DOW CORNING CORP, 67- Mem: Am Chem Soc. Res: Silicone resins and polymers. Mailing Add: Dow Corning Corp Midland MI 48640

BASS, STEVEN WILLIAM, b Philadelphia, Pa, Dec 9, 41; m 68; c 2. CLINICAL PHARMACOLOGY. Educ: Philadelphia Col Pharm & Sci, BS, 65; Thomas Jefferson Univ, PhD(pharmacol), 70. Prof Exp: Head toxicol sect, Wallace Labs, Div Wallace Pharmaceut, 70-71; asst to med dir clin pharmacol, Pfizer Labs, Div Pfizer Inc, 71-73; DIR CLIN DEVELOP CLIN PHARMACOL, CLIN RESOURCES, INC, DIV MEDCOM, INC, 73- Concurrent Pos: Fel pharmacol, Med Sch, Univ Pa, 70. Mem: Sigma Xi; Am Fedn Clin Res. Res: Clinical research in phases one, two and three of new drug development; analgesic, CNS, arthritis, cardiovascular, anti-infective and gastrointestinal. Mailing Add: 7 Pamela St Marlboro NJ 07746

BASS, VIRGINIA CARVEL, b Warrenton, Va, Jan 8, 34; m 64. ANALYTICAL CHEMISTRY. Educ: Sweet Briar Col, AB, 55; Tulane Univ, MS, 57; Univ Va, PhD(chem), 63. Prof Exp: Instr chem, Tulane Univ, 57-59; res assoc mat sci, Univ Va, 63-64, res assoc clin chem, univ hosp, 65-67; assoc prof chem, Longwood Col, 67-69; RES CHEMIST, FED BUR INVEST, 69- Mem: Am Chem Soc; fel Am Inst Chemists. Res: Analytical spectrophotometry and chromatography; toxicology and drug analysis; clinical chemistry. Mailing Add: FBI Lab Pennsylvania Ave at Ninth St NW Washington DC 20535

BASS, WILLIAM MARVIN, III, b Staunton, Va, Aug 30, 28; m 53; c 3. PHYSICAL ANTHROPOLOGY, ANTHROPOMETRICS. Educ: Univ Va, BA, 51; Univ Ky, MS, 56; Univ Pa, PhD(anthrop), 61. Prof Exp: Instr phys anthrop & anat, Grad Sch Med, Univ Pa, 56-60; instr anthrop, Univ Nebr, 60; from instr to prof, Univ Kans, 60-71; PROF ANTHROP & HEAD DEPT, UNIV TENN, KNOXVILLE, 71- Concurrent Pos: Res grant, Univ Kans, 61-68; NSF res grants, Smithsonian Inst, 62 & 63; Nat Park Serv grant, 63 & 66; NSF grants, 65, 67 & 69; Nat Geog grant, 68. Mem: Am Asn Phys Anthrop; Am Anthrop Asn; Soc Am Archaeol; Am Acad Forensic Sci. Res: Identification of human skeletal material. Mailing Add: Dept Anthrop 252 S Stadium Hall Univ of Tenn Knoxville TN 37916

BASSECHES, HAROLD, b Brooklyn, NY, Nov 27, 23; m 46; c 3. PHYSICAL CHEMISTRY. Educ: City Col New York, BS, 43; Ohio State Univ, PhD(phys chem), 51. Prof Exp: Mem tech staff, 52-59, supvr semiconductor mat group, Pa, 59-62, magnetic mat group, 62-67, head film technol dept, 67-70, HEAD THIN FILM MAT & TECHNOL DEPT, BELL TEL LABS, NJ, 70- Mem: Am Chem Soc; fel Am Inst Chem. Res: Preparation and processing of materials whose electric and magnetic properties are of use in modern solid state devices; development of thin film materials and processes for hybrid integrated circuits. Mailing Add: Thin Film Mats & Technol Dept Bell Tel Labs 555 Union Blvd Allentown PA 18103

BASSEL, ROBERT HAROLD, b Johnstown, Pa, Feb 16, 28; c 1. NUCLEAR PHYSICS. Educ: Univ Pittsburgh, BSc, 52, PhD(physics), 58. Prof Exp: Physicist, Oak Ridge Nat Lab, 58-66; assoc physicist, Brookhaven Nat Lab, 66-68; assoc prof physics, Univ Pittsburgh, 68-69; RES PHYSICIST, NAVAL RES LAB, 64- Mem: Fel Am Phys Soc; fel AAAS. Res: Nuclear and atomic reaction; evaluations of neutron induced reaction on cross sections. Mailing Add: Naval Res Lab Code 6660 Washington DC 20375

BASSETT, ALLEN MORDORF, b Brooklyn, NY, Feb 2, 25; m 49 & 71; c 4. GEMOLOGY, ECONOMIC GEOLOGY. Educ: Amherst Col, BA, 48; Columbia Univ, MA, 50, PhD(econ geol), 55. Prof Exp: Geologist, US Geol Surv, 51-58, geologist, Mojave Desert Borate Deposits, 52-58; asst prof geol, Ohio Wesleyan Univ, 58-61; from asst prof to prof geol, San Diego State Univ, 61-71; CONSULT GEOLOGIST, 71- Concurrent Pos: Consult geologist, Borax Consol Ltd, London, 55; dir, NSF Inst, 62 & 65; dir, Field Inst Brazil, 66; leader geol mapping, explor & gemstone studies, Himalaya Expeds, 70, 72, 74, 75 & 76; res assoc, Nat Ctr Sci Res, Paris, France, 75- Mem: Fel AAAS; fel Geol Soc Am; Soc Econ Geologists; Mineral Soc Am; fel Gemmological Asn Gt Brit. Res: Ore deposits, metallic and nonmetallic; Pleistocene geology of desert regions; structural geology and gemstone deposits of the Himalaya Mountains; gemstone inclusions. Mailing Add: PO Box 2222 La Jolla CA 92038

BASSETT, ALTON HERMAN, b Hartford, Conn, Nov 27, 30; m 56; c 2. TEXTILE CHEMISTRY. Educ: Middlebury Col, BA, 53. Prof Exp: Staff chemist, Am Viscose Corp, 55-58; proj dir sanit & surg prod, 58-60, dir prof develop, 60-69, asst dir res, 69-74, DIR RES, CHICOPEE MGR CO DIV, JOHNSON & JOHNSON, 74- Mem: AAAS; Am Asn Textile Technologists; Am Chem Soc. Res: Concentrated effort in development of new surgical dressings and applications; new textile constructions, nonwovens. Mailing Add: 73 Harriet Dr Princeton NJ 08540

BASSETT, ARTHUR LEON, b New York, NY, Feb 26, 35; m 60; c 1. CARDIOVASCULAR PHYSIOLOGY, PHARMACOLOGY. Educ: City Col New York, BS, 55; State Univ NY, PhD(physiol), 65. Prof Exp: Technician, Med Ctr, NY Univ, 56-58; jr scientist, State Univ NY Downstate Med Ctr, 58-60, asst physiol, 60-61 & 63-65; fel pharmacol, Col Physicians & Surgeons, Columbia Univ, 65-67, assoc, 67-68, asst prof pharmacol, 68-73; ASSOC PROF PHARMACOL & SURG, MED SCH, UNIV MIAMI, 73-, DIR GRAD AFFAIRS COMT, DEPT PHARMACOL, 75- Concurrent Pos: Polachek Found fel med res, 67-; Nat Heart Inst res career develop award, 69-73; grants, AMA, 69-72, Heart Asn Gtr Miami, 73- & Fla Heart Asn, 74- Mem: Biophys Soc; Am Soc Clin Pharmacol & Therapeut; Am Soc Pharmacol & Exp Therapeut; Am Physiol Soc; Soc Gen Physiol. Res: Excitatism-contraction coupling cardiac muscle; cardiac arrhythmias and antiarrhythmic drugs. Mailing Add: Univ of Miami Med Sch PO Box 520875 Miami FL 33152

BASSETT, CHARLES ANDREW LOOCKERMAN, b Crisfield, Md, Aug 4, 24; m 46; c 3. ORTHOPEDIC SURGERY, CELL PHYSIOLOGY. Educ: Columbia Univ, MD, 48, ScD(med), 55; Am Bd Orthop Surg, dipl. Prof Exp: From instr to assoc prof, 55-67, PROF ORTHOP SURG, COL PHYSICIANS & SURGEONS, COLUMBIA UNIV, 67- Concurrent Pos: Consult, Naval Med Res Inst, 53-54; assoc attend orthop surgeon, Presby Hosp, New York, 60-63, attend orthop surgeon, 63-; career scientist, New York City Health Res Coun, 61-71; consult, Food & Durg Admin; spec consult, NIH; mem comt skeletal syst, Nat Res Coun-Nat Acad Sci & NY State Rehab Hosp, West Haverstraw. Honors & Awards: Paralyzed Vet Am Award, 59; United Cerebral Palsy-Max Weinstein Award, 60; James Mather Smith Prize, 71. Mem: Orthop Res

Soc (secy-treas, 61-64, pres, 68-69); Am Acad Orthop Surg; Am Orthop Asn; Am Soc Cell Biol; fel Am Col Surg. Res: Bioelectric mechanisms in bone growth and physiology regenerating of central and peripheral nervous systems; tissue transplantations. Mailing Add: Col of Physicians & Surgeons Columbia Univ New York NY 10032

BASSETT, DAVID R, b Winston-Salem, NC, Jan 23, 39; m 61; c 3. PHYSICAL CHEMISTRY, COLLOID CHEMISTRY. Educ: Lafayette Col, AB, 61; Lehigh Univ, MS, 63, PhD(phys chem), 68. Prof Exp: RES CHEMIST, UNION CARBIDE CORP, 68- Mem: Am Chem Soc. Res: Surface chemistry; polymer colloids. Mailing Add: Tech Ctr Union Carbide Corp South Charleston WV 25303

BASSETT, EDWARD G, microbiology, enzymology, see 12th edition

BASSETT, EMMETT W, b Martinsville, Va, Jan 23, 21; m 50; c 3. MICROBIOLOGY. Educ: Tuskegee Inst, BS, 42; Univ Mass, MS, 50; Ohio State Univ, PhD(dairy technol), 54. Prof Exp: Res assoc microbiol, Col Physicians & Surgeons, Columbia Univ, 55-59, asst prof, 59-67; sr scientist, Ortho Res Found, NJ, 67-69; asst prof, 69-73, ASSOC PROF MICROBIOL, COL MED & DENT NJ, 73- Mem: Am Soc Biol Chem. Res: Aromatic synthesis in molds; studies on the combining region and biosynthesis of antibodies. Mailing Add: Dept of Microbiol Col of Med & Dent of NJ Newark NJ 07103

BASSETT, HENRY GORDON, b Newton, Mass, Nov 12, 24; m 49; c 3. STRATIGRAPHY. Educ: McGill Univ, BSc, 49, MSc, 50; Princeton Univ, AM & PhD(geol), 52. Prof Exp: Geologist, Shell Can Ltd, 52-72; HEAD, STRATIG SERV GROUP, SHELL DEVELOP CO, 72- Mem: Am Asn Petrol Geol; Can Asn Petrol Geol; fel Geol Asn Can. Res: Geology of northern and western Canada. Mailing Add: Shell Develop Co PO Box 481 Houston TX 77001

BASSETT, JAMES WILBUR, b Greenville, Tex, July 22, 23; m 49; c 6. ANIMAL SCIENCE. Educ: Tex A&M Univ, BS, 48, PhD(animal breeding, 65; Mont State Univ, MS, 57. Prof Exp: Asst county agt, Ark Agr Exten Serv, 48-50; asst prof wool technol, Mont State Univ, 51-63; PROF ANIMAL SCI, TEX A&M UNIV, 63- Mem: Am Soc Animal Sci. Res: Wool and mohair technology, production and marketing. Mailing Add: Dept of Animal Sci Tex A&M Univ College Station TX 77843

BASSETT, JOSEPH YARNALL, JR, b Asheville, NC, Aug 2, 27; m 53; c 2. ORGANIC CHEMISTRY. Educ: Univ NC, BS, 51, PhD(chem), 58. Prof Exp: Raw mat qual control analyst, Am Enka Corp, 53; lab asst, Univ NC, 53-56; develop chemist, Union Carbide Chem Co, WVa, 57-64; asst prof chem, WVa State Col, 62-64; asst prof, 64-69, PROF CHEM & CHMN DIV NATURAL SCI & MATH, WESTERN CAROLINA UNIV, 69- Mem: AAAS; Am Chem Soc. Res: Esterification; plasticizers; polyesters; vinyl resins; telomerization; oxidative chlorination of paraffins; organic synthesis by electrochemical reactions. Mailing Add: Western Carolina Univ PO Box 1524 Cullowhee NC 28723

BASSETT, LEWIS GORDON, b Mechanicville, NY, Sept 22, 02; m 24; c 2. ANALYTICAL CHEMISTRY. Educ: Rensselaer Polytech Inst, ChE, 23, PhD(chem), 35; Columbia Univ, AM, 31. Prof Exp: Instr math, 23-24, instr chem & chem eng, 24-38, from asst prof to prof anal chem, 38-73, EMER PROF ANAL CHEM, RENSSELAER POLYTECH INST, 73- Concurrent Pos: Sci ed, Int Conf Peaceful Uses of Atomic Energy, Geneva, 58-59. Mem: AAAS; Am Chem Soc. Res: Structure of metal chelates; solvent extraction of inorganic ions; photometric methods; analytical separations; radiation chemistry. Mailing Add: Dept of Anal Chem Rensselaer Polytech Inst Troy NY 12180

BASSETT, MARK JULIAN, b Washington, Ind, May 15, 40. PLANT BREEDING. Educ: Lake Forest Col, BA, 63; Univ Md, MS, 67, PhD(hort), 70. Prof Exp: ASST PROF VEG CROPS, UNIV FLA, 70- Mem: Am Soc Hort Sci; Am Genetic Asn. Res: Genetic studies and varietal improvement of snap beans, carrots, endive and lettuce. Mailing Add: Dept of Veg Crops Univ of Fla 3026 McCarty Hall Gainesville FL 32601

BASSETT, WILLIAM AKERS, b Brooklyn, NY, Aug 3, 31; m 62; c 3. MINERALOGY, GEOPHYSICS. Educ: Amherst Col, BA, 54; Columbia Univ, MA, 56, PhD(geol), 59. Prof Exp: Res assoc chem, Brookhaven Nat Lab, 60-61; from asst prof to assoc prof, 60-69, PROF GEOL, UNIV ROCHESTER, 69- Concurrent Pos: Res collabr, Brookhaven Nat Lab, 58-62; res assoc, Columbia Univ, 59-62; vis prof, Mass Inst Technol, 74. Mem: AAAS; Geol Soc Am; Mineral Soc Am; Am Geophys Union; Geochem Soc. Res: Structure and chemistry of sheet silicates; radioactive age dating of volcanic rocks and Pre-Cambrian rocks; behavior of solids at pressures and temperatures comparable with the earth's interior. Mailing Add: Dept of Geol Sci Univ of Rochester Rochester NY 14627

BASSETTE, RICHARD, b Washington, DC, May 1, 23; m 51; c 4. DAIRY SCIENCE. Educ: Univ Md, BS, 52, MS, 55, PhD(agr), 58. Prof Exp: Asst dairy husb, Univ Md, 53-58; asst prof, 58-64, ASSOC PROF DAIRY FOODS PROCESSING, KANS STATE UNIV, 64- Mem: Am Chem Soc; Am Dairy Sci Asn; Inst Food Technologists. Res: Flavor chemistry; gas chromatography; dairy chemistry; analysis of trace volatiles in biological fluids in relation to human and animal health. Mailing Add: Dept of Dairy & Poultry Sci Kans State Univ Manhattan KS 66502

BASSHAM, JAMES ALAN, b Sacramento, Calif, Nov 26, 22; m 56; c 5. BIOCHEMISTRY. Educ: Univ Calif, BS, 45, PhD(chem), 49. Prof Exp: RES CHEMIST, BIO-ORG CHEM GROUP, LAWRENCE BERKELEY LAB, UNIV CALIF, BERKELEY, 49-, ASSOC DIR CHEM BIODYNAMICS GROUP, LAB CHEM BIODYNAMICS, 60- Concurrent Pos: NSF sr fel, Oxford Univ, 56-57; lectr chem, Univ Calif, Berkeley, 57-59, adj prof chem, 72-; vis prof, dept biochem & biophys, Univ Hawaii, 68-69. Mem: Am Chem Soc; AAAS; Am Soc Biol Chemists; Am Soc Plant Physiol. Res: Regulation of metabolism in photosynthetic plant cells and in animal cells in tissue culture. Mailing Add: Bldg 3 Lawrence Berkeley Lab Univ of Calif Berkeley CA 94720

BASSI, SUKH D, b Kericho, Kenya; US citizen; m 71; c 1. DEVELOPMENTAL GENETICS. Educ: Knox Col, BS, 65; St Louis Univ, Mo, MS, 67, PhD(biol), 70. Prof Exp: Prog specialist biol, Clark Col, 70-71; asst prof, 71-74, ASSOC PROF BIOL, BENEDICTINE COL, 74- Concurrent Pos: Consult, Environ Coun Atchison, Kans, 71- Mem: Am Inst Biol Sci; Am Physiol Soc; Genetics Soc Am. Res: Interaction of juvenile hormone and molting hormone with carrier and receptor proteins in the large milkweed bug, Oncopeltus fasciatus. Mailing Add: Dept of Biol Benedictine Col Atchison KS 66002

BASSICHIS, WILLIAM, b Cleveland, Ohio, Aug 9, 37. THEORETICAL NUCLEAR PHYSICS. Educ: Mass Inst Technol, BS, 59; Case Inst Technol, MS, 61, PhD(physics), 63. Prof Exp: Vis scientist, Weizmann Inst Sci, 63-64 & Saclay Nuclear Res Ctr, France, 64-65; asst prof physics, Mass Inst Technol, 66-70; ASSOC PROF PHYSICS, TEX A&M UNIV, 70- Concurrent Pos: Consult, AEC, Lawrence Radiation Lab, Univ Calif, Livermore, 66- Res: Nuclear structure and reactions. Mailing Add: Dept of Physics Tex A&M Univ College Station TX 77843

BASSIN, ROBERT HARRIS, b Washington, DC, May 2, 38; m 61; c 2. MICROBIOLOGY. Educ: Princeton Univ, AB, 59; Rutgers Univ, PhD(microbiol), 65. Prof Exp: From res asst to res assoc microbiol, Rutgers Univ, 61-66; Imp Cancer Res Fund fel, London, Eng, 66-68; staff fel, 68-73, STAFF MICROBIOLOGIST & HEAD VIRAL BIOCHEM SECT, NAT CANCER INST, 73- Mem: AAAS; Soc Gen Microbiol; Tissue Cult Asn; Am Soc Microbiol; Am Asn Cancer Res. Res: RNA tumor virology; replication and transfomation by murine sarcoma and leukemia viruses in cell culture. Mailing Add: Nat Cancer Inst Bldg 41 Bethesda MD 20014

BASSINGTHWAIGHTE, JAMES B, b Toronto, Ont, Sept 10, 29; m 55; c 5. PHYSIOLOGY, BIOPHYSICS. Educ: Univ Toronto, BA, 51, MD, 55; Univ Minn, PhD(physiol), 64. Prof Exp: Intern, Toronto Gen Hosp, 55-56; gen practitioner, Ont, 56-57; house physician internal med, Hammersmith Hosp, London, 57-58; fel med & physiol, Mayo Grad Sch Med, Univ Minn, 58-61; teaching asst, Univ Minn, Minneapolis, 61-62; res asst, Mayo Grad Sch Med, Univ Minn, from asst prof to prof physiol, 67-75, assoc prof bioeng, 72-75, prof med, 75; PROF BIOENG & DIR CTR BIOENG, COL ENG & SCH MED, UNIV WASH, 75-, MEM BIOMATH GROUP FAC, 75- Concurrent Pos: Minn Heart Asn fel, 59-62; NIH career develop award, 64-74; vis prof, Pharmacol Inst, Univ Berne, 70-71; mem study sect, Nat Heart & Lung Inst, 70-74. Mem: AAAS; Am Heart Asn; Am Physiol Soc; Soc Comput Simulation; Microcirculatory Soc. Res: Cardiovascular and transport physiology; internal medicine; biomedical engineering; cardiology. Mailing Add: Ctr for Bioeng Univ of Wash Seattle WA 98195

BASSLER, GERALD CLAYTON, b Roaring Spring, Pa, July 7, 16; m 46. ORGANIC CHEMISTRY. Educ: Franklin & Marshall Col, BS, 38; Pa State Col, MS, 40, PhD(org chem), 43. Prof Exp: Instr org res, Pa State Col, 43-44, 45-47; res chemist, Heyden Chem Corp, NJ, 45; fel Mellon Inst, 47-54; sr org chemist, Stanford Res Inst, 54-69; MGR RES & TECH SERV, HILLS BROS COFFEE INC, 69- Mem: Am Chem Soc. Res: High explosives; penicillin and streptomycin recovery; penicillin precursors; new resins and plastics; hydrocarbon fuel stability; gas chromatography; adsorption spectroscopy. Mailing Add: Hills Bros Coffee Inc 2 Harrison St San Francisco CA 94119

BASSLER, RICHARD ALBERT, b New York, NY, Sept 24, 17; m 41; c 3. INFORMATION SCIENCE. Educ: Univ Colo, BS, 64; George Washington Univ, MS, 66; Laurence Univ, PhD(comput & educ), 74. Prof Exp: Dir, Audio Visual Servs, NAm Air Defense Command, 55-62; dir, Long-Range Automatic Data Processing Planning, Directorate Data Automation, US Air Force, 64-67; pres & chmn bd comput consult, ASSIST Corp, 67-69; ASSOC PROF & DIR, COMPUT SYST APPLN PROG, CTR TECHNOL & ADMIN, AM UNIV, 69- Concurrent Pos: Mem & vchmn, Data Processing Adv Bd, Northern Va Community Col, 75-; mem & adv, Eval Panel Non-Col Sponsored Educ, Am Coun Educ, Off Educ Credit, 75-76. Mem: Am Soc Info Sci; Soc Data Educr; Asn Comput Mach; Soc Mgt Info Syst; Data Processing Mgt Asn. Res: Examination of needs of computer users for technically proficient people in the future; the problem of educational institutions responding in a way to satisfy both sides of the equation; the way the current-day managers perceive their needs and sources of talent. Mailing Add: Ctr Technol & Admin Am Univ Washington DC 20016

BASSO, ELLEN BECKER, b New York, NY, Nov 8, 42; m 70. ANTHROPOLOGY. Educ: Hunter Col, BA, 63; Univ Chicago, MA, 65, PhD(anthrop), 69. Prof Exp: Instr anthrop, Univ Chicago, 69-70; ASST PROF ANTHROP, UNIV ARIZ, 71- Mem: AAAS; Am Anthrop Asn. Res: Symbolic systems; cognitive anthropology; kinship; social structure; language and culture. Mailing Add: Dept of Anthrop Univ of Ariz Tucson AZ 85725

BASSO, KEITH HAMILTON, b Asheville, NC, Mar 15, 40. ANTHROPOLOGY. Educ: Harvard Univ, BA, 62; Stanford Univ, MA, 65, PhD(anthrop), 67. Prof Exp: Assoc prof, 67-74, PROF ANTHROP, UNIV ARIZ, 74- Mem: Fel AAAS; fel Ling Soc Am; fel Am Anthrop Asn. Res: Southwestern ethnography, ethnographic semantics; religion. Mailing Add: Dept of Anthrop Univ of Ariz Tucson AZ 85721

BASSON, PHILIP WALTER, b New York, NY, June 29, 31; m 54; c 2. PALEOBOTANY, MARINE ECOLOGY. Educ: Eastern Mich Univ, BS, 59; Univ Mo, MA, 61, PhD(bot), 65. Prof Exp: Instr bot, Cornell Univ, 65-68; asst prof, 68-72, ASSOC PROF BIOL, AM UNIV BEIRUT, 72-, CHMN DEPT, 69- Concurrent Pos: Cornell Univ res grant, 67-68; sr consult, Arabian Am Oil Co, 75- Mem: Bot Soc Am; NY Acad Sci. Res: Plant macrofossils of Pennsylvanian Age; pollution monitoring and coastal zone management; determination of biotodes of Saudi Arabian Gulf Coast; calcareous algae of Lebanon. Mailing Add: Dept of Biol Am Univ of Beirut Beirut Lebanon.

BASTIAANS, GLENN JOHN, b Oak Park, Ill, Oct 25, 47; m 74. ANALYTICAL CHEMISTRY. Educ: Univ Ill, Urbana, BS, 69; Ind Univ, Bloomington, PhD(chem), 73. Prof Exp: Fel chem, Colo State Univ, 73-74; ASST PROF CHEM, GEORGETOWN UNIV, 75- Mem: Am Chem Soc; Soc Appl Spectros. Res: Study of the factors which limit analytical performance of atomic spectroscopy; development of instrumental techniques to solve problems of analysis; use of computers in chemical laboratories. Mailing Add: Dept of Chem Georgetown Univ Washington DC 20057

BASTIAN, JAMES W, b Indianapolis, Ind, Apr 17, 26; m 50; c 3. ENDOCRINOLOGY. Educ: Purdue Univ, PhD(biol sci), 54. Prof Exp: Pharmacologist, 54-67, HEAD DEPT PHARMACOL, ARMOUR PHARMACEUT CO, 67- Mem: Am Soc Pharmacol & Exp Therapeut. Res: Small animal screening; cardiovascular and anti-inflammatory pharmacology; platelet aggregation; calcium metabolism; calcitonin. Mailing Add: Dept of Pharmacol Armour Pharmaceut Co Kankakee IL 60901

BASTIAN, JOSEPH, b Mare Island, Calif, Feb 13, 44; m 66; c 3. NEUROBIOLOGY. Educ: Elmhurst Col, BS, 66; Univ Notre Dame, PhD(biol), 69. Prof Exp: Fel neurobiol, Univ Notre Dame, 69-70, Purdue Univ, 70-72; asst res neuroscientist, Sch Med, Univ Calif, San Diego, 72-74; ASST PROF ZOOL, UNIV OKLA, 74- Concurrent Pos: Fel, Nat Inst Neurol Dis & Stroke, 70, res grant, 75. Mem: AAAS; Soc Neurosci. Res: Neurophysiology of the electrosensory system in weakly electric fish and the role of the cerebellum in processing sensory information; mechanisms of the excitation-coupling process in skeletal muscle. Mailing Add: Dept of Zool 730 Van Vleet Oval Univ Okla Norman OK 73069

BASTIAN, ROBERT PAUL, analytical chemistry, physical chemistry, see 12th edition

BASTON, VIRGIL FOREST, physical chemistry, see 12th edition

BASTOS, MILTON LESSA, b Rio de Janeiro, Brazil, Mar 26, 27; m 53; c 5. TOXICOLOGY. Educ: Univ Rio de Janeiro, BS, 49; Univ Fluminense, Brazil, BS, 56,

PhD(toxicol), 65. Prof Exp: Chemist, Inst Legal Med, Brazil, 50-66; dir dept learning & res, Fed Univ Fluminense, 66-68, dir, Chem Inst, 68-69; consult toxicol, Lab Addictive Drugs, New York, 69-70; assoc chemist, Testing & Res Lab, NY State Narcotic Addiction Control Comn, 71-72; DIR TOXICOL, MED EXAMR OFF, NEW YORK, 73-; PROF TOXICOL, DEPT FORENSIC MED, NY UNIV, 73- Mem: Asn Off Racing Chem; Am Acad Forensic Sci; Int Asn Forensic Sci. Res: Distribution of methadone in different tissues according to the cause of death; metabolism as distribution of coccaine; development of analytical methods for toxicology. Mailing Add: Med Examr Off 520 First Ave New York NY 10016

BASTUSCHECK, CLIFFORD PAUL, b Washington, DC, Mar 23, 23; m 45; c 3. PHYSICS. Educ: Pa State Univ, BS, 44, MS, 50. Prof Exp: Instr physics, Pa State Univ, 45-50; sr physicist, 50-58, staff engr, 59-61, head solid state sect, 61-66, STAFF PHYSICIST, HRB-SINGER, INC, 66- Mem: Am Phys Soc. Res: Acoustics and ultrasonics; special optics; solid state imaging devices; infrared systems; remote sensing of environment. Mailing Add: 285 Ellen Ave State College PA 16801

BASU, ASIT PRAKAS, b Mar 17, 37; Indian citizen; m 66; c 2. STATISTICS. Educ: Univ Calcutta, BSc, 56, MSc, 58; Univ Minn, PhD(statist), 66. Prof Exp: Assoc lectr statist & math, Indian Inst Technol, Kharagpur, 60-62; asst statist, Rutgers Univ & Univ Minn, 62-66; asst prof, Univ Wis, 66-68; res staff mem, IBM Res Ctr, 68-70; asst prof indust eng, Northwestern Univ, 70-71; assoc prof math, Univ Pittsburgh, 71-74; PROF STATIST, UNIV MO-COLUMBIA, 74- Mem: Am Statist Asn; Inst Math Statist; fel Royal Statist Soc. Res: Nonparametric statistics; life testing and reliability; parametric and nonparametric inference in reliability. Mailing Add: Dept of Statist Univ of Mo Columbia MO 65201

BASU, DEBABRATA, b Dacca, Bangladesh, July 7, 24; Indian citizen; m; c 2. STATISTICS. Educ: Dacca Univ, BA, 45, MS, 46; Calcutta Univ, PhD(statist), 55. Prof Exp: From assoc prof to prof statist, Indian Statist Inst, Calcutta, 54-75; PROF STATIST, FLA STATE UNIV, 75- Concurrent Pos: Vis prof statist, Univ NC, Chapel Hill, 64-65 & Univ Chicago, 65-66; vis prof math, Am Univ Beirut, 64, Univ NMex, 68-69, Univ Manchester, Eng, 71-73 & Univ Western Australia, 75. Mem: Int Statist Inst. Res: Foundational questions in statistical inference. Mailing Add: Dept of Statist Fla State Univ Tallahassee FL 32304

BASU, PRASANTA KUMAR, b Mymensingh, Bangladesh, Apr 1, 22; m 48; c 1. OPHTHALMOLOGY. Educ: Univ Calcutta, BSc, 41, MB, 46, DOMS, 51. Prof Exp: Head ophthal, Ramakrishna Mission Hosp, Vrindaban, India, 51-59; assoc prof ophthal & dir ophthalmic res, 59-70, PROF OPHTHAL, UNIV TORONTO, 70-; SR OPHTHALMOLOGIST, TORONTO GEN HOSP, 59- Concurrent Pos: Colombo Plan fel, Univ Toronto, 55-56; res assoc, Univ Calif, San Francisco, 57; res assoc, Univ Toronto, 57-, clin teacher, 58-59; assoc, Med Res Coun Can, 65-; assoc med dir, Ont Div Eye Bank Can, 68- Mem: Asn Res Vision & Ophthal; assoc Can Ophthal Soc. Res: Ocular tissue transplantation; eye bank and tissue storage; ocular tumor; ocular penetration of steroids; tissue culture of eye tissues. Mailing Add: Rm 117 1 Spadina Crescent Toronto ON Can

BASU, SUBHASH CHANDRA, b Calcutta, India, May 28, 38; m 66. BIOCHEMISTRY. Educ: Univ Calcutta, BSc, 58, MSc, 60; Univ Mich, PhD(biochem), 66. Prof Exp: Lectr chem, Vidyasager Col, Univ Calcutta, 60-61; teaching asst biochem, Univ Mich, 61-64, teaching fel, 64-65, res asst, Rackham Arthritis Res Unit, 65-66; fel, Johns Hopkins Univ, 66-67, res assoc, 67-70; ASST PROF CHEM, UNIV NOTRE DAME, IND, 70- Concurrent Pos: Res fel, Helen Hay Whitney Found, NY, 67-70; vis scientist, Gothenburg Univ, 68; Kennedy Found travel fel, Sweden, 68. Mem: Soc Cryobiol. Res: Glycosphingolipids; glycolipids; neurochemistry; blood group glycosphingolipids. Mailing Add: Dept of Chem Univ of Notre Dame Notre Dame IN 46556

BATA, GEORGE L, b Budapest, Hungary, Nov 18, 24; Can citizen; m 50; c 1. POLYMER CHEMISTRY, POLYMER PHYSICS. Educ: Univ Budapest, BASc, 46, MSc, 47. Prof Exp: Mem new prod div, Arborite Co Ltd, 50-53 & Varcum Chem Co, 53-55; DIR TECHNOL, UNION CARBIDE CAN LTD, 55- Mem: Can Soc Chem Eng; Soc Plastics Eng; Chem Inst Can; Can Res Mgt Asn. Res: Polymer molecular weight distribution; polymer hydrodynamic volume; polydispersity versus physical properties of polyolefins; extrusion unit operation; polymer-human tissue interactions; muscle stimulation in vivo; environmental stress cracking phenomena. Mailing Add: 5624 Canterbury Ave Montreal PQ Can

BATAY-CSORBA, PETER ANDREW, b Budapest, Hungary, July 10, 45; Can; m 70. NUCLEAR PHYSICS. Educ: Mass Inst Technol, BS, 68; Calif Inst Technol, PhD(nuclear physics), 75. Prof Exp: Res asst nuclear physics, Kellogg Lab, Calif Inst Technol, 70-75; RES ASSOC NUCLEAR PHYSICS, UNIV COLO, BOULDER, 75- Mem: Sigma Xi; Am Phys Soc. Res: Nuclear structure and nuclear reactions. Mailing Add: Nuclear Physics Lab Univ of Colo Boulder CO 80302

BATCHELDER, ALAN COLEMAN, b Pasadena, Calif, Aug 4, 14; m 36 & 69; c 2. CHEMISTRY. Educ: Pomona Col, BA, 35; Univ Calif, PhD(biochem), 39; Golden Gate Col, LLB, 57. Prof Exp: Smith, Kline & French Co fel, Sch Med, Johns Hopkins Univ, 39-40; tutor biochem sci, Harvard Col & fel phys chem, Harvard Med Sch, 40-43; patent chemist, Hercules Powder Co, 43-45; dept head, Patent Div, Shell Develop Co, 46-60, mgr chem licensing, Licensing Div, 60-70, dir, Patent Div, 70-76; RETIRED. Res: Physical chemistry of amino acids and proteins, especially blood proteins; chemistry of ragweed pollen; protection and licensing of industrial property; patent law. Mailing Add: 29 Burlwood Dr San Francisco CA 94127

BATCHELDER, ARTHUR ROLAND, b Haverhill, Mass, Apr 17, 32; m 55; c 3. AGRONOMY. Educ: Univ Mass, BS, 54; Va Polytech Inst, MS, 66; Cornell Univ, PhD, 71. Prof Exp: SOIL SCIENTIST, AGR RES SERV, USDA, 57- Mem: Am Soc Agron; Soil Sci Soc Am; Int Soc Soil Sci. Res: Chemistry of agricultural waters and the eutrophication of farm ponds and reservoirs; influence of soil properties on ion and water uptake by plant roots. Mailing Add: Agr Res Serv USDA PO Box E Ft Collins CO 80522

BATCHELDER, ROBERT BRUCE, b Seattle, Wash, Nov 25, 23; m 48; c 3. GEOGRAPHY. Educ: Univ Wash, BA, 47; Northwestern Univ, MA, 49, PhD(geog), 51. Prof Exp: Lectr, Evanston Nat Col Educ, 48-49; asst prof, Stephen F Austin State Col, 51-53; asst prof geog, 53-70, PROF GEOG, BOSTON UNIV, 70- Concurrent Pos: Consult, Info Dynamics Inc; sci mem, Int Indian Ocean Exped, 63; prin investr grant contract, US Army Natick Labs, 64-66; grad sch grant, Boston Univ, 66- Mem: Asn Am Geogr; Am Geog Soc; Am Meteorol Soc. Res: Physical geography; physical climatology. Mailing Add: Dept of Geog Boston Univ Col of Lib Arts Boston MA 02215

BATCHELLER, OLIVER A, b Mattapoisett, Mass, May 26, 15; m 39, 54; c 2. HORTICULTURE. Educ: Ore State Col, BS, 36. Prof Exp: Asst county agt, Lane County, Ore, 36-37; nursery mgr, Calif Nursery Co, 37-41; PROF ORNAMENTAL HORT, CALIF STATE POLYTECH UNIV, POMONA, 46- Honors & Awards: Educ & Res Award, Am Florist Soc, 75. Res: Propagation; budding and grafting of Eucalyptus ficifolia. Mailing Add: Dept of Ornamental Hort Calif State Polytech Univ Pomona CA 91768

BATCHELOR, WILLIAM HENRY, b Palmerton, Pa, Feb 4, 21; m 48; c 3. MEDICINE, PHYSICAL CHEMISTRY. Educ: Harvard Univ, AB, 42, MD, 45. Prof Exp: Dir arthritis prog, 63-65, TRAINING & FELS OFFICER, NAT INST ARTHRITIS & METAB DIS, 65- Concurrent Pos: Res fel phys chem, Harvard Med Sch, 49-52; clin & res fel, Mass Gen Hosp, 54-63. Res: Impact of federal support programs on medical education. Mailing Add: Rm 620 Westwood Bldg Nat Inst Arthritis & Metab Dis Bethesda MD 20014

BATCHO, ANDREW DAVID, b Somerville, NJ, July 9, 34; m 70. ORGANIC CHEMISTRY. Educ: Rutgers Univ, BS, 55; Univ Calif, PhD(chem), 59. Prof Exp: Sr res chemist, Am Cyanamid Co, 58-61; SR RES CHEMIST, HOFFMANN-LA ROCHE, INC, 61- Mem: Am Chem Soc. Res: Synthetic organic chemistry; natural products. Mailing Add: Hoffmann-La Roche Inc Nutley NJ 07110

BATDORF, ROBERT LUDWIG, b Reading, Pa, Sept 18, 26; m 53; c 2. SOLID STATE SCIENCE. Educ: Albright Col, BA, 50; Univ Minn, PhD(phys chem), 55. Prof Exp: Mem tech staff, Am Tel & Tel Co, 55-65, supvr solid state device technol group, 65-68, dept head semiconductor mat & appl chem, 68-73, HEAD DEPT SEMICONDUCTOR TECHNOL, BELL LABS, 73- Mem: AAAS; Am Asn Crystal Growers; Am Chem Soc. Res: Diode and field effect transistor development; integrated circuit process technology, especially ion implantation techniques, diffusion, and chemical processprocessing; physical diagnostic analytical techniques such as scanning electron microscopy, electron beam and ion beam microprobes. Mailing Add: Bell Labs 2525 N 11th St Reading PA 19604

BATDORF, SAMUEL BURBRIDGE, b China, Mar 31, 14; m 40; c 2. PHYSICS, ENGINEERING MECHANICS. Educ: Univ Calif, AB, 34, MA, 36, PhD(theoret physics), 38. Prof Exp: Instr physics, Univ Utah, 38; from asst prof to assoc prof, Univ Nev, 38-43; physicist aeronaut res sci, Langley Lab, Nat Adv Comt Aeronaut, 43-51; adv physicist, Westinghouse Res Labs, 51-52; mgr develop, Mat Eng Dept, Westinghouse Elec Corp, 52-55, dir develop, Eng Hq Staff, 55-56; asst dir res & mgr electronics div, Lockheed Missile Syst Div, Lockheed Aircraft Corp, 56-57, dir weapon syst tech div, 57-58; chmn man-in-space proj, Inst Defense Anal, 58-59, chmn commun satellite proj, 59-60; mgr prod planning, Aeronutronic Div, Ford Motor Co, 60-61, dir res physics & electronics, 61-62; dir eff res, 62-64, dir appl mech & physics subdiv, 64-70, group dir appl technol concepts, 70-71, PRIN STAFF SCIENTIST, AEROSPACE CORP, 72- Concurrent Pos: Mem comt aircraft struct mat, Nat Adv Comt Aeronaut, 47-51; mem panel physics of solids, Res & Develop Bd, 49-51; adj prof, Univ Calif, Los Angeles, 73- Mem: Fel Am Phys Soc; assoc fel Am Inst Aeronaut & Astronaut; fel Am Soc Mech Engrs; sr mem Inst Elec & Electronics Engrs. Res: Mechanics; elasticity; theory of plates and shells; plasticity; magnetism; semiconductors; electronics; geophysics. Mailing Add: 28409 Seamount Dr Rancho Palos Verdes CA 90274

BATE, GEOFFREY, b Sheffield, Eng, Mar 30, 29; m 53; c 4. SOLID STATE PHYSICS. Educ: Univ Sheffield, BSc, 49, PhD(physics), 52. Prof Exp: Sci officer physics, Royal Naval Sci Serv, Eng, 52-56; res assoc, Univ BC, 56-57, asst prof, 57-59; staff physicist, Develop Lab, Data Systs Div, Int Bus Mach Corp, 59-60, adv physicist, 60-65, mgr advan rec technol, 63-65, SR PHYSICIST & MGR REC PHYSICS, IBM CORP, 65- Concurrent Pos: Adj prof, Syracuse Univ, 61- Mem: Am Phys Soc; Sigma Xi; sr mem Inst Elec & Electronics Engrs. Res: Ferromagnetism of high-coercivity materials; semiconductors and infrared photoconductors; physics of magnetic recording; magnetic properties of recording materials; high-frequency properties of soft-magnetic materials. Mailing Add: Sunshine Canyon Boulder CO 80302

BATE, GEORGE LEE, b Newton Falls, Ohio, Feb 15, 24; m 46; c 3. PHYSICS, APPLIED MATHEMATICS. Educ: Princeton Univ, AB, 45; Calif Inst Technol, MS, 46; Columbia Univ, PhD, 55. Prof Exp: Instr physics, Wheaton Col, 47-51; res asst, Columbia Univ, 51-55; from asst prof to assoc prof physics, Wheaton Col, 55-65; sr scientist, Isotopes, Inc, 65-67; prof physics, 67-69, chmn dept physics & math, 70-75, PROF PHYSICS & APPL MATH, WESTMONT COL, 69- Concurrent Pos: Res assoc & consult, Argonne Nat Lab, 56-65. Mem: AAAS; Geochem Soc; Am Asn Physics Teachers; Am Sci Affiliation. Res: Abundance of the elements by neutron activation analysis; induced fission of heavy elements. Mailing Add: Dept Physics & Math Westmont Col 955 LaPaz Rd Santa Barbara CA 93108

BATE, ROBERT THOMAS, b Denver, Colo, Apr 1, 31; m 51; c 5. SEMICONDUCTORS. Educ: Univ Colo, BS, 55; Ohio State Univ, MS, 57. Prof Exp: Physicist, Eng Lab, US Bur Standards, 55; res asst, Ohio State Univ, 56-57; physicist, Battelle Mem Inst, 57-64; physicist, 64-68, sr scientist, 68-72, BR MGR, CENT RES LABS, TEX INSTRUMENTS INC, DALLAS, 72- Mem: Fel Am Phys Soc. Res: Low temperature physics; electrical properties of semiconductors; semiconductor device research. Mailing Add: 512 Westshore Dr Richardson TX 75080

BATE, ROGER R, nuclear physics, computer science, see 12th edition

BATEMAN, BARRY LYNN, b Jacksonville, Tex, Sept 15, 43; m 63; c 2. COMPUTER SCIENCES. Educ: Tex A&M Univ, BA, 65, MS, 67, PhD(comput sci), 70. Prof Exp: From asst dir to actg dir comput ctr, Univ Southwestern La, 69-72, head dept comput sci, 70-72; CHMN DEPT COMPUT SCI, TEX TECH UNIV, 72- Concurrent Pos: Consult, Sigma Syst, 72- & Potomac Inst, 74- Mem: Asn Comput Mach; Soc Comput Simulation; Sigma Xi; Soc Indust & Appl Math; Math Asn Am. Res: Computer science education; computer center management; information storage and retrieval, data base management; simulation. Mailing Add: Dept of Comput Sci Tex Tech Univ PO Box 4420 Lubbock TX 79409

BATEMAN, DURWARD F, b Tyner, NC, May 28, 34; m 53; c 3. PLANT PATHOLOGY. Educ: NC State Col, BS, 56; Cornell Univ, MS, 58, PhD(plant path), 60. Prof Exp: From asst prof to assoc prof, 60-69, PROF PLANT PATH, CORNELL UNIV, 69-, CHMN DEPT, 70- Concurrent Pos: NIH spec fel, Dept Plant Path, Univ Calif, Davis, 67; vis prof plant path, NC State Univ, 74-75. Mem: AAAS; fel Am Phytopath Soc (vpres, 76); Am Soc Plant Physiologists; Int Soc Plant Path. Res: Ecology of soil borne pathogens; physiology of parasitism and disease and pathogen physiology; biochemical basis of pathogenic mechanisms and host reactions to infection; enzymatic decomposition of plant cell walls by phytopathogenic organisms. Mailing Add: Dept of Plant Path Cornell Univ Ithaca NY 14853

BATEMAN, FELICE DAVIDSON, b Springfield, Mass, Sept 2, 22; m 48; c 1. MATHEMATICS. Educ: Smith Col, AB, 43; Univ Mich, AM, 44, PhD(math), 50. Prof Exp: Instr math, NJ Col Women, Rutgers Univ, 47-49; instr, 54-55 & 57-58, ASST PROF MATH, UNIV ILL, URBANA-CHAMPAIGN, 58- Concurrent Pos: Mem sch math, Inst Advan Study, 49-50; vis lectr, Swarthmore Col, 61-62 & Sarah

Lawrence Col, 64-65. Res: Linear algebras with radical. Mailing Add: 108 Meadows Ct Urbana IL 61801

BATEMAN, GEORGE MONROE, chemistry, deceased

BATEMAN, JOHN HUGH, b East St Louis, Ill, Dec 21, 41; m 64; c 2. POLYMER CHEMISTRY, ORGANIC CHEMISTRY. Educ: Ind Univ, Bloomington, AB, 64; Univ Kans, PhD(org chem), 68. Prof Exp: SR RES CHEMIST, CIBA-GEIGY CORP, 68- Mem: Am Chem Soc. Res: Coatings; thermoset plastics; weatherable materials. Mailing Add: Plastics & Additives Res Ciba-Geigy Corp Ardsley NY 10502

BATEMAN, JOHN LAURENS, b Washington, DC, Mar 30, 26; m 50; c 5. RADIOBIOLOGY, ONCOLOGY. Educ: Mass Inst Technol, SB, 46; Johns Hopkins Univ, MD, 56. Prof Exp: Intern, Stamford Hosp, 56-57; resident internal med, Greenwich Hosp Asn, 57-58; res assoc, 58-61, res collabr, 61-67, SCIENTIST, MED RES CTR, BROOKHAVEN NAT LAB, 67- Concurrent Pos: Consult, St Charles & John T Mather Mem Hosps, Port Jefferson, NY, 66- Mem: Radiation Res Soc; NY Acad Sci. Res: Early and late effects of ionizing radiation on mammalian systems as functions of linear energy transfer and dose rate; bone marrow effects of maintenance chemotherapy in metastatic carcinoma. Mailing Add: 105 Beach Port Jefferson NY 11776

BATEMAN, JOSEPH R, b Chicago, Ill, Sept 3, 22. HEMATOLOGY, ONCOLOGY. Educ: George Washington Univ, MD, 47; Am Bd Internal Med, dipl, 57. Prof Exp: Intern med, Gorgas Hosp, Balboa, CZ, 47-48; fel, Vet Admin Hosp, Grand Junction, Colo, 54-56; resident, Vet Admin Hosps, Albuquerque, NMex, 58-61, hematologist, Oakland, Calif, 56-60; from instr to assoc prof med, 60-75, PROF MED, UNIV SOUTHERN CALIF, 75- Concurrent Pos: Fel hemat, Univ Southern Calif, 60-62; consult, Vet Admin Hosp, 64-; chmn, Western Cancer Study Group; chief med oncol div, Los Angeles County-Univ Southern Calif Med Ctr. Mem: Am Col Physicians; Am Soc Clin Oncol; Am Asn Cancer Res; Am Asn Cancer Educ; Am Fed Clin Res. Res: Clinical research; cancer chemotherapy. Mailing Add: Dept of Med Univ of Southern Calif Los Angeles CA 90033

BATEMAN, PAUL CHARLES, b Webster Groves, Mo, Sept 14, 10; m 43; c 2. PETROLOGY, STRUCTURAL GEOLOGY. Educ: Univ Calif, Los Angeles, AB, 36, PhD, 58. Prof Exp: Explor geophysicist, Int Geophys, Inc, 36-38; geologist, Develop Dept, E I du Pont de Nemours & Co, Inc, Del, 38-42; geologist, Mineral Deposits Br, 42-61 & Pac Coast Br, 61-66, chief, Field Geochem & Petrol Br, 66-70, GEOLOGIST, FIELD GEOCHEM & PETROL BR, US GEOL SURV, 70- Concurrent Pos: Vis prof, Univ Calif, Santa Cruz, 70; mem subcomn igneous rock nomenclature, Comn Petrol, Int Union Geol Sci. Mem: Geol Soc Am; Geochem Soc. Res: Constitution, structure and origin of batholiths with particular reference to the Sierra Nevada and related circum-Pacific batholiths. Mailing Add: US Geol Surv 345 Middlefield Rd Menlo Park CA 94025

BATEMAN, PAUL TREVIER, b Philadelphia, Pa, June 6, 19; m 48; c 1. MATHEMATICS. Educ: Univ Pa, AB, 39, AM, 40, PhD(math), 46. Prof Exp: Lectr statist, Bryn Mawr Col, 45-46; instr math, Yale Univ, 46-48; res, Off Naval Res Contract, Inst Advan Study, 48-50; from asst prof to assoc prof, 50-58, PROF MATH, UNIV ILL, URBANA-CHAMPAIGN, 58-, HEAD DEPT, 65- Concurrent Pos: NSF sr fel, Inst Advan Study, 56-57; researcher, Univ Pa, 61-62; mem math adv panel, NSF, 63-66; vis prof, City Univ New York, 64-65. Mem: Am Math Soc; Math Asn Am; London Math Soc; Indian Math Soc. Res: Number theory and related parts of algebra, analysis, and mathematics of computation. Mailing Add: Dept of Math Univ of Ill Urbana IL 61801

BATES, CHARLES, b Dayton, Ohio, May 4, 30; m 53; c 3. FOOD SCIENCE, FOOD TECHNOLOGY. Educ: Calif Inst Technol, BS, 51; Mass Inst Technol, PhD(food technol), 57. Prof Exp: Res assoc, Mass Inst Technol, 51-54; develop engr, Procter & Gamble Co, 57-72; TECH VPRES, AM MAIZE PROD CO, INC, 72- Mem: AAAS; Am Chem Soc; Inst Food Technologists; Am Asn Clin Chemists. Res: Process and product development of food products. Mailing Add: 760 Williams Dr Crown Point IN 46307

BATES, CHARLES CARPENTER, b Harrison, Ill, Nov 4, 18; m 42; c 3. OCEANOGRAPHY. Educ: DePauw Univ, BS, 39; Univ Calif, Los Angeles, MA, 44; Tex A&M Univ, PhD, 53. Prof Exp: Geophys trainee, Carter Oil Co, Okla, 39-41; spec asst to pres, Am Meteorol Soc, Ill, 45-46; oceanog technician sci surv, Bikini Atomic Bomb Test, Woods Hole Oceanog Inst, 46; assoc oceanogr, Div Oceanog, Hydrog Off, Navy Dept, 46-48, oceanogr & actg chief prog sect, 48-52, dep div dir, 52-57, tech coordr environ systs, Off Naval Res, 57-59 & Off Dept Chief Naval Opers, 59-60, chief, Vela Uniform Br, Advan Res Proj Agency, Off Secy Defense, 60-64, sci & tech dir, Naval Oceanog Off, 64-68; SCI ADV TO COMMANDANT, HQ, US COAST GUARD, 68- Concurrent Pos: Consult oceanogr & meteorologist, Bates & Glenn, Washington, DC, 47; consult meteorologist, A H Glenn & Assocs, La, 48-54; mem comt meteorol uses of satellites, Space Sci Bd, Nat Acad Sci; State Dept alt observer, Austral Season Insepction by US under Antarctic Treaty, 63-64; mem earth sci div, Nat Res Coun, 63-66, mem maritime transp res bd, 68-71; mem merchant marine coun, US Coast Guard, 68-71; co-chmn panel on marine facil, US-Japan Natural Resources Prog, 69- Honors & Awards: Meritorious Civilian Serv Award, US Navy, 51 & Superior Civilian Serv Award, 69; Silver Medal for Meritorious Achievement, US Dept Transp, 74. Mem: Am Meteorol Soc; Am Asn Petrol Geologists; Soc Explor Geophysicists (vpres, 65-66); Am Geophys Union. Res: Forecasting wave and surf conditions; meteorological engineering; delta formation; sea-ice; detection underground nuclear explosions; global budget of influx of petroleum hydrocarbons into the ocean from all sources. Mailing Add: GDS/62 US Coast Guard Hq Washington DC 20590

BATES, CLAYTON WILSON, JR, b New York, NY, Sept 5, 32; m 68; c 2. SOLID STATE PHYSICS. Educ: Manhattan Col, BS, 54; Polytech Inst Brooklyn, MS, 56; Washington Univ, PhD(physics), 66. Prof Exp: Solid-state physicist, Sylvania Elec Prod, NY, 55-57; sr engr physics, Varian Assoc, 66-72; ASSOC PROF MAT SCI, STANFORD UNIV, 72- Concurrent Pos: Vis prof reader, Imperial Col Sci & Technol, 68; consult, Varian Assoc, 72-; mem panel optical physics, Nat Acad Sci, 74-; consult & part owner, Diagnostics Info Inc, 75-; consult, Spectra-Physics, Inc, 75- Mem: Am Phys Soc; Optical Soc Am; sr mem Inst Elec & Electronic Engr; Soc Photo-Optical Instrumentation Engr; Royal Photog Soc Gt Brit. Res: Electrical and optical properties of crystalline and amorphous solids and surfaces; photoelectric emission; luminescence phosphors; electronically conducted glasses; interaction of intense radiation with matter. Mailing Add: Dept of Mat Sci & Eng Stanford Univ Stanford CA 94305

BATES, DARYL STOKESBURY, bacterial physiology, biochemistry, see 12th edition

BATES, DAVID JAMES, b Portland, Ore, Oct 22, 28; m 49; c 5. APPLIED PHYSICS. Educ: Ore State Univ, BS, 51, MS, 53; Stanford Univ, PhD(appl physics), 58. Prof Exp: Res assoc, W W Hansen Labs, Stanford Univ, 52-57; sect head power tube dept, Hughes Aircraft Co, 57-62; sect mgr, Phys Electronics Labs, 62-63; sr eng mgr, Wave Tube Dept, Varian Assocs, 63-65; sect head medium power res & develop, 65-73, MGR, POWER TUBE ENG DEPT, WATKINS-JOHNSON CO, 73- Mem: Inst Elec & Electronics Engrs. Res: Microwave amplifiers; electron linear accelerators; electron bombarded semiconductor amplifiers. Mailing Add: 820 La Verne Way Los Altos CA 94022

BATES, DAVID MARTIN, b Everett, Mass, May 31, 34; m 56; c 2. TAXONOMIC BOTANY. Educ: Cornell Univ, BS, 59; Univ Calif, Los Angeles, PhD(bot), 63. Prof Exp: From asst prof to assoc prof, 63-75, PROF BOT, CORNELL UNIV, 75-, DIR, L H BAILEY HORTORIUM, 69- Mem: AAAS; Soc Econ Botanists; Am Soc Plant Taxonomists; Inst Asn Plant Taxonomists; Int Orgn Biosyst. Res: Taxonomic and cytotaxonomic studies in the family Malvaceae; taxonomy of cultivated plants; ethnobotany. Mailing Add: L H Bailey Hortorium Cornell Univ Ithaca NY 14853

BATES, DONALD GEORGE, b Windsor, Ont, May 18, 33; m 57; c 2. HISTORY OF MEDICINE. Educ: Univ Western Ont, MD, 58, BA, 60; Johns Hopkins Univ, PhD, 75. Prof Exp: From instr to asst prof hist of med, Johns Hopkins Univ, 62-66; ASSOC PROF HIST OF MED & CHMN DEPT, McGILL UNIV, 66- Mem: Am Asn Hist Med; Hist Sci Soc; Soc Social Hist Med; Soc Med Anthrop. Mailing Add: Dept of Hist of Med McGill Univ Montreal PQ Can

BATES, FRANCIS LESLIE, b Marine City, Mich, May 17, 10. PHYSICAL CHEMISTRY. Educ: Univ Detroit, BS, 34, MS, 36; Iowa State Col, PhD(plant chem), 43. Prof Exp: Chemist, N J Schorn & Co, Detroit, 36-37; res asst, exp sta, Iowa State Col, 40-43; from instr to prof phys chem, Univ Detroit, 43-69; independent consult, 69-75; RETIRED. Mailing Add: 1171 SW 12th St Boca Raton FL 33432

BATES, GRACE ELIZABETH, b Albany, NY, Aug 13, 14. MATHEMATICS. Educ: Middlebury Col, BS, 35; Brown Univ, ScM, 38; Univ Ill, PhD(math), 46. Prof Exp: Teacher high sch, Vt, 35-36 & Pa, 38-43; instr math, Sweet Briar Col, 43-44; from instr to assoc prof, 46-56, PROF MATH, MT HOLYOKE COL, 56- Concurrent Pos: Mem, Comn Undergrad Prog Math, 70- Mem: Am Math Soc; Math Asn Am. Res: Modern algebra; free loops and nets and their generalizations; probability; mathematical statistics. Mailing Add: Dept of Math Mt Holyoke Col South Hadley MA 01075

BATES, HAROLD BRENNAN, JR, b Des Moines, Iowa, Feb 12, 35; m 64; c 3. ZOOLOGY. Educ: Drake Univ, Iowa, BA, 59, MA, 61; Iowa State Univ, PhD(zool), 70. Prof Exp: Instr biol, Burlington Col, 61-66; asst prof, 70-75, ASSOC PROF ZOOL, ALBANY STATE COL, 75- Mem: AAAS. Res: Effects of prostaglandin E(1) on liver polyploidy. Mailing Add: Dept of Biol Albany State Col Albany GA 31705

BATES, HARRY EUGENE, nonlinear optics, quantum electronics, see 12th edition

BATES, HENRY A, microbiology, see 12th edition

BATES, HOWARD FRANCIS, b Portland, Ore, May 27, 27; m 61. PHYSICS, ELECTRICAL ENGINEERING. Educ: Ore State Univ, BS, 50, MS, 56; Univ Alaska, PhD(geophys), 61. Prof Exp: Physicist, US Navy Electronics Lab, Calif, 51-52; physicist, Geophys Inst, Univ Alaska, 52-53, instr geophys, 57-61, assoc geophysicist, 61-62, assoc prof, 62-66; sr physicist, Stanford Res Inst, 66-70; PROF GEOPHYS, UNIV ALASKA, COLLEGE, 70-, PROF ELEC ENG, 75- Concurrent Pos: Mem comn III, Int Sci Radio Union, 62- Mem: Int Union Geod & Geophys; Int Asn Geomag & Aeronomy. Res: Aeronomy. Mailing Add: Geophys Inst Univ of Alaska College AK 99701

BATES, JAMES EDMUND, b Alliance, Ohio, Nov 19, 18; m 43; c 2. PHOTOGRAPHIC CHEMISTRY. Educ: Univ Pittsburgh, BS, 40. Prof Exp: Chemist, Res Dept, Ansco Div, Gen Aniline & Film Corp, 40-47, mgr processing develop lab, 47-60, prod res mgr black & white film, papers & chem, 60-63, assoc dir res develop, Photo & Reproduction Div, 63-65, res mgr black & white films, papers & chem, GAF Corp, 65-74, RES MGR COLOR PROD, GAF CORP, 74- Mem: Am Chem Soc; Soc Photog Scientists & Engrs; Royal Photog Soc Gt Brit. Res: Color films and their processing; black and white films and papers; emulsion research. Mailing Add: 10 Calgary Lane Binghamton NY 13901

BATES, JOHN BERTRAM, b Big Flats, NY, Sept 7, 14; m 44; c 2. COLLOID CHEMISTRY. Educ: Clarkson Col Technol, BS, 36; Rice Inst, MA, 39, PhD(colloid chem), 41. Prof Exp: Chemist, Corning Glass Works, NY, 36-37; res chemist, Sun Chem Corp, 41-42; Am Petrol Inst fel, Univ Mich, 42-44; leader, phys chem sect, Sun Chem Corp, 44-50, asst dir graphic arts labs, 50-56; vpres & tech dir, Ho-Par, Inc, 56-60; staff chemist, HRB-Singer, Inc, 60-62; TECH DIR, FLINT INK CORP, 62- Concurrent Pos: Mem steering comt, Nat Asn Printing Ink Mfrs Tech Inst, 75- Mem: AAAS; Am Chem Soc; Soc Rheol; Inter-Soc Color Coun; Soc Coating Technol. Res: Pigment and printing ink chemistry; x-ray diffraction; gas adsorption on pigments; rheology of pigment dispersions; iron cyanide chemistry; particle size methods; surface chemistry in pigment dispersions; spectrophotometry and spectrofluorometry. Mailing Add: Flint Ink Corp 25111 Glendale Detroit MI 48239

BATES, JOHN BRYANT, b Harlan, Ky, Mar 11, 42; m 63; c 2. CHEMICAL PHYSICS. Educ: Univ Ky, AB, 64, PhD(chem), 68. Prof Exp: Res assoc chem, Univ Md, College Park, 68-69; STAFF SCIENTIST, SOLID STATE DIV, OAK RIDGE NAT LAB, 69- Mem: Sigma Xi. Res: Solid state spectroscopy; vibrational spectroscopy, including experimental and theoretical aspects; structure and dynamics of inorganic and organic molecular crystals; radiation effects on insulation materials; matrix isolated species. Mailing Add: Solid State Div Oak Ridge Nat Lab PO Box X Oak Ridge TN 37830

BATES, JOSEPH H, b Little Rock, Ark, Sept 19, 33; m 55; c 4. MEDICINE, BACTERIOLOGY. Educ: Univ Ark, BS & MD, 57, MS, 63. Prof Exp: From instr to assoc prof med, 61-71, ASST PROF MICROBIOL, SCH MED, UNIV ARK, LITTLE ROCK, 63-, PROF MED, 71- Concurrent Pos: Fel infectious dis, 61-63; clin investr pulmonary dis, Vet Admin Hosp, Little Rock, 63-65, chief pulmonary & infectious dis serv, 65-67, chief med, 67- Mem: Am Thoracic Soc; Am Fedn Clin Res; Infectious Dis Soc Am. Res: Unclassified mycobacteria, mycobacteriophage and their interrelationships. Mailing Add: Vet Admin Hosp 300 E Roosevelt Rd Little Rock AR 72204

BATES, LLOYD M, b Westville, NS, Dec 25, 24; m 46; c 2. RADIOLOGICAL PHYSICS. Educ: Univ NB, BSc, 50; Univ Sask, MSc, 52; Johns Hopkins Univ, ScD(radiol physics), 66. Prof Exp: Physicist, Prov NB, 52-57; asst prof radiol, Sch Med, Univ Md, 57-59; from instr to asst prof radiol, Sch Pub Health, 59-74, radiation safety officer, Med Insts, 66-74, ASSOC PROF RADIOL SCI, SCH PUB HEALTH, JOHNS HOPKINS UNIV, 74- Concurrent Pos: On leave, sr staff physicist, Am Asn Physicists in Med, 74- Mem: Am Asn Physicists in Med; Am Col Radiol; Am Pub Health Asn; Health Physics Soc. Res: Evaluation of radiological imaging systems;

application of radioactive materials to physiological studies; design and development of radiotherapy apparatus; radiation control at medical installations. Mailing Add: AAPM Suite 307 6900 Wisconsin Ave Chevy Chase MD 20015

BATES, M NOBLE, b Schenectady, NY, Dec 19, 10; m 38; c 2. HISTOLOGY, EMBRYOLOGY. Educ: Hamilton Col, AB, 32; Oberlin Col, AM, 36; Cornell Univ, PhD(histol & embryol), 43. Prof Exp: Asst zool, Oberlin Col, 32-34, spec asst in chg zool mus, 34-35; asst histol & embryol, Med Col, Cornell Univ, 35-38, instr, 38-42; assoc, Jefferson Med Col, 42-49; asst prof, 49-57, ASSOC PROF ANAT, SCH MED, TEMPLE UNIV, 57- Concurrent Pos: Holbrook scholar, Marine Biol Lab, Woods Hole. Mem: AAAS; Am Asn Anatomists. Res: Nature of head glands of tropical fish; embryology of tongue; histogenesis of muscle; congenital malformations; preservation of embalmed anatomic material. Mailing Add: Dept of Anat Temple Univ Sch of Med Philadelphia PA 19140

BATES, MARGARET WESTBROOK, b Boston, Mass, Oct 5, 26. BIOCHEMISTRY, NUTRITION. Educ: Wellesley Col, BA, 48; Cornell Univ, MS, 50; Harvard Univ, DSc(nutrit), 54. Prof Exp: Fel nutrit, Harvard Univ, 54-55; USPHS fel med res, Univ Toronto, 55-57; from res assoc to res asst prof biochem & nutrit, 57-67, from res asst prof to res assoc prof, Lab Clin Sci, 67-73, RES ASSOC PROF PSYCHIAT, UNIV PITTSBURGH, 73- Mem: AAAS; Am Inst Nutrit; Am Chem Soc; Brit Biochem Soc. Res: Fat metabolism; ketone body metabolism; plasma transport of fatty acids. Mailing Add: Dept of Psychiat Univ of Pittsburgh Pittsburgh PA 15261

BATES, MARSTON, biology, deceased

BATES, PHILIP KNIGHT, b Cohasset, Mass, July 2, 02; m 29; c 3. RESEARCH ADMINISTRATION. Educ: Mass Inst Technol, BS, 42, PhD(bact), 29. Prof Exp: Asst bact, Mass Inst Technol, 24-25; teaching asst pub health, Med Sch, Tufts Col, 24-26; res assoc food bact, Mass Inst Technol, 27-28; head dept bact, Frigidaire Corp, Ohio, 28-32; res assoc, Mass Inst Technol, 33-35; bacteriologist, United Drug Co, Mass, 36-40; instr pharmacol, Sch Med, Boston Univ, 45-46; dir prod develop, Rexall Drug Co, 41-49; pres, Riker Labs, Inc, Calif, 49-52; gen mgr res, Carnation Co, 52-66; CONSULT, 67- Concurrent Pos: Mem gen comt foods & chmn comt dairy, oil & fat prod, Adv Bd Mil Personnel Supplies, Nat Acad Sci-Nat Res Coun, 65-69; ed, J Agr & Food Chem, 65- Mem: AAAS; Inst Food Technologists (treas, 45-53, pres, 54); Am Soc Microbiol; Am Chem Soc; Soc Chem Indust. Res: Administration of scientific laboratories; food product development. Mailing Add: 363 17th St Santa Monica CA 90402

BATES, RICHARD DOANE, JR, b Elizabeth, NJ, July 24, 44; m 71. PHYSICAL CHEMISTRY. Educ: Cornell Univ, BA, 66; Columbia Univ, MA, 67, PhD(chem), 71. Prof Exp: Preceptor chem, Columbia Univ, 67-68; ASST PROF CHEM, GEORGETOWN UNIV, 73- Mem: Sigma Xi; Am Chem Soc; Am Phys Soc; Am Asn Univ Profs. Res: Molecular dynamics of reacting and non-reacting chemical systems; vibrational energy transfer and the role of vibrational energy in chemical reactions; molecular interactions in solutions by magnetic double resonance. Mailing Add: Dept of Chem Georgetown Univ Washington DC 20057

BATES, RICHARD PIERCE, b Pennington, Tex, Jan 15, 26; m 52; c 1. AGRONOMY, PLANT BREEDING. Educ: Agr & Mech Col Tex, BS, 49, MS, 50; Univ Md, PhD(forage breeding), 53. Prof Exp: Technician, Seed Testing Lab, Agr & Mech Col Tex, 48-49, jr agronomist forage prod res, Substa 22, 50-51; asst agronomist, USDA, Md, 51-53; asst agronomist corn breeding & prod, Tex Substa 5, Agr & Mech Col Tex, 53-55; assoc plant breeder legume improv, 55-67, AGRONOMIST, SAMUEL ROBERS NOBLE FOUND, INC, 67- Mem: Am Soc Agron. Res: Forage crops, small grain and legume improvement and production. Mailing Add: Agr Div Samuel Roberts Noble Found Inc Ardmore OK 73401

BATES, ROBERT BROWN, b Huntington, NY, Dec 16, 33; m 55; c 3. ORGANIC CHEMISTRY. Educ: Rutgers Univ, BS, 54; Univ Wis, PhD(org chem), 57. Prof Exp: Res fel, Mass Inst Technol, 57-58; from instr to asst prof org chem, Univ Ill, 58-63; from asst prof to assoc prof , 63-69, PROF ORG CHEM, UNIV ARIZ, 69- Concurrent Pos: Sloan Found fel, 67-69. Mem: Am Chem Soc; Am Crystallog Asn. Res: Natural product structure, synthesis and biogenesis; carbanions. Mailing Add: 2212 N Frannea Dr Tucson AZ 85712

BATES, ROBERT ELLERY, b Providence, RI, Aug 21, 11; m 69. GEOMORPHOLOGY. Educ: Ind Univ, AB & AM, 32; Columbia Univ, PhD(geomorphol), 39. Prof Exp: Asst physiol, Columbia Univ, 34-36; instr geol, Western Reserve Univ, 36-39, asst prof geol & dean students, 39-41; assoc dean of men, Ind Univ, 41-45, dean of men, 45-46, asst dean students, 46-47; dir student affairs, Va Polytech Inst, 47-52; dean students, Colo State Univ, 52-62; DIR INSTNL STUDIES & PROF GEOL, CALIF STATE UNIV, HAYWARD, 63- Mailing Add: Dept of Geol Calif State Univ Hayward CA 94542

BATES, ROBERT GLENN, geology, deceased

BATES, ROBERT LATIMER, b Brookings, SDak, June 17, 12; m 35; c 2. ECONOMIC GEOLOGY. Educ: Cornell Univ, AB, 34; Univ Iowa, MS, 36, PhD(stratig), 38. Prof Exp: Asst, Dept Geol, Univ Iowa, 34-38; jr geologist, The Texas Co, 38-40; geologist, NMex Bur Mines, 41-45, chief oil & gas div, 45-47; from asst prof to assoc prof geol, Newark Col, Rutgers Univ, 47-51; assoc prof, 51-56, PROF GEOL, OHIO STATE UNIV, 56- Concurrent Pos: From instr to asst prof, NMex Sch Mines, 41-43; ed, J Nat Asn Geol Teachers, 60-64 & J Am Inst Prof Geologists, 69-70. Mem: Am Asn Petrol Geologists; fel Geol Soc Am; Nat Asn Geol Teachers (pres, 67-68); Am Inst Prof Geologists. Res: Economic geology of the nonmetallics. Mailing Add: Dept of Geol & Mineral Ohio State Univ Columbus OH 43210

BATES, ROBERT PARKER, food science, see 12th edition

BATES, ROBERT WESLEY, b Columbia, Iowa, Jan 31, 04; m 30; c 3. PHYSIOLOGICAL CHEMISTRY. Educ: Simpson Col, AB, 25; Univ Chicago, PhD(physiol chem), 31. Prof Exp: Asst prof chem, Simpson Col, 26-27; physiol chemist, E R Squibb & Sons, 27-29; res assoc, Sta for Exp Evolution, Carnegie Inst Technol, 31-32, investr, 33-41; biol chemist, Difco Labs, 41-42, foreman prod, 43-44; dept head, E R Squibb & Sons, 45-52; biochemist, Nat Inst Health, 52-75; RETIRED. Honors & Awards: Koch Award, Endocrine Soc. Mem: AAAS; Am Soc Biol Chem; Endocrine Soc; fel NY Acad Sci. Res: Chemistry and assay of enzymes; isolation and assay of hormones of the anterior pituitary, especially prolactin and thyrotropin; chemical methods for tryptophane and natural estrogens determination; hormonal induction of diabetes. Mailing Add: 1522 Mission Way Kino Springs AZ 85621

BATES, ROGER GORDON, b Cummington, Mass, May 20, 12; m 41; c 1. ANALYTICAL CHEMISTRY. Educ: Univ Mass, BS, 34; Duke Univ, AM, 36, PhD(phys chem), 37. Prof Exp: Asst chem, Duke Univ, 34-36; Sterling fel, Yale Univ, 37-39; chemist, Nat Bur Standards, 39-57, chief electrochem anal sect, 57-69, asst chief anal chem div, 58-67; PROF CHEM, UNIV FLA, 69- Concurrent Pos: Lectr, Trinity Col, DC, 47-49; USPHS fel, Univ Zurich, 53-54; chmn comn electrochem data, Int Union Pure & Appl Chem, 53-59, mem US nat comt, 59-65, comn electrochem, 59-67, comn symbols, terminology & units, 63-71, chmn comn electroanal chem, 71- Honors & Awards: Hillebrand Prize, 55; US Dept Com Except Serv Award, 57; Anal Chem Award, Am Chem Soc, 69. Mem: AAAS; Am Chem Soc; Electrochem Soc; Am Inst Chemists. Res: Electrode potentials; homogeneous equilibria; thermodynamics of electrolytic solutions; nonaqueous solutions; standardization of pH measurements, dissociation of weak acids and bases. Mailing Add: Dept of Chem Univ of Fla Gainesville FL 32611

BATES, THOMAS EDWARD, b Irricana, Alta, July 13, 26; m 53; c 2. SOIL SCIENCE. Educ: Ont Agr Col, BSA, 51; NC State Col, MS, 54; Iowa State Univ, PhD(soil fertil), 61. Prof Exp: Res officer agron, Tobacco Res Bd, Rhodesia & Nyasaland, 51-58; asst soil fertil, Iowa State Univ, 58-61; asst prof soil fertil, Ont Agr Col, 61-65; assoc prof soil fertil, 65-68, PROF SOIL SCI, UNIV GUELPH, 68- Concurrent Pos: Assoc ed, Soil Sci Soc Am J. Mem: Am Soc Agron; Agr Inst Can; Can Soil Sci Soc. Res: Soil fertility pertaining to fertilizer use and metal toxicities. Mailing Add: Dept of Land Resource Sci Univ of Guelph Guelph ON Can

BATES, THOMAS FULCHER, b Evanston, Ill, Jan 2, 17; m 42; c 3. MINERALOGY. Educ: Denison Univ, BA, 39; Columbia Univ, MS, 40, PhD(geol), 44. Hon Degrees: DSc, Denison Univ, 68. Prof Exp: From instr to assoc prof, 42-53, asst to vpres for res, 61-65, dir inst sci & eng, 63-65, asst dean grad sch, 64-65, vpres planning, 67-72, PROF MINERAL, PA STATE UNIV, 53- Concurrent Pos: Sci adv, Secy Interior, Washington, DC, 65-67. Mem: AAAS; fel Geol Soc Am; fel Mineral Soc Am; Am Geochem Soc. Res: X-ray and electron microscopic studies of clay and other minerals; origin of the Edwin clay, Ione, California; mineralogical investigation of fine-grained rocks; uraniferous black shales and lignites; clay minerals of Hawaii; application of earth sciences to land use and energy planning. Mailing Add: 212 Deike Bldg Pa State Univ University Park PA 16802

BATES, WILLIAM K, b Houston, Tex, Nov 16, 36; m 59; c 2. BIOCHEMISTRY. Educ: Rice Univ, BA, 59, PhD(biochem), 63. Prof Exp: USPHS trainee biol sci, Stanford Univ, 63-66; from asst prof to assoc prof, 66-74, PROF BIOL, UNIV NC, GREENSBORO, 74- Mem: AAAS; Am Chem Soc; Genetics Soc Am; Am Soc Microbiol. Res: Gene action; regulation in eucaryotic organisms; radiotracer applications; data processing and simulation of biological systems. Mailing Add: Dept of Biol Univ of NC Greensboro NC 27412

BATES, WILLIAM WANNAMAKER, JR, b Orangeburg, SC, Oct 12, 19; m 42; c 5. PHYSICAL CHEMISTRY. Educ: The Citadel, BS, 41; Duke Univ, PhD(phys chem), 51. Prof Exp: Res chemist, 50-53, asst to dir res, 53-58, assoc dir res, 58-64, DIR RES, RES CTR, LIGGETT & MYERS INC, 64- Mem: Am Chem Soc. Res: Dielectrics; analysis of chemical and physical properties of tobacco. Mailing Add: Res Ctr Liggett & Myers Inc Durham NC 27702

BATEY, HARRY HALLSTED, JR, b Grand Island, Nebr, Feb 1, 23; m 45; c 4. INORGANIC CHEMISTRY. Educ: Cornell Col, AB, 46; Ohio State Univ, MSc, 48, PhD(chem), 51. Prof Exp: From instr to assoc prof, 51-70, PROF CHEM, WASH STATE UNIV, 70- Concurrent Pos: UNESCO sci teaching expert, Iran, 68-69. Mem: AAAS; Am Chem Soc. Res: Inorganic nitrogen compounds; oxyhalides; gas chromatography of inorganic substances; non-aqueous solvents; science education. Mailing Add: Dept of Chem Wash State Univ Pullman WA 99163

BATH, DANIEL WHITE, cytology, genetics, see 12th edition

BATH, JAMES EDMOND, b Santa Ana, Calif, May 4, 38; m 58; c 3. ENTOMOLOGY, PLANT PATHOLOGY. Educ: Univ Calif, Davis, 60; Univ Wis, MS, 62, PhD(entom), 64. Prof Exp: From asst prof to assoc prof, 64-75, PROF ENTOM, MICH STATE UNIV, 75-, CHMN DEPT, 74- Mem: AAAS; Entom Soc Am; Am Phytopath Soc. Res: Mode of plant virus transmission by insects, especially circulative viruses and aphid vectors. Mailing Add: Dept of Entom Mich State Univ East Lansing MI 48824

BATHA, HOWARD DEAN, b Phillips, Wis, July 3, 25; m 48; c 4. PHYSICAL CHEMISTRY. Educ: Carroll Col, BA, 50; Univ Rochester, PhD, 54. Prof Exp: Group leader chem, Olin-Mathieson Chem Co, 54-56, sr res proj specialist, 56-58; res specialist, 58-63, mgr res br, 63-69, assoc dir res & develop, 69-74, DIR DEVELOP, CARBORUNDUM CO, 74- Mem: Am Chem Soc; Am Phys Soc; Electrochem Soc. Res: Free radical reactions; inorganic hydrides; high temperature reactions; mechanisms of crystal growth; silicon carbide. Mailing Add: Carborundum Co PO Box 337 Niagara Falls NY 14320

BATHER, ROY, b Eng, Mar 28, 25; m 50; c 2. VIROLOGY. Educ: Univ Toronto, BA, 49, MS, 50; Univ Edinburgh, PhD(biochem), 54. Prof Exp: Asst cancer res, Banting Inst, 49-52; asst, Cancer Campaign Unit, Poultry Res Ctr, Edinburgh, Scotland, 52-58; from asst prof to assoc prof cancer res, Sask Cancer & Med Res Ctr, Univ Sask, 58-70; sci adv, Drug Adv Bur, 70-71, RES SCIENTIST, VIRAL PROD DIV, BUR BIOLOGICS, DEPT NAT HEALTH & WELFARE, 71- Concurrent Pos: Brit Empire exchange fel, 54-55; Eleanor Roosevelt int cancer fel, Radium Inst, Orsay, France, 68-69. Mem: Tissue Cult Asn; Can Biochem Soc. Res: Biochemical approach to study of virus induced tumors; nucleic acids in tumor viruses and tissue culture; slow virus infections; role of measles virus in sub acute encephalitis. Mailing Add: Viral Prod Div Bur of Biologics Dept of Nat Health & Welfare Ottawa ON Can

BATHO, EDWARD HUBERT, b West Hoboken, NJ, Apr 13, 25. MATHEMATICS. Educ: Fordham Univ, BS, 50; Univ Wis, MS, 52, PhD(math), 55. Prof Exp: From instr to asst prof math, Univ Rochester, 54-59; assoc prof, 59-67, PROF MATH, UNIV NH, 67- Concurrent Pos: Mem sch math, Inst Advan Study, 58-59; NSF fel, 58-60; NSF fel, Sheffield Univ, 60; res fel math, Harvard Univ, 59-60. Mem: Am Math Soc; Math Asn Am. Res: Abstract algebra; ideal theory; structure of rings and algebras; lie algebras; algebraic geometry. Mailing Add: Dept of Math Univ of NH Durham NH 03824

BATHO, HAROLD FRANCIS, physics, deceased

BATKIN, STANLEY, b New York, NY, Nov 23, 12; m 52; c 3. NEUROLOGY, NEUROSURGERY. Educ: NY Univ, BS, 33; Univ Edinburgh, MD, 44; Am Bd Neurol Surg, dipl. Prof Exp: Clin tutor surg neurol, Univ Edinburgh, 44-46; vis neurologist, Dept Health, Scotland, 46-48; asst prof neurol, Med Sch, Univ Ark, 49-50; asst prof neurol & neurosurg, State Univ NY Upstate Med Ctr, 50-51, asst clin prof neurol & neurosurg & neurologist, Univ Hearing & Speech Ctr, 51-63; res assoc audiol, 63-66, res assoc physiol, 66-67, clin prof, 67-71, PROF SURG & PHYSIOL, SCH MED, UNIV HAWAII, MANOA, 71-, CHIEF DIV NEUROSURG, 67-, RES ASSOC DIV NEUROSURG ONCOL, CANCER CTR HAWAII, 73- Concurrent Pos: Chief neurosurg div, Vet Admin Hosp, Little Rock, Ark, 49-50; neurosurgeon, Syracuse Mem Hosp, Syracuse Univ Hosp, Couse-Irving Hosp, Syracuse Gen Hosp,

Midtown Hosp & Community Hosp of Syracuse, 51-63; neurol consult, NY State Educ Dept, Div Voc Rehab, 52-63; neurologist, Syracuse Cerebral Palsy Clin, Syracuse Univ Hearing & Speech Ctr, 52-63; lectr, Depts Spec Educ & Audiol, Syracuse Univ, 52-63; chief div neurol & neurosurg, Permanente Med Group, 63- Mem: AAAS; Am Acad Neurol; Asn Res Nerv & Ment Dis; Harvey Soc. Res: Neurophysiology; oncology. Mailing Add: Dept of Physiol Univ of Hawaii Sch of Med Honolulu HI 96822

BATLEY, FRANK, b Oldham, Eng, Dec 27, 20; m 46; c 3. RADIOLOGY. Educ: Univ Manchester, BSc, 41, MB, ChB, 44. Prof Exp: Registr radiation ther, Christie Hosp, Manchester, Eng, 45-51; asst, Ont Cancer Found, Hamilton, 52-56; assoc prof, Ohio State Univ Hosps, 56-57; dep dir, Ont Cancer Found, Kingston, 57-59; assoc prof radiol & chief radiation ther, State Univ NY Upstate Med Ctr, 59-67; PROF RADIOL & DIR RADIOTHER, OHIO STATE UNIV HOSPS, 67- Mem: Am Radium Soc; Am Col Radiol; Radiol Soc NAm. Res: Cancer therapy and statistical methods. Mailing Add: Ohio State Univ Hosp 410 W Tenth Ave Columbus OH 43210

BATLIN, ALEXANDER, b New York, NY, June 16, 14; m 43; c 2. BACTERIOLOGY. Educ: Johns Hopkins Univ, AB, 35; Univ Wis, MS, 49, PhD(bact), 55. Prof Exp: Mem Chem Corps, US Army, 41-62; CONSULT BACT & BIOCHEM, 62- Mem: AAAS; Am Chem Soc; Am Soc Microbiol; fel Am Acad Microbiol. Res: Bacterial physiology; intermediate metabolism; biomedical materials; water pollution. Mailing Add: 7601 Charleston Dr Bethesda MD 20034

BATOREWICZ, WADIM, b Poland, Nov 7, 34; US citizen; m 58. ORGANIC CHEMISTRY. Educ: Univ Minn, BCh, 60, PhD(org chem), 67. Prof Exp: Chemist, Minn Mining & Mfg Co, 60-62; SR RES CHEMIST, CHEM DIV, UNIROYAL, INC, 67- Mem: Am Chem Soc. Res: Chemistry of organophosphorus compounds; chemistry of polyurethanes; flame retardants for polymers; flammability of polymers. Mailing Add: Uniroyal Chem Div of Uniroyal Inc Elm St Naugatuck CT 06770

BATRA, GOPAL KRISHAN, b Lahore, Pakistan, Oct 10, 43; m 60; c 2. VIROLOGY, CANCER. Educ: Panjab Univ, India, BSc, 62, MSc, 66; Univ Ga, PhD(plant virol), 72. Prof Exp: Lectr biol, Govt Col, Mandi, India, 65-66; sr sci asst microbiol, Indian Drugs & Pharmaceut, Virbhadra, 66-68; res assoc virol, Univ Ga, 72-73; res assoc cell biol, 73-74, SR RES ASSOC VIROL, EMORY UNIV, 74- Mem: AAAS; Am Soc Microbiol; Soc Gen Microbiol; Sigma Xi. Res: Biochemical and biophysical properties of virus-infected cells; potential carcinogenesis of herpes and papova viral infection; viral latency and reactivation. Mailing Add: Dept of Pediat Sch of Med Emory Univ 69 Butler St SE Atlanta GA 30303

BATRA, INDER PAUL, b India, June 25, 42; m 70. SOLID STATE PHYSICS. Educ: Univ Delhi, BSc, 62, MSc, 64; Simon Fraser Univ, PhD(physics), 68. Prof Exp: Nat Res Coun Can fel, Simon Fraser Univ, 68-69; PROF SCIENTIST, INORG MAT SCI GROUP, RES LAB, IBM CORP, 69- Mem: Fel Am Phys Soc; Am Vacuum Soc. Res: Interband optical absorption; phonon assisted transitions in semiconductors; photoconductivity; stimulation scattering of laser light; theoretical studies of electronic transport; chemisorption and electronic structure of surfaces by molecular orbital techniques. Mailing Add: IBM Corp Res Lab Monterey &Cottle Rds San Jose CA 95193

BATRA, KARAM VIR, b Jhelum, Panjab, India, Jan 3, 32; US citizen; m 63; c 4. PHARMACOLOGY. Educ: Panjab Univ, India, BPharm, 52; Univ Minn, MS, 57; Univ Chicago, PhD(pharmacol), 63. Prof Exp: Gen trainee res mfg, Glaxo Labs Ltd, India, 52-54; med rep, Alamagamated Chem & Dyestuff Co Ltd, India, 54-55; res fel, Inst Med Res, Chicago Med Sch, 63-64; res biochemist, Vet Admin Hosp, Hines, Ill & Northwestern Univ, 64-66; res pharmacologist, IIT Res Inst, 66-68; from instr to asst prof pharmacol, Chicago Col Osteop Med, 68-73; mem staff, Haskell Lab Toxicol & Indust Med, E I du Pont de Nemours & Co, Inc, 73-75; ADJ ASSOC PROF PHARMACOL, UNIV DEL, 75- Concurrent Pos: Consult tumor res, Vet Hosp, Hines, 70- Mem: Am Soc Pharmacol & Exp Therapeut. Res: Metabolism and distribution of drugs; effect of drugs on cells in culture; radio-incorporation studies; biochemistry of carcinogenesis; chemical constitution and pharmacological-biological activity; lipoproteins; phytochemistry; cardiovascular pharmacology; preclinical investigations and screening; behavioral pharmacology and toxicology. Mailing Add: 386 Briar Lane Newark DE 19711

BATRA, LEKH RAJ, b Panjab, India, Nov 26, 29; nat US; m 60; c 2. MYCOLOGY. Educ: Panjab Univ, India, BSc, 52, MSc, 54; Cornell Univ, PhD(mycol), 58. Prof Exp: Res scholar mycol, Govt India, 54-55; lectr biol, Desh Bandhu Col, India, 55; instr plant path, Cornell Univ, 57-58; lectr biol, Swarthmore Col, 58-60; res assoc, Univ Kans, 60-62, from asst prof to assoc prof bot, 62-68; RES MYCOLOGIST, PLANT PROTECTION INST, AGR RES SERV, USDA, 68- Concurrent Pos: Directorate of Plant Quarantine, Govt India, 60; vis res fel, Fed Forest & Timber Mgt Res Ctr, Hamburg, Ger, 62. Mem: Mycol Soc Am; Bot Soc Am; NY Acad Sci. Res: Morphology and taxonomy Hemiascomycetes and Discomycetes; ambrosia fungi, cereal, cotton and sugar cane diseases; plant embryo culture. Mailing Add: Plant Protection Inst Agr Res Serv Plant Indust Sta Beltsville MD 20705

BATRA, PREM PARKASH, b Jhang Maghiana, India, Apr 1, 36; US citizen; m 61; c 1. BIOCHEMISTRY. Educ: Panjab Univ, India, BS, 55, MS, 58; Univ Ariz, PhD(agr chem), 61. Prof Exp: Res assoc biochem, Univ Ariz, 61-63; fels, Univ Utah, 63-64 & Johns Hopkins Univ, 64-65; biochemist, aging res lab & geront br NIH, Vet Admin Hosp, 65; from asst prof to prof biol, 65-75, PROF BIOL CHEM, SCH MED, WRIGHT STATE UNIV, 75- Concurrent Pos: NIH spec fel, Univ Ky, 73-74. Mem: Am Chem Soc; fel Am Soc Biol Chemists. Res: Biosynthesis of carotenoids in plants and microorganisms; mechanism of photosynthetic phosphorylation; enzymology of nucleic acids. Mailing Add: Dept of Biol Chem Wright State Univ Sch Med Dayton OH 45431

BATRA, SUZANNE WELLINGTON TUBBY, b New York, NY, Dec 15, 37; m 60; c 2. ENTOMOLOGY. Educ: Swarthmore Col, BA, 60; Univ Kans, PhD(entom), 64. Prof Exp: Res assoc entom, Univ Kans, 64-67; res, Utah State Univ, 68-69; consult, Apicult Res Br, 70-74, RES ENTOMOLOGIST, INSECT IDENTIFICATION & BENEFICIAL INSECT INTROD INST, AGR RES SERV, USDA, 74- Concurrent Pos: Grants, Sigma Xi & Am Philos Soc, 64-65; sr res officer, Punjab Agr Univ, 65; ed, J Kans Entom Soc, 66-67. Mem: AAAS; Entom Soc Am; Bee Res Asn; Weed Sci Soc Am. Res: Ecology, behavior of insects; evolution of social behavior; management of crop pollinators; insect-fungus symbioses; insect pathology; morphology and classification of Laboulbeniales, Discomycetes and Apoidea; biological control of weeds. Mailing Add: Beneficial Insect Introd Lab Bldg 417 Beltsville Agr Res Ctr USDA Beltsville MD 20705

BATSAKIS, JOHN G, b Petoskey, Mich, Aug 14, 29; m 57; c 3. MEDICINE, PATHOLOGY. Educ: Univ Mich, MD, 54. Prof Exp: Intern path, Univ Hosp, George Washington Univ, 54-55; resident, Univ Mich, 55-59; chief clin path, Walter Reed Gen Hosp, Washington, DC, 59-61; assoc pathologist, Bronson Methodist Hosp, Kalamazoo, Mich, 61-; from asst prof to assoc prof, 62-68, assoc dir clin labs, Univ Hosp, 64-70, PROF PATH, UNIV MICH, ANN ARBOR, 68-, CO-DIR CLIN LABS, UNIV HOSP, 70- Concurrent Pos: Consult, Vet Admin Hosp, Ann Arbor, 62- & Armed Forces, 72- Mem: Fel Col Am Path; Asn Mil Surg US; fel Am Col Physicians; Am Asn Clin Chemists; fel Am Soc Clin Pathologists. Res: Clinical chemistry; head and neck, gastrointestinal and thyroid pathology; clinical enzymology. Mailing Add: Dept of Path Univ of Mich Med Ctr Ann Arbor MI 48104

BATSEL, HENRY LEWIS, b Central City, Ky, Oct 15, 22; m 47; c 5. NEUROPHYSIOLOGY. Educ: Univ Ky, BS, 43, MS, 48; Vanderbilt Univ, PhD(anat), 55. Prof Exp: Physiologist, US Army Med Res Lab, 50-53, 55-61; instr anat, Vanderbilt Univ, 54-55; res assoc, 62-64, ADJ ASST PROF, UNIV CALIF, LOS ANGELES, 64-; RES PHYSIOLOGIST, VET ADMIN HOSP, LONG BEACH, 61- Concurrent Pos: Res scientist space med, Lockheed Missiles & Space Co, 64-69. Mem: Am Physiol Soc. Res: Temperature regulation; nervous regulation of respiration. Mailing Add: Med Res Prog 5901 E 7th St Vet Admin Hosp Long Beach CA 90801

BATSON, ALAN PERCY, b Birmingham, Eng, Sept 18, 32; m 57, 68; c 3. COMPUTER SCIENCE. Educ: Univ Birmingham, BSc, 53, PhD(physics), 56. Prof Exp: Fel, Univ Birmingham, 56-58; from instr to assoc prof physics, 58-67, PROF COMPUT SCI, UNIV VA, 67-, DIR COMPUT-SCI CTR, 62- Mem: Am Phys Soc; Asn Comput Mach. Res: Computer systems. Mailing Add: Comput-Sci Ctr Univ of Va Charlottesville VA 22903

BATSON, BLAIR EVERETT, b Hattiesburg, Miss, Oct 24, 20; m 54. PEDIATRICS. Educ: Vanderbilt Univ, BS, 41, MD, 44; Johns Hopkins Univ, MPH, 54. Prof Exp: Intern pediat, Vanderbilt Univ Hosp, 44-45; asst resident, Johns Hopkins Hosp, 45-46; asst resident, Vanderbilt Univ Hosp, 48-49, instr pediat, Sch Med, Vanderbilt Univ, 49-52; instr pediat, Med Sch & pub health adminr, Div Maternal & Child Health, Sch Hyg & Pub Health, Johns Hopkins Univ, 52-54, asst prof, 54-55; PROF PEDIAT & CHMN DEPT, SCH MED, UNIV MISS, 55- Concurrent Pos: Resident pediatrician, Vanderbilt Univ Hosp, 49-50; consult pediat & US Air Force examr, Am Bd Pediat. Mem: AMA; Am Acad Pediat; Am Pub Health Asn; Asn Am Med Cols; Am Pediat Soc. Res: Growth and development in children; immunization; handicapped children. Mailing Add: Univ of Miss Sch of Med Jackson MS 39216

BATSON, JACK DAVID, b Fairfield, Ala, Apr 9, 31. VERTEBRATE ECOLOGY. Educ: Univ Ala, BS, 53, MS, 57; Univ Ky, PhD(biol), 65. Prof Exp: Instr biol, Exten Div, Univ Ala, 58, Walker Jr Col, 58-61 & Birmingham-Southern Col, 61-62; asst prof, Delta State Col, 65-66 & Hiram Scott Col, 66-68; assoc prof, 68-75, PROF BIOL, GA COL MILLEDGEVILLE, 75- Mem: Nat Geog Soc; Sigma Xi. Res: Salamander taxonomy. Mailing Add: Dept of Biol Ga Col Milledgeville GA 31061

BATSON, LEWIS E, b Wiggins, Miss, Oct 31, 18; m 53; c 1. APPLIED MATHEMATICS. Educ: La State Univ, BS, 52, MS, 54; Univ Tex, PhD(math), 60. Prof Exp: Asst prof, 59-66, ASSOC PROF MATH, UNIV SOUTHWESTERN LA, 66- Mem: Math Asn Am. Res: Integral transforms. Mailing Add: Dept of Math Univ of Southwestern La Lafayette LA 70501

BATSON, MARGARET BAILLY, b New York, NY, July 13, 14; m 54. PSYCHIATRY, PEDIATRICS. Educ: Manhattanville Col, BA, 37; Columbia Univ, MA, 40, PhD(bact), 49; Univ Rochester, MD, 51. Prof Exp: Clin instr bact, Columbia Univ, 39-41; intern pediat, Johns Hopkins Univ Hosp, 51-52, asst resident, 53-54, instr, Sch Med, Univ & pediatrist-in-chg, Hosp, 54-55; asst resident pediat, Univ Minn, 52-53; asst prof, 55-59, ASSOC PROF PEDIAT, ASST PROF PSYCHIAT & CHIEF DIV HUMAN BEHAV, MED CTR, UNIV MISS, 59-, DIR INFANT & CHILD DEVELOP CLIN, 65-, ASST PROF MICROBIOL & ASSOC MEM GRAD FAC, 67- Res: Pediatric neurology; microbiology; immunology; human behavior. Mailing Add: Infant & Child Develop Clin Univ of Miss Med Ctr Jackson MS 39216

BATSON, OSCAR RANDOLPH, b Hattiesburg, Miss, Oct 26, 16; m 50; c 4. PEDIATRICS. Educ: Vanderbilt Univ, BA, 38, MD, 42; Am Bd Pediat, dipl. Prof Exp: From intern to resident pediat, Vanderbilt Hosp, 42-44, fel, 46-47; from instr to assoc prof, 47-59, actg dean sch med, 62-63, dean & dir med affairs, 63-72, vchancellor med affairs & vchancellor alumni & develop, 72-74, PROF PEDIAT, SCH MED, VANDERBILT UNIV, 59- Concurrent Pos: Mem med equip adv comt, NSF. Mem: Asn Am Med Cols; Am Acad Pediat; AMA; Am Pediat Soc; Asn Acad Health Ctrs. Res: Medical administration. Mailing Add: Vanderbilt Univ Sch of Med Nashville TN 37205

BATSON, WADE THOMAS, b Marietta, SC, May 7, 12; m 39; c 2. BOTANY. Educ: Furman Univ, BS, 34; Duke Univ, MA, 49, PhD, 52. Prof Exp: Teacher high sch, SC, 37-41, prin, 41-44; civil readjustment officer, 13th Naval Dist, 47; asst bot, Duke Univ, 48-50, instr, 50-52; asst prof biol, Univ SC, 52-54; asst chmn univ rels div, Oak Ridge Inst Nuclear Studies, 54-55; assoc prof, 55-59, actg head dept, 58-59, PROF BIOL, UNIV SC, 59- Mem: Am Bryol & Lichenological Soc. Res: Taxonomy and ecology. Mailing Add: Dept of Biol Univ of SC Columbia SC 29208

BATSON, WILLIAM EDWARD, JR, b Taylors, SC, Apr 3, 42; m 65; c 2. PLANT PATHOLOGY. Educ: Clemson Univ, BS, 65, MS, 66; Tex A&M Univ, PhD(plant path), 71. Prof Exp: Res assoc plant path, Tex A&M Univ, 67-71; ASST PROF PLANT PATH, MISS STATE UNIV, 71- Mem: Am Phytopath Soc. Res: Etiology and control of rhizoctonia soil rot of tomatoes. Mailing Add: Dept of Plant Path & Weed Sci Drawer PG Mississippi State MS 39762

BATT, CONRAD WILLIAM, b Buffalo, NY, May 25, 31. BIOCHEMISTRY. Educ: Cath Univ Am, BA, 54, MS, 56; Georgetown Univ, PhD(biochem), 59. Prof Exp: Instr chem, La Salle Acad, RI, 54-55 & De La Salle Col, Washington, DC, 55-59; from instr to assoc prof, 59-71, head dept, 64-71, PROF CHEM, MANHATTAN COL, 71- Concurrent Pos: NIH fel, Brookhaven Nat Lab, 63-64. Mem: AAAS; NY Acad Sci; Am Chem Soc; Am Inst Chemists. Res: Physical aspects of enzyme catalyzed reactions; kinetics and mechanisms of enzyme substrate interaction; thrombin. Mailing Add: Dept of Chem Manhattan Col Bronx NY 10471

BATT, ELLEN RAE, b New York, NY, Sept 24, 34. PHYSIOLOGY. Educ: Columbia Univ, AB, 56, MA, 59, PhD(zool), 67. Prof Exp: NIH fel, 67-70, res assoc, 70-73, ASST PROF PHYSIOL, COLUMBIA UNIV, 73- Mem: Am Soc Zoologists. Res: Transport through red blood cell, bacterial, and intestinal membranes; postnatal development of rodent intestine with respect to transport. Mailing Add: Dept of Physiol Columbia Univ New York NY 10032

BATT, RUSSELL HOWARD, b Worcester, Mass, July 27, 38; m 60; c 2. PHYSICAL CHEMISTRY. Educ: Univ Rochester, BS, 60; Univ Calif, Berkeley, PhD(chem), 65. Prof Exp: Asst prof chem, Wesleyan Univ, 64-66; res instr, Dartmouth Col, 66-68; asst prof, 68-73, ASSOC PROF CHEM, KENYON COL, 73- Mem: Am Inst Chemists. Res: Photoconductivity in molecular crystals; low temperature calorimetry; biology and chemistry of trace heavy metal environmental pollutants. Mailing Add: Dept of Chem Kenyon Col Gambier OH 43022

BATT, WILLIAM GEORGE, b Eng, Dec 1, 06; nat US; m 39, 61; c 12. BIOCHEMISTRY. Educ: Philadelphia Col Pharm, BS, 34, MS, 35, DSc(microchem), 38; Univ Del, BA, 60; Am Bd Clin Chem, dipl. Prof Exp: Asst microanalyst, Biochem Res Found, Franklin Inst, 37-40, head microanal lab, 40-53, asst dir, 53-55, dir, 55-66; prof, 66-73, EMER PROF CHEM, DEL STATE COL, 73- Concurrent Pos: Vis lectr, Wesley Col, Del, 74- Mem: AAAS; Am Phys Soc; Electrochem Soc; Sigma Xi; Am Chem Soc. Res: Microchemical procedures; chemical separation of radioisotopes; comparative study of microextraction procedures; x-ray induced ultraviolet radiation; biological effects of external radiation; enzyme studies; chemical studies ascites carcinoma cells. Mailing Add: 21 Townsend Rd Newark DE 19711

BATTAGLIA, CHARLES JOSEPH, physical chemistry, see 12th edition

BATTAGLIA, FREDERICK CAMILLO, b Weehawken, NJ, Feb 15, 32; m; c 2. PHYSIOLOGY, PEDIATRICS. Educ: Cornell Univ, BA, 53; Yale Univ, MD, 57. Prof Exp: USPHS fel biochem, Univ Cambridge, 59; Josiah Macy Found fel physiol, Sch Med, Yale Univ, 59-60; resident pediat, Johns Hopkins Univ Hosp, 60-62; mem lab perinatal physiol, NIH, PR, 62-64; asst prof pediat, Johns Hopkins Univ, 64-65; assoc prof, 65-69, dir div perinatal med, 70-74, PROF PEDIAT, OBSTET & GYNEC, MED CTR, UNIV COLO, DENVER, 69-, CHMN DEPT PEDIAT, 74- Concurrent Pos: Attend pediatrician, Colo Gen Hosp, 65- Mem: Soc Pediat Res; Soc Gynec Invest; Perinatal Res Soc; Am Acad Pediat. Res: Perinatal physiology; intrauterine growth retardation; biochemistry. Mailing Add: Dept of Pediat Univ of Colo Med Ctr Denver CO 80220

BATTAGLIA, SAMUEL THOMAS, information science, applied statistics, see 12th edition

BATTAILE, JULIAN, b Sept 26, 25; US citizen; m 58; c 2. BIOCHEMISTRY. Educ: La State Univ, BS, 47; Univ Ill, MS, 48; Ore State Univ, PhD(biochem), 60. Prof Exp: Instr chem, Southeast La Col, 47; asst prof, Southwestern La Inst, 48-50; chemist, Ethyl Corp, 51-54; instr chem, Ore State Univ, 57-58; NIH fel, Univ Calif, Davis, 60-62; from asst prof to assoc prof, 62-69, PROF CHEM, SOUTHERN ORE COL, 69- Mem: Am Chem Soc. Res: Biosynthesis of monoterpenes. Mailing Add: Dept of Chem Southern Ore Col Ashland OR 97520

BATTAN, LOUIS JOSEPH, b New York, NY, Feb 9, 23; m 52; c 2. METEOROLOGY. Educ: NY Univ, BS, 46; Univ Chicago, MS, 49, PhD(meteorol), 53. Prof Exp: Radar meteorologist, US Weather Bur, 47-51; res meteorologist, Univ Chicago, 51-58; assoc dir, Inst Atmospheric Physics, 58-73, PROF ATMOSPHERIC SCI, UNIV ARIZ, 58-, DIR, INST ATMOSPHERIC PHYSICS, 73- Concurrent Pos: Consult, US Weather Bur, NSF, US Air Force & US Army; mem comt atmospheric sci, Nat Acad Sci, 66-; mem US deleg, Cong World Meteorol Orgn, 67; mem US nat comt, Int Union Geod & Geophys. Honors & Awards: Meisinger Award, Am Meteorol Soc, Brooks Award, 71 & Second Half Century Award, 74. Mem: Fel AAAS (secy, Sect Atmospheric & Hydrospheric Sci); fel Am Meteorol Soc (pres, 66-67); fel Am Geophys Union; Nat Asn Sci Writers; Royal Meteorol Soc. Res: Radar meteorology; cloud physics; mesometeorology; tornadoes; weather modification. Mailing Add: Inst of Atmospheric Physics Univ of Ariz Tucson AZ 85721

BATTARBEE, HAROLD DOUGLAS, b Highlands, Tex, July 25, 40; m 65; c 1. ENDOCRINOLOGY, CARDIOVASCULAR PHYSIOLOGY. Educ: Univ Houston, BS, 66; Baylor Col Med, PhD(physiol), 71. Prof Exp: Instr, 71-72, ASST PROF PHYSIOL, SCH MED, LA STATE UNIV, SHREVEPORT, 72-, ASSOC MED, GRAD FAC, 73- Concurrent Pos: Edward Stiles res grant, La State Univ Sch Med, 73-74, La Heart Asn sr res grant-in-aid, 74-75; mem undergrad res comt, La Heart Asn, 74- Res: Thyroid physiology; liver carbohydrate metabolism; aldosterone secretion; hypertension. Mailing Add: Dept of Physiol PO Box 3932 La State Univ Med Ctr Shreveport LA 71130

BATTE, EDWARD G, b Ferris, Tex, Feb 4, 21; m 42; c 1. VETERINARY PARASITOLOGY. Educ: Agr & Mech Col Tex, BS, 42, MS, 49, DVM, 49. Prof Exp: Assoc parasitologist, Agr Exp Sta, Univ Fla, 49-51; res veterinarian, Calif Spray Chem Corp, 51-56; PROF VET SCI, NC STATE UNIV, 56- Mem: Am Vet Med Asn; Am Soc Parasitologists; Am Col Vet Toxicologists. Mailing Add: Dept of Vet Sci NC State Univ Raleigh NC 27607

BATTEN, ALAN HENRY, b Tankerton, Eng, Jan 21, 33; m 60; c 2. ASTRONOMY. Educ: Univ St Andrews, BSc, 55, DSc, 74; Univ Manchester, PhD(close binary systs), 58. Prof Exp: Res assoc astron, Univ Manchester, 58-61; sci officer, 61-70, ASSOC RES OFFICER, DOM ASTROPHYS OBSERV, 70- Concurrent Pos: Nat Res Coun Can fel, 59-61; lectr, Univ Victoria, 61-64; mem, Int Astron Union, 61-, Can nat comt, 64-70, organizing comts, comns 30-42, 67-, vpres comn 30, 73-; on leave, Vatican Observ, Castel Gandolfo, Italy, spring & summer, 70; guest investr, Inst Astron & Space Physics, Buenos Aires, 72. Mem: Am Astron Soc; Royal Astron Soc Can (vpres, 73-); fel Royal Astron Soc; Can Astron Soc (pres, 71-73). Res: Spectroscopic and photometric studies of close binary systems; radial velocities of stars. Mailing Add: Dom Astrophys Observ 5071 W Saanich Rd Victoria BC Can

BATTEN, CHARLES FRANCIS, b Manchester, NH, Apr 15, 42. PHYSICAL CHEMISTRY. Educ: St Anselms Col, AB, 64; Univ Notre Dame, MS, 66, PhD(chem), 71. Prof Exp: Fel chem, Univ Houston, 71-72, asst prof, 72-73; RES ASSOC CHEM, UNIV NEBR-LINCOLN, 75- Mem: Sigma Xi; Am Chem Soc; Am Soc Mass Spectrometry. Res: Photoelectron spectroscopy of organic and atmospheric molecules; photoionization mass spectrometry; energetics of ion reactions; analytic detectors based on photoionization. Mailing Add: Dept of Chem Univ of Nebr Lincoln NE 68588

BATTEN, EDMUND STANLEY, b Liberal, Kans, Aug 29, 32; m 61; c 3. METEOROLOGY, CLIMATOLOGY. Educ: Univ Calif, Los Angeles, BA, 58, MA, 61, PhD, 70. Prof Exp: Phys scientist, Dept Geophys & Astron, 59-69, PHYS SCIENTIST, PHYS SCI DEPT, RAND CORP, 69- Concurrent Pos: Mem comt exten of standard atmosphere, Task Group VI, 62; assoc prof geog, Calif State Univ, Northridge, 74- Mem: AAAS; Am Meteorol Soc; Am Geophys Union. Res: Structure and dynamics of the stratosphere, mesosphere, thermosphere and ionosphere; numerical modeling of the atmospheric general circulation; climatic dynamics. Mailing Add: Phys Sci Dept Rand Corp 1700 Main St Santa Monica CA 90406

BATTEN, GEORGE WASHINGTON, JR, b Houston, Tex, Sept 4, 37; m 61; c 1. MATHEMATICS. Educ: Rice Univ, BA, 59, MA, 61, PhD(math), 63. Prof Exp: Res assoc math, Univ Ill, 63-64; asst biomathematician, Univ Tex M D Anderson Hosp & Tumor Inst, 64-65, asst prof biomath, 65-66; ASST PROF MATH, UNIV HOUSTON, 66- Concurrent Pos: Vis lectr math, Rice Univ, 66. Mem: Am Math Soc; Soc Indust & Appl Math; Math Asn Am. Res: Probability theory; partial differential equations; numerical analysis. Mailing Add: Dept of Math Univ of Houston Houston TX 77004

BATTEN, ROGER LYMAN, b Hammond, Ind, June 22, 23. INVERTEBRATE PALEONTOLOGY. Educ: Univ Wyo, BA, 48; Columbia Univ, PhD(geol), 55. Prof Exp: Geologist, US Geol Surv, 54-55; from asst prof to assoc prof geol, Univ Wis, 55-62; assoc prof, 62-68, PROF GEOL, COLUMBIA UNIV, 68-, CUR, AM MUS NATURAL HIST, 67- Concurrent Pos: Assoc cur, Am Mus Natural Hist, 62-67. Mem: Paleont Soc; Syst Zool Soc; Soc Study Evolution. Res: Evolution and ecology of the primitive gastropods; Permian stratigraphy; recent marine ecology; population analysis. Mailing Add: Am Mus of Natural Hist New York NY 10024

BATTERMAN, BORIS WILLIAM, b New York, NY, Aug 25, 30; m 53; c 3. X-RAY CRYSTALLOGRAPHY, EXPERIMENTAL SOLID STATE PHYSICS. Educ: Mass Inst Technol, BS, 52, PhD(physics). Prof Exp: Elec res electrostatic generators, Nat Bur Standards, 51; lab asst soil solidification, Mass Inst Technol, 51-52, asst physics, 52-53; mem tech staff, Bell Tel Labs, 56-65; assoc prof, 65-68, PROF MAT SCI & APPL PHYSICS, CORNELL UNIV, 68-, DIR SCH APPL & ENG PHYSICS, 74- Concurrent Pos: Guggenheim & Fulbright-Hays fels, 71. Mem: AAAS; fel Am Phys Soc; Am Crystallog Asn. Res: X-ray and neutron diffraction applied to solid state physics problems; studies of thermal vibrations in crystals. Mailing Add: Sch of Appl & Eng Physics Cornell Univ Ithaca NY 14853

BATTERMAN, ROBERT COLEMAN, b Brooklyn, NY, Apr 12, 11; m 47; c 3. PHARMACOLOGY. Educ: NY Univ, BS, 31; MD, 35. Prof Exp: Instr therapeut, NY Univ, 39-50, asst prof med, 49-50; assoc prof physiol, pharmacol & med & chief arthritis sect, New York Med Col, Flower & Fifth Ave Hosps, 50-59; DIR & PRES, CLIN PHARMACOL RES INST, 59- Mem: Am Soc Pharmacol & Exp Therapeut; Am Chem Soc; Am Soc Clin Invest; Am Col Physicians; Soc Exp Biol & Med. Res: Clinical pharmacology; arthritis; cardiovascular disease. Mailing Add: Clin Pharmacol Res Inst 2123 Addison St Berkeley CA 94704

BATTERSBY, HAROLD RONALD ERIC, b Guildford, Eng, Nov 16, 22; US citizen; m 44. ANTHROPOLOGY, ANTHROPOLOGICAL LINGUISTICS. Educ: Univ Toronto, BA, 60; Ind Univ, PhD, 69. Prof Exp: ASSOC PROF ANTHROP, STATE UNIV NY COL GENESEO, 70- Mem: Fel Royal Asiatic Soc Gt Brit & Ireland; fel Royal Anthrop Inst Gt Brit & Ireland; fel Am Anthrop Asn; fel Royal Cent Asian Soc; fel Ling Soc Am. Res: Altaic cultures and linguistics; Uralic and Altaic studies. Mailing Add: Dept of Anthrop Blake D114 State Univ NY Col at Geneseo Geneseo NY 14454

BATTERSHELL, ROBERT DEAN, b Dover, Ohio, Jan 23, 31; m 58; c 4. PESTICIDE CHEMISTRY. Educ: Bowling Green State Univ, BS, 52; Purdue Univ, PhD, 60. Prof Exp: Chemist, Callery Chem Co, 52, 54-55; sr res chemist, Diamond Alkali Co, 59-68, GROUP LEADER, DIAMOND SHAMROCK CORP, 68- Mem: Am Chem Soc. Res: Synthesis of organic pesticides; chemistry of aliphatic and aromatic fluorine compounds; Hansch pi, rho, sigma analysis of chemically induced biological response. Mailing Add: Diamond Shamrock Corp T R Evans Res Ctr Box 348 Painesville OH 44077

BATTERTON, JOHN CLYDE, phycology, see 12th edition

BATTEY, JAMES F, nuclear physics, see 12th edition

BATTI, MARIO ALEX, b Bernardsville, NJ, May 10, 15; m 42; c 3. BOTANY. Educ: DePauw Univ, AB, 39. Prof Exp: Lab technician, 44-52, from asst res microbiologist to res microbiologist, 52-71,xSRxRESxSCIENTIS Miles Chem Co, 52-71, SR RES SCIENTIST, MILES LABS, INC, 71- Mem: Am Soc Microbiol; Soc Indust Microbiol. Res: Production of organic acids by fermentation, namely citric and itaconic acids. Mailing Add: Biosynthesis Res Lab Miles Labs Inc 1127 Myrtle St Elkhart IN 46514

BATTIFORA, HECTOR A, b Peru, Dec 11, 30; US citizen; m 59; c 3. PATHOLOGY. Educ: San Marcos Univ, Lima, BS, 50, MD, 57. Prof Exp: Assoc pathologist, Presby St Lukes Hosp, Chicago, 64-69; PROF PATH, MED SCH, NORTHWESTERN UNIV, CHICAGO, 69- Mem: Am Asn Path & Bact; Am Soc Exp Path; Soc Exp Biol & Med; Int Acad Path. Res: Experimental oncology; ultrastructure of human tumors and experimental glomerulonephritis. Mailing Add: Northwestern Univ Med Sch Chicago IL 60611

BATTIGELLI, MARIO C, b Florence, Italy, Dec 18, 27; US citizen; m 58; c 3. INDUSTRIAL MEDICINE. Educ: Univ Florence, MD, 51; Univ Pittsburgh, MPH, 57. Prof Exp: Resident indust med, Univ Milan, 52-53, instr, 54-55, assoc, 57-58; res assoc, Univ Pittsburgh, 59-61, asst prof occup med, 62-66; assoc prof, 66-69, assoc prof med, 69-74, PROF MED, SCH MED, UNIV NC, CHAPEL HILL, 74- Concurrent Pos: USPHS res grant, 61- Mem: AMA; Am Thoracic Soc; Indust Med Asn. Res: Environmental health; industrial toxicology; chest diseases; pulmonary physiology; inhalation experimental toxicology. Mailing Add: Univ of NC Sch of Med Chapel Hill NC 27515

BATTIN, RICHARD HORACE, mathematics, see 12th edition

BATTIN, WILLIAM T, b Hackensack, NJ, Aug 21, 27; m 51; c 3. ZOOLOGY. Educ: Swarthmore Col, BA, 50; Univ Minn, PhD(zool), 56. Prof Exp: From instr to asst prof biol, Wesleyan Univ, 55-58; assoc prof, Simpson Col, 58-59; from asst prof to prof, 59-74, DISTINGUISHED PROF BIOL, STATE UNIV NY BINGHAMTON, 74- Mem: AAAS; Am Soc Zool; Am Inst Biol Sci; Nat Asn Biol Teacher; Nat Sci Teachers Asn. Res: Physiology of nucleus and nuclear membrane; cytoplasmic DNA of amphibian oocytes; humanistic biology. Mailing Add: Dept of Biol State Univ of NY Binghamton NY 13901

BATTINO, RUBIN, b New York, NY, June 22, 31; m 60. PHYSICAL CHEMISTRY. Educ: Duke Univ, MA, 54, PhD(chem), 57. Prof Exp: Res chemist, Leeds & Northrup Co, 56-57; from instr to asst prof chem, Ill Inst Technol, 57-66; assoc prof, 66-69, PROF CHEM, WRIGHT STATE UNIV, 69- Concurrent Pos: Sect ed, Chem Abstr, 63- Mem: AAAS; Am Chem Soc. Res: Thermodynamics of solutions of nonelectrolytes; gas solubilities. Mailing Add: Dept of Chem Wright State Univ Dayton OH 45431

BATTIST, LEWIS, radiological physics, nuclear science, see 12th edition

BATTISTA, ARTHUR FRANCIS, b Can, Sept 7, 20; m 64. NEUROSURGERY, NEUROPHYSIOLOGY. Educ: McGill Univ, BSc, 43, MD, CM, 44; Univ Western Ont, MSc, 57; Am Bd Neurol Surg, dipl, 58; Hunter Col, MA, 69. Prof Exp: Intern, Royal Victoria Hosp, Montreal, 44-45; resident, NY Neurol Inst, 50-55; from instr to clin assoc prof, 56-64, assoc prof, 64-71, PROF NEUROSURG, MED SCH, NY UNIV, 71- Concurrent Pos: Am Physiol Soc Porter fel, Harvard Med Sch, 47-48, Milton fel, 48-49; from asst attend to assoc attend, NY Univ Hosp, 56-67, attend, 67-; asst vis surgeon, Bellevue Hosp, 57-; consult, Pub Health Marine Hosp, Staten Island, NY, 57- & Beekman-Downtown Hosp, 58-; attend, Vet Admin Hosp, New York, 58- Mem: Cong Neurol Surg; Am Asn Neurol Surg; Am Philos Asn; Asn Res Nervous &

Ment Dis; NY Acad Sci. Res: Neurophysiology of dyskinesias, hypothermia and pain; clinical neurosurgical practice. Mailing Add: NY Univ Med Sch New York NY 10016

BATTISTA, GUIDO WILLIAM, pharmacology, see 12th edition

BATTISTA, ORLANDO ALOYSIUS, b Cornwall, Ont, June 20, 17; nat US; m 45; c 2. CHEMISTRY. Educ: McGill Univ, BSc, 40. Hon Degrees: ScD, St Vincent Col, 55. Prof Exp: Res chemist, Am Viscose Corp, 40-51, sr res chemist, 52-53, leader pulping anal lab, 54, head anal group, 55-58, leader spec prod group, 58-59, spec prod sect, 59-60 & corp appl res sect, 60-61, mgr corp appl res dept, 61-63, asst dir res, Am Viscose Div, FMC Corp, 63, mgr interdisciplinary res, Chem Res & Develop Ctr, 63-65, asst dir cent res dept, 65-70; vpres sci & technol, Avicon, Inc, 70-74; CHMN & PRES, RES SERV CORP, 74- Concurrent Pos: Spec consult, Avicon, Inc, 74-; adj prof chem, Univ Tex, Arlington, 75- Honors & Awards: Chem Pioneer Ward, Am Inst Chemists, 69; Capt of Achievement, Am Acad Achievement, 71; James T Grady Award, Am Chem Soc, 73. Mem: Am Inst Chemists (pres-elect, 75-76, pres, 77); Am Chem Soc; Nat Asn Sci Writers; Am Med Writers Asn; fel NY Acad Sci. Res: Molecular weight of celluose; fine structure of cellulose; oxidation, acid hydrolysis and fractionation of cellulose; heat extruding viscose; washable crepe fabrics; waterproofing cellulose articles; molding hydroplastics; characterization and evaluation of wood pulps; colloidal macromolecular phenomena; microcrystal polymer science. Mailing Add: Res Serv Corp 5280 Trail Lake Dr Ft Worth TX 76133

BATTISTA, SAM P, b Orange, NJ, June 22, 24; m 50; c 3. PHARMACOLOGY, PHYSIOLOGY. Educ: St John's Univ, NY, BS, 50; Univ Wis, MS, 57; Boston Univ, PhD(pharmacol), 68. Prof Exp: Res chemist, Merck & Co, Inc, 51-53; CONSULT PHARMACOLOGIST, ARTHUR D LITTLE, INC, 58- Mem: Soc Cosmetic Chemists; NY Acad Sci; Am Asn Med Instrumentation; Am Soc Pharmacol & Exp Therapeut; Am Soc Testing & Mat. Res: Effects of pharmacological agents on ciliary function. Mailing Add: Arthur D Little Inc 35 Acorn Park Cambridge MA 02140

BATTISTE, MERLE ANDREW, b Mobile, Ala, July 22, 33; m 60; c 2. ORGANIC CHEMISTRY. Educ: The Citadel, BS, 54; La State Univ, MS, 56; Columbia Univ, PhD(org chem), 59. Prof Exp: Res assoc chem, Univ Calif, Los Angeles, 59-60; from asst prof to assoc prof, 61-70, PROF CHEM, UNIV FLA, 70-, CHMN ORG DIV, DEPT CHEM, 74- Concurrent Pos: Alfred P Sloan res fel, 67-69; Fulbright-Hays sr res scholar, Ger, 73-74. Mem: Am Chem Soc; The Chem Soc. Res: Non-benzenoid aromatic chemistry; multicharged aromatic ions; small-ring compounds; reaction mechanisms; stereochemistry and rearrangements of bridged polycyclic systems; organometallic chemistry; photochemical transformations. Mailing Add: Dept of Chem Univ of Fla Gainesville FL 32611

BATTISTO, JACK RICHARD, b Niagara Falls, NY, Sept 13, 22; m 50; c 2. IMMUNOLOGY, MICROBIOLOGY. Educ: Cornell Univ, BS, 49; Univ Mich, MS, 50, PhD(bact), 53. Prof Exp: Vis investr immunol, Rockefeller Inst Technol, 53-55; asst prof microbiol, Sch Med, Univ Ark, 55-57; Pop Coun sr fel, Weizmann Inst Sci, Israel, 63-64; from asst prof to prof microbiol & immunol, Albert Einstein Col Med, 67-74; SCI DIR, DEPT IMMUNOL, CLEVELAND CLIN FOUND, 74- Concurrent Pos: NIH fel, 53-55. Mem: AAAS; Am Asn Immunologists; Harvey Soc; NY Acad Sci. Res: Hypersensitivities to simple chemical compounds; immunological unresponsiveness to haptenes; naturally occurring delayed type iso-hypersensitivity; inherited and acquired serum differences; splenic influence on lymphoid tissue; regulation of immunological responses. Mailing Add: Dept of Immunol Res Div Cleveland Clin Found Cleveland OH 44106

BATTISTONE, GINO CHARLES, biochemistry, analytical chemistry, see 12th edition

BATTLE, ED LEN, b Wilton, Ala, Feb 22, 31; m 54; c 2. STATISTICAL MECHANICS, PLASMA PHYSICS. Educ: Auburn Univ, BS, 54; US Air Force Inst Technol, MS, 60; Univ Toronto, PhD(plasma physics), 67. Prof Exp: Lab technician, Coosa River Newprint Co, Ala, 49-50; assoc aircraft engr, Lockheed Aircraft Co, Ga, 54-55; US AIR FORCE, 55-, staff asst aircraft develop, Syst Prog Off, 56-58, mem staff, Ballistic Syst Div/Space Syst Div, Los Angeles & San Bernardino, 60-63, asst prof physics, Malstron AFB, Mont, 66-67, from asst prof to assoc prof, US Air Force Inst Technol, 67-71, dep head dept, 70-71, dir gen purpose & airlift aircraft studies, Hq, Pentagon, 71-75, DEP CHIEF STAFF/PLANS, HQ, FIRST STRATEGIC AIR DIV, 75- Mem: Am Phys Soc. Res: Development of plasma kinetic equations using many-body theory to reduce the Liouville equation; systems analysis. Mailing Add: First Strategic Air Div/XP Vandenberg AFB CA 93437

BATTLE, HELEN IRENE, b London, Ont, Aug 31, 03. DEVELOPMENTAL BIOLOGY, HUMAN GENETICS. Educ: Univ Western Ont, BA, 23, MA, 24; Univ Toronto, PhD(marine biol), 28. Hon Degrees: LLD, Univ Western Ont & DSc, Carleton Univ, 71. Prof Exp: Demonstr, 23-24, from instr to prof, 24-72, EMER PROF ZOOL, UNIV WESTERN ONT, 72- Honors & Awards: Centennial Medal, Can, 67. Mem: Hon mem Can Soc Zoologists; hon mem Nat Asn Biol Teachers; Can Asn Anatomists; Can Physiol Soc; Genetics Soc Can. Res: MacKinder's hereditary brachydactyly; fine structure trout blastoderm. Mailing Add: Dept of Zool Univ of Western Ont London ON Can

BATTLE, WARREN RICH, b Ellisburg, NJ, May 27, 19; m 44; c 4. AGRONOMY. Educ: Rutgers Univ, BS, 41, PhD(agron), 51. Prof Exp: Asst, Kans State Univ, 41-42; high sch instr, NJ, 44-47; instr farm crops, 47-48, res assoc, 48-51, from asst res specialist to assoc res specialist, 51-59, chmn dept farm crops, 61-63, chmn dept soils & crops, 63-71, PROF & RES SPECIALIST, RUTGERS UNIV, NEW BRUNSWICK, 59- Concurrent Pos: Fulbright sr fel, Univ Perugia, 54-55. Mem: AAAS; Am Soc Agron; Crop Sci Soc Am. Res: Plant genetics; soil and crop management. Mailing Add: Dept of Soils & Crops Rutgers Univ New Brunswick NJ 08903

BATTLES, MALCOLM HAYFORD, organic chemistry, see 12th edition

BATTLES, WILLIS RALPH, b Erie, Pa, Nov 12, 14; m 41; c 3. PETROLEUM CHEMISTRY. Educ: Univ Calif, Los Angeles, AB, 36. Prof Exp: Inspector, Shell Oil Co, Calif, 36-42; chem operator, Trojan Powder Co, Ohio, 42-43 & Dow Chem Co, Calif, 43-44; res chemist, Union Oil Co, 44-46; bus mgr & partner, Calif Car Bed Co, 46-51; chemist, Fletcher Oil Co, 51-57, chief chemist, 57-59; chemist, Charles Martin & Co, San Pedro, 59-60; CHEMIST, ATLANTIC RICHFIELD CO, WILMINGTON, 60- Mem: AAAS; Am Chem Soc; Am Inst Chemists. Res: Analytical methods research, especially for trace contaminants in petroleum products. Mailing Add: 560 S Helberta Ave Redondo Beach CA 90277

BATTLEY, EDWIN HALL, b Detroit, Mich, Jan 24, 25. MICROBIOLOGY, BIOCHEMISTRY. Educ: Harvard Col, BA, 49; Fla State Univ, MS, 51; Stanford Univ, PhD(biol), 56. Prof Exp: Asst, Lab Microbiol, Technol Univ Delft, 55-56 & Sch Med, Univ Wash, 56-57; instr biol chem, Seton Hall Col Med & Dent, 57-60; asst prof biol, Dartmouth Col, 60-62; ASSOC PROF BIOL, STATE UNIV NY STONY BROOK, 62- Mem: AAAS; Am Soc Microbiol; NY Acad Sci; Neth Soc Microbiol. Res: Biochemistry and physiology of microorganisms; thermodynamics of biological processes; general biology. Mailing Add: Div of Biol Sci State Univ of NY Stony Brook NY 11790

BATTON, ROBERT RALPH, b South Bend, Ind, Dec 14, 44; m 73; c 2. NEUROANATOMY. Educ: Ind Univ, AB, 67, PhD(anat), 75; Ariz State Univ, MS, 71. Prof Exp: FEL ANAT, COL PHYSICIANS & SURGEONS, COLUMBIA UNIV, 75- Res: Experimental neuroanatomy; quantitative neuroanatomy. Mailing Add: Dept of Anat Col of Phys & Surg Columbia Univ 630 W 168th St New York NY 10032

BATTS, BILLY STUART, b Raleigh, NC, July 14, 34; m 56; c 2. FISH BIOLOGY, ECOLOGY. Educ: NC State Univ, BS, 56, PhD, 70; Univ Wash, MS, 60. Prof Exp: Fisheries biologist, Univ Wash, 60-62; asst prof, 63-68, ASSOC PROF BIOL, LONGWOOD COL, 68- Mem: Am Fisheries Soc; Am Soc Ichthyologists & Herpetologists; Am Inst Biol Sci. Res: Lepidology of flounders; life history of skipjack tuna, especially age and growth, food habits, sexual maturity and reproduction. Mailing Add: Dept of Natural Sci Longwood Col Farmville VA 23901

BATTS, HENRY LEWIS, JR, b Macon, Ga, May 24, 22; m 45; c 3. ECOLOGY. Educ: Kalamazoo Col, AB, 43; Univ Mich, MS, 47, PhD(zool), 55. Hon Degrees: ScD, Western Mich Univ, 71. Prof Exp: Asst biol & instr ornith, Kalamazoo Col, 40-43; asst zool, Univ Mich, 46-48, teaching fel, 48-50; from instr to assoc prof, 50-59, PROF BIOL, KALAMAZOO COL, 59-; EXEC DIR, KALAMAZOO NATURE CTR, 61- Concurrent Pos: Instr, Exten Serv, Univ Mich, 53; ed, Bull Wilson Ornith Soc, 59-63; founding trustee, Environ Defense Fund, Inc, 67- Mem: AAAS; Am Soc Mammalogists; Am Ornithologists Union; Ecol Soc Am; Nat Audubon Soc; Wilson Ornith Soc (2nd vpres, 66-68, 1st vpres, 66-68, pres, 68-69). Res: Ecology of birds; nest activities of the American Goldfinch; bird nest identification; population and distribution of birds; pesticides; environmental degradation. Mailing Add: 2315 Angling Rd Kalamazoo MI 49008

BATY, JOSEPH A, physical chemistry, deceased

BATY, ROGER M, b Helena, Mont, Oct 2, 37; m 66; c 3. ANTHROPOLOGY. Educ: Mont State Univ, BA, 58; Stanford Univ, PhD, 70. Prof Exp: Pub sch teacher, Mont, 63-64; field dir overseas study progs, Exp Int Living, 64-66; consult, Intercult Dimension, 68-69, DIR INTERCULT DIMENSION, JOHNSTON COL, UNIV REDLANDS, 69- Mem: Am Anthrop Asn; Am Educ Res Asn; Soc Int Training, Educ & Res. Res: Training of teachers for work with children of culturally different backgrounds and for cross-cultural services; culture stress. Mailing Add: Intercult Dimension Johnston Col Univ of Redlands Redlands CA 92373

BATZA, EUGENE M, b Canton, Ohio, Mar 22, 17; m 46; c 2. AUDIOLOGY, SPEECH PATHOLOGY. Educ: Col Wooster, BA, 37; Northwestern Univ, 51, PhD(audiol, speech path), 56. Prof Exp: Asst prof audiol & speech path, Bowling Green State Univ, 54-57; chief coordr speech path, Bill Wilkerson Hearing & Speech Ctr, 57-62, dir training & res audiol & speech path, 62-63; vis prof, NTex State Univ, 63-64; PROF AUDIOL & SPEECH PATH & HEAD SECT CLIN, CLEVELAND CLIN EDUC FOUND, 64- Concurrent Pos: Assoc prof, Sch Med, Vanderbilt Univ, 57-63 & George Peabody Col, 57-63; prof, Tenn State Univ, 59-62; vis lectr, Western Reserve Univ, 64-65; adj prof, Cleveland State Univ, 73- Mem: Am Speech & Hearing Asn; Am Cong Rehab Med; Am Cleft Palate Asn; Int Asn Logopedics & Phoniatrics. Res: Disorders of vocal production and auditory perception. Mailing Add: Cleveland Clin Educ Found 2020 E 93rd St Cleveland OH 44106

BATZAR, KENNETH, b Brooklyn, NY, May 30, 38; m 62; c 1. POLLUTION CHEMISTRY, INORGANIC CHEMISTRY. Educ: Brooklyn Col, BS, 59, MA, 62; City Univ New York, PhD(inorg chem), 66. Prof Exp: SR RES CHEMIST, E I DU PONT DE NEMOURS & CO, NEWARK, 66- Mem: Am Chem Soc; Sigma Xi. Res: Coordination chemistry; solvent extraction; pigment technology; computer programming of chemical problems; physical studies of pigmentary systems; industrial wastewater treatment. Mailing Add: 20 Fuller Ave Piscataway NJ 08854

BATZEL, ROGER ELWOOD, b Idaho, Dec 1, 21; m 46; c 3. NUCLEAR CHEMISTRY. Educ: Univ Idaho, BS, 47; Univ Calif, PhD, 51. Prof Exp: Sr chemist, Calif Res & Develop Co, 51-53; assoc dir, 53-71, DIR, LAWRENCE LIVERMORE LAB, 71- Mem: Am Phys Soc. Res: High energy nuclear reactions; research and development. Mailing Add: Lawrence Livermore Lab PO Box 808 Livermore CA 94550

BATZER, HAROLD OTTO, b Gillett, Wis, Jan 22, 28; m 51; c 3. ENTOMOLOGY. Educ: Univ Minn, BS, 52, MS, 55, PhD, 65. Prof Exp: Entomologist, 54-64, insect ecologist, 64-70, PRIN INSECT ECOLOGIST, N CENT FOREST EXP STA, US FOREST SERV, 70- Concurrent Pos: Dep leader subj group S2.06-8, Int Union Forest Res Orgns, 71-; mem organizing comt forest entom session, Int Cong Entom, 75-76. Mem: Entom Soc Am; Soc Am Foresters. Res: Forest insects, particularly the role of insects in natural forest ecosystems; current emphasis on Malacosoma disstria in aspen forests. Mailing Add: NCent Forest Exp Sta Forest Serv USDA Folwell Ave St Paul MN 55108

BATZING, BARRY LEWIS, b Rochester, NY, May 6, 45; m 68; c 2. MICROBIOLOGY. Educ: Cornell Univ, BS, 67; Pa State Univ, MS, 69, PhD(microbiol), 71. Prof Exp: ASST PROF MICROBIOL, STATE UNIV NY COL CORTLAND, 73- Concurrent Pos: Investr, Immunol Carcinogenesis Sect, Biol Div, Oak Ridge Nat Lab-Grad Sch Biomed Sci, Univ Tenn, 71-73. Mem: Am Soc Microbiol; Sigma Xi; AAAS. Res: Physiology, ultrastructure, and ecological relationships of acetic acid bacteria and closely-related pseudomonads. Mailing Add: Dept of Biol Sci State Univ NY Col Cortland NY 13045

BATZLER, WILLIAM EMMET, b Appleton, Wis, Oct 6, 03; m 29; c 2. UNDERWATER ACOUSTICS. Educ: Univ Mich, AB, 26, MA, 29; Scripps Inst, Univ Calif, MS, 57. Prof Exp: Pub sch teacher, Mich, 26-43; mathematician & physicist, US Navy Electronics Lab, 46-49, PHYSICIST, US NAVAL UNDERSEA CTR, 69- Mem: AAAS; Acoust Soc Am. Res: Underwater acoustics; oceanographic factors affecting acoustic scattering. Mailing Add: 5385 Balboa Ave San Diego CA 92117

BAU, ROBERT, b Shanghai, China, Feb 10, 44; m 70. INORGANIC CHEMISTRY. Educ: Univ Hong Kong, BSc, 63; Univ Calif, PhD(chem), 68. Prof Exp: Res fel chem, Harvard Univ, 68-69; asst prof, 69-74, ASSOC PROF CHEM, UNIV SOUTHERN CALIF, 74- Concurrent Pos: Sloan Found res fel, 74-76; NIH res career develop award, 75-80. Mem: Am Chem Soc; The Chem Soc; Am Crystallog Asn. Res: Spectroscopic and crystallographic investigations of transition metal complexes; structural investigations of transition metal hydride compounds and metal nucleotide complexes. Mailing Add: Dept of Chem Univ of Southern Calif Los Angeles CA 90007

BAUBLIS, JOSEPH V, b Gardner, Mass, Jan 28, 31; m 56; c 4. PEDIATRICS,

VIROLOGY. Educ: Harvard Univ, AB, 52, McGill Univ, MD, CM, 56; Univ Mich, PhD(epidemiol), 67. Prof Exp: From instr to assoc prof path, Sch Med, 62-75, assoc prof pediat, 70-75, PROF PEDIAT & COMMUN DIS, SCH MED, UNIV MICH, ANN ARBOR, 75-, PROF PATH, 75-, LECTR EPIDEMIOL, SCH PUB HEALTH, 67- Mem: Am Pub Health Asn; Am Soc Microbiol; Soc Pediat Res. Res: Prenatal and perinatal infection as it may pertain to birth defects; slow virus disease of the central nervous system; diagnostic virology and pediatric infectious disease. Mailing Add: Univ of Mich Hosp Ann Arbor MI 48109

BAUCHWITZ, PETER SIEGBERT, b Halle, Ger, Sept 22, 20; US citizen; m 59; c 4. ORGANIC CHEMISTRY, PHYSICAL CHEMISTRY. Educ: Univ Akron, MS, 51; Univ Chicago, PhD(org chem), 56. Prof Exp: Develop engr, synthetic rubber govt labs, Univ Akron, 47-51; res chemist, Neoprene Plant, 56-62, RES CHEMIST, EXP STA, E I DU PONT DE NEMOURS & CO, 62- Mem: Am Chem Soc; Sigma Xi. Res: High polymers, rubbers, foams; intermediates; mechanisms of reaction; new processes and products; acetylene, free-radical, peroxide, analytical and urethane chemistry. Mailing Add: 1420 Athens Rd Wilmington DE 19803

BAUE, ARTHUR EDWARD, b St Louis, Mo, Oct 7, 29; m 56; c 3. SURGERY, CARDIOVASCULAR PHYSIOLOGY. Educ: Westminster Col, AB, 50; Harvard Univ, MD, 54; Am Bd Surg, dipl, 62; Am Bd Thoracic Surg, dipl, 63. Prof Exp: From intern to chief resident, Mass Gen Hosp, 54-61; fel surg, Harvard Med Sch, 61-62; sr registr thoracic surg, Sch Med, Bristol Univ, 62; asst prof surg, Univ Mo, 62-64; from asst prof to assoc prof, Univ Pa, 64-68; Edison prof surg, Sch Med, Washington Univ, 68-75; CHIEF CARDIOTHORACIC SURG, PROF SURG & CHMN DEPT, SCH MED, YALE UNIV, 75- Concurrent Pos: Markle scholar acad med, 63; USPHS res career develop award, 65-66; consult, Nat Bd Med Examr & Vet Admin, 64-; surgeon-in-chief & dir, Jewish Hosp St Louis, 68-75. Mem: Am Col Surg; AMA; Asn Am Med Cols; Am Physiol Soc; Soc Univ Surg. Res: Thoracic and cardiovascular surgery. Mailing Add: Yale Univ Sch of Med 333 Cedar St New Haven CT 06510

BAUER, ADELIA CATHERINE, b Wooster, Ohio, June 20, 16. PHYSIOLOGY. Educ: Russell Sage Col, BA, 39; Cornell Univ, MNS, 48; Univ Iowa, PhD(physiol), 57. Prof Exp: Med technologist, Samaritan Hosp, 39-43; lab asst biol, Rensselaer Polytech Inst, 45-46; jr nutritionist, Agr Exp Sta, Univ Hawaii, 48-50; asst nutrit, Nat Dairy Res Labs, Inc, NY, 51-54; instr physiol, Vassar Col, 57-59; RES PHYSIOLOGIST, LAB PHYS BIOL, NAT INST ARTHRITIS, METAB & DIGESTIVE DIS, 64- Concurrent Pos: NIH grant, Marine Biol Lab, Woods Hole, Mass, 59-64. Mem: AAAS; Am Chem Soc. Res: Muscle proteins; insect physiology; biochemistry. Mailing Add: Lab of Phys Biol Nat Inst Arth Metab & Dig Dis Bethesda MD 20014

BAUER, ALBERT WEBB, b Baltimore, Md, Apr 22, 28; m 53; c 2. ORGANIC CHEMISTRY. Educ: Gettysburg Col, AB, 49; Princeton Univ, MA, 51, PhD(chem), 53. Prof Exp: Res chemist, 52-61, DIV HEAD, E I DU PONT DE NEMOURS & CO, INC, 61- Mem: Am Chem Soc; Am Soc Heat, Refrig & Air-Conditioning Engrs. Res: Structure of starch; mechanism and new methods of dyeing synthetic fibers; new dyes; petroleum additives; diversification scouting; fluorocarbon application service and development. Mailing Add: 2615 Silverside Rd Wilmington DE 19810

BAUER, ARMAND, b Zeeland, NDak, Nov 29, 24; m 49; c 4. SOIL SCIENCE. Educ: NDak State Univ, BS, 50, MS, 55; Colo State Univ, PhD(soil sci), 64. Prof Exp: Asst soil scientist, NDak State Univ, 55-61; Dept Health, Educ & Welfare fel, 61-63; asst prof soil sci, 63-65, assoc prof soils, 65-69, PROF SOILS, N DAK STATE UNIV, 69- Mem: Am Soc Agron. Res: Soil fertility research with nitrogen, phosphorus, potassium and minor elements; fertility-moisture and fertility-temperature relationships. Mailing Add: Dept of Soils NDak State Univ Fargo ND 58102

BAUER, CARL AUGUST, b Marion Co, Kans, Nov 10, 16; m 41; c 3. ASTRONOMY. Educ: Univ Minn, BA, 42; Univ Chicago, MS, 44; Harvard Univ, PhD(astron), 49. Prof Exp: Instr, US Army Air Force, Univ Minn, 42; res asst, Yerkes Observ, Univ Chicago, 42-44; instr physics & math, reserve officers training corps, NDak Agr Col, 44; instr astron, Ind Univ, 45; instr & res assoc, observ, Univ Mich, 47-50 & Harvard Univ, 51; asst prof, 51-56, ASSOC PROF ASTRON, PA STATE UNIV, 56- Mem: AAAS; Am Astron Soc; Am Meteorol Soc; Int Astron Union. Res: Stellar and solar spectroscopy; meteoritics; production of helium by cosmic radiation in meteorites and other developments of the parent planet hypothesis; origin of comets. Mailing Add: 444 E McCormick Ave State College PA 16801

BAUER, CLIFFORD DAVID, b Washington, Iowa. BIOCHEMISTRY. Educ: Iowa Wesleyan Col, BS, 32; Univ Iowa, PhD, 40. Prof Exp: Tech dir, Whittier Labs, Chicago, 40-51; dir prod develop lab, spec prod div, Borden Co, 51-64; mem staff, Milwaukee Res Inst, 65-66; supvr food group, corn prod res, Anheuser-Busch, Inc, 66-75; RETIRED. Res: Food technology; nutrition. Mailing Add: 906 S Ninth Ave Washington IA 52353

BAUER, DAVID FRANCIS, b Lehighton, Pa, Apr 13, 40; m 65. MATHEMATICAL STATISTICS. Educ: East Stroudsburg State Col, BS, 63; Ohio Univ, MS, 65; Univ Conn, PhD(math statist), 70. Prof Exp: Instr math, Denison Univ, 65-66; assoc mem tech staff, Traffic Studies Ctr, Bell Tel Labs, NJ, 66-67; asst prof statist & comput sci, Univ Del, 70-74; ASST PROF MATH SCI, VA COMMONWEALTH UNIV, 74- Mem: Math Asn Am; Am Statist Asn. Res: Nonparametric methods and applications of statistics. Mailing Add: Dept of Math Sci Va Commonwealth Univ Richmond VA 23284

BAUER, DIETRICH CHARLES, b Elgin, Ill, July 1, 31; m 54. IMMUNOLOGY. Educ: Univ Ill, BS, 54; Mich State Univ, MS, 57, PhD(microbiol), 59. Prof Exp: Res asst microbiol, Mich State Univ, 57-59; fel immunol, Case Western Reserve Univ, 59-61; from asst prof to assoc prof microbiol, 61-69, PROF MICROBIOL, IND UNIV, INDIANAPOLIS, 69- Mem: AAAS; Am Soc Microbiol; Am Asn Immunol. Res: Virulence factors of Leptospirae; in vitro antibody synthesis; molecular forms of antibody; comparative reactivity of immunoglobulins; intracellular assembly of gamma globulin; suppression of immune response. Mailing Add: Dept of Microbiol Ind Univ Med Ctr Indianapolis IN 46207

BAUER, ELDON EUGENE, b St Louis, Mo, Dec 17, 17; m 40, 61; c 4. PHYSICAL CHEMISTRY, CHEMICAL ENGINEERING. Educ: McKendree Col, BS, 38; Univ Iowa, MS, 40, PhD(phys chem), 42. Prof Exp: Asst, Univ Iowa, 38-42; supvr res, develop & qual control, Eastman Kodak Co, 42-56; vpres mkt, Dynacolor Corp, Minn Mining & Mfg Co, 57-66; vpres & gen mgr, Graflex Inc, 67-68; PRES, A&R COLOR LABS, 68- Mem: Am Chem Soc; Soc Photog Scientists & Engrs; Photog Soc Am; Soc Motion Picture & TV Engrs. Res: Iodine monochloride systems; rheological properties of aluminum soap systems; color photographic processes and equipment. Mailing Add: 4199 W Henrietta Rd Rochester NY 14623

BAUER, ERNEST, b Vienna, Austria, Mar 24, 27; US citizen; m 59; c 2. ATMOSPHERIC PHYSICS. Educ: Cambridge Univ, BA, 47, PhD(theoret physics), 50. Prof Exp: Fel physics, Nat Res Coun Can, 50-52; res assoc, Inst Math Sci, NY Univ, 52-54, instr math, 53-54; asst prof physics, Univ NB, 54-55; adj asst prof, Col Eng, NY Univ, 55-56, res assoc, Inst Math Sci, 55-57, instr math, 56-57; prin scientist, Avco Res & Advan Develop, 57-59; staff scientist, Aeronutronic Div, Philco Corp, 59-65; RES STAFF MEM, SCI & TECHNOL DIV, INST DEFENSE ANAL, 65- Concurrent Pos: Lectr, Cath Univ Am, 68-69 & City Col New York, 70-71; exhibitioner, scholar & prizeman, Selwyn Col, Cambridge Univ. Mem: Fel Am Phys Soc; Am Inst Aeronaut & Astronaut; Am Geophys Union. Res: Atomic, molecular and radiation physics; physics of the earth's upper atmosphere; high temperature physics; atmospheric motions. Mailing Add: 8109 Fenway Rd Bethesda MD 20034

BAUER, ERNST GEORG, b Schoenberg, Ger, Feb 27, 28; US citizen; m 55; c 2. SURFACE PHYSICS. Educ: Univ Munich, MS, 53, PhD(physics), 55. Prof Exp: Asst, Phys Inst, Univ Munich, 54-58; br head, Naval Weapons Ctr, China Lake, Calif, 58-69; PROF, PHYS INST, CLAUSTHAL TECH UNIV, 69- Mem: Am Phys Soc; Am Vacuum Soc; Electron Micros Soc Am. Res: Surface science; electron and crystal physics. Mailing Add: Phys Inst Clausthal Tech Univ Leibnizstr 4 Clausthal-Zellerfeld Germany

BAUER, FRANCES BRAND, b New York, NY, July 5, 23; m 48. APPLIED MATHEMATICS. Educ: Brooklyn Col, AB, 43; Brown Univ, MS, 45, PhD(appl math), 48. Prof Exp: Res assoc appl math, Brown Univ, 45-48; res assoc aero eng structures, Polytech Inst Brooklyn, 49-50; sr mathematician, Reeves Instrument Corp, 50-51 & 51-61; mathematician, Bur Stand, Am Univ, 51-52; RES SCIENTIST ELASTICITY, FLUID DYNAMICS & COMPUT, COURANT INST MATH SCI, NY UNIV, 61- Mem: Am Math Soc; Asn Comput Mach. Res: Transonic flow; supercritical wing sections I; supercritical wing sections II. Mailing Add: 200 East End Ave New York NY 10028

BAUER, FRANCIS HARRY, b Reading Center, NY, Aug 5, 18; m 62. GEOGRAPHY. Educ: Univ Calif, Berkeley, AB, 49, MA, 52; Australian Nat Univ, PhD(geog), 60. Prof Exp: Lectr geog, Univ Calif, Riverside, 60-61; asst prof, San Diego State Col, 61-62; sr lectr & chmn dept, Univ Col Townsville, 62-65; from assoc prof to prof, Calif State Univ Hayward, 65-74; FIELD DIR, N AUSTRALIA RES UNIT, AUSTRALIAN NAT UNIV, 74- Concurrent Pos: Social Sci Res Coun Australia res grant, 63. Mem: Inst Australian Geog. Res: Historical geography of northern and central Australia, with emphasis on land settlement, land utilization and tropical agriculture; compilation of data on research in north Australia. Mailing Add: North Australia Res Unit PO Box 39448 Winnellie Northern Territory Australia

BAUER, FRANZ KARL, b Vienna, Austria, Jan 29, 17; US citizen; m 43; c 1. INTERNAL MEDICINE, NUCLEAR MEDICINE. Educ: La State Univ, MD, 41; Am Bd Internal Med, dipl, 50; Am Bd Nuclear Med, cert, 72. Prof Exp: Dir radioisotope unit, Vet Admin Ctr, Los Angeles, 51-56; assoc clin prof med & radiol & asst clin prof med & radiol, Univ Calif, Los Angeles, 51-56; assoc prof med, Sch Med, Univ Southern Calif, 56-60; prof med, Sch Med, Univ Calif, Los Angeles, 60-65; from assoc dean to dean sch med, 65-74, PROF MED, SCH MED, UNIV SOUTHERN CALIF, 65- Concurrent Pos: Coordr radioisotope res, Univ Southern Calif, 56-60; asst med dir, Los Angeles County Hosp, 56-60; chief med serv, Harbor Gen Hosp, 60-65; chief med serv, Rancho Los Amigos Hosp, 74- Mem: Am Col Physicians; AMA; Soc Exp Biol & Med; Am Fedn Clin Res. Res: Radioisotopes; metabolism; cardiovascular physiology. Mailing Add: Rancho Los Amigos Hosp 7601 E Imperial Hwy Downey CA 90242

BAUER, FREDERICK WILLIAM, b Wakefield, RI, Oct 22, 22; m 54; c 5. ORGANIC CHEMISTRY, RESEARCH ADMINISTRATION. Educ: Washington & Lee Univ, BS, 43; Princeton Univ, MA, 50, PhD(org chem), 52. Prof Exp: Jr asst res chemist, Tenn Eastman Corp, 43-44, asst res chemist, 46-47; technician, Textile Res Inst, 47, res chemist, 51-52; res chemist, Gen Labs, US Rubber Co, 52-54, org chemist, Textile Div, 54-56 & Res Ctr, 56-60; group leader, Cent Res Labs, 60-62, asst dir res admin, 62-64, dir res admin, 64-69, ASST MGR EMPLOYEE RELS, CORP RES & DEVELOP DIV, ALLIED CHEM CORP, 69- Mem: Am Chem Soc; Sigma Xi. Res: Structure of cellulose; rubber and polymers; textile finishing and chemical modifications; scientific personnel polymerization; polymer characterization; industrial hygiene, safety. Mailing Add: Corp Res & Develop Div Allied Chem Co PO Box 1021 R Morristown NJ 07960

BAUER, GUSTAV ERIC, b New York, NY, Jan 26, 35; m 57; c 2. ANATOMY, CYTOCHEMISTRY. Educ: Queens Col, NY, BS, 57; Western Reserve Univ, MA, 59; Univ Minn, PhD(anat), 63. Prof Exp: Jr chemist, Sperry-Rand Corp, 57; from instr to asst prof, 63-69, ASSOC PROF ANAT, UNIV MINN, MINNEAPOLIS, 69- Concurrent Pos: USPHS fel, 65-66; mem corp, Marine Biol Lab, Woods Hole. Mem: AAAS; Am Asn Anatomists. Res: Insulin and glucagon biosynthesis in teleost fish islets; isolation and characterization of subcellular particles; chemical carcinogenesis. Mailing Add: 262 Jackson Hall Univ of Minn Minneapolis MN 55455

BAUER, HANS FRED, b Los Angeles, Calif, Oct 23, 32; m 63; c 2. FUEL SCIENCE. Educ: Univ Calif, Los Angeles, BS, 54, PhD(inorg chem), 62. Prof Exp: Sr res engr, Rocketdyne Div, Rockwell Int, 61-67; group leader, Stauffer Chem Co, 67-68; SR CHEMIST, OXY RES, 68- Mem: Am Chem Soc. Res: Potash brine processes; phosphate extraction; solid waste pyrolysis; liquid synthetic fuels; coal tar chemistry; shale oil processing. Mailing Add: 22425 Robin Oaks Terr Diamond Bar CA 91765

BAUER, HENRY, b Minneapolis, Minn, Nov 3, 14; m 38. BACTERIOLOGY. Educ: Univ Minn, PhD(bact), 49; Am Bd Microbiol, dipl. Prof Exp: Bacteriologist, 38-41, supvr lab eval unit, 46-47, dep exec officer, 60-66, DIR DIV MED LABS, STATE DEPT HEALTH, MINN, 49- Concurrent Pos: Instr bact & immunol, Univ Minn, 47-48, lectr, Sch Pub Health, 49-; consult bacteriologist, Off Surgeon Gen, 51-54 & NIH, 57-67; mem heart dis control adv comt, USPHS, 63-65; mem adv comt, Commun Dis Ctr, 63-69. Honors & Awards: Award, Minn Med Asn, 63. Mem: Fel Am Acad Microbiol; Am Pub Health Asn; Asn State & Territorial Pub Health Lab Dirs. Res: Microbiology and virology; laboratory methodology. Mailing Add: Div of Med Labs Minn Dept of Health Minneapolis MN 55440

BAUER, HENRY HERMANN, b Vienna, Austria, Nov 16, 31; m 58; c 2. ELECTROCHEMISTRY, ANALYTICAL CHEMISTRY. Educ: Univ Sydney, BSc, 52, MSc, 53, PhD(chem), 56. Prof Exp: Res assoc chem, Univ Mich, 56-58; from lectr to sr lectr, Univ Sydney, 58-66; assoc prof, 66-69, PROF CHEM, UNIV KY, 69- Concurrent Pos: Fulbright travel award, 56-58; vis lectr chem & res scientist, Univ Mich, 65-66; vis prof, Univ Southampton, 72-73 & Japan Soc Promotion Sci, 74. Honors & Awards: Distinguished Achievement in Research, Univ Ky Res Found, 74. Mem: Am Chem Soc; Electrochem Soc; The Chem Soc; Australian Polarographic Soc (vpres, 64). Res: Polarography, particularly with alternating current; development of techniques and applications; mechanisms of electrode processes. Mailing Add: Dept of Chem Univ of Ky Lexington KY 40506

BAUER, JERE MARKLEE, b Ft Worth, Tex, Apr 24, 15; m 61; c 2. MEDICINE. Educ: Univ Tex, MD, 41, BA, 44. Prof Exp: Intern, Med Br, Univ Tex, 41-42; from asst resident to instr internal med, 42-46, instr & res asst endocrinol & metab, 46-48, from asst prof to assoc prof, 48-72, PROF INTERNAL MED, UNIV MICH, ANN ARBOR, 72- Mem: AAAS; AMA; fel Am Col Physicians; Am Diabetes Asn; Am Heart Asn. Res: Endocrine and metabolic diseases; metabolic aspects of cancer. Mailing Add: Univ of Mich Hosp Ann Arbor MI 48104

BAUER, LUDWIG, b Ger, July 27, 26; nat US; m 57; c 2. ORGANIC CHEMISTRY, MEDICINAL CHEMISTRY. Educ: Univ Sydney, BSc, 49, MSc, 50; Northwestern Univ, PhD(chem), 52. Prof Exp: Res assoc, Harvard Univ, 52-53, Columbia Univ, 53 & Univ Sydney, 53-54; res chemist, Elkin Chem Co, 55; from asst prof to assoc prof, 55-65, PROF CHEM, UNIV ILL MED CTR, CHICAGO, 65- Mem: Am Chem Soc; The Chem Soc. Res: Synthesis of potential medicinal agents, such as cancer chemotherapy, analgesics and antiradiation drugs; fundamental chemistry of heteroaromatic compounds, with an emphasis on the reactions of pyridine N-oxides and N-hydroxyuracils. Mailing Add: Dept of Chem Col of Pharm Univ of Chicago Med Ctr Chicago IL 60680

BAUER, MARK HENRY, b Easton, Pa, Feb 16, 12. PHYSIOLOGY. Educ: St Joseph's Col, Pa, 33, MS, 35; Woodstock Col, Md, PhL, 39, STL, 45; Princeton Univ, MA, 49, PhD(biol), 50. Prof Exp: Prof biol & chmn dept, St Joseph's Col, Pa, 53-58; res assoc physiol, Sch Med, 58-63, vpres for med ctr affairs, 63-69, ASSOC PROF BIOL, GEORGETOWN UNIV, 69- Mem: AAAS. Res: Cardiovascular physiology. Mailing Add: Dept of Biol Georgetown Univ Washington DC 20007

BAUER, MARVIN E, b Valparaiso, Ind, July 24, 43; m 69. AGRONOMY, REMOTE SENSING. Educ: Purdue Univ, BSA, 65, MS, 67; Univ Ill, PhD(crop physiol), 70. Prof Exp: RES AGRONOMIST, DEPT AGRON, PURDUE UNIV, 70-, PROG LEADER, CROP INVENTORY SYSTS & RES, LAB APPLN REMOTE SENSING, 70- Mem: Am Soc Agron; Crop Sci Soc Am; Am Soc Photogram. Res: Research and development of remote sensing for crop production inventories. Mailing Add: Lab Appln Remote sensing Purdue Univ 1220 Potter Dr West Lafayette IN 47906

BAUER, NEINZ, b Vienna, Austria, Nov 28, 14; US citizen; m 39; c 2. PATHOLOGY. Educ: Univ Vienna, 33-38; Emory Univ, MD, 51. Prof Exp: Nat Cancer Inst trainee, Emory Univ, 54-56, from assoc to assoc prof path, 56-61; clin assoc prof, 61-63, assoc prof, 63-65, PROF PATH, SCH MED, GEORGETOWN UNIV, 65- Concurrent Pos: Res assoc, Mt Sinai Hosp, NY, 61-64. Mem: Fel Col Am Path; Am Asn Path & Bact; NY Acad Sci; Am Soc Exp Path; Int Acad Path. Res: Infectious diseases; immunopathology; germ-free animal research; rheumatology. Mailing Add: Dept of Path Georgetown Univ Sch of Med Washington DC 20007

BAUER, RALPH HAROLD, physical organic chemistry, see 12th edition

BAUER, RICHARD G, b Kent, Ohio, Dec 9, 35; m 58; c 3. POLYMER CHEMISTRY, ORGANIC CHEMISTRY. Educ: Kent State Univ, BS, 56; Univ Akron, MS, 60, PhD(polymer sci), 66. Prof Exp: Res chemist, Gen Tire & Rubber Co, 56-60 & Air Reduction Chem & Carbide Co, 61-63; SR RES CHEMIST, GOODYEAR TIRE & RUBBER CO, AKRON, 63- Mem: AAAS; Am Chem Soc. Res: Preparation and characterization of polymers and organic chemicals related to the polymer field. Mailing Add: 1624 Chadwick Dr Kent OH 44240

BAUER, RICHARD M, b Appleton, Wis, Apr 30, 28; m 50; c 5. APPLIED PHYSICS. Educ: Lawrence Col, BS, 53. Prof Exp: From jr process engr to sr process engr, Marathon Corp, Wis, 55-58, sr process engr, Marathon Div, 58-60, group leader, 60-65, mgr indust prod develop, 65-67, mgr converting paperboard & spec prod res & develop, 67-68, assoc dir, 68-70, DIR RES & DEVELOP, NEENAH LAB, AM CAN CO, 70- Concurrent Pos: Marathon Div consult, Glamakote carton process to US & Europ countries, 61- Res: Conversion of plastics and paperboard to rigid food packages. Mailing Add: Neenah Lab Am Can Co 333 N Commercial St Neenah WI 54956

BAUER, ROBERT, b Grand Rapids, Mich, May 5, 26; m 51; c 3. CLINICAL BIOCHEMISTRY. Educ: Western Mich Univ, BS, 50, MS, 62. Prof Exp: Chemist, Socony Mobil Oil Co, Inc, 51-55; plastics chemist, Dow Chem Co, 55-58; chemist, Upjohn Co, 58-60; assoc res biochemist, 61-71, RES SCIENTIST, AMES CO DIV, MILES LABS, INC, 71- Mem: AAAS; Am Chem Soc. Res: Diagnostic aid and analytical aids for industry. Mailing Add: Ames Res Lab 1127 Myrtle St Elkhart IN 46514

BAUER, ROBERT OLIVER, b Chicago, Ill, Mar 2, 18; m 39; c 3. PHARMACOLOGY. Educ: Univ Mich, BS, 40; Wayne Univ, MS, 44, MD, 47; Am Bd Anesthesiol, dipl. Prof Exp: Intern, Wayne County Gen Hosp, Mich, 47-48; asst prof pharmacol, Sch Med, Boston Univ, 48-52; sr pharmacologist, Riker Lab, Inc, Los Angeles, 52-53; pharmacologist, Roswell Park Mem Inst, 55-58; from asst prof to assoc prof anesthesiol, 59-70, PROF ANESTHESIOL & PHARMACOL, CTR HEALTH SCI, UNIV CALIF, LOS ANGELES, 70- Concurrent Pos: Attend anesthesiologist, Ctr Health Sci, Univ Calif, Los Angeles, 58- Mem: AAAS; Am Soc Anesthesiol; Am Pharmaceut Asn; Am Soc Pharmacol & Exp Therapeut. Res: Analgesics; cardiovascular agents. Mailing Add: Dept of Anesthesiol & Pharmacol of Calif Ctr for Health Sci Los Angeles CA 90024

BAUER, ROBERT STEVEN, b Brooklyn, NY, Dec 8, 44; m 67; c 2. EXPERIMENTAL SOLID STATE PHYSICS, ELECTRONIC SPECTROSCOPY. Educ: Rensselaer Polytech Inst, BEE, 66; Stanford Univ, MS, 67, PhD(elec eng), 71. Prof Exp: MEM RES STAFF SURFACE & BULK ELECTRONIC STATES OF SOLIDS, XEROX PALO ALTO RES CTR, 70- Mem: Am Phys Soc; Am Vacuum Soc; Inst Elec & Electronics Engrs; Optical Soc Am; Sigma Xi. Res: Photoelectron and modulated optical spectroscopy studies of surface and bulk electronic states of crystalline and amorphous semiconductors, insulators and ionic conductors; chemical bonding, impurities/defects, interfaces, and lattice effects. Mailing Add: Gen Sci Lab Xerox Palo Alto Res Ctr 3333 Coyote Hill Rd Palo Alto CA 94304

BAUER, ROGER DUANE, b Oxford, Nebr, Jan 17, 32; m 56; c 3. BIOCHEMISTRY. Educ: Beloit Col, BS, 53; Kans State Univ, MS, 57, PhD(biochem), 60. Prof Exp: From asst prof to assoc prof, 59-69, chmn dept, 66-74, PROF CHEM, CALIF STATE UNIV, LONG BEACH, 69-, DEAN, SCH NATURAL SCI, 74- Concurrent Pos: Am Coun on Educ fel, 70-71. Mem: Am Chem Soc; Radiation Res Soc. Res: Protein dye interactions; metabolism and structure of nucleoproteins. Mailing Add: Dept of Chem Calif State Univ Long Beach CA 90840

BAUER, RONALD SHERMAN, b Huntington Park, Calif, Feb 24, 32; m 58; c 1. ORGANIC CHEMISTRY. Educ: Univ Calif, BS, 54, PhD(chem), 58. Prof Exp: CHEMIST, SHELL DEVELOP CO, 58- Mem: Am Chem Soc. Res: Synthesis and polymerization of halogenated allenes; polymerization of epoxides and related materials; process and product development on weatherable and high solids epoxy resins and coating systems. Mailing Add: Shell Develop Co PO Box 1380 Houston TX 77001

BAUER, RUDOLF WILHELM, b Rothenburg, Ger, Nov 28, 28; US citizen; m 58; c 3. EXPERIMENTAL PHYSICS. Educ: Amherst Col, BA, 52; Mass Inst Technol, PhD(physics), 59. Prof Exp: From instr to asst prof physics, Mass Inst Technol, 59-64; vis res physicist, 62-63, PHYSICIST, LAWRENCE LIVERMORE LAB, UNIV CALIF, 64- Mem: Am Phys Soc; Inst Elec & Electronics Engrs. Res: Radioactivity; beta and gamma ray spectroscopy; nuclear orientation, reactions and structure; neutron physics; plasma physics; diagnostics; accelerators; detectors. Mailing Add: L-221 Lawrence Livermore Lab Livermore CA 94550

BAUER, SIEGFRIED JOSEF, b Klagenfurt, Austria, Sept 13, 30; nat US; m 54; c 1. SPACE PHYSICS. Educ: Graz Univ, PhD(physics, geophys), 53. Prof Exp: Physicist, US Army Signal Res & Develop Lab, Ft Monmouth, NJ, 53-61; sr scientist & sect head space sci div, 61-65, head planetary ionospheres br, Lab Space Sci, 65-68, head ionospheric & radio physics br, 68-70, assoc chief lab planetary atmospheres, 70-75, ASSOC DIR SCI, GODDARD SPACE FLIGHT CTR, NASA, GREENBELT, 75- Concurrent Pos: Lectr, Dept Space Sci & Appl Physics, Cath Univ Am, 64-65; mem comns, Int Sci Radio Union. Honors & Awards: Except Sci Achievement Medal, NASA, 74. Mem: AAAS; Am Geophys Union; Int Asn Geomag & Aeronomy. Res: Physics of the upper atmosphere; planetary ionospheres; aeronomy; radio wave propagation. Mailing Add: 7 David Ct Silver Spring MD 20904

BAUER, SIMON HARVEY, b Kovno, Lithuania, Oct 12, 11; nat US; m 38; c 3. CHEMICAL KINETICS, STRUCTURAL CHEMISTRY. Educ: Univ Chicago, BS, 31, PhD(chem), 35. Prof Exp: Fel, Calif Inst Technol, 35-37; instr fuel technol, Pa State Col, 37-39; from instr to assoc prof, 39-50, PROF CHEM, CORNELL UNIV, 50- Concurrent Pos: Consult, Atlantic-Richfield Oil Co, Los Alamos Sci Labs & Naval Res Lab; Guggenheim Mem Found fel, 49; NSF sr fel, 62-63. Mem: Fel AAAS; fel Am Phys Soc; fel Am Inst Chemists; Fedn Am Scientists; Am Chem Soc. Res: Electron diffraction; compounds of boron; molecular spectra; rates of very rapid reactions as studied in shock tubes; chemical lasers. Mailing Add: Dept of Chem Cornell Univ Ithaca NY 14853

BAUER, STEWART THOMAS, b Chicago, Ill, Apr 25, 09; m 38; c 2. ORGANIC CHEMISTRY. Educ: Univ Ill, BS, 32; Univ Minn, MS, 34. Prof Exp: Res chemist, Armour Co, 35-41; sect head, USDA, 41-46 & Drackett Co, 46-54; chief chemist, 54-66, from asst tech dir to assoc tech dir, 66-71, TECH DIR, CROSBY CHEM, INC, 71- Mem: Am Chem Soc; Am Oil Chemists Soc. Res: Fatty acids and derivatives; terpenes and terpene polymers; tall oil products. Mailing Add: Res Lab Crosby Chem Inc PO Box 460 Picayune MS 39466

BAUER, THEODORE JAMES, b Iowa City, Iowa, Nov 18, 09; m 38; c 4. VENEREAL DISEASES. Educ: Univ Iowa, MD, 33, BS, 34. Prof Exp: Intern, US Marine Hosp, NY, 33-34; trainee, USPHS, 34-38, regional consult venereal dis control, San Francisco, 38-41, officer, Mo, 41-42, officer, Off Venereal Dis Control, Chicago Dept Health & med officer in chg, Chicago Intensive Treatment Control Ctr, 42-48, chief venereal dis div, Washington, DC, 48-52, chief commun dis ctr, 53-56, asst surgeon gen & dep chief, Bur State Servs, 56-59, chief, 59-62; med dir, 62-67, vpres res & develop, 64-67, SR VPRES RES & MED AFFAIRS, BECTON, DICKINSON & CO, 67- Concurrent Pos: Lectr, Sch Med, Georgetown Univ; vis fac mem, Sch Pub Health & Hyg, Johns Hopkins Univ; lectr, NJ Col Med & Dent, 62-; expert, Comn Venereal Infection & Treponematoses, WHO; mem, Surgeon Gen Adv Comt Community Health Serv, 62- & NJ Gov Coun Aging & Chronic Dis, 62- Honors & Awards: Distinguished Serv Award, USPHS, 62. Mem: Fel Am Pub Health Asn; fel Am Col Physicians; Am Venereal Dis Asn; Sci Res Soc Am. Mailing Add: Becton Dickinson & Co Stanley & Cornelia St Rutherford NJ 07070

BAUER, VICTOR JOHN, organic chemistry, see 12th edition

BAUER, WALTER, b Innsbruck, Austria, Mar 29, 35; US citizen; m 64. SOLID STATE PHYSICS. Educ: Univ Calif, Berkeley, AB, 57; Univ Ill, MS, 59, PhD(physics), 62. Prof Exp: Sr physicist, Atomics Int Div, NAm Aviation, Inc, 62-67, res specialist, 67-69; MEM TECH STAFF, SANDIA LABS, 69- Concurrent Pos: Exten instr, Univ Calif, Los Angeles, 63- Mem: AAAS; Am Phys Soc; Sigma Xi. Res: Defects in metals; electron radiation effects; inert gas migration in metals; ion implantation. Mailing Add: 911 Via Del Paz Livermore CA 94550

BAUER, WALTER F, applied mathematics, see 12th edition

BAUER, WALTER HERMANN, b Red Oak, Iowa, Sept 9, 07; m 34; c 3. PHYSICAL CHEMISTRY. Educ: Ore State Col, BS, 29; Univ Wis, PhD(phys chem), 33. Prof Exp: Asst instr chem, Univ Wis, 29-34; from instr to prof phys chem, 34-60, dean sch sci, 60-72, EMER DEAN SCH SCI, RENSSELAER POLYTECH INST, 72- Mem: AAAS; Soc Rheol; Am Chem Soc; Am Inst Chem; Am Soc Testing & Mat. Res: Kinetics of oxidation; rheology; chemistry of aluminum soaps; rheology of polymers. Mailing Add: 1625 Tibbits Ave Troy NY 12180

BAUER, WILLIAM, JR, b Philadelphia, Pa, Aug 23, 36; m 65. ORGANIC CHEMISTRY, CHEMICAL ENGINEERING. Educ: Univ Pa, BS, 58, PhD(chem), 62. Prof Exp: Chemist org synthesis, 62-70, lab head plastics synthesis, 70-73, sr res assoc, 73-74, PROJS LEADER, ROHM AND HAAS CO, 74- Mem: AAAS; Am Chem Soc; Soc Plastics Engr. Res: Process research and development for commercial manufacture of vinyl and related monomers. Mailing Add: 2046 Winthrop Rd Huntingdon Valley PA 19006

BAUER, WILLIAM EUGENE, b Greensburg, Pa, Jan 21, 33; m 68; c 1. ANALYTICAL CHEMISTRY. Educ: Miami Univ, BA, 54; Pa State Univ, PhD(chem), 59. Prof Exp: INSTR CHEM, LUCKNOW CHRISTIAN COL, INDIA, 59-, HEAD DEPT, 70- Concurrent Pos: Vis asst prof & asst ed, Newslett, Adv Coun Col Chem, Wabash Col, 64-65. Mem: Am Chem Soc; Indian Chem Soc. Res: Polarography; emission and absorption spectroscopy; instrumentation. Mailing Add: Lucknow Christian Col Gola Ganj Locknow-1 UP India

BAUER, WILLIAM ROBERT, biophysical chemistry, see 12th edition

BAUERLE, RONALD H, bNewport, Ky, Dec 16, 37; m 61; c 2. MOLECULAR BIOLOGY, MICROBIOLOGY. Educ: Villa Madonna Col, AB, 57; Univ Houston, MS, 59; Purdue Univ, PhD(microbiol), 63. Prof Exp: Instr microbiol, Purdue Univ, 61-62; res assoc molecular biol, Cold Spring Harbor Lab Quant Biol, 63-66; res assoc, div biol, Southwest Ctr Advan Studies, 66-69; ASSOC PROF BIOL, UNIV VA, 69- Concurrent Pos: Fel, NIH, 63-66, career develop award, 67-69, 71-76. Mem: AAAS; Genetics Soc Am; Am Soc Microbiol. Res: Regulation of gene function in bacteria; allosteric mechanisms of enzyme control; genetics of bacteria and fungi; mechanism of protein synthesis; molecular genetics; structure and evaluation of complex enzymes. Mailing Add: Dept of Biol Univ of Va Charlottesville VA 22903

BAUERMEISTER, HERMAN OTTO, b St Louis, Mo, Jan 18, 14; m 39. INDUSTRIAL ORGANIC CHEMISTRY, CHEMICAL ENGINEERING. Educ: Armour Inst Technol, BS, 37; Ill Inst Technol, MS, 41; DePaul Univ, JD, 46. Prof

Exp: Chemist, Commonwealth Edison Co, 37-40; patent examr insecticides & foods, US Patent Off, 40-42; res engr catalysis & synthetic rubber, Sinclair Refining Co, 42-46; PATENT ATTORNEY INORG CHEM & ELECTRONICS, MONSANTO CO, 46- Mem: Am Chem Soc; Am Bar Asn. Res: Heat transfer and distillation; catalysis; organic phosphorus chemistry; detergents; electronic devices. Mailing Add: 11750 Fawnridge Dr St Louis MO 63131

BAUERNFEIND, JACOB CHRISTOPHER, b North Branch, NY, Apr 30, 14; m 39; c 3. NUTRITION, BIOCHEMISTRY. Educ: Cornell Univ, BS, 36, MS, 39, PhD(biochem, nutrit, physiol), 40. Prof Exp: Asst nutrit, Cornell Univ, 36-40; nutritionist and res chemist, Hiram Walker and Sons, Inc, 40-44; chief appl nutrit, 44-55, dir food and agr prod develop, 55-61, dir agr res dept, 61-68; asst to vpres chem res, 68-71, NUTRIT RES COORDR, HOFFMANN-LA ROCHE INC, 72- Honors & Awards: Poultry Sci Res Award, 40; Food Technol Ind Achievement Award, 68. Mem: Sigma Xi; Am Chem Soc; Poultry Sci Asn; World's Poultry Sci Asn; Inst Food Technol. Res: General nutrition; biochemistry; nutrients in human nutrition; food technology; nutrition of the elderly; prevention of blindness of dietary origin; agricultural and food chemical research; experimental and practical feeding of farm animals. Mailing Add: Hoffmann-La Roche Inc Kingsland St Nutley NJ 07110

BAUGH, CHARLES M, b Fayetteville, NC, June 20, 31; m 52; c 4. BIOCHEMISTRY. Educ: Univ Chicago, SB, 58; Tulane Univ, PhD(biochem), 62. Prof Exp: Fel biochem, Tulane Univ, 62-63, from instr to asst prof, 63-65; asst prof pharmacol in med, Wash Univ, 65-67; assoc prof, 67-70, prof nutrit div, 70-73, PROF BIOCHEM & CHMN DEPT, COL MED, UNIV S ALA, 73- Concurrent Pos: External assessor, Nat Res Coun Australia, 76-77. Mem: Am Chem Soc; Am Inst Nutrit; Am Soc Biol Chem. Res: Biosynthesis of the pteridine nucleus, as found in the vitamins, folic acid and riboflavin; antimetabolites in nucleic acids. Mailing Add: Dept of Biochem Col of Med Univ of S Ala Mobile AL 36688

BAUGH, CLARENCE L, b Muskogee, Okla, Mar 31, 29; m 58; c 5. BACTERIAL PHYSIOLOGY. Educ: Univ Okla, BS, 54, MS, 56; Iowa State Univ, PhD(physiol bact), 61. Prof Exp: Instr bact, Iowa State Univ, 56-61; res microbiologist, US Army Biol Labs, Ft Detrick, Md, 61-64; res assoc, Merck Inst Therapeut Res, Pa, 64-71; ASSOC PROF BIOL, TEX TECH UNIV, 71- Mem: Am Soc Microbiol. Res: Carbon dioxide fixation; metabolic control mechanisms; tissue culture; intermediary metabolism; virology. Mailing Add: 6608 Orlando Ave Lubbock TX 79409

BAUGHMAN, DWIGHT JOE, b Lima, Ohio, Dec 20, 32; m 53; c 3. BIOPHYSICS. Educ: Ohio Northern Univ, BA, 53; Western Reserve Univ, MD, 57; Mass Inst Technol, PhD, 62. Prof Exp: Instr biol, Ohio Northern Univ, 53; NSF fel, Mass Inst Technol, 57-59, Nat Heart Inst fel, 59-62, res assoc, 62-69; sr scientist clin biochem, 69-71, Ortho Res fel, 71-74, DIR BIOCHEM, ORTHO RES FOUND, 74- Mem: Biophys Soc; fel NY Acad Sci; Hemat Soc; Int Soc Thrombosis & Haemostasis. Res: Blood coagulation; chemical kinetics; biochemical homeostasis; statistics; protein chemistry. Mailing Add: Ortho Res Found Raritan NJ 08869

BAUGHMAN, GLENN LAVERNE, b Dover, Pa, Apr 26, 31; m 56; c 2. ORGANIC POLYMER CHEMISTRY. Educ: Gettysburg Col, AB, 53; Pa State Univ, PhD(org chem), 61. Prof Exp: Chemist, Armstrong Cork Co, 55-57; res chemist, E I du Pont de Nemours & Co, 61-64; SR RES CHEMIST, CHEM DIV, PPG INDUSTS, BARBERTON, 64- Concurrent Pos: Instr, Barberton Tech Sch, 64-65. Mem: Am Chem Soc. Res: Unsaturated polyesters for thermosetting castings; monomer structure-polymer property correlations; organometallics chemistry; photochemistry. Mailing Add: 479 Dohner Dr Wadsworth OH 44281

BAUGHMAN, NEWTON MOORE, soil physics, see 12th edition

BAUGHMAN, RAY HENRY, b York, Pa, Jan 14, 43; m 69; c 2. CHEMICAL PHYSICS, POLYMER SCIENCE. Educ: Carnegie-Mellon Univ, BS, 64; Harvard Univ, MS, 66, PhD(mat sci), 71. Prof Exp: Staff scientist, 70-73, GROUP LEADER POLYMER SCI, MAT RES CTR, ALLIED CHEM CORP, 74- Mem: Am Phys Soc; Am Chem Soc. Res: Defect structure of solids, molecular dynamics and phase transformations; electrical, optical and mechanical properties of polymers; solid state reactions; molecular spectroscopy. Mailing Add: Mat Res Ctr Allied Chem Corp Morristown NJ 07960

BAUGHN, CHARLES (OTTO), (JR), b Bicknell, Ind, Feb 25, 21; m 44; c 3. BIOLOGY. Educ: Ind State Teachers Col, BSc, 47; Univ NC, MS, 49, PhD(parasitol, bact), 52. Prof Exp: Instr parasitol, Univ NC, 50-52; parasitologist, Stamford Res Labs, Am Cyanamid Co, 52-55, group leader bact, Lederle Labs, 55-61, mgr microbial dis sect, Agr Res Ctr, 61-64; asst dir res, Hess & Clark Div, Richardson-Merrell, Inc, 64-69; SR RES INVESTR, SQUIBB AGR RES CTR, 69- Mem: AAAS; Am Soc Microbiol; NY Acad Sci. Res: Chemotheraphy of microbial infections; resistance to infection. Mailing Add: 29 Maple Ave Flemington NJ 08822

BAUGUESS, CARL THOMAS, JR, b Lancaster Co, Pa, Sept 30, 28; m 65. PHARMACEUTICS. Educ: Univ NC, BS, 54, MS, 66; Univ Miss, PhD, 70. Prof Exp: Pharmacist, Jonesboro's Lee Drug Store, 54-57; instr & dir pharmaceut exten serv, Univ NC, 57-60, part-time instr, 60-64; pharm consult, NC Heart Asn, 64-67; asst prof pharmaceut, Northeast La State Univ, 67-68; ASSOC PROF PHARMACEUT, UNIV SC, 70-, DIR GRAD PROG, COL PHARM, 75- Honors & Awards: Lederle Award, 75. Mem: Am Pharmaceut Asn; Sigma Xi. Res: Investigations of new drugs for antitumor activity in mouse leukemia; pharmacokinetic studies of drug concentration changes after adminstration by various routes in larger animals; microencapsulation of basic drugs by spray dry techniques and coaservation in the development of novel dosage forms. Mailing Add: 129 W Circle Dr Lexington SC 29072

BAULD, NATHAN LOUIS, b Clarksburg, WVa, Dec 12, 34. ORGANIC CHEMISTRY. Educ: WVa Univ, BS, 56; Univ Ill, PhD(org chem), 59. Prof Exp: NSF fel, Harvard Univ, 59-60; res chemist, Rohm and Haas Co, Pa, 60-61; from instr to assoc prof, 61-73, PROF CHEM, UNIV TEX, AUSTIN, 73- Concurrent Pos: Sloan Found fel, 66-68. Mem: Am Chem Soc. Res: Physical and stereo chemistry; anion radicals; molecular orbital theory; aromatic systems. Mailing Add: Dept of Chem 213 Chem Bldg Univ of Tex Austin TX 78712

BAULKNIGHT, CHARLES WESLEY, b Concord, NC, Nov 15, 11; m 46; c 1. CHEMISTRY. Educ: J C Smith Univ, BS, 35; Am Inst Chemists, cert. Hon Degrees: DSc, J C Smith Univ, 59. Prof Exp: Asst, Univ Pa, 38-40; jr chemist, Chem Warfare Serv, Edgewood Arsenal, Md, 41-43; res chemist, Nat Defense Res Comt, Univ Pa, 44-45; res assoc, Mellon Inst, 45-46; phys chemist, Pitman-Dunn Lab, Frankford Arsenal, Philadelphia, 46-56 & missile & space vehicle dept, aerosci lab, Gen Elec Co, 56-63; RES SCIENTIST, GRUMMAN AEROSPACE CORP, 63- Concurrent Pos: Mem bd trustees & exec comt, J C Smith Univ, 65-; mem bd trustees & col pres adv coun, Dowling Col; mem bd trustees, Friends of Nassau County Mus. Mem: AAAS; fel Am Phys Soc; fel Am Inst Chemists; Am Chem Soc; Sigma Xi. Res: High temperature fluid properties of gases including transport phenomena and chemical kinetics; quantitative investigation of fatty acids via chromatography. Mailing Add: 19 Romscho St Bethpage NY 11714

BAUM, ALAN WILLIAM, organic chemistry, deceased

BAUM, ARTHUR ALOYSIUS, b Escanaba, Mich, July 10, 14; m 50; c 5. ORGANIC CHEMISTRY. Educ: Univ Notre Dame, BS, 36, MS, 37, PhD(org chem), 39. Prof Exp: SUPVR PROCESS DEVELOP, E I DU PONT DE NEMOURS & CO, 39- Mem: AAAS; Am Chem Soc. Res: Surface active agents; sulfur dyes; vulcanizing agents and accelerators for synthetic rubber; organic syntheses; rubber chemicals; dyes; intermediates. Mailing Add: 2311 Newport Gap Pike Wilmington DE 19808

BAUM, BERNARD, b Boston, Mass, Sept 21, 24; m 53; c 1. POLYMER CHEMISTRY. Educ: Lowell Technol Inst, BS, 47; Clark Univ, MS, 49, PhD(chem), 50. Prof Exp: Chemist, Sherwin-Williams Co, 50-53; develop assoc, Union Carbide Plastics Co, 53-62; group leader, Allied Chem Co, 62-63 & Borden Chem Co, 63-64; mgr prod develop, Cast Nylon Dept, Budd Co, 64-67; mgr, Chem & Mat Develop Lab, 67-69, mgr, Chem & Mat Div, 69-72, VPRES & MGR MAT RES & DEVELOP, DeBELL & RICHARDSON, 72- Mem: Commercial Develop Asn; Am Chem Soc; Air Pollution Control Asn; Soc Plastics Engr. Res: Physical organic; synthesis; polymer structure vs properties; coatings; elastomers; permanence properties of materials. Mailing Add: DeBell & Richardson Enfield CT 06082

BAUM, BERNARD R, b Paris, France, Feb 14, 37; m 61; c 1. TAXONOMY, BOTANY. Educ: Hebrew Univ, Jerusalem, MS, 63, PhD(tamarix taxon), 66. Prof Exp: Res scientist, Plant Res Inst, 67-73, SECT CHIEF, BIOSYSTEMATICS RES INST, CAN DEPT AGR, 73- Concurrent Pos: Asst ed, Can J Bot, 74- Mem: AAAS; Bot Soc France; Int Asn Plant Taxon; Can Bot Asn; Prof Inst Pub Serv Can. Res: Monograph of Tamarix and Avena; nomenclature; utilization of computers for taxonomy; taxometrics and statistics; evolution. Mailing Add: Biosystematics Res Inst Cent Exp Farm Ottawa ON Can

BAUM, BRUTON MURRY, b Brooklyn, NY, Dec 6, 34; m 65; c 1. ORGANIC CHEMISTRY, TEXTILE CHEMISTRY. Educ: Brooklyn Col, BS, 56; Univ Pittsburgh, PhD(org chem), 62. Prof Exp: RES CHEMIST, FMC CORP, 62- Mem: Am Chem Soc; Am Asn Textile Chemists & Colorists; Res: Preparation and bleaching of textiles; flame retardants. Mailing Add: 195 Clover Ln Princeton NJ 08540

BAUM, DAVID, b Saratoga Springs, NY, Aug 4, 27; m 59; c 4. PEDIATRIC CARDIOLOGY, BIOCHEMISTRY. Educ: Dartmouth Col, AB, 51; Cornell Univ, MD, 55. Prof Exp: Intern med, Kings County Hosp, NY, 55-56; resident pediat, New York Hosp-Cornell Med Ctr, 56-59; asst pediat cardiol, Johns Hopkins Hosp, 59-60 & Mayo Clin, 60-61; from instr to assoc prof pediat, Sch Med, Univ Wash, 62-71, asst prog dir, Clin Res Ctr, 63-70; ASSOC PROF PEDIAT & DIR DIV PEDIAT CARDIOL, SCH MED, STANFORD UNIV, 71- Concurrent Pos: Fel pediat cardiol, Johns Hopkins Hosp, 59-60 & Mayo Clin, 60-61; res trainee biochem, Sch Med, Univ Wash, 65-66; consult, Madigan Army Hosp, Ft Lewis & Rainier Sch, Buckley, Wash, 62-71 & Silas Hayes Army Hosp, Ft Ord, Calif. Mem: Soc Pediat Res; Am Acad Pediat; Soc Exp Biol & Med; Am Heart Asn; Am Physiol Soc. Res: Metabolic aspects of hypoxia as related to thermoregulation, exercise and growth in children with cardiovascular disease. Mailing Add: Dept of Pediat Stanford Univ Med Ctr Stanford CA 94305

BAUM, DENNIS WILLARD, b Allentown, Pa, Dec 29, 40; m 64; c 2. GAS DYNAMICS. Educ: Muhlenberg Col, BS, 63; Lehigh Univ, MS, 64, PhD(physics), 67. Prof Exp: Sr physicist gas dynamics, Physics Int Co, Inc, 67-72; MGR GAS DYNAMICS, ARTEC ASSOCS INC, 72- Mem: Sigma Xi. Res: Development and design of explosively-driven high energy-density devices such as pulsed magnetohydrodynamic generators, hypervelocity launchers and high performance shock-tubes. Mailing Add: 246 La Espiral Orinde CA 94545

BAUM, EDWARD JOSEPH, physical chemistry, see 12th edition

BAUM, GARY ALLEN, b New Richmond, Wis, Oct 2, 39; m 64; c 2. SOLID STATE PHYSICS. Educ: Wis State Univ, River Falls, BS, 61; Okla State Univ, MS, 64, PhD(physics), 69. Prof Exp: Res engr, Douglas Aircraft Co, Calif, 63-64; physicist, Electron Optics Group, Dow Chem Co, Colo, 64-65, sr physicist, 65-66; asst prof, 69-73, ASSOC PROF PHYSICS, INST PAPER CHEM, 73- Mem: Am Asn Physics Teachers; Tech Asn Pulp & Paper Indust. Res: Ferroelectric and antiferroelectric materials; vapor deposition, sputtering and accelerated ion techniques for thin film preparation; electrical and thermal properties of rutile; electrical properties of polymers. Mailing Add: 426 Fidelis Appleton WI 54911

BAUM, GEORGE, b Hungary, Jan 11, 33; US citizen; m 54; c 3. ORGANIC BIOCHEMISTRY. Educ: Case Inst Technol, BS, 54; Ohio State Univ, MS, 60, PhD, 66. Prof Exp: Engr solid fuel propellants, Wright Air Develop Command, 54-65; res engr synthetic hydraul fluids, Air Force Mat Lab, 55-65; RES CHEMIST, CORNING GLASS WORKS, 66- Concurrent Pos: Vis res assoc, Beth Israel Hosp, 75-76. Mem: NY Acad Sci; Am Chem Soc; Brit Chem Soc. Res: Organometallic synthesis; reinforced plastics; peptide chemistry; fluoroaromatic chemistry; ion-selective electrodes; immobilized enzymes; affinity chromatography; hormone secretion by tumor cells. Mailing Add: Corning Glass Works Sullivan Park Corning NY 14830

BAUM, GERALD ALLAN, polymer chemistry, see 12th edition

BAUM, GERALD L, b Milwaukee, Wis, Dec 19, 24; m 51; c 3. MEDICINE. Educ: Univ Wis, BS, 45, MD, 47. Prof Exp: Intern, Jewish Hosp, Cincinnati, 47-48, resident med, 48 & 49-51; res chest med, Bellevue Hosp, New York, 51-52; fel bronchology, St Luke's Hosp, Chicago, 52; fel med mycol, Jewish Hosp, Cincinnati, 56-58; chief pulmonary sect, Vet Admin Hosp, Cincinnati, 58-65; PROF MED, SCH MED, CASE WESTERN RESERVE UNIV, 65- Concurrent Pos: NIH fel, Tel Hashomer, Israel, 63-64 & Kupat Cholim award, Israel, 63-64; assoc prof, Univ Cincinnati, 64-65; chief pulmonary sect, Vet Admin Hosp, Cleveland, 65-73. Mem: Am Thoracic Soc; Am Col Physicians; Am Fedn Clin Res; Mycol Soc Am; Med Mycol Soc of the Americas. Res: Pulmonary diseases, particularly fungus infections; rehabilitation of patients with chronic pulmonary insufficiency and tuberculosis; laboratory aspects of mycology. Mailing Add: Case Western Reserve Univ Sch Med 2119 Abington Rd Cleveland OH 44106

BAUM, HARRY, b New York, NY, Oct 23, 15; m 44; c 2. ANALYTICAL CHEMISTRY. Educ: City Col New York, BS, 36; NY Univ, MS, 39. Prof Exp: Res chemist, Beth Israel Hosp, New York, 36-38; asst microchem, NY Univ, 38-39; res chemist, Gen Cigar Co, Pa, 40-41; supvr & plant chemist, US Rubber Co, Iowa, 41-43; supvr, Anal Res Lab, Publicker Indust, 46-47; supvr, Anal Develop Lab, 48-69, SUPT QUAL CONTROL, ROHM AND HAAS CO, 69- Concurrent Pos: Referee, Collab Int Pesticide Anal Coun, 69- Mem: AAAS; Am Chem Soc; Asn Off Anal Chemists; Am Soc Testing & Mat. Res: Analytical research and development. Mailing Add: 8210 Stockton Rd Elkins Park PA 19117

BAUM, JOHN, b New York, NY, June 2, 27; m 50; c 5. RHEUMATOLOGY, IMMUNOLOGY. Educ: NY Univ, BA, 49, MD, 54. Prof Exp: Instr med, Southwestern Med Sch, Univ Tex, 59-62, asst prof, 62-68; assoc prof med, 68-72, assoc prof prev med & community health, 69, assoc prof pediat, 70-72, PROF MED, PEDIAT, PREV MED & COMMUNITY HEALTH, SCH MED & DENT, UNIV ROCHESTER, 72-; DIR ARTHRITIS & CLIN IMMUNOL UNIT, MONROE COMMUNITY HOSP, ROCHESTER, 68- Concurrent Pos: Clin scholar, Arthritis Found, 65-70; dir arthritis clin, Parkland Mem Hosp, Dallas, Tex, 59-68, dir med clin, 65-67; consult rheumatol & co-dir arthritis clin, Scottish Rite Hosp Crippled Children, Dallas, 60-68; mem drug efficacy panel, Nat Acad Sci, 66 & DMSO panel, 72; sr assoc physician, Strong Mem Hosp, Rochester, NY, 68, dir pediat arthritis clin, 70-, physician & pediatrician, 72-; mem test comt rheumatol, Am Bd Internal Med, 70; mem merit rev bd immunol, Vet Admin, 72-; coordr prob area, Eval Therapeut in Arthritis, US-USSR Coop Pub Health & Med Sci, 74-; mem adv panel analgesics, sedatives & anti-inflammatory agts, US Pharmacopoeia, 75- Mem: Am Rheomatism Asn; Am Fedn Clin Res; Am Soc Human Genetics; Am Asn Immunologists; Heberden Soc. Res: Clinical studies of delayed hypersensitivity and mechanisms of inflammation in humans; chemotaxis of polymorphonuclear leukocytes in human disease; immunology of rheumatoid arthritis and systemic lupus erythematosus; drug studies in juvenile arthritis. Mailing Add: Sch of Med & Dent Univ of Rochester Rochester NY 14620

BAUM, JOHN DANIEL, b Pittsburgh, Pa, July 31, 18; m 48; c 3. MATHEMATICS. Educ: Yale Univ, BA, 41, MA, 49, PhD(math), 53. Prof Exp: Asst instr math, Yale Univ, 50-53; from instr to assoc prof, 53-63, PROF MATH, OBERLIN COL, 63- Concurrent Pos: NSF res grant, 55; sci fac fel, Princeton Univ, 59-60; hon vis prof, Birkbeck Col, Univ London, 66-67; vis hon prof, Univ New South Wales, 74. Mem: Am Math Soc; Math Asn Am; London Math Soc; Australian Math Soc. Res: Topology; topological dynamics. Mailing Add: 97 Parkwood Lane Oberlin OH 44074

BAUM, JOHN JOSEPH, b Canton, Ohio, Oct 12, 32; m 54; c 4. INSTRUMENTATION. Educ: Ohio Wesleyan Univ, BA, 56; Case Inst Technol, BS, 56; Univ Cincinnati, MS, 61. Prof Exp: Physicist, Aircraft Nuclear Propulsion Div, Gen ElecCo, 56-61; exec vpres & gen mgr, Molechem, Inc, Harshaw Chem Co, 61-66; mgr exp physics, Ill Inst Technol Res Inst, 66-67; assoc physicist, Argonne Nat Lab, 67-69; V PRES PROD DEVELOP, ANALOG TECHNOL CORP, PASADENA, 70- Concurrent Pos: Consult, Ill Inst Technol Res Inst & US Air Force, 67- Mem: Am Phys Soc. Res: Mechanisms of interactions of nuclear particles with semiconductor materials and their applications to solid state detector development; development of advanced analytical data systems.

BAUM, JOHN L, physics, see 12th edition

BAUM, JOHN WILLIAM, b Highland Park, Ill, Mar 22, 40; m 62; c 2. ORGANIC CHEMISTRY. Educ: Univ Minn, BA & BChem, 62; Wash Univ, PhD(org chem), 67. Prof Exp: Res chemist, MacMillan Bloedel Res Ltd, BC, 67-70; DIR CHEM DEVELOP, ZOECON CORP, 70- Mem: AAAS; Am Chem Soc. Res: Synthesis of indoles and alicyclic compounds; photochemistry of alicyclic ketones; chemistry of insect hormones and pheromones; experimental design in process development. Mailing Add: Zoecon Corp 975 California Ave Palo Alto CA 94304

BAUM, JOSEPH HERMAN, b Chicago, Ill, Sept 9, 27; m 70; c 2. PATHOLOGY. Educ: Roosevelt Univ, BS, 53; Northwestern Univ, PhD(path), 62. Prof Exp: Asst oncol, Chicago Med Sch, 53-54; fel path, Northwestern Univ, 62-63, instr, 63-66, asst prof path & dir student labs, 66-68; ASSOC PROF PATH, SCH MED, TEMPLE UNIV, 68-, ASST DEAN GRAD SCH, 73- Mem: AAAS; Am Soc Cell Biologists; Electron Micros Soc Am; Am Educ Res Asn; Am Soc Exp Path. Res: Cellular and molecular pathology; areas of oncology and cardiac pathology; biomedical education. Mailing Add: Dept of Path Temple Univ Sch of Med Philadelphia PA 19140

BAUM, LAWRENCE STEPHEN, b Scranton, Pa, Mar 3, 38; m 57; c 2. PLANT PHYSIOLOGY, CELL PHYSIOLOGY. Educ: Univ Ala, BS, 60, MS, 62, PhD(plant physiol), 65. Prof Exp: Asst prof biol, Exten Ctr, Univ Ala, 65-66; asst prof, 66-69, ASSOC PROF BIOL, NORTHEASTERN LA UNIV, 69- Mem: Bot Soc Am; Am Soc Plant Physiol. Res: Mineral nutrition in algae; effects of pesticides on micro-organisms in the soil. Mailing Add: Dept of Biol Northeast La Univ Monroe LA 71201

BAUM, LLOYD, b Ashton, Idaho, May 11, 23; m 53; c 2. DENTISTRY. Educ: Univ Ore, DMD, 46; Univ Mich, MS, 52. Prof Exp: Instr restorative dent, Sch Dent, Univ Southern Calif, 52-53; prof & clin dir, Sch Dent, Loma Linda Univ, 53-73; PROF DENT, SCH DENT MED, STATE UNIV NY STONY BROOK, 73- Mem: Am Acad Restorative Dent; Am Acad Crown & Bridge Prosthodont; fel Am Col Dent; Int Asn Dent Res. Res: Invention of precision intra-oral drilling device; development of a new dental filling gold. Mailing Add: Sch of Dent Med State Univ NY Stony Brook NY 11790

BAUM, MARTIN ALFRED, analytical chemistry, organic chemistry, see 12th edition

BAUM, MARTIN DAVID, b New York, NY, Jan 30, 41; m 64; c 2. PHOTOGRAPHIC CHEMISTRY. Educ: Mass Col Pharm, BS, 62, MS, 64; Univ Ill Med Ctr, PhD(org chem), 68. Prof Exp: Res chemist photochem, 68-70, res chemist dye synthesis, 70-71, mkt rep, 72-73, sr res chem, 74, RES SUPVR PHOTOG SYST, E I DU PONT DE NEMOURS & CO, INC, 74- Mem: Am Chem Soc. Res: Synthesis of dyes for use in photographic systems; development of products based on silver and non-silver photography. Mailing Add: Photoprod Dept PO Box 1009 E I du Pont de Nemours & Co Inc Rochester NY 14603

BAUM, O EUGENE, b Philadelphia, Pa, Oct 26, 16; m 47; c 2. PSYCHIATRY, PSYCHOANALYSIS. Educ: Univ Pa, BA, 37, MD, 43. Prof Exp: Dir resident psychiat training, 66-73, PROF PSYCHIAT, MED COL PA, 66-, COORDR UNDERGRAD PSYCHIAT, 73- Concurrent Pos: Pvt pract psychiat & psychoanal, 49-; mem, Comn Psychoanal Educ & Res, 72-74; consult, Norristown State Hosp. Mem: AAAS; Am Psychiat Asn; Am Psychoanal Asn; Asn Dirs Med Educ in Psychiat (secy, 75-76). Mailing Add: Dept of Psychiat Med Col of Pa Philadelphia PA 19129

BAUM, PARKER BRYANT, b Memphis, Tenn, Dec 21, 23; m 54; c 2. PHYSICAL CHEMISTRY, INORGANIC CHEMISTRY. Educ: Col of William & Mary, BS, 47; Univ Tenn, MS, 52; Univ NC, PhD(phys chem), 62. Prof Exp: Instr chem, Old Diminion Col, 47-50, asst prof, 52-56, assoc prof, 57-59, prof, 62-65; mem fac, 65-67, PROF CHEM, SKIDMORE COL, 67- Mem: AAAS; Am Chem Soc. Res: Thermodynamics; electromotive force measurements; exchange reactions using radiotracers. Mailing Add: Dept of Chem Skidmore Col Saratoga Springs NY 12866

BAUM, PAUL FRANK, b New York, NY, July 20, 36; m 61; c 2. MATHEMATICS. Educ: Harvard Univ, AB, 58; Princeton Univ, PhD(math), 63. Prof Exp: Instr math, Princeton Univ, 62-63; NSF fel, Oxford & Cambridge Univs, 63-64; fel, Inst Advan Study, 64-65; asst prof, Princeton Univ, 65-67; assoc prof, 67-72, PROF MATH, BROWN UNIV, 72- Mem: Am Math Soc. Res: Algebraic topology; lie groups. Mailing Add: Dept of Math Brown Univ Providence RI 02912

BAUM, PAUL M, b Brooklyn, NY, Feb 25, 35; m 62; c 1. EXPERIMENTAL PHYSICS. Educ: Columbia Univ, AB, 55; Univ Ill, MS, 57, PhD(physics), 62. Prof Exp: Res scientist, Gruman Aircraft Eng Corp, 62-63; ASST PROF PHYSICS, QUEENS COL, NY, 63- Mem: Am Asn Physics Teachers. Res: Study of properties of mesons using 300 mev betatron; energy levels of multiply ionized atoms. Mailing Add: Dept of Physics Queens Col Flushing NY 11367

BAUM, PETER JOSEPH, b Lennox, Calif, June 4, 43; wid; c 1. PHYSICS. Educ: Univ Calif, Santa Barbara, BA, 65; Univ Nev, Reno, MS, 67; Univ Calif, Riverside, PhD(physics), 71. Prof Exp: Res engr microelectronics, Autonetics Div, NAm Rockwell Corp, 67-68; physicist comput anal, Corona Lab, Naval Weapons Ctr, 68-70; res assoc solar-plasma physics, 70-74, Air Force Off Sci Res grant, 72-74, RES PHYSICIST SOLAR-PLASMA PHYSICS, UNIV CALIF, RIVERSIDE, 74- Concurrent Pos: US deleg, Int Atomic Energy Agency Fourth Conf Plasma Physics & Controlled Nuclear Fusion Res, USAEC, 71; consult, Sandia Labs, Albuquerque, NMex, 73-75; NSF grant, 75- Mem: Am Phys Soc. Res: Solar flares; laboratory study of magnetic field line reconnection; magnetic energy conversion; plasma radiation sources; controlled thermonuclear fusion. Mailing Add: Inst Geophys & Planetary Physics Univ of Calif Riverside CA 92507

BAUM, ROBERT HAROLD, b New York, NY, June 15, 36; m 57; c 3. BIOCHEMISTRY, MICROBIOLOGY. Educ: Cornell Univ, BS, 57; Univ Ill, PhD(biochem), 62. Prof Exp: Res asst, Univ Ill, 61-62; Am Cancer Soc fel, Oak Ridge Nat Lab, 62-64; asst prof biochem, State Univ NY Col Forestry, Syracuse Univ, 64-74; ASST PROF MICROBIOL, SCH PHARM, TEMPLE UNIV, 74- Mem: AAAS; Am Soc Microbiol; Am Chem Soc. Res: Metabolism of terpenoids by microorganisms; isolation and study of quinones of lactic acid bacteria. Mailing Add: Dept of Microbiol Sch of Pharm Temple Univ Philadelphia PA 19140

BAUM, SIEGMUND JACOB, b Vienna, Austria, Nov 14, 20; nat US; m 47; c 5. PHYSIOLOGY, RADIOBIOLOGY. Educ: Univ Calif, Los Angeles, BA, 49, MA, 50; Univ Calif, Berkeley, PhD(physiol), 59. Prof Exp: Physiologist, US Naval Radiol Defense Lab, 50-60; group leader physiol & radiobiol, Douglas Missile & Space Systs, 60-62; head cellular radiobiol div, 62-64, CHMN EXP PATH DEPT, ARMED FORCES RADIOBIOL RES INST, 64- Honors & Awards: Sci Award, US Naval Radiol Defense Lab, 60; Except Civilian Serv Award, Defense Nuclear Agency, 73. Mem: Am Physiol Soc; Radiation Res Soc; Int Soc Exp Hemat; Transplantation Soc. Res: Endocrinology; biological effects of radiation; erythrocyte precursor system; post-irradiation bone marrow therapy; experimental hematology; space radiobiology. Mailing Add: Exp Path Dept Armed Forces Radiobiol Res Inst Bethesda MD 20014

BAUM, STANLEY, b New York, NY, Dec 26, 29; m 58; c 3. RADIOLOGY. Educ: NY Univ, BA, 51; State Univ Utrecht, MD, 57; Univ Pa, cert med, 61. Prof Exp: Intern, Kings County Hosp Med Ctr, 57-58; Nat Cancer Inst resident radiol, Grad Hosp, Univ Pa, 58-61; fel cardiovasc radiol, Sch Med, Stanford Univ, 61, instr radiol, 62; from instr to prof, Univ Pa, 62-71; prof radiol, Harvard Med Sch, 71-75; PROF RADIOL & CHMN DEPT, SCH MED, UNIV PA, 75- CHIEF RADIOL, HOSP UNIV PA, 75- Concurrent Pos: Asst radiologist, Grad Hosp, Univ Pa, 62-; consult, Vet Admin Hosp, Wilmington, Del, 63-66 & Vet Admin Hosp, Philadelphia, 65-; asst radiologist, Presby-Univ Pa Med Ctr, 66-; radiologist, Mass Gen Hosp, 71- Mem: Asn Univ Radiol; Am Col Radiol; Radiol Soc NAm; Am Col Cardiol; Am Gastroenterol Asn. Res: Cardiovascular radiology; selective arteriography; pharmacologic control of portal hypertension. Mailing Add: Univ Hosp 3400 Spruce St Philadelphia PA 19014

BAUM, STEPHEN GRAHAM, b New York, NY, Apr 28, 37; m 61; c 1. INTERNAL MEDICINE, INFECTIOUS DISEASES. Educ: Cornell Univ, AB, 58; NY Univ, MD, 62. Prof Exp: From intern to resident med, Harvard Div, Boston City Hosp, 62-68; assoc, 68-69, asst prof med, 69-73, ASSOC PROF MED & CELL BIOL & DIR MD-PhD PROG, ALBERT EINSTEIN COL MED, 73-, CO-DIR INFECTIOUS DIS DIV, 75- Concurrent Pos: Nat Inst Allergy & Infectious Dis res assoc virol, 64-66; Med Res Coun spec fel, Nat Inst Med Res, London, 66-67; NY Health Res Coun career scientist, 68-; Am Cancer Soc fac res assoc. Mem: Infectious Dis Soc Am; Harvey Soc; Am Soc Microbiol. Res: Viral oncology. Mailing Add: Albert Einstein Col of Med 1300 Morris Park Ave Bronx NY 10461

BAUM, STUART J, b Brooklyn, NY, Mar 7, 39; m 64; c 2. INORGANIC CHEMISTRY, BIOORGANIC CHEMISTRY. Educ: Queens Col, BS, 60; Cornell Univ, MS, 63, PhD(molecular biol), 65. Prof Exp: Asst prof, 65-71, ASSOC PROF CHEM, STATE UNIV NY COL PLATTSBURGH, 71- Mem: Am Chem Soc. Res: Organic reaction mechanisms; organometallic chemistry; mechanisms of biochemical reactions; structure of coordination complexes. Mailing Add: Dept of Chem State Univ of NY Col Plattsburgh NY 12901

BAUM, WERNE A, b Giessen, Ger, Apr 10, 23; nat US; m 45; c 2. METEOROLOGY, CLIMATOLOGY. Educ: Univ Chicago, BS, 43, MS, 44, PhD(meteorol), 48. Hon Degrees: ScD, Mt St Joseph Col, RI, 71 & Univ RI, 74; DPA, Husson Col, 72. Prof Exp: From asst prof to prof meteorol, Fla State Univ, 49-58, head dept, 49-58, dir univ res, 57-58, dean grad sch & dir res, 58-60, dean faculties, 60-63; prof meteorol, dean faculties & vpres acad affairs, Univ Miami, 63-65; prof meteorol & vpres sci affairs, NY Univ, 65-67; dep adminr, Environ Sci Serv Admin, 67-68; prof physics & geog & pres, Univ RI, 68-73; PROF GEOG & UNIV CHANCELLOR, UNIV WIS-MILWAUKEE, 73- Concurrent Pos: Ed, J Meteorol, Am Meteorol Soc, 49-61, ed-in-chief, Periodicals, 57-61; mem comt climat adv to US Weather Bur, Nat Acad Sci, 55-58, chmn panel on educ, Comt Atmospheric Sci, 62-64; mem coun, Oak Ridge Inst Nuclear Studies, 58-62; mem exec reserve, US Weather Bur, 58-63, chmn adv comt educ & training, 64-66; trustee, Univ Corp Atmospheric Res, 59-63 & 65-67, corp secy, 63-67; mem adv panel on atmospheric sci prog, NSF, 63-67, chmn, 65-67; dir, Fund for Overseas Res Grants & Educ, 65-; mem sci adv coun, Tex Christian Univ Res Found, 67-, 73 & 75; US rep panel of experts on meteorol educ & training, World Meteorol Orgn, UN, 71-; mem, Nat Sea Grant Prog Adv Panel, 74-; life trustee, Univ RI Found, 75. Honors & Awards: Spec Citation, Am Meteorol Soc, 62, Charles Franklin Brooks Award, 75; Honors Medal, Fla Acad Sci, 64. Mem: Fel AAAS; fel Am Meteorol Soc (pres-elect, 76-77); Am Coun of Educ; fel Am Geophys Union. Res: Evaluation and analysis of foreign meteorological material; academic and scientific administration. Mailing Add: Univ of Wis Milwaukee WI 53201

BAUM, WERNER CHRISTIAN, b Naugatuck, Conn, July 4, 16; m 44; c 2. BIOLOGY. Educ: NY State Col Forestry, Syracuse Univ, BSc, 36; Rutgers Univ, MSc, 41, PhD(bot, plant physiol), 48. Prof Exp: Asst prof bot, Okla State Univ, 48-51; head dept bot microtechnol, Carolina Biol Supply Co, 51-52; asst prof biol, Albany Col Pharm, Union Univ, NY, 52-59; assoc prof, 59-64, PROF BIOL, STATE UNIV NY ALBANY, 64- Mem: AAAS; Bot Soc Am; Am Soc Plant Physiol; Am Inst Biol Sci. Res: Plant growth; phototaxis. Mailing Add: Dept of Biol State Univ of NY Albany NY 12203

BAUM, WILLIAM ALVIN, b Toledo, Ohio, Jan 18, 24; m 61. ASTRONOMY. Educ: Univ Rochester, BA, 43; Calif Inst Technol, MS, 45, PhD(physics), 50. Prof Exp: Physicist, Naval Res Lab, Washington, DC, 46-49; astronomer, Mt Wilson & Palomar Observs, 50-65; DIR PLANETARY RES CTR, LOWELL OBSERV, 65- Concurrent Pos: Adj prof, Ohio State Univ, 69- & Northern Ariz Univ, 73-; consult, Off Space Sci, NASA, 70-, sci team mem, Viking Mars Mission, 70- Mem: Am Astron Soc; Int Astron Union. Res: Photoelectric photometry; photoelectric image-receiving systems; optical instrument development; globular star clusters; stellar populations; magnitudes and redshifts of galaxies; planetary science; spacecraft instrumentation; observational cosmology; constancy of physical constants. Mailing Add: Lowell Observ PO Box 1269 Flagstaff AZ 86001

BAUM, WILLIAM STANHOPE, b Kent, Conn, May 2, 09; m 42; c 2. CANCER, INTERNAL MEDICINE. Educ: Yale Univ, BA, 31, MD, 36; Am Bd Internal Med, dipl. Prof Exp: From intern to resident pediat, Sch Med, Univ Rochester, 36-39; USPHS, 40-64, ward surgeon internal med, Hosps, Ellis Island, NY, 40-41, ward surgeon, Baltimore, Md, 41-42, ward surgeon & chief med serv, Boston, Mass, 43-44, chief med serv, 49-52, asst chief & dep chief med serv, Staten Island, NY, 47-49, med dir & asst chief field invests & demonstr br, Nat Cancer Inst, 52-57, dep med officer, Div Indian Health, 57-58, area med officer in chg, 58-64; physician, Stud Health Serv, Ariz State Univ, 64-65; internist, Kennedy Space Ctr, Fla, 65-68; physician, Ariz State Univ, 68-70; physician, 70-72, CHIEF MED CLINS, MARICOPA COUNTY GEN HOSP, 72-; HEALTH OFFICER, MARICOPA COUNTY HEALTH DEPT, PHOENIX, 70- Concurrent Pos: Res fel pediat, Sch Med, Univ Rochester, 39-40; med officer, Am Embassy, Berlin, Ger, 41-42; med officer aviation med, NIH, 42-43 & Indust Hyg Res Lab, 46-47. Mem: AMA; Aerospace Med Asn; fel Am Col Physicians. Res: Vitamin A distribution and storage; physiological effects of anoxia; cardiac function tests; environmental aspects of cancer. Mailing Add: Maricopa Co Gen Hosp 2601 E Roosevelt Phoenix AZ 85001

BAUMAN, BERNARD D, b Rochester, NY, June 15, 46; c 1. ORGANIC CHEMISTRY. Educ: Eastern Nazarene Col, BA, 68; State Univ NY Albany, PhD(org chem), 73. Prof Exp: Scholar org chem, Pa State Univ, 73-74; SR CHEMIST ORG CHEM, ROHM AND HAAS CO, 74- Mem: Am Chem Soc. Res: Synthetic and mechanistic organic chemistry. Mailing Add: Apt 1029 777 W Germantown Pike Plymouth Meeting PA 19462

BAUMAN, DAVID STANLEY, b Hobart, Okla, Apr 30, 38; m 69. MICROBIOLOGY. Educ: Univ Okla, BS, 61, MS, 64; WVa Univ, PhD(microbiol), 68. Prof Exp: Asst supvr microbiol, Norman Munic Hosp & lab consult, Prof Ctr Labs, 61-64; res assoc prev med, WVa Univ, 68-69; asst prof, 69-74, PROF COMMUNITY MED, UNIV KY, 74- Mem: AAAS; Am Soc Microbiol; Am Thoracic Soc. Res: Antigenic structure of Histoplasma capsulatum and mechanisms of immunity to mycotic diseases. Mailing Add: Dept of Community Med Univ of Ky Sch of Med Lexington KY 40506

BAUMAN, HOWARD EUGENE, b Woodworth, Wis, Mar 20, 25; m 48; c 2. FOOD SCIENCE. Educ: Univ Wis, BS, 49, MS, 51, PhD(bact), 53. Prof Exp: Head bact sect res & develop, Pillsbury Mills, Inc, 53-55, head biol res br, Cent res, Pillsbury Co, 55-60, assoc dir res, 60-67, dir corp res, 67-69, VPRES SCI & TECHNOL, PILLSBURY CO, MINNEAPOLIS, 69- Mem: AAAS; Am Soc Microbiol; Soc Indust Microbiol; Am Asn Cereal Chemists; Inst Food Technol. Res: Nutrition, microbiology. Mailing Add: 4580 Greenwood Dr Minnetonka MN 55343

BAUMAN, JOHN E, JR, b Kalamazoo, Mich, Jan 18, 33; m 64; c 3. INORGANIC CHEMISTRY. Educ: Univ Mich, BS, 55, MS, 60, PhD(chem), 61. Prof Exp: Res chemist, Midwest Res Inst, 55-58; asst prof, 61-65, assoc grad dean, 68-70, ASSOC PROF CHEM, UNIV MO-COLUMBIA, 65- Concurrent Pos: Mem staff, Scripps Inst Oceanog, Univ Calif, 70. Mem: Am Chem Soc; Sigma Xi; Calorimetry Conf. Res: Thermodynamics of inorganic coordination compounds; ionic reactions in solution; calorimetry. Mailing Add: Dept of Chem Univ Mo-Columbia Columbia MO 65201

BAUMAN, JOHN W, JR, b Stockton, Calif, Dec 17, 18; m 48; c 4. PHYSIOLOGY, BIOCHEMISTRY. Educ: Univ Southern Calif, AB, 48; Univ Calif, Berkeley, PhD(physiol), 55. Prof Exp: Instr physiol, Sch Med, NY Univ, 56-58, asst prof, 58-61; res scientist exp med, Bur Res, Princeton, NJ, 61-62, chief endocrinol sect, 62-70; ASSOC PROF PHYSIOL, NJ COL MED & DENT, 70- Concurrent Pos: Fel biol, Princeton Univ, 68-70; adj asst prof, Med Sch, NY Univ, 61-62; vis prof physiol, NJ Col Med & Dent, 68, 69. Mem: AAAS; Am Physiol Soc; Am Chem Soc; Endocrine Soc; Brit Soc Endocrinol. Res: Endocrine effects on renal function; mechanisms of antidiuresis and water diuresis; endocrine influence on cholesterol, bile and fat metabolism. Mailing Add: Dept of Physiol NJ Col of Med and Dent Newark NJ 07103

BAUMAN, LOYAL FREDERICK, b Nokomis, Ill, Oct 18, 20; m 47; c 5. PLANT BREEDING. Educ: Univ Ill, BS, 46, MS, 47, PhD(plant breeding), 50. Prof Exp: Asst corn breeding, Univ Ill, 47-53, res agronomist, Agr Res Serv, USDA, Ga, 53-59 & Ohio, 59; prof genetics, 59-70, PROG AGRON, PURDUE UNIV, WEST LAFAYETTE, 69- Mem: Am Soc Agron. Res: Corn breeding and genetics; cytoplasmic effects; quantitative inheritance. Mailing Add: Dept of Agron Agr Exp Sta Purdue Univ West Lafayette IN 47906

BAUMAN, NORMAN, b Brooklyn, NY, Aug 20, 32; m 57; c 3. IMMUNOLOGY, MEDICINE. Educ: Harvard Univ, AB, 53; NY Univ, MD, 57. Prof Exp: Intern, Bronx Munic Hosp, 57-58; fel rheumatology, Duke Univ, 60-61, asst res med, 61-62, fel rheumatology, 62-64, instr med, 62-64; SR SCIENTIST, LEDERLE LABS, AM CYANAMID CO, 64- Concurrent Pos: Res assoc, Nat Inst Arthritis & Metab Dis, 58-60. Mem: Math Asn Am; Am Rheumatism Asn. Res: Mechanism of drug action; complement; mediators of immune reactions; rheumatology. Mailing Add: Lederle Labs Am Cyanamid Co Pearl River NY 10965

BAUMAN, RICHARD GILBERT, b Warren, Ohio, Sept 14, 24; m 45; c 3. PHYSICAL CHEMISTRY. Educ: Case Inst Technol, BS, 44, MS, 48, PhD(phys chem), 51; Harvard Univ, advan mgt prog, 62. Prof Exp: Instr phys chem, Fenn Col, 46-47 & Case Inst Technol, 47-50; res chemist, 50-58, mgr tire res, 58-64, dir prod res, 64-72, DIR TIRE RES, B F GOODRICH CO, 73- Concurrent Pos: Chemist, Brookhaven Nat Lab, 53-54. Mem: Am Chem Soc. Res: Phase equilibria; polymerization reactions; physical properties of high polymers; polyelectrolytes; oxidation of high polymers; nuclear engineering; processing of reactor fuels; research administration; tire materials and construction. Mailing Add: B F Goodrich Co 6100 Oak Tree Blvd Cleveland OH 44131

BAUMAN, ROBERT ANDREW, b Rochester, NY, Apr 17, 23. ORGANIC CHEMISTRY. Educ: Univ Rochester, BS, 43; Univ Ill, PhD(chem), 46. Prof Exp: Spec res asst, Univ Ill, 43-46; res chemist, Gen Aniline & Film Corp, 46-48; asst, Univ Ill, 48-50; res chemist, Gen Aniline & Film Corp, 50-52; res chemist, Colgate-Palmolive-Peet Co, 52-53 & Colgate Palmolive Co, 53-63, res assoc, 63-68, SR RES ASSOC, COLGATE PALMOLIVE CO, 68- Mem: Am Chem Soc. Res: Organic synthesis; surfactants; oral chemotherapy. Mailing Add: 10 Landing Lane New Brunswick NJ 08901

BAUMAN, ROBERT POE, b Jackson, Mich, May 8, 28; m 49; c 4. CHEMICAL PHYSICS. Educ: Purdue Univ, BS, 49, MS, 51; Univ Pittsburgh, PhD(physics), 54. Prof Exp: Asst, Purdue Univ, 49-51 & Mellon Inst Indust Res, 51-54; from instr to assoc prof, 54-67, chmn dept, 67-73, PROF PHYSICS, UNIV ALA, BIRMINGHAM, 67-, DIR, PROJ ON TEACHING & LEARNING IN UNIV COL, 75- Mem: Am Chem Soc; Am Phys Soc; Am Asn Physics Teachers; Coblentz Soc (pres, 72-74). Res: Vibrational spectra and molecular structure; teaching-learning theory. Mailing Add: Dept of Physics Univ of Ala Birmingham AL 35233

BAUMAN, WILLIAM CARREL, b Cleveland, Ohio, Dec 7, 13; m 38; c 5. CHEMISTRY. Educ: Yale Univ, BS, 35, PhD(phys chem), 38. Prof Exp: Res chemist, 38-48, from asst dir to dir, Res Lab, 48-56, res dir, Chem Dept, 56-68, ASST DIR RES PHYS & INORG CHEM, DOW CHEM CO, 68- Mem: AAAS; Am Chem Soc. Res: Science; manufacture and uses of ion exchange resins; water treatment; reaction of atomic hydrogen with carbon tetrachloride. Mailing Add: Corp Res & Develop Dow Chem Co 2020 Bldg Midland MI 48640

BAUMANN, ARTHUR NICHOLAS, b Bogota, NJ, Dec 15, 22; m 47; c 2. INORGANIC CHEMISTRY, ANALYTICAL CHEMISTRY. Educ: Fla Southern Col, BS, 51. Prof Exp: Lab technician, phosphate rock div, Int Minerals & Chem Corp, Fla, 49-50; jr process engr, Davison Chem Co, 52; res chemist, Fla Exp Sta, 52-59, sr res chemist, tech div, 59-61, supvr tech serv, 61-64, chief chemist, 64-69, sr process engr, 69-74, MGR DEVELOP, AGR PROD DIV, INT MINERALS & CHEM CORP, 74- Mem: Am Chem Soc. Res: Production of alkali polyphosphates; defluorinated phosphate rock; anhydrous hydrogen fluoride; fluoride chemicals; potassium compounds; automation of wet chemical analytical methods for process control. Mailing Add: Agr Prod Div Int Minerals & Chem PO Box 867 Bartow FL 33830

BAUMANN, CARL AUGUST, b Milwaukee, Wis, Aug 10, 06. NUTRITIONAL BIOCHEMISTRY. Educ: Univ Wis, BS, 29, MS, 31, PhD(biochem), 33. Prof Exp: Asst biochem, Univ Wis-Madison, 29-34; Gen Educ Bd fel, Univ Heidelberg, Copenhagen Univ & Cambridge Univ, 34-36; res fel biochem, 36-38, from instr to assoc prof, 38-46, PROF BIOCHEM, UNIV WIS-MADISON, 46- Concurrent Pos: Consult, Off Sci Res & Develop, DC, 45-46; mem study sect nutrit, Am Cancer Soc; mem study sect nutrit & metab & nutrit, NIH, 48-60; mem nutrit surv, Interdept Comt Nutrit Nat Defense, NIH, Chile, 60; chief-of-party, Ford Found Basic Sci Proj, Agrarian Univ, Peru, 66-69; AID-Midwestern Univs Consortium Int Activities consult, Agr Univ, Indonesia, 74 & 75. Honors & Awards: Medal of Freedom & Cert of Merit, US Govt; Hon Mem, Order Brit Empire. Mem: AAAS; Am Chem Soc; Am Soc Biol Chem; Am Inst Nutrit. Res: Vitamins; nutrition; tumor development; skin and intestinal sterols; selenium metabolism. Mailing Add: Dept of Biochem Univ of Wis Madison WI 53706

BAUMANN, DONALD JOSEPH, b Mecosta, Mich, Nov 16, 15; m 45; c 10. INORGANIC CHEMISTRY. Educ: Univ Detroit, BS, 43; Creighton Univ, MS, 44; Iowa State Univ, PhD, 52. Prof Exp: Lab asst chem, Creighton Univ, 43-44, instr, 44-46; instr, Iowa State Univ, 46-61; asst prof, 51-59, chmn dept, 61-70, PROF CHEM, CREIGHTON UNIV, 59- Concurrent Pos: Mem indust waste control proj, Fed Water Pollution Control Admin, 70- Mem: Am Chem Soc. Res: Preparation of alkaline earthmetals; environmental chemistry. Mailing Add: Dept of Chem Creighton Univ Omaha NE 66131

BAUMANN, DUANE DENNIS, b Bloomington, Ill, Aug 13, 40; m 61; c 2. GEOGRAPHY, ENVIRONMENTAL MANAGEMENT. Educ: Ill State Univ, BS, 62, MS, 63; Clark Univ, PhD(geog), 68. Prof Exp: Instr geog, Univ RI, 64-65; asst prof, Ill State Univ, 65-66; asst prof, Southern Ill Univ, 67-69; res assoc, Clark Univ, 69-70; ASSOC PROF GEOG, SOUTHERN ILL UNIV, CARBONDALE, 70- Concurrent Pos: Clark Univ-NSF fel, Clark Univ, 69-70; mem formal partic comn man & environ, Int Union Geog & Geophys, 70-; consult, Comn Geog & Afro-Am, Asn Am Geogr, 71, res & appl nat needs, NSF, Univ Colo, 74 & US Corps Engrs, 74-75. Mem: AAAS. Res: Studies of human adjustment to natural hazards and decision-making in resource management, especially water resource management problems; research into the practicability of recycling renovated wastewater for municipal supply; factors related to human response to threat of tornado. Mailing Add: Dept of Geog Southern Ill Univ Carbondale IL 62901

BAUMANN, ELIZABETH WILSON, b Pueblo, Colo; m 54; c 1. ANALYTICAL CHEMISTRY. Educ: Brigham Young Univ, BA, 45; Univ Kans, PhD(chem), 53. Prof Exp: Chemist, Tex Co, 45-49; asst instr, Univ Kans, 49-51; res chemist, Rayon Res Lab, E I du Pont de Nemours & Co, 53-54; res assoc, Univ Kans, 54-55; chemist, 55-72, sr res chemist, 72-73, RES STAFF CHEMIST, SAVANNAH RIVER LAB, E I DU PONT DE NEMOURS & CO, 73- Mem: Am Chem Soc. Res: Ion exchange; water purification; ionic equilibria; ion selective electrodes. Mailing Add: Savannah River Lab E I du Pont de Nemours & Co Aiken SC 29801

BAUMANN, FREDERICK, b Los Angeles, Calif, Nov 26, 30; m 62; c 3. ANALYTICAL CHEMISTRY. Educ: Univ Calif, Los Angeles, BS, 52; Univ Wis, PhD(anal chem), 56. Prof Exp: Res chemist, Calif Res Corp, 56-64, sr res chemist, 64-65; mgr res, Varian Aerograph, 65-69, mgr data systs, 69-70, mgr liquid chromatography & res & eng, 70-72, MGR PROD DEVELOP, VARIAN INSTRUMENT DIV, 72- Concurrent Pos: Instr gas chromatography, Exten, Univ Calif, 64- Mem: Am Chem Soc. Res: Gas chromatography; liquid chromatography; laboratory automation; mass spectrometry; electrochemistry; microanalysis. Mailing Add: Varian Instrument Div 2700 Mitchell Dr Walnut Creek CA 94598

BAUMANN, GERT FRIEDRICH, b Leverkusen, Ger, Jan 28, 32; m 57; c 3. ORGANIC POLYMER CHEMISTRY, INDUSTRIAL ORGANIC CHEMISTRY. Educ: Univ Bonn, BS, 52; Stuttgart Tech Univ, MS, 54, PhD(org chem), 56. Prof Exp: Res chemist, 56-61, group leader, 61-71, SECT MGR, MOBAY CHEM CO, 71- Mem: Am Chem Soc. Res: Development and technical service of flexible and semiflexible urethane foam; new processes for continuous foam production; new urethane molding techniques for furniture and automotive industries; combustion modified urethane foams; market development for new urethane applications. Mailing Add: 1224 Satellite Circle Pittsburgh PA 15241

BAUMANN, JACOB BRUCE, b Fremont, Ohio, Sept 11, 32; m 57; c 3. ORGANIC CHEMISTRY. Educ: Amherst Col, BA, 54; Univ Mich, MS, 57, PhD(chem), 61. Prof Exp: From instr to asst prof chem, Defiance Col, 60-65; asst prof, 65-71, ASSOC PROF CHEM, PA STATE UNIV, SCHUYLKILL CAMPUS, 71- Mem: Am Chem Soc. Res: Aromatic nucleophilic substitution. Mailing Add: Dept Chem Schuylkill Campus Pa State Univ State Rte Schuylkill Haven PA 17972

BAUMANN, NORMAN PAUL, b Sylvan Grove, Kans, Nov 18, 27; m 54; c 1. NUCLEAR PHYSICS. Educ: Univ Kans, BS, 51, MS, 52, PhD(physics), 55. Prof

Exp: PHYSICIST, SAVANNAH RIVER LAB, E I DU PONT DE NEMOURS & CO, 55- Mem: Am Phys Soc; Am Nuclear Soc. Res: Reactor physics and design. Mailing Add: Exp Phys Div Savannah River Lab E I du Pont de Nemours & Co Aiken SC 29801

BAUMANN, RICHARD WILLIAM, b Castle Dale, Utah, July 24, 40; m 64; c 6. SYSTEMATIC ENTOMOLOGY, AQUATIC BIOLOGY. Educ: Univ Utah, BA, 65, MS, 67, PhD(aquatic entom), 70. Prof Exp: Res guest aquatic entom, Max Planck Limnol Inst, 70-71; asst prof limnol, Southwest Mo Univ, 71-72; assoc cur entom, Smithsonian Inst, 72-75; ASST PROF ENTOM, BRIGHAM YOUNG UNIV, 75- Concurrent Pos: Supvr res Moraca River, Yugoslavia, Off Limnol & Oceanog, Smithsonian Inst, 73-75; investr ecol, Bur Land Mgt, 74-75; adj prof, NTex State Univ, 74- Mem: Sigma Xi; Entom Soc Am; Am Entom Soc; NAm Benthol Soc. Res: Studies on the systematics, phylogeny and ecology of the stoneflies, Insecta plecoptera, of the northern hemisphere; ecological studies of aquatic insects in western North America. Mailing Add: 1617 W 1050 North Provo UT 84601

BAUMANN, THIEMA MARIE WOLF, b St Louis, Mo, May 30, 21; m 50; c 1. HEALTH SCIENCES, HUMAN GENETICS. Educ: Harris Col, AB, 42; Washington Univ, MA, 44; St Louis Univ, PhD(biol anat), 66. Prof Exp: Instr life sci, DePaul Hosp Sch Nursing, 47; instr math, Pub Schs, 48-58; instr genetics, Sch Dent, 61-67, asst prof, Grad Orthod Dept, 67-73, ASSOC CLIN PROF, GRAD ORTHOD DEPT, MED CTR, ST LOUIS UNIV, 73- Concurrent Pos: D I G res grants, 62-68; Orthod Educ & Res Found res grants, 67-69. Mem: AAAS; NY Acad Sci; Sigma Xi. Res: Microscopic neuroanatomy; genetic aspects of craniofacial growth; chromotolysis in the facial nucleus of irradiated and nonirradiated animals; therapeutic irradiation of force insulted oral tissues; effects of orthodontic procedures in Rhesus monkeys; biogenic amines; microscopic fluorescence; drug effects on tissues. Mailing Add: Grad Orthod Dept St Louis Univ Med Ctr St Louis MO 63104

BAUMANN, WINFRIED, b Tarutino, Rumania, Feb 13, 29; US citizen; m 59; c 2. LASERS. Educ: Univ Göttingen, Dipl, 56, Dr rer nat(phys chem), 59. Prof Exp: Res fel, Univ Calif, Berkeley, 59-60; sr scientist, Avco-Everett Res Lab, Mass, 60-65; sr scientist, Bell Aerospace Co Div, Textron, Buffalo, 65-75; TEACHER, BAD NENNDORF HIGH SCH, 75- Mem: Am Inst Aeronaut & Astronaut; Am Phys Soc; Combustion Inst; Bunsen Soc; Ger Soc Aeronaut & Astronaut. Res: Chemical lasers; isotope separation; computer science; shock tubes; spectroscopy; plasma physics. Mailing Add: Kamerstrasse 16 3052 Bad Nenndorf West Germany

BAUMANN, WOLFGANG JOSEF, b Crailsheim, Ger, May 26, 36; m 67; c 2. BIOCHEMISTRY, LIPID CHEMISTRY. Educ: Univ Stuttgart, BS, 57, MS, 61, PhD(org chem), 63. Prof Exp: Asst chem, Univ Stuttgart, 61-63; res fel, 63-66, res assoc, 66-67, from asst prof to assoc prof, 67-74, PROF CHEM, HORMEL INST, UNIV MINN, 74- Mem: Am Chem Soc; Soc Germ Chem; Am Oil Chemists Soc. Res: Synthesis and structrue of lipids; biochemistry of ether lipids and diol lipids; lipids of cancer cells; spectroscopy. Mailing Add: Univ of Minn Hormel Inst Austin MN 55912

BAUMBACH, DONALD OTTO, b Oil City, Pa, June 25, 26; m 62; c 1. PHYSICAL CHEMISTRY. Educ: Syracuse Univ, BS, 54; Pa State Univ, MS, 59, PhD(fuel technol), 62. Prof Exp: Chemist, Solvay Process Div, Allied Chem Corp, 54-56; res chemist, Lord Mfg Corp, 62-65; sr res chemist, Tex-US Chem Co, 65-69; ASST PROF CHEM, ST PAUL'S COL, VA, 70- Mem: Am Chem Soc. Res: Physical chemistry of polymers and carbons; physical and chemical properties of heterogeneous carbon systems and of heterogeneous polymer mixtures. Mailing Add: Dept of Chem Box 752 St Paul's Col Lawrenceville VA 23868

BAUMBER, JOHN SCOTT, b London, Eng, June 4, 37; m 61; c 4. PHYSIOLOGY. Educ: Univ Nottingham, BSc, 58; Queen's Univ, Ont, MSc, 60, PhD(physiol), 63, MD, 66. Prof Exp: Mo Heart Asn fel, 67-69; asst prof physiol, Univ Mo-Columbia, 69-70; asst prof, 70-72, ASSOC PROF PHYSIOL, UNIV CALGARY, 72-, ASST DEAN FAC MED, 74- Mem: Can Med Asn; Can Physiol Soc; Am Physiol Soc. Res: Cardiovascular, renal and endocrine physiology with emphasis on cardiodynamics, hemodynamics and heart failure; the renin-angiotensin-aldosterone system; hypoxia. Mailing Add: Div of Med Physiol Univ of Calgary Fac of Med Calgary AB Can

BAUMEISTER, CARL FREDERICK, b Dolliver, Iowa, May 15, 07; m 30; c 1. INTERNAL MEDICINE. Educ: Univ Chicago, BS, 30; Univ Iowa, MD, 33. Prof Exp: Intern & resident, Univ Hosps, Ind Univ & Univ Louisville, 33-37; physician, Council Bluffs Clin, 37-43; asst prof clin med, Col Med, Univ Ill, 43-65; clin asst prof med, Stritch Sch Med, Loyola Univ, Chicago, 70-73; ASST PROF CLIN MED, UNIV ILL COL MED, 73- Concurrent Pos: Head dept internal med, Suburban Med Ctr, Berwyn, 43; mem, Inst Med Chicago. Mem: AMA; Asn Am Med Cols; Am Heart Asn; Am Diabetes Asn; Am Med Writers Asn. Res: Role of the computer as a diagnostic aid in the surgical abdomen; mechanisms and prevention of type III sinus block headaches. Mailing Add: 120 S Delaplaine Rd Riverside IL 60546

BAUMEISTER, PHILIP WERNER, b Troy, Ohio, Mar 17, 29; m 52; c 3. PHYSICS. Educ: Stanford Univ, BS, 50; Univ Calif, Berkeley, PhD(physics), 59. Prof Exp: From asst prof to assoc prof, 59-72, PROF OPTICS, UNIV ROCHESTER, 72- Mem: Optical Soc Am; Am Vacuum Soc. Res: Optical interference coatings; optical instrumentation; optical properties of thin, solid films; optical filter design. Mailing Add: Inst of Optics Univ of Rochester Rochester NY 14627

BAUMEL, JULIAN JOSEPH, b Sanford, Fla, July 26, 22; m 45; c 3. ANATOMY. Educ: Univ Fla, BS, 47, MS, 50, PhD, 53. Prof Exp: From instr to assoc prof, 53-64, PROF ANAT, CREIGHTON UNIV, 64- Concurrent Pos: Chmn, Int Comt Avian Anat Nomenclature; pres, Nebr State Anat Bd. Mem: Am Asn Anatomists; Am Soc Zoologists; Am Ornith Union; World Asn Vet Anat. Res: Avian and human anatomy; vasculature; arthrology; peripheral nerves. Mailing Add: Dept of Anat Creighton Univ Sch of Med Omaha NE 68178

BAUMEL, PHILIP, b New York, NY, June 12, 32; m 56; c 3. HIGH ENERGY PHYSICS. Educ: City Col New York, BS, 53; Columbia Univ, PhD(physics), 60. Prof Exp: Asst prof physics, Fairleigh Dickinson Univ, 59-61; from instr to assoc prof, 61-72, PROF PHYSICS, CITY COL NEW YORK, 72-, ASST DEAN, 75- Mem: Am Asn Physics Teachers; Am Phys Soc. Res: Particle physics. Mailing Add: Dept of Physics City Col of New York New York NY 10031

BAUMGARDNER, JOHN HENRY, b Wellington, Tex, May 15, 16; m 41; c 4. ANIMAL NUTRITION. Educ: Tex Tech Col, BS, 39, MS, 40. Prof Exp: Tech adv nutrit, Educ Serv, Nat Cottonseed Prod Asn, Tex, 40-42; from asst prof to assoc prof, 45-61, PROF NUTRIT, TEX TECH UNIV, 61- Concurrent Pos: Consult Asian animal nutrit, US Feed Grains Coun, Washington, DC, 74-75, consult Europ animal nutrit, 75-76. Mem: Am Soc Animal Sci. Res: Infra-red head in the cooking of seed for food and feed; digestibility of sorghum grain by the ruminant. Mailing Add: Dept of Animal Sci Tex Tech Univ Lubbock TX 79409

BAUMGARDNER, KANDY DIANE, b Peoria, Ill, Sept 16, 46. GENETICS. Educ: Bradley Univ, BS, 68; Utah State Univ, PhD(zool), 74. Prof Exp: ASST PROF ZOOL, EASTERN ILL UNIV, 73- Mem: Am Inst Biol Sci. Res: The population biology of the lambdoid bacteriophages, particularly competition between lambda, phi-eighty and a hybrid type. Mailing Add: Dept of Zool Eastern Ill Univ Charlestown IL 61920

BAUMGARDNER, MARION F, b Wellington, Tex, Feb 6, 26; m 55; c 3. AGRONOMY. Educ: Tex Tech Col, BS, 50; Purdue Univ, MS, 55, PhD(soil fertil, plant nutrit), 64. Prof Exp: Lectr agron, Agr Inst, Allahabad, India, 50-53; admin secy int student conf, Nat Coun Churches, NY, 55-56; in-chg state soil testing opers, 58-61, from instr to assoc prof agron, 58-73, leader appln res progs, 66-70 & aerospace appln res progs, 70-72, PROF AGRON, PURDUE UNIV, WEST LAFAYETTE, 73-, LEADER EARTH SCI RES PROGS, LAB APPLN REMOTE SENSING, 72- Concurrent Pos: Purdue Univ rep, Ford Found Agr Develop Prog, Arg, 64-66; Purdue Res Found int travel grant lectr tour, 69; Am Soc Agron off rep, Triennial Cong Latin Am Asn Plant Scientists, Bogota, Colombia, 70; consult, Food & Agr Orgn, Arg, 70, Bulgaria, 71, 72 & Sudan, 74, Am Univ, Washington, DC & Lilly Endowment, Niger, 73 & Int Develop Res Centre, Ottawa, to Sudan, 75-76; ed spec publ 18, Aerospace Sci & Agr Develop, Am Soc Agron, 70; Klinck lectr, Can, 72; mem agr panel, Nat Acad Eng, 74; vis scientist, Int Inst Aerial Surv & Earth Sci, Enschede, Neth, 74-75; mem, Int Comt Remote Sensing for Develop, Nat Acad Sci, 75-76; chmn nat study panel remote sensing, Agr Res Inst. Mem: Am Soc Agron; Int Soil Sci Soc; Soil Sci Soc Am; Am Soc Photogram; Int Asn Ecol. Res: Applications of remote sensing technology; automatic identification, characterization and mapping of earth surface features; relationships between physicochemical and multispectral properties of soils. Mailing Add: Lab for Appl of Remote Sensing Purdue Univ 1220 Potter Dr West Lafayette IN 47906

BAUMGARDNER, RAY K, b Ridgway, Colo, Dec 26, 33; m 59; c 2. LIMNOLOGY. Educ: Ft Lewis Col, AB, 59; Adams State Col, BS, 61; Okla State Univ, MS, 62, PhD(zool), 66. Prof Exp: Asst prof aquatic biol, 65-69, assoc prof biol sci, 69-74, PROF BIOL SCI & HEAD DEPT, NORTHWESTERN STATE UNIV, 74- Mem: Am Soc Limnol & Oceanog. Res: Water pollution; primary productivity and its relation to water quality; effect of impoundments on streams. Mailing Add: Dept of Biol Northwestern State Univ Natchitoches LA 71457

BAUMGARDT, BILLY RAY, b Lafayette, Ind, Jan 17, 33; m 52; c 3. ANIMAL NUTRITION. Educ: Purdue Univ, BS, 55, MS, 56; Rutgers Univ, PhD(agr biochem), 59. Prof Exp: Instr dairy sci, Rutgers Univ, 56-59; from asst prof to assoc prof, Univ Wis-Madison, 59-67; PROF ANIMAL NUTRIT, PA STATE UNIV, UNIVERSITY PARK, 67-, HEAD DEPT, 70- Honors & Awards: Am Feed Mfgrs Asn Nutrit Res Award, 66. Mem: Am Dairy Sci Asn; Am Soc Animal Sci; Am Inst Nutrit. Res: Rumen physiology and biochemistry; food evaluation; regulation of food intake; relation of nutrition to body composition. Mailing Add: Dept of Animal Sci Pa State Univ University Park PA 16802

BAUMGART, JOHN KEPPLER, b Forest Park, Ill, Nov 10, 16. MATHEMATICS. Educ: Wheaton Col, Ill, BS, 38; Univ Mich, MA, 39. Prof Exp: Instr math, Cumberland Col, 39-42; instr radio, Air Corps Technol Schs, 42-43; assoc prof math & head dept, Elmhurst Col, 46-54; training officer, US Naval Res Lab, Washington, DC, 54-56; assoc prof math & head dept, Manchester Col, 56-65; PROF MATH, NORTH PARK COL, 65- Mem: AAAS; Am Math Soc; Math Asn Am. Res: Development of axiom systems in algebra; history of mathematics. Mailing Add: Dept of Math North Park Col 3225 Foster Ave Chicago IL 60625

BAUMGARTEN, ALEXANDER, b Warsaw, Poland, Nov 27, 35; Australian citizen; m 69; c 1. IMMUNOLOGY. Educ: Sydney Univ, MB, BS, 59; Univ New South Wales, PhD(immunopath), 69. Prof Exp: Resident med, Royal Prince Alfred Hosp, Sydney, Australia, 59; resident, Royal Adelaide Hosp, 60; registr, Alfred Hosp, Melbourne, 61; ASST PROF LAB MED, MED SCH, YALE UNIV, 70- Mem: Am Asn Immunologists; Am Soc Microbiol; NY Acad Sci. Res: Clinical and general immunology; protein polymorphism. Mailing Add: Dept of Lab Med Yale New Haven Hosp New Haven CT 06504

BAUMGARTEN, EDWIN, organic chemistry, operations research, see 12th edition

BAUMGARTEN, HENRY ERNEST, b Texas City, Tex, Feb 27, 21; m 48; c 4. SYNTHETIC ORGANIC CHEMISTRY. Educ: Rice Univ, BA, 43, MA, 44, PhD(org chem), 48. Prof Exp: Lab asst, Rice Univ, 41-43; asst, Off Rubber Reserve, Univ Ill, 48-49; from instr to prof, 49-64, actg chmn dept, 70-71, chmn dept, 71-75, FOUND PROF ORG CHEM, UNIV NEBR-LINCOLN, 64- Concurrent Pos: Guggenheim fel & vis assoc, Calif Inst Technol, 62-63; consult, div instnl progs, NSF, 63-67. Mem: AAAS; Am Chem Soc. Res: Heterocyclic compounds; reactions of organic nitrogen compounds; alpha lactams; organic electrochemistry. Mailing Add: Dept of Chem Univ of Nebr-Lincoln Lincoln NE 68508

BAUMGARTEN, REUBEN LAWRENCE, b New York, NY, Nov 19, 34; m 63; c 1. PHYSICAL ORGANIC CHEMISTRY. Educ: City Col New York, BS, 56; Univ Mich, MS, 58, PhD(chem), 62. Prof Exp: Instr chem, Hunter Col, 62-66, George N Shuster fel grant, 64-65; asst prof, 66-71, ASSOC PROF CHEM, LEHMAN COL, 71- Mem: AAAS; Am Chem Soc. Res: Organic chemistry of nitrogen; reaction of hydroxylamines and derivatives; nuclear magnetic resonance; magnetic isomerism; aromatic ring current; prochirality in nuclear magnetic resonance; qualitative detection of amines and glycoproteins. Mailing Add: Dept of Chem Herbert H Lehman Col Bedford Park Blvd W Bronx NY 10468

BAUMGARTEN, RONALD J, b New York, NY, May 7, 35. ORGANIC CHEMISTRY, ENVIRONMENTAL CHEMISTRY. Educ: Brooklyn Col, BS, 56; Johns Hopkins Univ, MA, 58, PhD(org chem), 62. Prof Exp: Fel, Brandeis Univ, 63-64; ASSOC PROF ORG CHEM, UNIV ILL, CHICAGO CIRCLE, 64- Concurrent Pos: Vis asst prof, Univ Ill, Urbana, 65-66; NSF fel, 70-72; vis scientist, State Univ Leiden, 71-72. Res: New approaches to deamination; photochemical reductions; new synthetic procedures; aldehyde and ketone syntheses. Mailing Add: Dept of Chem Univ Ill at Chicago Circle Chicago IL 60680

BAUMGARTEN, WERNER, b Berlin, Ger, Feb 14, 14; nat US; m 48; c 1. BIOCHEMISTRY. Educ: Univ Munich, Dipl, 37; Calif Inst Technol, PhD(org chem, biochem), 41. Prof Exp: Res chemist, Hiram Walker & Sons, 41-45, J T Baker Chem Co, NJ, 45-48 & Merck & Co, 48-49; res assoc, Sharp & Dohme, 49-56, res assoc, Merck Sharp & Dohme Res Labs, 56-65, sr res fel, 65-69, dir hemat, 69-71; sr info scientist, Res Labs, Hoffman-LaRoche, 71-73; asst dir clin res, Hoechst-Rossel Pharmaceut, 73-75; MED WRITER, ICI UNITED STATES INC, 75- Mem: Am Chem Soc; fel NY Acad Sci; Am Soc Biol Chem; Am Soc Hematol; Int Soc Hematol. Res: Sugar chemistry; heat of combustion of organic compounds; microbiological analysis of amino acids; determination of vitamins and carotenoids; isolation of natural products; fermentation; therapeutic enzymes; protein chemistry; blood coagulation. Mailing Add: 2420 Oaks Circle Huntingdon Valley PA 19006

BAUMGARTNER, FREDERICK MILTON, b Indianapolis, Ind, June 5, 10; m 36; c 4. WILDLIFE CONSERVATION. Educ: Butler Univ, AB, 31; Univ Kans, AM, 33; Cornell Univ, PhD(ornith), 37. Prof Exp: Ornithologist, State Dept Conserv, Mich, 36-39; assoc prof wildlife conserv, Okla State Univ, 39-64, assoc prof zool, 64-67; PROF CONSERV, UNIV WIS-STEVENS POINT, 67- Mem: Am Ornith Union; Wildlife Soc. Res: Ecology of the great horned owls and the bob-white quail; management of the bob-white quail; bird distribution in Oklahoma; methods of bird control. Mailing Add: Dept of Natural Resources Univ of Wis Stevens Pont WI 54481

BAUMGARTNER, FREDERICK NEIL, b Bluffton, Ohio, June 4, 20; m 44; c 3. CHEMISTRY. Educ: Miami Univ, AB, 42; Univ Ill, PhD(org chem), 48. Prof Exp: Res chemist, Esso Res & Eng Co, 48-57, sect head, 57-61, head mkt res chem div, Esso Int, Inc, 61-63, head mkt res & sales anal, Esso Chem Co, Inc, 63-65, head mkt res, Esso Chem SA, Belg, 65-68, sr mkt planning adv, Esso Chem Co, Inc, 68-70, MGR MKT RES, EXXON CHEM CO USA, 70- Mem: Am Chem Soc; Chem Mkt Res Asn; fel Am Inst Chemists. Res: Chemical market research, especially petrochemicals. Mailing Add: Exxon Chem Co USA PO Box 3272 Houston TX 77001

BAUMGARTNER, GEORGE JULIUS, b Atchison, Kans, June 27, 24; m 54; c 6. ORGANIC CHEMISTRY. Educ: St Benedict's Col, BSc, 45; Univ Notre Dame, PhD, 53. Prof Exp: Prof chem, Mt St Scholastica Col, 50-71, dean acad affairs, 68-71, PROF CHEM, BENEDICTINE COL, 71-, ASST DEAN, 73- Mem: Am Chem Soc. Res: Reactions of tetrahydrofurfuryl compounds. Mailing Add: Benedictine Col Atchison KS 66002

BAUMGARTNER, LEONA, b Chicago, Ill, Aug 18, 02; m 42; c 2. PEDIATRICS, PUBLIC HEALTH. Educ: Univ Kans, AB, 23, AM, 25; Yale Univ, PhD(immunol), 32, MD, 34; Am Bd Prev Med & Pub Health, dipl; Am Bd Pediat, dipl. Hon Degrees: LLD, Yale Univ, 70; ScD, NY Univ, 54, Russell Sage Col, 55, Smith Col, 56, Women's Med Col Pa, 66, Western Col Women, 66, Univ Mass, 66, MacMurray Col, 67, Univ Mich, 67, Med Col NY, 68, Clark Univ, 69; LHD, Keuka Col, 66; LLD, Oberlin Col, 65, Skidmore Col, 66. Prof Exp: Teacher high sch, Kans, 23-24; instr jr col, Kansas City, 25-26; instr bact, physiol & hyg, Univ Mont, 26-28; from intern to asst pediat, Cornell Univ & NY Hosp, 34-36; actg asst surgeon, USPHS, Washington, 36-37; med instr child & Sch Health, New York Dept Health, 37-38, dir pub health training, 38-39, dist health officer, 39-40, dir, Bur Child Hyg, 41-48, asst comnr in-chg maternal & child health serv, 48-53, comnr health, 54-62; asst adminr for tech coop & res, AID, 62-65; vis prof social med, Harvard Med Sch, 66-74; RETIRED. Concurrent Pos: Lectr, Columbia Univ, 39-42; instr, Med Col, Cornell Univ, 39-44, from asst prof to prof, 44-66; asst pediatrician, NY Hosp, 40-42, pediatrician, 42-56, assoc attend pediatrician, 56-74; adv, French Ministry Health, 45; assoc chief, US Children's Bur, 49-50, consult, 50-56; adv, Indian Minister Health, 55; exec dir, NY Found, 53; mem, US Off Exchange Mission, USSR, 58; mem, Nat Res Coun, Child Welfare League & Nat Health Coun; exec dir, Med Care & Educ Found, Inc, Tri-State Med Prog, 68-72; mem, Maternal & Child Health Expert Comn, WHO, 75-77. Honors & Awards: Lasker Award, Am Pub Health Asn, 59. Mem: Fel Am Pub Health Asn (pres, 58); fel Am Acad Pediat; Am Pediat Soc; Am Soc Microbiol. Res: Age and antibody formation; brucella infection; infant mortality; public health; medical and social history. Mailing Add: 1010 Memorial Dr Cambridge MA 02138

BAUMGARTNER, LUTHER LEROY, b Chatfield, Ohio, Sept 30, 13; m 40; c 3. BIOCHEMISTRY, ECOLOGY. Educ: Capital Univ, BSc, 35; Ohio State Univ, MA, 36, PhD(wildlife mgt), 40. Prof Exp: Res mammalogist, State Wildlife Res Unit, Ohio, 35-40; biologist, State Conserv Dept, Mich, 40-42; in charge biochem res lab, B F Goodrich Co, 45-53; DIR & OWNER, BAUMLANDA HORT RES LAB, 53- Concurrent Pos: Pres, Hort Prod Inc, 60-; ecologist, Hudson River Valley Comn, NY, 67-68; consult land use ecol. Mem: AAAS; Wildlife Soc; Am Soc Mammal; Am Soc Hort Sci; NY Acad Sci. Res: Fungicides; insecticides; weed killers; animal repellents; plant and animal ecology; animal food habits; agricultural chemicals evaluation; horticulture; wildlife and forest management; nursery management; livestock management in tropical areas; ecological factors in soybean production. Mailing Add: Rte 2 Box 44-R Newberry SC 29108

BAUMGARTNER, WERNER ANDREAS, b Poertschach, Austria, May 6, 35; m 68; c 2. PHYSICAL CHEMISTRY, BIOPHYSICAL CHEMISTRY. Educ: Univ New South Wales, BSc, 60, PhD(phys chem), 64. Prof Exp: Res asst phys chem, Commonwealth Sci & Indust Res Orgn, Australia, 55-56; res asst biophys chem, physics dept, St Vincent Hosp, Australia, 56-60; scholar, Univ New South Wales, 64-66; assoc scientist, Jet Propulsion Lab, Calif Inst Technol, 66; asst prof phys chem, Calif State Col, Long Beach, 67-70; DIR BIOPHYS CHEM, TRITIUM & DIAG LAB, NUCLEAR DYNAMICS, INC, 70- Concurrent Pos: Asst prof, Calif State Col, Long Beach, 67-68; assoc scientist, Scripps Inst Oceanog, 70 & immunochem lab, Vet Admin Hosp, Sepulveda, 70-71; res chemist, Vet Admin Hosp, Long Beach, 71-73 & Vet Admin Hosp, Wadsworth, 73- Mem: Am Chem Soc. Res: Theory of metal and enzyme catalysis; charge-transfer complexes in catalysis; immunochemistry and membrane processes; function of vitamin E and selenium; catalytic labeling of unstable biochemicals. Mailing Add: 5377 Appian Way Long Beach CA 90803

BAUMHOFF, MARTIN A, b Camino, Calif, Dec 22, 26; m 50; c 4. ANTHROPOLOGY. Educ: Univ Calif, AB, 54, PhD, 59. Prof Exp: From instr to assoc prof, 59-70, PROF ANTHROP, UNIV CALIF, DAVIS, 70- Mem: Am Anthrop Asn; Soc Am Archaeol. Res: Archaeology of western North America; statistical methods in anthropology; paleoecology. Mailing Add: Dept of Anthrop 331 Voorhies Hall Univ of Calif Davis CA 95616

BAUMHOVER, ALFRED HENRY, b Carroll, Iowa, June 17, 21; m 55; c 3. ECONOMIC ENTOMOLOGY. Educ: Iowa State Univ, BS, 49. Prof Exp: Area supvr, grasshopper control div, 49-51, med entomologist, insects affecting animals, res div, 51-58, med entomologist & prog adv, animal dis eradication div, 58-60, plant pest entomologist, entom res div, 60-61, entomologist in charge, screw worm res, 61-64, invest leader, tobacco insect invest, 64-72, RES LEADER, SOUTHERN REGION MID-ATLANTIC AREA, TOBACCO RES LAB, AGR RES SERV, USDA, 72- Honors & Awards: Order of Orange-Nassau, Netherlands, 55; Commendation, Netherlands Govt, 55; Sweepstakes Winner, Am Insect Photo Salon, Am Entom Soc, 57; Cert of Merit, USDA, 58, 59, Distinguished Serv Unit Award, 62 & 64; Man of Year Award, Progressive Farmer Mag, 59; Plaque Award, Southwest Animal Health Res Found, 65. Mem: AAAS; Am Asn Econ Entom. Res: Insect pest management with attractants, sterility and natural enemies. Mailing Add: Tobacco Res Lab USDA Rte 2 Box 16G Oxford NC 27565

BAUMILLER, ROBERT CAHILL, b Baltimore, Md, Apr 15, 31. GENETICS. Educ: Loyola Col, Md, BS, 53; St Louis Univ, PhL & PhD(biol), 61; Woodstock Col, Md, BT, 65. Prof Exp: Nat Found fel, Univ Wis, 61-62; asst med, Sch Med, Johns Hopkins Univ, 62-67; ASSOC PROF OBSTET & GYNEC & ASSOC PROF PEDIAT, SCH MED, GEORGETOWN UNIV, 67- Concurrent Pos: Prin investr, USPHS res grant, 64-67; AEC grant, 67-70; sr res fel, Kennedy Inst Study Reproduction & Bioethics; mem bd dir, Sex Info & Educ Coun US, 70-74; consult, Clin Pharmacol Assocs, 71-73; assoc ed, Linacre Quart, 71-; mem med adv bd, Hemophilia Soc, 74-; consult, Pope John XXIII Med-Moral Res Ctr, 75- Mem: AAAS; Genetics Soc Am; Am Soc Human Genetics; Soc Study Evolution; Environ Mutagen Soc. Res: Virus induced mutation both gene and chromosomal; effects of mutants in heterozygous condition; transformation of human leucocytes as affected by the menstrual cycle; human chromosome abnormalities. Mailing Add: Dept of Obstet & Gynec Georgetown Univ Sch of Med Washington DC 20007

BAUMSLAG, GILBERT, b Johannesburg, SAfrica, Apr 30, 33; m 59; c 2. MATHEMATICS. Educ: Univ Witwatersrand, BS, 53, Hons, 55; Univ Manchester, PhD(math), 58. Prof Exp: Lectr math, Univ Manchester, 58-59; instr, Princeton Univ, 59-60; from asst prof to assoc prof, Courant Inst Math Sci, NY Univ, 62-64; prof, City Univ New York, 64-68; mem, Inst Advan Study, 68-69; prof, Rice Univ, 69-73; DISTINGUISHED PROF MATH, CITY COL CITY UNIV NEW YORK, 73- Concurrent Pos: Sloan fel, Rice Univ. Mem: Am Math Soc; London Math Soc. Res: Group theory. Mailing Add: City Col City Univ New York Convent Ave & 138th St New York NY 10031

BAUMSTARK, JOHN SPANN, b Cape Girardeau, Mo, Dec 23, 27; m 52; c 3. BIOCHEMISTRY, MICROBIOLOGY. Educ: Southeast Mo State Col, BS, 51; Univ Mo, AM, 53, PhD(agr chem), 57. Prof Exp: Instr agr chem, Univ Mo, 55-57; sr scientist biochem microbiol, Res Dept, Mech Div, Gen Mills, Inc, 57-60, prin scientist, 60-61; from instr to asst prof obstet & gynec, Med Sch, Tufts Univ, 63-72; PROF & DIR RES OBSTET & GYNEC & ASSOC PROF BIOL CHEM, SCH MED, CREIGHTON UNIV, 72-, PROF PATH, 74- Concurrent Pos: Sr biochemist, St Margaret's Hosp, Boston, Mass, 61-72. Mem: Am Soc Exp Path; Am Soc Microbiol; Reticuloendothelial Soc; Am Chem Soc. Res: Proteolytic enzymes; proteins; amino acids; chromatography; experimental pathology. Mailing Add: Dept of Obstet & Gynec Creighton Univ Sch of Med Omaha NE 68108

BAUR, EBERHARD HEINZ, applied mathematics, applied statistics, see 12th edition

BAUR, FREDRIC JOHN, JR, b Toledo, Ohio, July 14, 18; m 43; c 3. PHYSIOLOGICAL CHEMISTRY, ORGANIC CHEMISTRY. Educ: Univ Toledo, BS, 39; Ohio State Univ, MS, 41, PhD(physiol chem), 43; cert com, Univ Cincinnati, 56. Prof Exp: Instr physiol chem, Ohio State Univ, 43-44; res chemist, 46-56, in charge flavor chem, Food Div, 56-60, in charge anal labs, 60-65, IN CHARGE FACTORY ANAL METHODS-SANIT, FOOD DIV, PROCTER & GAMBLE CO, 65- Mem: Am Chem Soc; Inst Food Technol; Am Oil Chemists Soc. Res: Interesterification, composition and structure of fats; synthesis of compounds containing fatty groups; acetin fats; chemistry and utility of flavors; analysis of foods and food raw materials. Mailing Add: Winton Hill Tech Ctr A3M36 Procter & Gamble Co Cincinnati OH 45224

BAUR, JOSEPH RALPH, b Indianapolis, Ind, May 27, 38; m 61; c 2. PLANT PHYSIOLOGY, BIOCHEMISTRY. Educ: Purdue Univ, BSc, 60, MSc, 63; Tex A&M Univ, PhD(plant physiol), 67. Prof Exp: PLANT PHYSIOLOGIST, SOUTHERN REGION, AGR RES SERV, USDA, 67- Mem: Am Soc Plant Physiol; Soc Exp Biol & Med; Weed Soc Am; Scand Soc Plant Physiol. Res: Plant metabolism; herbicide physiology and biochemistry; weed control; range land pasture establishment. Mailing Add: Dept Range Sci Agr Res Serv USDA Tex A&M Univ College Station TX 77843

BAUR, MARIO ELLIOTT, b Indianapolis, Ind, Aug 23, 34; m 60; c 5. PHYSICAL CHEMISTRY. Educ: Univ Chicago, AB, 53, MS, 55; Mass Inst Technol, PhD(chem), 59. Prof Exp: Asst chem, Mass Inst Technol, 55-56 & 58-59; NSF fel, State Univ Utrecht, 59-61; res assoc, Univ Calif, San Diego, 61-62; asst prof, 62-68, ASSOC PROF CHEM, UNIV CALIF, LOS ANGELES, 68- Mem: AAAS; Am Phys Soc; Chem The Chem Soc; Netherlands Phys Soc; Soc Rheology. Res: Statistical mechanics; optical and electrical properties; polymer physical chemistry; biophysics; paleochemistry. Mailing Add: Dept of Chem Univ of Calif Los Angeles CA 90024

BAUR, PAUL SCHUH, b Dyess, Ark, Nov 22, 38; m 63; c 3. CYTOLOGY. Educ: Tex A&M Univ, BS, 61, PhD(plant path), 70. Prof Exp: Adj prof biol, NTex State Univ, 70-72; staff mem plant path, Tex A&M Univ, 72-73; sr electromicros technician neurol, Baylor Col Med, 73; staff mem cell biol, 73-74, ASST PROF CELL BIOL, DEPT HUMAN BIOL CHEM & GENETICS & ASST PROF PHYSIOL & BIOPHYS, UNIV TEX MED BR GALVESTON, 74-; DIR SCANNING ELECTRON MICROS LAB, SHRINERS BURN INST, 74- Concurrent Pos: NTex State Univ grant, Botany Sect, Lunar Receiving Lab, Manned Spacecraft Ctr, NASA, Houston, 70-73. Mem: AAAS. Res: Scanning and transmission microscopy, including microchemical analysis, of mammalian cells in cultures; wound healing. Mailing Add: Dept of Human Biol Chem & Genet Univ of Tex Med Br Galveston TX 77550

BAUR, WERNER HEINZ, b Warsaw, Poland, Aug 2, 31; m 62; c 2. MINERALOGY, CRYSTALLOGRAPHY. Educ: Univ Göttingen, Dr rer nat, 56. Prof Exp: Sci officer, Univ Göttingen, 56-63, privat-dozent, 61; from asst prof to assoc prof mineral & crystallog, Univ Pittsburgh, 63-65; assoc prof, 65-68, PROF MINERAL & CRYSTALLOG, UNIV ILL, CHICAGO CIRCLE, 68-, HEAD DEPT GEOL SCI, 67- Concurrent Pos: Fel, Univ Berne, 57; vis assoc chemist, Brookhaven Nat Lab, 62-63; vis res prof, Univ Karlsruhe, 71-72; res assoc, Field Mus Natural Hist, Chicago, 74-; assoc ed, Am Mineralogist, 75-78. Mem: Fel Mineral Soc Am; Am Crystallog Asn; Ger Mineral Soc. Res: Crystal chemistry of minerals; crystal structure determination by x-ray and neutron diffraction; hydrogen bonding; computer simulation of crystal structures; empirical theories of bonding in crystals; predictive crystal chemistry. Mailing Add: Dept of Geol Sci Univ of Ill Box 4348 Chicago IL 60680

BAURER, THEODORE, b New York, NY, June 20, 24; m 47; c 2. PHYSICAL CHEMISTRY. Educ: Queens Col, NY, BS, 45; Syracuse Univ, PhD(chem), 53. Prof Exp: Asst, Syracuse Univ, 47-48 & Atomic Energy Comn, 51-53; res chemist, mineral beneficiation lab, Columbia Univ, 53-56; adv res engr, Sylvania Elec Prod, Inc & Sylvania Corning Nuclear Corp, 56-57; res phys chemist, Grumman Aircraft Eng Corp, 57-63; proj scientist, Gen Appl Sci Labs, Inc, 63-66; PROJ LEADER, CHEM SYSTS ANAL, VALLEY FORGE SPACE CTR, GEN ELEC CO, PHILADELPHIA, 66- Concurrent Pos: Guest scientist, free radicals res prog, Nat Bur Standards, 58-59; co-ed, Defense Nuclear Agency Reaction Rate Handbk. Mem: AAAS; Am Chem Soc; Am Phys Soc; Am Inst Aeronaut & Astronaut; Am Geophys Union. Res: Gas-phase kinetics; upper and lower atmosphere chemistry, including ablation, pollution, reentry and weapons effects; propulsion and combustion chemistry; free radicals chemistry; spacecraft internal and external environmental quality; space experiments definition; chemical systems analyses. Mailing Add: 751 Woodside Rd Rydal Jenkintown PA 19046

BAURIEDEL, WALLACE ROBERT, b Omaha, Nebr, Dec 11, 23; m 47; c 3. BIOCHEMISTRY. Educ: Iowa State Col, BS, 49, MS, 52, PhD(biochem), 54. Prof Exp: Res assoc, Vet Med Res Inst, Iowa State Col, 49-55, asst prof, 55-56; res biochemist, 56-64, sr res biochemist, 64-70, ASSOC SCIENTIST, AGR CHEM RES, DOW CHEM CO, 70- Mem: Am Chem Soc. Res: Vitamin chemistry and nutrition;

residue and metabolism of pesticides; metabolic fate of herbicides and pesticides. Mailing Add: 1404 Ashman Midland MI 48640

BAUSCH, AUGUSTUS FRANK, mathematics, deceased

BAUSERMAN, THOMAS, b Ronceverte, WVa, July 6, 15; m 46; c 1. MATHEMATICS. Educ: WVa Inst Technol, BS, 41; Wash State Univ, MA, 47; Univ Pittsburgh, PhD(math), 61. Prof Exp: Chemist, E I du Pont de Nemours & Co, WVa, 41; instr math, WVa Univ, 47-55; from asst prof to assoc prof, 55-61, PROF MATH, MARSHALL UNIV, 61-, CHMN DEPT, 63- Mem: Math Asn Am. Res: Geometry of the complex domain; introduction of modern mathematics programs in elementary and secondary schools; teacher instruction. Mailing Add: Dept of Math Marshall Univ Huntington WV 25701

BAUSHER, LARRY PAUL, b Reading, Pa, July 8, 39. ORGANIC CHEMISTRY. Educ: Franklin & Marshall Col, AB, 61; Univ Calif, Los Angeles, PhD(org chem), 67. Prof Exp: Res chemist, Hercules, Inc, 67-69; res fel chem, Wesleyan Univ, 69-71; RES FEL CHEM, YALE UNIV, 71- Mem: AAAS; Am Chem Soc; Asn Res Vision & Ophthal; Sigma Xi. Res: Physiology and pharmacology of aqueous humor dynamics. Mailing Add: 466 Middletown Ave No 4 New Haven CT 06513

BAUSLAUGH, PHILIP GARY, organic chemistry, photochemistry, see 12th edition

BAUSOR, SYDNEY CHARLES, b New York, NY, May 29, 10; m 43; c 3. BOTANY, MICROBIOLOGY. Educ: Columbia Univ, BA, 31, MA, 33, PhD(bot), 37. Prof Exp: Asst & lectr bot, Columbia Univ, 31-37; instr, Lehigh Univ, 37-40; res guest plant hormones, Univ Chicago, 40-41; instr morphol & plant physiol, Conn Col, 41-42; assoc examr physics & chem, US Civil Serv Comn, 42-43; microbiologist, Guayule Proj, USDA, 43-44; from assoc prof to prof bot & bact, Creighton Univ, 46-59; res assoc, dept surg, med ctr, Univ Ark, 59-60; PROF BIOL, CALIF STATE COL, PA, 60-, HEAD DEPT, 66- Res: Growth and cellular metabolism; fasciation; plant hormones. Mailing Add: 215 Fifth St California PA 15419

BAUSSUS-VON LUETZOW, HANS GERHARD, b Nortorf, Ger, Feb 13, 21; US citizen; m 52; c 3. MATHEMATICS. Educ: Univ Kiel, BS & MS, 49, DSc(math), 63. Prof Exp: Sci asst opers res, State Energy Admin, Ger, 49-51; asst prof math & physics, Zimmermann Col & Com Col, 52-56; unit chief math anal, Aeroballistics Lab, US Army Ballistic Missile Agency, 56-57, dep secy & dep br chief anal invest aerophys & geophys, 57-58, sci adv math statist, 59-60; sci adv earth & space sci, Aeroballistics Lab, Marshall Space Flight Ctr, 60-62; tech chief systs div, US Army Strategy & Tactics Anal Group, 62-63; prin scientist, Advan Progs, Gen Precision Co, 63-64; referent, Prog Coord Opers Res Budgeting, Off Fed Ministry Defense, Ger, 64-66; SR SCIENTIST, US ARMY TOPOG LABS, 66- Mem: Am Statist Asn; Am Geophys Union; Am Astronaut Soc; assoc fel Am Inst Aeronaut & Astronaut; NY Acad Sci. Res: Applied mathematics; dynamic meteorology; geodesy; operations research. Mailing Add: 8021 Garlot Dr Annandale VA 22003

BAUSUM, HOWARD THOMAS, b Kweilin, China, July 20, 33; US citizen; m 60; c 3. MICROBIOLOGY. Educ: Carson-Newman Col, BS, 54; Univ Tenn, MS, 56; Univ Tex, PhD(zool, genetics), 64. Prof Exp: Biologist, US Army Biol Labs, Ft Detrick, Md, 60-61; asst prof microbiol, Sch Med, Univ Kans, 64-65; asst prof genetics, Iowa State Univ, 65-68; microbiologist, US Army Biol Labs, 68-71; staff assoc, Comn Sci Educ, AAAS, 71-72; RES MICROBIOLOGIST, US ARMY MED BIOENG RES & DEVELOP LAB, 72- Mem: AAAS; Sigma Xi; Am Soc Microbiol; Genetics Soc Am; NY Acad Sci. Res: Environmental microbiology; microbiological aspects of waste management. Mailing Add: US Army Med Bioeng Res & Develop Lab Ft Detrick Frederick MD 21701

BAUTZ, LAURA PATRICIA, b Washington, DC, Sept 3, 40. ASTRONOMY, SCIENCE ADMINISTRATION. Educ: Vanderbilt Univ, BA, 61; Univ Wis, PhD(astron), 67. Prof Exp: From instr to asst prof astron, Northwestern Univ, Evanston, 65-72; prog dir, NSF, 72-73; assoc prof astron, Northwestern Univ, Evanston, 73-75; SR STAFF ASSOC, NSF, 75- Concurrent Pos: Trustee, Adler Planetarium, 73-; mem, Nat Coun Univ Res Adminr. Mem: Am Astron Soc; AAAS; Int Astron Union. Res: Evaluation of basic research programs; clusters of galaxies. Mailing Add: No 506 1325 18th St Washington DC 20036

BAVETTA, LUCIEN ANDREW, b Italy, June 4, 07; US citizen; m 33; c 2. BIOCHEMISTRY, NUTRITION. Educ: NY Univ, BS, 30; Univ Southern Calif, MS, 33, PhD(biochem), 42. Prof Exp: From instr to prof biochem & nutrit, Sch Dent, Univ Southern Calif, 48-74; prof biochem & nutrit, Grad Sch, 54-74; RETIRED. Concurrent Pos: Spec lectr, US Naval Dent Sch, 55-56; vis scientist, NIH, 55-56; USPHS res career award, 64-; NIH career award. Mem: Soc Exp Biol & Med (secy), 60-61); Int Asn Dent Res; Am Soc Biol Chemists; Inst Food Technologists; NY Acad Sci. Res: Biochemistry of collagen; developmental biology. Mailing Add: Ahmanson Ctr Univ of Southern Calif Sch Dent Los Angeles CA 90007

BAVISOTTO, VINCENT, b Buffalo, NY, Jan 21, 25; m 53; c 3. BIOCHEMISTRY. Educ: Univ Buffalo, BA, 48; Pa State Univ, MS, 50, PhD(biochem), 52. Prof Exp: Sr res chemist, Gen Biochem, Inc, 52-54; dir res, Paul Lewis Labs, Inc, 55-61, mgr res & develop, Paul Lewis Labs Div, Chas Pfizer & Co, Inc, 61-67; tech dir, Theodore Hamm Co, 67-73; ASSOC DIR, HEUBLIN INC, 73- Mem: Am Chem Soc; Am Soc Brewing Chem; Inst Food Technol; Master Brewers Asn Am; Am Soc Qual Control. Res: Basic and applied research on problems related to brewing, dairy, meat and enzyme manufacturing industries. Mailing Add: Heublin Inc 430 New Park Ave Hartford CT 06101

BAVLEY, ABRAHAM, b Boston, Mass, June 3, 15; m 46. CHEMISTRY. Educ: Tufts Col, BS, 36; Harvard Univ, MA & PhD(org chem), 40. Prof Exp: Chemist, Ansco Div, Gen Aniline & Film Co, Inc, NY, 40-43, leader colorformer group, 43-45; asst prof chem, Xavier Univ, Ohio, 45-46 & Gevaert Photo-Prod, Belg, 46-47; res chemist, Charles Pfizer & Co, Inc, NY, 47-51, res supvr, 51-58, mgr res, Greensborough Sect, 58-59; asst dir res, Gillette Safety Razor Co, Mass, 59-60; mgr res, Philip Morris, Inc, 60-65; vpres & tech dir, Marion Labs, Inc, Mo, 65-69; vpres & sci dir, Knoll Pharmaceut Co, 69-71; consult, 71-74; ASSOC, RES CORP, 74- Mem: Am Chem Soc; Optical Soc Am. Res: Organic synthesis; medicinal chemistry; chemistry of natural products. Mailing Add: 405 Lexington Ave New York NY 10017

BAWDEN, JAMES WYATT, b St Louis, Mo, Apr 23, 30; m 51; c 5. PHYSIOLOGY, DENTISTRY. Educ: Univ Iowa, DDS, 54, MS, 60, PhD(physiol), 61. Prof Exp: Pvt pract, 56-58; fel physiol, Univ Iowa, 58-61; from asst prof to assoc prof pedodontics, 61-65, asst dean res, 63-66, dean sch dent, 66-74, PROF PEDODONTICS, UNIV NC, CHAPEL HILL, 65- Concurrent Pos: NIH grant, 63-; vis prof, Sch Dent, Univ Lund, Sweden, 74-75. Mem: Int Asn Dent Res; Am Dent Asn; fel Int Col Dent; Am Acad Pedodontics; fel Am Col Dent. Res: Pedodontics; investigations in the placental transfer of calcium, fluoride, oxygen and carbon dioxide; the effect of maternal metabolic status on the metabolism of the fetus inutero; development of tooth enamel; trace elements. Mailing Add: Sch of Dent Univ of NC Chapel Hill NC 27514

BAWDEN, MONTE PAUL, b Salt Lake City, Utah, June 3, 43; m 62; c 4. PARASITOLOGY, IMMUNOLOGY. Educ: Univ Calif, Riverside, BA, 65; Rutgers Univ, New Brunswick, PhD(zool), 70. Prof Exp: Res asst, Rutgers Univ, 65-66, teaching asst, 68; fel, Dept Trop Pub Health, Sch Pub Health, Harvard Univ, 69-71; res assoc med parasitol, 71-73; PARASITOLOGIST, NAVY MED SERV CORPS, NAVAL MED RES INST, 73- Concurrent Pos: Rockefeller Found fel med parasitol, Sch Pub Health, Harvard Univ, 71-73. Mem: Am Soc Parasitologists; Am Soc Trop Med & Hyg; Soc Protozoologists; Royal Soc Trop Med & Hyg; Sigma Xi. Res: Malaria, trypanosomiasis, leishmaniasis and schistosomiasis; immunochemistry, immunology and serology. Mailing Add: Naval Med Res Inst Nat Naval Med Ctr Bethesda MD 20014

BAXEVANIS, JOHN JAMES, b Serres, Greece, Jan 12, 38; US citizen; m 70. GEOGRAPHY. Educ: City Univ New York, BA, 61; Ind Univ, Bloomington, MA, 63; Univ NC, Chapel Hill, PhD(geog), 68. Prof Exp: Asst prof geog, City Univ New York, 63-64; asst prof, Trenton State Col, 65-67; asst prof, Fla State Univ, 67-68; PROF GEOG, EAST STROUDSBURG STATE COL, 68-, DEAN, FAC SOCIAL SCI, 74- Concurrent Pos: Mem, Inst Balkan Studies, Greece, 63-; prof, Pa Consortium Int Educ, Austria, 72-73. Mem: Am Geog Soc Soc; Asn Am Geog. Res: Cultural geography of Europe; megalopolis; medieval city; Greece; population problems of underdeveloped countries; diffusion of technology; ekistics; urban growth; geography of wine. Mailing Add: 1947 Hillside Dr Stroudsburg PA 18360

BAXLEY, WILLIAM ALLISON, b Washington, DC, May 10, 33; m 56; c 3. INTERNAL MEDICINE. Educ: Duke Univ, BS, 55, MD, 62. Prof Exp: Intern med, Duke Hosp, Durham, NC, 62-63; med resident, Univ Wash, 63-64; asst med & res fel, Univ Wash & Vet Admin Hosp, 64-66; from instr to asst prof cardiol, Med Ctr, Univ Ala, 66-70; assoc prof med, Sch Med, Univ NMex, 70-72; ASSOC PROF MED, SCH MED, UNIV ALA, BIRMINGHAM, 72- Concurrent Pos: Mem coun clin cardiol, Am Heart Asn, 67. Mem: Am Fedn Clin Res; Am Heart Asn; fel Am Col Cardiol; fel Am Col Physicians. Res: Cardiovascular hemodynamics in patients with heart disease. Mailing Add: Dept of Med Univ of Ala Sch of Med Birmingham AL 35294

BAXMAN, HORACE ROY, b Lansing, Ill, Jan 27, 21; m 44; c 4. CHEMISTRY. Educ: Ind Univ, BS, 43; Cornell Univ, PhD(inorg chem), 47. Prof Exp: Chemist, Monsanto Chem Co, Oak Ridge, 47-48; assoc chemist, Argonne Nat Lab, 48-52; mem staff, 52-73; SECT LEADER, LOS ALAMOS SCI LAB, 73- Mem: Fel Am Inst Chemists; Am Chem Soc. Res: High Vacuum; boron compounds; spectrophotometry; unsymmetrical organoboron compounds; solvent extraction; fuel processing; uranium chemistry; process research on actinide elements; materials research on pyrolytic carbon. Mailing Add: Los Alamos Sci Lab Box 1663 Los Alamos NM 97544

BAXTER, ANN WEBSTER, b Evanston, Ill, July 25, 17; div; c 2. MICROBIAL PHYSIOLOGY. Educ: Rockford Col, BA, 38; Univ NC, Chapel Hill, PhD(bact immunol), 67. Prof Exp: Res technician physiol, sch med, Univ Louisville, 56-57; res technician bact, Am Sterilizer Co, Pa, 58-59; teacher, high sch, NC, 60-63; asst prof, 67-71, ASSOC PROF MICROBIOL, CLEMSON UNIV, 71- Honors & Awards: 75th Anniversary Bronze Medallion, Am Soc Microbiol, 74; Serv Plaque, 75. Mem: AAAS; Am Soc Microbiol; Brit Soc Gen Microbiol. Res: Bacterial growth and metabolism; factors affecting bacterial membrane transport; mechanisms of active transport of amino acids in bacteria; effects of environmental pollutants on survival of bacteria. Mailing Add: Dept of Microbiol Clemson Univ Clemson SC 29631

BAXTER, CLAUDE FREDERICK, b Hamburg, Ger, July 24, 23; nat US; m 52; c 3. PHYSIOLOGY, BIOCHEMISTRY. Educ: Univ Calif, Davis, PhD, 54. Prof Exp: Res physiologist, Univ Calif, Davis, 54; fel, McCollum-Pratt Inst, Johns Hopkins Univ, 54-56; SR RES BIOCHEMIST, DIV NEUROBIOL, CITY OF HOPE MED CTR, DUARTE, CALIF, 56-; CHIEF NEUROCHEM LABS, VET ADMIN HOSP, SEPULVEDA, 63- Concurrent Pos: Assoc prof physiol, Univ Calif, Los Angeles, 64-74, res prof psychiat, 74- Mem: AAAS; Int Soc Neurochem; Am Soc Biol Chemists; Am Chem Soc; NY Acad Sci. Res: Milk synthesis; sulfur metabolism; neurochemistry; metabolism and functional aspects of nitrogenous compounds in nervous system; neurotransmitter substances; protein synthesis; osmotic regulation in nervous system; nerve regeneration; aging of nervous system and convulsive disorders. Mailing Add: Neurochem Labs Vet Admin Hosp Sepulveda CA 91343

BAXTER, DAVID WOOD, physiology, see 12th edition

BAXTER, DONALD HENRY, b Schenectady, NY, Sept 6, 16; m 42; c 1. RADIOTHERAPY. Educ: Union Col, AB, 37; Albany Med Col, MD, 41. Prof Exp: Intern med, Albany Med Ctr Hosp, 41-42; resident radiol, Shreveport Charity Hosp, La, 46-47; radiologist, NLa Sanitarium, 47-49; from asst prof to assoc prof, 49-68, PROF RADIOL, ALBANY MED COL, 69- Concurrent Pos: Attend radiologist, Albany Med Ctr Hosp, 49- & Vet Admin Hosp, Albany, 51- Mem: AMA; Am Col Radiol. Mailing Add: Dept of Radiol Albany Med Ctr Hosp Albany NY 12208

BAXTER, DONALD WILLIAM, b Brockville, Ont, Aug 24, 26. MEDICINE. Educ: Queen's Univ, Ont, MD, CM, 51; McGill Univ, MSc, 53; Am Bd Psychiat & Neurol, dipl, 58. Prof Exp: From instr to asst prof, Sch Med, Univ Sask, 57-62; PROF NEUROL, FAC MED, McGILL UNIV, 63- Mem: Am Acad Neurol; Can Med Asn; Can Neurol Asn. Mailing Add: Div of Neurol Montreal Gen Hosp Montreal PQ Can

BAXTER, GENE FRANCIS, b Sanish, NDak, July 25, 22; m 49; c 3. POLYMER CHEMISTRY. Educ: Univ Wash, BS, 44. Prof Exp: Res chemist, Am Marietta Corp, 44-56, res group leader, 56-59, info coordr, 59-61, info coordr, Martin-Marietta Corp, 61-62; sr res scientist, Weyerhaeuser Co, 62-73; SR DEVELOP CHEMIST, GA-PAC CORP, 73- Mem: Am Chem Soc. Res: Polymers; adhesives; thermosetting resins; bonded wood products; plywood; laminated beams; construction materials; formaldehyde polymers; phenolic resins; aromatic amine reactions. Mailing Add: Chem Div Ga-Pac Corp 2883 Miller Rd Decatur GA 30032

BAXTER, GEORGE T, b Grover, Colo, Mar 19, 19; m 42; c 3. FISH BIOLOGY, LIMNOLOGY. Educ: Univ Wyo, BS & MS, 46; Univ Mich, PhD(zool), 52. Prof Exp: From instr to assoc prof, 47-59, actg head dept, 68-70, PROF ZOOL, UNIV WYO, 59- Mem: Am Fisheries Soc. Res: Limnology of alpine lakes; fishery management, primarily salmonids in the Rocky Mountains; ichthyology and herpetology of Wyoming. Mailing Add: Dept of Zool Univ of Wyo Laramie WY 82071

BAXTER, GLEN EARL, b Minneapolis, Minn, Mar 19, 30; m 54. MATHEMATICS. Educ: Univ Minn, BA, 51, MA, 52, PhD(math) 54. Prof Exp: Instr math, Univ Minn, 53-54 & Mass Inst Technol, 54-55; from instr to prof math, Univ Minn, 55-64; prof, Univ Calif, San Diego, 64-65; PROF MATH, PURDUE UNIV, 65- Concurrent Pos: Vis assoc prof, Stanford Univ, 60; vis prof, Aarhus Univ, 60-61; Fulbright res prof, 61-62. Mem: Inst Math Statist; Math Asn Am; Am Math Soc. Res: Probability theory; combinatorial analysis; Fourier analysis. Mailing Add: Dept of Math Purdue Univ Lafayette IN 47907

BAXTER, JAMES F, b Ptersburgh, Tenn, Aug 2, 16; m; c 3. POLYMER CHEMISTRY, TEXTILE CHEMISTRY. Educ: Middle Tenn State Col, BS, 40; Vanderbilt Univ, MS, 47, PhD(chem), 52. Prof Exp: Metallurgist, Tenn Coal Iron & RR Co Div, US Steel Corp, 40-41; asst prof physics, Middle Tenn State Col, 47-49; sr chemist, Rayon Div, 52-57, res chemist, Textile Fibers Dept, New Prod Div, 57-62, sr res chemist, 62-65, RES ASSOC, TEXTILE FIBERS DEPT, RES & DEVELOP LAB, E I DU PONT DE NEMOURS & CO, INC, 65- Mem: Tech Asn Pulp & Paper Indust. Res: Continuous filament non-woven fabric technology. Mailing Add: Textile Fibers Dept Res-Dev Lab E I du Pont de Nemours & Co Inc Old Hickory TN 37138

BAXTER, JAMES HUBERT, b Ashburn, Ga, Dec 19, 13; m 42; c 1. MEDICAL RESEARCH. Educ: Univ Ga, BS, 35; Vanderbilt Univ, MD, 41; Am Bd Internal Med, dipl, 49. Prof Exp: Res asst pharmacol, Med Sch, Vanderbilt Univ, 36-40; intern med, Johns Hopkins Univ, 41-42; asst resident med, Bowman Gray Sch Med, 42-44; res instr, Southwestern Med Sch, 44-46; Nat Res Coun Welch fel, Dept Biochem, Med Sch, Cornell Univ, 46-47, Hosp, Rockefeller Inst, New York, 47-48 & Dept Med, Johns Hopkins Univ Hosp, Baltimore, 48-50; SR STAFF MEM, LAB CELLULAR METAB, NAT HEART & LUNG INST, 50-; SURGEON-MED DIR, USPHS, 52- Mem: Am Soc Clin Invest; Soc Exp Biol & Med; Am Soc Pharmacol & Exp Therapeut; Am Heart Asn; AMA. Res: Metabolism; renal disease. Mailing Add: Lab of Cellular Metab Nat Heart & Lung Inst Bethesda MD 20014

BAXTER, JAMES WATSON, b Shamrock, Tex, Sept 9, 27. ECONOMIC GEOLOGY, MICROPALEONTOLOGY. Educ: Univ Ark, BS, 50, MS, 52; Univ Ill, PhD(geol), 58. Prof Exp: Asst mineral, Univ Ark, 50-51; res asst, Ark Inst Sci & Technol, 51-52; res asst, 52-54, asst geologist, 56-63, assoc geologist, 63-70, GEOLOGIST, ILL STATE GEOL SURV, 70- Mem: Geol Soc Am; Am Asn Petrol Geol; Soc Econ Paleont & Mineral; Geol Soc Belg. Res: Stratigraphy and sedimentation; geology of industrial minerals; fluorspar deposits; carbonate petrography; Mississippian biostratigraphy, foramini fera, calcareous algae. Mailing Add: Rm 300A Natural Resource Bldg Univ of Ill Urbana IL 61801

BAXTER, JOHN EDWARDS, b Meridian, Miss, Aug 29, 37; m 68; c 1. PHYSICAL CHEMISTRY, BIOCHEMISTRY. Educ: Millsaps Col, BS, 58; Vanderbilt Univ, MS, 60; Duke Univ, PhD(phys chem), 66. Prof Exp: Res assoc phys chem, Duke Univ, 66; asst prof physics & chem, NC Wesleyan Col, 66-71; ASST PROF BIOCHEM, CTR HEALTH SCI, UNIV TENN MED UNITS, MEMPHIS, 71- Mem: AAAS; Am Chem Soc. Res: Enzyme separation and characterization; utilization of self instructional methods in biochemistry education. Mailing Add: Dept of Biochem Univ of Tenn Med Units Memphis TN 38163

BAXTER, JOHN FRANKLIN, b New Castle, Pa, Dec 11, 09; m 40; c 3. CHEMISTRY. Educ: Bethany Col, WVa, AB, 32; Johns Hopkins Univ, PhD(chem), 42. Hon Degrees: ScD, Bethany Col, WVa, 60. Prof Exp: Teacher, High Sch, Ill, 33-35 & Ohio, 35-37; asst chem, Johns Hopkins Univ, 37-42; asst prof, Gettysburg Col, 42-43; lectr, Johns Hopkins Univ, 43-46; from assoc prof to prof, Washington & Lee Univ, 46-52; PROF CHEM, UNIV FLA, 52- Concurrent Pos: Instr chem, Loyola Col, Md, 39-42; teacher, Continental Classroom, Nat Broadcasting Co, 59-61; Ford Found sci & TV consult, Latin Am, 62- Honors & Awards: James T Grady Award, Am Chem Soc, 62. Mem: Fel AAAS; Am Chem Soc. Res: Physical inorganic chemistry; solution equilibria; acidity of hydrated beryllium ion; chemical education; television and film teaching. Mailing Add: Dept of Chem Univ of Fla Gainesville FL 32611

BAXTER, JOHN LEWIS, b San Diego, Calif, July 31, 25; m 47; c 2. FISHERIES MANAGEMENT. Educ: Univ Calif, Berkeley, AB, 51. Prof Exp: Marine biologist, fisheries lab, State Dept Fish & Game, Calif, 51-63, supvr, pelagic fish prog, 63-70, chief marine resources br, 70-71; spec asst to dir, Nat Marine Fisheries Serv, Washington, DC, 71-73; marine sport fish coordr, 73-74, SUPVR N OCEAN AREA, CALIF DEPT FISH & GAME, 74- Concurrent Pos: Res fel, Scripps Inst Oceanog; marine ed, Calif Fish & Game, 62-66; ed-in-chief, Fish Bull, 66-70; exec secy, Marine Fisheries Adv Comt, Dept of Commerce, 71-73. Mem: Am Inst Fishery Res Biol; Am Fisheries Soc. Res: Biology of sport and commercial fishes. Mailing Add: Calif Dept Fish & Game 411 Burgess Dr Menlo Park CA 94025

BAXTER, JOHN WALLACE, b Grover, Colo, Feb 4, 18; m 50. MYCOLOGY, PLANT PATHOLOGY. Educ: Univ Wyo, BS, 48; Purdue Univ, MS, 50, PhD, 52. Prof Exp: Asst prof plant path, Iowa State Univ, 52-56; assoc prof, 56-67, PROF BOT, UNIV WIS-MILWAUKEE, 67- Mem: Mycol Soc Am. Res: TAxonomy, life cycles and physiologic specialization of rust fungi. Mailing Add: Dept of Bot PO Box 413 Univ of Wis Milwaukee WI 53201

BAXTER, LUTHER WILLIS, JR, b Lawrenceburg, Ky, Nov 25, 24; m 48; c 3. PLANT PATHOLOGY. Educ: Eastern Ky State Col, BS, 50; La State Univ, MS, 52, PhD, 54. Prof Exp: Plant pathologist, Kaiser Aluminum & Chem Corp, 54-55; assoc plant pathologist, Clemson Univ, 55-58; prof plant path & head dept agr, Western Ky State Col, 58-66; assoc prof, 66-70, PROF PLANT PATH & PHYSIOL, CLEMSON UNIV, 70- Concurrent Pos: Consult, Kaiser Aluminum & Chem Corp, 56. Mem: Am Phytopath Soc. Res: Diseases of camellias; physiology of reproduction in fungi. Mailing Add: Dept of Plant Path & Physiol Clemson Univ Clemson SC 29631

BAXTER, NEAL EDWARD, b Bluffton, Ind, Sept 13, 08; m 52; c 5. INTERNAL MEDICINE, AEROSPACE MEDICINE. Educ: Ind Univ, AB, 32, MD, 35. Prof Exp: Intern, Indianapolis City Hosp, 35-36; physician gen pract, 36-54; MED DIR, WESTINGHOUSE ELEC CORP, 57- Concurrent Pos: Assoc, Woolery Clin, Bloomington, Ind, 36-37; specialist aerospace & internal med, pvt pract, 54-; dir found, Sch Bus, Ind Univ, 69-; aviation med examr, Fed Aviation Agency. Honors & Awards: John A Tamesia Award, Aerospace Med Asn, 65. Mem: Am Acad Gen Pract; Am Diabetes Asn; fel Am Acad Family Physicians; Aerospace Med Asn (pres, 65-66); Civil Aviation Med Asn (pres, 62). Res: Aviation physiology and oxygen equipment; syncopal reactions of anoxic subjects observed in the low pressure chamber; civil aviation medicine; medical aspects of business aviation. Mailing Add: 306 Kirkwood Ave Bloomington IN 47401

BAXTER, ROBERT WILSON, b CZ, June 7, 14; m 45; c 1. PALEOBOTANY. Educ: Wash Univ, AB, 37, MS, 47, PhD(paleobot), 49. Prof Exp: Asst instr bot, Univ Hawaii, 38-39; from asst prof to assoc prof, 49-63, chmn dept, 53-57, PROF BOT, UNIV KANS, 63- Concurrent Pos: Fulbright lectr, Univ Col WI, 55-56; NSF res award, 58, 61 & 63. Mem: Bot Soc Am; Int Soc Plant Morphol; Int Soc Plant Taxon. Res: Carboniferous flora of the central United States; cretaceous Dakota sandstone flora. Mailing Add: Dept of Bot Univ of Kans Lawrence KS 66044

BAXTER, ROSS M, b Erin, Ont, Oct 1, 18; m 46; c 3. PHARMACY. Educ: Univ Toronto, PhmB, 43; Univ Sask, BSc, 46; Univ Fla, MSc, 48, PhD, 51. Prof Exp: From asst prof to assoc prof, 51-60, PROF PHARMACOG, UNIV TORONTO, 60- Mem: AAAS; NY Acad Sci; Can Pharmaceut Asn; Acad Pharmaceut Sci; Asn Fac Pharm Can. Res: Medicinal chemistry. Mailing Add: Dept of Pharm Univ of Toronto Toronto ON Can

BAXTER, STUART DILLON, mathematics, see 12th edition

BAXTER, WARREN NESMITH, b Rutherford, NJ, Dec 29, 29; m 57; c 2. ORGANIC CHEMISTRY. Educ: Bates Col, BS, 50; Mass Inst Technol, PhD(chem), 53. Prof Exp: Asst, Mass Inst Technol, 51-53; chemist, E I du Pont de Nemours & Co, 53-55; fel, Univ Calif, Berkeley, 55-57; chemist, 57-61, sr res chemist, 61-66, SR SUPVR, E I DU PONT DE NEMOURS & CO, 66- Mem: Am Chem Soc. Res: Polyolefins; fluorocarbon polymers; nylon intermediates. Mailing Add: 2111 Liveoak Orange TX 77630

BAXTER, WILLARD ELLIS, b Chester, Pa, Dec 14, 29; m 54; c 3. MATHEMATICS. Educ: Ursinus Col, BS, 51; Univ Wis, MS, 52; Univ Pa, PhD, 56. Prof Exp: Asst prof math, Ohio Univ, 56-58; from asst prof to assoc prof, 58-67, PROF MATH, UNIV DEL, 67-, CHMN DEPT, 70- Mem: Am Math Soc; Math Asn Am. Res: Structure theory of rings. Mailing Add: Dept of Math Univ of Del Newark DE 19711

BAXTER, WILLIAM D, b Larned, Kans, Sept 13, 36; m 59; c 3. IMMUNOBIOLOGY. Educ: Phillips Univ, AB, 60; Univ Kans, PhD(zool), 66. Prof Exp: Asst prof, 66-71, ASSOC PROF BIOL, BOWLING GREEN STATE UNIV, 71- Concurrent Pos: Adj prof, Med Col Ohio, Toledo, 68- Mem: AAAS; Electron Micros Soc Am; Sigma Xi. Res: Immunology, especially early cellular events associated with the establishment of immunity; electron microscopy of animal tissues. Mailing Add: Dept of Biol Bowling Green State Univ Bowling Green OH 43402

BAXTER, WILLIAM JOHN, b Whinburgh, Eng, July 31, 35; m 60; c 3. METAL PHYSICS. Educ: Oxford Univ, BA, 57, DPhil(physics), 61. Prof Exp: Res physicist, Fulmer Res Inst, Eng, 60-61; res assoc internal friction, Cornell Univ, 61-63; assoc sr res physicist, 64-66, SR RES PHYSICIST, RES LABS, GEN MOTORS CORP, 66- Concurrent Pos: Fel, Cornell Univ, 61-63. Honors & Awards: Second Prize Photog Exhib, Am Soc Testing & Mat, 73. Mem: Am Soc Testing & Mat. Res: Exoelectron emission and photoelectron microscopy; fatigue and deformation of metals; oxidation of metals; internal friction. Mailing Add: Dept of Physics Res Labs Gen Motors Corp Warren MI 48090

BAXTER, WILLIAM LEROY, b San Diego, Calif, Dec 21, 29; m 58; c 2. VIROLOGY. Educ: Univ Calif, Los Angeles, AB, 56, PhD(microbiol), 61. Prof Exp: USPHS fel virol, Univ Calif, Los Angeles, 61-63; PROF MICROBIOL, SAN DIEGO STATE UNIV, 63- Concurrent Pos: Consult, clin lab, Sharp Mem Hosp, 63- & US Naval Hosp, 64- Mem: AAAS; Am Soc Microbiol; Am Inst Biol Sci; Electron Micros Soc Am; Sigma Xi. Res: Early steps of myxovirus infection; phase contrast and electron microscopy of virus-infected cell culture. Mailing Add: Dept of Microbiol San Diego State Univ San Diego CA 92182

BAXTER-GABBARD, KAREN LEE, b Terre Haute, Ind, Apr 27, 41. MOLECULAR BIOLOGY. Educ: Ind State Univ, AB, 63; Iowa State Univ, PhD(bact), 67. Prof Exp: Res scientist, Chas Pfizer & Co, 67-68; res assoc microbiol, Med Ctr, Ind Univ, Indianapolis, 68-71; INSTR LIFE SCI, CTR MED EDUC, IND STATE UNIV, TERRE HAUTE, 71- Concurrent Pos: NIH fel, 68. Mem: Am Soc Microbiol; NY Acad Sci. Res: Biochemical genetics; conformational responses of viral DNA with respect to salt concentration and viral proteins; mode of reproduction and infection of oncogenic viruses. Mailing Add: Dept of Microbiol Ctr Med Educ Ind State Univ Terre Haute IN 47809

BAY, DARRELL EDWARD, b Hays, Kans, Dec 22, 42; m 75; c 1. ENTOMOLOGY. Educ: Kans State Univ, BS, 64, MS, 67, PhD(entom), 74. Prof Exp: Entomologist, Walter Reed Army Inst Res, 68-71; ASST PROF ENTOM, TEX A&M UNIV, 74- Mem: Entom Soc Am. Res: Biology, ecology and control of livestock insects. Mailing Add: Dept of Entom Tex A&M Univ College Station TX 77843

BAY, ERNEST C, b Schenectady, NY, Aug 7, 29; m 63; c 2. MEDICAL ENTOMOLOGY, ECONOMIC ENTOMOLOGY. Educ: Cornell Univ, BS, 53, PhD(entom), 60. Prof Exp: Specialist entom, Univ Calif, Riverside, 60-61, asst entomologist biol control, 61-69, head div, 69-71; head dept entom, Univ Md, College Park, 71-75; SUPT & PROF ENTOM, WESTERN WASH RES & EXTEN CTR, WASH STATE UNIV, 75- Concurrent Pos: NIH grant, 61-65; WHO consult, Nicaraguan Ministry Health, 64, Far East, 67 & Nigeria, 72; secy gen, XV Int Cong Entom, 73-75. Mem: Entom Soc Am; Am Mosquito Control Asn; Am Soc Ichthyologists & Herpetologists. Res: Ecology and control of insects of medical and veterinary importance; ecology and control of chironomid midges; biological control of medically important insects. Mailing Add: Western Wash Res & Exten Ctr Wash State Univ Puyallup WA 98371

BAY, ROGER RUDOLPH, b La Crosse, Wis, Nov 27, 31; m 58; c 2. FORESTRY. Educ: Univ Idaho, BS, 53; Univ Minn, MF, 54, PhD(forestry), 67. Prof Exp: Forester, 54, res forester, 56-61, proj leader, NCent Forest Exp Sta, 61-70, from asst chief to chief, Br Watershed & Aquatic Habitat Res, Div Forest Environ Res, 70-73, asst to dep chief for res, 73-74, DIR INTERMOUNTAIN FOREST & RANGE EXP STA, FOREST SERV, USDA, 74- Mem: AAAS; Am Forestry Asn; Soc Am Foresters; Am Geophys Union; Soil Conserv Soc Am. Res: Watershed management research; forest hydrology. Mailing Add: Intermt Forest & Range Exp Sta Forest Serv USDA Ogden UT 84401

BAY, ZOLTAN LAJOS, b Gyulavari, Hungary, July 24, 00; m 47; c 3. QUANTUM PHYSICS. Educ: Univ Budapest, MS, 23, PhD(physics), 25. Prof Exp: Prof theoret physics, Univ Szeged, 30-36; prof atomic physics, Tech Univ Budapest, 38-48; res prof, George Washington Univ, 48-55; physicist, Nat Bur Standards, 55-73; SR RES SCIENTIST, AM UNIV, 73- Concurrent Pos: Dir, Tungsram Res Lab, 36-48; tech eng mgr, United Incandescent Lamp & Elec Co, 44-48. Mem: Fel Am Phys Soc; Hungarian Acad Sci. Res: Atomic and molecular spectroscopy; nuclear excited states, techniques and theory of fast coincidence experiments; ionization of matter by high energy radiations; optical masers; unified standardization of time and length. Mailing Add: 151 Quincy Sta Chevy Chase MD 20015

BAYAN, ARIS PAUL, b Istanbul, Turkey, Mar 13, 21; nat US; m 46; c 2. MICROBIOLOGY. Educ: Cornell Univ, BS, 42; Rutgers Univ, MS, 54. Prof Exp: Res asst, Squibb Inst Med Res, 46-52, res supvr, 52-57, res assoc, 57-59, sr res microbiologist, 59-69; asst mgr dept indust microbiol, 69-71, MGR MICROBIOL PROD, SCHERING CORP, 71- Mem: Am Soc Microbiol; Soc Indust Microbiol. Res: Microbiological biosynthetic processes. Mailing Add: Schering Corp 1011 Morris Ave Union NJ 07083

BAYER, DAVID E, b Grass Valley, Ore, Aug 1, 26. PLANT PHYSIOLOGY. Educ: Ore State Univ, BS, 51, MS, 53; Univ Wis, PhD(agron), 58. Prof Exp: Assoc agriculturalist, 59-62, asst botanist, 62-69, assoc botanist, 69-73, BOTANIST, UNIV CALIF, DAVIS, 73- Mem: Weed Soc Am. Res: Plant physiology, particularly pertaining to herbicides and their effects on plants. Mailing Add: Dept of Bot Univ of Calif Exp Sta Davis CA 95616

BAYER, DOUGLAS LESLIE, b Chicago, Ill, Feb 22, 45; m 68; c 1. COMPUTER SCIENCE. Educ: Knox Col, BA, 66; Mich State Univ, MS, 68, PhD(physics), 70. Prof Exp: Asst res specialist physics, Rutgers Univ, 71-72; MEM TECH STAFF COMPUT SCI, BELL LABS, 72- Mem: Asn Comput Mach. Res: Operating systems design. Mailing Add: Bell Labs 600 Mountain Ave 7C 207 Murray Hill NJ 07974

BAYER, FREDERICK MERKLE, b Asbury Park, NJ, Oct 31, 21. ZOOLOGY. Educ: Univ Miami, BS, 48; George Washington Univ, MS, 54, PhD(zool), 58. Prof Exp: Asst dir, State Mus, Fla, 42-46; asst marine invertebrates, Marine Lab, Univ Miami, 46-47; from asst cur to assoc cur, US Nat Mus, 47-61; assoc prof, 62-63, PROF, INST MARINE SCI, UNIV MIAMI, 64- Concurrent Pos: Mem, Bikini Sci Resurv Exped, US Navy, 47, Ifaluk Atoll Surv Team, Pac Sci Bd, Nat Res Coun-Nat Acad Sci, 53, Palau Island Exped, Vanderbilt Found, 55 & Gulf of Guinea Exped, Inst Marine Sci, Univ Miami, 64; ed, Bull Marine Sci of Gulf & Caribbean, 62- Mem: Soc Syst Zool; Am Paleont Soc; Am Malacol Union. Res: Biological oceanography; ecology of coral reefs; taxonomy of Octocorallia, Molusca, Pterobranchia. Mailing Add: Inst of Marine Sci Univ of Miami Coral Gables FL 33124

BAYER, HORST OTTO, b Stuttgart, Ger, Oct 21, 34; US citizen; m 55; c 2. ORGANIC CHEMISTRY. Educ: Rochester Inst Technol, BS, 57; Purdue Univ, PhD(org chem), 61. Prof Exp: Sr chemist, Pesticide Syntheses Group, 62-67, head res lab, 67-73, PROJ LEADER, RES LAB, ROHM AND HAAS CO, 73- Concurrent Pos: NSF-NATO fel, 61-62. Mem: Am Chem Soc. Res: Synthesis, structure and activity studies relating to herbicides, insecticides, fungicides and plant growth regulators; insecticide research and development. Mailing Add: Res Labs Rohm and Haas Co Springhouse PA 19477

BAYER, MARGRET HELENE JANSSEN, b Hamburg, Ger, July 8, 31; m 58; c 2. PLANT PHYSIOLOGY. Educ: Univ Hamburg, dipl, 58, PhD(bot, zool), 61. Prof Exp: Res assoc plant physiol, Inst Gen Bot & Bot Garden, 58-61; res assoc biol, 62-72, RES ASSOC MOLECULAR BIOL, INST CANCER RES, 73- Concurrent Pos: Guest prof biol, Univ Hamburg, 75. Mem: Am Inst Biol Sci; Am Soc Plant Physiologists. Res: Physiology and biochemistry of plant tumorgenesis; action of phytohormones on cell growth and development; studies on antibodies directed against surface antigens of virus-transformed cells. Mailing Add: Inst for Cancer Res 7701 Burholme Ave Philadelphia PA 19111

BAYER, RAYMOND P, chemistry, see 12th edition

BAYER, RICHARD EUGENE, b Milwaukee, Wis, Jan 11, 32; m 57; c 4. ANALYTICAL CHEMISTRY, INORGANIC CHEMISTRY. Educ: Carroll Col, Wis, BS, 54; Ind Univ, PhD(chem), 59. Prof Exp: Assoc prof, 58-66, PROF CHEM, CARROLL COL, WIS, 66-, CHMN DEPT, 67- Mem: Am Chem Soc. Res: Determination of stability constants of coordination compounds in non-aqueous solvents; rare earth chelates; atomic absorption; selective ion electrodes. Mailing Add: Dept of Chem Carroll Col Waukesha WI 53186

BAYER, ROBERT CLARK, b New York, NY, July 4, 44; m 67; c 3. AVIAN PHYSIOLOGY. Educ: Univ Vt, BS, 66, MS, 68; Mich State Univ, PhD(avian physiol), 72. Prof Exp: ASST PROF AVIAN PHYSIOL, UNIV MAINE, ORONO, 72- Mem: AAAS; Poultry Sci Asn; Am Soc Animal Sci; World Poultry Sci Asn. Res: Scanning and transmission electron microscopy as it pertains to gastrointestinal physiology and pathology; atherogenesis in avian species and physiology of environmental toxins; lobster nutrition. Mailing Add: 128 Hitchner Hall Univ of Maine Orono ME 04473

BAYER, SHIRLEY ANN, b Evansville, Ind, Aug 20, 40; m 73. NEUROANATOMY. Educ: St Mary-of-the-Woods Col, BA, 63; Calif State Univ, Fullerton, MA, 69; Purdue Univ, PhD(biol), 74. Prof Exp: Teacher biol, Guerin High Sch, Ill, 63-65 & Our Lady of Providence High Sch, Ind, 65-66; teacher & chmn sci dept, Marywood Sch, Calif, 66-70; ASST RES BIOLOGIST, PURDUE UNIV, 74- Mem: Soc Neurosci. Res: Ultrastructural and light microscopy studies on the development of the hippocampus in the rat brain using x-irradiation and 3H-thymidine autoradiography. Mailing Add: Dept of Biol Sci Purdue Univ West Lafayette IN 47907

BAYER, THOMAS NORTON, b Elyria, Ohio, Dec 23, 34; m 69. GEOLOGY, PALEONTOLOGY. Educ: Macalester Col, BA, 57; Univ Minn, MS, 60, PhD(geol), 65. Prof Exp: Instr geol, Macalester Col, 57-64; asst prof, 64-68, PROF EARTH SCI, WINONA STATE COL, 68- Concurrent Pos: Lectr, Macalester Col, 61-64; consult classroom prod, KTCA-TV, 62-63. Mem: AAAS; Geol Soc Am; Am Asn Petrol Geologists; Paleont Soc; Soc Econ Paleont & Mineral. Res: Paleoecology of lower Paleozoic invertebrate faunas. Mailing Add: Dept of Geol & Earth Sci Winona State Col Winona MN 55987

BAYES, ALFRED LEE, b Springfield, Mo, Nov 7, 14; m 42; c 2. INORGANIC CHEMISTRY. Educ: Univ Chicago, BS, 34, PhD(chem), 41. Prof Exp: Res chemist, Linde Air Prod Co, Tonawanda, NY, 35-38; asst chem, Univ Chicago, 38-41; asst chem, Res Lab, Linde Air Prod Co, 41-52, asst chem, Silicones Dept, 52-54, mgr develop, 54-56; asst dir res & develop, Silicones Div, Union Carbide Corp, 56-59; prod mgr, Gas Prod Dept, Linde Co, 59-60, admin asst to pres, 60-63, VPRES LINDE DIV, UNION CARBIDE CORP, 63- Mem: Soc Indust Chemists; Am Chem Soc; Am Oil Chem Soc. Res: Reactions of liquid ammonia; development of lubricating oil additives; preparation and purification of metal oxide powders; applications of industrial chemicals in food processing; research, product and process development of organo silicon chemicals. Mailing Add: Linde Div Union Carbide Corp 270 Park Ave New York NY 10017

BAYES, KYLE D, b Colfax, Wash, Mar 3, 35; m 61. PHYSICAL CHEMISTRY. Educ: Calif Inst Technol, BS, 56; Harvard Univ, PhD(chem), 59. Prof Exp: NSF fel, Univ Bonn, 59-60; from asst prof to assoc prof, 60-71, PROF CHEM, UNIV CALIF, LOS ANGELES, 71- Concurrent Pos: Sloan vis lectr, Harvard Univ, 66. Mem: AAAS; Am Phys Soc; Am Chem Soc. Res: Chemical kinetics and spectroscopy of gas phase reactions. Mailing Add: Dept of Chem Univ of Calif Los Angeles CA 90024

BAYFIELD, EDWARD GEOFFREY, b Annapolis Royal, NS, Jan 24, 00; nat US; m 37; c 2. CEREAL CHEMISTRY. Educ: Univ Alta, BSA, 23; McGill Univ, MSA, 24; Ohio State Univ, PhD, 31. Prof Exp: Agronomist & instr, Prov Sch Agr, Claresholm, Alta, 24-28; assoc agron, Ohio Exp Sta, 31-39; prof milling indust & head dept, Kans State Col, 39-45, cereal technologist, Exp Sta, 39-45; dir prod control & res, Standard Milling Col, Ill, 45-52; prof baking sci & mgt, 52-65, head dept baking sci & mgt & dir baking indust prog, 59-65, EMER PROF BUS & CONSULT, SCH BUS, FLA STATE UNIV, 65- Concurrent Pos: Nat Milling Co fel, Ohio Exp Sta & Ohio State Univ, 29-34; cereal technologist in charge fed soft wheat lab, USDA, 36-39. Mem: Fel AAAS; Am Asn Cereal Chemists; Am Soc Bakery Eng. Res: Milling and baking industries; wheat quality. Mailing Add: 1514 Golf Terr Tallahassee FL 32301

BAYHI, JOSEPH FRANKLIN, b New Orleans, La, Mar 20, 17; m 49; c 1. GEOPHYSICS. Educ: La State Univ, BS, 37. Prof Exp: Student engr, elec dept, Standard Oil Co, La, 37-39; asst operator geophys, Carter Oil Co, 39-41, res engr, 45-49, group leader geophys res, 49-54, sect head, 55-58; div mgr, Jersey Prod Res Co, 59-60; assoc chief geophys res, Humble Oil & Ref Co, 60-64; sect supvr, Esso Prod Res Co, 64-74, RES ADV, EXXON PROD RES CO, 74- Mem: Soc Explor Geophys. Res: Seismic exploration for oil; wave propagation theory; seismic source development and field techniques; instrumentation, data analysis and interpretation; electrical exploration methods, field techniques and data analysis; cathodic protection to casings and pipes. Mailing Add: Exxon Prod Res Co PO Box 2189 Houston TX 77001

BAYHURST, BARBARA P, b Vivian, La, Mar 31, 26; m 47; c 4. NUCLEAR CHEMISTRY, RADIO CHEMISTRY. Educ: La State Univ, BS, 46. Prof Exp: STAFF MEM RADIOCHEM, LOS ALAMOS SCI LAB, UNIV CALIF, 55- Mem: Am Chem Soc. Res: Absolute counting; cross-section measurements; atomic devices as research tools. Mailing Add: Los Alamos Sci Lab MS 514 Los Alamos NM 87545

BAYLESS, DAVID LEE, b Alliance, Ohio, June 12, 38; m 63; c 3. STATISTICS. Educ: Muskingum Col, AB, 60; Fla State Univ, MS, 63; Tex A&M Univ, PhD(statist), 68. Prof Exp: Asst, Fla State Univ, 61-63; assoc mem staff, Bell Tel Labs, NJ, 63-64; consult, Gen Food Tech Ctr, NY, 64-65; fel, Tex A&M Univ, 65-68; STATE PROG DEVELOP COORDR, RES TRIANGLE INST, 68- Concurrent Pos: Adj assoc prof statist, NC State Univ, 68-; asst to dir res, Nat Assessment of Educ Proj, 70. Mem: Am Statist Asn. Res: Industrial applications of statistics and computers specializing in sampling, regression model building, and developing large computer programs. Mailing Add: Res Triangle Inst PO Box 12194 Research Triangle Park NC 27709

BAYLESS, LAURENCE EMERY, b Richmond, Va, Aug 27, 38; m 62. BIOLOGY. Educ: Univ Tenn, AB, 61; Tulane Univ, MS, 62, PhD(biol), 66. Prof Exp: Asst prof, 66-70, ASSOC PROF BIOL, CONCORD COL, 70- Mem: Am Soc Ichthyol & Herpet; Soc Study Amphibians & Reptiles; Ecol Soc Am. Res: Population ecology and life history of amphibians. Mailing Add: Dept of Biol Concord Col Athens WV 24712

BAYLESS, PHILIP LEIGHTON, b Indianapolis, Ind, Feb 23, 28; m 49; c 3. CHEMISTRY. Educ: Oberlin Col, AB, 49; Duke Univ, PhD(chem), 54. Prof Exp: PROF CHEM & CHMN DEPT, WILMINGTON COL, 54- Concurrent Pos: Alexander von Humboldt fel, 63-64; fac fel, Nat Acad Sci, 70-71. Honors & Awards: Gustav Ohaus-NSTA Award, Nat Sci Teacher's Asn, 75. Mem: Am Chem Soc. Res: Mechanisms of chemical reactions; metal-organic compounds. Mailing Add: 71 Faculty Pl Wilmington OH 45177

BAYLEY, DONALD SPERRY, physics, see 12th edition

BAYLEY, HENRY SHAW, b Macclesfield, Eng, Aug 21, 38. BIOCHEMISTRY, NUTRITION. Educ: Univ Reading, BSc, 60; Univ Nottingham, PhD, 63; ARIC. Prof Exp: Asst lectr nutrit, Wye Col, Univ London, 63-65; asst prof, 65-68, ASSOC PROF NUTRIT, UNIV GUELPH, 68- Mem: Am Soc Animal Sci; Am Inst Nutrit; Can Soc Animal Prod; Nutrit Soc Can; Brit Nutrit Soc. Res: Energy metabolism: factors influencing digestion and absorption of major energy yielding nutrients, use of respiration calorimeter and substrate turnover rates as indicators of metabolic status in the developing animal. Mailing Add: Dept of Nutrit Univ of Guelph Guelph ON Can

BAYLEY, NED D, animal husbandry, see 12th edition

BAYLEY, RICHARD WILLIAM, geology, deceased

BAYLEY, STANLEY THOMAS, b Grays, Essex, Eng, Nov 5, 26; Can citizen; m 50; c 2. BIOPHYSICS, BIOCHEMISTRY. Educ: Univ London, BSc, 46, PhD(physics), 50. Prof Exp: Nuffield res fel biophys, King's Col, Univ London, 50-52; from asst res officer to sr res officer, Nat Res Coun Can, 52-67; chmn dept, 68-74, PROF BIOL, McMASTER UNIV, 67- Mem: Biophys Soc; Am Soc Biol Chemists; Can Biochem Soc; Can Soc Cell Biol. Res: Molecular biology; synthesis, structure and function of nucleic acids and proteins with particular reference to halophilic bacteria. Mailing Add: Dept of Biol McMaster Univ Hamilton ON Can

BAYLIFF, WILLIAM HENRY, b Annapolis, Md, Aug 29, 28; m 69. FISHERIES. Educ: Western Md Col, BA, 49; Univ Wash, MS, 54, PhD, 65. Prof Exp: Biologist, State Dept Fisheries, Wash, 52-54, 57-58; scientist, 58-65, SR SCIENTIST, INTER-AM TROP TUNA COMN, SCRIPPS INST OCEANOG, 65- Concurrent Pos: Marine fishery biologist, Food & Agr Orgn, UN, 67-68, consult, Develop Prog Fishery Proj, Callao, Peru, 71, convener, working party tuna tagging in Pac & Indian Oceans, Panel Experts Facilitation Tuna Res, 70-; mem ed comt, Fishery Bull, Nat Marine Fisheries Serv, 70- Honors & Awards: W F Thompson Award, Am Inst Fishery Res Biologists, 69. Mem: AAAS; Am Inst Fishery Res Biologists. Res: Biology and population dynamics of marine fishes. Mailing Add: Inter-Am Trop Tuna Comn Scripps Inst Oceanog La Jolla CA 92037

BAYLIN, GEORGE JAY, b Baltimore, Md, May 15, 11; m 38; c 3. RADIOLOGY. Educ: Johns Hopkins Univ, AB, 31; Duke Univ, MD, 37. Prof Exp: Instr, 41-42, assoc, 42-44, from asst prof to assoc prof, 44-50, PROF RADIOL, MED SCH, DUKE UNIV, 50-, PROF OTOLARYNGOL, 74- Mem: AAAS; fel Am Col Radiol; NY Acad Sci. Res: Mechanisms of pain in duodenal ulcer; x-ray aspects of small intestinal diseases; x-ray studies in newer aspects of kidney functions; studies of pulmonary blood flow, renal conditioning; renovascular hypertension; mastoid diseases. Mailing Add: Dept of Radiol Duke Univ Med Ctr Durham NC 27706

BAYLIS, JEFFREY ROWE, b Jackson, Mich, Nov 1, 45. ETHOLOGY, ICHTHYOLOGY. Educ: Univ Calif, Santa Barbara, BA, 68; Univ Calif, Berkeley, MA, 72, PhD(zool), 75. Prof Exp: Fel zool, Rockefeller Univ, 74-76; ASST PROF ZOOL, UNIV WIS-MADISON, 76- Mem: Animal Behav Soc; Sigma Xi; Soc Study Evolution; Am Soc Ichthyologists & Herpetologists; AAAS. Res: Analysis of the social communicatory signals of animals, with special regard to the evolutionary origins of the signals and the environmental pressures that shaped them. Mailing Add: Dept of Zool Univ of Wis Madison WI 53706

BAYLIS, JOHN ROBERT, JR, b Chicago, Ill, May 12, 27; m 54; c 2. GENETICS. Educ: Utah State Univ, BS, 51; Northwestern Univ, MS, 56; Fla State Univ, PhD(genetics), 66. Prof Exp: Biologist, Oak Ridge Nat Lab, 66-67; asst prof, 67-74, ASSOC PROF GENETICS, UNIV W FLA, 74- Mem: AAAS; Genetics Soc Am; Am Inst Biol Sci. Res: Fungal genetics; mutagenesis. Mailing Add: Dept of Biol Univ of WFla Pensacola FL 32504

BAYLISS, BERENICE, b Joliet, Ill, Feb 10, 02. BACTERIOLOGY. Educ: Mont State Col, BS, 49, MS, 51; Wash State Univ, PhD(bact), 64. Prof Exp: Teacher & prin, pub schs, Mont, 20-29; asst med technologist, Billings Clin, Mont, 37-43; supv technologist, Deaconess Hosp, 43-48; med technologist, student health serv, 49-51, from instr to asst prof bact, 51-65, from assoc prof to prof, 65-68, EMER PROF MICROBIOL, MONT STATE UNIV, 69- Mem: Am Soc Med Technologists. Res: Pathogenic bacteriology; immunohematology. Mailing Add: 440 Lewis Ave Billings MT 59102

BAYLISS, JOHN TEMPLE, b Richmond, Va, July 6, 39. PHYSICS. Educ: Bowdoin Col, BA, 61; Univ Va, PhD(physics), 67. Prof Exp: Asst prof physics, Va Commonwealth Univ, 67-73; CONSERV COORDR, VA ENERGY OFF, 75- Mem: Am Asn Physics Teachers. Res: Conservation techniques and state policy. Mailing Add: Va Energy Off 823 E Main St Richmond VA 23219

BAYLISS, PETER, b Eng; m 66. MINERALOGY, CRYSTALLOGRAPHY. Educ: Univ NSW, BE, 59, MSc, 62, PhD(geol), 67. Prof Exp: Asst prof, 67-70, ASSOC PROF GEOL, UNIV CALGARY, 70- Concurrent Pos: Abstractor, Mineral Abstr, 69; mem, X-ray Diffraction Powder File Index Comt, 70; assoc ed, Jour Mineral Asn Can, 75- Mem: Mineral Asn Can; Clay Minerals Soc; fel Mineral Soc Am. Res: Clay mineralogy in northern Canada; crystal structure analysis of sulphides. Mailing Add: Dept of Geol Univ of Calgary Calgary AB Can

BAYLOR, CHARLES, JR, b Baltimore, Md, Dec 5, 40; m 70; c 2. ORGANIC CHEMISTRY, PHOTOGRAPHIC CHEMISTRY. Educ: Morgan State Col, BS, 62; Utah State Univ, PhD(chem), 67. Prof Exp: RES CHEMIST, E I DU PONT DE NEMOURS & CO, INC, 66- Mem: Am Chem Soc. Res: Synthesis, analytical methods and structure determination of organic compounds; surfactant applications, polymers and emulsion polymerization. Mailing Add: Photo Prod Dept Exp Sta E I du Pont de Nemours & Co Inc Wilmington DE 19898

BAYLOR, CURTIS HORTON, b Marion, Va, Mar 20, 08; m 42; c 3. INTERNAL MEDICINE, OCCUPATIONAL MEDICINE. Educ: Emory & Henry Col, BS, 29; Johns Hopkins Univ, MD, 35; Am Bd Internal Med, dipl, 47. Prof Exp: From intern to resident internal med, Strong Mem Hosp, Rochester, NY, 35-38; adj prof, Am Univ Beirut, 38-41, assoc prof, 41-46; asst prof med, Univ Rochester, 46-47; prof internal med, Am Univ Beirut, 47-49; from asst med dir to med dir, Texaco, Inc, 49-74; RETIRED. Mailing Add: 205 Pelhamdale Ave Pelham NY 10803

BAYLOR, DENIS ARISTIDE, b Oskaloosa, Iowa, Jan 30, 40; m 65; c 2. NEUROPHYSIOLOGY. Educ: Knox Col, BA, 61; Yale Univ, MD, 65. Prof Exp: Fel neurophysiol, Sch Med, Yale Univ, 65-68; staff fel, Lab Neurophysiol, Nat Inst Neurol & Commun Disorders & Stroke, 68-70; USPHS spec fel, Physiol Lab, Cambridge Univ, 70-72; assoc prof physiol, Univ Colo Med Ctr, Denver, 72-74; assoc prof neurophysiol, Univ, 74-75, ASSOC PROF NEUROBIOL, SCH MED, STANFORD UNIV, 75- Honors & Awards: Sinsheimer Found Award Med Res, 75. Res: Generation and transmission of neural signals in the vertebrate retina. Mailing Add: Dept of Neurobiol Stanford Univ Sch Med Stanford CA 94305

BAYLOR, EDWARD RANDALL, b Uvalde, Tex, Jan 21, 14; m 42; c 2. MARINE ECOLOGY. Educ: Univ Ill, MS, 42; Princeton Univ, PhD(cellular physiol), 49. Prof Exp: Asst zool, Univ Tex, 38-39, Univ Ill, 41-42 & Princeton Univ, 46-47, 48-50; asst prof zool, Univ Mich, 50-58; assoc marine physiol, Woods Hole Oceanog Inst, 58-68; PROF MARINE SCI, STATE UNIV NY STONY BROOK, 68- Mem: Am Instrument Soc; Marine Technol Soc; Am Soc Limnol & Oceanog; Soc Gen Physiologists; Biophys Soc. Res: Invertebrate behavior; visual physiology; behavior of marine invertebrates resulting in their patchy distribution. Mailing Add: Dept of Marine Sci State Univ NY at Stony Brook Stony Brook NY 11790

BAYLOR, JOHN E, b Belvidere, NJ, Sept 16, 22; m 50; c 2. AGRONOMY. Educ: Rutgers Univ, BSc, 47, MSc, 48; Pa State Univ, PhD(agron), 58. Prof Exp: Asst exten specialist farm crops, Rutgers Univ, 48-49, assoc exten specialist, 49-55; assoc prof agron & assoc exten specialist, 57-65, PROF AGRON & EXTEN SPECIALIST, PA STATE UNIV, UNIVERSITY PARK, 65- Concurrent Pos: Consult, IRI Res Inst, Brazil, 64. Honors & Awards: Merit Award, Am Forage & Grassland Coun, 63, Medallion Award, 71. Mem: Am Soc Agron; Am Forage & Grassland Coun (pres, 67-70). Res: Forage crop production and management. Mailing Add: 106 Agr Admin Bldg Pa State Univ University Park PA 16802

BAYLOUNY, RAYMOND ANTHONY, b Paterson, NJ, Aug 11, 32; m 59; c 4. ORGANIC CHEMISTRY. Educ: Seton Hall Univ, BS, 54; Univ Md, MS, 58, PhD(org chem), 60. Prof Exp: Instr org chem & biochem, Brooklyn Col, 60-63; asst prof org chem, 63-67, chmn dept, 67-69, ASSOC PROF ORG CHEM, FAIRLEIGH DICKINSON UNIV, 67- Concurrent Pos: Sr fel, Princeton Univ, 69-70. Mem: Am Chem Soc; The Chem Soc. Res: Theoretical organic chemistry; biochemistry; thermal rearrangement of unsaturated esters; synthesis and structure analysis of natural products; newer methods of organic synthesis. Mailing Add: Dept of Chem Fairleigh Dickinson Univ Madison NJ 07940

BAYLY, GEORGE HENRY UNIACKE, b Toronto, Ont, Feb 13, 18; m 44; c 3. FORESTRY. Educ: Univ Toronto, BScF, 39, MScF, 52. Prof Exp: Chief, Div Reforestation, 53-57, asst dep minister, 57-66, dep minister, 66-71, secy of the treas bd, 71-74, DEP PROV SECY RESOURCES DEVELOP, DEPT LANDS & FORESTS, ONT, 74- Concurrent Pos: Chmn adv comt sci policy, Ont Govt. Mem: Can Inst Forestry. Res: Silviculture. Mailing Add: Parliament Bldgs Queen's Park Toronto ON Can

BAYLY, M BRIAN, b Northwood, Eng, Apr 16, 29; m 60; c 5. GEOLOGY. Educ: Cambridge Univ, BA, 52, MS & MSc, 62; Univ Chicago, PhD(geol), 62. Prof Exp: From asst prof to assoc prof, 62-74, PROF GEOL, RENSSELAER POLYTECH INST, 74- Res: Tectonics, structural geology, deformation processes, with emphasis on quantitative mechanical theories. Mailing Add: Dept of Geol Rensselaer Polytech Inst Troy NY 12181

BAYM, GORDON A, b New York, July 1, 35; m 58. THEORETICAL PHYSICS. Educ: Cornell Univ, BA, 56; Harvard Univ, AM, 57, PhD(physics), 60. Prof Exp: NSF fel, Inst Theoret Physics, Denmark, 60-62; asst res physicist, Univ Calif, Berkeley, 62-63; PROF PHYSICS, UNIV ILL, URBANA-CHAMPAIGN, 63- Concurrent Pos: A P Sloan res fel, 65-68. Mem: AAAS; fel Am Phys Soc; Am Astron Soc. Res: Low temperature physics; astrophysics; theory of many body systems. Mailing Add: Dept of Physics Univ of Ill Urbana IL 61801

BAYMAN, BENJAMIN, b New York, NY, Dec 12, 30; m 57; c 2. NUCLEAR PHYSICS. Educ: Cooper Union, BChE, 51; Univ Edinburgh, PhD(physics), 55. Prof Exp: Res fel theoret physics, Univ Edinburgh, 55-56; Ford Found fel theoret nuclear physics, Inst Theoret Physics, Denmark, 56-60; asst prof, Princeton Univ, 60-65; assoc prof physics, 65-68, PROF PHYSICS, UNIV MINN, MINNEAPOLIS, 68-, FEL, 75- Mem: Am Phys Soc. Res: Interpretation of experimental data on atomic nuclei obtained by nuclear reaction and radioactivity studies in terms of nuclear models. Mailing Add: Sch Phys & Astron Univ of Minn Minneapolis MN 55455

BAYMILLER, JOHN WILLIAM, chemistry, see 12th edition

BAYNE, CHARLES KENNETH, b Pittsburgh, Pa, Aug 22, 44. STATISTICS. Educ: Blackburn Col, BA, 66; Wash Univ, MS, 68; NC State Univ, PhD(statist), 74. Prof Exp: Mathematician, US Naval Ordnance Lab, 68-70; RES ASSOC STATIST, NUCLEAR DIV, UNION CARBIDE CORP, 74- Res: Experimental designs. Mailing Add: Bldg 9704-1 Nuclear Div Union Carbide Corp PO Box Y Oak Ridge TN 37830

BAYNE, CHRISTOPHER JEFFREY, b Trinidad, WI, Aug 31, 41; m 63; c 2. INVERTEBRATE PHYSIOLOGY. Educ: Univ Wales, BSc, 63, PhD(zool), 67. Prof Exp: Fel, Marine Sci Labs, Menai Bridge, Wales, 66-67; lectr marine zool, Univ Wales, 67-68; res assoc molluscan physiol, Univ Mich, 68-71; asst prof zool, 71-75, ASSOC PROF ZOOL, ORE STATE UNIV, 75- Mem: AAAS; Soc Invert Path; Am Inst Biol Sci; Am Soc Zool; Sigma Xi. Res: Molluscan immunology; cell culture; parasitology. Mailing Add: Dept of Zool Ore State Univ Corvallis OR 97331

BAYNE, DAVID ROBERGE, b Selma, Ala, Jan 29, 41; m 64. LIMNOLOGY. Educ: Tulane Univ, BA, 63; Auburn Univ, MS, 67, PhD(fisheries mgt), 70. Concurrent Pos: Asst prof biol, Ga Col, 70-71; ASST PROF FISHERIES & ALLIED AQUACULT, AUBURN UNIV, 72- Mem: Am Fisheries Soc; Wilderness Soc. Res: Physicochemical dynamics of large multipurpose reservoirs; fresh water plankton populations; effects of waste heat on aquatic communities. Mailing Add: Dept Fisheries & Allied Aquacult Auburn Univ Auburn AL 36830

BAYNE, GILBERT M, b Philadelphia, Pa, Mar 11, 21; m 45; c 4. PSYCHOPHARMACOLOGY, THERAPEUTICS. Educ: Ursinus Col, BS, 43; Univ Pa, MD, 47. Prof Exp: Intern, Hosp Univ Pa, Philadelphia, 47-48; asst med dir, 48-53, dir med res, 54-70, sr dir, 70, SR DIR LONG RANGE PLANNING MED RES, MERCK SHARP & DOHME RES LABS, 70- Concurrent Pos: Clin asst, Endocrinol Clin, Philadelphia Gen Hosp, 49-50, res asst, Med Serv, 50-52, asst vis physician, 52-53; res assoc, Dept Res Therapeut, Norristown State Hosp, Pa, 51-60; fel med, Dept Infectious Dis, Hosp Univ Pa, 51-52, asst instr, 52-53. Mem: NY Acad Sci; Am Heart Asn; Am Col Neuropsychopharmacol; Am Fedn Clin Res; AMA. Res: Medical research; mental health; infectious diseases. Mailing Add: Merck Sharp & Dohme Res Labs West Point PA 19486

BAYNE, HENRY GODWIN, b New York, NY, Dec 2, 25; m 58; c 3. BACTERIOLOGY. Educ: Brooklyn Col, BA, 49, MA, 54. Prof Exp: Sr technician microbiol, Western Reserve Univ, 50-51; health inspector, NY Dept Health, 53-55, jr bacteriologist, 55; MICROBIOLOGIST, WESTERN UTILIZATION RES, USDA, 55- Mem: AAAS; Am Soc Microbiol. Res: Investigation of biochemical and physiological processes of bacterial and other organisms found in agricultural products. Mailing Add: 246 Purdue Ave Berkeley CA 94708

BAYNTON, HAROLD WILBERT, b Brandon, Man, Sept 16, 20; nat US; m 47; c 3. METEOROLOGY. Educ: Univ Mich, MS, 57, MA, 59, PhD(diffusion), 63. Prof Exp: Weather forecaster, Meteorol Serv Can, 42-52, res meteorologist, 52-55; res assoc meteorol, Univ Mich, 55-58; res meteorologist, Systs Div, Bendix Aviation Corp, 58-63; climatologist, Martin Co, 63-64; METEOROLOGIST, NAT CTR FOR ATMOSPHERIC RES, BOULDER, 65- Concurrent Pos: Affil prof, Va Polytech Inst, 67-70. Mem: Am Meteorol Soc; Air Pollution Control Asn; fel Royal Meteorol Soc; Can Meteorol Soc. Res: Applications of probability and statistical methods to meteorology in the realm of applied meteorology; air pollution. Mailing Add: 415 Kiowa Pl Boulder CO 80303

BAYR, KLAUS J, b Gmunden, Austria, June 4, 39; m 68; c 1. GEOGRAPHY OF AFRICA, CULTURAL GEOGRAPHY. Educ: Graz Univ, MA(geog) & MA(phys educ), 66, PhD(geog, ethnog), 69. Prof Exp: Asst prof geog, Univ NH, 66-69; asst prof, 69-75, ASSOC PROF GEOG, KEENE STATE COL, 75- Mem: Austrian Asn Geog; Asn Am Geogr. Res: Settlement patterns in East Africa. Mailing Add: Dept of Social Sci Keene State Col Keene NH 03431

BAYRD, EDWIN DORRANCE, b Chicago, Ill, Nov 12, 17; m 42; c 5. MEDICINE. Educ: Dartmouth Col, AB, 39; Harvard Med Sch, MD, 42; Univ Minn, MS, 47. Prof Exp: From instr to assoc prof, 47-67, PROF MED, MAYO MED SCH, UNIV MINN, 67- Concurrent Pos: Fel trop med & parasitol, Tulane Univ, 43-44; fel med, Mayo Found, Univ Minn, 47; consult, Mayo Clin, Methodist & St Mary's Hosps, 47-; ed-in-chief, Mayo Clin Proc, 62-; Sir Norman Paul vis prof, Sydney Hosp, Australia, 63; chmn div hemat, Mayo Clin, 67-, pres staff, 69- Mem: AAAS; AMA; Am Fedn Clin Res; Am Soc Hemat; Int Soc Hemat. Res: Clinical and protein aspects of plasma cell disease, especially multiple myelome, macroglobulinemia and systematized amyloidosis. Mailing Add: Mayo Grad Sch of Med Univ of Minn Rochester MN 55902

BAYS, JAMES PHILIP, b West Frankfort, Ill, Feb 19, 41; m 63; c 2. BIO-ORGANIC CHEMISTRY, ORGANIC CHEMISTRY. Educ: Northwestern Univ, BS, 63; Univ Wis, PhD(org chem), 68. Prof Exp: NIH fel chem, Yale Univ, 68-70; ASST PROF CHEM, GRINNELL COL, 70- Concurrent Pos: ACM fac fel, Argonne Nat Lab, 71-72; asst prof biochem, Med Sch, Rush Univ, 74- Mem: Am Chem Soc; AAAS; Am Sci Affil. Res: Chemistry of dianions of phenylacetone and its derivatives; reactions of synthetic utility; synthesis of biologically important molecules. Mailing Add: Dept of Chem Grinnell Col Grinnell IA 50112

BAZER, FULLER WARREN, b Shreveport, La, Sept 2, 38; m 62; c 2. ANIMAL SCIENCE, REPRODUCTIVE PHYSIOLOGY. Educ: Centenary Col, BS, 60; La State Univ, Baton Rouge, MS, 63; NC State Univ, PhD(physiol), 69. Prof Exp: Asst prof, 68-74, ASSOC PROF ANIMAL SCI, UNIV FLA, 74- Mem: AAAS; Soc Study Reprod; Am Soc Animal Sci. Res: Uterine protein secretion of domestic animals as related to embryonic development and corpus luteum function. Mailing Add: Dept of Animal Physiol Univ of Fla Agr Exp Sta Gainesville FL 32601

BAZER, JACK, b New York, NY, Dec 23, 24; m 51; c 2. MATHEMATICS, PHYSICS. Educ: Cornell Univ, BA, 47; Columbia Univ, MA, 49; NY Univ, PhD(math), 53. Prof Exp: Jr res scientist, 51-53, res assoc, 53-58, sr scientist, 58, from asst prof to assoc prof, 58-65, PROF MATH, COURANT INST MATH SCI, NY UNIV, 65- Concurrent Pos: Consult, Grumman Aircraft Eng Corp, Long Island, 60; mem comn IV, Int Sci Radio Union. Mem: Am Math Soc. Res: Diffraction theory of scalar and vector fields; magnetogasdynamics; plasma physics; probability theory; partial differential equations. Mailing Add: Courant Inst of Math Sci 251 Mercer St New York NY 10012

BAZIN, MAURICE JACQUES, b Paris, France, Aug 17, 34; m 58; c 1. PHYSICS. Educ: Polytech Sch, Paris, dipl, 57; Stanford Univ, PhD(physics), 62. Prof Exp: Res asst physics, Stanford Univ, 59-61; res assoc, Lab LePrince-Ringuet, Paris, 62-64; from instr to asst prof, Princeton Univ, 64-68; ASSOC PROF PHYSICS, RUTGERS UNIV, 68- Mem: Am Phys Soc. Res: High energy physics; general relativity. Mailing Add: Dept of Physics Rutgers Univ New Brunswick NJ 08903

BAZINET, MAURICE L, b Haverhill, Mass, July 26, 18; m 53. CHEMISTRY. Educ: Univ Conn, BS, 49; Boston Univ, MS, 52. Prof Exp: Chemist, Nat Res Corp, 51-55 & Gulf Res & Develop Co, 55-56; CHEMIST, ARMY NATICK LABS, 56- Mem: AAAS; Am Soc Testing & Mat; Sigma Xi; Am Chem Soc. Res: Analysis of food flavors and aromas; design and development of analytical methods and techniques in food research. Mailing Add: 43 Cypress Rd Natick MA 01760

BAZLEY, NORMAN WILLIAM, b Windham, Conn, Jan 17, 33; m 56; c 4. APPLIED MATHEMATICS. Educ: Brown Univ, AB, 54; Univ Md, PhD(math), 59. Prof Exp: Mathematician, Nat Bur Standards, 57-59; asst res prof, Inst Fluid Dynamics & Appl Math, Univ Md, 59-60; mathematician, Nat Bur Standards, 60-62; chief math group, Inst Battelle, Switz, 62-72; PROF MATH, UNIV COLOGNE, 73- Concurrent Pos: Hon prof, Univ Mainz, 69- Mem: Am Math Soc. Res: Nonlinear functional analysis; approximation methods of mathematical physics. Mailing Add: Math Inst Univ of Cologne 5 Cologne 41 West Germany

BAZZANO, GAETANO, b Floridia, Italy, Feb 17, 32; m 69. INTERNAL MEDICINE, BIOCHEMISTRY. Educ: Univ Rome, MD, 56; Tulane Univ, PhD(biochem), 64. Prof Exp: Res asst med, Univ Rome, 57-58; res fel endocrinol, Sch Med, Tulane Univ, 58-59, instr med, 59-64; asst prof, 65-73, ASSOC PROF INTERNAL MED & BIOCHEM, SCH MED, ST LOUIS UNIV, 73-, DIR NUTRIT & METAB SECT, 66-; ASSOC PROF INTERNAL MED, PREV MED & NUTRIT, SCH MED, WASHINGTON UNIV, 73- Concurrent Pos: Fel metab & nutrit training prog, Sch Med, Tulane Univ, 59-64, fel, Div Metab & Nutrit, 64-65; med dir, St Louis City Hosp. Mem: NY Acad Sci; Am Chem Soc; Am Inst Chemists; Am Fedn Clin Res; Am Soc Clin Nutrit. Res: Vitamin B-12 and folic acid metabolism; glutamate feeding and cholesterol homeostasis; metabolism of cardiac glycosides. Mailing Add: St Louis City Hosp 1515 Lafayette Ave St Louis MO 63104

BAZZAZ, MAARIB BAKRI, b Baghdad, Iraq, Nov 27, 40; m 58; c 2. PLANT PHYSIOLOGY. Educ: Univ Ill, Urbana, BS, 61, MS, 63, PhD(biol), 72. Prof Exp: Instr plant physiol, Univ Baghdad, Col Sci, 64-66; res assoc photosynthesis, Inst Environ Studies, Univ Ill, 72-74; RES ASSOC PHOTOSYNTHESIS, DEPT HORT, UNIV ILL, URBANA, 74- Mem: Am Soc Plant Physiologists; Sigma Xi. Res: Maintenance and repair of the photosynthetic apparatus chloroplast in vitro. Mailing Add: Dept of Hort Univ of Ill Urbana IL 61801

BAZZELLE, WILLIAM EDWARD, b Hope, Ark, Dec 30, 43; m; c 1. ANALYTICAL CHEMISTRY, SPECTROSCOPY. Educ: Tex Southern Univ, BS, 64; Wayne State Univ, PhD(anal chem), 67. Prof Exp: Res chemist, Plastics Dept, Exp Sta, Wilmington, Del, 67-74, sr chemist, Houston, Tex, 74-75, ASST CHIEF CHEMIST, PLASTICS DEPT, E I DU PONT DE NEMOURS & CO, ORANGE, 75- Mem: Am Chem Soc. Res: Optical and infrared spectroscopy; inorganic analytical and electroanalytical chemistry; organic functional group analysis; gas chromatography; mass spectrometry. Mailing Add: E I du Pont de Nemours & Co PO Box 1089 Plastics Dept Orange TX 77630

BE, ALLAN WIE HWA, b Semarang, Indonesia, Feb 13, 31; m 53; c 2. GEOLOGY, BIOLOGY. Educ: Lehigh Univ, BA, 52; Columbia Univ, MA, 54, PhD(geol), 58. Prof Exp: SR RES ASSOC MARINE ECOL, LAMONT-DOHERTY GEOL OBSERV, COLUMBIA UNIV, 54-; ASSOC PROF BIOL, CITY COL NEW YORK, 70- Concurrent Pos: Lectr, Brooklyn & Queens Cols, 58-60; mem, Marine Biol Methods Panel, Nat Acad Sci, 62- & Plankton Adv Comt, Smithsonian Oceanog Sorting Ctr, 63-; convener zooplankton working group, Int Coun Explor Seas, 64-; res assoc, Am Mus Natural Hist. Mem: Am Soc Limnol & Oceanog; fel Geol Soc Am; Soc Syst Zool; Soc Econ Paleont & Mineral; Sigma Xi. Res: Ecology and paleoecology of planktonic foraminifera, pteropoda and coccolithophorids; geological and biological oceanography. Mailing Add: Lamont-Doherty Geol Observ Columbia Univ Palisades NY 10964

BEACH, BETTY LAURA, b Falls City, Nebr. FOOD SCIENCE, NUTRITION. Educ: Univ Nebr, Lincoln, BS, 58; Univ Wis-Madison, MS, 67, PhD(food sci & admin), 74. Prof Exp: Dietetic intern, USPHS, Staten Island, NY, 58-59, staff dietitian, Detroit, 59-62, clin dietitian, Staten Island, 62-63, chief food procurement & prod, 66-68, asst dir dietetics, 68-69; ASSOC PROF DIETETICS, UNIV TENN, KNOXVILLE, 69- Concurrent Pos: Nutrit Found Mary Swartz Rose fel, 71-72. Mem: Am Dietetic Asn; Soc Advan Food Serv Res; Food Systs Mgt Educ Coun. Res: Development of quantitative methods to control resources in food service systems; development of instructional strategies for training dietary personnel utilizing competency-based instruction and the mastery learning concept. Mailing Add: Col of Home Econ Univ of Tenn Knoxville TN 37916

BEACH, ELIOT FREDERICK, b Fairfield, Conn, May 26, 11; m 41; c 1. BIOCHEMISTRY. Educ: Brown Univ, BS, 33; Yale Univ, PhD(physiol chem), 37. Prof Exp: Asst biochem, Yale Univ, 34-37; res assoc metab & proteins, Children's Fund, Mich, 37-42, asst dir res lab, 46-48; res biochemist, 48-52, dir biochem lab, 52-70, ASST VPRES & DIR BIOCHEM RES, METROP LIFE INS CO, 70- Mem: AAAS; Am Soc Biol Chemists; Am Inst Nutrit; Am Chem Soc; Soc Exp Biol & Med. Res: Nutritive role of amino acids; amino acid composition of food and tissue proteins; surveys of nutritional status of populations by microchemical techniques; composition of serum proteins by electrophoretic techniques; alloxan diabetes; lipid metabolism; toxicology detecting drug therapy by urine testing. Mailing Add: Metrop Life Ins Co 1 Madison Ave New York NY 10010

BEACH, EUGENE HUFF, b Highland, Mich, Oct 9, 18; m 44; c 1. NUCLEAR PHYSICS, ENGINEERING MANAGEMENT. Educ: Univ Mich, BSE, 41, MS, 47, PhD(physics), 53. Prof Exp: Elec engr, Naval Ord Lab, 41-46, electronic scientist, Underwater Ord, 53-55, chief weapon mech div, 55-58, proj mgr, 58-59, chief underwater elec engr dept, 59-73, assoc head underwater weapons develop directorate, 73-75; HEAD ORD SYST DEVELOP DEPT, NAVAL SURFACE WEAPONS CTR, 75- Concurrent Pos: Instr, Univ Md & Mass Inst Technol; eng res assoc, Cyclotron Proj, Univ Mich, 50-52. Mem: Inst Elec & Electronics Engr; Am Phys Soc. Res: Angular distribution studies on phosphorous reactions. Mailing Add: 12201 Remington Dr Silver Spring MD 20902

BEACH, FRANK AMBROSE, b Emporia, Kans, Apr 13, 11; m 35; c 2. NEUROPSYCHOLOGY. Educ: Kans State Teachers Col, BS, 33, MS, 34; Univ Chicago, PhD, 40. Hon Degrees: DSc, McGill Univ, 66. Prof Exp: Asst neuropsychol, Harvard Univ, 35-36; asst cur, Dept Exp Biol, Am Mus Natural Hist, 36-42, cur & chmn dept animal behav, 42-46; prof psychol, Yale Univ, 46-52, Sterling prof, 52-58; PROF PSYCHOL, UNIV CALIF, BERKELEY, 58- Concurrent Pos: William James lectr, Harvard Univ, 52; Jacob Gimble lectr, Med Sch, Univ Calif, 52 & Univ Calif, Los Angeles, 53. Mem: Nat Acad Sci; Am Psychol Asn; fel Am Acad Arts & Sci; Soc Exp Psychol; Int Acad Sex Res (pres-elect, 75). Res: Species specific behavior in animals; role of the nervous system; role of hormones. Mailing Add: Dept of Psychol Univ of Calif Berkeley CA 94720

BEACH, GEORGE WINCHESTER, b Chicago, Ill, July 11, 13; m 40; c 3. CHEMISTRY. Educ: Northwestern Univ, BS, 37; Univ Chicago, PhD(biochem), 43. Prof Exp: Control chemist, Am Meat Inst, 33-35, res chemist, 35-37; chief chemist, R P Scherer Corp, Mich, 42-44, asst tech dir, 44-45; dir res, Libby, McNeill & Libby, 45-47, lab mgr, 47-62; pres, Stand Pharmacal Corp, 63-73; mem res dept, Am Hosp Asn, 73-75; V PRES, DATX CORP, 76- Mem: AAAS; Am Chem Soc; Am Pharmaceut Asn; Inst Food Technologists; fel Am Inst Chemists. Res: Food; drugs; nutrition; pharmaceutical manufacturing; biochemistry. Mailing Add: 303 E Ohio St Chicago IL 60611

BEACH, JAMES WILSON, b Dubuque, Iowa, Jan 29, 10; m 35; c 2. MATHEMATICS. Educ: Iowa State Col, BS, 31, MS, 36, PhD(math), 48. Prof Exp: Mem staff, Farley & Loetscher Mfg Co, 32-35; asst prof math & physics & head dept math, Univ Dubuque, 35-42; from instr to asst prof math, Iowa State Col, 42-48; assoc prof, NMex Sch Mines, 48-49; asst prof, Univ NMex, 49-53; assoc prof, 53-58, actg head dept, 64-67, PROF MATH, NORTHERN ILL UNIV, 58- Mem: Am Math Soc; Math Asn Am. Res: Hydrodynamics. Mailing Add: Dept of Math Northern Ill Univ DeKalb IL 60115

BEACH, JOHN YOUNGS, b Washington, DC, Nov 19, 12; m 38; c 2. PHYSICAL CHEMISTRY. Educ: Univ Calif, BS, 33; Calif Inst Technol, PhD(phys chem), 36. Prof Exp: Nat Res fel chem, Princeton Univ, 36-38, instr, 38-41; res chemist, Res Lab, 41-53, mgr anal res div, 53-72, SR ANAL ADV, CHEVRON RES CO, 72- Mem: Am Chem Soc. Res: Petroleum research; molecular spectroscopy; diffraction; mass spectroscopy. Mailing Add: Chevron Res Co Richmond CA 94802

BEACH, JOSEPH LAWRENCE, b Shelbyville, Ky, Jan 1, 43. RADIOLOGICAL PHYSICS. Educ: Univ Ky, BS, 64; Univ Wis, MS, 68, PhD(radiol sci), 72. Prof Exp: Fel neutron dosimetry, Univ Wis, 72-73; ASST PROF DOSIMETRY & MED PHYSICS, UNIV KANS, 73- Mem: Am Asn Physicists Med; Health Physics Soc; AAAS; Sigma Xi. Res: Beam dosimetry of fourteen million electron volt neutrons for radiotherapy, especially on collimator and patient scatter on depth dose and penumbra; internal dosimetry of plutonium in bone. Mailing Add: Dept of Radiation Biophys Univ of Kans Lawrence KS 66045

BEACH, LELAND KENNETH, b Holland, Mich, Mar 13, 13; m 39; c 3. CHEMISTRY. Educ: Hope Col, AB, 35; Wash Univ, MS, 37; Purdue Univ, PhD(org chem), 40. Prof Exp: Asst chem, Wash Univ, 35-37; from res chemist to sr chemist, 39-46, group leader, 46-56, res assoc, 57-74, CONSULT, EXXON RES & ENG CORP, 75- Concurrent Pos: Chem liaison in Europe, 56-59; consult, Plastics Educ Found, 74- Honors & Awards: Educ Serv Award, Plastics Inst Am, 75. Mem: Soc Plastics Engrs; Sigma Xi; Am Chem Soc; AAAS; NY Acad Sci. Res: Research on olefins, diolefins, ethanol; acrylonitrile; phthalic and maleic anhydride; acrolein; aromatics; toxic phosphorous compounds; high octane gasoline; metal prophyrins; high energy rocket fuels; isononyl alcohol and plasticizers. Mailing Add: 716 Saunders Ave Westfield NJ 07090

BEACH, LOUIS ANDREW, b Greenville, Ind, June 2, 25; m 56; c 4. EXPERIMENTAL NUCLEAR PHYSICS. Educ: Ind Univ, BS, 44, MS, 47, PhD(physics), 49. Prof Exp: Asst physics, Ind Univ, 46-49; res assoc nuclear physics, Cornell Univ, 49-51; nuclear physicist, 51-55, head nuclear reactions br, 55-65, head physics I sect, 65-71, HEAD NUCLEAR PHYSICS SECT, US NAVAL RES LAB, 71- Concurrent Pos: Lectr, Cath Univ Am, 60-66. Mem: AAAS; Am Phys Soc; Sigma Xi. Res: Disintegration studies of beta decay; interactions of Bremstrahlung radiation with matter; shielding of nuclear radiation; nuclear reactors; nuclear structure; studies of nuclear reactions and structure with cyclotron beams, particularly the study of few nucleon systems and light nuclei. Mailing Add: Code 6611 Naval Res Lab Washington DC 20375

BEACH, NANCY ANN, physical inorganic chemistry, see 12th edition

BEACH, NEIL WILLIAM, b Ann Arbor, Mich, Apr 11, 28; m 55; c 3. ZOOLOGY. Educ: Univ Mich, BS, 50, MS, 51, PhD(zool), 56. Prof Exp: Instr zool, Univ Mich, 56-57; asst prof biol, Lake Forest Col, 57-60; asst prof, 60-64, ASSOC PROF BIOL, GETTYSBURG COL, 64- Concurrent Pos: NSF grants, Mich Biol Sta, 55, Duke Marine Lab, 58 & 59; NSF sci fels, 60-62. Mem: Am Micros Soc; Soc Syst Zool; Ecol Soc Am; Am Soc Zool. Res: Invertebrate zoology and ecology; rotifera; distribution, life history and ecology of the oyster crab, Pinnotheres ostreum and pea crabs of Australia. Mailing Add: Dept of Biol Gettysburg Col Gettysburg PA 17325

BEACH, PAUL L, b Upper Sandusky, Ohio, Dec 11, 39; m 69. NUCLEAR PHYSICS. Educ: Capital Univ, BS, 61; Ohio Univ, MS, 64, PhD(nuclear physics), 66. Prof Exp: Res fel nuclear physics, Van de Graaff Lab, Ohio State Univ, 66-68; asst prof physics, 68-69, ASSOC PROF PHYSICS, CAPITAL UNIV, 69- Mem: AAAS; Am Asn Physics Teachers; Am Phys Soc. Mailing Add: Dept of Physics Capital Univ Columbus OH 43209

BEACHAM, LOWRIE MILLER, JR, b Atlanta, Ga, Oct 27, 11; m 39; c 3. CHEMISTRY. Educ: Univ SC, BS, 31. Prof Exp: Chemist, Pacific Mills, SC, 32-33; student chemist, Calco Chem Co, NJ, 33-34; food chemist, Food & Drug Admin, USDA, 34-41, asst chief food div, Fed Security Agency, 41-53, chief fruit & veg prod br, Dept Health, Educ & Welfare, 53-57, dep dir, Div Food, 57-64, dir div food standards & additives, 64-69, dir div food chem & technol, 69-72, dep dir off prod technol, 71-72, asst dir int stand, Bur Foods, 72-74; RETIRED. Concurrent Pos: Head, US delegs to Codex Alimentarius Comts on Food Labelling Fats & Oils, Fruit Juice, Cocoa & Chocolate Prod & Spec Dietary Foods; mem, US delegs to Comts on Processed Fruits & Veg, Quick Frozen Foods & Fishery Prod, 64-74; consult, Nat Canners Asn, Food & Agr Orgn UN & Nat Bur Stand, 75- Honors & Awards: US Dept Health, Educ & Welfare Superior Serv Award, 59. Mem: Inst Food Technologists; Asn Official Anal Chemists. Mailing Add: 2600 Valley Dr Alexandria VA 22302

BEACHAM, WOODARD DAVIS, b McComb, Miss, Apr 10, 11. OBSTETRICS & GYNECOLOGY. Educ: Univ Miss, BA, 32, BS, 33; Tulane Univ, MD, 35. Prof Exp: Asst biol, Univ Miss, 30-31; asst urol, Sch Med, La State Univ, 39-40; asst gynec & obstet, 40-41, from instr to assoc prof, 41-50, PROF CLIN GYNEC & OBSTET, TULANE UNIV, 50- Concurrent Pos: Mem active staff, Dept Obstet & Gynec, Southern Baptist Hosp, New Orleans, 40-, lectr, Nursing Sch, 42-, from secy to pres med staff, 60-61, chmn med staff exec comt, 63; lectr, Nursing Sch, Hotel Dieu Sister's Hosp, 42-53, consult, 53-; sr vis surgeon, Charity Hosp, New Orleans, 53- Honors & Awards: Musser Award, Tulane Univ, 35; Am Col Surg Med Rec Prize, 43; Distinguished Serv Award, Southern Med Asn, 75. Mem: Am Asn Med Cols; Asn Prof Gynec & Obstet; fel Am Gynec Soc; fel Am Asn Obstet & Gynec; Am Col Obstet & Gynec (pres, 51). Res: Abdominal pregnancy; tubal pregnancy; rupture of uterus; sterility studies; sickle cell disease and pregnancy; carcinoma in situ of the cervix; uterine and/or ovarian tumors weighing 25 pounds or more. Mailing Add: 4240 Magnolia at Gen Pershing New Orleans LA 70115

BEACHELL, HAROLD CHARLES, b Carlyle, Sask, Nov 20, 15; m 47; c 3. PHYSICAL CHEMISTRY. Educ: Queen's Univ, Ont, BA, 37; NY Univ, PhD(chem), 41. Prof Exp: Asst, NY Univ, 38-41; res chemist, E I du Pont de Nemours & Co, Inc, 41-46; from asst prof to assoc prof, 46-57, PROF CHEM, UNIV DEL, 57- Concurrent Pos: Fulbright res scholar to Ger, 57-58. Mem: Am Chem Soc; Chem Inst

Can. Res: Polymeric decomposition; ultraviolet and infrared spectroscopy; molecular structure studies; boron hydrides; nuclear magnetic resonance. Mailing Add: Dept of Chem Univ of Del Newark DE 19711

BEACHEM, MICHAEL THOMAS, b Newark, NJ, Nov 19, 23; m 47; c 5. ORGANIC CHEMISTRY. Educ: Seton Hall Univ, BS, 47; Rutgers Univ, MS, 50, PhD(chem), 54. Prof Exp: Asst chem, Rutgers Univ, 47-50; chemist, Merck & Co, 50-54; res chemist, 54-59, group leader, 59-65, sr res chemist, 65-72, PATENT LIAISON, AM CYANAMID CO, 72- Res: Textile finishes; cortical steroids; folic acids; sulfi sulfinic acids; sterochemistry; urea and melamine formaldehyde resins; fluorocompounds; marking and identification; chemiluminescence. Mailing Add: 50 Runyon Ave Somerset NJ 08873

BEACHER, ROBERT LINCOLN, soils, see 12th edition

BEACHLEY, ORVILLE THEODORE, b East Orange, NJ, Nov 8, 37; m 62; c 2. INORGANIC CHEMISTRY. Educ: Franklin & Marshall Col, BSc, 59; Cornell Univ, PhD(inorg chem), 62. Prof Exp: NIH fel chem, Univ Durham, 63-64; asst prof, Cornell Univ, 64-66; asst prof, 66-69, ASSOC PROF CHEM, STATE UNIV NY BUFFALO, 69- Mem: Am Chem Soc; The Chem Soc. Res: Preparative and physical inorganic chemistry with emphasis on organometallic, hydride and heterocyclic derivatives. of main group and transition elements. Mailing Add: Dept of Chem State Univ of NY Buffalo NY 14214

BEACHY, ROGER NEIL, b Plain City, Ohio, Oct 4, 44; m 67; c 1. PLANT VIROLOGY. Educ: Goshen Col, BA, 66; Mich State Univ, PhD(plant path), 73. Prof Exp: High sch teacher gen sci, United Church of Can Schs, Nfld, 67-68; res assoc plant virol, Univ Ariz, 73; RES ASSOC & NIH FEL PLANT VIROL, CORNELL UNIV, 73- Concurrent Pos: Nat Res Serv award, Div Allergy & Infectious Dis, NIH, 75-76. Mem: Am Phytopath Soc; Am Soc Plant Physiologists; AAAS; Int Asn Plant Tissue Cult. Res: Nature of plant virus, especially tobacco mosaic virus messenger RNAs and the mechanism which controls their genesis; effects of virus gene products on infected host cells. Mailing Add: Dept of Plant Path Cornell Univ Ithaca NY 14853

BEACOM, SEWARD ELMER, b East Rochester, NY, Sept 26, 12; m 55; c 1. ELECTROCHEMISTRY. Educ: Mt Union Col, BS, 34; Univ Mich, MS, 46; Univ Conn, PhD(chem), 54. Prof Exp: Teacher high schs, NY, 34-42; sr chemist, Chevrolet Motor Div, Gen Motors Corp, NY, 42-43; res anal chemist, Union Carbide & Carbon Res Labs, 43-44; instr chem, Univ Mich, 44-48; from asst prof to assoc prof chem, Cent Conn State Col, 48-57; res assoc, Univ Conn, 51-52; sr res chemist, Electrochem Dept, Res Labs, Gen Motors Corp, 57-62, head, 62-69, tech dir basic & appl sci, 69-74; RETIRED. Concurrent Pos: Vis prof, Pratt & Whitney Aircraft, East Hartford, Conn; Consult, Stanley Works, New Britain & Res Corp, NY, 55-56. Mem: Am Chem Soc; Am Electroplaters Soc; Electrochem Soc; Am Mgt Asn; Soc Automotive Eng. Res: Application of radiotracer techniques to fundamental studies of electrodeposition mechanisms; use of optical techniques to study diffusion layer characteristics in electrodeposition of metals. Mailing Add: Res Labs Gen Motors Corp Warren MI 48090

BEADLE, BUELL WESLEY, b Port Barre, La, Sept 9, 11; m 34; c 2. CHEMISTRY. Educ: Kans State Col, BS, 35, MS, 38; Purdue Univ, PhD(agr biochem), 42. Prof Exp: Asst chemist, Exp Sta, Kans State Col, 35-42; res chemist, Am Meat Inst, Chicago, 42-47, chemist in charge phys & anal chem, Am Meat Inst Found, 47-48; dir & vpres, George W Gooch Labs, 48-50; head tech div & comndr staff, Naval Ord Test Sta, China Lake, 50-51, head staff, 52-54; chmn div chem & chem eng, Midwest Res Inst, 56-57, mgr, 57-63; exec dir res & develop, 63-68, VPRES RES & DEVELOP, FARMLAND INDUSTS, INC, 68- Concurrent Pos: Asst, Purdue Univ, 40-42; from res assoc to asst prof pharmacol, Univ Chicago, 43-48. Mem: AAAS; Am Chem Soc; Am Soc Biol Chem; fel Am Inst Chemists; Am Oil Chemists Soc. Res: Physical and analytical methods; stability of vitamins; fat autoxidation, rancidity and antioxidants; spectrophotometric methods; high frequency electronic processing. Mailing Add: Farmland Indust Inc Res Ctr 103 W 26th Ave N Kansas City MO 64116

BEADLE, GEORGE WELLS, b Wahoo, Nebr, Oct 22, 03; m 28, 53; c 1. PLANT GENETICS. Educ: Univ Nebr, BS, 26, MS, 27; Cornell Univ, PhD(genetics), 31. Hon Degrees: Thirty-five from US & foreign cols & univs. in the US. Prof Exp: Asst, NY State Col Agr, Cornell Univ, 28-31; Nat Res Coun fel biol, Calif Inst Technol, 31-32, Inst fel, 32-35, instr, 35-36; asst prof genetics, Harvard Univ, 36-37; prof biol, Stanford Univ, 37-46; prof & chmn div, Calif Inst Technol, 46-61; pres, 61-68, PROF BIOL, UNIV CHICAGO, 61-, EMER PRES, 69- Concurrent Pos: Guest investr, Inst Biol, Paris, France, 35; Eastman vis prof, Oxford Univ, 58-59; trustee, Calif Inst Technol, 70-; hon trustee, Univ Chicago, 71. Honors & Awards: Nobel Prize in Med & Physiol, 58; Lasker Award, 50; Dyer Award, 51; Hansen Prize, 53; Albert Einstein Commemorative Award, 58; Am Cancer Soc Award, 59; Kimber Award, 60. Mem: Nat Acad Sci; AAAS (pres, 55); Genetics Soc Am (pres, 46); Am Philos Soc; Royal Danish Acad. Res: Cytology and genetics of maize; origin of Zea mays; physiological genetics of Drosophila; chemical genetics of Neurospora. Mailing Add: 5533 Dorchester Ave Chicago IL 60637

BEADLE, LESLIE DEWEY, b Millville, Wis, Dec 22, 09; m 43; c 2. MEDICAL ENTOMOLOGY. Educ: Wis State Univ, Platteville, BE, 31; Ohio Univ, MA, 37. Prof Exp: Instr biol, Wis State Univ, 38-43; jr entomologist, Upper Miss River Malaria Surv, 40-41; entomologist, USPHS, 43-45, entomologist water resources activities, 46-63, entomologist state aids activities, 63-69, med entomologist, Arbovirus Ecol Lab, 69-73; RETIRED. Mem: AAAS; Am Soc Trop Med & Hyg; Am Entom Soc; Sci Res Soc Am; Am Pub Helath Asn. Res: Bionomics of mosquitoes; vector control; mosquito-borne encephalitis. Mailing Add: Ctr for Dis Control USPHS Atlanta GA 30333

BEADLE, RALPH EUGENE, b Plentywood, Mont, Apr 26, 43; m 68; c 2. VETERINARY PHYSIOLOGY. Educ: Colo State Univ, DVM, 67; Univ Ga, PhD(vet physiol), 73. Prof Exp: Instr clins, Sch Vet Med, Univ Ga, 67-68, res assoc physiol, 68-71, asst prof clins, 72; resident, Sch Vet Med, Mich State Univ, 72-73, asst prof anesthesiol, 73-74; ASST PROF PHYSIOL, SCH VET MED, LA STATE UNIV, 74- Mem: Am Soc Vet Anesthesiologists; Am Soc Vet Physiologists & Pharmacologists. Res: Comparative aspects of chronic obstructive pulmonary disease and gas transport in the lungs. Mailing Add: Dept of Physiol & Pharmacol Sch Vet Med La State Univ Baton Rouge LA 70803

BEADLES, JOHN KENNETH, b Alva, Okla, Sept 22, 31; m 55; c 2. ICHTHYOLOGY, LIMNOLOGY. Educ: Northwest State Col, BS, 57; Okla State Univ, MS, 63, PhD(ichthyol, limnol), 66. Prof Exp: High sch teacher, 57-62; res asst, Okla State Univ, 63-64; asst prof biol, Ark State Univ, 65-66; asst prof zool, Haile Selassie Univ, 66-68; PROF BIOL & CHMN BIOL SCI DIV, ARK STATE UNIV, 70- Concurrent Pos: Proj dir, Environ Inventory & Assessment, US Army Corps Engrs, 72-74 & Environ Inventory, US Soil Conserv Serv, 74-75. Mem: Am Soc Ichthyologists & Herpetologists; Sigma Xi; Am Fishery Soc. Res: Vertebrate zoology; fish survey. Mailing Add: Drawer KK Ark State Univ State University AR 72467

BEAK, PETER, b Syracuse, NY, Jan 12, 36; m 59; c 2. ORGANIC CHEMISTRY. Educ: Harvard Univ, BA, 57; Iowa State Univ, PhD(org chem), 61. Prof Exp: From instr to assoc prof, 61-70, PROF ORG CHEM, UNIV ILL, URBANA-CHAMPAIGN, 70- Concurrent Pos: Sloan fel, 67-69; Guggenheim fel, 68-69. Res: New organic reactions; synthetic, structural and mechanistic organic chemistry. Mailing Add: 361b E C Univ of Ill Urbana IL 61801

BEAKLEY, JOHN W, microbiology, see 12th edition

BEAL, ERNEST O, b Lancaster, Ill, Mar 7, 28; m 46; c 3. AQUATIC BIOLOGY. Educ: NCent Col, BA, 50; Univ Iowa, MS, 53, PhD(bot), 55. Prof Exp: From asst prof to prof bot, NC State Univ, 54-68, actg head dept bot & bact, 63-64; HEAD DEPT BIOL, WESTERN KY UNIV, 68- Honors & Awards: Res Award, Asn Southeastern Biologists, 66. Mem: Bot Soc Am; Am Soc Plant Taxon; Ecol Soc Am; Am Inst Biol Sci; Int Asn Plant Taxon. Res: Systematic botany, especially ecology and taxonomy of marsh and aquatic plants. Mailing Add: Dept of Biol Western Ky Univ Bowling Green KY 42101

BEAL, JACK LEWIS, b Harper, Kans, July 7, 23; m 48; c 3. PHARMACOGNOSY. Educ: Univ Kans, BS, 48, MS, 50; Ohio State Univ, PhD(pharmacog), 52. Prof Exp: Instr pharmacy & pharmacog, Univ Kans, 49-50; from asst prof to assoc prof pharmacog, 52-63, PROF PHARMACOG, OHIO STATE UNIV, 63- Concurrent Pos: NSF fac fel, 58-59; mem revision comt, US Pharmacopoeia, 75- Mem: Am Asn Cols Pharm; Am Soc Pharmacog (pres, 62-63); fel Acad Pharmaceut Sci; Am Pharmaceut Asn. Res: Drugs of biological origin; drug plant research and cultivation; isolation of plant constituents; plant biochemistry; chemical microscopy. Mailing Add: Ohio State Univ Col of Pharm 500 W 12th Ave Columbus OH 43210

BEAL, JAMES BURTON, JR, b Galveston, Tex, Nov 22, 32; m 53; c 4. INORGANIC CHEMISTRY. Educ: Baylor Univ, BA, 53; Agr & Mech Col Tex, MS, 59, PhD(chem), 63. Prof Exp: Res chemist, Gen Chem Div, Allied Chem Corp, 62-63; assoc res dir, Chem Div, Ozark-Mahoning Co, 64-69; ASSOC PROF CHEM, UNIV MONTEVALLO, 69- Mem: Am Chem Soc; Sigma Xi. Mailing Add: Dept of Chem Univ of Montevallo Montevallo AL 35115

BEAL, JOHN ANTHONY, b Cleveland, Ohio, Mar 30, 45; m 69; c 3. ANATOMY, NEUROANATOMY. Educ: Xavier Univ, BS, 67; Univ Cincinnati, PhD(anat), 71. Prof Exp: ASST PROF ANAT, MED SCH, WAYNE STATE UNIV, 71- Res: Fine structure of the nervous system; synaptology; spinal cord. Mailing Add: Dept of Anat Wayne State Med Sch Detroit MI 48201

BEAL, JOHN MANN, b Miss, Nov 18, 15; m 43; c 3. SURGERY. Educ: Univ Chicago, BS, 37, MD, 41. Prof Exp: From intern to resident surgeon, NY Hosp, New York, 41-44 & 46-48; from instr to asst prof surg, Med Sch, Univ Calif, 49-53; assoc prof clin surg, Cornell Univ, 53-63; PROF SURG & CHMN DEPT, MED SCH, NORTHWESTERN UNIV, 63- Concurrent Pos: Instr, Med Col, Cornell Univ, 49; asst attend surgeon, NY Hosp, 49, attend surgeon, 53-63; chief surg serv, Wadsworth Gen Hosp, Los Angeles, 50-53; chmn dept surg, Chicago Wesley Mem Hosp, 63-69; chief div surg, Passavant Mem Hosp, Chicago, 63-73; consult, Vet Admin Res Hosp, 63-; chmn, Am Bd Surg, 70-71; chmn dept surg, Northwestern Mem Hosp, 73- Mem: Am Soc Clin Surg; Am Col Surg; Soc Univ Surg; Asn Am Med Cols; Am Surg Asn. Res: Surgical metabolism; gastrointestinal physiology. Mailing Add: 303 E Chicago Ave Chicago IL 60611

BEAL, PHILIP FRANKLIN, III, b Brewster, NY, Oct 2, 22; m 47; c 4. ORGANIC CHEMISTRY. Educ: Williams Col, AB, 43; Ohio State Univ, PhD(org chem), 49. Prof Exp: Jr chemist, Winthrop Chem Co, 43-44; res assoc, 50-66, head chem processes res & develop, 66-69, mgr, 69-70, GROUP MGR CHEM PROCESSES RES & DEVELOP, 70- Mem: Am Chem Soc. Res: Aliphatic diazo compounds; diazoketones; partial and total synthetic steroid work. Mailing Add: 3411 Lorraine Ave Kalamazoo MI 49008

BEAL, RICHARD SIDNEY, JR, b Victor, Colo, May 7, 16; m 41; c 2. ENTOMOLOGY. Educ: Univ Ariz, BS, 38; Univ Calif, PhD(entom), 52. Prof Exp: Asst prof biol, Westmont Col, 47-48; instr lib arts, San Francisco Baptist Col, 48-51; prof syst theol, Conservative Baptist Theol Sem, 51-56; entomologist, Agr Res Serv, USDA, 56-58; assoc prof entom, Ariz State Univ, 58-62; assoc dean arts & sci, 62-65, DEAN GRAD SCH, NORTHERN ARIZ UNIV, 65- Mem: Entom Soc Am; Soc Syst Zool; Am Entom Soc. Res: Biology and systematics of coleopterous family Dermestidae; biology of Arizona insects. Mailing Add: Box 4125 Northern Ariz Univ Flagstaff AZ 86001

BEAL, VIRGINIA ASTA, b Hull, Mass, Oct 31, 18. NUTRITION. Educ: Simmons Col, BS, 39; Harvard Univ, MPH, 45. Prof Exp: Nutritionist, Maternal & Child Health, Sch Pub Health, Harvard Univ, 39-46; instr physiol growth, Sch Med, Univ Colo, 48-54, asst prof human growth, 54-59, asst clin prof pediat, 59-71, nutritionist, Child Res Coun, 46-71; ASSOC PROF NUTRIT, UNIV MASS, AMHERST, 71- Mem: Fel Am Pub Health Asn; Am Inst Nutrit; Am Dietetic Asn; Soc Nutrit Educ; NY Acad Sci. Res: Nutritional intake during pregnancy and childhood and its relationship to physical and physiological growth and development; education. Mailing Add: Dept of Food Sci & Nutrit Univ of Mass Chenoweth Lab Amherst MA 01002

BEALE, JOHN HAMPDEN, organic chemistry, see 12th edition

BEALE, ROBERT SPENCER, analytical chemistry, see 12th edition

BEALE, WILLIAM L, horticulture, food science, see 12th edition

BEALES, FRANCIS WILLIAM, b Bristol, Eng, Feb 7, 19; Can citizen; m 54; c 3. GEOLOGY. Educ: Cambridge Univ, BA, 46; Univ Toronto, PhD(geol), 52. Prof Exp: Lectr geol, McMaster Univ, 47-51; assoc prof, 51-74, PROF GEOL, UNIV TORONTO, 74- Mem: Am Asn Petrol Geol; Soc Econ Paleont & Mineral; Geol Soc Am; Geol Asn Can; Int Asn Sedimentol. Res: Stratigraphy with special interest in limestones. Mailing Add: Dept of Geol Sci Univ of Toronto Toronto ON Can

BEALL, ARTHUR CHARLES, JR, b Atlanta, Ga, Aug 17, 29; m 49; c 2. MEDICINE. Educ: Emory Univ, BS, 50, MD, 53. Prof Exp: Intern surg, Barnes Hosp, St Louis, Mo, 53-54; resident, Methodist Hosp, Houston, Tex, 54-55; from instr to assoc prof, 59-71, PROF SURG, BAYLOR COL MED, 71- Concurrent Pos: Resident surg affil hosps, Baylor Col Med, 55-60, resident thoracic surg, 60-61; Houston Heart Asn fel, 60-61. Mem: Am Asn Thoracic Surg; Am Col Cardiol; Am Col Chest Physicians; Am Col Surgeons; Soc Thoracic Surg. Res: Cardiovascular surgery, especially open heart surgery; technics of extracorporeal circulation; pulmonary embolectomy; development of artificial heart valves. Mailing Add: Dept of Surg Baylor Col of Med Houston TX 77025

BEALL, DESMOND, b Germiston, SAfrica, Feb 2, 10; m 45; c 5. BIOCHEMISTRY. Educ: Univ BC, BA, 32; Univ Toronto, PhD(biochem), 35. Prof Exp: Asst, Connaught Labs, Univ Toronto, 34-35 & Nat Inst Med Res, London, 35-37; Beit Mem fel med res, Brit Post-Grad Med Sch, 37-39, lectr path, 39-41; res chemist, Ayerst, McKenna & Harrison, 41-49, asst dir res, 49-52, dir pharmaceut develop, Ayerst Labs, Inc, 53-67, vpres, 67-75; RETIRED. Concurrent Pos: Asst prof, Res Inst Endocrinol, McGill Univ, 47. Mem: Fel Chem Inst Can; The Chem Soc; Can Physiol Soc. Res: Sex hormones; steroids; vitamins; antibiotics; pharmaceutical product development. Mailing Add: 284 Lake St Rouses Point NY 12979

BEALL, FRANCIS CARROLL, b Baltimore, Md, Oct 3, 33; m 63; c 3. WOOD SCIENCE. Educ: Pa State Univ, BS, 64; State Univ NY Col Forestry, Syracuse Univ, MS, 66, PhD(wood physics), 68. Prof Exp: Res technologist, US Forest Prod Lab, 66-68; asst prof wood sci & technol, Pa State Univ, University Park, 68-73, assoc prof, 73-75; ASSOC PROF WOOD SCI, UNIV TORONTO, 75- Mem: Soc Wood Sci & Technol (secy-treas, 70-71); Forest Prod Res Soc; Tech Asn Pulp & Paper Indust; Am Soc Testing & Mat; Int Asn Wood Anatomists. Res: Wood carbonization; wood plastics composites; thermogravimetry of wood polymers; diffusion; paper physics; wood plasticization; dimensional stabilization; special methods of drying wood; temperature and humidity control instrumentation. Mailing Add: Fac of Forestry Univ Toronto 203 College St Toronto ON Can

BEALL, GEOFFREY, b Johannesburg, SAfrica, Feb 14, 08; nat US; m 34, 49; c 3. STATISTICS. Educ: Univ BC, BS, 31; Univ Ill, MS, 32; Univ London, PhD, 38. Prof Exp: Asst entomologist, Can Dept Agr, 28-44; statistician, Ont Dept Health, 44-45, Inst Paper Chem, Lawrence Col, 45-48, Preston Labs, 48-49 & Res Labs, Swift & Co, 49-50; head dept statistics, Univ Conn, 50-56; from statistician to vpres mfg, Gillette Safety Razor Co, 56-68; vis res statistician, Educ Testing Serv, 68-72; VIS PROF, NAT DEFENSE MED CTR, 72- Mem: Am Statist Asn; Biomet Soc. Res: Near random statistical distributions; transformations in statistical investigation; application of statistical methods in industry; experimental design. Mailing Add: Nat Defense Med Ctr PO Box 7432 Taipei Taiwan

BEALL, GEORGE HALSEY, b Montreal, Que, Oct 14, 35; m 62; c 3. GEOCHEMISTRY, CERAMICS. Educ: McGill Univ, BSc, 56, MSc, 58; Mass Inst Technol, PhD(geol), 62. Prof Exp: From sr geologist to res geologist, 62-65, res assoc geol, 65-66, MGR GLASS-CERAMIC RES DEPT, CORNING GLASS WORKS, 66- Mem: AAAS; Mineral Soc Am; fel Am Inst Chemists; Am Ceramic Soc. Res: Nucleation and growth of crystals in glass; glass-ceramics; synthetic mica; silica polymorph solid solutions; phase equilibria; petrology of basalt. Mailing Add: Res & Develop Dept Corning Glass Works Sullivan Pk Corning NY 14830

BEALL, HORACE ANSLEY, b Calif, Sept 25, 33; m 55; c 5. ELECTROOPTICS, ELECTRONIC ENGINEERING. Educ: Stanford Univ, BS & MSEE, 56. Prof Exp: Res engr autonetics, NAm Rockwell Corp, 56-60, group scientist, 61-67, mgr info sci & autonetics, 67-72, DIR INDUST ELECTRONICS, MICROELECTRONICS DIV, ROCKWELL INT CORP, 72- Mem: Inst Elec & Electronics Engr. Res: Photodetection. Mailing Add: 17742 Whitney Dr Santa Ana CA 92705

BEALL, ROBERT JOSEPH, b Washington, DC, May 19, 43; m 67; c 2. BIOCHEMISTRY. Educ: Albright Col, BS, 65; State Univ NY Buffalo, MA, 70, PhD(biol), 70. Prof Exp: Asst biol, State Univ NY Buffalo, 65-70; from instr to asst prof physiol, Sch Med, Case Western Reserve Univ, 70-74, from instr to asst prof, Sch Dent, 71-74; grants assoc, NIH, 74-75; DIR METAB PROG, EXTRAMURAL PROGS, NAT INST ARTHRITIS, METAB & DIGESTIVE DIS, 75- Mem: AAAS. Res: Molecular endocrinology, mechanism of action of adrenocorticotropic hormone; cyclic nucleotide synthesis and action, steriodogenesis, assay of adrenocorticotropic hormone in plasma, neoplastic endocrine tissue, genetic diseases, metabolic control. Mailing Add: Extramural Prog Westwood Bldg NIAMDD 5333 Westbard Ave Bethesda MD 20016

BEALMEAR, PATRICIA MARIA, b Dodge City, Kans, Oct 23, 29. EXPERIMENTAL PATHOLOGY, TRANSPLANTATION IMMUNOLOGY. Educ: Mt St Scholastica Col, BS, 49; Univ Notre Dame, PhD(biol), 65. Prof Exp: Chmn sci dept, high schs, Kans, 52-62, Mo, 62-63; asst prof biol, Mt St Scholastica Col, 65-66; res scientist microbiol, Univ Notre Dame, 66-71; asst prof exp biol, Baylor Col Med, 71-75; MEM STAFF, DEPT DERMAT, ROSWELL PARK MEM INST, 75- Mem: AAAS; Transplantation Soc; Radiation Res Soc; Am Soc Exp Path; Soc Exp Biol & Med. Res: Determination of the role of the thymus in the mammal; radiation pathology and bone marrow transplantation in germ-free mice. Mailing Add: Dept of Dermat Roswell Park Mem Inst Buffalo NY 14263

BEALOR, MARK DABNEY, b Shamokin, Pa, Sept 27, 21; m 51; c 1. ORGANIC CHEMISTRY. Educ: Univ Pa, BS, 49; Univ Ore, MA, 54, PhD(chem), 56. Prof Exp: Chemist, Sharp & Dohme Inc, 49-51; sr chemist, Niagara Div, Food Mach & Chem Co, Inc, 55-57; sr chemist, 57-68, RES SUPVR, S C JOHNSON & SON, INC, 68- Mem: Am Chem Soc; Soc Cosmetic Chem. Res: Chemical structure versus biological activity; carbonyl condensations; product development. Mailing Add: S C Johnson & Son Inc 15th & Howe Sts Racine WI 53403

BEALS, ALAN ROBIN, b Oakland, Calif, Jan 24, 28; c 2. CULTURAL ANTHROPOLOGY. Educ: Univ Calif, Los Angeles, BA, 48; Univ Calif, Berkeley, PhD(anthrop), 54. Prof Exp: From asst prof to assoc prof anthrop, Stanford Univ, 56-68; PROF ANTHROP, UNIV CALIF, RIVERSIDE, 68- Concurrent Pos: NSF fel, Mysore State, S India, 58-60 & Am Inst Indian Studies fel, 65-66; ed spec publ, Am Anthrop Asn, 63-65. Mem: AAAS; Am Anthrop Asn; Soc Asian Studies; Am Sociol Asn; Soc Appl Anthrop. Res: South Asia, conflict within groups or divisiveness, cultural change, population and ecology. Mailing Add: Dept of Anthrop Univ of Calif Riverside CA 92502

BEALS, EDWARD WESLEY, b Wichita, Kans, July 1, 33; m 67; c 1. ECOLOGY. Educ: Earlham Col, BA, 56; Univ Wis-Madison, MS, 58, PhD(bot), 61. Prof Exp: Instr biol, Am Univ Beirut, 61-62; from asst prof to assoc prof, Haile Selassie Univ, 62-65; asst prof bot & zool, Univ Wis Ctr Syst, 65-66; asst prof bot, 66-67, lectr zool, 67-68, asst prof, 68-69, assoc prof zool & bot, 69-73, PROF ZOOL & BOT, UNIV WIS-MADISON, 73- Mem: Ecol Soc Am; Am Ornithologists Union; Am Soc Naturalists; Soc Study Evolution; Cooper Ornith Soc. Res: Statistical analysis of ecological communities; interrelations between vegetation and animals, especially birds and mammals; factors in species diversity and niche specialization; plant taxonomy; allelopathy in plant communities. Mailing Add: Dept of Zool Univ of Wis Madison WI 53706

BEALS, ERNEST LESLIE, pharmaceutical chemistry, see 12th edition

BEALS, HAROLD OLIVER, b Mishawauka, Ind, July 4, 31; m 59; c 1. WOOD TECHNOLOGY, PALEOBOTANY. Educ: Purdue Univ, BSF, 55, MS, 57, PhD(bot), 60. Prof Exp: Secy, Evans Prod, Inc, Kans, 58-59; asst prof, 60-69, ASSOC PROF FORESTRY, AUBURN UNIV, 69- Concurrent Pos: Consult, G LeBlanc, Wis, 57-58 & Scherl & Roth, Ohio, 58. Mem: AAAS; Forest Prod Res Soc (secy-treas, 62-64). Res: Plant anatomy. Mailing Add: Dept of Forestry Auburn Univ Auburn AL 36380

BEALS, RALPH LEON, b Pasadena, Calif, July 19, 01; m 23; c 3. ANTHROPOLOGY, CULTURAL ANTHROPOLOGY. Educ: Univ Calif, Berkeley, AB, 26, PhD(anthrop), 30. Hon Degrees: LLD, Univ Calif, Los Angeles, 70. Prof Exp: Nat Res Coun fel biol sci, 30-32; Southwest Soc NY fel, 32-33; mus technician anthrop, Field Div Educ, Nat Park Serv, 33-35; lectr anthrop, Univ Calif, Berkeley, 35; from instr to prof, 36-69, EMER PROF ANTHROP, UNIV CALIF, LOS ANGELES, 69- Concurrent Pos: Dir, Latin Am Ethnic Studies, Smithsonian Inst, 42-43; fel, Ctr Advan Study Behav Sci, 55-56; John Simon Guggenheim Found fel, 58-59; prof, Univ Buenos Aires, 62; NSF res grant, 64-72. Mem: AAAS; Am Anthrop Asn (pres, 49-50); Am Sociol Asn; Am Ethnol Soc; Soc Am Archaeol. Res: Latin American societies with special emphasis on peasant communities; Indian-European relationships; socio-cultural change and economic problems. Mailing Add: Dept of Anthrop Univ of Calif Los Angeles CA 90024

BEALS, RICHARD WILLIAM, b Erie, Pa, May 28, 38; m 62; c 3. MATHEMATICAL ANALYSIS. Educ: Yale Univ, BA, 60, MA, 62, PhD(math), 64. Prof Exp: Instr math, Yale, 64-65; vis asst prof, 65-66, from asst prof to assoc prof, 66-71, PROF MATH, UNIV CHICAGO, 71- Mem: Am Math Soc; Math Asn Am. Res: Partial differential equations; linear operators. Mailing Add: Dept of Math Univ of Chicago Chicago IL 60637

BEALS, RODNEY K, b Portland, Ore, Jan 4, 31; m 56; c 3. ORTHOPEDIC SURGERY. Educ: Willamette Univ, BA, 53; Univ Ore, MD, 56; Am Bd Orthop Surg, dipl, 64. Prof Exp: From instr to asst prof, 61-67, ASSOC PROF ORTHOP SURG, MED SCH, UNIV ORE, 67- Mem: Am Acad Orthop Surg. Res: Clinical orthopedics. Mailing Add: Dept of Orthop Surg Univ Ore Med Sch Portland OR 97201

BEAM, CARL ADAMS, b Olympia, Wash, July 5, 20; m 52. MICROBIAL GENETICS, RADIOBIOLOGY. Educ: Brown Univ, BA, 42; Yale Univ, MS, 47, PhD(microbiol), 50. Prof Exp: Res assoc biophysics, Univ Calif, Berkeley, 50-56; asst prof, Yale Univ, 56-63; assoc prof biol, 63-65, PROF BIOL, BROOKLYN COL, 75- Concurrent Pos: NIH fel, 50-51, res grants & sr investr, 56-63. Mem: Radiation Res Soc; Biophys Soc; Genetics Soc Am; Soc Protozoologists. Res: Biological factors affecting radiation sensitivity; dinoflagellate genetics. Mailing Add: Dept of Biol Brooklyn Col Brooklyn NY 11226

BEAM, CHARLES FITZHUGH, JR, b New York, NY, Sept 24, 40; m 68; c 1. ORGANIC CHEMISTRY, POLYMER CHEMISTRY. Educ: City Col New York, BS, 63; Univ MD, College Park, PhD(org chem), 70. Prof Exp: Asst gen & phys chem, Univ Md, 63-65, org polymer chem, 65-68; res assoc, P M Gross Chem Lab, Duke Univ, 68-73; ASST PROF CHEM, NEWBERRY COL, 73- Mem: Am Chem Soc; fel The Chem Soc. Res: Organic chemistry of high polymers; organic synthesis with multiple anions. Mailing Add: Dept of Chem Newberry Col Newberry SC 29108

BEAM, JOHN E, b Honesdale, Pa, Oct 27, 31; m 61; c 3. FOOD SCIENCE, DAIRY SCIENCE. Educ: Pa State Univ, BS, 57; Univ Mass, MS, 59, PhD(food sci), 62. Prof Exp: Instr dairy mfg, Univ Mass, 57-59, res, 59-62; proj leader res & develop, Ross Labs, Abbott Labs, 62-64; in charge spec projs food res & develop, 64-67, mgr processed food res, 67-68, food res, 68, MGR PROD DEVELOP, BEECH-NUT INC, 68- Concurrent Pos: Mem lab comt, Wash Lab, Nat Canners Asn. Mem: AAAS; Inst Food Technol; Am Dairy Sci Asn; Sci Res Soc Am. Res: Dairy chemistry; gelling agents; infant nutrition; food product development; confections development; candy and chewing gum; cough and cold products; beverages. Mailing Add: Mgr of Prod Develop N Main St Beech-Nut Res & Develop Port Chester NY 10573

BEAM, JOHN EDGAR, theoretical nuclear physics, see 12th edition

BEAM, WILLIAM JAMES, elementary particle physics, cosmic ray physics, see 12th edition

BEAMAN, BLAINE LEE, b Portland, Ore, July 28, 42; m 65; c 1. MICROBIOLOGY. Educ: Utah State Univ, BS, 64; Univ Kans, PhD(microbiol), 68. Prof Exp: Asst prof med microbiol, Sch Med & Dent, Georgetown Univ, 70-75; ASST PROF MED MICROBIOL, UNIV CALIF SCH MED, DAVIS, 75- Concurrent Pos: Fel, Sch Med, NY Univ, 68-70. Mem: Am Soc Microbiol; AAAS. Res: Mechanisms of nocardial pathogenesis; induction, pathogenicity and biology of bacterial L-forms; the role of the alveolar macrophage in lung defense; infectious diseases of the lungs; bacterial cell wall biochemistry and bacterial ultrastructure. Mailing Add: Dept of Med Microbiol Univ of Calif Sch of Med Davis CA 95616

BEAMAN, JOHN HOMER, b Marion, NC, June 20, 29; m 58; c 2. SYSTEMATIC BOTANY. Educ: NC State Col, BS, 51; State Col Wash, MS, 53; Harvard Univ, PhD, 57. Prof Exp: From asst prof to assoc prof, 56-68, PROF BOT, MICH STATE UNIV, 68-, CURATOR, BEAL-DARLINGTON HERBARIUM, 56- Concurrent Pos: Nat Acad Sci-Nat Res Coun sr vis res assoc, Smithsonian Inst, 65-66; mem educ comt, Flora NAm Prog, 66-73; mem bd dirs, Orgn Trop Studies, 72- Mem: Bot Soc Am; Am Soc Plant Taxon; Int Asn Plant Taxon; Soc Econ Bot; Ecol Soc Am. Res: Biosystematic studies in alpine flora of Mexico and Central America; monographic studies in Astereae; Compositae of Mexico; computer applications in systematic biology. Mailing Add: Dept Bot & Plant Path Mich State Univ Plant Biol Lab East Lansing MI 48824

BEAMAN, TEOFILA, plant physiology, biochemistry, see 12th edition

BEAMER, PARKER REYNOLDS, b Centralia, Ill, July 27, 14; m 39; c 3. PATHOLOGY. Educ: Univ Ill, AB, 35, MS, 37, PhD(bact), 40; Washington Univ, MD, 43. Prof Exp: Asst bact, Univ Ill, 35-39; lectr, Sch Dent, Washington Univ, 41-42, asst path, Sch Med, 42-44, asst prof, 46-49; prof microbiol, dir dept & assoc prof path, Bowman Gray Sch Med, 49-53, assoc dean, 51-53; prof path, chmn dept & dir clin correlation & res, Sch Med, Ind Univ, 53-65; prof path, Sch Med, Univ Southern Calif, 65-69; chief pathologist, Porter Mem Hosp, 69-70; PROF PATH, CHICAGO MED SCH, 70- Concurrent Pos: Asst pathologist, Washington Univ Hosps, 46-49; attend physician, Vet Admin Hosp, Jefferson Barracks, 47-49; bacteriologist & assoc pathologist, NC Baptist Hosp, 49-53; consult, Vet Admin Hosp, Mt Home, Tenn, 51-53; ed, J Am Soc Clin Path, 56-; chief pathologist, Los Angeles County Gen Hosp, 65-69; life trustee, Am Bd Path. Mem: Am Asn Path & Bact; AMA; Am Soc Clin Path; Soc Exp Biol & Med; Col Am Pathologists. Res: Public health bacteriology; heat resistance of pathogenic microorganisms; Clostridium botulinum; chemotherapy in Salmonella and Shigella infections; pathogenic staphylococci; bacteriologic staining procedures; histoplasmosis; leptospirosis; pathology and bacteriology of miscellaneous unusual infectious diseases; relation of neoplasia to polluted water. Mailing Add: Dept of Path Univ Health Sci Chicago Med Sch Chicago IL 60612

BEAMER, PAUL DONALD, b Avonmore, Pa, Sept 21, 14; m 36; c 1. VETERINARY PATHOLOGY. Educ: Ohio State Univ, DVM, 41; Univ Ill, MS, 45, PhD, 51. Prof

Exp: Asst bacteriologist & pathologist, State Dept Agr, Ohio, 41-42; asst vet path & hyg, State Dept Agr, Ill, 42-44; war food asst, Univ Ill, 44-46, from asst prof to prof vet path & hyg, 46-60; adv to dean col vet med, Uttar Pradesh Agr Univ, India, 60-64; PROF VET PATH & HYG, UNIV ILL, URBANA-CHAMPAIGN, 64- Mem: Fel Am Vet Med Asn; Am Col Vet Path; Int Acad Path. Res: Infectious diseases of livestock; bovine tuberculosis; brucellosis; swine enteritis; Newcastle disease of poultry; lead poisoning in wild ducks. Mailing Add: Dept of Vet Path & Hyg Univ of Ill Urbana IL 61801

BEAMER, ROBERT LEWIS, b Pulaski, Va, June 9, 33; m 59; c 1. MEDICINAL CHEMISTRY. Educ: Med Col Va, BS, 55, MS, 57, PhD(chem), 59. Prof Exp: Asst chem, Med Col Va, 55-58; from asst prof to assoc prof pharmaceut chem, 59-69, assoc dean col pharm, 72-75, PROF MED CHEM, UNIV SC, 69- Concurrent Pos: Mead Johnson res grant, 62-63; NIH res grant, 64- Mem: AAAS; Am Chem Soc; Am Pharmaceut Asn. Res: Synthetic estrogenic agents; antimetabolites in cancer chemotherapy; catalytic hydrogenation; asymmetric synthesis; enzyme inhibition. Mailing Add: 1 Nob Hill Rd Columbia SC 29210

BEAMER, WILLIAM HOWARD, b Avonmore, Pa, July 5, 18; m 42; c 4. PHYSICAL CHEMISTRY, ENVIRONMENTAL SCIENCES. Educ: Col Wooster, BA, 39; Univ Md, MS, 41; Ohio State Univ, PhD(phys chem), 44. Prof Exp: Asst, Univ Md, 39-41; asst, Am Petrol Inst, Ohio State Univ, 41-44; res assoc, Los Alamos Sci Lab, 44-46, alternate group leader, 46; res chemist, Dow Chem Co, 46-51, div mgr, Rocky Flats Plant, Colo, 51, div leader, Spectros Lab, 51-55, asst dir chem physics res lab & head radiochem res lab, 55-68, dir radiochem res lab, 68-70; chmn & pres, Appl Radiation Corp, Calif, 70-75; MGR HEALTH & ENVIRON RES, WESTERN AREA, DOW CHEM CO, PITTSBURG, 75- Mem: Am Chem Soc; Am Nuclear Soc. Res: Radioactive isotopes; radiochemistry; radiation chemistry; nuclear fuel processing; health and environmental effects of chemical substances. Mailing Add: 233 Rutherford Dr Danville CA 94526

BEAMES, CALVIN G, JR, b Kingston, Okla, Oct 29, 30; m 52; c 3. PHYSIOLOGY. Educ: NMex Highlands Univ, AB, 55, MS, 56; Univ Okla, PhD(physiol, biochem), 61. Prof Exp: Asst physiol, Univ Okla, 55-56; teacher biol, Santa Fe High Sch, 56-57; asst physiol, Univ Okla, 57-60; trainee, Rice Univ, 60-61, NIH fel, 61-62; from asst prof to assoc prof, 62-70, PROF PHYSIOL, OKLA STATE UNIV, 70- Mem: AAAS; Am Soc Zool; Am Chem Soc; Am Physiol Soc; Am Soc Parasitol. Res: Carbohydrate metabolism and transport mechanisms of invertebrate organisms. Mailing Add: Dept of Physiol Sci Okla State Univ Stillwater OK 74074

BEAMES, R M, b Brisbane, Australia, Oct 12, 31; m 56; c 4. ANIMAL NUTRITION. Educ: Univ Queensland, BAgrSc, 54, MAgrSc, 62; McGill Univ, PhD(nutrit), 65. Prof Exp: Husb officer, Animal Res Inst Yeerongpilly, Queensland Dept Primary Industs, 54-65, sr husb officer, 65-68; asst prof animal sci, 69-71, ASSOC PROF ANIMAL SCI, FAC AGR SCI, UNIV BC, 71- Mem: Nutrit Soc Can; Can Soc Animal Prod; Brit Nutrit Soc; Am Soc Animal Sci; Australian Soc Animal Prod. Res: Supplementation of animals grazing poor quality pasture; nutritive value of grains for pigs; use of amino acids in pig rations. Mailing Add: Dept of Animal Sci Univ of BC Vancouver BC Can

BEAMESDERFER, JOHN WILLIAM, b Shaefferstown, Pa, Aug 24, 10; m 40; c 2. PHYSICAL CHEMISTRY. Educ: Gettysburg Col, BS, 32; Univ Mich, MS, 39, PhD(phys chem), 48. Prof Exp: Chemist, Gen Chem Co, Pa, 32-34; high sch instr, 34-38; fel, Univ Mich, 38-43; chief chemist, res & eng dept, Argus Inc, Mich, 43-47; from asst prof to assoc prof chem, 47-55, head dept, 52-67, PROF CHEM, UNIV MAINE, ORONO, 55- Mem: Fel AAAS; fel Am Inst Chemists; Am Chem Soc; Am Soc Eng Educ. Res: Wettability of solid surfaces; adsorption of vapors; selective adsorption from mixtures; properties of kaolin clay. Mailing Add: 285 Aubert Hall Univ of Maine Orono ME 04473

BEAMISH, FRED EARL, b Hanover, Ont, Oct 29, 01; m; c 2. ANALYTICAL CHEMISTRY. Educ: McMaster Univ, BA, 28, MA, 29. Hon Degrees: DSc, McMaster Univ, 62. Prof Exp: Asst, 28-31, lectr, 31-35, from asst prof to prof, 35-69, EMER PROF CHEM, UNIV TORONTO, 69- Honors & Awards: Fisher Sci Lect Award, 68. Mem: Am Chem Soc; Chem Inst Can; Fel Royal Soc Can. Res: Extraction from ores of the platinum group of metals; inorganic chemistry of the precious and semi-precious metals. Mailing Add: Dept of Chem Univ of Toronto Toronto ON Can

BEAMISH, FREDERICK WILLIAM HENRY, b Toronto, Ont, July 31, 35; m 58; c 2. ZOOLOGY. Educ: Univ Toronto, BA, 58, PhD(zool), 62. Prof Exp: Assoc scientist behav & physiol biol, Biol Sta, Fisheries Res Bd Can, 62-65; from asst prof to assoc prof, 65-72, PROF ZOOL, UNIV GUELPH, 72- Mem: Am Fisheries Soc; Can Soc Zool. Res: Respiratory metabolism in freshwater fish; muscular fatigue in marine fish; diurnal vertical migration by marine fish; physiology of migration. Mailing Add: Dept of Zool Univ of Guelph Guelph ON Can

BEAMISH, KATHERINE I, b Winnipeg, Man, June 25, 12. BOTANY, GENETICS. Educ: Univ BC, BSA, 49, MSA, 51; Univ Wis, PhD(bot & genetics), 54. Prof Exp: Res assoc cytogenetics, Univ Wis, 54-56; from instr to assoc prof biol & bot, 56-74, asst curator higher plant herbarium, 56-70, PROF BIOL & BOT, UNIV BC, 74-, CURATOR VASCULAR PLANT HERBARIUM, 70- Mem: AAAS; Bot Soc Am; Int Asn Plant Taxon; Genetics Soc Can; Can Bot Asn. Res: Cytotaxonomic studies in the flora of the Pacific Northwest. Mailing Add: Dept of Bot Univ of BC Vancouver BC Can

BEAMISH, ROBERT EARL, b Shoal Lake, Man, Sept 9, 16; m 43; c 3. CARDIOLOGY. Educ: Brandon Col, BA, 37; Univ Man, MD, 42, BSc, 44; FRCP(C), 50, Edinburgh, 61. Prof Exp: Demonstr med, Fac Med, Univ Man, 46-49, from lectr to assoc prof, 49-70; MED DIR, GREAT-WEST LIFE ASSURANCE CO, 70- Concurrent Pos: Cardiologist, Manitoba Clin, 46-70; asst physician, Winnipeg Gen Hosp, 47-57, assoc physician, 57-; Nuffield Dom traveling fel, Gt Brit, 47-48; mem bd dir, Can Heart Found, 63-75; regional rep, Coun Clin Cardiol, Am Heart Asn, Can, 63- Mem: Fel Am Col Cardiol; fel Am Col Physicians; Can Cardiovasc Soc (hon secy & treas, 57-, pres, 69-70); Inter-Am Soc Cardiol (vpres); Can Med Asn. Res: Clinical research in ischemic heart disease with special reference to the therapeutic use of anticoagulants. Mailing Add: Great-West Life Assurance Co 60 Osborne St N Winnipeg MB Can

BEAMS, HAROLD WILLIAM, b Belle Plaine, Kans, Aug 3, 03; m 35; c 2. ZOOLOGY. Educ: Fairmount Col, AB, 25; Northwestern Univ, AM, 26; Univ Wis, PhD(zool), 29. Prof Exp: DuPont fel histol & embryol, Dept Med, Univ Va, 29-30; from asst prof to prof, 30-74, EMER PROF ZOOL, UNIV IOWA, 74- Concurrent Pos: Res assoc, Argonne Nat Lab; Rockefeller traveling fel, 34-35; assoc ed, Microtomist's Vade-Mecum; mem, Corp Marine Biol Lab. Mem: AAAS; Am Soc Nat; Am Soc Zool (treas, 41-44); Am Micros Soc; Am Asn Anat. Res: Cytology; Golgi apparatus; mitochondria; polarity; sex chromosomes and spermatogenesis of man; salivary gland chromosomes; cytology of human ovary; colchicine studies upon cells; viscosity studies; effects of electric current upon polarity and growth of cells; studies with the electron microscopy on various cellular components. Mailing Add: Dept of Zoology Univ of Iowa Iowa City IA 52240

BEAMS, JESSE WAKEFIELD, b Belle Plaine, Kans, Dec 25, 98; m 31. PHYSICS. Educ: Fairmont Col, AB, 21; Univ Wis, MA, 22; Univ Va, PhD(physics), 25. Hon Degrees: ScD, Col of William & Mary, 41, Univ NC, 46, Washington & Lee Univ, 49, Fla Inst Technol, 71. Prof Exp: Instr physics & math, Ala Polytech Inst, 22-23; Nat Res Coun fel physics, Univ Va, 25-26; Nat Res Coun fel, Yale Univ, 26-27, instr, 27-28; from assoc prof to prof, 28-53, Smith prof, 53-69, chmn dept, 48-62, EMER PROF PHYSICS, UNIV VA, 69- Concurrent Pos: Off investr, Off Sci Res & Develop, 41-44; mem sci adv comt, Ballistic Res Lab, Ord Dept, Aberdeen Proving Ground, Md, 42-60; mem bd dir, Oak Ridge Inst Nuclear Studies, 48-54 & 60-; mem div physics, Nat Res Coun, 51-55; mem div physics, NSF, 52-54; mem gen adv comt, AEC, 54-60. Honors & Awards: Potts Medal, Franklin Inst, 42; Naval Ord Develop Award; Thomas Jefferson Award, 55; John Scott Award, Philadelphia, 56; Lewis Award, Am Philos Soc, 58; Award, Univ Wichita, 59; Award, Va Acad Sci, 63; Nat Medal of Sci, 67; Citation & Medal, AEC, 73. Mem: Nat Acad Sci; AAAS; Am Philos Soc; fel Am Phys Soc (pres, 58-59); Franklin Inst. Res: Ultra centrifuging; separation of isotopes and other substances by centrifuging; methods of acceleration of ions; production and use of high centrifugal fields; electrical breakdown in gases; Kerr cells; cavity oscillators; ram jets; tensile strength of metals; low temperature physics; constant speed rotation; magnetic balances; gravitation. Mailing Add: Dept of Physics Univ of Va Charlottesville VA 22903

BEAN, BRENT LEROY, b Rexburg, Idaho, June 13, 41; m 65; c 4. LASERS, QUANTUM OPTICS. Educ: Brigham Young Univ, BS, 66, MS, 68; NMex State Univ, PhD(physics), 72. Prof Exp: Res asst laser physics, Univ Laval, Quebec, 72-74; VIS ASST PROF PHYSICS, EMORY UNIV, 74-; CONSULT PHYSICS, ENG EXP STA, GA INST TECHNOL, 75- Mem: Optical Soc Am. Res: Development of the far infrared laser; use of the FIR laser and tuneable dye laser to measure the physical properties of biological and solid state samples. Mailing Add: Dept of Physics Emory Univ Atlanta GA 30322

BEAN, C THOMAS, JR, b Winchester, Ill, May 14, 20; m 44; c 2. POLYMER CHEMISTRY. Educ: Western Ill State Col, BEd, 42; Univ Kans, MA, 46, PhD(org chem), 49. Prof Exp: Anal control chemist, Hiram Walker & Sons, Inc, Ill, 42; asst instr chem, Univ Kans, 42-44; anal lab foreman, Tenn Eastman Corp & Clinton Eng Works, Tenn, 44-46; supvr asst instr qual anal, Univ Kans, 46-49; res chemist, J T Baker Co, 49-50; res chemist, 50-56, res supvr, 56-62, RES SECT MGR PLASTICS SECT EXISTING PROD RES, HOOKER CHEM CORP, 62-, MGR PROJ EVAL, 74- Mem: Am Chem Soc. Res: Vapor phase chlorination of hydrocarbons; insecticides; polyester resins; high temperature elastomers; corrosion control; epoxy resins; fire hazard reduction. Mailing Add: Hooker Chem Corp Res Ctr PO Box 8 Niagara Falls NY 14302

BEAN, CHARLES PALMER, b Buffalo, NY, Nov 27, 23; m 47; c 5. PHYSICS, BIOPHYSICS. Educ: Univ Buffalo, BA, 47; Univ Ill, AM, 49, PhD(physics), 52. Prof Exp: RES ASSOC, GEN ELEC RES & DEVELOP CTR, 51- Concurrent Pos: Consult, US Dept State, 57-58 & Nat Acad Sci, 59; adj assoc prof, Rensselaer Polytech Inst, 57-; vis scientist, Am Inst Physics, 58-; Coolidge fel, 71; guest investr, Rockefeller Univ, 73-; assoc ed, Biophys J, 75-; dir, Dudley Observ & Bellevue Res Found, 75- Honors & Awards: Indust Res Mag Award, 70. Mem: Nat Acad Sci; fel Am Phys Soc; Biophys Soc. Res: Solid state physics; membranes; neurophysiology. Mailing Add: Gen Elec Res & Develop Ctr PO Box 8 Schenectady NY 12301

BEAN, DANIEL JOSEPH, b Enosburg Falls, Vt, June 12, 34; m 58; c 3. LIMNOLOGY, PHYSIOLOGICAL ECOLOGY. Educ: Univ Vt, BA, 60, MS, 62; Univ RI, PhD(biol sci), 69. Prof Exp: Instr biol, Marist Col, 64-68; asst prof, 68-71, ASSOC PROF BIOL, ST MICHAEL'S COL (VT), 71- Concurrent Pos: Res assoc, Aquatec, Inc, Vt, 69-71. Mem: AAAS; Am Soc Limnol & Oceanog; Ecol Soc Am; Am Inst Biol Sci. Res: River zooplankton and botton fauna and the effect of thermal discharge upon their seasonal occurrence; gas production by lake sediments. Mailing Add: Dept of Biol St Michael's Col Winooski VT 05404

BEAN, GEORGE A, b Hempstead, NY, Apr 3, 33; m 57; c 2. PLANT PATHOLOGY. Educ: Cornell Univ, BS, 58; Univ Minn, MS, 60, PhD(plant path), 63. Prof Exp: Plant pathologist, Nat Capitol Region, US Dept Interior, 63-66; assoc prof bot, 66-72, ASSOC PROF PLANT PATH, UNIV MD, COLLEGE PARK, 72- Mem: Am Phytopath Soc. Res: Physiology and chemistry of plant disease causing fungi. Mailing Add: Dept of Biol Univ of Md College Park MD 20742

BEAN, GERRITT POST, b Amsterdam, NY, Apr 29, 29; m 56. ORGANIC CHEMISTRY. Educ: Northeastern Univ, BS, 52; Pa State Univ, PhD(org chem), 56. Prof Exp: Process develop chemist, Merck & Co, Inc, 56-57; res chemist, Althouse Chem Co, Inc, 57-60; asst prof chem, Douglass Col, Rutgers Univ, 60-66; vis prof, Univ East Anglia, 66-67; assoc prof, 67-71, PROF CHEM, WESTERN ILL UNIV, 71- Mem: AAAS; Am Chem Soc; The Chem Soc; fel Am Inst Chem. Res: Reactions and properties of 5-membered heterocycles; effect of substituents on reactions and properties of heterocycles; molecular orbital calculations. Mailing Add: Dept of Chem Western Ill Univ Macomb IL 61455

BEAN, JOHN LEWIS, b Village Mills, Tex, July 30, 33; m 53; c 2. ECONOMIC GEOGRAPHY, URBAN GEOGRAPHY. Educ: Univ Tex, Austin, BA, 59, MA, 62; Univ Pittsburgh, PhD(geog), 67. Prof Exp: Instr geog, Northwestern State Col, La, 63-65; asst prof, Univ Southern Miss, 65-67; asst prof, Tex Christian Univ, 67-70; ASSOC PROF GEOG, N TEX STATE UNIV, 70- Mem: Asn Am Geographers; Am Geog Soc. Res: Urban population growth, especially thresholds of population retardation and acceleration. Mailing Add: Dept of Geog N Tex State Univ Denton TX 76203

BEAN, JOHN WILLIAM, b Attercliffe, Ont, Sept 8, 01; nat US. PHYSIOLOGY. Educ: Univ Mich, AB, 24, MS, 25, PhD(physiol), 30, MD, 36. Prof Exp: Asst, 26-29, from instr to assoc prof, 29-44, actg chmn dept, 54-56, 62-63, PROF PHYSIOL, MED SCH, UNIV MICH, ANN ARBOR, 44- Concurrent Pos: Researcher, Physiol Inst, Fac Med Sci, Univ Buenos Aires, 41; mem med comt compressed air code, Dept Labor, NY, 50-58; mem physiol sect, Nat Bd Med Examr, 53-56. Mem: AAAS; Am Physiol Soc; Soc Exp Biol & Med; Undersea Med Soc. Res: Circulation and respiration; action of oxygen at atmospheric and high pressures; epileptiform convulsions and cerebral blood flow. Mailing Add: Dept of Physiol Univ of Mich Med Sch Ann Arbor MI 48104

BEAN, MICHAEL ARTHUR, b Alliance, Nebr, Sept 18, 40. PATHOLOGY, IMMUNOLOGY. Educ: Univ Colo, BA, 62, MD, 67. Prof Exp: Pathologist, Registry of Radiation Path, Armed Forces Inst Path, Washington, DC, 69; pathologist, Atomic Bomb Casualty Comn, Hiroshima, Japan, 69-70; assoc immunol, Sloan-Kettering Inst Cancer Res, 72-75, lab head tumor-host immunol, 74-75; SR INVESTR, VA MASON RES CTR, 75- Concurrent Pos: Asst attend path, Mem Hosp, New York, 73-75; asst prof biol, Sloan-Kettering Div, Grad Sch, Cornell Univ, 74-75; affil investr, Fred

Hutchinson Cancer Ctr, Seattle, 75- Mem: AAAS; Am Asn Cancer Res. Res: Tumor-host immunology. Mailing Add: Virginia Mason Res Ctr 1000 Seneca St Seattle WA 98101

BEAN, RALPH J, b Atlantic City, NJ, Aug 13, 33; m 59; c 3. MATHEMATICS. Educ: Univ Pittsburgh, BS, 57, MA, 60; Univ Md, PhD(math), 62. Prof Exp: Res scientist, Lewis Lab, NASA, 57-58; from instr to asst prof math, Univ Wis, 62-67; assoc prof, Univ Tenn, 67-71; PROF MATH, RICHARD STOCKTON STATE COL, 71- Concurrent Pos: Math Asn Am vis lectr, 63-; NSF res grants, 65-71. Mem: Am Math Soc; Math Asn Am; Nat Coun Teachers Math; Am Fedn Teachers. Res: Point set topology; Euclidian spaces and manifolds. Mailing Add: Dept of Math Richard Stockton State Col Pomona NJ 08240

BEAN, ROBERT JAY, b Dallas, Tex, Aug 9, 24. GEOPHYSICS. Educ: Southern Methodist Univ, BS, 48; Harvard Univ, AM, 50, PhD(geophys), 51. Prof Exp: Geophysicist, 51-68, SR STAFF GEOPHYSICIST, SHELL OIL CO, 68- Mem: Soc Explor Geophys. Res: Gravity and magnetic interpretation. Mailing Add: Shell Oil Co PO Box 481 Houston TX 77001

BEAN, ROBERT TAYLOR, b Cherokee, Iowa, Feb 8, 13; m 53; c 1. HYDROGEOLOGY, ENGINEERING GEOLOGY. Educ: Col Wooster, AB, 34; Ohio State Univ, MA, 42. Prof Exp: Asst geol, Stanford Univ, 46; geologist, US Geol Surv, 47; from jr to sr geologist eng geol, Calif Dept Water Resources, 47-56, supv eng geologist, 56-66; tech adv hydrogeol, UN, 66-71; CONSULT GEOLOGIST, 71- Concurrent Pos: Consult, Chile-Calif Prog, 64; lectr, Calif State Univ, Northridge & Calif State Univ, Los Angeles, 72- Mem: AAAS; Asn Eng Geologists (pres, 60-61); fel Geol Soc Am; Am Geophys Union; Soc Social Responsibility in Sci. Res: Water resources evaluation; international peace. Mailing Add: 2729 Willowhaven Dr La Crescenta CA 91214

BEAN, ROSS COLEMAN, b Thatcher, Ariz, Apr 22, 24; m 49; c 3. BIOCHEMISTRY. Educ: Univ Calif, BS, 46, PhD(comp biochem), 53; Stanford Univ, MS, 48. Prof Exp: Chemist, Western Regional Res Lab, 48-50; asst, Univ Calif, Berkeley, 51-53, jr biochemist, 53-55; jr biochemist, Univ Calif, Riverside, 55-56, asst prof biochem, 56-63; supvr biochem & physiol, Res Staff, 63-74, PRIN SCIENTIST, BIO-SCI STAFF, AERONUTRONIC DIV, PHILCO-FORD CORP, 74- Prof Exp: Consult, US Army Weapons Command, 67 & Off Saline Water, US Dept Interior, 69. Mem: AAAS; Biophys Soc; Am Soc Biol Chemists. Res: Photosynthesis; carbohydrate metabolism; membrane transport and electrophysiology. Mailing Add: Biosci 54 RL Aeronutronic Div Philco-Ford Corp Newport Beach CA 92663

BEAN, VERN ELLLIS, b La Grande, Ore, Jan 19, 37; m 61; c 4. HIGH PRESSURE PHYSICS. Educ: Brigham Young Univ, BS, 62, MS, 64, PhD(physics), 73. Prof Exp: Physicist, Lawrence Radiation Lab, Univ Calif, Berkeley, 64-67; PHYSICIST, NAT BUR STAND, 72- Res: Application of ultrasonic techniques to determine properties of materials at high pressures; development of techniques for the precise measurement of high pressure. Mailing Add: Nat Bur of Stand Pressure & Vacuum Sect Washington DC 20234

BEAN, WILLIAM BENNETT, b Manila, Philippines, Nov 8, 09; m 39; c 3. MEDICINE. Educ: Univ Va, BA, 32, MD, 35; Am Bd Internal Med, dipl, 47; Am Bd Nutrit, dipl, 51. Prof Exp: Student instr anat, Sch Med, Univ Va, 32-35; intern, Johns Hopkins Hosp, 35-36; teaching fel med, Thorndike Mem Lab, Harvard Univ, 36-37; sr med resident, Cincinnati Gen Hosp, 37-38; from instr to assoc prof med, Col Med, Univ Cincinnati, 38-48; prof med & head dept internal med, 48-70, SIR WILLIAM OSLER PROF MED, UNIV IOWA, 70-; DIR INST HUMANITIES IN MED, UNIV TEX MED BR GALVESTON, 74- Concurrent Pos: Asst res physician, Boston City Hosp, 36-37; fel nutrit, Col Med, Univ Cincinnati, 38-40; asst vis physician, Hillman Hosp, Ala, 40-42; asst attend physician, Cincinnati Gen Hosp, 41-46, clinician outpatient dept & attend physician, 46-48; asst ed, Nutrit Rev, 45-46; bk rev ed, Cincinnati J Med, 46-48; sr med consult, US Vet Admin, 47-; assoc ed, J Am Soc Clin Invest, 47-52; physician-in-chief, Univ Hosps, Iowa City, 48-70; assoc ed, Dis Chest, 51-61; Droessel Mem lectr, Milwaukee, 54; bk rev ed, Arch Internal Med, 55-62, ed-in-chief, 62-67; Harro Woltmann lectr, Ohio, 56, ann Marks Mem lectr, Col Chest Physicians, 57; consult ed, Mod Med, 64-67; ed-in-chief, Current Med Dig & consult ed med, Med Aspects of Human Sexuality, 67-; spec consult, Surgeon Gen, US Army. Honors & Awards: Horsley Mem Prize, Univ Va, 44; Gold Headed Cane Award, Univ Calif, 64. Mem: Fel AMA; Am Clin & Climat Asn (pres, 67); fel Am Col Chest Surg; master, Am Col Physicians; fel Am Med Writers Asn. Res: Nutrition; human physiology; skin changes in cirrhosis; vascular spiders and related lesions of the skin. Mailing Add: Inst of Humanities in Med Univ of Mex Med Br Galveston TX 77550

BEAN, WILLIAM CLIFTON, b Paris, Tex, Sept 10, 38. MATHEMATICS, AEROSPACE ENGINEERING. Educ: Univ Tex, Austin, BA, 59, MA, 60, PhD(math), 71. Prof Exp: Spec instr math, Univ Tex, Aus- tin, 60-63; aerospace technologist flight simulator software design, Flight Crew Support Div, 63-66, aerospace technologist, Mission Planning & Anal Div, 66-71, AEROSPACE ENG, ADVAN MISSION DESIGN BR, MISSION PLANNING & ANAL DIV, NASA-JOHNSON SPACECRAFT CTR, 71- Concurrent Pos: Asst instr, Univ Tex, Austin, 70-71. Honors & Awards: Apollo Achievement Award, NASA, 69. Mem: Am Inst Aeronaut & astronaut; Math Asn Am; Tensor Soc; NY Acad Sci. Res: Optimum interplanetary trajectory design; studies on extensor structures in the calculus of variations; space shuttle communication and data handling. Mailing Add: 734 Voyager Houston TX 77062

BEANE, DONALD GENE, b Aurora, Ill, Aug 25, 29; m 52; c 3. MATHEMATICS. Educ: Iowa Wesleyan Col, AB, 51; Univ Ill, MA, 58, PhD(math ed), 62; Ohio State Univ, MSc, 66. Prof Exp: Asst prof educ, 62-65; from asst prof to assoc prof math, 66-75, PROF MATH, COL WOOSTER, 75- Concurrent Pos: Teaching fel, Great Lakes Cols Asn, 75-76. Mem: Math Asn Am; Nat Coun Teachers Math. Res: Mathematical education; geometry; applications of linear mathematical models. Mailing Add: Dept of Math Col of Wooster Wooster OH 44691

BEANE, RICHARD EDWARD, b Buffalo, NY, July 12, 42; m 71. GEOCHEMISTRY, ECONOMIC GEOLOGY. Educ: Union Col, NY, BS, 64; Univ Ariz, MS, 68; Northwestern Univ, Evanston, PhD(geochem), 72. Prof Exp: Explor geologist, Phelps Dodge Corp, 67-68; ASST PROF GEOL, NMEX INST MINING & TECHNOL, 71- Concurrent Pos: Vis lectr geol, Univ Ariz, 76. Res: Aqueous solution-mineral interaction in geologic systems; mass transfer and hydrothermal alteration; geochemical environments of base-metal deposition and sources of materials; dump and in-situ leaching of copper. Mailing Add: Dept of Geosci NMex Inst of Mining & Technol Socorro NM 87801

BEAR, HERBERT S, JR, b Philadelphia, Pa, Mar 13, 29; m 51; c 2. MATHEMATICS. Educ: Univ Calif, Berkeley, BA, 50, PhD(math), 57. Prof Exp: Instr math, Univ Ore, 55-56; assoc, Univ Calif, Berkeley, 56-57; from instr to asst prof, Univ Wash, 57-62; assoc prof, Univ Calif, Santa Barbara, 62-67; prof, NMex State Univ, 67-69; chmn dept, 69-74, PROF MATH, UNIV HAWAII, MANOA, 69- Concurrent Pos: Vis asst prof, Princeton Univ, 59-60; vis assoc prof, Univ Calif, San Diego, 65-66. Mem: Am Math Soc; Math Asn Am. Res: Functional analysis. Mailing Add: Dept of Math Univ of Hawaii at Manoa Honolulu HI 96822

BEAR, JOHN L, b Lampasas, Tex, Mar 6, 34; m 60; c 3. INORGANIC CHEMISTRY. Educ: Southwest Tex State Col, BS, 55, MA, 56; Tex Tech Col, PhD(inorg chem), 60. Prof Exp: Fel inorg chem, Fla State Univ, 60-62, asst prof, 62-63; assoc prof, 63-72, PROF CHEM, UNIV HOUSTON, 72- Mem: Am Chem Soc. Res: Fast reactions in solution; thermodynamics of metal ion complex formation in solution; mixed solvent systems. Mailing Add: Dept of Chem Univ of Houston Houston TX 77004

BEAR, PHYLLIS DOROTHY, b New York, NY, Aug 7, 31. MICROBIAL GENETICS, MOLECULAR BIOLOGY. Educ: San Diego State Col, BS, 56; Univ Calif, Los Angeles, PhD(microbiol), 66. Prof Exp: Res technician marine genetics, Scripps Inst, Univ Calif, 56-62; assoc prof molecular biol, 68-74, PROF MOLECULAR BIOL, UNIV WYO, 74- Concurrent Pos: Carnegie fel, Carnegie Inst, 66-68. Mem: AAAS; Am Soc Microbiol. Res: Structure and function of microbial nucleic acid; regulation of gene expression in bacterial viruses; pathogenesis of Anaplasma marginale. Mailing Add: Dept of Microbiol Univ of Wyo Laramie WY 82071

BEAR, RICHARD SCOTT, b Miamisburg, Ohio, June 8, 08; m 31, 42, 72; c 2. MOLECULAR BIOLOGY, BIOPHYSICS. Educ: Princeton Univ, SB, 30; Univ Calif, Berkeley, PhD(chem), 33. Prof Exp: Nat Res Coun fel chem, Princeton Univ, 33-34; res assoc zool, Washington Univ, 34-38; asst prof chem, Iowa State Col, 38-41; from assoc prof to prof biol, Mass Inst Technol, 41-57; dean sci & humanities, Iowa State Univ, 57-61; dean grad sch, Boston Univ, 61-66, prof biol, 66-69; PROF ANAT, UNIV NC, CHAPEL HILL, 69- Concurrent Pos: Mem panel cytochem, Comt Growth, Nat Res Coun, 47-50; mem study sect, Biophys & Biophys Chem, NIH, 60-64; mem sci adv comt, Helen Hay Whitney Found, 60-64. Mem: AAAS; fel Am Acad Arts & Sci (vpres class II, 67-69); Am Crystallog Asn; Biophys Soc; Am Chem Soc. Res: X-ray diffraction and optical methods applied to the study of natural polymers and tissues. Mailing Add: Dept of Anat Univ of NC 111 Swing Bldg Chapel Hill NC 27514

BEARCE, DENNY N, b Pittsburgh, Pa, Nov 1, 34; m 57; c 2. GEOLOGY. Educ: Brown Univ, BA, 56; Mo Sch Mines, BS, 62, MS, 63; Univ Tenn, PhD(geol), 66. Prof Exp: Geologist, Mobile Oil Co, 63-64; asst prof geol, Eastern Ky Univ, 66-67; asst prof, 67-70, ASSOC PROF GEOL & CHMN DEPT, BIRMINGHAM-SOUTHERN COL, 70- Mem: Am Asn Petrol Geol. Res: Mapping and describing structure and stratigraphy of lower Cambrian and Precambrian rocks in Blue Ridge province of eastern Tennessee; structure and stratigraphy of Talladega Metamorphic Belt of east Alabama. Mailing Add: Dept of Geol Birmingham-Southern Col Birmingham AL 35204

BEARCE, WINFIELD HUTCHINSON, b Lewiston, Maine, Oct 20, 37; m 61. ORGANIC CHEMISTRY. Educ: Bowdoin Col, AB, 59; Lawrence Col, MS, 61, PhD, 64. Prof Exp: Sr res chemist, St Regis Paper Co, 64-66; ASSOC PROF CHEM, MO VALLEY COL, 66- Mem: Am Chem Soc. Res: Carbohydrate chemistry, particularly the stability, with respect to acid, of the glycosidic linkage in carbohydrate polymers; essential oils as a source of substrate for mechanistic studies. Mailing Add: Dept of Chem Mo Valley Col Marshall MO 65340

BEARD, BENJAMIN H, b Blair, Nebr, Jan 12, 18; m 45; c 1. PLANT BREEDING. Educ: Univ Mo, BS, 50, MS, 52; Univ Nebr, PhD(agron), 55. Prof Exp: Instr field crops, Univ Mo, 51-52; res geneticist, Agr Res Serv, USDA, Univ Minn, 55-60; RES GENETICIST, AGR RES SERV, USDA, UNIV CALIF, 60- Concurrent Pos: Res assoc agron, Univ Nebr, 51-53. Mem: Am Genetic Asn; Genetics Soc Am; Am Soc Agron. Res: Radiation genetics; cytogenetics; breeding and genetics of oilseed crops. Mailing Add: Dept of Agron & Range Sci Univ of Calif Davis CA 95616

BEARD, CHARLES IRVIN, b Ambridge, Pa, Nov 30, 16; m 48; c 2. RADIOPHYSICS. Educ: Carnegie Inst Technol, BS, 38; Mass Inst Technol, PhD(physics), 48. Prof Exp: Asst engr res labs, Westinghouse Elec Corp, Pa, 38-39; sr physicist, Field Res Labs, Magnolia Petrol Co, Tex, 48-50 & Appl Physics Lab, Johns Hopkins Univ, 50-56; sr eng specialist physics, Sylvania Electronic Defense Lab, 56-62; staff mem, Boeing Sci Res Labs, Wash, 62-71; RES PHYSICIST, NAVAL RES LAB, 71- Concurrent Pos: Chmn, US Comt II, Int Sci Radio Union, 70-73, mem US Nat Comt & assoc ed, Radio Sci, 73-75. Honors & Awards: Bolljahn Mem Award, Inst Elec & Electronics Engr, 62. Mem: AAAS; Am Phys Soc; fel Inst Elec & Electronics Engr. Res: Microwave spectroscopy; low frequency electromagnetic waves in conducting media; scattering of electromagnetic waves from random media, the ocean, atmospheric turbulence and internal waves. Mailing Add: Code 5369 Naval Res Lab Washington DC 20375

BEARD, CHARLES WALTER, b Tifton, Ga, Nov 16, 32; m 60; c 4. VETERINARY VIROLOGY. Educ: Univ Ga, DVM, 55; Univ Wis, MS, 64, PhD(vet sci), 65. Prof Exp: Mem staff, 65-72, DIR, SOUTHEAST POULTRY RES LAB, AGR RES SERV, USDA, 72- Mem: Am Vet Med Asn; Am Asn Avian Path. Res: Aerobiology; respiratory diseases; viruses; techniques of aerosol exposure; influence of route of administration on host response. Mailing Add: USDA Agr Res Serv 934 College Station Rd Athens GA 30601

BEARD, DAVID BREED, b Needham, Mass, Feb 1, 22; m 45; c 4. THEORETICAL PHYSICS. Educ: Hamilton Col, BS, 43; Cornell Univ, PhD(physics), 51. Prof Exp: Instr nuclear physics, Cath Univ Am, 50-51; instr theoret physics, Univ Conn, 51-53; from asst prof to prof physics, Univ Calif, 53-64; PROF PHYSICS & ASTRON & CHMN DEPT, UNIV KANS, 64- Concurrent Pos: Guggenheim fel & Fulbright scholar, Imp Col, Univ London, 65-66; consult, Naval Ord Lab, Cath Univ Am, 50-51, Pratt & Whitney Aircraft Corp, 52-53, AEC Proj, Univ Conn, 53-54, Lawrence Radiation Lab, 54-56 & 63-65, Lockheed Spacecraft & Missiles Res, 56-64, Sandia Corp, 59-68, Goddard Space Flight Ctr, 61-69 & US Atomic Energy Hq, 65-; NATO sr fel, 72. Mem: Fel AAAS; Am Geophys Union; fel Am Phys Soc; Fedn Am Scientists. Res: Meson theory of nuclear forces; level densities in heavy nuclei; beta-ray spectrometry; plasma physics; astro-physics; space physics; magnetospheric physics; quantum mechanics. Mailing Add: Dept of Physics & Astron Univ of Kans Lawrence KS 66044

BEARD, DAVID FRANKLIN, b Portage, Ohio, Aug 1, 12; m 36; c 3. PLANT BREEDING. Educ: Ohio State Univ, BS, 35, PhD(corn breeding), 40. Prof Exp: Exten agronomist, Ohio State Univ, 40-50; asst chief forage & range res br, Agr Res Serv, USDA, 50-51, chief, 51-58; dir res, 58-64, VPRES RES, WATERMAN-LOOMIS CO, 64- Mem: Fel Am Soc Agron; Crop Sci Soc Am; Am Genetic Asn. Res: Relative values of various testers of inbred lines of corn; procedures and methods to develop high yielding pest resistant varieties of alfalfa. Mailing Add: 10916 Bornedale Dr Adelphi MD 20783

BEARD, ELIZABETH L, b New Orleans, La, Apr 2, 32. PHYSIOLOGY. Educ: Tex Christian Univ, BA, 52, BS, 53, MS, 55; Tulane Univ, PhD(physiol), 61. Prof Exp: From instr to assoc prof, 55-68, PROF BIOL SCI, LOYOLA UNIV (LA), 68- Mem: AAAS; NY Acad Sci. Res: Blood proteolytic activity, particularly as related to stress in mammals and to atherosclerosis. Mailing Add: 6127 Garfield New Orleans LA 70118

BEARD, GEORGE B, b Marblehead, Mass, Feb 22, 24; m 55; c 5. EXPERIMENTAL NUCLEAR PHYSICS. Educ: Harvard Univ, AB, 47; Univ Mich, MS, 48, PhD(physics), 55. Prof Exp: Asst, Univ Mich, 50-51; from instr to asst prof physics, Mich State Univ, 54-60; assoc prof, 60-65, PROF PHYSICS, WAYNE STATE UNIV, 65-, CHMN DEPT, 73- Mem: Fel Am Phys Soc; Am Asn Physics Teachers. Res: Low energy nuclear spectroscopy; lifetimes of lowlying excited states; Mössbauer effect studies. Mailing Add: Dept of Physics Wayne State Univ Detroit MI 48202

BEARD, GEORGE VICTOR, physical chemistry, see 12th edition

BEARD, HELEN PEARL, b Welshfield, Ohio, Nov 3, 15. MATHEMATICAL ANALYSIS. Educ: Wilson Col, AB, 36; Univ Pa, AM, 37; Mass Inst Technol, PhD(math), 43. Prof Exp: Asst physics, Wilson Col, 40-41; from instr to asst prof math, Newcomb Col, Tulane Univ, 41-61; assoc prof, 61-69, PROF MATH, STATE UNIV NY BINGHAMTON, 69- Concurrent Pos: Vis asst prof, Statist Lab, Univ Calif, 53-54. Mem: Am Math Soc; Math Asn Am; Inst Math Statist. Res: Statistics and probability. Mailing Add: Dept of Math Sci State Univ NY Binghamton NY 13901

BEARD, JAMES B, b Piqua, Ohio, Sept 24, 35; m 55; c 2. PLANT PHYSIOLOGY. Educ: Ohio State Univ, BS, 57; Purdue Univ, MS, 59, PhD(environ plant physiol), 61. Prof Exp: From assoc prof to prof environ physiol grasses, Mich State Univ, 61-75; PROF ENVIRON PHYSIOL GRASSES, TEX A&M UNIV, 75- Concurrent Pos: NSF fel, 69-70. Mem: Am Soc Agron; Crop Sci Soc Am; Am Soc Plant Physiol. Res: Winterkill mechanisms of grasses; physiology of high temperature growth stoppage of grasses; plant micro-environment; biological decomposition of thatch; turfgrass culture; sod production and utilization. Mailing Add: Dept Soil & Crop Sci Tex A&M Univ College Station TX 77843

BEARD, JAMES DAVID, b Dayton, Ohio, July 21, 37; m 59; c 2. MEDICAL PHYSIOLOGY. Educ: DePauw Univ, BA, 59; Univ Tenn, PhD(physiol), 63. Prof Exp: Instr & fel physiol, Sch Med, Marquette Univ, 63-65; from instr to asst prof physiol & biophys, 65-74, ASSOC PROF PSYCHIAT, UNIV TENN CTR HEALTH SCI, MEMPHIS, 74- Concurrent Pos: Physiologist, Res Serv, Vet Admin Hosp, Wood, Wis, 63-65; Nat Acad Sci-Nat Comn Int Unit Physiol Sci travel grant, Tokyo, 65; dir alcohol & drug res ctr, Tenn Psychiat Hosp & Inst, 67-, dir clin labs, 67-, dir res, 72-; mem bd, Nat Coun Alcoholism. Mem: AAAS; Soc Neurosci; NY Acad Sci. Res: Pathophysiology of acute and chronic alcoholism and other drug ingestion; cardiovascular dynamics; fluid and electrolyte metabolism including trace metals; kidney, catecholamine metabolism; psychopharmacology, biological psychiatry. Mailing Add: Alcohol & Drug Res Ctr 865 Poplar Ave Memphis TN 38104

BEARD, JAMES MILLER, organic chemistry, physical chemistry, see 12th edition

BEARD, JEAN, b Cedar Falls, Iowa, Apr 12, 34. SCIENCE EDUCATION. Educ: State Univ Iowa, BA, 56; Univ Northern Iowa, MA, 60; Ore State Univ, PhD(sci educ), 69. Prof Exp: Teacher jr high sch, Minn, 56-59 & 60-61; instr prof educ & sci supvr, Mankato State Col, Wilson Campus Sch, 61-65; part-time instr sci educ, Ore State Univ, 65-68; ASSOC PROF NATURAL SCI, SAN JOSE STATE UNIV, 69- Mem: Nat Asn Biol Teachers; Nat Sci Teachers Asn; Nat Asn Res Sci Teaching. Res: Assessment of science achievements by students in general education science, kindergarten-14th grade, including science processes, science knowledge and attitudes about natural phenomena and organisms; rational general education about electrical energy alternatives. Mailing Add: Dept of Natural Sci San Jose State Univ San Jose CA 95192

BEARD, JOSEPH WILLIS, b Athens, La, Nov 5, 01; m. SURGERY. Educ: Univ Chicago, BS, 26; Vanderbilt Univ, MD, 29. Prof Exp: Intern, asst resident & res instr, Surg Serv, Vanderbilt Univ Hosp, 29-32; from asst to assoc path, Rockefeller Inst, 32-37; asst prof surg, Sch Med & Univ Hosp, Duke Univ, 37-42, from asst prof to prof exp surg, 37-63, assoc prof virol, 49-65, James B Duke Prof surg, 63-73; SR SCHOLAR, LIFE SCI RES LABS, 73- Mem: Am Soc Exp Path; Soc Exp Biol & Med; Am Soc Microbiol; Am Asn Immunol; Am Asn Cancer Res. Res: Surgical shock; chick embryo vaccine against equine encephalomyelitis; purification of animal viruses; influenza vaccines; virus-induced cancer. Mailing Add: Life Sci Res Labs 1509 1/2 49th St S St Petersburg FL 33705

BEARD, LUTHER STANFORD, b Langley, SC, Feb 21, 29; m 52; c 2. SYSTEMATIC BOTANY. Educ: Furman Univ, BS, 50; Univ NC, MA, 59, PhD(bot), 64. Prof Exp: Teacher high sch, SC, 55-58; instr biol, Furman Univ, 58-59; assoc prof, 61-67, PROF BIOL, CAMPBELL COL, NC, 67-, CHMN DEPT, 63- Mem: Am Soc Plant Taxon; Soc Study Evolution. Res: Taxonomy of the genus Schrankia and its relation to the genus Mimosa. Mailing Add: Box 366 Buies Creek NC 27506

BEARD, MARGARET ELZADA, b Washington, DC, Oct 17, 41. CELL BIOLOGY. Educ: Wellesley Col, BA, 63; Univ Mich, MS, 65, PhD(zool), 67. Prof Exp: Fel path, Albert Einstein Col Med, 67-69, asst prof, 69-70; asst prof res ophthal, Sch Med, Univ Ore, 72-75; ASST PROF BIOL, REED COL, 75- Mem: Am Soc Cell Biol; AAAS. Res: Investigations of the structure and function of peroxisomes. Mailing Add: Dept of Biol Reed Col Portland OR 97202

BEARD, OWEN WAYNE, b Wattensaw, Ark, Nov 15, 16; m 44; c 3. INTERNAL MEDICINE, CARDIOLOGY. Educ: Okla State Univ, 37-39; Univ Ark, BSM & MD, 43; Am Bd Internal Med, dipl, 50. Prof Exp: Instr med, Med Br, Univ Tex, 45-47; from asst prof to assoc prof, 47-53, assoc prof, 55-70, PROF MED, SCH MED, UNIV ARK, LITTLE ROCK, 70-, CHIEF GERIAT, SCH MED SCI, 75-; CHIEF GERIAT, VET ADMIN HOSP, LITTLE ROCK, 75- Concurrent Pos: Asst chief med, Vet Admin Hosp, Little Rock, 55-75. Mem: Am Heart Asn; Am Col Physicians; fel Am Col Cardiol; Geront Soc. Res: Cardiovascular disease. Mailing Add: Vet Admin Hosp 300 E Roosevelt Rd Little Rock AR 72206

BEARD, PERCY MORRIS, JR, b Opelika, Ala, Mar 31, 36; m 59; c 2. EXPERIMENTAL NUCLEAR PHYSICS. Educ: US Naval Acad, BS, 58; Duke Univ, PhD(nuclear physics), 64. Prof Exp: US Navy, 58-, res asst, Duke Univ, 59-64, studentship, Nuclear Power Prog, 64-66, SUBMARINE OFFICER, US NAVY, 66- Res: Low energy nuclear physics, especially research with thin windowless gas targets for high energy resolution neutron production. Mailing Add: 2319 S Lander Ln Charleston SC 29407

BEARD, RAIMON LEWIS, b Longmont, Colo, Apr 7, 12; m 41; c 2. ENTOMOLOGY. Educ: Wesleyan Univ, AB, 35; Yale Univ, PhD(zool), 39. Prof Exp: Asst biol, Yale Univ, 35-38; agr res scientist entom, Conn Agr Exp Sta, 39-46; tech aide, Insect Control Comt, Nat Res Coun, 46-47; agr res scientist entom, Conn Agr Exp Sta, 47-74; RETIRED. Concurrent Pos: Vis instr, Wesleyan Univ, 41-42; Fulbright sr res scholar, Commonwealth Sci & Indust Res Orgn, Australia, 53; tech assistance expert, Int Atomic Energy Agency, Thailand, 65-66. Mem: AAAS; Entom Soc Am; Am Soc Zool; Soc Invert Path; Entom Soc Can. Res: Insect ecology, physiology and pathology. Mailing Add: 864 Mountain Rd Cheshire CT 06410

BEARD, RODNEY RAU, b Guinda, Calif, Dec 27, 11; m 38; c 4. PREVENTIVE MEDICINE, ENVIRONMENTAL HEALTH. Educ: Stanford Univ, AB, 32, MD, 38; Harvard Univ, MPH, 40; Am Bd Prev Med, dipl, 49; Am Bd Indust Hyg, dipl, 62. Prof Exp: Clin lab technician, French Hosp, San Francisco, 33-37; intern, Gorgas Hosp, CZ, 37-38; asst resident pub health & prev med, Hosps & instr, Nursing Sch, Stanford Univ, 38-39; Rockefeller fel med sci, Sch Pub Health, Harvard Univ, 39-40; from instr to prof pub health & prev med, 40-69, exec, 49-69, dir rehab, 55-60, PROF FAMILY, COMMUNITY & PREV MED, SCH MED, STANFORD UNIV, 69- Concurrent Pos: Med dir, Pac Div, Pan Am Airways, 40-49; mem comn A, Comt Aviation Med, Nat Res Coun, 41-47, mem comt effects of atmospheric contaminants on human health & welfare, 68-70; clin prof, Sch Pub Health, Univ Calif, 52-64, lectr, 64-72; dept dir comn environ hyg, Armed Forces Epidemiol Bd, 54-56, 61-66, mem, 66-73; consult indust med, US Vet Admin, 54-67; consult, Off Surgeon Gen, US Dept Army, 54-; mem nat adv heart coun, NIH, 57-61, mem, Nat Adv Coun Environ Health Sci, 71-74, consult to dir, Nat Inst Environ Health Sci, 74-75; vis prof, Univ Milan, 60-61; trustee, Am Bd Prev Med, 61-70; mem, Nat Adv Coun Pub Health Training, 65-69; consult, Surgeon Gen, US Air Force, 66-69; mem tech adv comt, Calif Air Resources Bd, 69-72; mem subcomt atherosclerosis, Intersoc Comn Heart Dis Resources, 69- Mem: Fel Am Pub Health Asn; fel Am Acad Occup Med; Soc Occup & Environ Health; Asn Teachers Prev Med (pres, 58-59); fel Am Col Prev Med. Res: Occupational medicine; air pollution; epidemiology; behavior toxicology. Mailing Add: Dept of Family Commun & Prev Med Stanford Univ Sch of Med Stanford CA 94305

BEARD, WILLIAM CLARENCE, b Gallipolis, Ohio, June 9, 34; m 60. MINERALOGY. Educ: Ohio State Univ, BSc, 60, PhD(mineral), 65. Prof Exp: Res chemist, Linde Div, Union Carbide Corp, 65-67; asst prof, 67-69, ASSOC PROF GEOL, CLEVELAND STATE UNIV, 69- Mem: Mineral Soc Am; Am Ceramic Soc; Nat Asn Geol Teachers. Res: High temperature phase equilibrium of minerals and ceramics; crystal chemistry; hydrothermal synthesis; crystal growth; thermal analysis of minerals and ceramics. Mailing Add: Dept of Geol Sci Cleveland State Univ Cleveland OH 44115

BEARD, WILLIAM QUINBY, JR, b Beaufort, SC, Apr 10, 32; m 57; c 2. INDUSTRIAL CHEMISTRY. Educ: Univ NC, BS, 54; Duke Univ, PhD(org chem), 59. Prof Exp: Res fel chem, Duke Univ, 59-60; res chemist, Ethyl Corp, La, 60-68; SR RES CHEMIST, RES & DEVELOP LAB, CHEM DIV, VULCAN MAT CO, 68- Mem: Am Chem Soc. Res: Chlorinated hydrocarbons. Mailing Add: Res & Develop Lab Chem Div Vulcan Mat Co Box 545 Wichita KS 67202

BEARDEN, ALAN JOYCE, b Baltimore, Md, Nov 23, 31. BIOPHYSICS. Educ: Johns Hopkins Univ, AB, 50, PhD(physics), 59. Prof Exp: Instr physics, Univ Wis, Madison, 59-60; asst prof, Cornell Univ, 60-64; NIH spec fel biophysics, Univ Calif, San Diego, 64-66, asst prof chem, 66-68; res biophysicist & lectr, 69-72, RES BIOPHYSICIST & ADJ ASSOC PROF BIOPHYSICS, DONNER LAB & DIV MED PHYSICS, UNIV CALIF, BERKELEY, 73- Concurrent Pos: USPHS res career develop award, 70-75. Honors & Awards: Teaching Apparatus Award, Am Asn Physics Teachers, 62. Mem: Am Phys Soc; Biophys Soc; Am Soc Photobiol; AAAS. Res: Bioenergetics, photosynthesis, photobiology; structure and function of iron-sulfur proteins; physics of energy transduction processes in biology; applications of electron spin resonance and lasers to biophysics. Mailing Add: Donner Lab Univ of Calif Berkeley CA 94720

BEARDEN, HENRY JOE, b Starkville, Miss, May 12, 26; m 46; c 3. REPRODUCTIVE PHYSIOLOGY. Educ: Miss State Univ, BS, 50; Univ Tenn, MS, 51; Cornell Univ, PhD(animal breeding), 54. Prof Exp: Instr dairy, Univ Tenn, 51; asst prof dairy exten & res, Cornell Univ, 54-60; PROF DAIRY SCI & HEAD DEPT, MISS STATE UNIV, 60- Mem: Am Dairy Sci Asn; Soc Study Reproduction; US Animal Health Asn. Res: Animal physiology; dairy cattle genetics. Mailing Add: Dept of Dairy Sci Miss State Univ Drawer DD Mississippi State MS 39762

BEARDEN, JOYCE ALVIN, b Greenville, SC, Oct 19, 03; m 23; c 1. PHYSICS. Educ: Furman Univ, AB, 23, ScD, 51; fel, Univ Chicago, 25, PhD(physics), 26. Prof Exp: Asst, Univ Chicago, 23-25, instr, 26-29; assoc physics, 29-32, assoc prof, 32-39, physicist, Appl Physics Lab, 42-45, dir radiation lab, 43-57, chmn dept physics, 47-49, PROF PHYSICS, JOHNS HOPKINS UNIV, 39- Concurrent Pos: Consult, Nat Defense Res Comt, 40-42; physicist, Carnegie Inst, 41-42. Mem: AAAS; fel Am Phys Soc; Am Asn Physics Teachers; Inst Elec & Electronics Engrs. Res: Absolute x-ray intensities and electron distributions; the Compton effect; x-ray wavelengths by ruled and crystal gratings; refraction of x-rays; determination of charge on electron and other fundamental constants; x-ray wavelength standards. Mailing Add: Dept of Physics Johns Hopkins Univ Baltimore MD 21218

BEARDEN, WILLIAM HARLIE, b Kileen, Tex, Aug 18, 49; m 70; c 1. PHYSICAL ORGANIC CHEMISTRY. Educ: Centenary Col, BS, 71; Univ Houston, PhD(chem), 75. Prof Exp: NSF RES ASST CHEM, CALIF INST TECHNOL, 75- Mem: Am Chem Soc. Res: Structure determinations of organic compounds using nuclear magnetic resonance via nonclassical methods, including Lanthanide induced chemical shifts and dipolar couplings. Mailing Add: Gates & Crellin Lab Calif Inst Technol Pasadena CA 91109

BEARDMORE, WILLIAM BOONE, b Salem, Ohio, Aug 14, 25; m 48; c 2. MICROBIOLOGY. Educ: Ohio State Univ, BSc, 48, MSc, 50, PhD(bact), 53. Prof Exp: From asst to asst instr bact, Ohio State Univ, 48-53; assoc res microbiologist virol, 53-56, mgr biol control dept, 56-63, from res virologist to sr res virologist, 63-67, LAB DIR, PARKE, DAVIS & CO, 67- Mem: AAAS; Am Soc Microbiol. Res: Viruses and tissue cultures. Mailing Add: 213 Nesbit Lane Rochester MI 48063

BEARDSLEE, RONALD ALLEN, b Asheville, NC, Feb 1, 46; m 70. CLINICAL CHEMISTRY. Educ: Univ Calif, Santa Barbara, BS, 68, PhD(chem), 72. Prof Exp: Presidential intern protein res, Western Regional Res Ctr, USDA, 72-73; RES SCIENTIST CLIN CHEM, BIO-SCI LABS, 74- Mem: Am Chem Soc; Am Asn Clin Chem. Res: Development of new chemical and physical methods for the analysis of materials of clinical significance. Mailing Add: Bio-Sci Labs 7600 Tyrone Ave Van Nuys CA 91405

BEARDSLEY, DANIEL WALDO, b Moore Haven, Fla, Sept 17, 23; m 66; c 6. ANIMAL NUTRITION. Educ: Univ Fla, BSA, 47; Univ Ill, MS, 52, PhD(animal nutrit), 58. Prof Exp: Asst animal husb, Everglades Exp Sta, Inst Food & Agr Sci, Univ Fla, 49-50 & 53-56; assoc, Univ Ga, 58-63; assoc animal nutrit, Inst Food & Agr Sci, 65-68, PROF ANIMAL NUTRIT, INST FOOD & AGR SCI & DIR AGR RES

& EDUC CTR, UNIV FLA, 68- Mem: Am Soc Animal Sci. Res: Administration of vegetable, sugarcane and beef cattle research; beef cattle feeding; mineral nutrition; pasture and forage evaluation. Mailing Add: 1716 SE Ave J Belle Glade FL 33430

BEARDSLEY, GEORGE F, JR, physical oceanography, optics, see 12th edition

BEARDSLEY, GEORGE PETER, b New York, NY, Dec 29, 40; m 72. BIO-ORGANIC CHEMISTRY. Educ: Mass Inst Technol, BS, 67; Princeton Univ, PhD(chem), 71; Duke Univ, MD, 74. Prof Exp: Resident physician pediat, Yale New Haven Hosp, Yale Univ, 74-76; RES FEL PEDIAT HEMAT, CHILDREN'S HOSP MED CTR, HARVARD UNIV, 76- Mem: Am Chem Soc. Res: Bio-organic chemsitry of purines and ptoridinos. Mailing Add: Children's Hosp Med Ctr 300 Longwood Ave Boston MA 02115

BEARDSLEY, GRANT LINDLEY, JR, b Tampa, Fla, Jan 8, 33; m 58; c 4. FISH BIOLOGY. Educ: Davidson Col, BA, 54; Univ Miami, MS, 64, PhD(marine sci), 67. Prof Exp: FISHERY BIOLOGIST, NAT MARINE FISHERIES SERV, US DEPT COMMERCE, 67- Mem: Am Fisheries Soc; Am Soc Ichthyol & Herpet; Am Inst Fishery Res Biol. Res: Biology and ecology of commercially valuable marine fishes, principally tunas. Mailing Add: Southeast Fish Ctr 75 Virginia Beach Dr Miami FL 33149

BEARDSLEY, JOHN WYMAN, JR, b Los Angeles, Calif, Mar 25, 26; m 48; c 4. ENTOMOLOGY. Educ: Univ Calif, BS, 50; Univ Hawaii, MS & PhD, 63. Prof Exp: Asst entomologist, US Trust Territory of Pac Islands, 52-54; from asst to assoc entomologist, Exp Sta, Hawaiian Sugar Planters Asn, 54-72; assoc prof entom, Univ Hawaii, 63-66; assoc res entomologist, Univ Calif, Berkeley, 66-68; assoc prof entom & assoc entomologist, 68-72, PROF ENTOM & ENTOMOLOGIST, UNIV HAWAII, 72-; ENTOMOLOGIST, EXP STA, HAWAIIAN SUGAR PLANTERS ASN, 74- Mem: Am Entom Soc. Res: Taxonomy of scale insects and parasitic wasps; insect biology; sugar cane entomology. Mailing Add: Dept of Entom Univ of Hawaii Honolulu HI 96822

BEARDSLEY, RICHARD KING, b Cripple Creek, Colo, Dec 16, 18; m 42; c 3. CULTURAL ANTHROPOLOGY, ETHNOGRAPHY. Educ: Univ Calif, Berkeley, AB, 39, PhD(anthrop), 47. Prof Exp: From instr to assoc prof, 47-59, PROF ANTHROP, UNIV MICH, ANN ARBOR, 59- Concurrent Pos: Soc Sci Res Coun grant, Japan, 50; Ford Found foreign area res grants, Japan, 53-54, 64-65 & 70; bk rev ed, Am Anthrop Asn, 55-56; Guggenheim Found grant, Spain, 58-59; dir, Ctr Japanese Studies, 61-64 & 71-73; NSF grant, 73-74. Mem: AAAS; Am Anthrop Asn; Am Ethnol Soc; Japanese Soc Ethnol. Res: Japanese society, rural and modernizing; Soviet Asian ethnography; culture change and theories of modernization; anthropology of art. Mailing Add: Dept of Anthrop Univ of Mich 221 Angell Hall Ann Arbor MI 48104

BEARDSLEY, ROBERT CRUCE, b Jacksonville, Fla, Jan 28, 42; m 66; c 2. PHYSICAL OCEANOGRAPHY. Educ: Mass Inst Technol, BS, 64, PhD(oceanog), 68. Prof Exp: From asst prof to assoc prof oceanog, Mass Inst Technol, 67-75; ASSOC SCIENTIST, WOODS HOLE OCEANOG INST, 75- Mem: Am Geophys Union. Res: Geophysical fluid dynamics; coastal oceanography. Mailing Add: Dept of Phys Oceanog Rm 344 Clark Lab Woods Hole Oceanog Inst Woods Hole MA 02543

BEARDSLEY, ROBERT EUGENE, b Walton, NY, June 11, 23; m 48; c 3. MICROBIOLOGY, GENETICS. Educ: Manhattan Col, BS, 50; Columbia Univ, AM, 51, PhD(zool), 60. Prof Exp: From instr to assoc prof, 51-68, dir lab plant morphogenesis, 62-69, PROF BIOL, MANHATTAN COL, 68-, HEAD DEPT, 69- Concurrent Pos: Guggenheim fel, 66. Mem: AAAS; Am Inst Biol Sci; Am Soc Microbiol. Res: Mechanisms of tumor induction in crown gall. Mailing Add: Dept of Biol Manhattan Col New York NY 10471

BEARE, STEVEN DOUGLAS, b Detroit, Mich, May 31, 44; m 65. ORGANIC CHEMISTRY, PHYSICAL CHEMISTRY. Educ: Oakland Univ, BA, 65; Univ Ill, Urbana-Champaign, MS, 67, PhD(org chem), 69. Prof Exp: Teaching asst chem, Univ Ill, 65-67, res asst org chem, 67-69; RES CHEMIST, E I DU PONT DE NEMOURS & CO, 69- Mem: Am Chem Soc. Res: Determinations of optical purities and correlations of absolute configurations of alcohols, hydroxy acids, amino acids, amines, sulfoxides and phosphine oxides by nuclear magnetic resonance spectroscopy in chiral solvents. Mailing Add: Textile Res Lab E I du Pont de Nemours & Co Wilmington DE 19899

BEARE-ROGERS, JOYCE LOUISE, b Ont, Sept 8, 27; m 61; c 1. BIOCHEMISTRY, NUTRITION. Educ: Univ Toronto, BA, 51, MA, 52; Carleton Univ, Ont, PhD(biochem), 66. Prof Exp: Instr physiol, Vassar Col, 54-56; RES SCIENTIST, BUR NUTRIT SCI, FOOD DIRECTORATE, OTTAWA, 56- Mem: Am Inst Nutrit; Am Oil Chemists Soc; Can Inst Food Technol; Can Biochem Soc; Nutrit Soc Can. Res: Effects of dietary components on tissue phospholipids; nutritional aspects of long-chain fatty acids. Mailing Add: Bur of Nutrit Sci Food Directorate Ottawa ON Can

BEARINGER, VAN W, b Orchard, Nebr, Aug 30, 19; m 42; c 2. PHYSICS. Educ: Wayne State Univ, BS, 41; Iowa State Univ, PhD(physics), 50. Prof Exp: Asst, Iowa State Univ, 46-50; supvr res ctr, 50-51, from asst dir to dir, 51-61, gen mgr semiconductor prod div, 61-65, dir eng ord div, 65-66, vpres & gen mgr systs & res div, 66-72, VPRES SCI & ENG, HONEYWELL INC, 72- Mem: Am Phys Soc. Res: Solid state physics; properties of thin metals. Mailing Add: Honeywell Inc Honeywell Plaza Minneapolis MN 55408

BEARMAN, JACOB ELEAZER, b Minneapolis, Minn, June 28, 15; m 42; c 4. BIOMETRICS. Educ: Univ Minn, BA, 36, MA, 38, PhD(ma- th), 47. Prof Exp: Asst math, Univ Minn, 36-39; asst supt, Comt Sch Finance Surv, Minn State Dept Educ, 39-40; statistician, US Bur Census, 40-41; from instr to asst prof math, 46-52, from asst prof to assoc prof biostatist, 53-58, PROF BIOMET, UNIV MINN, MINNEAPOLIS, 58- Concurrent Pos: Mem, Surgeon Gen Adv Comt Epidemiol & Biomet, NIH, 58-62; statist consult, Surgeon Gen Comt Smoking & Health; statist consult, Univ Group Diabetes Prog, 60-; mem policy bd, Coronary Drug Proj, Nat Heart Inst, 63-; mem cancer chemother coop clin trials rev comt, Nat Cancer Inst, 66-68; vis prof, Tel-Aviv Univ, 68-69; mem biomet & epidemiol methodology adv comt, Food & Drug Admin, 70-; consult, Rockefeller Found; mem coun epidemiol, Am Heart Asn. Mem: Fel AAAS; fel Am Heart Asn; Math Asn Am; fel Am Statist Asn; fel Am Pub Health Asn. Res: Public health; methematical, industrial and medical statistics. Mailing Add: Dept of Biomet Sch of Pub Health Univ of Minn Minneapolis MN 55455

BEARMAN, RICHARD JOHN, b New York, NY, June 23, 29; m 61; c 2. PHYSICAL CHEMISTRY, CHEMICAL PHYSICS. Educ: Cornell Univ, AB, 51; Stanford Univ, PhD(chem), 56. Prof Exp: Asst, Yale Univ, 56-57; from asst prof to prof chem, Univ Kans, 57-71; PROF CHEM, UNIV NEW SOUTH WALES, 72- Concurrent Pos: Guggenheim fel, 62-63; vis reader, Univ New Eng, Australia, 68-69; hon fel, Res Sch Phys Sci, Australian Nat Univ, 68-69. Mem: Am Chem Soc; Am Phys Soc; fel Am Inst Chemists. Res: Statistical mechanics and thermodynamics of irreversible processes; membrane permeation; thermal diffusion in liquids; equilibrium theory of fluids; P-V-T measurements. Mailing Add: Fac of Military Studies Univ of New South Wales Duntroon Australia

BEARMAN, TONI CARBO, b Middletown, Conn, Nov 14, 42; m 70. INFORMATION SCIENCE. Educ: Brown Univ, AB, 69; Drexel Univ, MS, 73. Prof Exp: Subj specialist eng, Eng Libr, Univ Wash, 66-67; from libr asst biol sci to supvr phys sci, Phys Sci Libr, Brown Univ, 67-71; res asst curric develop, Grad Sch Libr Sci, Drexel Univ, 71-72; consult comput readable serv, Nat Fedn Abstracting & Indexing, World Inventory Abstracting & Indexing Serv, 73-74; proj coordr eng lit, NSF grant tech-oriented house jours, 72-73; EXEC DIR INFO SCI, NAT FEDN ABSTRACTING & INDEXING SERV, 74- Concurrent Pos: Mem Abstracting Bd, Int Coun Sci Unions US Nat Comt to Comt Int Sci & Tech Info Prog, 74-; mem adv comt, Conversion Sers, 74- Mem: AAAS; Am Soc Info Sci; Spec Libr Asn; Soc Study Sociol Sci. Res: Study of the scattering of scientific literature, particularly in interdisciplinary fields, including a study of the relationship between citations and indexing in discipline-oriented services. Mailing Add: Nat Fedn of Abstr & Index Serv 3401 Market St Philadelphia PA 19104

BEARN, ALEXANDER GORDON, b Cheam, Eng, Mar 29, 23; m 52; c 2. MEDICINE. Educ: Univ London, MB & BS, 46, MD, 51; Hon Degrees: Dr, Univ Rene Descartes, Paris. Prof Exp: Intern, Guy's Hosp, London, 46-47, serv, 47-49; intern & resident, Post-Grad Med Sch, Univ London, 49-51; from asst to assoc, Rockefeller Univ, 51-57, from assoc prof to prof, 57-66, from asst physician to sr physician, 51-66; PROF MED & CHMN DEPT, MED COL, CORNELL UNIV, 66-; PHYSICIAN-IN-CHIEF, NY HOSP, 66- Concurrent Pos: Adj prof, Rockefeller Univ, 66-; mem genetics training comt, USPHS. Mem: Nat Acad Sci; fel AAAS; Am Soc Biol Chemists; Harvey Soc (pres, 72); Soc Human Genetics (pres, 71). Res: Human genetics; internal medicine. Mailing Add: Cornell Univ Med Col 1300 York Ave New York NY 10021

BEARSE, ARTHUR EVERETT, b Attleboro, Mass, Sept 27, 11; m 37; c 4. ORGANIC CHEMISTRY. Educ: Univ Mass, BS, 33; Mass Inst Technol, PhD(org chem), 36. Prof Exp: Res chemist, Jackson Lab, E I du Pont de Nemours & Co, 36-38 & Arthur D Little, Inc, 38-39; res chemist, Battelle Mem Inst, 39-43, from asst supvr to supvr chem res, 43-53, div chief, 53-60, res adv, 60-64, assoc fel, 65-68, sr scientist, 69-76; RETIRED. Mem: Am Chem Soc. Res: Organic chemicals from coal; liquid propellants. Mailing Add: 2732 Northwest Blvd Columbus OH 43221

BEARSE, GORDON EVERETT, b Cambridge, Mass, June 13, 07; m 31; c 2. POULTRY SCIENCE. Educ: Univ Mass, BS, 28. Prof Exp: Asst poultry genetics, Agr Exp Sta, Univ Mass, 29; asst poultryman, poultry husb, Western Wash Res & Exten Ctr, 29-38, poultryman, 38-43, res poultryman, poultry sci, 43-45, poultryman, 46-53, poultry scientist, 53-63, poultry scientist, Dept Animal Sci, 63-72, EMER POULTRY SCIENTIST, WASH STATE UNIV, 72- Mem: Fel AAAS; fel Poultry Sci Asn (vpres, 57-59, pres, 59-60); World Poultry Sci Asn. Res: Effect of nutrition, genetic selection, managemental procedures and environment on performance of chickens and quality of their products. Mailing Add: Space 18 1018 Milwaukee Puyallup WA 98371

BEARSE, ROBERT CARLETON, b Hartford, Conn, May 22, 38; m 63; c 2. APPLIED PHYSICS. Educ: Rice Univ, MA, 62, PhD(physics), 64. Prof Exp: Res assoc physics, Argonne Nat Lab, 66-68, asst physicist, 68-69; from asst prof to assoc prof physics, 69-75, PROF PHYSICS, UNIV KANS, 75-, ASSOC DEAN RES ADMIN, 75- Concurrent Pos: Vis staff mem, Los Alamos Sci Lab, NMex, 72- Mem: Sigma Xi; AAAS; Am Phys Soc. Res: Trace element analysis in biological materials; applications of nuclear physics in other disciplines. Mailing Add: Dept of Physics Univ of Kans Lawrence KS 66045

BEARY, DEXTER F, b Battle Creek, Mich, Nov 1, 24; m 47; c 2. ANATOMY, BIOLOGY. Educ: Andrews Univ, AB, 51; Western Mich Univ, MA, 59; Loma Linda Univ, PhD(anat), 67. Prof Exp: Res asst pharmaceut, Upjohn Co, 51-59; instr biol, 59-64, CHMN DEPT BIOL, SOUTHWESTERN UNION COL, 64- Mem: Am Asn Anat; Am Inst Biol Sci. Res: Gross anatomy; histology; osteoporosis. Mailing Add: Dept of Biol Southwestern Union Col Keene TX 76059

BEASLEY, ANDREW BOWIE, b Upper Zion, Va, Sept 7, 31; div; c 2. ANATOMY, GENETICS. Educ: Va Polytech Inst, BS, 52, MS, 56; Med Col Pa, ScD(anat & genetics), 61. Prof Exp: Asst zool, Va Polytech Inst, 54-55; res asst genetics, Columbia Univ, 55-58; instr, 61-63, asst prof, 63-69, actg chmn dept, 71-73, ASSOC PROF ANAT, MED COL PA, 69-, ASST DEAN STUDENT AFFAIRS, 75- Mem: AAAS; Am Asn Anatomists; Am Soc Zoologists. Res: Developmental genetics and anatomy of central nervous system. Mailing Add: Dept of Anat Med Col of Pa Philadelphia PA 19129

BEASLEY, C A (BUD), b Twin Falls, Idaho, July 3, 32; m 56; c 4. PLANT PHYSIOLOGY, BIOCHEMISTRY. Educ: Univ Idaho, BS, 61; SDak State Univ, PhD(agron), 64. Prof Exp: Prod mgr herbicides & insecticides, Wasatch Chem Co, 63-64; plant physiologist, Stanford Res Inst, 64-70; RES PLANT PHYSIOLOGIST, UNIV CALIF, RIVERSIDE, 70- Mem: Am Soc Plant Physiol; Bot Soc Am; Scand Soc Plant Physiol; Int Asn Plant Tissue Cult. Res: Biochemical techniques and in vitro culture methods to study the chemical basis of differentiation and morphogenetic development of plants; plant hormones; cell wall biosynthesis. Mailing Add: Dept of Plant Sci Univ of Calif Riverside CA 92502

BEASLEY, CLARK WAYNE, b Pittsburg, Kans, June 6, 42; m 65. INVERTEBRATE ZOOLOGY. Educ: Kans State Col Pittsburg, BS, 64; Univ Okla, PhD(zool), 68. Prof Exp: Instr biol, Mo Southern State Col, 68-69; asst prof, 69-72, ASSOC PROF BIOL, McMURRY COL, 72-, CHMN DEPT, 73- Mem: AAAS; Am Inst Biol Sci. Res: Invertebrate taxonomy, especially Tardigrada. Mailing Add: Dept of Biol McMurry Col Abilene TX 79605

BEASLEY, CLOYD O, JR, b Florence, Ala, July 9, 33. PLASMA PHYSICS. Educ: Vanderbilt Univ, BA, 55, MA, 57; Univ Wis, PhD(physics), 62. Prof Exp: Appl sci rep physics, Int Bus Mach Corp, 57-58; PHYSICIST, THERMONUCLEAR DIV, OAK RIDGE NAT LAB, 62- Concurrent Pos: Res assoc, Culham Lab, UK, 66-67. Mem: Am Phys Soc. Res: Plasma micro-instability theory; micro-instabilities in mirror-confined plasmas; combined numerical-analytical approach to investigation of equilibria and stability on ToKa mathematics. Mailing Add: Oak Ridge Nat Lab 110 S Purdue Ave Oak Ridge TN 37830

BEASLEY, EDWARD EVANS, b Oakland, Calif, Mar 19, 24; m 49; c 1. PHYSICS. Educ: US Naval Acad, BS, 44; Univ Md, MS, 57, PhD(physics), 63. Prof Exp: Asst physics, Univ Md, 56-60; PROF PHYSICS & CHMN DEPT, GALLAUDET COL, 60- Mem: Am Phys Soc; Am Asn Physics Teachers. Res: Electron spin resonance in ultraviolet- irradiated aliphatic hydrocarbon glasses at liquid nitrogen temperature. Mailing Add: Dept of Physics Gallaudet Col Washington DC 20002

BEASLEY, JAMES DONALD, physics, see 12th edition

BEASLEY, JAMES GORDON, b Tela, Honduras, Nov 13, 28; US citizen; m 63; c 1. ORGANIC CHEMISTRY, MEDICINAL CHEMISTRY. Educ: Auburn Univ, BS, 51, MS, 55; Univ Va, PhD(org chem), 62. Prof Exp: Asst gen chem, Auburn Univ, 53-55; assoc prof chem, Memphis State Univ, 59-62; res assoc pharmaceut chem, Col Pharm, Univ Tenn, 62-63, asst prof pharmaceut & med chem, 63-65, assoc prof med chem, 65-71, dir chem enzym lab, 68-71; PROF CHEM, LAMBUTH COL, 71- Concurrent Pos: Asst, Univ Va, 59; co-investr, NSF grant, 64-66, co-prin investr, NSF res grant, 66-72; res grants, Marion Labs, 66-72 & A H Robins, 70- Mem: Am Chem Soc; Am Pharmaceut Soc; The Chem Soc. Res: Organic synthesis and biochemical evaluation of compounds with pharmacodynamic potential; correlation of molecular constitution and biochemical response; chemical enzymology. Mailing Add: Dept of Chem Lambuth Col Jackson TN 38301

BEASLEY, JOSEPH NOBLE, b Centerton, Ark, Mar 11, 24; m 56; c 1. Educ: Tex A&M Univ, DVM, 49, MS, 56; Univ Okla, PhD(med sci), 64. Prof Exp: Instr vet sci, Univ Ark, 49-51; instr vet med & surg, Tex A&M Univ, 51-52; pathologist, Ark Livestock Sanit Bd, 52-55; from asst prof to assoc prof vet path, Tex A&M Univ, 56-64; PROF ANIMAL SCI, UNIV ARK, FAYETTEVILLE, 68- Mem: Am Vet Med Asn; Am Col Vet Path; Int Acad Path; Conf Res Workers Animal Dis; Am Asn Avian Path. Res: Comparative pathology of psittacosis ornithosis; canine dirofilariasis. Mailing Add: Dept of Animal Sci Univ of Ark Fayetteville AR 72701

BEASLEY, MALCOLM ROY, b San Francisco, Calif, Jan 4, 40; m 62; c 3. LOW TEMPERATURE PHYSICS. Educ: Cornell Univ, BS, 62, PhD(physics), 68. Prof Exp: Res fel eng & appl physics, Harvard Univ, 67-69, from asst prof to assoc prof, 69-74; ASSOC PROF APPL PHYSICS & ELEC ENG, STANFORD UNIV, 74- Mem: Am Phys Soc. Res: Basic and applied superconductivity. Mailing Add: Dept of Appl Physics Stanford Univ Stanford CA 94305

BEASLEY, PHILIP GENE, b Harrisburg, Ill, Dec 22, 27; m 52. ANIMAL PHYSIOLOGY, PLANT PHYSIOLOGY. Educ: Washington Univ, AB, 49; Auburn Univ, MS, 62, PhD(zool), 67. Prof Exp: With Fed Bur Invest, 50-59; instr fisheries, Auburn Univ, 60-66; from asst prof to assoc prof biol, 66-71, PROF BIOL & CHMN DEPT, UNIV MONTEVALLO, 71- Mem: NY Acad Sci; AAAS. Mailing Add: Dept of Biol Univ of Montevallo Montevallo AL 35115

BEASLEY, THOMAS M, oceanography, radiochemistry, see 12th edition

BEASLEY, WILLIAM JOE, b Wharton, Tex, Sept 19, 43; m 65; c 1. MEDICAL MICROBIOLOGY, IMMUNOLOGY. Educ: Northwestern State Univ, La, BS, 65, MS, 67; Univ Ga, PhD(microbiol), 70. Prof Exp: Fel med microbiol, Col Med, Univ Fla, 70-72; instr, Fla Jr Col, Jacksonville, 72-73; microbiologist immunol, Naval Med Res Unit No One, 73-74; MICROBIOLOGIST IMMUNOL, NAVAL BIOMED RES LAB, 74- Mem: Am Soc Microbiol; Sigma Xi. Res: Immunobiology of infections induced by gram-negative organisms; pathogenic mechanisms of infections caused by gram-negative organisms; automated, rapid serological identification of bacteria. Mailing Add: Naval Biomed Res Lab Bldg 844 Naval Supply Ctr Oakland CA 94625

BEATLEY, JANICE CARSON, b Columbus, Ohio, Mar 18, 19. PLANT ECOLOGY. Educ: Ohio State Univ, BA, 40, MSc, 48, PhD(bot), 53. Prof Exp: High sch teacher, Ohio, 43-45; asst instr & asst, Ohio State Univ, 45-54, instr, 55-56, Muellhaupt scholar, 57-58; asst prof, ECarolina Col, 54-55 & NC State Col, 56-57; res assoc, NMex Highlands Univ, 59; asst res ecologist, Lab Nuclear Med & Radiation Biol, Univ Calif, Los Angeles, 60-67, assoc res ecologist, 67-72; ASSOC PROF BIOL SCI, UNIV CINCINNATI, 73- Concurrent Pos: Instr & asst prof, Univ Tenn, 52-60. Mem: Ecol Soc Am; Am Soc Plant Taxon. Res: Desert and radiation ecology; ecology of deciduous forest tree species and wintergreen herbs. Mailing Add: Dept of Biol Sci Univ of Cincinnati Cincinnati OH 45221

BEATON, ALBERT E, b Boston, Mass, Aug 9, 31; c 1. STATISTICS. Educ: State Teachers Col Boston, BS, 55; Harvard Univ, MEd, 56, EdD, 64. Prof Exp: Res asst, Harvard Univ, 56-57, managing dir, Littauer Statist Lab, 57-59, dir, Harvard Statist Lab, 59-62, actg mgr, Harvard Comput Ctr, 62, res assoc educ & IBM res fel, 62-64; adv statist & data anal, 64-70, DIR OFF DATA ANAL RES, EDUC TESTING SERV, 70-; VIS RES STATISTICIAN & VIS LECTR STATIST, PRINCETON UNIV, 66- Concurrent Pos: Res assoc, Educ Res Corp, 56-57; vis lectr, Sch Educ & Grad Sch, Boston Col, 60-61; pres, Albert E Beaton Assocs, Inc; consult, Cabot Corp, Ceir, Inc, Cides, Dominican Repub, Gen Elec Corp, Gillette Safety Razor Corp, Infrared Industs, Can Ministry of Youth, Nat Bur Econ Res, Prudential Life Ins Co, RCA Corp, S C Johnson & Co, Union Carbide Corp, US Res Corp, US Off Educ & US Off Econ Opportunity. Mem: Am Educ Res Asn; Am Psychol Asn; Am Sociol Asn; Am Statist Asn; Asn Comput Mach. Mailing Add: Off of Data Anal Res Educ Testing Serv Princeton NJ 08540

BEATON, GEORGE HECTOR, b Oshawa, Ont, Dec 20, 29; m 53; c 3. NUTRITION, BIOCHEMISTRY. Educ: Univ Toronto, BA, 52, MA, 53, PhD(nutrit), 55. Prof Exp: From asst prof to assoc prof nutrit, Sch Hyg, 55-63, prof & head dept, 63-75, actg dir, Sch Hyg, 74-75, PROF NUTRIT & FOOD SCI & CHMN DEPT, FAC MED, UNIV TORONTO, 75-, ACTG DEAN FAC FOOD SCI, 75- Concurrent Pos: WHO fel, Inst Nutrit Cent Am & Panama, 61; mem expert adv comt nutrit in pregnancy & lactation & expert adv panel nutrit, WHO, 64, mem expert comt nutrit, 70, mem expert group in vitamin requirements, Food & Agr Orgn-WHO, 67, mem expert group vitamin & mineral requirements, 69, mem expert comt protein & energy requirements, 71. Honors & Awards: Borden Award, Nutrit Soc Can, 68. Mem: Am Inst Nutrit; Nutrit Soc Can (secy, 57-60, vpres, 64-65, pres, 65-66); Can Physiol Soc. Res: Nutrition in pregnancy; biochemical assessment of nutritional status; intermediary metabolism and essential nutrients; public health applications of nutrition information; interpretation of nutrient requirements. Mailing Add: Dept of Nutrit & Food Sci Univ of Toronto Fac of Med Toronto ON Can

BEATON, JAMES DUNCAN, b Vancouver, BC, Aug 28, 30; m 52; c 3. SOIL FERTILITY. Educ: Univ BC, BSA, 51, MSA, 53; Utah State Univ, PhD(soils), 57. Prof Exp: Lab asst soil bact, Univ BC, 51-53, spec lectr, Dept Soil Sci, 56, instr, 57-59; phys chemist, Soil Sect, Exp Farm, Can Dept Agr, 59-61; soil scientist, Res & Develop Div, Consol Mining & Smelting Co Can, Ltd, 61-64, head soil sci res, Res & Corp Develop, 64-67; sr agronomist, Cominco Ltd, BC, 67-68; dir agr res, Sulphur Inst, 68-73; CHIEF AGRONOMIST, COMINCO LTD, ALTA, 73- Mem: Soil Sci Soc Am; Can Soc Soil Sci; Brit Soc Soil Sci; Int Soc Soil Sci; fel Am Soc Agron. Res: Soil fertility and reactions of fertilizers in soils; forest fertilization. Mailing Add: Cominco Ltd Suite 133 7330 Fisher St SE Calgary AB Can

BEATON, JOHN MCCALL, b Huntly, Scotland, June 21, 44; m 68. PSYCHOPHARMACOLOGY. Educ: Univ Aberdeen, BSc, 66, MSc, 69; Univ Ala, Birmingham, PhD(physiol), 73. Prof Exp: Sr res assoc psychol, Addiction Res Found, Toronto, 69-71; from res assoc to instr, 71-74, ASST PROF PSYCHIAT, UNIV ALA, BIRMINGHAM, 74-, PHARMACOL, 75- Mem: Brit Psychol Asn; Asn Psychophysiol Study Sleep; Soc Neurosci. Res: Psychopharmacological study of hallucinogenic agents using operant conditioning techniques; animal models of psychoses especially with relationship to schizophrenia; the effects of amino acid loadings on behavior. Mailing Add: Neurosci Prog Univ Sta Birmingham AL 35294

BEATON, JOHN ROGERSON, b Oshawa, Ont, Sept 7, 25; m 48; c 4. NUTRITION. Educ: Univ Toronto, BA, 49, MA, 50, PhD(nutrit), 52. Prof Exp: Res fel nutrit, Univ Toronto, 49-51, res assoc, 51-52, asst prof, 52-55, lectr food chem, 50-55; defence sci serv officer, Defence Res Bd, Can, 55-59, sect head, Defence Res Med Labs, 59-63; prof physiol, Univ Western Ont, 63-67; prof nutrit & chmn div, Univ Hawaii, 67-69; dean col human biol, Univ Wis-Green Bay, 69-70, dean cols, 70-75; DEAN COL HUMAN ECOL, UNIV MD, COLLEGE PARK, 75- Mem: Soc Exp Biol & Med; NY Acad Sci; Am Soc Biol Chemists; Am Inst Nutrit; Nutrit Soc Can. Res: Vitamin B-6; metabolic functions; amino acid metabolism in pregnancy and malignancy; physiology of cold exposure and hypothermia; metabolic effects of hormones; effects of exercise; fat mobilizing substances in urine; regulation of food intake; physiology and biochemistry of hyperphagia. Mailing Add: Univ of Md College Park MD 20742

BEATTIE, ALAN GILBERT, b Oakland, Calif, Apr 2, 34; m 61; c 3. SOLID STATE PHYSICS. Educ: Univ Calif, Berkeley, BA, 59; Univ Wash, MS, 60, PhD(physics), 65. Prof Exp: Res engr, Algae Res Proj, Univ Calif, Berkeley, 59-60; MEM STAFF, SANDIA CORP, 65- Mem: Am Phys Soc. Res: Electron structure of metals; ultrasonics; acoustic emission. Mailing Add: Orgn 9352 Sandia Labs Albuquerque NM 87115

BEATTIE, DIANA SCOTT, b Cranston, RI, Aug 11, 34; m 56; c 4. BIOCHEMISTRY. Educ: Swarthmore Col, BA, 56; Univ Pittsburgh, MS, 58, PhD(biochem), 61. Prof Exp: Res assoc biochem, Univ Pittsburgh, 61-68; asst prof, 68-70, assoc prof, 70-75, PROF BIOCHEM, MT SINAI SCH MED, 76- Mem: AAAS; Am Soc Biol Chemists; Am Soc Cell Biologists; Biophys Soc. Res: Mitochondrial metabolism, including various aspects of fatty acid metabolism and protein synthesis, enzymology and metabolic control mechanisms. Mailing Add: Dept of Biochem Mt Sinai Sch of Med New York NY 10029

BEATTIE, EDWARD J, b Philadelphia, Pa, June 30, 18; m 48; c 1. SURGERY. Educ: Princeton Univ, BA, 39; Harvard Univ, MD, 43; Am Bd Thoracic Surg, dipl, 51. Prof Exp: Resident surg, Peter Bent Brigham Hosp, Boston, 42-43, from asst resident surgeon to resident surgeon, 44-45, jr assoc surg, 45-47; fel, George Washington Univ, 47-48, asst prof & Markle scholar, 48-52; from asst prof to prof, Univ Ill, 52-65; PROF SURG, CORNELL UNIV, 65-; CHIEF THORACIC SURG, MEM HOSP, NEW YORK, 65-, CHMN DEPT SURG & CHIEF MED OFFICER, 66-, GEN DIR & CHIEF EXEC OFFICER, 74- Concurrent Pos: Instr, Harvard Univ, 45-47; Moseley travel fel from Harvard Univ, Postgrad Med Sch, Univ London, 46-47; consult, Gallinger Munic Hosp, Washington, DC, Walter Reed Army Hosp, Newton D Baker & Mt Alto Vet Admin Hosps, 48-52; consult, Chicago State Tuberc Sanitorium, Univ Ill Hosps & Hines Vet Admin Hosp, 52-65; thoracic surgeon, Presby-St Luke's Hosp, Chicago, 54-65; mem bd, Am Bd Thoracic Surg, 60-65, from vchmn to chmn, 65-69; mem, Sloan Kettering Inst, 66- Mem: Am Asn Thoracic Surg; Am Col Surg; Soc Vascular Surg; Soc Clin Surg; Am Surg Asn. Res: Thoracic surgery. Mailing Add: Mem Hosp 1275 York Ave New York NY 10021

BEATTIE, JAMES KENNETH, inorganic chemistry, see 12th edition

BEATTIE, JAMES MONROE, b Washington, DC, Feb 14, 21; m 42; c 2. HORTICULTURE. Educ: Univ Md, BS, 41; Cornell Univ, PhD(pomol), 48. Prof Exp: From asst prof to prof hort, Ohio Agr Exp Sta, 48-63, from asst dir to assoc dir, Ohio Agr Res & Develop Ctr, 63-73; DEAN COL AGR, DIR AGR EXP STA & DIR COOP EXTEN SERV, PA STATE UNIV, UNIVERSITY PARK, 73- Mem: Fel AAAS; fel Am Soc Hort Sci (pres, 69-70); Am Inst Biol Sci. Res: Nitrogen and mineral nutrition of fruit crops; reclamation of strip mine spoils; culture and management of fruit plantings. Mailing Add: 201 Agr Admin Bldg Pa State Univ University Park PA 16802

BEATTIE, ROBERT WALTER, b Owen Sound, Ont, Mar 12, 02; m 38. CHEMISTRY. Educ: Queen's Univ, Ont, BSc, 24, MSc, 25; Pa State Univ, PhD(chem), 32. Prof Exp: Demonstr chem eng & appl chem, Univ Toronto, 27-28; org res, Halowax Corp, 31-41; fel, Mellon Inst, 41-43; org res, Dominion Rubber Co, 43-75; RETIRED. Res: Metallic thorium; thorium alloys; halogenated naphthalenes; mercurated naphthalene; chlorination; extreme pressure lubricants; organic weed killers; insecticides. Mailing Add: 44 Stuart St Guelph ON Can

BEATTIE, THOMAS ROBERT, b Philadelphia, Pa, Aug 31, 40; m 61; c 3. ORGANIC CHEMISTRY. Educ: Univ Pa, BS, 61; Univ Wis, PhD(org chem), 65. Prof Exp: Sr res chemist, 66-73, RES FEL, MERCK & CO, INC, 73- Mem: Am Chem Soc. Res: Medicinal and synthetic organic chemistry. Mailing Add: 278 Jefferies Pl North Plainfield NJ 07060

BEATTIE, WILLARD HORATIO, b Oak Park, Ill, Mar 3, 27; m 54; c 3. PHYSICAL CHEMISTRY. Educ: Univ Chicago, BA, 51, MS, 54; Univ Minn, PhD(anal chem), 58. Prof Exp: Res chemist, Shell Chem Co, 58-62; sr res chemist, Beckman Instruments, Inc, 62; asst prof chem, Calif State Col, Long Beach, 62-67; MEM STAFF, LOS ALAMOS SCI LAB, 67- Concurrent Pos: Res prof, Univ Calif, Santa Barbara, 75-76. Mem: Am Chem Soc; Am Inst Chem. Res: Light scattering; chemical laser development; laser isotope separation; analytical and radiation chemistry of gases; instrumentation. Mailing Add: Group L-8 Los Alamos Sci Lab Box 1663 Los Alamos NM 87544

BEATTY, ALICE FERGUSON, b Dallas, Tex, Mar 5, 15; m 56. ZOOLOGY, ENTOMOLOGY. Educ: Southern Methodist Univ, BS, 36, MS, 40; La State Univ, PhD(entom), 55. Prof Exp: Asst prof biol, ETex State Col, 48-56; from asst prof to assoc prof zool, 56-74, PROF ZOOL, PA STATE UNIV, UNIVERSITY PARK, 74- Concurrent Pos: NSF grant, Mt Lake Biol Sta, Univ Va, 55; Danforth Found scholar, Pa State Univ, 55. Res: Odonata; morphology, systematics; biogeography, ecology, especially larvae of Neotropical Region. Mailing Add: Dept of Zool Pa State Univ University Park PA 16802

BEATTY, CHARLES LEE, b Kempton, Ind, Nov 7, 39; m 72; c 2. POLYMER PHYSICS, POLYMER SCIENCE. Educ: Purdue Univ, BS, 65; Case Western Reserve Univ, MS, 68; Univ Mass, MS, 71, PhD(polymer sci & eng), 72. Prof Exp: RES SCIENTIST POLYMER SCI, WEBSTER RES CTR, XEROX CORP, 71- Concurrent Pos: Assoc prof polymer sci, Univ Rochester, 76-77. Mem: Am Inst Chem Engrs; Mat Res Soc; Am Chem Soc; Am Phys Soc. Res: Solid-state properties of polymers ranging from diffusion to calorimetry; mechanical properties in the non-linear viscoelastic region. Mailing Add: Webster Res Lab Xerox Corp W-114 800 Phillips Rd Webster NY 14580

BEATTY, CLARISSA HAGER, b Colorado Springs, Colo, June 3, 19; m 43; c 2. PHYSIOLOGY. Educ: Sarah Lawrence Col, AB, 41; Columbia Univ, MA, 42, PhD(physiol), 45. Prof Exp: Asst physiol, Columbia Univ, 43-45, instr, 45-48; fel, Diabetic Res Found, Med Sch, Univ Ore, 48-53, from instr to asst prof, 53-62; assoc scientist, 61-67, SCIENTIST, ORE REGIONAL PRIMATE RES CTR, 67-; ASSOC PROF BIOCHEM, MED SCH, UNIV ORE, 62- Mem: Am Chem Soc; Am Physiol Soc; Soc Exp Biol & Med; Soc Study Reproduction. Res: Fetal muscle metabolism; smooth muscle metabolism. Mailing Add: Ore Regional Primate Res Ctr 505 NW 185th Ave Beaverton OR 97005

BEATTY, DAVID DELMAR, b Bellingham, Wash, Oct 10, 35; m 57; c 3. COMPARATIVE PHYSIOLOGY. Educ: Western Wash State Col, BA, 57; Univ Wyo, MSc, 59; Univ Ore, PhD(animal physiol), 64. Prof Exp: Instr zool, Univ Wyo, 59-60; asst prof, 64-70, ASSOC PROF ZOOL, UNIV ALTA, 70- Mem: AAAS; Am Soc Zool; Can Soc Zool. Res: Biochemistry of fish visual pigment; physiology of fish. Mailing Add: Dept of Zool Univ of Alta Edmonton AB Can

BEATTY, GEORGE FRANKLIN, b Cowan, Ind, June 9, 18. GEOGRAPHY OF ASIA. Educ: Ball State Univ, BS, 49; Univ Ill, MS, 51, PhD(geog), 58. Prof Exp: Asst prof geog & geol, Univ Tenn, Martin Br, 55-56; asst prof geog, Northern Ill Univ, 56-58; assoc prof coordr of sect, 58-65, chmn dept geog & geol, 65-73, PROF GEOG, BALL STATE UNIV, 65- Concurrent Pos: Fulbright fac res grant, Univ Calcutta, 63-64. Mem: Asn Am Geog; Am Meteorol Soc; Geog Soc India. Res: Regional development, with emphasis on water resources and agriculture and most work directed to Asia. Mailing Add: Dept of Geog & Geol Ball State Univ Muncie IN 47306

BEATTY, GLENN HURST, b Columbus, Ohio, Aug 2, 18; m 45; c 3. STATISTICS. Educ: Ohio State Univ, BA, 40; Iowa State Col, MS, 47. Prof Exp: Prin res statistician, 48-58, SR STATISTICIAN, BATTELLE MEM INST, 58- Mem: Am Statist Asn. Res: Experimental design and analysis in scientific research for government and industry. Mailing Add: Battelle Mem Inst 505 King Ave Columbus OH 43201

BEATTY, JAMES ROGER, b Iola, Kans, Apr 9, 17; m 45; c 4. PHYSICS. Educ: Kans State Teachers Col, BA, 43. Prof Exp: RES PHYSICIST, B F GOODRICH RES CTR, 43- Mem: Am Phys Soc; Am Chem Soc. Res: Physics of rubber elasticity; product performance versus rubber physical properties; compounding and evaluation of rubbers and test development. Mailing Add: Res Ctr B F Goodrich Co Brecksville OH 44141

BEATTY, JAMES WAYNE, JR, b Fargo, NDak, Sept 9, 34; m 59; c 3. PHYSICAL CHEMISTRY. Educ: NDak State Univ, BSc, 56; Mass Inst Technol, PhD(phys chem), 60. Prof Exp: From instr to asst prof physics, Colby Col, 60-63; from asst prof to assoc prof chem, 63-74, PROF CHEM, RIPON COL, 74- Mem: Am Chem Soc; Am Phys Soc. Res: Measurement of transport properties of gases, particularly diffusion coefficients; determination of force law constants from this data. Mailing Add: Dept of Chem Ripon Col Ripon WI 54971

BEATTY, JOHN JOSEPH, b Brooklyn, NY, Sept 5, 39. ANTHROPOLOGY. Educ: Brooklyn Col, BA, 65; Univ Okla, MS, 66; City Univ New York, PhD(anthrop), 72. Prof Exp: Instr, 71-72, ASST PROF ANTHROP, BROOKLYN COL, 72- Concurrent Pos: Consult, Off Native Am Progs, Dept Health, Educ & Welfare, 74-75. Mem: Fel NY Acad Sci; Am Anthrop Asn; Sigma Xi; AAAS; Animal Behav Soc. Res: Relationship between language and culture; cultural factors involved in sex behavior; acculturation and culture change. Mailing Add: 2983 Bedford Ave Brooklyn NY 11210

BEATTY, KENNETH WILSON, b Oklahoma City, Okla, Feb 22, 29; m 65; c 3. ZOOLOGY, LIMNOLOGY. Educ: Univ Calif, Davis, BS, 52, MA, 60, PhD(limnol), 68. Prof Exp: Lab technician animal husb, Univ Calif, Davis, 54-60; INSTR BIOL, COL SISKIYOUS, 60- Concurrent Pos: Biologist, US Geol Surv, 71-73. Mem: Am Soc Limnol & Oceanog; Ecol Soc Am; Am Inst Biol Sci; Nat Asn Biol Teachers; Am Nature Study Soc. Res: Ecology of new reservoirs; benthic ecology of alpine lakes; behavior and reproductive cycles of woodpeckers; mastitis studies under controlled management practices. Mailing Add: Dept of Biol Col of the Siskiyous Weed CA 96094

BEATTY, MARVIN THEODORE, b Bozeman, Mont, Mar 13, 28; m 56; c 4. SOILS, RESOURCE MANAGEMENT. Educ: Mont State Col, BS, 50; Univ Wis, PhD(soils), 55. Prof Exp: Soil scientist, Soil Conserv Serv, 50-56; from asst prof to assoc prof soils, 56-66, chmn environ resources unit, 70-72, PROF SOILS, UNIV WIS-MADISON, 66-, STATE PROG CHMN, NATURAL & ENVIRON RESOURCES, UNIV EXTEN, 72- Mem: Am Soc Agron; Soil Conserv Soc Am; Nat Univs Exten Asn. Res: Soil surveys and the interpretation of physical and chemical properties of soils as they affect sustained use of soils for agriculture; engineering; urban development; soil conservation; natural resource and environmental education programs. Mailing Add: 501 Exten Bldg Univ Exten Univ of Wis Madison WI 53706

BEATTY, MILLARD FILLMORE, JR, b Baltimore, Md, Nov 13, 30; m 51; c 3. CONTINUUM MECHANICS. Educ: Johns Hopkins Univ, BES, 59, PhD(mech), 64. Prof Exp: Instr eng physics, Loyola Col, Md, 59-60; instr mech, Johns Hopkins Univ, 60-63; asst prof continuum mech, Univ Del, 63-67; ASSOC PROF THEORET MECH, UNIV KY, 67- Mem: Soc Natural Philos; Math Asn Am; Am Asn Physics Teachers. Res: Non-linear elasticity and elastic stability theory; theory of couple-stresses; foundations of mechanics. Mailing Add: Dept of Eng Mech Univ of Ky Lexington KY 40506

BEATTY, OREN ALEXANDER, b Nobob, Ky, July 3, 01; m 33; c 9. MEDICAL ADMINISTRATION, PULMONARY DISEASES. Educ: Univ Louisville, MS, 30. Prof Exp: Med dir & supt, Richland Hosp, Mansfield, Ohio, 42-51, Hazelwood Sanatorium, Louisville, Ky, 51-62 & Richland Hosp, 62-72; tuberc controller Lorain County, Lorain County Tuberc Clin, 68-72; CONSULT, 72- Concurrent Pos: Asst prof, Sch Med, Univ Louisville, 52-62. Mem: AMA; fel Am Col Chest Physicians; fel Am Thoracic Soc. Res: Diseases of the chest; undulant fever in the respiratory tract; pulmonary cavities. Mailing Add: 3717 Hanover Rd Louisville KY 40207

BEATY, CHESTER BROOMELL, b Chicago, Ill, May 10, 25; m 47; c 2. GEOGRAPHY. Educ: La State Univ, BA, 48, MA, 40; Univ Calif, PhD(geog), 60. Prof Exp: Asst geog, La State Univ, 48-50 & Univ Calif, 53-56; res geogr, QmC, US Dept Army, 56-57; asst geog, Univ Calif, 58; from instr to prof, Univ Mont, 58-69, chmn dept, 63-69; PROF GEOG, UNIV LETHBRIDGE, 69- Mem: AAAS; Asn Am Geog; Geol Soc Am. Res: Geomorphology; desert land forms; mountain glaciation; Anglo-America geography; resource conservation. Mailing Add: Dept of Geography Univ of Lethbridge Lethbridge AB Can

BEATY, EARL CLAUDE, b Zeigler, Ill, Nov 6, 30; m 52; c 3. PHYSICS. Educ: Murray State Col, AB, 52; Washington Univ, PhD, 56. Prof Exp: PHYSICIST, NAT BUR STAND, 56-; FEL, JOINT INST LAB ASTROPHYS, UNIV COLO, BOULDER, 62- Concurrent Pos: Nat Res Coun fel, 56-57. Mem: Am Phys Soc. Res: Atomic physics. Mailing Add: Joint Inst for Lab Astrophys Univ of Colo Boulder CO 80302

BEATY, ELVIS ROY, b Ritz, Ark, Aug 10, 22; m; c 6. AGRONOMY. Educ: Okla State Univ, BS, 50; Miss State Univ, MS, 51, PhD(agron), 54. Prof Exp: From asst agronomist to assoc agronomist, 54-68, PROF AGRON, UNIV GA, 68- Mem: Am Soc Agron; Soc Range Mgt; Crop Sci Soc Am; Soil Conserv Soc Am. Res: Forage production and utilization; biomass accumulation; growth analysis. Mailing Add: Dept of Agron Univ of Ga Athens GA 30602

BEATY, MARJORIE HECKEL, b Buffalo, NY, Jan 21, 06; m 33; c 2. MATHEMATICS. Educ: Univ Rochester, AB, 28, MA, 29; Univ Colo, PhD(math), 39. Prof Exp: Asst math, Brown Univ, 29-31; instr, Univ SDak, 31-35; asst, Univ Colo, 35-36 & 37-38; actg head dept, 38-39, asst prof, 39-41 & 55-58, assoc prof, 58-61, actg dean women, 41, PROF MATH, UNIV S DAK, 61- Mem: Am Math Soc. Res: Complex roots of algebraic equations. Mailing Add: 314 Canby St Vermillion SD 57069

BEAUCHAMP, ERIC G, b Grenville, Que, Jan 6, 36; m 58; c 3. SOIL FERTILITY, PLANT NUTRITION. Educ: McGill Univ, BScAgr, 60, MSc, 62; Cornell Univ, PhD(soil sci), 65. Prof Exp: Res scientist, Res Br, Can Dept Agr, 65-67; asst prof soil sci, 67-70, ASSOC PROF SOIL SCI, UNIV GUELPH, 70- Mem: Am Soc Agron; Agr Inst Can; Can Soc Soil Sci; Int Soc Soil Sci; Can Soc Agron. Res: Response of field crops to fertilizers; soil nitrogen transformations. Mailing Add: Dept of Land Resource Sci Univ of Guelph Guelph ON Can

BEAUCHAMP, GARY KEITH, b Belvidere, Ill, Apr 5, 43; m 67; c 1. ANIMAL BEHAVIOR. Educ: Carleton Col, BA, 65; Univ Chicago, PhD(biopsychol), 71. Prof Exp: Fel chemosensation, 71-73, ASSOC MEM, MONELL CHEM SENSES CTR, UNIV PA, 73-, ASST PROF PSYCHOL, DEPT OTORHINOLARYNGOL & HUMAN COMMUN, MED SCH, 74- Mem: AAAS; Am Psychol Asn; Animal Behav Soc. Res: Investigation of the role of the chemical senses in regulating behavior and physiology in a variety of animal species, including humans. Mailing Add: Monell Chem Senses Ctr Univ of Pa 3500 Market St Philadelphia PA 19104

BEAUCHAMP, JESSE LEE, b Glendale, Calif, Nov 1, 42; m 64; c 1. PHYSICAL CHEMISTRY. Educ: Calif Inst Technol, BS, 64; Harvard Univ, PhD(chem), 67. Prof Exp: Noyes instr, 67-69, from asst prof to assoc prof, 69-74, PROF CHEM, CALIF INST TECHNOL, 74- Concurrent Pos: Alfred P Sloan Found fel, 68-70. Mem: Am Phys Soc; Am Chem Soc. Res: Gas phase ion chemistry; structures, properties and reactions of organic ions; chemical applications of ion cyclotron resonance spectroscopy. Mailing Add: A A Noyes Lab Calif Inst of Technol Pasadena CA 91109

BEAUCHAMP, JOHN J, b Nashville, Tenn, Sept 17, 37. STATISTICS, MATHEMATICS. Educ: Vanderbilt Univ, BA, 59, MAT, 60; Fla State Univ, MS, 63, PhD(statist), 66. Prof Exp: Asst prof math, Birmingham-Southern Col, 60-61; biomet trainee statist, Vanderbilt Univ, 61-62; statistician, Oak Ridge Nat Lab, 67-73; STATISTICIAN, COMPUT SCI DIV, NUCLEAR DIV, UNION CARBIDE CORP, 73- Mem: Biomet Soc; Am Statist Asn. Res: Nonlinear estimation; model building; canonical analysis. Mailing Add: Union Carbide Corp Nuclear Div PO Box Y Bldg 9704-1 Oak Ridge TN 37830

BEAUCHAMP, NICHOLAS ANTHONY, theoretical physics, systems analysis, see 12th edition

BEAUCHENE, ROY E, b Sioux City, Iowa, Sept 4, 25; m 52; c 4. NUTRITION, GERIATRICS. Educ: Morningside Col, BS, 51; Kans State Univ, MS, 52, PhD, 56. Prof Exp: Asst chem, Kans State Univ, 51-56; from asst prof to assoc prof human nutrit res, Tex Woman's Univ, 56-58, prof nutrit & biochem res, 58-63; NIH res scientist, Geront Br, Baltimore City Hosps, 63-68; assoc prof nutrit, 68-72, PROF NUTRIT, UNIV TENN, KNOXVILLE, 72- Concurrent Pos: Nutrit consult, Tenn Comn Aging, 72-; guest scholar, Kans State Univ, 73. Mem: Fel AAAS; Am Chem Soc; Geront Soc; Am Home Econ Asn; Sigma Xi. Res: Gerontology; bone density. Mailing Add: Dept of Nutrit Univ of Tenn Knoxville TN 37916

BEAUDET, PAUL R, theoretical physics, geophysics, see 12th edition

BEAUDET, ROBERT A, b Woonsocket, RI, Aug 18, 35; m 57; c 4. CHEMICAL PHYSICS, PHYSICAL CHEMISTRY. Educ: Worcester Polytech Inst, BS, 57; Harvard Univ, AM & PhD(phys chem), 62. Prof Exp: Res scientist, Jet Propulsion Lab, Calif Inst Technol, 61-63; from asst prof to assoc prof, 63-71, PROF CHEM, UNIV SOUTHERN CALIF, 71-, CHMN DEPT, 74- Concurrent Pos: Consult, Jet Propulsion Lab, Calif Inst Technol, 63-; NIH grant, Univ Southern Calif, 64; A P Sloan Found res fel, 67-71. Mem: Am Chem Soc; Am Phys Soc. Res: Microwave spectroscopy; rotational spectra; molecular structure; intra molecular interactions. Mailing Add: Dept of Chem Univ of Southern Calif Los Angeles CA 90007

BEAUDOIN, ALLAN ROGER, b New Britain, Conn, Aug 25, 27; m 50; c 4. EXPERIMENTAL EMBRYOLOGY. Educ: Univ Conn, BS, MS, 51; Univ Iowa, PhD(zool), 54. Prof Exp: Instr zool, Univ Iowa, 53-54; instr zool, Vassar Col, 54-55, instr physiol, 55-56; from instr to asst prof anat, Col Med, Univ Fla, 56-61; from asst prof to assoc prof, 61-69, PROF ANAT, MED SCH, UNIV MICH, ANN ARBOR, 69- Concurrent Pos: Vis res assoc prof, Karolinska Inst, Sweden, 67-68. Mem: AAAS; Am Soc Zoologists; Am Asn Anatomists; Teratology Soc; Soc Develop Biol. Res: Teratology. Mailing Add: Dept of Anat Univ of Mich 4614 Med Sci II Ann Arbor MI 48109

BEAUDOIN, JACQUES, b Lachine, Que, Aug 4, 39; m 65; c 1. VIROLOGY, BACTERIAL GENETICS. Educ: Univ Montreal, BA, 59, BScPharm, 63, MSc, 66; Univ Wis-Madison, PhD(bact), 70. Prof Exp: ASST PROF MICROBIOL & IMMUNOL, UNIV MONTREAL, 70- Mem: AAAS; Am Soc Microbiol; Can Soc Microbiol; Fr-Can Asn Advan Sci. Res: Genetics of Nocardia restrictus; nocardiophage; male-specific filamentous coliphages; coat proteins; coat protein mutants; replication; purification of haploid and diploid virions; serum inactivation; genetic mapping; control of morphogenesis; bacteriophages; molecular virology; viral genetics. Mailing Add: Dept of Microbiol & Immunol Univ of Montreal CP 6128 Montreal PQ Can

BEAUDOIN, RICHARD LAMPRON, zoology, parasitology, see 12th edition

BEAUDREAU, DAVID E, b Plummer, Idaho, May 30, 29; m 50; c 3. RESTORATIVE DENTISTRY, NEUROPHYSIOLOGY. Educ: Univ Wash, DDS, 54; Univ Pa, MSD, 65. Prof Exp: Instr, Dept Fixed Partial Dentures & Opers, Univ Wash, 56-58; pvt dent pract, 58-61; fel, Univ Pa, 61-63, from asst prof to assoc prof fixed partial dentures, 63-68, chmn dept, 63-68, dir periodont prosthesis, crown & bridge, 67-68; chmn dept, 68-73, PROF RESTORATIVE DENT, MED COL GA, 68-, ASSOC DEAN CURRICULUM, 73- Concurrent Pos: Golgi res grants, 66-71; consult, Vet

Admin Hosp, Philadelphia, 66-68 & Augusta, Ga, 68-; consult, US Army, Ft Benning, Ga, 68- & Ft Gordon, 70- Mem: Int Asn Dent Res. Res: Neurophysiology of occlusion and cyclic jaw movement. Mailing Add: Dept of Restorative Dent Med Col of Ga Augusta GA 30902

BEAUDREAU, GEORGE STANLEY, b Ferndale, Wash, Dec 2, 25; m 46; c 4. BIOCHEMISTRY. Educ: Wash State Col, BS, 49; Ore State Col, MS, 52, PhD, 54. Prof Exp: Fel biol sci, Purdue Univ, 54-56; res assoc, Duke Univ, 56-63; ASSOC PROF CHEM & AGR CHEM, ORE STATE UNIV, 63- Mem: Am Chem Soc; Am Asn Cancer Res. Res: Intermediary metabolism and enzymology; antibiotics; tumor virus; tissue culture. Mailing Add: Dept of Agr Chem Ore State Univ Corvallis OR 97330

BEAUDRY, JEAN ROMUALD, b Verdun, Que, Aug 30, 17; m 44; c 1. BIOSYSTEMATICS. Educ: Inst Agr d'Oka, LSA; McGill Univ, MSc; Univ Wis, PhD(genetics). Prof Exp: Asst agron, Macdonald Col, McGill Univ, 41-43, asst prof, 48-49; asst genetics, Univ Wis, 46-48; res assoc cytogenetics, 49-50; from asst prof to assoc prof biol, 50-60, PROF BIOL, UNIV MONTREAL, 60- Mem: Genetics Soc Can (pres, 62-63); French-Can Asn Advan Sci. Res: Cytogenetics, taxonomy and evolution of Solidago; cytotaxonomy of orthoptera; nature and concepts of species. Mailing Add: Dept of Biol Sci Univ of Montreal Montreal PQ Can

BEAUFAIT, WILLIAM RICHARD, forestry, see 12th edition

BEAUGE, LUIS ALBERTO, b Esparanza, Arg, Aug 24, 37; m 62; c 2. BIOPHYSICS. Educ: Nat Univ Cordoba, MD, 61, PhD(physiol), 64. Prof Exp: Asst surg, San Roque Hosp, Cordoba, Arg, 58-59, obstet, 59-60; asst surg, Nat Hosp Clin, Nat Univ Cordoba, 61-63, fel urol, Sc Med, 61-63, instr physiol, 61-65; fel pharmacol, NY Med Col, 65-66; fel biophys, Sch Med, Univ Md, Baltimore City, 66-68; asst prof biophys, Fac Chem Sci, Nat Univ Cordoba, 68-73; HEAD BIOPHYS, INST MED INVEST, MERCEDES & MARTIN FERREYRA, 71-; ASSOC PROF BIOPHYS, SCH MED, UNIV MD, BALTIMORE CITY, 73- Concurrent Pos: Fel, Nat Coun Sci & Tech Invest, Arg, 65-68; vis fel commoner, Trinity Col, Cambridge Univ, Eng, 75-76; mem corp, Marine Biol Lab, Woods Hole, Mass, 75- Mem: Nat Coun Sci & Tech Invest Arg; Biophys Soc; Arg Biophys Soc; Soc Gen Physiologists. Res: Mechanisms of cation transport through biological membranes. Mailing Add: Dept of Biophys Univ of Md Sch of Med Baltimore MD 21201

BEAULIEU, ALEXANDRE JACQUES, electrooptics, radiation physics, see 12th edition

BEAULIEU, J A E M, b Montreal, Que, Apr 17, 29; m 55; c 2. BACTERIOLOGY, IMMUNOLOGY. Educ: Univ Montreal, BSc, 51; McGill Univ, MSc, 53, PhD(bact, immunol), 55; Am Bd Microbiol, dipl. Prof Exp: Head dept bact, St Joan D'Arc Hosp, 53-56; lectr, Sch Med, Univ Ill, 56-57; from asst prof to assoc prof, 57-65, PROF BACT, FAC MED, UNIV OTTAWA, 65-, ASSOC DEAN, 71- Concurrent Pos: Head bact, Hosp Sacre Coeur, Hull, Que, 58-; Med Res Coun grants, 61-64; ed, Can Soc Microbiol News Bull, 64-67. Mem: Can Soc Microbiol. Res: Production of antibodies by monocytes in vitro; factors involved in monocytosis producing agent of Listeria; hospital infections. Mailing Add: Dept of Bact Univ of Ottawa Ottawa ON Can

BEAULIEU, JOHN DAVID, b Hanford, Wash, July 22, 44; m 67; c 2. STRATIGRAPHY, ENVIRONMENTAL GEOLOGY. Educ: Univ Wash, BS, 66; Stanford Univ, PhD(geol), 71. Prof Exp: Prof geol, Univ Ore, 69-70; GEOLOGIST, ORE DEPT GEOL & MINERAL INDUSTS, 70- Res: Cenozoic behavior of the San Andreas Fault in the Santa Cruz Mountains, California; stratigraphy of Oregon; geologic history and hazards of Oregon. Mailing Add: Dept of Geol & Mineral Industs 1069 State Off Bldg Portland OR 97201

BEAUMONT, RALPH HARRISON, JR, b Roxbury, NY, Nov 10, 23; m 50; c 6. CHEMISTRY. Educ: Colgate Univ, BA, 43; Rutgers Univ, PhD(chem), 53. Prof Exp: Asst, Rutgers Univ, 43-44, 46-59 & 51; res chemist spec prob sect, NB Lab, AEC, 49-51 & Allied Chem & Dye Corp, 51-52; asst dir res, Huyck Corp, 52-57, dir chem res, 57-63; tech dir, W F Fancourt & Co, 64-66, vpres res & develop, 66-70; PRES & DIR RES, BRIN-MONT CHEM INC, 68- Mem: Am Chem Soc; Am Asn Testile Chem & Colorists. Res: Analytical chemistry; inorganic chemistry; physical chemistry; kinetics and mechanism of reactions; textile chemistry; general textile auxiliaries; surfactants and rubber chemicals. Mailing Add: Brin-Mont Chem Inc 3921 Spring Garden St Greensboro NC 27407

BEAUMONT, RANDOLPH CAMPBELL, b Los Angeles, Calif, Oct 15, 41; m 66; c 2. INORGANIC CHEMISTRY. Educ: Whitman Col, BA, 63; Univ Idaho, PhD(inorg chem), 67. Prof Exp: ASST PROF CHEM, ALMA COL, MICH, 67- Mem: AAAS; Am Chem Soc. Res: Inorganic, physical and analytical chemistry; nitrites of period 2 elements. Mailing Add: Dept of Chem Alma Col Alma MI 48801

BEAUMONT, ROSS ALLEN, b Dallas, Tex, July 23, 14; m 40; c 2. MATHEMATICS. Educ: Univ Mich, AB, 36, MS, 37; Univ Ill, PhD(math), 40. Prof Exp: From instr to assoc prof, 40-54, PROF MATH, UNIV WASH, 54- Concurrent Pos: Advan of Educ Fund fel, Inst Advan Study, 54-55; ed, Pac J Math, 56-59 & 73-; vis prof, Univ Ariz, 73. Mem: Am Math Soc; Math Asn Am. Res: Theory of groups; ring theory; linear algebra. Mailing Add: Dept of Math Univ of Wash Seattle WA 98195

BEAUREGARD, LUDGER, b Montreal, Que, Mar 3, 20; m 46; c 4. GEOGRAPHY. Educ: Univ Paris, BA, 42; Univ Montreal, MA, 50, PhD(geog), 57. Prof Exp: Prof geog, Sch Advan Com Studies, 54-63; PROF GEOG, UNIV MONTREAL, 63- Mem: Can Asn Geog (pres, 68-69). Res: Urban geography; political geography; geography of Canada and Montreal. Mailing Add: 126 Bloomfield Montreal PQ Can

BEAUREGARD, RAYMOND A, b New Bedford, Mass, Feb 10, 43; m 64; c 3. ALGEBRA. Educ: Providence Col, AB, 64; Univ NH, MS, 66, PhD(algebra), 68. Prof Exp: Asst prof, 68-73, ASSOC PROF MATH, UNIV RI, 73- Mem: Am Math Soc; Math Asn Am. Res: Noncommutative rings; noncummutative integral domains, unique factorization; right LCM domains; principal right ideal domains; ring theory. Mailing Add: Dept of Math Univ of RI Kingston RI 02881

BEAVEN, MICHAEL ANTHONY, b London, Eng, Dec 4, 36; US citizen; m 64. PHARMACOLOGY, BIOCHEMISTRY. Educ: Chelsea Col Sci & Technol, London, BPharm, 59, PhD(med), 62. Prof Exp: Demonstr, Chelsea Col Sci & Technol, London, 59-62; vis fel, Lab Chem Pharmacol, 62-66, res pharmacologist, 66-68, sr investr, Exp Therapeut Br, 68-74, SR INVESTR, PULMONARY BR, NAT HEART & LUNG INST, 74- Concurrent Pos: USPHS vis fel, Nat Heart Inst, 62-63; Nat Heart Inst vis assoc, 64. Mem: Am Soc Pharmacol & Exp Therapeut; Brit Biochem Soc; The Chem Soc; Pharmaceut Soc Gt Brit. Res: Studies in the role of biogenic amines, and other tissue mediators, in physiological and pathological reactions with particular emphasis on possible applications in clinical medicine. Mailing Add: Pulmonary Br Rm 5N107 Bldg 10 Nat Heart and Lung Inst Bethesda MD 20014

BEAVER, ALBERT JOHN, soils, see 12th edition

BEAVER, DONALD LOYD, b Hayden, Colo, Sept 16, 43; m 66. VERTEBRATE ECOLOGY. Educ: Colo State Univ, BS, 65, MS, 67; Univ Calif, Berkeley, PhD(zool), 72. Prof Exp: ASST PROF ZOOL, MICH STATE UNIV, 72- Mem: Am Ornithologists Union; Cooper Ornith Soc; Ecol Soc Am; Sigma Xi. Res: Ecological and behavioral aspects of foraging in birds, including selection of foraging patch, optimal food choice and time and energy spent foraging. Mailing Add: Dept of Zool Mich State Univ East Lansing MI 48824

BEAVER, EARL RICHARD, b Newburgh, NY, Mar 2, 45; m 66. PHYSICAL CHEMISTRY. Educ: McMurry Col, BA, 66; Tex Tech Univ, 66-69, PhD(phys chem), 70. Prof Exp: Sr chemist, 69-73, process specialist, 73-74, GROUP SUPVR, MONSANTO CO, 74- Mem: Am Chem Soc; Am Inst Chem Engrs; Am Soc Testing & Mat. Res: Application of differential thermal analysis, thermogravimetric analysis and adsorption measurements to the solution of process problems in heterogeneous catalysis. Mailing Add: Process Technol Monsanto Co PO Box 1311 Texas City TX 77591

BEAVER, PAUL CHESTER, b Glenwood, Ind, Mar 10, 05; m 31; c 1. PARASITOLOGY, TROPICAL MEDICINE. Educ: Wabash Col, AB, 28; Univ Ill, MS, 29, PhD(zool), 35. Hon Degrees: DSc, Wabash Col, 63. Prof Exp: Asst zool, Univ Ill, Urbana, 28-29; instr, Univ Wyo, 29-31; asst, Univ Ill, 31-34; instr biol, Oak Park Jr Col, 34-37; asst prof, Lawrence Col, 37-42; biologist malaria control, Ga Dept Pub Health, 42-45; from asst prof to prof, 45-52, head dept parasitol, 56-71, WM VINCENT PROF TROP DIS & HYG, SCH MED & SCH PUB HEALTH & TROP MED, TULANE UNIV, 58- Concurrent Pos: Mem, Armed Forces Epidemiol Bd Comn Parasitic Dis, 53-73, dir, 67-73; mem microbiol fels rev panel, NIH, 60-63; mem, WHO Expert Panel on Parasitic Dis; mem, NIH Parasitic Dis Panel, US-Japan Coop Med Sci Prog, 65-69; mem bd sci counr, Nat Inst Allergy & Infectious Dis, 66-68; dir, Int Ctr Med Res, Tulane Univ, 67-; mem, Gorgas Mem Inst Trop & Prev Med Adv Sci Bd, 70-; ed, Am J Trop Med & Hyg, 60-66 & 72- Honors & Awards: Outstanding Civilian Serv Medal, US Dept Army, 73. Mem: Am Soc Trop Med & Hyg (vpres, 58, pres, 69); Am Soc Parasitol (pres, 68); fel Am Acad Microbiol; Am Micros Soc (vpres, 53); Am Pub Health Asn. Res: Occult and zoonotic helminthic infections; amebiasis; epidemiology of soil transmitted helminths. Mailing Add: Dept of Parasitol Tulane Univ Med Ctr New Orleans LA 70112

BEAVER, ROBERT JOHN, b Mt Carmel, Pa, Mar 27, 37; m 70. STATISTICS. Educ: Bloomsburg State Col, BS, 59; Bucknell Univ, MS, 64; Univ Fla, MStat, 66, PhD(statist), 70. Prof Exp: High sch teacher, Pa, 60-63; teaching asst math, Bucknell Univ, 63-64; asst statist, Univ Fla, 64-66, statist consult, Teacher Eval Proj, 66, dir comput-assisted instr proj, 66-67, instr statist, 67-68, teaching asst, 68-70; med statistician, Med Div, Oak Ridge Assoc Univs, 68; ASST PROF STATIST & STATISTICIAN, UNIV CALIF, RIVERSIDE, 70- Res: Model building as applied to problems of choice; paired and triple comparisons. Mailing Add: Dept of Statist Univ of Calif Riverside CA 92502

BEAVER, W DON, b Elkhart, Kans, May 7, 24; m 46; c 2. CHEMISTRY. Educ: Bethany Nazarene Col, AB, 46; Okla State Univ, MS, 53, PhD(chem), 55. Prof Exp: Teacher pub schs, Okla, 47-48; instr math & chem, Bethany Nazarene Col, 48-51; asst chem, Okla State Univ, 52-55; PROF CHEM, BETHANY NAZARENE COL, 55-, CHMN DIV NATURAL SCI, 74- Mem: AAAS; Am Chem Soc. Res: Use of borohydrides as reducing agents and coordination complexes of the transition metals in aqueous and non-aqueous solvents. Mailing Add: Dept of Chem Bethany Nazarene Col Bethany OK 73008

BEAVER, WILLIAM THOMAS, b Albany, NY, Jan 27, 33; m 61; c 3. CLINICAL PHARMACOLOGY. Educ: Princeton Univ, AB, 54; Cornell Univ, MD, 58. Prof Exp: Intern surg-med, Roosevelt Hosp, New York, 58-59; USPHS fel pharmacol, Med Col, Cornell Univ, 59-61, from instr to asst prof surg-med, 61-68; ASSOC PROF PHARMACOL & ANESTHESIOL, SCHS MED & DENT, GEORGETOWN UNIV, 68-, MEM GEN STAFF, DEPT ANESTHESIOL, UNIV HOSP, 69- Concurrent Pos: Clin asst, Mem Hosp, New York, 61-68; asst vis physician, Dept Med, James Ewing Hosp, 62-68; res assoc, Sloan-Kettering Inst Cancer Res, 63-68; attend physician, Calvary Hosp, 64-68; consult, Code Authority, Nat Asn Broadcasters, 67-, Food & Drug Admin, 69-, Fed Trade Comn, 70- & AMA Drug Eval, 71-; mem adv panel analgesics, sedatives & anti-inflammatory agts, US Pharmacopoeia, 71- Mem: AAAS; Am Soc Clin Pharmacol & Therapeut; Am Soc Pharmacol & Exp Therapeut. Res: Clinical pharmacology of analgesic drugs; design of clinical trials. Mailing Add: Dept of Pharmacol Georgetown Univ Sch of Med Washington DC 20007

BEAVERS, ALVIN HERMAN, b Okla, Jan 1, 13; m 46; c 3. AGRONOMY. Educ: NMex State Col, BS, 40; Univ Mo, MA, 48, PhD(soil sci), 50. Prof Exp: Soil scientist, Soil Conserv Serv, 41-43; asst, Univ Mo, 46-50; assoc prof soil physics, 50-63, PROF AGRON, UNIV ILL, URBANA-CHAMPAIGN, 63- Mem: Am Soc Agron. Res: Soil mineralogy; x-ray spectrographic and diffraction analyses; electrophoresis of clay minerals. Mailing Add: Dept of Agronomy Univ of Ill Urbana IL 61801

BEAVERS, DOROTHY (ANNE) JOHNSON, b Worcester, Mass, July 11, 27; m 52; c 1. ORGANIC CHEMISTRY. Educ: Clark Univ, BA, 49; Duke Univ, PhD(org chem), 55. Prof Exp: Res chemist, Am Cyanamid Co, 49-51; asst gen & org chem, Duke Univ, 51-53; sr res chemist, 54-61, SR RES ASSOC BIOCHEM & CLIN CHEM, EASTMAN KODAK CO, 61- Concurrent Pos: USPHS grant, 53-54. Mem: AAAS; Am Chem Soc; Soc Photog Scientists & Engrs. Res: Synthetic organic chemistry; surface active agents; aromatic cyclodehydration; chemical transfer; rapid processing of photographic emulsions; photographic paper products; biochemistry of mental illness and retardation; emulsion addenda; color reversal systems. Mailing Add: 70 Rainbow Dr Rochester NY 14622

BEAVERS, ELLINGTON MCHENRY, b Atlanta, Ga, Jan 29, 16; m 57. ORGANIC CHEMISTRY. Educ: Emory Univ, BS, 38, MS, 39, PhD(chem), Univ NC, 41. Prof Exp: Res chemist, 41-47, lab head, 47-59, res supvr, 57-61, asst dir res, 61-66, assoc dir res, 66-69, dir res, 69-73, vpres & mem bd dirs, 70-73, MEM EXEC COMT, ROHM AND HAAS CO, 71-, SR VPRES, 73- Mem: AAAS; Am Chem Soc; Am Inst Chemists; Am Inst Chem Engrs. Res: Agricultural chemicals; plastics; surfactants; enzymes; plasticizers. Mailing Add: 931 Coates Rd Meadow Brook PA 19046

BEAVERS, LEO EARICE, b Miller's Grove, Tex, May 7, 20; m 52; c 1. ORGANIC CHEMISTRY. Educ: Harvard Univ, AB, 50; Duke Univ, PhD, 55. Prof Exp: Res chemist, Am Cyanamid Co, 50-51; asst, Duke Univ, 51-52; res chemist, 54-60, HEAD LAB MKT RES & SR RES ASSOC, COLOR PHOTOG DIV, EASTMAN KODAK CO, 60- Mem: AAAS; Am Chem Soc; Soc Photog Scientists & Engrs; Soc Motion Picture & TV Engr. Res: Color photography. Mailing Add: 70 Rainbow Dr Rochester NY 14622

BEAVERS, WILLET I, b Billings, Mont, Nov 13, 33; m 59; c 2. ASTROPHYSICS,

ASTRONOMY. Educ: Univ Mo, BS, 55, MS, 59; Ind Univ, PhD(astrophys), 66. Prof Exp: Instr astron, Univ Mo, 63-65; asst prof, 65-69, ASSOC PROF PHYSICS, IOWA STATE UNIV, 69- Mem: Am Astron Soc; Optical Soc Am; Int Astron Union. Res: Observational and experimental astrophysics, design and construction of astronomical instruments. Mailing Add: Erwin W Fick Observ Iowa State Univ Ames IA 50010

BEBB, HERBERT BARRINGTON, b Wichita Falls, Tex, June 22, 35; m 58; c 2. SOLID STATE PHYSICS. Educ: Univ Okla, BS, 59; Syracuse Univ, MS, 64; Univ Rochester, PhD(optics), 65. Prof Exp: Assoc physicist, Fed Systs Div, Int Bus Mach Corp, 59-61, mem res staff, Res Lab, 61-62; res assoc, Inst Optics, Univ Rochester, 65-66; theoret physicist, Tex Instruments Inc, 66-69, mgr appl optics br, 69, dir advan technol lab, 69-74; MGR PROD DEVELOP DEPT, SYSTS DEVELOP DIV, XEROX CORP, 74- Concurrent Pos: Consult, Int Bus Mach Corp, 65. Mem: Am Phys Soc; Optical Soc Am; Math Asn Am. Res: Theoretical study of optical properties of solids; ultrasonic surface waves; liquid crystal phenomena; infrared sensors; semiconductor technology; systems engineering. Mailing Add: Xerox Corp 1341 W Mockingbird Lane Dallas TX 75247

BEBB, ROBERT LLOYD, b Columbus, Ohio, Jan 21, 13. ORGANIC CHEMISTRY. Educ: Ohio State Univ, AB, 34; Iowa State Col, PhD(org chem), 38. Prof Exp: Sr res org chemist, Firestone Tire & Rubber Co, 38-43, group leader synthetic rubber res, 43-51, group leader petrochem res, 51-55, mgr res & develop, Synthetic Rubber & Latex Div, 55-60, asst dir res div, 60-66, asst dir cent res, 66-67, assoc dir res, 67-71, res coordr, 71-75; RETIRED. Concurrent Pos: Vpres, Radiation Processing Inc, NY, 67-70. Mem: AAAS; Am Chem Soc; Ger Rubber Soc. Res: Synthetic rubber; polymerization; petrochemicals; organometallic compounds; metalation; organic synthesis. Mailing Add: 2374 Hume St NW North Canton OH 44720

BEBER, ADOLPH JOSEPH, chemistry, see 12th edition

BEBERNES, JERROLD WILLIAM, b Cotesfield, Nebr, Apr 7, 35; m 56; c 3. APPLIED MATHEMATICS. Educ: Univ Nebr, BS, 57, MA, 59, PhD(math), 62. Prof Exp: Mathematician, US Naval Ord Lab, Calif, 57; instr math, Univ Nebr, 59-60; from asst prof to assoc prof, 62-71, PROF MATH, UNIV COLO, BOULDER, 71- Mem: Am Math Soc; Math Asn Am. Res: Functional analysis and ordinary differential equations, especially differential inequalities and special boundary value problems. Mailing Add: Dept of Math Univ of Colo Boulder CO 80302

BEBOUT, DON GRAY, b Moneson, Pa, Jan 23, 31; m 52; c 3. SEDIMENTOLOGY. Educ: Mt Union Col, BS, 52; Univ Wis, MS, 54; Univ Kans, PhD(geol), 61. Prof Exp: Asst micropaleont, Univ Kans, 56-57 & 59-60; sr res specialist geol, Exxon Prod Res Co, Tex, 60-72; RES SCIENTIST GEOL, BUR ECON GEOL, UNIV TEX, AUSTIN, 72- Concurrent Pos: Lectr, Dept Geol Sci, Univ Tex, Austin, 75- Mem: Geol Soc Am; Am Asn Petrol Geologists; Soc Econ Paleontologists & Mineralogists. Res: Subsurface distribution of sedimentary facies along the Texas Gulf Coast; evaluating the potential of producing geothermal energy from Texas Gulf Coast geopressured reservoirs. Mailing Add: Bur of Econ Geol Univ of Tex Univ Sta Box X Austin TX 78712

BECHARA, IBRAHIM, b Bazzak, Syria, Jan 16, 43; US citizen; m 71; c 2. ORGANIC CHEMISTRY. Educ: Hobart Col, BS, 63; Univ Del, PhD(chem), 67. Prof Exp: RES CHEMIST, AIR PROD & CHEMS, 67- Mem: Am Chem Soc. Res: Synthetics of organic chemicals; homogenous catalysis of organic reactions, mainly isocyanate reactions. Mailing Add: 1159 Naaman's Creek Rd Boothwyn PA 19061

BECHER, PAUL, b Brooklyn, NY, Mar 24, 18; m 45; c 2. PHYSICAL CHEMISTRY. Educ: Polytech Inst Brooklyn, BS, 40, MS, 42, PhD(chem), 49. Prof Exp: Assoc prof chem, NGa Col, 48-51; sr proj chemist, Colgate-Palmolive Co, 51-57; res assoc, Atlas Chem Industs, Inc, 57-67, prin scientist, 67-70, mgr phys chem sect, Chem Res Dept, 70-, DIR SPECIALTY CHEM RES DEPT ICI US INC, 75- Mem: AAAS; Am Oil Chem Soc; Sigma Xi; Am Chem Soc; fel Am Inst Chemists. Res: Emulsions; surface chemistry. Mailing Add: Chem Res & Develop Lab ICI US Inc Wilmington DE 19897

BECHTEL, ROBERT CHRISTY, b Visalia, Calif, May 21, 24; m 48; c 2. ENTOMOLOGY. Educ: Univ Calif, BS, 50. Prof Exp: Res asst entom, Univ Calif, 54-57; SURV & SYST ENTOMOLOGIST, STATE DEPT AGR, NEV, 57- Concurrent Pos: Mem entom exped, Assocs Trop Biogeog, Mex, 53. Mem: Entom Soc Am; Soc Syst Zool; Am Entom Soc; Entom Soc Can. Res: Systematic entomology; Hymenoptera Sapygidae; Nevada insects. Mailing Add: Nev State Dept of Agr PO Box 11100 Reno NV 89510

BECHTEL, ROBERT D, b Chicago, Ill, Apr 2, 31; m 52; c 3. MATHEMATICS. Educ: McPherson Col, BS, 53; Kans State Univ, MS, 59; Purdue Univ, PhD(math educ), 63. Prof Exp: Teacher high schs, Kans, 55-58; asst prof math, Kans State Univ, 63-66; assoc prof, 66-71, PROF MATH EDUC, PURDUE UNIV, 71- Mem: Math Asn Am; Am Math Soc. Res: Mathematics education. Mailing Add: Div of Math Sci Purdue Univ Calumet Campus Hammond IN 46323

BECHTLE, GERALD FRANCIS, b Ottawa, Kans, Nov 4, 21. ORGANIC CHEMISTRY. Educ: Ottawa Univ (Kans), BS, 43; Univ Kans, MS, 50. Prof Exp: Sr chemist, Sherwin-Williams Co, 47-54; assoc chemist, Midwest Res Inst, 55-67; CHEMIST, COOK PAINT & VARNISH CO, 67- Res: Resin; paint; oils and fats. Mailing Add: Cook Paint & Varnish Co 1412 Knox North Kansas City MO 64116

BECHTLE, ROBERT M, b Valley City, NDak, July 27, 21. FOOD SCIENCE. Educ: Valley City State Col, BA, 47; Univ Ill, Chicago Circle, MS, 59; Kans State Univ, PhD(food sci), 70. Prof Exp: Res bacteriologist, US Vet Admin Hosps, Ft Snelling, Minn, 49-55; res contact, Sandoz Pharmaceut, NJ, 55-57; res adminr, Tice Lab, Univ Ill, Chicago Circle, 59-61; sanitarian & dir, Apple Orchard Develop Co, Ill, 61-65; microbiologist, Midwest Res Inst, Mo, 65-68; RES ASSOC DAIRY & POULTRY SCI, KANS STATE UNIV, 70- Mem: Am Dairy Sci Asn; Am Soc Microbiol; Inst Food Technologists; Am Chem Soc. Res: Tuberculosis bacteriology, antibiotic resistance development; potentiation of antibiotic activity; aerobiology; food microbiology and formulation; accelerated fermentation; dairy production research; analytical technics in animal nutrition. Mailing Add: Dept of Dairy & Poultry Sci Kans State Univ Manhattan KS 66502

BECHTOL, BRUCE EMERSON, b Oroville, Calif, Aug 30, 38; m 59; c 2. GEOGRAPHY. Educ: Calif State Univ, Chico, AB, 60; Univ Okla, MA, 65; Univ Ore, PhD(geog), 69. Prof Exp: Indust surv adv, AID, Guatemala, 67-68; from asst prof to assoc prof geog, 68-75, PROF GEOG, CALIF STATE UNIV, CHICO, 75- Mem: AAAS; Asn Am Geogr; Nat Coun Geog Educ. Res: Man-environment relationships; environmental planning and development; Latin America. Mailing Add: 31 Lawnwood Dr Chico CA 95926

BECHTOL, LAVON DEE, b Wabash, Ind, Feb 3, 18; m 38; c 2. MEDICINE, CHEMISTRY. Educ: Manchester Col, AB, 40; Purdue Univ, PhD(org chem), 44; Johns Hopkins Univ, MD, 47. Prof Exp: Asst chem, Purdue Univ, 40-41; pharmacologist, Eli Lilly & Co, Ind, 46; intern, Methodist Hosp, 47-48, resident, 48-49; asst med dir, Baxter Labs, Inc, 52-55; med dir, Ethicon, Inc, 55-61; dir clin res, Baxter Labs, Inc, 61-65; MEM STAFF, LILLY LAB CLIN RES, 65- Mem: AMA; Aerospace Med Asn; Am Chem Soc; Am Soc Clin Pharmacol & Therapeut; Am Med Writers' Asn. Res: Organic fluorides; endocrinology; clinical research; parenteral nutrition; antibiotics. Mailing Add: Lilly Lab for Clin Res Wishard Mem Hosp Indianapolis IN 46202

BECHTOLD, EDWIN WILLIAM, b Brooklyn, NY, Nov 22, 12; m 40; c 1. APPLIED OPTICS. Educ: Columbia Univ, BS, 34, MA, 48. Prof Exp: Practicing optometrist, NY, 34-43; physicist, Frankford Arsenal, Philadelphia, 43-44, optical physicist, 44-46; optical engr, Bendix Aviation Corp, NJ, 46; assoc entom, Columbia Univ, 46-49, assoc prof, 49-57, adj prof mech eng, 59-73; CONSULT OPTICAL ENGR, 71- Concurrent Pos: Lectr, Columbia Univ, 38-43 & Manhattan Col, 42-43. Mem: Assoc Optical Soc Am. Res: Higher order aberrations in optical systems; lens design. Mailing Add: 50 Papermill Rd Manhasset NY 11030

BECHTOLD, MAX FREDRICK, b North Manchester, Ind, Jan 3, 15; m 37; c 4. PHYSICAL CHEMISTRY. Educ: Manchester Col, BS, 35; Purdue Univ, PhD(phys chem), 39. Prof Exp: Chemist, 39-67, RES SUPVR, CENT RES DEPT, EXP STA, E I DU PONT DE NEMOURS & CO, INC, 67- Mem: AAAS; Sigma Xi; Am Chem Soc. Res: Inorganic polymerization kinetics; colloid chemistry of organic polymers; thin film techniques; solid state reactions in ceramics and metallic high temperature materials; refractory and magnetic powders; batteries, Rankin cycle engines; fluids; solar energy conversion. Mailing Add: Du Pont Exp Sta E I du Pont de Nemours & Co Inc Wilmington DE 19898

BECK, AARON TEMKIN, b Providence, RI, July 18, 21; m 50; c 4. PSYCHIATRY. Educ: Brown Univ, BA, 42; Yale Univ, MD, 46. Hon Degrees: MA, Univ Pa, 70. Prof Exp: Asst chief dept neuro-psychiat, Valley Forge Army Hosp, Va, 52-54; instr, 54-57, assoc, 57-58, from asst prof to assoc prof, 58-71, PROF PSYCHIAT, UNIV PA, 71- Concurrent Pos: Sect chief, Philadelphia Gen Hosp, 58-; consult, Vet Admin Hosp, Philadelphia, 67-; spec consult, Ctr Studies Suicide Prev, NIMH, 69-72, chmn, Task Force Suicide Prev in 70's, 69-70; trustee, Am Acad Psychoanal, 70- Mem: Psychiat Res Soc; Am Psychopath Asn; Am Col Psychiatrists; Asn Advan Behav Ther. Res: Depression; suicide; cognitive aspects of psychopathology; cognitive approaches to psychotherapy. Mailing Add: Stouffer Bldg Phila Gen Hosp Civic Ctr Blvd Philadelphia PA 19104

BECK, ALAN EDWARD, Can citizen. GEOPHYSICS. Educ: Univ London, BSc & BSc(physics), 51; Australian Nat Univ, PhD(geophys), 57. Prof Exp: Physicist, instruments develop sect, Brit Oxygen Co, Ltd, 51-52; scientist II, Nat Coal Bd Mining Res Estab, 56-57; Nat Res Coun Can fel, 57-58; from asst prof to assoc prof geophys, 58-65, actg head dept, 61-63, PROF GEOPHYS, UNIV WESTERN ONT, 65-, HEAD DEPT, 63- Concurrent Pos: Mem, Int Heat Flow Comt. Mem: Am Geophys Union; Soc Explor Geophys; Geol Asn Can; Can Geophys Union; Can Asn Physicists. Res: Terrestrial heat flow; energy balance of the earth; exploration methods. Mailing Add: Dept of Geophys Univ of Western Ont London ON Can

BECK, ALBERT J, b Nyack, NY, Aug 13, 35; m 59. ZOOLOGY. Educ: Univ Calif, Davis, AB, 57, MA, 61, PhD(zool), 66. Prof Exp: Assoc zool, Univ Calif, Davis, 63-64; asst res parasitologist, Inst Med Res, Malaysia, 65-69; asst res zoologist, Sch Pub Health, Univ Calif, Berkeley, 69-73; CONSULT & ANALYST, ECO-ANALYSTS, 73- Concurrent Pos: Med zoologist, Inst Med Res, Malaysia, 66-68; lectr geog, Calif State Univ, Chico, 75- Mem: AAAS; Am Soc Mammal; Wildlife Dis Asn; Malaysian Soc Trop Med & Parasitol. Res: Host-ectoparasite interactions; ecology of ectoparasites; epidemiology of zoonotic diseases; biological indicators; host-parasite relationships. Mailing Add: PO Box 1187 Chico CA 95926

BECK, ANATOLE, b New York, NY, Mar 19, 30; m 54; c 2. MATHEMATICS. Educ: Brooklyn Col, BA, 51; Yale Univ, MA, 53, PhD(math), 56. Prof Exp: Ford instr math, Williams Col, 55-56; Off Naval Res assoc, Tulane Univ, 56-57; traveling fel, Yale Univ, 57-58; from asst prof to assoc prof, 58-66, PROF MATH, UNIV WIS-MADISON, 66- Concurrent Pos: Vis scholar, Hebrew Univ, Israel, 64-65; vis prof, Univ Md, 71 & Technische Univ Munich, 73; assoc ed, Math Systs Theory, 67-; NSF sr fels, Univ Warwick, Univ London, Univ Erlangen & Hebrew Univ, 68-69; chair of math, London Sch Econ, Univ London, 73-75. Mem: Am Math Soc; Math Asn Am; London Math Soc; Am Fedn Teachers. Res: Probability of Banach spaces; topological dynamics; ergodic theory; measure theory. Mailing Add: Dept of Math Univ of Wis Madison WI 53706

BECK, BENNY LEE, b Burlington, Ind, June 13, 32; m 56; c 3. ANALYTICAL CHEMISTRY. Educ: Univ Wis, BS, 53, MS, 56, PhD(anal chem), 57. Prof Exp: From res chemist to sr res chemist, Humble Oil & Refining Co, 57-64; supv chemist, Anvil Points Oil Shale Res Ctr, Colo, 64-65; STAFF CHEMIST, CHEM PLANT LAB, EXXON CHEM CO, 65- Mem: Am Chem Soc; Am Soc Testing & Mat. Res: General petrochemical analysis. Mailing Add: Exxon Chem Co US 3600 Park St Baytown TX 77520

BECK, BRENDA E F, b Minneapolis, Minn, Mar 23, 40. ANTHROPOLOGY. Educ: Univ Chicago, BA, 62; Oxford Univ, dipl, 63, BLitt, 64, DPhil(anthrop), 68. Prof Exp: Lectr, Univ Paris, 67; lectr anthrop, Univ Chicago, 68-69; asst prof, 69-73, ASSOC PROF ANTHROP, UNIV BC, 73- Concurrent Pos: Res fel SAsian Studies, Univ Chicago, 68-69. Mem: Am Anthrop Asn; Royal Anthrop Inst Gt Brit & Ireland; Asn Social Anthrop of Gt Brit & Commonwealth. Res: South Asian studies; social traditions and world view of peasants, particularly the peasants of Tamilnadu, South India; folklore and ritual. Mailing Add: Dept of Anthrop Univ of BC Vancouver BC Can

BECK, CARL WELLINGTON, mineralogy, see 12th edition

BECK, CHARLES BEVERLEY, b Richmond, Va, Mar 26, 27; m 61; c 2. PLANT MORPHOLOGY. Educ: Univ Richmond, BA, 60; Cornell Univ, MS, 52, PhD(bot), 55. Prof Exp: Instr bot, Cornell Univ, 54-55; Cornell-Glasgow exchange fel, Glasgow Univ, 55-56; from instr to assoc prof, 56-65, PROF BOT, UNIV MICH, ANN ARBOR, 65-, CHMN DEPT, 71- Concurrent Pos: NSF sr fel, Univ Reading, 64; mem systs biol panel, NSF, 70. Mem: AAAS; Bot Soc Am; Am Inst Biol Sci. Res: Plant anatomy; Paleozoic paleobotany. Mailing Add: Dept of Bot Univ of Mich Ann Arbor MI 48104

BECK, CLIFFORD C, b Racine, Wis, May 15, 27; m 55; c 1. VETERINARY MEDICINE, PHARMACEUTICS. Educ: Mich State Univ, BS, 53, DVM, 54, MS, 59. Prof Exp: Asst prof vet med, Mich State Univ, 55-69, exten veterinarian, 63-69; assoc dir clin invest, 69-72, DIR ANIMAL HEALTH DEPT, PARKE, DAVIS & CO, 72- Concurrent Pos: Livestock health adv, Shepherd Mag, 60-; consult, Parke, Davis & Co, 62-69. Mem: Am Vet Med Asn; Animal Health Asn; Am Asn Exten Vet; Am Asn Zoo Vets; Indust Vet Asn. Res: Fetal immune response; ovine disease and parasite research; bovine reproductive diseases; development of new therapeutic agents and pharmaceuticals for livestock and companion animals. Mailing Add: Parke, Davis & Co 2800 Plymouth Rd Ann Arbor MI 48106

BECK, CURT WERNER, b Halle, Ger, Sept 10, 27; nat US; m 53; c 2. ORGANIC CHEMISTRY. Educ: Tufts Univ, BS, 51; Mass Inst Technol, PhD(org chem), 55. Prof Exp: Instr, Franklin Technol Inst, 51-56; asst prof, Robert Col, Istanbul, 56-57; lectr, 57-59, from asst prof to assoc prof, 59-66, PROF CHEM, VASSAR COL, 66- Concurrent Pos: Ed, Art & Archaeol Tech Abstracts; sect ed, Chem Abstracts. Mem: Am Chem Soc; Archaeol Inst Am; The Chem Soc; Soc Ger Chem; fel Int Inst Conserv Hist & Artistic Works. Res: Application of chemistry to archaeology; provenience analysis of amber artifacts. Mailing Add: Dept of Chem Vassar Col Poughkeepsie NY 12601

BECK, DAVID PAUL, b Wilmington, Del, Aug 3, 44; m 66; c 1. BIOCHEMISTRY. Educ: Princeton Univ, AB, 66; Johns Hopkins Univ, PhD(biochem), 71. Prof Exp: Helen Hay Whitney fel, 71; fel biochem, Harvard Univ, 71-74; RES ASSOC NEUROCHEM, MD PSYCHIAT RES CTR, BALTIMORE, 74- Mem: Am Soc Microbiol; Genetics Soc Am; Am Soc Cell Biol; AAAS. Res: Membrane structure and function; role of membranes in regulating cellular metabolism; substrate translocation processes. Mailing Add: Md Psychiat Res Ctr Box 3225 Baltimore MD 21228

BECK, DONALD EDWARD, b Logan, Iowa, June 12, 34. SURFACE PHYSICS. Educ: Univ Calif, Berkeley, BS, 56, MA, 58, PhD(physics), 65. Prof Exp: Vis lectr physics, St Andrews Univ, 66-67; asst prof, Univ Va, 67-72; asst prof, 72-75, ASSOC PROF PHYSICS, UNIV WIS-MILWAUKEE, 75- Mem: AAAS; Am Phys Soc; Sigma Xi. Res: Theoretical studies of the electronic properties of metal and semiconductor surfaces. Mailing Add: Dept of Physics Univ of Wis Milwaukee WI 53201

BECK, DONALD RICHARDSON, b Paterson, NJ, Mar 31, 40; m 68; c 1. QUANTUM CHEMISTRY, ATOMIC PHYSICS. Educ: Dickinson Col, BS, 62; Lehigh Univ, MS, 64, PhD(physics), 68. Prof Exp: Jr vis scientist physics, Joint Inst Lab Astrophys, Univ Colo, 68-69; res assoc chem, Yale Univ, 69-73, res assoc eng & appl sci, 73-74; ASST PROF CHEM, BELFER GRAD SCH SCI, YESHIVA UNIV, 74- Mem: Am Chem Soc; Am Phys Soc; Sigma Xi; AAAS. Res: Quantum theory of atoms, molecules and the solid state, including relativistic and correlation effects; properties; term, fine and hyperfine structure; transition probabilities; binding and Auger energies; autoionizing states. Mailing Add: Belfer Grad Sch of Sci Yeshiva Univ New York NY 10033

BECK, DORIS JEAN, b Blissfield, Mich. MICROBIAL GENETICS. Educ: Bowling Green State Univ, BS, 60; Mich State Univ, MS, 71, PhD(microbiol), 74. Prof Exp: Teacher sec educ chem, DeWitt Pub Schs, 61-63 & Holt Pub Schs, 63-68; asst microbiol, Mich State Univ, 68-71; ASST PROF MICROBIOL, BOWLING GREEN STATE UNIV, 74- Mem: Am Soc Microbiol; Sigma Xi. Res: Studies of platinum antitumor agents which cause mutations and induce filament formation in cells of Escherichia coli; recombination and excision repair of DNA damaged by platinum treatment. Mailing Add: 1301 Bourgogne Ave Bowling Green OH 43402

BECK, EDWARD C, b Spanish Fork, Utah, Feb 20, 18; m 44; c 4. PHYSIOLOGICAL PSYCHOLOGY. Educ: Brigham Young Univ, BA, 50; Univ Utah, PhD(psychol, neurophysiol), 54. Prof Exp: Res assoc physiol, Col Med, Univ Utah, 52-54; clin psychologist, 54-56, RES PSYCHOLOGIST & DIR NEUROPHYSIOL & PSYCHOPHYS LAB, VET ADMIN HOSP, 56- Concurrent Pos: Res instr psychiat, Col Med, Univ Utah, 55-58, asst res prof, 58-, neurol, 59-64, assoc res prof, 64-, asst res prof pharmacol, 60-, lectr psychol, 62-67, res prof psychol & chmn div med psychol, 67-; USPHS sr scientist, Neurobiol Res Unit, Nat Inst Hyg, France, 60-61. Mem: Am Psychol Asn; Interam Soc Psychol; Am Physiol Soc; Asn Res Nerv & Ment Dis; NY Acad Sci. Res: Electrophysiological correlates of behavior. Mailing Add: Dept of Psychol Utah State Univ Salt Lake City UT 84112

BECK, GAIL EDWIN, b Dunn Co, Wis, July 24, 23; m 50; c 3. FLORICULTURE, PLANT PHYSIOLOGY. Educ: Univ Mich, BS, 48, MS, 49; Univ Wis, PhD(floricult, plant physiol), 56. Prof Exp: Instr & exten specialist com floricult, 49-52, from instr to assoc prof floricult, 52-60, PROF FLORICULT, UNIV WIS-MADISON, 69- Mem: Am Soc Hort Sci; Am Hort Soc; Am Soc Plant Physiol; Soc Cryobiol. Res: Physiology of floriculture crops; plant stress physiology; plant-water relations. Mailing Add: Dept of Hort Univ of Wis Madison WI 53706

BECK, GLENN HANS, b Chester, Utah, July 27, 25; m 36; c 2. DAIRY HUSBANDRY. Educ: Univ Idaho, BS, 36; Kans State Col, MS, 38; Cornell Univ, PhD, 50. Prof Exp: Asst, Kans State Col, 36-37, instr dairy husb, 37-41; asst, Cornell Univ, 41-42; from asst prof to prof, Kans State Univ, 42-75, dir agr exp sta, 56-60, dean col agr, 60-75, vpres agr, 65-75; SR RES SPECIALIST, TECH ASST BUR, US AID, 75- Concurrent Pos: Head dairy dept, Univ Md, 53-56; provost, Ahmadu Bello Univ, Nigeria, 69-70; consult, Rockefeller Found, 71 & Overseas Liaison Comt, Am Coun Educ, 74- Mem: Am Soc Animal Sci; Am Dairy Sci Asn. Res: Artificial insemination; dairy cattle nutrition; mechanical milking. Mailing Add: Tech Assistance Bur US AID Washington DC 20523

BECK, HARRIS GRAYBILL, b Deland, Ill, June 24, 17; m 42; c 2. ORGANIC CHEMISTRY. Educ: Univ Chicago, BS, 39. Prof Exp: Chemist, Cent Processing Corp, 39-40; paint dir, Atlas Chem Industs Inc, 40-41, tech serv dir, 45-47, tech dir, 47-55; tech serv dir, Chicago Paint Lab, Glidden Co, 55-57, tech dir, 57-63, dir tech liaison, Int Group, 63-64, dir mkt & tech serv, 64-67, dir mkt & tech serv, Glidden-Durkee Div, 67-70, DIR DEVELOP, INT DEPT, GLIDDEN-DURKEE DIV, SCM CORP, 70- Res: Paint and plastics chemistry. Mailing Add: Int Dept SCM Glidden-Durkee Div 900 Union Com Bldg Cleveland OH 44114

BECK, HENRY NELSON, b Troy, Ohio, July 14, 27; m 54; c 2. POLYMER CHEMISTRY. Educ: Univ Mich, BS, 49, MS, 50, PhD(phys org chem), 57. Prof Exp: Chemist, Dow Corning Corp, Mich, 56-60, proj leader chem, 60-62, group leader, 62-63; res chemist, 63-64, SR RES CHEMIST, WESTERN DIV, DOW CHEM CO, 64- Mem: AAAS; Am Chem Soc; Sigma Xi. Res: Diazo and organosilicon compounds; silicones; polymer crystallization; nucleation; inorganic polymers; polymers; inorganic crystals. Mailing Add: 390 LaVista Rd Walnut Creek CA 94598

BECK, HENRY V, b Colby, Kans, July 5, 20; m 42; c 2. QUATERNARY GEOLOGY, HYDROGEOLOGY. Educ: Kans State Univ, BS, 46, MS, 49; Univ Kans, PhD(geol), 55. Prof Exp: Instr geol, Kans State Univ, 47-50; asst instr, Univ Kans, 50-52; from asst prof to assoc prof, 52-60, PROF GEOL, KANS STATE UNIV, 60- Mem: Am Asn Petrol Geol; Geol Soc Am; Sigma Xi; Nat Water Well Asn. Res: Geomorphology. Mailing Add: Dept of Geol Kans State Univ Manhattan KS 66502

BECK, IVAN THOMAS, b Budapest, Hungary, May 22, 24; nat Can; m 49; c 1. INTERNAL MEDICINE, GASTROENTEROLOGY. Educ: Univ Geneva, MD, 49; McGill Univ, dipl, 55, PhD(invest med), 63; FRCP(C). Prof Exp: Lectr pharmacol, McGill Univ, 49-52; actg head pharmacol, Univ Montreal, 54-56; res assoc, Royal Victoria Hosp, McGill Univ, 58-66, lectr, Dept Invest Med, 58-66, dir gastrointestinal lab, 60-66; assoc prof, 66-73, PROF MED, QUEENS UNIV, ONT, 73- Concurrent Pos: Assoc physician & gastroenterologist in chg, St Mary's Hosp, 60-66. Mem: Fel Am Col Physicians; Am Soc Pharmacol; Pharmacol Soc Can; Can Physiol Soc; Can Asn Gastroenterol (secy, 60, pres, 67). Res: Pharmacology; physiopathology and therapy of the gastrointestinal tract. Mailing Add: Dept of Med Queens Univ Kingston ON Can

BECK, JACOB WALTER, b Doylestown, Pa, Sept 4, 13; m 45; c 1. MEDICAL PARASITOLOGY. Educ: Pa State Col, BS, 36; Emory Univ, MS, 48; Rice Inst, PhD(parasitol), 50. Prof Exp: Asst prof bact & parasitol, Sch Med, Univ Ark, 51-52; in chg helminth unit, Parasitol & Mycol Sect, USPHS, 52-53; from asst prof to assoc prof parasitol, Sch Med, Univ Miami, 53-68; chmn dept allied health technol, 68-72, PROF ALLIED HEALTH STUDIES, MIAMI-DADE JR COL, 68- Concurrent Pos: China Med Bd fel to Cent Am, 55; vis prof parasitol, Sch Med, Marquette Univ. Mem: Am Soc Parasitol; Am Soc Trop Med & Hyg. Res: Immunology in trichinosis; chemotherapy of parasitic diseases. Mailing Add: Miami-Dade Med Ctr Campus NW Tenth Ave & 20th St Miami FL 33167

BECK, JAMES DONALD, b York Co, Pa, Dec 24, 40; m 62; c 2. INORGANIC CHEMISTRY. Educ: Kutztown State Col, BS, 62; Univ Del, PhD(chem), 69. Prof Exp: Teacher chem, Piscataway Twp High Sch, NJ, 62-64; asst, Univ Del, 64-68; asst prof chem, 68-69, interim head dept, 71, head dept, 71-72, ASSOC PROF CHEM, VA STATE COL, 69- Mem: Sigma Xi; Am Chem Soc; Nat Asn Sci Teachers; AAAS. Res: Calorimetry of Group IIIA halide complexes; uptake and degradation of selected herbicides in peanuts; uptake of cadmium by plants from sewage sludges. Mailing Add: Dept of Chem Va State Col Petersburg VA 23803

BECK, JAMES S, b Dallas, Tex, Dec 8, 31; m 62; c 3. BIOPHYSICS. Educ: Wash Univ, MD, 57; Univ Calif, Berkeley, PhD(biophys), 62. Prof Exp: Intern, Mt Zion Hosp & Med Ctr, San Francisco, 57-58; Nat Cancer Inst fel, Univ Calif, Berkeley, 58-62, assoc res biophysicist, Lawrence Radiation Lab, 62; asst prof physiol & lectr physics, Univ Minn, Minneapolis, 62-69; ASSOC PROF MED BIOPHYS, UNIV CALGARY, 69- Res: Mathematical biology; transport properties and structure of plasma membranes. Mailing Add: Div of Med Biophys Univ of Calgary Fac of Med Calgary AB Can

BECK, JAY VERN, b American Fork, Utah, Jan 15, 12; m 31; c 6. BIOCHEMISTRY, MICROBIOLOGY. Educ: Brigham Young Univ, AB, 33, AM, 36; Univ Calif, PhD(soil microbiol), 40. Prof Exp: Instr chem & math, High Sch, Utah, 35-36 & Dixie Jr Col, 36; technician plant nutrit, Univ Calif, 36-39; from jr chemist to assoc chemist, Food & Drug Admin, Fed Security Agency, 39-44; asst prof chem, Univ Idaho, 44-46; microbiologist, Pa Grade Crude Oil Asn, 46-47; from asst prof to assoc prof bact, Pa State Col, 47-51; PROF BACT, BRIGHAM YOUNG UNIV, 51- Concurrent Pos: Spec instr, Univ San Francisco, 38; Guggenheim Found fel, Univ Sheffield, 57-58; spec fel, NIH, 65- Mem: AAAS; Am Chem Soc; Am Soc Microbiol. Res: Uric acid as an index of filth in foods; minor element deficiency in plants; Utah sorgo syrup; purine fermentation by bacteria; purine determinations; microbial oxidation of sulfide minerals. Mailing Add: Dept of Microbiol Brigham Young Univ Provo UT 84601

BECK, JEANNE CRAWFORD, b Mt Pleasant, Pa, Mar 18, 43; m 66. BIOCHEMISTRY. Educ: Wilson Col, AB, 65; Johns Hopkins Univ, PhD(biochem), 69. Prof Exp: USPHS-Nat Cancer Inst fel, Sch Med, Johns Hopkins Univ, 69-71 & Biol Labs, Harvard Univ, 71-74; SR STAFF FEL, GERONT RES CTR, NAT INST AGING, 74- Mem: Am Chem Soc; Genetics Soc Am; Am Soc Cell Biol. Res: Role of membrane components in membrane structure and function; mechanism of transport of inorganic ions, amino acids, and sugars in kidney brush border; alteration of membrane function in aging tissues. Mailing Add: Geront Res Ctr Baltimore City Hosps Baltimore MD 21224

BECK, JOHN CHRISTIAN, b Audubon, Iowa, Jan 4, 24; m; c 1. INTERNAL MEDICINE. Educ: McGill Univ, BSc, 43, MD, CM, 47, MSc & dipl, 51; FRCP, 64. Prof Exp: Clin asst, Royal Victoria Hosp, Montreal, 52-54, asst physician, 55-57, assoc physician & chief endocrine-metab serv, 57-64; res assoc clin & lectr med & clin med, 54-55, from asst prof to assoc prof, 55-64, PROF MED, McGILL UNIV, 64-; PROF MED, UNIV CALIF, SAN FRANCISCO, 75- Concurrent Pos: Markel scholar, 54-59; res fel clin, McGill Univ, 55-57; dir univ clin & physician-in-chief, Royal Victoria Hosp, 64-74; dir, Robert Wood Johnson Clin Scholars Prog, San Francisco, 75- Mem: Endocrine Soc; Am Soc Clin Invest; Am Fedn Clin Res; fel Am Col Physicians; Int Soc Endocrinol (secy-gen). Res: Endocrinology; metabolism. Mailing Add: Clin Scholars Prog Suite 310 350 Parnassus San Francisco CA 94117

BECK, JOHN EDWIN, b Charleston, SC, Nov 17, 15; m 40; c 3. MEDICAL ADMINISTRATION. Educ: Ala Polytech Inst, BS, 35; Univ Va, MD, 39. Prof Exp: Physician, Decatur, Ga, 43-59; assoc med dir, Pfizer Labs, 59-62; assoc med dir, 62-63, assoc clin res dir, 64-66, CLIN RES DIR, SQUIBB INST MED RES, 67- Mem: AMA; Am Soc Microbiol. Res: Internal medicine; infectious diseases; clinical research. Mailing Add: Squibb Inst Med Res New Brunswick NJ 08903

BECK, JOHN LOUIS, b Newark, NJ, Aug 29, 31; m 73; c 3. ANALYTICAL CHEMISTRY. Educ: Seton Hall Univ, BS, 53, MSc, 67, PhD(anal chem), 69. Prof Exp: From chemist to sr res chemist, Merck & Co, Inc, 55-70, res fel & sect leader, 70-72; dir qual control, Am Hoechst Corp, Somerville, NJ, 72-74; VPRES QUAL CONTROL, J B WILLIAMS CO, INC, CRANFORD, NJ, 74- Mem: Am Pharmaceut Asn; Pharmaceut Mfrs Asn; Am Soc Qual Control; Am Soc Mass Spectroscopy; Am Chem Soc. Res: Mass spectroscopy, metabolism and structure determination with particular emphasis on the use of spectroscopy. Mailing Add: 536 Oak Ridge Rd Clark NJ 07066

BECK, JOHN R, b Las Vegas, NMex, Feb 26, 29; m 51; c 4. VERTEBRATE ECOLOGY, ECONOMIC BIOLOGY. Educ: Okla A&M Col, BS, 50; Okla State Univ, MS, 57. Prof Exp: Wildlife asst, King Ranch, Tex, 50-51; asst instr zool, Okla State Univ, 51-53; control agt, US Fish & Wildlife Serv, 53-54, res biologist, 55-57, control biologist, 57-65; instr physiol, Univ Tenn, 54-55; dir, Treasure Lake Job Corps Conserv Ctr, US Dept Interior, 65-67, state supvr div wildlife serv, Bur Sport Fisheries & Wildlife, 67-69; DIR QUAL CONTROL, BIO-SERV CORP, 69-, VPRES, 70- Concurrent Pos: Lectr, Ohio State Univ, 59-65; grain sanit consult, 59-; guest lectr, Bowling Green State Univ, 60-; chmn pesticide adv comt, Ferris State Col, 73- Mem: Wildlife Dis Asn; Wildlife Soc; fel Royal Soc Health; Am Soc Testing and Mat; Nat Pest Control Asn (dir (dir, 73-75). Res: Vertebrate pest control; development of avicides and control methods; zoonoses control; grain sanitation and quality control. Mailing Add: Bio-Serv Corp 1130 Livernois Troy MI 48084

BECK, JONATHON MOCK, b Lansing, Mich, Nov 11, 35. MATHEMATICS. Educ: Univ Mich, BS, 56, MA, 58; Columbia Univ, PhD(math), 64. Prof Exp: Instr nath, Univ Ill, 64-65; asst prof, Cornell Univ, 65-74; ASSOC PROF MATH, UNIV PR, 74- Concurrent Pos: Air Force Off Sci Res fel, 66-67. Mem: Am Math Soc. Res: Theory of categories, with applications in algebra and in algebraic topology. Mailing Add: Dept of Math Univ of PR Rio Piedras PR 00931

BECK, KARL MAURICE, b Belleville, Ill, June 16, 22; m 45; c 3. ORGANIC CHEMISTRY. Educ: Monmouth Col, BS, 43; Univ Ill, PhD(org chem), 48. Prof Exp: Asst chem, Univ Ill, 43, mem, Nat Defense Res Comt, Off Sci Res & Develop, 44-45, mem, Comt Med Res, 45-46; res chemist, Abbott Labs, 48-53, tech serv rep, 53-56, head tech serv, Chem Mkt Div, 57-62, sci employ, 62-64; vpres, Cyclamate Corp Am, 64-66; mgr prod planning & serv, 67-70, MGR CHEM DIV, NEW PROD RES, ABBOTT LABS, 70- Concurrent Pos: Tech ed, Food Prod Develop, 67- Mem: Am Chem Soc; Inst Food Technologists; Soc Soft Drink Technol; Soc Plastics Eng. Res: Synthetic organic chemistry; chemical warfare with poison gases; reduction of nitro carboxylic esters; piperazines; diuretics; phenolic Mannich bases; sweetening agents; food and beverage technology; food chemicals; plastics additives; chemical antimicrobial agents. Mailing Add: 224 E Sheridan Rd Lake Bluff IL 60044

BECK, KEITH RUSSELL, b Hudson, Mich, Apr 25, 44; m 68; c 2. SYNTHETIC ORGANIC CHEMISTRY. Educ: Adrian Col, BS, 65; Purdue Univ, PhD(chem), 70. Prof Exp: Vis asst prof & res assoc, Purdue Univ, 69-70; ASST PROF CHEM, ELMHURST COL, 70- Mem: Am Chem Soc. Res: Isolation, identification and synthesis of insect sex pheromones. Mailing Add: Dept of Chem Elmhurst Col Elmhurst IL 60126

BECK, LEONARD H, organic chemistry, see 12th edition

BECK, LLOYD, b Chatham, Ont, Nov 1, 22; m 50; c 4. PHARMACOLOGY. Educ: Univ Western Ont, BSc, 48, MSc, 50, PhD(med physiol), 53. Prof Exp: Res assoc pharmacol, Univ Mich, 53-54, from instr to prof, 54-68; prof & chmn dept, Univ NMex, 69-71; dir progs non-med use drugs, Dept Health & Welfare & sr sci adv basic res in Can, Ministry State for Sci & Technol, Ottawa, Ont, 71-72; prof pharmacol & head cardiovasc div, Sch Med, Univ NC, Chapel Hill, 72-73; PROF PHARMACOL & HEAD DEPT, SCH MED, UNIV MINN, DULUTH, 73- Concurrent Pos: Mem med adv comt, Coun High Blood Pressure Res, Am Heart Asn. Mem: Am Soc Pharmacol & Exp Therapeut; Pharmacol Soc Can; Int Soc Biochem Pharmacol. Res: Autonomic control of vasculature; shock; hypertension; autocoids; autoradiography. Mailing Add: Dept of Pharmacol Univ of Minn Sch of Med Duluth MN 55812

BECK, LLOYD WILLARD, b Batesville, Ind, Aug 10, 19; m 45; c 3. CHEMISTRY. Educ: DePauw Univ, AB, 41; Univ Wis, PhD(org chem), 44. Prof Exp: Res chemist, 44-55, assoc dir res div, 55-71, ASSOC DIR PROF & REGULATORY SERV DIV, PROCTER & GAMBLE CO, 71- Mem: AAAS; Am Chem Soc; Soc Invest Dermat; Am Inst Nutrit; Soc Toxicol. Res: Synthesis of compounds related to female sex hormones; synthetic lubricants; food chemistry; nutrition; metabolism; toxicology. Mailing Add: Ivorydale Tech Ctr Procter & Gamble Co Cincinnati OH 45217

BECK, LYLE VIBERT, b Lebanon, Ind, Apr 19, 06; m 40. PHARMACOLOGY. Educ: Wabash Col, AB, 28; Washington Univ, MS, 30; PhD(physiol), 33. Prof Exp: Asst physiol, Washington Univ, 28-30; asst, NY Univ, 30-31; asst, Univ Pittsburgh, 31-33; Lilly fel, NY Univ, 33-34; res biologist, Univ Pa, 34-35; Van Cott fel, Long Island Col Med, 35-36; Commonwealth fel, Sch Med, Univ Pa, 36-38; instr & assoc prof physiol, Hahnemann Med Col, 38-47; physiologist, Nat Cancer Inst, 47-50; assoc prof physiol & pharmacol, Sch Med, Univ Pittsburgh, 50-61; prof pharmacol, 61-76, EMER PROF PHARMACOL, SCH MED, IND UNIV, BLOOMINGTON, 76- Concurrent Pos: Mem corp, Marine Biol Lab. Mem: AAAS; Endocrine Soc; Am Pharmacol Soc; Am Physiol Soc; Am Chem Soc. Res: Kidney function; action of phlorizin; action of various substances on tumors and hosts; sulphur metabolism as affected by trauma, irradiation and diet; assay and metabolism of labeled and unlabeled insulins; luteinizing hormone and follicle-stimulating hormone Secretion. Mailing Add: Dept of Pharmacol 302 Myers Hall Ind Univ Sch of Med Bloomington IN 47401

BECK, MAE LUCILLE, b Buffalo, NY, Mar 27, 30. ORGANIC CHEMISTRY. Educ: Mich State Univ, BS, 51; Smith Col, AM, 55; Univ Pa, PhD(org chem), 60. Prof Exp: Polymer chemist, E I du Pont de Nemours & Co, 51-53; instr chem, Smith Col, 55-56; asst prof, 60-67, ASSOC PROF CHEM, SIMMONS COL, 67- Concurrent Pos: NIH grant, 62-64; res fel, Harvard Univ, 66-67. Mem: Am Chem Soc. Res: Organic sulfur chemistry; antimetabolites; organic natural products and drugs; environmental health. Mailing Add: Dept of Chem Simmons Col Boston MA 02115

BECK, MYRL EMIL, JR, b Redlands, Calif, May 13, 33; m 55; c 3. GEOPHYSICS. Educ: Stanford Univ, BA, 55, MS, 60; Univ Calif, Riverside, PhD(geol), 68. Prof Exp: Asst paleomagnetism, Dept Geophys, Stanford Univ, 61-62; geologist, Standard Oil Co Calif, 62; geologist, US Geol Surv, 62-66; assoc prof geol, 69-73, PROF GEOL, WESTERN WASH STATE COL, 73- Concurrent Pos: Res assoc, Univ Puget Sound, 72-; vis prof, Geophys Inst, Swiss Fed Inst Technol, Zurich, 73. Mem: Am Geophys Union; Soc Explor Geophys; Soc Terrestrial Magnetism & Elec Japan; Geol Soc Am; Int Asn Math Geologists. Res: Integration of paleomagnetism and other geophysical techniques for the solution of problems of regional geology. Mailing Add: Dept of Geol Western Wash State Col Bellingham WA 98225

BECK, PAUL EDWARD, b Lancaster, Pa, Jan 14, 37. ORGANIC CHEMISTRY. Educ: Franklin & Marshall Col, BS, 58; Duquesne Univ, PhD(org chem), 63. Prof Exp: Res chemist, E I du Pont de Nemours & Co, Inc, 63-66; assoc prof chem, 66-74, PROF CHEM, CLARION STATE COL, 74-, CHMN DEPT, 75- Mem: AAAS; Am Chem Soc; Am Inst Chemists; NY Acad Sci. Res: New synthetic methods; new polymers and polymer reactions; polymeric drug delivery systems. Mailing Add: Dept of Chem Clarion State Col Clarion PA 16214

BECK, PAUL W, b Crosby, Minn, Mar 28, 16; m 42; c 2. PHYSICAL CHEMISTRY. Educ: Univ Minn, BA, 40, MS, 42; Ill Inst Technol, PhD(chem), 52. Prof Exp: Res engr, Solar Mfg Corp, 43-46; res chemist, Sinclair Res Labs, 46-52; staff physicist, Philips Labs, Inc, 53-68; assoc prof, 68-74, PROF CHEM, WESTERN CONN STATE COL, 74- Mem: Am Chem Soc. Res: Magnetics; dielectrics; photochemistry. Mailing Add: Dept of Chem Western Conn State Col Danbury CT 06814

BECK, RAYMOND WARREN, b New Smyrna, Fla, Oct 20, 25; m 58; c 2. MICROBIOLOGY. Educ: Univ Fla, BS, 49, MS, 52; Univ Wis, PhD(bact), 56. Prof Exp: Asst bact, Univ Fla, 50-52 & Univ Wis, 52-55; from asst prof to assoc prof, 55-71, PROF BACT, UNIV TENN, KNOXVILLE, 71- Mem: Am Soc Microbiol. Res: Microbial physiology and biochemical virology; energy metabolism in bacteria; hydrogen production by photosynthetic bacteria; interference in Sindbis virus; and RNA synthesis by Sindbis virus. Mailing Add: Dept of Microbiol Univ of Tenn Knoxville TN 37916

BECK, ROBERT EDWARD, b Denver, Colo, June 7, 41; m 65; c 3. ALGEBRA, OPERATIONS RESEARCH. Educ: Harvey Mudd Col, BS, 63; Univ Pa, MA, 65, PhD(math), 69. Prof Exp: From instr to asst prof, 66-73, ASSOC PROF MATH, VILLANOVA UNIV, 73- Mem: Am Math Soc. Res: Computational methods in nonassociative algebras; algebraic models in differential equations, physics and biology. Mailing Add: Dept of Math Villanova Univ Villanova PA 19085

BECK, ROBERT NASON, b San Angelo, Tex, Mar 26, 28; m 58. NUCLEAR MEDICINE. Educ: Univ Chicago, AB, 54, BS, 55. Prof Exp: CHIEF SCIENTIST, ARGONNE CANCER RES HOSP, 57-; ASSOC PROF RADIOL SCI, UNIV CHICAGO, 67- Concurrent Pos: Consult, Int Atomic Energy Agency, 66-68; mem, Int Comn on Radiation Units, 68-; mem, Nat Coun on Radiation, Protection & Measurements, 70- Mem: Soc Nuclear Med; Am Asn Physicists in Med. Res: Development of an adequate theory of the process by which images can be formed of the distribution of radioactive material in a patient, in order to diagnose his disease. Mailing Add: Argonne Cancer Res Hosp 950 E 59th St Chicago IL 60637

BECK, ROLAND ARTHUR, b Mountain Iron, Minn, Apr 16, 13; m 34; c 2. INORGANIC CHEMISTRY, PHYSICAL CHEMISTRY. Educ: Maryville Col, BA, 34; Univ Minn, Minneapolis, MS, 39. Prof Exp: Instr, Pub High Schs, Minn, 34-41; res chemist, Texaco Res Labs, Texaco Inc, NY, 41-46, proj leader synthetic fuels, 46-48, supvr res, Texaco Res Lab, Calif, 49-60, dir, 60-68, mgr, Texaco Res Ctr, 68-75; SR STAFF MEM, ENERGY RES & DEVELOP ADMIN, 75- Concurrent Pos: Teaching asst, Univ Minn, Minneapolis, 38-39; conf leader, Mgt Develop Div, Calif Inst Technol, 56-68. Mem: AAAS; Am Inst Chem Eng; Am Chem Soc; Sigma Xi. Res: Coal gasification; synthetic fuels. Mailing Add: Energy Res & Develop Admin 20 Massachusetts Ave Washington DC 20028

BECK, RONALD RICHARD, b Tiltonsville, Ohio, Oct 8, 34; m 61; c 3. PHYSIOLOGY, ENDOCRINOLOGY. Educ: Univ Ohio, BS, 61; Ohio State Univ, MS, 67, PhD(physiol), 68. Prof Exp: From asst prof to assoc prof physiol, Sch Med, Ind Univ, Indianapolis, 68-75; ASSOC PROF PHYSIOL, SCH MED, UNIV SC, 75- Concurrent Pos: NIH grant, 72-74. Mem: Am Physiol Soc; Endocrine Soc. Res: Prostaglandins in blood pressure regulation. Mailing Add: Dept of Physiol & Pharmacol Univ of SC Sch of Med Columbia SC 29208

BECK, SIDNEY L, b New York, NY, Mar 28, 35; m 55; c 2. DEVELOPMENTAL GENETICS, TERATOLOGY. Educ: City Col New York, BS, 55; Univ Kans, MA, 57; Brown Univ, PhD(biol), 60. Prof Exp: Instr zool, Univ Mich, 60-61, USPHS fel, 61-64; NIH spec fel, Univ Col, London, 64-65, res assoc genetics, 65; from asst prof to assoc prof biol, Univ Toledo, 65-69; PROF BIOL & CHMN DEPT, WHEATON COL, MASS, 69- Concurrent Pos: Mem, Inst Lab Animal Resources Subcomt Genetic Standards, Nat Acad Sci, 66-69; biologist, US Environ Protection Agency, 75-76. Mem: AAAS; Am Soc Zoologists; Am Genetic Asn; Genetics Soc Am; Soc Develop Biol. Res: Mammalian genetics and development; teratogenesis, especially genetic contributions to differences in susceptibility and resistance to artificially induced maldevelopment; biological effects of environmental contaminants. Mailing Add: Dept of Biol Wheaton Col Norton MA 02766

BECK, SIDNEY M, b Pleasant Grove, Utah, Mar 10, 19; m 48; c 4. BACTERIOLOGY. Educ: Brigham Young Univ, AB, 41, MA, 48; Pa State Univ, PhD(bact), 51. Prof Exp: Chemist, US Bur Mines, 42-44; assoc prof bact, 51-74, PROF BIOL, UNIV IDAHO, 75- Mem: Am Chem Soc. Res: Bacterial purine metabolism and nutrition. Mailing Add: Dept of Bact Univ of Idaho Moscow ID 83843

BECK, STANLEY DWIGHT, b Portland, Ore, Oct 17, 19; m 43; c 4. ZOOLOGY. Educ: State Col Wash, BS, 42; Univ Wis, MS, 47, PhD(zool), 50. Prof Exp: From instr to prof, 48-69, W A HENRY PROF ENTOM, UNIV WIS-MADISON, 69- Mem: AAAS; Entom Soc Am; Phytochem Soc NAm. Res: Insect nutrition; metabolism and endocrinology; photoperiodism; development. Mailing Add: Dept of Entom Univ of Wis Madison WI 53706

BECK, WILLIAM CARL, b Chicago, Ill, Aug 24, 07; m 48; c 3. SURGERY. Educ: Northwestern Univ, BA, 28, MD, 32. Prof Exp: Intern, Robert Packer Hosp, Sayre, Pa, 32-33; resident surg, Univ Frankfurt, Ger, 33-35; assoc surg, Med Sch, Univ Ill, 36-42; from attend to chief, Surg Dept, Guthrie Clin & Robert Packer Hosp, Sayre, 46-72, PRES, GUTHRIE FOUND MED RES, 68- Concurrent Pos: Attend surgeon, St Joseph Hosp, Chicago, 35-42; assoc surgeon, Cook County Hosp, Chicago, 36-42; lectr surg, Sch Grad Med, Univ Pa, 49-70; clin prof surg, Hahnemann Med Col, 49-66. Mem: Am Surg Asn; Int Soc Surg; fel Am Col Surg; Royal Soc Med; Illum Eng Soc. Res: Hospital and surgical environment including lighting, ultraclean ventilation, noise control and design of hospitals; cancer control; delivery of health service. Mailing Add: Guthrie Found Guthrie Square Sayre PA 18840

BECK, WILLIAM J, b Fredericktown, Mo, Aug 2, 21; m 47; c 2. ENVIRONMENTAL HEALTH, BACTERIOLOGY. Educ: Southeast Mo State Col, BS, 47; Univ Mo-Columbia, MA, 50; Univ Mich, MPH, 59, PhD(environ health), 70. Prof Exp: Chief lab technician bacteriol, Vet Admin Hosp, 47-49; environ bacteriologist, Mo Div Health Labs, 50-58; food technologist, Washington, DC Health Dept, 60-61; chief Northwest Water Hyg Lab, 61-69, CHIEF ARCTIC HEALTH RES CTR, USPHS, 69- Concurrent Pos: Lectr, Sch Fisheries, Univ Wash, 65-69. Mem: Am Pub Health Asn; Nat Shellfisheries Asn; Conf State & Prov Pub Health Lab Dirs. Res: Public health aspects of milk, food, fresh and estuarine water; environmental effects of man's health and well being in the Arctic and sub-Arctic. Mailing Add: Arctic Health Res Ctr College AK 99701

BECK, WILLIAM NELSON, b Chicago, Ill, Dec 16, 23; m 48; c 3. MATHEMATICS, PHYSICS. Educ: Dakota Wesleyan Univ, BA, 46. Prof Exp: Instr physics, Dakota Wesleyan Univ, 46-49; electronics engr, Aircraft Div, Globe Corp, 51-53, chief field serv eng, 53-54; ASSOC PHYSICIST, ARGONNE NAT LAB, 54- Mem: Am Nuclear Soc. Res: Development and evaluation of nuclear reactor fuel materials; applications of ultrasonics in the field of nondestructive testing; neutron radiography techniques for irradiated materials. Mailing Add: Argonne Nat Lab EBR-II Div 9700 Cass Ave Argonne IL 60439

BECK, WILLIAM SAMSON, b Reading, Pa, Nov 7, 23; m; c 4. BIOCHEMISTRY, HEMATOLOGY. Educ: Univ Mich, BS, 43, MD, 46. Hon Degrees: AM, Harvard Univ, 70. Prof Exp: Clin instr, Med Sch, Univ Calif, Los Angeles, 50-53, instr, 53-55, asst prof med, 55-57; asst prof, 57-69, ASSOC PROF MED, HARVARD UNIV, 69- Concurrent Pos: Fel, NY Univ, 55-57; chief hemat & med sects, Atomic Energy Proj, Univ Calif, Los Angeles, 51-57; estab investr, Am Heart Asn, 55-; tutor biochem sci, Harvard Univ, 57-; chief hemat unit, Mass Gen Hosp, Boston, 57-; mem hemat study sect, NIH, 67-71; mem adv coun, Nat Inst Arthritis, Metab & Digestive Dis, 72-75. Honors & Awards: Wenner-Gren Prize, 55. Mem: AAAS; Am Chem Soc; Am Soc Biol Chem; Am Soc Clin Invest; Am Asn Cancer Res. Res: Vitamin B12; nucleic acid and bacterial metabolism; enzymology; blood cell biochemistry. Mailing Add: Hemat Res Lab Mass Gen Hosp Boston MA 02114

BECKEL, CHARLES LEROY, b Philadelphia, Pa, Feb 7, 28; m 58; c 4. THEORETICAL PHYSICS. Educ: Univ Scranton, BS, 48; Johns Hopkins Univ, PhD(physics), 54. Prof Exp: Asst, Johns Hopkins Univ, 49; asst, Sch Pharm, Univ

Md, 49-53; from asst prof to assoc prof physics, Georgetown Univ, 53-64; mem res staff, Inst Defense Anal, 64-66; assoc prof, 66-69, asst dean grad sch, 71-72, PROF PHYSICS, UNIV NMEX, 69- Concurrent Pos: Consult, Ballistics Res Lab, Aberdeen Proving Ground, Md, 55-57, Inst Defense Anal, 62-64 & 66-69, Dikewood Corp, 67-72 & 74- & Albuquerque Urban Observ, 69-71; Fulbright lectr, Univ Peshawar, 57-58 & Cheng Kung Univ, Taiwan, 63-64; actg dir, Inst social Res & Develop, Univ NMex, 72, actg vpres res, 72-73; vis prof theoret chem, Univ Oxford, 73. Mem: Am Phys Soc; Am Asn Physics Teachers. Res: Theoretical aspects of the structure of diatomic molecules; quantum mechanics; operations research; quantum biology; study of biomolecule conformation. Mailing Add: Dept of Physics & Astron Univ of NMex Albuquerque NM 87131

BECKEL, WILLIAM EDWIN, b Kingston, Ont, Apr 11, 26; m 53; c 3. ZOOLOGY. Educ: Queen's Univ (Ont), BA, 49, Univ Iowa, MSc, 53; Cornell Univ, PhD(entom), 55. Prof Exp: Head entom sect, Defense Res North Lab, 48-55; assoc entomologist physiol, Entom Div Can Dept Agr, 55-56; from asst prof to assoc prof zool, Univ Toronto, 56-63, prof & dean sci Scarborough Col, 64-68; acad vpres, 68-74, PROF BIOL SCI, UNIV LETHBRIDGE, 74- Mem: AAAS; Am Soc Zool; Entom Soc Can; Can Soc Zool. Res: Gross anatomy; histology and physiology of invertebrate animals; insects. Mailing Add: Univ of Lethbridge Lethbridge AB Can

BECKEN, BRADFORD ALBERT, b Providence, RI, Oct 5, 24; m 46; c 4. ACOUSTICS. Educ: US Naval Acad, BS, 46; US Naval Postgrad Sch, BS, 52; Univ Calif, Los Angeles, MS, 53, PhD(physics), 61. Prof Exp: Sonar prof officer, Oper Test & Eval Force, US Navy, 53-56, head surface ship sonar design br, Bur Ships, 56-58, sonar prog officer, US Navy Electronics Lab, 60-63, mgr ship sonar prog, Bur Ships, 63-65, head sonar adv develop br, Submarine Warfare Proj Off, 65-67; mgr syst eng lab, 67-68, prog mgr, 68-69, MGR ENG, SUBMARINE SIGNAL DIV, RAYTHEON CO, 70- Concurrent Pos: Consult, Airtronics Inc, Dulles Int Airport, DC, 67- Mem: Acoust Soc Am. Res: Directional distribution of ambient noise in the ocean; all aspects of the design, development, test, evaluation and operation of sonar systems for the detection of submarines. Mailing Add: Submarine Signal Div Raytheon Co Portsmouth RI 02871

BECKENBACH, EDWIN FORD, b Dallas, Tex, July 18, 06; m 33, 60; c 3. MATHEMATICAL ANALYSIS, GEOMETRY. Educ: Rice Univ, AB, 28, AM, 29, PhD(math), 31. Prof Exp: Nat res fel math, Princeton Univ, Ohio State Univ & Univ Chicago, 31-33; instr, Rice Univ, 33-40; asst prof, Univ Mich, 40-42; assoc prof, Univ Tex, 42-45; from assoc prof to prof, 45-74, EMER PROF MATH, UNIV CALIF, LOS ANGELES, 74- Concurrent Pos: Assoc ed, Am Math Monthly, 44-51; with Inst Numerical Anal, Nat Bur Standards, 48-49; mathematician, Rand Corp, 49-69; ed, Pac J Math, 51-; Guggenheim fel, 58-59; vis prof, Univ Del, 75-76. Mem: AAAS; Am Math Soc; Math Asn Am; Math Soc France; Indian Math Soc. Res: Meromorphic minimal surfaces; surfaces of negative curvature; convex, harmonic and subharmonic functions; complex variable theory; conformal mapping; potential theory. Mailing Add: Dept of Math Univ of Calif Los Angeles CA 90024

BECKENSTEIN, EDWARD, b New York, NY, Oct 21, 40. MATHEMATICAL ANALYSIS. Educ: Polytech Inst Brooklyn, BS, 62, MS, 64, PhD(math), 66. Prof Exp: From instr to asst prof math, Polytech Inst Brooklyn, 65-67; asst prof, St John's Univ, NY, 67-68; from asst prof to assoc prof, Polytech Inst Brooklyn, 68-72; ASSOC PROF MATH, ST JOHN'S UNIV, NY, 72- Res: Abstract algebra; functional analysis; theory of commutative Banach algebras. Mailing Add: Dept of Natural Sci St John's Univ Staten Island NY 10301

BECKER, AARON JAY, b Brooklyn, NY, Apr 28, 40; m 65; c 2. PHYSICAL CHEMISTRY, HIGH TEMPERATURE CHEMISTRY. Educ: Brooklyn Col, BS, 61; Univ Wash, MS, 64; Ill Inst Technol, PhD(chem), 71. Prof Exp: Scientist chem, IIT Res Inst, 65-68; fel, McMaster Univ, 71-73; SCIENTIST CHEM, ALCOA RES LAB, 73- Mem: Am Chem Soc; Sigma Xi. Res: Production of metals from ores; fused salt chemistry; utilization of natural resources; production of chemicals from coal. Mailing Add: Alcoa Res Labs Alcoa Center PA 15069

BECKER, ADRIAN ANTHONY, b Wien, Mo, Dec 21, 26; m 48; c 2. GEOPHYSICS. Educ: St Louis Univ, BS, 49. Prof Exp: Asst computer geophys explor, 49-50, res geophysicist, 50-51, from assoc res geophysicist to res geophysicist, 51-57, geophysicist, 57-59, actg asst regional geophysicist, 59-62, div geophysicist, 62-68, AREA GEOPHYSICIST, CONTINENTAL OIL CO, 68- Mem: Soc Explor Geophys. Res: Geophysical exploration methods. Mailing Add: 811 Thornwick Dr Houston TX 77024

BECKER, BARBARA, b Chicago, Ill, Jan 10, 32. BIOCHEMISTRY. Educ: Marymount Col, NY, BA, 54; Cath Univ, MA, 55; Georgetown Univ, PhD(chem), 66. Prof Exp: Instr chem, Marymount Jr Col, Va, 55-62; instr, Marymount Manhattan Col, 65-66; ASST PROF CHEM, MARYMOUNT COL, NY, 66-, DEAN STUDENTS, 68- Mem: AAAS; Am Chem Soc. Res: Fatty acid synthesis; enzyme purification. Mailing Add: Dept of Chem Marymount Col Tarrytown NY 10591

BECKER, BENJAMIN, b New York, NY, Apr 22, 16; m 51; c 2. MICROBIOLOGY, BIOCHEMISTRY. Educ: Rutgers Univ, BS, 37, MS, 62, PhD(microbiol), 65. Prof Exp: Asst prof microbiol, biochem & gen biol, Hamilton Col, 65-69; assoc prof cell biol, 69-75, PROF CELL BIOL, PURDUE UNIV, 75- Mem: Am Soc Microbiol; Am Chem Soc; Am Inst Biol Sci. Res: Actinomycetes; cell wall analyses; microbial transformations; asparaginase in leukemia; leprosy immunogens; new antifungal agents; antibiotics; nonspecific immunostimulaters. Mailing Add: Dept of Biol Purdue Univ Ft Wayne IN 46805

BECKER, BERNARD ABRAHAM, b Chicago, Ill, May 7, 20; m 44; c 4. TOXICOLOGY. Educ: Roosevelt Univ, BS, 41; Univ Southern Calif, MS, 51; Univ Iowa, PhD, 64. Prof Exp: Anal chemist, Armour Labs, 45-47, sr pharmacologist & head res prod control sect, 51-57; head res pharmacol sect, Strasenburgh Labs, 57-62; NIH spec fel, 63-64; from asst prof to prof pharmacol, Univ Iowa, 64-72; DIR TOXICOL, ABBOTT LABS, 72- Mem: AAAS; Soc Pharmacol & Exp Therapeut; Soc Exp Biol & Med; Soc Toxicol; NY Acad Sci. Res: Toxicology; drug induced teratology; drug evaluation. Mailing Add: Dept of Toxicol Abbott Labs D-468 North Chicago IL 60064

BECKER, CARL GEORGE, b Philadelphia, Pa, Mar 18, 36; m 61; c 3. PATHOLOGY. Educ: Yale Univ, BS, 57; Cornell Univ, MD, 61. Prof Exp: From intern to resident path, New York Hosp-Cornell Med Ctr, 61-66, asst prof, 66-68; pathologist, Naval Hosp. St Albans, 68-70; ASSOC PROF PATH & ASSOC PATHOLOGIST, NEW YORK HOSP-CORNELL MED CTR, 70- Concurrent Pos: USPHS trainee path, Med Col, Cornell Univ, 62-66; mem thrombosis coun, Am Heart Asn. Mem: Am Soc Exp Path; Harvey Soc; Am Asn Path & Bact; Sigma Xi. Res: Arteriosclerosis; rheumatic heart disease; renal disease; immunopathology; hemostasis and thrombosis; microbiology. Mailing Add: New York Hosp-Cornell Med Ctr 525 E 68th St New York NY 10021

BECKER, CHARLES BRUNNER, chemistry, deceased

BECKER, CHARLES EDWARD, b Logansport, Ind, Dec 5, 12. BIOCHEMISTRY. Educ: Ind Univ, AB, 47, AM, 48, PhD(biochem), 51; Am Bd Clin Chem, dipl. Prof Exp: Clin lab technician, Med Dept, US Army, 36-40; control chemist, Automotive Div, Studebaker Corp, 41-42; res biochemist, Endocrinol Dept, Upjohn Co, 51; res assoc biochem, Sch Pub Health, Harvard Univ, 52-55; res biochemist, Allied Sci Div, Ft Detrick, Md, 55-56; biochemist & chief med res lab, Vet Admin Hosp, Louisville, Ky, 56-57; res biochemist, Am Dent Asn, Ill, 58-62; assoc prof, 62-67, PROF DENT, UNIV DETROIT, 67- Mem: Am Chem Soc; fel Am Asn Clin Chemists; fel Am Inst Chemists; Health Physics Soc. Res: Intermediary metabolism and clinical chemistry. Mailing Add: Univ of Detroit Sch of Dent 2985 E Jefferson Ave Detroit MI 48207

BECKER, CHARLES HENRY, b Chicago, Ill, May 26, 14; m 40. PHARMACY. Educ: Univ Ill, BS, 37; Univ Fla, MS, 39, PhD(pharm), 40. Prof Exp: Res agron, Exp Sta, Univ Fla, 40; prof pharm & chem & head dept, Duquesne Univ, 40-47; assoc prof, 47-54, PROF PHARM, UNIV FLA, 54-, CHMN DEPT, 61-, ASST DEAN COL PHARM & GRAD STUDIES COORDR, 74- Concurrent Pos: Consult chemist, Pittsburgh, Pa, 41-47; chief chemist, Balch Flavor Co, 41-47. Res: Flavors for bottlers, foods and pharmacy; pharmaceutical emulsion; pharmaceutical product development and formulation. Mailing Add: Col of Pharm Univ of Fla Gainesville FL 32601

BECKER, CLARENCE DALE, b Albany, Ore, Aug 17, 30; m 62; c 4. FISHERIES. Educ: Ohio State Univ, BS, 53, MS, 55; Univ Wash, PhD(fisheries), 64. Prof Exp: Fisheries biologist, Fisheries Res Inst, Univ Wash, 55-59, res asst, Col Fisheries, 59-63, res assoc, 63-67; res scientist, 67-70, SR RES SCIENTIST, PAC NORTHWEST LABS, BATTELLE MEM INST, 70- Mem: Am Soc Parasitol; Am Fisheries Soc; Am Inst Fisheries Res Biol; Am Micros Soc; Helminth Soc Wash. Res: Fisheries biology; parasitology; haematozoa of fishes; limnology; freshwater ecology; environmental impacts; monitoring. Mailing Add: Ecosysts Dept Battelle-Northwest PO Box 999 Richland WA 99352

BECKER, CLIFFORD ANDREW L, inorganic chemistry, theoretical chemistry, see 12th edition

BECKER, DAVID ALVORD, b Syracuse, NY, Sept 12, 28; m 55; c 2. ZOOLOGY. Educ: Colo Col, AB, 54; NC State Col, MS, 57; Univ Nebr, PhD(zool), 66. Prof Exp: Asst dir labs & parasitologist, Div Labs, Nebr State Health Dept, 57-60; from instr to asst prof, 60-70, ASSOC PROF ZOOL, UNIV ARK, FAYETTEVILLE, 70- Mem: AAAS; Am Inst Biol Sci; Am Soc Parasitol; Am Soc Trop Med & Hyg. Res: Medical parasitology; helminth and copepod parasites of black basses; avian schistosomes. Mailing Add: Dept of Zool Univ of Ark Fayetteville AR 72701

BECKER, DAVID VICTOR, b New York, NY, May 24, 23; m 49; c 2. MEDICINE. Educ: Columbia Univ, AB, 43, MA, 44; NY Univ, MD, 48. Prof Exp: Rotating intern, Sinai Hosp, Baltimore, Md, 48-49; med intern, Maimonides Hosp, Brooklyn, 49-50; Runyan fel, Dept Clin Invest & Biophys, Sloan-Kettering Inst, New York, 50-52; chief radioisotope unit, Surg Res Unit, Brooke Army Med Ctr, Ft Sam Houston, Tex, 52-54; asst resident med serv, New York Hosp-Cornell Med Ctr, 54-55, radiologist, 57-62; from instr med & radiol to asst prof med, 57-61; assoc prof med, 61-75, PROF RADIOL, COL MED, CORNELL UNIV, 61-, PROF MED, 75-; DIR DIV NUCLEAR MED, NEW YORK HOSP-CORNELL MED CTR, 55- Concurrent Pos: Asst attend physician, New York Hosp-Cornell Med Ctr, 57-62, assoc attend physician, 62-74, attend radiologist, 71-; mem, Mayor's Tech Adv Comt Radiation, New York, 73-; mem med adv comt, Bur Radiol Health, NY State Dept Health, 73-; mem task force short lived radionuclides, Bur Radiol Health, Food & Drug Admin, 75-; consult, Pan-Am WHO, 75- Mem: Fel Am Col Physicians; Am Thyroid Asn; Soc Nuclear Med; Endocrine Soc; Am Fedn Clin Res. Res: Endocrinology; thyroid physiology and disease; nuclear medicine. Mailing Add: 525 E 68th St New York NY 10021

BECKER, DONALD A, b Valley City, NDak, July 27, 38; m 60; c 3. BOTANY, PLANT ECOLOGY. Educ: Valley City State Col, BS, 60; Univ NDak, MS, 65, PhD(ecol), 68. Prof Exp: High sch teacher, Wyo, 60-62; instr biol, Univ NDak, 67-68; from asst prof to assoc prof biol, Midland Lutheran Col, 68-75; ENVIRON SPECIALIST, MISSOURI RIVER BASIN COMN, OMAHA, 75- Concurrent Pos: NSF res fel, Okla State Univ, 70-71; Cottrell res grant, 75. Mem: Bot Soc Am; Soc Range Mgt; Sigma Xi; Am Inst Biol Sci. Res: Dispersal ecology; autoecology; abscission; nitrogen fixation in legumes. Mailing Add: 2116 Howard St Fremont NE 68025

BECKER, DONALD EUGENE, b Delavan, Ill, Feb 2, 23; m 49; c 5. ANIMAL NUTRITION, ANIMAL PHYSIOLOGY. Educ: Univ Ill, BS, 45, MS, 47; Cornell Univ, PhD(nutrit), 49. Prof Exp: Asst animal sci, Univ Ill, 45-47; asst animal husb, Cornell Univ, 47-49; assoc prof animal nutrit, Univ Tenn, 49-50; from asst prof to assoc prof animal nutrit, PROF ANIMAL NUTRIT, UNIV ILL, URBANA-CHAMPAIGN, 58-, HEAD DEPT ANIMAL SCI, 67- Concurrent Pos: Mem subcomt nutrient requirements of swine, Nat Res Coun; ed, J Animal Sci, 66-69. Honors & Awards: Am Feed Mfrs Award, 57. Mem: Fel AAAS; Am Soc Animal Sci (pres, 70-71); Animal Nutrit Res Coun; Poultry Sci Asn; Am Inst Nutrit. Res: Metabolic function of cobalt in ruminant nutrition; chemistry and morphology of ovine blood; protein and amino acid nutrition for pregnancy, lactation, growth and fattening; antibiotics and nonruminant nutrition; comparative value of carbohydrates and available energy values of feeds. Mailing Add: 328 Mumford Hall Univ of Ill Dept of Animal Sci Urbana IL 61801

BECKER, EDWARD BROOKS, b Emporia, Kans, Aug 12, 31; m 59; c 3. INORGANIC CHEMISTRY, ENVIRONMENTAL MANAGEMENT. Educ: Kans State Teachers Col, AB, 53; Univ Kans, PhD(inorg chem), 59. Prof Exp: Sr res chemist, Chem Div, Pittsburgh Plate Glass Co, 60-61; proj mgr, Alexandria Div, Am Mach & Foundry Co, 61-63; res chemist, Gulf Res & Develop Co, 64-70; DIR BUR AIR POLLUTION CONTROL & SOLID WASTE DISPOSAL, STATE OF WIS, 70- Concurrent Pos: Chmn, State & Territorial Air Pollution Prog Admin, 74. Mem: AAAS; Am Chem Soc; fel Am Inst Chemists. Res: High temperature plasma chemistry; air pollution control; solid waste management; nitrogen-phosphorus-potassium fertilizer chemistry; phase rule. Mailing Add: 1132 University Bay Dr Madison WI 53705

BECKER, EDWARD COULTON, b St Louis, Mo, Mar 15, 23; m 48; c 5. ENTOMOLOGY. Educ: Univ Mo, BS, 44; Univ Ill, MS, 50, PhD(entom), 52. Prof Exp: RES SCIENTIST, BIOSYST RES INST, AGR CAN, 52- Mem: Entom Soc Am; Entom Soc Can (treas, 61-). Res: Systematic entomology; systematics of Elateridae. Mailing Add: Biosystematics Res Inst Agr Can Ottawa ON Can

BECKER, EDWARD SAMUEL, b Bisbee, Ariz, Sept 8, 29; m 51; c 3. PULP TECHNOLOGY, POLLUTION CONTROL. Educ: Ore State Col, BS, 51, MS, 53, PhD(forest prod chem), 57. Prof Exp: Sect leader develop pulping, Rayonier, Inc, 57-64; sect leader, G L Pulp & Papermaking Union Camp, 64-65, dept mgr, 66-68; mgr tech develop, Columbia Cellulose Co, Ltd, 69-71, dir tech & environ control, 72; PRES, ECONOTECH SERV LTD, 72- Mem: Tech Asn Pulp & Paper Indust; Am Chem Soc; Air Pollution Control Asn; Chem Inst Can; Forest Prod Res Soc. Res: Technical consulting; research and development management; technical service; pulp and paper; pilot plant operation; management consulting. Mailing Add: Econotech Serv Ltd 852 Derwent Way Annacis Island New Westminster BC Can

BECKER, EDWIN DEMUTH, b Columbia, Pa, May 3, 30; m 53; c 2. PHYSICAL CHEMISTRY. Educ: Univ Rochester, BS, 52; Univ Calif, PhD(chem), 55. Prof Exp: Asst chem, Univ Calif, 53-54, instr, 55; phys chemist, NIH, 55-68, chief sect molecular biophys, 62-68, mem fac grad prog, 63-72, asst chief lab phys biol, Nat Inst Arthritis & Metab Dis, 68-72, CHIEF LAB CHEM PHYSICS, NAT INST ARTHRITIS, METAB & DIGESTIVE DIS, 72- Concurrent Pos: Prof lectr, Georgetown Univ, 58-; sr asst scientist, USPHS, 55-58; mem adv bd, Off Critical Tables, 67-69; mem, Joint Comt Atomic & Molecular Data, 67-; mem prog comt, Exp NMR Conf, 68-70, chmn, 69; mem comn molecular structure & spectros, Int Union Pure & Appl Chem, 73-, chmn comn, 75- Honors & Awards: Coblentz Mem Prize Chem Spectros, 66; Dept Health, Educ & Welfare Superior Serv Award, 74. Mem: AAAS; Coblentz Soc; Am Soc Appl Spectros; Am Chem Soc. Res: Molecular structure; infrared spectroscopy; nuclear magnetic resonance; hydrogen bonding; free radicals. Mailing Add: Bldg 2 Rm 120 NIH Bethesda MD 20014

BECKER, EDWIN NORBERT, b Ossian, Iowa, Aug 6, 22; m 50; c 5. PHYSICAL CHEMISTRY. Educ: Iowa State Univ, BS, 47; Univ Wis, PhD(phys chem), 53. Prof Exp: Asst prof chem, Col of St THomas, 53-55; from asst prof to assoc prof, 55-64, PROF CHEM, CALIF STATE COL, LONG BEACH, 64- Mem: AAAS; Am Chem Soc. Res: Reaction mechanisms. Mailing Add: Dept of Chem Calif State Col Long Beach CA 90801

BECKER, ELMER LEWIS, b Chicago, Ill, Feb 17, 18; m 46; c 3. IMMUNOLOGY. Educ: Univ Ill, MD, 45, PhD(biochem), 47. Prof Exp: Asst prof biochem, Col Med, Univ Ill, 47-49, asst prof bact, 50-51; chief dept immunochem, Walter Reed Army Inst Res, Washington, DC, 52-69; PROF PATH, HEALTH CTR, UNIV CONN, FARMINGTON, 70- Concurrent Pos: John Simon Guggenheim Mem Found fel, 69-70. Mem: AAAS; Am Acad Allergy; Am Asn Immunol. Res: Experimental allergy; antigen-activated enzyme systems; chemotaxis; lysosomal enzyme release; phagocytosis of neutrophils. Mailing Add: Dept of Path Univ of Conn Health Ctr Farmington CT 06032

BECKER, ERNEST I, b Cleveland, Ohio, Aug 18, 18; m 47; c 5. CHEMISTRY. Educ: Western Reserve Univ, BS, 41, MS, 43, PhD(chem), 46. Prof Exp: From instr to prof chem, Polytech Inst Brooklyn, 46-65; chmn dept, 65-67, chmn div natural sci, 65-70, PROF CHEM, UNIV MASS, HARBOR CAMPUS, 65-, CHMN DEPT, 69- Concurrent Pos: Consult to indust, 57- Mem: AAAS; Am Chem Soc; fel NY Acad Sci; The Chem Soc; Nat Sci Teachers Asn. Res: Synthesis reactions and spectra of tetracyclones; synthesis of enzyme models and spectra of fulvenes and other nonclassical aromatic hydrocarbons; organotin compounds; mechanism of organomagnesium reactions. Mailing Add: Dept of Chem Univ of Mass Harbor Campus Boston MA 02125

BECKER, ERNEST LOVELL, b Cincinnati, Ohio, Jan 13, 23; m 49; c 3. INTERNAL MEDICINE. Educ: Washington & Lee Univ, AB, 44; Univ Cincinnati, MD, 48; Am Bd Internal Med, dipl. Prof Exp: Asst dept pharmacol, Col Med, Univ Cincinnati, 46-47; res physician, Dept Med, Med Col Va, 49-51; investr, Mt Desert Biol Lab, Maine, 51-52; instr dept physiol, Col Med, NY Univ, 51-53; Markle Found scholar med sci, Dept Internal Med, Med Col Va, 55-57; from asst prof to assoc prof, 57-69, PROF MED, CORNELL UNIV, 69- Concurrent Pos: Lederle med fac award, 60-63; attend physician, NY Hosp, 69- Mem: AAAS; fel Am Col Physicians; Am Physiol Soc; Soc Exp Biol & Med; Am Med Clin Res. Res: Renal physiology and disease. Mailing Add: Dept of Med NY Hosp-Cornell Med Ctr New York NY 10021

BECKER, FREDERICK F, b New York, NY, July 23, 31; m 60. PATHOLOGY, ONCOLOGY. Educ: Columbia Univ, BA, 52; NY Univ, MD, 56. Prof Exp: Intern, Boston City Hosp, 56-57; asst & actg dir exp path, US Naval Med Res Inst, 60-61; from asst prof to assoc prof, 62-70, PROF PATH, SCH MED, NY UNIV, 70-; DIR PATH, BELLEVUE HOSP, 70- Concurrent Pos: Univ fel path, Sch Med, NY Univ, 57-60; career investr, NY Health Res Coun, 62-; consult, US Naval Hosp, St Albans, NY, 62- Res: Cell division. Mailing Add: Dept of Path NY Univ Sch of Med New York NY 10016

BECKER, GERALD ANTHONY, b Ossian, Iowa, May 30, 24; m 48; c 3. MATHEMATICS. Educ: Univ Iowa, BA, 50, MS, 51, PhD(math), 59. Prof Exp: High sch teacher, Iowa, 51-56; asst math, State Univ Iowa, 56-58; from asst prof to assoc prof, 58-65, PROF MATH, SAN DIEGO STATE UNIV, 65- Mem: Math Asn Am. Res: Mathematics curriculum; mathematical models in the social sciences. Mailing Add: Dept of Math San Diego State Univ San Diego CA 92182

BECKER, GERALD LEONARD, b San Diego, Calif, Oct 20, 40; m 65; c 2. BIOCHEMISTRY. Educ: Mass Inst Technol, BS, 62; Univ Chicago, MD, 66. Prof Exp: From intern to res assoc med, Beth Israel Hosp, Boston, 66-68; clin assoc geront br, Nat Inst Child Health & Human Develop, 68-70; fel instr physiol chem, Johns Hopkins Sch Med, 70-75; ASST PROF BIOCHEM, UNIV ALA, BIRMINGHAM, 75- Mem: AAAS. Res: Biochemistry of mitochondria and calcification. Mailing Add: Dept of Biochem Univ of Ala Univ Sta Birmingham AL 35294

BECKER, GORDON EDWARD, b St Louis, Mo, Sept 27, 20; m 56; c 2. SURFACE PHYSICS. Educ: Columbia Univ, BA, 42, AM, 43, PhD(physics), 46. Prof Exp: Asst physics, Columbia Univ, 42-44, res assoc, Radiation Lab, 43-46, instr physics, 46-47 & Stanford Univ, 47-51; sci staff, Hudson Labs, Columbia Univ, 51-53; MEM TECH STAFF, BELL TEL LABS, 53- Mem: Am Phys Soc; Inst Elec & Electronics Engrs. Res: Molecular beams. Mailing Add: Bell Tel Labs Murray Hill NJ 07974

BECKER, GWENETH (LESLIE), b Cedar Rapids, Iowa, Dec 11, 19; m 58; c 2. BIOLOGY. Educ: Univ Toronto, BA, 44; Smith Col, MA, 46; Radcliffe Col, PhD(bot), 56. Prof Exp: Res assoc genetics, Radiation Lab, Univ Calif, 49-59; INVESTR, ZOOL INST, UNIV MUNICH, 64- Res: Drosophila genetics. Mailing Add: Univ of Munich Inst of Zool Luisenstrasse 14 Munich Germany

BECKER, HARRY CARROLL, b Bisbee, Ariz, Oct 31, 13; m 40; c 1. ANALYTICAL CHEMISTRY. Educ: Univ Ill, BS, 36, MS, 37, PhD(anal chem), 40. Prof Exp: Anal res, 40-43, supvr, 44-62, RES ASSOC, TEXACO INC, 62- Mem: AAAS; Am Chem Soc; Sigma Xi. Res: Radiation chemistry; analysis of petroleum products; analytical studies with neutrons and energetic charged particles; scanning electron microscopy and electron microprobe. Mailing Add: Texaco Inc Beacon NY 12508

BECKER, HERMAN FREDERICK, b Ger, Jan 10, 07; US citizen; m 28; c 3. PALEOBOTANY. Educ: Brooklyn Col, BA, 47; Columbia Univ, MA, 52; Univ Mich, PhD, 56. Prof Exp: Horticulturist's asst, Brooklyn Bot Garden, 30-39; asst biol, Brooklyn Col, 39-54, instr, 56-58; res assoc paleobot & NSF grant, 58-65, cur, 65-75, EMER CUR, NY BOT GARDEN, 75- Concurrent Pos: Vis assoc prof, Univ Mich, 63-64; adj assoc prof, Columbia Univ, 67-; adj prof, Lehman Col, 69-. Mem: Bot Soc Am; Am Soc Plant Taxon; fel Geol Soc Am; Torrey Bot Club (pres, 71); Sigma Xi. Res: Tertiary paleobotany of southwestern Montana. Mailing Add: NY Bot Garden Bronx NY 10458

BECKER, JEFFREY MARVIN, b Baltimore, Md, July 13, 43; m 66; c 2. MICROBIOLOGY. Educ: Emory Univ, BA, 65; Ga State Univ, MS, 67; Univ Cincinnati, PhD(microbiol), 70. Prof Exp: Weizmann fel biophys, Weizmann Inst, Rehovot, Israel, 70-71, NATO-NSF fel, 71-72; asst prof, 72-75, ASSOC PROF MICROBIOL, UNIV TENN, KNOXVILLE, 75- Concurrent Pos: Consult, Oak Ridge Nat Lab, 74-; NIH res career develop award, 75-80. Mem: Am Soc Microbiol; AAAS; Am Asn Univ Prof. Res: Membrane transport; peptide transport and utilization in microorganisms and tissue culture. Mailing Add: Dept of Microbiol Univ of Tenn Knoxville TN 37916

BECKER, JERRY PAGE, b North Redwood, Minn, Mar 1, 37; m 59; c 3. MATHEMATICS. Educ: Univ Minn, Minneapolis, BS, 59; Univ Notre Dame, MS, 61; Stanford Univ, PhD(math educ), 67. Prof Exp: Pub sch teacher, Minn, 59-60 & 61-63; res asst math educ, Sch Educ, Stanford Univ, 63-64, asst, Sch Math Study Group, 64-67; from asst prof to assoc prof, Grad Sch Educ, Rutgers Univ, 67-75; VIS ASSOC PROF MATH SCI, NORTHERN ILL UNIV, 75- Concurrent Pos: Instr, NSF Inst, Univ San Francisco, 65-67 & Rutgers Univ, 67-68; consult, NSF Sci Educ Improv Prog, India, 67 & 68, supvr, 69 & 70, staff scientist math, Sci Liaison Staff, New Delhi, 71-72; mem US comn math instr, Nat Acad Sci, 73-77, cochmn panel US arrangements third int cong math educ, 74-76. Mem: Math Asn Am; Nat Coun Teachers Math; Int Study Group Math Learning; Brit Asn Teachers Math. Res: Mathematics curriculum development; teacher training; psychology of learning; achievement testing in mathematics; evaluation of mathematics education programs. Mailing Add: Dept of Math Sci Northern Ill Univ DeKalb IL 60115

BECKER, JOHN ANGUS, b New York, NY, Nov 25, 36. NUCLEAR PHYSICS. Educ: Queens Col (NY), BS, 57; Fla State Univ, PhD(physics), 62. Prof Exp: Asst physics, Fla State Univ, 58-62, res assoc, 62; res assoc, Brookhaven Nat Lab, 62-64; RES SCIENTIST, LOCKHEED PALO ALTO RES LABS, 64- Concurrent Pos: Sr fel, Oxford Univ, 71-72; assoc prof, Univ Minn, 74-75. Mem: Am Phys Soc. Res: Experimental and low energy nuclear physics; nuclear reactions initiated by electrostatic generator beams; particle detectors and instrumentation. Mailing Add: Lockheed Palo Alto Res Labs Palo Alto CA 94304

BECKER, JOHN W, b Baltimore, Md, Oct 19, 38; m 69; c 1. PHARMACEUTICAL CHEMISTRY. Educ: Univ Md, BS, 59, MS, 62; Univ Mich, PhD(pharm chem), 66. Prof Exp: Sr pharmaceut chemist, Pitman-Moore Div, Dow Chem Co, 67-68; SR PHARMACEUT CHEMIST, ELI LILLY & CO, 68- Mem: Am Pharmaceut Asn; Acad Pharmaceut Sci. Res: Dental enamel demineralization; drug physical stability; biopharmaceutics; drug availability. Mailing Add: 7450 Jewel Lane Indianapolis IN 46250

BECKER, JOSEPH F, b Mt Vernon, NY, Feb 26, 27; m 64. CHEMISTRY, BIOCHEMISTRY. Educ: Harvard Univ, AB, 50; Univ Del, MEd, 54; Columbia Univ, MS, 58, DEd, 62. Hon Degrees: JD, Seton Hall Univ, 73. Prof Exp: Teacher, High Sch, 51-58; assoc prof chem & biochem, 58-68, PROF CHEM, MONTCLAIR STATE COL, 68- Mem: AAAS; Am Chem Soc. Res: Biochemistry for non-science majors. Mailing Add: 9 Melrose Place Montclair NJ 07042

BECKER, JOSHUA A, b Philadelphia, Pa, Nov 28, 32; m 59; c 2. MEDICINE, RADIOLOGY. Educ: Temple Univ, AB, 53, MD, 57, MS, 61. Prof Exp: Intern, Philadelphia Gen Hosp, 58; resident radiol, Univ Hosp, Temple Univ, 61, instr, Sch Med, 63-65; from asst prof to assoc prof, Col Physicians & Surgeons, Columbia Univ, 65-70; PROF RADIOL, CHMN DEPT & DIR, STATE UNIV NY DOWNSTATE MED CTR, 70- Concurrent Pos: James Picker Found scholar radiol res, 64-65; from asst attend physician to assoc attend physician, Presby Hosp, 65-70; consult, Bronx Vet Admin Hosp, 67-70 & Brooklyn Vet Admin Hosp, 70- Mem: Am Col Radiol; Asn Univ Radiol. Mailing Add: Dept of Radiol State Univ NY Downstate Med Ctr Brooklyn NY 11203

BECKER, JOSPEH GERALD, b New York, NY, Mar 15, 27; m 62; c 2. MICROBIOLOGY. Educ: Univ Calif, Berkeley, BS, 49; Wash State Univ, MS, 53; Purdue Univ, PhD(microbiol), 56. Prof Exp: Asst bact, Wash State Univ, 53; res fel, Purdue Univ, 53-56; microbiologist biochem-enzyme res lab, Rohm and Haas Co, Pa, 56-59; dept head enzyme-fermentation lab, Nopco Chem Co, NJ, 59-62; enzyme-biochemist, Organon Pharmaceut Co, 62-64; SR MICROBIOLOGIST, COLGATE-PALMOLIVE RES CTR, 64- Mem: Soc Indust Microbiol; Am Soc Microbiol; Am Chem Soc. Res: Industrial and food microbiology; enzymes; fermentations; microbial physiology; microflora of the skin and of dental plaque; biocides. Mailing Add: Biol Sci Div Colgate-Palmolive Res Ctr Piscataway NJ 08854

BECKER, KENNETH LOUIS, b New York, NY, Mar 11, 31; m 54; c 2. INTERNAL MEDICINE, ENDOCRINOLOGY. Educ: Univ Mich, BA, 52; NY Med Col, MD, 56; Univ Minn, PhD(med, physiol), 64; Am Bd Internal Med, dipl, 66 & cert endocrinol & metab, 73. Prof Exp: Intern, Mt Sinai Hosp, NY, 56-57; resident internal med, Mayo Clin, 59-64; from asst prof to assoc prof med, George Washington Univ, 64-74; clin investr med & endocrinol, 64-65, CHIEF SECT METAB, VET ADMIN HOSP, WASHINGTON, DC, 65-; PROF MED, GEORGE WASHINGTON UNIV, 74-, DIR ENDOCRINOL, 74- Concurrent Pos: Consult, Wash Hosp Ctr, 64- Honors & Awards: Meritorious Res Award, Mayo Clin, 63. Mem: Fel Am Col Physicians; Am Fedn Clin Res; Am Soc Human Genetics; Am Soc Clin Pharmacol & Therapeut; Endocrine Soc. Res: Calcium metabolism; calcitonin; gynecomastia; gonadal-pituitary inter-relationships. Mailing Add: Vet Admin Hosp 50 Irving St NW 151-J Washington DC 20422

BECKER, KLAUS HERBERT, physical chemistry, see 12th edition

BECKER, LAWRENCE CHARLES, b Schenectady, NY, Aug 1, 34; m 57; c 2. NUCLEAR PHYSICS. Educ: Carleton Col, BA, 56; Yale Univ, BD, 59, MS, 60, PhD(physics), 64. Prof Exp: Asst prof, 63-70, ASSOC PROF PHYSICS, HIRAM COL, 70- Concurrent Pos: Res physicist, NASA Lewis Res Ctr, 69-70. Mem: AAAS; Am Phys Soc; Am Asn Physics Teachers. Res: Material studies utilizing the Mössbauer effect. Mailing Add: Dept of Physics Hiram Col Hiram OH 44234

BECKER, MARSHALL JOSEPH, b New York, NY, May 14, 38. ANTHROPOLOGY. Educ: Univ Pa, AB, 59, MA, 63, PhD(anthrop), 71. Prof Exp: Lectr anthrop, Beaver Col, 63; instr, Univ Toledo, 63-68; assoc prof, 68-72, PROF ANTHROP, WEST CHESTER STATE COL, 72- Mem: Am Anthrop Asn; Am Asn Phys Anthropologists; Soc Am Archaeol; Archaeol Inst Am. Res: Mesoamerican archaeology; archaeology and ethnohistory of the Delaware Indians; colonial ethnohistory and osteology; physical anthropology of Crete; Minoan trade routes, cultural and genetic contacts. Mailing Add: Dept of Anthrop West Chester State Col West Chester PA 19380

BECKER, MAURICE EDWIN, b Prattsville, NY, Sept 5, 21; m 47; c 4. BACTERIOLOGY. Educ: Cornell Univ, BS, 47, MS, 49; Purdue Univ, PhD(bact), 54. Prof Exp: Qual control, Food Processing Lab, Gen Foods Corp, 46; asst bact & food technol, NY Agr Exp Sta, Geneva, 47-49; instr bact, Fla State Univ, 49-51; bacteriologist, State Dept Health, Md, 54-60; CHIEF VIROL DIV, MICH DEPT PUB HEALTH, 60- Mem: Am Soc Microbiol; Tissue Cult Asn. Res: Flat sour spoilage of tomato juice; lysozyme; viral diagnostic procedures; tissue culture. Mailing Add: Mich Dept of Public Health Lansing MI 48914

BECKER, MILTON, b Chicago, Ill, July 13, 20; m 53; c 1. SOLID STATE PHYSICS. Educ: Univ Chicago, BS, 41; Purdue Univ, PhD(physics), 51. Prof Exp: Physicist, Electronics, Naval Res Labs, 42-47; physicist, semiconductors, Hughes Semiconductors, 51-58; PHYSICIST, SEMICONDUCTORS, CONTINENTAL DEVICE CORP, 58-, VPRES RELIABILITY, 61- Concurrent Pos: Consult, AEC Proj, 56-57; tech dir, Epidyne, Inc, 73-74. Mem: Am Phys Soc. Res: Microwave propagation in ionized media; optical properties of silicon and germanium in the infrared; electrical properties of semiconductor junctions. Mailing Add: 3100 Corda Dr Los Angeles CA 90049

BECKER, MILTON J, b Chicago, Ill, Feb 17, 25; m 50; c 3. BIOCHEMISTRY, BIOPHYSICS. Educ: Roosevelt Univ, BS, 50; Ill Inst Technol, MS, 54, PhD(biochem), 65. Prof Exp: Asst biochemist, Virus Res Lab, Dept Pub Health, Ill, 52-57; asst nucleoproteins, Dept Path, Presby-St Lukes Hosp, Chicago, 57-60; asst biochemist, Life Sci Div, Res Inst, Ill Inst Technol, 60-62, assoc biochemist, 62-64, res biochemist, 64-65; spec res fel biol, Lab Molecular Struct, Mass Inst Technol, 65-67; asst mem, Inst Biomed Res, AMA, 67-70; dir immunol & serol sect, Armour Pharmaceut Co, 70-73; DIR, IMMUNOLOGIC DIAG SERVS LTD, 73- Mem: Am Soc Cell Biol; Biophys Soc; Reticuloendothelial Soc; Soc Cryobiol. Res: Nucleic acid structure and metabolism; protein biosynthesis; physicochemical properties; antibody synthesis in vivo and in intro; photosynthesis; cell structure and function on molecular levels. Mailing Add: 2925 W Jerome St Chicago IL 60645

BECKER, NORWIN HOWARD, b New York, NY, Apr 23, 30; m 51; c 3. PATHOLOGY, HISTOCHEMISTRY. Educ: Cornell Univ, AB, 50, MNS, 52; State Univ NY, MD, 55. Prof Exp: Intern med, Montefiore Hosp, NY, 55-56; asst resident path, Long Island Jewish Hosp, 56-57; asst resident, Montefiore Hosp, 57-58, Nat Cancer Inst trainee, 58-59, from asst pathologist to assoc pathologist, 59-67; from asst prof to assoc prof, 64-74, PROF PATH, ALBERT EINSTEIN COL MED, 74-; ATTEND PATHOLOGIST, MONTEFIORE HOSP, 67- Mem: Am Asn Path & Bact; Am Soc Exp Path; Histochem Soc; Am Soc Cell Biologists; Am Soc Clin Pathologists. Res: Histochemistry and electron microscopy of the central nervous system; choroid plexus function; surgical pathology. Mailing Add: 39 Carlton Rd Monsey NY 10952

BECKER, RALPH SHERMAN, b Benton Harbor, Mich, Mar 14, 25; m 52; c 4. MOLECULAR SPECTROSCOPY. Educ: Univ Vt, BS, 49; Univ NH, MS, 50; Fla State Univ, PhD(chem), 55. Prof Exp: Dir phys & inorg res, Frederick S Bacon Labs, 52; from asst prof to assoc prof, 54-63, PROF CHEM, UNIV HOUSTON, 63- Concurrent Pos: Fulbright vis prof, Univ Barcelona, 62-63; consult, Weizmann Inst, 63; electronic transition in molecules, NSF, 64; consult, Phillips Petrol Co & Bryant Comput. Honors & Awards: Japanese Soc Promotion Sci Award, 71. Mem: AAAS; Am Chem Soc; Am Phys Soc. Res: Metal atom and ligand field perturbations; aromatic hydrocarbons; intercombinations of electronic states; photosynthesis; photochemistry; electron affinities; vision pigments; theoretical and experimental aspects of photoisomerization; chemical evolution; laser flash photolysis. Mailing Add: Dept of Chem Univ of Houston Houston TX 77004

BECKER, RANDOLPH ARMIN, b Bowler, Wis, Dec 19, 24; m 56; c 3. OPTICS, INSTRUMENTATION. Educ: Tex Col Arts & Indust, BS, 49, MS, 53. Prof Exp: Observing asst astron, Univ Chicago MacDonald Observ, Tex, 50; physicist, White Sands Missile Range, 50-57, chief optical res sect, Range Instrumentation Develop Div, 57-61; sr res engr space optics, Jet Propulsion Lab, Calif Inst Technol, 61-63, res specialist, 63-70, staff scientist, Space Photog Sect, 70-73; SR SCIENTIST, XEROX ELECTRO-OPTICAL SYSTS, 73- Res: Optical instrumentation; materials suitable for optical systems in interplanetary space; design and execution of television experiments on planetary space missions; deep-space mission design. Mailing Add: 616 E Mendocino Altadena CA 91001

BECKER, RICHARD LOGAN, b Dayton, Ohio, Dec 29, 29; m 54; c 2. THEORETICAL NUCLEAR PHYSICS. Educ: Harvard Univ, BA, 52; Yale Univ, MS, 53, PhD(physics), 57. Prof Exp: PHYSICIST, OAK RIDGE NAT LAB, 57- Concurrent Pos: Vis fel, Princeton Univ, 59-60; lectr & prof, Univ Tenn, 61- Mem: Fel Am Phys Soc. Res: Theoretical many-body physics; nuclear self-consistent field theory; nuclear structure theory; nuclear scattering theory. Mailing Add: Physics Div Oak Ridge Nat Lab Oak Ridge TN 37830

BECKER, RICHARD WILLIAM, b Buffalo, NY, Jan 21, 22; m 76; c 5. PHOTOGRAPHIC CHEMISTRY. Educ: Canisius Col, BS, 49; Univ Ill, MS, 50. Prof Exp: RES ASSOC PHOTOG CHEM, EASTMAN KODAK CO RES LABS, 50- Mem: Am Chem Soc; Soc Photog Scientists & Engrs. Res: Color image transfer chemistry. Mailing Add: Eastman Kodak Co Res Labs 1669 Lake Ave Rochester NY 14650

BECKER, ROBERT ADOLPH, b Tacoma, Wash, Feb 10, 13; m 44; c 5. SPACE PHYSICS. Educ: Col Puget Sound, BS, 35; Calif Inst Technol, MS, 37, PhD(physics), 41. Prof Exp: Asst physics, Col Puget Sound, 33-35; asst fel & asst nuclear physics, Calif Inst Technol, 35-41; asst physicist, Dept Terrestrial Magnetism, Carnegie Inst, 41; from asst physicist to physicist, Nat Bur Standards, 41-43; physicist, Radiation Lab, Univ Calif, 43; mem staff appl physics lab, Univ Wash, 43-45, sr physicist, 45; from asst prof to prof physics, Univ Ill, 46-60; dir space physics lab, 60-68, assoc gen mgr lab opers, 68-73, CONSULT PHYSICIST, AEROSPACE CORP, 73- Concurrent Pos: Guggenheim fel, 58-59. Mem: AAAS; Am Inst Aeronaut & Astronaut; Am Phys Soc; Am Astron Soc; Optical Soc Am. Res: Nuclear, space and atmospheric physics; astrophysics; proton and deuteron induced nuclear reactions; radioactive decay schemes from betatron produced radioactivities; element synthesis in the stars; structure of the upper atmosphere; planetary science. Mailing Add: PO Box 4609 Carmel CA 93921

BECKER, ROBERT HUGH, b Greenville, Pa, July 19, 34; m 63; c 6. ORGANIC CHEMISTRY. Educ: St Bonaventure Univ, BS, 56; Univ Notre Dame, PhD(org chem), 60. Prof Exp: Nat Insts Health res assoc org chem, NMex Highlands Univ, 59-61; instr chem, Lawrence Col, 61-63; from asst prof to assoc prof, Gannon Col, 63-75, dir sci res prog, 69-75; SR RES CHEMIST, CALSICAT DIV, MALLINCKRODT INC, 75- Concurrent Pos: Petrol Res Fund res grant, 63-65; consult, Calsicat Div, Mallinckrodt Inc, 67-75. Mem: Am Chem Soc; Am Oil Chem Soc. Res: Instrumental and qualitative organic analysis; spectroscopy; chromatography; catalysis. Mailing Add: Mallinckrodt Inc Calsicat Div 1707 Gaskell Ave Erie PA 16503

BECKER, ROBERT O, b Rivers Edge, NJ, May 31, 23; m 46; c 3. ORTHOPEDIC SURGERY, BIOPHYSICS. Educ: Gettysburg Col, BA, 46; NY Univ, MD, 48. Prof Exp: From instr to assoc prof, 56-66, PROF ORTHOP SURG, STATE UNIV NY UPSTATE MED CTR, 66-; CHIEF ORTHOP SECT, VET ADMIN HOSP, SYRACUSE, 56-, MED INVESTR, 72- Concurrent Pos: Mem spec study sect biomed eng, NIH; mem, Sanguine Study Comt, US Navy; assoc chief of staff for res, Vet Admin Hosp, Syracuse, 65-72. Honors & Awards: William S Middleton Award, 64. Mem: Orthop Res Soc; Inst Elec & Electronics Engrs; NY Acad Sci. Res: Biological solid state; organic semiconductors; biological control systems as applied to growth control; regenerative mechanisms and relationship to central nervous system to physical factors of environment. Mailing Add: Dept of Orthop Surg State Univ of NY Upstate Med Ctr Syracuse NY 13210

BECKER, ROBERT RICHARD, b Aitkin, Minn, Feb 16, 23; m 56; c 2. BIOCHEMISTRY. Educ: Univ NDak, BS, 48; Univ Wis, MS, 51, PhD(biochem), 52. Prof Exp: From instr to asst prof chem, Columbia Univ, 52-60; biochemist, Oak Ridge Nat Labs, 60-62; assoc prof chem, 62-67, PROF BIOCHEM, ORE STATE UNIV, 67- Concurrent Pos: Vis scientist, Dept Biol, Brookhaven Nat Labs, 68-69; mem, NIH Pathobiol Chem Study Sect, 75-77. Mem: AAAS; Am Soc Biol Chemists; Am Chem Soc. Res: Protein chemistry; polypeptides; theomophilic enzymes. Mailing Add: Dept of Biochem & Biophys Ore State Univ Corvallis OR 97331

BECKER, ROLAND FREDERICK, b Methuen, Mass, Aug 12, 12; m 36; c 2. BIOMECHANICS, ANATOMY. Educ: Mass State Col, BS, 35, MS, 37; Northwestern Univ, PhD(anat), 40. Prof Exp: Asst psychol, Mass State Col, 35-37; asst anat, Med Sch, Northwestern Univ, 38-40, instr, 40-44, assoc, 44-45, asst prof, 45-46; assoc prof, Sch Med, Univ Wash, 47, chmn dept, 47-48; dir neurol div, Daniel Baugh Inst Anat, Jefferson Med Col, 49-51; assoc prof anat, Duke Univ, 52-69; prof anat, 69-71, prof biomechanics & coordr res prog, 71-75, COORDR EDUC PROG, COL OSTEOP MED, MICH STATE UNIV, 75- Concurrent Pos: Consult, Eng Corps, US Army, 52-54 & NIH, 57-59; AID med educr, Thailand, 61-63; staff mem, Greenbrier Col Osteop Med, 74. Mem: Am Asn Anatomists. Res: Neuroanatomy and fetal physiology; behavioral development; neuromusculoskeletal system. Mailing Add: Dept of Biomechanics Mich State Univ Col Osteop Med East Lansing MI 48823

BECKER, SAMUEL WILLIAM, JR, b Rochester, Minn, Sept 11, 24; m 53; c 2. MEDICINE, DERMATOLOGY. Educ: Univ Ill, BS, 45, MD, 47; Univ Minn, MS, 51. Prof Exp: From clin instr to clin assoc prof, 53-73, CLIN PROF DERMAT, COL MED, UNIV ILL, 73- Res: Human pigmentary diseases; industrial dermatology. Mailing Add: Dept of Dermat Univ of Ill Col of Med Chicago IL 60680

BECKER, STEPHEN FRALEY, b Toledo, Ohio, Aug 13, 42; m 64. THEORETICAL PHYSICS. Educ: Miami Univ, AB, 65; Rutgers Univ, MS, 66, PhD(physics), 69. Prof Exp: ASST PROF PHYSICS, BUCKNELL UNIV, 69- Concurrent Pos: Instr, Lewisburg Fed Penitentiary, 71. Mem: Am Inst Physics; Am Phys Soc; Am Math Soc. Res: Three body problem; scattering theory; theoretical particle physics. Mailing Add: Dept of Physics Bucknell Univ Lewisburg PA 17837

BECKER, STEVEN ALLAN, b Anamosa, Iowa, Nov 23, 38; m 61; c 2. PLANT ANATOMY, MORPHOLOGY. Educ: Cornell Col, BA, 60; Univ Iowa, MS, 66, PhD(bot), 68. Prof Exp: Asst prof, 68-74, ASSOC PROF BOT, EASTERN ILL UNIV, 74- Mem: Bot Soc Am; Am Inst Biol Sci. Res: Developmental plant anatomy-morphology Mailing Add: Dept of Bot Eastern Ill Univ Charleston IL 61920

BECKER, ULRICH J, b Dortmund, Ger, Dec 17, 38; m 66; c 3. HIGH ENERGY PHYSICS. Educ: Univ Marburg, Vordiplom, 60, Univ Hamburg, Diplom, 64, PhD(physics), 68. Prof Exp: Sci employee exp physics, Deutschen Elektronen-Synchrotron, Hamburg, 64-68; res assoc high energy physics, 68-70, vis asst prof, 70-71, asst prof, 71-73, ASSOC PROF PHYSICS, MASS INST TECHNOL, 73- Concurrent Pos: Res coun mem, Deutschen Elektronen-Synchrotron, Hamburg, 70-71. Mem: Am Phys Soc. Res: High energy particle physics; discoverer of the new meson J (3.1) and study of the variety of these new particles, particularly vector mesons. Mailing Add: Lab for Nuclear Sci Bldg 44 Mass Inst of Technol Cambridge MA 02139

BECKER, VERYL E, b Glencoe, Minn, Sept 1, 39; m 61; c 2. PLANT PHYSIOLOGY. Educ: Gustavus Adolphus Col, BS, 61; SDak State Univ, MS, 63; Mich State Univ, PhD(plant physiol), 68. Prof Exp: Res assoc biochem, Ore State Univ, 67-69; asst prof biol, 69-75, ASSOC PROF BIOL, WAKE FOREST UNIV, 75- Mem: Am Soc Plant Physiol. Res: Cellular physiology; structure and function of cellular organelles; effects of the alkali metal ions on enzymes; nitrogen metabolism. Mailing Add: Dept of Biol Wake Forest Univ Winston-Salem NC 27109

BECKER, WALTER ALVIN, b Pittsburgh, Pa, Dec 1, 20; m 52; c 2. GENETICS. Educ: Stanford Univ, AB, 52; Univ Calif, Berkeley, MS, 55, PhD(genetics), 56. Prof Exp: Asst genetics, Univ Calif, 53-56; asst poultry scientist, Western Wash Exp Sta, Puyallup, 56-60; from asst prof to assoc prof animal sci, 60-65, assoc prof genetics, 65-68, actg chmn prog genetics, 67-68, PROF GENETICS, WASH STATE UNIV, 68- Concurrent Pos: Vis researcher, Inst Animal Genetics, Univ Edinburgh, 64-65. Mem: Fel AAAS; Genetics Soc Am; Poultry Sci Asn; Biomet Soc. Res: Animal breeding; genetics of cholestrol in chicken eggs; innovative methods of teaching; genetics of chicken body composition. Mailing Add: Dept of Animal Sci Wash State Univ Pullman WA 99163

BECKER, WARREN EARL, inorganic chemistry, see 12th edition

BECKER, WAYNE MARVIN, b Waukesha, Wis, May 29, 40; m 63; c 2. PLANT PHYSIOLOGY, BIOCHEMISTRY. Educ: Univ Wis-Madison, BS, 63, MS, 65, PhD(biochem), 67. Prof Exp: NATO fel molecular biol, Beatson Inst Cancer Res, Scotland, 67-68, NIH fel, 68-69; from asst prof to assoc prof bot, 69-75, PROF BOT, UNIV WIS-MADISON, 75- Concurrent Pos: Guggenheim res fel, Univ Edinburgh, 75-76. Mem: AAAS; Am Soc Plant Physiol; Am Soc Cell Biol; Brit Soc Cell Biol. Res: Regulation of eukaryotic gene expression; developmental biochemistry of seed germination; nucleic acid metabolism in plants. Mailing Add: Dept of Bot Univ of Wis Madison WI 53706

BECKER, WILLIAM BERNARD, b Brooklyn, NY, Sept 19, 09; m 36; c 1. ENTOMOLOGY. Educ: Syracuse Univ, BS, 34; Mass State Col, MS, 37, PhD(entom), 45. Prof Exp: Asst entom, Exp Sta, Mass State Col, 35-46, asst res prof, 46-47; from asst res prof to assoc res prof, 47-70, PROF ENTOM, UNIV MASS, AMHERS7, 70- Concurrent Pos: Mem, Northeastern Forest Pest Coun. Mem: Entom Soc Am; Soc Am Foresters; Entom Soc Can; Sigma Xi. Res: Biology and control of, or damage by, insect vectors of Dutch elm disease fungus, borers in unseasoned logs and certain other shade tree and forest insects. Mailing Add: Dept of Entom Univ of Mass Amherst MA 01002

BECKERBAUER, RICHARD, b Tilden, Nebr, Feb 25, 34; m 57; c 1. ORGANIC CHEMISTRY. Educ: Univ Nebr, BS, 59, MS, 60, PhD, 62. Prof Exp: Res chemist, 62-70, SR RES CHEMIST, PLASTIC PROD & RESINS DEPT, E I DU PONT DE NEMOURS & CO, 70- Mem: Am Chem Soc. Res: Copolymerization; block and graft polymers; emulsions; acrylic polymers. Mailing Add: Plastic Prod & Resins Dept E I du Pont de Nemours & Co Wilmington DE 19898

BECKERING, WILLIS, b Edgerton, Minn, Aug 21, 28; m 50; c 4. PHYSICAL INORGANIC CHEMISTRY. Educ: Calvin Col, BS, 50; DePauw Univ, MS, 52. Prof Exp: Asst, DePauw Univ, 50-52; Iowa State Col, 52-53, res asst, 53-55; RES CHEMIST, US BUR MINES, 55- Mem: Am Chem Soc; Soc Appl Spectros. Res: Theoretical and applied spectroscopy in the ultraviolet and infrared spectral regions; hydrogen bonding in phenolic compounds; x-ray fluorescent analysis; x-ray diffraction and electron microprobe. Mailing Add: 3010 Belmont Rd Grand Forks ND 58201

BECKERLE, JOHN C, b Mt Vernon, NY, Feb 14, 23; m 50; c 2. PHYSICAL OCEANOGRAPHY. Educ: Manhattan Col, BS, 48; Cath Univ Am, PhD(physics), 54. Prof Exp: Asst physics, Cath Univ Am, 48-51; physicist, Electronic Comput, Appl Physics Labs, Johns Hopkins Univ, 50; res assoc chem, Naval Ord Contract, Cath Univ Am, 51-53; mem tech staff, Bell Tel Labs, 53-60; mgr sonics res, Schlumberger Well Surv Corp, 60-64; ASSOC SCIENTIST, WOODS HOLE OCEANOG INST, 64- Mem: Acoust Soc Am. Res: Mathematical physics; electromagnetic and acoustic wave propagation; underwater sound; ocean sound velocity fluctuations; internal gravity waves; heat conduction in solids; sonic well logging; circulation in the Sargasso sea. Mailing Add: Dept of Geophysics Woods Hole Oceanog Inst Woods Hole MA 02543

BECKERLEY, JAMES GWAVAS, b Chicago, Ill, Feb 27, 15; m 39; c 1. PHYSICS. Educ: Stanford Univ, AB, 35, PhD(physics), 45. Prof Exp: Asst physics, Stanford Univ, 35-39; lectr, Univ Ga, 39-40, Judson Col, Rangoon, 40-42 & Columbia Univ, 42-45; metallurgist, Am Brake Shoe & Foundry Co, NJ, 45-46; physicist res div & dir tech adv, AEC, 47-49, classification, 49-54; prof lectr physics, George Washington Univ, 50-54; head eng physics dept, Schlumberger Well Serv Corp, 54-59; mem staff div sponsored res, Mass Inst Technol, 62-64; PRES, RADIOPTICS, INC, 64- Concurrent Pos: Lectr, Stevens Inst Technol, 47-49; consult, Md Gov Task Force Nuclear Power Plants, 69; mem planetology subcomt, Space Sci Steering Comt, NASA, 65-; ed, Ann Rev Nuclear Sci, 51-58; sr tech ed, Int Atomic Energy Comn, Austria, 60-62; ed, Am Nuclear Soc, 55-59. Mem: Am Nuclear Soc. Res: Electromagnetic radiation theory; neutron physics; nuclear power; laser materials. Mailing Add: Radioptics, Inc 10 Dupont St Plainview NY 11803

BECKERMAN, BARRY LEE, b New York, NY, Jan 12, 41; m 65; c 2. OPHTHALMOLOGY. Educ: Cornell Univ, AB, 61; NY Univ, MD, 65. Prof Exp: Intern surgery, Tufts New Eng Med Ctr, 65-66; Lt Comdr surgery, USPHS, 66-68; from resident to chief resident ophthal, NY Univ-Bellevue Med Ctr, 68-72; instr, Albert Einstein Col Med, 72-73; DIR ELECTROPHYSIOL, RETINA SERV, MONTEFIORE HOSP, BRONX, NY, 73- Concurrent Pos: Adj attend ophthal, Northern Westchester Hosp, Mt Kisco, NY, 73-; asst clin prof ophthal, Albert Einstein Col Med, Bronx, NY, 73- Mem: Asn Res Vision & Ophthal, Res: Clinical research in the diagnosis of ocular and especially retinal disorders, particularly development of diagnostic techniques and application to clinical disease. Mailing Add: 344 Main St Mt Kisco NY 10549

BECKERMAN, STEPHEN JOEL, b Washington, DC, Apr 9, 42. ANTHROPOLOGY. Educ: George Washington Univ, BA, 66; Univ NMex, PhD(anthrop), 75. Prof Exp: VIS INSTR ANTHROP, UNIV NMEX, 75- Mem: Am Anthrop Asn; Am Ethnol Asn. Res: Relationship of energy flow parameters to anthropologically interesting questions of cultural diversity, stability and evolution, focusing on tropical forest peoples. Mailing Add: Dept of Anthrop Univ of NMex Albuquerque NM 87131

BECKERS, JACQUES MAURICE, b Arnhem, Netherlands, Feb 14, 34; nat US; m 59; c 2. ASTROPHYSICS. Educ: Univ Utrecht, Drs, 59, DrAstron, 64. Prof Exp: Res physicist, High Altitude Observ, 62-64, Sacramento Peak Observ, 64-69 & High Altitude Observ, 69-70; RES PHYSICIST, SACRAMENTO PEAK OBSERV, 70- Honors & Awards: Henryk Arctowski Medal, Nat Acad Sci, 75. Mem: Fel AAAS; Optical Soc Am; Am Astron Soc; Netherlands Astron Soc; Int Astron Union. Res: Solar spectra; fine structures in the solar atmosphere; instrumentation for solar research; sunspots. Mailing Add: Sacramento Peak Observ Sunspot NM 88349

BECKERT, WILLIAM HENRY, b New York, NY, Sept 10, 20; m 48; c 2. CYTOLOGY, HISTOLOGY. Educ: St John's Univ, NY, BS, 49, MS, 51; NY Univ, PhD(cytol), 61. Prof Exp: From instr to assoc prof, 51-66, PROF BIOL, ST JOHN'S UNIV, NY, 66- Mem: AAAS; Am Soc Cell Biol; Am Soc Zoologists; NY Acad Sci; Tissue Cult Asn. Res: Sex chromatin body and drumstick in vertebrates and invertebrates; karyotypes of nonmammalian vertebrates; cell cultures. Mailing Add: 278 Spruce St West Hempstead NY 11552

BECKETT, JACK BROWN, b Hutsonville, Ill, Aug 10, 25; m 70; c 1. PLANT GENETICS. Educ: Univ Wash, BS, 50; Univ Wis, MS, 52; PhD(hort genetics), 54. Prof Exp: Asst hort, Univ Wis, 50-54; plant geneticist, USDA, Univ Ill, 54-63, PLANT GENETICIST, USDA, UNIV MO-COLUMBIA, 63-, ASST PROF AGRON, UNIV, 70- Concurrent Pos: Asst prof genetics, Univ Mo-Columbia, 67-70. Mem: AAAS; Am Soc Agron; Genetics Soc Am; Am Genetic Asn. Res: Genetics and cytogenetics of maize, with particular emphasis on development and use of A-B translocations. Mailing Add: Dept of Agron Univ of Mo Curtis Hall Columbia MO 65201

BECKETT, RALPH LAWRENCE, b Salt Lake City, Utah, June 6, 23; m 46; c 3. SPEECH PATHOLOGY. Educ: Univ Southern Calif, BA, 50, MA, 53, PhD(commun disorders), 68. Prof Exp: Asst prof speech, Los Angeles Harbor Col, 57-67; res fel, Univ Mo-Columbia, 68-69, asst prof speech path, 68-70; ASSOC PROF COMMUN DISORDERS, CALIF STATE UNIV, FULLERTON, 70- Concurrent Pos: Vis lectr, Lincoln Univ, 68-69; consult, Speech & Lang Inst, Orange County, Calif, 72- Mem: Am Speech & Hearing Asn. Res: Laryngeal physiology and pathology; physiological phonetics. Mailing Add: Dept of Speech Commun Calif State Univ Fullerton CA 92634

BECKFIELD, WILLIAM JOHN, b Pittsburgh, Pa, Aug 25, 20; m 45; c 8. PATHOLOGY. Educ: Allegheny Col, AB, 41; Univ Pa, MD, 44. Prof Exp: Asst prof path, Grad Sch Med, Univ Pa, 55-61, assoc prof, 61-62; PATHOLOGIST, LUTHER HOSP, 62- Concurrent Pos: Consult pathologist, Wyeth Labs, Am Home Prod Corp, 52-70; pathologist, Grad Hosp, Univ Pa, 54-56, vis lectr, Grad Sch Med, 63-67; pathologist, Philadelphia Gen Hosp, Pa, 56-62. Mem: Am Asn Path & Bact; Am Soc Exp Path; Am Soc Clin Path; Col Am Path; AMA. Res: Medical and surgical pathology; experimental pathology, toxicology and medicine. Mailing Add: Luther Hosp 310 Chestnut St Eau Claire WI 54701

BECKHORN, EDWARD JOHN, b Ithaca, NY; m; c 1. RESEARCH ADMINISTRATION, MICROBIOLOGY. Educ: Cornell Univ, BS, 47, MS, 48, PhD(biochem genetics), 50. Prof Exp: Res assoc, Dept Genetics, Carnegie Inst Wash, 50-52; dir microbiol res, Wallerstein Co Div, Baxter Labs, 52-57, from asst dir res to dir res, 57-70; mgr sci serv, Lehn & Fink Prod Co Div, Sterling Drug Co, 71-73; MGR SCI SERV, CUNNINGHAM & WALSH INC, 74- Mem: Asn Res Dirs; Am Soc Microbiol; Am Chem Soc; Soc Indust Microbiol; Am Asn Cereal Chemists. Res: Production and industrial or pharmaceutical applications of fermentation products, enzymes, antibiotics, organic acids; formulation, efficacy and toxicology of disinfectants, antimicrobials and consumer household chemical specialties; food processing with enzymes. Mailing Add: 465 Laurel Lane Kinnelon NJ 07405

BECKING, GEORGE C, b Mimico, Ont, Dec 2, 35; m 58; c 2. BIOCHEMISTRY. Educ: Queen's Univ (Ont), BSc, 58, MSc, 60, PhD(biochem), 62. Prof Exp: Fel biochem, Western Reserve Univ, 62-64; biochemist, Dow Chem Co, 64-65; mem staff res labs, Food & Drug Directorate, 65-74, MEM STAFF ENVIRON TOXICOL DIV, ENVIRON HEALTH DIRECTORATE, DEPT NAT HEALTH & WELFARE, 74- Mem: AAAS; Can Biochem Soc; Brit Biochem Soc. Res: Biochemical pharmacology and toxicology; factors affecting drug and steroid metabolism; drug interaction. Mailing Add: Environ Toxicol Div Environ Health Directorate Dept Nat Health & Welfare Ottawa ON Can

BECKING, RUDOLF (WILLEM), b Blora, Java, Indonesia, Oct 19, 22; m 52; c 3. ENVIRONMENTAL SCIENCES, NATURAL RESOURCES. Educ: Univ Wageningen, MF temperate forestry & MF trop forestry, 52; Univ Wash, PhD(forest mgt), 54. Prof Exp: Forest officer, Dutch Forest Serv, 54-56; asst prof, Univ NH, 56-57 & Pa State Univ, 57-58; assoc prof forestry, Auburn Univ, 58-60; from assoc prof to prof forestry, 60-70, PROF NATURAL RESOURCES, HUMBOLDT STATE UNIV, 70- Concurrent Pos: Mem, Sta Int Geobot Mediterranean & Alpine, France, 47- Mem: AAAS; Soc Am Foresters; Ecol Soc Am; Am Soc Photogram; Sigma Xi. Res: Phytosociology; forest ecology; photo interpretation; tropical rain forest dynamics; forest growth and yield. Mailing Add: Sch of Natural Resources Humboldt State Univ Arcata CA 95521

BECKJORD, PHILIP RAINS, b Manila, Philippines, Jan 9, 14; US citizen; m 37; c 4. PUBLIC HEALTH ADMINISTRATION. Educ: Univ Minn, BS, 35, MB, 37, MD, 38, MPH, 49; Am Bd Prev Med, dipl, 51; Johns Hopkins Univ, DrPH(epidemiol), 57. Prof Exp: Intern surg, Univ Minn Hosps, 37-38; gen pract, Minn, 38-40; mem sr staff pub health, US Mil Govt, US Army, Ger, 45-48, sr prev med officer, Hq US Forces, Austria, 49-52, sr prev med officer, Hq, First US Army, 52-53 & Div Prev Med, Walter Reed Army Inst Res, 53-57, dep chief prev med div & chief nutrit & civil affairs sect, Off Surgeon Gen, Washington, DC, 57-59, chief prof serv, Off Chief Med Officer, Korea, 59-60; PROF PUB HEALTH ADMIN & CHMN DEPT, TULANE UNIV, 60- Concurrent Pos: Mem, Armed Forces Disciplinary Control Bd, New York, 52-53; chief dept health pract, US Army Med Serv Grad Sch, 53-55; med mem, Spec Mil Intel Mission, Australasia & Far East, 54; instr, Johns Hopkins Univ, 55-56; dir div prev med & grad course prev med, Walter Reed Army Inst Res, 56-57; mem comt civil affairs, Strategic Div, Off Res Opers, Washington, DC, 57-59; mem pub health res study sect, NIH, 57-59; mem panel physiol, Comt Environ, Res & Develop Off, Dept Army, 58-59; mem curriculum comt med care, Asn Schs Pub Health, 60-61, chmn, 65-66; WHO fel, Guatemala, 62; mem, WHO Traveling Sem, Europe, 63; mem bd dirs, Social Welfare Planning Coun, New Orleans, 63-66 & 68-69; consult, USPHS Hosp, New Orleans, 63-; prog dir, Tulane Univ-USPHS, 63-, int health admin training, 63 & grad pub health nursing training, 62-64; chmn & coordr, Med Educ Nat Defense Comt, Tulane Univ, 64-65; chmn prof pract comt, New Orleans Hosp Coun, 64-65; dir gen prev med residency training prog, Tulane Univ, 64-; consult prev med, Touro Infirmary, 65-66; mem, USPHS Nat Community Health Res Training Comt, 65-68; Fulbright lectr, Univ Hamburg, 66-67; res assoc, Univ Edinburgh & Am Univ Beirut, 66-67; vis lectr, Univ Sarajevo & Univ Gothenburg, 67; pres & mem bd gov, Int Health Soc US, 65-69; consult, La State Health Dept, 68-, AID & New Orleans City Health Dept, 74-; pub health admin adv, Inst Int Educ, 73-; lectr, Kings Fund Col, London, 74. Mem: Fel Am Col Prev Med; fel Am Pub Health Asn; AMA; Asn Teachers Prev Med; Am Soc Trop Med & Hyg. Res: Epidemiology; disease control. Mailing Add: Dept of Health Serv Admin Tulane Sch Pub Health & Trop Med New Orleans LA 70112

BECKLER, DAVID (ZANDER), b Detroit, Mich, June 29, 18; m; c 3. SCIENCE ADMINISTRATION. Educ: Univ Rochester, BS, 39; George Washington Univ, JD, 43. Prof Exp: Patent attorney, Pennie, Davis, Marvin & Edmonds, 39-42; tech aide, Off Sci Res & Develop, 42-45; patent attorney, Eastman Kodak Co, 46; dep tech historian, Oper Crossroads, 46-47; chief tech intel br, Res & Develop Bd, Off Secy Defense, 47-49; mem int sci policy surv group, US Dept State, 49-50; exec dir comt atomic energy, Res & Develop Bd, 50-52; asst dir off indust develop, AEC, 52-53; exec officer, President's Sci Adv Comt, 53-73; ASST TO PRES, NAT ACAD SCI, 73- Concurrent Pos: Asst to spec asst to President for sci & technol, 57-62; asst to dir, Off Sci & Technol, 62-70, asst dir, 70- Res: Physical science administration. Mailing Add: 8709 Duvall St Fairfax VA 22030

BECKLOFF, GERALD LEE, b Hitchcock, Okla, Oct 26, 31; m 53; c 3. MEDICINE. Educ: Univ Kans, BS, 56, MS, 57; Univ Okla, MD, 61. Prof Exp: Asst med dir clin res, E R Squibb & Sons Div, Olin Mathieson Chem Corp, 62-64, assoc clin res dir, 64-68; dir clin pharmacol, 68-72, vpres res, 72-73, VPRES RES & DEVELOP, MARION LABS, INC, 73- Concurrent Pos: Asst instr, Med Sch, Univ Pa, 64-68; asst physician, Outpatient Dept, Pa Hosp, 64-68; res assoc, Truman Lab, Kansas City Gen Hosp, 68- Mem: AMA; Am Soc Clin Pharmacol & Therapeut; Am Soc Clin Oncol; Asn Psychophysiol Study Sleep. Res: Clinical research; drug evaluation in cancer chemotherapy. Mailing Add: Marion Labs Inc 10236 Bunker Ridge Rd Kansas City MO 64137

BECKMAN, ALEXANDER LYNN, b Los Angeles, Calif, June 9, 41; m 62; c 2. NEUROPHYSIOLOGY. Educ: Univ Calif, Los Angeles, BA, 64; Univ Calif, Santa Barbara, PhD(psychol), 68. Prof Exp: USPHS fel, Inst Neurol Sci, 68-70, ASST PROF PHYSIOL SCH MED, UNIV PA, 71- Concurrent Pos: Univ Pa Plant to Develop Scientists in Med Res fel, 70-73. Mem: Soc Neurosci; Am Physiol Soc. Res: Central nervous system mechanisms controlling body temperature; changes in central nervous system function during hibernation. Mailing Add: Dept of Physiol Univ of Pa Sch of Med Philadelphia PA 19174

BECKMAN, ARNOLD ORVILLE, b Cullom, Ill, Apr 10, 00; m 25; c 2. CHEMISTRY. Educ: Univ Ill, BS, 22, MS, 23; Calif Inst Technol, PhD(photochem), 28. Prof Exp: Res engr, Bell Tel Labs, 24-26; instr chem, Calif Inst Technol, 26-29, asst prof, 29-40; vpres, Nat Tech Labs, 37-39, pres, 39-50, Arnold O Beckman, Inc, 42-57, PRES & CHMN BD, BECKMAN INSTRUMENTS, INC, 50- Concurrent Pos: Pres, Helipot Corp, 44-58; consult, Orange County Air Pollution Control Dist; dir, Stanford Res Inst; trustee, Calif Inst Res Found & Calif Mus Found; vchmn adv comt, Los Angeles

Space Age Mus; chmn bd, Calif Inst Technol. Mem: Instrument Soc Am (pres, 52); Am Chem Soc; hon mem Am Inst Chem. Res: Applied chemistry; development of scientific instruments; photochemistry. Mailing Add: Beckman Instruments Inc 2500 Harbor Blvd Fullerton CA 92634

BECKMAN, CARL HARRY, b Cranston, RI, May 9, 23; m 45, 67; c 4. PLANT PATHOLOGY, HISTOPATHOLOGY. Educ: Univ RI, BS, 47; Univ Wis, PhD(plant path), 53. Prof Exp: Proj assoc, Univ Wis, 50-53; asst prof plant path, Univ RI, 53-58; plant pathologist, United Fruit Co, 58-62; assoc prof, 63-69, PROF PLANT PATH, UNIV RI, 69- Mem: Am Phytopath Soc; Phytochem Soc NAm. Res: Wilt diseases; physiology of parasitism; disease resistance; histochemistry. Mailing Add: Dept of Plant Path & Entom Univ of RI Kingston RI 02881

BECKMAN, DAVID LEE, b Dayton, Ohio, May 11, 39; m 69. PULMONARY PHYSIOLOGY. Educ: Ohio State Univ, BS, 62, MS, 64, PhD(physiol), 67. Prof Exp: Res assoc physiol, Ohio State Univ, 64-65; res assoc biosci, Univ Mich, 67-69, res physiologist, 70-71; asst prof anesthesiol & physiol, Wayne State Univ, 71-73; Hill prof physiol, Univ NDak, 73-76; PROF PHYSIOL, E CAROLINA UNIV, 76- Concurrent Pos: Partic, Biospace Training Prog, NASA & Univ Va, 67. Mem: AAAS; Am Physiol Soc; Soc Exp Biol & Med; Undersea Med Soc; Aerospace Med Asn. Res: Respiratory physiology; biomechanics; environmental physiology; trauma research; head and thoracic injury; lung mechanics; hyperbaric medicine; cardiovascular physiology; autonomic pharmacology. Mailing Add: Dept of Physiol ECarolina Univ Sch of Med Greenville NC 27834

BECKMAN, EDWARD LOUIS, b Kewanee, Ill, Dec 6, 16; m 45; c 1. PHYSIOLOGY. Educ: Northwestern Univ, MD, 43; Univ Southern Calif, MS, 50; Am Bd Prev Med, dipl. Prof Exp: Proj officer, Aero Med Equip Lab, Naval Air Materiel Ctr, Philadelphia, Pa, US Navy, 47, instr physiol, Sch Aviation Med, Pensacola, Fla, 48, chief dept & actg res dir, Aviation Med Acceleration Lab, Naval Air Develop Ctr, Johnsville, Pa, 49-53, dep dir, 59-61, exchange officer, Inst Aviation Med, Royal Air Force, Farnborough, Eng, 57-59, head environ stress div, Naval Med Res Inst, 61-65, dir dept physiol sci, 65-66, chief occup environ med off, 66-68, dir biomed res lab, Manned Spacecraft Ctr, NASA, 68-69; chief marine med div, Marine Biomed Inst, Univ Tex Med Br Galveston, 69-72; PROF PHYSIOL, BAYLOR SCH MED, 72-; PROF MARINE PHYSIOL, TEX A&M UNIV, 72- Concurrent Pos: Fel, Grad Sch Med, Univ Pa, 50-57. Mem: Fel Aerospace Med Asn; Am Physiol Soc; Undersea Med Soc. Res: Underwater physiology. Mailing Add: Moody Col of Marine Sci Ft Crockett Bldg 311 Galveston TX 77550

BECKMAN, FRANK SAMUEL, b New York, NY, Apr 10, 21; m 51; c 3. MATHEMATICS, COMPUTER SCIENCE. Educ: City Col New York, BS, 40; Columbia Univ, AM, 47, PhD(math), 65. Prof Exp: Civilian instr aircraft eng, US Army Air Corps, 41-44; asst prof math, Pratt Inst, 47-51; asst to mgr, Sci Comput Ctr, IBM Corp, 51-56, prod planning rep, 56-57, mgr appl prog systs, 57-59, mgr educ & data processing, Watson Sci Comput Lab, 59-60, from asst to assoc dir comput sci & educ systs res inst, 60-66, res mgr spec projs, 66-71; PROF COMPUT & INFO SCI & CHMN DEPT, BROOKLYN COL, 71- Concurrent Pos: Ed, SIAM Rev, 60-64; adj assoc prof, Columbia Univ, 65-71. Mem: Am Math Soc; Asn Comput Mach; Math Asn Am; Soc Indust & Appl Math. Res: Elliptical partial differential equations; education in computer-allied subjects; automata theory. Mailing Add: 16 Garwood Rd Fair Lawn NJ 07410

BECKMAN, JOSEPH ALFRED, b Macomb, Ill, Oct 30, 37; m 59; c 2. POLYMER CHEMISTRY. Educ: Western Ill Univ, AB, 60; Iowa State Univ, PhD(org chem), 65. Prof Exp: Res org chemist, 64-69, mgr org chem res, 69-71, MGR ELASTOMER SYNTHESIS RES, CENT RES LABS, FIRESTONE TIRE & RUBBER CO, 71- Mem: Am Chem Soc. Res: Synthesis, characterization, evaluation and application of organic and inorganic polymers. Mailing Add: 5646 Bonnie Lou Dr Akron OH 44319

BECKMAN, LEWIS DAVID, b Detroit, Mich, Jan 15, 40; m 68; c 2. VIROLOGY, MOLECULAR BIOLOGY. Educ: Wayne State Univ, BS, 63, PhD(biochem), 69. Prof Exp: Res asst biochem, Wayne State Univ, 63-69; res assoc molecular biol, Univ Tex, Dallas, 69-72, NIH res fel, 71-72; res assoc molecular genetics, Univ Mich, 72-73; ASST PROF VIROL, ALBANY MED COL, 73- Mem: Am Soc Microbiol. Res: Virus-cell interactions; transcription of viral RNA; structure and function of viral ribonucleoproteins. Mailing Add: Dept of Microbiol Albany Med Col Albany NY 12208

BECKMAN, WILLIAM, b Steelton, Pa, Sept 26, 09; m 34; c 2. INORGANIC CHEMISTRY, PHYSICAL CHEMISTRY. Educ: Youngstown Col, AB, 39; State Col Wash, MS, 47; Western Reserve Univ, PhD, 57. Prof Exp: Asst prof chem, Youngstown Col, 47-61; chmn div sci & math, Southwestern Col (Calif), 61-75, chmn dept chem, 64-75, chmn dept phys sci, 66-75; RETIRED. Mem: Am Chem Soc. Res: High temperature gelation phenomena of organophillic montmorillonite and mineral oil. Mailing Add: 817 Halecrest Dr Chula Vista CA 92010

BECKMAN, WILLIAM CURTIS, biology, see 12th edition

BECKMAN, WILLIAM P, plant breeding, genetics, see 12th edition

BECKMANN, ALBERT JULES, b New York, NY, Mar 2, 17; m 57; c 2. PEDIATRICS, PUBLIC HEALTH. Educ: Cornell Univ, BA, 38; Univ NC, MS, 41; Wake Forest Col, MD, 45. Prof Exp: Intern pediat, NC Baptist Hosp, 45-46; resident, Jewish Hosp, St Louis, 48-49; res assoc, Sch Pub Health, 49-51, asst prof pub health pract, 51-68, ADJ ASSOC PROF PUB HEALTH PRACT, SCH PUB HEALTH & ADMIN MED, COLUMBIA UNIV, 68- Concurrent Pos: Pediatrician, St John's Guild Floating Hosp, New York, 49-68, bd trustees, 70-, med dir, 72-; prin investr, NIH, 50-51; pvt pract, 51-; attend pediatrician, hosps, 53-; chief pediat, Franklin Gen Hosp, Valley Stream, NY, 63-, bd trustees, 63-, pres med bd, 74-; adj prof health sci, C W Post Col, Long Island Univ, 73-74; bd trustees, Nassau Acad Med, 74- Mem: Fel Am Pub Health Asn; AMA; Am Sch Health Asn; Royal Soc Health. Mailing Add: 111 Hempstead Ave Malverne NY 11565

BECKMANN, PETR, b Prague, Czech, Nov 13, 24; m 65. RADIOPHYSICS. Educ: Tech Univ Prague, MSc, 49, PhD(elec eng), 55; Czech Acad Sci, DSc(elec eng), 61. Prof Exp: Res engr, Res Inst Telecommun, Czech, 51-54; scientist, Geophys Inst, Czech Acad Sci, 55 & Inst Radio Eng & Electronics, 55-63; PROF ELEC ENG, UNIV COLO, BOULDER, 63- Mem: Fel Inst Elec & Electronics Engrs; assoc Opers Res Soc Am. Res: Wave propagation; probability theory; scattering by rough surfaces; random vector sums; computational linguistics. Mailing Add: Dept of Elec Eng Univ of Colo Boulder CO 80302

BECKNER, EVERET HESS, b Clayton, NMex, Feb 24, 35; m 55; c 3. PLASMA PHYSICS. Educ: Baylor Univ, BS, 56; Rice Univ, MA, 59, PhD(physics), 61. Prof Exp: Staff scientist, Lockheed Missiles & Space Co, 56-57; staff mem, 61-66, div supvr, 66-69, dept mgr, 69-74, DIR, SANDIA LABS, 74- Mem: Fel Am Phys Soc. Res: Plasma physics and magnetohydrodynamics; plasma spectroscopy; pulsed electron accelerators; beam-plasma interactions; electron beam fusion; laser fusion. Mailing Add: 809 Warm Sands Trail Albuquerque NM 87123

BECKWITH, JOHN BRUCE, b Spokane, Wash, Sept 18, 33; m 54; c 3. PATHOLOGY. Educ: Whitman Col, BA, 54; Univ Wash, MD, 58. Prof Exp: Resident pediat path, Children's Hosp, Los Angeles, Calif, 59-60, resident path, Cedars of Lebanon Hosp, Los Angeles, 60-62; chief resident, Children's Hosp, Los Angeles, Calif, 62-64; from instr to assoc prof path, 64-74, PROF PATH & PEDIAT, SCH MED, UNIV WASH, 74-; PATHOLOGIST, CHILDREN'S ORTHOP HOSP & MED CTR, 64- Concurrent Pos: NIH res grant, 66- Res: Sudden death in infancy; pathogenesis of tumors in children. Mailing Add: Dept of Path Children's Orthop Hosp & Med Ctr Seattle WA 98105

BECKWITH, JULIAN RUFFIN, b Petersburg, Va, Dec 28, 10; m 36; c 4. INTERNAL MEDICINE, CARDIOVASCULAR DISEASE. Educ: Univ Va, MD, 36. Prof Exp: Instr internal med, Univ Va, 40-41; chief dept, Chesapeake & Ohio Hosp, Clifton Forge, Va, 46-53; from asst to assoc prof, 53-62, PROF INTERNAL MED, SCH MED, UNIV VA, 62- Concurrent Pos: Fel, Coun Clin Cardiol, Am Heart Asn. Mem: AMA; Am Heart Asn; fel Am Col Physicians; fel Am Col Cardiol; Am Clin & Climat Asn. Res: Hypertension; electrocardiography. Mailing Add: Univ of Va Sch of Med Charlottesville VA 22903

BECKWITH, LEROY CHARLES, biology, ecology, see 12th edition

BECKWITH, MERTON MONROE, analytical chemistry, see 12th edition

BECKWITH, NEWELL PIERCE, b Omaha, Nebr, Jan 25, 15; m 39; c 4. ORGANIC CHEMISTRY. Educ: NDak State Col, BS, 36, DSc, 55. Prof Exp: Chemist, Chrysler Corp, 37-39; formulator & foreman, Rinshed-Mason Co, 40-41; chief auto res, 46-49, dir res, 49-52, vpres & tech dir, 52-54, vpres & gen mgr, Can, 54-67; vpres int mkt, TBI-Inmont Corp, 67-74, consult, Inmont SAfrica, 74-75, MGR, INMONT-VONAVAL, 75- Concurrent Pos: Secy, Paint Res Inst, 58-67; pres, Inmont Can Ltd, 67-74, Scarfe & Co, Brantford, Ont, 67-74 & Presstite, Georgetown, Ont, 68-74. Res: Accelerated tests for paint products; subtropical exposure work. Mailing Add: Inmont-Bonaval Bonn 53 West Germany

BECKWITH, RICHARD EDWARD, b San Jose, Calif, July 21, 27; m 69; c 3. STATISTICS, OPERATIONS RESEARCH. Educ: Stanford Univ, BS, 49, MS, 54; Purdue Univ, PhD, 59. Prof Exp: Statistician, US Fish & Wildlife Serv, 51-52; asst statist, Stanford Univ, 52-53; opers res, Case Inst Technol, 54-55; instr math, Purdue Univ, 55-58; sr res engr, Jet Propulsion Lab, Calif Inst Technol, 58-59; supvr opers res sect, Aeronutronic Div, Ford Motor Co, 59-62; assoc prof, Grad Sch Bus Admin, Univ Southern Calif, 62-68; prof quant methods & assoc dean, Sch Bus Admin, Ga State Univ, 68-69; PROF MGT SCI, GRAD SCH BUS ADMIN, TULANE UNIV, 69- Concurrent Pos: Lectr univ exten, Univ Calif, 59-63; consult, Kern County Land Co, 54 & Eli Lilly & Co, 57-58. Mem: Am Statist Asn; Opers Res Soc Am; Inst Mgt Sci; Am Inst Decision Sci. Res: Industrial operations research; applied statistics; mathematical programming; Monte Carlo method; queuing; reliability; management control systems; weapon systems analysis. Mailing Add: Grad Sch of Bus Admin Tulane Univ New Orleans LA 70118

BECKWITH, STEPHEN LYON, forestry, see 12th edition

BECNEL, IRWIN JOSEPH, b Taft, La, Aug 24, 09; m 39; c 1. ENTOMOLOGY. Educ: La State Univ, BS, 31, MS, 32. Prof Exp: Oil res, 33-37; res entomologist, La State Univ, 39-47; dir agr res, Freeport Sulphur Co, 47-69, asst vpres, 69-74; RETIRED. Mem: Entom Soc Am; Am Phytopath Soc; Soil Sci Soc Am; Am Hort Soc. Res: Economic entomology; cotton and citrus fruit insect control; sulphur in agriculture. Mailing Add: 311 Homestead Ave Metairie LA 70005

BECRAFT, GEORGE EARLE, b Sedro-Woolley, Wash, Oct 11, 22; m 48; c 4. GEOLOGY, INFORMATION SCIENCE. Educ: State Col Wash, BS, 49, MS, 50; Univ Wash, PhD(geol), 59. Prof Exp: Field asst, State Geol Surv, Wash, 48; instr geol, Field Camp, State Col Wash, 49, asst, 49-50; instr geol, Vassar Col, 50-51; geologist, 51-66, chief off tech reports, 66-72, CHIEF, OFF SCI PUBL, US GEOL SURV, 72- Concurrent Pos: Field asst, US Geol Surv, 50; instr, Spokane Exten, State Col Wash, 52-53. Mem: Fel Geol Soc Am; Soc Econ Geol; Asn Earth Sci Ed (pres, 75); AAAS. Res: Economic and regional geology; information science. Mailing Add: US Geol Surv Nat Ctr 904 Reston VA 22092

BEDELL, GEORGE NOBLE, b Harrisburg, Pa, May 1, 22; m 50, 70; c 9. INTERNAL MEDICINE. Educ: DePauw Univ, BA, 44; Univ Cincinnati, MD, 46. Prof Exp: Nat Heart Inst fel, 52-54; from asst prof to assoc prof, 54-68, PROF INTERNAL MED & DIR PULMONARY DIS DIV, UNIV IOWA, 68- Concurrent Pos: Nat Heart Inst spec fel, Univ Pa, 54-55; USPHS career develop award, 62; vis prof, Post-grad Sch Med, Univ London, 65-66. Mem: AAAS; Am Soc Clin Invest; Am Thoracic Soc; Am Col Chest Physicians; Am Fed Clin Res. Res: Pulmonary disease; clinical, cardiovascular and respiratory physiology. Mailing Add: Dept of Med Univ of Iowa Hosps Iowa City IA 52240

BEDELL, LOUIS ROBERT, b Niagara Falls, NY, Nov 14, 39; m 64; c 4. SURFACE PHYSICS. Educ: Auburn Univ, BS, 63; Brown Univ, MS, 68, PhD(physics), 71. Prof Exp: Res asst surface physics, Field Res Lab, Mobil Oil Co, 63-65; ASST PROF PHYSICS, NORTHEAST LA UNIV, 70- Mem: Am Phys Soc; Am Vacuum Soc; AAAS. Mailing Add: Dept of Physics Northeast La Univ Monroe LA 71201

BEDELL, THOMAS ERWIN, b Santa Cruz, Calif, Dec 21, 31; m 70; c 3. ANIMAL HUSBANDRY, RANGE MANAGEMENT. Educ: Calif State Polytech Col, BS, 53; Univ Calif, MS, 57; Ore State Univ, PhD, 66. Prof Exp: Jr specialist range mgt, Univ Calif, 57, asst agriculturist, Agr Exten, 57-63; from instr to asst prof range mgt, Ore State Univ, 66-70; exten range mgt specialist & assoc prof range mgt, Univ Wyo, 70-73; ASSOC PROF AGR, EXTEN & AREA LIVESTOCK EXTEN AGT, ORE STATE UNIV, 73- Res: Range management and improvement. Mailing Add: Rm 104 Courthouse Dallas OR 97338

BEDENBAUGH, ANGELA LEA OWEN, b Seguin, Tex, Oct 6, 39; m 61. CHEMISTRY. Educ: Univ Tex, BS, 61; Univ SC, PhD(chem), 67. Prof Exp: Lab instr, Univ Tex, 60-61; RES ASSOC CHEM, UNIV SOUTHERN MISS, 67- Mem: Am Chem Soc; NY Acad Sci; Sigma Xi. Res: Dissolving metal reductions. Mailing Add: Univ of Southern Miss Dept of Chem PO Box 466 Southern Sta Hattiesburg MS 39401

BEDENBAUGH, JOHN HOLCOMBE, b Newberry, SC, Apr 9, 31; m 61. ORGANIC CHEMISTRY. Educ: Newberry Col, BS, 53; Univ NC, MA, 57; Univ Tex, PhD(chem), 62. Prof Exp: Instr chem, Randolph-Macon Woman's Col, 56-58; assoc prof, Columbia Col, SC, 61-68; assoc prof, 68-74, PROF CHEM, UNIV SOUTHERN MISS, 74-, CHMN DEPT, 70- Mem: AAAS; Am Chem Soc; Nat Sci Teachers Asn. Res: Synthesis of heterocyclic compounds; chemical education; dissolving metal reductions; organic reagents for inorganic analysis. Mailing Add: Dept of Chem Univ of Southern Miss Hattiesburg MS 39401

BEDER, OSCAR EDWARD, b New York, NY, Aug 15, 14; m 42; c 2. PROSTHODONTICS. Educ: Rutgers Univ, BS, 36; Columbia Univ, DDS, 41. Prof Exp: Dent intern, Mem Hosp Treatment Cancer & Allied Dis, New York, 41-42; from asst to asst prof prosthodontics & maxillofacial prosthesis, Sch Dent, Columbia Univ, 42-52; assoc prof, 52-60, PROF PROSTHODONTICS & MAXILLOFACIAL PROSTHESIS, SCH DENT, UNIV WASH, 60- Concurrent Pos: Asst attend, Presby Hosp, New York, 44-52; consult, Vet Admin Regional Off, Brooklyn, 49-52 & Vet Admin Hosp, Seattle, Wash, 52-; attend dent staff, Children's Orthop Hosp, 52-54, dir dent serv, 54-64; consult, USPHS Hosp & Cleft Lip & Palate Rev Bd, Wash State Crippled Children's Serv, 54-; attend dent staff, King County Hosp Syst & Univ Hosp, 60. Mem: Am Cleft Palate Asn; Am Dent Asn; fel Royal Soc Health. Res: Maxillofacial prosthesis. Mailing Add: Dept of Prosthodontics Univ of Wash Sch of Dent Seattle WA 98195

BEDERKA, JOHN PAUL, JR, b Uniontown, Pa, Feb 18, 38; m 65; c 2. PHARMACOLOGY, ORGANIC CHEMISTRY. Educ: WVa Univ, BS, 59, MS, 61; Med Col Va, PhD(pharmacol), 67. Prof Exp: NIH trainee neuropharmacol, Univ Minn, Minneapolis, 67-68; spec staff Rockefeller Found visg prof pharmacol, Fac Sci, Bangkok, Thailand, 68-71; ASSOC PROF PHARMACOL-TOXICOL, OCCUP & ENVIRON MED, COL PHARM, UNIV ILL MED CTR, 71- Mem: Am Pharmaceut Asn; Am Chem Soc. Res: Metabolism of nitrogen heterocyclic compounds; cerebral structural, circulatory and electrical correlations; chemical bases of drug actions and inactivation; chemotherapy via nicotinamide and analogs; environmental pharmacology; teratology; toxicology. Mailing Add: Col of Pharm Univ of Ill at Med Ctr Chicago IL 60612

BEDERMAN, SANFORD HAROLD, b Cincinnati, Ohio, May 2, 32; m 60; c 1. GEOGRAPHY. Educ: Univ Ky, BA, 53; La State Univ, MA, 57; Univ Minn, PhD(geog), 73. Prof Exp: ASSOC PROF GEOG, GA STATE UNIV, 59- Concurrent Pos: NSF fac fel, London Sch Econ & Univ Ibadan, 65-66; Rockefeller Found res grant to work in Tanzania, Univ Minn, 71-72. Mem: Asn Am Geog; Soc Hist Discoveries; Nat Coun Geog Educ; Nigerian Geog Asn. Res: Geography of Africa; quality of life indicators in urban areas; decentralization of employment in metropolitan regions; geographical exploration and discovery. Mailing Add: Dept of Geog Ga State Univ Atlanta GA 30303

BEDERSON, BENJAMIN, b New York, NY, Nov 15, 21; m 56; c 4. EXPERIMENTAL PHYSICS. Educ: City Col New York, BSc, 46; Columbia Univ, MA, 48; NY Univ, PhD(physics), 50. Prof Exp: Res worker, Los Alamos Sci Lab, 44-45; mem staff, Res Lab Electronics, Mass Inst Technol, 50-51; asst prof physics, 52-57, assoc prof, 57-59, PROF PHYSICS, NY UNIV, 59-, CHMN DEPT, 73- Concurrent Pos: Chmn Int Conf Physics of Electronic & Atomic Collisions, 58 & 61; vis fel, Joint Inst Lab Astrophys, Nat Bur Stand, Univ Colo, 68-69; sci consult, Acad Press, NY; mem organizing comt, Int Conf Atomic Physics, 68; chmn atomic & molecular physics comt, Nat Acad Sci-Nat Res Coun, 70-; mem adv comt, Physics Div, Nat Sci Found, 72-75. Mem: Fel Am Phys Soc; AAAS. Res: Atomic collisions, structure and polarized beams; plasma physics; gaseous electronics. Mailing Add: Dept of Physics NY Univ New York NY 10003

BEDFORD, CLIFFORD LEVI, b Cheney, Wash, Feb 9, 12; m 41; c 4. FOOD SCIENCE. Educ: Univ Calif, BS, 32, MS, 34, PhD(bot), 41. Prof Exp: Asst, Univ Calif, 38-41; asst prof & asst horticulturist, State Col Wash, 41-49; res assoc prof hort, 49-59, prof, 59-60, PROF FOOD SCI, MICH STATE UNIV, 60- Mem: AAAS; Am Chem Soc; Inst Food Technologists. Res: Preservation of fruit and vegetable products; mycology; zymology. Mailing Add: Dept of Food Sci & Human Nutrit Mich State Univ East Lansing MI 48824

BEDFORD, JAMES WILLIAM, b Colfax, Wash, Dec 3, 42; m 65; c 1. WATER CHEMISTRY. Educ: Mich State Univ, BS, 65, MS, 67, PhD(entom), 70. Prof Exp: CHEMIST, DEPT NATURAL RESOURCES, STATE OF MICH, 70-, ASST DIR, MICH WATER RESOURCES LAB, 73- Mem: Am Fisheries Soc; Am Soc Limnol & Oceanog; Am Inst Biol Sci; AAAS. Res: Biological monitoring of pesticides; pesticide residue analysis; pesticide and heavy metal pollution; lamellibranch physiology. Mailing Add: Dept of Natural Resources 3500 N Logan Lansing MI 48906

BEDFORD, JOEL S, b Denver, Colo, Feb 14, 38; m 62; c 2. RADIOBIOLOGY, CHEMISTRY. Educ: Univ Colo, BA, 61, MS, 64; Oxford Univ, DPhil(radiobiol), 66. Prof Exp: Chemist med ctr, Univ Colo, 58-59, 61-62; instr, 66-67, James Picker Found res fel, 67-68, asst prof, 68-74, ASSOC PROF RADIOBIOL, COL MED, VANDERBILT UNIV, 74- Mem: Radiation Res Soc. Res: Influence of oxygen and dose rate on the survival of cultured mammalian cells exposed to ionizing radiation. Mailing Add: Dept of Radiol Vanderbilt Univ Hosp Nashville TN 37203

BEDFORD, JOHN MICHAEL, b Sheffield, Eng, May 21, 32. ANATOMY, PHYSIOLOGY. Educ: Cambridge Univ, BA, 55, MA & VetMB, 58; Univ London, PhD, 65. Prof Exp: Jr fel surg, Vet Sch, Bristol Univ, 58-59; res assoc physiol, Worcester Found, Mass, 59-61; asst prof, Royal Vet Col, Univ London, 61-66; scientist, Worcester Found, Mass, 66-67; from asst prof to assoc prof anat, Columbia Univ, 70-72; PROF REPRODUCTIVE BIOL & ANAT, MED COL, CORNELL UNIV, 72- Mem: Soc Study Reproduction; Am Asn Anatomists; Endocrine Soc; Brit Soc Study Fertil. Res: Reproductive physiology; sperm maturation in the male; capacitation of sperm in the female; fertilization; physiology of the ovum. Mailing Add: Dept of Obstet & Gynec Cornell Univ Med Col New York NY 10021

BEDGOOD, DALE RAY, b Saltillo, Tex, Aug 10, 32; m 59; c 2. MATHEMATICS, STATISTICS. Educ: ETex State Univ, BS, 54; Univ Ark, MA, 59; Okla State Univ, EdD(math), 64. Prof Exp: Asst math, Univ Ark, 58-59; instr, Northeast La State Col, 59-62, asst prof, 64-65, assoc prof & head dept, 65-67; PROF MATH & HEAD DEPT, E TEX STATE UNIV, 67- Concurrent Pos: Asst, Okla State Univ, 62-64; consult, NASA, Ala, 64-65, Cambridge Res Assoc, Mass, 66-67. Mem: Am Math Soc; Math Asn Am; Nat Coun Teachers Math. Res: Point-set topology; general analysis; teacher education and training at the elementary and secondary levels. Mailing Add: Dept of Math ETex State Univ Commerce TX 75428

BEDIENT, JACK DEWITT, b Ithaca, NY, Jan 30, 26; m 49; c 4. MATHEMATICS. Educ: Albion Col, BA, 48; Univ Colo, MS, 60, EdD(math educ), 66. Prof Exp: High sch teacher, Wash, 50-59; asst prof math, 63-71, ASSOC PROF MATH, ARIZ STATE UNIV, 71- Mem: Math Asn Am. Res: Mathematics education; training secondary school mathematics teachers. Mailing Add: Dept of Math Ariz State Univ Tempe AZ 85281

BEDIENT, PHILLIP E, b Foochow, China, Oct 15, 22; US citizen; m 43; c 2. MATHEMATICS. Educ: Park Col, AB, 43; Univ Mich, MA, 46, PhD(math), 59. Prof Exp: Instr math, Juniata Col, 50-52, asst prof, 53-55; instr, Univ Mich, 56-59; from asst prof to assoc prof, 59-68, PROF MATH, FRANKLIN & MARSHALL COL, 68-, CHMN DEPT MATH & ASTRON, 72- Mem: Am Math Soc; Math Asn Am. Res: Special functions; partial differential equations. Mailing Add: Dept of Math & Astron Franklin & Marshall Col Lancaster PA 17604

BEDINGER, CHARLES ARTHUR, JR, b Highlands, Tex, Apr 26, 42; m 70; c 1. ZOOLOGY. Educ: Tex A&M Univ, BS, 64, PhD(zool), 74; Sam Houston State Univ, MA, 67. Prof Exp: Res biologist, Res Found, Tex A&M Univ, 72-73; sr res biologist, 73-74, MGR BIOL RES, SOUTHWEST RES INST, 76- Mem: Am Soc Parasitol; Atlantic Estuarine Res Fedn; Sigma Xi. Res: Aquatic ecosystems research, including management of active projects in estuarine ecosystem dynamics, thermal pollution, fish parasitology, sewage treatment and molluscan ecology. Mailing Add: Southwest Res Inst 3600 Yoakum Blvd Houston TX 77006

BEDINGER, JOHN FRANKLIN, b Worsham, Va, Jan 26, 25; m 51; c 2. PHYSICS. Educ: Duke Univ, BA, 45; Univ Va, MA, 50. Prof Exp: Instr math & physics, St Helena Exten, Col William & Mary, 46-48; physicist, Geophys Res Directorate, Air Force Cambridge Res Ctr, 51-58; PHYSICIST, GCA CORP, 59- Mem: Am Phys Soc; Res Soc Am; Am Geophys Union; AAAS. Res: Upper atmospheric physics; airglow and aurora; ionosphere; rocket probing techniques; upper atmospheric composition and photo-chemistry through study of effects of ejecting small quantities of various elements and compounds from rockets. Mailing Add: Technol Div GCA Corp Bedford MA 01730

BEDKE, HAZEN H, meteorology, oceanography, see 12th edition

BEDNAR, JONNIE BEE, b Shiner, Tex, Aug 15, 41; m 65; c 2. MATHEMATICS, COMPUTER SCIENCE. Educ: Southwest Tex State Col, BS, 62; Univ Tex, Austin, MA, 64, PhD(math), 68. Prof Exp: Teaching asst math, Univ Tex, Austin, 62-65; engr sci II, Tracor, Inc, Tex, 65-68; asst prof math, Drexel Inst Technol, 68-69; asst prof, 69-72, ASSOC PROF MATH, UNIV TULSA, 72-, GRAD COORDR, 74-, CHMN DEPT MATH, 75- Mem: Am Math Soc; Soc Indust & Appl Math; Math Asn Am. Res: Ordered topological vector spaces: Choquet theory; numerical analysis. Mailing Add: Dept of Math Univ of Tulsa Tulsa OK 74104

BEDNAR, THOMAS W, plant physiology, biophysics, see 12th edition

BEDNARCYK, NORMAN EARLE, b Buffalo, NY, Oct 12, 38; m 63; c 3. FOOD SCIENCE. Educ: Mass Inst Technol, BS, 60, MS, 62; Rutgers Univ, PhD(food sci), 67. Prof Exp: Res chemist food prod develop, Lever Bros Co, 62-64; sr res chemist mix develop, Gen Mills, Inc, 67; head lipid chem, 67-72, head nutrit serv, 72-74, MGR SCI SERV, NABISCO, INC, 75- Mem: Inst Food Technologists; Am Oil Chemists' Soc; Nutrit Today Soc. Res: Nutritional chemistry and biochemistry as affected by the food supply. Mailing Add: 2111 Rte 208 Nabisco Inc Res Ctr Fair Lawn NJ 07410

BEDNARCZYK, LEONARD RONALD, toxicology, see 12th edition

BEDNAREK, ALEXANDER R, b Buffalo, NY, July 15, 33; m 54; c 4. MATHEMATICS. Educ: State Univ NY Albany, BS, 57; Univ Buffalo, MA, 59, PhD(math), 61. Prof Exp: Instr math, Univ Buffalo, 58-60; mathematician, Goodyear Aerospace Corp, Ohio, 61-62; asst prof math, Univ Akron, 62-63; from asst prof to assoc prof, 63-69, PROF MATH & CHMN DEPT, UNIV FLA, 69- Concurrent Pos: Consult info sci dept, Goodyear Aerospace Corp, 63- Mem: Am Math Soc; Math Asn Am; Polish Math Soc. Res: Topological algebra and the theory of relations, their applications to the mathematical theory of automata. Mailing Add: Dept of Math Univ of Fla Gainesville FL 32063

BEDNARSKI, THEODORE MARK, analytical chemistry, electrochemistry, see 12th edition

BEDNEKOFF, ALEXANDER G, b Seattle, Wash, June 13, 32; m 58; c 3. BIOCHEMISTRY. Educ: Univ Wash, BS, 54; Univ Wis, MS, 61, PhD(biochem), 62. Prof Exp: Asst prof biochem, SDak State Univ, 62-66; from asst prof to assoc prof, 66-73, PROF BIOCHEM, KANS STATE COL PITTSBURG, 73- Mem: Am Chem Soc; Sigma Xi; AAAS. Res: Blood group substances of cattle and sheep; urinary calculi. Mailing Add: Dept of Chem Kans State Col Pittsburg KS 66762

BEDNOWITZ, ALLAN LLOYD, b New York, NY, Oct 7, 39; m 66. X-RAY CRYSTALLOGRAPHY. Educ: Cooper Union, BChE, 60; Polytech Inst Brooklyn, MS, 63, PhD(chem physics), 66. Prof Exp: Fel x-ray crystallog, Brookhaven Nat Lab, 65-67; RES STAFF MEM X-RAY CRYSTALLOGRAPHY, THOMAS J WATSON RES CTR, IBM CORP, 67- Mem: AAAS; Am Crystallog Asn; Am Phys Soc; Res Soc Am. Res: Crystal structure analysis by direct methods; computer control of experiments in research laboratories; computer programming for problems in x-ray crystallography; laboratory automation. Mailing Add: Thomas J Watson Res Ctr IBM Corp PO Box 218 Yorktown Heights NY 10598

BEDO, DONALD ELRO, b Erie, Pa, Dec 25, 29. PHYSICS. Educ: Pa State Univ, BS, 51; Cornell Univ, PhD, 57. Prof Exp: Corning Glass Found fel, Cornell Univ, 56-58, res assoc physics, 58-60; RES PHYSICIST, US AIR FORCE CAMBRIDGE RES LABS, CRUU, 62- Mem: Am Phys Soc. Res: Soft x-ray and extreme ultraviolet spectroscopy; photoemission; electromagnetic theory. Mailing Add: US Air Force Cambridge Res Labs CRUU L G Hanscomb Field Bedford MA 01730

BEDOIT, WILLIAM CLARENCE, JR, b Chattanooga, Tenn, Apr 20, 22; m 43; c 2. POLYMER CHEMISTRY, PHYSICAL ORGANIC CHEMISTRY. Educ: Univ Tenn, BS, 47, MS, 48, PhD(phys org chem), 50. Prof Exp: Res chemist, Carbide & Carbon Chem Corp, 50-52 & Mallinckrodt Chem Works, 52-54; sr res chemist, Jefferson Chem Co, 54-57, mgr mkt develop, 57-70, mgr urethane mkt develop, 70-71; mkt mgr, Martin Sweets Co, 71-73, vpres, 73; PRES, UCT, INC, 73- Concurrent Pos: Instr, Morris Harvey Col, 51-52. Mem: Am Chem Soc; Soc Plastics Indust; fel Am Inst Chemists; Com Develop Asn. Res: Catalytic hydrogenation; chemical kinetics; distillation; organic preparations; market development; organic chemistry; petrochemicals; catalysis; all aspects of new urethane technology. Mailing Add: 2305 Stannye Dr Louisville KY 40222

BEDROSIAN, ALLEN J, analytical chemistry, soil chemistry, see 12th edition

BEDROSIAN, KARAKIAN, b Milford, Mass, June 29, 33; m 54; c 3. FOOD TECHNOLOGY. Educ: Univ Mass, BS, 54; Cornell Univ, MS, 56; Univ Ill, PhD(food technol), 58; Mich State Univ, MBA, 62. Prof Exp: Assoc res food technologist, Whirlpool Corp, 58-61, mgr food sci & technol, 61-62, dir res & develop, Tectrol Div, 62-63, dir res & develop, St Joseph Opers, 63-65; mgr new prod lab, Lever Bros Co, 65-66, tech mgr new prod & opers anal, 66-67; vpres res, DCA Food Indust, Inc, 67-71; PRES, BEDROSIAN & ASSOCS, 71- Mem: AAAS; Inst Food Technol; Soc Heating, Refrig & Air Conditioning Eng. Res: New product development; microwave processing; extruded foods; fruit and vegetable storage and distribution. Mailing Add: Sherwood Ct Alpine NJ 07620

BEDWELL, THOMAS HOWARD, b Forest, Idaho, Feb 17, 15; m 56. PHYSICS. Educ: Univ SDak, AB, 43, AM, 47, ME, 58; Univ Nebr, PhD(physics, educ), 66. Prof Exp: Instr physics, Univ SDak, 44-48; asst prof, Fla State Univ, 48-50; plant engr, Freeman Co, SDak, 50-53; chmn elec div, Southern State Col (SDak), 53-57; assoc prof physics, Univ SDak, 58-64; assoc prof, 64-68, PROF PHYSICS, NORTHERN ARIZ UNIV, 68- Concurrent Pos: Fel Univ Okla, 70-71; chmn, Am Physics Teachers Lect Series, 63-64; judge, Future Scientists of Am Found, Region VI, 58; environ monitoring consult, East SDak, Northern States Power, 63-64. Mem: Fel AAAS; Am Phys Soc; Inst Elec & Electronics Eng; Am Meteorol Soc; Am Geophys Union. Res: Environmental science; radio physics; infrared analysis of atmosphere. Mailing Add: Dept of Physics Northern Ariz Univ Box 6010 Flagstaff AZ 86001

BEE, ROBERT L, b Ithaca, NY, Dec 21, 38; m 60; c 2. ANTHROPOLOGY. Educ: Univ Kans, BA, 60, PhD(anthrop), 67; Univ Calif, Los Angeles, MA, 62. Prof Exp: Asst prof anthrop, 67-72, ASSOC PROF ANTHROP, UNIV CONN, 72- Mem: Fel Am Anthrop Asn; Soc Appl Anthrop. Res: Sociocultural change; American Indians, especially modern reservation communities; applied anthropology. Mailing Add: Dept of Anthrop Univ of Conn Storrs CT 06268

BEEBE, GEORGE WARREN, b Eau Claire, Wis, Sept 9, 36; m 59; c 3. PHOTOGRAPHIC CHEMISTRY. Educ: Wis State Univ, Eau Claire, BS, 58; Mich State Univ, PhD(tetrazoles), 64. Prof Exp: SR RES SPECIALIST, PHOTO PROD LAB, 3M CO, 63- Mem: Am Chem Soc; Soc Photog Scientists & Engrs. Res: Nitrogen heterocycles; condensed ring hydrocarbons; fluorescence, tetrazole synthesis; chemilumninescence; organic dyes; polyurethanes; aliphatic epoxides; resin curing systems; color photographic chemistry; photographic emulsion making, digestion and coating. Mailing Add: 3M Co Photo Prod Lab 3M Ctr 209-2C St Paul MN 55101

BEEBE, GILBERT WHEELER, b Mahwah, NJ, Apr 3, 12; m 33; c 4. MEDICAL STATISTICS. Educ: Dartmouth Col, AB, 33; Columbia Univ, MA, 38, PhD(sociol, statist), 42. Prof Exp: Statistician, Nat Comt Maternal Health, 34-41; mem tech staff, Milbank Mem Fund, 41-46; statistician, Div Med Sci, 46-58, DIR FOLLOW-UP AGENCY, NAT ACAD SCI-NAT RES COUN, 50- Concurrent Pos: Res assoc, Milbank Mem Fund, 39-41; statistician, Chief Reports & Anal Br, Control Div, Off Surgeon Gen, US Army, 43-46, consult, 46-50; consult, Comn Reorgn Exec Br, Hoover Comn, 48; chief epidemiol & statist dept, Atomic Bomb Casualty Comn, Japan, 58-60, 66-68 & 73-75, mem bd dirs & chief scientist, Radiation Effects Res Found, Japan, 75; mem subcomt somatic effects, Nat Acad Sci Adv Comt Biol Effects of Ionizing Radiations, 69-72; mem res training comt, Nat Inst Environ Sci, 70-73. Mem: AAAS; Am Epidemiol Soc; Am Acad Neurol; fel Am Statist Asn; Inst Math Statist. Res: Application of statistical methods to research in clinical medicine and epidemiology. Mailing Add: Nat Acad of Sci-Nat Res Coun 2101 Constitution Ave NW Washington DC 20418

BEEBE, RALPH ALONZO, b Monson, Mass, Mar 8, 98; m 28; c 3. SURFACE CHEMISTRY. Educ: Amherst Col, AB, 20; Princeton Univ, PhD(chem), 23. Prof Exp: From instr to prof, 23-66, EMER PROF CHEM & RESEARCHER, AMHERST COL, 66- Concurrent Pos: NIH prin investr, 60-75; vis prof, Carlton Univ, Can, 66-68; prof, Voorhees Col, 66-68. Mem: Fel AAAS; Am Chem Soc; fel Am Acad Arts & Sci. Res: Adsorption and catalysis; heat of adsorption; adsorption of polar and nonpolar gases on carbon surfaces; surface chemistry of bone mineral. Mailing Add: Dept of Chem Amherst Col Amherst MA 01002

BEEBE, RICHARD TOWNSEND, b Great Barrington, Mass, Jan 22, 02; m 32; c 3. INTERNAL MEDICINE. Educ: Princeton Univ, BS, 24; Johns Hopkins Univ, MD, 28; Am Bd Internal Med, dipl. Prof Exp: Res fel, Thorndike Mem Lab, 29-30; asst resident med, Johns Hopkins Univ, 30-32; assoc, 32-37, assoc prof, 37-48, head dept, 48, PROF MED, ALBANY MED COL, 48- Concurrent Pos: Physician-in-chief, Albany Hosp, 48-67, sr physician, 67- Mem: Assoc AMA; Am Soc Clin Invest; Am Clin & Climat Asn; fel Am Col Physicians. Mailing Add: 76 Schuyler Rd Loudonville NY 12211

BEEBE, THOMAS REED, organic chemistry, see 12th edition

BEECH, JOHN ALAN, b London, Eng, Sept 20, 27; m 63; c 3. CHEMISTRY, PHARMACOLOGY. Educ: Univ London, BSc, 53; Univ Md, PhD(pharmacol), 63. Prof Exp: Analyst, Glaxo Labs, Ltd, Eng, 53-55; analyst, Westminster Labs, Ltd, 55-57; spec lectr combustion res, Univ Toronto, 57-59; fel, Johns Hopkins Univ, 63-64; res assoc biomed res, Univ Md, Hosp, 64-65; head pharmacodyn, Strasenburgh Labs, Wallace & Tiernan Inc, NY, 65-69; assoc prof pharmacol, Fla A&M Univ, 69-74; DIR RES SUPPORT PROG, USV PHARMACEUT CO, REVLON, INC, NEW YORK, 74- Mem: Am Chem Soc; assoc mem Royal Inst Chem. Res: Plasma amino acids in health and disease; synthesis and purine and anticancer compounds and physiochemical theory of carcinogenesis. Mailing Add: 107 Wallace St Tuckahoe NY 10707

BEECHEM, HENRY A, chemistry, see 12th edition

BEECHER, GARY RICHARD, b Wilton, Wis, May 25, 39; m 62; c 3. NUTRITIONAL BIOCHEMISTRY. Educ: Univ Wis-Madison, BS, 61, MS, 63, PhD(biochem), 66. Prof Exp: Biochemist, US Army Med Res & Nutrit Lab, 66-68; asst prof biochem, Kans State Univ, 68-71; BIOCHEMIST, PROTEIN NUTRIT LAB, NUTRIT INST, AGR RES SERV, USDA, 71- Mem: AAAS; Am Chem Soc. Res: Physiological and biochemical responses of muscle to maturation, development, exercise and stress. Mailing Add: 16132 Kenny Rd Laurel MD 20810

BEECHER, HENRY KNOWLES, b Wichita, Kans, Feb 4, 04; m 34; c 3. MEDICINE. Educ: Univ Kans, AB, 26, AM, 27; Harvard Univ, MD, 32. Hon Degrees: MD, Univ Lund, 61; Dr, Univ Thessaloniki, 69. Prof Exp: Surg intern & asst resident, Mass Gen Hosp, 32-34, 35; Moseley traveling fel, Harvard Univ, 34-35; Dorr prof, 41-70, EMER DORR PROF RES ANESTHESIA, HARVARD MED SCH, 70-; ANESTHETIST IN CHIEF, MASS GEN HOSP, 36- Concurrent Pos: Numerous lectureships, 40-; consult, NIH, Surgeon Gen, US Army, US Air Force & USPHS; hon consult, US Navy; chmn subcomt anesthesia, Nat Res Coun; hon fel, Royal Col Surgeons, Ireland. Honors & Awards: Chevalier, Legion d'Honneur; Warren Prize, 31 & 37; Citation Distinguished Serv to Humanity, Univ Kans, 58; Hunter Mem Award, Am Therapeut Soc, 64; Knighthood, Queen Margrethe, Denmark, 74. Mem: AAAS; AMA; Am Soc Anesthesiol; Am Soc Clin Invest; Am Soc Pharmacol & Exp Therapeut. Res: Pharmacology of anesthesia; surgical shock; hypnotics; narcotics; measurement of subjective responses and influence of drugs thereon; the ethics of human experimentation. Mailing Add: Harvard Med Sch 10 Shattuck St Boston MA 02115

BEECHER, WILLIAM JOHN, b Chicago, Ill, May 23, 14. ORNITHOLOGY. Educ: Univ Chicago, BS, 47, MS, 48, PhD, 54. Prof Exp: Asst zool, Chicago Natural Hist Mus, 37-42, asst zool, Educ Dept, 46-54; asst zool, Conserv Dept, Cook County Forest Preserve Dist, 54-57; DIR, CHICAGO ACAD SCI, 58- Concurrent Pos: Mem biol comt, Ill Bd Higher Educ; vchmn, Ill Chap Nature Conserv; mem environ comt, Northeastern Ill Plan Comn; mem open lands proj. Honors & Awards: Nat Asn Biol Teachers Award. Mem: AAAS; Am Ornith Union; Wilson Ornith Soc; Cooper Ornith Soc; Soc Study Evolution. Res: Birds, anatomy and classification, ecology and migration; other vertebrates, comparative anatomy of the ear, migration and homing. Mailing Add: 2001 N Clark St Chicago IL 60614

BEECHLER, BARBARA JEAN, b Rockford, Ill, Dec 13, 28. MATHEMATICS. Educ: Univ Iowa, BA, 49, MS, 51, PhD(math), 55. Prof Exp: Instr math, Smith Col, 52-54; asst prof, Wilson Col, 55-58, assoc prof & chmn dept, 58-60; assoc prof, Wheaton Col (Mass), 60-67; assoc prof, 67-68, PROF MATH, PITZER COL, 68- Mem: Math Asn Am; Am Math Soc. Res: Arthmetic properties of integral domains; local rings. Mailing Add: Dept of Math Pitzer Col Claremont CA 91711

BEEDE, CHARLES HERBERT, b Chelsea, Mass, Dec 4, 34; m 57; c 1. ORGANIC CHEMISTRY. Educ: Northeastern Univ, BS, 57; Mass Inst Technol, PhD(org chem), 62. Prof Exp: Res chemist, Org & Polymer Chem, Hercules Powder Co, 61-63; RES CHEMIST, PRESSURE SENSITIVE ADHESIVES, JOHNSON & JOHNSON, 64- Mem: AAAS; Am Chem Soc; Sigma Xi. Res: Polymer synthesis, particularly vinyl, emulsion, stereoregular and condensation; organic synthesis; chemistry of small-ring compounds. Mailing Add: 19 Kings Rd East Brunswick NJ 08816

BEEHLER, ROGER EARL, b Rochester, Ind, May 8, 34; m 56; c 3. PHYSICS. Educ: Purdue Univ, BS, 56. Prof Exp: Physicist, Allison Div, Gen Motors Corp, Ind, 56-57; physicist, Nat Bur Standards, 57-60, proj leader atomic frequency standards, 60-65, asst sect chief, Atomic Time & Frequency Standards Sect, 65-66; physicist, Frequency & Time Div, Hewlett-Packard Co, 66-68; from asst chief to assoc chief time & frequency div, 68-75, CHIEF TIME & FREQUENCY SERV SEC, NAT BUR STANDARDS, 75- Honors & Awards: US Dept Com Silver Medal, 64. Res: Atomic beam frequency standards; time; frequency. Mailing Add: Time & Frequency Serv Sect Nat Bur of Standards Boulder CO 80302

BEEKMAN, BRUCE EDWARD, b Upland, Calif, Apr 18, 30; c 3. COMPARATIVE PHYSIOLOGY. Educ: San Diego State Col, BA, 52; Ind Univ, PhD, 65. Prof Exp: From asst prof to assoc prof physiol, 58-71, chmn dept biol, 72-75, PROF PHYSIOL, CALIF STATE UNIV, LONG BEACH, 71- Mem: AAAS; Am Soc Zoologists. Res: Physiology; endocrinology. Mailing Add: Dept of Biol Calif State Univ Long Beach CA 90840

BEEKMAN, JOHN ALFRED, b LaCrosse, Wis, July 14, 31; m 55; c 2. MATHEMATICS. Educ: Univ Iowa, BA, 53, MS, 57; Univ Minn, Minneapolis, PhD(math), 63. Prof Exp: From asst prof to assoc prof, 63-69, PROF MATH, BALL STATE UNIV, 69- Concurrent Pos: Vis prof statist, Univ Iowa, 69-70. Mem: Am Math Soc; Soc Actuaries; Inst Math Statist; Math Asn Am; Sigma Xi. Res: Probability theory; Gaussian Markov and collective risk stochastic processes; distributions of first exit times for compound Poisson and Ornstein-Uhlenbeck processes; applications in actuarial science, statistics, physics. Mailing Add: Dept of Math Sci Ball State Univ Muncie IN 47306

BEEKS, JOHN CHARLES, b Eagleville, Mo, July 7, 24; m 44; c 2. SOILS, FIELD CROPS. Educ: Univ Mo, BS, 48, MS, 55, EdD(agron, agr educ), 63. Prof Exp: Instr, Voc Schs, 48-58; instr crops, 58-61, from asst prof to assoc prof, 61-64, PROF CROPS & SOILS & CHMN DEPT AGR, NORTHWEST MO STATE UNIV, 64- Mem: Am Soc Agron; Soil Sci Soc Am. Res: Soil fertilization; soil water movement. Mailing Add: Dept of Agr Northwest Mo State Univ Maryville MO 64468

BEEL, JOHN ADDIS, b Butte, Mont, Sept 20, 21; m 44; c 2. ORGANIC CHEMISTRY. Educ: Mont State Col, BSc, 42; Iowa State Univ, PhD(org chem), 49. Prof Exp: Instr chem, Iowa State Univ, 45-49; asst prof & head dept, 49-52, assoc prof, 52-54, actg chmn sci div, actg assoc dean, PROF CHEM, UNIV NORTHERN COLO, 54- Honors & Awards: Meritorious Serv Award, Am Chem Soc, 69. Mem: Am Chem Soc. Res: Organometallic compounds; chemotherapy. Mailing Add: Dept of Chem Univ of Northern Colo Greeley CO 80631

BEELER, DONALD A, b Elmwood, NS, Aug 16, 31; m 53; c 4. BIOCHEMISTRY. Educ: McGill Univ, BSc, 53; Purdue Univ, MS, 59, PhD(biochem), 62. Prof Exp: Fel biochem, Univ Wis, 61-63; instr, 63-64, asst prof, 64-68, ASSOC PROF BIOCHEM, ALBANY MED COL, 68- Mem: Am Chem Soc; Am Oil Chem Soc. Res: Hepatic and biliary phospholipid and cholesterol metabolism; carotene biosynthesis; atherosclerosis. Mailing Add: Dept of Biochem Albany Med Col Union Univ Albany NY 12208

BEELER, FRED A, b Bluffton, Ind, Mar 8, 08; m 36. MATHEMATICS. Educ: Univ Alaska, BS, 31; Ind Univ, MA, 34; Univ Mich, PhD(math), 50. Prof Exp: Instr math, Univ Alaska, 30-33; asst, Univ Mich, 34-36; assoc prof, Hillsdale Col, 36-42; prof, 46-68, EMER PROF MATH, WESTERN MICH UNIV, 68- Concurrent Pos: Lectr qual control groups. Mem: Am Soc Qual Control; Math Asn Am; Am Watchmakers Inst. Res: Mathematical statistics; statistical quality control. Mailing Add: 1753 Greenlawn Kalamazoo MI 49007

BEELER, GEORGE W, JR, b West Point, NY, Oct 5, 38; m 65; c 1. BIOMEDICAL ENGINEERING. Educ: Princeton Univ, BSE, 60; Calif Inst Technol, MS, 61, PhD(elec eng), 65. Prof Exp: Assoc consult, Mayo Clin & instr, Mayo Med Sch, 67-68, CONSULT PHYSIOL & BIOPHYS, MAYO CLIN & ASST PROF, MAYO MED SCH, 69- Mem: AAAS; Inst Elec & Electronics Engr. Res: Analysis of biological systems; excitation-contraction coupling in cardiac muscle; application of computers to biomedical research and practice. Mailing Add: Dept of Physiol & Biophys Mayo Clin Rochester MN 55901

BEELER, JOE R, JR, b Beloit, Kans, Aug 13, 24; m 47; c 2. SOLID STATE PHYSICS. Educ: Univ Kans, PhD(physics), 55. Prof Exp: Staff mem, Sandia Corp, 52-55; sr physicist, Gen Elec Co, 55-57, prin physicist, 57-61, supvr solid state physics, 61-64, consult physicist, 64-67; PROF NUCLEAR ENG & MAT ENG, NC STATE UNIV, 67- Mem: Am Phys Soc. Res: Radiation damage in solids; computer simulation of collision chains and cascades; physics of defects in solids. Mailing Add: Dept of Nuclear & Mat Eng NC State Univ Raleigh NC 27607

BEELER, MYRTON FREEMAN, b Winthrop, Mass, Apr 27, 22. CLINICAL PATHOLOGY, CLINICAL CHEMISTRY. Educ: Harvard Univ, AB, 45; NY Med Col, MD, 49. Prof Exp: Resident path, Worcester City Hosp, Mass, 52-54; resident path, New Eng Deaconess Hosp, Boston, 54-56, asst pathologist, 56-58; dir labs, Ochsner Clin & Ochsner Found Hosp, 58-67; assoc prof, 67-71, PROF PATH, LA STATE UNIV MED CTR, NEW ORLEANS, 71-, DIR GRAD PROG CLIN CHEM, 70- Concurrent Pos: Assoc pathologist, Charity Hosp of La, New Orleans, 67-; consult clin path, Vet Admin Hosp, 71- Mem: AAAS; Am Soc Clin Path; Acad Clin Lab Physicians & Scientists. Res: Ceruloplasmin; tumor sterols and enzymes; instrument survey technics. Mailing Add: Dept of Path La State Univ Med Ctr New Orleans LA 70112

BEELER, NELSON FREDERICK, inorganic chemistry, see 12th edition

BEELIK, ANDREW, b Nizne Valice, Czech, Dec 12, 24; US citizen; m 51. ORGANIC CHEMISTRY. Educ: Univ Agrarian Sci, Hungary, dipl, 47; Univ Toronto, MSA, 52; McGill Univ, PhD(chem), 54. Prof Exp: Lectr org chem, Royal Mil Col, 54-55; from res chemist to sr res chemist, 55-62, res group leader, 62-72, RES SUPVR, OLYMPIC RES DIV, ITT RAYONIER, INC, 72- Mem: AAAS; Am Chem Soc. Res: Chemistry of wood constituents; cellulose acetate and viscose processes; instrumental analysis. Mailing Add: 2012 Walker Park Rd Shelton WA 98584

BEELMAN, ROBERT B, b Elyria, Ohio, May 16, 44; m 68. FOOD SCIENCE, ENOLOGY. Educ: Capital Univ, BS, 66; Ohio State Univ, MS, 67, PhD(food sci), 70. Prof Exp: Asst prof, 70-75, ASSOC PROF FOOD SCI, PA STATE UNIV, UNIVERSITY PARK, 75- Mem: Inst Food Technologists; Am Soc Enol. Res: Effects of processing on the quality of fruits, vegetables and mushrooms; evaluation of vinification practices on eastern United States table wines; stimulation of malo-lactic fermentation in eastern United States table wines. Mailing Add: Div of Food Sci Dept of Hort Pa State Univ University Park PA 16802

BEEM, JOHN KELLY, b Detroit, Mich, Jan 24, 42; m 64; c 1. GEOMETRY. Educ: Univ Southern Calif, AB, 63, MA, 65, PhD(math), 68. Prof Exp: Asst prof, 68-71, ASSOC PROF MATH, UNIV MO-COLUMBIA, 71- Concurrent Pos: Investr, NSF grant, 69-71. Mem: Am Math Soc; Math Asn Am. Res: Geometry of indefinite metric spaces and relativity. Mailing Add: 1106 Pannell Columbia MO 65201

BEEM, JOHN RAYMOND, b Greeley, Colo, Feb 22, 19; m 42; c 4. MEDICINE. Educ: Northwestern Univ, BS, 41; Univ Va, MD, 45. Prof Exp: Res fel med cardiol, Sch Med, Wash Univ, 51-53; instr med, Cardiol Sect & staff mem, Robinette Found, Sch Med, Univ Pa, 55-57; asst prof med, 57-66, dir hypertension-renal sect, 57-68, ASSOC PROF MED, HAHNEMANN MED COL & HOSP, 66-, STAFF MEM HOSP, 57-, DIR CLIN PHARMACOL SECT, 68- Concurrent Pos: Asst dir clin invest, Res Labs, Merck & Co, Inc, 53-59; dir clin res, Warner-Lambert Res Inst, 59-60, med dir, 60-63; assoc dir, Nat Heart Inst, 63-; mem coun arteriosclerosis, Am Heart Asn. Mem: Fel Am Col Cardiol; fel Am Col Chest Physicians; assoc Am Col Physicians; Aerospace Med Asn; Am Fedn Clin Res. Res: Internal medicine; cardiovascular diseases and research; hypertension; renal diseases; physiology; pharmacology; academic government; industrial research administration; systems management. Mailing Add: RR 2 Box 184 Perkasie PA 18944

BEEM, MARC O, b Chicago, Ill, June 25, 23; m 46; c 4. MEDICINE. Educ: Williams Col, BA, 45; Univ Chicago, MD, 48. Prof Exp: From asst prof to assoc prof, 54-64, PROF PEDIAT, UNIV CHICAGO, 64- Res: Infectious diseases; virology. Mailing Add: Dept of Pediat Box 286 Univ of Chicago Chicago IL 60637

BEEMAN, CURT PLETCHER, b Mt Vernon, Ohio, May 30, 44; m 66; c 2. ANALYTICAL CHEMISTRY. Educ: Univ Fla, BS, 67; Auburn Univ, MS, 69, PhD(chem), 71. Prof Exp: Res asst chem, Univ Alta, 71-73; RES ASSOC CHEM, MT HOLYOKE COL, 73- Mem: Am Chem Soc. Res: Nuclear magnetic resonance; proton magnetic resonance investigations of hydrogen bonding, proton exchange and phase equilibria; carbon magnetic resonance investigations in conformational analysis, carbonium ions and charge transfer complexes. Mailing Add: Carr Lab Mt Holyoke Col South Hadley MA 01075

BEEMAN, DAVID EDMUND, JR, b Sacramento, Calif, Dec 12, 38; m 65; c 1. THEORETICAL PHYSICS. Educ: Stanford Univ, BS, 61; Univ Calif, Los Angeles, MA, 63, PhD(physics), 67. Prof Exp: Res assoc theoret physics, Atomic Energy Res Estab, Eng, 67-69; asst prof, 69-74, ASSOC PROF PHYSICS, HARVEY MUDD COL, 69- Mem: AAAS; Am Phys Soc. Res: Theoretical solid state physics; microscopic theory of liquids. Mailing Add: Dept of Physics Harvey Mudd Col Claremont CA 91711

BEEMAN, ELIZABETH ANN, b Barre, Vt, Aug 12, 16. BEHAVIORAL BIOLOGY. Educ: Grinnell Col, AB, 38; Univ Chicago, MS, 39, PhD(behavior), 47; Mt Holyoke Col, MA, 41. Prof Exp: Asst, Univ Chicago, 41-44, asst biol sci, 44-45, from instr to asst prof, 45-51; mem sci fac, Sarah Lawrence Col, 51-60; assoc prof, 60-69, PROF BIOL SCI, MT HOLYOKE COL, 69- Concurrent Pos: NSF fac fel, 59-60; vis prof biobehav sci, Univ Conn, 74-75. Mem: AAAS; Am Soc Zool; Ecol Soc Am. Res: Behavior of mice, quail and chickens; distribution of cone cells in the retinas of various breeds of chickens. Mailing Add: Dept of Biol Sci Mt Holyoke Col South Hadley MA 01075

BEEMAN, ROBERT D, b Los Angeles, Calif, Mar 23, 32; m 68; c 3. MARINE ZOOLOGY, INVERTEBRATE ZOOLOGY. Educ: Humboldt State Col, BS, 54; Univ Idaho, MS, 56; Stanford Univ, PhD(marine biol), 66. Prof Exp: Instr zool, Humboldt State Col, 52-54; res fel, Univ Idaho, 54-56; instr zool, Humboldt State Col, 56-57; instr zool & dir, Mt San Antonio Col, 58-64; sci fac fel marine biol, Stanford Univ, 62-66; chmn dept marine biol, 67-69, PROF MARINE ZOOL, SAN FRANCISCO STATE UNIV, 66- Concurrent Pos: NSF res grants, 66-72. Mem: AAAS; Am Soc Zoologists; Inst Malacol; Soc Syst Zool; NY Acad Sci. Res: Migration ecology of elk; functional morphology of reproductive systems in anaspidean opisthobranch mollusks, including ultrastructure and capacitation of spermatozoa. Mailing Add: Dept of Marine Biol San Francisco State Univ San Francisco CA 94132

BEEMAN, WILLIAM WALDRON, b Detroit, Mich, Oct 21, 11; m 40; c 4. BIOPHYSICS. Educ: Univ Mich, BS, 37; Johns Hopkins Univ, PhD(physics), 40. Prof Exp: Res physicist, Gen Motors Corp, 40-41; from instr to assoc prof physics, 41-52, chmn biophys lab, 64-70, PROF PHYSICS, UNIV WIS-MADISON, 52- Concurrent Pos: Physicist, Argonne Nat Lab, Chicago, 46; consult, AEC, 46-52. Mem: Fel Am Phys Soc; Biophys Soc; Am Crystallog Asn. Res: X-ray absorption spectra of solids; small angle scattering of x-rays; structure of macromolecules. Mailing Add: Dept of Physics Univ of Wis Madison WI 53706

BEENKEN, MAY MARGARET, b Philadelphia, Pa, Oct 22, 01. MATHEMATICS. Educ: Univ Calif, Los Angeles, BEd, 23; Univ Chicago, AM, 26, PhD(math), 28. Prof Exp: Assoc math, Univ Calif, Los Angeles, 24-25; head dept, Wis State Col, Oshkosh, 28-47, dir div preprof educ, 44-47; prof math & chmn dept, 47-69, EMER PROF MATH, IMMACULATE HEART COL, 69- Mem: AAAS; Am Math Soc; Math Asn Am. Res: Projective differential geometry. Mailing Add: 1906 Park Ave Los Angeles CA 90026

BEER, ALAN E, b Milford, Ind, Apr 14, 37; m 59; c 4. OBSTETRICS & GYNECOLOGY, TRANSPLANTATION BIOLOGY. Educ: Ind Univ, BS, 59, MD, 62. Prof Exp: Intern, Methodist Grad Med Ctr, Ind, 62-63; surgeon, USPHS Div Indian Health, USPHS Indian Hosp, Tuba City, Ariz, 63-65; from resident to chief resident obstet & gynec, Hosp Univ Pa, 65-69; fel med genetics & obstet & gynec, Sch Med, Univ Pa, 68-70, asst prof, 70-71; asst prof, 71-73, ASSOC PROF CELL BIOL & OBSTET & GYNEC, UNIV TEX HEALTH SCI CTR, DALLAS, 73- Concurrent Pos: Smith Kline & French Foreign fel, Nigeria, 62-63. Honors & Awards: Carl Hartman Award, Am Fertil Soc, 70. Mem: AAAS; Transplantation Soc; Soc Study Reprod; Am Col Obstet & Gynec; Soc Gynec Invest. Res: Immunobiology of mammalian reproduction; immunological significance of the mammary gland; elicitation and expression of immunity in the uterus; transplantation immunobiology. Mailing Add: Dept of Cell Biol Univ of Tex Health Sci Ctr Dallas TX 75235

BEER, ALBERT CARL, b Mansfield, Ohio, Mar 7, 20; m 49; c 3. PHYSICS. Educ: Oberlin Col, AB, 41; Cornell Univ, PhD(physics), 44. Prof Exp: Asst, Nat Defense Res Comt proj, Cornell Univ, 42-44, res assoc, 44-45; physicist appl physics lab, Johns Hopkins Univ, 45-51; prin physicist, 51-52, asst chief div, 52-54, consult, 54-56, asst tech dir, 56-67, SR FEL, BATTELLE MEM INST, 67-; ADJ PROF, DEPT ELEC ENG, OHIO STATE UNIV, 69- Concurrent Pos: Chmn panel defects, ad hoc comt Characterization of Mat, Mat Adv Bd, Nat Acad Sci, 65-67. Mem: Fel Am Phys Soc; Inst Elec & Electronics Eng; Electrochem Soc. Res: Transport theory; semiconductors; semimetals; metals. Mailing Add: Battelle Mem Inst 505 King Ave Columbus OH 43201

BEER, CHARLES, b Le Sueur Co, Minn, Dec 18, 23. AGRICULT URAL ECONOMICS, BUSINESS MANAGEMENT. Educ: Univ Minn, St Paul, BS, 48; Mich State Univ, MS, 55, PhD(agr econ), 57. Prof Exp: Exten agent, Univ Minn, Anoka County, 49-53; asst prof agr econ, Mich State Univ, 57-62; prof, Univ Mo-Columbia, 62-64; specialist farm mgt, 64-66, asst dir, 66-68, DIR AGR PROD, EXTEN SERV, USDA, 68- Mem: Am Soc Farm Mgr & Rural Appraisers; Am Agr Econ Asn; Am Soc Agr Consult. Mailing Add: Agr Prod Exten Serv US Dept of Agr Washington DC 20250

BEER, GEORGE ATHERLEY, b Salmon Arm, BC, Jan 3, 35; m; c 1. NUCLEAR PHYSICS. Educ: Univ BC, BASc, 57, MSc, 59; Univ Sask, PhD(nuclear physics), 66. Prof Exp: Res officer reactor physics, Atomic Energy of Can Ltd, 59-61; Nat Res Coun fel nuclear physics, Inst Nuclear Physics, Darmstadt Tech Univ, 67-69; asst prof, 69-75, ASSOC PROF NUCLEAR PHYSICS, UNIV VICTORIA, 75- Concurrent Pos: Vis scientist, European Orgn Nuclear Res, 75-76. Mem: Can Asn Physicists. Res: Neutron gas scintillation spectroscopy; low-power reactor; nuclear structure investigations using inelastic electron scattering; intermediate energy physics and accelerator design. Mailing Add: Dept of Physics Univ of Victoria Victoria BC Can

BEER, JAMES ROBERT, wildlife management, zoology, deceased

BEER, MICHAEL, b Budapest, Hungary, Feb 20, 26; m 54; c 3. MOLECULAR BIOLOGY. Educ: Univ Toronto, BA, 49, MA, 50; Univ Manchester, PhD(phys chem), 53. Prof Exp: Instr physics, Univ Mich, 53-56; fel biophys, Nat Res Coun Can, 56-58; from asst prof to assoc prof, 58-71, PROF BIOPHYS, JOHNS HOPKINS UNIV, 71-, CHMN DEPT, 74- Concurrent Pos: Chmn biophys training comt, NIH, NIH spec fel, 67-68; mem, US Nat Comt of Int Union Pure & Appl Biophys, 74- Mem: Biophys Soc (pres, 75-76); Electron Micros Soc Am; AAAS. Res: Structure of biological macromolecules using optical methods; fine structure of cells; viruses and macromolecules as revealed by electron microscope. Mailing Add: Dept of Biophys Johns Hopkins Univ Baltimore MD 21218

BEER, REINHARD, b Berlin, Ger, Nov 5, 35; m 60. OPTICS, ASTRONOMY. Educ: Univ Manchester, BSc, 56, PhD(physics), 60. Prof Exp: Res asst physics, Univ Manchester, 56-60, sr res asst astron, 60-63; sr scientist, 63-67, group leader, 67-68, mem tech staff, 68-71, GROUP SUPVR INFRARED ASTRON, JET PROPULSION LAB, CALIF INST TECHNOL, 71- Concurrent Pos: Vis assoc prof astron, Univ Tex, Austin, 74. Honors & Awards: Except Sci Achievement Medal, NASA, 74. Mem: Optical Soc Am; Am Astron Soc. Res: Application of infrared interference spectroscopy to physics and astronomy; Fourier spectroscopy of planetary atmospheres. Mailing Add: Jet Propulsion Lab Calif Inst of Technol Pasadena CA 91103

BEER, ROBERT EDWARD, b Los Angeles, Calif, Apr 28, 18; m 41; c 1. ENTOMOLOGY. Educ: Univ Calif, BS, 47, MS, 48, PhD(entom), 50. Prof Exp: Res asst entom, Univ Calif, 47-49; from instr to assoc prof, 50-61, chmn dept, 61-68, PROF ENTOM, UNIV KANS, 61-, CHMN DEPT, 75- Concurrent Pos: From assoc state entomologist to state entomologist, Kans, 53-63; from assoc prof to prof, Biol Sta, Univ Mich, 57-69; prof, Biol Sta, Univ Minn, 71-73. Mem: AAAS; Entom Soc Am; Soc Study Evolution (treas). Res: Biology and taxonomy of mites and sawflies; agricultural entomology. Mailing Add: Dept of Entom Univ of Kans Lawrence KS 66044

BEER, STEVEN VINCENT, b Boston, Mass, July 19, 41; m 63; c 3. PLANT PATHOLOGY. Educ: Cornell Univ, BS, 65; Univ Calif, Davis, PhD(plant path), 69. Prof Exp: ASST PROF PLANT PATH, CORNELL UNIV, 69- Mem: Am Phytopath Soc; AAAS; Am Chem Soc. Res: Epidemiology, physiology and control of tree-fruit diseases with emphasis on fire blight; nature and control of post-harvest decay of fruit; mycotoxins produced by fruit pathogens. Mailing Add: Dept Plant Path Plant Sci Bldg Cornell Univ Ithaca NY 14853

BEER, SYLVAN ZAVI, b New York, NY, Feb 5, 29; m 52; c 2. PHYSICAL CHEMISTRY. Educ: Brooklyn Col, BS, 51; Polytech Inst Brooklyn, PhD(phys chem), 58. Prof Exp: Chemist, Hexagon Labs, 51-52; instr chem, New York Community Col, 55-57; Crucible Steel Co fel high temperature chem, Mellon Inst, 57-60; staff scientist, Crucible Steel Co Am, 60-62; sr engr, Dept Chem, Westinghouse Elec Corp, 62-64 & Solid State Phenomena Dept, 64-67; mgr res & develop, Special Metals Corp, 67-70; vpres technol, Technostruct Corp, 70-72; CONSULT, 72- Mem: AAAS; Am Chem Soc; Am Inst Chemists; Am Soc Metals; Fedn Am Scientists. Res: High temperature chemistry; electrochemistry; air and water pollution; sensor/transducer development, instrumentation, superionic conductors, energy storage. Mailing Add: 315 Scott Ave Syracuse NY 13224

BEERBOWER, JAMES RICHARD, b Ft Wayne, Ind, Apr 5, 27. PALEONTOLOGY. Educ: Univ Colo, BA, 49; Univ Chicago, PhD(paleozool), 54. Prof Exp: From instr to assoc prof geol, Lafayette Col, 53-64; prof, McMaster Univ, 64-69; PROF GEOL, STATE UNIV NY BINGHAMTON, 69- Mem: AAAS; Soc Vert Paleont; Soc Study Evolution; Paleont Soc; fel Geol Soc Am. Res: Paleoecology and evolution of vertebrate and invertebrate animals; vertebrate anatomy; sedimentology of alluvial deposits and marine shales. Mailing Add: Dept of Geol State Univ of NY Binghamton NY 13901

BEEREBOOM, JOHN JOSEPH, b Grand Rapids, Mich, Mar 6, 26; m 51; c 3. ORGANIC CHEMISTRY. Educ: Hope Col, BA, 49; Wayne State Univ, MS, 52, PhD(chem), 54. Prof Exp: Fel, Univ Manchester, 54-55; res chemist, 55-65, group supvr, 65-70, RES MGR, CENT RES, PFIZER INC, 70- Mem: Am Chem Soc. Res: Natural products; antibiotics; alkaloids; steroids; structure, isolation and synthesis; biologically active organic compounds; synthetic flavors; artificial sweeteners; polysaccharides; low-calorie food systems. Mailing Add: Pfizer Inc Groton CT 06344

BEERMAN, HERMAN, b Johnstown, Pa, Oct 13, 01; m 24. DERMATOLOGY. Educ: Univ Pa, AB, 23, MD, 27, ScD(med), 35; Am Bd Dermat, dipl, 35. Prof Exp: Field asst, Bur Entom, USDA, 25-26; intern, Mt Sinai Hosp, Philadelphia, 27-28; asst instr dermat & syphil, Sch Med, 29-30, instr, 30-36, assoc, 36-37, from asst prof to prof, 37-70, from asst prof to prof, Grad Sch Med, 40-70, chmn dept, 49-67, EMER PROF DERMAT, SCH MED, UNIV PA, 70- Concurrent Pos: Dermatologist, Hosp Univ Pa, 29-70, asst chief dermat clin, 38-70, asst dir inst study venereal dis, 38-54, assoc serol, Pepper Lab, 50, head dept dermat, Grad Hosp, 53-67; Abbott fel, Univ Pa, 32-46; consult dermatologist, USPHS, 37-70; asst dermatologist, Radium Clin, Philadelphia Gen Hosp, 38-40, dermatologist, 40-53, consult, 53-68, hon consult, 68-; dermatopathologist, Skin & Cancer Hosp, 48-54; treas & trustee, Inst Dermat Commun & Educ, 62-; consult, Grad Hosp, Univ Pa, 67-, Pa Hosp, 67- & Vet Admin Hosp, Coatesville, 67-; past assoc ed, Quart Rev Dermat & Quart Rev Syphilis; contrib ed, Am J Med Sci. Honors & Awards: Thomas Parran Award, Am Venereal Dis Asn, 74. Mem: Hon mem Am Dermat Asn (pres, 67-68); hon mem Am Acad Dermat (pres, 65-66); hon mem Soc Invest Dermat (pres, 47-58, secy-treas, 50-65); hon mem Am Soc Dermatopath (pres, 65-66); Asn Prof Dermat (pres, 67-68). Res: Dermatopathology; study of pathologic changes in patients; literary research. Mailing Add: 255 S 17th St Philadelphia PA 19103

BEERS, JOHN R, b Bridgeport, Conn, July 4, 33. BIOLOGICAL OCEANOGRAPHY. Educ: Bates Col, BS, 55; Univ NH, MS, 58; Harvard Univ, PhD(biol), 62. Prof Exp: Part-time instr biol, Univ NH, 57-58; marine biologist, Bermuda Biol Sta Res, Inc, 61-65, asst dir, 63-65; asst res zoologist, 65-71, ASSOC RES ZOOLOGIST, INST MARINE RESOURCES, UNIV CALIF, SAN DIEGO, 71-, LECTR, 69- Mem: AAAS; Am Soc Limnol & Oceanog. Res: Biological oceanography; invertebrate physiology; plankton ecology. Mailing Add: Inst of Marine Resources Univ of Calif at San Diego La Jolla CA 92093

BEERS, LINN YARDLEY, b Philadelphia, Pa, Apr 2, 13; m 45; c c 2. MOLECULAR SPECTROSCOPY. Educ: Yale Univ, BS, 34; Princeton Univ, MA, 37, PhD(physics), 41. Prof Exp: Asst physics, Princeton Univ, 40; instr, NY Univ, 40-41 & Smith Col, 41-42; staff mem radiation lab, Mass Inst Technol, 42-45, res assoc, 46; from asst prof to prof physics, NY Univ, 46-61; chief millimeter wave res sect, 61-62, chief radio standards physics div, 62-68, CONSULT, RADIO STANDARDS PHYSICS DIV, NAT BUR STANDARDS, 68- Concurrent Pos: Consult, Brookhaven Nat Lab, 47-58; Fulbright grant, Nat Standards Lab, Australia & vis lectr, Univ Sydney, 56; vis lectr, Univ Colo, Boulder, 63-64; adj prof, Univ Denver, 65-67; vis prof, Colo State Univ, 67-68 & Univ Colo, Denver, 69. Mem: Inst Elec & Electronics Engrs; Sigma Xi; Am Asn Physics Teachers; Am Phys Soc. Res: Direct determination of the charge of the beta particle; microwave and millimeter wave spectra of gases; stark effect; laser power and energy measurements. Mailing Add: 740 Willowbrook Rd Boulder CO 80302

BEERS, ROLAND FRANK, JR, b Brooklyn NY, Mar 31, 23; m 45; c 4. BIOCHEMISTRY. Educ: Dartmouth Med Sch, AB, 44; Univ Rochester, MD, 47; Mass Inst Technol, PhD(biochem), 51. Prof Exp: Am Cancer Soc fel, Mass Inst Technol, 50-52, res assoc, 51-52, asst prof phys biochem, 52-56; fel & instr physiol chem, Sch Med, Johns Hopkins Univ, 56-60, asst prof radiol, Sch Med & radiol sci, Sch Hyg & Pub Health, 60-61, assoc prof, 61-63, prof radiol sci, Sch Hyg & Pub Health, 62-73, dir div radiobiol, 61-73; V PRES RES AFFAIRS, MILES LABS, INC, 73- Concurrent Pos: Dir res, Childrens Hosp Inc, Md, 56-60; consult, Miles Labs, 57- & US Army Environ Hyg Agency, 62; adj prof biol, Notre Dame Univ, 74- Mem: Am Soc Biol Chemists; NY Acad Sci; Am Chem Soc; Am Soc Microbiol; Biophys Soc. Res: Enzymology, industrial microbiology, microbiology, molecular biology, allergy, medical physiology and biochemical engineering; polynucleotides; nucleic acid enzymology; hemeproteins; radiation biology; bone metabolism; industrial enzymes. Mailing Add: Miles Labs Inc 1127 Myrtle St Elkhart IN 46514

BEERS, THOMAS WESLEY, b Greensburg, Pa, Oct 23, 30; m 53; c 4. FOREST MENSURATION. Educ: Pa State Univ, BS, 55, MS, 56; Purdue Univ, PhD(forest mgt), 60. Prof Exp: From instr to assoc prof, 56-69, PROF FORESTRY, PURDUE UNIV, WEST LAFAYETTE, 69- Res: Forest mensuration techniques and instruments; inventory procedures and data processing methods, especially timber stand estimates by point sampling; forest management site productivity studies; computer programming of forestry problems. Mailing Add: Dept of Forestry Purdue Univ West Lafayette IN 47907

BEERSTECHER, ERNEST, JR, b Detroit, Mich, May 4, 19; m 63; c 4. BIOCHEMISTRY. Educ: Wayne Univ, BS, 40; Univ Tex, MA, 46, PhD(biochem), 48. Prof Exp: Asst exp med, Harper Hosp, Mich, 38-39; biochemist, Robison Labs, Inc, 40; toxicologist, Edgewood Arsenal, US Army, 40-42, res biochemist, 44-45; res biochemist, Biochem Inst, Univ Tex, 46-50; from asst prof to assoc prof, 50-55, PROF BIOCHEM, UNIV TEX DENT BR HOUSTON, 55- Concurrent Pos: Consult bacteriologist, Magnolia Petrol Co, Tex, 48-49. Mem: Am Chem Soc; Am Soc Biol Chemists. Res: Bioassay of drugs and metabolites; nutrition; endocrinology; bacterial metabolism; human metabolism; biosynthesis of amino acids; antibiotics; immunochemistry; dental biochemistry. Mailing Add: Dept of Biochem Univ of Tex Dent Br Houston TX 77025

BEESACK, PAUL RICHARD, b Hamilton, Ont, Aug 8, 24; m 49; c 2. MATHEMATICAL ANALYSIS. Educ: McMaster Univ, BA, 50, AM, 52; Wash Univ, PhD(math), 55. Prof Exp: Instr math, Wash Univ, 54-55; asst prof, McMaster Univ, 55-60; assoc prof, 60-65, PROF MATH, CARLETON UNIV, 65- Mem: Am Math Soc; Math Asn Am; Can Math Cong; Soc Indust & Appl Math. Res: Differential and integral inequalities. Mailing Add: Dept of Math Carleton Univ Ottawa ON Can

BEESCH, SAMUEL C, b Dayton, Ohio, Oct 29, 15; m 46; c 2. CHEMISTRY. Educ: Ohio State Univ, BA, 38, MSc, 48; Col Armed Forces, cert, 58. Prof Exp: Plant bacteriologist, Publicker Industs, 39-41, res bacteriologist, 41-44, head dept, 44-47, dir biol res, 47-50; asst head biol eng div, Chas Pfizer & Co, 50-56, head biol process develop, 56-59, tech supvr, Pfizer Int Corp, 59-61, biochem process coordr, 61-70, sr staff engr, 70-71; CONSULT BIOCHEM-BIOENG, 71- Mem: Am Chem Soc. Res: Industrial fermentation processes; anaerobes; antibiotics; solvents; vitamins; steroids; acids by fermentation; bioengineering. Mailing Add: 706 McQueen Rd Aberdeen NC 28315

BEESE, RONALD ELROY, b Milwaukee, Wis, May 19, 29; m 54; c 2. PHYSICAL CHEMISTRY, INORGANIC CHEMISTRY. Educ: Lawrence Col, BS, 51. Prof Exp: From res chemist to sr res chemist, 51-67, RES ASSOC, CANCO DIV, AM CAN CO, 67- Mem: Am Chem Soc; Nat Asn Corrosion Eng; Electrochem Soc. Res: Electrochemical corrosion research involving the evaluation of corrosion mechanisms related to products packaged in metallic containers. Mailing Add: Am Can Co 433 N Northwest Hwy Barrington IL 60010

BEESLEY, EDWARD MAURICE, b Belvidere, NJ, Jan 11, 15; m 40; c 3. MATHEMATICS. Educ: Lafayette Col, AB, 36; Brown Univ, ScM, 38, PhD(math), 43. Prof Exp: From instr to assoc prof, 40-55, actg head dept, 44-47, PROF MATH, UNIV NEV, RENO, 55-, CHMN DEPT, 47- Mem: Am Math Soc; Math Asn Am. Res: Theory of functions of a real variable. Mailing Add: Dept of Math Univ of Nev Reno NV 89507

BEESON, DONALD M, physical chemistry, see 12th edition

BEESON, EDWARD LEE, JR, b Bartow, Fla, Mar 9, 28; m 59; c 3. MOLECULAR SPECTROSCOPY. Educ: Emory Univ, BA, 49, MS, 50; Ga Inst Technol, PhD(physics), 60. Prof Exp: Instr physics, Chattanooga Univ, 50-51 & Ga Inst Technol, 54-59; from asst prof to assoc prof, 60-68, PROF PHYSICS, UNIV NEW ORLEANS, 68- Concurrent Pos: NIH res grant, 61-66. Mem: Am Phys Soc. Res: Microwave spectroscopy. Mailing Add: Dept of Physics Univ of New Orleans New Orleans LA 70122

BEESON, JAMES HAROLD, organic chemistry, analytical chemistry, see 12th edition

BEESON, PAUL BRUCE, b Livingston, Mont, Oct 18, 08; m 42; c 3. MEDICINE. Educ: McGill Univ, MDCM, 33. Hon Degrees: DSc, Emory Univ, 68, McGill Univ, 71, Albany Med Col, 75 & Yale Univ, 75. Prof Exp: Intern, Univ Pa Hosp, 33-35; from asst resident to resident, Rockefeller Inst Hosp, 37-39; resident med, Peter Bent Brigham Hosp, Boston, 39-40; chief physician, Harvard Univ Field Hosp, Unit, Eng, 41-42; from asst prof to prof med, Med Sch, Emory Univ, 42-52, chmn dept, 46-52; prof & chmn dept, Med Sch, Yale Univ, 52-65; Nuffield prof clin med, Oxford Univ, 65-74; PROF MED, UNIV WASH, 74-; DISTINGUISHED PHYSICIAN, US VET ADMIN, 74- Concurrent Pos: Physician-in-chief, Univ Serv, Grace-New Haven Community Hosp, Conn, 52-; med consult, West Haven Vet Admin Hosp, 52-; vis investr, Wright-Fleming Inst Microbiol Eng, 58-59. Mem: Nat Acad Sci; Am Soc Clin Invest; Soc Exp Biol & Med; Asn Am Physicians. Res: Infectious diseases; immunology; clinical medicine. Mailing Add: Seattle Vet Admin Hosp 4435 Beacon Ave S Seattle WA 98108

BEESON, WILLIAM JEAN, b Crawfordsville, Ind, June 26, 26; m 53; c 2. ANTHROPOLOGY. Educ: Univ Ill, BA, 51, MA, 52; Univ Ariz, PhD(anthrop), 66. Prof Exp: Res assoc archaeol, Univ Ill, 56-57; instr anthrop, 57-61, asst prof, 61-66, assoc prof, 66-71, PROF ANTHROP, CALIF STATE UNIV, SACRAMENTO, 71- Mem: Soc Am Archaeol; Am Anthrop Asn. Res: Archaeology, especially the Midwest, Southwest and California. Mailing Add: Dept of Anthrop Calif State Univ Sacramento CA 95819

BEESON, WILLIAM MALCOLM, b Meridian, Miss, Feb 3, 11; m 34; c 2. ANIMAL NUTRITION. Educ: Okla Agr & Mech Col, BS, 31; Univ Wis, MS, 32, PhD(animal nutrit), 35. Prof Exp: Asst animal husb, Univ Wis, 31-35; instr, Agr & Mech Col Tex, 35-36; from asst prof to assoc prof, Univ Idaho, 36-45; assoc prof, Univ Ariz, 37-38; assoc prof, 45-49, assoc prof animal sci, 49-59, PROF ANIMAL NUTRIT & LYNN DISTINGUISHED PROF AGR, PURDUE UNIV, WEST LAFAYETTE, 59-, HEAD ANIMAL NUTRIT RES, 45- Concurrent Pos: Chmn subcomt swine nutrit, Nat Res Coun, Nat Acad Sci, 50-64, chmn comt animal nutrit, 62-72; mem, US Nat Comt, Int Union Nutrit Sci, 62-68; Calcium Carbonate Co travel fel trace mineral res, 64. Honors & Awards: Am Feed Mfrs Asn Award, 52; Okla State Univ Res Award, 53; Distillers Feed Res Coun Award, 64; Morrison Award, 65. Mem: AAAS; fel Am Soc Animal Sci (past pres); Am Inst Nutrit; Am Dairy Sci Asn. Res: Animal nutrition research with cattle, sheep and swine. Mailing Add: Dept of Animal Sci Purdue Univ West Lafayette IN 47907

BEESTMAN, GEORGE BERNARD, b Hammond, Wis, July 17, 39; m 65; c 2. AGRICULTURAL CHEMISTRY, WEED SCIENCE. Educ: Wis State Univ, Riber Falls, BS, 61; Univ Wis, MS, 67, PhD(soil chem), 69. Prof Exp: Sr res chemist, Monsanto Co, 68-73, RES SPECIALIST, MONSANTO AGR PROD CO, 73- Mem: Am Soc Agron; Soil Sci Soc Am; Sigma Xi; AAAS; Weed Sci Soc Am. Res: Fertilizer nutrient imbalance effects on crop quality; influence of soil properties on insecticide uptake and translocation by plants; herbicide dissipation modes in soils; pesticide formulation. Mailing Add: Monsanto Agr Prod Co-T4C 800 N Lindbergh Blvd St Louis MO 63166

BEETCH, ELLSWORTH BENJAMIN, b Mankato, Minn, Jan 23; m 51; c 3. PHYSICAL CHEMISTRY. Educ: Mankato State Col, BS, 49; Kans State Univ, MS, 51, PhD(phys chem), 57. Prof Exp: Sr res chemist, Rahr Malting Co, 55-58; from asst prof to assoc prof, 58-66, PROF CHEM, MANKATO STATE COL, 66- Mem: Am Chem Soc. Res: Electrophoresis; interaction of proteins with small molecules; colorimetric determination of sulfur dioxide, radioisotopes; application of instruments to beer and wort color. Mailing Add: Dept of Chem Mankato State Col Mankato MN 56001

BEETHAM, WILLIAM PARKES, b Bellaire, Ohio, July 8, 02; m 25; c 3. OPHTHALMOLOGY. Educ: Ohio Wesleyan Univ, AB, 22; Harvard Univ, MD, 26. Prof Exp: From asst ophthal to asst clin prof, Harvard Sch Med, 33-63; consult surgeon, Mass Gen Hosp & Mass Eye & Ear Infirmary, 64-73; RETIRED. Concurrent Pos: Surgeon, Mass Gen Hosp & Mass Eye & Ear Infrimary, 39-64; ophthalmologist, Boston Lying-in-Hosp, New Eng Deaconess Hosp & New Eng Baptist Hosp, 39-; mem med adv bd, Selective Serv, 42-45. Mem: Am Ophthal Soc; Asn Res Vision & Ophthal; Am Acad Ophthal & Otolaryngol. Res: Diabetes. Mailing Add: 51 Annawan Rd Boston MA 02168

BEETLE, ALAN ACKERMAN, b Princeton, NJ, June 8, 13; m 40; c 3. AGRONOMY. Educ: Dartmouth Col, AB, 36; Univ Wyo, AM, 37; Univ Calif, PhD(bot), 41. Prof Exp: Instr agron, Univ Calif & jr agronomist, Exp Sta, 41-45, asst prof, Univ, 45-46; from asst prof to assoc prof, 46-55, PROF AGRON, UNIV WYO, 55- Mem: Bot Soc Am; Am Soc Agron; Am Soc Range Mgt. Res: Plant taxonomy; range management; vegetation survey; taxonomy of flowering plants, especially grasses. Mailing Add: Range Mgt Sect Univ of Wyo Laramie WY 82071

BEETON, ALFRED MERLE, b Denver, Colo, Aug 15, 27; m 45, 66; c 5. LIMNOLOGY. Educ: Univ Mich, BS, 52, MS, 54, PhD(zool), 58. Prof Exp: Instr biol, Wayne State Univ, 56-57; chief environ res prog, US Bur Com Fisheries, 57-66; PROF ZOOL & ASST DIR CTR GREAT LAKES STUDIES, UNIV WIS-MILWAUKEE, 66-, ASSOC DEAN GRAD SCH, 73- Concurrent Pos: Lectr civil eng, Univ Mich, 62-66; mem planning comt, Int Symp on Eutrophication, Nat Acad Sci, 65-68 & chmn panel on freshwater life & wildlife, Comt Water Qual Criteria, 70-72; consult, US Army Corps Engrs & Metrop Sanit Dist of Greater Chicago, 68-, Greater Cleveland Growth Asn, 70 & US Environ Protection Agency, 72-; chmn res adv coun, Wis Dept Natural Resources, 69-71. Honors & Awards: James W Moffett Publ Award, US Bur Com Fisheries Lab, Mich, 67. Mem: Am Soc Limnol & Oceanog (treas, 62-); Am Soc Zool; Int Soc Limnol; Int Asn Great Lakes Res. Res: Limnology of the Great Lakes; vertical migration and related behavior of planktonic Crustacea; limnology of man-made tropical lakes. Mailing Add: Ctr for Great Lakes Studies Univ of Wis Milwaukee WI 53201

BEEUWKES, REINIER, JR, applied mechanics, see 12th edition

BEEVERS, HARRY, b Durham, Eng, Jan 10, 24; nat US; m 49; c 1. PLANT PHYSIOLOGY, BIOCHEMISTRY. Educ: Univ Durham, BSc, 44, PhD, 47. Hon Degrees: DSc, Purdue Univ, 71 & Univ Newcastle, Eng, 74. Prof Exp: Asst plant physiol, Oxford Univ, 46-48, chief res asst, 48-50; vis prof, Purdue Univ, 50, from asst prof to prof, 51-69; PROF BIOL, UNIV CALIF, SANTA CRUZ, 69- Concurrent Pos: Demonstr & tutor, Oxford Univ, 46-50; Fulbright lectr, Australia, 62; NSF sr fel, 63-64. Honors & Awards: Stephen Hales Award, Am Soc Plant Physiol, 70. Mem: Nat Acad Sci; Am Soc Biol Chemists; Am Soc Plant Physiol (pres, 62); Brit Soc Exp Biol; fel Am Acad Arts & Sci. Res: Plant metabolism. Mailing Add: Div of Natural Sci Univ of Calif Santa Cruz CA 95060

BEEVERS, LEONARD, b Co Durham, Eng, Aug 7, 34; m 60. PLANT PHYSIOLOGY. Educ: Univ Durham, BSc, 58; Univ Wales, PhD(agr bot), 61. Prof Exp: Res assoc agron, Univ Ill, Urbana-Champaign, 61-63, from asst prof to assoc prof hort, 63-71; PROF BOT, UNIV OKLA, 71- Mem: Am Soc Plant Physiol; Brit Soc Exp Biol. Res: Intermediary nitrogen metabolism; mode of action of growth regulators; general plant physiology. Mailing Add: Dept of Bot & Microbiol Univ of Okla Norman OK 73069

BEFELER, BENJAMIN, b San Jose, Costa Rica, Dec 8, 39; US citizen; m 65; c 3. CARDIOVASCULAR DISEASE. Educ: Nat Univ Mex, MD, 63; Am Bd Internal Med, dipl, 74, cert cardiovasc dis, 75. Prof Exp: Intern, Span Hosp & Nat Univ Mex, 62; intern med, Mt Sinai Hosp, Cleveland, Ohio, 64, asst resident, 65; resident med, Vet Admin Hosp & Case Western Reserve Univ Hosp, 66-67, resident cardiol, 67-68; USPHS fel med, Med Sch, Univ Calif, San Francisco, 68-69; from instr to asst prof, 69-74, ASSOC PROF MED, SCH MED, UNIV MIAMI, 74-; CHIEF CARDIOVASC LAB, VET ADMIN HOSP, MIAMI, 71- Concurrent Pos: Preceptor clin diag, Med Sch, Case Western Reserve Univ, 66-68; staff cardiologist, Mt Sinai Hosp, Miami Beach, 69-71; staff physician, Cardiopulmonary Lab, Div Cardiol, Dept Internal Med, Mt Sinai Hosp Greater Miami, 69-71; attend physician, Univ Miami Hosps & Clins, 71- & Univ Miami Jackson Mem Hosp, 71-; civilian consult cardiol, US Air Force Hosp, Homestead, Fla, 72-; ed consult, Chest, 74; mem res & educ comt, Vet Admin Hosp, Miami, 74- Mem: Fel Am Col Physicians; fel Am Col Chest Physicians; fel Am Col Cardiol; fel Am Col Angiol; Am Heart Asn. Res: Mechanism of antiarrhythmias in patients with coronary artery disease; Wolff-Parkinson-White syndrome and the role of arrhythmias in determining patient prognosis. Mailing Add: 1201 NW 16th St Miami FL 33125

BEFU, HARUMI, b Los Angeles, Calif, Mar 20, 30; m 59; c 2. ANTHROPOLOGY. Educ: Univ Calif, Los Angeles, BA, 54; Univ Mich, MA, 56; Univ Wis, PhD(anthrop), 62. Prof Exp: Asst prof anthrop, Univ Nev, 61-62 & Univ Mo, 62-64; vis assoc prof, Univ Mich, 64-65; asst prof, 65-71, ASSOC PROF ANTHROP, STANFORD UNIV, 71- Concurrent Pos: Wenner-Gren Found anthrop res grant-in-aid, 62 & 64; NSF res grants, 66-67 & 69-74; Guggenheim fel, 72-73. Mem: Am Anthrop Asn; Asn Asian Studies; Japanese Soc Ethnol. Res: Methodology and theory; kinship; culture change; social exchange; social structure; Japanese culture. Mailing Add: Dept of Anthrop Stanford Univ Stanford CA 94305

BEG, MIRZA ABDUL BAQI, b Etawah, India, Sept 20, 34; m 58. THEORETICAL PHYSICS. Educ: D J Sind Govt Sci Col, Pakistan, BSc, 51; Univ Karachi, MSc, 54; Univ Pittsburgh, PhD(physics), 58. Prof Exp: Res fel math physics, Univ Birmingham, 58-60; res assoc theoret physics, Brookhaven Nat Lab, 60-62; mem sch math, Inst Advan Study, 62-64; from asst prof to assoc prof, 64-68, PROF PHYSICS, ROCKEFELLER UNIV, 68- Concurrent Pos: Consult, Brookhaven Nat Lab, 65-73, mem, High Energy Adv Comt, 75- Mem: Fel Am Phys Soc. Res: Field theory; physics of elementary particles. Mailing Add: Dept of Physics Rockefeller Univ New York NY 10021

BEGA, ROBERT V, b Milford, Mass, Aug 17, 28; m; c 1. PLANT PATHOLOGY. Educ: Univ Calif, BS, 53, PhD(path), 57. Prof Exp: Asst, 55-56, LECTR PLANT PATH, UNIV CALIF, BERKELEY, 56-, ASSOC, EXP STA, 57-; PLANT PATHOLOGIST, PAC SOUTHWEST FOREST & RANGE EXP STA, USDA, 56- Mem: Am Phytopath Soc. Res: Biology of rust fungi; root disease fungi; mycorrhiza; physiology of host-parasite relationship; aerobiology, microclimatology; biological control; air pollution. Mailing Add: Pac SW Forest & Range Exp Sta PO Box 245 Berkeley CA 94701

BEGALA, ARTHUR JAMES, b Newark, NJ, Sept 10, 40; m 66; c 2. PHYSICAL CHEMISTRY. Educ: Lehigh Univ, BS, 62; Rutgers Univ, PhD(chem), 71. Prof Exp: Chemist, E I du Pont de Nemours & Co, 62-63; res asst chem, Rutgers Univ, 64-68; SR RES CHEMIST, AM CYANAMID CO, 68- Mem: Am Chem Soc; Sigma Xi. Res: Synthesis and characterization of polyelectrolytes and investigation of various polymer interactions with specific metallic and oxide substrates. Mailing Add: 372 Nonopoge Rd Fairfield CT 06430

BEGANY, ALBERT JOHN, b Elizabeth, NJ, Feb 28, 13; m 42; c 3. PHARMACOLOGY. Educ: Univ Mo, AB, 40; Jefferson Med Col, MS, 57. Prof Exp: Control chemist, Merck & Co, Inc, NJ, 40-42; PHARMACOLOGIST & SUPVR TECH STAFF, PHARMACOL EVAL DIV, WYETH LABS, INC, 47- Mem: Am Chem Soc. Res: Pharmacology of synthetic anticoagulants, antihistamines, adrenolytic agents and antiemetics; hypothalamus-pituitary physiology; screening and evaluation of potential bronchodilator and anti-inflammatory agents; screening and evaluation of anti-PCA compounds. Mailing Add: Wyeth Labs Inc Box 8299 Philadelphia PA 19101

BEGG, CHARLES FREDERIC, b Des Moines, Iowa, Aug 8, 12; m 40; c 3. PATHOLOGY, HEMATOLOGY. Educ: Boston Univ, AB, 35; Harvard Univ, MD, 39. Prof Exp: Asst clin prof, 50-64, assoc prof, 64-69, CLIN PROF PATH, COLUMBIA UNIV, 69-; DIR LABS, ST LUKE'S HOSP, NEW YORK, 53- Concurrent Pos: Asst dir labs, St Luke's Hosp, 49-53. Mem: AAAS; Am Soc Clin Path; Soc Study Blood; Col Am Path; NY Acad Sci. Mailing Add: 86 Bayview Ave Port Washington NY 11050

BEGG, ROBERT WILLIAM, b Florenceville, NB, Dec 27, 14; m 43; c 4. BIOCHEMISTRY. Educ: Univ King's Col, BSc, 36; Dalhousie Univ, MSc, 38, MD, 42; Oxford Univ, DPhil(path), 50. Prof Exp: Asst prof biochem, Dalhousie Univ, 46-48, res assoc prof, 48-50; Nat Cancer Inst assoc prof med res, Univ Western Ont, 50-56, prof, 56-57; prof cancer res, head dept, lectr path & dir, Cancer & Med Res Inst, 57-62, prin, Saskatoon Campus, 67-74, dean med, 62-67, actg pres univ, 74-75, PROF CHEM PATH, UNIV SASK, 64-, PRES UNIV, 75- Concurrent Pos: Nat Cancer Inst Can fel, 48-50; dir, Sask Res Unit, Nat Cancer Inst Can, 57-62. Mem: AAAS; Can Biochem Soc. Res: Cancer research; tumor-host relations; lipid metabolism and enzymology. Mailing Add: Off of the Pres Univ of Sask Saskatoon SK Can

BEGGS, DAVID PHILLIP, physical chemistry, see 12th edition

BEGGS, WILLIAM H, b Ft Dodge, Iowa, Feb 19, 35; m 57; c 2. MEDICAL MICROBIOLOGY. Educ: Univ Minn, BA, 56; Kans State Univ, MS, 59; Univ Cincinnati, PhD(microbiol), 64; Nat Registry Microbiol, regist, 70. Prof Exp: Jr scientist, Univ Minn, 59-61; USPHS res fel, 64-65; MICROBIOLOGIST, BACT RES LAB, VET ADMIN HOSP, 65- Concurrent Pos: Res assoc, Dept Med, Med Sch, Univ Minn, 68- Mem: Am Chem Soc; Am Soc Microbiol. Res: Inhibitory actions of antibiotics and other antimicrobial drugs. Mailing Add: Bact Res Lab Vet Admin Hosp Minneapolis MN 55417

BEGHIAN, LEON E, b Istanbul, Turkey, July 25, 19; US citizen; m 64. NUCLEAR PHYSICS, NUCLEAR ENGINEERING. Educ: Oxford Univ, BA, 47, PhD, 51. Prof Exp: Staff mem, Clarendon Lab, Oxford Univ, 50-56; lectr nuclear eng, Mass Inst Technol, 56-65; prof nuclear sci & eng, 65-67, head dept nuclear eng, 67-68, head dept physics, 68-73, PROF PHYSICS, LOWELL TECHNOL INST, 68-, PROVOST, 73- Concurrent Pos: Chmn activation anal panel, Mass Gov Comt Law Enforcement; mem, Goals & Aims Comt, Mass Bd Higher Educ. Mem: Am Phys Soc. Res: Neutron kinetics. Mailing Add: Dept of Physics Lowell Technol Inst Lowell MA 01854

BEGIN, JOHN JOSEPH, b Versailles, Ohio, Apr 1, 21; m 43; c 2. POULTRY NUTRITION. Educ: Univ Ky, BS, 48; Purdue Univ, MS, 50; Pa State Univ, PhD, 60. Prof Exp: Res asst poultry, Purdue Univ, 48-50; asst field agent poultry improv, 50, exten poultry specialist, 50-51, asst poultry, 51-55, asst prof & poultry-husbandryman, 55-60, ASSOC PROF POULTRY SCI, UNIV KY, 60- Mem: Poultry Sci Asn. Res: Energy metabolism. Mailing Add: Dept of Animal Sci Univ of Ky Lexington KY 40506

BEGLE, EDWARD GRIFFITH, b Saginaw, Mich, Nov 27, 14; m 37; c 7. MATHEMATICS. Educ: Univ Mich, AB, 36, MA, 37; Princeton Univ, PhD(math), 40. Prof Exp: Asst math, Princeton Univ, 39-41; Nat Res Coun fel, Univ Mich, 41-42; from instr to assoc prof, Yale Univ, 42-61; PROF MATH, STANFORD UNIV, 61- Mem: AAAS; Am Math Soc (secy, 51-56); Math Asn Am. Res: Topology; locally connected spaces and generalized manifolds; homology and homotopy properties of spaces; mathematics education; curriculum evaluation; theories of mathematics learning. Mailing Add: Sch of Educ Stanford Univ Stanford CA 94305

BEGLEITER, HENRI, b Nimes, France, Sept 11, 35; US citizen; m 63; c 1. PSYCHOPHYSIOLOGY, NEUROPHYSIOLOGY. Educ: New Sch Social Res, PhD(psychophysiol), 67. Prof Exp: Res assoc psychophysiol, 64-66, asst prof psychiat & psychophysiol, 67-72, assoc prof psychiat, 72-76, DIR PSYCHOPHYSIOL LAB, SCH MED, STATE UNIV NY DOWNSTATE MED CTR, 66-, PROF PSYCHIAT, 76- Mem: AAAS; Am Psychol Asn; Soc Psychophysiol Res; NY Acad Sci. Res: Neurophysiological correlates of behavior. Mailing Add: Dept of Psychiat State Univ NY Downstate Med Ctr Brooklyn NY 11203

BEGUE, WILLIAM JOHN, b Chicago, Ill, June 15, 31; m 52; c 3. MICROBIAL PHYSIOLOGY, ANALYTICAL BIOCHEMISTRY. Educ: Col St Thomas, BS, 53; Univ Minn, MS, 60, PhD(microbiol), 63. Prof Exp: From instr to asst prof microbiol, Univ Ky Med Ctr, 63-67; sr scientist, 67-74, RES SCIENTIST ANAL MICROBIOL, LILLY RES LABS, DIV ELI LILLY & CO, 74- Mem: Am Soc Microbiol; AAAS; Am Inst Biol Sci. Res: Development of analytical methods for new compounds, mostly antibiotics as found in animal feed and tissues; microbial nutrition, physiology and metabolism; biochemistry; enzymology; animal nutrition; antibiotics. Mailing Add: Greenfield Lab Lilly Res Lab Div Eli Lilly & Co PO Box 708 Greenfield IN 46140

BEGUIN, FRED P, b Brussels, Belg, Oct 13, 09; US citizen; m 34; c 1. ACOUSTICS, PHYSICS. Educ: State Tech Col Brussels, BS, 31; State Sch Cinematography, Paris, BSc, 33; Univ of City of Paris, prof radio commun, 33; Nat Radio Commun Inst, Brussels, lic prof physics, 44; Colby Col, cert engr noise control, 71. Prof Exp: Patent analyst, French Thomson-CSF Co, Paris, 31-34; proj engr, Philips Res Labs, Holland, 34-44; tech dir recording & mfg, Decca Records Belg & France, 44-46; Europ dir, Motorola Corp Chicago, 46-50; audio consult, Electronics Div, Gen Elec Co, NY, 50-59; mgr electroacoustics res & develop, Am Optical Corp, 59-72; DIR HEARING & NOISE CONTROL CTR, HARRINGTON MEM HOSP, SOUTHBRIDGE, 72- Concurrent Pos: Chmn, TV Standardization Comn Belg, Ministry Commun, Brussels, 44-46; tech consult to US rep, Belgian Ministry Econ Affairs, 51-53; indust safety equip rep & mem noise & bioacoustics comts, Acoust Soc Am-US Am Standards Inst, 63-72; acoustical consult, Worcester Mem Auditorium, Mass, 65-67. Honors & Awards: Nat Sci Achievement Award, Audio Eng Soc, 70. Mem: Nat Soc Prof Engrs; sr mem Inst Elec & Electronics Engrs; fel Audio Eng Soc; Acoustical Soc Am; NY Acad Sci. Res: Engineering; architectural acoustics; sound recording and reproduction; psychological and physiological acoustics; bioacoustics; noise; shock; vibration; hearing; audiology; audiometry; noise-abatement; speech communications; environmental acoustics; measurement of real-ear attenuation of noise-protective devices. Mailing Add: RFD 1 Sturbridge MA 01566

BEGUN, GEORGE MURRAY, b Bedford, Mass, Aug 20, 21; m 48; c 4. PHYSICAL INORGANIC CHEMISTRY, CHEMICAL PHYSICS. Educ: Colo Col, BA, 43; Columbia Univ, MA, 44; Ohio State Univ, PhD(chem), 50. Prof Exp: Chemist, Electromagnetic Separation Plant, Tenn Eastman Corp, 44-46; res assoc photosurface proj, Ohio State Res Found, 50-51; CHEMIST, CHEM DIV, OAK RIDGE NAT LAB, 51- Mem: Am Chem Soc. Res: Electron impact studies; uranium chemistry; isotope separation and exchange; infrared, Raman and mass spectral studies; xenon and interhalogen compounds; fluorine and tritium chemistry. Mailing Add: 106 Colby Rd Oak Ridge TN 37830

BEHAL, FRANCIS JOSEPH, b Yoakum, Tex, Oct 10, 31; m 55; c 2. BIOCHEMISTRY. Educ: St Edward's Univ, BS, 53; Univ Tex, MA, 56, PhD(chem), 58. Prof Exp: From asst prof to assoc prof biochem, Med Col Ga, 58-68, dir grad div, 64-66, dir sch grad studies, 66-68, prof biochem & microbiol & dean sch grad studies, 68-71; PROF BIOCHEM & CHMN DEPT, SCH MED, TEX TECH UNIV, 71- Concurrent Pos: Mem, 10th Int Cancer Cong. Mem: Am Soc Biol Chemists; fel Am Acad Microbiol; fel Am Inst Chem; Soc Exp Biol & Med. Res: Comparative enzymology of mammalian and microbial peptidases. Mailing Add: Dept of Biochem Tex Tech Univ Sch of Med Lubbock TX 79409

BEHAN, MARK JOSEPH, b Denver, Colo, Jan 17, 31; m 54; c 3. PHYSIOLOGICAL ECOLOGY. Educ: Univ Denver, BA, 53; Univ Wyo, MS, 58; Univ Wash, PhD(plant physiol), 63. Prof Exp: Chmn wildlife biol degree progs, 70-74, PROF BOT, UNIV MONT, 61- Mem: Ecol Soc Am; Wildlife Soc. Res: Forest tree physiology, especially mineral nutrition and cycling; cycling of inorganic ions in ecosystems. Mailing Add: Dept of Bot Univ of Mont Missoula MT 59801

BEHAR, JOSE, b Trujillo, Peru, Aug 11, 33; US citizen; m 66; c 2. GASTROENTEROLOGY. Educ: San Marcos Univ, BS, 54; San Fernando Med Sch, MD, 61. Prof Exp: Intern med, Washington Hosp Ctr, 61-62; resident, Henry Ford Hosp, 62-64; fel, Boston City Hosp & Harvard Med Sch, 64-66; instr, Ind Univ, 66-67; instr, 68-69, ASST PROF MED, SCH MED, YALE UNIV, 69; CHIEF GASTROENTEROL SECT, VET ADMIN HOSP, WEST HAVEN, CONN, 69-

Mem: Am Gastroenterol Asn. Res: Gastrointestinal physiology with respect to gastrointestinal motility and sphincter function; pathogenesis, diagnosis and treatment of motility disorder of the gastrointestinatl tract. Mailing Add: Med Serv Vet Admin Hosp West Spring St West Haven CT 06516

BEHAR, MARJAM GOJCHLERNER, b Luck, Poland, Dec 8, 25; US citizen; m 51; c 3. ANALYTICAL BIOCHEMISTRY. Educ: Univ Havana, DSc, 50. Prof Exp: Res chemist, Va Smelting Co, Cuba, 50-53; control res chemist, Lab Geol, 53-54; instr high sch, Cuba, 54-55; chemist, Garden State Tanning Corp, Pa, 55-57; jr prod engr, Lansdale Div, Philco Corp, 58-61; res assoc anesthesiol, 62-70, RES SPECIALIST ANESTHESIOL, UNIV HOSP, UNIV PA, 70- Mem: Am Soc Anesthesiologists; Am Chem Soc; NY Acad Sci; Soc Appl Spectros; fel Am Inst Chemists. Res: Anesthesia research; effect of anesthetic agents on flood flows, metabolism and blood viscosity; radioimmuno assay. Mailing Add: Dept of Anesthesia Univ of Pa Hosp Philadelphia PA 19104

BEHARA, MINAKETAN, b Jamshedpur, India, June 15, 37; Can citizen; m 66; c 2. MATHEMATICS, MATHEMATICAL STATISTICS. Educ: Univ Saarland, DSc(math statist), 63. Prof Exp: Sci asst decision theory & sequential anal, Inst Europ Statist, Univ Saarland, 61-63; asst prof math, Univ Waterloo, 63-65, adj prof, 65-67; ASSOC PROF MATH, McMASTER UNIV, 67- Concurrent Pos: Adj asst prof, Univ Western Ont, 63-64; vis res prof, Inst Europ Statist, Univ Saarland, 64; vis prof, Univ Heidelberg, 67 & 72, actg chmn, 73; vis prof, Univ Karlsruhe, 68 & Univ Regensburg, 75. Mem: Am Math Soc; Inst Math Statist; Can Math Cong; fel Royal Statist Soc. Res: Information-theoretic contribution to statistical decision theory; entropy in coding and ergodic theories; probability theory. Mailing Add: Dept of Math McMaster Univ Hamilton ON Can

BEHBEHANI, ABBAS M, b Iran, July 23, 25; US citizen; m 58; c 3. VIROLOGY. Educ: Ind Univ, AB, 49; Univ Chicago, MS, 51; Univ Tex, PhD(virol), 55. Prof Exp: Assoc prof microbiol, Univ Tehran, 56-60; virologist, Med Ctr, Baylor Univ, 60-63, res asst prof virol, Col Med, 63-64; asst prof, 64-66, sr res assoc pediat, Sect Virus Res, 66-67, assoc prof path, Sch Med, 67-72, PROF PATH, SCH MED, UNIV KANS, 72- Concurrent Pos: Dir cent state health labs, Ministry Health, Tehran, 56-59; clin asst prof, Univ Tex Southwestern Med Sch Dallas, 61-64. Mem: AAAS; Am Soc Microbiol; Soc Exp Biol & Med. Res: Properties and characteristics of enteroviruses; use of organ culture for pathological studies of human disease; development of more rapid procedures in diagnostic virology. Mailing Add: Dept of Path & Oncol Univ of Kans Sch of Med Kansas City KS 66103

BEHER, WILLIAM TYERS, b Aurora, Ill, Dec 14, 22; m; c 5. BIOCHEMISTRY. Educ: NCent Col, BA, 44; Wayne State Univ, PhD(biochem), 50. Prof Exp: Res asst, 50-52, SR RES ASSOC BIOCHEM, EDSEL B FORD INST MED RES, 52- Concurrent Pos: Instr, Wayne State Univ, 52-60. Mem: AAAS; Am Chem Soc; Am Soc Biol Chem; Soc Exp Biol & Med; Am Heart Asn. Res: Sterol and bile acid metabolism and nutrition; steroid x-ray diffraction. Mailing Add: Edsel B Ford Inst for Med Res Henry Ford Hosp Detroit MI 48202

BEHFOROOZ, ALI, b May 24, 42; Iranian citizen; m 71; c 1. COMPUTER SCIENCE, EXPERIMENTAL STATISTICS. Educ: Univ Tehran, BS, 65, MS, 66; Mich State Univ, MS, 72, MS, 73, PhD(comput sci), 75. Prof Exp: Instr math, Univ Tehran & Nat Univ Iran, 65-70; res asst comput, Mich State Univ, 72-74; DIR INST RES & ASST PROF COMPUT, MOORHEAD STATE UNIV, 74- Concurrent Pos: Grant, Educ Radio & TV of Iran, 75. Mem: Asn Comput Mach; Soc Indust & Appl Math. Res: Cluster analysis and stochastic pattern recognition; formal languages and computer application in education. Mailing Add: Dir Inst Res Moorhead State Univ Moorhead MN 56560

BEHKI, RAM M, b Nawashahr, India, Mar 31, 32; m 60; c 3. MOLECULAR BIOLOGY. Educ: Panjab Univ, India, BSc, 52; Univ Nagpur, MSc, 54, PhD(biochem), 59. Prof Exp: Sr sci asst biochem, Nat Chem Lab, Coun Sci & Indust Res, India, 57-59; res fel, Stritch Med Sch, Loyola Univ, Ill, 59-60, instr, 60-61; Jane Coffin Childs Mem Fund fel, Nat Cancer Inst, Md, 61-62; res fel, Mass Gen Hosp, Boston, 62-63; res fel, Western Reserve Univ, 63-65; res officer, Microbiol Res Inst, 65-67, res scientist, Cell Biol Res Inst, 67-72, SR RES SCIENTIST, CHEM & BIOL RES INST, CAN DEPT AGR, 73- Concurrent Pos: Vis scientist, Ctr Study Nuclear Energy, Mol, Belg, 73-74. Mem: Am Chem Soc; Am Soc Plant Physiologists. Res: Nucleic acids and protein metabolism in normal and abnormal growth; plant cell culture; protoplast technology; genetic modification of plant cells by molecular biology techniques; nucleic acid isolation and characterization. Mailing Add: Chem & Biol Res Inst Can Dept of Agr Res Br Ottawa ON Can

BEHL, WISHVENDER K, b Dhariwal, India, Dec 26, 35. PHYSICAL CHEMISTRY, ELECTROCHEMISTRY. Educ: Univ Delhi, BSc & Hons, 55, MSc, 57, PhD(phys chem), 62. Prof Exp: jr res fel phys chem, Univ Delhi, 58-62; res assoc, NY Univ, 62-64 & Brookhaven Nat Lab, NY, 64-67; RES CHEMIST, US ARMY ELECTRONICS COMMAND, 67- Honors & Awards: Jr Sci Exhib Prize, Univ Delhi, 55. Mem: AAAS; fel Am Inst Chemists; Am Chem Soc; Electrochem Soc; NY Acad Sci. Res: Chemistry of molten salts, potential measurements polarography, chronopotentiometry, linear sweep voltammetry, thermodynamics, transference numbers and ionic mobilities measurements; diffusion coefficients; molten salt batteries; solid state galvanic cells; electroanalytical chemistry; lithium-inorganic electrolyte cell. Mailing Add: Power Sources Tech Area ET&D Lab US Army Elec Comnd Ft Monmouth NJ 07703

BEHLE, WILLIAM HARROUN, b Salt Lake City, Utah, May 13, 09; m 34; c 2. ORNITHOLOGY. Educ: Univ Utah, AB, 32, AM, 33; Univ Calif, PhD(zool), 37. Prof Exp: Asst mus vert zool, Univ Calif, 33-37; from instr to assoc prof biol, 37-50, head dept gen biol, 48-54, dir biol gen educ, 54-63, PROF BIOL, UNIV UTAH, 50- Mem: AAAS; Soc Study Evolution; Am Soc Mammal; Wildlife Soc; Am Ornithologists Union. Res: Geographic variation and distribution of birds; variation in Otocoris; birds of Utah; systematic ornithology. Mailing Add: Dept of Biol Univ of Utah Salt Lake City UT 84112

BEHLOW, ROBERT FRANK, b Cleveland, Ohio, Mar 16, 26; m 54; c 3. ANIMAL PARASITOLOGY. Educ: Ohio State Univ, DVM, 53. Prof Exp: Prof animal path & head state diag lab, Univ Ky, 54-64; gen mgr, Blue Chip Mills, Ill, 64; PROF ANIMAL SCI & EXTEN VET, NC STATE UNIV, 64- Concurrent Pos: Pres, NC Qual Enterprises Inc. Mem: Am Vet Med Asn. Res: Isolation of Histoplasma capsulatum from a calf and a pig; swine parasite control. Mailing Add: Dept of Animal Sci NC State Univ Raleigh NC 27609

BEHM, ROY, b Topeka, Kans, Feb 23, 30; m 54; c 2. ANALYTICAL CHEMISTRY. Educ: Univ Wash, BA, 55, PhD(chem), 62. Prof Exp: Teacher pub schs, Wash, 55-58; res chemist, Eastman Kodak Co, 62-63; assoc dean grad studies, 66-74, asst prof, 63-70, PROF CHEM, EASTERN WASH STATE COL, 70- Mem: AAAS; Am Chem Soc; Am Inst Chem. Res: Spectrophotometric studies of metal chelates; nonaqueous solvents; voltammetry. Mailing Add: Dept of Chem Eastern Wash State Col Cheney WA 99004

BEHME, RONALD JOHN, b Evansville, Ind, Apr 12, 38; m 63; c 2. GENETICS, MICROBIOLOGY. Educ: Univ Evansville, AB, 60; Ind Univ, PhD(genetics), 69. Prof Exp: Lectr zool & genetics, Ind Univ, 65-67; res assoc genetics, Univ Waterloo, 67-69; Can Med Res Coun fel, Univ Western Ont, 69-72; vis prof genetics, Univ Waterloo, 72-73; ASST PROF BACT, UNIV WESTERN ONT, 73- Mem: AAAS; Am Soc Microbiol; Can Soc Microbiol. Res: Paramecium genetics; DNA hybridization; symbionts; yeast relationships; nematode relationships; genetics of sporulation and membrane development in bacilli; genetics of antibiotic resistance in bacteria. Mailing Add: Dept of Bact Univ of Western Ont London ON Can

BEHMER, DAVID J, b Milwaukee, Wis, May 29, 41; m 63; c 2. FISHERIES. Educ: Wis State Col, Stevens Point, BS, 63; Iowa State Univ, MS, 65, PhD(fisheries biol), 66. Prof Exp: Lectr fisheries, Humboldt State Col, 66-67; asst prof biol, 67-72, ASSOC PROF BIOL, LAKE SUPERIOR STATE COL, 72- Mem: Am Fisheries Soc; Int Asn Gt Lakes Res. Res: Statistics; genetics. Mailing Add: Dept of Biol Lake Superior State Col Sault Ste Marie MI 49783

BEHN, ROBERT COLLINS, b Cleveland, Ohio, Dec 7, 13; m 56; c 1. METEOROLOGY. Educ: Ohio State Univ, BIE, 38; Univ Chicago, MS, 61. Prof Exp: RES METEOROLOGIST, BATTELLE MEM INST, 62- Mem: Am Meteorol Soc. Res: Physical meteorology. Mailing Add: 625 Glenmont Ave Columbus OH 43214

BEHNKE, JAMES RALPH, b Milwaukee, Wis, May 2, 43; m 66; c 1. FOOD SCIENCE. Educ: Univ Wis-Madison, BS, 66, MS, 68, PhD(food sci), 71. Prof Exp: From group leader to sr group leader prod develop, Quaker Oats Co, 71-73, sect mgr, 73-74; MGR PROD DEVELOP, PILLSBURY CO, 74- Mem: Inst Food Technologists; Am Asn Cereal Chemists. Res: Identification, development and process engineering of consumer marketed and packaged food products. Mailing Add: Pillsbury Res & Develop Labs 311 Second St SE Minneapolis MN 55414

BEHNKE, ROBERT J, b Stamford, Conn, Dec 30, 29; m 63; c 2. ICHTHYOLOGY. Educ: Univ Conn, BA, 57; Univ Calif, Berkeley, MA, 60, PhD(zool), 65. Prof Exp: Am Acad Sci exchange scholar, USSR, 64-65; asst prof zool, Univ Calif, Berkeley, 66; ASST PROF FISHERIES, COLO STATE UNIV, 66- Mem: Am Fisheries Soc; Am Soc Ichthyol & Herpet. Res: Systematics of family, Salmonidae and systematics of North American freshwater fishes. Mailing Add: Colo Coop Fisheries Unit Colo State Univ Ft Collins CO 80521

BEHNKE, ROY HERBERT, b Chicago, Ill, Feb 24, 21; m 44; c 4. INTERNAL MEDICINE. Educ: Hanover Col, AB, 43; Ind Univ, MD, 46; Am Bd Internal Med, dipl, 56. Prof Exp: Markle scholar, Sch Med, Ind Univ, Indianapolis, 52-57, from instr to prof med, 60-72; PROF INTERNAL MED & CHMN DEPT, COL MED, UNIV S FLA, TAMPA, 72- Concurrent Pos: Chief med, Indianapolis Vet Admin Hosp, 57-72; chmn, Vet Admin Pulmonary Dis Res Comt, 62-68, mem, Vet Admin Coop Studies Res Comt, 66-70; chmn, Inter-Soc Comn Heart Dis Res, 69- Mem: Am Fedn Clin Res; Am Heart Asn; Am Thoracic Soc; AMA; Am Col Physicians. Res: Cardiopulmonary disease. Mailing Add: Dept of Internal Med Univ of SFla Col of Med Tampa FL 33620

BEHNKE, WALTER ERIC, organic chemistry, see 12th edition

BEHNKE, WILLIAM DAVID, b Pasadena, Calif, Jan 15, 41; m 62; c 3. BIOLOGICAL CHEMISTRY, BIOPHYSICS. Educ: Univ Calif, Berkeley, AB, 63; Univ Wash, PhD(biochem), 68. Prof Exp: Fel, Harvard Med Sch, 68-72; asst prof chem, Univ SC, 72-74; ASST PROF BIOCHEM, COL MED, UNIV CINCINNATI, 74- Res: Molecular interaction of lectins with tumor cell membranes; metalloprotein structure and function. Mailing Add: Dept of Biol Chem Univ of Cincinnati Col of Med Cincinnati OH 45267

BEHNKEN, DONALD WASHINGTON, b New York, NY, May 18, 24; m 53; c 4. EXPERIMENTAL STATISTICS. Educ: Dartmouth Col, AB, 45; Yale Univ, BS, 45; Columbia Univ, MBA, 49; NC State Col, PhD(statist), 59. Prof Exp: Physicist, Gen Elec Co, 46-47; cost analyst, E I du Pont de Nemours & Co, 49-50, head cost acct, 50-52, physicist, 52-56; consult statistician, 59-61, HEAD MATH ANAL, AM CYANAMID CO, STAMFORD, 61- Concurrent Pos: Statistician in residence, Univ Wis, 69-70. Mem: Am Statist Asn; Inst Math Statist; Int Asn Statist in Phys Sci. Res: Experimental designs for estimating linear and nonlinear parameters; application of statistical methods and probability models to scientific research problems; time series. Mailing Add: Allen Rd Norwalk CT 06851

BEHOF, ANTHONY F, JR, b Chicago, Ill, Apr 30, 37. PHYSICS. Educ: DePaul Univ, BS, 59; Univ Notre Dame, PhD(nuclear physics), 65. Prof Exp: Instr physics, Univ Notre Dame, 65; res physicist & Nat Res Coun-Naval Res Lab fel, US Naval Res Lab, 65-67; ASST PROF PHYSICS, DEPAUL UNIV, 67- Mem: AAAS; Am Phys Soc; Am Asn Physics Teachers. Res: Experimental nuclear physics, especially low energy nuclear reactions; nuclear structure physics. Mailing Add: Dept of Physics DePaul Univ 1215 W Fullerton Ave Chicago IL 60614

BEHR, ELDON AUGUST, b Minneapolis, Minn, July 29, 18; m 50; c 2. WOOD SCIENCE, WOOD TECHNOLOGY. Educ: Univ Minn, BS, 40, PhD(agr biochem), 48. Prof Exp: Asst chemist, Am Creosoting Co, Ky, 40-42; wood technologist, Air Forces, US Army, NJ, 42-43; tech dir res, Chapman Chem Co, 47-50, vpres & mgr tech dept, 50-59; assoc prof forestry, 59-67, PROF FORESTRY, MICH STATE UNIV, 67- Mem: Forest Prod Res Soc; Am Soc Testing & Mat; Am Wood Preservers' Asn; Int Res Group Wood Preservation; Brit Wood Preserving Asn. Res: Evaluation of wood preservatives; fibre preservatives; natural durability of wood; termite resistance of wood; test methods for decay of wood products. Mailing Add: Dept of Forestry Mich State Univ East Lansing MI 48824

BEHR, INGA, b Plovdiv, Bulgaria, Sept 8, 23; US citizen; m 51; c 2. CHEMISTRY. Educ: Univ Geneva, MS, 47, PhD(anal chem), 50. Prof Exp: Asst, Univ Geneva, 48-50; chemist, Weizmann Inst, 50-51; res chemist, Israel AEC, 52-58; res chemist, Bio-Sci Lab, Calif, 59; NIH grant, City of Hope, 60-62; instr gen chem & qual anal, 62-66, asst prof, 66-68, ASSOC PROF GEN CHEM & QUAL ANAL, PASADENA CITY COL, 68- Concurrent Pos: Asst prof, Los Angeles State Col, 63-64. Mem: Sr mem Am Chem Soc. Res: Clinical biochemistry; nature of urochrome pigments; marine biology. Mailing Add: 1619 Huntington Dr South Pasadena CA 91030

BEHR, LYELL CHRISTIAN, b Minneapolis, Minn, May 4, 16; m 54; c 4. ORGANIC CHEMISTRY. Educ: Univ Minn, BChem, 37; Univ Ill, PhD(org chem), 41. Prof Exp: Lab asst chem, Univ Ill, 37-40; Hormel fel, Univ Minn, 41-42; chemist, Chem Warfare Serv, Columbia Univ, 42-43; res chemist, E I du Pont de Nemours & Co, Del, 43-47; from asst prof to assoc prof chem, 47-54, head dept, 63-64, PROF CHEM, MISS STATE UNIV, 54-, DEAN COL ARTS & SCI, 64- Mem: AAAS; Am

Chem Soc; The Chem Soc. Res: Heterocyclic and synthetic organic chemistry; azoxy compounds. Mailing Add: Box AS Mississippi State MS 39762

BEHRE, CHARLES HENRY, JR, b Atlanta, Ga, Mar 16, 96; m 21. GEOLOGY. Educ: Univ Chicago, BS, 18, PhD(geol), 25. Hon Degrees: DSc, Franklin & Marshall Col, 51. Prof Exp: Geologist, Wis Geol Surv, 16-17; asst zool, Univ Chicago, 18-20, asst geol, 20-21; instr, Lehigh Univ, 21-23; asst prof, Univ Cincinnati, 24-30; from assoc prof to prof econ geol, Northwestern Univ, 30-41, chmn dept, 33-37; pres, Behre, Dolbear & Co, 64-69, vpres, 69-75; prof, 41-64, chmn dept, 56-59, EMER PROF, COLUMBIA UNIV, 64-; CONSULT GEOLOGIST, BEHRE, DOLBEAR & CO, 46- Concurrent Pos: Guggenheim Mem fel, 37-38; from asst geologist to sr geologist, US Geol Surv, 21-45; coop geologist, Pa Geol Surv, 22-; exchange lectr, Wash Univ, St Louis, 33 & Univ Cincinnati, 34; geol adv, Burma, Haiti & Algerian Govts; chmn comt geog explor, Joint Res & Develop Bd, 45-46; mem earth sci panel, NSF, 45-48. Honors & Awards: Miner Award, Am Asn Geol Teachers; Posepny Medal, Czech Acad Sci. Mem: Fel AAAS; fel Geol Soc Am; Am Asn Petrol Geol; fel Mineral Soc Am; Soc Econ Geologists (secy, 42-46, pres, 67-68). Res: Geology of ore deposits especially lead and zinc deposits; non-metallic deposits; ground water; regional geology of Pennsylvania, Idaho, Nevada, Colorado, Mexico and Galapagos Islands; contact metamorphism; structural geology as applied to mineral deposits; mineral economics and world affairs. Mailing Add: 330 Christie Heights St Leonia NJ 07605

BEHREND, DONALD FRASER, b Manchester, Conn, Aug 30, 31; m 57; c 3. WILDLIFE MANAGEMENT, ENVIRONMENTAL SCIENCES. Educ: Univ Conn, BS, 58, MS, 60; State Univ NY Col Forestry, Syracuse Univ, PhD(forest zool), 66. Prof Exp: Asst forestry & wildlife mgt, Univ Conn, 58-60; forest game mgt specialist, Ohio Div Wildlife, 60; from res asst to res assoc forest zool, State Univ NY Col Forestry, Syracuse, 60-67; asst prof wildlife resources, Univ Maine, 67-68; dir wildlife res, Archer & Anna Huntington Wildlife Forest Sta, 68-73, actg dean grad studies, 73, EXEC DIR INST ENVIRON PROG AFFAIRS & ASST VPRES RES, STATE UNIV NY COL ENVIRON SCI & FORESTRY, SYRACUSE, 73- Concurrent Pos: Leader deer res, Maine Dept Inland Fisheries & Game, 67-68. Mem: Am Inst Biol Sci; Soc Am Foresters; Wildlife Soc; Sigma Xi. Res: Ecology, behavior and management of forest wildlife with emphasis on snowshoe hares, white-tailed deer and songbirds; environmental research administration. Mailing Add: State Univ of NY Col of Environ Sci & Forestry Syracuse NY 13210

BEHRENDS, RALPH EUGENE, b Chicago, Ill, May 20, 26; m 61; c 2. ELEMENTARY PARTICLE PHYSICS. Educ: US Naval Acad, BS, 47; Univ Calif, Los Angeles, PhD(physics), 56. Prof Exp: Instr physics, Univ Calif, Los Angeles, 56-57; asst physicist, Brookhaven Nat Lab, 57-59; NSF fel physics, Inst Advan Study, 59-60; res assoc, Univ Pa, 60-61; from asst prof to assoc prof, 61-66, PROF PHYSICS, BELFER GRAD SCH SCI, YESHIVA UNIV, 66- Mem: Am Phys Soc. Res: Electromagnetic corrections to decay processes; symmetries of strong and weak interactions. Mailing Add: Dept of Physics Grad Sch of Sci Yeshiva Univ New York NY 10033

BEHRENDT, JOHN CHARLES, b Stevens Point, Wis, May 18, 32; m 61; c 2. MARINE GEOPHYSICS. Educ: Univ Wis, BS, 54, MS, 56, PhD(geophys), 61. Prof Exp: Asst seismologist, Antarctic geophys, Arctic Inst NAm, 56-58; proj assoc, Univ Wis, 58-64; geophysicist, Colo, 64-72, geologist in charge, Off Marine Geol, Woods Hole Off, 72-74, CHIEF BR ATLANTIC-GULF OF MEX GEOL, OFF MARINE GEOL, US GEOL SURV, 74- Mem: Fel Royal Astron Soc; Am Geophys Union; Soc Explor Geophys; Geol Soc Am; Glaciol Soc. Res: Aeromagneitc, aeroradioactivity, gravity and seismic investigations on West African continental margin, Antarctica and western United States; aeromagnetic, multichannel, seismic and gravity research of the transition North America to the Atlantic Ocean. Mailing Add: Off of Marine Geol US Geol Surv Woods Hole MA 02543

BEHRENS, EARL WILLIAM, b Albany, NY, Nov 4, 35; m 66; c 2. MARINE GEOLOGY. Educ: Cornell Univ, BA, 56; Univ Mich, MS, 58; Rice Univ, PhD(geol), 63. Prof Exp: Asst prof geol, 65-71, RES SCIENTIST ASSOC, INST MARINE SCI, UNIV TEX, 61-, ASSOC PROF GEOL, 71- Concurrent Pos: Prin investr, NSF grant, 70-72 & US Army Coastal Eng Res Ctr, 72-73, 74-75, panelist, Shoreline Erosion Adv Panel, US Army Corps Engrs, 74-79. Mem: AAAS; Am Geophys Union; Geol Soc Am; Soc Econ Paleont & Mineral. Res: Holocene stratigraphy and sedimentation; carbonate sedimentary petrology; sea level fluctuations; tidal inlet sedimentation and hydraulics; beach processes. Mailing Add: Inst of Marine Sci Univ of Tex Port Aransas TX 78373

BEHRENS, ERNST WILHELM, b Piraeus, Greece, Nov 10, 31; m 62; c 2. POLYMER PHYSICS. Educ: Univ Göttingen, BS, 54, teacher's dipl & MS, 57, PhD(solid state physics), 61. Prof Exp: Physicist, Siemens Co, Ger, 61-66; scientist, Lockheed-Ga Co, 66-69; physicist, 69-73, SR RES SCIENTIST, ARMSTRONG CORK CO, 73- Concurrent Pos: Europ AEC res fel magnetic resonance br, Grenoble Nuclear Res Ctr, France, 61. Mem: AAAS; Am Phys Soc; Am Acad Mech. Res: Physical properties of high polymers and composite materials. Mailing Add: Res & Develop Ctr Armstrong Cork 2500 Columbia Ave Lancaster PA 17604

BEHRENS, HERBERT CHARLES, b Cedarburg, Wis, Dec 16, 04; m 30; c 3. OPHTHALMOLOGY. Educ: Univ Wis, BA, 27; Northwestern Univ, MD, 31; Univ Pa, MSc, 35. Prof Exp: Asst clin prof ophthal, Loma Linda Univ, 41, 48-52, assoc prof, 52-57; from assoc prof to prof, 57-71, EMER PROF OPHTHAL, UNIV SOUTHERN CALIF, 71- Mem: Am Acad Ophthal & Otolaryngol. Mailing Add: 6712 S Friends Ave Whittier CA 90601

BEHRENS, HERBERT ERNEST, b Milwaukee, Wis, Nov 9, 15; m 41; c 5. PHYSICS, PHOTOGRAPHY. Educ: Ill State Norm Univ, BEd, 37; Univ Wis, MS, 47. Prof Exp: Teacher high sch, Ill, 37-41; RES PHYSICIST PHOTO-PROD DEPT, E I DU PONT DE NEMOURS & CO, PARLIN, NJ, 47- Mem: Am Phys Soc; Soc Photog Sci & Eng. Res: Photographic materials and processes. Mailing Add: 31 Sheridan Ave Metuchen NJ 08840

BEHRENS, MILDRED ESTHER, b Geneva, NY, Apr 26, 22. ZOOLOGY. Educ: Univ Rochester, BS, 50; Syracuse Univ, PhD(zool), 63. Prof Exp: INVESTR VISION RES, MASONIC MED RES LAB, 63- Mem: Am Physiol Soc; Am Soc Zool; Asn Res Vision & Ophthal; Soc Gen Physiol. Res: Electrophysiology of the planarian photoreceptor and the Limulus lateral eye. Mailing Add: Masonic Med Res Lab Bleecker St Utica NY 13501

BEHRENS, OTTO KARL, b Evansville, Ind, Aug 22, 11; m 35; c 2. BIOCHEMISTRY. Educ: DePauw Univ, AB, 32; Univ Ill, AM, 33, PhD(biochem), 35. Prof Exp: Asst, George Wash Univ, 35-37 & Rockefeller Inst Med Res, 37-39; Lalor Found fel, Cambridge Univ, 39 & med col, Cornell Univ, 39-40; res biochemist, 40-46, head bio-org chem res dept, 47-49, immuno-chem res, 49-51, dir biochem res div, 52-63, biopharmacol res div, 63-66, dir res chem, 66-68, ASSOC DIR RES, ELI LILLY & CO, 68- Mem: AAAS; Am Chem Soc; Am Soc Biol Chem; Soc Exp Biol & Med; NY Acad Sci. Res: Chemistry and metabolism of amino acids; peptides and proteins; biosynthesis of pencillins; protein hormones; antibiotics. Mailing Add: Eli Lilly & Co 307 E McCarty St Indianapolis IN 46206

BEHRENS, RICHARD, b Zenda, Wis, Nov 14, 21; m 50; c 2. WEED SCIENCE. Educ: Univ Wis, BS, 49, MS, 50, PhD(agron), 52. Prof Exp: Plant physiologist agr res serv, USDA, 52-58; assoc prof agron, 58-63, chmn plant physiol fac, 68-71, PROF AGRON, UNIV MINN, MINNEAPOLIS, 63- Honors & Awards: Publ Award, Weed Sci Soc Am, 62. Mem: Am Soc Plant Physiol; fel Weed Sci Soc Am (pres, 67). Res: Basic and applied aspects of weed control research; absorption, translocation and mode of action of herbicides; effect of environment on herbicidal action; effect of environment on weed competition. Mailing Add: 1490 W County Rd C-2 St Paul MN 55113

BEHRENS, RUDOLF ADOLF, b Rochester, NY, Aug 14, 28; m 51; c 1. RUBBER CHEMISTRY. Educ: Rutgers Univ, BS, 50. Prof Exp: Proj engr, Int Latex Corp, 55-56; chemist, Hewitt-Robins Inc, 56-59; res chemist, Tex-US Chem Co, 59-65; SR RES CHEMIST, AM CYANAMID CO, 65- Mem: Am Chem Soc. Res: Specialty elastomer rheology, vulcanization, reinforcement and resistance to special environments; also, vulcanization accelerators for general purpose and specialty elastomers. Mailing Add: Am Cyanamid Co Bound Brook NJ 08805

BEHRENTS, ROLF GORDON, b Galesburg, Ill, July 21, 47. HUMAN DEVELOPMENT, ORTHODONTICS. Educ: St Olaf Col, BA, 69; Meharry Med Col, DDS, 73; Case Western Reserve Univ, MS, 75. Prof Exp: FEL CRANIOFACIAL RES, CTR HUMAN GROWTH & DEVELOP, UNIV MICH, 75- Concurrent Pos: Pvt pract, 76. Honors & Awards: Milo Hellman Res Award, Am Asn Orthodontists, 76. Mem: Am Dent Asn; Am Asn Orthodontists. Res: Study of craniofacial growth in general with particular emphasis on the mechanisms controlling facial skeletal morphology in normal and abnormally growing humans. Mailing Add: Ctr for Human Growth & Develop Univ of Mich 1111 E Catherine St Ann Arbor MI 48109

BEHRINGER, ROBERT ERNEST, b Springfield, Mass, May 18, 31; m 56; c 5. LASERS. Educ: Worcester Polytech Inst, BS, 53; Univ Calif, MA, 55, PhD(physics), 58. Prof Exp: Physicist, Gen Elec Co, 53; asst physics, Univ Calif, 53-58; assoc physicist, Int Bus Mach Corp, 58-60; consult, Atomics Int Div, NAm Aviation, Inc, 60-63; PHYSICIST, US OFF NAVAL RES, 63- Concurrent Pos: From asst prof to assoc prof physics, San Fernando Valley State Col, 60-65; consult, Marquardt Corp, 61-63. Mem: Am Phys Soc; Am Asn Physics Teachers; Sigma Xi. Res: Theoretical investigation in the field of ferro-magnetism; nuclear resonance; metallic conductivity; electrooptics; laser physics. Mailing Add: Off of Naval Res 1030 E Green St Pasadena CA 91106

BEHRISCH, HANS WERNER, b Vienna, Austria, Nov 26, 41; Can citizen; m 67; c 2. BIOCHEMISTRY, COMPARATIVE PHYSIOLOGY. Educ: Univ BC, BSc, 64, PhD(comp biochem), 69; Ore State Univ, MA, 66. Prof Exp: Asst prof, 69-73, ASSOC PROF BIOCHEM, INST ARCTIC BIOL, UNIV ALASKA, 73- Concurrent Pos: Vis prof, Inst Zoophysiol, Univ Innsbruck, 74-75. Mem: AAAS; Can Physiol Soc; Am Soc Biol Chemists; Am Physiol Soc. Res: Molecular mechanisms of adaptation to extreme environments and environmental change; regulation of metabolism on three levels of organization: whole organism, organ tissue and enzymic. Mailing Add: Inst of Arctic Biol Univ of Alaska Fairbanks AK 99701

BEHRLE, FRANKLIN C, b Ansonia, Conn, June 4, 22; m 45; c 5. PEDIATRICS. Educ: Dartmouth Col, AB, 44; Yale Univ, MD, 46. Prof Exp: From instr to assoc prof pediat, Sch Med, Univ Kans, 51-61; asst dean, 67-73, PROF PEDIAT, NJ MED SCH, 61-, CHMN DEPT, 65- Mem: Soc Pediat Res; fel Am Acad Pediat; AMA. Res: Respiratory problems of newborn. Mailing Add: Martland Hosp Newark NJ 07107

BEHRMAN, ABRAHAM SIDNEY, b Covington, Ky, Dec 15, 92. WATER CHEMISTRY, INDUSTRIAL CHEMISTRY. Educ: Univ Ky, BS, 14. Prof Exp: Mem staff educ dept, Philippines, 15-16, mem staff, Bur Sci, 16-17; vpres & chem dir, Infilco, Inc, Chicago, 19-42; vpres & dir res, Velsicol Corp, 44-46; CHEM CONSULT, 47- Mem: AAAS; Am Chem Soc. Res: Water purification; ion exchange materials and processes; activated carbon; siliceous gels and gel catalysts; halide preparation; halogen recovery. Mailing Add: 4206 N Broadway Chicago IL 60613

BEHRMAN, EDWARD JOSEPH, b New York, NY, Dec 13, 30; m 53; c 3. BIOCHEMISTRY. Educ: Yale Univ, BS, 52; Univ Calif, PhD(biochem), 57. Prof Exp: Fel biochem, Cancer Res Inst, Nat Cancer Inst, 57-60, res assoc, 60-64; asst prof res, Brown Univ, 64-65; from asst prof to assoc prof, 65-69, PROF BIOCHEM, OHIO STATE UNIV, 69- Concurrent Pos: Res fel biochem, Harvard Med Sch, 59-61, res assoc, 61-64, tutor, Harvard Col, 61-64. Mem: The Chem Soc; Am Chem Soc. Res: Peroxydisulfate oxidations; reaction mechanisms; nucleic acid chemistry; sugar phosphates. Mailing Add: Dept of Biochem Ohio State Univ Columbus OH 43210

BEHRMAN, HAROLD R, b Vidora, Sask, Nov 26, 39; m 59; c 3. PHYSIOLOGY, BIOCHEMISTRY. Educ: Univ Man, BSc, 62, MSc, 65; NC State Univ, PhD(physiol), 67. Prof Exp: Res asst sensory & digestive physiol, NC State Univ, 64-67; Can Med Res Coun res fel reproductive endocrinol, Harvard Med Sch, 67-70, assoc, 70-71, asst prof physiol, 71-72; dir dept reproductive biol, Merck Inst Therapeut Res, NJ, 72-75; ASSOC PROF OBSTET & GYNEC & PHARMACOL & DIR REPRODUCTIVE BIOL SECT, SCH MED, YALE UNIV, 76- Concurrent Pos: Lalor fel, Harvard Med Sch, 71-73. Mem: Can Fedn Biol Sci; Soc Exp Biol & Med; Endocrine Soc; Soc Endocrinol; Can Physiol Soc. Res: Hormone and prostaglandin interrelationships in endocrine systems. Mailing Add: Dept of Obstet & Gynec Yale Univ Sch of Med New Haven CT 06510

BEHRMAN, RICHARD ELLIOT, b Philadelphia, Pa, Dec 13, 31; m 54; c 4. PEDIATRICS, PHYSIOLOGY. Educ: Amherst Col, BA, 53; Harvard Univ, LLB, 56; Univ Rochester, MD, 60. Prof Exp: Intern pediat, Johns Hopkins Hosp, 60-61, resident pediat & mem staff, Lab Obstet Physiol, 63-65; scientist, Lab Perinatal Physiol, Nat Inst Neurol Dis & Blindness, 61-63, sect chief physiol & biochem & actg lab chief, 62-63; from asst prof to assoc prof pediat, Med Sch, Univ Ore & from assoc scientist to scientist, Ore Regional Primate Res Ctr, 65-68, physician in-chg nursery serv, Med Sch & chmn dept perinatal physiol, Res Ctr, 65-68; prof pediat, Col Med, Univ Ill, 68-71; prof, 71-74, CARPENTIER PROF PEDIAT, COL PHYSICIANS & SURGEONS, COLUMBIA UNIV, 74-, CHMN DEPT, 71-; DIR PEDIAT SERV, BABIES HOSP, 71- Concurrent Pos: Whipple scholar, Univ Rochester & Johns Hopkins Hosp, 60-61; Wyeth fel pediat, 63-65; grants, Med Res Found, Ore, 65-66, United Health Found, 66-67, Nat Inst Child Health & Human Develop, 66-69, Pharmaceut Mfrs Asn Found, 70-72 & Robert Wood Johnson Found, 73-76; examr, Am Bd Pediat, 74-77, mem sub-bd neonatal-perinatal med, 75-77; mem bd material, child & family health res, Nat Res Coun, 74- Res: Fetal, newborn and placental physiology; transfers of gases and solutes across the placenta; reproductive physiology; water and electrolyte balance; acid-base adjustments; membrane transport; bilirubin metabolism; protein binding. Mailing Add: Dept of Pediat Col of Physicians & Surgeons Columbia Univ New York NY 10032

BEHRMAN, SAMUEL J, b Worcester, SAfrica, Sept 10, 20; US citizen; m 56; c 2. OBSTETRICS & GYNECOLOGY. Educ: Univ Cape Town, MB, ChB, 44; Univ Mich, MS, 49; FRCOG, 60. Prof Exp: From instr to assoc prof obstet & gynec, 48-56, PROF OBSTET & GYNEC, COORDR POSTGRAD MED & RES LECTR PUB HEALTH, UNIV MICH, ANN ARBOR, 56-, LECTR FAMILY PLANNING & DIR CTR RES REPRODUCTIVE BIOL, 66- Concurrent Pos: Res grants, Macy Found, 48-62 & NIH, 64-67; consult, Ypsilanti State Hosp, Wayne County Gen Hosp & Sinai Hosp, 56-; ed, Int J Fertil, 58- & jour, Int Fertil Asn, 62; guest prof, Univ London, 64; consult maternal health comt, AMA, 64-; mem sci adv bd, Human Life Found; consult, US AID; pres, Int Fedn Fertil Socs, 71-74; mem task force immunol, WHO, 72-; mem adv bd, Food & Drug Admin, 73- Mem: Fel Am Asn Obstet & Gynec; Am Fertil Soc; fel Am Col Surgeons; Am Col Obstet & Gynec; Royal Soc Med. Res: Reproductive physiology. Mailing Add: Ctr for Res in Reproductive Biol Univ of Mich Ann Arbor MI 48104

BEHRMANN, ELEANOR MITTS, b Williamstown, Ky, May 24, 17; m 46; c 2. CHEMISTRY, ORGANIC BIOCHEMISTRY. Educ: Univ Ky, BS, 38, MS, 39; Iowa State Univ, PhD, 43. Prof Exp: Chemist, Shell Develop Co, 43-46; instr chem, Harvard Univ, 46-48; instr, 57-64, ASST PROF CHEM, UNIV CINCINNATI, 64- Mem: Am Chem Soc. Res: Glucose amine chemistry; phenolic resins; organic synthesis. Mailing Add: Dept of Chem Univ of Cincinnati Cincinnati OH 45521

BEHUN, JOHN DAVID, organic chemistry, see 12th edition

BEICHL, GEORGE JOHN, b Philadelphia, Pa, Aug 20, 18; m 51; c 4. PHYSICAL CHEMISTRY, INORGANIC CHEMISTRY. Educ: St Joseph's Col, BS, 39; Univ Pa, MS, 42, PhD, 53. Prof Exp: Instr chem, St Joseph's Col, 40-44; asst, AEC, Los Alamos, 46; from asst prof to assoc prof, 47-56, PROF CHEM, ST JOSEPH'S COL, 56-, CHMN DEPT, 66- Concurrent Pos: Sr res investr, Univ Pa, 53-54; NSF fel, Munich, Ger, 58-59. Mem: AAAS; Am Chem Soc; The Chem Soc. Res: Hydrides; boron and copper compounds; kinetics. Mailing Add: Dept of Chem St Joseph's Col Philadelphia PA 19131

BEIDLEMAN, JAMES C, b Wilkes-Barre, Pa, Nov 13, 36; m; c 2. ALGEBRA. Educ: Bucknell Univ, BA, 58, MS, 59; Pa State Univ, PhD(math), 64. Prof Exp: Asst prof math, 64-67, ASSOC PROF MATH, UNIV KY, 67- Concurrent Pos: NSF grant on near-rings, 66-68, grant on groups & near-rings, 68-70. Mem: Am Math Soc; London Math Soc. Res: Schunk classes; Fitting classes; formations of finite solvable groups. Mailing Add: Dept of Math Univ of Ky Lexington KY 40506

BEIDLEMAN, RICHARD GOOCH, b Grand Forks, NDak, June 3, 23; m 46; c 3. ECOLOGY, HISTORY OF SCIENCE. Educ: Univ Colo, BA, 47, MA, 48, PhD(zool), 54. Prof Exp: Asst univ mus, Univ Colo, 46-48; asst prof zool, Colo State Univ, 48-56; asst prof biol, Univ Colo, 56-57; from asst prof to assoc prof, 57-64, PROF BIOL, COLO COL, 64- Concurrent Pos: Ford Found Fund Adv Ed grant, 54-55; Am Inst Biol Sci vis biologist prog comnr, Comn Undergrad Educ Biol Sci, NSF; mem consult bur steering comt, Biol Sci Curric Study. Honors & Awards: Romco Environ Award, 71. Mem: Fel AAAS; Am Soc Zool; Ecol Soc Am; Am Soc Mammal; Am Soc Ichthyol & Herpet. Res: Species association groups amoung birds; vertebrate ecology of Western biotic communities; winter bird population studies; small mammal population studies; significance of the American frontier on natural science; history of American biology. Mailing Add: Dept of Biol Colo Col Colorado Springs CO 80903

BEIDLER, LLOYD M, b Allentown Pa, Jan 17, 22; m 46; c 6. PHYSIOLOGY, BIOPHYSICS. Educ: Muhlenberg Col, BS, 43; Johns Hopkins Univ, PhD(biophysics), 51. Prof Exp: Physicist, radiation lab, Johns Hopkins Univ, 44-45; from asst prof to prof physiol, 50-74, PROF BIOL SCI, FLA STATE UNIV, 74- Concurrent Pos: Bowditch lectr, 59; Nat lectr, Sigma Xi, 61; Tanner lectr, Inst Food Technol Sci Lectr, 68; mem, Inst Food Technol Sci Lectr Panel, 66-68; mem comt human factors & psychophysiol, Adv Bd Mil Personnel Supplies, Nat Res Coun-Nat Acad Sci. Mem: AAAS; Am Soc Gen Physiol; Optical Soc Am; Am Physiol Soc; Biophys Soc. Res: physiological properties of chemoreceptors; mechanisms of taste stimulation; olfaction. Mailing Add: Dept of Biol Sci Fla State Univ Tallahassee FL 32306

BEIERLE, JOHN W, microbiology, biochemistry, see 12th edition

BEIERWAITES, WILLIAM HENRY, b Saginaw, Mich, Nov 23, 16; m 42; c 3. MEDICINE. Educ: Univ Mich, AB, 36, MD, 41; Am Bd Internal Med, dipl. Prof Exp: From intern to asst resident med, Cleveland City Hosp, 41-43; resident, 44, from instr to assoc prof, 45-59, PROF MED, UNIV HOSP, UNIV MICH, ANN ARBOR, 59-, DIR NUCLEAR MED, 52- Concurrent Pos: Guggenheim fel, 66-67; Commonwealth Fund fel, 67; mem adv comt, IAE Comn, 74- Mem: Fel Am Col Physicians; Am Thyroid Asn (vpres, 64); Am Fedn Clin Res (pres, 54); Soc Nuclear Med (pres-elect, 64, pres, 65); Am Cancer Soc. Res: Internal nuclear medicine; diagnosis and treatment of cancer using radionuclide labeled compounds. Mailing Add: W-5614 Univ Hosp 1313 E Ann St Ann Arbor MI 48104

BEIGELMAN, PAUL MAURICE, b Los Angeles, Calif, July 21, 24; m 53; c 1. MEDICINE. Educ: Univ Southern Calif, MD, 48; Am Bd Internal Med, dipl, 55. Prof Exp: Intern med, Los Angeles County Hosp, 47-48; asst resident, Stanford Univ, Hosps, 48-49; resident path, Cedars of Lebanon Hosp, 49-50; asst resident med, Peter Bent Brigham Hosp, 50-51, asst, 53-55; res asst anat & clin asst prof med, Sch Med, Univ Calif, Los Angeles, 55-56; sr physician, Sepulveda Vet Admin Hosp, 56; from asst prof to assoc prof med, 56-73, assoc prof physiol, 63-64, assoc prof pharmacol, 69-73, PROF MED & PHARMACOL, SCH MED, UNIV SOUTHERN CALIF, 73- Concurrent Pos: Res fel, Harvard Med Sch, 53-55 & Dazian Found, 54-55; attend physician, Los Angeles County Hosp, 56-; consult physician, Wadsworth Vet Admin Hosp, 62-66 & San Fernando Vet Admin Hosp, 66-72; attend physician, Rancho Los Amigos Hosp, Downey, 74- Mem: Endocrine Soc; Am Physiol Soc; Am Fedn Clin Res; fel Am Col Physicians; Soc Exp Biol & Med. Res: Bacteriology; antibiotics; epidemiology; virology; collagen diseases; endocrinology; metabolism; diabetes mellitus; electrophysiology. Mailing Add: Dept Pharmacol Sch Med Univ of Southern Calif Los Angeles CA 90033

BEIGLER, MYRON ARNOLD, b Detroit, Mich, Nov 8, 26. NUTRITON, RESEARCH ADMINISTRATION. Educ: NY Univ, BS, 50, BA, 53. Prof Exp: Int develop mgr, Amino Acid Div, Ajinomoto Corp, 66-69; vpres, Vivonex Corp, 69-71; gen mgr, Vivonex Div, Norwich Pharmacal, 71-72; VPRES & DIR, INST AGRISCI & NUTRIT, RES DIV, SYNTEX CORP, 72- Mem: Am Inst Nutrit; Am Soc Animal Sci; Inst Food Technologists; Am Asn Cereal Chemists; AAAS. Res: Therapeutic nutrition directed toward development of oral and intravenous products for use by patients with various metabolic and physiological problems; use of alpha-keto amino acids in nutrition of patients with renal and/or hepatic disease. Mailing Add: Syntex Res 3401 Hillview Ave Palo Alto CA 94304

BEIL, GARY MILTON, b Clinton, Iowa, Dec 27, 38; m 60; c 2. PLANT BREEDING, STATISTICS. Educ: Iowa State Univ, BS, 60, MS, 63, PhD(plant breeding), 65. Prof Exp: Res assoc agron, Iowa State Univ, 60-65; plant breeder, Caladino Farm Seed, Inc, 65-67; DIR NORTHEASTERN EXP STA, DeKALB AgRES, INC, 67- Mem: Am Soc Agron; Crop Sci Soc Am; Am Entom Asn. Res: Research and development on the genetics of maize. Mailing Add: Northeastern Exp Sta DeKalb AgRes Inc 1440 Okemos Rd Mason MI 48854

BEILBY, ALVIN LESTER, b Watsonville, Calif, Sept 17, 32; m 58; c 2. ANALYTICAL CHEMISTRY. Educ: San Jose State Col, BA, 54; Univ Wash, PhD(chem), 58. Prof Exp: Instr chem, 58-60, from asst prof to assoc prof, 60-72, PROF CHEM & CHMN DEPT, POMONA COL, 72- Concurrent Pos: Petrol Res Fund fac award, Univ Ill, 64-65; guest worker & res chemist, Nat Bur Standards, 71-72. Mem: Am Chem Soc; Sigma Xi. Res: Electroanalytical chemistry; nature of carbon electrodes; trace metal analysis; chemical education. Mailing Add: Seaver Chem Lab Pomona Col Claremont CA 91711

BEILER, ADAM CLARKE, b Berlin, Ger, July 8, 12; US citizen; m 38; c 2. PHYSICS. Educ: Allegheny Col, BS, 33; Univ Ill, MS, 36, PhD(physics), 41. Prof Exp: Instr physics, Univ Ark, 40-41; mat engr, 41-48, mgr magnetics sect, 48-52, magnetic & semiconductor appln eng, 52-55, mgr, Magnetic Eng Dept, 55-62, chief scientist, Aerospace Elec Div, 62-69, ADV ENGR, SYSTS DEVELOP DIV, WESTINGHOUSE ELEC CORP, 69- Mem: Am Phys Soc; Am Soc Testing & Mat; Inst Elec & Electronics Engrs. Res: Electrical discharges in gases; molecular physics; magnetic materials; solid state physics; metallurgy; materials science; solid state microwaves. Mailing Add: MS 496 Systs Develop Div Westinghouse Elec Corp Box 746 Baltimore MD 21203

BEILER, JAY MORTON, biochemistry, see 12th edition

BEILER, THEODORE WISEMAN, b Meadville, Pa, Apr 29, 24; m 51; c 1. ORGANIC CHEMISTRY. Educ: Allegheny Col, BS, 48; Harvard Univ, MA, 50, PhD(chem), 52. Prof Exp: Sr asst scientist, USPHS, 51-53; from asst prof to assoc prof chem, Stetson Univ, 53-62; Fulbright lectr, Univ Panjam, WPakistan, 62-63; prof, 63-68, WILLIAM KENAN PROF CHEM, STETSON UNIV, 68-, CHMN DEPT, 63- Concurrent Pos: Vis scientist, NIH, 58. Mem: Am Chem Soc. Res: Syntheses of alicyclic systems; transformations of amino acids. Mailing Add: Dept of Chem Stetson Univ DeLand FL 32730

BEILFUSS, ERWIN ROLAND, b Milwaukee, Wis, May 7, 20; m 46; c 2. PARASITOLOGY. Educ: Carroll Col, BA, 48; Univ Wis, MS, 50; Okla State Univ, PhD, 56. Prof Exp: Asst microbiol, Pabst Res Lab, 45-46; asst fishery biol, Univ Wis, 48-49; instr biol, Park Col, 49-51; asst prof, Macalester Col, 53-56, asst prof, 57-76, chmn dept, 70-73, EMER ASST PROF BIOL, HUMBOLDT STATE UNIV, 76- Mem: Am Soc Parasitol; Am Micros Soc; Wildlife Dis Asn. Res: Parasitic helminths. Mailing Add: PO Box 3444 Eureka CA 95501

BEIMER, ROBERT GLENN, analytical organic chemistry, see 12th edition

BEIN, DONALD, b New York, NY, Dec 13, 34; m 60; c 2. MATHEMATICS. Educ: Brooklyn Col, BS, 57; NY Univ, PhD(math), 64. Prof Exp: From asst prof to assoc prof, 64-74, PROF MATH, FAIRLEIGH DICKINSON UNIV, 74- Mem: Am Math Soc; Math Asn Am; Soc Indust & Appl Math. Res: Parabolic partial differential equations. Mailing Add: Dept of Math Fairleigh Dickinson Univ Teaneck NJ 07666

BEINDORFF, ARTHUR BAKER, b Omaha, Nebr, Apr 16, 25; m 46; c 3. POLYMER CHEMISTRY. Educ: Univ Nebr, BS, 47, MA, 49, PhD(biochem), 52. Prof Exp: From chemist to sr chemist, Chemstrand Corp, Monsanto Co, NC, 52-58, group leader nylon polymer res, 58-59, head personnel sect res ctr, 59-62, mgr patent liaison, 62-67, assoc dir patent liaison, Textiles Div, NY, 67-68, tech dir polyester, Ala, 68-70, dir tire technol, 70-74, DIR APPAREL FIBERS TECHNOL, MONSANTO TEXTILES CO, FLA, 74- Concurrent Pos: Chmn dept chem, Athens Col, 52-55; prof, St Bernard Col, 54-56. Mem: Am Chem Soc; Soc Automotive Eng; Am Asn Textiles Technol. Res: Exploratory organic polymer chemistry; nylon; polyester; steel. Mailing Add: Monsanto Textiles Co Box 12830 Pensacola FL 32575

BEINEKE, LOWELL WAYNE, b Decatur, Ind, Nov 20, 39; m 67; c 2. MATHEMATICS. Educ: Purdue Univ, BS, 61; Univ Mich, PhD(math), 65. Prof Exp: Res asst math inst soc res, Univ Mich, 62-63; from asst prof to assoc prof, 65-71, PROF MATH, PURDUE UNIV, FT WAYNE, 71- Concurrent Pos: Res asst, Univ Col, Univ London, 66-67; consult inst soc res, Univ Mich, 65; tutor, Oxford Univ, 73-74. Mem: Am Math Soc; Math Asn Am. Res: Linear graph theory; network theory; mathematical models in the social sciences; combinatorial analysis. Mailing Add: Dept of Math Purdue Univ Ft Wayne IN 46805

BEINEKE, THOMAS ANDREW, b Cincinnati, Ohio, Aug 2, 39; m 66; c 3. PHYSICAL CHEMISTRY. Educ: Ohio Univ, BS, 61; Calif Inst Technol, PhD(phys chem), 66. Prof Exp: Instr chem, Iowa State Univ & asst chemist, Ames Lab, AEC, 65-68; asst prof chem, Monmouth Col, Ill, 68-71; PUB HEALTH LAB EVALUATOR, STATE OF ILL DEPT PUB HEALTH, 71- Mem: Am Chem Soc; Am Crystallog Asn. Res: Analysis of crystal and molecular structure by x-ray diffraction. Mailing Add: 542 Stange St Springfield IL 62704

BEINEKE, WALTER FRANK, b Indianapolis, Ind, Mar 7, 38; m 61; c 2. FOREST GENETICS. Educ: Purdue Univ, BS, 60; Duke Univ, MF, 61; NC State Univ, PhD(forest genetics), 66. Prof Exp: ASSOC PROF FORESTRY, PURDUE UNIV, 64- Mem: Soc Am Foresters. Res: The genetic improvement of black walnut, Juglans nigra, including selection, vegetative propagation, progeny testing, flowering, and breeding. Mailing Add: Dept of Forestry Purdue Univ West Lafayette IN 47906

BEINERT, HELMUT, b Lahr, Ger, Nov 17, 13; nat US; m 44; c 4. BIOCHEMISTRY, ENZYMOLOGY. Educ: Univ Leipzig, Dr rer nat(chem), 43. Prof Exp: Res assoc biol chem, Kaiser Wilhelm Inst Med Res, Ger, 43-45; biochemist, Air Force Aeromed Ctr, Ger, 46 & US Air Force Sch Aviation Med, 47-50; res assoc, Inst Enzyme Res, 51-52, from asst prof to assoc prof, 52-62, PROF ENZYME CHEM, INST ENZYME RES, UNIV WIS-MADISON, 62-, PROF BIOCHEM, 67-, CHMN SECT III, 58- Mem: Am Chem Soc; Am Soc Biol Chem. Res: Enzymes and coenzymes of biological oxidation. Mailing Add: Inst for Enzyme Res Univ of Wis Madison WI 53706

BEINFEST, SIDNEY, b Brooklyn, NY, Oct 29, 17; m 42; c 2. ORGANIC CHEMISTRY. Educ: Brooklyn Col, BA, 38; Univ Ark, MS, 39; Polytech Inst Brooklyn, PhD(chem), 49. Prof Exp: Dir res & develop, Premo Pharmaceut Labs, Inc, 41-45; vpres in charge opers, Berkeley Chem Corp, 46-63; tech vpres, Millmaster Chem Corp, 63-64, vpres technol, Millmaster Onyx Chem Corp, Jersey City, 64-74, VPRES CORP TECH, MILLMASTER ONYX CORP, NY, 74- Mem: AAAS; Am Chem Soc; Sci Res Soc Am; Am Inst Chem Eng. Res: Cyanine dyes; pyrazine derivatives; barbiturates; anti-malarial drugs; muscle relaxant; tranquillizing drugs. Mailing Add: Millmaster Onyx Corp 99 Park Ave New York NY 10016

BEINFIELD, WILLIAM HARVEY, b St Louis, Mo, Apr 8, 18; m 50; c 2. INTERNAL MEDICINE, PHARMACOLOGY. Educ: Univ Wis, BA, 40; Columbia Univ, MD, 43; Am Bd Internal Med, dipl, 52. Prof Exp: From asst instr to instr med, 50-54, instr physiol, 50-52, assoc, 52-54, from assoc prof to prof physiol, 56-64, assoc prof pharmacol, 56-60, ASST PROF MED, NEW YORK MED COL, 56-, PROF PHARMACOL, 60- Concurrent Pos: Adv drug & formulary comt, Dept Hosps, NY. Mem: Fel Am Col Physicians; Am Heart Asn; Am Thoracic Soc; Am Geriat Soc; fel NY Acad Sci. Res: Cardiopulmonary physiology and pharmacology. Mailing Add: Dept of Pharmacol Basic Sci Bldg New York Med Col Valhalla NY 10595

BEINHART, ERNEST GEORGE, JR, b Quincy, Fla, July 21, 25; m 48; c 5. PLANT PHYSIOLOGY, AGRONOMY. Educ: Pa State Univ, BS, 49; NC State Univ, MS, 51; Duke Univ, PhD(bot), 59. Prof Exp: Asst chem, NC State Univ, 49-50, asst agron, 50-51; res biochemist, R J Reynolds Tobacco Co, 54-56; res plant physiologist, Agr Res Serv, USDA, 56-69; prof crop physiol, Univ Nebr Mission in Colombia, 69-72; prof agron, Univ Fla Mission in El Salvador, 72-73; CONSULT, 73- Concurrent Pos: Lectr, Clemson Univ, 65-69; govt Japan guest researcher, Nat Inst Animal Indust, Japan, 67. Mem: Am Soc Plant Physiol; Am Soc Agron. Res: Studies of plant responses to environmental stress; application of research to agricultural problems of developing nations. Mailing Add: 220 N Clemson Ave Clemson SC 29631

BEINING, PAUL R, b Pittsburgh, Pa, Feb 2, 23. MICROBIOLOGY, BIOCHEMISTRY. Educ: Spring Hill Col, BS, 49; Cath Univ Am, MS, 52, PhD(microbiol), 62. Prof Exp: Instr biol, Univ Scranton, 49-51; instr, St Joseph's Col, Pa, 62-63, asst prof, 63-66; assoc prof & chmn dept, Wheeling Col, 66-67; assoc prof, 67-75, PROF BIOL, UNIV SCRANTON, 75- Concurrent Pos: Researcher, Georgetown Univ, 69-71; guest researcher, NIH. Mem: AAAS; Am Soc Microbiol. Res: Biochemical and serological characteristics of in-vivo and in-vitro cultivated Staphyloccus aureus; increased protection with active immunization with influenza virus; polyphasic taxonomy of marine and non-marine Micrococci; ultrastructure and biochemical analysis of staphylococcal membrane systems. Mailing Add: Dept of Biol Univ of Scranton Scranton PA 18510

BEIQUE, RENE ALEXANDRE, b Cornerbrook, Nfld, Nov 29, 25; m 51; c 3. RADIATION PHYSICS. Educ: Univ Montreal, BA, 47, BSc, 50, MSc, 51; Mass Inst Technol, PhD(physics), 58. Prof Exp: Jr physicist, High Voltage Lab, Mass Inst Technol, 52-53; physicist, Montreal Gen Hosp, 58-68; assoc prof, 69-71, PROF RADIOL, UNIV MONTREAL, 71-; PHYSICIST, NOTRE DAME HOSP, 68- Concurrent Pos: Lectr, McGill Univ, 60-69, asst prof, 69-; consult, Montreal Children's Hosp, 60- Mem: Soc Nuclear Med; Can Asn Physicists; Can Asn Radiol. Res: Physics of radiology, especially diagnostic radiology. Mailing Add: Sch of Med Univ of Montreal Montreal PQ Can

BEIRNE, BRYAN PATRICK, b Wexford, Ireland, Jan 22, 18; m 48; c 2. ECONOMIC ENTOMOLOGY. Educ: Univ Dublin, PhD(entom), 40, MSc, 41, MA, 42. Prof Exp: Asst lectr zool, Univ Dublin, 42-43, asst dir mus zool & comp anat, 42-49, lectr entom, 43-49; sr entomologist res br, Can Dept Agr, 49-55, dir, Res Inst, Belleville, Ont, 55-67; PROF PEST MGT & DIR PESTOLOGY CTR, SIMON FRASER UNIV, 67- Mem: Fel Can Soc Zool; Am Entom Soc; Entom Soc Can; Royal Entom Soc London; Royal Irish Acad. Res: Biological control; pest management; ecology of agricultural insects; insect taxonomy. Mailing Add: Dept of Biol Sci Simon Fraser Univ Burnaby BC Can

BEIRNE, GILBERT ARTHUR, b San Francisco, Calif, Aug 28, 20. DERMATOLOGY. Educ: Univ Calif, AB, 41; Creighton Univ, MD, 45. Prof Exp: Clin instr dermat, Univ Calif, San Francisco, 52-56, from asst clin prof to assoc clin prof, 56-75. Concurrent Pos: Asst attend dermat, San Francisco Gen Hosp. Mem: Fel Am Col Physicians; Am Acad Allergy; Am Acad Dermat. Res: Chemosurgery; development, treatment and prognosis in cutaneous carcinoma; individual cancer resistance. Mailing Add: 1515 Trousdale Dr Burlingame CA 94010

BEISCHER, DIETRICH EBERHARD, b Ravensburg, Ger, Dec 10, 08; m 41; c 3. PHYSICAL CHEMISTRY. Educ: Stuttgart Tech Univ, DrIng(chem), 32. Prof Exp: Dozent colloid chem, Kaiser Wilhelm Inst & Univ Berlin, 33-40; prof inorg chem, Ger Univ Strasbourg, France, 41-45; prof phys chem, Kaiser Wilhelm Inst, 46; prof colloid chem, Naval Aerospace Med Inst, 47-67, chief chem sci div, Res Dept, 67-75; RETIRED. Mem: AAAS; Am Chem Soc; Aerospace Med Asn; Electron Micros Soc Am. Res: Colloid chemistry; electron microscopy; tracer research; aviation physiology. Mailing Add: Res Dept Naval Aerospace Med Inst Pensacola FL 32512

BEISEL, CLIFFORD GORDON, b Tustin, Calif, Aug 31, 16; m; c 2. FOOD SCIENCE. Educ: Univ Southern Calif, BS, 37. Prof Exp: Chief chemist, Citrus Juice & Flavor Co, Calif, 37-41; prod mgr, Panama Coca Cola Bottling Co, 41-42; asst dir res, Fla Citrus Canners Coop, 46-50; dir res & develop, Real Gold Citrus Prod, 50-53; mgr develop dept, Orange Prod Div, Sunkist Growers, Inc, 53-54, head dept, 54-60, tech dir, 60-67, prod mgr, 67-74, MGR PROD RES & DEVELOP DIV, SUNKIST RES CTR, 74- Mem: Inst Food Technol. Res: Citrus products manufacture, pharmaceutical and peel derivatives; microbiology of citrus products. Mailing Add: Sunkist Res Ctr 760 E Sunkist St Ontario CA 91761

BEISEL, WILLIAM R, b Philadelphia, Pa, Apr 8, 23; m 49; c 5. INFECTIOUS DISEASES, METABOLISM. Educ: Muhlenberg Col, BA, 46; Ind Univ, Indianapolis, MD, 48; Am Bd Internal Med, dipl, 55. Prof Exp: Intern, Fitsimmons Army Hosp, Denver, Colo, US Army, 48-49, resident med, Letterman Gen Hosp, San Francisco, 49-52, asst chief dept med, 21st Army Sta Hosp, Korea, 53-54, chief dept med, US Army Hosp, Ft Leonard Wood, Mo, 54-56, basic sci yr, Walter Reed Army Inst Res, DC, 56-57, chief dept metab, 57-60, chief phys sci div, Walter Reed Med Unit, Ft Detrick, Md, 62-68; SCI ADV, US ARMY MED RES INST INFECTIOUS DIS, 69- Concurrent Pos: Fel metab, Sch Med, Univ Calif, San Francisco, 61-62; chief med sect 4, Walter Reed Army Hosp, DC & asst prof, Sch Med, Georgetown Univ, 57-60; mem study sect pharmacol & exp therapeut, NIH, 59-60, study sect endocrinol, 64-68; consult metab dis to Surgeon Gen, US Army, 66-; assoc prof, Sch Med, Univ Md, 67-; mem infective agents res eval comt, Vet Admin, 69-71; chmn subcomt nutrit & infection, Nat Res Coun. Honors & Awards: Hoff Gold Medal, Water Reed Army Inst Res, 57; Award & Commendation, Army Sci Conf, 66; Stitt Award, Asn Mil Surgeons, 68. Mem: Fel Am Col Physicians; Am Inst Nutrit; Infectious Dis Soc Am; Endocrine Soc. Res: Broad aspects of endocrinology and metabolism, especially as they pertain to and help regulate the responses of a host to the stress of infectious illness. Mailing Add: US Army Med Res Inst Infectious Dis Frederick MD 21701

BEISER, HELEN R, b Chicago, Ill, Nov 15, 14. PSYCHOANALYSIS, CHILD PSYCHIATRY. Educ: Univ Ariz, BSEd, 35; Univ Ill, MS & MD, 41. Prof Exp: From instr to asst prof, 52-62, ASSOC PROF PSYCHIAT, UNIV ILL COL MED, 62-; CHMN CHILD ANAL TRAINING, INST PSYCHOANAL, 69- Concurrent Pos: Fel staff & consult, Inst Juv Res, 48-; med dir, North Shore Ment Health Clin, 53-54. Mem: Am Psychoanal Asn; Am Acad Child Psychiat; Am Psychiat Asn. Res: Evaluation and measurement in child diagnosis and treatment; personality and achievement in medical students; process of supervision in psychotherapy; psychological meaning of games. Mailing Add: 180 N Michigan Ave Chicago IL 60601

BEISER, MORTON, b Regina, Sask, Nov 16, 36; m 61; c 3. PSYCHIATRY. Educ: Univ BC, MD, 60. Prof Exp: Intern, Montreal Gen Hosp, 60-61; resident psychiat med ctr, Duke Univ, 61-64; lectr behav sci, 66-67, asst prof, 67-70, ASSOC PROF BEHAV SCI, SCH PUB HEALTH, HARVARD UNIV, 70- Concurrent Pos: NIMH res fel med ctr, Duke Univ, 64; res fel med ctr, Cornell Univ, 65-66; asst psychiatrist, Mass Gen Hosp, 70-; consult, Ment Health Br, Indian Health Serv, 72-; prin investr grants, Dept Ment Health, Commonwealth of Mass, 72- & Indian Health Serv, Dept Health, Educ & Welfare, 72; staff psychiatrist, Tufts New Eng Med Ctr, 72- & Children's Hosp Med Ctr, Boston, 74- Honors & Awards: Master of Res of First Degree, Nat Inst Health & Med Res, France, 72. Mem: AAAS; Am Psychiat Asn; Am Psychopath Asn; hon mem Royal Soc Med, Belg. Res: Psychiatric epidemiology; cross-cultural psychiatry; epidemiology of childhood psychiatric disorders. Mailing Add: Harvard Univ Sch of Pub Health 677 Huntington Ave Boston MA 02115

BEISER, SAM MEYER, microbiology, deceased

BEISHLAG, GEORGE ALBERT, b Syracuse, NY, Mar 4, 07; m 36. GEOGRAPHY. Educ: Wayne State Univ, AB, 30; Clark Univ, MA, 37; Univ Md, PhD(geog), 53. Prof Exp: Teacher pub sch, Detroit, Mich, 31-42; intel specialist, Off Strategic Serv, US Dept State & Cent Intel Agency, 42-49; adj geogr, Am Univ, 49-53; chmn dept, 65, PROF GEOG, TOWSON STATE COL, 54- Concurrent Pos: Maps consult, Secretariat UN Librn, 46; lectr geog, Naval Post Grad Intel Sch, Washington, DC, 46; res geogr, Rural Land Classification Proj, Govt PR, 51; consult, Joseph Mealey Assocs, 71- Mem: Asn Am Geog; Nat Coun Geog Educ; Int Geog Union; Am Geog Soc; Latin Am Studies Asn. Res: North and South Americas; Arctic; Maryland. Mailing Add: Dept of Geog Towson State Col Baltimore MD 21204

BEISHLINE, ROBERT RAYMOND, b Ogden, Utah, Nov 28, 30; m 59; c 1. PHYSICAL ORGANIC CHEMISTRY. Educ: Brigham Young Univ, BS, 55, MS, 57; Pa State Univ, PhD(chem), 62. Prof Exp: Res chemist, Am Cyanamid Co, 62-63; asst prof chem, State Univ NY Albany, 63, Res Found fac res fel, 64; assoc prof, 66-74, PROF CHEM, WEBER STATE COL, 66- Mem: Am Chem Soc. Res: Chemical kinetics of organic reactions; reaction mechanisms; linear free energy; relationships, amine oxide chemistry. Mailing Add: Dept of Chem Weber State Col Ogden UT 84403

BEISLER, JOHN ALBERT, b Hackensack, NJ, May 18, 37; m 68; c 2. MEDICINAL CHEMISTRY. Educ: Fairleigh Dickinson Univ, BS, 60; Rutgers Univ, PhD(org chem), 64. Prof Exp: Staff fel steroid chem, Nat Inst Arthritis & Metab Dis, 66-71; sr investr med chem, Microbiol Assoc Whittaker Corp, 71-74; RES CHEMIST MED CHEM, NAT CANCER INST, 74- Concurrent Pos: Lectr, Am Univ, Washington, DC, 73. Mem: Am Chem Soc; The Chem Soc; AAAS; Am Crystallog Asn. Res: Rational design and synthesis of drug molecules having a potential chemotherapeutic value in the treatment of cancer in humans. Mailing Add: Bldg 37 Rm 6D19 Nat Inst Health Bethesda MD 20014

BEISNER, HENRY MICHAELS, physics, see 12th edition

BEISPIEL, MYRON, b New York, NY, Nov 21, 31; m 54; c 3. ORGANIC CHEMISTRY. Educ: Brooklyn Col, BS, 53; NY Univ, PhD(chem), 59. Prof Exp: Instr chem, NY Univ, 57-59; with Nuodex Prod Co, Div Heyden Newport Chem Co, 59-61; with Sci Design Co, Inc, 61-63; CHMN SCI DEPT, RANNEY SCH, 63- Concurrent Pos: Dir environ health lab, Woodbridge Twp, NJ; consult, Cent Jersey Regional Air Pollution Control Agency, NJ & Middlesex County Health Dept, 73- Mem: Am Chem Soc; Brit Anal Chem Soc. Res: Organic and inorganic synthesis; analysis techniques, especially gas chromatography; photo chemistry; auto and catalytic oxidation; air and water pollution problems; analytical microscopy; environmental health and impact studies; public health. Mailing Add: 11 Gayle St New Monmouth NJ 07748

BEISSER, ARNOLD RAY, b Santa Ana, Calif, Oct 5, 25; m 53. PSYCHIATRY. Educ: Stanford Univ, AB, 48, MD, 50. Prof Exp: Intern, Charity Hosp La, 49-50; resident, Metrop State Hosp, 53-55; resident, Los Angeles County Gen Hosp, 53-54; staff psychiatrist, Pasadena Guid Clin, 56; chief psychiatrist outpatient & aftercare dept, Metrop State Hosp, Dept Ment Hyg, State of Calif, 56-57, coordr residency training, 57-58, chief prof educ, 58-65, DIR CTR TRAINING IN COMMUNITY PSYCHIAT, DEPT MENT HYG, STATE OF CALIF, 65- Concurrent Pos: Consult, Metrop State Hosp, 65-, Los Angeles County Ment Health Dept, 66-, Ment Health Develop Comn Los Angeles County, 66-, Family Serv Los Angeles, 70- & Psychiat Residency Training Prog, Olive View Hosp, 70-; mem fac eng mgt sem, Grad Sch Bus Admin, Univ Calif, Los Angeles, 66-73, assoc clin prof psychiat, Sch Med, 67-73, lectr community ment health, Sch Pub Health, 67-73; mem training fac, Gestalt Ther Inst Los Angeles, 69-; teaching consult, Sch Med, Univ Southern Calif, 70- Honors & Awards: Physician of the Year Award, Calif Governor's Comt Employ of Handicapped, 71. Mem: AAAS; fel Am Psychiat Asn. Res: Development of human potential; community service systems; consultation; sports and leisure; psychotherapy. Mailing Add: Ctr for Training in Community Psychiat Dept of Ment Hyg State of Calif 11665 W Olympic Blvd Suite 200 Los Angeles CA 90064

BEISSNER, ROBERT EDWARD, b San Antonio, Tex, Oct 27, 33; m 54; c 6. THEORETICAL SOLID STATE PHYSICS. Educ: St Mary's Univ, BS, 55; Tex Christian Univ, MA, 60, PhD(physics), 65. Prof Exp: Nuclear engr, Westinghouse Elec Corp, 55-56 & Gen Dynamics Corp, 56-69; PHYSICIST, SOUTHWEST RES INST, 69- Mem: Am Phys Soc; AAAS. Res: Theory of defects in metals; applications of scattering theory in nondestructive evaluation. Mailing Add: Southwest Res Inst PO Drawer 28510 San Antonio TX 78284

BEISTEL, DONALD W, b Sunbury, Pa, Feb 29, 36; m 63; c 1. PHYSICAL CHEMISTRY. Educ: Bucknell Univ, BS, 58; Univ Del, PhD(chem), 63. Prof Exp: Mem tech staff, Directorate Res & Develop, US Army Missile Command, Redstone Arsenal, Ala, 63-64; asst prof chem, Franklin & Marshall Col, 64-66 & Marshall Univ, 66-68; asst prof, 68-69, ASSOC PROF CHEM, UNIV MO-ROLLA, 69- Mem: AAAS; Am Chem Soc; Soc Appl Spectros. Res: Elucidation of molecular structure and electronic configurations of polynuclear aromatic compounds by application of nuclear magnetic resonance; applications of computer methods to chemistry; molecular orbital calculations on chemical carcinogens. Mailing Add: Dept of Chem Univ of Mo Rolla MO 65401

BEISWANGER, JOHN PAUL G, chemistry, see 12th edition

BEITCH, BARBARA ROSE, biology, physiology, see 12th edition

BEITCH, IRWIN, b Brooklyn, NY, Nov 28, 37; m 63; c 2. EMBRYOLOGY, CYTOLOGY. Educ: Univ Richmond, BS, 60, MS, 62; Univ Va, PhD(biol), 68. Prof Exp: NIH trainee ophthal res, Col Physicians & Surgeons, Columbia Univ, 67-69; asst prof, 69-71, ASSOC PROF BIOL, QUINNIPIAC COL, 71- Mem: AAAS; Am Soc Zool; Am Inst Biol Sci; NY Acad Sci. Res: Electron microscopy of keratinization and feather development; electron microscopy of eye tissues; histology; histochemistry. Mailing Add: Dept of Biol Quinnipiac Col Hamden CT 06518

BEITCHMAN, BURTON DAVID, b Philadelphia, Pa, May 1, 26; m 56; c 2. ORGANIC CHEMISTRY. Educ: Temple Univ, BA, 48; Rutgers Univ, MS, 52, PhD(org chem), 54; Widener Col, MBA, 73. Prof Exp: Chemist, Publicker Industs, Inc, 48-49; asst res specialist, Rutgers Univ, 54-56; chemist, Nat Bur Standards, 57-60; chemist, 60-66, SR RES CHEMIST, HOUDRY LABS, AIR PROD & CHEM, INC, 66- Mem: Am Chem Soc; Catalysis Soc. Res: New products and process research and development; fluorochemicals; oxidation reactions, homogeneous and heterogeneous catalysis; organic synthesis; isocyanates; polyurethanes; chemical additives; waste gas adsorbents; peroxide decomposition accelerators. Mailing Add: 134 Parkview Dr Springfield PA 19064

BEITEL, GEORGE A, physics, see 12th edition

BEITER, MARION, b Buffalo, NY, Aug 23, 07. MATHEMATICS. Educ: Canisius Col, AB, 44; St Bonaventure Col, MS, 48; Cath Univ Am, PhD(math), 60. Prof Exp: Teacher parochial schs, NY, Wash & Ohio, 25-39; prin & teacher, NY, Ohio, WVa, 39-49; prin, Charleston Cath High Sch, WVa, 49-52; PROF MATH & CHMN DEPT, ROSARY HILL COL, 52- Mem: Am Math Soc; Math Asn Am. Res: Magnitude of coefficients of cyclotomic polynomial. Mailing Add: Dept of Math Rosary Hill Col 4380 Main St Buffalo NY 14226

BEITINGER, THOMAS LEE, b Prairie du Chein, Wis, Mar 4, 45; m 67; c 2. PHYSIOLOGICAL ECOLOGY. Educ: Hamline Univ, BS, 67; Univ RI, MS, 69; Univ Wis-Madison, PhD(zool), 74. Prof Exp: ECOLOGIST, GREAT LAKES THERMAL EFFECTS PROG, ARGONNE NAT LAB, 74- Mem: Am Fisheries Soc; Sigma Xi. Res: Behavioral and physiological responses of animals to environmental variables; temperature preference, avoidance and regulation of ectothermic organisms. Mailing Add: Radiol & Environ Res Div Argonne Nat Lab 9700 S Cass Ave Argonne IL 60439

BEITINS, INESE ZINTA, b Riga, Latvia. PEDIATRIC ENDOCRINOLOGY. Educ: Univ Toronto, MD, 62; FRCP(C), 67. Prof Exp: Rotating intern med & surg, Toronto Western Hosp, 62-63; resident pediat, Hosp Sick Children, Toronto, 63-65; fel pediat, Johns Hopkins Hosp, 65-66; resident path & med, Hosp Sick Children & Toronto Gen Hosp, 66-67; staff physician pediat, Hosp Sick Children, 67-68; fel pediat endocrinol, Johns Hopkins Univ Hosp, 68-71; Busswell fel & res asst prof pediat, State Univ NY Buffalo, 72-73; ASST PROF PEDIAT, HARVARD MED SCH, 73- Mem: Lawson Wilkins Pediat Endocrine Soc; Am Fedn Clin Res; Endocrine Soc; Royal Col Med London. Res: Adrenal steroid measurement and their role in hypertension, maternal fetal interrelationships and disorders such as congenital virilizing adrenal hyperplasia; Gonadotropin heterogeneity, subunits and biological versus immunological activity. Mailing Add: Vincent Res Hosp Mass Gen Hosp 32 Fruit St Boston MA 02114

BEITZ, DONALD CLARENCE, b Stewardson, Ill, Mar 30, 40; m 63; c 2. NUTRITIONAL BIOCHEMISTRY. Educ: Univ Ill, BS, 62, MS, 63; Mich State Univ, PhD(biochem, dairy sci), 67. Prof Exp: ASSOC PROF ANIMAL SCI & BIOCHEM, IOWA STATE UNIV, 67- Mem: AAAS; Am Dairy Sci Asn; Am Soc Animal Sci; Am Inst Nutrit. Res: Fatty acid and cholesterol synthesis and lipolysis in cattle and goats; cholesterol absorption and gluconeogenesis in young ruminants; lactate metabolism in pigs; etiology of milk fever. Mailing Add: Dept of Animal Sci Iowa State Univ Ames IA 50011

BEITZEL, RICHARD EARL, b Hiles, Wis, Feb 13, 27; m 50; c 3. ORGANIC CHEMISTRY, SCIENCE EDUCATION. Educ: Minot State Col, BS, 50; Colo State Col Educ, MA, 53; Univ Iowa, PhD(sci educ), 63. Prof Exp: Teacher jr high sch, Colo, 52-53; teacher, Iowa, 54-57; instr chem, 57-58, asst prof, 58-63, from assoc prof to prof, 63-68, VPRES ACAD AFFAIRS, BEMIDJI STATE COL, 68- Mem: Nat Asn Res Sci Teaching; Am Chem Soc; Nat Sci Teachers Asn. Res: Teaching chemistry in the elementary school. Mailing Add: Bemidji State Col Bemidji MN 56601

BEIZER, LAWRENCE H, b Scranton, Pa, Feb 17, 09; m 51; c 3. INTERNAL MEDICINE, HEMATOLOGY. Educ: Univ Pa, BS, 30; Harvard Med Sch, MD, 34; Univ Minn, MS, 42. Prof Exp: Intern, Philadelphia Gen Hosp, 34-36; resident clin path, Hosp Univ Pa, 36-37; fel, Mayo Clin, 37-43; instr med, 46-49, assoc, 49-50, from asst prof to assoc prof, 50-65, dir div hemat, 50-70, PROF CLIN MED, SCH MED, UNIV PA, 65-, DIR DIV ONCOL, GRAD HOSP, 70- Concurrent Pos: Assoc med, Woman's Med Col, 46-48, from clin asst prof to clin assoc prof, 48-56, consult, Hosp, 47-56; consult, Valley Forge Army Hosp, Phoenixville, 47-74, Vet Hosp, Philadelphia, 53-56, Walson Army Hosp, Ft Dix, NJ, 60-63 & Vet Hosp, Wilmington, Del, 62; physician, Presby-Univ Pa Med Ctr, 70- Mem: Fel Am Col Physicians; Am Soc Hemat. Res: Clinical hematology, especially bone marrow aspiration biopsy and mutiple myeloma. Mailing Add: Suite 302 Grad Med Bldg 419 S 19th St Philadelphia PA 19146

BEJNAR, WALDEMERE, b Hamtramck, Mich, Feb 7, 20; m 46; c 5. GEOLOGY. Educ: Univ Mich, BS, 43, MA, 47; Univ Ariz, PhD(geol), 50; Univ NMex, MS, 65. Prof Exp: Geologist, US Dept Reclamation, 50 & Foreign Sect, US Geol Surv, Nigeria, 50-52; asst prof geol, NMex Inst Mining & Technol, 52-55; consult geologist, 55-58; assoc sponsor, Dale Carnegie & Assocs, 58-63; PROF GEOL & GEOG, NMEX HIGHLANDS UNIV, 65- Concurrent Pos: Pres, Mira Uranium Corp, 56 & Waldemere Bejnar & Assocs, Inc, 59-; consult geologist, 58- Mem: Am Inst Prof Geol. Res: Ore deposition; ground water; landslides. Mailing Add: Div of Geol NMex Highlands Univ Las Vegas NM 87701

BEKEFI, GEORGE, b Prague, Czech, Mar 14, 25; m 61. PHYSICS. Educ: London Univ, BSc, 48; McGill Univ, MSc, 50, PhD(physics), 52. Prof Exp: Res assoc, McGill Univ, 52-55, asst prof physics, 55-57; res assoc plasma physics res lab electronics, 57-61, from asst prof to assoc prof physics, 61-68, PROF PHYSICS, MASS INST TECHNOL, 68- Mem: Am Asn Physics Teachers; fel Am Phys Soc. Res: Microwave antennae and propagation; plasma physics. Mailing Add: Dept Physics Mass Inst Technol Rm 36-213 Massachusetts Ave Cambridge MA 02139

BEKENSTEIN, JACOB DAVID, b Mexico City, Mex, May 1, 47; US citizen. ASTROPHYSICS. Educ: Polytech Inst Brooklyn, BS & MS, 69; Princeton Univ, PhD(physics), 72. Prof Exp: Res assoc, Ctr Relativity Theory, Dept Physics, Univ Tex, Austin, 72-74; SR LECTR, DEPT PHYSICS, BEN-GURION UNIV OF NEGEV, ISRAEL, 74- Mem: Israeli Phys Soc. Res: Physical properties of collapsed stars; methods of detection of astrophysical black holes; variability of particle properties; radiation mechanism of quasars. Mailing Add: Dept of Physics Ben-Gurion Univ of the Negev Beer Sheva Israel

BEKOFF, ANNE LAURENS, b Denver, Colo, May 19, 47; m 70. NEUROEMBRYOLOGY. Educ: Smith Col, BA, 69; Wash Univ, St Louis, Mo, PhD(neurobiol & develop), 74. Prof Exp: Res assoc neurobiol, Dept Environ Pop & Organismic Biol, 74-75, FEL NEUROPHYSIOL, DEPT PHYSIOL, UNIV COLO MED CTR, 75- Mem: Soc Neurosci; Am Soc Zoologists; AAAS; Am Polar Soc; US Antarctic Res Prog. Res: Analyzing the mechanisms underlying the development of specific neural connections in the motor system of the chick embryo and the development of coordinated movement. Mailing Add: Dept of Physiol Univ of Colo Med Ctr Denver CO 80220

BEKOFF, MARC, b Brooklyn, NY, Sept 6, 45; m 70. ANIMAL BEHAVIOR. Educ: Washington Univ, AB, 67, PhD(animal behav), 72; Hofstra Univ, MA, 68. Prof Exp: Asst prof biol, Univ Mo, St Louis, 73-74; ASST PROF BIOL, UNIV COLO, BOULDER, 74- Mem: Animal Behav Soc; Am Soc Zoologists; Am Soc Mammalogists; US Antarctic Res Prog; Sigma Xi. Res: Sociobiology of mammals, especially canids; mathematical models of social interaction; animal social development; communication processes. Mailing Add: Dept of Environ Pop & Org Biol Ethol Group Univ of Colo Boulder CO 80302

BELAIR, ERNEST JOSEPH, b Alburg, Vt, July 1, 27; m 51; c 5. PHYSIOLOGY. Educ: Univ Vt, BA, 50; St Michael's Col, Vt, MA, 51; Univ Ottawa, PhD(physiol), 66. Prof Exp: Teacher, Mooers Cent Sch, NY, 53-54; from instr to asst prof biol, St Michael's Col, Vt, 54-62; part-time instr physiol, Univ Ottawa, 62-65; sect head cardiovasc pharmacol, Pharmaceut Div, Pennwalt Corp, NY, 65-73; SR PHARMACOLOGIST, PRE-CLIN RES, ROHM AND HAAS CO, 73- Mem: Am Soc Soc; AAAS; NY Acad Sci. Res: Cardiovascular and renal pharmacology and physiology; gastrointestinal pharmacology; general pharmacology. Mailing Add: Pre-Clin Res Dept Rohm and Haas Co Spring House PA 19477

BELAMARICH, FRANK ALEXANDER, b Newark, NJ, Jan 2, 27; m 56; c 4. CELL BIOLOGY. Educ: Montclair State Col, BA, 57; Harvard Univ, MA, 60, PhD(biol), 62. Prof Exp: Res assoc biochem, Sch Med, Univ Buffalo, 61-63; from asst prof to assoc prof, 63-72, PROF BIOL, BOSTON UNIV, 72- Concurrent Pos: Mem corp, Marine Biol Lab, 66- Mem: Int Soc Thrombosis & Haemostasis; AAAS; Am Soc Zool. Res: Mechanism of aggregation of hemostatic cells. Mailing Add: Dept of Biol Boston Univ Boston MA 02215

BELAND, GARY LAVERN, b Los Angeles, Calif, May 18, 42; m 69; c 1. ECONOMIC ENTOMOLOGY. Educ: Kearney State Col, BA, 65; Univ Nebr, Lincoln, MS, 68, PhD(entom), 72. Prof Exp: Instr entom, Univ Nebr, Lincoln, 69-72; RES ENTOMOLOGIST, AGR RES SERV, USDA, 72- Mem: Entom Soc Am; Sigma Xi. Res: Host plant resistance; biology and ecology of insect pest on soybeans and corn in the Midsouth. Mailing Add: US Delta States Agr Res Ctr USDA PO Box 225 Stoneville MS 38776

BELAND, JACQUES (ROBERT), b Cabano, Que, May 5, 23. GEOLOGY. Educ: Laval Univ, BA, 44, BSc, 48, MSc, 49; Princeton Univ, PhD(geol), 53. Prof Exp: Geologist, Geol Surv Br, Que Dept Nat Resources, 53-62; assoc prof, 62-66, chmn dept, 66-75, PROF GEOL, UNIV MONTREAL, 66- Concurrent Pos: Geol Soc fel. Mem: Geol Asn Can; Mineral Asn Can; Can Inst Mining & Metall. Res: Structural geology; tectonics. Mailing Add: 204 Brookfield Ave Ville Mont-Royal Montreal PQ Can

BELAND, RENE, b Can, Apr 26, 18; m 45; c 7. GEOLOGY, MINERALOGY. Educ: Laval Univ, BA, 37; Queen's Univ (Ont), BSc, 42; Univ Toronto, PhD(geol), 46. Prof Exp: Assoc prof geol & mineral, 46-54, PROF STRUCT GEOL, LAVAL UNIV, 54- Concurrent Pos: Field geologist, Dept of Mines, Que, 45- Mem: Soc Econ Geol; Geol Asn Can; Mineral Asn Can. Res: Hydrosynthesis of sulphide and sulphosalt minerals; petrographic and petrologic problems in connection with ore genesis. Mailing Add: Dept of Geol Laval Univ Quebec PQ Can

BELANGER, DAVID GERALD, b Rockville, Conn, Dec 8, 44; m 71; c 1. SYSTEMS THEORY. Educ: Union Col, BS, 66; Case Western Reserve Univ, MS, 68, PhD(math), 71. Prof Exp: Asst prof, 71-75, ASSOC PROF MATH, UNIV S ALA, 75- Concurrent Pos: Mathematician, US Army Corps Engrs, 73- Mem: Soc Indust & Appl Math; Math Asn Am. Res: Stability of dynamical systems. Mailing Add: 6229 Parkwood Dr Mobile AL 36608

BELANGER, LEONARD FRANCIS, b Montreal, Que, Mar 11, 11; m 38; c 2. HISTOLOGY, HISTOCHEMISTRY. Educ: Univ Montreal, BA, 31, MD, 37; Harvard Univ, MA, 40. Prof Exp: Asst prof histol & embryol, Univ Montreal, 41; res assoc, McGill Univ, 45-46; assoc prof, 46-47, PROF HISTOL & EMBRYOL & HEAD DEPT, UNIV OTTAWA, 47- Concurrent Pos: Vis prof, Univ BC & lectr, Cornell Univ, 63-64. Honors & Awards: Dow Award & Parizeau Medal, 68; Steindler Award, Orthopaedic Res Soc, 72. Mem: AAAS; Am Asn Anat; Histochem Soc; Can Physiol Soc (treas, 51-54); Royal Soc Can. Res: Techniques of autoradiography; mechanism of minerization and growth of cartilage, bone and teeth; arteriosclerosis; fluorosis; inner ear; hard tissue; histophysiology; bone implantation. Mailing Add: Dept of Histol Fac of Med Univ of Ottawa 275 Nicholas St Ottawa ON Can

BELANGER, PATRICE CHARLES, organic chemistry, see 12th edition

BELANGER, PIERRE ANDRE, b Pointe-au-Pic, Que, Nov 1, 41; m 66; c 1. LASERS. Educ: Univ Laval, BSc, 66, DSc(optics), 71. Prof Exp: Res & develop specialist lasers, Gen-Tec, Inc, 71-72; res asst, 70-71, ASST PROF PHYSICS, LAVAL UNIV, 72- Mem: Optical Soc Am; Can Asn Physicists. Res: Generation of very high peak power laser pulse; optics of high peak power laser beam. Mailing Add: Lab for Res Optics & Laser Dept of Physics Laval Univ Quebec PQ Can

BELASCO, IRVIN JOSEPH, biological chemistry, see 12th edition

BELCASTRO, PATRICK FRANK, US citizen; m 63; c 2. PHARMACY. Educ: Duquesne Univ, BS, 42; Purdue Univ, MS, 51, PhD(pharm), 53. Prof Exp: Instr pharm & chem, Duquesne Univ, 46-49; asst prof pharm, Ohio State Univ, 53-54; ASST PROF PHARM, PURDUE UNIV, WEST LAFAYETTE, 54- Mem: Am Pharmaceut Asn; Am Inst Hist Pharm; Fine Particle Soc; Am Soc Hosp Pharmacists. Res: Effects of x-irradiation on pharmacological action of certain drugs. Mailing Add: Sch of Pharm & Pharmacal Sci Purdue Univ West Lafayette IN 47907

BELCHER, JANE COLBURN, b New York, NY, June 11, 10. ZOOLOGY, EVOLUTION. Educ: Colby Col, AB, 32; Columbia Univ, MA, 33; Univ Mo, PhD(zool), 40. Prof Exp: Instr biol, Colby Col, 33-36; instr zool, Hunter Col, 37; asst anat & zool, Univ Mo, 38-40; instr biol, 40-45, asst prof, 45-51, from assoc prof to prof, 51-72, Dorys McConnell Duberg prof ecol, 72-75, EMER DORYS McCONNELL DUBERG PROF ECOL, SWEET BRIAR COL, 75- Mem: AAAS; Am Soc Zool; NY Acad Sci. Res: Cytology and endocrinology of reproductive cycle; Virginia invertebrates; seasonally and experimentally induced changes in the reproductive tract of the female bat Myotis grisescens; evolution. Mailing Add: PO Box BA Sweet Briar VA 24595

BELCHER, ROBERT ORANGE, b Williamsburg, Ky, June 29, 18; m 38; c 2. BOTANY. Educ: Berea Col, AB, 38; Univ Mich, MS, 47, PhD(bot), 55. Prof Exp: Asst biol, Purdue Univ, 38-40; fel bot, Univ Mich, 46; head dept, 58-66, from asst prof to assoc prof, 46-57, PROF BIOL, EASTERN MICH UNIV, 57- Honors & Awards: US Typhus Comn medal, 46. Mem: AAAS; Bot Soc Am; Am Soc Plant Taxon. Res: Systematic botany of pantropical composite weeds; erechthitoid species of Senecio in Australasia. Mailing Add: PO Box 242 Ypsilanti MI 48197

BELDEN, DON ALEXANDER, JR, b Akron, Ohio, July 13, 26; m 52; c 2. ZOOPHYSIOLOGY. Educ: Middlebury Col, BA, 50; Williams Col, MA, 52; Wash State Univ, PhD(zool), 58. Prof Exp: Asst zool, Middlebury Col, 49-50; asst biol, Williams Col, 50-52; res fel zool, Wash State Univ, 52-56; asst prof, 56-65, ASSOC PROF ZOOL, UNIV DENVER, 65- Res: Cellular biology; histochemistry. Mailing Add: 2060 E Amherst Ave Denver CO 80210

BELDEN, EVERETT LEE, b Bridgeport, Nebr, Aug 14, 38; m 61; c 2. VETERINARY IMMUNOLOGY, VETERINARY MICROBIOLOGY. Educ: Univ Wyo, BS, 60, MS, 62; Univ Calif, Davis, PhD(microbiol), 71. Prof Exp: Instr microbiol, 62-65, ASST PROF IMMUNOL, UNIV WYO, 70- Mem: Am Soc Microbiol; Sigma Xi. Res: Humoral and cellular immune response in the bovine; infectious reproductive diseases in cattle and sheep; chlamydial diseases in domestic animals. Mailing Add: Div of Microbiol & Vet Med Univ of Wyo Laramie WY 82071

BELDIN, RICHARD ALLEN, statistics, mathematical biology, see 12th edition

BELDING, HARWOOD SEYMOUR, physiology, deceased

BELDING, RALPH CEDRIC, b Mich, June 2, 15; m 38; c 4. POULTRY PATHOLOGY. Educ: Mich State Col, DVM, 46, MS, 54; Ohio State Univ, PhD, 58. Prof Exp: Veterinarian, Fed Regional Poultry Res Lab, USDA, 48; RESEARCHER MICROBIOL & PUB HEALTH, MICH STATE UNIV, 49-, ASSOC PROF MICROBIOL & PUB HEALTH, 74- Mem: Am Vet Med Asn. Res: Pathogenic bacteriology; hemodialysis. Mailing Add: Dept of Microbiol & Pub Health Mich State Univ East Lansing MI 48823

BELEW, JOHN SEYMOUR, b Waco, Tex, Nov 3, 20; m 44; c 2. ORGANIC CHEMISTRY. Educ: Baylor Univ, BS, 41; Univ Wichita, MS, 47; Univ Wis, PhD(chem), 51. Prof Exp: Res assoc org chem, Brown Univ, 51-53; actg asst prof chem, Univ Va, 53-56; from asst prof to assoc prof, 56-63, PROF CHEM, BAYLOR UNIV, 63-, DEAN COL ARTS & SCI, 74- Mem: AAAS; Am Chem Soc; Am Inst Chem; The Chem Soc. Res: Reactions of ozone; heterocycle and o-quinone syntheses; oxidations with transition metal oxides. Mailing Add: Col of Arts & Sci Baylor Univ Waco TX 76703

BELFORD, GENEVA GROSZ, b Washington, DC, May 18, 32; m 54. APPLIED MATHEMATICS. Educ: Univ Pa, BA, 53; Univ Calif, Berkeley, MA, 54; Univ Ill, PhD(comput theory), 60. Prof Exp: Res assoc chem, Univ Ill, Urbana-Champaign, 59-61; NIH fel, 61-64; asst prof math, 64-72, RES ASST PROF, CTR ADVAN COMPUT, UNIV ILL, URBANA-CHAMPAIGN, 72- Mem: Soc Indust & Appl Math. Res: Approximation theory and approximate solution of differential and integral equations; mathematical problems in data management and computer networks. Mailing Add: Ctr for Advan Comput Univ of Ill Urbana IL 61801

BELFORD, JOHN F, b Birmingham, Eng, July 10, 26; US citizen; m 52; c 3. ULTRASOUND, BIOMEDICAL ENGINEERING. Educ: Mass Inst Technol, BS, 52; John Carroll Univ, SM, 62. Prof Exp: Mem tech staff piezoelec, Bell Tel Labs, Inc, 52-59; staff engr acoust, Clevite Corp, 59-62; physicist ultrasonics, Branson Instruments, Inc, 62-63; dir res med physics, Hoffrel Instruments, Inc, 63-65, vpres, 65-66; eng mgr, Andersen Labs, Inc, 66-69; VPRES, HOFFREL INSTRUMENTS, INC, 69- Mem: Inst Elec & Electronics Eng; Acoust Soc Am; Am Inst Ultrasound Med; Asn Advan Med Instrumentation; Int Fedn Med & Biol Eng. Res: Piezoelectricity; acoustics; ultrasonics; elasticity; medical and biological physics and engineering; medical imaging. Mailing Add: 39 Highland Rd Westport CT 06880

BELFORD, JULIUS, b Brooklyn, NY, Feb 3, 20; m 55; c 3. PHARMACOLOGY. Educ: Brooklyn Col, BA, 40; Long Island Univ, BS, 43; Yale Univ, PhD(pharm), 49. Prof Exp: From asst prof to assoc prof pharm, 49-69, asst to dean, 58-61, actg chmn dept pharmacol, 71-72, PROF PHARM, STATE UNIV NY DOWNSTATE MED CTR, 69- Concurrent Pos: Exchange prof, St Mary's Hosp Med Sch, London, 52-53; USPHS spec fel, Mario Negri Inst Pharmacol Res, Milan, Italy, 65. Mem: AAAS; Am Soc Pharmacol; Harvey Soc; Asn Am Med Cols. Res: Pharmacology of drugs affecting the cardiovascular and respiratory systems; experimental epilepsy and anticonvulsants; medical education. Mailing Add: Dept of Pharmacol State Univ NY Downstate Med Ctr Brooklyn NY 11203

BELFORD, RUE LINN, b St Louis, Mo, Dec 13, 31; m 54. PHYSICAL CHEMISTRY, INORGANIC CHEMISTRY. Educ: Univ Ill, BS, 53; Univ Calif, Berkeley, PhD(chem), 55. Prof Exp: Chemist, Univ Ill, 52; asst, Univ Calif, 53, chemist radiation lab, 53-54; instr, 55-57, asst prof, 57-63, ASSOC PROF PHYS CHEM, UNIV ILL, 63- Concurrent Pos: Sloan fel, 61-63; mem NIH biophys & biophys chem fel comt, 65-69, sr fel, 66-67. Mem: AAAS; Am Phys Soc; Am Chem Soc; Soc Appl Spectros. Res: Spectra and bonding of transition-metal chelate compounds; electron paramagnetic resonance and nuclear quadrupole coupling; molecules excited by hypersonic shock waves. Mailing Add: Dept of Chem Univ of Ill Urbana IL 61801

BELILES, ROBERT PRYOR, b Louisville, Ky, Dec 21, 32; m 58; c 2. TOXICOLOGY. Educ: Univ Louisville, BA, 54, MS, 58; Iowa State Univ, PhD(pharmacol), 62. Prof Exp: Instr pharmacol, Iowa State Univ, 61-62; instr radiation biol, Univ Rochester, 62-64; pharmacologist, Woodard Res Corp, 64, chief environ toxicol div, 64-67; dir toxicol, Lakeside Labs, 67-75; ASSOC DIR PHARM TOXICOL, LITTON BIONETICS, LITTON INDUST, INC, 75- Concurrent Pos: Mem, Nat Agr Chem Comt Wildlife Pesticides, 65-67; asst clin prof pharmacol, Med Col Wis, 68-75. Mem: Teratology Soc; Am Indust Hyg Asn; Soc Toxicol; NY Acad Sci; Am Asn Lab Animal Sci. Res: Toxocology of environmental contaminants; iron compounds; behavioral and reproductive toxicology. Mailing Add: Litton Bionetics 101 W Jefferson St Falls Church VA 22046

BELINFANTE, FREDERIK J, b The Hague, Holland, Jan 6, 13; nat US; m 37; c 3. PHYSICS. Educ: State Univ Leiden, BA, 33, MA, 36, PhD(theoret physics), 39. Prof Exp: Asst theoret physics, State Univ Leiden, 36-46, lectr, 45-46; assoc prof, Univ BC, 46-48; assoc prof, 48-51, PROF THEORET PHYSICS, PURDUE UNIV, WEST LAFAYETTE, 51- Mem: Fel Am Phys Soc; Neth Phys Soc. Res: General-relativistic quantum field theory; hidden-variables theories; foundations of quantum theory. Mailing Add: Dept of Physics Purdue Univ West Lafayette IN 47907

BELINFANTE, JOHAN G F, b Leiden, Neth, Feb 2, 40; US citizen. APPLIED MATHEMATICS, PHYSICS. Educ: Purdue Univ, BS, 58; Princeton Univ, PhD(physics), 61. Prof Exp: NSF fel physics, Calif Inst Technol, 61-62; res asst, Univ Pa, 62-64; from asst prof to assoc prof, Carnegie-Mellon Univ, 64-73; ASSOC PROF MATH, GA INST TECHNOL, 73- Mem: Am Math Soc; Am Phys Soc; Math Asn Am; Soc Indust & Appl Math. Res: Quantum mechanics; lie algebras. Mailing Add: Sch of Math Ga Inst of Technol Atlanta GA 30332

BELITSKUS, DAVID, b Cuddy, Pa, Dec 24, 38; m 68; c 1. PHYSICAL INORGANIC CHEMISTRY. Educ: Duquesne Univ, BS, 61; Univ Pittsburgh, PhD(phys chem), 64. Prof Exp: Res engr, 64-67, sr res scientist, 67-75, SCI ASSOC, ALCOA LABS, ALUMINUM CO AM, 75- Mem: Am Chem Soc; Sigma Xi. Res: X-ray crystal structure analyses; oxidation of light metal alloys; batteries and fuel cells; properties of lubricants and corrosion inhibitors; metallothermic reductions; metal matrix composites; carbon properties. Mailing Add: 2678 Elaine Dr Lower Burrell PA 15068

BELJAN, JOHN RICHARD, b Detroit, Mich, May 26, 30; m 52; c 3. AEROSPACE MEDICINE, BIOMEDICAL ENGINEERING. Educ: Univ Mich, BS, 51, MD, 54. Prof Exp: Clin instr surg, Sch Med, Univ Mich, 56-59; dir med serv, Stuart Co, Calif, 65; from asst prof to assoc prof surg, Sch Med, Univ Calif, Davis, 66-74, asst to dean, 70-71, asst dean, 71-74, from asst prof to assoc prof eng, Col Eng, 68-74; PROF SURG & DEAN SCH MED, PROF BIOMED ENG, COL SCI & ENG & VPROVOST, WRIGHT STATE UNIV, 74- Concurrent Pos: Mem ground support team, Dept Defense Task Force for Proj Gemini, 64-65; consult, Atlas Chem Industs, Inc, Del, 69-73 & Dept Health, Educ & Welfare, 72- Mem: Aerospace Med Asn; Am Col Surgeons; Am Inst Aeronaut & Astronaut; Inst Elec & Electronics Engrs; Royal Soc Med. Res: General surgery; human performance in the aerospace environment; orbital flight effects on calcium kinetics and fracture healing; telemetry. Mailing Add: Sch of Med Wright State Univ Dayton OH 45431

BELKIN, BARRY, b Philadelphia, Pa, May 4, 40; m 64; c 2. OPERATIONS RESEARCH. Educ: Mass Inst Technol, BS, 62; Univ Pa, MA, 64; Cornell Univ, PhD(math), 68. Prof Exp: SR ASSOC OPERS RES/MATH, DANIEL H WAGNER, ASSOCS, 67- Honors & Awards: Rist Prize, Mil Opers Res Soc, 75. Mem: Inst Math Statist; Soc Indust & Appl Math. Res: First passage problems; inventory requirements problems; theory of random walk. Mailing Add: Daniel H Wagner Assocs Station Square One Paoli PA 19301

BELKIN, DANIEL ARTHUR, b Los Angeles, Calif, July 23, 34; m 64. COMPARATIVE PHYSIOLOGY, ANIMAL BEHAVIOR. Educ: Univ Calif, Los Angeles, BA, 56; Univ Fla, PhD(zool), 61. Prof Exp: Asst biol, Col Med, Univ Fla, 56-58, lectr, 58-60, res fel physiol, 61-64, res assoc biol, 64-65, instr , 65-66, asst prof physiol, 66-71; PROF PHYSIOL & DIR, BAYT SHAHIN FIELD STA PHYSIOL & BEHAV ECOL, 71- Mem: Am Soc Zool; Am Soc Ichthyol & Herpet; Am Physiol Soc. Res: Comparative physiology and behavior of extremes of adaptation; diving physiology; ecological physiology of reptiles; environmental physiology; predator-prey ethology; physiology of starvation. Mailing Add: Bayt Shahin Field Sta PO Box 422 Alpaugh CA 93201

BELKIN, JOHN NICHOLAS, b Petrograd, Russia, Oct 24, 13; nat US; m 38, 50, 70; c 7. ENTOMOLOGY, TAXONOMY. Educ: Cornell Univ, BS, 38, PhD(med entom), 46. Prof Exp: Asst entom, Cornell Univ, 38-40, instr, 40-42; jr entomologist, Tenn Valley Authority, Wilson Dam, 42; asst, Cornell Univ, 46; asst specialist, Rutgers Univ, 46; assoc prof biol & head dept, Assoc Cols Upper NY, 46-49; from asst prof entom & asst entomologist to assoc prof & assoc entomologist, 49-58, prof & entomologist, 58-62, PROF ZOOL, UNIV CALIF, LOS ANGELES, 62- Mem: Fel AAAS; Soc Syst Zool. Res: Zoogeography, phylogeny and ecology of mosquitoes. Mailing Add: Dept of Biol Univ of Calif Los Angeles CA 90024

BELKNAP, HERBERT JOHN, b Edwards, Miss, Jan 1, 21; m 52; c 3. ANALYTICAL CHEMISTRY. Educ: Univ Miss, BS, 42, MS, 53; Univ Tex, PhD, 55. Prof Exp: Indust res chemist, E I du Pont de Nemours & Co, 55-58; indust anal chemist, 58-67, ENVIRON CONTROL MGR, TEX INSTRUMENTS INC, 67- Mem: Am Chem Soc; Air Pollution Control Asn; Water Pollution Control Fedn. Res: Chemical methods of analysis; atmospheric monitoring; waste water treatment and reuse. Mailing Add: Tex Instruments Inc PO Box 5474 MS161 Dallas TX 75222

BELKNAP, ROBERT WAYNE, b Omaha, Nebr, Jan 13, 24; m 53; c 2. PHYSIOLOGY. Educ: Creighton Univ, BS, 49, MS, 51; Univ Calif, PhD, 58. Prof Exp: Jr res physiologist, Univ Calif, 58-60; asst prof, 60-64, ASSOC PROF BIOL, CREIGHTON UNIV, 64- Concurrent Pos: Instr, Vet Admin Hosp, Omaha, 62- Mem: AAAS; NY Acad Sci; Am Soc Zool; Soc Nuclear Med; Am Statist Asn. Res: Mammalian physiology and environmental physiology; tissue metabolism; heat and cold exposure; hypothermia. Mailing Add: Dept of Biol Creighton Univ Omaha NE 68131

BELL, ALEXANDER GRAHAM, b Durham, Ont, Mar 11, 29. HUMAN GENETICS. Educ: Univ Toronto, MD, 53, MA, 62. Prof Exp: Jr intern med, Toronto East Gen Hosp, Ont, 53-54; gen practitioner, Sudbury Clin, Ont, 54-55 & Guelph Med Group, 55-57; sr house officer, West-Kent Hosps Asn, Maidstone, Kent & Addenbrooke's Hosp, Cambridge, UK, 57-58; health officer, Dept Health, Ont & Govt Parliament Bldg, Toronto, 58-59; res fel human genetics, Hosp Sick Children & Univ Toronto, 60-62; res assoc, Biol Div, Oak Ridge Nat Lab, Tenn, 62-63; ASST PROF PATH & ZOOL, UNIV TORONTO, 64-, RES ASSOC MED & ASST PROF PEDIAT, 68- Concurrent Pos: Queen Elizabeth II scientist, 65- Mem: Am Soc Human Genetics; Can Med Asn; Genetics Soc Can. Res: Sex chromatin survey; Toronto newborns; clinical and medical genetics; irradiation-induced chromosome abnormalities in human leucocyte cultures; irradiation-induced polyploidy and endoreduplication in human leucocyte cultures. Mailing Add: Dept of Zool Univ of Toronto Toronto ON Can

BELL, ALFRED LEE LOOMIS, JR, b Englewood, NJ, Jan 12, 23; m 47; c 4. MEDICINE. Educ: Harvard Univ, MD, 47; Am Bd Internal Med, dipl, 58; Am Bd Cardiovasc Dis, dipl, 64. Prof Exp: From intern to asst resident, St Luke's Hosp, 47-50; NIH fel, USPHS. 50-51; mem fac internal med, Sch Aviation Med, Randolph AFB, 51-53; physician-in-charge, 53-64, from asst attend physician to assoc attend physician, 55-64, from asst cardiologist to assoc cardiologist, 55-65, assoc attend cardiologist, 65-69, DIR, CARDIOPULMONARY LAB, ST LUKE'S HOSP, 64-, ATTEND PHYSICIAN, HOSP, 64-, ATTEND CARDIOLOGIST, 69-, CHIEF DIV PULMONARY DIS, 71- Concurrent Pos: Instr, Col Physicians & Surgeons, Columbia Univ, 53-58, assoc, 58-64, asst prof, 64-74, assoc prof, 74-; fel, Coun Clin Cardiol, Am Heart Asn. Mem: Fel Am Thoracic Soc; fel Am Col Physicians; fel Am Col Chest Physicians; fel NY Acad Med; Int Cardiovasc Soc. Res: Cardiopulmonary physiology. Mailing Add: St Luke's Hosp 421 W 113th St New York NY 10025

BELL, ALOIS ADRIAN, b Bloomfield, Nebr, Jan 25, 34; m 58; c 2. PLANT PATHOLOGY. Educ: Univ Nebr, BSc, 55, MSc, 58, PhD(bot), 61. Prof Exp: Asst prof bot, Univ Md, 61-65; res plant pathologist plant indust sta, 65-70, DIR NAT COTTON PATH RES LAB, CROPS RES DIV, AGR RES SERV, USDA, 70- Concurrent Pos: Grad Sch Res Coun grant, Univ Md, 62; USPHS res grant, 63-66. Mem: AAAS; Am Phytopath Soc; Am Soc Plant Physiol. Res: Physiology of fungi and diseased plant tissue; biochemistry and physiology of disease resistance and fungal pathogenesis of cotton. Mailing Add: Cotton Path Lab Crops Res Div Agr Res Serv USDA PO Drawer JF College Station TX 77840

BELL, ANTHONY E, b Ramsgate, Eng, Aug 23, 37; m 63; c 1. SURFACE PHYSICS. Educ: Univ London, BS, 59, PhD(phys chem), 62. Prof Exp: Res assoc, Univ Chicago, 63-65; physicist, Field Emission Corp, 65-69; ASSOC PROF CHEM, LINFIELD COL, 69- Mem: Am Chem Soc. Res: Adsorption studies; field emission energy distribution; field ion microscopy of adsorbed organic molecules; catalytic oxidation of methanol to formaldehyde; adsorption properties of carbon; LEED-auger surface analysis; miniature biophysics transducers. Mailing Add: Linfield Res Inst Linfield Ave McMinnville OR 97128

BELL, ANTHONY J, physical organic chemistry, see 12th edition

BELL, AUDRA EARL, b Providence, Ky, May 9, 18; m 41; c 3. GENETICS. Educ: Univ Ky, BS, 39; La State Univ, MS, 41; Iowa State Col, PhD(genetics), 48. Prof Exp: Asst genetics, Iowa State Col, 41-42, 46-48; from asst prof to assoc prof, 48-54, PROF ANIMAL SCI, PURDUE UNIV, WEST LAFAYETTE, 54- Concurrent Pos: Guggenheim fel, 64-65. Mem: AAAS; Am Genetics Asn; Am Soc Animal Sci; Genetics Soc Am; Am Soc Naturalists. Res: Population genetics in poultry, drosophilia and tribolium; genetics of disease resistance. Mailing Add: Dept of Animal Sci Purdue Univ West Lafayette IN 47907

BELL, BARBARA, b Evanston, Ill, Apr 1, 22. SOLAR PHYSICS, CLIMATOLOGY. Educ: Radcliffe Col, AB, 44, MA, 49, PhD(astron), 51. Prof Exp: Res asst, Harvard Observ, 48-57, sci asst to dir, 57-71, astronr, Harvard Col Observ, 71-73, ASTRONR, CTR ASTROPHYSICS, 73- Mem: Am Astron Soc; AAAS; Int Astron Union; Archaeol Inst Am. Res: Spectroscopy; solar-terrestrial correlations; modern climatology, post-glacial climate fluctuations and their possible role in ancient history, especially of Egypt and the Nile. Mailing Add: Ctr for Astrophys 60 Garden St Cambridge MA 02138

BELL, BRUCE MCCONNELL, b Appleton, Wis, Oct 14, 41; m 62; c 1. PALEOBIOLOGY. Educ: Earlham Col, BA, 63; Univ Cincinnati, MS, 66, PhD(paleobiol), 70. Prof Exp: CURATOR PALEONT, NY STATE MUS, 69- Mem: AAAS; Paleont Soc; Paleont Res Inst; Brit Paleont Asn. Res: Paleobiology of the Echinodermata, including anatomy and ontogeny, evolution and phylogeny, functional morphology, paleoecology and population dynamics; edrioasteroidea. Mailing Add: Dept of Educ 973 EBA Educ Bldg NY State Mus Albany NY 12234

BELL, CARL, F, b Otsego, Ohio, Nov 24, 33; m 61. PLANT PATHOLOGY. Educ: Muskingum Col, BS, 55; Miami Univ, MS, 58; Ohio State Univ, PhD(plant path), 61. Prof Exp: Assoc prof, 61-69, PROF BIOL, SHEPHERD COL, 69- Mem: AAAS; Am Phytopath Soc; Am Bot Soc. Res: Physiology of fungus parasitism. Mailing Add: Dept of Sci & Math Shepherd Col Shepherdstown WV 25443

BELL, CECIL COOPER, JR, b Bedford, Va, May 15, 22; m 47; c 4. SURGERY. Educ: Randolph-Macon Col, BS, 42; Med Col Va, MD, 45; Am Bd Surg, dipl. Prof Exp: From assoc instr to assoc prof surg, Med Col Va, 57-73; PROF SURG & ASST DEAN, EASTERN VA MED SCH, 73-; CHIEF OF STAFF, VET ADMIN HOSP, 73- Concurrent Pos: Asst chief surg, Vet Admin Hosp, Richmond, Va, 60-73, assoc chief staff res, 62-73; pres, Nat Asn Vet Admin Chiefs of Staff, 74-75, mem, Vet Admin Cent Off Med Sch Assistance Rev Comt, 74-76 & chmn, Vet Admin Cent Off Med Res Adv Comt, 75-76. Mem: Asn Am Med Cols; Am Col Surgeons; Am Gastroenterol Asn. Res: Steroid metabolism research; liver and adrenal steroid metabolism; controlling mechanisms of gallstone formation. Mailing Add: Vet Admin Hosp Hampton VA 23667

BELL, CHARLES BERNARD, JR, b New Orleans, La, Aug 20, 28; m 53; c 4. MATHEMATICS, STATISTICS. Educ: Xavier Univ, La, BS, 47; Univ Notre Dame, MS, 48, PhD, 53. Prof Exp: Asst math, Univ Notre Dame, 49-51; res engr, Douglas Aircraft Co, 51-55; asst prof math & physics, Xavier Univ, La, 55-57; instr math & res assoc statist, Stanford Univ, 57-58; from asst prof to assoc prof math, San Diego State Col, 58-66; prof, Case Western Reserve Univ, 66-68; prof statist, Univ Mich, Ann Arbor, 68-74; PROF MATH, TULANE UNIV, 74- Concurrent Pos: NSF res grants, 57-; consult, 58- Mem: Inst Math Statist; Am Statist Asn; Am Math Soc. Res: Structure of stochastic processes; statistical completeness; theory of distribution-free statistics; statistical applications in engineering and biology. Mailing Add: Dept of Math Tulane Univ New Orleans LA 70118

BELL, CHARLES E, JR, b Norwood, Mass, Aug 8, 31; m 55; c 4. PHYSICAL ORGANIC CHEMISTRY. Educ: Univ Va, BS, 54, MS, 55, PhD(chem), 61. Prof Exp: Chemist, Prod Res Div, Esso Res & Eng Co, Standard Oil, NJ, 60-62; from asst prof to assoc prof chem, 62-67, asst dean grad stud, 72-75, PROF CHEM, OLD DOM UNIV, 67-, ASSOC DEAN GRAD STUDIES, 75- Mem: AAAS; Am Chem Soc. Res: Kinetics and mechanisms of organic reactions; analysis of chemical composition of jellyfish. Mailing Add: Dept of Chem Old Dom Univ Norfolk VA 23508

BELL, CHARLES LEIGHTON, physical chemistry, see 12th edition

BELL, CHARLES W, b Cheyenne, Wyo, Sept 30, 31; m 51; c 6. PLANT PHYSIOLOGY. Educ: Univ Wyo, BA, 53; Univ Wash, MS, 57; Wash State Univ, PhD(bot), 62. Prof Exp: PROF NATURAL SCI & BIOL, SAN JOSE STATE UNIV, 62- Concurrent Pos: Partic, Biol Sci Curric Study Inst Prep Col Teachers, Univ Colo, 63. Mem: AAAS; Am Inst Biol Sci. Res: Plant mineral nutrition; computer modeling and simulation of biological systems. Mailing Add: Dept of Natural Sci San Jose State Univ San Jose CA 95192

BELL, CHRISTOPHER KEITH, economic geology, see 12th edition

BELL, CLARA G, b Roumania, Oct 16, 37; US citizen; m 64; c 3. IMMUNOLOGY, IMMUNOGENETICS. Educ: Hebrew Univ Jerusalem, BSc, 59, MSc, 60, PhD(immunol), 64. Prof Exp: Fel immunol, Univ Sydney, 64, res fel immunochem, 66-68; USPHS res assoc immunol, 69-70, asst prof, 70-73, ASSOC PROF IMMUNOL, UNIV ILL, CHICAGO, 73- Concurrent Pos: On leave, NIH career develop award & assoc prof, Karolinska Inst Tumor Biol, Stockholm, Sweden, 75- Honors & Awards: Boris Pragel Gold Medal, NY Acad Sci, 72. Mem: Am Asn Immunologists; AAAS; Am Med Writers Asn; NY Acad Sci. Res: Functional analysis of lymphocyte sub-populations in terms of membrane markers and specificities, DNA, immunoglobulin, antibody synthesis and cellular responsiveness in immunity and surveillance. Mailing Add: 4951 N Kimball Ave Chicago IL 60625

BELL, CLYDE RITCHIE, b Cincinnati, Ohio, Apr 10, 21; m 43. BOTANY. Educ: Univ NC, AB, 47, MA, 49; Univ Calif, Berkeley, PhD(bot), 53. Prof Exp: Instr bot, Univ Ill, 53-55; from asst prof to assoc prof, 55-66, PROF BOT, UNIV NC, CHAPEL HILL, 66- Concurrent Pos: Dir, NC Bot Garden, 66- Mem: Bot Soc Am (treas, 72-76); Am Inst Biol Sci; Am Soc Plant Taxonomists (secy, 59-62); Int Asn Plant Taxonomists; fel AAAS. Res: Systematic botany; plant evolution; pollination biology; flora of the Southeastern United States. Mailing Add: Dept of Bot Coker Hall Univ of NC Chapel Hill NC 27514

BELL, CURTIS PORTER, b Augusta, Ga, July 26, 34; m 61; c 1. MATHEMATICS. Educ: Wofford Col, BS, 55; Univ Ga, MA, 60, PhD(math), 63. Prof Exp: Instr math & physics, Limestone Col, 55-56; asst math, Univ Ga, 62-63; asst prof, 63-67, ASSOC PROF MATH, WOFFORD COL, 67- Mem: Am Math Soc; Math Asn Am. Res: Point set topology. Mailing Add: Dept of Math Wofford Col Spartanburg SC 29301

BELL, DAVID, mathematical analysis, see 12th edition

BELL, DOROTHY MAYS, b Denton, Tex, Dec 9, 09; m 26; c 2. SPEECH PATHOLOGY. Educ: Univ Tex, BS, 32; Univ Denver, MA, 54, PhD(speech path, spec educ), 58. Prof Exp: DIR SPEECH & HEARING CLIN, TEX CHRISTIAN UNIV, 51-, HEAD DIV SPEECH & HEARING PATH, 66- Concurrent Pos: Res grants, Ohio Univ, Univ Colo, Univ Denver, Univ Wichita & Cent Mo State Col; consult, Ft Worth Soc Crippled Children & Adults, Inc, 65- & Ft Worth Child Study Clin, 66-; educ consult div, Mgt Serv Assocs. Mem: Am Speech & Hearing Asn; Coun Except Children; Am Educ Res Asn. Res: Language development; articulation learning disabilities. Mailing Add: 2314 Fairmount Ave Ft Worth TX 76110

BELL, DURHAM KEITH, plant pathology, soil microbiology, see 12th edition

BELL, EDWIN LEWIS, II, b Danville, Pa, May 13, 26; m 50; c 3. HERPETOLOGY. Educ: Bucknell Univ, BS, 48; Pa State Univ, MS, 50; Univ Ill, PhD(zool), Univ Ill, 54. Prof Exp: Asst prof biol, Moravian Col, 52-54; from asst prof to assoc prof, 54-65, PROF BIOL & CHMN DEPT, ALBRIGHT COL, 65- Concurrent Pos: Dir biol sci curric study teachers inserv inst, NSF, 66-67. Mem: AAAS; Nat Asn Biol Teachers; Am Soc Ichthyol & Herpet; Herpetologists League; Asn Adv Health Prof. Res: Herpetology of Huntingdon County, Pennsylvania; study of lizard Sceloporus occidentalis; sexing of Drosophila larvae; use of Drosophila in unique situations in genetics lab. Mailing Add: Dept of Biol Albright Col Reading PA 19604

BELL, ELY EUGENE, physics, deceased

BELL, ERNEST PERCY, JR, organic chemistry, see 12th edition

BELL, EUGENE, b New York, NY, Oct 20, 18; m 38; c 2. DEVELOPMENTAL BIOLOGY. Educ: NY Univ, BA, 48; Univ RI, MSc, 51; Brown Univ, PhD(biol), 54. Prof Exp: Res assoc & USPHS fel, Brown Univ, 54-56; from asst prof to assoc prof, 56-67, PROF BIOL, MASS INST TECHNOL, 67- Concurrent Pos: Consult, Mass Gen Hosp, 60-; prog dir, Develop Biol Films, Educ Develop Ctr; mem develop biol panel, NSF, 63-64. Mem: AAAS; Am Soc Zool; Soc Develop Biol; Biophys Soc; Int Soc Develop Biol. Res: Developmental genetics; regulation of nucleic acid and protein synthesis in differentiating cells. Mailing Add: Dept of Biol Mass Inst Technol Cambridge MA 02139

BELL, FRANK F, b Savannah, Tenn, Nov 6, 15; m 40; c 3. AGRONOMY. Educ: Univ Tenn, BS, 37, MS, 49; Iowa State Univ, PhD(agron), 56. Prof Exp: From instr to assoc prof, 46-61, PROF AGRON, UNIV TENN, KNOXVILLE, 61- Concurrent Pos: Vis prof, Iowa State Univ, 61, NC State Univ, 67 & Univ Ky, 71; US rep, Int Soil Cong, Adelaide, Australia, 68; agron consult, US AID Prog, India, 70. Honors & Awards: Distinguished Teaching Award, Univ Tenn, 67. Mem: AAAS; Am Soc Agron; Soil Sci Soc Am; Int Soc Soil Sci. Res: Soil productivity; management research. Mailing Add: Dept of Plant & Soil Sci Univ of Tenn Knoxville TN 37916

BELL, FRANK HEATON, b Clarksville, Pa, July 21, 19; m 45; c 3. PLANT NEMATOLOGY. Educ: Waynesburg Col, BS, 40; Univ WVa, MS, 42; Ohio State Univ, PhD(plant path), 49. Prof Exp: Asst instr bot, Ohio State Univ, 46-49; agr res scientist, Camp Detrick, 49-52; res specialist plant path, Point IV, Bolivia, 52-56; rep, Rohm and Haas Co, Peru, 56-65 & Far East, 65-70, MEM STAFF, ROHM AND HAAS ESPANA SA, 70- Mem: Am Phytopath Soc; Am Inst Biol Sci; Am Ornith Union. Res: Plant disease control; insect control; weed control. Mailing Add: Rohm and Haas Espana SA Provenza 216 Barcelona 11 Spain

BELL, FRED E, b Knoxville, Tenn, Feb 12, 25; m 50; c 1. BIOCHEMISTRY. Educ: Emory Univ, AB, 50, PhD(biochem), 57. Prof Exp: Instr biochem, Sch Med, Univ Va, 56-57; USPHS fel, Harvard Univ & Mass Gen Hosp, 57-58; from instr to asst prof, 58-72, ASSOC PROF BIOCHEM, SCH MED, UNIV NC, CHAPEL HILL, 72- Mem: AAAS; Am Chem Soc; Brit Biochem Soc. Res: Protein biosynthesis; nucleic acids; enzymes. Mailing Add: Dept of Biochem Univ of NC Sch of Med Chapel Hill NC 27514

BELL, GEORGE IRVING, b Ill, Aug 4, 26; m 56; c 2. REACTOR PHYSICS, THEORETICAL BIOLOGY. Educ: Harvard Univ, AB, 47; Cornell Univ, PhD(physics), 51. Prof Exp: Asst, Cornell Univ, 47-49; mem staff, 51-69, ASSOC DIV LEADER THEORET PHYSICS DIV, LOS ALAMOS SCI LAB, 70-, GROUP LEADER THEORET BIOL & BIOPHYS, 74- Concurrent Pos: Vis lectr, Harvard Univ, 62-63; vis prof, Med Ctr, Univ Colo, 70- Honors & Awards: Cert of Merit, Am Nuclear Soc, 66. Mem: Fel AAAS; fel Am Phys Soc; fel Am Nuclear Soc; Biophys Soc. Res: Theoretical physics and immunology; nuclear reactor theory; mathematical models in biophysics. Mailing Add: T Div Los Alamos Sci Lab PO Box 1663 Los Alamos NM 87545

BELL, GORDON RUSSELL, b Vancouver, BC, Mar 27, 24; m 48; c 4. MICROBIOLOGY. Educ: Univ BC, BSA, 46; Univ Iowa, MSc, 54; Univ Western Ont, PhD(bact), 55. Prof Exp: Asst immunol & bact, Univ BC, 46-47; instr, Univ Iowa, 51-52; asst scientist res inst, Can Dept Agr, Ont, 52-57; from assoc scientist to scientist biol sta, Fisheries Res Bd Can, 57-75, HEAD FISH HEALTH PROG, PAC BIOL STA, FISHERIES & MARINE SERV, 75- Concurrent Pos: Co-chmn, Can Comt Fish Dis, 70- Mem: Can Soc Microbiol; fel Brit Inst Biol. Res: Metabolism of sulfur oxidizing bacteria; degradation of phenoxyacetic acid herbicides by soil bacteria; tissue transaminases of salmon; physiology of salmon; fish diseases; microscopy and aquatic microbiology; cellular defense mechanisms in salmon. Mailing Add: Biol Sta Fisheries & Marine Serv Dept of Environ Nanaimo BC Can

BELL, GRAYDON DEE, b Paducah, Ky, May 5, 23; m 54; c 3. PHYSICS, ASTROPHYSICS. Educ: Univ Ky, BS, 49; Calif Inst Technol, MS, 51, PhD(physics), 57. Prof Exp: Asst prof physics, Robert Col, Istanbul, 51-54; res asst, Calif Inst Technol, 54-56, res fel, 56-57; from asst prof to assoc prof, 57-65, PROF PHYSICS, HARVEY MUDD COL, 65-, CHMN DEPT, 71- Concurrent Pos: NSF res grants, 60-70, fac fel, Nat Res Coun Can, 70-71; physicist, Nat Bur Standards, 63-64. Mem: AAAS; Am Phys Soc; Am Astron Soc; Am Asn Physics Teachers. Res: Experimental measurement of f-values for spectral lines of heavy elements. Mailing Add: Dept of Physics Harvey Mudd Col Claremont CA 91711

BELL, HAROLD, b New York, NY, June 25, 32; m 56; c 4. MATHEMATICS. Educ: Univ Miami, BS, 58, MS, 59; Tulane Univ, PhD(math), 64. Prof Exp: From asst prof to assoc prof, 64-74, PROF MATH, UNIV CINCINNATI, 74- Mailing Add: Dept of Math Univ of Cincinnati Cincinnati OH 45221

BELL, HAROLD E, b Taber, Alta, Aug 7, 26; m 51; c 3. CLINICAL PATHOLOGY. Educ: Univ Alta, BSc, 47, MD, 49. Prof Exp: Lectr clin path, Univ Alta, 59-60, asst dir lab serv, Univ Hosp, 60; res asst lab med, Univ Minn, 61-62; asst dir lab serv, Univ Hosp Fac Med, 62-66, assoc prof, 66-71, PROF CLIN PATH, UNIV ALTA, 71-, CHMN DEPT LAB MED, HOSP, 73- Mem: Col Am Path; Can Asn Med Biochem; Am Soc Clin Path; Can Med Asn; Can Asn Path. Res: Clinical biochemistry; separation of isozymes of lactic dehydrogenase; structure of immune globulins. Mailing Add: Dept of Clin Path Univ of Alta Edmonton AB Can

BELL, HAROLD MORTON, b Monticello, Ky, May 7, 40; m 60); c 2. ORGANIC CHEMISTRY. Educ: Eastern Ky Univ, BBS, 60; Purdue Univ, PhD(chem), 64. Prof Exp: Asst prof, 66-75, ASSOC PROF CHEM, VA POLYTECH INST & STATE UNIV, 75- Mem: Am Chem Soc. Res: Organophosphorus chemistry; metal hydride reductions; nuclear magnetic resonance; computer-assisted instruction. Mailing Add: Dept of Chem Va Polytech Inst & State Univ Blacksburg VA 24061

BELL, HARRY K, horticulture, plant physiology, see 12th edition

BELL, HOWARD E, b Albany, NY, Jan 20, 37; m 61; c 2. MATHEMATICS. Educ: Union Univ, NY, BS, 58; Univ Wis, MS, 59, PhD(math), 61. Prof Exp: Asst prof math, Union Univ, NY, 61-63 & State Univ NY Binghamton, 63-66; assoc prof, Union Univ, NY, 66-67; ASSOC PROF MATH, BROCK UNIV, 67-, CHMN DEPT, 75- Mem: Am Math Soc; Math Asn Am; Can Math Cong. Res: Algebra, especially rings and near-rings. Mailing Add: Dept of Math Brock Univ St Catharines ON Can

BELL, IAN, b Chorley, Eng, Sept 29, 32; m 56; c 3. INDUSTRIAL ORGANIC CHEMISTRY. Educ: Univ Manchester, BSc, 53, PhD(org chem), 56. Prof Exp: Tech officer org chem, Imp Chem Indust, Ltd, Eng, 58-60; from res chemist to dir res, Nease Chem Co, Inc, 60-67; chemist, Winthrop Labs, Eng, 67-68; VPRES RES & DEVELOP, NEASE CHEM CO, INC, 68- Mem: Am Chem Soc; The Chem Soc. Res: Acetylenic chemistry; steroid structure determination; oxo process; aromatic and aliphatic intermediates; hydrotropes. Mailing Add: Nease Chem Co Inc PO Box 221 State College PA 16801

BELL, JAMES HENRY, b London, Eng, July 10, 17; m 41. ALGEBRA. Educ: Univ Western Ont, BA, 37; Univ Wis, AM, 38, PhD(math), 41. Prof Exp: Asst, Univ Wis, 39-41; sr res assoc, Nat Res Coun Can, 41-42, jr radio res engr, 42-45; asst prof math, Univ Man, 45-46; from asst prof to assoc prof, Mich State Col, 46-53; tech dir mace proj & dir guid & navig, AC Spark Plug Div, 58-63, dir reliability, 63-67, dir reliability, AC Electronics Div, 67-70, dir customer serv, Delco Electronics Div, 70-73, MGR ENG MILWAUKEE OPERS, DELCO ELECTRONICS DIV, GEN MOTORS CORP, 73- Mem: Am Math Soc; Math Asn Am. Res: Mathematics; matrix theory; radar, ultra-high frequency antennae and propagation; systems analysis; radar, missiles and inertial guidance. Mailing Add: 5620 S Kurtz Rd Hales Corner WI 53130

BELL, JAMES MILTON, b Portsmouth, Va, Nov 5, 21. PSYCHIATRY. Educ: Meharry Med Col, MD, 47; Am Bd Psychiat & Neurol, dipl & cert psychiat & child psychiat. Prof Exp: Intern, Harlem Hosp, NY, 47-48; from asst physician to clin dir, Lakin State Hosp, 48-51; fel psychiat, Menninger Sch Psychiat, 53-56; asst sect chief, Children's Unit, Topeka State Hosp, Kans, 56-58; clin asst prof, 59-68, CLIN ASSOC PROF PSYCHIAT, ALBANY MED COL, 68-; CLIN DIR & PSYCHIATRIST, BERKSHIRE FARM CTR & SERV FOR YOUTH, 59- Concurrent Pos: Resident psychiat, Winter Vet Admin Hosp, 53-56; civilian consult, Irvin Army Hosp, Kans, 57-58; fel child psychiat, Menninger Found, 57-58; chief consult psychiatrist, Albany Home for Children, 59-; asst to dispensary psychiatrist, Albany Med Ctr Clin, 60-; trainee consult, Albany Child Guid Ctr, 60-; mem instrnl staff, Frederick A Moran Inst Delinq & Crime, St Lawrence Univ, 65-70; mem & vchmn subcomt returning vet, NY State Post Vietnam Planning Comt, 68-; consult, Astor Home for Children, NY; mem, Gov State Wide Comt for 1970 White House Conf Children & Youth; mem bd dirs & exec comt, Guild Farm, Mass; deleg, White House Conf Children & Youth; consult on adolescence & mem med adv bd, NY State Div for Youth. Mem: Fel AAAS; Group Advan Psychiat; dipl mem Pan-Am Med Asn; Insts Relig & Health; AMA. Res: Child psychiatry; delinquency. Mailing Add: Hudsonview Old Post Rd North-Croton-on-Hudson NY 10520

BELL, JAMES RICHARD, b San Francisco, Calif, Apr 24, 42; m 65; c 2. COMPUTER SCIENCE, RESEARCH ADMINISTRATION. Educ: Dartmouth Col, BA, 64; Stanford Univ, MS, 66, PhD(computer sci), 68. Prof Exp: Programmer, Bell Tel Labs, 60-63 & Kiewit Comput Ctr, Dartmouth Col, 64; mkt staff, IBM Corp, NY, 65; programmer, Control Data Corp, 66; instr, Stanford Univ, 68; res engr, Stanford Res Inst, 66-69; MGR RES & DEVELOP, DIGITAL EQUIP CORP, 69- Concurrent Pos: Mem fac comput sci, Northeastern Univ, 71-; ed adv bd, Computer Languages, Pergamon Press, 74- Mem: Asn Comput Mach, 63. Res: Computer programming languages; algorithms; computer design. Mailing Add: Digital Equip Corp Maynard MA 01754

BELL, JERRY ALAN, b Davenport, Iowa, June 28, 36; m 61; c 2. PHYSICAL CHEMISTRY. Educ: Harvard Univ, AB, 58, PhD(chem), 62. Prof Exp: Res assoc chem, Brandeis Univ, 61-62; asst prof, Univ Calif, Riverside, 62-67; assoc prof, 67-72, PROF CHEM, SIMMONS COL, 72- Mem: Am Inst Physics; Am Chem Soc; The Chem Soc. Res: Kinetics of fast reactions; methylene reactions; flash photolysis; photochemistry. Mailing Add: Dept of Chem Simmons Col Boston MA 02115

BELL, JIMMY TODD, b Hazlehurst, Ga, Dec 17, 38; m 61; c 2. PHYSICAL CHEMISTRY. Educ: Berry Col, AB, 60; Univ Miss, PhD(phys chem), 63. Prof Exp: Res chemist chem technol div, 63-74, GROUP LEADER, OAK RIDGE NAT LAB, 74- Mem: AAAS; Am Chem Soc; Am Nuclear Soc. Res: High temperature and pressure spectroscopy of transuranium element aqueous solutions; photochemistry of actinides; tritium equilibrias; tritium permeation. Mailing Add: Box X 4500 South Bldg Oak Ridge Nat Lab Oak Ridge TN 37830

BELL, JOHN BARR, JR, b Richmond, Va, June 17, 19; m 45; c 3. ORGANIC CHEMISTRY. Educ: Worcester Polytech Inst, BS, 41; NY Univ, MS, 44, PhD(chem), 47. Prof Exp: Res chemist, Nat Oil Prod Co, 43-45, Jefferson Chem Co, Inc, 46-53 & Mkt Res Div, 53-55; sr mkt analyst, Commercial Develop Div, 55-59, supvr mkt surv sect, 59-61, asst mgr mkt res, 61-63, mgr, 63-73, DIR PROJ DEVELOP, LATIN AM, AM CYANAMID CO, 73- Mem: Am Chem Soc; Chem Mkt Res Asn; Com Chem Develop Asn; Soc Chem Indust. Res: Chemical market research; market development. Mailing Add: Cyanamid Latin Am-Asia Div Berdan Ave Wayne NJ 07470

BELL, JOHN CLARENCE, b Chicago, Ill, Dec 25, 15; m 44; c 3. MATHEMATICS. Educ: Chicago Norm Col, dipl, 36; Univ Ill, BS, 37, MS, 38, PhD(math), 42. Prof Exp: Instr math, SDak State Col, 38-40; asst, Univ Ill, 41-42, instr, 42-44; res engr, 44-50, asst supvr, 50-57, consult math physics div, 57-63, res assoc exp physics div, 63-71, FEL, APPL MATH & MECH DIV, BATTELLE MEM INST, 71- Mem: Am Math Soc. Res: Differential geometry; stress analysis; heat conduction; fluid mechanics; ballistics; lubrication theory. Mailing Add: Battelle Mem Inst 505 King Ave Columbus OH 43201

BELL, JOHN FREDERICK, b Ashland, Ore, Jan 7, 24; m 50; c 4. FOREST MENSURATION, FOREST MANAGEMENT. Educ: Ore State Univ, BS, 49; Duke Univ, MF, 51; Univ Mich, PhD, 70. Prof Exp: Tech asst, Ore State Forestry Dept, 49-50, inventory forester, 51-53, unit forester, 53-55, supvr forest inventory, 55-59; from asst prof to assoc prof, 59-72, PROF FOREST MENSURATION, SCH FORESTRY, ORE STATE UNIV, 72- Mem: Soc Am Foresters. Res: Forest measurements and sampling. Mailing Add: Sch of Forestry Ore State Univ Corvallis OR 97331

BELL, JOHN FREDERICK, b Plato, Minn, Feb 18, 15; m 47; c 2. BACTERIOLOGY. Educ: Univ Minn, BA, 36, PhD(bact), 42; Wayne Univ, MD, 45; Am Bd Med Microbiol, dipl. Prof Exp: Bacteriologist, Minn State Dept Conserv, 34-39; res assoc, Sch Animal Path, Univ Pa, 39-40; bacteriologist, Minn State Dept Conserv, 40-41; instr & res assoc path, Wayne Univ, 41-43; intern, US Marine Hosp, Seattle, 45-46; commissioned officer, 46-50, sr surgeon, 52-57, MED DIR, ROCKY MOUNTAIN LAB, NIH, 57- Concurrent Pos: Guggenheim fel, 61 & 62; mem, Pan Am Health Orgn, Buenos Aires, 67-69. Mem: Am Acad Microbiol; Wildlife Dis Asn; Am Pub Health Asn; AMA. Res: Wildlife disease; equine encephalitis; botulism; rabies; parasitism. Mailing Add: Rocky Mountain Lab Nat Inst Health Hamilton MT 59840

BELL, JOHN MILTON, b Islay, Alta, Jan 16, 22; m 44; c 5. ANIMAL NUTRITION. Educ: Univ Alta, BScA, 43; McGill Univ, MSc, 45; Cornell Univ, PhD, 48. Prof Exp: From asst prof to assoc prof animal husb, 48-54, prof & head dept, 54-75, ASSOC DEAN RES COL AGR, UNIV SASK, 75- Concurrent Pos: Chmn comt animal nutrit, Can Dept Agr; chmn subcomt lab animal nutrit, Nat Res Coun-Nat Acad Sci; mem bd gov, Int Develop Res Ctr, 75- Honors & Awards: Borden Award, 62; First Agr Laureate, Can, 70; Order of Can, 72. Mem: Am Soc Animal Sci; Am Inst Nutrit; Can Soc Animal Prod (pres, 52); fel Agr Inst Can; Nutrit Soc Can. Res: Swine nutrition; dairy calf nutrition; toxic factors in rapeseed; environmental temperature effects on animals. Mailing Add: Col of Agr Room 223 John Mitchell Bldg Univ of Sask Saskatoon SK Can

BELL, JOHN PERKINS, b San Francisco, Calif, Jan 24, 40; m 65; c 2. BIOCHEMISTRY, BIOLOGY. Educ: Univ Wash, BS, 62; Univ Wis, Madison, PhD(org chem), 67. Prof Exp: Biochemist, Stanford Res Inst, 67-69; biochemist, Syntex Res Div, 69-75, BIOCHEMIST, SYNTEX RES, 75- Mem: AAAS; Am Chem Soc; Sigma Xi. Res: RNA and DNA polymerases; nucleotides as probes of biochemical mechanisms and as drugs; synthesis and biological properties of polynucleotides; development of new drugs and immunoassays; identification of drug metabolites. Mailing Add: Syntex Res 3401 Hillview Palo Alto CA 94304

BELL, JOHN SEBASTIAN, geology, see 12th edition

BELL, JOHN THOMAS, JR, b Columbus, Ga, Aug 30, 26; m 48; c 3. VETERINARY ANATOMY. Educ: Univ Ga, DVM, 52; Univ Minn, PhD, 56. Prof Exp: Instr vet histol & embryol, Univ Minn, 52-56; asst prof, Iowa State Univ, 56-59, assoc prof vet anat, Mich State Univ, 59-62; PROF ANAT & HISTOL & HEAD DEPT, SCH VET MED, UNIV GA, 62- Honors & Awards: Norden Lab Distinguished Teacher of Yr Award, 70. Mem: Am Vet Med Asn; Am Asn Vet Anat (pres, 69-70); Am Asn Anat, Conf Res Workers Animal Dis; Am Asn Vet Inst. Res: Comparative hematology; histology and diseases of marine animals; histochemistry of hemopoietic tissues. Mailing Add: Sch of Vet Med Univ of Ga Athens GA 30601

BELL, KATHERINE LAPSLEY, b Columbus, Ohio, May 9, 45. PLANT ECOLOGY. Educ: Agnes Scott Col, BA, 66; Wake Forest Univ, MA, 68; Univ Alta, PhD(bot), 74. Prof Exp: Fel plant ecol, Univ Alta, 73-75; ASST PROF BIOL, UNIV NEV, LAS VEGAS, 75- Mem: Ecol Soc Am; Sigma Xi. Res: Physiological ecology of arctic and alpine plants. Mailing Add: Dept of Biol Sci Univ of Nev Las Vegas NV 89154

BELL, LLOYD FRANKLIN, operations research, statistics, see 12th edition

BELL, MALCOLM RICE, b New Britain, Conn, Aug 22, 28; m 60; c 3. ORGANIC CHEMISTRY. Educ: Rensselaer Polytech Inst, BS, 50; Cornell Univ, PhD(chem), 55. Prof Exp: Develop chemist, Allied Chem & Dye Corp, 50-51; asst, Cornell Univ, 51-55; res assoc, 55-61, res chemist, group leader, 61-65, sect head, 65-68, ASSOC DIR CHEM, STERLING-WINTHROP RES INST, 68- Mem: Am Chem Soc. Res: Gliotoxin; synthetic analgesic agents; synthetic therapeutic agents. Mailing Add: Sterling-Winthrop Res Inst Rensselaer NY 12144

BELL, MARCUS ARTHUR MONEY, b Victoria, BC, Mar 1, 35; m 59; c 3. PLANT ECOLOGY, ETHNOBOTANY. Educ: Univ BC, BSF, 57, PhD(plant ecol), 64; Yale Univ, MF, 58. Prof Exp: Lectr plant & human ecology, Bobot, morphol & taxon, 61-63, instr, 63-64, asst prof bot, 64-70, ASSOC PROF BIOL, UNIV VICTORIA, 70-, CURATOR, HERBARIUM BOT COLLECTIONS, 63- Concurrent Pos: Nat Res Coun Can res grants, 65-, travel grant, Van Dusen fel & Ger Acad Exchange Serv grant, 65; Netherlands Educ Dept grant & Brit Coun grant, 66. Mem: AAAS; Ecol Soc Am; Can Bot Asn; Int Asn Plant Taxon; Int Asn Plant Ecol. Res: Forest ecology; classification and analysis of plant communities and environments; forest productivity, environmental information; ecological aspects of waste disposal; ethnobotany of British Columbia native groups. Mailing Add: Dept of Biol Univ of Victoria Victoria BC Can

BELL, MARION RANDOLPH, b Starkville, Miss, May 27, 40; m 59; c 2. INSECT PATHOLOGY. Educ: Miss State Univ, BS, 63, MS, 68, PhD(entom), 75. Prof Exp: Biol lab technician, 63-68, RES ENTOMOLOGIST, WESTERN COTTON RES LAB, AGR RES SERV, USDA, 68- Mem: Soc Invert Path; Entom Soc Am. Res: Improving the effectiveness of using insect pathogens as biological control agents of cotton pests; mass production of insect pathogens; study of pathogens affecting insect mass rearing facilities. Mailing Add: Western Cotton Res Lab USDA 4135 E Broadway Rd Phoenix AZ 85040

BELL, MARVIN CARL, b Centertown, Ky, Nov 24, 21; m 48; c 2. ANIMAL NUTRITION. Educ: Univ Ky, BS, 47, MS, 49; Okla Agr & Mech Col, PhD(animal nutrit), 52. Prof Exp: Res asst animal husb, Univ Ky, 47-48; asst, Okla Agr & Mech Col, 48-51; assoc prof animal husb & assoc animal husbandman, 51-65, from assoc scientist to scientist, AEC, 58-74, PROF ANIMAL SCI, UNIV TENN, KNOXVILLE, 65- Concurrent Pos: Mem staff travel lectr prog, Oak Ridge Inst Nuclear Studies, 59-; part-time loan to Oak Ridge Nat Lab Civil Defense Res, 65-68; Nat Feed Ingredients Asn fel, 69. Honors & Awards: Gamma Sigma Delta res award, 70. Mem: AAAS; Am Inst Nutrit; Am Soc Animal Sci; Nutrit Today Soc; Animal Nutrit Res Coun. Res: Digestion and metabolism studies with cattle and sheep; urea and stilbestrol feeding to ruminants; fescue poisoning investigations; feeding value of wood molasses; radioisotope tracers in nutrition studies with cattle, sheep, swine and rats; irradiation effects on metabolism in large animals; mineral metabolism. Mailing Add: Dept of Animal Sci Univ of Tenn Knoxville TN 37916

BELL, MARVIN DRAKE, b Tulsa, Okla, Jan 14, 29; m 50; c 4. SOLID STATE PHYSICS. Educ: Okla State Univ, BS, 55, MS, 56, PhD(physics), 64. Prof Exp: Asst chemist, Phillips Petrol Co, 52-54, res physicist, 56-58; res assoc physics, Okla State Univ, 64-65; assoc prof, Western Ky Univ, 65-67; PROF PHYSICS & CHMN DEPT, CENT MO STATE UNIV, 67- Mem: Am Phys Soc; Optical Soc Am; Am Asn Physics Teachers. Res: Electron spin resonance in semiconducting diamonds; photoconductivity; photovoltaic effect; electrical properties of diamonds; holography; coherent optics. Mailing Add: Dept of Physics Cent Mo State Univ Warrensburg MO 64093

BELL, MARY, b Providence, RI, May 21, 37. CELL BIOLOGY, HUMAN ANATOMY. Educ: Brown Univ, AB, 58; Yale Univ, PhD(anat), 65. Prof Exp: Asst scientist, Dept Cutaneous Biol & Electron Micros, Ore Regional Primate Res Ctr, 64-71, assoc scientist, 71-74; ASST PROF, DEPT ENVIRON HEALTH & DEPT ANAT, COL MED, UNIV CINCINNATI, 74- Concurrent Pos: Assoc prof, Div Dermat, Sch Med, Univ Ore, 70-74. Mem: Am Soc Cell Biol; Am Asn Anatomists. Res: Ultrastructure of skin; effects of polychlorinated biphenyl compounds in epithelia. Mailing Add: Dept Environ Health Col of Med Univ of Cincinnati Cincinnati OH 45267

BELL, MAURICE EVAN, b New Castle, Ohio, Sept 10, 10; m 43; c 1. GEOPHYSICS. Educ: Kenyon Col, ScB, 32; Mass Inst Technol, PhD(physics), 37. Prof Exp: Res engr, Westinghouse Elec & Mfg Co, 37-41; res assoc underwater sound, Harvard Univ, 41-42; res assoc opers res, Nat Defense Res Comt, Off Sci Res & Develop & US Navy, 42-47; sci liaison officer, Off Naval Res, London, 48, dep sci dir, 49-50, sci dir, 51-53; eng mgr, Sylvania Missile Systs Lab, Sylvania Elec Prod, Inc, 54, mgr solid state br, Physics Lab, 54-56; asst dean res, Col Earth & Mineral Sci, 56-72, PROF GEOPHYS & DIR EARTH & MINERAL SCI EXP STA, PA STATE UNIV, UNIVERSITY PARK, 56-, ASSOC DEAN RES, COL EARTH & MINERAL SCI, 72- Mem: AAAS; Am Phys Soc; Opers Res Soc. Res: Ionization probabilities; dielectric properties of solids; high vacuum lubricants; operations research; missile systems; semiconductor physics. Mailing Add: 118 Mineral Sci Bldg Pa State Univ University Park PA 16802

BELL, MAX EWART, b Milton, Iowa, Nov 19, 27; m 52; c 3. PLANT MORPHOLOGY. Educ: Parsons Col, BS, 50; Iowa State Col, MS, 52, PhD, 54. Prof Exp: Asst bot, Iowa State Col, 50-54; PROF BOT, NORTHEAST MO STATE UNIV, 54- Mem: Bot Soc Am. Res: Corn embryology and anatomy. Mailing Add: Dept of Bot Northeast Mo State Univ Kirksville MO 63501

BELL, NORMAN H, b Gainesville, Ga, Feb 11, 31; m 59, 68, 72; c 3. ENDOCRINOLOGY, METABOLISM. Educ: Emory Univ, AB, 51; Duke Univ, MD, 55. Prof Exp: From intern to asst resident med, Duke Univ Med Ctr, 55-57; clin assoc, Nat Inst Allergy & Infectious Dis, 57-59; staff investr endocrinol, Nat Heart Inst, 59-63; assoc med, Med Sch, Northwestern Univ, 63-65, asst prof, 65-68; assoc prof, 68-71, PROF MED PHARMACOL & DIR CLIN RES CTR, SCH MED, IND UNIV, INDIANAPOLIS, 71- Concurrent Pos: Attend physician, Res Vet Admin Hosp, 64-68; chief sect metab, Indianapolis Vet Admin Hosp, 68-74, assoc chief staff, 71- Mem: AAAS; Am Fedn Clin Res; Am Soc Clin Invest; Endocrine Soc; Soc Exp Biol & Med. Res: Calcium metabolism. Mailing Add: Vet Admin Hosp 1481 W Tenth St Indianapolis IN 46202

BELL, OLIVER E, JR, b Laurel, Miss, Nov 29, 36; m 57; c 2. BIOCHEMISTRY. Educ: Auburn Univ, BS, 59, MS, 62, PhD(biochem), 65. Prof Exp: Trainee biochem, Sansum Clin Res Found, Calif, 65-66; instr, 66-67, ASST PROF BIOCHEM, SCH MED, UNIV MISS, 67- Mem: AAAS. Res: Phospholipid metabolism in normal and neoplastic tissue; metabolism of analogs of natural phospholipids; lipid metabolism in brain tissue. Mailing Add: Dept of Biochem Univ of Miss Med Ctr Jackson MS 39216

BELL, PAUL HADLEY, b Cornerville, Ohio, Aug 3, 14; m 41; c 5. BIOCHEMISTRY. Educ: Marietta Col, AB, 36; Pa State Col, MS, 38, PhD(phys chem), 40. Hon Degrees: DSc, Marietta Col, 60. Prof Exp: Asst, Pa State Col, 37-40; res chemist, 40-56, head biochem res dept, 56-69, RES FEL BIOCHEM, LEDERLE LABS, AM CYANAMID CO, 69- Mem: Am Chem Soc; fel NY Acad Sci. Res: Mechanism of drug action; purification and chemistry of antibiotics and pituitary hormones; enzyme kinetics; metabolism of drugs; intermediate metabolism; steroid action; fibrinolytic enzymes; immunological control systems; transfer factor. Mailing Add: Lederle Labs Am Cyanamid Co Pearl River NY 10965

BELL, PERSA RAYMOND, b Ft Wayne, Ind, Apr 24, 13; m 41, 67; c 1. NUCLEAR MEDICINE. Educ: Howard Col, BSc, 36, DSc, 54. Prof Exp: Mem staff, Nat Defense Res Comt Proj, Univ Chicago, 40-41; mem staff radiation lab, Mass Inst Technol, 41-46; physicist, Oak Ridge Nat Lab, 46-67; chief lunar & earth sci div, NASA Manned Spacecraft Ctr, Houston, 67-70, CHIEF LUNAR & EARTH SCI DIV, NASA MANNED SPACECRAFT CTR MAN PROJ, OAK RIDGE NAT LAB, 70- Honors & Awards: Longstreth Medal, Franklin Inst. Mem: Am Phys Soc; Am Geophys Union; Soc Nuclear Med. Res: Electronic and physics instrumentation; nuclear physics; nuclear medical instrumentation; planetary research. Mailing Add: Bldg 9201-2 Oak Ridge Nat Lab Oak Ridge TN 37830

BELL, PETER M, b New York, NY, Jan 3, 34; m 59; c 4. GEOCHEMISTRY, GEOPHYSICS. Educ: St Lawrence Univ, BS, 56; Univ Cincinnati, MS, 59; Harvard Univ, PhD(geophys sci), 63. Prof Exp: Fel, 63-64, GEOPHYSICIST, CARNEGIE INST, 64- Mem: Geol Soc Am; fel Mineral Soc Am; fel Am Geophys Union. Res: High pressure-temperature experimental geochemistry and geophysics. Mailing Add: Carnegie Inst of Wash Geophys Lab 2801 Upton St NW Washington DC 20008

BELL, RAYMOND FRANK, b Carbondale, WVa, June 9, 11; m 42; c 1. APPLIED MATHEMATICS, MATHEMATICAL ANALYSIS. Educ: WVa Inst Technol, AB, 33; Univ Mich, MS, 38. Prof Exp: Teacher high sch, WVa, 33-39; chmn dept math, York Col, 39-40; instr, Eastern Wash Col Educ, 40-42; prof physics, NDak State Teachers Col, Valley City, 44-46; from asst prof to assoc prof math, Eastern Wash State Col, 47-54, chmn dept, 51-58, assoc prof, 60-67; assoc prof, 67-74, PROF MATH, WVA INST TECHNOL, 74- Concurrent Pos: Asst ground sch instr aerial navig, US Naval Dept, Pensacola, 42-44; fel math, Univ Mich, 45-46, lectr, 58-60; mem, Int Math Cong. Mem: Am Math Soc; Math Asn Am; Soc Indust & Appl Math. Res: Mathematical physics; linear differential equations; integral transforms; head conduction; partial differential equations. Mailing Add: PO Box 140 Glen Ferris WV 25090

BELL, RAYMOND MARTIN, b Weatherly, Pa, Mar 21, 07; m 42; c 3. PHYSICS. Educ: Dickinson Col, AB, 28; Syracuse Univ, AM, 30; Pa State Col, PhD(physics), 37. Prof Exp: Asst physics, Syracuse Univ, 28-30 & Pa State Col, 30-37; from instr to assoc prof, 37-46, prof & chmn dept, 46-75, EMER PROF PHYSICS, WASHINGTON & JEFFERSON COL, 75- Mem: AAAS; Am Phys Soc; Am Asn Physics Teachers. Res: Raman spectra; propagation of radio waves; Allison magnetooptic effect; dielectric measurements of porous materials; barium titanate. Mailing Add: 413 Burton Ave Washington PA 15301

BELL, REUBEN HAYS, organic chemistry, carbohydrate chemistry, see 12th edition

BELL, RICHARD DENNIS, b Prague, Okla, Apr 21, 37; m 61. PHYSIOLOGY. Educ: Univ Okla, BS, 60, MS, 65, PhD(physiol), 68; Okla State Univ, BS, 61. Prof Exp: RES ASST PROF UROL & ASST PROF PHYSIOL, BIOPHYS & ALLIED HEALTH EDUC, HEALTH SCI CTR, UNIV OKLA, 68- Concurrent Pos: USPHS res grant, 69-71. Mem: NY Acad Sci; Am Physiol Soc; Sigma Xi. Res: Renal physiology; body fluids and electrolytes; blood and lymph vascular systems of the kidney and their relationships to renal function. Mailing Add: Res 151F Vet Admin Hosp Oklahoma City OK 73104

BELL, RICHARD OMAN, b Havre, Mont, Feb 16, 33; m 57; c 4. SOLID STATE PHYSICS. Educ: Mont State Univ, BS, 55; Univ Calif, Los Angeles, MS, 58; Boston Univ, PhD(physics), 68. Prof Exp: Mem tech staff, Hughes Res Labs, Calif, 56-58; staff scientist, Raytheon Co, Mass, 60-62; sr scientist, Tyco Labs, 62-66; sr scientist, Borders Electronic Res, 66-67; group leader electronic mat, Tyco Labs, Waltham, 67-74, PRIN SCIENTIST, MOBIL TYCO SOLAR ENERGY CORP, 75- Honors & Awards: Distinguished Tech Achievement Award, Tyco Labs, 70. Mem: AAAS; Am Phys Soc; Inst Elec & Electronics Eng. Res: Electronic materials, especially semiconductors and for use in solar cells and nuclear radiation detectors, and crystal growth and evaluation of the physical and electronic properties of these materials. Mailing Add: 24 Austin Rd Sudbury MA 01776

BELL, ROBERT EDWARD, clinical pathology, hematology, deceased

BELL, ROBERT EDWARD, b Eng, Nov 29, 18; m 47; c 1. NUCLEAR PHYSICS. Educ: Univ BC, BA, 39, MA, 41; McGill Univ, PhD(physics), 48. Hon Degrees: DSc, Univ NB, 71, Univ Laval, 73; LLD, Univ Toronto, 73. Prof Exp: Physicist, Nat Res Coun Can, 41-45; res physicist, Atomic Energy Proj, Chalk River, Ont, 45-56; assoc prof physics, 56-60, dir, Foster Radiation Lab, 60-69, vdean phys sci, 64-67, dean grad studies & res, 69-70, RUTHERFORD PROF PHYSICS, McGILL UNIV, 60-, PRIN & VCHANCELLOR, 70- Mem: Fel Am Phys Soc; Can Asn Physicists (pres, 65-66); fel Royal Soc Can; fel Royal Soc London. Res: Beta and gamma ray spectroscopy; high speed coincidence techniques; proton induced nuclear reactions; positron annihilation and positronium; delayed proton radioactivity; teaching and administration. Mailing Add: McGill Univ PO Box 6070 Sta A Montreal PQ Can

BELL, ROBERT EUGENE, b Marion, Ohio, June 16, 14; m 38; c 2. ARCHAEOLOGY, PHYSICAL ANTHROPOLOGY. Educ: Univ NMex, BA, 40; Univ Chicago, MA, 43, PhD(anthrop), 47. Prof Exp: From asst prof to prof anthrop, 47-67, chmn dept, 47-55, GEORGE LYNN CROSS RES PROF ANTHROP, UNIV OKLA, 67-, HEAD CUR SOCIAL SCI DIV, STOVALL MUS, 47- Concurrent Pos: Wenner-Gren Found res grant, 55-56; ed elect, Soc Am Archaeol, 65-66, ed, Am Antiq, 66-70. Mem: AAAS; Soc Am Archaeol; fel Am Anthrop Asn; Am Asn Phys Anthrop. Res: Archaeology of Southern Plains; early man in the new world. Mailing Add: Dept of Anthrop Univ of Okla Norman OK 73069

BELL, ROBERT GALE, b Robinson, Ill, Oct 5, 37; m 66. BIOCHEMISTRY. Educ: Bradley Univ, AB, 59; St Louis Univ, PhD(biochem), 64. Prof Exp: NIH fel, Southern Ill Univ, 64-67; res assoc, St Louis Univ, 67-68; asst prof biochem, Univ RI, 68-70; asst prof chem, Univ Dayton, 70-71; asst prof biochem, 71-74, ASSOC PROF BIOCHEM, UNIV RI, 74- Concurrent Pos: Nat Heart & Lung Inst res grants, 72-78. Res: Vitamin K, anticoagulants, blood clotting. Mailing Add: Dept of Biochem Univ of RI Kingston RI 02881

BELL, ROBERT GRAHAM, b Aberdeen, Scotland, Mar 17, 43; Can citizen; m 68; c 3. AGRICULTURAL MICROBIOLOGY. Educ: Univ Durham, BSc, 65; Cambridge Univ, PhD(microbiol), 68. Prof Exp: RES SCIENTIST, POLLUTION CONTROL PROG, RES BR, CAN DEPT AGR, 73- Mem: Int Asn Water Pollution Res. Res: Treatment, disposal and utilization of agricultural wastes, particularly livestock wastes. Mailing Add: Soil Sci Sect Agr Res Sta Lethbridge AB Can

BELL, ROBERT JOHN, b Lewisburg, WVa, Dec 18, 34; m 57; c 2. SPECTROSCOPY, SOLID STATE PHYSICS. Educ: Va Polytech Inst, BS, 56, PhD(solid state physics), 63; Rice Inst, MA, 58. Prof Exp: Asst instr physics, Va Polytech Inst, 62-63; sr res mem, Southwest Res Inst, 63-65; from asst prof to assoc prof, 65-72, actg chmn dept, 70-71, PROF PHYSICS, UNIV MO-ROLLA, 72- Concurrent Pos: Asst prof, Trinity Univ, Tex, 64-65; vis scientist, Max Planck Inst Solid State Physics, Stuttgart, Ger, 71-72. Honors & Awards: Res Award, Univ Mo-Rolla, 71. Mem: Am Phys Soc. Res: Submillimeter specttroscopy studies of semiconductors, dielectrics and superconductors at room and cryogenic temperatures; surface physics and chemistry. Mailing Add: Dept of Physics Univ of Mo Rolla MO 65401

BELL, ROBERT LLOYD, b McKeesport, Pa, Sept 3, 23; m 51; c 2. MEDICINE, NEUROSURGERY. Educ: Washington & Jefferson Col, BS, 43; Univ Pittsburgh, MD, 47. Prof Exp: Asst instr neurosurg, State Univ NY Downstate Med Ctr, 54-55, from instr to assoc prof, 55-61; chief neurosurg, US Vet Admin Ctr, Wadsworth, Kans, 61-67; CHIEF NEUROSURG, VET ADMIN HOSP, 67- Mem: Asn Neurol Surg; Soc Nuclear Med; fel Am Col Surg; Cong Neurol Surg. Res: Nuclear instrumentation; brain tumor localization. Mailing Add: 51 S 12th Ave Coatesville PA 19320

BELL, ROBERT MAURICE, b Lincoln, Nebr, Mar 24, 44; m 66; c 2. BIOCHEMISTRY, MOLECULAR BIOLOGY. Educ: Univ Nebr, Lincoln, BS, 66; Univ Calif, Berkeley, PhD(biochem), 70. Prof Exp: NSF fel, Wash Univ, 70-72; ASST PROF BIOCHEM, MED CTR, DUKE UNIV, 72- Concurrent Pos: NSF grant, Med Ctr, Duke Univ, 73-74, NIH grant, 74-, Nat Found-March of Dimes grant, 74-; estab investr, Am Heart Asn, 74. Mem: Am Soc Microbiol. Res: Regulation of membrane phospholipid and triacylglycerol synthesis; enzymology of lipid synthesis. Mailing Add: Dept of Biochem Duke Univ Med Ctr Durham NC 27710

BELL, ROGER ALISTAIR, b Walton-on-Thames, Eng, Sept 16, 35; m 60; c 2. ASTROPHYSICS. Educ: Univ Melbourne, BSc, 57; Australian Nat Univ, PhD(astron), 62. Prof Exp: Lectr physics, Univ Adelaide, 62-63; asst prof, 63-69, ASSOC PROF ASTRON, UNIV MD, 69- Mem: Am Astron Soc; Royal Astron Soc; Int Astron Union. Res: Stellar atmospheres, particularly cepheid variables, white dwarfs and narrow band photometry. Mailing Add: Astron Prog Univ of Md College Park MD 20742

BELL, ROMA RAINES, b Clark, SDak, July 21, 44; m 71. NUTRITION. Educ: SDak State Univ, BS, 66; Univ Calif, PhD(nutrit), 70. Prof Exp: Res assoc nutrit, 70-74,

ASST PROF NUTRIT, UNIV ILL, URBANA-CHAMPAIGN, 74- Mem: AAAS; Am Chem Soc. Res: Nutritional surveys; mineral metabolism; calcium, phosphorus and selenium. Mailing Add: Human Resources & Family Studies Univ of Ill Urbana-Champaign Urbana IL 61801

BELL, RONDAL E, b Kennett, Mo, Dec 29, 33; m 54; c 2. CELL PHYSIOLOGY. Educ: William Jewell Col, BA, 55; Univ NMex, MS, 60; Univ Miss, PhD(immunol, path), 71. Prof Exp: Asst biol, Univ NMex, 57-60; PROF BIOL, MILLSAPS COL, 60- Concurrent Pos: Dir, NSF grants, Univ Colo, 64-66. Mem: AAAS; Am Soc Microbiol; Sigma Xi; NY Acad Sci. Res: Biochemical and immunological aspects of hormonally induced polyarteritis nodosa. Mailing Add: Dept of Biol Millsaps Col Jackson MS 39210

BELL, ROSEMOND KAY, b Bessemer, Ala, Oct 20, 03; m 32; c 1. METALLURGICAL ANALYTICAL CHEMISTRY, METALLURGICAL CHEMISTRY. Educ: Cumberland Univ, BA, 25; George Peabody Col, BS, 27; Vanderbilt Univ, MS, 29. Prof Exp: Teacher sci high sch, Tenn, 26-27; teacher, WVa, 27-28; asst instr chem, Vanderbilt Univ, 29-30; RES CHEMIST, INST MAT RES, NAT BUR STANDARDS, 30- Honors & Awards: Silver Medal, US Dept Com. Mem: AAAS; Am Chem Soc; Am Soc Test & Mat. Res: Metallurgical analyses; nonferrous metals. Mailing Add: Inst for Mat Res Nat Bur of Standards Washington DC 20234

BELL, ROSS TAYLOR, b Apr 23, 29; m 57. SYSTEMATIC ZOOLOGY. Educ: Univ Ill, BA, 49, MS, 50, PhD(entom), 53. Prof Exp: Instr, 55-58, from asst prof to assoc prof, 58-72, PROF ZOOL, UNIV VT, 73- Mem: Entom Soc Am; Soc Study Evolution. Res: Systematics and morphology of the carabid beetles. Mailing Add: Dept of Zool Univ of Vt Burlington VT 05401

BELL, RUSSELL A, b Christchurch, NZ, Feb 3, 35; m 62; c 2. ORGANIC CHEMISTRY. Educ: Univ Victoria, NZ, MSc, 58; Univ Wis, MS, 60; Stanford Univ, PhD(chem), 63. Prof Exp: Res assoc org chem, Univ Mich, 62-63; fel, Nat Res Coun Can, 63-64; from asst prof to assoc prof, 64-73, PROF ORG CHEM, McMASTER UNIV, 73- Mem: Am Chem Soc; The Chem Soc. Res: Total synthesis of organic natural products related to macrolide antibiotics; nuclear magnetic resonance; nuclear overhauser effect. Mailing Add: Dept of Chem McMaster Univ Hamilton ON Can

BELL, SAMUEL DENNIS, JR, b New York, NY, Nov 5, 20; m 44; c 4. MICROBIOLOGY. Educ: Princeton Univ, AB, 43; Columbia Univ, MD, 45; Harvard Univ, MPH, 53. Prof Exp: Res fel microbiol, 53-55, from asst prof to assoc prof, 55-71, asst physician, Univ Health Serv, 67-75, ASSOC PHYSICIAN, UNIV HEALTH SERV, HARVARD UNIV, 75-, CONSULT MICROBIOLOGIST, 67- Concurrent Pos: Sect ed, Biol Abstr. Mem: Soc Trop Med & Hyg. Res: Virus and rickettsial diseases; typhus; trachoma; adenoviruses. Mailing Add: Harvard Univ Health Serv Fourth Floor 75 Mt Auburn St Cambridge MA 02138

BELL, SANDRA LUCILLE, b Dupo, Ill, Dec 20, 35. BOTANY, CYTOGENETICS. Educ: Eastern Ill Univ, BS, 57; Univ Chicago, PhD(bot), 60. Prof Exp: Instr bot, 60-62, asst prof, 62-64, ASSOC PROF BOT, UNIV TENN, 64- Concurrent Pos: Consult biol div, Oak Ridge Nat Lab, 62-69; res assoc, Nat Inst Agron, Paris, France, 66-67. Mem: AAAS; Am Soc Plant Physiol; Am Soc Cell Biol. Res: Chromosome cytology of higher plants; chemical effects on chromosomes. Mailing Add: Dept of Botany Univ of Tenn Knoxville TN 37916

BELL, STANLEY C, b Philadelphia, Pa, June 23, 31; m 67; c 2. ORGANIC CHEMISTRY, MEDICINAL CHEMISTRY. Educ: Univ Pa, BA, 52; Temple Univ, MA, 54, PhD(phyrimidines), 59. Prof Exp: Res assoc medicinal chem, Merck Sharp & Dohme, 54-59; sr res chemist, 59-61, group leader, 61-68, MGR, MEDICINAL CHEM SECT, WYETH LABS DIV, AM HOME PROD CORP, 68- Mem: Am Chem Soc. Res: Synthesis of new organic compounds as potential medicinal products; investigation of novel chemical reactions. Mailing Add: Wyeth Labs Div Am Home Prod Corp King of Prussia Rd & Lancaster Ave Radnor PA 19087

BELL, STOUGHTON, b Waltham, Mass, Dec 20, 23; m 49; c 4. MATHEMATICS, COMPUTER SCIENCE. Educ: Univ Calif, AB, 50, MA, 53, PhD, 55. Prof Exp: Researcher math, Univ Calif, 52-55; staff mem math res dept, Sandia Corp, 55-64, supvr systs anal div, 64-66; assoc prof math, 67-71, PROF MATH & COMPUT SCI, UNIV NMEX, 71-, DIR COMPUT CTR, 67- Concurrent Pos: Vis lectr, Univ NMex, 56-66; nat lectr, Asn Comput Mach, Inc, 73-74. Mem: AAAS; Soc Indust & Appl Math; Am Statist Asn; Am Math Soc; Math Asn Am. Res: Rarefied gas dynamics; ordinary differential equations; operations research; weapons systems analysis; computing systems. Mailing Add: Comput Ctr Univ NMex 2701 Campus Blvd NE Albuquerque NM 87131

BELL, THADDEUS GIBSON, b Leominster, Mass, Feb 6, 23; m 50; c 2. ACOUSTICS. Educ: Yale Univ, BS, 45. Prof Exp: PHYSICIST ACOUST, US NAVAL UNDERWATER SOUND LAB, 47- Honors & Awards: Solberg Award, Am Soc Naval Engrs, 68. Mem: Fel Acoust Soc Am; Am Soc Naval Engrs. Res: Sonar design and analysis; underwater sound propagation; operations research; military oceanography. Mailing Add: Naval Underwater Systs Ctr New London CT 06320

BELL, THOMAS DONALD, b Hagerman, Idaho, Oct 24, 10; m 35; c 2. ANIMAL HUSBANDRY. Educ: Univ Idaho, BS, 32, MS, 36. Univ Wis, PhD(animal breeding), 39. Prof Exp: Asst animal husbandman, NMex Col, 39-42; animal husbandman substa 17, Exp Sta, Agr & Mech Col, Univ Tex, 42-43; prof animal husb & chmn br, Utah State Agr Col, 43-50; in charge sheep invests, Kans State Col, 50-57; prof animal husb & head dept, 57-70, PROF ANIMAL INDUSTS & HEAD DEPT, UNIV IDAHO, 70- Mem: Fel Am Genetic Asn; fel Am Soc Animal Sci (vpres, 63). Res: Sheep feeding; breeding and genetics. Mailing Add: Dept of Animal Indust Univ of Idaho Moscow ID 83843

BELL, THOMAS NORMAN, b Runcorn, Eng, Apr 18, 32; m 58. PHYSICAL CHEMISTRY. Educ: Univ Durham, BSc, 53, PhD(kinetics), 56. Prof Exp: Fel, Cambridge Univ, 56-57; fel, Univ Manchester, 57-58; sr lectr chem, Univ Adelaide, 58-67; assoc prof, 67-75, chmn dept, 70-75, PROF CHEM, SIMON FRASER UNIV, 75- Concurrent Pos: Nat Res Coun Can fel, 63-64. Mem: The Chem Soc; fel Chem Inst Can. Res: Free radical reactions in gas phase; reaction mechanisms. Mailing Add: Dept of Chem Simon Fraser Univ Vancouver BC Can

BELL, VERNON LEE, JR, b Omaha, Nebr, June 2, 27; m 49; c 2. POLYMER CHEMISTRY, ORGANIC CHEMISTRY. Educ: Doane Col, AB, 50; Ala Polytech Inst, MS, 53; Univ Nebr, Lincoln, PhD(chem), 58. Prof Exp: Water anal chemist, US Geol Surv, 51; fiber res chemist, Chemstrand Corp, Ala, 53-55; water anal chemist, US Geol Surv, 55-56; res chemist, Film Dept, Exp Sta, E I du Pont de Nemours & Co, Inc, Del, 58-61 & Mylar Res Lab, Ohio, 62-63; polymer chemist, 63-68, head polymeric mat sect, 68-70, head nonmetallic mat br, 70-74, STAFF CHEMIST, LANGLEY RES CTR, NASA, 74- Concurrent Pos: Liaison mem comt aerospace struct adhesives, Nat Mat Adv Bd, 72-74. Mem: Am Chem Soc. Res: Aromatic organophosphorus chemistry; benzacridines; Zeigler catalysis; polyhydrocarbons; polytrienes and Diels-Alder adducts; aromatic polyesters, synthesis and radiation effects; synthesis and properties of thermostable aromatic-heterocyclic polymers; polyimides; polyimidazopyrrolones; adhesives; composites. Mailing Add: 27 York Point Dr Seaford VA 23696

BELL, WALLACE G, b Lander, Wyo, Oct 28, 22; m 49; c 2. GEOLOGY. Educ: Univ Wyo, BA, 48, PhD(geol), 55. Prof Exp: Sr geologist explor, Gulf Oil Corp, 55-57; asst prof geol, Univ Mo, 57-59; res geologist, Gulf Res & Develop Co, 59-67; supvr tech data syst, Mene Grande Oil Co, 67-71; dir geol div, Lucius Pitkin, Inc, 71-75; CONSULT GEOLOGIST, 75- Mem: Geol Soc Am; Am Asn Petrol Geol; Am Inst Mining, Metall & Petrol Engrs. Res: Data processing; structural geology; stratigraphy; sedimentation. Mailing Add: 591 Rambling Rd Grand Junction CO 81501

BELL, WARREN NAPIER, b Winnipeg, Man, May 8, 21; m 50; c 2. HEMATOLOGY. Educ: Univ Man, MD, LM, CC, 44; Univ Pa, MSc, 52, DSc, 55. Prof Exp: Assoc med, Univ Pa, 41-53, Damon Runyan scholar, 52-53, mem hemat staff, Hosp, 51-53; Storey Fund fel, Cambridge Univ, 53-54; ASSOC PROF MED & PROF CLIN LAB SCI, UNIV MISS & DIR LABS, UNIV HOSP, 54- Honors & Awards: Chown Prize, Univ Man, 44. Mem: AMA. Res: Mechanisms of coagulation and hemolysis. Mailing Add: 3928 Eastwood Dr Jackson MS 39211

BELL, WAYNE ELLIOT, physical chemistry, see 12th edition

BELL, WAYNE HARRELL, b Washington, DC, May 23, 45; m 67; c 2. MARINE MICROBIOLOGY. Educ: Univ Miami, BS, 67; Harvard Univ, AM, 69, PhD(biol), 76. Prof Exp: Assoc, 69-71, INSTR BIOL, MIDDLEBURY COL, 71- Concurrent Pos: Res asst, Marine Biol Lab, Woods Hole, Mass, 69-70. Mem: AAAS; Am Soc Limnol & Oceanog; Am Soc Microbiol. Res: Interactions between marine bacteria and phytoplankton; bacterial chemotaxis; ecological studies using continuous cultures. Mailing Add: Dept of Biol Middlebury Col Middlebury VT 05753

BELL, WILLIAM CHARLES, geology, see 12th edition

BELL, WILLIAM E, b Fairmont, WVa, July 31, 29; m 58; c 2. PEDIATRIC NEUROLOGY. Educ: Univ WVa, AB, 51, BS & MS, 53; Med Col Va, MD, 55; Am Bd Psychiat & Neurol, dipl neurol, 63; Am Bd Pediat, dipl, 69. Prof Exp: Instr neurol, Univ Iowa, 62-64, asst prof pediat & neurol, 64; asst prof pediat, Univ Tex, 64-65; asst from asst prof to assoc prof, 65-72, PROF PEDIAT & NEUROL, UNIV IOWA, 72- Mem: Fel Am Acad Neurol; fel Am Acad Pediat. Res: Infections and metabolic disorders of the nervous system in childhood. Mailing Add: Dept of Pediat Univ of Iowa Iowa City IA 52240

BELL, WILLIAM EARL, atomic physics, see 12th edition

BELL, WILLIAM HARRISON, b St Louis, Mo, Mar 28, 27; m 65; c 4. ORAL SURGERY. Educ: St Louis Univ, BS, 50, DDS, 54; Am Bd Oral Surg, dipl. Prof Exp: Intern oral surg, Metrop Hosp, New York, 54-55; resident, Univ Tex M D Anderson Hosp & Tumor Inst, Houston, 55-56; teaching fel, Univ Tex Dent Br Houston, 56-57, asst prof, Dent Sci Inst, 57-66, asst mem inst, 66-72; ASSOC PROF ORAL SURG, SOUTHWESTERN MED SCH, UNIV TEX HEALTH SCI CTR DALLAS, 72- Concurrent Pos: Consult, Vet Admin Hosp, Dallas, 61-; pvt pract, 61- Mem: Am Dent Asn; Am Soc Oral Surg; Am Equilibration Soc; Int Asn Dent Res. Res: Bone graft and bone graft substitutes; developmental and acquired jaw deformities; surgical-orthodontics and bone healing; bone physiology. Mailing Add: Dept of Oral Surg Southwestern Med Sch Dallas TX 75235

BELL, WILLIAM ROBERT, JR, b Greece, NY; m 66; c 2. INTERNAL MEDICINE, HEMATOLOGY. Educ: Univ Notre Dame, BS, 57; George Washington Univ, MS, 60, PhD(pharmacol), 61; Harvard Med Sch, MD, 63; Univ London, HRA, 68. Prof Exp: Consult, Lab Physiol, Nat Cancer Inst, 59-61; intern med, Johns Hopkins Hosp, 63-64, asst resident, 66-67; clin assoc, NIH, 64-66; vis scientist, Royal Postgrad Med Sch London, Hammersmith Hosp, 67-68; sr resident med, Harvard Med Ctr-Peter Bent Brigham Hosp, 68-69; asst prof, 70-74, ASSOC PROF MED & RADIOL, JOHNS HOPKINS UNIV, 74- Concurrent Pos: Admin officer, Nat Cancer Inst, 65-66, exec co-dir lymphoma task force, 64-66; consult, Twyford Labs, London, 67-68. Mem: Am Fedn Clin Res. Res: Blood coagulation; thrombotic disorders. Mailing Add: Dept of Hemat Johns Hopkins Hosp Baltimore MD 21205

BELL, WILLIS H, botany, see 12th edition

BELL, WILSON BRYAN, b Surry Co, Va, May 16, 13; m 40; c 2. VETERINARY SCIENCE. Educ: Va Polytech Inst, BS, 34, MS, 35, PhD(animal path), 52; Cornell Univ, DVM, 39. Prof Exp: Teaching fel biol, Va Polytech Inst, 34-35; asst bact, NY State Vet Col, Cornell Univ, 35-39; assoc vet sci, Col Agr & Exp Sta, Univ Calif, 38-46; assoc prof animal path, 46-47, prof, 47-73, assoc dir agr exp sta, 54-73, dean col agr, 62-68, DIR UNIV DEVELOP, VA POLYTECH INST & STATE UNIV, 68- Concurrent Pos: Consult, Univ Tenn, AEC, & Agr Res Lab, Oak Ridge; mem subcomt on animal response, Nat Acad Sci Comt on Supersonic Transport-Sonic Boom, 67-; vet med rev comt, NIH, 69- Mem: AAAS; Am Vet Med Asn. Res: Veterinary medicine and bacteriology; animal pathology. Mailing Add: Dept of Animal Sci Va Polytech Inst & State Univ Blacksburg VA 24060

BELLABARBA, DIEGO, b Rome, Italy, Aug 13, 35; Can citizen; m 63; c 3. ENDOCRINOLOGY. Educ: Univ Rome, MD, 59. Prof Exp: Asst physician, Univ Hosp, Rome, Italy, 60-64; clin res fel med, Mass Gen Hosp, Boston, 64-65; res assoc, Bronx Vet Admin Hosp, NY, 66-68; asst prof, 68-71, ASSOC PROF MED, MED SCH, UNIV SHERBROOKE, 71- Concurrent Pos: Int comt on exchange persons Fulbright scholar, 64-65; Med Res Coun Can scholar, 69-74; Que Health Res Coun res scholar, 75-78. Mem: AAAS; Am Fedn Clin Res; Endocrine Soc; Am Thyroid Asn; Can Soc Clin Invest. Res: Ontogeny of hormone receptors; metabolism of triiodothyronine in health and disease. Mailing Add: Fac of Med Univ Hosp Ctr Sherbrooke PQ Can

BELLAH, ROBERT GLENN, b Nocona, Tex, June 21, 27; m 57; c 4. ENVIRONMENTAL BIOLOGY. Educ: McPherson Col, BS, 54; Kans State Univ, MS, 57, PhD(bot), 69. Prof Exp: Teacher pub sch, Kans, 55; from instr to assoc prof, 57-74, PROF BIOL, BETHANY COL, KANS, 74- Mem: Ecol Soc Am; Am Inst Biol Sci. Res: Social behavior of vertebrates, especially domestic chickens; plant ecology; forest successions on flood plains of rivers. Mailing Add: Dept of Biol Bethany Col Lindsborg KS 67456

BELLAIRE, FRANK ROLLAND, b Grand Rapids, Mich, Dec 21, 09; m 37; c 4. METEOROLOGY. Educ: Univ Hawaii, AB, 36; Univ Mich, BS, 41, MA, 39, MS, 57. Prof Exp: Teacher pub sch, Mich, 39-41; res asst meteorol, Mass Inst Technol, 41-42; meteorologist, US Naval Electronics Lab, 46-51 & 52-54; res meteorologist & lectr, Univ Mich, 54-67; assoc prof earth sci, Millersville State Col, 67-75; RETIRED.

Mem: Am Meteorol Soc. Res: Micrometeorology; radio propagation. Mailing Add: 1111 Richmond Rd Lancaster PA 17603

BELLAK, LEOPOLD, bVienna, Austria, June 22, 16; nat US; m; c 2. PSYCHIATRY, PSYCHOANALYSIS. Educ: Hamerling Gym, Vienna, matura, 35; Boston Univ, MA, 39; NY Med Col, MD, 44. Prof Exp: Instr psychol, Cambridge Univ & Univ Boston Ctr Adult Educ, 39-42; instr, City Col New York, 42-44; from intern to med officer, St Elizabeths Hosp, DC, 44-46; instr psychol, George Washington Univ & USDA Grad Sch, 45-46; assoc psychiat, NY Med Col, 46-50, clin asst prof, 50-56; vis clin prof psychiat, 65-75, CLIN PROF PSYCHIAT, ALBERT EINSTEIN COL MED, NY UNIV, 75-, RES PROF PSYCHOL POSTDOCTORAL PROG PSYCHOTHER, 75-; VIS PROF PSYCHIAT & BEHAV SCI, SCH MED, GEORGE WASHINGTON UNIV, 75- Concurrent Pos: Dir dept psychiat, City Hosp, Elmhurst, 58-64; NIMH grant proj dir trouble shooting clin, Elmhurst Hosp; NIMH prin investr schizophrenia res proj; chief psychiat consult, Altro Health & Rehab Serv, Inc, 47-58. Mem: Fel Am Psychiat Asn; Am Psychoanal Asn; fel Am Psychol Asn; Soc Proj Tech & Personality Assessment (pres, Soc Proj Tech, 58); fel Royal Soc Med. Res: Projective techniques; manic-depressive psychosis; schizophrenia; ego function assessment; geriatric psychiatry. Mailing Add: 22 Rockwood Dr Larchmont NY 10538

BELLAMA, JON MICHAEL, inorganic chemistry, see 12th edition

BELLAMY, DAVID, b Rochester, NY, July 17, 26; m 58; c 4. PHYSICS. Educ: Yale Univ, BS, 49; Univ Chicago, MBA, 69. Prof Exp: Chief engr, Fenwal Labs, Inc, 50-60, biomed engr, Baxter Labs, Inc, 60-65; dir biomed eng, Travenol Labs, Inc, 65-75; VPRES TECH ADMIN, BAXTER LABS, INC, 75- Mem: Soc Cryobiol; NY Acad Sci; Asn Advan Med Instrumentation. Res: Blood processing equipment; hospital devices. Mailing Add: Baxter Labs Inc One Baxter Pkwy Deerfield IL 60015

BELLAMY, DAVID P, b Bristol, VA, Apr 25, 44; m 65; c 1. TOPOLOGY. Educ: King Col, AB, 64; Mich State Univ, MS, PhD(math), 68. Prof Exp: Asst prof, 68-73, ASSOC PROF MATH, UNIV DEL, 73- Mem: Am Math Soc; Math Asn Am. Res: Point set topology; continua theory. Mailing Add: Dept of Math Univ of Del Newark DE 19711

BELLAMY, RAYMOND EDWARD, b Worcester, Mass, Dec 8, 12; m 36; c 5. BIOLOGY. Educ: Univ Fla, BS, 35, MS, 36, PhD(biol), 47; Univ Mich, MA, 41. Prof Exp: Instr, Dodd Jr Col, La, 36-37; field biologist div malaria invest, State Dept Pub Health, Ga, 37-42; actg asst prof biol, Univ Fla, 47-48; sr asst scientist, Commun Dis Ctr, USPHS, Ga, 48-51, scientist dir, Bakersfield Field Sta, 51-67; res scientist Res Inst, 67-72, RES SCIENTIST RES STA, CAN DEPT AGR, 72- Mem: AAAS; Am Soc Trop Med & Hyg; Am Mosquito Control Asn; Entom Soc Can; Can Soc Zool. Res: Biology of mosquitoes; natural history of arthropod-borne virus encephalitides; biology and reproduction of mosquitoes. Mailing Add: Can Agr Res Sta 107 Sci Crescent Univ of Sask Campus Saskatoon SK Can

BELLAMY, WINTHROP DEXTER, b Philadelphia, Pa, Oct 7, 15; m 47; c 2. MICROBIOLOGY, BIOCHEMISTRY. Educ: Cornell Univ, BS, 38, PhD(biochem & microbiol), 45. Prof Exp: Teaching asst bact, Cornell Univ, 41-43, instr, 43-45; res assoc microbiol, Sterling Winthrop Chem Co, 45-48; RES BIOCHEMIST, RES & DEVELOP CTR, GEN ELEC CO, 49- Mem: Fel AAAS; fel NY Acad Sci; Am Soc Microbiol; fel Am Acad Microbiol; Am Chem Soc. Res: Microbial degradation of organic wastes; single cell protein production from cellulosic wastes; use of thermophilic microorganisms for waste conversion. Mailing Add: Corp Res & Develop Gen Elec Co Schenectady NY 12345

BELLANTI, JOSEPH A, b Buffalo, NY, Nov 21, 34; m 58; c 7. PEDIATRICS, MICROBIOLOGY. Educ: Univ Buffalo, MD, 58; Am Bd Pediat, cert, 64; Am Bd Allergy & Immunol, dipl, 74. Prof Exp: Intern, Millard Fillmore Hosp, Buffalo, 58-59; resident pediat, Children's Hosp Buffalo, 59-61; NIH spec trainee immunol, J Hillis Miller Health Ctr, Univ Fla, 61-62; res virologist, Dept Virus Dis, Walter Reed Army Inst Res, Washington, DC, 62-64; from asst prof to assoc prof, 63-70, PROF PEDIAT & MICROBIOL, SCH MED, GEORGETOWN UNIV, 70-, DIR CTR INTERDISCIPLINARY STUDIES IMMUNOL, 75- Concurrent Pos: Mead Johnson grant pediat res, 64; mem growth & develop comt, NIH, 70-75; mem med adv comt, Nat Kidney Found, 71-; chmn, Infectious Dis Comt, 72; ed-in-chief, Pediat Res, 74-; dir, Am Bd Allergy & Immunol, 75- Honors & Awards: William Peck Sci Res Award, 66; Sci Exhibit Award, Am Acad Clin Pathologists & Col Am Path, 66; E Mead Johnson Award, Am Acad Pediat, 70. Mem: AMA; fel Am Acad Pediat; fel Am Acad Allergy; fel Am Acad Allergists; fel Am Asn Clin Immunol & Allergy. Res: Immunologic aspects of facultatively slow virus infections; biochemical changes in human polymorphonuclear leukocytes during maturation. Mailing Add: Dept of Pediat Georgetown Univ Sch of Med Washington DC 20007

BELLEAU, BERNARD ROLAND, b Montreal, Que, Mar 15, 25; m 50; c 2. ORGANIC CHEMISTRY, BIOCHEMICAL PHARMACOLOGY. Educ: McGill Univ, PhD(biochem), 50. Prof Exp: Asst steroid chem, Sloan-Kettering Inst, 50-52; asst prof biochem, Laval Univ, 52-53; fel Case Inst Technol, 53-54; res chemist chemotheraphy, Reed & Carnick, 54-55; asst prof biochem, Laval Univ, 55-58; from assoc prof to prof, Univ Ottawa, 58-71; PROF CHEM, MCGILL UNIV, 71- Concurrent Pos: Consult, Bristol Labs, Inc, consult res dir, 66- Honors & Awards: Merck Sharpe & Dohme Award, 62; Parizean medal, 70. Mem: Am Chem Soc; Soc Biochem Can; Can Inst Chem; fel Royal Soc Can. Res: Alkaloid chemistry and drug metabolism; enzyme and drug mechanisms. Mailing Add: Dept of Chem McGill Univ PO Box 6070 Sta A Montreal PQ Can

BELLENOT, STEVEN F, b Glendale, Calif, Aug 4, 48; m 75. MATHEMATICAL ANALYSIS. Educ: Harvey Mudd Col, BS, 70; Claremont Grad Sch, PhD(math), 74. Prof Exp: ASST PROF MATH, FLA STATE UNIV, 74- Mem: Am Math Soc; Math Asn Am; AAAS. Res: Functional analysis; geometric and topological structure of locally convex topological vector spaces, especially Banach spaces and Schwartz spaces. Mailing Add: Dept of Math Fla State Univ Tallahassee FL 32306

BELLER, BARRY M, b New York, NY, Dec 12, 35; m 58; c 2. CARDIOLOGY, INTERNAL MEDICINE. Educ: Columbia Univ, AB, 56, MD, 60. Prof Exp: Intern, Univ Chicago Hosps, 60-61, from asst resident to sr resident internal med, 61-64, USPHS res trainee cardiol, 63-64; assoc med, Albert Einstein Col Med, 66-67, asst prof, 67-68; assoc prof med & head sect cardiovasc dis, Med Sch, Univ Tex, San Antonio, 68-73; ASSOC PROF MED, YESHIVA UNIV, 73- Concurrent Pos: Consult, USPHS Surv, 62; consult tech rev comt artificial heart prog, Nat Heart Inst-NIH, 69; mem coun clin cardiol, Am Heart Asn, 69, fel, 70. Mem: AAAS; Am Fedn Clin Res; Am Heart Asn; fel Am Col Cardiol. Res: Myocardial protein synthesis; hemodynamics; electrocardiography. Mailing Add: Dept of Med Yeshiva Univ Bronx NY 10461

BELLER, FRITZ K, b Munich, Ger, May 17, 24; US citizen; m 48; c 2. OBSTETRICS & GYNECOLOGY. Educ: Univ Berlin, Dr med, 48; Univ Giessen, MedSciD, 55. Prof Exp: Docent obstet & gynec, Univ Giessen, 52-56; docent, Univ Tübingen, 56-58, head physics, 58-61, prof, 61; vis assoc prof, Sch Med, NY Univ, 61-63, from assoc prof to prof, 63-72; PROF OBSTET & GYNEC & CHMN DEPT, UNIV MÜNSTER, 72- Mem: Fel Am Col Obstet & Gynec; Am Col Surg; Am Soc Exp Path; Soc Gynec Invest; fel Royal Soc Med. Res: Hematological problems in obstetrics; intravascular coagulation and proteolysis; gynecological oncology. Mailing Add: Von-Esmarch-Str 125 Münster 4400 West Germany

BELLER, MARTIN LEONARD, b New York, NY, Apr 30, 24; m 47; c 3. ORTHOPEDIC SURGERY. Educ: Columbia Univ, AB, 44, MD, 46; Am Bd Orthop Surg, dipl, 55. Prof Exp: Intern, Mt Sinai Hosp, New York, 46-47; resident orthop surg, Hosp Joint Dis, 49-52; ASST CHIEF, PHILADELPHIA GEN HOSP, 52-; ASSOC PROF ORTHOP SURG, SCH MED, UNIV PA, 72- Concurrent Pos: Attend orthop surgeon, Albert Einstein Med Ctr, 60-70, chmn dept orthop surg, Daroff Div, 70-; attend orthop surgeon, Hosp Univ Pa, 63- Mem: AAAS; fel Geront Soc; fel Am Geriat Soc; AMA; fel Am Col Surgeons. Res: Radioactive phosphorus and other isotopes in evaluating circulation of bone; x-ray densitometry of bone with aluminum step wedge; bone metabolism; osteoporosis; Paget's disease. Mailing Add: 1936 Spruce St Philadelphia PA 19103

BELLES, FRANK EDWARD, b Cleveland, Ohio, Feb 28, 23; m 46; c 3. FUEL SCIENCE. Educ: Western Reserve Univ, BS, 47; Case Inst Technol, MS, 52. Prof Exp: Aeronaut res scientist, NASA Lewis Res Ctr, 47-57, head gas dynamics sect, 57-64, head kinetics sect, 64-71, chief propulsion chem br, 71-72, dir aerospace safety res & data inst, 72-74; MGR LABS RES & DEVELOP, AM GAS ASN, 74- Mem: Am Chem Soc; Combustion Inst; Am Soc Gas Engrs; AAAS. Res: Flame quenching and fire extinguishing; high-temperature reactions; gaseous detonations; rocket propellant safety; combustion research and development. Mailing Add: 29308 Wolf Rd Bay Village OH 44140

BELLET, EUGENE MARSHALL, b Hollywood, Calif, July 3, 40; m 68; c 5. PESTICIDE CHEMISTRY. Educ: Univ Calif, Riverside, BS, 63, MS, 69, PhD(chem & entom), 71. Prof Exp: Staff physicist residue chem, Univ Calif, Riverside, 65-68, environ sci fel, 69-71; res fel, Univ Calif, Berkeley, 71-73; chemist, Off Pesticide Prog, Environ Protection Agency, 73-74; mgr prod regist div, Weslaco Tech Ctr, Ansul Co, 74-76; DIR RES & DEVELOP, KALO LAB, 76- Mem: Am Chem Soc; Entom Soc Am; Sigma Xi. Res: Synthesis, chemistry and toxicology of organophosphorus compounds and bicyclic phosphates; chemistry of organoarsenicals; synthesis and development of new pesticides. Mailing Add: 9233 Ward Pkwy Kansas City MO 64114

BELLET, RICHARD JOSEPH, b East Orange, NJ, Oct 26, 27; m 55; c 4. POLYMER CHEMISTRY. Educ: Princeton Univ, AB, 49; Columbia Univ, AM, 51, PhD(org chem), 56. Prof Exp: Chemist, Berkeley Chem Corp, 49-51; sr res scientist, Hooker Chem Co, 56-57; sr res chemist, Tex-US Chem Co, Texaco, Inc & US Rubber Co, 57-61, group leader, 61-63; sr res chemist, Plastics Div, 63-65, tech supvr, 65-66, mgr, 66-70, CORP RES COORDR, PLASTICS DIV, ALLIED CHEM CORP, 70- Mem: Am Chem Soc. Res: Fine organics; pharmaceuticals; ultraviolet stabilizers; butadiene chemistry; antioxidants; emulsion polymerization; synthetic rubber; polymer characterization, development and commercialization; impact and condensation polymers; polyamides; applications research. Mailing Add: 416 Morris Ave Mountain Lakes NJ 07046

BELLETIRE, JOHN LEWIS, b Chicago, Ill, Aug 23, 43. MEDICINAL CHEMISTRY. Educ: Univ Chicago, SB, 65; Northwestern Univ, PhD(org chem), 72. Prof Exp: RES SCIENTIST MED CHEM, PFIZER INC, 74- Mem: Am Chem Soc; The Chem Soc. Res: Medicinal chemistry of diabetes; organic synthesis of novel heterocycles; anthracyclinones; heterolytic fragmentation reactions; photochemistry. Mailing Add: Pfizer Inc Cent Res Eastern Point Rd Groton CT 06340

BELLHORN, MARGARET BURNS, b Sharon, Pa, Nov 26, 39. BIOCHEMISTRY. Educ: Allegheny Col, BS, 61; Yale Univ, MS, 62; Albert Einstein Col Med, PhD(biochem), 71. Prof Exp: From instr ophthal to instr biochem, 72-75, ASST PROF OPHTHAL, ALBERT EINSTEIN COL MED, 73-, ASST PROF BIOCHEM, 75- Mem: Am Chem Soc; Asn Res Vision Ophthal. Res: Biochemical and morphological analysis of spontaneous and induced retinal degenerations and transport of substances through the retinal vessels; secondary ion mass spectrometry as a tool for chemical analysis of biological systems. Mailing Add: Albert Einstein Col of Med Montefiore Hosp & Med Ctr Bronx NY 10467

BELLI, LUIGI BRUTUS, public health, microbiology, see 12th edition

BELLIN, ALBERT IRVING, applied mechanics, see 12th edition

BELLIN, JUDITH SCHRYVER, b Holland, Aug 6, 24; nat US; m 47; c 3. PHYSICAL CHEMISTRY, PHOTOBIOLOGY. Educ: Hunter Col, BS, 45; Polytech Inst Brooklyn, MS, 47, PhD, 57. Prof Exp: Assoc dept cell physiol, Rockefeller Inst Med Res, 56-58; NIH fel, 58-62, res asst prof chem, 62-69, assoc prof, 69-74, PROF BIOCHEM & DIV HEAD, POLYTECH INST BROOKLYN, 74- Mem: AAAS; Am Chem Soc; NY Acad Sci. Res: Photochemistry; application of computers to chemical instrumentation. Mailing Add: Dept of Chem 333 Jay St Polytech Inst NY Brooklyn NY 11201

BELLIN, STUART ARNOLD, biochemistry, see 12th edition

BELLINA, JOSEPH JAMES, JR, b Orange, NJ, Mar 27, 40; m 62; c 5. SURFACE PHYSICS. Educ: Univ Notre Dame, BS, 61, PhD(physics), 66. Prof Exp: Res assoc physics, Barus Surface Physics Lab, Brown Univ, 66-68; res scientist, McDonnell Douglas Corp, 68-74; vis asst prof, Southern Ill Univ-Edwardsville, 74-75; ASST PROF PHYSICS, ST MARY'S COL, 75- Mem: Am Phys Soc; Am Vacuum Soc. Res: Low energy electron diffraction; field emission microscopy; vacuum technology; auger spectroscopy; low energy electron diffraction; surface chemistry. Mailing Add: 402 N Esther South Bend IN 46617

BELLINA, RUSSELL FRANK, b Newark, NJ, June 24, 42; m 66; c 2. ORGANIC CHEMISTRY. Educ: Fairleigh Dickinson Univ, BS, 64; Seton Hall Univ, MS, 66, PhD(org chem), 68. Prof Exp: Res chemist, 68-73, res supvr synthesis chem, 73-74, LAB ADMINR, E I DU PONT DE NEMOURS & CO, INC, 74- Mem: Am Chem Soc. Res: Agricultural chemistry, synthesis of biologically active chemicals designed for use as insecticides, fungicides, bactericides, herbicides, and plant growth modifiers. Mailing Add: Du Pont Exp Sta Biochem Dept Bldg 324 Wilmington DE 19898

BELLINGER, LARRY LEE, b Oakland, Calif, May 12, 47; m 73; c 1. ANIMAL PHYSIOLOGY. Educ: Univ Calif, Davis, BS, 69, PhD(physiol), 74. Prof Exp: NIH FEL PHYSIOL, STATE UNIV NY BUFFALO, 74- Mem: Sigma Xi. Res: The study of neurophysiological and neuroendocrine parameters involved in food intake control. Mailing Add: Dept of Surg State Univ of NY Buffalo NY 14215

BELLINGER, PETER F, b New Haven, Conn, June 15, 21; m 53. ENTOMOLOGY.

Educ: Yale Univ, BA, 42, PhD, 52. Prof Exp: Lectr zool, Univ Col WI, 52-56; instr, Yale Univ, 56-58; from asst prof to assoc prof, 58-65, PROF BIOL, CALIF STATE UNIV, NORTHRIDGE, 65- Mem: Entom Soc Am; Lepidop Soc. Res: Collembola. Mailing Add: Dept Biol Calif State Univ 18111 Nordoff State Northridge CA 91324

BELLINO, FRANCIS LEONARD, b Elizabeth, NJ, Dec 7, 38; m 63; c 8. STEROID BIOCHEMISTRY, MOLECULAR BIOLOGY. Educ: St Francis Col, Pa, BA, 60; Univ Notre Dame, BS, 61; Univ Pittsburgh, MS, 65; State Univ NY Buffalo, PhD(biophys sci), 71. Prof Exp: Assoc engr, Westinghouse Elec Corp, 61-65; instr physics, St Francis Col, Pa, 65-67; ASSOC RES SCIENTIST, MED FOUND BUFFALO, 71- Mem: Inst Elec & Electronic Engrs; Biophys Soc; AAAS. Res: Enzymology and physiological regulation of estrogen biosynthesis in human placenta and female. Mailing Add: 835 Casey Rd East Amherst NY 14051

BELLINO, VITO VICTOR, b Elizabeth, NJ, May 31, 17; m 42; c 1. ORGANIC CHEMISTRY. Educ: Univ Notre Dame, BS, 39, PhD(chem), 51; Columbia Univ, MA, 48; Temple Univ, LLB, 57. Prof Exp: From chemist to chief chemist, Reilly Tar & Chem Corp, 39-48; lab supvr, Barrett Div, Allied Chem Corp, 50-58; mgr patent adv sect, Wyeth Labs, 58-70; chief patent counsel, 70-72, ASST GEN COUNSEL PATENTS & TRADEMARKS, AM HOME PROD CORP, 73- Mem: AAAS; Am Chem Soc. Res: Pharmaceuticals; organic synthesis; coal tar chemicals. Mailing Add: Legal Dept Am Home Prod Corp 685 Third Ave New York NY 10017

BELLIS, EDWARD DAVID, b Ridley Park, Pa, June 28, 27. ZOOLOGY. Educ: Pa State Univ, BS, 51, Univ Okla, MS, 53, Univ Minn, PhD(zool), 57. Prof Exp: Asst zool, Univ Okla, 51-53; teaching asst, Univ Minn, 53-57; instr, Univ Ga, 57; from asst prof to assoc prof, 58-71, PROF ZOOL, PA STATE UNIV, 71- Mem: AAAS; Am Soc Ichthyol & Herpet; Ecol Soc Am; Soc Study Evolution; Am Soc Nat. Res: Herpetology; speciation; general ecology. Mailing Add: Dept of Biol McAllister Bldg Pa State Univ University Park PA 16802

BELLIS, ERNEST ANTHONY, b Liverpool, Eng, Aug 31, 08; nat US; m 39; c 2. ORGANIC CHEMISTRY. Educ: Marquette Univ, MS, 42. Prof Exp: Prof chem, Wis State Col, 37-56; prof, 56-74, EMER PROF CHEM, UNIV WIS-MILWAUKEE, 74- Mem: AAAS; Am Chem Soc. Res: Science education in elementary and secondary schools. Mailing Add: Dept of Chem Univ of Wis 3203 N Downer Ave Milwaukee WI 53201

BELLIS, HAROLD E, b Middletown, Conn, Sept 18, 30; m 54; c 2. PHYSICAL CHEMISTRY, MATHEMATICS. Educ: Cent Conn State Col, BSc, 51; Univ Conn, PhD(chem), 57. Prof Exp: Res chemist metall, Electrochem Dept, 57-69, staff scientist, 69-76, STAFF SCIENTIST, INDUST CHEM DEPT, E I DU PONT DE NEMOURS & CO, INC, 76- Mem: Am Chem Soc; Am Soc Metals; Am Electroplaters Soc. Res: High temperature chemistry and metallurgy of molten metal systems; molecular spectroscpy; electroless plating. Mailing Add: Chestnut Run Labs Indust Chem Dept E I du Pont de Nemours & Co Wilmington DE 19898

BELLIS, VINCENT J, JR, b Penn Yan, NY, Jan 18, 38; m 58; c 2. AQUATIC ECOLOGY. Educ: NC State Col, BS, 60, MS, 63; Univ Western Ont, PhD(bot), 66. Prof Exp: ASSOC PROF BIOL, E CAROLINA UNIV, 66- Mem: AAAS; Phycol Soc Am; Am Inst Biol Sci; Brit Phycol Soc; Int Phycol Soc. Res: General estuarine and coastal ecology; effects of development on natural ecosystems of eastern North Carolina. Mailing Add: Dept of Biol E Carolina Univ Greenville NC 27834

BELLMAN, RICHARD ERNEST, b New York, NY, Aug 26, 20; m 63; c 2. MATHEMATICS. Educ: Brooklyn Col, BA, 41; Univ Wis, MA, 43; Princeton Univ, PhD(math), 46. Prof Exp: Instr electronics, Truax Field, Wis, 42-43; math, Princeton Univ, 43-44; math physicist, Radio & Sound Lab, USN, San Diego, 44-45 & Los Alamos, NMex, USA, 45-46; fine instr math, Princeton Univ, 46-47, res assoc & asst prof, 47-48; assoc prof, Stanford Univ, 48-52; mathematician, Rand Corp, 52-65; PROF MATH ELEC ENG & MED, UNIV SOUTHERN CALIF, 65-, PROF MED, SCH MED, 74- Concurrent Pos: Assoc, Ctr Study Dem Inst, 69- Mem: Am Math Soc. Res: Analytic number theory; linear and non-linear differential equations ordinary and partial; stochastic processes; dynamic programming; invariant imbedding; quasilinearization; mathematical biosciences. Mailing Add: Dept of Math Univ of Southern Calif Los Angeles CA 90007

BELLMER, ELIZABETH HENRY, b New York, Sept 30, 27. BOTANY, ZOOLOGY. Educ: Trinity Col (DC), AB, 59; Cath Univ Am, MS, 62, PhD(bot), 68. Prof Exp: Instr, 59-65, asst prof, 65-71, ASSOC PROF BIOL, TRINITY COL (DC), 71- Mem: AAAS; Am Genetic Asn; Am Inst Biol Sci. Res: Time of embryonic fusion of the malleus and incus of the guinea pig; distribution, variation and chromosome number in the Appalachian shale barren endemic Eriogonum Allenii Watson. Mailing Add: Dept of Biol Trinity Col Washington DC 20017

BELLO, CARMEN T, b Philadelphia, Pa, July 19, 16; m 44; c 2. INTERNAL MEDICINE, PHARMACOLOGY. Educ: Temple Univ, BS, 37, MD, 43, MS, 49. Prof Exp: Instr pharmacol & med, 50-53, assoc pharmacol, 53-56, assoc med, 53-58, from asst prof to assoc prof med & pharmacol, 58-64, clin prof med, 64-66, PROF PHARMACOL, HOSP & SCH MED, TEMPLE UNIV, 64-, PROF MED, 66- Mem: Am Fedn Clin Res; Am Therapeut Soc; Am Soc Pharmacol & Exp Therapeut. Res: Hypertension; evaluation of varying hemodynamic patterns of essential hypertension. Mailing Add: Temple Univ Hos & Sch of Med Philadelphia PA 19140

BELLO, JAKE, b Detroit, Mich, Feb 22, 28; m 57. PHYSICAL BIOCHEMISTRY. Educ: Wayne Univ, BS, 48, PhD(chem), 52. Prof Exp: Fel org chem, Purdue Univ, 52-53; fel protein chem, Calif Inst Technol, 53-56; chemist, Eastman Kodak Co, NY, 56-58; sr res assoc protein chem, Polytech Inst Brooklyn, 58-59; assoc scientist, 60-67, PRIN SCIENTIST, ROSWELL PARK MEM INST, 67-, DIR GRAD STUDIES CHEM, ROSWELL PARK DIV, GRAD SCH, STATE UNIV NY BUFFALO, 73- Concurrent Pos: Mem comt biophys training, NIH. Mem: AAAS; Am Chem Soc. Res: Protein chemistry and structure; mechanism of gelation of gelatin; polymers; halogenation; nitro compounds. Mailing Add: Roswell Park Mem Inst 666 Elm St Buffalo NY 14263

BELLO, LEONARD JOHN, b Philadelphia, Pa, Mar 5, 37; m 59; c 4. BIOCHEMISTRY. Educ: Temple Univ, BA, 58; Johns Hopkins Univ, PhD(biochem), 62. Prof Exp: NIH fel microbiol, Sch Med, 62-64, assoc, 65, ASST PROF MICROBIOL, SCH VET MED, UNIV PA, 65- Res: Biochemical virology; regulation of genetic expression; enzymology; nucleic acid metabolism. Mailing Add: Dept of Microbiol Univ of Pa Sch Vet Med Philadelphia PA 19104

BELLOLI, ROBERT CHARLES, b St Louis, Mo, Nov 3, 42; m 68; c 2. ORGANIC CHEMISTRY, ORGANOMETALLIC CHEMISTRY. Educ: St Louis Univ, BS in Chem, 64; Univ Calif, Berkeley, PhD(org chem), 68. Prof Exp: Teaching asst chem, Univ Calif, Berkeley, 64-65; asst prof, 68-72, ASSOC PROF CHEM, CALIF STATE UNIV, FULLERTON, 72- Concurrent Pos: Res Corp grant, 69-70; Am Chem Soc-Petrol Res Fund grant, 72-74. Mem: AAAS; Am Chem Soc; The Chem Soc. Res: Physical-organic chemistry; organometallic chemistry, especially tin and mercury. Mailing Add: Dept of Chem Calif State Univ Fullerton CA 92634

BELLONE, CLIFFORD JOHN, b San Francisco, Calif, Feb 9, 41; m 63; c 5. IMMUNOBIOLOGY. Educ: Univ Notre Dame, BS, 63, PhD(microbiol), 71. Prof Exp: Fel immunol, Univ Calif, San Francisco, 71-74; ASST PROF IMMUNOL, SCH MED, ST LOUIS UNIV, 74- Res: Use of synthetic antigens to study the nature of the immune recognition unit at both the T and B cell level. Mailing Add: Dept of Microbiol St Louis Univ Sch of Med St Louis MO 63104

BELLUCE, LAWRENCE P, b Chester, Pa, June 15, 32; m 55; c 2. MATHEMATICS. Educ: Univ Calif, Berkeley, BA, 58; Univ Calif, Los Angeles, MA, 61, PhD(math), 64. Prof Exp: Asst prof, math, Univ Calif, Riverside, 63-67; ASSOC PROF MATH, UNIV BC, 67- Res: Functional analysis; mathematical logic; theory of rings. Mailing Add: Dept of Math Univ of BC Vancouver BC Can

BELLVILLE, JOHN WELDON, b Wauseon, Ohio, Aug 7, 26; m 56; c 4. MEDICAL RESEARCH. Educ: Cornell Univ, AB, 48, MD, 52; Am Bd Anesthesiol, dipl. Prof Exp: Intern, Minneapolis Gen Hosp, 52-53; asst in surg, Med Col, Cornell Univ, 53-55; Bowen-Harlow Brooks scholar, 55; fel, Sloan-Kettering Inst Cancer Res, 55-56; res instr surg, Med Col, Cornell Univ, 56-59, asst prof anesthesiol in surg, 59; from assoc prof to prof anesthesia, Stanford Univ, 60-71; PROF ANESTHESIOL, SCH MED, UNIV CALIF, LOS ANGELES, 72- Concurrent Pos: Resident anesthesiol, New York Hosp, 53-55; clin asst anesthesiol serv, Mem Hosp & James Ewing Hosp, 55-59, asst attend, 59-60; asst clinician, Sloan-Kettering Inst Cancer Res, 56-58, asst div exp surg, 58-60, assoc, 60; consult, Arthur D Little Co, Inc, 70-74. Honors & Awards: Borden Res Award, 52. Mem: AAAS; Am Chem Soc; Am Soc Pharmacol & Exp Therapeut; fel Am Col Anesthesiol; Sigma Xi. Res: Anesthesiology; clinical pharmacology; control of respiration; instrumentation. Mailing Add: Dept of Anesthesiol Univ of Calif Sch of Med Los Angeles CA 90024

BELLWARD, GAIL DIANNE, b Brock, Sask, May 27, 39; div. BIOCHEMICAL PHARMACOLOGY. Educ: Univ BC, BSP, 60, MSP, 63, PhD(pharmacol), 66. Prof Exp: Instr med chem & pharmacol, 60-61, lectr pharmacol, 66-67, asst prof, 67-73, ASSOC PROF PHARMACOL, UNIV BC, 73- Concurrent Pos: Fel med, Sch Med, Emory Univ, 68-69. Mem: AAAS; NY Acad Sci; Can Asn Res Toxicol; Pharmacol Soc Can; Am Soc Pharmacol & Exp Therapeut. Res: Metabolic pharmacology; drug metabolism and interactions; toxicology; enzymology. Mailing Add: Fac of Pharm Univ of BC Vancouver BC Can

BELLY, ROBERT T, b Trenton, NJ, Feb 19, 45. MICROBIAL ECOLOGY. Educ: Rutgers Univ, AB, 66; Pa State Univ, PhD(microbiol), 70. Prof Exp: Res assoc microbiol, Ind Univ, 70-71 & Univ Wis, 71-73; SR RES CHEMIST MICROBIOL, EASTMAN KODAK CO, 73- Concurrent Pos: NIH fel, Univ Wis, 72-73. Mem: Am Soc Microbiol; Soc Indust Microbiol; AAAS; Sigma Xi. Res: Microbial degradation and biological pollution control. Mailing Add: Res Labs Bldg B2 Eastman Kodak Co Rochester NY 14650

BELMAN, SIDNEY, b New York, NY; m 51; c 2. BIOCHEMISTRY. Educ: City Col New York, BS, 48; Polytech Inst Brooklyn, MS, 52; NY Univ, PhD(biochem), 58. Prof Exp: Instr indust med, 58-61, asst prof environ med, 61-67, ASSOC PROF ENVIRON MED, NY UNIV, 67- Mem: AAAS; NY Acad Sci; Environ Mutagen Soc; Am Chem Soc; Am Asn Cancer Res. Res: Immunochemistry; cancer. Mailing Add: NY Univ Dept of Environ Med 550 First Ave New York NY 10016

BELMONT, ARTHUR DAVID, b Calgary, Alta, Jan 6, 20; nat US; m 53; c 2. METEOROLOGY. Educ: Univ Calif, AB, 42; Calif Inst Technol, MS, 43; McGill Univ, PhD(meteorol), 56. Prof Exp: Meteorologist, US Weather Bur, Ger, 47, officer in-chg, Am Annex, US Danish Weather Sta, Greenland, 47-48; meteorologist, US Air Force Weather Serv, Washington, DC, 49; res meteorologist, Univ Calif, Los Angeles, 49-52 & 53-54; res assoc, McGill Univ, 54-57; prin scientist, Gen Mills, Inc, 57-64; DIR METEOROL RES, CONTROL DATA CORP, 64- Concurrent Pos: Secy, Int Comn Meteorol Upper Atmosphere, 60-75, pres, 75-79. Mem: Am Meteorol Soc; Am Geophys Union; Royal Meteorol Soc. Res: Stratospheric meteorology, especially of polar regions; ozone; general circulation of upper atmosphere; variability of atmosphere; interpretation and applications of satellite data. Mailing Add: 3216 36th Ave NE Minneapolis MN 55418

BELMONT, EMANUEL, radiochemistry, see 12th edition

BELMONT, HERMAN S, b Philadelphia, Pa, Mar 13, 20; m 46; c 3. PSYCHIATRY, PSYCHOANALYSIS. Educ: Univ Pa, AB, 40, MD, 43. Prof Exp: Rockefeller fel psychiat, Pa Hosp Ment & Nerv Dis, Inst Pa Hosp, 44; instr sch med, Univ Pa, 47-48, assoc, 48-52; assoc prof & sr attend child psychiat, Hahnemann Med Col & Hosp, 52-60, clin prof, 60-63; dir child anal training, Inst Philadelphia Asn Psychoanal, 57-63; PROF MENT HEALTH SCI & DIR DIV CHILD PSYCHIAT, HAHNEMANN MED COL & HOSP, 63-, DEP CHMN, 72-; SUPVR & TRAINING ANALYST CHILD & ADULT PSYCHOANAL, INST PHILADELPHIA ASN PSYCHOANAL, 56- Concurrent Pos: Mem, Govt Comprehensive Ment Health Planning Adv Comt, 64-66; mem adv comt child psychiat planning, Comnr Ment Health, State of Pa, 64-; mem task force II, Joint Comn Ment Health Children, 67- Mem: Am Psychoanal Asn; Int Psychoanal Asn; Am Orthopsychiat Asn; Am Psychiat Asn; Am Acad Child Psychiat. Res: Undergraduate and graduate training; administration of a program in child psychiatry integrating individual psychodynamics and social applications; application of individual psychodynamic knowledge to preventive work with children. Mailing Add: 245 N Broad St Philadelphia PA 19107

BELMONTE, ALBERT ANTHONY, b Lynn, Mass, July 17, 44; m 67; c 2. PHARMACEUTICS. Educ: Northeastern Univ, BS, 67, MS, 69; Univ Conn, PhD(pharmaceut), 72. Prof Exp: Res asst pharm, Univ Conn, 70-72; ASST PROF PHARMACEUT SCH PHARM, AUBURN UNIV, 72- Concurrent Pos: Consult, Natchaug Hosp, 70-72; Auburn Univ Res Found fel, 73-74. Honors & Awards: Cert of Recognition, Am Pharmaceut Asn, 67; Award, McKesson & Robbins, 67; Lederle Pharm Fac Award, 76. Mem: AAAS; Am Pharmaceut Asn; Acad Pharmaceut Sci; Am Asn Cols Pharm. Res: Membrane phenomena and model systems; surface chemistry of carcinogens; electrical phenomena at interfaces; dosage form design and their parameters. Mailing Add: Sch of Pharm Auburn Univ Auburn AL 36830

BELMONTE, ROCCO GEORGE, b New York, NY, July 6, 15. BIOLOGY. Educ: Georgetown Univ, AB, 40; Woodstock Col, Md, PhL, 41, STL, 48; Fordham Univ, MS, 44; Cath Univ Am, PhD(biol), 54. Prof Exp: Instr biol, Loyola Col, Md, 41-43 & Fordham Univ, 49-51; asst prof, Canisius Col, 54-61 & Fordham Univ, 61-62; asst prof, 62-64, ASSOC PROF BIOL, ST PETER'S COL, NJ, 64- Mem: AAAS; Am Asn Jesuit Sci; Am Inst Biol Sci. Res: Effects of adrenal extracts on blood-forming organs in amphibians; effects of inhibiting agents on mitosis in Allium cepa. Mailing Add: Dept of Biol St Peter's Col Jersey City NJ 07306

BELNER, ROBERT JOSEPH, b Cleveland, Ohio, Feb 16, 26; m 47; c 2. PHYSICAL

CHEMISTRY, POLYMER CHEMISTRY. Educ: Case Inst Technol, BS, 46, MS, 48, PhD(phys chem), 55. Prof Exp: Assoc prof, Cleveland Col Mortuary Sci, 46-47; asst, Case Inst Technol, 46-47, res assoc, 47-55; sr res chemist, Cent Res, Gen Tire & Rubber Co, 55-59, develop engr, Chem Div, 59-60, tech mgr, New Prod Dept, 60-62, tech dir, 62-64, asst tech dir, Chem-Plastics Div, 64-65, prod mgr rigid plastics, 65-67, mgr, 67; vpres eng & develop, EMC Plastics Div, Amerace Corp, Ind, 67-69; vpres corp & dir, Arvey Corp, 69-70, PRES, LAMCOTE DIV, ARVEY CORP, 70-, PRES, METALS DIV & DISPLAY DIV, 73- Mem: AAAS; Am Chem Soc; Soc Plastics Eng; Am Mgt Asn. Res: High polymers; polymerization techniques; stereospecific polymer synthesis; heterogeneous catalysis; polymer rheology; plastic fabrication; film conversion; plated plastics. Mailing Add: Arvey Corp 3450 N Kimball Ave Chicago IL 60618

BELOHLAV, LEO RUDOLF, b Vienna, Austria, May 31, 25; US citizen; m 46; c 1. RESEARCH ADMINISTRATION. Educ: Univ Vienna, Dr phil(org chem), 53. Prof Exp: Res asst chem, Univ Vienna, 52-56; res assoc org chem, Purdue Univ, 56-58; dir res, Great Lakes Chem Co, 58-64; sr staff assoc govt res, 64-66, mgr tech eval, Celanese Res Co, 66-68, DIR TECH EVAL, CELANESE RES CO, CELANESE CORP, 68- Concurrent Pos: Res fel, Harvard Univ, 55-56; consult, Donau-Pharmazie, Austria, 51-55. Mem: Am Chem Soc. Res: Natural products; halogen compounds; high temperature polymers; research on research. Mailing Add: 12 Tanglewood Lane Berkeley Heights NJ 07922

BELON, ALBERT EDWARD, b Gap, Hautes Alpes, France, May 2, 30; US citizen; m 55; c 3. AERONOMY, REMOTE SENSING. Educ: Univ Alaska, BS, 52; Univ Calif, Los Angeles, MA, 54. Prof Exp: Physicist geophys, Int Geophys Inc, 53-54; res geophysicist, Univ Calif, Los Angeles, 54-56; instr, Geophys Inst, Univ Alaska, 56-61, assoc prof physics, 61-69; prof dir solar-terrestrial res, NSF, 68-70; PROF PHYSICS, GEOPHYS INST, UNIV ALASKA, 69- Concurrent Pos: Consult, Los Alamos Sci Lab, 67-; mem, Inter-Union Comn Solar-Terrestrial Physics, 68-70 & Comt Polar Res, Panel on Upper Atmospheric Physics, Nat Acad Sci, 73- Honors & Awards: Meritorious Serv Award, NSF, 70; Except Sci Achievement Medal, NASA, 74. Mem: Am Geophys Union; Int Union Geod & Geophys. Res: Spectrophotometry of upper atmospheric phenomena; auroral morphology and magnetospheric convection; applications of satellite and aircraft imagery to Alaskan resources surveys. Mailing Add: Geophys Inst Univ of Alaska Fairbanks AK 99701

BELOTE, THOMAS ALEXANDER, physics, see 12th edition

BELSER, WILLIAM LUTHER, JR, b Hershey, Pa, May 28, 25; m 49. GENETICS. Educ: Lafayette Col, BA, 48; Yale Univ, PhD(microbiol), 55. Prof Exp: USPHS fel marine genetics, Scripps Inst, 55-57, asst res biologist, 57-61, asst prof microbiol, 61-66, ASSOC PROF MICROBIOL, UNIV CALIF, RIVERSIDE, 66- Mem: Am Soc Microbiol; Brit Soc Gen Microbiol. Res: Regulation of metabolic flow in bacteria with emphasis on tryptophan and pyrimidine biosynthesis; comparative biochemistry in the Enterobacteriaceae; enzyme formation and function. Mailing Add: Div of Life Sci Univ of Calif Riverside CA 92502

BELSERENE, EMILIA PISANI, b New Rochelle, NY, Dec 12, 22; m 44; c 3, ASTRONOMY. Educ: Smith Col, AB, 43; Columbia Univ, AM, 47, PhD(astron), 52. Prof Exp: Asst, Lick Observ, 43-44; assoc in astron, Columbia Univ, 54-73; MEM FAC PHYSICS & ASTRON, HERBERT H LEHMAN COL, CITY UNIV NY, 73- Concurrent Pos: Lectr, Hunter Col, 52, 56-68; adj assoc prof, Lehman Col, 68- Mem: Am Astron Soc. Res: Photographic photometry; globular clusters. Mailing Add: Dept of Physics & Astron Herbert H Lehman Col City Univ NY Bronx NY 10468

BELSHAW, CYRIL SHIRLEY, b Waddington, NZ, Dec 3, 21; Can citizen; m 43; c 2. ANTHROPOLOGY. Educ: Univ NZ, BA, 43, MA, 45; Univ London, PhD(social anthrop), 49. Prof Exp: Dist officer, Brit Solomon Islands Protectorate, 43-46; sr res fel, Australian Nat Univ, 49-53; from asst to assoc prof anthrop, Univ BC, 53-62, dir, UN Training Centre, 62-63, head dept anthrop & sociol, 68-74, PROF ANTHROP, UNIV BC, 63- Concurrent Pos: Consult, S Pac Comn, 53; co-dir, Indian Res Proj, Dept Citizenship & Immigration, 54-56; US Social Sci fac res grant, Tri-Inst Pac Prog, Bishop Mus, Yale Univ & Univ Hawaii, 58-59; consult capital in peasant soc, UNESCO, 62, consult community develop, UN Bur Social Affairs, 63, mem, UN Econ & Social Coun Working Group Eval Texh Asst Thailand, 65, Guggenheim Mem fel, UN Res Inst Social Develop, Geneva, 65-66, consult contrib anthrop develop, UNESCO, 70, Can del, Gen Conf, 72 & 74; vpres, Social Sci Res Coun Can, 69-70, chmn comt scholarly rels, 70-72; coun mem Can, Int Union Anthrop & Ethnol Sci, 70-; Can Coun fel, Switz, 72-73; ed, Current Anthrop, 74- Mem: Int Social Sci Coun; Am Anthrop Asn; Can Sociol & Anthrop Asn; Soc Int Develop; Royal Soc Can. Res: Anthropology and public policy; economic anthropology; social and political organization; social science policy; international cooperation; social change Oceania, Europe and Asia; higher education. Mailing Add: Dept of Anthrop & Sociol Univ of BC Vancouver BC Can

BELSHE, JOHN FRANCIS, b Marshall, Mo, Feb 6, 35; m 57; c 2. ZOOLOGY, ECOLOGY. Educ: Cent Mo State Col, BS, 57; Univ Miami, MS, 61, PhD(zool), 66. Prof Exp: Instr biol, Miami-Dade Jr Col, 61-63, asst prof, 63-64; from asst prof to assoc prof, 64-74, PROF BIOL, CENT MO STATE COL, 74- Mem: AAAS; Am Inst Biol Sic; Am Fisheries Soc; Am Soc Zool. Res: Ichthyology; aquatic biology; population studies of Tribolium; environmental education. Mailing Add: Dept of Biol Cent Mo State Col Warrensburg MO 64093

BELSKY, MELVIN MYRON, b Brooklyn, NY, Apr 26, 26; m 52; c 1. BIOLOGY. Educ: Brooklyn Col, BS, 49, MA, 51; Univ Pa, PhD(bot), 55. Prof Exp: From asst prof to assoc prof, 62-68, PROF BIOL, BROOKLYN COL, 68- Mem: Am Soc Microbiol; Soc Plant Physiologists; Brit Soc Gen Microbiol; AAAS. Res: Intermediary metabolism; marine mycology; hormone action; enzymology; membrane transport. Mailing Add: Dept of Biol Brooklyn Col Brooklyn NY 11210

BELSKY, THEODORE, b New Brunswick, NJ, Dec 24, 30; m 52; c 1. PUBLIC HEALTH, ANALYTICAL CHEMISTRY. Educ: Univ Calif, Berkeley, BS, 60, PhD(biophys), 66. Prof Exp: Biophysicist, Space Sci Lab, Univ Calif, Berkeley, 63-66; asst res geochemist, Space Sci Lab, Univ Calif, Los Angeles, 66-68; asst chemist, Western Regional Lab, USDA, Calif, 69; asst pub health chemist, 69-74, PUB HEALTH CHEMIST, CALIF STATE DEPT HEALTH, 74- Mem: Am Pub Health Asn; AAAS; Am Chem Soc. Res: Organic microanalysis of air pollutants and occupational contaminants. Mailing Add: Calif State Dept of Health 2151 Berkeley Way Berkeley CA 94704

BELSON, HENRY S, b Boston, Mass, Apr 9, 25; m 56. PHYSICS. Educ: Univ NH, BS, 49; Rutgers Univ, Phd(physics), 55. Prof Exp: Prin scientist, Univac Div, Sperry Rand Corp, 55-63; res physicist, 63-70, CHIEF, MAGNETISM & METALL DIV, US NAVAL ORD LAB, 70- Concurrent Pos: Lectr, Univ Pa, 56-59, res assoc, 59-60. Mem: Am Phys Soc; Am Asn Physics Teachers; Fedn Am Sci. Res: Magnetic materials and measurements; ferromagnetic resonance; magnetostriction and magnetic anistropy measurements. Mailing Add: US Naval Ord Lab White Oak Silver Spring MD 20910

BELT, CHARLES BANKS, JR, b New York, NY, Dec 11, 31; m 57; c 5. ECONOMIC GEOLOGY. Educ: Williams Col, BS, 53; Columbia Univ, MA, 55, PhD(econ geol), 59. Prof Exp: Asst explor geologist, Mineracao Hannaco, Brazil, 58-59; geologist, Bear Creek Coord Unit, 59-60; res assoc eng exp sta, Univ Utah, 61; asst prof geol & geol eng inst technol, 61-66, ASSOC PROF GEOL, ST LOUIS UNIV, 66- Mem: Geol Soc Am; Inst Mining, Metall & Petrol Eng; Sigma Xi. Res: Origin of hydrothermal ore deposits; relation between igneous intrusions and ore deposition; atomic absorption spectrophotometry and its use in the analysis of rocks and minerals for trace and major chemical elements; environmental geology; fluvial constriction. Mailing Add: Dept of Earth & Atmospheric Sci St Louis Univ St Louis MO 63156

BELT, EDWARD SCUDDER, b New York, NY, Aug 4, 33; m 61; c 4. SEDIMENTOLOGY. Educ: Williams Col, BA, 55; Harvard Univ, AM, 57; Yale Univ, MA, 59, PhD(geol), 63. Prof Exp: Asst prof geol, Villanova Univ, 62-66; asst prof, 66-70, ASSOC PROF GEOL, AMHERST COL, 70-, CHMN DEPT, 71- Concurrent Pos: NSF grant, 64-67; Am Philos Soc & Nat Geog Soc grant, 72-73. Mem: Am Inst Mining, Metall & Petrol Eng; Am Asn Petrol Geol; Soc Econ Paleont & Mineral; Geol Soc Am; Int Asn Sedimentology. Res: Stratigraphic relationships, facies and depositional environments and regional tectonics of Carboniferous strata in eastern Canada and Scotland; sedimentology of various paleoenvironments; coral zonation and reef development in modern Caribbean reefs. Mailing Add: Dept of Geol Amherst Col Amherst MA 01002

BELT, GEORGE HARLEY, JR, b Washington, DC, Jan 28, 38; m 60; c 2. MICROMETEOROLOGY, FOREST HYDROLOGY. Educ: NC State Col, BS, 60; Yale Univ, MF, 62; Duke Univ, DF, 68. Prof Exp: Asst prof forestry, 65-68, proj leader, Water Resources Res Inst, 65-75, ASSOC PROF FORESTRY, UNIV IDAHO, 68- Concurrent Pos: Coop proj leader, US Forest Serv, USDA, Idaho, 66- Mem: Soc Am Foresters; Am Meteorol Soc. Res: Forest meteorology, particularly predictive modeling of physical and biological processes influenced by land management. Mailing Add: Col of Forestry Univ of Idaho Moscow ID 83843

BELT, ROGER FRANCIS, b Springfield, Ohio, Mar 20, 29; m 56; c 3. CRYSTALLOGRAPHY. Educ: Ohio State Univ, BS, 50; Duquesne Univ, MS, 52; Univ Iowa, PhD(chem), 56. Prof Exp: Chemist, E I du Pont de Nemours & Co, 53; sr chemist, B F Goodrich Co, 56-61; sr scientist, Harshaw Chem Co, 61-64; staff scientist, 64-71, RES DIR, AIRTRON DIV, LITTON INDUSTS, INC, 71- Mem: AAAS; Am Chem Soc; Am Crystallog Asn; fel Am Inst Chem; Am Inst Physics. Res: Solid state physics; x-ray crystallography, crystal growth; morphology; physical properties; perfection. Mailing Add: Airtron Div of Litton Indust Inc 200 E Hanover Ave Morris Plains NJ 07950

BELT, WARNER DUANE, b Bellaire, Ohio, Dec 29, 25; m 50; c 4. ANATOMY, CYTOLOGY. Educ: Bethany Col, BS, 48; Ohio State Univ, MSc, 50, PhD(anat), 55. Prof Exp: Asst instr anat, Ohio State Univ, 51-53, instr, 53-55; res assoc, Univ Claif, Los Angeles, 55-56; asst prof, Emory Univ, 56-59; from asst prof to assoc prof, Med Col Va, 59-64; assoc prof, 64-67, PROF ANAT, SCH MED, TUFTS UNIV, 67-, CHMN DEPT, 73- Mem: Am Asn Anat; Electron Micros Soc Am; Histochem Soc. Res: Fine structural cytology; structure of endocrine and exocrine secretory cells; fine structure of steroids secretory organs; cytoplasmic interrelationships. Mailing Add: Dept of Anat Tufts Univ Sch of Med Boston MA 02111

BELTHUIS, LYDA CAROL, b Parkersburg, Iowa, May 29, 13. GEOGRAPHY. Educ: Northern Iowa Univ, BA, 39; Univ Northern Colo, MA, 43; Univ Mich, PhD(geog), 47. Prof Exp: Teacher hist & geog, Iowa Pub Schs, Burlington, 39-43, prin, 43-45; from asst prof to assoc prof, 47-55, head dept, 55-65, PROF GEOG, UNIV MINN, DULUTH, 55- Concurrent Pos: Univ Minn sabbatical leave, 53-54, res leaves, 59, 63, 66 & 71. Mem: Asn Am Geog; Am Geog Soc; Nat Coun Geog Educ; Can Geog Soc; Royal Geog Soc Australia; Res: Historical geography, irrigation power-snowy mountains; Murrumbidgee and Coleambally irrigation areas; Australia. Mailing Add: Dept of Geog Univ of Minn Duluth MN 55812

BELTON, MICHAEL J S, b Bognor Regis, Eng, Sept 29, 34; m 61; c 2. ASTRONOMY. Educ: Univ St Andrews, BSc, 59; Univ Calif, Berkeley, PhD(astron), 64. Prof Exp: ASSOC ASTRONR, PLANETARY SCI DIV, KITT PEAK NAT OBSERV, 64- Mem: AAAS; Am Geophys Union; Am Meteorol Soc; Am Astron Soc. Res: Planetary atmospheres and spectroscopy; space photography. Mailing Add: Kitt Peak Nat Observ 950 N Cherry Tucson AZ 85704

BELTRAMI, EDWARD J, b New York, NY, Apr 21, 34; m 62. MATHEMATICS. Educ: Polytech Inst Brooklyn, BS, 56; NY Univ, MS, 59; Adelphi Univ, PhD(math), 62. Prof Exp: Res mathematician, Grumman Aircraft Eng Corp, 56-62; res mathematician, Saclant Anti-Submarine Warfare Res Ctr, 62-63 & Grumman Aircraft Eng Corp, 63-66; PROF APPL MATH & URBAN SCI, STATE UNIV NY STONY BROOK, 66- Mem: Opers Res Soc; Soc Indust & Appl Math; Math Asn Am. Res: Mathematical optimization techniques and abstract analysis, with emphasis on optical resource allocation. Mailing Add: Dept of Appl Math State Univ NY Stony Brook NY 11790

BELTRAN, ENRIQUE, b Mexico, DF, Apr 26, 03; m 35; c 2. ZOOLOGY. Educ: Nat Univ Mex, BSc, 21, DSc(protozool), 26. Hon Degrees: MSc, Nat Univ Mex, 39; DSc, Univ Havana, 57 & Univ Michoacan, 64. Prof Exp: Instr bot, Nat Univ Mex, 21-23, asst prof zool, 24-28, prof, 30-64; DIR, MEX INST NATURAL RESOURCES, 53- Concurrent Pos: Microbiologist, Mex Dept Agr, 23-25; dir, Marine Biol Lab, Veracruz, 26-27 & Inst Biotecnico, 34-35; Guggenheim fel, 32-33; tech adv, Dept Educ, 35-36, chief sec educ, 37-38; chief lab protozool, Inst Pub Health & Trop Dis, 39-52; prof protozool, Nat Polytech Inst, Mex, 40-64. Honors & Awards: Officer, Palmes Academiques; Legion d'honneur; Commandeur Merite Agricole, Repub Francaise, 60; John C Phillips Medal, Int Union Conserv Nature, 66. Mem: Mex Soc Hist Sci & Technol (pres, 64-); AAAS; Am Micros Soc (vpres, 45); hon mem Soc Am Foresters; Mex Natural Hist Soc (pres). Res: Biology; general zoology; hydrobiology; protozoology; history of biology; natural resources. Mailing Add: Dr Vertiz 724 Mexico DF Mexico

BELTRAN, ERNESTO G, food technology, food chemistry, see 12th edition

BELTZ, ALEX D, b Watertown, SDak, Oct 9, 25; m 62. PHYSIOLOGY. Educ: Walla Walla Col, BA, 49, MA, 57; Western Wash State Col, BAEd, 51; Mich State Univ, PhD(physiol), 66. Prof Exp: Asst prof biol, George Fox Col, 56-59; res asst insect physiol, Entom Res Sta, USDA, 61-62; asst prof physiol & biol, Univ Redlands, 62-63; instr, Eastern Mich Univ, 63-65; ASSOC PROF BIOL, ARMSTRONG STATE COL, 67- Mem: AAAS; Am Soc Zool. Res: Sex pheromones of the housefly; neonatal rat physiology and heart rate-temperature relationships. Mailing Add: Dept of Biol 11935 Abercorn St Armstrong State Col Savannah GA 31406

BELTZ, LEROY DUANE, b Pierce, Nebr, Apr 25, 24; m 44; c 4. PHARMACEUTICAL CHEMISTRY. Educ: Univ Nebr, BSc, 51; Univ Conn, PhD(pharmaceut chem), 56. Prof Exp: Asst pharmacol & pharmaceut chem, Univ Nebr, 51-52; instr pharm, Univ Conn, 52-56; asst prof, Univ Fla, 56-58; assoc prof, Ferris State Col, 58-61, prof, 61-66; PROF PHARM, OHIO NORTHERN UNIV, 74-, DEAN COL PHARM, 66- Mem: Am Pharmaceut Asn. Res: Synthetic organic pharmaceuticals with bacteriostatic or bacteriocidal properties. Mailing Add: Col of Pharm Ohio Northern Univ Ada OH 45810

BELTZ, RICHARD EDWARD, b Loma Linda, Calif, June 20, 29; m 51; c 2. BIOCHEMISTRY. Educ: Walla Walla Col, BS, 50; Univ Southern Calif, MA, 51, PhD(biochem, nutrit), 56. Prof Exp: From instr to assoc prof, 56-70, PROF BIOCHEM, SCH MED, LOMA LINDA UNIV, 70- Concurrent Pos: USPHS sr res fel & career develop award, 60-64. Mem: AAAS; Am Chem Soc. Res: Nucleic acids; control of DNA biosynthesis; metabolism of deoxynucleotides and deoxynucleosides; liver regeneration. Mailing Add: Dept of Biochem Loma Linda Univ Sch Med Loma Linda CA 92354

BELYEA, HELEN R, b St John, NB, Feb 11, 13. GEOLOGY. Educ: Dalhousie Univ, BA, 34, MS, 36; Northwestern Univ, PhD(geol), 39. Prof Exp: Geologist, 45-61, sr geologist, 61-68, res scientist, 68-69, SR RES SCIENTIST, GEOL SURV CAN, 69- Honors & Awards: Officer, Order of Can, 76. Mem: Soc Econ Paleont & Mineral; Int Asn Sedimentol; fel Royal Soc Can; Geol Asn Can. Res: Devonian stratigraphy of Western Canada. Mailing Add: Inst Sediment & Petrol Geol 3303 33rd St NW Calgary AB Can

BELZER, FOLKERT O, b Soerabaja, Indonesia, Oct 5, 30; US citizen; m; c 3. SURGERY. Educ: Colby Col, MA, 53; Boston Univ, MS, 54, MD, 58. Prof Exp: Instr surg, Med Sch, Univ Ore, 63-64, asst resident surgeon, 64; sr lectr surg, Guy's Hosp, Eng, 64-66; asst prof surg & chief surg outpatient clin, 66-69, asst chief transplantation, 69, assoc prof surg, 69-73, PROF SURG, MED CTR, UNIV CALIF, SAN FRANCISCO, 73-, CO-DIR TRANSPLANT SERV, 69- Concurrent Pos: Attend surgeon, Moffitt Hosp & San Francisco Gen Hosp, 69- Res: General surgery; transplantation. Mailing Add: Dept of Surg Univ of Calif Med Ctr San Francisco CA 94122

BELZILE, RENE, b Kapuskasing, Ont, Mar 6, 30; m 57; c 3. ANIMAL NUTRITION. Educ: Univ Ottawa, BA, 51; McGill Univ, BScAgr, 55; Purdue Univ, MSc, 57; Univ Sask, PhD(animal nutrit), 63. Prof Exp: Lectr biochem, Ont Agr Col, 57-60; from asst prof to assoc prof animal nutrit, 63-72, PROF ANIMAL NUTRIT, LAVAL UNIV, 72-, CHMN DEPT ANIMAL SCI, 74- Concurrent Pos: Feedstuffs Act Can res grant, 65-67; asst ed, Can J Animal Sci, 75-77. Mem: Nutrit Soc Can; Can Soc Animal Sci. Res: Protein and amino acid nutrition for poultry; non-protein nitrogen for dairy cattle; soybean meal for mink; light control for raising mink. Mailing Add: Fac of Agr & Foods Laval Univ Quebec PQ Can

BEMAN, FLOYD L, organic chemistry, see 12th edition

BEMENT, ROBERT EARL, b Denver, Colo, Jan 24, 18; m 39; c 2. RANGE SCIENCE. Educ: Colo State Univ, BS, 40, MF, 47, PhD, 68; US Army Command & Gen Staff Col, dipl, 62. Prof Exp: Instr range forage plants, Colo State Univ, 47-48; range conservationist soil conserv serv, USDA, 48, soil conservationist, 48-55, res range conservationist cent plains exp range, 55-61, res range conservationist in charge, 61-65, res range scientist in charge, 65-73, range scientist agr res serv, 66-69, range scientist in charge forage & range res br crops res lab, Colo State Univ, 69-73; CONSULT, 73- Concurrent Pos: Spec lectr col forestry, Colo State Univ, 59-; agr res serv rep, Tech Comt Seedling Estab Range Plants, Great Plains Coun, 60-, comt chmn, 62-63, agr res serv rep, Range & Livestock Mgt Comt, 70-, consult, Food & Agr Orgn, UN, Iceland, 73-78; consult, Int Prog Ecol Mgt of Arid and Semi-arid Rangelands in Africa and Near East, 74-75. Mem: Am Soc Range Mgt; Ecol Soc Am; Am Soc Animal Sci. Res: Relationship of season and degree of grazing to vegetation, gains per animal and animal gain per acre; artificial range revegetation, including role of temperature and moisture as modified by mulches. Mailing Add: PO Box 524 Mancos CO 81328

BE MENT, SPENCER L, b Detroit, Mich, Apr 1, 37; m 62; c 3. ELECTRONICS, NEUROPHYSIOLOGY. Educ: Univ Mich, BSE, 60, MSE, 62, PhD(bioeng), 67. Prof Exp: From instr to asst prof, 63-71, ASSOC PROF ELEC ENG, UNIV MICH, 71- Mem: AAAS; Inst Elec & Electronics Engrs; Soc Neurosci; Biomed Eng Soc; Sigma Xi. Res: Communications and systems theory applied to information coding and transmission by nervous systems; passive and active electrical properties of biological neural tissue. Mailing Add: Univ Mich Bioelec Sci Lab 1124 G G Brown Bldg Ann Arbor MI 48109

BEMILLER, JAMES NOBLE, b Evansville, Ind, Apr 7, 33; m 60; c 2. BIOCHEMISTRY, CARBOHYDRATE CHEMISTRY. Educ: Purdue Univ, BS, 54, MS, 56, PhD(biochem), 59. Prof Exp: Asst biochem, Purdue Univ, 54-59, asst prof, 59-61; from asst prof to assoc prof chem & biochem, 61-68, actg chmn dept, 66-67, PROF CHEM & BIOCHEM, SOUTHERN ILL UNIV, CARBONDALE, 68-, SCH MED, 71- Mem: Fel AAAS; Am Chem Soc (chmn div carbohydrate chem, 65-66); Am Asn Cereal Chem; Sigma Xi; Soc Complex Carbohydrates. Res: Biochemistry of aging; carbohydrate chemistry; cell surface carbohydrates; polysaccharides as industrial gums. Mailing Add: Dept of Chem & Biochem Southern Ill Univ Carbondale IL 62901

BEMIS, CURTIS ELLIOT, JR, b Boston, Mass, Feb 18, 40; m 62; c 3. NUCLEAR PHYSICS, NUCLEAR CHEMISTRY. Educ: Univ NH, BS, 61; Mass Inst Technol, PhD(nuclear chem), 65. Prof Exp: Res asst nuclear chem, Mass Inst Technol, 61-64; Sweden-Am Found & Fulbright-Hayes fels nuclear spectros, Nobel Inst Physics, Stockholm, Sweden, 64-65; researcher, 65-66, RES STAFF MEM NUCLEAR SPECTROS, OAK RIDGE NAT LAB, 66- Mem: AAAS; fel Am Phys Soc. Res: Nuclear spectroscopy and low energy nuclear physics; heavy ion physics; transuranium element nuclear and atomic structure physics and health physics; production and identification of new transuranium elements and isotopes. Mailing Add: Chem Div Bldg 5505 Oak Ridge Nat Lab Oak Ridge TN 37830

BEMIS, WILLIAM PUTNAM, b Calif, Nov 21, 22; m 50; c 3. GENETICS. Educ: Univ Calif, MS, 50; Mich State Univ, PhD(bot, hort), 52. Prof Exp: Asst prof hort, Univ Ill, 52-59; PROF HORT, UNIV ARIZ, 59- Mem: Am Soc Hort Sci; Am Genetic Asn; Am Soc Naturalists; Soc Econ Bot. Res: Cucurbita genetics; ecological adaptation of xerophytic crops. Mailing Add: Dept of Plant Sci Univ of Ariz Tucson AZ 85721

BEMMELS, WILLIAM DAVID, b Winona, Minn, Oct 25, 13; m 37; c 2. PLASMA PHYSICS. Educ: Colo Col, AB, 34; Syracuse Univ, AM, 36; Univ Colo, PhD(physics), 41. Hon Degrees: MHL, Ottawa Univ, Kans, 70. Prof Exp: Asst physics, Syracuse Univ, 34-36; asst, Univ Colo, 36-40; prof physics & math, Lambuth Col, 40-41; dean eng studies, 41-49, dean, Univ, 49-68, PROF PHYSICS & MATH, OTTAWA UNIV, KANS, 41- Mem: Am Phys Soc; Math Asn Am; Am Asn Physics Teachers. Res: Design of audio oscillators; conduction in gases at high pressures; effects of pressure and temperature upon mobilities of ions in air and commercial nitrogen at high pressures. Mailing Add: Dept of Physics & Math Ottawa Univ Ottawa KS 66067

BEMRICK, WILLIAM JOSEPH, b Superior, Wis, Jan 19, 27; m 67; c 6. PARASITOLOGY. Educ: Superior State Col, BS, 50; Univ Wis, MS, 55, PhD(zool), 57. Prof Exp: Res asst zool & parasitol, Univ Wis, 53-57, teaching asst, 57; asst prof vet med parasitol, 57-67, ASSOC PROF VET MED PARASITOL, UNIV MINN, ST PAUL, 67- Concurrent Pos: Presiding officer, Annual Midwestern Conf Parasitologists, 74. Mem: Am Soc Parasitol. Res: Host parasite relationships; parasitic immunology; electron microscopy of parasites. Mailing Add: Dept of Biol Col of Vet Med Univ of Minn St Paul MN 55101

BEMSKI, GEORGE, b Warsaw, Poland, May 20, 23; nat US; m 48; c 2. BIOPHYSICS. Educ: Univ Calif, BA, 49, PhD(physics), 53. Prof Exp: Mem tech staff, Bell Tel Labs, 53-61; mem tech staff, Brazilian Ctr Phys Res, 62-64; mem tech staff, Albert Einstein Col Med, 64-67; MEM STAFF, VENEZUELAN INST SCI INVESTS, 67- Concurrent Pos: Adj prof, Newark Col, Eng, 55-57; prof, Cent Univ Venezuela, 67- Mem: Am Phys Soc; Biophys Soc. Res: Semiconductor physics; phenomena of recombination of carriers; paramagnetic resonance as applied to semiconductors. Mailing Add: Venezuelan Inst for Sci Res Apartado 1827 Caracas Venezuela

BEN, MANUEL, b Syracuse, NY, July 30, 16; m 40; c 2. PHYSICAL CHEMISTRY, PHYSICS. Educ: Univ Mich, BS, 39. Prof Exp: Asst bearing metallurgist bearing dept, AC Spark Plug Div, 40-41, bearing chemist, 41-45, sr res chemist mfg develop eng dept, 45-53, sr res chemist electrochem dept, GM Res Labs, GM Tech Ctr, Warren, 53-54, supvr, 54-66, supvr chem eng dept, GM Mfg Develop, 66-72, sr design engr, Chevrolet Motor Div, 72-74, STAFF DEVELOP ENGR MFG DEVELOP, GEN MOTORS CORP, 74- Honors & Awards: C H Proctor Mem Award, 64. Mem: Am Electroplaters Soc (vpres, 59-62, pres, 62-63, past pres, 63-64); Electrochem Soc; Brit Inst Metal Finishing; NY Acad Sci. Res: Electrodeposition from aqueous solutions; deposition of metals by chemical reduction and chemical and electrochemical surface treatments; air and water pollution; energy. Mailing Add: 13430 Rosemary Blvd Oak Park MI 48237

BEN, MAX, b Utica, NY, Oct 12, 26; m 52; c 4. PHARMACOLOGY, PHYSIOLOGY. Educ: Syracuse Univ, AB, 51; Princeton Univ, MA, 53, PhD(biol, physiol), 54. Prof Exp: Lab asst gen biol, Utica Col, 50-51; asst biol & biochem, Princeton Univ, 51-53, asst endocrinol to Prof Swingel, 53-54; assoc surg res, Harrison, Dept Surg Res, Med Sch, Univ Pa, 54-56 & Merck, Sharp & Dohme, 56-58; sect head, Endocrine Res & Eval, Squibb Inst, 58; sr scientist, Warner-Lambert Res Inst, 58-62; sci dir pharmacol & co-dir mkt, Hazelton Labs, Inc, 62-64, sr pharmacologist & sci dir develop, 64-65; UN tech assistance expert pharmacol & adv to Nat Coun Res & Develop, Prime Minister's Off, Israel, 65-68; exec secy, Nat Biol Cong, 69-70; EXEC VPRES & DIR, HEALTH SCI ASSOCS, BETHESDA, MD, 70-; PRES, MEDISERV INT, 74- Concurrent Pos: Consult pharmacol & toxicol, NIH contract, 58. Mem: AAAS; Am Physiol Soc; Am Soc Pharmacol & Exp Therapeut. Mailing Add: Mediserv Int 8150 Leesburg Pike Vienna VA 22180

BEN, VICTOR RALPH, organic chemistry, see 12th edition

BENACERRAF, BARUJ, b Caracas, Venezuela, Oct 29, 20; US citizen; m 43; c 1. IMMUNOLOGY, EXPERIMENTAL PATHOLOGY. Educ: Univ Paris, Lic es lett, 40; Columbia Univ, BS, 42; Med Col Va, MD, 45. Prof Exp: Res fel immunol, Med Sch, Columbia Univ, 47-49; sr researcher, Nat Ctr Sci Res, France, 49-56; prof path, Sch Med, NY Univ, 56-68; chief lab immunol, Nat Inst Allergy & Infectious Dis, 68-70; FABYAN PROF COMP PATH & CHMN DEPT PATH, HARVARD MED SCH, 70- Concurrent Pos: Mem immunol study sect, NIH, 64-68; adv immunol, WHO, 65-; mem sci adv bd, Mass Gen Hosp, 71-74; pres, Fedn Am Socs Exp Biol, 74-75; assoc ed, Am J Path, Lab Invest, J Exp Med & J Immunol; trustee & mem sci adv bd, Trudeau Found; mem bd gov, Weizmann Inst Sci, 75- Mem: Nat Acad Sci; Am Asn Immunologists (pres, 73-74); Am Asn Path & Bact; Am Soc Exp Path; Soc Exp Biol & Med. Res: Antibody synthesis and structure; hypersensitivity; immunopathology; immunogenetics. Mailing Add: Dept of Path Harvard Med Sch Boston MA 02115

BENACH, JORGE L, b Havana, Cuba, June 27, 45; US citizen; m 68; c 2. MEDICAL ENTOMOLOGY. Educ: Upsala Col, BA, 66; Seton Hall Univ, MS, 68; Rutgers Univ, PhD(entom), 71. Prof Exp: Med entomologist, Bur Acute Commun Dis Control, Albany, 71-72, SR MED ENTOMOLOGIST, WHITE PLAINS REGIONAL OFF, NY STATE DEPT HEALTH, 72- Mem: AAAS; Am Asn Trop Med & Hyg; NY Acad Sci; Am Mosquito Control Asn. Res: Characterization and identification of Rickettsial pathogens and the epidemiology of Rickettsial diseases in New York. Mailing Add: NY State Dept Health 901 N Broadway White Plains NY 10603

BENADE, ARTHUR HENRY, b Chicago, Ill, Jan 2, 25; m 48; c 2. ACOUSTICS. Educ: Wash Univ, AB, 48, PhD(physics), 52. Prof Exp: Instr, 52-62, PROF PHYSICS, CASE WESTERN RESERVE UNIV, 62- Concurrent Pos: Vis prof, Indian Inst Technol, Kanpur, 64-65 & Univ Mich, 74. Mem: Am Phys Soc; fel Acoust Soc Am (vpres-elect, 73-74, vpres, 74-75); AAAS; Am Asn Physics Teachers. Res: Musical acoustics; statistical properties of room acoustics; musical perception. Mailing Add: Dept of Physics Case Western Reserve Univ Cleveland OH 44106

BENADE, LEONARD E, b Evansville, Ind, Nov 13, 44; m 75. CANCER, OCCUPATIONAL HEALTH. Educ: Univ Va, BA, 66; George Washington Univ, MPh & PhD, 71. Prof Exp: Guest worker biochem, Nat Cancer Inst, 68-71; chemist, Cent Intel Agency, 71-73; sr biochemist, Enviro Control, Inc, 74-75; BIOMED CONSULT, JRB ASSOCS, INC, 75- Concurrent Pos: Lectr biol, Northern Va Community Col, 74-; consult Mitre Corp, 75. Mem: AAAS. Res: Biochemistry, metabolism and chemotherapy of neoplasia, specifically ascorbate toxicity in Ehrlich ascites carcinoma cells; cancer research. Mailing Add: JRB Assoc Inc 1651 Old Meadow Rd McLean VA 22101

BENARD, MARK, b Pittsburgh, Pa, Jan 4, 44; m 73. ALGEBRA. Educ: Tulane Univ, BA, 65; Yale Univ, MPhil, 67, PhD(math), 69. Prof Exp: Instr math, Yale Univ, 69-70; asst prof, 70-75, ASSOC PROF MATH, TULANE UNIV, 75- Mem: Am Math Soc; Math Asn Am. Res: Finite group theory and representation theory, with emphasis on the Schur index. Mailing Add: Dept of Math Tulane Univ New Orleans LA 70118

BENARDE, MELVIN ALBERT, b Brooklyn, NY, June 15, 23; m 51; c 3. EPIDEMIOLOGY, PUBLIC HEALTH. Educ: St John's Univ, NY, BS, 48; Univ Mo, MS, 50; Mich State Univ, PhD(bact), 54. Prof Exp: Res asst virus res, Pub Health Res Inst, New York, 50-51; food specialist, Off Naval Res, Naval Supply Depot, Bayonne, 54-55; asst prof, Seafood Processing Lab, Univ Md, 55-61; assoc prof, Bioeng Lab, Dept Civil Eng, Rutgers Univ, 61-67; assoc prof commun med, 67-74, PROF

COMMUN MED, HAHNEMANN MED COL, 74- Concurrent Pos: WHO fel, Sch Hyg & Trop Med, Univ London, 63. Mem: Fel Am Pub Health Asn; Am Soc Microbiol; Asn Teachers Prev Med; fel Royal Soc Health. Res: Nosocomial infections; microbial metabolism of synthetic detergents; effects of synthetic chemicals on tissue cultures; germicidal effects of chlorine compounds; evaluation of medical care delivery; environmental factors in pancreatic cancer. Mailing Add: Dept of Com Med & Environ Health Hahnemann Med Col Philadelphia PA 19102

BENBOW, RALPH LAWRENCE, b Watertown, SDak, Apr 8, 42; m 68; c 2. SOLID STATE PHYSICS. Educ: SDak Sch Mines & Technol, BS, 64, MS, 67; Iowa State Univ, PhD(physics), 75. Prof Exp: Physicist, Harry Diamond Labs, 64-69; RES ASSOC PHYSICS, NORTHERN ILL UNIV, 74- Mem: Am Phys Soc. Res: Vacuum ultraviolet photo-emission spectroscopy of layered semiconductors, interfaces of thin films with layered semiconductors, metals and chemisorbed gases on metals using the unique properties of synchrotron radiation. Mailing Add: Synchrotron Radiation Ctr Univ of Wis PO Box 6 Stoughton WI 53589

BENCOSME, SERGIO ARTURO, b Montecristi, Dominican Repub, Apr 27, 20; Can citizen; m; c 5. PATHOLOGY. Educ: Univ Montreal, MD, 47; McGill Univ, MSc, 48, PhD(path), 50. Prof Exp: Asst prosector, Royal Victoria Hosp, Montreal, 50-51; asst prof path, Univ Ottawa, 51-53; assoc prof, Queen's Univ, Ont, 53-57; asst res pathologist, Univ Calif, Los Angeles, 57-59; assoc prof, 59-65, PROF PATH, QUEEN'S UNIV, ONT, 65- Concurrent Pos: Can Life Ins Officer's Asn fel, 53-56; Can Cancer Soc traveling fel, US, 54; Ont Cancer Treatment & Res Found traveling fel, Univ Calif, Los Angeles, 56, lectr, 57; actg dir, Hotel Dieu Hosp, Ont, 57; pres, Burton Soc Electron Micros, 59; mem coun basic sci, Am Heart Asn, 62-; dir inst biomed, Univ Pedro Henriquez Urena, Santo Domingo, 73. Mem: Am Soc Cell Biol; Am Asn Anat; Am Asn Path & Bact; Am Soc Exp Path; Soc Exp Biol & Med. Res: Pancreatic islets; kidney pathology; myocardium pathology; cancer of the breast. Mailing Add: Dept of Path Queen's Univ Kingston ON Can

BENDA, GERD THOMAS ALFRED, b Berlin, Ger, Nov 19, 27; nat US; m 56; c 2. PLANT PHYSIOLOGY. Educ: Princeton Univ, AB, 46; Yale Univ, MS, 49, PhD(bot), 51. Prof Exp: Instr plant physiol, Yale Univ, 52-53; vol worker dept plant path, Rothamsted Exp Sta, 53-54; USPHS res fel virus lab, Univ Calif, 54-56, jr res biologist, 56-57; from asst prof to assoc prof biol, Univ Notre Dame, 57-64; Rockefeller Found grant virol sect, Inst Agron, Brazil, 64-65; PLANT PHYSIOLOGIST, AGR RES SERV, USDA, 66- Mem: Bot Soc Am; Am Phytopath Soc; Am Soc Cell Biol; Asn Trop Biol; Am Soc Sugar Cane Technol. Res: Physiology of plant virus diseases. Mailing Add: US Sugar Cane Field Sta Box 470 Houma LA 70360

BENDA, STEPAN VACLAV, b Puclice, Czech, Feb 24, 11; nat US; m 39; c 3. BIOPHYSICS, PHYSICAL CHEMISTRY. Educ: Prague Tech Univ, EngD, 34, PhD, 36. Prof Exp: Asst biophys, Eng Sch, Prague Tech Univ, 34-35; chem officer phys chem, Mil Acad, Pardubice, 35-37; exten specialist biophys & biophys chem, Res Dept, Ministry of Agr, Prague, 37-48; commentator sci scripts, Brit Broadcasting Corp, London, Radio Vaticano, Free Europe & Voice of Am, 48-52; asst prof natural sci, Del State Col, 55-57, assoc prof, 57-58, prof, 58-64; PROF PHYSICS & HEAD DEPT, HAMPTON INST, 64- Concurrent Pos: NSF fel, Cornell Univ, 58 & Stanford Univ, 60; lectr, Univs Ger, Austria & France; Fulbright scholar, 63-64. Mem: Am Asn Physics Teachers; Economet Soc; Opers Res Soc Am. Res: Physics for engineering; biological effects of radiation; amino acids; growth plants. Mailing Add: Dept of Physics Hampton Inst Hampton VA 23368

BENDALL, VICTOR IVOR, b London, Eng, Dec 14, 35; m 66; c 2. ORGANIC CHEMISTRY. Educ: Univ London, BSc, 57; Bucknell Univ, MS, 60; Brown Univ, PhD(chem), 64. Prof Exp: Res chemist, Brit-Thompson-Houston Co, Eng, 57-58; res assoc, Univ Chicago, 63-64; asst prof chem, George Peabody Col, 64-66; assoc prof, 66-69, PROF CHEM, EASTERN KY UNIV, 69- Mem: Am Chem Soc; The Chem Soc. Res: Small and medium alicyclic compounds; carbenes and reactive intermediates. Mailing Add: Dept of Chem Eastern Ky Univ Richmond KY 40475

BENDANA, FRANK, plant physiology, plant biochemistry, see 12th edition

BEN DANIEL, DAVID J, b Philadelphia, Pa, Nov 10, 31; m 57; c 2. PHYSICS, RESEARCH ADMINISTRATION. Educ: Univ Pa, BA, 52, MS, 53; Mass Inst Technol, PhD(nuclear eng), 60. Prof Exp: Physicist, Lawrence Radiation Lab, Univ Calif, Livermore, 59-60; physicist, Res Lab, 60-68, mgr advan progs, 68-70, MGR TECH VENTURES OPER, GEN ELEC CO, 70- Concurrent Pos: Adj assoc prof, Rensselaer Polytech Inst, 61-67; adj prof, State Univ NY Albany, 68-69, vis prof, 70-; vis fel & lectr, Harvard Bus Sch, 69-70. Mem: AAAS; Am Phys Soc. Res: Plasma physics, especially thermonuclear containment. Mailing Add: Gen Elec Res & Develop Ctr Box 8 Schenectady NY 12301

BENDAT, JULIUS SAMUEL, b Chicago, Ill, Oct 26, 23; m 47; c 2. RANDOM DATA ANALYSIS, APPLIED MATHEMATICS. Educ: Univ Calif, AB, 44; Calif Inst Technol, MS, 48; Univ Southern Calif, PhD(math), 53. Prof Exp: Res asst physics radiation lab, Univ Calif, 42-45; asst appl mech, Calif Inst Technol, 46-48; asst prof aeronaut eng col aeronaut, Univ Southern Calif, 48-49, lectr math, 49-58; res engr, Northrop Aircraft, Inc, 53-55; mem sr staff, Thompson-Ramo-Wooldridge, Inc, 55-62; pres, Measurement Anal Corp, 62-70; CONSULT, 70- Concurrent Pos: Asst, Univ Calif, 44-46; res physicist, Calif Inst Technol, 47-48. Mem: Am Math Soc; Math Asn Am; Soc Indust & Appl Math. Res: Abstract analysis; prediction and filter theory; system engineering; mathematical physics; information theory and circuit theory; random noise theory; vibration and acoustics research. Mailing Add: J S Bendat Co 833 Moraga Dr Los Angeles CA 90049

BENDEL, WARREN LEE, b Ravenna, Ohio, Jan 19, 25; m 56; c 1. NUCLEAR PHYSICS. Educ: Kent State Univ, BS, 46; Univ Ill, MS, 47, PhD(physics), 53. Prof Exp: Lab asst physics, Kent State Univ, 46; asst, Univ Ill, 46-50; res asst, Los Alamos Sci Lab, 49; res asst, Univ Ill, 50 & 52-53; PHYSICIST, US NAVAL RES LAB, 53- Concurrent Pos: Vis fac mem, Univ Md, 71-72. Mem: Am Phys Soc; Sigma Xi. Res: Nuclear spectroscopy; photonuclear reactions; electron accelerators; electron scattering. Mailing Add: Code 6632 Radiation Technol Div US Naval Res Lab Washington DC 20375

BENDELL, JAMES FRANCIS SIDNEY, vertebrate zoology, see 12th edition

BENDELOW, VICTOR MARTIN, b Sheffield, Eng, Mar 15, 20; Can citizen; m 45; c 1. CEREAL CHEMISTRY. Educ: Univ Man, BSc, 57, MSc, 63, PhD, 70. Prof Exp: CEREAL CHEMIST, CAN DEPT AGR, 48- Mem: Master Brewers Asn Am; Am Asn Cereal Chemists. Mailing Add: Can Dept of Agr Res Sta 25 Dafoe Rd Winnipeg MB Can

BENDER, ALLAN DOUGLAS, b Iowa City, Iowa, May 26, 36; m 57; c 4. PHARMACOLOGY. Educ: Williams Col, BA, 57; Univ Iowa, MS, 59; Jefferson Med Col, PhD(physiol), 62. Prof Exp: Asst physiol, Univ Iowa, 58-59; asst, Lankenau Hosp, Philadelphia, 59-61; asst prof biol, West Chester State Col, 61-62; sr scientist, 62-64, group leader, 64-66, asst sect head, 66-67, coordr long range planning, 67, assoc dir biomed info & long range planning, 67, dir sci info, 67-71, VPRES RES & DEVELOP PLANNING & OPERS, RES & DEVELOP DIV, SMITH KLINE & FRENCH LABS, 71- Concurrent Pos: Vis lectr, Jefferson Med Col, 62-68. Mem: AAAS; fel Am Col Cardiol; Am Chem Soc; fel Geront Soc; fel Am Inst Chem. Res: Cardiovascular physiology and pharmacology; pharmacologic aspects of aging; drug development; data processing; management science; long range and project planning. Mailing Add: 401 Thayer Rd Swarthmore PA 19081

BENDER, CARL MARTIN, b Brooklyn, NY, Jan 18, 43; m 66; c 1. HIGH ENERGY PHYSICS, MATHEMATICAL PHYSICS. Educ: Cornell Univ, AB, 64; Harvard Univ, AM, 65, PhD(theoret physics), 69. Prof Exp: Mem physics res, Inst Advan Study, 69-70; asst prof math, 70-73, ASSOC PROF MATH, MASS INST TECHNOL, 73- Mem: Am Inst Physics. Res: Quantum field theory; convergence of perturbation theory; massless field theory; infinite dimensional field theory; Pade theory. Mailing Add: Room 2-379 Dept of Math Mass Inst of Technol Cambridge MA 02139

BENDER, CHARLES F, b Fortuna, Calif, June 15, 41; m 66. THEORETICAL ATOMIC PHYSICS, MOLECULAR PHYSICS. Educ: Univ of the Pac, BS, 63, MS, 64; Univ Wash, PhD(chem), 68. Prof Exp: Fel, Battelle Mem Inst, 68-69; chem consult, 70-71; chemist, 71-72, leader compatability group, 72-75, LEADER THEORET ATOMIC & MOLECULAR PHYSICS GROUP, LAWRENCE LIVERMORE LAB, UNIV CALIF, 75- Concurrent Pos: Instr dept appl sci, Univ Calif, Davis, 73- Mem: AAAS. Res: Development of theoretical and computational methods for calculation of electron proppproperties of small chemical systems. Mailing Add: L215 Lawrence Livermore Lab Livermore CA 94550

BENDER, DANIEL FRANK, b Cincinnati, Ohio, Mar 22, 36; m 59; c 7. PHYSICAL ORGANIC CHEMISTRY, ANALYTICAL CHEMISTRY. Educ: Xavier Univ, Ohio, BS, 59, MS, 63; Univ Cincinnati, PhD(phys org chem), 67. Prof Exp: Chemist, Nat Air Pollution Control Admin, USPHS, US Dept Health, Educ & Welfare, 60-61, res chemist, 61-67, res chemist environ control admin, Bur Solid Waste Mgt, 67-69, supvr res chemist, Res Serv, Chem Studies Group, 69-73; RES CHEMIST WATER POLLUTION, ENVIRON PROTECTION AGENCY, 73- Concurrent Pos: Pyrolysis session moderator, Nat Indust Solid Wastes Mgr Conf, 70, pyrolysis session chmn, Am Chem Soc Solid Waste Chem Symp, 70; adj asst prof chem, Raymond Walters Br, Univ Cincinnati, 73- Mem: Am Chem Soc; The Chem Soc; Sigma Xi. Res: Structure-activity relationships; reaction mechanisms; organic molecular orbital theory; absorption and fluorescence spectroscopy; column and thin-layer chromatography; ultramicroorganic analysis of environmental pollutants; pyrolytic synthesis; water pollution; analytical methodology research; environmental research. Mailing Add: 9536 Lansford Dr Cincinnati OH 45242

BENDER, DAVID FULMER, b Reno, Nev, Feb 10, 13; m 35; c 3. PHYSICS. Educ: Calif Inst Technol, BS, 33, MS, 34, PhD(physics), 37. Prof Exp: Instr physics, Univ La, 37-42; supply asst prof, Vanderbilt Univ, 42-43; asst prof, Fisk Univ, 43-46; prof, Whittier Col, 46-74; MEM STAFF, JET PROPULSION LAB, 74- Concurrent Pos: Prin scientist, Space & Info Syst Div, NAm Rockwell Corp, 61-74. Mem: Am Inst Aeronaut & Astronaut; Am Astronaut Soc; Am Asn Physics Teachers. Res: Astronautics; space physics; inertial guidance; celestial mechanics. Mailing Add: Jet Propulsion Lab 156/220 4800 Oak Grove Dr Pasadena CA 91103

BENDER, EDWARD ANTON, b Brooklyn, NY, Feb 7, 42; div; c 3. MATHEMATICS. Educ: Calif Inst Technol, BS, 63, PhD(math), 66. Prof Exp: Instr math, Harvard Univ, 66-68; res mathematician, Inst Defense Anal, 68-74; PROF MATH, UNIV CALIF, SAN DIEGO, 74- Mem: Am Math Soc. Res: Combinatorics. Mailing Add: Dept of Math Univ of Calif San Diego La Jolla CA 92037

BENDER, GEORGE ALMON, b Osseo, Wis, Sept 26, 04; m 22, 48; c 6. PHARMACY. Educ: SDak State Univ, PhG, 23, DSc, 58; Phila Col Pharm & Sci, PhM, 45. Prof Exp: Pharmacist, Watertown, SDak, 24-29; ed, Northwest Druggist, St Paul, 29-33 & J Nat Asn Retail Druggists, Chicago, 33-45; serv ed, Am Druggist, New York, 46; ed, Mod Pharm, Parke Davis & Co, Detroit, 47-58, dir instnl adv, 58-69, historian, 68-69; asst to exec secy, Nat Asn Retail Druggists, Chicago, 69-70; prof pharm & admin aide to dean col pharm, 70-74, EMER PROF PHARM, UNIV ARIZ, 74- Mem: Am Pharmaceut Asn; Am Inst Hist Pharm (pres, 75-77); Can Pharmaceut Asn. Res: History of the health sciences. Mailing Add: Col of Pharm 123 Univ of Ariz Tucson AZ 85721

BENDER, HARVEY ALAN, b Cleveland, Ohio, June 5, 33; m 56; c 3. GENETICS, ENTOMOLOGY. Educ: Case Western Reserve Univ, AB, 54; Northwestern Univ, MS, 57, PhD(biol sci), 59. Prof Exp: USPHS fel, Univ Calif, 59-60; from asst prof to assoc prof, 60-69, PROF BIOL, UNIV NOTRE DAME, 69- MEM SR STAFF RADIATION LAB, 64- Concurrent Pos: Vis prof in-serv inst, Purdue Univ, 62; Gosney fel & USPHS spec fel, Calif Inst Technol, 65-66. Mem: Fel AAAS; Am Genetic Asn; Am Soc Zool; Am Inst Biol Sci; NY Acad Sci. Res: Development and physiological genetics; cytochemistry; histochemistry. Mailing Add: Dept of Biol Univ of Notre Dame Notre Dame IN 46556

BENDER, HOWARD LEONARD, b Williamstown, WVa, Sept 14, 93; m 18; c 2. PLASTICS CHEMISTRY. Educ: Marietta Col, AB, 17, AM, 18; Case Inst, BS, 19; Columbia Univ, PhD(chem), 23. Hon Degrees: DSc, Marietta Col, 53. Prof Exp: Res chemist, Dow Chem Co, 19-21; instr chem, Marietta Col, 21; res chemist, Union Carbide Plastics Co, 22-51, asst dir res, 51-58, consult, 58-66; INDEPENDENT CONSULT PLASTICS, 66- Honors & Awards: Hyatt Award, 53. Mem: Am Chem Soc. Res: Thermosetting resins and plastics; synthetic resins; acetylene gas reactions; polymers; phenolformaldehyde resins. Mailing Add: 159 Peyton St Winchester VA 22601

BENDER, HOWARD SANFORD, b Brooklyn, NY, Aug 29, 35; m 61; c 2. POLYMER CHEMISTRY, ORGANIC CHEMISTRY. Educ: State Univ NY Buffalo, BS, 57; Bucknell Univ, MS, 60; Univ Del, PhD(org chem), 62. Prof Exp: Res chemist cent res labs, Interchem Corp, 62-65; from assoc sr res chemist to sr res chemist, 65-71, SUPVR COATINGS RES, RES LAB, GEN MOTORS CORP, 71- Honors & Awards: Roon Award, 70. Mem: Am Chem Soc; Fedn Socs Paint Technol. Res: Surface coatings; paint; organic synthesis; rheology; small ring compounds; heterocycles; chemotherapy; aziridine chemistry. Mailing Add: Polymers Dept Gen Motors Res Warren MI 48090

BENDER, JAMES ARTHUR, b Webster, SDak, Nov 27, 23; m 48; c 3. APPLIED PHYSICS. Educ: Univ SDak, BA, 47, MA, 48; Rensselaer Polytech Inst, MS, 68. Prof Exp: Instr physics, Univ SDak, 48-49; res physicist, Snow, Ice & Permafrost Res Estab, 50-56, chief snow & ice basic res br, 56-58, chief basic res br, 58-61; chief res div, US Army Cold Regions Res & Eng Lab, 61-68; PHYS SCI ADMINR, HQ ARMY MAT COMMAND, OFF DEP FOR LABS, 68- Concurrent Pos: Int chmn ground ice, Int Union Geod & Geophys, 60-; spec mem ad hoc comt int hydrol prog report, Nat Acad Sci, 62-63, mem status & needs for US prog in hydrol, 63-64, mem panel glaciol, comt polar res, 62-64, ex-officio mem, 64-, chmn ad hoc comt spacecraft appln glaciol, panel glaciol & comt polar res, 65-; mem planning comt, Int Conf Permafrost, Purdue Univ, 63. Mem: Soc Rheol; Am Geophys Union; Am Polar Soc; Glaciol Soc. Res: Properties of snow, ice and frozen ground operations in cold regions; solid-earth physics; operations research and systems analysis; operation of laboratories; ballistics; armaments. Mailing Add: Hq Army Mat Command Off of Deputy for Labs Washington DC 20315

BENDER, LEONARD FRANKLIN, b Philadelphia, Pa, Oct 2, 25; m 48; c 4. MEDICINE. Educ: Jefferson Med Col, MD, 48; Univ Minn, MS, 52; Am Bd Phys Med & Rehab, dipl. Prof Exp: From asst prof to prof phys med & rehab, Med Sch, Univ Mich, Ann Arbor, 54-75; PROF PHYS MED & & CHMN DEPT, SCH MED, WAYNE STATE UNIV, 75-; VPRES & MED DIR, REHAB INST, DETROIT, 75- Concurrent Pos: Consult, Wayne County Gen Hosp, Saginaw Community Hosp & Detroit Cerebral Palsy Ctr. Mem: AMA; Am Asn Electromyog & Electrodiag; Am Rheumatism Asn; Am Cong Rehab Med; Am Acad Phys Med & Rehab (pres, 74). Res: Ultrasound; rheumatoid arthritis; rehabilitation; electromyography; prosthetics; orthetics. Mailing Add: Dept of Phys Med & Rehab Wayne State Univ Sch of Med Detroit MI 48202

BENDER, MARGARET MCLEAN, b Easthampton, Mass, May 14, 16; m 43. ORGANIC CHEMISTRY. Educ: Mt Holyoke Col, BA, 37, MA, 39; Yale Univ, PhD(org chem), 41. Prof Exp: Instr chem, Conn Col, 41-42; proj assoc, Yale Univ, 42-43; proj assoc, 43-45, lectr exten div, 51-69, PROJ ASSOC RADIOCARBON DATING, UNIV WIS-MADISON, 63- Res: Radiocarbon dating; carbon isotope ratios of plants. Mailing Add: Dept of Meteorol Univ of Wis 1225 W Dayton St Madison WI 53706

BENDER, MAURICE, management, biochemistry, see 12th edition

BENDER, MAX, b Boston, Mass, Oct 23, 14; m 47; c 2. PHYSICAL CHEMISTRY, COLLOID CHEMISTRY. Educ: Northeastern Univ, SB, 36; Mass Inst Technol, SM, 37; NY Univ, PhD(chem), 50. Prof Exp: Charge of lab, Manton-Gaulin Mfg Co, Inc, 38-40; sr res chemist & group leader colloid & phys chem, Interchem Corp, 41-48; sr res chemist, Am Cyanamid Co, 50-61; from asst prof to assoc prof chem, 61-69, PROF CHEM, FAIRLEIGH DICKINSON UNIV, 69- Concurrent Pos: Consult. Mem: AAAS; fel NY Acad Sci; Am Chem Soc; Am Inst Chem. Res: Electrokinetics; dispersions; emulsions; rheology; interfactial chemistry; sorption; diffusion; gelation; membranes. Mailing Add: 16 S Woodland Ave East Brunswick NJ 08816

BENDER, MERRILL ARTHUR, b Cleveland, Ohio, Dec 7, 23; m 44; c 6. MEDICINE. Educ: Middlebury Col, AB, 44; Harvard Med Sch, MD, 48. Prof Exp: Radiation therapist, US Naval Hosp, 52-53; assoc chief radiologist cancer res, 53-55, chief radioisotope res, 55-58, CHIEF DEPT NUCLEAR MED, ROSWELL PARK MEM INST, 58- Concurrent Pos: Chmn div nuclear med, State Univ NY Buffalo, 56-; mem adv comt med uses isotopes & mem adv comt isotopes & radiation develop, AEC; mem, Nat Coun Radiation Protection & Measurements; chmn radiopharmaceut adv comt, Food & Drug Admin. Mem: Soc Nuclear Med (pres, 67-68). Res: Nuclear medicine. Mailing Add: Dept of Nuclear Med Roswell Park Mem Inst Buffalo NY 14203

BENDER, MICHAEL A, b New York, NY, July 25, 29; m 50, 69; c 2. CYTOGENETICS. Educ: Univ Wash, BS, 52; Johns Hopkins Univ, PhD(biol), 56. Prof Exp: Nat Cancer Inst fel biol, Johns Hopkins Univ, 56-58; biologist, Oak Ridge Nat Lab, 58-69; assoc prof radiol, Sch Med, Vanderbilt Univ, 69-71; geneticist, US AEC, 71-73; vis prof radiol, Sch Med, Johns Hopkins Univ, 73-75; SR SCIENTIST, BROOKHAVEN NAT LAB, 75- Mem: Fel AAAS; Am Soc Cell Biol; Radiation Res Soc; Am Soc Photobiol; Environ Mutagen Soc. Res: Eukaryotic chromosome structure; mechanisms of mutagenesis; environmental mutagenesis. Mailing Add: Med Dept Brookhaven Nat Lab Upton NJ 11973

BENDER, MICHAEL E, b Spring Valley, Ill, Feb 18, 39; m 61; c 2. POLLUTION BIOLOGY. Educ: Southern Ill Univ, BA, 61; Mich State Univ, MS, 62; Rutgers Univ, PhD(environ sci), 68. Prof Exp: Biologist, USPHS, 62-64; asst prof environ health, Univ Mich, 68-70; sr marine biologist & chmn dept ecol & pollution, 70-73, ASST DIR & HEAD DIV ENVIRON SCI & ENG, VA INST MARINE SCI, 73- Concurrent Pos: Assoc prof, Sch Marine Sci, Col William & Mary & Dept Marine Sci, Univ Va, 70-; vpres, Environ Control Technol Corp, Mich. Mem: Am Soc Limnol & Oceanog; Ecol Soc Am; Am Fisheries Soc; Water Pollution Control Fedn; Int Asn Theoret & Appl Limnol. Res: Water quality; bioassays; coastal zone management; wetlands; information transfer to management organizations. Mailing Add: Dept of Ecol & Pollution Va Inst of Marine Sci Gloucester Point VA 23062

BENDER, MORRIS BORIS, b Russia, June 8, 05; nat US; m 36; c 5. NEUROLOGY. Educ: Univ Pa, BS, 27, MD, 31. Prof Exp: Hon res fel physiol, Yale Univ, 36-37; adj neurologist, Mt Sinai Hosp, 37-46, assoc neurologist, 46-51; prof, 66-68, HENRY P & GEORGETTE GOLDSCHMIDT PROF NEUROL, MT SINAI SCH MED, 68-, CHMN DEPT, 66-; DIR & CHIEF NEUROL, MT SINAI HOSP, 51- Concurrent Pos: Asst, Sch Med, NY Univ, 38-39, head lab exp neurol, 38-42, from instr to assoc prof, 39-51, prof, Univ & attend neurologist, Hosp, 51-66; consult neurologist, US Vet Admin, 46; assoc attend neurologist, Bellevue Hosp, 46-51, vis neurologist & psychiatrist, 49-51, dir neurol serv, 51-66; clin prof, Col Physicians & Surgeons, Columbia Univ, 51-66. Honors & Awards: Southern Cross Brazil, 67. Mem: Am Neurol Asn (pres, 72-73); Am Psychol Asn; fel Am Col Physicians; Am Physiol Soc; Asn Res Nerv & Ment Dis. Res: Vestibular-oculomotor system; perception. Mailing Add: Dept of Neurol Mt Sinai Sch of Med New York NY 10029

BENDER, MYRON LEE, b St Louis, Mo, May 20, 24; m 52; c 3. CHEMISTRY. Educ: Purdue Univ, BS, 44, PhD(chem), 48. Hon Degrees: DSc, Purdue Univ, 69. Prof Exp: Fel, Harvard Univ, 48-49; AEC fel, Univ Chicago, 49-50; instr chem, Univ Conn, 50-51; from instr to assoc prof, Ill Inst Technol, 51-60; assoc prof, 60-62, PROF CHEM, NORTHWESTERN UNIV, EVANSTON, 62-, PROF BIOCHEM, 74- Concurrent Pos: Sloan res fel, 59-63; mem biochem study sect, NIH. Honors & Awards: Midwest Award, Am Chem Soc, 72. Mem: Nat Acad Sci; AAAS; Am Chem Soc; Am Soc Biol Chem. Res: Organic and enzymatic reaction mechanisms; mechanisms of catalysis; isotope effects. Mailing Add: Dept of Chem Northwestern Univ Evanston IL 60201

BENDER, NORMAN CHARLES, b Buffalo, NY, Dec 22, 95; m 24; c 3. PEDIATRICS. Educ: Univ Mich, BS, 18, MD, 19; Am Bd Pediat, dipl, 34. Prof Exp: ASST PROF PEDIAT, MED SCH, STATE UNIV NY BUFFALO. Concurrent Pos: Pediatrician, Children's Hosp & Deaconess Hosp; consult pediatrician var hosps. Mem: Am Acad Pediat; AMA; World Med Asn. Mailing Add: 306 Woodbridge Ave Buffalo NY 14214

BENDER, PAUL A, b Buffalo, NY, Jan 18, 31; m 55. SOLID STATE PHYSICS. Educ: Grinnell Col, BA, 52; Wash Univ, MA, 55; Univ Colo, PhD(physics), 60. Prof Exp: Asst prof physics, Drury Col, 55-57; asst prof, 60-71, ASSOC PROF PHYSICS, WASH STATE UNIV, 71- Mem: AAAS; Am Phys Soc; Am Asn Physics Teachers; Am Geophys Union. Res: Cosmic rays; nuclear spin resonance of solids; electron spin resonance in igneous rocks. Mailing Add: Dept of Physics Wash State Univ Pullman WA 99163

BENDER, PAUL ELLIOT, b Long Beach, NY, Dec 11, 42; m 65; c 2. SYNTHETIC ORGANIC CHEMISTRY. Educ: State Univ NY Stony Brook, BS, 63; Univ Ill, Urbana, MS, 66, PhD(org chem), 69. Prof Exp: SR INVESTR MED CHEM, SMITH KLINE & FRENCH LABS, 69- Mem: Am Chem Soc. Res: Design and synthesis of amidines, cannabinoids and various heterocyclic compounds to investigate their potential for medicinal use. Mailing Add: Smith Kline & French Labs 1500 Spring Garden St Philadelphia PA 19101

BENDER, PAUL J, b Mansfield, Ohio, Nov 20, 17; m 43. PHYSICAL CHEMISTRY. Educ: Yale Univ, BS, 39, PhD(phys chem), 42. Prof Exp: Phys chemist, Off Sci Res & Develop proj, Yale Univ, 42; instr phys chem, 42-45, from asst prof to assoc prof, 45-55, prof, 55-73, PROF CHEM, UNIV WIS-MADISON, 74- Mem: Am Chem Soc. Res: Thermodynamic properties of solutions; Raman spectroscopy; vapor pressures for liquids; nuclear magnetic resonance spectroscopy. Mailing Add: Dept of Chem Univ of Wis Madison WI 53706

BENDER, PETER LEOPOLD, b New York, NY, Oct 18, 30; m 53; c 3. PHYSICS. Educ: Rutgers Univ, BS, 51; Princeton Univ, PhD(physics), 56, MA, 57. Prof Exp: PHYSICIST, NAT BUR STAND, 56- Concurrent Pos: Mem joint inst lab astrophys, Nat Bur Stand & Univ Colo, 62-, chmn, 69-70; adj prof, Univ Colo, 64- Honors & Awards: Gold Medal, US Dept Com, 59; Samuel Wesley Stratton award, Nat Bur Stand, 62. Mem: Fel AAAS; Am Astron Soc; Am Geophys Union; fel Optical Soc Am; fel Am Phys Soc. Res: Atomic physics; geophysics; astronomy. Mailing Add: Joint Inst for Astrophys Nat Bur Stand Boulder CO 80302

BENDER, PHILLIP R, b Milwaukee, Wis, Sept 20, 27; m 56; c 5. MATHEMATICS. Educ: Purdue Univ, BS, 51; Marquette Univ, MS, 59; Iowa State Univ, PhD(math), 66. Prof Exp: Qual control engr, A-P Controls Corp, 51-55; instr math, Milwaukee Sch Eng, 50-60; instr, 60-63, asst prof, 66-74, ASSOC PROF & ASST CHMN DEPT, MARQUETTE UNIV, 74- Mem: Math Asn Am; Soc Indust & Appl Math. Res: Differential equations; asymptotic behavior of solutions; numberical solutions. Mailing Add: Dept of Math Marquette Univ Milwaukee WI 53233

BENDER, REINHOLD, b Berlin, Ger, Oct 31, 37; US citizen; m 67; c 2. ORGANIC CHEMISTRY. Educ: Univ Munich, BS, 59, MS, 64, PhD(tetrazole chem), 67. Prof Exp: SR CHEMIST CHEM DEVELOP, WYETH LABS INC, 69- Res: Semisynthetic penicillins and cephalosporins. Mailing Add: Beverly Dr Kennett Square PA 19348

BENDER, ROGER H, mathematics, see 12th edition

BENDER, ROGER STILLMAN, b Cresson, Pa, Aug 20, 15; m 60; c 2. NUCLEAR PHYSICS, RADIATION DOSIMETRY. Educ: Yale Univ, BS, 36; Harvard Univ, PhD(nuclear physics), 45. Prof Exp: Res assoc nuclear physics, Univ Wis, 45-48; assoc prof, Univ Pittsburgh, 48-57; physicist, Oak Ridge Nat Lab, 57-66; PROF PHYSICS, THE CITADEL, 66- Concurrent Pos: Consult, Med Col Va, 67- Mem: Am Inst Physics; Am Phys Soc; Am Acad Sci; Am Asn Physics Prof. Res: Nuclear structure; accelerator development; instrumentation used to study nuclear structure; radiation; dosimetry in radiation biological experiments. Mailing Add: Dept of Physics The Citadel Charleston SC 29409

BENDER, WALTER LOUIS, b Oak Park, Ill, Nov 12, 26; m 59; c 3. FOREST MANAGEMENT. Educ: Univ Mich, BSF, 49, MF, 50. Prof Exp: Property mgr asst, Univ Mich, 44 & 47-49; forester, Ahonen Lumber Co, 49-50; div supt, Firestone Plantations Co, Africa, 50-52; consult, Liberia Mining Co, 52-53; forester, Pomery & McGowin Forest Consults, 53; forestry adv, US Govt, Haiti, 53-57, timber mgt specialist, Colombian Govt, 57-61; mgt rep, 61-62, mgr Far East sales, 62-65, for projs mgr, 65-69, dir admin & planning, 70, for planning mgr, 71-72, SPEC PROJ MGR, WEYERHAEUSER CO, 72- Mem: Soc Am Foresters; Int Soc Trop Foresters; Am Tech Asn Pulp & Paper Indust; Colombian Asn Forest Eng; Mex Tech Asn Pulp & Paper Indust. Res: Tropical forest management; utilization of tropical hardwoods for pulp and paper. Mailing Add: 805 N C St Tacoma WA 98403

BENDER, WILLIAM, b Philadelphia, Pa, Oct 4, 00. PHYSICS. Educ: Univ Colo, BS, 25, MS, 27; Yale Univ, PhD(physics), 31. Prof Exp: Consult physicist, 31-35; res physicist, E I du Pont de Nemours & Co, 35-40 & Visking Corp, 40-46; consult physicist, 46-48; prof physics & head dept, Univ SDak, 48-58; consult staff physicist, Bendix Corp, 58-60; prof, 61-68, EMER PROF PHYSICS, WESTERN WASH STATE COL, 68- Mem: AAAS; Am Phys Soc; Math Asn Am; Am Asn Physics Teachers. Res: Engineering physics; outer space research; relativity theory; scale coordinate geometry and physics; macro quantum theory. Mailing Add: Dept of Physics & Astron Western Wash State Col Bellingham WA 98225

BENDERSKY, MARTIN, b New York, NY, Apr 15, 45; m 71; c 1. TOPOLOGY. Educ: City Col New York, BS, 66; Univ Calif, Berkeley, PhD(math), 71. Prof Exp: Asst math, Aarhus, Denmark, 71-72; vis lectr, 72-73, ASST PROF MATH, UNIV WASH, 73- Res: Study of the Adams-Novikov spectral sequence, particularly the description of an unstable version and an appropriate Curtis-Kan algebra. Mailing Add: Dept of Math Univ of Wash Seattle WA 98120

BENDET, IRWIN (JACOB), b New York, NY, May 9, 27; m 60; c 2. BIOPHYSICS. Educ: City Col New York, BS, 49; Univ Mich, MA, 50; Univ Calif, PhD(biophys), 54. Prof Exp: Res asst, Univ Calif, 52-53; res assoc, 54-58, from asst prof to assoc prof, 58-66, PROF BIOPHYS, UNIV PITTSBURGH, 66- Concurrent Pos: Assoc ed jour, Biophys Soc, 70-73. Mem: AAAS; Biophys Soc; Electron Micros Soc Am; NY Acad Sci. Res: Electron microscopy; ultracentrifugation; birefringence; ultraviolet dichroism; biophysics of viruses, proteins, nucleic acid and sperm. Mailing Add: Dept of Biophys & Microbiol Univ of Pittsburgh Pittsburgh PA 15260

BENDICH, AARON, b New York, NY, June 18, 17; m 40; c 2. BIOCHEMISTRY. Educ: City Col New York, BS, 39; Columbia Univ, PhD(biochem), 46. Prof Exp: Asst, Columbia Univ, 40-43, mem res staff, Off Sci Res & Develop, 43-46, res assoc, univ, 46-47; fel org biochem studies nucleic acids, 47-48, from asst to assoc, 48-60, from asst prof to assoc prof biochem, 53-61, synthesis sect, 54-58, org chem sect, 58-69, chmn post-doctoral studies comn, 60, CHIEF DIV CELL BIOCHEM, 69-, PROF BIOCHEM, 61-, MEM, 60- Concurrent Pos: Instr, City Col New York, 46-48, 52, 54-55, 59; fel, Am-Swiss Found Sci Interchange, 56; mem sci adv bd, St Jude Children's Hosp, 66-; ed, Arch Biochem & Biophys. Honors & Awards: R Thornton Wilson Award, 60; Alfred P Sloan Found award cancer res, 64; Res Career Award, NIH, 64-66. Mem: AAAS; Am Chem Soc; Am Soc Biol Chem; Harvey Soc; NY Acad Sci. Res: Repair of genetic disease; chemical basis of malignant change. Mailing Add: Sloan-Kettering Inst Cancer Res 410 E 68th St New York NY 10021

BENDITT, EARL PHILIP, b Philadelphia, Pa, Apr 15, 16; m 45; c 3. PATHOLOGY. Educ: Swarthmore Col, BA, 37; Harvard Univ, MD, 41. Prof Exp: From instr to assoc prof path, Univ Clins, Univ Chicago, 44-57; PROF PATH & CHMN DEPT, SCH MED, UNIV WASH, 57- Concurrent Pos: Asst dir res, La Rabida Sanitarium, Ill, 50-56; vis scientist & Commonwealth Fund fel, Sir William Dunn Sch Path, Oxford, 65; coun mem, Nat Inst Environ Health Sci, 70-73; consult, Vet Admin & USPHS. Mem: Nat Acad Sci; Am Soc Exp Path (vpres, 74-75, pres, 75-76); Soc Exp Biol & Med; Histochem Soc (vpres, 62-63, pres, 63-64); Am Asn Path & Bact. Res: Cell injury; inflammation; wound healing; atherosclerosis and heart diseases. Mailing Add: Dept of Path Univ of Wash Sch of Med Seattle WA 98195

BENDIX, SELINA (WEINBAUM), b Pasadena, Calif, Feb 16, 30; m 53; c 3. ENVIRONMENTAL MANAGEMENT. Educ: Univ Calif, Los Angeles, BS, 50; Univ Calif, PhD(zool), 57. Prof Exp: Asst plant physiol, Earhart Lab, Calif Inst Technol, 50-51; asst res physicochem biol, Dept Zool, Univ Calif, 51-54; jr res biologist, Lab Comp Biol, Kaiser Found Res Inst, 57-59, asst res biologist, 59-64; asst prof biol, San Francisco State Col, 64-65; lectr, Mills Col, 65-69; sci ed, Freedom News, 70-72; RES DIR, BENDIX RES-ENVIRON CONSULT, 72-; ENVIRON REV OFFICER, CITY & CO OF SAN FRANCISCO, DEPT CITY PLANNING, 74- Concurrent Pos: Consult, Berkwood Sch, 59-70; mem, Calif Atty Gen Task Force Environ Probs of San Francisco Bay Area. Mem: AAAS; Asn Environ Prof; Scientists' Inst Pub Info; Int Asn Ecol; Asn Women in Sci. Res: Algology; cellular physiology; genetics; effect of climatic factors on growth; human ecology; heavy metal pollution; environmental impact assessment; environmental policy and decision-making process. Mailing Add: 100 Larkin St San Francisco CA 94102

BENDIXEN, LEO E, b Mills, Utah, Oct 21, 23; m 49; c 4. PLANT PHYSIOLOGY. Educ: Utah State Univ, BS, 53; Univ Calif, Davis, MS, 57, PhD(plant physiol), 60. Prof Exp: Sr lab technician agron, Univ Calif, Davis, 55-61; from asst prof to assoc prof, 61-70, assoc prof bot, 67-70, PROF AGRON & BOT, OHIO STATE UNIV, 70- Concurrent Pos: Sr fel, East-West Ctr, 75-76. Mem: Am Soc Plant Physiol; Weed Sci Soc Am; Am Soc Agron; Crop Sci Soc Am. Res: Physiological and biochemical aspects of herbicides; physiological and ecological aspects of plant reproduction and growth; physiological and anatomical aspects of plagiotropic growth. Mailing Add: Dept of Agron Col of Agr Ohio State Univ 1885 Neil Ave Columbus OH 43210

BENDLER, HARRY MORRISON, physics, astronomy, see 12th edition

BEN-DOR, SHMUEL, b Tel-Aviv, Israel, Sept 9, 34. ANTHROPOLOGY. Educ: Hebrew Univ, Israel, BA, 57; Univ Minn, Minneapolis, MA, 62, PhD(anthrop), 65. Prof Exp: Teacher, Talmud-Torah Minneapolis, Minn, 58-62; teaching asst anthrop, Univ Minn, Minneapolis, 62, teaching assoc, 64-65; vis lectr, Mem Univ, 63-64; asst prof, Wayne State Univ, 65-69; ASST PROF ANTHROP, CASE WESTERN RESERVE UNIV, 69- Mem: Am Ethnol Soc. Res: Social structure; Arctic, especially the Eskimos; East Europeam Jewish society. Mailing Add: Dept of Anthrop Case Western Reserve Univ Cleveland OH 44106

BENDOW, BERNARD, b Portland, Maine, Apr 30, 42. SOLID STATE PHYSICS. Educ: Yeshiva Univ, BA, 64; NY Univ, PhD(physics), 69. Prof Exp: Res physicist, Univ Calif, San Diego, 69-71; PHYSICIST, AIR FORCE CAMBRIDGE RES LABS, 71- Honors & Awards: Award Tech Achievement, US Air Force, 72; O'Day Mem Award, Air Force Cambridge Res Labs, 73. Mem: Am Phys Soc; Optical Soc Am. Res: Optical properties of laser materials and optical phenomena related to the interaction of laser light with solids. Mailing Add: Air Force Cambridge Res Labs Hanscom AFB MA 01731

BENDT, PHILIP JOSEPH, b Syracuse, NY, Dec 21, 19. CRYOGENICS, EXPERIMENTAL NUCLEAR PHYSICS. Educ: Mass Inst Technol, BS, 42; Columbia Univ, PhD(physics), 51. Prof Exp: Res physicist, Clinton Lab, Monsanto Chem Co, 46-47; res asst, Columbia Univ, 48-51; STAFF MEM, LOS ALAMOS SCI LAB, 51- Mem: Am Phys Soc. Res: Superfluid hydrodynamics; polarized proton targets; electron paramagnetic resonance; neutron scattering and fission experiments. Mailing Add: Los Alamos Sci Lab Los Alamos NM 87545

BENDURE, RAYMOND LEE, b Middletown, Ohio, July 1, 43; m 65; c 2. PHYSICAL CHEMISTRY, SURFACE CHEMISTRY. Educ: Purdue Univ, BS, 65; Iowa State Univ, PhD(phys chem), 68. Prof Exp: Res chemist surface chem, Miami Valley Labs, 68-72, group leader paper prod div, 72-73, SECT HEAD PAPER PROD DIV, PROCTER & GAMBLE CO, 73- Mem: Am Chem Soc. Res: Capillary ripples propagation; surface and interfacial tension determination; applications of surface and colloid chemistry to foams, emulsions, wetting, adhesion, and phase behavior; absorbent structures process development. Mailing Add: Procter & Gamble Tech Ctr 6200 Center Hill Rd Cincinnati OH 45224

BENDURE, ROBERT J, b Steubenville, Ohio, May 2, 20; m 42; c 2. INORGANIC CHEMISTRY, POLLUTION CHEMISTRY. Educ: Miami Univ, BA, 42. Prof Exp: Chem analyst, 42-44 & 46-47, sr analyst, 47-49, from chemist to sr chemist, 49-61, supv chemist, 61-68, mgr chem labs, 68-70, DIR CHEM RES, ARMCO STEEL CORP, 70- Honors & Awards: Lundell Bright Mem Award, Am Soc Testing & Mat, 72. Mem: AAAS; Am Chem Soc; Am Soc Testing & Mat. Res: Inorganic analytical chemistry using both classical and instrumental methods; ferrous and nonferrous metals analysis; inclusions; intermetallic compounds and dissolved gases in steel; air and water pollution control; corrosion. Mailing Add: Res Ctr Armco Steel Corp Middletown OH 45042

BENEDEK, GEORGE BERNARD, b New York, NY, Dec 1, 28; m 55; c 2. PHYSICS. Educ: Rensselaer Polytech Inst, BS, 49; Harvard Univ, MA, 51, PhD(physics), 54. Prof Exp: Mem staff, Joint Harvard-Lincoln Lab, Mass Inst Technol Proj, 53-55; res fel, Harvard Univ, 55-57, lectr solid state physics, 57-58, asst prof appl physics, 58-61; assoc prof, 61-65, PROF PHYSICS, MASS INST TECHNOL, 65- Concurrent Pos: Guggenheim fel, 60; prof fel, Atomic Energy Res Estab, Harwell, Eng, 67; mem gov bd, Am Inst Physics, 71-74. Mem: Fel Am Phys Soc. Res: Magnetic resonance; high pressure physics; solid state physics; critical phenomena and light scattering with lasers; biological physics. Mailing Add: Dept of Physics Mass Inst of Technol Cambridge MA 02139

BENEDEK, TIBOR, medical mycology, deceased

BENEDETTI, JACQUELINE KAY, b Chicago, Ill, July 22, 48. BIOSTATISTICS. Educ: Univ Calif, Los Angeles, BA, 70; Univ Wash, PhD(biomath), 74. Prof Exp: NAT INST GEN MED SCI FEL, HEALTH SCI COMPUT FACIL & DEPT BIOMATH, SCH MED, UNIV CALIF, LOS ANGELES, 74- Mem: Inst Math Statist; Am Statist Asn; Biomet Soc. Res: Analysis of categorical data, including multidimensional contingency table analysis, and measures of association for two-way tables; clinical trials statistics. Mailing Add: Dept of Biomath Univ of Calif Sch of Med Los Angeles CA 90024

BENEDETTO, FRANK ARISTIDE, b Macon, Ga, Mar 30, 14. PHYSICS. Educ: St Louis Univ, AB, 36; Fordham Univ, MS, 40, PhD(cosmic rays), 46. Prof Exp: Asst prof physics & math, Spring Hill Col, 38-39; from asst prof to assoc prof, 47-58, chmn dept, 54-67, PROF PHYSICS & MATH, LOYOLA UNIV, LA, 58-, SECY BD DIRS & EXEC ASST TO PRES & VPRES COMMUN, 66- Concurrent Pos: Am Philos Soc grants, Fordham Univ & Loyola Univ. Mem: Am Phys Soc; Am Geophys Union; Seismol Soc Am; Am Asn Physics Teachers. Res: Mesotron intensity variations with atmospheric temperature changes; mesotron production in the atmosphere; terrestrial ionization balance; radium protection. Mailing Add: Dept of Physics Loyola Univ New Orleans LA 70118

BENEDETTO, JOHN, b Melrose, Mass, July 16, 39; m 68; c 1. ANALYTICAL MATHEMATICS. Educ: Boston Col, BA, 60; Harvard Univ, MA, 62; Univ Toronto, PhD(math), 64. Prof Exp: Assoc mem tech staff, Radio Corp Am, Burlington, 61-62, mem tech staff, 62-64; teaching fel math, Univ Toronto, 62-64; asst prof, NY Univ, 64-65; res assoc, Inst Fluid Dynamics & Appl Math, 65-66, asst prof, 66-68, ASSOC PROF MATH, UNIV MD, COLLEGE PARK, 68- Concurrent Pos: Assoc mathematician, Int Bus Mach Corp, Mass, 64-; prof, Scuola Normale Superiore, Pisa, 70-71. Mem: Am Math Soc; Math Asn Am. Res: Theory of generalized function; Laplace and Fourier type transforms and topological vector spaces. Mailing Add: Dept of Math Univ of Md College Park MD 20742

BENEDICT, ALBERT ALFRED, b Pasadena, Calif, Nov 26, 21; m 47; c 1. IMMUNOCHEMISTRY. Educ: Univ Calif, AB, 48, MA, 50, PhD(bact), 52. Prof Exp: Res assoc bact, Univ Calif, 49-52; res assoc virol med br, Univ Tex, 52-53, asst prof prev med & pub health, 53-57; assoc prof bact, Univ Kans, 57-63; PROF MICROBIOL & CHMN DEPT, UNIV HAWAII, 63- Concurrent Pos: WHO consult, 71. Mem: AAAS; Am Soc Microbiol; Soc Exp Biol & Med; Am Asn Immunol. Res: Avian immunoglobulins; structure and synthesis of immunoglobulins. Mailing Add: Dept of Microbiol Univ of Hawaii Honolulu HI 96822

BENEDICT, CHAUNCEY, b Lake Placid, NY, June 10, 30; m 54; c 1. PLANT BIOCHEMISTRY. Educ: Cornell Univ, BS, 54, MS, 56; Purdue Univ, PhD(plant biochem), 60. Prof Exp: Res assoc biochem, Brookhaven Nat Lab, 60-61; fel & asst prof biol, Dartmouth Med Sch, 61-62; asst prof biochem, Wayne State Univ, 62-66; from assoc prof to prof plant sci, 66-74, PROF PLANT PHYSIOL, TEX A&M UNIV, 74- Concurrent Pos: Mem subcomt biol chem, Nat Res Coun, 64-67. Mem: Am Soc Plant Physiol. Res: Plant metabolism and carotene syntehsis. Mailing Add: Dept of Plant Sci Tex A&M Univ College Station TX 77843

BENEDICT, DONALD LEE, b Galesburg, Ill, July 5, 16; m 44; c 3. PHYSICS, RESEARCH ADMINISTRATION. Educ: Knox Col, BA, 38; Univ Wis, PhD(physics), 43. Prof Exp: Asst, Univ Wis, 38-41; asst sect S-1, Off Sci Res & Develop, 41-43; res physicist, Sylvania Elec Prods, Inc, NY, 43-45; res fel electronics, Harvard Univ, 45-49; consult, Raytheon Mfg Co, Mass, 47-49; assoc chmn eng res div, Stanford Res Inst, 49-54, dir phys sci div, 54-59, dir Europ Off, Switz, 59-60, dir phys & biol sci div, 60-61, assoc dir res, 61-63, dir poulter Res Labs, 63-64, exec dir phys & indust sci, 64-66; pres, Ore Grad Ctr, 66-69; personal serv contracts, US Agency Int Develop, 70-71, UN Develop Prog, 72, Nat Sci Found, 73-74 & World Bank, 74-75, PERSONAL SERV CONTRACT, NAT SCI FOUND, 76- Mem: AAAS; Am Phys Soc; Am Math Soc; Inst Elec & Electronics Engrs; Sigma Xi. Res: Solid state physics, applied mathematics; planning, developing and administering science-based applied research programs and graduate level educational programs. Mailing Add: Court St Chelsea VT 05038

BENEDICT, GEORGE FREDERICK, b Los Angeles, Calif, Mar 17, 45; m 67; c 2. ASTRONOMY. Educ: Univ Mich, BS(astron) & BS(physics), 67; Northwestern Univ, MS, 69, PhD(astron), 72. Prof Exp: ASTRONOMER, UNIV TEX, AUSTIN, 72- Concurrent Pos: Consult, Boller & Chivens Div, Perkin Elmer Corp, 73- Mem: Am Astron Soc. Res: Far ultraviolet spectroscopy of early-type stars; galaxy photometry: digital image processing and analysis. Mailing Add: Dept of Astron Univ of Tex Austin TX 78712

BENEDICT, HARRIS MILLER, b Cincinnati, Ohio, Aug 13, 07; m 48. POLLUTION BIOLOGY. Educ: Univ Cincinnati, AB, 29, AM, 30; Univ Chicago, PhD(plant physiol), 32. Prof Exp: Asst hort, Purdue Univ, 33-34; asst plant physiologist in charge soil conserv serv nursery, Bur Plant indust, USDA, 34-36, assoc physiologist in charge forage crops invests, 36-42 & spec guayule res proj, Salinas, 42-45, plant physiologist, 45-46, sr plant physiologist, Div Rubber Invest, Bur Plant Indust, 47-49; sr plant physiologist, Isotopes Div, AEC, 49-50; sr plant physiologist, Stanford Res Inst, 50-59, mgr agr res ctr, 59-61, supvr plant sci, 61-64, mgr plant biol sect, 64-66, staff scientist, 66-72; CONSULT AIR POLLUTION, 72- Concurrent Pos: Off del, USDA Fourth Int Grassland Cong, Wales, 37; plant ecologist, Forest & Range Exp Sta, US Forest Serv, Colo, 46-47; sr plant physiologist, Stanford Res Inst, 47. Mem: AAAS; Am Soc Plant Physiol; Am Soc Agron; Air Pollution Control Asn. Res: Seed germination; weed control; physiology of range plants; photoperiodism; physiology of rubber plants; growth regulators; effect of ultraviolet radiation on growth and on the calcium and phosphorus contents of plants; response of vegetation to air pollution; mineral nutrition. Mailing Add: PO Box 393 South Pasadena CA 91030

BENEDICT, IRVIN J, b Sayre, Okla, Nov 24, 24; m 47; c 3. BIOLOGY, PUBLIC HEALTH. Educ: Univ Tex, BS, 51, EdD(health educ), 68; Trinity Univ, Tex, MS, 54. Prof Exp: Teacher, San Antonio Independent Sch Dist, 52-54; head sanit div, City Water Bd, San Antonio, 54-59; PROF BIOL & HEALTH SCI, SAN ANTONIO COL, 59-, ASSOC DEAN, 75- Concurrent Pos: Consutl, City Water Bd, San Antonio, 59-60; consult sanit, Raba & Assoc, Consult Engr, 74- Mem: Sigma Xi. Res: Utilization of molecular membrane techniques in field situations for public water supply sampling; communicable disease content in public school textbooks; evapotranspiration applications to disposal of domestic wastes. Mailing Add: 16403 Hidden View San Antonio TX 78232

BENEDICT, JAMES HAROLD, b Floral Park, NY, Mar 9, 22; m 48; c 4. BIOCHEMISTRY. Educ: NY Univ, AB, 44; Univ Pittsburgh, PhD(chem), 50. Prof Exp: Asst chem, Univ Pittsburgh, 46-47; CHEMIST, PROCTER & GAMBLE CO, 49- Mem: Fel AAAS; Am Chem Soc; Am Oil Chem Soc. Res: Fat metabolism; development of semimicro analytical methods; trace analyses; toxicology. Mailing Add: Sharon Woods Tech Ctr Procter & Gamble Co Cincinnati OH 45241

BENEDICT, JEAN DAVIDSON, b Portsmouth Va; div. BIOCHEMISTRY. Educ: Hunter Col, AB, 41; Mass Inst Technol, MA, 47. Prof Exp: Res asst biochem, Harvard univ, 47-48; res asst biochem, Pub Health Res Inst, New York, 48-54; chemist, 54-58, prog analyst metab, 58-62, scientist adminr, 62-70, ophthal, Nat Eye Inst, 70-73, SCIENTIST ADMINR NEUROL, NAT INST NEUROL & COMMUN DIS & STROKE, NIH, 73- Mem: Am Chem Soc. Res: Stroke and trauma. Mailing Add: 4912 Bangor Dr Kensington MD 20795

BENEDICT, JOHN HOWARD, JR, b San Antonio, Tex, Mar 31, 44; m 65; c 1. ENTOMOLOGY. Educ: Calif State Univ, Los Angeles, BA, 69; Univ Calif, Davis, PhD(entom), 75. Prof Exp: RES ENTOMOLOGIST VI, ROCKEFELLER FOUND-

UNIV CALIF, DAVIS, 75- Mem: Entom Soc Am; Entom Soc Can; Pac Coast Entom Soc. Res: Cooperative research projects in host plant resistance with plant breeders and entomologists throughout the United States and Mexico to identify sources of resistance in cotton to lygus bugs and other pest species. Mailing Add: Dept of Entom Univ of Calif Davis CA 95616

BENEDICT, JOSEPH T, b Chicago, Ill, May 21, 20; m 57. INORGANIC CHEMISTRY, ACADEMIC ADMINISTRATION. Educ: Univ Chicago, BS, 43; Mass Inst Technol, PhD(inorg chem), 50. Prof Exp: Asst, Mass Inst Technol, 44-50; res assoc chem, Columbia Univ, 50-53; res chemist, Nickel Processing Corp, Div Int Nickel Co, 53-57; control mgr electronic chem, Merck & Co, Inc, 57-58; mgr tech develop, Int Div, Stauffer Chem Co, 58-63; from asst prof to assoc prof, 63-71, PROF CHEM, FAIRLEIGH DICKINSON UNIV, 71-, DIR MBA PROG CHEM MKT & ECON, PHARMACEUT MKT & ECON, 74- Mem: Am Chem Soc. Res: Fused salt technology; chemical-pharmaceutical marketing and economics. Mailing Add: Dept of Chem Fairleigh Dickinson Univ Teaneck NJ 07666

BENEDICT, PETER CARL, b Vienna, Austria, Aug 14, 17; nat US; m 41; c 2. GEOLOGY. Educ: Columbia Univ, BS, 49; Univ Zurich, PhD(geol), 52. Prof Exp: Geologist, Bear Creek Mining Co, 52-53; mineralogist, Crane Co, 53-54; geologist, Calif Explor Co, 55-58; ASSOC PROF EARTH SCI, STATE UNIV NY ALBANY, 58- Res: Structural geology and tectonics. Mailing Add: Dept of Geol State Univ of NY Albany NY 12203

BENEDICT, ROBERT CURTIS, b Bridgeport, Conn, May 17, 32; m 58; c 3. BIOCHEMISTRY. Educ: Univ Conn, BS, 53; Cornell Univ, MS, 55; Univ Pa, PhD(biochem), 65. Prof Exp: Pa Plan Med Res fel, 64-66; RES CHEMIST, EASTERN UTILIZATION RES & DEVELOP DIV, USDA, . WYNDMOOR, 66- Mem: AAAS; Am Chem Soc; Am Soc Animal Sci; Inst Food Technologists. Res: Dairy chemistry; biochemistry of reproduction; tobacco smoke investigations; gradient centrifugation of intact cells; meat science. Mailing Add: 342 Forest Ave Willow Grove PA 19090

BENEDICT, ROBERT GLENN, b Iona Co, Mich, July 16, 11; m 45; c 5. MICROBIOLOGY. Educ: Mich State Univ, BS, 36; Va Polytech Inst, MS, 38; Univ Wis, PhD(agr bact), 42. Prof Exp: Asst gen bact, Va Polytech Inst, 36-38; res asst agr bact, Univ Wis, 38-41, instr, 41-42; cur, Northern Regional Res Lab, USDA, 42-46, investr, 45-53; investr classified res, Chem Corps, US Army, 52-56; in charge microbial technol polymers unit, Northern Regional Lab, USDA, 56-58 & new prod explor & res unit, 58-60; res asst prof. pharmacog, Col Pharm, 60-63, actg asst prof, 63-64, res assoc prof civil eng, Univ, 73, RES ASSOC PROF PHARMACOG, COL PHARM, UNIV WASH, 64- Concurrent Pos: Consult & expert witness, Antibiotic Litigations, Fed Trade Comn, 59-60 & Patent Litigations, Am Cyanamid Co, Can, Europe & Mid E, 63-; gen med res grants, NIH, 60-68, 69-72; vis assoc prof microbiol, Univ Puget Sound, 73-74. Honors & Awards: Superior Serv Award, USDA, 50. Mem: Am Soc Pharmacog; Am Soc Microbiol. Res: Mushroom toxins and secondary metabolites of possible interest in medicine; taxonomy of microorganisms in disposal of sewage and solid wastes. Mailing Add: 4229 NE 104th Pl Seattle WA 98125

BENEDICT, WILLIAM SIDNEY, b Lake Linden, Mich, July 4, 09; m 36; c 1. CHEMICAL PHYSICS. Educ: Cornell Univ, AB, 28, AM, 29; Mass Inst Technol, PhD, 33. Prof Exp: Nat Res Coun fel chem, Princeton Univ, 33-35; res chemist, Gen Chem Co, 36-42; phys chemist, Carnegie Inst Geophys Lab, 42-46; sr physicist, Nat Bur Stand, 46-52; res contract dir, Johns Hopkins Univ, 52-67; PROF, INST MOLECULAR PHYSICS, UNIV MD, 67- Mem: AAAS; Optical Soc Am; Am Phys Soc. Res: Structure of polyatomic molecules; infrared spectroscopy; planetary atmospheres. Mailing Add: 4935 Massachusetts Ave NW Washington DC 20016

BENEDICT, WINFRED GERALD, b Wallaceburg, Ont, Mar 18, 19; m 43; c 4. PLANT PATHOLOGY. Educ: Ont Agr Col, BSA, 49; Univ Toronto, PhD(plant path & mycol), 52; Univ Leeds, MPhil, 71. Prof Exp: Asst-assoc plant pathologist, Can Dept Agr, 52-57; assoc prof, 57-66, PROF BIOL, UNIV WINDSOR, 66- Concurrent Pos: Vis prof, Univ Leeds, 69-70. Mem: Am Phytopath Soc; Can Phytopath Soc; fel Linnean Soc London. Res: Diseases of forage and field crops; physiology of plant infection; mycology, myxomycetes; effect of light intensity and wavelength on early stages of leaf spot infections. Mailing Add: Dept of Biol Univ of Windsor Windsor ON Can

BENEDICTY, MARIO, b Trieste, Italy, July 16, 22; m 47; c 2. MATHEMATICS. Educ: Univ Rome, MathD, 46, Libera Docenza, 52. Prof Exp: Asst prof anal geom, Univ Rome, 48-51, assoc prof algebraic geom, 51-58; assoc prof, 57-60, chmn dept, 63-69, PROF MATH, UNIV PITTSBURGH, 60- Concurrent Pos: Co-worker math, Dizionario Enciclopedico dell Enciclopedia Italiana, contribr, 60; vis prof, Univ BC, 60-62. Mem: AAAS; Am Math Soc; Math Asn Am; NY Acad Sci; Can Math Cong. Res: Algebraic geometry and related fields; modern algebra; projective geometries; computer-assisted instruction; geometry. Mailing Add: Dept of Math Univ of Pittsburgh Pittsburgh PA 15260

BENEKE, EVERETT SMITH, b Greensboro, NC, July 6, 18. BOTANY, MICROBIOLOGY. Educ: Miami Univ, AB, 40; Ohio State Univ, MS, 41; Univ Ill, PhD(bot), 48. Prof Exp: From asst prof to assoc prof, 48-58, PROF BOT & MICROBIOL, DEPT BOT & PLANT PATH, MICH STATE UNIV, 58-, DEPT MICROBIOL & PUB HEALTH, 67- Concurrent Pos: WHO fel, Indonesia, 69. Honors & Awards: Sigma Xi Jr Award for Distinguished Res, Mich State Univ Chap, 58; Distinguished Serv Plaque, Am Asn Bioanalysts, 67. Mem: Am Soc Microbiol; Soc Indust Microbiol; Int Asn Microbiol Socs; hon mem Mycol Soc Am (actg secy-treas, 54, secy-treas, 56-59, vpres, 60, pres, 62); Fel AAAS. Res: Saprolegniales and other related water molds; mycology; horticulture; medical mycology. Mailing Add: Dept of Microbiol & Pub Health Mich State Univ East Lansing MI 48823

BENENSON, ABRAM SALMON, b Napanoch, NY, Jan 22, 14; m 39; c 4. MEDICINE. Educ: Cornell Univ, AB, 33, MD, 37; Am Bd Path, Am Bd Prev Med & Am Bd Med Microbiol, dipl. Prof Exp: Rotating intern, Queens Gen Hosp, New York, 37-39; psychiat intern, Bellevue Hosp, 39; asst & chief lab serv, Tripler Gen Hosp, Honolulu, 41-42; asst virus lab, Army Med Dept Res & Grad Sch, US Dept Army, Washington, DC, 46-47, instr path & bact, Med Sch Tufts Col, 47-49, prof mil sci & tactic, 47-49, dir, Second Army Area Med Lab, Ft George G Meade, Md, 49-52, dir, Trop Res Med Lab, San Juan, PR, 52-54, dir exp med, Camp Detrick, Md, 54-55, dir div immunol, Walter Reed Army Inst Res, 56-60, dir div commun dis & immunol, 60-62; dir, Pakistan-SEATO Cholera Res Lab, 62-66; prof prev med & microbiol, Jefferson Med Col, 66-69; PROF COMMUNITY MED & CHMN DEPT, COL MED, UNIV KY, 69- Concurrent Pos: From assoc mem to mem comn immunization, Armed Forces Epidemiol Bd, 56-72, dir, 67-72; mem study group on typhoid vaccines, WHO & expert comt on cholera, 62-73, mem expert adv comt bact dis, 73-, consult, Expert Comt Biol Stand, 58 & Expert Comt on Smallpox, 64-; mem bd sci counsr, Div Biol Stand, NIH, 67-71, chmn, 69-71, chmn cholera adv comt, 68-72; ed, Control of Commun Dis of Man, Am Pub Health Asn, 70-75; mem, Cholera Panel, US-Japan Coop Sci Prog, 73- Honors & Awards: First Award, Army Sci Conf, 57; Legion of Merit, 62. Mem: AAAS; Am Pub Health Asn; Am Epidemiol Soc (secy-treas); AMA; Am Asn Immunol. Res: Immunology; virology; epidemiology; clinical pathology; infectious diseases. Mailing Add: Dept of Community Med Univ of Ky Col of Med Lexington KY 40506

BENENSON, DAVID MAURICE, b Brooklyn, NY, Jan 22, 27; m 55; c 2. PLASMA PHYSICS. Educ: Mass Inst Technol, BS, 50; Calif Inst Technol, MS, 53, PhD(aeronaut), 57. Prof Exp: Proj engr, Calif Inst Technol, 50-53; res engr, Res Labs, Westinghouse Elec Corp, 57-63; assoc prof, 63-67, PROF PLASMA & GAS DYNAMICS, STATE UNIV NY BUFFALO, 67- Concurrent Pos: Instr, Carnegie Inst Technol, 58-62; investr, NSF, 64-; consult, Gen Elec Co, 74- Inst Elec & Electronic Engrs; Am Soc Mech Engrs. Res: Plasma and gas dynamics; turbomachinery; boundary layer analysis; flow problem; heat transfer; aerodynamic noise; energy conversion; power distribution. Mailing Add: Fac of Eng & Appl Sci State Univ of NY at Buffalo Amherst NY 14226

BENENSON, RAYMOND ELLIOTT, b New York, NY, Dec 12, 25; m 56. NUCLEAR PHYSICS. Educ: Mass Inst Technol, BS, 46; Univ Chicago, MS, 49; Univ Wis, PhD(physics), 55. Prof Exp: Assoc physicist, Brookhaven Nat Lab, 54-56; physicist, Wright Air Develop Ctr, Brookhaven, 56; instr physics, City Col New York, 56-60, from asst prof to assoc prof, 60-67; assoc prof, 67-70, PROF PHYSICS, STATE UNIV NY ALBANY, 70- Concurrent Pos: Res assoc, Pegram Lab, Columbia Univ, 56-; co-worker, Phys Inst Basel, 62-63. Res: Fast neutron and gamma ray spectroscopy in conjunction with Van de Graaff accelerators. Mailing Add: Dept of Physics State Univ of NY 1400 Washington Ave Albany NY 12203

BENENSON, WALTER, b New York, NY, Apr 27, 36; m 69; c 1. NUCLEAR PHYSICS. Educ: Yale Univ, BS, 57; Univ Wis, MS, 59, PhD(physics), 62. Prof Exp: Res assoc nuclear physics, Nuclear Res Inst, Strasbourg, 62-63; from asst prof to assoc prof, 63-72, PROF NUCLEAR PHYSICS, MICH STATE UNIV, 72- Concurrent Pos: Vis fel, Australian Nat Univ, 68; vis prof, Univ Grenoble, 70; Nat Acad Sci fel, Poland & Czech, 74. Mem: Am Phys Soc. Res: Polarization of fast neutrons; electromagnetic transitions in light nuclei; mass measurements of nuclei far from stability; direct nuclear reactions; use of computers in nuclear physics. Mailing Add: Cyclotron Lab Mich State Univ East Lansing MI 48823

BENEPAL, PARSHOTAM S, b Bhagomajra-Punjab, India, Feb 25, 33; m 54; c 3. PLANT GENETICS, PLANT PHYSIOLOGY. Educ: Punjab, India, BS, 53, MS, 58; Kans State Univ, PhD(plant genetics, hort), 67. Prof Exp: Res asst, Regional Res Lab, USDA, Kans State Univ, 62-65; PROF LIFE SCI & CHMN DEPT, VA STATE COL, 67- Concurrent Pos: USDA Agr Res Serv grant, 68-75. Mem: AAAS; Am Inst Biol Sci; Am Genetic Asn; Am Soc Plant Physiol; fel Royal Hort Soc. Res: Plant sciences; genetics and biochemistry of insect and disease resistance; breeding for quality and agronomic and physiological research on crop plants. Mailing Add: 1106 Duke of Gloucester Colonial Heights VA 23834

BENERITO, RUTH ROGAN, b New Orleans, La, Jan 12, 16; m 50. PHYSICAL CHEMISTRY. Educ: Sophie Newcomb Mem Col, BS, 35; Tulane Univ, MS, 38; Univ Chicago, PhD(chem), 48. Prof Exp: Instr chem, Randolph-Macon Women's Col, 40-43; asst prof, Tulane Univ, 43-53; phys chemist, 53-58, head colloidal chem invest, Cotton Chem Lab, 58-61, HEAD ORGANO-PHYS INVEST, COTTON CHEM REACTIONS LAB, SOUTHERN REGIONAL RES LAB, USDA, 61- Concurrent Pos: Adj prof, Med Sch & Grad Sch, Tulane Univ. Honors & Awards: Distinguished Serv Award, USDA, 64, 70; Fed Woman Award, 68, Southern Chemist Award, 68; Garvan Medalist, 70; Southwest Regional Award, Am Chem Soc, 72. Mem: AAAS; Am Chem Soc; Sigma Xi; Am Oil Chemists Soc; fel Am Inst Chemists. Res: Thermodynamics of electrolytic solutions; kinetics and reaction mechanisms; cellulose reactions; surface phenomena; equilibrium; x-ray diffraction. Mailing Add: 4733 Marigny St New Orleans LA 70122

BENES, ELINOR SIMSON, b Yreka, Calif, Oct 5, 24; m 45; c 4. MICROSCOPIC ANATOMY, VERTEBRATE ZOOLOGY. Educ: San Jose State Col, AB, 45; Ariz State Univ, MS, 60; Univ Calif, Davis, PhD(zool), 66. Prof Exp: NIH fel, Univ Calif, Davis, 66-67; instr, 67, asst prof, 69-73, ASSOC PROF BIOL SCI, CALIF STATE UNIV, SACRAMENTO, 73- Mem: AAAS; Am Inst Biol Sci; Am Soc Ichthyol & Herpet; Soc Study Amphibians & Reptiles; Am Soc Zool. Res: Behavioral response to color; effects of light on microanatomy; growth and metabolism of lizards; feeding and maintenance of lizards. Mailing Add: Dept of Biol Sci Calif State Univ Sacramento CA 95819

BENES, VACLAV EDVARD, b Brussels, Belgium, July 24, 30; nat US; m 51; c 2. MATHEMATICS. Educ: Harvard Univ, BA, 50; Princeton Univ, PhD(logic), 53. Prof Exp: Instr logic & philos, Princeton Univ, 52-53; MEM TECH STAFF, BELL TEL LABS, INC, 53- Mem: Am Math Soc; Math Asn Am; Soc Indust & Appl Math; Inst Math Statist. Res: Probability; stochastic processes; logic; set theory; linear systems; circuit theory; functional equations; information theory; theory of queues and congestion. Mailing Add: Bell Tel Labs Inc Murray Hill NJ 07974

BENESCH, REINHOLD, b Bielsko, Pyland, Aug 13, 19; nat US; m 46. BIOCHEMISTRY. Educ: Univ Leeds, BSc, 41, MSc, 45; Northwestern Univ, PhD(biochem), 50. Prof Exp: Asst lectr, Univ Reading, 42-43; instr biochem, Johns Hopkins Univ, 47-48; from instr to asst prof, Northwestern Univ, 50-52; asst prof, Univ Iowa, 52-55; USPHS spec res fel, Enzyme Inst, Univ Wis, 55-56; Am Heart Asn estab investr, Marine Biol Lab, Woods Hole Oceanog Inst, 56-62; assoc prof, 60-66, PROF BIOCHEM, COL PHYSICIANS & SURGEONS, COLUMBIA UNIV, 66- Concurrent Pos: Lalor fel, Marine Biol Lab, Woods Hole Oceanog Inst, 51; NIH res career award, 62- Mem: AAAS; Am Chem Soc; Am Soc Biol Chemists. Res: General biochemistry; calcification; nutrition; protein chemistry; electrochemistry; polarography; sulfur compounds; structure function relationships in hemoglobin. Mailing Add: Col Physicians & Surgeons Columbia Univ New York NY 10032

BENESCH, RUTH ERICA, b Paris, France, Feb 25, 25; US citizen; m 46; c 2. BIOCHEMISTRY. Educ: Univ London, BS, 46; Northwestern Univ, PhD(biochem), 51. Prof Exp: Demonstr chem, Univ Reading, 46-47; res assoc biochem, Johns Hopkins Univ, 47-48; fel, State Univ Iowa, 52; fel, Enzyme Inst, Univ Wis, 55; independent investr, Marine Biol Lab, Woods Hole Oceanog Inst, 56-60; res assoc, 60-64, asst prof, 64-72, ASSOC PROF BIOCHEM, COL PHYSICIANS & SURGEONS, COLUMBIA UNIV, 72- Mem: Am Chem Soc; Am Soc Biol Chemists; Am Soc Hemat. Res: Physical biochemistry; protein chemistry; enzymology; structure-function relationships in hemoglobin; hematology. Mailing Add: Col Physicians & Surgeons Columbia Univ New York NY 10032

BENESCH, SAMUEL ELI, b Baltimore, Md, Aug 19, 24; m 50; c 2. SYSTEMS THEORY. Educ: Johns Hopkins Univ, BA, 50; Univ Ill, MA, 51, PhD(math), 53. Prof Exp: Res supvr exterior ballistics, Calif Inst Technol, 53-62; staff lab dept mgr space systs, TRW Systs Group, 62-70; exec dir psychol, Los Angeles Ctr Group Psychother, 70-71; dir training crisis intervention, Childrens Hosp, Los Angeles, 71-

72; STAFF SCIENTIST DEFENSE SYSTS, TRW SYSTS GROUP, REDONDO BEACH, 72- Mem: AAAS; Am Math Soc; Am Soc Psychical Res. Res: Systems models of consciousness relating psychological phenomena to functions of complex variables. Mailing Add: 8895 Appian Way Hollywood CA 90046

BENESCH, WILLIAM MILTON, b Baltimore, Md, Apr 22, 22; m 46; c 3. MOLECULAR PHYSICS, MOLECULAR SPECTROSCOPY. Educ: Lehigh Univ, BA, 42; Johns Hopkins Univ, MA, 49, PhD(physics), 52. Prof Exp: Jr instr physics, Johns Hopkins Univ, 42-44; mem, Comn Relief Belg Fel, Univ Liege, 52-53; asst prof, Univ Pittsburgh, 53-60; res fel, Weizmann Inst Sci, 60; assoc prof, 62-66, PROF PHYSICS, UNIV MD, COLLEGE PARK, 66-, DIR INST MOLECULAR PHYSICS, 73- Concurrent Pos: Consult, Argonne Nat Lab, 74- Mem: Am Phys Soc; Optical Soc Am; Soc Appl Spectros. Res: Infrared spectroscopy; solar and atmospheric spectroscopy; molecular interactions; atomic and molecular structure. Mailing Add: Inst for Molecular Physics Univ of Md College Park MD 20742

BENESI, HANS ARPAD, chemistry, see 12th edition

BENET, LESLIE Z, b Cincinnati, Ohio, May 17, 37; m 60; c 2. BIOPHARMACEUTICS. Educ: Univ Mich, AB, 59, BS, 60, MS, 62; Univ Calif, PhD(pharmaceut chem), 65. Prof Exp: Asst prof pharm, Wash State Univ, 65-69; asst prof, 69-71, ASSOC PROF PHARM & PHARMACEUT CHEM, MED CTR, UNIV CALIF, SAN FRANCISCO, 71-, VCHMN DEPT PHARM, SCH PHARM, 73- Concurrent Pos: Consult, Merck, Sharp & Dohme Res Labs, 72-; assoc ed, J Pharmacokinetics & Biopharmaceut, 73- Mem: AAAS; Am Pharmaceut Asn; fel Acad Pharmaceut Sci. Res: Pharmacokinetics relating to the intestinal absorption and renal transport of drugs; modeling processes; drug bioavailability as a function of the route of administration and multiparticulate dissolution. Mailing Add: Sch of Pharm Univ of Calif San Francisco CA 94143

BENET, SULA, US citizen. CULTURAL ANTHROPOLOGY. Educ: Univ Warsaw, dipl, 35; Columbia Univ, PhD(anthrop), 44. Prof Exp: PROF ANTHROP, HUNTER COL, 44- Mem: Fel NY Acad Sci; fel Am Anthrop Asn; Am Ethnol Soc; Polish Inst Sci. Res: Effects of culture on longevity. Mailing Add: 315 Central Park W New York NY 10025

BENEVENGA, NORLIN JAY, b San Francisco, Calif, Feb 16, 34; m 56; c 3. BIOCHEMISTRY, NUTRITION. Educ: Univ Calif, Davis, BS, 59, EE & MS, 60, PhD(nutrit), 65. Prof Exp: Res assoc amino acid metab, Mass Inst Technol, 65; proj assoc, 65-66, from asst prof to assoc prof, 66-74, PROF MEAT & ANIMAL SCI & NUTRIT SCI, UNIV WIS-MADISON, 74- Concurrent Pos: Mem, Nat Res Coun Sub-Comt Lab Animal Nutrit, 74. Mem: AAAS; Am Nutrit Soc; Am Inst Nutrit; Am Soc Animal Sci; Brit Biochem Soc. Res: Nutritional and metabolic effects of amino acid excesses and imbalances and digestion and metabolism of lactose. Mailing Add: Dept of Meat & Animal Sci Univ of Wis 905 University Ave Madison WI 53706

BENEVENTO, LOUIS ANTHONY, b Waterbury, Conn, Nov 17, 40. NEUROPHYSIOLOGY, NEUROANATOMY. Educ: Rensselaer Polytech Inst, BS, 62, MS, 64; Univ Md, PhD(physiol), 68. Prof Exp: Nat Inst Neurol Dis & Stroke res fel neurosci, Brown Univ, 68-70, resident fel, 69; spec fel neurophysiol, Max Planck Inst Psychiat, Ger, 70-71; asst prof physiol, Sch Med, Univ Va, 71-72; ASSOC PROF ANAT & OPHTHAL, COL MED, UNIV ILL MED CTR, 72- Mem: AAAS; NY Acad Sci; Inst Elec & Electronics Engrs; Am Acad Arts & Sci; Am Anat Asn. Res: Anatomical and physiological correlates of the central nervous system-neocortex. Mailing Add: Col of Med Univ of Ill at the Med Ctr Chicago IL 60680

BENFER, ROBERT ALFRED, b San Antonio, Tex, Mar 1, 39; m 61; c 1. ANTHROPOLOGY. Educ: Univ Tex, Austin, BA, 63, MA, 64, PhD(anthrop), 68. Prof Exp: Asst prof anthrop, Univ Wis-Milwaukee, 67-69; asst prof, Univ Mo-Columbia, 69-71, chmn dept, 71-75, ASSOC PROF ANTHROP, UNIV MO-COLUMBIA, 71- Concurrent Pos: Grant, Univ Mo-Columbia Res Coun, 70 & 72; co-prin investr, NSF Grant, 71-72. Mailing Add: Dept of Anthrop Univ of Mo-Columbia Columbia MO 65201

BENFEY, BRUNO GEORG, b Düsseldorf, Ger, Oct 9, 17; m 49; c 3. PHARMACOLOGY, THERAPEUTICS. Educ: Univ Hamburg, MD, 48; Univ Göttingen, dipl chem, 50. Prof Exp: Asst org chem, Univ Göttingen, 48-50; asst, Second Clin Internal Med, Univ Hamburg, 50-51; sci asst, Inst Physiol Chem, Univ Göttingen, 51; USPHS fel physiol chem, Yale Univ, 51-52; lectr, 52-54, asst prof, 54-60, ASSOC PROF PHARMACOL, McGILL UNIV, 60- Mem: Am Soc Pharmacol & Exp Therapeut; Pharmacol Soc Can; Can Physiol Soc. Res: Molecular weight of heparin; isolation of pikromycin from actinomycetes; enzymatic studies with amino acids and peptides; chemistry of pituitary hormones; pharmacology and biochemistry of sympathomimetic and adrenergic blocking agents. Mailing Add: Dept of Pharmacol McGill Univ Montreal PQ Can

BENFEY, OTTO THEODOR, b Berlin, Ger, Oct 31, 25; nat US; m 46; c 4. PHYSICAL ORGANIC CHEMISTRY, HISTORY OF SCIENCE. Educ: Univ London, BSc, 45, PhD(org chem), 47. Prof Exp: From asst prof to assoc prof chem, Haverford Col, 48-56; assoc prof, Earlham Col, 56-59, Prof, 59-73; ED, CHEMISTRY, 64-; DANA PROF CHEM & HIST OF SCI, GUILFORD COL, 73-, CHMN CHEM DEPT, 73- Concurrent Pos: Res fel, Harvard Univ, 55-56; lectr, Univ Dublin, 61, Univ Sao Paulo, 63 & Univ Santiago, Chile, 66; ed adv, Index Chemicus, 64-68; mem Adv Coun Col Chem, 65-68; Danforth Found E Harris Harbison fel, 67-68; Fulbright-Hays res & study award, Kwansei-Gakuin Univ, Japan, 70-71. Honors & Awards: Col Teacher Award, Mfg Chem Asn, 67. Mem: Am Chem Soc; Hist Sci Soc; Soc Hist Technol. Res: Reaction kinetics; mechanism of organic reactions; solvent and salt effects in unimolecular solvolysis; history of chemistry; philosophy of science; spectral studies of strong acid solutions; Oriental science. Mailing Add: 801 Woodbrook Dr Greensboro NC 27410

BENFIELD, JOHN R, b Vienna, Austria, June 24, 31; US citizen; m 63; c 3. SURGERY. Educ: Columbia Univ, AB, 52; Univ Chicago, MD, 55; Am Bd Surg, dipl, 63; Am Bd Thoracic Surg, dipl, 65. Prof Exp: Instr surg, Sch Med, Univ Chicago, 61-64; asst prof, Univ Wis, 64-67; assoc prof surg, Sch Med, Univ Calif-Harbor Gen Hosp, 67-71; PROF SURG, UNIV CALIF, LOS ANGELES, 72- Concurrent Pos: Consult, US Naval Hosp, San Diego & US Naval Hosp, Camp Pendelton, Calif. Mem: Am Surg Asn; fel Am Col Surg; Soc Univ Surg; Am Asn Thoracic Surg. Res: Lung cancer; lung transplantation; obesity. Mailing Add: Harbor Gen Hosp 1000 W Carson St Torrance CA 90509

BENFORADO, JOSEPH MARK, b New York, NY, June 20, 21; m 48; c 6. CLINICAL PHARMACOLOGY. Educ: City Col New York, BS, 41; Columbia Univ, MA, 42; State Univ NY, MD, 51. Prof Exp: Intern, Boston City Hosp, 51-52; res fel pharmacol, Harvard Med Sch, 52-53, instr, 53-57, assoc, 57-58; Life Ins med res fel, Oxford Univ, 57-58; assoc prof pharmacol, Sch Med, State Univ NY Buffalo, 58-67; ASSOC CLIN PROF MED, SCH MED, UNIV WIS-MADISON, 67- Mem: AAAS; Am Soc Pharmacol. Res: Clinical medicine; drug abuse. Mailing Add: Univ of Wis Sch of Med Madison WI 53706

BENGELSDORF, IRVING SWEM, b Chicago, Ill, Oct 23, 22; m 49; c 3. SCIENCE WRITING. Educ: Univ Ill, BS, 43; Univ Chicago, MS, 48, PhD(chem), 51. Prof Exp: Res fel chem, Calif Inst Technol, 51-52; instr, Univ Calif, Los Angeles, 52-54 & Gen Elec Res Lab, 54-59; group leader org chem res, Texus Res Ctr, 59-60; sr scientist, US Borax Res Corp, 60-63; sci ed, Los Angeles Times, 63-70; LECTR SCI COMMUN & DIR, CALIF INST TECHNOL, 71- Concurrent Pos: Sr lectr, Sch Jour, Univ Southern Calif, 66, 68 & Dept Chem, 71, 73, 74; sr lectr, Dept Chem, Univ Calif, Los Angeles, 67, 68, 71; sci columnist & contrib ed, Enterprise Sci Serv, Newspaper Enterprise Asn, 71-74. Honors & Awards: Jean M Kline Mem Award, Am Cancer Soc, 65; James T Grady Award, Am Chem Soc, 67; Westinghouse Writing Award, AAAS, 67, 69; Claude Bernard Sci Jour Award, Nat Soc Med Res, 68. Mem: AAAS; Am Chem Soc; Chem Soc London; NY Acad Sci. Res: Explanation of principles and philosophy of science and technology to nonscientists. Mailing Add: 256 S Arden Blvd Los Angeles CA 90004

BENGTSON, ROGER D, b Wausa, Nebr, Apr 29, 41; m 63; c 2. PLASMA PHYSICS, ATOMIC PHYSICS. Educ: Univ Nebr, BS, 62; Va Polytech Inst, MS, 64; Univ Md, PhD, 68. Prof Exp: Fac assoc, 68-70, asst prof, 70-75, ASSOC PROF PHYSICS, UNIV TEX, AUSTIN, 75- Mem: Am Phys Soc; Sigma Xi. Res: Plasma spectroscopy. Mailing Add: Dept of Physics Univ of Tex Austin TX 78712

BENHAM, GRAHAM HARVEY, b Hove, Eng, Mar 8, 12; nat US; m; c 3. BIOCHEMISTRY. Educ: Univ London, BSc, 35, PhD(biochem), 38. Prof Exp: Demonstr biochem, Physiol Labs, Univ Col, Univ London, 35-38; Commonwealth Fund fel, Univ Wis, 38-40; chemist, Ogilvie Flour Mills, Winnipeg, 40-41 & Montreal, 41-42; asst prof agr chem, MacDonald Col, McGill Univ, 42-45; assoc prof biochem, Ill Inst Technol, 45-50; supvr, Armour Res Found, 50-55; dir res & process develop, Am Agr Chem Co, 55-63; res adv, Kabul Univ, Afghanistan, 63-67; asst dir off res serv, NY Univ, 68-69; dir grant develop, 69-72, PROF NUTRIT SCI, UNIV WIS-GREEN BAY, 70- Concurrent Pos: Consult, Food & Nutrit Group, US AID, 69. Mem: AAAS; Am Chem Soc; Am Inst Chem; NY Acad Sci; Sigma Xi. Res: Nutrition and food analysis; fertilizers and phosphates; nutritional biochemistry; problems of world hunger; higher education and research in Afghanistan; salt as a vehicle for nutrients. Mailing Add: Dept of Nutrit Sci Univ of Wis Green Bay WI 54302

BENHAM, ROSS STEPHEN, b Calgary, Alta, Feb 13, 11; m 40; c 2. MICROBIOLOGY. Educ: Univ Chicago, SM & PhD(microbiol), 57; Am Bd Microbiol, dipl. Prof Exp: Dir clin lab servs, Muskoka Hosp, 39-48; dir, Clin Microbiol Labs, Hosps & Clins, Univ Chicago, 48-69, asst prof med, 57-69, res assoc, Dept Microbiol, 59-69, assoc prof med & clin microbiol, 67-69; CHIEF MICROBIOLOGIST, PROF SERV CORP, 69- Concurrent Pos: Abbot fel, Univ Chicago, 66; hon mem staff, St Joseph Mercy Hosp, Pontiac, 69- Mem: AAAS; fel Am Pub Health Asn; Am Soc Microbiol; Am Soc Med Technol; NY Acad Sci. Res: Clinical microbiology; immunology; bacterial and protozoan physiology; epidemiology. Mailing Add: St Joseph Mercy Hosp 900 Woodward Ave Pontiac MI 48053

BENIAMS, HERMAN NORIGAARD, physical biochemistry, see 12th edition

BENICA, WILLIAM STEINHART, b Logansport, Ind, Sept 9, 21; m 44; c 8. PHARMACY. Educ: Rutgers Univ, BSc, 42; Univ Minn, MSc, 43, PhD(pharmaceut chem), 48. Prof Exp: Develop chemist, Schering Corp, NJ, 48, supvr pharmaceut develop dept, 48-63; tech dir, 63-68, VPRES CONTROL, WILLIAM H RORER, INC, 68- Concurrent Pos: Asst prof, Med Col Va, 48. Res: Quality control. Mailing Add: William H Rorer Inc 500 Virginia Dr Ft Washington PA 19034

BENIGNI, JOSEPH D, organic chemistry, see 12th edition

BENINGTON, FREDERICK, b Chelsea, Mass, m40;xc m 40; c 2. ORGANIC CHEMISTRY, MATHEMATICS. Educ: Tufts Univ, BSc, 39. Prof Exp: Res chemist, Innis Speiden Co, 39-41 & Niacet Chem Corp, 41-45; physicist, Corning Glass Works, 45-47; res engr, Battelle Mem Inst, 47-49, sr res assoc, 49-52; assoc prof, 62-71, PROF MED CHEM, MED CTR, UNIV ALA, 71- Concurrent Pos: Instr, Niagara Univ, 42-44; consult chemist, 49- Mem: Am Chem Soc; Nat Speleol Soc. Res: Thermodynamics and kinetics of gas reactions; free radical reactions; organometallic reactions; syntehsis of psychotomimetic compounds. Mailing Add: Neurosci Prog Dept of Psychiat Univ of Ala Med Ctr Birmingham AL 35294

BENIOFF, PAUL, b Pasadena, Calif, May 1, 30; m 59; c 3. PHYSICS. Educ: Univ Calif, Berkeley, PhD(chem), 59. Prof Exp: Nuclear chemist, Lab Electronics, Tracerlab, Inc, 52-54; Weizmann fel, Weizmann Inst, Israel, 60-61; Ford grant, Inst Theoret Physics, Copenhagen Univ, 61; THEORET CHEMIST, ARGONNE NAT LAB, 61- Mem: AAAS; Am Phys Soc. Res: Molecular structure; relationship between foundations of physics and mathematics. Mailing Add: Argonne Nat Lab 9700 S Cass Ave Argonne IL 60439

BENIRSCHKE, KURT, b Glückstadt, Ger, May 26, 24; nat US; m 52; c 3. PATHOLOGY. Educ: Univ Hamburg, MD, 48. Prof Exp: Assoc path, Harvard Med Sch, 57-60; prof & chmn dept, Dartmouth Med Sch, 60-70; PROF REPRODUCTIVE BIOL & PATH, UNIV CALIF, SAN DIEGO, 70-; DIR RES, SAN DIEGO ZOO, 75- Mem: Col Am Path; Am Asn Path & Bact; AMA. Res: Pathology and endocrinology of human fetus and placenta; gemellology; mammalian hybrids. Mailing Add: Dept of Obstet Univ of Calif at San Diego La Jolla CA 92037

BENISEK, WILLIAM FRANK, b Los Angeles, Calif, Oct 12, 38; m 60; c 2. BIOCHEMISTRY. Educ: Calif Inst Technol, BS, 60; Columbia Univ, MA, 61; Univ Calif, Berkeley, PhD(biochem), 66. Prof Exp: USPHS fel biophys, Yale Univ, 66-68; biochemist, Sch Med, Stanford Univ, 68-69; asst prof, 69-75, ASSOC PROF BIOCHEM, SCH MED, UNIV CALIF, DAVIS, 75- Mem: Am Soc Biol Chem; Am Chem Soc. Res: Protein chemistry; structures and conformations of protein molecules; chemical mechanisms of enzyme-catalyzed reactions. Mailing Add: Dept of Biol Chem Univ of Calif Sch of Med Davis CA 95616

BENISON, BETTY BRYANT, b Irvine, Ky, Aug 25, 38; m 59; c 1. PHYSICAL EDUCATION, BEHAVIORAL PHYSIOLOGY. Educ: La State Univ, Baton Rouge, BS, 55; Univ Mich, Ann Arbor, MA, 58; Univ NMex, PhD(health, phys educ), 68. Prof Exp: Instr phys educ, C E Byrd High Sch, Shreveport, La, 55-56 & Miss State Col Women, 56-57; asst, Univ Mich, Ann Arbor, 57-58; asst prof, Stephen F Austin Univ, 58-59, NTex State Univ, 59-64 & Univ NMex, 64-69; PROF BIOMECH & PHYSIOL, TEX CHRISTIAN UNIV, 69- Concurrent Pos: Mem, Nat Coun Outdoor Educ, Am Alliance Health, Phys Educ & Recreation, 73-; consult col educ adv bd, Am Heart Asn, Tarrant County, 74-75; Tex Christian Univ grant, 74-75; consult adapted & corrective ther, Standford Convalescent Ctrs, 74-; Am Heart Asn grant, 75; Tex Christian Univ fel, Cooper Aerobics Ctr, Dallas, Tex, 75-76; NIH-NASA grant, Southwest Med Sch, Dallas, 76; sabbatical mem staff, Aerobics Ctr, Dallas, 76. Mem:

Am Pub Health Asn; Am Col Sports Med; Int Soc Sports Psychol; Soc Behav Kinesiology; Am Heart Asn. Res: Proper electrode placement in stress testing; mechanical differentials in sports and rhythmic activities; comparative studies of cardiac impairments; systematic development and evaluation of high-level physical fitness programs. Mailing Add: 6412 San Juan Ave Ft Worth TX 76133

BEN-ISRAEL, ADI, b Rio de Janeiro, Brazil, Nov 6, 33; m 64; c 2. APPLIED MATHEMATICS. Educ: Israel Inst Technol, BSc, 55, DiplIng, 56, MSc, 59; Northwestern Univ, PhD(eng sci), 62. Prof Exp: Vis asst prof statist, Carnegie Inst Technol, 62-63; sr lectr math, Israel Inst Technol, 63-65; assoc prof systs eng, Univ Ill, Chicago Circle, 65-66; from assoc prof to prof eng sci, Northwestern Univ, 66-70; prof appl math & chmn dept, Technion-Israel Inst Technol, 70-75; PROF, INDUST ENG DEPT, NORTHWESTERN UNIV, 75- Mem: Am Math Soc. Res: Matrix theory; numerical analysis; mathematical progamming; operations research; engineering design. Mailing Add: Dept of Indust Eng Northwestern Univ Evanston IL 60201

BENITEZ, ALLEN, b Honolulu, Hawaii, July 30, 23; m 55; c 4. ORGANIC CHEMISTRY. Educ: Stanford Univ, BS, 51, MS, 55. Prof Exp: Asst chem, Carnegie Inst Dept Plant Biol, 52-54; CHEMIST, STANFORD RES INST, 55- Mem: Am Chem Soc. Res: Chlorophylls; synthesis; nucleosides; organic analysis. Mailing Add: Stanford Res Inst 333 Ravenswood Ave Menlo Park CA 94025

BENJAMIN, CHESTER RAY, b Alliance, Ohio, Jan 23, 23; m 47; c 1. MYCOLOGY. Educ: Mt Union Col, BS, 48; Univ Iowa, MS, 54, PhD(bot), 55. Prof Exp: Chemist res & develop div, Babcock-Wilcox Co, 48-51; asst bot, Univ Iowa, 51-53, asst mycol, 53-55; mycologist, cult collection invest, fermentation lab, Northern Utilization Res & Develop Div, Agr Res Serv, 55-60, prin mycologist, cur nat fungus collections & leader mycol invests, Crops Protection Res Br, Plant Sci Res Div, 60-71, liaison officer assigned to Dept of State, Int Progs Div, 71-74, ASST DIR INT PROGS DIV, AGR RES SERV, USDA, 75- Concurrent Pos: Hon res assoc, Nat Res Coun, 61-67 & Smithsonian Inst, 61-; chmn toxic micro-organisms panel, Joint US-Japan Coop Develop Natural Resources Prog, 64-69, collabr, 69-; mem, US Nat Comt, Int Union Biol Sci, 65-71, secy, 69-71; bd trustees, Am Type Cult Collection, 68-74, secy-treas, 69-71, vchmn, 71-72; mem spec comt fungi & lichens, Tenth & 11th Int Bot Cong, 64-72; chmn US deleg, US-Japan Joint Panel Conf Toxic Micro-Organisms, 66-68; alt US deleg, 17th Gen Assembly, Int Union Biol Sci, 70; mem US deleg to 16th, 17th & 18th session of conf, Food & Agr Orgn UN, 71, 73 & 75, alt US deleg to regional conf for Latin Am, 72, Asia & Far East, 72, mem US deleg to coun, 57th & 58th sessions, 71, 60th, 61st & 62nd sessions, 73 & 65th & 67th sessions, 75, conf mem, Staff Pension Comt, 71-73. Honors & Awards: Plaque Award, Japanese Toxic Micro-Organisms Panel, US-Japan Coop Natural Resources Prog, 69. Mem: Fel AAAS; Mycol Soc Am (pres, 66-67); Am Phytopath Soc; Bot Soc Am; Brit Mycol Soc. Res: Classification of the ascomycetes and mucorales; nomenclature of fungi; international agriculture and organization affairs related to agriculture. Mailing Add: Int Progs Div ARS USDA Fed Bldg Hyattsville MD 20782

BENJAMIN, DANIEL MARSHALL, b Minneapolis, Minn, Sept 18, 16; m 41; c 2. FOREST ENTOMOLOGY. Educ: Univ Minn, BS, 39, MS, 46, PhD(entom), 50. Prof Exp: Asst entom, Univ Minn, 39-42; asst forester, US Dept Agr, 47-52, asst sta leader, 52-54; from asst prof to assoc prof, 54-65, PROF ENTOM, UNIV WIS-MADISON, 65- Concurrent Pos: Vis scientist, EAfrican Agr Forestry Res Orgn, 65-67; prog coordr & vis prof, Midwestern Univ Consortium for Int Activities, Indonesian Higher Agr Educ Proj, Inst Pertanian, Java, MUCIA/USAID, 73-75. Mem: Entom Soc Am; Entom Soc Can. Res: Biological-ecological aspects of forest insects and control through management; population dynamics of neodiprion sawflies; ecology of insects of tea. Mailing Add: Dept of Entom Univ of Wis Madison WI 53706

BENJAMIN, FRED BERTHOLD, b Darmstadt, Ger, Oct 24, 12; nat US. PHYSIOLOGY. Educ: Bonn Univ, DMD, 35; Univ Ill, MS, 49; Loyola Univ, PhD(physiol), 53. Prof Exp: Pvt pract dent, Kashmir, India, 36-46; res assoc physiol, Univ Ill, 49-53; asst prof, Univ Pa, 53-60; sr res coordr, Life Sci Lab, Repub Aviation Corp, 60-64; chief eval br, Off Space Med, NASA, 64-66, staff asst, Apollo Med Support, 66-70; SR RES PHYSIOLOGIST, RES INST, NAT HWY TRAFFIC SAFETY ADMIN, 70- Concurrent Pos: Prof lectr, Sch Med, George Washington Univ, 64-66. Mem: AAAS; Soc Exp Biol & Med; Am Physiol Soc; Aerospace Med Asn. Res: Pain sensation; reaction to heat; effects of alcohol and narcotic drugs. Mailing Add: Dept of Transp N43-40 Washington DC 20590

BENJAMIN, HIRAM BERNARD, b Austria, July 4, 01; nat US; m 27. PHYSIOLOGY, ANATOMY. Educ: Marquette Univ, MD, 30, MSc, 49. Prof Exp: Intern, Milwaukee County Hosp, Wis, 31, resident, 31-38; pathologist, St Anthony Hosp, 39; assoc prof, 49-74, ADJ PROF ANAT, MED COL WIS, 74-; CHIEF SURG, EVANGEL DEACONESS HOSP, 54- Mem: Am Col Angiol; Am Col Chest Physicians; AMA; Am Chem Soc; Asn Mil Surg US. Res: Gastrointestinal, blood expanders. Mailing Add: Dept of Anat Med Col of Wis Milwaukee WI 53233

BENJAMIN, L, b Tonypandy, South Wales, Mar 17, 32; m 54; c 4. SURFACE CHEMISTRY. Educ: Kings Col, Univ London, BS & dipl, 53, PhD(phys chem), 57; Imp Col, dipl, 58. Prof Exp: Res assoc phys chem, Imp Col, Univ London, 56-58; res fels, Nat Res Coun Can, 58-60; res chemist, Ivorydale Tech Ctr, 60-66, SECT HEAD, MIAMI VALLEY LAB, PROCTER & GAMBLE CO, 66- Concurrent Pos: UK Atomic Energy Comn asst, Imp Col, Univ London, 56-58. Res: Fundamental investigations of solution behavior; ionic, organic and metal solutions; thermodynamics; applications in fuel cell and detergency fields; surface chemistry. Mailing Add: 414 Beech Tree Dr Cincinnati OH 45224

BENJAMIN, PHILIP PALAMOOTTIL, b Eraviperur, India, Sept 5, 32; m 60; c 4. NUCLEAR CHEMISTRY, NUCLEAR MEDICINE. Educ: Univ Madras, BSc, 52; St John's Col, India, MSc, 55; McGill Univ, PhD(radiochem), 65. Prof Exp: Lectr chem, Ewing Christian Col, Univ Allahabad, 55-60, asst prof, 65-66; ASST PROF RADIOL & RADIOCHEM, CASE WESTERN RESERVE UNIV, 66- Concurrent Pos: Res grants, Squibb Inst Med Sci, 68-69; Am Cancer Soc res grant, 69-; consult, Squibb Inst Med Res & Abbott Labs. Mem: AAAS; Soc Nuclear Med; fel Am Inst Chem; NY Acad Sci; Chem Inst Can. Res: Radiochemical investigations of nuclear fission products; distribution of nuclear charge in fission; applications of radiochemistry to nuclear medicine. Mailing Add: Div of Radiation Biol Case Western Reserve Univ Cleveland OH 44106

BENJAMIN, RICHARD KEITH, b Argenta, Ill, Apr 9, 22; m 46; c 2. MYCOLOGY. Educ: Univ Ill, BS, 47, MS, 49, PhD(bot), 51. Prof Exp: Nat Res Coun fel bot, Harvard Univ, 51-52; from asst prof to assoc prof, 52-62, PROF BOT, CLAREMONT GRAD SCH, 62-; MYCOLOGIST, RANCHO SANTA ANA BOT GARDEN, 52- Concurrent Pos: Ed-in-chief, Mycol Soc Am, 71-75. Mem: Bot Soc Am; Mycol Soc Am (secy-treas, 60-62, vpres, 63, pres, 65); Brit Mycol Soc. Res: Laboulbeniales; Mucorales. Mailing Add: Dept of Bot Claremont Grad Sch Claremont CA 91711

BENJAMIN, RICHARD WALTER, b Albany, NY, Dec 8, 35; m 61; c 1. NUCLEAR PHYSICS. Educ: Lamar State Col, BS, 58; Southern Methodist Univ, MS, 61; Univ Tex, Austin, PhD(physics), 65. Prof Exp: Designer eng, Los Alamos Sci Lab, 58-59; res scientist physics, Tex Nuclear Corp, 61-66; res assoc, Swiss Fed Inst Technol, 66-68; RES PHYSICIST, SAVANNAH RIVER LAB, E I DU PONT DE NEMOURS & CO, 68- Concurrent Pos: Lectr, Paine Col, 69-73. Mem: Am Phys Soc; Am Nuclear Soc; AAAS; Swiss Phys Soc; Europ Phys Soc. Res: Neutron cross section measurement, evaluation and testing, particularly of the actinide nuclides. Mailing Add: Savannah River Lab E I du Pont de Nemours & Co Aiken SC 29801

BENJAMIN, ROBERT FREDRIC, b Washington, DC, Jan 19, 45; m 68. EXPERIMENTAL PHYSICS. Educ: Cornell Univ, BS, 67; Mass Inst Technol, PhD(physics), 73. Prof Exp: STAFF PHYSICIST LASER FUSION, LOS ALAMOS SCI LAB, 73- Mem: Am Phys Soc; AAAS; Sigma Xi. Res: Laser-plasma interactions, x-ray diagnostics and short pulse phenomena. Mailing Add: Los Alamos Sci Lab Los Alamos NM 87545

BENJAMIN, ROBERT MYLES, b Bronxville, NY, Aug 13, 27; m 60; c 4. NEUROPHYSIOLOGY. Educ: St Lawrence Univ, BS, 49; Brown Univ, MS, 51, PhD(psychol), 53. Prof Exp: USPHS fel, 54-56, from instr to assoc prof physiol, 56-65, PROF NEUROPHYSIOL & PHYSIOL, SCH MED, UNIV WIS-MADISON, 65- Concurrent Pos: Mem neurol study sect, NIH, 67-71. Mem: Am Physiol Soc; Am Asn Anatomists; Soc Neurosci. Res: Sensory neurophysiology. Mailing Add: Sch of Med Univ of Wis Madison WI 53706

BENJAMIN, ROBERT STEPHEN, b Brooklyn, NY, Apr 20, 43; m 65; c 2. CLINICAL PHARMACOLOGY. Educ: Williams Col, AB, 64; NY Univ, MD, 68. Prof Exp: Sr clin assoc, Baltimore Cancer Res Ctr, Nat Cancer Inst, 72-73; from asst prof med to physician specialist, Los Angeles County-Univ Southern Calif Med Ctr, 73-74; ASST INTERNIST & ASST PROF CLIN PHARMACOL, UNIV TEX SYST CANCER CTR, M D ANDERSON HOSP & TUMOR INST, 74- Concurrent Pos: Clin assoc, Nat Cancer Inst, Baltimore Cancer Res Ctr, 70-72; mem, NIH, Nat Cancer Inst Clin Trials Comt, 75-, Am Cancer Soc jr fac clin fel, 75. Honors & Awards: J D Lane Award, USPHS, 73. Mem: Am Fedn Clin Res; Am Soc Clin Oncol; Am Asn Cancer Res. Res: Clinical trials in medical oncology; clinical pharmacology of chemotherapeutic agents; particular interest in clinical and pharmacologic studies of anthracycline antibiotics. Mailing Add: Univ of Tex Syst Cancer Ctr M D Anderson Hosp & Tumor Inst Houston TX 77030

BENJAMIN, ROLAND JOHN, b Williamsfield, Ill, May 18, 28; m 50; c 1. APPLIED MECHANICS, OPTICS. Educ: Univ Ill, BS, 50, MS, 51, PhD(theoret & appl mech), 55. Prof Exp: Sr struct engr, Downey Div, NAm Aviation, Inc, 52-56; sr engr, Technol Ctr Div, Cook Elec Co, 56-58, staff engr, 58-62, dir eng, Aerospace Sect, 62-64, mgr, 64-67; dir optical eng, 67-70, dir eng, 70-74, ASST VPRES OPTICS, CONSUMER PHOTO PROD, BELL & HOWELL CO, 74- Res: Aircraft and missile structural analysis; development of supersonic and hypersonic recovery systems; photographic optics, cine and still cameras and projectors; plastic optics development. Mailing Add: Consumer Photo Prod Bell & Howell 7100 McCormick Rd Chicago IL 60645

BENJAMINI, ELIEZER, b Tel-Aviv, Israel, Feb 8, 29; US citizen; m 53; c 2. IMMUNOLOGY. Educ: Univ Calif, Berkeley, BS, 52, MS, 54, PhD(insect toxicol), 58. Prof Exp: Lab technician insect toxicol, Univ Calif, Berkeley, 52-55; toxicologist, Nat Cancers Res Asn Lab, Calif, 54-57; jr specialist, Citrus Exp Sta, Univ Calif, Riverside, 57-58; from asst res scientist to assoc res scientist, Kaiser Found Res Inst, San Francisco, 59-66, res scientist & asst dir lab med entom, 66-70; PROF MED MICROBIOL, SCH MED, UNIV CALIF, DAVIS, 70- Concurrent Pos: Prin investr, NIH res grants, 61- Mem: Am Chem Soc; Entom Soc Am; Am Asn Immunol. Res: Immune response to antigenic determinants and its manipulation; characterization and synthesis of antigenic determinants of proteins with particular emphasis on the relationship between structure and immunological activity. Mailing Add: Dept of Med Microbiol Univ of Calif Sch of Med Davis CA 95616

BENJAMINOV, BENJAMIN S, b Sofia, Bulgaria, Mar 21, 23; US citizen; m 47; c 2. ORGANIC CHEMISTRY. Educ: Univ Kans, BA, 52; Allegheny Col, MS, 58; Weizmann Inst, PhD(org chem), 64. Prof Exp: Instr chem, Univ Mass, 53-54 & Rockford Col, 54-56; assoc prof & head dept, Alliance Col, 56-59; assoc prof org chem, 59-66, PROF CHEM, ROSE-HULMAN INST TECHNOL, 66- Concurrent Pos: Nat Cancer Inst fel, 62-64; Fulbright grant & vis prof, Univ Strasbourg, 70-71. Mem: AAAS; Am Chem Soc; The Chem Soc; Sigma Xi; Israel Chem Soc. Res: Chemistry of natural products, especially terpenes and steroids; reaction mechanisms; stereochemical, elucidative and structural problems; coordination compounds; bicyclic and small ring compounds; biosynthetic and synthetic investigations; lipid chemistry. Mailing Add: Dept of Chem Rose-Hulman Inst of Technol Terre Haute IN 47803

BENJAMINS, JOYCE ANN, b Bay City, Mich, June 1, 41; m 65; c 2. NEUROCHEMISTRY. Educ: Albion Col, BA, 63; Univ Mich, PhD(biochem), 67. Prof Exp: NIH grant pediat & genetics, Sch Med, Stanford Univ, 67-68; res assoc neurol, Sch Med, Johns Hopkins Univ, 68-69, from instr to asst prof, 69-73; asst prof biochem & res scientist, Biol Sci Res Ctr, Sch Med, Univ NC, Chapel Hill, 73-75; ASST PROF NEUROL, SCH MED, WAYNE STATE UNIV, 75- Mem: Soc Neurosci; Am Soc Neurochem. Res: Developmental neurochemistry; lipid synthesis and assembly of membranes in developing nervous system; cell isolation; myelination in organ cultures. Mailing Add: Harper Hosp Dept of Neurol 3990 John R Detroit MI 48201

BENJAMINSON, MORRIS AARON, b Bronx, NY, Aug 6, 30; m 58; c 2. MICROBIOLOGY, DEVELOPMENTAL GENETICS. Educ: Long Island Univ, BS, 51; NY Univ, MS, 61, PhD(biol), 67. Prof Exp: Med technician, First Med Field Lab, US Army, 52-54; sr biol technician, Sloan-Kettering Inst Cancer Res, 54-55; med technician microbiol, Vet Admin Hosp, New York, 56-59; res asst, Margaret M Caspary Inst Vet Res, 59-61; res assoc, Bronx-Lebanon Hosp Ctr, 61-64; sr task leader, Naval Appl Sci Lab, 64-69; asst prof, Dent Sch, NY Univ, 69-74; ASSOC PROF ALLIED HEALTH & COORDR MED TECHNOL, YORK COL, NY, 74- Concurrent Pos: US Navy contract, NY Univ, 69-71, NSF grant, 74-75; adj asst prof, City Col New York, 69-; consult, Dept Air Resources, City of New York, 71- Mem: AAAS; Am Soc Microbiol; NY Acad Sci. Res: Clinical microbiology; cytochemistry; automation; aerobiology. Mailing Add: Dept of Allied Health Sci York Col Jamaica NY 11451

BEN-JONATHAN, NIRA, b Holon, Israel, Nov 23, 40. NEUROENDOCRINOLOGY, REPRODUCTIVE PHYSIOLOGY. Educ: Univ Tel-Aviv, BSc, 67; Univ Ill, Urbana, MSc, 69, PhD(physiol), 72. Prof Exp: Teacher chem, Alpha High Sch, Tel-Aviv, 66-67; from teaching asst to res asst physiol, Univ Ill, Urbana, 68-71; fel endocrinol, Univ Tex Southwestern Med Sch, 72-74, asst prof physiol, 74-75; ASST PROF PHYSIOL, SCH MED, IND UNIV, '76- Concurrent Pos: Fel, Pop Coun, 72. Mem: Am Endocrine Soc; Am Physiol Soc. Res: Neuronal-endocrine interrelations; hypothalamic catecholamines and releasing hormones and the control of tropic hormones secretion by the anterior pituitary gland. Mailing Add: Dept of Physiol Ind Univ Sch of Med Indianapolis IN 46202

BENKESER, ROBERT ANTHONY, b Cincinnati, Ohio, Feb 16, 20; m 46; c 5. ORGANIC CHEMISTRY. Educ: Xavier Univ, Ohio, BS, 42; Univ Detroit, MS, 44; Iowa State Col, PhD(org chem), 47. Prof Exp: From asst prof to assoc prof, 46-54, PROF CHEM, PURDUE UNIV, WEST LAFAYETTE, 54-, HEAD DEPT, 74- Honors & Awards: F S Kipping Award, Am Chem Soc, 69. Mem: Am Chem Soc; The Chem Soc. Res: Organometallics and synthetic organic chemistry; reactions of organosilicon compounds; reductions in amine solvents; metallations and reactions of benzylic Grignard systems. Mailing Add: Dept of Chem Purdue Univ West Lafayette IN 47907

BENMAMAN, JOSEPH DAVID, b Tetuan, Morocco, Dec 12, 24; US citizen; m 60; c 2. BIOPHARMACEUTICS. Educ: Univ Madrid, Lic pharm, 54, PhD(pharmaceut chem), 60. Prof Exp: Asst prof phys chem, Univ Madrid, 54-60; prof phys chem & chmn dept, Univ Oriente, Venezuela, 60-63; res chemist, Sch Med, Univ Calif, San Francisco, 63-65, res chemist, Sch Pharm, 65-68; from asst prof to assoc prof phys chem, 68-74, PROF PHYS CHEM & BIOPHARMACEUT, COL PHARM, MED UNIV OF SC, 74- Mem: Am Chem Soc; Am Pharmaceut Asn; Acad Pharmaceut Sci; Pan-Am Fedn Pharm & Biochem (secy, Sci Div, 72-73). Res: Methods of separation and analysis of biological materials; kinetics and mechanisms of reactions; drug stability; kinetics of absorption, distribution, metabolism and excretion of drugs; clinical pharmacokinetics; drug design. Mailing Add: Col of Pharm Med Univ of S C Charleston SC 29401

BENN, WALTER R, b Chicago, Ill, Jan 29, 27; m 52; c 3. ORGANIC CHEMISTRY. Educ: Univ Chicago, BS, 50; Univ Wis, PhD, 56. Prof Exp: SR RES INVESTR, G D SEARLE & CO, 55- Res: Synthetic and naturally occurring steroids. Mailing Add: 605 Westgate Rd Deerfield IL 60015

BENNE, ERWIN JOHN, b Morrowville, Kans, May 21, 02; m 28; c 2. BIOCHEMISTRY. Educ: Kans State Univ, BS, 28, MS, 31, PhD(chem), 37. Prof Exp: Instr high schs, Kans, 28-30; instr chem, Kans State Col, 30-38; res asst, 38-43, res assoc, 43-46, prof res, 46-59, prof, 59-71, EMER PROF CHEM, EXP STA, MICH STATE UNIV, 71- Mem: AAAS; Am Chem Soc; Am Soc Agron; Asn Off Anal Chem. Res: Development of analytical methods; effects of minor elements and soil conditions on plant growth; nutritive value of animal feeds and forages. Mailing Add: 201 N Lakeview Ave Sturgis MI 49091

BENNER, BLAIR RICHARD, b Braddock, Pa, Mar 29, 47; m 74. CHEMICAL METALLURGY. Educ: Pa State Univ, BS, 69; Stanford Univ, MS, 71. Prof Exp: Jr metallurgist, NMex State Bur Mines & Mineral Resources, 71-73; RES METALLURGIST EXTRACTIVE METALL, DEEPSEA VENTURES INC, TENNECO CO, 73- Mem: Am Inst Mining, Metall & Petrol Engr. Res: Hydrometallurgical processing of deep ocean nodules; pyrolysis of metal chlorides. Mailing Add: Deepsea Ventures Inc Gloucester Point VA 23062

BENNER, DAVID BRIGHT, genetics, see 12th edition

BENNER, ERNEST JACK, b Leavenworth, Wash, Mar 4, 31; m 57; c 2. MEDICINE. Educ: Cent Wash Col Educ, BA, 53; Univ Ore, MD, 63. Prof Exp: Instr med, Med Sch, Univ Ore, 63-64; asst & res fel, Univ Wash, 64-65; from asst prof to assoc prof, Med Sch, Univ Ore, 65-68; assoc prof, Univ Calif, Davis, 68-72; CONSULT INTERNIST, 72- Mem: Fel Am Col Physicians; Am Soc Microbiol; Am Fedn Clin Res. Res: Infectious diseases; clinical pharmacology of antibiotics. Mailing Add: Med Group Scripps Dr Sacramento CA 95825

BENNER, FRANK CLARENCE, physical chemistry, see 12th edition

BENNER, GERELD S, inorganic chemistry, see 12th edition

BENNER, VELMA, b Grove City, Ill; m 23; c 1. GEOGRAPHY. Educ: Washington Univ, BS, 47, AM, 50. Prof Exp: Instr geog, Washington Univ, 50-53, univ col, 53-54, lectr, 54-59, asst prof, 59-64; assoc prof geog, Meramec Community Col, 64-75; RETIRED. Concurrent Pos: Mem, Int Geog Union. Mem: AAAS; Asn Am Geogr; Am Geog Soc; Nat Coun Geog Educ. Res: Economic and urban geography. Mailing Add: 275 N Union Blvd St Louis MO 63108

BENNET, DOROTHEA, b Honolulu, Hawaii, Dec 27, 29. GENETICS. Educ: Barnard Col, BA, 51; Columbia Univ, PhD(zool), 56. Prof Exp: Res assoc genetics, Columbia Univ, 57-62; from asst prof to prof anat, Med Col, Cornell Univ, 62-76; MEM, MEMORIAL SLOAN-KETTERING CANCER CTR, 76- Mem: Genetics Soc Am; Soc Develop Biol. Res: Effects of mutant genes on development in mammals. Mailing Add: Mem Sloan-Kettering Cancer Ctr New York NY 10021

BENNET, WILLIAM BAKER, organic chemistry, see 12th edition

BENNETT, ALAN JEROME, b Philadelphia, Pa, June 13, 41; m 63; c 3. SOLID STATE PHYSICS. Educ: Univ Pa, BA, 62; Univ Chicago, MS, 63, PhD(physics), 65. Prof Exp: NSF fel, Cambridge Univ, 66; physicist, 66-74, MGR SOLID STATE COMMUN BR, GEN ELEC RES & DEVELOP CTR, 75- Concurrent Pos: Vis assoc prof, Cornell Univ, 70 & Technion, 72. Mem: Am Phys Soc; Inst Elec & Electronics Engrs; AAAS. Res: Superconductivity; surface physics; tunnelling; band structure; microwave devices; ultrasonic imaging. Mailing Add: Gen Elec Res & Develop Ctr PO Box 8 Schenectady NY 12301

BENNETT, ALLAN IGO, JR, physics, see 12th edition

BENNETT, ALLISON CARR, b Mountain Park, Okla, Mar 10, 38; m 65; c 2. SOIL CHEMISTRY, SOIL FERTILITY. Educ: Okla State Univ, BS, 60, MS, 63; Auburn Univ, PhD(soil chem), 69. Prof Exp: ASST PROF AGRON & SOILS, AGR EXP STA, AUBURN UNIV, 69- Mem: Am Soc Agron; Soil Sci Soc Am. Res: Chemical thermodynamics of soil solution phosphorus and other dilute aqueous solution phosphorus; toxicity of herbicides in relation to soil chemical properties. Mailing Add: Route 3 Box 247 Opelika AL 36801

BENNETT, ARTHUR LAWRENCE, b Oconto, Wis, June 25, 05; m 32; c 2. PHYSIOLOGY. Educ: Lawrence Col, AB, 27; Univ Chicago, PhD(physiol), 33; Rush Med Col, MD, 37. Prof Exp: Adj prof, Sch Med, Univ Tex, 30-31; asst prof physiol & pharmacol, 34-41, prof, 41-73, asst chmn dept physiol, 54-67, chmn dept, 67-70, EMER PROF PHYSIOL & PHARMACOL, COL MED, UNIV NEBR MED CTR, OMAHA, 73- Honors & Awards: Freer Prize & Rush Medal. Mem: AAAS; Am Physiol Soc; Am Asn Electromyog & Electrodiag. Res: Electrophysiology of nerve and muscle. Mailing Add: Dept of Physiol Univ of Nebr Med Ctr Omaha NE 68105

BENNETT, BURTON GEORGE, b Omaha, Nebr, July 7, 39; m 65; c 2. ENVIRONMENTAL SCIENCES. Educ: Ore State Univ, BA & BS, 62; Univ Wash, MS, 63. Prof Exp: Physicist radiation physics, 64-70, PHYS SCIENTIST ENVIRON RADIOACTIVITY, HEALTH & SAFETY LAB, US ENERGY RES & DEVELOP ADMIN, 70- Concurrent Pos: Adv, US Deleg to UN Sci Comt on Effects Atomic Radiation, 72, consult, Secretariat, 75- Res: Fallout strontium 90 in diet and man; environmental levels and dose to man from fallout tritium and plutonium; collective dose from nuclear fuel cycle operations. Mailing Add: Health & Safety Lab USERDA 376 Hudson St New York NY 10014

BENNETT, CARL ALLEN, b Winfield, Pa, Nov 22, 21; m 44; c 2. STATISTICS. Educ: Bucknell Univ, AB, 40, MA, 41; Univ Mich, AM, 42, PhD(math), 52. Prof Exp: Jr chemist, Chem & Metall Lab, Univ Chicago, 44; sr chemist & qual control supvr, Tenn Eastman Corp, 44-46; from chief statist & head appl math to mgr appl math dept, Hanford Labs, Gen Elec Co, 47-65; mgr appl math dept, 65-68, mgr systs & electronics div, 68-70, sr staff scientist, 70-71, STAFF SCIENTIST, HUMAN AFFAIRS RES CTRS, PAC NORTHWEST LABS, BATTELLE MEM INST, 71- Concurrent Pos: Res assoc math, Princeton Univ, 50. Mem: Fel AAAS; fel Am Statist Asn; fel Am Soc Qual Control; Biomet Soc (pres Western NAm Region, 62-63); Inst Math Statist. Res: Application of statistical and mathematical techniques to industrial problems; order statistics; nonparametric methods; variance component analysis. Mailing Add: 11121 SE 59th St Bellevue WA 98006

BENNETT, CARLYLE WILSON, b Narrows, Ky, June 8, 95. PLANT PATHOLOGY. Educ: Univ Ky, BS, 17; Mich State Col, MS, 19; Univ Wis, PhD, 26. Prof Exp: Asst bot, Mich State Col, 17-19, from instr to asst prof, 19-28; res assoc, Ohio Exp Sta, 28-29; pathologist, Div Sugar Plant Invest, Bur Plant Indust, 29-43, pathologist, Bur Plant Indust, Soils & Agr Eng, 43-45, sr pathologist, 45-46, prin pathologist, Agr Res Serv, 46, supt, 53-64, plant pathologist, 64-66, COLLABR, US AGR RES STA, US DEPT AGR, 65- Mem: AAAS; Bot Soc Am; Am Phytopath Soc; Am Soc Sugar Beet Technol. Res: Virus diseases of plants. Mailing Add: US Agr Res Sta US Dept of Agr Box 5098 Salinas CA 93901

BENNETT, CARROLL G, b Richmond, Va, Dec 8, 33; m 56; c 2. DENTISTRY. Educ: Randolph-Macon Col, BS, 55; Med Col Va, DDS, 59, MS, 62; Am Bd Pedodont, dipl, 67. Prof Exp: Instr dent anat & pedodontics, Sch Dent, WVa Univ, 62-63, from asst prof to prof pedodontics & chmn dept, 63-73, asst dean student affairs, 72-73; prof pedodontics & chmn div, 73-74, PROF PEDIAT DENT & CHMN DEPT, COL DENT, UNIV FLA, 74- Mem: Am Dent Asn; Am Soc Dent for Children; Int Asn Dent Res. Res: Pedodontics; growth, development and pulpal reactions in children's dentistry. Mailing Add: Dept of Pediat Dent Univ of Fla Col of Dent Box J426 Gainesville FL 32601

BENNETT, CECIL JACKSON, b Eau Claire, Wis, Oct 4, 27; m 51; c 2. ANIMAL GENETICS, BEHAVIORAL GENETICS. Educ: Univ Wis, BS, 49, PhD(zool & genetics), 59; Univ Wash, MA, 53. Prof Exp: Lab maintenance man zool, Univ Okla, 53-55; res asst genetics, Univ Wis, 55-57; from asst prof to assoc prof, 57-75, PROF BIOL, NORTHERN ILL UNIV, 75- Mem: AAAS; Soc Study Evolution; Genetics Soc Am; Am Genetics Asn; Am Inst Biol Sci. Res: Population genetics; gene action; behavior in Drosophila. Mailing Add: Dept of Biol Northern Ill Univ DeKalb IL 60115

BENNETT, CHARLES FRANKLIN, b Oakland, Calif, Apr 10, 26; m 47; c. 1. BIOGEOGRAPHY, CULTURAL GEOGRAPHY. Educ: Univ Calif, Los Angeles, BA, 55, PhD(geog), 60. Prof Exp: From instr to assoc prof geog, 59-69, PROF BIOGEOG, UNIV CALIF, LOS ANGELES, 69- Concurrent Pos: Consult, Tex Instruments Corp, 63-64 & AEC, 65-67; Off Naval Res fel res in Panama, Univ Calif, Los Angeles, 63; res assoc, Smithsonian Trop Res Inst, 66-; Guggenheim fel res in Spain, 70-71. Mem: Fel AAAS; Ecol Soc Am; Brit Ecol Soc; Am Geog Soc; Am Soc Mammalogists. Res: Ecology of human modified ecosystems; ecology of the humid tropics; ecology of agricultural systems; conservation of resources in underdeveloped countries. Mailing Add: Dept of Geog Univ of Calif Los Angeles CA 90024

BENNETT, CLARENCE EDWIN, b Providence, RI, May 23, 02; m 28; c 2. PHYSICS. Educ: Brown Univ, PhB, 23, MS, 24, PhD(physics), 30. Prof Exp: Asst physics, Brown Univ, 23-24, instr, 24-31; instr, Mass Inst Technol, 31-34; from asst prof to prof, 34-70, from actg head dept to head dept, 37-67, EMER PROF PHYSICS & CONSULT, UNIV MAINE, ORONO, 70- Concurrent Pos: Mem, Nat Acad Sci-Nat Res Coun adv panel, Nat Bur Stand, 60-63. Mem: Fel AAAS; fel Am Phys Soc; Optical Soc Am; Am Asn Physics Teachers; Am Soc Eng Educ. Res: Optical dispersion of gases at high pressure; capillary electrometer; application of linear displacement interferometer to measurement of dispersion and index of refraction of gases as functions of pressure; molar refraction versus density for carbon dioxide. Mailing Add: 65 Forest Ave Orono ME 04473

BENNETT, CLIFTON FRANCIS, b Tillamook, Ore, July 27, 25; m 56; c 3. WOOD CHEMISTRY. Educ: Lewis & Clark Col, 49; Ore State Univ, MS, 52; McGill Univ, PhD(wood chem), 56. Prof Exp: Res chemist, Weyerhaeuser Co, 51-53; RES CHEMIST, CROWN ZELLERBACH CORP, 55- Concurrent Pos: Nat dir, US Jr CofC, 58-59; instr, Portland Continuation Ctr, Ore State Syst Higher Educ, 63-65; adv, Chem Explorer Post 404, 67-; Nat Acad Sci exchange scientist, Czech & Romania, 72-73. Mem: Am Chem Soc. Res: Cellulose reactions; substitutions and fractionations; organic sulfur compounds; lignin; organic chemical synthesis; pulp and paper chemistry. Mailing Add: 1423 NE Sixth Ave Camas WA 98607

BENNETT, DAVID ARTHUR, b Cleveland, Ohio, Dec 9, 42; c 2. BIOINORGANIC CHEMISTRY. Educ: Muskingum Col, BS, 64; Northwestern Univ, MAT, 65; Cornell Univ, PhD(chem), 73. Prof Exp: Instr sci, Am Sch, Switz, 65-67; res assoc chem, Purdue Univ, 72-74; ASST PROF CHEM, MIDDLEBURY COL, 74- Mem: Am Chem Soc. Res: Chemical and biochemical studies of porphyrins and porphyrin containing biological molecules. Mailing Add: Dept of Chem Middlebury Col Middlebury VT 05753

BENNETT, DAVID GORDON, b Winston-Salem, NC, Sept 26, 41; m 60; c 1. POPULATION GEOGRAPHY. Educ: E Carolina Univ, AB, 63; Mich State Univ, MA, 65, PhD(pop geog), 68. Prof Exp: Instr geog, La State Univ, New Orleans, 64-65; asst prof, Ark State Univ, 66-67; asst prof, 67-75, ASSOC PROF GEOG, UNIV N C, GREENSBORO, 75- Concurrent Pos: Res asst geog, Map Div, Libr Cong, Washington, DC; mem, Pop Reference Bur. Mem: Asn Am Geog; Pop Asn Am; Nat Audubon Soc; Nat Geog Soc. Res: Population characteristics and movements; trade areas; social problems. Mailing Add: Dept of Geog Univ of N C Greensboro NC 27412

BENNETT, DON C, b Salt Lake City, Utah, June 28, 25; m 48; c 4. GEOGRAPHY. Educ: Univ Utah, AB, 40; Syracuse Univ, MA, 51, PhD(geog), 57. Prof Exp: Instr geog, Syracuse Univ, 56-57, asst prof, 57-59; asst prof, 59-63, ASSOC PROF GEOG, IND UNIV, BLOOMINGTON, 63- Concurrent Pos: NSF res grant, 62-; Fulbright Hays res grant, Philippines, 63-64. Mem: Asn Am Geog; Am Geog Soc; Regional Sci Asn; Am Inst Planners; Pop Asn Am. Res: Urban form, function, density and growth; Southeast Asia. Mailing Add: Dept of Geog Ind Univ Bloomington IN 47401

BENNETT, DONALD RAYMOND, b Mishawaka, Ind, Feb 16, 26; m 47; c 2. PHARMACOLOGY. Educ: Univ Mich, BS, 49, MS, 51, MD, 55, PhD, 58. Prof Exp: From asst prof to assoc prof pharmacol, Univ Mich, 57-65; MGR, BIOSCI RES LAB, DOW CORNING CORP, 65- Mem: Am Soc Pharmacol & Exp Therapeut; Soc Toxicol. Res: Organosilicon pharmacology and toxicology. Mailing Add: Biosci Res Lab Dow Corning Corp Midland MI 48640

BENNETT, DWIGHT G, JR, b Pittsburgh, Pa, Aug 27, 35; m 59; c 2. VETERINARY MEDICINE. Educ: Univ Ill, BS, 57, DVM, 59; Univ Wis, MS, 63, PhD(vet sci), 64. Prof Exp: Private practice, 59; asst vet sci, Univ Wis, 62-63, USPHS fel, 63-64; from asst prof to assoc prof vet parasitol, Purdue Univ, 64-73, prof vet parasitol & large animal clin, 73-75; PROF CLIN SCI, COLO STATE UNIV, 75- Concurrent Pos: Mem, Nat Bd Vet Med Examrs, 71- Mem: Am Vet Med Asn; Am Asn Vet Parasitol; World Asn Advan Vet Parasitol; Am Asn Equine Practitioners. Res: Chemotherapy and immunology of nematode parasites of ruminants; chemotherapy of parasites of horses. Mailing Add: Vet Hosp Col of Vet Med Colo State Univ Ft Collins CO 80521

BENNETT, E MAXINE, b Beaver City, Nebr, July 14, 15. OTOLARYNGOLOGY. Educ: Hastings Col, AB, 36; Univ Nebr, MD, 42; Am Bd Otolaryngol, dipl, 49. Prof Exp: Med dir, Wis Bur Handicapped Children, 50-53; assoc prof surg, 53-63, chmn div otolaryngol, 59-68, PROF SURG, SCH MED, UNIV WIS-MADISON, 63- Concurrent Pos: Consult, Vet Admin Hosp, Madison, 60-70; vis prof, Columbia Univ, 69. Mem: Am Acad Opthal & Otolaryngol; Am Laryngol, Rhinol & Otol Soc. Res: Medical education, particularly as it relates to clinical application, graduate and undergraduate levels; otology and audiology, particularly in the preschool age, nonverbal child. Mailing Add: Dept of Surg Univ of Wis Sch of Med Madison WI 53706

BENNETT, EDGAR F, b Colebrook, NH, JUly 29, 29; m 70. REACTOR PHYSICS. Educ: Univ NH, BS, 51, MS, 53; Princeton Univ, PhD(physics), 57. Prof Exp: PHYSICIST, DIV REACTOR PHYSICS, ARGONNE NAT LAB, 57- Mem: Am Phys Soc; Am Nuclear Soc. Res: Radiation detection; radiation instrumentation. Mailing Add: D-316 Argonne Nat Lab 9700 S Cass Ave Argonne IL 60439

BENNETT, EDWARD LEIGH, b Hood River, Ore, Nov 20, 21; m 54; c 3. BIOCHEMISTRY, NEUROBIOLOGY. Educ: Reed Col, BA, 43; Calif Inst Technol, PhD, 49. Prof Exp: Asst chem, Calif Inst Technol, 42-49; res chemist, Radiation Lab, Univ Calif, Berkeley, 49-51; Am Cancer Inst fel, Inst Cytophysiol, Copenhagen Univ, 51-52; RES CHEMIST, LAWRENCE BERKELEY LAB, UNIV CALIF, BERKELEY, 52- Mem: Am Chem Soc; Am Soc Biol Chem; NY Acad Sci; Am Soc Neurochem: Soc Neurosci. Res: Nucleic acid metabolism; biochemical psychology; biochemistry & memory. Mailing Add: Lab of Chem Biodynamics Lawrence Berkeley Lab Univ Calif Berkeley CA 94720

BENNETT, EDWARD OWEN, b St Louis, Mo, Mar 16, 26; m 47; c 2. BACTERIOLOGY. Educ: Univ Houston, BS, 49; Univ Iowa, MS, 51; Baylor Univ, PhD, 58. Prof Exp: From asst prof to assoc prof, 51-63, chmn dept bact, 64-67, assoc dean col arts & sci, 67-73, PROF BACT, UNIV HOUSTON, 63- Mem: Am Soc Microbiol; Soc Indust Microbiol; Brit Soc Gen Microbiol. Res: Antimicrobial agents. Mailing Add: Dept of Biol Univ of Houston Houston TX 77004

BENNETT, ELBERT WHITE, b Texarkana, Tex, Jan 24, 29; m 62; c 3. EXPERIMENTAL NUCLEAR PHYSICS. Educ: Univ Tex, BS, 51, MA, 52, PhD(physics), 57. Prof Exp: Staff mem, 56-63, assoc group leader, 63-67, MEM STAFF, LOS ALAMOS SCI LAB, 67- Mem: Am Phys Soc. Res: Neutron scattering; diagnostic and effects measurements of nuclear explosions. Mailing Add: 263 Dos Brazos Los Alamos NM 87544

BENNETT, EMMETT, b Goshen, Ohio, Sept 16, 04; m 34; c 2. BIOCHEMISTRY. Educ: Ohio State Univ, BS, 29; Univ Mass, MS, 34; Pa State Univ, PhD, 50. Prof Exp: Asst, Exp Sta, 30-36, asst res prof chem, 36-53, res prof, 53-62, res prof forestry, 62-68, EMER RES PROF CHEM, UNIV MASS, AMHERST, 68- Prof Exp: Lectr, Univ Malawi, 66-68. Mem: Am Chem Soc; Am Soc Plant Physiol. Res: Chemistry of plant products. Mailing Add: 384A Northampton Rd Amherst MA 01002

BENNETT, FREDERICK DEWEY, b Miles City, Mont, June 2, 17; m 39; c 4. FLUID MECHANICS. Educ: Oberlin Col, AB, 37; Pa State Col, MS, 39, PhD(physics), 41. Prof Exp: Asst physics, Pa State Col, 38-41; from instr to asst prof, Univ NH, 41-43; physicist, Spec Projs Lab, AMC, Wright Field, Ohio, 43-46; from asst prof to assoc prof elec eng, Univ Ill, 46-48; physicist, Ballistics Res Lab, Aberdeen Proving Ground, 48-63, chief exterior ballistics div, 63-70; CONSULT, 70- Concurrent Pos: Fel, Dept Mech & Mat Sci, Johns Hopkins Univ, 71- Mem: AAAS; Am Phys Soc; Am Geophys Union. Res: Physics of hydrogen palladium system; aircraft antennas; diffusion of hydrogen through metals; optical methods in analysis of airflow; exploding wires; volcanic ash formation. Mailing Add: 139 Broadway Bel Air MD 21014

BENNETT, FREDERICK WILLIAM, b Commerce, Ga, Apr 13, 00; m 21, 63; c 5. DAIRY BACTERIOLOGY. Educ: Univ Ga, BSA, 19; Iowa State Col, MS, 32, PhD, 50. Prof Exp: Assoc prof, 20-67, EMER ASSOC PROF DAIRYING, UNIV GA, 67- Concurrent Pos: Dairy prod grader & inspector, US Dept Agr, 41-55; mem bd dirs, Learning Found Int, Inc, 69-71. Mem: Fel AAAS; Am Dairy Sci Asn; Am Soc Microbiol. Res: Bacteriophage and chemistry of agriculture, especially dairying; studies of streptococcus, staphylococci and viricidal agents; Escherichia coli in dairy products and reaction between lactic streptococcal bacteriophage and animals. Mailing Add: 410 University Dr Athens GA 30601

BENNETT, GARY LEE, b Twin Falls, Idaho, Jan 17, 40; m 61; c 1. NUCLEAR PHYSICS. Educ: Univ Idaho, BS, 62, MNucSc, 66; Wash State Univ, PhD(nuclear physics), 70. Prof Exp: Physicist, Atomic Energy Div, Phillips Petrol Co, 62-66; res asst physics, Wash State Univ, 67-70; nuclear physicist, Space Nuclear Propulsion Off, NASA-Lewis Res Ctr, 70-71; flight safety officer, Space Nuclear Syst Div, Ener Res & Develop Admin, 71-74; TECH ASST, REACTOR SAFETY RES DIV, US NUCLEAR REGULATORY COMN, 74- Mem: Am Phys Soc; Am Asn Physics Teachers; Am Inst Aeronaut & Astronaut. Res: Nuclear reactor kinetics; low energy nuclear reactions; nuclear astrophysics; nuclear rocket propulsion; radiation shielding. Mailing Add: Reactor Safety Res Div Nuclear Regulatory Comn Washington DC 20555

BENNETT, GEORGE EDWARD, organic chemistry, see 12th edition

BENNETT, GEORGE WILLIAM, aquatic biology, see 12th edition

BENNETT, GERALD WILLIAM, b Hempstead, NY, June 15, 33; m 55; c 6. RADIATION PHYSICS, MEDICAL PHYSICS. Educ: Brooklyn Polytech Inst, BME, 55; Hofstra Univ, MA, 62; State Univ NY Stony Brook, PhD(physics), 68. Prof Exp: Jr engr, NY Naval Shipyard, 56; mech engr, Fairchild Engine Div, NY, 56-59; opers engr, Cosmotron Div, Brookhaven Nat Lab, 59-64, develop engr, 64-68; physicist, Ger Electron Synchrotron, Hamburg, 68-69; PHYSICIST, BROOKHAVEN NAT LAB, 69- Concurrent Pos: Prof, State Univ NY Stony Brook, Clin Campus, 75-; prin invest, Nat Cancer Inst Res Grant, 75-77. Mem: AAAS; Am Phys Soc. Res: Particle accelerator development; nuclear structure physics; production and shielding of radiation from high energy particles; high energy particle beam instrumentation; clinical applications of heavy charged particles. Mailing Add: Brookhaven Nat Lab Upton NY 11973

BENNETT, GORDON FRASER, b Ootacamund, India, Aug 20, 30; Can citizen; m 59. ENTOMOLOGY, PROTOZOOLOGY. Educ: Univ Toronto, BA, 53, MA, 54, PhD(entom), 57. Prof Exp: Res fel parasitol, Ont Res Found, 57-68; ASSOC PROF BIOL, MEM UNIV NFLD, 68- Concurrent Pos: Vis scientist, NIH, 63-65. Mem: Am Soc Parasitol; Entom Soc Am; Wildlife Dis Asn; Can Soc Zool; Malaysian Soc Parasitol & Trop Med. Res: Biology and taxonomy of myiasis-producing Diptera; biology of vectors of hematozoa; biology and taxonomy of hematozoa. Mailing Add: Dept of Biol Mem Univ of Nfld St Johns NF Can

BENNETT, HAROLD EARL, b Missoula, Mont, Feb 25, 29; m 52. OPTICAL PHYSICS. Educ: Univ Mont, BA, 51; Pa State Univ, MS, 53, PhD(physics), 55. Prof Exp: Physicist, Nat Bur Standards, 53 & Wright Air Develop Ctr, 55-56; PHYSICIST, US NAVAL WEAPONS CTR, 56-, HEAD PHYS OPTICS BR, 60-, ASSOC HEAD PHYS DIV, 71- Concurrent Pos: Fel Ord Sci, Naval Weapons Ctr, 72. Honors & Awards: L T E Thompson Award, Naval Weapons Ctr, 74. Mem: Fel Optical Soc Am; Am Phys Soc. Res: Optical properties of solids; solid state physics; optical instrumentation; thin films; laser components. Mailing Add: Code 6018 Michelson Lab US Naval Weapons Ctr China Lake CA 93555

BENNETT, HARRY, b New York, NY, May 28, 95; m 21; c 2. INDUSTRIAL CHEMISTRY. Educ: NY Univ, BS, 17. Prof Exp: Mem staff, E I du Pont de Nemours & Co, 18-19; pres, Glyco Prod Co, 22-58 & Cheminform Inst, 58-62; PRES, BENNETT-ROSENDAHL CO, INC, 62-, DIR, B R LAB, 70- Concurrent Pos: Ed, Chem Publ Co, 33- Mem: Fel Am Inst Chemists; Am Chem Soc; Soc Indust Chem; Am Soc Testing & Mat; Asn Textile Chemists & Colorists. Res: Chemical specialty formulation, including emulsions, synthetic and commercial waxes, polishers, water-soluble resins, anti-tack materials, emulsifiers, waterproofing; chemical trade marks. Mailing Add: 714 W 51st St Miami Beach FL 33140

BENNETT, HARRY JACKSON, b Bernice, La, Aug 1, 04; m 31; c 3. ZOOLOGY. Educ: La State Univ, BS, 26; Univ Ill, MS, 28, PhD(parasitol), 35. Prof Exp: Instr, 29-32 & 33-35, from asst prof to assoc prof, 35-45, asst to dean col arts & sci, 39-41, dir marine lab, 46-56, dir sci training progs, 69-74, assoc dean grad sch & coordr res, 73-74, PROF ZOOL, LA STATE UNIV, BATON ROUGE, 46- Concurrent Pos: Mem, US Schistosmoiasis Comn, 45-46; scientist, Div Trop Dis, USPHS Hosp, 47-49; dir indust res lab, 56-; chmn, Asn Acads Sci, 57-65, secy-treas acad conf, 62-65, pres, 68; dir, Vis Scientist Prog, 58-65; biol consult, Gulf South Res Inst & Raytheon Co, 75- Honors & Awards: La State Univ Found Distinguished Fac Fel, 68. Mem: Fel AAAS; Am Soc Zool; Am Micros Soc; Am Soc Limnol & Oceanog; Am Soc Parasitologists. Res: Parasitology; classification, life cycles and taxonomy of trematodes; marine zoology. Mailing Add: Dept of Zool & Physiol La State Univ Baton Rouge LA 70803

BENNETT, HENRY STANLEY, b Tottori, Japan, Dec 22, 10; US citizen; m 35; c 4. ANATOMY, CELL BIOLOGY. Educ: Oberlin Col, AB, 32; Harvard Univ, MD, 36. Hon Degrees: DSc, Monmouth Col, 62. Prof Exp: Nat Res Coun fel & univ res fel anat, Harvard Med Sch, 37-39, instr anat & pharmacol, 39-41, assoc anat, 41-48; prof & head dept, Sch Med, Univ Wash, 48-60; prof biophys & dean div biol sci, Sch Med, Univ Chicago, 61-65, prof anat, 61-69, Robert R Bensley prof biol & med sci, 66-69; PROF ANAT & CHMN DEPT, SARAH GRAHAM KENAN PROF BIOL & MED SCI & DIR LAB REPRODUCTIVE BIOL, UNIV NC, CHAPEL HILL, 69- Concurrent Pos: Asst prof cytol, Mass Inst Technol, 45-48; Eastman Mem lectr, Univ Rochester, 58; Phillips lectr, Haverford Col, 59; mem bd trustees, Salk Inst Biol Studies, 63-; mem, US-Japan Comt Sci Coop, 64-, US co-chmn, 69- Honors & Awards: Order of Sacred Treas, Second Class, Japanese Govt, 74. Mem: Am Asn Anat (pres, 60-61); Asn Am Physicians; Am Soc Cell Biol; Am Acad Arts & Sci; Am Physiol Soc. Res: Histochemistry and cytochemistry of sulfhydryl groups; cell ultrastructure, molecular structure and dynamic behavior of membranes; development of methods for structural analysis of cells and tissues. Mailing Add: Div of Health Affairs Univ of NC Dept of Anat Chapel Hill NC 27514

BENNETT, HERALD DURWARD, b Akron, Ohio, June 17, 16; m 45; c 4. PLANT MORPHOLOGY, CYTOLOGY. Educ: WVa Univ, AB, 40, MS, 41; Univ Iowa, PhD(bot), 59. Prof Exp: Instr bot, Univ Iowa, 46-49; from assoc prof morphol & cytol to prof bot, 49-74, PROF BIOL, W VA UNIV, 74- Mem: Bot Soc Am; Int Phycol Soc. Res: Algae as they are affected by factors of their natural environment. Mailing Add: Dept of Biol WVa Univ Morgantown WV 26506

BENNETT, HERBERT STANTON, b Quincy, Mass, May 18, 36. PHYSICS. Educ: Harvard Univ, AB, 58, PhD(physics), 64; Univ Md, MS, 60. Prof Exp: Res assoc physics, U K Atomic Energy Res Estab, Harwell, Eng, 64-65; res assoc, Univ Ill, Urbana, 65-66; SOLID STATE PHYSICIST, INST FOR MAT RES, NAT BUR STAND, 66-, CHIEF SOLID STATE MAT SECT, 72- Concurrent Pos: Sci technol fel, Dept Com, 71-72. Honors & Awards: Md Outstanding Young Scientist Award, 70; Civil Serv Comn Educ for Pub Mgt Award, 71. Mem: Am Phys Soc; Inst Elec & Electronic Engrs; Am Soc Testing & Mat. Res: Magneto-optical effects in solids; second order phase transitions; imperfections in solids; interactions of intense laser beams with matter; nonlinear optics; infrared and optical absorption in materials. Mailing Add: Nat Bur of Stand Bldg 223 Rm A259 Washington DC 20234

BENNETT, HUGH DEVERAUX, b Brooklyn, NY, Jan 19, 18; m 41. MEDICINE. Educ: Univ Chicago, BS, 40, MD, 42. Prof Exp: Instr med, Med Sch, Northwestern Univ, 51-53; from asst prof to assoc prof, Col Med, Baylor Univ, 53-62; PROF MED & ASSOC DEAN, HAHNEMANN MED COL, 62- Concurrent Pos: Asst chief med serv, Vet Admin Hosp, Ill, 51-53, chief med serv, Houston, 53-62; attend physician, Philadelphia Gen Hosp, 62-; vis prof, Free Univ Lille, France, 75- Mem: Fel Am Col Physicians; AMA; Am Geriat Soc. Res: Hepatic and gastrointestinal diseases. Mailing Add: Dept of Med Hahnemann Med Col Philadelphia PA 19102

BENNETT, IAN CECIL, b Bebington, Eng, Aug 2, 31; US citizen; m 60; c 2. DENTISTRY. Educ: Univ Liverpool, BDS, 56; Univ Toronto, DDS, 59; Univ Wash, MSD, 64. Prof Exp: Assoc prof pedodontics, Dalhousie Univ, 63-65; asst prof, Univ Ky, 65-68, dir med ctr commun, 67-68; assoc prof pedodontics & assoc dean, NJ Dent Sch, 68-69, PROF PEDODONTICS & DEAN NJ DENT SCH, COL MED & DENT NJ, 69- Mem: AAAS; Am Dent Asn; Am Acad Pedodontics; Am Soc Dent for Children; Can Dent Asn. Res: Calcification of bone and tooth substance; effect of tetracycline on calcification and on organ culture of bone. Mailing Add: NJ Dent Sch Col of Med & Dent of NJ 100 Bergen St Newark NJ 07103

BENNETT, IVAN FRANK, b Hartford, Conn, Sept 6, 19; m 44; c 2. PSYCHIATRY. Educ: Trinity Col, BS, 41; Jefferson Med Col, Thomas Jefferson Univ, MD, 44. Prof

Exp: Instr psychiat, Sch Med, Univ Pa, 54-56, guest lectr, Sch Dent, 54-56; clin asst prof, Sch Med, Georgetown Univ, 56-58; from asst prof to assoc prof, 58-72, PROF PSYCHIAT, SCH MED, IND UNIV, INDIANAPOLIS, 72-; SR PHYSICIAN, LILLY LAB CLIN RES, ELI LILLY & CO, 63- Concurrent Pos: Asst physician, State Hosp, Harrisburg, Pa, 48-50; asst chief acute intensive treatment serv & chief physiol treatment sect, Vet Admin Hosp, Coatesville, Pa, 50-56; chief psychiat res, Psychiat & Neurol Serv, Dept Med & Surg, Vet Admin, Washington, DC, 56-58; clin investr, Lilly Lab Clin Res, Eli Lilly & Co, Indianapolis, 58-63; dir, Lilly Psychiat Clin, Marion County Gen Hosp, Indianapolis, 59- Mem: AMA; fel Am Col Physicians; fel Am Psychiat Asn; fel Am Col Neuropsychopharmacol; Asn Res Nerv & Ment Dis. Res: Physiological and pharmacological therapies in psychiatry. Mailing Add: 8452 Green Braes North Dr Indianapolis IN 46234

BENNETT, IVAN LOVERIDGE, JR, b Washington, DC, Mar 4, 22; m 44; c 4. PATHOLOGY. Educ: Emory Univ, AB, 43, MD, 46; Am Bd Internal Med, dipl, 54. Prof Exp: Intern, Grady Mem Hosp, Atlanta, 46-47; vis investr bact, Naval Med Res Inst, Bethesda, Md, 47-49; asst & fel path, Johns Hopkins Univ, 49-50; asst resident med, Duke Univ Hosp, 50-51; asst, Emory Univ, 51-52; asst prof internal med, Yale Univ, 52-54; from assoc prof to prof med, Johns Hopkins Univ, 54-58, Baxley prof path & dir dept, 58-66; from dep dir to actg dir, Off Sci & Technol, Exec Off of the President, 66-69; dir med ctr, 69-73, VPRES HEALTH AFFAIRS, NY UNIV, 69-, PROF MED & DEAN SCH MED, 70-, PROVOST MED CTR, 73- Concurrent Pos: Asst pathologist, Johns Hopkins Hosp, 49-50, physician & consult bact, 54-58, pathologist-in-chief, 58-68; chief resident, Grady Mem Hosp, 51-52; assoc physician, Grace-New Haven Hosp, 52-54; attend physician, West Haven Vet Admin Hosp, 53-54; consult med, Baltimore City Hosps, 54-58, Loch Raven Vet Admin Hosp, 55-58 & Clin Ctr, USPHS, Bethesda, 55-58; consult path, Baltimore City Hosps, 58-66. Spec consult, Surgeon-Gen, US Army, 56-; mem comn epidemiol surv, Epidemiol Bd, Armed Forces, 56; consult, Commun Dis Ctr, USPHS, 57-; mem bd sci adv, Armed Forces Inst Path, 62-; mem exec comt, Div Med Sci, Nat Res Coun, 64-; chmn bd sci counr, Nat Inst Dent Res, 66-; chmn path training comt, Nat Inst Gen Med Sci, 66-; mem, President's Sci Adv Comt, 66-69; mem, US Japan Coop Med Sci Comt, NIH, 66-, chmn, 72, consult, Rev of Training Grants Prog, 70; mem, Bd Med, Nat Acad Sci, 67-70, mem, Int Environ Progs Comt, 70-74, consult, Space Sci Bd Rev of NASA Life Sci Prog, 70, mem comt on US int develop inst, Bd Sci & Technol for Int Develop, 70, mem, Bd Sci & Technol for Int Develop & Bd Human Resources, 70-; mem, President's Comt on Pop, 68-69, chmn panel chem & biol weapons, President's Sci Adv Comt, 69, chmn panel on biol & med sci, 69 & mem, President's Comt on the Nat Medal of Sci, 70-; mem adv panel sci & technol, Comt Sci & Astronaut, US House Rep, 69-; chmn ad hoc adv comt for rev of testing safety at Edgewood Arsenal, Md & Ft McClellan, Ala, Dept of Army, 69; consult, Carnegie Endowment for Int Peace, 70-; mem, Life Sci Adv Comt of NASA, 71-; chmn, Adv Panel on Presidential Prizes for Innovation, 72; mem ad hoc comt on sci & tech advice to Fed Govt, Nat Acad Sci, 73-; mem, Defense Sci Bd, Dept of Defense & consult, Off Dir, Defense Res & Eng, 74-77; mem, Comt on Joint Study of US-USSR Systs for Fundamental Res, Nat Acad Sci, 74-; mem, Dean Search Comt Uniformed Serv Univ of Health Sci, Dept of Defense, Washington, DC, 74; consult, Nat Security Coun, 73-; mem joint planning comt on world food, health & pop, Nat Acad Sci & Sci & Technol Policy Off, 74-; consult, Off Technol Assessment, Cong of US, 74-; mem, Panel to Rev US-USSR Sci Exchanges & Rels of BISE under CIR, Nat Acad Sci-Nat Res Coun, 75-; mem, Adv Group on Contrib of Technol to Econ Strength, STPO, NSF Directorate for Sci, Technol & Int Affairs, Washington, DC, 75- Mem: Inst of Med of Nat Acad Sci; Am Fedn Clin Res (pres, 57-58); Am Soc Clin Invest; Soc Exp Biol & Med; Am Soc Microbiol. Res: Infectious diseases; pathogenesis infection; mechanism of fever. Mailing Add: NY Univ Med Ctr 550 First Ave New York NY 10016

BENNETT, IVEN, b Chico, Calif, June 24, 21; m 45; c 1. CLIMATOLOGY, GEOGRAPHY. Educ: Chico State Col, AB, 48; Univ Nebr, MS, 50; Boston Univ, PhD(geog), 62. Prof Exp: Instr geog, Rutgers Univ, 50-53; res assoc, Am Geog Soc, 53-55; geog, US Army Qm Res & Develop Ctr, 55-62; head desert & trop lab, US Army Natick Labs, 62-66; PROF GEOG, UNIV NMEX, 66- Mem: AAAS; Asn Am Geog; Am Meteorol Soc; Am Geog Soc; Int Solar Energy Soc. Res: Climatology and geography of Southwestern United States. Mailing Add: Dept of Geog Univ of NMex Albuquerque NM 87131

BENNETT, JAMES AUSTIN, b Taber, Alta, Jan 29, 15; nat US; m 40; c 5. ANIMAL BREEDING. Educ: Utah State Univ, BS, 40, MS, 41; Univ Minn, PhD(animal breeding), 57. Prof Exp: Livestock asst, Dom Dept Agr Can, 41-45; asst prof, 45-50, PROF ANIMAL HUSB & HEAD DEPT, UTAH STATE UNIV, 50- Mem: AAAS; Am Soc Animal Sci; Am Genetics Asn. Res: Breeding phases of beef cattle, sheep and swine. Mailing Add: Dept of Animal Sci Utah State Univ Logan UT 84321

BENNETT, JAMES GORDY, JR, b Washington, DC, Aug 29, 32; m 56; c 2. ORGANIC CHEMISTRY. Educ: NY State Col Teachers, Albany, BS, 54; Rensselaer Polytech Inst, PhD(chem), 59. Prof Exp: Assoc res chemist, Parke Davis & Co, Mich, 59-60; sr res chemist, Huyck Felt Co, NY, 60-63; res chemist, Chem Develop Opers, 63-68, RES CHEMIST, PLASTICS DEPT, GEN ELEC CO, 68- Mem: Am Chem Soc. Res: Organic and polymer synthesis; oxidative coupling chemistry; cyclopropyl chemistry. Mailing Add: RD Feura Bush Rd Delmar NY 12054

BENNETT, JAMES HALLAM, b Galveston, Tex, Nov 2, 38; m 59; c 4. MATHEMATICS, COMPUTER SCIENCE. Educ: Cornell Univ, BA, 59; Princeton Univ, MA, 61, PhD(math), 62. Prof Exp: Hildebrant res instr math, Univ Mich, 62-64, asst prof, 64-66; managing dir, Appl Logic Res Inst, 68-71, INDUST MGR, APPL LOGIC CORP, 71-, SR SCIENTIST, 61- Mem: AAAS; Am Math Soc; Asn Symbolic Logic; Asn Comput Mach. Res: Mathematical theory and recursive function theory; computer theorem proving; time-sharing systems; information retrieval. Mailing Add: Appl Log Corp 900 State Rd Princeton NJ 08540

BENNETT, JAMES MARVIN, b St Louis, Mo, June 28, 39; m 66. ENVIRONMENTAL MANAGEMENT. Educ: Wash Univ, BS, 63, PhD(bot), 68. Prof Exp: Asst prof biol, NY Univ, 68-71; dir environ affairs, 71-75, DIR ENVIRON & INDUST AFFAIRS, JOS SCHLITZ BREWING CO, 75- Concurrent Pos: Lectr, New Sch Soc Res, 69-71; consult, W A Benjamin Publ, 70-71; comnr, Comn Undergrad Educ Biol Sci, 70-71; adj asst prof, NY Univ, 71. Mem: AAAS; Am Inst Biol Sci. Res: Management of environmental control procedures; ecology; air pollution and water pollution control; solid waste management; resource recovery; economic and social benefits. Mailing Add: Jos Schlitz Brewing Co PO Box 614 Milwaukee WI 53201

BENNETT, JAMES PETER, b Chicago, Ill, Aug 25, 44. CROP PHYSIOLOGY, PHYSIOLOGICAL ECOLOGY. Educ: Washington Univ, BA, 66; Univ Mich, Ann Arbor, MA, 69; Univ BC, PhD(plant sci), 75. Prof Exp: Sr scientist ecol, Hudson River Valley Comn, Tarrytown, NY, 69-70; instr, Pratt Inst, 70-71; ASST PROF & OLERICULT CROP ECOL, DEPT VEG CROPS, UNIV CALIF, DAVIS, 75- Mem: AAAS; Am Inst Biol Sci; Am Soc Hort Sci; Crop Sci Soc Am. Res: Plant competition, density, spacing, productivity; whole plant development and growth analysis; allometry; yield component compensation; biostatistics; vegetable and field crops; air pollution effects on vegetation. Mailing Add: Dept of Veg Crops Univ of Calif Davis CA 95616

BENNETT, JEAN MCPHERSON, b Kensington, Md, May 9, 30; m 52. OPTICAL PHYSICS. Educ: Mt Holyoke Col, BA, 51; Pa State Univ, MS, 53, PhD(physics), 55. Prof Exp: Physicist, Nat Bur Standards, 51, 53 & Wright Air Develop Ctr, 55-56; PHYSICIST, NAVAL WEAPONS CTR, 56-, RES SCIENTIST, 60- Mem: Fel Optical Soc Am. Res: Optical properties of solids; solid state physics; interferometry; thin films. Mailing Add: Code 6018 Michelson Lab Naval Weapons Ctr China Lake CA 93555

BENNETT, JESSE HARLAND, b Lehi, Utah, June 21, 36; m 58; c 3. PLANT PHYSIOLOGY. Educ: Utah State Univ, BS, 61; Univ Utah, PhD(bot) & cert environ toxicol, 69. Prof Exp: USPHS trainee environ toxicol, Ctr Environ Biol, Univ Utah, 65-68, assoc plant physiol, Air Pollution Lab, 68; NIH fel & res assoc bioenergetics, Ctr Biol Natural Syst, Wash Univ, 69-70; assoc plant physiol, Air Pollution Lab, Univ Utah, 70-74; PLANT PHYSIOLOGIST, PLANT STRESS LAB, PLANT PHYSIOL INST, AGR RES SERV, USDA, 74- Concurrent Pos: Air pollution consult & researcher, Ajax Presses, Utah Power & Light Co, 71-73; sci reviewer, Environ Qual, Nat Acad Sci, Coun Environ Qual, NSF, Energy Res & Develop Admin & USDA, 74-75. Mem: Bot Soc Am; Am Soc Plant Physiologists. Res: Physiological/metabolic effects, modes of action and control of environmental stresses on plant photosynthesis, bioenergetics and growth and fruiting processes. Mailing Add: Plant Physiol Inst Agr Res Serv Plant Stress Lab USDA Beltsville MD 20705

BENNETT, JOAN WENNSTROM, b Brooklyn, NY, Sept 15, 42; m 66; c 2. GENETICS, MYCOLOGY. Educ: Upsala Col, BS, 63; Univ Chicago, MS, 64, PhD(bot), 67. Prof Exp: Res assoc, NSF fel, Univ Chicago, 67-68; Nat Res Coun fel, 68-70; ASST PROF BIOL, TULANE UNIV, 70- Concurrent Pos: NSF fel, 70-71. Mem: AAAS; Torrey Bot Club; Bot Soc Am; Genetics Soc Am; Am Soc Microbiol; Mycol Soc Am. Res: Parasexual cycle; aflatoxin biosynthesis; genetics of secondary metabolism. Mailing Add: Dept of Biol Tulane Univ New Orleans LA 70118

BENNETT, JOE CLAUDE, b Birmingham, Ala, Dec 12, 33; m 58; c 3. IMMUNOLOGY. Educ: Howard Col, AB, 54; Harvard Med Sch, MD, 58; Am Bd Internal Med, dipl, 68. Prof Exp: Intern med, Med Ctr, Univ Ala, 58-59, asst resident, 59-60; Arthritis & Rheumatism Found fel, 60-62; res assoc molecular biol, NIH, 62-64; sr res fel biol chem, Calif Inst Technol, 64-65; assoc prof med & microbiol & asst dir div clin immunol & rheumatology, 66-70, PROF MED & MICROBIOL, CHMN DEPT MED & DIR DIV CLIN IMMUNOL & RHEUMATOLOGY, MED CTR, UNIV ALA, BIRMINGHAM, 70- Concurrent Pos: Clin fel rheumatology, Mass Gen Hosp, Harvard Med Sch, 60-61, res fel, 61-62; Markle scholar acad med, 65-70; Nat Inst Gen Med Sci res career develop award, 65-75; mem study sect for training in rheumatology, NIH, 69-72, chmn, 70-72, mem & chmn study sect for allergy & immunol, 75-78. Mem: Genetics Soc Am; Am Rheumatism Asn (secy-treas, 74-76); Am Asn Immunologists; Am Soc Clin Invest; Am Asn Physicians. Res: Genetic determinants of protein structure; structural aspects of immunoglobulins; molecular basis of disease states; immunochemistry; cell surface structure. Mailing Add: Dept of Med Univ of Ala Sch of Med Birmingham AL 35233

BENNETT, JOHN FRANCIS, b Palo Alto, Calif, Jan 13, 25; m 64; c 1. SCIENCE EDUCATION, EVOLUTIONARY BIOLOGY. Educ: Stanford Univ, AB, 46, PhD(biol), 62. Prof Exp: Physicist, US Navy Electronics Lab, 50-51; asst biol, Hopkins Marine Sta, Calif, 53-54; asst, Stanford Univ, 54-56, actg instr, 57-60; NIH fel microbiol, 61-63; asst prof hist & philos sci, Univ Pa, 63-69; vis scholar biol sci, Stanford Univ, 69-71; proj dir natural sci, Sullivan Assocs, 71-74; staff mem, Encycl Britannica, 75; LECTR EMBRYOL, HUMAN ECOL & GRAD BIOL, SAN JOSE STATE UNIV, 76- Concurrent Pos: Fulbright guest prof, Inst Statist, Univ Vienna, 67-68. Mem: AAAS; Sigma Xi. Res: Macroevolution theory; human ecology; integration of knowledge. Mailing Add: 2323 Sharon Rd Menlo Park CA 94025

BENNETT, JOHN HENRY, b Cincinnati, Ohio, July 21, 35; m 58; c 3. APPLIED MATHEMATICS. Educ: Harvard Univ, AB, 57, SM, 58, PhD(appl math), 62. Prof Exp: Sr mathematician, Bettis Atomic Power Lab, Pa, 62-64; sr mathematician, United Aircraft Res Labs, 64, supvr appl math, 67-68, chief math anal sect, 68-72, mgr info systs dept, Sikorsky Aircraft Div, United Technologies Corp, 72-73, DIR DATA PROCESSING, UNITED TECHNOLOGIES CORP, 73- Concurrent Pos: Lectr, Carnegie Inst Technol, 63-64 & Trinity Col (Conn), 65-72. Res: Numerical solution of the neutron transport equation and the Navier-Stokes equations. Mailing Add: United Technologies Corp Hartford CT 06101

BENNETT, JOHN M, b Boston, Mass, Apr 24, 33; m 57; c 3. INTERNAL MEDICINE, HEMATOLOGY. Educ: Harvard Univ, AB, 55; Boston Univ, MD, 59. Prof Exp: Instr med, Harvard Med Sch, 65-66; head morphol & histochem sect, Clin Path Dept, NIH, 66-68; asst prof med, Sch Med, Tufts Univ, 68-69; asst prof, 69-71, ASSOC PROF MED, SCH MED, UNIV ROCHESTER, 71-, ASSOC DIR CLIN ONCOL, UNIV ROCHESTER CANCER CTR, 74- Concurrent Pos: Dir outpatient labs, Boston City Hosp, 68-69; mem lymphoma task force, Nat Cancer Inst, 68-74; dir hemat & med oncol, Highland Hosp, Rochester, 69-74; head med oncol unit, Strong Mem Hosp, 74- Mem: AAAS; Am Soc Hemat; Am Fedn Clin Res. Res: Diagnosis by cytochemical techniques and treatment of malignant disorders of the hematopoietic system. Mailing Add: Univ of Rochester Cancer Ctr 601 Elmwood Ave Rochester NY 14642

BENNETT, JOHN PHILLIP, b Portsmouth, Eng, June 12, 31; m 56; c 2. REPRODUCTIVE PHYSIOLOGY, ENDOCRINOLOGY. Educ: Univ London, BSc, 53, Hons, 54; Cambridge Univ, PhD(biol), 63. Prof Exp: Exp off animal reproduction, Agr Res Coun Unit Reproductive Physiol & Biochem, Cambridge Univ, 56-63; head endocrinol, Brit Drug Houses Ltd, Eng, 63-67, dep to head biol res, 66-67; head dept reproductive physiol, Syntex Res Div, Syntex Corp, 67-72, dir res admin, 70-72, CORP DIR ADMIN & DIR AQUACULT DIV, SYNTEX INC, 72-, VPRES, 72- Concurrent Pos: Mem pop control, WHO, 66-; Gt Brit deleg, World Cong Fertil & Steril, Stockholm, 66. Mem: Am Fertil Soc; Soc Study Reproduction; Brit Soc Endocrinol; Brit Soc Study Fertil; World Maricult Soc. Res: Applied aspects of reproduction, leading to the development of new methods of contraception in the human and to unique methods of increasing reproduction in animals; aquaculture in crustaceans. Mailing Add: Syntex Corp Hillview Ave Palo Alto CA 94304

BENNETT, JOHN RICHARD, b Detroit, Mich, Nov 9, 45; m 68; c 3. OCEANOGRAPHY. Educ: Univ Wis, Madison, BS, 68, MS, 70, PhD(meteorol), 72. Prof Exp: Scientist oceanog, Nat Oceanic & Atmospheric Admin, 72-74; ASST PROF PHYS OCEANOG, MASS INST TECHNOL, 74- Mem: Am Geophys Union; AAAS. Res: Numerical modeling of lake and ocean circulations. Mailing Add: Rm 54-1316 Mass Inst Technol Cambridge MA 02139

BENNETT, JOHN WILLIAM, b Milwaukee, Wis, July 18, 15; m 40; c 2. CULTURAL ANTHROPOLOGY, ECOLOGY. Educ: Beloit Col, BA, 37; Univ Chicago, MA, 41,

PhD(anthrop), 46. Prof Exp: Instr & asst sociol & anthrop, Ohio State Univ, 46-49; chief pub opinion & sociol res div, Japan Occup, 49-51; asst prof sociol & anthrop, Ohio State Univ, 52-59; chmn dept anthrop, 67-70, 74-75, PROF ANTHROP, WASH UNIV, 59- Concurrent Pos: Grants, Off Naval Res, 52-54, Soc Sci Res Coun, 52-54, NSF, 62-64, 74-76 & Agr Develop Coun & Wenner-Gren Found, 66; mem, NIMH Soc Sci Rev Comt, 68-72; mem, Technol & Human Affairs Prog, Wash Univ; sr fel, Ctr Biol Natural Systs, Wash Univ; mem, Ann Rev Anthrop educ bd, 71-75; mem, Conf on Japanese Industrialization, 71-73; grant, NIMH, 71-72; mem, UNESCO Man-Biosphere Prog, 73- Mem: Am Ethnol Soc (pres, 70-72); Asn Asian Studies. Res: Economic-developmental and ecological studies of contemporary agrarian peoples, North America, Far East, Latin America; modern Japanese society; cooperative and communal societies. Mailing Add: Dept of Anthrop Wash Univ St Louis MO 63130

BENNETT, KENNETH A, b Butler, Okla, Oct 3, 35; m 59; c 2. BIOLOGICAL ANTHROPOLOGY, HUMAN GENETICS. Educ: Univ Tex, BA, 61; Univ Ariz, MA, 66, PhD(anthrop), 67. Prof Exp: Asst prof anthrop, Univ Ore, 67-70; assoc prof, 70-75, PROF ANTHROP, UNIV WIS-MADISON, 75- Mem: AAAS; Am Asn Phys Anthrop; Soc Study Evolution; Am Eugenics Soc; Am Inst Biol Sci. Res: Osteology of prehistoric human populations; evolutionary genetics, systematics and taxonomy; electrophoretic techniques in human populations. Mailing Add: Dept of Anthropology Univ of Wis Madison WI 53706

BENNETT, KIMBERLY D, b San Mateo, Calif, July 23, 47; m 70. NEUROCYTOLOGY. Educ: Univ Wash, BSc, 69; Univ Calif, Los Angeles, PhD(anat), 74. Prof Exp: INSTR ANAT, NORTHWESTERN UNIV, 74- Concurrent Pos: Fel anat, Northwestern Univ, 74- Mem: Am Soc Cell Biol. Res: Structure and biochemistry of glycoproteins and glycolipids of neuronal plasma membrane. Mailing Add: Dept of Anat Northwestern Univ Sch of Med Chicago IL 60611

BENNETT, LARRY E, b, San Diego, Calif, Feb 29, 40. INORGANIC CHEMISTRY. Educ: San Diego State Col, BS, 62; Stanford Univ, PhD(chem), 65. Prof Exp: NSF fel chem, Columbia Univ, 65-66; asst prof, Univ Fla, 66-70; assoc prof, 70-74, PROF CHEM, SAN DIEGO STATE COL, 74- Mem: Am Chem Soc. Res: Coordination complexes of transition metals; mechanisms of electron transfer and substitution processes; design and synthesis of complexes for investigation of mechanism and possible new modes of reaction. Mailing Add: Dept of Chem San Diego State Col San Diego CA 92115

BENNETT, LAWRENCE HERMAN, b Brooklyn, NY, Oct 17, 30; m 53; c 3. PHYSICS, MATERIALS SCIENCE. Educ: Brooklyn Col, BA, 51; Univ Md, MS, 55; Rutgers Univ, PhD(physics), 58. Prof Exp: Physicist, US Naval Ord Lab, 51-53; physicist, Metal Physics Sect, 58-63, CHIEF, ALLOY PHYSICS SECT, NAT BUR STANDARDS, 63- Concurrent Pos: Lectr, Grad Sch, Univ Md, 58-61, assoc prof, 61- Honors & Awards: Gold Medal, Dept of Commerce, 71; Burgess Mem Award, Am Soc Metals, 75. Mem: Fel Am Phys Soc; Am Soc Metals; Am Inst Mining, Metall & Petrol Eng; Am Soc Test & Mat. Res: Nuclear magnetic resonance; Mössbauer effect; alloy theory; magnetic properties of materials; catalysis. Mailing Add: Nat Bur of Standards Washington DC 20234

BENNETT, LEE COTTON, JR, b Philadelphia, Pa, Mar 14, 33; m 53; c 3. MARINE GEOPHYSICS. Educ: Haverford Col, BA, 55; Temple Univ, MS, 58; Bryn Mawr Col, PhD, 66. Prof Exp: Asst solid state physics, Franklin Inst Labs, Pa, 55-58, res physicist, 58-59; asst geophys, Woods Hole Oceanog Inst, 63-66; asst prof oceanog, Univ Wash, 66-73, asst prof geophys, 68-73; GEOPHYSICIST, L C B CONSULTS, 73- Concurrent Pos: Actg asst prof, Univ Wash, 64; instr, Shoreline Community Col, 75-76. Mem: Am Geophys Union; Soc Explor Geophys; Acoust Soc Am. Res: Continuous seismic profiling; seismic absorption in unconsolidated sediments; shoreline processes. Mailing Add: L C B Consults 2032 S Newton Seattle WA 98112

BENNETT, LEONARD LEE, JR, b Savannah, Ga, Nov 10, 20; m 49; c 3. BIOCHEMICAL PHARMACOLOGY. Educ: Vanderbilt Univ, AB, 42, MS, 43; Univ NC, PhD(org chem), 49. Prof Exp: Asst, Vanderbilt Univ, 42-43; instr, Univ Ga, 43-44; asst & fel, Univ NC, 45-47; res assoc, 45, res chemist, 48-56, head biochem div, 56-64, DIR BIOCHEM RES, BIOCHEM DIV, SOUTHERN RES INST, 64- Concurrent Pos: Mem adv comt res ther cancer, Am Cancer Soc, 56-58 & 60-63; mem chemother study sect, NIH, 64-68; mem ed bd, Molecular Pharmacol, 68- & Cancer Res, 74- Mem: AAAS; Am Asn Cancer Res; Am Soc Biol Chemists. Res: Biochemistry of cancer; mechanisms of drug action; biochemistry of purines and pyrimidines. Mailing Add: Southern Res Inst 2000 Ninth Ave S Birmingham AL 35205

BENNETT, LESLIE R, b Denver, Colo, Feb 13, 18; m 48; c 4. RADIOLOGY. Educ: Univ Calif, AB, 40; Univ Rochester, MD, 43. Prof Exp: Intern, Strong Mem Hosp, Rochester, NY, 44; asst radiation biol & pediat, Univ Rochester, 46-49; asst clin prof radiol, 49-52, actg chief div radiobiol, Atomic Energy Proj, 50-52, from asst prof to assoc prof radiol, 52-60, PROF RADIOL, SCH MED, UNIV CALIF, LOS ANGELES, 60-, CHIEF CLIN RADIOISOTOPE SERV, 52- Mem: Am Roentgen Ray Soc; Am Soc Exp Path; Radiation Res Soc; Soc Nuclear Med. Res: Radiobiology. Mailing Add: Dept of Radiol Univ of Calif Ctr Health Sci Los Angeles CA 90024

BENNETT, LLOYD M, b Columbus, Ind, Nov 22, 28; m 56; c 2. SCIENCE EDUCATION. Educ: Ball State Teachers Col, AB, 50; Butler Univ, MS, 52; Univ Tex, MA, 58; Fla State Univ, PhD(marine biol, sci educ), 63. Prof Exp: Elem sch teacher, Ind, 50-51; res chemist, Firestone Tire & Rubber Co, Ind, 51-53; high sch teacher, Ind, 53-55; instr & asst biol, Univ Tex, 55-57; teacher & chmn dept sci, Tex, 57-59; assoc prof biol, Amarillo Col, 59-60; prof biol & chem & dir eve div, Indian River Jr Col, 60-61; asst physiol, Fla State Univ, 61-63; asst prof educ & sci educ, 63-66, assoc prof sci educ, 66-71, univ grants, 63-70, PROF SCI EDUC, TEX WOMAN'S UNIV, 71- Concurrent Pos: Hogg Found ment health grant, Univ Tex & Hogg Found, 64; co-researcher, Hogg Found grant, 65-66; UNESCO consult elem sci & math educ, Govt Philippines, 74. Mem: AAAS; Nat Asn Res Sci Teaching; Nat Asn Biol Teachers; Nat Sci Teachers Asn; Sch Sci & Math Asn. Res: Devising, writing, standardizing and implementing new curricular programs in science in junior high school, including laboratory experiences; research in science for preschool level; preschool curricular developmental research in science. Mailing Add: PO Box 22846 Tex Woman's Univ Sta Denton TX 76204

BENNETT, LONNIE TRUMAN, b Gorum, La, June 9, 33; m 55; c 4. MATHEMATICS, STATISTICS. Educ: Northwest State Col (La), BS, 55; La State Univ, MA, 60; Okla State Univ, PhD(statist), 66. Prof Exp: Mathematician, Western Geophys Co Am, 55; teacher pub sch, 57-59; instr math, La State Univ, 60-62; asst prof, 66-69, ASSOC PROF MATH, NORTHEAST LA STATE UNIV, 69- Concurrent Pos: Consult, Marshall Space Flight Ctr, NASA, Ala, 66-67. Mem: Math Asn Am; Am Statist Asn. Res: Experimental design. Mailing Add: Dept of Math Northeast La Univ Monroe LA 71201

BENNETT, MARVIN HERBERT, b New York, NY, May 3, 31; m 63; c 4. NEUROANATOMY, NEUROPHYSIOLOGY. Educ: Univ Calif, Los Angeles, BS, 58, PhD(physiol), 67. Prof Exp: Sr electronic technician, Dept Anat, Univ Calif, Los Angeles, 58-61; instr anat, La State Univ Med Ctr, New Orleans, 67-71; ASST PROF NEUROSURG, MED SCH, UNIV PITTSBURGH, 71- Mem: AAAS; Am Asn Anatomists; Soc Neurosci; assoc Am Physiol Soc. Res: Electrophysiological and behavioral investigations of sensory systems organization and function. Mailing Add: Dept of Neurol Surg Univ of Pittsburgh Med Sch Pittsburgh PA 15261

BENNETT, MARY KATHERINE, b Waterbury, Conn, Jan 30, 40. MATHEMATICS. Educ: Albertus Magnus Col, BS, 61; Univ Mass, MA, 65, PhD(math), 66. Prof Exp: Teacher high sch, Conn, 61-63; asst prof math, Univ Mass, Amherst, 66-68; John Wesley Young res instr, Dartmouth Col, 68-70; asst prof, 70-72, ASSOC PROF MATH, UNIV MASS, AMHERST, 72- Mem: Am Math Soc; Math Asn Am. Res: Lattice theory, particularly lattices of convex geometries and orthomodular lattices; category theory. Mailing Add: Dept of Math Univ of Mass Amherst MA 01002

BENNETT, MICHAEL, b Goose Creek, Tex, Feb 11, 36; m 60; c 4. EXPERIMENTAL PATHOLOGY, HEMATOLOGY. Educ: Baylor Univ, MD, 61. Prof Exp: Intern, Cleveland Metrop Gen Hosp, Ohio, 62; resident path, Philadelphia Gen Hosp, Pa, 63; biologist, Oak Ridge Nat Labs, 63-65; mem staff, Roswell Park Mem Inst, 65-68, assoc cancer res scientist, 68-72; ASSOC PROF PATH, SCH MED, BOSTON UNIV, 72- Concurrent Pos: Resident path, Buffalo Gen Hosp, 69-71. Mem: Am Soc Exp Path. Res: Function and morphology of hemopoietic stem cells; cellular immunology; bone marrow transplantation. Mailing Add: Dept of Path Boston Univ Sch of Med Boston MA 02118

BENNETT, MICHAEL J, b Cleveland, Ohio, May 9, 36; m 56; c 2. NUCLEAR PHYSICS. Educ: Ohio State Univ, BSc, 60; Univ NMex, MSc, 63; Fla State Univ, PhD(physics), 67. Prof Exp: STAFF MEM, PHYSICS DIV, LOS ALAMOS SCI LAB, 60- Mem: Am Phys Soc. Res: Nuclear structure. Mailing Add: Physics Div Los Alamos Sci Lab Los Alamos NM 87544

BENNETT, MICHAEL VANDER LAAN, b Madison, Wis, Jan 7, 31; m 63. NEUROPHYSIOLOGY. Educ: Yale Univ, BS, 52; Oxford Univ, DPhil(physiol), 57. Prof Exp: Res worker, Col Physicians & Surgeons, Columbia Univ, 57-58, res assoc, 58-59, from asst prof to assoc prof neurol, 59-67; prof anat, 67-74, PROF NEUROSCI & DIR DIV CELLULAR NEUROBIOL, ALBERT EINSTEIN COL MED, 74- Concurrent Pos: Mem & trustee, Marine Biol Lab; Nat Neurol Res Found fel, 58-60; NIH sr res fel, 60-62; USPHS career develop award, 62- Mem: Am Physiol Soc; Biophys Soc; fel NY Acad Sci; Am Soc Cell Biol; Soc Neurosci. Res: Electrobiology; synaptic transmission; neural organization. Mailing Add: Albert Einstein Col of Med Yeshiva Univ Bronx NY 10461

BENNETT, MIRIAM FRANCES, b Milwaukee, Wis, May 17, 28. BIOLOGICAL RHYTHMS, COMPARATIVE PHYSIOLOGY. Educ: Carleton Col, BA, 50; Mt Holyoke Col, MA, 52; Northwestern Univ, PhD(biol), 54. Prof Exp: Asst zool, Mt Holyoke Col, 50-52; asst biol, Northwestern Univ, 52-53; from instr to prof, Sweet Briar Col, 54-73, chmn dept, 64-73; DANA PROF BIOL & CHMN DEPT, COLBY COL, 73- Concurrent Pos: Guest investr, Inst Zool, Univ Munich, 60-61 & 68; NSF fel, 61; mem, Marine Biol Lab, Woods Hole; trustee, Kents Hill Sch, 75-; dir, Res Inst of Gulf of Maine, 75- Mem: Fel AAAS; Ecol Soc Am; Am Zool Soc; Am Micros Soc. Res: Biological rhythmicity; invertebrate and amphibian hormones. Mailing Add: Dept of Biol Colby Col Waterville ME 04901

BENNETT, ORUS LAMAR, agronomy, plant physiology, see 12th edition

BENNETT, OVELL FRANCIS, b Middleboro, Mass, Nov 5, 29; m 55; c 2. ORGANIC CHEMISTRY. Educ: State Teachers Col, Mass, BS, 53; Boston Col, MS, 55; Pa State Univ, PhD(chem), 58. Prof Exp: Asst, Boston Col, 53-55; asst, Pa State Univ, 55-56; res chemist, E I du Pont de Nemours & Co, NJ, 58-61; asst prof chem, 61-64, ASSOC PROF CHEM, BOSTON COL, 64- Mem: AAAS; Am Chem Soc. Res: Organic mechanisms; organosulfur chemistry. Mailing Add: Dept of Chem Boston Col Chestnut Hill MA 02167

BENNETT, PETER BRIAN, b Portsmouth, Eng, June 12, 31; m 56; c 2. ANESTHESIOLOGY, PHYSIOLOGY. Educ: Univ London, BSc, 51; Univ Southampton, PhD(physiol, biochem), 64. Prof Exp: Asst head surg sect, Royal Navy Physiol Lab, Alverstoke, Eng, 53-56, head inert gas narcosis sect, 56-66; head pressure physiol, Can Defence & Civil Inst Environ Res, 66-68; head pressure physiol, Royal Navy Physiol Lab, 68-72; prof biomed eng, 72-75, PROF ANESTHESIOL, MED CTR, DUKE UNIV, 72-, DIR RES ANESTHESIOL, 73-, CO-DIR F G HALL ENVIRON RES LAB, 74-, ASSOC PROF PHYSIOL-PHARMACOL, 75- Concurrent Pos: Mem comt decompression sickness, Nat Inst Safety & Occup Health, 72-74. Mem: AAAS; Brit Inst Biol; Undersea Med Soc (pres, 75-76); Soc Neurosci; Aerospace Med Asn. Res: Inert gas narcosis; mechanisms of anesthesia; high pressure nervous syndrome; decompression sickness; oxygen toxicity; diving physiology and medicine; pharmacology. Mailing Add: F G Hall Environ Res Lab Duke Univ Med Ctr Durham NC 27710

BENNETT, PETER HOWARD, b Farnworth, Eng, June 21, 37; m 63; c 1. EPIDEMIOLOGY, MEDICINE. Educ: Univ Manchester, BSc, 58, MB, ChB, 61. Prof Exp: House physician, Royal Infirmary, Manchester, Eng, 62, house surgeon, 62-63; house phsyician, Postgrad Med Sch, London, Eng, 63-64; house physician, Nat Hosp Nerv Dis, Queen Square, 64; vis assoc arthritis, NIH, 64-68, assoc chief epidemiol, Clin Field Studies Unit, 68-70, CHIEF EPIDEMIOL & FIELD STUDIES BR, NAT INST ARTHRITIS, METAB & DIGESTIVE DIS, PHOENIX, ARIZ, 70- Concurrent Pos: Vis fel arthritis, NIH, 63; res fel, Postgrad Med Sch, London, 64; assoc, Col Med, Univ Ariz, 68- Mem: AAAS; Am Pub Health Asn; Brit Med Asn; Am Rheumatism Asn; Am Diabetes Asn. Res: Clinical rheumatology; epidemiologic studies of arthritis and diabetes. Mailing Add: Nat Inst Arthritis Metab Dig Dis 1440 E Indian School Rd Phoenix AZ 85014

BENNETT, RALPH BLOUNT, mathematics, see 12th edition

BENNETT, RALPH DECKER, b Williamson, NY, June 30, 00; m 34; c 2. PHYSICS. Educ: Union Col, NY, BS, 21, MS, 23; Univ Chicago, PhD(physics), 25. Hon Degrees: ScD, Union Col, NY, 45. Prof Exp: Instr math, Union Col, NY, 21-23, asst prof physics, 25-26; Nat Res Coun fel, Princeton Univ, 26-27; Calif Inst Technol, 27-28; res assoc, Univ Chicago, 28-31; assoc prof elec measurements, Mass Inst Technol, 31-37, prof, 37-46; tech dir, Naval Ord Lab, 44-54; mgr tech dept, Knolls Atomic Power Lab, 54-56; mgr Vallecitos Atomic Lab, Gen Elec Co, 56-61; vpres, Martin Co, 61-67; CONSULT, 67- Concurrent Pos: Life trustee & vchmn bd, Union Col, NY; chmn adv bd oceanog, US Navy, mem sci adv comt, Naval Oceanog Off, mem adv bd, Naval Postgrad Sch; chmn sci adv comt, US Coast Guard; mem, Atomic Indust Forum, 63-69. Honors & Awards: Navy Distinguished Civilian Serv Medal; Off, Order Brit Empire; Legion of Merit; Meritorious Pub Serv Award, US Coast Guard, 75. Mem: Fel Am Phys Soc; fel Inst Elec & Electronics Eng; Am Nuclear Soc. Res: X-rays in relation to quantum theory; dielectrics-relation of properties of atomic structure; cosmic ray instrumentation and measurement; electronics; research management and administration. Mailing Add: 204 San Rafael Ave Belvedere CA 94920

BENNETT, RALPH EDGAR, b Grant, Mich, Feb 17, 09; m 33; c 1. BOTANY, MICROBIOLOGY. Educ: Univ Mich, AB, 33, MA, 34, PhD(mycol), 40. Prof Exp: Instr bot, Univ Mich, 40-41; res supvr fermentation, Joseph Seagram & Sons, 42-43; res microbiologist, Commercial Solvents Corp, 43-50, assoc dir, 50-52; sr microbiologist, Univ Pa, 52-55; head dept microbial-biochem, Squibb Inst Med Res, 55-61, head dept chemother, 61-64, res assoc sci info, 64-68, dir sci info, 68-73; CONSULT, OMNI RES CO, 74- Concurrent Pos: Consult, Squibb Inst Med Res, 52-55 & Nat Dairies Res Labs, 52-55. Mem: Am Chem Soc; Mycol Soc Am; Am Soc Microbiol; Am Acad Microbiol. Res: Cytology of fungi; physiology of microorganisms; industrial fermentations; microbiology of production and use of antibiotics; enzymes of microorganisms; information systems to aid biological and medical research. Mailing Add: 616 Ewing St Princeton NJ 08540

BENNETT, RAYMOND DUDLEY, b Meriden, Conn, May 9, 31; m 61; c 2. AGRICULTURAL BIOCHEMISTRY, NATURAL PRODUCTS CHEMISTRY. Educ: Univ Conn, BS, 53; Purdue Univ, MS, 55, PhD(pharm chem), 58. Prof Exp: Fel org chem, Univ Va, 58-60; res chemist, Nat Insts Health, 60-63; res chemist, Calif Inst Technol, 64-69, RES CHEMIST, FRUIT & VEG CHEM LAB, AGR RES SERV, US DEPT AGR, 69- Mem: Am Chem Soc; AAAS; Phytochem Soc NAm. Res: Structure determination of citrus constituents; nuclear magnetic resonance; biochemistry of citrus fruits. Mailing Add: Fruit & Veg Chem Lab US Dept of Agr 263 S Chester Ave Pasadena CA 91106

BENNETT, RICHARD BOND, b Grove City, Pa, Nov 28, 32; m 61; c 2. ORGANIC CHEMISTRY. Educ: Grove City Col, BS, 54; Ohio State Univ, MSc, 55, PhD(carbohydrate chem), 64. Prof Exp: Res chemist, Gulf Res & Develop Co, 58; asst prof chem, Millikin Univ, 61-63; from asst prof to assoc prof, 67-74, chmn dept, 69-71, PROF CHEM, THIEL COL, 74- Mem: Am Chem Soc. Res: Aldonamide hydrolysis; ketose synthesis. Mailing Add: Dept of Chem Thiel Col Greenville PA 16125

BENNETT, RICHARD HENRY, b Hart Co, Ky, Mar 4, 44; m 65; c 2. PETROLEUM CHEMISTRY. Educ: Western Ky Univ, BS, 65; Vanderbilt Univ, PhD(org chem), 69. Prof Exp: Res chemist petrol chem, Texaco Res Lab, 69-75; RES CHEMIST LIQUID CHROMATOGRAPHY, JEFFERSON CHEM CO, 75- Mem: Am Chem Soc; Sigma Xi. Res: Improvement of refinery processes; isolation and identification of trace organics present in refinery waste water streams; high pressure liquid chromatography. Mailing Add: Jefferson Chem Co 7114 N Lamar Blvd Austin TX 78753

BENNETT, RICHARD THOMAS, b Trenton, NJ, Jan 7, 30; m 58; c 4. PLASTICS CHEMISTRY. Educ: Yale Univ, BS, 52; Rutgers Univ, MS, 54, PhD(org chem), 56. Prof Exp: Res chemist polymeric films, E I du Pont de Nemours & Co, 56-58, tech rep, Appln Res, 58-62; tech dir packaging, Am Bag & Paper Corp, 62-64; assoc paper coating, hot melt & solvents, 64-66, tech supvr, Plastics Div, 66-67, prod mgr, 67-71, bus mgr, 71-73, asst to pres, 73-74, GEN MGR SPECIALTY CHEM DIV, ALLIED CHEM CORP, 74- Mem: Am Chem Soc. Res: Films; polymeric systems; adhesives; high energy electron irradiation. Mailing Add: Allied Chem Corp 2829 Glendale Ave Toledo OH 43614

BENNETT, ROBERT BOWEN, b Tillamook, Ore, Mar 28, 27; m 49; c 4. PHYSICS. Educ: Willamette Univ, BA, 50, Univ Ore, MA, 54, PhD(physics), 58. Prof Exp: From asst prof to assoc prof, physics, 57-65; UNESCO sr lectr, Univ Col Rhodesia & Nyasaland, 65-66, lectr, Univ Zambia, 66-67; ASSOC PROF PHYSICS, CENT WASH STATE COL, 67- Concurrent Pos: Staff physicist, Comn Col Physics, Univ Md, College Park, 68-69. Mem: Am Asn Physics Teachers; Am Phys Soc. Res: Collision process in atomic systems; pressure broadening of spectral lines. Mailing Add: Dept of Physics Cent Wash State Col Ellensburg WA 98926

BENNETT, ROBERT LEO, b Wilkinsburg, Pa, Dec 18, 11; m 37; c 2. PHYSICAL MEDICINE. Educ: Univ Pittsburgh, BS, 34, MD, 36; Univ Minn, MS, 40. Hon Degrees: DSc, Univ Pittsburgh, 60. Prof Exp: Asst prof phys med, Univ Wis, 40-41; dir phys med, 41-58, asst dir, 48-53, med dir, 53-58, exec dir, 58-74, EMER MED DIR & DIR PROF EDUC, GA WARM SPRINGS HOSP, 74-; PROF PHYS MED, MED SCH, EMORY UNIV, 45- Concurrent Pos: Consult, Area 3, Vet Admin, 45-65; chmn dept phys med, Med Sch, Emory Univ, 46-66; secy-treas, Am Bd Phys Med & Rehab, 47-53, chmn, 53-62; med dir, Atlanta Cerebral Palsy Sch Clin, 53-54, trustee, 54-; mem med adv comt, Nat Found, 59-72; mem eval comt prosthetics res & develop, Nat Acad Sci-Nat Res Coun, 61-64; mem, Am Rehab Found, 61-; Horowitz vis prof, Inst Phys Med & Rehab, NY Univ, 62; mem med res study sect, Voc Rehab Admin, 63-66; med dir, Ga Rehab Ctr, 64-73; vchmn, Joint Comn Undergrad Educ Phys Med & Rehab, 64-69; mem, Presidential Task Force Phys Handicapped, 69-70; mem phys ther adv comt, Ga State Col, 69-; clin prof, Sch Med, Med Col Ga, 70- & Col Med, Med Univ SC, 70-; mem hon staff, All Children's Hosp, St Petersburg, Fla; consult, Crippled Children's Div, State of Ga; mem, Int Rehab Med. Honors & Awards: Citation, President's Comt Nat Employ Phys Handicapped Week, 54; Rehab Award, Ga Div Voc Rehab, 66. Mem: Am Acad Cerebral Palsy; Am Rheumatism Asn; Int Fedn Phys Med; hon mem Peruvian Soc Phys Med & Rehab; hon mem Ger Soc Phys Med. Res: Methods of care of convalescent poliomyelitis; electromyographic studies of denervated muscle tissue; apparatus for support of weakened bodily segments. Mailing Add: Ga Warm Springs Hosp Warm Springs GA 31830

BENNETT, ROBERT PUTNAM, b Hartford, Conn, Dec 11, 32; m 57; c 4. ORGANIC CHEMISTRY. Educ: Trinity Col, Conn, BS, 55, MS, 57; Case Inst Technol, PhD(org chem), 60. Prof Exp: Fel chem, Pa State Univ, 60-61; res chemist Am Cyanamid Co, 61-66, sr res chemist, 66-69; lab dir, 69-74, dir chem res & develop, 74, ASST VPRES RES & DEVELOP, APOLLO CHEM CORP, 75- Mem: Am Chem Soc; Am Inst Chem; Am Soc Test & Mat; Nat Asn Corrosion Eng. Res: Catalytic reactions; combustion reactions; catalysis petroleum chemistry; air pollution chemistry and control. Mailing Add: 35 S Jefferson Rd Whippany NJ 07981

BENNETT, STELMAN EMERSON, entomology, deceased

BENNETT, STEPHEN LAWRENCE, chemistry, see 12th edition

BENNETT, STEWART, b New York, NY, Feb 14, 33; m 56; c 2. PHYSICS, Educ: Cornell Univ, AB, 53, PhD(nuclear physics), 61. Prof Exp: Res physicist anti-submarine warfare, Opers Res Inc, 60-61; staff scientist plasma physics, Res & Adv Develop Div, Avco Corp, 61-63, chief sect, 63-69; MGR SPEC PROJ, POLAROID CORP, 69- Mem: Am Phys Soc; Am Inst Aeronaut & Astronaut. Res: Plasma and discharge physics; cosmic rays. Mailing Add: Polaroid Corp Cambridge MA 02183

BENNETT, THEDA (ELIZABETH), physiology, see 12th edition

BENNETT, TRAVIS H, b El Paso, Tex, Nov 16, 31; m 59; c 2. MICROBIOLOGY. Educ: Tex Western Col, BA, 53; Univ Tex, MA, 62. Prof Exp: Res bacteriologist, William S Merrell Co, 61-66; from assoc bacteriologist to bacteriologist, 66-75, ASST SR MICROBIOLOGIST, ELI LILLY & CO, 75- Res: Antimicrobial research and development methods using in vitro and in vivo technics; host resistance factors; diagnostic bacteriology; vaccine development and assay studies; antimicrobial resistance studies and bacterial genetics. Mailing Add: Box 708 Dept G 709 Vet Res Eli Lilly & Co Greenfield IN 46140

BENNETT, WILLARD HARRISON, b Findlay, Ohio, June 13, 03. PLASMA PHYSICS. Educ: Ohio State Univ, AB, 24; Univ Wis, MS, 26; Univ Mich, PhD(physics), 28. Prof Exp: Instr physics, Univ Mich, 26-28; Nat Res Coun fel, Calif Inst Technol, 28-30; instr, Ohio State Univ, 30-33, asst prof, 33-38; dir res, Electronic Res Corp, 38-41; appl res, Inst Textile Technol, 45-46; sect chief, Nat Bur Stand, 46-50; prof physics, Univ Ark, 50-51; physicist, US Naval Res Lab, 51-61; BURLINGTON PROF PHYSICS, NC STATE UNIV, 61- Concurrent Pos: Consult, Los Alamos Sci Lab, 55- Mem: Fel Am Phys Soc. Res: Pinch effect; infrared spectra of symmetric top molecules; negative ion source; tandem accelerator; nonmagnetic mass spectrometer; radiation belts in laboratory tube; principles of nuclear fusion using intense relativistic electron beams; nuclear fusion. Mailing Add: Dept of Physics NC State Univ Raleigh NC 27607

BENNETT, WILLIAM EARL, b Eskridge, Kans, Dec 29, 23; m 52. INORGANIC CHEMISTRY. Educ: Sterling Col, BS, 47, Univ Kans, PhD(chem), 51. Prof Exp: Res assoc chem, Univ Chicago, 51-52 & Mass Inst Technol, 52-53; asst prof, 53-59, ASSOC PROF INORG CHEM, UNIV IOWA, 59- Mem: Am Chem Soc. Res: Coordination compounds; ion exchange. Mailing Add: Dept of Chem Univ of Iowa Iowa City IA 52240

BENNETT, WILLIAM ERNEST b Salters, SC, Feb 2, 28; m 53; c 1. IMMUNOPATHOLOGY. Educ: Lincoln Univ, Pa, AB, 50; Temple Univ, MS, 55; Univ Pa, PhD(med microbiol), 60. Prof Exp: Instr-asst prof microbiol, Meharry Med Col, 60-63; NIH fel investr cellular biol, Rockefeller Univ, 63-65; chief immunopath sect, Med Lab Div, US Biol Lab, 66-69; microbiologist sci admin, Div Res Grants, NIH, 69-70, health sci adminr, Div Physician & Health Professions Educ, 70-74; CHIEF INST RESOURCES BR SCI ADMIN, DIV MED, HEALTH RESOURCES ADMIN, DEPT HEALTH, EDUC & WELFARE, 74- Concurrent Pos: Consult, Oak Ridge Nat Lab, 62-65; NIH grants assoc, 69-70. Mem: Soc Exp Path; Soc Reticuloendothelial Syst; Sigma Xi; Am Soc Microbiol; Fedn Am Soc Exp Biol. Res: Mechanisms of cellular immunity; ultrastructural cytochemistry; biochemistry and physiology of macrophages; federal science administration and medical school resource management. Mailing Add: Rm 4B-55 Bldg 31 NIH 9000 Rockville Pike Bethesda MD 20014

BENNETT, WILLIAM FREDERICK, b Plainview, Ark, Jan 23, 27; m 50; c 3. AGRONOMY, PLANT NUTRITION. Educ: Okla State Univ, BS, 50; Iowa State Col, MS, 52, PhD, 58. Prof Exp: Exten area agronomist, Iowa State Col, 51-54, exten agronomist, 54-57; exten soil chemist, Tex A&M Univ, 57-63; chief agronomist, Elcor Chem Corp, 63-68; PROF AGRON, TEX TECH UNIV, 68-, ASST DEAN AGR SCI, 70- Concurrent Pos: Consult, King Ranch, Arg. Mem: AAAS; Am Soc Agron; Soil Sci Soc Am; Sigma Xi. Res: Soil fertility and fertilizer use on agronomic and vegetable crops; plant analysis and soil testing; developing diagnostic tools. Mailing Add: Office of the Dean Tex Tech Univ PO Box 4169 Lubbock TX 79409

BENNETT, WILLIAM HUNTER, b Taber, Alta, Nov 5, 10; nat US; m 50; c 4. AGRONOMY, Educ: Utah State Univ, BS, 36, MSc, 48; Univ Wis, PhD, 57. Prof Exp: From asst county agr agent to county agr agent, 37-42, exten agronomist, 46-47, from asst prof agron to prof agron, 47-69, prof soils & meteorol, 69-73, dir exten serv, 64-73, from asst dir to actg dir, 56-58, actg dean, Col Agr, 58, EMER PROF PLANT SCI & EMER DIR UNIV EXTEN, UTAH STATE UNIV, 73-, HEAD DEPT, 69- Mem: Am Soc Agron. Res: Forage yield and quality; miscellaneous legume trials; grass seed production; pasture mixture investigation. Mailing Add: Col of Agr Utah State Univ Logan UT 84321

BENNETT, WILLIAM RALPH, JR, b NJ, Jan 30, 30; m 52; c 3. PHYSICS. Educ: Princeton Univ, BA, 51; Columbia Univ, PhD(physics), 59. Hon Degrees: MA, Yale Univ, 65. Prof Exp: Asst physics, Radiation Lab, Columbia Univ, 52-54, asst, Cyclotron Group, 54-57; instr, Yale Univ, 57-59, res assoc, 59; mem tech staff, Bell Tel Labs, 59-62; from assoc prof to prof physics & appl sci, 62-70, CHARLES BALDWIN SAWYER PROF ENG & APPL SCI & PROF PHYSICS, YALE UNIV, 70- Concurrent Pos: Consult, Tech Res Group, Inc, 62-67; vis scientist, Am Inst Physics Prog, 63-64; Sloan Found fel, 63-65; consult, Inst Defense Anal, 63-70; mem adv panel, Nat Acad Sci to Nat Bur Standards, 63-69, chmn, 66; consult, Army Res Off, Durham, 65- & CBS Labs, 66-68; John S Guggenheim Found fel, 67; guest of Soviet Acad Sci, 67 & 69; consult & mem bd dirs, Laser Sci Corp, 68-69; mem lab adv bd for res, Naval Res Adv Comt, 68-76. Honors & Awards: Morris N Liebmann Award, Inst Elec & Electronics Eng, 65. Mem; Fel Am Phys Soc; fel Optical Soc Am; fel Inst Elec & Electronics Eng. Res: Atomic and molecular physics; positronium; optical pumping; inelastic collisions; radiative lifetimes; excitation processes; optical and rf spectroscopy; gas lasers. Mailing Add: Dunham Lab Yale Univ New Haven CT 06520

BENNETT, WORD BROWN, JR, b Nashville, Tenn, Oct 24, 15; m 39; c 4. NATURAL PRODUCTS CHEMISTRY. Educ: Vanderbilt Univ, MS, 40. Prof Exp: Chief chemist, 40-56, res dir, 56-66, VPRES RES & DEVELOP, US TOBACCO CO, 66- Mem: Am Chem Soc. Res: Dark tobacco. Mailing Add: 2700 Overhill Dr Nashville TN 37214

BENNICK, ALFRED HAROLD, astronomy, see 12th edition

BENNING, CALVIN JAMES, b Chicago, Ill, Aug 8, 25; m 52; c 5. ORGANIC CHEMISTRY. Educ: Univ Notre Dame, BSc, 50; Ohio State Univ, PhD(chem), 53. Prof Exp: Chemist res & develop, Hudson Foam Plastics, 53-55 & M W Kellogg Co, 55-58; res chemist, W R Grace & Co, 58-62, res supvr plastic develop, 62-64, mgr plastic develop sect, 64-67, res mgr, Corp Res Lab, 67-68; mgr new technol res, 68-70, ASST DIR, CORP RES CTR, INT PAPER CO, 70- Mem: Am Chem Soc. Res: Analytical microscopy; polymer science and technology; paper and coating research; cellulose and polymer composites; materials for health and medical devices. Mailing Add: Corp Res Ctr Int Paper Co PO Box 797 Tuxedo Park NY 10987

BENNINGHOFF, WILLIAM SHIFFER, b Ft Wayne, Ind, Mar 23, 18; m 69; c 2. PLANT ECOLOGY. Educ: Harvard Univ, SB, 40, MA, 42, PhD(plant geog), 48. Prof Exp: Asst bot mus, Harvard Univ, 38-40, lab instr, Univ exten, 46-48; botanist, US Geol Surv, 48-56, chief Alaska terrain & permafrost sect, 52-56; assoc prof, 56-60, assoc prof biol sta, 57, prof, 61, 63, 66, asst dir bot gardens, 65-66, PROF BOT, UNIV MICH, 60- Concurrent Pos: Mem panel biol & med sci, Nat Acad Sci-Nat Res Coun Comt Polar Res, 62-, chmn, 66-; mem, Sci Comt Antarctic Res working group in biol, 68-; dir, US Aerobiol prog Int Biol Prog, 67-, chmn, Int Aerobiol Working Group, 68-; mem, US Nat Comt of Int Union Biol Sci, 75. Mem: Fel AAAS; Am Geog Soc; Am Soc Limnol & Oceanog; Arctic Inst NAm; Int Soc Plant Geog & Ecol.

Res: Spore dispersal; historical phytogeography; phytocoenology; Arctic and subarctic ecology. Mailing Add: Dept of Bot Univ of Mich Ann Arbor MI 48104

BENNINGTON, JAMES LYNNE, b Evanston, Ill, Apr 29, 35; m 59; c 2. PATHOLOGY. Educ: Univ Chicago, MD, 59, MS, 62. Prof Exp: Intern, Presby-St Lukes Hosp, Chicago, 59-60; resident path, Kaiser Found Hosp, Oakland, Calif, 62-64, pathologist-in-chief, 69-70; assoc prof path, Sch Med, Univ Wash, 64-69; CLIN ASSOC PROF PATH, MED SCH, UNIV CALIF, SAN FRANCISCO, 70-, CHMN DEPT PATH & CLIN LAB, CHILDREN'S HOSP SAN FRANCISCO & ADULT MED CTR, 70- Concurrent Pos: Pathologist & consult, Child Develop Ctr, Oakland, Calif, 62-64; chief & dir labs, King County Hosp, Wash, 66-69. Mem: Am Soc Clin Path; Col Am Path; Int Acad Path. Res: Oncology; cellular kinetics of normal and neoplastic tissue. Mailing Add: Dept of Path Children's Hosp 3700 California St San Francisco CA 94119

BENNINGTON, KENNETH OLIVER, b Forsyth, Mont, Oct 13, 16; m 69. GEOCHEMISTRY. Educ: Mont State Univ, BS, 47, BA, 49; State Col Wash, MS, 51; Univ Chicago, PhD(geol), 60. Prof Exp: Res assoc geol, Univ Chicago, 57-58; crystallographer, Univ Wash, 59-63, res assoc prof geophys, 64-68; RES CHEMIST, THERMODYN LAB, ALBANY METALL RES CTR, BUR MINES, US DEPT INTERIOR, 68- Mem: Geochem Soc; Am Geophys Union. Res: Compositional changes and mineralogical reactions attributable to mechanical stress; thermodynamic properties of silicate minerals. Mailing Add: Thermodyn Lab Albany Metall Res Ctr Bur of Mines US Dept Interior Albany OR 97321

BENNINGTON, NEVILLE LYNNE, b Canton, Ohio, Aug 8, 06; m 30; c 2. ZOOLOGY. Educ: Col of Wooster, AB, 28; Northwestern Univ, AM, 30, PhD(zool), 34. Prof Exp: Instr zool, Northwestern Univ, 30-36; instr biol, Col of Wooster, 36-37; from assoc prof to prof bot & zool, Beloit Col, 37-62; asst comnr prof educ, State Univ NY, 62-66; div dir pre-col sci educ, NSF, 66-68; coordr fac res & projs, Wis State Univ, Oshkosh, 68-76; RETIRED. Concurrent Pos: Biologist in charge fisheries div, Wis Conserv Dept, 46; sect head insts div, NSF, 59-61. Mem: AAAS; Am Inst Biol Sci; assoc Am Soc Zool. Res: Cytology and reproductive rhythms of teleosts; lake survey and lake management; fish migration. Mailing Add: 833 Windward Ct Oshkosh WI 54901

BENNION, MARION, b Murray, Utah, Sept 23, 25. NUTRITION, FOOD SCIENCE. Educ: Utah State Univ, BS, 47; Columbia Univ, MA, 49; Univ Wis, PhD(food sci, biochem), 56. Prof Exp: Instr food & nutrit, Idaho State Univ, 49-50; instr, Brigham Young Univ, 52-53, from asst prof to assoc prof, 55-61; vis asst prof, Univ Calif, Davis, 61-62; head dept, 62-69, PROF FOOD & NUTRIT, BRIGHAM YOUNG UNIV, 62- Mem: AAAS; Inst Food Technol; Am Dietetic Asn; Am Inst Nutrit. Res: Decomposition of fats during frying; trace elements in human nutrition. Mailing Add: Dept of Food Sci & Nutrit Brigham Young Univ Provo UT 84602

BENNISON, ALLAN P, b Stockton, Calif, Mar 8, 18; m 41; c 3. GEOLOGY. Educ: Univ Calif, AB, 40. Prof Exp: Fel geol, Antioch Col, 40-42; photogrammet engr, US Geol Surv, 42- 45; asst chief geologist, Tri-Pet Corp, Columbia, 45-49; staff stratigrapher, Sinclair Oil & Gas Co, Okla, 49-68; GEOL CONSULT, 68- Concurrent Pos: Lectr, Univ Tulsa, 64; deleg, Int Geol Cong, Prague, Czech, 68-; compiler geol hwy map ser, Am Asn Petrol Geol, 72- Mem: AAAS; Am Asn Petrol Geol; Soc Econ Paleont & Mineral; Geol Soc Am. Res: Cretaceous stratigraphy of California; stratigraphy of Oklahoma, Colombia, Somalia and Mexico; tectonics of Alaska, Wyoming and Utah. Mailing Add: 804 Beacon Bldg Tulsa OK 74103

BENNISON, BERTRAND EARL, b Boston, Mass, Apr 18, 15; m 43; c 4. MEDICINE. Educ: Mass Inst Technol, SB, 37; Harvard Univ, MD, 41; Univ Pittsburgh, MPH, 54. Prof Exp: Med intern, US Marine Hosp, NY, 41-42; med officer, USPHS, 42-54; asst dir med res div, Esso Res & Eng Co, 54-59; asst dir res, Ortho Pharmaceut Corp, NJ, 60-66; prof biol sci & head dept, Drexel Univ, 66-71; DIR, LEON COUNTY HEALTH DEPT, FLA, 71- Concurrent Pos: Adj prof, Div Reproductive Biol, Hahnemann Med Col; vis prof, Dept Urban & Regional Planning, Fla State Univ, 72-75. Mem: AAAS; Am Pub Health Asn. Res: Preventive medicine. Mailing Add: Leon County Health Dept Tallahassee FL 32304

BENNUN, ALFREDO, b Buenos Aires, Arg, July 9, 34; m 61; c 1. BIOCHEMISTRY, BIOPHYSICS. Educ: Nat Univ Cordoba, BS, 54, MS, 57, PhD(biochem, pharm), 63. Prof Exp: Lab instr, Dept Psychiat, Sch Med, Nat Univ Cordoba, 56-57, instr histol & embryol, 57-59; instr biol chem, Univ Buenos Aires, 62-63; fel biochem, Weizmann Inst, 63-64; res assoc, Duke Univ, 64-65; NIH fel, New York, 65-66 & spec fel, Cornell Univ, 66-67; lectr prof, Univ PR, 67-68, asst prof, 68-69; ASSOC PROF BIOCHEM, NEWARK COL ARTS & SCI, RUTGERS UNIV, 69- Concurrent Pos: Resident, Clin Path Lab, Ment Health Hosp of State Inst Health, Cordoba, Arg, 56-57; res grants, NIH, 68-70, Res Corp, 69-72, Rutgers Res Fund, 69-73 & Charles & Johanna Busch Mem Fund, 74-76; mem comt bioenergetics, Int Union Pure & Appl Biophys. Mem: Am Chem Soc; Am Soc Plant Physiol; Genetics Soc Am; Biophys Soc; NY Acad Sci. Res: Phosphorus metabolism; enzymology; cellular respiration; oxidative and photophosphorylation; hormone receptors and enzymatic regulation; theoretical bioenergetics. Mailing Add: Newark Col Arts & Sci Rutgers Univ Newark NJ 07102

BENOIST, JEAN, b Lyon, France, Nov 24, 29; m; c 3. PHYSICAL ANTHROPOLOGY. Educ: Univ Lyon, MD, 55; Univ Paris, DSc, 64. Prof Exp: Assoc prof, 62-67, PROF ANTHROP, UNIV MONTREAL, 67- Concurrent Pos: Mem coun, Int Asn Human Biol. Res: Human genetics of race crossing; microevolution, especially in small populations. Mailing Add: Dept of Anthrop Univ of Montreal Montreal PQ Can

BENOIT, GEORGE JULIEN, JR, b Marshalltown, Iowa, Oct 2, 17; m 42; c 3. BIOCHEMISTRY. Educ: Univ Wash, BS, 37, PhD(chem), 42. Prof Exp: Res chemist, Exp Sta, Hercules Powder Co, 42-45, res chemist, Calif Res Corp, 45-66, res assoc, 66-67, SR RES ASSOC, CHEVRON RES CORP, 67- Mem: Am Chem Soc. Res: Lubricating oil additives. Mailing Add: Chevron Res Corp Richmond CA 94802

BENOIT, GEORGE RAYMOND, soil physics, see 12th edition

BENOIT, JEAN CLAUDE, b Montreal, Que, Oct 6, 26; m 54; c 3. BIOLOGY, MICROBIOLOGY. Educ: Univ Montreal, BS, 49, MS, 51, PhD(bact), 55. Prof Exp: Res asst bact, Inst Microbiol & Hyg, 53-70, PROF MICROBIOL, FAC MED, UNIV MONTREAL, 70- Mem: Can Soc Microbiol. Res: Biology of mycobacteria. Mailing Add: Dept of Microbiol Fac of Med Univ of Montreal Montreal PQ Can

BENOIT, JOHN WILLIAM, computer science, see 12th edition

BENOIT, PAUL, b Montreal, Que, Aug 17, 31; m 56, 71; c 1. INSECT TAXONOMY. Educ: Univ Montreal, BScA, 54; Laval Univ, MSc, 58; Cornell Univ, PhD(insect taxon), 62. Prof Exp: Entomologist, Que Prov Dept Agr, 55-56; RES SCIENTIST, CAN DEPT ENVIRON, 56- Mem: Soc Syst Zool; Entom Soc Am; Entom Soc Can. Res: Forest insect pests survey, research and management. Mailing Add: Can Forestry Serv CP 3800 Quebec PQ Can

BENOIT, PETER WELLS, b Boston, Mass, Dec 9, 39; m 65; c 1. DENTISTRY, ANATOMY. Educ: Tufts Univ, BS, 61, DMD, 65, MS, 67, PhD(anat), 71. Prof Exp: Consult pharmacol, Astra Pharmaceut Prod Inc, Mass, 71-72; ASST PROF ORAL BIOL, SCH DENT, EMORY UNIV, 72- Concurrent Pos: USPHS fel oral path, Tufts Univ, Boston, 65-67, instnl training grant anat, 67-71. Mem: Int Asn Den Res; Am Acad Oral Path. Res: Dental research. Mailing Add: Emory Univ Sch of Dent Clifton Rd Atlanta GA 30322

BENOIT, PHILIPPE STANISLAS, b Fitchburg, Mass, Nov 2, 40. PHARMACOLOGY, PHYSIOLOGY. Educ: Mass Col Pharm, BS, 62, MS, 64, PhD(pharmacol), 70. Prof Exp: From instr to asst prof physiol & pharmacol, Mass Col Pharm, 65-71; asst prof, 71-74, ASSOC PROF PHARMACOL, UNIV ILL MED CTR, 74-, COORDR PHARMACOL, 73- Concurrent Pos: Grant, Univ Ill Med Ctr, 72-73; lectr gen & ocular pharmacol, Ill Col Optom, 72-74, adj assoc prof, 74-, consult biol sci, 73-; consult, Nat Bd Examr Optom, 72-74; consult prescription res, R A Gosselin & Co, 72-74; Acad Pharmaceut Sci rep to Nat Soc Med Res, 72-; mem bd dirs, Vision Conserv Inst Inc, 74- Mem: Am Pharmaceut Asn; Acad Pharmaceut Sci; Am Soc Pharmacog; Am Asn Cols Pharm. Res: Search and screening for new analgesic and antiflammatory agents; drug side-effects; mechanisms of hypotensive drugs; antispasmodics-acetylcholine receptor; drug induced glaucoma. Mailing Add: Dept of Pharmacog & Pharmacol Univ of Ill at the Med Ctr Chicago IL 60612

BENOIT, ROBERT LUCIEN, inorganic chemistry, see 12th edition

BENOITON, NORMAND LEO, b Somerset, Man, Sept 30, 32; m 73. BIOCHEMISTRY. Educ: Loyola Col, Can, BSc, 53; Univ Montreal, MSc, 55, PhD(biochem), 56. Prof Exp: Vis scientist, Nat Insts Health, 56-58; Imp Chem Industs res fel, Univ Exeter, 58-60; assoc res officer, Nat Res Coun Can, 60-61; from asst prof to assoc prof, 61-70, PROF BIOCHEM, UNIV OTTAWA, 70- Concurrent Pos: Assoc, Med Res Coun Can, 61- Mem: AAAS; Am Chem Soc; Chem Inst Can; Can Biochem Soc; Brit Chem Soc. Res: N-Methylamino acid chemistry and biochemistry; peptide chemistry. Mailing Add: Dept of Biochem Univ of Ottawa Ottawa ON Can

BENOKRAITIS, VITALIUS, b Keturkaimis, Lithuania, July 25, 41; m 67; c 1. NUMERICAL ANALYSIS. Educ: Kent State Univ, BS, 64; Univ Ill, MS, 66; Univ Tex, PhD(comput sci), 74. Prof Exp: Asst prof comput sci, Va Commonwealth Univ, 74-75; MATHEMATICIAN, BALLISTIC RES LABS, US ARMY, 75- Mem: Soc Indust & Appl Math; Asn Comput Mach; Am Math Soc. Res: Iterative solution of large linear systems; numerical solution of partial differential equations; mathematical modeling. Mailing Add: 613 Shamrock Rd Bel Air MD 21014

BENOLKEN, ROBERT MARSHALL, b St Paul, Minn, May 11, 32; m 57; c 6. CELL PHYSIOLOGY. Educ: Marquette Univ, BS, 54; Johns Hopkins Univ, PhD(biophys), 59. Prof Exp: Asst physics, Marquette Univ, 53-54; asst biophys, Johns Hopkins Univ, 54-56, res assoc, 56-59; from asst prof to assoc prof, Univ Minn, Minneapolis, 59-68; PROF NEURAL SCI, GRAD SCH BIOMED SCI, UNIV TEX, 68- Mem: Soc Gen Physiologists; Biophys Soc. Res: Cellular physiology; physiology of visual sense cells; electrophysiology. Mailing Add: Univ Tex Grad Sch Biomed Sci 6420 Lamar Fleming Houston TX 77025

BEN-PORAT, TAMAR, b Worms, Ger, Sept 4, 29; US citizen; m 59; c 2. VIROLOGY, BIOCHEMISTRY. Educ: Hebrew Univ Jerusalem, MS, 54; Univ Ill, PhD(microbiol), 59. Prof Exp: Assoc mem staff virol, Albert Einstein Med Ctr, 58-70; res assoc prof microbiol, Sch Med, Temple Univ, 70-72; ASSOC PROF MICROBIOL, SCH MED, VANDERBILT UNIV, 72- Concurrent Pos: Mem rev comt, Cancer Res Ctr, Nat Cancer Inst. Mem: Am Soc Microbiol. Res: Analysis of virus growth at the molecular level. Mailing Add: Dept of Microbiol Vanderbilt Univ Sch of Med Nashville TN 37232

BENSADOUN, ANDRE, b Fes, Morocco, Nov 10, 31; US citizen; m 59; c 2. NUTRITION, PHYSIOLOGY. Educ: Univ Bordeaux, BS, 51; Nat Sch Advan Agron, France, BSEng, 56; Univ Toulouse, MS, 56; Cornell Univ, PhD(nutrit), 60. Prof Exp: Res nutritionist, Cornell Univ, 59-65; asst prof nutrit, Univ Ill, Urbana, 65-67; ASSOC PROF PHYSIOL, CORNELL UNIV, 67- Mem: Am Inst Nutrit; Soc Exp Biol & Med. Res: Lipid metabolism; lipid transport; lipoproteins; lipases. Mailing Add: Div of Nutrit Sci Cornell Univ Ithaca NY 14850

BENSCH, KLAUS GEORGE, b Miedar, Ger, Sept 1, 28; nat US; m 55; c 3. HUMAN PATHOLOGY. Educ: Univ Erlangen, MD, 53; Am Bd Path, dipl, 59. Prof Exp: From instr to assoc prof path, Sch Med, Yale Univ, 58-68; PROF PATH, SCH MED, STANFORD UNIV, 68- Mem: Am Asn Path & Bact. Res: Cellular metabolism, proteins and nucleic acids. Mailing Add: Dept of Path Stanford Univ Sch of Med Stanford CA 94305

BENSEL, JOHN PHILLIP, b Glen Ridge, NJ, Oct 30, 45; m 70. SOLID STATE PHYSICS. Educ: Stevens Inst Technol, BS, 67; Univ Pa, PhD(physics), 73. Prof Exp: Res assoc physics, Univ Ill, Urbana, 73-76; ASST PROF ELECTRONICS TECHNOL, SOUTH SIDE VA COMMUNITY COL, 76- Mem: Am Phys Soc. Res: Development of instrumentation for study of fluctuation phenomena in solids; measurement of electrical noise spectra near phase transitions and search for anamolies related to charge density wave formation. Mailing Add: Southside Va Community Col Alberta VA 23821

BENSELER, ROLF WILHELM, b San Jose, Calif, Sept 24, 32; m 61. DENDROLOGY. Educ: Univ Calif, Berkeley, BS, 57, PhD(bot), 68; Yale Univ, MF, 58. Prof Exp: Jr specialist, Sch Forestry, Univ Calif, 58-60; instr biol, Modesto Jr Col, 61-63; lectr, San Jose State Col, 68; from asst prof to assoc prof, 68-75, PROF BIOL SCI, CALIF STATE UNIV, HAYWARD, 75- Mem: Bot Soc Am; Wilderness Soc. Res: Reproductive biology of woody plants. Mailing Add: Dept of Biol Sci Calif State Univ Hayward CA 94542

BENSEN, DAVID WARREN, b Paterson, NJ, Feb 6, 28; m 54; c 4. SOIL CHEMISTRY. Educ: Rutgers Univ, BSA, 54, MS, 55, PhD(soil chem), 58. Prof Exp: Chemist, Hanford Atomic Prod Oper, 58-63; sr res scientist, Isotopes, Inc, 63-66; tech mgr gov contracts, US Naval Radiol Defense Lab, 66-67; res analyst, Res & Eng Off, Civil Defense, Off Secy Army, 67-73, DEP DIR HAZARD EVAL & VULNERABILITY REDUCTION DIV, DEFENSE CIVIL PREPAREDNESS AGENCY, US DEPT DEFENSE, 73- Mem: Sigma Xi; Health Physics Soc. Res: Soil chemistry; soil-plant relationships; soil fertility; plant nutrition; radioactive waste disposal; radionuclide reactions with soils and minerals; fallout contamination inthe food chain; fallout phenomenology; radiological hazard assessment; radiological countermeasures and protection. Mailing Add: 4624 Sunflower Dr Rockville MD 20853

BENSEN, JACK F, b Chicago, Ill, Jan 4, 23; c 3. SPEECH & HEARING SCIENCES. Educ: Univ Miami, BA, 49; Univ WVa, MA, 51; Univ Fla, PhD, 61. Prof Exp: Speech therapist, Univ Fla, 51-53; dean men, Dana Col, 54-55; PROF SPEECH, UNIV MIAMI, 55-, DIR, SPEECH & HEARING CLIN, 68- Concurrent Pos: Mem staff, S Fla Cleft Palate Clin, 56-58; consult, United Cerebral Palsy, 65-68; mem bd dirs, Miami Hearing & Speech Ctr, 65-68. Mem: Am Speech & Hearing Asn; Am Cleft Palate Asn. Res: Cleft palate; speech analysis of the cerebral palsied. Mailing Add: Speech & Hearing Clin Univ of Miami Coral Gables FL 33124

BENSEND, DWIGHT WINFRED, b Turtle Lake, Wis, Mar 3, 13; m 41; c 3. WOOD SCIENCE, WOOD TECHNOLOGY. Educ: Univ Minn, BS, 37, PhD(silvicult, statist), 42. Prof Exp: Instr forestry, Univ Minn, 38-42, technologist, Forest Prod Lab, Wis, 42-44; assoc prof wood technol, Utah State Univ, 44-47; prof, Iowa State Univ, 47-61; prof forestry, Univ Ky-US Agency Int Develop contract team at Indonesia, 61-62; PROF WOOD TECHNOL, IOWA STATE UNIV, 62- Honors & Awards: Distinguished Serv Award, Forest Prod Res Soc, 72; Frudden Award, 74. Mem: Soc Wood Sci & Technol (secy-treas, 62-63); Forest Prod Res Soc (treas, Midwest Sect, 70-75); Soc Am Foresters; Int Asn Wood Anat. Res: Wood anatomy and physical properties; tension wood development and properties in cottonwood and soft maple; variation in specific gravity and fiber length in pine. Mailing Add: Dept of Forestry Iowa State Univ Ames IA 50011

BENSINGER, DAVID AUGUST, b St Louis, Mo, May 14, 26; div; c 2. DENTISTRY, PERIODONTOLOGY. Educ: Wash Univ, BA, 44; St Louis Univ, DDS, 48. Prof Exp: From instr to asst prof, 49-55, asst dean, 69-71, ASSOC PROF DENT MED, SCH DENT, WASH UNIV, 55-, ASSOC DEAN, 71- Concurrent Pos: Consult, US Air Force, 59- Mem: Fel Am Col Dent; Am Acad Periodont; Am Soc Periodont; Int Asn Dent Res. Res: Oral diagnosis; time-lapse alterations in periodontal disease. Mailing Add: Sch of Dent Wash Univ St Louis MO 63110

BENSINGER, JAMES ROBERT, b Washington, DC, Aug 20, 41; m 66; c 2. ELEMENTARY PARTICLE PHYSICS. Educ: Bucknell Univ, BS, 63; Univ Wis, PhD(physics), 70. Prof Exp: Asst prof physics, Univ Pa, 70-74; ASST PROF PHYSICS, BRANDEIS UNIV, 74- Mem: Am Phys Soc. Res: One pion production and Boson spectroscopy using bubble chamber techniques. Mailing Add: Dept of Physics Brandeis Univ Waltham MA 02154

BENSLEY, EDWARD HORTON, b Toronto, Ont, Dec 10, 06; m 44. HISTORY OF MEDICINE. Educ: Univ Toronto, BA, 27, MD, 30; FRCP(C). Hon Degrees: DSc, Acadia Univ, 64. Prof Exp: Vdean fac med, 61-67, PROF EXP MED, McGILL UNIV, 65-, LECTR HIST MED, 68- Concurrent Pos: Hon consult, Montreal Gen Hosp, 62- Mem: Fel Am Col Physicians; fel Chem Inst Can. Mailing Add: Fac of Med McGill Univ Montreal PQ Can

BENSON, ANDREW ALM, b Modesto, Calif, Sept 24, 17; m 42; c 3. BIOCHEMISTRY, PLANT PHYSIOLOGY. Educ: Univ Calif, Berkeley, BS, 39; Calif Inst Technol, PhD(org chem), 42. Hon Degrees: DPhil, Univ Oslo, 65. Prof Exp: Instr chem, Univ Calif, Berkeley, 42-43, asst dir, Bio-Org Group, Radiation Lab, 46-55; res assoc, Stanford Univ, 44-45; asst, Calif Inst Technol, 45-46; from assoc prof to prof agr & biol chem, Pa State Univ, 55-61; prof in residence physiol chem & biophys & res biochemist, Lab Nuclear Med, 61-62, chmn dept marine biol, 65-70, assoc dir inst, 66-70, chmn marine biol res div, 67-70, actg chmn, 70-71, PROF BIOL, SCRIPPS INST, 62-, DIR PHYSIOL RES LAB, 70- Concurrent Pos: Fulbright lectr, Agr Col Norway, 51-52; consult, Molecular Biol Prog, NSF, 61-63. Honors & Awards: Sugar Res Found Award, 50; Lawrence Mem Award, US Atomic Energy Comn, 62; Stephen Hales Award, Am Soc Plant Physiol, 72. Mem: Nat Acad Sci; Am Soc Biol Chem; Am Chem Soc; Am Soc Plant Physiol. Res: Thyroxine analog synthesis; path of carbon in photosynthesis; tracer methodology; neutron activation analysis; radiochromatography; hot-atom chemistry; lipid biochemistry; biological membrane surfactants; phosphonic and sulfonic acid metabolism; wax metabolism; coral metabolism. Mailing Add: Scripps Inst of Oceanog Univ of Calif San Diego CA 92093

BENSON, BARRETT WENDELL, b Brattleboro, Vt, Sept 4, 39; m 57; c 4. ORGANIC CHEMISTRY. Educ: Middlebury Col, AB, 61; Univ Vt, PhD(org chem), 65. Prof Exp: Asst prof chem, Fresno State Col, 65-67; assoc prof, 67-72, PROF CHEM & CHMN DEPT, BLOOMSBURG STATE COL, 72- Concurrent Pos: NSF res grants, 68-; vis prof, Dartmouth Col, 74. Mem: Am Chem Soc. Res: Organic structure determination; organic reaction mechanisms. Mailing Add: Dept of Chem Bloomsburg State Col Bloomsburg PA 17815

BENSON, BRENT W, b Chicago, Ill, Apr 10, 41; m 62; c 2. BIOPHYSICS, RADIATION CHEMISTRY. Educ: Knox Col, Ill, BA, 63; Pa State Univ, MS, 65, PhD(biophys), 69. Prof Exp: Asst prof physics, Southern Ill Univ, 69-72; ASSOC PROF PHYSICS, LEHIGH UNIV, 72- Mem: AAAS; Biophys Soc; Am Phys Soc; Radiation Res Soc. Res: Electron spin resonance studies of radiation induced free radicals in important biological molecules; biological and molecular effects due to decay of incorporated radioisotopes; drug-membrane interactions. Mailing Add: Physics Bldg 16 Lehigh Univ Bethlehem PA 18015

BENSON, BRUCE BUZZELL, b Choteau, Mont, Feb 22, 22; m 50. CHEMICAL PHYSICS, OCEANOGRAPHY. Educ: Amherst Col, BA, 43; Yale Univ, MS, 45, PhD(physics), 47. Prof Exp: Asst physics, Yale Univ, 43-44, instr, 44-46, asst nuclear physics, 46-47; from instr to assoc prof, 47-60, PROF PHYSICS, AMHERST COL, 60- Concurrent Pos: Guggenheim fel, 58-59; assoc, Woods Hole Oceanog Inst, 57-67; consult, Earth Sci Sect, NSF, 64-67, Oceanog Sect, 67-68, IDOE Sect, 73-74; assoc ed, J Marine Chem, 74- Mem: AAAS; Am Geophys Union; Geochem Soc; Am Phys Soc; Am Asn Physics Teachers. Res: Mass spectrometry; oceanography; geophysics; geochemistry; solution of gases in water. Mailing Add: Dept of Physics Amherst Col Amherst MA 01002

BENSON, CARL SIDNEY, b Minneapolis, Minn, June 23, 27; m 55; c 3. GLACIOLOGY. Educ: Univ Minn, BA, 50, MS, 56; Calif Inst Technol, PhD(geol), 60. Prof Exp: Physicist, Snow, Ice & Permafrost Res Estab, 52-56; asst prof geophys res, 60-61, assoc res geophysicist, 61-62, assoc prof, 62-69, chmn dept geol, 69-73, PROF GEOPHYS INST, UNIV ALASKA, 69- Concurrent Pos: Mem Pac Northwest region comt & glaciers comt, Am Geophys Union, 64-67; glaciol panel, Comt Polar Res, Nat Res Coun-Nat Acad Sci, 64-67; mem, Gov Environ Adv Bd Alaska, 75- Mem: AAAS; Am Geophys Union; Meteorol Soc; fel Arctic Inst NAm; Glaciol Soc; Geol Soc Am. Res: Low temperature air pollution; freezing process in turbulent streams; glacier mass balance. Mailing Add: Geophys Inst Univ of Alaska Fairbanks AK 99701

BENSON, CHARLES EVERETT, b Dayton, Ohio, Dec 15, 37; m 60; c 1. BIOCHEMICAL GENETICS, MEDICAL MICROBIOLOGY. Educ: Franklin Col, AB, 60; Miami Univ, MS, 64; Wake Forest Univ, PhD(microbiol), 69. Prof Exp: Teacher gen sci, Cornell Heights Elem Sch, 60-61; res asst cardiac physiol, Dept Res, Miami Valley Hosp, 63-65; trainee, 69-71, scholar, Pa Plan Develop Scientists Med Res, 71-74, res assoc biochem genetics, 73-75, ASST PROF MICROBIOL, SCH ALLIED MED PROF & SCH MED, UNIV PA, 75- Mem: Am Soc Microbiol; NY Acad Sci; Soc Gen Microbiol; Sigma Xi; Am Soc Allied Health Prof. Res: Interrelationship of gene and enzyme function in purine metabolism of enteric bacteria and cultured diploid fibroblasts and the mechanism of salmonella pathogenesis. Mailing Add: Dept of Med Technol Sch of Allied Med Prof Univ Pa Philadelphia PA 19174

BENSON, CYRIL BROWNLOW, b Eng, Feb 10, 26; m 56; c 2. PHYSICS. Educ: Cambridge Univ, BA, 50; Univ Toronto, MA, 53, PhD(physics), 55. Prof Exp: Asst prof, 56-63, ASSOC PROF PHYSICS, UNIV OTTAWA, ONT, 63- Res: Low temperature physics; properties of liquid helium. Mailing Add: Dept of Physics Univ of Ottawa Ottawa ON Can

BENSON, DEAN CLIFTON, b Hazelton, NDak, Oct 25, 18; m 43; c 2. MATHEMATICS. Educ: Sioux Falls Col, BA, 41; Iowa State Col, MS, 47, PhD(math), 54. Prof Exp: Aeronaut res photographer, Nat Adv Comt Aeronaut, Langley Field, 44-45; asst prof math, Sioux Falls Col, 47-50; instr, Iowa State col, 50-54; asst prof, SDak Sch Mines & Tech, 54-56; asst prof, Chico State Col, 56-60; PROF MATH, S DAK SCH MINES & TECHNOL, 60-, CHMN DEPT, 65- Mem: Am Math Soc; Math Asn Am. Res: Analysis; differential equations; complex variables; vector analysis. Mailing Add: Dept of Math SDak Sch of Mines & Technol Rapid City SD 57701

BENSON, DON GEHR, JR, developmental physiology, see 12th edition

BENSON, DONALD CHARLES, b Modesto, Calif, June 6, 27; m 54; c 2. MATHEMATICS. Educ: Pomona Col, BA, 50; Stanford Univ, MS, 53, PhD(math), 54. Prof Exp: Instr math, Princeton Univ, 54-55; asst prof, Carnegie Inst Technol, 55-57; from asst prof to assoc prof, 57-68, PROF MATH, UNIV CALIF, DAVIS, 68- Mem: Am Math Soc; Math Asn Am. Res: Analysis Mailing Add: Dept of Math Univ of Calif Davis CA 95616

BENSON, DONALD WARREN, b Jamestown, NY, Aug 17, 21; m 46; c 3. ANESTHESIOLOGY. Educ: Univ Chicago, BS, 48, MD, 50, PhD(pharmacol), 57. Prof Exp: From instr to asst prof anesthesiol, Univ Chicago, 53-56; from assoc prof to prof, Johns Hopkins Univ, 56-75; PROF ANESTHESIOL & CHMN DEPT, UNIV CHICAGO, 75- Mem: Am Soc Anesthesiol; AMA; Int Anesthesia Res Soc. Res: Respiratory physiology; mechanical respirators; hypothermia. Mailing Add: Dept of Anesthesiol Univ of Chicago Chicago IL 60637

BENSON, EDMUND WALTER, b Woburn, Mass, July 27, 38. INORGANIC CHEMISTRY. Educ: Univ NH, BS, 61; Univ Tenn, MS, 63, PhD(chem), 67. Prof Exp: Res assoc chem, Univ Ill, 66-67; asst prof, 67-72, ASSOC PROF CHEM, CENT MICH UNIV, 72- Mem: AAAS; Am Chem Soc; Sigma Xi. Res: Chemistry of the rare earth elements; water analysis and quality. Mailing Add: Dept of Chem Cent Mich Univ Mt Pleasant MI 48858

BENSON, ELLIS STARBRANCH, b Honan, China, Oct 28, 19; m 47; c 3. PATHOLOGY. Educ: Augustana Col, BA, 41; Univ Minn, MD, 45. Prof Exp: From instr to asst prof clin lab med, 45-57, from asst dir to dir clin labs, 53-66, asst prof path & clin lab med, 57-61, PROF LAB MED & BIOCHEM, UNIV MINN, MINNEAPOLIS, 61-, HEAD LAB MED, 66-, HEAD PATH, 73- Concurrent Pos: USPHS sr res fel, 57-; consult, Vet Admin, 52- & USPHS, 64-72; trustee, Am Bd Path, 71- Mem: Am Soc Clin Path; Am Soc Biol Chem; Am Asn Path & Bact; Biophys Soc; Am Soc Cell Biol. Res: Chemistry and structure of heart muscle; hemoglobin and red blood cell structure and function. Mailing Add: 4134 Coffman Ln Minneapolis MN 55406

BENSON, ERNEST PHILLIP, JR, b New Castle, Pa, Feb 21, 36; m 60. INORGANIC CHEMISTRY. Educ: Geneva Col, BS, 58; Mich State Univ, PhD(inorg chem), 63. Prof Exp: Res fel, Pa State Univ, 63-64; asst prof chem, Fairleigh Dickinson, Florham-Madison Campus, 64-66; asst prof, 66-74, ASSOC PROF CHEM, IDAHO STATE UNIV, 74- Mem: Am Chem Soc. Res: Electron transfer reactions; preparation of coordination compounds; reactions of oxyanions. Mailing Add: Dept of Chem Idaho State Univ Pocatello ID 83201

BENSON, FREDERIC RUPERT, b Cape Girardeau, Mo, Sept 21, 15; m 39; c 1. ORGANIC CHEMISTRY, INFORMATION SCIENCE. Educ: Wesleyan Univ, AB, 36; NY Univ, MS, 38, PhD(org chem), 47. Prof Exp: Res chemist, Hambrock Chem Corp, NJ, 37; instr chem, Bergen Jr Col, 38-41; res chemist, Picatinny Arsenal, US War Ord Dept, NJ, 41-45; chief res chemist, Res Lab, Remington Rand, Inc, 46-54; sr res chemist, Metalectro Corp, 54-56; supvr info br, Chem Res Dept, Atlas Chem Industs, Inc, 54-56, supvr info br, Chem Res Dept, 56-59, mgr info sect, 59-70, MGR INFO SERV, ICI UNITED STATES INC, 70- Mem: AAAS; Am Chem Soc; Drug Info Asn. Res: Organic nitrogen chemistry; tetrazoles; triazoles; hydrazine chemistry; polyols; surfactants; automation of chemical information; information research. Mailing Add: ICI United States Inc Chem Res & Develop Lab Wilmington DE 19897

BENSON, GEOFFREY WADDINGTON, physical chemistry, polymer chemistry, see 12th edition

BENSON, GEORGE CAMPBELL, b Toronto, Ont, July 25, 19; m 46; c 2. PHYSICAL CHEMISTRY. Educ: Univ Toronto, BA, 42, MA, 43, PhD(chem), 45. Prof Exp: Res chemist, Chem Div, Nat Res Coun Can, 45-49; mem solid state physics group, Bristol Univ, 49-50; RES CHEMIST, CHEM DIV, NAT RES COUN CAN, 51- Honors & Awards: Bronze Medal, Brit Asn, 42. Mem: Am Phys Soc; Am Chem Soc; Royal Soc Can; Chem Inst Can. Res: Thermodynamic studies of mixtures of nonelectrolytes. Mailing Add: Div of Chem Nat Res Coun Can Ottawa ON Can

BENSON, GILBERT THOMAS, b Los Angeles, Calif, Oct 3, 29; m 57; c 1. GEOLOGY. Educ: Stanford Univ, BS, 52, MS, 53; Yale Univ, PhD(geol), 63. Prof Exp: Geologist, Texaco, Inc, 53-58; asst prof geol, Univ Ore, 62-68; asst prof geol, 68-74, ASSOC PROF EARTH SCI, PORTLAND STATE UNIV, 74- Mem: Geol Soc Am; Am Asn Petrol Geol. Res: Structural, Areal, and petroleum geology. Mailing Add: Dept of Geol Portland State Univ Portland OR 97207

BENSON, HARRIET, b Kansas City, Mo, May 17, 41. ORGANIC CHEMISTRY, SCIENTIFIC BIBLIOGRAPHY. Educ: Wellesley Col, BA, 63; Univ Kans, PhD(org chem), 67. Prof Exp: Lit specialist, Shell Develop Co, 63-64; fel, State Univ Groningen, 67-69; LIT SPECIALIST, ALZA CORP, 69- Mem: AAAS; Am Chem Soc; Am Med Writers Asn; Drug Info Asn. Res: Conformational factors in free radical and photochemical reactions; scientific literature related to pharmaceutical sciences including medicinal, pharmaceutical, organic, physical and polymer chemistry, clinical and veterinary medicine and toxicology; technical writing and editing; Federal Drug Administration regulations. Mailing Add: 724 Arastradero Rd Palo Alto CA 94306

BENSON, HERBERT, b Yonkers, NY, Apr 24, 35; m 62; c 2. CARDIOLOGY. Educ:

Wesleyan Univ, BA, 57; Harvard Med Sch, MD, 61. Prof Exp: Instr physiol & med, 69, asst prof med, 70-72, ASSOC PROF MED, HARVARD MED SCH, 72-; ASSOC PHYSICIAN MED & DIR HYPERTENSION SECT, BETH ISRAEL HOSP, 74- Concurrent Pos: USPHS fel cardiol, Thorndike Mem Lab, Harvard Med Sch, 65-67, Med Found grant behav physiol, 67-69; prog dir, Gen Clin Res Ctr, Boston City Hosp, 72-74 & Gen Clin Res Ctr, Beth Israel Hosp, 74-; expert consult, Spec Action Off Drug Abuse Prev, Exec Off of Pres, Washington, DC, 74. Mem: AAAS; Am Fedn Clin Res; Am Physiol Soc; Am Psychosom Soc; fel Am Col Cardiol. Res: Behavioral aspects of cardiovascular disease and other diseases related to so-called stress; the counterpart of the fight or flight response—the relaxation response. Mailing Add: Beth Israel Hosp Dept of Med 330 Brookline Ave Boston MA 02215

BENSON, HERBERT LINNE, JR, b Kansas City, Mo, Apr 26, 34; m 57; c 3. PHYSICAL CHEMISTRY. Educ: Bethany Col, Kans, BS, 56; Univ Wis, PhD(radiation chem), 61. Prof Exp: Res chemist, 61-66, SR RES CHEMIST, SHELL OIL CO, DEER PARK, 66- Mem: Am Chem Soc. Res: Radiation chemistry of organic compounds; petroleum chemistry. Mailing Add: 310 Parliament Houston TX 77034

BENSON, JAMES MILLER, b Dayton, Tenn, Dec 31, 07; m 38; c 3. PHYSICS. Educ: E Tenn State Col, BS, 29; Univ Tenn, MS, 32, PhD, 50. Prof Exp: Head hydrodyn anal, Langley Aeronaut Lab, Nat Adv Comt Aeronaut, 40-45, head instrument res, 45-48; instr physics, Univ Tenn, 48-50; vpres & dir res, 50-72, CONSULT, TELEDYNE HASTINGS-RAYDIST, INC, 72- Mem: Inst Elec & Electronics Eng; Am Inst Navig; Am Vacuum Soc. Res: Vacuum techniques; electronics; radio propagation; instrumentation and controls. Mailing Add: PO Box 1275 Hampton VA 23661

BENSON, JOHN ALEXANDER, JR, b Manchester, Conn, July 23, 21; m 47; c 4. GASTROENTEROLOGY. Educ: Wesleyan Univ, BA, 43; Harvard Med Sch, MD, 46. Prof Exp: Instr med, Harvard Med Sch, 56-59; assoc prof med, 59-65, head div gastroenterol, 59-75, PROF MED, MED SCH, UNIV ORE, 65- Concurrent Pos: Attend in med, Vet Admin Hosps, Boston, 58-59 & Portland, Ore, 60-75; consult gastroenterol, Vancouver, Wash, 60-; mem subspecialty bd gastroenterol, Am Bd Internal Med, 61-65; mem, Am Bd Internal Med, 69-, secy-treas, 72-75, pres, 75-; consult, Madigan Gen Army Hosp. Hosp. Mem: Am Gastroenterol Asn (secy, 70-73, vpres, 75-); Am Col Physicians; AMA; Am Clin & Climat Asn; Am Soc Internal Med. Res: Normal and abnormal intestinal absorption. Mailing Add: Am Bd of Internal Med 200 SW Market St Portland OR 97201

BENSON, JOHN EDWARD, b Merchantville, NJ, Oct 21, 24. SURFACE CHEMISTRY. Educ: Pa State Col, BS, 50; Princeton Univ, MA, 50, PhD(chem), 57. Prof Exp: Res asst, Forrestal Res Ctr, 55-56; asst prof chem, Pa State Univ, 56-61 & Gettysburg Col, 61-64; assoc prof & head dept, 64-66, prof & head dept, 66-71, DANA PROF CHEM, DICKINSON COL, 71- Concurrent Pos: NSF sci fac fel, Stanford Univ, 64-65. Mem: AAAS; Am Chem Soc. Res: Surface phenomena; adsorption; catalysis; chemical history. Mailing Add: Dept of Chem Dickinson Col Carlisle PA 17013

BENSON, KATHERINE ADAMS, b Roanoke, Va, June 29, 38; m 62; c 2. DEVELOPMENTAL PHYSIOLOGY. Educ: Col William & Mary, BS, 60; Univ Va, PhD(biol), 65. Prof Exp: Asst prof zool, La State Univ, 65-67; ASSOC PROF BIOL, RADFORD COL, 67- Mem: Am Soc Zool. Res: Mechanisms of histolysis of insect tissues during metamorphosis; reorientation of physiological mechanisms in developing cells of insects and the lower vertebrates. Mailing Add: Dept of Biol Radford Col Radford VA 24141

BENSON, LOIS MARY, b Hudson, NY, Apr 17, 31; m 71. VIROLOGY. Educ: Albany Col Pharm, Union Univ, NY, BS, 52; Baylor Univ, MS, 65. Prof Exp: From jr bacteriologist to sr bacteriologist, Div Labs & Res, NY State Dept Health, 53-63; lab asst virol, Col Med, Baylor Univ, 63-65; SR BACTERIOLOGIST, DIV LABS & RES, NY STATE DEPT HEALTH, 65- Concurrent Pos: Health Res, Inc res grants, 65-68; specialist microbiologist, pub health & med lab microbiologist, Am Acad Microbiol, 75. Mem: Am Soc Microbiol; Reticuloendothelial Soc; Am Soc Clin Path. Res: Diagnostic virology; tissue culture; immunology; pathogenesis of disease due to lymphocytic choriomeningitis virus; cytomegalovirus; tumor viruses; rabies virus; mouse genetics; parabiosis. Mailing Add: Griffin Lab NY State Dept Health Div Labs & Res New Scotland Ave Albany NY 12201

BENSON, LOREN ALLEN, b St Louis, Mo, Mar 19, 32; m 58; c 3. PHYSICAL CHEMISTRY. Educ: Wash Univ, AB, 52; St Louis Univ, PhD, 59. Prof Exp: Oper analyst, Inst Defense Anal, 58-65; mem staff, Lambda Corp, 65-67; chief scientist, Keystone Comput Assocs, 67-70; CHIEF SCIENTIST, DATRAN CORP, 70- Mem: AAAS; Opers Res Soc Am; Asn Comput Mach. Res: Computer science; operations research; communications analysis. Mailing Add: 8719 Bradgate Ct Alexandria VA 22308

BENSON, LYMAN DAVID, b Kelseyville, Calif, May 4, 09; m 31; c 2. PLANT TAXONOMY. Educ: Stanford Univ, AB, 30, AM, 31, PhD(biol), 40. Prof Exp: Instr, Bakersfield Jr Col, 31-38; from instr to asst prof bot, Univ Ariz, 38-44, asst botanist, Exp Sta, 38-44; from assoc prof to prof, 44-74, head dept, 44-74, Wig distinguished prof, 63 & 74, EMER PROF BOT, POMONA COL, 74- Mem: Fel AAAS; Bot Soc Am; Am Soc Naturalists; Soc Study Evolution; Am Soc Plant Taxon (secy-treas, 49, pres, 60). Res: Classification and distribution of Ranunculaceae, Grameae, Cactaceae and Prosopis for North America; vascular flora of western North America; taxonomy and distribution of North American species of Ranunculus; taxonomic and general botanical textbooks; evolution of the floras of North America. Mailing Add: Dept of Bot Pomona Col Claremont CA 91711

BENSON, NELS R, b Chicago, Ill, Feb 20, 14; m 38; c 5. HORTICULTURE, SOIL SCIENCE. Educ: Univ Fla, BS, 36, MS, 38; State Col Wash, PhD(agron), 50. Prof Exp: Asst, Agr Exp Sta, Univ Fla, 38-39; jr soil scientist, Soil Conserv Serv, USDA, Wis & Mo, 41-43, asst soil scientist, Mich & Mo, 43-46; asst soil scientist, 46-51, assoc soil scientist & assoc horticulturist, 51-58, SOIL SCIENTIST & HORTICULTURIST, TREE FRUIT RES CTR, WASH STATE UNIV, 58- Mem: Soil Sci Soc Am; Am Soc Agron; Am Soc Hort Sci. Res: Effect of pesticide residues in soils on fruit trees, especially arsenic; use of herbicides in orchards; calcium nutrition and bitter pit of apple; use of preplant soil fumigation on growth of fruit trees; zinc uptake by apple trees with and without micorrhiza. Mailing Add: Tree Fruit Res Ctr Wash State Univ 1100 N Western Wenatchee WA 98801

BENSON, NORMAN G, b Berlin, Conn, May 7, 23; m 51; c 5. BIOLOGY, LIMNOLOGY. Educ: Univ Maine, BS, 48; Univ Mich, MA, 51, PhD(fisheries), 53. Prof Exp: Fishery res biologist, US Fish & Wildlife Serv, 57-61, chief, N Cent Res Invest, US Bur Sport Fisheries & Wildlife, 61-75, FISH & WILDLIFE ADMINR & TEAM LEADER NAT STREAM ALTERATION TEAM, US FISH & WILDLIFE SERV, 75- Mem: Am Fisheries Soc; Am Soc Limnol & Oceanog; Ecol Soc Am; fel Am Inst Fishery Res Biol. Res: Trout stream ecology; fish population dynamics; large reservoir limnology and fish populations; water management. Mailing Add: 1304 Woodhill Rd Columbia MO 65201

BENSON, PETER HOWARD, b Martinez, Calif, Aug 25, 35; m 58; c 2. ECOLOGY, ENVIRONMENTAL BIOLOGY. Educ: Univ Calif, Berkeley, BA, 57; San Francisco State Col, MA, 60; Univ Southern Calif, PhD(biol), 73. Prof Exp: Res asst bact genetics, Stanford Univ, 60-61; teaching asst biol, Univ Southern Calif, 61-62, res asst midwater ecol, 63-65, fel biol res, Allan Hancock Found, 65-68; instr marine ecol, Univ Calif, Los Angeles, 67-68; from scientist to sr scientist, 68-73, MGR DEPT MARINE BIOL, LOCKHEED AIRCRAFT SERVICE CO, 73- Concurrent Pos: Specialist, Antarctic oceanol prog, 65-67; coop study of Kuroshio, 68-69. Mem: Am Inst Biol Sci; Marine Technol Soc; Water Pollution Control Asn; Phycol Soc Am. Mem: Experimental studies on marine fouling organisms; general ecology of marine invertebrates of California; environmental monitoring including studies on the effects of water quality parameters upon benthic organisms, particularly diatom populations. Mailing Add: Lockheed Ocean Lab 3380 N Harbor San Diego CA 92101

BENSON, RALPH CRISWELL, b St Louis, Mo, Apr 19, 11; m 37; c 3. OBSTETRICS & GYNECOLOGY. Educ: Lehigh Univ, BA, 32, Johns Hopkins Univ, MD, 36. Prof Exp: From asst prof to assoc prof obstet & gynec, Med Sch, Univ Calif, 47-56; PROF OBSTET & GYNEC & CHMN DEPT, MED SCH, UNIV ORE, 56- Mem: Am Gynec Soc; fel Am Col Obstet & Gynec. Res: Clinical research gynecologic cancer; obstetrical complications, especially abortion; hormonal aspects of gynecology; perinatal complications. Mailing Add: 1960 SW Vista Ave Portland OR 97201

BENSON, RICHARD EDWARD, b Racine, Wis, May 8, 20; m 42; c 2. ANALYTICAL CHEMISTRY. Educ: Ariz State Teachers Col, BA, 42; Univ Nebr, MA, 44, PhD(org chem), 46. Prof Exp: Lab asst, Univ Nebr, 42-46; res chemist, Chem Dept, 46-56, res supvr, 57-67, ASSOC RES DIR, CENT RES DEPT, EXP STA, E I DU PONT DE NEMOURS & CO, INC, 67- Mem: Am Chem Soc. Res: Synthetic organic chemistry; allene; ferrocene. Mailing Add: Cent Res Dept Exp Sta EI du Pont de Nemours & Co Wilmington DE 19899

BENSON, RICHARD HALL, b Huntington, WVa, May 19, 29; m 57. PALEONTOLOGY, OCEANOGRAPHY. Educ: Marshall Col, BS, 51; Univ Ill, MS, 53, PhD(geol), 55; Univ Leicester, MSc, 71. Prof Exp: Asst, State Geol Surv, Ill, 51-53, asst geologist, 55-56; from asst prof to assoc prof geol, 55-64, PROF GEOL, UNIV KANS, 64-; RES PALEOBIOLOGIST & CUR, SMITHSONIAN INST, 64- Honors & Awards: Nat Mus Nat Hist Dirs Award, Smithsonian Inst, 74. Mem: Paleont Soc; Soc Syst Zool; Paleont Asn Gt Brit; AAAS; Int Paleont Asn. Res: Micropaleontology; Recent and Cenozoic Ostracoda; deep-sea environmental studies; paleoecology; Cenozoic stratigraphy. Mailing Add: Dept of Paleobiol Smithsonian Inst Mus Natural Hist Washington DC 20560

BENSON, RICHARD NORMAN, b Sioux City, Iowa, Nov 11, 35; m 59; c 1. MICROPALEONTOLOGY. Educ: Augustana Col, Ill, BA, 58; Univ Minn, Minneapolis, PhD(geol), 66. Prof Exp: Geologist, Humble Oil & Refining Co, 66-69; asst prof geol, Augustana Col, Ill, 69-75; GEOLOGIST, DEL GEOL SURV, 75- Concurrent Pos: Geologist, Water Resources Div, US Geol Surv, 74- Mem: AAAS; Geol Soc Am; Am Asn Petrol Geol; Am Quaternary Asn. Res: Radiolarian and foraminiferal biostratigraphy and paleoecology; stratigraphy of the Atlantic Coastal Plain and continental shelf; paleoecology of marine sedimentary rocks; silica in sediments and sedimentary rocks; petroleum geology. Mailing Add: Del Geol Surv Univ of Del Newark DE 19711

BENSON, ROBERT FRANKLIN, b Cumberland, Md, Jan 22, 41; m 63; c 1. ORGANIC CHEMISTRY. Educ: Wva Univ, BS, 63; Univ SC, MS, 65; Rensselaer Polytech Inst, PhD(org chem), 73. Prof Exp: Assoc res chemist, Sterling-Winthrop Res Inst, 65-73; SR RES CHEMIST, GAF RES LAB, 75- Concurrent Pos: Assoc, Rensselaer Polytech Inst, 73-75. Mem: Am Chem Soc. Res: Research, development and production of azo dyestuffs. Mailing Add: GAF Corp Riverside Ave Rensselaer NY 12144

BENSON, ROBERT FREDERICK, b Minneapolis, Minn, Mar 16, 35; m 58; c 3. IONOSPHERIC PHYSICS. Educ: Univ Minn, BS, 56, MS, 59; Univ Alaska, PhD(geophys), 63. Prof Exp: Int Geophys Year scientist, Arctic Inst NAm, 56-58; asst geophysicist, Geophys Inst, Univ Alaska, 59-63; asst prof astron, Univ Minn, 63-64; GEOPHYSICIST, GODDARD SPACE FLIGHT CTR, NASA, 64- Concurrent Pos: Nat Acad Sci-Nat Res Coun resident res assoc, 64-65. Mem: Am Polar Soc; Am Geophys Union. Res: Radio star scintillations; ionospheric cross modulation; plasma waves, instabilities and non-linear phenomena in the topside ionosphere; ionospheric electron temperature. Mailing Add: NASA-Goddard Space Flight Ctr Greenbelt MD 20771

BENSON, ROBERT HAYNES, b Hanover, NH, 24; m 50; c 3. ANIMAL BREEDING. Educ: Univ NH, BS, 49; WVa Univ, MS, 51; Univ Wis, PhD, 55. Prof Exp: Dairy husbandman, Univ Wis, 53-55; EXTEN DAIRYMAN, UNIV CONN, 55- Concurrent Pos: Mem coord group, Nat Dairy Herd Improv Asn Prog, 68-71. Mem: Am Dairy Sci Asn. Res: Type classification of dairy cattle; environmental influences and their effect on milk production; dairy cattle breeding. Mailing Add: Dept of Animal Indust Univ of Conn Storrs CT 06268

BENSON, ROBERT LELAND, b Tucson, Ariz, Oct 27, 41; m 66. ENTOMOLOGY, ENZYMOLOGY. Educ: Pomona Col, BA, 63; Univ Ill, Urbana-Champaign, PhD(entom), 69. Prof Exp: Staff fel, Geront Res Ctr, NIH, Md, 68-70; ASST PROF ENTOM, WASH STATE UNIV, 70- Mem: AAAS; Am Inst Biol Sci; Am Soc Zoologists; Entom Soc Am. Res: Synthesis of aminosugars and chitin in insects; properties of enzymes involved in aminosugar metabolism; hormonal control metabolic processes; ultrastructural localization of disaccharidases in mammalian intestine. Mailing Add: Dept of Entom Wash State Univ Pullman WA 99163

BENSON, ROYAL H, b Galveston, Tex, Oct 25, 25; m 54; c 4. ORGANIC CHEMISTRY, ANALYTICAL CHEMISTRY. Educ: Univ Houston, BS, 48, MS, 56. Prof Exp: Lab instr chem, Univ Houston, 47-48; res chemist, M D Anderson Hosp & Tumor Inst, 48-50; res chemist, Radioisotope Unit, Vet Admin, 50-56; res specialist, 56-70, SR RES SPECIALIST, MONSANTO CO, 70- Mem: Am Chem Soc; Am Nuclear Soc. Res: Chemical and analytical research using radioisotopes; radioactivity assay techniques, liquid scintillation and gamma spectrometry. Mailing Add: Monsanto Co PO Box 1311 Texas City TX 77591

BENSON, SIDNEY WILLIAM, b New York, NY, Sept 26, 18; m 55, 70; c 8. PHYSICAL CHEMISTRY. Educ: Columbia Univ, BA, 38; Harvard Univ, MA & PhD(phys chem), 41. Prof Exp: Res assoc, Harvard Univ, 41; instr chem, City Col New York, 42-43; res chemist, Manhattan Proj, Kellex Corp, 43; from asst prof to prof chem, Univ Southern Calif, 43-63, dir chem physics prog, 62-63; CHMN DEPT KINETICS & THERMOCHEM, STANFORD RES INST, 63- Concurrent Pos: Res assoc, Nat Defense Res Coun, 44-45; Guggenheim & Fulbright fel, France, 50-51; NSF sr fel, 57-58 & 71-72; vis prof, Univ Calif, Los Angeles, 59; vis lectr, Univ Ill,

59; Glidden lectr, Purdue Univ, 61; consult, Goodyear Tire & Rubber Co, 57-62 & 70-75, Douglas Aircraft Co, 58-70, Jet Propulsion Labs, 61-65 & Aerospace Labs, 61-; mem int ed bd, Elsevier Publ Co, 65-; ed-in-chief, Int J Chem Kinetics, 67-; mem adv coun, Gordon Res Conf, 68-71; mem comt selection postdoctoral res fels, NSF, 69; mem org solvents adv comt, Nat Air Pollution Control Admin, 69-72; chem eval panel phys chem, Nat Bur Stand, 69-72; mem Comt on Data for Sci & Technol & chmn task group chem kinetics, Int Coun Sci Unions, 69-; mem, Comn Motor Vehicle Exhausts, Nat Acad Sci, 71-74 & Sci Adv Panel, State of Calif Air Resources Bd, 72-; chmn div phys chem, Am Chem Soc, 73-75. Mem: Fel AAAS; fel Am Phys Soc; Am Chem Soc. Res: Chemical lasers; photochemistry; kinetics; theory of liquid structure; catalysis adsorption; statistical mechanics; free radicals thermochemistry. Mailing Add: Dept of Kinetics & Thermochem Stanford Res Inst Menlo Park CA 94025

BENSON, WALTER RODERICK, b Chicago, Ill, Oct 16, 29; m 57; c 3. PHARMACEUTICAL CHEMISTRY, ANALYTICAL CHEMISTRY. Educ: Univ Ill, BS, 51; Univ Colo, PhD(chem), 58. Prof Exp: Chemist, Griffith Labs, Ill, 53-54; asst, Univ Colo, 54-57; fel, Purdue Univ, 58-59; asst prof chem, Colo State Univ, 59-63; sect head, Pesticides Br, Div Food Chem, 63-69, chief, Phys Chem Br, Div Drug Chem, 69-74, BR CHIEF, INSTRUMENTAL APPLN RES BR, DIV DRUG CHEM, BUR DRUGS, FOOD & DRUG ADMIN, 74- Concurrent Pos: Prof lectr, USDA Grad Sch, 65 & Am Univ, 67. Honors & Awards: Merit Award, Food & Drug Admin, 70. Mem: AAAS; Am Chem Soc; Asn Off Anal Chem; Sigma Xi; assoc mem Am Pharmaceut Asn. Res: Electron spin immunoassay; radioactive drug analysis; synthetic organic chemistry; bridgehead rearrangements; fluorination; organic mechanisms; carbamate pesticides; oximes; agricultural and pharmaceutical chemistry; photochemistry. Mailing Add: Instrumental Appln Res Br Div Drug Chem Bur Drugs Food & Drug Admin HFD 420 Washington DC 20204

BENSON, WALTER RUSSELL, b Tamaqua, Pa, July 27, 20; m 50; c 4. PATHOLOGY, MEDICINE. Educ: Duke Univ, MD, 44. Prof Exp: Instr path, Sch Med, Duke Univ, 52-54, assoc, 54-55; asst prof, Sch Med, Univ Louisville, 55-56; from asst prof to assoc prof, 56-67, PROF PATH, MED SCH, UNIV NC, CHAPEL HILL, 67- Mem: Am Asn Path & Bact; Am Soc Exp Path; Soc Exp Biol & Med; Am Soc Clin Path; Int Acad Path. Res: Cardiovascular pulmonary diseases; protein metabolism. Mailing Add: Dept of Path Univ of NC Chapel Hill NC 27515

BENSON, WILBUR MAXWELL, b Honeyford, NDak, May 9, 15; m 46; c 2. PHARMACOLOGY, TOXICOLOGY. Educ: NDak State Univ, BS, 39; Univ Fla, MS, 41; Univ Wis, MD, 48, PhD(physiol), 49. Prof Exp: Asst pharmacologist, Sharp & Dohme, Inc, 41-43; asst physiol, Sch Med, Univ Wis, 44-46; med intern, Calif Hosp, 49-50; sr pharmacologist, Hoffmann-La Roche, Inc, 50-53, vchmn, Dept Pharmacol, 53-57; Hill prof pharmacol, Sch Med, Univ Minn, 57-60; dir neuropharmacol & psychopharmacol, Mead Johnson Res Ctr, 60-62; chmn, Dept Pharmacol, Stanford Res Inst, 62-69; biomed consult, 70-74; ASSOC DIR CLIN RES, BOEHRINGER INGELHEIM LTD, 75- Concurrent Pos: USPHS fel, 48-49; mem adv comt preclin psycho-pharmacol, NIMH, 61-63 & Nat Formulary, 62-64. Mem: AAAS; Am Soc Pharmacol & Exp Therapeut; Am Pharmaceut Asn; Am Med Asn; Am Soc Clin Pharmacol & Therapeut. Res: Assay of cardioactive agents; inhibition of cholinesterase; phthalylsulfathiazole; quinuclidinol esters and morphinans; pharmacology and toxicology of sulfonamides, thiosemicarbazones, hydrazides, morphinans, psychotherapeutic and cardiovascular drugs. Mailing Add: Box 718 Scarsdale NY 10583

BENSON, WILLIAM EDWARD BARNES, b West Haven, Conn, May 15, 19; m 44; c 3. GEOLOGY. Educ: Yale Univ, BA, 40, MS, 42, PhD(geol), 52. Prof Exp: Geologist, Conn State Geol & Nat Hist Surv, 40-41; geologist, US Geol Surv, 42-54, actg chief gen geol br, 53-54; exec secy div earth sci, Nat Res Coun, 54-55; chief geologist, Manidon Mining Inc, 55-56; chief geologist, 56-75, CHIEF SCIENTIST EARTH SCI DIV, NSF, 75- Concurrent Pos: From prog dir to head earth sci sect, NSF, 57-75. Mem: AAAS; Am Geophys Union; Geol Soc Am. Res: Marine geology; geomorphology; glacial geology; tertiary stratigraphy; economic geology. Mailing Add: Nat Sci Found 1800 G St NW Washington DC 20550

BENSON, WILLIAM HOWARD, b Halethorpe, Md, Jan 30, 02; m 27; c 1. MATHEMATICS. Educ: US Naval Acad, BS, 25. Prof Exp: From asst prof to assoc prof math, 55-70, registr, 58-66, EMER ASSOC PROF MATH, DICKINSON COL, 70- Res: Probability; statistics. Mailing Add: 234 Walnut St Carlisle PA 17013

BENSTON, MARGARET LOWE, b Longview, Wash, Oct 1, 37; div. PHYSICAL CHEMISTRY. Educ: Willamette, BA, 59; Univ Wash, PhD(phys chem), 64. Prof Exp: Proj assoc theoret chem, Univ Wis, 64-66; ASST PROF CHEM, SIMON FRASER UNIV, 66- Mem: Am Phys Soc. Res: Theoretical studies in molecular and atomic quantum mechanics. Mailing Add: Dept of Chem Simon Fraser Univ Burnaby BC Can

BENSUSAN, HOWARD BERNARD, b Boston, Mass, July 22, 22; m 51; c 2. BIOLOGICAL CHEMISTRY. Educ: Mass Inst Technol, SB, 45; Purdue Univ, MS, 48; Boston Univ, PhD(biochem), 53. Prof Exp: Biochemist, Sch Med, Boston Univ, 49-53, instr, 53; asst prof physiol res, Sch Med, Univ Md, 53-56; from asst prof to assoc prof, 56-75, PROF BIOCHEM, SCH MED, CASE WESTERN RESERVE UNIV, 75- Mem: AAAS; Am Chem Soc; Am Soc Biol Chem; Am Asn Univ Prof. Res: Connective tissue proteins: their structure, mechanism of crosslinking, function, and time-dependent alterations; platelet binding and mechanisms of their function. Mailing Add: Dept Biochem Sch Med Case Western Reserve Univ Cleveland OH 44106

BENT, DONALD FREDERICK, b Clinton, Mass, Nov, 16, 25; m 51; c 3. MICROBIOLOGY. Educ: Univ NH, BS, 48, MS, 53; Univ Md, PhD(bact), 57. Prof Exp: Spec staff asst, Am Coun Ed, 56; instr bact & immunol, Harvard Med Sch, 56-59; owner & dir, Health Serv Lab, 60-74; VPRES, BIOSPHERIC CONSULT INT, INC, 74- Concurrent Pos: Mem staff, Biol Labs, Mass Dept Pub Health, 56-59; mem fac, Colby Jr Col Women, 56-69; vpres, Bio-tronics Res, Inc, 61-66; Fulbright lectr, Med Fac, Univ Malaya, 66-67; dir clin microbiol div, Metrop Path Lab, Inc, NJ, 69-71. Mem: Am Soc Microbiol; Water Pollution Control Fedn. Res: Clinical and environmental microbiology; bacterial toxins; immunology; diagnostic and automated microbiology. Mailing Add: Box 32 New LondonNH 03257

BENT, FORREST CONSER, cytogenetics, see 12th edition

BENT, HENRY ALBERT, b Cambridge, Mass, Dec 21, 26; m 59; c 2. PHYSICAL CHEMISTRY. Educ: Oberlin Col, AB, 49; Univ Calif, PhD(chem), 52. Prof Exp: Asst, Univ Calif, 49-51; instr phys chem, Univ Conn, 52-55; res fel, Univ Minn, Minneapolis, 55-57, from asst prof to prof inorg chem, 57-69; PROF CHEM, NC STATE UNIV, 69- Concurrent Pos: Lectr, Gulbenkian Inst Advan Study, Portugal, 70, 71 & 72; consult ed, Oxford Univ Press; Chautaugua course dir, AAAS, 72, 73 & 74; comt mem comt prof training, Am Chem Soc, 72- & nat counr div chem educ, 75- Mem: Fel AAAS; Am Chem Soc; Am Phys Soc. Res: Thermodynamics; molecular structure; intermolecular interactions; localized molecular orbitals; applications of chemical principles to environmental problems and exercise physiology; history of chemistry; chemical education. Mailing Add: Dept of Chem Dabney Hall NC State Univ Raleigh NC 27607

BENT, RICHARD LINCOLN, b Rochester, NY, Oct 25, 17. ORGANIC CHEMISTRY. Educ: Cambridge Univ, BA, 39. Prof Exp: Chemist, 41-48, res chemist, 48-71, RES ASSOC, COLOR PHOTOG DIV, EASTMAN KODAK CO, 71- Mem: Am Chem Soc; Soc Photog Sci & Eng. Res: Synthesis of photographic developing agents. Mailing Add: 37 Park View Rochester NY 14613

BENT, ROBERT DEMO, b Cambridge, Mass, Dec 22, 28; m 56; c 3. PHYSICS. Educ: Oberlin Col, AB, 50; Rice Inst, AM, 52, PhD(physics), 54. Prof Exp: Res assoc physics, Rice Inst, 54-55; res assoc, Columbia Univ, 55-58; from asst prof to assoc prof, 58-66, PROF PHYSICS, IND UNIV, BLOOMINGTON, 66- Concurrent Pos: Guggenheim fel, Oxford & Atomic Energy Authority Res Estab, Harwell, Eng, 62-63. Mem: Fel Am Phys Soc. Res: Intermediate energy nuclear physics; nuclear astrophysics. Mailing Add: Dept of Physics Ind Univ Bloomington IN 47401

BENTALL, RAY, b Grand Rapids, Mich, June 28, 17; m 38; c 2. GEOLOGY. Educ: Univ Mich, BS, 38. Prof Exp: Geologist, State Div Geol, Tenn, 40-43; geologist, US Geol Surv, Tenn, 43-45 & Nebr, 45-67, hydrologist, Nebr, 67-73; HYDROLOGIST, CONSERV & SURV DIV, INST AGR & NATURAL RESOURCES, UNIV NEBR, LINCOLN, 74- Res: Hydrology. Mailing Add: Rm 113 Nebr Hall 901 N 17th St Lincoln NE 68508

BENTE, PAUL FREDERICK, JR, general chemistry see 12th edition

BENTFELD, MARY J, zoology, see 12th edition

BENTLEY, CHARLES FRED, b Cambridge, Mass, Mar 14, 14; m 43; c 2. SOIL SCIENCE. Educ: Univ Alta, BSc, 39, MSc, 41; Univ Minn, PhD(soils), 45. Prof Exp: Teacher, 33-36; asst soils, Univ Alta, 40-41; instr, Univ Minn, 42-43; from instr to asst prof, Univ Sask, 43-46; from asst prof to prof, Univ Alta, 46-68, from assoc dean agr to dean agr, 57-68; agr adv, External Aid Off, Ottawa, 68-69; PROF SOIL SCI, UNIV ALTA, 69- Concurrent Pos: Colombo Plan specialist, Ceylon, 52-53; Food & Agr Orgn consult, Thailand, 62; leader, Can Agr Task Force to India, 67-; consult, Can Int Develop Agency, 68-; mem bd gov, Int Develop Res Ctr, Can; mem & chmn gov bd, Int Crops Res Inst, Semi-Arid Tropics, Hyderabad, India; pres, Int Soil Sci Soc & Chmn Organizing Comt 11th Cong, 74-78. Mem: Fel AAAS; fel Agr Inst Can (past pres); Can Soil Sci Soc; Am Soc Agron; Int Soil Sci Soc (pres, 74-78). Res: Soil management and fertility; soil genesis and classification; international development; the population problem. Mailing Add: Dept of Soil Sci Univ of Alta Edmonton AB Can

BENTLEY, CHARLES RAYMOND, b Rochester, NY, Dec 23, 29; m 64; c 2. GEOPHYSICS, GLACIOLOGY. Educ: Yale Univ, BS, 50; Columbia Univ, PhD(geophys), 59. Prof Exp: Asst, Columbia Univ, 52-55, geophysicist, 55-56; traverse co-leader seismologist, Int Geophys Year, Antarctic Prog, Arctic Inst NAm, 56-58, traverse leader, 58-59; proj assoc geophys, 59-61, from asst prof to assoc prof, 61-68, PROF GEOPHYS, UNIV WIS, MADISON, 68- Concurrent Pos: NSF sr fel, Mass Inst Technol, 68-69; secy sci comt Antarctic Res, Working Group Solid Earth Geophys, 75-; mem sci coun, Int Antarctic Glaciological Proj. Honors & Awards: Bellingshausen-Lazarev Medal, Acad Sci, USSR, 71. Mem: AAAS; Seismol Soc Am; Am Geophys Union; Geol Soc Am; Soc Explor Geophys. Res: Geophysical studies on polar ice caps and glaciological applications of geophysical methods; magnetotellurics; theoretical elastic wave propagation studies. Mailing Add: Geophys & Polar Res Ctr Weeks Hall Univ of Wis Madison WI 53706

BENTLEY, CLEO L, b Dermott, Ark, June 6, 19; m 42; c 3. VERTEBRATE ANATOMY, PARASITOLOGY. Educ: Agr, Mech & Norm Col, Ark, BS, 41; Univ Wis, MS, 50, PhD(zool), 56. Prof Exp: High sch teacher, 41-43; training officer, Vet Admin, 46-47; instr, 50-54, PROF BIOL, AGR, MECH & NORM COL, ARK, 56- Mem: Nat Inst Sci. Res: Effect of radio-active materials on animals infected with Eimeria Tenella. Mailing Add: Dept of Biol Agr Mech & Norm Col Pine Bluff AR 71602

BENTLEY, DONALD LYON, b Los Angeles, Calif, Apr 25, 35; m 57; c 3. MATHEMATICAL STATISTICS. Educ: Stanford Univ, BS, 57, MS, 58, PhD(statist), 61. Prof Exp: Asst prof math statist, Colo State Univ, 61-64; from asst prof to assoc prof, 64-74, PROF MATH, POMONA COL, 74- Concurrent Pos: Prin investr, NSF grant, 62-64; NIH grant, 68-70; NSF fac fel, 70-71; consult, Allergan Pharmaceut Corp, 65- Mem: Inst Math Statist; Am Statist Asn; Biomet Soc; Royal Statist Soc. Res: Mathematical statistics and stochastic processes and their application to biological and social phenomena. Mailing Add: Dept of Math Pomona Col Claremont CA 91711

BENTLEY, FLOYD EDWARD, organic chemistry, industrial engineering, see 12th edition

BENTLEY, HERSCHEL LAMAR, b Slyacauga, Ala, July 14, 39; m 59; c 4. MATHEMATICS. Educ: Univ Ariz, BS, 61, MS, 63; Rensselaer Polytech Inst, PhD(math), 65. Prof Exp: Eng trainee, Hughes Aircraft Co, 59-60; asst, Univ Ariz, 61-63; asst prof math, Rensselaer Polytech Inst, 65-66, Univ NMex, 66-69 & Bucknell Univ, 69-71; assoc prof, 71-75, PROF MATH, UNIV TOLEDO, 75- Mem: Math Asn Am; Am Math Soc. Res: General topology. Mailing Add: Dept of Math Univ of Toledo Toledo OH 43606

BENTLEY, J PETER, b Oldham, Eng, Sept 15, 31; US citizen; m 57; c 2. BIOCHEMISTRY. Educ: Univ Ore, MS, 61, PhD(biochem, physiol), 63. Prof Exp: Res asst, Harvard Med Sch, 57-59; res asst, 59-62, res assoc, 62-64, instr exp biol, 64-65, asst prof exp biol & biochem, 65-68, ASSOC PROF BIOCHEM, SCH MED, UNIV ORE, 68- Mem: Am Soc Biol Chem; Soc Complex Carbohydrates. Res: Biosynthesis of connective tissue components; collagen and proteoglycans. Mailing Add: Univ of Ore Med Sch Portland OR 97201

BENTLEY, KENNETH CHESSAR, b Montreal, Que, Sept 22, 35; m 61; c 2. ORAL SURGERY. Educ: McGill Univ, DDS, 58, MD, CM, 62. Prof Exp: Jr rotating intern, Montreal Gen Hosp, 62-63, jr asst resident, 63-64; partic prog oral surg, Bellevue Hosp, New York, 64-66, chief resident, 65-66; from asst prof to assoc prof, 66-75, PROF ORAL SURG, McGILL UNIV, 75-, CHMN DEPT ORAL SURG & DIR DIV SURG & ORAL MED, 68-; DENT SURGEON-IN-CHIEF, MONTREAL GEN HOSP, 70- Mailing Add: Montreal Gen Hosp Montreal PQ Can

BENTLEY, KENTON EARL, b Detroit, Mich, June 1, 27; m 53. ANALYTICAL CHEMISTRY, REMOTE SENSING. Educ: Univ Mich, BS, 50; Univ NMex, PhD(anal chem), 59. Prof Exp: Lab asst electrochem, Univ Mich, 49-50; phys chemist, Res Dept, Consol Electrodynamics Corp, 56-57; consult chemist, 57-59; asst prof chem, Am Univ Beirut, 59-61; res scientist, Lockheed-Calif Co, Lockheed

Aircraft Corp, 62-63; res specialist space sci, Jet Propulsion Lab, Calif Inst Technol, 63-65; res scientist, Electrochem Lab, Hughes Aircraft Co, 65-67; asst dept mgr, Space Environ Dept, Manned Spacecraft Ctr, 67, mgr space physics dept, 67-70, mgr earth resources dept, 68-70, DIR SCI & APPLN BR, AEROSPACE SYSTS DIV, LOCKHEED ELECTRONICS CO, INC, 70-, MGR IRAN EARTH RESOURCES PROG, 75- Concurrent Pos: Rockefeller Found res grant, 59-61. Mem: AAAS; Am Chem Soc; Combustion Inst; Am Astronaut Soc. Res: Instrumental methods of chemical analysis; chemistry of the lower valence states of transition metals; electrochemical studies; earth resources/remote sensing program operations and management. Mailing Add: Sci & Appln Br Aerospace Syst Div Lockheed Electronics Co 16811 El Camino Real Houston TX 77058

BENTLEY, MICHAEL DAVID, b Jacksonville, Fla, Feb 7, 39; m 63; c 1. ORGANIC CHEMISTRY. Educ: Auburn Univ, BSChem, 63, MS, 65; Univ Tex, Austin, PhD(org chem), 69. Prof Exp: NIH fel, Univ Calif, Berkeley, 68-69; asst prof chem, 69-74, ASSOC PROF CHEM, UNIV MAINE, 74- Mem: Am Chem Soc. Res: Mechanisms of carbonium ion and carbanion reactions; chemistry of sulfamic acid derivatives; molecular orbital correlations. Mailing Add: Dept of Chem Univ of Maine Orono ME 04473

BENTLEY, ORVILLE GEORGE, b Midland, SDak, Mar 6, 18; m 42; c 2. NUTRITIONAL BIOCHEMISTRY, ANIMAL NUTRITION. Educ: SDak State Col, BS, 42; Univ Wis, MS, 47, PhD(biochem), 50. Hon Degrees: DS, SDak State Univ, 74. Prof Exp: Asst, Univ Wis, 46-50; from asst prof to prof animal sci, Ohio Univ, Exp Sta, 50-58; dean agr & dir exp sta, SDak State Univ, 58-65; DEAN COL AGR, UNIV ILL, URBANA-CHAMPAIGN, 65- Concurrent Pos: Mem bd dirs, Midwest Univs Consortium Int Activities, Inc; bd dirs, Farm Found; bd trustees, Am Univ Beirut; mem, Agr Res Policy Adv Comt; chmn, Ill State Rural Develop Adv Coun. Honors & Awards: Am Feed Mfrs Award, 58. Mem: Am Chem Soc; fel Am Soc Animal Sci; Am Inst Nutrit. Res: Large animal nutrition; metabolism studies with animal and bacteria; microbiological assays; studies on nutrition and metabolism of rumen microflora. Mailing Add: Col of Agr Univ of Ill Urbana IL 61801

BENTLEY, PATRICK E, b Dec 25, 36; US citizen; m 60; c 2. PHARMACOLOGY. Educ: Univ Mo, AB, 60; St Louis Univ, MS, 64; Univ Louisville, PhD(pharmacol), 68; Kirksville Col Osteop Med, DO, 73. Prof Exp: Asst chemist, Anheuser-Busch, Inc, Mo, 60-61; res asst biol, St Louis Univ, 61-63; high sch teacher, Mo, 63-64; instr pharmacol, Kirksville Col Osteop Med, 67-74; ASSOC PROF PHARMACOL & CHMN DEPT, GREENBRIER COL OSTEOP MED, 74- Concurrent Pos: Consult, Still-Hildrith Hosp & Kirksville Osteop Hosp, 68-71; intern, Kirksville Osteop Hosp, 73-74. Mem: AAAS; NY Acad Sci. Res: Renal autonomic pharmacology in the conscious dog; toxicity of radiopaque diagnostic agent; characterization of large molecules; behavioral physiology and pharmacology of vertebrate and invertebrate. Mailing Add: Dept of Pharmacol Greenbrier Col of Osteop Med Lewisburg WV 24901

BENTLEY, PETER JOHN, b Perth, Australia, Jan 13, 30; m 54; c 2. PHYSIOLOGY, PHARMACOLOGY. Educ: Univ Western Australia, BSc, 52, PhD(physiol), 60. Prof Exp: Res asst pharmacol, Univ Col, Univ London, 53-54; asst physiol, Univ Western Australia, 55-58, lectr, 58-60; res fel pharmacol, Bristol Univ, 61-62, lectr, 62-65; res assoc zool, Duke Univ, 65-66, assoc prof pharmacol, 66-68; PROF PHARMACOL & OPHTHAL, MT SINAI SCH MED, 68- Mem: Am Zool Soc; Endocrine Soc; Am Physiol Soc; Brit Soc Endocrinol. Res: Endocrinology; zoology. Mailing Add: Mt Sinai Sch of Med New York NY 10029

BENTLEY, RICHARD FOSTER, b Detroit, Mich, Jan 28, 45. NUCLEAR PHYSICS. Educ: Williams Col, BA, 67; Univ Colo, MS, 69, PhD(physics), 72. Prof Exp: STAFF SCIENTIST PHYSICS, LOS ALAMOS SCI LAB, 72- Mem: Am Phys Soc. Res: Low and medium energy nuclear physics; experiment, accelerators and theory. Mailing Add: PO Box 1663 MS 808 Los Alamos NM 87545

BENTLEY, RONALD, b Derby, Eng, Mar 10, 22; US citizen; m 48; c 3. BIOCHEMISTRY. Educ: Univ London, BSc, 43, PhD(org chem), 45. Hon Degrees: DSc, Univ London, 65. Prof Exp: Commonwealth Fund fel, Col Physicians & Surgeons, Columbia Univ, 46-47, res assoc biochem, 51-52; mem sci staff, Nat Inst Med Res, London, 48-51; from asst prof to assoc prof biochem, 53-60, PROF BIOCHEM, UNIV PITTSBURGH, 60-, CHMN DEPT, 72- Concurrent Pos: Guggenheim fel, Inst Biochem, Univ Lund, 64; mem microbial chem study sect, NIH, 72- Mem: AAAS; Am Soc Microbiol; Am Soc Biol Chemists; Am Soc Chem; The Chem Soc. Res: Biochemistry of microorganisms; carbohydrate biochemistry; biosynthesis of aromatic compounds; stereochemical implications in biology. Mailing Add: 730 Crabtree Hall Univ of Pittsburgh Dept of Biochem Pittsburgh PA 15261

BENTON, ALLEN HAYDON, b Ira, NY, Sept 4, 21; m 47; c 3. ZOOLOGY. Educ: Cornell Univ, BS, 48, MS, 49, PhD(vert zool), 52. Prof Exp: From instr to assoc prof biol, State Univ NY Albany, 49-62; prof, 62-73, chmn dept, 66-69, DISTINGUISHED TEACHING PROF BIOL, STATE UNIV NY COL FREDONIA, 73- Concurrent Pos: Vis prof biol, Concord Col, 69-70. Mem: AAAS; Am Ornith Union; Am Soc Mammal; Am Inst Biol Sci; Ecol Soc Am. Res: Bird distribution in New York State; life histories of small mammals; taxonomy, distribution and host relationships of American Siphonaptera. Mailing Add: Dept of Biol State Univ NY Fredonia NY 14063

BENTON, ALLEN WILLIAM, b Greenwich, Conn, June 8, 31; m 54; c 3. BIOCHEMISTRY, PHYSIOLOGY. Educ: Univ Conn, BS, 58, MS, 62; Cornell Univ, PhD(entom), 65. Prof Exp: Soil scientist, Soil Conserv Serv, USDA, 54-62; asst prof entom & apicult, Rutgers Univ, 65-66; asst prof, 66-69, ASSOC PROF ENTOM, PA STATE UNIV, 69- Concurrent Pos: Consult, Center Labs, 65-, Pharmacia, 75-, Good Samaritan Hosp, Johns Hopkins Univ, 75- & Buffalo Gen Hosp, 74- Mem: Entom Soc Am; Bee Res Asn. Res: Venoms, biochemistry and physiological effects upon animals; biochemistry of allergens. Mailing Add: Dept of Entom Rm 6 Armsby Bldg Pa State Univ University Park PA 16802

BENTON, ARTHUR LESTER, b New York, NY, Oct 16, 09; m 39; c 3. NEUROPSYCHOLOGY. Educ: Oberlin Col, AB, 31, AM, 33; Columbia Univ, PhD, 35. Prof Exp: Assoc prof med psychol, Sch Med, Univ Louisville, 46-48; PROF NEUROL & PSYCHOL, UNIV IOWA, 48- Concurrent Pos: USPHS res grant, Univ Iowa, 54-, USPHS spec fel neuropsychol, 58-59; consult, NIH, 61-; vis prof, Fac Med, Univ Milan, 64, Free Univ, Amsterdam, 71, Dept Neurol, Univ Helsinki, 74 & Tokyo Metrop Inst Geront, 74. Mem: Am Psychol Asn; Am Neurol Asn; Am Acad Neurol; Int Neuropsychol Soc (pres, 70-72); Am Orthopsychiat Asn (pres, 64-65). Res: Behavioral disabilities associated with brain disease. Mailing Add: Dept of Neurol Univ of Iowa Hosps Iowa City IA 52242

BENTON, BYRL E, b Armstrong, Iowa, Sept 4, 12; m 39; c 2. State Univ, BSPharm, 35, MSPharm, 39; Univ Ill, PhD(pharmacol), 37. Prof Exp: Pharmacist, Hermanson Drug Co, 35-37; asst pharm, SDak State Univ, 37-40; from instr to assoc prof, Univ Ill, 40-49; DEAN, COL PHARM, DRAKE UNIV, 49- Concurrent Pos: Prescription ed, Nat Asn Retail Druggists J, 45-48; sci ed, La Farmacia Mod, 46-51; mem, Am Bd Dipl Pharm. Mem: AAAS; Am Pharmaceut Asn. Res: Cardiac research involving digitalis; compressed tablet research, disintegration. Mailing Add: Drake Univ Col of Pharm Des Moines IA 50311

BENTON, CHARLES HERBERT, (JR), organic chemistry, see 12th edition

BENTON, DUANE ALLEN, b Waterloo, Iowa, June 1, 31; m 55; c 3. BIOCHEMISTRY. Educ: Mich State Univ, BS, 52; Univ Wis, MS, 54, PhD(biochem), 56. Prof Exp: Asst prof biochem, Okla State Univ, 56-59; cancer res scientist, Roswell Park Mem Inst, 59-63; assoc prof animal nutrit, Cornell Univ, 63-66; head biochem invest, USDA, 66-70; DIR NUTRIT RES, ROSS LABS, 70- Mem: AAAS; Am Inst Nutrit; Am Chem Soc. Res: Amino acid interrelationships in the nutrition of animals; dietary interrelations and their effects on metabolism. Mailing Add: Ross Labs 625 Cleveland Ave Columbus OH 43216

BENTON, DUANE MARSHALL, b Savannah, Ohio, Nov 1, 33; m 60; c 2. ENVIRONMENTAL CHEMISTRY. Educ: Ashland Col, AB, 55; Univ Utah, BS, 58; Ohio State Univ, MS, 64; George Washington Univ, MS, 67. Prof Exp: Asst chem, Ohio State Univ, 55-57; asst anal res chemist, Hess & Clark Div, Richardson-Merrell Inc, 58-59; instr physics, math & chem, Ashland Col, 59-60 & 62-64, from asst prof to assoc prof phys sci, 64-73; ENVIRON PHYS SCIENTIST, DEPT OF ARMY HQ, ENVIRON OFF, WASHINGTON, DC, 73- Concurrent Pos: AEC res grant, 64-67. Mem: Am Meteorol Soc. Res: Photosynthesis. Mailing Add:

BENTON, EDWARD ROWELL, b Milwaukee, Wis, Jan 20, 34; m 58; c 3. GEOPHYSICS. Educ: Harvard Univ, AB, 56, AM, 57, PhD(appl math), 61. Prof Exp: Asst sr engr appl mech group, Arthur D Little, Inc, 60-62; lectr math, Univ Manchester, 62-63; staff scientist, Nat Ctr Atmospheric Res, 63-65, asst dir adv study prog, 67-69; asst prof, Univ Colo, Boulder, 65-67, prof & chmn astro-geophys, 69-74; CONSULT, ABERDEEN PROVING GROUNDS, BALLISTIC RES LABS, US ARMY, 74- Concurrent Pos: Consult, Arthur D Little, Inc, 62-63; Environ Sci Serv Admin res grant, 66-67; NSF res grant, 70-; spec asst univ rels, Univ Corp Atmospheric Res, 75-77. Mem: AAAS; Am Geophys Union; Am Phys Soc; Sigma Xi. Res: Magnetohydrodynamics; rotating fluids; applied mathematics; boundary layer theory; turbulence theory; dynamo theory of geomagnetism. Mailing Add: 780 Juniper Ave Boulder CO 80302

BENTON, EUGENE VLADIMIR, b Simferopol, Russia, July 23, 37; US citizen; m 61; c 2. RADIOLOGICAL PHYSICS. Educ: San Jose State Col, BA, 58, MA, 60; Stanford Univ, PhD(physics), 68. Prof Exp: Instr physics, San Jose State Col, 58-60; from jr investr to investr, US Naval Radiol Defense Lab, 61-69; res prof, 69-73, PROF PHYSICS, UNIV SAN FRANCISCO, 73- Concurrent Pos: Consult, Lawrence Berkeley Lab, Univ Calif, 69- Honors & Awards: Gold Medal Sci Achievement, Naval Radiol Defense Lab, 69. Mem: AAAS; Am Phys Soc. Res: Medical physics; nuclear photographic emulsions; radiation dosimetry; materials science; solid state physics; radiation effects in solids; dielectric nuclear particle track detectors. Mailing Add: Dept of Physics Univ of San Francisco San Francisco CA 94117

BENTON, FRANCIS LEE, b Moxahala, Ohio, Feb 12, 12; m 40; c 3. ORGANIC CHEMISTRY. Educ: Ohio State Univ, BChE, 35, MS, 36, PhD(chem), 40. Prof Exp: Instr chem, Univ Notre Dame, 40-42, from asst prof to assoc prof, 42-52; res chemist, Armour Lab, 52-59; assoc prof chem, 59-62, PROF CHEM, ST MARY'S COL, IND, 62- Mem: Am Chem Soc. Res: General organic synthesis; isolation of naturally occurring substances of biological origin and investigation of their molecular structure. Mailing Add: Dept of Chem St Mary's Col Notre Dame IN 46556

BENTON, GEORGE STOCK, b Oak Park, Ill, Sept 24, 17; m 45; c 4. METEOROLOGY. Educ: Univ Chicago, SB, 42, PhD(meteorol), 47. Prof Exp: From asst to asst prof meteorol, Univ Chicago, 42-48; civil eng, 48-52, assoc prof, 52-57, actg chmn dept civil eng, 58-60, chmn dept mech, 60-66, chmn dept earth & planetary sci, 69-70, dean fac arts & sci, 70-72, PROF METEOROL, JOHNS HOPKINS UNIV, 57- Concurrent Pos: Dir, Insts Environ Res, Environ Sci Serv Admin, Boulder, Colo, 66-69; trustee, Univ Corp Atmospheric Res; vpres, Homewood Divs, 72- Mem: Am Soc Civil Eng; fel Am Meteorol Soc; fel Am Geophys Union; fel Am Acad Arts & Sci. Res: Flow of stratified and rotating fluid; general circulation of the atmosphere; hydrometeorology; hydrology. Mailing Add: Div of Arts & Sci Johns Hopkins Univ Baltimore MD 21218

BENTON, JOHN WILLIAM, JR, b Enterprise, Ala, July 3, 30; m 55; c 3. PEDIATRICS, NEUROLOGY. Educ: Univ Ala, BS, 51; Med Col Ala, MD, 55; Am Bd Pediat, dipl, 62; Am Bd Psychiat & Neurol, dipl, 69. Prof Exp: Intern pediat, Univ Ala Hosp, Birmingham, 55-56; resident, Univ Utah, 56-57 & Univ Minn, Minneapolis, 59-60; instr, 60-62, asst prof pediat & neurol, 62-66, dir, Ctr Develop & Learning Disorders, 62-68, assoc prof pediat, 66-69, interim chmn dept, 68-69, PROF PEDIAT & CHMN DEPT, SCH MED & DIR CHILDREN'S HOSP, UNIV ALA, BIRMINGHAM, 69-, ASSOC PROF, DIV NEUROL, SCH MED, 66- Concurrent Pos: NIH trainee neurol, Med Col, Univ Ala, 60-62; fel, Mass Gen Hosp, Harvard Med Sch, 64-66. Mem: Am Pediat Soc; Am Acad Neurol; Am Acad Pediat. Res: Acute ataxia; acute encephalopathy of undetermined origin; convulsive disorders. Mailing Add: Children's Hosp Univ of Ala Sch of Med Birmingham AL 35233

BENTON, JOSEPH GEORGE, b New York, NY, Jan 20, 15; c 2. MEDICINE. Educ: Brooklyn Col, AB, 37; NY Univ, MS, 38; PhD, 40, MD, 45. Prof Exp: Instr anat, NY Univ, 36-41, asst physiol, Col Med, 41-42; lectr, Exten Div, Hunter Col, 42-46; instr pharmacol, Col Med, NY Univ, 47-48, assoc prof phys med & rehab, 49-58, consult, Inst Rehab Med, Bellevue Med Ctr & attend physician, Univ Hosp, 50-58; PROF REHAB MED & CHMN DEPT, COL MED, STATE UNIV NY DOWNSTATE MED CTR, 58-, DEAN COL HEALTH RELATED PROF, 66- Concurrent Pos: Fel, Med Col, Cornell Univ, 46-47; fel therapeut, Col Med, NY Univ, 47-49; fel, Med Div, Goldwater Mem Hosp, 48-50; from intern to vis physician, Bellevue Hosp, New York, 45-58; asst physician, Beth Israel Hosp & Hosp Joint Dis, 46-48 & NY Univ Clins & Irvington House, 47; assoc vis physician, Med Div, Goldwater Mem Hosp, 51-52; assoc prof & dir educ & res, Col Med, NY Univ, 52; dir, NY Regional Respiratory & Rehab Ctr, 54-58; chief rehab med dept, State Univ Hosp, 58- & Kings County Hosp Ctr, Brooklyn, 58-; consult, Brooklyn Vet Admin Hosp, Long Island Col & Long Island Jewish Hosps, Maimonides Hosp, Brooklyn, Nassau County Dept Health, Brooklyn-Cumberland Med Ctr, Methodist Hosp & Bur Indian Affairs, US Dept Interior; mem, Am Bd Phys Med & Rehab; chmn, Comn Educ Phys Med & Rehab, NY; spec consult health sci to pres, Long Island Univ; mem adv coun med educ to pres, York Col, NY. Mem: Am Asn Anat; Am Soc Pharmacol & Exp Therapeut; Geont Soc; Am Col Physicians; Asn Am Med Cols. Res: Experimental morphology; renal and hepatic physiology; clinical pharmacology; thermal environment and thyroid-adrenal apparatus; rehabilitation medicine and cardiovascular disease; training programs in rehabilitation foch for physicians and allied health personnel. Mailing Add: State Univ NY Downstate Med Ctr 450 Clarkson Ave Brooklyn NY 11203

BENTON, KENNETH CURTIS, b Whitinsville, Mass, Sept 15, 41. POLYMER

CHEMISTRY. Educ: Worcester Polytech Inst, BS, 63; Univ Akron, PhD(polymer chem), 69. Prof Exp: Sr res chemist, Copolymer Rubber & Chem Corp, 67-70, group leader polymers, CPC Int, Inc, 70-74; SR RES CHEMIST, STAND OIL CO, OHIO, 74- Mem: Am Chem Soc. Res: Synthesis of acrylonitrile barrier resins for packaging applications. Mailing Add: Stand Oil Res & Eng Dept 4440 Warrensville Center Rd Warrensville Heights OH 44128

BENTON, ROBERT S, b Moxahala, Ohio, June 30, 14; m 42; c 3. ANATOMY. Educ: Univ Mich, AB, 37; Univ Chicago, AM, 40; Duke Univ, PhD(anat), 60. Prof Exp: Asst prof anat, Med Col SC, 49-60; assoc prof, 60-69, PROF ANAT, MED CTR, UNIV KY, 69- Mem: Am Asn Anat; Am Asn Phys Anthrop. Res: Gross anatomy; comparative anatomy of primates. Mailing Add: Dept of Anat Univ of Ky Med Ctr Lexington KY 40506

BENTON, STEPHEN ANTHONY, b San Francisco, Calif, Dec 1, 41; m 64; c 1. OPTICAL PHYSICS. Educ: Mass Inst Technol, BS, 63; Harvard Univ, MS, 65, PhD(appl physics), 68. Prof Exp: Asst prof appl physics, Div Eng & Appl Physics, Harvard Univ, 68-73; SR SCIENTIST, RES LABS, POLAROID CORP, 61- Concurrent Pos: Vis prof physics, Univ Reading, Eng, 72-73. Mem: Optical Soc Am; Soc Photographic Scientists & Engrs; Sigma Xi. Res: Image communication systems. Mailing Add: Res Labs Polaroid Corp Cambridge MA 02139

BENTON, WILLIAM J, b Franklin, Va, June 28, 33; m 55; c 2. VETERINARY VIROLOGY. Educ: Univ Ga, DVM, 56; Univ Del, MS, 59, PhD(biol sci), 66. Prof Exp: Res assoc poultry path & virol, 56-66, asst prof, 66-67, actg chmn dept animal sci & agr biochem, 67-68, prof & chmn dept, 68-71, asst dean col agr sci & asst dir agr exp sta, 71-74, PROF ANIMAL SCI, UNIV DEL, 71-, ASSOC DEAN COL AGR SCI & ASSOC DIR AGR EXP STA, 74- Mem: Am Vet Med Asn; Am Asn Avian Path. Res: Application of pathology, serology, immunology and virology in the study of avian diseases such as infectious laryngotracheitis, infectious synovitus, avian leukosis, mycoplasma gallisepticum infection, Newcastle disease, infectious bronchitis and infectious bursal disease. Mailing Add: Col of Agr Sci Univ of Del Newark DE 19711

BENTRUDE, WESLEY GEORGE, b Waterloo, Iowa, Mar 13, 35; m 55; c 2. ORGANIC CHEMISTRY. Educ: Iowa State Univ, BS, 57; Univ Ill, PhD(org chem), 61. Prof Exp: Res chemist, Celanese Chem Co, 61-63; res assoc, Univ Pittsburgh, 63-64; asst prof chem, 64-67, ASSOC PROF CHEM, UNIV UTAH, 67- Mem: Am Chem Soc; Am Sci Affiliation. Res: Organophosphorus chemistry; mechanisms of free-radical reactions; physical organic chemistry. Mailing Add: Dept of Chemistry Univ of Utah Salt Lake City UT 84112

BENTS, ULRICH HENRY, physics, see 12th edition

BENTZ, ALAN PAUL, organic chemistry, see 12th edition

BENTZ, RALPH WAGNER, b Pennsylvania, Mar 9, 19; m 50; c 2. PHYSICAL CHEMISTRY, ORGANIC CHEMISTRY. Educ: Albright Col, BS, 43; Lehigh Univ, MS, 44, PhD(chem), 48. Prof Exp: Lab technician, A Wilhelm Co, 36-43; dept supvr, Manhattan Proj, 44-45; sr chemist, Tenn Eastman Co, 48-53; mkt analyst, Eastman Chem Prod, Inc, 53-66, SR MKT ANALYST, EASTMAN CHEM PROD, INC, 66- Mem: Am Chem Soc; Chem Mkt Res Asn. Res: Petrochemicals; plastics; packaging. Mailing Add: 4511 Preston Pl Kingsport TN 37664

BENUA, RICHARD SQUIER, b Bexley, Ohio, Aug 11, 21; m 54; c 4. NUCLEAR MEDICINE. Educ: Western Reserve Univ, BS, 43; Johns Hopkins Univ, MD, 46; Univ Minn, MS, 52. Prof Exp: Asst prof med, Cornell Univ, 56-66; dir nuclear med serv & assoc prof internal med & radiol, Med Br, Univ Tex, 66-70; CHIEF NUCLEAR MED SERV, MEM HOSP CANCER & ALLIED DIS, 70-; ASSOC PROF MED & RADIOL, MED SCH, CORNELL UNIV, 70- Concurrent Pos: Fel med, Mayo Clin, 50-53; fel surg, Western Reserve Univ, 53-55; asst pituitary-thyroid sect, Sloan-Kettering Inst Cancer Res, 55-60, assoc, 60-66, assoc div biophys, 64-66, assoc mem, 70-; clin asst, James Ewing Hosp, 55-59, asst attend, 59-63, assoc vis physician, 63-66, co-chief endocrine clin, 62-66; clin asst, Mem Hosp, 55-59, asst attend, 59-63, assoc attend, 63-66, attend, 70- Mem: Am Thyroid Asn; Endocrine Soc; Fedn Clin Res; Soc Nuclear Med. Res: Internal medicine; thyroid gland. Mailing Add: Nuclear Med Serv Mem Hosp New York NY 10021

BENUCK, IRWIN, b Chicago, Ill, Sept 5, 48; m 75. PSYCHOPHYSIOLOGY, GENETICS. Educ: Loyola Univ, Chicago, BS, 70; Ill Inst Technol, MS, 73, PhD(psychol), 74. Prof Exp: Res asst biol, Northwest Inst Med Res, 68-73; grad asst psychol, Ill Inst Technol, 72-74; GENETIC COUNR, NORTH HEALTH DIS GENETIC COUNS PROG, 74- Concurrent Pos: Adj asst prof psychol, Brenau Col, 74- Mem: Sigma Xi; Am Soc Human Genetics; Behav Genetics Soc; AAAS. Res: Neuroendocrinology and behavior; neural and sensory aspects of sexual, aggressive, agonistic and maternal behavior; biochemical correlates of abnormal behavior, and genetics and behavior. Mailing Add: Apt 17D 2601 Thompson Bridge Rd Gainesville GA 30501

BENUCK, MYRON, b Chicago, Ill, July 7, 34; m 68; c 2. NEUROCHEMISTRY. Educ: Univ Chicago, BA & BS, 57, MS, 62; Univ Ill, PhD(physiol), 66. Prof Exp: Fel neurochem, Albert Einstein Col Med, 65-67; SR RES SCIENTIST BIOCHEM, NY STATE RES INST NEUROCHEM & DRUG ADDICTION, 67- Mem: Am Soc Neurochem; Inst Soc Neurochem. Res: Brain metabolism, specifically degradation of proteins and peptides by brain enzymes. Mailing Add: NY State Res Inst for Neurochem Wards Island New York NY 10035

BENUMOF, REUBEN, b New York, NY, Nov 30, 12; m 36; c 2. ENGINEERING PHYSICS. Educ: City Col New York, BS, 33, MS, 37; NY Univ, PhD(physics), 45. Prof Exp: Elec engr, Fed Power Comn, 37-38; lectr physics, City Col New York, 45-56; assoc prof, Stevens Inst Technol, 51-56; PROF PHYSICS & HEAD DEPT, STATEN ISLAND COMMUNITY COL, 56- Concurrent Pos: Res partic, Oak Ridge Nat Lab, 56-58; fel, NSF, 59-61, dir optical pumping equip proj, 63-65; dir nuclear reactor kinetics res proj, 65-70, dir physics eng, 71-; fel, State Univ NY, 72; vis prof, Univ Zambia, 73. Mem: AAAS; Am Phys Soc; Am Asn Physics Teachers; Am Soc Eng Educ. Res: Classical theory of electricity and magnetism; nuclear reactor physics; mathematical modeling of biological and physical systems. Mailing Add: 21-15 34th Ave Long Island City NY 11106

BENVENISTE, JACOB, b Portland, Ore, Dec 21, 21; m 44; c 3. NUCLEAR PHYSICS. Educ: Reed Col, BA, 43; Univ Calif, PhD(physics), 52. Prof Exp: Radio engr, Naval Res Lab, 43-46; asst physics, Reed Col, 46; asst physics, Univ Calif, 47-50, physicist, Lawrence Radiation Lab, 50-63; physicist, Aerospace Corp, 63-65, sr staff scientist advan concepts, 65, dir nuclear effects subdiv, 65-68; dir res, Physics Int Co, 68-72, vpres, 69-72; SR STAFF SCIENTIST, AEROSPACE CORP, 72- Concurrent Pos: Mem, Nuclear Cross Sect Adv Group, 56-63; deleg, Geneva Conf, 58. Mem: AAAS; Am Phys Soc. Res: Solid state physics; radiation damage and effects; instrumentation; laser gyroscopes. Mailing Add: 1200 S Catalina 211 Redondo Beach CA 90277

BENYSHEK, LARRY L, b Concordia, Kans, Feb 26, 47; m 66; c 2. ANIMAL BREEDING. Educ: Kans State Univ, BS, 69; Va Polytech Inst & State Univ, MS, 71, PhD(animal breeding), 73. Prof Exp: Dir res & educ beef cattle performance prog, NAm Limousin Found, 73-74; asst prof agr, Ft Hays Kans State Col, 74-76; ASST PROF BEEF CATTLE BREEDING, UNIV GA, 76- Concurrent Pos: Consult, NAm Limousin Found, 73- Mem: Sigma Xi; Am Soc Animal Sci; Am Genetic Asn. Res: Estimation of genetic parameters in new breeds of cattle; crossing of newly introduced breeds of cattle for increased commercial production; selection based on other than current measures of growth in beef cattle. Mailing Add: Dept of Animal & Dairy Sci Livestock-Poultry Bldg Univ of Ga Athens GA 30602

BENZ, EDMUND WOODWARD, b Nashville, Tenn, May 8, 11; m 45; c 4. SURGERY. Educ: Vanderbilt Univ, AB, 37, MD, 40. Prof Exp: Res asst physiol, Sch Med, 38-40, surg training, Hosp, 40-45, resident surgeon, 44-45, asst prof clin surg, Sch Med, 52-68, ASSOC CLIN PROF SURG, SCH MED, VANDERBILT UNIV, 68- Concurrent Pos: Pvt pract, 45-; mem surg staff, Hosps; chmn cancer study group, Tenn/Mid-South Regional Med Prog, 68-72. Mem: AAAS; AMA; Am Col Surg. Res: Traumatic shock; wound healing and temperature; control of respiration; action potentials in peripheral nerve, splenectomy in hemophilia. Mailing Add: 422 Med Arts Bldg 1211 21st Ave S Nashville TN 37212

BENZ, EDWARD JOHN, b Pittsburgh, Pa, June 11, 23; m 45; c 4. CLINICAL PATHOLOGY, MICROBIOLOGY. Educ: Univ Pittsburgh, 43, MD, 46; Univ Minn, MS, 52. Prof Exp: DIR LABS & PATHOLOGIST, ST LUKE'S HOSP, 53- Concurrent Pos: Consult, Palmerton Hosp; adj prof microbiol, Lehigh Univ, 65-; chmn med adv comt labs, Pa Dept Health, 74- Mem: Am Asn Path & Bact; Int Acad Path; Col Am Pathologists; Am Soc Clin Pathologists. Mailing Add: St Luke's Hosp Ostrum St Bethlehem PA 18015

BENZ, GEORGE WILLIAM, b Ulm, Ger, June 26, 22; US citizen; m 47; c 5. ORGANIC CHEMISTRY. Educ: Dartmouth Col, AB, 43; Univ Tex, MA, 46, PhD(org chem), 50. Prof Exp: Chemist, Winthrop Chem Co, 43-44, Heyden Chem Co, 49-52, Callery Chem Co, 52 & Polak's Frutal Works, 52-55; PRES, ULBECO, INC, 55- Concurrent Pos: Prof, Ulster County Community Col, Stone Ridge, NY, 69-, chmn dept phys sci, 71- Mem: Am Chem Soc. Res: Organic synthesis; organometallic compounds; perfume and flavor chemicals. Mailing Add: 5 Sheryl St RD 7 Kingston NY 12401

BENZ, WOLFGANG, b Heilbronn, Ger, Mar 16, 32; m 60; c 1. MASS SPECTROMETRY. Educ: Univ Heidelberg, dipl chem, 60, Dr rer nat (chem), 62. Prof Exp: Res assoc chem, Mass Inst Technol, 62-63; asst, Inst Org Chem, Univ Heidelberg, 63-64; chemist, Badische Anilin & Sodafabrik, 64-68; SR CHEMIST, HOFFMANN-LA ROCHE INC, 68- Mem: Am Chem Soc; Ger Chem Soc. Res: Organic mass spectrometry; mechanisms of mass spectral fragmentations; analytical use of mass spectrometry; use of computers for data processing in mass spectrometry. Mailing Add: Hoffmann-La Roche Inc Nutley NJ 07110

BENZER, SEYMOUR, b New York, NY, Oct 15, 21; m 42; c 2. BEHAVIORAL GENETICS. Educ: Brooklyn Col, BA, 42; Purdue Univ, MS, 43, PhD(physics), 47. Hon Degrees: DSc, Purdue Univ, 68 & Columbia Univ, 74. Prof Exp: From instr to asst prof physics, Purdue Univ, 45-48; biophysicist, Oak Ridge Nat Lab, 48-49; res fel biophys, Calif Inst Technol, 49-51; Fulbright res scholar, Pasteur Inst, Paris, 51-52; from asst prof to Stuart distinguished prof biophys, Purdue Univ, 53-67; prof biol, 67-75, BOSWELL PROF NEUROSCI, CALIF INST TECHNOL, 75- Concurrent Pos: NSF sr res fel, Cambridge Univ, 57-58. Honors & Awards: Honor Award, Brooklyn Col, 56; Sigma Xi Award, Purdue Univ, 57; Howard Taylor Ricketts Award, Chicago, 61; Gold Medal Sci Award, Alumni Asn, City Col New York, 62; Gairdner Award, Can, 64; McCoy Award, Purdue Univ, 65; Lasker Award, 71; T Duckett Jones Award, 75. Mem: Nat Acad Sci; fel AAAS; Am Acad Arts & Sci; Am Philos Soc; Biophys Soc Am. Res: Molecular genetics; bacterial viruses; genetics and physiology of the nervous system and behavior; behavioral and developmental genetics and neurophysiology of Drosophila. Mailing Add: Div of Biol Calif Inst of Technol Pasadena CA 91125

BENZING, DAVID H, b Chicago, Ill, Oct 13, 37; m 62. BIOLOGY. Educ: Miami Univ, BA, 59; Univ Mich, MS, 62, PhD(bot), 65. Prof Exp: ASSOC PROF BIOL, OBERLIN COL, 65- Mem: NSF sci fac fel, Univ SFla, 71-72; Mem: AAAS; Bot Soc Am; Am Soc Enology & Viticulture. Res: Adaptive biology of vascular epiphytes; nutrition, water balance and factors controlling host preferences. Mailing Add: Dept of Biol Oberlin Col Oberlin OH 44074

BENZING, GEORGE, III, b Hamilton, Ohio, Nov 18, 26. PEDIATRIC CARDIOLOGY, PHYSIOLOGY. Educ: Univ Cincinnati, BS, 51, MD, 58; Am Bd Pediat, dipl, 63, cert pediat cardiol, 64. Prof Exp: From asst to assoc prof, 64-72, PROF PEDIAT, COL MED, UNIV CINCINNATI, 72-; ATTEND CARDIOLOGIST, CHIREN'S HOSP, 70- Concurrent Pos: Assoc attend cardiologist, Children's Hosp, 66-70. Mem: Am Heart Asn. Res: Myocardial performance as evaluated by ventricular function curves and cardiopulmonary bypass. Mailing Add: Div of Cardiol Children's Hosp Cincinnai OH 45229

BENZINGER, HAROLD EDWARD, JR, b New York, NY, July 9, 40; m 64. MATHEMATICS. Educ: Mass Inst Technol, BS, 62; Syracuse Univ, PhD(math), 67. Prof Exp: Asst prof, 67-75, ASSOC PROF MATH, UNIV ILL, URBANA-CHAMPAIGN, 75- Mem: Am Math Soc; Math Asn Am. Res: Boundary value problems for ordinary linear differential equations; boundary value problems; Fourier analysis. Mailing Add: Dept of Math 273 Altgeld Hall Univ of Ill Urbana IL 61801

BENZINGER, JAMES ROBERT, b Buffalo, NY, Nov 26, 22; m 48; c 3. PLASTICS CHEMISTRY. Educ: Canisius Col, BS, 48, MS, 54. Prof Exp: Plant chemist, 48-53, res chemist, 53-56, group leader, 56-58, mgr plastics res & develop, 58-67, MGR ADVAN DEVELOP, SPAULDING FIBRE CO, INC, 67- Mem: Am Chem Soc; Soc Plastics Eng. Res: Product and process development of thermosetting laminates and filamant wound structures for the electrical insulation industry; chemistry of high polymers and adhesives, including phenolics, epoxies, melamines, polyurethanes and polyesters. Mailing Add: Spaulding Fibre Co Inc Tonawanda NY 14127

BENZINGER, ROLF HANS, b Rostock, Ger, Dec 4, 35; m 62; c 1. BIOCHEMICAL GENETICS. Educ: Johns Hopkins Univ, BA, 56, PhD(biochem genetics), 61. Prof Exp: Ger Res Asn fel, Max Planck Inst Biochem, 61-65; State of Geneva assistantship, Lab Biochem Genetics, Geneva, 65-67; asst prof biol, 67-70, ASSOC PROF BIOL, UNIV VA, 70- Concurrent Pos: Ctr Advan Studies res grant, Univ Va, 67-68; NIH res grants, 68-76; NIH res career develop award, 71-76. Res: Application of infectious nucleic acid assays; origin and fate of multilength filamentous phage DNA; injection of T5 phage DNA and control over host processes. Mailing Add: Dept of Biol Gilmer Hall Univ of Va Charlottesville VA 22903

BENZINGER, WILLIAM DONALD, b Pittsburgh, Pa, Feb 6, 40; m 62; c 2. INDUSTRIAL CHEMISTRY. Educ: Univ Notre Dame, BS, 61; Pa State Univ, PhD(inorg chem), 67. Prof Exp: Res chemist, Res Labs, US Army Edgewood Arsenal, Md, 67-68; SR RES CHEMIST, PENNWALT CORP, 69- Mem: Am Chem Soc. Res: Coordination chemistry; inorganic polymers; fiber reinforced composites; coatings; electrochemistry; industrial processes; membrane technology; ultrafiltration; waste treatment. Mailing Add: Pennwalt Corp 900 First Ave King of Prussia PA 19406

BEN-ZVI, EPHRAIM, b Otynia, Poland, Nov 14, 22; nat US; m 50; c 1. PHYSICAL CHEMISTRY. Educ: Israel Inst Technol, BS, 52, ChemE, 53; Univ Calif, Davis, PhD(chem), 60. Prof Exp: Res fel chem, Calif Inst Technol, 60-61; from asst prof to assoc prof, 61-71, PROF CHEM, IMMACULATE HEART COL, 71- Mem: Am Chem Soc. Res: Kinetics of redox reactions in aqueous solutions; coordination compounds; reactions of the oxy-acids of phosphorus; ESR studies of trapped free radicals. Mailing Add: Dept of Chem Immaculate Heart Col Los Angeles CA 90027

BEOUGHER, ELTON EARL, b Gove, Kans, Mar 22, 40; m 60; c 2. MATHEMATICS. Educ: Ft Hays Kans State Col, BS, 61, MA, 64; Univ Mich, PhD(math educ), 68. Prof Exp: Teacher, Winona Consol Schs, Kans, 61-64 & Garden City Schs, 64-65; asst prof, 68-72, ASSOC PROF MATH, FT HAYS KANS STATE COL, 72-, CHMN DEPT, 73- Mem: Math Asn Am. Res: Mathematical education and curriculum; psychology of learning mathematics. Mailing Add: Dept of Math Ft Hays Kans State Col Hays KS 67601

BERAHA, LOUIS, b New York, NY, June 7, 26; m 54; c 2. PLANT PATHOLOGY. Educ: Cornell Univ, BS, 49; Univ Ill, MS, 51, PhD(plant path), 54. Prof Exp: Plant pathologist, Agr Res Serv, 54-55, biol sci br, Agr Mkt Serv, 55-65, invests leader, Mkt Qual Res Div, Agr Res Serv, 65-71, RES LEADER & LOCATION LEADER, MKT PATH LAB, N CENT REGION, AGR RES SERV, USDA, 71- Concurrent Pos: Res assoc, Univ Chicago, 62- Mem: Am Phytopath Soc. Res: Post harvest diseases and control; radiation preservation of food; physiology of parasitism; fungal genetics. Mailing Add: Agr Res Serv USDA Rm 183 536 S Clark St Chicago IL 60605

BERAN, DONALD WILMER, b Wheatland, Wyo, Aug 15, 35; m 56; c 1. METEOROLOGY. Educ: Utah State Univ, BSc, 58; Colo State Univ, MSc, 66; Univ Melbourne, PhD(meteorol), 70. Prof Exp: Engr-meteorologist, Martin Co, 62-64; res asst meteorol, Colo State Univ, 64-66; sr meteorologist, Allied Res Assocs, 66-67; res meteorologist, 70-74, SUPVRY RES METEOROLOGIST, WAVE PROPAGATION LAB, ENVIRON RES LAB, NAT OCEANIC & ATMOSPHERIC ADMIN, 74- Mem: Am Meteorol Soc; Sigma Xi; Acoust Soc Am. Res: Satellite meteorology; lee waves and mountain winds; clear air turbulence and its detection; development of acoustic sounding methods for remote sensing of boundary layer phenomena. Mailing Add: Environ Res Lab R45x8 Nat Oceanic & Atmospheric Admin Boulder CO 80302

BERAN, GEORGE WESLEY, b Riceville, Iowa, May 22, 28; m 54; c 3. VETERINARY PUBLIC HEALTH. Educ: Iowa State Univ, DVM, 54; Kans Univ, PhD(med microbiol), 59. Hon Degrees: LHD, Silliman Univ, Philippines, 73. Prof Exp: Epidemic intel serv officer, USPHS, 54-56; prof microbiol, Silliman Univ, Philippines, 60-73, dir, Van Houweling Lab, 61-73; PROF VET MICROBIOL, COL VET MED, IOWA STATE UNIV, 73- Concurrent Pos: Consult, Diamond Labs, 69, 74 & Wildlife Vaccines, Inc, 74; dir, Negros Oriental Prov Lab, Philippines, 70-73; consult vet pub health, WHO, UN, 71, consult rabies, 71-76. Honors & Awards: Adopted Son, Prov of Negros Oriental, Philippines, 73; Recognition Award, Secy of Health, Philippines, 73. Mem: Fel Royal Soc Health; Am Asn Food Hyg Veterinarians; Asn Teachers Vet Pub Health; Am Vet Med Asn. Res: Epidemiology and immunology of rabies; epidemiology of pseudorabies; epidemiology of California encephalitis; microbiology of anaerobic waste handling systems. Mailing Add: Vet Microbiol & Prev Med Iowa State Univ Col of Vet Med Ames IA 50011

BERAN, JO ALLAN, b Odell, Nebr, Aug 24, 42; m 64; c 2. ENVIRONMENTAL CHEMISTRY. Educ: Hastings Col, BA, 64; Univ Kans, PhD(chem), 68. Prof Exp: Asst prof, 68-71, ASSOC PROF CHEM, TEX A&I UNIV, 71-, CHMN DEPT, 75- Mem: Am Chem Soc. Res: Gas phase studies of ions produced by photoionization; investigations in chemical education. Mailing Add: 503 Alexander Kingsville TX 78363

BERAN, MARK JAY, b New York, NY, Aug 19, 30; m 53; c 3. PHYSICS, ENGINEERING SCIENCE. Educ: Mass Inst Technol, SB, 52; Harvard Univ, SM, 53, PhD, 55. Prof Exp: Hydrodynamicist, Hydrodyn Lab, Mass Inst Technol, 51-55; group leader math, Waltham Airborne Systs Lab, Radio Corp Am, 55-56; res physicist, Air Force Cambridge Res Ctr, 56-59 & Tech Opers, 59-61; from assoc prof to prof mech eng, Univ Pa, 61-74; PROF ENG, TEL-AVIV UNIV, 74- Concurrent Pos: Fulbright scholar, Dept Electronics, Weizmann Inst Sci, Israel, 67-68. Mem: Fel Optical Soc Am; Acoust Soc Am; Am Phys Soc. Res: Underwater acoustics; applied physics; fluid mechanics; electromagnetic theory; plasma physics; statistical continuum theory; coherence theory. Mailing Add: ch of Eng Tel-Aviv Univ Tel-Aviv Israel

BERANBAUM, SAMUEL LOUIS, b Toronto, Ont, Can, Apr 20, 15; US citizen; m 49; c 3. RADIOLOGY. Educ: Univ Toronto, BA, 37, MD, 40. Prof Exp: Intern, Mt Sinai Hosp, Cleveland, 40-42; resident radiol, Postgrad Hosp, New York, 42-45; instr, Columbia Univ, 45-48; from asst clin prof to assoc clin prof, 48-60, CLIN PROF RADIOL, NY UNIV, 60-, ATTEND RADIOLOGIST, UNIV HOSP, BELLEVUE MED CTR, 55- Concurrent Pos: Assoc attend radiologist, Postgrad Hosp, New York, 45-48; dir radiol, St Barnabas Hosp for Chronic Dis, 45-70; consult, St Lukes Hosp, 70-; consult radiology, Vet Admin Hosp. Mem: AAAS; fel Am Col Radiol; Radiol Soc NAm; Am Roentgen Ray Soc; Asn Gastro-Intestinal Radiologists. Res: Special procedures in roentgen diagnostics; fluoroscopy and radiography of the gastro-intestinal tract; gastric volvulus; elusive abdominal tumors. Mailing Add: 121 E 60th St New York NY 10022

BERANEK, LEO LEROY, b Solon, Iowa, Sept 15, 14; m 41; c 2. ACOUSTICS, COMMUNICATIONS. Educ: Cornell Col, AB,36; Harvard Univ, MS, 37, ScD(physics), 40. Hon Degrees: ScD, Cornell Col, 46; DEng, Worcester Polytech Inst, 71. Prof Exp: From instr to asst prof physics & communication eng, Harvard Univ, 40-43, dir electro-acoustic lab, 43-46 & systs res lab, 45-46; Guggenheim Mem Found fel, Harvard Univ & Mass Inst Technol, 46-47; assoc prof communications eng, Mass Inst Technol, 47-58, tech dir acoustics lab, 47-53; PRES, BOSTON BROADCASTERS, INC, 63- Concurrent Pos: Pres, Bolt, Beranek & Newman, Inc, 53-69, dir, 53-, chief scientist, 69-; trustee, Cornell Col, 56-71; lectr, Mass Inst Technol, 58-; chmn bd, Mueller-BBN GmbH, Munich, 62-; mem, Aeronaut & Space Eng Bd, Nat Acad Eng, 65-, comn pub eng policy, 63-70, mem coun, 69-71; vchmn, Mass Comn Ocean Mgt; mem vis comt, Dept Soc Rels & Psychol, Harvard Univ, 64-70, dept biol, 71-, adv comt mgt develop progs, Grad Sch Bus Admin, 66-70, advan mgt prog, 65. Honors & Awards: Biennial Award, Acoust Soc Am, 44; President's Cert Merit, 48; Wallace Clement Sabine Award, Acoust Soc Am, 61; Gold Medal, Audio Eng Soc, 71; Gold Medal, Acoust Soc Am, 75. Mem: Nat Acad Eng; fel AAAS; fel Acoust Soc Am (vpres, 49-50, pres, 54-55); fel Audio Eng Soc (pres, 67-68); fel Am Acad Arts & Sci. Res: Architectural acoustics; electromechanico-acoustical devices; acoustic measurements; noise control techniques; technology assessment. Mailing Add: 7 Ledgewood Rd Winchester MA 01890

BERARD, ANTHONY D, JR, b Lynn, Mass, Sept 23, 42; m 66. TOPOLOGY. Educ: The Citadel, BS, 64; Case Western Reserve Univ, MA, 67, PhD(math), 68. Prof Exp: Asst prof math, US Air Force Inst Technol, 68-72; ASSOC PROF MATH, KING'S COL, 72- Mem: Am Math Soc; Math Asn Am. Res: Characterization of metric spaces by the use of their midsets. Mailing Add: Dept of Math King's Col Wilkes-Barre PA 18702

BERARD, COSTAN WILLIAM, b Cranford, NJ, Dec 23, 32; m 58; c 2. MEDICINE, PATHOLOGY. Educ: Princeton Univ, AB, 55; Harvard Univ, MD, 59. Prof Exp: Intern surg, Univ Rochester, 59-60; res assoc, Walter Reed Inst Res, 60-62, dept chief, 62-63; resident, 63-66, STAFF PATHOLOGIST, NAT CANCER INST, 66- Mem: AMA; Am Soc Exp Path. Res: Metabolic response to cancer with particular reference to immunology; metabolic response to trauma with particular reference to wound healing. Mailing Add: Path Anat Br NIH Lab of Path Bethesda MD 20014

BERARD, RAYMOND ANDRE, physical chemistry, polymer chemistry, see 12th edition

BERARDI, LEAH CASTILLON, b Gulfport, Miss, Oct 14, 21; m 54. PROTEIN CHEMISTRY. Educ: Tulane Univ, BS, 41. Prof Exp: Chemist, Censorship Lab, US War Dept, 43-45; RES CHEMIST, SOUTHERN REGIONAL RES CTR, USDA, 45- Mem: Sigma Xi; Inst Food Technologists; Am Oil Chemists Soc; Am Asn Cereal Chemists; Phytochem Soc NAm. Res: Proteins of cottonseed and other oilseeds; preparation of edible protein products from oilseeds; elucidation of chemical, physical and structural characteristics of oilseed protein products. Mailing Add: Southern Regional Res Ctr USDA PO Box 19687 New Orleans LA 70179

BERARDINELLI, FRANK MICHAEL, b Newark, NJ, June 6, 20; m 57; c 2. POLYMER CHEMISTRY. Educ: Seton Hall Univ, BS, 43. Prof Exp: Chemist, 43-47, res chemist, 47-49, sr res chemist, 49-60, group leader polyacetal res & develop, 60-74, RES SUPVR, CELANESE RES CO, 74- Mem: Am Chem Soc; Sigma Xi. Res: Synthesis and processing of cellulosic plastics; textile finishes and plasticizer synthesis and evaluation; polyethelene film studies; emulsion, condensation and ionic polymerization of polymers; polyacetals; composites; high temperature polymers. Mailing Add: Celanese Corp of Am Box 1000 Summit NJ 07901

BERBEE, JOHN GERARD, b Hamilton, Can, Oct 12, 25; m 50; c 3. PLANT PATHOLOGY, FORESTRY. Educ: Univ Toronto, BScF, 49; Yale Univ, MF, 50; Univ Wis, PhD(plant path), 54. Prof Exp: Asst forest biologist, Can Dept Agr Sci Serv, 54-57; assoc prof plant path, 57-69, PROF PLANT PATH & FORESTRY, UNIV WIS-MADISON, 69- Concurrent Pos: Sr lectr plant sci, Univ Ife, Nigeria, 64-67. Mem: Am Phytopath Soc; Soc Am Foresters. Res: Disease of forest trees and breeding for disease resistance; forest tree virology; tropical plant pathology. Mailing Add: Dept of Plant Path Univ of Wis Madison WI 53706

BERBERIAN, STERLING KHAZAG, b Waukegan, Ill, Jan 15, 26; m 61; c 2. MATHEMATICAL ANALYSIS. Educ: Mich State Univ, BS, 48, MS, 50; Univ Chicago, PhD(math), 55. Prof Exp: Instr math, Fisk Univ, 50, Southern Ill Univ, 51-52 & Univ Ill, 52-53; asst prof, Mich State Univ, 55-57; from asst prof to prof, Univ Iowa, 57-66; ed, Math Rev, Ann Arbor, Mich, 66-68; PROF MATH, UNIV TEX, AUSTIN, 68- Concurrent Pos: Vis prof, Ind Univ, Bloomington, 70-71. Mem: Am Math Soc; Math Asn Am. Res: Topological algebraic structures; Hilbert space; integration theory; representations of locally compact groups; operator algebras. Mailing Add: Dept of Math Univ of Tex Austin TX 78712

BERBLINGER, KLAUS WILLIAM, b Zurich, Switz, Sept 8, 10; US citizen; m 74; c 2. PSYCHIATRY, PSYCHOSOMATIC MEDICINE. Educ: Univ Munich, MD, 34; Univ Berne, MD, 36. Prof Exp: Instr psychiat, Sch Med, Duke Univ, 49-51; from asst prof to assoc prof, Med Sch, Univ Md, 51-58; assoc prof, 58-65, PROF PSYCHIAT, SCH MED, UNIV CALIF, SAN FRANCISCO, 65- Concurrent Pos: Fel radiol, Univ Berne, 36; fel tuberc, Swiss Res Inst, Zurich, 37; fel psychiat & neurol, Univ Clin Neuropathy, Zurich, 38; grant psychiat & psychosom med, Duke Hosp, 49-50. Mem: AAAS; corresp fel Ger Neuropsychiat Soc; Int Soc Social Psychiat; Am Psychiat Asn; Am Col Psychiat. Res: Psychopathology; psychotherapy; psychosomatic medicine; problem of hysteria. Mailing Add: Dept of Psychiat Univ of Calif San Francisco CA 94143

BERCAW, JAMES ROBERT, b Canton, Ohio, Aug 10, 23; m 44; c 4. TEXTILE PRODUCTS. Educ: William Jewell Col, AB, 48; Ohio State Univ, PhD(chem), 54. Prof Exp: Res chemist, 54-57, from res supvr to sr supvr, 57-63, tech supt, 63-65, indust tech mgr, 65-66, asst nylon prod mgr, 66-68, asst tech serv mgr, 68-74, MGR PLANT SERV & TECHNOL, TEXTILE FIBERS DEPT, E I DU PONT DE NEMOURS & CO, INC, 74- Mem: AAAS; Am Chem Soc; Am Burn Asn. Res: Man-made fiber research, development, evaluation and production. Mailing Add: Textile Fibers Dept E I du Pont de Nemours & Co Wilmington DE 19898

BERCH, JULIAN, b Winnipeg, Man, Oct 13, 16; nat US; m 44; c 4. CHEMISTRY. Educ: Univ Wash, BS, 38. Prof Exp: Chemist, Fish & Wildlife Serv, 39-40; chemist, Nat Bur Stands, 42-43; res assoc, Textile Res Found, 43-44 & Harris Res Labs, 45-67; RES SUPVR, GILLETTE RES INST, 67- Mem: AAAS; Am Chem Soc; Am Asn Textile Chem & Colorists. Res: Surface active agents and detergents; technology of non-woven fabrics; mechanism of wool felting and shrinkage control; wool chemistry; textile application of rubber lattices and permanent-press finishes; water pollution. Mailing Add: 2100 Washington Ave Silver Spring MD 20910

BERCHTOLD, GLENN ALLEN, b Pekin, Ill, July 1, 32; m 59; c 3. ORGANIC CHEMISTRY. Educ: Univ Ill, BS, 54; Univ Ind, PhD(org chem), 59. Prof Exp: From instr to assoc prof, 60-69, PROF CHEM, MASS INST TECHNOL, 69-, CHMN DEPT, 71- Mem: Am Chem Soc. Res: Organic synthesis. Mailing Add: Dept of Chem Mass Inst Technol Cambridge MA 02139

BERCOS, JAMES, mathematical statistics, operations research, see 12th edition

BERCOV, RONALD DAVID, b Edmonton, Alta, Dec 14, 37; m 65; c 2. ALGEBRA. Educ: Univ Alta, BSc, 59; Calif Inst Technol, PhD(math), 62. Prof Exp: Res assoc math, Cornell Univ, 62-63; from asst prof to assoc prof, 63-74, PROF MATH, UNIV ALTA, 74- Concurrent Pos: Vis assoc prof, Univ Wash, 69; ed, Can Math Bull, 70. Mem: Am Math Soc; Math Asn Am; Can Math Cong. Res: Group theory, primarily in the field of permutation groups. Mailing Add: Dept of Math Univ of Alta Edmonton AB Can

BERCZ, JENO PETER, organic chemistry, inorganic chemistry, see 12th edition

BERCZI, ISTVAN, b Bekes, Hungary, Nov 12, 38; Can citizen; m 67; c 2. IMMUNOLOGY. Educ: Budapest Vet Sch, Hungary, DVM, 62; Univ Man,

PhD(immunol), 72. Prof Exp: Res scientist microbiol, Vet Med Res Inst, Hungarian Acad Sci, Budapest, 62-67; ASST PROF IMMUNOL, UNIV MAN, 72- Mem: Am Asn Immunologists; Am Asn Cancer Res; Transplantation Soc; Can Soc Immunol. Res: Cancer immunology; regulation of the immune response. Mailing Add: Dept of Immunol Univ of Man Fac of Med Winnipeg MB Can

BERDAHL, JAMES MAYNARD, b Sioux City, Iowa, Apr 12, 32; m 56; c 2. ORGANIC CHEMISTRY. Educ: Antioch Col, BS, 55; Wayne State Univ, MS, 58. Prof Exp: From assoc scientist to sr scientist, Mead Johnson & Co, 59-68; develop chemist, Gen Elec Co, 68-69; PROCESS DEVELOP CHEMIST, ABBOTT LABS, 69- Mem: Am Chem Soc. Res: Sympathomimetic amines; antidepressants; condensation polymers; production support. Mailing Add: 4020 Harper Ave Gurnee IL 60031

BERDAN, JEAN MILTON, b New Haven, Conn, May 9, 16. INVERTEBRATE PALEONTOLOGY. Educ: Vassar Col, AB, 37; Yale Univ, MS, 43, PhD(geol), 49. Prof Exp: Geologist, Ground Water, 42-46, GEOLOGIST, PALEONT & STRATIG, US GEOL SURV, 49- Mem: Geol Soc Am; Am Paleont Soc; Brit Paleont Asn. Res: Lower Paleozoic ostracode faunas. Mailing Add: US Geol Surv Rm E303 US Nat Mus Washington DC 20244

BERDANIER, CAROLYN DAWSON, b East Brunswick, NJ, Nov 14, 36; m 57; c 3. NUTRITIONAL BIOCHEMISTRY. Educ: Pa State Univ, BS, 58; Rutgers Univ, MS, 63, PhD(nutrit), 66. Prof Exp: Res asst home econ, Douglass Col, 62-63; NIH fel animal sci, Rutgers Univ, 66-67; res nutritionist, Human Nutrit Res Div, Agr Res Serv, USDA, 68-75; RES ASST PROF, DEPTS MED & BIOCHEM, COL MED, UNIV NEBR, 75- Concurrent Pos: Asst prof nutrit, Univ Md, 70-75. Mem: AAAS; fel Am Inst Chemists; NY Acad Sci; Am Inst Nutrit; Soc Exp Biol & Med. Res: Hormones and metabolic control mechanisms; nutritient-genetic interactions; carbohydrate nutrition; diet-drug interactions; obesity; diabetes; lipemia. Mailing Add: 1621 Brent Blvd Lincoln NE 68520

BERDANIER, CHARLES REESE, JR, b Knoxville, Pa, Dec 10, 28; m 57; c 3. SOIL MORPHOLOGY, MINERALOGY. Educ: Mansfield State Col, BS, 53; Pa State Univ, BS, 56, MS, 58; Rutgers Univ, PhD(soil genesis & morphol), 67. Prof Exp: Res asst soils, Pa State Univ, 56; soil scientist, US Soil Conserv Serv, NJ, 61-67; soil scientist, World Soil Geog Unit, 67-69, res soil scientist, Soil Surv Lab, 69-71, soil scientist, Soil Surv Invests, Soil Surv Lab, USDA, 71-73, SOIL SCIENTIST, NAT SOIL SURV LAB, USDA, 73- Mem: Soil Sci Soc Am; Int Soc Soil Sci; Am Soc Agron; Crop Sci Soc Am; Soil Conserv Soc Am. Res: Fundamental processes of soil genesis and observations of resultant soil types. Mailing Add: 1621 Brent Blvd Lincoln NE 68520

BERDICK, MURRAY, b New Rochelle, NY, June 27, 20; m 47; c 1. POLYMER CHEMISTRY. Educ: George Washington Univ, BS, 42; Polytech Inst Brooklyn, MS, 49, PhD(polymer chem), 54. Prof Exp: Asst, Am Electroplaters Soc, Bur Stand, Washington, DC, 40-42; test engr, Gen Elec Co, NY, 42-43, chemist, 43-46; proj leader, Evans Res & Develop Corp, 46-51, coord res, 53-60, vpres & dir res, 60-61; res mgr, 62-64, dir, Clinton Labs, 64-69, dir res labs, 69-70, dir appl res, 71-75, DIR REGULATORY AFFAIRS, CHESEBROUGH-POND'S INC, 75- Concurrent Pos: Chmn, Food & Drug Admin-Toilet Goods Asn Sci Liaison Comt; chmn, Inter-Indust Color Comt, 71- Honors & Awards: CIBS Award, Cosmetic, Toiletry & Fragrance Asn, Inc, 72. Mem: Am Chem Soc; Soc Chem Indust; Am Inst Chemists; NY Acad Sci; Soc Invest Dermat. Res: Crosslinking in copolymerization; keratin; thixotropic compositions; interactions of proteins and polyelectrolytes; polymerization; unsaturated polyesters; fiber physics; properties of high polymers; cosmetic and pharmaceutical products; sustained release drugs; transepidermal moisture loss. Mailing Add: Chesebrough-Pond's Inc Res Labs Trumbull Indust Park Trumbull CT 06611

BERDJIS, CHARLES CHOAIB, b Kashan, Iran, July 1, 18; US citizen; m 41; c 2. PATHOLOGY, RADIOBIOLOGY. Educ: Univ Geneva, MD, 40, cert path, 45; Univ Paris, MD, 47. Prof Exp: From intern to resident & asst, Cantonal Hosp, Med Sch, Univ Geneva, 40-45; asst med, Anticancer Ctr, Hotel-Dieu Hosp & Inst Cancer, Paris, France, 46-49; chief, Anticancer Ctr, Tehran, Iran, 50-53; from lectr to asst prof, Sch Med, Univ Calif, San Francisco, 53-57; US Army, 57-, chief path to commanding officer, 4th Army Med Lab, Brooke Army Med Ctr, Tex, 57-59, sr pathologist, Walter Reed Army Med Unit, Ft Detrick, Md, 59-63, chief exp med & biomed depts, Med Res Directorate & Res Labs, Edgewood Arsenal, 64-69, dep comdr, Shape Int Hosp, Belg, 69-70, RES PATHOLOGIST RADIOBIOL & CHIEF EXP NEPHROLOGY, ARMED FORCES INST PATH, US ARMY, 70- Concurrent Pos: Consult, Univ Tehran, 50-53; chief path, AEC, 53-57; mem, Int Cong Radiation. Mem: AMA; Am Soc Exp Path; Radiation Res Soc; Asn Mil Surg US; Am Chem Soc. Res: Cancerology; endocrinology; nutrition; infectious disease; immunology; epidemiology; toxicology. Mailing Add: Armed Forces Inst of Path Washington DC 20305

BERDON, JOHN KENNETH, b Natchez, Miss, Oct 27, 29. PERIODONTICS. Educ: Loyola Univ, La, BS, 51, DDS, 55; Baylor Univ, BSD, 61, MSD, 63. Prof Exp: Asst prof periodont, chmn sect & chmn postgrad periodont, Sch Dent, Univ Detroit, 63-66, chmn postgrad studies, 64-66; prof periodont & chmn dept, Sch Dent, WVa Univ, 66-72; prof & chmn dept, Col Dent Med, Med Univ SC, 72-74; PROF PERIODONT & ASST DEAN STUDENT PROGS, SCH DENT, UNIV MISS, 74- Concurrent Pos: Consult, Vet Admin, WVa, 67- Mem: Am Acad Periodont; Am Dent Asn; Am Acad Oral Med; Brit Soc Periodont; Int Asn Dent Res. Res: Dental hygiene. Mailing Add: Univ of Miss Sch of Dent Univ Med Ctr 2500 N State St Jackson MS 39216

BEREDJICK, NICKY, b Sofia, Bulgaria, Oct 6, 29; m 57. ORGANIC CHEMISTRY, POLYMER CHEMISTRY. Educ: Hebrew Univ, MS, 54; Syracuse Univ, PhD(chem), 57. Prof Exp: Res chemist, Plastic Film Res & Develop Dept, Visking Co Div, Union Carbide Co, 57-58; proj chemist, Res Dept, Stand Oil Co Ind, 58-62; prin res scientist, Ford Motor Co, 62-64; sr tech affairs officer, UN, 64, from chief petrochem & textile industs sect, Div Indust Technol to sr indust develop officer, UN Indust Develop Orgn, 64-72, dep dir, UN Relief Oper Bangladesh, UN Hq New York, 72-74, dir admin & opers, UN Emergency Oper, 74-75, OFFICER-IN-CHG, UN EMERGENCY OPER, 75- Mem: Am Chem Soc. Res: Synthesis and mechanism of stereo-specific polymerization and copolymerizations; coorelations of structure with physical properties in high molecular weight systems; transfer and adaptation of technology to developing countries. Mailing Add: UN Secretariat Bldg UN Plaza New York NY 10017

BEREMAN, ROBERT DEANE, b Clinton, Ind, Oct 2, 43; m 65. INORGANIC CHEMISTRY, BIOINORGANIC CHEMISTRY. Educ: Butler Univ, BS, 65; Mich State Univ, PhD(inorg chem), 69. Prof Exp: Res asst chem, Butler Univ, 61-65; asst inorg chem, Mich State Univ, 65-69; NSF fel phys-inorg chem, Univ Ill, Urbana, 69-70; ASSOC PROF INORG CHEM, STATE UNIV NY BUFFALO, 70- Concurrent Pos: Dreyfus Found fel, 75- Mem: Am Chem Soc. Res: Preparation and characterization by magnetic resonance techniques—nuclear magnetic, electron spin and nuclear quadrupole resonance—of theoretically interesting transition metal complexes; magnetic resonance spectroscopy; bioinorganic model complexes; copper enzymes. Mailing Add: Dept of Chem State Univ of NY Buffalo NY 14214

BEREN, SHELDON KUCIEL, b Marietta, Ohio, Oct 7, 22; m 46, 68; c 7. CHEMISTRY. Educ: Harvard Univ, BS, 44. Prof Exp: Chem engr, AEC, 44-46; dir res & prod, Marco Chem Co, Tex, 46-47; mem staff, Okmar Oil Co, 48-70; CHMN, BEREN CORP, 70- Res: Oils, fats, waxes and soaps. Mailing Add: 2160 First of Denver Plaza 633 17th St Denver CO 80202

BERENBAUM, MORRIS BENJAMIN, b Chicago, Ill, Dec 19, 24; m 46; c 3. POLYMER CHEMISTRY, ORGANIC CHEMISTRY. Educ: City Col New York, BChE, 44; Polytech Inst Brooklyn, PhD(org chem), 51. Prof Exp: Res engr, Stand Brands, Inc, 44, jr engr, Process Res on Food Prods, 44-45; chemist, Stauffer Chem Corp, 45-48, res chemist, 51-60; supvr res sect, Synthetic Rubber, Thiokol Chem Corp, 51-58, mgr res dept, 58-60, dir res dept, 60-62, tech dir, 62-69; tech dir, Res & Develop, Specialty Chem Div, 69-71, VPRES, ALLIED CHEM CORP, 71- Mem: Am Chem Soc; Soc Plastics Eng. Res: Organic sulfur compounds; alkyl polysulfide polymers; urethanes and hydrocarbon polymers; fluorocarbons; specialty chemicals. Mailing Add: Allied Chem Corp Specialty Chem Div PO Box 1087R Morristown NJ 07960

BERENBERG, WILLIAM, b Haverhill, Mass, Oct 29, 15; m 40; c 3. MEDICINE. Educ: Harvard Univ, AB, 36; Boston Univ, MD, 40. Prof Exp: Assoc prof, 53-70, PROF PEDIAT, HARVARD MED SCH, 70-; ASSOC PHYSICIAN IN CHIEF, CHILDRENS HOSP BOSTON, 70- Concurrent Pos: Consult, Mass Gen Hosp, 50-; chief med serv, Childrens Hosp Boston, 53-70; mem, Mass Cerebral Palsy Comn; chmn res coun, Nat United Cerebral Palsy. Honors & Awards: Jacobi Award, 49; Weinstein Award, United Cerebral Palsy, 69; Presidential Medal Merit, Ecuador, 70. Mem: Soc Pediat Res; Am Pediat Soc; Am Acad Pediat; Am Acad Cerebral Palsy; Am Acad Neurol. Res: Handicapped children; pulmonary disease. Mailing Add: 50 Beresford Rd Chestnut Hill MA 02167

BERENBOM, MAX, b Saskatoon, Sask, Sept 4, 19; nat US; m 47; c 2. BIOCHEMISTRY. Educ: Univ Sask, BA, 41, MA, 43; Univ Toronto, PhD(biochem), 47. Prof Exp: Sr res fel, Nat Cancer Inst, 47-49, spec res fel, 49-50, vis scientist, 50-51; assoc oncol, Med Ctr, Univ Kans, 51-54, asst prof oncol & biochem, 54-57; DIR BIOCHEM, MENORAH MED CTR, 57- Mem: Am Chem Soc; fel AAAS; Am Asn Clin Chem. Res: Intermediary metabolism; enzymes and tissue components in normal and abnormal states; radiation effects; carcinogenesis. Mailing Add: Menorah Med Ctr Kansas City MO 64110

BERENDS, ERNEST A, inorganic chemistry, physical chemistry, see 12th edition

BERENDSEN, PETER BARNEY, b Los Angeles, Calif, July 14, 37; m 62; c 3. ANATOMY, CELL BIOLOGY. Educ: St Marys Col, Calif, BS, 60; George Washington Univ, MS, 65, PhD(anat), 72. Prof Exp: Res asst hemat, Armed Forces Inst Path, 60-65; res assoc path, Univ Mich, 65-67; guest worker physiol, NIH, 70-72, res physiologist, 72; instr anat, 72-73, ASST PROF ANAT, COL MED & DENT NJ, 73- Mem: Electron Micros Soc Am; Am Asn Anat; Sigma Xi. Res: Intestinal absorption; lipid metabolism; reticuloendothelial and endothelial function. Mailing Add: Col of Med & Dent of NJ 100 Bergen St Newark NJ 07103

BERENDT, RAYMOND DONALD, b Milwaukee, Wis, Mar 30, 20; m 44; c 2. ACOUSTICS. Educ: Univ Scranton, BS, 49. Prof Exp: Instr math, Temple Univ, 50; physicist, Nat Bur Stand, 50-75; ACOUST CONSULT, 75- Honors & Awards: Silver Medal Award, Dept Com, 69; Award of Merit, Pollution Eng Mag, 72; Spec Achievement Award, Nat Bur Stand, 72. Mem: Acoust Soc Am; Am Soc Testing & Mat. Res: Architectural acoustics; applied acoustics; noise control engineering; noise pollution; building noise control. Mailing Add: 6015 Sherborn Lane Springfield VA 22152

BERENDT, RICHARD FREDERICK, bacteriology, see 12th edition

BERENDZEN, RICHARD EARL, b Walters, Okla, Sept 6, 38; m 64; c 2. ASTRONOMY, HISTORY OF SCIENCE. Educ: Mass Inst Technol, BS, 61; Harvard Univ, MA, 67, PhD, 68. Prof Exp: Res scientist, Geophys Corp Am, 59-64; fel, Harvard Univ, 61-64; from lectr to prof astron, Boston Univ, 65-73, actg chmn dept, 71-72; DEAN COL ARTS & SCI, AM UNIV, WASHINGTON, DC, 74- Concurrent Pos: Res scientist, Astronaut Div, Ling-Temco-Vought Corp, 62-63; staff mem astron ed, Proj Physics, Harvard Univ, 65-66; mem comn teaching astron & comn hist sci of Int Astron Union; consult, Educ Coop, Natick, Mass; adv, NAm Fedn Planetarium Educators; ed, J Col Sci Teaching, Nat Sci Teachers Asn; mem Astron Surv Comt, Nat Acad Sci; astron adv bd, Ctr Hist & Philos Physics, Am Inst Physics; invited partic sem, Int Union Pure & Appl Physics, 70; arranger & chmn, Int Conf Educ & Hist Mod Astron, Am Astron Soc-NY Acad Sci; dir, Nat Sci Found prog, Case Studies Proj Develop Mod Astron; arranger & chmn, Conf Life Beyond Earth & Mind of Man, NASA, Boston, 72; consult, Off Acad Affairs, Am Coun Educ & Space Sci Bd, Nat Acad Sci, 73-74; mem energy task force, Wash Bd Trade, 74-75 & comts col sci teaching & publ of Nat Sci Teachers Asn, 74-; evaluator, Mid States Asn Cols, 75. Mem: Fel AAAS; Am Astron Soc; Am Hist Sci Soc; Am Asn Higher Educ; Nat Sci Teachers Asn. Res: Astronomy education; sociology of science; science education and planning; academic administration. Mailing Add: The Am Univ Washington DC 20016

BERENS, ALAN PAUL, b Cincinnati, Ohio, June 15, 34; m 57; c 5. MATHEMATICAL STATISTICS. Educ: Univ Dayton, BS, 55; Purdue Univ, MS, 57, PhD(math statist), 63. Prof Exp: Instr math, Purdue Univ, 58-62; head statist dept, Technol Inc, Ohio, 62-68; vpres, Beta Industs, 68-69; RES STATISTICIAN, UNIV DAYTON RES INST, 69- Mem: Inst Math Statist; Am Statist Asn. Res: Application of mathematical statistics and probability to physical science research; reliability. Mailing Add: Univ of Dayton Res Inst Dayton OH 45409

BERENS, ALAN ROBERT, b Oak Park, Ill, Sept 28, 25; m 49; c 2. POLYMER SCIENCE. Educ: Harvard Univ, AB, 47; Case Western Reserve Univ, MS, 49, PhD(chem), 51. Prof Exp: Res chemist, 50-60, from res assoc to sr res assoc, 60-71, RES FEL POLYMERIZATION, B F GOODRICH CO, 71- Mem: Am Chem Soc. Res: Vinyl polymerization; polymer structure and properties; diffusion in polymers. Mailing Add: Res Ctr B F Goodrich Co Brecksville OH 44141

BERENSON, GERALD SANDERS, b Bogalusa, La, Sept 19, 22; m 51; c 4. INTERNAL MEDICINE. Educ: Tulane Univ, BS, 43, MS, 45. Prof Exp: Asst med, Sch Med, Tulane Univ, 48-49, instr, 49-52; USPHS fel pediat, Univ Chicago, 52-54; from asst prof to assoc prof, 54-63, PROF MED, SCH MED, LA STATE UNIV MED CTR, NEW ORLEANS, 63-, DIR SPECIALIZED CTR RES IN ARTERIOSCLEROSIS, 72- Mem: Am Chem Soc; Soc Exp Biol & Med; Am Soc Biol Chem; fel Am Col Physicians; fel Am Col Cardiol. Res: Cardiology; atherosclerosis; biochemistry of connective tissues. Mailing Add: Dept of Med La State Univ Med Ctr New Orleans LA 70112

BERENSON, LEWIS JAY, b Boston, Mass, Oct 8, 32; m 56; c 4. MATHEMATICS EDUCATION. Educ: Yeshiva Col, BA, 53, Yeshiva Univ, MS, 54; Columbia Univ, PhD(math educ), 61. Prof Exp: Teacher parochial & pub schs, NY, 54-59; instr math & math educ, Hofstra Univ, 59-62; from assoc prof to prof math, Nassau Community Col, 62-72; ASSOC PROF TECHNOL EDUC, CTR TECHNOL EDUC, TEL AVIV UNIV, 72- Concurrent Pos: NSF fac fel, Courant Inst Math Sci, NY Univ, 64-65; lectr, Grad Dept Educ, Brooklyn Col, 67-68; vis prof, Dept Math & Sci Educ, Belfer Grad Sch Sci, Yeshiva Univ, 68-69; mem, Comt In-Serv Educ Teachers of Math & Comt Adaptation of Elem Sch Math to Needs of Cult Deprived & guest lectr, Ministry of Educ & Cult Israel, 69-70. Mem: Math Asn Am; Nat Coun Teachers Math. Res: Mathematics curricula for vocational schools; mathematics for the college liberal arts student; mathematical preparation of elementary school teachers. Mailing Add: Ctr Technol Educ Tel Aviv Univ 52 Golomb St Holon Israel

BERENT, STANLEY, b Norfolk, Va, Mar 10, 41; c 2. CLINICAL PSYCHOLOGY, NEUROSCIENCE. Educ: Old Dom Univ, BS, 66; Va Commonwealth Univ, MS, 68; Rutgers Univ, PhD(clin psychol), 72. Prof Exp: Instr psychol, Va Commonwealth Univ, 67-68; res asst psychiat, Rutgers Med Sch, 68-71; intern psychol, NIHM, St Elizabeths Hosp, 71-72; ASST PROF PSYCHIAT, MED SCH, UNIV VA, 72-, ASST PROF, DEPT PSYCHOL, 75- Concurrent Pos: Consult, NIMH, St Elizabeths Hosp, 72-75. Mem: Am Psychol Asn; Soc Neurosci; Int Soc Neuropsychol. Mailing Add: Box 203 Univ of Va Med Ctr Charlottesville VA 22901

BERENYI, NICHOLAS MIKLOS, soil science, see 12th edition

BERES, JOHN JOSEPH, b Beaver Falls, Pa, May 1, 47; m 75. PHYSICAL ORGANIC CHEMISTRY. Educ: Geneva Col, BS, 69; Carnegie-Mellon Univ, MS, 73, PhD(phys chem), 75. Prof Exp: FEL POLYMERS, UNIV MASS, 75- Mem: Am Chem Soc; Am Crystallog Asn; Am Sci Asn; Sigma Xi. Res: X-ray structural investigations of polymers; structure-property relationships in polymers, especially polyurethanes and polyphosphazenes. Mailing Add: Army Mat & Mech Res Ctr Watertown MA 02172

BERES, WILLIAM PHILIP, b Peabody, Mass, Jan 8, 36; m 66; c 3. NUCLEAR PHYSICS. Educ: Mass Inst Technol, BS, 59, PhD(physics), 64. Prof Exp: Asst physics, Mass Inst Technol, 60-64; res physicist, GCA Corp, 64; res assoc physics, Univ Md, 64-66; asst prof, Duke Univ, 66-69; assoc prof, 69-75, PROF PHYSICS, WAYNE STATE UNIV, 75- Mem: Am Phys Soc; Am Asn Physics Teachers. Res: Theory of nuclear structure and reactions; microscopic nuclear calculations; quasiparticles; local and non-local potentials; photonuclear reactions. Mailing Add: Dept of Physics Wayne State Univ Detroit MI 48202

BERESFORD, WILLIAM ANTHONY, b London, Eng, Aug 13, 36; m 57; c 3. HISTOLOGY, NEUROANATOMY. Educ: Oxford Univ, BA, 59, DPhil(histol), 63. Prof Exp: Lectr histol, Univ Liverpool, 63-65; asst prof anat, Am Univ Beirut, 65-68; asst prof, 68-70, ASSOC PROF ANAT, WVA UNIV, 70- Mem: Am Asn Anat; Anat Soc Gt Brit & Ireland; Ger Anat Soc; Int Asn Dent Res. Res: Factors influencing fracture repair; ectopic bone and cartilage; effects of vitamin A and steroid hormones on bone and cartilage; bone induction. Mailing Add: Dept of Anat WVa Univ Med Ctr Morgantown WV 26506

BERESNIEWCZ, ALEKSANDER, b Lithuania, Apr 5, 27; nat US; m 54; c 3. PHYSICAL CHEMISTRY. Educ: Marquette Univ, BS, 51; Univ Ill, PhD(chem), 54. Prof Exp: Asst, Univ Ill, 51-53; res chemist, 54-68, RES ASSOC, E I DU PONT DE NEMOURS & CO, INC, 68- Res: Physical chemistry of high polymers; relation between the structure of polymers and their physical properties. Mailing Add: 2501 Elmdale Lane Wilmington DE 19810

BERESTON, EUGENE SYDNEY, b Baltimore, Md, Feb 21, 14; m 42; c 3. DERMATOLOGY, MYCOLOGY. Educ: Johns Hopkins Univ, AB, 33; Univ Md, MD, 37; Univ Pa, MSc, 45, DSc(dermat), 55. Prof Exp: Intern, Conemaugh Valley Mem Hosp, Johnstown, Pa, 37-38 & Mercy Hosp, Baltimore, 38-39; asst dermat, Skin & Cancer Unit, NY Postgrad Hosp, Columbia Univ, 40-41; resident, Montefiore Hosp, NY, 40-41; asst dermatologist, Outpatient Dept, Univ Hosp, 46-50, instr, Sch Med, 46-47, assoc, 47-52, from asst prof to assoc prof dermat, 52-72, PROF MED IN DERMAT, SCH MED, UNIV MD, BALTIMORE CITY, 72-, DERMATOLOGIST, UNIV HOSP, 50- Concurrent Pos: Asst, Sch Med, Johns Hopkins Univ, 46-50, instr, 50-60, chief fungus lab, Hosp, 46-51, dermatologist, 50-60; attend dermatologist, Sinai Hosp, 47-; dermatologist, Mercy Hosp, 50-67, chief dermatologist, 67-; asst investr, US Dept Army Fungus Res Proj, 51-57; consult dermatologist, US Vet Admin Hosp, Baltimore & Ft Howard; consult dermatologist, Spring Grove State Hosp, 47- Mem: Fel AMA; fel Am Col Physicians; fel Am Acad Dermat; fel Royal Soc Health. Res: Aspergillus infection of nails; nutritional requirements of microsporum group of fungi and Trichophyton Tonsurans; fluorescent compound in fungus infected hairs. Mailing Add: 22 E Eager St Baltimore MD 21202

BERETS, DONALD JOSEPH, b New York, NY, July 6, 26; m 56; c 2. PHYSICAL CHEMISTRY. Educ: Harvard Univ, AB, 46, AM, 47, PhD(chem), 49. Prof Exp: Res chemist, 49-54, group leader, 54-60, sect mgr, Energy Conversion Res, 60-67, mgr mat sci sect, 67-73, MGR CATALYST RES, CHEM RES DIV, AM CYANAMID CO, 73- Concurrent Pos: Assoc, Mass Inst Technol, 49. Mem: Am Chem Soc; Am Phys Soc; AAAS. Res: New product development; catalysis; surface chemistry; pigments; solid state physics and chemistry. Mailing Add: Chem Res Div Am Cyanamid Co Stamford CT 06904

BERETSKY, IRWIN, b New York, NY, Jan 7, 35; m 56; c 4. BIOMEDICAL ENGINEERING, BIOACOUSTICS. Educ: City Univ New York, BChE, 57; Rensselaer Polytech Inst, MSME, 64; State Univ NY Upstate Med Ctr, MD, 68. Prof Exp: Engr, Combustion Eng Inc, 56-58, Alco Prod Inc, 58-60 & Gen Elec Missile & Space Vehicles, 60-61; analyst heat transfer & fluid dynamics, Mechanical Tech Inc, 61-64; dir med eng, Syracuse Univ Res Corp, 65-68; DIR BIOMED RES, TECHNICON INSTRUMENTS CORP, 71-; ASST PROF PATH, NY MED COL, 73-; ASST PROF MED, ALBERT EINSTEIN COL MED, 76- Concurrent Pos: Chief emergency serv, Ramapo Gen Hosp, 74-; fac mem, Am Heart Asn NY, 74-; prin investr, Nat Heart & Lung Inst Res Contract Ultrasound Diag Atherosclerosis, 75- Mem: Am Inst Ultrasound in Med; Am Col Emergency Physicians; Critical Care Soc; AMA; Am Acad Family Pract. Res: Research and development of diagnostic ultrasound for medical use, particularly research concerned with improvements in the diagnostic capability and a program devoted to tissue characterization. Mailing Add: Technicon Instruments Corp 511 Benedict Ave Tarrytown NY 10591

BEREZNEY, RONALD, b New York, NY, Dec 25, 43. CELL BIOLOGY, BIOCHEMISTRY. Educ: Fairleigh Dickinson Univ, BS, 66; Purdue Univ, PhD(membrane biochem), 71. Prof Exp: Res asst molecular biol, Inst Cancer Res, Col Physician & Surgeons, Columbia Univ, 65-66; fel cell biol, Univ Freiburg, Ger, 71-72; res assoc pharmacol, Sch Med, Johns Hopkins Univ, 72-75; ASST PROF CELL & MOLECULAR BIOL, STATE UNIV NY BUFFALO, 75- Mem: Am Soc Cell Biol. Res: Isolation and analysis of mammalian nuclei, nuclear membranes and the nuclear protein matrix in normal and cancer cells; DNA replication, RNA transport; relationship of the nuclear membrane to endoplasmic reticulum. Mailing Add: Div of Cell & Molecular Biol State Univ of NY Buffalo NY 14214

BERG, ARTHUR R, b Clay Center, Kans, Feb 9, 37; m 61; c 3. PLANT MORPHOLOGY. Educ: Tex Tech Univ, BS, 60; Univ Calif, Davis, PhD(bot), 66. Prof Exp: Plant physiologist, Pac Southwest Forest & Range Exp Sta, US Forest Serv, 66-72; LECTR BOT, UNIV ABERDEEN, 72- Mem: Bot Soc Am; Soc Develop Biol. Res: Plant morphogenesis; growth regulator physiology; developmental anatomy and morphology. Mailing Add: Dept of Bot Univ Aberdeen St Machar's Dr Aberdeen Scotland

BERG, CLAIRE M, b Mt Vernon, NY, Apr 24, 37. GENETICS. Educ: Cornell Univ, BS, 59; Univ Chicago, MS, 62; Columbia Univ, PhD(genetics), 66. Prof Exp: NATO fel, Med Res Coun-Microbiol Gen Res Unit, Hammersmith Hosp, London, Eng, 66-67; Jane Coffin Childs fel, Molecular Biol Lab, Univ Geneva, Switz, 67-68; asst prof, 68-72, ASSOC PROF BIOL, UNIV CONN, 72- Mem: Genetics Soc Am; Am Soc Microbiol; Fedn Am Scientists; AAAS. Mailing Add: Biol Sci Group Univ of Conn Storrs CT 06268

BERG, CLARENCE PETER, b Mead, Nebr, Aug 30, 00; m 27; c 3. BIOCHEMISTRY. Educ: Augustana Col, AB, 24; Univ Ill, MA, 25, PhD(biochem), 29. Hon Degrees: LLD, Augustana Col, 60. Prof Exp: From instr to prof, 29-68, EMER PROF BIOCHEM, UNIV IOWA, 68- Concurrent Pos: Vis prof, Med Sch, Univ Tenn, 48 & 68 & State Univ NY Col Med Buffalo, 70. Honors & Awards: Iowa Award, Am Chem Soc, 63. Mem: AAAS; Am Soc Biol Chem; Am Chem Soc; fel Am Inst Nutrit; Soc Exp Biol & Med. Res: Resolution of amino acids; influence of optical isomerism on the metabolism of the amino acids and their use for maintenance and growth; intermediary metabolism of tryptophan, methionine, lysine and histidine. Mailing Add: Univ Iowa 4-432 Basic Sci Bldg Dept of Biochem Iowa City IA 52242

BERG, CLIFFORD OSBURN, b Stoughton, Wis, Aug 9, 12; m 40; c 2. FRESH WATER BIOLOGY. Educ: Luther Col, AB, 34; Univ Mich, MS, 39, PhD(zool), 49. Hon Degrees: ScD, Luther Col, 70. Prof Exp: From asst prof to assoc prof, Ohio Wesleyan Univ, 47-53; assoc prof entom, 53-59, PROF ENTOM, CORNELL UNIV, 59- Concurrent Pos: NIH grant, 54-67; NSF grants, 59-60 & 62-; Fulbright scholar, Brazil, 67; Guggenheim fel, SAm, 66 & 67; resident ecologist, Smithsonian Inst, 70-71. Mem: Ecol Soc Am; Entom Soc Am; Am Soc Limnol & Oceanog; Am Micros Soc; Entom Soc Can. Res: Life cycles and ecological relationships of aquatic invertebrates; biology and ecology of snail-killing flies. Mailing Add: Dept of Entom Cornell Univ Ithaca NY 14853

BERG, CLYDE C, b Meriden, Kans, Nov 2, 36; m 58; c 3. PLANT BREEDING, PLANT GENETICS. Educ: Kans State Univ, BS, 58; Okla State Univ, MS, 61; Wash State Univ, PhD(genetics), 65. Prof Exp: Asst prof agron & genetics & asst agronomist, Wash State Univ, 65-66; RES GENETICIST, AGR RES SERV, USDA, 66- Mem: AAAS; Am Soc Agron; Crop Sci Soc Am; Genetics Soc Am; Genetics Soc Can. Res: Genetics, cytogenetics and breeding of Festuca arundinacea, Festuca pratnesis, Lolium perenne, Lolium multiforum and Dactylis glomerata; Bromus inermis; Phleumpratense. Mailing Add: US Regional Pasture Res Lab University Park PA 16802

BERG, DANA B, b Chicago, Ill, July 8, 21; m 52; c 1. SCIENCE WRITING. Educ: Ill Inst Technol, BS, 43. Prof Exp: Res & develop engr, Linde Air Prods Lab, NY, 43-48; from asst ed to ed, Putman Pub Co, 48-66, chmn ed bd, 66-67, staff consult, 67-68; SR TECH ED, ARGONNE NAT LAB, 68- Mem: Am Inst Chem Eng; Am Nuclear Soc; Soc Tech Communication. Res: Technical writing; editing; staff administration. Mailing Add: 1629 77th Ave Elmwood Park IL 60635

BERG, DWIGHT HILLIS, b Montrose, Pa, Jan 6, 16; m 43; c 2. PLANT MORPHOLOGY. Educ: Mansfield State Teachers Col, BS, 47; Cornell Univ, MS, 48, PhD(plant morphol), 50. Prof Exp: Teacher pub schs, Pa, 37-42 & 46-47; asst bot, Cornell Univ, 49-50; assoc prof biol, 50-64, PROF BIOL, HIRAM COL, 64- Mem: AAAS; Bot Soc Am. Res: Morphology of the gametophytes; fertilization and proembryo of Araucaria Bidwilli. Mailing Add: Dept of Biol Hiram Col Hiram OH 44234

BERG, EDUARD, b Trier, Ger, Nov 9, 28. GEOPHYSICS. Educ: Univ Saarlandes, dipl phys, 53, Dr rer nat(physics), 55. Prof Exp: Mem, Inst Sci Res Cent Africa, Congo, 55-63; from assoc prof to prof geophys & Geol, Univ Alaska, 63-72, dir seismol prog, 63-72; GEOPHYSICIST, HAWAII INST GEOPHYS, 72- Concurrent Pos: Res mem fac, Univ Bonn, 57 & Univ Calif, Berkeley, 61; seismol consult, Mining Union Upper Katanga, 60-61; head seismic & volcanic depts & time serv, Inst Sci Res Cent Africa, 59-; vis prof, Hawaii Inst Geophys, 71-72. Res: Seismology; volcanology. Mailing Add: Hawaii Inst Geophys Univ of Hawaii Honolulu HI 96822

BERG, EDWARD, parasitology, see 12th edition

BERG, EUGENE WALTER, b Dade City, Fla, Nov 10, 26; m 47; c 3. ANALYTICAL CHEMISTRY. Educ: Miss Col, BS, 49; Univ Tex, PhD(chem), 52. Prof Exp: From asst prof to assoc prof, 52-63, PROF CHEM, LA STATE UNIV, 63-, CHMN DEPT, 73- Mem: Am Chem Soc; assoc Int Union Pure & Appl Chem. Res: Ion exchange; chromatography; separation techniques; solvent extraction; beta-dikotane chelates. Mailing Add: Dept of Chem La State Univ Baton Rouge LA 70803

BERG, GENE ARTHUR, b Red Wing, Minn, Apr 29, 46; m 67; c 1. GEOMETRY. Educ: Augsburg Col, BA, 68; Colo State Univ, MS, 73, PhD(math), 75. Prof Exp: ASST PROF MATH, VA COMMONWEALTH UNIV, 75- Mem: Math Asn Am; Am Math Soc. Res: An enumeration problem in finite geometries with application to coding theory. Mailing Add: Dept of Math Sci Va Commonwealth Univ Richmond VA 23184

BERG, GEORGE G, b Warsaw, Poland, May 27, 19; nat US; m 52; c 3. TOXICOLOGY, CYTOCHEMISTRY. Educ: Temple Univ, BA, 42; Columbia Univ, MS, 47, PhD(zool), 54. Prof Exp: Res assoc, Sch Med, Georgetown Univ, 49-51; res collabr, Dept Biol, Brookhaven Nat Lab, 54-55; lectr gen biol & embryol, Univ Sch, 55-63, instr pharmacol, 56-64, asst prof, 60-64, ASSOC PROF RADIATION BIOL, SCH MED, UNIV ROCHESTER, 64- Concurrent Pos: USPHS grant, 56-62. Mem: Fedn Am Sci; Histochem Soc. Res: Membrane transport; polyphosphates; phosphohydrolases; environmental toxicity. Mailing Add: Dept Radiation Biol & Biophys Univ of Rochester Sch of Med Rochester NY 14642

BERG, GERALD, b New York, NY, Nov 3, 28; m 55; c 3. VIROLOGY, MICROBIOLOGY. Educ: Utica Col, AB, 51; Syracuse Univ, MS, 52, PhD(bact), 55. Prof Exp: Asst, Upstate Med Ctr, State Univ NY, 53-54; res fel virol, Children's Hosp Res Found, Cincinnati, Ohio, 55; researcher, USPHS, 55-64, chief virus dis studies, 64-67, chief virol, Cincinnati Water Res Lab, Fed Water Qual Admin, 67-74; CHIEF BIOL METHODS BR, ENVIRON MONITORING & SUPPORT LAB, US

ENVIRON PROTECTION AGENCY, 74- Concurrent Pos: Adj prof, Univ Cincinnati, 74- Mem: AAAS; fel Am Acad Microbiol; Am Soc Microbiologists; Am Pub Health Asn. Res: Medical bacteriology; virus multiplication, chemotherapy, isolation and identification; epidemiology of virus dissemination; kinetic virucidal and bactericidal studies; statistical studies with viruses in cell cultures; viruses in renovated, waste, and other waters; viruses in solids associated with various waters. Mailing Add: Environ Monitoring & Support Lab US Environ Protection Agency Cincinnati OH 45268

BERG, HENRY CLAY, b Brooklyn, NY, Apr 23, 29; m 51; c 1. GEOLOGY. Educ: Brooklyn Col, BA, 51; Harvard Univ, AM, 56. Prof Exp: Geologist, 56-74, chief, Menlo Tech Reports Unit, 65-67, MGR ALASKAN MINERAL RESOURCE ASSESSMENT PROG, US GEOL SURV, 74- Mem: AAAS; fel Geol Soc Am; fel Geol Asn Can. Res: Areal geology of southeastern Alaska; mineral resources of Alaska. Mailing Add: Br of Alaskan Geol US Geol Surv 345 Middlefield Rd Menlo Park CA 94025

BERG, HOWARD CURTIS, b Iowa City, Iowa, Mar 16, 34; m 64; c 3. BIOPHYSICS. Educ: Calif Inst Technol, BS, 56; Harvard Univ, AM, 60, PhD(chem physics), 64. Prof Exp: Asst prof biol, Harvard Univ, 66-69, assoc prof biochem, 69-70, chmn bd tutors biochem sci, 66-70; assoc prof molecular, cellular & develop biol, 70-74, PROF MOLECULAR, CELLULAR & DEVELOP BIOL, UNIV COLO, BOULDER, 74- Concurrent Pos: Vis scientist, Univ Wis, 70. Mem: AAAS; Am Phys Soc; Am Chem Soc; NY Acad Sci; Am Soc Microbiologists. Res: Chemical structure of cell membranes; methods for separating macromolecules or small particles by mass; motility and chemotaxis of bacteria; spin-exchange in the atomic hydrogen maser. Mailing Add: Dept of Biol Univ of Colo Boulder CO 80302

BERG, IRA DAVID, b New York, NY, Nov 27, 31; m 62; c 2. MATHEMATICS. Educ: Univ Pa, BS, 53; Lehigh Univ, MS, 59, PhD(math), 62. Prof Exp: Lectr math, Yale Univ, 62-64; from asst prof to assoc prof, Univ Ill, Urbana, 64-70; assoc prof, Queen's Univ, Ont, 70-71; ASSOC PROF MATH, UNIV ILL, URBANA, 71- Mem: Am Math Soc. Res: Functional and harmonic analysis; summability. Mailing Add: Dept of Math Altgeld Hall Univ of Ill Urbana IL 61801

BERG, JAMES IRVING, b Minneapolis, Minn, May 5, 40; m 66; c 2. SOLID STATE SCIENCE. Educ: Univ Minn, BS, 62; Ohio State Univ, PhD(physics), 69. Prof Exp: Prin res scientist physics, Honeywell Corp Res Ctr, 69-71; sr physicist, Graphics Res & Develop Ctr, Addressograph Multigraph Corp, 72-74; res assoc mat sci, Case Western Reserve Univ, 74-76; MEM STAFF, OWENS-CORNING FIBERGLAS CORP, TECH CTR, 76- Mem: Am Phys Soc. Res: Compound semiconductors; photoconducting films; amorphous materials; infrared spectroscopy. Mailing Add: Tech Ctr Owens-Corning Fiberglas Corp Granville OH 43023

BERG, JEFFREY HOWARD, b New York, NY, Feb 6, 43; m 66; c 2. ORGANIC CHEMISTRY. Educ: Yeshiva Col, BA, 64; NY Univ, PhD(org chem), 69, MBA, 73. Prof Exp: Teaching asst chem, NY Univ, 64-69; sr chemist, Gen Foods Corp, 69-73; qual control supvr & chief chemist, I Rokeach & Sons, Inc, 73-74; SR PROJ SCIENTIST, JOHNSON & JOHNSON, 74- Mem: Am Chem Soc; Inst Food Technologists; Inst Elec & Electronics Engrs. Res: Chlorinated anthraquinodimethanes and related compounds; crowded anthracenes; food chemistry with emphasis on flavors and carbohydrates; orthopaedic and textile new product development; biomedical devices. Mailing Add: Johnson & Johnson Res US Hwy 1 New Brunswick NJ 08903

BERG, JOHN RICHARD, b Chippewa Falls, Wis, Apr 24, 32; m 56; c 4. PHYSICAL CHEMISTRY, INORGANIC CHEMISTRY. Educ: Col St Thomas, BS, 54; Iowa State Univ, PhD(phys chem, physics), 61. Prof Exp: Asst chem, Iowa State Univ, 54-61; sr chemist, 61-66, supvr, 66-69, MGR, MINN MINING & MFG CO, 69- Mem: Am Chem Soc. Res: Photoconductivity; solid state chemistry; preparative inorganic chemistry; imaging chemistry; optics; data processing. Mailing Add: 2305 N Western Ave St Paul MN 55113

BERG, JOHN ROBERT, b Chicago, Ill, Feb 18, 15; m 40; c 1. GEOLOGY, CHEMISTRY. Educ: Augustana Col, AB, 38; Univ Iowa, MS, 40, PhD(geol), 42. Prof Exp: Asst mineral, petrol & econ geol, Univ Iowa, 39-42; supvr mil explosives, E I du Pont de Nemours & Co, 42-44; sub-surface geologist, Shell Oil Co, Kans, 44-46; assoc prof geol, 46-48, head dept, 53-63, dean univ col, 62-69, PROF GEOL, WICHITA STATE UNIV, 48- Concurrent Pos: Mem bd dirs & secy-treas, Petrol Resources Fund, Inc, 69- Mem: Am Chem Soc; Geol Soc Am; Am Asn Petrol Geologists; Am Geophys Union; Nat Asn Geol Teachers (pres, 56-58). Res: Petrology of alkaline igneous rocks; pre-Cambrian stratigraphy and petrology; petroleum exploratory statistics; geological education. Mailing Add: Dept of Geol Wichita State Univ Wichita KS 67208

BERG, JOSEPH WILBUR, JR, b Essington, Pa, Oct 6, 20; m 50; c 3. GEOPHYSICS. Educ: Univ Ga, BS, 47; Pa State Univ, MS, 52, PhD(geophys), 54. Prof Exp: Instr, Armstrong Col (Ga), 47-48; asst prof, Univ Tulsa, 54-55; assoc prof geophys, Univ Utah, 55-60; prof, Ore State Univ, 60-66; EXEC SECY DIV EARTH SCI, NAT ACAD SCI, 66- Concurrent Pos: Geophysicist, Inst Defense Anal, 60-61; vis prof, Cornell Univ, 69-70. Mem: Soc Explor Geophys (vpres, 75); Seismol Soc Am (pres, 68); Am Geophys Union; Earthquake Eng Res Inst; Geol Soc Am. Res: Theoretical seismology; earth structure as determined by seismic, gravitational and magnetic methods; heat flow. Mailing Add: Nat Acad of Sci 2101 Constitution Ave NW Washington DC 20418

BERG, LAWRENCE RAYMOND, b Hoquiam, Wash, Nov 18, 16; m 40; c 2. POULTRY NUTRITION. Educ: Wash State Univ, BS, 39, PhD(animal sci), 54; Kans State Univ, MS, 41. Prof Exp: Asst exten agent, Wash State Univ, 40-41; asst prof poultry, Univ Idaho, 41-42; from asst poultry scientist to assoc poultry scientist, 42-62, POULTRY SCIENTIST, WESTERN WASH RES & EXTEN CTR, WASH STATE UNIV, 62- Mem: Poultry Sci Asn; Am Inst Nutrit. Res: Factors affecting egg quality; nutrient balance for growth; zinc metabolism; vanadium toxicity; energy utilization. Mailing Add: Western Wash Res & Exten Ctr Puyallup WA 98371

BERG, MARIE HIRSCH, b Mannheim, Ger, Mar 20, 09; nat US; m 35; c 1. BIOCHEMISTRY. Educ: Univ Heidelberg, PhD(chem), 34. Prof Exp: Prod chemist, Fat & Oil Indust, Ger, 34-36; asst to Prof G Bredig, Univ Karlsruhe, 36-37; Lambert Pharmacal Co fel, Colgate-Palmolive-Peet Co fel & res assoc, Northwestern Univ, 41-47; chief chemist, Dept Path, St Luke's Hosp, Chicago, Ill, 47-48; biochemist, Dept Dermat, Univ Mich, 48-51; mem staff, Wayne County Gen Hosp, Eloise, Mich, 51-52; mem dept biol, Hamline Univ, 52-59 & dept chem, 55-59; res assoc, Sch Med, Univ Minn, 53-57 & dept agr biochem, 59-60; prof chem & chmn div natural sci & math, Northwestern Col, Minn, 60-67; lectr sci, 67-74, ADV FIRE PROTECTION PROG, METROP COMMUNITY COL, 74- Mem: Fel AAAS; fel Am Sci Affil; Am Chem Soc; Nat Asn Fire Sci & Admin. Res: Science education; porphyrines and lead poisoning; fire science. Mailing Add: Dept Natural & Math Sci Metrop Community Col 50 Willow St Minneapolis MN 55403

BERG, OLGA ARONOWITZ, b New York, NY, Feb 7, 23; m 52; c 3. ENDOCRINOLOGY, ENVIRONMENTAL SCIENCES. Educ: Brooklyn Col, BA, 44; NY Univ, MS, 47, PhD, 50. Prof Exp: Asst, NY Zool Soc, 49-52; res assoc zool, Barnard Col, Columbia Univ, 52-55; LECTR, UNIV SCH, UNIV ROCHESTER, 55-, PRES ROCHESTER COMT SCI INFO, 74- Concurrent Pos: Am Cancer Soc fel, Columbia Univ, 52-53; guest biologist, Brookhaven Nat Lab, 54- Mem: AAAS. Res: Fish genetics and endocrinology. Mailing Add: 109 Southern Pkwy Rochester NY 14618

BERG, PAUL, b New York, NY, June 30, 26; m 47; c 1. BIOCHEMISTRY. Educ: Pa State Univ, BS, 48; Western Reserve Univ, PhD, 52. Prof Exp: From asst to assoc prof microbiol, Sch Med, Wash Univ, 55-59; from assoc prof to prof, 59-69, chmn dept, 69-74, WILLSON PROF BIOCHEM, MED CTR, STANFORD UNIV, 70- Concurrent Pos: Am Cancer Soc res fel, Inst Cytophysiol, Copenhagen & Sch Med, Wash Univ, 52-54 & scholar, 54-57; mem study sect physiol chem, NIH, 62-66; chmn, Gordon Conf Nucleic Acids, 66-75; mem gen med basic sci comt & biochem comt, Nat Bd Med Examr, 68-69; mem adv bd, Jane Coffin Childs Found, 71. Honors & Awards: Eli Lilly Award, 59. Mem: Inst of Med of Nat Acad Sci; fel AAAS; Am Acad Arts & Sci; Am Soc Biol Chem (pres); Am Chem Soc. Res: Microbial and animal biochemistry. Mailing Add: Dept of Biochem Stanford Univ Med Ctr Stanford CA 94305

BERG, PAUL WALTER, b New York, NY, Mar 18, 25; m 46; c 2. MATHEMATICS. Educ: NY Univ, BA, 47, PhD(math), 53. Prof Exp: Asst math, NY Univ, 47-51; instr, Rutgers Univ, 51-53; Nat Res Coun fel, 53-54; actg asst prof math, 54-55, from asst prof to assoc prof, 55-67, PROF MATH, STANFORD UNIV, 67-, VCHMN DEPT, 74- Mem: Am Math Soc; Math Asn Am. Res: Partial differential equations; calculus of variations; applied mathematics. Mailing Add: Dept of Math Stanford Univ Stanford CA 94305

BERG, RICHARD ALLEN, b Chicago, Ill, June 12, 42; m 68. ASTRONOMY. Educ: Univ Ill, Urbana, BSc, 64; Univ Va, MA, 66, PhD(astron), 70. Prof Exp: Phys scientist astron & cartog, Aeronaut Chart & Info Ctr, St Louis, Mo, 64-70; fel, Dept Physics, Univ Del, 70-72, asst prof astron, 72-73; ASST PROF ASTRON, UNIV ROCHESTER, 73- Concurrent Pos: Consult, Rand Corp, 71- Mem: Am Astron Soc; Int Astron Union. Res: Spectroscopy and photometry of spectrum variable stars; high-speed photometry of transient events in cataclysmic variables; computer applications in astronomical observing. Mailing Add: Dept of Physics & Astron Univ of Rochester Rochester NY 14627

BERG, RICHARD BLAKE, b Portland, Ore, Mar 7, 37; m 65; c 2. GEOLOGY. Educ: Beloit Col, BS, 59; Univ Mont, PhD(geol), 64. Prof Exp: Instr geol, State Univ NY Col Plattsburgh, 64-66; ECON GEOLOGIST, MONT BUR MINES & GEOL, 66- Mem: AAAS; Geol Soc Am; Am Inst Prof Geologists; Clay Minerals Soc; Mineral Soc Am. Res: Igneous and metamorphic petrology; geochemistry; mineralogy; economic geology; geology of industrial minerals in Montana. Mailing Add: Mont Bur of Mines & Geol Butte MT 59701

BERG, ROBERT ALVIN, physical chemistry, see 12th edition

BERG, ROBERT R, b St Paul, Minn, May 28, 24; m 46; c 3. GEOLOGY. Educ: Univ Minn, BA, 48, PhD(geol), 51. Prof Exp: Geologist, oil & gas explor, Calif Co, 51-56; div geologist, Cosden Petrol Corp, 57-58; consult geologist, Berg & Wasson, 59-66; PROF GEOL & HEAD DEPT, TEX A&M UNIV, 67-, DIR RES, 72- Mem: Fel Geol Soc Am; Am Asn Petrol Geol; Am Inst Prof Geol (pres, 71). Res: Stratigraphy; geology of petroleum. Mailing Add: Dept of Geol Tex A&M Univ College Station TX 77843

BERG, ROBERT W, b Welch, Minn, Apr 9, 17; m 44; c 3. POULTRY GENETICS. Educ: Univ Minn, BS, 41, MS, 50, PhD(poultry husb, animal breeding), 53. Prof Exp: Inspector, Minn Poultry Improve Bd, 45-48; asst geneticist, Western Coop Hatcheries, 53; geneticist & hatchery mgr, Jerome Turkey Hatchery, Inc, 53-58; POULTRY EXTEN SPECIALIST, INST AGR, UNIV MINN, ST PAUL, 58- Mem: Poultry Sci Asn; World Poultry Sci Asn. Res: Poultry husbandry and genetics; effects of light intensity on growth of turkeys in a total confinement program. Mailing Add: Dept of Animal Sci Inst of Agr Univ of Minn St Paul MN 55101

BERG, ROY TORGNY, b Millicent, Alta, Apr 8, 27; m 51; c 4. ANIMAL GENETICS. Educ: Univ Alta, BSc, 50; Univ Minn, MS, 54, PhD(animal breeding), 55. Prof Exp: Lectr animal husb, 50-52, from asst prof animal sci to assoc prof, 55-63, PROF ANIMAL SCI, UNIV ALTA, 63- Concurrent Pos: Nat Res Coun sr res fel, 64-65. Mem: AAAS; Am Soc Animal Sci; Am Meat Sci Asn; Agr Inst Can; Genetics Soc Can. Res: Animal breeding; genetics. Mailing Add: Dept of Animal Sci Univ of Alta Edmonton AB Can

BERG, SELWYN S, b New York, NY, Mar 26, 31; m 54; c 3. APPLIED PHYSICS. Educ: San Diego State Col, BS, 54; Univ Wash, MS, 56; Cornell Univ, PhD(physics), 64. Prof Exp: Scientist reactor physics, Westinghouse Atomic Power, 56-58; reactor supvr, Cornell Univ, 61-62; sr scientist nuclear reactors, Von Karman Ctr, Aerojet, 64-65; sr scientist physiol physics, City of Hope & Aerojet, 65-66; prin physicist, SDS Data Systs, Pomona, 66-67; physicist, Conductron-Mo, 67-69; pres, Med Screening Systs Inc, 69-70; TECH DIR, BIO-MED ENG CONSULT, 70- Concurrent Pos: Vis scientist, Brookhaven Nat Lab, 57-58; consult, City of Hope Nat Res Ctr, 64-65; biophysicist, 67- Mem: Am Phys Soc; Am Nuclear Soc; Am Asn Physicists in Med; Inst Elec & Electronics Engrs. Res: Optical spectroscopy; geometrical optics; theoretical reactor physics; experimental reactor physics; re-entry physics; fluid flow; physiological physics; ion optics; electrooptics; coherent optics; biomedical engineering and electronics. Mailing Add: 1048 Lake Forest Dr Claremont CA 91711

BERG, WILLIAM ALBERT, b Sterling, Colo, Jan 12, 30; m 60; c 4. SOIL CONSERVATION. Educ: Colo State Univ, BS, 53, MS, 57; NC State Univ, PhD(soils), 60. Prof Exp: Soil scientist, US Forest Serv, 60-68; ASSOC PROF AGRON, COLO STATE UNIV, 68- Mem: Am Soc Agron; Am Soc Range Mgt; Soil Conserv Soc Am. Res: Vegetative stabilization of disturbed lands and mine wastes; application of soils information in range and forest management. Mailing Add: Dept of Agron Colo State Univ Ft Collins CO 80523

BERG, WILLIAM DARRAGH, mathematics, see 12th edition

BERG, WILLIAM EUGENE, b Round Mountain, Nev, Dec 6, 18; m 47. BIOLOGY. Educ: Calif Inst Technol, BS, 39, MS, 40; Stanford Univ, PhD(biol), 46. Prof Exp: Asst biol, Stanford Univ, 41-43, res assoc, 43; res assoc, Univ Southern Calif, 43; from instr to prof biol, 47-70, PROF ZOOL, UNIV CALIF, BERKELEY, 70-, VCHMN DEPT, 73- Mem: Am Soc Zoologists. Res: Physiological embryology; aviation physiology; human exercise physiology. Mailing Add: Dept of Zool Univ of Calif Berkeley CA 94720

BERG, WINFRED EMIL, b Fredericksburg, Tex, Mar 4, 17; m 41; c 4. SCIENCE

ADMINISTRATION. Educ: US Naval Acad, BS, 39; Mass Inst Technol, SB, 49. Prof Exp: Prog officer sci admin, Naval Res Lab, US Navy, 49-52, sr prog officer, 55-60, dir astronaut develop div, Off Chief Naval Opers, 60-62, asst chief naval res, Off Naval Res, 62-65; sr staff asst, Nat Aeronaut & Space Coun, Exec Off of President, 65-71; SR STAFF OFFICER, BD ENERGY STUDIES, NAT ACAD SCI-NAT RES COUN, 72- Concurrent Pos: Mem ocean dynamics adv comt, NASA, 75- Honors & Awards: Legion of Merit, US Navy, 60. Mem: Am Inst Aeronaut & Astronaut. Mailing Add: 2059 Huntington Ave Alexandria VA 22303

BERGDALL, IRENE FLOY, b Cissna Park, Ill, July 7, 10. MATHEMATICS. Educ: Huntington Col, AB, 31; Univ Mich, AM, 42. Prof Exp: Teacher pub sch, Ill, 31-42; assoc prof math, Huntington Col, 42-75; RETIRED. Res: Geometry; algebra; modern mathematics. Mailing Add: Dept of Math Huntington Col Huntington IN 46750

BERGDOLL, MERLIN SCOTT, b Petersburg, WVa, Sept 23, 16; m 42; c 3. BIOCHEMISTRY. Educ: WVa Univ, BS, 40; Purdue Univ, PhD(agr biochem), 46. Prof Exp: Asst chemist, Purdue Univ, 42-46; instr, Biochem & Food Res Inst, Univ Chicago, 46-50, asst prof, 50-56, assoc prof, Food Res Inst, 56-66; prof, Dept Foods & Nutrit, 66-68, PROF, FOOD RES INST, UNIV WIS, MADISON, 66- & DEPT FOOD SCI, 68- Mem: Am Chem Soc; Am Soc Microbiol; Inst Food Technol; Am Soc Biol Chem. Res: Production, purification, physico-chemical properties, structure, immunology and analysis of the staphylococcal enterotoxins. Mailing Add: Food Res Inst Univ of Wis MadisonWI 53706

BERGE, DOUGLAS G, b Montevideo, Minn, Sept 19, 38; m 66. ANALYTICAL CHEMISTRY. Educ: Mankato State Col, BS & BA, 61; Univ Iowa, PhD(anal chem), 65. Prof Exp: Asst prof chem, Adrian Col, 65-66; asst prof, 66-70, ASSOC PROF CHEM, UNIV WIS-OSHKOSH, 70- Mem: Am Chem Soc. Res: Organic reagents; equilibrium studies. Mailing Add: Dept of Chem Univ of Wis Oshkosh WI 54901

BERGE, JOHN WILLISTON, b Madison, Wis, July 29, 30; m 56; c 3. POLYMER CHEMISTRY. Educ: Univ Wis, BS, 51, PhD(chem), 59. Prof Exp: Chemist, E I du Pont de Nemours & Co, Inc, 51-52, res chemist, 58-63; SR CHEMIST, S C JOHNSON & SONS, INC, 63- Mem: Am Chem Soc; Soc Rheol. Res: Dynamical mechanical properties of high polymers; synthetic polymers and resins for polishes; light scattering; ultra centrifugation. Mailing Add: 1529 Crabapple Dr Racine WI 53405

BERGE, JON PETER, b Madison, Wis, Feb 4, 35; m 59; c 2. EXPERIMENTAL HIGH ENERGY PHYSICS. Educ: Univ Calif, BA, 56, MA, 58, PhD(physics), 63. Prof Exp: Res assoc physics, Lawrence Radiation Lab, Univ Calif, Berkeley, 56-67; res officer, Univ Oxford, 67-70; physicist, Ctr Europ Nuclear Res, Geneva, 70-74; PHYSICIST, FERMI NAT ACCELERATOR LAB, 74- Mem: Am Phys Soc; AAAS. Res: Bubble chamber data analysis computer programming. Mailing Add: Fermi Nat Accelerator Lab Batavia IL 60510

BERGE, KENNETH G, b Wahkon, Minn, Feb 9, 26; m 48; c 3. INTERNAL MEDICINE. Educ: Univ Minn, BA, 48, BS, 49, BMed, 51, DMed, 52, MS, 55. Prof Exp: Asst med, Mayo Clin, 55, consult, 55-70; from instr to assoc prof, Mayo Grad Sch Med, Univ Minn, 57-74, PROF MED MAYO MED SCH, 74-; HEAD SECT MED, MAYO CLIN, 70- Concurrent Pos: Consult & vchmn steering comt coronary drug proj, Nat Heart & Lung Inst, 62-, mem epidemiol & biomet adv comt, 70-72, mem policy adv bd, Hypertension Detection & Followup Prog, 74-, mem policy bd, Aspirin Myocardial Infarction Study, 75-; fel coun arteriosclerosis & epidemiol, Am Heart Asn; mem sect arteriosclerosis & ischaemic heart dis, Int Soc Cardiol; mem bd dir & exec comt, Rochester Methodist Hosp, 70- Honors & Awards: Billings Silver Medal, AMA, 57. Mem: Fel Am Col Physicians; AMA; Am Fedn Clin Res. Res: Cardiovascular research, especially epidemiology, atherosclerosis and hypertension. Mailing Add: Mayo Clin 200 First St SW Rochester MN 55901

BERGE, TRUMAN KENT, b Erskine, Minn, Sept 2, 22; m 48; c 5. PHYSICS, COMPUTER SCIENCE. Educ: US Mil Acad, BS, 46; Purdue Univ, MS, 52; grad, Air War Col, 67. Prof Exp: Reconnaissance pilot, US Air Force, US, Japan & Korea, 46-50, instr physics, US Mil Acad, 52-54 & asst prof, 54-55, res adminstr, Europ Off, Air Res & Develop Command, Belg, 55-58, asst prof physics, US Air Force Acad, 58-60, assoc prof, acad counr & exec, Basic Sci Div, 60-62, physicist, Eastern Test Range, Patrick Air Force Base, Fla, 62-64, physicist & chief, Support Instrumentation Dir, Nat Range Div, 64-65, dept dir tactical opers, Korea, 65, dep comdr fighter group, Minn, 65-66, reconnaissance pilot, 67-68, div chief comput div, Air Force Hq, Washington, DC, 68-74; PROG ANALYST, NAT ENERGY INFO CTR, FED ENERGY ADMIN, WASHINGTON, DC, 75- Concurrent Pos: Founder & exec dir, Fed Comput Performance Eval & Simulation Ctr, Washington, DC, 72-74. Mem: AAAS; Am Phys Soc; Am Asn Physics Teachers; Brit Phys Soc. Res: Nuclear reactions; administration of geophysics and physics research; underwater acoustics; satellite, spacecraft and missile test range instrumentation; management science; simulation; energy information. Mailing Add: 4610 Mansfield Manor Dr Washington DC 20022

BERGE, TRYGVE OBERT, b Stoughton, Wis, Aug 19, 09; m 31; c 5. VIROLOGY, RICKETTSIAL DISEASES. Educ: Univ Wis-Madison, BA & MS, 35; Univ Calif, Berkeley, PhD(bact), 51. Prof Exp: Instr bact, NDak Agr Col, 36-40; asst bacteriologist, Rocky Mountain Lab, Mont, 41-42; chief bact & serol sect, Torney Gen Hosp, Calif, 42-44; officer, US Typhus Comn, 44-46; chief viral & rickettsial br, 406th Med Gen Lab, Tokyo, 46-49; chief immunol br, 6th Army Area Med Lab, Calif, 51-56; chief virol div, US Army Med Unit, Md, 56-61; chief virol br, Armed Forces Inst Path, 61-64; cur collection animal viruses & rickettsiae, Am Type Cult Collection, 64-71; CONSULT, CTR DIS CONTROL, USPHS, 71- Concurrent Pos: Consult, Armed Forces Inst Path, 65-75, Delta Regional Primate Res Ctr, Covington, La, 66-70 & Nat Cancer Inst, 67-70; ed, Catalogue of Arthropod-Borne Viruses of World, 68-75; chmn subcomt info exchange, Am Comt Arthropod-Borne Viruses, 68-75. Honors & Awards: US Typhus Comn Medal, 45. Mem: Am Soc Trop Med & Hyg; Am Asn Immunol; fel Am Acad Microbiol. Res: Arthropod-borne and respiratory disease viruses; rickettsial diseases; preservation and characterization of human and animal viruses. Mailing Add: USPHS Ctr for Dis Control PO Box 2087 Ft Collins CO 80521

BERGEAUX, PHILLIP JAMES, b Eunice, La, May 1, 18; m 42; c 2. AGRONOMY, SOIL CHEMISTRY. Educ: Univ Southwestern La, BS, 40; Univ Ga, MS, 60, EdD(adult educ), 74. Prof Exp: Agronomist, Tenn Corp, 47-56; asst agronomist, 56-62, ASSOC AGRONOMIST, UNIV GA, 62- Concurrent Pos: Consult, Lowe & Stevens Advert Agency, 63-69. Honors & Awards: Nat County Agts Asn Distinguished Serv Award, 74. Mem: Am Soc Agron. Res: Use of radioactive phosphates to determine extent of absorption of phosphates from fertilizer applied phosphates and phosphates contained by the soil in an available form; interaction of phosphates absorbed by Irish potato plants with zinc, aluminum and iron; soil fertility. Mailing Add: Exten Agron Dept Univ of Ga Coop Exten Serv Athens GA 30601

BERGELAND, MARTIN E, b Madison, Minn, Oct 14, 35; m 59; c 2. VETERINARY PATHOLOGY. Educ: Univ Minn, BS, 57, DVM, 59, PhD(vet path), 65. Prof Exp: From instr to assoc prof vet path, Univ Minn, St Paul, 59-70; prof vet path & dir vet diag med, Univ Ill, Urbana-Champaign, 70-73; PROF VET SCI, SDAK STATE UNIV, 73- Mem: Am Col Vet Path; Am Vet Med Asn; Am Asn Avian Path; Nat Conf Vet Diagnosticians. Res: Pathology of clostridial infections of animals. Mailing Add: Dept of Vet Sci SDak State Univ Brookings SD 57006

BERGEN, CATHERINE MARY, b Garden City, NY, Jan 16, 12. SCIENCE EDUCATION. Educ: Wellesley Col, AB, 33; Columbia Univ, AM, 35, PhD(sci educ), 42. Prof Exp: Sci consult pub sch, NY, 34-37; asst physics, Hofstra Col, 39-40; teacher elem sci, Lincoln Sch, Teachers Col, Columbia Univ, 40-43; assoc prof, 41-54, PROF SCI, JERSEY CITY STATE COL, 54- Mem: AAAS; Am Phys Soc; Nat Sci Teachers Asn; Am Asn Physics Teachers. Res: Sources of science information used by children; analysis of science material published for laymen; use of mathematics in science. Mailing Add: 39 Duncan Ave Jersey City NJ 07304

BERGEN DELMAR WESLEY, b Coldwater, Kans, Dec 28, 31; m 52; c 4. NUCLEAR PHYSICS. Educ: Greenville Col, AB, 53; Univ Kans, MA, 55; Univ NMex, PhD(physics), 67. Prof Exp: Engr, Boeing Airplane Co, 56-57; staff mem physics, 57-71, asst to assoc dir weapons, 71-73, DIR WEAPONS PROG OFF, LOS ALAMOS SCI LAB, UNIV CALIF, 73- Res: Neutron cross section measurements; study of the fission reaction; gamma ray spectroscopy. Mailing Add: Los Alamos Sci Lab Box 1663 Los Alamos NM 87544

BERGEN, DONNA CATHERINE, b Crawfordsville, Ind, Mar 17, 45. NEUROLOGY. Educ: Vassar Col, BA, 67; Univ Ill, MD, 71. Prof Exp: Intern med, Evanston Hosp, Ill, 71-72; resident neurol, 72-75, STAFF NEUROLOGIST, RUSH MED COL, 75- Mem: Am Acad Neurol; Am Epilepsy Soc. Res: Properties of visually evoked cerebral potential in patients with multiple sclerosis and other disorders of visual pathways. Mailing Add: Rush Med Col 1753 W Congress Pkwy Chicago IL 60612

BERGEN, JOHN RICHARD, b Clifton, NJ, May 12, 21; m 47; c 3. PHYSIOLOGY. Educ: Colgate Univ, AB, 42; Univ Vt, MS, 48; Tufts Col, PhD(physiol), 52. Prof Exp: Mem staff, 52-62, SR SCIENTIST, WORCESTER FOUND EXP BIOL, 62- Mem: AAAS; Am Physiol Soc; Am Soc Biol Psychiat; Am Col Neuropsychopharmacol; Am Soc Neurochem; Soc Neurosci; NY Acad Sci. Res: Neurosphysiology; neuropharmacology; psychopharmacology. Mailing Add: 222 Maple Ave Shrewsbury MA 01545

BERGEN, JOHN VANDERVEER, b New York, NY, Nov 3, 34; m 58; c 2. PHARMACY, MEDICINAL CHEMISTRY. Educ: Philadelphia Col Pharm, BS, 56; Univ Wis, PhD(pharmaceut, org chem), 61. Prof Exp: Asst, Univ Wis, 57-60; from asst prof to prof pharmaceut chem, Idaho State Univ, 60-68; lectr pharmaceut, Univ Kans, 68-69; from assoc dir to dir, Nat Formulary, Am Pharmaceut Asn, 69-75; PRES, PHILADELPHIA COL PHARM & SCI, 75- Concurrent Pos: Dean col pharm, Idaho State Univ, 63-68, dir med arts, 64-68; mem, Pharm Rev Comt, USPHS, 68-72, US Adopted Names Coun, 70-72, Expert Adv Panel Int Pharmacopeia & Pharmaceut Prep, WHO, 74- & US Pharmacopeia Comt Rev, 75- Mem: Am Chem Soc; Am Pharmaceut Asn; Am Asn Cols Pharm; fel Acad Pharmaceut Sci. Res: Drug standards; correlation of molecular structure and thermodynamic parameters with pharmacological activity. Mailing Add: Philadelphia Col of Pharm & Sci 43rd St & Kingsessing Mall Philadelphia PA 19104

BERGEN, JOHN VICTOR, b Edinburgh, Scotland, Aug 11, 30; US citizen; m 57; c 4. GEOGRAPHY. Educ: Col Wooster, BA, 52; Ind Univ, MA, 58, PhD(geog), 64. Prof Exp: Asst prof geog, Ferris State Col, 60-66; asst prof geog, Western Ill Univ, 66-68; assoc prof geog, 68-74, PROF GEOG, WESTERN ILL UNIV, 74-, MAP LIBRN, 68- Mem: Asn Am Geog; Nat Coun Geog Educ; Spec Libr Asn. Res: Rural and urban settlement patterns; historical bibliographies of geography in higher education. Mailing Add: Dept of Geog Western Ill Univ Macomb IL 61455

BERGEN, ROBERT LUDLUM, JR, b Islip, NY, Oct 29, 29; m 51; c 4. POLYMER CHEMISTRY. Educ: Williams Col, AB, 51; Cornell Univ, MS, 53, PhD(phys chem), 55. Prof Exp: Sr res chemist chem div, US Rubber Co, 55-60, sr res specialist, 60-69; mgr textile & plastic res, Res Ctr, 69-72, mgr Paracril res & develop, 72-74, MGR SYNTHETIC RUBBER RES & DEVELOP, CHEM DIV, UNIROYAL INC, 74- Concurrent Pos: Instr, New Haven Col, 64-67, adj prof, 67-69. Mem: Fel AAAS; Am Chem Soc; Soc Rheol; Sigma Xi. Res: Physical chemistry and physics of polymers; rheology; statistical experimentation; physics of polymer blends; environmental stress cracking of plastics; fiber spinning; fibers for pollution control; polymer flammability; rubber chemistry; synthetic elastomers for specialty applications such as oil and heat resistance. Mailing Add: 79 Lebanon Rd Bethany CT 06525

BERGEN, STANLEY S, JR, b Princeton, NJ, May 2, 29; wid; m 65; c 5. INTERNAL MEDICINE. Educ: Princeton Univ, AB, 51; Columbia Univ, MD, 55. Prof Exp: Med dir, St Luke's Hosp, Greenwich, Conn, 62-64; actg chief med serv, Ft Jay Army Hosp, New York, NY, 61-62; dir med, Cumberland Hosp, 64-68; chief commun med, Brooklyn-Cumberland Med Ctr, 68-70; sr vpres, New York City Health & Hosps Corp, 70-71; PROF MED, RUTGERS MED SCH & NJ MED SCH & PRES, COL MED & DENT NJ, 71- Concurrent Pos: Zabriskie fel, St Luke's Hosp, New York, 59-60; Hartford Found grant, 63-65; assoc prof med, State Univ NY Downstate Med Ctr, 64-71. Mem: NY Acad Sci; fel Am Col Physicians; fel Am Col Nutrit; Asn Am Med Cols; Am Pub Health Asn. Res: Carbohydrate and hepatic metabolism; endocrinology; development of new health care delivery system. Mailing Add: Col of Med & Dent of NJ 100 Bergen St Newark NJ 07103

BERGEN, WERNER GERHARD, b Warstade, WGer, Apr 23, 43; US citizen; m 66; c 1. ANIMAL NUTRITION. Educ: Ohio State Univ, BSc, 64, MSc, 66, PhD(protein nutrit), 67. Prof Exp: ASSOC PROF ANIMAL NUTRIT, MICH STATE UNIV, 67- Mem: Am Soc Animal Sci; Am Inst Nutrit; Fedn Am Socs Exp Biol; Soc Exp Biol & Med. Res: Ruminant nutrition; plasma amino acid and protein metabolism; protein quality of single cell proetins; protein synthesis efficiency in anaerobic bacteria; amino acid transport; protein digestion; muscle biology. Mailing Add: Dept Animal Husb Col Agr Mich State Univ East Lansing MI 48824

BERGENBACK, RICHARD EDWARD, b Allentown, Pa, Oct 23, 26; m 47; c 3. GEOLOGY. Educ: Lafayette Col, AB, 48; Lehigh Univ, MS, 50; Pa State Univ, PhD, 64. Prof Exp: Asst, Lehigh Univ, 48-50; geologist, Cities Serv Res & Develop Co, 57-58; geologist-in-chg fuels br, Tenn Valley Authority, 58-60; teacher, Baltimore County, Md, 60-61; asst, Pa State Univ, 61-64, res assoc, 64-65; asst prof geol, Southern Conn State Col, 65-66; assoc prof geol & geog & chmn dept, Howard Univ, 66-68; ASSOC PROF GEOL, UNIV TENN, CHATTANOOGA, 68- Concurrent Pos: Instr eve col, Univ Chattanooga, 58-59. Mem: Geol Soc Am; Am Asn Petrol Geologists. Res: Petrology and geochemistry of sedimentary rocks, especially carbonate rocks; regional stratigraphy, especially coal-bearing rocks. Mailing Add: 2242 Peterson Dr Chattanooga TN 37421

BERGENDAHL, MAXIMILIAN HILMAR, b Reading, Pa, Apr 24, 21; m 45; c 1.

ECONOMIC GEOLOGY. Educ: Brown Univ, AB, 50. Prof Exp: Geologist, US Geol Surv, 51-67; SR GEOLOGIST, AMAX EXPLOR INC, 67- Mem: Am Asn Petrol Geol; Soc Econ Geol; Am Inst Prof Geol. Res: Geology and stratigraphy of Florida phosphate deposits; late Cenozoic stratigraphy of south-central Florida; uranium deposits of Wyoming; stratigraphy of northeastern Wyoming; pre-Cambrian geology; ore deposits of central Colorado; gold deposits of the United States; metals exploration of western United States. Mailing Add: AMAX Explor Inc Box C Belmar Sta Denver CO 80226

BERGER, ABRAHAM WILLIAM, physical chemistry, see 12th edition

BERGER, ANDREW JOHN, b Warren, Ohio, Aug 30, 15; div; c 2. ANATOMY, ORNITHOLOGY. Educ: Oberlin Col, AB, 39; Univ Mich, AM, 47, PhD, 50. Prof Exp: From instr to assoc prof anat, Med Sch, Univ Mich, 50-64; chmn dept zool, 65-71, PROF ZOOL, UNIV HAWAII, 64- Concurrent Pos: Asst ed, Wilson Bull, Wilson Ornith Soc, 50-51 & The Auk, Am Ornithologists' Union, 53-54; Guggenheim fel, 63; Fulbright lectr, Univ Baroda, 64-65; Univ Hawaii Res Admin grant, 65; NSF grants, 66-69, 70-75. Mem: Am Asn Anatomists; fel Am Ornithologists' Union; Wilson Ornith Soc (second vpres, 71-73, first vpres, 73-75, pres, 75-); Cooper Ornith Soc; fel AAAS. Res: Human anatomy; avian anatomy and life history. Mailing Add: Dept of Zool Univ of Hawaii Honolulu HI 96822

BERGER, ARTHUR, b Chicago, Ill, Feb 17, 16; m 51; c 3. ORGANIC CHEMISTRY. Educ: Univ Ill, BS, 37, MS, 38, PhD(chem), 40. Prof Exp: Fel plastic intermediates, Northwestern Univ, 42-44; sr chemist, Amino Acids & Proteins, A E Staley Mfg Co, 44-52; GROUP LEADER, SYNTHETIC ORG PHARMACEUTS, BAXTER LABS INC, 52- Mem: AAAS; Am Chem Soc; Int Soc Heterocyclic Chem; Coblentz Soc. Res: Synthetic pharmaceuticals; organic analysis; information retrieval. Mailing Add: Baxter Labs Inc 6301 Lincoln Morton Grove IL 60053

BERGER, BERNARD, organic chemistry, see 12th edition

BERGER, BEVERLY JANE, b Morristown, NJ, Apr 28, 39. POPULATION BIOLOGY. Educ: Univ NMex, BA, 61, MEdSci, 65, MS, 67; Univ Calif, Davis, PhD(genetics), 71. Prof Exp: NIH fel, Sch Pub Health, Univ Calif, Berkeley, 71-73; MATH BIOLOGIST, LAWRENCE LIVERMORE LAB, UNIV CALIF, 73- Mem: Genetics Soc Am; AAAS. Res: Effects of non-nuclear pollutants on man and vegetation; biological systems for energy storage, production, and conservation. Mailing Add: Lawrence Livermore Lab PO Box 808 Livermore CA 94550

BERGER, CARL, physical chemistry, see 12th edition

BERGER, DANIEL RICHARD, b Oakland, Calif, Nov 3, 33; m 58. ORGANIC CHEMISTRY. Educ: Stanford Univ, BS, 54; Northwestern Univ, PhD(chem), 58. Prof Exp: Instr chem, Northwestern Univ, 57-58; chemist, Am Cyanamid Co, 58-59; chemist, 59-64, tech serv mgr, Org Chem Div, 64-74, PROD DEVELOP MGR, ORG CHEM DIV, RICHARDSON CO, 74- Concurrent Pos: Lectr eve div, Northwestern Univ, 59-68. Mem: Am Oil Chem Soc; Am Chem Soc. Res: Rearrangements of alpha-halogenated ethers; mechanism; ethers; surfactant chemistry. Mailing Add: Richardson Co 2700 Lake St Melrose Park IL 60160

BERGER, EDMOND LOUIS, b Salem, Mass, Dec 5, 39; m 64; c 2. HIGH ENERGY PHYSICS. Educ: Mass Inst Technol, BS, 61; Princeton Univ, PhD(physics), 65. Prof Exp: Asst prof physics, Dartmouth Col, 65-68; res physicist, Lawrence Radiation Lab, 68-69; asst physicist, 69-70; ASSOC PHYSICIST, ARGONNE NAT LAB, 70- Concurrent Pos: Vis scientist theory div, CERN, Geneva, Switz, 72-74. Mem: Fel Am Phys Soc. Res: Strong interactions of elementary particles; phenomenological models of quasi-two-body and multiparticle production. Mailing Add: High Energy Physics Bldg 362 Argonne Nat Lab Argonne IL 60439

BERGER, EDWARD MICHAEL, b New York, NY, May 2, 44; m 67; c 3. GENETICS. Educ: Hunter Col, BA, 65; Syracuse Univ, MS, 67, PhD(genetics), 69. Prof Exp: NIH fel develop biol, Harvard Univ, 69-70, Univ Chicago, 70-71 & State Univ NY Albany, 71-75; ASST PROF BIOL, DARTMOUTH COL, 75- Mem: AAAS; Genetics Soc Am; Soc Zool; Am Soc Naturalists. Mailing Add: Dept of Biol Dartmouth Col Hanover NH 03755

BERGER, EUGENE Y, b Philadelphia, Pa, Dec 11, 19; m 51; c 2. MEDICINE. Educ: Lafayette Col, AB, 40; NY Univ, MD, 44. Prof Exp: Res resident, Goldwater Mem Hosp, 46-48, asst med, Col Med, 49-51, from instr to asst prof, 51-61, ASSOC PROF MED, COL MED, NY UNIV, 61- Concurrent Pos: Res assoc, Goldwater Mem Hosp, NY Univ, 53-55, assoc dir res serv, 55-71; mem, Am Bd Internal Med. Mem: AAAS; Am Soc Clin Invest; Biophys Soc; Soc Exp Biol & Med; Am Physiol Soc. Res: Clinical physiology—renal, salt, water balance, body composition, gastrointestinal, ion exchange, edematous states, endocrine effects, mercurials; isotopes; mechanisms of anesthesia, water structure. Mailing Add: 126 Ritchie Dr Yonkers NY 10705

BERGER, FRANK MILAN, b Pilsen, Czech, June 25, 13; nat US; m; c 2. MEDICAL RESEARCH, PHARMACOLOGY. Educ: Prague Tech Univ, MD, 37. Hon Degrees: DSc, Philadelphia Col Pharm & Sci, 66. Prof Exp: Chief res, Monsall Hosp, Manchester, 41-43; chief bacteriologist, W Riding of Yorkshire, 43-45; head pharm & bact, British Drug House, 45-47; asst prof pediat, Univ Rochester, 47-49; dir res, Wallace Labs, 49-58, pres, 58-73; PROF PSYCHIAT & CO-DIR PSYCHOPHARMACOL DOCTORAL PROG, UNIV LOUISVILLE SCH MED, 74- Concurrent Pos: Mem adv coun biol, Princeton Univ, prof lectr, 69-76; mem ad hoc study group clin pharmacol, Walter Reed Army Med Ctr, Washington, DC, 74- Mem: Fel AAAS; fel NY Acad Sci; Am Soc Pharmacol & Exp Therapeut; Am Soc Microbiol; fel Royal Soc Med. Res: Muscle relaxants mephenesin, carisoprodol; tranquilizers; meprobamate; non-specific immunity, protodyne. Mailing Add: 190 East 72nd St New York NY 10021

BERGER, FRANKLIN GORDON, b Providence, RI, Sept 26, 47. BIOCHEMISTRY, MOLECULAR BIOLOGY. Educ: State Univ NY Buffalo, BA, 69; Purdue Univ, PhD(biochem), 74. Prof Exp: Res asst biochem, Purdue Univ, 69-74; RES ASSOC BIOCHEM, CORNELL UNIV, 74- Concurrent Pos: NIH fel, 75- Mem: Am Soc Microbiol. Res: Regulation of gene expression; in vitro analysis of transcriptional and translational processes. Mailing Add: Biochem Molecular & Cell Biol Cornell Univ Ithaca NY 14853

BERGER, HAROLD, b New York, NY, Aug 9, 27; m 51; c 3. PARASITOLOGY. Educ: NY Univ, BA, 52, MS, 57, PhD(biol), 62. Prof Exp: Technician cancer res, Montefiore Hosp, NY, 52-53; biologist physiol, Pearl River, NY, 53-57, res parasitologist, Princeton, NJ, 57-75, SR RES PARASITOLOGIST, AM CYANAMID CO, 75- Concurrent Pos: NIH fel parasitol, Univ Mex, 62-63. Mem: Am Soc Parasitologists. Res: Testing and development of compounds in domestic animals in order to discover and evaluate materials of possible utility in the treatment against parasitic helminth infection. Mailing Add: Am Cyanamid Co PO Box 400 Princeton NJ 08540

BERGER, HAROLD, b Syracuse, NY, Oct 7, 26; m 52; c 5. RADIOLOGICAL PHYSICS. Educ: Syracuse Univ, BS, 49, MS, 51. Prof Exp: Physicist x-ray dept advan develop lab, Gen Elec Co, 50-59; sr physicist solid state devices div, Battelle Mem Inst, 59-60; assoc physicist metall div, Argonne Nat Lab, 60-70, sr physicist & group leader nondestructive testing mat sci div, 70-73; PHYSICIST REACTOR RADIATION DIV, NAT BUR STANDS, 73-, PROG MGR NONDESTRUCTIVE EVAL, 75- Concurrent Pos: Vis scientist, Ctr Nuclear Studies, Grenoble, 68-69; tech ed, Mat Eval, Am Soc Nondestructive Testing, 69-, Mehl hon lectr fel, 73-75. Honors & Awards: Achievement Award, Am Soc Nondestructive Test, 67; Radiation Indust Award, Am Nuclear Soc, 74. Mem: Am Phys Soc; Am Soc Nondestructive Test; Sigma Xi; fel Am Nuclear Soc; Am Soc Test & Mat. Res: X-ray and light detection techniques; ultrasonic imaging methods for nondestructive evaluation; proton radiography; neutron radiography; development and standardization of nondestructive evaluation techniques. Mailing Add: 9832 Canal Rd Gaithersburg MD 20760

BERGER, HILLARD, b Chicago, Ill, Dec 10, 38; m 57; c 2. MOLECULAR GENETICS. Educ: Univ Calif, Los Angeles, BA, 61; Vanderbilt Univ, PhD(molecular biol), 65. Prof Exp: Trainee biol, Stanford Univ, 65-67; asst prof, 67-71, ASSOC PROF BIOL, JOHNS HOPKINS UNIV, 71- Mem: Genetics Soc Am; Am Chem Soc; Am Soc Microbiologists. Res: Bacteriophage genetics-mechanisms of recombination in phage. Mailing Add: Dept of Biol Johns Hopkins Univ Baltimore MD 21218

BERGER, HOWARD MARTIN, aeronautics, see 12th edition

BERGER, JACQUES, b New York, NY, Apr 14, 34; m 54. PROTOZOOLOGY, INVERTEBRATE ZOOLOGY. Educ: Pa State Univ, BSc, 55; Univ Ill, MS, 58, PhD(zool), 64. Prof Exp: Instr zool, Univ Ill, 62-63 & Duke Univ, 63-64; asst prof, NC State Univ, 64-65; asst prof, 65-68, ASSOC PROF ZOOL, UNIV TORONTO, 68- Concurrent Pos: NSF fel marine biol, Duke Univ, 64-65; sr fel, Massey Col, Univ Toronto, 73- Mem: Soc Protozool; Am Micros Soc; Am Soc Zool; Asn Fr Speaking Protistologists. Res: Multivariate morphometrics of Tetrahymena, Paramecium and Euplotes; biology of Ciliate protozoa; morphology, morphogenesis, systematics and bionomics of echinoid inhabiting Ciliates. Mailing Add: Dept of Zool Univ of Toronto Toronto ON Can

BERGER, JAMES DENNIS, b Spokane, Wash, July 16, 42; m 66; c 1. CELL BIOLOGY. Educ: Ind Univ, AB, 64, MA, 65, PhD(zool), 69. Prof Exp: ASSOC PROF ZOOL, UNIV BC, 70- Mem: Soc Protozoologists; Can Soc Cell Biol. Res: Genetic and physiological basis of regulation of DNA synthesis and other cell cyle events in Paramecium. Mailing Add: Dept of Zool Univ of BC 2075 Wesbrook Pl Vancouver BC Can

BERGER, JAMES EDWARD, b Pinson Fork, Ky, Sept 10, 35; m 56; c 3. PHARMACOLOGY. Educ: Univ Cincinnati, BS, 59, MS, 61; Univ Fla, PhD(pharmacol), 67. Prof Exp: Res asst bact, Surg Bact Res Lab, Cincinnati Gen Hosp, Ohio, 60-62, res assoc, 62-64; asst prof, 67-71, ASSOC PROF PHARMACOL, COL PHARM, BUTLER UNIV, 71- Concurrent Pos: Mead-Johnson grant, 70-71; consult pharmacologist, Vet Admin Hosp, 69- Mem: AAAS; Am Pharmaceut Asn. Res: Staphylococcal toxin purification and production; action of quinidine on the heart as related to alpha and beta adrenergic receptors in the tissue. Mailing Add: Dept of Pharmacol Butler Univ Indianapolis IN 46208

BERGER, JAY MANTON, b New York, NY, Oct 29, 27; m 48; c 3. THEORETICAL PHYSICS, COMPUTER SCIENCE. Educ: Univ Mich, BS, 47; Columbia Univ, AM, 48; Case Inst Technol, PhD(physics), 53. Prof Exp: Instr physics, Fenn Col, 48-51; res assoc, Princeton Univ, 52-56; sr engr, Adv Systs Develop Div, 56-71, prod assurance adv, Data Processing Group, 71-72, ADV, LITIGATION ANAL, INT BUS MACH CORP, 72- Concurrent Pos: Consult, Radiation Lab, Univ Calif, 54 & Lockheed Missile Systems, 55. Mem: Am Phys Soc. Res: Noise and information theory; applied mathematics; theoretical nuclear physics; plasma physics; astrophysics; field theory; systems design. Mailing Add: 57 Fuller Rd Briarcliff Manor NY 10510

BERGER, JERRY EUGENE, b Jefferson City, Mo, Aug 8, 33; m 54; c 4. ATMOSPHERIC CHEMISTRY. Educ: Westminster Col, BA, 55; Univ Ky, MS, 57, PhD(phys chem), 59. Prof Exp: Res chemist, 59-65, sr res chemist, 65-67, group leader, 66-67, sr chemist, 67-70, spec analyst, 70-75, SR STAFF CHEMIST, SHELL OIL CO, 75- Mem: Am Chem Soc; Soc Automotive Engrs. Res: Environmental conservation; air conservation. Mailing Add: Res & Develop Dept Shell Oil Co One Shell Plaza Houston TX 77002

BERGER, JOEL GILBERT, b Brooklyn, NY, June 3, 37; m 62; c 1. MEDICINAL CHEMISTRY. Educ: City Col New York, BS, 57; NY Univ, MS, 60, PhD(org chem), 62. Prof Exp: Chemist, Esso Res & Eng Co, 62-64 & Wallace & Tiernan, Inc, 64-65; res assoc, Princeton Univ, 65-66; sr med chemist, 66-73, GROUP LEADER, ENDO LABS, INC, 73- Mem: Am Chem Soc; NY Acad Sci. Res: Reaction mechanisms; chemistry of pyrroles and indoles; synthesis of drugs affecting the central nervous system. Mailing Add: Endo Labs Inc 1000 Stewart Ave Garden City NY 11533

BERGER, JULIUS, b Hamilton, Ont, Nov 28, 13; nat US; m 36; c 2. BIOCHEMISTRY. Educ: McMaster Univ, BA, 33, MA, 34; Univ Wis, MS, 36, PhD(biochem), 38. Prof Exp: Asst biochem & bact, Univ Wis, 35-37, asst, 38-40; assoc bot, Conn Col, 40-43; sr res chemist, 43-64, HEAD DEPT MICROBIOL, HOFFMANN-LA ROCHE, INC, 64- Mem: AAAS; Am Chem Soc; Am Soc Microbiol; Am Acad Microbiol; NY Acad Sci. Res: Production of new antibiotics; microbial transformations of steroids, carbohydrates and other organic compounds; screening for new antibiotics; microbial transformations of organic compounds. Mailing Add: Dept of Microbiol Hoffmann-La Roche Inc Nutley NJ 07110

BERGER, KENNETH WALTER, b Evansville, Ind, Mar 22, 24; m 46; c 4. AUDIOLOGY. Educ: Evansville Col, BA, 49; Ind State Univ, MA, 50; Southern Ill Univ, MS, 60, PhD, 62. Prof Exp: Speech & hearing therapist, Carmi Pub Schs, Ill, 55-61; asst audiol, Southern Ill Univ, 61-62; dir audiol, 62-72, MEM FAC AUDIOL, KENT STATE UNIV, 72- Mem: Acoust Soc Am; Am Speech & Hearing Asn. Res: Psycho-physics of sound; hearing loss in children; audiometric testing of discrimination for speech. Mailing Add: Dept of Audiol Kent State Univ Kent OH 44242

BERGER, KERMIT CARL, soils, see 12th edition

BERGER, LAWRENCE, b New York, NY, Apr 6, 26; m 50; c 4. MEDICINE. Educ: NY Univ, BA, 47; Chicago Med Sch, MD & MB, 51; Am Bd Internal Med, dipl. Prof Exp: Intern, Mt Sinai Hosp, 51-52, resident med, 54-56; resident, Montefiore Hosp, 52-53; NIH res fel, Med Col, Cornell Univ, 53-54; asst, Goldwater Mem Hosp, 56-57, asst attend, 56-58; res fel, 57-69; clin asst physician, 57-63, asst attend physician, 63-72, RES ASSOC MED, MT SINAI HOSP, 69-, ASSOC ATTEND, 72-, ASSOC CLIN PROF MED, MT SINAI SCH MED, 72- Mem: AAAS; AMA; Am Fedn Clin Res; fel Am Col Physicians; fel NY Acad Med. Res: Renal function; physiology. Mailing Add: 119 E 84th St New York NY 10028

BERGER, LEO, b New York, NY, Jan 30, 18; m 41; c 5. MEDICINAL CHEMISTRY. Educ: Long Island Univ, BS, 38; NY Univ, MS, 41. Prof Exp: Jr control chemist, Fruit Industs, Ltd, NY, 38-39; from res chemist to sr res chemist, 41-69, res fel, 69-74, RES GROUP CHIEF, HOFFMANN-LA ROCHE, INC, 74- Mem: AAAS; Am Chem Soc; Sigma Xi; Inflammation Res Asn. Res: Synthetic vitamins; natural and synthetic glycosides; analgesics; central nervous system drugs; anticoagulants; antirheumatics. Mailing Add: Hoffmann-La Roche Inc Roche Park Nutley NJ 07110

BERGER, LESLIE RALPH, b London, Eng, Dec 18, 28; nat; m 54; c 2. MICROBIAL PHYSIOLOGY. Educ: Univ Cincinnati, BS, 50; Univ Wash, MS, 53; Univ Calif, PhD(microbiol), 57. Prof Exp: Res microbiologist, Scripps Inst Oceanog, 57-59; asst biochemist, Univ Calif, Davis, 59-60; from asst prof to assoc prof microbiol, 60-69, PROF MICROBIOL, UNIV HAWAII, 69- Res: Physiology of growth at increased hydrostatic pressure; cytology and physiology of photosynthetic organisms at increased hydrostatic pressure; kinetics of allosteric enzymes at extreme physical environments. Mailing Add: Dept of Microbiol 2538 The Mall Univ of Hawaii Honolulu HI 96822

BERGER, LUC, b Morges, Switz, May 2, 33. SOLID STATE PHYSICS. Educ: Univ Lausanne, BS, 55, PhD(physics), 60. Prof Exp: Swiss Nat Sci Found fel physics, 60-61, from instr to assoc prof, 61-74, PROF PHYSICS, CARNEGIE-MELLON UNIV, 74- Mem: Am Phys Soc; Swiss Phys Soc. Res: Electronic structure of transition metals and of transition metal compounds; ferromagnetism and antiferromagnetism; transport processes in ferromagnets. Mailing Add: Dept of Physics Carnegie-Mellon Univ Pittsburgh PA 15213

BERGER, MARTIN, b New York, NY, May 23, 26; m 47; c 3. PHYSICS. Educ: Columbia Univ, BS, 49. Prof Exp: Physicist, Tire Div, US Rubber Co, 50-55; engr, Res Labs, Chrysler Corp, 55-56; physicist, Chem Res Div, Esso Res & Eng Co, 56-60, sr physicist, Ctr Basic Res, 60-62, res assoc polymers, 62-64, sect head elastomer res, Enjay Polymer Lab, 64-68, dir phys & eng sci, Corp Res Lab, 68-75, DIR GOVT RES LABS, EXXON RES & ENG CO, 75- Res: Polymer structure and properties, their relationship and methods of changing structure to improve properties. Mailing Add: Corp Res Lab Exxon Res & Eng Co PO Box 45 Linden NJ 07036

BERGER, MARTIN JACOB, b Vienna, Austria, July 12, 22; nat US; m; c 3. MATHEMATICAL PHYSICS. Educ: Univ Chicago, PhD, 51. Prof Exp: Fel statist, Univ Chicago, 51-52; PHYSICIST, NAT BUR STAND, 52- Concurrent Pos: Mem comt nuclear sci, Nat Acad Sci-Nat Res Coun. Mem: Am Phys Soc; Radiation Res Soc. Res: Radiation transport theory; Monte Carlo methods; atmospheric and auroral physics; nuclear medicine; shielding. Mailing Add: Nat Bur Stand Ctr Radiation Res Washington DC 20234

BERGER, MELVYN STUART, b Brooklyn, NY, Aug 23, 39; c 1. MATHEMATICS. Educ: Univ Toronto, BA, 61; Yale Univ, MA, 63, PhD(math), 64. Prof Exp: From asst prof to assoc prof math, Univ Minn, Minneapolis, 64-69; assoc prof, 69-74, PROF MATH, BELFER GRAD SCH, YESHIVA UNIV, 74- Concurrent Pos: Vis mem, Courant Inst, 66-68; vis mem, Inst Advan Study, Princeton Univ, 72-73. Mem: Am Math Soc; London Math Soc. Res: Nonlinear functional analysis; partial differential equations, mathematical physics and differential geometry. Mailing Add: Dept Math Belfer Grad Sch Yeshiva Univ New York NY 10033

BERGER, NEIL EVERETT, b New York, NY, Oct 8, 42; m 75. APPLIED MATHEMATICS. Educ: Columbia Univ, BS, 63; NY Univ, MS, 65, PhD(appl math), 68. Prof Exp: Asst prof, 68-75, ASSOC PROF MATH, UNIV ILL, CHICAGO CIRCLE, 75- Honors & Awards: Monroe Martin Prize, Inst Fluid Dynamics & Appl Math, Univ Md, 74. Mem: Soc Indust & Appl Math; AAAS; Am Math Soc; Math Asn Am. Res: Partial differential equations of elasticity; fluid dynamics; asymptotic expansions. Mailing Add: Dept of Math Univ of Ill Chicago Circle Box 4348 Chicago IL 60680

BERGER, PHILIP JEFFREY, b Newark, NJ, June 28, 43; m 65; c 2. ANIMAL BREEDING, STATISTICAL ANALYSIS. Educ: Del Valley Col, BS, 65; Ohio State Univ, MS, 67, PhD(animal breeding), 70. Prof Exp: Res & teaching asst animal breeding, Ohio State Univ, 65-70, fel, 71; ASST PROF ANIMAL BREEDING, IOWA STATE UNIV, 72-, ASSOC MEM GRAD FAC, 74- Mem: Am Soc Animal Sci; Am Dairy Sci Asn; Biomet Soc; Sigma Xi. Res: Development of computing procedures for best linear unbiased prediction of breeding values, predictability of correlated responses in breeding populations, economic evaluation of productive traits and recovery of selection response. Mailing Add: Dept of Animal Sci Iowa State Univ 36 Kildee Hall Ames IA 50011

BERGER, RAINER, b Graz, Austria, July 3, 30; US citizen; m 59; c 2. ANTHROPOLOGY, GEOGRAPHY. Educ: Univ Ill, PhD(chem), 60. Prof Exp: Staff scientist geochem, Convair Sci Res Dept, 60-62; sr res scientist, Lockheed Aircraft Corp, 62-63; from asst prof to assoc prof, 63-73, PROF ANTHROP, GEOG & GEOPHYS, UNIV CALIF, LOS ANGELES, 73- Concurrent Pos: Consult, Follow-on prog to Apollo, NAm Aviation Inc/NASA, 64-65; NASA fel, 66-67; NSF fel, 66-68; Guggenheim fel, 68-69. Honors & Awards: Distinguished Serv Award, Univ Calif, Los Angeles, 68. Mem: AAAS; Am Anthrop Asn; Geochem Soc; Am Chem Soc; Soc Am Archaeol. Res: Scientific chronometric and tracing methods, especially radiocarbon dating in archaeology, geography and history and in environmental studies in geo- and cosmochemistry and geophysics. Mailing Add: Inst of Geophys Univ of Calif Los Angeles CA 90024

BERGER, RICHARD DONALD, b Macungie, Pa, Jan 15, 34; m 62; c 2. PLANT PATHOLOGY. Educ: Kutztown State Col, 55; Univ Wis, PhD(plant path), 62. Prof Exp: Res fel, Univ Wis, 62-63; plant pathologist, Pa State Univ, 63-66; PROF PLANT PATH, UNIV FLA, 66- Concurrent Pos: Assoc ed, J Am Phytopath Soc, 76-; res award, Fla Fruit & Veg Asn, 72. Honors & Awards: Campbell Award, Am Phytopath Soc & Campbell Soup Co, 74. Mem: Am Phytopath Soc. Res: Epidemiology of vegetable diseases; Cercospora and Helminthosporium biology; chemical control; disease modeling; computer simulation of epidemics. Mailing Add: JFAS Univ of Fla Gainesville FL 32611

BERGER, RICHARD S, b Brooklyn, NY, Apr 28, 29; m 54; c 2. ORGANIC POLYMER CHEMISTRY. Educ: Stanford Univ, BS, 50; Univ Wis, PhD(org chem), 54. Prof Exp: Fel, Univ Minn, 54; chemist, Shell Develop Co, 56-65; mgr fiber chem br, Res Div, 65-70, dir res projs, 70-73, DIR LONG RANGE PLANNING, PHILLIPS FIBERS CORP, PHILLIPS PETROL CO, 73- Mem: Am Chem Soc; Am Mgt Asn. Res: Synthetic polymer chemistry; polymer and fiber stabilization; chemical modification of polyamide, polyester and polyolefin polymers and fibers. Mailing Add: Res Dept Phillips Fibers Corp PO Box 66 Greenville SC 29602

BERGER, ROBERT LEWIS, b Omaha, Nebr, Sept 2, 25; m 50; c 4. BIOPHYSICS, INSTRUMENTATION. Educ: Colo Agr & Mech Col, BS, 50; Pa State Univ, MS, 53, PhD(physics), 56. Prof Exp: Instr physics, Park Col, 50-51; fel, Brit-Am Exchange, Am Cancer Soc, Cambridge Univ, 56; asst prof physics, Utah State Univ, 57-60, assoc prof, 60-62; MEM STAFF, LAB TECH DEVELOP, NAT HEART & LUNG INST, 62- Concurrent Pos: Vis scientist, Dept Chem, Univ Calif, San Diego, 69-71; ed, Nat Bur Stand publ, 70-72. Mem: Fel AAAS; fel Chem Soc Gt Brit; Soc Gen Physiol; Am Phys Soc; Biophys Soc. Res: Enzyme kinetics and mechanisms; spectroscopy, fluid dynamics and computers; microcalorimetry; dye laser spectroscopy. Mailing Add: Nat Heart & Lung Inst Bldg 10 Rm 5D18 Bethesda MD 20014

BERGER, ROBERT S, b Tours, Tex, Jan 2, 33; m 59; c 4. INSECT TOXICOLOGY. Educ: Tex A&M Univ, BS, 54, MS, 57; Cornell Univ, PhD(biochem), 61. Prof Exp: Entomologist agr res serv, USDA, 61-63; assoc prof insect toxicol, 63-70, PROF ZOOL & ENTOM, AUBURN UNIV, 70- Mem: AAAS; Entom Soc Am; Am Chem Soc. Res: Insecticide chemistry; insect attractants. Mailing Add: Dept of Zool & Entom Auburn Univ Auburn AL 36830

BERGER, S EDMUND, b Osijek, Yugoslavia, Nov 13, 22; nat US; m 47; c 2. ORGANIC CHEMISTRY. Educ: Univ Rome, ChD(org chem), 48. Prof Exp: Asst org chem, Univ Rome, 48-49; res chemist, Dept Legal Med, Harvard Med Sch, 49-51; res chemist, 51-63, RES SUPVR, ALLIED CHEM CORP, 63- Mem: Am Chem Soc. Res: Polyurethane chemistry; polyesters; diisocyanates; food acidulants; surfactants; liquid ion exchange; exploratory organic synthesis; catalysis. Mailing Add: 298 Grayton Rd Tonawanda NY 14150

BERGER, SELMAN A, b Brooklyn, NY, Aug 31, 42; m 67; c 2. ANALYTICAL CHEMISTRY, ENVIRONMENTAL CHEMISTRY. Educ: Brooklyn Col, BS, 64; Univ Conn, MS, 67, PhD(anal chem), 69. Prof Exp: Teaching asst chem, Univ Conn, 64-69; postdoctoral fel, Dalhousie Univ, Halifax, NS, 69-70; sr chemist, Toxicol Lab, Bur Labs, NYC Dept Health, 70-71; ASST PROF CHEM, JOHN JAY COL, CITY UNIV NEW YORK, 71-, ASST TO DEPT CHMN, 72- Concurrent Pos: Doctoral fac chem, City Univ New York, Grad Ctr, 73- Mem: Am Chem Soc; Sigma Xi. Res: Spectroscopic methods for trace analysis; flame methods; complexation and solvent extraction methods. Mailing Add: Dept Sci John Jay Col Crim Just City Univ NY 445 W 59th St New York NY 10019

BERGER, SHELBY LOUISE, b New York, NY, Jan 5, 41; m 73. BIOCHEMISTRY, CELL BIOLOGY. Educ: Bryn Mawr Col, AB, 62; Harvard Univ, PhD(biophys), 68. Prof Exp: Fel biochem, Nat Inst Arthritis & Metab Dis, 68-70, staff fel, 70-71, sr staff fel biochem & cell biol, Nat Inst Dent Res, 72-74 & Nat Cancer Inst, 74-76, RES CHEMIST BIOCHEM & CELL BIOL, NAT CANCER INST, NIH, 76- Res: Control of the metabolism of messenger ribonucleic acid and heterogeneous nuclear ribonucleic acid in physiologically resting, growing and malignant human lymphocytes. Mailing Add: Nat Cancer Inst Nat Insts of Health Bethesda MD 20014

BERGER, SHELDON, b Chicago, Ill, Nov 12, 28; m 50; c 4. MEDICINE, ENDOCRINOLOGY. Educ: Univ Ill, BA, 49, BS, 51, MD, 53. Prof Exp: Physician in charge, Radioisotope Lab, 59-63, asst dir dept metab & endocrine res, 60-63; assoc dir dept res & educ, Evanston Hosp, 63-65, coordr res labs, 65-69; asst prof, 65-68, ASSOC PROF MED, NORTHWESTERN UNIV, CHICAGO, 68- Concurrent Pos: NIH trainee metab & endocrine, Michael Reese Hosp, Chicago, 58-59; NIH trainee thyroid res methods, Beth Israel Hosp, Boston, 59. Mem: Am Col Physicians; Endocrine Soc. Res: Diabetes; general metabolism; thyroid physiology; internal medicine. Mailing Add: 707 N Fairbanks Ct Chicago IL 60611

BERGER, STEVEN BARRY, b New York, NY, Dec 29, 46. PHYSICS. Educ: Mass Inst Technol, SB, 67, PhD(physics), 73. Prof Exp: Res assoc physics, Boston Univ, 73-74; FEL PHYSICS, MASS INST TECHNOL, 74- Concurrent Pos: Fight-for-Sight fel, NY, 74-76; asst ed, Am J Physics, 75- Mem: Sigma Xi; Am Phys Soc; Am Optical Soc; AAAS. Res: Elementary particle physics; biophysics of vision. Mailing Add: Mass Inst of Technol Rm 20B-136 Cambridge MA 02139

BERGER, WILLIAM J, b Arnold, Pa, Nov 10, 21; m 65. MATHEMATICS, COMPUTER SCIENCE. Educ: Carnegie Inst Technol, BS, 50, MS, 51, PhD(math), 54. Prof Exp: Mathematician & engr, Radio Corp Am, Fla, 55; mathematician, Hastings-Raydist Corp, Va, 56-57; res engr & mathematician, Convair-Astronaut, Gen Dynamics Corp, Calif, 57-58; res mathematician, Lockheed Aircraft Corp, 58-60; res engr, Aeronutronic Div, Ford Motor Co, 60; consult mathematician, Gen Elec Co, Va, 60-65; assoc prof math, Howard Univ, 65-69; PROF MATH, DC TEACHERS COL, 69- Mem: Math Asn Am. Res: Scientific computing; astronautics; functional analysis; geometry. Mailing Add: Dept of Math DC Teachers Col 1100 Harvard St Washington DC 20009

BERGER, WOLFGANG HELMUT, b Erlangen, Ger, 1937. OCEANOGRAPHY, MICROPALEONTOLOGY. Educ: Univ Erlangen, cand geol, 61; Univ Colo, Boulder, MSc, 63; Univ Calif, San Diego, PhD(oceanog), 68. Prof Exp: Asst res oceanogr, Scripps Inst Oceanog, 68-70; asst geol, Univ Kiel, 70-71; asst prof, 71-74, ASSOC PROF OCEANOG, SCRIPPS INST OCEANOG, 74- Concurrent Pos: Lectr oceanog, Calif State Univ, San Diego, 68-70; consult, Oceanog Div, NSF, 73-75 & deep sea drilling proj, Joint Oceanog Inst Deep Earth Sampling, 74- Mem: AAAS; Geol Soc Am; Am Geophys Union; Am Soc Limnol & Oceanog; Soc Econ Paleontologists & Mineralogists. Res: Ecology and paleoecology of planktonic foraminifera and coccolithophores; biogenous sedimentation in the deep ocean; paleo-oceanography. Mailing Add: Scripps Inst of Oceanog Univ of Calif San Diego La Jolla CA 92037

BERGERON, CLYDE J, JR, b New Orleans, La, July 2, 32; m 56; c 3. PHYSICS. Educ: Loyola Univ, BS, 55; La State Univ, PhD, 59. Prof Exp: Res asst, Loyola Univ, 54-55; from asst to assoc prof, 55-68, PROF PHYSICS, UNIV NEW ORLEANS, 68-, CHMN DEPT, 74- Concurrent Pos: Oak Ridge Nat Lab equip subcontract, 63-; Res Corp grant, 67-68; consult, Electronuclear Div, Oak Ridge Nat Lab, 61 & 62-66. Mem: Am Phys Soc; Sigma Xi; Am Asn Physics Teachers. Res: Solid state low temperature physics; physics of metals, especially superconductivity. Mailing Add: Dept of Physics Univ of New Orleans Lake Front New Orleans LA 70122

BERGERON, GEORGES ALBERT, b Quebec, Que, Oct 11, 16; m 45; c 4. PHSYIOLOGY. Educ: Laval Univ, BA, 37, MD, 42; FRCPS(C), 47. Prof Exp: Resident med, Hotel-Dieu de Quebec, 43; from lectr to asst prof, 43-51, PROF PHYSIOL, MED SCH, LAVAL UNIV, 51-, VDEAN, 64-, HEALTH SCI DIR, 70- Concurrent Pos: Fel, Western Reserve Univ, 44; Markle Found scholar, 50-55. Mem: AAAS; Am Physiol Soc; Can Physiol Soc. Res: Circulation; medical education. Mailing Add: Dept of Physiol Laval Univ Med Sch Quebec PQ Can

BERGERON, JOHN ALBERT, b Cumberland, RI, June 13, 29; m 52; c 3. CYTOLOGY. Educ: Brown Univ, BA, 51; Cornell Univ, PhD(zool), 56. Prof Exp: Asst hist embryol, Cornell Univ, 51-55; assoc physiologist, Brookhaven Nat Lab, 55-61, physiologist, 61-64; MEM STAFF, GEN ELEC RES LAB, 64-, RES SCIENTIST, RES & DEVELOP CTR, 64-, MGR DIAGNOSTIC PROJS, CORP RES & DEVELOP, 70- Concurrent Pos: Mem, Biol Stain Comn. Mem: Soc Gen Physiol; Am Soc Microbiol; Am Soc Cell Biol; Am Soc Plant Physiol; Soc Cryobiol. Res: Electromicroscopy; sub-cellular fractions; physicochemical aspects of the behavior of dyes and photosynthetic pigments; hemostatic mechanisms. Mailing Add: Res & Develop Ctr Bldg K-1 Gen Elec Res Lab PO Box 8 Schenectady NY 12301

BERGERON, KENNETH DONALD, b Jacksonville, Fla, May 20, 46; m 68. PLASMA PHYSICS, STATISTICAL MECHANICS. Educ: Brown Univ, ScB, 68; Brandeis Univ, MS, 69, PhD(physics), 75. Prof Exp: PHYSICIST, PLASMA THEORY DIV, SANDIA LABS, 74- Res: Ion emission in diodes; fluid kinetic theory. Mailing Add: Div 5241 Sandia Labs Albuquerque NM 87115

BERGERON, MICHEL, b Alma, Que. PHYSIOLOGY, NEPHROLOGY. Educ: Laval Univ, BA, 53, MD, 59; McGill Univ, MSc, 64. Prof Exp: Intern internal med, Huntington Mem Hosp, 59-60; resident, Lahey Clin, 61-62; Med Res Coun Can res fel, Nuclear Res Ctr, France, 64-67, scholar, 67-72; from asst prof to assoc prof physiol, 67-75, PROF PHYSIOL, UNIV MONTREAL, 75- Mem: AAAS; Am Soc Nephrology; Can Soc Physiol; Can Soc Nephrology; Fr Soc Electron Micros. Res: Amino acid transport; micropuncture techniques; electron microscopy; radioautography, renal physiology; membrane transport. Mailing Add: Dept of Physiol Univ of Montreal Montreal PQ Can

BERGES, DAVID ALAN, b Evansville, Ind, Aug 5, 41; m 65; c 3. ORGANIC CHEMISTRY, MEDICINAL CHEMISTRY. Educ: Evansville Col, BA, 63; Ind Univ, PhD(chem), 67. Prof Exp: Assoc org chem, Nat Ctr Sci Res, Inst Chem Natural Substances, France, 67-68; ASSOC SR INVESTR, SMITH KLINE & FRENCH LABS, 68- Mem: Am Chem Soc; The Chem Soc. Res: Isolation, structure elucidation and synthesis of natural products; development of new organo-chemical synthetic methods; conformational analysis; biological properties, chemistry and synthesis of betalactam antibiotics. Mailing Add: Smith Kline & French Labs 1500 Spring Garden St Philadelphia PA 19101

BERGESON, GLENN BERNARD, b Park City, Utah, Jan 18, 26; m 48; c 4. PLANT NEMATOLOGY. Educ: Utah State Univ, BS, 48; Univ Calif, PhD(entom), 58. Prof Exp: Jr nematologist, Shell Agr Lab, 48-53; sr lab tech plant nematol, Univ Calif, 54-57; asst prof, 58-74, ASSOC PROF BOT & PLANT PATH, PURDUE UNIV, 74- Mem: Am Phytopath Soc; Soc Nematol. Res: Control methods for plant parasitic nematodes; plant disease complexes caused by nematodes and microorganisms. Mailing Add: Dept of Bot Purdue Univ West Lafayette IN 47906

BERGESON, HAVEN ELDRED, b Logan, Utah, Dec 22, 33; m 57; c 6. HIGH ENERGY PHYSICS. Educ: Univ Utah, BS, 58, PhD(physics), 62. Prof Exp: Physicist, Space Sci Lab, Gen Elec Co, 61-63; asst prof physics, 63-64, asst res prof, 64-68, assoc prof, 68-75, PROF PHYSICS, UNIV UTAH, 75- Mem: Am Phys Soc. Res: Cosmic ray physics; cosmic ray anisotropies; high energy neutrino interactions; muon production; ultra high energy collisions; extensive air showers; air scintillation. Mailing Add: Dept of Physics Univ of Utah Salt Lake City UT 84112

BERGGREN, MICHAEL J, b Menominee, Mich, Feb 22, 39. RADIOLOGICAL PHYSICS. Educ: Univ Mich, BS, 61; Stanford Univ, PhD(physics), 69. Prof Exp: Teaching asst physics, Stanford Univ, 61-63, res asst, 63-69; asst prof, Mankato State Col, 69-73; vis scientist, 74-75, RES FEL, BIOPHYS SCI UNIT, MAYO FOUND, MAYO CLINIC, 75- Concurrent Pos: Res asst, Kaman Nuclear, Colo, 62. Mem: Am Phys Soc. Res: Computerized tomography from x-ray and radioisotope sources; element selective imaging with x-ray sources. Mailing Add: Biophys Sci Unit Med Sci Bldg Mayo Found Rochester MN 55901

BERGGREN, RONALD B, b Staten Island, NY, June 13, 31; m 54; c 2. SURGERY. Educ: Johns Hopkins Univ, AB, 53; Univ Pa, MD, 57; Am Bd Surg, dipl; Am Bd Plastic Surg, dipl. Prof Exp: From asst instr to instr, Univ Pa, 58-65; from asst prof to assoc prof, 65-73, PROF SURG, OHIO STATE UNIV, 73-, DIR PLASTIC SURG, COL MED, UNIV HOSPS & CHILDRENS HOSP, COLUMBUS, 65- Mem: AAAS; fel Am Col Surg; Am Soc Plastic & Reconstruct Surg; Am Soc Surg of Trauma; Am Asn Plastic Surg. Res: Low temperature preservation of tissues and organs; clinical problems in reconstruction of traumatic deformities of the face. Mailing Add: Dept of Surg Ohio State Univ Col of Med Columbus OH 43210

BERGGREN, WILLIAM ALFRED, b New York, NY, Jan 15, 31; m 54; c 4. GEOLOGY, MICROPALEONTOLOGY. Educ: Dickinson Col, BA, 52; Univ Houston, MSc, 57; Univ Stockholm, PhD(geol), 60, DSc(geol), 62. Prof Exp: Micropaleontologist, Shell Oil Co, 55-57; res assoc micropaleont, Univ Stockholm, 57-62, Doktorand stipendiat, 60-62; res paleontologist, Oasis Oil Co, Libya, 62-65; SR SCIENTIST, WOODS HOLE OCEANOG INST, 65- Concurrent Pos: Fel, Princeton Univ, 60-61; Am Asn Petrol Geol grant-in-aid, 61-62; Sigma Xi grant-in-aid, 62; exchange student, Soviet-Am Cult Exchange Prog, 62; Rumanian-Am Acad Sci Exchange Prog, 69; mem adv panel, Joint Oceanog Deep Earth Sampling Prog Atlantic, Inst Comn Deep Sea Stratig, 67-68; working group on biostratig zonation of Cretaceous & Cenozoic for correlation in marine geol, Comn Stratig, Int Union Geol Sci, 68- Mem: Fel Geol Soc Am; Am Asn Petrol Geol; Soc Explor Paleont & Mineral; Swiss Geol Soc; Geol Soc France. Res: Studies in Mesozoic and Cenozoic planktonic Foraminifera, including their evolutionary development and world-wide stratigraphic correlation based on their occurrence. Mailing Add: Dept of Geophys & Geol Woods Hole Oceanog Inst Woods Hole MA 02543

BERGH, ARPAD A, b Hungary, Apr 26, 30; US citizen; m 56; c 3. PHYSICAL CHEMISTRY. Educ: Univ Szeged, MS, 52; Univ Pa, PhD(phys chem), 59. Prof Exp: Res assoc, Geophys Inst, Hungary, 52-56; mem tech staff, Bell Tel Labs, 59-63, supvr planar transistor tech, 64-69, HEAD DEPT, BELL LABS, 69- Mem: Inst Elec & Electronics Engrs; Electrochem Soc. Res: Reaction kinetics; homogeneous reactions and catalysis in gases and aqueous solutions; physical chemistry of electron devices; processing, operation and failure mechanisms of semiconductor devices; opto-electronic devices including III-V semiconductor materials and the physics and fabrication of LED displays and opto-isolators. Mailing Add: Rm 2D-332 Bell Labs 600 Mountain Ave Murray Hill NJ 07974

BERGH, BERTHOLD ORPHIE, b Sask, Jan 30, 25; m 48; c 2. GENETICS. Educ: Univ Sask, BS, 50; Ohio State Univ, MSc, 51; Univ Calif, PhD(genetics), 56. Prof Exp: Res officer, Fruit Breeding, Can Dept Agr, 51-52; asst, Genetics Dept, 52-56, asst geneticist, Citrus Exp Sta, 56-67, GENETICS SPECIALIST, CITRUS EXP STA, UNIV CALIF, RIVERSIDE, 67- Mem: AAAS; Bot Soc Am; Am Genetics Asn; Am Soc Hort Sci. Res: Basic genetic studies of the tomato and the pepper; commercial improvement of the avocado. Mailing Add: 3040 Cimarron Rd Riverside CA 92506

BERGHOEFER, FRED G, b Chicago, Ill, June 7, 21; m 50; c 3. OPERATIONS RESEARCH. Educ: Rose Polytech, BS, 43; Univ Chicago, SM, 49. Prof Exp: Radar engr, 43-45, analyst, Opers Eval Group, 50-66, DIV DIR, CTR NAVAL ANALYSES, OPERS EVAL GROUP, NAVAL RES LAB, 66- Mem: AAAS; Opers Res Soc Am; Inst Elec & Electronics Engrs. Res: Operations analysis of naval operations, especially antisubmarine warfare and air warfare aspects. Mailing Add: 2720 N Fillmore St Arlington VA 22207

BERGIN, MARION JOSEPH, b Lavoye, Wyo, May 2, 27; m 58; c 3. GEOLOGY. Educ: Univ Wyo, BS, 51. Prof Exp: GEOLOGIST, US GEOL SURV, 51- Mem: Geol Soc Am. Res: Geology of mineral fuels. Mailing Add: US Geol Surv 956 Nat Ctr Reston VA 22092

BERGLUND, DONNA LOU, b Geneva, Ill, Jan 4, 45. INORGANIC CHEMISTRY. Educ: Cent Methodist Col, BA, 65; Ohio State Univ, PhD(inorg chem), 69. Prof Exp: Vis res fel, Ohio State Univ, summer 69; asst prof inorg & phys chem, Goucher Col, 69-72; ASST PROF CHEM, COL OF WOOSTER, 72- Concurrent Pos: Consult, Great Lakes Cols Asn, 73-77 & mem bd dirs, 74-77. Mem: Am Chem Soc; Am Asn Univ Profs. Res: Transition metal complexes containing cationic phosphine, potentially tridentate phosphine, and phosphine oxide ligands; coordination chemistry—synthesis and structural characterization of transition metal complexes containing ligands such as nitrosyl and halide species. Mailing Add: Dept of Chem Col of Wooster Wooster OH 44691

BERGLUND, ERIK, b Boston, Mass, June 9, 24; m 61. PULMONARY DISEASES. Educ: Karolinska Inst, Sweden, ML, 49, MD, 55. Prof Exp: Res assoc physiol, Sch Pub Health, Harvard Univ, 51-57; asst prof, Gothenburg Univ, 57-63; assoc prof exp surg, Karolinska Inst, Stockholm, 63-72; PHYSICIAN IN CHIEF, RENSTROMSKA HOSP, 74- Mem: Swedish Cardiol Soc. Res: Cardiovascular and pulmonary physiology. Mailing Add: Renstromska Hosp Gothenburg Sweden

BERGLUND, JOHN VERNE, b Johnsonburg, Pa, Apr 23, 39; m 61; c 2. SILVICULTURE, FOREST ECOLOGY. Educ: Pa State Univ, BS, 62, MS, 64; State Univ NY Col Forestry, Syracuse, PhD(silvicult & ecol), 68. Prof Exp: Sr programmer, Data Processing Assocs, 64; prof res asst silvicult, 65-68, asst prof, 68-71, ASSOC PROF SILVICULT, STATE UNIV NY COL ENVIRON SCI & FORESTRY, 71- Mem: Soc Am Foresters; Ecol Soc Am. Res: Forest site evaluation; site requirements of forest tree species; computer applications in forestry; applications of multivariate statistics in forest research. Mailing Add: Dept of Silviculture SUNY Col Environ Sci & Forestry Syracuse NY 13210

BERGMAN, ABRAHAM, b Seattle, Wash, May 11, 32; m 57; c 1. MEDICINE, PEDIATRICS. Educ: Reed Col, BA, 54; Western Reserve Univ, MD, 58; Am Bd Pediat, dipl, 63. Prof Exp: Intern, Children's Hosp Med Ctr, Boston, 58-59, jr asst resident, 59-60; exchange registr, Pediat Unit, St Mary's Hosp, London, Eng, 60-61; res fel pediat, State Univ NY Upstate Med Ctr, 61-63; ASSOC PROF PEDIAT, SCH MED & ASSOC PROF HEALTH SERV, SCH PUB HEALTH & COMMUNITY MED, UNIV WASH, 64-; DIR OUTPATIENT SERV, CHIREN'S ORTHOP HOSP & MED CTR, 64- Concurrent Pos: Mem coop fac, Grad Sch Pub Affairs, Univ Wash; consult, Nat Health Serv Corps, Dept Health, Educ & Welfare; consult Indian health, US Senate Comt Interior & Insular Affairs; pres, Nat Found Sudden Infant Death. Mem: Fel Am Acad Pediat. Res: Epidemiology; public policy in health; medical care; injury prevention. Mailing Add: Children's Orthop Hosp & Med Ctr Seattle WA 98105

BERGMAN, ELLIOT, b Brooklyn, NY, Feb 2, 30; m 52; c 3. ORGANIC CHEMISTRY. Educ: Brooklyn Col, BS, 51; Cornell Univ, PhD(org chem), 55. Prof Exp: Instr org chem, Univ Calif, 55-56; supvr org chem, Agr Res Labs, Shell Develop Co, 56-67, head org chem div, Woodstock Agr Res Ctr, Shell Res Ltd, Eng, 67-70, asst to mgr res & develop, Agr Div, Shell Chem Co, 70-71, proj mgr consumer prod, 71-74, supv org chem, Agr Res Labs, 65-67; vpres res & develop, 74-75, VPRES DEM & INT MKT, RPR INC, 75- Concurrent Pos: Off Naval Res asst, Cornell Univ, 51-53. Mem: Am Chem Soc. Res: Flourine, organometallic, organophosphorus, polymer and agricultural chemistry; consumer products; cholesteric liquid crystal materials for consumer, industrial and medical products (visual displays). Mailing Add: 6700 Sierra Lane Dublin CA 94566

BERGMAN, EMMETT NORLIN, b Slayton, Minn, May 6, 29; m 53; c 4. PHYSIOLOGY. Educ: Univ Minn, BS, 50, DVM & MS, 53, PhD(physiol), 59. Prof Exp: Instr physiol, Univ Minn, 50-53; vet lab off, Walter Reed Army Inst Res, 53-55; from instr to asst prof physiol, Univ Minn, 55-61; from assoc prof to prof, 61-74, PROF VET PHYSIOL & COORDR ELECTIVE PROGS, STATE UNIV NY VET MED, CORNELL UNIV, 74- Concurrent Pos: NSF sr fel, 63-64; mem metab study sect, Div Res Grants, NIH, 66-70; vis prof physiol, Med Ctr, Univ Calif, San Francisco, 69-70; consult, Coun Grad Schs US. Mem: AAAS; Am Physiol Soc; Am Vet Med Asn; Conf Res Workers Animal Dis; NY Acad Sci. Res: Ketone body, carbohydrate and fatty acid metabolism; electrolyte metabolism; ruminant metabolism; visceral blood flow; liver metabolism. Mailing Add: Dept Physiol Cornell Univ State Univ NY Vet Med Col Ithaca NY 14850

BERGMAN, ERNEST L, b Munich, Ger, July 12, 22; nat US; m 48. HORTICULTURE, PLANT NUTRITION. Educ: Kanton Landwirtsch, Schwand, Switz, dipl landwirt, 41; Ore State Col, BS, 55; Mich State Univ, MS, 56, PhD(hort), 58. Prof Exp: Res asst veg crops, Kanton Gartenbau Schule, Oeschberg, Switz, 45; pomologist, Schweizerisch Obstbauzentrale, 46; pvt enterprise, 47-53; from asst prof to assoc prof, 58-71, PROF PLANT NUTRIT, PA STATE UNIV, 71- Honors & Awards: Kenneth Post Award, Am Soc Hort Sci, 68. Mem: Fel AAAS; Am Soc Hort Sci; Am Soc Plant Physiol; Am Soc Agron; Int Soc Hort Sci. Res: Mineral nutrition of horticultural plants, vegetables and fruits; physiological, pathological, and virus disorders in connection with nutrient deficiency or toxicity problems. Mailing Add: Dept of Hort Pa State Univ University Park PA 16802

BERGMAN, GUNNAR (BROR), physical chemistry, see 12th edition

BERGMAN, HAROLD LEE, b Sault St Marie, Mich, July 8, 41; m 68; c 2. PHYSIOLOGICAL ECOLOGY. Educ: Eastern Mich Univ, BA, 68, MS, 71; Mich State Univ, PhD(fisheries biol), 73. Prof Exp: Fishery biologist, Great Lakes Fishery Lab, US Fish & Wildlife Serv, Ann Arbor, Mich, 68-71; res asst fishery biol, Dept Fisheries & Wildlife, Mich State Univ, East Lansing, 71-73, res assoc environ physiol, Dept Physiol, 74; res assoc environ impact, Environ Sci Div, Oak Ridge Nat Lab, Oak Ridge, Tenn, 74-75; ASST PROF ZOOL & PHYSIOL, UNIV WYO, 75- Concurrent Pos: Ed, Black Thunder Study, Wyo Environ Inst, 75. Mem: Sigma Xi; Am Fisheries Soc; AAAS; NAm Benthological Soc. Res: Physiological ecology of fishes and the effects of environmental purturbation on aquatic animals. Mailing Add: Dept of Zool & Physiol Univ of Wyo Laramie WY 82071

BERGMAN, HYMAN CHAIM, b Latvia, Oct 10, 05; nat US; m 37; c 2. PHARMACOLOGY. Educ: Univ Calif, Los Angeles, AB, 28; Univ Southern Calif, PhD(physiol), 37. Prof Exp: Jr chemist, Bur Animal Indust, USDA, 28-29; biochemist, Scripps Metab Clin, 29-33 & Hormone Assay Lab, 37-38; pharmacologist, Wilson Labs, 38-43; res assoc, Cedars Lebanon Hosp, Los Angeles, 43-48; dir, Joffe Labs, 48-51, Primorganics, Inc, 51-57 & Bergman Labs, 57-62; sr chemist, NAm Aviation, Inc, 62-69; SR CHEMIST, BERGMAN LABS, 69- Mem: AAAS; Am Chem Soc; Soc Exp Biol & Med; NY Acad Sci. Res: Carbohydrate metabolism; hormones; anterior pituitary; adrenal cortex; pharmacology of tissue extracts; mechanism of shock; cardiovascular physiology. Mailing Add: 2006 Chariton St Los Angeles CA 90034

BERGMAN, JOHN GEORGE, JR, b Menominee, Mich, Oct 20, 35; m 66. INORGANIC CHEMISTRY. Educ: Wayne State Univ, AB, 59, MS, 62; Mass Inst Technol, PhD(chem), 65. Prof Exp: High sch teacher, Mich, 59-61; res assoc surface chem, Res Lab Electronics, Mass Inst Technol, 65-66; MEM TECH STAFF CRYSTAL CHEM, BELL TEL LABS, 66- Mem: AAAS; Am Chem Soc; Am Phys Soc; Am Crystallog Asn. Res: Structural inorganic chemistry; crystallography; nonlinear optics; magnetochemistry; crystal growth. Mailing Add: Rm 4E-404 Bell Tel Labs Holmdel NJ 07733

BERGMAN, KENNETH HARRIS, b Plainfield, NJ, Dec 11, 35. METEOROLOGY. Educ: Rutgers Univ, BA, 58; Univ Wash, PhD(atmospheric sci), 69. Prof Exp: Design engr, Semiconductor Div, Radio Corp Am, 58-59; asst, Dept Atmospheric Sci, Univ Wash, 60-65; asst prof meteorol, San Jose State Col, 65-70; fel, Tex A&M Univ, 70-71; res meteorologist, Nat Hurricane Res Lab, 71-73, RES METEOROLOGIST, NAT METEOROL CTR, NAVAL WEAPONS STA, NAT OCEANIC & ATMOSPHERIC AGENCY, 73- Mem: Am Meteorol Soc; Am Geophys Union. Res: Mass exchange between stratosphere and troposphere; dynamics and dynamic stability of convective atmospheric vortices; atmospheric transport and diffusion in the 1000 meter boundary layer; dynamic meteorology, numerical modeling and weather prediction; numerical objective analysis of synoptic data. Mailing Add: Develop Div W322 NOAA NWS Nat Meteorol Ctr Washington DC 20233

BERGMAN, MOE, b Brooklyn, NY, Mar 28, 16; m 38; c 1. AUDIOLOGY. Educ: Univ Ill, AB, 37; Columbia Univ, MA, 39, EdD(audiol), 49. Prof Exp: Dir spec educ, Peekskill Union Free Schs, 38-43; chief audiologist, NY Regional Off, Vet Admin, 46-53; prof lect audiol, Col Physicians & Surgeons & Teachers Col, Columbia Univ, 48-58; dir speech & hearing ctr, 58-69, dir commun sci prog, 69-72, prof audiol, Hunter Col, 53-76; PROF AUDIOL, SCH MED, TEL-AVIV UNIV, ISRAEL, 76- Concurrent Pos: Consult, Albert Einstein Col Med, 62-; consult, Indust Home for Blind, 59-64 & New York Dept Health, 60-; exec officer, PhD Prog Speech, City Univ New York, 65-69; vis prof, Med Sch, Tel-Aviv Univ, 68-69 & 71. Mem: AAAS; Am Speech & Hearing Asn; Acoust Soc Am. Res: Binaural and stereophonic hearing; audiology in audiosurgery; auditory perception in blind persons; development of audiology services; hearing in primitive peoples; presbycucis; aging speech perception. Mailing Add: 10 Wissotzky St Tel-Aviv Israel

BERGMAN, NORMAN, b Seattle, Wash, Oct 14, 26; m 52; c 2. ANESTHESIOLOGY. Educ: Reed Col, BA, 49; Univ Ore, MD, 51. Prof Exp: Assoc anesthesiol, Columbia Univ, 54-58; asst attend, Presby Hosp, New York, 54-58; from asst prof to prof, Col Med, Univ Utah, 58-70; chief anesthesiol serv, Vet Admin Hosp, 58-70; PROF ANESTHESIOL & CHMN DEPT, MED SCH, UNIV ORE & DIR ANESTHESIA SERV, MED SCH HOSPS, 70- Concurrent Pos: Vis res assoc, Royal Col Surgeons & Post-Grad Med Sch, Univ London & Hammersmith Hosp, 63-64; Nat Heart Inst spec fel, 63-64. Mem: Am Soc Anesthesiol; AMA. Res: Respiratory changes during anesthesia; physiology of artificial respiration. Mailing Add: 7050 SW Ridgemont Portland OR 97225

BERGMAN, ROBERT GEORGE, b Chicago, Ill, May 23, 42; m 65. ORGANIC CHEMISTRY, ORGANOMETALLIC CHEMISTRY. Educ: Carleton Col, BA, 63; Univ Wis, PhD(chem), 66. Prof Exp: Arthur Amos Noyes instr chem, 67-69, from asst prof to assoc prof, 69-73, PROF CHEM, CALIF INST TECHNOL, 73- Concurrent Pos: NATO fel chem, Columbia Univ, 66-67; Am Chem Soc Petrol Res Fund grant, 67-72; Res Corp grant, 68-70; Alfred P Sloan Found fel, 70-72; NIH & NSF grants; Dreyfus Found Teacher-Scholar award, 70-75. Mem: Am Chem Soc. Res: Synthesis and reaction mechanisms of organic and transition metal organometallic compounds; catalysis; synthesis of nonbenzenoid aromatic molecules. Mailing Add: Dept of Chem Calif Inst of Technol Pasadena CA 91109

BERGMAN, ROBERT KAYE, b Great Falls, Mont, Apr 5, 34; m 57; c 6. IMMUNOLOGY. Educ: Mont State Univ, BS, 56, MS, 57; Univ Mo, PhD(environ physiol), 63. Prof Exp: Instr dairy cattle nutrit, Dairy Indust Dept, Mont State Univ, 58-60; res asst environ physiol, Dairy Husb Dept, Univ Mo, 60-63; staff fel, 63-66, OFFICER IMMUNOL, ROCKY MOUNTAIN LAB, NAT INST ALLERGY & INFECTIOUS DIS, 66- Res: Biological activities of substances in Bordetella pertussis that enhance hypersensitivity reactions such as histamine sensitivity, anaphylaxis, leucocytosis and experimental allergic encephalomyelitis in experimental animals. Mailing Add: Nat Inst Allergy & Infectious Dis Rocky Mountain Lab Hamilton MT 59840

BERGMAN, RONALD ARLY, b Chicago, Ill, June 21, 27; m 54; c 4. ANATOMY, PHYSIOLOGY. Educ: Univ Ill, BS, 50, MS, 53, PhD(physiol, zool & chem), 55. Prof Exp: Nat Found Infantile Paralysis fel, Karolinska Inst, Sweden, 55-56; instr epidemiol, Sch Hyg & Pub Health, Johns Hopkins Univ, 56-57, res assoc anat, Sch Med, 57-59, from asst prof to assoc prof, 68-74; PROF HUMAN MORPHOL, SCH MED, AM UNIV BEIRUT, 74- Concurrent Pos: Vis assoc prof anat, Am Univ Beirut, 69. Mem: Am Asn Anatomists; Am Soc Cell Biol; Electron Micros Soc Am. Res: Comparative and cell biology; structure, function and pathology of striated and smooth muscle and myogenesis. Mailing Add: Sch of Med Am Univ of Beirut Beirut Lebanon

BERGMAN, RUSSEL THEODORE, b Hancock, Mich, Nov 14, 04; m 32; c 2. MEDICINE. Educ: Pac Union Col, BS, 29; Loma Linda Univ, MD, 30. Prof Exp: CLIN PROF UROL, SCH MED, LOMA LINDA UNIV, 48- Concurrent Pos: Clin prof, Univ Calif, Irvine; mem staff, White Mem Med Ctr, 33-; sr attend surgeon, Glendale Adventist Hosp, 33, Los Angeles County-Univ Southern Calif Med Ctr, 33; & Behrens Mem Hosp, 40- Mem: Am Med Asn; fel Am Col Surg; Int Col Surg; Am Urol Asn; Int Soc Urol. Res: Motion picture studies of surgical anatomy. Mailing Add: Sch of Med Dept of Urol Loma Linda Univ Loma Linda CA 92354

BERGMAN, STEFAN, b Czestochowa, Poland; US citizen; m 50. MATHEMATICS. Educ: Univ Berlin, PhD(math), 33. Prof Exp: Privatdocent & lectr math, Univ Berlin, 31-33; prof, Univs Tomsk & Tiflis, USSR, 34-37; lectr, Mass Inst Technol, 39-40 & 51-52; lectr, Yeshiva Col, 40-41, Brown Univ, 41-45 & Harvard Univ, 45-51; prof, 52-74, EMER PROF MATH, STANFORD UNIV, 74- Mem: Am Math Soc; Soc Indust & Appl Math; Am Acad Arts & Sci. Res: Pure and applied mathematics; fluid mechanics; theory of functions of several complex variables; partial differential equations; theory of compressible fluids. Mailing Add: Dept of Math Stanford Univ Stanford CA 94305

BERGMANN, ERIC ARNOLD, b Cincinnati, Ohio, May 5, 09; m 45; c 2. ANALYTICAL CHEMISTRY, POLYMER CHEMISTRY. Educ: Yale Univ, PhB, 31; Univ Cincinnati, MA, 33, PhD(chem), 35. Prof Exp: Control chemist, Andrew Jergens Co, 35-36; res chemist, Procter & Gamble Co, 36-40; res chemist, Texaco Inc, 40-74; RETIRED. Mem: Am Chem Soc. Res: Properties of greases; oxidation of fatty oils; thermochemistry. Mailing Add: Osborne Hill Rd Fishkill NY 12524

BERGMANN, ERNEST EISENHARDT, b New York, Nov 2, 42; m 67; c 3. LASERS. Educ: Columbia Univ, AB, 64; Princeton Univ, MA, 66, PhD(physics), 69. Prof Exp: Asst prof, 69-74, ASSOC PROF PHYSICS, LEHIGH UNIV, 74- Mem: Am Phys Soc; Optical Soc Am. Res: Quantum mechanics; theory and experiment in gas lasers and optical resonators. Mailing Add: Dept of Physics Bldg 16 Lehigh Univ Bethlehem PA 18015

BERGMANN, FRED HEINZ, b Feuchtwagen, Ger, Jan 26, 28; nat; m 66; c 2. GENETICS. Educ: Mass Inst Technol, BS, 50, MS, 51; Univ Wis, PhD(biochem), 57. Prof Exp: Jr res chemist, Ethicon Suture Labs, 51-53; USPHS fel, Microbiol Dept, Med Sch, Washington Univ, 57-59; fel biol, Brandeis Univ, 59-61; res biochemist, NIH, 61-66; sci adminr, Res Grants Br, 66-72; DIR GENETICS PROG, NAT INST GEN MED SCI, 72- Honors & Awards: USPHS Superior Serv Award, 74. Mem: Am Soc Human Genetics; Genetics Soc Am. Res: Carboxyl activation of amino acids; protein synthesis; science policy. Mailing Add: Genetics Prog Rm 918 Westwood Bldg Nat Inst Gen Med Sci NIH Bethesda MD 20014

BERGMANN, JOHN FRANCIS, b Los Angeles, Calif, Oct 8, 28. GEOGRAPHY. Educ: Univ Calif, Los Angeles, AB, 50, PhD, 59; Univ Tex, MA, 52. Prof Exp: Assoc geog, Univ Calif, Los Angeles, 56-57; asst prof, Southern Methodist Univ, 57-60; asst prof, 60-63, ASSOC PROF GEOG, UNIV ALTA, 63- Mem: Am Geog Soc; Asn Am Geog; Can Asn Geog. Res: Cultural and historical geography in Middle America; cultural geography of Cacao in aboriginal America and its commercialization in early Guatemala; settlement and land use in Patagonia. Mailing Add: Dept of Geog Univ of Alta Edmonton AB Can

BERGMANN, LOUIS LAWRENCE, b Vienna, Austria, Jan 23, 07; nat US; m 38. ANATOMY. Educ: Univ Vienna, MD, 32. Prof Exp: Demonstr anat, Univ Vienna, 29-31, asst, 31-34; asst path, Nat Med Col, Shanghai, 34-35; vol intern, Vienna Hosps, 35-37; pvt pract, 37-38; from assoc prof to prof anat, Med Sch, Middlesex Univ, 39-44; from instr to prof, Col Med, NY Univ, 44-64; PROF ANAT, NEW YORK MED COL, 64- Mem: Am Asn Anat; NY Acad Sci. Res: Neuroanatomy; vascular anatomy and pathology; gross anatomy; neurohistology; blood vessels of human ganglia and changes in their vascular pattern associated with age; parkinsonism. Mailing Add: Dept of Anat New York Med Col New York NY 10029

BERGMANN, OTTO, b Vienna, Austria, Feb 7, 25; m 57; c 2. THEORETICAL PHYSICS. Educ: Univ Vienna, PhD(physics), 49. Prof Exp: Scholar, Dublin Inst Adv Studies, 51-52; sr res fel, Dept Math Physics, Univ Adelaide, 52-55; sr res fel physics, Univ New Eng, Australia, 56-58; res physicist, Res Inst Adv Study, Univ Baltimore, 58-60; assoc prof physics, Univ Ala, 60-62; assoc prof, 62-67, PROF PHYSICS, GEORGE WASHINGTON UNIV, 67- Concurrent Pos: Vis prof, Univ Ala, 58-59; vis prof, Inst Theoret Physics, Univ Vienna, Austria, 73-74. Mem: Am Phys Soc. Res: Classical and quantized field theories; theory of relativity. Mailing Add: 1039 19th St S Arlington VA 22202

BERGMANN, PETER GABRIEL, b Berlin, Ger, Mar 24, 15; US citizen; m 36; c 2. THEORETICAL PHYSICS. Educ: Prague Univ, PhD(theoret physics), 36. Prof Exp: Mem & asst, Sch Math, Inst Adv Study, 36-41; asst prof physics, Black Mountain Col, 41-42 & Lehigh Univ, 42-44; staff mem & asst dir sonar anal group, Div War Res, Columbia Oceanog Inst, Woods Hole, 44-47; assoc prof, 47-50, PROF PHYSICS, SYRACUSE UNIV, 50- Concurrent Pos: Adj prof, Polytech Inst Brooklyn, 47-57; vis prof, Brandeis Univ, 57 & King's Col, Univ London, 58; vis prof, Yeshiva Univ, 59-63, prof & chmn dept, 63-64 & vis prof, Belfer Grad Sch Sci, 70-; lectr, Marburg Univ, 55, Univ Stockholm, 58 & Int Sch Cosmology & Gravitation, Int Ctr Sci Culture, Erice, Italy, 75; mem, Int Comt Gen Relativity & Gravitation. Honors & Awards: Boris Pregel Award, NY Acad Sci, 70. Mem: AAAS; Fedn Am Sci (vchmn, 61, chmn, 64); Am Phys Soc; Am Math Soc; Europ Phys Soc. Res: Relativistic field theories, including quantization; wave propagation and scattering; electron optics; tactosols; stochastic problems; irreversible processes. Mailing Add: Dept of Physics Syracuse Univ Syracuse NY 13210

BERGMARK, WILLIAM R, b Mankato, Minn, Oct 28, 40; m 63. ORGANIC CHEMISTRY. Educ: St Olaf Col, BA, 62; Mass Inst Technol, PhD(org chem), 66. Prof Exp: Res chemist, Am Cyanamid Co, 66-67; res assoc org chem, State Univ NY Buffalo, 67-68; asst prof, 68-73, PROF ORG CHEM, ITHACA COL, 73- Concurrent Pos: Res grants, Petrol Res Fund & Res Corp, 69- Mem: Am Chem Soc. Res: Photochemistry; small ring compounds; chemiluminescence. Mailing Add: Dept of Chem Ithaca Col Ithaca NY 14850

BERGNA, HORACIO ENRIQUE, b La Plata, Arg, Feb 1, 24; nat US; m 51; c 1. PHYSICAL CHEMISTRY, COLLOID CHEMISTRY. Educ: La Plata Nat Univ, Lic, 48, Dr Chem, 50. Prof Exp: Chemist colloid sci, Lab Testing Mat & Tech Res, 44-50; Fr Govt fel, Univ Paris, 51; US Govt & Orgn Am States fels, Mass Inst Technol, 52, mem staff surface chem, 52-56; sr res chemist, 56-74, STAFF CHEMIST, E I DU PONT DE NEMOURS & CO, 74- Concurrent Pos: Asst anal chem, La Plata Nat Univ, 44-46. Honors & Awards: Croix de Chevalier dans l'ordre des Palmes Academiques, French Govt, 70. Mem: AAAS; Am Chem Soc; Sigma Xi; NY Acad Sci; Arg Chem Asn. Res: Physical chemistry of inorganic colloids. Mailing Add: 34 Vining Lane Westhaven Wilmington DE 19807

BERGNER, PER-ERIK EMIL, b Orebro, Sweden, Apr 13, 32; m 56; c 3. THEORETICAL BIOLOGY. Educ: Univ Uppsala, MA, 56; Karolinska Inst, Sweden, BM, 60, Res MD, 62. Prof Exp: Res asst med physics, Karolinska Inst, Sweden, 58-61; sr scientist rd traffic, Swedish Rd Inst, Stockholm, 61-62; asst prof med physics, Karolinska Inst, Sweden, 62-64; vis scientist, Theoret Biol, Med Div, Oak Ridge Assoc Univs, 64-65, res scientist, 65-69; CHIEF INVESTR, RES CTR, ROCKLAND STATE HOSP, ORANGEBURG NY, 69- Concurrent Pos: With Swedish Res Inst Nat Defense, 57-58. Mem: Soc Math Biol. Res: Generalized tracer kinetics; cell population dynamics; dose-response theory. Mailing Add: Dept of Theoret Biokinetics Res Ctr Rockland State Hosp Orangeburg NY 10962

BERGNES, MANUEL, b New York, NY, Mar 21, 15; m 41; c 1. MEDICINE. Educ: Wagner Col, BS, 37; Long Island Col Med, MD, 41. Prof Exp: Intern, Holy Name Hosp, 41-42; resident path, Kings County Hosp, 45-48; asst path, Long Island Col Med, 46-48; PATHOLOGIST, PHOENIXVILLE HOSP, 49-; PATHOLOGIST, SACRED HEART HOSP, 50-; ASST PROF PATH, MED COL PA, 59- Concurrent Pos: Mem med adv bd, Nat Pituitary Agency. Mem: Am Soc Clin Path; Col Am Path. Res: Pathology. Mailing Add: Dept of Path Sacred Heart Hosp Norristown PA 19401

BERGOFSKY, EDWARD HAROLD, b Baltimore, Md, June 18, 27. PHYSIOLOGY. Educ: Univ Md, BS, 48, MD, 52. Prof Exp: Intern med, Mt Sinai Hosp, NY, 52-53; asst resident, Bellevue Hosp, 53-54; from asst resident to resident, Mt Sinai Hosp, 54-57; Polachek Found fel, NY, 57-59; NY Heart Asn sr res fel, 59-62; from asst prof to prof physiol, Sch Med, NY Univ, 62-74; PROF MED & HEAD PULMONARY DIS SECT, STATE UNIV NY STONY BROOK, 74- Concurrent Pos: Res fel physiol, Presby Hosp, Columbia Univ, 55-56; instr, Columbia Univ, 61-62. Mem: AAAS; Am Soc Clin Invest; Am Physiol Soc; Am Fedn Clin Res; NY Acad Sci. Res: Respiratory and circulatory physiology; regulation of pulmonary and systemic circulations;

distribution of oxygen and carbon dioxide in tissues; mechanical performance and biophysics of muscles of respiration. Mailing Add: Sch of Med State Univ NY Stony Brook NY 11790

BERGOMI, ANGELO, b Milan, Italy, Oct 22, 33; US citizen. ORGANIC CHEMISTRY. Educ: Univ Milan, Dr(org chem), 60; Univ Birmingham, MSc, 65. Prof Exp: Chemist org chem, Montecatini-Edison, 61-69; SR RES CHEMIST ORG CHEM, GOODYEAR TIRE & RUBBER CO, 69- Mem: Am Chem Soc. Res: Catalytic isomerization of hydrocarbons; synthesis of rubber chemicals. Mailing Add: Goodyear Tire & Rubber Co 142 Goodyear Blvd Akron OH 44316

BERGON, LLOYD, b Brooklyn, NY, Dec 7, 44. EXPERIMENTAL PATHOLOGY. Educ: Brooklyn Col, BA, 65; State Univ NY Buffalo, PhD(exp path), 74. Prof Exp: AM HEART ASN FEL HYPERTENSION, PENROSE RES LAB, 74- Res: Isolation and characterization of a pituitary protein noted in rats bred based on their resistance to salt-induced hypertension. Mailing Add: Penrose Res Lab 34th & Girard Ave Philadelphia PA 19104

BERGQUIST, HARLAN RICHARD, b Shell Lake, Wis, Nov 21, 08; m 53. GEOLOGY. Educ: Univ Minn, BA, 31, MA, 36, PhD(geol), 38. Prof Exp: Instr, Univ Minn, 36-38; geologist, Standard Oil Co Tex & Calif Co, 38-39; asst prof geol & phys geog, Univ Miss, 39-40; geologist & micropaleontologist, State Geol Surv, Miss, 40-42; assoc prof geol & phys geog, Univ Miss, 42; geologist, 42-60, RES GEOLOGIST, US GEOL SURV, 60- Mem: Fel Geol Soc Am; Am Paleont Soc. Res: Biostratigraphy of American Cretaceous and Jurassic foraminifera. Mailing Add: US Geol Surv Stop 970 Reston VA 22092

BERGQUIST, JAMES WILLIAM, b Ottumwa, Iowa, Apr 23, 28; m 51. COMPUTER SCIENCES, APPLIED MATHEMATICS. Educ: Iowa State Univ, BS, 50; Univ Southern Calif, MS, 55, PhD(math), 63. Prof Exp: Math analyst, Lockheed Aircraft Corp, 51-54; design engr, Gilfillan Bros Electronics, 55-57; mathematician, 58-68, SCI REP, IBM CORP, 68- Concurrent Pos: IBM consult, Calif Inst Technol, vis assoc, 63-68. Mem: Am Math Soc; Soc Indust & Appl Math; Am Inst Aeronaut & Astronaut. Res: Combinatorial analysis; mathematical theory of vision; computers and computation; symbolic manipulation. Mailing Add: 4705 Daleridge Rd La Canada CA 91011

BERGREN, WILLIAM RAYMOND, bTacoma, Wash, July 25, 10; m 41; c 2. BIO-ORGANIC CHEMISTRY. Educ: Calif Inst Technol, BS, 32, PhD(bio-org chem), 41. Prof Exp: Asst animal physiol, Calif Inst Technol, 35-37; develop chemist, Fisheries Indust, 40-41; dir res, Truesdail Labs, 41-42; from asst prof to assoc prof biochem, 47-67, PROF BIOCHEM, UNIV SOUTHERN CALIF, 67- Concurrent Pos: Head biochem res lab, Children's Hosp, Los Angeles, 51-73, dir res, 73- Mem: AAAS; Am Asn Clin Chem; Brit Biochem Soc. Res: Biochemistry of genetic diseases. Mailing Add: 4650 Sunset Blvd Los Angeles CA 90027

BERGS, VICTOR VISVALDIS, b Kuldiga, Latvia, Dec 20, 23; US citizen; m 60; c 1. VIROLOGY. Educ: Boston Univ, AM, 55; Univ Pa, PhD(med microbiol), 58. Prof Exp: Res asst virol, Children's Med Ctr, Boston, 54-55; res assoc, Children's Hosp, Philadelphia, 58; asst prof, Inst Microbiol, Rutgers Univ, 58-62; res virologist, Stanford Res Inst, 63-65; from asst prof to assoc prof microbiol, Sch Med, Univ Miami, 65-73; ASSOC VIROLOGIST & MGR, LIFE SCI, INC, 73- Concurrent Pos: Asst dir, Variety Children's Res Found, 66-70. Mem: Am Asn Cancer Res; Am Asn Immunol; Am Soc Microbiol; Soc Exp Biol & Med; fel Am Acad Microbiol. Res: Oncogenic and endogenous viruses; diagnosis and chemotherapy of virus-induced diseases; tissue culture. Mailing Add: 14829 Crown Dr Largo FL 33540

BERGSAGEL, DANIEL EGIL, b Outlook, Sask, Apr 25, 25; m 50; c 4. INTERNAL MEDICINE, ONCOLOGY. Educ: Univ Man, MD, 49; Oxford Univ, DPhil, 55; Am Bd Internal Med, dipl, 59. Prof Exp: Asst resident, Winnipeg Gen Hosp, 50-51 & Salt Lake County Gen Hosp, Utah, 51-52; assoc internist, Univ Tex M D Anderson Hosp, 55-65; PROF MED, UNIV TORONTO, 65-; CHIEF OF MED, PRINCESS MARGARET HOSP, 65- Mem: AMA; Am Fedn Clin Res; Int Soc Hemat. Res: Tumor growth; cancer chemotherapy. Mailing Add: Princess Margaret Hosp 500 Sherbourne St Toronto ON Can

BERGSMA, DANIEL, b Wallington, NJ, Apr 4, 09; m 37; c 2. PREVENTIVE MEDICINE. Educ: Oberlin Col, AB, 32; Yale Univ, MD, 36; Univ Mich, MPH, 46. Prof Exp: Chief bur venereal dis control, State Dept Health, NJ, 40-42, dep dir health, 46-47, pub health consult, 47-48, state comnr health, 48-59; assoc dir med care, 59-64, dir med dept, 64-69, VPRES PROF EDUC, NAT FOUND-MARCH OF DIMES, 69- Concurrent Pos: Mem, Interstate Sanit Comn, 51-59; adv to US deleg, World Health Assembly, Geneva, 56; head party, Health Surv Nicaragua, Int Coop Admin, 58. Mem: AAAS; AMA; fel Am Pub Health Asn; Am Soc Pub Admin; Asn State & Territorial Health Off (pres, 56). Res: Administrative methodology. Mailing Add: Nat Found-March of Dimes 1275 Mamaroneck Ave White Plains NY 10605

BERGSTRESSER, KENNETH A, b Lower Saucon Township, Pa, May 25, 12; m 39; c 3. BIOLOGY, ZOOLOGY. Educ: Albright Col, BS, 34; Univ Pittsburgh, MS, 37; Lehigh Univ, PhD(cell biol), 74. Prof Exp: Asst prof biol, Beaver Col, 37-41; field dir, Am Red Cross, 43-45; assoc prof biol, Moravian Col, 46-52; chief bacteriologist, R K Laros Co, 52-58; assoc prof biol, 58-71, PROF BIOL, MORAVIAN COL, 71- Mem: AAAS; Am Soc Microbiol; Sigma Xi; Am Soc Zoologists. Res: Bacteriology; industrial fermentations; cytology. Mailing Add: Dept of Biol Moravian Col Bethlehem PA 18018

BERGSTRESSER, PAUL RICHARD, b Ottawa, Kans, Aug 24, 41; m 69; c 2. DERMATOLOGY. Educ: Sch Med, Stanford Univ, MD, 68. Prof Exp: Intern med, Sch Med, Univ NMex, 68-69; resident dermatol, Sch Med, Stanford Univ, 69-70; resident, 72-74, res assoc, 74-75, ASST PROF DERMATOL, SCH MED, UNIV MIAMI, 75- Mem: Soc Invest Dermatol; AAAS. Res: Computer assisted diagnoses in medicine; control mechanisms in epidermal proliferation. Mailing Add: 427 Aragon Ave Coral Gables FL 33134

BERGSTRESSER, THOMAS KARL, b Ottawa, Kans, Oct 8, 38; m 66. THEORETICAL SOLID STATE PHYSICS. Educ: Calif Inst Technol, BS, 60; Univ Calif, Berkeley, PhD(physics), 66. Prof Exp: NSF fel, Cavendish Lab, Cambridge Univ, 66-67; res assoc, James Franck Inst, Univ Chicago, 67-69; asst prof physics, Univ Wis, Madison, 69-73; VIS ASST PROF PHYSICS, CLARK UNIV, WORCESTER, MASS, 73- Mem: Am Phys Soc. Res: Band structure and optical properties of semiconductors and insulators; magnetism and critical phenomena. Mailing Add: Dept of Physics Clark Univ Worcester MA 01610

BERGSTROM, CLARENCE GEORGE, b Chicago, Ill, Oct 23, 25; m 48; c 3. ORGANIC CHEMISTRY. Educ: Ill Inst Technol, PhD(chem), 51. Prof Exp: RES CHEMIST, G D SEARLE & CO, 50- Mem: Am Chem Soc. Res: Steroids; synthetic organic chemistry; medicinal chemistry. Mailing Add: 6945 N Osceola Ave Chicago IL 60631

BERGSTROM, DAVID WALLACE, b Oak Park, Ill, May 7, 17; m 45; c 2. ZOOLOGY. Educ: N Park Col, AA, 39; Cent YMCA Col, BS, 41; Univ NMex, MS, 51. Prof Exp: Instr biol, N Park Col, 45-51; Denforth fel, 56-57, coordr biol sci, Middletown Campus, 67-73, asst prof zool, 51-68, ASSOC PROF ZOOL & PHYSIOL, MIDDLETOWN & OXFORD CAMPUSES, MIAMI UNIV, 68- Mem: AAAS; Am Asn Higher Educ. Res: Invertebrate systematics and ecology; human biology; preparation of college teachers. Mailing Add: Dept of Zool Middletown Campus Miami Univ Middletown OH 45042

BERGSTROM, ROBERT CHARLES, b Newcastle, Wyo, Mar 23, 22; m 48; c 2. VETERINARY PARASITOLOGY. Educ: Colo State Univ, BS, 50; Univ Wyo, MS, 56, PhD(zool), 64. Prof Exp: Surv supvr, Bur Entom & Plant Quarantine, USDA, 50-51; instr high sch, Wyo & Mont, 51-57; technologist, Dept Entom & Parasitol, Univ Wyo, 57-58; instr high sch, Wyo, 58-61; teaching asst zool, 61-62, PARASITOLOGIST, DIV MICROBIOL & VET MED, UNIV WYO, 62- Concurrent Pos: NIH fel, Med Sch, Mex, 66. Mem: Am Soc Parasitologists; World Asn Advan Vet Parasitol. Res: Helminths of domestic animals, especially research physiological responses of the host animal to nematode parasite infections. Mailing Add: Div of Microbiol & Vet Med Univ of Wyo Laramie WY 82070

BERGSTROM, ROBERT EDWARD, b Rock Island, Ill, Mar 27, 23; m 46; c 3. GEOLOGY. Educ: Augustana Col (Ill), AB, 47; Univ Wis, MS, 50, PhD(geol), 53. Prof Exp: Instr geol, Beloit Col, 50-51; integrated lib studies, Univ Wis, 52-53; asst geologist, Ill Geol Surv, 53-55, assoc geologist, 55-59, geologist, 59-61; proj geologist, Kuwait, Parsons Corp, 61-63; geologist & head groundwater geol & geophys explor sect, 63-74; PRIN GEOLOGIST, ILL GEOL SURV, 74- Concurrent Pos: Mem comt geol sci, Nat Acad Sci. Mem: Geol Soc Am. Res: Ground water geology; geology of Kuwait; geology of waste disposal. Mailing Add: Ill Geol Surv Urbana IL 61801

BERGSTROM, STIG MAGNUS, b Skövde, Sweden, June 12, 35. GEOLOGY, INVERTEBRATE PALEONTOLOGY. Educ: Lund Univ, Sweden, Fil Kand, 59, PhD(geol), 61. Prof Exp: Amanuensis geol, Lund Univ, 58-60; res asst, Ohio State Univ, 60-61; asst prof & lectr, Lund Univ, 62-68; from asst prof & cur to assoc prof & cur, 68-72, PROF GEOL, OHIO STATE UNIV, 72- Concurrent Pos: Scand-Am Found fel, 64. Mem: Geol Soc Am; Paleont Soc; Paleont Asn; Am Asn Petrol Geologists; Bot Soc Am. Res: Stratigraphy, fossils and geologic history of the lower Paleozoic of North America and Northwestern Europe. Mailing Add: Dept of Geol & Mineral Ohio State Univ 125 S Oval Mall Columbus OH 43210

BERGSTROM, WILLIAM H, b Bay City, Mich, Jan 1, 21; m 44; c 4. PEDIATRICS. Educ: Amherst Col, BA, 42; Univ Rochester, MD, 45. Prof Exp: Assoc prof, 55-65, PROF PEDIAT, STATE UNIV NY UPSTATE MED CTR, 55- Mem: AAAS; Am Pediat Soc; Soc Pediat Res; Soc Exp Biol & Med. Res: Electrolyte metabolism; blood-bone equilibria. Mailing Add: Dept of Pediat State Univ NY Upstate Med Ctr Syracuse NY 13210

BERGTROM, GERALD, b New York, NY, Feb 14, 45. INSECT PHYSIOLOGY. Educ: City Col New York, BSc, 67; Brandeis Univ, PhD(biol), 73. Prof Exp: Fel insect develop, Univ Conn, 73-75; ASST PROF BIOL, VANDERBILT UNIV, 75- Mem: AAAS; Am Inst Biol Sci; Am Soc Zoologists. Res: Changes in macromolecular synthesis during insect development and in response to insect growth hormones in organ and tissue culture, using the hemoglobins of chironomus as a model system. Mailing Add: Dept of Gen Biol Vanderbilt Univ Nashville TN 37235

BERHENKE, LUTHER FREDERICK, organic chemistry, see 12th edition

BERING, EDGAR ANDREW, JR, b Salt Lake City, Utah, Feb 18, 17; m 44; c 3. MEDICINE, NEUROSURGERY. Educ: Univ Utah, AB, 37; Harvard Univ, MD, 41. Prof Exp: Surg house officer, Boston City Hosp, 41-42; spec res assoc phys chem, Med Sch, Harvard Univ, 42; asst neurosurg, spec res assoc & demonstr anat, NY Med Col & Flower & Fifth Ave Hosps, 46-48; Moseley traveling fel from Harvard Med Sch, Nat Hosp Queens Sq, London, 48-49; resident neurosurg, Children's Hosp, Boston, 49-50, asst neurosurgeon, 50-54; dir neurosurg res lab, Children's Med Ctr, 52-63, assoc neurosurgeon, 54-64; actg chief prog anal, Nat Inst Neurol Dis & Stroke, 64-65, spec asst to dir, 65-71, chief spec proj br, 71-74; ASSOC CLIN PROF NEUROSURG, SCH MED, GEORGETOWN UNIV, 69- Concurrent Pos: Resident, Peter Bent Brigham Hosp, Boston, 49-50, jr assoc, 50; res fel surg, Harvard Med Sch, 50-52; Cushing fel, 50-51; Nat Res Coun sr fel, 50-52; from asst to asst clin prof, Harvard Med Sch, 52-64; mem, Am Bd Neurol Surg, 52; from consult to sr consult, Lemuel Shattuck Hosp, Boston, 54-64; attend neurosurgeon, West Roxbury Vet Admin Hosp, 55-64; vis lectr, Univ Calif, Los Angeles, 58; vis scientist, Nat Inst Neurol Dis & Stroke, 63-64. Mem: AAAS; Am Asn Neurol Surg; Neurosurg Soc Am; Am Acad Neurol; Soc Neurosci. Res: Metabolism and physiology of the nervous system, cerebrospinal fluid; problems relating to the surgery of the nervous system; biophysics. Mailing Add: PO Box 1498 Easton MD 21601

BERINGER, FREDERICK MARSHALL, b New York, NY, May 8, 20; m 44; c 1. ORGANIC CHEMISTRY. Educ: Harvard Univ, BS, 41; Columbia Univ, MS, 44, PhD(org chem), 47. Prof Exp: Org res chemist, Barrett Div, Allied Chem & Dye Corp, NJ, 41-43; asst, Rockefeller Inst, 47; from instr to assoc prof org chem, 48-59, head dept chem, 64-68, dean sci, 68-70, dean arts & sci, 70-74, PROF ORG CHEM, POLYTECH INST NEW YORK, 59-, ASSOC PROVOST, 74- Concurrent Pos: Vis assoc prof, Yale Univ, 58-59; vis scholar, Columbia Univ, 63-64; ed, Monogr, Am Chem Soc, 63-72. Mem: AAAS; Am Chem Soc; The Chem Soc. Res: Organic synthesis; reaction mechanisms; kinetics; polyvalent iodine. Mailing Add: Polytech Inst of New York 333 Jay St Brooklyn NY 11201

BERINGER, ROBERT, b Pittsburgh, Pa, Oct 14, 17; m 42; c 3. PHYSICS. Educ: Washington & Jefferson Col, BS, 39; Yale Univ, PhD(physics), 42. Prof Exp: Asst radiation lab, Mass Inst Technol, 42-46; asst prof physics, 46-50, from assoc prof to PROF PHYSICS, YALE UNIV, 50-, DIR HEAVY ION ACCELERATOR LAB. Mem: Am Phys Soc. Res: Counting techniques; microwave physics; electronics; nuclear particle accelerators. Mailing Add: Sloane Physics Lab Yale Univ New Haven CT 06520

BERISFORD, ROBERT, physical chemistry, inorganic chemistry, see 12th edition

BERK, ABRAHAM ALBERT, b New York, NY, Aug 1, 07; m 31; c 3. CHEMISTRY. Educ: Cooper Union, BS, 28; NY Univ, MS, 31. Prof Exp: Mem staff, 31-41, assoc chemist, 41-45, chemist, 46-51, CHIEF INDUST WATER BR, US BUR MINES, 51- Honors & Awards: Hecht Award, Am Soc Testing & Mat, 56; Distinguished Serv Award, Dept Interior, 66; Res Cert, Am Soc Mech Engrs, 68; Distinguished Serv Cert, Am Chem Soc, 70. Mem: Am Chem Soc; Am Soc Mech Engrs. Res: Corrosion of metals, including piping, boilers, air preheaters, tanks and valves; analytical methods in water; water-formed deposits; recovery of potassium salts from polyhalite; waterborne industrial wastes; boiler feedwater treatment; fuels utilization. Mailing Add: US Bur of Mines College Park MD 20740

BERK, BERNARD, b Brooklyn, NY, Dec 23, 15; m 40; c 1. CHEMISTRY. Educ: Ind Univ, BA, 37, MA, 38, PhD(org chem), 41; NY Univ, MChE, 47. Prof Exp: Jr chemist, Ind Ord Works, 41-42; res chemist, Lawrence Richard Bruce, Conn, 42-46; res & develop chemist & head dept, Tech Serv Lab, 46-66, mgr mfg, Overseas Plants, 53-55, assoc tech dir, Overseas Div, 55-58, dir steroid prep lab, Squibb Inst Med Res, 58-66, dir res prod lab, 66-70, MGR, ANTIBIOTIC MFG PLANNING, E R SQUIBB & SONS, 70- Mem: Assoc Am Chem Soc; Am Inst Chem Eng; assoc NY Acad Sci. Res: Organic synthesis; antibiotics; wool dyeing processes; vapor phase catalytic reduction over intermetallic catalysts; crystallization of penicillin salts. Mailing Add: ER Squibb & Sons Princeton NJ 08540

BERK, HAROLD, b Minneapolis, Minn, July 27, 17; m 42; c 3. HISTOLOGY. Educ: Northwestern Univ, DDS, 41; Am Bd Pedodont, dipl, 51. Prof Exp: Chief dent clin, Forsyth Dent Infirmary Children, 42-46; asst prof, 46-72, ASSOC PROF ORAL PEDIAT, SCH DENT MED, TUFTS UNIV, 72-, ASST PROF RESTORATIVE DENT, FORSYTH DENT CTR, 73- Concurrent Pos: Lectr, Forsyth Sch Dent Hygienists, 43-; consult, Childrens Hosp, Washington, DC, Beth Israel Hosp, 44-, Floating Hosp, New Eng Med Ctr, 45- & USPHS Hosp, Brighton, 50-55; dent surgeon, NIH, 55-57. Honors & Awards: Dipl d'Honneur, Int Film Cong, Paris, 62. Mem: Fel Am Acad Pedodont; fel Am Acad Dent Handicapped; fel Col Dent; fel Int Col Dent; Am Soc Dent for Children. Res: Dental pulp, materials and caries. Mailing Add: 1249 Beacon St Brookline MA 02146

BERK, HERBERT L, plasma physics, see 12th edition

BERK, JACK EDWARD, b Philadelphia, Pa, Nov 24, 11; m 37; c 2. MEDICINE. Educ: Univ Pa, BA, 32, MSc, 39, DSc(med), 43; Jefferson Med Col, MD, 36; Am Bd Internal Med & Am Gastroenterol, dipl, 43. Prof Exp: Instr gastroenterol, Grad Sch Med, Univ Pa, 41-46; asst prof, Med Sch, Temple Univ, 46-54, res assoc & asst dir, Fels Res Inst, 46-54; assoc prof clin med, Wayne State Univ, 54-62, clin prof, 62-63; prof med & chmn dept, 63-73, CHMN DIV GASTROENTEROL, UNIV CALIF, IRVINE, 63- Concurrent Pos: Consult, Surgen Gen, US Army, 47-, Cedars of Lebanon, Mt Sinai & White Mem Hosps, Los Angeles, Mem Hosp, Long Beach, 63- & Long Beach Vet Admin Hosp, 63-; dir dept med, Sinai Hosp, Detroit, Mich, 54-63; vis lectr, Grad Sch Med, Univ Pa, 61-; chief gastroenterol sect, Tilton & Rhoads Gen Hosps. Honors & Awards: Distinguished Serv Award, Mich State Med Sco, 59; Schindler Award, Am Soc Gastrointestinal Endoscopy, 66; Rorer Award, Am Col Gastroenterol, 70 & 74. Mem: Am Soc Gastrointestinal Endoscopy (pres, 58); Bockus Int Soc Gastroenterol (pres, 67-71); fel Am Col Physicians; Am Gastroenterol Asn; Am Col Gastroenterol (pres-elect, 74). Res: Acidity of the first part of duodenum in ulcer; hepatic function tests and the status of the liver in varying disorders; radiographic demonstration of the extra-hepatic bile ducts; radiographic visualization of the liver and spleen; characterization of serum amylase; tetracycline fluorescence in gastric lesions. Mailing Add: Dept of Med Univ of Calif Irvine CA 92664

BERK, KENNETH N, b Takoma Park, Md, Sept 24, 38; m 68. MATHEMATICS. Educ: Carnegie Inst Technol, BS, 60; Univ Minn, PhD(math), 65. Prof Exp: Asst prof math, Northwestern Univ, 65-68; asst prof, Northeastern Ill State Col, 68-69; ASSOC PROF MATH, ILL STATE UNIV, 69- Mem: Inst Math Statist; Soc Indust & Appl Math; Am Statist Asn. Res: Probability; statistics; time series. Mailing Add: Dept of Math Ill State Univ Normal IL 61761

BERK, RICHARD SAMUEL, b Chicago, Ill, Oct 7, 28. MICROBIOLOGY, BIOCHEMISTRY. Educ: Roosevelt Univ, BS, 50; Univ Minn, MS, 55; Univ Chicago, PhD(microbiol), 58. Prof Exp: Researcher, Univ Calif, Los Angeles, 58-61; sr res biochemist, Magna Prods, Calif, 61-62; assoc prof microbiol, 62-69, PROF MICROBIOL, COL MED, WAYNE STATE UNIV, 69- Mem: Am Chem Soc; Am Soc Microbiol; Soc Exp Biol & Med. Res: Bacteriocins; endotoxins; sulfur metabolism; bacterial cytology; electron microscopy; pathological study of exotoxins. Mailing Add: Col of Med Wayne State Univ Detroit MI 48201

BERK, ROBERT HAROLD, mathematical statistics, see 12th edition

BERK, ROBERT NORTON, b Pittsburgh, Pa, Sept 3, 30; m 56; c 3. RADIOLOGY. Educ: Univ Pittsburgh, BS, 51, MD, 55. Prof Exp: From instr to asst prof radiol, Sch Med, Univ Pittsburgh, 61-65; chief radiol, Passavant Hosp, Pittsburgh, 66-68; assoc prof, Sch Med, Univ Calif, San Diego, 68-74; PROF RADIOL & CHMN DEPT, UNIV TEX HEALTH SCI CTR, DALLAS, 74- Concurrent Pos: Consult radiol, Balboa Naval Hosp, San Diego & Vet Admin Hosp, La Jolla, 70-74; dir radiol, Parkland Mem Hosp, Dallas, 74-; consult radiol, Vet Admin Hosp, Dallas, 74- & Baylor Univ Med Ctr, 74-; chief radiol serv, Children's Med Ctr, Dallas, 74- Mem: Asn Univ Radiologists; Am Col Radiol; Radiol Soc NAm; Am Roentgen Ray Soc; Soc Gastrointestinal Radiologists. Res: Pharmacokinetics of radiographic contrast materials for cholangiography and cholecystography. Mailing Add: Univ Tex Health Sci Ctr 5323 Harry Hines Blvd Dallas TX 75235

BERK, TOBY STEVEN, b Chicago, Ill, Jan 15, 44; m 65. COMPUTER SCIENCES. Educ: Univ Mich, BSE, 65; Purdue Univ, MS, 68, PhD(comput sci), 72. Prof Exp: Instr comput sci, Ind Univ-Purdue Univ, Indianapolis, 70-72; asst prof, 72-75, ASSOC PROF COMPUT SCI, FLA INT UNIV, 75- Concurrent Pos: Consult, Dade County Bd Educ, 73 & Metrop Mus, 74; reviewer, Acad Press, 76. Mem: Asn Comput Mach; Inst Elec & Electronic Engrs. Res: Computer graphics, minicomputers, graphic languages. Mailing Add: Dept of Math Sci Fla Int Univ Miami FL 33199

BERKE, HARRY L, b New York, NY, July 17, 10; m 45; c 3. BIOCHEMISTRY, PHARMACOLOGY. Educ: City Col NY, BS, 32; Univ Rochester, MS, 48, PhD(biochem), 55. Prof Exp: With Civil Serv, New York, 39-42; res asst, Manhattan Proj, US AEC, Univ Rochester, 43-45, asst chief physiol sect, 45-46, scientist, 46-55, from instr to asst prof radiation biol & toxicol, Univ, 56-63; ASSOC PROF OCCUP & ENVIRON HEALTH, WAYNE STATE UNIV, 63- Mem: Fel AAAS; Am Chem Soc; Am Indust Hyg Asn; fel Am Inst Chem; Health Physics Soc. Res: Inhalation toxicology; radioactive rare earths and polonium; effects of radiation, radioisotopes and uranium on the blastogenic response in lymphocytes. Mailing Add: Dept Occup & Environ Health Wayne State Univ Sch of Med Detroit MI 48202

BERKEBILE, CHARLES ALAN, b Jamaica, NY, Mar 4, 38; m 60; c 1. MINERALOGY, CRYSTALLOGRAPHY. Educ: Allegheny Col, BS, 60; Boston Univ, AM, 61, PhD(geol), 64. Prof Exp: Mem staff crystal growth, Lab for Insulation Res, Crystal Physics Sect, Mass Inst Technol, 63-64; asst prof earth sci, Southampton Col, Long Island, 64-67; mem res staff, Tech Staffs Div, Corning Glass Works, 67-69; assoc prof geol & chmn dept, 69-75, PROF GEOL & MARINE SCI, SOUTHAMPTON COL, LONG ISLAND UNIV, 75- Concurrent Pos: Vis assoc scientist & res collab, Crystal Struct Anal Group, Chem Dept, Brookhaven Nat Lab, 66-67. Mem: Geol Soc Am; Mineral Soc Am; Am Crystallog Asn; Nat Asn Geol Teachers. Res: Ground-water resources; marine and environmental geology. Mailing Add: Sci Div Southampton Col Long Island Univ 239 Montauk Hwy Southampton NY 11968

BERKEBILE, JAMES MARCUS, chemistry, see 12th edition

BERKELEY, EDMUND, b Red Hill, Va, Aug 10, 12; m 36; c 3. BOTANY. Educ: Univ Va, BS, 41, MS, 46; Univ NC, PhD(bot), 51. Prof Exp: Teacher high sch, Va, 46-47; from instr to asst prof biol, Washington & Lee Univ, 47-49; prof & head dept, Washington Col, Md, 51-52; from asst prof to assoc prof, Univ of the South, 52-57; asst adminstr, Univ Ctr Va, 57-59; from asst prof to assoc prof biol, Univ NC, Greensboro, 60-67, actg head dept, 61-63; chmn div natural sci, Cent Va Community Col, 67-70, prof biol, 67-72; prof biol, Piedmont Va Community Col, 72-74; RETIRED. Mem: AAAS; Bot Soc Am. Res: Development of the mitotic spindle in the giant amoeba; floral morphology; biographical studies in the history of American botany. Mailing Add: 42 Canterbury Rd Bellair Charlottesville VA 22901

BERKELHAMMER, GERALD, b Newark, NJ, Feb 3, 31; m 54; c 3. ORGANIC CHEMISTRY. Educ: Brown Univ, AB, 52; Univ Wash, PhD(org chem), 57. Prof Exp: Res chemist, 57-60, group leader, 60-70, MGR ORG SYNTHESIS, AM CYANAMID CO, 70- Mem: Am Chem Soc; The Chem Soc. Res: Agricultural and medicinal chemistry. Mailing Add: Agr Div Am Cyanamid Co Box 400 Princeton NJ 08540

BERKELMAN, KARL, b Lewiston, Maine, June 7, 33; m 59; c 3. EXPERIMENTAL HIGH ENERGY PHYSICS. Educ: Univ Rochester, BS, 55; Cornell Univ, PhD(physics), 59. Prof Exp: Instr & res assoc physics, 59-60, from asst prof to assoc prof, 61-67, PROF PHYSICS, LAB NUCLEAR STUDIES, CORNELL UNIV, 67- Concurrent Pos: NSF fels, Ital Nat Synchrotron Lab, Frascati, 60-61; European Orgn Nuclear Res, Geneva, Switz, 60-61 & 67-68; sci staff mem, Ger Electron Synchrotron, Hamburg, 74-75. Mem: Am Phys Soc. Res: Meson photoproduction experiments; high energy electron scattering experiments; phenomenological theory of high energy particle reactions; electron-positron colliding beam experiments. Mailing Add: Lab of Nuclear Studies Cornell Univ Ithaca NY 14850

BERKENKAMP, BILL BRODIE, b National City, Calif, Jan 30, 31; m 52; c 3. PLANT PATHOLOGY. Educ: Ariz State Univ, BS, 55; Univ Ariz, MS, 58, PhD(plant path), 62. Prof Exp: RES OFFICER, CAN DEPT AGR, 62- Concurrent Pos: Mem subcomt plant dis, Nat Res Coun Can, 63- Mem: Am Phytopath Soc; Can Phytopath Soc. Res: Diseases of field crops; mechanisms of resistance. Mailing Add: Res Br Res Sta Can Dept Agr Lacombe AB Can

BERKES, JOHN STEPHAN, b Buzias, Rumania, Sept 5, 40; US citizen; m 69; c 1. SURFACE SCIENCE. Educ: Mich Technol Univ, BS, 62; Pa State Univ, MS, 64, PhD(solid state sci), 68. Prof Exp: Scientist mat sci, 68-73, mgr, 73-74, MGR MATERIAL SCI, XEROX CORP, 74- Mem: Am Ceramic Soc; Am Vacuum Soc; Sigma Xi. Res: Structure, thermal properties, viscous flow, bulk and surface stability of amorphous chalcogenide photoconducting materials; thin films technology. Mailing Add: Xerox Corp Webster Res Ctr W 114 800 Phillips Rd Webster NY 14580

BERKEY, DENNIS ALAN, b Woodburn, Ore, Mar 3, 44; m 65; c 2. SEED PHYSIOLOGY. Educ: Ore State Univ, BS, 68; Miss State Univ, MS, 73, PhD(agron), 74. Prof Exp: SEED PHYSIOLOGIST, ASGROW SEED CO, UPJOHN CO, 75- Mem: Soc Com Seed Technologists; Am Soc Agron; Crop Sci Soc Am. Res: Seed physiology and pathology relating to seed quality, crop performance and seed testing. Mailing Add: Asgrow Seed Co 634 Lincoln Way E Ames IA 50010

BERKEY, EDGAR, b Mexico City, Mex, Sept 1, 40; US citizen; m 63; c 1. PHYSICAL CHEMISTRY, NUCLEAR SCIENCE. Educ: Stanford Univ, BS, 62; Cornell Univ, PhD(nuclear sci), 67. Prof Exp: Sr res scientist, 67-71, MGR LIQUID METAL TECHNOL PHYS CHEM, WESTINGHOUSE RES LABS, 71- Concurrent Pos: Gen chmn, Am Nuclear Soc, Int Conf Liquid Metal Technol Energy Prod, 75-76. Mem: Am Nuclear Soc. Res: Liquid metal technology for advanced energy production systems like fast breeder reactor, fusion reactor, and high power density electromechanical machines; including instrumentation development, chemistry, materials interactions and system definition. Mailing Add: Westinghouse Res Labs 1310 Beulah Rd Pittsburgh PA 15235

BERKEY, GORDON BRUCE, b DuBois, Pa, June 1, 42; m 64; c 2. ASTROPHYSICS, PHYSICS. Educ: Cornell Univ, AB, 64; Purdue Univ, MS, 67, PhD(physics), 69. Prof Exp: ASST PROF PHYSICS, MINOT STATE COL, 69- Mem: AAAS; Am Phys Soc. Res: Propagation of galactic cosmic-ray electrons; production of x-rays and gamma rays by cosmic-ray electrons. Mailing Add: Dept of Physics Minot State Col Minot ND 58701

BERKEY, REYNOLD A, organic chemistry, see 12th edition

BERKHEIMER, HENRY EDWARD, b Williamsport, Pa, Oct 13, 29; div; c 3. RUBBER CHEMISTRY. Educ: Dickinson Col, AB, 51; Bucknell Univ, MS, 53; Pa State Univ, PhD(org chem), 58. Prof Exp: Res fel, Harvard Univ, 58-59; res chemist, Jackson Lab, E I du Pont de Nemours & Co, Inc, NJ, 59-62; sr res chemist, A E Staley Mfg Co, 62-63; res chemist, Elastomer Chem Dept, Exp Sta, 63-69, RES CHEMIST, ELASTOMERS LAB, E I DU PONT DE NEMOURS & CO, INC, CHESTNUT RUN, 69- Res: Polymer synthesis and mechanisms of polymerization; latex technology. Mailing Add: Elastomers Lab Chestnut Run E I duPont deNemours & Co PO Box 406 Wilmington DE 19898

BERKHEISER, SAMUEL WILLIAM, b Ashland, Pa, July 29, 22; m 46; c 1. CLINICAL PATHOLOGY. Educ: Western Reserve Univ, BS, 43, MD, 46. Prof Exp: Intern, Geisinger Mem Hosp, Danville, Pa, 46-47; res path & bact, Albany Hosp & Med Sch, NY, 49-50; path & clin path, Youngstown Hosp Asn, Ohio, 50-52, asst pathologist, 52; from asst pathologist to assoc pathologist, Guthrie Clin, Robert Packer Hosp, Sayre, 52-55; assoc pathologist, 55-69, DIR LABS, HARRISBURG POLYCLIN HOSP, PA, 69- Concurrent Pos: Spec study, Armed Forces Inst Path, 52; consult, Vet Admin Hosp, Lebanon, Pa. Mem: Col Am Pathologists; Am Soc Clin Pathologists; AMA; Aerospace Med Asn. Res: Nevi and malignant melanomas of skin; tumors of thyroid and parotid gland; cysts of mediastinum and diaphragm; hypothyroidism; metabolic craniopathy; problems in diagnosis of skin lesions; epithelial hyperplasia of lung associated with cortisone and thromboembolism. Mailing Add: Dept of Labs Harrisburg Polyclin Hosp Harrisburg PA 17105

BERKHOFF, GERMAN ADOLFO, b Valdivia, Chile, Apr 17, 32; m 57; c 3. VETERINARY MICROBIOLOGY. Educ: Sch Vet Med, Santiago, Chile, DVM, 56; Cornell Univ, PhD(vet microbiol), 73. Prof Exp: Res vet animal virol, Inst Bact Chile, 56-65; asst prof avian dis, Sch Vet Med, Santiago, Chile, 65-69; asst, NY State Vet Col, Cornell Univ, 69-73; ASST PROF VET MICROBIOL, SCH VET MED, PURDUE UNIV, 73- Mem: Am Asn Avian Pathologists; Am Soc Microbiol; Conf Res Workers Animal Dis. Res: Study of diseases caused by anaerobic bacteria in animals. Mailing Add: Microbiol Path & Pub Health Purdue Univ Sch Vet Med West Lafayette IN 47907

BERKHOUT, AART W J, b Dinteloord, Netherlands, June 13, 39; m 66; c 2.

GEOPHYSICS, GEOLOGY. Educ: Technol Univ Delft, MSc, 64; Queen's Univ (Ont), PhD(geol, geophys), 68. Prof Exp: Res scientist, Tulsa Res Ctr, Sinclair Oil Corp, 67-69; res geophysicist, Prod Res Lab, Atlantic Richfield Co, 69-71; assoc geophysicist, 71-74, GEOPHYS SPECIALIST, MOBIL OIL CORP, 74- Mem: Soc Explor Geophysicists; Royal Netherlands Geol & Mining Soc; Netherlands Royal Inst Engrs; Europ Asn Explor Geophys. Res: Interpretation of potential field data, especially gravity and magnetic observations, in terms of geological prospects for oil or mineral accumulation; regional gravity field of the Canadian Arctic. Mailing Add: Mobil Oil Corp ESC Box 900 Dallas TX 75221

BERKLAND, TERRILL RAYMOND, b Mason City, Iowa, Oct 17, 41; m 67. SCIENCE EDUCATION. Educ: Loras Col, BS, 64; Drake Univ, MA, 66; Univ Iowa, PhD(sci educ), 73. Prof Exp: Instr gen sci, 66-70, ASST PROF EARTH SCI & EDUC, CENT MO STATE UNIV, 72- Res: Understanding of science process and interest in science teaching among prospective elementary teachers. Mailing Add: Dept of Earth Sci Cent Mo State Univ Warrensburg MO 64093

BERKLEY, DAVID A, b Brooklyn, NY, Apr 28, 40; m 73. ACOUSTICS. Educ: Cornell Univ, BEE, 61, PhD(appl physics), 66. Prof Exp: Res assoc electron & med physics, Chalmers Tech Sweden, 66-68; MEM TECH STAFF ACOUST RES, BELL TEL LABS, 68- Mem: Acoust Soc Am; Optical Soc Am. Res: Acoustics signal processing and electroacoustics; speech and hearing. Mailing Add: Bell Tel Labs Rm 2D537 Murray Hill NJ 07974

BERKMAN, ANTON HELMER, botany, deceased

BERKMAN, JAMES ISRAEL, b Cambridge, Mass, Nov 14, 13; m 42; c 2. MEDICINE, PATHOLOGY. Educ: Harvard Univ, AB, 35, MA, 36; NY Univ, MD, 40. Prof Exp: From asst pathologist to assoc pathologist, Montefiore Hosp, NY, 48-53; CHMN DEPT LABS, LONG ISLAND JEWISH-HILLSIDE MED CTR, 53-; PROF PATH, STATE UNIV NY STONY BROOK, 71- Concurrent Pos: Attend pathologist, Montefiore Hosp, NY, 63-; pathologist in chief, Queens Hosp Ctr, 64-; prof lectr, State Univ NY Downstate Med Ctr; consult path, Creedmoor State Hosp, 75- Mem: Am Asn Path & Bact; Int Acad Path; Col Am Pathologists; Am Soc Clin Path; Am Soc Nephrology. Res: Vascular disease in diabetes; renal disease. Mailing Add: LI Jewish-Hillside Med Ctr New Hyde Park NY 11040

BERKMAN, MICHAEL G, b Poland, Apr 4, 17; nat US; m 41; c 2. CHEMISTRY. Educ: Univ Chicago, BS, 37, PhD(org chem), 41; De Paul Univ, JD, 58; John Marshall Law Sch, MPL, 62. Prof Exp: Res chemist, Deavitt Labs, Univ Chicago, 39-41, Am Can Co, Ill, 41-42 & Argonne Nat Lab, 46-51; chief chemist, Colburn Labs, Inc, 51-59; patent atty, Mann, Brown & McWilliams, 59-63; PATENT ATTY, KEGAN, KEGAN & BERKMAN, 63- Concurrent Pos: Patent consult. Mem: Am Chem Soc. Res: Industrial research and development; synthetic coatings; halogenation; atomic energy; heavy metal complexes; chlorination of hydrocarbons; food technology; federal food and drug laws. Mailing Add: 939 Glenview Rd Glenview IL 60025

BERKMAN, ROBERT NELSON, b Flint, Mich, July 22, 28; m 51; c 3. VETERINARY PATHOLOGY. Educ: Mich State Univ, BS, 50, MS, 51, BS, 54, DVM, 56, PhD(path), 58. Prof Exp: Instr vet path, Mich State Univ, 56-58; sr scientist path & pharmacologist, Eli Lilly & Co, Ind, 58-59, head vet res, 59-68; dir res, 68-74, SR DIR RES, SCHERING CORP, 74- Concurrent Pos: Path consult, fission prod lab, Univ Mich, 57-58. Res: Pharmacological and pathological studies of drugs for use in veterinary medicine. Mailing Add: Schering Corp Kenilworth NJ 07016

BERKMAN, SAM, b Chicago, Ill, Apr 7, 16; m 43; c 3. BACTERIOLOGY. Educ: Univ Chicago, BS, 38, MS, 39, PhD(bact), 42. Prof Exp: Asst bact, Univ Chicago, 38-42 & 42-43; res bacteriologist, US Dept War, Camp Detrick, Md, 46-48; PRES, BIO-SCI LABS, 48- Concurrent Pos: Consult, Chocolate Prod Co, Chicago, 39-43. Mem: AAAS; Am Soc Microbiol; Soc Exp Biol & Med; Am Asn Clin Chemists; NY Acad Sci. Res: Physiology of bacteria; mode of action of chemotherapeutic agents and of antibiotics; accessory growth factor requirements of the members of the genus Pasteurella; clinical biochemistry. Mailing Add: Bio-Sci Lab 7600 Tyrone Ave Van Nuys CA 91405

BERKNER, KLAUS HANS, b Dessau, Ger, May 2, 38; US citizen; m 60; c 2. PHYSICS. Educ: Mass Inst Technol, SB, 60; Univ Calif, Berkeley, PhD(physics), 64. Prof Exp: PHYSICIST, LAWRENCE BERKELEY LAB, UNIV CALIF, BERKELEY, 64- Concurrent Pos: NSF fel, UK Atomic Energy Authority, Culham Lab, Eng, 65-66. Mem: Am Phys Soc; Sigma Xi. Res: Atomic collisions; plasma physics; controlled fusion research. Mailing Add: Lawrence Berkeley Lab Univ of Calif Berkeley CA 94720

BERKOFF, CHARLES EDWARD, b London, Eng, Sept 29, 32; m 61; c 3. ORGANIC CHEMISTRY, MEDICINAL CHEMISTRY. Educ: Univ London, BSc, 56, PhD(org chem) & dipl, Imp Col, Univ London, 59; FRIC, 72. Prof Exp: Jr chemist, Brit Drug Houses, London, 49-54; Monsanto bursary, 56-59; Fulbright travel scholar, 59-60; fel, Univ Southampton, 60-61; asst to dir res, Nicholas Res Inst, Eng, 61-62; asst dir chem res & develop, Biorex Labs, 62; group leader org chem, Wyeth Labs, Am Home Prod Co, Pa, 63-64; sr scientist, Sci Info Dept, 64-66, group leader, 66-70, asst dir chem support, 70-71, mgr technol assessment & long range planning, Planning & Opers Dept, 71-74, MGR ORG CHEM DEPT, RES & DEVELOP DIV, SMITH KLINE & FRENCH LABS, 74- Concurrent Pos: Res assoc, Johns Hopkins Univ, 59-60; mem adv coun, Smithsonian Sci Info Exchange, 74- Mem: AAAS; Am Chem Soc; Royal Inst Chem; fel Am Inst Chemists; Am Mgt Asn. Res: Terpenoids, steroids and other natural products; small ring and heterocyclic compounds, especially those of medicinal interest; chemistry and biochemistry of insect hormones. Mailing Add: Res & Develop Div Smith Kline & French Labs 1500 Spring Garden St Philadelphia PA 19101

BERKOFF, ROBERT BERNARD, b Calgary, Alta, July 7, 12; m 40; c 2. CHEMISTRY. Educ: Univ Alta, BSc, 40. Prof Exp: Plant chemist refinery lab, Alta, 40-46, sr chemist, Ont, 46-50, chief chemist, Man, 50-53, tech supt refinery opers anal, 53-56, chief chemist, refinery lab, Que, 56-57 & Ont, 57, coordr tech labs res, 57-67, MGR OPERS DIV, IMP OIL ENTERPRISES , LTD, 67- Mem: Fel Chem Inst Can. Res: Administration of research technical support laboratories; environmental quality. Mailing Add: Imp Oil Enterprises Ltd Res Dept PO Box 3022 Sarnia ON Can

BERKOFSKY, LOUIS, meteorology, see 12th edition

BERKOVITZ, LEONARD DAVID, b Chicago, Ill, Jan 24, 24; m 53; c 2. MATHEMATICS. Educ: Univ Chicago, PhD(math), 51. Prof Exp: AEC fel, Stanford Univ, 51-52; res fel, Calif Inst Technol, 52-54; mathematician, Rand Corp, 54-62; PROF MATH, PURDUE UNIV, 62-, HEAD DEPT, 75- Concurrent Pos: Ed, J Control, Soc Indust & Appl Math. Mem: Am Math Soc; Math Asn Am; Soc Indust &

Appl Math. Res: Calculus of variations; optimal control; game theory. Mailing Add: Div of Math Sci Purdue Univ West Lafayette IN 47907

BERKOWITZ, AMI EMANUEL, b Brooklyn, NY, Sept 13, 26; m 53; c 3. MAGNETISM. Educ: Univ Pa, PhD(physics), 53. Prof Exp: Sr physicist, Res & Develop Lab, Franklin Inst, 46-59, head magnetics br, 57-59; mem res staff, Thomas J Watson Res Ctr, Int Bus Mach Corp, 60-66, mgr mat develop group, Advan Memory Dept, 66-68; STAFF MEM, GEN ELEC RES & DEVELOP CTR, 68- Mem: Am Phys Soc. Res: Magnetic properties of fine particles, alloys and oxides; structure of metals and oxides. Mailing Add: Gen Elec Res & Develop Ctr The Knolls Schenectady NY 12301

BERKOWITZ, BARRY ALAN, b Brookline, Mass, Dec 29, 42; m 63; c 2. PHARMACOLOGY. Educ: Northeastern Univ, BS, 64; Univ Calif, PhD(pharmacol), 68. Prof Exp: Fel, 68-70, res assoc, 70-71, ASSOC MEM PHARMACOL, ROCHE INST MOLECULAR BIOL, 71- Concurrent Pos: Vis asst prof pharmacol, Med Col, Cornell Univ, 71- Mem: AAAS; NY Acad Sci; Am Soc Pharmacol & Exp Therapeut. Res: Neurochemistry and pharmacology of blood vessels; pharmacology of narcotics and narcotic antagonist analgesics; biogenic amines; cardiovascular disease. Mailing Add: Dept of Physiol Chem Roche Inst of Molecular Biol Nutley NJ 07110

BERKOWITZ, DAVID B, b Hazleton, Pa, Feb 9, 36; m 61; c 1. BIOCHEMISTRY, MOLECULAR BIOLOGY. Educ: Temple Univ, BS, 58, MS, 60; Univ Wis, PhD(biochem), 65. Prof Exp: NSF fel bact genetics, NIH, 65-67; ASST PROF BIOCHEM, UNIV PA, 67- Mem: AAAS; Genetics Soc Am; Am Soc Microbiol. Res: Genetics. Mailing Add: Dept of Biochem Univ of Pa Med Sch Philadelphia PA 19104

BERKOWITZ, HARRY LEO, b New York, NY, Mar 10, 37; m 60; c 2. RADIATION PHYSICS, THEORETICAL PHYSICS. Educ: Adelphi Col, AB, 59; Columbia Univ, MA, 61; Stevens Inst Technol, PhD(physics), 72. Prof Exp: Lectr physics lab, Columbia Univ, 60-61; RES PHYSICIST, US ARMY ELECTRONICS TECHNOL DEVELOP LAB, 61- Mem: Am Phys Soc. Res: Use of nuclear reactions for determining dopant profiles in silicon. Mailing Add: Electron Technol & Devices Lab AMSEL-TL-EN Ft Monmouth NJ 07703

BERKOWITZ, HARRY WILLIAM, mathematics, see 12th edition

BERKOWITZ, JEROME, b Brooklyn, NY, Oct 2, 28; m 54; c 2. MATHEMATICS. Educ: NY Univ, PhD(math), 53. Prof Exp: Mathematician, Reeves Instrument Corp, 48; res asst, 50-56, from asst prof to assoc prof math, 56-65, PROF, COURANT INST MATH SCI, NY UNIV, 65- Mem: Am Math Soc; Math Asn Am. Res: Functional analysis; magnetohydrodynamics. Mailing Add: Courant Inst of Math Sci NY Univ 251 Mercer St New York NY 10012

BERKOWITZ, JOAN B, b Brooklyn, NY, Mar 13, 31; m 59; c 1. PHYSICAL CHEMISTRY. Educ: Swarthmore Col, BA, 52; Univ Ill, PhD(phys chem), 55. Prof Exp: NSF fel, Yale Univ, 55-57; PHYS CHEMIST, ARTHUR D LITTLE, INC, 57- Mem: Am Chem Soc; Electrochem Soc; Am Phys Soc. Res: Thermodynamics of inorganic systems at high temperatures; oxidation of refractory metals and alloys; electrochemistry in flames; mass spectrometry; solid-gas interactions; inorganic coating technology; heterogeneous catalytic recombination; high temperature vaporization. Mailing Add: Arthur D Little Inc 15 Acorn Park Boston MA 02140

BERKOWITZ, JOSEPH, b Czech, Apr 22, 30; US citizen; m 58; c 2. PHYSICAL CHEMISTRY. Educ: NY Univ, BE, 51; Harvard Univ, PhD(phys chem), 55. Prof Exp: Jr chem engr, Brookhaven Nat Lab, 51-52; res assoc physics, Univ Chicago, 55-57; physicist, 57-73, SR PHYSICIST, ARGONNE NAT LAB, 73- Concurrent Pos: Vis asst prof, Univ Ill, 59-60; Guggenheim Found fel phys inst, Univ Freiburg, 65-66. Mem: Am Chem Soc; Am Phys Soc; Am Soc Mass Spectrometry. Res: Molecular structure; chemical kinetics; high temperature thermodynamics; mass spectrometry; photoionization; photoelectron spectroscopy. Mailing Add: Physics Div Bldg 203 Argonne Nat Lab Argonne IL 60439

BERKOWITZ, LEWIS MAURICE, b New York, NY, Apr 12, 31; m 68; c 1. ORGANIC CHEMISTRY. Educ: City Col New York, BS, 52; Columbia Univ, AM, 53, PhD(chem), 57. Prof Exp: Chemist, Engelhard Indust Inc, 57-58, US Vitamin & Pharmaceut Indust Corp, 59-60 & Gen Foods Corp, 60-63; CHEMIST, US ARMY, EDGEWOOD ARSENAL, 63- Mem: Am Chem Soc. Res: Structure-activity relationship of chemotherapeutic and physiologically active compounds; organic synthesis; enzyme immobilization. Mailing Add: US Army Edgewood Arsenal MD 21010

BERKOWITZ, SIDNEY, b Perth Amboy, NJ, Aug 1, 21; m 45; c 2. INDUSTRIAL ORGANIC CHEMISTRY. Educ: Rutgers Univ, BA, 65. Prof Exp: Res chemist, 58-70, SR RES CHEMIST, FMC CORP, PRINCETON, 70- Mem: Catalyst Soc. Res: Heterocyclic chemistry with emphasis on the reactions of isocyanic acid and many of its polymeric derivatives, both linear and cyclic; high temperature catalytic and hot tube oxidations. Mailing Add: 310 S Fourth Ave Highland Park NJ 08904

BERKOWITZ, STEVEN ARLEN, b Newark, NJ, Nov 3, 46; m 70. BIOPHYSICAL CHEMISTRY. Educ: Farleigh Dickinson Univ, BS, 68; NY Univ, PhD(biochem), 75. Prof Exp: Fel, Yale Univ, 75- Res: The physical chemistry of macromolecules and associated processes involved in creating supramolecular structures such as viruses and microtubules. Mailing Add: Dept of Biol Yale Univ New Haven CT 06520

BERKSON, ASTRID J, mathematics, see 12th edition

BERKSON, BURTON MARTIN, biology, mycology, see 12th edition

BERKSON, DAVID M, b Chicago, Ill, Oct 16, 28; m 57; c 3. MEDICINE. Educ: Univ Ill, BS, 49; Northwestern Univ, Chicago, MD, 53. Prof Exp: ASST DIR HEART DIS CONTROL PROG, CHICAGO BD HEALTH, 61-; ASSOC PROF COMMUNITY HEALTH & PREV MED, MED SCH, NORTHWESTERN UNIV, CHICAGO, 70- Concurrent Pos: Assoc attend physician, Michael Reese Hosp, Chicago, 61-; head sect cardiovasc dis, St Joseph Hosp, Chicago, 63- Mem: AAAS; Am Heart Asn; AMA; fel Am Col Physicians; fel Am Col Cardiol. Res: Cardiovascular diseases; epidemiology of hypertensive and atherosclerotic coronary heart disease. Mailing Add: 104 S Michigan Ave Chicago IL 60603

BERKSON, EARL ROBERT, b Chicago, Ill, June 6, 34; m 60; c 3. MATHEMATICS. Educ: Univ Calif, Los Angeles, BS, 56, MA, 57; Univ Chicago, PhD(math), 60. Prof Exp: From instr to asst prof math, Univ Calif, Los Angeles, 60-66; from asst prof to assoc prof, 66-73, PROF MATH, UNIV ILL, URBANA, 73- Concurrent Pos: Vis asst prof, Univ Calif, Berkeley, 64-65. Mem: Am Math Soc. Res: Operator theory; spectral theory; complex analysis. Mailing Add: Dept of Math Univ of Ill Urbana IL 61801

BERKSON, HAROLD, b Easton, Pa, Oct 30, 29; m 58; c 2. ENVIRONMENTAL BIOLOGY. Educ: Rutgers Univ, BA, 51; Amherst Col, MA, 53; Univ Calif, San

Diego, PhD(environ physiol), 63. 64. Prof Exp: Asst comp anat, Amherst Col, 51-53; asst eng scientist, NY Univ, 54-55; biochem technician, M D Anderson Hosp & Tumor Inst, Univ Tex, 55-56; asst biol oceanogr, Narragansett Marine Lab, Univ R I, 56-58; asst physiol & ecol, Scripps Inst, Univ Calif, 58-63; marine biologist, Fed Water Pollution Control Admin, 63-68; chief biol & ecol sci sect, Estuarine & Oceanog Progs Br, 68-70; specialist environ policy, Cong Res Serv, Libr of Congress, 70-72; SR ENVIRON SPECIALIST, NUCLEAR REGULATORY COMN, 72- Mem: AAAS; Am Soc Limnol & Oceanog; Ecol Soc Am; Am Fisheries Soc; Int Asn Fish, Game & Conserv Comnrs. Res: Estuarine ecology; adaptation to the environment; prolonged and deep-diving by air-breathing vertebrates; environmental mamagement; environmental impact assessment. Mailing Add: US Nuclear Regulatory Comn Site Safety & Environ Assessment Washington DC 20555

BERKUT, MICHAEL KALEN, b NY, June 30, 15; m 38; c 2. BIOLOGICAL CHEMISTRY. Educ: NC State Col, BS, 41; Univ NC, PhD(biol chem), 53. Prof Exp: Chemist, NC State Lab Hyg, 41; instr chem, NC State Col, 42; from instr to assoc prof biol chem, 47-74, ASSOC PROF BIOCHEM, NUTRIT & PHYSIOL, UNIV NC, CHAPEL HILL, 74- Concurrent Pos: Res biochemist, US AEC, 62-64. Mem: AAAS; Am Chem Soc; Am Asn Clin Chemists. Res: Mineral metabolism; mechanism of biological calcification; neural and humoral influences on hematopoiesis. Mailing Add: Dept of Biochem Univ of NC Chapel Hill NC 27514

BERKY, JOHN JAMES, b Billings, Mont, Nov 19, 24; m 50; c 3. MEDICAL MICROBIOLOGY. Educ: NDak State Univ, BS, 49; Pa State Univ, MS, 51, PhD(microbial physiol), 54; Nat Registry Microbiol, registered, 63; Am Acad Microbiol, cert specialist microbiol, Pub Health & Medical Lab Microbiol, 70, Food, Dairy & Sanit Microbiol, 75. Prof Exp: Food res microbiologist, Armour Res Div, Chicago, 54-56, asst sect chief bacteriol, 56-58; asst chief, Lab Div, US Army Biol Opers, Pine Bluff, Ark, 58-59, chief, Biol Lab Div, 59-71; SR INVESTR, VIROL & CELL CULT LAB, MICROBIOL & IMMUNOL DIV, NAT CTR FOR TOXICOL RES, 71- Mem: AAAS; Am Soc Microbiol; Am Chem Soc; NY Acad Sic; Sigma Xi. Res: Microbial metabolism, food research microbiology, microbial processes fermentations by-products purification and assay; cell culture systems in toxicology, toxicity testing, carcinogenesis; virus-chemical interactions; contract administration and laboratory management. Mailing Add: Nat Ctr for Toxicol Res Jefferson AR 72079

BERL, SOLL, b New York, NY, June 12, 18; m 56; c 4. BIOCHEMISTRY, NEUROCHEMISTRY. Educ: St John's Univ, BS, 40; Univ Wis, MS, 43; Western Reserve Univ, MD, 40. Prof Exp: Intern, Long Island Col Hosp, 50-51; resident psychiat, Bellevue Med Ctr, NY Univ, 51-54; NIH fel, 56; sr res scientist neurochem, NY Psychiat Inst, 57-61, assoc res scientist, 61-62; res assoc, 62-64, asst prof, 64-70, ASSOC PROF NEUROCHEM, COL PHYSICIANS & SURGEONS, COLUMBIA UNIV, 70- Concurrent Pos: Vet Admin training fel, 51-54; asst attend, Vanderbilt Clin, Columbia-Presby Med Ctr, 57-64; NIH res career develop award, 62. Honors & Awards: Lucy G Moses Prize, 70. Mem: AAAS; Am Psychiat Asn; Acad Psychoanal; Am Soc Biol Chemists; Am Chem Soc. Res: Amino acid and protein metabolism of the central nervous system in relation to structure and function; psychiatry. Mailing Add: 630 W 168th St New York NY 10032

BERLAD, ABRAHAM LEON, b New York, NY, Sept 20, 21; m 49; c 3. CHEMICAL PHYSICS. Educ: Brooklyn Col, BA, 43; Ohio State Univ, PhD(physics), 50. Prof Exp: Aeronaut res scientist, Nat Adv Comt Aeronaut, 51-54, head combustion fundamentals sect, 54-56; sr staff scientist, Convair Sci Res Lab, 56-63; sr staff scientist, Dept Aeronaut Sci, Univ Calif, Berkeley, 63-64; mem staff, Physics Dept, Gen Res Corp, 64-66; PROF ENG, STATE UNIV NY STONY BROOK, 66- Concurrent Pos: Consult, AEC, 73-75 & Energy Res & Develop Agency, 75-; vis prof, Univ Calif, San Diego & Hebrew Univ Jerusalem, 74; vis sr scientist, Brookhaven Nat Lab, 75. Mem: Am Phys Soc; Combustion Inst. Res: Irreversible processes; unstable media; crystallization and combustion theory; radiation phenomena; stability of reaction processes; atmospheric and environmental rate processes; thermokinetic systems. Mailing Add: Dept of Mech Col of Eng State Univ of NY Stony Brook NY 11790

BERLANDI, FRANCIS J, analytical chemistry, nuclear chemistry, see 12th edition

BERLEANT-SCHILLER, RIVA, b Buffalo, NY, Nov 19, 35; m 58; c 3. ANTHROPOLOGY. Educ: State Univ NY Buffalo, BA, 56; Long Island Univ, MS, 67; State Univ NY Stony Brook, PhD(anthrop), 74. Prof Exp: Librn, Nassau County Mus Natural Hist, NY, 67-69; ASST PROF ANTHROP & SOCIOL, QUEENSBOROUGH COMMUNITY COL, 73- Mem: Fel Am Anthrop Asn. Res: Strategies of household survival among Caribbean subsistence producers; relation between popular images and scientific concepts of prehistory and prehistoric people since the nineteenth century. Mailing Add: 25 Highfield Rd Glen Cove NY 11542

BERLEY, DAVID, b Brooklyn, NY, Mar 25, 30; m 62; c 2. PHYSICS. Educ: Union Col, NY, BS, 51; Cornell Univ, PhD(physics), 57. Prof Exp: Res assoc, Columbia Univ, 57-59 & 62; Louis de Broglie fel, Col de France, 59-61; physicist, Brookhaven Nat Lab, 62-74, head exp div, Accelerator Dept, 69-74; PROG DIR, NAT SCI FOUND, 74- Concurrent Pos: Fulbright fel, 59 & 60. Mem: AAAS; fel Am Phys Soc; Sigma Xi. Res: Elementary particle physics; weak interactions; decay of muons, hyperons and kaons; strong interactions; resonances and their quantum numbers; management of elementary particle physics programs. Mailing Add: Nat Sci Found Washington DC 20550

BERLIN, BRENT, b Pampa, Tex, Dec 20, 36; m 57; c 2. ANTHROPOLOGY. Educ: Univ Okla, BA, 59; Stanford Univ, MA, 60, PhD(anthrop), 64. Prof Exp: Instr soc anthrop, Harvard Univ, 65-66; asst prof anthrop, 66-71, ASSOC PROF ANTHROP, UNIV CALIF, BERKELEY, 71- Mem: Am Anthrop Asn; Ling Soc Am. Res: Linguistic anthropology, especially ethnoscience; ethnography of the Tzeltal; ethnobotany. Mailing Add: Dept of Anthrop 233 Kroeber Hall Univ of Calif Berkeley CA 94720

BERLIN, BYRON SANFORD, b Detroit, Mich, Mar 19, 21; m 46; c 5. VIROLOGY. Educ: Wayne State Univ, BA, 42, MD, 45; Am Bd Med Microbiol, dipl, 68. Prof Exp: Coman fel, Univ Chicago, 50-52, resident med, 52-54; res assoc, Sch Pub Health, Univ Mich, 54-59, asst prof, 59-67; ASSOC PROF MED & PATH & DIR CLIN VIROL LAB, MED SCH, NORTHWESTERN UNIV, CHICAGO, 67- Concurrent Pos: Mem, Am Bd Med Microbiol. Mem: Am Asn Immunol; Soc Exp Biol & Med; Radiation Res Soc. Res: Immunology; infection. Mailing Add: Med Sch Northwestern Univ Chicago IL 60611

BERLIN, CHARLES I, b New York, NY, Dec 26, 33; m 58; c 4. AUDIOLOGY. Educ: NY Univ, BS, 53; Univ Wis, MA, 54; Univ Pittsburgh, PhD(hearing & speech), 58. Prof Exp: Audiologist & speech pathologist, US Vet Admin Hosp, Calif, 59-61; fel med audiol, Johns Hopkins Hosp, 62-63, asst prof otolaryngol, Johns Hopkins Hosp, 63-67; assoc prof, 67-70, PROF OTOLARYNGOL, MED SCH, LA STATE UNIV MED CTR, NEW ORLEANS, 70-, DIR, KRESGE HEARING RES LAB OF THE SOUTH, 68- Concurrent Pos: Nat Inst Neurol Dis & Blindness res career develop award, 63-67. Mem: AAAS; Acoust Soc Am; Am Speech & Hearing Asn. Res: Hearing and speech sciences; temporal lobe function; dichotic listening; communication in pathological states in humans and animals; speech disorders; hearing disorders. Mailing Add: Dept of Otorhinolaryngol La State Univ Med Sch New Orleans LA 70119

BERLIN, ELLIOTT, b Baltimore, Md, Jan 4, 34; m 56; c 5. BIOCHEMISTRY. Educ: Johns Hopkins Univ, BS, 57, MA, 59; Univ Md, College Park, PhD(biochem), 66. Prof Exp: RES CHEMIST, AGR RES SERV, USDA, 62- Concurrent Pos: Lectr, Catonsville Community Col, 65-68; asst prof, Univ Md, College Park, 69-70; instr, Bowie State Col, 71-72; adj prof, Southeastern Univ, 73-74. Mem: AAAS; Am Chem Soc; Biophys Soc. Res: Nutritional factors affecting physical chemical properties and biological function of biomembranes; membrane fluidity; lipoprotein chemistry; water binding by proteins and other membranes. Mailing Add: Nutrit Inst Agr Res Ctr Agr Res Serv USDA Beltsville MD 20705

BERLIN, GRAYDON LENNIS, b St Petersburg, Pa, May 21, 43; m 64; c 1. PHYSICAL GEOGRAPHY. Educ: Clarion State Col, BS, 65; Ariz State Univ, MA, 67; Univ Tenn, PhD(geog), 70. Prof Exp: Asst prof & res assoc geog, Fla Atlantic Univ, 68-69; ASST PROF GEOG, NORTHERN ARIZ UNIV, 69- Concurrent Pos: Res assoc, US Geol Surv, 70- Mem: Am Soc Photogram; Asn Am Geogr; Nat Res Soc NAm. Res: Environmental impacts and modifications; remote sensing of natural resources; fossil landscapes; urban climatology; land use. Mailing Add: Dept of Geog Northern Ariz Univ Flagstaff AZ 86001

BERLIN, IRVING NORMAN, b Chicago, Ill, May 31, 17; m 43; c 3. PSYCHIATRY. Educ: Univ Calif, Los Angeles, BA, 39; Univ Calif, San Francisco, MD, 43. Prof Exp: Lectr psychiat, Sch Med, Univ Calif, Los Angeles, 50-52, clin instr, 52-54, asst res psychiatrist, 55-56, asst clin prof psychiat, 56-60, assoc clin prof, 60-65; prof psychiat & pediat & head div child psychiat, Sch Med, Univ Wash, 65-75; PROF PSYCHIAT & PEDIAT & HEAD CHILD PSYCHIAT, UNIV CALIF, DAVIS, 75- Concurrent Pos: Fel child psychiat, Langley Porter Inst, 46-50; consult, McAuley Neuropsychiat Inst, Calif, 60-, Ctr Training Community Psychiat, 61- & Training Br, NIMH, 74-; Grant Found grant, 73-75. Mem: Fel Am Orthopsychiat Asn; Am Psychiat Asn; fel Am Pub Health Asn; fel Royal Soc Health; Am Acad Child Psychiat (pres, 75). Res: Childhood psychosis; mental health consultation and community psychiatry; training in mental health specialties; early intervention and prevention child mental illness. Mailing Add: Dept of Psychiat Sch of Med Sacramento Med Ctr Ment Health Clin Bldg Rm 27 Sacramento CA 95817

BERLIN, JERRY D, b Trenton, Mo, Aug 28, 34; m 58; c 3. CELL BIOLOGY. Educ: Univ Mo, BS, 60, MA, 61; Iowa State Univ, PhD(cell biol), 64. Prof Exp: Fel, Iowa State Univ, 64; biol scientist, Gen Elec Co, 64-65; sr res scientist, Pac Northwest Lab, Battelle Mem Inst, 65-68; assoc prof biol, 68-73, PROF BIOL, TEX TECH UNIV, 73-, ADJ PROF ANAT, SCH MED, 74- Concurrent Pos: Grants, AEC, Cotton Producers Inst & Cotton Inc, 73- Mem: AAAS; Am Micros Soc; Am Soc Cell Biol; Bot Soc Am; Electron Micros Soc Am. Res: Fine structure of host-parasite interface between higher plants and their fungal parasites; centrioles and flagella in fungi; ultrastructure of intracellular protein movement in poikilotherm livers; Golgi complex; effects of irradiation on cells; formation of cell walls in higher plants. Mailing Add: Dept of Biol Tex Tech Univ Lubbock TX 79409

BERLIN, KENNETH DARRELL, b Quincy, Ill, June 12, 33; m 58; c 2. ORGANIC CHEMISTRY, MEDICINAL CHEMISTRY. Educ: NCent Col, Ill, BA, 55; Univ Ill, PhD(org chem), 58. Prof Exp: Fel chem, Univ Fla, 58-60; from asst prof to prof, 60-71, REGENTS PROF CHEM, OKLA STATE UNIV, 71- Concurrent Pos: App mem res rev panel NSF postdoctoral fac fels, 69 & predoctoral fac fels, 70; mem med chem study sect B, NIH, 69-73; mem ed bd, J Phosphorus Chem, 70- Honors & Awards: Award in Res, Sigma Xi, 69. Mem: Sr mem Am Chem Soc; The Chem Soc. Res: Organophosphorus chemistry; heterocyclic medicinal compounds; nuclear magnetic resonance spectroscopy; organometallics; synthetic organic chemistry. Mailing Add: Dept of Chem Okla State Univ Stillwater OK 74074

BERLIN, NATHANIEL ISAAC, b New York, NY, July 4, 20; m 53; c 2. MEDICAL RESEARCH. Educ: Western Reserve Univ, BS, 42; Long Island Col Med, MD, 45; Univ Calif, PhD(med physics), 49. Prof Exp: Intern, Kings County Hosp, Brooklyn, 45-46, res pathologist, 46-47; fel, Univ Calif, 49-50, res assoc, 50-51, instr, 51, lectr & res assoc, 51-52, lectr & assoc res med physicist, 52-53; Nat Heart Inst spec res fel, Nat Inst Med Res, London, 53-54; med officer, Anal Br, Effects Div, Hqs Armed Forces Spec Weapons Proj, 54-56; head metab serv, Gen Med Br, NIH, 56-59, chief, 59-61, clin dir, 61-71, sci dir gen labs & clin, Nat Cancer Inst, 69-72, dir div cancer biol & diag, 72-75; GENEVIEVE B TEUTON PROF MED & DIR, CANCER CTR, NORTHWESTERN UNIV, CHICAGO, 75- Concurrent Pos: Mem med staff, Highland-Alameda County Hosp, Calif, 54-60; consult, US Naval Hosp, 55-65, Radiation Lab, Univ Calif, 55-57 & Defense Atomic Support Agency, Dept Defense, 57-59; assoc ed, Cancer Res; mem adv comt blood dis & resources, Nat Heart & Lung Inst. Honors & Awards: Superior Serv Award, Dept Health, Educ & Welfare. Mem: AAAS; Am Soc Clin Invest; Soc Exp Biol & Med; Am Physiol Soc; Am Asn Cancer Res. Res: Cobalt metabolism; erythropiesis; use of radioactive isotopes in biological and medical research; cancer research. Mailing Add: Northwestern Univ Med Sch 303 E Chicago Ave Chicago IL 60611

BERLINCOURT, TED GIBBS, b Fremont, Ohio, Oct 29, 25; m 53; c 1. PHYSICS. Educ: Case Inst Technol, BS, 49; Yale Univ, MS, 50, PhD(physics), 53. Prof Exp: Lab asst, Yale Univ, 49-50, res asst, 50-51; proj physicist, Naval Res Lab, 52-55; supvr electronic properties unit, Atomics Int Div, NAm Aviation, Inc, 55-64, group leader electronic properties, NAm Rockwell Sci Ctr, 64-65, assoc dir, 65-70, chief scientist, NAm Rockwell Microelectronics Co, 69-70; prof physics & chmn dept, Colo State Univ, 70-72; DIR, PHYS SCI DIV, OFF NAVAL RES, 72- Concurrent Pos: Mem instrumentation panel, Nat Acad Sci-Nat Res Coun Physics Surv, 71, Navy liaison rep, Div Phys Sci, Nat Res Coun, 72- Mem: AAAS; fel Am Phys Soc; Sigma Xi; Inst Elec & Electronic Engrs. Res: Superconductivity; Fermi surfaces; de Haas-van Alphen effect; galvano-magnetic effects; magnetism; cryogenics; pulsed magnetic fields; metal-oxide-semiconductor large-scale-integration technology; scintillation counters; x-ray diffraction; research and development management and policy. Mailing Add: Off of Naval Res 800 N Quincy St Arlington VA 22217

BERLIND, ALLAN, b New York, NY, Dec 24, 42; m 68. ZOOLOGY. Educ: Swarthmore Col, BA, 64; Harvard Univ, MA, 65, PhD(biol), 69. Prof Exp: NIH fel, Univ Calif, Berkeley, 69-71; ASST PROF BIOL, WESLEYAN UNIV, 71- Mem: Am Soc Zoologists. Res: Neurophysiology; neuroendocrinology; mechanisms of release of neurosecretory peptides. Mailing Add: Dept of Biol Wesleyan Univ Middletown CT 06457

BERLINER, ERNST, b Katowice, Ger, Feb 18, 15; nat US; m 47; c 1. ORGANIC CHEMISTRY. Educ: Harvard Univ, MA, 40, PhD(org chem), 43. Prof Exp: Pvt asst antimalarials, Harvard Univ, 43-44; lectr, 44-45, from asst prof to assoc prof, 45-53,

PROF ORG CHEM, BRYN MAWR COL, 53- Concurrent Pos: Guggenheim fel, 62; mem ed bd, J Org Chem, 63-68. Honors & Awards: Award, Am Chem Soc, 71. Mem: AAAS; Am Chem Soc; The Chem Soc. Res: Physical-organic aspects of aromatic chemistry; hyperconjugation; aromatic substitution; isotope effects; polynuclear aromatic reactivities; halogenation of acetylenes. Mailing Add: Dept of Chem Bryn Mawr Col Bryn Mawr PA 19010

BERLINER, FRANCES (BONDHUS), b Oskaloosa, Iowa, Nov 21, 21; m 47; c 1. ORGANIC CHEMISTRY. Educ: William Penn Col, BA, 43; Bryn Mawr Col, MA, 44, PhD(org chem), 47. Prof Exp: Demonstr, 46-50, from instr to asst prof, 50-57, LECTR CHEM, BRYN MAWR COL, 57- Mem: Am Chem Soc. Res: Aromatic chemistry. Mailing Add: Dept of Chem Bryn Mawr Col Bryn Mawr PA 19010

BERLINER, KURT, b Charlottenburg, Ger, Aug 9, 99; nat US; m 55. CARDIOLOGY. Educ: Univ Berlin, MD, 22; Am Bd Internal Med, dipl, 38. Prof Exp: CONSULT CARDIOLOGIST, SYDENHAM HOSP, 36-; ASSOC PROF MED, NY MED COL, 47- Concurrent Pos: Assoc attend physician, Flower & Fifth Ave, Metrop & Bird S Coler Hosps, 47-; fel, Coun Clin Cardiol, Am Heart Asn. Mem: Fel Am Col Physicians; Am Heart Asn; AMA. Res: Electrocardiography. Mailing Add: 1235 Park Ave New York NY 10028

BERLINER, LAWRENCE J, b Los Angeles, Calif, Sept 18, 41; m 67. BIOPHYSICAL CHEMISTRY. Educ: Univ Calif, Los Angeles, BS, 63; Stanford Univ, PhD(phys chem), 67. Prof Exp: Res chemist, Sun Oil Co, 62 & Chevron Res Corp, Stand Oil Co, Calif, 63; Brit Heart Found-Am Heart Asn fel molecular biophysics, Oxford Univ, 67-68; Am Heart Asn fel phys chem, Stanford Univ, 68-69; asst prof chem, 69-75, ASSOC PROF CHEM, OHIO STATE UNIV, 75- Concurrent Pos: Estab investr, Am Heart Asn, 75-80. Mem: Am Chem Soc; Biophys Soc. Res: Structure-function relations in biomolecules by magnetic resonance methods; small molecule binding; conformational changes; x-ray structural investigations; applications of magnetic resonance to biology. Mailing Add: Dept of Chem Ohio State Univ Columbus OH 43210

BERLINER, MARTHA D, b Antwerp, Belg, Nov 18, 28; US citizen; m 52; c 2. MICROBIOLOGY, MYCOLOGY. Educ: Hunter Col, BA, 49; Univ Mich, MS, 50; Columbia Univ, PhD(mycol), 53. Prof Exp: Jr scientist cytol, Sloan Kettering Inst, 50-51; sr scientist microbiol, Res & Advan Develop Div, Med Sci Dept, Avco Corp, 60-62, sr staff scientist space microbiol, 62-65; from instr to asst prof, 65-71, ASSOC PROF BIOL, SIMMONS COL, 71- Concurrent Pos: Res assoc, Sch Med, Tufts Univ, 53-54; instr, Lynn Hosp, Mass, 53-62; res assoc, Dept Microbiol, Sch Pub Health, Harvard Univ, 65-75; ed consult, Charles River Breeding Labs, 75- Mem: Am Mycol Soc; Am Soc Microbiol; NY Acad Sci; Int Soc Human & Animal Mycol; Med Mycol Soc Americas. Res: Medical mycology; microscopy; biological rhythms; radiation and environmental affects; protoplasts and wall deficient variants; biological parameter of pathogenicity in fungi; protoplasts of fungi and algae. Mailing Add: Dept of Biol Simmons Col 300 The Fenway Boston MA 02115

BERLINER, ROBERT WILLIAM, b New York, NY, Mar 10, 15; m 41; c 4. PHYSIOLOGY, INTERNAL MEDICINE. Educ: Yale Univ, BS, 36; Columbia Univ, MD, 39. Hon Degrees: DSc, Med Col Wis & Yale Univ, 73. Prof Exp: Intern, Presby Hosp, NY, 39-41; resident physician, Goldwater Mem Hosp, 42-43; asst med, Col Med, NY Univ, 43-44, instr, 44-47; asst prof med, Columbia Univ, 47-50; chief, Lab Kidney & Electrolyte Metab, Nat Heart Inst, 50-62, dir intramural res, 54-68; dir labs & clins, NIH, 68-69, dep dir for sci, 69-73; PROF PHYSIOL & MED & DEAN, SCH MED, YALE UNIV, 73- Concurrent Pos: Res fel, Goldwater Mem Hosp, NY, 43-44, res asst, 44-47; res assoc, Dept Hosps, NY, 47-50; lectr, Sch Med, George Washington Univ, 51-73; Am Soc Clin Invest rep, Nat Res Coun, 57-60; prof lectr, Sch Med & Dent, Georgetown Univ, 64-73. Honors & Awards: Distinguished Serv Award, Dept Health, Educ & Welfare, 62; Homer W Smith Award, 65; Distinguished Achievement Award, Mod Med, 69; Res Achievement Award, Am Heart Asn, 70. Mem: Inst of Med of Nat Acad Sci; AAAS (vpres, 72); Am Physiol Soc (pres, 67); Am Soc Clin Invest (pres, 59); Am Soc Nephrol (pres, 68). Res: Physiology of kidney; electrolyte transport. Mailing Add: Yale Univ Sch of Med 333 Cedar St New Haven CT 06510

BERLINGHIERI, JOEL CARL, b Boston, Mass, Dec 18, 42; m 71. Educ: Boston Col, BS, 64; Univ Rochester, MS, 66, PhD(physics), 70. Prof Exp: Asst prof physics, Colgate Univ, 69-71; ASST PROF PHYSICS, THE CITADEL, 71- Concurrent Pos: Res assoc, Univ Houston, 73-75. Mem: Am Phys Soc; Am Asn Physics Teachers; Acoust Soc Am; Optical Soc Am. Res: Kaon interactions at high energies; acousto-optic interactions. Mailing Add: Dept of Physics The Citadel Charleston SC 29409

BERLINGUET, LOUIS, b Trois-Rivieres, Que, June 20, 26; m 50; c 3. ORGANIC BIOCHEMISTRY. Educ: Univ Montreal, BSc, 47; Laval Univ, PhD(chem), 50. Prof Exp: From asst prof to prof biochem, Sch Med, Laval Univ, 47-68, asst dir, 58-63, chmn dept, 63-68; VPRES RES, UNIV QUE, 68- Honors & Awards: Sci Award, Prov Que, 50. Mem: Am Chem Soc; NY Acad Sci; Can Biochem Soc (pres, 64); fel Chem Inst Can. Res: Synthesis of unnatural amino acids; chemotherapy of cancer with antimetabolites; chemistry of proteins and peptides. Mailing Add: Univ of Quebec 2700 rue Eistein Ste Foy Quebec PQ Can

BERLINROOD, MARTIN, b New York, NY, Aug 20, 43; m 66; c 2. DEVELOPMENTAL BIOLOGY. Educ: City Col New York, BS, 64; Univ Tex, Austin, PhD(biol), 69. Prof Exp: Res assoc biol sci, State Univ NY Albany, 68-70; ASST PROF BIOL SCI, GOUCHER COL, 70- Mem: AAAS; Am Soc Zoologists. Res: Axoplasmic transport in embryonic nerve fibers; axonal outgrowth and guidance in amphibian embryos. Mailing Add: Dept of Biol Sci Goucher Col Towson MD 21204

BERLMAN, ISADORE B, b St Louis, Mo, Jan 13, 22; m 53; c 2. PHYSICS. Educ: Wash Univ, PhD(physics), 50. Prof Exp: Assoc physicist radiol physics, Argonne Nat Lab, 50-71; ASSOC PROF PHYSICS, HEBREW UNIV, JERUSALEM, 71- Mem: Am Phys Soc. Res: Energy levels of the light nuclei; neutron detection; nuclear physics; luminescence; radiation damage. Mailing Add: 21 Radak Jerusalem Israel

BERLOW, STANLEY, b New York, NY, June 16, 21; m 47; c 6. PEDIATRICS. Educ: Univ Mich, BA, 40; Harvard Univ, MA, 41, MD, 50. Prof Exp: Intern pediat, Mass Gen Hosp, 50-51, chief resident, 53-54; resident, Babies Hosp, NY, 51-53; from instr to asst prof, Med Sch, Marquette Univ, 54-65; assoc prof pediat, Chicago Med Sch & med dir dysfunctioning child prog, Michael Reese Hosp & Med Ctr, Chicago, 65-69; ASSOC PROF PEDIAT, MED SCH, UNIV WIS-MADISON, 71- Res: Biochemistry of mental retardation. Mailing Add: Waisman Ment Retardation Ctr Univ of Wis Madison WI 53706

BERLOWITZ, LAURENCE JACK, b New York, NY, Oct 20, 34; m 54; c 2. DEVELOPMENTAL BIOLOGY, CELL BIOLOGY. Educ: Univ Calif, Berkeley, AB, 54, PhD(genetics), 65; Univ Calif, Los Angeles, MA, 58. Prof Exp: Mem tech staff, Thompson-Ramo-Wooldridge Corp, 58-60; human factors scientist, Western Develop Lab, Philco Corp, 60-61; instr biol sci, Chabot Col, 61-64; res fel med res coun epigenetics res group, Inst Animal Genetics, Edinburgh, Scotland, 65-66; from asst prof to assoc prof biol, State Univ NY Buffalo, 66-75, co-chmn dept, 68-69; PROG DIR GENETIC BIOL, NAT SCI FOUND, 75- Concurrent Pos: NIH spec fel, Univ Nijmegen, Neth, 72-73, grants assoc, 74-75. Honors & Awards: John Belling Prize in Genetics, 70. Mem: AAAS; Am Soc Cell Biol; Genetics Soc Am; Brit Soc Cell Biol; Soc Develop Biol. Res: Molecular and ultrastructural changes in the nucleus during differentiation. Mailing Add: Nat Sci Found Washington DC 20550

BERLYN, GRAEME PIERCE, b Chicago, Ill, Sept 6, 33; m 58; c 2. PLANT ANATOMY, TREE PHYSIOLOGY. Educ: Iowa State Univ, BS, 56, PhD(anat), 60. Prof Exp: From instr to asst prof wood anat, 60-67, ASSOC PROF WOOD ANAT, SCH FORESTRY & ENVIRON STUDIES, YALE UNIV, 67- Concurrent Pos: Res collabr, Brookhaven Nat Lab, 63- Mem: Soc Wood Sci & Technol; Am Soc Plant Physiol; Int Asn Wood Anatomists; Bot Soc Am. Res: Quantitative cytochemistry; electron microscopy; reaction wood; developmental and experimental anatomy of pine embryos and seedlings; cell wall structure; microtechnical methods; effect of enzymatic hydrolysis on cell wall structure and free space; variation in DNA content of nuclei from populations of Pinus rigida. Mailing Add: Sch of Forestry & Environ Studies Yale Univ 370 Prospect New Haven CT 06511

BERLYN, MARY BERRY, b Iowa Co, Iowa, Dec 28, 38; m 58; c 2. GENETICS. Educ: Iowa State Univ, BS, 59; Yale Univ, PhD(genetics), 66. Prof Exp: Res staff biologist, Yale Univ, 67-71, res assoc, 71-73; STAFF SCIENTIST, CONN AGR EXP STA, 74- Mem: AAAS; Genetics Soc Am. Res: Somatic cell genetics of plants; RNA polymerase mutants of Ecoli; histidine and polyaromatic biosynthetic pathways in prokaryotes and eukaryotes. Mailing Add: Conn Agr Exp Sta 123 Huntington St New Haven CT 06504

BERMAN, ALAN, b Brooklyn, NY, Nov 2, 25; m 62; c 5. PHYSICS, PHYSICAL OCEANOGRAPHY. Educ: Columbia Univ, AB, 47, PhD(physics), 52. Prof Exp: Res physicist, Hudson Labs, Columbia Univ, 52-57, assoc dir, 57-63, dir, 63-67; DIR RES, US NAVAL RES LAB, 67- Honors & Awards: Super Civilian Serv Award, Dept Navy, 69; Distinguished Civilian Serv Award, Dept Defense, 73. Mem: Fel Am Phys Soc; fel Acoust Soc Am. Res: Atomic beams. Mailing Add: Res Dept US Naval Res Lab Washington DC 20390

BERMAN, ALEX, b New York, NY, Feb 7, 14; m 43. HISTORY OF PHARMACY. Educ: Fordham Univ, BS, 47; Univ Wis, MS, 51, PhD(hist of pharm & sci), 54. Prof Exp: Pharmacist, Vet Admin, 48-49; asst hist of pharm, Alumni Res Found, Univ Wis, 50-53; staff assoc, Am Pharmaceut Asn, 53-54; pharmacist, Univ Mich Hosp, 54-55; asst prof hist of pharm, Univ Wis, 55-56; asst prof pharm, Univ Mich, 56-61; assoc prof, Univ Tex, 61-68; prof hist & social studies in pharm, 68-74, EMER PROF HIST & HIST STUDIES IN PHARM, COL PHARM, UNIV CINCINNATI, 75- Concurrent Pos: Guggenheim Mem Found fel, 58-59; NSF grant, 62-64. Honors & Awards: Kremers Award, Am Inst Hist of Pharm, 63. Mem: Am Pharmaceut Asn; Am Soc Hosp Pharmacists; Am Asn Hist of Med; Am Inst Hist of Pharm (pres, 67-69). Res: Nineteenth century pharmaceutical and medical Americana and French pharmacy; history of hospital pharmacy. Mailing Add: 222 Senator Pl Cincinnati OH 45220

BERMAN, ALVIN LEONARD, b Baltimore, Md, July 19, 24; m 49; c 3. NEUROANATOMY, NEUROPHYSIOLOGY. Educ: Johns Hopkins Univ, AB, 45, PhD(physiol), 57. Prof Exp: USPHS fel neuroanat, Johns Hopkins Hosp, 57-59, USPHS fel psychiat, 59-60; asst prof physiol, Univ Md, 60-61; from asst prof to assoc prof anat & neurophysiol, 61-69, PROF NEUROPHYSIOL, DIR MULTI-DISCIPLINARY LABS & COORDR INTERDISCIPLINARY CURRICULUM, UNIV WIS-MADISON, 69- Mem: Am Asn Anat; Am Physiol Soc; Am Acad Neurol; Asn Res Nerv & Ment Dis; Soc Neurosci. Res: Structure and function of mammalian central nervous system; cytoarchitecture of the the brain stem, thalamus and basal telencephalon of the cat; electrophysiology of somatic and auditory areas of cat cerebral cortex. Mailing Add: Dept of Neurophysiol Univ of Wis Madison WI 53706

BERMAN, ARTHUR IRWIN, b New York, NY, Jan 1, 25; m 57; c 6. SCIENCE EDUCATION. Educ: Brooklyn Col, AB, 45; Stanford Univ, MS, 49, PhD(physics), 54. Prof Exp: Instr, Western Elec Co, 44-45; mem staff, Los Alamos Sci Lab, 49-52; res assoc high energy physics lab, Stanford Univ, 54; proj physicist, Pratt & Whitney Aircraft, 54-56; from assoc prof to prof physics, Rensselaer Polytech, Hartford Grad Ctr, 56-69; sr res fel, Inst Studies Higher Educ, Copenhagen Univ, 70-74; ED CONSULT, 69-; VIS PROF, ENERGY LAB, TECH UNIV DENMARK, 74- Concurrent Pos: Consult, Smithsonian Astrophys Observ, 59-61, UNESCO, 72 & Orgn Econ Co-op & Develop, 74; lectr, NASA, 61-69; Fulbright prof, Tech Univ Denmark, 62-63 & 67-68. Mem: AAAS; Am Asn Physics Teachers; Am Inst Aeronaut & Astronaut. Res: X-ray astronomy; photonuclear reactions; seminar/autolecture learning system; media-activated learning groups. Mailing Add: Energy Lab Tech Univ Denmark Copenhagen Denmark

BERMAN, BARRY L, b Chicago, Ill, Mar 8, 36; m 63; c 2. NUCLEAR PHYSICS. Educ: Harvard Univ, BA, 57; Univ Ill, MS, 59, PhD(physics), 63. Prof Exp: NUCLEAR PHYSICIST, LAWRENCE LIVERMORE LAB, 63- Concurrent Pos: Vis assoc prof, Dept Physics & Electron Accelerator Lab, Yale Univ, 69-70; vis prof, Univ Toronto, 70; mem subcomt photonuclear reactions, US Nuclear Data Comt, 73, mem subcomt basic sci, 73-75. Mem: Fel Am Phys Soc. Res: Photonuclear reactions; neutron physics; nuclear astrophysics; applications. Mailing Add: Lawrence Livermore Lab Livermore CA 94550

BERMAN, BARRY L, b Baltimore, Md, Oct 9, 42; m 64; c 1. LOW TEMPERATURE PHYSICS. Educ: Case Inst Technol, BS, 64; Univ Wis, MS, 66, PhD(physics), 69. Prof Exp: Asst prof physics, Bellarmine-Ursuline Col, 69-70; asst prof, 70-75, ASSOC PROF PHYSICS, WELLS COL, 75- Mem: Am Asn Physics Teachers; Am Phys Soc. Res: Specific heats; thermal conductivities; dielectric constants; fluid dynamics; biophysics; use of the computer in undergraduate science instruction. Mailing Add: Dept of Physics Wells Col Aurora NY 13026

BERMAN, DAVID ALBERT, b Rochester, NY, Nov 4, 17; m 45; c 2. PHARMACOLOGY. Educ: Univ Southern Calif, BS, 40, MS, 48, PhD(pharmacol), 51. Prof Exp: Life Ins Med res fel, 51-52; from instr to assoc prof, 52-63, PROF PHARMACOL & TOXICOL, UNIV SOUTHERN CALIF, 63- Concurrent Pos: USPHS spec fel, 60-61. Res: Cardiac metabolism in relation to drug action. Mailing Add: Dept of Pharmacol Sch Med Univ of Southern Calif Los Angeles CA 90033

BERMAN, DAVID ALVIN, b Milwaukee, Wis, Apr 13, 24; m 70. INORGANIC CHEMISTRY, CORROSION. Educ: Univ Wis, BS, 48; Univ Mich, MS, 50, PhD(inorg chem), 57. Prof Exp: Instr, Wis State Teachers Col, Oshkosh, 50; asst prof chem, Exten Div, Univ Wis, 55-63 & Carroll Col, Wis, 64-65; chemist, Naval Air Eng Ctr, 67-70, CHEMIST, NAVAL AIR DEVELOP CTR, 70- Mem: AAAS; Am Chem

Soc; Nat Asn Corrosion Eng. Res: Corrosion; electrochemistry. Mailing Add: Air Vehicle Technol Dept Naval Air Develop Ctr Warminster PA 18974

BERMAN, DAVID S, b Los Angeles, Calif, Jan 10, 40; m 68. VERTEBRATE PALEONTOLOGY. Educ: Univ Calif, Los Angeles, BA, 62, MA, 65, PhD(zool), 69. Prof Exp: ASST CUR VERT PALEONT, CARNEGIE MUS NATURAL HIST, 70- Mem: Paleont Soc; Am Soc Ichthyologists & Herpetologists; Soc Study Evolution. Res: Paleontology of late Paleozoic vertebrates. Mailing Add: Carnegie Mus of Natural Hist 4400 Forbes Ave Pittsburgh PA 15213

BERMAN, DAVID THEODORE, b Brooklyn, NY, June 14, 20; m 44; c 2. VETERINARY SCIENCE. Educ: Brooklyn Col, BA, 41; Cornell Univ, DVM, 44; Univ Wis, MS, 46, PhD, 49. Prof Exp: Res asst, 44-46, from instr to assoc prof, 46-57, chmn dept, 64-68, assoc dean grad sch, 69-75, PROF VET SCI, UNIV WIS-MADISON, 57- Concurrent Pos: Coop agent, USDA, 46-57; mem, Conf Res Workers Animal Dis NAm, 47-; consult, Nat Brucellosis Comt, 58-; grad sch G I Haight traveling fel, State Serum Inst, Copenhagen, Denmark & Vet Lab, Weybridge, Eng, 63; expert comt brucellosis, WHO, 63-70, mem, 70-; mem bd sci counr, Nat Inst Allergy & Infectious Dis, 74-78. Mem: AAAS; Am Soc Microbiol; Am Vet Med Asn; US Animal Health Asn; Am Asn Immunol. Res: Immunology and bacteriology of brucellosis; microbial variation; staphylococcal infection; hypersensitivity in experimental arthritis; antigenic analysis mycobacteria; chlamydia; immune response mucosae. Mailing Add: Dept of Vet Sci Univ of Wis Madison WI 53706

BERMAN, DONALD, b Zanesville, Ohio, Sept 12, 25; m 54; c 3. VIROLOGY, WATER POLLUTION. Educ: Ohio State Univ, BS, 50, MS, 54. Prof Exp: Bacteriologist, Sect Feeds & Fertilizers, Div Plant Indust, State Dept Agr, Ohio, 54-63; res microbiologist, Robert A Taft Sanit Eng Ctr, US Dept Health, Educ & Welfare, 63-70; MICROBIOLOGIST, ENVIRON PROTECTION AGENCY, 70- Mem: Am Soc Microbiol; AAAS; Water Pollution Control Fedn; Sigma Xi. Res: Isolation of small quantities of virus from large volumes of fluid; kinetic studies of virus inactivation, evaluation and mode of action of virucides; effect of pH on viruses; recovery of viruses from solids. Mailing Add: Environ Monitoring & Support Lab Environ Protection Agency Cincinnati OH 45268

BERMAN, ELEANOR, b Duluth, Minn, Oct 5, 21. PHARMACOLOGY, BIOCHEMISTRY. Educ: Wis State Col, Superior, BS, 42; Univ Minn, Minneapolis, MS, 50, PhD(pharmacol), 57. Prof Exp: Teacher pub schs, Mich, 42-45 & Wis, 45-46; med technologist, St Mary's Hosp, Duluth, Minn, 46-48; clin biochemist, Ill Masonic Hosp, Chicago, 53-66; TOXICOLOGIST, HEKTOEN INST, COOK COUNTY HOSP, 66- Mem: Am Asn Clin Chem; fel Am Acad Forensic Sci; Am Chem Soc; fel Am Inst Chem; Am Soc Testing & Mat. Res: Clinical toxicology; lead poisoning and trace metals in clinical chemistry. Mailing Add: Hektoen Inst Cook County Hosp Chicago IL 60612

BERMAN, ELIZABETH ALEXANDRA, b Minneapolis, Minn, Feb 27, 37; m 57; c 2. ALGEBRA. Educ: Univ Minn, BA, 58, MA, 59; Univ Mo-Kans City, PhD(math), 70. Prof Exp: Asst prof math, Rockhurst Col, 70-75; LECTR MATH, UNIV MO-KANS CITY, 76- Concurrent Pos: Consult, Acad Press Inc, 73, Addison-Wesley Publ Co, 75. Mem: Am Math Soc; Math Asn Am; Nat Coun Teachers Math. Res: Polynomial identity rings; developmental mathematics. Mailing Add: 3016 W 73 St Prairie Village KS 66208

BERMAN, ELLIOT, b Quincy, Mass, Jan 13, 30; m 53; c 3. ORGANIC CHEMISTRY. Educ: Brown Univ, ScB, 51; Boston Univ, PhD(org chem), 56. Prof Exp: Sr chemist, Fundamental Res Dept, Nat Cash Register Co, 55-57, head org sect, 57-59; head chem sect, Res Div, Itek Corp, 59-61, mgr chem dept, 61-64, tech dir res, 64-65, dir, Lexington Res Labs, 65-67, dir photosensitive mat, 67-69; consult, 69-73; pres & chmn, Solar Power Corp, 73-75; DIR CTR ENERGY STUDIES, BOSTON UNIV, 75- Concurrent Pos: Instr, Univ Dayton, 56-57; asst prof, Univ Cincinnati, 56-58. Mem: AAAS; Am Chem Soc; Soc Photog Sci & Eng; Solar Energy Soc; The Chem Soc. Res: Reversible photochemical systems; kinetics and mechanisms of reactions at solid interfaces; photochemistry; photophysics; solar energy utilization; photography. Mailing Add: 67 Dewson Rd Quincy MA 02169

BERMAN, GERALD, b Can, Nov 12, 24; US citizen; m 48; c 4. MATHEMATICS. Educ: Univ Toronto, BA, 47, MA, 48, PhD(math), 50. Prof Exp: From instr to assoc prof math, Ill Inst Technol, 50-59; PROF MATH, UNIV WATERLOO, 59-, CHMN DEPT COMBINATORICS & OPTIMIZATION, 66- Concurrent Pos: Consult, Inst Air Weapons Res, Chicago, 57-59, Martin-Marietta Corp, Colo, 59-63 & Defense Res Bd Can, 64-68. Mem: Am Math Soc; Math Asn Am; Soc Indust & Appl Math; Can Math Cong. Res: Applied mathematics; cominatorics; optimization. Mailing Add: 306 Shakespeare Dr Waterloo ON Can

BERMAN, HELEN MIRIAM, b Chicago, Ill, May 19, 43. BIOLOGICAL STRUCTURE. Educ: Columbia Univ, AB, 64; Univ Pittsburgh, PhD(crystallog), 67. Prof Exp: NIH traineeship biochem crystallog, Univ Pittsburgh, 67-69; res assoc molecular struct, 69-73, ASST MEM, INST CANCER RES, 73- Mem: AAAS; Am Crystallog Asn; Biophys Soc; Am Chem Soc. Res: Crystal structures of nucleic acid components and proteins; structural interactions between nucleic acids and proteins. Mailing Add: Inst for Cancer Res 7701 Burholme Ave Philadelphia PA 19111

BERMAN, HERBERT IRVING, organic chemistry, see 12th edition

BERMAN, HERBERT JOSHUA, b Boston, Mass, Oct 20, 24. PHYSIOLOGY. Educ: Univ RI, BS, 45; Boston Univ, AM, 48, PhD(physiol), 53. Prof Exp: Instr zool, Univ Mass, 48-49; asst, 52-53, res assoc biol, 54-61, from asst res prof to assoc res prof, 61-69, PROF BIOL, BOSTON UNIV, 69- Mem: Radiation Res Soc; Am Soc Hemat; Microcirc Soc (secy-treas, 59-); Am Soc Zool; Am Physiol Soc. Res: Physiology of circulation; small blood vessels; hypertension; thromboembolism and blood coagulation; ecaluation of plasma expanders; physiology of aging. Mailing Add: Dept of Biol Boston Univ Boston MA 02215

BERMAN, HORACE AARON, b Brooklyn, NY, Nov 21, 15; m 39; c 3. APPLIED CHEMISTRY. Educ: Columbia Univ, BS, 35; ChE, 36. Prof Exp: Chemist, Union Chem Corp, NJ, 36; chem engr, Herstein Labs, Inc, NY, 37; draftsman, Westminster Tire Corp, 38-40; chemist San Francisco field off, Nat Bur Stand, 40-57, head paint lab, 45-57, DC, 57-68; RES CHEMIST, FED HWY ADMIN, 68- Concurrent Pos: Teacher, Drew Sch, San Francisco, 45; Emerson Inst, 59-61; abstractor, Chem Abstracts, 58-72; auditor, Calorimetry Conf, 63. Mem: Am Chem Soc; Am Soc Test & Mat. Res: Thermochemistry and phase equilibria of portland cement constituents; analytical chemistry; paint and varnish technology; synthetic resins; colorimetry of paint surfaces. Mailing Add: Off of Res Mat Div Fed Hwy Admin Washington DC 20590

BERMAN, IRWIN, b New York, NY, Dec 4, 24; m 51; c 2. ANATOMY, HEMATOLOGY. Educ: Seton Hall Univ, BS, 49; NY Univ, MS, 51, PhD(physiol), 55. Prof Exp: Asst radiobiol, Brookhaven Nat Lab, 55-56; res assoc, Sch Med, Stanford Univ, 56-59; asst prof, 59-62, ASSOC PROF ANAT, SCH MED, UNIV MIAMI, 62- Concurrent Pos: Am Cancer Soc fel, Brookhaven Nat Lab, 55-56 & Sch Med, Stanford Univ, 56-57; NIH res fel, 57-58; USPHS spec res fel, Harvard Med Sch, 63-64. Mem: AAAS; Am Soc Cell Biol; Am Asn Anat; Radiation Res Soc. Res: Electron microscopy; radiation biology; oncology. Mailing Add: Dept of Biol Struct Univ of Miami Sch of Med Miami FL 33152

BERMAN, JEROME RICHARD, b New York, NY, Aug 7, 20; m 46; c 1. INTERNAL MEDICINE. Educ: Univ Cincinnati, BA, 41, MD, 44. Prof Exp: Fel, 47-50, from instr to assoc prof med, 48-67, clin prof, 67-74, PROF MED, UNIV CINCINNATI, 74-; ATTEND PHYSICIAN, HOLMES HOSP, CINCINNATI GEN HOSP, 74- Mem: AAAS; Am Fedn Clin Res; Am Asn Study Liver Dis; fel Am Col Physicians; fel Am Col Gastroenterol. Res: Gastroenterology. Mailing Add: Univ of Cincinnati Med Ctr 321 Bethesda Ave Cincinnati OH 45267

BERMAN, JOEL DAVID, b Minneapolis, Minn, Jan 1, 43. MATHEMATICS. Educ: Univ Minn, Minneapolis, BA, 65; Univ Wash, PhD(math), 70. Prof Exp: Asst prof, 70-75, ASSOC PROF MATH, UNIV ILL, CHICAGO CIRCLE, 75- Mem: Am Math Soc; Math Asn Am. Res: Lattice theory and universal algebra. Mailing Add: Dept of Math Univ of Ill Chicago Circle Chicago IL 60680

BERMAN, JULIAN L, b Minneapolis, Minn, July 8, 35; m 59; c 2. PEDIATRICS, GENETICS. Educ: Univ Minn, Minneapolis, BA, 56, BS, 57, MD, 60. Prof Exp: Intern, Mt Zion Hosp Med Ctr, San Francisco, 60-61; resident pediat, Children's Mem Hosp, Northwestern Univ, 61-63, chief resident, 62-63, fel biochem genetics, 63-66, assoc pediat, 66-67; dir genetics sect, 68-75, dir med educ, 70-75, ATTENDING PHYSICIAN, COOK COUNTY HOSP, 67-; PROF PEDIAT, CHICAGO MED SCH, 72-, CHMN DEPT PEDIAT & ACTG ASSOC DEAN GRAD EDUC, 75- Concurrent Pos: Lectr & dir PKU clin, Med Ctr Hosp, Univ Ill, Chicago, 70- Mem: Am Acad Pediat; Am Fedn Clin Res; Am Soc Human Genetics; Soc Pediat Res. Res: Phenylketonuria and its variant states. Mailing Add: Dept of Pediat Chicago Med Sch Chicago IL 60612

BERMAN, KENNETH SIDNEY, b Boston, Mass, July 29, 36; m 65; c 3. ORAL BIOLOGY. Educ: Harvard Univ, AB, 58, DMD, 62. Prof Exp: Nat Inst Dent Res fel oral biol, Harvard Univ, 62-65; asst prof, Case Western Reserve Univ, 65-68; ASST PROF RESTORATIVE DENT, TUFTS UNIV, 68- Concurrent Pos: Nat Inst Dent Res career develop award, 66-68. Mem: AAAS; Am Soc Microbiol; Int Asn Dent Res. Res: Oral microbiological aspects of dental caries and periodontal disease. Mailing Add: 136 Harrison Ave Boston MA 02111

BERMAN, LAWRENCE URETZ, b Chicago, Ill, Aug 4, 19; m 50; c 2. ORGANIC CHEMISTRY. Educ: Univ Ill, Urbana, BS, 41; Northwestern Univ, MS, 47. Prof Exp: Asst chem org chem dept, IIT Res Inst, 48-56, assoc chemist, 56, res chemist, 56-65; res chemist bus equip group, Explor Res Dept, Bell & Howell Co, 65-68; RES CHEMIST, RES & DEVELOP DIV, KRAFTCO CORP, 68- Mem: AAAS; Am Chem Soc; Am Inst Chem. Res: Synthesis of new organic compounds, including monomers and polymers; development of organic reaction processes by new techniques and commercial scale-up. Mailing Add: 8025 N Hamlin Ave Skokie IL 60076

BERMAN, LOUIS, b London, Eng, Mar 21, 03; nat US; m 34; c 1. ASTRONOMY. Educ: Univ Minn, AB, 25, AM, 27; Univ Calif, PhD(astrophys), 29. Prof Exp: Asst, Univ Minn Observ, 25-27; instr astron, Carleton Col, 29-31; instr astron & math, San Mateo Col, 31-35 & City Col San Francisco, 35-68; LECTR ASTRON, UNIV SAN FRANCISCO, 68- Mem: AAAS; Am Astron Soc. Res: Comet positions and orbits; spectroscopic and visual binaries; variable stars; nebular and stellar spectroscopy; galactic rotation. Mailing Add: 1020 Laguna Ave Burlingame CA 94010

BERMAN, M LAWRENCE, b Stanford, Conn, July 13, 29; m. ANESTHESIOLOGY, CLINICAL PHARMACOLOGY. Educ: Univ Conn, BS, 51; Univ Wash, MS, 54, PhD, 56; Univ NC, MD, 64. Prof Exp: Toxicologist & asst chief environ health lab, Air Materiel Command, Wright-Patterson AFB, 56-57, prin investr pharmacol-biochem, Aerospace Med Lab, Wright Air Develop Ctr, 57-60; resident anesthesia, Sch Med, Univ NC, Chapel Hill, 65-67; from asst prof to assoc prof, Sch Med, Northwestern Univ, 67-74; PROF ANESTHESIA, SCH MED, VANDERBILT UNIV, 74- Mem: Am Soc Clin & Exp Pharmacol; Int Anesthesia Res Soc; Am Soc Clin Res; Am Soc Anesthesiol; Asn Univ Anesthetists. Res: Man in space program; humoral response to stress; catabolism of catechol amines; psychotropic drugs; drug metabolism. Mailing Add: Sch of Med Vanderbilt Univ Nashville TN 37205

BERMAN, MONES, b Lithuania, Aug 20, 20; nat US; m 45; c 2. BIOMATHEMATICS. Educ: Cooper Union, BEE, 50; Polytech Inst Brooklyn, PhD(physics), 57. Prof Exp: Engr res asst electronics, Sloan Kettering Inst, Cornell Univ, 46-50, asst biophysics, 53-58, asst prof, Med Sch, 55-58; biophysicist theoret biol, Nat Inst Arthritis & Metab Dis, 58-72, CHIEF LAB THORET BIOL, NAT CANCER INST, NIH, 72- Honors & Awards: Super Serv Award, Dept Health, Educ & Welfare, 74. Mem: AAAS; Biophys Soc; Soc Math Biol. Res: Theoretical biology; kinetics; metabolic systems; mathematical biology; computers; radiation physics. Mailing Add: Lab Theoret Biol Nat Cancer Inst NIH Bethesda MD 20014

BERMAN, PAUL RONALD, b New York, NY, Feb 5, 45. ATOMIC PHYSICS, LASERS. Educ: Rensselaer Polytech Inst, BS, 65; Yale Univ, MS, 66, PhD(physics), 69. Prof Exp: Instr physics, Yale Univ, 69-71; asst prof, 71-75, ASSOC PROF PHYSICS, NY UNIV, 75- Mem: Am Phys Soc; Am Asn Physics Teachers; AAAS. Res: Theoretical atomic and laser physics with special emphasis on collision effects. Mailing Add: Dept of Physics NY Univ 4 Wash Pl New York NY 10003

BERMAN, REUBEN, b Minneapolis, Minn, Feb 8, 08; m 31; c 6. CARDIOLOGY. Educ: Univ Minn, Minneapolis, BA, 29, MD, 33. Prof Exp: Pvt pract med, 37-; CLIN PROF MED, UNIV MINN, MINNEAPOLIS, 63- Concurrent Pos: Prin investr, Coronary Drug Proj, Mt Sinai Hosp, Minneapolis. Mem: Am Col Physicians; Am Col Cardiol; Am Heart Asn. Mailing Add: 5620 Edgewater Blvd Minneapolis MN 55417

BERMAN, ROSE LOUISE, b New York, NY. BIOLOGY, NATURAL SCIENCES. Educ: Hunter Col, BA, 33. Prof Exp: Asst, Mt Sinai Hosp, NY, 34-42; biologist & bacteriologist, Chemists Club, Col Physicians & Surgeons, Columbia Univ, 44, Ortho Res Found, NJ, 44-45 & Dr E Friedheim Res Lab, NY, 45-47; biol supvr, Food Res Lab, 47-48; biologist & bacteriologist, Dr E Friedheim Res Lab, NY, 48-49; DIR & OWNER, CLIN LAB, 49- Concurrent Pos: Pres, Clin Lab Dirs NY State, 67- Mem: AAAS; fel Am Inst Chem; Am Asn Clin Chem; NY Acad Sci. Res: Medical bioanalysis; endocrinology; chemotherapy; rat hyperemia pregnancy test discoverer. Mailing Add: Clin Lab 1780 Broadway New York NY 10019

BERMAN, SAM MORRIS, b Worcester, Mass, Feb 4, 33; m 57. THEORETICAL PHYSICS. Educ: Univ Miami, Fla, BS, 54, MS, 55; Calif Inst Technol, PhD(physics), 59. Prof Exp: Instr, Calif Inst Technol, 55-59; NSF fel, Univ Copenhagen & Bohr Inst, Denmark, 59-64; assoc prof, Linear Accelerator Ctr, 64-69, PROF, LINEAR

ACCELERATOR CTR, STANFORD UNIV, 69- Concurrent Pos: Consult, IBM, 58-59; consult, Space Tech Labs, 59- Res: Quantum theory of fields; quantum electrodynamics and related high energy physics; plasma physics and hydromagnetics. Mailing Add: Linear Accelerator Ctr Stanford Univ Stanford CA 94305

BERMAN, SANFORD, b Syracuse, NY, Apr 22, 27; m 50; c 2. MICROBIOLOGY. Educ: Syracuse Univ, BS, 49, MS, 51, PhD(microbiol), 54. Prof Exp: Bacteriologist, Ralph M Parsons Co, Md, 53-55; bacteriologist, 55-60, res bacteriologist, 60-62, CHIEF MICROBIOL SECT, DEPT BIOLOGICS RES, WALTER REED ARMY INST RES, 62- Honors & Awards: Super Accomplishment Award, Off of Surgeon Gen, US Dept Army, 57; Super Performance Award, Dept Army, 66 & 70. Mem: AAAS; Am Soc Microbiol; Soc Cryobiol. Res: Methods of production and evaluation of biologics, particularly in the dry state, such as bacterial enzymes, bacterial, viral and rickettsial vaccines. Mailing Add: Dept of Biol Res Walter Reed Army Inst of Res Washington DC 20012

BERMAN, SIMEON MOSES, b Rochester, NY, Mar 28, 35; m 55; c 6. MATHEMATICAL STATISTICS. Educ: City Col New York, BA, 56; Columbia Univ, MA, 58, PhD(math statist), 61. Prof Exp: Lectr math, City Col New York, 57-60; res assoc math statist, Columbia Univ, 60-61, asst prof, 61-65; ASSOC PROF MATH, COURANT INST MATH SCI, NY UNIV, 65- Mem: Inst Math Statist; Math Asn Am. Res: Probability theory, especially stochastic processes. Mailing Add: Courant Inst of Math Sci NY Univ 251 Mercer St New York NY 10012

BERMES, BORIS JOHN, b Merchantville, NJ, Mar 2, 26; m 55; c 3. HYDROGEOLOGY. Educ: Colo Sch Mines, Geol Eng, 50; Univ Utah, MS, 54. Prof Exp: Geologist, US Geol Surv, 50-54; exploitation engr, Shell Oil Co, 54-56; hydraulic engr, US Dept Interior, 56-73; SR SCIENTIST WATER RESOURCES, GERAGHTY & MILLER INC, TAMPA, 73- Concurrent Pos: Analog model specialist, UN Develop Prog, 69; tech officer, Food & Agr Orgn UN, 69-71. Mem: Int Asn Hydrogeologists. Res: Application of electronic technology to resource appraisal and management for purposes of exploitation and/or environmental protection. Mailing Add: 11 Oak Rd St Augustine FL 32084

BERMES, EDWARD WILLIAM, JR, b Chicago, Ill, Aug 20, 32; m 57; c 5. CLINICAL CHEMISTRY. Educ: St Mary's Col, Minn, BS, 54; Loyola Univ Ill, MS, 56, PhD(biochem), 59. Prof Exp: Asst biochem, Loyola Univ, Ill, 54-56; chief biochemist, Hektoen Inst Med Res, Cook County Hosp, Chicago, 58-60; clin instr, 59-65, from asst prof to assoc prof, 66-74, PROF BIOCHEM & PATH, STRITCH SCH MED, LOYOLA UNIV CHICAGO, 74-, DIR CLIN CHEM, HOSP, 69-, ASSOC DIR, SCH MED TECHNOL, 70- Concurrent Pos: Chief biochemist, W Suburban Hosp, Chicago, 60-62; dir biochem, St Francis Hosp, 62-68; ed, Chicago Clin Chemist, 69- Mem: AAAS; Am Chem Soc; Am Asn Clin Chem; Am Asn Clin Path; NY Acad Sci. Res: Enzymology. Mailing Add: Dept of Path Loyola Univ Stritch Sch Med Maywood IL 60153

BERMON, STUART, b Philadelphia, Pa, Dec 2, 36; m 61; c 1. SOLID STATE PHYSICS. Educ: Univ Pa, BA, 58; Univ Ill, MS, 59, PhD(physics), 64. Prof Exp: Staff mem physics, Lincoln Lab, Mass Inst Technol, 64-66; asst prof eng, Brown Univ, 66-73; ASSOC PROF PHYSICS, CITY COLLEGE NEW YORK, 73- Mem: Am Phys Soc. Res: Electron tunneling; superconductivity; properties of thin films; transport in amorphous semiconductors; cyclotron resonance in semiconductors and metals. Mailing Add: Dept of Physics City College of New York New York NY 10031

BERMUDEZ, VICTOR MANUEL, b New York, NY, Jan 24, 47. SOLID STATE SCIENCE. Educ: Mass Inst Technol, BS, 67; Princeton Univ, MA, 69, PhD(phys chem), 76. Prof Exp: RES PHYSICIST, NAVAL RES LAB, 72- Mem: Sigma Xi; Am Phys Soc. Res: Surface chemistry and surface physics of insulating materials. Mailing Add: Code 6440 Naval Res Lab Washington DC 20375

BERN, HOWARD ALAN, b Montreal, Que, Jan 30, 20; US citizen; m 46; c 2. ENDOCRINOLOGY, ZOOLOGY. Educ: Univ Calif, Los Angeles, AB, 41, MA, 42, PhD(zool), 48. Prof Exp: Teaching asst zool, Univ Calif, Los Angeles, 41-42, 46, Nat Res Coun fel, 46-48; from instr to assoc prof, Univ Calif, Berkeley, 48-61; assoc res endocrinologist, 56-61, RES ENDOCRINOLOGIST, CANCER RES LAB, 61-, PROF ZOOL, UNIV CALIF, BERKELEY, 61- Concurrent Pos: Guggenheim fel, 51-52; NSF sr fel, 58-59, 65-66; fel, Ctr Adv Study in Behav Sci, Stanford Univ, 60; res prof, Miller Inst Basic Res Sci, 61; chmn, northern grad group endocrinol, Univ Calif, 61-; vis prof, Bristol Univ, 65-66, Triandrum, Kerala, India, 67 & Univ Tokyo, 71; mem neuroendocrinol panel, Int Brain Res Orgn; mem adv comt carcinogenesis, Nat Cancer Inst; Eli Lilly lectureship, Endocrine Soc, 75. Mem: Am Soc Zool (pres, 67); Endocrine Soc; Am Asn Cancer Res; Soc Exp Biol & Med; Am Asn Anat. Res: Comparative endocrinology, especially of prolactin; endocrinology of perinatal carcinogenesis, especially mammary gland; hormones and neoplastic growth; neurosecretion and comparative neuroendocrinology. Mailing Add: Dept of Zool Univ of Calif Berkeley CA 94720

BERNABEI, AUSTIN M, b New Rochelle, NY, July 5, 28. NUCLEAR PHYSICS. Educ: Manhattan Col, BCE, 50; Cath Univ Am, MS, 53; NY Univ, MNucE, 59, PhD(physics), 64. Prof Exp: ASSOC PROF PHYSICS, MANHATTAN COL, 56- Mem: Am Phys Soc; Am Asn Physics Teachers. Res: Neutron resonance analysis, particularly the Doppler distortion effect and Mössbauer effect. Mailing Add: Dept of Physics Manhattan Col Bronx NY 10471

BERNABEI, STEFANO, b San Martino in Rio, Italy, Feb 29, 44. PLASMA PHYSICS. Educ: Univ Milan, PhD(physics), 69. Prof Exp: Res asst physics, Univ Milan, 69-71; res assoc, 71-74, RES STAFF MEM PLASMA PHYSICS LAB, PRINCETON UNIV, 74- Concurrent Pos: Nat Coun Res Italy fel, Plasma Physics Lab, Princeton Univ, 70-71. Res: Plasma heating with radio frequency waves; particularly heating at the lower hybrid frequency in toroidal machines. Mailing Add: Plasma Physics Lab J Forrestal Campus PO Box 451 Princeton NJ 08540

BERNADY, KAREL FRANCIS, b Baltimore, Md, Feb 1, 41; m 64; c 2. PHARMACEUTICAL CHEMISTRY. Educ: Loyola Col, Md, BS, 63; Fordham Univ, PhD(org chem), 68. Prof Exp: Res chemist org chem, Lederle Labs, 67-75, GROUP LEADER, AM CYANAMID CO, 75- Mem: Am Chem Soc; NY Acad Sci; AAAS. Res: Chemistry of natural products; process research on pharmaceuticals and fine chemicals. Mailing Add: 381 Gemini Dr S Somerville NJ 08876

BERNAL, ERNESTO, b June 22, 21; m 45; c 4. VETERINARY PATHOLOGY. Educ: Nat Univ Colombia, DMV, 45. Prof Exp: Veterinarian pub health, Colombian Govt, 45-53; prof path, Univ Caldas, Colombia, 54-56; pathologist, Univ Miami, 57-69; pathologist, Chas Pfizer Med Res Inst, 69-71; pathologist, Squibb Inst Med Res, 72-74; PATHOLOGIST, GULF SOUTH RES INST, 74- Mem: Int Acad Path; Am Vet Med Asn; Environ Mutagen Soc; Soc Pharmacol & Environ Pathologists. Res: Chemical and environmental-induced cancer in laboratory animals; experimental development pathology; teratology and mutagenesis; experimental toxicology. Mailing Add: Gulf South Res Inst Box 1177 New Iberia LA 70560

BERNAL, IVAN, b Barranquilla, Colombia, Mar 28, 31; m 57. CHEMISTRY. Educ: Clarkson Col, BS, 54; Univ Va, MS, 56; Columbia Univ, PhD(chem), 63. Prof Exp: Staff mem chem, RCA Labs, NJ, 56-59; Harvard Corp fel, 63-64; asst prof chem, State Univ NY Stony Brook, 64-67; chemist, Brookhaven Nat Lab, 67-73; assoc prof, 73-75, PROF CHEM & ASSOC CHMN DEPT, UNIV HOUSTON, 75- Honors & Awards: US Sr Scientist Award, Alexander von Humboldt Found, 75. Mem: Am Chem Soc; Am Crystallog Asn. Res: Problems of valence states and of stereochemistry of transition metal ions; spectroscopic and x-ray crystallographic work. Mailing Add: Dept of Chem Univ of Houston Houston TX 77004

BERNAL G, ENRIQUE, b Barranquilla, Colombia, July 30, 38; US citizen; m 61; c 2. OPTICAL PHYSICS, LASERS. Educ: Col St Thomas, BS, 60; Univ Minn, MS, 63. Prof Exp: Res asst & res scientist, 63-66, prin res scientist, 66-71, sr prin res scientist, 71-75, STAFF SCIENTIST LASERS & OPTICAL MAT, CORP RES CTR, HONEYWELL INC, 75- Honors & Awards: H W Sweatt Award, Honeywell Inc, 75. Mem: Inst Elec & Electronic Engrs; Am Phys Soc. Res: Properties of transparent infrared materials and the effect of laser irradiation on optical properties; optical materials, interferometry and effects of laser irradiation of opaque solids. Mailing Add: Honeywell Corp Res Ctr 10701 Lyndale Ave S Bloomington MN 55420

BERNAL-LLANAS, ENRIQUE, b Mexico, DF, May 26, 25; m 54; c 4. PHARMACOLOGY, MICROBIOLOGY. Educ: Nat Univ Mex, BA, 42; Nat Polytech Inst, Mex, MSc, 48. Prof Exp: Res med bact, Trop Dis Inst, 45-49; asst prof & lectr, Nat Polytech Inst, Mex, 49; researcher, Inst Behring, 52-57; prod exec, Cyanamid of Mex, 57-65; plant mgr, Bristol Labs, Mex, 66-72; TECH DIR, BRISTOL LABS & MEAD JOHNSON LABS, MEX, 73- Mem: AAAS; Am Soc Microbiol; Soc Mex Microbiol. Res: Biochemistry of influenza virus; biochemistry and immunology of antibiotic-resistant strains of bacteria; antibiotics; research and development of antibiotic products. Mailing Add: Insurgentes Sur 3825 Tlalpan Mexico DF Mexico

BERNARD, CAMILLE STEPHEN, b Wellington, PEI, Apr 16, 27; nat; m 51; c 2. ANIMAL BREEDING. Educ: Laval Univ, BSc, 49; Univ Wis, MSc, 50, PhD(genetics), 54. Prof Exp: Res officer animal husb div, Exp Farm, 53-68, DIR RES STA, CAN AGR, 68- Res: Animal breeding and feeding. Mailing Add: Res Sta Can Agr Lennoxville PQ Can

BERNARD, DAVY LEE, b Vermillion Parish, La, July 16, 36; m 60; c 2. NUCLEAR PHYSICS. Educ: Univ Southwestern La, BS, 58; Rice Univ, MA, 60, PhD(physics), 65. Prof Exp: Asst prof physics, Univ Va, 65-68; assoc prof, 68-74, PROF PHYSICS, UNIV SOUTHWESTERN LA, 74- Concurrent Pos: Res Corp grant, 70. Mem: Am Phys Soc. Res: Atomic physics, especially energy loss and charge exchange of ions passing through matter; neutron physics, particularly measurement of neutron scattering cross sections. Mailing Add: Dept of Physics Box 1210 Univ of Southwestern La Lafayette LA 70501

BERNARD, GARY DALE, b Everett, Wash, Feb 19, 38; m 64; c 4. VISION, PHYSIOLOGICAL OPTICS. Educ: Univ Wash, BSEE, 59, MSEE, 60, PhD(elec eng), 64. Prof Exp: From instr to asst prof elec eng, Mass Inst Technol, 65-68; asst prof ophthal, eng & appl sci, Sch Med, 68-71, ASSOC PROF OPHTHAL, VISUAL SCI, ENG & APPL SCI, YALE UNIV, 71- Concurrent Pos: Ford Found fel, Mass Inst Technol, 65-66; res career develop award, Nat Eye Inst, 71-75. Mem: AAAS; Asn Res Vision & Ophthal; Optical Soc Am; Biophys Soc; Lepidopterists Soc. Res: Optical structure and function of eyes, intra-ocular color filters; optics of the compound eye; processing of visual information; interaction of electromagnetic waves with dielectric structures; Diptera and Lepidoptera. Mailing Add: 80 Dawes Ave Hamden CT 06517

BERNARD, HARVEY RUSSELL, b New York, NY, June 12, 40; m 62; c 2. ANTHROPOLOGY. Educ: Queens Col, BA, 61; Univ Ill, MA, 63, PhD(cult anthrop), 68. Prof Exp: From lectr to assoc prof anthrop, Wash State Univ, 66-72; res assoc marine affairs, Scripps Inst Oceanog, 72; ASSOC PROF ANTHROP, W VA UNIV, 72-, RES ASSOC, REGIONAL RES INST, 74- Concurrent Pos: NSF field sch grants, 67-70; Fulbright res scholar & lectr, Univ Athens, 69-70; sci collabr, Democritos Nuclear Res Ctr, Athens, 69-70; assoc dir, Social Res Ctr, Wash State Univ, 71-72; Am Philos Soc grant study Am Indian languages, 71, 72 & 75. Mem: Fel AAAS; fel Am Anthrop Asn; fel Soc Appl Anthrop. Res: Anthropological linguistics; Greece and Mexico; technology and cultural change; marine peoples; adjustment problems of repatriated migrant workers in Greece; numerical analysis of closed social groups. Mailing Add: Dept of Sociol & Anthrop WVa Univ Morgantown WV 26505

BERNARD, JOHN MILFORD, b Duluth, Minn, May 22, 33; m 63; c 1. PLANT ECOLOGY. Educ: Univ Minn, Duluth, BS, 58; Rutgers Univ, MS, 60, PhD(plant ecol), 63. Prof Exp: Tech asst biol, Rutgers Univ, 58-62; instr, Franklin & Marshall Col, 62-64; from asst prof to assoc prof, 64-72, PROF BIOL, ITHACA COL, 72-, CHMN DEPT, 73- Concurrent Pos: Finger Lakes sci grant, 66. Mem: Ecol Soc Am; Torrey Bot Club. Res: Structure and function of wetland ecosystems. Mailing Add: Dept of Biol Ithaca Col Ithaca NY 14850

BERNARD, JOSEPH LIONEL, b Plattsburgh, NY, May 13, 28; m 50; c 3. ANALYTICAL CHEMISTRY, INSTRUMENTATION. Educ: St Michael's Col, BS, 49. Prof Exp: Res chemist, E I du Pont de Nemours & Co, 53-65; field lab supvr, 65-70, supvr, Customer Training Sect, 70-74, TECH MKT SPECIALIST, PERKIN-ELMER CORP, 74- Mem: Am Chem Soc; Soc Appl Spectros. Res: Physical and chemical characterization of research compounds through gas chromatography, thin layer chromatography, infrared spectroscopy, x-ray diffraction and fluorescence; radiochemistry; emission spectroscopy; thermogravimetric analysis; derivation of new analytical techniques; physico-chemical studies and characterization of viscosity and rheology. Mailing Add: Perkin-Elmer Corp 1625 E Edinger Ave Santa Ana CA 92705

BERNARD, MARIE (WITTE), b Richmond, Ind, Sept 19, 14. BIOLOGY. Educ: Marian Col, BS, 41; Fordham Univ, MS, 45, PhD(biol), 47. Prof Exp: From instr to assoc prof, 47-63, PROF BIOL, MARIAN COL, 63- Mem: Am Inst Biol Sci; Nat Asn Biol Teachers. Res: Cytology of plants; a comparative cytological study of three species of the Chenopodiaceae. Mailing Add: Dept of Biol Sci Marian Col Indianapolis IN 46222

BERNARD, PATRICK SPITALETTA, b Jersey City, NJ, Jan 3, 31; m 61; c 2. PHYSIOLOGY, BIOCHEMISTRY. Educ: Univ RI, BS, 57. Prof Exp: Supvr pharmacol & neuropharmacol, Ciba Pharmaceut Col, 57-71, SR RES SCIENTIST NEUROPHARMACOL, CIBA-GEIGY PHARMACEUT CO, 71- Mem: AAAS. Res: Neuropharmacology; neurophysiology; effects of drugs on the central and peripheral nervous system. Mailing Add:

BERNARD, RICHARD FERNAND, b Lewiston, Maine, Aug 19, 34; m 59. ZOOLOGY. Educ: Univ Maine, BA, 56; Mich State Univ, MS, 59, PhD(zool), 62.

Prof Exp: Assoc prof zool, Wis State Univ, Superior, 62-69; assoc prof biol, 69-73, PROF BIOL, QUINNIPIAC COL, 73- Concurrent Pos: Bd Regents Instnl study grant, 63-64. Mem: Am Ornith Union; Am Soc Mammalogists; Nat Audubon Soc. Res: Effects of insecticides on wildlife. Mailing Add: Dept of Biol Quinnipiac Col Hamden CT 06517

BERNARD, RICHARD LAWSON, b Detroit, Mich, Aug 12, 26; m 52; c 4. PLANT BREEDING. Educ: Ohio State Univ, BS, 49, MS, 50; NC State Col, PhD(field crops), 60. Prof Exp: Res instr agron, NC State Col, 54; res agronomist, 54-61, RES GENETICIST, USDA REGIONAL SOYBEAN LAB, UNIV ILL, URBANA-CHAMPAIGN, 61- Mem: Am Soc Agron; Am Genetic Asn; Soc Study Evolution. Res: Qualitative genetics, germplasm and breeding of soybeans. Mailing Add: US Regional Soybean Lab 178 Davenport Hall Univ of Ill Urbana IL 61801

BERNARD, RICHARD RYERSON, b Jacksonville, Fla, Dec 24, 17; m 43; c 3. MATHEMATICS. Educ: Univ Va, PhD(math), 49. Prof Exp: Instr math, Univ Va, 48-49; from instr to asst prof, Yale Univ, 49-55; assoc prof, 55-59, PROF MATH, DAVIDSON COL, 59- Mem: Math Asn Am. Mailing Add: Davidson Col Box 656 Davidson NC 28096

BERNARD, RUDY ANDREW, b New York, NY, May 31, 30; m 60; c 3. NEUROPHYSIOLOGY. Educ: Univ Montreal, BA, 53; Cornell Univ, MNS, 60, PhD(animal physiol), 62. Prof Exp: Asst nutrit, Cornell Univ, 58-60; fel neurophysiol, Univ Wis, 62-64; asst prof physiol, State Univ NY Downstate Med Ctr, 64-66 & Rockefeller Univ, 66-69; assoc prof, 69-75, PROF PHYSIOL, MICH STATE UNIV, 75- Mem: AAAS; Am Physiol Soc; Soc Neurosci. Res: Anatomy of central taste pathways; electrophysiology of peripheral and central taste mechanisms; behavioral studies of taste preferences; taste and hypertension. Mailing Add: 112 Giltner Hall Mich State Univ East Lansing MI 48824

BERNARD, SELDEN ROBERT, b Scobey, Mont, Nov 10, 25; div; c 3. MATHEMATICAL BIOLOGY. Educ: Univ Denver, BS, 48; Univ Chicago, PhD(math biol), 60. Prof Exp: Asst chemist, Kaiser Aluminum & Chem Co, 49; fel radiol physics, Oak Ridge Nat Lab, 49-50; health physicist, Union Carbide Nuclear Co, 50-54; res health physicist, 54-57, MEM RES STAFF, OAK RIDGE NAT LAB, 60- Mem: AAAS; Health Physics Soc. Res: Mathematical biology and health physics; mathematical studies of biological phenomena, especially blood-gas equilibria and compartmental models for metabolism of inorganic ions and molecules; estimation of radiation exposure to organs and tissues from internally contained radionuclides. Mailing Add: Health Physics Div Oak Ridge Nat Lab Oak Ridge TN 37830

BERNARD, WALTER JOSEPH, b Manchester, NH, Dec 25, 23; m 48; c 4. PHYSICAL CHEMISTRY. Educ: Univ NH, BS, 50, MS, 51; Mass Inst Technol, PhD(chem), 54. Prof Exp: Assoc dir eng, 64-68, dir res & develop, 68-71, SR SCIENTIST, SPRAGUE ELEC CO, 71- Mem: Electrochem Soc; Am Chem Soc. Res: Inorganic fluorides; silicon chemistry; oxide films on metals; dielectrics. Mailing Add: Res & Develop Ctr Sprague Elec Co North Adams MA 01247

BERNARD, WILLIAM HICKMAN, b New Orleans, La, July 7, 32; div; c 4. PHYSICS. Educ: Tulane Univ, BS, 54, PhD(physics), 63. Prof Exp: Asst prof physics, 62-69, PROF PHYSICS, LA TECH UNIV, 69- Mem: AAAS; Am Asn Physics Teachers; Am Crystallog Asn. Res: X-ray crystallography. Mailing Add: Dept of Physics La Tech Univ Ruston LA 71271

BERNARDI, DOMINIC JOSEPH, organic chemistry, see 12th edition

BERNARDI, SALVATORE DANTE, b Italy, Apr 12, 19; nat US; m 41; c 2. MATHEMATICS. Educ: Yale Univ, BS, 41; Springfield Col, MEd, 43; Univ Notre Dame, MS, 45; NY Univ, PhD(math), 51. Prof Exp: Instr math, Springfield Col, 41-43, Univ Notre Dame, 43-45 & Lehigh Univ, 45-56; ASSOC PROF MATH, NY UNIV, 46- Mem: Am Math Soc; Am Soc Eng Educ; Math Asn Am. Res: Theory of Schlicht functions. Mailing Add: Dept of Math NY Univ-Courant 251 Mercer St New York NY 10012

BERNARDIN, JOHN EMILE, b Santa Monica, Calif, Feb 28, 37; m 58; c 2. PHYSICAL BIOCHEMISTRY. Educ: Univ Calif, Riverside, BA, 63; Univ Ore, PhD(phys biochem), 70. Prof Exp: Chemist, 63-70, RES CHEMIST, WESTERN REGIONAL RES CTR, AGR RES SERV, USDA, 70- Mem: Am Chem Soc; Am Asn Cereal Chemists. Res: Rheology of biopolymers. Mailing Add: Western Regional Res Ctr USDA 800 Buchanan St Albany CA 94710

BERNARDIN, LEO J, b Lawrence, Mass, June 4, 30; m 56; c 4. PAPER CHEMISTRY. Educ: Tufts Univ, BS, 52; Inst Paper Chem, MS, 54, PhD(wood chem), 58. Prof Exp: Res scientist, Marathon Div, Am Can Co, 57-59; res scientist, 59-66, SR RES SCIENTIST, RES & ENG, KIMBERLY-CLARK CORP, 66- Mem: Am Chem Soc; Tech Soc Pulp & Paper Indust. Res: Wood and cellulose chemistry; prehydrolysis of wood; hemicellulose chemistry; synthetic fiber paper; chemical modification of cellulose. Mailing Add: Res & Eng Kimberly-Clark Corp Neenah WI 54956

BERNARDIS, LEE L, b Graz, Austria, Sept 18, 26; US citizen; m 58; c 1. PHYSIOLOGY. Educ: Graz Univ, DrPhil, 49; Univ Western Ont, PhD(physiol, biochem), 61. Prof Exp: Res tech liver dis, Hosp for Sick Children, Toronto, Ont, 53-56; res mem meat processing & diet res, Can Packers, Ltd, 56-57; res asst physiol, Univ Western Ont, 57-61; res assoc exp path, 61-63, res asst prof, 63-69, res assoc prof, 69-73, RES PROF SURG, STATE UNIV NY BUFFALO, 73- Concurrent Pos: Proj dir, Dept Surg, State Univ NY Buffalo & dir, Neuroendocrine-Neurovisceral Res Unit, 73- Mem: Am Asn Univ Prof; Fedn Am Scientists; Am Inst Nutrit; Int Brain Res Orgn; Int Soc Neuroendocrinol. Res: Hypothalamic regulation of growth, endocrines and metabolism. Mailing Add: Dept of Surg State Univ NY Clin Ctr 462 Grider Buffalo NY 14215

BERNARDO, PETER D, b Sagamore, Pa, Aug 21, 37; m 61; c 3. PHYSICAL PHARMACY. Educ: Duquesne Univ, BS, 59; Univ Mich, MS, 63, PhD(pharmaceut chem), 66. Prof Exp: Asst res pharmacist, Parke, Davis & Co, 60-61; SR PHARMACEUT CHEMIST, SMITH KLINE & FRENCH LABS, 66- Mem: Am Pharmaceut Asn; Acad Pharmaceut Sci; Am Chem Soc. Res: Dissolution rate and crystal growth of polymorphic compounds; product development of solid dosage forms for human and animal health; preformulation research; exploratory studies on new drug delivery systems. Mailing Add: Smith Kline & French Labs 1500 Spring Garden St Philadelphia PA 19101

BERNASCONI, CLAUDE FRANCOIS, b Zurich, Switz, Feb 17, 39; m 63; c 2. PHYSICAL ORGANIC CHEMISTRY. Educ: Swiss Fed Inst Technol, dipl, 63, PhD(chem), 65. Prof Exp: Res asst chem, Swiss Fed Inst Technol, 65-66 & Max Planck Inst Phys Chem, 66-67; asst prof, 67-72, ASSOC PROF CHEM, UNIV CALIF, SANTA CRUZ, 72- Concurrent Pos: Res grants, Petrol Res Fund, 69-71, 72-74 & 75-77, Res Corp, 70, Alfred P Sloan Found, 71-73 & NSF, 71-77. Mem: AAAS; Am Chem Soc; The Chem Soc. Res: Mechanisms of nucleophilic aromatic substitution reactions; fast reaction kinetics. Mailing Add: Div Natural Sci Univ of Calif Santa Cruz CA 95064

BERNAT, THOMAS PHILLIP, b Mansfield, Ohio, Apr 19, 44. LOW TEMPERATURE PHYSICS. Educ: Claremont Men's Col, BA, 66; Brandeis Univ, PhD(physics), 72. Prof Exp: Res assoc physics, Brandeis Univ, 71-72; vis asst prof, 72-75, ASST PROF PHYSICS, LA STATE UNIV, BATON ROUGE, 75- Res: Cryogenic instrumentation applied to gravitational radiation detection; properties of liquid and solid helium isotopic mixtures; cryogenic instrumentation applied to surface studies. Mailing Add: Dept of Physics & Astron La State Univ Baton Rouge LA 70803

BERNAU, SIMON J, b Wanganui, NZ, June 12, 37; m 59; c 2. MATHEMATICS. Educ: Univ Canterbury, BSc, 58, MS, 59; Cambridge Univ, BA, 61, PhD(math), 64. Prof Exp: Lectr math, Univ Canterbury, 64-65, sr lectr, 65-66; prof, Univ Otago, NZ, 66-69; ASSOC PROF MATH, UNIV TEX, AUSTIN, 69- Mem: Am Math Soc; London Math Soc; Australian Math Soc. Res: Lattice ordered algebraic systems; operator theory; functional analysis. Mailing Add: Dept of Math Univ of Tex Austin TX 78712

BERNAYS, PETER MICHAEL, b New York, NY, July 19, 18; m 47; c 3. PHYSICAL INORGANIC CHEMISTRY, INFORMATION SCIENCE. Educ: Mass Inst Technol, BS, 39; Univ Ill, MS, 40, PhD(anal chem), 42. Prof Exp: From instr to asst prof chem, Ill Inst Technol, 46-50; assoc prof, Southwestern La Inst, 50-54; asst ed, 54-57, assoc ed, 58, head assignment dept, 59-61, asst to dir res & develop, 61-65, personnel asst, 65-66, sr assoc ed, 66-69, sr ed, 69-71, ASST MGR PHYS, ANAL & INORG DEPT, CHEM ABSTR, 71- Concurrent Pos: Res Corp Cotrell grant, 48-50. Mem: AAAS; Am Chem Soc. Res: X-ray diffraction; inorganic compounds; chemistry of scandium; structural properties of lime; chemical documentation. Mailing Add: Chem Abstr Serv Ohio State Univ Columbus OH 43210

BERNDT, ALAN FREDRIC, b New York, NY, Mar 14, 32; m 57; c 3. PHYSICAL CHEMISTRY, DENTAL RESEARCH. Educ: Cooper Union, BChE, 53; Calif Inst Technol, PhD(phys chem), 57. Prof Exp: Asst, Calif Inst Technol, 54-57; chemist, Monsanto Chem Co, 57-58; mem tech staff, Radio Corp of Am Labs, 58-60; chemist, Argonne Nat Lab, 60-65; from asst prof to assoc prof, 65-71, PROF CHEM, UNIV MO-ST LOUIS, 71-, ASST DEAN GRAD SCH & ASST DIR RES, 75- Mem: Am Chem Soc; Am Crystallog Asn; fel Am Inst Chemists; Int Asn Dent Res. Res: Crystallography; x-ray diffraction; dental fluoride chemistry. Mailing Add: Dept of Chem Univ of Mo St Louis MO 63121

BERNDT, BRUCE CARL, b St Joseph, Mich, Mar 13, 39; m 63; c 1. MATHEMATICS. Educ: Albion Col, AB, 61; Univ Wis-Madison, MS, 63, PhD(math), 66. Prof Exp: Asst lectr math, Glasgow Univ, 66-67; from asst prof to assoc prof, 67-75, PROF MATH, UNIV ILL, URBANA-CHAMPAIGN, 75- Concurrent Pos: Mem, Inst Advan Study, Princeton, NJ, 73-74. Mem: Am Math Soc; Math Asn Am. Res: Analytic number theory; classical analysis; complex integration. Mailing Add: Dept of Math 273 Altgeld Hall Univ of Ill Urbana IL 61801

BERNDT, DONALD CARL, b Toledo, Ohio, Apr 11, 35; m 60; c 2. PHYSICAL ORGANIC CHEMISTRY. Educ: Ohio State Univ, BSc, 57, PhD(org chem), 61. Prof Exp: NSF fel org chem, Ohio State Univ, 62; from asst prof to assoc prof chem, 62-74, PROF CHEM, WESTERN MICH UNIV, 74- Mem: Am Chem Soc. Res: Elucidation of reaction mechanisms in organic chemistry, including the study of medium, catalytic and structural influences. Mailing Add: Dept of Chem Western Mich Univ Kalamazoo MI 49001

BERNDT, EDWARD WALTON, organic chemistry, see 12th edition

BERNDT, WILLIAM O, b St Joseph, Mo, May 33; m 54; c 5. PHARMACOLOGY, TOXICOLOGY. Educ: Creighton Univ, BS, 54; Univ Buffalo, PhD(pharmacol), 59. Prof Exp: Fel pharmacol, Sch Med, Univ Buffalo, 59; from instr to assoc prof, Dartmouth Med Sch, 59-74; PROF PHARMACOL & TOXICOL & CHMN DEPT, MED CTR, UNIV MISS, 74- Concurrent Pos: Estab investr, Am Heart Asn, 64-69. Mem: Am Chem Soc; Am Soc Pharmacol & Exp Therapeut; Am Soc Nephrol. Res: Electrolyte and organic acid transport; renal physiology and pharmacology; metabolic aspects of transport processes; diffusion processes. Mailing Add: Dept of Pharmacol & Toxicol Univ of Miss Med Ctr Jackson MS 39216

BERNDTSON, WILLIAM EVERETT, b Middletown, Conn, Oct 1, 44; m 69; c 3. REPRODUCTIVE PHYSIOLOGY. Educ: Univ Conn, BS, 66; Cornell Univ, PhD(reprod physiol), 71. Prof Exp: Trainee reprod, Okla State Univ, 71-72; res assoc, 72-75, ASST PROF REPROD, COLO STATE UNIV, 76- Mem: Soc Study Reprod; Am Soc Animal Sci. Res: Cryopreservation of mammalian spermatozoa for artificial insemination; influence of hormones, drugs and exteroceptive stimuli on quantitative aspects of spermatogenesis. Mailing Add: Animal Reprod Lab Colo State Univ Ft Collins CO 80523

BERNE, BRUCE J, b New York, NY, Mar 8, 40; m 61. CHEMICAL PHYSICS, PHYSICAL CHEMISTRY. Educ: Brooklyn Col, BS, 61; Univ Chicago, PhD(chem physics), 64. Prof Exp: NATO fel, 66; mem chem fac, 66-69, assoc prof chem, 69-72, PROF CHEM, COLUMBIA UNIV, 72- Concurrent Pos: Alfred P Sloan Found fel, 67-70; Guggenheim fel, 71-72; mem panel, Nat Resource Comput Chem, Nat Acad Sci-Nat Res Coun. Mem: Am Phys Soc; AAAS. Res: Equilibrium and non-equilibrium statistical mechanics, structure and dynamics of the condensed phases of matter; laser light scattering. Mailing Add: Dept of Chem Columbia Univ New York NY 10027

BERNE, ROBERT MATTHEW, b Yonkers, NY, Apr 22, 18; m 44; c 4. MEDICAL PHYSIOLOGY. Educ: Univ NC, AB, 39; Harvard Med Sch, MD, 43. Hon Degrees: DSc, Med Col Ohio, 73. Prof Exp: From asst prof to prof physiol, Case Western Reserve Univ, 52-66, asst prof med, 57-66; CHARLES SLAUGHTER PROF PHYSIOL & CHMN DEPT, UNIV VA MED CTR, 66- Concurrent Pos: Ed, Circulation Res, 70-75; mem panel, Heart & Blood Vessel Dis Task Force, Nat Heart & Lung Inst, 72-, mem proj comt, Heart & Lung Prog, 75-77; mem panel, Heart & Vascular Dis Nat Res & Demonstration Rev Comt, NIH, 73-74; mem selection comt, Ciba Found Award Hypertension, 75-77; mem basic sci sect, Am Heart Asn. Honors & Awards: Carl J Wiggers Award, Am Physiol Soc, 75; Asn Chmn Depts Physiol Teaching Award, 76. Mem: Am Physiol Soc (pres, 72-73); Asn Chmn Depts Physiol (pres, 70-71); Am Soc Clin Invest; Microcirculatory Soc; Am Heart Asn. Res: Cardiovascular physiology, particularly the chemical factors involved in the local regulation of organ blood flow. Mailing Add: Dept of Physiol Jordan Hall Univ of Va Med Ctr Charlottesville VA 22901

BERNECKER, RICHARD RUDOLPH, b Allentown, Pa, Feb 23, 36; m 59; c 4. PHYSICAL CHEMISTRY. Educ: Muhlenberg Col, BS, 57; Cornell Univ, PhD(phys chem), 62. Prof Exp: Res chemist, Standard Oil, Ind, 61-64, Los Alamos Sci Lab, Univ Calif, 64-69; RES CHEMIST EXPLOSIVES, NAVAL SURFACE WEAPONS

CTR, MD, 69- Mem: Combustion Inst; Am Chem Soc. Res: Sensitivity and detonability of explosives and propellants; elucidation and characterization of mechanism of transition from deflagration to detonation in explosives and propellants. Mailing Add: 204 Pewter Lane Silver Spring MD 20904

BERNEIS, HANS LUDWIG, b Heidelberg, Ger, Jan 20, 14; nat US; m 42; c 2. PHYSICAL CHEMISTRY, ORGANIC CHEMISTRY. Educ: Royal Univ, Naples, Dr Chem, 38; Wayne Univ, PhD(chem), 49. Prof Exp: Petrol chemist, Keystone Oil Res Co & Petrol Specialties, 40-46; instr chem, Wayne Univ, 48-49; assoc prof, NMex Western Col, 49-50; AEC fel, Ind Univ, 50-52; CHEMIST, DEPT PATENT LAW, UPJOHN CO, 52- Mem: Am Chem Soc. Res: Reaction mechanisms; kinetics; steroid chemistry. Mailing Add: Upjohn Co Patent Law Dept 301 Henrietta St Kalamazoo MI 49001

BERNEKING, ARMOUR DALE, physical chemistry, see 12th edition

BERNER, DAVID LEO, b Alton, Ill, Feb 3, 38; m 61; c 1. BIOCHEMISTRY. Educ: St Louis Univ, BS, 60; Iowa State Univ, MS, 64, PhD(biochem), 68. Prof Exp: Jr chemist, Ames Lab Atomic Res, Iowa State Univ, 60-64, res asst biochem, Dept Food Technol, 64-68; SR RES CHEMIST LIPID CHEM, CAMPBELL INST FOOD RES, CAMPBELL SOUP CO, CAMDEN, 68- Mem: Am Chem Soc; Am Inst Biol Sci; Am Oil Chemists Soc. Res: Methods development in analytical chemistry; lipase activity and specificity; autoxidation mechanisms; flavor stability; natural and synthetic antioxidants; fats and oils applications in heat processed and frozen foods. Mailing Add: Camden NJ

BERNER, FREDERIC WALDEMAR, industrial chemistry, see 12th edition

BERNER, LEO DEWITTE, JR, b Pasadena, Calif, Feb 11, 22; m 47; c 2. BIOLOGICAL OCEANOGRAPHY. Educ: Pomona Col, BA, 43; Scripps Inst Oceanog, Univ Calif, MS, 52; Univ Calif, PhD(oceanog), 57. Prof Exp: Asst, Pomona Col, 48-49; res biologist, Scripps Inst Oceanog, Univ Calif, 52-56, jr res biologist, 58-61; fisheries res biologist, US Fish & Wildlife Serv, 57-58; assoc prog dir div sci personnel & educ, NSF, 61-65; res scientist, 65-66, assoc prof oceanog, 66-72, asst dean, 67-71, ASSOC DEAN GRAD COL, TEX A&M UNIV, 71-, PROF OCEANOG, 72- Concurrent Pos: Biologist, US Navy Oper Wigwam, 55. Mem: Fel AAAS; Am Soc Limnol & Oceanog; Ecol Soc Am; Marine Biol Asn UK. Res: Plankton zeography and ecology; pelagic tunicates; taxonomy of tunicata. Mailing Add: 1108 Neal Pickett Dr College Station TX 77840

BERNER, LEWIS, b Savannah, Ga, Sept 30, 15; m 45; c 2. ENTOMOLOGY. Educ: Univ Fla, BS, 37, MS, 39, PhD(zool), 41. Prof Exp: From asst prof to assoc prof biol, 46-54, actg dir div biol sci, 70-73, dir, 73-75, PROF BIOL, UNIV FLA, 54-, HEAD BIOL SCI, 59- Concurrent Pos: NIH grant, 53-60; consult entomologist, Gold Coast, Brit WAfrica, 50 & Nyasaland, Brit Cent Africa, 52; ed, Fla Entomologist, 50-63; prof entom, Lake Itasca Biol Sta, Univ Minn, 58-62, 68-69; USPHS grant, 63-66. Mem: Fel Entom Soc. Res: Ecology and taxonomy of Ephemeroptera; biology; parasitology; mayflies of Southeast. Mailing Add: Dept of Zool Univ of Fla Gainesville FL 32601

BERNER, ROBERT A, b Erie, Pa, Nov 25, 35; m 59; c 3. MARINE GEOCHEMISTRY. Educ: Univ Mich, BS, 57, MS, 58; Harvard Univ, PhD(geol), 62. Prof Exp: Sverdup fel oceanog, Scripps Inst, 62-63; from asst prof to assoc prof geophys sci, Univ Chicago, 63-65; assoc prof geol & geophys, 65-71, PROF GEOL & GEOPHYS, YALE UNIV, 71- Concurrent Pos: Alfred P Sloan Found fel phys sci, 68-72; Guggenheim fel, Guggenheim Found, 72; assoc ed, Am J Sci. Honors & Awards: Award, Mineral Soc Am, 71. Mem: Fel Geol Soc Am; Geochem Soc; Mineral Soc Am. Res: Kinetics of geochemical processes, chemical diagenesis; mineral-water equilibira; biogeochemistry. Mailing Add: Dept of Geol & Geophys Yale Univ New Haven CT 06520

BERNERS, EDGAR DAVIS, b Milwaukee, Wis, Aug 22, 27; m 60; c 4. EXPERIMENTAL NUCLEAR PHYSICS. Educ: Col Holy Cross, BS, 49; Univ Wis, PhD(physics), 57. Prof Exp: From instr to asst prof physics, Marquette Univ, 57-67; asst fac fel, 67-70, ASSOC FAC FEL, UNIV NOTRE DAME, 70- Mem: Am Phys Soc; Am Asn Physics Teachers. Res: Nuclear physics; electrostatic accelerators. Mailing Add: Dept of Physics Univ of Notre Dame Notre Dame IN 46556

BERNETTI, RAFFAELE, b Florence, Italy, Aug 24, 32; m 60; c 3. ORGANIC CHEMISTRY, ANALYTICAL CHEMISTRY. Educ: Univ Pisa, MS, 55; Univ Pa, PhD(org chem), 59; Univ Chicago, MBA, 70. Prof Exp: Lab dir, Milan Lab, 60-65, sect head org chem, Moffett Tech Ctr, ARGO, 65-67, ASST DIR PROD METHODOLOGY, MOFFETT TECH CTR, ARGO, CPC INT, 67- Concurrent Pos: Instr, DePaul Univ, 70. Mem: Am Chem Soc; fel Am Inst Chem. Res: Heterocyclic chemistry; carbohydrate chemistry; starch and dextrose production; dextrose, levulose sweeteners; analytical methods for carbohydrate and food products; project management and coordination. Mailing Add: CPC Int, CPC Develop Int Plaza Englewood Cliffs NJ 07632

BERNEY, CHARLES V, b Walla Walla, Wash, July 11, 31; m 56; c 3. MOLECULAR SPECTROSCOPY. Educ: Whitman Col, BA, 53; Univ Wash, PhD(phys chem), 62. Prof Exp: Res fel, Mellon Inst, 62-65; asst prof chem, Univ NH, 65-72; sr res assoc, Air Force Rocket Propulsion Lab, 72-73; SR RES ASSOC, NUCLEAR ENG DEPT, MASS INST TECHNOL, 73- Mem: AAAS; Am Phys Soc; Am Chem Soc. Res: Molecular structure; spectroscopy; application of neutron scattering to molecular spectroscopy. Mailing Add: NW13-221 Nuclear Eng Dept Mass Inst Technol Cambridge MA 02139

BERNEY, STEVEN, b Ger, Dec 31, 35; US citizen; m 61; c 3. RHEUMATOLOGY, IMMUNOLOGY. Educ: Rutgers Univ, BA, 58; State Univ NY Upstate Med Ctr, MD, 62. Prof Exp: Intern & resident internal med, Ohio State Univ Hosp, 62-63 & 65-67; USPHS fel rheumatic dis & immunol, NY Univ, 67-71, Arthritis Found fel, 70-71; ASST PROF MED, HEALTH SCI CTR, TEMPLE UNIV, 71- Mem: Am Rheumatism Asn; Reticuloendothelial Soc; Am Fedn Clin Res. Res: Lymphocyte physiology and immunology; isolation and purification of membrane antigen receptors; circulatory behavior of lymphocytes; mechanism of immunologic depression in aging. Mailing Add: Dept of Med Temple Univ Health Sci Ctr Philadelphia PA 19140

BERNFELD, PETER, b Leipzig, Ger, June 1, 12; nat US; m 40; c 2. ENZYMOLOGY, BIOCHEMISTRY. Educ: Univ Leipzig, MS, 35; Univ Geneva, PhD(chem), 37. Prof Exp: From res fel to chief chemist org chem, Univ Geneva, 37-49, privat docent, 47-49; from asst prof to assoc prof biochem & nutrit & biochemist, Cancer Res Control Unit, Tufts Univ, 49-57; VPRES & DIR RES, BIO-RES INST & BIO-RES CONSULTS, 57- Honors & Awards: Werner Medal, Swiss Chem Soc, 48. Mem: AAAS; Am Chem Soc; Am Soc Biol Chem; Soc Exp Biol & Med; Am Asn Cancer Res. Res: Proteins; enzymes; polysaccharides; immobilized enzymes and antigens; studies on serum glycoproteins; humoral antigens prognosticating susceptibility to mammary cancer in women and mice; inhalation toxicity of smoke including tobacco smoke. Mailing Add: Bio-Res Inst 9 Commercial Ave Cambridge MA 02141

BERNHAGEN, RALPH JOHN, b Toledo, Ohio, Aug 2, 10; m 40; c 2. GEOLOGY. Educ: Ohio State Univ, BA, 37, MA, 39. Prof Exp: Jr paleontologist, Shell Oil Co, Tex, 39-41; eng geologist, Lockwood, Andrews & Duller Eng Co, 41; chief geologist, Ohio Water Resources Bd, 41-52; asst chief, Ohio Div, Geol Surv, 52-57; div chief & state geologist, Ohio Geol Surv, 57-68; chief water planning sect, 68-72, ADMINR SHORELAND MGT SECT, OHIO DEPT NATURAL RESOURCES, 72- Mem: Fel Geol Soc Am; Am Asn Petrol Geologists; Asn Am State Geologists; Am Inst Prof Geologists. Res: Stratigraphy of Ohio and the Appalachian Basin; ground water and petroleum geology. Mailing Add: 5916 Linworth Rd Worthington OH 43085

BERNHARD, RICHARD ALLAN, b Pittsburgh, Pa, Oct 29, 23; m 53; c 2. ORGANIC CHEMISTRY. Educ: Stanford Univ, BS, 50; Calif Inst Technol, PhD(chem), 55. Prof Exp: Res chemist, Lemon Prods, Adv Bd, Calif, 55-57; from asst prof to assoc prof food sci & technol, 57-70, PROF FOOD SCI & TECHNOL, UNIV CALIF, DAVIS, 70- Honors & Awards: Asgrow Award of Merit, 65. Mem: Am Chem Soc; Inst Food Technol. Res: Chemistry and biochemistry of terpenes and sulfur compounds in plants; chemotaxonomy; analytical methods for determination of micro-constituents in materials of biological origin; flavor chemistry; theory of gas chromatographic processes. Mailing Add: Dept of Food Sci & Technol Univ of Calif Davis CA 95616

BERNHARD, VICTOR MONTWID, b Milwaukee, Wis, June 29, 27; m 60; c 3. MEDICINE. Educ: Northwestern Univ, BS, 47, MD, 51. Prof Exp: From clin instr to clin asst prof surg, Med Col Wis, 59-64; NIH fel cardiovasc surg, Col Med, Baylor Univ, 65; assoc prof surg, 66-74, PROF SURG, MED COL WIS, 74- Mem: Soc Vascular Surg; AMA; Am Col Surg; Int Cardiovasc Soc. Res: Vascular diseases; vascular surgery. Mailing Add: 8700 W Wisconsin Ave Milwaukee WI 53226

BERNHARD, WILLIAM ALLEN, b Philadelphia, Pa, Oct 9, 42; m 65; c 2. BIOPHYSICS. Educ: Union Col, BS, 64; Pa State Univ, MS, 66, PhD(biophysics), 68. Prof Exp: Asst biophysics, Pa State Univ, 64-68; res fel biol & med, Argonne Nat Lab, 68-70; ASST PROF RADIATION BIOL & BIOPHYS, UNIV ROCHESTER, 70- Concurrent Pos: Career develop award, NIH, 72, res grant award, 75; vis prof biol, Univ Regensburg, 75-76; assoc ed, Radiation Res, 75-78. Mem: Biophys Soc; Radiation Res Soc. Res: Effects of ionizing radiation on the primary structure of nucleic acids; electron spin resonance spectroscopy. Mailing Add: Dept of Radiation Biol & Biophys Univ of Rochester Rochester NY 14642

BERNHARDT, HARVEY ADOLPHUS, physical chemistry, see 12th edition

BERNHARDT, ROBERT L, III, b Salisbury, NC, Apr 28, 39; m 61; c 1. MATHEMATICS. Educ: Univ NC, Chapel Hill, BS, 61, MA, 64; Univ Ore, PhD(math), 68. Prof Exp: Asst prof math, Univ NC, Greensboro, 68-74; ASST PROF MATH, CHICAGO STATE UNIV, 74- Mem: Am Math Soc; Math Asn Am. Res: Ring theory; torsion theory of modules. Mailing Add: Dept of Math Chicago State Univ Chicago IL 60628

BERNHART, ARTHUR, b Chicago, Ill, May 14, 08; m 39; c 2. MATHEMATICS. Educ: Olivet Col, AB, 30; Univ Mich, MS, 31, PhD(physics), 34. Prof Exp: Prof math & physics, Ottawa Univ, 34-38; asst prof math, Bucknell Univ, 38-43; from asst prof to prof, Univ Okla, 43-73. Concurrent Pos: Math Asn Am vis lectr, 63-71. Mem: Am Math Soc; Math Asn Am. Res: Four color problem; non-commutative algebra; theory of games; pursuit curves; axiomatic geometry; mathematical linguistics. Mailing Add: 821 S Berry Rd Norman OK 73069

BERNHEIM, FREDERICK, b West End, NJ, Aug 18, 05; m 28; c 1. PHARMACOLOGY. Educ: Harvard Univ, AB, 25; Cambridge Univ, PhD(biochem), 28. Prof Exp: Nat Res Coun fel, 29-30; from asst prof to assoc prof physiol, 30-45, PROF PHARMACOL, SCH MED, DUKE UNIV, 45- Mem: Am Soc Biol Chem; Am Soc Pharmacol & Exp Therapeut; NY Acad Sci. Res: Oxidative processes in the animal body; biochemical aspects of pharmacology. Mailing Add: Dept of Pharmacol Duke Med Ctr Durham NC 27706

BERNHEIM, MARY LILIAS CHRISTIAN, b Gloucester, Eng, June 28, 02; nat US; m 28; c 1. BIOCHEMISTRY. Educ: Cambridge Univ, BA, 25, MA, 28, PhD(biochem), 29. Prof Exp: From instr to assoc prof, 30-65, PROF BIOCHEM, SCH MED, DUKE UNIV, 65- Res: Enzymes effecting the oxidation or reduction of small molecules containing nitrogen. Mailing Add: Dept of Biochem Duke Med Ctr Durham NC 27710

BERNHEIM, ROBERT A, b Hackensack, NJ, June 8, 33; div; c 1. PHYSICAL CHEMISTRY, CHEMICAL PHYSICS. Educ: Brown Univ, BS, 55; Harvard Univ, MA, 57; Univ Ill, PhD(chem), 59. Prof Exp: Res fel chem, Columbia Univ, 59-61; from asst prof to assoc prof, 61-69, PROF CHEM, PA STATE UNIV, UNIVERSITY PARK, 69- Concurrent Pos: NSF sr fel, 67-68; Guggenheim fel, 74-75. Mem: AAAS; Am Chem Soc; Am Phys Soc. Res: Nuclear magnetic and electron paramagnetic resonance; optical pumping; nuclear spin relaxation; nuclear electric quadrupole interactions; laser spectroscopy. Mailing Add: 152 Davey Lab Pa State Univ University Park PA 16802

BERNHEIMER, ALAN WEYL, b Philadelphia, Pa, Dec 9, 13; m 42; c 1. MICROBIOLOGY, TOXINOLOGY. Educ: Temple Univ, BS, 35, AM, 37; Univ Pa, PhD(med sci), 42. Prof Exp: Asst biol, Temple Univ, 35-37; instr bact, Pa State Col Optom, 37-39; chmn basic med sci, 69-74, from instr to prof bact, 41-58, PROF MICROBIOL, SCH MED, NY UNIV, 58- Concurrent Pos: Mem, Marine Biol Lab; consult, Surgeon Gen, 57-; NIH res career award, 62-; trustee, Cold Spring Harbor Lab Quant Biol, 63-68. Honors & Awards: Lily Award, 48. Mem: Am Soc Microbiol; fel Am Acad Microbiol; Am Micros Soc; Mineral Soc Am; Am Asn Immunol. Res: Extracellular toxins and enzymes of bacteria; streptococci and staphylococci; cell membranes. Mailing Add: Dept of Microbiol NY Univ Sch of Med New York NY 10016

BERNHEIMER, HARRIET P, b New York, NY, May 27, 19; m 42; c 1. MICROBIAL GENETICS. Educ: Hunter Col, BA, 39; NY Univ, MS, 41, PhD(microbiol), 50. Prof Exp: Asst biochem, Col Physicians & Surgeons, Columbia Univ, 50-53; from instr to asst prof med, 58-72, ASSOC PROF MED, STATE UNIV NY DOWNSTATE MED CTR,72- Concurrent Pos: NIH grant, 63-70; Health Res Coun NY career scientist award, 64-70; Irma T Hirschl trust career scientist award, 73-77. Mem: AAAS; Genetics Soc Am; Am Soc Microbiol; Harvey Soc. Res: Genetic control of capsule production, studied through the bacterial transformation system; biosynthesis of capsular polysaccharide in pneumococcus; pneumococcus bacteriophages. Mailing Add: State Univ of NY Downstate Med Ctr Dept of Med 450 Clarkson Ave Brooklyn NY 11203

BERNHOLZ, WILLIAM FRANCIS, b New York, NY, Jan 26, 24; m 49; c 3. LIPID CHEMISTRY, SURFACE CHEMISTRY. Educ: Manhattan Col, BS, 45; Fordham Univ, MS, 48. Prof Exp: Assoc technologist, Cent Labs, Gen Foods Corp, 48-57; develop chemist, Avon Prod, Inc, 57-62; res chemist, Colgate-Palmolive Co, Inc, 62; chief chemist textile lab, Drew Chem Corp, 63-66, mgr indust res lab, Drew Div, Pac

Veg Oil Corp, 66-75; DIR INT TECH SERV, PVO INT, INC, 75- Mem: Am Chem Soc; Am Oil Chemists Soc. Res: Chemistry of coffee, gelatin; gel forming proteins and carbohydrates; caramelization of sugars; browning reaction; enzyme isolation; colorimetric, chromatographic and high vacuum techniques; product development; convenience foods and cosmetics; aerosol formulation and packaging; textile lubricant and antistat specialties. Mailing Add: 11 Ledge Rd Wayne NJ 07470

BERNI, RALPH JOHN, b New Orleans, La, Nov 1, 31; m 57; c 3. ANALYTICAL CHEMISTRY. Educ: La State Univ, BS, 54; Tulane Univ, MS, 61, PhD(inorg chem), 66. Prof Exp: Res assoc chemist, 55-65, res chemist, 65-74, RES LEADER, SOUTHERN REGIONAL RES CTR, USDA, 75- Mem: Am Chem Soc; Am Inst Chemists; Sigma Xi. Res: Etherification and esterification of cellulose, including reaction mechanisms; metal ion complexes of urea derivatives; transition elements; instrumental analysis. Mailing Add: 645 Aris Ave Metairie LA 70005

BERNICH, ERIKA HANDLER, x-ray crystallography, physical chemistry, see 12th edition

BERNICK, SOL, b St Paul, Minn, Sept 29, 15; m 44; c 3. ANATOMY, ORAL PATHOLOGY. Educ: Univ Minn, BS, 39; Univ Southern Calif, MS, 47, PhD, 55. Prof Exp: Instr histol & path, Sch Dent, 47-50, asst prof, 50-53, res asst anat, Sch Med, 54-59; sr res assoc, 59-66, PROF ANAT, SCH DENT, UNIV SOUTHERN CALIF, 66- Mem: AAAS; Am Acad Oral Path; Int Asn Dent Res; Reticuloendothelial Soc; Am Asn Anatomists. Res: Innervation, electron microscopy and histochemistry of teeth; effect of nutrition upon the bone and teeth; reticuloendothelial system; fat metabolism; aging of kidney, lung, teeth, bones. Mailing Add: Dept of Anat Univ Southern Calif Sch Med Los Angeles CA 90033

BERNIER, BERNARD, b Riviere-du-Loup, Que, May 2, 26; m 53; c 6. SOIL SCIENCE. Educ: Laval Univ, BA, 48, BSc, 52, MSc, 53; Oxford Univ, PhD(soil sci), 56. Prof Exp: PROF FOREST SOILS, LAVAL UNIV, 56-, HEAD DEPT ECOL, 66- Mem: Int Soc Soil Sic. Res: Forest soils fertility. Mailing Add: Fac of Forestry Laval Univ Quebec PQ Can

BERNIER, CLAUDE, b St Boniface, Man, June 18, 31; m 58; c 3. PHYTOPATHOLOGY. Educ: Univ Man, BA, 53, BSA, 57, MSc, 61; Univ Minn, PhD(plant path), 64. Prof Exp: Res officer veg crops, Can Dept Agr, 57-61; from asst prof to assoc prof phytopath, 65-74, PROF PHYTOPATH, UNIV MAN, 74- Mem: Am Phytopath Soc; Can Phytopath Soc. Res: Resistance to ergot; biology of the ergot fungus; diseases of Vicia Faba. Mailing Add: Dept of Plant Sci Univ of Man Winnipeg MB Can

BERNIER, EDWARD JOSEPH, b Chicopee, Mass, Mar 7, 22; m 43; c 2. TEXTILE PHYSICS, APPLIED PHYSICS. Educ: Brown Univ, AB, 43, ScM, 47. Prof Exp: Physicist, Bur Mines, US Dept Interior, 43-44; physicist, Harvey-Wells Electronics, Inc, 44-49; CHIEF PHYSICIST, BELDING HEMINWAY CO, INC, 49- Mem: Textile Res Inst; Am Phys Soc; Soc Rheol; Inst Elec & Electronics Eng; Am Soc Test & Mat. Res: Management of research, development and technical service with emphasis on sewing threads and industrial yarns and cords of all commercial fibers for all market applications; properties and processing textile materials and composites. Mailing Add: R R 2 Box 118 Woodstock CT 06281

BERNIER, GLORIA A, b Newton Falls, NY, Mar 31, 24. MATHEMATICAL STATISTICS, EXPERIMENTAL STATISTICS. Educ: St Lawrence Univ, BS, 45. Prof Exp: Eng asst, Gen Elec Co, 45-48; SR RES MATHEMATICIAN, ST REGIS PAPER CO, 48- Mem: Tech Asn Pulp & Paper Indust; Am Forestry Asn. Res: Study of mechanisms responsible for hydration of cellulose pulp; investigation of the use of synthetic fibers as reinforcing agents in paper. Mailing Add: St Regis Paper Co West Nyack NY 10994

BERNIER, JOSEPH LEROY, b Chicago, Ill, Apr 5, 09; m 36; c 2. ORAL PATHOLOGY. Educ: Univ Ill, DDS, 32, MS, 34; Am Bd Oral Path, dipl, 48; Am Bd Periodont, dipl; FRCS, 49; Am Bd Oral Med, dipl, 57; FRCM, 61. Prof Exp: Instr oral path, Univ Ill, 32-34; mem staff, Walter Reed Gen Hosp, US Army, 34-38, chief oral path br, Armed Forces Inst Path, 38-39, chief dept dent lab, CZ, 39-41, chief oral path br, Armed Forces Inst Path, 41-42, dent coordr, 15th Hosp Ctr, 42-43, dent surgeon, Camp Polk, La, 43, dir dent div, McCloskey Gen Hosp, Temple, Tex, 43-44, instr, Ft Lewis, Wash, 44, dent surgeon, 254th Gen Hosp, 44-45, chief oral path div, Armed Forces Inst Path, 45-60, asst surgeon gen & chief, Army Dent Corps, 60-67; PROF ORAL PATH & CHMN DEPT, GEORGETOWN UNIV, 45- Concurrent Pos: Concurrent Pos: Mem comt dent, Nat Res Coun & Nat Adv Dent Health Coun, 54; pres, Am Bd Oral Path, 59-60; prof lectr, Jefferson Med Col, 59-; spec lectr, Dent Sch, Fairleigh Dickinson Univ, 61-; chmn dent res adv comt, US Dept Army. Honors & Awards: Cert of Achievement, US Dept Army, 49 & 52; Seaman Award, Asn Mil Surgeons of US, 52; Leadership Award, Sch Dent Med, Tufts Univ, 55; Award, Hinman Clin, Ga, 55; Callahan Gold Medal Award, Ohio State Dent Asn, 55; Award of Merit, Sch Dent, Georgetown Univ, 57; Gold Medal & Cert, Pierre Fauchard Acad, 61; Gold Medal Award, RI State Dent Soc, 65; Plaque, Baylor Univ, 65; Miller Medal, Am Acad Oral Med, 67; William J Gies Award, Am Acad Periodont, 69; Outstanding Dentists Award, Nat Libr Med. Mem: Fel Am Col Dent; Am Soc Oral Surg; Am Acad Oral Path (pres, 59-60); Am Acad Periodont (pres, 59-60); fel Int Acad Oral Path (secy). Res: Pathology of periodontitis; etiology of oral cancer; atomic effects on oral structures; pathology of dental pulp diseases. Mailing Add: Dept of Oral Path Georgetown Univ Sch of Dent Washington DC 20007

BERNIER, PAUL EMILE, b St Michel, Que, Oct 22, 11; US citizen; m 40. POULTRY GENETICS. Educ: Laval Univ, BSA, 32; Univ Calif, PhD(genetics), 47. Prof Exp: Poultry husbandman, Ecole Superieure d'Agr, Que, 28-45, lectr poultry husb, 32-45, lectr genetics, 39-45; chief inspector nat poultry breeding prog, Can Dept Agr, Ont, 45-47; assoc prof, poultry genetics, 47-55, PROF POULTRY GENETICS, ORE STATE UNIV, 55- Honors & Awards: Res Prize, Poultry Sci Asn, 52. Mem: AAAS; Biomet Soc; Poultry Sci Asn; Genetics Soc Am; Teratology Soc. Res: Genetics of domestic fowl and teratology. Mailing Add: Dept of Poultry Sci Ore State Univ Corvallis OR 97331

BERNING, JEAN ACKERMAN, b Buffalo, NY, June 13, 26; m 49. COMPUTER SCIENCES. Educ: Univ Buffalo, BA, 48; Johns Hopkins Univ, MA, 51. Prof Exp: Comput scientist, Bausch & Lomb Inc, 51-57; mathematician, US Civil Serv, US Army Electronics Res & Develop Lab, Ft Belvior, 57-63; MGR COMPUT SECT, LIBBEY-OWENS-FORD CO, 63- Mailing Add: Libbey-Owens-Ford Co 1701 E Broadway Toledo OH 43605

BERNING, PETER H, b Port Washington, NY, Dec 16, 24; m 49. MATHEMATICS. Educ: Johns Hopkins Univ, AB, 48, MA, 51. Prof Exp: Mathematician, Bausch & Lomb Optical Co, 51-57 & Eng Res & Develop Labs, Ft Belvoir, 57-63; mathematician, 63-67, ASST DIR RES, LIBBEY-OWENS-FORD CO, 67- Mem: Optical Soc Am. Res: Thin film optics, especially metal optics; thin films and surface physics. Mailing Add: 401 E Wayne St Maumee OH 43537

BERNING, WARREN WALT, b Cincinnati, Ohio, July 29, 20; m 46; c 2. PHYSICS, METEOROLOGY. Educ: Univ Cincinnati, BA, 42; Calif Inst Technol, MS, 46. Prof Exp: Aerodynamicist, McDonnell Corp, St Louis, 46 & Curtiss-Wright Corp, Columbus, 46-47; meteorologist, Ballistic Res Lab, Aberdeen Proving Ground, 47-50, physicist, 50-67; dep chief radiation directorate, Defense Atomic Support Agency, 67-71, ASST TO DEP DIR THEORET RES, DEFENSE NUCLEAR AGENCY, 71- Mem: AAAS; assoc fel Am Inst Aeronaut & Astronaut; Am Meteorol Soc; Am Geophys Union; fel Am Phys Soc. Res: Electromagnetic propagation through atmosphere; lower level meteorology; upper atmosphere physics. Mailing Add: Defense Nuclear Agency Washington DC 20305

BERNKOPF, MICHAEL, b Boston, Mass, Jan 11, 27; m 65. MATHEMATICS. Educ: Dartmouth Col, BA, 49; Columbia Univ, MA, 58; NY Univ, PhD(hist math), 65. Prof Exp: Asst prof math, Fairleigh Dickinson Univ, 65-66; assoc prof, 66-70, PROF MATH, PACE UNIV, 70- Mem: Am Math Soc; Math Asn Am. Res: History of modern mathematics, particularly analysis. Mailing Add: Dept of Math Pace Univ New York NY 10038

BERNLOHR, ROBERT WILLIAM, b Columbus, Ohio, Apr 20, 33; m 55; c 4. BIOCHEMISTRY. Educ: Capital Univ, BS, 55; Ohio State Univ, PhD(biochem), 58. Prof Exp: Fel biol, Oak Ridge Nat Lab, 58-60; asst prof biochem, Ohio State Univ, 60-62; asst prof microbiol, Univ Minn, 63-64, from assoc prof to prof microbiol & biochem, 65-74; prof microbiol & head dept, 74-75, PROF BIOCHEM & BIOPHYS & HEAD DEPT, PA STATE UNIV, 75- Concurrent Pos: USPHS res career develop award, 62-72; mem adv panel metab biol prog, NSF, 68-71. Mem: AAAS; Am Soc Biol Chem; Biochem Soc; Am Soc Microbiol. Res: Bacterial metabolism; enzymology; regultion; spore formation. Mailing Add: Dept of Biochem & Biophys 108 Althouse Pa State Univ University Park PA 16802

BERNOFSKY, CARL, b Brooklyn, NY, Nov 22, 33; m 57; c 2. BIOCHEMISTRY. Educ: Brooklyn Col, BS, 55; Univ Kans, PhD(biochem), 63. Prof Exp: Res asst biochem, Am Meat Inst Found, 56-58; res fel, Case Western Reserve Univ, 62-67; asst prof biochem, Mayo Grad Sch Med, Univ Minn, 67-73, assoc prof biochem, Mayo Med Sch, 73-75; VIS ASSOC PROF BIOCHEM, SCH MED, TULANE UNIV, 75- Concurrent Pos: Consult, Mayo Clin & Found, 67-74; NSF grants, 71-73 & 74-76; NIH grant, 73-76. Mem: AAAS; Am Chem Soc; NY Acad Sci; Am Soc Biol Chem. Res: Enzymology; pyridine nucleotides; mitochondrial metabolism; neoplasia; regulation of cellular growth. Mailing Add: Dept of Biochem Tulane Univ Sch Med New Orleans LA 70112

BERNS, DONALD SHELDON, b Bronx, NY, June 27, 34; m 56; c 3. PHYSICAL CHEMISTRY, BIOPHYSICS. Educ: Wilkes Col, BS, 55; Univ Pa, PhD(phys chem), 59. Prof Exp: Calif Res Corp grant chem, Yale Univ, 59-60 & Rockefeller Found grant, 60-61; resident res assoc chem, Argonne Nat Lab, 61-62; sr res scientist phys chem, Div Labs & Res, NY State Dept Health, Albany, 62-67, assoc res scientist, 67-71; asst prof, 62-67, ASSOC PROF BIOCHEM, ALBANY MED COL, 67-; DIR PHYS CHEM LAB, DIV LABS & RES, NY STATE DEPT HEALTH, 68-, PRIN RES SCIENTIST PHYS CHEM, 71- Concurrent Pos: NSF grant, Div Labs & Res, NY State Dept Health, 62-; adj prof chem, Rensselaer Polytech Inst, 71-; Environ Protection Agency spec sr fel & vis prof physiol & microbiol chem, Hadassah Med Sch, Hebrew Univ, Israel, 72-73; NIH grant, 75- Mem: Am Chem Soc; Biophys Soc; Am Soc Biol Chem; Am Soc Photobiol. Res: Energy transduction and transfer in model membrane systems; protein structure and function, particularly algal biliproteins, protein-lipid interaction and membrane assembly; physical chemical properties of biological macromolecules, polyelectrolytes and electrolytes. Mailing Add: Div of Labs & Res NY State Dept Health Albany NY 12201

BERNS, KENNETH, b Cleveland, Ohio, June 14, 38; m 64. BIOCHEMISTRY. Educ: Johns Hopkins Univ, AB, 60, PhD(biol), 64, MD, 66. Prof Exp: Mem house staff, Johns Hopkins Univ Hosp, 66-67; staff assoc, Nat Inst Arthritis & Metab Dis, 67-68, staff assoc, Nat Inst Allergy & Infectious Dis, 68-70; asst prof microbiol, 70-74, ASSOC PROF MICROBIOL, SCH MED, JOHNS HOPKINS UNIV, 74-, ASST PROF PEDIAT, 70-, DIR, YR ONE PROG, 73- Res: DNA structure; structure of animal virus chromosomes. Mailing Add: Dept of Microbiol Sch of Med Johns Hopkins Univ Baltimore MD 21205

BERNS, MICHAEL W, b Burlington, Vt, Dec 1, 42; m 63; c 2. DEVELOPMENTAL BIOLOGY, CELL BIOLOGY. Educ: Cornell Univ, BS, 64, MS, 66, PhD(biol), 68. Prof Exp: Assoc dir laser biol, Pasadena Found Med Res Calif, 69-70; asst prof zool, Univ Mich, 70-72; ASSOC PROF DEVELOP BIOL & CELL BIOL, UNIV CALIF, IRVINE, 72-, CHMN DEPT, 75- Concurrent Pos: NSF res grant laser microbeam effects on chromosomes, 70, 72; NIH-Heart & Lung Inst grant, 71-76; co-prin investr laser microbeam studies on mitosis, NSF, 73-75; co-dir carcinogenesis training grant, NIH-Nat Cancer Inst, 75-80; NSF equip grant, 75. Mem: AAAS; Tissue Cult Asn; Am Soc Cell Biol; Am Soc Photobiol; Soc Develop Biol. Res: Laser microbeam studies on chromosomes; nucleoli and mitochondria of tissue culture cells; studies on mitosis of cells; cellular and embryonic development. Mailing Add: Dept of Develop Biol & Cell Biol Univ of Calif Irvine CA 92664

BERNSOHN, JOSEPH, b New York, NY, Mar 21, 14; m 39; c 1. NEUROSCIENCES. Educ: Brooklyn Col, BS, 40; Univ Ill, MS, 41, PhD, 44. Prof Exp: Res biochemist, Rutgers Univ, 44-46; chief biochemist, Bur Labs, Conn, 46-49; biochemist, Cancer Res Found, 49-50; CHIEF NEUROPSYCHIAT RES LAB, VET ADMIN HOSP, ILL, 50- Concurrent Pos: Asst prof biochem, Med Sch, Northwestern Univ, 54-64, assoc prof, 64-; prof biochem & biophys, pharmacol & exp therapeut, Stritch Sch Med, Loyola Univ, Ill, 69- Mem: AAAS; Am Chem Soc; assoc Am Acad Neurol Surg. Res: Brain metabolism, mechanism of drug action; tranquilizing and energizing drugs; lipid metabolism in brain; enzymes in brain; psychopharmacology. Mailing Add: Neuropsychiat Res Lab Vet Admin Hosp Hines IL 60141

BERNSTEIN, ABRAM BERNARD, b New York, NY, July 11, 35; m 62; c 2. METEOROLOGY. Educ: City Col NY, BS, 56; Pa State Univ, MS, 59. Prof Exp: Res meteorologist, Nat Oceanic & Atmospheric Admin, 58-73; STAFF SCIENTIST, NAT ADV COMT OCEANS & ATMOSPHERE, 73- Mem: Am Meteorol Soc; AAAS. Mailing Add: 1205 Holly St NW Washington DC 20012

BERNSTEIN, ALAN, b Brooklyn, NY, Nov 21, 26; m 48; c 3. VIROLOGY. Educ: Philadelphia Col Pharm, BS, 51; Univ Pa, PhD(med microbiol), 54. Prof Exp: Bacteriologist, Philadelphia Childrens Hosp, 48-51, res asst, 51-53; bacteriologist, virus & rickettsia sect, Communicable Dis Ctr, USPHS, 53-58; bacteriologist, 58-68, MANAGING DIR, WYETH LABS, MARIETTA, 68- Mem: AAAS; Am Soc Microbiol; NY Acad Sci. Res: Production of bacterial and viral vaccines; control testing of biologicals; serology; epidemiology; diagnostic virology. Mailing Add: 306 Cornell Ave Lancaster PA 17603

BERNSTEIN, ALECK, b London, Eng, June 19, 22; US citizen; m 54; c 5. MICROBIOLOGY, GENETICS. Educ: Univ London, MB, BS, 49, dipl bact, 52,

MD, 53. Prof Exp: From jr bacteriologist to sr bacteriologist, Pub Health Lab Serv, London, 49-62, consult bacteriologist, 62-63, dep dir enteric reference lab, 60-63; ASSOC PROF MICROBIOL, MED COL WIS, 63- Concurrent Pos: Mem exec comt, Int Comt Enteric Phage Typing, 65. Mem: AAAS; Am Soc Microbiol; Genetics Soc Am; Brit Soc Gen Microbiol; Royal Col Path. Res: Bacteriophage genetics, particularly the genetics of Vi-phage II of Salmonella typhi; diagnostic microbiology, particularly bacteriophage typing of enteric pathogens; colicinogeny and colicin activity. Mailing Add: Dept of Microbiol Med Col of Wis Milwaukee WI 53233

BERNSTEIN, ALLEN RICHARD, b Chicago, Ill, Feb 14, 41. MATHEMATICS. Educ: Calif Inst Technol, BS, 62; Univ Calif, Los Angeles, MA, 64, PhD(math), 65. Prof Exp: Asst prof math, Univ Wis-Madison, 65-68; assoc prof, 68-74, PROF MATH, UNIV MD, COLLEGE PARK, 74- Mem: Am Math Soc; Asn Symbolic Logic. Res: Mathematical logic and its applications; analysis. Mailing Add: Dept of Math Univ of Md College Park MD 20742

BERNSTEIN, BARBARA ELAINE, b Washington, DC, Sept 26, 48; m 72. MATHEMATICS, PSYCHIATRY. Educ: Univ Chicago, BA, 70; Univ Md, MEd, 71, PhD(human develop), 73. Prof Exp: INSTR MATH & STATIST, BOWIE STATE COL, 73- Mem: Am Math Soc; Math Asn Am; Asn Women in Math. Res: Analysis and empirical validation of effective methods for communicating mathematical concepts; psychiatric and physiological factors which influence learning; psychoanalytic interpretations of classroom interactions and other behaviors relevant to learning. Mailing Add: 7809 Chestnut Ave Bowie MD 20715

BERNSTEIN, BARRY, b New York, NY, Nov 20, 30; m 54; c 2. APPLIED MATHEMATICS. Educ: City Col, BS, 51; Ind Univ, MA, 54, PhD(math), 56. Prof Exp: Mathematician, US Naval Res Lab, 51-53, 56-61 & Nat Bur Stand, 61-66; PROF MATH, ILL INST TECHNOL, 66-, ACTG DIR CTR APPL MATH, 70- Concurrent Pos: Lectr, Univ Md, 56-57, 64- & Catholic Univ, 58-64; consult, Nat Bur Stand, 68- Mem: Am Math Soc; Soc Rheol; Soc Indust & Appl Math. Res: Rheology; continuum mechanics; thermodynamics; mathematical analysis; gas dynamics; biomechanics. Mailing Add: Dept of Math Ill Inst of Technol Chicago IL 60616

BERNSTEIN, BENJAMIN TOBIAS, b Brooklyn, NY, Apr 4, 34; m 56; c 2. SOLID STATE PHYSICS, PHYSICAL CHEMISTRY. Educ: Brooklyn Col, BS, 57; Iowa State Univ, PhD(phys chem, physics), 59. Prof Exp: Res assoc metall, Iowa State Univ, 59-60; mem tech staff solid state physics sect, Union Carbide Res Inst, 60-62; supvr solid state physics, Res Div, Am-Stand, 62-63, mgr physics & electronics, 63-66; dir eng, Tech Prod Div, Singer Co, 66-70; V PRES AUTOMATED SYSTS DIV, UNION SPEC MACH CO, 70- Mem: AAAS; fel Am Inst Chemists; Am Phys Soc. Res: Ultrasonics; elasticity of solids; energy band theory; relaxation phenomena; transport properties; x-ray diffraction. Mailing Add: 1428 Sheridan Rd Highland Park IL 60035

BERNSTEIN, BURTON, b Brooklyn, NY, Dec 11, 20; m 61; c 2. LASERS, OPTICAL PHYSICS. Educ: Brooklyn Col, BA, 42; Columbia Univ, MA, 49; Mass Inst Technol, PhD(physics), 54. Prof Exp: Res scientist res lab, Philco Corp, 54-57; staff scientist, United Nuclear Corp, 57-58; res physicist, Gen Precision Labs, 58-61; staff physicist, Loral Electronics Corp, 61-62; sr staff scientist, Repub Aviation Corp, 62-64; mgr crystal physics & laser develop lab, Airtron Div Litton Industs, 64-66; assoc prof physics, 66-69, PROF PHYSICS, STATE UNIV NY COL NEW PALTZ, 69- Mem: Am Phys Soc. Res: Semiconductor surface physics; solid state laser materials. Mailing Add: Dept of Physics State Univ of NY Col New Paltz NY 12561

BERNSTEIN, CAROL, b Paterson, NJ, Mar 20, 41; m 62; c 2. MOLECULAR BIOLOGY. Educ: Univ Chicago, BS, 61; 61; Yale Univ, MS, 64; Univ Calif, Davis, PhD(genetics), 67. Prof Exp: NIH fel zool, Univ Calif, Davis, 68-69; RES ASSOC MICROBIOL, COL MED, UNIV ARIZ, 68- Mem: Genetics Soc Am; Am Soc Microbiol; Biophys Soc. Res: Structure and replication of DNA of microorganisms; mechanisms of chemical mutagenesis. Mailing Add: Dept of Microbiol Univ of Ariz Col of Med Tucson AZ 85724

BERNSTEIN, DAVID, b Minsk, Russia, Oct 20, 10; nat US; m 37; c 2. OTORHINOLARYNGOLOGY. Educ: NY Univ, BS, 30, MD, 35; Am Bd Otolaryngol, dipl, 43; Am Bd Surg, dipl, 48. Prof Exp: From asst clin prof to assoc clin prof otolaryngol, Postgrad Med Sch, 50-67, CLIN PROF OTORHINOL & OTORHINOL PLASTIC SURG, MED CTR, NY UNIV, 67-, MEM PLANNING COMT, DEPT OTORHINOLARYNGOL, 69- Concurrent Pos: Dir otorhinolaryngol & otorhinolaryngol plastic surg & mem exec comt, Jacques Loewe Found Hosp; attend otolaryngologist & otorhinolaryngol plastic surgeon, NY Vet Admin Hosp, 67-70, consult otorhinolaryngol plastic surg, 70-; chief otolaryngol & otolaryngol plastic surg serv, Maimonides Med Ctr, Brooklyn, 69-; mem exec comt, Joint Conf Comn & secy, Med Bd, Metrop Jewish Geriat Ctr, 69-, pres exec comt, Med Bd, 73- Mem: Fel Am Acad Facial Plastic & Reconstruct Surg; fel Am Col Chest Physicians; fel Am Acad Ophthal & Otolaryngol; fel NY Acad Med; fel Int Col Surg. Res: Ear, nose, throat and facial plastic surgery. Mailing Add: 1342 51st St Brooklyn NY 11219

BERNSTEIN, DAVID M, physics, see 12th edition

BERNSTEIN, DOROTHY LEWIS, b Chicago, Ill, Apr 11, 14. MATHEMATICS. Educ: Univ Wis, BA & MA, 34; Brown Univ, PhD(math), 39. Prof Exp: Instr math, Mt Holyoke Col, 37-40 & Univ Wis, 41-42; res assoc, Univ Calif, 42-43; from instr to prof, Univ Rochester, 43-59; chmn math dept, 60-70, PROF MATH, GOUCHER COL, 59- Mem: Am Math Soc; Math Asn Am; Soc Indust & Appl Math. Res: Partial differential equations; double Laplace transforms. Mailing Add: Dept of Math Goucher Col Baltimore MD 21204

BERNSTEIN, ELAINE KATZ, b Baltimore, Md, May 14, 22; m 55; c 2. BIOCHEMISTRY, SCIENCE WRITING. Educ: Goucher Col, BA, 41; Oberlin Col, MA, 42. Prof Exp: Asst scientist, Metall Lab, Univ Chicago, 43-46; assoc scientist, Argonne Nat Lab, 46-58, consult clin chem, 58-60; sci info specialist, John Crerar Libr, Chicago, 63-70; writer & ed med & clin chem, 70-74; SCI WRITER & ED, 74- Mem: AAAS; Sigma Xi; Fedn Am Scientists; Scientists in Pub Interest. Res: Clinical chemistry; liver function; toxicology; radiation-induced biochemical changes; leukemia. Mailing Add: 955 Wildwood Lane Highland Park IL 60035

BERNSTEIN, ELLIOT R, b New York, NY, Apr 14, 41; m 65; c 2. CHEMICAL PHYSICS, SPECTROSCOPY. Educ: Princeton Univ, BA, 63; Calif Inst Technol, PhD(chem physics), 67. Prof Exp: Res assoc paramagnetic resonance, Univ Chicago, 67-69; asst prof chem, Princeton Univ, 69-75; ASSOC PROF CHEM, COLO STATE UNIV, 75- Concurrent Pos: Consult, Los Alamos Nat Lab, 75- Mem: Am Phys Soc. Res: Optical, Raman and magnetic resonance spectroscopy of inorganic and organic molecular crystals; exchange interactions, magnetic ordering and exciton structure in molecular solids; structural, magnetic, ferro- and antiferroelectric phase transitions in solids. Mailing Add: Dept of Chem Colo State Univ Ft Collins CO 80523

BERNSTEIN, EMIL OSCAR, b New York, NY, Dec 16, 29; m 51; c 2. CELL BIOLOGY. Educ: Syracuse Univ, AB, 51, MS, 53; Univ Calif, Los Angeles, PhD(zool), 56. Prof Exp: Am Cancer Soc fel, Univ Calif, Los Angeles, 56-57; from instr to assoc prof zool, Univ Conn, 57-65; assoc prof, Univ Md, 65-68; sr res assoc, 68-73, PRIN RES ASSOC, GILLETTE RES INST, 73- Concurrent Pos: NSF grant, 59-, fel biol inst, Carsberg Found, Copenhagen, 63-64. Mem: AAAS; Am Soc Cell Biol; Soc Gen Physiol; Electron Micros Soc Am. Res: Cell physiology; growth and division of cells; metabolic problems; problem of obligate autotrophy; scanning electron microscopy; skin morphology and physiology; ultrastructure; cell biology and cell physiology. Mailing Add: Biomed Sci Div Gillette Res Inst 1413 Res Blvd Rockville MD 20850

BERNSTEIN, EUGENE F, b New York, NY, Oct 9, 30; m 54; c 3. SURGERY. Educ: State Univ NY Downstate Med Ctr, MD, 54; Univ Minn, MS, 61, PhD(surg), 64; Am Bd Thoracic Surg, dipl, 68. Prof Exp: From instr to assoc prof surg, Univ Minn, 63-69; PROF SURG, UNIV CALIF, SAN DIEGO, 69- Concurrent Pos: Markle scholar acad med, 63-68; consult, Gulf-Gen Atomic, Naval Hosp, San Diego & Medtronic, Inc, Minn. Mem: AAAS; fel Am Col Surg; Am Soc Artificial Internal Organs (secy-treas, 68-70, pres-elect, 70, pres, 71); Asn Acad Surg; Soc Exp Biol & Med. Res: Cardiovascular surgery; cardiovascular physiology; prolonged mechanical circulatory support. Mailing Add: Dept of Surg Univ Hosp of San Diego County San Diego CA 92103

BERNSTEIN, EUGENE H, b New York, NY; m 52; c 3. BIOCHEMISTRY, VIROLOGY. Educ: US Merchant Marine Acad, BS, 47; Rutgers Univ, BS, 51, MS, 52, PhD(biochem, microbiol), 55. Prof Exp: USPHS fel, Sloan-Kettering Inst Cancer Res, New York, 55-56; biochemist, Colgate-Palmolive Co Dent Res, NJ, 56-58, sect head, 58-59; OWNER & DIR, UNIV LABS, INC, 59- Concurrent Pos: Vis investr, Inst Microbiol, Rutgers Univ, 59-68. Mem: AAAS; Am Chem Soc; NY Acad Sci; Am Soc Microbiologists; Int Leukemia Soc. Res: Enzymology; virology, especially avian tumor and leukemia viruses; toxicology; microbiology; nutrition; carcinogenesis; cancer chemotherapy. Mailing Add: Univ Labs Inc 810 N Second Ave Highland Park NJ 08904

BERNSTEIN, EUGENE MERLE, b Baltimore, Md, Feb 13, 31; m 60; c 2. NUCLEAR PHYSICS. Educ: Duke Univ, BS, 53, MA, 54, PhD(physics), 56. Prof Exp: Instr & res assoc physics, Duke Univ, 56-57; instr & res assoc physics, Univ Wis, 57-59, lectr & res assoc, 60-61; NSF fel, Inst Theoret Physics, Copenhagen, 59-60; from asst prof to assoc prof physics, Univ Tex, 61-65; vis staff mem, Los Alamos Sci Lab, 65-67; prof physics, Univ Tex, 67-68; PROF PHYSICS, WESTERN MICH UNIV, 68- Mem: Am Phys Soc; NY Acad Sci. Res: Low energy nuclear physics. Mailing Add: Dept of Physics Western Mich Univ Kalamazoo MI 49001

BERNSTEIN, GERALD SANFORD, b Trenton, NJ, July 4, 28; m 52; c 1. REPRODUCTIVE BIOLOGY, OBSTETRICS & GYNECOLOGY. Educ: Temple Univ, BA, 50; Univ Mass, MA, 52; Univ Del, PhD(chem), 55; Univ Southern Calif, MD, 62. Prof Exp: Pop Coun fel, 55-56, USPHS fel, 56-58; res fel pharmacol, Sch Med, Univ Southern Calif, 59-62, intern, Los Angeles County-Univ Southern Calif Med Ctr, 62-63, resident physician obstet & gynec, 63-67; from instr to asst prof, 67-71, ASSOC PROF OBSTET & GYNEC, UNIV SOUTHERN CALIF, 71- Concurrent Pos: Dir family planning serv, Los Angeles County-Univ Southern Calif Med Ctr, 68-; chmn bd dir, Los Angeles Regional Family Planning Coun, 68-; med dir, Los Angeles Affil Planned Parenthood, 68- Mem: AAAS; NY Acad Sci; Am Fertil Soc; Am Soc Study Reproduction. Res: Andrology; family planning; physiology of sperm. Mailing Add: Woman's Hosp Los Angeles County Univ of Southern Calif Med Ctr Los Angeles CA 90033

BERNSTEIN, HAROLD JOSEPH, b Toronto, Ont, Aug 26, 14; m 48; c 3. PHYSICAL CHEMISTRY. Educ: Univ Toronto, BA, 35, MA, 36, PhD(chem), 38. Prof Exp: Demonstr, Univ Toronto, 35-38; 1851 Exhib scholar, Copenhagen, Denmark, 38-40; assoc res chemist, 46-52, sr res chemist, 52-57, PRIN RES CHEMIST, NAT RES COUN CAN, 57- Concurrent Pos: Bourke lectr, 60. Mem: Am Chem Soc; Am Phys Soc; The Chem Soc; fel Royal Soc Can; fel Chem Inst Can. Res: Molecular spectroscopy; non-bonded interactions; nuclear magnetic resonance. Mailing Add: Div of Chem Nat Res Coun Can Ottawa ON Can

BERNSTEIN, HARRIS, b Brooklyn, NY, Dec 12, 34; m 62; c 2. GENETICS. Educ: Purdue Univ, BS, 56; Calif Inst Technol, PhD(genetics), 61. Prof Exp: USPHS res fel genetics, Yale Univ, 61-63; asst prof, Univ Calif, Davis, 63-68; assoc prof, 68-74, PROF MICROBIOL, COL MED, UNIV ARIZ, 74- Concurrent Pos: NSF res grant, 64-76. Mem: Genetics Soc Am; Am Soc Microbiol; Fedn Am Scientists. Res: Bacteriophage genetics; mechanisms of recombination; molecular morphogenesis; DNA replication; DNA repair and mutation. Mailing Add: Dept of Microbiol Univ of Ariz Col of Med Tucson AZ 85724

BERNSTEIN, HERBERT JACOB, b Brooklyn, NY, Sept 26, 44; m 68; c 2. COMPUTING. Educ: Washington Square Col, NY Univ, BA, 64, MS, 65, PhD(math), 68. Prof Exp: Assoc res scientist, AEC Comput & Appl Math Ctr, NY Univ, 68-70; sci programmer analyst, 70-74, ASSOC SCIENTIST, CHEM DEPT, BROOKHAVEN NAT LAB, ASSOC UNIV INC, 74- Mem: Am Math Soc; Am Crystallog Asn; Soc Indust Appl Math; Math Asn Am. Res: Applications of computers to chemistry, especially to crystallography. Mailing Add: Chem Dept Bldg 555A Brookhaven Nat Lab Upton NY 11973

BERNSTEIN, IRA BORAH, b Bronx, NY, Nov 8, 24; m 55; c 2. THEORETICAL PHYSICS. Educ: City Col NY, BChE, 44; NY Univ, PhD(physics), 50. Prof Exp: Res physicist, Westinghouse Res Labs, 50-54; sr res physicist, Plasma Physics Lab, Princeton Univ, 54-64; PROF APPL SCI, YALE UNIV, 64- Concurrent Pos: Fulbright scholar, 62-63; Guggenheim fel, 69. Mem: Fel Am Phys Soc. Res: Theoretical plasma physics. Mailing Add: Dept of Eng & Appl Sci Yale Univ New Haven CT 06520

BERNSTEIN, IRWIN S, b New York, NY, Mar 9, 34; m 65. MATHEMATICS. Educ: City Col New York, BS, 55; Mass Inst Technol, SM, 56, PhD(math), 59. Prof Exp: Sr staff scientist, Avco Res & Advan Develop, 59-63 & Aerospace Res Ctr, Gen Precision, Inc, 63-68; asst prof math, 68-74, ASSOC PROF MATH, CITY COL NEW YORK, 74- Mem: Am Math Soc. Res: Uniqueness of solutions of elliptic partial differential equations; numerical solutions of partial differential equations; analytical celestial mechanics. Mailing Add: Dept Math City Col New York Convent Ave & 139th St New York NY 10031

BERNSTEIN, IRWIN SAMUEL, b Brooklyn, NY, July 11, 33; c 4. PRIMATOLOGY. Educ: Cornell Univ, BA, 54; Univ Chicago, MA, 55, PhD(psychol), 59. Prof Exp: Res fel & res assoc psychobiol, 60-67, SOCIOBIOLOGIST, YERKES REGIONAL PRIMATE RES CTR, 67-; PROF PSYCHOL, ZOOL & ANTHROP, UNIV GA, 68- Concurrent Pos: NSF, NIH, NIMH, Walter Reed Army Inst Res & Wenner Gren Found Anthrop Res grants & contracts. Mem: AAAS; Animal Behav Soc; Int Primatol Soc; Int Soc Study Aggression; Am Psychol Asn. Res: Primate social behavior; principles of group organization; communication and comparative patterns of activities in both field and captive group studies; hybridization; endocrine correlates of stress and aggression; infant socialization. Mailing Add: Dept of Psychol Univ of Ga Athens GA 30602

BERNSTEIN, ISADORE A, b Clarksburg, WVa, Dec 23, 19; m 42; c 2. BIOCHEMISTRY. Educ: Johns Hopkins Univ, AB, 41; Western Reserve Univ, PhD(biochem), 52. Prof Exp: Res assoc biochem, Western Reserve Univ, 51-52, sr instr, 52-53; res assoc, Inst Indust Health, 53-56 & 59-70, from instr to assoc prof bi biol chem, Univ, 54-61, from assoc prof to prof indust health, 61-67, PROF BIOL CHEM, UNIV MICH, ANN ARBOR, 68-, PROF ENVIRON & INDUST HEALTH & RES SCIENTIST, INST ENVIRON & INDUST HEALTH, 70- Concurrent Pos: USPHS spec fel, Inst Protein Res, Osaka Univ, 63-64; WHO travel fel, Western Europe, 72; consult, Vet Admin Hosp, Allen Park, Mich, 61-70 & 72- Honors & Awards: Taub Int Mem Award Res Psoriasis, Baylor Col Med, 59. Mem: Fel AAAS; Am Soc Biol Chem; Am Soc Microbiol; Am Soc Cell Biol; Soc Invest Dermat (vpres, 73-74). Res: Cutaneous metabolism, keratinization, neoplasia and differentiation; mitochondrial metabolism and regeneration; biochemical mechanisms of accommodation. Mailing Add: Sch of Pub Health Univ of Mich Ann Arbor MI 48104

BERNSTEIN, ISIDOR MAYER, b Chicago, Ill, Nov 2, 96; m 23; c 2. CHEMISTRY. Prof Exp: Chemist, Heyl Labs, Inc, New York, 20-21 & Sigmund Ullman Co, 21-24; chief chemist, Stand Printing Ink Co, Ohio, 30-36 & Southern Div, Int Printing Ink Co, 36-40; dir res, H D Roosen Co, NY, 40-46; res assoc & dir nat printing ink res inst, Lehigh Univ, 46-48; tech dir, Gotham Ink & Color Co, 48-55; pres, Flexo-Gravure Inks, Inc, 55-63; CHEM CONSULT, 63- Concurrent Pos: Lectr, NY Univ, 50- Mem: Fel Am Inst Chem; NY Acad Sci. Res: Drying oil polymerization; molecular weight determination; pigment identification and analysis; polymer fractionation; synthetic drying oils; pigment dispersion; photomicrography and photomicrographical analysis; microscope illumination and image acuity; creativity in ancient Egyptian mathematics. Mailing Add: 445 A Cheshire Ct Lakewood NJ 08701

BERNSTEIN, JACK, b Utica, NY, Oct 24, 16; m 40; c 1. MEDICINAL CHEMISTRY. Educ: Cornell Univ, AB, 37, PhD(org chem), 41. Prof Exp: Org res chemist, 41-60, ASST DIR ORG CHEM DEPT, SQUIBB INST MED RES, 60- Mem: Am Chem Soc. Res: Benzyl fluorides; antimalarial agents; chemistry of penicillin; chemotherapy of tuberculosis; synthetic drugs, including antibacterial and antifungal agents; tranquilizing agents; radiopaque agents. Mailing Add: Org Chem Dept PO Box 4000 Squibb Inst for Med Res Princeton NJ 08540

BERNSTEIN, JAY, b New York, NY, May 14, 27; m 57; c 2. PATHOLOGY. Educ: Columbia Univ, BA, 48; State Univ NY, MD, 52. Prof Exp: Instr path, Sch Med, Wayne State Univ, 58-60, asst prof pediat, 60-62; from asst prof to assoc prof path, Albert Einstein Col Med, 62-69; DIR ANAT PATH, WILLIAM BEAUMONT HOSP, 69- Concurrent Pos: Attend pathologist, Children's Hosp Mich, Detroit, 57-62; sr res assoc, Child Res Ctr, Wayne State Univ, 60-62; contrib ed path, J Pediat, 68-; vis assoc prof path, Albert Einstein Col Med, 69-74, vis prof, 74-; partic pathologist, Int Study Kidney Dis in Childhood, 68-; mem ed bd, Perspectives in Pediat Path, 71-; mem sci adv bd, Mich Kidney Dis Found, 72-; consult pathologist, Children's Hosp of Mich, 75- Mem: Am Asn Pathologists & Bacteriologists; Am Soc Nephrology; Am Pediat Soc; Biophys Soc. Res: Neonatal pathology and congenital abnormalities; renal disease-pathogenesis and ultrastructure. Mailing Add: William Beaumont Hosp Dept Anat Path 3601 W 13 Mile Rd Royal Oak MI 48072

BERNSTEIN, JERALD JACK, b Brooklyn, NY, Mar 30, 34; m 57; c 1. NEUROANATOMY, EXPERIMENTAL NERUOLOGY. Educ: Hunter Col, BA, 55; Univ Mich, MS, 57, PhD, 59. Prof Exp: Asst fisheries, Univ Mich, 56-58, asst zool, Mus Zool, 58-59, asst psychol, 58-59; assoc prof biol, Clarion State Col, 59-60; res biologist, Lab Neuroanat Sci, NIH, 60-65; asst prof anat, Ctr Neurobiol Sci, 65-69, ASSOC PROF NEUROSCI & OPHTHAL, CTR NEUROBIOL SCI & CTR HUMAN PROSTHETIC RES, COL MED, UNIV FLA, 69- Mem: AAAS; Soc Neurosci; Am Soc Cell Biol; Am Asn Anat; Am Physiol Soc. Res: Physiological and neurological basis of central nervous system regeneration. Mailing Add: Dept of Neurosci Univ of Fla Col of Med Gainesville FL 32601

BERNSTEIN, JEREMY, b Rochester, NY, Dec 31, 29. THEORETICAL PHYSICS. Educ: Harvard Univ, BA, 51, MA, 53, PhD(physics), 55. Prof Exp: Res assoc physics, Cyclotron Lab, Harvard Univ, 55-57; mem, Inst Advan Study, 57-59; assoc, Brookhaven Nat Lab, 60-62; assoc prof, NY Univ, 62-67; PROF PHYSICS, STEVENS INST TECHNOL, 67- Concurrent Pos: NSF fel, 59-61. Honors & Awards: Westinghouse-AAAS Writing Prize, 64; Am Inst Physics-US Steel Found Sci Writing Prize, 70. Mem: Fel Am Phys Soc. Res: Elementary particles and weak interactions. Mailing Add: Dept of Physics Stevens Inst of Technol Hoboken NJ 07030

BERNSTEIN, KENNETH, b New York, NY, Oct 15, 45; m 67; c 1. MOLECULAR GENETICS, VIROLOGY. Educ: Dartmouth Col, AB, 66; Ind Univ, Bloomington, PhD(microbiol, genetics), 70. Prof Exp: Asst prof biol, Williams Col, 70-74; ASST PROF BIOL, STATE UNIV NY, FREDONIA, 74- Mem: AAAS; Genetics Soc Am. Res: Control phenomena operating in bacterial growth and in viral infection of bacterial and animal cells; control of RNA synthesis in Escherichia coli. Mailing Add: Dept of Biol State Univ of NY 99 Washington Ave Fredonia NY 14063

BERNSTEIN, LEON, b Skoudas, Lithuania, Dec 27, 14; m 45; c 1. NUMBER THEORY. Educ: Kaunas State Univ, BA, 39; Vilnius State Univ, MA, 41, PhD(math), 45. Hon Degrees: LHD, Northern Mich Univ, 72. Prof Exp: Asst prof math, Vilnius State Univ, 40-41, lectr, 44-45, 49-57; lectr, Hebrew Univ, Tel-Aviv, 58-64 & Tel-Aviv Univ, 64-66; vis prof math, Northern Mich Univ, 67 & Syracuse Univ, 67-68; VIS PROF MATH, ILL INST TECHNOL, 68- Concurrent Pos: Guest lectr, Univ Hamburg, 64; guest prof, Univ Cologne, 65, Univ Munich & Univ Göttingen, 67. Mem: Am Math Soc; Math Soc Am; Sigma Xi. Res: Disclosure of many infinite classes of algebraic number fields of any degree greater than or equal to 3 whose basis, when properly constructed, yields periodic Jacobi algorithms; theory of functions; units; zeros of arithmatic functions. Mailing Add: Dept of Math Ill Inst of Technol Chicago IL 60616

BERNSTEIN, LEON, b London, Eng, Aug 15, 16; US citizen; m 43. PHYSIOLOGY, PUBLIC HEALTH ADMINISTRATION. Educ: Univ London, BSc, 39, PhD(physiol), 57; MRCS & LRCP, 49. Prof Exp: Demonstr histol & pharmacol, London Hosp Med Col, 39-40, demonstr physiol, 46-49, lectr, 49-51, sr lectr, 51-57; chief physiol res lab, Vet Admin Hosp, Baltimore, Md, 57-61, actg assoc chief staff, 59-61; chief physiol res lab, Vet Admin Hosp, San Francisco, Calif, 61-65; actg chief prog proj br, Nat Heart Inst, 66-67; chief off sci eval, 67-71, dir health systs res & develop serv, 71-74, SPEC ASST TO CHIEF MED DIR, US VET ADMIN, 74- Concurrent Pos: Asst prof, Johns Hopkins Univ, 58-61; assoc prof, Sch Med, Univ Md, 60-61; assoc clin prof, Sch Med & mem consult staff, Cardiovasc Res Inst, Univ Calif, San Francisco, 61-65; prof, Sch Med, George Washington Univ, 70-; consult, Nat Comt Careers in Med Lab, 72-73 & Bur Health Manpower Educ, NIH, 74- Mem: Am Physiol Soc; Brit Physiol Soc; AAAS. Res: Physiology of respiration; health services research and development; management sciences; program planning and evaluation for ubiquitous health care delivery systems. Mailing Add: Dept of Med & Surg Vet Admin (105) Washington DC 20420

BERNSTEIN, LEON, plant physiology, see 12th edition

BERNSTEIN, LESLIE, b Poland, Apr 26, 24; US citizen; m 57. OTOLARYNGOLOGY, MAXILLOFACIAL SURGERY. Educ: Univ Witwatersrand, BDS, 47, MB, BCh, 54; Royal Col Physicians & Surgeons, Eng, dipl laryngol & otol, 59; Am Bd Otolaryngol, dipl, 63. Prof Exp: From asst prof to assoc prof otolaryngol, Col Med, Univ Iowa, 60-69, chmn sect maxillofacial-plastic surg, 69, prof otolaryngol & maxillofacial surg, 69-72; PROF OTORHINOLARYNGOL & CHMN DEPT, MED SCH, UNIV CALIF, DAVIS, 72- Honors & Awards: Harris P Mosher Award, 68; Award of Merit, Am Acad Ophthal & Otolaryngol, 70; Ira Tresley Res Award, 72. Mem: Am Broncho-Esophagol Asn; fel Am Acad Ophthal & Otolaryngol; fel Am Acad Facial Plastic & Reconstruct Surg; fel Am Laryngol, Rhinol & Otol Soc; Int Asn Maxillofacial Surg (vpres). Res: Ear; Larynx; broncho-esophagology; cleft palate. Mailing Add: Dept of Otorhinolaryngol Univ of Calif Med Sch Davis CA 95616

BERNSTEIN, LIONEL M, b Chicago, Ill, Sept 10, 23; m 52; c 3. MEDICINE. Educ: Univ Ill, BS, 44, MD, 45, MS, 51, PhD, 54. Prof Exp: Res assoc med & clin sci, Univ Ill, 52-53, instr med & physiol, 53-54; chief metab res div, Med Nutrit Lab, Fitzsimons Army Hosp, Denver, Colo, 54-55; physician sr grade, Vet Admin Hosp, Sepulveda, Calif, 55-56; chief gastroenterol sect, Vet Admin Hosp, Hines, Ill, 56-57, assoc chief staff, 57-62; chief med serv, Vet Admin West Side Hosp, Chicago, 62-67; dir res serv, Vet Admin Cent Off, 67-70; assoc dir extramural progs, Nat Inst Arthritis & Metab Dis, 70-73; dir off prog opers, 73-74, SPEC ASST, OFF ASST SECY HEALTH, DEPT HEALTH, EDUC & WELFARE, 75- Concurrent Pos: Prof, Univ Ill, 66-70. Mem: AMA; Am Gastroenterol Asn; Am Fedn Clin Res; fel Am Col Physicians. Res: Renal physiology of clinical disease states; hemodynamics of hypertension and gastroenterology. Mailing Add: Off of Asst Secy Health Dept of Health Educ & Welfare Washington DC 20201

BERNSTEIN, MARVIN HARRY, b Los Angeles, Calif, Dec 30, 43; m 66; c 2. COMPARATIVE PHYSIOLOGY. Educ: Univ Calif, Los Angeles, BA, 65, MA, 66, PhD(physiol), 70. Prof Exp: NIH trainee cardiovasc physiol, Univ Calif, Los Angeles, 70-71; from temp instr to temp asst prof zool, Duke Univ, 71-73; ASST PROF BIOL, NMEX STATE UNIV, 73- Mem: Am Physiol Soc; Am Soc Zoologists; Am Inst Biol Sci; AAAS. Res: Respiration, gas transport, cardiovascular function and temperature regulation in birds and mammals; during rest and locomotion, growth and adulthood, at sea level and high altitude. Mailing Add: Dept of Biol NMex State Univ Box 3AF Las Cruces NM 88003

BERNSTEIN, MAURICE HARRY, b St Louis, Mo, Apr 3, 23; m 67; c 1. ANATOMY, CELL BIOLOGY. Educ: Wash Univ, AB, 47, PhD(zool), 50. Prof Exp: Am Cancer Soc-Nat Res Coun fels, Univ Mo-Columbia, 50-51 & Univ Calif, Berkeley, 51-52; jr staff mem genetics, Carnegie Inst Wash, 52-54; asst res biologist, Virus Lab, Univ Calif, Berkeley, 54-56; res physiologist, Res Inst, Mt Sinai Hosp, Los Angeles, Calif, 56-57; asst res anatomist, Sch Med, Univ Calif, Los Angeles, 57-58; from asst prof to assoc prof, 58-68, PROF ANAT, SCH MED, WAYNE STATE UNIV, 68- Concurrent Pos: Lalor Found fel, Marine Biol Lab, Woods Hole, Mass, 59; consult, Vet Admin, 65-70 & NIH, 67; vis prof anat & reprod biol, Univ Hawaii, 71-72. Mem: AAAS; Am Asn Anat; Am Soc Cell Biol; Asn Res Vision & Ophthal; Soc Study Reprod. Res: Enzyme localizations and functional sperm morphology; fine structure of retina, cytochemistry of photoreceptor function; role of pigment epithelium in visual function. Mailing Add: Dept of Anat Wayne State Univ Sch Med Detroit MI 48201

BERNSTEIN, RICHARD BARRY, b Long Island, NY, Oct 31, 23; m 48; c 4. PHYSICAL CHEMISTRY. Educ: Columbia Univ, AB, 43, AM, 44, PhD(chem), 48. Prof Exp: Chemist, SAM Labs, Columbia Univ, 42-44 & Manhattan Dist, 44-46; asst prof chem, Ill Inst Technol, 48-53; from asst prof to prof, Univ Mich, 53-63; prof, Univ Wis-Madison, 63-66, W W Daniells prof, 66-73; W T DOHERTY PROF CHEM, UNIV TEX, AUSTIN, 73- Concurrent Pos: Sloan fel, 56-60; NSF sr fel, 60-61; chmn off chem & chem technol, Nat Res Coun, 74-76. Mem: Nat Acad Sci; AAAS; Am Chem Soc; fel Am Phys Soc; fel Am Acad Arts & Sci. Res: Chemical kinetics; molecular interactions by the molecular beam scattering technique. Mailing Add: Dept of Chem Univ of Tex Austin TX 78712

BERNSTEIN, SELDON EDWIN, b Bangor, Maine, Jan 3, 26; m 50, 59; c 4. MEDICAL GENETICS. Educ: Univ Maine, BA, 49, MA, 52; Brown Univ, PhD(biol), 56. Prof Exp: Asst, Univ Maine, 49-52; asst, Brown Univ, 52-54; assoc staff scientist, 56-58, staff scientist, 58-67, SR STAFF SCIENTIST, JACKSON LAB, 67-, ASST DIR, 64- Concurrent Pos: Mem bd dirs, Col of the Atlantic, 69-, chmn, 72-; lectr zool, Univ Maine, 71- Mem: Genetics Soc Am; Soc Exp Hemat; Am Genetic Asn; Am Soc Hemat; NY Acad Sci. Res: Physiological genetics of hematological mutants; therapy of radiation sickness; ways of bypassing the histocompatibility barrier; experimental hematology. Mailing Add: Jackson Lab Bar Harbor ME 04609

BERNSTEIN, SEYMOUR, b Newark, NJ, Aug 12, 15; m 55; c 1. ORGANIC CHEMISTRY. Educ: Princeton Univ, AB, 36, AM, 38, PhD(chem), 39. Prof Exp: Res asst & assoc chem, Princeton Univ 39-43; RES CHEMIST & GROUP LEADER, LEDERLE LABS, AM CYANAMID CO, 43- Mem: Am Chem Soc; Am Soc Biol Chemists; Endocrine Soc. Res: Natural products specializing in steroids. Mailing Add: Lederle Labs Am Cyanamid Co Pearl River NY 10965

BERNSTEIN, SEYMOUR, b Chicago, Ill, Feb 20, 09; m 38; c 2. PARTICLE PHYSICS. Educ: Univ Ill, BS, 30; Univ Chicago, MS, 36, PhD(physics), 39. Prof Exp: Engr, City Bur Eng, Chicago, 30-34; instr physics, Austin Col, 39-42; res assoc, Manhattan Dist Nuclear Energy Lab, Univ Chicago, 42-44; res physicist, Oak Ridge Nat Lab, 44-47, sect head, 47-52, chief physicist, 52-64; PROF PHYSICS, UNIV ILL, CHICAGO CIRCLE, 64- Concurrent Pos: Mem comt neutron stand, Nat Res Coun, 46-48; US adv, Int Conf Peaceful Uses Atomic Energy, Geneva, Switz, 55; lectr, Univ Tenn, 48-55; vis prof, Israel Inst Technol, 55-56; prof, Univ Miami, 61-62. Mem: Fel Am Phys Soc; AAAS. Res: X-rays; neutron physics; nuclear physics; atomic energy; Mossbauer effect; elementary particles; experimental particle physics at high energies. Mailing Add: Dept of Physics Univ of Ill PO Box 4348 Chicago IL 60680

BERNSTEIN, SHELDON, b Milwaukee, Wis, Mar 23, 27; m 48; c 3. PHYSIOLOGICAL CHEMISTRY. Educ: Univ Wis, BS, 49, PhD(physiol chem), 52. Prof Exp: Asst physiol chem, Univ Wis, 50-52; res biochemist, Upjohn Co, 52-53; vpres & dir res, 53-65, PRES & DIR RES, MILBREW INC, 65- Concurrent Pos: Pres, Bader & Bernstein, Inc, 61- Mem: AAAS; Am Chem Soc; Am Soc Microbiol; fel Am Inst Chemists; Inst Food Technologists. Res: Yeast and yeast products; enzymes; microbiological nutrients; nucleic acids; amino acids; vitamins; custom fermentations and fermentation research; food supplements; dairy products; whey fractionation. Mailing Add: Milbrew Inc 330 S Mill St Juneau WI 53039

BERNSTEIN, SIDNEY, b New York, NY, June 10, 18; m 41; c 3. MICROBIOLOGY. Educ: City Col New York, BS, 39; Iowa State Col, MS, 42; Yale Univ, PhD(microbiol), 60. Prof Exp: Res microbiologist, Vet Admin Hosp, Sunnymount, NY, 46-53 & West Haven, Conn, 53-69; assoc dean, Sch Allied Health & Natural Sci, 72-, PROF BIOL, QUINNIPAC COL, 69- Concurrent Pos: NIH grant, 58-63; lectr microbiol, Sch Med, Yale Univ, 60- Mem: Am Soc Microbiol; Am Pub Health Asn; Am Thoracic Soc. Res: Chemotherapy of tuberculosis; etiology and pathogenesis of chronic pulmonary disease; effect of steroids on chronic pulmonary processes; metabolism of the autotrophic hydrogen bacteria, Hydrogenomonas ruhlandii. Mailing Add: Sch Allied Health & Natural Sci Quinnipac Col Hamden CT 06518

BERNSTEIN, SOL, b West New York, NJ, Feb 3, 27; m 63; c 1. MEDICINE. Educ: Univ Southern Calif, AB, 52, MD, 56. Prof Exp: Resident internal med, Los Angeles County Hosp, Calif, 57-60; asst prof, 62-67, ASSOC PROF MED, UNIV SOUTHERN CALIF, 67-, MED DIR, LOS ANGELES COUNTY-UNIV SOUTHERN CALIF MED CTR, 74- Concurrent Pos: Fel cardiol, St Vincents Hosp, Los Angeles, 58-59; asst dir dept med, Los Angeles County-Univ Southern Calif Med Ctr, 66-72, chief prof serv, Gen Hosp, 72-74. Mem: Fel Am Col Cardiol; Am Heart Asn; fel Am Col Physicians; Am Diabetes Asn; Am Fedn Clin Res. Res: Cardiology; endocarditis, myocarditis and pulmonary embolus; infectious diseases; peripheral vascular disease; metabolism; lactic acidosis; diabetes. Mailing Add: Sch of Med Univ of Southern Calif Los Angeles CA 90033

BERNSTEIN, STANLEY CARL, b New York, NY, May 19, 37; m 60; c 2. PHYSICAL ORGANIC CHEMISTRY. Educ: Queens Col, NY, BS, 58; Univ Mich, MS, 61, PhD(chem), 63. Prof Exp: Res assoc chem, Ohio State Univ, 62-64, instr, 63-64; asst prof, Wright State Univ, 64-70; ASST PROF CHEM, ANTIOCH COL, 70- Mem: Am Chem Soc. Res: Mechanism and stereochemistry of organic reactions; organosilicon chemistry; chemical education. Mailing Add: Dept of Chem Antioch Col Yellow Springs OH 45387

BERNSTEIN, STANLEY H, b Brooklyn, NY, Sept 10, 24. INTERNAL MEDICINE, IMMUNOLOGY. Educ: Col William & Mary, BS, 45; NY Univ, MD, 48. Prof Exp: Vis prof, All India Med Inst, 64; assoc prof med, Univ Conn, 68-74; ASSOC PROF CLIN MED, SCH MED, UNIV MIAMI, 74- Concurrent Pos: Dir med serv, Mt Sinai Hosp, 68-74. Mem: Am Fedn Clin Res; fel Am Col Physicians; Am Heart Asn; AMA; fel NY Acad Med. Res: Pathogenesis of rheumatic fever and acute glomerulonephritis following streptococcal infection; the relation of streptococci to pharyngeal infections; myoglobin-antimyoglobin reactions in cardiac muscle after various injuries and its determination by immunofluorescent techniques. Mailing Add: 3800 S Ocean Dr Hollywood FL 33019

BERNSTEIN, STEPHEN, b Rochester, NY, Nov 14, 33; m 60. NEUROPSYCHOLOGY. Educ: Princeton Univ, AB, 55; Univ Wis, MA, 59, PhD(psychol), 62. Prof Exp: NIMH fels, Zurich, Switz, 62-64 & Paris, France, 64-65; trainee ment health training prog, Brain Res Inst, 65-66, ASST PROF PSYCHIAT, UNIV CALIF, LOS ANGELES, 66- Concurrent Pos: NIMH career develop award, 66- Mem: AAAS; Bee Res Asn; Sigma Xi; NY Acad Sci. Res: Brain research; emotional development and complex learning in rhesus monkeys; neural and sensory mechanisms controlling learned and fixed behavior in the ant. Mailing Add: Dept of Psychiat Med Ctr Univ of Calif Los Angeles CA 90024

BERNSTEIN, WILLIAM CARL, b Stillwater, Minn, Apr 12, 04; m 33; c 3. CANCER. Educ: Univ Minn, BS, 25, MD, 28; Am Bd Colon & Rectal Surg, dipl. Prof Exp: Attend proctologist, Univ Hosps, 46-58, assoc prof, Univ, 54-59, clin prof, 59-72, dir div proctol, 58-72, EMER CLIN PROF SURG, UNIV MINN, MINNEAPOLIS, 72- Concurrent Pos: Chief proctol surv, US Vet Hosp, Minneapolis, 46 & Ancker Hosp, St Paul, 46- Mem: Fel Am Col Surg; fel Am Proctol Soc. Res: Cancer of rectum and colon. Mailing Add: 740 River Dr St Paul MN 55116

BERNSTORF, EARL CRANSTON, b Judsonia, Ark, June 17, 21; m 44; c 4. ANATOMY. Educ: Taylor Univ, BS, 44; Ind Univ, MA, 48, PhD(human anat), 50. Prof Exp: Teacher high schs, Ind, 44-46; from instr to asst prof anat, Hahnemann Med Col, 50-58; Ind Univ-AID contract prof, Postgrad Med Ctr, Karachi, Pakistan, 58-63; vis assoc prof, Med Ctr, Ind Univ, 63-64; ASSOC PROF ANAT, MEHARRY MED COL, 64- Mem: Endocrine Soc; Am Asn Anat; Pakistan Asn Advan Sci. Res: Histology; neuroanatomy; endocrinology. Mailing Add: Dept of Anat Meharry Med Col Nashville TN 37208

BERNTHAL, FREDERICK MICHAEL, b Sheridan, Wyo, Jan 10, 43; m 69; c 1. NUCLEAR CHEMISTRY, NUCLEAR PHYSICS. Educ: Valparaiso Univ, BS, 64; Univ Calif, Berkeley, PhD(chem), 69. Prof Exp: Res staff chemist, Heavy Ion Accelerator Lab, Yale Univ, 69-70; asst prof, 70-75, ASSOC PROF CHEM & PHYSICS, MICH STATE UNIV, 75- Mem: Am Chem Soc; Am Phys Soc. Res: Nuclear spectroscopy and structure; models and spectroscopy of deformed nuclei; behavior of nuclei at very high spin; nuclear decay; applications of nuclear technology. Mailing Add: Dept of Chem Mich State Univ East Lansing MI 48824

BERNTON, HARRY SAUL, b Jan 15, 84; m 16; c 2. ALLERGY. Educ: Harvard Univ, AB, 04, MD, 08. Prof Exp: Dir, Bender Lab, NY, 09-14; pathologist, State Bd Health, RI, 14-18; prof path, George Washington Univ, 19-22; from assoc prof to prof hyg & prev med, Georgetown Univ, 24-48; prof clin med, Col Med, Howard Univ, 48-73; RETIRED. Concurrent Pos: Instr path, Harvard Univ, 16-18; spec expert, USPHS, 22-24; clin specialist allergy, Bur Agr & Indust Chem, USDA, 36-53 & Agr Res Serv, 53-71. Mem: AAAS; fel Am Acad Allergy (pres, 28); fel Am Col Allergists; fel Am Col Physicians; Am Asn Path & Bact. Res: Hygiene of hay fever and asthma; protein sensitization. Mailing Add: 4000 Cathedral Ave NW Washington DC 20016

BERNTSEN, ROBERT ANDYV, b Northfield, Minn, Feb 22, 17; m 43; c 2. INORGANIC CHEMISTRY. Educ: St Olaf Col, BA, 39; NY Univ, MS, 41; Purdue Univ, PhD(chem), 49. Prof Exp: Instr chem, St Olaf Col, 41-43; head dept, Minot State Teachers Col, 43-46; from asst prof to assoc prof, 48-57, PROF CHEM, AUGUSTANA COL, 57-, HEAD DEPT, 68- Mem: Am Chem Soc. Res: Oxidation of carbohydrates with oxides of nitrogen. Mailing Add: Dept of Chem Augustana Col Rock Island IL 61201

BERNTZEN, ALLEN KEITH, parasitology, biochemistry, see 12th edition

BEROZA, MORTON, b New Haven, Conn, Mar 7, 17; m 46; c 2. ANALYTICAL CHEMISTRY. Educ: George Washington Univ, BS, 43; Georgetown Univ, MS, 46, PhD(org chem, biochem), 50. Prof Exp: Sci asst drugs, Food & Drug Admin, 39-42; chemist & chem engr plastics, US Naval Ord Lab, 46-48; RES CHEMIST, USDA, BELTSVILLE, 48- Honors & Awards: Hillebrand Prize, 63; Award in Chromatography & Electrophoresis, Am Chem Soc, 69; Harvey W Wiley Award, 70; Gold Medal Award, Synthetic Org Chem Mfrs Asn, 73. Mem: AAAS; Am Chem Soc; Entom Soc Am; Soc Exp Biol & Med; Am Inst Chemists. Res: Insecticidal alkaloids of plants, isolation and chemical structure; insecticidal synergists of natural origin; synthesis of insect repellents, attractants and insecticides; spectroscopy; chromatography; pesticide residue analysis; analytical micromethodology. Mailing Add: 821 Malta Lane Silver Spring MD 20901

BERQUIST, KENNETH R, b Rhame, NDak, Feb 21, 18; m 42; c 7. MICROBIOLOGY. Educ: Univ SDak, BA, 41, MA, 48; Wash State Univ, PhD(bact), 64. Prof Exp: Sr bacteriologist lab div, SDak State Health Dept, 41-43 & 46-48; instr bact med sch, Univ SDak, 48-50; asst dir pub health bact, Lab Sect, Wash State Dept Health, 50-60, head immunol-virol sect, Div Labs, 63-65; RES MICROBIOLOGIST, CTR DIS CONTROL, PHOENIX LABS, USPHS, 65- Mem: Am Soc Microbiol; Tissue Cult Asn; Asn State & Territorial Pub Health Lab Dirs. Res: Immunology; virology; fluorescent antibody techniques. Mailing Add: Phoenix Labs Ctr Dis Cont USPHS 4420 N Seventh St Phx Field Sta Phoenix AZ 85014

BERR, CHARLES ERNEST, organic chemistry, see 12th edition

BERRA, TIM MARTIN, b St Louis, Mo, Aug 31, 43; m 67. ICHTHYOLOGY, ZOOGEOGRAPHY. Educ: St Louis Univ, BS, 65; Tulane Univ, MS, 67, PhD(biol), 69. Prof Exp: NIH trainee environ biol, Tulane Univ, 66-69; Fulbright scholar zool, Australian Nat Univ, 69-70, demonstr, 70; sr tutor biol, Univ Papua New Guinea, 70-71; asst prof, 72-76, ASSOC PROF ZOOL, MANSFIELD REGIONAL CAMPUS, OHIO STATE UNIV, 76- Mem: Am Soc Ichthyologists & Herpetologists; Am Fisheries Soc; Soc Syst Zool. Res: Ecology, taxonomy, zoogeography, movements and home range of freshwater fishes; bibliography of naturalists, history of biology and natural history of Tropics, Australia and New Guinea. Mailing Add: Dept of Zool Ohio State Univ Mansfield OH 44906

BERREMAN, DWIGHT WINTON, b Salem, Ore, May 30, 28; m 54; c 3. PHYSICS. Educ: Univ Ore, BS, 50; Calif Inst Technol, MS, 52, PhD(physics), 55. Prof Exp: Physicist x-ray optical design, Appl Res Labs, Calif, 55; physicist infrared & visual optics, Stanford Res Inst, 55-56, consult, 56-57; asst prof physics, Univ Ore, 56-61; MEM TECH STAFF, BELL LABS, 61- Mem: Am Phys Soc; Optical Soc Am. Res: Physical optics in films, crystals and liquid crystals. Mailing Add: Bell Labs Murray Hill NJ 07974

BERREMAN, GERALD DUANE, b Portland, Ore, Sept 2, 30; m 52; c 3. ANTHROPOLOGY. Educ: Univ Ore, BA, 52, MA, 53; Cornell Univ, PhD(anthrop), 59. Prof Exp: From asst prof to assoc prof anthrop, Univ Calif, Berkeley, 59-66, PROF ANTHROP, UNIV CALIF, BERKELEY, 66- Mem: Am Anthrop Asn; Soc Appl Anthrop; Am Ethnol Soc; Asn Asian Studies. Res: Culture change; intergroup relations; social stratification; community study; applied anthropology; Aleutian Islands, India, especially Himalayan area; research methods. Mailing Add: Dept of Anthrop 232 Kroeber Hall Univ of Calif Berkeley CA 94720

BERREND, ROBERT E, b Wausau, Wis, July 12, 25; m 57; c 2. INVERTEBRATE ZOOLOGY. Educ: Univ Wis, BA, 49, MA, 52, PhD(protozool), 58. Prof Exp: Instr zool, Mont State Univ, 58-60; asst prof biol, 60-64, ASSOC PROF BIOL, SAN FRANCISCO STATE UNIV, 64- Mem: AAAS. Res: Distribution of and locomotion in testaceans; estuarine ecology. Mailing Add: Dept of Biol San Francisco State Univ 1600 Holloway San Francisco CA 94132

BERRETH, JULIUS R, b Artas, SDak, Aug 15, 29; m 58; c 3. NUCLEAR CHEMISTRY, PHYSICAL CHEMISTRY. Educ: Cent Wash Col, BA, 52; Wash State Univ, MS, 59. Prof Exp: Chemist, Scott Paper Co, 56-57; chemist, Phillips Petrol Co, 57-59, res chemist, Atomic Energy Div, 59-68; res chemist & group leader nuclear chem, Aerojet Nuclear Co, 68-73; GROUP LEADER, ALLIED CHEM CORP, 73- Mem: Am Chem Soc. Res: Preparations and separations of radioactive elements and determination of integral and differential cross-sections; neutron activation analysis on biological and inorganic samples; nuclear waste management. Mailing Add: Allied Chem Corp 550 Second St Idaho Falls ID 83401

BERRETT, DELWYN GREEN, b Menan, Idaho, July 27, 35; m 63; c 5. ORNITHOLOGY. Educ: Brigham Young Univ, BS, 57, MS, 58; La State Univ, PhD(ornith), 62. Prof Exp: Mem staff zool, Ricks Col, 62-64; ASSOC PROF ZOOL, BRIGHAM YOUNG UNIV, HAWAII CAMPUS, 64- Mem: Am Ornith Union; Cooper Ornith Soc; Wilson Ornith Soc. Res: Taxonomy, classification and distribution of birds. Mailing Add: Dept of Biol Brigham Young Univ Laie HI 96762

BERRETTONI, JULIO NARCISO, statistics, see 12th edition

BERRI, MANUEL PHILLIP, b San Antonio, Tex, July 12, 31. MATHEMATICS. Educ: Rockhurst Col, BS, 52; Univ Notre Dame, MS, 56; Univ Calif, Los Angeles, PhD(math), 61. Prof Exp: Instr math, Loyola Univ, Calif, 57-59; instr, Univ Calif, Los Angeles, 59-61, lectr, 61-62; asst prof, Tulane Univ, 62-66; assoc prof math, 66-74, actg chmn dept, 67-68, PROF MATH, UNIV NEW ORLEANS, 74- Mem: AAAS; Am Math Soc; Math Asn Am. Res: Point-set topology; mathematics education. Mailing Add: Dept of Math Univ of New Orleans New Orleans LA 70122

BERRIER, HARRY HILBOURN, b Norborne, Mo, July 6, 17; m 50. VETERINARY PATHOLOGY. Educ: Univ Mo, BSc, 41, MSc, 60; Kans State Univ, DVM, 45. Prof Exp: Instr pub schs, Mo, 41-42; practicing veterinarian, Mo, 45-46; ASSOC PROF VET PATH, COL VET MED, UNIV MO-COLUMBIA, 48- Concurrent Pos: NSF fel, Vet Path Div, US Armed Forces Inst Path, 63-64. Mem: Am Vet Med Asn; Am Soc Vet Clin path; fel Am Col Vet Toxicol. Res: Ovine pregnancy diseases; veterinary medicine diagnostic aids; canine surgical mouth speculum. Mailing Add: Col of Vet Med Univ of Mo Columbia MO 65201

BERRIER, JOHN VINCENT, b Columbus, Ohio, Apr 10, 45; m 72; c 1. SYNTHETIC ORGANIC CHEMISTRY. Educ: Bucknell Univ, BS & MS, 67; Princeton Univ, MA, 69, PhD(chem), 72. Prof Exp: ASSOC PROF CHEM, ANNE ARUNDEL COMMUNITY COL, 71- Mem: Am Chem Soc; Am Asn Univ Prof. Res: Synthesis of heterocyclic compounds. Mailing Add: Sci Div Anne Arundel Community Col Arnold MD 21012

BERRILL, MICHAEL, b Montreal, Que, Apr 18, 44; m 69. ANIMAL BEHAVIOR. Educ: McGill Univ, BSc, 64; Univ Hawaii, MS, 65; Princeton Univ, PhD(animal behav), 68. Prof Exp: Asst prof biol, 68-74, ASSOC PROF BIOL, TRENT UNIV, 74- Res: Biology of the Stauromedusae; behavior of synaptid Holothuria; embryonic and larval development of behavior of invertebrates, especially mysid, amphipod, decapod crustaceans and gastropod mollusks. Mailing Add: Dept of Biol Trent Univ Peterborough ON Can

BERRILL, NORMAN JOHN, b Bristol, Eng, Apr 28, 03. DEVELOPMENTAL BIOLOGY. Educ: Bristol Univ, BSc, 24; Univ London, PhD, 28, DSc, 32. Hon Degrees: LLD, Univ Windsor, Ont, 67; DSc, Univ BC, 72 & McGill Univ, 73. Prof Exp: Asst zool, Univ Col, Univ London, 25-27; lectr comp physiol, Med Sch, Univ

Leeds, 27-28; from asst prof to assoc prof zool, McGill Univ, 28-46, Strathcona prof, 46-65, in charge dept, 37-40, chmn dept, 40-47; Guggenheim fel, 64-65; lectr, Univ Hawaii, 64-65 & Swarthmore Col, 67-68; RETIRED. Mem: Royal Soc Can; fel Royal Soc; Marine Biol Asn UK; Am Soc Zool; Soc Develop Biol. Res: Embryology of ascidians; regeneration in annelids; tunicate development; origin of vertebrates; asexual development in tunicates and hydroids; experimental morphogenesis in tunicates, polychaets and hydroids. Mailing Add: 410 Swarthmore Ave Swarthmore PA 19081

BERRIS, BARNET, b Toronto, Ont, Apr 23, 21; m 44; c 3. INTERNAL MEDICINE. Educ: Univ Toronto, BA, 41, MD, 44; Univ Minn, MS, 50; FRCPC, 52. Prof Exp: Instr med, Univ Minn, 49-50; clin teacher, 50-57, assoc, 57-65, assoc prof, 65-67, PROF MED, UNIV TORONTO, 67- Mem: Fel Am Col Physicians. Res: Clinical medicine. Mailing Add: Dept of Med New Mt Sinai Hosp Toronto ON Can

BERRY, ANDREW CAMPBELL, b Somerville, Mass, Nov 23, 06. MATHEMATICS. Educ: Harvard Univ, AB, 25, AM, 26, PhD, 29. Prof Exp: Instr math, Harvard Univ, 26-27, 28-29; Nat Res fel, Brown Univ, 29-31 & Princeton Univ, 31; from instr to asst prof, Columbia Univ, 31-41; assoc prof, 41-43, Child prof, 43-58, Colman prof, 58-75, EMER PROF MATH, LAWRENCE UNIV, 75- Mem: Math Asn Am. Res: Analysis; probability; Berry-Esseen Theorem on normal law in probability. Mailing Add: 123 N Green Bay Rd Appleton WI 54911

BERRY, BRIAN JOE LOBLEY, b Sedgley, UK, Feb 16, 34; US citizen; m 58; c 3. URBAN GEOGRAPHY, URBAN RESEARCH & DEVELOPMENT. Educ: Univ London, BSc, 55; Univ Wash, MA, 56, PhD(geog), 58. Prof Exp: From asst prof to prof geog, Univ Chicago, 58-72, Irving B Harris prof urban geog, 72-76, chmn training progs urban studies, 63-70, chmn dept geog & dir ctr urban studies, 74-76; WILLIAMS PROF CITY & REGIONAL PLANNING & DIR LAB FOR COMPUT GRAPHICS & SPATIAL ANAL, HARVARD UNIV, 76- Honors & Awards: Meritorious Contrib to Geog Award, Asn Am Geogr, 68; Hon Fel, Urban Land Inst, 75. Mem: Nat Acad Sci; Am Geog Soc; Asn Am Geogr; Regional Sci Asn; Am Statist Asn. Res: Policy-oriented urban research, with particular concern for deliberate change in spatial systems. Mailing Add: PO Box 190 Park Forest IL 60466

BERRY, CHARLES A, b Rogers, Ark, Sept 17, 23; m 44; c 3. AEROSPACE MEDICINE. Educ: Univ Calif, Berkeley, AB, 45; Univ Calif, San Francisco, MD, 47; Harvard Univ, MPH, 56. Prof Exp: Intern, City-County Hosp, San Francisco, 47-48; gen pract med, 48-51; resident aviation med training prog, US Air Force Sch Aerospace Med, 51-52, surgeon, Albrook AFB, CZ, 52-55, resident aviation med training prog, US Air Force Sch Aerospace Med, 55-56, asst chief dept aviation med, 59-62; chief med opers off, Manned Spacecraft Ctr, NASA, 62-66, dir med res & opers, 66-74; PRES, UNIV TEX HEALTH SCI CTR HOUSTON, 74- Concurrent Pos: Vis lectr, Sch Pub Health, Harvard Univ; clin assoc prof prev med, Ohio State Univ; clin prof, Sch Pub Health, Univ Tex; prof & chmn dept aerospace med, Univ Tex Med Br, 67-74; mem bioastronaut comt & awards & fels comt, Int Astronaut Fedn; mem, Orbiting Int Lab, Int Acad Astronaut; mem bd, Earth Awareness Found Int. Honors & Awards: Tuttle Award, Aerospace Med Asn, 61, Bauer Founders Award, 66; Cert of Achievement, Surgeon Gen US Air Force, 62; Honor Citation, AMA, 62; Montgomery Award, Soc Aerospace Prof, 63; NASA Exceptional Serv Medal, 65; Gold Medal, Czech Acad Med, 66; Boynton Award, Am Astronaut Soc, 66; Gold Medal, Am Col Chest Physicians, 66; John Jeffries Award, 66; Hubertus Strughold Award, 67; Day Mem Medal, Acad Natural Sci, 68; Physician Mission Award, Carlo Erba Found, Italy, 69; Silver Medal, Port Ministry of Health & Welfare, 69; Guggenheim Int Astronaut Award, 69; NASA Exceptional Sci Achievement Medal, 69; Med Serv Award, Am Col Surgeons, 70; Herman Oberth Award, 74; Cedars of Lebanon Award, 74. Mem: Fel Aerospace Med Asn (pres, 70); fel Am Col Prev Med (vpres aerospace med); fel Am Col Physicians; fel Am Astronaut Soc; fel Am Col Chest Physicians. Res: Medical support of manned spaceflight; dysbarism; clinical aviation medicine; medical selection. Mailing Add: Off of the Pres Univ of Tex Health Sci Ctr Houston TX 77025

BERRY, CHARLES ARTHUR, b Ketchum, Idaho, Aug 19, 29; m 54; c 1. PHARMACOLOGY. Educ: Univ Idaho, BS, 51; Idaho State Univ, BS, 54; Univ Iowa, MS, 59, PhD(pharmacol), 61. Prof Exp: Instr pharmacol, Univ Iowa, 61-62; fel, Stanford Univ, 62-64; asst prof, 64-68, ASSOC PROF PHARMACOL, NORTHWESTERN UNIV, 68- Mem: AAAS; Am Soc Pharmacol & Exp Therapeut. Res: Problems of attention, learning, memory, motivation and animal behavior; investigation of drugs and neural systems interactions; definition of cortical-subcortical relationships. Mailing Add: Dept of Pharmacol Northwestern Univ Med Sch Chicago IL 60611

BERRY, CHARLES RICHARD, b Morgantown, WVa, Mar 8, 27; m 48; c 2. PLANT PATHOLOGY. Educ: Glenville State Col, AB, 49; WVa Univ, MS, 55, PhD(plant path), 58. Prof Exp: Plant pathologist, Div Forest Dis Res, 57-62, PLANT PATHOLOGIST, DIV FOREST PROTECTION RES, SOUTHEASTERN FOREST EXP STA, USDA, 66-, PROJ LEADER AIR POLLUTION, 62- Mem: Mycol Soc Am; Am Phytopath Soc. Res: Reclamation; air pollution; forest disease. Mailing Add: USDA Southeastern Forest Exp Sta Forestry Sci Lab Athens GA 30601

BERRY, CHESTER RIDLON, b Boston, Mass, Aug 15, 19; m 40; c 3. SOLID STATE PHYSICS. Educ: Dartmouth Col, AB, 40, AM, 42; Cornell Univ, PhD(exp physics), 46. Prof Exp: Tech supvr, Manhattan Dist Proj, Oak Ridge, 44-45; res physicist, Eastman Kodak Co, 46-75; RETIRED. Mem: AAAS; fel Am Phys Soc; Am Crystallog Asn; Am Asn Physics Teachers; fel Soc Photog Sci & Eng. Res: Electron and x-ray diffraction; theory of photographic processes; imperfections in crystals; crystal growth; scattering and absorption of light. Mailing Add: 37 Heritage Dr South Orleans MA 02662

BERRY, CLARK GREEN b Ilion, NY, Sept 28, 08; m 36; c 2. CHEMISTRY. Educ: Clarkson Col Technol, BS, 30. Prof Exp: Chemist, Atmospheric Nitrogen Corp, 30-31 & Skenandoa Rayon Corp, 33-37, control supvr, 37-38; res chemist, Inst Paper Chem, 38-42 & Skenandoa Rayon Corp, 42-46; chief chemist, Del Rayon Co, 46-52; lab dir, New Bedford Rayon Co, 52-53, chief chemist & dir res, 53-66; SUPVRY CHEMIST & BR HEAD, BUR ENGRAVING & PRINTING, US TREASURY DEPT, 66- Mem: Am Chem Soc; Tech Asn Pulp & Paper Indust. Res: Viscose rayon; cellulose; synthetic fibers. Mailing Add: Tech Serv Div Bur Engrav & Print US Treasury Dept 14th & C St Washington DC 20226

BERRY, CLYDE MARVIN, b Posey, Ill, June 18, 13; m 40; c 3. INDUSTRIAL HYGIENE, CHEMICAL ENGINEERING. Educ: McKendree Col, BS, 33; Univ Ill, MS, 36; Univ Iowa, MSChE, 40, PhD(indust hyg), 41. Prof Exp: Indust hyg engr, USPHS, 41-48 & Esso Standard Oil Co, 48-55; ASSOC DIR, INST AGR MED, UNIV IOWA, 55- Concurrent Pos: Mem adv comt, Div Occup Health, USPHS, 57-60, consult, 60-, mem adv comt, Div Accident Prev, 60-63, consult, 63-; consult, Nat Comn Community Health Serv, 64- Mem: Fel Am Pub Health Asn; Am Indust Hyg Asn (pres, 67); Am Conf Govt Indust Hygienists; Am Col Health Asn; Am Soc Safety Engr. Res: Bacterial quality of air; industrial carcinogens; charged aerosols. Mailing Add: Oakdale Campus Univ of Iowa Oakdale IA 52319

BERRY, DAISILEE H, b Honolulu, Hawaii. PEDIATRICS, HEMATOLOGY. Educ: Univ Ark, BS, 49, MD, 59; Western Reserve Univ, MN, 52. Prof Exp: Intern resident pediat, Univ Ark, 59-61; resident fel, Washington Univ, 61-63; fel hemat oncol, Univ Ark, 63-64, asst prof pediat hemat oncol, 64-67; fel hemat, Duke Univ, 67-69; asst prof pediat hemat oncol, 69-70, ASSOC PROF PEDIAT HEMAT ONCOL, UNIV ARK, LITTLE ROCK, 70- Mem: Am Soc Hemat; Am Soc Clin Oncol; Acad Pediat. Res: White cell separation and enzymology; clinical manifestations of sickle cell anemia. Mailing Add: Ark Childrens Hosp 804 Wolfe St Little Rock AR 72001

BERRY, DAVID A, b Marietta, Ohio, Jan 13, 22; m 53; c 1. ORGANIC CHEMISTRY. Educ: Marietta Col, AB, 47. Prof Exp: Org res chemist, Marietta Dyestuffs, Ohio, 43-45 & Diamond Alkali Co, 46-47; proj leader, 47-60, asst chief polymer sect, 60-68, chief org & polymer div, 68-70, chief org chem div, 70-73, MGR ANAL & PHYS CHEM SECT, BATTELLE MEM INST, 73- Mem: AAAS; Am Chem Soc. Res: Chlorinated paraffins; dyestuff intermediates; iodonium compounds; drying oils from paraffin wax; stabilization of chlorinated paraffins; preparation of azo dyestuffs; polyesters and alkyd resins; thermally-stable polymers; rosin derivatives. Mailing Add: Org Chem Div Battell Mem Inst 505 King Ave Columbus OH 43201

BERRY, DWIGHT BEECHER, b Troy, Ohio, Dec 5, 37; m 62; c 3. PHYSICS. Educ: Manchester Col, BA, 59; Ind Univ, Bloomington, MS, 62; Mich State Univ, PhD(physics), 69. Prof Exp: Teaching asst physics, Ind Univ, Bloomington, 60-62; instr, Manchester Col, 62-64; teaching asst, Mich State Univ, 64-65, res asst, 65-69; asst prof, 69-71, ASSOC PROF PHYSICS, MANCHESTER COL, 71- Mem: Am Asn Physics Teachers; Am Phys Soc. Res: Nuclear physics gamma ray spectroscopy in the neutron deficient region. Mailing Add: Dept of Physics Manchester Col North Manchester IN 46962

BERRY, EDWIN X, b San Francisco, Calif, June 20, 35; m 57 & 73; c 3. ATMOSPHERIC PHYSICS. Educ: Calif Inst Technol, BS, 57; Dartmouth Col, MS, 60; Univ Nev, PhD(physics), 65. Prof Exp: Res asst ionospheric physics, Thayer Sch Eng, Dartmouth Col, 59-60; radio propagation engr, Advan Commun Eng Div, Cook Elec Co, 60-61; from res asst to res assoc atmospheric physics, Desert Res Inst, Univ Nev, Reno, 62-72; prog mgr weather modification, 73-74, STAFF & PHYS SCIENTIST, WESTERN PROJ OFF, NAT SCI FOUND, 74- Concurrent Pos: Mem comt on land use planning, 72; chmn Gov's adv comt weather modification, State of Nev, Am Meterol Soc, 74. Mem: AAAS; Am Meteorol Soc; Am Geophys Union. Res: Cloud physics. Mailing Add: 566 Acacia Ave San Bruno CA 94066

BERRY, FREDERICK ALMET FULGHUM, b Tulsa, Okla, Apr 12, 29. GEOLOGY. Educ: Stanford Univ, BS, 49, MS, 51, PhD(geol), 59. Prof Exp: Asst, A I Levorsen, Consult Geologist, 53-54; geologist fuels br, US Geol Surv, 54-56; dir & mgr geol res, Petrol Res Corp, Colo, 54-61; vis prof geol, 60-61, ASSOC PROF GEOL & GEOPHYS, UNIV CALIF, BERKELEY, 61- Concurrent Pos: Consult, US Govt, 53-54. Mem: AAAS; Soc Explor Geophys; Am Asn Petrol Geologists; Geol Soc Am; Am Geophys Union; Geochem Soc. Res: Geophysics and geochemistry of fluids; geology of petroleum; physical stratigraphy; regional tectonics. Mailing Add: Dept of Geol & Geophys Univ of Calif Berkeley CA 94720

BERRY, FREDERICK HAMER, b Staunton, Va, Jan 3, 21; m 44; c 1. FOREST PATHOLOGY. Educ: Duke Univ, BS, 43, MF, 44; Univ Md, PhD, 60. Prof Exp: Forest asst, Duke Univ, 44; silviculturist spruce & fir, Northeastern Forest Exp Sta, Maine, 45-47; forester, Nat Forest Admin, US Forest Serv, Ariz, 47-49; plant pathologist, Crops Res Div, USDA, Md, 49-60; forest pathologist, US Forest Serv, Cent States Forest Exp Sta, 60-66, FOREST PATHOLOGIST, US FOREST SERV, NORTHEASTERN FOREST EXP STA, 66- Mem: Soc Am Foresters; Am Phytopath Soc. Res: Diseases of forest and shade trees. Mailing Add: US Forest Serv Insect & Dis Lab Delaware OH 43015

BERRY, GAIL WRUBLE, b Kalamazoo, Mich, Nov 7, 39. PSYCHIATRY. Educ: Kalamazoo Col, AB, 60; NY Univ Sch Med, MD, 64. Prof Exp: Intern pediat, Kings County Hosp, Brooklyn, NY, 64-65; resident psychiat, NY Med Col, Metrop Hosp, NY, 65-68; staff psychiatrist, Beth Israel Med Ctr, NY, 68-74; from clin instr psychiat to clin assoc, Mt Sinai Sch Med, NY, 68-75; CHIEF PSYCHIAT INPATIENT SERV, BETH ISRAEL MED CTR, NY, 74-, ASST CLIN PROF PSYCHIAT, MT SINAI SCH MED, NY, 75- Concurrent Pos: Assoc ed, Academy, Am Acad Psychoanal, 76- Mem: Am Psychiat Asn; Am Acad Psychoanal; Am Soc Adolescent Psychiat. Res: Psychoanalytic research into aspects of incest and into psychology of women. Mailing Add: 1225 Park Ave New York NY 10028

BERRY, GEORGE WILLARD, b Poolville, NY, Feb 22, 15; m 46; c 2. GEOLOGY. Educ: Colgate Univ, AB, 36; Cornell Univ, MS, 38, PhD(struct geol), 41. Prof Exp: From asst to sr asst geol, Cornell Univ, 36-38, instr, 38-41; geol scout, Texas Co, Tex, 41-42; geologist, Fla, 46 & Wyo, 46-48; dist geologist, Sun Oil Co, Wyo, 48-52, asst div geologist, Colo, 52-61; geologist, Tex, 61-64 & Colo, 64-70; res geologist, Cordero Mining Co, 70-72; CONSULT GEOLOGIST, 72- Mem: AAAS; Am Asn Petrol Geologists; Am Inst Mining, Metall & Petrol Engrs; Geol Soc Am; Soc Econ Geol. Res: Rocky Mountain geology; geothermal energy. Mailing Add: 600 Spruce St Boulder CO 80302

BERRY, GEORGE WILLIAM, b Parker, Colo, Feb 6, 07; m 30; c 5. CHEMISTRY. Educ: Colo State Col, BSc, 30; Univ Nebr, MSc, 32, PhD(phys chem), 34. Prof Exp: Instr org chem, Univ Nebr, 34-35; res chemist, Socony-Vacuum Oil Co, Kans, 35-45; res chemist, Flintkote Co, 45-56, mgr built-up roofing dept, 56-63, tech mgr, Archit Prod Div, 63-66; sr res assoc, Johns-Manville Corp, 66-72; SR ASSOC, ROBT M STAFFORD INC, 72- Mem: Am Chem Soc; fel Am Inst Chem. Res: Utilization of asphalt in building materials; petroleum refining; surface tension of dilute solutions of sodium palmitate. Mailing Add: PO Box 11075 Charlotte NC 28209

BERRY, GUY C, b Greene Co, Ill, May 11, 35; m 57; c 3. POLYMER CHEMISTRY. Educ: Univ Mich, BS, 57, MS, 58, PhD(chem eng), 60. Prof Exp: Fel polymer sci, Mellon Inst, 60-67; sr fel & assoc prof chem & polymer sci, 67-73, PROF CHEM & POLYMER SCI, CARNEGIE MELLON UNIV, 73- Mem: Am Chem Soc; Soc Rheol. Res: Physical chemistry of polymers; solution viscosity; light scattering; rheology. Mailing Add: Dept of Chem Carnegie Mellon Univ Pittsburgh PA 15213

BERRY, HENRY GORDON, b Huddersfield, Eng, July 25, 40; m 68; c 4. ATOMIC PHYSICS. Educ: Oxford Univ, BA, 62; Univ Wis-Madison, MS, 63, PhD(physics), 67. Prof Exp: Instr physics, Univ Wis-Madison, 67-68; fel physics, Univ Ariz, 68-69; guest researcher, Res Inst, Stockholm, Sweden, 69-70; maitre de conf, Univ Lyon, France, 70-72; ASST PROF PHYSICS, UNIV CHICAGO, 72- Mem: Am Phys Soc. Res: Beam foil spectroscopy; studies of radiative lifetimes, transition energies, spectral analysis and hyperfine structures of heavy ions; heavy ion collisions in solids and gases. Mailing Add: Dept of Physics Univ of Chicago 1110 E 58th St Chicago IL 60637

BERRY, HERBERT WEAVER, b Syracuse, NY, Dec 17, 13; m 42; c 2. PHYSICS. Educ: Syracuse Univ, AB, 37, AM, 39; Washington Univ, PhD(physics), 42. Prof Exp:

Instr, Case Inst Technol, 42-43; res physicist, Radiation Lab, Univ Calif, 43-45; asst prof physics, Univ Okla, 46; from asst prof to assoc prof, 46-54, PROF PHYSICS, SYRACUSE UNIV, 54- Mem: Am Phys Soc; Am Asn Physics Teachers. Res: Plasma physics; ionization of gases; scattering in gases. Mailing Add: Dept of Physics Syracuse Univ Syracuse NY 13210

BERRY, JAMES FREDERICK, b Baltimore, Md, Nov 11, 27; m 52; c 4. BIOCHEMISTRY. Educ: Johns Hopkins Univ, BA, 49; Univ Rochester, PhD(biochem), 53. Prof Exp: Lectr biochem, Univ Western Ont, 55-56; instr physiol chem, Johns Hopkins Univ, 59-61; assoc prof, 61-66, PROF NEUROL, MED SCH, UNIV MINN, MINNEAPOLIS, 66- Concurrent Pos: USPHS fel biochem, Univ Western Ont, 53-55, sr res fel, 53-56; mult sclerosis fel, Agr Res Coun Inst Animal Physiol, Babraham, Eng, 56-57; fel physiol chem, Johns Hopkins Univ, 58-59; assoc dir biochem res, Sinai Hosp, Baltimore, 57-61. Mem: Fel AAAS; Am Soc Biol Chemists; Am Oil Chemists' Soc; Am Chem Soc; Am Soc Neurochem. Res: Biochemistry of degenerating and regenerating nerve; lipid and fatty acid composition and biosynthesis in nervous tissue; biosynthesis of choline esters; alcohol and acetaldehyde metabolism. Mailing Add: Dept of Neurol Univ of Minn Med Sch Minneapolis MN 55455

BERRY, JAMES WESLEY, b Rankin, Ill, Mar 23, 26; m 47; c 5. ORGANIC CHEMISTRY. Educ: Augustana Col, AB, 49; Univ Ill, PhD(org chem), 53. Prof Exp: Res chemist, Rayonier, Inc, 53-55; asst prof chem, Whitworth Col, Wash, 55-56; PROF BIOCHEM, UNIV ARIZ, 56- Mem: Am Chem Soc. Res: Polysaccharides. Mailing Add: Dept of Agr Biochem Univ of Ariz Tucson AZ 85721

BERRY, JAMES WILLIAM, b Bristol, Va, Dec 7, 35; m 70. INVERTEBRATE ECOLOGY. Educ: ETenn State Univ, BS, 57; Va Polytech Inst, MS, 58; Duke Univ, PhD(zool), 66. Prof Exp: From instr to asst prof, 65-70, ASSOC PROF ZOOL, BUTLER UNIV, 70- Concurrent Pos: USPHS fel, 66-67. Mem: Ecol Soc Am; Am Inst Biol Sci. Res: Spider ecology in North Carolina piedmont; spiders of the Florida Everglades; spider distribution on Pacific atolls; chlorine toxicity on fish. Mailing Add: Dept of Zool Butler Univ Indianapolis IN 46208

BERRY, JAMES WILLIAM, JR, horticulture, see 12th edition

BERRY, JEWELL EDWARD, b Independence, Mo, Feb 7, 26; m 54; c 3. PARASITOLOGY. Educ: Fisk Univ, BA, 51, MA, 53; Univ Notre Dame, PhD(biol), 56. Prof Exp: ASSOC PROF BIOL, PRAIRIE VIEW AGR & MECH COL, 56- Mem: Am Soc Zoologists. Res: Trematode morphology; taxonomy and life-cycles; invertebrate history. Mailing Add: Dept of Biol Prairie View Agr & Mech Col Prairie View TX 77445

BERRY, JOE GENE, b Sayre, Okla, Feb 19, 44; m 65; c 1. FOOD SCIENCE, POULTRY SCIENCE. Educ: Okla State Univ, BS, 65, MS, 67; Kans State Univ, PhD(food sci), 70. Prof Exp: ASSOC PROF ANIMAL SCI, PURDUE UNIV, 70- Mem: Inst Food Technologists; Poultry Sci Asn. Res: Prevention of egg shell damage in production, processing and retailing; poultry products technology. Mailing Add: Poultry Bldg Purdue Univ West Lafayette IN 47906

BERRY, JOHN WILLIAM, b Indianapolis, Ind, Jan 21, 16; m 43; c 3. ANALYTICAL CHEMISTRY. Educ: Ind Univ, AB, 37; Univ Iowa, MS, 41, PhD(anal chem), 43. Prof Exp: Anal spectrographer, Dow Chem Co, 37-38; chem physicist, 42-72, SR RES CHEMIST, AM CYANAMID CO, 72- Res: Analytical and physical chemistry; analytical spectrographic analysis; flame photometry; analytical and industrial instrumentation. Mailing Add: Sci Serv Dept Am Cyanamid Co 1937 W Main St Stamford CT 06904

BERRY, KEITH O, b Ft Collins, Colo, Aug 6, 38; m 60; c 2. INORGANIC CHEMISTRY, ANALYTICAL CHEMISTRY. Educ: Colo State Col, BA, 60; Iowa State Univ, PhD(inorg chem), 66. Prof Exp: Instr, 65-70, ASSOC PROF CHEM, UNIV PUGET SOUND, 70- Mem: AAAS; Am Chem Soc; Forensic Sci Soc. Res: Transition metal complexes with oxygen-donor ligands; microcrystalline detection of trace elements. Mailing Add: Dept of Chem Univ of Puget Sound Tacoma WA 98416

BERRY, LEONARD, b Malmesbury, Eng, June 5, 30; m 66; c 4. PHYSICAL GEOGRAPHY, RESOURCE MANAGEMENT. Educ: Bristol Univ, BSc, 51, MSc, 56, PhD(geog), 69. Prof Exp: Asst lectr geog, Univ Hong Kong, 54-57; from lectr to sr lectr, Univ Khartoum, 57-65; prof, Univ Col, Dar es Salaam, 65-69, dean fac arts & soc sci, 67-69; dir resource mgt, Univ Dar es Salaam, 69-71; actg dir grad sch geog, 71-72, PROF GEOG, CLARK UNIV, 69-, DIR INT DEVELOP PROG, 74-, DEAN GRAD SCH, 75- Concurrent Pos: Consult, Int Develop Res Ctr, Ottawa, 72; mem comt to Indonesia on natural resources planning, Nat Acad Sci, 72, comt environ implications of US mat policy, 72-73 & comt on remote sensing for develop, 75. Mem: Royal Geog Soc; Inst Brit Geogr; Asn Am Geogr. Res: Tropical geomorphology; rural water development; regional planning; natural resource planning; environmental problems and development. Mailing Add: Grad Sch of Geog Clark Univ Worcester MA 01610

BERRY, LEONARD GASCOIGNE, b Toronto, Ont, Aug 17, 14; m 41; c 2. MINERALOGY. Educ: Univ Toronto, BA, 37, MA, 38, PhD(mineral), 41. Prof Exp: Asst, Geol Surv, Can, 35-36; asst, Ont Dept Mines, 36-37, geologist, 38-40; engr optical shops, Res Enterprises, Ltd, 40-44; lectr mineral, 44-46, from asst prof to prof, 46-67, MILLER MEM RES PROF, QUEEN'S UNIV, ONT, 67- Concurrent Pos: Asst mineral, Univ Toronto, 37-40; mem, Can nat Comt Crystallog, 48-70; Guggenheim fel, 53; assoc ed, Powder diffraction File, 56-69, ed, 69-; ed, Can Mineralogist, 57- Honors & Awards: Willet G Miller Medal, Royal Soc Can, 63. Mem: Fel Mineral Soc (vpres, 63, pres, 64); fel Am Geol Soc; Am Crystallog Asn; Mineral Asn Can (vpres, 58-59); fel Royal Soc Can. Res: Morphological and structural crystallography of minerals; crystal structure; x-ray diffraction and general mineralogical studies. Mailing Add: Dept of Geol Sci Queen's Univ Kingston ON Can

BERRY, LEONIDAS HARRIS, b Woodsdale, NC, July 20, 02; m 37, 59; c 1. MEDICINE. Educ: Wilberforce Univ, BS, 24; Univ Chicago, SB, 25, MD, 29; Univ Ill, MS, 33; Am Bd Internal Med, dipl, 46; Am Bd Gastroenterol, dipl, 46. Hon Degrees: ScD, Wilberforce Univ, 45. Prof Exp: From jr attend physician to assoc attend physician, Provident Hosp, 33-43, chmn div digestive dis, 34-60, from vchmn to chmn dept med, 43-48; PROF MED, COOK COUNTY GRAD SCH MED, 46- Concurrent Pos: Sr attend physician, Provident Hosp, 43-; mem dept med, Michael Reese Hosp, 46-63, sr attend Physician, 64; from clin asst prof to clin assoc prof, Sch Med, Univ Ill, 52-; consult gastroenterologist, Women's & Children's Hosp, Chicago, 56- & Alexian Bros Hosp, 61-; int lectr, Cult Affairs Div, US Dept State, Africa, Asia & Europe, 65, 66 & 70; mem bd trustees, Cook County Grad Sch Med; mem nat adv coun, Fed Regional Med Progs Versus Heart Dis, Cancer & Stroke; mem nat tech adv comts, Medicare Prog; spec dep, Prof Community Affairs, Cook County Hosp. Honors & Awards: Distinguished Serv Award, Nat Med Asn, 58. Mem: Nat Med Asn (1st vpres, 59, pres elect, 64, pres, 65); Am Gastroenterol Asn; fel Am Col Gastroenterol; AMA; fel Am Col Physicians. Res: Racial, sociological and pathological aspects of tuberculosis; techniques of gastroscopy; gastro-biopsy instrument; therapy of chronic gastritis and peptic ulcer; gastric cancer; medical history; narcotic rehabilitation. Mailing Add: Suite 303 2600 S Michigan Blvd Chicago IL 60616

BERRY, LEVETTE JOE, b North Birmingham, Ala, June 17, 10; m 34; c 1. MICROBIOLOGY. Educ: Southwest Tex Teachers Col, BS, 30; Univ Tex, PhD(physiol), 39. Prof Exp: Instr zool, Univ Tex, 39-40; asst prof biol, Bryn Mawr Col, 40-43, from asst prof to prof, 45-70, secy fac, 64-69, Found Microbiol lectr, 65-66, actg provost, 69-70; chmn dept microbiol, 70-75, PROF MICROBIOL, UNIV TEX, AUSTIN, 70- Concurrent Pos: Res assoc, Hillman Hosp, Ala, 43-45; investr biochem inst, Univ Tex, 48-49 & Andean Inst Biol, Peru, 55-56; consult, Chem Corps, US Army, NIH & Vet Admin; ed, J Bact, 64-68; adv comn bact & mycol lunar receiving lab, NASA, 69-71; mem int fel rev comn, NIH, 70-73, chmn, 71-72; Am Inst Biol Sci adv group microbiol to Off Naval Res, 70- Mem: AAAS; Am Acad Microbiol; Soc Exp Biol & Med; Am Soc Microbiol; Reticuloendothelial Soc (vpres, 65, pres, 67-69). Res: Medical microbiology; host response to infection; immunology; physiological chemistry; stress physiology; ribosomal vaccines and cellular immunity. Mailing Add: Dept of Microbiol Univ of Tex Austin TX 78712

BERRY, LOUIS MILTON, b La Grange, Tex, Oct 22, 18; m 61; c 3. CHEMISTRY. Educ: NTex State Univ, BS, 41. Prof Exp: Chemist, Gulf Oil Co, 41-43, Los Alamos Sci Lab, Univ Calif, 43-45 & Tex Co, 45-48; pvt pract retail merchandise, 48-51; staff mem chem & lubrication, 52-58, sect supvr chem, lubrication, electrochem & coatings, 58-60, div supvr chem, lubrication, electrochem, coatings & ceramics, 60-65, dept mgr chem, lubrication, electrochem, coatings, ceramics, plastics & metall, 65-69, DIR MAT & PROCESSES, SANDIA CORP, 69- Mem: Am Soc Metals; Am Soc Testing & Mat; Am Inst Chemists. Mailing Add: Sandia Labs Albuquerque NM 87115

BERRY, MAXWELL (RUFUS), b Atlanta, Ga, June 7, 10; m 34; c 4. INTERNAL MEDICINE. Educ: Cornell Univ, AB, 31, MD, 35; Univ Minn, PhD(med), 42. Prof Exp: Bursar, Knickerbocker Found, Cornell Univ, 33-35; intern med, Bellevue Hosp, NY, 35-37, res physician, 37-38; assoc med, Hosp, Med Col Va, 42, asst prof, 43-44; assoc, Sch Med, Emory Univ, 44-49; DIR CANCER CLIN, ST JOSEPH'S INFIRMARY, 47-, DIR BERRY CLIN, 48- Concurrent Pos: Consult, Grady Hosp, 44-49; pres, Northwest Hosp Corp, 66-70; chmn bd dirs, West Paces Ferry Hosp, 66-70; chmn bd trustees, Annandale at Suwanee, Inc, 66-73, pres, 73- Mem: Fel Am Col Physicians; fel Am Col Gastroenterol (pres, 65-66). Res: Tomac oxygen nebulizer; plethysmolymograph; physiology of circulation and respiration; studies on etiology of duodenal ulcer; clinical gastroenterology. Mailing Add: 3250 Howell Mill Rd NW Atlanta GA 30327

BERRY, MICHAEL JAMES, b Chicago, Ill, July 17, 47; m 67; c 2. PHYSICAL CHEMISTRY, CHEMICAL PHYSICS. Educ: Univ Mich, BS, 67; Univ Calif, Berkeley, PhD(chem), 70. Prof Exp: ASST PROF CHEM, UNIV WIS-MADISON, 70- Mem: Am Phys Soc; Am Chem Soc. Res: Chemical laser studies of the energy partitioning into the products of gas-phase bimolecular, unimolecular and photochemical reactions. Mailing Add: Dept of Chem Univ of Wis 1101 University Ave Madison WI 53706

BERRY, MICHAEL JOHN, b Southport, Eng, Oct 5, 40; Can citizen. SEISMOLOGY. Educ: Univ Toronto, BSc, 61, MA, 62, PhD(seismol), 65. Prof Exp: Res asst inst geophys & planetary sci, Univ Calif, Los Angeles, 65-66; RES SCIENTIST, SEISMOL DIV, EARTH PHYSICS BR, GOVT CAN, 67- Mem: Seismol Soc Am; Am Geophys Union; Soc Explor Geophys. Res: Studies of the Earth's crust and upper mantle using reflection and refraction seismology and the dispersion of seismic surface waves. Mailing Add: Seismol Div Earth Physics Br Ottawa ON Can

BERRY, MYRON GARLAND, b Franklin, NH, May 24, 19; m 48; c 3. PHYSICAL CHEMISTRY. Educ: Colby Col, BA, 40; Harvard Univ, MA, 42; Syracuse Univ, PhD(phys chem), 51. Prof Exp: Teacher, Dept Phys Sci, Urbana Jr Col, 46-49; asst prof chem, Ohio Wesleyan, 51-55; from asst prof to assoc prof, 55-62, admin asst, 62-69, PROF CHEM, MICH TECHNOL UNIV, 62- Mem: Am Chem Soc. Res: Surface chemistry; photochemistry. Mailing Add: Dept of Chem Mich Technol Univ Houghton MI 49931

BERRY, PAUL JOSEPH, physical chemistry, see 12th edition

BERRY, PAUL MCCLELLAN, b Oklahoma City, Okla, Aug 4, 33; m 61; c 3. APPLIED MATHEMATICS. Educ: Univ Okla, BS, 54, MA, 61, PhD, 68. Prof Exp: Petrol lab dir, Creole Petrol Corp, 56-59; asst math, Univ Okla, 59-64; res technologist, Mobile Oil Corp, 64-68; res assoc, 68-70, MGR DRILLING & PROD MECH RES, MOBIL RES & DEVELOP CORP, 70- Concurrent Pos: Trainee cardiovasc res, NIH, 61-63; vis indust prof comput sci, Southern Methodist Univ, 70- Mem: Math Asn Am; Am Math Soc; Soc Indust & Appl Math. Res: Mathematical modeling of drilling and oilfield production operations. Mailing Add: Mobil Res & Develop Corp 2123 Matagorda Dallas TX 75232

BERRY, RALPH EUGENE, b Gering, Nebr, June 14, 40; m 66; c 2. ENTOMOLOGY. Educ: Colo State Univ, BS, 63, MS, 65; Kans State Univ, PhD(entom), 68. Prof Exp: Entomologist, Pesticide Regulation Div, Agr Res Serv, USDA, 65; asst prof, 68-74, ASSOC PROF ENTOM, ORE STATE UNIV, 74- Mem: Entom Soc Am. Res: Applied entomology; insect migration; biology and control of soil arthropods; ecology of soil insects; investigations of non-insecticidal control of insect pests; pest population ecology. Mailing Add: Dept of Entom Ore State Univ Corvallis OR 97331

BERRY, RAYMOND ORVIL, b Chico, Tex, Oct 28, 02; m 29; c 2. ANIMAL CYTOLOGY. Educ: NTex Teachers Col, BS, 29; Agr & Mech Col, Tex, MS, 32; Johns Hopkins Univ, PhD(cytol), 39. Prof Exp: Prof biol, Blinn Jr Col, Tex, 29-31; instr & asst prof, Agr & Mech Col Tex, 31-36, assoc geneticist exp sta, 38-47, asassoc prof physiol, exp sta & col, 47-52, prof animal husb, 52-60; dir, Wortham Res Lab, 60-70 & Inst Reprod, Prairie View Agr & Mech Col, 70-76; RETIRED. Mem: AAAS; assoc Am Soc Zool; Genetics Soc Am; Am Soc Animal Sci; Soc Exp Biol & Med. Res: Infertility problems in cattle; cytology of sheep, goats, horses, mules and jacks; physiology of reproduction in sheep and cattle; physiology of mammalian sperm; cytology of maturation process of artificially ovulated ova. Mailing Add: PO Box 986 Hempstead TX 77445

BERRY, RICHARD EMERSON, b Washington, NJ, Nov 11, 33; m 54; c 2. SOLID STATE PHYSICS. Educ: Lafayette Col, BS, 54; Princeton Univ, MA, 56, PhD(physics), 58. Prof Exp: Researcher, Gen Elec Co, 57-58; asst prof physics, Lafayette Col, 58-62; assoc prof, Tex Technol Col, 62-65; chmn dept, 65-74, PROF PHYSICS, INDIANA UNIV PA, 74- Mem: Am Phys Soc. Res: Electron spin resonance in solid state; theory of electricity and magnetism; electromagnetic theory and unidentified flying objects effects. Mailing Add: Dept of Physics Indiana Univ of Pa Indiana PA 15701

BERRY, RICHARD G, b Bethel, Conn, Jan 29, 16; m 42; c 5. NEUROPATHOLOGY.

Educ: Wesleyan Univ, BA, 37; Albany Med Col, MD, 42. Prof Exp: Asst psychol, Wesleyan Univ, 38; intern, US Naval Hosp, Newport, RI, 42-43; res neurol, Jefferson Hosp, Philadelphia, 46-47; staff neurologist, US Naval Hosp, 47-50; instr, US Naval Med Sch, 50-53; assoc prof, 54-59, PROF NEUROL, JEFFERSON MED COL, 59-, DIR NEUROPATH LAB, 54- Concurrent Pos: Fel neuropath, US Armed Forces Inst Path, 53-54; clin asst prof, Sch Med & staff neurologist, Univ Hosp, Georgetown Univ, 50-54; consult, Vet Admin Hosps, Lebanon, Pa, 55- & Coatesville, 58-, NJ State Hosp, Ancora, 59-, US Naval Hosp & Eastern Pa Psychiat Inst, Philadelphia, 60- Mem: AAAS; Am Neurol Asn; Am Acad Neurol; Am Asn Neuropath; Asn Res Nerv & Ment Dis. Res: Cerebral vascular disease. Mailing Add: Dept of Neurol Jefferson Med Col Philadelphia PA 19107

BERRY, RICHARD LEE, b Shelby, Ohio, July 28, 42; m 64; c 2. MEDICAL ENTOMOLOGY, INSECT TAXONOMY. Educ: Univ Notre Dame, BS, 64; Tulane Univ, MS, 67; Ohio State Univ, PhD(entom), 70. Prof Exp: MED ENTOMOLOGIST, OHIO DEPT HEALTH, VECTOR BORNE DIS UNIT, 70- Mem: Entom Soc Am; Am Mosquito Control Asn; Am Soc Trop Med & Hyg; Sigma Xi. Res: Epidemiology of mosquito borne encephalitis viruses; Arborvirus vector potential of mosquitoes; biology and ecology of mosquitoes; biological control of mosquitoes; taxonomy of Tenebrionidae. Mailing Add: Ohio Dept of Health Vector Borne Dis Unit PO Box 2568 Columbus OH 43216

BERRY, RICHARD STEPHEN, b Denver, Colo, Apr 9, 31; m 55; c 3. PHYSICAL CHEMISTRY. Educ: Harvard Univ, AB, 52, AM, 54, PhD(chem), 56. Prof Exp: Instr chem, Harvard Univ, 56-57 & Univ Mich, 57-60; asst prof, Yale Univ, 60-64; assoc prof, 64-67, PROF CHEM, UNIV CHICAGO, 67- Concurrent Pos: Alfred P Sloan fel, 62-66; guest prof, Copenhagen Univ, 67; Arthur D Little prof, Mass Inst Technol, 68; sect lectr, Int Union Pure & Appl Chem Cong, Sydney, 69; Phillips lectr, Haverford Col, 69; Seydel-Wooley lectr, Ga Inst Technol, 71; Guggenheim fel, 73; dir, Bull Atomic Scientists, 75-76. Mem: Fel Am Phys Soc; Am Chem Soc. Res: Atomic-molecular processes; spectroscopy; resource management. Mailing Add: Dept of Chem Univ of Chicago Chicago IL 60637

BERRY, RICHARD WALLACE, b Seattle, Wash, July 4, 14; m 52; c 2. FORESTRY. Educ: Ore State Univ, BS, 41, MS, 61, PhD(forest mgt), 62. Prof Exp: Forester fire protection, State Bd Forestry, Ore, 41-45, forest conserv, 45-49, res dir forest res, 49-57; res dir, Ore Forest Lands Res Ctr, 57-59; from asst prof to assoc prof forestry, 61-72, PROF FORESTRY, NORTHERN ARIZ UNIV, 72- Mem: Soc Am Foresters. Res: Arid land ecology; late Wisconsin Glacial Period to present geochronology. Mailing Add: Dept of Forestry Northern Ariz Univ Flagstaff AZ 86001

BERRY, RICHARD WARREN, b Quincy, Mass, June 21, 33; m 58; c 3. MARINE GEOLOGY, MARINE GEOCHEMISTRY. Educ: Lafayette Col, BS, 55; Wash Univ, MA, 57, PhD(geochem), 63. Prof Exp: Instr geol, Trinity Col, Conn, 59-61; from asst prof to assoc prof, 61-71, PROF GEOL, SAN DIEGO STATE UNIV, 71-, ASSOC DIR CTR MARINE STUDIES, 75- Concurrent Pos: Fulbright prof, Univ Baghdad, 65-66; Royal Norweg Coun Indust & Sci Res fel, Univ Oslo, 68-69; consult, Nat Coun Educ Geol Sci. Mem: AAAS; fel Am Inst Chem; Mineral Soc Am; Geol Soc Am; Nat Asn Geol Teachers. Res: Mineralogy and geochemistry of clay sized ocean bottom sediments and their relationship to cationic concentrations in sea water; distribution of unconsolidated quaternary sediments on the continental shelf. Mailing Add: Dept of Geol Sci San Diego State Univ San Diego CA 92182

BERRY, ROBERT EDDY, b East Prairie, Mo, Jan 23, 30; m 51; c 3. BIOCHEMISTRY. Educ: Vanderbilt Univ, AB, 51; Univ Mo, MS, 57, PhD(agr biochem), 59. Prof Exp: From asst to instr agr biochem, Univ Mo, 55-59; appl res chemist, Nestle Co, Ohio, 59-63; invests head, US Fruit & Veg Prod Lab, 63-72, DIR, US SUBTROP PROD LAB, 72- Mem: AAAS; Am Chem Soc; Inst Food Technol. Res: New citrus products; pollution abatement; space foods; analysis of natural products; development of food processing, especially methods of dehydration; chemical changes in foods processing and storage; chemistry of citrus; nucleotides and nucleic acids. Mailing Add: US Subtrop Prod Lab PO Box 1909 Winter Haven FL 33880

BERRY, ROBERT JOHN, b Belleville, Ont, Dec 11, 29; m 55; c 2. SOLID STATE PHYSICS. Educ: Queen's Univ, Ont, BSc, 51, MSc, 52; Ottawa Univ, PhD(solid state physics), 72. Prof Exp: Assoc res officer, 52-69, SR RES OFFICER, HEAT & SOLID STATE PHYSICS, NAT RES COUN CAN, 69- Mem: Can Asn Physicists. Res: Temperature standards and scales; resistance thermometry; electrical restivity of metals. Mailing Add: Physics Div Nat Res Coun Ottawa ON Can

BERRY, ROBERT WADE, b Granbury, Tex, July 21, 30; m 55; c 3. PLANT PATHOLOGY. Educ: Tex A&M Univ, 52, MS, 60; Univ Wis, PhD(plant path), 63. Prof Exp: County agr agent, 54-58, AREA PLANT PATHOLOGIST, TEX AGR EXTEN SERV, 63- Mem: Am Phytopath Soc. Res: Disease control in plants. Mailing Add: Tex Agr Ext Serv Rte 3 Box 213AA Lubbock TX 79401

BERRY, ROBERT WALTER, b Atlanta, Ga, Oct 27, 28; m 54; c 2. INORGANIC CHEMISTRY, PHYSICAL CHEMISTRY. Educ: Clemson Agr Col, BS, 50; Mich State Univ, PhD(chem), 56. Prof Exp: Mem tech staff, Bell Tel Labs, 56-67, HEAD THIN FILM TECHNOL DEPT, BELL LABS, 67- Mem: Am Vacuum Soc; Inst Elec & Electronic Engrs; Sigma Xi. Res: Thin metallic and dielectric films as applied to electrical components. Mailing Add: Bell Labs 555 Union Blvd Allentown PA 18103

BERRY, ROBERT WAYNE, b Gilmer, Tex, Sept 14, 44. NEUROBIOLOGY. Educ: Calif Tech Inst, BS, 67; Univ Ore, MS, 68, PhD(biol), 70. Prof Exp: ASST PROF ANAT, SCH MED, NORTHWESTERN UNIV, CHICAGO, 73- Concurrent Pos: USPHS fel neuropath, Albert Einstein Col Med, 70-71; Nat Inst Neurol Dis & Stroke fel, Calif Inst Technol, 71-72. Mem: Soc Neurosci; Soc Gen Physiol. Res: Neuronal protein and RNA synthesis. Mailing Add: Dept of Anat Northwestern Univ Med Sch Chicago IL 60611

BERRY, ROY ALFRED, JR, b New Hebron, Miss, Dec 11, 33; m 57; c 2. ORGANIC CHEMISTRY. Educ: Miss Col, BS, 56; Univ NC, PhD(org chem), 62. Prof Exp: NSF res fel org chem, Univ Fla, 61-62; asst prof, 62-69, PROF CHEM, MILLSAPS COL, 69- Concurrent Pos: R J Reynolds res fel, 59-60; Petrol Res Fund asst, 60-61. Mem: Am Chem Soc. Res: The basicities of the ferrocenylazobenzenes halogenation decarboxylation. Mailing Add: Dept of Chem Millsaps Col Jackson MS 39210

BERRY, SPENCER JULIAN, b Quincy, Mass, May 25, 33; m 57; c 3. INSECT PHYSIOLOGY. Educ: Williams Col, BA, 55; Wesleyan Univ, MA, 57; Western Reserve Univ, PhD(biol), 65. Prof Exp: Technician biochem, Harvard Univ Huntington Labs, Mass Gen Hosp, 57-58; asst prof biol, 64-71, ASSOC PROF BIOL, WESLEYAN UNIV, 71- Res: Physiology of insect development, particularly at the cellular and subcellular level. Mailing Add: Dept of Biol Atwater Sci Ctr Wesleyan Univ Middletown CT 06457

BERRY, STANLEY Z, b NY, May 10, 30; m 59; c 2. PLANT BREEDING. Educ: Cornell Univ, BS, 52; Univ NH, MS, 53; Univ Calif, PhD(plant path), 56. Prof Exp: Plant pathologist, USDA, 56-60; plant breeder, Campbell Soup Co, 60-67; ASSOC PROF HORT, OHIO STATE UNIV & OHIO AGR RES & DEVELOP CTR, 67- Mem: Am Phytopath Soc; Am Soc Hort Sci. Res: Plant breeding and the utilization of disease resistance; tomato and vegetable breeding. Mailing Add: Dept of Hort Ohio Agr Res & Develop Ctr Wooster OH 44691

BERRY, WILLIAM BENJAMIN NEWELL, b Boston, Mass, Sept 1, 31; m 61; c 1. PALEOBIOLOGY. Educ: Harvard Univ, AB, 53, AM, 55; Yale Univ, PhD(geol), 57. Prof Exp: Asst prof geol, Univ Houston, 57-58; vis asst prof paleont, 58-60, from asst prof to assoc prof, 60-68, assoc dir mus paleont, 63-69, vchmn dept paleont, Univ, 67-69, PROF PALEONT, UNIV CALIF, BERKELEY, 68-, CHMN DEPT, 75-, CUR PALEOZOIC INVERT, MUS PALEONT, 60-, DIR, 75- Concurrent Pos: Guggenheim fel, 66-67. Mem: Paleont Soc; Norweg Geol Soc. Res: Graptolites; paleozoic biostratigraphy; carbonate petrography; community ecology. Mailing Add: Dept of Paleont Univ of Calif Berkeley CA 94720

BERRY, WILLIAM FRANCIS, b Patterson, NJ, May 2, 20; m 49; c 2. PETROGRAPHY, GEOLOGY. Educ: Univ Mass, BS, 50, MS, 52; Pa State Univ, PhD(geol), 63. Prof Exp: Asst geol, Univ Mass, 50-51, teaching fel, 51-52; res assoc, Pa State Univ, 52-61; proj coordr coal res, Bituminous Coal Res Inc, 61-68; PRES, W F BERRY ASSOCS, 68- Concurrent Pos: Mem, Int Comn Coal Petrog, Int Comt Petrog Nomenclature & Int Comt Petrog Stand, 61- Honors & Awards: Am Iron & Steel Inst Award, 61. Mem: AAAS; Geol Soc Am; Am Asn Petrol Geol; Am Chem Soc; Am Inst Mining, Metall & Petrol Eng. Res: Coal petrography as related to the production of metallurgical coke and coal combustion, classification, preparation and other uses. Mailing Add: 5134 Scenic Dr Murraysville PA 15618

BERRY, WILLIAM LEE, b Auburn, NY, Sept 1, 27; m 50; c 2. ORGANIC CHEMISTRY. Educ: Cornell Univ, BA, 51; Univ Mich, PhD(chem), 56. Prof Exp: Res chemist, 55-57, group leader, 57-60, dir org pigments res & develop, 60-63, prod mgr, Pigments Div, 63-67, from asst tech dir to tech dir, 67-71, dept mgr color pigments, 71-73, DEPT MGR PLASTICS, AM CYANAMID CO, 73- Mem: Am Chem Soc. Res: Synthetic organic chemistry; colored hetero and carbocyclic systems and their utilization; high performance thermoplastic and thermoset resins. Mailing Add: Indust Chem & Plastics Div Am Cyanamid Co Wayne NJ 07470

BERRYHILL, DAVID LEE, b Council Bluffs, Iowa, Mar 9, 44; m 68. BACTERIOLOGY. Educ: Simpson Col, BA, 66; Iowa State Univ, MS, 69, PhD(bact), 71. Prof Exp: Asst prof, 71-75, ASSOC PROF BACT, NDAK STATE UNIV, 75- Mem: Am Pub Health Asn; Am Soc Microbiol; Can Soc Microbiologists; Genetics Soc Am; Sigma Xi. Res: Bacterial genetics; bacteriophage. Mailing Add: Dept of Bact NDak State Univ Fargo ND 58102

BERRYHILL, VIRGINIA FARMER, b Columbia, SC, Sept 4, 35; m 59; div; c 2. CELL PHYSIOLOGY, CHEMICAL EMBRYOLOGY. Educ: Univ SC, BS, 57, MS, 59. Prof Exp: Asst prof biol, Lenoir-Rhyne Col, 63-64; from asst prof to assoc prof, Wesleyan Col, Ga, 64-70; SCI TEACHER, JAMES H HAMMOND ACAD, 70- Mem: AAAS. Res: Biochemical cytology and embryology; pharmalogical testing on embryos for micro or macro anatomical or detectable biochemical anomalies and stimuli to precocious development. Mailing Add: 3934 Montgomery Ave Columbia SC 29205

BERRYHILL, WALTER REECE, b Charlotte, NC, Oct 14, 00; m 30; c 2. MEDICINE. Educ: Univ NC, BA, 21; Harvard Univ, MD, 27. Hon Degrees: ScD, Davidson Col, 56. Prof Exp: Intern, Boston City Hosp, 27-29; actg assoc prof physiol, Univ NC, 29-30; resident med, Univ Hosps, Cleveland, Ohio, 30-31; instr, Western Reserve Univ, 31-33; from asst prof to prof, 33-68, from asst dean to dean, 38-64, dir div educ & res in community care, 65-69, sr consult, Div Educ & Res in Community Med Care, 71-75, EMER DEAN SCH MED, UNIV NC, CHAPEL HILL, 64-, EMER SARAH GRAHAM KENAN PROF MED, 68- Honors & Awards: O Max Gardner Award, Bd Trustees, Univ NC & Distinguished Citizen Award, Gov of NC, 64. Mem: AAAS; fel AMA; fel Am Col Physicians; Asn Am Med Cols. Res: Internal medicine and medical education. Mailing Add: PO Box 866 Chapel Hill NC 27514

BERRYMAN, ALAN ANDREW, b Tanganyika, Africa, Jan 5, 37; m 68. ENTOMOLOGY. Educ: Univ London, BSc, 59; Univ Calif, Berkeley, MS, 61, PhD(entom), 65. Prof Exp: Asst prof entom, 64-69, ASSOC PROF ENTOM, WASH STATE UNIV, 69- Mem: Entom Soc Am; Entom Soc Can. Res: Population dynamics of bark beetles; mathematical and computer models of population dynamics and sterile male control theory; resistance of conifers to insect and fungus invasion. Mailing Add: Dept of Entom Wash State Univ Pullman WA 99163

BERRYMAN, GEORGE HUGH, b South Shields, Eng, Apr 3, 14; nat US; m 39; c 3. NUTRITION, RESEARCH ADMINISTRATION. Educ: Univ Scranton, BS, 35; Pa State Col, MS, 36; Univ Minn, PhD(biochem, human nutrit), 41; Univ Chicago, MD, 50. Prof Exp: Asst animal nutrit, Univ Ill, 37-38; chief div food & nutrit, Army Med Sch, Washington, DC, 41-44; commanding officer & dir res, Med Nutrit Lab, Univ Chicago, 44-46; head nutrit br, Qm Food & Container Inst, Chicago, 48-49; rotating intern, USPHS Hosp, Staten Island, 50-51; assoc, 51-54, head clin invest, 54-58, dir med sci proj, 58-60, dir clin develop, 60-64, med dir, 65-70, V PRES MED AFFAIRS, DEPT MED, ABBOTT LABS, 70-; CLIN ASSOC PROF MED, SCH MED, UNIV ILL, 62- Concurrent Pos: Clin asst prof med, Univ Ill, 53-62. Mem: AAAS; Am Med Asn; Am Soc Clin Nutrit; Am Acad Allergy; fel Am Col Physicians. Res: Nutritional status; appraisal; relation of food; food composition; military nutrition; metabolic disease; allergy; drug reactions. Mailing Add: Col of Med Univ of Ill 1737 W Polk St Chicago IL 60680

BERRYMAN, JACK HOLMES, b Salt Lake City, Utah, July 28, 21; m 41; c 2. RESOURCE MANAGEMENT, FISH AND GAME MANAGEMENT. Educ: Univ Utah, BS, 47, MS, 48. Prof Exp: Proj leader big game, Utah State Dept Fish & Game, 47-48, actg coordr wildlife, 48-50; asst regional supvr, US Fish & Wildlife Serv, NMex, 50-53, Minn, 53-59; assoc prof wildlife, Utah State Univ, 59-65; chief div wildlife serv, Bur Sport Fisheries & Wildlife, US Dept Interior, 65-74; actg dep assoc dir, US Fish & Wildlife Serv, 74, CHIEF DIV TECH ASSISTANCE, US FISH & WILDLIFE SERV, WASHINGTON, DC, 74- Concurrent Pos: Consult, Off Secy Interior, 62-64 & Gov Comt State Recreation Plan, 64-; del, White House Conf Conserv, 62. Honors & Awards: Minn Award, 60. Mem: Wildlife Soc (vpres, 60-62, pres, 64); NY Acad Sci; Am Fisheries Soc; Sigma Xi. Res: Relationships of land and resource use; energy developments; population requirements to fish and wildlife resource management and use. Mailing Add: 10503 Linfield St Fairfax VA 22030

BERS, LIPMAN, b Riga, Latvia, May 22, 14; m 38; c 2. MATHEMATICS. Educ: Univ Prague, Dr rer nat(math), 38. Prof Exp: Asst dynamics & mech, Brown Univ, 42-43, res instr, 43-44, sr res mathematician, 44-45; from asst prof to assoc prof math, Syracuse Univ, 45-51; mem staff, Inst Advan Study, 49-51; vis prof, NY Univ, 51-53, prof, 53-64, chmn dept grad sch, 59-64; PROF MATH, COLUMBIA UNIV, 64- Concurrent Pos: Guggenheim fel & Fulbright award, 59-60; ed, Trans, Am Math Soc,

59-64; with Nat Bur Standards & Nat Adv Comt Aeronaut. Mem: Nat Acad Sci; fel Am Acad Arts & Sci; Am Math Soc (vpres, 64-65). Res: Complex function theory and its generalizations; partial differential equations; gas dynamics. Mailing Add: Dept of Math Columbia Univ New York NY 10027

BERSHADER, DANIEL, b New York, NY, Mar 14, 23; c 2. AEROPHYSICS. Educ: Brooklyn Col, AB, 42; Princeton Univ, MA, 46, PhD(physics), 48. Prof Exp: Instr physics, Princeton Univ, 43-44; flight res electronic engr, Bell Aircraft Corp, NY, 44; physicist, Naval Ord Lab, Washington, DC, 44-45; instr physics, Palmer Phys Lab, Princeton Univ, 48-49; res assoc, Univ Md, 49-51, assoc res prof, 51-52; res assoc & assoc prof, Princeton Univ, 52-56; assoc prof, 56-64, PROF AEROPHYSICS, STANFORD UNIV, 64-, VCHMN DEPT, 74- Concurrent Pos: Mgr gas dynamics, Lockheed Res Lab, 56-64; distinguished vis prof, Syracuse Univ, 69. Mem: Fel Am Phys Soc; fel Am Inst Aeronaut & Astronaut; Am Soc Eng Educ; Am Asn Physics Teachers. Res: Kinetic processes in high speed air flow; optical methods in fluid dynamics; plasma dynamics; physics of planetary entry. Mailing Add: Dept of Aeronaut & Astronaut Stanford Univ Stanford CA 94305

BERSOHN, MALCOLM, b New York, NY, May 13, 25; m 64. SYNTHETIC ORGANIC CHEMISTRY, COMPUTER SCIENCES. Educ: Harvard Univ, BS, 43; Columbia Univ, MA, 57, PhD(chem), 60. Prof Exp: Res scientist spectros, Calif Res Corp, 60-62; asst prof paramagnetic resonance, 62-66, ASSOC PROF CHEM, UNIV TORONTO, 66- Concurrent Pos: Guggenheim fel, Comput Sci Dept, Stanford Univ, 69-71. Mem: Asn for Comput Mach. Res: Application of computer science to synthetic organic chemistry. Mailing Add: Dept of Chem Univ of Toronto Toronto ON Can

BERSOHN, RICHARD, theoretical chemistry, see 12th edition

BERSON, JEROME ABRAHAM, b Sanford, Fla, May 10, 24; m 46; c 3. ORGANIC CHEMISTRY. Educ: City Col New York, BS, 44; Columbia Univ, AM, 47, PhD(chem), 49. Prof Exp: Asst chemist, Hoffmann-La Roche, Inc, 44; lab asst, Columbia Univ, 46-49; Nat Res Coun fel, Harvard Univ, 49-50; from asst prof to prof chem, Univ Southern Calif, 50-63; prof, Univ Wis, 63-69; chmn dept, 71-74, PROF CHEM, YALE UNIV, 69- Concurrent Pos: Sloan fel, 57-61; NSF sr fel, 59-60; vis prof, Univ Calif, Los Angeles, 62, Univ Cologne, 65 & Univ Western Ont, 67; consult, Goodyear Tire & Rubber Co, 65-74; mem med chem study sect, NIH, 69-73; Sherman Fairchild distinguished scholar, Calif Inst Technol, 74-75; mem adv comt, Assembly Math & Phys Sci, Nat Res Coun, 75- Mem: Nat Acad Sci; Am Chem Soc; The Chem Soc; Am Acad Arts & Sci. Res: Synthetic and theoretical organic chemistry; mechanisms of organic reactions. Mailing Add: Dept of Chem Yale Univ New Haven CT 06520

BERSON, ROBERT CHAMBLISS, b Brownsville, Tenn, Sept 12, 12; m 40; c 2. MEDICINE. Educ: Vanderbilt Univ, BS, 34, MD, 37. Prof Exp: Intern surg serv, Vanderbilt Univ Hosp, 37-38, instr clin med, Sch Med, 42-47, asst prof & asst dean sch med, 52-55; asst resident, Med Serv, Baltimore City Hosp, Md, 38-39; asst prof med, Col Med, Univ Ill, 48-52; prof, dean & vpres univ, Med Col Ala, 55-62; dean, STex Med Sch, Univ Tex, 62-64; exec dir, Asn Am Med Cols, 65-69; CHIEF STAFF, BALTIMORE VET ADMIN HOSP, 70- Concurrent Pos: Asn dir, Surv Med Educ, 49-52. Mem: AMA; Asn Am Med Cols (pres, 63-64). Res: Medical education. Mailing Add: 3900 Loch Raven Blvd Baltimore MD 21218

BERSON, SOLOMON AARON, internal medicine, deceased

BERSTED, BRUCE HOWARD, b Chicago, Ill, Sept 25, 40; m 68; c 2. POLYMER PHYSICS. Educ: Beloit Col, BA, 63; Univ Ill, MS, 65, PhD(phys chem), 69. Prof Exp: RES CHEMIST, AMOCO CHEM CORP, 69- Concurrent Pos: App to write test procedure vapor pressure osmometry, Am Soc Testing & Mat, 75- Mem: Am Phys Soc. Res: Methods of molecular weight determination; polymer rheology and its relation to molecular weight distribution; dynamic mechanical properties of polymers and composites. Mailing Add: Amoco Chem Corp Res Ctr Bldg 503 Rm 2306 Naperville IL 60540

BERSTEIN, BARRY, b New York, NY, Nov 20, 30; m 54; c 2. APPLIED MATHEMATICS, MECHANICS. Educ: City Col New York, BS, 51; Ind Univ, MA, 54, PhD(math), 56. Prof Exp: Mathematician, Naval Res Lab, 51-52, 56-61; Nat Bur Stand, 61-66; PROF MATH, ILL INST TECHNOL, 66- Concurrent Pos: Vis assoc prof, Purdue Univ, 65-66; consult, E I du Pont de Nemours & Co, 68- & Nat Bur Stand, 69- Mem: Am Math Soc; Soc Natural Philos; Soc Rheol; Soc Eng Sci. Res: Rheology; differential equations; viscoelasticity; hypoelasticity; thermodynamics; blood tracer methods. Mailing Add: Dept of Math Ill Inst of Technol Chicago IL 60616

BERSTEIN, GREGOR, colloid chemistry, see 12th edition

BERSTEIN, IRVING AARON, chemistry, see 12th edition

BERSTEIN, ISRAEL, b Brichany, USSR, June 23, 26; m 70. MATHEMATICS. Educ: Univ Bucharest, MS, 54; Bucharest Inst Math, PhD(math), 58. Prof Exp: Res assoc math, Bucharest Inst Math, 54-58, sr res assoc, 58-59, calculator, 58-61; lectr, Israel Inst Technol, 61-62; from asst prof to assoc prof, 62-67, PROF MATH, CORNELL UNIV, 67- Mem: Am Math Soc; Israel Math Union. Res: Algebraic topology, especially homotopy theory. Mailing Add: Dept of Math Cornell Univ Ithaca NY 14850

BERT, MARK HENRY, b Lima, Peru, May 1, 16; US citizen; m 60; c 2. NUTRITIONAL BIOCHEMISTRY, STATISTICS. Educ: Nat Col Agr & Vet Sci, Lima, BS, 39; Univ Ill, Urbana, MS, 48, PhD(nutrit biochem), 55. Prof Exp: Specialist in flax fiber processing, Dept Agr, Peru, 40-41; tech dir, Desfibradora de Lino, SA, 41-43; Inter-Am Trade scholar flax fiber processing firms, Ore, 44-46; consult var indust firms & Dept Agr, Peru, 46-47; res biochemist lab, Corn Prod Co, Mass, 55-58, dir res lab, 58-63, biochemist in-chg biochem sect, NJ, 63-65; asst prof nutrit biochem, 65-68, asst prof nutrit & food & head dept, 68-70, ASSOC PROF NUTRIT & FOOD, UNIV MASS, AMHERST, 70- Concurrent Pos: Univ Mass res grant, 66-68. Mem: Animal Nutrit Res Coun; Am Chem Soc; Am Statist Asn; Inst Food Technologists; fel Am Inst Chemists. Res: Amino acid metabolism; food irradiation effects on protein quality, vitamins, enzyme activity, lipids; nutritive value of algae; vitamin metabolism; statistical design and analysis; computer applications to nutrition; atherosclerosis; radioisotopic techniques. Mailing Add: Dept of Nutrit & Food Skinner Hall Univ of Mass Amherst MA 01002

BERTA, DOMINIC ANDREW, b New Kensington, Pa, Sept 20, 40; m 62; c 2. PLASTIC CHEMISTRY. Educ: St Vincent Col, Pa, BS, 63; WVa Univ, MS, 64, PhD(phys chem), 66. Prof Exp: Res assoc, Univ SC, 66-67; SR RES CHEMIST PHYS CHEM, HERCULES INC, 67- Res: Development of packaging film specifically related to polypropylene and oriented polypropylene film. Mailing Add: Hercules Res Ctr Hercules Inc Hercules Rd Wilmington DE 19899

BERTA, MICHAEL A, physical chemistry, nuclear chemistry, see 12th edition

BERTALANFFY, FELIX D, b Vienna, Austria, Feb 20, 26; nat Can; m 54. ANATOMY, HISTOLOGY. Educ: McGill Univ, MSc, 51, PhD(anat), 54. Prof Exp: Res asst, McGill Univ, 53-54; from asst prof to assoc prof, 55-64, PROF ANAT, UNIV MAN, 64- Mem: Am Asn Cancer Res; Am Asn Anatomists; Can Asn Anatomists; Can Soc Cytol; fel Royal Micros Soc. Res: Cancer and cancer research; histochemistry and cytochemistry; exfoliative cytology; cytodynamics of normal and cancerous tissues; chemotherapeutic agents; respiratory system; fluorescence microscopy. Mailing Add: Dept of Anat Univ of Man Fac of Med & Dent Winnipeg MB Can

BERTANI, GIUSEPPE, b Como, Italy, Oct 23, 23; nat US; m 54; c 2. MOLECULAR GENETICS. Educ: Univ Milan, DrNatSc, 45. Prof Exp: Res fel biol, Zool Sta, Naples, 45-46; asst zool, Univ Milan, 46-47 & Univ Zurich, 47-48; res fel genetics, Carnegie Inst, Cold Spring Harbor, NY, 48-49; res assoc bact, Ind Univ, 49-50 & Univ Ill, 50-54; sr res fel biol, Calif Inst Technol, 54-57; assoc prof med microbiol, Med Sch, Univ Southern Calif, 57-60; vis prof microbial genetics, 60-64, PROF MICROBIAL GENETICS, KAROLINSKA INST, SWEDEN, 64- Mem: AAAS; Genetics Soc Am; Am Soc Microbiol; Brit Soc Gen Microbiol; Europ Molecular Biol Orgn. Res: Genetics of bacteria and bacterial viruses. Mailing Add: Karolinska Inst 10401 Stockholm Sweden

BERTANI, LAURA MARIE, b New York, NY, Aug 4, 41. BIOCHEMISTRY, ORGANIC CHEMISTRY. Educ: Col New Rochelle, BA, 63; Fordham Univ, MS, 65, PhD(chem), 68. Prof Exp: Instr, 68-71, res assoc, 71-73, RES ASST PROF MED, MT SINAI SCH MED, 71- Mem: AAAS; Am Chem Soc; Am Inst Chemists. Res: Catecholamine metabolism in normal and disease states. Mailing Add: Mt Sinai Sch of Med Fifth Ave at 100th St New York NY 10029

BERTANI, LILLIAN ELIZABETH, b Ind, July 9, 31; m 54; c 2. MICROBIOLOGY. Educ: Univ Mich, BS, 53; Calif Inst Technol, PhD(virol), 57. Prof Exp: Res assoc med microbiol, Univ Southern Calif, 57-60; res assoc microbial genetics, Karolinska Inst, Sweden, 61-65; Med Res Coun fel, Swedish Med Res Coun, 66-75; DOCENT MED MICROBIOL, KAROLINSKA INST, 75- Concurrent Pos: USPHS fel, 60-61; docent microbiol, Univ Stockholm, 65. Res: Temperate bacteriophages; lysogeny; virology. Mailing Add: Karolinska Inst 10401 Stockholm Sweden

BERTAUT, EDGARD FRANCIS, b Chicago, Ill, May 23, 31; m 57; c 5. COMPUTER SCIENCE. Educ: Loyola Univ, Ill, BSc, 53; Carnegie Inst Technol, MS, 56, PhD(inorg chem), 58. Prof Exp: Instr chem, Carnegie Inst Technol, 57-58; asst prof, Pa State Univ, 58-60, fel mineral sci, 60-62; asst prof chem, Univ Detroit, 62-68; assoc prof, 68-70, prof natural sci & chmn div, 70-71, PROF COMPUT SCI & CHMN PROG, FED CITY COL, 74- Mem: Am Chem Soc; Am Ceramic Soc; Asn Comput Mach; AAAS. Res: High temperature spectra; televised instruction; application of computers to chemistry; computer graphics; operating systems. Mailing Add: 929 E St NW Washington DC 20004

BERTELL, ROSALIE, b Buffalo, NY, Apr 4, 29. Educ: D'Youville Col, BA, 51; Cath Univ Am, MA, 59, PhD(math), 66. Prof Exp: Teacher Bishop O'Hern High Sch, NY, 56-57; asst math, Cath Univ Am, 57-58; instr, Sacred Heart Jr Col, Pa, 58-62, assoc prof, 65-68; coordr & teacher, D'Youville Acad, Diocese Atlanta, 68-69; assoc prof math, D'Youville Col, 69-73; SR CANCER RES SCIENTIST, ROSWELL PARK MEM INST, 73-; ASST RES PROF, GRAD SCH, STATE UNIV NY, BUFFALO, 74- Concurrent Pos: Cancer res grant & cancer res scientist, Roswell Park Mem Inst, 70-73, sr res sci consult, 75- Mem: Biomet Soc; Hasting Inst Soc, Ethics & Life Sci; NY Acad Sci. Res: Mathematical statistics; analysis; measure theory; aging effect in humans associated with exposure to ionizing radiation; updating relative risk methodology for biomedical applications; life-style and chronic diseases. Mailing Add: Roswell Park Mem Inst Buffalo NY 14263

BERTELLI, DOMENICK J, organic chemistry, see 12th edition

BERTELSON, ROBERT CALVIN, b Milwaukee, Wis, Nov 5, 31; m 60; c 2. SYNTHETIC ORGANIC CHEMISTRY. Educ: Univ Wis, BS, 52; Mass Inst Technol, PhD(org chem), 57. Prof Exp: Sr res chemist, Nat Cash Register Co, 57-61, group leader, 61-74; OWNER, CHROMA CHEM, 74- Concurrent Pos: Sr res assoc, Nat Acad Sci-Nat Res Coun, 72-74. Mem: Am Chem Soc. Res: Organic synthesis; organic photochemistry; non-silver photographic systems; photochromism. Mailing Add: 5312 Bliss Pl Dayton OH 45440

BERTHIAUME, LAURENT, b Montreal, Que, July 29, 41; m 67; c 3. VIROLOGY. Educ: Univ Montreal, BSc, 65, MSc, 69, PhD(microbiol), 72. Prof Exp: HEAD ELECTRON MICROS LAB, ARMAND FRAPPIER INST, 66-, ASST PROF VIROL, 74- Concurrent Pos: Mem, Bact Virus Subcomt, Int Comt Taxon Viruses, 76- Mem: Can Soc Microbiologists; Micros Soc Can. Res: Oncogenic cell transformation by herpes simplex and/or latent viruses; rapid diagnosis of viral infections by electron microscopy techniques. Mailing Add: 531 Des Prairies Blvd Ville de Laval Quebec PQ Can

BERTHIAUME, PIERRE, algebra, see 12th edition

BERTHOLD, ROBERT, JR, b Paterson, NJ, Aug 2, 41; m 66; c 1. APICULTURE, ENTOMOLOGY. Educ: Juniata Col, BS, 63; Rutgers Univ, MS, 65; Pa State Univ, PhD(entom), 68. Prof Exp: Res asst entom, Rutgers Univ, 63-66 & Pa State Univ, 66-68; ASST PROF BIOL, DEL VALLEY COL, 68- Concurrent Pos: Apiary inspector, Pa Dept Agr. Mem: Entom Soc Am; Am Entom Soc. Res: Insect behavior; German cockroach; honey bee; development of new honey products including marketing studies. Mailing Add: Dept of Biol Delaware Valley Col Doylestown PA 18901

BERTHOLF, DENNIS E, b Harper, Kans, Aug 19, 41; m 62; c 3. MATHEMATICS. Educ: Univ Kans, BS, 63; NMex State Univ, MA, 65, PhD(math), 68. Prof Exp: Asst prof math, 68-74, ASSOC PROF MATH, OKLA STATE UNIV, 74- Mem: Am Math Soc; Math Asn Am. Res: Abelian group theory; homological algebra. Mailing Add: Dept of Math & Statist Okla State Univ Stillwater OK 74074

BERTHOLF, LLOYD BERNARD, b Marion, Ohio, Oct 19, 15; m 39. METEOROLOGY, OCEANOGRAPHY. Educ: Univ Minn, BS, 38. Prof Exp: Apprentice meteorologist, Pan Am World Airways, 39-40, asst meteorologist, 40-42, sect meteorologist, Brazil, 42-46, asst meteorologist, 46-51; meteorologist, 51-54, supvry oceanogr, 54-71, dir oceanog surv dept, Phys Sci Adminr, 71-74, DIR MAT MGT DIRECTORATE, LOGISTIC MGT ADMINR, US NAVAL OCEANOG OFF, 74- Honors & Awards: Sustained Super Performance, US Naval Oceanog Off, 73. Res: Military applications of oceanography; administering, coordination and planning collection; evaluation, analysis and processing of oceanographic survey data. Mailing Add: Naval Oceanog Off Washington DC 20373

BERTHOLF, LLOYD MILLARD, b Kechi, Kans, Dec 15, 99; m 21; c 2.

APICULTURE. Educ: Southwestern Col, Kans, AB, 21, LLD, 59; Johns Hopkins Univ, AM, 25, PhD(zool, physiol), 28; DHL, Ewha Woman's Univ, Korea, 70. Hon Degrees: DHumanities, Ill Wesleyan Univ, 75. Prof Exp: Instr biol, NC Col Women, 22-24; prof, Western Md Col, 24-48, dean freshmen, 33-39, dean fac, 39-48; prof biol & dean col, Col Pac, 48-56, acad vpres, 57-58; pres, 58-68, EMER PRES, ILL WESLEYAN UNIV, 68-; ACTG EXEC DIR, MID-ILL AREA WIDE HEALTH PLANNING CORP, 71- Concurrent Pos: Nat Res fel, Ger, 30-31; pres, Cent States Col Asn, 68-69; consult, Ewha Woman's Univ, Korea, 69-70. Mem: AAAS; Am Soc Zoologists; Entom Soc Am; Am Inst Biol Sci. Res: Molts, digestion of carbohydrates and reactions to light in honey bees; physiological and stimulative effect of ultraviolet in insects; effects of insecticides on honey bees; metamorphosis in tunicates. Mailing Add: 307 Phoenix Ave Bloomington IL 61701

BERTHRONG, MORGAN, b Aurora Hills, Va, July 17, 18; m 43; c 6. MEDICINE. Educ: Harvard Med Sch, MD, 43. Prof Exp: Resident path, Johns Hopkins Univ & Hosp, 46-50, from instr to assoc prof, 49-53; prof & head dept, 59-61, CLIN PROF PATH, SCH MED, UNIV COLO, 61-, PATHOLOGIST, PENROSE HOSP, COLORADO SPRINGS, 54-, DIR LABS, 67- Concurrent Pos: Consult, Ft Carson Army Hosp, 55-59 & Fitzsimmons Army Hosp, 64-; vis prof path & actg head dept, Sch Med, Stanford Univ, 66-67. Mem: Am Asn Path & Bact; Am Soc Clin Path; Am Thoracic Soc; AMA; Col Am Path. Res: Gneral anatomic pathology, especially diseases of lungs, liver, kidneys and congenital heart; experiments with tuberculosis, hypersensitivity and so-called collagen vascular diseases. Mailing Add: Dept of Path Penrose Hosp Colorado Springs CO 80907

BERTIE, JOHN E, b London, Eng, Mar 24, 36; Can citizen. PHYSICAL CHEMISTRY, CHEMICAL PHYSICS. Educ: Univ London, PhD(phys chem), 60. Prof Exp: Fel chem, Div Appl Chem, Nat Res Coun Can, 60-62, asst res off, 62-65, assoc res off, 65-67; assoc prof, 67-75, PROF CHEM, UNIV ALTA, 75- Mem: Am Phys Soc; Optical Soc Am; Chem Inst Can; Spectros Soc Can. Res: Vibrational spectra of crystals; use of vibrational spectroscopy, x-ray diffraction and dielectric relaxation to study high pressure phases of molecular crystals. Mailing Add: Dept of Chem Univ of Alta Edmonton AB Can

BERTIN, ERNEST PETER, b Seattle, Wash, Nov 2, 20. PHYSICAL CHEMISTRY, INORGANIC CHEMISTRY. Educ: Gonzaga Univ, 44, MA, 45; Alma Col, STD(theol), 52; Univ Notre Dame, PhD(chem), 57. Prof Exp: Teacher prep sch, Seattle, 45-48; assoc prof chem, 57-67, PROF CHEM, SEATTLE UNIV, 67- Concurrent Pos: Fel radiation lab, Univ Notre Dame, 62-63; res appointee radiation chem, Hanford Proj, Wash, 70-71. Mem: Am Chem Soc. Res: Radiation chemistry. Mailing Add: Dept of Chem Seattle Univ Seattle WA 98122

BERTIN, EUGENE P, b Williamsport, Pa, Oct 29, 21. ANALYTICAL CHEMISTRY, INORGANIC CHEMISTRY. Educ: Univ Ill, BS, 48, MS, 49, PhD(chem), 52. Prof Exp: Electronic technician, Sylvania Elec Prod, Inc, 43-46; instr chem, Univ Ill, 52-53; design & develop engr, Electron Tube Div, RCA Corp, 53-64, sr engr, Com Receiving Tube & Semiconductor Div, 64-69, MEM TECH STAFF, RCA LABS, 69- Mem: AAAS; Am Inst Chem; Am Soc Testing & Mat; Microbeam Anal Soc; Am Chem Soc. Res: X-ray spectrometry; electron-probe microanalysis. Mailing Add: Rm E-120 RCA Labs Princeton NJ 08540

BERTIN, HENRY JOHN, JR, b San Francisco, Calif, Oct 30, 25. PHYSICAL ORGANIC CHEMISTRY. Educ: Univ Calif, BS, 50; Stanford Univ, PhD(chem), 57. Prof Exp: Tech asst chem, Hanford Atomic Energy Plant, Gen Elec Co, 50-53; instr, 57-59, from asst prof to assoc prof, 59-68, PROF CHEM, SAN FRANCISCO STATE COL, 68- Mem: Am Chem Soc. Res: Mechanistic studies on carbonyl addition reactions of aryl-substituted tetralones and indanones; infrared spectroscopy. Mailing Add: Dept of Chem San Francisco State Col 1600 Holloway Ave San Francisco CA 94132

BERTINI, HUGO W, b Chicago, Ill, Dec 12, 26; m 54; c 3. PHYSICS. Educ: Northwestern Univ, MS, 51; Univ Tenn, PhD(physics), 62. Prof Exp: PHYSICIST, OAK RIDGE NAT LAB, 53- Mem: Am Phys Soc; Sigma Xi. Res: Nuclear reactor physics; high energy nuclear reactions. Mailing Add: PO Box X Oak Ridge Nat Lab Oak Ridge TN 37830

BERTINO, JOSEPH R, b Port Chester, NY, Aug 16, 30; m 56; c 4. PHARMACOLOGY, MEDICINE. Educ: State Univ NY Downstate Med Ctr, MD, 54. Hon Degrees: MA, Yale Univ, 69. Prof Exp: Intern, Grad Hosp, Univ Pa, 54-55; resident internal med, Vet Admin Hosp, Philadelphia, 55-56; asst prof pharmacol, 61-64, assoc prof, 64-69, PROF MED & PHARMACOL, SCH MED, YALE UNIV, 69- Concurrent Pos: USPHS res fel hemat & biochem, Sch Med, Univ Wash, 58-61, career develop award, Nat Cancer Inst, 64-74; consult, Nat Serv Ctr, 64- Mem: Am Soc Hemat; Am Asn Cancer Res; Am Fedn Clin Res; Am Soc Clin Invest; Am Soc Biol Chemists. Res: Leukocyte enzymes; folic acid and vitamin B12 metabolism; cancer chemotherapy. Mailing Add: Dept of Pharmacol Yale Univ Sch of Med New Haven CT 06510

BERTINUSON, TORVALD ARTHUR, b Alexander, NDak, Sept 12, 23; m 47; c 6. AGRONOMY. Educ: Univ Denver, BS, 50, MS, 51. Prof Exp: Instr agron, Univ Mass, 51-54; instr pub sch, Washington, DC, 54-55; dir res, Cullman Bros Tobacco Co, Inc, 55-64; DIR AGR RES, CONSOL CIGAR CORP, 64- Mem: Am Soc Agron; Crop Sci Soc Am; Soil Sci Soc Am. Res: Fertility studies on shade tobacco soils; maturity studies of shade tobacco; wrapper tobacco breeding; exploration of new wrapper tobacco areas; technical assistance cigar tobacco area Latin America. Mailing Add: Consol Cigar Corp 131 Oak St Glastonbury CT 06033

BERTKE, ELDRIDGE MELVIN, histology, pathology, see 12th edition

BERTLES, JOHN F, b Spokane, Wash, June 8, 25; m 48; c 3. INTERNAL MEDICINE, HEMATOLOGY. Educ: Yale Univ, BS, 45; Harvard Med Sch, MD, 52; Am Bd Internal Med, dipl, 61. Prof Exp: Intern & asst resident med, Presby Hosp, New York, 52-55; instr, Harvard Med Sch, 59-61; DIR HEMAT DIV & CHIEF HEMAT CLIN, ST LUKE'S HOSP CTR, 62-; PROF MED, COL PHYSICIANS & SURGEONS, COLUMBIA UNIV, 74- Concurrent Pos: USPHS res fel, Sch Med & Dent, Univ Rochester, 55-56; USPHS res fel, 55-57; res fel, Harvard Med Sch, 56-59; from asst attend physician to attend physician, St Luke's Hosp Ctr, 62-; from asst clin prof to assoc clin prof med, Col Physicians & Surgeons, Columbia Univ, 62-, assoc prof, 71-74. Mem: Am Soc Clin Invest; Am Physiol Soc; fel Am Col Physicians; Am Soc Hemat; Am Fedn Clin Res. Res: Human hemoglobinopathies. Mailing Add: Hemat Div St Luke's Hosp Ctr New York NY 10025

BERTOCCI, UGO, electrochemistry, physical chemistry, see 12th edition

BERTOLACINI, RALPH JAMES, b Pawtucket, RI, Aug 8, 25; m 53; c 3. INORGANIC CHEMISTRY. Educ: Univ RI, BS, 49; Mich State Univ, MS, 51. Prof Exp: Asst, Univ RI, 48-49 & Mich State Univ, 49-51; sr res scientist, Am Oil Co, 51-68, PROJ MGR & RES ASSOC, RES & DEVELOP DIV, AMOCO OIL, 68- Mem: Am Chem Soc. Res: Catalysis; analytical chemistry. Mailing Add: Amoco Oil Res Dept PO Box 400 Naperville IL 60540

BERTOLINI, MAURICE JOSEPH, biochemistry, organic chemistry, see 12th edition

BERTON, JOHN ANDREW, b Villa Park, Ill, June 22, 30; m 52; c 5. GEOMETRY. Educ: Univ Ill, AB, 55, MA, 57, PhD(geom), 64. Prof Exp: From instr to asst prof math, Ind State Col, 59-64; assoc prof, Ripon Col, 64-67; chmn dept, 67-75, PROF MATH, OHIO NORTHERN UNIV, 67- Mem: AAAS; Am Math Soc; Math Asn Am; Soc Indust & Appl Math. Res: Differential geometry. Mailing Add: Dept of Math Ohio Northern Univ Ada OH 45810

BERTON, WILLIAM MORRIS, b Fresno, Calif, Feb 8, 24; m 49; c 3. PATHOLOGY. Educ: Univ Calif, MD, 49. Prof Exp: Intern, US Naval Hosp, Mass, 49-50; res path, Methodist Hosp, Ind, 50-51; from intern to resident, Duke Univ Hosp, 51-54, instr, Sch Med, Duke Univ, 52-54; head lab serv & chief res, US Naval Hosp, Camp Pendleton, 55-56; from asst prof to prof path, Univ Tenn, 56-68; PROF PATH, UNIV NEBR MED CTR, OMAHA, 69- Mem: Asn Am Med Cols; Col Am Path; Am Soc Clin Path; Am Soc Exp Path; NY Acad Sci. Res: Chromatographic methods as applied to pathology and bacteriology. Mailing Add: 13136 Mason Ave Omaha NE 68154

BERTONCINI, PETER JOSEPH, theoretical chemistry, see 12th edition

BERTONI, HENRY LOUIS, b Chicago, Ill, Nov 15, 38; c 2. ELECTROPHYSICS. Educ: Northwestern Univ, BS, 60; Polytech Inst Brooklyn, MS, 62, PhD(electrophys), 67. Prof Exp: From instr to assoc prof, 66-75, PROF ELECTROPHYS, POLYTECH INST NEW YORK, 75- Concurrent Pos: Consult, Magi, Inc, NY, 69-, Anderson Labs, Conn, 72-74, Rockwell Int, Calif, 75- & Panametrics, Inc, Mass, 76. Mem: Inst Elec & Electronics Engrs; Int Sci Radio Union; Acoustical Soc Am; Sigma Xi. Res: Electromagnetic, elastic and acoustic wave propagation and scattering, as applied to ultrasonic nondestructive evaluation, acoustic surface wave devices, integrated optics and radar signatures. Mailing Add: Polytech Inst of New York 333 Jay St Brooklyn NY 11201

BERTONIERE, NOELIE RITA, b New Orleans, La, Oct 17, 36. ORGANIC CHEMISTRY. Educ: St Mary's Dominican Col, BS, 59; Univ New Orleans, PhD(org chem), 71. Prof Exp: Res asst biochem, Sch Med, Tulane Univ, 59-60; RES CHEMIST TEXTILES & FOOD, SOUTHERN REGIONAL RES CTR, AGR RES SERV, USDA, 60- Mem: Am Chem Soc. Res: Cellulose chemistry; durable press cotton fabric; flame resistant cotton fabric; photochemistry of small ring heterocycles and carbenes. Mailing Add: Southern Regional Res Ctr PO Box 19687 New Orleans LA 70179

BERTOZZI, EUGENE R, b Lowell, Mass, Sept 26, 15; m 43; c 3. POLYMER CHEMISTRY. Educ: Worcester Polytech Inst, BS, 38; Brooklyn Polytech Inst, MS, 50. Prof Exp: Chemist res & develop, Benzol Prod Co, 38-42; mgr develop, Dept Res & Develop, 42-64, asst tech dir res & develop, 64-68, dir advan technol, 68-72, PRIN SCIENTIST, CHEM DIV, THIOKOL CORP, 72- Mem: Am Chem Soc; Sigma Xi. Res: Process development of organic polysulfide polymers; new polymer products development in condensation and hydrocarbon polymers. Mailing Add: 2115 Stackhouse Dr Yardley PA 19067

BERTOZZI, WILLIAM, b Framingham, Mass, June 9, 31; m 56. PHYSICS. Educ: Mass Inst Technol, SB, 53, PhD(physics), 58. Prof Exp: Staff mem, Div Sponsored Res, Lab Nuclear Sci, 57-58, from instr to assoc prof, 58-68, PROF PHYSICS, MASS INST TECHNOL, 68- Mem: Am Phys Soc; Am Asn Physics Teachers. Res: Nuclear physics. Mailing Add: Dept of Physics Mass Inst of Technol Cambridge MA 02139

BERTRAM, EWART GEORGE, b Can, Apr 10, 23; m 49; c 2. NEUROANATOMY. Educ: Univ Western Ont, BA, 46, MSc, 49; Univ Buffalo, PhD(anat), 54. Prof Exp: Asst anat sch med, Univ Buffalo, 50-53; instr sch med, Marquette Univ, 53-56, asst prof, 56-60; assoc prof fac med, Univ Alta, 60-67; PROF ANAT, UNIV TORONTO, 67- Mem: Am Asn Anatomists; Am Acad Neurol; Can Asn Anatomists. Res: Interthalamic and intrathalamic connections as shown by Golgi studies; sex determination by chromatin studies; changes in ribonucleic acid in neurons following prolonged activity; chemical basis of Golgi method; neuroanal connections and capillary-neuronal relationship as shown by modified Golgi studies. Mailing Add: Dept of Anat Univ of Toronto Toronto ON Can

BERTRAM, LEON LEROY, b Corry, Pa, Jan 8, 17; m 44; c 2. CELLULOSE CHEMISTRY. Educ: Pa State Univ, BS, 41. Prof Exp: Chemist, 41-45, develop supvr, 46-50, chief chemist, 50, oper supvr, 50-51, asst mgr, 51-54, supt carboxymethyl cellulose dept, 55-69, ASST PLANT MGR, HOPEWELL PLANT, HERCULES, INC, 69- Mem: Am Chem Soc. Res: Cellulose chemistry. Mailing Add: Hopewell Plant Hercules Inc Hopewell VA 23860

BERTRAMSON, BERTRAM RODNEY, b Potter, Nebr, Jan 25, 14; m 38; c 3. AGRONOMY. Educ: Univ Nebr, BS, 37, MS, 38; Ore State Univ, PhD(soils), 41. Prof Exp: From asst soil physics to asst instr soils, Univ Nebr, 37-38; asst soil fertil, Ore State Col, 38-41; instr soils, Soil Lab, Colo State Col, 41-42; asst prof soil chem, Univ Wis, 46; assoc soil chemist, Dept Agron, Purdue Univ, 46-49; chmn dept agron, 49-67, DIR RESIDENT INSTR, COL AGR, WASH STATE UNIV, 67- Mem: Fel AAAS; fel Am Soc Agron (vpres, 60, pres, 61); Crop Sci Soc Am; Soil Sci Soc Am; Soil Conserv Soc Am. Res: Soil physics; physical properties of the soil as affected by manuring and fertilizers; irrigation efficiency; phosphorus analyses; soil fertility; comparative efficiency of organic phosphorus and of superphosphate; research on potash, manganese, boron, sulfur and magnesium. Mailing Add: Col of Agr Wash State Univ Pullman WA 99163

BERTRAN, CARLOS ENRIQUE, b Santurce, PR, July 4, 26; m 52; c 1. CARDIOLOGY. Educ: Cornell Univ, BA, 45, MD, 48; Am Bd Internal Med, dipl, 55. Prof Exp: Clin asst med, Postgrad Med Sch, NY Univ, 51-52; assoc med, 52-53, asst prof, 53-59, ASST PROF CLIN MED, SCH MED & SCH DENT, UNIV PR, SAN JUAN, 59- Concurrent Pos: Dir cardiac clin, Rio Piedras Munic Hosp, 59-66; dir cardiac clin, Presby Hosp, San Juan, 66-68, dir coronary care unit, 68-, chief of staff, 69-70. Mem: AMA; Am Col Physicians; Am Heart Asn; assoc Am Col Cardiol. Res: Arrhythmias. Mailing Add: Ashford Med Ctr San Juan PR 00907

BERTRAND, ANSON RABB, b Purmela, Tex, Aug 19, 23; m 46; c 2. SOIL PHYSICS. Educ: Agr & Mech Col Tex, BS, 47, Univ Ill, MS, 49; Purdue Univ, PhD(soil physics), 55. Prof Exp: Voc agr teacher, High Sch, Tex, 47; from instr to assoc prof agron, Purdue Univ, 49-61; dir, Southern Piedmont Soil Conserv Res Ctr, Ga, 61-64; chief southern br, Soil & Water Conserv Res Div, Agr Res Serv, USDA, 64-67; head dept agron, Univ Ga, 67-71; DEAN COL AGR SCI, TEX TECH UNIV, 71- Mem: Am Soc Agron; Soil Sci Soc Am. Res: Gaseous diffusion in soil and plant root growth and distribution. Mailing Add: Col of Agr Sci Tex Tech Univ Lubbock TX 79409

BERTRAND, FOREST, b St Pie-de-Guire, Que, May 31, 18; m 45; c 4. AGRICULTURE. Educ: Laval Univ, BA, 39, BSc, 43; McGill Univ, MSc, 45; Cornell Univ, PhD, 56. Prof Exp: Res officer, Can Dept Agr, 45-60; tech adv, 60-62, head res div, 62-67, dir res & educ, 67-73, DIR GEN RES & EDUC, QUE DEPT AGR, 73- Concurrent Pos: Secy, Que Agr Res Coun, 60-63, chmn, 63-72, mem, 72- Mem: AAAS; Am Soc Plant Physiol; Am Soc Hort Sci; Can Soc Plant Physiol; Can Soc Hort Sci. Res: Horticulture; vegetable crop and small fruits production and breeding. Mailing Add: 1570 Chemin Ste Foy Quebec PQ Can

BERTRAND, GARY LANE, b Lake Charles, La, Sept 25, 35; m 64; c 2. PHYSICAL CHEMISTRY. Educ: McNeese State Col, BS, 57; Tulane Univ, PhD(phys chem), 64. Prof Exp: Fel chem, Carnegie Inst Technol, 64-65, asst prof, 65-66; asst prof, 66-68, ASSOC PROF CHEM, UNIV MO-ROLLA, 68- Mem: AAAS; Am Chem Soc. Res: Investigations of solvent effects and deuterium isotope effects on the thermodynamic properties of dissolved species; solution calorimetry and dilatometry. Mailing Add: Dept of Chem Univ of Mo Rolla MO 65401

BERTRAND, HELEN ANNE, b Wilmington, Del, Nov 24, 39. BIOCHEMISTRY. Educ: Univ Del, BA, 61; NC State Univ, PhD(biochem), 70. Prof Exp: Res assoc biol, Univ Del, 61-63; res technician plant biol, Rockefeller Univ, 63-64; from instr to asst prof biochem, Med Col Pa, 71-75; ASST PROF PHYSIOL, UNIV TEX HEALTH SCI CTR SAN ANTONIO, 75- Concurrent Pos: NIH training grant, Med Col Pa, 70-71. Mem: Am Chem Soc; Sigma Xi. Res: Lipid metabolism; membrane biochemistry. Mailing Add: Dept of Physiol Univ of Tex Health Sci Ctr San Antonio TX 78289

BERTRAND, JOSEPH AARON, b Lake Charles, La, Mar 20, 33; m 57; c 1. INORGANIC CHEMISTRY. Educ: McNeese State Col, BS, 55; Tulane Univ, MS, 56, PhD(inorg chem), 61. Prof Exp: Instr chem, McNeese State Col, 56-59; res assoc, Mass Inst Technol, 61-62; from asst prof to assoc prof, 62-70, PROF CHEM, GA INST TECHNOL, 70-, DIR CHEM, 74- Concurrent Pos: Alfred P Sloan res fel, 66-68; vis prof, Univ Kans, 68. Mem: Am Chem Soc; Am Crystallog Asn. Res: Structure of transition metal complexes. Mailing Add: Sch of Chem Ga Inst of Technol Atlanta GA 30332

BERTRAND, JOSEPH E, b Kaplan, La, June 20, 24; m 51; c 4. ANIMAL NUTRITION. Educ: Southwestern La Inst, BS, 48; La State Univ, MS, 52, PhD(ruminant nutrition), 60. Prof Exp: Res assoc animal indust, La State Univ, 52-56; asst prof, Ark State Col, 56-58; ruminant nutritionist, Com Solvents Corp, New York, 60-66; ASSOC PROF ANIMAL SCI, AGR RES CTR, UNIV FLA, 66- Mem: Am Soc Animal Sci; Poultry Sci Asn; Am Dairy Sci Asn. Res: Ruminant nutrition; beef cattle production, especially growing and developing light-weight calves to desired feedlot weights and finishing cattle for market, emphasizing maximum use of locally grown roughages and concentrates. Mailing Add: Agr Res Ctr Univ of Fla Jay FL 32565

BERTRAND, KENNETH JOHN, b Green Bay, Wis, Jan 29, 10; m 37. PHYSICAL GEOGRAPHY, GEOGRAPHY OF POLAR REGIONS. Educ: Univ Wis, BS, 32, PhD(geog), 40. Prof Exp: Teaching asst geog, Syracuse Univ, 32-33; teaching asst, Univ Wis, 33-35, instr, Exten Div, 35-37; from instr to assoc prof, Okla State Univ, 37-43; regional geogr, US Bd Geog Names, 43-44, chief res div, 43-46, asst dir, 46; from asst prof to prof, 46-75, EMER PROF GEOG, CATH UNIV AM, 75- Concurrent Pos: Mem, Adv Comt Antarctic Names, US Bd Geog Names, 47-73, chmn, 62-73. Honors & Awards: Pub Serv Award, US Dept of the Interior, 73; Citation Meritorious Contributions to Field of Geog, Asn Am Geog, 74. Mem: Asn Am Geog; Am Geog Soc; Arctic Inst NAm; Glaciol Soc. Res: History of Antarctic exploration and mapping. Mailing Add: 6808 40th Ave Univ Park Hyattsville MD 20782

BERTRAND, RENE ROBERT, b Manchester, NH, Feb 5, 36; m 62. PHYSICAL CHEMISTRY, CHEMICAL ENGINEERING. Educ: Worcester Polytech Inst, BS, 57; Mass Inst Technol, PhD(phys chem), 62. Prof Exp: Chemist, Esso Res & Eng Co, Stand Oil Co, 61-66; prog mgr res & develop, Am Cryobenics, Inc, 66-68; sr res chemist, 68-73, ENG ASSOC, GOVT LAB, EXXON RES & ENG CO, 73- Mem: Am Chem Soc. Res: Fluidized bed combustion of coal; enviromental aspects of coal conversion processes. Mailing Add: Govt Res Lab Exxon Res & Eng Co PO Box 8 Linden NJ 07036

BERTSCH, CHARLES RUDOLPH, b Long Island, NY, June 13, 31; m 66; c 2. INORGANIC CHEMISTRY. Educ: Syracuse Univ, BS, 52; Pa State Univ, PhD(chem), 55. Prof Exp: Res chemist, US Naval Propellants Plant, 55-56, Olin Mathieson Chem Corp, 57-58 & Pennsalt Chem Corp, 58-61; sr prod ed, 61-69, mgr ed prod, 69-73, HEAD ED PROCESSING DEPT, AM CHEM SOC, 73- Mem: Am Chem Soc. Res: Editing and production of chemical journals. Mailing Add: R D 1 Kinternersville PA 18930

BERTSCH, GEORGE FREDERICK, b Oswego, NY, Nov 5, 42; m 64; c 1. NUCLEAR PHYSICS. Educ: Swarthmore Col, BA, 62; Princeton Univ, PhD(physics), 65. Prof Exp: Fel physics, Niels Bohr Inst, Copenhagen, Denmark, 65-66; from instr to asst prof, Princeton Univ, 66-69; asst prof, Mass Inst Technol, 69-70; asst prof, 70-74, PROF PHYSICS, MICH STATE UNIV, 74- Concurrent Pos: Alfred P Sloan Found fel, 69-71. Mem: AAAS. Res: Theoretical studies of nuclear structure based on shell model; qualitative and quantitative understanding of experimental results. Mailing Add: Dept of Physics Mich State Univ East Lansing MI 48823

BERTSCH, WALTER FRANK, b San Diego, Calif, Sept 26, 34; m 61; c 2. BIOPHYSICS, PLANT PHYSIOLOGY. Educ: Pomona Col, BA, 56; Yale Univ, MS, 57, PhD(bot), 59. Prof Exp: NSF fel plant physiol, Sorbonne & Lab du Phytotron, France, 59-61; biologist, Oak Ridge Nat Lab, 61-65; NIH spec fel biochem, Univ Cambridge, 65-66; assoc prof biol sci, 66-72, PROF BIOL SCI, HUNTER COL, 72- Concurrent Pos: Consult, Clint McDade & Sons, Inc, 62- Mem: Bot Soc Am; Am Soc Plant Physiol. Res: Photosynthesis; solid state aspects of energy conversion; photobiology; control of growth by light in plants; psychophysics; reception of audio messages by humans. Mailing Add: Dept of Biol Sci Hunter Col 695 Park Ave New York NY 10021

BERUBE, GENE ROLAND, b Malden, Mass, Nov 16, 37; m 59; c 3. PHYSICAL BIOCHEMISTRY. Educ: Univ Mass, Amherst, BS, 59. Prof Exp: Chemist, Environ Res Inst, US Army Natick Labs, 61-63; dir anal labs, Astra Pharmaceut Prod Inc, 63-67; group leader chem instrumentation, 67-72, GROUP LEADER APPL RES, CHESEBROUGH POND'S INC, 72- Mem: Instrument Soc Am; Am Asn Advan Med Instrumentation; Am Chem Soc; Soc Cosmetic Chemists; Royal Soc Health. Res: Physical chemical definition of hair, skin and fingernails and the alterations of these properties by cosmetics. Mailing Add: 349 Sharon Dr Cheshire Ct 06410

BERWICK, MARTIN ALFRED, organic chemistry, see 12th edition

BES, DANIEL RAUL, nuclear physics, see 12th edition

BESANCON, ROBERT MARTIN, b Missoula, Mont, July 29, 10; m 47. OPTICAL PHYSICS, COSMIC RAY PHYSICS. Educ: Univ Mont, BA, 31; Univ Ill, MA, 33. Prof Exp: Instr math & physics, Northern Mont Col, 36-37; from instr to asst prof physics, Univ Ill, 38-42, 45-55; from physicist to phys sci adminr, US Air Force Mat Lab, 55-75; ED, ENCYCL PHYSICS, 66- Mem: AAAS; Am Phys Soc; Am Asn Physics Teachers. Res: Reflectance; transmittance; luminescence. Mailing Add: 515 Grand Ave Dayton OH 45405

BESARAB, ANATOLE, b Hamburg, Ger, May 9, 43; US citizen; m 70; c 2. NEPHROLOGY, PHYSIOLOGY. Educ: Univ Pa, BS, 65, MD, 69. Prof Exp: From intern to resident, Pa Hosp, 69-72; fel nephrol, Boston City Hosp & Beth Israel Hosps, Harvard Med Sch, 72-75; ASST PROF MED & PHYSIOL, THOMAS JEFFERSON UNIV, 75- Concurrent Pos: USPHS spec fel, 72-74. Mem: Am Soc Nephrol; Int Soc Nephrol; Am Soc Artificial Internal Organs; Am Fedn Clin Res. Res: Hormonal modulation of renal function with emphases on intrarenal hormones on regulating glomerular filtration rate; interrelation of parathyroid hormone and vitamin D on mineral excretion. Mailing Add: Dept of Physiol Thomas Jefferson Univ Philadelphia PA 19107

BESCH, EMERSON LOUIS, b Hammond, Ind, June 9, 28; m 55; c 4. ENVIRONMENTAL PHYSIOLOGY. Educ: Southwestern Tex State Col, BS, 52, MA, 55; Univ Calif, Davis, PhD(physiol), 64. Prof Exp: Instr biol, Southwest Tex State Col, 54-55; res asst animal physiol, Univ Calif, Davis, 60-64, res physiologist & lectr animal physiol, 64-67; from assoc prof to prof physiol & mech eng, Kans State Univ, 67-74; PROF & ASSOC DEAN COL VET MED, UNIV FLA, 74- Concurrent Pos: NIH trainee, 61-64; chmn comt lab animal housing, Inst Lab Animal Resources, Nat Acad Sci, 75-; Am Asn Lab Animal Sci res award, 75. Mem: AAAS; Aerospace Med Asn; Am Physiol Soc; Am Asn Lab Animal Sci; Soc Neurosci. Res: Aviation physiology; effects of mechanical forces on biological systems; physiological effects of man-machine relationships; adaptation to high altitude; comparative physiology of environmental stresses. Mailing Add: Col of Vet Med Box J-12S JHMHC Univ of Fla Gainesville FL 32610

BESCH, EVERETT DICKMAN, b Hammond, Ind, May 4, 24; m 51; c 5. VETERINARY PARASITOLOGY, VETERINARY PUBLIC HEALTH. Educ: Agr & Mech Col Tex, DVM, 54; Univ Minn, Minneapolis, MPH, 56; Okla State Univ, PhD(vet parasitol), 63. Prof Exp: Instr vet bact, Agr & Mech Col Tex, 54; instr vet parasitol, Univ Minn, Minneapolis, 54-56; asst prof, Okla State Univ, 56-64, prof & Head dept, 64-68; DEAN SCH VET MED, LA STATE UNIV, 68- Mem: Am Vet Med Asn; Am Asn Vet Parasitol; Conf Pub Health Vets; Am Vet Med Col (secy, 74-76). Res: Host-parasite relationships of ruminant parasites; responses of immature stages of ruminant parasites to their free living and parasitic environments. Mailing Add: Sch of Vet Med Audubon Hall La State Univ Baton Rouge LA 70803

BESCH, HENRY ROLAND, JR, b San Antonio, Tex, Sept 12, 42; m 64; c 1. PHARMACOLOGY, CARDIOLOGY. Educ: Ohio State Univ, BSc, 64, PhD(pharmacol, biophys), 67. Prof Exp: Instr steroid biochem, Ohio State Univ, 67-68; asst prof pharmacol, 71-73, asst prof med biophys, 72-73, ASSOC PROF PHARMACOL & MED, SCH MED, IND UNIV, INDIANAPOLIS, 73- Concurrent Pos: USPHS fel, Baylor Col Med, 68-70; Med Res Coun fel, Royal Free Hosp Sch Med, London, Eng, 71-72, vis sr lectr, 73-74; res assoc, Krannert Inst Cardiol, 74-; Showalter awardee, 75- Mem: Am Soc Pharmacol & Exp Therapeut; Brit Biochem Soc; NY Acad Sci; Int Study Group Res Cardiac Metab; Am Heart Asn. Res: Cardiac pharmacology; subcellular mechanisms of inotropic and chronotropic drugs. Mailing Add: Dept of Pharmacol Ind Univ Sch of Med Indianapolis IN 46202

BESCH, PAIGE KEITH, b San Antonio, Tex, June 23, 31; m 57; c 2. BIOCHEMICAL PHARMACOLOGY. Educ: Trinity Univ, BS, 54, MS, 55; Ohio State Univ, PhD(physiol, pharmacol), 60. Prof Exp: Biochemist-in-chg metab sect, Surg Res Unit, Brooke Army Med Ctr, Ft Sam Houston, Tex, 53-54; sr biochemist, Clin Lab, Robert B Green Mem Hosp, San Antonio, 54-55; sr res asst to chmn dept endocrinol, Southwest Found Res & Educ, 55-58; asst, Dept Physiol, Col Med, Ohio State Univ, 58-60, res assoc, Steroid Res Lab, 60, instr pharmacol, Univ, 60-62, co-dir med sch & dir hosp, 60-66, asst prof physiol, obstet & gynec, Med Sch, 62-66, asst prof pharmacol, 63-66, assoc prof obstet & gynec, Univ Hosp, 66-68; assoc prof, 68-73, PROF OBSTET & GYNEC, BAYLOR COL MED, 73- Concurrent Pos: Res assoc, Columbus Psychiat Inst & Hosp, Ohio, 58-59; dir steroid res lab, Grant Hosp Res & Educ Found, 61-; dir labs, Midwest Found Res & Educ, 64-68; sr consult, Med Res Consult, Inc, 62-68. Mem: AAAS; Am Asn Clin Chemists; Sigma Xi; sr mem Am Chem Soc; Soc Exp Biol & Med. Res: Steroid metabolism; pharmacological effects of drugs; ability to alter steroids; fertility and sterility; chemical contraception. Mailing Add: Dept of Obstet & Gynec Baylor Col of Med Houston TX 77025

BESCHEL, ROLAND ERNEST, plant ecology, deceased

BESHINSKE, RAYMOND JOSEPH, computer science, see 12th edition

BESIC, FRANK CHARLES, b Chicago, Ill, July 13, 08; m 36; c 2. DENTISTRY. Educ: Univ Ill, DDS, 34. Prof Exp: Instr & assoc oper dent, Univ Ill, 35-46; chief dent clin, Forsyth Dent Infirmary, 46-47; from res assoc to assoc prof dent, 49-68, prof dent surg, 68-73, EMER PROF DENT SURG, UNIV CHICAGO, 73-, RES ASSOC, 73- Concurrent Pos: Mem dent sect, Nat Res Coun, 51-54. Mem: Am Dent Asn; Int Asn Dent Res. Res: Dental caries. Mailing Add: Dept of Dent Surg Univ of Chicago Chicago IL 60637

BESKID, GEORGE, b Erie, Pa, Mar 20, 29; m 54; c 3. BACTERIOLOGY. Educ: Univ Buffalo, BA, 50; Syracuse Univ, MS, 54, PhD(bact), 59. Prof Exp: Asst bact, Syracuse Univ, 57-59; res assoc microbiol, Hahnemann Med Col & Hosp, 59-61, instr microbiol & surg, 61-63; sr bacteriologist, 63-71, RES GROUP CHIEF, HOFFMANN-LA ROCHE, INC, 71- Mem: AAAS; Am Soc Microbiol; AMA. Res: Bacterial sporulation; endotoxin shock; surgical infections; amylase producing bacteria; chemotherapy. Mailing Add: Dept of Chemother Hoffmann-La Roche Inc Nutley NJ 07110

BESOZZI, ALFIO JOSEPH, b New York, NY, Dec 2, 21; m; c 4. ORGANIC CHEMISTRY. Educ: NY Univ, AB, 43, PhD(org chem), 51. Prof Exp: Jr pharmaceut chemist, Winthrop Chem Co Div, Sterling Drug, Inc, 43-44; proj leader & consult chem, Evans Res & Develop Corp, 51-56; sr res chemist, 56-67, res group head, 67-74, INDUST HYG CORRDR, PETRO-TEX CHEM CORP, 74- Mem: Am Chem Soc; Catalysis Soc. Res: Petrochemical research including heterogeneous catalysis, polymerization and synthesis; pharmaceutical chemsitry. Mailing Add: Petro-Tex Chem Corp 8600 Park Place Houston TX 77017

BESS, HENRY ALVER, b Newville, Ala, Jan 21, 07; m 35; c 2. ENTOMOLOGY. Educ: Ala Polytech Inst, BS, 27; Univ Fla, MS, 31; Ohio State Univ, PhD(entom), 34. Prof Exp: Teacher, Ala, 27-29; asst, Univ Fla, 29-31 & Ohio State Univ, 31-34; Nat Res Coun fel, Citrus Exp Sta, Univ Calif, 35-36; from asst entomologist to entomologist, Div Forest Insect Invests, Bur Entom & Plant Quarantine, USDA, 36-

48; prof entom & head dept, 48-58, sr prof entom & sr entomologist, 58-72, EMER SR PROF ENTOM, UNIV HAWAII, 72- Concurrent Pos: Fulbright scholar, Univ Ceylon, 54-55 & Kenya, 60-61; NSF grants, 58-61 & 63-68; NSF res grant, res assoc & vis prof, Nat Chung Hsing Univ, 73-74. Mem: AAAS; Entom Soc Am; Ecol Soc Am. Res: Fluctuations in insect populations; immunity of insects to insect parasites; biology of insect pests and their parasites; ecology and biological control of insect pests. Mailing Add: Dept of Entom Univ of Hawaii Honolulu HI 96822

BESS, JOHN CLIFFORD, b Tucson, Ariz, Mar 10, 37. AUDIOLOGY, SPEECH PATHOLOGY. Educ: Univ Ariz, BA, 61; Tulane Univ, MS, 65; Univ Okla, PhD(audiol), 69. Prof Exp: AUDIOL COORDR, VET ADMIN HOSP, ATLANTA & ASST PROF AUDIOL, DIV ALLIED HEALTH SCI, SCH MED, EMORY UNIV, 69- Concurrent Pos: Vet Admin fel, Vet Admin Hosp, Atlanta, 72-77; consult, Northwest Ga Speech & Hearing Ctr, 69-73 & US Fed Penitentiary Hosp, Atlanta & Doctor's Mem Hosp, 70-; instr, Ga State Univ, 70-; consult task force speech path & audiol serv needs in prisons, Am Speech & Hearing Asn, 72-73, chmn comt manpower needs & utilization, 73- Mem: Acoust Soc Am; Am Speech & Hearing Asn; Coun Except Children. Res: Electroencephalographic audiology; laterality effects in audition. Mailing Add: Vet Admin Hosp Atlanta 1670 Clairmont Rd NE Decatur GA 30033

BESSAC, FRANK BAGNALL, b Lodi, Calif, Jan 13, 22; m 53; c 2. ANTHROPOLOGY. Educ: Col Pac, BA, 46; Univ Calif, Berkeley, MA, 57; Univ Wis, PhD(anthrop), 63. Prof Exp: Exec officer, China Relief Mission, US State Dept, Paotou, 48; dir Mongol proj, Econ Coop Admin, China, 48-49; Fulbright scholar, Inner Mongolia, 49-50; vis lectr anthrop, Univ Tex, 58-59; from lectr to asst prof, Lawrence Univ, 59-64; Agr Develop Coun fel, Taiwan, 64-65; assoc prof, 65-70, PROF ANTHROP, UNIV MONT, 70- Concurrent Pos: Nat Inst Ment Health fel, 68-69. Mem: Fel Am Anthrop Asn; Am Ethnol Soc; Asn Asian Studies; Mongolia Soc. Res: East Asian and central Asian ethnology; Chinese social change; theoretical problems connected with demarcating units of study such as ethnic groups; relations of social organizations and interethnic relations; usefulness and validity of culture area concept. Mailing Add: 410 Traynor Dr Missoula MT 59801

BESSE, ARTHUR L, b Cambridge, Mass, Sept 26, 19; m 43; c 3. EXPERIMENTAL PHYSICS. Educ: Harvard Univ, BS, 42. Prof Exp: Res assoc electronics & acoust underwater sound lab, Harvard Univ, 42-47; from res scientist, Cryovac Div, Dewey & Almy Chem Co to dir, W R Grace & Co, 47-64; res assoc plasma physics & optics, Stevens Inst Technol, 64-68; PHYSICIST, US AIR FORCE CAMBRIDGE RES LAB, HANSCOM FIELD, 68- Concurrent Pos: Naval Ord develop award, 45. Mem: Optical Soc Am. Res: Plasma physics; optics; supersonic flow; space physics. Mailing Add: 30 Lantern Lane Weston MA 02193

BESSE, JOHN C, b Rayne, La, June 23, 33; m 62; c 3. PHARMACOLOGY, PHYSIOLOGY. Educ: Univ Southwestern La, BSc, 56; Vanderbilt Univ, PhD(pharmacol), 66. Prof Exp: Fel pharmacol, State Univ NY Downstate Med Ctr, 65-67; ASST PROF PHARMACOL, MED CTR, UNIV ALA, BIRMINGHAM, 67- Res: Autonomics and cardiovascular pharmacology; mechanism of action of drugs and hormones in smooth muscle at the level of receptors and membranes; cardiovascular physiology; biophysics of circulation and its correlation to drug activity on the circulation. Mailing Add: Univ of Ala Med Ctr 1919 Seventh Ave S Birmingham AL 35233

BESSERMAN, MARION, b Jamaica, NY, Mar 11, 21. PHYSICAL CHEMISTRY. Educ: Univ Wash, PhD(chem), 53. Prof Exp: Instr chem, Reed Col, 51-52; from instr to asst prof, 52-57, ASSOC PROF CHEM, WESTERN WASH STATE COL, 57- Res: Colloidal electrolytes; teaching of general chemistry. Mailing Add: Dept of Chem Western Wash State Col Bellingham WA 98225

BESSETTE, FRANCE MARIE, b Magog, Que, Can, June 26, 44. MOLECULAR BIOPHYSICS. Educ: Univ Sherbrooke, BSc, 66, PhD(phys chem), 69. Prof Exp: Phys sci specialist, Ministry Health, Que, 70; prof phys chem, Col Gen & Voc Educ, Prov Que, Can, Lionel-Groulx, 71; ASST PROF BIOPHYS, UNIV SHERBROOKE, 71- Mem: Biophys Soc; Spectros Soc Can. Res: Lipid-protein interaction in membranes; characterization of molecular parameters by optical techniques of functional proteins reincorporated into artificial lipid vesicles. Mailing Add: Dept of Biophys Fac of Med Univ of Sherbrooke Sherbrooke PQ Can

BESSETTE, RUSSELL ROMULUS, b New Bedford, Mass, July 21, 40; m 64; c 2. ANALYTICAL CHEMISTRY. Educ: Univ Mass, Amherst, MS, 65, PhD, 67. Prof Exp: Air Force Off Sci Res res asst, Univ Mass, Amherst, 65-67; sr res chemist, Res Ctr, Olin Corp, 67-68; asst prof, 68-71, ASSOC PROF CHEM, SOUTHEAST ERN MASS UNIV, 71- Mem: Am Chem Soc. Res: Electrochemical investigations in nonaqueous media; electrochemistry of tungsten bronzes. Mailing Add: Dept of Chem Southeastern Mass Univ North Dartmouth MA 02747

BESSEY, OTTO ARTHUR, b Niotaze, Kans, Oct 7, 04; m 29; c 2. BIOCHEMISTRY. Educ: Univ Mont, AB, 28; Univ Pittsburgh, PhD(biochem, nutrit), 32. Prof Exp: Asst, Univ Pittsburgh, 28-33 & Columbia Univ, 34; clin biochemist, Margaret Hague Hosp, NJ, 34; res fel path, Med Sch, Harvard Univ, 34-36, res assoc, 36-40, biol chem, 37-39, assoc, 39-42; chief, Div Physiol & Nutrit, Pub Health Res Inst, NY, 41-48, dir, 42-45; prof biol chem & head dept, Col Med, Univ Ill, 48-52; prof biochem & nutrit & chmn ddpt, Sch Med, Univ Tex, 52, prof, 52-56; assoc chief, Environ Protection Div, US Army Qm Res & Eng Ctr, Mass, 56-60; res & training grants adminr, Div Occup Health, USPHS, 60-69; HEALTH SCIENTIST ADMINR, NAT INST ENVIRON HEALTH SCI, NIH, 67-, ASSOC DIR EXTRAMURAL PROGS, 73- Concurrent Pos: Mem sci adv comt, Nat Vitamin Found, 46-49; mem comt dent, Nat Res Coun, 51-; pres, Am Bd Clin Chem, 49-55; secy & treas, Am Bd Nutrit, 48-58. Mem: Am Soc Biol Chem (secy, 46-49); Am Chem Soc; Am Inst Nutrit; Harvey Soc. Res: Chemistry and physiology of nutrition; biochemistry of metabolism; pathology of deficiency diseases; environmental stress. Mailing Add: NIH Bethesda MD 20014

BESSEY, PAUL MACK, b Cudahy, Wis, Mar 23, 25; m 47; c 5. HORTICULTURE. Educ: Univ Wis, BS, 49, MS, 51; Mich State Univ, PhD(hort), 57. Prof Exp: Asst veg crops & hort, Univ Wis, 49-51; instr & asst hort, Univ Maine, 51-54; asst hort, Mich State Univ, 54-57; asst hort, 57-65, ASSOC PROF HORT & ASSOC HORTICULTURIST, UNIV ARIZ, 65- Mem: Am Soc Hort Sci; Inst Food Technologists. Res: Postharvest physiology of fruits and vegetables, potato breeding, vegetable flavor in relation to market quality standards. Mailing Add: Dept of Hort & Landscape Archit Univ of Ariz Tucson AZ 85721

BESSEY, ROBERT JOHN, b East Lansing, Mich, Apr 8, 16; m 45; c 2. PHYSICS. Educ: Mich State Univ, BS, 37; Univ Mich, MS, 38, PhD(physics), 43. Prof Exp: Asst physics, Univ Mich, 38-42; instr, Albion Col, 42; from instr to asst prof, Univ, Idaho, 42-47; asst prof, Univ Okla, 47-53; from asst prof to assoc prof, 53-58, PROF PHYSICS, UNIV WYO, 58- Concurrent Pos: Vis scientist, Sonnenborg Observ, Netherlands, 68-69, 76. Mem: Am Phys Soc; Am Asn Physics Teachers; Am Astron Soc. Res: Theory of plasma oscillations; physics of solar chromosphere. Mailing Add: Dept of Physics Univ of Wyo Laramie WY 82070

BESSEY, WILLIAM HIGGINS, b East Lansing, Mich, Mar 18, 13; m 45; c 2. SOLID STATE PHYSICS. Educ: Univ Chicago, BS, 34; Carnegie Inst Technol, MS, 35, DSc, 40. Prof Exp: Asst, Carnegie Inst Technol, 35-39; instr physics, SDak Sch Mines, 39-40 & NC State Col, 40-42; from instr to asst prof, Carnegie Inst Technol, 42-52; assoc prof, Sch Mines & Metal, Univ Mo, 52-56; PROF PHYSICS & HEAD DEPT, BUTLER UNIV, 56- Mem: AAAS; Am Phys Soc; Am Asn Physics Teachers; Sigma Xi. Res: Diffraction of molecular rays; high speed testing of metals; shaped charges; electrical properties of evaporated metal films; thermal conductivity of ionic crystals. Mailing Add: Dept of Physics Butler Univ Indianapolis IN 46208

BESSLER, STUART ALAN, b Minneapolis, Minn, Oct 9, 31; m 54; c 3. MATHEMATICAL STATISTICS. Educ: Univ Minn, BS & BA, 54, MA, 58; Stanford Univ, PhD(statist), 60. Prof Exp: Adv res engr, Sylvania Elec Defense Labs, 59-60; res analyst, Gen Tel & Elec Labs, 60-61; opers res analyst, C-E-I-R, Inc, 61-63; partner, Decision Studies Group, 63-68; DIR DEPT BIOSTATIST, SYNTEX RES, SYNTEX CORP, 68- Res: Mathematical statistics and its applications to operations research and medicine; decision theory; design of experiments. Mailing Add: Syntex Corp 3401 Hillview Palo Alto CA 94304

BESSMAN, ALICE NEUMAN, b Washington, DC, Nov 7, 22; m 45; c 2. INTERNAL MEDICINE. Educ: Smith Col, BA, 43; George Washington Univ, MD, 49; Am Bd Internal Med, dipl, 59. Prof Exp: Fel med, DC Gen Hosp, George Washington Univ, 51-52; Nat Found Infantile Paralysis fel pediat, Mass Gen Hosp, 52-53; fel med, DC Gen Hosp, George Washington Univ, 53-54; staff physician, Group Health Asn, Inc, Washington, DC, 54-56; vis physician, Dept Med, Baltimore City Hosps, Johns Hopkins Univ, 56-62, hosp physician, 62-67, asst chief chronic & community med, 67-68; ASSOC PROF MED, SCH MED, UNIV SOUTHERN CALIF, 68-; CHIEF DIABETES SERV & ASST MED DIR, RANCHO LOS AMIGOS HOSP, 68- Concurrent Pos: Dept Health, Educ & Welfare Social Rehab Serv grant, 65- Mem: Am Fedn Clin Res; AMA; Am Col Physicians. Res: Significance of elevated serum ammonia levels in gastrointestinal bleeding and hepatic failure; ketone and lactate metabolism in diabetes mellitus; delivery of out-patient care via nurse practitioners. Mailing Add: Rancho Los Amigos Hosp 7601 E Imperial Hwy Downey CA 90242

BESSMAN, MAURICE JULES, b Newark, NJ, July 31, 28; m 52; c 4. BIOCHEMISTRY. Educ: Harvard Univ, AB, 49; Tufts Univ, MS, 52, PhD(biochem), 56. Prof Exp: Fel, Nat Cancer Inst, USPHS, 56-58; instr, Sch Med, Wash Univ, 58; from asst prof to assoc prof, 58-66, PROF BIOL, McCOLLUM-PRATT INST, JOHNS HOPKINS UNIV, 66- Mem: AAAS; Am Soc Biol Chem; Am Chem Soc. Res: Enzymology of nucleic acid metabolism. Mailing Add: Dept Biol McCollum-Pratt Inst Johns Hopkins Univ Baltimore MD 21218

BESSMAN, SAMUEL PAUL, b NJ, Feb 3, 21; m 45; c 2. PEDIATRICS, BIOCHEMISTRY. Educ: Wash Univ, MD, 44. Prof Exp: Intern, St Louis Children's Hosp, 44-46; pathologist, USPHS, US Marine Hosp, Va, 46-47; fel biochem, Children's Hosp, Washington, DC, 47-48, dir biochem res, 48-54; assoc prof pediat, Sch Med, Univ Md, 54-60, assoc prof biochem, 58-68, prof pediat res, 60-68; PROF PHARMACOL & CHMN DEPT, SCH MED, UNIV SOUTHERN CALIF, 68-, PROF PEDIAT, 69- Concurrent Pos: Fel, Psychiat Inst, NY, 48; assoc prof clin pediat, Sch Med, George Washington Univ, 52; fel, Harvard Univ, 52-53; ed, Biochem Med; assoc ed, Anal Biochem. Mem: Soc Pediat Res; Am Chem Soc; Am Acad Pediat; Am Soc Biol Chemists. Res: Biochemistry of disease; mechanism of mental symptoms; carbohydrate diseases. Mailing Add: Dept of Pharmacol Univ of Southern Calif Los Angeles CA 90033

BESSO, JOSEPH AUGUSTUS, JR, b Haverhill, Mass, Oct 2, 42; m 69; c 1. NEUROBIOLOGY. Educ: Univ Vt, BA, 69, PhD(physiol), 73. Prof Exp: NIH fel neurobiol, Albert Einstein Col Med, 73-74; ASST PROF NEUROBIOL & CHMN DEPT, TEX COL OSTEOP MED, N TEX STATE UNIV, 74- Mem: Soc Neurosci; Sigma Xi; Am Soc Zoologists. Res: Electrophysiology and ultrastructural characteristics of synaptic transmission in animals. Mailing Add: Dept of Neurobiol Tex Col of Osteop Med Denton TX 76203

BESSO, MICHAEL M, b Brooklyn, NY, Mar 30, 30; m 55; c 2. ORGANIC CHEMISTRY. Educ: Lowell Technol Inst, BSc, 50; Lehigh Univ, MS, 57, PhD(org chem), 59. Prof Exp: Chemist, Pinatel Piece Dye Works, Can, 50-51 & Nat Lead Co, 53-55; res chemist, Chas Pfizer & Co, 59-65; sr res chemist, 65-68, group leader Fibers, Plastics & Coatings, 69-74, MGR POLYMER CHEM, FLAMMABILITY & POWDER RES, CELANESE RES CORP, 74- Mem: Am Chem Soc; Fiber Soc. Res: Polymer flammability and flame retardant fibers; plastics, cellulose and polyester fiber technology; polymer synthesis; powder coating technology; increasing productivity of research; regenerated protein fibers; fermentation chemicals. Mailing Add: Celanese Res Co PO Box 1000 Summit NJ 07901

BEST, ALAN C G, b Kuala, Lumpur, Malaya, Oct 8, 39. GEOGRAPHY. Educ: Univ Nebr, BA, 60; Mich State Univ, MA, 62, PhD(geog), 65. Prof Exp: Asst prof geog, Eastern Mich Univ, 64-67; asst prof, Boston Univ, 67-71, ASSOC PROF GEOG, BOSTON UNIV, 71- Mem: Asn Am Geogr; Am Geog Soc. Res: Political and population geography, especially Sub-Saharan Africa; viability of Swaziland, Botswana, Lesotho and other newly independent states. Mailing Add: Dept of Geog Boston Univ Boston MA 02215

BEST, AUDREY NANCE, b Pittsburgh, Pa, Apr 23, 21. BACTERIOLOGY, BIOCHEMISTRY. Educ: Pa State Univ, BS, 43; Univ Ga, PhD(bact), 66. Prof Exp: Jr fel, Mellon Inst, 43-52; res asst, Biol Div, 53-66, BIOCHEMIST BIOL DIV, OAK RIDGE NAT LAB, 66- Mem: Am Soc Microbiol. Res: Protein synthesis; microbial physiology; transfer ribonuclear acid metabolism; methodology of cancer detection. Mailing Add: Biol Div Oak Ridge Nat Lab Oak Ridge TN 37830

BEST, CHARLES HERBERT, b West Pembroke, Maine, Feb 27, 99; m 24; c 2. PHYSIOLOGY. Educ: Univ Toronto, BA, 21, MA, 22, MD, 25; Univ London, DSc, 28; FRCP(E); FRCPS(C). Hon Degrees: Twenty-five from US & foreign univs, 45-72. Prof Exp: Res mem, Connaught Labs, Univ Toronto, 22-32, assoc dir, 32-41, asst prof & res assoc, Banting & Best Dept Med Res, 24-41, actg head dept physiol hyg, Sch Hyg, 29-41, prof physiol, Univ, 29-74, dir dept, 29-65, dir, Banting & Best Dept Med Res, 41-67, EMER DIR DEPT PHYSIOL, UNIV TORONTO, 65-, EMER PROF PHYSIOL, 74-, EMER DIR BANTING & BEST DEPT MED RES, 67- Concurrent Pos: Co-discoverer of insulin with F G Banting, 21; sci dir int health div, Rockefeller Found, 41-48; chmn conf liver injury, Josiah Macy Jr Found, 47-53; hon pres, Int Diabetes Fedn, 49-; pres, Int Union Physiol Sci, 53-; mem sci adv comt, Nutrit Found, 50-64, trustee, 64-; hon dir, Muscular Dystrophy Asn Can, 64-; mem bd sci dirs, Jackson Mem Lab, 55-; mem adv comt med res, WHO, 63- Reeve prize, Univ Toronto, 23, J J Mackenzie & Ellen Mickle fels, 25, Charles Mickle fel & lect, 39, Banting Mem lect, 44; Addison lect, Guy's Hosp, London, 56; Mitchell lect, Queen's Univ, Belfast, 57; Oslerian oration, Univ London, 57; Woodyat Mem lect, Univ

Chicago, 59; 150th anniversary lect, Karolinska Inst, Sweden, 60; Joslin Mem lect, Boston Univ, 62; lectr, Col of France, 62; Grow Mem lectr, US Air Force, 64; George W Corner lect, Univ Rochester, 70. Honors & Awards: Comdr, Order of Brit Empire, 44; Legion of Merit, 47; Liberty Cross, Norway, 47; Comdr, Order of Crown, Belg, 48; Knight Comdr, Mil & Hospitaller Order St Lazarus of Jerusalem, 64; Companion, Order of Can, 67; Companion of Honour from HM Queen Elizabeth II, 71; Brazil Sci Biennial Award, 71; Gairdner Found Int Award, 71. F N G Starr Gold Medal, Can Med Asn, 36; Baly Medal, Royal Col Physicians, 39; Gold Medal, Can Pharmaceut Mfrs Asn, 48; Banting Medal, Am Diabetes Asn, 49; Flavelle Medal, Royal Soc Can, 50; Coronation Medal, 53; Phillips Mem Medal, Am Col Physicians, 53; Queen Elizabeth of the Belgians Gold Medal, 56; Medal, Royal Neth Acad Sci & Lett, 58; Dale Medal, Soc Endocrinol, London, 59; Silver Medal, Czech Soc Phys Med, 62; La Grande Medaille d'Argent, Paris, 62; Humanitarian Award, Can B'nai B'rith, 63; Centennial Medal, 67. Mem: For assoc Nat Acad Sci; Am Physiol Soc; Am Soc Biol Chem; Am Philos Soc; Am Diabetes Asn (past pres). Res: Insulin; histamine; carbohydrate and fat metabolism; choline; heparin. Mailing Add: Rm 211 Charles H Best Inst Univ of Toronto 112 College St Toronto ON Can

BEST, EDGAR ALLAN, b Enumclaw, Wash, Aug 17, 25; m 50; c 1. MARINE BIOLOGY. Educ: Univ Wash, BS, 53. Prof Exp: Fishery biologist, Ore Fish Comn, 53; marine biologist, Calif Dept Fish & Game, 54-57, proj leader, 57-64, supvr Point Arguello surv, 64-65; SUPVR SMALL FISH INVEST, INT PAC HALIBUT COMN, UNIV WASH, 65- Concurrent Pos: Mem int trawl fisheries comt, US State Dept, 59-65; res assoc ichthyol, Santa Barbara Mus Natural Hist, 65-68. Mem: Am Inst Fishery Res Biologists; Am Soc Ichthyologists & Herpetologists. Res: Marine biology and ecology; ecology; life history of marine fishes. Mailing Add: PO Box 5009 Univ Sta Seattle WA 98105

BEST, EDWARD WILLSON, b Windsor, Ont, Apr 15, 27; m 51; c 4. GEOLOGY. Educ: Univ Western Ont, BSc, 49; Univ Wis, PhD(geol), 53. Prof Exp: Geologist, Ohio Oil Co, 53-55; geologist, Triad Oil Co, Ltd, 55-58, chief geologist, 58-64, explor mgr, 64-70; VPRES, BP CANADA LTD, 70- Mem: Am Asn Petrol Geologists; Paleont Soc. Res: Stratigraphy; structure; petroleum geology. Mailing Add: BP Canada Ltd 335 Eighth Ave SW Calgary AB Can

BEST, GARY KEITH, b Weatherford, Okla, Oct 8, 38; m 62. MICROBIOLOGY. Educ: Southwestern State Col, Okla, BS, 60; Okla State Univ, PhD(microbiol), 65. Prof Exp: Res microbiologist, Agr Res Serv, USDA, 65-68; asst prof microbiol, 68-74, ASSOC PROF MICROBIOL, MED COL GA, 74- Mem: Am Soc Microbiol. Res: Bacterial cell wall synthesis; effect of antibiotics on microorganisms; physiological basis for psychrophily. Mailing Add: Dept of Microbiol Med Col of Ga Augusta GA 30904

BEST, GEORGE HAROLD, b Chicago, Ill, Aug 4, 20; m 44; c 3. PHYSICS. Educ: Purdue Univ, BS, 42; Northwestern Univ, MS, 48, PhD(physics), 49. Prof Exp: MEM STAFF & GROUP LEADER, LOS ALAMOS SCI LAB, UNIV CALIF, 49- Mem: Am Phys Soc; Mil Opers Res Soc. Res: Nuclear physics; radioactivity; nuclear weapons effects; nuclear reactors. Mailing Add: Los Alamos Sci Lab Univ of Calif Box 1663 MS632 Los Alamos NM 87545

BEST, JAY BOYD, b Tulsa, Okla, Nov 21, 21. PHYSIOLOGY, BIOPHYSICS. Educ: Univ Tex, AB, 47; Univ Chicago, PhD(physiol), 53. Prof Exp: Res assoc psychol, Univ Chicago, 53-55; biophysicist, Walter Reed Army Inst Res, 55-60; assoc prof physiol, Col Med, Univ Ill, 60-63; PROF PHYSIOL & BIOPHYS, COLO STATE UNIV, 63- Concurrent Pos: NIH grant, 61-67; res fel, Univ Chicago, 63-66; NASA grants, 64-67. Mem: Fel AAAS; Am Soc Zoologists. Res: Mathematical biophysics; molecular mechanisms of memory, differentiation and aging; protopsychology. Mailing Add: Dept of Physiol & Biophys Colo State Univ Ft Collins CO 80521

BEST, LAVAR, b Salt Lake City, Utah, Mar 28, 31; m 54; c 4. BIOACOUSTICS. Educ: Univ Utah, BS, 53, MS, 57, PhD(audiol), 64. Prof Exp: Instr res, Med Ctr, Univ Colo, 64-66; asst prof speech & hearing sci, 67-71, ASSOC PROF SPEECH & HEARING SCI, UNIV DENVER, 71- Mem: NY Acad Sci. Res: Cortical processing of auditory stimuli. Mailing Add: Dept of Speech Path & Audiol Univ of Denver Denver CO 80210

BEST, MAURICE MCDONALD, JR, b Louisville, Ky, Feb 8, 20; m 49; c 3. MEDICINE. Educ: Univ Louisville, MD, 45; Am Bd Internal Med, dipl, 52; Am Bd Cardiovasc Dis, dipl, 56. Prof Exp: Intern, Methodist Hosp, Indianapolis, 45-46, resident, 46-47; resident, Louisville Gen Hosp, 47-49; from instr to assoc prof, 49-70, PROF MED, SCH MED, UNIV LOUISVILLE, 70- Concurrent Pos: Consult physician, Vet Admin Hosp, Louisville; mem coun arteriosclerosis, Am Heart Asn. Mem: AAAS; Am Col Physicians; Am Fedn Clin Res; Am Heart Asn; NY Acad Sci. Res: Cardiovascular disease, especially arteriosclerosis; medical research. Mailing Add: Med Dent Res Bldg Univ of Louisville Sch of Med Louisville KY 40202

BEST, MELVYN EDWARD, high energy physics, nuclear physics, see 12th edition

BEST, PHILIP ERNEST, b Perth, Australia, July 28, 38; m 60; c 3. SOLID STATE PHYSICS, SURFACE PHYSICS. Educ: Univ Western Australia, BSc, 59, PhD(physics), 63. Prof Exp: Instr & res assoc physics, Lab Atomic & Solid State Physics, Cornell Univ, 63-65; exp physicist, Res Labs, United Aircraft Corp, 65-70; fel, Inst Mat Sci, 70-71, ASSOC PROF PHYSICS, UNIV CONN, 71- Mem: Am Phys Soc; Inst Elec & Electronics Engrs; Brit Inst Physics & Phys Soc; Australian Inst Physics. Res: Interaction of slow electrons and x-rays with matter; electronic excitations at surfaces and in the bulk of solids; electronic properties of small molecules. Mailing Add: Inst of Mat Sci U-136 Univ of Conn Storrs CT 06268

BEST, RAYMOND VICTOR, b BC, Can, Sept 13, 15; m 50; c 3. GEOLOGY. Educ: Univ BC, MASc, 52; Princeton Univ, PhD, 59. Prof Exp: Lectr geol, McMaster Univ, 54-60; asst prof, 60-67, ASSOC PROF GEOL, UNIV BC, 67- Mem: Paleont Soc. Res: Taxonomy of lower Cambrian olenellid trilobites. Mailing Add: Dept of Geol Univ of BC Vancouver BC Can

BEST, RICHARD JAMES, physical chemistry, see 12th edition

BEST, TROY LEE, b Fort Sumner, NMex, Aug 30, 45; c 2. VERTEBRATE ECOLOGY, SYSTEMATIC ZOOLOGY. Educ: Eastern NMex Univ, BS, 67; Univ Okla, MS, 71, PhD(zool), 76. Prof.Exp: Preparator zool, Stovall Mus Sci & Hist, Univ Okla, 68-71, res asst, Dept Zool, 71-74; ASST PROF VERT ZOOL, DEPT BIOL, NORTHEASTERN UNIV, 74- Concurrent Pos: Index ed, Syst Zool, 72- Mem: Am Soc Mammalogists; Ecol Soc Am; Soc Syst Zool; Am Ornithologists Union; Wildlife Soc. Res: Assessment of interrelationships between variation in ecological and morphological characteristics of mammals; mammalian systematics utilizing univariate and multivariate statistical techniques; effect of character selection on data analyses. Mailing Add: Dept of Biol Northeastern Univ 360 Huntington Ave Boston MA 02115

BEST, WILLIAM ROBERT, b Chicago, Ill, July 14, 22; m 44; c 2. HEMATOLOGY, BIOSTATISTICS. Educ: Univ Ill, BS, 45, MD, 47, MS, 51; Am Bd Internal Med, dipl, 57, cert hemat, 72. Prof Exp: From clin asst to assoc prof med, 48-70, asst dir aeromed lab, 64-67, PROF MED, UNIV ILL MED CTR, 70-, ASSOC DEAN, A LINCOLN SCH MED, 72-, CHIEF STAFF, UNIV ILL HOSP, 75- Concurrent Pos: Attend physician, Ill Res & Educ Hosp, 53-; consult, Vet Admin Hosp, Hines, 53-, chief midwest res support ctr, 67-72; consult, Grant Hosp, 54-67. Mem: AAAS; Am Fedn Clin Res; Am Soc Hemat; Int Soc Hemat; Am Statist Asn. Res: Statistical methods; cooperative clinical trials; computers in medicine. Mailing Add: Univ of Ill Hosp Univ of Ill Med Ctr Chicago IL 60612

BESTER, JOHN (FRANCIS), b Cargill, Ont, Apr 17, 22; m 46; c 3. PHARMACOLOGY. Educ: Univ Sask, BSc, 49; Ohio State Univ, MSc, 50, PhD(pharmacol), 52. Prof Exp: Asst prof pharmacol, Ferris Inst, 52-53; from asst prof to assoc prof pharm, Sch Pharm, Univ Southern Calif, 53-68; HEALTH SCIENTIST ADMINR, NAT INST OCCUP SAFETY & HEALTH, OFF EXTRAMURAL ACTIV, 68- Mem: AAAS; Am Pharmaceut Asn. Res: Intravenous anesthetics; diuretics; protein bound iodine metabolism; muscle relaxants. Mailing Add: Nat Inst Occup Safety & Health Off of Extramural Activ Rockville MD 20852

BESTUL, ALDEN BEECHER, b Forestburg, SDak, Sept 1, 21; m 47; c 4. PHYSICAL CHEMISTRY. Educ: St Olaf Col, BA, 42; Pa State Col, MS, 44; Univ Md, PhD(chem), 54. Prof Exp: Res asst cryogen, Pa State Col, 44-45; res thermochemist, Stamford Res Labs, Am Cyanamid Co, 46-48; res physicist, Nat Bur Standards, 48-70; GEN PHYS SCIENTIST, OFF SPEC STUDIES, NAT OCEANIC & ATMOSPHERIC ADMIN, 70- Concurrent Pos: Com Sci & Technol fel, 69-70. Mem: AAAS; Am Chem Soc; Am Phys Soc; Soc Rheology; Am Ceramic Soc; Brit Soc Glass Technol. Res: Molecular structure in the vitreous state; rheology of polymers; thermochemistry of organic compounds; research and development evaluation; environmental policy analsis; vitreous state; polymer rheology; thermodynamics and thermochemistry. Mailing Add: 9400 Overlea Dr Rockville MD 20850

BETAK, JOHN FREDERICK, urban geography, environmental psychology, see 12th edition

BETCHOV, ROBERT, b Switz, Oct 12, 19; nat US; m 48; c 2 2. PHYSICS. Educ: Col de Geneve, BA, 39; Univ Geneva, lic sci math, 41, lic sci physics, 43; Univ Berne, PhD(physics), 46. Prof Exp: Physicist, Swiss Fed Govt, 43-46; chief asst, Dutch Inst Technol, Technol Univ Delft, 47-49; asst prof, Univ Md, 49-54; res assoc, George Washington Univ, 54-55; lectr, Johns Hopkins Univ, 55-57; scientist, Douglas Aircraft, Calif, 57-59, Space Technol Lab, 59-60 & Aerospace Corp, 60-67; prof, Univ Berne, 67-68; PROF AEROSPACE & MECH ENG, UNIV NOTRE DAME, 68- Concurrent Pos: Lectr, Univ Calif, Los Angeles, 57-68. Mem: AAAS; Am Phys Soc. Res: Fluid dynamics; electronics; applied mathematics. Mailing Add: Dept of Aerospace & Mech Eng Univ of Notre Dame Notre Dame IN 46556

BETH, ERIC WALTER, b Vienna, Austria, June 7, 12; nat US. PHYSICS. Educ: Univ Vienna, PhD(physics), 34. Prof Exp: Asst physics, Univ Vienna, 36-37; res fel, Univ Calif, 37-38, asst, 38; asst physics, Reed Col, 39-40; vis instr physics & math, Mills Col, 40-41; instr physics, Ill Inst Technol, 41-42; spec consult, Electronic Res Labs, Mass, 46-53; mem sci staff, Melpar, Inc, 53-54; physicist, Union Switch & Signal Div, Westinghouse Air Brake Co, 54-62; ASST PROF PHYSICS, STATE UNIV NY BUFFALO, 62- Res: Wave mechanics; quantum theory of the non-idea gas; upper air properties. Mailing Add: Dept of Physics State Univ of NY Buffalo NY 14214

BETHE, HANS ALBRECHT, b Strasbourg, Ger, July 2, 06; nat US; m 39; c 2. THEORETICAL PHYSICS. Educ: Univ Munich, PhD(physics), 28. Hon Degrees: DSc, Polytech Inst Brooklyn, 50, Univ Denver, 52, Univ Chicago, 53, Univ Birmingham, England, 56, Harvard Univ, 58, Univ Munich & Tech Univ Munich, 68, Univ Seoul & Univ Delhi, 69. Prof Exp: Instr physics, Univ Frankfurt, 28-29 & Univ Stuttgart, 29; lectr, Univ Munich, 30-33 & Univ Manchester, England, 33-34; fel, Univ Bristol, 34-35; asst prof, 35-37, prof, 37-74, JOHN WENDELL ANDERSON PROF PHYSICS, CORNELL UNIV, 74- Concurrent Pos: Rockefeller Found Int Educ Bd fel, Cambridge Univ & Univ Rome, 30-32; staff mem radiation lab, Mass Inst Technol, 42-43; chief theoret physics div, Los Alamos Sci Lab, NMex, 43-46; vis prof, Columbia Univ, 41-48; consult, Los Alamos Sci Lab, 47-, Atomic Power Develop Assocs, 53- & Avco Res Lab, 55-; vis prof, Cambridge Univ, 55-56; consult, President's Sci Adv Comt, 56-59; mem, US Delegation to Discussions on Discontinuance of Nuclear Weapons Tests, Geneva, 58-59; vis prof, Calif Inst Technol, 64. Honors & Awards: Nobel Prize in Physics, 67; Morrison Prize, NY Acad Sci, 38, 40; US Medal Merit, 46; Draper Medal, Nat Acad Sci, 48; Planck Medal, 55; Enrico Fermi Prize, US AEC, 61. Mem: Nat Acad Sci; fel Am Phys Soc (pres, 54); Am Astron Soc; Am Philos Soc; Royal Soc London. Res: Quantum theory of atoms; theory of metals; quantum theory of collisions; theory of atomic nuclei; meson theory; energy production in stars; quantum electrodynamics; shock wave theory. Mailing Add: Dept of Physics Cornell Univ 318 Newman Lab Ithaca NY 14853

BETHEIL, JOSEPH JAY, b New York, NY, Apr 16, 24; m 48; c 2. BIOCHEMISTRY. Educ: City Col New York, BS, 44; Univ Wis, MS, 47, PhD(biochem), 49. Prof Exp: Res asst, Dept Chem, Off Sci Res & Develop Proj, Columbia Univ, 44-45; from instr to asst prof biochem, Sch Med & Dent, Univ Rochester, 50-55; asst prof, 55-58, ASSOC PROF BIOCHEM, ALBERT EINSTEIN COL MED, YESHIVA UNIV, 58- Concurrent Pos: USPHS fel, Inst Enzyme Res, Univ Wis, 49-50, spec fel, Nat Inst Med Res, London, 64-65. Mem: AAAS; Am Soc Biol Chemists; Am Inst Nutrit; Harvey Soc; Am Chem Soc. Res: Protein chemistry and structure; intermediary metabolism; study of cellular aging; regulation of protein biosynthesis; nutrition. Mailing Add: Albert Einstein Col of Med Yeshiva Univ New York NY 10461

BETHEL, EDWARD LEE, b Strawn, Tex, Feb 18, 26. MATHEMATICS. Educ: Southern Methodist Univ, BA, 50; NTex State Univ, MA, 57; Univ Tex, PhD(math), 67. Prof Exp: Radar instr, US Civil Serv, Keesler AFB, Miss, 50-51; systs engr, Chance Vought Aircraft Corp, Tex, 51-53; aerophys engr, Gen Dynamics/Convair, 57; instr math, Univ Tex, 57-63 & Knox Col, Ill, 63-64; assoc prof, Clemson Univ, 64-67; assoc prof, 67-74, PROF MATH, KENT STATE UNIV, 74- Mem: Am Math Soc; Math Asn Am. Res: General topology and convex sets in metric spaces. Mailing Add: Dept of Math Kent State Univ Kent OH 44240

BETHEL, JAMES SAMUEL, b New Westminster, BC, Aug 13, 15; m 41; c 3. WOOD TECHNOLOGY. Educ: Univ Wash, BSF, 37; Duke Univ, MF, 39, DF, 47. Prof Exp: Instr forestry, Pa State Col, 39-41; asst prof, Va Polytech Inst, 41-42; asst mgr, Tidewater Plywood Co, Ga, 46-48, prod mgr, 48-49; assoc prof wood technol, NC State Col, 49-50, prof & dir wood prod lab, 50-58, head wood prod dept & actg dean, Grad Sch, 58-59; head spec proj, Sci Educ Sect, NSF, 59-62; assoc dean grad sch & col forest resources, 62-64, PROF FORESTRY, UNIV WASH, 62-, DEAN COL FOREST RESOURCES, 64- Concurrent Pos: UN consult, Yugoslavia, 52; adv, Econ Develop Admin, PR, 56; consult, US Dept Health, Educ & Welfare, 56-, Gov Res Triangle Comt, NC, 57-59 & NSF, 62-; mem exec comt & dir, Orgn Trop Studies, 63-; vpres, 64-65, pres, 66-67; ed, J Soc Am Foresters; consult, President's Coun Environ

Qual; mem comt effects herbicides Vietnam, Nat Acad Sci, 71-72; consult, Comt Eng Policy, Nat Acad Eng, 72-73; bd mem, Bd Agr & Renewable Resources, Nat Acad Sci, 73-75, chmn, Task Force on Educ Agr & Renewable Resources & Comt on Renewable Resources for Indust Mat, 74-75. Mem: Fel AAAS; Soc Am Foresters; Forest Prod Res Soc (secy-treas, 52-56); Soc Wood Sci & Technol (pres, 58-59, 60-61); Ecol Soc Am. Res: Wood science, especially anatomy and morphology; relationship of environment to structure of wood in forest trees; wood moisture relationships. Mailing Add: Col of Forest Resources Univ of Wash Seattle WA 98195

BETHKE, GEORGE WILLIAM, JR, b Green Bay, Wis, July 17, 30; m 63; c 2. APPLIED PHYSICS. Educ: Univ Wis, BS, 52; Harvard Univ, MA, 54, PhD(phys chem), 56. Prof Exp: CONSULT SCIENTIST, GEN ELEC CO, 56- Mem: Optical Soc Am. Res: Research and instrumentation in fields of spectroscopy and optics, including extreme ultraviolet; lasers, lidar, light scattering, atmospheric measurements and plasma diagnostics. Mailing Add: Valley Forge Space Ctr Gen Elec Co PO Box 8555 Philadelphia PA 19101

BETHKE, PHILIP MARTIN, b Chicago, Ill, Mar 22, 30; m 55; c 7. GEOLOGY. Educ: Amherst Col, BA, 52; Columbia Univ, MA, 54, PhD(geol), 57. Prof Exp: Asst geol, Columbia Univ, 52-53, res asst, 53-54; asst prof, Mo Sch Mines, 55-59; GEOLOGIST, US GEOL SURV, 57- Mem: Geol Soc Am; Geochem Soc; Mineral Soc Am (treas, 73-74); Soc Econ Geol. Res: Chemistry of ore deposition; mineralogy of ore minerals; minor elements in ore deposits; time-space relations in ore deposits; phase equilibria of ore minerals. Mailing Add: 4220 Franklin St Kensington MD 20795

BETHLAHMY, NEDAVIA, b Tel-Aviv, Palestine, July 23, 18; US citizen; m 43; c 2. FOREST HYDROLOGY. Educ: Pa State Col, BS, 39; Yale Univ, MF, 40; Cornell Univ, PhD, 56. Prof Exp: Forester northeastern forest exp sta, US Forest Serv, 46-55, res forester, Pac Northwest Forest & Range Exp Sta, 56-63, proj leader, Intermountain Forest & Range Exp Sta, 63-66; watershed mgr expert, Food & Agr Orgn, UN, Taiwan, 66-68; res forester, 68-69, PRIN FOREST HYDROLOGIST, INTERMOUNTAIN FOREST & RANGE EXP STA, US FOREST SERV, 69- Concurrent Pos: Adj prof forest hydrol, Univ Idaho, 69-76. Mem: AAAS; Soc Am Foresters; Am Geophys Union. Res: Hydrology of wildlands, with special emphasis on effects of major wild fires; watershed management. Mailing Add: Intermntn Forest & Range Exp Sta US Forest Serv 1221 S Main St Moscow ID 83843

BETHUNE, JOHN EDMUND, b Mar 23, 27; Can citizen; m 48; c 2. ENDOCRINOLOGY. Educ: Acadia Univ, BA, 47, BSc, 48; Dalhousie Univ, MD, 53; FRCPS(C), 57. Prof Exp: Asst prof med, Dalhousie Univ, 58-61; from asst prof to assoc prof, 61-70, head sect endocrinol & dir clin res ctr, 67-72, PROF MED, UNIV SOUTHERN CALIF, 70-, CHMN DEPT, 72- Concurrent Pos: Consult, Los Angeles Vet Admin Hosp, 68 & Sepulveda Vet Admin Hosp, 70. Mem: Fel Am Col Physicians; Endocrine Soc; Am Fedn Clin Res; Am Soc Nephrology; Am Physiol Soc. Res: Adrenal hormone activity, with special reference to calcium and bone metabolism. Mailing Add: Sch of Med Dept of Med Univ of Southern Calif Los Angeles CA 90033

BETHUNE, JOHN LEMUEL, b Baddeck, NS, July 22, 25; m 58. PHYSICAL CHEMISTRY, BIOPHYSICS. Educ: Acadia Univ, BSc, 47; Clark Univ, PhD(chem), 61. Prof Exp: Prof chemist, Can Breweries, Ltd, 48-58; asst prof chem, Clark Univ, 61; res assoc, 61-64, assoc, 64-66, asst prof, 66-69, ASSOC PROF BIOCHEM, HARVARD MED SCH, 69- Res: Effect of chemical reactions upon the interpretation of transport experiments. Mailing Add: Harvard Med Sch 721 Huntington Ave Boston MA 02115

BETHUNE, VICTOR GEORGE, organic chemistry, biochemistry, see 12th edition

BETSO, STEPHEN RICHARD, b Brooklyn, NY, May 28, 45; m 68. ANALYTICAL CHEMISTRY, POLYMER CHEMISTRY. Educ: St Johns Univ, NY, BS, 66; Ohio State Univ, MS, 70, PhD(chem), 71. Prof Exp: Res fel anal chem, Univ Ga, 71-73; ANAL CHEMIST, DOW CHEM CO, 73- Mem: Am Chem Soc; Sigma Xi. Res: Development of new analytical methodology; application of analytical technology to chemical production; polymer analytical chemistry; electroanalytical and thermal analytical chemistry. Mailing Add: Dow Chem Co Anal Labs 574 Bldg Midland MI 48640

BETT, HILLYARD DOBSON, b Dundee, Scotland, May 8, 13; m 40; c 2. BIOCHEMISTRY. Educ: Univ Toronto, BA, 38, MA, 39, PhD(physiol hyg), 44. Prof Exp: Asst chem, Univ Toronto, 38-39; tech asst, 39-41, res asst, 41-46, res assoc, 46-60, RES MEM, CONNAUGHT MED RES LABS, 60- Res: Preparation of pituitary hormones; preparation of liver extracts, vitamin B-12 and effect of these extracts on bone marrow; investigation of procedures for the purification of insulin; development of large scale processes. Mailing Add: Connaught Med Res Labs 1755 Steeles Ave Willowdale ON Can

BETT, JOHN ALEXANDER STUART, b Dehra Dun, India, Nov 19, 33; m 62; c 1. PHYSICAL CHEMISTRY. Educ: Univ St Andrews, BSc, 56; Ill Inst Technol, PhD(chem), 62. Prof Exp: Fel, McGill Univ, 62-64; fel chem, Mellon Inst, 64-68; RES PROJ ENGR, POWER SYSTS DIV, UNITED TECHNOLOGIES CORP, 68- Res: Heterogeneous catalysis; electrocatalysis; surface science. Mailing Add: Power Systs Div United Technologies Corp Middletown CT 06457

BETTELHEIM, FREDERICK A, b Gyor, Hungary, June 3, 23; nat US; m 47; c 1. PHYSICAL CHEMISTRY. Educ: Cornell Univ, BS, 53; Univ Calif, MS, 54, PhD(phys chem), 56. Prof Exp: Asst chemist, Agr Exp Sta, Israel, 47-51; analyst, NY Agr Exp Sta, 52-53; asst, Univ Calif, Davis, 53-56; instr, Univ Mass, 56-57; from asst prof to assoc prof, 57-64, PROF PHYS CHEM, ADELPHI UNIV, 64- Concurrent Pos: Lalor Found fel, 58; mem adv bd mil personnel supplies, Nat Acad Sci, 67-71. Mem: Am Chem Soc; NY Acad Sci; Fedn Am Socs Exp Biol; Asn Res Vision & Ophthal. Res: Physical chemistry of high polymers; mucopolysaccharides; glycoproteins; silicons; x-ray diffraction, birefringence; light scattering high vacuum vapor sorption; infra-red dichroism; molecular, supermolecular structures in cornea, lens, vitreous of the eye. Mailing Add: Dept of Chem Adelphi Univ Garden City NY 11530

BETTEN, CORNELIUS, JR, b Lake Forest, Ill, May 10, 11; m 41; c 3. CHEMISTRY. Educ: Cornell Univ, AB, 31, PhD(chem eng), 36. Prof Exp: Instr anal chem, Cornell Univ, 30-36; chem engr, Procter & Gamble Co, 36-37; asst chief chemist, Champion Papers, Inc, 37-43, chief chemist, US Plywood-Champion Papers Inc, 45-46, RES CHEMIST, CHAMPION INT CORP, 46- Concurrent Pos: Instr schs, Ohio, 39-43. Mem: Am Chem Soc; Am Tech Asn Pulp & Paper Indust. Res: Production of paper pulp and development and utilization of by-products. Mailing Add: Champion Int Corp Hamilton OH 45013

BETTENCOURT, JOSEPH S, JR, b Cambridge, Mass, Mar 5, 40; m 63; c 3. PARASITOLOGY. Educ: Suffolk Univ, AB, 62; Univ NH, MS, 65, PhD(zool), 76. Prof Exp: Instr, 65-67, ASST PROF BIOL, MARIST COL, 69- Mem: Am Soc Parasitologists; AAAS; Sigma Xi. Res: Marine fish and fresh water fish hematozoa; life cycles of Cryptobia, host specificity and vector transmission. Mailing Add: Dept of Biol Marist Col Poughkeepsie NY 12601

BETTERTON, HARRY O, b Hammond, Ind, Oct 17, 34; m 66; c 2. MICROBIAL PHYSIOLOGY, BIOCHEMISTRY. Educ: Ball State Teachers Col, BS, 60; Southern Ill Univ, MS, 63, PhD(microbiol), 68. Prof Exp: Asst prof biol, microbiol & physiol, Ball State Univ, 68-69; res assoc biochem, Ind Univ, Bloomington, 69-70; ASST PROF MICROBIOL, MED CTR, UNIV ARK, LITTLE ROCK, 70- Concurrent Pos: NSF grant, Ball State Univ, 68-69; fel, Ind Univ, Bloomington, 69-70. Mem: AAAS; Am Soc Microbiol. Res: Biochemical genetics; mechanisms of drug inhibition; subcellular molecular biology. Mailing Add: Dept of Microbiol Univ of Ark Med Ctr Little Rock AR 72201

BETTERTON, JESSE OATMAN, JR, b Omaha, Nebr, Aug 2, 20; m 49; c 3. PHYSICS. Educ: Lehigh Univ, BS, 42; Oxford Univ, DPhil, 51. Prof Exp: Metallurgist, Dow Chem Co, 42-47 & Oak Ridge Nat Lab, 51-68; ASSOC PROF PHYSICS, UNIV NEW ORLEANS, 68- Mem: Am Soc Metals; Am Inst Mining, Metall & Petrol Engrs; Am Phys Soc. Res: Low temperature specific heats; alloy properties; high field superconductivity; thermodynamic properties; phase equilibria; zone purification of the early transition elements and their alloys. Mailing Add: Dept of Physics Univ of New Orleans New Orleans LA 70122

BETTICE, JOHN ALLEN, b Columbus, Ind, Nov 27, 42; m 67. MAMMALIAN PHYSIOLOGY. Educ: Univ Dayton, BS, 65; Johns Hopkins Univ, PhD(physiol), 72. Prof Exp: ASST PROF PHYSIOL, SCH MED, CASE WESTERN RESERVE UNIV, 74- Concurrent Pos: Fel sch med, Johns Hopkins Univ, 72-73; fel, Univ Kans Med Ctr, 73-74. Mem: AAAS. Res: Mechanisms of the physiological reactions which buffer changes in acid-base balance of the body and the effects of such inbalances on body functions and cellular metabolism. Mailing Add: Dept of Physiol 2119 Abington Rd Case Western Reserve Univ Cleveland OH 44106

BETTINGER, DONALD JOHN, b Cincinnati, Ohio, Aug 20, 27; m 51; c 4. INORGANIC CHEMISTRY. Educ: Miami Univ, BS, 47; Univ Cincinnati, MS, 50; Univ NC, PhD(chem), 55. Prof Exp: Asst prof chem, Davidson Col, 50-51 & WVa Univ, 54-58; from asst prof to assoc prof, Denison Univ, 58-63; chmn dept chem, 63-74, head div math & natural sci, 67-74, PROF CHEM, OHIO NORTHERN UNIV, 63- Mem: Am Chem Soc; Am Inst Chemists. Res: Preparative inorganic chemistry; inorganic aggregation and polymerization phenomena; structure and stereochemistry of coordination compounds; chemical education. Mailing Add: Dept of Chem Ohio Northern Univ Ada OH 45810

BETTINGER, RICHARD THOMAS, b Dayton, Ohio, Aug 3, 31; m 52; c 7. SPACE PHYSICS. Educ: Syracuse Univ, BS, 55; Univ Md, PhD(physics), 64. Prof Exp: Asst prof physics, Univ Md, College Park, 64-74; MEM STAFF, BETCO ELECTRONICS, 74- Mem: Am Geophys Union. Res: In situ probe systems for the measurement of ionospheric parameters; vertical ozone distribution measured from satellites. Mailing Add: Betco Electronics 15504 Old Columbia Pk Burtonsville MD 20730

BETTIS, JERRY RAY, b Chandler, Okla, Dec 13, 42; m 61; c 3. OPTICAL PHSYICS. Educ: Univ Okla, BSEP, 64; Air Force Inst Technol. MS, 67, PhD(optical physics), 75. Prof Exp: Res physicist laser triggered high voltage switching, 67-71, RES PHYSICIST OPTICAL PHYSICS, AIR FORCE WEAPONS LAB, KIRTLAND AFB, N MEX, 73- Res: Laser induced damage to transparent dielectrics and dielectric thin film coating; laser triggered switching of multimegavolt systems with sub-nanosecond precision. Mailing Add: Air Force Weapons Lab/LRE Kirtland AFB NM 87117

BETTLER, PHILIP C, b Burlington, Iowa, Aug 25, 17; m 56. PHYSICS. Educ: Univ Okla, BA, 40, MS, 46; Ore State Univ, PhD, 58. Prof Exp: Asst, Univ Okla, 40-42; mem staff radar modulators, Radiation Lab, Mass Inst Technol, 42-45; asst cyclotron design, Radiation Lab, Univ Calif for Univ Rochester, 46-47; asst physicist, Microwave Lab, Univ Calif, 47-51; physicist geophys inst, Univ Alaska, 51-54; res physicist, Linfield Res Inst, Ore, 56-61; res assoc, 61-65, assoc prof physics, 65-69, PROF PHYSICS, UNIV NEV, RENO, 69- Mem: AAAS; Am Phys Soc; Am Asn Physics Teachers. Res: Field electron emission; field ion emission; solid state physics and surface physics; aurora, upper atmospheric physics. Mailing Add: Dept of Physics Univ of Nev Reno NV 89507

BETTMAN, JEROME WOLF, b San Francisco, Calif, June 22, 09; m 35; c 2. MEDICINE, OPHTHALMOLOGY. Educ: Univ Calif, AB, 31, MD, 35. Prof Exp: From asst resident to resident ophthal, 35-37, from assoc instr to assoc clin prof, 37-57, CLIN PROF SURG, MED SCH, STANFORD UNIV, 57-, EXEC DIR BASIC SCI COURSE IN OPHTHAL, 70-; CLIN PROF OPHTHAL, MED SCH, UNIV CALIF, SAN FRANCISCO, 64- Concurrent Pos: Edward Jackson lectr, Univ Colo, 52; consult, Marine Hosp, USPHS & San Francisco Hosp, 57-59; chief ophthal, Presby Med Ctr, 59-66; mem comt sci assemblies, NIH-Calif Med Asn Sci Bd. Honors & Awards: Award of Honor, Am Acad Ophthal & Otolaryngol, 58. Mem: Asn Res Vision & Ophthal. Res: Factors influencing the blood volume of the choroid and the retina; toxic cataracts. Mailing Add: 3910 Sand Hill Rd Woodside CA 94062

BETTMAN, MAX, b Ger, Nov 15, 25; nat US; m 57; c 1. PHYSICAL CHEMISTRY. Educ: Reed Col, BA, 49; Calif Inst Technol, PhD(chem), 52. Prof Exp: Res chemist, Calif Res & Develop Co, 52-53 & Union Carbide Corp, 53-64; RES CHEMIST, SCI LAB, FORD MOTOR CO, 64- Mem: Am Chem Soc. Res: Solid state chemistry; heterogeneous catalysis. Mailing Add: 28328 E Kalong Circle Southfield MI 48076

BETTONVILLE, PAUL JOHN, b St Louis, Mo, July 19, 18; m 50; c 2. PHARMACOLOGY. Educ: St Louis Univ, MD, 42. Prof Exp: Resident internal med, Univ Hosps, 46-48, instr pharmacol & internal med, Sch Med, 48-59, ASST PROF CLIN MED, SCH MED, ST LOUIS UNIV, 59- Mem: AAAS; AMA. Res: Chemical carcinogenesis. Mailing Add: 6331 Pershing St Louis MO 63130

BETTS, DONALD DRYSDALE, b Montreal, Que, May 16, 29; m 54; c 4. THEORETICAL PHYSICS. Educ: Dalhousie Univ, BSc, 50, MSc, 52; McGill Univ, PhD(math physics), 55. Prof Exp: Nat Res Coun Can fel, 55-56, from asst prof to assoc prof physics, 56-66, PROF PHYSICS, UNIV ALTA, 66-, DIR, THEORET PHYSICS INST, 72- Concurrent Pos: NATO sci fel, King's Col, London, 63-64, Nuffield fel, 70-71; mem, Int Comn Thermodyn & Statist Mech; chmn comn thermodyn & statist mech, Int Union Pure & Appl Physics, 72-75; vis prof chem & physics, Cornell Univ, 75. Mem: Am Phys Soc; Can Asn Physicists (vpres, 68-69, pres, 69-70). Res: Theoretical solid state and statistical physics; mathematical methods of physics. Mailing Add: Dept of Physics Univ of Alta Edmonton AB Can

BETTS, HENRY BROGNARD, b New Rochelle, NY, May 25, 28; m 70.

REHABILITATION MEDICINE. Educ: Princeton Univ, BA, 50; Univ Va, MD, 54. Prof Exp: Staff physiatrist phys med & rehab, 63-64, assoc med dir, 64-65, MED DIR, REHAB INST CHICAGO, 65-, V PRES, 69-; PROF REHAB MED & CHMN DEPT, MED SCH, NORTHWESTERN UNIV, CHICAGO, 69- Concurrent Pos: Consult staff mem, Northwestern Mem Hosp, 67-; dir, Res & Training Ctr 20; mem gov comt, Employ of Handicapped. Honors & Awards: Meritorious Serv Citation, US Pres Comt Employ of Handicapped, 65. Mem: Am Cong Rehab Med (vpres, 71-, pres, 75-76); Am Acad Phys Med & Rehab. Res: Research of prosthetics and orthotics; neuromuscular studies; sociological research. Mailing Add: Rehab Inst of Chicago 345 E Superior Chicago IL 60611

BETTS, ROBERT HOLLADAY, b Calgary, Alta, Dec 30, 19; m 47; c 1. PHYSICAL CHEMISTRY. Educ: Univ Alta, BSc, 42, MSc, 43; McGill Univ, PhD(chem), 45. Prof Exp: Res chemist, Nat Res Coun Can, 45-52; res chemist, Atomic Energy Can, Ltd, 52-55, head res chem br, 55-63, asst dir chem & metall div, 63-66; PROF CHEM & HEAD DEPT, UNIV MAN, 66- Mem: Chem Inst Can; The Chem Soc. Res: Physical chemistry of aqueous solutions; reaction kinetics; radiation chemistry; radio and nuclear chemistry. Mailing Add: Dept of Chem Univ of Man Winnipeg MB Can

BETTS, SHERMAN WILCOX, meteorology, oceanography, see 12th edition

BETTWY, MARY LEON, b Altoona, Pa, Aug 7, 28. INORGANIC CHEMISTRY. Educ: Seton Hill Col, BA, 56; Univ Notre Dame, MS, 58, PhD(chem), 59. Prof Exp: From asst prof to assoc prof, 59-70, chmn dept, 64-71, PROF CHEM, SETON HILL COL, 70- Mem: Am Chem Soc; Soc Appl Spectros. Res: Infrared absorption spectral studies of coordination compounds. Mailing Add: Dept of Chem Seton Hill Col Greensburg PA 15601

BETZ, BARBARA JEAN, b Boscobel, Wis, Jan 25, 09; m 41; c 2. PSYCHIATRY. Educ: Mt Holyoke Col, AB, 31; Johns Hopkins Univ, SM, 33, MD, 38; Am Bd Psychiat & Neurol, dipl. Prof Exp: Instr med, Johns Hopkins Univ, 42-46, from asst prof to assoc prof phychiat, 46-65; assoc prof col med, Cornell Univ, 65-68, prof, 68-69; clin prof, Dartmouth Med Sch, 69-72; adj prof psychiat, Cornell Univ Col Med, 72-75; MEM FAC PSYCHIAT, UNIV CALIF, LOS ANGELES, 75- Concurrent Pos: Asst psychiatrist, Johns Hopkins Hosp, 42-46, psychiatrist, 46-65; dir psychiat educ, Spring Grove State Hosp, 62-65; coordr psychiat educ, New York Hosp-Cornell Med Ctr, Westchester Div, 65-69; dir psychiat educ & res, Brattleboro Retreat, 69-70; pvt pract, 70-75; chief ment health unit, Sepulveda Vet Admin Hosp, 75- Mem: Asn Advan Psychother; Am Asn Social Psychiat; AMA; fel Am Psychiat Asn. Res: Psychotherapy of schizophrenia. Mailing Add: 906 Iliff St Pacific Palisades CA 90272

BETZ, DANIEL OLIVER, JR, b Ada, Ohio, Sept 15, 16; m 45; c 3. NEMATOLOGY, PLANT PATHOLOGY. Educ: Ohio State Univ, BSc & MS, 47. Prof Exp: Agent golden nematode control, USDA, 48-49, supvr, 49-56, agriculturist, Plant Pest Control Br, 56-60, nematologist, Pesticides Regulation Div, Environ Protection Agency, 60-65, asst chief staff off, 65-74; RETIRED. Honors & Awards: USDA Cert Merit, 63. Res: Nematode identification; laboratory supervision; conduct of nationwide system nematode surveys; administration of federal insecticide, fungicide and rodenticide act, specializing in nematocides and fungicides. Mailing Add: 8405 Wagon Wheel Rd Alexandria VA 22309

BETZ, EBON ELBERT, b Springport, Mich, Sept 3, 14; m 43; c 4. MATHEMATICS. Educ: Albion Col, AB, 34; Univ Mich, AM, 35; Univ Pa, PhD(math), 39. Prof Exp: Asst instr math, Univ Pa, 35-38; instr, Haverford Col, 39-41; from instr to assoc prof, 41-57, PROF MATH, US NAVAL ACAD, 57- Mem: Am Math Soc. Res: Accessibility and separation by simple closed curves; topology. Mailing Add: Dept of Math US Naval Acad Annapolis MD 21402

BETZ, GABRIEL POHL, b Mendota, Ill, Mar 2, 11; m 47; c 2. GEOGRAPHY. Educ: Univ Ill, BS, 38, MS, 39; Syracuse Univ, PhD(geog), 51. Prof Exp: Asst prof geog, Univ Buffalo, 51-55; chmn dept geog & earth sci, 57-64, dean, Sch Arts & Sci, 62-68, PROF GEOG, CALIFORNIA STATE COL, PA, 55- Mem: Asn Am Geogr; Nat Coun Geog Educ; Am Geog Soc. Res: Geography of Anglo America; economic geography. Mailing Add: Dept of Geog & Earth Sci California State Col California PA 15419

BETZ, GEORGE, b Asherville, Kans, Apr 27, 34; m 57; c 3. OBSTETRICS & GYNECOLOGY, ENDOCRINOLOGY. Educ: Kans State Col, BS, 56; Univ Kans, MD, 60, PhD(biochem), 68. Prof Exp: Resident gen surg, Denver Gen Hosp, 61-62; resident obstet & gynec, Univ Kans, 62-65, instr, 65-68; asst prof, 68-73, ASSOC PROF OBSTET & GYNEC, UNIV COLO, DENVER, 73-, RES ASSOC BIOCHEM, 71- Mem: Endocrine Soc; Soc Gynec Invest. Res: Biochemistry of reproduction with primary emphasis on steroid converting enzymes; endocrinology of reproduction. Mailing Add: Dept of Obstet & Gynec Univ of Colo Med Ctr Denver CO 80220

BETZ, JOHN VIANNEY, b Philadelphia, Pa, Apr 5, 37; m 62; c 2. BACTERIOLOGY, VIROLOGY. Educ: St Joseph's Col, Pa, BSc, 58; St Bonaventure Univ, PhD(microbiol), 63. Prof Exp: Fel microbiol, Ind Univ, 62-63, instr bact, 63; asst prof, 63-69, ASSOC PROF BOT & BACT, UNIV S FLA, 69- Mem: Am Soc Microbiol. Res: Interactions of host bacteria and bacteriophages; nature of lysogeny; genetic factors affecting lysogenesis; mechanism of flagellar movement. Mailing Add: Dept of Bot & Bact Univ of SFla Tampa FL 33620

BETZ, NORMAN LEO, b Baton Rouge, La, Jan 23, 38; m 60; c 5. BEHAVIORAL PHYSIOLOGY, FOOD BIOCHEMISTRY. Educ: La State Univ, BA, 61, MS, 63, PhD(food sci), 66. Prof Exp: Res assoc, Agr Res Serv, Entom Res Div, USDA, 62-66; chief field messing br/instr, US Army Quartermaster Sch, 66-68; assoc sr scientist, Mallinckrodt Chem Works, 68-72; SCIENTIST, RALSTON PURINA CO, 72- Concurrent Pos: Exec ed, Cereal Foods World, Am Asn Cereal Chemists, 74-; exec vpres, Physicians Nutrit Serv Inc, 75- Mem: Inst Food Technologists; Am Asn Cereal Chemists. Res: Primary responsibility for human and animal palatability programs in areas of taste, odor, color, texture and appetite control. Mailing Add: Ralston Purina Co 900 Checkerboard Sq St Louis MO 63188

BETZ, ROBERT F, b Chicago, Ill, Jan 25, 23; m 51; c 3. BIOCHEMISTRY, ECOLOGY. Educ: Ill Inst Technol, BS, 48, MS, 52, PhD(biochem), 55. Prof Exp: Instr biochem & microbiol, Ill Inst Technol, 50-52; assoc prof biol, 55-68, PROF BIOL SCI, NORTHEASTERN ILL STATE COL, 68- Res: Biochemistry and physiology of Myxomycetes, especially their enzymes; regeneration of prairies in abandoned cemeteries; autecology and conservation of prairie plants. Mailing Add: Dept of Biol Sci Northeastern Ill State Col Bryn Mawr at St Louis Ave Chicago IL 60625

BETZ, THOMAS WILLIAM, b Sarnia, Ont, Sept 19, 36; US citizen; m 68. DEVELOPMENTAL ENDOCRINOLOGY. Educ: Univ Mo, BA, 58, MA, 60; Univ Ill, PhD(zool), 65. Prof Exp: Asst prof biol, Carleton Univ, Ont, 65-67; NIH fel biochem, Emory Univ, 67-68; asst prof biol, 68-71, ASSOC PROF BIOL, CARLETON UNIV, ONT, 71- Concurrent Pos: Res grants, Nat Res Coun Can & Ont Dept Univ Affairs, 65-67; partic, Laurentian Hormone Conf, 70; NATO fel zool, Univ Gothenburg, 72-73. Mem: Int Soc Develop Biologists; Europ Soc Comp Endocrin; AAAS; Am Soc Zoologists; Am Inst Biol Scientists. Res: Developmental endocrinology of the spleen, duodenum, neural retina, adenohypophysis, yolk sac, hatching and growth of chicken embryos. Mailing Add: Dept of Biol Carleton Univ Colonel By Dr Ottawa ON Can

BEUCHAT, LARRY RAY, b Meadville, Pa, July 23, 43. FOOD MICROBIOLOGY. Educ: Pa State Univ, BS, 65; Mich State Univ, MS, 67, PhD(food sci), 70. Prof Exp: Group leader res & develop, Quaker Oats Co, 70-72; asst prof, 72-76, ASSOC PROF FOOD MICROBIOL, UNIV GA, 76- Mem: Inst Food Technologists; Am Soc Microbiol; Soc Appl Bact; Int Asn Milk, Food & Environ Sanitarians; Mycol Soc Am. Res: Environmental factors affecting growth and death of Vibrio parahaemolyticus; chemical changes occurring in peanuts fermented or infected with fungi; conditions for growth of microbial pathogens on pecan nuts. Mailing Add: Dept of Food Sci Univ of Ga Agr Exp Sta Experiment GA 30212

BEUG, MICHAEL WILLIAM, b Austin, Tex, May 18, 44; m 68. ORGANIC CHEMISTRY, ENVIRONMENTAL CHEMISTRY. Educ: Harvey Mudd Col, BS, 66; Univ Wash, PhD(chem), 71. Prof Exp: Asst prof chem, Harvey Mudd Col, 71-72; MEM FAC CHEM, EVERGREEN STATE COL, 72- Mem: AAAS; Am Chem Soc. Res: The effects of pesticides, polychlorinated biphenyls, heavy metals and petroleum on terrestrial and aquatic ecosystems; the detection of pollutants by chromatography and gas chromatography; infra red spectroscopy. Mailing Add: Dept of Chem Evergreen State Col Olympia WA 98505

BEUHLER, ROBERT JAMES, JR, b Lake Charles, La, Feb 21, 42; m 65; c 2. PHYSICAL CHEMISTRY. Educ: Univ Mich, BS, 63; Univ Wis, Madison, PhD(phys chem), 68. Prof Exp: Res assoc, Argonne Nat Lab, 68-70; res assoc, 70-72, asst chemist, 72-75, CHEMIST, BROOKHAVEN NAT LAB, 75- Res: Crossed molecular beam kinetics; focusing and orientation of molecules in molecular beams; ion-molecule reactions; ion impact phenomena; biological mass spectrometry. Mailing Add: Chem Dept Brookhaven Nat Lab Upton NY 11973

BEUK, JACK FRANK, b Ely, Minn, July 17, 17; m 46; c 3. BIOCHEMISTRY, ENZYMOLOGY. Educ: Univ Minn, BS, 41; Rutgers Univ, MS, 43. Prof Exp: Res chemist biochem, 43-65, MGR ENZYMES, SWIFT & CO, 65- Honors & Awards: Nat Indust Achievement Award, Inst Food Technol, 61. Mem: Inst Food Technol. Res: Development of new enzyme formulas for tenderizing of meat by both ante and post mortem techniques; research and adaptation of new enzyme sources for antemortem tenderization of meat. Mailing Add: 6 S Madison St Hinsdale IL 60521

BEUS, STANLEY S, b Salt Lake City, Utah, July 31, 30; m 53; c 5. INVERTEBRATE PALEONTOLOGY, STRATIGRAPHY. Educ: Utah State Univ, BS, 57, MS, 58; Univ Calif, Los Angeles, PhD(geol), 63. Prof Exp: From asst prof to assoc prof geol, 62-70, chmn dept geol & geog, 66-69, PROF GEOL, NORTHERN ARIZ UNIV, 70- Concurrent Pos: Paleontologist, Mus Northern Ariz, 63-; co-ed, J Paleont, 74- Mem: Geol Soc Am; Paleont Soc. Res: Paleozoic stratigraphy. Mailing Add: Dept of Geol & Geog Box 6030 Northern Ariz Univ Flagstaff AZ 86001

BEUTE, MARVIN KENNETH, b Jenison, Mich, Mar 3, 35; m 55; c 2. PLANT PATHOLOGY. Educ: Calvin Col, BA, 63; Mich State Univ, PhD(plant path), 68. Prof Exp: ASSOC PROF PLANT PATH, NC STATE UNIV, 68- Mem: AAAS; Am Phytopath Soc. Res: Ecology of soil borne pathogens; disease complexes; diseases of the peanut. Mailing Add: Dept of Plant Path NC State Univ Raleigh NC 27609

BEUTER, JOHN H, b Chicago, Ill, Dec 24, 35; m 60; c 2. FOREST ECONOMICS. Educ: Mich State Univ, BS, 57, MS, 58; Iowa State Univ, PhD(forestry econ), 66. Prof Exp: Resource analyst, US Forest Serv, 61-63, economist & proj leader, 65-68; economist, Mason, Bruce & Girard Consult Foresters, 68-70; ASSOC PROF FOREST MGT, ORE STATE UNIV, 70- Mem: Soc Am Foresters; Forest Prod Res Soc; Am Forestry Asn. Res: Economics of forest management and forest products marketing and utilization; general resource economics; operations research. Mailing Add: Dept of Forest Econ Ore State Univ Corvallis OR 97331

BEUTHIN, FREDERIC C, b Detroit, Mich, Sept 22, 40; m 63; c 2. PHARMACOLOGY. Educ: Wayne State Univ, BS, 65, MS, 68; Purdue Univ, PhD(pharmacol), 71. Prof Exp: ASST PROF PHARMACOL & PHARM, IDAHO STATE UNIV, 70- Mem: AAAS; NY Acad Sci. Res: Biochemical pharmacology; neurochemical correlates of drug response; role of central nervous system transmitters in hypertension. Mailing Add: Dept of Pharmacol Idaho State Univ Pocatello ID 83209

BEUTLER, ERNEST, b Berlin, Ger, Sept 30, 28; US citizen; m 50; c 4. MEDICINE. Educ: Univ Chicago, PhB, 46, BS, 48, MD, 50. Prof Exp: From instr to asst prof med, Univ Chicago, 55-59, attend physician, Univ Clins, 55-59; CHMN DIV MED, CITY OF HOPE MED CTR, 59- Concurrent Pos: Assoc clin prof med, Univ Southern Calif, 60-64, clin prof, 64-65 & 67-; clin prof, Univ Calif-Calif Col Med, 65-67; mem hemat study sect, NIH, 70-74. Honors & Awards: Gairdner Found Award, 75. Mem: Nat Acad Sci; Asn Am Physicians; Am Soc Human Genetics; Am Soc Hemat. Res: Red cell biochemistry and physiology; biochemical genetics; hematology. Mailing Add: City of Hope Med Ctr 1500 E Duarte Rd Duarte CA 91010

BEUTNER, ERNST HERMAN, b Berlin, Ger, Aug 27, 23; nat US; m 49; c 4. MICROBIOLOGY. Educ: Univ Pa, PhD, 51. Prof Exp: Res supvr, Sias Res Labs, Brooks Hosp, 51-55; res assoc, Harvard Sch Dent Med, 55-56; asst prof bact & immunol, 62-68, PROF MICROBIOL, STATE UNIV NY BUFFALO, 68- Concurrent Pos: Mem subcomt fluorescent antibodies, Nat Comt Clin Lab Standards; consult, Vet Admin Hosp, Buffalo. Honors & Awards: Rocha Lima Award, 67. Mem: Hon mem Polish Dermat Soc; Soc Exp Biol & Med; Am Soc Immunol; NY Acad Sci; Int Col Exp Dermat. Res: Light and electron microscopy of bacteria; antibiotic susceptibility of bacteria to nitrofurans; tissue immunology of pituitary and salivary glands; immunopathology of thyroiditis; systemic lupus erythematosus; myasthenia gravis; pemphigus and bullous pemphigoid. Mailing Add: Sch of Med State Univ of NY Buffalo NY 14214

BEUZEVILLE, CARLOS F, b Lima, Peru, Aug 17, 26; m 56. PEDIATRICS. Educ: San Marcos Univ, Lima, BSc, 44, BM & MD, 53. Prof Exp: Asst, Clin Lab, Hosp Loayza, 54-55; res asst biol, Nat Inst Andean Biol, 55-56; Int Educ Exchange Serv fel, Bryn Mawr Col, 56-58; assoc prof biochem, Univ San Agustin, Peru, 58-60; vis asst prof physiol, Albert Einstein Col Med, 60-69; from intern to resident pediat, Montefiore Hosp & Med Ctr, 69-71; RESIDENT PEDIAT, LONG ISLAND JEWISH HOSP & MED CTR, 71- Mem: Fel Am Acad Pediat. Res: Metabolism of antidiuretic hormones. Mailing Add: 640 W Boston Post Rd Mamaroneck NY 10543

BEVAK, JOSEPH PERRY, b Detroit, Mich, Mar 8, 29. PHYSICAL CHEMISTRY. Educ: Wayne State Univ, BS, 50; Mass Inst Technol, PhD(chem), 55. Prof Exp: From instr to assoc prof, 56-68, from actg head dept to head dept, 58-69, PROF CHEM, SIENA COL, 68- Mem: Am Chem Soc; Am Phys Soc. Res: Thermodynamics of solutions. Mailing Add: Dept of Chem Siena Col Loudonville NY 12211

BEVAN, DONALD EDWARD, b Seattle, Wash, Feb 23, 21. FISH BIOLOGY. Educ: Univ Wash, BS, 48, PhD(fisheries), 59. Prof Exp: Sci asst, Fisheries Res Inst, 47-48, res assoc, 48, proj leader, Kodiak Island Res, 48-58, proj supvr pink salmon res, 58-60, res assoc prof, Col Fisheries, 60-64, assoc prof, 64-66, assoc dean, Col Fisheries, 65-69, dir comput ctr, 68-69, PROF, COL FISHERIES, UNIV WASH, 66-, ASST VPRES RES, 69- & adj prof, Inst Marine Studies, 73- Concurrent Pos: Consult fisheries indust, 48-, pulp & paper indust, 62-; fel, Moscow State Univ, 59-60. Mem: AAAS; Am Inst Fishery Res Biologists; Am Soc Ichthyologists & Herpetologists. Res: Population dynamics; biometrics; computer science; Soviet fisheries. Mailing Add: 201 Admin Bldg Univ of Wash Seattle WA 98195

BEVAN, JOHN ACTON, b London, Eng, Apr 24, 30; m 56; c 4. PHARMACOLOGY. Educ: Univ London, BSc, 50, MB & BS, 53. Prof Exp: Demonstr pharmacol, St Bartholemew's Med Col, 50-52; from asst prof to assoc prof, 57-67, PROF PHARMACOL, UNIV CALIF, LOS ANGELES, 67-, ACTG CHMN DEPT, 65- Concurrent Pos: Ed, Blood Vessel. Mem: Am Soc Pharmacol & Exp Therapeut; Am Physiol Soc. Res: Pharmacology of the cardiovascular system. Mailing Add: Univ of Calif Sch of Med Los Angeles CA 90024

BEVANS, MARGARET, b Flushing, NY, Oct 2, 10. PATHOLOGY. Educ: Trinity Col, DC, BA, 31; Columbia Univ, MD, 35. Prof Exp: Asst pathologist, Bellevue Hosp, 38-43; pathologist, Goldwater Mem Hosp, 43-46; from asst prof to assoc prof path, Col Physicians & Surgeons, Columbia Univ, 47-57; DIR LABS, NEW YORK INFIRMARY, 57- Concurrent Pos: Dir labs, Midtown Hosp, 43-; pathologist, Columbia Res Serv, Goldwater, 47-56, consult, 58; lectr path, Col Physicians & Surgeons, Columbia Univ, 58-61; consult, New York Vet Admin Hosp, 58-64; mem coun arteriosclerosis, Am Heart Asn. Mem: AMA; Am Soc Exp Path; Am Asn Pathologists & Bacteriologists; Int Acad Path. Res: Arteriosclerosis; experimental nephritis; bile metabolism; cirrhosis. Mailing Add: New York Infirmary E 15th St & Stuyvesant Sq New York NY 10003

BEVC, VLADISLAV, b Ljubljana, Yugoslavia, Apr 9, 32; US citizen; m 61; c 2. ELECTROMAGNETICS, PLASMA PHYSICS. Educ: Univ Calif, Berkeley, BSc, 57, MSc, 58, PhD(microwave electronics), 61. Prof Exp: Nat Acad Sci-Nat Res Coun & Air Force Off Sci Res res fel, Univ Eng Lab & St Catherine's Col, Oxford Univ, 62-63; assoc prof elec eng, Naval Postgrad Sch, 63-66; mem tech staff, Aerospace Corp, 65-69; physicist, Lawrence Livermore Lab, Univ Calif, 69-73; PRES, BEVC ENG INC, 73- Concurrent Pos: Lectr & sr lectr, Univ Col, Univ Exten & Community Serv Div, Univ Southern Calif, 67- Mem: Am Phys Soc; Am Soc Eng Educ; Nat Soc Prof Engrs. Res: High energy electron beam dynamics, controlled thermonuclear fusion; electromagnetic theory; wave propagation in plasmas; electron optics. Mailing Add: Bevc Eng Inc 51 Hardester Ct Danville CA 94526

BEVELANDER, GERRIT, b West Sayville, NY, Apr 6, 05; m 35. HISTOLOGY, ANATOMY. Educ: Hope Col, AB, 26; Univ Mich, AM, 28; Johns Hopkins Univ, PhD(zool), 32. Prof Exp: Assoc, US Bur Fisheries, 28-29; instr biol, Union Univ, NY, 31-33; res assoc, Col Dent, NY Univ, 33-34, from instr to assoc prof anat, Univ, 34-47, prof histol, 47-62; prof, 62-72, EMER PROF HISTOL, UNIV TEX DENT BR HOUSTON, 72-; CONSULT, MONTEREY ABALONE FARMS, 73- Concurrent Pos: Consult, USPHS, 57-61 & 63-67; mem staff, Marine Biol Lab, Woods Hole & Bermuda Biol Sta. Mem: Am Soc Zoologists; Soc Exp Biol & Med; Am Asn Anatomists; fel NY Acad Sci; Int Asn Dent Res (pres). Res: Comparative histology of fishes; experimental approach to problems in calcification; histochemistry; integumentary derivatives; dynamic aspects of calcification; marine biology; electron microscope studies relating to mineralization. Mailing Add: PO Box 2656 Carmel CA 93921

BEVER, ARLEY TUNIS, (JR), b Bristol, Colo, Aug 25, 22; m 46; c 3. BIOCHEMISTRY. Educ: Colo Agr & Mech Col, BS, 48; Cornell Univ, MS, 51, PhD(biochem), 52. Prof Exp: Asst, Cornell Univ, 51-52; res fel biol, Harvard Univ, 52-55; from asst prof to assoc prof biochem, Sch Med, Univ Okla, 55-63; spec asst to assoc dir res grants, Off of Dir, NIH, 63-64, assoc chief res anal & eval div res grants, 64-68; dep planning dir, 68-70, head off budget, prog & planning, 70-73, DEP PLANNING DIR, NSF, 73- Mem: AAAS; Am Chem Soc; NY Acad Sci; Am Inst Chemists. Res: Oxidative enzymes; enzyme hormone interrelationships; uterine metabolism; skin biochemistry; science information systems; research analysis; science planning and policy. Mailing Add: NSF 1800 G St NW Washington DC 20550

BEVER, CHRISTOPHER THEODORE, b Munich, Ger, Mar 12, 19; nat US; m 44; c 4. PSYCHIATRY. Educ: Harvard Univ, BA, 40, MD, 43. Prof Exp: Intern, Hartford Hosp, Conn, 44; resident psychiat, St Elizabeths Hosp, Washington, DC, 47-48, psychiatrist, 48-50; psychiatrist, Washington Inst Ment Hyg, 50-51; dir, Montgomery County Ment Hyg Clin, 51-54; assoc prof psychiat, Sch Med, Univ NC, 54-56; MEM FAC, WASHINGTON SCH PSYCHIAT, 56-, CHMN PSYCHOTHER PRECEPTORSHIP & SEM PROG, 66-, MEM BD DIRS, 74- Concurrent Pos: Clin instr, Sch Med, Georgetown Univ, 49-51; instr, Washington Psychoanal Inst, 54-61, teaching analyst, 61-; dir psychiat outpatient ctr, NC Mem Hosp, 54-56; consult, Family Serv Agency, 56-58, St Elizabeths Hosp, Washington, DC, 59-61 & 73-74 & Walter Reed Army Hosp, 71-; assoc psychiat, Med Sch, George Washington Univ, 57-60, from asst clin prof to assoc clin prof, 60-74, clin prof, 74-; pres, DC Inst Ment Hyg, 66-68, consult, 66-72; pres, Community Psychiat Clin, 73-75; trustee, William Alanson White Psychiat Found, 75- Mem: AMA; fel Am Psychiat Asn; Am Psychoanal Asn; fel Am Orthopsychiat Asn; fel Acad Psychoanal. Res: Continuing education in psychotherapy. Mailing Add: Suite 303 2141 K St NW Washington DC 20037

BEVER, ENID L, b Lakewood, Ohio, July 7, 18. BIOCHEMISTRY. Educ: Pa State Univ, BS, 39; Univ Pa, MS, 40; Tex Women's Univ, PhD, 54. Prof Exp: Asst, Dept Res Med, Univ Pa, 40-43; instr sci, Stevens Sch, 43-46; instr chem, Centenary Jr Col, 46-48; instr, Pa State Univ, 48-51, instr nutrit res, 51-52; asst prof, Tex Women's Univ, 52-55; from assoc prof to prof chem, Milwaukee-Downer Col, 55-64, chmn dept 55-64, chmn div natural sci, 57-64; prof chem, Lawrence Univ, 64-68; chmn dept chem, Converse Col, 68-73; RETIRED. Mem: AAAS; Am Chem Soc; fel Am Inst Chemists; NY Acad Sci. Res: Vitamin A and protein interrelationships in animal nutrition; chemistry education; lipid chemistry and metabolism of the tocopherols. Mailing Add: PO Box 344 Etowah NC 28729

BEVER, JAMES EDWARD, b Bellingham, Wash, July 7, 20; m 46; c 1. GEOLOGY. Educ: Wash State Univ, BS, 42; Univ Mich, MS, 49, PhD(mineral), 54. Prof Exp: Asst purchasing agent, Traub Mfg Co, 47-48; instr mineral, Univ Mich, 50-54, instr conserv & acad counr, 52-54; from asst prof to assoc prof geol, 54-65, PROF GEOL, MIAMI UNIV, 65- Concurrent Pos: Consult, Oliver Mining Co, 53-54, Encyclop Americana, 56-57, Nat Res Coun, 57-58 & Gen Elec Co, 65-; vpres & dir, Timberline Minerals, Inc, 71- Mem: AAAS; Geol Soc Am; Mineral Soc Am; Mineral Asn Can. Res: Field geology; petrography and petrology; mineral resources. Mailing Add: Dept of Geol Miami Univ Oxford OH 45056

BEVER, WAYNE MELVILLE, b Lewiston, Idaho, Mar 5, 04; m 31; c 2. PLANT PATHOLOGY. Educ: Univ Idaho, BS, 27, MS, 28; Univ Wis, PhD(plant path & agron), 40. Prof Exp: Jr pathologist div cereal crops & dis, Bur Plant Indust, USDA, 27-38, from asst pathologist to pathologist, 39-53, pathologist cereal crops sect, Field Crops Res Br, Agr Res Serv, 53-66; prof plant path & head dept, 57-72, EMER PROF PLANT PATH, UNIV ILL, URBANA-CHAMPAIGN, 72- Mem: Fel AAAS; Am Phytopath Soc. Res: Cereal Crops. Mailing Add: 609 S Russell Champaign IL 61820

BEVERAGE, DAVID GAVIN, mathematics, see 12th edition

BEVERIDGE, DAVID L, b Coshocton, Ohio, Jan 29, 38; m 64; c 2. PHYSICAL CHEMISTRY. Educ: Col Wooster, BA, 59; Univ Cincinnati, PhD(phys chem), 65. Prof Exp: Asst operating chemist, Monsanto Res Corp, Ohio, 60-62; USPHS fel, Ctr Appl Wave Mech, Paris, 65-66 & Carnegie-Mellon Univ, 66-68; asst prof pharmacol, Mt Sinai Sch Med, 68-70; assoc prof, 70-74, PROF CHEM, HUNTER COL, 74- Concurrent Pos: NIH res career develop award, 72-77. Mem: AAAS; Am Chem Soc. Res: Quantum chemistry and biochemistry. Mailing Add: Dept of Chem Hunter Col New York NY 10021

BEVERIDGE, JAMES MACDONALD RICHARDSON, b Dunfermline, Scotland, Aug 17, 12; nat US; m 40; c 7. BIOCHEMISTRY. Educ: Acadia Univ, BSc, 37; Univ Toronto, PhD(biochem), 40; Univ Western Ont, MD, 50. Hon Degrees: DSc, Acadia Univ, 62; LLD, Mt Allison Univ, 66. Prof Exp: Asst, Dept Med Res, Banting Inst, 40; assoc biochemist, Pac Fisheries Exp Sta, 44-46; lectr path chem, Univ Western Ont, 46-50; Craine prof biochem & head dept, Queen's Univ, Ont, 50-64, dean sch grad studies, 60-64; PRES, ACADIA UNIV, 64- Mem: Royal Soc Can. Res: Analysis of protein hydrolysates; choline and fatty liver problem; dietary liver necrosis and lipid and bile acid metabolism. Mailing Add: Off of the Pres Acadia Univ Wolfville NS Can

BEVERIDGE, THOMAS ROBINSON, b Sandwich, Ill, June 30, 18; m 46; c 2. GEOLOGY. Educ: Monmouth Col, BS, 39; Univ Iowa, MS, 47, PhD(geol), 49. Prof Exp: Planning engr, NAm Aviation, Inc, 42-43; geologist & draftsman, Iowa Geol Surv, 45-46; asst geol, Univ Iowa, 46-47, instr, 47-49; geologist, Mo Geol Surv, 49-55, state geologist, 55-64; chmn dept geol & geophys, 65-71, PROF GEOL, UNIV MO-ROLLA, 64- Mem: Geol Soc Am. Res: Mississippian stratigraphy; structural geology of Missouri; engineering geology. Mailing Add: Dept of Geol Univ of Mo Rolla MO 65401

BEVERLEY-BURTON, MARY, b Abergavenny, Wales, June 10, 30; m 57, 68; c 3. ZOOLOGY. Educ: Univ Wales, BSc, 53; Univ London, PhD(parasitol), & dipl, Imp Col, 58. Prof Exp: Exp officer pest control, Scotland Dept Agr, Glasgow, 53-54, sci officer seed potato cert, Seed Testing Sta, 54-55; asst lectr parasitol, Imp Col, Univ London, 55-58; Nuffield res fel, Univ Col Rhodesia & Nyasaland, 58-60; asst ed helminth, Commonwealth Agr Bur, Eng, 67-68; asst prof zool, 68-72, ASSOC PROF ZOOL, UNIV GUELPH, 73- Mem: Am Soc Parasitol; Can Soc Wildlife & Fishery Biol; Can Soc Zool; Brit Soc Parasitol; Zool Soc London. Res: Helminth taxonomy; electron microscopy; biology of parasitic helminth. Mailing Add: Dept of Zool Univ of Guelph Guelph ON Can

BEVERUNG, WARREN NEIL, b New Orleans, La, Sept 3, 41; m 64; c 2. MEDICINAL CHEMISTRY. Educ: La State Univ, New Orleans, BS, 64, PhD(org chem), 68. Prof Exp: Asst org chem, La State Univ, New Orleans, 64-68; instr & fel, Univ Ill, Chicago Circle, 68-69; SR RES CHEMIST, ORG DIV, BRISTOL LABS, 69- Mem: Am Chem Soc. Res: Total synthesis of 9-azasteroids; general photochemical processes; synthesis of new biologically active heterocyclic systems. Mailing Add: Med Chem Dept Bristol Labs PO Box 657 Syracuse NY 13201

BEVILL, RARDON DIXON, III, b Winnfield, La, Feb 23, 38; m 61; c 3. BIOCHEMISTRY, CLINICAL CHEMISTRY. Educ: Washington & Lee Univ, BS, 60; Univ Minn, PhD(biochem), 65. Prof Exp: NSF fel, 67-68; asst prof molecular biol, Albert Einstein Col Med, 68-70; dir res applns, Medi-Comput Corp, 70-73; DIR RES APPLNS, SHEARSON HAYDEN STONE, INC, 73- Concurrent Pos: USPHS career develop award, 68-70. Mem: AAAS; Am Soc Biol Chem; Am Chem Soc; Am Soc Microbiol. Res: Biosynthesis, structure and function of polysaccharides; enzymatic methods of analysis of biological materials. Mailing Add: Shearson Hayden Stone Inc 85 Church St New Haven CT 06510

BEVILL, RICHARD F, JR, b Christopher, Ill, May 18, 34; m 53; c 2. VETERINARY PHARMACOLOGY. Educ: Univ Ill, DVM, 64, PhD(pharmacol), 72. Prof Exp: ASST PROF PHARMACOL, COL VET MED, UNIV ILL, URBANA, 72- Res: Pharmacokinetics of antibacterial drugs following their administration to domestic animals; relationships between plasma, urine and tissue residues are stressed. Mailing Add: Col of Vet Med Univ of Ill Urbana IL 61801

BEVIN, A GRISWOLD, b New Haven, Conn, Dec 7, 35; m 58; c 5. PLASTIC SURGERY, RECONSTRUCTIVE SURGERY. Educ: Wesleyan Univ, AB, 56; Yale Univ, MD, 60; Am Bd Surg, dipl, 67; Am Bd Plastic Surg, dipl, 71. Prof Exp: Instr surg, Sch Med, Yale Univ, 66-68; instr plastic surg, 68-69, ASSOC PROF PLASTIC SURG & CHIEF DIV PLASTIC & RECONSTRUCT SURG, SCH MED, UNIV NC, CHAPEL HILL, 69- Concurrent Pos: Consult, Watts Hosp, Durham, NC, 69- Mem: AAAS; NY Acad Sci; Am Asn Surg of Trauma; Am Soc Surg of Hand; Am Asn Plastic Surgeons. Res: Wound healing; surgery of the hand; maxilla facial trauma. Mailing Add: Div Plastic & Reconstruct Surg Univ of NC Med Sch Chapel Hill NC 27514

BEVINGTON, PHILIP RAYMOND, b New York, NY, July 31, 33; m 54; c 2. NUCLEAR PHYSICS. Educ: Harvard Univ, AB, 54; Duke Univ, PhD(physics), 60. Prof Exp: Instr physics, Duke Univ, 60-61, res assoc, 60-63, asst prof, 61-63; asst prof, Stanford Univ, 63-68; ASSOC PROF PHYSICS, CASE WESTERN RESERVE UNIV, 68- Mem: Am Phys Soc. Res: Research in nuclear structure through investigation of nuclear reactions attainable with Van de Graaff accelerators; nucleon-nucleon interactions at medium energies. Mailing Add: Dept of Physics Case Western Reserve Univ Cleveland OH 44106

BEVIS, JEAN HARWELL, b Miami, Fla, Dec 6, 39; m 60; c 2. ALGEBRA, SYSTEMS THEORY. Educ: Univ Fla, BS, 61, MS, 62, PhD(math), 65. Prof Exp: Asst prof math, Va Polytech Inst, 65-69; assoc prof, 69-73, PROF MATH, GA STATE UNIV, 73- Mem: Am Math Soc; Math Asn Am; Soc Indust & Appl Math; Asn Comput Mach. Res: Lattices; incidence matrices; graphs; automata; category theory; algebraic representation of properties of graphs. Mailing Add: Dept of Math Ga State Univ Atlanta GA 30303

BEVOLO, ALBERT JOSEPH, b St Louis, Mo, Oct 20, 40; m 66; c 2. SOLID STATE PHYSICS. Educ: St Louis Univ, BS, 62; MS, 64, PhD(physics), 70. Prof Exp: Fel physics, St Louis Univ, 70-72; Presidential intern, Ames Lab, US AEC, 72-73, ASST PHYSICIST, AMES LAB, ENERGY RESOURCE & DEVELOP ADMIN, 73- Mem: Am Phys Soc. Res: Low temperature specific heat of tungsten bronzes; IV-VI semicomputers and soft mode superconductors; electrochemical energy conversion. Mailing Add: Ames Lab Energy Resource & Develop Admin Ames IA 50011

BEWICK, HOWARD ALBANY, inorganic chemistry, see 12th edition

BEWLEY, GLENN CARL, b Middletown, Ohio, July 19, 42; m 65; c 3. DEVELOPMENTAL GENETICS. Educ: Miami Univ, BSEd, 65, MA, 67; Univ NC, Chapel Hill, PhD(genetics), 74. Prof Exp: Aerospace physiologist, US Naval Reserve, 67-70; NIH fel dept biol, Univ Va, 74-75; ASST PROF GENETICS, NC STATE UNIV, 75- Mem: Genetics Soc Am. Res: Regulation of gene function; genetic and biochemical analyses of isozymes in Drosophila; genetic organization and fine structure of the eukaryotic genome. Mailing Add: Dept of Genetics Gardner Hall NC State Univ Raleigh NC 27607

BEWLEY, JOHN DEREK, b Preston, Eng, Dec 11, 43; m 66; c 2. PLANT PHYSIOLOGY, BIOCHEMISTRY. Educ: Queen Elizabeth Col, Univ London, BSc, 65, PhD(plant physiol), 68. Prof Exp: Fel plant biochem, Inst Cancer Res, 68-70; asst prof biol, 70-73; ASSOC PROF BIOL, UNIV CALGARY, 73- Mem: Am Soc Plant Physiol; Can Soc Plant Physiol; Brit Soc Exp Biol; Scand Soc Plant Physiol; Japanese Soc Plant Physiol. Res: Dormancy and survival mechanisms in plants, biochemical and physiological aspects; biochemical ecology. Mailing Add: Dept of Biol Univ of Calgary Calgary AB Can

BEYCHOK, SHERMAN, b New York, NY, Sept 10, 31; m 50; c 1. BIOCHEMISTRY, IMMUNOBIOLOGY. Educ: City Col New York, BS, 52; NY Univ, MS, 55, PhD(chem), 57. Prof Exp: Instr biochem, NY Med Col, 52-56; guest lectr & mem staff, Div Sponsored Res, Mass Inst Technol, 56-60 & Children's Cancer Res Found, Harvard Med Sch, 60-61; asst prof biochem, Col Physicians & Surgeons, 62-67, assoc prof biophys & chmn subcomt biophys, 67-68, PROF BIOL SCI & CHEM, COLUMBIA UNIV, 68-, CHMN DEPT BIOL SCI, 73- Concurrent Pos: Trustee, Cold Spring Harbor Lab, 75- Mem: Am Chem Soc; Am Soc Biol Chem; Harvey Soc. Res: Structure and solution properties of biological macromolecules and related polymers; conformation of proteins; immunoglobulins; membrane proteins. Mailing Add: Dept of Biol Sci Columbia Univ New York NY 10027

BEYEA, JAN EDGAR, b Englewood, NJ, Dec 16, 39; m 65; c 2. NUCLEAR PHYSICS. Educ: Amherst Col, BA, 62; Columbia Univ, PhD(physics), 68. Prof Exp: Res assoc physics, Columbia Univ, 68-70; ASST PROF PHYSICS, HOLY CROSS COL, 70- Mem: Am Phys Soc; Sigma Xi. Res: Radiation damage; nuclear shell model; safety of nuclear power. Mailing Add: Dept of Physics Holy Cross Col Worcester MA 01610

BEYER, ARTHUR FREDERICK, b Toledo, Ohio. PALEOBOTANY. Educ: Ohio Univ, BSc, 43; Ohio State Univ, MSc, 45; Univ Cincinnati, PhD(bot), 50. Prof Exp: Instr bot, Western Reserve Univ, 47-50; assoc prof, 50, PROF BOT, MIDWESTERN STATE UNIV, 51-, CHMN DEPT BIOL, 60- Mem: Bot Soc Am; Sigma Xi. Res: Tamarix gallica morphology; ecology; paleoxylotomy. Mailing Add: Dept of Biol Midwestern State Univ Wichita Falls TX 76308

BEYER, EDGAR HERMAN, b Melrose Park, Ill, Apr 27, 31; m 54; c 3. PLANT BREEDING, PLANT GENETICS. Educ: Univ Ill, BS, 58; Purdue Univ, MS, 62, PhD(plant breeding & genetics), 64. Prof Exp: Asst prof forage breeding, Univ Md, College Park, 63-66; res dir, Farm Seed Res Corp, 66-74; VEG & FLOWER SEED PROD MGR, FERRY-MORSE SEED CO, 74- Concurrent Pos: Mem, Nat Cert Alfalfa Variety Rev Bd, 69-72. Mem: Am Soc Agron; Crop Sci Soc Am; Sigma Xi. Res: Testing performance of forage crop varieties; combining ability studies with alfalfa single crosses; improvement of Trifolium pratense by interspecific hybridization; alfalfa breeding and development. Mailing Add: 1807 Kirklyn Dr San Jose CA 93124

BEYER, ELMO MONROE, JR, b Corpus Christi, Tex, Mar 22, 41; m 64; c 1. PLANT PHYSIOLOGY, PLANT CHEMISTRY. Educ: Tex Technol Col, BS, 63; Tex A&M Univ, PhD(plant physiol), 69. Prof Exp: Res biologist, Cent Res Dept, 69-74, RES SUPVR, CENT RES DEPT, E I DU PONT DE NEMOURS & CO, INC, 74- Mem: AAAS; Am Soc Plant Physiologists; Am Inst Biol Scientist. Res: Phytohormones, especially mechanism of ethylene action; effect of ethylene on translocation; plant growth regulators, utility and mechanism of action. Mailing Add: Cent Res Dept Exp Sta E I du Pont de Nemours & Co Inc Wilmington DE 19898

BEYER, GEORGE LEIDY, b Philadelphia, Pa, Jan 12, 19; m 46; c 3. POLYMER CHEMISTRY. Educ: Juniata Col, BS, 41; Rutgers Univ, PhD(anal chem), 45. Prof Exp: Lab asst, Rutgers Univ, 41-43, instr chem, 43-44; phys chemist, 45-55, RES ASSOC RES LABS, EASTMAN KODAK CO, 55- Mem: Am Chem Soc. Res: Polymer molecular characterization; light-scattering; exclusion chromatography. Mailing Add: Eastman Kodak Co Kodak Park Rochester NY 14650

BEYER, JACQUELYN L, b Mitchell, SDak, July 11, 24. GEOGRAPHY. Educ: Univ Colo, BA, 44, MA, 54; Univ Chicago, PhD(geog), 57. Prof Exp: Instr geog, Chicago Teacher's Col, 56; asst prof, Mont State Univ, 57-60; lectr, Univ Cape Town, 60-64; asst prof geog, Rutgers Univ, Newark, 64-70; assoc prof geog, 70-74, PROF GEOG & CHMN DEPT GEOG & ENVIRON STUDIES, UNIV COLO, COLORADO SPRINGS, 74- Concurrent Pos: Vis lectr, Univ Tex, 58-59. Mem: Am Geog Soc; Asn Am Geogr; Nat Coun Geog Educ; African Studies Asn; Am Water Resources Asn. Res: Resource management, conservation and perception; economic development and cultural change, especially in Africa. Mailing Add: Dept of Geog Univ of Colo Colorado Springs CO 80907

BEYER, KARL HENRY, JR, b Henderson, Ky, June 19, 14; m 40; c 2. PHARMACOLOGY, PHYSIOLOGY. Educ: Western Ky State Col, BS, 35; Univ Wis, PhM, 37, PhD(physiol), 40, MD, 43. Prof Exp: Instr chem, Western Ky State Col, 35-36; instr physiol, Med Sch, Univ Wis, 39-43; from asst dir to dir pharmacol res, Sharp & Dohme, 43-50, asst dir res, 50-56, dir res, Merck Inst Therapeut Res, 56-58, pres, 61-66, sr vpres res, Merck, Sharp & Dohme Res Labs, 66-73; RETIRED. Concurrent Pos: Chmn med chem sect, Gordon Res Conf, 56; vpres life sci, Merck, Sharp & Dohme Res Labs, 58-66; lectr, Swed Govt, 62, Sch Med, Howard Univ, 64, Inst Pharmacol, Free Univ Berlin, 66, Sch Med, Univ Wis, 67, Med Sch, Temple Univ, Grad Med Sch, Univ Pa, Jefferson Med Col & Med Col Pa; mem drug res bd, Nat Acad Sci, 64-70; treas bd trustees, Biol Abstr, 65-69; mem int comt, Third Int Pharmacol Cong, Sao Paulo, 66 & mem int adv comt, Fourth Int Cong, Basle, 69; resident fel, Col Physicians Philadelphia; mem coun on circulation, Am Heart Asn; vis prof pharmacol, Hershey Med Ctr, Pa State Univ & Med Sch, Vanderbilt Univ. Honors & Awards: Merck Sci Award, 59; Gairdner Found Award, 64; Mod Pioneers in Creative Indust Award, Nat Asn Mfrs, 65; Mod Med Award Distinguished Achievement, 67; Found Award in Pharmacodynamics, Am Pharmaceut Asn, 67; Cert of Distinction, Am Therapeut Soc, 68; Lasker Award, 75. Mem: Fel AAAS; fel Am Col Physicians; Am Physiol Soc; Am Soc Pharmacol & Exp Therapeut (secy, 59-61, pres elect, 63-64, pres, 64-65, past-pres, 65-66); Am Chem Soc. Res: Autonomic, renal and chemical pharmacology; pharmacodynamics and toxicology. Mailing Add: PO Box 276 Gwynedd Valley PA 19437

BEYER, LOUIS MARTIN, b Paducah, Ky, Nov 7, 39; m 59; c 2. NUCLEAR PHYSICS. Educ: Murray State Univ, BS, 62; Mich State Univ, MS, 63, PhD(nuclear physics), 67. Prof Exp: From asst prof to assoc prof, 67-75, PROF PHYSICS, MURRAY STATE UNIV, 75- Mem: Am Asn Physics Teachers. Res: Low energy nuclear structure by methods of beta and gamma ray spectroscopy; atomic structure and lifetime by beam-foil spectroscopic means; x-ray produced by heavy charged particle reactions. Mailing Add: Dept of Physics Murray State Univ Murray KY 42071

BEYER, ROBERT EDWARD, b Englewood, NJ, Feb 20, 28; m 54; c 3. BIOCHEMISTRY. Educ: Univ Conn, AB, 50, MSc, 52; Brown Univ, PhD(biol), 54. Prof Exp: Asst biol, Brown Univ, 51-53; USPHS fel, Wenner-Gren Inst, Stockholm, 54-56; instr physiol, Sch Med, Tufts Univ, 56-58, asst prof, 58-62; asst prof enzyme chem, Enzyme Inst, Univ Wis-Madison, 62-65; assoc prof zool, 65-69, PROF ZOOL, UNIV MICH, 69- Concurrent Pos: USPHS sr res fel, Sch Med, Tufts Univ, 58-59; NIH res career develop award, Tufts Univ, 59-62 & Univ Wis, 62-65. Mem: AAAS; Am Physiol Soc; Am Soc Biol Chemists; Am Soc Cell Biol; Biophys Soc; Can Physiol Soc. Res: Hormonal control of metabolic systems; mechanism of oxidative phosphorylation and its physiological control; cold acclimation. Mailing Add: Dept Cellular & Molecular Biol Lab Chem Biol Univ of Mich Ann Arbor MI 48104

BEYER, ROBERT THOMAS, b Harrisburg, Pa, Jan 27, 20; m 44; c 4. ACOUSTICS. Educ: Hofstra Univ, AB, 42; Cornell Univ, PhD, 45. Prof Exp: Asst physics, Cornell Univ, 42-45; from instr to assoc prof physics, 45-58, exec officer, 66-68, chmn dept, 68-74, PROF PHYSICS, BROWN UNIV, 68- Concurrent Pos: Advan Educ Fund fel, Univ Calif, Los Angeles, 53-54; consult Russian trans, Am Inst Physics, 55-, Raytheon, 61-71 & Off Naval Res, 74-75. Mem: Fel Am Phys Soc; fel Acoust Soc Am (vpres, 61-62, pres, 68-69); fel Inst Elec & Electronics Engrs. Res: Underwater sound; acoustic relaxation times; ultrasonic absorption in liquids and solids; nonlinear acoustics; liquid state; physics in the Soviet Union. Mailing Add: 132 Cushman Ave East Providence RI 02914

BEYER, WENDELL T, b Van Nuys, Calif, Nov 26, 39; m 64. COMPUTER SCIENCE. Educ: Univ Ore, BA, 62, MA, 64; Mass Inst Technol, PhD(math), 69. Prof Exp: Mem tech staff, Bell Tel Labs, 64-65; asst prof comput sci, 69-74, SR SYSTS PROGRAMMER, UNIV ORE, 74- Mem: Asn Comput Mach. Res: Programming language design. Mailing Add: Comput Ctr Univ of Ore Eugene OR 97403

BEYER, WILLIAM A, b Tyrone, Pa, Nov 9, 24; m 55; c 2. MATHEMATICS. Educ: Pa State Univ, BS, 49, PhD(math), 59; Univ Ill, Urbana, MS, 50. Prof Exp: Teaching asst math, Pa State Univ, 52-53, instr, 54-55 & 57-59; teaching asst, Queen's Univ, Ont, 53-54; mathematician, Gen Elec Co, Ohio, 55-57; STAFF MEM, LOS ALAMOS SCI LAB, 60- Mem: Am Math Soc; Math Asn Am; Soc Indust & Appl Math. Res: Mathematical biology; probability; Laplace transforms; mechanics; fractional dimension theory; biological mathematics. Mailing Add: Los Alamos Sci Lab Los Alamos NM 87544

BEYER, WILLIAM HYMAN, b Akron, Ohio, Mar 8, 30; m 59; c 3. MATHEMATICAL STATISTICS, MATHEMATICS. Educ: Univ Akron, BS, 52; Va Polytech Inst, MS, 54, PhD(statist), 61. Prof Exp: Group leader reliability & qual control, Goodyear Aerospace Corp, 53-57; corp staff statistician, Gen Tire & Rubber Co, 57-58; asst prof math, Va Polytech Inst, 58-61; from asst prof to assoc prof math & statist, 61-66, head dept math, 69-74, PROF MATH & STATIST, UNIV AKRON, 66- Concurrent Pos: Instr, Univ Akron, 54-56; partic, NSF In-Serv Inst Sec Sch Teachers Math & lectr, Univ Akron, 63-66; ed statist, Chem Rubber Co, Ohio, 66- Mem: Am Statist Asn; Inst Math Statist; Biomet Soc; Math Asn Am. Res: Experimental design; analysis of variance. Mailing Add: Dept of Math Univ of Akron 302 E Buchtel Ave Akron OH 44304

BEYER, WILLIAM W, b Twin Falls, Idaho, May 26, 30; m 50; c 3. FOOD SCIENCE. Educ: Univ Wis, BS, 53, MS, 57, PhD(food technol), 60. Prof Exp: Scientist prod develop refrig foods, 58-62, tech mgr prod develop refrig foods, 62-64, mgr res & develop refrig foods, 64-69, dir food technol, 69-72, DIR REFRIG FOODS RES & DEVELOP, PILLSBURY CO, 72- Mem: AAAS; Inst Food Technologists. Res: Effect of fungicides and insecticides on the physico-chemical properties of cherries; refrigerated foods product development. Mailing Add: 311 Second St SE Minneapolis MN 55414

BEYERLEIN, FLOYD HILBERT, b Frankenmuth, Mich, Apr 15, 42; m 65; c 2. ANALYTICAL CHEMISTRY. Educ: Mich State Univ, BS, 64, MS, 67, PhD(anal chem), 70. Prof Exp: RES CHEMIST ANAL CHEM, S C JOHNSON & SON, INC, 70- Mem: Am Chem Soc. Res: Methods development in support of product research; update analysis techniques already available. Mailing Add: S C Johnson & Son Inc 1525 Howe St Racine WI 53403

BEYERS, ROBERT JOHN, b Long Beach, Calif, Sept 13, 33; m 54; c 4. ECOLOGY, AQUATIC BIOLOGY. Educ: Univ Miami, BS, 54; Univ Tex, PhD(zool), 62. Prof Exp: NSF fel ecol, Inst Marine Sci, Univ Tex, 61-62, res assoc scientist, 62-64, head ecol prog, 63-64; from asst prof to assoc prof zool, Lab Radiation Ecol, Univ Ga, 64-74; dir, Savannah River Ecol Lab, 67-74; PROF BIOL SCI & CHMN DEPT, UNIV S ALA, 74- Concurrent Pos: Co-prin investr, AEC contract, 67-74; Fed Water Qual Admin grant, 70-74; vpres, Echo Environ Consults, Inc, 75- Mem: AAAS; Am Soc Limnol & Oceanog; Ecol Soc Am; Inst Soc Limnol. Res: Limnology; marine science; microcosm techniques; measurement of carbon dioxide metabolism in aquatic organisms and ecosystems; effects of radiation and pollution on aquatic ecosystems; fish behavior. Mailing Add: Dept of Biol Sci Univ of SAla Mobile AL 36688

BEYERS, WILLIAM BJORN, b Seattle, Wash, Mar 24, 40; m 68. ECONOMIC GEOGRAPHY, REGIONAL ECONOMICS. Educ: Univ Wash, BA, 62, PhD(geog), 67. Prof Exp: Asst prof geog, 67-75, ASSOC PROF GEOG, UNIV WASH, 75- Concurrent Pos: Vis asst prof regional sci, Harvard Univ, 71; Environ Protection Agency grant, Harvard Univ & Cornell Univ, 71-72; vis asst prof urban planning, Cornell Univ, 71-72. Mem: AAAS; Asn Am Geog; Regional Sci Asn; Am Econ Asn. Res: Regional development; location theory; input-output analysis. Mailing Add: Dept of Geog Univ of Wash Seattle WA 98195

BEYLER, ARTHUR LEWIS, b Wyanet, Ill, Jan 31, 22; m 43; c 2. ENDOCRINOLOGY. Educ: DePauw Univ, AB, 47, MA, 48; Univ Calif, Los Angeles, PhD(zool), 52. Prof Exp: Lectr zool, DePauw Univ, 48; head endocrinol res, 52-63, ASST DIR BIOL DIV, STERLING-WINTHROP RES INST, 63- Mem: AAAS; Endocrine Soc; NY Acad Sci; Soc Study Reproduction. Res: Hormonal interactions; hormone-enzyme interrelationships; factors influencing enzyme reactions; drug actions on carbohydrate, fat and protein metabolism. Mailing Add: Sterling-Winthrop Res Inst Rensselaer NY 12144

BEYLER, ROGER ELDON, b Nappanee, Ind, May 20, 22; m 44; c 3. ORGANIC CHEMISTRY. Educ: NCent Col, BA, 44; Univ Ill, MS, 47, PhD(chem), 49. Prof Exp: Res chemist, Merck & Co, Inc, 49-59; dean, Col Lib Arts & Sci, 66-73, dean, Col Lib Arts, 73-74, PROF CHEM, SOUTHERN ILL UNIV, CARBONDALE, 59- Concurrent Pos: Orgn Econ Coop & Develop sci fel, Univ 64. Mem: Am Chem Soc. Res: Synthesis of the alkaloid sparteine; adrenal steroid total synthesis; steroid synthesis. Mailing Add: Dept of Chem & Biochem Southern Ill Univ Carbondale IL 62901

BEYSTER, JOHN R, b Detroit, Mich, July 26, 24; m 55; c 3. NUCLEAR PHYSICS. Educ: Univ Mich, BSE(math) & BSE(physics), 45, MS, 47, PhD(physics), 50. Prof Exp: Sr scientist, Physics Dept, Westinghouse Atomic Power Div, 50-51; mem staff, Los Alamos Sci Lab, 51-57 & Proj Linac, Gen Atomic Div, Gen Dynamics Corp, 57-59; PRES & BD CHMN, SCI APPLN INC, 69- Concurrent Pos: Mem comt radiation sources, Nat Acad Sci, 61-64. Mem: Am Phys Soc; Am Nuclear Soc. Res: Accelerator physics; nuclear and reactor physics investigations of neutron cross sections and differential neutron spectra in the energy range from subthermal to fourteen million volts using accelerators and research reactors. Mailing Add: Sci Appln Inc PO Box 2351 La Jolla CA 92038

BEZDEK, JAMES CHRISTIAN, b Harrisburg, Pa, Oct 22, 39; m 63; c 4. APPLIED MATHEMATICS. Educ: Univ Nev, Reno, BSCE, 69; Cornell Univ, PhD(appl math), 73. Prof Exp: Instr eng mech, Cornell Univ, 72; asst prof math, State Univ NY Col Oneonta, 73-74; ASST PROF MATH, MARQUETTE UNIV, 74- Mem: Math Asn Am; Soc Indust & Appl Math; Pattern Recognition Soc. Res: Pattern recognition, cluster analysis and unsupervised learning using fuzzy sets and graphs; applied probability and ordinary differential equations. Mailing Add: Dept of Math Marquette Univ Milwaukee WI 53233

BEZDICEK, DAVID FRED, b Jackson, Minn, Sept 18, 38; m 62; c 2. SOIL MICROBIOLOGY. Educ: SDak State Univ, BS, 60; Univ Minn, MS, 64, PhD(soil sci), 67. Prof Exp: Field supvr, Calif Packing Corp, 60-61; asst prof agron, Univ Md, College Park, 67-74; ASSOC PROF SOILS, WASH STATE UNIV, 74- Mem: Am Soc Agron. Res: Waste disposal on land; symbiotic nitrogen fixation; phytotoxicity-crop residues; mine reclamation. Mailing Add: Dept of Agron & Soils Wash State Univ Pullman WA 99163

BEZKOROVAINY, ANATOLY, b Riga, Latvia, Feb 11, 35; US citizen; m 64; c 2. BIOCHEMISTRY. Educ: Univ Chicago, BS, 56; Univ Ill, MS, 58, PhD(biochem), 60. Prof Exp: Res assoc biochem, Oak Ridge Nat Lab, 60-61; res chemist, Nat Animal Dis Lab, USDA, Iowa, 61-62; from asst biochemist to assoc biochemist, Presby-St Luke's Hosp, 62-70, assoc prof, 70-73, PROF BIOCHEM & SR BIOCHEMIST, RUSH-PRESBY-ST LUKE'S MED CTR, 73- Concurrent Pos: From asst prof to assoc prof biochem, Univ Ill Col Med, 62-70, res prof, Dept Path, 74- Mem: Am Soc Biol Chemists; Am Chem Soc; Brit Biochem Soc; Am Inst Chemists. Res: Protein chemistry; glycoproteins; iron metabolism; biochemistry of milk; comparative biochemistry of proteins. Mailing Add: Dept of Biochem Rush-Presby-St Luke's Med Ctr Chicago IL 60612

BEZMAN, RICHARD DAVID, b Pittsburgh, Pa, Oct 2, 46; m 68. PHYSICAL CHEMISTRY. Educ: Pa State Univ, BSc, 67; Harvard Univ, PhD(chem), 72. Prof Exp: Mem tech staff, GTE Lab, Inc, 73-75; STAFF CHEMIST, LINDE DIV, UNION CARBIDE CORP, 75- Mem: Am Chem Soc; Electrochem Soc. Res: Heterogeneous catalysis; electrochemistry; solid state chemistry. Mailing Add: Tarrytown Tech Ctr Union Carbide Corp Tarrytown NY 10591

BEZNAK, MARGARET, b Budapest, Hungary, May 10, 14; nat Can; m 36. PHYSIOLOGY. Educ: Univ Budapest, MD, 39. Prof Exp: Demonstr physiol, Univ Budapest, 34-37, lectr, 37-40, asst prof, 40-46; res assoc, Hungary Biol Res Inst, Tihany, 46-48; from sr lectr to assoc prof, 53-60, head dept, 60-69, PROF PHYSIOL, UNIV OTTAWA, 60-, V DEAN FAC MED, 69- Concurrent Pos: Med Res Coun Gt Brit grant, Univ Birmingham, 49-53. Mem: Am Physiol Soc; Can Physiol Soc; Brit Physiol Soc. Res: Cardiac hypertrophy; correlation between size and work of heart and effect of endocrine glands on these processes. Mailing Add: Fac of Med Univ of Ottawa Ottawa ON Can

BEZUSZKA, STANLEY JOHN, b Wilna, Poland, Jan 26, 14; nat US. MATHEMATICAL PHYSICS. Educ: Boston Col, AB, 39, AM, 40, MS, 42; Weston Col, STL, 47; Brown Univ, PhD(physics), 53. Prof Exp: From instr to asst prof physics, Boston Col, 41-43; instr math, Weston Col, 43-45; from asst prof to assoc prof, 53-69, chmn dept, 53-67, PROF MATH, BOSTON COL, 69-, DIR MATH INST, 58- Concurrent Pos: Instr, Polaroid Corp, 57-61, mem educ policies comn, 63-66. Mem: AAAS; Am Math Soc; Math Asn Am; Acoustical Soc Am; Am Phys Soc. Res: Theoretical physics; scattering of ultrasonic waves; matrix theory; vector analysis and mathematical physics. Mailing Add: Dept of Math Boston Col Chestnut Hill MA 02167

BHAGAT, BUDH DEV, b India, Jan 1, 26. PHYSIOLOGY, PHARMACOLOGY. Educ: Univ London, PhD(pharmacol), 61. Prof Exp: Asst prof pharmacol, Med Sch, Howard Univ, 64-66; asst prof, NY Med Col, 66-68; assoc prof, 68-71, PROF PHYSIOL & PHARMACOL, SCH MED, ST LOUIS UNIV, 71- Concurrent Pos: Fel pharmacol, Med Sch, Univ Wis, 62 & Med Sch, Univ Minn, 63; mem adv bd neurosci res, Acad Press, 71- Res: Autonomic nervous system; neurotransmitter and cardiovascular studies; factors affecting release uptake, storage synthesis, metabolism and replenishment of catecholamines. Mailing Add: Dept of Physiol St Louis Univ Sch of Med St Louis MO 63104

BHAGAT, SATINDAR M, b Jammu, India, July 19, 33; m 64. PHYSICS. Educ: Univ Jammu & Kashmir, India, BS, 50; Univ Delhi, MSc, 53, PhD(physics), 56. Prof Exp: Lectr physics, Univ Delhi, 55-57; Govt of India res fel, Clarendon Lab, Oxford Univ, 57-60; res assoc, Carnegie Inst Technol, 60-62, instr, 61-62; asst prof, 62-67, ASSOC PROF PHYSICS, UNIV MD, COLLEGE PARK, 67- Res: Thermodynamic and hydrodynamic properties of liquid helium; properties of magnetic systems at low temperatures and other problems in solid state physics. Mailing Add: Dept of Physics Univ of Md College Park MD 20742

BHAGAVAN, HEMMIGE, b Hemmige, India, Oct 23, 34. NUTRITION, BIOCHEMISTRY. Educ: Univ Mysore, BSc, 53; Indian Inst Sci, Bangalore, AIISc, 58; Univ Ill, Urbana-Champaign, PhD(nutrit, biochem), 63. Prof Exp: Res asst biochem & nutrit, Cent Food Tech Res Inst, Univ Mysore, India, 57-58, jr sci asst biochem, 58-59; res assoc biol sci, Purdue Univ, 63-64; CHIEF BIOCHEM LAB, RES INST, ST JOSEPH HOSP, 64- Mem: Am Inst Nutrit; Fedn Am Socs Exp Biol; NY Acad Sci; Brit Biochem Soc. Res: Metabolism and function of vitamins; biochemical effects of vitamin deficiencies; central nervous system function. Mailing Add: Res Inst St Joseph Hosp Lancaster PA 17604

BHAGAVAN, NADHIPURAM V, b Mysore, India, Oct 5, 31; m 62; c 2. CLINICAL BIOCHEMISTRY. Educ: Univ Mysore, BSc, 51; Univ Bombay, MSc, 55; Univ Calif, PhD(pharmaceut chem), 60. Prof Exp: Asst res biochemist, Med Ctr, Univ Calif, San Francisco, 61-65; asst biochemist, 65-70, asst prof anat, 66-70, assoc prof biochem & med technol, 70-72, PROF BIOCHEM & MED TECHNOL, MED SCH, UNIV HAWAII, MANOA, 72- Concurrent Pos: Assoc prof biol & chem, Hawaii Loa Col, 69-70; consult biochemist, Kaiser Found Hosp, Honolulu, 72- Mem: Am Chem Soc. Res: Immunochemistry; biochemical and immunochemical studies on normal and malignant cell nuclei. Mailing Add: Depts Biochem & Med Technol Univ of Hawaii at Manoa Honolulu HI 96822

BHAKAR, BALRAM SINGH, b Wardha, India, Jan 1, 37; m 68; c 1. THEORETICAL NUCLEAR PHYSICS. Educ: Agra Univ, BSc, 57; Aligarh Muslim Univ, India, MSc, 60; Univ Delhi, PhD(physics), 65. Prof Exp: Res assoc physics, Bonner Nuclear Lab, Rice Univ, 66-68 & Univ Sussex, 68-69; ASST PROF PHYSICS, UNIV MANITOBA, 69- Mem: Am Phys Soc. Res: 3-nucleon system using separable potential and Faddeev theory and study of nuclear matter; 4-nucleon system. Mailing Add: Dept of Physics Univ of Manitoba Winnipeg MB Can

BHALLA, CHANDER P, b Hariana, India, Sept 15, 32; US citizen; m 62; c 3. ATOMIC PHYSICS. Educ: Punjab Univ, India, BS, 52, Hons, 54, MS, 55; Univ Tenn, PhD(physics), 60. Prof Exp: Asst physics, Univ Tenn, 56-60; sr scientist, Westinghouse Elec Corp, Pa, 60-64; from asst prof to assoc prof physics, Univ Ala, 64-66; assoc prof, 66-72, PROF PHYSICS, KANS STATE UNIV, 72- Concurrent Pos: Consult, Oak Ridge Nat Lab, 58-60, Nat Bur Standards, DC, 62-64, space div, Northrop Corp, Ala, 65-69 & Argonne Nat Lab, 66-69; vis prof, FOM Inst Atomic & Molecular Physics, Amsterdam, Netherlands, 73-74. Mem: Am Nuclear Soc; fel Am Phys Soc. Res: Heavy ion interactions; Auger effect; radiative transitions; internal conversion processes; inelastic and elastic energy loss of heavy ions; atomic structure. Mailing Add: Dept of Physics Kans State Univ Manhattan KS 66506

BHALLA, RANBIR J R SINGH, b Anandpur, India, July 24, 43; m 68; c 2. SOLID STATE SCIENCE. Educ: Univ Jabalpur, BS, 61, MS, 63, PhD(appl physics), 69; Pa State Univ, PhD(solid state sci), 70. Prof Exp: Lectr physics, Govt Sci Col, Univ Jabalpur, 64-66; res physicist inorg mats sci, Cent Res Lab, Am Cyanamid Co, 70-72; SR RES ENGR INORG MATS SCI, LAMP DIV, WESTINGHOUSE ELEC CORP, 72- Mem: Electrochem Soc; Electron Microprobe Soc; Illum Eng Soc. Res: Preparation, properties and characterization of inorganic materials; display materials, including cathodo and photochromics and phosphors. Mailing Add: Dept 8212 Westinghouse Elec Corp 1 Westinghouse Plaza Bloomfield NJ 07003

BHALLA, SATISH CHANDER, b Kasauli, India, July 10, 34; m 67. BIOLOGY, GENETICS. Educ: Panjab Univ, India, MS, 56; Univ Kans, MA, 63; Univ Notre Dame, PhD(biol), 66. Prof Exp: Res asst mosquito behav, Malaria Inst India, Delhi, 56-61; asst prof biol & genetics, Millersville State Col, 67-68; asst prof, 69-74, ASSOC PROF VECTOR GENETICS, UNIV MD, BALTIMORE, 74- Mem: Entom Soc Am; Am Mosquito Control Asn; Genetics Soc Can. Res: Mosquito behavior; vector genetics; insecticides; housefly and mosquito genetics. Mailing Add: Dept of Med Entom Univ of Md Sch of Med Baltimore MD 21201

BHALLA, VINOD KUMAR, b Lahore, India, Aug 4, 40; m 66; c 2. ENDOCRINOLOGY. Educ: Agra Univ, BS, 62, MS, 64, PhD(natural prod), 68. Prof Exp: Fel org chem, Univ Ga, 68-69, res assoc biochem & reproductive physiol, 69-72; res assoc biochem & endocrinol, Emory Univ, 72-74; ASST PROF ENDOCRINOL, MED COL GA, 74- Mem: Endocrine Soc; Soc Study Reproduction; Am Chem Soc; Soc Complex Carbohydrates. Res: Mechanism of action of gonadotropin in testicular functions. Mailing Add: Dept of Endocrinol Med Col of Ga Augusta GA 30902

BHANDARKAR, DILEEP PANDURANG, b Bombay, India, July 16, 49; m 73; c 1. COMPUTER SCIENCE. Educ: Indian Inst Technol, Bombay, BTechnol, 70; Carnegie-Mellon Univ, MS, 71, PhD(elec eng), 73. Prof Exp: MEM TECH STAFF COMPUT SCI, TEX INSTRUMENTS INC, 73- Mem: Inst Elec & Electronic Engrs; Asn Comput Mach. Res: Computer performance evaluation; computer architecture; fault tolerant computing. Mailing Add: Tex Instruments Inc PO Box 5936 MS 132 Dallas TX 75222

BHANGOO, MAHENDRA SINGH, b Ludhiana, India, Apr 15, 31; m 44; c 2. SOIL FERTILITY, PLANT NUTRITION. Educ: Agra Univ, BSc, 50; Univ Calif, Los Angeles, MSc, 54; Kans State Univ, PhD(soil fertil), 56. Prof Exp: Instr soils, Kans State Univ, 55-57; agronomist, Standard Fruit Co, Cent Am, 57-59, asst to res dir banana prod, La, 60-63; chemist, Cane Sugar Refining Res Proj Inc, 64-66; assoc prof agr, 66-69, PROF AGRON, UNIV ARK, PINE BLUFF, 69- Mem: Am Soc Agron; Soil Sci Soc Am; Sigma Xi. Res: Soil fertility; plant nutrition and irrigation requirement of bananas, cereals and legumes; phosphate chemistry in cane sugar refining process; soybean nutrient requirement, root development and irrigation. Mailing Add: Dept of Agr Univ of Ark Pine Bluff AR 71601

BHAPKAR, VASANT PRABHAKAR, b India, Apr 8, 31; m 61; c 3. MATHEMATICAL STATISTICS. Educ: Univ Bombay, BSc, 51, MSc, 53; Univ NC, PhD(math statist), 59. Prof Exp: Lectr statist, Univ Poona, 54-60, reader, 60-68, prof, 68-72; assoc prof statist, Univ Ky, 69-73, PROF STATIST, UNIV KY, 73- Concurrent Pos: Vis assoc prof statist, Univ NC, 64-66. Mem: Inst Math Statist; fel Am Statist Asn. Res: Categorical data, non-parametric methods in statistics; multivariate analysis; statistical inference. Mailing Add: Dept of Statist Univ Of Ky Lexington KY 40506

BHARADWAJ, PREM DATTA, b Gorakhpur, India, May 20, 31; m 49; c 4. NUCLEAR PHYSICS. Educ: NREC Col, India, BSc, 50; Agra Univ, MSc, 52; State Univ NY Buffalo, PhD, 64. Prof Exp: Asst prof physics, BR Col, India, 52-54; lectr physics, GPI Col, 54-56 & Govt Col, Meerut, 56-59; asst prof physics, BR Col, 59-60; from asst prof to assoc prof, 62-66, PROF PHYSICS, NIAGARA UNIV, 66- Res: High energy physics; theoretical physics. Mailing Add: Dept of Physics Niagara Univ Niagara Falls NY 14109

BHARATI, AGEHANANDA, b Vienna, Austria, Apr 20, 23; US citizen. CULTURAL ANTHROPOLOGY, SOUTH ASIAN STUDIES. Prof Exp: Lectr Ger & philos, Delhi Univ, 51; reader philos, Banaras Hindu Univ, 51-54; guest prof, Nalanda Inst Buddhist Res, India, 54-55; vis prof comp relig, Royal Mahamukuta Buddhist Acad, Bangkok, Thailand, 55-56; vis prof Indian philos, Univ Tokyo & Kyoto Univ, 56-57; res assoc Indian studies, Univ Wash, 57-61; asst prof anthrop, Syracuse Univ, 61-64, assoc prof, 64-68, PROF ANTHROP, SYRACUSE UNIV, 68-, CHMN DEPT, 71- Concurrent Pos: Sr res fel, NIMH, 64; guest prof, Univ Hawaii & Univ Wash, 65; Rose Morgan distinguished vis prof, Univ Kans, 70; sr fel, Am Inst Ceylonese Studies, Philadelphia, 70-71. Mem: Fel Am Anthrop Asn; fel Royal Anthrop Inst Gt Brit & Ireland; Asn Asian Studies; Am Oriental Soc; Soc Sci Study Relig. Res: South Asian culture and society; South Asian minorities in Africa; anthropology of ritual and belief systems; value orientations; anthropological linguistics, especially patterns of linguistic dissimulation in modern India; contemporary Buddhism in Ceylon. Mailing Add: Dept of Anthrop Syracuse Univ 500 University Pl Syracuse NY 13210

BHARGAVA, HEMENDRA NATH, b Delhi, India, Sept 30, 42; m 71. BIOCHEMICAL PHARMACOLOGY. Educ: Banaras Hindu Univ, India, BPharm, 63, MPharm, 65; Univ Calif, San Francisco, PhD(pharm chem), 69. Prof Exp: Fel pharmacol, Univ Calif, San Francisco, 69-72, res pharmacologist, 72-75, lectr, 74-75; ASST PROF PHARMACOL, UNIV ILL MED CTR, 75- Mem: Am Soc Pharmacol & Exp Therapeut. Res: Biochemical mechanisms in the central PhD(natural prod), 68. Prof Exp: Fel org chem, Univ Ga, 68-69, res assoc biochem & reproductive physiol, 69-72; res assoc biochem & endocrinol, Emory Univ, 72-74; ASST PROF ENDOCRINOL, MED COL GA, 74- Mem: Endocrine Soc; Soc Study Reproduction; Am Chem Soc; Soc Complex Carbohydrates. Res: Mechanism of action of gonadotropin in testicular functions. Mailing Add: Dept of Endocrinol Med Col of Ga Augusta GA 30902

BHARGAVA, RAMESHWAR NATH, b Allahabad, India, Dec 25, 39; m 65; c 2. PHYSICS, SOLID STATE PHYSICS. Educ: Univ Allahabad, BS, 57, MS, 59; Columbia Univ, PhD(physics), 66. Prof Exp: Res asst physics, Watson Labs, Int Bus Mach Corp, 62-66, consult, Watson Res Ctr, 66-67; mem tech staff, Bell Tel Labs, 67-70; mem tech staff, Philips Labs Div, 70-74, SR PROG LEADER, PHILIPS LABS DIV, N AM PHILIPS CORP, 74- Mem: Am Phys Soc; Inst Elec & Electronic Engrs. Res: Galvanomagnetic properties of semimetals and semiconductors; optical properties of semiconductors primarily in gallium arsenide and gallium phosphide; device work in gallium phosphide diodes; deep states and nonradiative processes in semiconductors. Mailing Add: Lab Div NAm Philips Corp 345 Scarborough Rd Briarcliff Manor NY 10510

BHARGAVA, TRILOKI NATH, b Lucknow, India, Aug 21, 33; US citizen. APPLIED STATISTICS. Educ: Lucknow Univ, BSc, 52, MSc, 54; Mich State Univ, PhD(math & statist), 62. Prof Exp: Lectr math & statist, Khalsa Col, India, 54-55; lectr, Gujerat Univ, India, 55-57; asst prof, Agra Univ, 57-58; res asst statist, Mich State Univ, 59-61; from asst prof to assoc prof math & statist, Kent State Univ, 62-67, PROF MATH & STATIST, KENT STATE UNIV, 67- Concurrent Pos: NASA grant, Kent State Univ, 63-67, NSF grant, 64, Environ Protection Agency res grant, 71-; res assoc, Ctr Urban Regionalism, Kent State Univ, 70- Mem: Fel Royal Statist Soc; Am Math Soc; Am Statist Soc; Inst Math Statist; Math Asn Am. Res: Applied probability and statistics; graph theory; binary systems; ecological systems. Mailing Add: Dept of Math Kent State Univ Kent OH 44242

BHARTENDU, b Banda, India, Dec 15, 35; m 64; c 2. ATMOSPHERIC PHYSICS. Educ: Univ Allahabad, BSc, 55, MSc, 58; Univ Sask, PhD(atmospheric physics), 64. Prof Exp: Res asst physics, Univ Sask, 60-64; fel, NMex Inst Mining & Technol, 65-66; RES SCIENTIST, ATMOSPHERIC ENVIRON SERV, DEPT ENVIRON, CAN, 66- Concurrent Pos: Fel, Univ Sask, 66; mem working group joint comt atmospheric elec, Int Asn Meteorol & Atmospheric Physics & Int Asn Geomagnetism & Aeronomy, 67-71; mem subcomn II, Int Comn Atmospheric Elec, Int Asn Meteorol & Atmospheric Physics, 71-; mem, Can Comt Atmospheric Elec, 72-75, chmn, 75-; mem, Can Nat Comt, Int Union Radio Sci, 74- Mem: Am Geophys Union; Am Meteorol Soc; Can Meteorol Soc. Res: Atmospheric electricity including its relationship with meteorology, fair weather atmospheric electricity, biological effects of atmospheric electricity; atmospheric acoustics, including thunder and pressure waves from nuclear and chemical explosions. Mailing Add: Atmospheric Environ Serv Dept of Environ 4905 Dufferin St Downsview ON Can

BHARUCHA, KEKI RUSTOMJI, b Bombay, India, Feb 4, 28. ORGANIC CHEMISTRY. Educ: Univ Bombay, BSc, 46, MSc, 49; Univ London, PhD(org chem) & DIC, 52. Prof Exp: Res assoc, Univ Toronto, 52-53; SR RES CHEMIST, SR SCIENTIST & GROUP LEADER, RES LABS, CAN PACKERS, LTD, 53- Mem: Am Chem Soc; fel The Chem Soc. Res: Chemistry of steroids; synthesis of analogs of bioactive compounds; antiviral agents. Mailing Add: Res & Develop Labs Can Packers Ltd 2211 St Clair Ave W Toronto ON Can

BHARUCHA, NANA R, b Bombay, India, Oct 20, 26. ELECTROCHEMISTRY, SURFACE CHEMISTRY. Educ: Univ Bombay, BSc, 46, MSc, 48, PhD(chem), 50; Univ Manchester, PhD(chem technol), 53. Prof Exp: Res chemist, Paint Res Sta, Eng, 54-60; sr investr chem, Brit Non-Ferrous Metals Res Asn, 60-66, head appl chem div, 66-68; group leader, 68-69, head electrochem & corrosion dept, 69-70, MGR, CHEM RES DIV, NORANDA RES CTR, 70- Mem: Fel Brit Inst Metal Finishing; Electrochem Soc; Soc Chem Indust; Am Chem Soc; Inst Corrosion Sci & Technol. Mem: Corrosion; organic coatings; metal finishing; lubrication; chemical metallurgy. Mailing Add: Noranda Res Ctr 240 Hymus Blvd Pointe Claire PQ Can

BHARUCHA-REID, ALBERT TURNER, b Hampton, Va, Nov 13, 27; m 54; c 2. APPLIED MATHEMATICS. Educ: Iowa State Univ, BS, 49. Prof Exp: Asst, mat biol, Univ Chicago, 50-53 & math statist, Columbia Univ, 53-54, res assoc, 54-55; asst res statistician, Univ Calif, Berkeley, 55-56; from instr to asst prof math, Univ Ore, 56-61; assoc prof, 61-65, PROF MATH, WAYNE STATE UNIV, 65-, DIR CTR RES IN PROBABILITY, 67- Concurrent Pos: Prin investr, US Air Force res grant, 54-55; co-prin investr, US Army res off grant, 56-62; res fel, math inst, Polish Acad Sci, 58-59; co-prin investr, NSF grant, 62-64; vis prof, Inst Math Sci, Univ Madras, India, 63-64 & math res ctr, Univ Wis, Madison, 66-67; prin investr, NIH res grant, 66-69 & NSF grant, 69-71; prof, Ga Inst Technol, 73-74. Mem: AAAS; Am Math Soc; Soc Indust & Appl Math; Inst Math Statist; Int Asn Math Geol (vpres, 72-76). Res: Markov processes and their applications; mathematical biology; probabilistic analysis and its applications; semigroups of operators. Mailing Add: Dept of Math Wayne State Univ Detroit MI 48202

BHASIN, MADAN M, b Lahore, India, June 23, 38; m 61; c 2. PHYSICAL CHEMISTRY, SURFACE CHEMISTRY. Educ: Univ Delhi, BSc, 58; Notre Dame Univ, PhD(phys chem), 64. Prof Exp: Teaching asst phys chem, Ind Univ, 59-60; proj chemist, 63-69, PROJ SCIENTIST, UNION CARBIDE CORP, 69- Mem: Am Chem Soc; NAm Catalysis Soc. Res: Heterogenous catalysis; spectroscopic investigations of catalyst surfaces; diffusion phenomenon; diffusion through membranes. Mailing Add: Union Carbide Corp Res & Develop Dept PO Box 8361 South Charleston WV 25303

BHASKAR, SURINDAR NATH, b Rasul, India, Jan 7, 23; nat US; m 50; c 3. PATHOLOGY. Educ: Punjab Univ, BDS, 42; Northwestern Univ, DDS, 46; Univ Ill, MS, 48, PhD(anat), 51; Am Bd Oral Path & Am Bd Oral Med, dipl. Prof Exp: Instr oral path & histol, Col Dent, Univ Ill, 51-52, assoc prof path, 52-55; US Army, 55-, chief oral tumors br, Armed Forces Inst Path, 55-60, chief dept oral path, 60-70, dir, Army Inst Dent Res, Walter Reed Army Med Ctr, 70-73, dir personnel, Off Surgeon Gen, 73-75, ASST SURGEON GEN & CHIEF DENT CORPS, US ARMY, 75- Concurrent Pos: Prof, Sch Dent & Med, Georgetown Univ. Mem: Fel AAAS; fel Am Acad Oral Path; Am Acad Oral Med; Am Dent Asn; Int Acad Oral Path. Res: General and oral pathology; human and experimental tumors of salivary glands; oral tumors and oral diseases. Mailing Add: Asst Surg Gen & Chief Dent Corps US Army Pentagon Washington DC 20310

BHASKARAN, GOVINDAN, b Mavelikara, India, Feb 12, 35; m 62; c 2. INSECT PHYSIOLOGY, DEVELOPMENTAL BIOLOGY. Educ: Univ Kerala, BSc, 55, MSc, 57; Univ Bombay, PhD(zool), 62. Prof Exp: Asst res officer, Indian Coun Med Res, Haffkine Inst, Bombay, 61; sci officer, Bhabha Atomic Res Ctr, Trombay, 61-68; res assoc entom, Univ Ill, Urbana-Champaign, 68-70; sr scientist, Biol Dept, Zoecon Corp, 70-75; SR SCIENTIST, INST DEVELOP BIOL, TEX A&M UNIV, 73-, ASSOC PROF BIOL, 75- Mem: AAAS; Soc Develop Biol; Radiation Res Soc. Res: Insect physiology and development; developmental biology of imaginal discs in the Diptera. Mailing Add: Inst of Develop Biol Tex A&M Univ College Station TX 77843

BHAT, CLARITA CSAKY, organic chemistry, see 12th edition

BHAT, MULKI RADHAKRISHNA, b Mulki, India, May 7, 30; m 67. NUCLEAR PHYSICS. Educ: Univ Bombay, BSc, 51; Univ Poona, MSc, 54; Ohio State Univ, PhD(nuclear physics), 61. Prof Exp: Nat Res Coun Can fel, 61-63; res assoc neutron physics, 64-66, from asst physicist to assoc physicist, 66-73, PHYSICIST, BROOKHAVEN NAT LAB, 73- Mem: AAAS; Am Phys Soc. Res: Neutron physics; measurement of resonance parameters and neutron capture gamma rays; nuclear spectroscopy; neutron cross-section evaluation. Mailing Add: Brookhaven Nat Lab Upton NY 11973

BHAT, UGGAPPAKODI NARAYAN, b Vittal, India, Nov 17, 33; m 59; c 2. OPERATIONS RESEARCH, STATISTICS. Educ: Univ Madras, BA, 53, BT, 54; Karnatak Univ, India, MA, 58; Univ Western Australia, PhD(math statist), 64. Prof Exp: Teacher high sch, India, 54-56; lectr statist, Karnatak Univ, India, 58-61; asst math, Univ Western Australia, 61-64, temp lectr, 64-65; asst prof statist, Mich State Univ, 65-66; from asst prof to assoc prof opers res, Case Western Reserve Univ, 66-69; from assoc prof to prof comput sci & opers res, 69-75, head dept, 72-74, PROF INDUST ENG & OPERS RES & HEAD DEPT, SOUTHERN METHODIST UNIV, 75- Concurrent Pos: Assoc ed, Opsearch, 68-74, Opers Res, 68- & Mgt Sci, 69-74. Mem: Am Statist Asn; Inst Math Statist; Opers Res Soc Am; Inst Mgt Sci; Am Inst Indust Engrs. Res: Queueing theory; probabilistic models for computer and information systems; applied probability; stochastic processes. Mailing Add: Dept of Indust Eng & Opers Res Southern Methodist Univ Dallas TX 75275

BHAT, VENKATRAMANA KAKEKOCHI, b Padre, India, July 26, 33; US citizen; m 67; c 2. PHARMACEUTICAL CHEMISTRY, MEDICINAL CHEMISTRY. Educ: Univ Madras, BSc, 55, BPharm, 58; Univ Wash, PhD(pharmaceut chem), 63. Prof Exp: Res assoc carbohydrate chem, Georgetown Univ, 63-67; pharmaceut chemist, Gulf South Res Inst, 67-71; CONSULT CHEMIST, 71- Mem: AAAS; Am Chem Soc; Am Inst Chemists; NY Acad Sci; Sigma Xi. Res: Synthesis of sulfur containing compounds related to Ephedrine; synthesis of nucleosides, carbohydrates and antimalarials; analytical biochemistry, especially analysis of drugs, herbicides and pesticides using gas chromatography. Mailing Add: 8002 53rd Ave W Everett WA 98203

BHATIA, ANAND K, b WPakistan, Jan 26, 34; m 63; c 1. ATOMIC PHYSICS. Educ: Univ Delhi, BSc, 53, MSc, 55; Univ Md, PhD(physics), 62. Prof Exp: Asst prof physics, Wesleyan Univ, 62-63; res assoc, Nat Acad Sci, 63-65; AEROSPACE TECHNOLOGIST, GODDARD SPACE FLIGHT CTR, NASA, 65- Res: High energy physics; molecular physics. Mailing Add: Goddard Space Flight Ctr Code 602 NASA-Greenbelt MD 20771

BHATIA, AVADH BEHARI, b Barabanki, India, Aug 16, 21; m 50; c 2. THEORETICAL PHYSICS, SOLID STATE PHYSICS. Educ: Univ Allahabad, BSc, 40, MSc, 42, DPhil, 46; Univ Liverpool, PhD(theoret physics), 51. Prof Exp: Lectr physics, Univ Allahabad, 44-47; prof theoret physics, Phys Res Lab, India, 50-52; Imp Chem Indust fel, Univ Edinburgh, 52-53; Nat Res Coun Can fel, 53-55; from asst prof to assoc prof physics, 55-60, PROF PHYSICS, UNIV ALTA, 60- Mem: Am Phys Soc; Acoust Soc Am; Can Asn Physicists; fel Royal Soc Can; fel Brit Inst Physics & Phys Soc. Res: Nuclear scattering and nuclear reactions; transport phenomena in and thermodynamic properties of metals and alloys; ultrasonics. Mailing Add: Dept of Physics Univ of Alta Edmonton AB Can

BHATIA, KISHAN, b Poona, India, Mar 23, 36. ORGANIC CHEMISTRY, ANALYTICAL CHEMISTRY. Educ: Univ Poona, BSc, 58, MSc, 60; Univ Ark, PhD(org chem), 65. Prof Exp: Demonstr chem, Fergusson Col, Univ Poona, 59-60; jr res fel org chem, Nat Chem Lab, India, 60-61; asst, Univ Ark, 61-65; sr res chemist, US Steel Corp, 65-70; fel chem, Mellon Inst, Carnegie-Mellon Univ, 71-75; CHROMATOGRAPHY SECT LEADER, ALLIED CHEM CORP, 75- Mem: Am Chem Soc. Res: Mechanism of organic reactions; application of chromatographic, spectroscopic and isotope techniques to the study of problems in organic chemistry. Mailing Add: Allied Chem Corp CRL PO Box 1021-R Morristown NJ 07960

BHATIA, NAM PARSHAD, b Lahore, India, Aug 24, 32; m 62; c 3. MATHEMATICS. Educ: Agra Univ, BSc, 52, MSc, 54 & 56; Dresden Tech Univ, Dr rer nat(math), 61. Prof Exp: Lectr math, REI Degree Col, Agra Univ, 55-56; asst prof math, Birla Col, India, 56-58; aspirant, Dresden Tech Univ, 58-61; asst prof math, Birla Col, India, 61-62; vis mathematician, Res Inst Advan Studies, div Martin Co, Md, 62-63; from asst prof to assoc prof math, Case Western Reserve Univ, 63-68; vis assoc prof, dept math & inst fluid dynamics & appl math, Univ Md, College Park, 68-69, PROF MATH, DIV MATH, UNIV MD, BALTIMORE COUNTY & INST FLUID DYNAMICS & APPL MATH, COLLEGE PARK, 69- Concurrent Pos: NSF res grant, 64-; ed, Math Systs Theory. Mem: Am Math Soc; Math Asn Am; Soc Indust & Appl Math; Ger Soc Appl Math & Mech. Res: Dynamical and semi-dynamical systems; theory and application of ordinary differential equations; control theory; stability theory. Mailing Add: Div Math Univ Md Baltimore Co 5401 Wilkens Ave Baltimore MD 21228

BHATIA, SHYAM SUNDER, b Rawalpindi, Pakistan, July 7, 24; US citizen; m 50; c 2. ECONOMIC GEOGRAPHY, GEOGRAPHY OF SOUTH ASIA. Educ: Univ Panjab, BSc, 43, MA, 47; Univ Kans, PhD(geog), 59. Prof Exp: Lectr geog, Univ Col Panjab, New Delhi, 48-56; reader, Univ Delhi, 59-66; assoc prof, Univ Wis-Oshkosh, 66-70, PROF GEOG, UNIV WIS-OSHKOSH, 70- Concurrent Pos: Reader, Indian Sch Int Studies, 60-64. Mem: Asn Am Geogr; Am Geog Soc; Asn Asian Studies. Res: Agricultural geography and spatial analysis of change. Mailing Add: Dept of Geog Univ of Wis Oshkosh WI 54901

BHATIA, SUSHIL, b Lyallpur, India, July 12, 43; m 72; c 2. POLYMER CHEMISTRY. Educ: Univ Delhi, BSc, 64, MSc, 66; Liege Univ, Belgium, DrSc(polymer chem), 71. Prof Exp: Sci collabr, Liege Univ, Belgium, 71-72; res mgr adhesives, Morgan Adhesives Co, 72-76; RES SECT HEAD CONSUMER PROD, DENNISON MFG CO, 75- Mem: Royal Inst Chem; Brit Plastics Inst. Res: Pressure-sensitive adhesives; removable and permanent systems for consumer and industrial applications; head transfer inks; selection and modification of inks, printing technique and transfer to different kinds of fabrics. Mailing Add: Dennison Mfg Co 300 Howard St Framingham MA 01701

BHATIA, VISHNU NARAIN, b Lucknow, India, Aug 2, 24; nat US; m 51; c 2. PHARMACY. Educ: Benares Hindu Univ, BPharm, 45; Univ Iowa, MS, 49, PhD(pharm, pharmacol), 51. Prof Exp: Chief chemist, Mathur & Manzoor, Ltd, India, 45-46; asst prof pharmacog, Benares Hindu Univ, 46-47; head pharm div, CIPLA, Ltd, 51-52; from asst prof to assoc prof pharm, Col Pharm, 52-62, PROF PHARM, COL PHARM, WASH STATE UNIV, 62-, DIR HONORS PROG, 65- Concurrent Pos: Pres, Nat Collegiate Honors Coun, 67-68; dir, Int Progs, 71- Mem: Am Pharmaceut Asn. Res: Industrial pharmacy. Mailing Add: Honors Prog Wash State Univ Pullman WA 99163

BHATNAGAR, AJAY SAHAI, b Muree, India, Sept 26, 42; m 70. ORGANIC CHEMISTRY, REPRODUCTIVE ENDOCRINOLOGY. Educ: Cambridge Univ, BA, 63, MA, 67; Univ Basel, PhD(org chem), 67. Prof Exp: Res assoc physiol, 68-70, ASST PROF OBSTET & GYNEC, MED COL VA, 70-, ASST PROF BIOCHEM, 73- Concurrent Pos: Fel org chem, Univ Basel, 67. Mem: AAAS; NY Acad Sci; Am Chem Soc; Swiss Chem Soc. Res: Analytical methodology for the estimation of steroids; steroid metabolism in the female; endocrinology of pregnancy. Mailing Add: Dept of Obstet & Gynec Med Col of Va Richmond VA 23298

BHATNAGAR, ANIL KUMAR, b Jhansi, India, Jan 8, 42; m 65; c 1. SOLID STATE PHYSICS, LOW TEMPERATURE PHYSICS. Educ: Punjab Univ, India, BS, 58; Univ Allahabad, MSc, 62; Univ Md, PhD(solid state physics), 68. Prof Exp: Res assoc solid state physics, State Univ NY Stony Brook, 67-68; from asst prof to assoc prof solid state physics, St John's Univ, 68-74; ASSOC PROF PHYSICS, FORDHAM UNIV, 74- Concurrent Pos: Res collabr, Brookhaven Nat Lab, 68-73; Res Corp grant, 69-70. Mem: Am Phys Soc. Res: Superconductivity, thin films, transport properties of solids fluctuations in superconductors. Mailing Add: Dept of Physics Fordham Univ Bronx NY 10458

BHATNAGAR, DINECH C, b Lahore, WPakistan, Apr 14, 34; m 62; c 2. INORGANIC CHEMISTRY, ANALYTICAL CHEMISTRY. Educ: Univ Delhi, BSc, 53, MSc, 55; Wayne State Univ, PhD(chem), 63. Prof Exp: Res asst chem, Nat Chem Lab, India, 56-59; instr, Detroit Inst Technol, 62-63; fel, Univ Ariz, 63-65; asst prof, La Verne Col, 65-67; master, Sch Mines, Haileybury, Ont, Can, 67-68; MASTER CHEM, ALGONQUIN COL, 68- Concurrent Pos: Consult, Garett Res Corp, Calif, 65. Mem: Am Chem Soc; Chem Inst Can. Res: Optical rotary dispersion of inorganic complex ions; luminescence of metal complexes. Mailing Add: Dept of Chem Algonquin Col 200 Lees Ave Ottawa ON Can

BHATNAGAR, KUNWAR PRASAD, b Gwalior, India, Mar 21, 34; m 61; c 2. ANATOMY, NEUROANATOMY. Educ: Agra Univ, BSc, 56; Vikram Univ, India, MSc, 58; State Univ NY Buffalo, PhD(anat), 72. Prof Exp: From lectr to asst prof zool, Madhya Pradesh Educ Serv, India, 58-67; teaching asst anat, State Univ NY Buffalo, 68-72; ASST PROF ANAT, SCH MED, UNIV LOUISVILLE, 72- Mem: AAAS; Am Soc Mammal; Am Asn Anatomists; Am Soc Zool; Anat Soc India. Res: Mammalian olfaction; rhinencephalon in Chiroptera; sensory physiology. Mailing Add: Health Sci Ctr Dept of Anat Univ of Louisville Sch of Med Louisville KY 40201

BHATNAGAR, MAHESH KUMAR, b Allahabad, India, July 4, 36; m 67; c 1. VETERINARY ANATOMY. Educ: Vikram Univ, India, BVSc & AH, 59; Univ Guelph, PhD(cell biol), 68. Prof Exp: Res asst genetics, Indian Vet Res Inst, India, 59-63; asst prof biomed sci, Ont Vet Col, 68-73, ASSOC PROF BIOMED SCI, ONT VET COL, UNIV GUELPH, 73- Concurrent Pos: Med Res Coun Can fel, 69-70. Mem: Can Asn Anatomists; Am Asn Vet Anatomists; World Asn Vet Anatomists; Pan Am Asn Anatomists. Res: Experimental histopathology; cancer research; cell biology. Mailing Add: Dept of Biomed Sci Ont Vet Col Univ of Guelph Guelph ON Can

BHATNAGAR, RAJENDRA SAHAI, b Lucknow, India, Mar 10, 36; m 66; c 2. BIOCHEMISTRY. Educ: Agra Univ, BS, 54, MS, 56; Duke Univ, MS, 63, PhD(biochem), 64. Prof Exp: Lectr, Govt Col, Rupar, 57-58; tech asst, Tech Develop Dept, Ministry Indust, 58-60; asst instr med & biochem, Sch Med, Univ Pa & Philadelphia Gen Hosp, 65-67; sr res assoc biochem, Med Sch, Northwestern Univ, 67-68; assoc prof, 68-74, PROF BIOCHEM, SCH DENT, UNIV CALIF, SAN FRANCISCO, 74- Concurrent Pos: Fel, Vet Admin Hosp, Hines, Ill, 64-65 & Sch Med, Univ Pa & Philadelphia Gen Hosp, 65-67; res career develop award, 69; vis prof biophys & biochem, Indian Inst Sci, Bangalore, India, 72-73. Mem: AAAS; Am Chem Soc; Am Soc Biol Chemists; NY Acad Sci; fel Am Inst Chemists. Res: Biology of connective tissue, especially the biosynthesis and regulation of collagen. Mailing Add: Sch of Dent Univ of Calif San Francisco CA 94143

BHATNAGAR, RANBIR KRISHNA, b India; US citizen. PHARMACOLOGY. Educ: Univ Lucknow, BSc, 54; Agra Univ, BVSc, 58; Mich State Univ, MS, 63, PhD(pharmacol), 71. Prof Exp: Veterinarian, Govt Uttar Pradesh, India, 58-59; res assoc physiol, Col Vet Sci, Mathura, India, 59-61; res assoc pharmacol, Univ Chicago, 65-67; res assoc, 71-72, ASST PROF PHARMACOL, UNIV IOWA, 72- Mem: Am Soc Pharmacol & Exp Therapeut; Soc Neurosci. Res: Neurochemistry and neuropharmacology with particular emphasis on central nervous system neurotransmitters; factors which influence the growth and differentiation of neurons in central nervous system. Mailing Add: 2310 Basic Sci Bldg Univ of Iowa Dept of Pharmacol Iowa City IA 52242

BHATTACHARJEE, JNANENDRA K, b Sylhet, Bangladesh, Feb 1, 36. MICROBIAL GENETICS. Educ: MC Col, Sylhet, BS, 57; Univ Dacca, MS, 59; Southern Ill Univ, PhD(microbiol), 66. Prof Exp: Res assoc microbiol, Albert Einstein Med Ctr, Pa, 65-66, asst mem, 66-68; assoc prof, 68-73, PROF MICROBIOL, MIAMI UNIV, 73- Concurrent Pos: NSF res grant, 69, 71, 75; Lilly Res Found grant, 70, 72; S & H Found res grant, 74. Honors & Awards: President's Award, President of Pakistan, 60. Mem: AAAS; Genetics Soc Am; Am Soc Microbiol. Res: Genetics of yeast; biosynthetic mechanism of lysine and other related amino acids in yeast; regulation of gene action in eucaryotic organisms; single-cell protein. Mailing Add: Dept of Microbiol Miami Univ Oxford OH 45056

BHATTACHARJI, SOMDEV, b Calcutta, India, Apr 23, 32; US citizen; m 62; c 3. STRUCTURAL GEOLOGY. Educ: Univ Calcutta, BS, 50, Hons, 51; Indian Sch Mines, MS, 54; Univ Chicago, MS, 57, PhD(geol), 59. Prof Exp: Instr ling, Univ Chicago, 59-60, res assoc & lectr S Asian lang, 60-61, asst prof, 61-62; fel geol, Nat Res Coun Can, 62-64; from instr to assoc prof geol, 64-72, dep chmn dept, 69-72, PROF GEOL, BROOKLYN COL, CITY UNIV NEW YORK, 72- Concurrent Pos: NSF grants, 66-68, 70-72 & 74-75; City Univ of New York fac res grant, 73-76; mem, Int Geodynamic Comt, 72. Mem: AAAS; fel Geol Soc Am; Am Geophys Union; Geol Soc Edinburgh, Scotland; Int Soc Paleontology. Res: Tectonophysics; petrogenesis; languages. Mailing Add: Brooklyn Col City Univ New York Brooklyn NY 11210

BHATTACHARYA, AMAR NATH, b Calcutta, India, Oct 1, 34; m 66; c 1. PHARMACOLOGY, PHYSIOLOGY. Educ: Bengal Vet Col, India, BVetS, 57; Ohio State Univ, MS, 63, PhD(pharmacol), 67. Prof Exp: State vet, Directorate Vet Serv, Govt WBengal, India, 57-59; demonstr, Dept Clin Vet Med & Pharmacol, Bengal Vet Col, 59-61; res asst pharmacol, Col Med, Ohio State Univ, 62-67, res assoc pharmacol & med, 68; asst prof, 70-73, ASSOC PROF PHARMACOL, COL PHARM, OHIO NORTHERN UNIV, 74- Concurrent Pos: Ford Found fel physiol, Sch Med, Univ Pittsburgh, 68-70. Mem: Soc Study Reprod; Brit Soc Study Fertil; Indian Sci Cong Asn. Res: Neuroendocrine control mechanisms of corticotropin and gonadotropins, role of central catecholamines and serotonin; gonad-anterior pituitary feedback interrelationship in subhuman primates, studies with castration, steroids and psychotropic drugs; prostaglandins. Mailing Add: Dept of Pharmacol Ohio Northern Univ Ada OH 45810

BHATTACHARYA, PRADEEP KUMAR, b Dacca, Brit India, Jan 12, 40. PLANT PHYSIOLOGY, PLANT PATHOLOGY. Educ: Banaras Hindu Univ, BSc, 57, MSc, 59; Univ Sask, PhD(biol), 66. Prof Exp: Lectr bot, MLK Degree Col, Gorakhpur Univ, India, 59-61; res assoc biol & plant physiol, Univ Sask, 66-67; fel bot, Univ Western Ont, 67-69; fel plant path, Univ Wis-Madison, 69-70; asst prof biol, Rockford Col, 70-73; ASSOC PROF BIOL & CHMN DEPT, IND UNIV NORTHWEST, 73- Concurrent Pos: Nat Res Coun Can assoc, Dept Biol, Univ Sask, 66-67; Nat Res Coun Can fel, Dept Bot, Univ Western Ont, 67-69; NIH fel, Dept Plant Path, Univ Wis-Madison, 69-70. Mem: Am Inst Biol Scientists; Am Soc Cell Biol; Can Soc Plant Physiol; Can Soc Cell Biol; Sigma Xi. Res: Physiology of host-parasite relations with particular reference to obligate parasitism; cytochemical and related biochemical events in plant-parasite interaction studies in wheat rust, barley mildew and root rot of cabbage. Mailing Add: Dept of Biol Ind Univ Northwest 3400 Broadway Gary IN 46408

BHATTACHARYA, RABINDRA NATH, b Barisal, EPakistan, Jan 11, 37; m 67. MATHEMATICAL STATISTICS. Educ: Univ Calcutta, BSc, 56, MSc, 59; Univ Chicago, PhD(statist), 67. Prof Exp: Res off statist, River Res Inst, Calcutta, 61; lectr, Kalyani Agr Univ, 61-64; asst prof, Univ Calif, Berkeley, 67-72; ASSOC PROF MATH, UNIV ARIZ, 72- Mem: Inst Math Statist; Am Math Soc. Res: Central limit theorems; markov processes; statistical mechanics. Mailing Add: Dept of Math Univ of Ariz Tucson AZ 85721

BHATTACHARYA, RAMENDRA KUMAR, b India, Mar 29, 31. MATHEMATICS. Educ: Univ Calcutta, BSc, 51, MSc, 53; Stanford Univ, PhD(math), 64. Prof Exp: Lectr math, Taki Govt Col, India, 54-57 & Jadavpur Univ, 57-59; asst prof math, Univ Ariz, 64-67 & Southern Ill Univ, Carbondale, 67-69; ASSOC PROF MATH, PAC UNIV, 69- Mem: Am Math Soc; Math Asn Am; Calcutta Math Soc. Res: Theory of affine connections; system analysis. Mailing Add: Dept of Math Pac Univ Forest Grove OR 97116

BHATTACHARYYA, BIBHUTI BHUSHAN, b Bhatpara, India, Aug 1, 35; m 62. MATHEMATICAL STATISTICS. Educ: Univ Calcutta, BSc, 53, MSc, 55; Univ London, PhD(economet), 59. Prof Exp: Res scholar statist, Indian Statist Inst, 56; vis statistician, NC State Col, 59-61; res fel economet, Inst Econ Growth, 62; lectr math & statist, Univ Toronto, 62-63; asst prof, 63-69, ASSOC PROF STATIST, NC STATE UNIV, 69- Mem: Inst Math Statist. Res: Econometrics; mathematical programming and its application to national planning as well as production planning of individual firms; sequential estimation problems. Mailing Add: Dept of Statist PO Box 5126 NC State Univ Raleigh NC 27607

BHATTACHARYYA, GOURI KANTA, b Hooghly, WBengal, India; Jan 12, 40; m 62; c 1. STATISTICS, MATHEMATICAL STATISTICS. Educ: Univ Calcutta, BSc, 58, MSc, 60; Univ Calif, Berkeley, PhD(statist), 66. Prof Exp: Lectr statist, R K Mission, Narendrapur, India, 61; res officer, River Res Inst, Calcutta, 61-63; asst prof, Univ Calif, Berkeley, 66; asst/prof,/Univ from asst prof to assoc prof statist, 66-75, PROF STATIST, UNIV WIS-MADISON, 75- Concurrent Pos: Consult sch med, Univ Wis, 67-; prof, Indian Inst Mgt, Ahmedabad, 69-70; vis prof, Indian Statist Inst, Calcutta, 75-76. Mem: Inst Math Statist; fel Royal Statist Soc. Res: Statistical inference, life testing and reliability studies; nonparametric methods. Mailing Add: Dept of Statist Univ of Wis 1210 W Dayton Madison WI 53706

BHATTACHARYYA, MARYKA HORSTING, b Glen Ridge, NJ, Sept 17, 43; m 71. BIOCHEMISTRY. Educ: Tufts Univ, BS, 65; Univ Wis, PhD(biochem), 70. Prof Exp: Fel biochem, Univ Wis, 70-71, res assoc, 71-74; ASST BIOCHEMIST, ARGONNE NAT LAB, 74- Res: Biochemical interactions of plutonium with the body; development of therapeutic means for its removal in cases of accidental exposure. Mailing Add: Argonne Nat Lab 9700 Cass Ave Argonne IL 60439

BHATTACHARYYA, PRANAB K, b India, Aug 9, 38; m 69; c 1. PHYSICAL CHEMISTRY, ANALYTICAL CHEMISTRY. Educ: Calcutta Univ, BS, 59, MS, 62; Columbia Univ, PhD(phys chem), 71. Prof Exp: Res assoc nuclear magnetic resonance chem dept, Columbia Univ, 71-74; SR SCIENTIST & SUPVR SPECTROS TECHNOL, ANAL RES SECT, HOFFMANN-LA ROCHE INC, 74- Mem: NY Acad Sci; Am Chem Soc; Am Phys Soc; Int Soc Magnetic Resonance. Res: Quantitative analysis and characterization of drug molecules and complex organic compounds using the techniques of pulsed Fourier transform nuclear magnetic resonance and molecular spectroscopy; nuclear magnetic resonance studies of biological molecules. Mailing Add: Anal Res Sect Bldg 86 Hoffmann-La Roche Inc Nutley NJ 07110

BHATTI, WAQAR HAMID, b Sohawa, WPakistan, Nov 22, 31; m 64. PHYTOCHEMISTRY, HEMATOLOGY. Educ: Univ Panjab, BSc, 55, MSc, 57; Philadelphia Col Pharm, PhD(pharmacog), 63. Prof Exp: Lectr pharmacog, Univ Panjab, Pakistan, 57; lectr bot, Univ Peshawar, 58-59; asst biol, Philadelphia Col Pharm, 59-63; asst prof pharmacog, Sch Pharm, NDak State Univ, 63-67; res fel internal med, Univ Iowa Hosp, 67-68; res assoc, Sch Pharm, Univ Pittsburgh, 68-69; asst prof, 69-74, ASSOC PROF PHARMACOG, BUTLER UNIV, 74- Concurrent Pos: Res assoc, Einstein Med Ctr, 62-63 & Vet Admin, 63-; teaching fel pharmacog, NSF Inst Prog High Sch Students & Teachers, 64, prin investr, NSF grant, 64- Mem: AAAS; Am Pharmaceut Asn; Am Soc Pharmacog. Res: The chemical investigation of both higher and lower plants and the isolation of their active principles which may serve as new drugs useful in the prophylaxis or treatment of cancer or cardiac and hemolytic disorders. Mailing Add: Col of Pharm Butler Univ 46th at Sunset Ave Indianapolis IN 46207

BHAVNAGRI, VISPI PESHOTAN, pharmacy, biopharmaceutics, see 12th edition

BHAVNANI, BHAGU R, b Hyderabad, Pakistan, Feb 16, 36. BIOCHEMISTRY, PHARMACEUTICAL CHEMISTRY. Educ: Univ Bombay, BS, 58 & 60, MS, 62; Mass Col Pharm, PhD(biochem), 66. Prof Exp: Res asst steroid biochem, Worcester Found Exp Biol, 62-66; lectr, McGill Univ, 68-69, asst prof investigative med, 69-75; assoc scientist, Royal Victoria Hosp, Montreal, 68-75; ASST PROF OBSTET & GYNEC & CLIN BIOCHEM, UNIV TORONTO & STAFF SCIENTIST, ST

MICHAEL'S HOSP, TORONTO, 75- Concurrent Pos: Fel biochem, McGill Univ, 66-68; Med Res Coun Can res grant, 70-75. Mem: AAAS; Endocrine Soc; NY Acad Sci; Can Soc Endocrinol & Metab. Res: Genetic aspects of cortisol metabolism in the guinea pig; steroid formation and metabolism during pregnancy in humans; formation of ring B unsaturated estrogens in the pregnant mare; developmental aspects of fetal glycogen metabolism. Mailing Add: Dept of Obstet & Gynec St Michael's Hosp Toronto ON Can

BHIWANDKER, NUTAN C, b Bombay, India, Mar 7, 34; m 60; c 2. INORGANIC CHEMISTRY, CERAMICS. Educ: Univ Bombay, BSc, 54; Univ Ottawa, MSc, 60; McGill Univ, PhD(inorg chem), 66. Prof Exp: Chemist, B P Refinery Ltd, Aden, 55-58; mem res staff, Gulton Industs, 64-66, Can Tech Tape, 66-68 & Ampex Corp, 68-72; mem res staff, C & Z circuits, 72-73, MGR TECH CERAMICS, DATA MAGNETICS CORP, 73- Mem: Am Chem Soc; Am Ceramic Soc. Res: Ferrites; magneto-chemistry of transition metal compounds; electronic and magnetic ceramics; magnetic memory and logic; pressure sensitive adhesives. Mailing Add: Data Magnetics Corp 355 Maple Ave Torrance CA 90505

BHUSSRY, BALDEV RAJ, b Sialkot, Pakistan, Feb 8, 28; m 56. ANATOMY, DENTISTRY. Educ: Sir C E M Dent Col, India, BDS, 49; Univ Rochester, MS, 53, PhD(anat), 56. Prof Exp: Clin asst, Univ Bombay, 49-50; instr anat physiol, Eastman Dent Dispensary, 55-56; instr dent res & anat, Univ Rochester, 56-57; from asst prof to assoc prof anat, 57-64, dir res training, Sch Dent, 57-65, PROF ANAT, SCH MED & DENT, GEORGETOWN UNIV, 64-, CHMN DEPT, 63- Concurrent Pos: USPHS career develop award, 58-63; mem dent educ rev comt, NIH, 68- Mem: Am Asn Anatomists; fel Am Acad Oral Path; Int Asn Dent Res; NY Acad Sci. Res: Structure of human enamel; effect of fluorides on tooth development; role of nutritional deficiencies on oral and dental structures. Mailing Add: Dept of Anat Georgetown Univ Sch Med & Dent Washington DC 20007

BHUVANESWARAN, CHIDAMBARAM, b India, Nov 10, 34. BIOCHEMISTRY. Educ: Univ Bombay, BSc, 55 & 57, PhD(biochem), 63. Prof Exp: Jr sci officer, Cent Food Tech Res Inst, Mysore, India, 60-64; sci officer, Atomic Energy Estab, Bombay, 64-65; res assoc biochem, Ore State Univ, 65-68; RES ASSOC BIOCHEM, FAC MED, UNIV MAN, 68- Mem: AAAS; Am Chem Soc. Res: Electron transport system, especially succinic dehydrogenase. Mailing Add: Dept of Biochem Univ of Man Winnipeg MB Can

BHUYAN, BIJOY KUMAR, b Calcutta, India, Aug 30, 30; m 55; c 2. BIOCHEMISTRY. Educ: Utkal Univ, India, BSc, 48; Univ Calcutta, MSc, 51; Univ Wis, MS, 54, PhD(biochem), 56. Prof Exp: Res assoc & fel biochem, Univ Wis, 56-57; fel, Nat Res Coun Can, 57-58; sr sci officer, Hindustan Antibiotics, India, 58-59; res assoc microbiol, 60-63, RES ASSOC BIOCHEM, UPJOHN CO, 63- Mem: Tissue Cult Asn; Am Asn Cancer Res. Res: Penicillin production; anti-tumor antibiotics; tissue culture; cell cycle. Mailing Add: Upjohn Co Kalamazoo MI 49001

BI, LE-KHAC, b Hue, Viet Nam, Jan 28, 40. POLYMER CHEMISTRY. Educ: Univ Saigon, BS, 63; Univ Wis, Madison, MS, 71; Univ Akron, PhD(polymer sci), 75. Prof Exp: RES SCIENTIST POLYMER SYNTHESIS, ARCO/POLYMERS, INC, 75- Res: Preparation, properties and morphology of block copolymers. Mailing Add: Arco/Polymers Inc 440 College Park Dr Monroeville PA 15146

BIAGAS, WILFRED MICHAEL, b Opelousas, La, Sept 22, 23; m 57; c 2. PHYSICAL CHEMISTRY. Educ: Univ Chicago, MS, 52; St Louis Univ, PhD(phys chem), 58. Prof Exp: Res assoc physics sect, Knolls Atomic Power Lab, Gen Elec Co, 57-58; engr reactor metall, Bettis Atomic Power Lab, Westinghouse Elec Corp, 58-61; fel chem, Univ Pittsburgh, 61-62; physicist cent res lab, Crucible Steel Co Am, Pa, 62-65; assoc prof chem, California State Col, Pa, 65-66; assoc prof phys sci, 66-69, chmn dept, 66-74, PROF PHYS SCI, POINT PARK COL, 69-, DIR ALUMNI AFFAIRS & ASSOC DIR GOVT GRANTS, 75- Mem: Am Chem Soc; Am Crystallog Asn. Res: Crystal structures of metal-organic complexes; metal physics; defect structures. Mailing Add: Dept of Phys Sci Point Park Col 201 Wood St Pittsburgh PA 15222

BIAGETTI, RICHARD VICTOR, b Woonsocket, RI, Jan 13, 40; m 64; c 2. ELECTROCHEMISTRY, INORGANIC CHEMISTRY. Educ: Providence Col, BS, 60, MS, 62; Univ NH, PhD(inorg chem), 66. Prof Exp: Mem staff, 65-69, SUPVR ELECTROCHEM, BELL TEL LABS, 69- Mem: Am Chem Soc; Electrochem Soc. Res: Coordination chemistry of nonaqueous transition metal nitrates; failure mechanism analyses and development of improved lead-acid battery systems. Mailing Add: Dept of Electrochem Bell Labs Rm 1E-245 600 Mountain Ave New Providence NJ 07974

BIAGGI, VIRGILIO, JR, b Mayaguez, PR, Dec 25, 13; m 40; c 3. ORNITHOLOGY. Educ: Univ PR, BSA, 40; Agr & Mech Col Tex, MSc, 42; Ohio State Univ, PhD, 49. Prof Exp: Asst instr biol, 37-40, instr zool, 41-44, assoc prof, 46-49, PROF ZOOL, UNIV PR, MAYAGUEZ, 49-, DIR DEPT BIOL, 58- Concurrent Pos: Guggenheim fel, 57; asst to dean arts & sci, Univ PR, Mayaguez, 60-62. Mem: AAAS; Wilson Ornith Soc; Am Ornith Union; Wildlife Soc. Res: Studies of life histories of birds of Puerto Rico. Mailing Add: Dept of Biol Univ of PR Mayaguez PR 00708

BIAGLOW, JOHN E, b Cleveland, Ohio, Apr 1, 37; m 60; c 4. BIOCHEMISTRY. Educ: John Carroll Univ, BS, 59; Loyola Univ, Chicago, MS, 61, PhD(biochem), 63. Prof Exp: NIH fel biochem, 63-65, ASST PROF BIOCHEM, CASE WESTERN RESERVE UNIV, 65-, ASST PROF RADIOL, 67- Concurrent Pos: Pres, Custom Biochem Co; life sci consult. Mem: Am Chem Soc; NY Acad Sci. Res: Enzymology; kinetics; control and organization of enzyme systems; oncology; physiological controls; oxygen utilization. Mailing Add: Dept of Radiol Sch of Med Case Western Reserve Univ 2119 Abington Rd Cleveland OH 44106

BIALAS, WAYNE FRANCIS, b Middletown, NY, Aug 5, 49. OPERATIONS RESEARCH, STATISTICS. Educ: Clarkson Col Technol, BS, 71; Cornell Univ, MS, 74, PhD(oper res), 75. Prof Exp: Res assoc & instr, Col Eng, Cornell Univ, 75; ASST PROF INDUST ENG, STATE UNIV NY BUFFALO, 76- Mem: Am Statist Asn; Asn Comput Mach; Inst Math Statist; Math Asn Am; Oper Res Soc Am. Res: Mathematical models of mine acid drainage; asymptotic properties of probability measures; economic models for flood prone regions. Mailing Add: Dept of Indust Eng State Univ of NY Bell Hall Amherst NY 14260

BIALY, JERZY JOZEF, organic chemistry, see 12th edition

BIANCHI, CARMINE PAUL, b Newark, NJ, Apr 9, 27; m 57; c 3. CELL PHYSIOLOGY. Educ: Columbia Univ, AB, 50; Rutgers Univ, MS, 53, PhD(biochem, physiol), 56. Prof Exp: Vis scientist & res assoc, NIH, 58-60; asst mem, Inst Muscle Dis, New York, 60-61; from assoc to assoc prof physiol, 61-68, PROF PHARMACOL, SCH MED, UNIV PA, 68- Concurrent Pos: Res fel pub health, NIH, 56-58. Mem: Biophys Soc; Am Soc Zoologists; Soc Gen Physiol; Am Physiol Soc. Res: Role of mono- and divalent cations in controlling cell membrane function and metabolism. Mailing Add: Dept of Pharm Univ of Pa Sch of Med Philadelphia PA 19104

BIANCHI, DONALD ERNEST, b Santa Cruz, Calif, Nov 22, 33; m 56; c 3. PHYSIOLOGY, MYCOLOGY. Educ: Stanford Univ, AB, 55, AM, 56; Univ Mich, PhD, 59. Prof Exp: Asst prof bot, San Fernando Valley State Col, 59-63, assoc prof biol, 63-66, PROF BIOL, CALIF STATE UNIV, NORTHRIDGE, 66-, DEAN SCH SCI & MATH, 73- Concurrent Pos: NSF sci fac fel, 65-66; sr researcher, Univ Geneva, 70. Mem: AAAS; Bot Soc Am; Am Soc Microbiol; Am Soc Plant Physiol. Res: Physiology of fungi. Mailing Add: Dean's Off Calif State Univ Northridge CA 91324

BIANCHI, ROBERT GEORGE, b Chicago, Ill, Mar 20, 25; m 50; c 4. PHARMACOLOGY. Educ: Franklin & Marshall Col, BS, 45. Prof Exp: Jr investr biol res, 50-60, SR INVESTR PHARMACOL, SEARLE LABS, 60- Mem: AAAS. Res: Gastrointestinal pharmacology; compounds useful in the treatment of gastrointestinal diseases. Mailing Add: Searle Labs PO Box 5110 Chicago IL 60680

BIANCHI, WILLIAM C, b Los Angeles, Calif, June 27, 29; m 53; c 5. GROUNDWATER HYDROLOGY. Educ: Univ Calif, Davis, BS, 52, PhD(soil sci), 58. Prof Exp: Asst prof soil sci, Univ Nev, 56-59; res soil scientist, Agr Res Serv, USDA, 59-67, RES LEADER WATER MGT, AGR RES SERV, USDA, 67- Mem: Am Soc Agron; Soil Sci Soc Am; Am Water Works Asn. Res: Agricultural drainage and irrigation; ground water hydrology, water flow through soils, especially technology for artificially recharging urban and agricultural well fields. Mailing Add: 5768 N Eighth Fresno CA 93710

BIANCHINE, JOSEPH RAYMOND, b Albany, NY, Sept 7, 29; m 56. PHARMACOLOGY. Educ: St Bernardine of Siena Col, BS, 51; Albany Med Col, PhD(pharmacol), 59; State Univ NY Albany, MD, 60. Prof Exp: Intern & asst resident physician, Johns Hopkins Hosp, 60-62; mem staff, Cancer Control Prog, USPHS, 62-64; chief resident, Baltimore City Hosp, 64-65; from instr to assoc prof med, Johns Hopkins Univ, 65-73, asst pharm & exp therapeut, 70-73; chmn dept pharmacol, Tex Tech Univ, 73-74; PROF PHARMACOL & CHMN DEPT & PROF MED, SCH MED, OHIO STATE UNIV, 74- Concurrent Pos: Asst chief med, Baltimore City Hosps, 71-73. Res: Clinical pharmacology; neurochemistry; internal medicine. Mailing Add: Dept of Pharmacol Ohio State Univ Sch of Med Columbus OH 43210

BIANCO, DONALD R, b Des Moines, Iowa, May 28, 18; m 42; c 3. PHYSICS. Educ: Iowa State Col, BS, 45; Univ Iowa, MS, 49; Univ Md, PhD, 54. Prof Exp: Asst physics, Univ Iowa, 47-49, Hydraul Lab, 46; assoc physicist, Appl Physics Lab, Johns Hopkins Univ, 49-54, sr physicist, 57-66; res assoc, Collins Radio, 54-57; PRIN ENGR & GROUP SUPVR, BENDIX AEROSPACE SYSTS, BENDIX CORP, 66- Concurrent Pos: Instr, Montgomery Jr Col, 57-61. Mem: AAAS; Inst Elec & Electronics Engrs. Res: Microwave spectroscopy; pore structure adsorbents; microwave and ionospheric physics; satellite design and development; space science instrument development. Mailing Add: 8253 N Antcliff Rd Howell MI 48843

BIANCULLI, JOSEPH AMEDEO, chemistry, see 12th edition

BIAS, WILMA B, b Muskogee, Okla, Dec 23, 28; m 47; c 3. HUMAN GENETICS. Educ: Univ Okla, BS, 49; Johns Hopkins Univ, PhD(genetics), 63. Prof Exp: Fel med genetics, Sch Med, Johns Hopkins Univ, 63-64; from instr to asst prof med, 64-73, asst prof epidemiol, Sch Hyg, 67-73, ASSOC PROF MED, SCH MED & ASSOC PROF EPIDEMIOL, SCH HYG, JOHNS HOPKINS UNIV, 73-, ASST PROF SURG, SCH MED, 69- Mem: AAAS; Am Soc Human Genetics. Res: Biochemical genetics; immunogenetics. Mailing Add: Dept of Surg Johns Hopkins Univ Sch Med Baltimore MD 21205

BIAVATI, BRUCE J, radiological physics, see 12th edition

BIBB, WILLIAM ROBERT, b Salisbury, NC, May 28, 32; m 55; c 2. IMMUNOLOGY. Educ: Univ NC, BS, 57, MS, 59, PhD(bact), 62. Prof Exp: Res assoc virol, virus lab, Sch Med, Univ NC, Chapel Hill, 62-63, USPHS trainee, 63-64, asst prof bact, 64-65; mem med res br, Div Biol & Med, US Atomic Energy Comn, Washington, DC, 65-71, tech asst to comnr, 71-73, CHIEF RES & DEVELOP, US ENERGY & DEVELOP ADMIN, OAK RIDGE, 73- Mem: Am Soc Microbiol; Electron Micros Soc Am; Reticuloendothelial Soc; Am Nuclear Soc. Res: Metabolism; amino acid transport; enzyme synthesis; bacterial cell structure; amino acid decarboxylases; organ and tissue transplantation; effects of low-dose radiation; radiation protection. Mailing Add: PO Box 861 Oak Ridge TN 37830

BIBBO, MARLUCE, b Sao Carlos, Brazil, July 14, 39. CYTOPATHOLOGY. Educ: Dr Alvaro Guiao Inst Educ, Brazil, BS, 57; Univ Sao Paulo, MD, 63, DSc, 68. Prof Exp: Resident obstet & gynec, Univ Sao Paulo, 63-65, trainee exfoliative cytol, 65, trainee nucleic acids, 65-66, instr morphol, Fac Med, 65-68, asst prof morphol & obstet & gynec & chief cytol serv, 68-70; res assoc, 69-70, ASST PROF OBSTET & GYNEC, UNIV CHICAGO, 70- Mem: Int Acad Cytol; Am Soc Cytol; fel Asn Clin Sci; World Asn Gynec Cancer Prev. Res: Exfoliative and experimental cytology; computerized cell evaluations; pattern recognition; endocrinologic cell assessments. Mailing Add: Dept of Obstet & Gynec Univ of Chicago Chicago IL 60637

BIBEAU, ARMAND A, b US, July 11, 24; m 50; c 1. ZOOLOGY. Educ: St Anselm's Col, AB, 50; Univ NH, MS, 51; Univ Conn, PhD, 55. Prof Exp: Cytologist, St Francis Hosp, Conn, 53-54; USPHS fel, Harvard Univ, 55; head dept biol, 59-74, PROF BIOL, ST ANSELM'S COL, 55- Mem: AAAS; Soc Protozool. Res: Parasitology; protozoology; exfoliative cytology. Mailing Add: Dept of Biol St Anselm's Col Manchester NH 03102

BIBEL, DAVID JAN, b San Francisco, Calif, Apr 6, 45; m 70; c 1. MEDICAL MICROBIOLOGY, MICROBIAL ECOLOGY. Educ: Univ Calif, Berkeley, AB, 67, CPhil, 71, PhD(immunol), 72. Prof Exp: Microbiologist, 72-73, BACTERIOLOGIST, LETTERMAN ARMY INST RES, 73- Concurrent Pos: Lectr, Sch Pub Health, Univ Calif, Berkeley, 75 & Ctr Health Studies, Antioch Col W, 75- Mem: Am Soc Microbiol; AAAS; Fedn Am Scientists. Res: Microbiology and ecology of human skin; morphology and growth of L-forms and cell wall defective bacteria; host-parasite relationships; history of medical microbiology and immunology. Mailing Add: 787 Euclid Ave Berkeley CA 94708

BIBER, MARGARET CLARE BOADLE, b Melbourne, Australia, Jan 18, 43; m 69; c 1. NEUROPHARMACOLOGY. Educ: Univ London, BSc, 64; Oxford Univ, DPhil(biochem, pharmacol), 67. Prof Exp: Res assoc pharmacol, Sch Med, Yale Univ, 68-69, from instr to asst prof, 69-75; ASSOC PROF PHYSIOL, MED COL VA, 75- Concurrent Pos: A B Coxe Mem fel pharmacol, Sch Med, Yale Univ, 67-68. Mem: Am Soc Pharmacol & Exp Therapeut; Soc Neurosci; Am Soc Neurochem; Marine Biol Asn UK. Res: Studies on the regulation of neurotransmitter formation, particularly of catecholamines and serotonin, by ongoing nervous activity, drugs and hormones. Mailing Add: Dept of Physiol Med Col of Va Richmond VA 23298

BIBERMAN, LUCIEN MORTON, b Philadelphia, Pa, May 31, 19; m; c 3. PHYSICS. Educ: Rensselaer Polytech Inst, BS, 40. Prof Exp: Phys chemist, Congoleum-Nairn, Inc, NJ, 41-42; physicist, US Navy, SC, 42-44 & US Naval Ord Testing Sta, 44-57; dir syst div, Midway Labs, Univ Chicago, 57-59, assoc dir, Labs Appl Sci, 59-63; MEM SR RES STAFF, INST DEFENSE ANAL, 63- Concurrent Pos: Consult, US Naval Ord Testing Sta, 57-60, Univ Mich, 59 & Off Secy Defense, 62-; adj prof, Univ RI, 69-, vis prof, 71-72. Mem: AAAS; fel Optical Soc Am; fel Inst Elec & Electronics Engrs. Res: Military applications of optics and electronics with emphasis on infrared and ultraviolet techniques. Mailing Add: Inst Defense Anal 400 Army Navy Dr Arlington VA 22202

BIBERSTEIN, ERNEST LUDWIG, b Breslau, Ger, Nov 11, 22; nat; m 49; c 4. VETERINARY BACTERIOLOGY. Educ: Univ Ill, BS, 47; Cornell Univ, DVM, 51, MS, 54, PhD(vet bact), 55; Am Col Vet Microbiologists, dipl. Prof Exp: Actg asst prof path, NY State Vet Col, Cornell Univ, 55-56; asst prof vet med, 56-60, assoc prof, 60-66, chmn dept, 69-74, PROF MICROBIOL, UNIV CALIF, DAVIS, 66- Concurrent Pos: NIH spec fel, 63-64 & 68-69. Mem: Am Vet Med Asn; Am Soc Microbiol; fel Am Acad Microbiol; Path Soc Gt Brit & Ireland. Res: Infectious diseases of animals; pasteurellosis; haemophilus infections; clinical microbiology. Mailing Add: Dept of Vet Microbiol Univ of Calif Davis CA 95616

BIBLER, NED EUGENE, b Bucyrus, Ohio, July 25, 37; m 63; c 3. RADIATION CHEMISTRY. Educ: Denison Univ, BSc, 59; Ohio State Univ, MSc, 62, PhD(chem), 65. Prof Exp: Res chemist, 65-73, STAFF CHEMIST RADIATION CHEM, SAVANNAH RIVER LAB, E I DU PONT DE NEMOURS & CO, 73- Mem: Am Chem Soc. Res: Radiation chemistry associated with radioactive isotope production and nuclear fuel reprocessing. Mailing Add: Savannah River Lab E I du Pont de Nemours & Co Aiken SC 29801

BIBLIS, EVANGELOS J, b Thessaloniki, Greece, Apr 8, 29; US citizen; m 61; c 1. WOOD SCIENCE & TECHNOLOGY. Educ: Univ Thessaloniki, BS, 53; Yale Univ, MF, 61, PhD(forestry), 65. Prof Exp: Qual control engr, Kaman Aircraft Corp, 61-63; PROF WOOD TECHNOL, AUBURN UNIV, 65- Mem: Soc Wood Sci & Technol; Forest Prod Res Soc. Res: Mechanical behavior of wood and wood structures. Mailing Add: Dept of Forestry Auburn Univ Auburn AL 36830

BICAK, LADDIE JOHN, science education, biology, see 12th edition

BICE, CLAUDE WESLEY, b Kansas City, Mo, May 3, 18; m 41; c 2. AGRICULTURAL BIOCHEMISTRY. Educ: Ft Hays Kans State Col, BS, 40, MS, 43; Univ Minn, PhD(biochem), 50. Prof Exp: Lab aide northern regional res lab, Bur Agr & Indust Chem, USDA, 41-42, res chemist, 43-45; res leader, R T French Co, 50-52, mgr tech res, 52-70, asst res dir, 70-74; PRIN RES SCIENTIST, FRITO-LAY, INC, 74- Concurrent Pos: Mem study panel food protection, Nat Res Coun-Nat Acad Sci, 72-74. Mem: Soc Nutrit Educ; Am Asn Cereal Chemists; Inst Food Technologists. Res: Nutrition, food safety, starch, cereals, dehydrated potates and snack technology. Mailing Add: Frito-Lay Inc 900 N Loop 12 Irving TX 75061

BICE, DAVID EARL, b Cornville, Ariz, Apr 8, 38; m 60; c 4. IMMUNOLOGY. Educ: Utah State Univ, BS, 62; Univ Ariz, MS, 64; La State Univ, PhD(trop med), 68. Prof Exp: La State Univ res assoc, Int Ctr Med Res & Training, San Jose, Costa Rica, 64-68, from instr to asst prof med & microbiol, La State Univ Med Ctr, New Orleans, 68-75; ASSOC SCIENTIST, INHALATION TOXICOL RES INST, LOVELACE FOUND, 75- Concurrent Pos: Res fels immunol, La State Univ, 68-69 & Harvard Med Sch & Tufts Univ, 70-71; scientist, Dept Med, La State Univ Div of Charity Hosp, 68-75. Mem: Am Asn Immunol. Res: Cellular immunology in relation to human immunodeficiencies, cancer and lung diseases. Mailing Add: Inhalation Toxicol Res Inst PO Box 5890 Dept Path Albuquerque NM 87115

BICEK, EDWARD JOHN, chemistry, see 12th edition

BICHARD, J W, b Toronto, Ont, May 9, 32. SOLID STATE PHYSICS. Educ: Univ Toronto, BASc, 55; Univ Notre Dame, PhD(physics), 60. Prof Exp: Asst prof physics, 60-70, ASSOC PROF PHYSICS, UNIV BC, 70- Concurrent Pos: Nat Res Coun Can grant, 64-65. Mem: Can Asn Physicists. Res: Nuclear spectroscopy. Mailing Add: Dept of Physics Univ of BC Vancouver BC Can

BICHSEL, HANS, b Basel, Switz, Sept 2, 24; nat US; m 59; c 2. RADIATION PHYSICS. Educ: Univ Basel, PhD(physics), 51. Prof Exp: Exchange fel physics, Princeton Univ, 51-52, res asst physics, 52-54; lectr optics, Univ Basel, 54-55; res assoc physics, Rice Inst, 55-57; asst prof physics, Univ Wash, 57-59; from asst prof to assoc prof, Univ Southern Calif, 59-68; assoc prof, Univ Calif, Berkeley, 68-69; assoc prof radiol, 69-72, PROF RADIOL, UNIV WASH, 72- Concurrent Pos: Mem subcomt penetration charged particles in matter, Nat Acad Sci-Nat Res Coun, 62-; NSF grant, 62-64; consult, Pac Northwest Labs, Battelle Mem Inst, 66-69 & Int Comn Radiation Units & Measurements, 70- Mem: Am Phys Soc; Am Asn Physics Teachers; Swiss Soc Physics & Natural Hist. Res: Interaction of radiation with matter, neutron physics, use of radiation in medicine and biology; cancer therapy with fast neutron beams; neutron radiation dosimetry. Mailing Add: Dept of Radiol Univ Hosp Seattle WA 98195

BICHTELER, KLAUS RICHARD, b Leipzig, Ger, Mar 15, 38; m 69; c 1. MATHEMATICS. Educ: Univ Hamburg, dipl, 62, PhD(gen relativity), 65. Prof Exp: Asst math, Univ Heidelberg, 65-68; asst prof, Southwest Ctr Advan Studies, 66-69; asst prof, 69-71, ASSOC PROF MATH, UNIV TEX, AUSTIN, 71- Mem: Am Math Soc. Res: General relativity; representation theory of topological groups and algebras. Mailing Add: Dept of Math Univ of Tex Austin TX 78712

BICK, GEORGE HERMAN, b Neptune, La, Sept 23, 14; m 45; c 2. ENTOMOLOGY, ZOOLOGY. Educ: Tulane Univ, BS, 36, MS, 38; Cornell Univ, PhD(entom), 47. Prof Exp: Inspector, Food & Drug Admin, USDA, La, 38-41; field biologist, State Dept Conserv, La, 41-42; asst prof, Univ Miss, 47-48 & Tulane Univ, 48-56; assoc prof, Univ Southwestern La, 56-59; prof, Clarion State Col, 59-60; PROF BIOL, ST MARY'S COL, IND, 60- Mem: Fel AAAS; Entom Soc Am. Res: Biology of Odonata. Mailing Add: Dept of Biol St Mary's Col Notre Dame IN 46556

BICK, KENNETH F, b Janesville, Wis, Feb 14, 32; m 58; c 3. GEOLOGY. Educ: Yale Univ, BS, 54, MS, 56, PhD(geol), 58. Prof Exp: Asst prof geol, Washington & Lee Univ, 58-61; assoc prof, 61-66, chmn dept geol, 62-68, PROF GEOL, COL WILLIAM & MARY, 66- Mem: AAAS; Geol Soc Am; Am Asn Petrol Geologists; Am Geophys Union. Res: Structural and stratigraphic geology; application of computers to geological problems. Mailing Add: Dept of Geol Col of William & Mary Williamsburg VA 23185

BICK, RODGER LEE, b San Francisco, Calif, May 21, 42. HEMATOLOGY, ONCOLOGY. Educ: Univ Calif, Irvine, MD, 70. Prof Exp: From intern to resident internal med, Kern Gen Hosp, Bakersfield, Calif, 70-72; fel, 72-73, MEM FAC HEME-ONCOL, UNIV CALIF CTR HEALTH SCI, LOS ANGELES, 73-; MED DIR, BAY AREA HEMAT ONCOL RES LABS, 73- Concurrent Pos: Dir med educ & hemat-oncol, Kern Gen Hosp, 73-74; attend hematologist, Santa Monica Med Ctr, Calif, 74-, dir oncol, 75-; attend hematologist, St John's Health Sci Ctr, Santa Monica & Univ Calif Ctr Health Sci, Los Angeles, 74-; mem thrombosis coun, Am Heart Asn. Mem: Int Soc Thrombosis & Hemostasis; Am Soc Hemat; Am Cancer Soc. Res: Hemostasis and blood coagulation; alterations of hemostasis associated with cardiopulmonary bypass; hypercoagulability and thrombosis, acquired antithrombin-III deficiency; alterations of hemostasis associated with malignancy, nature of disseminated intravascular coagulation. Mailing Add: Bay Area Hemat Oncol Res Labs 1260 15th St Santa Monica CA 90404

BICK, THEODORE A, b Brooklyn, NY, Dec 18, 30; m 56; c 4. MATHEMATICS. Educ: Union Col, BS, 58; Univ Rochester, MS, 60, PhD(ergodic theory), 64. Prof Exp: Instr math, Hobart & William Smith Cols, 61-64, from asst prof to assoc prof, 64-66; ASSOC PROF MATH, UNION COL, NY, 66- Mem: Am Math Soc; Math Asn Am. Res: Ergodic theory; functional analysis; real variable theory; measure and integration. Mailing Add: Dept of Math Union Col Schenectady NY 12308

BICKEL, EDWIN DAVID, b Louisville, Ky, Nov 11, 41; m 67; c 2. MALACOLOGY, ENVIRONMENTAL GEOLOGY. Educ: Univ Louisville, AB, 63, MSc, 65; Ohio State Univ, PhD(geol), 70. Prof Exp: Asst prof geol, 70-73, DIR COOP EDUC, MINOT STATE COL, 73- Concurrent Pos: Work-group researcher, Northern Great Plains Resources Prog, 74-75; contracted malacologist, Off Endangered Species, US Fish & Wildlife Serv, 74-; environ consult, 75-; proj prin investr, NDak Regional Environ Assessment Prog, 76- Mem: Am Asn Petrol Geologists; Am Malacol Union; Paleont Soc; Sigma Xi; AAAS. Res: Systematics and ecology of living and fossil non-marine Mollusca; Cretaceous-Tertiary molluscan biostratigraphy, coal stratigraphy; socio-economic and environmental impacts of energy development. Mailing Add: 1904 Sixth St NW Minot ND 58701

BICKEL, PETER J, b Bucharest, Romania, Sept 21, 40; US citizen; m 64; c 2. MATHEMATICAL STATISTICS. Educ: Univ Calif, Berkeley, AB, 60, MA, 61, PhD(statist), 63. Prof Exp: From asst prof to assoc prof statist, 63-70, PROF STATIST, UNIV CALIF, BERKELEY, 70- Concurrent Pos: Lectr, Imp Col, Univ London, 65-66; J S Guggenheim Found fel, 70. Mem: Fel Inst Math Statist; Am Statist Asn; Royal Statist Soc. Res: Nonparametric statistics; sequential analysis; robustness. Mailing Add: Dept of Statist Univ of Calif Berkeley CA 94720

BICKEL, ROBERT JOHN, b Louisville, Ky, Nov 8, 16; m 42. MATHEMATICS. Educ: Univ Louisville, AB, 37; Northwestern Univ, MA, 41; Univ Pittsburgh, PhD(math), 60. Prof Exp: Teacher, pub schs, Ky, 37-41; from instr to assoc prof, 46-60, actg head dept, 68-69, 73-74, PROF MATH, DREXEL UNIV, 60-, ASSOC HEAD DEPT, 63- Mem: Math Asn Am; Soc Indust & Appl Math (treas, 55-62). Res: Analysis; summability; divergent series. Mailing Add: Dept of Math Drexel Univ Philadelphia PA 19104

BICKEL, THOMAS FULCHER, b Detroit, Mich, Nov 20, 37; m 64. ALGEBRA. Educ: Univ Mich, BS, 59, MA, 60, PhD(math), 65. Prof Exp: Instr math, Mass Inst Technol, 65-67; asst prof, 67-73, ASSOC PROF MATH, DARTMOUTH COL, 73- Mem: Am Math Soc. Res: Finite group theory; theory of finite groups of Lie type; permutation groups. Mailing Add: Dept of Math Dartmouth Col Hanover NH 03755

BICKEL, WILLIAM SAMUEL, b Ottsville, Pa, June 8, 37. EXPERIMENTAL PHYSICS, SPECTROSCOPY. Educ: Pa State Univ, BS, 59, PhD(physics), 65. Prof Exp: Res assoc physics, 65, from asst prof to assoc prof, 65-75, PROF PHYSICS, UNIV ARIZ, 75- Mem: Am Phys Soc. Res: Spectroscopic diagnostics; atomic physics; spectroscopy of fast excited ions; measurement of mean lives of excited ions; vacuum ultraviolet and visible spectroscopy; biophysics; time-resolved polarized light scattering from biological macromolecules. Mailing Add: Dept of Physics Univ of Ariz Tucson AZ 85721

BICKERDIKE, ERNEST LAWRENCE, b Bakersfield, Calif, Mar 17, 08; m 42; c 2. ANALYTICAL CHEMISTRY, INORGANIC CHEMISTRY. Educ: Univ Southern Calif, BA, 30, MA, 32, PhD(chem), 37. Prof Exp: Instr sci, Taft Jr Col, 35-38; instr sci, 38-44, from instr to prof chem, 44-75, EMER PROF CHEM, UNIV CALIF, SANTA BARBARA, 75- Mem: AAAS; Am Chem Soc. Res: Rare earths. Mailing Add: Dept of Chem Univ of Calif Santa Barbara CA 93106

BICKERMAN, HYLAN A, b New York, NY, Oct 26, 13; m 41; c 2. RESPIRATORY PHYSIOLOGY, CLINICAL PHARMACOLOGY. Educ: Columbia Univ, BA, 34, MA, 35; NY Univ, MD, 39. Prof Exp: CHIEF ASTHMA-EMPHYSEMA CLIN, COLUMBIA-PRESBY MED CTR, 61- Concurrent Pos: Res fel med & chest dis, Columbia Univ Res Serv, Goldwater Mem Hosp, New York, 42-47, vis physician, 57-68; chief med serv, Cushing Gen Hosp, Boston, 45-46; asst clin prof, Col Physicians & Surgeons, Columbia Univ, 57-58, assoc clin prof, 58-; consult, St Barnabas Hosp, New York, 65-71 & Brookhaven Mem Hosp, Suffolk County, NY, 68-; dir respiratory lab & inhalation ther dept, Francis Delafield Hosp, New York, 68-75; panelist on emphysema, Int Cong Chest Dis, Lausanne, Switz, 70; asst attend physician, Presby Hosp, New York, 70-; mem adv panel, Food & Drug Admin; dir pulmonary function labs, Doctors Hosp, New York. Honors & Awards: Golden Tree of Life Award, Am Asn Inhalation Therapists, 67. Mem: AAAS; fel AMA; fel Am Col Chest Physicians; fel Am Col Physicians; Am Thoracic Soc. Res: Respiratory physiology and management of patients with chronic obstructive lung disease, especially pharmacologic therapy; development of various modalities of inhalation therapy including aerosols and pressure breathing. Mailing Add: Col of Physicians & Surgeons Columbia Univ New York NY 10027

BICKERSTAFF, ARTHUR, forestry, see 12th edition

BICKERSTAFF, THOMAS ALTON, b Tishomingo, Miss, Sept 5, 04; m 33; c 3. MATHEMATICS. Educ: Univ Miss, BA, 28, MA, 29; Univ Mich, PhD(math), 48. Prof Exp: From instr to asst prof, 29-46, registr, 36-46, prof & chmn dept, 46-72, EMER PROF MATH, UNIV MISS, 72- Concurrent Pos: Instr math, V-12 trainees, US Navy, 44-45. Mem: Am Math Soc; Math Asn Am; Inst Math Statist. Res: Mathematical and order statistics; certain order probabilities in nonparametric sampling. Mailing Add: Dept of Math Univ of Miss University MS 38677

BICKERTON, ROBERT KEITH, b East Liverpool, Ohio, Oct 1, 34; m 57; c 2. PHARMACOLOGY. Educ: Univ Pittsburgh, BS, 56, MS, 58, PhD(pharmacol), 60. Prof Exp: Sr res pharmacologist, 60-62, unit leader, 62-63, from asst chief to chief sect pharmacol, 63-66, dir pharmacometrics div, 66-68, dir res, 68-69, vpres res, 69-73, V PRES SCI AFFAIRS, NORWICH PHARMACAL CO, 73- Mem: Am Pharmaceut Asn; Am Soc Pharmacol & Exp Therapeut. Res: Mechanism of action and development of antihypertensive drugs. Mailing Add: Norwich Pharmacal Co PO Box 191 Norwich NY 13815

BICKFORD, ARTHUR ALTON, veterinary pathology, see 12th edition

BICKFORD, CHARLES ALLEN, b Gorham, NH, May 25, 05; m 32; c 2. FOREST BIOMETRY. Educ: Dartmouth Col, BS, 25; Univ Idaho, MS, 31. Prof Exp: Asst off blister rust control, US Forest Serv, Wash, 29-31, jr forester, Southern Forest Exp Sta, 31-35; from asst silviculturist to silviculturist, 36-46, forester, Northeastern Forest Exp Sta, 46-47, statistician, 47-63; from assoc prof to prof statist, 63-71, EMER PROF STATIST, STATE UNIV NY COL ENVIRON SCI & FORESTRY, 71- Mem: Fel AAAS; Soc Am Foresters; Am Statist Asn; Biomet Soc. Res: Silviculture of southern pines; use of fire in southern pine stands; forest mensuration; statistical analysis in relation to forest research; methods of thinning; natural regeneration. Mailing Add: 34 Compton Bristol TN 37620

BICKFORD, LAWRENCE RICHARDSON, b Elmira, NY, Nov 24, 21; m 43; c 3. SOLID STATE PHYSICS. Educ: Alfred Univ, BS, 43; Mass Inst Technol, PhD(physics), 49. Prof Exp: Assoc prof physics, State Univ NY Col Ceramics, Alfred, 49-54; sr physicist res lab, 54-63, dir gen sci, Res Ctr, 62-63, dir res lab, Tokyo, Japan, 63-65, dir memory & mat res, Res Lab, NY, 65-70, mgr mat sci & technol, 70-73, RES STAFF MEM, IBM RES LAB, 73- Mem: Fel Am Phys Soc; Inst Elec & Electronics Engrs. Res: Ferrites; titanates; glass; organic conductors; measuring and explaining contact effects which interfere with the accurate determination of conductivity in anisotropic organic conductors. Mailing Add: K01/281 IBM Res Lab 5600 Cottle Rd San Jose CA 95193

BICKFORD, MARION EUGENE, JR, b Memphis, Tenn, Aug 30, 32; m 54; c 3. PETROLOGY, GEOCHEMISTRY. Educ: Carleton Col, BA, 54; Univ Ill, MS, 58, PhD(geol), 60. Prof Exp: Asst prof geol, San Fernando Valley State Col, 60-63; asst res geophysicist, Inst Geophys, Univ Calif, Los Angeles, 63-64; from asst prof to assoc prof geol, 64-73, PROF GEOL, UNIV KANS, 73- Mem: Mineral Soc Am; Geol Soc Am; Am Geophys Union; Geochem Soc. Res: Geochronology of the Precambrian of southeastern Missouri; geochemistry of silicic extrusives and of migmatites. Mailing Add: Dept of Geol Univ of Kans Lawrence KS 66045

BICKFORD, REGINALD G, b Brewood, Eng, Jan 20, 13; nat US; m 45; c 2. NEUROSCIENCES. Educ: Cambridge Univ, BA, MD & BCh, 36; FRCP, 71. Prof Exp: Mayo res assoc, Univ Minn, 46-48, from assoc prof to prof physiol, 53-69; PROF NEUROSCI, SCH MED & HEAD EEG LAB, UNIV HOSP, UNIV CALIF, SAN DIEGO, 69- Concurrent Pos: Med Res Coun fel, Univ London, 37-40; consult EEG, Mayo Clin, 46-69; mem neurol study sect, USPHS, 60-64, mem comput study sect, 64-68, mem adv comt epilepsies, 64-70, mem clin pract comt deleg to Soviet Union, 68 & mem sci info prog adv comt, Nat Inst Neurol Dis & Stroke, 70- Mem: Am EEG Soc (pres, 56); Am Physiol Soc; Int League Against Epilepsy; Am Neurol Asn. Res: Electrical activity of the brain in man and animals; experimentally produced changes and their relation to behavioral and psychological effects; computer analysis and automation of the electroencephalogram. Mailing Add: Dept of Neurosci Univ of Calif Sch of Med La Jolla CA 92093

BICKING, JOHN BEEH, b Reading, Pa, Feb 11, 20; m 49; c 2. ORGANIC CHEMISTRY. Educ: Drexel Inst, BS, 42; Univ Del, MS, 48, PhD(chem), 53. Prof Exp: Control chemist, Gen Chem Div, Allied Chem & Dye Corp, 46-47; SR RES ASSOC, MERCK, SHARP & DOHME RES LABS DIV, MERCK & CO, INC, 48- Mem: Am Chem Soc. Res: Synthetic organic chemistry. Mailing Add: 622 Salford Ave Lansdale PA 19446

BICKLEY, HARMON C, b Detroit, Mich, May 1, 30; m 56, 68; c 4. PATHOLOGY, DENTISTRY. Educ: Univ Mich, DDS, 59; Univ Rochester, PhD(exp path), 64. Prof Exp: USPHS trainee, Univ Rochester, 59-63; from asst prof to assoc prof path & chmn dept oral biol, Col Dent, Univ Ky, 63-73; ASSOC PROF ORAL BIOL, COL DENT, MED CTR, UNIV IOWA, 73- Concurrent Pos: USPHS res grant & res career develop award, 64-67. Mem: Am Dent Asn; Int Asn Dent Res; Tissue Cult Asn. Res: Effects of lathyrogens on cultured connective tissue; immunopathology; psychophysiologic disease mechanisms. Mailing Add: Dept Oral Biol Col Dent Univ Iowa Med Ctr Iowa City IA 52242

BICKLEY, WILLIAM ELBERT, b Knoxville, Tenn, Jan 20, 14; m 41; c 4. ENTOMOLOGY. Educ: Univ Tenn, BS, 34, MS, 36; Univ Md, PhD(entom), 40. Prof Exp: Teacher, pub schs, Tenn, 35-37; teaching fel entom, Univ Md, 37-40, instr, 40-42; from asst entomologist to sr asst sanitarian, malaria control in war areas, USPHS, 42-46; asst prof biol, Univ Richmond, 46-49; assoc prof, 49-57, head dept, 57-71, PROF ENTOM, UNIV MD, 57- Concurrent Pos: Entomologist, State Dept Pub Health, Va, 47-49; ed, Mosquito News, 73- Mem: Entom Soc Am; Mosquito Control Asn (pres, 61); Am Soc Trop Med & Hyg; Coun Biol Ed; Am Heartworm Soc. Res: Chrysopidae; insect morphology; Japanese beetle; mosquitoes; vegetable insects; alfalfa weevil. Mailing Add: Dept of Entom Univ of Md College Park MD 20742

BICKMORE, JOHN TARRY, b Logan, Utah, Jan 15, 28; m 53; c 3. APPLIED PHYSICS, ENGINEERING PHYSICS. Educ: Idaho State Col, BS, 50; Univ Rochester, PhD(biophys), 56. Prof Exp: Res assoc, Univ Rochester, 52-55; res physicist, 55-75, SR SCIENTIST, XEROX CORP, 75- Mem: AAAS; Soc Photog Sci & Eng. Res: Chemical effects of ionizing radiation; xerographic process research, particularly the mechanisms, materials properties and failure modes in the development process. Mailing Add: 1849 Blossom Rd Rochester NY 14625

BICKNELL, EDWARD J, b Kansas City, Mo, Jan 23, 28; m 52. COMPARATIVE PATHOLOGY, CLINICAL PATHOLOGY. Educ: Univ Mo, BA, 48; Kans State Univ, MS, 51, DVM, 60; Mich State Univ, PhD(path), 65. Prof Exp: Instr path, Mich State Univ, 60-64; asst prof, Kans State Univ, 65-66; assoc prof, Iowa State Univ, 66-67; assoc prof vet sci, SDak State Univ, 68-72; ANIMAL PATHOLOGIST & EXTEN VETERINARIAN, AGR EXP STA, UNIV ARIZ, 72- Res: Pathology of feedlot cattle disease; pathology of pesticide toxicosis in wild fowl. Mailing Add: Agr Exp Sta Univ of Ariz Tucson AZ 85721

BICOFF, JUAN PEDRO, b Buenos Aires, Arg, Oct 23, 26; US citizen; m 55; c 3. PEDIATRIC CARDIOLOGY. Educ: Nat Col Buenos Aires, BA, 43. Prof Exp: Res assoc, Cook County Children's Hosp, 60-62; lectr, Cath Univ Cordoba, 63-64; asst prof, 65-68, ASSOC PROF PEDIAT CARDIOL, CHICAGO MED SCH, 68-; DIR PEDIAT EDUC & CHIEF PEDIAT CARDIOL, ST FRANCIS HOSP, 74- Concurrent Pos: Fel pediat cardiol, Cook County Children's Hosp, Ill, 59-60; prof pediat, Loyola Univ. Mem: Fel Am Col Cardiol; fel Am Acad Pediat; Arg Cardiol Soc; Arg Pediat Soc; fel Am Heart Asn. Res: Natural history of ventricular septal defects; artificial placenta and other systems for the infant supportive perfusion. Mailing Add: Dept of Pediat Cardiol St Francis Hosp Evanston IL 60202

BIDANI, NIRMALA DEVI, population geography, geography of south asia & asia, see 12th edition

BIDDINGTON, WILLIAM ROBERT, b Piedmont, WVa, Mar 30, 25; m 47; c 1. DENTISTRY. Educ: Univ Md, DDS, 48. Prof Exp: From instr to assoc prof, Dent Sch, Univ Md, 48-59; PROF ENDODONTICS & HEAD DEPT, SCH DENT, W VA UNIV MED CTR, 59-, DEAN SCH DENT, 68- Mem: Am Dent Asn; Int Asn Dent Res; Am Col Dent; Am Asn Endodontists. Res: Endodontics; periodontics. Mailing Add: Off of Dean WVa Univ Sch of Dent Morgantown WV 26505

BIDDLE, JOHN CHARLES, b South Bend, Ind, Aug 20, 32; m 51; c 4. MATHEMATICS EDUCATION. Educ: Ind Univ, Bloomington, BS, 56, MAT, 57, EdD, 66. Prof Exp: Instr sci & math, Flint Community Col, 57-63; instr math, Sch Educ, sch educ, Ind Univ, Bloomington, 63-66; asst prof, Cent Mich Univ, 66-68; asst prof math & dir comput ctr, Col Petrol & Minerals, Dhahran, Saudi Arabia, 68-69; assoc prof math & dir comput ctr, Davis & Elkins Col, 69-70; DIR INSTNL RES & AUTOMATIC DATA PROCESSING, CALIF STATE COL, BAKERSFIELD, 70- Concurrent Pos: Ed, Book Shelf sect, The Pentagon, 68-70. Mem: Math Asn Am. Res: Computer applications. Mailing Add: Calif State Col 9001 Stockdale Highway Bakersfield CA 93309

BIDDLE, RICHARD ALBERT, b Philadelphia, Pa, July 16, 30; m 55; c 4. INORGANIC CHEMISTRY. Educ: Pa State Univ, BS, 52; Iowa State Univ, MS, 55. Prof Exp: Instr & res assoc chem, Iowa State Univ, 52-56; res chemist, Elkton Div, 57-60, head inorg & propellant chem group, Res Dept, 60-66, SCIENTIST RES DEPT, ELKTON DIV, THIOKOL CHEM CORP, 66- Mem: AAAS; Am Chem Soc. Res: Inorganic chemical syntheses with emphasis on boron hydrides and high vacuum techniques; thermal decomposition of perchlorates and kinetics of solid phase reactions; space simulation studies; thermoanalytical techniques; hygrometry. Mailing Add: 36 Merry Rd Newark DE 19713

BIDDULPH, LOWELL GEORGE, b Ogden, Utah, Apr 14, 06; m 31; c 5. GEOLOGY. Educ: Brigham Young Univ, AB, 38; Univ Mich, MA, 39; Univ Utah, EdD, 50. Prof Exp: Teacher high sch, Idaho, 28-37; prof geol & dean, 38-72, EMER DEAN STUDENTS, RICKS COL, 72- Mem: AAAS. Res: Geysers and hot springs; birds of Yellowstone National Park; geology of Teton Mountains, Wyoming. Mailing Add: Ricks Col Rexburg ID 83440

BIDDULPH, ORLIN, b Hooper, Utah, Jan 27, 08; m 42; c 2. BOTANY. Educ: Brigham Young Univ, BS, 29, MS, 33; Univ Chicago, PhD(bot), 34. Prof Exp: Asst prof bot & head dept, Univ SDak, 34-37; from asst prof to prof, 37-68, actg chmn dept biophys, 68-73, EMER PROF BOT & BIOPHYS, WASH STATE UNIV, 73- Concurrent Pos: Nat Res fel, 41; consult, GE Hanford Eng Works, 49-50; Guggenheim fel, 59. Mem: AAAS; Am Soc Plant Physiol; Bot Soc Am. Res: Photoperiodism; mineral nutrition; translocation; radiation effects; radioactive tracers. Mailing Add: Dept of Botany Wash State Univ Pullman WA 99163

BIDE, RICHARD W, b Calgary, Alta. PATHOLOGICAL CHEMISTRY. Educ: Univ Alta, BSc, 59, MSc, 61; Aberdeen Univ, PhD(biochem), 64. Prof Exp: Nat Res Coun Can fel radiobiol, Atomic Energy Can, Ltd, 64-66; NAT RES COUN CAN FEL BIOCHEM, ANIMAL DIS RES INST, CAN DEPT AGR, 66- Mem: Am Soc Animal Sci; Can Soc Clin Chem; Brit Biochem Soc. Res: Enzymology; radiobiology; veterinary pathological chemistry; biomedical profiling. Mailing Add: Animal Dis Res Inst Can Dept of Agr PO Box 640 Lethbridge AB Can

BIDELMAN, WILLIAM PENDRY, b Los Angeles, Calif, Sept 25, 18; m 40; c 4. ASTRONOMY. Educ: Harvard Univ, SB, 40; Univ Chicago, PhD(astron), 43. Prof Exp: Asst, Yerkes Observ, Univ Chicago, 41-43; physicist, ballistic res lab, Aberdeen Proving Ground, US Ord Dept, 43-45; from instr astron to asst prof astrophys, Univ Chicago, 45-53; from asst astronr to assoc astronr, Lick Observ, Univ Calif, 53-62; prof astron, Univ Mich, 62-69 & Univ Tex, 69-70; PROF ASTRON & CHMN DEPT, CASE WESTERN RESERVE UNIV & DIR WARNER & SWASEY OBSERV, 70- Mem: Am Astron Soc; Int Astron Union. Res: Spectral classification; observational astrophysics; astronomical data. Mailing Add: Warner & Swasey Observ 1975 Taylor Rd East Cleveland OH 44112

BIDINOSTI, DINO RONALD, b Winnipeg, Man, Mar 27, 33; m 57; c 4. PHYSICAL CHEMISTRY. Educ: Univ Man, BSc, 55, MSc, 56; McMaster Univ, PhD(chem), 59. Prof Exp: Defense Res Bd Can fel phys chem, Univ Ottawa, 59-60; res assoc & Air Res & Develop Command fel, Cornell Univ, 60-61; asst prof, 61-66, ASSOC PROF PHYS CHEM, UNIV WESTERN ONT, 66- Mem: Chem Inst Can; The Chem Soc. Res: Application of mass spectrometry to chemical systems; thermochemistry. Mailing Add: Dept of Chem Univ of Western Ont London ON Can

BIDLACK, DONALD EUGENE, b Oakwood, Ohio, Apr 16, 32; m 53; c 3. VETERINARY MEDICINE. Educ: Ohio State Univ, BSc, 54, DVM, 57. Prof Exp: Pvt pract, Ind, 57-65; vet, Eaton Labs, Norwich Pharmacal Co, 65-73; RES VET, SYNTEX RES, 73- Mem: Am Vet Med Asn; Am Asn Swine Practr. Res: Industrial pharmaceutical research and development. Mailing Add: 21650 Fitzgerald Dr Cupertino CA 95014

BIDLACK, VERNE CLAUDE, JR, b South Bend, Ind, Mar 5, 23; m 46; c 4. ORGANIC CHEMISTRY. Educ: Univ Mich, BS, 44; Pa State Univ, MS, 48, PhD(chem), 50. Prof Exp: Res chemist, E I du Pont de Nemours & Co, 50-51; chief org chemist pharmaceuts, Henry K Wampole Co, Inc, 51-55; sr com develop engr, Archer-Daniels-Midland Co, 55-62; mgr mkt & prod develop, Atlantic Refining Co, 62, chem mkt res, 62-64; dir long planning, Chem Group, W R Grace & Co, 64-66; DEVELOP ENGR & PURCHASING AGENT, FMC CORP, 66- Mem: Am Chem Soc; Chem Mkt Res Asn; Soc Plastics Indust. Res: Fatty acids and vegetable oils; natural and synthetic resins; protective coatings; adhesives; medicinal chemicals; plastics; fertilizers; plastic films. Mailing Add: 407 Cheltena Ave Jenkintown PA 19046

BIDLEMAN, TERRY FRANK, b Chicago, Ill, Feb 17, 42. ANALYTICAL CHEMISTRY, ENVIRONMENTAL CHEMISTRY. Educ: Ohio Univ, BS, 64; Univ Minn, PhD(anal chem), 70. Prof Exp: Fel chem, Dalhousie Univ, 70-72; res assoc food & resource chem, Univ RI, 72-75; ASST PROF CHEM, UNIV SC, 75- Mem: Am Chem Soc. Res: Pesticide residue analysis; transport of pesticides in the environment; chemical aspects of air, sea transfer; solvent extraction equilibria; equilibria in natural water systems. Mailing Add: Dept of Chem Univ of SC Columbia SC 29208

BIDLINGMAYER, WILLIAM LESTER, b Cleveland, Ohio, July 7, 20; m 50; c 2. ENTOMOLOGY. Educ: Univ Fla, BSA, 49, MS, 52. Prof Exp: Entomologist, Fla State Bd Health, 50-51; asst sanitarian, tech develop lab, Commun Dis Ctr, USPHS, 51-55; ENTOMOLOGIST, FLA MED ENTOM LAB, DIV OF HEALTH, 55- Mem: Am Mosquito Control Asn; Entom Soc Am. Res: Biology and ecology of mosquitoes and sandflies; sampling and population dynamics of mosquito populations. Mailing Add: Fla Med Entom Lab Div of Health PO Box 520 Vero Beach FL 32960

BIDLINGMEYER, BRIAN ARTHUR, b Dallas, Tex, Aug 8, 44; m 72; c 1. ANALYTICAL CHEMISTRY. Educ: Kenyon Col, AB, 66; Purdue Univ, PhD(anal chem), 71. Prof Exp: Res chemist, Stand Oil Co Ind, 71-74; MGR ORG BIOSCI

RES, WATERS ASSOCS, 75- Mem: Am Chem Soc; Am Soc Testing & Mat. Res: Application of liquid chromatography to difficult separations in organic and biological research, specifically pesticide residues, long-chain fatty acids and lipids; data handling, manipulation and interpretation in chromatography. Mailing Add: Waters Assocs 34 Maple St Milford MA 01757

BIDNEY, DAVID, b Sept 25, 08; US citizen; m 40; c 2. HISTORY OF ANTHROPOLOGY, ETHNOLOGY. Educ: Univ Toronto, BA, 28, MA, 29; Yale Univ, PhD(philos), 32. Prof Exp: Sterling fel, Yale Univ, 39-40; Res assoc anthrop, Wenner-Gren Found Anthrop Res, 41-50; from assoc prof to prof anthrop & philos, 50-64, prof anthrop & philos educ, 65-73, EMER PROF ANTHROP & PHILOS EDUC, IND UNIV, BLOOMINGTON, 74- Concurrent Pos: Guggenheim fel, NY, 50-51; Ford Found fel, Israel, 64-65. Mem: Fel Am Anthrop Asn; Royal Anthrop Inst. Res: History and theory of anthropology; ethics, religion, mythology and phenomenology; ethnological education. Mailing Add: 5946 N New Jersey St Indianapolis IN 46220

BIDWELL, LEONARD NATHAN, b Camden, NJ, Dec 20, 34; m 63; c 2. MATHEMATICS. Educ: Univ Pa, BA, 56, MA, 57, PhD(math), 60. Prof Exp: Vis asst prof math, Haverford Col, 60-61; mathematician, Gen Elec Co, 61-62; asst prof math, 62-67, ASSOC PROF MATH, RUTGERS UNIV, CAMDEN, 67- Mem: Am Math Soc; Math Asn Am. Res: Topology; functional analysis. Mailing Add: Dept of Math Rutgers Univ Camden NJ 08102

BIDWELL, ORVILLE WILLARD, b Whitehouse, Ohio, Jan 14, 18; m 44; c 2. SOIL MORPHOLOGY, SOIL CLASSIFICATION. Educ: Oberlin Col, AB, 40; Ohio State Univ, BSc, 42, PhD(agron), 49. Prof Exp: Asst agron, Ohio State Univ, 46-49; from asst prof to assoc prof, 50-60, actg head dept agron, 70-71, PROF SOILS, KANS STATE UNIV, 60- Concurrent Pos: Vis prof, Ahmadu Bello Univ, Nigeria, 73; chmn ed bd, J Soil & Water Conserv, 75- Honors & Awards: Merit Award, Soil Conserv Soc Am, 63. Mem: Fel AAAS; fel Am Soc Agron; Soil Sci Soc Am; fel Soil Conserv Soc Am. Res: Soil classification and development. Mailing Add: Dept of Agron Col of Agr Kans State Univ Manhattan KS 66506

BIDWELL, ROGER GRAFTON SHELFORD, b Halifax, NS, June 8, 27; m 50; c 4. PLANT PHYSIOLOGY. Educ: Dalhousie Univ, BS, 47; Queen's Univ, Ont, BA, 50, MA, 51, PhD, 54. Prof Exp: Res officer bact, Defense Res Bd, 52-56; botanist, Atlantic Regional Lab, Nat Res Coun, 56-59; assoc prof bot, Univ Toronto, 59-65; prof biol, Case Western Reserve Univ, 65-69, chmn dept, 67-69; PROF BIOL, QUEEN'S UNIV, ONT, 69- Concurrent Pos: Secy, Biol Coun Can, 72- Mem: AAAS; Can Soc Plant Physiol (secy-treas, 63-65, vpres, 71-72, pres, 72-73); fel Royal Soc Can. Res: Intermediary metabolism in plants; carbohydrates and amino acids; process of photosynthesis; metabolism of marine algae. Mailing Add: Dept of Biol Queen's Univ Kingston ON Can

BIEBEL, PAUL JOSEPH, b Belleville, Ill, Feb 26, 28; m 51; c 6. PHYCOLOGY. Educ: Univ Notre Dame, BS, 49; St Louis Univ, MS, 55; Ind Univ, PhD(phycol), 63. Prof Exp: Instr biol, Spring Hill Col, 59-63; from asst prof to assoc prof, 63-74, PROF BIOL, DICKINSON COL, 74- Mem: Bot Soc Am; Phycol Soc Am; Am Bryol & Lichenological Soc; Int Phycol Soc. Res: Life cycles, morphology, genetics and morphogenesis of algae, especially saccoderm desmids. Mailing Add: Dept of Biol Dickinson Col Carlisle PA 17013

BIEBER, ALLAN LEROY, b Mott, NDak, Aug 14, 34; m 58; c 2. BIOCHEMISTRY. Educ: NDak State Univ, BS, 56, MS, 58; Ore State Univ, PhD(biochem), 62. Prof Exp: NIH training grant, Sch Med, Yale Univ, 61-63; from asst prof to assoc prof, 63-75, PROF CHEM, ARIZ STATE UNIV, 75- Mem: Am Chem Soc. Res: Metabolism of purines; purine analogs and their respective nucleotides; biochemistry of snake venoms and toxins. Mailing Add: Dept of Chem Ariz State Univ Tempe AZ 85281

BIEBER, GENE LAWRENCE, b Lafayette, Ind, Aug 15, 36; m 69. AGRONOMY, PLANT PHYSIOLOGY. Educ: Purdue Univ, BS, 58; Kans State Univ, MS, 60; Auburn Univ, PhD(agron), 67. Prof Exp: Asst agronomist, Miss State Univ, 67-71; asst prof agr, Southeast Mo State Univ, 71-74; with Daviess County Exten Serv, Purdue Univ, Lafayette, 74-75; MEM FAC, DEPT AGR, SOUTHEAST MO STATE UNIV, 75- Mem: Am Soc Agron; Crop Sci Soc Am. Res: Roadside turf research; crops variety testing. Mailing Add: Dept of Agr Southeast Mo State Univ Cape Girardeau MO 63701

BIEBER, IRVING, b New York, NY, Apr 4, 08; m 37. MEDICINE, PSYCHIATRY. Educ: NY Univ, BS, 27, MD, 30. Prof Exp: Asst neurophysiol, Sch Med, Yale Univ, 33; asst clin prof neurol, 47-49, CLIN PROF PSYCHIAT, COL MED, NY UNIV, 46- Concurrent Pos: Attend psychiatrist, Flower & Fifth Ave Hosps, 46-; res dir, Post-Grad Inst Psychother, 49-54; res neuropsychiatrist, Mem Ctr, 54-63; assoc, Sloane Kettering Inst, 56-63; attend psychiatrist, Metrop Hosp. Mem: AMA; fel Am Psychiat Asn; fel Acad Psychoanal; fel NY Acad Med. Res: Psychoanalysis of homosexuals; olfaction in sexual development. Mailing Add: 132 E 72nd St New York NY 10021

BIEBER, LORAN LAMOINE, b Mott, NDak, Apr, 33; m 55; c 3. BIOCHEMISTRY. Educ: NDak State Univ, BS, 55, MS, 56; Ore State Univ, PhD(biochem), 63. Prof Exp: Asst agr chem, NDak State Univ, 56-57; lab technician, Hosp Lab, Ft Carson, Colo, 57-58 & Chem Ctr, Edgewood, Md, 58-59; PROF BIOCHEM, MICH STATE UNIV, 59- Concurrent Pos: NIH fel biochem, Univ Calif, Los Angeles, 63-65; NIH fel, Mich State Univ, 66-74 & NSF fel, 67-71; vis prof cell physiol, Wenner-Gren Inst, Stockholm, 74. Mem: AAAS; Am Soc Biol Chemists; Am Chem Soc; Entom Soc Am. Res: Functions of carnitine; mode of actions of polyene antibiotics. Mailing Add: Dept of Biochem Mich State Univ East Lansing MI 48824

BIEBER, SAMUEL, b US, Feb 5, 26; m 49; c 2. DEVELOPMENTAL BIOLOGY. Educ: NY Univ, BA, 44, MS, 48, PhD(vert zool), 52. Prof Exp: Teaching fel, Washington Sq Col, NY Univ, 48-51, res fel, 51-52; sr biologist, res labs, Burroughs Wellcome & Co, 52-62; prof biol, assoc dean sci & assoc dean grad sch, Long Island Univ, 62-64, assoc dean grad fac & spec adv to provost for sci, 64-66, dean, Conolly Col, 66-69, from adj asst prof to adj assoc prof grad sch, 57-62; campus dean, 69-71, PROVOST TEANECK-HACKENSACK CAMPUS, FAIRLEIGH DICKINSON UNIV, 71- Concurrent Pos: Sci collabr, New York Aquarium, 52- Mem: AAAS; Am Chem Soc; Am Soc Zoologists; Soc Develop Biol; Soc Exp Biol & Med. Res: Renotropic effects of steroids; antimetabolites and embryogenesis and regeneration; nucleic acid metabolism during gametogenesis and embryogenesis; experimental cancer chemotherapy; chemical suppression of the immune response. Mailing Add: Fairleigh Dickinson Univ 1000 River Rd Teaneck NJ 07666

BIEBER, THEODORE IMMANUEL, b Zurich, Switz, July 6, 25. ORGANIC CHEMISTRY, BIOCHEMISTRY. Educ: NY Univ, BA, 45, MS, 46, PhD(chem), 51. Prof Exp: Asst chem, NY Univ, 47-51; asst prof, Coe Col, 51-52, Adelphi Col, 53-56 & Ga Inst Technol, 56-57; from assoc prof to prof, Univ Miss, 57-63; PROF CHEM, FLA ATLANTIC UNIV, 63- Mem: AAAS; Am Chem Soc; NY Acad Sci; The Chem Soc. Res: Organophosphorus and organosulfur chemistry; stereochemistry; amide synthesis; iodine-containing dyes; heterocycles; dehydrogenation; respiratory chain phosphorylation; metallocene chemistry. Mailing Add: Dept of Chem Fla Atlantic Univ Boca Raton FL 33432

BIEBERLY, FRANK GEARHART, b Spearville, Kans, Feb 29, 12; m 39; c 7. AGRONOMY. Educ: Kans State Univ, BS, 38, MS, 49. Prof Exp: Instr, Paxico High Sch, 38-41; county agent, Morris County, Kans, 41-42 & Hamilton County, 42-46; PROF AGRON & EXTEN AGRONOMIST, KANS STATE UNIV, 46- Mem: Am Soc Agron. Res: Soil management, including tillage and fertility, in the low rainfall areas of the great plains; moisture conservation for crop production and soil stability; safe and effective use of pesticides. Mailing Add: Dept of Agron Waters Hall Kans State Univ Manhattan KS 66502

BIEBERMAN, ROBERT ARTHUR, b Rock Island, Ill, Apr 3, 23; m 44; c 2. PETROLEUM GEOLOGY. Educ: Ind Univ, AB, 48, AM, 50. Prof Exp: PETROLEUM GEOLOGIST, NMEX BUR MINES, 50- Mem: Am Asn Petrol Geologists. Res: Subsurface geology. Mailing Add: 601 Fitch Ave Socorro NM 87801

BIEBUYCK, DANIEL P, b Deinze, Belg, Oct 1, 25; m 50; c 7. ANTHROPOLOGY. Educ: Univ Ghent, BA, 46, Lic es Philos et Lett, 48, PhD(anthrop, classics), 54. Prof Exp: Res fels, Inst Sci Res Cent Africa, 49-57; prof, Univ Lovanium, Leopoldville, 57-61; prof, Univ Del, 61-64; prof, Univ Calif, Los Angeles, 64-66, H RODNEY SHARP PROF ANTHROP, UNIV DEL, 66- Concurrent Pos: Vis lectr, Univ Liege, 56-58; vis lectr, Univ London, 60; cur African collections, Univ Calif, Los Angeles, 65-66; fac res grant & African Studies Ctr res grant, Univ Calif, Los Angeles, 66-67; Joint Comt African Studies res grant, Soc Sci Coun, 67-68; fac res grant, Univ Del, 68 & J T Last pvt res grant, 69-70; interim dir Black studies, Univ Del, 70-71; assoc sem, Columbia Univ, 70-; Nat Endowment Humanities res grant, 71-73. Honors & Awards: Laureate Concours, Royal Belg Acad Overseas Sci, 55. Mem: Fel Am Anthrop Asn; fel African Studies Asn; fel Int African Inst; fel Royal Belg Acad Overseas Sci; fel Int Inst Differing Civilizations. Res: Central African ethnography; African art; African oral literature; African systems of thought; African land tenure. Mailing Add: 271 W Main St Newark DE 19711

BIEDEBACH, MARK CONRAD, b Pasadena, Calif, Apr 21, 32; m 64; c 2. BIOPHYSICS. Educ: Univ Southern Calif, BE, 56, MS, 58; Univ Calif, Los Angeles, PhD(biophys), 64. Prof Exp: NIH res fel biol systs, Calif Inst Technol, 64-65; Nat Inst Neurol Dis & Blindness res fel neurophysiol, Lab Cellular Neurophysiol, Paris, France, 65-66; ASSOC PROF BIOL, CALIF STATE UNIV, LONG BEACH, 67- Mem: NY Acad Sci. Res: Biophysics of invertebrate nervous system-integration as determined by electrophysiological stimulation and recording. Mailing Add: Dept of Biol Calif State Univ 6101 E Seventh St Long Beach CA 90840

BIEDENHARN, LAWRENCE CHRISTIAN, JR, b Vicksburg, Miss, Nov 18, 22; m 50; c 2. NUCLEAR PHYSICS. Educ: Mass Inst Technol, BS, 44, PhD(physics), 49. Prof Exp: From res asst to res assoc, Mass Inst Technol, 48-50; physicist, Oak Ridge Nat Lab, 50-52; asst prof physics, Yale Univ, 52-54; assoc prof, Rice Inst, 54-61; PROF PHYSICS, DUKE UNIV, 61- Concurrent Pos: Sr Fulbright fel, 57-58; Guggenheim fel, 58; NSF sr fel, 64-65; consult, Los Alamos Sci Lab. Mem: AAAS; Fedn Am Scientists; fel Am Phys Soc; Am Asn Physics Teachers; Am Math Soc. Res: Theoretical physics; nuclear reactions. Mailing Add: Dept of Physics Duke Univ Durham NC 27706

BIEDERMAN, EDWIN WILLIAMS, JR, b Stamford, Conn, June 30, 30; m 58; c 4. GEOLOGY, MINERALOGY. Educ: Cornell Univ, BA, 52; Pa State Univ, PhD(mineral), 58. Prof Exp: Res geologist & tech group leader, Geochem & Sedimentology, Cities Serv Res & Develop, Co, 58-68, res planner, 68-72; ASST DIR, PA TECH ASSISTANCE PROG, PA STATE UNIV, 72- Concurrent Pos: Mem bd adv, Micropaleont Press, Am Mus Natural Hist, 70- Mem: Am Asn Petrol Geologists; Soc Econ Paleontologists & Mineralogists; Geochem Soc. Res: Sedimentary petrology; recent sediments; geochemistry; origin of oil; geological statistics; economic geology; photomicrography; general and physical geology; electron microscopy. Mailing Add: 501 Keller Bldg Pa State Univ University Park PA 16802

BIEDLER, JUNE LEE, b New York, NY, June 24, 25. CELL GENETICS. Educ: Vassar Col, AB, 47; Columbia Univ, MA, 54; Cornell Univ, PhD(biol), 59. Prof Exp: Exchange investr, Inst Gustave-Roussy, France, 59-60; res fel, 59-60, res assoc, 60-62, assoc, 62-72, sect head, 66-72, ASSOC MEM, SLOAN-KETTERING INST CANCER RES, 72-, LAB HEAD, 75-; ASSOC PROF BIOL, SLOAN-KETTERING DIV, GRAD SCH MED SCI, CORNELL UNIV, 73- Concurrent Pos: From instr to asst prof, Sloan-Kettering Div, Grad Sch Med Sci, Cornell Univ, 62-73; USPHS res career develop award, 63. Mem: Fel AAAS; Am Asn Cancer Res; Am Soc Cell Biol; Genetics Soc Am; NY Acad Sci. Res: Cytogenetics and somatic cell genetics; tumor biology; chromosome structure-function relationships; drug resistance of mammalian cells. Mailing Add: Sloan-Kettering Inst Cancer Res Lab of Cell & Biochem Genetics Rye NY 10580

BIEFELD, LAWRENCE PAUL, chemistry, see 12th edition

BIEFELD, PAUL FRANKLIN, b Brownsville, Pa, Nov 5, 25; m 50; c 3. PHYSICAL CHEMISTRY. Educ: Denison Univ, BS, 48; Mich State Univ, MS, 51. Prof Exp: Asst, Mich State Univ, 48-51 & chem res infrared spectros, State Hwy Dept, Mich, 51-56; chemist, characterization lab, Newark, 56, mgr anal lab, Granville, 56-71, SR SCIENTIST, DEPT CHEM TECHNOL, TECH CTR, OWENS-CORNING FIBERGLAS CORP, 71- Mem: Am Chem Soc; Soc Appl Spectros; Coblentz Soc. Res: Application of instrumental analysis in the fields of glass and organic plastics and resins. Mailing Add: Dept of Chem Technol Owens-Corning Fiberglas Corp Tech Ctr Granville OH 43023

BIEFER, GREGORY JAMES, b Montreal, Que, Sept 12, 21; m 52; c 2. PHYSICAL CHEMISTRY. Educ: McGill Univ, BSc, 49, PhD(phys chem), 52. Prof Exp: Chemist, E I du Pont de Nemours & Co, 52-56 & AEC Can, 56-60; CHEMIST, CAN DEPT ENERGY, MINES & RESOURCES, 60- Mem: Nat Asn Corrosion Eng; Chem Inst Can; Prof Inst Pub Serv Can. Res: Metal corrosion and anti-corrosion. Mailing Add: 85 Beaver Ridge Ottawa ON Can

BIEGELSEN, DAVID K, b St Louis, Mo, Oct 18, 43; m 66. EXPERIMENTAL SOLID STATE PHYSICS. Educ: Yale Univ, BA, 65; Wash Univ, MA & PhD(physics), 70. Prof Exp: MEM RES STAFF PHYSICS, XEROX PALO ALTO RES CTR, 70- Mem: Am Phys Soc; Inst Elec & Electronics Engrs. Res: Using acousto-optic probe of photoelastic properties of solids. Mailing Add: Xerox Palo Alto Res Ctr 3333 Coyote Hill Rd Palo Alto CA 94304

BIEGEN, JOSEPH ROBERT, b Northport, NY, June 13, 38; SURFACE PHYSICS. Educ: Clarkson Col Technol, BChE, 59, BS, 60; MS, 67, PhD(physics), 71. Prof Exp: Asst prof physics & math, Broome Community Col, 60-67; TEACHER PHYSICS & MATH, POTSDAM CENT SCH, 71- Concurrent Pos: Res assoc, Clarkson Col Technol, 71-74. Mem: Inst Colloid & Surface Sci. Res: Thermodesorption of adsorbed gases and surface reactions and mechanics during adsorption and desorption; microbalance techniques for the study of surface reactions. Mailing Add: 15 Cedar St Potsdam NY 13676

BIEHL, ARTHUR TREW, b New York, NY, May 18, 24; m 50; c 2. PHYSICS. Educ: Ill Inst Technol, BS, 45; Calif Inst Technol, MS, 47, PhD(physics), 49. Prof Exp: Asst, Calif Inst Technol, 47-48, asst cosmic ray exp, 48-49, res fel physics, 49-50; res engr atomic energy res dept, NAm Aviation, Inc, 50-52; physicist radiation lab, Univ Calif, 52-55, lectr nuclear eng, 55-57; vpres & tech dir, Aerojet-Gen Nucleonics & tech dir, Aerojet-Gen Corp, Gen Tire & Rubber Co, 57-59; spec asst to dir defense res & eng, Off Secy Defense, 59-60; pres, MB Assocs, 60-66; assoc dir, Lawrence Radiation Lab, Univ Calif, 66-71; VPRES, RES & DEVELOP ASSOCS, 71- Mem: Am Phys Soc. Res: Radar maintenance; cosmic ray and nuclear physics; statistical studies of cosmic rays at high altitudes; atomic energy; solid propellent rocket research. Mailing Add: R & D Assocs PO Box 9695 Marina del Rey CA 90291

BIEHL, EDWARD ROBERT, b Pittsburgh, Pa, July 14, 32; m 55; c 3. PHYSICAL CHEMISTRY, ORGANIC CHEMISTRY. Educ: Univ Pittsburgh, BS, 58, PhD(chem), 61. Prof Exp: Sr res chemist, Monsanto Res Corp, 61-62; from asst prof to assoc prof, 62-73, PROF CHEM, SOUTHERN METHODIST UNIV, 73- Mem: Am Chem Soc. Res: Benzyne chemistry. Mailing Add: Dept of Chem Southern Methodist Univ Dallas TX 75222

BIEHL, JOSEPH PARK, b Berkeley, Calif, June 14, 22; m 49; c 4. NEUROLOGY. Educ: Stanford Univ, AB, 43, MD, 46. Prof Exp: Intern internal med, Boston City Hosp, 45-46; sr asst res physician, Cincinnati Gen Hosp, 48-49, res neurologist, 49-50; from instr to assoc prof, 50-72, PROF NEUROL, COL MED, UNIV CINCINNATI, 72- Concurrent Pos: Fel med, Col Med, Univ Cincinnati, 50-52; consult, Vet Admin Hosp; attend neurologist, Cincinnati Gen Hosp, Drake Mem Hosp & Longview State Hosp. Mem: Asn Res Nerv & Ment Dis. Res: Clinical neurology; electroencephalography. Mailing Add: Dept of Neurol Univ of Cincinnati Col of Med Cincinnati OH 45221

BIEHLER, SHAWN, b Jersey City, NJ, May 20, 37; m 60; c 2. GEOPHYSICS. Educ: Princeton Univ, BSE, 58; Calif Inst Technol, MS, 61, PhD(geophys), 64. Prof Exp: Res fel geophys, Calif Inst Technol, 64-66; asst prof, Mass Inst Technol, 66-70; ASSOC PROF GEOPHYS, UNIV CALIF, RIVERSIDE, 70- Mem: AAAS; Am Geophys Union; Am Soc Explor Geophys; Seismol Soc Am; Europ Asn Explor Geophys. Res: Application of geophysics to geothermal areas; relationship of gravity anomalies to geologic structure. Mailing Add: Dept of Geol Sci Univ of Calif Riverside CA 92505

BIEHN, WILLIAM LAWRENCE, plant pathology, see 12th edition

BIEL, ERWIN REINHOLD, climatology, meteorology, deceased

BIEL, JOHN HANS, b Namslau, Ger, Sept 30, 20; nat US; m 47. CHEMISTRY. Educ: Dartmouth Col, AB, 42; Univ Mich, MS, 45, PhD(org chem), 47. Prof Exp: Sr res chemist, Lakeside Labs, Inc, 47-50, chief med chem div, 50-54, chief chem div, 54-59, dir labs, 59-62; vpres & dir res & develop, Aldrich Chem Co, Inc, Wis, 62-69; mgr dept gen pharmacol, 69-71, dir div pharmacol & med chem, 71-73; VPRES EXP THER, ABBOTT LABS, 73- Concurrent Pos: Asst prof col med, Univ Ill, 59-62; clin prof pharmacol, Med Col, Wis, 62- Honors & Awards: Res Achievement Award, Am Pharmaceut Asn Found, 66. Mem: Am Chem Soc; Am Soc Pharmacol & Exp Therapeut; Am Col Neuropsychopharmacol; Col Int Neuropsychopharmacol. Res: Dopaminergic agents; tetrahydrocannabinols; chemotherapy; heterocyclic and olefinic chemistry; hydrazines; acetylenes; antispasmodics bronchodilators; psychotherapeutics; metabolic and medicinal agents; catecholamine releasing agents; cardiovascular and peptic ulcer drugs; structure-activity correlations in the fields of antidepressant agents; monoamine oxidase and enzyme inhibitors. Mailing Add: Abbott Labs North Chicago IL 60064

BIELER, BARRIE HILL, b Pasadena, Calif, June 17, 29; m 55; c 3. MINERALOGY. Educ: Calif Inst Technol, BS, 51, MS, 52; Pa State Univ, PhD(mineral), 55. Prof Exp: Field asst reconnaissance geol mapping, US Geol Surv, 52; res asst mineral, Pa State Univ, 52-55; geologist, US Geol Surv, 55-57; ceramist, 57-67, res ceramist, West Div, 67-70, sr res ceramist, 70-74, DEVELOP SPECIALIST, WESTERN DIV, DOW CHEM CO, 74- Concurrent Pos: Assoc prof gen sci, John F Kennedy Univ, 68- Mem: Am Ceramic Soc. Res: Geothermal exploration; ceramics and refractories; slags in basic oxygen surface steelmaking; melting and fining of soda-lime glass. Mailing Add:

BIELLIER, HAROLD VICTOR, b Bois D'Arc, Mo, Jan 22, 21; m 42; c 2. PHYSIOLOGY. Educ: Univ Mo, BS, 43, PhD, 55. Prof Exp: Asst prof mil sci & tactics, 50-53, assoc prof, 53-69, PROF POULTRY HUSB, UNIV MO-COLUMBIA, 69- Concurrent Pos: Assoc ed, Poultry Sci, 67-75. Mem: Fel AAAS; Poultry Sci Asn; Sigma Xi. Res: Endocrine and reproductive physiology of domestic poultry. Mailing Add: Dept of Poultry Husb Univ of Mo-Columbia Columbia MO 65201

BIELSKI, BENON H J, b Poland, June 12, 27; US citizen; m 69. RADIATION CHEMISTRY. Educ: Harvard Univ, BA, 51; Columbia Univ, MA, 56, PhD(chem), 57. Prof Exp: Asst prof chem, Univ Fla, 57-58; res assoc, 58-60, assoc chemist, 60-63, CHEMIST, BROOKHAVEN NAT LAB, 64- Concurrent Pos: Vis prof, State Univ NY Downstate Med Ctr, 71- Mem: Am Chem Soc. Res: Radiation chemistry of inorganic and biological systems; chemistry of free radicals; low temperature biochemistry. Mailing Add: Dept of Biochem 450 Clarkson Ave State Univ NY Downstate Med Ctr Brooklyn NY 11203

BIELY, JACOB, b Bogopol, Russia, Jan 2, 03; m 29; c 4. POULTRY NUTRITION. Educ: Univ BC, BSA, 26, MSA, 30; Kans State Col, MS, 29. Hon Degrees: DSc, Univ BC, 70. Prof Exp: Res asst, 27-42, from asst prof to prof, 43-68, head dept, 52-68, RES PROF, UNIV BC, 69- Mem: Fel AAAS; fel Agr Inst Can; fel Poultry Sci Asn; fel Royal Soc Can. Res: Nutritive value of animal protein concentrates; vitamin and amino acid requirements of growing chickens; fat utilization by growing and laying birds; stress and nutritive requirements of chickens; effect of environment on egg production; cholesterol absorption; nutritive value of wheat; utilization of rapeseed. Mailing Add: Dept of Poultry Sci Univ of BC Vancouver BC Can

BIEMANN, KLAUS, b Innsbruck, Austria, Nov 2, 26; m 56; c 2. ORGANIC CHEMISTRY. Educ: Univ Innsbruck, PhD(chem), 51. Prof Exp: Instr chem, Univ Innsbruck, 51-55; res assoc, 55-57, from instr to assoc prof, 57-63, PROF CHEM, MASS INST TECHNOL, 55- Mem: Am Chem Soc; Am Acad Arts & Sci; Austrian Chem Soc; hon mem Chem Soc Belg. Res: Mass spectroscopy of organic compounds. Mailing Add: Dept of Chem Mass Inst of Technol Cambridge MA 02139

BIEMPICA, LUIS, b Orense, Spain, Aug 14, 25; US citizen; m 53; c 2. GASTROENTEROLOGY, PATHOLOGY. Educ: Univ Buenos Aires, MD, 52, cert nutrit, 56, PhD(med), 62. Prof Exp: Resident med, Bronx Munic Hosp Ctr, 66-67 & resident path, 68-71; asst prof path, Albert Einstein Col Med, 64-73, assoc med, 64-70, ASST PROF MED, ALBERT EINSTEIN COL MED, YESHIVA UNIV, 70-, ASSOC PROF PATH, 73- Concurrent Pos: Arg Res Coun fel, Albert Einstein Col Med, 61-62, USPHS res grant, 67-70. Honors & Awards: Riopedre Prize, Arg Asn Advan Sci, 69. Mem: AAAS; Am Soc Cell Biologists; Histochem Soc; Am Asn Study Liver Dis; Am Asn Pathologists & Bacteriologists. Res: Liver diseases; cytochemistry; enzymology; ultrastructure. Mailing Add: Dept of Path Albert Einstein Col of Med Bronx NY 10461

BIEN, GEORGE SUNG-NIEN, b An-Ching, China, July 2, 03; nat US; m 38; c 4. PHYSICAL CHEMISTRY. Educ: Shanghai Col, BS, 26; Brown Univ, MS, 32, PhD(phys chem), 34. Prof Exp: Res chemist, Golden Sea Res Inst Chem Indust, China, 34-37; prof, Kuang-Si Univ, 37-40; sub-mgr & supt, Lanchow Chem Works, 40-44; prof, Hua Chung Univ, 44-49; assoc prof, Sterling Col, 50-51; CHEM SPECIALIST, SCRIPPS INST, UNIV CALIF, 51- Mem: AAAS; Am Chem Soc; fel Am Inst Chem. Res: Conductivity of electrolytic solutions; industrial utilization of bittern and the extraction of potassium chloride therefrom; chemical properties of recent marine sediments; genesis of clay minerals; C-14 dating and study in ocean waters. Mailing Add: Scripps Inst Oceanog 1166 Ritter Univ of Calif La Jolla CA 92037

BIEN, PAUL BEH NIEN, b Anhwei, China, July 2, 03; nat US; m 37; c 3. PHYSICAL CHEMISTRY. Educ: Shanghai Baptist Col, China, BS, 25; Brown Univ, PhB, 28, ScM, 29, PhD(phys chem), 32. Prof Exp: Actg prof phys chem, Wuhan Univ, 33; prof phys & anal chem, Anhwei Univ, China, 33-34; assoc prof phys & gen chem, Hopei Inst Technol, 34-37; assoc prof, Lingnam Univ, 38-40; prof & actg dean col sci, Cheeloo Univ, 44-46; prof phys & org chem, Shanghai Univ, 46-49; sr sci master physics & chem, Queen's Col, Hongkong, 49-55; res fel phys chem, Ind Univ, 55-56; assoc prof gen chem & thermodyn, Youngstown Univ, 56-60; res chemist, Oak Ridge Nat Lab, 60-68; ASSOC PROF CHEM, FURMAN UNIV, 68- Mem: AAAS; Sci Res Soc Am; Am Chem Soc; fel Am Inst Chem. Res: Physical chemistry of electrolytic systems; conductivity of electrolytes in nonaqueous solvents; viscosity; electromotive force; transference; molten salts; isopiestic measurements in aqueous solutions. Mailing Add: Dept of Chem Furman Univ Greenville SC 29613

BIENENSTOCK, ARTHUR IRWIN, b New York, NY, Mar 20, 35; m 57; c 3. SOLID STATE PHYSICS, X-RAY CRYSTALLOGRAPHY. Educ: Polytech Inst Brooklyn, BS, 55, MS, 57; Harvard Univ, PhD(appl physics), 62. Prof Exp: NSF fel, Atomic Energy Res Estab, Eng, 62-63; asst prof appl physics, Harvard Univ, 62-67; assoc prof mat sci, 67-72, PROF MAT SCI & ENG, STANFORD UNIV, 72-, VPROVOST FAC AFFAIRS, 72- Concurrent Pos: Mem ad hoc comt, Nat Acad Sci-Nat Res Coun, 69-70. Mem: Am Crystallog Asn; fel Am Phys Soc. Res: Semiconductor band theory; structure and properties of imperfectly crystallized and amorphous materials; amorphous semiconductor devices; determination of atomic arrangements in amorphous materials and of the relationships of these arrangements to electrical, optical and thermal properties. Mailing Add: Dept of Mat Sci Stanford Univ Stanford CA 94305

BIENFANG, PAUL KENNETH, b Watertown, Wis, Apr 14, 48; m 73. BIOLOGICAL OCEANOGRAPHY. Educ: Univ Hawaii, BS, 71, MS, 74. Prof Exp: Res asst oceanog, Hawaii Inst Marine Biol, 69-73; res asst, 73-74, SR SCIENTIST OCEANOG, OCEANIC INST, 73- Concurrent Pos: Oceanog consult, Environ Consult Inc, 69-74, Oceanic Inst, 72, Sunn, Low, Tom & Hara Inc, 75 & Oceanic Eng Dept, Univ Hawaii, 76. Mem: Am Soc Limnol & Oceanog. Res: Phytoplankton ecology, particularly nutritional control of sinking rates of phytoplankton and dynamics of substrate limited growth in phytoplankton; thermal pollution assessment, especially effects on phytoplankton biomass and productivity. Mailing Add: Oceanic Inst Waimanalo HI 96795

BIENIARZ, JOSEPH, b Kety, Poland, Dec 13, 12; m 45; c 4. OBSTETRICS & GYNECOLOGY. Educ: Jagiellonian Univ, Cand Med, 35, MD, 37. Prof Exp: From asst prof to assoc prof obstet & gynec, Med Acad, Gdansk, 53-57; assoc prof uterine physiol, Univ Montevideo, 58-67; DIR LAB UTERINE PHYSIOL, MICHAEL REESE HOSP & MED CTR, 68- Concurrent Pos: Brit Coun bursary, Univ Edinburgh, 57-58; Rockefeller Found res grant, Inst Advan Studies, Montevideo, 58-59, vis res prof, 59, NIH grant, 64-65 & 66-68; NIH grant, Karolinska Inst, 66; exchange visitor, Dept Anat, Univ Ill, Chicago Circle, 58; partic, Pan-Am Health Orgn Adv Comt Med Res, Washington, DC, 69; assoc prof, Univ Chicago, 69, prof, 71. Honors & Awards: Cross of Braves & Polonia Restituta, Comn Underground Army, Poland, 45; Pub Health Award for Distinguished Work, Minister Health, Poland, 53; Best Sci Expos Prize, World Cong Gynec & Obstet, Arg, 64. Mem: AAAS; fel NY Acad Sci; Am Col Obstetricians & Gynecologists; fel Latin Am Asn Invest Human Reprod; Int Soc Res Reprod. Res: Computer surveillance of labor; systemic and placental circulation in the pregnant woman; factors inhibiting uterine contractility; fetal distress; uterine physiology. Mailing Add: Dept of Obstet & Gynec Michael Reese Hosp & Med Ctr Chicago IL 60616

BIENVENU, RENE JOSEPH, JR, b Colfax, La, Mar 19, 23; m 48; c 3. MICROBIOLOGY. Educ: La State Univ, BS, 44, MS, 49; Univ Tex, PhD(bact), 57. Prof Exp: Bacteriologist, Confederate Mem Hosp, 49-50; from asst prof to assoc prof, 50-62, head dept, 60-69, PROF MICROBIOL, NORTHWESTERN STATE UNIV, 62-, DEAN COL SCI & TECHNOL, 68- Concurrent Pos: Consult, Natchitoches Parish Hosp, 58- Mem: Am Soc Microbiol. Res: Immunology; antigenic aspects of Brucella species. Mailing Add: Col of Sci & Technol Northwestern State Univ Natchitoches LA 71457

BIENVENUE, GORDON RAYMOND, b Fall River, Mass, Oct 11, 46; m 68; c 2. AUDIOLOGY, PSYCHOACOUSTICS. Educ: Univ Mass, Amherst, BA, 68; Mich State Univ, MA, 69; Pa State Univ, PhD(audiol), 75; Am Speech & Hearing Asn, Clin Cert Audiol. Prof Exp: Chief audiol, Brooke Gen Hosp, 69-71; asst audiol, 71-75, RES ASST AUDIOL & PSYCHOACOUST, PA STATE UNIV, 75- Concurrent Pos: Instr audiol, US Army Med Field Serv Sch, San Antonio, Tex, 69-71; lectr audiol, Brooke Gen Hosp, 69-71; consult, Indust Audiol, 71- Mem: Am Speech & Hearing Asn; Sigma Xi; Acoust Soc Am; Am Audiol Soc; Mil Audiol & Speech Pathol Soc. Res: Investigation of psychoacoustic and physiological acoustic phenomena with particular emphasis on hearing conservation and hearing diagnostic procedures especially in the study of the effects of high level noise. Mailing Add: Environ Acoust Lab 110 Moore Bldg Pa State Univ University Park PA 16802

BIENZ, DARREL RUDOLPH, b Bern, Idaho, Apr 1, 26; m 50; c 5. PLANT BREEDING. Educ: Univ Idaho, BS, 50; Cornell Univ, PhD(plant breeding), 55. Prof Exp: Asst prof hort, Univ Idaho, 54-58, horticulturist, Agr Res Serv, USDA & Univ Idaho, 58-59; from asst prof to assoc prof, 59-71, PROF HORT, WASH STATE UNIV, 71- Concurrent Pos: Fulbright exchange prof, Turkey, 67-68. Mem: Am Soc Hort Sci; Potato Asn Am; Am Genetic Asn. Res: Vegetable breeding and incompatability; general vegetable research. Mailing Add: Dept of Hort Wash State Univ Pullman WA 99163

BIER, MILAN, b Vukovar, Yugoslavia, Dec 7, 20; nat US; m 52. BIOPHYSICS. Educ: Univ Geneva, License Sci Chim, 46; Fordham Univ, PhD(biochem), 50. Prof Exp: Asst prof chem, Fordham Univ, 50-62; VIS RES PROF CHEM, UNIV ARIZ, 62-; RES BIOCHEMIST, VET ADMIN HOSP, 62- Concurrent Pos: Head chem res lab, Inst Appl Biol, 52-60; consult, Univs Space Res Asn, 73- Mem: Am Chem Soc; Am Soc Biol Chem; Am Soc Artificial Internal Organs; Am Inst Aeronautics & Astronaut. Res: Biomedical engineering, electrophoresis, membrane processes and artificial kidney technology; protein and enzyme biophysics, plasma fractionation, applied immunology. Mailing Add: Vet Admin Hosp Tucson AZ 85723

BIERANOWSKI, LEONARD PAUL, mycology, see 12th edition

BIERENBAUM, MARVIN L, b Philadelphia, Pa, Aug 30, 26; m 51; c 2. CARDIOLOGY, NUTRITION. Educ: Rutgers Univ, BS, 47; Hahnemann Med Col, MD, 53. Prof Exp: Intern med, Beth Israel Hosp, Newark, NJ, 53-54; resident med & cardiol, Vet Admin Hosp, Brooklyn, 54-57; DIR ATHEROSCLEROSIS RES PROJ, ST VINCENT'S HOSP, 57- Concurrent Pos: Prog coordr heart dis control, NJ Dept Health, 57-; prin investr, NIH grant, 60-; consult, Fairleigh Dickinson Univ, 62-; mem coun arteriosclerosis, Am Heart Asn, 64; clin assoc prof med, NJ Col Med & Dent, 67- Mem: AMA; fel Am Col Physicians; Am Inst Nutrit; Am Soc Clin Invest. Res: Nutritional management in prevention and therapy of arteriosclerotic coronary heart disease. Mailing Add: Atherosclerosis Res Ctr St Vincent's Hosp Montclair NJ 07042

BIERER, BERT WORMAN, b Philadelphia, Pa, June 2, 11; m 31; c 7. POULTRY PATHOLOGY. Educ: Univ Pa, DVM, 34. Prof Exp: Vet inspector, Baltimore City Health Dept, 36-41; livestock inspector, USDA, 42-47, asst lab dir poultry path, 47-57, lab dir, 57-73, vet, SC Agr Exp Sta, 57-62, poultry scientist, 62-64, PROF POULTRY SCI, CLEMSON UNIV, 64- Concurrent Pos: Poultry pathologist, Del State Diag Lab, 43-44; USDA Hatch Act grants, 57- Mem: Am Vet Med Asn; Poultry Sci Asn; Am Asn Avian Path. Res: Poultry disease diagnosis, research covering various areas in field avian diseases and normal values; development of attenuated fowl cholera vaccine used in drinking water. Mailing Add: 5552 Sylvan Dr Columbia SC 29206

BIERHORST, DAVID WILLIAM, b New Orleans, La, Oct 17, 24; m 45; c 2. PLANT MORPHOLOGY. Educ: Tulane Univ, BS, 47, MS, 49; Univ Minn, PhD(bot), 52. Prof Exp: Instr bot, Univ Minn, 52-53; asst prof biol, Univ Va, 53-55; from asst prof to assoc prof bot, Cornell Univ, 55-68; PROF BOT, UNIV MASS, AMHERST, 68- Mem: Bot Soc Am; Int Soc Plant Morphologists (vpres, 70-75); Am Fern Soc (vpres, 73-75). Res: Developmental anatomy; plant anatomy; morphology. Mailing Add: Dept of Bot Univ of Mass Amherst MA 01002

BIERI, JOHN GENTHER, b Norfolk, Va, May 24, 20; m 43; c 3. BIOCHEMISTRY, NUTRITION. Educ: Antioch Col, BA, 43; Pa State Col, MS, 44; Univ Minn, PhD(biochem), 49. Prof Exp: Nutritionist, Naval Med Res Inst, 44-46; instr biochem, Univ Minn, 48-49; assoc prof, Med Br, Univ Tex, 49-55; BIOCHEMIST, NAT INST ARTHRITIS, METAB & DIGESTIVE DIS, 55- Mem: AAAS; Am Soc Biol Chemists; Am Chem Soc; Soc Exp Biol & Med; Am Inst Nutrit. Res: Nutritional biochemistry; metabolism of vitamin E and vitamin A; polyunsaturated fatty acids; selenium. Mailing Add: Rm 5N-102 Bldg 10 Nat Inst Arthritis Metab & Digestive Dis Bethesda MD 20014

BIERI, ROBERT, b Washington, DC, Feb 7, 26; c 3. OCEANOGRAPHY, ECOLOGY. Educ: Antioch Col, BS, 49; Univ Calif, MS, 53, PhD(oceanog), 58. Prof Exp: Res biologist, Scripps Inst, Univ Calif, 54-55; oceanogr, Lamont Geol Observ, 55-57; assoc dir personnel, 57-59, asst prof biol, 59-67, chmn dept biol, 59-63, ASSOC PROF BIOL, ANTIOCH COL, 67-, CHMN ENVIRON STUDIES CTR, 68- Mem: Am Soc Limnol & Oceanog; Marine Biol Asn UK. Res: Zooplankton, Chaetognatha; Neuston; sea surface community life histories, food web and behavior; forest community succession. Mailing Add: Environ Studies Ctr Antioch Col Yellow Springs OH 45387

BIERLEIN, THEO KARL, b Ansbach, Ger, Feb 6, 24; nat US; m 49; c 6. PHYSICAL CHEMISTRY. Educ: Univ Wash, BS, 45, PhD(chem), 50. Prof Exp: Chemist, Gen Elec Co, 50-52, engr, 52-55, sr scientist, 55-63; mgr phys metall unit, Reactor & Mat Technol Dept, Pac Northwest Labs, Battelle Mem Inst, 63-65, MGR DAMAGE ANAL, MAT DEPT, WESTINGHOUSE-HANFORD, 65- Mem: Am Chem Soc; Am Soc Metals; Electron Micros Soc Am; Am Nuclear Soc. Res: Relationship between microstructure and properties of metals and alloys; deformation and fracture mechanism; effect of reactor irradiation on structure of fuel and its cladding. Mailing Add: Westinghouse-Hanford 326 Bldg 300 Area PO Box 1970 Richland WA 99352

BIERLY, EUGENE WENDELL, b Pittston, Pa, Sept 11, 31; m 53; c 3. METEOROLOGY, SCIENCE ADMINISTRATION. Educ: Univ Pa, AB, 53; US Naval Postgrad Sch, cert, 54; Univ Mich, MS, 57, PhD, 68. Prof Exp: Asst, dept civil eng, meteorol labs, Univ Mich, 56-60, asst res meteorologist, dept eng mech, 60-63, lectr, 61-63; meteorologist, US Atomic Energy Comn, 63-66; prog dir meteorol, 66-71, coordr global atmospheric res prog, 71-74, head, Off Climate Dynamics, 74-75, HEAD, CLIMATE DYNAMICS RES SECT, NSF, 75- Concurrent Pos: Consult, Reactor Develop Co, 61-62 & Pac Missile Range, 62-63; Cong fel, 70-71. Mem: AAAS; fel Am Meteorol Soc; Air Pollution Control Asn; Royal Meteorol Soc; Am Polit Sci Asn. Res: Air pollution; diffusion; lake breezes; atmospheric tracers; science management; environmental problems and science policy. Mailing Add: NSF 1800 G St NW Washington DC 20550

BIERLY, JAMES N, JR, b Lewisberg, Pa, May 6, 22; m 47. PHYSICS. Educ: Kutztown State Teachers Col, BS, 43; Bucknell Univ, MS, 47; Temple Univ, PhD(physics), 61; Am Bd Radiol, dipl, 55. Prof Exp: Instr physics, Bucknell Univ, 47; physicist radiol dept, Jefferson Med Col, 51-55; physicist, Vet Admin Hosp, 55-56; res physicist, Franklin Inst, 56-62; from asst prof to assoc prof physics, 62-73, PROF PHYSICS, PHILADELPHIA COL PHARM & SCI, 73- Concurrent Pos: Consult, Misericordia Hosp. Mem: Am Phys Soc; Sigma Xi; Am Nuclear Soc; Soc Nuclear Med; Am Col Radiol. Res: Thermoelectricity in semiconductors; consulting work in medical radiation physics. Mailing Add: 4627 Hazel Ave Philadelphia PA 19143

BIERLY, MAHLON ZWINGLI, JR, b Philadelphia, Pa, Apr 24, 22; m 44; c 2. MEDICINE. Educ: Franklin & Marshall Col, BS, 43; Jefferson Med Col, MD, 46; Am Bd Pediat, dipl, 55. Prof Exp: Resident & asst chief, Philadelphia Hosp Contagious Dis, 49-50; resident pediat, Children's Hosp Philadelphia, 51-52; pvt pract, 52-53; MEM STAFF, MED DEPT, WYETH LABS, 53- Mem: Am Pub Health Asn; Am Col Prev Med; Am Acad Pediat. Res: Biological agents; vaccines, toxoids and antisera. Mailing Add: Wyeth Labs Box 8299 Philadelphia PA 19101

BIERMAN, ARTHUR, b Vienna, Austria, Oct 14, 25; US citizen; m 54; c 2. THEORETICAL PHYSICS. Educ: Univ Chicago, PhD(math, biol), 54; Columbia Univ, MA, 57. Prof Exp: Nat Found Infantile Paralysis fel, 54-55; from instr to asst prof physics, City Col New York, 58-61; sr res scientist, Lockheed-Calif Co, 62-63; assoc prof physics, Los Angeles State Col, 63; from asst prof to assoc prof, 64-69, PROF PHYSICS, CITY COL, CITY UNIV NY, 69-, ACTG ASSOC PROVOST, 72- Concurrent Pos: Consult, Lockheed-Calif Co, 62 & Lockheed Ga Co, 65-66; US Atomic Energy Comn grant, 64-71. Mem: Am Phys Soc. Res: Theory of molecular excitons in large molecules. Mailing Add: 137 W 78th St New York NY 10024

BIERMAN, DON EDWARD, b Kowel, Poland, July 24, 31; US citizen; m 55. TRANSPORTATION GEOGRAPHY, POLITICAL GEOGRAPHY. Educ: George Washington Univ, BA, 63, MA, 66; Mich State Univ, PhD(geog), 70. Prof Exp: Instr geog, Univ Louisville, 66-70, asst prof geog, 70-74, ASSOC PROF GEOG, UNIV LOUISVILLE, 74-, CHMN DEPT, 72- Concurrent Pos: Vis scholar, Transp Ctr, Northwestern Univ, 71-72; mem, Transp Res Forum; tutor, Univ Oxford, 73-74. Mem: Asn Am Geogr; Am Soc Traffic & Transp. Res: Urban transportation system analysis; economic development, transportation. Mailing Add: 2134 Lowell Ave Louisville KY 40205

BIERMAN, EDWIN LAWRENCE, b Far Rockaway, NY, Sept 17, 30; m 56; c 2. INTERNAL MEDICINE, METABOLISM. Educ: Brooklyn Col, AB, 51; Cornell Univ, MD, 55. Prof Exp: Intern, NY Hosp, 55-56, asst physician, Diabetes Study Group, 56-57; from asst chief to chief metab res div, US Army Med Res & Nutrit Lab, Fitzsimons Army Hosp, Denver, 57-59; asst resident, NY Hosp, 59-60, outpatient physician, 60-62; from asst prof to assoc prof, 62-68, PROF MED, SCH MED, UNIV WASH, 68-; CHIEF DIV METAB, ENDOCRINOL & GERONT, VET ADMIN HOSP, 62- Concurrent Pos: Asst physician, Rockefeller Inst Hosp, NY, 56-57, assoc physician & asst prof, Rockefeller Inst, 60-62. Mem: Endocrine Soc; Am Diabetes Asn; Asn Am Physicians; Am Soc Clin Invest; Am Col Physicians. Res: Lipid and carbohydrate metabolism, with particular emphasis on diabetes and hyperlipemia; gerontology. Mailing Add: Sch of Med Univ of Wash Seattle WA 98195

BIERMAN, HOWARD RICHARD, b Newark, NJ, Jan 27, 15; m; c 3. MEDICINE. Educ: Wash Univ, BSc & MD, 39; Am Bd Internal Med, dipl. Prof Exp: From jr to sr intern, Barnes Hosp, 40, resident house officer, 41; clin physiologist & prin clin investr, Nat Cancer Inst, 46-53; chief clin sect, Lab Exp Oncol & assoc clin prof med, Sch Med, Univ Calif, 47-53; dir, Hosp Tumors & Allied Dis, City of Hope Med Ctr, 53-56, dir, Hosp Blood Dis, 56-59, chmn dept internal med & sci dir, 53-59, med dir, 56-59; DIR, INST CANCER & BLOOD RES, BEVERLY HILLS, 59-; CLIN PROF MED, SCH MED, LOMA LINDA UNIV, 69- Concurrent Pos: Chief resident, St Louis Isolation Hosp, Contagious Dis, 40; spec consult, Nat Cancer Inst, 53-; emer attend physician, Los Angeles County Hosp. Mem: Fel Am Col Physicians; fel NY Acad Sci; fel Am Col Angiol; fel Int Soc Hemat; fel Am Soc Clin Oncol. Res: Hematology and oncology; chemotherapy and clinical pharmacology; bone marrow metabolism; biophysical characteristics of tumors; aviation medicine; barometrics and acceleration; nutrition and vitamins; leukocyte kinetics. Mailing Add: 152 N Robertson Blvd Beverly Hills CA 90211

BIERMANN, ALAN WALES, b Newport News, Va, Feb 5, 39; m 68; c 1. COMPUTER SCIENCE. Educ: Ohio State Univ, BEE & MSc, 61; Univ Calif, Berkeley, PhD(comput sci), 68. Prof Exp: Asst prof comput sci, San Fernando Valley State Col, 62-64 & 68-69; res assoc, Stanford Univ, 69-71; asst prof, Ohio State Univ, 71-73; ASST PROF COMPUT SCI, DUKE UNIV, 74- Mem: Sigma Xi; Inst Elec & Electronics Engrs; Asn Comput Mach. Res: Learning and inference theory; automatic program synthesis. Mailing Add: Dept of Comput Sci Duke Univ Durham NC 27706

BIERMANN, WENDELL J, b St Louis, Mo, June 2, 22; m 49; c 7. PHYSICAL CHEMISTRY. Educ: St Louis Univ, BS, 44; Univ Wis, PhD(chem), 50. Prof Exp: Asst, Univ Wis, 46-50; from asst prof to prof phys chem, Univ Man, 50-63; sr res chemist, Carrier Res & Develop Co, 63-66, chief chemist res div, 66-75, PROJ MGR, ENERGY SYSTS DIV, CARRIER CORP, 75- Res: Electrochemistry and solution thermodynamics; analytical chemistry; air pollution abatement methods; metallurgy; materials specification; thick film technology; new energy sources; solar energy utilization. Mailing Add: Energy Systs Div Carrier Corp Carrier Tower Syracuse NY 13201

BIERON, JOSEPH F, b Buffalo, NY, Oct 19, 37; m 59; c 5. ORGANIC CHEMISTRY. Educ: Canisius Col, BS, 59, MS, 61; State Univ NY Buffalo, PhD(chem), 65. Prof Exp: Asst prof chem, 66-71, ASSOC PROF CHEM, CANISIUS COL, 71-, DEAN COL ARTS & SCI, 71- Mem: Am Chem Soc. Res: Oxidation of polymeric materials and confirmational analysis of organic molecules by nuclear magnetic resonance spectroscopy. Mailing Add: Dept of Chem Canisius Col 2001 Main St Buffalo NY 14208

BIERSDORF, WILLIAM RICHARD, b Salem, Ore, Sept 27, 25; m 66; c 2. VISUAL PHYSIOLOGY, OPHTHALMOLOGY. Educ: Wash State Univ, BS, 50, MS, 51; Univ Wis, PhD(psychol), 54. Prof Exp: Exp psychologist vision, Walter Reed Army Inst Res, 54-67, asst chief dept sensory psychol, 62-66, actg chief dept psychophysiol, 66-67; ASSOC PROF OPHTHAL, OHIO STATE UNIV, 67-, ASSOC PROF BIOPHYS, 70-, PROJ SUPVR VISION RES, INST RES IN VISION, 67- Concurrent Pos: US Army res grant vision, Ohio State Univ, 67-68, USPHS res grant, 68-76; Am Psychol Asn del, Inter-Soc Color Coun, 66-75; adv, Nat Res Coun-Armed Forces Comt Vision, 68-; mem visual sci study sect, Div Res Grants, NIH, 71-75. Mem: Fel AAAS; fel Am Psychol Asn; Int Soc Clin Electroretinography; Asn Res Vision & Ophthal; Psychonomic Soc. Res: Electrophysiology of human vision. Mailing Add: Inst for Res in Vision 1314 Kinnear Rd Columbus OH 43212

BIERSMITH, EDWARD L, b Kansas City, Mo, Feb 26, 42; m 68; c 1. ORGANIC CHEMISTRY, BIO-ORGANIC CHEMISTRY. Educ: Rockhurst Col, AB, 63; Univ Kans, PhD(org chem), 69. Prof Exp: Asst chem, Rockhurst Col, 62-63; asst org chem, Univ Kans, 63-69; res assoc, Univ Okla, 69-70; ASST PROF CHEM, NORTHEAST LA UNIV, 70- Mem: Am Chem Soc; The Chem Soc. Res: Photolysis of 2-cyclopropyl cyclohexanone: a three-carbon photochemical ring expansion; total synthesis of bicyclo heptan-2-one; isolation and identification of sterols from marine life, especially Renilla, Acrapora palmata; reactions of cyclic amides. Mailing Add: Northeastern La Univ Monroe LA 71201

BIERWAGEN, MAX EUGENE, b Flagler, Colo, May 15, 25; m 47; c 3. PHARMACOLOGY. Educ: Concordia Teachers Col, BS, 47; Univ Colo, PhD(pharmacol), 56. Prof Exp: Instr pharmacol, Sch Med, Univ Colo, 56-57; asst dir pharmacol res, 63-65, sr res pharmacologist, 57-65, dir pharmacol res, 65-75, ASST DIR RES PLANNING & LICENSING, BRISTOL LABS, INC, 75- Mem: AAAS; Am Chem Soc; NY Acad Sci; Sigma Xi. Res: Pharmacodynamics; inflammation; allergy and immunology; experimental hypertension; enzymology. Mailing Add: Res Admin Bristol Labs Inc Syracuse NY 13201

BIES, DAVID ALAN, b Los Angeles, Calif, Aug 15, 25; m 54; c 2. ACOUSTICS. Educ: Univ Calif, Los Angeles, BA, 48, MA, 51, PhD, 53. Prof Exp: Sr consult, Bolt, Beranek & Newman, Inc, 53-72; sr res fel, 72-75, READER MECH ENG, UNIV ADELAIDE, SOUTH AUSTRALIA, 76- Mem: AAAS; Acoustical Soc Am; Australian Acoustical Soc; Inst for Noise Control Eng. Res: Sound propagation in ducts; uses of anechoic and reverberant rooms; uses of holography for study of vibration and sound radiation; general problems of noise control. Mailing Add: Dept of Mech Eng Univ of Adelaide Adelaide 5001 Australia

BIESELE, FERDINAND CHARLES, b New Braunfels, Tex, June 14, 12; m 36; c 3. MATHEMATICS. Educ: Univ Tex, AB, 32, AM, 33, PhD(math), 41. Prof Exp: Teacher high schs, Tex, 33-35; instr math, San Antonio Jr Col, 35-37; teacher high schs, Tex, 38-39; from instr to assoc prof, 41-56, PROF MATH, UNIV UTAH, 56- Mem: Am Math Soc; Math Asn Am. Res: Group theory; abstract algebra. Mailing Add: Dept of Math Univ of Utah Salt Lake City UT 84112

BIESELE, JOHN JULIUS, b Waco, Tex, Mar 24, 18; m 43; c 3. CYTOLOGY. Educ: Univ Tex, AB, 39, PhD(zool), 42. Prof Exp: Int Cancer Res Found fel, 42 & 43-44; res assoc genetics, Carnegie Inst Technol, 44-46; asst, Sloan-Kettering Inst Cancer Res, 46, res fel, 47, head cell growth sect, 47-58, assoc, 47-55, mem, 55-58; asst prof anat, Med Col, Cornell Univ, 50-52, assoc prof biol, Grad Sch Med Sci, Sloan-Kettering Div, 52-55, prof biol, 55-58; prof zool, 58-73, MEM GRAD FAC, UNIV TEX, AUSTIN, 58-, DIR GENETICS FOUND, 59- & PROF ZOOL & EDUC, 73- Concurrent Pos: Instr, Univ Pa, 43-44; res assoc, Mass Inst Technol, 46-47; mem cell biol study sect, NIH, 58-63; assoc scientist, Sloan-Kettering Inst Cancer Res, 59-; counr, Cancirco, 62-; assoc chmn, Conf Advan Sci & Math Teaching, Tex, 65, chmn, 66; mem discussion group on chem carcinogenesis, Nat Cancer Inst, 67, cancer res training comt, 69-72, res career awards, 62-77. Mem: Fel AAAS; Am Soc Naturalists; Am Soc Cell Biol; Am Asn Cancer Res; Environ Mutagen Soc. Res: Electron microscopy; cytochemistry of cancerous and normal mammalian tissue; effects of differentiation, ageing, hormones and drugs on cell chemistry and morphology; tumor chemotherapy; tissue culture; antimetabolites and mitotic poisons. Mailing Add: Dept of Zoology Univ of Tex Austin TX 78712

BIESTER, JOHN LOUIS, b Aurora, Ill, Aug 29, 18; m 47; c 2. CHEMISTRY, ACADEMIC ADMINISTRATION. Educ: Beloit Col, BA, 41; Syracuse Univ, MS, 43, PhD, 59. Prof Exp: Asst, Syracuse Univ, 42-43; res chemist, Standard Oil Co, Ind, 43; asst Syracuse Univ, 46-47; asst prof, 48-58, assoc dir field placement, 64-69, dir, 69-75, ASSOC PROF CHEM, BELOIT COL, 58-, DIR FIELD PLACEMENT & CAREER COUN, 75- Concurrent Pos: Assoc prog dir, acad yr insts, NSF, 61-62, consult, sci personnel & educ div, 62-69. Mem: AAAS; fel Am Inst Chemists. Res: Science education; preparation of science teachers. Mailing Add: Off of Field Placement Beloit Col Beloit WI 53511

BIESTERFELDT, HERMAN JOHN, JR, b New York, NY, Jan 31, 37. MATHEMATICS. Educ: Pa State Univ, BS, 57, PhD(math), 63. Prof Exp: Instr math, Pa State Univ, 59-60; asst prof, Lebanon Valley Col, 62-63; asst prof, SDak Sch Mines & Technol, 63-65, Univ Wis, 65-67 & Drexel Inst Technol, 67-68; ASSOC PROF MATH, SOUTHERN ILL UNIV, 68- Concurrent Pos: Vis asst prof, Univ Mass, 66-67. Mem: Am Math Soc; Math Asn Am. Res: Vibration theory; convergence structure and general topology; polynomial theory for solutions to various differential equations; nonstandard analysis; K-theory. Mailing Add: Dept of Math Southern Ill Univ Carbondale IL 62901

BIETZ, JEROLD ALLEN, b Mayville, NDak, Feb 22, 42; m 63; c 1. PROTEIN CHEMISTRY. Educ: Mayville State Col, BS, 63; Univ NDak, MS, 66. Prof Exp: RES CHEMIST FOOD PROTEINS, NORTHERN REGIONAL RES CTR, AGR RES SERV, USDA, PEORIA, 66- Mem: AAAS; Am Chem Soc; Am Asn Cereal Chemists. Res: Concerning isolation, characterization and comparison of wheat gluten proteins relates structure to functionality, quality and genetic background through peptide characterization, electrophoresis and amino acid sequence analysis. Mailing Add: Northern Regional Res Ctr USDA 1815 N University Peoria IL 61604

BIEVER, KENNETH DUANE, b Hot Springs, SDak, Jan 8, 40; m 60; c 2. ENTOMOLOGY. Educ: SDak State Univ, BS, 62; Univ Calif, Riverside, PhD(entom), 67. Prof Exp: RES ENTOMOLOGIST, BIOL CONTROL INSECTS RES LAB, AGR RES SERV, USDA, 66- Mem: Entom Soc Am; Soc Invert Path; Int Orgn Biol Control; Am Mosquito Control Asn. Res: Influence of climate and weather on insects; biological control of insects; medical entomology; insect pathology. Mailing Add: Biol Control of Insects Res Lab USDA PO Box A Columbia MO 65201

BIGAT, TEVFIK KAYA, physical chemistry, organic chemistry, see 12th edition

BIGBEE, DANIEL E, b Walters, Okla, Apr 16, 30; m 50; c 5. POULTRY SCIENCE. Educ: Okla State Univ, BS, 56, MS, 58; Mich State Univ, PhD(poultry sci), 62. Prof Exp: Asst prof poultry sci, Okla State Univ, 62-66; asst prof, 66-70, ASSOC PROF POULTRY SCI, UNIV MD, COLLEGE PARK, 70- Mem: Poultry Sci Asn; World Poultry Sci Asn. Res: Food science pertaining to poultry products. Mailing Add: Dept of Poultry Sci Univ of Md College Park MD 20740

BIGELEISEN, JACOB, b Paterson, NJ, May 2, 19; m 45; c 3. PHYSICAL CHEMISTRY. Educ: NY Univ, AB, 39; State Col Wash, MS, 41; Univ Calif, PhD(chem), 43. Prof Exp: Asst chem, Univ Calif, 41-43; res scientist, SAM Labs, Manhattan Proj, Columbia Univ, 43-45; res assoc, cryogenic lab, Ohio State Univ, 45-46; fel, Inst Nuclear Studies, Univ Chicago, 46-48; from assoc chemist to sr chemist, Brookhaven Nat Lab, 48-68; prof chem, 68-73, chmn dept, 70-75, TRACY H HARRIS PROF CHEM, UNIV ROCHESTER, RIVER CAMPUS, 73- Concurrent Pos: Vis prof, Cornell Univ, 53-54; NSF fel & hon vis prof, Swiss Fed Inst Technol, 62-63; Gilbert N Lewis lectr, 63; John Simon Guggenheim fel, 74-75. Honors & Awards: Am Chem Soc Award, 58; E O Lawrence Mem Award, 64. Mem: Nat Acad Sci; Am Acad Arts & Sci; Am Chem Soc; Am Phys Soc. Res: Dissociation of strong electrolytes; thermodynamics of electrolytes; acids and bases; semi-quinones; photochemistry in rigid media; color of organic compounds; low temperature; spectroscopy; use of isotopes as tracers; chemistry of isotopes. Mailing Add: Dept of Chem Univ of Rochester River Campus Rochester NY 14627

BIGELOW, CHARLES C, b Edmonton, Alta, Apr 25, 28; m 53; c 2. CHEMISTRY. Educ: Univ Toronto, BASc, 53; McMaster Univ, MSc, 55, PhD(chem), 57. Prof Exp: Nat Res Coun Can fel protein chem, Carlsberg Lab, Copenhagen, 57-59; assoc, Sloan-Kettering Inst, 59-62; from asst prof to assoc prof chem, Univ Alta, 62-65; vis prof, Fla State Univ, 65; from assoc prof to prof biochem, Univ Western Ont, 65-74; PROF & HEAD DEPT BIOCHEM, MEM UNIV NEWFOUNDLAND, 74- Concurrent Pos: Res grants, Nat Res Coun Can, 63-69, Med Res Coun Can, 65-; vis prof biochem, Univ Toronto, 73-74. Mem: Fel Chem Inst Can; Am Chem Soc; Am Soc Biol Chem; Can Biochem Soc. Res: Physical chemistry of proteins. Mailing Add: Dept of Biochem Mem Univ of Newfoundland St John's NF Can

BIGELOW, HOWARD ELSON, b Greenfield, Mass, June 28, 23; m 56. BOTANY. Educ: Oberlin Col, AB, 49, MA, 51; Univ Mich, PhD(bot), 56. Prof Exp: Attache de recherche, Univ Montreal, 56-57; from instr to assoc prof bot, 57-70, PROF BOT, UNIV MASS, AMHERST, 70- Mem: AAAS; Mycol Soc Am (vpres, 74-75, pres-elect, 75-76); Bot Soc Am; Am Inst Biol Sci; Ger Soc Mycol. Res: Mycology; taxonomy and ecology of fleshy fungi, especially the Agaricales. Mailing Add: Dept of Bot Univ of Mass Amherst MA 01002

BIGELOW, MARGARET ELIZABETH BARR, b Elkhorn, Man, Apr 16, 23; m 56. MYCOLOGY. Educ: Univ BC, BA, 50, MA, 52; Univ Mich, PhD(bot), 56. Prof Exp: Nat Res Coun fel, Univ Montreal, 56-57; from instr to asst prof bot, 57-70, ASSOC PROF BOT, UNIV MASS, 70- Mem: Mycol Soc Am (ed-in-chief, Mycologia, 76-); Am Inst Biol Sci; Int Asn Plant Taxonomists. Res: Systematics and ecology of pyrenomycetous ascomycetes. Mailing Add: Dept of Bot Univ of Mass Amherst MA 01002

BIGELOW, MAURICE HUBBARD, b New Haven, Conn, Oct 25, 03; m 26; c 1. MARINE BIOLOGY. Educ: Northeastern Univ, BS, 24; Univ Pittsburgh, PhD(chem), 33. Prof Exp: Chemist, Lever Bros Soap Co, Mass, 21-25; head dept sci, Am Col, Greece, 25-28; instr chem, Univ Pittsburgh, 29-33; indust fel, Mellon Inst, 33-35; dir tech serv, Plaskon Div, Libby-Owens-Ford Glass Co, 35-36; vpres, Allied Chem Corp, 56-63; head dept sci, Port Charlotte Adult Educ Ctr, 67-70; PRES, CHARLOTTE BICHEM RES LAB, INC, 70- Mem: Am Chem Soc; Am Soc Testing & Mat. Res: X-ray spectroscopy of glass; plastics; acetylene and carbon monoxide chemistry; marine biology of the Gulf of Mexico. Mailing Add: 115 Port Charlotte Village Port Charlotte FL 33952

BIGELOW, MELVIN JEROME, b Bangor, Mich, Mar 20, 26; m 58; c 2. ORGANIC CHEMISTRY. Educ: Western Mich Univ, BS, 47; Northwestern Univ, PhD(org chem), 50. Prof Exp: Instr chem, Mich Col Mining & Technol, 50-51; asst prof, Union Col, NY, 51-54; res chemist, Olin Mathieson Co, 54-55; assoc prof chem & head dept, Drury Col, 55-58; sci intel anal, Off Naval Intel, 58-59; assoc prof chem, Cent Mich Univ, 59-61; assoc prof, 61-64, PROF CHEM, IDAHO STATE UNIV, 64- Mem: AAAS; Am Chem Soc. Res: Steroid stereochemistry; chemical education. Mailing Add: 73 Fordham St Pocatello ID 83201

BIGELOW, NOLTON H, b Providence, RI, Feb 8, 18; m 42; c 2. PATHOLOGY, FORENSIC MEDICINE. Educ: Yale Univ, AB, 40; Cornell Univ, MD, 43. Prof Exp: Assoc prof path, Albany Med Col, 52-58; dir labs, St Barnabas Hosp, New York, 59-62; instr, Sch Legal Med, Harvard Univ, 64-67; dir lab, New Eng Hosp, 64-68; DIR LAB, SANCTA MARIA HOSP, CAMBRIDGE, 69- Concurrent Pos: Consult, NY Hosp, 44-45, RI Hosp, 47-48 & Cancer Res Inst, New Eng Deaconess Hosp, 64-; assoc med exam, Norfolk County, Mass, 68. Mem: Am Asn Path & Bact; fel Col Am Path. Res: Pharmacology, problems of muscle pain and spasm; analgesic agents such as procaine; oncology and forensic studies. Mailing Add: 53 Hill Rd Belmont MA 02178

BIGELOW, ROBERT SIDNEY, b Canning, NS, Apr 26, 18; m 46; c 4. ZOOLOGY. Educ: McGill Univ, BSc, 50, PhD(entom), 54. Prof Exp: Tech officer syst entom, Can Dept Agr, 52-53; from asst prof to assoc prof entom, McGill Univ, 53-62; READER ZOOL, UNIV CANTERBURY, NZ, 62- Concurrent Pos: Nat Res Coun Can fel, Oxford Univ, 54-55; mem, Panel Human Aggressiveness, UNESCO Conf, Paris, 70. Mem: AAAS; Soc Study Evolution; Am Soc Naturalists; Australian Asn Advan Sci; Entom Soc NZ. Res: Evolutionary genetics; human evolution; anthroplogy; history; sociology. Mailing Add: Dept of Zool Univ of Canterbury Christchurch New Zealand

BIGELOW, WILBUR CHARLES, b Wyoming Co, Pa, Mar 18, 23; m 50; c 2. PHYSICAL CHEMISTRY. Educ: Pa State Univ, BS, 44; Univ Mich, MS, 48, PhD(phys chem), 52. Prof Exp: Res chemist, US Naval Res Lab, Washington, DC, 44-46; res assoc res inst, 51-56, asst prof sci, 56-59, assoc prof chem & metall eng, 59-62, PROF CHEM & METALL ENG, UNIV MICH, 62- Mem: Am Chem Soc; Electron Micros Soc Am; Am Soc Metals; Electron Probe Anal Soc Am; Int Metallog Soc. Res: Electron microscopy, diffraction and microprobe analysis; microstructures of metals and ceramics systems and their relationships to composition and physical properties. Mailing Add: Dept of Mat & Metall Eng Univ of Mich Ann Arbor MI 48104

BIGGER, CYNTHIA ANITA HOPWOOD, b Sheffield, Ala, Mar 3, 42; m 63; c 1. MOLECULAR BIOLOGY. Educ: La State Univ, Baton Rouge, BS, 66, MS, 69, PhD(microbiol), 71. Prof Exp: NIH fel, Dept Molecular Biol, Univ Edinburgh, 71-72; vis instr res, Dept Chem, 73, res assoc, Dept Microbiol, 74, RES ASSOC, SCH VET MED, LA STATE UNIV, BATON ROUGE, 74- Mem: Am Soc Microbiol; Am Chem Soc. Res: Binding of polycyclic aromatic hydrocarbons and their metabolites to cytoplasmic binding proteins and nuclear DNA and the relationship of these interactions to chemical carcinogenesis. Mailing Add: Physiol Pharmacol & Toxicol La State Univ Sch Vet Med Baton Rouge LA 70803

BIGGER, J THOMAS, JR, b Cambridge, Mass, Jan 17, 35. CARDIOLOGY, PHARMACOLOGY. Educ: Emory Univ, AB, 55; Med Col Ga, MD, 60. Prof Exp: Intern & asst resident, Columbia Univ Med Div, Bellevue Hosp, 60-62; asst resident & resident cardiol, Presby Hosp, New York, 63-65; instr, 66-67, assoc, 67-68, from asst prof to assoc prof med, 68-72, assoc prof, 72-75, PROF MED & PHARMACOL, COL PHYSICIANS & SURGEONS, COLUMBIA UNIV, 75- Concurrent Pos: Vis fel cardiol, Col Physicians & Surgeons, Columbia Univ, 65-66, fel, 66-67; NY Heart Asn fel, 65-68; Nat Heart & Lung Inst res career develop award, 72-77; asst physician, Presby Hosp, New York, 65-68, from asst attend physician to assoc attend physician, 68-75, attend physician, 75-; sr investr, NY Heart Asn, 68-72; assoc ed, Circulation, 70- Mem: AAAS; Am Heart Asn; Am Physiol Soc; Am Soc Pharmacol & Exp Therapeut; Am Fedn Clin Res. Res: Cardiac electrophysiology and arrhythmias; clinical and laboratory evaluation of cardioactive drugs, particularly cardiac antiarrhythmic drugs. Mailing Add: Dept of Med & Pharmacol Columbia Univ New York NY 10032

BIGGER, THEODORE C, b York, SC, Oct 10, 12; m 36; c 2. SOIL SCIENCE. Educ: Clemson Univ, BS, 34; Va Polytech Inst & State Univ, MS, 48; Mich State Univ, PhD(soil sci), 54. Prof Exp: Asst prof agron, Univ Ga, 48-50; assoc prof, Tenn Polytech Inst, 53-57; assoc prof, Univ Ark, 57-59; assoc prof, 59-64, head dept, 59-68, PROF AGR, MIDDLE TENN STATE UNIV, 64- Concurrent Pos: Mem, Water Resources Conf, NSF, 63. Mem: Am Soc Agron; Soil Sci Soc Am; Soil Conserv Soc Am. Res: Relationship of quick soil tests, green tissue tests and yield with applied fertilizer, especially phosphorus and potassium. Mailing Add: Dept of Agr Middle Tenn State Univ Murfreesboro TN 37130

BIGGERS, CHARLES JAMES, b Gastonia, NC, Feb 7, 35; m 58; c 3. BIOLOGY, GENETICS. Educ: Wake Forest Col, BS, 57; Appalachian State Teachers Col, MA, 59; Univ SC, PhD(biol), 69. Prof Exp: Instr biol, Col of Orlando, 59-61, asst prof, 62-66, chmn dept, 61-66; exten instr, Univ SC, 66-69; ASSOC PROF BIOL, MEMPHIS STATE UNIV, 69- Mem: AAAS; Am Genetic Asn; Genetics Soc Am; Am Soc Mammalogists. Res: Genetic studies of serum protein polymorphisms in Peromyscus polionotus, with primary interest in transferrin. Mailing Add: 2473 Lynnfield Rd Memphis TN 38117

BIGGERS, JOHN DENNIS, b Gt Brit, Aug 18, 23; m 48; c 3. PHYSIOLOGY. Educ:

Univ London, BSc(vet sci) & BSc(spec), 46, PhD(physiol), 52, DSc, 65. Prof Exp: Student demonstr physiol, Royal Vet Col, Univ London, 44-45, demonstr, 45-47; asst lectr, Univ Sheffield, 47-48; lectr vet physiol, Univ Sydney, 48-53, sr lectr, 53-55; sr lectr physiol, Royal Vet Col, Univ London, 55-59; assoc mem, Wistar Inst & assoc prof, Sch Vet Med, Univ Pa, 59-61, King Ranch res prof reproductive physiol, 61-65; prof pop dynamics & assoc prof obstet & gynec, Johns Hopkins Univ, 66-72; PROF PHYSIOL & MEM LAB HUMAN REPRODUCTION & REPRODUCTIVE BIOL, HARVARD MED SCH, 72- Concurrent Pos: Vis scientist, Strangeways Res Lab, Eng, 54-55; Commonwealth fel, St John's Col, Cambridge Univ, 54-55; Damon Runyon Mem Fund grant, 60-62; NIH grant, 60-65 & 67-; Pop Coun grant, 60-; Lalor Found grant, 61; NSF grant, 61-64; adv, WHO, 64 & 71-; mem study sect reproductive biol, NIH, 67-71, adv, 69-; mem, ICRO Comt, 68-; Ford Found grant, 69-70; ed, Biol of Reproduction, 70-74; Upjohn lectr, Am Fertil Soc, 73; Ford Found adv, 74-; prog proj dir, NIH Grant, 74-; mem corp, Marine Biol Lab, Woods Hole. Mem: Am Statist Asn; Am Soc Cell Biol; Soc Study Reproduction (pres, 68-69); fel Royal Col Vet Surg; fel Royal Statist Soc. Res: Endocrinology of reproduction; developmental biology; bioassay organ and embryo culture; variation in experimental animals; reproduction in wild animals. Mailing Add: Dept of Physiol Harvard Med Sch Boston MA 02115

BIGGERSTAFF, JOHN A, b Berea, Ky, Sept 10, 31. PHYSICS. Educ: Univ Ky, BS & MS, 53, PhD(physics), 61. Prof Exp: Physicist, eng lab, US Army Signal Corps, 55; sr engr nucleonics, Martin Co, 55-57; res assoc nuclear physics, Univ Ky, 61-63; PHYSICIST, OAK RIDGE NAT LAB, 62- Res: Low energy nuclear spectroscopy; instrumentation for nuclear spectroscopy. Mailing Add: Physics Div Oak Ridge Nat Lab 6000 Bldg X-10 Oak Ridge TN 37830

BIGGERSTAFF, ROBERT HUGGINS, b Richmond, Ky, June 1, 27; m 50; c 2. PHYSICAL ANTHROPOLOGY, ORTHODONTICS. Educ: Howard Univ, BS, 51, DDS, 55; Univ Pa, MS, 66, PhD(anthrop), 69; cert orthod, 73. Prof Exp: NIH trainee, Univ Pa, 64-66, res investr, 67-69; asst prof, 69-73, ASSOC PROF ORTHOD, UNIV KY, 73- Mem: Am Asn Phys Anthrop; Am Anthrop Asn; Am Dent Asn; Am Asn Orthod. Res: Analysis of morphological and mensurational variations in the post-canine dentition; the metric description of occlusion; analysis of arch form; environmental variables which can cause deviations in tooth form. Mailing Add: Dept of Orthod Univ of Ky Col of Dent Lexington KY 40506

BIGGERSTAFF, WARREN RICHARD, b Folsom, NMex, May 2, 18; m 42; c 3. ORGANIC CHEMISTRY. Educ: Willamette Univ, BA, 40; Ore State Col, MS, 42; Univ Wis, PhD(org chem), 48. Prof Exp: Asst chem, Ore State Col, 40-42; asst, Univ Wis, 42-44, instr org chem, 47-48; PROF CHEM, FRESNO STATE COL, 58- Concurrent Pos: Chmn dept chem, Fresno State Col, 61-66, assoc vpres acad planning, 69-70; vis assoc, Sloan-Kettering Inst Cancer Res, 56-57; vis prof, Univ Lund, 67. Mem: Am Chem Soc. Res: Synthetic hormone substitutes; diethylstilbestrol and analogs related to progesterone and desoxycorticosterone; ring D metabolites of the estrogens; unsaturated thiolactones. Mailing Add: Dept of Chem Fresno State Col Fresno CA 93710

BIGGINS, JOHN, b Sheffield, Eng, Mar 30, 36; m 63. PLANT PHYSIOLOGY, PLANT BIOCHEMISTRY. Educ: Univ London, BSc, 60; Univ Calif, Berkeley, PhD(plant physiol), 65. Prof Exp: Asst prof bot, Univ Pa, 65-70; ASSOC PROF BIOL, BROWN UNIV, 70- Mem: Am Soc Photobiol; Am Soc Plant Physiol. Res: Correlation of structure and biochemical function of energy transducing systems; mechanism of photosynthesis and respiration in higher plants and algae; marine microbiology; phytoplankton physiology. Mailing Add: Dept of Biol & Med Sci Brown Univ Providence RI 02912

BIGGS, DAVID FREDERICK, b London, Eng, Jan 3, 39; m 62; c 2. PHARMACOLOGY. Educ: Univ Nottingham, BPharm, 61; King's Col, Univ London, 67. Prof Exp: Head dept pharmacol, Res Labs, May & Baker Ltd, Eng, 63-69; asst prof biopharm, 69-71, ASSOC PROF BIOPHARM, UNIV ALTA, 71- Mem: Brit Inst Biol; Brit Pharmacol Soc; Pharmaceut Soc Gt Brit. Res: Pharmacological properties of novel chemical compounds; relationship between structure and activity among neuromuscular blocking agents; general anaesthetics; pharmaceutical toxicology. Mailing Add: Fac of Pharm & Pharmaceut Sci Univ of Alta Edmonton AB Can

BIGGS, DONALD LEE, b Italy, Tex, Sept 25, 20; m 52. GEOLOGY. Educ: Univ Mo, AB, 49, AM, 51; Univ Ill, PhD(geol), 57. Prof Exp: Instr geol, Univ Mo, 52-54; asst geologist, State Geol Surv, Ill, 54-56; from asst prof to assoc prof, 56-66, PROF GEOL, IOWA STATE UNIV, 66- Mem: Geochem Soc; Geol Soc Am. Res: Mineralogy; sedimentary petrology; geochemistry. Mailing Add: Dept of Earth Sci Iowa State Univ Ames IA 50010

BIGGS, FRANK, b Langdon, Mo, Dec 16, 27; m 64; c 7. MATHEMATICAL PHYSICS. Educ: Univ Ark, BS, 56, MS, 57, PhD(physics), 65. Prof Exp: STAFF MEM, SANDIA LAB, 56-, RES PHYSICIST, 65- Concurrent Pos: Physicist, Sandia Lab, 57-65, consult, 60-65; prof, Univ NMex, 75- Mem: AAAS; Am Phys Soc. Res: X-ray cross sections and instrumentation; analysis of experimental data; applied mathematics; energy studies. Mailing Add: 3515 Monte Vista NE Albuquerque NM 87106

BIGGS, HOMER GATES, b Greene, NY, Feb 16, 30; m 52; c 2. BIOCHEMISTRY. Educ: State Univ NY Binghamton, BA, 51; Univ Iowa, MS, 54, PhD(biochem), 56. Prof Exp: Instr & asst dir clin chem, Univ Tenn, 56-59, asst prof, Div Chem & Clin Chem, Dept Med Labs, 59-63; assoc prof path, Med Col & dir clin chem, Univ Hosp, Univ Ala, 63-70, dep dir educ, Dept Clin Path, 69-70; dir clin chem, Med Ctr, Ind Univ, Indianapolis, 70-74, prof clin path, 70-75; V PRES & TECH DIR, BIOZYME LABS, INC, 75- Mem: Am Chem Soc; Am Asn Clin Chemists; NY Acad Sci. Res: Serum proteins and enzymes. Mailing Add: Biozyme Labs Inc 2409 W State St Olean NY 14760

BIGGS, MAX WILLIAM, industrial medicine, see 12th edition

BIGGS, R B, b Baltimore, Md, Feb 27, 37; m 60; c 1. MARINE GEOLOGY. Educ: Lehigh Univ, BA, 59, MS, 61, PhD(geol), 63. Prof Exp: Asst geol, Lehigh Univ, 60-62; res assoc, Chesapeake Biol Lab, 62-64; from asst prof to assoc prof geol, 64-70, ASSOC PROF, COL MARINE STUDIES, UNIV DEL, 70-, ASST DEAN, 72- Mem: Sigma Xi. Res: Marine geology, especially the geochemistry of modern sediments. Mailing Add: Col of Marine Studies Univ of Del Newark DE 19711

BIGGS, ROBERT HILTON, b US, May 5, 31; m 53; c 2. PLANT PHYSIOLOGY. Educ: ECarolina Univ, BS, 53; Purdue Univ, MS, 55, PhD(plant physiol), 58. Prof Exp: Asst biochemist, 58-65, assoc prof fruit crops & assoc biochemist, 65-69, PROF FRUIT CROPS & BIOCHEMIST, UNIV FLA, 69- Mem: Am Soc Plant Physiol; Am Soc Hort Sci. Res: Abscission; photoperiodism; dormancy. Mailing Add: Dept of Fruit Crops Univ of Fla Gainesville FL 32601

BIGGS, WALTER CLARK, JR, b Wilmington, NC, June 17, 31; m 63; c 2. VERTEBRATE ZOOLOGY, ANIMAL BEHAVIOR. Educ: Eastern Carolina Univ, BS, 53; Tex A&M Univ, MS, 60; NC State Univ, PhD(zool), 69. Prof Exp: Asst prof biol, 60-68, ASSOC PROF BIOL, UNIV NC, WILMINGTON, 69- Mem: Animal Behav Soc; Nat Parks Asn. Res: Social behavior and organization of animals, especially mother-young interactions among mammals; suckling behavior of domestic and feral swine compared in a variety of environmental regimes. Mailing Add: Dept of Biol Univ of NC Wilmington NC 28401

BIGGS, WALTER GALE, meteorology, see 12th edition

BIGHAM, CLIFFORD BRUCE, b Kamloops, BC, June 12, 28; m 54; c 3. NUCLEAR PHYSICS. Educ: Queen's Univ, Ont, BS, 51, MS, 52; Univ Liverpool, PhD(nuclear physics), 54. Prof Exp: From assoc res officer to sr res officer reactor physics, 54-66, SR RES OFFICER ACCELERATOR PHYSICS BR, ATOMIC ENERGY CAN, LTD, 66- Mem: Am Nuclear Soc; assoc Can Asn Physicist. Res: Experimental reactor physics; nuclear physics data pertaining to and parameters of nuclear reactor lattices; RF power for linear accelerators; superconducting cyclotron design. Mailing Add: Physics Div Atomic Energy of Can Ltd Chalk River ON Can

BIGHAM, ERIC CLEVELAND, b Kannapolis, NC, Mar 26, 47; m 70. MEDICINAL CHEMISTRY. Educ: NC State Univ, BS, 69; Princeton Univ, MA, 71, PhD(chem), 75. Prof Exp: RES SCIENTIST MED CHEM, CENT RES DIV, PFIZER INC, 73- Mem: Am Chem Soc; Sigma Xi. Res: Synthesis and semi-synthesis of antibiotics, drug design, new synthetic methods and other aspects of medicinal chemistry. Mailing Add: Pfizer Cent Res Eastern Point Rd Groton CT 06340

BIGHLEY, LYLE DELEVAN, b Albert Lea, Minn, Nov 8, 36; m 59; c 1. PHYSICAL PHARMACY. Educ: Univ Minn, Minneapolis, BS, 60; Univ Iowa, MS, 63, PhD, 66. Prof Exp: Assoc res pharmacist, Parke Davis & Co, Mich, 66-68; sr chemist, Cent Res Labs, Minn Mining & Mfg Co, Minn, 68-69; asst prof, 69-73, ASSOC PROF PHARM, UNIV IOWA, 73- Mem: Am Pharmaceut Asn; Acad Pharmaceut Sci. Res: Studies concerning the influence of physical and chemical properties of a drug and its dosage form on the availability of the drug to the body. Mailing Add: Dept of Pharm Univ of Iowa Col of Pharmacol Iowa City IA 52240

BIGLER, RODNEY ERROL, b Pocatello, Idaho, Mar 15, 41. NUCLEAR PHYSICS, BIOPHYSICS. Educ: Portland State Univ, BS, 66; Univ Tex, Austin, PhD(nuclear physics), 71. Prof Exp: Res assoc, 71-73, ASSOC BIOPHYS, MEM SLOAN-KETTERING CANCER CTR, 73-; ASST PROF RADIATION BIOL, SLOAN-KETTERING DIV, CORNELL UNIV, 74- Concurrent Pos: Res collabr, Brookhaven Nat Lab, 73- Mem: AAAS; Am Asn Physics Teachers; Am Phys Soc; Am Asn Physicists in Med; Soc Nuclear Med. Res: New methods to diagnose, treat and evaluate the results of therapy directed toward malignant disease. Mailing Add: Mem Sloan-Kettering Cancer Ctr Biophys Lab 1275 York Ave New York NY 10021

BIGLER, STUART GRAZIER, b Johnstown, Pa, Oct 21, 27; m 48; c 5. METEOROLOGY. Educ: Pa State Univ, BS, 52; Agr & Mech Col Tex, MS, 57. Prof Exp: Assoc, Ill State Water Surv, 52-55; from res assoc to res scientist III, 55-59; meteorologist & head radar sferics unit, US Weather Bur, 59-64, supvr radiation systs sect, 64-70; DIR, ALASKA REGION, NAT WEATHER SERV, 70- Concurrent Pos: Chmn, working group uses radar in meteorol, World Meteorol Orgn, 63-65; US Dept Commerce sci & tech training fel, 65-66; chief sounding systs br, data acquisition div, Weather Bur Environ Sci Serv Admin, 67- Honors & Awards: Spec Award, Am Meteorol Soc, 57; Dept Commerce Gold Medal, 70. Mem: Am Meteorol Soc. Res: Radar meteorology and cloud physics. Mailing Add: Alaska Region Nat Weather Serv Anchorage AK 99501

BIGLER, WILLIAM NORMAN, b Oakland, Calif, Aug 29, 37; m 61; c 3. BIOCHEMISTRY. Educ: Univ Calif, Berkeley, AB, 60; San Jose State Col, MS, 63; Univ Colo, Boulder, PhD(biochem), 68. Prof Exp: Instr chem, Univ Colo, Boulder, 66-67; NIH fel, Univ Calif, Los Angeles, 68-70, actg asst prof biochem, 70; asst prof, 70-75, ASSOC PROF CHEM, ROCHESTER INST TECHNOL, 75- Mem: AAAS; Am Chem Soc. Res: Biochemistry of cell division, regulatory enzymes, effects of ionizing radiation and slime molds. Mailing Add: Rochester Inst of Technol 1 Lomb Memorial Dr Rochester NY 14623

BIGLEY, NANCY JANE, b Sewickley, Pa, Feb 1, 32. BACTERIOLOGY, IMMUNOLOGY. Educ: Pa State Univ, BS, 53; Ohio State Univ, MSc, 55, PhD(bact, immunol), 57. Prof Exp: Res assoc immunol, Ohio State Univ, 57-65, asst prof microbiol, 65-68, assoc prof, Fac Microbiol & Cellular Biol, 68-69; from assoc prof to prof microbiol, Chicago Med Sch, 69-76; PROF MICROBIOL, SCH MED, WRIGHT STATE UNIV, 76- Mem: AAAS; Am Soc Microbiol; Am Asn Immunologists; Reticuloendothelial Soc. Res: Viral alterations of tissue components; autoimmunization; Rh antigen structural aspects; nucleic acid antigens; RNA-protein subfractions of Salmonella typhimurium involved in protective immunity; tumor-specific DNA-protein antigens. Mailing Add: Dept of Microbiol Wright State Univ Sch of Med Dayton OH 45431

BIGLEY, ROBERT HARRY, b Auburn, Wash, Aug 11, 29; m 59; c 2. INTERNAL MEDICINE, GENETICS. Educ: Univ Wash, BS, 51; Univ Ore, MD, 53. Prof Exp: From instr to assoc prof med, 60-75, PROF MED & MED GENETICS, SCH MED, HEALTH SCI CTR, UNIV ORE, 75- Concurrent Pos: USPHS res fel coagulation, Med Sch, Univ Ore, 59-60. Mem: Am Soc Hemat; Int Soc Hemat; Am Soc Human Genetics. Res: Genetic disease in man including characterization of altered primary gene products. Mailing Add: Dept of Med Univ of Ore Health Sci Ctr Portland OR 97201

BIGLIERI, EDWARD GEORGE, b San Francisco, Calif, Jan 17, 25; m 53; c 3. INTERNAL MEDICINE, ENDOCRINOLOGY. Educ: Univ San Francisco, BS, 48; Univ Calif, San Francisco, MD, 52. Prof Exp: Intern, Med Ctr, Univ Calif, San Francisco, 52-53, resident, 53-54; resident, Vet Admin Hosp, San Francisco, 54-56; clin assoc, Nat Heart Inst, 56-58; asst res physician, Metab Unit, 58-61, from asst prof to assoc prof, 62-71, PROF MED, MED CTR, UNIV CALIF, SAN FRANCISCO, 71-; CHIEF, ENDOCRINE METAB DIV, SAN FRANCISCO GEN HOSP, 62-, PROG DIR, CLIN STUDY CTR, 64- Concurrent Pos: Vis scientist, Prince Henry's Hosp, Monash Univ, Australia, 67; consult, Clin Invest Ctr, Oak Knoll Naval Hosp, Calif, 67-; mem coun high blood pressure res, Am Heart Asn. Mem: Asn Am Physicians; fel Am Col Physicians; Am Soc Clin Invest; Endocrine Soc. Res: Adrenal; mineralocorticoid hormones. Mailing Add: Clin Study Ctr San Francisco Gen Hosp San Francisco CA 94110

BIGNALL, KEITH E, b Marquette, Mich, Feb 18, 32; m 58. NEUROPHYSIOLOGY. Educ: Univ Mich, BSEE, 58, MS, 61, PhD(physiol), 63. Prof Exp: Asst prof, 64-72, ASSOC PROF PHYSIOL, UNIV ROCHESTER, 72- Concurrent Pos: NSF fel, 63-64. Mem: Am Physiol Soc; Soc Neurosci; NY Acad Sci; Int Brain Res Orgn. Res: Sensory system physiology; physiology of thalamocortical systems; temperature regulation; neurobehavioral development. Mailing Add: Dept of Physiol Univ of Rochester Sch of Med Rochester NY 14642

BIHLER, IVAN, b Osijek, Yugoslavia, Aug 12, 24; Can citizen; m 48. PHARMACOLOGY, BIOCHEMISTRY. Educ: Hebrew Univ, Jerusalem, MSc, 54, PhD(biochem), 57. Prof Exp: Asst biochem, Hebrew Univ, Jerusalem, 53-58; from asst prof to assoc prof, 63-67, PROF PHARMACOL & THERAPEUT, UNIV MAN, 67- Concurrent Pos: Fel pharmacol, Univ Rochester, 58-59; fel biochem, Washington Univ, 59-61; Imp Chem Indust res fel, Univ Edinburgh, 61-63; med res assoc, Med Res Coun, Can, 64. Mem: Am Soc Pharmacol & Exp Therapeut; Pharmacol Soc Can; NY Acad Sci; Can Biochem Soc; Brit Biochem Soc. Res: Mechanisms of cell membrane permeability and its regulation; membrane transport of sugars in intestine and muscle; effect of drugs and hormones. Mailing Add: Dept of Pharmacol & Therapeut Univ of Man Winnipeg MB Can

BIK, M J J, geomorphology, quaternary geology, see 12th edition

BIKALES, NORBERT M, b Berlin, Ger, Jan 7, 29; nat US; m 51; c 2. ORGANIC CHEMISTRY, POLYMER CHEMISTRY. Educ: City Col, BS, 51; Polytech Inst Brooklyn, MS, 56, PhD(chem), 61. Prof Exp: Res chemist, Am Cyanamid Co, 51-62; tech dir, Gaylord Assoc, Inc, 62-65; CHEM CONSULT, 65-; PROF CHEM & DIR CONTINUING EDUC SCI, RUTGERS UNIV, NEW BRUNSWICK, 73- Concurrent Pos: Ed, Encycl Polymer Sci & Technol, 62-; polymer workshop dir, Fairleigh Dickinson Univ & Ctr Prof Advan, 66-68; part-time prof, Upsala Col & consult to trustees, Columbia Univ & Strickman Found, 67-73; mem comt lead paints, Nat Acad Sci, 72-74; assoc mem comn macromolecules, Int Union Pure & Appl Chem, 74- Mem: AAAS; Am Inst Chem; Soc Plastics Eng; Am Chem Soc; NY Acad Sci. Res: Polymer synthesis and properties; chemical modification of natural polymers; water-soluble polymers; monomer synthesis; petrochemicals; textile, rubber and mining chemicals; tobacco filtration; pollution control. Mailing Add: Wright Chem Lab Rutgers Univ New Brunswick NJ 08903

BIKERMAN, JACOB JOSEPH, b Odessa, Russia, Oct 26, 98; nat US; m 33; c 2. PHYSICAL CHEMISTRY. Educ: Univ St Petersburg, Russia, MSc, 21. Prof Exp: Head, Munic Milk Testing Lab, St Petersburg, 20-21; from staff mem to asst ed, Ger Chem Soc, Berlin, 24-35; dir res, Glass Fibers, Ltd, Scotland, 39-41; sr physicist, Metal Box Co, Ltd, Eng, 41-44; leader adhesives res, Printing & Allied Trades Res Asn, 44-45; asst dir tech info, Merck & Co, Inc, 46-51; sr chemist, Yardney Labs, Inc, 51-55; supvr adhesives lab, Mass Inst Technol, 56-64; sr res assoc, Horizons Inc, 64-70; ADJ PROF CASE WESTERN RESERVE UNIV, 74- Concurrent Pos: With Brooklyn Polytech Inst, 52. Mem: Am Chem Soc; Soc Rheol. Res: Physics and chemistry of surfaces, particularly electrokinetics, foams, friction and adhesion. Mailing Add: 15810 Van Aken Blvd Cleveland OH 44120

BIKERMAN, MICHAEL, b Berlin, Ger, July 30, 34; US citizen; m 56; c 3. GEOCHRONOLOGY. Educ: Queens Col, NY, BS, 54; NMex Inst Mining & Technol, BS, 56; Univ Ariz, MS, 62, PhD(geol), 65. Prof Exp: Instr sci & eng, Ft Lewis Agr & Mech Col, 56-57; assayer, Holly Minerals, Cinnabar Mine, Idaho, 57; instr sci, Boise Jr Col, 58-60; asst geochem, geochronology lab, Univ Ariz, 63-65; asst prof geol, Wichita State Univ, 65-67; asst prof, 67-71, ASSOC PROF EARTH & PLANETARY SCI, UNIV PITTSBURGH, 71- Mem: Geol Soc Am; Geochem Soc. Res: Age determination and correlation of volcanic rocks, their petrology, structural relations and plutonic antecedents; isotope geology and geochronology of metamorphic terranes. Mailing Add: Dept of Earth & Planetary Sci Univ of Pittsburgh Pittsburgh PA 15260

BIKIN, HENRY, b Chicago, Ill, Oct 7, 18; m 43; c 2. PHARMACEUTICAL CHEMISTRY. Educ: Purdue Univ, BS, 40, MS, 48, PhD(pharmacy), 50. Prof Exp: Hosp pharmacist, Univ Hosp, Univ Mich, 40-41; pharmaceut chemist, Burroughs Wellcome & Co, 41-43; lab asst, Purdue Univ, 47-48; pharmaceut chemist, Eli Lilly & Co, 49-57; mgr pharmaceut develop & prod, Corvel, Inc, 57-67; RES SCIENTIST, ELI LILLY & CO, 67- Mem: AAAS; fel Am Inst Chem; Am Chem Soc; Am Pharmaceut Asn. Res: Incompatibilities; stabilities of liquid products; tablet products; formulations; stabilities; manufacturing trouble shooter. Mailing Add: 5231 Nob Lane Indianapolis IN 46226

BIKLE, DANIEL DAVID, b Harrisburg, Pa, Apr 25, 44; m 65; c 1. INTERNAL MEDICINE, BIOCHEMISTRY. Educ: Harvard Univ, AB, 65; Univ Pa, MD, 69, PhD(biochem), 74. Prof Exp: Clin fel internal med, Peter Bent Brigham Hosp, 69-71; RES INTERNIST INTERNAL MED & BIOCHEM, LETTERMAN ARMY INST RES, 74- Res: Mechanism of action of Vitamin D. Mailing Add: Dept of Med Letterman Army Inst of Res Presidio of San Francisco CA 94129

BIL, MILOS SIDNEY, b Heralec, Czech, Sept 10, 11; US citizen; m 42. ORGANIC CHEMISTRY, INORGANIC CHEMISTRY. Educ: Charles Univ, Prague, Dr rer nat, 36. Prof Exp: Res chemist & asst to plant mgr, United Chem & Metall Works, Czech, 36-38, chem engr, 38-40, plant mgr, 40-45, mgr org div, 45-51, ed-in-chief, tech instr, 52-56, head appl res labs, 56-58; corresp, Czech chem indust, Frankfurt, Ger, 59; RES & DEVELOP CHEMIST, CLAIROL, INC, 59- Concurrent Pos: Mem, Comt Develop Synthetic Dyes, Czech, 55-58. Mem: Am Chem Soc; Soc Chem Indust. Res: Synthesis and manufacture of organic chemicals, especially dye intermediates, dyes, insecticides, herbicides and aromatic fluorine intermediates. Mailing Add: 2 Blachley Rd Stamford CT 06902

BILAN, M VICTOR, b WUkraine, June 6, 22; US citizen; m 57; c 3. FOREST PHYSIOLOGY, ECOLOGY. Educ: Univ Munich, BS, 49; Duke Univ, MF, 54, DF, 57. Prof Exp: Res asst forest biol, Duke Univ, 53-57; from asst prof to assoc prof forestry, Stephen F Austin State Col, 57-66, PROF TREE PHYSIOL & ECOL, STEPHEN F AUSTIN STATE UNIV, 67- Concurrent Pos: NSF travel grant, Austria, 61, res grant, 62-66. Mem: Soc Am Foresters; Ecol Soc Am; Sigma Xi; Am Inst Biol Sci. Res: Effect of environment on root growth; root-shoot growth correlation; seed production; growth and development of southern pines; moisture relations in trees, drought resistance. Mailing Add: Sch of Forestry Stephen F Austin State Univ Nacogdoches TX 75961

BILANIUK, LARISSA TETIANA, b Ukraine, July 15, 41; US citizen; m 64; c 2. RADIOLOGY. Educ: Wayne State Univ, BA, 61, MD, 65. Prof Exp: Intern med, Philadelphia Gen Hosp, 65-66; resident, 66-67 & 68-71, assoc, 72-74, ASST PROF RADIOL, HOSP UNIV PA, 74- Concurrent Pos: Res fel, Cancer Res Ctr, Heidelburg, Ger, 67-68; NIH fel, 68-69 & 70-71; clin fel, Am Cancer Inst, 69-70; fel, Armed Forces Inst Path, Washington, DC, 70; Rothschild Ophthal Found fel, Paris, 72. Honors & Awards: Physicians Recognition Award, AMA, 69 & 73. Mem: Radiol Soc NAm; Am Col Radiol; Am Roentgen Ray Soc; Asn Univ Radiologists. Res: Computerized axial tomography in the diagnosis and treatment of brain tumors and cerebrovascular disease. Mailing Add: Dept of Radiol Hosp of Univ of Pa Philadelphia PA 19104

BILANIUK, OLEXA-MYRON, b Ukraine, Dec 15, 26; US citizen; m 64; c 2. NUCLEAR PHYSICS. Educ: Cath Univ Louvain, Cand Eng, 49; Univ Mich, BSE, 52, MS, 53, AM, 54, PhD(physics), 57. Prof Exp: Instr, Univ Mich, 57-58; asst prof physics, Univ Rochester, 59-64; assoc prof, 64-70, PROF PHYSICS, SWARTHMORE COL, 70- Concurrent Pos: Vis scientist, Arg Atomic Energy Comn, 62-63; NSF sci fac fel, Max Planck Inst, Heidelberg, 67-68 & Univ Paris, Orsay, 72. Mem: Am Phys Soc; Am Asn Physics Teachers; Europ Phys Soc; Fr Soc Physics; Sigma Xi. Res: Experimental and theoretical nuclear spectroscopy; nuclear reaction mechanisms and final state interactions; possibility of existence of supraluminal particles. Mailing Add: Dept of Physics Swarthmore Col Swarthmore PA 19081

BILBAO, MARCIA KEPLER, b Rochester, Minn, Jan 14, 31; m 54; c 3. RADIOLOGY. Educ: Univ Minn, Minneapolis, BS, 52; Columbia Univ, MD, 57; Am Bd Radiol, dipl, 62. Prof Exp: From instr to assoc prof, 61-69, PROF RADIOL, MED SCH, UNIV ORE, 69- Concurrent Pos: Consult, Kaiser Permanaente Clin, 61-63. Honors & Awards: Janeway Prize, Col Physicians & Surgeons, Columbia Univ, 57; Award of Merit, Am Women's Med Asn, 57; Aesculaplus Award, Ore Med Asn, 67; Cert of Merit, AMA, 68. Res: Diagnostic radiology; gas contrast radiography; gastrointestinal radiography. Mailing Add: Dept of Radiol Univ of Ore Med Sch Portland OR 97201

BILBRUCK, JAMES DONALD, plant pathology, see 12th edition

BILEAU, CLAIRE OF THE SAVIOR, b Woonsocket, RI, Sept 9, 11. PHYSIOLOGY. Educ: River Col, BS, 41; Cath Univ Am, MS, 45, PhD(physiol), 57. Prof Exp: From instr to assoc prof, 37-52, PROF BIOL, RIVER COL, 53-, CHMN DEPT, 45- Concurrent Pos: NIH res grant, 57-62. Mem: Soc Protozoologists; AAAS; Nat Asn Biol Teachers. Res: Endocrine physiology of cold-blooded vertebrates; thyroid and pituitary of the turtle; histology; cytology; cytochemistry. Mailing Add: Dept of Biol River Col Nashua NH 03060

BILELLO, MICHAEL ANTHONY, b NY, Oct 24, 24; m 51; c 5. METEOROLOGY, CLIMATOLOGY. Educ: Univ Wash, BS, 50; McGill Univ, MSc, 72. Prof Exp: Meterologist, hydrol sect, US Weather Bur, 51-54 & northern hemisphere map unit, 54-55, nat weather anal ctr, 55-56; meteorologist, snow, ice & permafrost res estab, Corps Engrs, 56-61 & cold regions res & eng lab, 61-68, RES METEOROLOGIST, SNOW & ICE BR, COLD REGIONS RES & ENG LAB, US ARMY, 68- Concurrent Pos: Vis instr, Dartmouth Col, 64, instr glaciol sem, 68. Honors & Awards: US Army Civilian Qual Increase Award, 67; Ann Grad Award, Can Asn Geogr, 73; US-Latin Am Coop Sci Award, NSF, 75. Mem: Am Meteorol Soc; Glaciol Soc; Can Geog Soc. Res: Meteorology and climatology in association with physical properties of snow cover and formation growth; decay of sea, lake and river ice; special projects on cold regions environmental research. Mailing Add: 12 Spencer Rd Hanover NH 03755

BILES, CHARLES MORGAN, b Yakima, Wash, Nov 29, 39; m 63; c 3. MATHEMATICS. Educ: St Martin's Col, BS, 61; Univ Ariz, MS, 64; Univ NH, PhD(math), 69. Prof Exp: Instr math & chem, St Martin's Col, 64-66; ASSOC PROF MATH, HUMBOLDT STATE UNIV, 69- Mem: Am Math Soc; Math Asn Am. Res: Rings of continuous functions, especially the topics of structure spaces of these rings; extension theory and Wallman-type compactifications. Mailing Add: Dept of Math Humboldt State Univ Arcata CA 95521

BILES, JOHN ALEXANDER, b Del Norte, Colo, May 4, 23; m 43; c 2. PHARMACEUTICAL CHEMISTRY. Educ: Univ Colo, BS, 44, PhD(chem), 49. Prof Exp: Asst chem, Univ Colo, 44-47, instr pharm, 47-48, asst chem, 48-49; prof pharmaceut chem, Midwestern Univ, 49-50; asst prof pharm, Ohio State Univ, 50-52; from asst prof to prof pharmaceut chem, 57-69, PROF BIOMED CHEM, UNIV SOUTHERN CALIF, 69-, DEAN, SCH PHARM, 68- Honors & Awards: Fink Gold Medal, 45. Mem: Am Pharmaceut Asn; Am Chem Soc. Res: Chemical microscopy. Mailing Add: Sch of Pharm Univ of Southern Calif Los Angeles CA 90007

BILETCH, HARRY, b Brooklyn, NY, Feb 9, 22; m 49; c 4. ORGANIC POLYMER CHEMISTRY. Educ: City Col New York, BS, 44; Polytech Inst Brooklyn, MS, 51, PhD(chem), 53. Prof Exp: Chemist, bur food & drugs, Dept Health, New York, 46-48; asst, Polytech Inst Brooklyn, 49-51; sr res fel, Off Naval Res, 51-52; sr res chemist, Interchem Corp, 52-54; fel, Polytech Inst Brooklyn, 54-55; group leader, basic polymer res group, Continental Can Co, 55-59 & polymer res, Foster Grant Co, 59-61; res dir, Great Am Plastics Co, 61-62; mgr, chem dept, Am Polymer & Chem Corp, 62-64 & res & develop, Terrell Corp, 64-69; DIR, RES & DEVELOP DEPT, SOLAR CHEM CORP, 69- Mem: AAAS; Soc Plastics Engrs; Am Chem Soc; Am Soc Testing & Mat. Res: Organic synthesis azonitriles; kinetics; mechanisms of organic reaction; monomer synthesis; co-polymerization of new monomers; high temperature and fluorinated polymers; semiconductors; heterocycles; polymer evaluations; x-ray and spectroscopic analysis; synthesis of high impact polymers. Mailing Add: Res & Develop Dept Solar Chem Corp Solar Park Leominster MA 01453

BILIMORIA, MINOO HORMASJI, b Barkhera, India, Aug 24, 30; m 67; c 2. MICROBIOLOGY, BIOCHEMISTRY. Educ: Univ Bombay, BSc, 52, MSc, 54; Bangalore Univ, PhD(microbiol), 63. Prof Exp: Microbiologist & chemist, Geoffrey Manners & Co Ltd, India, 54-58; res asst microbiol, Indian Inst Sci, Bangalore, 60-63; fel biochem, Sch Med, Duke Univ, 64-67; BIOCHEMIST, IMP TOBACCO CO CAN, LTD, 68- Mem: Can Biochem Soc; Indian Soc Biol Chem; Can Soc Microbiologists; NY Acad Sci; Environ Mutagen Soc. Res: Thermophilic bacilli, pectolytic enzymes from a variety of microorganisms; microsomal electron transport; L-asparaginases; biological effects of tobacco smoke. Mailing Add: Res Dept Imperial Tobacco Ltd PO Box 6500 Montreal PQ Can

BILLEN, DANIEL, b New York, NY, Nov 27, 24; m 51; c 3. MICROBIOLOGY. Educ: Cornell Univ, BS, 48; Univ Tenn, MS, 49, PhD(microbiol), 51. Prof Exp: Instr microbial metab, Univ Tenn, 50-51; res biologist radiobiol, Oak Ridge Nat Lab, 51-57; assoc prof, Post Grad Sch Med, Univ Tex, 57-60, biologist & prof biol & chief sect radiation biol, Univ Tex M D Anderson Hosp & Tumor Inst, 60-66; prof radiation biol & mem dept microbiol & radiol, Col Med, Univ Fla, 66-73; PROF BIOL & DIR, UNIV TENN-OAK RIDGE GRAD SCH BIOMED SCI, OAK RIDGE NAT LAB, 74- Concurrent Pos: Lectr, Univ Tenn, 52; prog dir metab biol prog, NSF, 60-61, consult, Career Develop Rev Br, NIH, 66-, consult radiation study sect, 69-73; ed, Microbios J; & Cytobios J; hon ed, Photochem & Photobiol J, 71-73. Mem: Am Soc Microbiol; Am Acad Microbiol; Radiation Res Soc; Am Soc Cell Biol. Res: Microbial physiology, biochemistry and cytology; radiobiology; microbiological genetics; tissue culture. Mailing Add: Univ Tenn-Oak Ridge Grad Sch Biomed Sci Biol Div Oak Ridge Nat Lab PO Box Y Oak Ridge TN 37830

BILLENSTIEN, DOROTHY CORINNE, b Easton, Pa, Feb 27, 21. ANATOMY. Educ: Boston Univ, BS, 42; Univ Colo, MS, 51, PhD(anat), 56. Prof Exp: From instr to asst prof anat, Albert Einstein Col Med, 56-61; asst prof, 61-64, ASSOC PROF ANAT, COLO STATE UNIV, 64- Mem: Am Asn Anatomists; Am Soc Zoologists. Res: Neurosecretion; neuroendocrinology. Mailing Add: Dept of Anat Colo State Univ Ft Collins CO 80521

BILLERA, LOUIS JOSEPH, b New York, NY, Apr 12, 43; m 64; c 2.

MATHEMATICS, OPERATIONS RESEARCH. Educ: Rensselaer Polytech Inst, BS, 64; City Univ New York, MA, 67, PhD(math), 68. Prof Exp: Asst prof opers res, 68-73, ASSOC PROF OPERS RES & MATH, CORNELL UNIV, 73- Concurrent Pos: NSF fel math, Hebrew Univ, Israel, 69; vis res assoc, Brandeis Univ, 74-75. Mem: Am Math Soc; Math Asn Am. Res: N-person game theory; combinatorial mathematics. Mailing Add: Dept of Opers Res Upson Hall Cornell Univ Ithaca NY 14850

BILLERBECK, FRED WILLIAM, JR, b Chicago, Ill, Sept 20, 28; m 61; c 4. FOOD SCIENCE. Educ: Purdue Univ, BS, 51, MS, 53, PhD(agr), 59. Prof Exp: Food technician radiation, Qm Food & Container Inst Armed Forces, 58-60; gen mgr food processing, Westfield Sommers Foods, Inc, 60-63; mgr prod develop baby foods, 63-67, RES MGR PROD DEVELOP, GERBER PRODS CO, 67- Mem: Inst Food Technologists. Res: Food technology and bacteriology; development of baby foods, special food formulations, cereals and general canned foods; thermal, dehydration and radiation in food preservation; horticulture; frozen foods; snack products. Mailing Add: Res Dept Gerber Prods Co Fremont MI 49412

BILLHARTZ, WILLIAM (HENRY), optics, see 12th edition

BILLIAR, REINHART BILLIE, b Crete, Nebr, July 2, 36; m 65; c 2. REPRODUCTIVE ENDOCRINOLOGY, BIOCHEMISTRY. Educ: Kans State Univ, BSc, 58; Univ Utah, PhD(biochem), 64. Prof Exp: Fel biochem, Harvard Med Sch, 63-65; from instr to asst prof, 65-75, ASSOC PROF REPRODUCTIVE BIOL, CASE WESTERN RESERVE UNIV, 75- Mem: AAAS; Am Endocrine Soc; Soc Exp Biol & Med; Soc Gynec Invest. Res: In vivo and in vitro metabolism of progesterone; hormonal control of uterine protein secretion; hypothalamic-steroid interactions and control of gonado-tropin secretion. Mailing Add: Dept of Reproductive Biol Case Western Reserve Univ Cleveland OH 44106

BILLICA, HARRY ROBERT, b Spokane, Wash, July 14, 19; m 46; c 3. CHEMISTRY. Educ: Univ NC, BS, 41; Univ Wis, PhD(org chem), 48. Prof Exp: Shift supvr exp sta, E I du Pont de Nemours & Co, 41-42, chemist, 42; chemist, US Naval Res Lab, Washington, DC, 42-45; res chemist, Exp Sta, E I du Pont de Nemours & Co, Del, 47-50, group leader tech sect, Nylon Plant, Tenn, 50-51, tech supvr, Dacron Polyester Fiber Plant, NC, 51-59, RES MGR, FIBER SURFACE RES SECT, TEXTILE FIBERS DEPT, E I DU PONT DE NEMOURS & CO, 59- Concurrent Pos: Fiber Soc lectr, 70-71. Mem: AAAS; Am Chem Soc; Fiber Soc (pres, 76); NY Acad Sci. Res: Chemical warfare; protective clothing; high pressure catalysis; hydrogenation; polymerization; catalyst development; synthetic fibers; physics and chemistry of fiber and film surfaces; polymer morphology. Mailing Add: E I du Pont de Nemours & Co PO Box 800 Kinston NC 28501

BILLIG, FRANKLIN A, b Los Angeles, Calif, Feb 11, 23; m 57; c 1. ORGANIC CHEMISTRY, SCIENCE EDUCATION. Educ: Univ Southern Calif, AB, 54. Prof Exp: Res chemist, Am Potash & Chem Corp, 54-58, sr res chemist, 58-64; SUPVR CHEM LABS, UNIV SOUTHERN CALIF, 64-, SAFETY OFFICER, DEPT CHEM, 73- Mem: Fel Am Inst Chemists; Am Chem Soc; fel AAAS; The Chem Soc. Res: Organometallics; science education techniques; theoretical organic chemistry. Mailing Add: Dept of Chem Univ of Southern Calif Los Angeles CA 90007

BILLIG, OTTO, b Vienna, Austria, Aug 10, 10; nat US; m 43; c 1. PSYCHIATRY. Educ: Univ Vienna, MD, 37. Prof Exp: Res neuropsychiat, Highland Hosp, Duke Univ, 39-41, assoc, 41-45, instr, Med Sch, 42-43, assoc, 43-46, asst prof, 46-47; from asst prof to assoc prof psychiat, 48-69, PROF CLIN PSYCHIAT, SCH MED, VANDERBILT UNIV, 69- Concurrent Pos: Dir Asheville exten, Rehab Clin, Duke Univ, 41-47, clin dir, 45-46, sr psychiatrist, Hosp, 46-47; vis psychiatrist & psychiatrist-in-chief, Out-Patient Dept, Vanderbilt Univ Hosp, 48-60; lectr, Sch Social Work, Univ Tenn, 49-68; consult, Vet Admin Hosps, Roanoke, Va, 49-56, Tenn, 50- & US Army, 50-56; chmn, Metro Bd Health, 63-73. Mem: AMA; Am Psychiat Asn; Soc Personality Assessment; World Fedn Ment Health; Int Soc Art & Psychopath. Res: Cultural approach to dynamic psychiatry; emotional aspects of art; schizophrenic thought disturbances. Mailing Add: Vanderbilt Univ Sch of Med Nashville TN 37232

BILLIGHEIMER, CLAUDE ELIAS, b Breslau, Ger, Nov 25, 30; Can citizen; m 55. PURE MATHEMATICS, MATHEMATICAL ANALYSIS. Univ Melbourne, BA & BSc, 53, MA, 58; Univ Toronto, PhD(math), 66. Prof Exp: Lectr math, Royal Mil Col, Canberra, 55-57; exp officer statist, Commonwealth Sci & Indust Res Orgn, Australia, 58; res fel math, Canberra Univ Col, 59; lectr, Australian Nat Univ, 60-61; res fel, Univ Toronto, 62, lectr, 62-66; sr lectr, Univ Melbourne, 66; asst prof, 66-70, ASSOC PROF MATH, McMASTER UNIV, 70- Concurrent Pos: Nat Res Coun Can grant, 69-76; mem bd gov, Maimonides Col, 72-76. Mem: Sigma Xi; Am Math Soc; Can Math Cong; Australian Math Soc. Res: Differential equations; nonlinear differential equations; boundary problems of linear differential and difference equations; linear operators in Hilbert space; spectral theory; physics; theoretical physics; fluid flow and aerodynamics. Mailing Add: Dept of Math McMaster Univ Hamilton ON Can

BILLINGHAM, EDWARD J, JR, b Lebanon, Pa, Dec 6, 34; m 58. ANALYTICAL CHEMISTRY. Educ: Lebanon Valley Col, BS, 56; Pa State Univ, PhD(chem), 61. Prof Exp: Assoc prof chem & chmn dept, Thiel Col, 60-65; from asst prof to assoc prof, 65-71, PROF CHEM, UNIV NEV, LAS VEGAS, 71- Mem: AAAS; Am Chem Soc. Res: Thermometric precipitation processes in both aqueous and fused salt media, including determination of thermodynamic properties as well as analytical implications; trace metal determinations. Mailing Add: Dept of Chem Univ of Nev Las Vegas NV 89154

BILLINGHAM, JOHN, b Worcester, Eng, Mar 18, 30; US citizen; m 56; c 2. AEROSPACE MEDICINE, EXOBIOLOGY. Educ: Oxford Univ, BA, 51, MA, BM & BCh, 54. Prof Exp: Med res officer aviation physiol, Royal Air Force Inst Aviation Med, Farnborough, Eng, 56-63; br chief environ physiol & space med, NASA Manned Spacecraft Ctr, Houston, Tex, 63-65, div chief biotechnol & aerospace med, Ames Res Ctr, 66-75, CHIEF BIOCOSMOLOGY DIV & CHIEF PROG OFF FOR SEARCH FOR EXTRATERRESTRIAL INTELLIGENCE, NASA AMES RES CTR, MOFFETT FIELD, 76- Concurrent Pos: Staff mem marine sci coun, Exec Off of the President, 69. Mem: Aerospace Med Asn. Res: Evolution of and search for extraterrestrial intelligent life; aerospace medicine and physiology; biotechnology; bioengineering. Mailing Add: 55 Golden Oak Dr Portola Valley CA 94025

BILLINGHAM, RUPERT EVERETT, b Warminster, Eng, Oct 15, 21; m 51; c 3. CELL BIOLOGY, IMMUNOLOGY. Educ: Oxford Univ, BA, 43, MA, 47, DPhil(zool), 50, DSc(zool), 57. Hon Degrees: DSc, Trinity Col, Conn, 65. Prof Exp: Lectr zool, Univ Birmingham, 47-51; hon res assoc, Univ Col, Univ London, 51-57; prof zool, Univ Pa, 57-71, prof med genetics & chmn dept, Sch Med, 65-71, dir, Henry Phipps Inst Med Genetics, 65-71; PROF CELL BIOL & CHMN DEPT, UNIV TEX HEALTH SCI CTR, DALLAS, 71- Honors & Awards: Alvarenga Prize, Am Col Physicians, 63; Hon Award, Am Asn Plastic Surgeons, 64. Mem: Fel Royal Soc; fel Am Acad Arts & Sci; Am Asn Immunologists; Soc Develop Biol; Am Fertil Soc. Res: Biology and immunology of transplantation; immunobiology of mammalian reproduction; wound healing and regeneration. Mailing Add: Dept of Cell Biol Univ of Tex Health Sci Ctr Dallas TX 75235

BILLINGS, BRUCE HADLEY, b Chicago, Ill, July 6, 15; m 75; c 4. OPTICS. Educ: Harvard Univ, AB, 36, MA, 37; Johns Hopkins Univ, PhD(physics), 43. Prof Exp: Mem radiol safety sect atomic bomb test, Bikini, 46; dir res, Baird-Atomic, Inc, 47-63; exec vpres, 55-58, asst dir defense res & eng, US Dept Defense, 59-60; dir, Baird-Atomic Holland, 60-62; vpres & gen mgr lab opers, Aerospace Corp, 63-69; spec asst to ambassador for sci & technol, Taipei, Taiwan, 69-72; VPRES, AEROSPACE CORP, 73- Concurrent Pos: Deleg, Marseille, Conf Thin Films, France, 49 & 63; assoc ed jour, Optical Soc Am, 50; mem staff, Ealing Corp & Diffraction, Ltd, 60; mem, Sci Adv Bd, US Air Force, 62-; Am comnr, Joint Comn Rural Reconstruct, Taipei, Taiwan; mem UN adv comt, Appln Sci & Technol, 72-; vpres, Int Comn Optics, 73-75. Honors & Awards: Brilliant Star, Govt Repub China, 72. Mem: Fel Am Acad Arts & Sci (secy, 55-59); Am Phys Soc; Acoust Soc Am; Optical Soc Am. Res: Infrared receivers and crystal optics; development of tunable and fixed narrowband filters; Fourier transform spectroscopy; energy conversion. Mailing Add: 955 L'Enfant Plaza SW Suite 4040 Washington DC 20024

BILLINGS, CHARLES EDGAR, JR, b Boston, Mass, June 15, 29; m 55; c 1. AEROSPACE MEDICINE, ENVIRONMENTAL PHYSIOLOGY. Educ: NY Univ, MD, 53; Ohio State Univ, MSc, 60. Prof Exp: Resident physician, Univ Vt, 57-58; instr prev med, Ohio State Univ, 60-62, from asst prof to prof prev med, physiol & aviation, 62-73; RES MED OFFICER, NASA-AMES RES CTR, 73- Concurrent Pos: Consult, Webb Assocs, 60-73; med consult, Beckett Aviation Corp, 62-73 & US Army, 64-; tech consult, Fed Aviation Agency, 63-66, mem med adv comt, 66-70; mem comt hearing, acoust & biomech, Nat Acad Sci-Nat Res Coun, 65-69 & 75-; clin prof prev med, physiol & aviation, Ohio State Univ, 73- Mem: Am Physiol Soc; fel Aerospace Med Asn; fel Am Col Prev Med. Res: Effects of environmental stress on man's ability to perform in the work environment, particularly the flight environment. Mailing Add: Man-Machine Integration Br LTI:239-3 NASA-Ames Res Ctr Moffett Field CA 94035

BILLINGS, DONALD EARL, b Colo, Sept 18, 14; m 45. ASTROPHYSICS. Educ: Univ Colo, BA, 35, MA, 39, PhD(physics), 49. Prof Exp: Teacher pub sch, Colo, 35-37; asst physics, Univ Colo, 39-41; physicist mine warfare, US Naval Yard, SC, 42-45; instr physics, Univ Colo, 46-49; asst prof, La State Univ, 49-51; res physicist high altitude observ, 51-62, PROF ASTRO-GEOPHYS, UNIV COLO, 62- Mem: Fel AAAS; Am Phys Soc; Am Astron Soc; Am Asn Physics Teachers; fel Royal Astron Soc. Res: Physical characteristics of the solar corona; solar flares; energy states of atomic nuclei; viscosity of molten glass; ferromagnetism. Mailing Add: Dept of Astro-Geophys Univ of Colo Boulder CO 80309

BILLINGS, FREDERIC TREMAINE, b Pittsburgh, Pa, Feb 22, 12; m 42; c 3. INTERNAL MEDICINE, MEDICAL EDUCATION. Educ: Princeton Univ, BA, 33; Oxford Univ, BSc, 36; Johns Hopkins Univ, MD, 38. Prof Exp: From instr to assoc prof, 39-64, dean med students, 61-68, assoc dean med ctr prog develop, 68-75, CLIN PROF MED, MED SCH, VANDERBILT UNIV, 64- Concurrent Pos: Prof, Meharry Med Col, 51-, head dept med, 51-57, trustee, 60-; trustee, Princeton Univ, 56-60 & The Choate Sch, 67-; dir, Nat Med Fels, Inc, 57-64 & Am Asn Rhodes Scholars, 58-62. Mem: Am Clin & Climat Asn (secy-treas, 58-68, pres, 69); master Am Col Physicians; fel AMA; Asn Am Physicians. Res: Cardiology and infectious diseases. Mailing Add: Med Arts Bldg Nashville TN 37212

BILLINGS, JAMES JENKINS, physics, see 12th edition

BILLINGS, MARLAND PRATT, b Boston, Mass, Mar 11, 02; m 38; c 2. GEOLOGY. Educ: Harvard Univ, AB, 23, AM, 25, PhD(geol), 27. Hon Degrees: DSc, Wash Univ, 60 & Univ NH, 66. Prof Exp: From asst to instr geol, Harvard Univ, 22-28; assoc geol, Bryn Mawr Col, 28-29, assoc prof, 29-30; from asst prof to prof, 30-72, chmn div geol sci, 46-51, EMER PROF GEOL, HARVARD UNIV, 72- Concurrent Pos: Lectr, Am Asn Petrol Geologists, 48; consult, Pa Turnpike, 38 & Metrop Dist Comn, 58- Mem: Nat Acad Sci; AAAS (vpres, 46-47); Am Acad Arts & Sci; fel Geol Soc Am (vpres, 51, pres, 59); fel Mineral Soc Am. Res: Structural geology; petrology; New England geology; engineering geology. Mailing Add: Dept of Geol Sci Harvard Geol Mus 24 Oxford St Cambridge MA 02138

BILLINGS, WILLIAM DWIGHT, b Washington, DC, Dec 29, 10; m 58. BOTANY, ECOLOGY. Educ: Butler Univ, AB, 33; Duke Univ, AM, 35, PhD(bot), 36. Hon Degrees: DSc, Butler Univ, 55. Prof Exp: Instr bot, Univ Tenn, 36-37; from instr to prof biol, Univ Nev, 38-52, head dept, 50-52; from assoc prof to prof, 52-67, JAMES B DUKE PROF BOT, DUKE UNIV, 67- Concurrent Pos: Ed, Ecol, Ecol Soc Am, 51-56 & Ecol Monogr, 68, 69; Fulbright res scholar, Univ NZ, 59. Honors & Awards: Cert of Merit, Bot Soc Am, 60; Mercer Award, Ecol Soc Am, 62. Mem: AAAS; Am Soc Naturalists; Soc Study Evolution; Bot Soc Am; Ecol Soc Am (vpres, 60). Res: Effect of geologic substratum on plant growth and distribution; desert, arctic and alpine ecology; physiological ecology; ecology of ecological races; environment. Mailing Add: Dept of Bot Duke Univ Durham NC 27706

BILLINGSLEY, LAWRENCE WINSTON, b Montreal, Que, May 16, 09; m 38; c 4. BIOCHEMISTRY. Educ: McGill Univ, BSc, 32, MSc, 33, PhD(biochem), 37. Prof Exp: Demonstr biochem, McGill Univ, 33-37; lectr biol, Sir George Williams Col, 37-38; res biologist, Nat Res Coun Can, 38-42, biol ed, Can J Res, 38-40, secy med res comt, 40-42; ed, Can Chem & Process Indust, 45-46; chief tech info & res librn, Ayerst, McKenna & Harrison, Ltd, 46-49; admin officer, Med Res Labs, Defence Res Bd, Dept Nat Defence, Ottawa, 49-52, sci ed, 52-68; assoc ed, Fisheries Res Bd, Can, 68-74; RETIRED. Mem: Am Fisheries Soc. Res: Carotene and vitamin A; metabolism of small animals; vitamin content of foods. Mailing Add: 95 Renfrew Ave Ottawa ON Can

BILLINGSLEY, PATRICK PAUL, b Sioux Falls, SDak, May 3, 25; m 53; c 5. MATHEMATICS. Educ: US Naval Acad, BS, 48; Princeton Univ, MA, 52, PhD(math), 55. Prof Exp: NSF fel, Princeton Univ, 57-58; asst prof statist, 58-62, assoc prof math & statist, 62-67, PROF MATH & STATIST, UNIV CHICAGO, 67- Concurrent Pos: Vis prof, Copenhagen Univ, 64-65; ed, Ann Probability. Mem: Am Math Soc; fel Inst Math Statist. Res: Probability theory; stochastic process theory. Mailing Add: Dept of Statist Univ of Chicago Chicago IL 60637

BILLINGSLEY, SUSAN V, history of science, see 12th edition

BILLMAN, JOHN HENRY, b Brooklyn, NY, Feb 8, 12; m 37; c 2. ORGANIC CHEMISTRY. Educ: Univ Va, BS, 34; Princeton Univ, AM, 35, PhD(chem), 37. Prof Exp: Instr org & inorg chem, Univ Ill, 37-39; from instr to assoc prof, 39-58, PROF ORG CHEM, IND UNIV, BLOOMINGTON, 58- Concurrent Pos: Vis lectr, Yale Univ, 46-47; prof, Univ Del, 58; consult, US Dept Health, Educ & Welfare, Am Viscose Corp & New Castle Prod; mem Pharmacol & Endocrinol Fel Panel; civilian investr, Off Sci Res & Develop. Mem: Am Chem Soc. Res: Pharmaceutical chemistry;

organic synthesis; organic analytical reagents; compounds for use as plant stimulants; antitumor and antiviral agents; catalysis; amino acids. Mailing Add: Dept of Chem Ind Univ Bloomington IN 47401

BILLMAN, KENNETH WILLIAM, b Covington, Ky, Jan 9, 33; m 54; c 4. LASERS. Educ: Thomas More Col, AB, 55; Univ Cincinnati, MS, 58, PhD(physics), 59. Prof Exp: Instr physics, Univ Cincinnati, 57-59; from instr to asst prof, Mass Inst Technol, 59-67; staff physicist, Electronics Res Ctr, 67-70, Ames Res Ctr, 70-72, ASST CHIEF PHYS GAS-DYNAMICS & LASERS BR, AMES RES CTR, NASA, 72- Concurrent Pos: Adj prof physics, Colo State Univ, 75-76. Mem: Am Phys Soc; assoc Inst Fundamental Studies. Res: Development of ultraviolet and x-ray lasers; laser matter interaction studies, especially inverse bremstrahlung absorption; laser induced chemistry and isotope separation; harmonic conversion; laser energy conversion; supersonic electric discharge lasers. Mailing Add: NASA Ames Res Ctr MS 230-3 Moffett Field CA 94035

BILLMEYER, FRED WALLACE, JR, b Chattanooga, Tenn, Aug 24, 19; m 51; c 3. CHEMISTRY. Educ: Calif Inst Technol, BSc, 41; Cornell Univ, PhD(phys chem), 45. Prof Exp: Res chemist, Plastics Dept, E I du Pont de Nemours & Co, 45-57, res assoc, 57-64; PROF CHEM, RENSSELAER POLYTECH INST, 64- Concurrent Pos: Prof, Univ Del, 52-64. Mem: AAAS; Am Chem Soc; Inter-Soc Color Coun (pres, 69-70, secy, 70-); fel Optical Soc Am; fel Am Phys Soc. Res: Molecular structure of polymers; color science and technology; optical properties of plastics. Mailing Add: 2121 Union St Schenectady NY 12309

BILLO, EDWARD JOSEPH, b Brantford, Ont, Aug 3, 38; m 61; c 3. INORGANIC CHEMISTRY, ANALYTICAL CHEMISTRY. Educ: McMaster Univ, BSc, 61, MSc, 63, PhD(chem), 67. Prof Exp: Res assoc, Purdue Univ, 67-69; asst prof, 69-74, ASSOC PROF CHEM, BOSTON COL, 75- Mem: Am Chem Soc. Res: Solution equilibria of transition metal chelates; kinetics of fast reactions of metal chelates; metal complexes of peptides. Mailing Add: Dept of Chem Boston Col Chestnut Hill MA 02167

BILLOTTI, JOSEPH EUGENE, b Baltimore, Md, May 7, 32. MATHEMATICS. Educ: Fordham Univ, AB, 56, MAT, 58, MA, 60; Woodstock Col, PhL, 57, STL, 64; Brown Univ, PhD(appl math), 69. Prof Exp: NASA res asst, Brown Univ, 65-68; asst prof, 69-75, ASSOC PROF MATH & CHMN DEPT, LE MOYNE COL, NY, 75- Concurrent Pos: Vis prof, dept sci affairs, Orgn Am States & Invest Ctr, Nat Polytech Inst, Mex, 72-73. Mem: Am Math Soc; Math Asn Am; Soc Indust & Appl Math. Res: Ordinary and functional differential equations; global analysis; dissipative systems for periodic functional differential equations. Mailing Add: Dept of Math Le Moyne Col Syracuse NY 13214

BILLS, ALAN MORRIS, b Sutton-in-Ashfield, Eng, Jan 11, 41; US citizen; m 66; c 3. CARBOHYDRATE CHEMISTRY, ORGANIC CHEMISTRY. Educ: Lake Forest Col, BA, 61; Inst Paper Chem, MS, 63, PhD(paper chem), 67. Prof Exp: Res chemist, 66-74, SR RES CHEMIST, WESTVACO CORP, CHARLESTON, SC, 74- Mem: Am Chem Soc; Tech Asn Pulp & Paper Indust. Res: The Koenigs-Knorr condensation of glycosyl halides and simple sugars; pulp and paper chemistry and technology; chemistry of natural byproducts from wood, particularly tall oil. Mailing Add: 116 President Circle Summerville SC 29483

BILLS, CHARLES EVERETT, biochemistry, deceased

BILLS, CHARLES WAYNE, b Meeker, Colo, Nov 6, 24; m 47; c 3. INORGANIC CHEMISTRY, ORGANIC CHEMISTRY. Educ: Colo Agr & Mech Col, BS, 47; Univ Colo, MS, 52, PhD(chem), 54. Prof Exp: Analyst, Los Alamos Sci Lab, Univ Calif, 47, org chem & radiochem synthesis, 48-50; asst, Univ Colo, 50-52, qual anal, 53; res chemist corrosion, Prod Res Dept, Stanolind Oil & Gas Co, 54-55 & Mallinckrodt Chem Works, Mo, 55-56; chief geochem & geophys res & develop br, US AEC, Colo, 56-58, nuclear chemist, Chem Separations Br, Hanford Opers Off, Wash, 58-59, dep dir health & safety div, Nat Reactor Testing Sta, Idaho Opers, 59-62, dir nuclear technol div, 62-74, ASST MGR PROD & TECH SUPPORT, IDAHO OPERS, ENERGY RES & DEVELOP ADMIN, 74- Concurrent Pos: Princeton Univ fel pub affairs, 62-63; ex-officio dir, Eastern Idaho Nuclear Indust Coun, 66- & Intermountain Sci Experience Ctr, Inc, 73- Mem: Am Chem Soc; Am Nuclear Soc. Res: Aluminum stearates; geochemistry of uranium; processing nuclear power reactor fuels; transuranic and fission product recovery; nuclear reactor operations, safety and technology; radioactive waste management. Mailing Add: 1090 E 21st St Idaho Falls ID 83401

BILLS, DANIEL GRANVILLE, b Wenatchee, Wash, Sept 8, 24; m 45; c 3. PHYSICS. Educ: Wash State Univ, BS, 49, MS, 51; Harvard Univ, PhD(physics), 57. Prof Exp: Instr physics, Wash State Univ, 55-57, asst prof, 57-60; pres, 54-69, CHMN, BD DIRS, GRANVILLE-PHILLIPS CO, 54- Mem: Am Phys Soc; Am Vacuum Soc. Res: Surface physics; physical electronics; ultrahigh vacuum. Mailing Add: Granville-Phillips Co 5675 E Arapahoe Boulder CO 80303

BILLS, DONALD DUANE, b Hillsboro, Ore, Dec 4, 32; m 56; c 1. FOOD SCIENCE. Educ: Ore State Univ, BS, 59, MS, 64, PhD(food sci), 66. Prof Exp: Asst prof, 65-68, ASSOC PROF FOOD SCI & TECHNOL, ORE STATE UNIV, 68- Mem: AAAS; Am Chem Soc; Am Dairy Sci Asn; Inst Food Technologists. Res: Trace components in foods, including flavor constituents, pesticides and carcinogens. Mailing Add: Dept of Food Sci & Technol Ore State Univ Corvallis OR 97331

BILLS, JAMES LAVAR, b Murray, Utah, Aug 4, 35; m 58; c 3. INORGANIC CHEMISTRY. Educ: Univ Utah, BS, 58; Mass Inst Technol, PhD(inorg chem), 63. Prof Exp: Asst prof, 62-68, ASSOC PROF CHEM, BRIGHAM YOUNG UNIV, 68- Mem: Am Chem Soc. Res: Thermochemistry; coordination compounds. Mailing Add: 1720 W 1400 North Provo UT 84601

BILLS, JOHN LAWRENCE, b Moore, Idaho, July 11, 20; m 46; c 7. CHEMISTRY. Educ: Stanford Univ, AB, 42, PhD(chem), 47. Prof Exp: Res chemist, Union Oil Co, Calif, 44-46, 47-49, group leader, 49-52; instr chem, Stanford Univ, 46-47; mkt develop engr, Brea Chem, Inc, 52-57; mgr mkt res, Am Potash & Chem Corp, 57-64; MGR CORP DEVELOP, KERR-McGEE CORP, 64- Mem: Am Chem Soc. Res: Hydroforming and desulfurization catalysts; organic synthesis of isoquinoline derivatives; oxidation and combustion of petroleum products as a synthetic tool; acetylene chemistry, hydrogen cyanide; phthalic acids; polymers; fertilizers and related products. Mailing Add: Kerr-McGee Corp Oklahoma City OK 73102

BILLS, ROBERT F, b Harvey, Ill, Jan 10, 31; m 54; c 3. CYTOLOGY, ELECTRON MICROSCOPY. Educ: Univ Ill, BS, 54, MS, 58, PhD(bot), 60. Prof Exp: Asst bot, Univ Ill, 56-57, Electron Micros Lab, 57-60; res assoc biol, Mass Inst Technol, 60-61; from asst prof to assoc prof, 61-70, PROF BIOL, UNIV SOUTHERN CALIF, 70-, DIR ELECTRON MICROS LAB, ALLAN HANCOCK FOUND, 61- Concurrent Pos: NIH fel, Mass Inst Technol, 60-61; vis prof, Path Inst & Air Hyg Inst, Univ Düsseldorf, 68-69. Mem: AAAS; NY Acad Sci; Electron Micros Soc Am; Am Soc Cell Biol. Res: Cell ultrastructure, electron microscopy; cytologic effects of toxic air pollutants on lung cells; connective tissues of lung; development and aging of cells. Mailing Add: Electron Micros Lab Allan Hancock Found USC Los Angeles CA 90007

BILLUPS, NORMAN FREDERICK, b Portland, Ore, Oct 15, 34; m 57; c 2. PHARMACY, PHARMACEUTICAL CHEMISTRY. Educ: Ore State Univ, BS, 58, MS, 61, PhD(phys pharm), 63. Prof Exp: Instr pharm, Ore State Univ, 58-59, asst, 59-62, NIH res fel phys pharm, 62-63; assoc prof, 63-71, PROF PHYS PHARM, UNIV KY, 71- Concurrent Pos: Fel, Am Found Pharmaceut Educ, 62-63; consult, Blue Cross-Blue Shield & Mkt Measures. Honors & Awards: Lyman Award, Am Asn Cols Pharm, 71; Res Achievement Award, Am Soc Hosp Pharmacists, 75. Mem: Am Pharmaceut Asn; Acad Pharmaceut Sci; Acad Gen Pract Pharm; Am Asn Cols Pharm; AAAS. Res: Various dosage forms of pharmaceutical products, especially disintegrating agents and ointment technology; drug release from selected ointment bases; kinetics of drug absorption and distribution; protein binding of drugs; complexation; percutaneous adsorption; socio-economics of pharmacy. Mailing Add: Col of Pharm Univ of Ky Lexington KY 40506

BILLUPS, W EDWARD, b Huntington, WVa, Apr 7, 39; m 66; c 1. ORGANIC CHEMISTRY. Educ: Marshall Univ, BS, 61, MS, 65; Pa State Univ, PhD(chem), 70. Prof Exp: Res chemist, Union Carbide Corp, WVa, 61-68; ASST PROF CHEM, RICE UNIV, 70-, ASSOC, RICHARDSON COL, 72- Mem: Am Chem Soc; The Chem Soc. Res: Chemistry of small ring compounds, reactive intermediates, transition metal catalysis. Mailing Add: Dept of Chem Rice Univ Houston TX 77001

BILLY, BAHE, agricultural chemistry, range management, see 12th edition

BILODEAU, GERALD GUSTAVE, b Waterville, Maine, Nov 2, 29; m 56; c 6. MATHEMATICAL ANALYSIS. Educ: Univ Maine, BA, 50; Harvard Univ, AM, 51, PhD(math), 59. Prof Exp: Sr mathematician, Westinghouse Atomic Power Div, Pa, 55-59; adv res engr, Sylvania Electronics Systs, 59-60; from asst prof to assoc prof, 60-71, PROF MATH, BOSTON COL, 71- Mem: Am Math Soc; Math Asn Am; Sigma Xi; Soc Indust & Appl Math. Res: Orthogonal functions; integral transforms; partial differential equations. Mailing Add: 200 Harvard Circle Newtonville MA 02160

BILOW, NORMAN, b Chicago, Ill, Sept 9, 28; m 54; c 3. CHEMISTRY. Educ: Roosevelt Univ, BS, 49; Univ Chicago, MS, 52, PhD(chem), 56. Prof Exp: Res chemist, Emulsol Corp, 49-52; asst, Univ Chicago, 55-56; res chemist, Dow Chem Co, 56-59; staff scientist, 59-65, sr staff chemist, 65-69, head chem synthesis group, 67-69, head polymer & chem technol sect, Aerospace Group, 69-70, polymer & phys chem sect, 70-73, SR SCIENTIST, ADVAN TECHNOL LAB, HUGHES AIRCRAFT CO, 73- Honors & Awards: Indust Res Mag Award, 70 & 74. Mem: Am Chem Soc; Sigma Xi; fel Am Inst Chemists; NY Acad Sci. Res: Stereospecific polyolefins; polyphenylenes; phenolics; ablative resins; free radicals; photochemistry; diazo compounds; surfactants; polyurethanes; intumescent paints; electrical insulation; polyferrocenes; poly-aromatic-heterocyclic polymers; high temperature polymers. Mailing Add: 16685 Calneva Dr Encino CA 90056

BILPUCH, EDWARD GEORGE, b Connellsville, Pa, Feb 10, 27; m 52. PHYSICS. Educ: Univ NC, BS, 50, MS, 52, PhD, 56. Prof Exp: Res assoc, 56-62, from asst prof to assoc prof, 62-71, PROF PHYSICS, DUKE UNIV, 71-, DEP DIR, TRIANGLE UNIVS NUCLEAR LAB, 68- Mem: Fel Am Phys Soc. Res: High resolution neutron and charged particle cross section measurements; fine structure of isobaric analogue states. Mailing Add: Dept of Physics Duke Univ Durham NC 27706

BILTONEN, RODNEY LINCOLN, b Sudbury, Ont, Aug 24, 37; US citizen; m 60; c 2. BIOPHYSICAL CHEMISTRY, BIOCHEMISTRY. Educ: Harvard Univ, AB, 59; Univ Minn, PhD(phys chem), 65. Prof Exp: NIH fel, 65-66; asst prof phys chem, 66-72, ASSOC PROF PHARMACOL & BIOCHEM, SCH MED, JOHNS HOPKINS UNIV, 72- Mem: Am Chem Soc; AAAS; Am Soc Biol Chemists. Res: Structure and thermodynamics of macromolecules of biological significance; thermodynamics of ligand binding to macromolecules; application of calorimetric techniques to the study of biological systems; thermodynamic aspects of enzyme catalysis. Mailing Add: Dept of Physiol Chem Johns Hopkins Univ Sch of Med Baltimore MD 21205

BILYEU, RUSSELL GENE, b Krum, Tex, Jan 22, 30; m 53; c 4. MATHEMATICAL ANALYSIS. Educ: NTex State Univ, BS, 52, MS, 57; Univ Kans, PhD(math), 60. Prof Exp: Engr systs eng, Chance Vought Aircraft, Inc, 52-57; from asst prof to assoc prof, 60-70, PROF MATH, N TEX STATE UNIV, 70- Mem: Am Math Soc; Math Asn Am. Res: Normed linear spaces. Mailing Add: 1709 Westchester Denton TX 76201

BIMBER, RUSSELL MORROW, b Warren, Pa, Mar 26, 29; m 51; c 3. ORGANIC CHEMISTRY. Educ: Antioch Col, BS, 52; Western Reserve Univ, MS, 62. Prof Exp: Chemist, Diamond Alkali Co, 52-62, SR CHEMIST, DIAMOND SHAMROCK CORP, 62- Mem: AAAS; Am Chem Soc. Res: Organic syntheses; chlorinations; Diels-Alder reactions; chemistry of chloral; bench scale process research and development. Mailing Add: 10471 Prouty Rd Painesville OH 44077

BINCER, ADAM MARIAN, b Krakow, Poland, Apr 25, 30; US citizen; m; c 2. THEORETICAL PHYSICS. Educ: Mass Inst Technol, SB, 53, PhD(physics), 56. Prof Exp: Res assoc physics, Brookhaven Nat Lab, 56-58; asst res, Univ Calif, Berkeley, 58-60; from asst prof to assoc prof, 60-68, PROF PHYSICS, UNIV WIS, MADISON, 68- Concurrent Pos: Fulbright scholar, Univ Sao Paulo, Brazil, 65. Mem: Am Phys Soc. Res: Theory of fundamental particles; analyticity properties of scattering amplitudes; group theory. Mailing Add: Dept of Physics Univ of Wis Madison WI 53706

BINDER, CHARLES R, fuel science, organic chemistry, see 12th edition

BINDER, DANIEL, b New York, NY, Feb 20, 27; m 56. PHYSICS. Educ: City Col New York, BS, 47; Yale Univ, PhD(physics), 50. Prof Exp: Physicist, Oak Ridge Nat Lab, 50-60; SR STAFF SCIENTIST, HUGHES AIRCRAFT CO, 60- Mem: Am Phys Soc. Res: Radioactivity; radiation damage and dosimetry; transient effects on electronics. Mailing Add: 30004 Via Borica Rancho Palos Verdes CA 90274

BINDER, FRANKLIN LEWIS, b Bristol, Pa, Nov 14, 45; m 65; c 2. MICROBIAL PHYSIOLOGY, MYCOLOGY. Educ: Ind Univ Pa, BA, 67; WVa Univ, MS, 69, PhD(microbiol), 71. Prof Exp: Res asst, WVa Univ, 67-71; ASST PROF MICROBIOL, MARSHALL UNIV, 71- Concurrent Pos: Instr microbiol, Ohio Univ, Portsmouth, 73-75; fac res grant, Marshall Univ, 75 & 76. Honors & Awards: Res Award, Sigma Xi, 75. Mem: Am Soc Microbiol; Mycol Soc Am; AAAS; Sigma Xi. Res: Physiological and biochemical studies with the haustorial mycoparasite Tieghemiomyces Parasiticus in axenic culture; transport mechanism in fungi. Mailing Add: Dept of Biol Sci Marshall Univ Huntington WV 25701

BINDER, HENRY JOSEPH, b New York, NY, Dec 5, 36; m 61; c 2. INTERNAL

MEDICINE, GASTROENTEROLOGY. Educ: Dartmouth Col, AB, 57; NY Univ, MD, 61. Prof Exp: Instr med, Yale Univ, 65-66; clin instr, Univ Calif, San Francisco, 67-68; asst prof, Univ Chicago, 68-69; asst prof, 69-72, ASSOC PROF MED, YALE UNIV, 72- Mem: Am Fedn Clin Res: Am Gastroenterol Asn. Res: Gastrointestinal physiology; intestinal transport. Mailing Add: Dept of Internal Med Yale Univ New Haven CT 06510

BINDER, LAURENCE OSCAR, JR, organic chemistry, see 12th edition

BINDRA, JASJIT SINGH, b Rawalpindi, India, Oct 20, 42; m 70. ORGANIC CHEMISTRY. Educ: Agra Univ, MSc, 64, PhD(chem), 68. Prof Exp: Scientist med chem, Cent Drug Res Inst, India, 68-69; res assoc chem, Ind Univ, Bloomington, 69-71; res chemist, 71-74, SR RES CHEMIST, PFIZER, INC, 74- Mem: Am Chem Soc; The Chem Soc. Res: Organic synthesis; medicinal agents; molecular pharmacology. Mailing Add: Med Res Labs Pfizer Inc Groton CT 06340

BINDSCHADLER, ERNEST, organic chemistry, see 12th edition

BINFORD, CHAPMAN HUNTER, b Darlington Heights, Va, Oct 3, 00; m; c 2. PATHOLOGY. Educ: Hampden-Sydney Col, BA, 23; Med Col Va, MD, 29; Am Bd Path, dipl. Hon Degrees: DSc, Hampden-Sydney Col, 62. Prof Exp: Med officer, US Marine Hosp, Va, 30-31; res worker, USPHS Cancer Lab, Harvard Univ, 31-32 & Leprosy Invest Sta, Honolulu, 33-36; res worker res path, NIH, 36-37; pathologist, US Marine Hosp, Detroit, 37-42, New Orleans, 42-45 & Baltimore, 45-51; chief infectious dis path, 51-55, chief lab invest & registry of leprosy, 55-60, chief geog path div, 60-63, CHIEF LEPROSY BR, ARMED FORCES INST PATH, WASHINGTON, DC, 63-; RES PATHOLOGIST, LEONARD WOOD MEM, 60- Concurrent Pos: USPHS fel, Cancer Lab, Harvard Univ, 31-32; med dir, Leonard Wood Mem, 63-72. Honors & Awards: Ward Burdick Award, Am Soc Clin Path, 68; Damien Dutton Award, Leprosy, 71; Maude Abbott Lectr, Int Acad Path, 73. Mem: Fel AMA; fel Am Col Path; fel Am Soc Clin Path; Am Asn Pathologists & Bacteriologists; Int Acad Path (pres, 58-59). Res: Leprosy and fungus diseases. Mailing Add: 6046 23rd St N Arlington VA 22205

BINFORD, JESSE STONE, JR, b Freeport, Tex, Nov 1, 28; m 55; c 2. PHYSICAL CHEMISTRY. Educ: Rice Univ, BA, 50, MA, 52; Univ Utah, PhD(chem), 55. Prof Exp: Instr chem, Univ Tex, 55-58; from asst prof to assoc prof, Col Pac, 58-61; assoc prof, 61-71, PROF CHEM, UNIV S FLA, 72- Concurrent Pos: Fulbright prof chem & chmn dept, Nat Univ Honduras, 68-69; consult, Fla Consort, AID, Honduras, 69. Mem: AAAS; Am Chem Soc; Calorimetry Conf. Res: Solution calorimetry; biochemical calorimetry; calorimetry of bacterial growth. Mailing Add: 1905 E 111th Ave Tampa FL 33612

BINFORD, LAURENCE CHARLES, b Chicago, Ill, Jan 11, 35. VERTEBRATE ZOOLOGY, ORNITHOLOGY. Educ: Univ Mich, BS, 57; La State Univ, PhD(vert zool), 68. Prof Exp: Instr zool, La State Univ, 65-66; asst cur, 68-73, ASSOC CUR BIRDS & MAMMALS & CHMN DEPT, CALIF ACAD SCI, 73- Concurrent Pos: Vpres bd dirs, Point Reyes Bird Observ, 69-72, pres bd dirs, 72-; fel, Calif Acad Sci, 72; mem ed bd, Western Birds, 73- Mem: Elective mem Am Ornith Union; Cooper Ornith Soc; Wilson Ornith Soc. Res: Ornithology, especially avian taxonomy; ornithogeography, particularly Mexico. Mailing Add: Calif Acad Sci Golden Gate Park San Francisco CA 94118

BINFORD, LEWIS R, b Norfolk, Va, Nov 21, 30; m 52; c 2. ANTHROPOLOGY, ARCHAEOLOGY. Educ: Univ NC, BA, 57; Univ Mich, MA, 58, PhD(anthrop), 64. Prof Exp: Res assoc archaeol, Mus Anthrop, Univ Mich, 58-59, cur, 60-61; asst prof anthrop, Univ Chicago, 61-65; asst prof, Univ Calif, Santa Barbara, 65-66; assoc prof, Univ Calif, Los Angeles, 66-70; ASSOC PROF ANTRHOP, UNIV N MEX, 70- Concurrent Pos: Consult, State Mich, 59. Mem: AAAS; fel Soc Am Archaeol; fel Am Anthrop Asn. Res: Study of evolutionary processes through the use of archaeological data; archaeological theory and method. Mailing Add: Dept of Anthrop Univ of NMex Albuquerque NM 87106

BINFORD, SALLY R, archeology, physical anthropology, see 12th edition

BING, ARTHUR, b Springfield, Mass, Apr 18, 16; m 54; c 1. PLANT PHYSIOLOGY. Educ: Univ Conn, BS, 39; Cornell Univ, PhD, 49. Prof Exp: Asst bot, 46-49, PROF FLORICULT, ORNAMENTAL RES LAB, CORNELL UNIV, 49- Mem: Am Soc Plant Physiologists; Weed Sci Soc Am; Am Soc Hort Sci; Int Soc Hort Sci; Sigma Xi. Res: Weed control and factors influencing flowering of ornamentals. Mailing Add: Cornell Ornamental Res Lab Riverhead NY 11901

BING, DAVID H, b East Cleveland, Ohio, Aug 3, 38; m 61; c 3. IMMUNOLOGY, PROTEIN CHEMISTRY. Educ: Wesleyan Univ, BA, 60; Case Western Reserve Univ, PhD(immunol), 66. Prof Exp: Asst prof microbiol & pub health, Mich State Univ, 68-72, assoc prof microbiol & pub health & human develop, 72-73; SR INVESTR, CTR BLOOD RES, BOSTON, 73- Concurrent Pos: Am Cancer Soc fel immunochem, Univ Calif, Berkeley, 66-68; foreign res worker, Nat Inst Health & Med Res, 70-71; res assoc, Div Immunol, Childrens Hosp Res Ctr, Boston, 72-73; prin res assoc, Dept Biol Chem, Harvard Med Sch, 73- Mem: Am Chem Soc; Am Asn Immunologists. Mailing Add: Ctr for Blood Res 800 Huntington Ave Boston MA 02115

BING, GEORGE FRANKLIN, b Barberton, Ohio, Dec 16, 24; m 51; c 5. THEORETICAL PHYSICS. Educ: Oberlin Col, BA, 48; Case Inst Technol, MS, 51, PhD(theoret physics), 54. Prof Exp: Physicist, Battelle Mem Inst, 51-52; instr, Case Inst Technol, 52-54; mem staff, Lawrence Radiation Lab, Univ Calif, 54-61; asst dir nuclear test detection off, advan res proj agency, US Dept of Defense, DC, 61-63; mem staff, Lawrence Radiation Lab, Univ Calif, 63-67; sci adv to SACEUR, Supreme Hq Allied Powers Europe, 67-70; MEM STAFF, LAWRENCE LIVERMORE LAB, UNIV CALIF, 70- Mem: Am Phys Soc. Res: Low energy nuclear physics; Monte Carlo calculations; nuclear photo-effect and weapons physics; hydrodynamics; controlled thermonuclear process. Mailing Add: 4128 Colgate Way Livermore CA 94550

BING, KURT, b Cologne, Ger, Apr 30, 14; nat US; m 52; c 2. MATHEMATICS. Educ: Hebrew Univ, Jerusalem, MSc, 46; Harvard Univ, PhD(math), 53. Prof Exp: Assoc math, Univ Calif, 52-53; from asst prof to assoc prof, 53-62, PROF MATH, RENSSELAER POLYTECH INST, 62- Mem: Am Math Soc; Math Asn Am; Asn Symbolic Logic. Res: Mathematical logic; foundations of mathematics. Mailing Add: Dept of Math Rensselaer Polytech Inst Troy NY 12181

BING, R H, b Oakwood, Tex, Oct 20, 14; m; c 4. MATHEMATICS. Educ: Southwest Tex State Teachers Col, BS, 35; Univ Tex, MEd, 38, PhD(math), 45. Prof Exp: From instr to asst prof pure math, Univ Tex, 43-47; from asst prof to prof math, Univ Wis, Madison, 47-73; PROF MATH, UNIV TEX, AUSTIN, 73- Concurrent Pos: Vis prof, Univ Va, 49-50 & Univ Tex, 71-72; mem, Inst Advan Study, 57-58, 62-63, 67; chmn conf bd math sci, Nat Acad Sci, mem nat sci bd, 68-75. Honors & Awards: Distinguished Serv Award, Math Asn Am, 74. Mem: Nat Acad Sci; AAAS (vpres, 59); Am Math Soc; Math Asn Am (pres, 63-64). Res: Topology; continuous curves; metric spaces; Euclidean spaces. Mailing Add: Dept of Math Univ of Tex Austin TX 78712

BING, RICHARD JOHN, b Nurenburg, Bavaria, Oct 12, 09; nat US; m 38; c 4. INTERNAL MEDICINE. Educ: Univ Munich, MD, 34; Univ Bern, MD, 35. Prof Exp: Instr physiol, Col Physicians & Surgeons, Columbia Univ, 39-41; instr, NY Univ, 41-43; instr med, Med Sch, Johns Hopkins Univ, 43-44, asst prof surg, 45-47, assoc prof surg & asst prof med, 47-51; prof exp med & clin physiol, Med Col Ala, 51-56; prof med, Washington Univ & dir univ med serv, Vet Admin Hosp, 56-59; prof med & chmn dept, Col Med, Wayne State Univ, 59-70; DIR EXP CARDIOL & INTRAMURAL MED, HUNTINGTON MEM HOSP, PASADENA, 70-; PROF MED, UNIV SOUTHERN CALIF, 70-; RES ASSOC, CALIF INST TECHNOL, 70- Concurrent Pos: Harvey lectr, Royal Soc Med, 54-55. Honors & Awards: Res Accomplishment Award, Am Heart Asn, 74. Mem: Am Soc Clin Invest; Am Physiol Soc; Harvey Soc; Soc Exp Biol & Med; Asn Am Physicians. Res: Physiology and biochemistry of heart muscle; clinical cardiology and internal medicine. Mailing Add: Dept Exp Cardiol & Intramurl Med Huntington Mem Hosp Pasadena CA 91105

BINGEL, AUDREY SUSANNA, b Bronx, NY, Jan 24, 42. PHARMACOLOGY, REPRODUCTIVE ENDOCRINOLOGY. Educ: Hunter Col, AB, 63; Univ Ill, Chicago, PhD, 68. Prof Exp: ASSOC PROF PHARMACOL, COL PHARM, UNIV ILL, 67- Mem: Am Inst Biol Sci; Am Soc Pharmacog; Endocrine Soc; Acad Pharmaceut Sci; Soc Study Reprod. Res: Antifertility screening; ovulation timing. Mailing Add: Dept of Pharmacog & Pharmacol Univ of Ill Col of Pharm Chicago IL 60612

BINGENHEIMER, LEVI EDWIN, JR, b Braggadocio, Mo, Aug 13, 21; m 43; c 4. PHARMACEUTICAL CHEMISTRY, ANALYTICAL CHEMISTRY. Educ: Univ Tenn, BS, 41; Purdue Univ, MS, 48, PhD(pharmaceut chem), 50. Prof Exp: Asst prof pharm & pharmaceut chem, St Louis Col Pharm, 49-52, assoc prof, 52-54; assoc prof pharmaceut chem, Col Pharm, Univ Tenn, 54-61, actg head dept, 59-61; sr res assoc, Dorsey Labs, Wander Co, 61-62, mgr anal res & qual control, 62-67; mgr qual control, Neisler Labs, 67-69; DIR QUAL CONTROL, PHARMACEUT PROD DIV, MALLINCKRODT, INC, 69- Mem: Am Pharmaceut Asn; Am Chem Soc; Am Soc Qual Control. Res: Analytical chemistry of pharmaceutical materials. Mailing Add: Mallinckrodt Inc Res & Dev Lab Mallinckrodt & Second St St Louis MO 63147

BINGGELI, RICHARD LEE, b Sioux Falls, SDak, May 16, 37; m 62; c 3. NEUROPHYSIOLOGY. Educ: Univ Calif, Los Angeles, AB, 59, PhD(anat), 64. Prof Exp: From instr to asst prof, 64-70, ASSOC PROF ANAT, SCH MED, UNIV SOUTHERN CALIF, 70- Concurrent Pos: Consult, Astropower, Inc, Douglas Aircraft Co, Inc, 63-66. Res: Neurophysiological and behavioral studies of the avian visual system. Mailing Add: Dept of Anat Univ of Southern Calif Sch Med Los Angeles CA 90033

BINGHAM, CARLETON DILLE, b Washington, DC, Mar 25, 29; m 58; c 5. ANALYTICAL CHEMISTRY, RADIOCHEMISTRY. Educ: San Diego State Col, AB, 50; Univ Calif, Los Angeles, PhD(phys chem), 59. Prof Exp: Radiol engr div radiation safety, Univ Calif, Los Angeles, 53-54, sr radiol engr, 54-59; sr res chemist atomics int div, NAm Aviation, Inc, 59-61, supvr anal chem, 61-68; proj engr fast breeder reactor chem, NAm Rockwell Corp, 68-71; DIR, NEW BRUNSWICK LAB, US ENERGY RES & DEVELOP ADMIN, 71- Concurrent Pos: Alt US rep, Int Stand Org Tech Comm 85—Nuclear Energy, 76- Mem: AAAS; Am Chem Soc; Am Nuclear Soc; Health Physics Soc; Am Soc Test & Mat. Res: Analytical chemistry of nuclear materials, nondestructive assay, nuclear materials safeguards measurements, reference standards for nuclear materials. Mailing Add: PO Box 150 New Brunswick NJ 08903

BINGHAM, CARROL R, b Fallston, NC, May 22, 38; m 61; c 3. NUCLEAR PHYSICS. Educ: NC State Univ, BSNE, 60, MS, 62; Univ Tenn, PhD(nuclear physics), 65. Prof Exp: Res assoc nuclear spectros, Oak Ridge Nat Lab, 65-66; asst prof, 66-71, ASSOC PROF PHYSICS, UNIV TENN, KNOXVILLE, 71- Concurrent Pos: Consult, Union Carbide Corp, 66- Mem: Am Phys Soc. Res: Nuclear spectroscopy with single-nucleon and two-nucleon transfer reactions; experimental investigation of two-nucleon and three-nucleon systems; nuclear spectroscopy from decay properties of short-lived nuclei. Mailing Add: Dept of Physics Univ of Tenn Knoxville TN 37916

BINGHAM, CHRISTOPHER, b New York, NY, Apr 16, 37; m 67. STATISTICS. Educ: Yale Univ, BA, 58, MA, 60, PhD(math), 64. Prof Exp: Res assoc math, Princeton Univ, 64-66; asst prof statist, Univ Chicago, 66-72; ASSOC PROF APPL STATIST, UNIV MINN, ST PAUL, 72- Mem: Biomet Soc; Soc Indust & Appl Math; Am Statist Asn; Inst Math Statist; Royal Statist Soc. Res: Probability distributions of directions; statistical climatology; time series analysis. Mailing Add: Dept of Appl Statist Univ of Minn St Paul MN 55108

BINGHAM, EDGAR, b Boone, NC, June 23, 21; m 44; c 1. GEOGRAPHY. Educ: Univ Tenn, BA & MS, 48; Ohio State Univ, PhD(geog), 54. Prof Exp: Lectr geog, Univ Tenn, Knoxville, 50-51; res geogr, Res & Develop Labs, US Army, Natick, Mass, 51-53; registr, dir admis & asst dean, Emory & Henry Col, 55-65, from asst prof to assoc prof, 53-71, PROF GEOG, EMORY & HENRY COL, 71- Concurrent Pos: Lectr, Exten Div, Univ Md, 51-52 & Univ Va, 53-60; consult, US Army Res & Develop Labs, Natick, 67-68; lectr geog, Univ Va, 69 & Mary Washington Col, 71; vchmn, Wash County Planning Comn, Va, 72; mem, Mt Rogers Regional Planning Comn, Va, 72. Mem: Asn Am Geogrs; Asn Asian Studies. Res: Appalachian studies; south Asia studies; physical geography and physiography. Mailing Add: Dept of Geog Emory & Henry Col Emory VA 24327

BINGHAM, EDWIN THEODORE, b Ogden, Utah, Nov 4, 36; m 62; c 2. PLANT BREEDING. Educ: Utah State Univ, BS, 59, MS, 61; Cornell Univ, PhD(genetics), 64. Prof Exp: NIH fel genetics, Univ Minn, 64-65; ASST PROG AGRON, UNIV WIS-MADISON, 65- Res: Genetics and breeding of alfalfa; genetics of autotetraploid maize; germ plasma transfer in alfalfa; developmental genetics of the alfalfa leaf; cytogenetics of alfalfa. Mailing Add: Dept of Agron Univ of Wis-Madison Madison WI 53706

BINGHAM, FELTON WELLS, b Greenville, Miss, Aug 18, 35; m 59; c 4. EXPERIMENTAL ATOMIC PHYSICS. Educ: Tulane Univ, BS, 57; Ind Univ, MS, 59, PhD(physics), 62. Prof Exp: Assoc physics, Ind Univ, 57-59, res asst, 59-62; res assoc, Univ Ill, 62-64; MEM TECH STAFF, SANDIA LABS, 64- Mem: Am Phys Soc. Res: Atomic and molecular collision and radiation processes of importance in high-power gas lasers; collisions between heavy ions and atoms of energies above 10 kiloelectron volts. Mailing Add: Sandia Labs Div 5216 Albuquerque NM 87115

BINGHAM, FRANK THOMAS, b Pasadena, Calif, Jan 30, 21; m 46; c 1. SOIL CHEMISTRY. Educ: Univ Calif, BS, 43, PhD(chem), 51. Prof Exp: Prin lab tech soil chem, 46-51, asst chemist soils & plant nutrit, Citrus Exp Sta, Riverside, 51-59, assoc chemist & assoc prof soil sci, 59-69, PROF SOIL SCI & CHEM, DEPT SOIL SCI & AGR ENG, UNIV CALIF, RIVERSIDE, 69- Res: Soil chemistry and fertility as they relate to nutrition; interaction of phosphorous and the micronutrients. Mailing Add: Dept of Soil Sci Univ of Calif Riverside CA 92502

BINGHAM, HARRY H, JR, b Chicago, Ill, May 25, 31; m 60; c 3. PHYSICS. Educ: Princeton Univ, AB, 52; Calif Inst Technol, PhD(physics), 60. Prof Exp: Physicist, Electro-Optical Systs, Inc, Calif, 59-60 & Polytech Sch, Paris, 60-62; Ford Found fel physics, Europ Orgn Nuclear Res, Switz, 62-64; from asst prof to assoc prof, 64-72, PROF PHYSICS, UNIV CALIF, BERKELEY, 72- Mem: Fel Am Phys Soc. Res: Elementary particle physics; bubble chambers; particle beams; computers; particle decays; coherent pi- and k- nucleus interactions; polarized photoproduction, k-pi scattering; interactions of high energy pi, k, p, neutrino; gamma beams in H, D heavy liquid. Mailing Add: Dept of Physics Univ of Calif Berkeley CA 94720

BINGHAM, NELSON ELDRED, b Edinburg, Ohio, June 7, 01; m 34; c 4. BIOLOGY. Educ: Ohio State Univ, BS, 23; Columbia Univ, MA, 30, PhD(sci educ), 38. Prof Exp: Prin & teacher pub sch, Ohio, 23-29; instr sci, NJ State Teachers Col, 29-32; sci teacher, Lincoln Sch, Teachers Col, Columbia Univ, 32-44; dir lab sch & assoc prof, Temple Univ, 44-45; assoc prof sci educ, Northwestern Univ, 45-50; prof, 50-71, EMER PROF SCI EDUC, UNIV FLA, 71- Concurrent Pos: Consult, Coronet Instr Films, 52-; Sci Fair Work Confs mem, La, SC, 57 & PR, 58-; mem, Teachers Col, Columbia Univ Contract Team, India, 62-64; ed, Sci Educ, 69- Mem: AAAS; Nat Asn Biol Teachers; Nat Asn Res Sci Teaching (pres, 47); Nat Sci Teachers Asn. Res: Preparation of science teachers; teaching sciences to educationally deprived youth. Mailing Add: 1718 NW 10th Ave Gainesville FL 32605

BINGHAM, PETER, electronics, optics, see 12th edition

BINGHAM, RICHARD CHARLES, b Middlebury, Vt, May 23, 44; m 72. ORGANIC CHEMISTRY. Educ: Univ Vt, BS, 66; Princeton Univ, PhD(chem), 70. Prof Exp: Fel chem, Univ Tex, Austin, 71-72; RES CHEMIST, E I DU PONT DE NEMOURS & CO, INC, 73- Mem: Am Chem Soc. Res: Synthesis of novel dyes and pigments; electronic spectra of organic molecules; interpretation of the structures of molecules; theoretical organic chemistry. Mailing Add: Pigments Dept Exp Sta E I du Pont de Nemours & Co Wilmington DE 19898

BINGHAM, ROBERT GRAHAM, physics, astronomy, deceased

BINGHAM, ROBERT J, b Blackfoot, Idaho, Feb 17, 32; m 59; c 5. BIOCHEMISTRY, FOOD CHEMISTRY. Educ: Utah State Univ, BS, 59; Univ Wis, MS, 62, PhD(biochem, food sci), 64. Prof Exp: Health physicist, Phillips Petrol Co, AEC, Idaho, 59; res asst food sci, Univ Wis, 59-63; asst prof food chem, NC State Univ, 63-68; head agr-prod res, Beatrice Foods Co, 68-70; vpres, Nutrico, 70-74; VPRES, PROMARKCO, 74- Concurrent Pos: Dir, Banfield of Tulsa, 73- & Prime Western, Inc, 75- Mem: Fel Am Inst Chem; Am Chem Soc; Inst Food Technol; Am Dairy Sci Asn. Res: Nutritional properties and means of protecting or enhancing nutritional value during processing and storage; utilization of by-products of food industry for high quality animal feeds. Mailing Add: 5505 E 64th St Tulsa OK 74136

BINGHAM, ROBERT LODEWIJK, b Amsterdam, Netherlands, Sept 13, 30; US citizen; m 60; c 3. NUCLEAR PHYSICS, PLASMA PHYSICS. Educ: Williams Col, BA, 52; Harvard Univ, MA, 53; Columbia Univ, PhD(physics), 59. Prof Exp: Res asst meson physics, Nevis Cyclotron Lab, Columbia Univ, 55-59; res assoc plasma physics, Princeton Univ, 59-63; res staff mem, Gen Elec Co, NY, 63-68, tech counsr large lamp dept, Ohio, 68-69; Sloan exec fel grad sch bus, Stanford Univ, 69-70; gen mgr, tektran, Air Prods & Chem Inc, 70-71; consult, R L Bingham, 72-73; COORDR PLANS DIV CONTROLLED THERMONUCLEAR RES, US ENERGY RES & DEVELOP ADMIN, 74- Mem: Am Phys Soc. Res: Gaseous discharges; controlled thermonuclear fusion; particle detectors and analyzers; energy conversion. Mailing Add: 6909 East Ave Chevy Chase MD 20015

BINGHAM, SAMUEL WAYNE, b Fallston, NC, Apr 6, 29; m 55; c 3. PLANT PHYSIOLOGY, WEED SCIENCE. Educ: NC State Col, BS, 54, MS, 56; La State Univ, PhD(weed sci), 60. Prof Exp: Res agronomist, Crops Div, USDA, 56-58; asst specialist, La State Univ, 60-61; assoc prof, 61-72, PROF PLANT PHYSIOL, VA POLYTECH INST & STATE UNIV, 72- Mem: Weed Sci Soc Am; Am Soc Hort Sci; Int Turfgrass Soc. Res: Weed control in turf, penetration, translocation and fate of herbicides in plants; effects of herbicides on physiological process in plants; fate of herbicides in surface water. Mailing Add: Dept of Plant Physiol Va Polytech Inst & State Univ Blacksburg VA 24061

BINGLER, EDWARD CHARLES, b Philadelphia, Pa, Nov 4, 35; m 54; c 3. GEOLOGY. Educ: Lehigh Univ, BA, 59; NMex Inst Mining & Technol, MS, 61; Univ Tex, PhD, 64. Prof Exp: Geologist, NMex Bur Mines & Mineral Resources, 64-67; asst prof geol & geol eng, SDak Sch Mines & Technol, 67-69; ASSOC MINING GEOLOGIST, NEV BUR MINES, UNIV NEV, RENO, 69- Mem: Fel Geol Soc Am; Soc Econ Geol. Res: Structural geology and petrology; field geology. Mailing Add: Nev Bur Mines Univ of Nev Reno NV 89507

BINHAMMER, ROBERT T, b Watertown, Wis, Apr 28, 29; m 52; c 3. ANATOMY. Educ: Kalamazoo Col, BA, 51; Univ Tex, MA, 53, PhD(anat), 55. Prof Exp: From instr to assoc prof, 55-70, asst dean, 67-75, PROF ANAT, COL MED, UNIV CINCINNATI, 70-, ASSOC DEAN, 75- Mem: Assoc mem Radiation Res Soc; Am Asn Anatomists. Res: Radiation biology; parabiosis in physiological studies; pituitary physiology; cytology. Mailing Add: Dept of Anat Univ of Cincinnati Col of Med Cincinnati OH 45219

BINKERD, EVAN FRANCIS, b Lynch, Nebr, Mar 30, 19; m 42. FOOD SCIENCE. Educ: Iowa State Univ, BS, 42. Prof Exp: Prod chemist, Nat Aniline Div, Allied Chem & Dye Corp, NY, 42; res chemist, 42-43 & 46-48, head fatty acid derivatives res sect, 48-50, admin asst to vpres res div, 50-56, assoc tech dir food res div, 56-59, asst tech dir, 59-63, lab mgr, 63-64, mgr food res div, 64-67, dir res, 67-71, V PRES RES, ARMOUR & CO, 71- Concurrent Pos: Mem, Nat Livestock & Meat Bd & Food Sci Adv Comt, Col Agr, Univ Ill; partic & mem panel, White House Conf Food, Nutrit & Health, 69; adv, US Govt Int Codex Comn, Geneva, 71 & Rome, 72 & 74. Honors & Awards: Distinguished Achievement Award, Iowa State Univ, 74. Mem: AAAS; Am Chem Soc; Am Meat Sci Asn; Indust Res Inst; fel Inst Food Technologists. Mailing Add: Armour & Co 15101 N Scottsdale Rd Scottsdale AZ 85260

BINKLEY, FRANCIS, b Scotland, Ill, July 11, 15; m 40; c 4. BIOCHEMISTRY, IMMUNOCHEMISTRY. Educ: Univ Ill, BS, 38; Univ Mich, MS, 39; Cornell Univ, PhD(biochem), 42. Prof Exp: Asst, Univ Mich, 38-39, Cornell Univ, 39-42 & Rockefeller Inst, 42-44; assoc prof, Sch Med, Univ Utah, 46-51; PROF BIOCHEM, EMORY UNIV, 51- Concurrent Pos: Consult, Vet Admin, 48- Mem: AAAS; Am Chem Soc; Am Soc Biol Chem; Am Soc Exp Biol & Med; Harvey Soc. Res: Metabolism; enzyme systems; renal enzymes and renal function; metabolism of nervous tissue. Mailing Add: Dept of Biochem Emory Univ Atlanta GA 30322

BINKLEY, ROGER WENDELL, b Newark, NJ, Mar 9, 41; m 62; c 2. ORGANIC CHEMISTRY, PHOTOCHEMISTRY. Educ: Drew Univ, BA, 62; Univ Wis, PhD(org chem), 66. Prof Exp: From asst prof to assoc prof, 66-74, PROF CHEM, CLEVELAND STATE UNIV, 74- Concurrent Pos: Res collab grant, 68-69. Mem: Am Chem Soc. Res: Photochemistry of carbohydrates and of saturated nitrogen systems. Mailing Add: Dept of Chem Cleveland State Univ Cleveland OH 44115

BINKLEY, STEPHEN BENNETT, b Scotland, Ill, Jan 24, 10; m 35; c 1. BIOCHEMISTRY. Educ: Eureka Col, BS, 34; Univ Nebr, MS, 36, PhD(org chem), 38.Prof Exp: Mem sr res staff, Parke, Davis & Co, 38-39; sr instr biochem, Sch Med, St Louis Univ, 40-42; mem sr res staff, Parke, Davis & Co, 42-45; asst dir res, Bristol Labs, Inc, 45-49; prof biol chem, Col Med, Univ Ill, 49-72, dean grad col, 69-72; HEAD MATH SCI DIV, EUREKA COL, 72- Mem: AAAS; Am Chem Soc; Am Soc Biol Chemists; NY Acad Sci. Res: Vitamins; hormones; antibiotics; medicinal chemistry; vitamin K; barbiturates; penicillin; vitamin B; antihistaminics; analgesics. Mailing Add: RR 2 Eureka IL 61530

BINKLEY, SUE ANN, b Dayton, Ohio, May 19, 44; c 1. BIOLOGICAL RHYTHMS, ENDOCRINOLOGY. Educ: Univ Colo, BA, 66; Univ Tex, Austin, PhD(zool), 71. Prof Exp: NIH fel physiol, Nat Inst Child Health & Human Develop, 71-72; asst prof biol, 73-76, ASSOC PROF BIOL, TEMPLE UNIV, 76- Mem: Endocrine Soc; Am Soc Zoologists; Inst Soc Chronobiol. Res: Biological basis of circadian rhythms in vertebrates and in the function of the pineal gland. Mailing Add: Dept of Biol Temple Univ Philadelphia PA 19122

BINKLEY, WENDELL WILFRED, organic chemistry, see 12th edition

BINN, LEONARD NORMAN, b Lithuania, Nov 6, 27; nat US; m 53; c 3. MEDICAL BACTERIOLOGY. Educ: NY Univ, BA, 49; Univ Mich, MS, 51, PhD(bact), 55; Am Bd Microbiologists, dipl. Prof Exp: MED BACTERIOLOGIST, VIROL, WALTER REED ARMY INST RES, 55- Mem: Am Soc Microbiol; US Animal Health Asn; Soc Trop Med & Hyg. Res: Bacterial viruses; stable smallpox vaccine; immunology of adenoviruses; immunity in arthropod-borne virus diseases; laboratory animal virus diseases and canine viruses. Mailing Add: Div Vet Med Walter Reed Army Inst Res Washington DC 20012

BINNELL, JAMES MONROE, b Chicago, Ill, Apr 5, 22; m 49. ORGANIC CHEMISTRY, CHEMICAL ENGINEERING. Educ: Northwestern Univ, BS, 44, BChE, 48, MS, 49; Univ Fla, PhD(chem), 55. Prof Exp: Asst dir res, Snow Crop, Fla, 51-55; tech dir, Orange Crystals, Inc, 55-57; DIR RES, TROPICANA PROD, INC, 57- Concurrent Pos: Vpres, Prof Eng Serv, Inc, 55- Mem: AAAS; Am Chem Soc; Am Inst Chem Eng; fel Am Inst Chemists. Res: Citrus products and by-products. Mailing Add: Tropicana Prod Inc PO Box 338 Bradenton FL 33505

BINNENDIJK, LEENDERT, b Gouda, Holland, May 13, 13; m 44; c 2. ASTRONOMY. Educ: Univ Leiden, MS, 41, Dr, 47. Prof Exp: Asst astron, Univ Leiden Observ, 41-46; teacher high sch, Holland, 43-46; res assoc, Sproul Observ, Swarthmore Col, 47-50; dir, Goodsell Observ, Northfield, Minn, 50-53; assoc prof, 53-60, PROF ASTRON, UNIV PA, 60- Mem: Am Astron Soc; Int Astron Union. Res: Photographic and photoelectric photometry; positional astronomy; astrophysics; stellar statistics. Mailing Add: Dept of Astron Univ of Pa Philadelphia PA 19104

BINNING, LARRY KEITH, b Fond du Lac, Wis, Aug 29, 42; m 64; c 1. HORTICULTURE. Educ: Univ Wis, BS, 65; Mich State Univ, MS, 67, PhD(herbicide physiol, crop sci), 69. Prof Exp: Asst prof, 69-73, ASSOC PROF HORT, UNIV WIS-MADISON, 73- Mem: Am Soc Agron; Weed Sci Soc Am. Res: Herbicide use and physiology. Mailing Add: Dept of Hort Univ of Wis Madison WI 53706

BINNS, WALTER ROBERT, b Williamsburg, Kans, Nov 7, 40; m 62; c 1. LASERS, COSMIC RAY PHYSICS. Educ: Univ Ottawa, BS, 62; Colo State Univ, MS, 66, PhD(physics), 69. Prof Exp: SCIENTIST, RES LABS, McDONNELL DOUGLAS CORP, 69- Mem: Am Phys Soc. Res: Research on pulsed visible and infrared laser systems and research to determine the abundances of very heavy cosmic rays in the primary cosmic radiation. Mailing Add: Res Labs McDonnell Douglas Corp St Louis MO 63166

BINNS, WAYNE, biological science, chemistry, see 12th edition

BINTZ, GARY LUTHER, b Manchester, Iowa, Feb 25, 41. COMPARATIVE PHYSIOLOGY, ENVIRONMENTAL PHYSIOLOGY. Educ: Cornell Col, BA, 63; Univ NMex, MS, 66, PhD(zool), 68. Prof Exp: Asst prof physiol, 68-74, ASSOC PROF BIOL, EASTERN MONT STATE COL, 74- Res: Water metabolism, including the effects of negative water balance on whole body and tissue levels; studies on several rodents, especially ground squirrels capable of hibernation. Mailing Add: Sci Bldg Eastern Mont State Col Billings MT 59101

BINZ, CARL MICHAEL, b Elgin, Ill, June 29, 47; m 75. ANALYTICAL CHEMISTRY. Educ: Loras Col, BS, 69; Purdue Univ, MS, 71, PhD(chem), 74. Prof Exp: Res assoc, Purdue Univ, 74-75; INSTR ANAL CHEM, LORAS COL, 75- Res: Determination of trace materials in air and water and their accumulated effects. Mailing Add: Dept of Chem Box 933 Loras Col Dubuque IA 52001

BIOLSI, LOUIS, JR, b Port Jefferson, NY, Aug 21, 40; m 62; c 2. THEORETICAL CHEMISTRY. Educ: State Univ NY Albany, BS, 61, MS, 63; Rensselaer Polytech Inst, PhD(theoret chem), 67. Prof Exp: Proj assoc, theoret chem inst, Univ Wis, 66-67; res assoc theoret chem, Brown Univ, 68; asst prof, 68-71, ASSOC PROF CHEM, UNIV MO-ROLLA, 71- Mem: Am Phys Soc; Am Chem Soc. Res: Scattering and kinetic theory of polyatomic molecules and the specification on nonequilibrium states. Mailing Add: Dept of Chem Univ of Mo-Rolla Rolla MO 65401

BIONDI, FRANK JOSEPH, b Bethlehem, Pa, Sept 22, 14; m 43; c 3. CHEMISTRY, ELECTRONICS. Educ: Lehigh Univ, BS, 36; Columbia Univ, MS, 40. Prof Exp: Chemist chem res & develop, Bell Tel Labs, Inc, 36-62, DIR, ELECTRON DEVICE TECH LAB, BELL LABS, INC, MURRAY HILL, NJ, 62- Concurrent Pos: Consult Mat Adv Bd, Nat Acad, 57-59. Mem: Am Chem Soc; Am Soc Testing & Mat; Electrochem Soc. Res: Wood preservation; plastics; finishes; powder metallurgy; metals heat treating; electron tube and semiconductor device materials and processing; electrochemical batteries. Mailing Add: Bell Labs Inc Murray Hill NJ 07971

BIONDI, MANFRED ANTHONY, b Carlstadt, NJ, Mar 5, 24; m 52; c 2. ATOMIC PHYSICS, AERONOMY. Educ: Mass Inst Technol, SB, 44, PhD(physics), 49. Prof Exp: Res assoc physics, Mass Inst Technol, 46-49; from res physicist to adv physicist, Westinghouse Res Lab, 49-57, mgr physics dept, 57-60; PROF PHYSICS, UNIV PITTSBURGH, 60- Concurrent Pos: Mem adv panel physics, NSF; vis fel, Joint Inst Lab Astrophys, 72-73. Mem: Am Phys Soc; Am Geophys Union. Res: Interactions and reactions involving electrons, ions and excited atoms; plasma physics; electromagnetic properties of metals at liquid helium temperatures; airglow studies. Mailing Add: Dept of Physics Univ of Pittsburgh Pittsburgh PA 15260

BIONDO, FRANK X, b Brooklyn, NY, Mar 25, 27. MICROBIOLOGY, MYCOLOGY. Educ: Temple Univ, BA, 48; Athenaeum Ohio, MS, 51, PhD(exp med), 55. Prof Exp: Res dir bovine mastitis, Agr Res Inst, NJ, 53-54; instr microbiol, NY Med Col, 56-60; asst prof microbiol, Long Island Univ, 60-68 & C W Post Col, 68-72, ASSOC PROF HEALTH SCI, C W POST COL, LONG ISLAND UNIV, 72- Concurrent Pos: Part time med technologist, Ital Hosp, New York, 44-60; La State Univ fel parasitol & trop med, PR Dominican Repub & Haiti, 58. Mem: AAAS; Am Soc Microbiol; Am Soc Trop Med & Hyg. Res: Medical mycology; tranplantable tumors; tissue extracts and their effect on Micrococcus pyogenes var aureus infections in mice and transplantable tumors; biochemistry of dematiaceous fungi and Candida species. Mailing Add: Dept of Biol C W Post Col Long Island Univ Greenvale NY 11548

BIORDI, JOAN CONCETTA, b Ellwood City, Pa, June 10, 40. PHYSICAL CHEMISTRY. Educ: Chatham Col, BS, 62; Carnegie-Mellon Univ, MS, 65, PhD, 66. Prof Exp: NATO res fel spectros, Cambridge Univ, 66-67; fel gas phase kinetics, Radiation Res Lab, Carnegie-Mellon Univ, 67-69; PHYS CHEMIST, US BUR MINES, 69- Mem: Combustion Inst. Res: Photochemically induced hot atom reactions; discharge flow systems for studying reactions of atoms; radical reactions using the sector method; molecular beam-mass spectrometry applied to flame microstructure and flame inhibition. Mailing Add: Safety Res Ctr USBM 4800 Forbes Ave Pittsburgh PA 15213

BIRCH, ALBERT FRANCIS, b Washington, DC, Aug 22, 03; m 33; c 3. GEOPHYSICS. Educ: Harvard Univ, BS, 24, MA, 29, PhD(physics), 32. Hon Degrees: DSc, Univ Chicago, 70. Prof Exp: Engr, NY Tel Co, 24-26; instr & tutor physics, 30-32, res assoc geophysics, 32-37, from asst prof to prof, 37-74, EMER PROF GEOL, HARVARD UNIV, 74- Concurrent Pos: Staff mem radiation lab, Mass Inst Technol, 42. Honors & Awards: Day Medal, Geol Soc Am, 50; William Bowie Medal, Am Geophys Union, 60; Legion of Merit Award, Am Acad Arts & Sci; Nat Medal of Sci, 67; Vetlesen Prize, 69; Penrose Medal, Geol Soc Am, 69; Gold Medal, Royal Astron Soc Gt Brit, 73. Mem: Nat Acad Sci; fel Am Phys Soc; fel Geol Soc Am (pres, 64); Seismol Soc Am; Am Geophys Union. Res: Properties of materials at high pressures and high temperature; geothermal studies; elasticity. Mailing Add: Hoffman Lab Harvard Univ Cambridge MA 02138

BIRCH, HERBERT G, medicine, psychology, deceased

BIRCH, HOMER JAMES, b Struthers, Ohio, June 1, 18; m 44; c 2. ANALYTICAL CHEMISTRY. Educ: Youngstown Col, BS, 41; Univ Ill, MS, 49, PhD(anal chem), 52. Prof Exp: Res chemist, E I du Pont de Nemours & Co, 52-56; proj group leader, Callery Chem Co, 56-59; ASSOC RES CONSULT, US STEEL CORP RES CTR, 59- Mem: AAAS; Am Chem Soc; Electron Micros Soc Am; Am Soc Metals. Res: Electron microscopy and diffraction; x-ray diffraction; physical methods of analysis. Mailing Add: US Steel Corp Res Ctr MS 15 125 Jamsen Ln Monroeville PA 15146

BIRCH, MARTIN CHRISTOPHER, b Crewe, Eng, July 14, 44; m 70. ENTOMOLOGY. Educ: Oxford Univ, BA, 66, DPhil(entom), 69. Prof Exp: Res asst org chem, Dyson Perrins Lab, Oxford Univ, 69-70; res asst entom, Univ Calif, Berkeley, 70-73; ASST PROF ENTOM, UNIV CALIF, DAVIS, 73- Concurrent Pos: Royal Entom Soc fel, 65- Mem: Royal Entom Soc; Entom Soc Am; Sigma Xi. Res: Insect behavior and physiology, particularly chemical communication; behavioral mechanisms in pheromone communication of scolytidae and lepidoptera and applications to forest and agricultural management. Mailing Add: Dept of Entom Univ of Calif Davis CA 95616

BIRCH, ROBERT LEE, b Sharon, Pa, Nov 8, 14; m 56; c 2. INVERTEBRATE ZOOLOGY. Educ: Westminster Col, BS, 38; Pa State Col, MS, 48. Prof Exp: Instr zool, 48-65, ASST PROF BIOL, W VA UNIV, 65-, DIR TERRA ALTA BIOL STA, 64- Concurrent Pos: Partic, Orgn Inland Biol Field Sta. Mem: AAAS; Am Soc Parasitol; Am Ornith Union; Wilson Ornith Soc; Am Inst Biol Sci. Mailing Add: Dept of Biol W Va Univ Morgantown WV 26506

BIRCHAK, JAMES ROBERT, b Latrobe, Pa, Mar 20, 39; m 63; c 2. EXPERIMENTAL SOLID STATE PHYSICS. Educ: Carnegie Inst Technol, BS, 61; Rice Univ, MA, 64, PhD(physics), 66; Wayne State Univ, MBA, 71. Prof Exp: Assoc sr res physicist, Gen Motors Res Lab, 66-71; SR RES PHYSICIST, SOUTHWEST RES INST, 71- Mem: AAAS; Am Phys Soc; Inst Elec & Electronics Engrs; Sigma Xi; Am Soc Nondestructive Testing. Res: Irreversible magnetization of superconductors; magnet sources of internal friction; ultrasonics; electric and magnetic properties of metals. Mailing Add: Instrumentation Res Dept Southwest Res Inst San Antonio TX 78284

BIRCHARD, RALPH EDWIN, b Oelwein, Iowa, Apr 17, 14; m 40; c 4. URBAN GEOGRAPHY, GEOGRAPHY OF AFRICA, EUROPE. Educ: Univ Northern Iowa, BA, 37; Univ Ill, MS, 38; Univ Iowa, PhD(geog), 54. Prof Exp: From instr to assoc prof geog, Okla State Univ, 46-68; ASSOC PROF GEOG, E CAROLINA UNIV, 68- Concurrent Pos: Prof geog, Haile Selassie Univ, 65-67. Mem: Asn Am Geog. Res: The main business districts of Addis Ababa. Mailing Add: Dept of Geog Box 2723 E Carolina Univ Greenville NC 27834

BIRCHENALL, CHARLES ERNEST, b Coatesville, Pa, Feb 19, 22; m 72; c 3. CHEMISTRY. Educ: Temple Univ, AB, 43; Princeton Univ, MA, 45, PhD(chem), 46. Prof Exp: Asst, Manhattan Proj, Princeton Univ, 43-46; mem staff, Metall Res Lab, Carnegie Inst Technol, 46-52, asst prof metall eng, 51-52; assoc prof chem, Princeton Univ, 52-60; dean sch grad studies, 64-67, DISTINGUISHED PROF METALL, UNIV DEL, 60- Concurrent Pos: Sci fac fel, NSF, Imp Col, Univ London, 58-59; metallurgist, Lawrence Radiation Lab, Univ Calif, Berkeley, 67-68; mem, Panel Magnetohydrodynamics, Off Sci & Technol, 68-69. Mem: Fel AAAS; fel Am Soc Metals; Am Inst Mining, Metall & Petrol Engrs. Res: Diffusion in metals, oxides and sulfides; corrosion. Mailing Add: Dept of Chem Eng Univ of Del Newark DE 19711

BIRCHFIELD, GENE EDWARD, b Bartlesville, Okla, Apr 14, 28; m 59; c 3. ATMOSPHERIC PHYSICS, GEOPHYSICS. Educ: Univ Chicago, AB, 52, MS, 55, PhD(geophys sci), 62. Prof Exp: Asst meteorol, Univ Chicago, 56-62; NSF fel geophys sci, Univ Stockholm & Mass Inst Technol, 62-63; assoc prof, 63-74, PROF ENG & GEOL SCI, NORTHWESTERN UNIV, 74- Concurrent Pos: Vis scientist, Inst Oceanog Sci, Wormley, Surrey, Eng, 72-73. Mem: Am Meteorol Soc; Am Geophys Union. Res: Geophysical fluid dynamics; physical limnology. Mailing Add: Dept of Eng Sci Technol Inst Northwestern Univ Evanston IL 60201

BIRCHFIELD, WRAY, b Ware Shoals, SC, July 2, 20; m 45; c 3. PLANT PATHOLOGY. Educ: Univ Fla, BS, 49, MS, 51; La Sa State Univ, PhD(plant path), 54. Prof Exp: Lab asst plant path, Univ Fla, 49-50; asst, La State Univ, 51-54; phytonematologist, Citrus Exp Sta, Fla State Plant Bd, 54-59; res plant pathologist, 59-72, NEMATOLOGIST, USDA, LA STATE UNIV, 72- Mem: Am Phytopath Soc. Res: Diseases of plants caused by nematodes; host-parasite relations of nematode diseases; soil microbiology and fumigation studies of soils; fungal and nematode complexes in relation to root diseases of plants; taxonomy of nematodes, fungi and plants. Mailing Add: Dept of Bot & Plant Path La State Univ Baton Rouge LA 70803

BIRD, CHARLES DURHAM, b Norman, Okla, July 7, 32; Can citizen; m 57; c 3. BOTANY. Educ: Univ Man, BS, 56; Okla State Univ, MS, 58, PhD(bot), 60. Prof Exp: Nat Res Coun Can fel, 60-62; asst prof bot, Univ Alta, 62-65; assoc prof, 67-74, admin off dept, 66-69, PROF BOT, UNIV CALGARY, 74-, ACTG CUR HERBARIUM, 62- Concurrent Pos: Assoc ed bot, Can Field Naturalist, 74- Mem: Am Bryol & Lichenological Soc; Am Soc Mammal; Am Soc Plant Taxon; Brit Bryol Soc; Int Asn Lichenology. Res: Floristics; ecology and taxonomy of plants, especially of lichens and bryophytes. Mailing Add: Dept of Biol Univ of Calgary Calgary AB Can

BIRD, CHARLES EDWARD, b Kingston, Ont, Oct 19, 31; m 56; c 2. ENDOCRINOLOGY, METABOLISM. Educ: Queen's Univ, Ont, MD & CM, 56; McGill Univ, PhD(exp med), 67. Prof Exp: Clin asst med, Toronto Gen Hosp, Can, 61-63; clin asst, Royal Victoria Hosp, Can, 63-65; from asst prof to assoc prof, 65-75, PROF MED, QUEEN'S UNIV, ONT, 75- Mem: Can Med Asn; Can Soc Clin Invest; Endocrine Soc. Res: Biological disposition of progesterone; progesterone metabolism in the human fetus; androgen metabolism in carcinoma of the prostate; androgen and estrogen metabolism in carcinoma of the breast. Mailing Add: Dept of Med Queen's Univ Kingston ON Can

BIRD, CHARLES NORMAN, b Rolla, Mo, Dec 27, 29. ORGANIC CHEMISTRY. Educ: Col St Thomas, BS, 51; Univ Md, PhD(org chem), 57. Prof Exp: Coordr polyester develop, 56-71, supvr graphic arts prod, 71-72, SUPVR PHOTOGRAPHIC BASES, MINN MINING & MFG CO, 3M CTR, ST PAUL, 72- Mem: Am Chem Soc. Res: Pyrolysis of esters; amides and lactones; organic synthesis; adhesives; rubber chemicals; polyester films; photographic films and chemicals. Mailing Add: Afton MN 55001

BIRD, DONAL WILLIAM, poultry husbandry, see 12th edition

BIRD, EMERSON WHEAT, b Philadelphia, Pa, Aug 20, 01. CLINICAL CHEMISTRY. Educ: Pa State Col, BS, 23; Iowa State Col, PhD(chem), 29. Prof Exp: From asst chem to instr chem, 23-28, asst prof dairy indust, 28-37, from assoc prof to prof biochem, dairy & food indust, 37-72, CHEMIST CLIN LABS, IOWA STATE UNIV, 74- Concurrent Pos: Researcher food & container inst, US Army Qm Corps Proj, 46-49; mem adv bd Qm res & develop comt dairy, oil & fat prods, Nat Acad & Nat Res Coun, 57- Honors & Awards: Borden Award, 52. Mem: Am Chem Soc; Am Dairy Sci Asn; Am Inst Food Technol. Res: Application of chemistry to manufacture of dairy products; butter composition control device; investigations concerning clinical chemical methods. Mailing Add: 2138 Hughes Ave Ames IA 50010

BIRD, FLOYD WESTON, analytical chemistry, see 12th edition

BIRD, FRANCIS HOWE, b Battle Creek, Mich, Apr 22, 13; m 37; c 3. NUTRITION. Educ: Univ Mich, BS, 36; Univ Calif, Berkeley, PhD, 48. Prof Exp: Sr lab technician, Univ Calif, 41-47; feed researcher, Eastern States Farmers Exchange, Inc, 47-61; MEM FAC ANIMAL & VET SCI, UNIV MAINE, ORONO, 61- Concurrent Pos: Vis scientist, Poultry Res Ctr, Agr Res Coun, Scotland, 70; mem exec comt, Animal Nutrit Res Coun, 50-53, 62-64, secy-treas, 52-53. Mem: AAAS; Poultry Sci Asn; Am Inst Nutrit; Brit Nutrit Soc. Res: Poultry nutrition and physiology. Mailing Add: Hitchner Hall Univ of Maine Orono ME 04473

BIRD, GEORGE RICHMOND, b Bismark, NDak, Jan 25, 25; m 48; c 3. PHOTOGRAPHIC CHEMISTRY. Educ: Harvard Univ, AB, 49, AM, 52, PhD(phys chem), 53. Prof Exp: Nat Res Coun fel, Radiation Lab, Columbia Univ, 52-53; asst prof chem, Rice Inst, 53-58; res chemist, Polaroid Corp, 58-61, mgr phys chem lab, 61-69; dir sch, 71-74, PROF CHEM, SCH CHEM, RUTGERS UNIV, 69- Concurrent Pos: Guggenheim fel, Photog Inst, Fed Tech Sch, Zurich, 74-75. Mem, Am Phys Soc; Am Chem Soc; fel Optical Soc Am; fel Soc Photog Sci & Eng. Res: Spectroscopy; basic photographic science. Mailing Add: Sch of Chem Rutgers Univ New Brunswick NJ 08903

BIRD, GEORGE W, b Newton, Mass, June 16, 39; m 67. NEMATOLOGY, PLANT PATHOLOGY. Educ: Rutgers Univ, BS, 61, MS, 63; Cornell Univ, PhD(plant path), 67. Prof Exp: Res scientist, Res Br, Can Dept Agr, 66-68; from asst prof to assoc prof, Dept Plant Path, Univ Ga, 68-73; ASSOC PROF ENTOM, MICH STATE UNIV, 73- Mem: Am Phytopath Soc; Soc Nematol (secy, 71-74). Res: Phytonematology; biology nematodes. Mailing Add: Dept of Entom Mich State Univ East Lansing MI 48824

BIRD, HARVEY HAROLD, b Montclair, NJ, Aug 25, 34; m 57; c 2. PLASMA PHYSICS, GEOPHYSICS. Educ: Johns Hopkins Univ, BA, 56; Calif Inst Technol, MS, 59; Stevens Inst Technol, PhD(physics), 69. Prof Exp: Mem tech staff, Space Technol Labs, Thompson-Ramo-Woolridge, Calif, 59-62; from instr to asst prof, 62-70, ASSOC PROF PHYSICS, FARLEIGH DICKINSON UNIV, 70- Concurrent Pos: US del, Int Asn Geomagnetism & Aeronomy, Madrid, Spain, 69. Mem: Am Phys Soc; Am Geophys Union. Res: Physics of the earth's magnetosphere; instability theory in plasmas. Mailing Add: Dept of Physics Fairleigh Dickinson Univ Rutherford NJ 07070

BIRD, HERBERT RODERICK, b Madison, Wis, Jan 7, 12; m 37; c 4. ANIMAL NUTRITION. Educ: Univ Wis, BS, 33, MS, 35, PhD(biochem), 38. Prof Exp: Asst agr chem, Univ Wis, 33-38; assoc prof poultry nutrit, Univ Md, 38-44; biochemist, Bur Animal Indust, USDA, 44-48, in-chg poultry invests, 48-53; chmn dept poultry sci, 53-64 & 66-71, PROF POULTRY SCI, UNIV WIS-MADISON, 53- Concurrent Pos: Prof & chief party, Wis Contract, US AID, Univ Rio Grande do Sul, Brazil, 64-66; prof & coodr, Midwest Univs Consort Int Activities-AID-Indonesian Agr Higher Educ Proj, 71-73. Honors & Awards: Am Feed Mfrs Award, 48; Newman Int Award, 49; USDA Award, 49; Borden Award, 53. Mem: AAAS; Am Chem Soc; Poultry Sci Asn (pres, 57-58); World Poultry Sci Asn (vpres, 62-66); Am Inst Nutrit. Res: Poultry nutrition; vitamins; antibiotics; growth stimulants; amino acids. Mailing Add: Dept of Poultry Sci Univ of Wis-Madison Madison WI 53706

BIRD, JOHN BRIAN, b Birmingham, Eng, Aug 28, 23; m 47; c 3. GEOGRAPHY. Educ: Cambridge Univ, BA, 47, MA, 49. Prof Exp: Lectr geog, Univ Toronto, 47-50; from asst prof to assoc prof, 50-62, chmn dept, 67-74, PROF GEOG, McGILL UNIV, 62- Concurrent Pos: Leader, Arctic field parties, Can Dept Mines & Tech Survs, 48, 50, 52, 54 & 56; leader arctic expeds, McGill Univ, 58; mem, Can Nat Comt, Int Geog Union, 60-68, chmn, 60-64; mem, Assoc Comt Quaternary Res, Nat Res Coun Can, 66-; chmn, Organizing Comt, Int Geog Cong, 68- Mem: Am Geog Soc; Asn Am Geogrs; fel Arctic Inst NAm; Can Asn Geogrs (pres, 58-59); Brit

Glaciol Soc. Res: Physiography; historical geography; arctic terrain analysis. Mailing Add: Dept of Geog McGill Univ Montreal PQ Can

BIRD, JOHN MALCOMB, b Newark, NJ, Dec 27, 31; m 57; c 2. GEOLOGY. Educ: Union Col, NY, BS, 55; Rensselaer Polytech Inst, MS, 59, PhD(geol), 62. Prof Exp: From instr to prof geol, State Univ NY Albany, 61-72; chmn dept geol sci, 69-72; PROF GEOL, COL ENG, CORNELL UNIV, 72- Concurrent Pos: Res grants, Geol Soc Am, 62-, NY State Geol Surv, 63-, NSF, 64, 68, 72 & 73, Nat Acad Sci Day Fund, 69 & Petrol Res Found, 75; mem, NY State Mus & Sci Serv, 63-67; Nat Acad Sci exchange vis scientist, Polish Acad Sci, 67; res assoc, Lamont-Doherty Geol Observ, Columbia Univ, 70, sr res assoc, 70-72; distunguished vis scientist, Am Geol Inst, 71-; assoc ed, J Geophys Res, 71-73; chmn, Appalachian Working Group, US Geodynamics Comt, 71-73. Mem: AAAS; fel Geol Soc Am; Am Geophys Union; fel Can Geol Soc; Sigma Xi. Res: Neotectonics of nuclear power plant sites; genesis of ore deposits; Appalachian geology; lithosphere paleotectonics; continent evolution; evolution of mountain belts; ophiolite tectonics; petrology of ultramatic rocks. Mailing Add: Dept of Geol Sci Cornell Univ Ithaca NY 14850

BIRD, JOHN WILLIAM CLYDE, b Erie, Pa, Nov 10, 32; m 54; c 2. PHYSIOLOGY, BIOCHEMISTRY. Educ: Univ Colo, BA, 58, MA, 59; Univ Iowa, PhD(physiol), 61. Prof Exp: From asst prof to assoc prof physiol & biochem, 61-71, PROF PHYSIOL & BIOCHEM, RUTGERS UNIV, 71- Concurrent Pos: Nat Acad Sci France, 64; Fulbright prof biochem, Cairo Univ, 64-65; NIH career develop award, 66-71; Fulbright res scholar, Belg, 69-70. Mem: AAAS; Am Physiol Soc; Fedn Clin Res. Res: Normal and dystrophic muscle metabolism; enzymology; lysosomes. Mailing Add: Dept of Physiol Rutgers Univ New Brunswick NJ 08903

BIRD, JOSEPH FRANCIS, b Scranton, Pa, Feb 17, 30; m 55; c 1. THEORETICAL PHYSICS. Educ: Univ Scranton, AB, 51; Cornell Univ, PhD(theoret physics), 58. Prof Exp: Sr staff mem, 58-62, PRIN STAFF MEM, THEORET PHYSICS, APPL PHYSICS LAB, JOHNS HOPKINS UNIV, 62- Mem: Am Phys Soc. Res: Astrophysics, cosmogony, star formation theory; combustion, acoustic instability in solid fuel rockets; vision, psychophysical and physiological analysis; neural noise; ocean magnetism. Mailing Add: Appl Physics Lab Johns Hopkins Univ Laurel MD 20810

BIRD, JOSEPH GORDON, b Waycross, Ga, Oct 6, 15; m 39; c 3. CLINICAL PHARMACOLOGY. Educ: Fla Southern Col, BS, 37; Univ Md, MD, 42, PhD(pharmacol), 49. Prof Exp: Rotating intern, Univ Hosp, Baltimore, Md, 42-43; asst resident med, 43-44 & 46-47; instr pharmacol, Sch Med, Univ Md, 47-54, asst med, 50-54; dir, Div Clin Res, Sterling-Winthrop Res Inst, 54-67, dir exp med, 67-70; asst dir, 70-72, ASSOC DIR CLIN PHARMACOL, CIBA-GEIGY CORP, 73- Concurrent Pos: Staff physician, Univ Hosp & Md Gen Hosp, Baltimore, 50-54; med dir, Radio Div, Bendix Corp, 49-54. Mem: AAAS; Am Col Allergists; assoc Am Chem Soc; Am Soc Clin Pharmacol & Exp Therapeut; NY Acad Sci. Res: Pharmaceutical research; clinical trials of new drugs. Mailing Add: Ciba-Geigy Corp 556 Morris Ave Summit NJ 07901

BIRD, JUNIUS BOUTON, b Rye, NY, Sept 21, 07; m 34; c 3. ANTHROPOLOGY. Hon Degrees: DSc, Wesleyan Univ, 58. Prof Exp: Archaeologist, Pa State Mus, 29; field asst archaeol, 31-37, from asst cur to cur, 37-72, EMER CUR ARCHAEOL, AM MUS NATURAL HIST, 72- Concurrent Pos: Consult, Mus Primitive Art, NY, 56-; trustee, Textile Mus, Washington, DC, 57- Honors & Awards: Viking Fund Award, 57. Mem: Soc Am Archaeol (pres, 61-62); NY Acad Sci (corresp secy, 51-55). Res: American archaeology, especially of South America; prehistoric fabrics. Mailing Add: Am Mus of Natural Hist 79th St & Central Park W New York NY 10024

BIRD, LUTHER SMITH, b Greenville, SC, Nov 25, 21; m 47; c 2. PLANT PATHOLOGY. Educ: Clemson Univ, BS, 48; Tex A&M Univ, MS, 50, PhD(genetics), 55. Prof Exp: From instr to assoc prof, 50-65, PROF PLANT PATH, TEX A&M UNIV, 65- Mem: Crop Sci Soc Am; Am Phytopath Soc; Am Soc Agron; Am Genetic Asn. Res: Breeding for and genetics of multiple disease resistance and escape in cotton; breeding for environmental neutrality; seed quality and cold tolerance in cotton; nature of resistance and escape mechanisms. Mailing Add: Dept of Plant Sci Tex A&M Univ College Station TX 77843

BIRD, MARION TAYLOR, b Carterville, Ill, Oct 7, 04; m 30; c 4. MATHEMATICS. Educ: Ill Wesleyan Univ, BS, 28; Univ Ill, AM, 29, PhD(math), 34. Prof Exp: Asst math, Univ Ill, 29-34; from asst prof to assoc prof math & astron, Southwestern Col, Kans, 34-36; instr math, Univ Wis, 36-37; from instr to assoc prof math, Utah State Col, 37-46; instr, Naval Radio Training Sta, US Army Air Force prog & army specialized training prog, 43-45; asst prof, Allegheny Col, 46-47; from asst prof to prof, 47-71, assoc head dept, 58-59, EMER PROF MATH, SAN JOSE STATE UNIV, 71- Mem: Math Asn Am. Res: Difference equations; integral function theory. Mailing Add: 45 Pala Ave San Jose CA 95127

BIRD, NANCY L, zoology, parasitology, see 12th edition

BIRD, RICHARD PUTNAM, b Durango, Colo, July 19, 38; m 62; c 3. BIOPHYSICS. Educ: Univ Colo, BS, 60, MS, 62; Univ Calif, Berkeley, MBioradiol, 68, PhD(biophys), 72. Prof Exp: Res asst physics, Univ Colo, 61-62; health physicist, Idaho Opers Off, 62-64; phys sci adminr, San Francisco Opers Off, US AEC, 64-66; radiation technician, Off Environ Health & Safety, 67; res assoc physics, Kans State Univ, 71-74; RES ASSOC PHYSICS, COLUMBIA UNIV, 74- Mem: Radiation Res Soc. Res: Biological effect of radiation of different quality, particularly accelerator-produced radiations, using eucharyotic cell systems. Mailing Add: Brookhaven Nat Lab Bldg 902B Upton NY 11973

BIRD, ROBERT EARL, b San Antonio, Tex, Nov 4, 43; m 64. MOLECULAR BIOLOGY. Educ: Kans State Univ, BS, 65, PhD(genetics), 69. Prof Exp: Investr Molecular biol, Biol Div, Oak Ridge Nat Lab, 69-71; res dir, Dept Molecular Biol, Univ Geneva, Switz, 71-74; SR STAFF FEL, LAB MOLECULAR BIOL, NAT INST ALLERGY, METAB & DIGESTIVE DIS, NIH, 74- Mem: Genetics Soc Am; Am Soc Microbiol. Res: Genetics, microbiology and biochemistry as related to the regulation of chromosome replication and cell division; control and mechanism of DNA replication in vivo and in vitro. Mailing Add: Bldg 2 Rm 202 Nat Insts of Health Bethesda MD 20014

BIRD, ROBERT MONTGOMERY, b Charlottesville, Va, Feb 1, 15. CLINICAL MEDICINE. Educ: Univ Va, BS, 37, MD, 39; Am Bd Internal Med, dipl. Prof Exp: Intern med & asst res physician, New York Hosp, 39-42; asst med, Med Col, Cornell Univ, 40-42, res assoc physiol, 46-47, from instr to asst prof, 47-50, instr med, 47-52; assoc prof, Sch Med, Univ Okla, 52-61, assoc prof physiol, 56-62, prof med, 61-74, prof physiol, 62-74, assoc dean planning & develop, 65-70, dean, Sch Med, 70-74; DIR, LISTER HILL NAT CTR BIOMED COMMUN, NAT LIBR MED, 74- Concurrent Pos: Asst physician to outpatients, New York Hosp, 46-47, physician, 48-52. Mem: AAAS; Am Col Physicians; Am Physiol Soc; Harvey Soc; Am Clin & Climat Asn. Res: Hematology; metabolism of bone marrow and leukemic cells; genetic linkage; coagulation research; clinical studies. Mailing Add: Nat Libr of Med 8600 Rockville Pike Bethesda MD 20014

BIRD, SAMUEL OSCAR, II, b Charleston, WVa, Feb 26, 34; m 55. VERTEBRATE PALEONTOLOGY. Educ: Marshall Col, BS, 55; Univ Wis, MS, 58; Univ NC, PhD(geol), 62. Prof Exp: Asst prof geol & geog, 62-67, ASSOC PROF GEOL, MARY WASHINGTON COL, 67-, CHMN DEPT, 70- Concurrent Pos: Exten grant for res partic, Col Teachers Prog, 65-67; proj dir, Co SIP, NSF, 70. Mem: AAAS; Am Geol Inst; Soc Syst Zool; Am Paleont Soc. Res: Quantitative systematics; molluscan ecology and community structure; quantitative approaches to stratigraphic correlation. Mailing Add: Dept of Geol Mary Washington Col Fredericksburg VA 22402

BIRD, THOMAS JOSEPH, b Scranton, Pa, Nov 15, 27; m 52; c 5. MEDICAL MICROBIOLOGY. Educ: Univ Scranton, BS, 51; Univ Pa, MS, 54, PhD(med microbiol), 56. Prof Exp: Instr microbiol, Med Sch, Northwestern Univ, 56-58; instr, Stritch Sch Med, Loyola Univ, Ill, 58-59, asst prof, 59-60; res microbiologist, 64-72, CHIEF MICROBIOLOGIST, HINES VET ADMIN HOSP, HINES, ILL, 72- Concurrent Pos: Consult, Ill Dept Pub Health, 65-; vis prof microbiol, Univ Chicago Med Sch, 70- Mem: AAAS; Am Soc Microbiol; Am Genetics Soc; NY Acad Sci; Brit Soc Gen Microbiol. Res: Virus relationships to host; bacteriophages; lysogeny; tissue culture; drug effects on host virus system; pseudomonas infections; pyocins. Mailing Add: Hines Vet Admin Hosp Hines IL 60141

BIRDEN, JOHN HARLAN, b Sheridan, Ind, Nov 23, 18; m 43; c 4. INORGANIC CHEMISTRY, PHYSICS. Educ: Ind Cent Col, BS, 42. Prof Exp: Res radiochemist, 44-66, sr res chemist, 66-70, RES SPECIALIST, MOUND LAB, MONSANTO CHEM CO, 70- Res: Nuclear research calorimetry as applied to the design of high accuracy calorimeter systems, for the evaluation of radioactive heat standards, and for inventory and process control of nuclear materials. Mailing Add: 6013 Jasmine Dr Dayton OH 45449

BIRDSALL, CLAIR MALLERY, b Susquehanna, Pa, Jan 12, 15; m 38. ANALYTICAL CHEMISTRY. Educ: Syracuse Univ, BS, 36; Wesleyan Univ, AM, 38; Yale Univ, PhD(phys chem), 42. Prof Exp: Asst chem, Wesleyan Univ, 36-38; instr, Gilbert Sch, Conn, 38-39; asst chem, Yale Univ, 39-42, instr anal chem, 42-44; staff supvr, Carbide & Carbon Co, Tenn, 44-46; res physicist, Linde Air Prod Co, 46-58, sr res chemist, 58-62, sr develop chemist, Union Carbide Consumer Prod Div, 62-65, HEAD ANAL GROUP, UNION CARBIDE CONSUMER PROD DIV, 65- Mem: Am Soc Testing & Mat. Res: Investigation of surfaces; study of the physical properties of fibres and fibre masses; preparation of colloidal sols; ionization constants from cell electro motive force; quantitative analysis; infrared analysis; Geiger counters; mass spectroscopy; properties of ozone; gas purities. Mailing Add: 28 Macy Rd Briarcliff Manor NY 10510

BIRDSALL, HENRY ALFRED, b New York, NY, Dec 21, 14; m 40; c 8. PHYSICAL CHEMISTRY. Educ: Fordham Univ, AB, 36; Columbia Univ, AM, 40. Prof Exp: Mem tech staff, Bell Tel Labs, Inc, 36-61; staff rep, Am Tel & Tel Co, 61-66, SR STAFF ENGR, WESTERN ELEC CO, AM TEL & TEL CO, 66- Concurrent Pos: Instr, Sch Eng, Cooper Union Univ, 41-45; US del, Int Orgn Standardization Paper, Board & Pulps. Mem: Am Chem Soc; Am Soc Testing & Mat; Tech Asn Pulp & Paper Indust. Res: Chemical methods of analysis; electrical insulation; fibrous materials; relationship between paper and print quality; development of international standards for paper and pulp. Mailing Add: Western Elec Co 50 Lawrence Rd Springfield NJ 07081

BIRDSALL, JOHN J, b Algoma, Wis, July 14, 30; m 54; c 4. FOOD TECHNOLOGY, BIOCHEMISTRY. Educ: Univ Wis, BS, 52, MS, 53, PhD(dairy & food indust), 56. Prof Exp: Food technologist, Wis Alumni Res Found Inst, Inc, 56-58, asst dir lab proj, 58-60, asst dir labs, 60-69, vpres, 69-74, mgr mkt res div, 73-74; SCI DIR, AM MEAT INST, 74- Concurrent Pos: Chmn study panel pesticides, Agr Res Inst & chmn indust comt indust liaison panel, Food Protection Comt, 71; chmn, Coun Indust Liaison Panel, Food & Nutrit Bd, 72. Mem: Am Inst Nutrit; Inst Food Technol; Am Chem Soc; Am Pub Health Asn; Am Soc Microbiol. Res: Nutritional and environmental health problems as related to composition and quality of foods; pesticide residues; food additives and naturally occuring toxic factors. Mailing Add: Am Meat Inst PO Box 3556 Washington DC 20007

BIRDSALL, WILLIAM JOHN, b Waterbury, Conn, Oct 28, 44; m 69; c 1. INORGANIC CHEMISTRY. Educ: Univ Maine, BA, 66; Pa State Univ, PhD(chem), 71. Prof Exp: ASST PROF CHEM, ALBRIGHT COL, 71- Concurrent Pos: Fel, Pa State Univ, 71; consult, Hershey Foods Corp, 72-73. Mem: AAAS; Am Chem Soc; Sigma Xi. Res: Synthesis and structure determination of phosphorus compounds; isolation of natural products with emphasis on inorganic constituents. Mailing Add: Dept of Chem Albright Col Reading PA 19604

BIRDSELL, DALE CARL, b Spokane, Wash, Feb 9, 40; m 62; c 2. MICROBIOLOGY, PHYSIOLOGY. Educ: Wash State Univ, BS, 62, MS, 64; Univ Calif, Riverside, PhD(biol), 67. Prof Exp: Res assoc microbiol, Scripps Clin & Res Found, 67-68; asst prof, Sch Med, Univ Louisville, 69-70; asst prof, Sch Dent, Loyola Univ, 70-74; ASST PROF BASIC DENT SCI, UNIV FLA, 74- Concurrent Pos: NIH fel microbiol, Scripps Clin & Res Found, 68-69. Mem: AAAS; Am Soc Microbiol; Electron Micros Soc Am. Res: Structure and function of the synthesis of macromolecular components of cell surfaces; membrane genesis in microorganisms. Mailing Add: Dept of Basic Dent Sci Univ of Fla Miller Health Ctr Gainesville FL 32610

BIRDSELL, JOSEPH BENJAMIN, b South Bend, Ind, Mar 20, 08; m 44. PHYSICAL ANTHROPOLOGY. Educ: Mass Inst Technol, BS, 31; Harvard Univ, PhD(phys anthrop), 41. Prof Exp: Instr anthrop, State Col Wash, 41-43; from asst prof to assoc prof, 47-70, PROF ANTHROP, UNIV CALIF, LOS ANGELES, 70- Concurrent Pos: Co-leader, Harvard-Adelaide Exped, Australia, 38-39, E C Australian fieldwork, 54. Mem: Am Anthrop Asn; Soc Am Archaeol; Am Soc Human Genetics; Soc Study Evolution; Am Asn Phys Anthrop. Res: Human population genetics; early man; trihybrid origin of the Australian aborigines. Mailing Add: Dept of Anthrop Univ of Calif Los Angeles CA 90024

BIRDSEY, MONROE ROBERTS, b Middletown, Conn, Mar 15, 22; div; c 3. BOTANY. Educ: Univ Miami, AB, 46; Columbia Univ, AM, 47; Univ Calif, PhD(bot), 55. Prof Exp: Asst, Univ Miami, 48-50 & Univ Calif, 51-54; instr bot, Oakland Jr Col, 55; from instr to asst prof, Miami Univ, 55-63; assoc prof, 63-67, PROF BOT, MIAMI-DADE COMMUNITY COL, 67- Mem: Bot Soc Am; Int Asn Plant Taxon. Res: Taxonomy and anatomy of Araceae and Cycadales. Mailing Add: Dept of Biol S Campus Miami-Dade Community Col Miami FL 33176

BIRDSONG, MCLEMORE, b Suffolk, Va, Dec 11, 11; m 41; c 3. PEDIATRICS. Educ: Univ Va, MD, 37. Prof Exp: Intern, Univ Hosp, Univ Va, 37-38, resident, 38-39; house officer & resident pediat, Childrens & Infants Hosp, Boston, 39-41; from asst prof to assoc prof, 41-55, actg chmn dept, 60-64, PROF PEDIAT, SCH MED,

UNIV VA, 55- Mem: Am Acad Pediat; AMA; Asn Am Med Cols. Res: Infectious diseases; rheumatic fever. Mailing Add: Dept of Pediat Univ of Va Sch of Med Charlottesville VA 22903

BIRDSONG, RAY STUART, b Naples, Fla, June 18, 35; m 61; c 2. BIOLOGY, ICHTHYOLOGY. Educ: Fla State Univ, BS, 62, MS, 63; Univ Miami, PhD(marine sci), 69. Prof Exp: Asst prof biol, 68-73, ASSOC PROF BIOL & OCEANOG, OLD DOM UNIV, 73- Concurrent Pos: Asst prog dir, Syst Biol & Ecol Progs, NSF, 70-71. Mem: Am Soc Ichthyol & Herpet; Soc Syst Zool. Res: Ecology, distribution and classification of fishes, especially marine fishes; functional morphology of fishes in general. Mailing Add: Dept of Biol Old Dom Univ Norfolk VA 23508

BIRDWHISTELL, RALPH KENTON, b Columbus, Ohio, May 11, 24; m 43; c 3. PHYSICAL INORGANIC CHEMISTRY. Educ: Ohio State Univ, BSc, 49; Univ Kans, PhD(chem), 53. Prof Exp: Fel, Mich State Univ, 53-54, asst prof, 54-58; assoc prof chem, Butler Univ, 63-67; PROF CHEM & CHMN DEPT, UNIV W FLA, 67- Concurrent Pos: Vis lectr, Univ Ill, 62-63. Mem: Am Chem Soc; Sigma Xi; AAAS; NY Acad Sci. Res: Synthesis and thermodynamic properties of inorganic complex compounds; species occurring in both aqueous and nonaqueous media. Mailing Add: Dept of Chem Univ of West Fla Pensacola FL 32503

BIRECKA, HELENA M, b Poland, May 13, 21; m 48; c 1. PLANT PHYSIOLOGY, BIOCHEMSITRY. Educ: Univ Perm, MS, 44; Timiriazev Acad, Moscow, PhD(plant physiol), 48. Prof Exp: Asst prof agr chem, Agr Univ, Warsaw, 49-51, assoc prof plant physiol, 53-61, prof, 61-68; assoc prof agr chem, Univ Poznan, 51-53; consult, Int Atomic Energy Agency, Vienna, 68-69; res assoc plant physiol, Yale Univ, 69-70; assoc prof plant physiol & biochem, 70-74, PROF BIOSCI, UNION COL, NY, 74- Concurrent Pos: Partic, Int Bot Cong, Montreal, 59, Edinburgh, 64; head plant metab lab, Polish Acad Sci, 60-68, chmn, Comt Use Isotopes & Nuclear Energy in Agr & Biol Sci, 62-68; prof, Isotope Lab, Inst Plant Cultivation, Warsaw, 61-68; Food & Agr Orgn fel, 64. Mem: Polish Bot Soc; Polish Biochem Soc; Am Soc Plant Physiol; Am Soc Agron. Res: Mineral nutrition of plants; alkaloid biosynthesis and metabolism; photosynthesis; long distance translocation in plants; enzyme biosynthesis and activity as related to hormone action. Mailing Add: Dept of Biol Sci Union Col Schenectady NY 12308

BIRELY, JOHN H, b Glen Ridge, NJ, Oct 17, 39; m 68. CHEMICAL PHYSICS. Educ: Yale Univ, BS, 61; Univ Calif, Berkeley, MS, 63; Harvard Univ, PhD(phys chem), 67. Prof Exp: Asst phys chem, Univ Calif, Berkeley, 61-63 & Harvard Univ, 63-66; NIH fel, Cambridge Univ, 66-67; asst prof chem, Univ Calif, Los Angeles, 67-69; mem tech staff, Aerospace Corp, 69-74; GROUP LEADER, LOS ALAMOS SCI LAB, 74- Mem: Am Phys Soc; Am Chem Soc. Res: Atomic and molecular collision physics; molecular spectroscopy; gas phase chemical kinetics and energy transfer; laser photochemistry. Mailing Add: Los Alamos Sci Lab PO Box 1663 Los Alamos NM 87545

BIRGE, ANN CHAMBERLAIN, b San Francisco, Calif, Jan 20, 25; m 48; c 3. RADIATION BIOPHYSICS. Educ: Vassar Col, AB, 46; Radcliffe Col, AM, 47, PhD(physics), 51. Prof Exp: Physicist, Donner Lab, Univ Calif, Berkeley, 51-59, consult, 59-62, lectr, Eng & Sci Exten, 63-65; assoc prof, 65-75, PROF PHYSICS, CALIF STATE UNIV, HAYWARD, 75- Mem: AAAS; Fedn Am Scientists; Am Asn Physics Teachers. Mailing Add: 1 Greenwood Common Berkeley CA 94708

BIRGE, RAYMOND THAYER, b Brooklyn, NY, Mar 13, 87; m 13; c 2. PHYSICS. Educ: Univ Wis, AB, 09, AM, 10, PhD, 14. Hon Degrees: LLD, Univ Calif, 55. Prof Exp: From instr to asst prof physics, Syracuse Univ, 13-18; from instr to prof, 18-55, chmn dept, 33-55, EMER PROF PHYSICS, UNIV CALIF, 55- Concurrent Pos: Mem div phys sci & chmn comt phys constants, Nat Res Coun, 30-37. Mem: Nat Acad Sci; AAAS; fel Am Phys Soc (vpres, 54, pres, 55); fel Optical Soc Am; Am Asn Physics Teachers. Res: Spectroscopy; band and line series relations in spectra; carbon and oxygen isotopes; probable values of general physical constants; statistics; history of physics. Mailing Add: Dept of Phys Univ of Calif Berkeley CA 95709

BIRGE, ROBERT RICHARDS, b Washington, DC, Aug 10, 46; m 68; c 1. CHEMICAL PHYSICS. Educ: Yale Univ, BS, 68; Wesleyan Univ, PhD(chem), 72. Prof Exp: Researcher chem air pollution res div, Environics Br, Kirtland Air Force Base, NMex, 72-73; res fel, Harvard Univ, 73-75; ASST PROF CHEM, UNIV CALIF, RIVERSIDE, 75- Concurrent Pos: Consult atmospheric spectros, Air Pollution Res Div, Environics Br, Kirtland Air Force Base, NMex, 73-75; mem & sci consult, Bice Comn Atmospheric Pollution, US Senate, 73-75; NIH fel, 74-75. Honors & Awards: Nat Sci Award, Am Cyanamid Co, 64. Mem: Am Chem Soc; Am Phys Soc. Res: Applications of high resolution tunable dye laser spectroscopy and quantum theory to the study of electronically excited states; photophysical properties of the visual chromophores; solvent effects on molecular properties. Mailing Add: Dept of Chem Pierce Hall Univ of Calif Riverside CA 92502

BIRGE, ROBERT WALSH, b Berkeley, Calif, Jan 30, 24; m 48; c 3. HIGH ENERGY PHYSICS. Educ: Univ Calif, AB, 45; Harvard Univ, AM, 47, PhD(physics), 50. Prof Exp: Jr physicist, 42-45, PHYSICIST, LAWRENCE RADIATION LAB, UNIV CALIF, BERKELEY, 50-, LECTR, 58-, ASSOC DIR PHYSICS, 73- Concurrent Pos: Orgn Europ Econ Coop sr vis fel, 60; Guggenheim fel, 61; consult, Arms Control & Disarmament Agency, 63; NATO sr fel sci, 71; vis scientist, Europ Orgn Nuclear Res, 71-72; mem sci policy comt, Stanford Linear Accelerator Ctr, 70-74. Mem: Am Phys Soc; Italian Phys Soc. Res: Experimental high energy particle physics. Mailing Add: Lawrence Radiation Lab Univ of Calif Berkeley CA 94720

BIRGE, WESLEY JOE, b Pomeroy, Wash, Mar 11, 29; m 69. ZOOLOGY. Educ: Eastern Wash State Col, BA, 51; Ore State Col, MS, 53, PhD(zool), 55. Prof Exp: Asst zool, Ore State Col, 52-53; instr, Univ Ill, 55-58, from asst prof to assoc prof, 58-62; assoc prof, Univ Minn, Morris, 62-68; ASSOC PROF BIOL SCI, UNIV KY, LEXINGTON, 68- Mem: AAAS; Am Soc Zoologists; Soc Develop Biol; Am Soc Cell Biol; Histochem Soc. Res: Developmental biology; cellular biology; experimental neuroembryology; cytochemistry; electron microscopy, histology. Mailing Add: Dept of Zool Univ of Ky Lexington KY 40506

BIRGENEAU, ROBERT JOSEPH, b Toronto, Ont, Mar 25, 42; m 66; c 4. SOLID STATE PHYSICS. Educ: Univ Toronto, BSc, 63; Yale Univ, PhD(physics), 66. Prof Exp: Instr physics, Yale Univ, 66-67; fel, Oxford Univ, 67-68; mem tech staff, Bell Lab Inc, 68-75, head res dept, 75; PROF PHYSICS, MASS INST TECHNOL, 75- Concurrent Pos: Guest scientist, AEK Res Estab, Riso, Denmark, 71; mem solid state sci adv panel, Nat Res Coun, 74-78; guest scientist, Brookhaven Nat Lab, 68- Mem: Fel Am Phys Soc; AAAS. Res: Neutron and x-ray scattering spectroscopy of condensed matter, especially near magnetic and structural phase transitions. Mailing Add: Dept of Physics Mass Inst of Technol Cambridge MA 02139

BIRITZ, HELMUT, b Vienna, Austria, Oct 4, 40; m 66. THEORETICAL PHYSICS. Educ: Univ Vienna, PhD(theoret physics), 62. Prof Exp: Res assoc theoret physics, Univ Vienna, 62-63 & Max Planck Inst Physics & Astrophys, 63-67; asst prof, 68-74, ASSOC PROF THEORET PHYSICS, GA INST TECHNOL, 74- Mem: Am Phys Soc. Res: Theoretical high energy physics. Mailing Add: Sch of Physics Ga Inst of Technol Atlanta GA 30332

BIRIUK, GEORGE, b Sdolbunov, Russia, Apr 1, 28. MATHEMATICS. Educ: Ind Univ, MA, 57; Univ Calif, Los Angeles, PhD(math), 63. Prof Exp: From asst prof to assoc prof, 62-70, PROF MATH, CALIF STATE UNIV, NORTHRIDGE, 70- Concurrent Pos: Assoc ed, Math Rev, 67-68. Mem: Am Math Soc. Res: Functional analysis, especially self adjoint and normal operators in Hilbert spaces; applications to differential operators. Mailing Add: Dept of Math Calif State Univ Northridge CA 91324

BIRK, JAMES PETER, b Cold Spring, Minn, Aug 21, 41. INORGANIC CHEMISTRY. Educ: St John's Univ, Minn, BA, 63; Iowa State Univ, PhD(phys chem), 67. Prof Exp: Res assoc inorg chem, Univ Chicago, 67-68; asst prof, Univ Pa, 68-73; ASSOC PROF CHEM, ARIZ STATE UNIV, 73- Mem: Am Chem Soc. Res: Kinetics and mechanisms of reactions of transition metal complexes in solution; oxidation-reduction reactions; substitution reactions; linkage isomerism. Mailing Add: Dept of Chem Ariz State Univ Tempe AZ 85281

BIRKE, RONALD LEWIS, b St Louis, Mo, Jan 4, 39; m 62; c 2. ELECTROCHEMISTRY, ANALYTICAL CHEMISTRY. Educ: Univ NC, BS, 61; Mass Inst Technol, PhD(anal chem), 65. Prof Exp: Res asst chem, Mass Inst Technol, 62-63, 65; vis fel, Free Univ Brussels, 65-66; instr, Harvard Univ, 66-69; from asst prof to assoc prof, Univ South Fla, 69-74; ASSOC PROF CHEM, CITY COL NEW YORK, 74- Concurrent Pos: William F Milton Fund grant, Harvard Univ, 66-67; consult, US Army Electronic Components Commmand, NJ, 68 & US Army Harry Diamond Labs, DC, 69; NIH grant, 70-73 & 75- Mem: AAAS; Am Chem Soc; Electrochem Soc; NY Acad Sci. Res: Thermodynamics and kinetics of electrochemical reactions in solution and interfaces; principles of electrochemical measurements; electroanalytical chemical analysis; biological redox processes and electrochemiluminescence. Mailing Add: Dept of Chem City Col of New York New York NY 10031

BIRKELAND, CHARLES JOHN, b Warwick, NDak, Apr 16, 16; m 41; c 3. HORTICULTURE. Educ: Mich State Col, BS, 39; Kans State Col, MS, 41; Univ Ill, PhD(hort), 47. Prof Exp: Res asst, Kans State Col, 39-41, asst horticulturist, 41-46; from asst to asst prof, 46-49, actg head dept, 49-50, PROF HORT & HEAD DEPT, UNIV ILL, URBANA-CHAMPAIGN, 50- Mem: AAAS; fel Am Soc Hort Sci; fel Am Pomol Soc; Am Genetic Asn; Bot Soc Am. Res: Anatomy and photosynthesis of fruit-tree leaves; internal structure of apple leaves of varieties, species and hybrids, with special reference to growth and fruitfulness; fruit breeding. Mailing Add: Dept of Hort Univ of Ill Urbana IL 61801

BIRKELAND, JORGEN MAURICE, b Fergus Falls, Minn, Nov 15, 98; m 28; c 2. BACTERIOLOGY. Educ: NDak State Agr Col, BS, 27, MS, 28; Univ Chicago, PhD(bact), 33. Prof Exp: Bacteriologist & chemist, Chicago Bd Educ, 28-30; Nat Res Coun fel, Rothamsted Exp Sta, Harpenden, Eng, 33-34; from asst prof to prof, 35-69, EMER PROF BACT, OHIO STATE UNIV, 69- Concurrent Pos: Sci attache, US Dept State, Stockholm, 53-54. Mem: AAAS; Am Soc Microbiol; Soc Exp Biol & Med. Res: Chemotherapy; tuberculosis; immunity; bacterial variation and physiology; serological studies on plant viruses. Mailing Add: 299 Piedmont Rd Columbus OH 43214

BIRKELAND, JORGEN WYATT, plasma physics, spectroscopy, see 12th edition

BIRKELAND, PETER WESSEL, b Seattle, Wash, Sept 19, 34; m 59; c 1. GEOLOGY, SOIL SCIENCE. Educ: Univ Wash, BS, 58; Stanford Univ, PhD(geol), 62. Prof Exp: Asst prof soil morphol, Univ Calif, Berkeley, 62-67; assoc prof, 67-72, PROF GEOL SCI, UNIV COLO, BOULDER, 72- Mem: Geol Soc Am; Soil Sci Soc Am; Am Quaternary Asn. Res: Geomorphology; Pleistocene geology; soil morphology and genesis; soil stratigraphy and mineralogmineralogy; soil development rates in alpine and arctic environments. Mailing Add: Dept of Geol Sci Univ of Colo Boulder CO 80304

BIRKELAND, STEPHEN P, b Chicago, Ill, Mar 8, 32; m 55; c 4. SOLID STATE CHEMISTRY. Educ: Univ NMex, PhD(chem), 59. Prof Exp: Sr chemist photog sci, 61-64, res specialist, 64-70, prod develop supvr, 70-72, MGR RES & DEVELOP, 3M CO, 72- Concurrent Pos: NIH fel fel, Univ Wis, 59-60 & Stanford Univ, 60-61. Res: Data recording, electron beam recording and imaging and display technology. Mailing Add: Visual Prod Div 3M Co 3M Ctr St Paul MN 55101

BIRKENHAUER, ROBERT JOSEPH, b Toledo, Ohio, Feb 7, 16; m 41; c 2. TEXTILE CHEMISTRY. Educ: Univ Detroit, BS, 37; Univ Notre Dame, MS, 39, PhD(phys chem), 41. Prof Exp: Chemist, Gulf Refining Co, Tex, 37-38; asst, Univ Notre Dame, 38-41; res chemist, process develop viscose rayon, 41-44, prod develop, 44-46, res group leader equip develop, 46, res supvr, 46-47, plant res supvr, Rayon Dept, 47-53, Textile Fibers Dept, 47-53, process supt rayon, 53-58, dacron, 58-60, mgr tech serv, 60-62, ASST TO PROD MGR DACRON, TEXTILE FIBERS DEPT, E I DU PONT DE NEMOURS & CO, 62- Mem: Am Chem Soc. Res: Process and development of viscose rayon and of dacron and polyester fibers. Mailing Add: Textile Fibers Dept E I du Pont de Nemours & Co Wilmington DE 19898

BIRKENHOLZ, DALE EUGENE. BIOLOGY. Educ: Iowa State Univ, BS, 56; Southern Ill Univ, MA, 58; Univ Fla, PhD(biol), 62. Prof Exp: Instr biol, Univ Fla, 61-62; assoc prof biol sci, 62-70, PROF ECOL, ILL STATE UNIV, 70- Mem: AAAS; Am Soc Mammalogists; Ecol Soc Am; Am Soc Zoologists; Am Ornith Union. Res: Ecology and behavior of small mammals; ornithology. Mailing Add: Dept of Biol Sci Ill State Univ Normal IL 61761

BIRKETT, FRANK ELLIOT, b Columbia, Tenn, Mar 4, 11; m 45, 58; c 1. CHEMISTRY. Educ: Univ Ill, BS, 33, MS, 35. Prof Exp: CHEMIST, PRATT & LAMBERT, INC, 35- Mem: Am Chem Soc; Fedn Socs Paint Technol. Res: Protective and decorative coatings. Mailing Add: 228 Paramount Pkwy Buffalo NY 14223

BIRKETT, JAMES DAVIS, b Norwalk, Conn, Sept 30, 36; m 60; c 3. PHYSICAL CHEMISTRY. Educ: Bowdoin Col, AB, 58; Yale Univ, MS, 60, PhD(chem), 63. Prof Exp: CHEMIST RES & DEVELOP, ARTHUR D LITTLE, INC, 62- Mem: AAAS; Am Inst Chem; Am Chem Soc; Electrochem Soc. Res: Electrochemistry of energy conversion systems; membrane permeation systems, including artificial internal organs. Mailing Add: Oxbow Rd Lincoln MA 01773

BIRKETT, RAY EDWARD, nuclear physics, atomic physics, see 12th edition

BIRKHEAD, PAUL KENNETH, b Poplar Bluff, Mo, Aug 23, 28. GEOLOGY, INVERTEBRATE PALEONTOLOGY. Educ: Univ Mo, AB, 51, MA, 60; Univ NC, PhD(geol), 65. Prof Exp: Cartogr, US Geol Surv, 53-58; from instr to assoc prof, 63-

74, PROF GEOL, CLEMSON UNIV, 74- Mem: Paleont Res Inst; Soc Econ Paleont & Mineral; Geol Soc Am; Nat Asn Geol Teachers; Am Inst Prof Geologists. Res: Photogrammetry; field mapping; Stromatoporoidea. Mailing Add: Dept of Chem & Geol Clemson Univ Clemson SC 29631

BIRKHOFF, GARRETT, b Princeton, NJ, Jan 10, 11; m 38; c 3. MATHEMATICS. Educ: Harvard Univ, AB, 32. Hon Degrees: Dr, Nat Univ Mex, 51, Univ Lille, 59 & Case Inst Technol, 64. Prof Exp: From instr to assoc prof, 36-47, PROF MATH, HARVARD UNIV, 47- Concurrent Pos: Guggenheim fel, 48; consult, Los Alamos Sci Lab, 51-, Gen Motors Corp, 59-, Argonne Nat Lab, 63-, Brookhaven Nat Lab, 64- & Rand Corp, 65-; chmn, Conf Bd Math Sci, 69-70. Mem: Nat Acad Sci; Soc Indust & Appl Math (pres, 67-68); Math Asn Am (vpres, 70-71); Am Math Soc(vpres, 58-59); Am Acad Arts & Sci (vpres, 66-67). Res: Modern algebra; fluid mechanics; numerical methods; reactor theory. Mailing Add: 45 Fayerweather St Cambridge MA 02138

BIRKHOFF, ROBERT D, b Chicago, Ill, Jan 29, 25; m 45. EXPERIMENTAL SOLID STATE PHYSICS. Educ: Mass Inst Technol, BS, 45; Northwestern Univ, PhD(physics), 49. Prof Exp: Consult, health physics div, 50-55, physicist, 55-67, sect chief, radiation physics sect, 67-74, MEM RES STAFF, HEALTH PHYSICS DIV, OAK RIDGE NAT LAB, 74-; PROF PHYSICS, UNIV TENN, 67- Concurrent Pos: Assoc prof physics, Univ Tenn, 49-55. Mem: Am Phys Soc; Health Physics Soc. Res: Beta ray spectroscopy; interaction of radiation with matter; health physics; electronic and optical properties of liquids; electronic structure of cylinders. Mailing Add: Health Physics Div Oak Ridge Nat Lab Oak Ridge TN 37830

BIRKS, LAVERNE STANFIELD, b Rockford, Ill, Feb 4, 19; m 42; c 3. PHYSICS. Educ: Univ Ill, BS, 42; Univ Md, MS, 51. Prof Exp: Physicist, Optical Physics Div, 42-49, sect head x-ray anal, 49-58, BR HEAD X-RAY OPTICS, OPTICAL PHYSICS DIV, US NAVAL RES LAB, 58- Concurrent Pos: Chmn, Comn V.4, Int Union Pure & Appl Physics, 75; consult x-ray anal. Honors & Awards: Spectros Soc Award, 62; Sigma Xi Award, 62; Gold Medal, Spectros Soc, 67; E O Hulbert Award, Naval Res Lab, 70. Mem: Am Phys Soc; Micros Soc; Sigma Xi; Am Soc Testing & Mat; Electron Probe Anal Soc Am (pres, 68); Soc Appl Spectros. Res: X-ray and electron optics. Mailing Add: Naval Res Lab Mat Sci Div Washington DC 20375

BIRKS, RICHARD IRWIN, b Montreal, Que, Aug 12, 24; m 52; c 2. PHYSIOLOGY. Educ: McGill Univ, BA, 49, MSc, 54, PhD(physiol), 57. Prof Exp: Nat Res Coun Can fel biophys, Univ Col, Univ London, 57-59; from asst prof to assoc prof, 59-69, PROF PHYSIOL, McGILL UNIV, 69- Concurrent Pos: Med res assoc, Med Res Coun Can, 61-, mem, Subcomt Physiol, Pharmacol & Anat, 64-67. Mem: Brit Physiol Soc. Res: Mechanism of transmitter turnover in peripheral nervous systems; mechanism of excitation-contraction coupling in muscle. Mailing Add: Dept of Physiol McGill Univ Montreal PQ Can

BIRKY, CARL WILLIAM, JR, b Urbana, Ill, June 5, 37; m 60; c 1. GENETICS. Educ: Ind Univ, BA, 59, PhD(embryol), 63. Prof Exp: From instr to asst prof zool, Univ Calif, Berkeley, 63-70; assoc prof, 70-76, PROF GENETICS, OHIO STATE UNIV, 76- Concurrent Pos: NIH fel, 69, Mem: Am Inst Biol Sci; Am Soc Zool; Soc Develop Biol; Int Soc Develop Biol; Genetics Soc Am. Res: Non-chromosomal genetics, especially genetics of mitochondria and chloroplasts; developmental polymorphism in rotifers. Mailing Add: Dept of Genetics Ohio State Univ Columbus OH 43210

BIRKY, MERRITT MERLE, b Kouts, Ind, June 17, 36; m 58; c 3. PHYSICAL CHEMISTRY, INHALATION TOXICOLOGY. Educ: Goshen Col, BA, 58; Univ Va, PhD(chem), 61. Prof Exp: Nat Res Coun fel thermodyn, Heat & Thermodyn Sect, 61-62, res chemist, Atomic Physics Sect, 62-69, Fire Ctr, 69-73, ACTG CHIEF TOXICOL COMBUSTION PROD, CTR FIRE RES, NAT BUR STANDARDS, 75- Concurrent Pos: Vis assoc prof, Dept Mat Sci, Univ Utah, 73-74. Mem: AAAS. Res: Analysis and toxicity of combustion products and environmental pollutants, including the effects of these chemicals on biological systems; recycling of consumable resources and the use of nonpolluting energy source. Mailing Add: Ctr for Fire Res Nat Bur of Standards Washington DC 20234

BIRMAN, JOAN SYLVIA, b New York, NY, May 30, 27; m 50; c 3. MATHEMATICS. Educ: Barnard Col, Columbia Univ, BA, 48, Columbia Univ, MA, 50; NY Univ, PhD(math), 68. Prof Exp: Asst prof math, Stevens Inst Technol, 68-73; PROF MATH, COLUMBIA UNIV, 73- Concurrent Pos: Vis asst prof, Princeton Univ, 71. Mem: Am Math Soc. Res: Group theory; topology; Riemann surfaces. Mailing Add: Dept of Math Columbia Univ New York NY 10027

BIRMAN, JOSEPH HAROLD, b West Hartford, Conn, June 2, 24; m 45; c 2. GEOLOGY. Educ: Brown Univ, AB, 48; Calif Inst Technol, MSc, 50; Univ Calif, Los Angeles, PhD, 57. Prof Exp: Chmn dept, 50-70, PROF GEOL, OCCIDENTAL COL, 50- Concurrent Pos: NSF grant glacial geol studies, Turkey, 63; consult, Mining & Oil Cos, 52- Mem: Geol Soc Am; Am Inst Mining, Metall & Petrol Engrs. Res: Minerals exploration and development; glacial geology; shallow earth temperatures in ground water exploration. Mailing Add: Dept of Geol Occidental Col Los Angeles CA 90041

BIRMAN, JOSEPH LEON, b New York, NY, May 21, 27; m 50; c 3. THEORETICAL PHYSICS. Educ: City Col New York, BS, 47; Columbia Univ, MA, 50, PhD(theoret chem), 52. Hon Degrees: Dr es Sci, Univ Rennes, France, 74. Prof Exp: Asst, Columbia Univ, 48-52; from physicist to head luminescence sect, Gen Tel & Electronics Res Labs, NY, 52-62; from assoc prof to prof physics, NY Univ, 62-74; HENRY SEMAT PROF PHYSICS, CITY COL NEW YORK, 74- Concurrent Pos: Mary Amanda Wood vis lectr, Univ Pa, 60; vis prof, Univ Paris, 69-70. Mem: Fel Am Phys Soc. Res: Solid state and many-body quantum theory; group theory. Mailing Add: Dept of Physics CUNY 138 St Convent Ave New York NY 10031

BIRMINGHAM, EUGENE, animal science, meat science, see 12th edition

BIRMINGHAM, MARION KRANTZ, b Munich, Ger, Nov 2, 17; Can citizen; m 42; c 2. BIOCHEMISTRY. Educ: Bennington Col, BA, 41; McGill Univ, MSc, 46, PhD(exp med), 49. Prof Exp: Res biochemist, Geront Unit, 49-54, RES BIOCHEMIST, THERAPEUT RES UNIT, ALLAN MEM INST PSYCHIAT, 54-; PROF PSYCHIAT, McGILL UNIV, 72- Concurrent Pos: Demonstr, McGill Univ, 53-61, lectr, 54-61, hon lectr biochem, 59, from asst prof to assoc prof psychiat, 61-72; 12th Ann Eduardo Brown Menendez lectr; med res assoc, Med Res Coun Can. Mem: Am Soc Biol Chemists; Endocrine Soc; Can Physiol Soc. Res: Adrenal steroid biogenesis and its control. Mailing Add: Allan Mem Inst of Psychiat 1033 Pine Ave W Montreal PQ Can

BIRNBAUM, ALLAN, mathematical statistics, see 12th edition

BIRNBAUM, DAVID, b New York, NY, Mar 30, 40; m 65; c 3. ELEMENTARY PARTICLE PHYSICS. Educ: Cornell Univ, BEngPhys, 61; Univ Rochester, MA, 63, PhD(physics), 67. Prof Exp: Res assoc physics, Univ Rochester, 67; res physicist, Carnegie-Mellon Univ, 67-69, asst prof physics, 69-71; ASST PROF PHYSICS, HAMLINE UNIV, 71- Mem: Am Phys Soc; Am Asn Physics Teachers. Res: Experimental particle physics; applications of computers to experimental physics; physics education; astronomy. Mailing Add: Dept of Physics Hamline Univ St Paul MN 55104

BIRNBAUM, EDWARD ROBERT, b Brooklyn, NY, Oct 28, 43; m 68; c 3. INORGANIC CHEMISTRY, BIOINORGANIC CHEMISTRY. Educ: Brooklyn Col, BS, 64; Univ Ill, Urbana, MS, 66, PhD(inorg chem), 68. Prof Exp: Asst prof, 68-73, ASSOC PROF INORG CHEM, NMEX STATE UNIV, 73- Mem: AAAS; Am Chem Soc; The Chem Soc. Res: Calorimetric and nuclear magnetic resonance studies of inorganic and biological complexes of the lanthanide ions. Mailing Add: Dept of Chem NMex State Univ Las Cruces NM 88003

BIRNBAUM, ERNEST RODMAN, b Newark, NJ, Oct 4, 33. INORGANIC CHEMISTRY. Educ: Univ Calif, Berkeley, BA, 55; Univ Southern Calif, MS, 58; Univ Fla, PhD(chem), 61. Prof Exp: Sr res engr, Rocketdyne Div, NAm Aviation, Inc, 61-62; fel chem, Univ Tex, 62-63; from instr to asst prof, 63-71, ASSOC PROF CHEM, ST JOHN'S UNIV, NY, 71- Mem: Am Chem Soc; The Chem Soc. Res: Inorganic reactions in nonaqueous solvents; kinetics and mechanisms of inorganic reactions; Lewis acid-base reactions; hydrido complexes of transition metals. Mailing Add: Dept of Chem St John's Univ Jamaica NY 11432

BIRNBAUM, GEORGE. PHYSICS. Educ: Brooklyn Col, 41; George Washington Univ, MS, 49, PhD(physics), 56. Prof Exp: Head microwave dielectric measurement group, Nat Bur Standards, Washington, DC, 46-54, head microwave frequency & spectros sect, Boulder Labs, 54-56; sr scientist & head quantum physics sect, Hughes Res Labs, Calif, 56-62; GROUP LEADER SPECTROS SCI CTR, NAM ROCKWELL CORP, 62- Concurrent Pos: Mem USA comn I & II, Int Radio Sci Union; mem & chmn, Nat Acad Sci adv panel to radio standards lab, Nat Bur Standards, 65-68. Mem: Fel Am Phys Soc; fel Inst Elec & Electronics Engrs. Res: Microwave and far infrared spectroscopy, masers, and lasers. Mailing Add: NAm Rockwell Corp Sci Ctr PO Box 1085 Thousand Oaks CA 91360

BIRNBAUM, GEORGE I, b Cracow, Poland, July 14, 31; Can citizen; m 67; c 2. STRUCTURAL CHEMISTRY. Educ: Columbia Univ, BS, 54, MA, 55, PhD(org chem), 61. Prof Exp: Res worker x-ray crystallog, Columbia Univ, 59-61, res assoc, 61-64; fel, Nat Res Coun Can, 64-65; NIH spec fel, Glasgow Univ, 65-66; ASSOC RES OFFICER, NAT RES COUN CAN, 66- Mem: Am Chem Soc; Am Crystallog Asn; NY Acad Sci. Res: X-ray structure analyses of biologically significant organic molecules. Mailing Add: Div of Biol Sci Nat Res Coun Sussex Dr Ottawa ON Can

BIRNBAUM, HERMANN, b Gera, Ger, Apr 30, 05; nat; m 44; c 1. ORGANIC CHEMISTRY. Educ: Univ Leipzig, PhD(chem), 34. Prof Exp: Asst to Dr Berl, Carnegie Inst Technol, 35-36; instr chem, Duquesne Univ, 36-37; chemist, res & qual control, Hachmeister, Inc, H J Heinz Co, 37-70; CONSULT, EMULSIFIERS, 70- Mem: Am Chem Soc; Am Oil Chemists Soc; Inst Food Technologists. Res: Oils and fats; food emulsifiers; cereal chemistry and molecular distillation. Mailing Add: 5701 Munhall Rd Pittsburgh PA 15217

BIRNBAUM, JEROME, b Brooklyn, NY, June 22, 39; m 62; c 2. MICROBIOLOGY, BIOCHEMISTRY. Educ: Brooklyn Col, BS, 61; Univ Cincinnati, MS, 64, PhD(microbiol), 66. Prof Exp: Res asst physiol, Inst Muscle Dis, 61-63; sr res microbiologist, fermentation dept, Merck Sharp & Dohme Res Labs, 66-69, assoc dir dept, 69-71, dir dept, 71-74, SR DIR BASIC MICROBIOL DEPT, MERCK INST THERAPEUT RES, MERCK & CO, INC, 74- Mem: AAAS; Am Soc Microbiol; Am Chem Soc; Soc Indust Microbiol; Am Fedn Scientists. Res: Metabolic and genetic regulation in microbial synthesis of antibiotics, amino acids, antiviral compounds and vitamins. Mailing Add: Dept of Basic Microbiol Merck Inst Therapeut Res Rahway NJ 07065

BIRNBAUM, LINDA SILBER, b Passaic, NJ, Dec 21, 46; m 67; c 2. MOLECULAR BIOLOGY. Educ: Univ Rochester, AB, 67; Univ Ill, Urbana, MS, 69, PhD(microbiol), 72. Prof Exp: Asst prof microbiol, Univ Ill, Urbana, 72; fel biochem, Univ Mass, 73-74; asst prof biol, Kirkland, 74-75; RES ASSOC BIOCHEM, MASONIC MED RES LAB, 75- Mem: Am Soc Microbiol; AAAS; Am Inst Biol Sci. Res: Biochemistry of drug metabolism and carcinogenesis in liver and extra-hepatic tissues from organism of different ages. Mailing Add: Masonic Med Res Lab 2150 Bleecker St Utica NY 13503

BIRNBAUM, MILTON, b Brooklyn, NY, Nov 27, 20; m 57; c 2. LASERS. Educ: Brooklyn Col, AB, 42; Univ Md, MS, 48, PhD(physics), 53. Prof Exp: Nuclear physicist, Naval Res Lab, DC, 46-53; head dept physics, Bulova Res & Develop Co, 53-55; assoc prof, Polytech Inst Brooklyn, 55-61; sect head, 61-70, HEAD QUANTUM OPTICS DEPT, AEROSPACE CORP, 70- Concurrent Pos: Consult, Brookhaven Nat Lab, 57-58, Repub Aviation Corp, 60 & Loral Electronics Corp, 60-61. Mem: Fel Am Phys Soc; Optical Soc Am; Inst Elec & Electronics Engrs; fel Brit Inst Physics. Res: Accelerator design; neutron binding energies; electron paramagnetic resonance; laser physics; optics; semiconductors. Mailing Add: 4904 Elkridge Dr Palos Verdes Peninsula CA 90274

BIRNBAUM, NATHAN, chemistry, deceased

BIRNBAUM, SANFORD MILTON, b Plainville, Conn, Mar 4, 15; m 46; c 1. BIOCHEMISTRY. Educ: Univ Conn, BS, 37; Univ Cincinnati, PhD(biochem), 50. Prof Exp: Mem staff, Biochem Lab, Nat Cancer Inst, 50-63; exec secy, Biochem Study Sect, Div Res Grants, NIH, 63-74; SPONSORED RES ADMINR, BRANDEIS UNIV, 74- Mem: AAAS; Am Chem Soc; Am Soc Biol Chemists. Res: Biochemistry of cancer; enzymology; nutrition; science administration. Mailing Add: 5 Hawthorne Village Concord MA 01742

BIRNBAUM, SIDNEY, b New York, NY, Sept 22, 28; m 54; c 3. MATHEMATICS. Educ: NY Univ BA, 48, MS, 49; Univ Colo, PhD(math), 65. Prof Exp: Engr, Martin Co, 53-56, res scientist, 56-65; assoc prof math, Univ SC, 65-70, actg head dept, 67-68; prof math & chmn dept, 70-73, ASSOC DEAN SCH SCI, CALIF STATE POLYTECH UNIV, POMONA, 73- Concurrent Pos: Consult, Sandia Corp, NMex, 66- Mem: AAAS; Am Math Soc; Math Asn Am; Soc Indust & Appl Math. Res: Functional analysis; spectral theory; probability; aeroelasticity. Mailing Add: Sch of Sci Calif State Polytech Univ Pomona CA 91766

BIRNBAUM, ZYGMUNT WILLIAM, b Lwow, Poland, Oct 18, 03; nat US; m 40; c 2. MATHEMATICS, STATISTICS. Educ: Univ Lwow, LLM, 25, PhD(math), 29. Prof Exp: Teacher math, Gymnasium, Poland, 26-29; chief actuary, Life Ins Co Phoenix, Poland, 31-36; asst biomet, NY Univ, 37-39; from asst prof to assoc prof, 39-50, PROF MATH, UNIV WASH, 50-, DIR LAB STATIST RES, 48- Concurrent Pos: Vis prof, Stanford Univ, 51-52, Guggenheim fel, 60-61; consult, Boeing Co, 56- & US Dept Health, Educ & Welfare, 64-; ed, Ann Math Statist, 67-70. Mem: Am Math Soc; fel Inst Math Statist (pres, 63-64); Math Asn Am; fel Am Statist Asn; fel Polish Inst Actuaries. Res: Inequalities; analytic functions; probabilitv. Mailing Add: Dept of Math Univ of Wash Seattle WA 98195

BIRNBAUMER, LUTZ, b Vienna, Austria, Feb 6, 39; Arg citizen; m 65. BIOCHEMISTRY, ENDOCRINOLOGY. Educ: Univ Buenos Aires, BA, 62, BS, 64, PhD(glycogen metab), 66. Prof Exp: Nat Inst Arthritis, Metab & Digestive Dis fel, 68-69; staff fel endocrinol, Nat Inst Arthritis, Metab & Digestive Dis, 69-71; ASSOC PROF PHYSIOL, SCH MED, NORTHWESTERN UNIV, CHICAGO, 71- Concurrent Pos: Nat Inst Child Health & Human Develop res grant, Northwestern Univ, 71- Mem: Endocrine Soc; Am Soc Biol Chemists. Res: Cellular, biochemical and molecular bases of hormone action; role of cyclic nucleotides in cellular regulation; plasma membrane biochemistry; reproductive physiology of the ovary. Mailing Add: Dept of Physiol Northwestern Univ Sch of Med Chicago IL 60611

BIRNBOIM, HYMAN CHAIM, b Winnipeg, Man, Feb 29, 36; m 59; c 2. MOLECULAR BIOLOGY. Educ: Univ Man, MD, 60, MSc, 63. Prof Exp: Intern, St Boniface Hosp, Winnipeg, 60-61, asst resident med, 61-62; fel biochem, Nat Cancer Inst Can, 62-63, Med Res Coun, Can, 63-67; ASSOC RES OFFICER, ATOMIC ENERGY CAN, LTD, 67- Concurrent Pos: Assoc ed, Can J Biochem, 73-; sr lectr, Dept Biochem, Univ Ottawa, 75- Mem: Can Biochem Soc; AAAS. Res: Organization of genetic information in DNA of higher organisms. Mailing Add: Atomic Energy of Can Ltd Chalk River ON Can

BIRNBOIM, MEYER HAROLD, b Kamentz-Letewsk, Poland, Sept 14, 30; Can citizen; m 57; c 1. PHYSICS, RHEOLOGY. Educ: Univ BC, BA, 52, MA, 55; Univ Wis, PhD(physics), 61. Prof Exp: Scientist, Grain Res Lab, Bd Grain Comnr, Govt Can, 54-56; fel physics, Mellon Inst, 60-69; ASSOC PROF MECH, RENSSELAER POLYTECH INST, 69- Concurrent Pos: NIH grant, 64-67. Mem: Soc Rheol Soc; Soc Natural Philos. Res: Continuum physics; viscoelastic properties of dilute polymer solutions, polymer gels, crystalline polymers and metal crystals. Mailing Add: Dept of Mech Rensselaer Polytech Inst Troy NY 12181

BIRNEY, DION SCOTT, JR, b Washington, DC, Apr 23, 26; m 55; c 3. ASTRONOMY. Educ: Yale Univ, BS, 50; Georgetown Univ, MA, 56, PhD(astron), 61. Prof Exp: Engr, Eng Res Corp, Md, 50-52, Am Instrument Co, 52-54 & Stone Straw Corp, DC, 54-56; asst astron, Georgetown Univ, 56-60; from instr to asst prof astron, Univ Va, 60-68; assoc prof, 68-74, PROF ASTRON, WELLESLEY COL, 74- Mem: Am Astron Soc. Res: Photometry of Cepheid variables and eclipsing binaries. Mailing Add: Dept of Astron Wellesley Col Wellesley MA 02181

BIRNEY, ELMER CLEA, b Satanta, Kans, Mar 26, 40; m 61; c 2. MAMMALOGY, SYSTEMATICS Educ: Ft Hays Kans State Col, BS, 63, MS, 65; Univ Kans, PhD(zool), 70. Prof Exp: Instr biol, Kearney State Col, 65-66; asst prof, 70-74, ASSOC PROF ECOL, UNIV MINN, MINNEAPOLIS, 74-, CUR MAMMAL, JAMES FORD BELL MUS NATURAL HIST, 70- Mem: Ecol Soc Am; Am Soc Mammal; Soc Study Evolution; Soc Syst Zool; Am Soc Naturalists. Res: Mammalian systematics and evolution; ecology and behavior of mammals. Mailing Add: J F Bell Mus of Nat Hist Univ of Minn Minneapolis MN 55455

BIRNIE, JAMES HOPE, endocrinology, deceased

BIRNIE, RICHARD WILLIAMS, b Boston, Mass, Dec 8, 44; m 73. GEOLOGY. Educ: Dartmouth Col, AB, 68, AM, 71; Harvard Univ, PhD(geol), 75. Prof Exp: ASST PROF GEOL, DARTMOUTH COL, 75- Concurrent Pos: Vis staff mem, Los Alamos Sci Lab, 75- Mem: Geol Soc Am; Mineral Soc Am; Soc Econ Geologists; Explorers Club. Res: Infrared radiation thermometry of volcanoes; remote sensing applied to prospecting for ore deposits; sulfide mineralogy and crystal chemistry. Mailing Add: Dept of Earth Sci Dartmouth Col Hanover NH 03755

BIROS, FRANCIS JOHN, organic chemistry, physical chemistry, see 12th edition

BIRSS, FRASER WILLIAM, b Limerick, Sask, Dec 9, 32; m 57; c 3. THEORETICAL CHEMISTRY. Educ: Univ Sask, BA, 53, MA, 54; Oxford Univ, DPhil(chem), 56. Prof Exp: Fel theoret chem, Oxford Univ, 56-58 & Univ Rochester, 58-59; from asst prof to assoc prof, 59-71, PROF CHEM, UNIV ALTA, 71- Res: Molecular orbital theory; self consistent field theory. Mailing Add: Dept of Chem Univ of Alta Edmonton AB Can

BIRSTEIN, SEYMOUR J, b Brooklyn, NY, May 1, 27; m 52; c 1. SURFACE CHEMISTRY, CLOUD PHYSICS. Educ: NY Univ, BA, 47; Mont State Col, MS, 49. Prof Exp: Res chemist res labs, Air Reduction Co, Inc, 49-50; res chemist, 51-59, proj scientist, 59-68, CHIEF AEROSOL INTERACTION BR, AIR FORCE CAMBRIDGE RES LABS, 68- Mem: Am Chem Soc; Am Meteorol Soc; Electron Micros Soc Am; Sigma Xi. Res: Application of aerosols in controlling the aerospace environment; basic research and its application to meteorological and weather modification processes. Mailing Add: 85 East India Row Boston MA 02110

BIRSTEN, OSCAR GEORGE, chemistry, see 12th edition

BIRT, ARTHUR ROBERT, b Winnipeg, Man, May 19, 06; m 33; c 3. MEDICINE. Educ: Univ Man, MD, 30; Royal Col Physicians & Surgeons, Can, cert dermat & syphilol, 43, FRCPS(C), 65. Prof Exp: From lectr med to asst prof dermat, 43-63, ASSOC PROF DERMAT & CHMN DEPT, UNIV MAN, 64- Mem: Am Acad Dermat; Am Dermat Asn; Can Dermat Asn (secy, 52-57, pres, 62). Res: Medical mycology and cause of cutaneous flushing in human mast cell disease; photodermatitis in North American Indians. Mailing Add: 714 Med Arts Bldg Winnipeg MB Can

BIRTCH, ALAN GRANT, b La Porte, Ind, Feb 3, 32; m 54; c 4. SURGERY. Educ: Johns Hopkins Univ, BA, 54; Johns Hopkins Med Sch, MD, 58. Prof Exp: Resident surg, Peter Bent Brigham Hosp, 58-63, asst, 63-66; instr surg, Harvard Med Sch, 66-67, assoc, 68-70, asst prof, 70-72; PROF SURG & ASST CHMN, SCH MED, SOUTHERN ILL UNIV, 72-, CHMN GEN SURG, 72-, CHIEF SECT TRANSPLANTATION, 72- Concurrent Pos: Res fel surg, Harvard Med Sch, 63-64; Arthur-Tracey Cabot fel surg, Peter Bent Brigham Hosp, 64-65, chief resident surg, 64-65; res fel physiol, Harvard Med Sch, 65-66; consult surg, Robert Breck Brigham Hosp, 68-72. Mem: Am Surg Asn; Transplantation Soc; Am Col Surgeons; Cent Surg Asn; Soc Univ Surgeons. Res: Clinical and basic transplantation immunology. Mailing Add: Southern Ill Univ Sch of Med PO Box 3926 Springfield IL 62708

BIRTEL, FRANK T, b New Orleans, La, Apr 4, 32; m 64; c 2. MATHEMATICS. Educ: Loyola Univ, La, BS, 52; Univ Notre Dame, MS, 53, PhD(math), 60. Prof Exp: Sr mathematician, US Naval Nuclear Power Sch, 55-57; instr math, Conn Col, 56-57; asst prof, Ohio State Univ, 60-62; from asst prof to assoc prof, 62-67, PROF MATH, TULANE UNIV, 67- Concurrent Pos: Lectr, res assoc & Off Naval Res fel, Yale Univ, 61-62; vis prof, Cath Univ Nijmegen, 68-69. Mem: Math Asn Am. Res: Uniform algebras; several complex variables, harmonic analysis, topological rings and algebras. Mailing Add: Dept of Math Tulane Univ New Orleans LA 70118

BIRTWISTLE, JAMES S, organic chemistry, see 12th edition

BIRUM, GAIL HUBERT, organic chemistry, see 12th edition

BIRZIS, LUCY, b Lithuania, June 19, 19; US citizen. NEUROPHYSIOLOGY. Educ: Univ Pa, AB, 40; Univ Idaho, MS, 43; Univ Calif, Los Angeles, PhD(physiol), 55. Prof Exp: Physicist, Bell Tel Labs, NY, 43-45; ed, Chem Publ Co, 45-46; sr lab technician, Univ Calif, Los Angeles, 51-52, res physiologist, 52-56; instr pharmacol, Sch Med, Univ Fla, 56-59; physiologist, Stanford Res Inst, 60-66; RES ASSOC NEUROSURG, MED SCH, STANFORD UNIV, 66- Concurrent Pos: NIH grants, 61-66 & 62-67. Res: Central regulation of body temperature; electrical correlates of brain function; mechanisms of centrally active drugs; local cerebral blood flow. Mailing Add: Div of Neurosurg Stanford Univ Med Sch Palo Alto CA 94304

BIS, RICHARD F, b Woonsocket, RI, Mar 9, 35; m 57; c 2. SOLID STATE PHYSICS. Educ: Worcester Polytech Inst, BS, 57; Univ Md, MS, 61, PhD(physics), 69. Prof Exp: PHYSICIST, US NAVAL ORD LAB, 57- Concurrent Pos: Instr, Univ Md, 64- Res: Optical and electrical properties of tin tellurium and its alloys with lead; bulk and film samples. Mailing Add: US Naval Surface Weapons Ctr White Oak Silver Spring MD 20910

BISAILLON, ANDRE, b Montreal, Que, Aug 9, 43. VETERINARY ANATOMY, MAMMALOGY. Educ: Col Bourget, BA, 65; Univ Montreal, DMV, 69, MScV, 73. Prof Exp: Instr gross vet anat, Univ Sask, 71-73; ASST PROF GROSS VET ANAT, UNIV MONTREAL, 73- Mem: Am Asn Vet Anatomists; World Asn Vet Anatomists; Am Soc Mammalogists; Am Soc Zoologists. Res: Morphology of aquatic mammals, mainly pinnipeds; study of aquatic locomotion of pinnipeds and gross anatomy of the blood vessels of semi-aquatic and aquatic mammals. Mailing Add: Fac of Vet Med Univ of Montreal St Hyacinthe PQ Can

BISALPUTRA, THANA, b Bangkok, Thailand, Jan 6, 32; m 64. BOTANY. Educ: Univ New Eng, Australia, BSc, 59, MSc, 61; Univ Calif, Davis, PhD(bot), 64. Prof Exp: Demonstr bot, Univ New Eng, Australia, 59-61; res asst bot, Univ Calif, Davis, 61-64; from asst prof to assoc prof, 64-75, PROF BOT, UNIV BC, 75- Res: Light and electron microscopic study of the vascular anatomy of higher plants and the ultrastructure of algae. Mailing Add: Dept of Bot Univ of BC Vancouver BC Can

BISCAYE, PIERRE EGINTON, b New York, NY, Nov 24, 35; m 58; c 4. MARINE GEOCHEMISTRY. Educ: Wheaton Col, Ill, BS, 57; Yale Univ, PhD(geochem), 64. Prof Exp: Geochemist, Jersey Prod Res Co, Okla, 64-65; scientist, Isotopes, Inc, NJ, 65-67; RES ASSOC, LAMONT-DOHERTY GEOL OBSERV, COLUMBIA UNIV, 67- Concurrent Pos: Nat Acad Sci-Nat Res Coun resident res assoc, Inst Space Studies, NY, 67-69. Mem: AAAS; Am Geophys Union; Clay Minerals Soc; Am Geochem Soc; Geol Soc Am. Res: Isotope-, major- and trace-element geochemistry, and mineralogy to study atmospheric dusts, suspended particulates, deep-sea and continental shelf sediments; sources, sinks and transport mechanisms of marine sedimentation. Mailing Add: Lamont-Doherty Geol Observ Palisades NY 10964

BISCHOFF, ERIC RICHARD, b Joliet, Ill, June 16, 38; m 61; c 2. DEVELOPMENTAL BIOLOGY, CELL BIOLOGY. Educ: Knox Col, Ill, AB, 60; Wash Univ, PhD(zool), 66. Prof Exp: Fel anat, Univ Pa, 65-68, instr, med sch, 68-69; ASST PROF ANAT, MED SCH, WASH UNIV, 69- Mem: Am Soc Cell Biol; Am Soc Zoologists. Res: Differentiation of cultured skeletal muscle cells; role of cell division during development; effect of pyrimidine analogues on differentiated functions. Mailing Add: Dept of Anat Sch of Med Wash Univ St Louis MO 63110

BISCHOFF, FRITZ EMIL, b Milwaukee, Wis, May 21, 99. ORGANIC CHEMISTRY, BIOCHEMISTRY. Educ: Univ Wis, BS, 20, MS, 22, PhD(org chem), 24; Am Bd Clin Chem, dipl, 51. Prof Exp: Plant supvr & res chemist, Newport Co, 20-21; res chemist, Jackson Lab, E I du Pont de Nemours & Co, 24-25; chief chemist, 25-39, DIR RES, RES INST, POTTER METAB LAB, COTTAGE HOSP, SANTA BARBARA, CALIF, 39- Concurrent Pos: Adv trustee, Int Cancer Res Found, 46-48. Mem: Am Chem Soc; Am Soc Biol Chemists; Am Asn Clin Chemists; Am Inst Nutrit; Am Asn Cancer Res. Res: Contact catalysis; organic synthesis; chemistry and physiology of protein hormones; physiology of steroid hormones; hormones and vitamins in cancer; red cell estronase; solid state carcinogenesis; alkyl titanates. Mailing Add: Santa Barbara Cottage Hosp 320 W Pueblo Santa Barbara CA 93105

BISCHOFF, HARRY WILLIAM, b Evansville, Ind, May 15, 22; m 47; c 5. ZOOLOGY, BOTANY. Educ: Evansville Col, BA, 49; Vanderbilt Univ, MS, 52; Univ Tex, PhD(phycol), 63. Prof Exp: From instr to assoc prof, 50-65, dean men, 53-54, PROF BIOL, TEX LUTHERAN COL, 65- Concurrent Pos: NSF fel, 59-60. Mem: Bot Soc Am; Phycol Soc Am. Res: Phycology; culturing and systematics. Mailing Add: Dept of Biol Tex Lutheran Col Seguin TX 78155

BISCHOFF, JAMES LOUDEN, b Los Angeles, Calif, Mar 20, 40; m 65; c 2. GEOCHEMISTRY. Educ: Occidental Col, AB, 62; Univ Calif, Berkeley, PhD(geochem), 66. Prof Exp: Fel marine geochem, Woods Hole Oceanog Inst, 66-67, asst scientist, 67-69; from asst prof to assoc prof geol sci, Univ Southern Calif, 69-74; CONSULT PROF, DEPT GEOL, STANFORD UNIV & GEOLOGIST, MARINE GEOL BR, US GEOL SURV, 74- Mem: AAAS; Geochem Soc. Res: Chemistry pf water-rock interaction; metal deposits in the oceans. Mailing Add: Br of Marine Geol US Geol Surv Menlo Park CA 94025

BISEL, HARRY FERREE, b Manor, Pa, June 17, 18; m 54; c 3. CANCER. Educ: Univ Pittsburgh, BS, 39, MD, 42. Prof Exp: Intern, Med Ctr, Univ Pittsburgh, 42-43; staff physician, Eastern State Hosp, Knoxville, 48; res physician, Knoxville Gen Hosp, 50-51; resident, Mem Sloan-Kettering Cancer Ctr, 51-53; asst cancer coord, Sch Med, Univ Pittsburgh, 53-63, asst prof med, 54-63; consult clin oncol, 63-64, HEAD DIV CLIN ONCOL, MAYO CLIN, 64-; ASSOC PROF MED, MAYO GRAD SCH MED, UNIV MINN, 70- Concurrent Pos: Asst med, Boston City Hosp, 49-50; lectr, Grad Sch Pub Health & Sch Med, Univ Pittsburgh, 53-56; consult & exec secy, Clin Studies Panel, Cancer Chemother Nat Serv Ctr, 55-56; consult, Nat Cancer Inst, 56- Mem: Am Cancer Soc; AMA; Am Fedn Clin Res; Am Asn Cancer Res; Am Soc Clin Oncol (pres, 65). Res: Cancer chemotherapy. Mailing Add: Sunny Slopes Rochester MN 55901

BISER, ERWIN, mathematics, see 12th edition

BISGARD, GERALD EDWIN, b Denver, Colo, Aug 4, 37; m 61; c 3. VETERINARY PHYSIOLOGY. Educ: Colo State Univ, BS, 59, DVM, 62; Purdue Univ, MS, 67; Univ Wis, PhD(physiol), 71. Prof Exp: From instr to asst prof vet clins, Purdue Univ, 62-69; asst prof, 71-74, ASSOC PROF VET PHYSIOL, DEPT VET SCI, UNIV WIS-MADISON, 74- Mem: Am Physiol Soc; Am Vet Med Asn; Acad Vet Cardiol; Am Heart Asn. Res: Control of respiration at high altitude, role of peripheral and central chemoreceptors; pulmonary circulation and myocardial function. Mailing Add: Dept of Vet Sci Univ of Wis 1655 Linden Dr Madison WI 53706

BISHARA, RAFIK HANNA, b Cairo, Egypt, Mar 21, 41; US citizen; m 68; c 1. ANALYTICAL CHEMISTRY. Educ: Cairo Univ, BSc, 62; Butler Univ, MSc, 70; Purdue Univ, PhD(bionucleonics), 72. Prof Exp: Phamaceut chemist & head biochem unit, Chem Indust Develop Co, Egypt, 62-67; anal chemist, 67-69, sr anal chemist,

71-75, RES SCIENTIST, ELI LILLY & CO, 76- Mem: Am Chem Soc; Am Pharmaceut Asn; Acad Pharmaceut Sci. Res: Analytical methods, specifications, reference standards; investigational new drug and new drug applications for new drug substances; corresponding precursors and degradation products; chromatographic techniques and compilation of analytical data. Mailing Add: Dept M 769 Eli Lilly & Co PO Box 618 Indianapolis IN 46206

BISHARA, SAMIR EDWARD, b Cairo, Egypt, Oct 31, 35; m 75. ORTHODONTICS. Educ: Univ Alexandria, BChD, 57, dipl, 66; Univ Iowa, MS, 70, DDS, 72. Prof Exp: From intern to mem med staff dent, Moassat Hosp, Alexandria, Egypt, 57-68; asst prof orthod, 70-73, ASSOC PROF ORTHOD, UNIV IOWA, 73- Concurrent Pos: Fel, Guggenheim Dent Clin, 59-60; consult, Am Cleft Palate Asn, 75-76. Mem: Int Asn Dent Res; Am Asn Orthodontists; Am Cleft Palate Asn; Int Dent Fedn; AAAS. Res: Changes in facial and dental relations in normal and cleft populations as related to age and/or management of the cleft; post-treatment changes in orthodontics. Mailing Add: 1014 Penkridge Dr Iowa City IA 52240

BISHIR, JOHN WILLIAM, b Joplin, Mo, Sept 23, 33; m 54; c 4. APPLIED MATHEMATICS. Educ: Univ Mo, AB, 55; State Univ Iowa, MS, 57; NC State Univ, PhD(statist), 61. Prof Exp: From instr to asst prof math, NC State Univ, 57-62; asst prof statist, Fla State Univ, 62-63; from asst prof to assoc prof, 63-68, PROF MATH, NC STATE UNIV, 68- Mem: Math Asn Am; Inst Math Statist. Res: Mathematical modeling in forest genetics. Mailing Add: Dept of Math NC State Univ Raleigh NC 27607

BISHNOI, UDAI RAM, b Punjab, India, Aug 15, 40; m 63; c 1. AGRONOMY. Educ: Rajasthan Univ, India, BSc, 61; Punjab Agr Univ, MSc, 63; Miss State Univ, PhD(agron), 71. Prof Exp: Asst prof agron, Rajasthan Univ, 63-67; asst seed technol, Miss State Univ, 67-71; seed physiologist, Hulsey Seed Testing Lab, Decatur, Ga, 71-72; ASSOC PROF PLANT SCI, ALA A&M UNIV, HUNTSVILLE, 72- Mem: Am Soc Agron; Am Soc Crop Sci. Res: Evaluation of existing varieties of triticale for growth, yield and forage quality; seed production, testing, physiology and processing. Mailing Add: PO Box 149 Ala A&M Univ Huntsville AL 35762

BISHOP, ALLEN DAVID, JR, b Meridian, Miss, Sept 5, 38; m 62; c 2. INORGANIC CHEMISTRY, PHYSICAL CHEMISTRY. Educ: Millspas Col, BS, 60; La State Univ, Baton Rouge, MS, 63; Univ Houston, PhD(inorg chem), 67. Prof Exp: Chemist, Res & Develop Lab, Texaco, Inc, Tex, 63-65; fel chem, Univ Houston, 67; assoc prof, 67-74, PROF CHEM, MILLSAPS COL, 74- Mem: Am Chem Soc; Geochem Soc; Am Soc Oceanog. Res: Polymerization of silicic acid under conditions similar to geologic conditions, effect of various ions on the polymerization process; complexes of oxomolybdenum V, complexes of molybdenum in the plus-5 oxidation state in an attempt to stabilize this oxidation state; thixotropic properties of clays, effect of various organic chelating agents on the thixotropic properties of clays. Mailing Add: Dept of Chem Millsaps Col Box 15463 Jackson MS 39210

BISHOP, AMASA STONE, nuclear physics, see 12th edition

BISHOP, B A, b Cleburne, Tex, Sept 14, 31; m 64; c 2. GEOLOGY. Educ: Tex Christian Univ, BA, 55, MA, 57; Univ Tex, PhD(geol), 66. Prof Exp: Explor geologist, Mene Grande Oil Co, 58-59 & Tenneco Oil Co, 64-67; ASST PROF GEOL, E CAROLINA UNIV, 67- Mem: Am Asn Petrol Geologists; Soc Econ Paleontologists & Mineralogists; Geol Soc Am; AAAS; Sigma Xi. Res: Stratigraphy; sedimentary petrography; description, classification and interpretation of stratified sedimentary rocks; petrography and paleoecology of carbonate sediments and carbonate rocks. Mailing Add: Dept of Geol ECarolina Univ Greenville NC 27834

BISHOP, BEVERLY PETTERSON, b Corning, NY, Oct 19, 22; m 44; c 1. PHYSIOLOGY. Educ: Syracuse Univ, BA, 44; Univ Rochester, MA, 46; State Univ NY Buffalo, PhD(physiol), 58. Prof Exp: Asst psychol, Univ Rochester, 44-46; teacher, Ohio Pub Sch, 46-48; instr bus admin, State Univ NY Buffalo, 49-50; asst physiol, Glasgow Univ, 56-57; from instr to assoc prof, 58-75, PROF PHYSIOL, STATE UNIV NY BUFFALO, 75- Mem: Am Physiol Soc; Soc Neurosci; Am Cong Rehab Med; Am Asn Electromyog & Electrodiag. Res: Neurophysiology; neural regulation of respiration and control of motor systems. Mailing Add: Dept of Physiol Sch of Med State Univ of NY Buffalo NY 14214

BISHOP, CHARLES ALDRICH, b Kingston, Ont, Dec 13, 35; m 68. ETHNOLOGY, ANTHROPOLOGY. Educ: Univ Toronto, BA, 61, MA, 63; State Univ NY Buffalo, PhD(anthrop), 69. Prof Exp: Asst prof anthrop, Fla State Univ, 67-69; asst prof, Eastern NMex Univ, 69-70; assoc prof, 70-74, PROF ANTHROP, STATE UNIV NY COL OSWEGO, 74- Concurrent Pos: Assoc, Royal Int Mus, Toronto, 69- Mem: Fel Am Anthrop Asn; Am Ethnol Soc; Soc Am Ethnohist; Sigma Xi; assoc Current Anthrop. Res: Ethnology, culture change, ethnohistory, cultural ecology, geographical emphasis; North American Indians, subarctic, Ojibwa and Cree. Mailing Add: Dept of Anthrop State Univ NY Col Oswego NY 13126

BISHOP, CHARLES ANTHONY, b Rochester, NY, May 28, 34; m 57; c 6. ORGANIC CHEMISTRY, PHYSICAL CHEMISTRY. Educ: Rochester Inst Technol, BS, 57; Iowa State Univ, PhD(org chem), 61. Prof Exp: Res chemist, 61-63, sr res chemist, 63-68, lab head, 68-74, SR LAB HEAD, EASTMAN KODAK CO, 75- Mem: Soc Photog Sci & Eng; Am Chem Soc. Res: Displacement and elimination reactions; quinonoid compounds; photographic chemistry, dyes. Mailing Add: Res Labs Eastman Kodak Co Rochester NY 14650

BISHOP, CHARLES FRANKLIN, b Doylestown, Pa, June 29, 18; m 43; c 4. PLANT PATHOLOGY. Educ: Goshen Col, BA, 40; Univ WVa, MS, 42, PhD(plant path), 48. Prof Exp: Exten specialist plant path & entom, Univ WVa, 44-56; prof agr, 56-63, chmn div natural sci, 68-73, PROF BIOL & HEAD DEPT, GOSHEN COL, 63- Concurrent Pos: Vis prof, Univ S Fla, 73 & Univ Fla, 75-76; open fac fel, Lilly Endowment Fund, 75. Mem: Am Entom Soc; Am Phytpath Soc. Res: Purpletop of potatoes. Mailing Add: Dept of Biol Goshen College Goshen IN 46526

BISHOP, CHARLES JOHNSON, b Sask, Can, Jan 6, 20; m 51; c 1. HORTICULTURE. Educ: Acadia Univ, BSc, 41; Harvard Univ, AM & PhD(cytogenetics), 47. Prof Exp: Assoc prof genetics & agr res officer, Acadia Univ, 47-52, res prof & supt exp sta, Kentville, NS, 52-58; dir res sta, Summerland, BC, 58-59; assoc dir prog, 59-64, RES COORDR HORT, RES BR, CAN DEPT AGR, 64- Mem: Fel Am Soc Hort Sci; Am Pomol Soc (vpres, 57); Genetics Soc Can (pres, 58); Can Soc Hort Sci (pres, 57); fel Royal Soc Can. Res: X-ray and thermal neutron induced mutations in apples; polyploidy and scab resistance in apples; fruit and vegetable breeding; male sterility in tomatoes; editorial practices in plant science reporting; coordination of horticultural research. Mailing Add: Cent Exp Farm Can Dept of Agr Ottawa ON Can

BISHOP, CHARLES JOSEPH, b Gary, Ind, June 22, 41; m 63; c 2. NUCLEAR CHEMISTRY, NUCLEAR PHYSICS. Educ: Purdue Univ, BS, 63; Univ Wash, PhD(nuclear chem), 69. Prof Exp: Res assoc nuclear physics, Univ Wash, 63-69; res engr, Electrical Subsysts Dept, Boeing Aerospace Co, 69-75, MGR NUCLEAR PROJS, NUCLEAR SYST PROJS DEPT, BOEING ENG & CONSTRUCTION CO, 75- Concurrent Pos: Mem, Radioactive Waste Mgt Subcomt, Atomic Indust Forum, 75- Mem: Am Inst Aeronaut & Astronaut. Res: Energy conversion, especially nuclear power, solar energy; photovoltaics. Mailing Add: Boeing Eng & Construct PO Box 3707 Seattle WA 98124

BISHOP, CHARLES (WILLIAM), b Elmira, NY, June 30, 20; m 44; c 1. BIOCHEMISTRY. Educ: Syracuse Univ, BS, 42, MS, 44; Univ Rochester, PhD(biochem), 46; Am Bd Clin Chem, dipl. Prof Exp: Asst chem, Syracuse Univ, 42-44; res chemist, Manhattan Proj, Univ Rochester, 44-46; instr chem, Kent State Univ, 46-47; asst prof, 47-64, ASSOC PROF BIOCHEM, SCH MED, STATE UNIV NY BUFFALO, 64-; HEAD BIOCHEM LAB, BUFFALO GEN HOSP, 67- Concurrent Pos: NIH spec res fel, Glasgow Univ, 55-56; supvr med res lab, Buffalo Gen Hosp, 47-51, asst lab dir, Chronic Dis Res Inst, 51-55; founder & dir, Blood Info Serv. Mem: AAAS; Am Chem Soc; Am Soc Biol Chemists; Am Asn Clin Chemists; Soc Exp Biol & Med. Res: Red cell metabolism; carbohydrate and purine metabolism; blood storage systems; clinical chemistry. Mailing Add: Biochem Lab Buffalo Gen Hosp Buffalo NY 14203

BISHOP, CLAUDE TITUS, b Liverpool, NS, May 13, 25; m 52; c 1. IMMUNOCHEMISTRY, CARBOHYDRATE CHEMISTRY. Educ: Acadia Univ, BSc, 45, BA, 46; McGill Univ, PhD(chem), 49. Prof Exp: Asst res officer, Cardohydrate Chem Sect, Div Biosci, 49-55, assoc res officer, 55-60, sr res officer, 60-67, prin res officer, 67-70, asst dir biochem lab, 70-72, ASSOC DIR BIOL SCI DIV, NAT RES COUN CAN, 72- Concurrent Pos: Ed-in-chief, Can J Res, 71- Mem: Can Biochem Soc; Chem Inst Can; Am Chem Soc. Res: Carbohydrate chemistry; structures and immunochemistry of microbial polysaccharide antigens. Mailing Add: Div of Biol Sci NRC 100 Sussex Dr Ottawa ON Can

BISHOP, DAVID C, b Aurora, Ill, June 19, 38; m 64; c 3. IMMUNOLOGY, IMMUNOCHEMISTRY. Educ: Univ Dayton, BS, 60; Marquette Univ, MS, 65, PhD(immunol), 68. Prof Exp: SR RES SCIENTIST IMMUNOL, ORTHO DIAG, INC, 70- Concurrent Pos: Am Cancer Soc res fels, Harvard Med Sch, 68-69 & Inst Microbiol, Rutgers Univ, 69-70. Mem: AAAS; Am Asn Clin Chemists; Sigma Xi; Reticuloendothelial Soc; NY Acad Sci. Res: Cellular immunology, including in vitro models for cell mediated immunity, target cell cytotoxicity and macrophage function; models for autoimmune, anti-inflammatory and anticancer therapy. Mailing Add: Ortho Diag Inc Rte 202 Raritan NJ 08869

BISHOP, DAVID HUGH LANGLER, b London, Eng, Dec 31, 37; m 71; c 3. MOLECULAR BIOLOGY, VIROLOGY. Educ: Univ Liverpool, BSc, 59, PhD(biochem), 62. Prof Exp: Fel biochem, Nat Ctr Sci Res, Gif-sur-Yvette, France, 62-63; res assoc, Univ Edinburgh, 63-66 & Univ Ill, 66-69; asst prof molecular biol, Columbia Univ, 69-70; assoc prof, Rutgers Univ, 71-75, prof virol, 75; PROF MICROBIOL & SR SCIENTIST, COMPREHENSIVE CANCER CTR, UNIV ALA MED CTR, BIRMINGHAM, 75- Concurrent Pos: Mem, Virus Cancer Prog Sci Rev Comt A, 75-77. Mem: Am Soc Microbiol. Res: Molecular aspects of the replication of RNA viruses, viral genetics and host-virus interactions. Mailing Add: Dept of Microbiol Univ of Ala Med Ctr Birmingham AL 35294

BISHOP, DAVID MICHAEL, b London, Eng, Sept 19, 36. THEORETICAL CHEMISTRY. Educ: Univ London, BSc, 57, PhD(chem), 60. Prof Exp: Asst prof chem, Carnegie Inst Technol, 62-63; from asst prof to assoc prof, 63-72, PROF CHEM, UNIV OTTAWA, 72- Concurrent Pos: Am Chem Soc fel, 60-62. Res: Molecular quantum chemistry. Mailing Add: Dept of Chem Univ of Ottawa Ottawa ON Can

BISHOP, DAVID WAKEFIELD, b Philadelphia, Pa, May 23, 12; m 39; c 2. PHYSIOLOGY. Educ: Swarthmore Col, AB, 34; Univ Pa, PhD(zool, physiol), 42. Prof Exp: Asst biol, Univ Colo, 34-35; instr zool, Univ Pa, 35-41; res assoc physiol, Swarthmore Col, 42; asst prof zool, Univ Colo, 46-47; asst prof, Univ Ill, 47-48; prof physiol, Univ Mass, 48-51; vis prof zool, Calif Inst Technol, 51-52; mem physiol staff, Carnegie Inst, Washington, 52-67; vis prof, Cornell Univ, 67-68; PROF PHYSIOL, MED COL OHIO, 68- Concurrent Pos: Field zoologist, Acad Nat Sci, Philadelphia, 37; agent, US Fish & Wildlife Serv, 42; vis investr, Barro Colo Lab, Panama; mem, Marine Biol Lab, Woods Hole; vis fel, Max Planck Inst, Heidelberg, 56-57; NSF consult, 63-; vpres, Int Physiol Cong, 68. Mem: Am Zool Soc; Am Soc Gen Physiol (secy, 61-63, pres-elect, 64); Soc Study Develop & Growth; Soc Exp Biol & Med; Soc Study Reproduction. Res: Reproductive physiology; germ cell cytology; sperm metabolism; flagellation; differentiation; autoallergy. Mailing Add: Dept of Physiol Med Col of Ohio PO 6190 Toledo OH 43614

BISHOP, EDWIN VANDEWATER, b Jamaica, NY, May 17, 35; m 57; c 1. PHYSICS. Educ: Swarthmore Col, BA, 58; Yale Univ, MS, 60, PhD(physics), 66. Prof Exp: Res staff astronr, Yale Univ, 66-69; ASST PROF ASTRON & PHYSICS, YOUNGSTOWN STATE UNIV, 69- Mem: Am Phys Soc. Res: Solar-system physics; thermodynamic properties, particularly temperature-profile histories of the major planets' interiors, especially Jupiter's. Mailing Add: Dept of Physics & Astron Youngstown State Univ Youngstown OH 44503

BISHOP, ERRETT A, b Newton, Kans, July 14, 28; m 56; c 3. MATHEMATICS. Educ: Univ Chicago, BS, 48, MS, 50, PhD(math), 55. Prof Exp: From instr to prof math, Univ Calif, Berkeley, 54-65; Miller prof, 64-65, PROF MATH, UNIV CALIF, SAN DIEGO, 65- Concurrent Pos: Sloan fel, 59-62; fel, Inst Advan Study, 61-62. Mem: Am Math Soc. Res: Theory of functions of several complex variables; theory of uniform algebras and functional analysis. Mailing Add: Dept of Math Univ of Calif at San Diego La Jolla CA 92037

BISHOP, EVERETT LASSITER, JR, b Savannah, Ga, Aug 5, 16; m 40; c 4. BIOLOGY. Educ: Emory Univ, AB, 38, MS, 39; Univ Iowa, PhD(zool), 42. Prof Exp: Instr biol, Armstrong Jr Col, 42-44; asst sanitarian, Carter Lab, USPHS, 44-46; from asst prof to assoc prof, 46-54, PROF BIOL, UNIV ALA, 54- Mem: AAAS; Am Soc Prof Biologists; Am Micros Soc. Res: Regeration in protozoa; ultracentrifugation upon invertebrates; cytology of protozoa; effects of insecticdes on aquatic life. Mailing Add: Dept of Biol Univ of Ala Box 1927 University AL 35486

BISHOP, FREDERIC LENDALL, b Chicago, Ill, Mar 6, 09; m 31; c 2. PHYSICS. Educ: Univ Pittsburgh, BS, 30, MS, 34, PhD(physics), 42. Prof Exp: Technician, Am Tel & Tel Co, 31-32; asst, Univ Pittsburgh, 32-33; technician, Am Window Glass Co, 33-42, dir res, 42-53; asst tech dir, Kimble Glass Co, 53-62; tech adv, Consumer & Tech Prod Div, Owens Ill Glass Co, Ohio, 62-64, mgr mat develop, Owens-Ill, Inc, 64-70; CONSULT GLASS FOR ELECTRONICS, 70- Concurrent Pos: Eve lectr, Univ Pittsburgh, 36-47; ed, Laser J. Mem: Fel Am Ceramic Soc; Brit Soc Glass Technol. Res: Velocity of sound in gases; physical properties of glass; flow of glass in tanks; manufacture of safety glass; chemistry of glass; light; spectroscopy; heat; surface of sheet glass. Mailing Add: 28823 Thornhill Dr Sun City CA 92381

BISHOP, GALE ARDEN, b Jamestown, NDak, Dec 10, 42; m 64; c 2. PALEONTOLOGY. Educ: Sdak Sch of Mines, BS, 65, MS, 67; Univ Tex, Austin, PhD(geol), 71. Prof Exp: ASST PROF GEOL, GA SOUTHERN COL, 71- Mem: Sigma Xi; Geol Soc Am; Paleont Soc. Res: Fossil crabs; preservation processes; functional morphology. Mailing Add: Dept of Geol Ga Southern Col Statesboro GA 30458

BISHOP, GEORGE HOLMAN, physiology, deceased

BISHOP, GUY WILLIAM, b Medford, Ore, May 16, 26; m 48; c 3. ENTOMOLOGY. Educ: Ore State Col, BS, 51, MS, 53; State Col Wash, PhD(entom), 58. Prof Exp: Asst entom, Ore State Col, 48-52; asst, State Col Wash, 54-56; from asst entomologist to assoc entomologist, 57-74, assoc prof, 68-74, PROF ENTOM & ENTOMOLOGIST, UNIV IDAHO, 74- Mem: Am Entom Soc. Res: Virus vectors of plants; taxonomy of Cucujidae, especially Cryptolestes. Mailing Add: Dept of Entom Univ of Idaho Moscow ID 83843

BISHOP, JACK BELMONT, b Seymour, Tex, Aug 26, 43; m 71; c 2. GENETICS, TOXICOLOGY. Educ: McMurry Col, BA, 67; La State Univ, Baton Rouge, MS, 70, PhD(genetics), 74. Prof Exp: Res geneticist, Bee Breeding & Stock Ctr Res Lab, Agr Res Serv, USDA, 72-75; RES GENETICIST, NAT CTR TOXICOL RES, DEPT HEALTH EDUC & WELFARE, FOOD & DRUG ADMIN, 75- Mem: Environ Mutagens Soc; Genetics Soc Am; Am Genetics Asn; AAAS. Res: Chemical mutagenesis including studies of chemically induced meiotic recombination in the mouse and drosophila; studies in dosimetry and spermatogenic and cell stage response specificity with chemical mutagens in experimental organisms such as the mouse, drosophila and the honeybee. Mailing Add: Div of Mutagenesis Nat Ctr for Toxicol Res Jefferson AR 72079

BISHOP, JACK GARLAND, b Ft Worth, Tex, Sept 12, 19; m 46; c 1. PHYSIOLOGY. Educ: NTex State Col, BS, 46, MS, 49; Ind Univ, PhD(physiol), 55. Prof Exp: Instr, NTex State Col, 49-51; asst physiol, Ind Univ, 51-53, res assoc, 53-54; from asst prof to prof, 54-69, chmn dept, 56-69, dir res, 61-69, actg dir grad studies, 68-69, ASSOC DEAN GRAD SCH, COL DENT, BAYLOR UNIV, 70- Mem: AAAS; Am Physiol Soc; Int Asn Dent Res. Res: Cardiovascular physiology. Mailing Add: Baylor Col of Dent 800 Hall St Dallas TX 75226

BISHOP, JACK LYNN, b Abilene, Kans, Nov 15, 29; m 58; c 2. ENTOMOLOGY. Educ: Kans State Univ, BS, 56, MS, 57, PhD, 59. Prof Exp: From asst prof to assoc prof entom, Va Polytech Inst & State Univ, 58-67; entomologist, Shell Develop Co, Calif, 67-70, TOXICOLOGIST-ENVIRONMENTALIST, SHELL CHEM CO, 70- Mem: Entom Soc Am. Res: Forage entomology and insect toxicology. Mailing Add: Pesticide Regulation Dept Shell Chem Co 2401 Crow Canyon Rd San Ramon CA 94583

BISHOP, JAMES MARTIN, b Dodge City, Kans, Mar 13, 36; m 64; c 2. PHYSICS. Educ: Kans State Teachers Col, AB, 58; Univ Wis, MS, 60, PhD(physics), 67. Prof Exp: Instr physics, Ohio Univ, 66-67, asst prof, 67-72; FAC FEL, DEPT PHYSICS, UNIV NOTRE DAME, 72- Mem: AAAS; Am Phys Soc; Asn Comput Mach. Res: Elementary particle physics; computer applications in physics. Mailing Add: Dept of Physics Univ of Notre Dame Notre Dame IN 46556

BISHOP, JAY LYMAN, b Salt Lake City, Utah, July 7, 32; m 58; c 8. METALLURGICAL CHEMISTRY, ORGANIC CHEMISTRY. Educ: Univ Utah, BS, 53, PhD(org chem), 62. Prof Exp: Res assoc org chem, Ariz State Univ, 62-63, instr & res assoc chem, 64-67; sr chemist, Ciba Pharmaceut Corp, 67-71; chief chemist & metallurgist, Assoc Smelters Int, 72-73 & United Refinery Inc, 73-75; CHIEF CHEMIST & METALLURGIST, US NAT METALS INC, 75- Concurrent Pos: Pres, Bishop Mfg Co, 71-, consulting chemist, 72-; lectr, Univ Utah, 60-62; traveling sci lectr, Ariz Acad Sci, 62-64; sr pres, 481st Quorum of 70, 70-71. Mem: AAAS; Am Chem Soc; Sigma Xi. Res: Heterocyclics; polymers; amine-formaldehyde alkylation; industrial processes; chemotherapeutic agents; precious metals refining and fabricating. Mailing Add: 11 West 900 North Bountiful UT 84010

BISHOP, JOHN MICHAEL, b York, Pa, Feb 22, 36; m 59. VIROLOGY, BIOCHEMISTRY. Educ: Gettysburg Col, AB, 57; Harvard Univ, MD, 62. Prof Exp: Intern internal med, Mass Gen Hosp, Boston, 62-63, resident, 63-64; res assoc virol, NIH, 64-66, sr investr, 66-68; from asst prof to assoc prof, 68-72, PROF MICROBIOL, MED CTR, UNIV CALIF, SAN FRANCISCO, 72- Concurrent Pos: Res grants, NIH, 68-, Cancer Res Coord Comt, Univ Calif, 68- & Calif Div, Am Cancer Soc, 69- Mem: AAAS; Am Soc Biol Chemists. Res: Biochemistry of animal viruses; replication of nucleic acids; viral oncogenesis; molecular genetics. Mailing Add: Dept of Microbiol Univ of Calif Med Ctr San Francisco CA 94143

BISHOP, JOHN RUSSELL, b Pa, Dec 6, 20; m 45; c 4. AGRICULTURAL CHEMISTRY. Educ: Ursinus Col, BS, 42. Prof Exp: Res chemist, 42-60, chief chemist org synthesis, 60-63, mgr, Agr Chem Lab, 63-68, dir, 68-75, DIR RES & DEVELOP, AMCHEM PROD, INC, PA, 75- Mem: Am Chem Soc; Weed Sci Soc Am. Res: Organic chemical research pertaining to agricultural chemicals; preparation of chemicals for use as herbicides, plant hormones, plant growth regulators; defoliators; formulation of herbicides and plant growth regulators. Mailing Add: 2407 Fairview Ave Hatfield PA 19440

BISHOP, JOHN WATSON, b Glenridge, NJ, Mar 21, 38; m 62; c 2. ECOLOGY, AQUATIC BIOLOGY. Educ: Rutgers Univ, BA, 60; Cornell Univ, MS, 62, PhD(limnol), 66. Prof Exp: Asst prof biol, Univ Del, 64-66; asst prof, 66-71, ASSOC PROF BIOL, UNIV RICHMOND, 71- Concurrent Pos: Univ res found grant, Univ Del, 65-66; Off Water Resources grant, 68-70. Mem: Am Soc Limnol & Oceanog; Ecol Soc Am. Res: Zooplankton ecology; trophic relationships in aquatic ecosystems. Mailing Add: Dept of Biol Univ of Richmond Richmond VA 23173

BISHOP, JOHN WILLIAM, b Jefferson Co, Ind, June 12, 16; m 39; c 2. PETROLEUM CHEMISTRY. Educ: DePauw Univ, BA, 38; Western Reserve Univ, MA, 39, PhD(org chem), 42. Prof Exp: Chemist, Tidewater Oil Co, 42-44, group leader, res dept, 44-56, mgr opers anal, 56-68; MGR ECON & EVAL, GETTY OIL CO, 68- Mem: Am Chem Soc; Am Nuclear Soc; Sigma Xi; AAAS. Res: Hydrocarbon isomerization; lubricants; additives for lubricants; electronic computer applications in petroleum industry; planning and economic evaluations; nuclear power technology. Mailing Add: Getty Oil Co 3810 Wilshire Blvd Los Angeles CA 90054

BISHOP, LEWIS GRAHAM, comparative physiology, neurophysiology, see 12th edition

BISHOP, MARGARET S, b Lewiston, Mich, June 21, 06; m 37; c 2. GEOLOGY. Educ: Univ Mich, AB, 29, MS, 31, PhD(geol), 33. Prof Exp: Geologist, Pure Oil Co, 29-30, asst to chief geologist, 33-38, consult geologist, 38-53; from asst prof to prof, 53-71, EMER PROF GEOL, UNIV HOUSTON, 71- Mem: Geol Soc Am; Am Asn Petrol Geologists. Res: Subsurface geology. Mailing Add: PO Box 2567 Texas City TX 77590

BISHOP, MARILYN FRANCES, b Sacramento, Calif, Jan 19, 50. THEORETICAL SOLID STATE PHYSICS. Educ: Univ Calif, Irvine, BA, 71 & 72, MA, 73, PhD(physics), 76. Prof Exp: Res asst physics, Univ Calif, Irvine, 72-76; RES ASSOC PHYSICS, PURDUE UNIV, 76- Mem: Am Phys Soc. Res: Optical properties of dielectrics; surface electromagnetic waves; electronic properties of crystals, especially metals. Mailing Add: Dept of Physics Purdue Univ West Lafayette IN 47907

BISHOP, MURIEL BOYD, b Billingsley, Ala, Oct 7, 28; m 60; c 1. BIOCHEMISTRY, ORGANIC CHEMISTRY. Educ: Huntingdon Col, BA, 52; Emory Univ, MS, 55; Mich State Univ, PhD(chem), 58. Prof Exp: NSF grant pharmacol, Yale Univ, 58-59; asst prof, 60-74, ASSOC PROF CHEM, CLEMSON UNIV, 74- Mem: Am Chem Soc. Res: Nucleic acid chemistry, especially enzymes associated with pyrimidine biosynthesis; structure of nucleic acids. Mailing Add: Dept of Chem Clemson Univ Clemson SC 29631

BISHOP, NORMAN IVAN, b Silverton, Colo, June 29, 28; m 48; c 4. PLANT PHYSIOLOGY. Educ: Univ Utah, BS, 51, MS, 52, PhD(physiol), 55. Prof Exp: Res asst biol, Univ Utah, 52-55; res assoc biochem, Univ Chicago, 55-57, asst prof, 57-60; assoc prof, Fla State Univ, 60-63; assoc prof, 63-65, PROF PLANT PHYSIOL, ORE STATE UNIV, 65- Concurrent Pos: Lalor Found fel, 56; Guggenheim fel, 71. Mem: AAAS; Am Soc Plant Physiol; Biophys Soc; Am Soc Biol Chemists; Am Soc Photobiol. Res: Photosynthesis and metabolism of microorganisms; photochemistry; general biochemistry. Mailing Add: Dept of Bot Ore State Univ Corvallis OR 97331

BISHOP, RICHARD LAWRENCE, b Lake Odessa, Mich, Aug 12, 31; m 54; c 5. GEOMETRY. Educ: Case Inst Technol, BS, 54; Mass Inst Technol, PhD(math), 59. Prof Exp: PROF MATH, UNIV ILL, URBANA-CHAMPAIGN, 59- Mem: Am Math Soc; Math Asn Am. Res: Relations between the topological and analytical properties of riemannian manifolds. Mailing Add: Dept of Math Univ of Ill Urbana IL 61801

BISHOP, RICHARD S, cytology, biochemistry, see 12th edition

BISHOP, ROBERT FREDERICK, b Somerset, NS, Apr 24, 13; m 37; c 3. SOIL SCIENCE, PLANT NUTRITION. Educ: Acadia Univ, BSc, 34; McGill Univ, MSc, 45; Mich State Univ, PhD(soil sci), 53. Prof Exp: Sch prin, 34-38; agr asst, Can Dept Agr, 38-45, agr scientist, 45-50, res officer, 50-75; RETIRED. Concurrent Pos: Mem, Atlantic Fertil Comt. Mem: Can Soc Soil Sci; Int Soc Soil Sci. Res: Soil fertility and plant nutrition, especially of cereals, field crops and vegetables. Mailing Add: 18 Forest Hill Kentville NS Can

BISHOP, SANFORD PARSONS, b Springfield, Vt, Aug 28, 36; m 57; c 3. COMPARATIVE PATHOLOGY, CARDIOVASCULAR DISEASES. Educ: NY State Vet Col Cornell Univ, DVM, 60; Ohio State Univ, MSc, 65, PhD(vet path), 68; Am Col Vet Pathologists, dipl; Am Col Vet Internal Med, dipl & cert cardiol. Prof Exp: Pvt pract vet med, NY, 60-62; instr cardiol, Sch Vet Med, Univ Pa, 62-63; from asst prof to assoc prof path, Col Med & Dept Vet Path, Ohio State Univ, 68-75; PROF PATH, UNIV ALA, BIRMINGHAM, 75- Concurrent Pos: NIH res career develop award, 72-75; mem, Int Study Group Res Cardiac Metab. Mem: AAAS; Am Asn Pathologists & Bacteriologists; Am Vet Med Asn; Am Heart Asn; Int Acad Path. Res: Comparative cardiovascular pathology; hypertrophy; congestive heart failure; atherosclerosis; myocardial infarction. Mailing Add: Dept of Path Univ of Ala University Sta Birmingham AL 35294

BISHOP, STEPHEN HURST, b Philadelphia, Pa, June 22, 36; m 62; c 3. BIOCHEMISTRY, COMPARATIVE PHYSIOLOGY. Educ: Gettysburg Col, BA, 58; Duke Univ, MA, 60; Rice Univ, PhD(biol), 64. Prof Exp: ASST PROF BIOCHEM, BAYLOR COL MED, 67- Concurrent Pos: Fel biochem, Univ Kans, 64-67. Mem: AAAS; Am Soc Limnol & Oceanog; Am Chem Soc; Am Soc Biol Chem; Am Soc Zool. Res: Phosphoglycoprotein structure-function; nitrogen metabolism. Mailing Add: Dept of Biochem Baylor Col of Med Houston TX 77030

BISHOP, THOMAS PARKER, b Richland, Ga, July 22, 36; m 62; c 1. SOLID STATE PHYSICS. Educ: Carson-Newman Col, BS, 59; Emory Univ, MS, 63; Clemson Univ, PhD, 68. Prof Exp: Asst prof, 67-72, ASSOC PROF PHYSICS, GA SOUTHERN COL, 72- Mem: Am Asn Physics Teachers; Am Phys Soc. Res: Color centers in solids; electron paramagnetic resonance; health physics. Mailing Add: Dept of Physics Ga Southern Col Statesboro GA 30459

BISHOP, VERNON SPILMAN, b McPherson, Kans, Oct 1, 35; m; c 3. PHYSIOLOGY, BIOPHYSICS. Educ: Miss Col, BS, 58; Univ Kans, MS, 60; Univ Miss, PhD(physiol, biophys), 64. Prof Exp: Radiol safety officer & asst prof nuclear eng, Tex A&M Univ, 60-61 & 64-65, asst prof biol, 64-65; fel, Cardiovasc Res Inst, Univ Calif, 65-66; res physiologist & chief weightlessness sect, Biodyn Br, US Air Force Sch Aerospace Med, 66-68; assoc prof pharmacol, Univ Tex Med Sch, San Antonio, 68-73; prof, 73-74. Concurrent Pos: Adv, Atomic Energy Lab, Ceylon, 60; consult, Tex A&M Univ Syst, 60-61. Mem: Soc Nuclear Med; Am Physiol Soc. Res: Control factors involved in the regulation of the circulatory system, particularly the failing and stressed heart. Mailing Add: Dept of Pharmacol Univ of Tex Health Sci Ctr San Antonio TX 78284

BISHOP, WILLIAM P, b Lakewood, Ohio, Jan 18, 40; m 63. NUCLEAR SCIENCE. Educ: Col Wooster, BA, 62; Ohio State Univ, PhD(phys chem), 67. Prof Exp: Vis res assoc chem, Ohio State Univ, 67-69; mem staff nuclear fuel cycle, Sandia Labs, 69-75; CHIEF WASTE MGT BR, US NUCLEAR REGULATORY COMN, 75- Mem: AAAS; Am Nuclear Soc; Sigma Xi. Res: Radiation chemistry, radiation dosimetry; underground testing; nuclear fuel cycle research; nuclear waste management. Mailing Add: Waste Mgt Br MF US Nucl Regulatory Comn Washington DC 20555

BISHOP, WILLIAM RICHARD, b San Francisco, Calif, Sept 19, 26; m 67. MEDICAL EDUCATION. Educ: Univ Ore, BS, 50, MS, 52, PhD(biomed sci), 60, PhD(environ health educ), 62. Prof Exp: Head cytol res labs, Col Med, Univ Ore, 62-63; head div sci & math, Univ Dubuque, 63-64; dir res microanal, Med Res Found, Eli Lilly & Co, 64-67; prof toxicol-pharmaceut, Univ Tenn Med Unit, 67-70; HEAD CYTOTOXICOL & ELECTRON MICROS, MAT SCI TOXICOL DIV, AMA, 70- Concurrent Pos: Consult, Balzers High Vacuum, Inc & NASA, 66-, Vet Admin Hosp, Int Path, Univ Tenn Med Units & St Jude Children's Res Hosp, 67- & Campbell Orthop Clin, Memphis, Tenn, 68-; secy, Residency Rev Comts, AMA, 70- Res: Survey medical education and training programs; allied health; testing and evaluation of biomaterials for medical and paramedical applications; tissue culture; sorption; electron microscopy; freeze-replication; environmental health; allied medical professions; medical and post-graduate medical education; continuing medical education of health team. Mailing Add: Div of Med Educ Am Med Asn 535 N Dearborn Chicago IL 60610

BISHOP, YVONNE M, b Eng, Jan 12, 25; US citizen. BIOSTATISTICS. Educ: Univ

London, BA, 47; Harvard Univ, MSc, 61, PhD(statist), 67. Prof Exp: Math ed, Ministry Agr & Fisheries, Eng, 49-51; ed asst J Iron & Steel Inst, 51-52; sr ed asst, Brit Coal Utilization Res Asn, 52-53; statistician, Pac Biol Sta, BC, 53-54, Int Pac Salmon Fisheries Comn, Can, 54-56, Inter-Am Trop Tuna Comn, 56-57 & St Paul's Rehab Ctr for Newly Blinded Adults, 58-61; instr statist, 61-63, asst prof appl biostatist, 63-71, ASSOC PROF BIOSTATIST, SCH PUB HEALTH, HARVARD UNIV, 71- Concurrent Pos: Head, Div Biostatist & Data Processing, Sidney Farber Career Ctr, 66-75. Mem: Biomet Soc; Am Statist Asn; Am Pub Health Asn. Res: Statistics applied to general public health problems, particularly cancer; development of statistical methodology, particularly handling discrete multivariate data. Mailing Add: Harvard Sch of Pub Health 677 Huntington Ave Boston MA 02115

BISMANIS, JEKABS EDWARDS, b Svitene, Latvia, Dec 28, 11; m 37; c 3. BACTERIOLOGY. Educ: Latvian Univ, MD, 35; Med Coun Can, LMCC, 50; Royal Col Physicians Can, fel, 73. Prof Exp: Asst bact, Latvian Univ, 35-39; in charge, Pasteur Inst, 39-44; in charge bact lab, Dept Health, Riga, Latvia, 39-44; practicing physician, Ger, 44-48; lectr bact, Univ Ottawa, 48-51; med bacteriologist, Lab Pub Health, Regina, Sask, 51-52; from asst prof to assoc prof, 52-74, PROF BACT, UNIV BC, 74- Mem: Can Soc Microbiol; Genetics Soc Can. Res: Virology; activation of latent infections by immunosuppression. Mailing Add: Dept of Microbiol Univ of BC Vancouver BC Can

BISNO, ALAN LESTER, b Memphis, Tenn, Sept 28, 36; m 63; c 2. INTERNAL MEDICINE, INFECTIOUS DISEASES. Educ: Princeton Univ, AB, 58; Wash Univ, MD, 62. Prof Exp: Intern internal med, Vanderbilt Univ Hosp, 62-63, resident, 63-65; med epidemiologist, Ctr Dis Control, USPHS, 65-68; from asst prof to assoc prof, 69-74, PROF INFECTIOUS DIS, UNIV TENN, MEMPHIS, 74-, CHIEF SECT, 71- Concurrent Pos: NIH training grant infectious dis, Med Sch, Univ Tenn, Memphis, 68-69; Am Col Physicians teaching & res scholar, 69-72. Mem: Infectious Dis Soc Am; fel Am Col Physicians; Cent Soc Clin Res; Am Fedn Clin Res. Res: Host-parasite relations in infectious diseases and clinical epidemiology as it relates to infectious diseases; special interest and emphasis in streptococcal diseases and their sequelae. Mailing Add: Div of Infectious Dis Univ of Tenn Dept of Med Memphis TN 38163

BISQUE, RAMON EDWARD, b Stambaugh, Mich, Sept 1, 31; m 54; c 5. GEOCHEMISTRY. Educ: St Norbert Col, BS, 53; Iowa State Col, MS, 56 & 57, PhD(geochem), 59. Prof Exp: From asst prof to assoc prof geochem, 59-69, PROF CHEM & HEAD DEPT, COLO SCH MINES, 69- Concurrent Pos: Consult, Arthur D Little, Inc, Mass, 62-64 & Am Geol Inst, DC, 64-66; co-founder, Earth Sci, Inc, Colo, 62; assoc dir, Earth Sci Curric Proj, Colo Sch Mines, 65, dir, 65- Mem: Am Chem Soc; Geol Soc Am; Am Geochem Soc. Res: Geochemistry and analytical chemistry as applied to mineral exploration, benefication and exploitation; curriculum development in earth science. Mailing Add: Dept of Chem Colo Sch of Mines Golden CO 80401

BISSELL, CHARLES LYNN, b Cumberland, Iowa, Mar 22, 39; m 59; c 3. PHYSICAL CHEMISTRY. Educ: Tarkio Col, BA, 61; Iowa State Univ, MS, 63; Tex A&M Univ, PhD(phys chem), 65. Prof Exp: From asst prof to assoc prof, 65-74, PROF CHEM, NORTHWESTERN STATE UNIV, LA, 74- Mem: Am Chem Soc. Res: Electrochemistry of fused salts with conductivity and electrochemical cells as the primary concern. Mailing Add: Dept of Chem Northwestern State Univ Natchitoches LA 71457

BISSELL, EUGENE RICHARD, b San Francisco, Calif, June 24, 28; m 63. ORGANIC CHEMISTRY. Educ: Univ Calif, BS, 49; Mass Inst Technol, PhD(org chem), 53. Prof Exp: CHEMIST, LAWRENCE LIVERMORE LAB, UNIV CALIF, 53- Mem: Am Chem Soc; AAAS. Res: Organic synthesis; fluorine and polymer chemistry; synthesis of biologically active materials. Mailing Add: 101 Via Lucia Alamo CA 94507

BISSELL, GROSVENOR WILLSE, b Buffalo, NY, Mar 17, 15; m 43. INTERNAL MEDICINE. Educ: Univ Buffalo, MD, 39. Prof Exp: Intern, Buffalo Gen Hosp, 39-40, asst res med & pediat, 40-41, chief res med, 41-42; assoc, Thorndike Mem Lab, 43-44; instr internal med, Sch Med, Univ Buffalo, 44-46, assoc, 46-48, asst prof, 48-51; asst med dir, Armour Labs, 51-52; asst prof internal med, Sch Med, Univ Buffalo, 52-56, assoc prof med, 56-62; chief med serv, Vet Admin Hosp, Buffalo, 52-62; chief endocrine sect & assoc chief staff for res, Sunmount Vet Admin Hosp, Tupper Lake, NY, 62-65; from assoc prof to prof internal med, Sch Med, Wayne State Univ, 65-76; & chief med serv, Vet Admin Hosp, Allen Park, 65-76; CHIEF MED SERV, VET ADMIN HOSP, SAGINAW, 76- Concurrent Pos: Res fel, Thorndike Mem Lab, 42-43; instr clin path, Sch Med, Harvard Univ & vis physician, Boston City Hosp, 43-44; asst internal med, E J Meyer Mem Hosp, 44-51, attend, 52-62; chmn, Vet Admin Coop Study Oral Hypoglycemic Agents in Treatment of Diabetes Mellitus, 51-64; consult endocrinol, Tri County Mem Hosp, 53-62. Mem: AAAS; AMA; Am Diabetes Asn; Am Thyroid Asn; Edocrine Soc. Res: Endocrinology; metabolism. Mailing Add: Vet Admin Hosp Saginaw MI 48602

BISSELL, HAROLD JOSEPH, b Springville, Utah, Feb 9, 13; m 40; c 3. GEOLOGY. Educ: Brigham Young Univ, BS, 34; Univ Iowa, MS, 36, PhD(sedimentation), 48. Prof Exp: Teacher pub sch, Utah, 37-38; instr geol, Brigham Young Univ, 38-40; jr geologist, US Army Corps Engrs, Miss, 41-42; engr, Geneva Steel Co, Utah, 42-43; chmn dept geol geog, Br Agr Col, Utah State Agr Col, 43-44; asst geologist, US Geol Surv, Utah, 44-45; from asst prof to assoc prof, 46-52, chmn dept, 54-56, PROF GEOL, BRIGHAM YOUNG UNIV, 52- Concurrent Pos: Assoc geologist, US Geol Surv, Utah, 47-50; vis prof, Univ Wash, 56-66 & Univ NC, Chapel Hill, 66-67. Mem: Am Asn Petrol Geologists; fel Geol Soc Am; Paleont Soc; Soc Econ Paleontol & Mineral; Int Asn Sedimentologists. Res: Lake Bonneville sedimentation; Pennsylvanian Fusulinidae of Utah; eolian deposits of Utah; Pleistocene sedimentation in southern Utah Valley, Utah; lower Triassic southern Nevada; sedimentary petrographic carbonates; Permo-Triassic eastern Great Basin. Mailing Add: Dept of Geol Brigham Young Univ Provo UT 84602

BISSELL, MINA JAHAN, b Tehran, Iran, May 14, 40; m 67; c 2. CELL BIOLOGY. Educ: Radcliffe Col, AB, 63; Harvard Univ, MA, 65, PhD(molecular genetics), 69. Prof Exp: Fel bact physiol, Dept Microbiol, Harvard Univ, 69-70; fel cancer res & virol, Dept Molecular Biol, 70-72, SR BIOCHEMIST CANCER RES, LAWRENCE BERKELEY LAB, UNIV CALIF, BERKELEY, 72- Concurrent Pos: Fel, Milton Fund Med Res, Harvard Univ, 69-70; Am Cancer Soc fel, Univ Calif, Berkeley, 70-72. Mem: AAAS; Am Soc Cell Biol; Asn Women in Sci; Tissue Culture Asn; Fedn Am Scientists. Res: Biochemistry of cancer cells; metabolic regulation in differentiated, nondifferentiated and virus tranformed cells in culture; mechanism of action of antitumor drugs with emphasis on rifamycin derivatives. Mailing Add: Chem Biodyn Lawrence Berkeley Lab Univ of Calif Berkeley CA 94720

BISSELL, ROBERT, b Anamosa, Iowa, Dec 18, 39. THEORETICAL CHEMISTRY, ORGANIC CHEMISTRY. Educ: Northwestern Univ, BA, 61; Columbia Univ, MA, 62; Duke Univ, PhD(chem), 68. Prof Exp: Instr chem, Duke Univ, 66-67; res assoc, Cornell Univ, 67-68; asst prof, Hampden-Sydney Col, 68-69 & Longwood Col, 69-74; ASST PROG ADMINR, PETROL RES FUND, DEPT RES GRANTS & AWARDS, AM CHEM SOC, 74- Mem: Am Chem Soc. Res: Applications of semi-empirical theories to the study of organic reactions. Mailing Add: Dept of Res Grants & Awards Am Chem Soc 1155-16th St NW Washington DC 20036

BISSETT, DONALD LYNN, b North Adams, Mass, July 4, 49; m 73. BIOCHEMISTRY. Educ: Univ Conn, BA, 71; Mich State Univ, PhD(biochem), 75. Prof Exp: RES SCIENTIST BIOCHEM, MIAMI VALLEY LAB, PROCTER & GAMBLE CO, 75- Mem: Am Soc Microbiol. Res: Skin biochemistry. Mailing Add: Miami Valley Lab Procter & Gamble Co Ross OH 45247

BISSETT, MARJORIE LOUISE, b Miami, Ariz, Sept 21, 25. MICROBIOLOGY. Educ: Univ Calif, Los Angeles, AB, 49; Univ Mich, PhD(epidemiol sci), 65. Prof Exp: Microbiologist, Calif State Dept Pub Health, 49-55, asst microbiologist, 55-61; trainee epidemiol sci, Univ Mich, 61-65; RES MICROBIOLOGIST, CALIF STATE DEPT PUB HEALTH, 65- Concurrent Pos: Mem, Diag Prod Adv Comt, Food & Drug Admin, 72-76; Am Acad Microbiol fel, 73. Mem: Am Pub Health Asn; Am Soc Microbiol; NY Acad Sci; Conf State & Prov Pub Health Lab Dirs. Res: Clinical and public health microbiology; microbiology and laboratory diagnosis of Neisseria gonorrheae, Yersinia sporans, Enterobacteriaceae and other medically important organisms. Mailing Add: Microbial Dis Lab 2151 Berkeley Way Berkeley CA 94709

BISSETT, ORVILLE R, b Saginaw, Mich, Nov 30, 18; m 44; c 4. PLANT PHYSIOLOGY, FORESTRY. Educ: Mich State Univ, BS, 48, MF, 49; Univ Mass, PhD(bot), 65. Prof Exp: Forester, Mass Dept Natural Resources, 49-53 & New Eng Silvicult Serv, 53-55; teacher math & sci, Althol High Sch, 55-57; logging contractor & consult forester, 57-63; asst bot, Univ Mass, 63-65; asst prof, 65-70, ASSOC PROF BIOL, CENT CONN STATE COL, 70- Mem: Soc Am Foresters; Am Soc Plant Physiol; Scandinavian Soc Plant Physiol. Res: Mineral nutrition of white pine. Mailing Add: Dept of Biol Cent Conn State Col New Britain CT 06050

BISSEY, LUTHER TRAUGER, b New Britain, Pa, Aug 2, 12; m 40. CHEMISTRY, PETROLEUM ENGINEERING. Educ: Pa State Univ, BS, 34, MS, 42. Prof Exp: Analyst, Mineral Indust Exp Sta, Col Mineral Indust, 36-38, instr, Petrol & Natural Gas Dept, 38-50, from asst prof to assoc prof, 50-73, EMER ASSOC PROF, PETROL & NATURAL GAS DEPT, PA STATE UNIV, 73- Mem: Am Gas Asn. Res: Water-flooding; household refuse incineration; natural gas pipeline deposits; gas flow in porous media; gas reservoir mechanics; instrumentation and communications. Mailing Add: 930 N Atherton St State College PA 16801

BISSHOPP, FREDERIC EDWARD, b Beloit, Wis, Oct 2, 34; m 58; c 1. APPLIED MATHEMATICS. Educ: Ill Inst Technol, BS, 54; Univ Chicago, MS, 56, PhD(physics), 59. Prof Exp: Asst prof, 61-65, ASSOC PROF MATH, BROWN UNIV, 65- Res: Fluid mechanics; electromagnetic theory; partial differential equations. Mailing Add: Div of Appl Math Brown Univ Providence RI 02912

BISSING, DONALD EUGENE, b Hays, Kans, Sept 10, 34; m 54; c 2. ORGANIC CHEMISTRY. Educ: Fort Hays State Col, BA, 59; Univ Kans, PhD(org chem), 62. Prof Exp: Mgr process develop, Monsanto Co, 62-74; DIR CHEM ENG RES & DEVELOP CHEM, BASF WYANDOTTE CORP, 74- Mem: Am Chem Soc; Res & Eng Soc Am. Res: Chemistry of agricultural products, polyether polyols, graft polyols; environmental control chemistry and engineering. Mailing Add: BSAF Wyandotte Corp 1609 Biddle Ave Wyandotte MI 48192

BISSINGER, BARNARD HINKLE, b Lancaster, Pa, Jan 27, 18; m 50; c 2. MATHEMATICS, ENGINEERING. Educ: Franklin & Marshall Col, BS, 38; Syracuse Univ, MA, 40; Cornell Univ, PhD(math), 43. Prof Exp: Instr math, Cornell Univ, 43; asst prof, Mich State Col, 44; opers res analyst, US Army Air Force, 44 & Chennault Flying Tigers, China, 45; gen mgr, Athletic Shoe Factory, 46-52; Lehman Prof math & chmn dept, Lebanon Valley Col, 53-68; prof on leave, 68-69, DIR MATH SCI, PA STATE UNIV, CAPITOL CAMPUS, 70- Concurrent Pos: Assoc actuary, Pa State Univ, 68; NSF sci fac fel, Princeton Univ, 58-59. Mem: Am Math Soc; Math Asn Am; fel Royal Statist Soc. Res: Operations research; mathematical analysis; functional analogues of continued fractions; group physical mortality. Mailing Add: Dept of Math Sci Pa State Univ Capitol Campus Middletown PA 17057

BISSINGER, WILLIAM ELLIS, organic chemistry, see 12th edition

BISSON, PETER ANDRE, b Dover, Del, Aug 28, 45. AQUATIC BIOLOGY. Educ: Univ Calif, Santa Barbara, BA, 67; Ore State Univ, MS, 69, PhD(fisheries), 75. Prof Exp: SR SCIENTIST AQUATIC BIOL, WEYERHAEUSER CO, 75- Mem: Ecol Soc Am. Res: Biology of streams; nutrient relationships; ecology of fish populations. Mailing Add: Weyerhaeuser Co PO Box 188 Longview WA 98632

BISSONNETTE, HOWARD LOUIS, b Detroit, Mich, Aug 28, 27; m 57; c 3. PLANT PATHOLOGY. Educ: Col St Thomas, BS, 52; Univ Minn, MS, 58, PhD, 64. Prof Exp: Res asst, Univ Minn, 53-56, plant pathologist, USDA, 57-62; exten plant pathologist, NDak State Univ, 62-68; PROF PLANT PATH & PLANT PATHOLOGIST, AGR EXTEN, UNIV MINN, MINNEAPOLIS, 68- Mem: Am Phytopath Soc; Mycol Soc Am; Am Soc Sugar Beet Technol. Res: Soil mycroflora; root rot diseases of sugar beets and peas; general pathology; potato diseases; aerial application of fungicides. Mailing Add: 3456 Milton St St Paul MN 55112

BISSONNETTE, JOHN MAURICE, b Montreal, Can, Feb 11, 39; m 70; c 3. OBSTETRICS & GYNECOLOGY. Educ: Loyola Col Montreal, BA, 60; McGill Univ, MD CM, 64. Prof Exp: Rotating intern, Montreal Gen Hosp, 64-65, asst resident, 65-66; resident obstet & gynec, Johns Hopkins Hosp, Baltimore, 66-69; fel pulmonary physiol, Sch Pub Health, Johns Hopkins Univ, 69-70; asst prof obstet & gynec, Univ Calif, Irvine, 70-73; asst prof, 73-74, ASSOC PROF OBSTET & GYNEC, UNIV ORE HEALTH SCI CTR, 74- Mem: Am Physiol Soc; Perinatal Res Soc. Res: Respiratory and cardiovascular physiology of the mammalian placenta, especially transport of respiratory gasses across the placenta, transport of water and nonelectrolytes across the placenta and regulation of capillary volume in the placenta. Mailing Add: Dept of Obstet & Gynec Univ of Ore Health Sci Ctr Portland OR 97201

BISSOT, THOMAS CHARLES, b Grand Rapids, Mich, Apr 20, 30; m 51; c 8. INORGANIC CHEMISTRY, PHYSICAL CHEMISTRY. Educ: Aquinas Col, BS, 52; Univ Mich, MS, 53, PhD(chem), 56. Prof Exp: Res chemist, 55-71, RES ASSOC, E I DU PONT DE NEMOURS & CO, 71- Mem: Am Chem Soc. Res: Boron hydrides; high vacuum techniques; catalysis; polymer and surface chemistry. Mailing Add: Deer Run Rd RD 3 Little Baltimore Newark DE 19711

BISWAL, NILAMBAR, b Khamar, Orissa, India, Feb 20, 34; m 67; c 3. VIROLOGY. Educ: Punjab Col Vet Med, BVSc & AH, 58; Mich State Univ, MS, 63, PhD(microbiol), 65. Prof Exp: Asst res virologist, Virus Lab, Univ Calif, Berkeley, 65-67; fel virol, 67-68, asst prof, 68-72, ASSOC PROF VIROL, BAYLOR COL MED,

HOUSTON, 72- Mem: Am Soc Microbiol; AAAS; Sigma Xi. Res: Basic mechanism of biosynthesis of herpes virus DNA; analysis of structure and replication of herpes virus DNA. Mailing Add: Dept of Virol Baylor Col of Med Houston TX 77025

BISWAS, ASIT KUMAR, b Balasore, India, Feb 25, 39. ENVIRONMENTAL MANAGEMENT. Educ: Indian Inst Technol, Kharagpur, BTech, 60, MTech, 61; Univ Strathclyde, PhD(water management), 67. Prof Exp: Asst civil engr, Ward, Ashcroft & Parkman, UK, 61-62; res fel hydraul eng, Loughborough Univ Technol, Eng, 62-63; lectr water resources eng, Univ Strathclyde, 63-67; vis prof water resource planning, Queen's Univ, Ont, 67-68; sr res officer, Dept Energy, Mines & Resources, 68-70; DIR DEPT ENVIRON, OTTAWA, CAN, 70- Concurrent Pos: Vis prof, Univ Ottawa, 68-70; sr consult, UN Environ Prog, 74-; Rockefeller Found Int Relations fel, 74. Honors & Awards: Walter L Huber Res Medal, Am Soc Civil Engrs, 74. Mem: Am Soc Civil Engrs; Int Water Resource Asn; Int Asn Hydraulic Res. Res: Resources management; interrelationship between population, food, energy, raw materials and environment; climatic changes and their effects on world food production, mathematical modelling for environmental and resources management. Mailing Add: Planning & Finance Serv Dept of Environ Ottawa ON Can

BISWAS, NRIPENDRA NATH, b Calcutta, India, Aug 1, 30; m 59; c 2. PHYSICS. Educ: Univ Calcutta, BSc, 49, MSc, 51, DrPhil(physics), 55. Prof Exp: Res asst cosmic rays, Bose Res Inst, Calcutta, 52-55; res scholar physics, Max Planck Inst Physics, Göttingen, 55-57; res assoc, Physics Inst, Univ Bologna, 57-58; vis scientist high energy physics, Lawrence Radiation Lab, Univ Calif, Berkeley, 58-60; res scientist bubble chamber anal, Max Planck Inst Physics, Munich, 61-65; asst prof, 65-66, ASSOC PROF PHYSICS, UNIV NOTRE DAME, IND, 66- Res: Elementary particle physics, theoretical and experimental; study of cosmic rays. Mailing Add: Dept of Physics Univ of Notre Dame Notre Dame IN 46556

BISWAS, PROSANTO K, b Calcutta, India, Mar 1, 34; US citizen; m 62; c 1. HORTICULTURE, PLANT PHYSIOLOGY. Educ: Univ Calcutta, BS, 58; Univ Mo, MS, 59, PhD(hort), 62. Prof Exp: Asst prof hort & head dept, 62-69, PROF HORT & CHMN DEPT PLANT & SOIL SCI, 69- Concurrent Pos: Chmn, Macon County Bd Educ. Mem: Am Soc Plant Physiol; Am Soc Hort Sci; Weed Sci Soc Am. Res: Physiological responses of plants to different growth regulators; metanolism and mode of action of herbicides by plants. Mailing Add: Dept of Plant Sci Tuskegee Inst Tuskegee AL 36088

BISWAS, SHIB D, b Agra, India, Feb 1, 40; m 69. CHEMISTRY, BIOCHEMISTRY. Educ: Univ Allahabad, BSc, 57, MSc, 60, PhD(chem), 64. Prof Exp: Fel chem, Univ Cambridge, 64; fel biochem, 64-65, lectr, 65-67 & Sch Dent Hyg, 65, ASST PROF BIOCHEM & ORAL BIOL, UNIV MANITOBA, 68- Res: Lignin complexes in metal ions; etiology of perio-dontal and caries diseases processes in relation to biochemical standpoint; application of radiochemicals in the study of metabolic pathways in oral bacteria. Mailing Add: Dent Col Univ of Manitoba Winnipeg MB Can

BISWELL, HAROLD HUBERT, b Fayette, Mo, Nov 8, 05; m 39; c 1. ECOLOGY, FORESTRY. Educ: Cent Col, Mo, AB, 30; Univ Nebr, AM, 32, PhD(bot), 34. Prof Exp: Asst bot, Univ Nebr, 34; jr range examr, Calif Forest & Range Exp Sta, US Forest Serv, 34-36, asst conservationist, 36-37, San Joaquin Exp Range, 36-40, from assoc forest ecologist to sr forest ecologist, 40-47, in-chg range res, 40-47; from assoc prof forestry & assoc forest ecologist, Exp Sta to prof & forest ecologist, 47-72, EMER PROF FORESTRY, UNIV CALIF, BERKELEY & EMER FOREST ECOLOGIST, EXP STA, 73- Mem: AAAS; Am Soc Range Mgt; Soc Am Foresters; Ecol Soc Am. Res: Range ecology and management; effects of environment upon root development of deciduous forest trees; fire effects and use in land management; harmonizing fire and forest management; fire ecology. Mailing Add: 28 The Crescent Berkeley CA 94708

BITANCOURT, AGESILAU ANTONIO, b Manaos, Brazil, June 6, 99. PLANT PATHOLOGY. Educ: Sorbonne, DSc, 55. Prof Exp: Asst phytopath, Biol Inst, Rio de Janeiro, 20-26; prof bot & phytopath, Higher Sch Agr, Univ Sao Paulo, 26-28; dir, Agr Exp Sta, Rio de Janeiro, 28-31; phytopathologist, Biol Inst Sao Paulo, 31-69, dir div biol-veg, 45-49, dir gen, 49-53, biol-veg, 53-69; RETIRED. Concurrent Pos: Guggenheim fel, 41-42; adv, Brazilian Res Coun, 69- Mem: Brazilian Acad Sci; fel Am Phytopath Soc. Res: Plant cancer; plant growth hormones. Mailing Add: Inst Biologico Caixa postal 7119 Sao Paulo Brazil

BITCOVER, EZRA HAROLD, b New York, NY, Jan 17, 20; m 44; c 4. LEATHER CHEMISTRY. Educ: Rutgers Univ, BS, 41; Univ Mass, PhD(agron), 50. Prof Exp: Jr chemist, Tenn Valley Authority, 41-43; res chemist, Prophylactic Brush Co, 43-46; soils chemist, Citrus Exp Sta, Univ Fla, 48; anal res supvr, Lindsay Chem Co, 50-52; res chemist food processing, Miner Labs, 53-56; RES BIOCHEMIST, EASTERN UTILIZATION RES DIV, USDA, 56- Mem: AAAS; Am Leather Chem Asn; Am Chem Soc; Am Inst Chemists (secy, 58-63). Res: Nitrogen chemicals; processing of vegetables. Mailing Add: 8237 Michener Ave Philadelphia PA 19150

BITHER, TOM ALLEN, JR, b Berkeley, Calif, Oct 8, 17; m 42; c 3. CHEMISTRY. Educ: Univ Calif, BS, 39; Yale Univ, PhD(chem), 42. Prof Exp: RES CHEMIST, E I DU PONT DE NEMOURS & CO, 42- Mem: AAAS; Am Chem Soc. Res: Chemical kinetics of exchange reactions; vinyl ionic and condensation polymerization; inorganic solid state chemistry; reactions at high pressure; heterogeneous catalysis. Mailing Add: Cent Res & Develop Lab E I du Pont de Nemours & Co Wilmington DE 19898

BITLER, WILLIAM REYNOLDS, b Nyack, NY, Nov 25, 27; m 54. SOLID STATE PHYSICS. Educ: Carnegie Inst Technol, BS, 53, MS, 55, PhD(physics), 57. Prof Exp: Asst prof metall eng, Carnegie Inst Technol, 56-62; assoc prof, 62-69, PROF METALL, PA STATE UNIV, 69-, CHMN METALL SECT, 69- Concurrent Pos: Consult, Allegheny-Ludlum Steel Corp, 57- Mem: Am Phys Soc; Am Soc Metals; Am Inst Mining, Metall & Petrol Engrs. Res: Physical metallurgy; kinetics; statistical thermodynamics; magnetism; optical properties. Mailing Add: 221 Mineral Indust Bldg Pa State Univ University Park PA 16802

BITMAN, JOEL, b Elizabeth, NJ, Nov 29, 26; m 50; c 3. BIOCHEMISTRY. Educ: Cornell Univ, AB, 46; Univ Minn, MS, 48, PhD(physiol chem), 50. Prof Exp: Asst physiol chem, Univ Minn, 47-50, res fel, 50; chemist, Div Nutrit & Physiol, Dairy Husb Res Br, USDA, 51-55; res assoc, Columbia Univ, 55-56; chemist & leader physiol invests, Dairy Cattle Res Br, 56-74, CHIEF BIOCHEM LAB, ANIMAL PHYSIOL & GENETICS INST, AGR RES CTR-EAST, USDA, 74- Res: Chemistry and metabolism of sex hormones; determination of estrogens in blood; vitamin A deficiency and cerebrospinal fluid mechanics; body temperature during the estrous cycle; chemical composition of the mammary gland during mastitis; mechanism of estrogen action in the uterus; DDT effect on carbonic anhydrase and egg-shell thickness; estrogenic effects of DDT. Mailing Add: Biochem Lab Animal Physiol Inst ARC-East USDA Beltsville MD 20705

BITONTI, JOHN, b Belle-Vernon, Pa, Jan 3, 34; m 58; c 1. SPEECH PATHOLOGY, AUDIOLOGY. Educ: Calif State Col, Pa, BS, 60; WVa Univ, MA, 61, EdD(speech path), 69. Prof Exp: Speech clinician, Westmoreland County Schs, Pa, 60-62; from asst prof to assoc prof, 62-69, PROF SPEECH PATH & CHMN DEPT, CALIF STATE COL, PA, 69- Concurrent Pos: Consult, WVa Pub Schs, 66, Fayette County Pub Schs, Pa, 67 & United Health Serv, 68-70; chmn, Prof Training Comt, Pa Speech & Hearing Asn, 68-70, mem exec coun, 68- Honors & Awards: Honors of Asn, Pa Speech & Hearing Asn, 75. Mem: Am Speech & Hearing Asn. Mailing Add: Dept of Speech Path & Audiol Calif State Col California PA 15419

BITTAR, EVELYN EDWARD, b Jaffa, Israel, Oct 12, 28; m 61; c 4. PHYSIOLOGY. Educ: Colby Col, BA, 51; Yale Univ, MD, 55. Prof Exp: Instr med, Med Sch, George Washington Univ, 59-61; chief med officer geriat, St Elizabeth's Hosp, Washington, DC, 61-63; Fulbright lectr nephrology, Univ Damascus, 63-64; vis scientist cell physiol, Bristol Univ, 64-65; Wellcome Trust investr, Cambridge Univ, 67-68; from vis assoc prof to assoc prof, 68-72, PROF PHYSIOL, UNIV WIS-MADISON, 72- Concurrent Pos: NIH fel, DC Gen Hosp, 60-61; Wellcome Trust fel, Oxford Univ, 64-66. Honors & Awards: Osler Medal, Am Asn Hist Med; Order of Merit, Govt of Syria, 64. Mem: Brit Biochem Soc; Brit Biophys Soc; Brit Soc Exp Biol; Am Physiol Soc; Am Biophys Soc. Res: Cell physiology with special reference to membrane transport and ion transport in single cell preparations; mode of action of hormones and drugs. Mailing Add: Dept of Physiol Univ of Wis Madison WI 53706

BITTER, GARY G, b Hoisington, Kans, Feb 2, 40; m 62; c 3. MATHEMATICS EDUCATION, COMPUTER EDUCATION. Educ: Kans State Univ, BS, 62; Kans State Teachers Col, MA, 65; Univ Denver, PhD(comput & math educ), 70. Prof Exp: Teacher math & sci, Derby High Sch, 62-65; teacher math, Forsythe Jr High Sch, 65-66; instr math, Washburn Univ, 66-67; instr comput sci & math, Colo Col, 67-70; asst prof, 70-74, ASSOC PROF MATH EDUC, ARIZ STATE UNIV, 74- Concurrent Pos: Lectr, Univ Colo, 68-69; consult, Kaman Nuclear, 67-70; supvr student teachers & comput educ res assoc, Univ Denver, 69-70. Mem: Asn Educ Data Systs; Nat Coun Teachers Math; Math Asn Am. Res: Incorporation of computer into elementary and secondary school curriculum; futuristic mathematics curricula; mathematics laboratory; metric education and the application of the hand held calculator in the school. Mailing Add: Col of Educ Ariz State Univ Tempe AZ 85281

BITTER, HAROLD LOUIS, b Baltimore, Md, Aug 24, 18; m 46; c 3. PHYSIOLOGY, TOXICOLOGY. Educ: Univ Md, BS, 49, MS, 51; Univ Rochester, PhD(physiol), 58. Prof Exp: Biologist, Agr Res Ctr, USDA, Md, 49-51; aviation physiologist, Sch Aerospace Med, US Air Force, 56-61, aviation physiologist for technol, Wright-Patterson AFB, Ohio, 61-63, chief biol systs sect, Sch Aerospace Med, 63-65, chief physiol chem, 65-68, assoc chief aviation physiologist in Air Force, 68-73; RETIRED. Mem: AAAS; Aerospace Med Asn; Human Factors Soc; NY Acad Sci. Res: Ventilation-perfusion aspects of pressure breathing; aviation physiology; physiological problems in space flight; toxicological problems involving renal, cardiovascular, pulmonary and metabolic research as related to manned space flight. Mailing Add: 4110 Valleyfield Dr San Antonio TX 78222

BITTERS, WILLARD PAUL, b Eau Claire, Wis, June 4, 15; m 40; c 2. HORTICULTURE. Educ: St Norbert Col, BA, 37; Univ Wis, MA, 40, PhD(plant physiol), 42. Prof Exp: Lab asst, Univ Wis, 39-40; res asst, 40-42; asst horticulturist, Univ Ariz, 42-26; from asst horticulturist to assoc horticulturist, 46-58, HORTICULTURIST & PROF HORT, CITRUS EXP STA & AGR RES CTR, UNIV CALIF, RIVERSIDE, 58- Res: Effect of rootstocks on long term yields; fruit quality and size; effect of rootstocks in relation to disease resistance; clonal varieties of citrus; ornamental citrus; dwarfing rootstocks. Mailing Add: 1185 La Subida Ct Riverside CA 92507

BITTINGER, MARVIN LOWELL, b Akron, Ohio, Aug 9, 41; m 65; c 1. MATHEMATICS. Educ: Manchester Col, BA, 63; Ohio State Univ, BS, 65; Purdue Univ, PhD(math educ), 68. Prof Exp: Asst prof math, Ind Univ-Purdue Univ, Indianapolis, 68-74, ASSOC PROF MATH, PURDUE UNIV, WEST LAFAYETTE, 74- Concurrent Pos: Consult, Addison-Wesley Publ Co, 70- Mem: Math Asn Am. Res: Mathematics education; trigonometry; logic and proof. Mailing Add: Dept of Math Purdue Univ West Lafayette IN 47907

BITTLE, JAMES LONG, b Norristown, Pa, Mar 7, 27; m 54; c 2. VETERINARY MICROBIOLOGY, VIROLOGY. Educ: Univ Calif, BS, 51, DVM, 53; Am Col Vet Microbiol, dipl. Prof Exp: Virologist, Virus Res Lab, Pitman-Moore Co, 58-60; supt tissue cult prod, head dept polio & virus vaccine testing & mgr biol testing, Lederle Labs, NY, 60-65; dir virus res, Pitman-Moore Co, Ind, 65-66, head dept infectious dis, Pitman-Moore Div, Dow Chem Co, 67-69, DIR RES, PITMAN-MOORE, INC, 69- Mem: Am Vet Med Asn; Soc Exp Biol & Med. Res: Viral infectious diseases; vaccine development; production; testing. Mailing Add: Pitman-Moore Inc PO Box 344 Washington Crossing NJ 08560

BITTLE, WILLIAM ELMER, b Mansfield, Ohio, May 11, 26; m 49; c 3. ETHNOLOGY, ANTHROPOLOGICAL LINGUISTICS. Educ: Univ Calif, Los Angeles, BA, 49, MA, 50, PhD(anthrop), 56. Prof Exp: Rockefeller Found grant study lang & symbolism, Univ Mich, Ann Arbor, 51-52; from instr to assoc prof ling & anthrop, 52-65, PROF ANTHROP, UNIV OKLA, 65- Concurrent Pos: Cur ethnol, Univ Okla Mus, 68-; mem bd dirs, Am Indian Theater Ensemble, 71- Res: North American Indians and their language. Mailing Add: Dept of Anthrop Univ of Okla 455 W Lindsey Norman OK 73069

BITTMAN, LORAN R, b New York, NY, Dec 24, 16; m 55; c 2. APPLIED PHYSICS. Educ: NY Univ, BA, 42, MS, 52, PhD(physics), 53. Prof Exp: Physicist, Westinghouse Elec Corp, 42-49; asst, NY Univ, 49-52; proj engr, Sperry-Gyroscope Co, 52-59; prin staff scientist, Martin Co, 59-64; res dir & prof elec eng, Univ Admin, Genesys, Univ Fla, 64-69; prin engr, Missile Systs Div, Raytheon Co, 69-71; consult, 71-73, DIR STATIST, MASS DEPT PUB WELFARE, BOSTON, 73- Mem: Fel AAAS; Am Inst Physicists; sr mem Inst Elec & Electronics Engrs. Res: Development of operations management techniques; electro-optics and holography; physical electronics research in gas and vacuum tubes leading to microwave tubes; energy conversion devices; high resolution recording techniques; laser application; infrared detectors and optical systems. Mailing Add: 39 Blake Rd Lexington MA 02173

BITTMAN, ROBERT, b New York, NY, Mar 19, 42. BIO-ORGANIC CHEMISTRY. Educ: Queens Col, NY, BS, 62; Univ Calif, Berkeley, PhD(chem), 65. Prof Exp: NSF fel chem, Max Planck Inst Phys Chem, 65-66; asst prof, 66-70, ASSOC PROF CHEM, QUEENS COL, NY, 71- Concurrent Pos: Am Cancer Soc res grant, 67-71; asst secy, Org Reactions, 68, secy, 69- Mem: NY Acad Sci. Res: Fast reactions in solution; mechanisms of biochemical and organic reactions. Mailing Add: Dept of Chem Queens Col City Univ New York Flushing NY 11367

BITTNER, BURT JAMES, b Ft Collins, Colo, Feb 8, 21; m 46; c 4. ELECTROMAGNETICS. Educ: Colo State Univ, BSEE, 43. Prof Exp: Jr engr, Submarine Signal Co, 46-48; div supvr, Sandia Corp, 48-56; vpres & res dir, Gulton Indust Div, 57-58; sr staff scientist, Kaman Sci Corp, 58-74; GROUP LEADER, INTERSTELLAR, LTD, 65-, DIR, 74-; RES PHYSICIST, UNIV COLO,

COLORADO SPRINGS, 74- Concurrent Pos: Task engr, Pac Nuclear Tests, 62; secy & pres, Air Acad Bd Educ, 63-75; mem, Group Six, Int Radio Consult Comt, 64-; consult, Army Res Off, Duke Univ, 68-; consult, Nat Ctr Atmospheric Res, 74; sr consult, Kaman Sci Corp & Kaman Aerospace, 74-; consult, Shock Tube Lab, Univ NMex. Mem: Inst Elec & Electronics Engrs; Prof Soc Protective Design. Res: Antennas; nuclear electromagnetic plasma propagation; solar furnace magnetohydrodynamics; energy-economics education. Mailing Add: 11415 Hungate Rd Black Forest CO 80908

BITTNER, JOHN WILLIAM, b Iowa, Mar 6, 26; m 64; c 2. NUCLEAR PHYSICS. Educ: Univ Western Ont, BSc, 48, MS, 50, PhD, 54. Prof Exp: Assoc physicist, 54-63, PHYSICIST, BROOKHAVEN NAT LAB, 63- Mem: Am Phys Soc; Inst Elec & Electronics Engrs. Res: Nuclear accelerators; scanning transmission electron microscope. Mailing Add: Brookhaven Nat Lab Bldg 463 Upton NY 11973

BITTON, GABRIEL, b Marrakech, Morocco, Sept 8, 40; m 70; c 1. ENVIRONMENTAL BIOLOGY. Educ: Univ Toulouse, BS & Agr Eng, 65; Laval Univ, MS, 67; Hebrew Univ, PhD(microbiol), 73. Prof Exp: Teacher biol, Acad Quebec, 66-68; res fel appl microbiol, Harvard Univ, 72-74; ASST PROF ENVIRON MICROBIOL, UNIV FLA, 74- Mem: Am Soc Microbiol; Fr Can Asn Advan Sci. Res: Microbiology of surfaces; environmental virology; flocculation of microorganisms; clay minerals and microbial ecology. Mailing Add: Dept of Environ Eng Sci 210 A P Black Hall Univ of Fla Gainesville FL 32611

BITZ, MIRIAM L, b Aberdeen, SDak, Oct 24, 21. GEOCHEMISTRY. Educ: Ore State Univ, BS, 42; Dominican Col, Calif, MA, 51; Fordham Univ, MS, 63, PhD(chem), 67. Prof Exp: Teacher parochial sch, Ore, 42-50, prin, 50-55; assoc prof phys sci, Mt Angel Col, 55-60; res chemist, Univ Calif, San Diego, 66-70; PROF PHYS SCI, MT ANGEL COL, 70- Mem: Am Chem Soc; Geochem Soc; Am Geol Soc; NY Acad Sci. Res: Organic geochemistry; ozonolysis of polymer-like material in coal, ancient rocks and meteorites; gas chromatography; mass spectrometry. Mailing Add: 3548 45th Ave NE Salem OR 97303

BITZER, CARL WILFRID, b Johnson City, Tenn, Sept 14, 33; m 56; c 2. MATHEMATICS. Educ: Duke Univ, BS, 55; Univ NC, Chapel Hill, MA, 66, PhD(math), 68. Prof Exp: Programmer, US Naval Ord Test Sta, Calif, 56-63; asst prof math, Univ NC, Greensboro, 67-73; COMPUT PROGRAMMER, McDERMOTT HUDSON ENG, 74- Mem: Am Math Soc. Res: Stieltjes integral equations. Mailing Add: J Ray McDermott 1010 Common St New Orleans LA 70112

BITZER, MORRIS JAY, b Huntington, Ind, Mar 3, 36; m 56; c 3. AGRONOMY. Educ: Purdue Univ, BS, 58, MS, 65, PhD(plant breeding, genetics), 68. Prof Exp: Teacher, high schs, Ind, 59-63; res asst hybrid wheat, Purdue Univ, 63-68; asst prof agron, Univ Ga, 68-72; EXTEN SPECIALIST GRAIN CROPS, UNIV KY, 72- Mem: Am Soc Agron; Crop Sci Soc Am. Res: Hybrid vigor and gene action in hybrid wheat; pollen dispersal and seed set for hybrid wheat production; effect of nitrogen on yields of dwarf wheats; vernalization of soft winter wheats. Mailing Add: Dept of Agron Univ of Ky Lexington KY 40506

BIVENS, RICHARD LOWELL, b Denver, Ill, Aug 19, 39; m 63; c 2. PHYSICAL CHEMISTRY. Educ: Monmouth Col, Ill, BA, 61; Case Western Reserve Univ, MS, 63, PhD(chem), 69. Prof Exp: Asst prof, 65-72, ASSOC PROF CHEM, ALLEGHENY COL, 72- Mem: Am Chem Soc. Res: Thermodynamics and statistical mechanics of phase transitions. Mailing Add: Dept of Chem Allegheny Col Meadville PA 16335

BIVER, CARL JOHN, JR, b Owensboro, Ky, July 1, 32; m 58; c 4. APPLIED PHYSICS, MATERIALS SCIENCE. Educ: Univ Notre Dame, BS, 54, MS, 56. Prof Exp: Asst scientist metall, Argonne Nat Lab, 56-59; from physicist to sr physicist, X-ray Dept, 59-66, SR PHYSICIST, NEUTRON DEVICES DEPT, GEN ELEC CO, 66- Honors & Awards: Gen Elec Co Inventors Award, 72. Mem: Am Vacuum Soc; AAAS; Am Soc Metals. Res: Gaseous permeation phenomena; high pressure physics; thermoelectric devices physics; photoconductivity, heat transfer, metallurgical and chemical kinetics; mathematical modeling. Mailing Add: 4 Westwood Lane Belleair Clearwater FL 33516

BIXBY, JOHN N, b Appleton, Wis, Dec 25, 14; m 38; c 3. FOOD CHEMISTRY, BIOCHEMISTRY. Educ: Univ Wis, BS, 37, MS, 53, Prof Exp: Biochemist, Wander Co, Ill, 37-51, res mgr foods, 54-63; asst dairy chem, Univ Wis, 51-54; tech dir, Crystal Dairy Prod, Inc, 63-64 & H C Christians Co, 64-70; SR SCIENTIST, LAND O'LAKES, INC, 70- Mem: Am Dairy Sci Asn; Am Inst Food Technologists; Am Chem Soc. Res: Human nutrition and biochemistry as related to the dairy and food industry; new and existing food product development; food and dairy processing techniques and sanitation. Mailing Add: Land O'Lakes Inc 614 McKinley Pl Minneapolis MN 55413

BIXBY, ROBERT EUGENE, b Oakland, Calif, Sept 14, 45; m 66; c 2. OPERATIONS RESEARCH. Educ: Univ Calif, Berkeley, BS, 68; Cornell Univ, MS, 70, PhD(opers res), 72. Prof Exp: ASST PROF MATH, UNIV KY, 72- Concurrent Pos: Vis prof, Math Res Ctr, Univ Wis, 74-75. Mem: Am Math Soc. Res: Games matroids and graphs; combinatorial optimization, matroid theory and linear programming. Mailing Add: Dept of Math Univ of Ky Lexington KY 40506

BIXBY, WILLIAM ELLIS, b Rockford, Ill, Dec 5, 20; m 46; c 2. PHYSICS. Educ: Northern Ill State Col, BSE, 44. Prof Exp: Asst instr physics, Wabash Col, 43-45; asst, Ohio State Univ, 45-46; res engr, Battelle Mem Inst, 46-59; chief res engr, Micro-data Div, Bell & Howell Co, 59-67, dir bus equip group, Advan Develop Dept, 67-71; SR SCIENTIST, CHEM RES DEPT, A B DICK CO, 71- Honors & Awards: Kosar Mem Award, Soc Photog Scientists & Engrs, 72. Mem: Soc Photog Scientists & Engrs; Soc Motion Picture & TV Engrs; Nat Microfilm Asn. Res: Photoconductivity in thin films; vacuum deposition; electrostatics, xerography micro-imaging. Mailing Add: A B Dick Co 5700 W Touhy Ave Chicago IL 60648

BIXLER, A L M, b Upton, Mass, Aug 26, 09; m 37; c 3. PAPER CHEMISTRY. Educ: Pa State Univ, BS, 33; Lawrence Col, MS, 35, PhD, 37. Prof Exp: Res & develop engr, Riegel Paper Corp, 37-46, dir res & develop, 46-48, asst supt, Milford Mill, 48, dir qual control, 49, supt, Warren Mill, 50-55, mgr, Upper Mills, 55-56 & Paper Prod, Carolina Div, 56-58, mfg servs, 59; vpres mfg, Hamilton Paper Co, 59-61; vpres mfg & mgr, Miguon Mill, Weyerhaeuser Co, 61-62, group leader res ctr paper div, 63-71, scientist new technol invest & eval, 71-75; CONSULT, PULP & PAPER INDUST, 75- Mem: Tech Asn Pulp & Paper Indust. Res: New technology investigation; engineering, operating and product problems; dry forming. Mailing Add: Apt 9 2269 Glencoe Hills Dr Ann Arbor MI 48104

BIXLER, DAVID, b Chicago, Ill, Jan 7, 29; m 51; c 5. GENETICS, DENTISTRY. Educ: Ind Univ, AB, 50, PhD(zool), 56, DDS, 59. Prof Exp: Asst prof dent sci, 59-66, asst prof med genetics, 66-70, assoc prof dent sci & med genetics, 70-74, PROF MED GENETICS & ORAL-FACIAL GENETICS & CHMN DEPT ORAL-FACIAL GENETICS, SCHS DENT & MED, IND UNIV, INDIANAPOLIS, 74- Concurrent Pos: USPHS fel, Sch Dent, Ind Univ, Indianapolis, 56-58; USPHS career develop award, 67-72. Mem: AAAS; Soc Human Genetics; Int Asn Dent Res. Res: Histopathology of salivary glands; genetics of cleft lip and palate; hereditary anomalies of oral structures. Mailing Add: Ind Univ Sch of Dent 1121 W Michigan St Indianapolis IN 46202

BIXLER, DEAN A, b Gretna, Nebr, Oct 24, 19; m 41; c 2. TEXTILE CHEMISTRY. Educ: Univ Nebr, Lincoln, BS, 40, MS, 46; Purdue Univ, PhD(agr biol), 53. Prof Exp: DIR RES, CONE MILLS CORP, 69- Mailing Add: 4108 Redwine Greensboro NC 27410

BIXLER, HARRIS JACOB, b Harrisburg, Pa, Dec 14, 31; m 56; c 2. PHYSICS, CHEMISTRY. Educ: Mass Inst Technol, SB, 53, SM, 58, ScD(chem eng), 60; Univ Toronto, MASc, 57. Prof Exp: Res assoc chem eng, Mass Inst Technol, 60, asst prof, 60-64; vpres res & dir, Amicron Corp, 64-70; vpres & tech dir, 70-71, exec vpres, 71-75, PRES & DIR, MARINE COLLOIDS, 75- Concurrent Pos: Consult, Butcher Polish Co, 60-64. Mem: AAAS; Am Chem Soc; Am Inst Chem Engrs. Res: Structure-polymer relationships in high polymers; surface chemistry; polymer and small molecule crystallization; rheology of disperse systems; semipermeability of polymers; biomedical engineering and process engineering. Mailing Add: Marine Colloids PO Box 308 Rockland ME 04841

BIXLER, JOHN WILSON, b Eau Claire, Wis, July 27, 37. ANALYTICAL CHEMISTRY. Educ: Lakehead Col, BS, 59; Univ Minn, PhD(anal chem), 63. Prof Exp: Instr chem, Lake Forest Col, 63-65, asst prof, 65-69; ASSOC PROF CHEM, STATE UNIV NY COL BROCKPORT, 69- Concurrent Pos: Vis prof, Purdue Univ, 75-76. Mem: Am Chem Soc. Res: Electroanalytical chemistry; chemical instrumentation. Mailing Add: Dept of Chem State Univ of NY Col Brockport NY 14420

BIXLER, MILO EVERETT, chemistry, see 12th edition

BIZZELL, OSCAR MCARTHUR, b Newton Grove, NC, Sept 26, 21; m 43; c 3. NUCLEAR CHEMISTRY. Educ: Univ NC, BS, 42. Prof Exp: Chemist anal div, US Rubber Co, MC, 42-43; chemist chem div, Oak Ridge Nat Lab, 46-48; adv field serv br, 48-53, asst chief radiol safety br, 53-54, chief tech develop br, 54-56, chief isotopes sect div civilian appl, 56-58, chief isotopes appl br, US Atomic Energy Comn, 62-72; mem, Presidential Prizes Staff, Exec Off of the President, 72-73; NUCLEAR CONSULT, 73- Mem: Am Chem Soc; Am Nuclear Soc. Res: Isotopes applications; radiochemistry; activation analysis. Mailing Add: 16501 Walnut Hill Rd Gaithersburg MD 20760

BJARNGARD, BENGT E, b N Akarp, Sweden, Oct 2, 34; m 58; c 3. RADIOLOGICAL PHYSICS. Educ: Univ Lund, Fil Mag, 58, Fil Lic(radiation physics), 62. Prof Exp: Asst radiation physics, Univ Lund, 59-61; res physicist, Atomic Energy Co, Sweden, 61-65; dir radiation physics, Controls for Radiation, Inc, 65-68; asst prof, 68-74, ASSOC PROF RADIOTHER, HARVARD MED SCH, 74-, DIR PHYSICS, 68- Concurrent Pos: Lectr, Harvard Sch Pub Health. Mem: Am Asn Physicists in Med; Health Physics Soc. Res: Dosimetry; medical radiological physics. Mailing Add: Dept of Radiother Harvard Med Sch Boston MA 02115

BJERKNES, JACOB (AALL BONNEVIE), meteorology, deceased

BJORK, CARL KENNETH, SR, b Burlington, Iowa, Mar 16, 26; m 47; c 7. AGRICULTURAL CHEMISTRY. Educ: Augustana Col, AB, 47; DePauw Univ, MS, 50; Univ Ky, PhD(inorg chem), 53. Prof Exp: Res chemist, Int Mining & Chem Co, 53-54; res & develop engr, 54-59, patent agt & group leader, 59-70, MGR BIOPROD SECT, PATENT DEPT, DOW CHEM CO, 70- Mem: Am Chem Soc; Sigma Xi. Res: Patent law; patent application and prosecution. Mailing Add: 2712 Lambros Dr Midland MI 48640

BJORK, PHILIP R, b Wyandotte, Mich, Sept 14, 40; m 64; c 2. VERTEBRATE PALEONTOLOGY. Educ: Univ Mich, BS, 62, PhD(geol), 68; SDak Sch Mines & Technol, MS, 64. Prof Exp: From asst prof to assoc prof geol, Univ Wis-Stevens Point, 68-75; ASSOC PROF GEOL, S DAK SCH MINES & TECHNOL & DIR MUS GEOL, 75- Mem: Soc Vert Paleont; Am Paleont Soc; AAAS; Am Soc Naturalists; Am Quaternary Asn. Res: Blancan carnivora in North America; Oligocene faunas of North America; Blancan faunas of North America. Mailing Add: Mus of Geol S Dak Sch of Mines & Technol Rapid City SD 57701

BJORKHOLM, PAUL J, b Milwaukee, Wis, May 27, 42; m 67; c 2. ASTROPHYSICS. Educ: Princeton Univ, BA, 64; Univ Wis, MA, 65, PhD(low energy nuclear physics), 69. Prof Exp: Sr scientist x-ray astron & lunar geol, 67-73, SR STAFF SCIENTIST, AM SCI & ENG, INC, 73- Mem: AAAS; Am Phys Soc; Am Astron Soc. Res: X-ray astronomy; design and development of x-ray imaging techniques; autoradiographic cervical cancer screening techniques; lunar geophysics; remote sensing of elemental composition, natural and induced radioactivity. Mailing Add: Am Sci & Eng Inc 955 Massachussetts Ave Cambridge MA 02142

BJORKLUND, ELAINE M, b Ogden, Utah, May 8, 28; m. GEOGRAPHY. Educ: Univ Utah, BS, 49; Univ Wash, MA, 51; Univ Chicago, PhD(geog), 55. Prof Exp: Geogr, Res & Planning Sect, US Bur Census, 55-56; from asst prof to assoc prof geog, Vassar Col, 56-66; assoc prof, 67-71, PROF GEOG, UNIV WESTERN ONT, 71- Mem: Asn Am Geogrs; Soc Woman Geogrs. Res: Human, cultural and urban geography; areal functional organization. Mailing Add: Dept of Geog Univ of Western Ont London ON Can

BJORKLUND, GORDON HERBERT, organic chemistry, see 12th edition

BJORKLUND, RICHARD GUY, b Milwaukee, Wis, Feb 10, 28; m 54; c 5. ECOLOGY, CONSERVATION. Educ: Mont State Univ, BS, 51, MS, 53; Univ Mich, PhD, 58. Prof Exp: Res asst, Univ Mich, 54-55; from instr to assoc prof biol, 57-68, chmn dept, 64-70, PROF BIOL, BRADLEY UNIV, 68-, DEAN LIB ARTS & SCI, 73- Concurrent Pos: Asst blister rust control, US Forest Serv, 48; aide, Inst Fish Res, 53-54. Mem: Am Inst Biol Sci; Am Fisheries Soc; Wilson Ornith Soc. Res: Environmental biology; temperature sense; photoperiodism; endocrine functions; aquatic biology; ecology of herons; population ecology and reproductive behavior of other birds. Mailing Add: Lib Arts & Sci Bradley Univ Peoria IL 61625

BJORKLUND, RUSSELL FOSTER, biophysics, physics, see 12th edition

BJÖRKMAN, OLLE, b Jönköping, Sweden, July 29, 33; m 55; c 2. PHYSIOLOGICAL ECOLOGY, PHOTOBIOLOGY. Educ: Univ Stockholm, MS, 57; Univ Uppsala, PhD, 60. Prof Exp: Asst scientist, Dept Genetics & Plant Breeding, Royal Agr Col Sweden & Univ Uppsala, 57-61, res fel, Swedish Natural Sci Res Coun, 61-63; res fel, 64-65, STAFF BIOLOGIST, DEPT PLANT BIOL, CARNEGIE INST WASH, CALIF, 66- Concurrent Pos: Assoc courtesy prof biol, Stanford Univ, 67-74, consult prof biol, 75-; mem fac, Univ Uppsala. Mem: AAAS; Am Soc Plant Physiologists; Am Soc Photobiol; Scand Soc Plant Physiol. Res: Physiological and biochemical mechanisms of plant response and adaptation to ecologically diverse environments; environmental and biological control of photosynthesis. Mailing Add: Dept of Plant Biol Carnegie Inst of Wash Stanford CA 94305

BJORKSTEN, JOHAN AUGUSTUS, b Tammerfors, Finland, May 27, 07; nat US; m 61; c 4. CHEMISTRY. Educ: Univ Helsinki, MS, 27, PhD(protein chem), 31. Prof Exp: Int Educ Bd fel, Univ Minn, 31-32; res chemist, Fleton Chem Co, 33-34, chief chemist, 34-35; res chemist, Pepsodent Co, 35, in-chg develop, 36; chief chemist, Ditto, Inc, 36-41; chem dir, Quaker Chem Prod Corp, Pa, 41-44; CHMN, BD DIRS, BJORKSTEN RES LABS, 44- Concurrent Pos: Vpres, ABC Packaging Mach Corp, Fla, 40- & Bee Chem Co, Ill, 44-57; pres, Bjorksten Res Found, 53-; dir, Griffolyn Col, Inc, Tex, 56- Mem: AAAS; Am Chem Soc; Am Asn Cereal Chemists; Am Asn Textile Chemists & Colorists; fel Am Inst Chemists (pres, 62-63). Res: Synthetic resins and plastics; paper and fibrous materials; special coating compositions; proteins; industrial problems; chemical gerontology. Mailing Add: Bjorksten Res Labs Box 9444 Madison WI 53715

BJORNDAL, ARNE MAGNE, b Ulstein, Norway, Aug 19, 16; m 52; c 3. DENTISTRY. Educ: Volda State Col, BS, 39; Univ Oslo, DDS, 47; Univ Iowa, MS & DDS, 56; Am Bd Endodont, dipl, 64. Prof Exp: Instr pedodont, 48-52; from instr to assoc prof, 55-62, PROF OPERATIVE DENT & ENDODONT, COL DENT, UNIV IOWA, 62- Concurrent Pos: Mem, Coun Med TV, 65- Mem: AAAS; Am Asn Endodont; Int Asn Dent Res; Norweg Dent Asn. Res: Endodontics in relation to hemophiliacs and blood dyscrasias. Mailing Add: Col of Dent Univ of Iowa Iowa City IA 52240

BJORNERUD, EGIL KRISTOFFER, b Hamar, Norway, May 3, 25; nat US; m 59; c 4. PHYSICS. Educ: Univ Wash, BS, 49; Calif Inst Technol, PhD(physics), 54. Prof Exp: Staff mem physics dept, Rand Corp, Calif, 54-55; reactor dept, Inst Atomic Energy, Norway, 55-56; res assoc, Gen Atomic Div, Gen Dynamics Corp, 56-59, asst export mgr, 59-62; assoc dir sci div, Nat Eng Sci Co, Calif, 62-64; chief nuclear technol aerospace div, 64-67, DIR LASER DEVELOP, RES DIV, BOEING CO, 70- Concurrent Pos: Consult, Atomic Energy Comn, India, 59-60. Mem: Am Phys Soc; Am Nuclear Soc; Am Inst Aeronaut & Astronaut; fel Royal Astron Soc. Res: Cloud-chamber studies of elementary particle interactions; experimental studies of flame emission spectra; neutron diffusion and criticality experiments; reactor physics and nuclear technology. Mailing Add: 3821 E Prospect Seattle WA 98102

BJORNSON, AUGUST SVEN, b Reykjavik, Iceland, May 10, 22; nat US; m 44; c 3. ORGANIC CHEMISTRY. Educ: Univ Wis, BSc, 44, MSc, 45, Univ Kans, PhD(chem), 48. Prof Exp: Res chemist, 48-51, group leader, 51-53, supvr, 53-54, supvr develop dept, 54-56, res mgr, 56-60, res mgr new prod develop, 60-62, mgr develop planning, 62-66, MGR PATENTS & CONTRACTS, E I DU PONT DE NEMOURS & CO, INC, 66- Mem: AAAS; Am Chem Soc. Res: Negotiations and administration of technical agreements. Mailing Add: E I du Pont de Nemours & Co Inc Wilmington DE 19801

BJUGSTAD, ARDELL JEROME, b Sheldon, NDak, Apr 28, 33; m 59; c 2. RANGE SCIENCE. Educ: NDak State Univ, BS, 59, PhD(range mgt), 65. Prof Exp: Res asst range res, NDak State Univ, 59-63; range conservationist, Cent States Forest Res Ctr, 63-64, range scientist, NCent Forest Exp Sta, 64-68, proj leader, 68-71, asst br chief, 71-73, PROJ LEADER, ROCKY MOUNTAIN FOREST & RANGE EXP STA, US FOREST SERV, 73- Mem: Am Inst Biol Sci; Am Soc Range Mgt; Soc Am Foresters; Wildlife Soc; Nat Wildlife Fedn. Res: Range and wildlife habitat ecology and management; watershed management; silviculture. Mailing Add: USDA Forest Serv Forest Res Lab SDSM&T Rapid City SD 57701

BLACHLY, PAUL H, b Portland, Ore, Dec 4, 29; m 56; c 6. PSYCHIATRY. Educ: Reed Col, BA, 50; Univ Ore, MD & MS, 55. Prof Exp: Dep chief addict serv, USPHS Hosp, Ft Worth, Tex, 59-61; from asst prof to assoc prof, 61-68, PROF PSYCHIAT, MED SCH, UNIV ORE, 68- Concurrent Pos: Consult, Vet Admin Hosp, Portland, 64- Mem: Am Psychiat Asn; AMA. Res: Psychophysiologic aspects of cardiovascular disease; psychopharmacology; electroconvulsive therapy; suicide; drug addiction. Mailing Add: Dept of Psychiat Univ of Ore Med Sch Portland OR 97201

BLACHMAN, ARTHUR GILBERT, b Cleveland, Ohio, June 29, 26; m 53; c 3. PHYSICS. Educ: Western Reserve Univ, BS, 50; Ohio State Univ, MS, 56; NY Univ, PhD(physics), 66. Prof Exp: Physicist res dept, Erie Resistor Corp, Pa, 50-51; physicist digital comput lab, Mass Inst Technol, 51-52; teacher high sch, Colo, 52-54; res staff mem, Watson Lab, New York, 56-65, RES STAFF MEM RES CTR, IBM CORP, YORKTOWN HEIGHTS, 65- Mem: AAAS; Am Phys Soc; Am Vacuum Soc. Res: Semiconductor physics; hyperfine structure of atomic metastable states; structure and properties of sputtered thin films. Mailing Add: 215 Cedar Dr E Briarcliff Manor NY 10510

BLACHMAN, NELSON MERLE, b Cleveland, Ohio, Oct 27, 23; m 53; c 2. PHYSICS. Educ: Case Inst Technol, BS, 43; Harvard Univ, AM & PhD(eng sci, appl physics), 47. Prof Exp: Spec res assoc, Underwater Sound Lab, Harvard Univ, 43-45, res assoc, Ctr Commun Res, 45-46, Cruft Lab, 46; assoc scientist, Brookhaven Nat Lab, 47-51; physicist, Comput Br, Off Naval Res, 51-53, math br, 53-54, SR SCIENTIST, WESTERN DIV, SYLVANIA ELECTRONIC SYSTS GROUP, GEN TEL & ELECTRONICS CORP, 54- Concurrent Pos: Lectr, Univ Md, 51-52; mem, Comn VI, Int Sci Radio Union, 57-; sci liaison officer, London Br, Off Naval Res, 58-60; lectr, Eng Exten, Univ Calif, 61-63; lectr, Stanford Univ, 67. Mem: Fel AAAS; Inst Math Statist; fel Inst Elec & Electronics Engrs; Soc Indust & Appl Math; fel Inst Elec Engrs London, Res: Statistical communication theory; information theory. Mailing Add: 443 Ferne Ave Palo Alto CA 94306

BLACK, ALEX, b Richmond, Ky, Nov 16, 06; m 29. ANIMAL NUTRITION. Educ: Univ Ky, AB, 29; Pa State Univ, MS, 33; Univ Rochester, PhD(nutrit), 38. Prof Exp: Asst, 29-34, instr, 34-38, from asst prof to prof, 38-69, asst dir agr exp sta, 53-60, assoc dir, 60-69, EMER PROF ANIMAL NUTRIT, PA STATE UNIV, 69- Mem: Am Soc Animal Sci; Am Inst Nutrit; NY Acad Sci. Res: Human and animal nutrition; energy metabolism; protein, fat and mineral metabolism. Mailing Add: 126 W Mitchell Ave State College PA 16801

BLACK, ARTHUR HERMAN, b Toledo, Ohio, Aug 16, 19; m 45; c 3. ANALYTICAL CHEMISTRY. Educ: Univ Toledo, BS, 41, MS, 48; Univ Mich, MS, 55. Prof Exp: From instr to assoc prof chem, 46-61, asst dean col arts & sci, 61-64; dean men, 64-68, assoc res found; ASSOC DEAN COL ARTS & SCI, UNIV TOLEDO, 68-, PROF CHEM, 70- Mem: AAAS; NY Acad Sci; Am Chem Soc. Res: Volumetric and colorimetric methods involving various chelating agents. Mailing Add: 3209 Kylemore Rd Toledo OH 43606

BLACK, ARTHUR LEO, b Redlands, Calif, Dec 1, 22; m 45; c 3. BIOCHEMISTRY. Educ: Univ Calif, BS, 48, PhD(comp physiol), 51. Prof Exp: From instr to assoc prof, 51-62, PROF BIOCHEM, UNIV CALIF, DAVIS, 62- Honors & Awards: Borden Award, Am Inst Nutrit, 63. Mem: AAAS; Am Soc Biol Chemists; Am Inst Nutrit; Am Physiol Soc. Res: Gluconeogenesis and its control; biosynthesis of amino acids in animals; study of intermediary metabolism in intact animals, especially ruminants. Mailing Add: 891 Linden Lane Davis CA 95616

BLACK, BENJAMIN MARDEN, b Salt Lake City, Utah, July 4, 10; m 42. SURGERY. Educ: Stanford Univ, AB, 31, AM, 33, MD, 36; Univ Minn, MS, 40; Am Bd Surg, dipl, 42. Prof Exp: Instr anat, Stanford Univ, 31-32; asst, 39-41, mem sect surg, Mayo Clin, 42-74, from instr to prof, Mayo Grad Sch Med, 43-74, EMER PROF SURG, MAYO MED SCH, 74- Res: Diseases and surgery of neck and abdomen. Mailing Add: 10918 Amber Trail Sun City AZ 85351

BLACK, BILLY C, II, b Beatrice, Ala, Feb 1, 37; m 61; c 1. BIOCHEMISTRY, ANALYTICAL CHEMISTRY. Educ: Tuskegee Inst, BS, 60; Iowa State Univ, MS, 62, PhD(biochem), 64. Prof Exp: Res asst biochem, Iowa State Univ, 60-64; PROF CHEM, ALBANY STATE COL, 64-, CHMN DEPT, 66- Concurrent Pos: Chmn div sci & math, 69-70. Mem: Am Chem Soc; Am Oil Chem Soc; Int Food Technologists; Am Inst Chemists. Res: Fatty acid metabolism; molecular dynamics; glyceride structure. Mailing Add: Albany State Col PO Box 440 Albany GA 31705

BLACK, BRUCE ALLEN, b Price, Utah, Feb 24, 38; m 55; c 5. GEOLOGY. Educ: Brigham Young Univ, BS, 64, MS, 65; Univ Wis-Madison, PhD(geol), 69. Prof Exp: Res geologist, res & develop div, 68-70, mgt analyst, Exec Dept, 70-74, DIST GEOLOGIST, PHILLIPS PETROL CO, 74- Res: Stratigraphy and structure in the southern Wasatch Mountains, Utah, and sedimentary structures in the Ouachita Mountains of Oklahoma and Arkansas; sedimentology; marine geology and mineral resources; geothermal and uranium exploration geology. Mailing Add: 9400 La Playa Ave NE Albuquerque NM 87111

BLACK, CARL (ELLSWORTH), b Jacksonville, Ill, May 19, 20; m 43; c 2. POLYMER SCIENCE. Educ: Ill Col, AB, 40; Northwestern Univ, PhD(chem), 43. Hon Degrees: DSc, Ill Col, 68. Prof Exp: Res chemist, Naval Res Lab, Anacostia Sta, Washington, DC, 43-45; res chemist, 45-52, res mgr, 52-56, dir, Carothers Res Lab, 56-60, prod mgr, Textile Fibers Dept, 60-67, tech dir, 67-69, dir, Qiana Div, 69-75, DIR, QIANA, LYCRA & HOSIERY FIBERS DIV, TEXTILE FIBERS DEPT, E I DU PONT DE NEMOURS & CO, 75- Mem: Am Chem Soc. Res: X-ray crystallography; structure and physical properties of high polymers; corrosion of metals; piezoelectric crystals; infrared absorption spectra; synthetic fibers. Mailing Add: Textile Fibers Dept E I du Pont de Nemours & Co Wilmington DE 19898

BLACK, CHARLES ALLEN, b Lone Tree, Iowa, Jan 22, 16; m 39; c 3. SOIL FERTILITY. Educ: Colo State Univ, BS, 37; Iowa State Univ, MS, 38, PhD(soil fertil), 42. Prof Exp: Asst land classifier, Resettlement Admin, 37; jr soil surveyor, Soil Conserv Serv, USDA, 38; instr soils, 39-42, res assoc, 43-44, from res asst prof to res assoc prof, 44-49, prof, 49-67, DISTINGUISHED PROF SOILS, IOWA STATE UNIV, 67- Concurrent Pos: Vis prof, Cornell Univ, 55-56; Kearney Found lectr, Univ Calif, 60; NSF fel, 64-65; consult, USDA, 64; chmn bd dirs, Coun Agr Sci & Technol, 72, pres, 73, exec vpres, 74- Honors & Awards: Soil Sci Award, Am Soc Agron, 57. Mem: AAAS; fel Am Soc Agron (pres, 70-71); Soil Sci Soc Am (pres, 62); fel Am Inst Chemists; Int Soc Soil Sci. Res: Soil phosphorus and soil-plant relationships; selection by sequential elimination based on concordant judgments. Mailing Add: Dept of Agron Iowa State Univ Ames IA 50011

BLACK, CLANTON CANDLER, JR, b Tampa, Fla, Nov 27, 31; m 52; c 3. BIOCHEMISTRY. Educ: Univ Fla, BSA, 53, MSA, 57, PhD(agron), 60. Prof Exp: NIH fel biochem, Cornell Univ, 60-62; fel, Kettering Res Lab, 62-63, staff scientist, 63-67; asst prof biol, Antioch Col, 63-67; from assoc prof to prof biochem, 67-74, PROF BOT & HEAD DEPT, UNIV GA, 74- Mem: Am Soc Biol Chemists; Am Soc Plant Physiologists. Res: Photosynthesis; enzymology; electron transport; photophosphorylation. Mailing Add: Dept of Biochem GSRC Univ of Ga Athens GA 30601

BLACK, CRAIG C, b Peking, China, May 28, 32; US citizen; m 54, 67; c 2. VERTEBRATE PALEONTOLOGY. Educ: Amherst Col, AB, 54, MA, 57; Harvard Univ, PhD(biol), 62. Prof Exp: Assoc cur vert paleont, Carnegie Mus, 60-63, cur, 63-70; assoc prof ecol & systs, Mus Natural Hist, Univ Kans, 70-72; dir mus & prof Geosci, Tex Tech Univ, 72-75; DIR, CARNEGIE MUS NATURAL HIST, 75- Concurrent Pos: Mem, Mus Panel, Nat Endowment Arts, 74-77. Mem: Soc Vert Paleont (pres, 70-71); fel Geol Soc Am; Soc Study Evolution; fel Linnean Soc London; AAAS. Res: Tertiary mammals; paleoecology and biogeography; rodent evolution. Mailing Add: Carnegie Mus of Nat Hist 4400 Forbes Ave Pittsburgh PA 15213

BLACK, CRAIG PATRICK, b Greeley, Colo, Apr 25, 46; m 69. ANIMAL PHYSIOLOGY, ANIMAL ECOLOGY. Educ: Tufts Univ, BS, 68; Dartmouth Col, PhD(biol), 75. Prof Exp: Lectr, Dept Biol Sci, 74-75, NIH FEL, DEPT PHYSIOL, SCH MED, DARTMOUTH COL, 75- Concurrent Pos: Vis asst prof, Dept Biol Sci, Dartmouth Col, 75-76. Mem: Ecol Soc Am; Cooper Ornith Soc; Am Ornithologists Union. Res: Avian energetics; avian respiration physiology; high altitude physiology. Mailing Add: Dept of Physiol Dartmouth Col Sch of Med Hanover NH 03755

BLACK, DAVID CHARLES, b Waterloo, Iowa, May 14, 43; m 67. THEORETICAL ASTROPHYSICS, METEORITICS. Educ: Univ Minn, BS, 65, MS, 67, PhD(physics), 70. Prof Exp: Nat Acad Sci fel, 70-72, RES SCIENTIST THEORET ASTROPHYS, NASA AMES RES CTR, 72- Res: Theoretical studies of the formation and evolution of stars and planetary systems; interpretation of rare gas isotopic data from meteorites and lunar samples. Mailing Add: Theoret Studies Br Space Sci Div NASA Ames Res Ctr Moffett Field CA 94035

BLACK, DONALD K, b Ely, Eng, Feb 23, 36. ORGANIC CHEMISTRY. Educ: Univ London, BSc, 60, PhD(org chem), 63. Prof Exp: Res chemist, Middlesex Hosp Med Sch, London, 64-66; Nat Acad Sci resident res assoc org chem, Naval Stores Res Sta, USDA, 66-68; RES CHEMIST, RES CTR, HERCULES, INC, 68- Mem: The Chem Soc. Res: Diterpene, resin acid and pesticide chemistry. Mailing Add: Res Ctr Hercules Inc Wilmington DE 19899

BLACK, DONALD LEE, b Lexington, NC, Oct 10, 43; m 73; c 1. PHYSICAL CHEMISTRY, ELECTRON MICROSCOPY. Educ: Wake Forest Univ, BS, 66; Univ Mich, MS, 70, PhD(phys chem), 71. Prof Exp: Teaching fel chem, Univ Mich, 67-68; SR RES CHEMIST, EASTMAN KODAK CO RES LABS, 70- Mem: Am Chem Soc; Sigma Xi. Res: Colloid stability studies of photographic components using electron and optical microscopic techniques; dye fading mechanisms in photographic chemistry; surface chemistry studies; molecular spectroscopy and structure. Mailing Add: Eastman Kodak Co Res Labs 343 State St Rochester NY 14650

BLACK, DONALD LEIGHTON, b Portland, Maine, Aug 3, 28; m 50; c 3. ANIMAL PHYSIOLOGY. Educ: Univ Maine, BS, 54; Cornell Univ, MS, 57, PhD(animal physiol), 59. Prof Exp: Assoc prof animal physiol, 59-68, PROF VET & ANIMAL SCI, UNIV MASS, AMHERST, 68- Res: Oviduct physiology; physiology of ovarian function. Mailing Add: Dept of Animal Sci Univ of Mass Amherst MA 01002

BLACK, DONALD MILLER, chemistry, see 12th edition

BLACK, EMILIE A, b New Haven, Conn; m 45; c 1. MEDICAL ADMINISTRATION, PEDIATRICS. Educ: George Washington Univ, BS, 42, MD, 45. Prof Exp: Pvt pediat pract, 49-66; med officer pediat, DC Dept Pub Health, 63-68; prog adminr clin sci, 68-72, asst chief RGB, 72-73, dep dir sci prog, 74-75, DIR CLIN & PHYSIOL SCI PROG, NAT INST GEN MED SCI, 76- Mem: Int Asn Study Pain; Int Soc Burn Injuries; fel Am Asn Surg Trauma; Am Burn Asn; Am Trauma Soc. Res: Pathophysiology of severe burns and trauma; basic physiologic and metabolic changes associated with anesthetic agents; nature and control of pain, mode of action of anesthetic agents. Mailing Add: Nat Inst of Gen Med Sci 5333 Westbard Ave Bethesda MD 20016

BLACK, EMMETT RUSSELL, JR, b Kosciusko, Miss, Oct 1, 44; m 68; c 1. ENTOMOLOGY. Educ: Miss State Univ, BS, 67, MS, 69, PhD(entom), 72; Am Registry Prof Entomologists, cert. Prof Exp: Res asst entom, Southern Grain Insects, USDA, Miss State Univ, 66-69, res asst, Dept Entom, 69-72; CONSULT ENTOMOLOGIST, 72- Mem: Entom Soc Am. Res: Pest management of agricultural insects. Mailing Add: PO Box 922 Indianola MS 38751

BLACK, FRANCIS LEE, b Taipei, Formosa, Jan 2, 26; Can citizen; m 47; c 3. VIROLOGY. Educ: Univ BC, BA, 47, MA, 49; Univ Calif, PhD(biochem), 52. Prof Exp: Fel virus chem, Rockefeller Found, 52-54; chemist viruses, Lab Hyg, Dept Nat Health & Welfare, Can, 54-55; res assoc & lectr, 55-61, from asst prof to assoc prof, 61-73, PROF EPIDEMIOL, SCH MED, YALE UNIV, 73- Mem: Am Asn Immunol; Soc Exp Biol & Med; Am Epidemiol Soc; Sigma Xi. Res: Chemistry and functions of viruses; viral genetics; measles virus; infectious disease patterns in isolated populations; South American Indians. Mailing Add: Dept of Epidemiol Yale Univ Sch of Med New Haven CT 06510

BLACK, GRAHAM, b Sheffield, Eng, July 21, 38. PHYSICAL CHEMISTRY. Educ: Univ Sheffield, BSc, 58, PhD(chem), 61. Prof Exp: Res fel chem, Univ Calif, Riverside, 61-62; PHYSICIST, STANFORD RES INST, 63- Res: Charge transfer, atom recombination and light absorption as means of producing excited atoms and molecules; physical and chemical properties of these excited species. Mailing Add: Stanford Res Inst Menlo Park CA 94025

BLACK, HOMER SELTON, b Port Arthur, Tex, Sept 9, 35; m 68. CANCER. Educ: Agr & Mech Col Tex, BS, 56; Sam Houston State Teachers Col, MEd, 60; La State Univ, PhD(bot), 64. Prof Exp: Res chemist, Seed Protein Pioneering Res Lab, Southern Utilization Res & Develop Div, USDA, 64-65; res fel biochem, Univ Tex-M D Anderson Hosp & Tumor Inst, 65-66; asst prof biol, Sam Houston State Univ, 66-68; res asst prof dermat, 68-74, adj asst prof biochem, 69-74, ASSOC RES PROF DERMAT, BAYLOR COL MED, 74-; PHYSIOLOGIST, PHOTOBIOL LAB, VET ADMIN HOSP, HOUSTON, TEX, 68- Concurrent Pos: Nat Acad Sci-Nat Res Coun res assoc, 64- Res: Role of toxins in plant disease; physiology of diseased plant tissues; lipid metabolism of skin; chemical carcinogenesis. Mailing Add: 811 Briar Park Dr Houston TX 77042

BLACK, HOWARD CHARLES, b Warsaw, Ind, Sept 20, 12; m 38. ORGANIC CHEMISTRY. Educ: DePauw Univ, AB, 34; Univ Ill, AM, 36, PhD(biochem), 38. Prof Exp: Res chemist, Swift & Co, 37-41, in chg fat & oil res div, Res Labs, 41-50, asst dir res, 50-54, assoc dir, 54-63, dir, 63-70, gen mgr sci res, 70-73; RETIRED. Mem: Am Chem Soc; Am Oil Chem Soc (pres, 57). Res: Amino acid metabolism; fats and oils; autoxidation; antioxidants; emulsifying agents; industrial oils; fatty acids; drying oils; oil seed extraction; solvent extraction of oils; soaps and detergents; adhesives; proteins; plastics. Mailing Add: 8 Hinckley Circle Belle Vista AR 72717

BLACK, HUGH ELIAS, b Mindemoya, Ont, Oct 30, 38; m 61; c 3. VETERINARY PATHOLOGY. Educ: Ont Vet Col, DVM, 63; Ohio State Univ, MSc, 69, PhD(vet path), 72. Prof Exp: Pvt pract vet med, Wellesley Vet Clin, Wellesley, Ont, 63-67; path resident vet path, Ohio State Univ, 67-72; RES PATHOLOGIST, PROCTER & GAMBLE CO, 72- Concurrent Pos: Adj asst prof path, Dept Vet Pathobiol, Ohio State Univ, 74- Mem: Am Col Vet Pathologists; Int Acad Path; Am Vet Med Asn; Can Vet Med Asn. Res: Parathyroid gland ultrastructure and secretory activity; calcium homeostatic mechanisms; pathophysiology of skeletal disease. Mailing Add: Procter & Gamble Co R&D Miami Valley Labs PO Box 39175 Cincinnati OH 45239

BLACK, JACK, b Glasgow, Scotland, Aug 19, 16; US citizen; m 50; c 2. PHARMACOLOGY. Educ: Glasgow Univ, MB, ChB, 39. Prof Exp: Resident, Newark Gen Hosp, Eng, 40; resident contagious dis, Belvedere Hosp, Glasgow, Scotland, 46-47 & Monsall Hosp, Manchester, Eng, 47; asst med officer health, dis, tuberc & pub health, Southport, Eng, 47-48; med researcher tuberc, Valley View Hosp, Paterson, NJ, 48-49; pvt practitioner, New York, 49-55; physician, Clin Invest, Med Res Div, Schering Corp, 56-60, dir clin pharmacol, 60-65, assoc dir med res, 65-66; dir biol res, 66-74, VPRES EXP THERAPEUT, WARNER LAMBERT RES INST, 74- Mem: Am Fedn Clin Res; NY Acad Sci; Am Soc Microbiol; AMA; Am Col Clin Pharmacol & Chemother. Res: Clinical pharmacology; metabolism, toxicology and clinical screening of all classes of potentially useful new chemical structures. Mailing Add: 2 Hamilton Rd Morristown NJ 07960

BLACK, JAMES FRANCIS, b Butte, Mont, Jan 15, 19; m 55; c 2. PHYSICAL CHEMISTRY. Educ: Univ Calif, BS, 40; Princeton Univ, MA & PhD(chem), 43. Prof Exp: Asst, Princeton Univ, 40-42; sr res assoc, Esso Res & Eng Co, 43-53, proj leader, 53-56, res assoc, 56-60, sr res assoc, 60-67, SCI ADV, EXXON RES & ENG CO, 67- Concurrent Pos: Chmn air pollution res adv comt, Coord Res Coun, 71- Mem: Fel AAAS; Am Chem Soc; Am Nuclear Soc; Am Meteorol Soc; Am Inst Chem. Res: Nuclear chemistry; catalysis; radioisotope applications; polymerization; liquid phase oxidation, free radical chemistry; air pollution chemistry; mathematical modeling of dispersion of air pollutants; stimulation of convective rainfall. Mailing Add: Exxon Res & Eng Co PO Box 51 Linden NJ 07036

BLACK, JEFFREY HOWARD, b Prairie City, Ore, Feb 11, 43; m 66; c 2. HERPETOLOGY, ECOLOGY. Educ: Ore State Univ, BS, 65; Univ Mont, MS, 70; Univ Okla, PhD(zool), 73. Prof Exp: Asst biol, Univ Mont, 65-67; asst zool, Univ Okla, 70-72; ASST PROF BIOL, OKLA BAPTIST UNIV, 72- Concurrent Pos: Adv biol, Cent Okla Grotto, Nat Speleol Soc, 69-; res assoc herpet, J Willis Stovall Mus, Univ Okla, 70-; mem res coun, Okla City Zoo, 75; res assoc herpet, Okla Biol Surv, 75; actg chmn biol, Okla Baptist Univ, 75. Mem: Herpetologist's League; Am Soc Ichthyologists & Herpetologists; Soc Study Amphibians & Reptiles; Sigma Xi; Nat Speleol Soc. Res: Amphibians of Oklahoma; cavelife of Oklahoma; ecology of temporary pools. Mailing Add: Dept of Biol Okla Baptist Univ Shawnee OK 74801

BLACK, JESSIE KATE, b Hogansville, Ga. DEVELOPMENTAL BIOLOGY. Educ: Tuskegee Inst, BS, 66; Atlanta Univ, MS, 72, PhD(develop biol), 75. Prof Exp: High sch biol teacher, 68-71; FEL BIOL, EMORY UNIV, 76- Mem: Soc Develop Biol; Am Zoologist Soc; Soc Study Reproduction. Res: Study of insulin effects on chick heart cells to test the effects of insulin on embryonic heart cell surfaces. Mailing Add: PO Box 92179 Atlanta GA 30314

BLACK, JOE BERNARD, b Noxapater, Miss, Nov 1, 33; m 55; c 1. INVERTEBRATE ZOOLOGY. Educ: Miss Col, BS, 55; Tulane Univ, MS, 57, PhD(zool), 63. Prof Exp: Asst prof biol, La Col, 57-60; assoc prof, Miss Col, 62-64; assoc prof, 64-69, PROF BIOL, McNEESE STATE UNIV, 69- Mem: Am Soc Zool; Int Soc Astacology; Am Sci Affil. Res: Development, taxonomy, ecology and genetics of crawfishes. Mailing Add: Dept of Biol McNeese State Univ Lake Charles LA 70601

BLACK, JOHN ALEXANDER, b Kettle, Scotland, June 15, 40. BIOCHEMISTRY. Educ: Glasgow Univ, BSc, 61, PhD(biochem), 64. Prof Exp: Res assoc biochem, Univ BC, 64-67; asst prof, 67-72, ASSOC PROF BIOCHEM & MED GENETICS, MED SCH, UNIV ORE, 72- Res: Structural, functional and genetic relationships of proteins. Mailing Add: Dept of Biochem & Med Genetics Univ of Ore Med Sch Portland OR 97201

BLACK, JOHN B, b Ann Arbor, Mich, Apr 11, 39; m 65; c 1. ENDOCRINOLOGY, ANATOMY. Educ: Mercer Univ, AB, 61; Med Col Ga, PhD(endocrinol), 69. Prof Exp: Asst prof, 68-75, PROF BIOL & ACTG CHMN DEPT, AUGUSTA COL, 75- Mem: Sigma Xi. Res: Reproductive endocrine physiology; effect of androgens on hypothalamic-hypophyseal-ovarian axis; separation and storage of sperm fractions. Mailing Add: Dept of Biol Augusta Col Augusta GA 30904

BLACK, JOHN DAVID, b Winslow, Ark, July 2, 08; m; c 2. VERTEBRATE ZOOLOGY. Educ: Univ Kans, BA, 35; Ind Univ, MA, 37; Univ Mich, PhD(zool), 40. Prof Exp: Instr zool, Univ Ark, 37-38; prof sci, Anderson Col, Ind, 40-41 & 45-47; asst prof zool, Ala Polytech Inst, 41-43; biologist, State Conserv Comn, Wis, 43-45; assoc prof zool, Eastern Ill State Col, 47-48; prof, 48-74, EMER PROF ZOOL, NORTHEAST MO STATE TEACHERS COL, 48- Concurrent Pos: Sr sci fac fel, NSF, 59. Res: Taxonomy of Mississippi Valley fishes; fisheries biology; vertebrate distribution and ecology; biological conservation. Mailing Add: 1408 E Parkview Kirksville MO 63501

BLACK, JOHN EARLE, b Montreal, Que, Oct 26, 35; m 65; c 1. SOLID STATE PHYSICS. Educ: McGill Univ, BEng, 59; Queen's Univ (Ont), MSc, 61; Univ Sask, PhD(auroral physics), 67. Prof Exp: ASSOC PROF PHYSICS, BROCK UNIV, 74- Mem: Am Asn Physics Teachers. Res: Radio astronomy; auroral morphology; electrical conductivity of metals. Mailing Add: Dept of Physics Brock Univ St Catharines ON Can

BLACK, JOHN LARRY, b McKenzie, Tenn, Apr 16, 42; m 63; c 1. EMBRYOLOGY, BIOCHEMISTRY. Educ: Univ Tenn, BS, 64, MS, 65, PhD(animal sci), 67. Prof Exp: Asst prof biol, Union Univ, Tenn, 67-68; asst prof chem, Jackson State Community Col, 68-70; chmn div sci & math, 70-73, PROF BIOL, BETHEL COL, 70- Mem: AAAS; Am Soc Animal Sci. Res: Ovarian morphology of prenatal pig and neonatal pig. Mailing Add: Dept of Biol Bethel Col McKenzie TN 38201

BLACK, JOHN WILSON, b Veedersburg, Ind, Feb 9, 06; m 36; c 4. SPEECH & HEARING SCIENCE. Educ: Wabash Col, AB, 27; Univ Iowa, MA, 30, PhD, 35. Prof Exp: Prof speech, Adrian Col, 27-35 & Kenyon Col, 36-49; PROF SPEECH, OHIO STATE UNIV, 49- Concurrent Pos: Dir voice commun proj, Off Naval Res, 47-73; mem, Armed Forces Comt Hearing & Bio-acoust, Nat Res Coun, 53-59; ed, Speech Monographs, 58-59; Regents prof, Ohio State Univ, 66-; Fulbright fel, 54-55; NSF fel, Italy, 61. Honors & Awards: President's Award Merit, 49. Mem: Fel AAAS; fel Am Acoust Soc; Am Psychol Asn; fel Am Speech & Hearing Asn (vpres, 64); Speech Commun Asn (vpres, 65, pres, 66). Res: Statistics of language; intelligibility; spectro-analysis; auditory responses. Mailing Add: 1400 Lincoln Rd Columbus OH 43212

BLACK, LOWELL LYNN, b Norman, Ark, Oct 9, 38; m 60; c 1. PLANT PATHOLOGY. Educ: Univ Ark, BS, 60, MS, 62; Univ Wis, PhD(plant path), 65. Prof Exp: Asst prof plant path, WVa Univ, 65-68; asst prof, 68-70, ASSOC PROF PLANT PATH, LA STATE UNIV, 70- Mem: AAAS; Am Phytopath Soc; Am Soc Plant Physiologists; Am Inst Biol Sci. Res: Physiology of diseased plants; virus diseases; diseases of vegetables. Mailing Add: Dept of Plant Path La State Univ Baton Rouge LA 70803

BLACK, MARTIN LUTHER, b Knoxville, Tenn, May 5, 21; m 44; c 2. SYNTHETIC ORGANIC CHEMISTRY. Educ: Univ Tenn, BS, 44, PhD(chem), 49. Prof Exp: Lab asst org chem, 44-46, patent assoc legal dept, 49-50, jr res chemist, 50-53, SR RES CHEMIST, PARKE, DAVIS & CO, 53- Mem: Am Chem Soc. Res: Chemotherapy of tuberculosis; atherosclerosis; antifertility agents; synthetic cephalosporins. Mailing Add: Parke-Davis & Co Ann Arbor MI 48106

BLACK, OTIS DEITZ, b Etowah, Tenn, June 5, 13; m 36; c 2. INDUSTRIAL CHEMISTRY. Educ: Ohio State Univ, AB, 38, MS, 39, PhD(phys chem), 42. Prof Exp: Asst chem, Ohio State Univ, 38-42; res engr, Camden Plant, Radio Corp Am, 42-68, RES ENGR, CAMDEN PLANT, RCA CORP, 68- Res: Activity coefficients; resins and plastics; mycology; printed circuits; industrial chemical safety; emission spectroscopy; thin films; magnetic films; solder metallurgy; analytical instrumentation.

BLACK, OWEN, JR, biochemistry, see 12th edition

BLACK, PAUL H, b Boston, Mass, Mar 11, 30; m 62; c 3. MEDICINE, VIROLOGY. Educ: Dartmouth Col, AB, 52; Columbia Univ, MD, 56. Prof Exp: Intern med, Mass Gen Hosp, 56-57, asst resident, 57-58, clin & res fel, 58-60, resident, 60-61; sr surgeon, NIH, 61-67, consult, Nat Inst Allergy & Infectious Dis, 64-67; asst prof, 67-70, ASSOC PROF MED, HARVARD MED SCH, 70- Concurrent Pos: Assoc physician, Mass Gen Hosp, 67-70. Mem: Am Soc Clin Invest; NY Acad Sci. Res: Oncogenic viruses and the mechanisms by which these viruses induce malignancy. Mailing Add: Dept of Infectious Dis Mass Gen Hosp Boston MA 02114

BLACK, PERRY, b Montreal, Que, Oct 2, 30; m 63; c 3. NEUROSURGERY, NEUROPHYSIOLOGY. Educ: McGill Univ, BSc, 51, MD, CM, 56; Am Bd Neurol Surg, dipl, 66. Prof Exp: Intern & asst resident med & gen surg, Jewish Gen Hosp, Montreal, 56-58; asst resident neurol, Montreal Neurol Inst, 58-59; resident neurosurg, Johns Hopkins Hosp, 59-63, from instr to asst prof, Sch Med, 64-69, asst prof psychiat, 67-70, ASSOC PROF NEUROL SURG, SCH MED, JOHNS HOPKINS UNIV, 69-, ASSOC PROF PSYCHIAT, 70-, NEUROSURGEON, 64- Concurrent Pos: NIH fel physiol, Johns Hopkins Univ, 61-62; dir lab neurol sci, Friends Med Sci Res Ctr, Baltimore, Md, 64-; vis neurosurgeon, Baltimore City Hosps, 65-; neurosurg consult, Columbia Hosp & Clin Found, Md, 69- & NCharles Gen Hosp, Baltimore, 75-; neurosurgeon, Good Samaritan Hosp, 70-; mem neurol A study sect, NIH, 73-77; state coordr, Epilepsy Found Am, 73- Mem: Cong Neurol Surg; Am Asn Neurol Surg; Am Asn Neuropath; Soc Neurosci; Am Neurol Asn. Res: Neurosurgery, particularly head injury in children; experimental study of recovery of motor function; clinical and experimental epilepsy; frontal and temporal lobe functions in learning and memory. Mailing Add: Dept of Neurol Surg Johns Hopkins Hosp Baltimore MD 21205

BLACK, PETER ELLIOTT, b New York, NY, Nov 19, 34; m 56; c 4. WATERSHED MANAGEMENT. Educ: Univ Mich, BS, 56, MF, 58; Colo State Univ, PhD(watershed mgt), 61. Prof Exp: Res forester, Coweeta Hydrol Lab, Southeastern Forest & Range Exp Sta, US Forest Serv, 56-59; asst coop watershed mgt unit, Colo State Univ, 59-60; asst prof forestry & watershed mgt, Humboldt State Col, 61-65; PROF WATERSHED MGT, STATE UNIV NY COL ENVIRON SCI & FORESTRY, 65- Concurrent Pos: Watershed mgt consult, Nat Park Serv, Washington, DC, 63-64; prin, Impact Consults, Syracuse, 74- Mem: Am Geophys Union; Am Soil Conserv Soc; Am Forestry Asn; Am Water Resources Asn. Res: Forest influences; wildland hydrology; water law; water resource economics. Mailing Add: State Univ NY Col Environ Sci & Forestry Syracuse NY 13210

BLACK, RICHARD BLACKBURN, geographical exploration, see 12th edition

BLACK, RICHARD GLYNN, b Walkerville, Ont, Aug 16, 28; m 54; c 3. MEDICAL SCIENCE, ANESTHESIOLOGY. Educ: Univ Toronto, BASc, 54, MD, 60. Prof Exp: Instr physiol, Univ Toronto, 60-62; asst prof neurosurg, 64-68, ASSOC PROF ANESTHESIOL, UNIV WASH, 68- Mem: AAAS; Sigma Xi; NY Acad Sci; Asn Univ Anesthetists. Res: Neurophysiology of central pain states and the clinical management of chronic pain problems with emphasis on multidisciplinary attack on the clinical aspects of chronic pain. Mailing Add: Dept of Anesthesiol Sch of Med Univ of Wash Mail Stop RN-10 Seattle WA 98195

BLACK, RICHARD H, b Deerfield, Wis, Apr 24, 25. APPLIED MATHEMATICS. Educ: Univ Wis, BS, 47, MS, 51, PhD(appl math), 63. Prof Exp: Instr math, Univ Wis-Milwaukee, 56-63, asst prof math & dir, Comput Ctr, 63-68; ASSOC PROF MATH, CLEVELAND STATE UNIV, 68- Mem: Asn Comput Mach. Res: Pattern recognition; computer science. Mailing Add: Dept of Math Cleveland State Univ Cleveland OH 44115

BLACK, ROBERT CORL, b Philadelphia, Pa, June 16, 41; m 63; c 1. BOTANY, PLANT PHYSIOLOGY. Educ: Pa State Univ, BS, 63, MS, 65, PhD(bot), 70. Prof Exp: Food technologist, Gen Foods Corp, 65-67; ASST PROF BIOL, PA STATE UNIV, DELAWARE COUNTY CAMPUS, 70- Mem: AAAS; Am Soc Plant Physiologists; Am Inst Biol Sci. Res: Mechanism and control of indole acetic acid biosynthesis in plants and plant tissue cultures. Mailing Add: Dept of Biol Pa State Univ Delaware Co Campus Media PA 19063

BLACK, ROBERT EARL LEE, b Cassville, Mo, Nov 20, 28; m 53; c 3. EMBRYOLOGY. Educ: William Jewell Col, BA, 51; Univ Wash, PhD(zool), 57. Prof Exp: USPHS fel, Calif Inst Technol, 57-59; assoc prof biol, 59-65, PROF BIOL & MARINE SCI, COL WILLIAM & MARY, 65- Mem: AAAS; Am Soc Zool. Res: Embryology of marine invertebrates. Mailing Add: Dept of Biol Col of William & Mary Williamsburg VA 23185

BLACK, ROBERT FOSTER, b Dayton, Ohio, Feb 1, 18; m; c 2. GEOLOGY. Educ: Col Wooster, BA, 40; Syracuse Univ, MA, 42; Johns Hopkins Univ, PhD(geol), 53. Prof Exp: Asst geol, Syracuse Univ, 40-42; asst, Calif Inst Technol, 42-43; geologist, US Geol Surv, 43-59; prof geol, Univ Wis-Madison, 59-70; PROF GEOL, UNIV CONN, 70- Concurrent Pos: Geologist, Roosevelt Wildlife Conserv Dept, 41-42, chief Alaska Terrain & Permafrost Sect, 46-49, res geologist, 50-53; with US Geol Surv, foreign geol, Mex, 54-56; assoc prof, Univ Wis-Madison, 56-59; mem, Nat Acad Sci Study Group on Permafrost, 72-; consult, NSF, 73- Mem: Fel AAAS; fel Geol Soc Am; Am Soc Photogram; fel Arctic Inst NAm; Am Geophys Union. Res: Glacial and quaternary geology of polar regions, north-central and northeast United States, including present and past permafrost. Mailing Add: Dept of Geol Univ of Conn Storrs CT 06268

BLACK, RODNEY ELMER, b Pawhuska, Okla, Sept 21, 16; m 58; c 1. PHYSICAL CHEMISTRY. Educ: Okla Agr & Mech Col, BS, 38; Univ Wis, MS, 40, PhD(chem), 42. Prof Exp: Lab asst, Agr Res Dept, Exp Sta, Okla Agr & Mech Col, 36-38; asst, Univ Wis, 38-41; res chemist, Barrett Div, Allied Chem Dye Corp, 42-43; sr res chemist, Phillips Petrol Co, 43-47; prof chem & head dept, Morningside Col, 47-51; chmn div phys sci, 53-54, ASSOC PROF CHEM, UNIV KY, 51- Concurrent Pos: Assoc, Cornell Univ, 58-59. Mem: AAAS; Am Chem Soc; Electrochem Soc. Res: Electrochemistry; physical properties of organic compounds; purification and identification of compounds; microscopy; electrodeposition of alloys; crystallization; heterocyclic and organic syntheses; polarography; microscopy; electrodeposition of tungsten alloys. Mailing Add: Dept of Chem Univ of Ky Lexington KY 40506

BLACK, ROGER LEWIS, b Syracuse, NY, Feb 12, 24; m 45; c 3. INTERNAL MEDICINE. Educ: Syracuse Univ, MD, 46. Prof Exp: Asst med, Johns Hopkins Univ, 54-57; clin fel, Mass Gen Hosp, 55; sr investr metab dis, Nat Inst Arthritis, 55-63, asst to dir labs & clins, NIH, 64-65, ASSOC DIR CLIN CTR, NIH, 65- Concurrent Pos: Co-dir rheumatol serv, DC Gen Hosp, 58-64; clin asst prof med, Sch Med, Georgetown Univ, 58, clin prof med, 71- Mem: Am Rheumatism Asn. Res: Clinical research; rheumatic diseases; research administration. Mailing Add: 1413 Bradley Ave Rockville MD 20851

BLACK, SAMUEL HAROLD, b Lebanon, Pa, May 1, 30; m 61; c 2. MICROBIOLOGY. Educ: Lebanon Valley Col, BS, 52; Univ Mich, MS, 58, PhD(microbiol), 60. Prof Exp: NSF fel, Delft Technol Univ, 60-61; instr microbiol, Univ Mich, Ann Arbor, 61-62; from asst prof to assoc prof, Baylor Col Med, 62-71; from assoc prof to prof microbiol & pub health, Mich State Univ, 71-75; PROF MICROBIOL & HEAD DEPT, COL MED, TEX A&M UNIV, 75- Concurrent Pos: Lectr, Univ Mich-Flint, 61-62 & Univ Houston, 64-66; guest prof, Swiss Fed Inst Technol, 69-70. Mem: AAAS; Am Soc Microbiol; fel Am Acad Microbiol; Am Soc Cell Biologists; Soc Invert Path. Res: Medical microbiology; microbial physiology; cytology of morphogenesis in prokaryotic and eukaryotic cells; biology of Bacillus species, of Hanaenula species, and of Treponema pallidum. Mailing Add: Col of Med Tex A&M Univ College Station TX 77843

BLACK, SAMUEL P W, b Barbourville, Ky, Dec 19, 16; m 44; c 4. MEDICINE. Educ: Yale Univ, BS, 40; Johns Hopkins Univ, MD, 43. Prof Exp: Instr & asst prof, Yale Univ, 50-55; assoc prof, 55-60, PROF SURG, UNIV MO-COLUMBIA, 60- Mem: Am Col Surg; Asn Res Nerv & Ment Dis; Am Asn Neurol Surg. Res: Intraaneurysmal hemodynamics; cerebellar physiology. Mailing Add: Med Ctr Univ of Mo Columbia MO 65202

BLACK, SIMON, b Deerfield, Wis, Aug 9, 17; m 44; c 3. CHEMISTRY. Educ: Univ Wis, BS, 38, MS, 41, PhD(biochem), 42. Prof Exp: Asst pharmacol, Univ Chicago, 42-45, res assoc, 45-46, from instr to asst prof biochem, Dept Med, 46-51; fel, Mass Gen Hosp, 51-52; chemist, 52-58, CHIEF SECT BIOCHEM OF AMINO ACIDS, NIH, 58- Concurrent Pos: Consult, Off Sci Res & Develop. Mem: AAAS; Am Soc Biol Chemists; Am Chem Soc. Res: Biochemistry of amino acids; enzymes of amino acid metabolism and activation; origin of life and the genetic code. Mailing Add: NIH Bldg 4 Bethesda MD 20014

BLACK, TRUMAN D, b Houston, Tex, Sept 30, 37; m 57; c 1. SOLID STATE PHYSICS. Educ: Univ Houston, BS, 59; Rice Univ, MA, 62, PhD(physics), 64. Prof Exp: Mem tech staff, Tex Instruments, 64-65; asst prof physics, 65-69, ASSOC PROF PHYSICS, UNIV TEX, ARLINGTON, 69- Mem: Am Phys Soc; Am Asn Physics Teachers; AAAS. Res: Spin lattice interactions in paramagnetic solids; low temperature magnetic effects; laser optics; holography. Mailing Add: 814 Pin Oak Lane Arlington TX 76012

BLACK, VIRGINIA H, b Detroit, Mich, Sept 24, 41; m 63; c 2. CELL BIOLOGY, DEVELOPMENTAL BIOLOGY. Educ: Kalamazoo Col, AB, 63; Sacramento State Col, MA, 66; Stanford Univ, PhD(anat), 68. Prof Exp: Instr anat, 67-68, from instr to asst prof cell biol, 68-73, ASSOC PROF CELL BIOL, SCH MED, NY UNIV, 73- Concurrent Pos: USPHS res grant, 69-78. Mem: AAAS; Am Asn Anat; NY Acad Sci; Soc Develop Biol; Am Soc Cell Biologists. Res: Embryonic development of the reproductive system and adrenal; cytodifferentiation and functional modulation of steroid-secreting cells; influence of steroids on reproductive system development. Mailing Add: Dept of Cell Biol NY Univ Sch of Med New York NY 10016

BLACK, WALLACE GORDON, b Arlington, Mass, Feb 24, 22; m 49; c 5. REPRODUCTIVE PHYSIOLOGY. Educ: Univ Wis, BS, 48, MS, 49, PhD(genetics), 52. Prof Exp: Instr reproductive physiol, Univ Wis, 52-53; asst prof animal husb, Univ Nev, 53-54; assoc prof, 54-58, PROF ANIMAL SCI, UNIV MASS, AMHERST, 58- Mem: AAAS; Am Inst Biol Sci; Am Asn Lab Animal Sci; Asn Animal Tech Educrs. Res: Embryo transplantation; uterine defense mechanisms; role of progesterone in maintenance of pregnancy and estrous cycle control; laboratory animal technology. Mailing Add: 204 Paige Lab Vet An Sci Univ of Mass Amherst MA 01002

BLACK, WAYNE EDWARD, b East St Louis, Ill, Sept 15, 35. PUBLIC HEALTH. Educ: Ill Col, BS, 57; Univ St Louis, PhD, 64; Nat Registry Clin Chem, cert. Prof Exp: Chief radiol health res, 63-64, chief radiol health res & lab serv, 64-67, DIR LABS, ST LOUIS COUNTY HEALTH DEPT, 67-, ASST DIR DEPT COMMUNITY HEALTH & MED CARE, 73- Concurrent Pos: Lectr physiol chem, Washington Univ, 66-70, instr dept physiol chem, Sch Dent, 69- Mem: Fel Am Inst Chemists; AAAS; Am Chem Soc; Am Asn Clin Chem. Res: Laboratory work, including environmental health. Mailing Add: Dept Commun Health & Med Care 801 S Brentwood Blvd Clayton MO 63105

BLACK, WILLIAM, b Philadelphia, Pa, June 14, 35; m 62. PHYSIOLOGICAL PSYCHOLOGY. Educ: Bucknell Univ, BA, 57; Western Reserve Univ, MS, 59, PhD(psychol), 62. Prof Exp: Res psychologist, Walter Reed Army Inst Res, 61-64; asst prof psychol, Purdue Univ, 64-70; RES ASSOC, INST PSYCHIAT RES, SCH MED, IND UNIV, INDIANAPOLIS, 70- Mem: AAAS; Am Psychol Asn; Soc Neurosci. Res: Neurophysiological mechanisms underlying motivational processes and neural substates of learning. Mailing Add: Inst of Psychiat Res 1100 W Michigan Indianapolis IN 46202

BLACK, WILLIAM BRUCE, b Indianapolis, Ind, Feb 25, 23; m 45; c 2. ORGANIC CHEMISTRY, POLYMER CHEMISTRY. Educ: Univ Va, BA, 50, MS, 53, PhD(org chem), 54. Prof Exp: Res chemist, Chemstrand Res Ctr, 54-61, group leader, 61-62, Chemstrand Co div, Monsanto Co, 62-65, mgr polymer sci, 65-69, SR GROUP LEADER, MONSANTO TECH CTR, CHEMSTRAND CO DIV, MONSANTO CO, 69- Mem: AAAS; Am Chem Soc; NY Acad Sci; Fiber Soc. Res: Synthesis, spinning and evaluation of novel polymers and fibers, especially very high modulus, very high strength aromatic fibers; the attainment of unusually high strength and high modulus fibers from conventional polymers. Mailing Add: Monsanto Tech Ctr PO Box 12830 Pensacola FL 32575

BLACK, WILLIAM CARMAN, chemistry, see 12th edition

BLACK, WILLIAM CARTER, JR, b San Diego, Calif, Feb 25, 39. LOW TEMPERATURE PHYSICS. Educ: Pomona Col, BA, 61; Univ Ill, Urbana, MS, 63, PhD(physics), 67. Prof Exp: Asst physics, Univ Ill, Urbana, 61-67; asst prof, Univ Calif, San Diego, 67-70; dir & gen mgr, 71-72, PRES, S H E CORP, 72- Mem: Am Phys Soc. Res: Superconducting devices and systems; advanced helium cryostats and refrigeration methods; electronic instrumentation for cryogenics. Mailing Add: 1191 Avenida Amantea La Jolla CA 92037

BLACK, WILLIAM CORMACK, b Denver, Colo, May 18, 01; m 25; c 1. PATHOLOGY. Educ: Univ Colo, MD, 25. Prof Exp: Instr clin path, 25-30, from instr to prof path, 30-45, clin prof, 46-73, EMER CLIN PROF PATH, UNIV COLO MED CTR, DENVER, 73- Concurrent Pos: Pathologist, St Luke's Hosp, 45-74. Mem: AAAS; fel Am Col Path. Res: Tuberculosis; allergy; oncology; mycology; tropical medicine. Mailing Add: Dept of Path Univ of Colo Med Ctr Denver CO 80203

BLACK, WILLIAM FRANCIS, b Winnipeg, Man, Sept 30, 27; m 57; c 3. ECOLOGY, INVERTEBRATE ZOOLOGY. Educ: Univ Man, BSc, 48; McGill Univ, PhD(zool), 56. Prof Exp: Demonstr zool, McGill Univ, 49-55; lectr biol, Sir George William Univ, 56-57, asst from asst prof to assoc prof, 57-67; TEACHER ZOOL, BADDECK RURAL HIGH SCH, 67- Concurrent Pos: Consult scientist, Cape Breton Develop Corp, Marine Farming Res Div, 72- Mem: Fel AAAS; Am Ornithologists Union; Sigma Xi. Res: Marine biology; ornithology; ecology of shellfish in the Bras D'or Lakes, Cape Breton, Canada. Mailing Add: Baddeck Rural High Sch Victoria County Baddeck NS Can

BLACK, WILLIAM JAMES, b Winnipeg, Man, May 8, 39. BIOCHEMISTRY, ENZYMOLOGY. Educ: Univ Man, BS, 60, MS, 65, PhD(biochem), 69. Prof Exp: Res asst biochem, Col Med, Univ Man, 60-63, res assoc, 65-66; RES FEL MOLECULAR BIOL, ALBERT EINSTEIN COL MED, 69- Concurrent Pos: Muscular Dystrophy Asn Can fel, 69-72; deleg, NATO Advan Study Conf, Protein Struct, Venice, Italy, 70. Honors & Awards: E L Drewry Award & Medal, Univ Man, 70. Mem: AAAS; Can Biochem Soc. Res: Cathecolamine metabolism; muscle metabolism with emphasis on the regulatory enzymes, especially glycogen phosphorylase, glycerophosphate dehydrogenase and Fructose-l, 6-diphosphatase; relationship between protein structure and function; enzyme kinetics. Mailing Add: Dept of Molecular Biol Albert Einstein Col of Med Bronx NY 10461

BLACK, WILLIAM RICHARD, b Brownsville, Pa, June 8, 42; m 63; c 3. TRANSPORTATION GEOGRAPHY. Educ: Calif State Col, Pa, BSc, 64; Univ Iowa, MA, 66, PhD(geog), 69. Prof Exp: Asst prof geog, Miami Univ, 68-69; asst prof, 69-72, ASSOC PROF GEOG, IND UNIV, BLOOMINGTON, 72-, DIR, CTR URBAN

& REGIONAL ANAL, 73- Concurrent Pos: Tech dir, Ind Nat Transp Needs Study, Div Com, State of Ind, 71; ed, Regional Sci Perspectives, 72-74; tech dir, Bloomington Mass Transit Tech Study, City of Bloomington-Urban Mass Transp Admin US, 73; adj prof civil eng, Purdue Univ, Lafayette, 73; mem, Hwy Res Bd, Nat Acad Sci-Nat Res Coun; proj dir, Coun State Govts, Ky, 74-75; dir rail planning, Pub Serv Comn Ind, 75- Mem: Asn Am Geogr; Regional Sci Asn. Res: Measurement of mobility; analysis of variations in the gravity models distance exponent; factorial flow analysis; urban travel behavior; regional transport network analysis. Mailing Add: Dept of Geog Ind Univ Bloomington IN 47401

BLACKADAR, ALFRED KIMBALL, b Newburyport, Mass, July 6, 20; m 46; c 3. METEOROLOGY. Educ: Princeton Univ, AB, 42; NY Univ, PhD(meteorol), 50. Prof Exp: Res assoc prof meteorol, NY Univ, 46-56; assoc prof, 56-60, PROF METEOROL, PA STATE UNIV, 60-, HEAD DEPT, 67- Concurrent Pos: Ed, Meteorol Monogr, 61-65; mem, Nat Acad Sci-US Nat Comt Global Atmospheric Res Prog, 68-; mem, NSF Adv Panel Atmospheric Sci, 68-; mem comt basic sci adv to US Army, Nat Acad Sci-Nat Res Coun, 70-73, mem exec comt, Earth Sci Div. Honors & Awards: Charles F Brooks Award, Am Meteorol Soc, 69; Humboldt Found Sr Scientist Award, 73. Mem: Fel AAAS; fel Am Meteorol Soc (secy, 65-69, pres, 71-72); fel Am Geophys Union. Res: Atmospheric energy transformation; atmospheric turbulence. Mailing Add: 805 W Foster Ave State College PA 16801

BLACKADAR, BRUCE EVAN, b Nyack, NY, Oct 22, 48. MATHEMATICS. Educ: Princeton Univ, AB, 70; Univ Calif, Berkeley, MA, 74, PhD(math), 75. Prof Exp: ASST PROF MATH, UNIV NEV, RENO, 75- Mem: Am Math Soc. Res: Tensor products and crossed products of von Neumann algebras; induced representations of locally compact groups; representations of products and extensions of groups. Mailing Add: Dept of Math Univ of Nev Reno NV 89507

BLACKADAR, ROBERT GORDON, b Ottawa, Ont, Mar 18, 28. GEOLOGY. Educ: Univ Toronto, BA, 50, MA, 51, PhD(geol), 54. Prof Exp: GEOLOGIST, GEOL SURV CAN, 53-, CHIEF SCI ED, 70- Concurrent Pos: Sci ed, Geol Surv Can, 65-70. Mem: Geol Soc Am. Res: Precambrian geology of Arctic regions; geography and geology of the Arctic. Mailing Add: Geol Surv Bldg 601 Booth St Ottawa ON Can

BLACKARD, CLYDE ERHARDT, b Indianapolis, Ind, Nov 28, 32; m 57; c 3. SURGERY, UROLOGY. Educ: Butler Univ, BS, 54; Ind Univ, MD, 57. Prof Exp: Intern, Univ Nebr Hosp, Omaha, 57-58; resident gen surg, Creighton Univ, 58-59; resident urol, Univ Louisville Hosps, 59-62; from instr to assoc prof, 62-74, PROF UROL SURG, HEALTH SCI CTR, UNIV MINN, MINNEAPOLIS, 62-; CHIEF UROL, VET ADMIN HOSP, 69- Concurrent Pos: Consult urol, Gillette State Hosp for Crippled Children, 65-; asst chief urol, Vet Admin Hosp, Minneapolis, 67-69; consult urol, Nicollet Clin, Minneapolis, 72- Mem: AMA; fel Am Col Surg; Am Urol Asn; Soc Univ Urol. Res: Carcinoma of the prostate. Mailing Add: Sect of Urol Vet Admin Hosp Minneapolis MN 55417

BLACKARD, WILLIAM GRIFFITH, b Baltimore, Md, July 14, 33; m 60; c 3. INTERNAL MEDICINE, ENDOCRINOLOGY. Educ: Duke Univ, MD, 57; Am Bd Internal Med, dipl, 65. Prof Exp: Intern med, New York Hosp-Cornell Med Ctr, 57-58, resident, 58-59; fel endocrinol & metab, Med Ctr, Duke Univ, 59-60, resident med, 60-61, fel endocrinol & metab, 63-64; from instr to asst prof med, 64-68, ASSOC PROF MED, SCH MED, LA STATE UNIV, NEW ORLEANS, 68- Concurrent Pos: Markle scholar acad med, 68-73. Mem: Am Fedn Clin Res; Endocrine Soc; fel Am Physicians; Am Soc Clin Invest; Am Diabetes Asn. Res: Carbohydrate and lipid metabolism; insulin and growth hormone homeostesis; diabetes; calcium and phosphorous metabolism. Mailing Add: La State Univ Med Ctr New Orleans LA 70112

BLACKBOURN, ANTHONY, b Orpington, Eng, June 6, 38; Can citizen; m 69. ECONOMIC GEOGRAPHY. Educ: Univ London, BSc, 59; Univ Ga, MA, 61; Univ Toronto, PhD(geog), 68. Prof Exp: Lectr, 63-67, asst prof, 67-71, ASSOC PROF GEOG, UNIV WINDSOR, 71- Mem: Asn Am Geogrs; Regional Sci Asn. Res: Industrial geography; transportation. Mailing Add: Dept of Geog Univ of Windsor Windsor ON Can

BLACKBURN, ARCHIE BARNARD, b Austin, Tex, July 6, 38; m 67; c 2. PSYCHIATRY. Educ: Baylor Univ, BA, 60, MS & MD, 65; Am Bd Psychiat & Neurol, dipl psychiat, 74. Prof Exp: Resident psychiatrist, Vet Admin Hosp & Univ Tex Southwestern Med Sch Dallas, 69-70 & Vet Admin Hosp & Univ Okla, 72; asst prof, Health Sci Ctr, Univ Okla, 72-73; ASST PROF PSYCHIAT, BAYLOR COL MED, 73-; STAFF PSYCHIATRIST, VET ADMIN HOSP, HOUSTON, 73- Concurrent Pos: Staff psychiatrist & med dir, Drug Abuse Treatment Unit, Vet Admin Hosp, Oklahoma City, 72- Mem: AAAS; Am Psychiat Asn; Asn Psychophysiol Study Sleep. Res: Clinical hypodynamics; biorhythms; methadone maintenance; application of research methodology in clinical disorders of sleep. Mailing Add: Vet Admin Hosp Houston TX 77031

BLACKBURN, BENJAMIN (COLEMAN), b Medina, NY, May 2, 08. BOTANY, DENDROLOGY. Educ: Cornell Univ, BS, 29; Rutgers Univ, PhD(bot), 49. Prof Exp: Instr hort, Rutgers Univ, 36-42; adj prof bot, 51-72, EMER PROF BOT, DREW UNIV, 72- Concurrent Pos: Adminr, Willowwood Arboretum, Rutgers Univ, 67- Mem: Int Asn Plant Taxon; Int Dendrol Soc; Sigma Xi. Res: Collecting and growing plants not sufficiently known, both native and exotic species, selections, mutations or other derivatives, desirable for study in contexts of botany, horticulture and forestry. Mailing Add: Willowwood Arboretum Rutgers Univ Box 125 Gladstone NJ 07934

BLACKBURN, DALE WARREN, b La Porte, Ind, Sept 12, 26; m 52; c 2. PHARMACEUTICAL CHEMISTRY. Educ: Purdue Univ, BS, 48, MS, 51, PhD(pharmaceut chem), 54. Prof Exp: Anal chemist, Whitehall Pharmacal Co, 50; instr pharmaceut chem, Purdue Univ, 51-52, asst prof, 54-56; group leader, 56-66, ASST DIR ORG CHEM SECT, SMITH KLINE CORP, 66- Mem: Am Chem Soc; Am Pharmaceut Asn. Res: Organic and radiochemical synthesis; high pressure reactions; instrumental analysis; separation and purification; radiological defense. Mailing Add: Smith Kline Corp 1500 Spring Garden Philadelphia PA 19101

BLACKBURN, EDWARD VICTOR, b Skegness, Eng, Sept 18, 44. ORGANIC CHEMISTRY. Educ: Univ London, BSc, 66; Univ Nottingham, PhD(org chem), 69. Prof Exp: Fel, 69-70, lectr, 70-72, ASST PROF CHEM, ST JEAN UNIV COL, UNIV ALBERTA, 72- Mem: The Chem Soc; assoc Royal Inst Chem; Am Chem Soc; Chem Inst Can. Res: Photochemistry of olefins with particular reference to stilbene analogues; polar effects in radical reactions. Mailing Add: Dept of Chem Col Universitaire St Jean Edmonton AB Can

BLACKBURN, HENRY WEBSTER, JR, b Miami, Fla, Mar 22, 25; m 51; c 3. EPIDEMIOLOGY. Educ: Univ Miami, BS, 47; Tulane Univ, MD, 48; Univ Minn, MS, 57. Prof Exp: Intern, Chicago Wesley Mem, 48-49; intern, Am Hosp of Paris, 49-50; med officer-in-chg, USPHS, Austria & Ger, 50-53; chief resident med, Ancker Hosp, Minn, 56; asst prof, 56-61, assoc prof, 61-68, PROF PHYSIOL HYG, SCH PUB HEALTH, UNIV MINN, MINNEAPOLIS, 68-, DIR PHYSIOL HYG, SCH PUB HEALTH & PROF MED, SCH MED, 72- Concurrent Pos: Med dir, Underwriting Mutual Serv Ins Co, Minn, 56-; consult, Univ Group Diabetes Prog, 61-; temporary consult, WHO, 61, 64, 65, 70 & 71; chmn coun epidemiol, Am Heart Asn, 71-73; med consult, Retail Credit Co, 69-72; vis prof, Univ Geneva, 70-71; dir electrocardiographic ctr & mem steering comt, Coronary Drug Proj, Nat Heart & Lung Inst, nat vchmn multiple risk factor intervention trial, 72-75, mem adv comt, Cardiol Br, 75- Mem: AMA; Am Heart Asn; fel Am Col Cardiol; fel Am Epidemiol Soc. Res: Epidemiology of cardiovascular disease; electrocardiography; internal medicine; insurance medicine. Mailing Add: Lab of Physiol Hyg Univ of Minn Minneapolis MN 55455

BLACKBURN, JOHN FRANCIS, b Chicago, Ill, June 11, 03; m 43. PHYSICS. Educ: Univ Chicago, BS, 26; Calif Inst Technol, PhD(physics), 32. Prof Exp: Pumpman, Neale-Rainbow Light Corp, Calif, 27; engr, Robert C Burt, Calif, 28; geophysicist, Superior Oil Co, 30; physicist, Consol Steel Co, 33; chief engr, Lansing Mfg Co, 36-38 & F F Ferreira, Mex, 39; prod engr, Vultee Aircraft Corp, Calif, 40; staff mem, Radiation Lab, Mass Inst Technol, 42-46, res engr, Dynamic Anal & Control Lab, 46-57, res engr, Lincoln Lab, 52-56, res assoc mech eng, 54-57; asst to vpres eng & res, Raytheon Co, 57-60; staff consult to pres, Aerospace Corp, 60-68; INDEPENDENT TECH & MGT CONSULT, 68- Concurrent Pos: Chief, US Mutual Security Agency Surv Group, Ital Electrotech Indust, 52-53. Honors & Awards: Mach Design Div Award, Am Soc Mech Engrs, 53. Mem: Am Chem Soc; Inst Elec & Electronics Engrs; Acoust Soc Am; Am Soc Mech Engrs. Res: Radio wave propagation; high voltage x-rays; sound recording and reproduction; radio and electronic equipment design; radar; fire control; electronic and mechanical computers; servomechanism; hydraulic power control mechanism; machine design; electronic components; management. Mailing Add: 414 S Cliffwood Ave Los Angeles CA 90049

BLACKBURN, JOHN GILL, b Lake Charles, La, June 25, 35; m 56; c 2. PHYSIOLOGY. Educ: Tulane Univ, BS, 59, PhD(physiol), 65. Prof Exp: From instr to asst prof, 64-75, ASSOC PROF PHYSIOL, MED UNIV SC, 75- Concurrent Pos: USPHS res grant, 65-68, grant co-investr, 73- Res: Neurophysiological regulation of feeding activity; regulation of the swallowing reflex; electrophysiology of spinal injury. Mailing Add: Dept of Physiol Med Univ of SC Charleston SC 29401

BLACKBURN, MAURICE, b Melbourne, Australia, Oct 27, 15; nat US; m 66; c 5. ZOOLOGY. Educ: Univ Melbourne, DSc(zool). Prof Exp: Res officer marine biol, Australian Commonwealth Sci & Indust Res Orgn, 37-56; prof zool, Univ Hawaii, 56-57; assoc res biologist, 57-60, RES BIOLOGIST, SCRIPPS INST, UNIV CALIF, 60- Mem: Fel AAAS; Am Soc Limnol & Oceanog; fel Am Inst Fishery Res Biologists. Res: Biological oceanography. Mailing Add: Scripps Inst of Oceanog Univ of Calif La Jolla CA 92093

BLACKBURN, PAUL EDWARD, b West Branch, Iowa, May 24, 24; m 48; c 2. HIGH TEMPERATURE CHEMISTRY. Educ: Ohio Wesleyan Univ, BA, 48; Ohio State Univ, PhD(chem), 54. Prof Exp: Res assoc, Ohio State Univ, 48-51; res chemist, Westinghouse Res Labs, 54-61; proj dir, Arthur D Little, Inc, 61-65; ASSOC CHEMIST, ARGONNE NAT LAB, 65- Mem: AAAS; Am Phys Soc; Am Chem Soc; Electrochem Soc; NY Acad Sci. Res: High temperature physical chemistry; vapor pressures; thermodynamics; crystal structures; gas-solid reactions. Mailing Add: CEN-205 Argonne Nat Lab Argonne IL 60439

BLACKBURN, THOMAS HENRY, b Newton, NC, June 1, 23; m 44; c 2. MATHEMATICS. Educ: Lenoir-Rhyne Col, AB, 45; Western Reserve Univ, MA, 48. Prof Exp: From instr to asst prof, 46-55, PROF MATH & HEAD DEPT, LENOIR-RHYNE COL, 58- Mem: Math Asn Am. Res: Functions of a complex variable. Mailing Add: Dept of Math Lenoir-Rhyne Col Hickory NC 28601

BLACKBURN, THOMAS ROY, b St Louis, Mo, Nov 30, 36; m 60; c 3. ANALYTICAL CHEMISTRY, PHYSICAL CHEMISTRY. Educ: Carleton Col, BA, 58; Harvard Univ, MA, 59, PhD(anal chem), 62. Prof Exp: From instr to asst prof chem, Carleton Col, 62-64; asst prof, Wellesley Col, 64-67; assoc prof, 67-72, PROF CHEM, HOBART & WILLIAM SMITH COLS, 72- Concurrent Pos: NSF basic res grant, 64-66. Mem: AAAS. Res: Electrochemistry and electrode reactions; acid-base electrochemistry through the hydrogen electrode; geochemistry of aqueous solutions. Mailing Add: Dept of Chem Hobart & William Smith Cols Geneva NY 14456

BLACKBURN, WALTER EVANS, chemistry, deceased

BLACKBURN, WILBERT HOWARD, b Loa, Utah, Mar 23, 41; m 65; c 3. WATERSHED MANAGEMENT. Educ: Brigham Young Univ, BS, 65; Univ Nev, Reno, MS, 67, PhD, 73. Prof Exp: Lab instr bot, Brigham Young Univ, 63-65; res aide range ecol, Intermountain Forest & Range Exp Sta, 64-65; res asst, Univ Nev, Reno, 65-66, range ecologist, 66-74, asst prof range & watershed mgt, 73-75; ASSOC PROF WATERSHED MGT, TEX A&M UNIV, 75- Mem: Am Soc Range Mgt. Res: Pinyon-juniper invasion, condition and trend; vegetation and soil relationships; range hydrology; infiltration studies of selected vegetation; soil units; range and forest practices and water quality. Mailing Add: Range Sci Dept Tex A&M Univ College Station TX 77843

BLACKBURN, WILL R, b Durant, Okla, Nov 4, 36. PATHOLOGY, TROPICAL MEDICINE. Educ: Univ Okla, BS, 57; Tulane Univ, MD, 61. Prof Exp: Intern med, Col Physicians & Surgeons, Columbia Univ, 61-62, resident path, 62-64, instr, 63-64; resident & instr, Univ Colo, 64-66; from asst prof to assoc prof, Col Med, Pa State Univ, 69-73; PROF PATH, COL MED, UNIV S ALA, 73- Concurrent Pos: Fel, Univ Colo, 64-66; res pathologist, Div Geog Path, Armed Forces Inst Path, 66-68. Mem: AAAS; Am Asn Path & Bact; Soc Invert Path; Soc Pediat Res; Am Soc Exp Biol. Res: Human nutrition; protozoology; developmental biology; pediatric pathology; cellular immunology; electron microscopy; parasitology. Mailing Add: Dept of Path Univ of S Ala Mobile AL 36617

BLACKBURN, WILLIAM HOWARD, b Ottawa, Ont, July 30, 41; m 64; c 2. PETROLOGY. Educ: St Francis Xavier Univ, BSc, 64; Mass Inst Technol, PhD(geol), 67. Prof Exp: Asst prof, 67-71, ASSOC PROF GEOL & GEOCHEM, UNIV KY, 71- Mem: Mineral Soc Am; Am Geophys Union; Geol Asn Can; Mineral Asn Can. Res: Equilibration of metamorphic rocks; mobility of chemical elements during metamorphism and kinetics of metamorphic reactions; variation of major and minor element chemistry of granites; tectonics and basalt geochemistry. Mailing Add: Dept of Geol Univ of Ky Lexington KY 40506

BLACK-CLEWORTH, PATRICIA ANN, b Los Angeles, Calif, May 10, 41. ETHOLOGY, NEUROPHYSIOLOGY. Educ: Stanford Univ, AB, 63; Univ Calif, Los Angeles, MA, 65, PhD(zool), 69. Prof Exp: NIMH training grant neurophysiol & ethology, Dept Anat & Space Biol Lab, 69-71, ASST RES ANATOMIST NEUROPHYSIOL, MENT RETARDATION RES CTR, NEUROPSYCHIAT INST, UNIV CALIF, LOS ANGELES, 72- Mem: Am Soc Ichthyologists & Herpetologists; Animal Behav Soc; Soc Neurosci. Res: Neurophysiological mechanisms of classical conditioning; chronic neurophysiological recordings; social communication in mammals and fishes; human ethology. Mailing Add: Lab of Neurophysiol Neuropsychiat Inst Univ of Calif Los Angeles CA 90024

BLACKERBY, BRUCE ALFRED, b Colon, Repub Panama, Nov 14, 36; US citizen; m 57; c 4. GEOLOGY. Educ: Univ Calif, Riverside, AB , 59; Univ Calif, Los Angeles, PhD(geol), 65. Prof Exp: From asst prof to assoc prof, 63-71, PROF GEOL, FRESNO STATE COL, 71-, CHMN DEPT, 69- Concurrent Pos: NSF instrl grant, 64-65; geologist, US Geol Surv, 65- Mem: Geol Soc Am. Res: Tertiary and Mesozoic volcanics; regional geology; geology of Sierra Nevada, especially batholith. Mailing Add: Dept of Geol Fresno State Col Fresno CA 93726

BLACKETT, DONALD WATSON, b Boston, Mass, May 2, 26; m 51. MATHEMATICS. Educ: Harvard Univ, AB, 47; Princeton Univ, MA, 48, PhD(math), 50. Prof Exp: Instr math, Princeton Univ, 50-51, res assoc, 51-53; from asst prof to assoc prof, 53-62, PROF MATH, BOSTON UNIV, 62- Mem: Am Math Soc; Math Asn Am. Res: Algebra; operations research. Mailing Add: 97 Eliot Ave West Newton MA 02165

BLACKETT, SHIRLEY ALLART, b Greenport, NY, Jan 24, 28; m 51. MATHEMATICS. Educ: Univ Rochester, AB, 48; Pa State Univ, EdM, 51. Prof Exp: Test specialist, Math Test Develop Dept, Educ Testing Serv, 49-52; from instr to asst prof, 53-64, ASSOC PROF MATH, NORTHEASTERN UNIV, 64- Mem: Math Asn Am. Res: Testing and educational measurement; operations research. Mailing Add: Off 505 UR Bldg Northeastern Univ Boston MA 02115

BLACKFORD, VIRGINIA LEE, b Washington, DC, Apr 10, 13. VIROLOGY. Educ: Am Univ, BS, 49; George Washington Univ, MD, 51, PhD(bact), 57. Prof Exp: Bacteriologist, Walter Reed Army Med Ctr, DC, 52-55; bacteriologist, Naval Med Res Inst, Md, 55-59; asst prof microbiol, Col Med, Howard Univ, 59-61; exec secy, Med Res Study Sect, Off Voc Rehab, Dept Health, Educ & Welfare, 61-63; SCIENTIST ADMINR, NAT HEART & LUNG INST, 63- Mem: NY Acad Sci; Am Soc Microbiol; Tissue Cult Asn; Asn Am Med Cols; Mineral Soc Am. Res: Antibiotics and chemotherapy; fungicidal assay by growth experiments; antibiotic production by coliform bacteria; host-parasite relationship of viruses; metabolism and nutrition of rickettsiae. Mailing Add: 4858 Battery Ln Bethesda MD 20014

BLACKHAM, ANGUS UDELL, b East Ely, Nev, Apr 16, 26; m 46; c 6. PHYSICAL ORGANIC CHEMISTRY. Educ: Brigham Young Univ, AB, 49; Univ Cincinnati, MS, 50, PhD, 52. Prof Exp: From asst prof to assoc prof, 52-60, PROF CHEM, BRIGHAM YOUNG UNIV, 60- Prof Exp: Res chemist, US Indust Chem Co, Ohio, 59-60 & Deering Milliken Res Corp, 74-75. Mem: Am Chem Soc. Res: Electroorganic chemistry; azoalkanes and aliphatic azines; transition metal complexes; catalytic reforming of hydrocarbons. Mailing Add: Dept of Chem Brigham Young Univ Provo UT 84601

BLACKHURST, HOMER TENNYSON, b Cass, WVa, Feb 16, 12; m 43; c 3. HORTICULTURE. Educ: WVa State Teachers Col, AB, 35; Agr & Mech Col Tex, MS, 40, PhD(genetics), 47. Prof Exp: Tech asst cytol, 38-47, HORTICULTURIST, EXP STA, TEX A&M UNIV, 47-, PROF HORT, 55- Mem: Assoc Am Genetic Asn; assoc Am Soc Hort Sci. Res: Cytogenetics of Rosa, Citrus and Rubus; breeding of southern peas, snapbeans and crucifers; cytogenetic studies of Rosa Rubiginosa and its hybrids. Mailing Add: 302 Plant Sci Bldg Tex A&M Univ College Station TX 77843

BLACKLER, ANTONIE W C, b Portsmouth, Eng, Oct 19, 31; m 70. DEVELOPMENTAL BIOLOGY. Educ: Univ London, BSc, 53, PhD(embryol), 56. Prof Exp: Sr biologist, Loughborough Col Technol, Eng, 56-57; asst lectr zool, Queen's Univ, Belfast, 57-59; Brit Empire Cancer Campaign res asst embryol, Oxford Univ, 59-61; extraordinary prof zool, Univ Geneva, 61-64; assoc prof, 64-73, PROF ZOOL, CORNELL UNIV, 73- Mem: Soc Develop Biol; Swiss Zool Soc; Int Soc Develop Biol. Res: Origin and differentiation of amphibian sex cells; nucleo-cytoplasmic interactions in amphibian development. Mailing Add: Sect on Genetics, Develp & Physiol Cornell Univ Ithaca NY 14853

BLACKLOW, NEIL RICHARD, b Cambridge, Mass, Feb 26, 38; m 63; c 2. VIROLOGY. Educ: Harvard Univ, BA, 59; Columbia Univ, MD, 63. Prof Exp: Resident med, Beth Israel Hosp, Harvard Univ, 63-65; res virologist, 65-68; clin fel med, Mass Gen Hosp, 68-69; NIH sr scientist, Harvard Univ, 69-71; from asst prof to assoc prof med, Sch Med, Boston Univ, 71-76; PROF MED & MICROBIOL, SCH MED, UNIV MASS, 76- Mem: AAAS; Infectious Dis Soc Am; Am Soc Microbiol; Am Asn Immunol; NY Acad Sci. Res: Defective viruses; clinical virology; clinical infectious disease. Mailing Add: Univ of Mass Med Ctr 55 Lake Ave N Worcester MA 01605

BLACKMAN, CARL F, JR, b Annapolis, Md, Mar 16, 41; m 67. MOLECULAR BIOLOGY, PHYSICAL CHEMISTRY. Educ: Colgate Univ, AB, 63; Pa State Univ, MS, 67, PhD(biophys), 69. Prof Exp: Fel, Brookhaven Nat Lab, 68-70; biologist, Bur Radiol Health, 70; BIOLOGIST, ENVIRON PROTECTION AGENCY, 70- Mem: Am Soc Photobiol; NY Acad Sci; Am Inst Biol Sci; Biophys Soc; Geront Soc. Res: Interactions of non-ionizing electromagnetic radiation—DC to 300 GHz—with biological systems at the molecular and cellular level, particularly interactions which alter genetic processes. Mailing Add: Environ Protection Agency MD-72 Research Triangle Park NC 27711

BLACKMAN, JEROME, b New York, NY, Apr 22, 28; m 55. MATHEMATICS. Educ: Mass Inst Technol, BS, 48; Cornell Univ, PhD(math), 51. Prof Exp: From instr to asst prof, 52-59, ASSOC PROF MATH, SYRACUSE UNIV, 59- Res: Mathematical analysis, geometric quantization. Mailing Add: Dept of Math Syracuse Univ Syracuse NY 13210

BLACKMAN, LESLIE EVERETT, chemistry, see 12th edition

BLACKMAN, SAMUEL WILLIAM, b New York, NY, Aug 25, 13; m 39; c 3. ORGANIC CHEMISTRY. Educ: Cornell Univ, AB, 34; NY Univ, MS, 37; Polytech Inst Brooklyn, PhD(chem), 60. Prof Exp: Dir control, Premo Pharmaceut Labs, Inc, 35-42; microanalyst, Hoffmann-La Roche, 42-44; sr chemist, Burroughs Wellcome & Co, Inc, 44-70; ASSOC PROF CHEM, YESHIVA COL, 60- Mem: Am Chem Soc. Res: Polynuclear condensed heterocycles. Mailing Add: 1349 Lexington Ave New York NY 10028

BLACKMON, BOBBY GLENN, b Rodessa, La, May 17, 40; m 58; c 3. FOREST SOILS. Educ: La Tech Univ, BS, 62; Duke Univ, MF, 63; La State Univ, PhD(forest soils), 69. Prof Exp: Res asst soil fertility, Univ Ark, Stuttgart, 63-65; soil scientist, 67-72, PROJ LEADER FOREST SOILS, SOUTHERN HARDWOODS LABS, USDA FOREST SERV, STONEVILLE, MISS, 72- Mem: Soil Sci Soc Am; Am Soc Agron; Soc Am Foresters. Res: Forest soil fertility and tree nutrition including nutrient cycling, fertilization and nutrient changes in forest soils. Mailing Add: US Forest Serv Southern Hardwoods Lab Box 227 Stoneville MS 38776

BLACKMON, CLINTON RALPH, b Timmonsville, SC, Aug 13, 19; m 45; c 5. PLANT BREEDING. Educ: Clemson Col, BS, 41; Univ Mass, MS, 49; Rutgers Univ, PhD(farm crops), 55. Prof Exp: Instr bot, Univ Mass, 48; assoc prof agron, Nat Agr Col, 48-55, head dept, 49-55; from asst prof to assoc prof, Univ Maine, 56-62; prof & agr adv, Panama, 62-64; assoc prof plant sci, 64-71, chmn dept hort, 69-75, PROF PLANT SCI, DELAWARE VALLEY COL SCI & AGR, 71-, DEAN, 75- Concurrent Pos: Partic, Comn Undergrad Educ Biol Sci Teaching Conf, NSF, Washington, DC, 68 & Biol Inst, Williams Col, 68. Mem: Am Soc Hort Sci; Am Genetic Asn; Am Soc Agron. Res: Plant breeding techniques; fungi spores in the plant and animal environment; varietal improvement; pollen viability in economic plants. Mailing Add: Delaware Valley Col of Sci & Agr Doylestown PA 18901

BLACKMON, CYRIL WELLS, b Timmonsville, SC, Apr 12, 23; m 53; c 3. PLANT PATHOLOGY. Educ: Va Polytech Inst & State Univ, BS, 49; Trinity Col, Tex, MS, 53; Tex A&M Univ, PhD, 61. Prof Exp: Instr pub sch, SC, 49-50; asst mgr, Farmers Co-op Exchange, Farm Supply Store, 50; microbiologist, Southwest Found Res & Educ, 53-54; operator plant prod sulfuric acid, Consol Chem Co, Tex, 54-55; instr hort, Northeast La State Col, 55-56; asst plant path, Exp Sta, Univ Tex, 56-57; plant pathologist, Agr Res Serv, USDA, 57-62, Pesticide Regulation Div, 62-65; asst prof plant path, 65-68, bot & bact, 68-74, ASST PROF PLANT PHYSIOL & PATH, EDISTO EXP STA, CLEMSON UNIV, 74- Mem: Am Phytopath Soc; Am Soc Hort Sci; Sigma Xi. Res: Diseases of cotton, mainly physiology of resistance of cotton varieties to bacterial blight of cotton, Xanthomonas malvacearum; soybean diseases. Mailing Add: Edisto Exp Sta Clemson Univ Blackville SC 29817

BLACKMON, JOHN R, b Canton, Ohio, Jan 28, 30. INTERNAL MEDICINE. Educ: Mt Union Col, BS, 52; Western Reserve Univ, MD, 56; Am Bd Internal Med, dipl, 63; Am Bd Cardiovasc Dis, dipl, 70. Prof Exp: Intern, Univ Hosps, Cleveland, 56-57, resident, 57-59, fel cardiol, 59-60; from instr to asst prof cardiol, 62-69, ASSOC PROF MED, SCH MED, UNIV WASH, 69-, DIR HOSP CARDIOL LABS, 64- Mem: AMA; Am Fedn Clin Res. Res: Total and regional blood flow in normal and cardiac subjects. Mailing Add: Univ of Wash Sch of Med Seattle WA 98144

BLACKMON, MAURICE LEE, b Beaumont, Tex, Aug 24, 40; m 70. ELEMENTARY PARTICLE PHYSICS. Educ: Lamar State Col, BS, 62; Mass Inst Technol, PhD(physics), 67. Prof Exp: Appointee physics, Argonne Nat Lab, 67-69; asst prof, Syracuse Univ, 69-74; MEM STAFF, NAT CTR ATMOSPHERIC RES, 74- Mem: Am Phys Soc. Res: Theory of elementary particles. Mailing Add: Nat Ctr Atmospheric Res PO Box 1470 Boulder CO 80302

BLACKMORE, RAYMOND HORNER, food technology, see 12th edition

BLACKMORE, ROBERT VALENTINE, b Provost, Alta, July 10, 20; m 43; c 3. MICROBIOLOGY, BIOCHEMISTRY. Educ: Univ Alta, DDS, 43; Univ Rochester, PhD(microbiol), 63; FRCD(C), 69. Prof Exp: Pvt pract, 46-48; sessional demonstr oper dent, Univ Alta, 48-50, lectr, 51-54, asst prof, 54-57; Colgate res fel dent, Univ Rochester, 57-58, Nat Res Coun Can fel, 58-62; PROF MICROBIOL & DENT RES, UNIV ALTA, 63- Concurrent Pos: Mem assoc comt dent res, Nat Res Coun Can, 63-66, exec mem, 66-67; mem, Med Res Coun Can, 67-69, mem grants comts for microbiol & infectious dis & for dent sci, 67-72. Res: Virology, especially autointerference by herpes simplex virus; biology, especially purification and characterization of bacterial endonuclease. Mailing Add: Fac of Dent Univ of Alta Edmonton AB Can

BLACKMORE, WILLIAM PETER, b Maryfield, Sask, May 28, 20; nat US; m 48. PHARMACOLOGY. Educ: Univ Sask, BS, 48; Univ Ill, MS, 49, PhD(pharmacol), 53; Univ Tex, MD, 62. Prof Exp: High sch prin, Sask, 43-44; dir mfg pharm, Res Hosp, Univ Ill, 50-51; from asst prof to assoc prof pharm, Univ Tex Southwestern Med Sch Dallas, 53-62; from asst dir to dir clin pharmacol, 62-70, DIR DIV CLIN RES, STERLING-WINTHROP RES INST, 70- Mem: AAAS; Am Col Clin Pharmacol & Chemother; Am Soc Pharmacol & Exp Therapeut; Am Heart Asn; Soc Exp Biol & Med. Res: Renal and electrolyte pharmacology and physiology; clinical research. Mailing Add: Div of Clin Res Sterling-Winthrop Res Inst Rensselaer NY 12144

BLACK-SCHAFFER, BERNARD, b New York, NY, Nov 22, 10; m 39; c 2. PATHOLOGY. Educ: NY Univ, BSc, 32; Univ Vienna, MD, 37. Prof Exp: Asst path, Sch Med, Yale Univ, 40-42; asst prof, Med Col Va, 42-45; assoc prof, Sch Med, Duke Univ, 45-50; from assoc prof to prof, Col Med, Univ Cincinnati, 52-68; PROF PATH, IND UNIV, BLOOMINGTON, 68-, DIR COMBINED MED DEGREE PROG, 69- Concurrent Pos: Pathologist, Atomic Bomb Casualty Comn, Nat Res Coun, Japan, 50-52. Mem: AAAS; Am Asn Path & Bact. Res: Cardiovascular research. Mailing Add: Combined Med Degree Prog Ind Univ Bloomington IN 47401

BLACKSHEAR, GERTRUDE LIEBL, b Sheboygan, Wis, Dec 16, 22; m 48; c 6. PHYSIOLOGY. Educ: Univ Wis, BS, 44, MD, 47; Univ Minn, PhD(physiol), 75. Prof Exp: Teaching assoc physiol, 67-75, INSTR PHYSIOL, UNIV MINN, 75- Res: Study of capillary permeability, particularly in the heart with emphasis on changes due to physical or chemical agents. Mailing Add: Dept of Physiol Univ of Minn Minneapolis MN 55455

BLACKSTEAD, HOWARD ALLAN, b Minot, NDak, Feb 24, 40; m 64. MAGNETIC RESONANCE, LOW TEMPERATURE PHYSICS. Educ: NDak State Univ, BS, 62; Dartmouth Col, MA, 64; Rice Univ, PhD(physics), 67. Prof Exp: Res assoc physics, Univ Ill, Urbana-Champaign, 67-69; asst prof, 69-75, ASSOC PROF PHYSICS, UNIV NOTRE DAME, 75- Mem: Am Phys Soc. Res: Superfluid helium flow properties; ferromagnetic resonance in the heavy rare earth metals; nuclear quadrupole resonance; phonon spectroscopy. Mailing Add: Dept of Physics Univ of Notre Dame Notre Dame IN 46556

BLACKSTOCK, DAVID THEOBALD, b Austin, Tex, Feb 13, 30; m 55; c 4. ACOUSTICS. Educ: Univ Tex, BS, 52, MA, 53; Harvard Univ, PhD(appl physics), 60. Prof Exp: Physicist, Bioacoust Br, Wright Air Develop Ctr, Ohio, 54-56; sr res asst, Acoust Res Lab, Harvard Univ, 60; sr res staff mem, Nonlinear Acoust, Res Dept, Gen Dynamics/Electronics NY, 60-63; assoc prof elec eng, Univ Rochester, 63-70; FAC RES SCIENTIST, APPL RES LABS, UNIV TEX, AUSTIN, 70- Concurrent Pos: Mem hearing, bioacoust & biodynamics comt, Nat Acad Sci-Nat Res Coun, 63-70. Mem: Fel Acoust Soc Am; Am Asn Physics Teachers. Res: Nonlinear acoustics; wave motion. Mailing Add: Appl Res Labs Univ of Tex Austin TX 78712

BLACKSTONE, DONALD LEROY, b Chinook, Mont, June 16, 09; m 36; c 4. GEOLOGY. Educ: Univ Wash, BS, 31; Univ Mont, MS, 34; Princeton Univ, PhD(geol), 36. Prof Exp: Geologist, Carter Oil Co, Okla, asst prof geol, Univ Mo, 39-42; geologist, Carter Oil Co, Okla, 42-45; asst prof geol, Univ Mo, 45-46; from assoc prof to prof, 46-74, head dept, 62-68, EMER PROF GEOL, UNIV WYO, 74- Concurrent Pos: Wyo state geologist & dir, Wyo Geol Surv, 66-68. Mem: Am Asn Petrol Geologists; fel Geol Soc Am; Am Geophys Union. Res: Structural geology; relationship of Precambrian structure to later deformation; structural pattern in south central Wyoming; structural geology of Pryor Mountains, Montana. Mailing Add: Dept of Geol Univ of Wyo Laramie WY 82070

BLACKWELDER, RICHARD ELIOT, b Madison, Wis, Jan 29, 09; m 35. INVERTEBRATE ZOOLOGY. Educ: Stanford Univ, AB, 31, PhD(zool), 34. Prof Exp: Bacon traveling scholar, Smithsonian Inst, 35-38; asst cur entom, Am Mus Natural Hist, 38-40; from asst cur to assoc cur insects, US Nat Mus, 40-54; assoc prof, St John Fisher Col, 56-58; assoc prof, 58-65, PROF ZOOL, SOUTHERN ILL UNIV, CARBONDALE, 65- Concurrent Pos: Mem expeds, Panama, 31, WI, 35-37; agt, White-Fringed Beetle Identification Unit, USDA, 38; ed, Air-Track Mfg Co, 43-45. Mem: AAAS; Soc Syst Zool (secy-treas, 49-60, pres, 61); Brazilian Entom Soc. Res: Principles of biology; morphology, classification, bibliography and nomenclature; coleopterous family Staphylinidae; invertebrates. Mailing Add: Dept of Zool Southern Ill Univ Carbondale IL 62901

BLACKWELL, BARRY M, b Birmingham, Eng, July 5, 34; m 58; c 3. PSYCHOPHARMACOLOGY. Educ: Cambridge Univ, BA, 57, MB, BChir, 60, MA & MD, 66; Univ London, DPM, 67. Prof Exp: Intern med & surg, Guys Hosps, Cambridge, 60-61; resident neurol, Whittington Hosp, London, 61-62; resident psychiat, Maudsley Hosp & Inst Psychiat, 62-68; asst prof psychiat, Univ Cincinnati, 68-70, prof psychiat & assoc prof pharmacol, 70-75; PROF PSYCHIAT & PHARMACOL & CHMN DEPT PSYCHIAT, SCH MED, WRIGHT STATE UNIV, 75-; GROUP DIR CLIN PSYCHIAT RES, MERRELL-NAT CO, 68- Mem: Royal Col Psychiat; Am Col Neuropsychopharmacol. Res: Interactions between foodstuffs and monoamine oxidase inhibitors; various aspects of clinical and psychopharmacology research; clinical research.

BLACKWELL, DAVID (HAROLD), b Centralia, Ill, Apr 24, 19; m 44; c 7. MATHEMATICS, STATISTICS. Educ: Univ Ill, AB, 38, AM, 39, PhD(math), 41. Prof Exp: Rosenwald fel, Inst for Advan Study, 41-42; asst statistician, Off Price Admin, 42; instr math & physics, Southern Agr & Mech Col, 42-43; instr math, Clark Univ, 43-44; prof, Howard Univ, 44-54; prof statist, Univ Calif, Berkeley, 54-73, PROF STATIST & MATH & DIR STUDY CTR UNIV CALIF, UNITED KINGDOM & IRELAND, 73- Concurrent Pos: Res fel, Brown Univ, 43. Mem: AAAS; Am Math Soc; fel Inst Math Statist. Res: Markoff chains; sequential analysis. Mailing Add: Study Ctr Univ of Calif 26 Dover St London W1X 4DX England

BLACKWELL, FLOYD ORIS, b Mayfield, Ohio, Feb 27, 25; m 51; c 4. ENVIRONMENTAL HEALTH. Educ: Wash State Univ, BS, 50; Univ Mass, MS, 54; Univ Calif, Berkeley, MPH, 65, DrPH, 69; Am Acad dipl. Prof Exp: Sanitarian, Benton-Franklin Dist Health Dept, 50-53; health & sanit adv, US AID, Pakistan, 54-56, sr sanit adv, 56-59; asst prof environ & pub health, Am Univ Beirut,59-64; assoc prof, Rutgers Univ, 67-71; assoc prof, Col Med, Univ Vt, 71-74; ASSOC PROF ENVIRON & PUB HEALTH, E CAROLINA UNIV, 74- Concurrent Pos: Mem pub health rev comt, Bur Health Manpower Educ, NIH, 71-73; mem bd dirs, Am Acad Sanitarians, 72- Honors & Awards: H L Nicholas Award, NJ Sanitarians Assoc, Inc, 70; Cert of Merit, Nat Environ Health Asn, 70. Mem: Am Pub Health Asn; Nat Environ Health Asn (pres, 76). Res: Environmental control of communicable diseases, particularly food borne diseases, considering biological, microbiological, physical and social aspects in complex interrelationships; ecology of human well-being. Mailing Add: 1210 E Rock Spring Rd Greenville NC 27834

BLACKWELL, HAROLD RICHARD, b Harrisburg, Pa, Jan 16, 21; m 43; c 2. VISION. Educ: Haverford Col, BS, 41; Brown Univ, MA, 42; Univ Mich, PhD(psychol), 47. Prof Exp: Res psychologist, Polaroid Corp, 43; res assoc, L C Tiffany Found, New York, 43-45; dir vision res labs, Univ Mich, 46-58; DIR INST RES IN VISION, OHIO STATE UNIV, 58-, PROF PHYSIOL OPTICS & RES PROF OPHTHALMOL, 58-, PROF BIOPHYS, 65- Concurrent Pos: Exec secy vision comt, Nat Res Coun, US Armed Forces, 45-55. Honors & Awards: Lomb Medal, Optical Soc Am, 50; Gold Medal, Illuminating Eng Soc, 72. Mem: Optical Soc Am; Asn Res Vision & Ophthal; Am Acad Optom; Psychonomic Soc; Illuminating Eng Soc. Res: Psychophysics and psychophysiology of vision; psychophysical methodology and sensory theory; physiological optics. Mailing Add: Inst for Res in Vision Ohio State Univ Res Ctr Columbus OH 43212

BLACKWELL, JOHN, b Sheffield, Eng, Jan 15, 42; m 65; c 1. BIOPHYSICS, POLYMER SCIENCE. Educ: Univ Leeds, BSc, 63, PhD(biophys), 67. Prof Exp: Fel chem, State Univ NY Col Forestry, Syracuse Univ, 67-69; asst prof, 69-74, ASSOC PROF MACROMOLECULAR SCI, CASE WESTERN RESERVE UNIV, 74- Concurrent Pos: NIH res career develop award, 73-77. Mem: Biophys Soc; Soc Complex Carbohydrates; AAAS; Am Chem Soc; Am Phys Soc. Res: Structure and interactions of biological macromolecules; x-ray crystallography and infrared and Raman spectroscopy of polysaccharides; circular dichroism studies of polysaccharide-protein interactions; polyurethane structures. Mailing Add: Dept of Macromolec Sci Case Western Reserve Univ Cleveland OH 44120

BLACKWELL, JOHN HENRY, b Melbourne, Australia, July 9, 21; m 46; c 4. APPLIED MATHEMATICS. Educ: Univ Melbourne, BSc, 41; Univ Western Ont, MSc, 47, PhD(physics), 52. Prof Exp: Demonstr physics, 46-47, instr, 47-49, lectr, 49-52, from asst prof to prof, 52-62, appl math, 62-63, sr prof, 63-67, PROF & HEAD DEPT APPL MATH, UNIV WESTERN ONT, 67- Concurrent Pos: Vis res fel, Australian Nat Univ, 55-56; vis Commonwealth fel, Oxford Univ, 64-65; mem, Can Nat Comt, Int Union Theoret & Appl Math. Honors & Awards: Can Efficiency Decoration, 67. Mem: Can Asn Physicists; Can Math Cong; Can Soc Mech Eng. Res: Heat flow and diffusion theory; fluid dynamics; hydromagnetics. Mailing Add: Dept of Appl Math Univ of Western Ont London ON Can

BLACKWELL, LAWRENCE A, b Houston, Tex, Nov 23, 31; m 57; c 2. ACADEMIC ADMINISTRATION, SAFETY ENGINEERING. Educ: Rice Inst, BA, 54; Duke Univ, PhD(physics), 58. Prof Exp: Engr microwave, Tex Instruments, 58-62; vpres res & develop, Microwave Physics Inc, 62-66; ACAD ADMINR, RICE UNIV, 66- Mem: Am Soc Safety Engr. Mailing Add: 5331 Forest Haven Houston TX 77066

BLACKWELL, LEO HERMAN, b Austin, Tex, June 2, 34; m 63; c 3. PHYSIOLOGY. Educ: Univ Tex, BA, 55; ETex State Col, MA, 60; Mich State Univ, PhD(physiol), 64. Prof Exp: Res technician physics, Univ Tex M C Anderson Hosp & Tumor Inst Houston, 56-59, res assoc, 59-60; asst biol, ETex State Col, 59; asst physiol, Mich State Univ, 62-63; resident assoc, Argonne Nat Lab, 63-64; radiation safety officer, 64-68, actg chmn div radiation biol, 67, asst prof radiation biol & clin physiol, 64-67, assoc prof radiation biol & clin physiol, 67-74, asst to vchancellor acad affairs, 74-75, ASST PROF PHYSIOL & BIOPHYS & PROF RADIATION ONCOL, UNIV TENN CTR HEALTH SCI, MEMPHIS, 75- Concurrent Pos: Consult, City of Memphis Hosps, 67- Mem: AAAS; Radiation Res Soc; Am Inst Biol Sci; Health Physics Soc. Res: Cell proliferation; biological effects of radiation; radiation dosimetry. Mailing Add: Ctr for the Health Sci Univ of Tenn Memphis TN 38163

BLACKWELL, PAUL K, II, b Miami, Fla, July 10, 31; m 49; c 3. COMPUTER SCIENCE, MATHEMATICS. Educ: Univ Chicago, BA, 52; Syracuse Univ, MS, 61, PhD(math), 68. Prof Exp: Assoc prof math, Milwaukee Sch Eng, 55-57; analyst, Systs Develop Corp, 57-58; analyst defense systs, Gen Elec Co, NY, 58-61, consult, 61-63, analyst defense systs, 63-65; mathematician, Syracuse Univ Res Corp, 65-68; ASSOC PROF COMPUT SCI, UNIV MO-COLUMBIA, 68- Mem: Am Math Soc; Asn Comput Mach; Inst Elec & Electronics Engrs; Math Asn Am; Soc Indust & Appl Math. Res: Logic functions; finite Fourier transform; combinatorics. Mailing Add: Dept of Comput Sci Univ of Mo Columbia MO 65201

BLACKWELL, RICHARD QUENTIN, b Wichita, Kans, Sept 5, 18; m 45, 60; c 4. BIOCHEMISTRY. Educ: Univ Wichita, AB, 42; Northwestern Univ, PhD(biochem), 49. Prof Exp: Lectr chem, Northwestern Univ, 49-50, from instr to assoc prof, 50-61; head dept biochem, US Naval Med Res Unit 2, Taipei, Taiwan, 58-74, MEM DIV CANCER BIOL & DIAG, NAT CANCER INST, 74- Mem: AAAS; Am Chem Soc. Res: Clinical biochemistry; human hemoglobin variants; nutrition. Mailing Add: Div of Cancer Biol & Diag Nat Cancer Inst Bethesda MD 20014

BLACKWELL, ROBERT JERRY, b Clovis, NMex, Dec 31, 25; m 55; c 3. PHYSICS. Educ: Tex Christian Univ, BA, 47; Univ NC, PhD(physics), 53. Prof Exp: Instr physics, Univ NC, 47-52, asst prof, 53-54; mem prod res div, Humble Oil & Ref Co, 54-64 & Esso Prod Res Co, 64-74, SR RES ADV, EXXON CORP, 74- Mem: Am Phys Soc; Am Asn Physics Teachers; Am Inst Chem Eng; Am Inst Min, Metall & Petrol Eng. Res: Cosmic rays; flow through porous media; petroleum production. Mailing Add: 8904 Memorial Houston TX 77024

BLACKWELL, ROBERT LEIGHTON, b Seneca, NMex, Nov 24, 24; m 50; c 1. ANIMAL GENETICS. Educ: NMex Col, BS, 49; Ore State Col, MS, 51; Cornell Univ, PhD(animal genetics), 53. Prof Exp: From asst prof to assoc prof animal husb, NMex State Univ, 53-58; dir, Sheep Exp Sta, Agr Res Serv, USDA, 59-66; PROF ANIMAL & RANGE SCI & HEAD DEPT, MONT STATE UNIV, 66- Res: Genetics and breeding of domestic animals. Mailing Add: Dept of Animal & Range Sci Mont State Univ Bozeman MT 59715

BLACKWOOD, ALLISTER CLARK, b Calgary, Alta, Nov 22, 15; m 43, c 3. MICROBIOLOGY. Educ: Univ Alta, BSc, 42, MSc, 44; Univ Wis, PhD, 49. Prof Exp: Bacteriologist, Nat Res Coun, 44-46 & Div Appl Biol, Prairie Regional Lab, 48-57; chmn dept, 57-68, PROF MICROBIOL, MACDONALD COL, McGILL UNIV, 57-, DEAN & VPRIN, 72- Concurrent Pos: Ed, Can J Microbiol, 70- Mem: Am Soc Microbiol; Can Soc Microbiol (1st vpres, 63-64, pres, 64-65); Soc Indust Microbiol; fel Royal Soc Can; Brit Soc Appl Bact. Res: Bacterial physiology; metabolism of yeasts; metabolism of phenolics; water pollution. Mailing Add: Fac of Agr Macdonald Col McGill Univ Ste Anne de Bellevue PQ Can

BLACKWOOD, CARLTON E, b Jamaica, WI; US citizen; m 53; c 4. BIOCHEMISTRY. Educ: Long Island Univ, BS, 51; NY Univ, MS, 59, PhD(biol chem), 62. Prof Exp: Res asst biochem, Funk Found Med Res, 53-59; res assoc, Columbia Univ, 59-67; ASSOC PROF BIOL, IONA COL, 67- Concurrent Pos: Res assoc, Columbia Univ, 59- & St Barnabas Hosp, Bronx, NY, 67-; lectr biol, Hunter Col, 63-68. Mem: Sigma Xi; Am Soc Cell Biol; Tissue Cult Soc. Res: Cancer immunology; immunology of aging; aging as related to connective tissues; aging and emphysema. Mailing Add: Dept of Biol Iona Col 715 North Ave New Rochelle NY 10801

BLACKWOOD, CHESLEY M, b Nfld, Jan 2, 30; m 56; c 2. BIOCHEMISTRY, FOOD TECHNOLOGY. Educ: Dalhousie Univ, BSc, 50; Univ Toronto, MA, 54; Univ Wash, PhD(biochem), 61. Prof Exp: Assoc scientist, Fisheries Res Bd Can, Nfld, 61-62; chief inspection br, Maritimes Region, Can Dept Fisheries, 62-66, asst dir, 66-70, dir inspection br, Fisheries & Marine Serv, Dept Environ, 70-75, DIR INSPECTION & TECHNOL BR, ENVIRON CAN, FISHERIES & MARINE SERV, 75- Res: Product development fisheries; application of data from basic research in solving technical problems at fish processing plant level. Mailing Add: Fisheries & Marine Serv 9th Floor 580 Booth St Ottawa ON Can

BLACKWOOD, ROBERT KEITH, b Edmonton, Alta, Feb 17, 30; nat US; m 53; c 2. ORGANIC CHEMISTRY. Educ: Univ Alta, BSc, 51; Purdue Univ, PhD(org chem), 55. Prof Exp: Chem lab demonstr, Univ Alta, 48-51; teaching asst, Purdue Univ, 51-54; res chemist, 55-68, MGR PROCESS RES, PFIZER, INC, 68- Mem: Am Chem Soc. Res: Pydoxin syntheses; aliphatic nitro compounds; chemistry of the tetracycline antibiotics; medicinal process research and development. Mailing Add: Pfizer Inc Groton CT 06340

BLACKWOOD, UNABELLE BOGGS, b Looneyville, WVa, Aug 5, 29; m 56; c 2. NUTRITION. Educ: Berea Col, BA, 52; Columbia Univ, MA, 56; Univ Md, PhD(poultry nutrit), 58. Prof Exp: Asst res prof, Univ Md, 57-59 & Ohio State Univ, 59-62; lectr, Univ Bridgeport, 63-67 & Univ Akron, 69-75; NUTRIT CONSULT, 73- Mem: AAAS; Soc Exp Biol & Med; NY Acad Sci; Int Acad Prev Med. Res: Orthomolecular nutritional treatment of mental, emotional and other physical illnesses. Mailing Add: 1025 Amelia Ave Akron OH 44302

BLADES, ARTHUR TAYLOR, b Milton, Ont, July 20, 26; m 50; c 3. PHYSICAL CHEMISTRY. Educ: Univ Western Ont, BSc, 48; Univ Wis, PhD(phys chem), 52. Prof Exp: Photochemist, Nat Res Coun Can, 52-54; fel nuclear chem, McMaster Univ, 54-55; RES CHEMIST, RES COUN ALTA, 55- Concurrent Pos: Mem sci adv comt, Environ Conserv Authority, Alta, 74- Mem: Chem Inst Can. Res: Chemical kinetics; kinetic isotope effects in gas phase reactions; chemistry in flames. Mailing Add: 11315 87th Ave Edmonton AB Can

BLADES, CHARLES ERNEST, b Milton, Ont, Feb 6, 20; US citizen; m 46; c 3. ORGANIC CHEMISTRY. Educ: Univ Alta, BSc, 47; Univ Wis, PhD(chem), 53. Prof Exp: Asst, Univ Wis, 48-50 & 51-53; res chemist, Air Reduction Labs, 53-57, sect head, 57-58, mgr tech sales lab, 59-61, mgr develop lab, 61-69; dir res, Airco Chem & Plastics Co, 69-71; RES COORDR, AIR PROD & CHEM CO, 71- Res: Acetylenics; vinyl monomers; polymers; paper coatings; paints; textiles; water soluble resins; plastics. Mailing Add: Air Products & Chemicals Co 1 Possumtown Rd Piscataway NJ 08854

BLADES, JOHN DIETERLE, b Cincinnati, Ohio, Nov 1, 24; m 49; c 3. PHYSICS. Educ: Western Md Col, BS, 49; Univ Cincinnati, MS, 51, PhD(physics), 54. Prof Exp: Sr staff scientist, Burroughs Corp, 54-65; mgr solid state physics lab, Franklin Inst Res Labs, 65-71; MGR SOLID STATE MAT, GRAPHIC PROD DEVELOP LAB, ADDRESSOGRAPH-MULTIGRAPH CORP, CLEVELAND, 71- Mem: AAAS; Am Phys Soc. Res: Electrodynamics of solid state systems; magnetism; microwave techniques; mechanics; physics of the thin film and particulate state. Mailing Add: 1110 Sheerbrook Dr Chagrin Falls OH 44022

BLAEDEL, WALTER JOHN, b New York, NY, May 26, 16; m 42; c 3. ANALYTICAL CHEMISTRY. Educ: Univ Calif, Los Angeles, BA, 38, MA, 39; Stanford Univ, PhD(chem), 42. Prof Exp: Instr chem, Northwestern Univ, 41-42, res assoc, Nat Defense Res Comt, 42-44; res assoc, Manhattan Proj, Univ Chicago, 44-46;

res assoc, Radiation Lab, Univ Calif, 46-47; from asst prof to assoc prof chem, 47-57, PROF CHEM, UNIV WIS-MADISON, 57- Concurrent Pos: Mem comt postdoctoral fels in chem, Nat Acad Sci, 63-66; mem anal chem adv comt, Nat Bur Stand-Nat Acad Sci, 67-69; mem comt anal chem, Nat Res Coun, 75- Mem: Am Chem Soc; Sigma Xi. Res: Analysis in membrane systems; separations; electrochemistry; ion exchange. Mailing Add: Dept of Chem Univ of Wis Madison WI 53706

BLAESE, ROBERT MICHAEL, b Minneapolis, Minn, Feb 16, 39; m 62; c 2. IMMUNOLOGY, PEDIATRICS. Educ: Gustavus Adolphus Col, BS, 61; Univ Minn, MD, 64. Prof Exp: Intern, Parkland Hosp, Dallas, 64-65; resident pediat, Hosp, Univ Minn, 65-66; clin assoc immunol, 66-68, sr investr, 68-73, HEAD CELLULAR IMMUNOL SECT, NAT CANCER INST, 73- Mem: Am Soc Clin Invest; Am Asn Immunologists; Soc Pediat Res: Am Fedn Clin Res; AAAS. Res: Fundamental mechanisms of host defense by studying patients with immunodeficiency diseases, allergy, malignancy and autoimmunity; cellular immunology; macrophage function; cytotoxicity; antibody formation; neonatal immunology and phagocytosis. Mailing Add: Nat Cancer Inst Bethesda MD 20014

BLAGA, AUREL, organic chemistry, see 12th edition

BLAGER, FLORENCE BERMAN, b Wheeling, WVa, July 2, 28; m 73. SPEECH PATHOLOGY. Educ: Ohio Univ, BA, 50; Univ Denver, MA, 66, PhD(speech path), 70. Prof Exp: Teacher Eng & speech, High Sch, New York, 62-65; admin asst, Videotape Self-Confrontation Res Proj, Speech & Hearing Ctr, Univ Denver, 69-70; CHIEF SPEECH PATH & AUDIOL, JOHN F KENNEDY CHILD DEVELOP CTR, UNIV COLO MED CTR, DENVER, 70- Concurrent Pos: Asst prof, Dept Otolaryngol, Univ Colo Med Ctr, Denver & Dept Speech Path & Audiol, Univ Colo, Boulder; adj asst prof, Dept Speech Path & Audiol, Univ Northern Colo, Colo State Univ & Univ Denver; consult speech path, Children's Diag Ctr & Day Care Ctr, Univ Colo Med Ctr, Denver & Phenylketonuria Clin, John F Kennedy Child Develop Ctr. Mem: Am Speech & Hearing Asn. Res: Communication methodology. Mailing Add: John F Kennedy Child Develop Ctr Univ of Colo Med Ctr Denver CO 80220

BLAGG, JOHN CREIGHTON LEE, b Mt Sterling, Iowa, Aug 8, 09; m 37; c 2. CHEMISTRY. Educ: Iowa Wesleyan Col, BS, 31; Columbia Univ, PhD(phys chem), 36. Hon Degrees: DSc, Iowa Wesleyan Col, 61. Prof Exp: Lab asst, Columbia Univ, 31-36; res chemist, Pfizer, Inc, New York, 36-39, head prod dept, 39-51, asst supt prod, 51-58, prod mgr, 58-63, plant mgr, Brooklyn, 63-72; RETIRED. Mem: Am Chem Soc. Res: Rate of thermal decomposition of deuterium iodide; chemical engineering; pharmacy. Mailing Add: 27 Glendale Rd Park Ridge NJ 07656

BLAHA, ELI WILLIAM, b Collinsville, Ill, Oct 30, 27; m 48; c 2. ORGANIC CHEMISTRY. Educ: Univ Ill, BS, 51; Univ Iowa, PhD(org chem), 54. Prof Exp: Chemist, Standard Oil Co, Ind, 54-60, proj chemist, Am Oil Co Div, 60-64, sr proj chemist, Amoco Chem Corp, 64-70, group leader polystyrene process develop, 70-73, RES SUPVR, AMOCO CHEM CORP, STANDARD OIL CO, INC, 73- Mem: Am Chem Soc; Soc Plastics Engrs. Res: Discovery, development and production of motor oil additives, including detergents, dispersants, oxidation inhibitors and antiwear agents; research and development of improved polystyrene resins and new condensation polymers. Mailing Add: Amoco Chem Corp PO Box 400 Naperville IL 60540

BLAHA, GORDON C, b Chicago, Ill, July 21, 34. ANATOMY, HISTOLOGY. Educ: Northern Ill Univ, BS, 57; Univ Ill, MS, 61, PhD(anat), 63. Prof Exp: Instr anat, Univ Ill Col Med, 62-63; from instr to asst prof, 63-72, ASSOC PROF ANAT, COL MED, UNIV CINCINNATI, 72- Mem: AAAS; Am Asn Anatomists; Soc Study Reproduction. Res: Reproductive physiology; senescent decline in reproductive function; senescence in golden hamster. Mailing Add: Dept of Anat Univ of Cincinnati Col of Med Cincinnati OH 45267

BLAHA, STEPHEN, b North Tarrytown, NY, Aug 20, 44; m 68; c 2. ELEMENTARY PARTICLE PHYSICS. Educ: Univ Notre Dame, BS, 66; Rockefeller Univ, PhD(physics), 71. Prof Exp: Res assoc elem particle physics, Univ Wash, 71-73; res assoc & instr, Cornell Univ, 73-75; ASST PROF ELEM PARTICLE PHYSICS, SYRACUSE UNIV, 75- Mem: Am Phys Soc; Sigma Xi. Res: Models of the substructure of elementary particles. Mailing Add: Dept of Physics Syracuse Univ Syracuse NY 13210

BLAHD, WILLIAM HENRY, b Cleveland, Ohio, May 11, 21; m 45; c 3. NUCLEAR MEDICINE, INTERNAL MEDICINE. Educ: Tulane Univ, MD, 45; Am Bd Internal Med, dipl, 53; Am Bd Nuclear Med, dipl, 72. Prof Exp: Ward officer metab res, Vet Admin, 51-52, asst chief radioisotope serv, 52-56; CHIEF NUCLEAR MED SERV, WADSWORTH VET ADMIN HOSP, 56- Concurrent Pos: Prof med, Univ Calif, Los Angeles, 66- Mem: Fel Am Col Physicians; AMA; Health Physics Soc; Soc Exp Biol & Med; Soc Nuclear Med. Res: Development of nuclear medicine techniques and study of body composition. Mailing Add: Nuclear Med Serv Wadsworth Vet Admin Hosp Los Angeles CA 90073

BLAIN, DANIEL, b Kashing, China, Dec 17, 98; m 36; c 1. PSYCHIATRY. Educ: Washington & Lee Univ, BA, 21; Vanderbilt Univ, MD, 29; Am Bd Neurol & Psychiat, cert. Hon Degrees: LLD, LaSalle Col, 68; DrS, Washington & Lee Univ, 72. Prof Exp: House officer, Peter Bent Brigham Hosp & Boston City Hosp, 31; Austen Riggs Found fel, Stockbridge, Mass, 32; res psychiatrist, Blythewood Sanitarium, Greenwich, Conn, 37; assoc dir, Tratelja Farms Sanitarium, Lake George, NY, 38-42; chief psychiat div, Vet Admin, 45-48; prof psychiat & chmn dept, Georgetown Univ, 47-48, prof clin psychiat, 48-58; prof psychiat, 58-68, EMER PROF CLIN PSYCHIAT, UNIV PA, PHILADELPHIA, 68- Concurrent Pos: Practicing psychiatrist, 32-42; consult, Vet Admin, 48-58; state dir ment hyg, Calif, 59-63; dir psychiat planning & develop, Pa Hosp, 63-66, life visitor, 67-; dir, Philadelphia State Hosp, 66-70. Mem: Fel AMA; Am Psychiat Asn (pres, 64-65); fel Am Col Physicians; Am Psychopath Asn; Asn Res Nerv & Ment Dis. Res: Blood of fowl; spinal fluid; administrative and social psychiatry. Mailing Add: 2100 Clarkson Ave Philadelphia PA 19144

BLAINE, LAMDIN ROBERT, b New York, NY, Mar 2, 21; m 49; c 3. PHYSICS. Educ: Gettysburg Col, BA, 49; Univ Md, MS, 63. Prof Exp: Pub sch teacher, Pa, 50-51; physicist, Ballistics Res Lab, Aberdeen, Md, 51-52; physicist infrared, Nat Bur Standards, 52-66; PHYSICIST-AEROSPACE TECHNOLOGIST, EARTH OBSERV SYSTS DIV, SENSOR DEVELOP BR, GODDARD SPACE FLIGHT CTR, 66- Mem: AAAS; Optical Soc Am. Res: Molecular absorption and emission spectroscopy in the mean and far infrared; design of associated instrumentation. Mailing Add: Earth Observ Systs Div Goddard Space Flight Ctr Code 652 Greenbelt MD 20771

BLAINE, ROGER LEE, analytical chemistry, see 12th edition

BLAIR, A JAMES, JR, b Philadelphia, Pa, Jan 26, 23; m 64; c 3. INTERNAL MEDICINE, ENDOCRINOLOGY. Educ: Amherst Col, AB, 46; Cornell Univ, MD, 51; McGill Univ, PhD(exp med), 61. Prof Exp: Intern path, Grace-New Haven Community Hosp, Conn, 51-52; intern med, Univ Hosps, Minneapolis, Minn, 52-53; resident, Univ Hosps, Columbus, Ohio, 53-54; resident, Royal Victoria Hosp, Montreal, Que, 55-56; from instr to asst prof med, Univ Wis, 60-65; Swedish Med Res Coun res fel endocrinol, Karolinska Inst, Sweden, 65-66; asst prof med, 66-69, ASSOC PROF MED, UNIV MICH, ANN ARBOR, 69-; DIR ENDOCRINE SERV, WAYNE COUNTY GEN HOSP, ELOISE, 69- Mem: Endocrine Soc; Am Fedn Clin Res: Am Diabetes Asn; fel Am Col Physicians; Am Physiol Soc. Res: Metabolism of adrenocortical and progestational hormones. Mailing Add: Dept of Med Wayne County Gen Hosp Eloise MI 48152

BLAIR, ALAN HUNTLEY, b Vancouver, BC, Oct 20, 33; m 60; c 2. BIOCHEMISTRY. Educ: Univ BC, BSc, 56, MSc, 57; Univ Calif, Berkeley, PhD(biochem), 64. Prof Exp: Damon Runyan Cancer Res Fund fel, Biophys Res Lab, Harvard Med Sch, 64-65; assoc staff med, Peter Bent Brigham Hosp, Boston, Mass, 65; asst prof, 66-69, ASSOC PROF BIOCHEM, DALHOUSIE UNIV, 69- Mem: Can Biochem Soc; Am Chem Soc. Res: Enzymology; mechanism of action of alcohol and aldehyde dehydrogenases. Mailing Add: Dept of Biochem Dalhousie Univ Halifax NS Can

BLAIR, ALBERT PATRICK, b Dayton, Tex, Nov 3, 13; m 36. ZOOLOGY. Educ: Univ Tulsa, BS, 36; Ind Univ, PhD(zool), 40. Prof Exp: Nat Res Found fel, 40-41; cur, NY Zool Soc, 42; asst cur, Dept Animal Behav, Am Mus Natural Hist, 43-47; from asst prof to assoc prof, 47-56, PROF ZOOL, UNIV TULSA, 56- Mem: Am Soc Zoologists; Genetics Soc Am; Am Soc Ichthyologists & Herpetologists. Res: Population genetics; geographical variation; sex behavior of toads. Mailing Add: Dept of Life Sci Univ of Tulsa Tulsa OK 74104

BLAIR, ALEXANDER MARSHALL, b Windsor, Ont, May 24, 32. GEOGRAPHY. Educ: Univ Western Ont, BA, 54, MA, 61; Univ Ill, Urbana, PhD(geog), 65. Prof Exp: Demonstr geog, Univ Western Ont, 59-61, part-time instr & reader, Exten Div, 60-61; asst, Univ Ill, Urbana, 62-63, Bur Community Planning, 63; lectr, 64-65, asst prof, 65-69, actg chmn dept, 69-71, ASSOC PROF GEOG, YORK UNIV, ONT, 69- Honors & Awards: Mackintosh Prize, Univ Western Ont, 60. Mem: Asn Am Geog; Can Asn Geog. Res: Non-mrtallic mineral industries in Southern Ontario. Mailing Add: Dept of Geog York Univ Fac Arts & Sci Toronto ON Can

BLAIR, ALLEN G, b Mora, Minn, Dec 14, 26; m 57; c 4. RESEARCH ADMINISTRATION. Educ: Hamline Univ, BS, 49; Univ Minn, MS, 54; Univ Pittsburgh, PhD(physics), 60. Prof Exp: Instr physics, St Cloud State Col, 53-54, Augsburg Col, 54-56 & St Cloud State Col, 56-57; MEM PHYSICS STAFF, LOS ALAMOS SCI LAB, 60- Concurrent Pos: NATO fel, Saclay, France, 66-67; prog mgr, AEC, 73-74. Mem: Am Phys Soc; Am Asn Physics Teachers. Res: Nuclear and nonnuclear energy conversion research development. Mailing Add: Los Alamos Sci Lab Los Alamos NM 87545

BLAIR, ANDREW DRYDEN, JR, b Toronto, Ont, Dec 11, 39; m 66; c 2. PHARMACEUTICAL CHEMISTRY. Educ: Univ BC, BSP, 64; Univ Wash, PhD(pharm chem), 72. Prof Exp: RES ASSOC MED, UNIV WASH, 72- Mem: AAAS; Am Soc Microbiol. Res: Research and development of analytical procedures for selected drugs and the study of the pharmocokinetics of these drugs in patients with normal and impaired renal function. Mailing Add: Harborview Med Ctr Univ of Wash 325 9th Ave Room 2C-33 Seattle WA 98104

BLAIR, BARBARA ANN, b Gastonia, NC, Oct 21, 26. BIOCHEMISTRY. Educ: Agnes Scott Col, BA, 48; Univ Tenn, MS, 53, PhD(chem), 56. Prof Exp: Res assoc, Med Sch, Univ Buffalo, 56-57 & Med Sch, Univ Va, 57-61; asst prof chem, Wilson Col, 61-62; asst prof, 62-67, asst dean, 69-74, ASSOC PROF CHEM, SWEET BRIAR COL, 67-, DEAN, 74- Concurrent Pos: Vis lectr & actg head chem dept, Women's Christian Col, Madras, India, 68-69. Mem: Fel AAAS. Mailing Add: Sweet Briar Col Sweet Briar VA 24595

BLAIR, BILLIE D, b Melbourne, Ark, May 4, 27; m 49; c 2. ENTOMOLOGY. Educ: Univ Ark, BS, 51, MS, 60; Ohio State Univ, PhD(entom), 64. Prof Exp: Instr agr, Ohio Coop Exten Serv, 54-60; specialist entom, Univ, 60-63, ENTOMOLOGIST, OHIO AGR RES & DEVELOP CTR & OHIO COOP EXTEN SERV, OHIO STATE UNIV, 63-, PROF ENTOM, UNIV, 66- Mem: Entom Soc Am. Res: Soybean insects. Mailing Add: Dept of Entom Ohio State Univ Columbus OH 43210

BLAIR, BYRON OLIVER, b Crawford, Kans, Nov 1, 20; m 46. CROP ECOLOGY. Educ: Ft Hays Kans State Col, BS, 47, MS, 48; Cornell Univ, PhD(agron), 54. Prof Exp: Range conservationist, Southwestern Forest & Range Sta, Ariz, 48-51; from crop ecologist to assoc prof, 54-72, PROF AGRON, PURDUE UNIV, WEST LAFAYETTE, 72- Honors & Awards: Crop & Soils Award Agr Jour, 65. Mem: Fel Am Soc Agron. Res: Microclimatology; crop physiology. Mailing Add: Dept of Agron Purdue Univ West Lafayette IN 47906

BLAIR, CHARLES BARKLEY, JR, b Loudon, Tenn, July 19, 17; m 44; c 2. CYTOLOGY, COMPARATIVE ANATOMY. Educ: Maryville Col, Tenn, AB, 38; Univ NC, PhD(zool), 51. Prof Exp: Pub sch teacher, NC, 38-46; lab asst zool, Univ NC, 46-47, instr, 47-51; from assoc prof to prof biol, Birmingham Southern Col, 51-60, chmn dept, 53-60; dir gross anat, Emory Univ, 60-68; ASSOC PROF ANAT & RADIOL, COL VET MED, UNIV GA, 68- Concurrent Pos: Vis prof, Emory Univ, 58-59; res partic, Oak Ridge Inst Nuclear Studies, 60. Mem: AAAS; assoc Am Vet Med Asn. Res: Electron microscopy, cytochemistry and histochemistry of normal, aging and diet deficient tissues; gross anatomical anomalies traceable to embryonic malformations or anomalous developmental processes; embryology; armadillology. Mailing Add: Dept of Anat & Radiol Univ of Ga Col of Vet Med Athens GA 30602

BLAIR, CHARLES EUGENE, b New York, NY, Dec 30, 49. OPERATIONS RESEARCH. Educ: Mass Inst Technol, BS, 71; Carnegie-Mellon Univ, MA, 72, PhD(math), 75. Prof Exp: Vis asst prof, Dept Math, NC State Univ, 75-76; ASST PROF, DEPT BUS ADMIN, UNIV ILL, URBANA, 76- Mem: Am Math Soc. Res: Integer programming, semi-infinite programming, combinatorics. Mailing Add: Dept of Bus Admin Univ of Ill Urbana IL 61801

BLAIR, CHARLES MELVIN, JR, b Vernon, Tex, Oct 24, 10; m 36; c 2. PHYSICAL CHEMISTRY, ORGANIC CHEMISTRY. Educ: Rice Univ, AB, 31, AM, 32; Calif Inst Technol, PhD(chem), 35. Prof Exp: Chemist, Petrolite Corp, 35-43, dir res, 43-53, pres, 53-64; vchancellor, Washington Univ, 64-67; vpres & tech dir, 67-70, CHMN BD DIRS, MAGNA CORP, 70- Concurrent Pos: Mem bd visitors, Chapman Col, 74- Mem: AAAS; Am Chem Soc; Nat Asn Corrosion Eng. Res: Colloid and surface chemistry; demulsifiers; corrosion inhibitors; chemical processing of petroleum. Mailing Add: Magna Corp 11808 S Bloomfield Ave Santa Fe Springs CA 90670

BLAIR, DONALD GEORGE RALPH, b Lloydminster, Sask, Nov 5, 32; m 59; c 2. BIOCHEMISTRY. Educ: Univ Alta, BSc, 55, MSc, 56; Univ Wis, PhD, 61. Prof Exp: Lectr cancer res, 61-63, lectr biochem, 63-67, from asst prof to assoc prof cancer res, 63-74, ASSOC PROF CANCER RES, UNIV SASK, 74- Concurrent Pos: Assoc prof,

Dept Path, Univ Wash, 74-75. Mem: AAAS; Am Soc Biol Chemists; Am Asn Cancer Res; Can Biochem Soc. Res: Cell biology; rat intestinal sucrase; mechanisms and regulation of the biosynthesis of pyrimidine nucleotides; mechanisms of the resistance of mammalian cells to purine and pyrimidine analogues; RNA synthesis in neoplastic cells; RNA polymerase. Mailing Add: Dept of Cancer Res Univ of Sask Saskatoon SK Can

BLAIR, EMIL, b Satu-mare, Rumania, Oct 20, 23; US citizen. THORACIC SURGERY, CARDIOVASCULAR SURGERY. Educ: Med Col Ga, MD, 46. Prof Exp: Instr med & NIH fel, Med Ctr, Duke Univ, 52-54; Halsted fel surg, Univ Colo Med Ctr, Denver, 56-57; asst prof surg, Sch Med, Univ Md, Baltimore, 60-66; prof cardiothoracic surg, Col Med, Univ Vt, 66-69; PROF SURG, UNIV COLO MED CTR, DENVER, 69- Concurrent Pos: Vis prof, Univ Uppsala, 65-66; consult, Cent Off, Vet Admin, 65-68. Honors & Awards: Hektoen Award, AMA, 55. Mem: Am Asn Thoracic Surg; Am Col Surg; Asn Advan Med Instrumentation; Am Soc Artificial Internal Organs; Int Cardiovasc Soc. Res: Hypothermia; hyperbaric oxygen; carciovascular pulmonary physiology and surgery; shock; biomechanics of trauma; emergency medical services; education. Mailing Add: Dept Surg Gen Rose Mem Hosp 1050 Clermont Denver CO 80220

BLAIR, ETCYL HOWELL, b Wynona, Okla, Oct 16, 22; m 49; c 3. ORGANIC CHEMISTRY. Educ: Southwestern Col, Kans, AB, 47; Kans State Col, MS, 49, PhD(chem), 52. Hon Degrees: DSc, Southwestern Col, Kans, 74. Prof Exp: Instr org chem, Southwestern Col, Kans, 47-48; res asst, Kans State Col, 48-51; res chemist, 51-56, group leader, 56-65, div leader, 65-67, dir res, Britton Lab, 67-68, mgr res & develop, Agr Dept, 68-71, dir res & develop, Ag-Org Dept, 71-73, DIR HEALTH & ENVIRON RES, US AREA RES & DEVELOP, DOW CHEM CO, 73- Concurrent Pos: Environ health consult, Hydrosci Assocs, Inc, 75- Mem: AAAS; Am Chem Soc; NY Acad Sci; Sigma Xi. Res: Isolation and characterization of natural products; synthesis of biological active compounds; organic phosphorus compounds. Mailing Add: US Area Res & Develop Dow Chem Co 2020 Dow Center Midland MI 48640

BLAIR, EUGENE BAXTER, b Childress, Tex, Mar 7, 22; m 46; c 2. MEDICAL BACTERIOLOGY. Educ: Tex Technol Col, BA, 48; Univ Tex, MA, 50, PhD(bact), 55. Prof Exp: Chief serol sect, 4th Army Med Lab, US Army, 50-52, bacteriologist, 97th Gen Hosp Lab, Frankfurt am Main, Ger, 54-57, chief diag sect, Dept Bact, Walter Reed Army Inst Res, 57-63, asst chief dept bact, 63, asst chief microbiol div, Army Med Res & Nutrit Lab, 63-66, chief dept bact, 9th Med Lab, Repub SVietnam, 66-67, asst chief microbiol div, 67-68, chief microbiol div, Army Med Res & Nutrit Lab, 68-72, chief microbiol sect, Clin Invest Serv, Fitzsimmons Army Med Ctr, 72-75; RETIRED. Concurrent Pos: Mem lab comt, Vet Admin-Armed Forces Coop Study Tuberc, 69-72. Mem: Am Soc Microbiol. Res: Microbiology of cholera; diagnostic and epidemiological methods for study of staphylococcal disease; microbiology of tuberculosis. Mailing Add: 1955 Oswego St Aurora CO 80010

BLAIR, GEORGE RICHARD, b San Bernardino, Calif, Aug 17, 20; m 51; c 2. INORGANIC CHEMISTRY. Educ: Univ Redlands, BA, 42. Prof Exp: Chief anal chemist, Metal Hydrides, Inc, Mass, 43-46; anal group leader, Mineral Eng Lab, Am Cyanamid Co, 50-52; chief res chemist, McMillan Lab, Inc, Mass, 52-59; head space mat group, 59-68, HEAD CHEM PROCESS, ELECTRON DYNAMICS DIV, HUGHES AIRCRAFT CO, 68- Honors & Awards: NASA Award for Contribution to Surveyor Prog. Mem: Am Chem Soc; Am Ceramic Soc; Sigma Xi; Electrochem Soc. Res: High vacuum studies; clean surfaces; instrumental analysis. Mailing Add: Hughes Aircraft Co Elec Dyn Div 3100 Lomita Blvd Torrance CA 90509

BLAIR, GRACE, b Marshall, Ark, Feb 23, 15. CHEMISTRY. Educ: Univ Ark, BA, 33, MA, 34, BSMT, 60; Ohio State Univ, PhD(chem), 47. Prof Exp: Teacher high sch, Ark, 35-36; aide, Nat Bur Stands, Washington, DC, 38-41, jr physicist, 42-43, jr chemist, 43-44; chemist, Southern Regional Res Lab, Bur Agr & Indust Eng, USDA, 48-51; instr chem, State Univ NY, 51-53; assoc, Washington Univ, 53-55; res assoc, Med Ctr, Univ Ala, 55-58; asst prof, Little Rock Univ, 60-61; chemist, Clin Lab, Ark Baptist Hosp, 61-63; res & writing, 63-65; assoc prof chem, Ark Polytech Col, 65-68; chemist, Sigma Chem Co, 69-73. Mem: Am Chem Soc; NY Acad Sci. Res: Thermal expansion and density of liquids and solids; alcoholic fermentation; action of amylases; isolation of plant and animal constituents; synthesis and structure of carbohydrate derivatives; oxidation of cellulose; function of vitamin C; clinical chemistry. Mailing Add: 102 N School Ave Fayetteville AR 72701

BLAIR, JAMES BRYAN, b Waynesburg, Pa, May 26, 44; m 66; c 2. BIOCHEMISTRY. Educ: WVa Univ, BS, 66; Univ Va, PhD(biochem), 70. Prof Exp: Lab asst chem, WVa Univ, 64-66; fel, Inst Enzyme Res, Univ Wis, 69-72; ASST PROF BIOCHEM, WVA UNIV, 72- Concurrent Pos: NIH fel, 70. Mem: AAAS; Am Chem Soc; Sigma Xi. Res: Regulation of carbohydrate and lipid metabolism in mammals; kinetic and regulatory properties of hepatic pyruvate kinase. Mailing Add: Dept of Biochem Med Ctr WVa Univ Morgantown WV 26506

BLAIR, JAMES EDWARD, b Cedar Rapids, Iowa, Dec 6, 35; m 61; c 2. PHYSICAL CHEMISTRY. Educ: Rockhurst Col, BS, 58; Univ Wis, PhD(chem), 63. Prof Exp: Res chemist, Photo Prod Dept, E I du Pont de Nemours & Co, 63-65; asst prof chem, 65-68, ASSOC PROF CHEM, MILTON COL, 68- Mem: Am Chem Soc. Res: Sedimentation studies of polymer solutions; proteins and DNA. Mailing Add: 2210 Ravenswood Rd Madison WI 53711

BLAIR, JAMES STUART, b Monmouth, Ill, Dec 18, 97; m 21; c 2. CHEMISTRY. Educ: Univ Kans, AB, 19, AM, 20; Stanford Univ, PhD(org chem), 24. Prof Exp: Res chemist, Fixed Nitrogen Res Lab, USDA, 20-23; instr chem, Ore State Col, 24-25 & Stanford Univ, 25-27; chemist, Res Dept, Am Can Co, 33-57, from res assoc to sr res assoc, 57-62; consult, 63-65; VIS RES PROF CHEM, STETSON UNIV, 65- Mem: AAAS; Am Chem Soc; Inst Food Technol. Res: Determination of pH in and application of pH control of foods; chemistry of flavorful components and of other quality factors in foods. Mailing Add: 35 N University Circle DeLand FL 32720

BLAIR, JOHN DENNIS, b Elma, Wash, Nov 16, 07; m 47; c 2. ORTHOPEDIC SURGERY. Educ: Stanford Univ, AB, 28; Univ Ore, MD, 32. Prof Exp: Resident orthop, Med Sch & Hosps, Univ Iowa, 38-41, asst, Med Sch, 40-41; asst prof orthop surg, Med Sch, Baylor Univ, 51; asst clin prof surg, Med Sch, Stanford Univ, 53-58; mem staff, Walter Reed Army Hosp, DC, 58-61; CHIEF ORTHOP & REHAB SERVS, FAIRMONT HOSP, 61- Concurrent Pos: Asst clin prof orthop surg, Med Sch, Univ Calif, 63-75. Mem: Am Acad Orthop Surg; Am Col Surg. Mailing Add: Fairmont Hosp 15400 Foothill Blvd San Leandro CA 94578

BLAIR, JOHN MORRIS, b Russellville, Ark, May 24, 19; m 47; c 2. EXPERIMENTAL NUCLEAR PHYSICS. Educ: Okla Agr & Mech Col, BS, 40; Univ Wis, PhM, 42; Univ Minn, PhD(physics), 47. Prof Exp: Jr scientist, Los Alamos Sci Lab, NMex, 43-45; res assoc physics, Univ Minn, 47-48; assoc scientist, Argonne Nat Lab, 48-50; from asst prof to assoc prof, 50-62, PROF PHYSICS, UNIV MINN, MINNEAPOLIS, 62- Mem: Am Phys Soc. Res: Nuclear physics; development and use of Van de Graaf machines; reaction involving heavy ions. Mailing Add: Sch of Physics Univ of Minn Minneapolis MN 55455

BLAIR, JOHN SANBORN, b Madison, Wis, Apr 28, 23; m 51; c 3. PHYSICS. Educ: Yale Univ, BS, 43; Univ Ill, MA, 49, PhD(physics), 51. Prof Exp: Jr physicist, Manhattan Proj, 44-45; res assoc, Univ Ill, 51-52; from instr to assoc prof, 52-61, PROF PHYSICS, UNIV WASH, 61- Concurrent Pos: Res assoc, Princeton Univ, 57-58; NSF sr fel, Inst Theoret Physics, Copenhagen, 61-62. Mem: Am Phys Soc. Res: Theoretical nuclear physics. Mailing Add: Dept of Physics Univ of Wash Seattle WA 98105

BLAIR, MCCLELLAN GORDON, b Pittsburgh, Pa, Oct 20, 38; m 60; c 4. SOLID STATE PHYSICS. Educ: Yale Univ, BS, 60; Univ Rochester, MA, 62, PhD(physics), 67. Prof Exp: Res asst, Univ Rochester, 60-67; physicist, Energy Conversion Div, Nuclear Mat & Equip Co, Pa, 67-71, mgr systs develop, 71-72; ARTIFICIAL HEART PROG MGR, ARCO MED PROD CO, 72- Concurrent Pos: Contribr, Int Conf Thermal Conductivity, 67, speaker, 68. Mem: Am Phys Soc; Asn Advan Med Instrumentation. Res: Biomaterials; biomedical devices; microwave ultrasonics; phonon-phonon interactions; nuclear magnetic resonance; thermal conductivity; vapor deposition of thin films; optical pumping; lasers; nuclear safety; health physics. Mailing Add: Arco Med Prod Box 546 Leechburg PA 15656

BLAIR, MURRAY REID, JR, b Somerville, Mass, July 13, 28; m 51. PHARMACOLOGY. Educ: Tufts Univ, BS, 49, PhD(pharmacol), 53. Prof Exp: From instr to asst prof pharmacol, Tufts Univ, 53-58; vis asst prof, State Univ NY Buffalo, 56-57, asst prof, 58-60; consult pharmacologist, Arthur D Little, Inc, 60-61; assoc prof pharmacol, Med Col Va, 61-66, from asst dean to assoc dean sch med, 61-66; DIR RES, ASTRA PHARMACEUT PROD, INC, 66-, ASSOC DIR SCI AFFAIRS, 74- Concurrent Pos: USPHS fel, Tufts Univ, 53-54. Mem: Am Soc Pharmacol & Exp Therapeut; NY Acad Sci. Res: Pharmacology and physiology of the smooth muscle of the gastrointestinal tract; biological action of polypeptids; action of atropine; medical education. Mailing Add: Astra Pharmaceut Prod Inc 7 112 Neponset St Worcester MA 01606

BLAIR, PAUL V, b Kimball, Nebr, Nov 11, 29. BIOCHEMISTRY. Educ: Utah State Univ, BS, 55; Purdue Univ, PhD(genetics), 61. Prof Exp: Fel biochem, Inst Enzyme Res, Univ Wis, 61-64, asst prof, Univ, 64-66; asst prof, 66-69, ASSOC PROF BIOCHEM, SCH MED, UNIV IND, INDIANAPOLIS, 69- Mem: Am Chem Soc. Res: Mitochondrial metabolism; correlation of metabolic functions with ultrastructure. Mailing Add: Dept of Biochem Ind Univ Sch of Med Indianapolis IN 46205

BLAIR, PHYLLIS BEEBE, b Buffalo, NY, May 17, 31; m 55. IMMUNOLOGY. Educ: Cornell Univ, BS, 53; Univ Calif, Berkeley, PhD(zool), 58. Prof Exp: Jr res zoologist, Cancer Res Genetics Lab, Univ Calif, Berkeley, 59-61, asst res zoologist, 61-62; res biologist, Vet Admin Hosp, Long Beach, Calif, 62-63; asst res zoologist, Dept Bact, 63-67, assoc prof immunol, 67-72, PROF IMMUNOL, UNIV CALIF, BERKELEY, 72- Mem: AAAS; Transplantation Soc; Am Asn Cancer Res; Am Asn Immunologists. Res: Mouse mammary tumors; developmental aspects and role of tumor virus; immunologic comparison of viruses and virus-host interactions; tissue transplantation. Mailing Add: Dept of Bact & Immunol Univ of Calif Berkeley CA 94720

BLAIR, ROBERT LOUIE, b Bath, Ill, Aug 30, 27; div; c 4. MATHEMATICS. Educ: Univ Iowa, BA, 49, MS, 50, PhD(math), 52. Prof Exp: Asst physics, Univ Iowa, 47-49 & 49-50, instr math, 50-52; NSF fel, Univ Chicago, 52-53; from instr to asst prof math, Univ Calif, Davis, 52-55; asst prof, Mich State Univ, 55-57; from asst prof to assoc prof, Univ Ore, 57-60; from assoc prof to prof, Purdue Univ, 60-67; PROF MATH, OHIO UNIV, 67- Mem: Am Math Soc; Math Asn Am. Res: Ring theory; lattice theory; algebraic systems of continuous functions; general topology. Mailing Add: Dept of Math Ohio Univ Athens OH 45701

BLAIR, ROBERT MARKS, b Bonner Springs, Kans, Sept 5, 25; m 50; c 2. PLANT ECOLOGY. Educ: Univ Mich, BSF, 51, MWM, 52. Prof Exp: Dist biologist, Mont Dept Fish & Game, 52-56; RES ECOLOGIST, SOUTHERN FOREST EXP STA, US FOREST SERV, 56- Mem: Wildlife Soc; Soc Range Mgt. Res: Ecological relationship between forest management practices and wildlife habitat. Mailing Add: 7600 SFA Station Nacogdoches TX 75961

BLAIR, ROBERT PAUL, b Pittsburgh, Pa, Jan 28, 31. ORGANIC CHEMISTRY, INORGANIC CHEMISTRY. Educ: Washington & Jefferson Col, AB, 53; Bucknell Univ, MA, 55; Univ NC, PhD(org chem), 61. Prof Exp: Sr res chemist, Monsanto Chem Co, 61-62; from asst prof to assoc prof, 62-70, PROF CHEM, IND INST TECHNOL, 70- Honors & Awards: Kekiongan Feather Award, Ind Inst Technol, 70. Mem: Am Chem Soc. Res: Electrophilic substitution of metal chelates; use of optically active chelates to determine mechanism of reaction. Mailing Add: Dept of Chem Ind Inst of Technol Ft Wayne IN 46803

BLAIR, ROBERT WILLIAM, b Kansas City, Mo, Apr 20, 17; m 41; c 1. PETROLEUM GEOLOGY. Educ: Univ Colo, BA, 38. Prof Exp: Recorder, US Geol Surv, 38-39; engr, Gravity Serv Co, 39; geologist, US Geol Surv, 46-47; geologist, Continental Oil Co, 47-49, div geologist, 49-54; chief surface geologist, Sahara Petrol Co, Egypt, 54-57; asst to regional geologist, Continental Oil Co, 57-61; chief geologist, San Jacinto Oil & Gas Co, 61-62; geologist, Continental Oil Co, 62-67; consult geologist, 67-69; geologist, Geophoto Serv, Inc, 69-70; CONSULT GEOLOGIST, 71- Mem: Am Asn Petrol Geologists; fel Geol Soc Am. Res: Domestic and foreign petroleum exploration. Mailing Add: 1280 Elm St Denver CO 80220

BLAIR, ROGER L, b Sterling, Ill, Sept 6, 41; m 63; c 2. FOREST GENETICS. Educ: Univ Ill, BSF, 64; Yale Univ, MF, 65; NC State Univ, PhD(forest genetics), 70. Prof Exp: Biometrician, Int Paper Co, Ga, 68-70, res forester, 70-75; FORESTRY RES DIR, POTLATCH CORP, 75- Concurrent Pos: Adj asst prof, NC State Univ, 71- Mem: Sigma Xi; Soc Am Foresters; Tech Asn Pulp & Paper Indust. Res: Forest tree breeding; forest disease resistance; administration of forest research. Mailing Add: Potlatch Corp PO Box 1016 Lewiston ID 83501

BLAIR, WILLIAM FRANKLIN, b Dayton, Tex, June 25, 12; m 33. VERTEBRATE BIOLOGY. Educ: Univ Tulsa, BS, 34; Univ Fla, MS, 35; Univ Mich, PhD(zool), 38. Prof Exp: Asst, Mammal Div, Mus Zool, Univ Mich, 35-37, res assoc, Lab Vert Biol, 37-46; from asst prof to assoc prof zool, 46-54; res scientist genetics found, 51-54, PROF ZOOL, UNIV TEX, AUSTIN, 55-, DIR RES VERT SPECIATION, 53- Concurrent Pos: Mem adv comt, Div Biol & Med, NSF, 66-69; chmn US nat comt, Int Biol Prog, 68-72, vpres spec comt, 69-74; mem arid lands res comt, AEC, 68-; bd adv to Cong Ad Hoc Comt Environ, 69-; mem int environ progs comt, Nat Acad Sci-Nat Res Coun, 70- Honors & Awards: Am Inst Biol Sci Distinguished Serv Award, 75. Mem: Am Soc Ichthyologists & Herpetologists (vpres, 56); AAAS; Ecol Soc Am (vpres, 56, pres, 63); Soc Study Evolution (treas, 55-57, vpres, 58, pres, 62); Am Inst Biol Sci (vpres, 71, pres, 72). Res: Population genetics; ecology and speciation of mammals, reptiles and amphibians; mating call as an isolation mechanism in anurans;

population dynamics of terrestrial vertebrates; biogeography of North and South America. Mailing Add: Dept of Zool Univ of Tex PO Box 7366 Austin TX 78712

BLAIS, JEAN ROBERT, forest entomology, see 12th edition

BLAIS, NORMAND C, b Springfield, Mass, Jan 4, 26; m 51; c 3. CHEMICAL PHYSICS. Educ: Union Col, BS, 52; Yale Univ, MS, 53, PhD(physics), 56. Prof Exp: MEM RES STAFF, LOS ALAMOS SCI LAB, UNIV CALIF, 56- Mem: Am Phys Soc. Res: Collision interactions between atoms or molecules; measuring transport properties of gases; reactive scattering in molecular beams; theoretical calculations. Mailing Add: 56 Coyote Los Alamos NM 87544

BLAIS, ROGER A, b Shawinigan, Que, Feb 4, 26; m 50; c 2. ECONOMIC GEOLOGY. Educ: Laval Univ, BASc, 49, MSc, 50; Univ Toronto, PhD(geol), 54. Prof Exp: Geologist, Que Dept Natural Resources, 53-56; chief develop engr, Iron Ore Co, Can, 56-61; from assoc prof to prof econ geol, 61-70, assoc dean res, 70-71, DIR RES SERV, POLYTECH SCH, UNIV MONTREAL, 71- Concurrent Pos: Consult geologist various Can mining co, 61-; mem senate, Univ Montreal, 68-70; dir, Eldorado Nuclear Ltd, 68-; chmn nat study group solid earth sci in Can, Sci Coun Can, 68-70; mem earth sci comt, Nat Res Coun Can, 67-69, chmn, 69; mem exec comt, Nat Adv Comt Res Geol Sci, 68- Honors & Awards: Gold Medal, Can Inst Mining & Metall, 67, Barlow Mem Gold Medal, 62. Mem: Geol Asn Can (pres, 69-70); Can Inst Mining & Metall (past pres); Mineral Asn Can; Soc Econ Geol, Res: Applied geostatistics; applications of the theory of regionalized variables to ore reserve estimation, grade control and mineral exploration; origin of iron ores; geology of base metal sulphide deposits and their relations to volcanic and carbonate rocks; metallogenesis of vein type uranium mineralization. Mailing Add: Box 6079 Sta A Montreal PQ Can

BLAIS, ROGER NATHANIEL, b Duluth, Minn, Oct 3, 44; m 71. ATMOSPHERIC PHYSICS, REMOTE SENSING. Educ: Univ Minn, BA, 66; Univ Okla, PhD(exp physics), 71. Prof Exp: Teaching asst, Univ Okla, 66-67; res asst, Univ Okla, 67-71; instr physics, Westark Community Col, 71-72; ASST PROF PHYSICS & GEOPHYS SCI, OLD DOM UNIV, 72- Mem: AAAS; Am Phys Soc; Sigma Xi. Res: Smoke plume diffusion modeling; atmospheric electricity; electrical breakdown wave propagation; remote sensing of air pollution; airborne particle trace analysis. Mailing Add: Dept of Physics & Geophys Sci Old Dom Univ Norfolk VA 23508

BLAISDELL, BAALIS EDWIN, b Lynn, Mass, May 11, 11; m 40; c 2. APPLIED MATHEMATICS. Educ: Mass Inst Technol, BS, 32, PhD(phys chem), 35. Prof Exp: Res assoc, Mass Inst Technol, 35-42; res chemist, Linde Air Prod Co, 42-46 & Pioneering Res Lab, Textile Fibers Dept, E I du Pont de Nemours & Co, 46-53; prof chemp Siena Col, 53-54; assoc prof, 54-61, head dept, 61-68, PROF MATH, JUNIATA COL, 61- Mem: Math Asn Am; Am Math Soc; Soc Indust & Appl Math. Res: Round-off error; photochemistry of dyes; random packing, automated gas chromatograph-mass spectrometer-computer systems. Mailing Add: Dept of Math Juniata Col Huntingdon PA 16653

BLAISDELL, JAMES PERSHING, b Holbrook, Idaho, June 29, 18; m 41; c 5. RANGE ECOLOGY. Educ: Utah State Univ, BS, 39; Univ Idaho, MS, 42; Univ Minn, PhD(bot), 56. Prof Exp: Field asst, Grazing Serv, US Dept Interior, 37-39; agr aid, US Forest Serv, 40-42, range conservationist, Soil Conserv Serv, 46, range conservationist, US Forest Serv, 46-66, ASST DIR INTERMOUNTAIN FOREST & RANGE EXP STA, US FOREST SERV, 66- Mem: Soc Range Mgt; Ecol Soc Am. Res: Management of livestock and big game ranges; effects of fire on range vegetation; plant growth in relation to climate. Mailing Add: US Forest Serv Bldg 507 25th St Ogden UT 84401

BLAISDELL, RICHARD KEKUNI, b Honolulu, Hawaii, Mar 11, 25; m 62; c 2. INTERNAL MEDICINE, HEMATOLOGY. Educ: Univ Redlands, AB, 45; Univ Chicago, MD, 47. Prof Exp: Intern med, Johns Hopkins Hosp, 48-49; asst resident, Charity Hosp, New Orleans, 49-50; instr path, Duke Univ, 54-55; resident med, Univ Chicago, 55-57, instr, 57-58; res assoc, Atomic Bomb Casualty Comn, 59-61; asst prof, Univ Chicago, 61-66; chmn dept, 66-70, PROF MED, SCH MED, UNIV HAWAII, MANOA, 66-, CHIEF DIV HEMAT, 70- Mem: AAAS; Am Fedn Clin Res; Am Soc Hemat; Am Asn Hist Med. Res: Hemoproliferative disorders. Mailing Add: Dept of Med Univ of Hawaii Sch of Med Honolulu HI 96816

BLAIVAS, MURRAY A, b St Louis, Mo, Feb 26, 17; m 41; c 3. CLINICAL CHEMISTRY, HEMATOLOGY. Prof Exp: Dir, Kings County Res Labs, Inc, 37-69; GEN MGR, CLIN LABS DIV, HOFFMANN-LA ROCHE, INC, 69- Mem: NY Acad Sci; fel Am Inst Chemists. Res: Computerization in clinical chemistry. Mailing Add: Clin Labs Div Hoffmann-La Roche Inc 340 Kingsland St Nutley NJ 07110

BLAKE, CARL BLOCH, chemistry, see 12th edition

BLAKE, CARL THOMAS, b Jacksonville, NC, Apr 14, 26; m 49; c 2. AGRONOMY. Educ: NC State Col, BS, 52, MS, 57; Pa State Univ, PhD(agron), 63. Prof Exp: Asst county agr agt, NC Agr Exten Serv, 52-53; spec acct, Carolina Tel & Tel Co, 53-54; AGRON EXTEN SPECIALIST TURF MGT, NC AGR EXTEN SERV, 54-60, 61-, EXTEN PROF CROP SCI, NC STATE UNIV, 61- Mem: Am Soc Agron. Res: Ecological and physiological aspects of plant communities, particularly production of cultivated species. Mailing Add: 139 Williams Hall NC State Univ Raleigh NC 27607

BLAKE, DANIEL BRYAN, b New York, NY, Apr 4, 39; m 66. PALEOBIOLOGY. Educ: Univ Ill, Urbana, BS, 60; Mich State Univ, MS, 62; Univ Calif, Berkeley, PhD(paleont), 66. Prof Exp: Actg instr paleont, Univ Calif, Berkeley, 66-67; asst prof, 67-72, ASSOC PROF GEOL, UNIV ILL, URBANA, 72- Mem: Soc Econ Paleontologists & Mineralogists; Paleont Soc; Soc Syst Zool; Int Palaeont Union; Brit Palaeont Asn. Res: Paleozoic Bryozoa; Cenozoic and modern asteroids. Mailing Add: Dept of Geol Univ of Ill Urbana IL 61801

BLAKE, DANIEL MELVIN, b Miami, Fla, Oct 23, 43; m 65. INORGANIC CHEMISTRY, ORGANOMETALLIC CHEMISTRY. Educ: Colo State Univ, BS, 65; Wash State Univ, PhD(inorg chem), 69. Prof Exp: Vis asst prof chem, Harvey Mudd Col, 69-70; asst prof, 70-74, ASSOC PROF CHEM, UNIV TEX, ARLINGTON, 74- Mem: Am Chem Soc; The Chem Soc. Res: Coordination chemistry; homogeneous catalysis; platinum metals chemistry; organotransition metal chemistry; photochemistry. Mailing Add: Dept of Chem Univ of Tex Arlington TX 76019

BLAKE, DAVID ANDREW, b Baltimore, Md, Aug 26, 41; m 63; c 2. PHARMACOLOGY, TERATOLOGY. Educ: Univ Md, Baltimore City, BS, 63, PhD(pharmacol), 66. Prof Exp: Res assoc, Lab Chem Pharmacol, Nat Heart Inst, 66-67; from asst prof pharmacol to assoc prof pharmacol & toxicol, Sch Pharm, Univ Md, Baltimore City, 67-73, actg head dept, 69-70, chmn dept, 70-73; ASST PROF GYNEC-OBSTET & PHARMACOL, SCH MED, JOHNS HOPKINS UNIV, 73- Concurrent Pos: Mem drug abuse adv comn, Food & Drug Admin. Mem: Teratology Soc; Soc Toxicol; Am Soc Pharmacol & Exp Therapeut. Res: Drug disposition; chemoteratogenicity; biochemical toxicology; biotransformation of volatile fluorocarbons; hepatotoxicity of halothane; mechanism of action of thalidomide; long acting devices for drug release. Mailing Add: Depts Gynec-Obstet & Pharmacol Johns Hopkins Univ Sch of Med Baltimore MD 21205

BLAKE, DORIS HOLMES, b Stoughton, Mass, Jan 11, 92; m 18; c 1. ENTOMOLOGY. Educ: Boston Univ, AB, 13; Radcliffe Col, AM, 17. Prof Exp: Intern, Boston Psychopathic Hosp, 16-17; sci asst, Bur Entom, USDA, 19-22, jr entomologist, 22-28, asst entomologist, 28-34, collabr, 34-60; RES ASSOC, DEPT ENTOM, SMITHSONIAN INST, 60- Res: Taxonomy of beetles; Chrysomelidae. Mailing Add: 3416 Glebe Rd N Arlington VA 22207

BLAKE, EMMET REID, b Abbeville, SC, Nov 29, 08; m 47; c 2. ORNITHOLOGY. Educ: Presby Col, SC, AS, 28; Univ Pittsburgh, MS, 33. Hon Degrees: DSc, Presby Col, SC, 66. Prof Exp: Asst zool, Univ Pittsburgh, 31-32; tech asst, Nat Geog Soc exped, Brazil & Venezuela, 30-31; leader, Mandel-Field Mus exped, Orinoco River, 31-32; ornithologist, Field Mus exped, Guatemala, 33-34; leader, Carnegie Mus exped, Brit Honduras, 35; asst ornithologist, 35-37, asst cur ornith, 37-47, assoc cur, 47-55, cur, Div Birds, 55-73, EMER CUR, DIV BIRDS, FIELD MUS NATURAL HIST, 73- Concurrent Pos: Mem Field Mus Natural Hist zool exped, Brit Guiana & Brazil, 37-38, Brit Guiana exped, 38-39, Southwestern zool exped, 41 & Mex field studies, 53; leader, Conover Peru Exped, 58. Mem: Fel Am Ornith Union; Cooper Ornith Soc; Wilson Ornith Soc. Res: Origin, distribution and variation of neotropical birds. Mailing Add: Field Mus of Natural Hist Roosevelt Rd & Lake Shore Dr Chicago IL 60605

BLAKE, FRANCIS GILMAN, b New York, NY, Apr 27, 17; m 39; c 2. RESEARCH ADMINISTRATION. Educ: Harvard Univ, SB, 38, AM, 40, MES, 48, PhD, 49. Prof Exp: Asst, Harvard Univ, 40-41, spec res assoc, 41-44; assoc scientist, Univ Calif, Los Alamos Labs, NMex, 44-46; asst acoust res lab, Harvard Univ, 46-49, res fel & lectr, 49-50; res physicist, Explor Div, Chevron Oil Field Res Co, 50-52, sr res physicist, 52-54, supvr geophys res sect, 54-62, mgr explor res div, 63-66, sr res scientist, 66-71; tech asst to dir, Off Sci & Technol, Exec Off of President, 71-73; sr policy analyst, Sci & Technol Policy Off, NSF, 73-76; ASST DIR POLICY & PLANNING, DIV GEOTHERMAL ENERGY, US ENERGY RES & DEVELOP ADMIN, 76- Mem: AAAS; Soc Explor Geophys; Am Geophys Union; Marine Technol Soc (vpres, 71-74). Res: Science policy; research management; energy policy and planning. Mailing Add: Div of Geothermal Energy US Energy Res & Develop Admin Washington DC 20545

BLAKE, GEORGE HENRY, JR, b Faunsdale, Ala, June 21, 22; m 46; c 4. ENTOMOLOGY. Educ: Auburn Univ, BS, 47, MS, 49; Univ Ill, PhD, 58. Prof Exp: Assoc entomologist, 50-64, assoc prof, 64-65, PROF ENTOM & ZOOL, AUBURN UNIV, 65- Mem: Entom Soc Am. Res: Teaching and graduate research. Mailing Add: Dept of Zool & Entom Auburn Univ Auburn AL 36830

BLAKE, GEORGE MARSTON, b Los Angeles, Calif, Jan 21, 32; m 52; c 4. FORESTRY. Educ: Univ Idaho, BS, 57; Univ Minn, MS, 59, PhD, 64. Prof Exp: Asst, Univ Minn, 57-62; from asst prof silvicult to assoc prof forestry, 62-72, PROF FORESTRY, UNIV MONT, 72-, RES COORDR FORESTRY, 73- Mem: AAAS; Am Soc Foresters. Res: Genetic variation in quaking aspen; hybridization of subalpine and western larch. Mailing Add: Sch of Forestry Univ of Mont Missoula MT 59801

BLAKE, GEORGE ROWLAND, b Provo, Utah, Mar 14, 18; m 41; c 4. SOIL PHYSICS. Educ: Brigham Young Univ, AB, 43; Ohio State Univ, PhD, 49. Prof Exp: Asst prof soil physics & asst res specialist, Rutgers Univ, 49-55; assoc prof soil physics, 55-60, PROF SOIL PHYSICS, UNIV MINN, ST PAUL, 61- Concurrent Pos: Coop agt, USDA, 49-55; NSF sr fel, Ger, 62-63; consult, Ford Found, Chile, 67; Fulbright guest prof & lectr, Univ Hohenheim, Stuttgart, Ger, 70-71; guest prof & consult, Univ Kesthely, Hungary, 74. Mem: Soil Sci Soc Am; fel Am Soc Agron; Soil Conserv Soc Am; Int Soc Soil Sci. Res: Soil structure; tillage; in situ water relations. Mailing Add: Dept of Soil Sci Univ of Minn St Paul MN 55108

BLAKE, J BERNARD, b New York, NY, Dec 14, 35; m 60; c 2. ASTROPHYSICS, SPACE PHYSICS. Educ: Univ Ill, BS, 57, MS, 58, PhD(physics), 62. Prof Exp: Res assoc physics, Univ Ill, 62; DEPT HEAD SPACE PARTICLES & FIELDS DEPT, SPACE PHYSICS LAB, AEROSPACE CORP, 62- Mem: Am Phys Soc; Am Geophys Union; Am Astron Soc. Res: Low energy nuclear physics; magnetospheric and cosmic-ray physics; nucleosynthesis; nuclear physics. Mailing Add: Space Physics Lab Aerospace Corp PO Box 92957 Los Angeles CA 90009

BLAKE, JAMES A, b Fresno, Calif, Sept 24, 41, m 63; c 2. INVERTEBRATE ZOOLOGY, MARINE BIOLOGY. Educ: Fresno State Col, BA, 63, MA, 65; Univ Maine, PhD(zool), 69. Prof Exp: Res assoc marine biol, Ira C Darling Ctr, Univ Maine, 68-69; asst prof marine biol, 69-73, actg dir, Pac Marine Sta, 73, ASSOC PROF MARINE BIOL, PAC MARINE STA, UNIV OF THE PAC, 73- Concurrent Pos: Res assoc, NSF grant cruise to eastern Can Arctic, 68-69. Mem: Am Soc Zoologists; Am Soc Limnol & Oceanog; Int Oceanog Found; Am Micros Soc; Sigma Xi. Res: Systematics, reproduction and larval development of polychaetes. Mailing Add: Pac Marine Sta Univ of the Pac Dillon Beach CA 94929

BLAKE, JAMES J, b New York, NY, June 10, 37. ORGANIC CHEMISTRY, BIOCHEMISTRY. Educ: Mass Inst Technol, BS, 58; Univ Calif, PhD(chem), 65. Prof Exp: Develop engr, rheology, Aerojet Gen Corp, 58-60; ASST RES BIOCHEMIST, HORMONE RES LAB, UNIV CALIF, SAN FRANCISCO, 65- Concurrent Pos: NIH fel, 65-67. Mem: AAAS; Am Chem Soc. Res: Synthesis of peptides related to pituitary hormones. Mailing Add: Hormone Res Lab Univ of Calif San Francisco CA 94122

BLAKE, JAMES NEAL, b Great Bend, Kans, July 18, 33; m 63. SPEECH & HEARING SCIENCES. Educ: Kans State Univ, BS, 55, MS, 60; Univ Southern Miss, PhD(speech & hearing sci), 68. Prof Exp: Speech therapist, Charlotte-Mecklenburg Schs, 60-64; asst prof speech & hearing, Western Carolina Univ, 66-70; ASSOC PROF AUDIOL & SPEECH PATH, UNIV LOUISVILLE, 70- Concurrent Pos: Consult, Rauch Ctr Retarded Children, 70- Mem: Ling Soc Am; Speech Asn Am; Am Speech & Hearing Asn; Am Psychol Asn; Coun Except Children. Res: Relationship between normal and abnormal verbal behavior and various cognitive states; effects of variable berbal loadings on cognitive. Mailing Add: Audiol Ctr Univ of Louisville Sch of Educ Louisville KY 40208

BLAKE, JOHN WILSON, b New York, NY, Sept 11, 33; m 54; c 6. AQUATIC BIOLOGY, POLLUTION ECOLOGY. Educ: Mass Inst Technol, BS, 55; Univ NC, MA, 58, PhD(marine ecol), 61. Prof Exp: Asst marine ecol, Univ NC, 55-60; biologist, Aquacultural Res Corp, 60-62, dir res marine ecol, 62-65; sr marine biologist, William F Clapp Labs, Inc, Battelle Mem Inst, 65-69; TECH DIR ECOL, ENVIRON SYSTS CTR, RAYTHEON CO, 69- Concurrent Pos: Chemist, Liggett & Myers Tobacco Co, NC, 56-57; vis investr radioecol, Bur Commercial Fisheries Radiobiol Lab, NC, 60-61; res assoc, Div Estuarine Sci, Acad Natural Sci

Philadelphia, Pa, 61-; consult, Marine Resources Prog, Univ RI, 63-65. Mem: AAAS; Am Inst Biol Sci; Ecol Soc Am; Am Soc Limnol & Oceanog; Atlantic Estuarine Res Soc (secy-treas, 63-64). Res: Physiological ecology of marine mollusks; culture of mollusk larvae; farming of the sea; pollution of coastal areas; commercial shellfish populations. Mailing Add: Systems Ctr Raytheon Environ 421 Old State Rd Berwyn PA 19312

BLAKE, JOSEPH THOMAS, b Provo, Utah, Jan 24, 19; m 41; c 6. VETERINARY MEDICINE. Educ: Brigham Young Univ, BS, 49; Iowa State Univ, MS, 50, PhD, 55, DVM, 56. Prof Exp: Instr animal husb, Ariz State Univ, 50-51; instr dairy husb, Iowa State Univ, 51-53; from asst prof to assoc prof vet sci, 56-69, PROF VET SCI, UTAH STATE UNIV, 69- Mem: Am Vet Med Asn; Am Soc Vet Physiol & Pharmacol; Am Soc Animal Sci; Am Asn Vet Med Cols. Res: Physiology and nutrition in relation to veterinary medicine. Mailing Add: Dept of Vet Sci Utah State Univ Logan UT 84321

BLAKE, JULES, b New York, NY, July 7, 24; m 49; c 3. ORGANIC CHEMISTRY. Educ: Univ Pa, BS, 49, MS, 51, PhD(org chem), 54. Prof Exp: Res chemist, Marshall Lab, E I du Pont de Nemours & Co, Philadelphia, 54-59, res supvr, 60-65, develop supvr, 65-66; res dir, Indust Chem Div, Mallinckrodt Chem Works, 66-69, dir res & develop & gen mgr res & develop div, Chem Group, 69-71; V PRES RES & DEVELOP, KENDALL CO, 71- Concurrent Pos: Rep to Indust Res Inst; mem, Nat Acad Sci-Nat Res Coun adv panel, Anal Chem Div, Nat Bur Stand. Mem: Am Chem Soc; The Chem Soc. Res: Organic and polymer synthesis; evaluation and commerical development. Mailing Add: Kendall Co 225 Franklin St Boston MA 02110

BLAKE, JULIAN GASKILL, b Macon, Ga, Jan 23, 45. QUANTUM OPTICS, EXPERIMENTAL PHYSICS. Educ: Amherst Col, AB, 66; Harvard Univ, MAT, 69, PhD(appl physics), 73. Prof Exp: LECTR PHYSICS, HARVARD UNIV, 73-, DIR, GORDON McKAY LAB, DIV ENG & APPL PHYSICS, 75- Mem: Am Asn Physics Teachers; Am Phys Soc. Res: Photoelectron counting statistics; experimental optical studies of solids. Mailing Add: Gordon McKay Lab Harvard Univ 9 Oxford St Cambridge MA 02138

BLAKE, LAMONT VINCENT, b Somerville, Mass, Nov 7, 13; m 38, 57; c 3. MICROWAVE PHYSICS, RADIO ENGINEERING. Educ: Mass State Col, BS, 35; Univ Md, MS, 50. Prof Exp: Radio interference investr, Ark Power & Light Co, 37-40; physicist, Naval Res Lab, 40-67, consult physicist, Radar Div, 67-69, head radar geophys br, 69-72; SR SCIENTIST, TECHNOL SERV CORP, SILVER SPRING, MD, 72- Concurrent Pos: Free lance tech writer, 60- Honors & Awards: Civilian Serv Award, US Navy, 51, Superior Civilian Serv Award, 72; Sci Res Soc Am Appl Sci Award, 63. Mem: AAAS; Am Phys Soc; Sigma Xi; Inst Elec & Electronics Engrs; Am Geophys Union. Res: Design of radar systems and components; radar maximum range theory; atmospheric absorption of radio waves; radio noise theory. Mailing Add: 7814 Oaklawn Dr Alexandria VA 22306

BLAKE, LOUIS HARVEY, b Lynn, Mass, Apr 26, 43. MATHEMATICS. Educ: Tufts Univ, BS, 64; Univ Md, College Park, PhD(math), 69. Prof Exp: Asst prof math, Northern Ill Univ, 69-70, ASST PROF MATH, WORCESTER POLYTECH INST, 70- Mem: Am Math Soc. Res: Stochastic processes. Mailing Add: Dept of Math Worcester Polytech Inst Worcester MA 01609

BLAKE, MARTIN IRVING, b Paterson, NJ, Oct 20, 23; m 48; c 4. PHARMACY. Educ: Brooklyn Col Pharm, BS, 47; Rutgers Univ, MS, 50; Ohio State Univ, PhD(pharm), 51. Prof Exp: Asst prof pharmaceut chem, Duquesne Univ, 51-55; prof, NDak State Univ, 55-59; resident res assoc, Argonne Nat Lab, 59-60; PROF PHARM & HEAD DEPT, COL PHARM, UNIV ILL MED CTR, 60- Concurrent Pos: Consult & fac assoc, Chem Div, Argonne Nat Lab, 60-; consult pharm serv, Westside Vet Admin Hosp, Chicago, 65-66 & Hines Vet Admin Hosp, Ill, 66-; mem rev comt, US Pharmacopeia, 70- Mem: Sigma Xi; Am Pharmaceut Asn; Am Asn Cols Pharm; AAAS; Am Chem Soc. Res: Nonaqueous titrimetry; volatile oil and drug analysis; isotope effects; biosynthesis; ion exchange. Mailing Add: Dept of Pharm Univ of Ill at the Med Ctr Chicago IL 60612

BLAKE, MILTON CLARK, JR, b San Francisco, Calif, Feb 20, 32; m 59. GEOLOGY. Educ: Univ Calif, Berkeley, AB, 58; Stanford Univ, PhD(geol), 65. Prof Exp: Jr geologist, 58-64, GEOLOGIST, BR WESTERN ENVIRON GEOL, US GEOL SURV, 64- Mem: Geol Soc Am; Mineral Soc Am. Res: Igneous and metamorphic petrology; low grade metamorphic rocks of Western United States; Blueschist-facies rocks of the world. Mailing Add: Br of Western Environ Geol US Geol Surv 345 Middlefield Rd Menlo Park CA 94025

BLAKE, OLIVER DUNCAN, b Gloucester, Mass, July 28, 10; m 39; c 5. GEOLOGY. Educ: Antioch Col, AB, 46; Ohio State Univ, PhD(geol), 52. Prof Exp: Instr geol, Marietta Col, 46-48; lectr, Col Wooster, 49-50; instr, Ohio State Univ, 50-51; from asst prof to assoc prof, Mont Sch Mines, 51-53; geologist-adminr, Mont Oil & Gas Conserv Comn, 57-62; assoc prof geol, Univ Redlands, 62-64; ASSOC PROF GEOL, W VALLEY COL, 64- Concurrent Pos: Geologist, Mont Bur Mines & Geol. Mem: AAAS; Geol Soc Am; assoc Am Asn Petrol Geologists. Res: Stratigraphy of Paleozoic bryozoa. Mailing Add: Dept of Geol W Valley Col Campbell CA 95008

BLAKE, RICHARD L, b Berkeley Springs, WVa, Mar 8, 37; m 63. ASTROPHYSICS. Educ: Rensselaer Polytech Inst, 59; Univ Colo, PhD(astro-geophys), 68. Prof Exp: Physicist, Naval Res Lab, 59-63; asst prof astron & astrophys, Enrico Fermi Inst Nuclear Studies, Univ Chicago, 68-74; STAFF PHYSICIST, LOS ALAMOS SCI LAB, 74- Mem: Am Astron Soc; Am Phys Soc; Am Geophys Union; Soc Appl Spectros. Res: X-ray spectroscopy of ions and atoms; x-ray physics; solar physics; x-ray astronomy. Mailing Add: Apt 9 3790 Gold St Los Alamos NM 87544

BLAKE, ROBERT GEORGE, b Cornell, Ill, May 4, 06; m 29; c 1. MATHEMATICS. Educ: Univ Fla, AB, 38, MA, 45, PhD(math), 53. Prof Exp: Teacher pub schs, Fla, 26-43; instr, 43-49, asst prof, 49-55, ASSOC PROF MATH, UNIV FLA, 55- Mem: Am Math Soc; Math Asn Am; Soc Indust & Appl Math. Res: Distributions and generalized transforms; partially ordered algebraic systems. Mailing Add: Dept of Math Univ of Fla Gainesville FL 32603

BLAKE, ROBERT J, b Brooklyn, NY, Aug 22, 28; m 55; c 4. PHYSICAL CHEMISTRY, CHEMICAL ENGINEERING. Educ: Rensselaer Polytech Inst, BChE, 51. Prof Exp: Proj scientist gas purification, 56-64, group leader, 65-71, ASSOC DIR RES & DEVELOP SEPARATION RES, UNION CARBIDE CORP, TARRYTOWN, NY, 71- Mem: Am Chem Soc. Res: Gas purification and separations; removal of hydrogen sulfide and carbon dioxide from synthetic gas, natural gas and refinery gas; removal of sulfur dioxide from stock gas. Mailing Add: 3278 Curry St Yorktown Heights NY 01598

BLAKE, ROBERT L, b Claremont, NH, Mar 21, 33; m 61; c 2. BIOCHEMISTRY, PHARMACOLOGY. Educ: Bates Col, BS, 55; Univ Rochester, MS, 57; Univ Calif, San Francisco, PhD(biochem), 62. Prof Exp: Res assoc pharmacol, Univ Rochester, 55-57; asst biochem, Univ Calif, Berkeley & San Francisco, 57-59; USPHS fel, McArdle Mem Lab, Univ Wis-Madison, 62-64; res chemist, Med Ctr, Univ Calif, San Francisco, 64-65, Cardiovasc Res Inst fel, 65-66; assoc staff scientist, 66-70, STAFF SCIENTIST, JACKSON LAB, 70- Concurrent Pos: NSF travel fel, NATO Advan Study Inst, Bergen, Norway, 68. Mem: AAAS; NY Acad Sci; Genetics Soc Am; Brit Biochem Soc. Res: Biochemical control mechanisms of mammalian genetic expression; mitochondrial biogenesis, hormonal induction of enzymes; inborn errors of metabolism; pharmacogenetics. Mailing Add: Jackson Lab Bar Harbor ME 04609

BLAKE, ROLAND CHARLES, b Howland, Maine, Mar 8, 20; m 51; c 2. HORTICULTURE. Educ: Univ Maine, BS, 49; Univ Minn, PhD(hort, plant breeding), 54. Prof Exp: Asst horticulturist, Northwestern Wash Exp Sta, State Col Wash, 54-57; horticulturist, Southern Ore Br Exp Sta, Agr Res Serv, USDA, 57-59; res horticulturist, Plant Indust Dept, Crops Res Div, Univ Southern Ill, Carbondale, 59-73; RES HORTICULTURIST, DEPT HORT, OHIO AGR RES & DEVELOP CTR, AGR RES SERV, USDA, 73- Mem: Am Soc Hort Sci. Res: Small fruit breeding; tree fruit breeding; breeding pear for good quality and fire blight resistance. Mailing Add: Dept of Hort Ohio Agr Res & Develop Ctr Wooster OH 44691

BLAKE, ROLLAND LAWS, b Minneapolis, Minn, Jan 16, 24; m 53; c 3. GEOLOGY. Educ: Univ Minn, BS, 50, MS, 51, PhD(geol), 58. Prof Exp: Geologist, Cleveland-Cliffs Iron Co, Mich, 51-52, Minn, 52-54; petrol & mineral res geologist, 58-71, SUPVRY GEOLOGIST, US BUR MINES, 71- Mem: Am Crystallog Asn; Mineral Soc Am; Soc Econ Geol. Res: Mineralogy; iron silicates; carbonates; iron oxides associated with iron ore deposits; manganese minerals; petrology; pre-Cambrian in Canadian shield area. Mailing Add: 6701 Southdale Rd Minneapolis MN 55435

BLAKE, THOMAS MATHEWS, b Sheffield, Ala, Aug 4, 20. CARDIOLOGY. Educ: Univ Ala, BA, 41; Vanderbilt Univ, MD, 44; Am Bd Internal Med, dipl. Prof Exp: Intern med, Vanderbilt Univ Hosp, 44-45, intern path, 47-48, asst resident med, 48-49, fel cardiovasc res, Med Sch, 50-52, instr, 52-54, instr clin med, 54-55; from asst prof to assoc prof, 55-70, PROF MED, SCH MED, UNIV MISS, 70- Concurrent Pos: Asst resident, Strong Mem Hosp, Rochester, NY, 49-50; chief clin physiol sect, Res Lab, Vet Admin Hosp, 54-55; mem coun arteriosclerosis, Am Heart Asn. Mem: AMA; Am Fedn Clin Res: fel Am Col Cardiol; fel Am Col Chest Physicians; fel Am Col Physicians. Res: Electrocardiography; cardiovascular physiology. Mailing Add: Dept of Med Univ of Miss Sch of Med Jackson MS 39216

BLAKE, WESTON, JR, b Boston, Mass, Feb 26, 30; m 60; c 2. GLACIAL GEOLOGY. Educ: Dartmouth Col, AB, 51; McGill Univ, MSc, 53; Ohio State Univ, PhD(geol), 62; Univ Stockholm, Fil Lic(geog), 64, Fil Dr(phys geog), 75. Prof Exp: PLEISTOCENE GEOLOGIST, GEOL SURV CAN, 62-; RES SCIENTIST & HEAD PALEOECOL & GEO-CHRONOL SECT, TERRAIN SCI DIV, 69- Concurrent Pos: Mem subcomt glaciers, Nat Res Coun Can, 66-72; mem Can nat comt, Int Geol Correlation Prog, 74-; assoc ed, Can J Earth Sci, 75- Mem: Geol Asn Can; Arctic Inst NAm; Glaciol Soc; Geol Soc Finland; Swedish Soc Anthrop & Geog. Res: Glacial history and geomorphological processes in arctic regions; radiocarbon dating; pleistocene marine faunas. Mailing Add: Geol Surv of Can 601 Booth St Ottawa ON Can

BLAKE, WILLIAM DEWEY, b Summit, NJ, June 27, 18; m 42; c 2. PHYSIOLOGY. Educ: Dartmouth Col, AB, 40; Harvard Med Sch, MD, 43. Prof Exp: Intern med, Presby Hosp, NY, 44, asst resident, 47, res fel, 48-49; from instr to asst prof physiol, Sch Med, Yale Univ, 49-52; assoc prof, Univ Ore, 52-60; PROF PHYSIOL & HEAD DEPT, UNIV MD, BALTIMORE CITY, 60- Mem: AAAS; Am Physiol Soc; Am Soc Clin Invest; Soc Neurosci; Am Soc Nephrology. Res: Mammalian physiology; neuroendocrine control of kidney. Mailing Add: Dept of Physiol Univ of Md Sch of Med Baltimore MD 21201

BLAKELEY, PHILLIP EARL, b Canwood, Sask, Apr 3, 22; m 45; c 3. ENTOMOLOGY. Educ: Univ Sask, BS, 45; Univ Alta, MSc, 54. Prof Exp: Entomologist, Crop Insect Sect, 46-57, INFO OFFICER FIELD CROPS, CAN AGR RES STA, LETHBRIDGE, 57- Mem: Entom Soc Can; Agr Inst Can; Can Phytopath Soc. Res: Liaison between research officers and extension services; entomology and plant pathology concerning field crops. Mailing Add: Can Agr Res Sta Lethbridge AB Can

BLAKELY, LAWRENCE MACE, b Los Angeles, Calif, Nov 12, 34; m 60; c 2. PLANT PHYSIOLOGY. Educ: Mont State Univ, BA, 56, MA, 58; Cornell Univ, PhD(plant physiol), 63. Prof Exp: Asst bot, Mont State Univ, 56-58; asst plant physiol, Cornell Univ, 58-62, res assoc, 62-63, instr, 63; PROF BOT, CALIF STATE POLYTECH UNIV, POMONA, 63- Mem: AAAS; Bot Soc Am; Am Soc Plant Physiologists; Am Inst Biol Sci; Tissue Cult Asn. Res: Physiology of plant growth; tissue culture of plant cells; variation of plant cells in tissue culture; growth and development of plant cells in tissue culture. Mailing Add: Dept of Biol Sci Calif State Polytech Univ Pomona CA 91768

BLAKELY, ROBERT FRASER, b Newark, NJ, Apr 1, 21; m 43; c 2. GEOPHYSICS. Educ: Miami Univ, AB, 46, MA, 48. Prof Exp: Instr physics, Miami Univ, 47-49; spectrographer, Dept Conserv, State Geol Surv, Ind, 49-53, geophysicist, 53-74; ASSOC PROF GEOPHYS, IND UNIV, BLOOMINGTON, 74- Mem: AAAS; Soc Explor Geophys; Am Geophys Union; Am Meteorol Soc; Asn Comput Mach. Res: Spectrographic determination of major constituents in limestones; physical properties of Indiana sediments; computers in geology; solid earth geophysics and meteorology. Mailing Add: 116 Meadowbrook Ave Bloomington IN 47401

BLAKEMAN, CRAWFORD HARRIS, JR, b Middlesboro, Ky, Sept 8, 46; m 67; c 2. ANTHROPOLOGY. Educ: Univ Ky, BS, 68, MA, 72; Southern Ill Univ, Carbondale, PhD(anthrop), 74. Prof Exp: ASST PROF ANTHROP, MISS STATE UNIV, 74- Concurrent Pos: Proj dir archaeol surv & excavations, Nat Park Serv & Corps Engrs, 74- Mem: AAAS; Am Anthrop Asn; Sigma Xi. Res: Prehistoric adaptations of human groups to various environmental settings in the Southeastern United States, involving an analysis of settlement patterns, subsistence patterns and processes of cultural change in the prehistory of the region. Mailing Add: Dept of Anthrop PO Drawer GN Mississippi State MS 39762

BLAKEMORE, GEORGE JEFFERSON, JR, b Philadelphia, Pa, Aug 12, 24; m 71. APPLIED STATISTICS, OPERATIONS RESEARCH. Educ: George Washington Univ, BA, 49, MA, 53. Prof Exp: Opers analyst, Opers Res Off, Johns Hopkins Univ, 50-57; proj statistician, Arinc Res Corp, 57-63; prin reliability engr, Honeywell, Inc, 63-68; RES SCIENTIST, DRAPER LAB, MASS INST TECHNOL, 68- Mem: Opers Res Soc Am; Am Statist Asn; Am Soc Qual Control. Res: Systems-cost effectiveness; reliability engineering; quality control; acceptance sampling; life testing. Mailing Add: Draper Lab Mass Inst of Technol 68 Albany St Cambridge MA 02139

BLAKEMORE, JOHN SYDNEY, b London, Eng, May 25, 27; m 53; c 2. SOLID STATE PHYSICS. Educ: Univ London, BSc, 48, PhD(physics), 51. Prof Exp: Res physicist, Stand Telecommun Labs, London, 50-54; res fel, Univ BC, 54-56; staff scientist, Honeywell, Inc, 62-63; sr staff scientist, Lockheed Res Labs, 63-64; PROF

PHYSICS, FLA ATLANTIC UNIV, 64-, DEAN SCI, 75- Concurrent Pos: Fulbright-Hays sr fel, Univ New South Wales, 71. Mem: Fel Am Phys Soc; Inst Elec & Electronics Engrs; fel Brit Inst Physics. Res: Semiconductors; bulk properties of semiconducting materials; photoconductivity, recombination and trapping; statistics of charge carriers. Mailing Add: Col of Sci Fla Atlantic Univ Boca Raton FL 33432

BLAKEMORE, WILLIAM STEPHEN, b Stockdale, Pa, June 22, 20; m 49; c 6. SURGERY. Educ: Washington & Jefferson Col, BS, 42; Univ Pa, MD, 45; Am Bd Surg, dipl, 52; Am Bd Thoracic Surg, dipl, 55. Prof Exp: Intern, Hosp Univ Pa, 45-46, asst resident, 48-51; asst instr, Sch Med, Univ Pa, 48-51, instr, 51-52, assoc, 52-54, from asst prof to prof, 54-73, asst chief surg clin, Grad Hosp, 55-62, from asst dir to assoc dir, Harrison Dept Surg Res, Univ, 56-73, from J William White asst prof to J William White assoc prof surg res, 56-62, chmn dept surg, 62-69, surgeon-in-chief, Grad Hosp, 62-73, chmn dept grad surg, 69-73; PROF SURG & CHMN DEPT, MED COL OHIO, 73-, CHIEF SURG, HOSP, 73- Concurrent Pos: Fel, Harrison Dept Surg, Univ Pa, 48-51, res fel, Runyan Clin, 50-51; scholar, Am Cancer Soc, 52-57; Ravdin traveling fel, 53-54; vis fel, Karolinska Hosp, Sweden, 53-54; mem study sect pharmacol, NIH, 47, mem study sect surg, 65-69; a attend physician, Philadelphia Gen Hosp, 52-73; assoc surg, Hosp Univ Pa, 52-73; chief surgeon, Emergency Am Med Team to Algeria, 62; mem comt blood & transfusion probs, Nat Res Coun Dean's Comt, Vet Admin Hosp, Wilmington, Del, 62-68; mem cent adv comt, Coun Cardiovasc Surg, Am Heart Asn, 66-73; life trustee, Washington & Jefferson Col. Mem: AAAS; Asn Cancer Res; Am Asn Thoracic Surg; Am Col Surg; AMA. Res: Cardiopulmonary physiology; cancer immunology. Mailing Add: Med Col of Ohio Box 6190 Toledo OH 43614

BLAKER, JOHN WARREN, b Wilkes-Barre, Pa, May 25, 34; m 57; c 2. OPTICS. Educ: Wilkes Col, BS, 55; Mass Inst Technol, PhD(chem), 58. Prof Exp: Asst prof physics, Fairleigh Dickinson Univ, 58-61; mem staff, G C Dewey Co, 61-62 & John Wiley & Sons, Inc, 62-64; PROF PHYSICS, VASSAR COL, 64- Mem: AAAS; Am Phys Soc; Am Chem Soc; Am Asn Physics Teachers. Res: Optics and solid state theory. Mailing Add: Dept of Physics Vassar Col Poughkeepsie NY 12601

BLAKER, ROBERT GORDON, b Webster, Mass, Sept 19, 30; m 53; c 2. LABORATORY MEDICINE, IMMUNOLOGY. Educ: Univ Calif, Berkeley, BA, 55, MA, 57; Univ Pa, PhD(med microbiol), 58. Prof Exp: Res microbiologist, Lister Inst, London, Eng, 58-59; instr, Woman's Med Col, 59-62; instr microbiol & immunol, State Univ NY Downstate Med Ctr, 62-68; from asst prof to assoc prof path, Med Ctr, Duke Univ, 68-71; DIR MICROBIOL DEPT, BIO-SCI LABS, 71- Concurrent Pos: Microbiologist, Durham Vet Admin Hosp, NC, 68-71. Mem: Am Soc Microbiol; Am Soc Human Genetics; Infectious Dis Soc Am. Res: Tumor immunology; human genetics and birth defects. Mailing Add: Bio-Sci Labs 7600 Tyrone Ave Van Nuys CA 91405

BLAKER, ROBERT HOCKMAN, b Meadow Bridge, WVa, Apr 6, 20; m 45. PHYSICAL CHEMISTRY, INFORMATION SCIENCE. Educ: Berea Col, AB, 42; Calif Inst Technol, PhD(chem), 49. Prof Exp: Instr chem, Berea Col, 43-44 & Off Naval Res, Calif Inst Technol, 46-49; Imp Chem Industs fel, Univ Liverpool, 49-50; res chemist, Jackson Lab, 50-53, res supvr, Petrol Lab, 53-56, asst dir lab, 56-60, res mgr, Orchem Dept, 60-64, head patent serv, 64-70, MGR CENT PATENT & REPORT SERV, SECRETARY'S DEPT, E I DU PONT DE NEMOURS & CO, INC, 70- Honors & Awards: Am Dyestuff Reporter Award, 53. Mem: Am Chem Soc; Am Soc Info Sci. Res: Size and shape of high polymer molecules; theory of dyeing synthetic fibers; fuels and lubricants for internal combustion engines. Mailing Add: Ashland Rd Hockessin DE 19707

BLAKERS, ALBERT LAURENCE, b Perth, Australia, Jan 2, 17; m 46; c 4. MATHEMATICS. Educ: Univ Western Australia, BSc, 39; Princeton Univ, PhD(topol), 47. Prof Exp: Lectr math, Univ Western Australia, 39 & Scotch Col, Australia, 40; res physicist, Nat Res Coun Can, 42-45; res fel homotopy theory, Princeton Univ, 47-48, instr math, 47-49; from asst prof to assoc prof, Lehigh Univ, 49-52; PROF MATH & CHMN DEPT, WESTERN AUSTRALIA UNIV, 52- Concurrent Pos: Instr, Rutgers Univ, 49; vis prof, Princeton Univ, 58-59 & Dartmouth Col, 72-73; sr res assoc, Stanford Univ, 65-66; mem nat adv comt math, Australia & Australian Subcomt, Int Comn Math Instr; dir nat math summer sch, Australian Nat Univ, 69- Mem: Am Math Soc; Math Asn Am; Australian Asn Math Teachers (pres, 70 & 71); Australian Math Soc. Res: Relations between homology and homotopy sequences; topology; mathematics education. Mailing Add: Dept of Math Univ of Western Australia Perth 6009 Australia

BLAKESLEE, A EUGENE, b Sayre, Pa, June 20, 28; m 54; c 2. PHYSICAL CHEMISTRY. Educ: Pa State Univ, BS, 50; Cornell Univ, PhD(phys chem), 55. Prof Exp: Investr photoconductor develop, NJ Zinc Co, 55-57; mem tech staff semiconductor mat, Bell Tel Labs, 57-60; engr, Advan Semiconductor Lab, Gen Elec Co, 60-61; chemist, Components Div Lab, 62-64 & World Trade Lab, 64-65, CHEMIST, RES DIV, IBM CORP, 66- Mem: Electrochem Soc. Res: X-ray crystallography; photoconductive zinc oxide powder; electrophotography; vapor epitaxial growth and diffusion in semiconductors: III-V compound semiconductor materials and devices. Mailing Add: Watson Res Ctr IBM Corp Yorktown Heights NY 10598

BLAKESLEE, ALTON LAUREN, b Dallas, Tex, June 27, 13; m 37; c 2. SCIENCE WRITING. Educ: Columbia Univ, BA, 35. Prof Exp: Reporter, Jour-Every Eve, Del, 35-39; staff mem, 39-42, foreign news desk, 42-46, sci writer, 46-69, SCI ED, ASSOC PRESS, 69- Concurrent Pos: Pres, Am Tentative Soc, 73- Honors & Awards: Westinghouse Award, AAAS, 52; Polk Award, 52; Lasker Award, 54, 62 & 64; Bronze Medal, Am Heart Asn, 54, Blakeslee Award, 63 & 64; Grady Medal, Am Chem Soc, 59; Distinguished Serv Award for Gen Reporting, Sigma Delta Chi, 65; Honor Award for Distinguished Serv in Jour, Univ Mo, 66; Sci Writers Award, Am Dent Asn, 67; Robert T Morse Award, Am Psychiat Asn, 73. Mem: Nat Asn Sci Writers. Mailing Add: Assoc Press 50 Rockefeller Plaza New York NY 10020

BLAKESLEE, DENNIS LAUREN, b Wilmington, Del, July 2, 38; m 68; c 1. IMMUNOBIOLOGY. Educ: Middlebury Col, BA, 59; Columbia Univ, MS, 60; Univ Wis, MS, 67, PhD(genetics), 70. Prof Exp: Trainee, Dept Path, Sch Med, Univ Pa, 70-72; fel, 72-74, RES SCHOLAR & ASST PROF, NAT CANCER INST CAN, DEPT PATH, QUEENS UNIV, 74- Mem: AAAS; Genetics Soc Am. Res: Physical and immunochemical investigations of aspects of the host-tumor relationship. Mailing Add: Dept of Path Richardson Lab Queens Univ Kingston ON Can

BLAKESLEE, GEORGE M, b Philadelphia, Pa, Aug 30, 46; m 67; c 2. FOREST PATHOLOGY. Educ: Albright Col, BS; Duke Univ, MF, 70, PhD(forest path), 75. Prof Exp: ASST PROF FOREST PATH, UNIV VT, 75- Mem: Am Phytopath Soc; Soc Am Foresters; Sigma Xi. Res: Forest ecosystem response to the stress of soil compaction, with emphasis on soil microbe populations, mycorrhizal fungi and soil-borne pathogens of forest trees. Mailing Add: Dept of Forestry Sch of Natural Resources Univ Vt Burlington VT 05401

BLAKESLEE, THEODORE EDWIN, b Sandusky, Ohio, June 24, 18; m 45; c 2. ENTOMOLOGY. Educ: Ohio State Univ, BSc, 41; Univ Md, MSc, 52. Prof Exp: US Army, 48-73, entomologist, Okinawa, 48-50, chief entom div, Second Army Area Med Lab, 51-54, 406th Med Gen Lab, 54-57, Environ Health Lab, 57-59 & Sixth Army Med Lab, 59-73; RETIRED. Mem: Entom Soc Am; Am Mosquito Control Asn. Res: Mosquito bionomics; insecticide resistance. Mailing Add: 618 Woodbine Dr San Rafael CA 94903

BLAKLEY, EDWIN RAYMOND, b Sask, Can, Nov 24, 19; m 47; c 5. MICROBIAL BIOCHEMISTRY. Educ: Univ Sask, BSA, 49, MSc, 50; Univ Minn, PhD(agr biochem), 54. Prof Exp: SR RES OFFICER, NAT RES COUN CAN, 54- Mem: Can Soc Microbiol; Chem Inst Can. Res: Microbial degradation of aromatic, alicyclic and related halogenated compounds; microbial transformation of organic compounds; enzymes. Mailing Add: Prairie Regional Lab Nat Res Coun Saskatoon SK Can

BLAKLEY, GEORGE ROBERT, b Chicago, Ill, May 6, 32; m 57; c 3. MATHEMATICS, MATHEMATICAL BIOLOGY. Educ: Univ Md, PhD(math), 60. Prof Exp: Off Naval Res res assoc, Cornell Univ, 60-61; Nat Acad Sci-Nat Res Coun fel, Harvard Univ, 61-62; asst prof math, Univ Ill, 62-66; assoc prof, State Univ NY Buffalo, 66-70; PROF MATH & HEAD DEPT, TEX A&M UNIV, 70- Mem: AAAS; Am Math Soc; Math Asn Am; Soc Indust & Appl Math; Am Inst Biol Sci. Res: Linear algebra; nonlinear operators; population genetics; mathematical ecology. Mailing Add: Dept of Math Tex A&M Univ College Station TX 77843

BLAKLEY, RAYMOND L, b Christchurch, NZ, May 14, 26; m; c 4. BIOCHEMISTRY. Educ: Canterbury Univ Col, BSc, 46, MSc, 47; Univ Otago, NZ, PhD(biochem), 51; Australian Nat Univ, DSc, 65. Prof Exp: Actg head dept biochem, Australian Nat Univ, 65-66; PROF BIOCHEM, COL MED, UNIV IOWA, 68- Concurrent Pos: Vis prof, Univ Calif, 62-63; prof fel, Australian Nat Univ, 65-68. Mem: Am Soc Biol Chemists. Res: Folic acid metabolism; role of cobamide coenzymes in ribonucleotide reduction; mechanism and metabolic interactions; structure of allosteric sites; suppressors of lymphocyte transformation. Mailing Add: Dept of Biochem Univ of Iowa Col of Med Iowa City IA 52240

BLALOCK, JAMES CAREY, analytical chemistry, deceased

BLALOCK, THOMAS JACKS, b Clinton, SC, Mar 27, 10; m 35; c 2. INORGANIC CHEMISTRY. Educ: Presby Col, SC, BS, 31; Univ NC, MA, 48. Prof Exp: From instr to asst prof, 46-72, EMER ASST PROF CHEM, NC STATE UNIV, 72- Mem: Am Chem Soc. Res: General and qualitative chemistry. Mailing Add: Dept of Chem NC State Univ Raleigh NC 27607

BLAMBERG, DONALD LEE, b Baltimore, Md, Sept 11, 32; m 63; c 3. NUTRITION, PHYSIOLOGY. Educ: Univ Md, BS, 54, MS, 56, PhD(nutrit, biochem), 60. Prof Exp: Res nutritionist, Quaker Oats Co, 60-61; res asst poultry, Univ Md, 62-66; asst prof, 66-72, ASST TO DIR, LIFE SCI & AGR EXP STA, UNIV MAINE, ORONO, 72- Mem: AAAS; Poultry Sci Asn. Res: Role of zinc in the reproduction of the domestic fowl; influence of dietary amino acid balance on voluntary food consumption of the chick. Mailing Add: 106 Winslow Hall Univ of Maine Orono ME 04473

BLANC, EDWARD JOSEPH, organic chemistry, see 12th edition

BLANC, FRANCIS LOUIS, b Knoxville, Tenn, Aug 22, 16; m 38; c 2. ENTOMOLOGY. Educ: Univ Calif, BS, 38. Prof Exp: Inspector entom & plant quarantine, 40-43 & 45-47, syst entomologist, 47-58, supvr insect pest detection & surv, 58-71, PROG SUPVR PEST PREVENTION, STATE DEPT AGR, CALIF, 71- Mem: Entom Soc Am. Res: Systematics of Diptera, Tephritidae. Mailing Add: 5309 Spilman Ave Sacramento CA 95819

BLANC, JOSEPH, b Cernauti, Romania, Mar 12, 30; US citizen; m 49; c 3. SOLID STATE SCIENCE, PHYSICAL CHEMISTRY. Educ: Columbia Col, AB, 54; Columbia Univ, MA, 55, PhD(phys chem), 59. Prof Exp: Res asst infrared spectros, Columbia Univ, 56-57 & 58-59; MEM TECH STAFF, RCA LABS, 59- Concurrent Pos: Res assoc, Lab d'Optique Electronique CNRS, Toulouse, France, 74-75. Honors & Awards: RCA Labs Achievement Award, 67-71. Mem: AAAS; Am Phys Soc; Am Chem Soc. Res: Photochemistry of indigoid dyes; chemical physics of defects in solids and in the spectroscopy of the lanthanide ions; study of defects in semiconductors, especially via transmission electron microscopy. Mailing Add: 471 Walnut Lane Princeton NJ 08540

BLANC, ROBERT PARMELEE, b Redlands, Calif, July 2, 30; m 52; c 3. GEOLOGY. Educ: Univ Calif, Los Angeles, BA, 53, MA, 58. Prof Exp: Res geologist, Richfield Oil Corp, 59-63, field div geologist, Colo, 63-64, explor supvr, Diversification Div, Calif, 64-65; staff geologist, Tidewater Oil Co, 65-68; minerals explor mgr, 68-70, MINERALS EXPLOR OPERS MGR, GETTY OIL CO, 70- Mem: Am Asn Petrol Geologists; Geol Soc Am; Am Inst Mining, Metall & Petrol Engrs; Soc Econ Geologists. Res: Economic geology of non-metallic mineral deposits; minerals exploration. Mailing Add: Minerals Explor Dept Getty Oil Co 3810 Wilshire Blvd Los Angeles CA 90005

BLANC, WILLIAM ANDRE, b Geneva, Switz, Sept 28, 22; nat US; m 54; c 1. PATHOLOGY. Educ: Univ Geneva, BA, 40, MD, 47, MedScD, 52; Am Bd Path, dipl. Prof Exp: Resident path, Univ Geneva, 47-50, assoc, 50-53; fel pediat path, Harvard Univ, 53-54; instr, 54-57, assoc prof, 61-66, PROF PATH, COL PHYSICIANS & SURGEONS, COLUMBIA UNIV, 66- Concurrent Pos: From asst attend pathologist to assoc attend pathologist, Columbia-Presby Med Ctr, 57-67, attend pathologist, 67-; dir path, Babies Hosp; consult, USPHS. Honors & Awards: Laureate, Fac Med, Univ Geneva, 51. Mem: AAAS; Am Pediat Soc; fel Am Soc Clin Path; Int Acad Path; AMA; fel Col Am Path. Res: Pathology in pediatrics and obstetrics; placenta diseases of prematurity; fetal medicine; morphometrics of cardiovascular adaptation at birth. Mailing Add: Col of Physician & Surgeons Columbia Univ New York NY 10003

BLANCH, GERTRUDE, b Kolno, Poland, Feb 2, 97; US citizen. MATHEMATICS. Educ: NY Univ, BS, 32; Cornell Univ, PhD(math), 35. Prof Exp: Mathematician, Math Tables Proj, NY, 38-42, Nat Bur Standards, 42-48 & Inst Numerical Anal, Calif, 48-54; sr mathematician, Aerospace Res Labs, Wright-Patterson AFB, 54-67; RES & WRITING, 67- Honors & Awards: Fed Women's Award, 64. Mem: Am Math Soc; Math Asn Am; Soc Indust & Appl Math. Res: Numerical analysis; Mathieu functions. Mailing Add: 3625 First Ave Apt 14 San Diego CA 92103

BLANCHAER, MARCEL CORNEILLE, b Antwerp, Belg, Apr 4, 21; nat Can; m 41; c 2. BIOCHEMISTRY. Educ: Queen's Univ, Ont, BA, 45, MD & CM, 46. Prof Exp: Assoc prof physiol, 48-64, PROF BIOCHEM & HEAD DEPT, UNIV MAN, 64- Concurrent Pos: Dir biochem, St Boniface Hosp, 54-64; Markle scholar, 48-53. Mem: Can Biochem Soc; Can Soc Clin Chem. Res: Muscle biochemistry. Mailing Add: Dept of Biochem Univ of Manitoba Winnipeg MB Can

BLANCHAR, ROBERT W, b Sherburn, Minn, June 30, 37; m 58; c 1. SOIL CHEMISTRY, BIOCHEMISTRY. Educ: Macalester Col, BA, 58; Univ Minn, MS, 61, PhD(soil sci), 64. Prof Exp: Res agronomist soil chem, Int Minerals & Chem Corp, 64-68; PROF AGRON, UNIV MO-COLUMBIA, 68- Mem: Am Soc Agron. Res: Polyphosphate chemistry in soils; ion products of iron, aluminum and manganese in soils; arsenic transformations in soils; chemical transformations in coal mine spoils; chemistry of Pragipan soils. Mailing Add: Dept of Agron Univ of Mo Columbia MO 65201

BLANCHARD, CONVERSE HERRICK, b Boston, Mass, Sept 25, 23; m 46; c 4. NUCLEAR PHYSICS. Educ: Harvard Univ, AB, 45; Univ Wis, PhD(physics), 50. Prof Exp: Nuclear physicist, Nat Bur Standards, 50-53; assoc prof physics, Pa State Univ, 53-61; assoc prof, 61-63, PROF PHYSICS, UNIV WIS-MADISON, 63- Mem: Fel Am Phys Soc. Res: Theoretical nuclear and quantum physics. Mailing Add: Dept of Physics Univ of Wis Madison WI 53706

BLANCHARD, DUNCAN CROMWELL, b Winterhaven, Fla, Oct 8, 24; m 57; c 3. MARINE SCIENCES. Educ: Tufts Univ, BNavalSc, 45, BS, 47; Pa State Univ, MS, 51; Mass Inst Technol, PhD(meteorol), 61. Prof Exp: Res assoc, Res Lab, Gen Elec Co, 48-49; assoc scientist, Woods Hole Oceanog Inst, 51-68; SR RES ASSOC, ATMOSPHERIC SCI RES CTR, STATE UNIV NY ALBANY, 68- Concurrent Pos: Mem joint panel air/sea interaction, Nat Acad Sci, 63-65; mem steering comt, Ocean Sci Comt Study Ocean Dumping, Nat Acad Sci-Nat Res Coun, 74-76. Mem: AAAS; Sigma Xi; Am Meteorol Soc. Res: Aerosol production by the sea, bubble production in the sea, and flux of charged particles across the surface; ejection of dissolved organic matter and bacteria from water surfaces into the atmosphere. Mailing Add: Atmospheric Sci Res Ctr State Univ of NY Albany NY 12203

BLANCHARD, FRANK NELSON, b Ann Arbor, Mich, May 24, 31; m 53; c 4. GEOLOGY. Educ: Univ Mich, AB, 53, MS, 54, PhD(geol), 60. Prof Exp: Asst prof, 58-69, ASSOC PROF GEOL, UNIV FLA, 69- Mem: Mineral Soc Am; Nat Asn Geol Teachers; Mineral Asn Can. Res: Mineralogy; crystallography; petrology; petrography; x-ray analysis. Mailing Add: Dept of Geol Univ of Fla Gainesville FL 32601

BLANCHARD, FRED AYRES, b Cleveland, Ohio, May 27, 23; m 48; c 3. BIOPHYSICS. Educ: La State Univ, BS, 44; Univ Cincinnati, MS, 48, PhD(physics), 51. Prof Exp: RADIOCHEMIST TRACERS, DOW CHEM CO, 51- Mem: AAAS; Am Chem Soc. Res: Applications of radioisotopes to biological, chemical and industrial tracer problems; liquid scintillation counting; computer processing of counting data; low level counting; environmental science. Mailing Add: Environ Sci Res Lab Bldg 1602 Dow Chem Co Midland MI 48640

BLANCHARD, GORDON CARLTON, b Monroe, NH, Dec 5, 32; m 53; c 2. MICROBIOLOGY. Educ: Univ Vt, BS, 56, MS, 58; Syracuse Univ, PhD(microbiol), 61. Prof Exp: Sr scientist microbiol, Melpar, Inc Div, Westinghouse Air Brake Co, 61-62, br head, 62-67; RES MICROBIOLOGIST, VET ADMIN HOSP, 67- Mem: AAAS; Am Soc Microbiol. Res: Characterization of ragweed pollen allergens and study of hypersensitive reactions; automated methods for microbiology. Mailing Add: Vet Admin Hosp 150 S Huntington Ave Boston MA 02130

BLANCHARD, JONATHAN EWART, b Truro, NS, Mar 22, 21; m 58; c 2. GEOPHYSICS. Educ: Dalhousie Univ, BSc, 40; Univ Toronto, MA, 47, PhD(geophys), 52. Prof Exp: Dir, 49-66, vpres, 66-68, PRES, NS RES FOUND CORP, 68- Concurrent Pos: From asst prof to prof geophys, Dalhousie Univ, 52-56. Mem: Soc Explor Geophysicists; Am Geophys Union; Can Asn Physicists; Can Inst Mining & Metall; fel Royal Soc Can. Mailing Add: NS Res Found Corp PO Box 790 100 Fenwick St Dartmouth NS Can

BLANCHARD, RICHARD LEE, b Hutchinson, Kans, May 2, 33; m 55; c 2. RADIOCHEMISTRY, ECOLOGY. Educ: Ft Hays Kans State Col, BS, 55; Vanderbilt Univ, MS, 58; Washington Univ, PhD(chem), 63. Prof Exp: Chemist, Div Radiol Health, USPHS, Oak Ridge Nat Lab, 56-58, health phyisicst, Utah, 58-59, chemist, Robert A Taft Sanit Eng Ctr, 62-64, chg physics & bioassay group, Radiol Health Res Activities, 64-68, dep chief radiol eng lab, Bur Radiol Health, Environ Control Admin, Ohio, 68-74; DIR RADIOCHEM & NUCLEAR ENG FACIL, US ENVIRON PROTECTION AGENCY, 74- Concurrent Pos: Adj prof nuclear eng, Univ Cincinnati, 68-; mem N 13 subcomt, Am Nat Standards Inst, 74- Mem: Health Physics Soc; Am Inst Chemists; Sigma Xi. Res: Radiation dose to man from naturally occurring radionuclides and from radionuclides in the environment of nuclear power reactors; movement and concentration of radionuclides in aquatic environments. Mailing Add: Radiochem & Nuclear Eng Facil US Environ Protection Agency Cincinnati OH 45268

BLANCHE, ERNEST EVRED, b Passaic, NJ, Oct 22, 12; m 38; c 2. MATHEMATICS, STATISTICS. Educ: Bucknell Univ, AB & AM, 38; Univ Ill, PhD(math), 41. Prof Exp: Asst math, Univ Ill, 38-41; instr math & statist, Mich State Univ, 41-42; statist dir, Curtiss-Wright Corp, NY, 42-43, head math & statist lab, 43-44; prin statistician, Foreign Econ Admin, 44-45; instr, US Army Univ, Italy, 45, prin admin analyst, Army Serv Forces, 46-47; chief statist br, Res & Develop Div, Gen Staff, US War Dept, 47-48; chief statistician, Logistics Div, US Dept Army, 48-54, vpres for res, Frederick Res Corp, Bethesda, Md, 54-55; pres & chief res scientist, Blanche & Assocs, Inc, Kensington, 55-70; chmn bd trustees, Capitol Inst Technol, 70-74, actg pres, 71-72; PRES, ERNEST E BLANCHE CONSULTS, CHEVY CHASE, 74- Concurrent Pos: Army mem subpanel math & statist, Comt Basic Phys Sci, Res & Develop Bd, 49; consult logistics div, US Dept Army, 54-; adj prof math & statist, Am Univ, 46-66; prof math & statist & trustee, Capitol Inst Technol, 69-70. Mem: Am Math Soc; Math Asn Am; Am Statist Asn; Inst Math Statist. Res: Probability; choice and chance; work simplification and work measurement; logistics; computer research and applications; system design; programming engineering computations; traffic projections. Mailing Add: 14818 Carrolton Rd Manor Club Rockville MD 20853

BLANCHET, WALDO W E, b New Orleans, La, Aug 6, 10; m 43; c 2. SCIENCE EDUCATION. Educ: Talladega Col, AB, 31; Univ Mich, MS, 36, PhD(sci educ), 46. Prof Exp: Head sci dept, Ft Valley Normal & Indust Sch, 32-35, head dept & dean sch, 36-38; prof sci & admin dean, 39-66, pres, 66-73, EMER PRES, FT VALLEY STATE COL, 73-; CONSULT SCI EDUC & HIGHER EDUC, 73- Concurrent Pos: Consult & lectr, NSF sci insts, Atlanta Univ & Albany State Col; consult spec proj, Tenn Valley Authority, 63-64; mem, Nat Adv Coun on Educ Disadvantaged Children, 70- Mem: AAAS; Nat Inst Sci; Nat Asn Res Sci Teaching. Res: Curriculum problems of general science education on college level. Mailing Add: 110 Lamar St Ft Valley GA 31030

BLANCK, HARVEY F, JR, b North Ridgeville, Ohio, Apr 4, 32; m 60; c 2. PHYSICAL CHEMISTRY, ANALYTICAL CHEMISTRY. Educ: Miami Univ, BA, 54; Ohio State Univ, MS, 62, PhD(chem), 67. Prof Exp: PROF CHEM, AUSTIN PEAY STATE UNIV, 64- Mem: Am Chem Soc. Mailing Add: Dept of Chem Austin Peay State Univ Clarksville TN 37040

BLANCO, VICTOR MANUEL, b Guayama, PR, Mar 10, 18; m 43, 71; c 3. ASTRONOMY. Educ: Univ Chicago, BS, 43; Univ Calif, MA, 47, PhD(astron), 49. Hon Degrees: DSc, Univ PR, 72. Prof Exp: From asst prof to assoc prof physics, Univ PR, 49-51; from instr to prof astron, Case Inst Technol, 51-65; dir astrometry & astrophys div, US Naval Observ, Washington, DC, 65-67; DIR, CERRO TOLOLO INTER-AM OBSERV, 67- Concurrent Pos: Mem comn galactic struct, Int Astron Union. Mem: Am Astron Soc (vpres, 72-75). Res: Galactic structure and Magellanic clouds research. Mailing Add: Cerro Tololo Inter-Am Observ Casilla 63-D La Serena Chile

BLAND, BRIAN HERBERT, b Calgary, Alta, July 10, 43; m 64; c 2. NEUROBIOLOGY, NEUROPSYCHOLOGY. Educ: Univ Calgary, BSc, 66, MSc, 68; Univ Western Ont, PhD(neuropsychol), 71. Prof Exp: NATO sci fel, Inst Neurophysiol, Oslo, Norway, 71-73; asst prof, Univ Sask, 73-74; ASST PROF NEUROPSYCHOL, UNIV CALGARY, 74- Mem: Can Physiol Soc. Res: Electrophysiological and pharmacological investigation of hippocampal formation function. Mailing Add: Dept of Psychol Univ of Calgary Calgary AB Can

BLAND, CHARLES E, b Raleigh, NC, Aug 17, 43; m 63; c 1. MYCOLOGY. Educ: Univ NC, Chapel Hill, AB, 64, PhD(mycol), 69. Prof Exp: ASST PROF BIOL, E CAROLINA UNIV, 69- Mem: AAAS; Mycol Soc Am; Electron Micros Soc Am; Am Soc Cell Biol. Res: Fungi; parasitic in marine crustaceans. Mailing Add: Dept of Biol ECarolina Univ Greenville NC 27834

BLAND, CLIFFORD J, b London, Eng, June 22, 36; m 63; c 2. COSMIC RAY PHYSICS, RADIATION PHYSICS. Educ: Univ London, BSc, 58, PhD(physics), 61; Royal Col Sci, ARCS, 58; Imp Col, Univ London, DIC, 61. Prof Exp: Res fel physics, Lab Cosmic Physics, La Paz, 62-63; res fel, Inst Physics, Milan, 64-68; ASSOC PROF PHYSICS, UNIV CALGARY, 68- Concurrent Pos: Lectr, La Paz, 62-63 & Milan, 64-68; consult, Nat Inst Nuclear Physics, Italy, 64-68; Nat Res Coun Can res grants, 68-75. Mem: Am Geophys Union. Res: Geomagnetic effects; primary electrons; modulation theory; terrestrial radioactivity. Mailing Add: Dept of Physics Univ of Calgary Calgary AB Can

BLAND, HESTER BETH, b Sullivan, Ind, June 22, 06. HEALTH SCIENCES. Educ: Ind State Univ, BS, 42; Butler Univ, MS, 49; Ind Univ, HSD, 56. Hon Degrees: LLD, Ind State Univ, 73. Prof Exp: Supvr & teacher pub schs, Ind, 30-41, teacher, 42-47; supvr plant protection, Allison Div, Gen Motors Corp, 41-42; consult health educ, Ind State Bd Health, 47-72; ADJ PROF, IND STATE UNIV, 72- Concurrent Pos: Lectr, Div Allied Sci, Sch Med, Ind Univ, 50-72 & Dept Health Educ, 58-; dir, Am Sch Health Asn & Pharmaceut Mfrs Asn Drug Educ Curric Proj, 69-70; consult ed, J Sch Health, 74- Honors & Awards: Ind State Univ Alumni Distinguished Serv Award, 62; Am Asn Health, Phys Educ & Recreation Distinguished Serv Award, 67; Distinguished Serv Award, Am Sch Health Asn, 70, William A Howe Award, 75. Mem: Fel Am Sch Health Asn (pres, 72-73); Am Asn Health, Phys Educ & Recreation. Res: Critical areas of health education, especially venereal disease, drug misuse, smoking and alcohol; school and public health. Mailing Add: 2511 Parkwood Indianapolis IN 46224

BLAND, JEFFREY S, b Peoria, Ill, Mar 21, 46; m; c 2. BIO-ORGANIC CHEMISTRY. Educ: Univ Calif, Irvine, BS, 67; Univ Ore, PhD(chem), 71. Prof Exp: ASST PROF CHEM, UNIV PUGET SOUND, 71- Concurrent Pos: Vis prof, Univ Hawaii, 74; sr res fel, Univ Ore, 74 & 75; clin lab dir, Medicomp Inc, 75. Mem: AAAS; Am Chem Soc; Am Asn Clin Chemists. Res: Investigation of the effects of biological antioxidants upon the rate of photohemolysis of erythrocytes; examination of mechanisms of photodynamic processes and relation to membrane structure. Mailing Add: Dept of Chem Univ of Puget Sound Tacoma WA 98416

BLAND, JOHN EDWARD, b Cleveland, Ohio, Dec 5, 42; m 66; c 2. ARCHAEOLOGY, BIOLOGICAL ANTHROPOLOGY. Educ: Case Inst Technol, BS, 64, MA, 67; Univ Mass, PhD(anthrop), 70. Prof Exp: Asst prof, 69-73, ASSOC PROF ANTHROP, CLEVELAND STATE UNIV, 73- Mem: Soc Am Archaeol; fel Am Anthrop Asn. Res: Cultural paleoecological adaptational systems of the Ohio Valley; primate anatomy, emphasis upon primate osteology and musculature. Mailing Add: Dept of Anthrop Cleveland State Univ Cleveland OH 44115

BLAND, JOHN (HARDESTY), b Globe, Ariz, Nov 7, 17; m 44; c 4. MEDICINE. Educ: Earlham Col, AB, 40; Jefferson Med Col, MD, 44; Am Bd Internal Med, dipl, 52. Prof Exp: Resident, Burlington County Hosp & Pa Hosp, Philadelphia, 44-46; chief cardiovasc sect, Sta Hosp, Ft Hood, Tex, 46-48; res fel cardiol, 48-49, from instr to asst prof, 49-55, dir rheumatism res unit, 50-69, ASSOC PROF MED, COL MED, UNIV VT, 55- Concurrent Pos: Fel rheumatic dis, Mass Gen Hosp & New Eng Ctr Hosp, Boston, 50; hon res fel rheumatol, Univ Manchester, 58-59; vis sci worker, NIH, 72-73. Mem: AAAS; fel Am Col Physicians; Am Fedn Clin Res; NY Acad Sci; Am Rheumatism Asn. Res: Rheumatic disease; connective tissue metabolism. Mailing Add: Dept of Med Univ of Vt Col of Med Burlington VT 05401

BLAND, RICHARD P, b Oklahoma City, Okla, Nov 23, 28; m 53. MATHEMATICAL STATISTICS. Educ: Univ Okla, BS, 52; Univ NC, PhD(math statist), 61. Prof Exp: Anal engr, Chance Vought Aircraft, Inc, 53-56; res asst statist, Univ NC, 60-61; asst prof, Johns Hopkins Univ, 61-63; asst prof, 63-72, ASSOC PROF STATIST, SOUTHERN METHODIST UNIV, 72- Mem: Am Statist Asn; Am Inst Math Statist. Res: Statistical decision theory and statistical inference. Mailing Add: Dept of Statist Southern Methodist Univ Dallas TX 75222

BLAND, ROBERT GARY, b New York, NY, Feb 25, 48; m 74. OPERATIONS RESEARCH, APPLIED MATHEMATICS. Educ: Cornell Univ, BS, 69, MS, 72, PhD(oper res), 74. Prof Exp: ASST PROF MATH SCI, STATE UNIV NY BINGHAMTON, 74- Concurrent Pos: Res assoc oper res, Ctr Oper Res & Econometrics & lectr appl sci, Cath Univ Louvain, 75-76. Mem: Oper Res Soc Am. Res: Mathematical programming; combinatorial optimization; matroids; graph theory; networks. Mailing Add: Dept of Math Sci State Univ of NY Binghamton NY 13901

BLAND, ROGER GLADWIN, b Los Angeles, Calif, Dec 28, 39; m 64. ENTOMOLOGY. Educ: Univ Calif, Davis, BS, 61; Ore State Univ, MS, 64; Univ Ariz, PhD(entom), 67. Prof Exp: Asst prof biol, 67-71, univ res grants, 68-72, ASSOC PROF BIOL, CENT MICH UNIV, 71- Mem: AAAS; Entom Soc Am; Am Inst Biol Sci. Res: Nutrition, behavior and ecology of insects. Mailing Add: Dept of Biol Cent Mich Univ Mt Pleasant MI 48858

BLANDAU, RICHARD JULIUS, b Erie, Pa, Aug 5, 11; m 37; c 1. EXPERIMENTAL EMBRYOLOGY. Educ: Linfield Col, AB, 35; Brown Univ, PhD(biol), 39; Univ Rochester, MD, 48. Prof Exp: Instr biol, embryol & histol, Brown Univ, 39-41; instr anat, Harvard Med Sch, 41-42; from instr to asst prof, Sch Med, Univ Rochester, 42-49; assoc prof, 49-51, from asst dean to assoc dean sch med, 55-64, PROF ANAT, SCH MED, UNIV WASH, 51- Concurrent Pos: Borden res award, Sch Med, Univ Rochester, 48; consult, Nat Primate Ctr, Calif & WHO. Honors & Awards: Isidor Rubin Award, 52; Ortho Res Award, Am Fertil Soc, 56 & Ortho Medal, 69; Vienna

Film Festival Award, 59; Barren Found Medalist, 69. Res: Significance of ovum and spermatozoan age on development; germ cells; ovulation; reproductive physiology. Mailing Add: Sch of Med Univ of Wash Seattle WA 98195

BLANDER, MILTON, b Brooklyn, NY, Nov 1, 27; m 60; c 3. PHYSICAL CHEMISTRY. Educ: Brooklyn Col, BS, 50; Yale Univ, PhD(phys chem), 53. Prof Exp: Res assoc, Cornell Univ, 53-55; sr chemist, Oak Ridge Nat Lab, 55-62; mem tech staff, NAm Rockwell Sci Ctr, 62-67, group leader, 67-71; GROUP LEADER, ARGONNE NAT LAB, 71- Mem: Fel AAAS; Am Chem Soc; fel Meteoritical Soc; Am Geophys Union. Res: Meteoritics; geochemistry; nucleation phenomena; solution chemistry; molten salts; high temperature vapors; thermodynamics. Mailing Add: Argonne Nat Lab 9700 S Cass Ave Argonne IL 60439

BLANDFORD, ROBERT ROY, b Columbus, Ga, Jan 1, 37; m 66; c 1. PHYSICAL OCEANOGRAPHY, SEISMOLOGY. Educ: Calif Inst Technol, BS, 59, PhD(geophys), 64. Prof Exp: Res scientist, NY Univ, 64-66; DIR RES, SEISMIC DATA LAB, GEOTECH, TELEDYNE, INC, 66- Concurrent Pos: Consult, Rand Corp, 57-66. Mem: Am Geophys Union. Res: Rossby waves; thermocline; internal waves; Gulf Stream dynamics; signal detection; arrays; time-series. Mailing Add: Geotech Seismic Data Lab 300 N Washington Alexandria VA 22313

BLANEY, DONALD JOHN, b Cincinnati, Ohio, May 18, 26; m 57; c 1. BIOCHEMISTRY. Educ: Xavier Univ, Ohio, BS, 49; Univ Iowa, PhD(biochem), 53; Univ Mich, MD, 57. Prof Exp: Res asst, Univ Mich, 53-55; intern, Philadelphia Gen Hosp, 57-58; fel dermat, Cincinnati Gen Hosp, 58-61, instr, 61-63; assoc prof, 65-67, ASSOC CLIN PROF DERMAT, UNIV CINCINNATI, 67-; ASST PROF DERMAT, CINCINNATI GEN HOSP, 63- Mem: AAAS; Am Chem Soc. Res: Immunology. Mailing Add: Dept of Dermat Cincinnati Gen Hosp Cincinnati OH 45219

BLANK, ALBERT ABRAHAM, b New York, NY, Nov 29, 24; m 50; c 4. MATHEMATICAL PHYSICS, VISION. Educ: Brooklyn Col, AB, 44; Brown Univ, MS, 46; NY Univ, PhD(math), 51. Prof Exp: Math analyst physiol optics, Columbia Univ, 51-52; assoc info & control systs, Control Systs Lab, Univ Ill, 53-54; asst prof math, Univ Tenn, 54-59; from assoc prof to prof, Courant Inst Math Sci, NY Univ, 59-69; PROF MATH, CARNEGIE-MELLON UNIV, 69- Concurrent Pos: Scientist, Magneto-Fluid Dynamics Div, Courant Inst Math Sci, NY Univ, 57-69; assoc ed, Am Math Monthly. Mem: Fel AAAS; Am Phys Soc; Am Math Soc; Math Asn Am. Res: Partial differential equations and applications; binocular space perception; magnetofluid dynamics; studies of stability and diffusion in connection with problems in controlling thermonuclear reactions. Mailing Add: Dept of Math Carnegie-Mellon Univ Pittsburgh PA 15213

BLANK, BENJAMIN, b Philadelphia, Pa, May 12, 31; m 58; c 3. ORGANIC CHEMISTRY, MEDICINAL CHEMISTRY. Educ: Temple Univ, BA, 53, PhD(chem), 58. Prof Exp: Chemist, Muscular Dystrophy Asn, Vet Admin Hosp, Philadelphia, 56; sr med chemist, 58-74, SR INVESTR, SMITH KLINE & FRENCH LABS, 74- Concurrent Pos: Biochemist, Johnson Res Found, Sch Med, Univ Pa, 69-70. Mem: AAAS; NY Acad Sci; Am Chem Soc; Am Inst Chemists; Sigma Xi. Res: Thyromimetics; peptides; nonsteroidal anti-inflammatories; inhibitors of steroidal biosynthesis and gluconeogenesis; cephalosporins. Mailing Add: 502 Bruce Rd Trevose PA 19047

BLANK, CARL HERBERT, b Toledo, Ohio, Mar 16, 27; m 50; c 1. PUBLIC HEALTH, MICROBIOLOGY. Educ: Univ Toledo, BS, 50; Utah State Univ, MS, 57; Univ NC, MPH, 65, DrPH(lab practice & admin), 67. Prof Exp: Jr microbiologist, Div Labs, Utah State Health Dept, 51-54, sr microbiologist, 54-58, actg asst dir, Div Labs, 58-62, asst dir, 62-64; dep dir, Bur Labs, Utah State Div Health, 67-72; CHIEF EXAM & DOCUMENTATION BR, CTR DIS CONTROL, 72- Concurrent Pos: Clin instr, Univ Utah, 68-70; consult, Ctr Dis Control, 71. Mem: Am Pub Health Asn; Am Soc Microbiol; Conf State & Territorial Pub Health Lab Dirs. Res: Incidence of Q fever; improved methods in isolation and identification of actinomyces species; toxo plasmosis; fluorescent antibody techniques; laboratory evaluation and improvement programs. Mailing Add: Ctr for Dis Control 1600 Clifton Rd Atlanta GA 30333

BLANK, CHARLES ANTHONY, b Brooklyn, NY, Apr 15, 22; m 53; c 3. PHYSICAL INORGANIC CHEMISTRY, AERONOMY. Educ: Brooklyn Col, BA, 46, MA, 49; Syracuse Univ, PhD(phys inorg chem), 54. Prof Exp: Lab technician res, Int Flavors & Fragrances, Inc, 41-43; asst chem, Syracuse Univ, 47-48; res chemist, Film Dept, E I du Pont de Nemours & Co, 51-52, Bell Tel Labs, 54-56 & Int Bus Mach Corp, 56-59; supvr tech staff, Phys Sci Dept, Melpar, Inc, Va, 59-60; PHYS SCIENTIST, US DEPT DEFENSE, 60- Mem: Am Chem Soc; fel Am Inst Chemists; Am Geophys Soc; NY Acad Sci. Res: Coordination compounds; organic and inorganic synthesis; solid state chemistry; chemistry of electronic devices and materials; physical instrumentation; geophysics; nuclear weapon effects; chemistry and physics of ionosphere; research administration. Mailing Add: 5110 Sideburn Rd Fairfax VA 22030

BLANK, FRITZ, b Sept 29, 14; US citizen; m 38; c 3. MEDICAL MYCOLOGY. Educ: Swiss Fed Inst Technol, Dipl Ing Agr, 38, DrScNat(bot), 39, DrScTech(org chem), 49. Prof Exp: Res worker, Pharmaceut Indust, 45-50; med mycologist & asst prof bact & immunol, McGill Univ, 51-62; assoc prof microbiol, 62-63, PROF MED MYCOL, SCH MED, TEMPLE UNIV, 63- Mem: Am Soc Microbiol; Am Chem Soc; Can Soc Microbiol; Int Soc Human & Animal Mycol; Swiss Bot Soc. Res: Chemistry and physiology of pathogenic fungi. Mailing Add: Skin & Cancer Hosp Temple Univ Sch of Med Philadelphia PA 19140

BLANK, HARVEY, b Chicago, Ill, June 21, 18; m 50; c 1. DERMATOLOGY. Educ: Univ Chicago, BS, 38, MD, 42; Am Bd Dermat & Syphil, dipl, 48. Prof Exp: Intern, Harper Hosp, Mich, 42-43; fel dermat, Univ Pa, 46, assoc, Grad Sch Med, 47-52; assoc med dir, Squibb Inst Med Res, 51-56; PROF DERMAT & MED, SCH MED, UNIV MIAMI, 56-, CHMN DEPT DERMAT, 60- Concurrent Pos: Mem res dept, Children's Hosp, Philadelphia, 46-56; Nat Res Coun fel, Univ Pa, 47-48; consult, Surgeon-Gen, US Dept Army, 48-; assoc dermat, Col Physicians & Surgeons, Columbia Univ, 51-55; mem adv comt Qm res & develop, Nat Res Coun, 52-; chief ed, Arch Dermat, 62-64; chmn comn cutaneous dis, Armed Forces Epidemiol Bd, 62-; consult, Vet Admin; mem dermat training grant comt, NIH. Mem: Soc Invest Dermat; Am Acad Dermat; Am Dermat Asn; fel Am Col Physicians; fel AMA. Res: Virus diseases of the skin; mycology; experimental cytology; investigative and clinical dermatology. Mailing Add: Dept of Dermat Univ of Miami Sch of Med Miami FL 33152

BLANK, HORACE RICHARD, JR, b New York, NY, Apr 27, 29; m 60; c 3. REGIONAL GEOLOGY. Educ: Southwestern Univ, Tex, BS, 48; Univ Wash, MS, 50, PhD(geol), 59. Prof Exp: Geophysicist electromagnetic & magnetic prospecting, NJ Zinc Co, Pa. 51; geol engr iron deposits, Columbia Iron Mining Co, US Steel Corp, 56-57; geologist, Regional Geophys Br, US Geol Surv, 61-68; ASSOC PROF GEOL & GEOPHYS, CTR VOLCANOL, UNIV ORE, 68- Concurrent Pos: NSF res fel, 59-60; res fel, Victoria Univ, NZ, 60-61. Mem: Soc Explor Geophys; Geol Soc Am; Am Geophys Union. Res: Regional geophysics; volcanology; field geology and metamorphic petrology; geology of iron deposits; ash flow fields and calderas; electromagnetic and magnetic prospecting; aeromagnetic and gravity surveys and interpretation; volcanology. Mailing Add: Ctr for Volcanology Univ of Ore Eugene OR 97403

BLANK, JOHN MOULTON, b Massillon, Ohio, Mar 20, 23; m 46; c 4. PHYSICS. Educ: Col Wooster, AB, 47; Mass Inst Technol, SM, 50, PhD(physics), 52. Prof Exp: Physicist, 52-67, RES PHYSICIST, GEN ELEC CO, 67- Res: Solid state physics; ferromagnetism; ferrites; ferroelectrics; thin films; semiconductors; light emitting diodes. Mailing Add: 2972 Washington Blvd Cleveland Heights OH 44118

BLANK, MARTIN, b US, Feb 28, 33; m 55; c 2. PHYSIOLOGY, BIOPHYSICS. Educ: City Col New York, BS, 54; Columbia Univ, PhD(phys chem), 57; Univ Cambridge, PhD(colloid sci), 59. Prof Exp: From instr to asst prof, 59-68, ASSOC PROF PHYSIOL, COL PHYSICIANS & SURGEONS, COLUMBIA UNIV, 68- Concurrent Pos: NIH fels, 60-70; res chemist, Unilever Res Lab, Eng, 64 & 67, Netherlands, 69; vis scientist, Weizmann Inst, 67, Univ Calif, 68 & Hebrew, 70; liaison scientist, Off Naval Res, London, 74-75. Mem: AAAS; Am Chem Soc; Biophys Soc; Electrochem Soc. Res: Physical chemistry of membrane models; permeability and rheology of monolayers and bilayers; electrical effects at interfaces and membranes; transport mechanisms in natural membranes, such as NakATPase, sperm cell, lung surfactant system; theoretical aspects of membrane processes. Mailing Add: Col of Physicians & Surgeons Columbia Univ New York NY 10032

BLANK, ROBERT EUGENE, b Marion, Ohio, Sept 9, 19; m 43; c 3. ORGANIC CHEMISTRY. Educ: Western Reserve Univ, AB, 41, MS, 51. Prof Exp: Chemist vinyl stabilizers, Sherwin-Williams Lab, Western Reserve Univ, 41-42, chemist oil alcoholysis, 46-52; chemist specialty coatings, 52-54, group leader spec projs, 54-57, dir can coating lab, Tin Can Div, 57-66, dir auxiliaries res, 66-69, dir inorganic res & develop, Tech Dept, Chem Div, 69-70, prod progs mgr, 70-71, DIR RES, TECH DEPT, CHEM DIV, SHERWIN-WILLIAMS CO, 71- Mem: Am Chem Soc; Indust Res Inst; Sigma Xi. Mailing Add: Chem Div Sherwin Williams Co Box 6520 Cleveland OH 44101

BLANK, ROBERT H, b USA, Feb 14, 26; m 49; c 2. BIOLOGY. Educ: NY Univ, BA, 51. Prof Exp: Biochemist fermentation, 51-66, mgr microbiol res serv, 66-70, MGR BIOTHERAPEUT RES SERV, LEDERLE LABS, AM CYANAMID CO, 70- Mem: Health Physics Soc; Soc Res Adminr. Res: Fermentation biochemistry; antibiotics and microbiological transformation of steroids. Mailing Add: Lederle Labs Middletown Rd Pearl River NY 10965

BLANK, ZVI, b Tel-Aviv, Israel. SOLID STATE CHEMISTRY. Educ: Israel Inst Technol, BS, 60; NY Univ, MChE, 66, PhD(chem eng), 70. Prof Exp: Asst res scientist crystal growth, NY Univ, 65-69; staff engr solid state chem, Mat Res Ctr, Allied Chem Corp, 69-75; RES SCIENTIST DISPLAY TECHNOL, CORP RES & DEVELOP LAB, SINGER CO, 75- Mem: Am Chem Soc; Am Asn Crystal Growers; Int Solar Energy Soc. Res: Liquid crystals; electrochemical materials; acousto-optic materials; solid state chemistry related to energy storage and conversion; metal hydrides; crystal growth; structure-properties relationships in inorganics and in polymers. Mailing Add: Singer Co Corp Res & Develop Lab 286 Eldridge Rd Fairfield NJ 07006

BLANKE, BERTRAM CHARLES, b Chicago, Ill, Nov 12, 17; m 60. PHYSICAL CHEMISTRY. Educ: Univ Ill, BS, 38, MS, 40. Prof Exp: Res chemist, Citrus Prod Co, Ill, 40-41; res sect mgr, Mound Lab, 45-70, qual assurance engr, AEC, 70-73, NUCLEAR OPERS OFFICER, ENERGY RES & DEVELOP ADMIN, DAYTON AREA OFF, MOUND LAB, MONSANTO CO, 73- Concurrent Pos: Instr, Indust Col Armed Forces. Mem: Am Chem Soc. Res: Radiochemistry; heavy metal alpha emitters separation and properties; physical properties of actinides. Mailing Add: Mound Lab Miamisburg OH 45342

BLANKE, ROBERT VERNON, b Leavenworth, Kans, Dec 3, 24; m 50; c 3. TOXICOLOGY. Educ: Northwestern Univ, BS, 49; Univ Ill, MS, 53, PhD(pharmacol), 58. Prof Exp: Res asst, Univ Ill, 50-53, instr, 53-57; asst toxicologist, Med Examr Off, Md, 57-60; toxicologist, Ill, 60-63; toxicologist, Med Examr Off, Commonwealth of Va, 63-72; PROF PATH, PHARMACOL & LEGAL MED, MED COL VA, 72- Concurrent Pos: Sr chemist, Cook County Coroner's Lab, Chicago, 50-53; assoc prof, Med Col Va, 63-; consult, Vet Admin Hosp, Richmond, 64-70, 74-; mem toxicol study sect, Div Res Grants, NIH, 70-74. Mem: AAAS; Am Acad Forensic Sci; Am Chem Soc; NY Acad Sci; Soc Toxicol. Res: Analytical toxicology; trace metal metabolism; general toxicology. Mailing Add: Div of Clin Path Med Col Va Box 696 MCV Sta Richmond VA 23298

BLANKENBECLER, RICHARD, b Kingsport, Tenn, Feb 4, 33; m 53; c 2. THEORETICAL PHYSICS. Educ: Miami Univ, AB, 54; Stanford Univ, PhD(physics), 58. Prof Exp: NSF fel physics, Princeton Univ, 58-63, assoc prof, 63-67; prof, Univ Calif, Santa Barbara, 67-69; PROF, STANFORD LINEAR ACCELERATOR CTR, STANFORD UNIV, 69- Mem: Am Phys Soc. Res: Particle physics and theoretical high energy physics. Mailing Add: Stanford Linear Accelerator Ctr Stanford Univ PO Box 4349 Stanford CA 94305

BLANKENHORN, DAVID HENRY, b Cleveland, Ohio, Nov 16, 24; m 48; c 4. CARDIOLOGY. Educ: Univ Cincinnati, MD, 47. Prof Exp: Res asst, Rockefeller Inst, 52-54; chief resident internal med, Cincinnati Gen Hosp, 54-55; instr med, Univ Cincinnati, 55-57; from asst prof to assoc prof, 57-67, PROF MED, UNIV SOUTHERN CALIF, 67-, HEAD CARDIOL SECT, 63- Mem: Am Fedn Clin Res; Am Soc Exp Biol; Am Soc Clin Invest. Res: Lipid metabolism; vascular disease. Mailing Add: Dept of Med Univ of Southern Calif Los Angeles CA 90033

BLANKENHORN, PAUL RICHARD, b Shenandoah, Pa, Apr 16, 44; m 68; c 2. WOOD SCIENCE & TECHNOLOGY, MATERIALS ENGINEERING. Educ: Pa State Univ, BS, 66, MS, 68, PhD(wood sci, mat sci), 72. Prof Exp: Eng aide aerospace eng, Johnsville Naval Air Sta, 66-67; res asst nuclear eng, Breazeale Nuclear Reactor, 67-68, res asst wood sci, Sch Forest Resources, 70-72, res assoc mat sci, Pa Transp Inst, 72-75, ASST PROF WOOD TECHNOL, SCH FOREST RESOURCES, PA STATE UNIV, 75- Honors & Awards: Wood Award, 2nd, Forest Prod Res Soc, 72. Mem: Forest Prod Res Soc; Soc Wood Sci & Technol; Am Carbon Soc. Res: Physical and mechanical properties of natural occurring polymers and polymer base composites. Mailing Add: 106 Horseshoe Circle Pennsylvania Furnace PA 16865

BLANKENSHIP, FLOYD ALLEN, b Atlanta, Ga, Nov 21, 30. PHYSICAL CHEMISTRY. Educ: Univ Ga, BS, 57; Univ Ill, PhD(chem), 62. Prof Exp: Chemist, Celanese Chem Corp, 57; res chemist, Rohm & Haas Co, 62-66; asst prof, 66-68, ASSOC PROF CHEM, TOWSON STATE COL, 68- Mem: Am Chem Soc. Res: Computer utilization in chemistry; electronic structures of transition metal complexes;

properties of non-ionic surfactants. Mailing Add: Dept of Chem Towson State Col Baltimore MD 21204

BLANKENSHIP, FORREST (FARLEY), b Gatesville, Tex, Oct 14, 13; m 34; c 2. PHYSICAL CHEMISTRY. Educ: Univ Tex, BA, 32, MA, 33, PhD(chem), 43. Prof Exp: Prof chem & physics, Paris Jr Col, 34-42; from asst prof to assoc prof chem, Univ Okla, 43-51; prin chemist, Oak Ridge Nat Lab, 51-58, sect chief reactor chem div, 58-62, sr sci adv, 62-73; RETIRED. Mem: Fel Am Chem Soc. Res: Hydrates of hydrocarbons; chemistry of molten salts; high temperature nuclear reactor materials; fission product behavior. Mailing Add: 627 Lake Shore Dr Kingston TN 37763

BLANKENSHIP, JAMES EMERY, b Sherman, Tex, Mar 19, 41; m 30; c 2. NEUROPHYSIOLOGY. Educ: Austin Col, BA, 63; Yale Univ, MS, 65, PhD(neurophysiol), 67. Prof Exp: Fel physiol, Med Sch, NY Univ, 67-69; NIH fel neurophysiol, NIMH, 69-70; ASST PROF PHYSIOL & MEM STAFF, DIV COMP MARINE NEUROBIOL, MARINE BIOMED INST, UNIV TEX MED BR GALVESTON, 70- Concurrent Pos: NIH res grant, Nat Inst Neurol Dis & Stroke, 70, 72, 73; Off Naval Res contract, 73. Mem: AAAS; Soc Neurosci; Am Physiol Soc. Res: Cellular electrophysiology in Aplysia studying neural basis of sexual behavior; comparative neurobiology, synaptic and interneuronal properties; influence of hyperbaric nitrogen tensions on neurons. Mailing Add: Marine Biomed Inst Univ of Tex Med Br Galveston TX 77550

BLANKENSHIP, JAMES LYNN, b Knoxville, Tenn, Mar 26, 31; m 51; c 4. EXPERIMENTAL SOLID STATE PHYSICS, SEMICONDUCTORS. Educ: Univ Tenn, BS, 54, MS, 55, PhD(physics), 73. Prof Exp: Circuit develop engr, 55-58, res engr, 58-65, educ assignment, 65-67, physicist, 68-72, SR RES STAFF MEM, OAK RIDGE NAT LAB, 72- Mem: Am Phys Soc. Res: Nuclear medicine image processing. Mailing Add: Oak Ridge Nat Lab Bldg 3500 Oak Ridge TN 37831

BLANKENSHIP, JAMES R, atmospheric physics, see 12th edition

BLANKENSHIP, JAMES W, b Lafayette, Ind, Jan 15, 28; m 51; c 3. BIOCHEMISTRY. Educ: Southern Missionary Col, BA, 51; Univ Ark, MS, 53; Univ Wyo, PhD(biochem), 69. Prof Exp: Instr biochem, Univ Wyo, 67-68, asst prof, 69-70; asst prof, 70-74, ASSOC PROF NUTRIT, SCH ALLIED HEALTH PROFESSIONS, LOMA LINDA UNIV, 74- Mem: Am Chem Soc; Am Oil Chem Soc. Res: Lipid metabolism in insects; minor lipids and their effects on normal and abnormal tissue development. Mailing Add: Dept of Nutrit Sch Allied Health Professions Loma Linda Univ Loma Linda CA 92354

BLANKENSHIP, JAMES WILLIAM, b San Diego, Calif, June 8, 43; m 68; c 2. PHARMACOLOGY. Educ: Tex A&M Univ, BS, 65, MS, 67; Univ Utah, PhD(pharmacol), 72. Prof Exp: Res assoc cell biol, Med Col Ga, 72-74; ASST PROF PHARMACOL, MED UNIV SC, 74- Res: Effects of drugs on the cell nucleus; processes controlling metabolism of nucleic acids; nuclear proteins and polyamines. Mailing Add: Dept of Pharmacol Med Univ of SC Charleston SC 29401

BLANKENSHIP, LYTLE HOUSTON, b Campbellton, Tex, Mar 1, 27; m 54; c 4. WILDLIFE RESEARCH. Educ: Tex A&M Univ, BS, 50; Univ Minn, MS, 52; Mich State Univ, PhD(physiol), 57. Prof Exp: Biologist, Game Div, Mich Dept Conserv, 54-56; res biologist, Minn Div Game & Fish, 56-61; wildlife biologist, Wildlife Res Br, Bur Sport Fisheries & Wildlife, US Fish & Wildlife Serv, 61-69; res scientist, Tex A&M Univ-E African Agr & Forestry Res Orgn, 69-72; assoc prof, 72-75, PROF, AGR EXP STA, TEX A&M UNIV, 75- Mem: Wildlife Soc; Wildlife Dis Asn; E Afica Wildlife Soc; Wildlife Protection & Conserv Soc S Africa; S African Wildlife Mgt Asn. Res: Ecology, behavior and diseases of doves, waterfowl, woodcock, moose and deer; physiology of deer; ecology, reproduction and meat production of large game animals in East Africa; range nutrition of plants and animals. Mailing Add: Agr Exp Sta Tex A&M Univ Box 1051 Uvalde TX 78801

BLANKENSTEIN, WILLIAM E, b Natchez, Miss, June 15, 26. ORGANIC CHEMISTRY. Educ: Auburn Univ, BS, 49, MS, 50; Univ Tex, PhD(org chem), 57. Prof Exp: Asst polymer chemist, Tenn Eastman Co, 50-53; res chemist, Carothers Res Lab, E I du Pont de Nemours & Co, 56-57; res chemist, Esso Res Lab, Standard Oil Co, 57-59; res chemist, Dacron Res Lab, 59-64, RES CHEMIST, TEXTILE RES LAB, E I DU PONT DE NEMOURS & CO, 64- Mem: Am Chem Soc. Res: Textile and polymer chemistry in industry; halogenated acetylenic and olefinic ethers and alcohols; quinoline compounds. Mailing Add: 1108 Arundel Dr Wilmington DE 19808

BLANKESPOOR, HARVEY DALE, b Boyden, Iowa, Oct 15, 39; m 64; c 2. INVERTEBRATE ZOOLOGY, PARASITOLOGY. Educ: Westmar Col, BA, 63; Iowa State Univ, MS, 67, PhD(invert zool), 70. Prof Exp: Asst prof biol, Univ Northern Iowa, 70-71 & Trinity Christian Col, 71-72; ASST PROF ZOOL, UNIV MICH, ANN ARBOR, 72- Honors & Awards: Chester A Herrick Award, Midwest Conf Parasitologists, 70. Mem: Am Soc parasitol; Am Micros Soc; Am Soc Trop Med & Hyg. Res: Host-parasite relationships of parasitic flatworms. Mailing Add: Mus of Zool Univ of Mich Ann Arbor MI 48109

BLANKESPOOR, RONALD LEE, b Hull, Iowa, Feb 22, 46; m 69; c 1. ORGANIC CHEMISTRY. Educ: Dordt Col, AB, 68; Iowa State Univ, PhD(org chem), 71. Prof Exp: Asst prof chem, Univ Wis-Oshkosh, 71-73; ASST PROF CHEM, WAKE FOREST UNIV, 73- Mem: Am Chem Soc; Sigma Xi. Res: Electron spin resonance studies of hydrocarbon radical anions. Mailing Add: Dept of Chem Wake Forest Univ Winston-Salem NC 27109

BLANKINSHIP, WILLIAM AUBREY, b McDowell, Va, Aug 18, 20; m 40; c 2. APPLIED MATHEMATICS. Educ: Univ Va, BA, 41; Princeton Univ, MA, 49, PhD(math), 49. Prof Exp: Jr scientist, US Rubber Co, 42-43; asst math, Univ Ill, 43-44; supvr mathematician, 54-74, CONSULT MATH & SPEECH ACOUST, NAT SECURITY AGENCY, 74- Mem: Am Math Soc. Res: Modern algebra; statistics and probability; computing systems; techniques for bandwidth compression for human speech. Mailing Add: 137 Spa View Ave Annapolis MD 21401

BLANKLEY, CLIFTON JOHN, b Chicago, Ill, Apr 21, 42. MEDICINAL CHEMISTRY. Educ: Stanford Univ, BS, 63; Mass Inst Technol, PhD(org chem), 67. Prof Exp: SR RES CHEMIST, PARKE DAVIS & CO, 67- Mem: Am Chem Soc; AAAS. Res: Synthesis of potential anti-infl-mmatory anti-inflammatory agents, anti-bacterial agents and anti-tumor agents; study of quantitative structure activity relationships. Mailing Add: Parke Davis & Co 2800 Plymouth Rd Ann Arbor MI 48106

BLANKS, JANET MARIE (CLARENBACH), b Berkeley, Calif, Sept 25, 44; m 67. NEUROCYTOLOGY. Educ: Humboldt State Col, BA, 66; Univ Calif, Los Angeles, PhD(anat), 73. Prof Exp: Feldman fel neurocytol, Jules Stein Eye Inst, Univ Calif, Los Angeles, 73-74; VIS SCIENTIST, MAX PLANCK INST BRAIN RES, 74- Mem: Asn Res Vision & Ophthal. Res: Correlation of biochemical and morphological changes which occur during synaptogenesis in the mammalian central nervous system; research involves light and electron microscopic autoradiography and Golgi light microscopy. Mailing Add: Max Planck Inst Brain Res Deutschordenstrasse 46 6 Frankfurt/M-Niederrad West Germany

BLANN, H MARSHALL, b Los Angeles, Calif, Aug 22, 35; m 59; c 1. NUCLEAR CHEMISTRY. Educ: Univ Calif, Los Angeles, BS, 57; Univ Calif, Berkeley, PhD(chem), 60. Prof Exp: From instr to assoc prof, 60-74, PROF CHEM, UNIV ROCHESTER, 74- Concurrent Pos: Res Corp grant, 60- Mem: Am Phys Soc. Res: Nuclear reaction mechanisms at moderate excitation energy; nuclear fission. Mailing Add: Dept of Chem Univ of Rochester Rochester NY 14627

BLANPIED, GEORGE DAVID, b Ridgewood, NJ, June 29, 30; m 52; c 3. POMOLOGY, PHYSIOLOGY. Educ: Dartmouth Col, BA, 52; Cornell Univ, MS, 55; Mich State Univ, PhD(pomol), 59. Prof Exp: Asst prof, 55-57 & 59-64, ASSOC PROF POMOL, CORNELL UNIV, 65- Concurrent Pos: Vis prof, Univ Dublin, 65-66 & Agr Can, 74-75. Mem: Am Soc Hort Sci; Am Soc Plant Physiologists. Res: Biological and commercial aspects of fruit maturity, storage, handling and marketing. Mailing Add: Dept of Pomol Cornell Univ Ithaca NY 14850

BLANPIED, WILLIAM ANTOINE, b Rochester, NY, May 11, 33; m 59; c 1. SCIENCE POLICY, HISTORY OF SCIENCE. Educ: Yale Univ, BS, 55; Princeton Univ, PhD(physics), 59. Prof Exp: Instr physics, Princeton Univ, 58-59; NSF fel, Synchrotron Lab, Frascati, Italy, 59-60; instr, Yale Univ, 60-62, asst prof, 62-66; assoc prof, Case Western Reserve Univ, 66-69; staff scientist,. NSF, New Delhi, India, 69-71; assoc prof, Case Western Reserve Univ, 71-72; sr res fel, Harvard Univ, 72-74; HEAD DIV PUB SECTOR PROGS, AAAS, 74- Concurrent Pos: Yale Univ jr fac fel, Cambridge Electron Accelerator, Harvard Univ, 63-64. Mem: Am Phys Soc; Am Hist Sci Soc; AAAS. Res: Science curriculum development; science and public policy; public understanding of science; scientific freedom and responsibility; history of science in Asia. Mailing Add: AAAS 1776 Massachussetts Ave NW Washington DC 20036

BLANQUET, RICHARD STEVEN, b Brooklyn, NY, Jan 17, 40; m 67. INVERTEBRATE PHYSIOLOGY, BIOCHEMISTRY. Educ: City Col New York, BS, 61; Duke Univ, PhD(zool), 66. Prof Exp: Res assoc zool, Lab Quant Biol, Univ Miami, 65-67; instr physiol, 67-69, asst prof, Dept Biol, 69-73, ASSOC PROF BIOL, GEORGETOWN UNIV, 73- Mem: AAAS; Am Soc Zool. Res: Physiology and biochemistry of lower invertebrates; problems of coelenterate nematocyst discharge and chemical nature of capsule and enclosed toxin; adaptation to environmental stress. Mailing Add: Dept of Biol Georgetown Univ Washington DC 20057

BLANTON, CHARLES DEWITT, JR, b Kings Mountain, NC, Jan 11, 37; m 59; c 1. ORGANIC CHEMISTRY, MEDICINAL CHEMISTRY. Educ: Western Carolina Col, BS, 59; Univ Miss, PhD(org chem), 63. Prof Exp: Fel, Dept Chem, Ind Univ, 63-64; assoc res prof pharmaceut & med chem, Sch Pharm, Auburn Univ, 64-68; assoc prof med chem, 68-75, PROF MED CHEM, SCH PHARM, UNIV GA, 75- Mem: AAAS; Am Chem Soc; Am Pharmaceut Asn; Int Soc Heterocyclic Chem; The Chem Soc. Res: Synthesis and study of structural-activity relationships of folic acids; heteroaromatic compounds; Mannich bases. Mailing Add: Sch of Pharm Univ of Ga Athens GA 30601

BLANTON, FRANKLIN SYLVESTER, b Jones Mill, Ala, Dec 12, 02; m 43; c 3. MEDICAL ENTOMOLOGY. Educ: Cornell Univ, PhD, 51. Prof Exp: Entomologist, Bur Entom & Plant Quarantine, USDA, Med Dept, US Dept Army, 42-56; prof, 56-74, EMER PROF ENTOM, UNIV FLA, 74- Mem: AAAS; fel Entom Soc Am. Res: Taxonomy of Diptera of medical importance; virus transmission; control of insects by chemicals; thermal treatments; ecology of yellow fever mosquitoes. Mailing Add: Dept of Entom Univ of Fla Gainesville FL 32601

BLANTON, JACKSON ORIN, b Atlanta, Ga, Oct 28, 39; m 62; c 2. PHYSICAL OCEANOGRAPHY, LIMNOLOGY. Educ: Univ Fla, BSCE, 62; Ore State Univ, MS, 64, PhD(oceanog), 68. Prof Exp: Consult coastal eng, Marine Adv Inc, 64-65; res assoc, Univ Fla, 65-66; asst prof phys oceanog, Marine Lab, Duke Univ, 68-70; res scientist phys limnol, Can Ctr Inland Waters, 70-74; PROG MGR PHYS OCEANOG, US ENERGY RES & DEVELOP ADMIN, 74- Concurrent Pos: Consult phys oceanog, Res Triangle Inst, NC, 68-69. Mem: Am Geophys Union; Oceanog Soc Japan; Int Asn Great Lakes Res; Int Soc Limnol. Res: Energy dissipation in lakes and estuaries; processes of upwelling and water mass exchanges; maintenance and dissipation of thermoclines or frontal zones in oceans and lakes; physical nearshore processes. Mailing Add: Div of Biomed & Environ Res US Energy Res & Develop Admin Washington DC 20545

BLANTON, MADISON VAN, b Wallins Creek, Ky, Oct 14, 37; m 62; c 2. BIOCHEMISTRY, BIOLOGY. Educ: ETenn State Univ, BS, 61, MA, 63; Va Polytech Inst, PhD(biochem), 69. Prof Exp: Sr res chemist, Anheuser-Busch, Inc, 68-71; res chemist, Royal Bond Inc, 72; SR FOOD CHEMIST, RALSTON PURINA CO, 72- Res: Protein chemistry, isolation and chemical modification of proteins for industrial and food use; nutrition as applied to protein supplements to foods. Mailing Add: Ralston Purina Co 900 Checkerboard Plaza St Louis MO 63188

BLANTON, PATRICIA LOUISE, b Clarksville, Tex, July 9, 41. GROSS ANATOMY, PERIODONTICS. Educ: Hardin-Simmons Univ, BA, 62; Baylor Univ, MS, 64, PhD(anat), 67; Baylor Dent Col, DDS, 74, cert periodont, 75. Prof Exp: From asst prof to assoc prof gross anat, 67-75, PROF GROSS ANAT & PERIODONT, BAYLOR DENT COL, 75- Concurrent Pos: Mem, Nat Adv Coun Health Prof Educ, NIH, 73-74. Honors & Awards: First Place Award, Student Clinicians, Am Dent Asn, 72. Mem: Am Asn Anatomists; Am Acad Periodont; Int Soc Electromyographic Kinesiology; Am Dent Asn; Sigma Xi. Res: Electromyographic evaluation of the human head and neck musculature; masticatory under normal conditions and relative to temporomandibular joint dysfunction and occlusal discrepancies and adjustments. Mailing Add: Dept of Gross Anat Baylor Dent Col Dallas TX 75226

BLANTON, RICHARD EDWARD, b Cheyenne, Wyo, Nov 16, 43. ANTHROPOLOGY. Educ: Univ Mich, BA, 66, MA, 67, PhD(anthrop), 70. Prof Exp: Asst prof anthrop, Rice Univ, 70-72; ASST PROF ANTHROP, HUNTER COL, 72- Concurrent Pos: NSF grants archaeol, Oaxaca, Mex, 71- Mem: AAAS; Soc Am Archaeol; Am Anthrop Asn; Ecol Soc Am; World Pop Soc. Res: Cultural evolution in prehispanic Oaxaca, Mexico, with special emphasis on role of population processes in culture change and adaptation. Mailing Add: Dept of Anthrop Hunter Col 695 Park Ave New York NY 10021

BLANTON, WILLIAM GEORGE, b Bowie, Tex, Jan 9, 30; m 61; c 1. BIOLOGICAL OCEANOGRAPHY. Educ: Tex Wesleyan Col, BS, 58; NTex State Univ, MS, 59; Tex A&M Univ, PhD(biol oceanog), 66. Prof Exp: Res scientist, Off Naval Res Proj, Inst Marine Sci, Univ Tex, 59-61; dir corrosion res, Nutro Prod Corp, Tex, 62-65; Dow Chem Co co-investr, Off Saline Waters grant, Tex A&M Univ, 65-66; ASSOC PROF BIOL, TEX WESLEYAN COL, 66- Concurrent Pos: Dow Chem Co co-investr, Environ Studies Group, Tex Christian Univ, 67-69; co-investr, Ecol Study Cedar Bayou, US Steel Corp, 69; prin investr, Ecol Study Lavaca Bay, Alcoa Aluminum grant, 70-71; group leader, Environ Response Team, Oil & Hazardous Chem Div, Environ Protection Agency, 71-73; staff scientist, Biologische Anstalt Helgoland, 73-74. Mem: AAAS; Am Soc Microbiol; Am Soc Oceanog. Res: Microbial ecology; cell physiology; intermediate metabolism of marine bacteria. Mailing Add: Sci Div Tex Wesleyan Col Ft Worth TX 76105

BLASDELL, ROBERT FERRIS, b Kuala Lumpur, Malaya, May 12, 29; US citizen; m 63; c 2. SYSTEMATIC BOTANY, PLANT MORPHOLOGY. Educ: Ohio Wesleyan Univ, BA, 51; Univ Mich, MA, 54, PhD(bot), 59. Prof Exp: Asst prof biol, La State Univ, 59-64; asst prof, 64-69, ASSOC PROF BIOL, CANISIUS COL, 69- Mem: Am Fern Soc; Torrey Bot Club; Am Soc Plant Taxonomists; Bot Soc Am; Int Asn Plant Taxonomists. Res: Taxonomy and morphology of the vascular plants; fern genus Paesia. Mailing Add: Dept of Biol Canisius Col Buffalo NY 14208

BLASE, ERNEST FREDERICK, physical chemistry, radiation physics, see 12th edition

BLASER, HENRY WESTON, b Pitman, NJ, May 1, 10; m 48; c 2. BOTANY. Educ: Temple Univ, BS, 31, AM, 33; Cornell Univ, PhD(bot), 40. Prof Exp: From asst to instr biol, Temple Univ, 31-38; from asst to instr bot, Cornell Univ, 38-46; from asst prof to prof, 46-75, EMER PROF BOT, UNIV WASH, 75- Concurrent Pos: With USDA. Mem: AAAS; Bot Soc Am; Int Soc Plant Morphol. Res: Comparative morphology; anatomy of vascular plants. Mailing Add: Dept of Bot Univ of Wash Seattle WA 98105

BLASER, ROBERT U, b Akron, Ohio, Sept 1, 16; m 40; c 1. RESEARCH ADMINISTRATION, RESOURCE MANAGEMENT. Educ: Univ Akron, BS, 37. Prof Exp: Res engr, 37-47, chief eng physics sect, 47-51, chief nuclear eng sect, 49-51, asst supt mat dept, Res Ctr, 51-61, staff asst, 61-64, chief plans, Res & Develop Div, 64-72, TECH MGR ACCT, BABCOCK & WILCOX CO, 72- Honors & Awards: Charles B Dudley Medal, 57. Mem: Am Phys Soc; Am Soc Mech Engrs; Am Nuclear Soc; Soc Res Adminrs; Metric Asn. Res: Steam generating equipment; nuclear power; materials research; environmental testing; nuclear materials for fuel elements and structural parts; forecasting and planning of research, development, and allocation of resources. Mailing Add: 115 Vincent Blvd Alliance OH 44601

BLASER, ROY EMIL, b Duncan, Nebr, Mar 3, 12; m 35; c 1. AGRONOMY. Educ: Univ Nebr, BS, 34; Rutgers Univ, MS, 36; NC State Col, PhD(agron), 47. Prof Exp: Asst, Rutgers Univ, 34-35, instr, 35-36, asst, 36-37; from asst agronomist to agronomist, Exp Sta, Univ Fla, 37-46; actg prof agron, Cornell Univ, 46-47, prof, 47-49; prof, 49-72, UNIV PROF AGRON, VA POLYTECH INST & STATE UNIV, 72- Honors & Awards: Medallion Award, Am Forage & Grassland Coun. Mem: Am Soc Agron (pres, 70); Crop Sci Soc Am; Am Forage & Grassland Coun; Soil Sci Soc Am; Am Dairy Asn. Res: Physiology and ecology of production and utilization of grasses and legumes for pasture, silage, hay and turf; interrelations of plants, animals and climatic factors; management of turfgrass and highway cover; effects of de-iceing salts on soil and plants. Mailing Add: Dept of Agron Va Polytech Inst & State Univ Blacksburg VA 24601

BLASIE, J KENT, b Flint, Mich, Apr 30, 43; m 65. BIOPHYSICS, PHYSICS. Educ: Univ Mich, BS, 64, PhD(biophys), 68. Prof Exp: Asst prof, 68-72, ASSOC PROF BIOPHYS, UNIV PA, 72- Concurrent Pos: NIH fel, 64-68, career develop award, 71-76; assoc biophysicist, Brookhaven Nat Lab, 73- Mem: Biophys Soc. Res: Structural and dynamical study of natural and artificial biological membrane systems via x-ray and neutron diffraction and nuclear magnetic resonance and the forces responsible for their structure and interaction. Mailing Add: Dept of Biochem & Biophys Univ of Pa Philadelphia PA 19104

BLASING, TERENCE JACK, b Waukesha, Wis, Dec 16, 43; m 67; c 2. CLIMATOLOGY. Educ: Univ Wis-Madison, BS, 66, MS, 68, PhD(meteorol), 75. Prof Exp: RES ASSOC DENDROCHRONOLOGY, LAB TREE-RING RES, UNIV ARIZ, 71- Mem: Am Meteorol Soc; Asn Am Geogrs. Res: Large scale climatic variations and long range forecasting; statistical meteorology; paleoclimatology; bioclimatology and agricultural climatology. Mailing Add: Lab of Tree-Ring Res Univ of Ariz Tucson AZ 85721

BLASINGHAM, MARY CYNTHIA, b Indianapolis, Ind, Aug 27, 48. PHYSIOLOGY. Educ: Duke Univ, BA, 70; Ind Univ, MS, 72, PhD(physiol), 76. Prof Exp: Asst, 71-76, RES ASSOC PHYSIOL, IND UNIV MED CTR, 76- Res: Metabolism of prostaglandin E by the lung during hemorrhagic shock; renal, pulmonary and cardiovascular physiology. Mailing Add: Dept of Physiol Ind Univ Med Ctr 1100 W Michigan St Indianapolis IN 46202

BLASS, ELLIOTT MARTIN, b New York, NY, Sept 10, 40; m 76; c 2. PSYCHOPHYSIOLOGY, ANIMAL BEHAVIOR. Educ: Brooklyn Col, BS, 63; Univ Conn, MS, 64; Univ Va, PhD(psychol), 67. Prof Exp: Fel zool, Univ Pa, 67-69; asst prof, 69-74, ASSOC PROF PSYCHOL, JOHNS HOPKINS UNIV, 74- Concurrent Pos: Consult ed, J Comp & Physiol Psychol, 74- Mem: AAAS; Neurosci Soc; Psychonomic Soc; Develop Psychobiol Soc. Res: Development of ingestive behavior in infants, neurology and physiology of ingestion; maternal-infant relationships. Mailing Add: Dept of Psychol Johns Hopkins Univ Baltimore MD 21218

BLASS, GERHARD ALOIS, b Chemnitz, Ger, Mar 12, 16; nat US; m 45; c 5. THEORETICAL PHYSICS. Educ: Univ Leipzig, PhD(theoret physics). Prof Exp: Sci collabr solid state theory, Siemens & Halske, Berlin, 43-45; lectr math & physics, OhmPolytechnicum, Nuremberg, 46-49; asst prof physics, Col St Thomas, 49-51; from asst prof to assoc prof, 51-61, PROF MATH PHYS & ASTROPHYS, UNIV DETROIT, 61- Concurrent Pos: Chmn dept physics, Univ Detroit, 62-69. Mem: AAAS; Am Phys Soc; Am Asn Physics Teachers; Math Asn Am. Res: Field theory; solid state physics; philosophy of science. Mailing Add: Dept of Physics Univ of Detroit Detroit MI 48221

BLASS, JOHN P, b Vienna, Austria, Feb 21, 37; US citizen; m 60; c 2. NEUROPSYCHIATRY, BIOCHEMISTRY. Educ: Harvard Univ, AB, 58; Univ London, PhD(biochem), 60; Columbia Univ, MD, 65. Prof Exp: Intern, Mass Gen Hosp, 65-66, asst resident med, 66-67; res assoc, Molecular Dis Br, Nat Heart Inst, 67-70; ASST PROF BIOCHEM & PSYCHIAT, SCH MED, UNIV CALIF, LOS ANGELES, 70- Mem: Am Soc Biol Chemists; Int Brain Res Orgn; Int Soc Neurochem; Am Fedn Clin Res; Soc Neurosci. Res: Metabolic disorders affecting the nervous system. Mailing Add: Ment Retardation Ctr Neuropsychiat Inst Univ of Calif Los Angeles CA 90024

BLASS, WILLIAM ERROL, b Minneapolis, Minn, Aug 5, 37; m 59; c 3. MOLECULAR SPECTROSCOPY, COMPUTER SCIENCE. Educ: St Mary's Col, Minn, BA, 59; Mich State Univ, PhD(physics), 63. Prof Exp: Asst prof physics, St Mary's Col, Minn, 63-67; asst prof, 67-71, ASSOC PROF PHYSICS & ASTRON, UNIV TENN, KNOXVILLE, 71- Concurrent Pos: Co-prin investr, AEC res contract, 64-67; prin investr, Univ Tenn-NASA grant, 68-70; mem sr sci staff, NASA grant, 69- Mem: Am Phys Soc; Am Asn Physics Teachers. Res: Theoretical and experimental molecular structure and spectroscopy, especially accidental and essential resonance phenomena; application of real time computers in physics; methods of data analysis in the batch processor environment. Mailing Add: Dept of Physics & Astron Univ of Tenn Knoxville TN 37916

BLATCHFORD, JOHN KERSLAKE, b Chicago, Ill, Sept 6, 25; m 60; c 1. ORGANIC CHEMISTRY. Educ: Univ Colo, BA, 49, MA, 56; Univ Cincinnati, PhD(org chem), 63. Prof Exp: Res chemist, Chicago Copper & Chem Co, 50-51; chief chemist, Edcan Labs, 57-61; res chemist, 63-68, staff chemist, Res & Eng Ctr, 68-71, STAFF SCIENTIST, WHIRLPOOL CORP, 71- Mem: AAAS; Am Chem Soc; Sigma Xi; Int Soc Hybrid Microelectronics. Res: Organometallic compounds; polycyclic-aromatic hydrocarbons; organic synthesis; cyclopropanes; carbene chemistry; snake venoms; detergency; thick film hybrid electronics; polymer synthesis; chromatography. Mailing Add: Whirlpool Corp Res & Eng Ctr Monte Rd Benton Harbor MI 49022

BLATCHLEY, DONALD E, b Chicago, Ill, Feb 5, 33; m 57; c 3. NUCLEAR PHYSICS. Educ: DePauw Univ, BA, 55; Univ Ill, MS, 57; Ind Univ, PhD(physics), 64. Prof Exp: Assoc scientist, Bettis Atomic Power Div, Westinghouse Elec Corp, 57-59; res assoc nuclear physics, Univ Iowa, 64-65 & Argonne Nat Lab, 65-67; MEM TECH STAFF, AEROSPACE CORP, 67-; FEL, JOHNSTON COL, UNIV REDLANDS, 70- Res: Low energy nuclear physics. Mailing Add: Dept of Physics Johnston Col Univ Redlands Redlands CA 92373

BLATSTEIN, IRA M, b Hackensak, NJ, Apr 18, 44; m 68; c 2. UNDERWATER ACOUSTICS. Educ: Drexel Univ, BS, 67; Cath Univ Am, MS, 71, PhD(physics), 73. Prof Exp: RES PHYSICIST UNDERWATER EXPLOSION EFFECTS, WHITE OAK LAB, NAVAL SURFACE WEAPONS CTR, 67- Mem: Am Phys Soc; Acoustical Soc Am. Res: Long range propagation of underwater explosion energy; normal mode theory; ray theory and modified ray theory; ocean basin reverberation. Mailing Add: Code WR-14 White Oak Lab Naval Surface Weapons Ctr Silver Spring MD 20910

BLATT, ALBERT HAROLD, b Cincinnati, Ohio, Jan 9, 03; m 35; c 1. ORGANIC CHEMISTRY. Educ: Harvard Univ, BS, 23, AM, 25, PhD(chem), 26. Prof Exp: Assoc, Harvard Univ, 26-28; asst to mgr, Columbus-McKinnon Chain Co, 28-30; mem fac, Univ Buffalo, 30-32; assoc prof chem, Howard Univ, 32-38; from assoc prof to prof, 39-75, chmn dept, 61-68, EMER PROF CHEM, QUEENS COL, NY, 75-; PROF CHEM, FLA INST TECHNOL, 75- Concurrent Pos: Sci liaison officer, London Mission, Off Sci Res & Develop, 44-45, tech aide, Div 8, 45. Mem: Am Chem Soc. Mailing Add: Dept of Chem Fla Inst of Technol Melbourne FL 32901

BLATT, ELIZABETH KEMPSKE, b Baltimore, Md, Dec 2, 36; m 58; c 3. MEDICAL PHYSIOLOGY, NEUROENDOCRINOLOGY. Educ: Goucher Col, BA, 56; Vassar Col, MA, 58; Union Univ, PhD(life sci & syst), 71. Prof Exp: Instr physiol, Mt Holyoke Col, 58-59; res asst zool, Kans State Univ, 64-65; from instr to asst prof biol, Concord Col, 67-73; ASSOC PROF PHYSIOL & CHMN DEPT, WVA SCH OSTEOP MED, 74- Concurrent Pos: Fac fel, US Dept Health, Educ & Welfare, 69-70; grant-in-aid, Sigma Xi, 70; consult health planning, Appalachian Regional Comn, 73-74. Mem: AAAS; Fedn Am Scientists; Asn Women Sci; Sigma Xi; assoc mem Am Osteop Asn. Res: Influence of pineal on reproductive rhythmicity; electrophysiological correlates of melatonin activity. Mailing Add: WVa Sch of Osteop Med 400 N Lee St Lewisburg WV 24901

BLATT, FRANK JOACHIM, b Vienna, May 1, 24; nat US; m 46; c 2. PHYSICS. Educ: Mass Inst Technol, BSEE, 46, MSEE, 48; Univ Wash, PhD(physics), 53. Prof Exp: Asst elec eng, Mass Inst Technol, 46-48; assoc, Univ Wash, 48-50, asst physics, 50-53; res assoc, Univ Ill, 53-55, res asst prof, 55-56; from asst prof to assoc prof, 56-61, chmn dept, 69-73, PROF PHYSICS, MICH STATE UNIV, 61- Concurrent Pos: Consult, Naval Res Lab, 55-59, Eng Res Inst, Univ Mich, 58-59 & Boeing Aircraft Co, 59; NSF fel, Oxford Univ, 59-60; consult, Argonne Nat Lab, 61-64, E I du Pont de Nemours & Co, Inc, 62- & Argonne Nat Lab, 67-75; guest prof, Swiss Fed Inst Technol, 63-64; mem prog rev comt, Solid State Sci Div, Argonne Nat Lab, 66-72. Mem: Am Phys Soc; Biophys Soc. Res: Physics of solids; transport theory; thermoelectric and thermomagnetic effects. Mailing Add: Dept of Physics Mich State Univ East Lansing MI 48823

BLATT, HARVEY, b New York, Sept 11, 31; m 56; c 3. PETROLOGY. Educ: Ohio State Univ, BS, 52; Univ Tex, MA, 58; Univ Calif, Los Angeles, PhD(geol), 63. Prof Exp: From instr to assoc prof geol, Univ Houston, 62-68; ASSOC PROF GEOL, UNIV OKLA, 68- Concurrent Pos: Fulbright sr fel, 74-75. Mem: Geol Soc Am; Soc Econ Paleont & Mineral; Int Asn Sedimentol. Res: Sedimentary petrology; petrography and genesis of sandstones, limestones, dolomites and cherts; provenance studies; petrographic stratigraphy; particle size distributions of sediments. Mailing Add: Sch of Geol & Geophys Univ of Okla Norman OK 73069

BLATT, IRVING MYRON, b New York, NY, May 31, 25; m 49; c 3. OTOLARYNGOLOGY. Educ: Cornell Univ, AB, 49, MD, 52; Univ Mich, MS, 56. Prof Exp: Instr otolaryngol, Med Sch, Univ Mich, 57-58; from asst prof to assoc prof otorhinol, Sch Med, La State Univ, 60-65, prof & head dept, 65-70, dir dept educ & res, 63-65; NAT CONSULT OTOLARYNGOL, SURGEON GEN, US AIR FORCE, 68- Concurrent Pos: Consult, US Vet Admin Hosp, 64- & Tex Tech Univ Sch Med. Honors & Awards: Am Acad Ophthal & Otolaryngol Res Award, 59. Mem: Fel Am Acad Ophthal & Otolaryngol; fel Am Fedn Clin Res; fel Am Col Surg; fel Am Laryngol, Rhinol & Otol Soc. Res: Otorhinolaryngology; biochemistry and mineralogy of salivary gland disease; vestibular pathophysiology; temporal bone; cochlear pathology. Mailing Add: 410 W Second St Thibodaux LA 70301

BLATT, JEREMIAH LION, b Ft Worth, Tex, Oct 27, 20; m 59; c 3. MOLECULAR BIOLOGY, SCIENCE EDUCATION. Educ: Univ Calif, AB, 43; Univ Calif, Los Angeles, PhD(biochem), 55. Prof Exp: Assoc, Agr Exp Sta, Univ Calif, 45-47; res assoc & instr biochem, Univ Chicago, 55-57; asst prof physiol, Mt Holyoke Col, 57-59; asst prof biophys, Kans State Univ, 59-67; PROF & CHMN DIV NATURAL SCI, CONCORD COL, 67- Mem: AAAS; Am Chem Soc. Res: Theory of chromatography; protein structure and biosynthesis; information transfer in biological systems; general education in natural science. Mailing Add: Div of Natural Sci Concord Col Athens WV 24712

BLATT, JOEL HERMAN, b Washington, DC, Mar 30, 38; m 68. SPECTROSCOPY, INSTRUMENTATION. Educ: Harvard Univ, BA, 59; Univ Ala, Huntsville, MS, 67; Univ Ala, Tuscaloosa, PhD(physics), 70. Prof Exp: Physicist, Missile Support Command Calibration Ctr, Redstone Arsenal, US Army, 62-64, res physicist, Missile Command Propulsion Lab, 64-66; sr scientist, Hayes Int Corp, Ala, 66-67; sr res asst physics, Univ Ala, Huntsville, 68-70; asst prof, 70-74, ASSOC PROF PHYSICS, FLA INST TECHNOL, 74- Mem: Am Asn Physics Teachers; Nat Speleol Soc; Optical Soc Am. Res: Mass spectrometry of organic radicals; optical spectroscopy and fluorometry; radiometry; general instrumentation in optical, oceanographic and astronomy areas. Mailing Add: Dept of Physics & Space Sci Fla Inst of Technol Melbourne FL 32901

BLATT, JOEL MARTIN, b Brooklyn, NY, Oct 28, 42. BIOCHEMISTRY. Educ: Brooklyn Col, BS, 63; Univ Wis-Madison, PhD(physiol chem), 68. Prof Exp: NIH fel, Purdue Univ, West Lafayette, 68-70, res assoc molecular biol, 70-72; ASST PROF BIOCHEM, MEHARRY MED COL, 72- Concurrent Pos: NSF res grant, Meharry Med Col, 74-75. Mem: AAAS; Am Chem Soc; Am Inst Biol Sci; Am Soc Microbiol. Res: Interlocking control mechanisms for gene expression and enzyme activity; regulation of branched-chain amino acid biosynthesis in bacteria. Mailing Add: Dept of Biochem & Nutrit Meharry Med Col Nashville TN 37208

BLATT, S LESLIE, b Philadelphia, Pa, June 10, 35; m 59; c 3. NUCLEAR PHYSICS. Educ: Princeton Univ, AB, 57; Stanford Univ, MS, 59, PhD(physics), 65. Prof Exp: Res assoc physics, Van de Graaff Accelerator Lab, 64-66, from asst prof to assoc prof, 66-75, PROF PHYSICS, OHIO STATE UNIV, 75- Concurrent Pos: Vis researcher, Nuclear Res Ctr, Strasbourg, France, 72-73. Mem: Am Phys Soc; Inst Elec & Electronics Eng. Res: Studies of charged particle radiative capture reactions and other nuclear reactions; gamma-ray detector systems; design and applications of nuclear instrumentation and computer-based data acquisition systems. Mailing Add: Dept of Physics Ohio State Univ Columbus OH 43210

BLATT, SYLVIA, b New York, NY, July 10, 18. CLINICAL CHEMISTRY, SCIENCE ADMINISTRATION. Educ: Hunter Col, AB, 38; City Col New York, MS, 40. Prof Exp: Asst chemist, City Col New York, 43-46 & Morrisania City Hosp, 46-49; chemist, Bd of Water Supply Lab, New York, 49-53; supvr biochem unit, 53-60, asst chief div supv clin labs & blood banks, 60-62, chief div lab field serv, 62-71, ASST DIR LAB IMPROV, BUR LABS, NEW YORK CITY DEPT HEALTH, 71- Concurrent Pos: Consult, Prof Exam Serv, Am Pub Health Asn, 65- Mem: Am Asn Clin Chemists; Am Chem Soc; Am Pub Health Asn. Res: Clinical chemistry, especially the technical and administrative aspects; quality control and proficiency testing programs in clinical chemistry; laboratory improvement programs. Mailing Add: New York City Dept of Health 455 First Ave New York NY 10016

BLATTEIS, CLARK MARTIN, b Berlin, Ger, June 25, 32; nat US; m 58; c 3. PHYSIOLOGY. Educ: Rutgers Univ, BA, 54; Univ Iowa, MSc, 55, PhD(physiol), 57. Prof Exp: Asst physiol, Univ Iowa, 54-55; res physiologist, US Army Res Inst Environ Med, 63-66; assoc prof, 55-74, PROF PHYSIOL & BIOPHYS, CTR FOR HEALTH SCI, UNIV TENN, MEMPHIS, 74- Concurrent Pos: USPHS res fel, Peruvian Inst Andean Biol, 61-62 & Nuffield Inst Med Res, Oxford, 62-63; Orgn Am States exchange sci award, 68; vis prof physiol Sch Med, Univ Guanajuato, 68-; Fulbright-Hays sr scholar, 75. Mem: AAAS; Am Heart Asn; Am Physiol Soc; Soc Exp Biol & Med; hon mem Peruvian Soc Physiol Sci. Res: Energy metabolism; temperature regulation; altitude. Mailing Add: Dept of Physiol & Biophys Univ of Tenn Ctr for Health Sci Memphis TN 38163

BLATTI, STANLEY PARRIS, b Kasson, Minn, Apr 28, 42; m 62; c 2. BIOCHEMISTRY, ANIMAL VIROLOGY. Educ: Univ Minn, Minneapolis, BA, 64; Mich State Univ, PhD(biochem), 68. Prof Exp: NIH fels animal virol, Albert Einstein Col Med, 68-69; NIH fels molecular biol, Sch Med, Univ Calif, San Francisco, 69-71; ASST PROF BIOCHEM, UNIV TEX MED SCH HOUSTON, 71- Mem: Am Chem Soc. Res: Molecular basis of transcription and its regulation in eucaryotic cells; multiple forms of eukaryotic DNA-dependent RNA polymerase; eukaryotic sigma factors. Mailing Add: Dept of Biochem & Molecular Biol Univ of Tex Med Sch Houston TX 77025

BLATTLER, DELBERT PAUL, b Wheeling, WVa, May 24, 38; m; c 4. CLINICAL BIOCHEMISTRY. Educ: Univ Calif, Berkeley, AB, 61; Univ Ore, PhD(chem), 67; Am Bd Clin Chem, cert, 75. Prof Exp: Fel chem, Okla State Univ, 68, fel physics, 69; chief clin & metab lab, Idaho State Sch & Hosp, 69-71; chief chemist, Prov Lab of Sask, 71-73; ASST DIR, BIO-SCI LAB, VAN NUYS, 74- Mem: Am Asn Clin Chemists; Can Soc Clin Chemists; Am Soc Microbiol; Am Chem Soc; Soc Study Social Biol. Res: Clinical chemistry; enzymology; protein chemistry. Mailing Add: Bio-Sci Labs 7600 Tyrone Ave Van Nuys CA 91405

BLATTNER, FREDERICK RUSSELL, b St Louis, Mo, Oct 6, 40; m 71. BIOPHYSICS. Educ: Oberlin Col, BA, 62; Johns Hopkins Univ, PhD(biophys), 68. Prof Exp: RES ASSOC BIOPHYS, McARDLE LAB, UNIV WIS-MADISON, 68- Mailing Add: Lab of Genetics Univ of Wis Madison WI 53706

BLATTNER, MEERA MCCUAIG, b Chicago, Ill, Aug 14, 30; m 56; c 3. COMPUTER SCIENCES. Educ: Univ Chicago, BA, 52; Univ Southern Calif, MA, 66; Univ Calif, Los Angeles, PhD(eng), 73. Prof Exp: Asst prof math, Los Angeles Harbor Col, 66-73; instr comput sci, Univ Mass & res fel comput sci, Harvard Univ, 73-74; ASST PROF MATH SCI, RICE UNIV, 74- Mem: Asn Comput Mach; Am Math Soc; Inst Elec & Electronics Engrs; Soc Indust & Appl Math. Res: Formal languages and automata theory and theoretical computer science. Mailing Add: Dept of Math Sci Rice Univ Houston TX 77001

BLATTNER, RUSSELL JOHN, b St Louis, Mo, July 3, 08; m 39; c 2. PEDIATRICS. Educ: Univ Wash, AB, 29, MD, 33. Prof Exp: Path intern, Barnes Hosp, St Louis, 33-34; jr resident physician, St Louis Children's Hosp, 34-35, from asst resident physician to resident physician, 35-37; house physician, Princess Elizabeth of York Hosp, London, 37; from instr to assoc prof pediat, Sch Med, Univ Wash, 37-47; prof, 47-68, DISTINGUISHED SERV PROF PEDIAT & CHMN DEPT, BAYLOR COL MED, 68- Mem: Fel Am Acad Pediat; Am Pediat Soc; Soc Pediat Res; Am Soc Trop Med & Hyg. Res: Infectious disease with emphasis on virus epidemiology; congenital malformations. Mailing Add: Dept of Pediat Baylor Col of Med Houston TX 77025

BLATZ, PAUL E, b Pittsburgh, Pa, Aug 29, 23; m 65; c 5. BIOCHEMISTRY. Educ: Southern Methodist Univ, BS, 51; Univ Tex, PhD(chem), 55. Prof Exp: Sr res chemist org synthesis, Dow Chem Co, 55-59; res assoc polymer chemist, Socony Mobil Oil Co, 59-63; assoc prof chem, NMex Highlands Univ, 63-64; from assoc prof to prof, Univ Wyo, 64-71; chmn dept, 71-76, PROF CHEM, UNIV MO-KANSAS CITY, 71- Mem: AAAS; Am Chem Soc; Biophys Soc; Am Asn Biol Chemists. Res: Chemistry of vision, particularly the spectroscopy and chemistry of visual pigments, retinal and related polyenes. Mailing Add: Dept of Chem Univ of Mo 5100 Rockhill Kansas City MO 64110

BLATZ, PAUL J, physical chemistry, see 12th edition

BLAU, EDMUND JUSTIN, b New York, NY, June 2, 15; m 46; c 3. INFORMATION SCIENCE, SCIENTIFIC BIBLIOGRAPHY. Educ: Cornell Univ, BChem, 35; Univ Chicago, SM, 36; Ohio State Univ, PhD(phys chem), 53. Prof Exp: Teacher, Col Prep Sch, New Rochelle, NY, 40-41; chemist, Ind Ord Works, 41-42; phys chemist, Nat Bur Standards, 46-48 & 53-56; mem staff appl physics lab, 56-63, BIBLIOGRAPHER, APPL PHYSICS LAB, JOHNS HOPKINS UNIV, 63-, LIBR SYSTS ANALYST, 68- Mem: Am Soc Info Sci. Res: Infrared and Raman spectroscopy; molecular structure; bibliography. Mailing Add: Johns Hopkins Appl Physics Lab Johns Hopkins Rd Laurel MD 20810

BLAU, HAROLD, b New York, NY, July 4, 35; m 60; c 2. ANTHROPOLOGY. Educ: State Univ NY, BS, 56; Brooklyn Col, MA, 64; New Sch Social Res, PhD(anthrop), 69. Prof Exp: From instr to asst prof, 64-70, ASSOC PROF ANTHROP, NEW YORK CITY COMMUNITY COL, 70- Concurrent Pos: Res ethnologist, NY State Mus, 64-68; prof anthrop, New Sch Social Res, 64-72; asst prof nutrit res, New York Med Col, 69-70. Mem: Fel Am Anthrop Asn; Am Ethnol Asn; Am Folklore Soc; Soc Ethnohist. Res: American Indians; Iroquois and Onondaga; religion; mythology; nutrition; pediatrics; medical urban anthropology. Mailing Add: Dept of Anthrop New York City Commun Col Brooklyn NY 11201

BLAU, HENRY HESS, b Dayton, Ohio, May 16, 97; m 24; c 3. PHYSICAL INORGANIC CHEMISTRY. Educ: Carnegie Inst Technol, BS, 19, ChemE, 28; Mass Inst Technol, MS, 20; Univ Pittsburgh, PhD(phys chem), 35. Prof Exp: Macbeth-Evans Glass Co, Charleroi, Pa, 20-27, res dir, 27-36; res & mfg exec, Corning Glass Works, 36-41; vpres in chg res & mfg, Fed Glass Co, 41-63; prof glass sci, 44-68, EMER PROF GLASS SCI, OHIO STATE UNIV, 68- Concurrent Pos: Vpres, Fed Paper Bd Co, 58-63; mem bd dirs, Fed Glass Co, 48-63; mem ceramic data comt, Nat Res Coun, 54-55, chmn, 55-58, ceramic chem comt, 58-; adv ceramic comt, Int Critical Tables, 58-; Am rep, Int Glass Cong; consult. Honors & Awards: Toledo Award, 62; US Dept Defense Award, 46; Alumni Merit Award, Carnegie Inst Technol; S B Meyer Award, Am Ceramic Soc, 56, Bleininger Award, 64. Mem: Fel AAAS; Am Chem Soc; Nat Inst Ceramic Eng; Am Inst Chem Eng; fel Am Ceramic Soc (treas, 63-64). Res: Glass technology; refractories; opal glasses; heat and chemical resisting glasses; color filters; glass furnaces; glass blocks; infrared transmitting glasses and filters; heat treatment of glasses; glass structure; nucleation and crystallization of glasses; kinetics of glass formation. Mailing Add: 2841 S Dorchester Rd Columbus OH 43221

BLAU, HENRY HESS, JR, b Speers, Pa, Feb 16, 30; m 58; c 3. ATMOSPHERIC PHYSICS. Educ: Yale Univ, BS, 51; Ohio State Univ, MSc, 52, PhD(physics), 55. Prof Exp: Physicist, Arthur D Little, Inc, 55-66, leader physics sect, Res & Develop Div, 62-66; physicist, Space Sci Lab & group leader geophys, Missile & Space Div, Gen Elec Co, 66-67; physicist, Arthur D Little, Inc, 67-71; V PRES & DIR, ENVIRON RES & TECHNOL, INC, 71- Concurrent Pos: Trustee & sci dir, Mt Washington Observ, 70- Mem: Fel Optical Soc Am; Am Phys Soc; AAAS; Sigma Xi. Res: Optical physics; molecular spectroscopy; air quality sensors. Mailing Add: 48 Highland Circle Wayland MA 01778

BLAU, JULIAN HERMAN, b New York, NY, Dec 23, 17; m 48; c 2. MATHEMATICS. Educ: City Col New York, BS, 38; NY Univ, MS, 39; Univ NC, PhD(math), 48. Prof Exp: Instr math, Mass Inst Technol, 47-49; asst prof, Pa State Univ, 49-52; assoc prof, 52-62, PROF MATH, ANTIOCH COL, 62- Concurrent Pos: NSF fac fel, 66-67. Mem: Am Math Soc; Math Asn Am. Res: Mathematics in social science; theory of social choice. Mailing Add: Dept of Math Antioch Col Yellow Springs OH 45387

BLAU, LAWRENCE MARTIN, b New York, NY, Jan 6, 38; m 59; c 2. MEDICAL PHYSICS. Educ: Princeton Univ, BA, 59; Univ Rochester, MA, 63, PhD(physics), 65. Prof Exp: Res assoc physics, Columbia Univ, 65-69; ASSOC SCIENTIST, HOSP SPEC SURG & ASST PROF PHYSICS IN RADIOL, MED COL, CORNELL UNIV, 69-, DIR COMPUT SERV, HOSP SPEC SURG, 73- Mem: Soc Nuclear Med; Am Phys Soc; Am Asn Physics Teachers. Res: Investigation and application of physics, mathematics and computers in nuclear medicine; application of computer technology in hospitals. Mailing Add: Hosp for Spec Surg Res 535 E 70th St New York NY 10021

BLAU, MONTE, b New York, NY, June 17, 26; m 46. NUCLEAR MEDICINE. Educ: Polytech Inst Brooklyn, BS, 48; Univ Wis, PhD(chem), 52. Prof Exp: Res asst geochronomet lab, Yale Univ, 52-53; mem staff, Radioisotope Lab, Montefiore Hosp, 53-54; mem staff dept biochem res, Roswell Park Mem Inst, 54-57, mem staff dept nuclear med, 57-75; RES PROF NUCLEAR MED & BIOPHYSICS, STATE UNIV NY BUFFALO, 70-, CHMN DEPT NUCLEAR MED, 75- Mem: Am Asn Cancer Res; Am Soc Nuclear Med (pres, 72-73); Am Chem Soc. Res: Cancer; radiopharmaceuticals; radioisotope instrumentation. Mailing Add: Sch of Med State Univ of NY Buffalo NY 14214

BLAUFOX, MORTON D, b New York, NY, July 19, 34; m 59; c 3. INTERNAL MEDICINE, NUCLEAR MEDICINE. Educ: State Univ NY, MD, 59; Univ Minn, PhD(med), 64. Prof Exp: Asst, Peter Bent Brigham Hosp, 64-66; instr med & dir sect nuclear med, 66-68, asst prof radiol, 66-71, asst prof med, 68-72, ASSOC PROF RADIOL, ALBERT EINSTEIN COL MED, 71-, ASSOC PROF MED, 72- Concurrent Pos: Am Heart Asn advan res fel, 64-66; scholar, Harvard Col & res fel, Harvard Med Sch, 64-66; asst attend physician, Bronx Munic Hosp Ctr, 66-71, assoc attend physician, 71-72, attend physician, 72-; mem sect renal dis, Coun Circulation, mem, High Blood Pressure Res Coun & Coun Cardiovasc Radiol, Am Heart Asn; mem panel radiopharmaceut, Comt on Scope, US Pharmacopeia; mem hypertension adv comt, New York Dept Health. Honors & Awards: Edgar Cigelman Award, 52. Mem: Am Fedn Clin Res; Soc Nuclear Med; Am Soc Nephrology; Am Soc Artificial Internal Organs; Int Soc Nephrology. Res: Nephrology. Mailing Add: Albert Einstein Col of Med Yeshiva Univ Bronx NY 10461

BLAUG, SEYMOUR MORTON, b New York, Dec 21, 23; m 53; c 4. PHARMACY. Educ: Columbia Univ, BS, 50, MS, 52; Univ Iowa, PhD(pharmacy), 55. Educ: From instr to prof pharm, Univ Iowa, 54-74; PROF PHARM & DEAN SCH PHARM, UNIV NC, CHAPEL HILL, 74- Concurrent Pos: Mem, US Pharmacopeia Rev Comt, 60-70. Honors & Awards: Borden Award, 49; Abbott Res Award, 69. Mem: Am Pharmaceut Asn; NY Acad Sci; fel Acad Pharmaceut Sci. Res: Physical-chemical phenomena that affect pharmaceutical systems; stability studies; flow properties of pharmaceutical systems; formulation; emulsion. Mailing Add: Sch of Pharm Beard Hall Univ of NC Chapel Hill NC 27514

BLAUMANIS, OTIS RUDOLF, b Riga, Latvia, Mar 22, 39; US citizen. PHYSIOLOGY, BIOPHYSICS. Educ: Johns Hopkins Univ, BA, 65, PhD(physiol, biophys), 70. Prof Exp: ASST PROF PHYSIOL, SCH MED, UNIV MD, BALTIMORE CITY, 70- Mem: AAAS; NY Acad Sci; Soc Neurosci. Res: Physiology of the central nervous system; cerebrovascular disease; structure and function of the cerebral cortex; pathophysiology of epilepsy; biomedical instrumentation. Mailing Add: 660 W Redwood St Baltimore MD 21201

BLAUNSTEIN, ROBERT P, b New York, NY, Nov 9, 39; m 62; c 2. RESEARCH ADMINISTRATION. Educ: City Col New York, BS, 61; Western Reserve Univ, MS, 64; Univ Tenn, PhD(physics), 68. Prof Exp: Res scientist radiation physics, Oak Ridge Nat Lab, 68; asst prof physics, Univ Tenn, Knoxville, 68-73; TECH REP, US ENERGY RES & DEVELOP ADMIN, 73- Concurrent Pos: Consult, Oak Ridge Nat Lab, 68- Mem: Am Phys Soc; Radiation Res Soc; Sigma Xi. Res: Interaction of low energy electrons with polyatomic molecules; electron attachment and scattering processes; interaction of ultra-violet light with polyatomic molecules; excimer formation inter- and intra-molecular energy transfer. Mailing Add: Div of Biomed & Environ Res US Energy Res & Develop Admin Washington DC 20545

BLAUSTEIN, BERNARD DANIEL, b Philadelphia, Pa, Apr 26, 29; m 54; c 3. PHYSICAL CHEMISTRY. Educ: Univ Pa, BSChE, 50; Johns Hopkins Univ, MA, 51, PhD(chem), 57. Prof Exp: Jr instr chem, Johns Hopkins Univ, 50-56; instr, Sch Pharm, Univ Md, 55-56; asst technologist, Res Ctr, US Steel Corp, 56-58; supvry phys chemist, Bur Mines, US Dept Interior, 58-62, res chemist, 62-68, supvry res chemist, 68-75; STAFF SCIENTIST, PITTSBURGH ENERGY RES CTR, US ENERGY RES & DEVELOP ADMIN, 75- Concurrent Pos: Instr, Univ Pittsburgh, 57-70, adj assoc prof, 70- Mem: AAAS; Am Chem Soc. Res: Separation and identification of organic mixtures; chemistry of coal; production of chemicals from coal. Mailing Add: Pittsburgh Energy Res Ctr US Energy Res & Develop Admin Pittsburgh PA 15213

BLAUSTEIN, ERNEST HERMAN, b Boston, Mass, June 29, 21; m 42; c 2. MICROBIOLOGY. Educ: Boston Col, AB, 51; Mass Inst Technol, MPH, 42; Boston Univ, PhD(microbiol), 52. Prof Exp: From instr to assoc prof, 46-57, chmn dept sci, 64-69, sci adv, Col Lib Arts, 69-73, PROF BIOL, BOSTON UNIV, 57-, ASSOC DEAN, DIV GEN EDUC, COL LIB ARTS, 69-, COORDR, SIX YR LIB ARTS MED EDUC PROG, 73- Concurrent Pos: Spec res fel, NIH, 60-61. Mem: AAAS; fel Am Acad Microbiol; Am Soc Microbiol; Am Pub Health Asn; Tissue Cult Asn. Res: Virology; tissue culture; medical microbiology. Mailing Add: Dept of Biol Boston Univ Boston MA 02215

BLAUT, JAMES MORRIS, b New York, NY, Oct 20, 27; m 64. AGRICULTURAL GEOGRAPHY. Educ: Univ Chicago, PhB, 48, BSc, 50; Imp Col Trop Agr, BWI, cert, 49-50; La State Univ, MSc, 54, PhD, 58. Prof Exp: Instr geog, Univ Malaya, 51-53; from instr to asst prof, Yale Univ, 56-61; field dir inter-Am prog adv social sci training, Pan-Am Union-Orgn Am States, Puerto Rico, 61-63; consult agr res, Govt Venezuela, 63-64; consult, social sci res, UNESCO, Dominican Repub, 64-65; dir Caribbean Res Inst, Col Virgin Islands, 65-66; prof geog, Clark Univ, 66-74; PROF GEOG GRAD COL, UNIV ILL, CHICAGO CIRCLE, 74- Concurrent Pos: NSF & Am Coun Learned Socs res & travel grants; dir, Singapore Agr Surv, 51-53; vis asst prof, Dept Agr Econ, Cornell Univ, 60; dir, High Sch Geog Inst, Virgin Islands, 65, Brazil, 68; cor mem, Humid Tropics Comn, Int Geog Union, 59-, specialist regional planning, Orgn Am States-UN Mission to Peru, 59; participant, Pan-Am Union Seminar on Plantations, PR, 57; del tech conf, Int Union Conserv, Greece, 58. Mem: AAAS; Am Anthrop Asn; Asn Am Geog; Am Geog Soc; Soc Ethnomusicol. Res: Cultural geography and economics of peasant agriculture, especially in Malaysia and the Caribbean; philosophy of social science; cultural ecology; environmental cognition. Mailing Add: Grad Col Box 4348 Univ of Ill at Chicago Circle Chicago IL 60680

BLAW, MICHAEL ERVIN, b Gary, Ind, Nov 14, 27; m 50; c 4. PEDIATRIC NEUROLOGY. Educ: Univ Chicago, PhB, 49, MD, 54; Am Bd Psychiat & Neurol, dipl. Prof Exp: Resident pediat, Univ Chicago, 54-56; resident neurol, Univ Mich, 56-59; instr pediat, Univ Chicago, 59-60; from asst prof to assoc prof pediat neurol, Univ Minn, Minneapolis, 60-68; PROF PEDIAT NEUROL, UNIV TEX SOUTHWESTERN MED SCH DALLAS, 68- Concurrent Pos: Nat Inst Neurol Dis & Blindness spec trainee grant, 58 & pediat neurol training grant, 65; NIH grant, 64; consult, Scottish Rite Hosp, Dallas; mem bd trustees, United Cerebral Palsy Asn Dallas County. Mem: AAAS; Int Child Neurol Asn (vpres); Child Neurol Soc; fel Am Acad Neurol. Res: Clinical studies related to neurological disease in childhood. Mailing Add: Dept of Neurol Univ of Tex Southwestern Med Sch Dallas TX 75235

BLAY, JORGE ALBERT, b Buenos Aires, Arg, Mar 25, 32; m 65; c 2. ORGANIC CHEMISTRY. Educ: Eng Technol Inst, Arg, BS, 56; Univ Buenos Aires, MS, 58, PhD(anal chem), 60. Prof Exp: Analyst, City Water Works Lab, Arg, 54-55; res chemist, AEC, 55-60; res chemist, 60-63, sr res chemist, 63-68, RES ASSOC, CELANESE CHEM CO, 68- Concurrent Pos: Asst inorg & radio chem, Univ Buenos Aires, 57-60. Mem: Am Chem Soc; Arg Chem Asn. Res: Analytical chemistry; uranium processing; radiochemistry; electroanalysis; polarography; analytical instrumentation; computer control of data units; catalytic oxidation. Mailing Add: Celanese Chem Co PO Box 9077 Corpus Christi TX 78408

BLAYDES, DAVID FAIRCHILD, b Columbus, Ohio, Aug 17, 34; m 61; c 1. PLANT PHYSIOLOGY, BIOCHEMISTRY. Educ: Ohio State Univ, BS, 56; Univ Wis, MS, 57; Ind Univ, PhD(bot), 62. Prof Exp: Res assoc, Mich State Univ, 62-64, NIH fel, 64-65; asst prof, 65-70, ASSOC PROF BIOL, W VA UNIV, 70- Mem: Bot Soc Am; Am Soc Plant Physiologists; NY Acad Sci; Scand Soc Plant Physiologists; Japanese Soc Plant Physiologists. Res: Physiology, biochemistry and molecular biology of growth and development; tissue culture; natural products. Mailing Add: Dept of Biol W Va Univ Morgantown WV 26506

BLAYLOCK, LYNN GAIL, b Fromberg, Mont, Jan 23, 22; m 50; c 3. ANIMAL NUTRITION. Educ: Colo Agr & Mech Col, BS, 48; Agr & Mech Col Tex, MS, 50, PhD(biochem & nutrit), 53. Prof Exp: Asst nutrit, Agr & Mech Col Tex, 48-52; asst prof poultry nutrit, Wash State Col, 52-54 & Colo Agr & Mech Col, 54-55; dir animal nutrit & res, Supersweet Feed Div, Int Milling Co, 55-67, mgr nutrit res & animal technol, 67-69, assoc dir res, 69-71, DIR ANIMAL NUTRIT RES, SUPERSWEET RES FARM, INT MULTIFOODS CORP, COURTLAND, 71- Mem: Am Inst Nutrit; Poultry Sci Asn; Am Soc Animal Sci; Am Dairy Sci Asn. Res: Nutritional requirements of poultry, swine and cattle; interrelationships of environmental temperature and humidity on the physiological response of animals on various diets; food product development for animal products. Mailing Add: 1024 S Payne St New Ulm MN 56073

BLAYLOCK, W KENNETH, b Bristol, Va, Sept 6, 31; m; c 3. DERMATOLOGY. Educ: King Col, BS, 53; Med Col Va, MS, 58. Prof Exp: Intern, Duke Hosp, Durham, NC, 58-59, resident dermat, 59-61 & internal med, 61-62; from asst prof to assoc prof med, 64-67, PROF DERMAT, MED COL VA, 67-, ASST DEAN GRAD MED EDUC, 72- Concurrent Pos: Dermat rep, Fed Drug Admin, 74-; chmn workshop comt eczema, NIH, 74-; mem resident eval comt, Am Asn Prof Dermat, 74- Mem: Am Acad Dermat; Am Dermat Asn; Soc Invest Dermat; Am Asn Prof Dermatologists. Res: Allergy and immunology. Mailing Add: Med Col of Va Box 164 Richmond VA 23298

BLAZEVIC, DONNA JEAN, b Wyandotte, Mich, Nov 8, 31. MEDICAL MICROBIOLOGY. Educ: Univ Mich, BS, 54, MPH, 63. Prof Exp: Res assoc virol, Dept Microbiol, Univ Mich, 63-65; from instr to assoc prof, 65-75, PROF CLIN MICROBIOL, DEPT LAB MED, UNIV MINN, MINNEAPOLIS, 75-, LECTR, DEPT MICROBIOL, 70-, ASSOC DIR DIAG MICROBIOL, UNIV MINN HOSPS, 71- Concurrent Pos: Mem subcomt microbiol reagents, Nat Comt Clin Lab Stand, 72-; trustee, Am Soc Med Technol Educ & Res Fund, Inc, 73-78. Mem: Am Soc Microbiol; Am Soc Med Technol; AAAS; Acad Clin Lab Physicians & Scientists. Res: Development of rapid tests for diagnosis of infectious disease and identification of microorganisms of medical importance. Mailing Add: Dept of Lab Med & Path Univ of Minn Box 198 Mayo Minneapolis MN 55455

BLAZKOVEC, ANDREW A, b Kewaunee, Wis, Dec 9, 36. IMMUNOBIOLOGY. Educ: Univ Wis, BS, 58, MS, 61, PhD(biol), 63. Prof Exp: NIH fel immunol, Swiss Nat Res Inst, 63-64; res assoc, Swiss Nat Res Inst & Univ London, 64-66; asst prof, 67-71, ASSOC PROF MED MICROBIOL, UNIV WIS-MADISON, 71- Mem: Am Asn Immunologists; AAAS. Res: Role and significance of delayed-type cutaneous hypersensitivity in acquired cellular resistance to infectious agents. Mailing Add: Dept of Med Microbiol Med Sch Univ of Wis Madison WI 53706

BLAZQUEZ Y SERVIN, CARLOS HUMBERTO, b Mexico City, Mex, Oct 9, 26; nat US; m 65; c 2. PLANT PATHOLOGY. Educ: Univ Calif, BS, 55; Univ Fla, MS, 57, PhD(plant path), 59. Prof Exp: Jr plant pathologist virus diseases, Bur Plant Path, Calif State Dept Agr, 55; plant pathologist, Firestone Rubber Plantations, Brazil, 59-61; asst plant pathologist, United Fruit Co, Honduras, 62-63; plant pathologist, Dow Chem Co, Mich, 64; plant pathologist citrus res, Univ West Indies, 64-66; asst prof plant path & asst plant pathologist, 66-72, ASSOC PROF PLANT PATH & ASSOC PLANT PATHOLOGIST, AGR RES CTR, UNIV FLA, 72- Concurrent Pos: Hon consult, Univ West Indies, 67; proj leader aerial appl grant, Univ Fla, 69-70; partic, Plant Protection Cong, Paris, 70; prin investr, Inst Food & Agr Sci, Univ Fla-NASA ERTS-B Prog, 73- Honors & Awards: Achievement Award, Chevron Chem Co, Fla, 70. Mem: Mycol Soc Am; Am Phytopath Soc; Asn Appl Biol; Am Chem Soc; Am Asn Hort Sci. Res: Pesticide residue analysis; remote sensing; low volume and ultra low volume aerial application methods for disease control; effect of fungicide degradation on biological activity. Mailing Add: Agr Res Ctr Univ of Fla Route 1 Box 2G Immokalee FL 33934

BLAZYK, JACK, b New York, NY, Sept 22, 47; m 72; c 2. BIOCHEMISTRY. Educ: Hamilton Col, AB, 69; Brown Univ, PhD(biochem), 75. Prof Exp: Vis asst prof, Univ SC, 74-75; ASST PROF CHEM, OHIO UNIV, 75- Concurrent Pos: Fel gen med div, NIH, 75. Mem: AAAS; Am Chem Soc. Res: An investigation of the structure and function of biological membranes and the study of membrane-bound enzymes. Mailing Add: Dept of Chem Ohio Univ Athens OH 45701

BLEAK, ALVIN THOMAS, b Logandale, Nev, Jan 10, 19; m 43; c 2. RANGE MANAGEMENT, PLANT ECOLOGY. Educ: Univ Utah, BS, 41; Utah State Univ, MS, 70. Prof Exp: Shift engr, Chem Control Labs, Utah Ord Plant, Remington Arms Co, Inc, 42-43; res range conservationist, Intermountain Forest & Range Exp Sta, US Forest Serv, 46-54, RES RANGE CONSERVATIONIST, CROPS RES DIV, AGR RES SERV, USDA, 54- Mem: Ecol Soc Am; Am Soc Range Mgt; Am Soc Agron; Crops Sci Soc Am; Am Grassland Coun. Res: Revegetation and management of rangelands including selection of grasses and legumes; plant response to environmental factors; methods of establishment; fertilization and control of undesirable plants. Mailing Add: UMC 63 Utah State Univ Logan UT 84322

BLEAKNEY, JOHN SHERMAN, b Corning, NY, Jan 14, 28; m 52; c 2. MALACOLOGY. Educ: Acadia Univ, BSc, 49, MSc, 51; McGill Univ, PhD(zool), 56. Prof Exp: Cur ichthyol & herpet, Nat Mus Can, 52-58; assoc prof biol, 58-61, PROF BIOL, ACADIA UNIV, 61- Concurrent Pos: Nat Res Coun Can sr res fel, Eng & Denmark, 67-68. Mem: Am Soc Ichthyologists & Herpetologists; Am Inst Biol Sci; Am Malacol Union; Malacol Soc London; Marine Biol Asn UK. Res: Herpetology and nudibranch molluscs of Eastern Canada; fauna of the Minas Basin, Nova Scotia. Mailing Add: Dept of Biol Acadia Univ Wolfville NS Can

BLEASDALE, JAMES LEWIS, b Meadville, Pa, May 29, 21; m 44; c 2. ORGANIC CHEMISTRY. Educ: Allegheny Col, BS, 44; Univ Minn, PhD(org chem), 50. Prof Exp: Asst, Univ Minn, 46-50; res chemist, 50-68, RES ASSOC, E I DU PONT DE NEMOURS & CO, 68- Mem: Am Chem Soc. Res: Physical character of textile polymers and fibers. Mailing Add: E I du Pont de Nemours & Co Waynesboro VA 22980

BLECHARCZYK, WALTER JOSEPH, b New Bedford, Mass, Aug 19, 21; m 51; c 5. BIOMEDICAL ENGINEERING. Educ: RI State Col, BS, 43. Prof Exp: Rubber chemist, 47-53, from asst chief chemist to chief chemist, 53-70, TECH DIR, DAVOL, INC, 70- Mem: Am Chem Soc; Am Soc Testing & Mat; Soc Plastics Eng. Res: Technology of rubber and rubber latex; synthetic polymers; plastics; manufacture of surgical and sundry products. Mailing Add: Davol Inc 100 Sochanossett Crossroad Cranston RI 02920

BLECHER, MARVIN, b New York, NY, Sept 12, 40; m 62; c 2. NUCLEAR PHYSICS, ELEMENTARY PARTICLE PHYSICS. Educ: Columbia Univ, BA, 62; Univ Ill, Urbana-Champaign, PhD(physics), 67. Prof Exp: ASST PROF PHYSICS, VA POLYTECH INST & STATE UNIV, 68- Mem: Am Phys Soc. Mailing Add: Dept of Physics Va Polytech Inst Blacksburg VA 24060

BLECHER, MELVIN, b Rahway, NJ, July 19, 22; m 50, 63, 74; c 2. BIOCHEMISTRY. Educ: Rutgers Univ, BS, 49; Univ Pa, PhD(biochem), 54. Prof Exp: Asst instr biochem, Sch Med, Univ Pa, 49-51; res assoc, Col Physicians & Surgeons, Columbia Univ, 54-56; from instr to asst prof, Albert Einstein Col Med, 56-61; assoc prof, 61-68, PROF BIOCHEM, SCHS MED & DENT & GRAD SCH, GEORGETOWN UNIV, 68- Concurrent Pos: Vis prof, Med Clin, Univ Rome; vis prof biochem, Univ Paris. Mem: AAAS; Am Soc Biol Chemists; Am Soc Pharmacol & Exp Therapeut; Soc Exp Biol & Med; AMA. Res: Mechanism of action of hormones; regulatory mechanisms in metabolism of cyclic nucleotides. Mailing Add: Dept of Biochem Georgetown Univ Sch of Med & Dent Washington DC 20007

BLECHMAN, HARRY, b Brooklyn, NY, Aug 22, 18; m 43; c 2. MICROBIOLOGY. Educ: City Col New York, BS, 38; NY Univ, DDS, 51; Hunter Col, MA, 59; Am Bd Endodont, dipl, 68. Prof Exp: Clin bacteriologist, St John's Long Island Col Hosp & Rockaway Beach Hosp, 38-41; clin bacteriologist, Tilton Gen Hosp, 41-43; sr bacteriologist, Harlem Hosp, 46-47; chmn dept microbiol, 51-54, from asst prof to prof, 51-68, chmn dept microbiol, prev med & hyg, 54-68, asst dean col dent, 64-68, dean, 68-75, PROF ENDODONT & CHMN DEPT, COL DENT, NY UNIV, 75- Concurrent Pos: Exec secy, Guggenheim Inst Dent Res, 54-64; mem, Nat Bd Dent Examr; pres, Am Bd Endodont, 75; consult, Vet Admin Hosps, USPHS & NY State Narcotic Addiction Comn. Mem: AAAS; fel Am Col Dent; Am Soc Microbiol; Sigma Xi; Harvey Soc. Res: Oral microbiology; antibiosis and interactions among oral microorganisms in dental caries and periodontal disease. Mailing Add: NY Univ Col of Dent 421 First Ave New York NY 10010

BLECHNER, JACK NORMAN, b New York, NY, Jan 7, 33; m 57; c 3. OBSTETRICS & GYNECOLOGY. Educ: Columbia Col, BA, 54; Yale Univ, MD, 57; Am Bd Obstet & Gynec, dipl. Prof Exp: Intern, Bronx Munic Hosp Ctr, 57-58; Josiah Macy, Jr Found fel physiol, Yale Univ, 58-60; asst resident obstet & gynec, Columbia Presby Med Ctr, 60-63, chief resident, 63-65; from asst prof to assoc prof, Col Med, Univ Fla, 65-70; PROF OBSTET & GYNEC & HEAD DEPT, SCH MED, UNIV CONN, 70- Concurrent Pos: Univ fel, Yale Univ, 62-63; Am Cancer Soc fel, 64; chief dept obstet & gynec, New Britain Gen Hosp, 70-; consult, Vet Admin Hosp, 70-, St Francis Hosp, 70-, Mt Sinai Hosp, Hartford, Conn, 70-, Conn State Family Planning Coun, 71-, Hartford Hosp, 72- & Bristol Hosp, 73-; examr, Am Bd Obstet & Gynec. Mem: Am Asn Obstetricians & Gynecologists; Am Gynec Soc; Asn Profs Gynec & Obstet; Soc Gynec Invest; Perinatal Res Soc. Res: Fetal physiology; membrane transfer; intrauterine transfusion and experimental intrauterine surgery; respiratory gas exchange. Mailing Add: Dept of Obstet & Gynec Univ of Conn Sch of Med Farmington CT 06032

BLECK, RAINER, b Danzig, Ger, Nov 4, 39; m 66; c 2. METEOROLOGY. Educ: Free Univ Berlin, MS, 64; Pa State Univ, PhD(meteorol), 68. Prof Exp: Res asst meteorol, Inst Theoret Meteorol, Free Univ Berlin, 65; scientist, Nat Ctr Atmospheric Res, 67-75; ASSOC PROF METEOROL, UNIV MIAMI, 75- Res: Weather analysis and prediction using isentropic coordinates; mountain lee waves; cloud physics. Mailing Add: Rosenstiel Sch of Marine & Atmospheric Sci Univ of Miami Coral Gables FL 33124

BLECKE, RONALD GENE, organic chemistry, polymer chemistry, see 12th edition

BLECKER, HARRY HERMAN, b Philadelphia, Pa, Apr 10, 27; m 51; c 4. ORGANIC CHEMISTRY. Educ: Bucknell Univ, BS, 51; Rutgers Univ, MS, 54, PhD(chem), 55. Prof Exp: Res assoc chem, Univ Mich, 54-56; asst prof, Bucknell Univ, 56-57; from asst prof to assoc prof, 57-67, PROF CHEM, UNIV MICH, FLINT, 67-, CHMN DEPT, 65- Mem: Am Chem Soc. Res: Reaction mechanisms; bioanalytical chemistry. Mailing Add: Univ of Mich 1321 E Court St Flint MI 48503

BLEDSOE, JAMES O, JR, b New Bern, NC, Mar 22, 38; m 62; c 2. ORGANIC CHEMISTRY. Educ: Univ NC, Chapel Hill, BS, 60; Univ Okla, PhD(org chem), 65. Prof Exp: Proj leader org synthesis, Int Flavors & Fragrances, Inc, 65-66; res chemist, 66-71, MGR NEW PROD RES, ORG CHEM GROUP, GLIDDEN DURKEE DIV, SCM CORP, 71- Mem: Am Chem Soc. Res: Synthesis of natural products, particularly terpenes, especially for uses in the perfumery and flavor fields. Mailing Add: 8750 Burkhall St Jacksonville FL 32211

BLEDSOE, JOHN D, b Republic, Mo, Apr 23, 19; m 39; c 4. GEOPHYSICS, ENGINEERING. Educ: Southwest Mo State Col, BA, 39; Univ Chicago, cert meteorol, 43; St Louis Univ, MSc, 51, PhD(geophys, eng), 54. Prof Exp: Res meteorologist, Thunderstorm Proj, US Weather Bur, 46-47; instr meteorol & eng, Parks Col Aeronaut Technol, St Louis Univ, 51-S2; engr, Missile Div, McDonnell Aircraft Corp, Mo, 52-53; asst proj engr, Ord Div, Universal Match Corp, 53-54; proj engr, Aerojet-Gen Corp, Calif, 54-59; group scientist, Autonetics Res Ctr, NAm Rockwell, 59-63; proj mgr geophys & lunar explor, Earth Sci Div, Teledyne, Inc, 63-66; lab mgr, Northrop Space Labs, Northrop Corp, 66-67, Lunar Receiving Lab & Joint Venture, Brown & Root-Northrop Corp, Manned Spacecraft Ctr, Houston, Tex, 67-69; sr scientist & dept mgr, Syst Sci Ctr, Comput Sci Corp, Falls Church, 69-74; PRIN ENGR SCIENTIST, A A MATHEWS INC, 74- Concurrent Pos: Mem & chmn panel on systs eval, Comt Rapid Excavation, Nat Acad Sci-Nat Acad Eng, 67-68; US del & chmn task group on res & develop, Int Adv Conf on Tunneling, Orgn Econ Coop & Develop, 70. Mem: Soc Min Eng. Res: Systems analysis of earth science engineering applications, especially tunneling & mining through the development of mathematical models simulating the process; development of procedures and equipment for lunar geophysical exploration and the processing of lunar geological samples and data. Mailing Add: A A Mathews Inc 11900 Parklawn Dr Rockville MD 20852

BLEDSOE, WOODROW WILSON, b Maysville, Okla, Nov 12, 21; m 44; c 3. MATHEMATICS. Educ: Univ Utah, BS, 48; Univ Calif, MA, 50, PhD(math), 53. Prof Exp: Lectr math, Univ Calif, 51-52; mathematician & mgr systs anal dept, Sandia Corp, 53-60; mathematician & pres, Panoramic Res, Inc, Calif, 60-66; actg chmn dept math, 67-69, chmn, 73-75, PROF MATH, UNIV TEX, AUSTIN, 66- Concurrent Pos: Vis prof, Mass Inst Technol, 70-71. Mem: Am Math Soc; Asn Comput Mach; Inst Elec & Electronics Engrs. Res: Real variables; topological product measures; systems analysis; artificial intelligence; automatic theorem proving. Mailing Add: 3002 Willowood Circle Austin TX 78703

BLEECKER, MARGIT, b Vienna, Austria, Nov 20, 43; US citizen; m 64. NEUROCHEMISTRY, NEUROENDOCRINOLOGY. Educ: NY Univ, BA, 64; State Univ NY Downstate Med, Ctr, PhD(anat), 70. Prof Exp: Instr anat, Sch Med, Johns Hopkins Univ, 70-72; INSTR ANAT, SCH MED, TEMPLE UNIV, 72- Mem: Am Soc Neurochem; Int Soc Psychoneuroendocrinol; Am Asn Anat; Soc Neurosci. Res: In vitro studies on Taurine in the rat brain. Mailing Add: Dept of Anat Temple Univ Sch of Med Philadelphia PA 19122

BLEECKER, WALTER M, genetics, biochemistry, see 12th edition

BLEFKO, ROBERT L, b July 5, 32; US citizen. TOPOLOGY. Educ: Pa State Univ, PhD(math), 65. Prof Exp: Asst prof math, Univ RI, 65-68; ASSOC PROF MATH, WESTERN MICH UNIV, 68- Mem: Am Math Soc; Math Asn Am. Res: General topology, specializing in continuous function theory. Mailing Add: Dept of Math Western Mich Univ Kalamazoo MI 49008

BLEI, IRA, b Brooklyn, NY, Jan 19, 31; m 52; c 2. BIOPHYSICAL CHEMISTRY. Educ: Brooklyn Col, BS, 52, MA, 54; Rutgers Univ, PhD(biophys chem), 57. Prof Exp: Instr chem, Rutgers Univ, 53-57; prin res chemist, Lever Bros Co Res Ctr, 57-61; supvr biophys & biodetection sect, Melpar, Inc, 61-66, mgr environ sci lab, 66-67; from asst prof to assoc prof, 67-74, PROF BIOL, RICHMOND COL, NY, 74- Mem: AAAS; Am Chem Soc; NY Acad Sci; Fedn Am Scientists. Res: Detection of extra-terrestrial life; optical rotatory dispersion; polarization of fluorescence; physical chemistry of non-aqueous phosphatide solutions; biological differentiation of sodium and potassium; mechanism of action of psychoactive drugs. Mailing Add: 130 Stuyvesant Pl Staten Island NY 10301

BLEIBERG, MARVIN JAY, b Brooklyn, NY, Feb 19, 28; m 60; c 2. PHARMACOLOGY. Educ: Col William & Mary, BS, 49; Med Col Va, PhD(pharmacol), 57. Prof Exp: USPHS trainee steroid biochem, Worcester Found Exp Biol, 57-58; from instr to asst prof pharmacol, Jefferson Med Col, 58-61; pharmacologists, Food & Drug Admin, 61-62; sr scientist res div, Melpar, Inc, 62-64; pharmacologist, Div Environ Toxicol, Woodard Res Corp, 64-68, dir div pharmacol, 68-70, sect head pharmacol, Spec Projs Dept, 70-75; PHARMACOLOGIST & TOXICOLOGIST, PETITION REV, DIV TOXICOL, BUR FOODS, US FOOD & DRUG ADMIN, WASHINGTON, DC, 75- Concurrent Pos: Vis lectr, Howard Univ. Mem: AAAS; Biomet Soc; Am Soc Pharmacol; Am Chem Soc; Soc Toxicol. Res: Toxicology; biochemical and autonomic nervous system pharmacology; behavioral pharmacology. Mailing Add: 3613 Old Post Rd Fairfax VA 22030

BLEIBTREU, HERMANN KARL, b Darmstadt, Ger, Aug 31, 33; US citizen; m 63; c 2. PHYSICAL ANTHROPOLOGY, ANTHROPOLOGY. Educ: Harvard Univ, BA, 56, PhD, 64. Prof Exp: Tutor dept anthrop, Harvard Univ, 61; res asst biol, Case-Western Reserve Univ, 62; asst prof anthrop, Univ Calif, Los Angeles, 64-66; from asst prof to assoc prof, Univ Ariz, 66-71, prof anthrop & dean col lib arts, 71-75; DIR, MUS NORTHERN ARIZ, 75- Concurrent Pos: NIH grant pop characteristics Pima-Papago, Univ Ariz, 67-68, NSF grants, 68 & 69-; mem bd, Prescott Col, 71- Mem: Fel AAAS; Am Asn Phys Anthrop; fel Am Anthrop Asn; Am Eugenics Soc. Res: Human microevolution; multivariate methods in anthropometrics. Mailing Add: Mus of Northern Ariz Ft Valley Rd Flagstaff AZ 86001

BLEICH, HERMANN EWALD, b Frankfurt, Ger, Dec 10, 38. PHYSICAL BIOCHEMISTRY. Educ: Hannover Tech Inst, diplom, 66; Columbia Univ, PhD(physics), 73. Prof Exp: Res assoc chem, Rockefeller Univ, 71-73, asst prof, 73-74; RES ASSOC BIOCHEM, UNIV CONN HEALTH CTR, 74- Mem: AAAS. Res: Nuclear magnetic resonance study of biological oligopeptides; fundamentals of nuclear magnetic resonance and instrumental development. Mailing Add: Dept of Biochem Univ of Conn Health Ctr Farmington CT 06032

BLEICHER, MICHAEL NATHANIEL, b Cleveland, Ohio, Oct 2, 35; m 57; c 3. MATHEMATICS. Educ: Calif Inst Technol, BS, 57; Tulane Univ, MS, 59, PhD(math), 61; Univ Warsaw, PhD(math), 61. Prof Exp: NSF fel math, Univ Calif, Berkeley, 61-62; from asst prof to assoc prof, 62-69, chmn dept, 73-75, PROF MATH, UNIV WIS-MADISON, 69- Mem: Am Math Soc; Math Asn Am; Polish Math Soc; Sigma Xi. Res: Geometry of numbers; convexity; number theory; foundations of mathematics; universal algebra. Mailing Add: Dept of Math Univ of Wis Madison WI 53705

BLEICHER, SHELDON JOSEPH, b New York, NY, Apr 9, 31. INTERNAL MEDICINE, METABOLISM. Educ: NY Univ, BA, 51; Western Ill Univ, MS, 52; State Univ NY Downstate Med Ctr, MD, 56. Prof Exp: Res fel med, Thorndike Mem Lab, Boston City Hosp & Harvard Med Sch, 60-63; dir metab res unit, Jewish Hosp & Med Ctr Brooklyn, 63-69; from instr to assoc prof, 63-75, PROF MED, STATE UNIV NY DOWNSTATE MED CTR, 75-; CHIEF DIV ENDOCRINOL & METAB & ATTEND PHYSICIAN, JEWISH HOSP & MED CTR BROOKLYN, 69- Concurrent Pos: USPHS fel, 61-63; NIH res career develop award; consult-expert in med use of radioisotopes, Int Atomic Energy Agency. Mem: Am Col Physicians; Am Fedn Clin Res: Am Diabetes Asn; AMA; Endocrine Soc. Res: Diabetes; hypoglycemia; carbohydrate metabolism in pregnancy; fat metabolism. Mailing Add: 114 Magnolia Lane Roslyn Heights NY 11577

BLEICK, WILLARD EVAN, b Newark, NJ, Dec 8, 07; m 36; c 3. MATHEMATICAL PHYSICS. Educ: Stevens Inst Technol, ME, 29; Johns Hopkins Univ, PhD(appl math), 33. Prof Exp: Mem, Sch Math Inst Advan Study, 33-34 & 35-36; actuarial student, Prudential Ins Co, 36-38; engr, Gibbs & Cox, Inc, 38-40; instr math, Cooper Union, 39-40; from instr to asst prof, US Naval Acad, 40-46, assoc prof math & mech, 46-50, PROF MATH & MECH, NAVAL POSTGRAD SCH, 50- Mem: Am Asn Physics Teachers. Res: Quantum mechanics; stress analysis; crystal lattice energies; digital computers; missile dynamics; control optimization. Mailing Add: Dept of Math Naval Postgrad Sch Monterey CA 93940

BLEIDNER, WILLIAM EGIDIUS, b Staten Island, NY, Sept 8, 21; m 47; c 1. ANALYTICAL BIOCHEMISTRY. Educ: Univ Rochester, BS, 43, MS, 45; Univ Minn, PhD(biochem), 52. Prof Exp: Lab instr quant anal, Univ Rochester, 43-44; teaching asst phys chem, Univ Minn, 45-46, lectr quant anal, Dept Biochem, 46-51; sr chemist, Grasselli Chem Exp Sta, 52-53, sr res investr, 53-56, res scientist, 56-57, res scientist, Stine Lab, 57-61, res assoc, 61-65, RES SUPVR, EXP STA, INDUST & BIOCHEM, E I DU PONT DE NEMOURS & CO, INC, 65- Mem: Am Chem Soc; Sigma Xi. Res: All phases of chromatography; trace analysis; product quality. Mailing Add: 4 Beech Ave Brookland Terr Wilmington DE 19805

BLEIER, WILLIAM JOSEPH, b Jourdanton, Tex, July 8, 47; m 73. REPRODUCTIVE BIOLOGY. Educ: Univ Tex, Austin, BA, 69; Tex Tech Univ, MS, 71, PhD(zool), 75. Prof Exp: ASST PROF ZOOL, NDAK STATE UNIV, 75- Mem: Sigma Xi; Am Soc Mammalogists. Res: Factors controlling embryonic development in mammals. Mailing Add: Dept of Zool NDak State Univ Fargo ND 58102

BLEIL, CARL EDWARD, b Detroit, Mich, Oct 7, 23; wid; c 4. PHYSICS. Educ: Mich State Univ, BS, 47; Univ Okla, MS, 50, PhD(physics), 53. Prof Exp: Asst, Univ Okla, 51-53; SUPVRY RES PHYSICIST, RES LABS, GEN MOTORS CORP, 53- Concurrent Pos: Mem solid state sci panel, SSS Comt, Nat Res Coun, 74- Mem: AAAS; Am Phys Soc. Res: Compound semiconductors; surface physics; crystal growth; coherent optics. Mailing Add: Gen Motors Corp Res Labs 12 Mile & Mound Rds Warren MI 48090

BLEIL, DAVID FRANKLIN, b Detroit, Mich, Dec 4, 08; m 39; c 3. PHYSICS. Educ: Univ Mich, BS, 34, MS, 37; Mich State Col, PhD(physics), 48. Prof Exp: From asst to instr physics, Mich State Col, 36-43; res physicist, Naval Ord Lab, 43-51, chief solid state div, 51-52, chief physics dept, 51-58, assoc tech dir res, Naval Surace Weapons Ctr-White Oak Lab, 58-73; RETIRED. Concurrent Pos: Navy sponsored sabbactical year, Oxford Univ, 67; govt consult; mem fac, Univ Md. Honors & Awards: Superior Civilian Serv Award, US Dept Navy, 63; Navy Distinguished Civilian Serv Award, Secy of Navy. Mem: Fel Am Phys Soc; Am Asn Physics Teachers; Sigma Xi. Res: Geophysical prospecting; induced polarization; resistivity methods; magnetism; solid state. Mailing Add: Rte 2 Box 251-B Hedgesville WV 25427

BLEISTEIN, NORMAN, b New York, NY, June 29, 39; m 61; c 2. APPLIED MATHEMATICS, MATHEMATICAL ANALYSIS. Educ: Brooklyn Col, BS, 60; NY Univ, MS, 61, PhD(math), 65. Prof Exp: Res assoc math, Courant Inst Math Sci, NY Univ, 65-66; asst prof, Mass Inst Technol, 66-69; assoc prof, 69-74, PROF MATH, UNIV DENVER, 74- Concurrent Pos: Assoc ed, Soc Indust & Appl Math J, 73- Mem: Soc Indust & Appl Math. Res: Asymptotic methods; geometrical theory of diffraction; inverse problems in acoustics; electromagnetics; seismology. Mailing Add: Dept of Math Univ of Denver Denver CO 80210

BLEIWEIS, ARNOLD SHELDON, b Brooklyn, NY, Aug 8, 37; m 67. MICROBIOLOGY, BIOCHEMISTRY. Educ: Brooklyn Col, BS, 58; Pa State Univ, MS, 60, PhD(microbiol), 64. Prof Exp: Lectr biol, Brooklyn Col, 58; asst microbiol, Pa State Univ, 58-63; fel epidemiol, Med Sch, Washington Univ, 64-65; asst biochem, Albert Einstein Med Ctr, 66-67; ASST PROF BACT, UNIV FLA, 67-, ASSOC PROF MICROBIOL, 72-, ASSOC PROF IMMUNOL & MED MICROBIOL, COL MED, 72-, ASSOC MICROBIOLOGIST, INST FOOD & AGR SCI, 74- Concurrent Pos: NIH training grant epidemiol, 64-65, fel, 66-67, res grant sr investr, 69-; Eli Lilly res gift, 69-; consult, Nat Inst Dent Res, 70- Mem: AAAS; Am Soc Microbiol. Res: Structural and antigenic aspects of streptococcal cell walls; cell wall biosynthesis; bacterial cytology; immunologic studies of carcinogenic streptococci. Mailing Add: Dept of Microbiol Univ of Fla Gainesville FL 32601

BLEJER, HECTOR P, b Caracas, Venezuela, Dec 17, 33; US citizen. OCCUPATIONAL MEDICINE, TOXICOLOGY. Educ: McGill Univ, BSc, 56, MD,

CM, 58; Univ Toronto, dipl indust health, 63; Am Bd Prev Med, dipl, 68. Prof Exp: Pvt pract, Ont, 63-65; head occup health Southern Calif, Bur Occup Health & Environ Epidemiol, State Calif Dept Pub Health, 65-74; DEP DIR, DIV FIELD STUDIES CLIN INVEST, NAT INST OCCUP SAFETY & HEALTH, 75- Concurrent Pos: Lectr, Air Pollution Control Inst, Univ Southern Calif, 69-72, asst clin prof, Sch Med, 70-; attend staffman, Los Angeles County-Univ Southern Calif Med Ctr, 70-; consult, Nat Acad Sci, 73- Mem: Am Thoracic Soc; fel Am Occup Med Asn; Am Pub Health Asn; fel Am Col Prev Med; Soc Occup & Environ Health. Res: Environmental medicine, epidemiology, toxicology and cancerigenesis; occupational and environmental pollution control research. Mailing Add: Nat Inst Occup Safety & Health Rm 523 US Post Off Bldg Cincinnati OH 45202

BLEM, CHARLES R, b Dunkirk, Ohio, Apr 18, 43; m 65; c 2. VERTEBRATE ECOLOGY. Educ: Ohio Univ, BS, 65; Univ Ill, Urbana, MS, 68, PhD(zool), 69. Prof Exp: ASSOC PROF BIOL, VA COMMONWEALTH UNIV, 69- Mem: AAAS; Ecol Soc Am; Am Soc Ichthyologists & Herpetologists; Am Soc Mammal; Am Ornith Union; Cooper Ornith Soc. Res: Energetics and general ecology of vertebrates, especially birds. Mailing Add: Dept of Biol Va Commonwealth Univ Richmond VA 23284

BLEND, MICHAEL J, b Mobile, Ala, Dec 30, 42; m 70. REPRODUCTIVE PHYSIOLOGY, TOXICOLOGY. Educ: Univ Scranton, BS, 65; Univ Nebr, Omaha, MS, 67; Cornell Univ, PhD(animal physiol), 70. Prof Exp: Asst prof biol, Univ Detroit, 70-73; DIR RADIOASSAY, DIV NUCLEAR MED, MT CARMEL MERCY HOSP, MICH, 73- Mem: AAAS; Soc Study Reproduction. Res: Role or mode of action of steroid hormones on the canine prostate; steroid hormone and pesticide interactions in the mammalian organism. Mailing Add: Div of Nuclear Med Mt Carmel Mercy Hosp Detroit MI 48235

BLENDEN, DONALD, b Webster Groves, Mo, Aug 13, 29; m 51; c 2. VETERINARY PUBLIC HEALTH. Educ: Univ Mo, BS, 51, MA, 53, DVM, 56; Am Bd Vet Pub Health, dipl. Prof Exp: Med technician, Boone County Hosp, 52-56; clin vet, Pvt Pract, 56-58; PROF VET MICROBIOL, UNIV MO-COLUMBIA, 58- Mem: Fel Am Pub Health Asn; Am Vet Med Asn; Am Soc Microbiol. Res: Comparative medicine; epidemiology and microbiology of zoonoses; leptospirosis; listeriosis; rabies; enteric infections. Mailing Add: Univ of Mo Col of Vet Med Columbia MO 65201

BLESS, ROBERT CHARLES, b Ithaca, NY, Dec 3, 27. ASTRONOMY. Educ: Univ Fla, BS, 47; Cornell Univ, MS, 55; Univ Mich, PhD(astron), 58. Prof Exp: Physicist, Naval Res Lab, 47-48; lectr astron, Univ Mich, 58; res assoc, 58-61, from asst prof to assoc prof, 61-69, PROF ASTRON, UNIV WIS-MADISON, 69-, CHMN DEPT, 72- Concurrent Pos: Nat Res Coun sr fel, Goddard Space Flight Ctr, 67-68. Mem: Am Astron Soc. Res: Photoelectric spectrophotometry; space astronomy. Mailing Add: Washburn Observ Univ of Wis Madison WI 53706

BLESSEL, KENNETH WAYNE, b Tullahoma, Tenn, Sept 9, 43; m 66; c 2. ANALYTICAL CHEMISTRY. Educ: Gannon Col, BA, 66; Univ Mass, MS, 70, PhD(anal chem), 72. Prof Exp: GROUP LEADER ANAL RES, HOFFMANN-LA ROCHE INC, 72- Mem: Am Soc Testing & Mat; Am Soc Qual Control. Res: Development of new and improved methods of analysis for new drug candidates. Mailing Add: 340 Kingsland St Nutley NJ 07110

BLETHEN, SANDRA LEE, biochemistry, see 12th edition

BLETNER, JAMES KARL, b Mason, WVa, July 29, 12; m 35; c 2. POULTRY HUSBANDRY. Educ: WVa Univ, BS, 34, MS, 50; Ohio State Univ, PhD, 58. Prof Exp: Asst exten poultryman, Agr Exp Sta, WVa Univ, 35-38, asst prof poultry husb, Univ & asst poultry husbandman, Agr Exp Sta, 46-55; instr poultry husb, Ohio State Univ, 55-58; assoc prof, 58-71, PROF POULTRY HUSB, UNIV TENN, KNOXVILLE, 71- Concurrent Pos: Feed serviceman, Quaker Oats Co, 38-44 & 46. Mem: AAAS; Am Inst Nutrit; Poultry Sci Asn; World Poultry Sci Asn. Res: Poultry nutrition and management. Mailing Add: Dept of Animal Sci Univ of Tenn Knoxville TN 37916

BLETZINGER, JOHN CALVIN, b Coraopolis, Pa, Jan 22, 14; m 41; c 4. CHEMISTRY. Educ: Allegheny Col, AB, 36; Lawrence Univ, MS, 38, PhD, 40. Prof Exp: Lab asst, Allegheny Col, 34-36; chief chemist, Sterling Pulp & Paper Co, 40-42; process engr, Eau Claire Ord Plant, US Rubber Co, 42-44; res chemist, War Prod Div, 44-45, res coordr, 45-53, supt sanit prod lab, 53-66, SR RES ASSOC, KIMBERLY-CLARK CORP, 66- Mem: AAAS; assoc Tech Asn Pulp & Paper Indust; assoc Am Chem Soc; NY Acad Sci; Soc Indust Microbiol. Res: Chemical structure of cellulose as related to papermaking properties; development of cellulosic specialties and processes particularly in the field of catamenial protection and other related sanitary products. Mailing Add: Res Ctr Kimberly Clark Corp Neenah WI 54956

BLEUER, NED KERMIT, b Oak Park, Ill, Nov 21, 43; m 65; c 2. GLACIAL GEOLOGY. Educ: Univ Wis-Madison, BS, 65, PhD(geol), 70; Univ Ill, MS, 67. Prof Exp: GLACIAL GEOLOGIST, IND GEOL SURV, BLOOMINGTON, IND, 68- Mem: Am Quaternary Asn. Res: Glacial stratigraphy and history of Indiana. Mailing Add: Ind Geol Surv 611 N Walnut Grove Bloomington IN 47401

BLEULER, ERNST, b Kusnacht, Switz, Jan 4, 16; nat US; m 45; c 3. NUCLEAR PHYSICS. Educ: Swiss Fed Inst Technol, dipl, 38, Dr Sc Nat, 42. Prof Exp: Asst physics, Swiss Fed Inst Technol, 38-47, privat dozent, 46-47; vis prof nuclear physics, Purdue Univ, 47-48, from assoc prof to prof, 48-64; PROF NUCLEAR PHYSICS, PA STATE UNIV, UNIVERSITY PARK, 64- Concurrent Pos: Guggenheim fel, Europ Orgn Nuclear Res, 61-62. Mem: AAAS; Am Phys Soc; Am Asn Physics Teachers; Swiss Phys Soc. Res: Nuclear electron scattering; technique of absorption measurements; disintegration energies of light nuclei; alpha particle reactions; electronic instrumentation. Mailing Add: Dept of Physics Pa State Univ University Park PA 16802

BLEVINS, CHARLES EDWARD, b Ritzville, Wash, Sept 5, 24; m 48; c 2. ANATOMY. Educ: Stanford Univ, BA, 47, MA, 48; Univ Calif, San Francisco, PhD(anat), 61. Prof Exp: Instr anat, physiol & zool, Glendale Col, 49-51; instr anat physiol, Contra Costa Col, 51-58, chmn dept life sci, 61-62; res anatomist, Div Otolaryngol, Univ Calif, San Francisco, 59-60; from instr to asst prof anat, Univ Wash, 62-65; asst prof, Baylor Col Med, 65-67; assoc prof, Med Sch, Northwestern Univ, 67-74; CHMN DEPT ANAT, SCH MED, IND UNIV, INDIANAPOLIS, 74- Mem: AAAS; Am Asn Anat. Res: Gross and microscopic anatomy; histology; histochemistry and ultrastructure of muscles, nerves, connective tissue and joints of the middle ear and larynx. Mailing Add: Dept of Anat Ind Univ Sch of Med Indianapolis IN 46202

BLEVINS, DALE GLENN, b Ozark, Mo, Aug 29, 43; m 67; c 1. PLANT NUTRITION. Educ: Southwest Mo State Univ, BS, 65; Univ Mo, MS, 67; Univ Ky, PhD(plant physiol), 72. Prof Exp: Res asst biochem, Dept Bot, Ore State Univ, 72-74; ASST PROF PLANT NUTRIT, UNIV MD, COLLEGE PARK, 74- Mem: Am Soc Plant Physiologists; Australian Soc Plant Physiologists. Res: Determination of the role of calcium in the infection of legume roots by rhizobia; studies on the role of malate in nitrate uptake and reduction with regard to potassium mobility in plants. Mailing Add: Dept of Bot Univ of Md College Park MD 20742

BLEVINS, GILBERT SANDERS, b Smyth Co, Va, Oct 31, 27; m 50; c 2. EXPERIMENTAL PHYSICS. Educ: Va Polytech Inst, BS, 53; Duke Univ, PhD(physics), 57. Prof Exp: Asst, Duke Univ, 53-57; sr res physicist, Nat Cash Register Co, 57-58; tech consult, Sperry Farragut Co, 58-60; mgr data acquisition systs, Guided Missile Range Div, Pan Am World Airways, 60-62; assoc dir appl physics, Surface Div, Westinghouse Elec Corp, 62-65; consult, US Dept Defense, 65-69; TECH DIR, US NAVY SECURITY GROUP, 69- Mem: Am Phys Soc. Res: Systems engineering; applied microwave and solid state physics. Mailing Add: 13300 Briarwood Dr Laurel MD 20810

BLEVINS, MAURICE EVERETT, b Franklinville, NJ, Mar 30, 28; m 52; c 1. APPLIED PHYSICS, COMPUTER SCIENCES. Educ: Duke Univ, BS, 52, PhD(physics), 58. Prof Exp: Asst physics, Duke Univ, 52-56, res assoc, 56-59; prof, Wofford Col, 59-64; fac mem, Spartanburg Tech Educ Ctr, 64-72, head electronics dept, 66-72; SR RES ENGR, STEEL MEDDLE MFG CO, GREENVILLE, 72- Concurrent Pos: Consult, 68- Mem: Am Phys Soc; Am Asn Physics Teachers. Res: Low temperature physics; high energy physics; superconductivity; textile technology; electronics. Mailing Add: 98 Canterbury Rd Spartanburg SC 29302

BLEVINS, RAYMOND DEAN, b Butler, Tenn, Apr 3, 39; m 62; c 4. PHYSIOLOGY, BIOCHEMISTRY. Educ: ETenn State Univ, BS, 60, MA, 61; Univ Tenn, PhD(physiol, biochem), 71. Prof Exp: Teacher & chmn dept sci, Va High Sch, 61-62; assoc prof biol, King Col, 62-69, chmn dept, 62-64; asst prof human physiol & anat, Univ Tenn, Knoxville, 69-71; ASSOC PROF HUMAN PHYSIOL & ANAT, E TENN STATE UNIV, 71- Concurrent Pos: Consult, Biol Div, Oak Ridge Nat Lab, 74- Mem: AAAS; Am Soc Zoologists; Am Inst Biol Sci; Am Chem Soc. Res: Effects of specific chemicals on the nucleic acids, protein and lipids on human cells grown in culture. Mailing Add: 1502 Mushogean Dr Johnson City TN 37601

BLEVINS, ROBERT L, b Finely, Ky, Oct 24, 31; m 55; c 2. SOIL SCIENCE. Educ: Univ Ky, BS, 52, MS, 58; Ohio State Univ, PhD(agron), 67. Prof Exp: Soil scientist, Soil Conserv Serv, 58-61; mem staff, Eastern Ky Resource Develop Prog, 61-62; asst prof soil surv, Univ Ky, 62-65; res assoc agron, Ohio State Univ, 65-67; ASST PROF AGRON, UNIV KY, 67- Mem: Am Soc Agron; Soil Sci Soc Am. Res: Forest soils; soil classification; genesis and land use, especially problems involving soil management. Mailing Add: Dept of Agron Univ of Ky Lexington KY 40506

BLEVIS, BERTRAM CHARLES, b Toronto, Ont, Feb 26, 32; m 59; c 3. PHYSICS. Educ: Univ Toronto, BA, 53, MA, 54, PhD(physics), 56. Prof Exp: Sci officer radio physics, Defence Res Telecommun Estab, 56-57, group leader ultra high frequency propagation, 57-62, sci officer satellite commun, 62-66; sci liaison officer, Can Defence Res Staff, Washington, DC, 66-69; leader satellite commun sect, Commun Res Ctr, 69-71, mgr space systs prog, 71-75, DIR INT ARRANGEMENTS, DEPT OF COMMUN, 75- Concurrent Pos: Mem Can Nat Orgn, Int Radio Consult Comt, 61-74, deleg, Int Telecommun Union, 63-; mem Can comn II, Inst Sci Radio Union, 65- Mem: Sr mem Inst Elec & Electronics Engrs; Can Asn Physicists. Res: Gaseous discharge electronics; lunar radar; radio auroral backscattering; radio propagation, particularly application to space communications; satellite communications and broadcasting applications and systems studies. Mailing Add: Dept of Commun 300 Slater St Ottawa ON Can

BLEWETT, CHARLES WILLIAM, b Norfolk, Va, Aug 15, 43; m 64; c 3. ORGANIC CHEMISTRY. Educ: Thomas More Col, AB, 63; Univ Cincinnati, PhD(org chem), 69. Prof Exp: Chemist, 68-73, GROUP LEADER, EMERY INDUSTS, INC, 73- Mem: Am Chem Soc; Soc Cosmetic Chemists; Am Oil Chemists Soc. Res: Fatty acid synthesis and applications; organometallic chemistry; oxo chemistry. Mailing Add: Emery Industs Inc 4900 Este Ave Cincinnati OH 45232

BLEWETT, JOHN PAUL, b Toronto, Ont, Apr 12, 10; nat US; m 36. PHYSICS. Educ: Univ Toronto, BS, 32, MA, 33; Princeton Univ, PhD(physics), 36. Prof Exp: Royal Soc Can fel, Cambridge Univ, 36-37; res physicist, Gen Elec Co, 37-46; from scientist to sr scientist, 47-60, dep chmn accelerator dept, 60-72, SPEC ASST TO DIR FOR ENERGY, BROOKHAVEN NAT LAB, 73- Concurrent Pos: Ed, Particle Accelerators, 70- Mem: Fel Am Phys Soc; fel Inst Elec & Electronics Engrs; fel NY Acad Sci. Res: X-ray crystal analysis; mass spectrography; nuclear physics; oxide coated cathodes; microwaves; design of high energy electron and proton accelerators. Mailing Add: 650 Old Harbor Rd New Suffolk NY 11956

BLEWETT, MYRTLE HILDRED, b Toronto, Ont, May 28, 11; nat US; m 36; div. PHYSICS. Educ: Univ Toronto, BA, 35. Prof Exp: Engr appl physics, Gen Elec Co, 42-47; physicist, Brookhaven Nat Lab, 47-64; sr physicist, Argonne Nat Lab, Ill, 64-69; SR PHYSICIST, EUROP ORGN NUCLEAR RES, 69- Mem: Am Phys Soc. Res: Particle accelerators; high energy physics. Mailing Add: ISR Div Europ Orgn for Nuclear Res 1211 Geneva 23 Switzerland

BLEWITT, HARRY LYON, organic chemistry, biochemistry, see 12th edition

BLEWITT, THOMAS HUGH, b Cleveland, Ohio, Feb 8, 21; m 43; c 2. SOLID STATE PHYSICS. Educ: Case Inst Technol, BS, 42; NY Univ, BS, 44; Carnegie Inst Technol, DSc(physics), 50. Prof Exp: Sr physicist, Oak Ridge Nat Lab, 50-61; SR PHYSICIST, MAT SCI DIV, ARGONNE NAT LAB, 61-; PROF MAT ENG, UNIV ILL, CHICAGO CIRCLE, 65- Mem: Fel Am Phys Soc. Res: Irradiation effects; metal physics. Mailing Add: Mat Sci Div Argonne Nat Lab Argonne IL 60439

BLEYMAN, LEA KANNER, b Halle, Ger, Nov 9, 36; US citizen; c 1. GENETICS, CELL BIOLOGY. Educ: Brandeis Univ, BA, 58; Columbia Univ, MA, 60; Ind Univ, Bloomington, PhD(genetics of ciliates), 66. Prof Exp: Res assoc zool, Univ Ill, Urbana, 64-69; res assoc, Labs for Reproductive Biol, Univ NC, Chapel Hill, 69-72; ASSOC PROF BIOL, BARUCH COL, CITY UNIV NEW YORK, 73- Mem: AAAS; Genetics Soc Am; NY Acad Sci; Am Soc Cell Biol; Soc Protozoologists. Res: Genetics of ciliates, especially inheritance and determination of mating type specificity and expression and nuclear-cytoplasmic relations in cellular development. Mailing Add: Box 349 17 Lexington Ave New York NY 10010

BLEYMAN, MICHAEL ALAN, b New York, NY, Aug 25, 37; m 58; c 1. MOLECULAR GENETICS. Educ: Brooklyn Col, BA, 59; Ind Univ, Bloomington, MA, 64; Univ Ill, Urbana-Champaign, PhD(biol), 68. Prof Exp: Trainee dept microbiol, Univ Ill, Urbana-Champaign, 69; ASST PROF ZOOL & BIOCHEM, UNIV NC, CHAPEL HILL, 69- Mem: Am Soc Microbiol; Genetics Soc Am; Soc Protozool. Res: Transcriptional mapping; transcription and translation of nucleic acids; evolution of genetic code; regulation and control of cell metabolism at the molecular level. Mailing Add: Dept of Zool Univ of NC Chapel Hill NC 27514

BLICHER, ADOLPH, b Warsaw, Poland; nat US; m 34; c 1. SOLID STATE SCIENCE. Educ: Polytech Inst Poland, DSc(tech sci), 38. Prof Exp: Dir eng, Polish Broadcasting Co, 44-46; sr physicist, Radio Receptor Co, NY, 51-55; mgr spec prod develop & applns dept, RCA Corp, 55-56, mgr comput devices develop dept, Electronic Components & Devices, 56-62, mgr spec prod develop & applns dept, 63-64, mgr advan devices & applns dept, Solid State Technol Ctr, 64-73, consult, RCA Labs, Princeton, 73-74; CONSULT, 74- Mem: Sr mem Inst Elec & Electronics Engrs. Res: Solid state physics and electronics. Mailing Add: 345 Judges Lane North Plainfield NJ 07063

BLICK, JAMES DONALD, b Santa Monica, Calif, July 19, 22. GEOGRAPHY. Educ: Univ Calif, Los Angeles, AB, 47, MA, 50, PhD(geog), 56. Prof Exp: Asst prof geog, Univ Md, 56-57; instr, Univ Idaho, 57-58; instr geog & geol, Fullerton Jr Col, Calif, 58-60; asst prof, Univ of the Pac, 60-63, assoc prof geog, 63-66; asst prof, 66-69, ASSOC PROF GEOG, SAN DIEGO STATE UNIV, 69- Concurrent Pos: Secy-treas, Calif Coun Geog Educ, 61-64; NSF grant, Univ Fla & Univ Madrid, 65-66. Mem: Asn Am Geog; Am Geog Soc. Res: Geography of California; desert urbanization; urban expansion and house types. Mailing Add: Dept of Geog San Diego State Univ San Diego CA 92182

BLICKENSDERFER, PETER W, b Cincinnati, Ohio, Nov 28, 32; m 56; c 4. ANALYTICAL CHEMISTRY. Educ: Col Wooster, BA, 54; Univ Mo, PhD(phys chem), 63. Prof Exp: Instr chem, Univ Mo-Rolla, 61-62; assoc res chemist, Res Labs, Whirlpool Corp, 62-70, sr res chemist, 70-71; head dept, 72-75, ASSOC PROF CHEM, KEARNEY STATE COL, 71- Mem: Am Chem Soc. Res: Adsorption of gases on solids; surface chemistry of silicate minerals; high vacuum techniques; analytical methods development for trace inorganic constituents. Mailing Add: 1202 W 31st St Kearney NE 68847

BLICKENSTAFF, CARL CURTISS, entomology, see 12th edition

BLICKENSTAFF, ROBERT THERON, b North Manchester, Ind, July 3, 21; m 44; c 2. ORGANIC CHEMISTRY. Educ: Manchester Col, AB, 44; Purdue Univ, MS, 46, PhD(org chem), 48. Prof Exp: Res chemist, Procter & Gamble Co, 48-55; res assoc, Med Units, Univ Tenn, Memphis, 55-58; ASSOC PROF BIOCHEM, SCH MED, IND UNIV, INDIANAPOLIS, 64- Concurrent Pos: Res chemist, Vet Admin Hosp, Indianapolis, 58-70, exec secy res, 70-73. Mem: Am Chem Soc; Sigma Xi; Am Inst Chemists. Res: Steroids. Mailing Add: Vet Admin Hosp 1481 W Tenth St Indianapolis IN 46202

BLICKLE, ROBERT LOUIS, b Ironton, Ohio, Nov 12, 13; m 47; c 1. ENTOMOLOGY. Educ: Ohio State Univ, BS, 37, PhD(entom), 42; Univ NH, MS, 39. Prof Exp: Asst, Exp Sta, La State Univ, 37; asst chem & entom, Univ NH, 38-41; asst zool, Ohio State Univ, 41-42; from asst prof to assoc prof, 46-57, PROF ENTOM, UNIV NH, 57-, ACTG CHMN DEPT, 69- Mem: Entom Soc Am. Res: Synergistic action of chemicals on organic insecticides; taxonomy of Diptera, Hymenoptera and Trichoptera; study of parasites of Synanthropic flies. Mailing Add: Dept of Entom Univ of NH Durham NH 03824

BLIESE, JOHN CARL WILLIAM, b Waterloo, Iowa, Mar 10, 13; m 39; c 2. ZOOLOGY. Educ: Univ Northern Iowa, BA, 35; Columbia Univ, MA, 36; Iowa State Univ, PhD(zool), 53. Prof Exp: High sch instr, Iowa, 36-41; sci supvr, Campus Sch, Univ Northern Iowa, 41-42; civilian instr, US Air Force, 42-44 & Tenn Eastman Corp, 44-45; sci supvr, Campus Sch, Univ Northern Iowa, 45-47; instr biol, Cornell Col, 47-49; instr zool, Iowa State Univ, 49-53; assoc prof, 53-70, head dept, 62-66, PROF BIOL, KEARNEY STATE COL, 70- Mem: Am Sci Affil; Am Inst Biol Sci; Nat Asn Biol Teachers; Am Ornithologists Union. Res: Photoperiodism of aquarium plants; roosting of bronzed grackles and associated birds; diurnal ecology of sandhill cranes. Mailing Add: Dept of Biol Kearney State Col Kearney NE 68847

BLIESNER, WILLIAM CLARK, b Twin Falls, Idaho, Apr 27, 35; m 57; c 4. PAPER & PULP TECHNOLOGY, CHEMICAL ENGINEERING. Educ: Univ Idaho, BS, 57; Lawrence Univ, MS, 61, PhD(pulp & paper technol), 63. Prof Exp: Assoc staff tech dir, Potlach Forests, Inc, Idaho, 63-64; res engr, Res & Develop Div, Beloit Corp, Wis, 64-66; res assoc, Res & Develop Div, Consol Papers, Inc, 66-71, lab & staff mgr, 68-71; dir res & develop, 71-74, VPRES RES & DEVELOP, APPLETON MILLS, 75- Honors & Awards: George Olmsted Award, Am Paper Inst, 70. Mem: Tech Asn Pulp & Paper Indust. Res: Advanced technology in the area of wet pressing of paper, properties and functions of wet press fabrics. Mailing Add: Appleton Mills PO Box 1899 Appleton WI 54911

BLINCOE, CLIFTON (ROBERT), b Odessa, Mo, Nov 21, 26; m 49; c 3. PHYSICAL BIOCHEMISTRY. Educ: Univ Mo, BS, 47, MA, 48, PhD, 55. Prof Exp: Instr dairy husb, Univ Mo, 50-55, asst prof, 55-56; asst prof & asst res chemist, 56-60, assoc prof biochem, 60-69, PROF BIOCHEM, UNIV NEV, RENO, 69-, RES CHEMIST, 60- Concurrent Pos: Vis prof US-Ireland scholar exchange prog, Trinity Col, Dublin, 69-70; Nat Feed Ingredients Asn travel fel, 74. Mem: AAAS; Am Chem Soc; Am Soc Animal Sci; NY Acad Sci; Brit Soc Chem Indust. Res: Physical biochemistry of domestic animals as studied with radiotracers. Mailing Add: Div of Biochem Univ of Nev Reno NV 89507

BLINDER, SEYMOUR MICHAEL, b New York, NY, Mar 11, 32. THEORETICAL CHEMISTRY. Educ: Cornell Univ, AB, 53; Harvard Univ, AM, 55, PhD(chem physics), 59. Prof Exp: Physicist, Appl Physics Lab, Johns Hopkins Univ, 58-61; asst prof, Carnegie Inst Technol, 61-62; res assoc, Harvard Univ, 62-63; from asst prof to assoc prof, 63-70, PROF CHEM, UNIV MICH, 70- Concurrent Pos: Guggenheim fel, Univ Col, Univ London, 65-66; NSF sr fel, Ctr de Mecanique Ondulatoire Appliquee, Paris, 70-71 & Math Inst, Oxford, 70-71. Mem: Am Phys Soc. Res: Theoretical chemistry; applications of quantum mechanics to atomic and molecular problems. Mailing Add: Dept of Chem Univ of Mich Ann Arbor MI 48104

BLINKS, JOHN ROGERS, b New York, NY, Mar 21, 31; m 53; c 3. PHARMACOLOGY. Educ: Stanford Univ, AB, 51; Harvard Univ, MD, 55. Prof Exp: Med house officer, Peter Bent Brigham Hosp, Boston, 55-56; res assoc, Nat Heart Inst, Bethesda, Md, 56-58; instr pharmacol, Harvard Med Sch, 58-61, assoc, 61-64, asst prof, 64-68; assoc prof, Mayo Grad Sch Med, 68-73, PROF PHARMACOL, MAYO MED SCH, UNIV MINN, 73-; HEAD DEPT PHARMACOL, MAYO FOUND, 68- Concurrent Pos: Markle scholar med sci, 61-66; estab investr, Am Heart Asn, 65-70; hon res asst, Univ Col, Univ London, 62-63; mem prog proj comt, Nat Heart & Lung Inst, 68-71; field ed, J Pharmacol & Exp Therapeut, 69-71; vis lectr, Dept Physiol, Univ Auckland, 74; mem res comt, Am Heart Asn, 75-79; mem cardiol adv comt, Nat Heart & Lung Inst, 75-79. Mem: Biophys Soc; Cardiac Muscle Soc; Am Physiol Soc; Am Soc Pharmacol & Exp Therapeut; Soc Gen Physiologists. Res: Physiology and pharmacology of heart muscle. Mailing Add: Dept of Pharmacol Mayo Found Rochester MN 55901

BLINKS, LAWRENCE ROGERS, b Michigan City, Ind, Apr 22, 00; m 28; c 1. PLANT PHYSIOLOGY. Educ: Harvard Univ, BS, 23, MA, 25, PhD(gen physiol), 26. Prof Exp: Asst, Div Gen Physiol, Rockefeller Inst, 26-29, assoc, 29-33; assoc prof plant physiol, 33-36, prof biol, 36-65, dir, Hopkins Marine Sta, 43-65, EMER PROF BIOL, STANFORD UNIV, 65- Concurrent Pos: Guggenheim fel, Fla, Cold Spring Harbor & Woods Hole, 39-40, Rockefeller Inst & Nobel Inst, 49; with Bikini Sci Resurv, 47; Am Scand fel, 49; ed, J Gen Physiol, 51-62; consult, NSF, 51-65, asst dir, 54-55; ed, Biol Bull, 52-55; Fulbright scholar, Cambridge Univ, 57; ed, Botanica Marine, 59-; with Te Vega Exped, Borneo & Solomon Islands, 65; prof, Univ Calif, Santa Cruz, 65-73, emer prof, 73-; with Alpha Helix, Great Barrier Reef, 66. Honors & Awards: Hales Award, Am Soc Plant Physiol, 52. Mem: Nat Acad Sci; AAAS (vpres, 55); fel Am Acad Arts & Sci; Am Soc Plant Physiol; Soc Gen Physiol (pres, 52). Res: Electrical phenomena in single cells; physiology of algae; photosynthesis; permeability; salt accumulation. Mailing Add: Hopkins Marine Sta Stanford Univ Pacific Grove CA 93950

BLINN, DEAN WARD, b Carroll, Iowa, May 30, 41; m 65; c 2. PHYCOLOGY, AQUATIC ECOLOGY. Educ: Simpson Col, BA, 64; Univ Mont, MA, 66; Univ BC, PhD(phycol), 69. Prof Exp: Asst prof biol, Univ NDak, 69-71; ASST PROF BIOL, NORTHERN ARIZ UNIV, 71- Concurrent Pos: Dir, Ariz Pub Serv, 74-; dir, Nat Park Serv, 75- Res: Ecology, nutrition and life histories of freshwater species of algae; effects of thermal pollution on algae; effect of heavy metals on algae. Mailing Add: Dept of Biol Sci Northern Ariz Univ Flagstaff AZ 86001

BLINN, ELLIOTT L, b Pittsburgh, Pa, Sept 25, 40; m 69. INORGANIC CHEMISTRY. Educ: Univ Pittsburgh, BS, 62; Ohio State Univ, MS, 64, PhD(inorg chem), 67. Prof Exp: Asst chem, Ohio State Univ, 61-62; fel, State Univ NY Buffalo, 67-68; asst prof, 68-74, ASSOC PROF CHEM, BOWLING GREEN STATE UNIV, 74- Mem: Am Chem Soc; Brit Chem Soc. Res: Coordination chemistry; reaction mechanism of metal complexes. Mailing Add: Dept of Chem Bowling Green State Univ Bowling Green OH 43402

BLINN, ROGER C, b Long Beach, Calif, Jan 17, 21; m 47; c 4. PESTICIDE CHEMISTRY. Educ: Univ Calif, Los Angeles, AB, 43. Prof Exp: Chemist, Richfield Oil Corp, 43-46; chemist pesticide res, Univ Calif, Riverside, 46-63; res chemist, 63-69, GROUP LEADER, METAB GROUP, AM CYANAMID CO, 69- Mem: Fel Am Chem Soc; Entom Soc Am. Res: Development of analytical methods for pesticide residues. Mailing Add: Agr Div Am Cyanamid Co PO Box 400 Princeton NJ 08540

BLINN, WALTER CRAIG, b Belleville, Ill, May 16, 30; m 57; c 2. SCIENCE EDUCATION, NATURAL SCIENCE. Educ: Ill State Univ, BS, 51; Okla State Univ, MS, 58; Northwestern Univ, PhD(biol), 61. Prof Exp: Instr biol, Northwestern Univ, 60-61; from instr to assoc prof, 61-69, PROF NATURAL SCI, MICH STATE UNIV, 69- Concurrent Pos: Consult, Welch Sci Co, 61-64; NSF res grant, 63-65. Mem: AAAS. Res: History and philosophy of science; metaphysical foundations of scientific thought. Mailing Add: Dept of Natural Sci Univ Col Mich State Univ East Lansing MI 48824

BLINT, RICHARD JOSEPH, b Fairmont, Minn, Feb 23, 45; m 66; c 3. CHEMICAL PHYSICS. Educ: St Mary's Col, Minn, BA, 67; Calif Inst Technol, PhD(chem), 72. Prof Exp: Presidential internship, Brookhaven Nat Lab, 72-73; assoc chem, Nat Res Coun, 73-74; RES ASSOC PHYSICS, UNIV ILL, 74- Mem: Am Inst Physics. Res: Molecular quantum mechanics is utilized to investigate electronic excitation in ion-atom systems; surface adsorption and catalytic activity are studied through the energetics of small molecular aggregates. Mailing Add: Dept of Physics & Mat Res Lab Univ of Ill Urbana IL 61801

BLISCHKE, WALLACE ROBERT, b Oak Park, Ill, Apr 20, 34; m 60; c 1. STATISTICAL ANALYSIS. Educ: Elmhurst Col, BS, 56; Cornell Univ, MS, 58, PhD(statist), 61. Prof Exp: Res fel biomath, NC State Col, 61-62; mem tech staff, Appl Math Dept, TRW Space Tech Labs, 62-64; sr statistician & proj dir, C-E-I-R, Inc, 64-69; CONSULT STATIST ANAL, 69-; ASSOC PROF, SCH BUS, UNIV SOUTHERN CALIF, 72- Mem: Inst Math Statist; Am Statist Asn; Biomet Soc; Inst Mgt Sci. Res: Mathematical statistics, especially estimation, mixtures of distributions and discrete distributions; applied statistics, especially experimental design, sample surveys, regression and related data analysis techniques. Mailing Add: Dept of Quantitative Bus Anal Univ of Southern Calif Los Angeles CA 90007

BLISS, ARTHUR DEAN, b Waterbury, Conn, Jan 7, 27; m 57; c 2. ORGANIC CHEMISTRY. Educ: Yale Univ, BS, 49, PhD(org chem), 55. Prof Exp: Res chemist, Olin Mathieson Chem Corp, 55-57, sr res chemist, 57-63, res assoc, 63-69; sect head nylon res & develop, 69-74, MGR, RES LAB, FOSTER GRANT CO, INC, 74- Mem: Am Chem Soc. Res: Synthesis and rearrangements of amino-alcohols; preparation of boron hydrides and organic thiophosphates; preparation and reaction of aliphatic sultams; condensation polymerization; nylon 6; nylon 12; copolyamides; addition polymerization; polystyrene; acrylonitrile butadiene styrene. Mailing Add: Prospect Hill Rd Harvard MA 01451

BLISS, CHESTER ITTNER, b Springfield, Ohio, Feb 1, 99. BIOMETRICS. Educ: Ohio State Univ, AB, 21; Columbia Univ, AM, 22, PhD(zool), 26. Prof Exp: From assoc entomologist to entomologist, Bur Entom, USDA, 26-33; res student, Galton Lab, Univ Col, Univ London, 33-35; mem, Inst Plant Protection, Leningrad, 35-37; lectr biomet, Yale Univ, 42-63, sr res assoc & lectr statist, 63-67, res assoc, 68-75; RETIRED. Concurrent Pos: Consult biometrician, 38-; biometrician, Conn Agr Exp Sta, 40-70 & Storrs Exp Sta, 40-53; mem revision comt, US Pharmacopoeia, 50-60; mem, Gonville & Caius Col, Cambridge Univ, 53-54. Mem: Biomet Soc (secy, 47-55, treas, 51-56, pres, 62-63, vpres, 64); fel Am Statist Asn (vpres, 43 & 47); fel Inst Math Statist; hon fel Royal Statist Soc; Int Statist Inst. Res: Statistical methods in biology; biological assay; experimental design in biological, agricultural and medical research. Mailing Add: 597 Prospect St New Haven CT 06511

BLISS, DOROTHY CRANDALL, b Westerly, RI, Feb 20, 16. ECOLOGY. Educ: Univ RI, BS, 36, MS, 38; Univ Tenn, PhD, 57. Prof Exp: Teacher high sch, RI, 36-37; instr sci & math, Greenbrier Jr Col, 41-43, Southern Sem & Jr Col, 43-47; instr bot, Univ Wyo, 47-49; from asst prof to assoc prof, 49-68, PROF BIOL, RANDOLPH-MACON WOMAN'S COL, 68- Concurrent Pos: Southern Fel Fund scholar, 55-56. Mem: Ecol Soc Am; Am Inst Biol Sci; Sigma Xi. Res: Ferns of Rhode Island; ecological problem in the spruce-fir region of the Smoky Mountains. Mailing Add: Dept of Biol Randolph-Macon Womans Col Box 278 Lynchburg VA 24504

BLISS, DOROTHY ELIZABETH, b Cranston, RI, Feb 13, 16. INVERTEBRATE PHYSIOLOGY. Educ: Brown Univ, AB, 37, ScM, 42; Radcliffe Col, PhD, 52. Hon Degrees: ScD, Brown Univ, 72. Prof Exp: Teacher, Milton Acad, Mass, 42-47 & 48-49; res fel biol, Harvard Univ, 52-55; res asst prof anat, Albert Einstein Col Med, 56-64, vis res assoc prof, 64-66; from asst cur to cur living invert, 56-74, CHAIRWOMAN & CUR DEPT FOSSIL & LIVING INVERT, AM MUS NATURAL HIST, 74- Concurrent Pos: NSF res grant, 57-; adj prof, City Univ New York, 71- Mem: Fel AAAS; Am Soc Zool; Am Inst Biol Sci. Res: Invertebrate zoology; neuroendocrinology. Mailing Add: Am Mus of Natural Hist Cent Park W at 79th St New York NY 10024

BLISS, EUGENE LAWRENCE, b Pittsburgh, Pa, Feb 18, 18; m 47; c 3. MEDICINE, PSYCHIATRY. Educ: Yale Univ, BA, 39; NY Med Col, MD, 43. Prof Exp: Lectr psychiat, Col Med, Univ Utah, 50-51, asst prof, 51-53; asst clin prof, Sch Med, Yale Univ, 53-56; assoc prof, 56-61, FOUND FUND PROF PSYCHIAT, COL MED, UNIV UTAH, 61-, CHMN DEPT, 70- Concurrent Pos: Consult, Vet Admin Hosp, Salt Lake City, 50-53 & 56- Honors & Awards: Moses Award, 55. Mem: Am Psychosom Soc; Am Psychiat Asn; AMA. Res: Neurochemistry; neuroendocrinology. Mailing Add: Dept of Psychiat Univ of Utah Salt Lake City UT 84112

BLISS, FREDRICK ALLEN, b Inavale, Nebr, Dec 5, 38; m 63; c 2. PLANT GENETICS, PLANT BREEDING. Educ: BSc, Univ Nebr, 60; Univ Wis, PhD(hort, genetics), 65. Prof Exp: Fel genetics, Univ Minn, St Paul, 65-66; asst prof, 66-71, ASSOC PROF HORT, UNIV WIS-MADISON, 71- Concurrent Pos: Lectr, Univ Ife, Nigeria, 66-68; mem vis fac, Inst Agron & Plant Genetics, Univ Göttingen, 68-69. Mem: Am Soc Hort Sci; Crop Sci Soc Am. Res: Plant breeding and genetical research dealing with Phaseolus vulgaris; quantitative genetics studies; genetic control of protein synthesis; edible legumes. Mailing Add: Dept of Hort Univ of Wis Madison WI 53705

BLISS, HORACE HOPKINS, chemistry, see 12th edition

BLISS, LAURA, b Davenport, Iowa, Apr 10, 16. CHEMISTRY. Educ: Iowa State Col, BS, 38, PhD(enzyme chem), 43. Prof Exp: Patent chemist, Hercules Powder Co, Del, 43-45; from instr to asst prof, 45-57, ASSOC PROF CHEM, RANDOLPH-MACON WOMAN'S COL, 57- Concurrent Pos: NSF sci fac fel, Inst Enzyme Res, 61-62; researcher, Dept Biochem, Univ Queensland, 68-69. Mem: AAAS; Am Chem Soc. Res: Enzyme chemistry; phosphorylase; hydrolysis of beta-D-glucosides; phosphorylase of waxy maize; acetylesterases. Mailing Add: Dept of Chem Randolph Macon Woman's Col Lynchburg VA 24503

BLISS, LAWRENCE CARROLL, b Cleveland, Ohio, Nov 29, 29; m 52; c 2. BOTANY. Educ: Kent State Univ, BS, 51, MS, 53; Duke Univ, PhD(bot), 56. Prof Exp: Instr biol, Bowling Green State Univ, 56-57; from instr to prof bot, Univ Ill, Urbana, 57-68; PROF BOT & DIR CONTROLLED ENVIRON FACIL, UNIV ALTA, 68- Concurrent Pos: Fulbright res scholar, NZ, 63-64; assoc ed, Oecologia, 69-75; dir Can tundra terrestrial productivity study, Int Biol Prog, 70-75; assoc ed, Arctic & Alpine Res; mem environ protection bd, Can Arctic Gas Pipeline Ltd, 71-; mem consultative comt, Nat Mus Can, 75-; dir revegetation modeling study, Alta Oil Sands, 75- Mem: AAAS; Am Inst Biol Sci; Arctic Inst NAm; Ecol Soc Am; Can Bot Asn. Res: Arctic ecosystem and manipulation studies; plant production; plant-soil relationships; microenvironmental studies; eco-physiology, including plant growth chamber research; tundra and boreal forest ecology. Mailing Add: Dept of Bot Univ of Alta Edmonton AB Can

BLISS, MYRON, JR, b Jacksonville, Fla, Aug 20, 43; m 65; c 3. ENTOMOLOGY. Educ: Ohio State Univ, BS, 65; Pa State Univ, MS, 67, PhD(entom), 70. Prof Exp: Res specialist entom, Ciba-Geigy Corp, Ardsley, NY & Greensboro, NC, 70-74, tech planning coordr, 74; PROD SUPVR INSECTICIDES & FUNGICIDES, DIAMOND SHAMROCK CORP, CLEVELAND, 74- Mem: Entom Soc Am; AAAS. Res: Development of insecticides and fungicides for agricultural use. Mailing Add: 5957 Chestnut Hills Dr Parma OH 44129

BLISSITT, CHARLES W, b College Park, Ga, Aug 1, 32; m 59; c 3. PHARMACY. Educ: Southern Col Pharm, BS, 54; Univ Fla, PhD(pharm), 58. Prof Exp: From asst prof to assoc prof pharm, WVa Univ, 58-64; Dean, St Louis Col Pharm, 64-70; DEAN COL PHARM, UNIV OKLA, 70-, PROF PHARM, 72- Res: Pharmaceutical delivery systems; drug abuse education. Mailing Add: Col of Pharm Univ of Okla 625 Elm Norman OK 73069

BLITCH, LEE WESLEY, b Cairo, Ga, Jan 31, 02; m 31; c 2. CHEMISTRY. Educ: Emory Univ, BS, 22; Johns Hopkins Univ, PhD(chem), 25. Prof Exp: From instr to asst prof, 25-27, asst prof, Valdosta div, 34-42, from assoc prof to prof, 42-74, EMER PROF CHEM, 74- Mem: Am Chem Soc. Mailing Add: Dept of Chem Emory Univ Atlanta GA 30322

BLITZ, RUTH R, b Boston, Mass, Nov 27, 35; m 63. BACTERIOLOGY, IMMUNOLOGY. Educ: Brandeis Univ, BA, 57; Univ Calif, Berkeley, MA, 59, PhD(bact), 65. Prof Exp: Instr immunochem, Univ Calif, Berkeley, 62-63; asst prof biol, Calif State Col Hayward, 64-65; from asst prof to assoc prof, 65-73, PROF BIOL, SONOMA STATE COL, 73- Res: Comparative immunology, especially immune mechanisms of lower animals and phylogenetic development of specific immune responses. Mailing Add: Dept of Biol Calif State Col Sonoma Rohnert Park CA 94928

BLITZER, LEON, b New York, NY, Dec 13, 15; m 42; c 2. CELESTIAL MECHANICS. Educ: Univ Ariz, BS, 38, MS, 39; Calif Inst Technol, PhD(physics), 43. Prof Exp: Instr physics, Calif Inst Technol, 43-45, res assoc, Off Sci Res & Develop Proj, 42-46; from asst prof to assoc prof physics, 46-50, PROF PHYSICS, UNIV ARIZ, 50- Concurrent Pos: Consult, Space Technol Labs, 55-70 & Jet Propulsion Lab, Calif Inst Technol, 71-75. Mem: AAAS; Am Phys Soc; Am Asn Physics Teachers. Res: Exterior ballistics of rockets; stellar spectroscopy; spectroscopy; satellite orbit perturbations; satellite and planetary perturbation theory. Mailing Add: Dept of Physics Univ of Ariz Tucson AZ 85721

BLITZSTEIN, WILLIAM, b Philadelphia, Pa, Sept 3, 20; m 44; c 2. ASTRONOMY. Educ: Univ Pa, PhD, 50. Prof Exp: Physicist with Dr J Razek, 41-47; res engr, Franklin Inst Labs, 47-52; electronic scientist, Frankford Arsenal, 52-54; res assoc, 47-54, asst prof astron & elec eng, 54-60, assoc prof astron, 60-64, PROF ASTRON, UNIV PA, 64- Concurrent Pos: Consult, Princeton Observ, 48-50, Decker Corp, Radio Corp Am & Spitz Labs, 58- Mem: Int Astron Union; Am Astron Soc; Optical Soc Am; Royal Astron Soc. Res: Electronic instrumentation; astronomical photoelectric photometry; eclipsing binaries; celestial mechanics. Mailing Add: Dept of Astron Univ of Pa Philadelphia PA 19104

BLIVAISS, BEN BURTON, b Chicago, Ill, July 4, 17; m 46; c 3. PHYSIOLOGY. Educ: Univ Chicago, BS, 38, MS, 40, PhD(zool), 46. Prof Exp: Lab asst biol, Wright Jr Col, 38-41; lab asst zool, Univ Chicago, 43-44, asst zool, Toxicity Lab, 45-46, Dept Zool, 46; instr physiol, Univ Ill, 46-47; guest investr zool, Univ Chicago, 47-48; from instr to assoc prof, 48-66, PROF PHYSIOL, CHICAGO MED SCH, 66-, COORDR, BASIC MED SCI EDUC, 75- Mem: AAAS; Aerospace Med Asn; Am Soc Zoologists; Endocrine Soc; Soc Exp Biol & Med. Res: Cryptorchidism; thyroid-gonad relations in fowl; mating behavior of fowl; thiouracil on pigment; pharmacology of respiratory poisons; testicular tumors; adrenal-gonad relations; steroid hormone metabolism; experimental arthritis; aerospace physiology; nutritional effects on implantation and gestation; experimental carcinogenesis. Mailing Add: Dept Physiol & Biophys Chicago Med Sch Chicago IL 60612

BLIVEN, CHARLES WATSON, pharmacy, see 12th edition

BLIVEN, FLOYD E, JR, b Erie, Pa, May 20, 21; m 50; c 6. ORTHOPEDIC SURGERY. Educ: Univ Rochester, AB & MD, 42; Am Bd Orthop Surg, dipl, 56. Prof Exp: From instr to asst prof, Sch Med, Univ Rochester, 52-56; assoc prof, 56-62, PROF ORTHOP SURG, MED COL GA, 62-; CHIEF ORTHOP, TALMADGE MEM HOSP, AUGUSTA, 56- Concurrent Pos: Consult, Battery State Hosp Tuberc, 56-, Vet Admin Hosp, Augusta, 56-, Crippled Children Serv, Ga Dept Pub Health, 60-, Milledgeville State Hosp, 63- & US Army Hosp Spec Treatment Ctr, Ft Gordon, 65- Mem: Am Acad Orthop Surg; Orthop Res Soc; Am Col Surg; AMA. Res: Infection of bone and joint; circulation and developmental studies of the hip; disability evaluation. Mailing Add: Dept of Surg Med Col of Ga Augusta GA 30902

BLIZARD, DAVID ARTHUR, b Hastings, Eng, Mar 26, 42; m 68; c 1. BEHAVIORAL BIOLOGY. Educ: Univ Wales, BA, 65, PhD(psychol), 68. Prof Exp: Fel behav genetics, Jackson Lab, 68-70; guest investr physiol psychol, Rockefeller Univ, 70-72; ASST PROF NEUROL, MED CTR, NY UNIV, 72- Concurrent Pos: Adj asst prof, Rockefeller Univ, 72- Mem: Behav Genetics Asn; Asn Study Animal Behav; Brit Psychol Soc. Res: Role of hormones and biogenic amines in etiology of sex and genetic influences on behavior; psychosomatic illness. Mailing Add: Milbank Labs 340 E 24th St New York NY 10010

BLIZARD, ROBERT BROOKS, b New York, NY, Dec 23, 23. ACOUSTICS. Educ: Princeton Univ, AB, 43; Mass Inst Technol, PhD(physics), 49. Prof Exp: Mem staff, Sperry Gyroscope Co, 49-52; mem staff, Schlumberger Well Surv Corp, 52-59; res scientist, Martin-Marietta Denver Div, 59-75; MEM STAFF, SCHLUMBERGER TECHNOL CORP, 75- Res: Communication theory; inertial instrumentation; optics; acoustic logging of oil wells. Mailing Add: Schlumberger-Doll Res Ctr PO Box 307 Ridgefield CT 06877

BLIZARD-COX, JANE BERGGREN, b Rochester, NY, Dec 13, 24; m 72. PHYSICS. Educ: Univ Rochester, AB, 45; Mass Inst Technol, PhD(physics), 49. Prof Exp: Lectr physics, Hofstra Col, 49-51, asst prof, 51-52; lectr, Univ Conn, 53-54; physicist, NEng Inst Med Res, 54-56; lectr, Vassar Col, 55-56; asst prof, Tex & Anderson Hosp, 56-60; asst res scientist, Martin-Marietta Corp, 60-63; res scientist, Physics Eng & Chem Corp, 64-65; res physicist, Denver Res Inst, Univ Denver, 65-70; researcher, Earth Sci Curric Prog, 71-73, TUTOR PHYSICS, DORM TUTORING PROG, UNIV COLO, 72- Mem: AAAS. Res: Space, nuclear and solar physics; solar activity. Mailing Add: 827 16th St Boulder CO 80302

BLIZNAKOV, EMILE GEORGE, b Kamen, Bulgaria, July 28, 26; m 54. MICROBIOLOGY, IMMUNOLOGY. Educ: Acad Med, Sofia, MD, 53. Prof Exp: Dir, Regional Sanit-Epidemiol Sta, Pirdop, Bulgaria, 53-55; staff scientist-microbiologist, Res Inst Epidemiol & Microbiol, Sofia, 55-59; SR STAFF SCIENTIST, NEW ENG INST MED RES, 61- Concurrent Pos: Fel, Res Inst Epidemiol & Microbiol, Moscow, USSR, 58-59. Mem: AAAS; Am Soc Microbiol; Am Soc Neurochem; Reticuloendothelial Soc; fel Royal Soc Trop Med & Hyg. Res: Administration of research; specific and nonspecific resistance to experimental bacterial, parasitic and viral infections and tumors; phagocytosis of particles by the reticuloendothelial system; mediators and their metabolism in allergy and shock; experimental and clinical screening for drugs with anti-cancer and anti-infectious effect. Mailing Add: New Eng Inst for Med Res PO Box 308 Ridgefield CT 06877

BLIZZARD, RICHARD REESE, SR, b Cincinnati, Ohio, Nov 16, 20; m 43; c 3. ORGANIC CHEMISTRY, MATHEMATICS. Educ: Univ Cincinnati, BA, 47. Prof Exp: Chemist, Carthage Mills, Inc, 47-49, Ander Chem Co, 49-60 & Ander Chem Corp Div, Columbian Carbon Co, 60-62; chemist, Levey Div, Cities Serv Co, 62-65, lab mgr printing ink, 65-67, mgr coatings & dispersions labs, 67-74; DIV SAFETY SUPVR, PRINTING INK DIV, BORDEN CHEM, DIV BORDEN INC, 74- Res: Research and development of alkyds, resins, varnishes, vehicles, conventional and specialty printing inks, protective coating systems, primers, special pigment dispersions and aerosol coatings. Mailing Add: 281 Bonham Rd Cincinnati OH 45215

BLIZZARD, ROBERT M, b East St Louis, Ill, June 20, 24. PEDIATRICS. Educ: Northwestern Univ, BS, 48, MD, 52; Am Bd Pediat, dipl. Prof Exp: Rotating intern, Iowa Methodist Hosp & Raymond Blank Mem Hosp Children, Des Moines, Iowa, 52-53; from asst resident physician to resident physician, Raymond Blank Mem Hosp Children, 53-55; fel pediat endocrinol & pediatrician, Harriet Lane Home Children, Johns Hopkins Hosp, 55-57; asst prof pediat & med, Ohio State Univ, 57-59, assoc prof pediat, 59-60, assoc, Div Endocrinol & Metab, Univ Hosp, 57-60; from assoc prof to prof pediat, Johns Hopkins Hosp, 60-73, chief div pediat endocrinol, 60-73, actg chmn dept pediat, 72-73; CHMN DEPT PEDIAT, SCH MED, UNIV VA, 73- Concurrent Pos: Dir, Nat Pituitary Agency, 63-67; pres, Human Growth Found. Honors & Awards: Ayerst Award, Endocrine Soc. Mem: Am Fedn Clin Res; Soc Pediat Res; Endocrine Soc; Am Pediat Soc; Am Thyroid Asn. Mailing Add: Dept of Pediat Univ of Va Sch of Med Charlottesville VA 22903

BLOCH, AARON NIXON, b Chicago, Ill, Feb 28, 42; m 67; c 2. CHEMICAL PHYSICS, SOLID STATE CHEMISTRY. Educ: Yale Univ, BS, 63; Univ Chicago, PhD(chem physics), 68. Prof Exp: Res assoc chem, Mass Inst Technol, 68-69; asst prof, 69-74, ASSOC PROF CHEM, JOHNS HOPKINS UNIV, 74- Concurrent Pos: Consult, Bell Labs, 71-73; fac visitor, Thomas J Watson Res Labs, IBM Corp, 74-75; A P Sloan Found fel, 74-76. Mem: Am Phys Soc. Res: Organic conductors; electronic processes in nearly one-dimensional structures; metal-insulator transitions; pseudopotential theory of chemical trends in solids; chemical physics of surfaces. Mailing Add: Dept of Chem Johns Hopkins Univ Baltimore MD 21218

BLOCH, ALAN, b New York, NY, Nov 28, 15; m; c 2. PHYSICS, MATHEMATICS. Educ: Swarthmore Col, BA, 38; Oberlin Col, MA, 39. Prof Exp: Teaching fel physics, Iowa State Col, 40-41; head comput develop sect, Arma Corp, 46-49; chief develop engr, Audio Instrument Co, Inc, 51-55; head syst anal dept, GPL Div, Gen Precision Inc, 55-60, staff consult to vpres res & eng, 60-61, regional mkt mgr, Librascope Div, 61-63, staff physicist, 63-65; staff analyst, Gyrodyne Co Am, Inc, 65-70; SCI CONSULT & WRITER, 70- Mem: AAAS; Am Phys Soc; Asn Comput Mach; Audio Eng Soc; NY Acad Sci. Res: Hard and social sciences, particularly system analysis and synthesi synthesis; interdisciplinary problem solving. Mailing Add: 333 E 30th St 14-J New York NY 10016

BLOCH, ALEXANDER, b Freiburg, Ger, Oct 6, 23; US citizen. BIOCHEMICAL PHARMACOLOGY, CANCER. Educ: City Col NY, BS, 54; Long Island Univ, MS, 58; Cornell Univ, PhD(microbiol, biochem), 62. Prof Exp: Asst biol, Long Island Univ, 57-58; asst microbiol, Sloan-Kettering Inst, 58-62; from scientist to assoc scientist, 62-74, PRIN SCIENTIST, CANCER RES, ROSWELL PARK MEM INST, 74- Concurrent Pos: Res prof, Niagara Univ, 68- & Canisius Col, 68-; assoc res prof, State Univ NY Buffalo, 68-75, res prof, 75- Mem: Soc Exp Biol & Med; fel NY Acad Sci; AAAS; Am Chem Soc; Am Asn Cancer Res. Res: Experimental cancer chemotherapy; biochemical pharmacology of cyclic nucleotides; mode of action of purine and pyrimidine analogs. Mailing Add: Roswell Park Mem Inst 666 Elm St Buffalo NY 14203

BLOCH, ALFRED, b Gailingen, Ger, Aug 8, 03; nat US; m; c 2. BIOCHEMISTRY. Educ: Univ Zurich, PhD(org chem), 27. Prof Exp: Asst instr chem, Univ Zurich, 27-28; dir lab, M E Bircher, MD, Switz, 28-33; owner clin chem lab, 33-38; microchemist, Biochem Res Found, Del, 39-42; assoc dir res, Ethicon, Inc, 43-68; INDUST CONSULT, 68- Concurrent Pos: Vis investr, Col Eng, Rutgers Univ, 74- Mem: AAAS; Am Chem Soc; Am Asn Clin Chemists; NY Acad Sci. Res: Fibrous proteins, especially collagen; tanning; wound healing; proteolytic enzymes; preparation of immobilized enzymes. Mailing Add: 9 Bloomfield Ave Somerset NJ 08873

BLOCH, BERNARD LOUIS, nuclear physics, see 12th edition

BLOCH, DAVID PAUL, b Chicago, Ill, Feb 10, 26; m 52; c 3. CYTOLOGY. Educ: Northwestern Univ, BS, 48; Univ Wis, PhD(bot), 52. Prof Exp: Asst bot, Univ Wis, 49-52; res assoc cytol, Col Physicians & Surgeons, Columbia Univ, 53-56; asst prof zool, Univ Calif, Los Angeles, 56-61; assoc prof, 61-73, PROF BOT, UNIV TEX, AUSTIN, 73- Concurrent Pos: Am Cancer Soc fel, 53-55; Damon Runyon Found fel, 55-56; USPHS & NSF grants, 57-; Guggenheim fel, 64-65; USPHS res career develop award, 64- Mem: AAAS; Histochem Soc; Am Soc Cell Biol; Bot Soc Am; Soc Develop Biol. Res: Cell biology; genetics and cytochemistry; nucleic acid and histone metabolism. Mailing Add: Dept of Bot Univ of Tex Austin TX 78712

BLOCH, EDWARD HENRY, b Berlin, Ger, Feb 1, 14; US citizen. ANATOMY. Educ: Univ Chicago, BSc, 39, PhD(microphysiol), 49; Univ Tenn, MD, 45. Prof Exp: Asst, Univ Tenn, 40-44; intern, Michael Reese Hosp, Chicago, 46; assoc, Univ Chicago, 46-48; from asst prof to assoc prof, 53-71, PROF ANAT, CASE WESTERN RESERVE UNIV, 71- Concurrent Pos: Estab investr, Am Heart Asn, 50-55. Mem: Microcirc Soc; Am Asn Immunol; Soc Rheol; Am Heart Asn; Am Asn Anatomists. Res: Physics and chemistry of red cell coatings; scanning microspectrophotometry of organs in vivo; microvascular physiology and pathology of the liver, spleen, gastrointestinal tract, muscle and lung; high speed cinephotography; systems analysis of organs; instrumentation for the microscopic study of living organs in situ. Mailing Add: Dept of Anat Case Western Reserve Univ Cleveland OH 44106

BLOCH, ERIC, b Munich, Ger, Apr 4, 28; nat US; m 61; c 2. BIOCHEMISTRY. Educ: City Col New York, BS, 48; Univ Tex, AM, 50, PhD(biochem), 53. Prof Exp: Asst, Univ Tex, 50-52; res biochemist, Worcester Found Exp Biol, 52-57; res assoc, Children's Cancer Res Found, 57-58; asst prof, 58-65, ASSOC PROF BIOCHEM, OBSTET & GYNEC, ALBERT EINSTEIN COL MED, 65- Concurrent Pos: Asst, Med Sch, Harvard Univ, 57; lectr, steroid training prog, Worcester Found Exp Biol, 57, 58; res career develop award, 60-69. Mem: Fel AAAS; Am Soc Biol Chemists; Endocrine Soc; Soc Study Reproduction; Soc Gynec Invest. Res: Fetal adrenal and testicular functional development; developmental and reproductive endocrinology; steroid hormone function. Mailing Add: Dept of Biochem Gynec & Obstet Albert Einstein Col of Med Bronx NY 10461

BLOCH, FELIX, b Zurich, Switz, Oct 23, 05; nat US; m 40; c 4. PHYSICS. Educ: Univ Leipzig, PhD, 28. Prof Exp: Lorentz-Fonds fel, Univ Utrecht, 30; Orsted-Fonds fel, Copenhagen Univ, 31; privat-docent, Univ Leipzig, 32; Int Educ Bd fel, Univ Rome, 33; prof physics, Stanford Univ, 34-41; investr, Manhattan Dist, 41-44; radio res lab, Harvard Univ, 44-45; prof, 45-74, EMER PROF PHYSICS, STANFORD UNIV, 74- Concurrent Pos: Lectr, Inst Henri Poincare, Paris, 33. Honors & Awards: Nobel Prize, physics, 52. Mem: Nat Acad Sci; fel Am Phys Soc; fel Am Acad Sci. Res: Electron theory of metals; quantum theory of ferromagnetism; stopping power of matter; x-ray phenomena; quantum-electrodynamics; neutrons; polarization of neutrons; magnetic moments of nuclei; superconductivity. Mailing Add: Dept of Physics Stanford Univ Stanford CA 94305

BLOCH, HERMAN SAMUEL, b Chicago, Ill, June 15, 12; m 40; c 3. PETROLEUM CHEMISTRY. Educ: Univ Chicago, BS, 33, PhD(org chem), 36. Prof Exp: Res chemist, Universal Oil Prod Co, 36-39, res group leader, 39-45, coordr chem res div, 45-55, dep dir refining res, 55-61, assoc dir process res, 61-64 & res, 64-73, DIR CATALYSIS RES, UOP, INC, 73- Concurrent Pos: Vpres, Chicago Tech Socs Coun, 47-49; chmn comt phys sci, Ill Bd Higher Educ, 69-70. Honors & Awards: Award of Merit, Chicago Tech Socs Coun, 57; Honor Scroll Award, Am Inst Chem, 57; Eugene J Houdry Award Appl Catalysis, Catalysis Soc, 71; E V Murphree Award Indust & Eng Chem, Am Chem Soc, 74. Mem: Nat Acad Sci; fel AAAS (vpres chem, 70); hon mem Am Inst Chem; Am Chem Soc; Soc Chem Indust. Res: Catalytic reforming, cracking and isomerization of hydrocarbons; chemicals from petroleum; synthetic resins; oils; paints; organic solvents; petroleum refining catalysis; automotive exhaust catalysts. Mailing Add: 9700 Kedvale Ave Skokie IL 60076

BLOCH, INGRAM, b Louisville, Ky, Aug 27, 20; m 45, 60; c 2. THEORETICAL PHYSICS. Educ: Harvard Univ, BA, 40; Univ Chicago, MS, 41, PhD(physics), 46. Prof Exp: Asst math biophys & lab asst physics, Univ Chicago, 41-42, asst instr electronics eng, sci & mgt war training, 42-43, jr scientist, Metall Lab, 43-44; asst scientist, Los Alamos Sci Lab, Univ Calif, 44-46; assoc nuclear physics, Univ Wis, 46-47; asst, Yale Univ, 47-48; from asst prof to assoc prof physics, 48-57, PROF PHYSICS, VANDERBILT UNIV, 57- Mem: Fel Am Phys Soc; AAAS; Am Asn Physics Teachers. Res: Quantum theory; theory of particles; relativity. Mailing Add: Dept of Physics Vanderbilt Univ Nashville TN 37235

BLOCH, KONRAD EMIL, b Neisse, Ger, Jan 21, 12; nat US; m. BIOLOGICAL CHEMISTRY. Educ: Tech Hochschule, Munich, Chem E, 34; Columbia Univ, PhD(biochem), 38. Prof Exp: Instr biochem, Columbia Univ, 41-44, assoc, 44-46; from asst prof to prof, Univ Chicago, 46-54; HIGGINS PROF BIOCHEM, HARVARD UNIV, 54- Concurrent Pos: Mem panel cmt growth, Nat Res Coun, 50-; Guggenheim fel, Tech Hochschule, Zurich, 53. Honors & Awards: Nobel Prize med physiol, 64; Fritsche Award, Am Chem Soc, 64. Mem: Nat Acad Sci; fel Am Acad Arts & Sci; Am Chem Soc; Am Soc Biol Chem (pres, 67); Am Philos Soc. Res: Structure, biosynthesis and function of lipids. Mailing Add: Dept of Chem Harvard Univ Cambridge MA 02138

BLOCH, SYLVAN C, b Vicksburg, Miss, Nov 9, 31; m 58; c 2. PHYSICS. Educ: US Coast Guard Acad, BS, 54; Univ Miami, MS, 59; Fla State Univ, PhD(physics), 62. Prof Exp: Physicist microwave spectros, Martin Co, Fla, 59; pres, Recon, Inc, 61, res dir, 61-63; asst prof physics, Fla State Univ, 63; from asst prof to assoc prof, 63-70, PROF PHYSICS, UNIV S FLA, 70- Concurrent Pos: Consult, Recon, Inc, 63- Mem: Am Phys Soc; Inst Elec & Electronics Eng; Am Asn Physics Teachers. Res: Theoretical and experimental plasma physics; atmospheric physics. Mailing Add: Dept of Physics Univ of S Fla Tampa FL 33620

BLOCHER, JOHN MILTON, JR, b Baltimore, Md, Jan 6, 19; m 41; c 5. PHYSICAL CHEMISTRY. Educ: Baldwin-Wallace Col, BS, 40; Ohio State Univ, PhD(phys chem), 46. Prof Exp: Asst chem, Ohio State Univ, 40-43, 45, 46-47; instr physics, Capital Univ, 43-44; jr chemist, Tenn Eastman Corp, 44-45; res engr, 46-50, asst supvr, 50-58, div chief, 58-69, FEL, BATTELLE-COLUMBUS LABS, 69- Mem: Am Chem Soc; Electro-chem Soc; Am Nuclear Soc. Res: Thermodynamics of halide vapor equilibria; mass spectrometry; photovoltaic effect; high purity metals; refractory coatings; chemical vapor deposition. Mailing Add: Mat Proc & Fabrication Dept Battelle-Columbus Labs 505 King Ave Columbus OH 43201

BLOCK, ARTHUR MCBRIDE, b Newark, NJ, June 26, 38; m 67; c 2. ENVIRONMENTAL CHEMISTRY, QUANTUM CHEMISTRY. Educ: Cornell Univ, AB, 61; Rutgers Univ, PhD(phys chem), 67. Prof Exp: Lab asst chem, Rutgers Univ, 62-67; lectr phys chem, Univ PR, 67-68, asst prof, 68-72; scientist I terrestrial ecol prog, 72-75, SCIENTIST II TERRESTRIAL ECOL PROG, PR NUCLEAR CTR, 75- Concurrent Pos: Asst prof sch pub health, Univ PR, 75- Mem: Am Chem Soc; Sigma Xi; Int Soc Quantum Biol; Quantum Chem Prog Exchange. Res: Chemistry of plant growth regulators; elemental transport in tropical rain forests; structure and electronic distribution of single electron reduced organic systems. Mailing Add: PR Nuclear Ctr Caparra Heights Sta San Juan PR 00935

BLOCK, BARRY, nuclear physics, geophysics, see 12th edition

BLOCK, BURTON PETER, b Saginaw, Mich, Sept 3, 23; m 44; c 3. INORGANIC CHEMISTRY. Educ: Harvard Univ, BS, 44; Univ Tenn, MS, 45; Univ Ill, PhD(chem), 49. Prof Exp: Jr chemist, Tenn Eastman Corp, 44; instr chem, Univ Chicago, 49-52; asst prof, Pa State Univ, 52-57; sr chemist, 57-58, proj leader, 58-59, group leader, 59-70, SR SCIENTIST, PENNWALT CORP, 70- Concurrent Pos: Mem staff, Manhattan Proj, 44. Honors & Awards: Am Chem Soc Award, 68. Mem: Am Chem Soc. Res: Preparation, stability and structure of coordination compounds; chemistry of less familiar elements; inorganic polymers; chemical nomenclature; inorganic antitumor agents. Mailing Add: Tech Div Pennwalt Corp King of Prussia PA 19406

BLOCK, DOUGLAS ALFRED, b Rockford, Ill, July 12, 21; m 60; c 2. GEOLOGY, STRATIGRAPHY. Educ: Wheaton Col, Ill, AB, 43; Univ Iowa, MS, 52; Univ NDak, PhD(Pleistocene geol), 65. Prof Exp: From instr to assoc prof geol, Wheaton Col, Ill, 45-67; prof geol, 67-71, PROF GEOL & EARTH SCI, ROCK VALLEY COL, 71- Mem: Geol Soc Am; fel Am Sci Affil; Nat Asn Geol Teachers. Res: Glacial geology and geomorphology; sedimentary petrology; Pleistocene stratigraphy. Mailing Add: Dept of Geology Rock Valley Col Rockford IL 61111

BLOCK, DUANE LLEWELLYN, b Madison, Wis, Dec 27, 26; m 49; c 2. OCCUPATIONAL MEDICINE. Educ: Univ Wis, BS, 49, MD, 51; Am Bd Prev Med, dipl & cert occup med, 61. Prof Exp: Plant physician, Cadillac Div, Gen Motors Corp, 52-54, med dir, Gen Motors Tech Ctr, 54-55; physician-in-chg, Rouge Med Serv, 55-70, MED DIR, FORD MOTOR CO, 70- Concurrent Pos: Mem vis staff, William Beaumont Hosp, 60-; clin asst prof occup & environ health, Med Sch, Wayne State Univ, 60-; lectr occup med, Sch Pub Health, Univ Mich, 60-; chmn bd, Occup Health Inst, 72-; mem, Mich Task Force Prev Health Serv & Mich Occup Health Stand Comn, 75-; mem, Am Bd Prev Med, 75-; vchmn air pollution res adv comt, Coord Res Coun, 75- Honors & Awards: Meritorious Serv Award, Am Occup Med Asn, 72, William S Knudsen Award, 75. Mem: Am Occup Med Asn (pres, 69-70); fel Am Acad Occup Med; Int Asn Occup Health. Res: Health effects of automotive emissions; effects of the work environment on worker health. Mailing Add: Ford Motor Co The American Rd Dearborn MI 48121

BLOCK, ERIC, b New York, NY, Jan 25, 42; m 66; c 2. ORGANIC CHEMISTRY. Educ: Queens Col, NY, BS, 62; Harvard Univ, AM, 64, PhD(chem), 67. Prof Exp: Fel, Harvard Univ, 67; asst prof, 67-73, ASSOC PROF CHEM, UNIV MO-ST LOUIS, 73- Concurrent Pos: Vis assoc prof chem, Harvard Univ, 74. Mem: Am Chem Soc; The Chem Soc. Res: Organic sulfur chemistry; organic photochemistry; novel methods of organic synthesis. Mailing Add: Dept of Chem Univ of Mo St Louis MO 63121

BLOCK, FRED BERT, medicinal chemistry, see 12th edition

BLOCK, GEORGE E, b Joliet, Ill, Sept 16, 26; m; c 3. SURGERY. Educ: Univ Mich, MD, 51, MS, 58; Am Bd Surg, dipl, 59. Prof Exp: Intern, Univ Hosp, Univ Mich, 51-52, resident, 51-53 & 55-58, from instr to asst prof surg, 58-60; from asst prof to assoc prof, 60-67, PROF SURG, UNIV CHICAGO, 67- Concurrent Pos: Coordr clin oncol, Univ Chiago; attend surgeon, Cook County Hosp; consult, US Naval Hosp, Great Lakes, Ill; Am Cancer Soc fel, Ben May Lab Cancer Res, Univ Chicago, 55. Honors & Awards: McClintock Award, Univ Chicago, 65. Mem: Fel Am Col Surgeons; Am Surg Asn; Soc Surg Alimentary Tract; Soc Head & Neck Surg; Int Soc Surg. Res: Inflammatory diseases of the bowel; thyroid cancer; cancer of the colon. Mailing Add: Dept of Surg Univ of Chicago Hosp & Clin Chicago IL 60637

BLOCK, HENRY DAVID, b New York, NY, Feb 22, 20; m 46; c 1. APPLIED MATHEMATICS. Educ: City Col New York, BS, 40, BCE, 43; Iowa State Univ, MS, 47, PhD(math), 49. Prof Exp: Asst prof math, Iowa State Col, 49-53; asst prof, Univ Minn, 53-55; from instr to assoc prof, 55-61, PROF APPL MATH, CORNELL UNIV, 61- Concurrent Pos: Vis prof appl electrophys, Univ Calif, San Diego, 66-67; Air Force Off Sci Res grant res in robots, 66-70; chmn, Gordon Res Conf Biomath, 70; John Simon Guggenheim Found fel biomath, 70-71; ed, J Comput, Soc Indust & Appl Math, 71-74 & J Appl Math, 75-78; vis prof, Dept Info Sci, Univ Calif, Santa Cruz, 76. Mem: Am Math Soc; Math Asn Am; Inst Elec & Electronics Eng; Asn Comput Mach; Artificial Intelligence Soc Brit. Res: Mechanics; mathematical biology; computers; robots; artificial intelligence; adaptive and evolutionary systems; control systems; numerical and functional analysis; pattern recognition; optimization; mathematical economics, statistics and physics. Mailing Add: Dept of Theoret & Appl Mech Thurston Hall Cornell Univ Ithaca NY 14853

BLOCK, HENRY WILLIAM, b Newark, NJ, Sept 9, 41; m 68. MATHEMATICAL STATISTICS. Educ: Carnegie-Mellon Univ, BS, 63; Ohio State Univ, MS, 64, PhD(math), 68. Prof Exp: Asst math, Ohio State Univ, 63-68; asst prof, 68-75, ASSOC PROF MATH, UNIV PITTSBURGH, 75- Concurrent Pos: Vis asst prof, Rensselaer Polytech Inst, 74-75; vis assoc prof, Univ Calif, Berkeley, 75-76. Mem: Am Math Soc; Inst Math Statist; Math Asn Am; Am Statist Asn. Res: Reliability theory; statistical distribution theory; infinitely divisible laws. Mailing Add: Dept of Math Univ of Pittsburgh Pittsburgh PA 15232

BLOCK, JACOB, b Brooklyn, NY, Mar 8, 36; m 63; c 2. WATER CHEMISTRY. Educ: Brooklyn Col, BS, 56; Case Inst Technol, PhD(anal chem), 61. Prof Exp: Sr res analyst, Olin Mathieson Chem Corp, 61-65; res chemist, 65-67, SR RES CHEMIST, W R GRACE & CO, 68- Mem: Am Inst Chem; Nat Water Supply Improv Asn. Res: Desalination; polyelectrolytes for scale prevention; ion exchange; waste water renovation. Mailing Add: Washington Res Ctr W R Grace & Co Clarksville MD 21029

BLOCK, JEROME BERNARD, b New York, NY, Jan 7, 31; m 56; c 3. ONCOLOGY, IMMUNOLOGY. Educ: Stanford Univ, BA, 52; NY Univ, MD, 56. Prof Exp: Intern med, Mass Mem Hosp, Boston, 56-57, NIH fel, 57-59; resident, Univ Wash, 59-60; clin assoc oncol, Nat Cancer Inst, 60-64, staff physician, 64-66, CHIEF, BALTIMORE RES CTR, NAT CANCER INST, 66- Concurrent Pos: Instr med,

George Washington Univ, 62-63. Honors & Awards: Karger Int Prize Biochem of Leukemia, 64. Mem: AAAS; Am Asn Cancer Res; Am Fedn Clin Res; Am Soc Hemat. Res: Chemotherapy; internal medicine. Mailing Add: 3100 Wyman Park Dr Baltimore MD 21211

BLOCK, JOHN HARVEY, b Yakima, Wash, May 11, 38; m 64; c 3. PHARMACEUTICAL CHEMISTRY. Educ: Wash State Univ, BS & BPharm, 61, MS, 63; Univ Wis, PhD(pharmaceut chem), 66. Prof Exp: Asst prof, 66-71, ASSOC PROF PHARMACEUT CHEM, ORE STATE UNIV, 72- Concurrent Pos: NIH spec fel & vis scholar, Stanford Univ, 72-73. Mem: Acad Pharmaceut Sci; Am Pharmaceut Asn; Am Chem Soc. Res: Medicinal chemistry. Mailing Add: Sch of Pharm Ore State Univ Corvallis OR 97331

BLOCK, LAWRENCE HOWARD, b Baltimore, Md, Nov 24, 41. CLINICAL PHARMACOLOGY. Educ: Univ Md, BS, 62, MS, 66, PhD(pharmaceut), 69. Prof Exp: Asst prof pharm, Univ Pittsburgh, 68-70; from asst prof to assoc prof, 70-75, PROF PHARMACEUT CHEM & PHARMACEUT, SCH OF PHARM, DUQUESNE UNIV, 75-, DIR, CLIN PHARMACOKINETICS RES LAB, 75- Concurrent Pos: Secy-treas & dir, Pesa Res, Inc, Pa, 70- Mem: AAAS; Acad Pharmaceut Sci; Am Pharmaceut Asn; Soc Cosmetic Chem; NY Acad Sci. Res: Mass transport; biopharmaceutics and pharmacokinetics; dosage form design; drug interactions. Mailing Add: Dept Pharmaceut Chem & Pharmaceut Sch of Pharm Duquesne Univ Pittsburgh PA 15219

BLOCK, LEROY PELTON, chemistry, see 12th edition

BLOCK, MARTIN M, b Newark, NJ, Nov 29, 25; m 49; c 2. PHYSICS. Educ: Columbia Univ, BS, 47, MA, 48, PhD(physics), 51. Prof Exp: Instr math, Newark Col Eng, 47-48; asst physics, Columbia Univ, 47-51, scientist, 51; assoc, Duke Univ, 51, from asst prof to assoc prof, 52-61; chmn dept, 61-68, PROF PHSYICS, NORTHWESTERN UNIV, 61- Concurrent Pos: Guggenheim fel, 58-59. Mem: Am Phys Soc; Italian Physics Soc. Res: High energy physics; liquid helium bubble chamber; neutrino physics. Mailing Add: Dept of Physics Northwestern Univ Evanston IL 60201

BLOCK, MATTHEW HAROLD, b Brooklyn, NY, Dec 13, 15; m 42; c 4. EMBRYOLOGY, HEMATOLOGY. Educ: City Col New York, BS, 37; Univ Mich, MS, 38; Univ Chicago, PhD(anat), 41, MD, 43. Prof Exp: Asst anat, Univ Chicago, 38-41, 42, asst med, 43, intern, Clins, 43-44; assoc physician, Metal Proj, Manhattan Eng Dist, 43-45; asst med, Univ Chicago, 47, USPHS sr fel, 48-50, asst prof, 50-53; assoc prof, 54-61, PROF MED, MED SCH, UNIV COLO, 61- Concurrent Pos: Med consult, AEC, Argonne Nat Lab, 47-53; assoc ed, J Lab & Clin Med, 51-53; exchange prof, Univ Ulm, Ger, 68; hemat consult, Atomic Bomb Casualty Comn, Japan, 75. Honors & Awards: Capps Prize, Inst Med, Chicago, 46; Distinguished Serv Award, Univ Chicago, 64. Mem: AMA; Am Soc Clin Invest; Am Asn Cancer Res; Asn Exp Biol & Med; Int Soc Hemat. Res: Irradiation histopathology; pathologic mechanisms in therapy of hematopoietic tumors, especially nitrogen mustards and radioisotopes; embryology and histopathology of blood forming organs; pathology of cold injury. Mailing Add: Univ of Colo Med Ctr 4200 E Ninth Ave Denver CO 80220

BLOCK, MICHAEL JOSEPH, b Chicago, Ill, June 29, 42; m 70; c 1. ORGANIC CHEMISTRY, AGRICULTURAL CHEMISTRY. Educ: Univ Mich, BS, 64; Harvard Univ, MA, 65, PhD(chem), 69. Prof Exp: RES CHEMIST, UNION OIL RES CTR, 68- Mem: AAAS; Am Chem Soc. Res: Physical-organic and inorganic chemistry; petroleum coke and graphite research. Mailing Add: Union Oil Res Ctr PO Box 76 Brea CA 92621

BLOCK, PAUL, JR, b New York, NY, May 11, 11; m 48; c 3. ORGANIC CHEMISTRY. Educ: Yale Univ, AB, 33; Columbia Univ, PhD(org chem), 43. Prof Exp: Fel, Mellon Inst, 43-44; RES PROF CHEM, UNIV TOLEDO, 45- Concurrent Pos: Hon fel, Med Sch, Yale Univ, 48-49; chmn bd trustees, Med Col Ohio at Toledo, 64-70; mem, Nat Adv Health Manpower Coun, 67-70. Mem: Fel NY Acad Sci. Res: Organic chemistry of iodine; synthetic problems related to thyroxine. Mailing Add: The Blade 541 Superior St Toledo OH 43660

BLOCK, RICHARD BLANCHARD, plasma physics, nuclear physics, see 12th edition

BLOCK, RICHARD EARL, b Chicago, Ill, Mar 8, 31; m 64; c 1. MATHEMATICS. Educ: Univ Chicago, PhD(math), 56. Prof Exp: Instr math, Univ Ind, 56-58; Off Naval Res res assoc, Yale Univ, 58-59; Bateman res fel, Calif Inst Technol, 59-60, asst prof, 60-64; assoc prof, Univ Ill, 64-68; PROF MATH, UNIV CALIF, RIVERSIDE, 68- Concurrent Pos: Res assoc, Yale Univ, 67. Mem: Am Math Soc. Res: Algebra; Lie algebras; associative and non-associative ring theory; derivations; combinatorial designs. Mailing Add: Dept of Math Univ of Calif Riverside CA 92502

BLOCK, ROBERT CHARLES, b Newark, NJ, Feb 11, 29; m 52; c 2. NUCLEAR PHYSICS, NUCLEAR ENGINEERING. Educ: Newark Col Eng, BS, 50; Columbia Univ, MA, 53; Duke Univ, PhD, 56. Prof Exp: Physicist, Oak Ridge Nat Lab, 55-66; PROF NUCLEAR SCI, RENSSELAER POLYTECH INST, 66- Concurrent Pos: Lectr, vis scientists prog, Am Inst Physics & Am Asn Physics Teachers, 58-; exchange scientist, Atomic Energy Authority Res Estab, Harwell, Eng, 62-63; guest lectr, Rensselaer Polytech Inst, 65-66; vis prof, Kyoto Univ, Japan, 73-74; consult, Knolls Atomic Power Lab; mem, Nuclear Cross-sect Adv Comt, US AEC. Mem: Am Phys Soc; Am Nuclear Soc; AAAS. Res: Basic and applied neutron physics research. Mailing Add: Dept of Nuclear Sci Rensselaer Polytech Inst Troy NY 12181

BLOCK, RONALD EDWARD, b Charleston, SC, Apr 29, 41; m 65. PHYSICAL CHEMISTRY, BIOPHYSICAL CHEMISTRY. Educ: Col Charleston, BS, 63; Clemson Univ, MS, 66, PhD(phys chem), 69. Prof Exp: Fel chem, Univ Fla, 69-70; res scientist, 72-73, ASSOC SCIENTIST BIOCHEM NUCLEAR MAGNETIC RESONANCE STUDIES, PAPANICOLAOU CANCER RES INST, 74- Concurrent Pos: Vis asst prof chem, Univ Fla, 70; adj asst prof neurol, Sch Med, Univ Miami, 74-76, adj assoc prof, 76-; consult nuclear magnetic resonance proj, Nat Cancer Inst, 75- Mem: Am Chem Soc; AAAS. Res: Application of nuclear magnetic resonance spectroscopy to studies of the structure of polymers, biological membranes, polypeptides and tissue components. Mailing Add: Papanicolaou Cancer Res Inst 1155 NW 14th St PO Box 236188 Miami FL 33123

BLOCK, STANLEY, b Baltimore, Md, Feb 14, 26; m 50; c 3. PHYSICAL CHEMISTRY, CRYSTALLOGRAPHY. Educ: Johns Hopkins Univ, MA, 52, PhD(chem), 55. Prof Exp: Chemist, Crown Cork & Seal Co, Ind, 43-50; physicist solid state, 54-67, CHIEF CRYSTALLOG SECT, NAT BUR STANDARDS, 67- Concurrent Pos: Lectr, Am Univ, 58 & Howard Univ, 67; UNESCO tech asst expert in x-ray crystallog, 59-61. Honors & Awards: S B Meyer, Jr Award, Am Ceramic Soc, 59; Gold Medal Award, US Dept Com, 74. Mem: Am Chem Soc; Am Crystallog Asn. Res: High pressure, inorganic chemistry. Mailing Add: Crystallog Sect Rm A221 Bldg 223 Nat Bur of Standards Washington DC 20234

BLOCK, STANLEY L, b Cincinnati, Ohio, Oct 10, 23; m 47; c 2. PSYCHIATRY. Educ: Univ Cincinnati, BS, 50, MD, 47; Chicago Inst Psychoanal, cert, 64. Prof Exp: From instr to asst prof, 53-64, assoc clin prof, 64-71, assoc prof, 71-74, PROF PSYCHIAT, UNIV CINCINNATI, 75- Concurrent Pos: Consult, Vet Admin Hosp, 57- & Jewish Family Serv, 62-; lectr, Hebrew Union Col, 62-; vis assoc prof, Med Ctr, Univ Ky, 64-69, vis prof, 69-; dir dept psychiat, Jewish Hosp, Cincinnati, Ohio, 67- Mem: Fel Am Psychiat Asn; Am Psychoanal Asn; Am Asn Hist Med; Am Acad Sci; Hist Sci Soc. Res: Psychoanalysis; history of psychiatry; group psychotherapy. Mailing Add: 3216 Burnet Ave Cincinnati OH 45229

BLOCK, WALTER DAVID, b Dayton, Ohio, Oct 16, 11; m 41; c 2. BIOCHEMISTRY, NUTRITION. Educ: Univ Dayton, BS, 33; Univ Mich, MS, 34, PhD(biochem), 38. Prof Exp: From instr to asst prof biochem, Univ Mich, Ann Arbor, 38-48; pvt consult, 48-52; assoc prof biochem, 52-66, PROF NUTRIT, UNIV MICH, ANN ARBOR, 66- Concurrent Pos: Res consult, Caylor-Nickel Res Found, Bluffton, Ind, 53- Mem: AAAS; Soc Exp Biol & Med; Am Inst Nutrit; Am Soc Biol Chemists. Res: Lipid metabolism; amino acid metabolism; protein research. Mailing Add: 1335 Glendaloch Circle Ann Arbor MI 48104

BLOCKER, HENRY DERRICK, b Waterboro, SC, May 18, 32; m 58; c 2. SYSTEMATIC ENTOMOLOGY. Educ: Clemson Univ, BS, 54, MS, 58; NC State Univ, PhD(entom), 65. Prof Exp: Asst prof, 65-71, ASSOC PROF ENTOM, KANS STATE UNIV, 71- Mem: Entom Soc Am. Res: Taxonomy and biology of leafhoppers. Mailing Add: Dept of Entom Kans State Univ Manhattan KS 66506

BLOCKER, TRUMAN GRAVES, JR, b West Point, Miss, Apr 17, 09; m 38; c 4. SURGERY. Educ: Austin Col, BA, 29; Univ Tex, MD, 33. Prof Exp: Rotating intern, Grad Hosp, Univ Pa, 33-35; res surgeon, Med Br, Univ Tex, 35-36; instr surg, Col Physicians & Surgeons, Columbia Univ, 36; asst prof, Sch Med, Univ Tex, 36-42, assoc prof, 46, prof plastic & maxillofacial surg & dir spec surg unit & grad div, 46, med dir, Univ Br Hosps, 50-53, dean clin fac, 53-56, chmn dept surg, 60-65, exec dir, Med Br, 64-67, pres, 67-74, PROF SURG, UNIV TEX MED BR GALVESTON, 46-, EMER PRES, 74- Concurrent Pos: Attend surgeon & plastic surgeon, John Sealy Hosp, 36-; consult, Vet Admin, 46-; Surgeon Gen; commanding officer, 807th US Army Hosp Unit, 56-64; mem med panel, Asst Secy Defense for Res & Eng, Nat Res Coun comt on trauma & surg study sect, NIH. Mem: Fel Am Col Surg; fel Asn Mil Surg US; fel Am Asn Plastic Surg; fel Am Soc Plastic & Reconstruct Surg; fel Am Surg Asn. Res: Application of plastic techniques to general surgery; protein therapy and burns; tissue culture with epithelium; speech training for cleft palate children; lymphatic cannulation. Mailing Add: Dept of Surg Univ of Tex Med Br Galveston TX 77550

BLOCKSTEIN, WILLIAM LEONARD, b Irwin, Pa, Nov 11, 25; m 53; c 3. HEALTH SCIENCES, ACADEMIC ADMINISTRATION. Educ: Univ Pittsburgh, BS, 50, MS, 53, PhD(pharm), 59. Prof Exp: Asst, Sch Pharm, Univ Pittsburgh, 52-53, admin asst to dean & instr pharm, 53-58; asst to dean & assoc prof, Sch Pharm, Wayne State Univ, 59-64; assoc prof, 64-67, PROF PHARM, SCH PHARM, UNIV WIS-MADISON, 67-, CLIN PROF PREV MED, MED SCH, 72-, CHMN HEALTH SCI UNIT, UNIV EXTEN, 69- Concurrent Pos: Vis scientist & lectr, Am Asn Cols Pharm, 69-72. Honors & Awards: Distinguished Serv Award, Mich State Pharmaceut Asn, 64 & Univ Exten, Univ Wis-Madison, 69; Meritorious Award, Col Pharm, Wayne State Univ, 70. Mem: Fel Am Col Apothecaries; fel Acad Pharmaceut Sci. Res: Continuing education in pharmacy; education and educational evaluation in health sciences; health care delivery systems. Mailing Add: Univ Exten Rm 415 Univ of Wis Madison WI 53701

BLODGETT, FREDERIC MAURICE, b Bucksport, Maine, July 3, 20; m; c 5. PEDIATRICS. Educ: Bowdoin Col, BS, 42; Yale Univ, MD, 45; Am Bd Pediat, dipl, 55. Prof Exp: Instr pediat, Harvard Med Sch, 52-54, assoc, 54-57; from asst prof to assoc prof, Sch Med, Yale Univ, 57-66; PROF PEDIAT, MED COL WIS, 66- Concurrent Pos: Chmn comt instr clin years, Sch Med, Yale Univ, 63-65. Honors & Awards: Children's Bur Award, Interchange Experts Prog, Int Health Res Act, 66. Mem: Fel Am Acad Pediat. Res: General care of children. Mailing Add: Milwaukee Children's Hosp Milwaukee WI 53233

BLODGETT, ROBERT BELL, b Milwaukee, Wis, Feb 2, 16; m 39; c 2. CHEMISTRY. Educ: Northwestern Univ, BS, 37; Univ Wis, PhD(phys chem), 40. Prof Exp: Res chemist, Rayon Dept, E I du Pont de Nemours & Co, 40-46, res supvr cellophane res, 46-48; chief chemist, Robbins Mills, Inc, 48-54; res mgr, Okonite Co, NJ, 54-61, dir res, 61-66, vpres res, 66-74, CHIEF ENGR, POWER CABLE WIRE & CABLE DIV, ANACONDA CO, 74- Mem: AAAS; Am Chem Soc; Inst Elec & Electronics Eng. Res: High-voltage insulation; cable technology; high-tenacity rayon for fabric uses; plasticization and characterization of high polymers; fabric finishing and dyeing. Mailing Add: Power Cable Wire & Cable Div Anaconda Co Greenwich Off Park 3 Greenwich CT 06830

BLODI, FREDERICK CHRISTOPHER, b Vienna, Austria, Jan 11, 17; nat US; m 46; c 2. MEDICINE. Educ: Univ Vienna, MD, 40. Prof Exp: Instr ophthal, Univ Vienna, 46-47; res fel, Inst Ophthal, Columbia Univ, 47-52; assoc prof, 52-66, PROF OPHTHAL, UNIV IOWA, 66-, HEAD DEPT, 67- Mem: Am Ophthal Soc; Asn Res Vision & Ophthal; Am Acad Ophthal & Otolaryngol. Res: Morphologic pathology of the eye; electromyography of the eye muscles. Mailing Add: Dept of Ophthal Univ Hosp Univ of Iowa Iowa City IA 52240

BLOEDEL, JAMES R, b Minneapolis, Minn, Apr 7, 40; c 3. NEUROPHYSIOLOGY. Educ: St Olaf Col, BA, 62; Univ Minn, Minneapolis, PhD(physiol), 67, MD, 69. Prof Exp: Asst Prof, 68-73, ASSOC PROF NEUROSURG & PHYSIOL, UNIV MINN, MINNEAPOLIS, 73-, DIR NEUROSURG RES, 72- Concurrent Pos: NIH awards, Univ Minn, Minneapolis, 70-76; vis scientist, Lab Neurophysiol, Good Samaritan Hosp, Portland, Ore, 72. Mem: Am Physiol Soc; Soc Neurosci. Res: Cerebellar physiology; pathophysiology of pain. Mailing Add: Dept of Neurosurg Univ of Minn Minneapolis MN 55455

BLOEDORN, FERNANDO GERMANO, radiotherapy, deceased

BLOEMBERGEN, NICOLAAS, b Dordrecht, Neth, Mar 11, 20; nat US; m 50; c 3. QUANTUM OPTICS. Educ: State Univ Utrecht, Phil Can, 41, Phys Drs, 43; State Univ Leiden, Dr Phil(physics), 48. Hon Degrees: AM, Harvard Univ, 51. Prof Exp: Asst physics, Harvard Univ, 46-47; assoc, State Univ Leiden, 47-48; jr fel, Soc Fels, 49-51, from assoc prof to Gordon McKay Prof appl physics, 51-74, RUMFORD PROF PHYSICS, HARVARD UNIV, 74- Honors & Awards: Buckley Prize, Am Phys Soc, 58; Liebmann Prize, Inst Elec & Electronics Eng, 59; Ballantine Medal, Franklin Inst, 61; Nat Medal of Sci, President of US, 74. Mem: Nat Acad Sci; fel Am Phys Soc; fel Inst Elec & Electronics Eng; Am Acad Arts & Sci; fel Optical Soc Am. Res: Nuclear and electronic magnetic resonance; solid state masers; nonlinear optics. Mailing Add: Pierce Hall Harvard Univ Cambridge MA 02138

BLOEMER, WILLIAM LOUIS, b Covington, Ky, Nov 26, 47; m 69. PHYSICAL

CHEMISTRY, CHEMICAL INSTRUMENTATION. Educ: Thomas More Col, BA, 67; Univ Ky, PhD(chem), 72. Prof Exp: Teaching asst chem, Univ Ky, 68-73; ASST PROF CHEM, SANGAMON STATE UNIV, 73- Concurrent Pos: Adj prof chem, Sch Med, Southern Ill Univ, 73- Mem: Am Chem Soc. Res: Molecular orbital calculations; educational applications of computers. Mailing Add: Sangamon State Univ Springfield IL 62708

BLOHM, THOMAS ROBERT, b South Bend, Ind, Oct 28, 20; m 48; c 1. BIOCHEMISTRY. Educ: Univ Notre Dame, BS, 42; Northwestern Univ, PhD(biochem), 48. Prof Exp: From instr to assoc prof biochem, Dent Br, Univ Tex, 48-52; head sect, Pharmacol Dept, 52-56, head dept biochem, 56-61, PRIN RES SCIENTIST, MERRELL-NATIONAL LABS DIV, RICHARDSON-MERRELL INC, 61- Concurrent Pos: Fel, Coun Arteriosclerosis, Am Heart Asn. Mem: Fel AAAS; Am Soc Biol Chem; fel Am Inst Chem; Int Soc Biochem Pharmacol; Am Chem Soc. Res: Biochemical pharmacology, lipid metabolism; atherisclerosis; metabolic diseases. Mailing Add: Merrell-Nat Labs Div Richardson-Merrell Inc 110 E Amity Rd Cincinnati OH 45215

BLOIS, MARSDEN SCOTT, JR, b San Antonio, Tex, Jan 5, 19; m 41; c 5. DERMATOLOGY. Educ: US Naval Acad, BS, 41; Stanford Univ, MS, 51, PhD(physics), 52, MD, 59. Prof Exp: Researcher & adminr, Nat Bur Standards, 52-53; researcher & adminr, Naval Ord Lab, 53; with microwave lab & physics dept, Stanford Univ, 53-64, actg dir, Biophys Lab, 61-64, assoc prof dermat, 64-69; PROF DERMAT & MED INFO SCI, MED CTR, UNIV CALIF, SAN FRANCISCO, 69- Honors & Awards: Gold Award, Am Acad Dermat, 68. Mem: AAAS; Am Phys Soc; Biophys Soc; NY Acad Sci. Res: Thin metallic films; magnetic resonance methods in biomedical research; biological free radicals; melanin and melanoma; photobiology; computer applications in medicine; medical information science. Mailing Add: Apt 401 2120 Pacific Ave San Francisco CA 94115

BLOK, JOHAN, chemical physics, see 12th edition

BLOM, CHRISTIAN JAMES, b Berkley, Calif, Dec 3, 28; m 53; c 3. GEOLOGY. Educ: Calif Inst Technol, BS, 50; Innsbruck Univ, PhD(geol), 53. Prof Exp: Geologist, Standard Oil Co Calif, 55-64; geologist, Occidental Petrol Corp, 64-65, explor mgr & vpres, Occidental of Libya, Inc, 66-67, sr vpres & resident mgr, 67-69, sr vpres & coordr Libyan affairs, 69, EXPLOR MGR EASTERN HEMISPHERE, OCCIDENTAL PETROL CORP, 69- Mem: Geol Soc Am; Am Asn Petrol Geol; Am Inst Mining, Metall & Petrol Eng. Res: International petroleum exploration. Mailing Add: 203 Fairway Dr Bakersfield CA 93309

BLOM, GASTON EUGENE, b Brooklyn, NY, Mar 29, 20; m 46; c 4. CHILD PSYCHIATRY, EDUCATIONAL PSYCHOLOGY. Educ: Colgate Univ, BA, 41; Harvard Univ, MD, 44; Am Bd Psychiat & Neurol, dipl & cert psychiat, 55; Boston Psychoanal Inst, dipl psychoanal, 56, dipl child anal, 58; Am Bd Psychiat & Neurol, cert child psychiat, 59. Prof Exp: Resident psychiat, USPHS Hosp, Ft Worth, Tex, 46-47; vet resident, Univ Mich Hosp, Ann Arbor, 47-48; fel child psychiat, Mass Gen Hosp, Boston, 48-49 & Judge Baker Guid Ctr, 49-50; asst psychiat, Harvard Med Sch, 51-53, instr, 54-58; assoc prof, Med Sch & chief, Child Psychiat Div, Univ Colo Med Ctr, Denver, 58-63, dir, Day Care Ctr, 63-70, actg chmn, Dept Psychiat, 70-71, prof psychiat, Med Sch, 63-75, prof educ, Denver Ctr, 66-75; PROF PSYCHIAT & ELEMENTARY & SPEC EDUC, MICH STATE UNIV, 75- Concurrent Pos: Dir, Child Psychiat Unit, Mass Gen Hosp, 56-58; consult, Ment Health Study Sect, Div Res Grants, NIH, 60-64, mem, Conf Social Competence & Clin Pract, Nat Inst Ment Health, 65; mem, Conf on Adaptation to Change, Found Fund for Res in Psychiat, 68; consult, Nat Adv Comn on Dyslexia & Related Reading Dis, Dept Health, Educ & Welfare, 68-69 & Denver Pub Schs, 69-75; training & supvry analyst, Denver Psychoanal Inst, 69-75 & supvry child analyst, 70-75; mem, Conf on Child Psychiat in Gen Psychiat Training, 71; mem vis staff, Gentofte Hosp, Copenhagen, Denmark, 72-73; med consult, US Gen Acct Off, 74-75. Mem: Fel Am Psychiat Asn; Am Psychoanal Asn; fel Am Acad Child Psychiat; Coun Except Children; Int Reading Asn. Res: Psycho-educational principles and techniques; factors in reading achievement and disability; content analysis of primers in the United States and foreign countries; cross national studies in reading; interdisciplinary approaches to school related problems and issues. Mailing Add: Dept of Psychiat B115 WFee Hall Mich State Univ East Lansing MI 48824

BLOMBERG, RICHARD NELSON, b Can, Oct 21, 24; nat US; m 49; c 2. ORGANIC CHEMISTRY. Educ: Cornell Univ, BA, 46; Stanford Univ, MS, 47. Prof Exp: Assoc pharmaceut, Wyeth, Inc, 47-50; assoc, 50-67, RES ASSOC TEXTILE FIBERS, EXP STA, E I DU PONT DE NEMOURS & CO, 67- Mem: Am Chem Soc. Res: Textile fibers. Mailing Add: 3 Deer Run Dr Wilmington DE 19807

BLOMGREN, GEORGE EARL, b Chicago, Ill, Apr 15, 31; m 56; c 2. PHYSICAL CHEMISTRY. Educ: Northwestern Univ, BS, 52; Univ Wash, PhD(chem), 56. Prof Exp: Boese fel, Columbia Univ, 56-57; res chemist, Nat Carbon Co, 57-62, group leader, Consumer Prod Div, 62-73, res group leader, Battery Prod Div, 73-75, SR TECHNOL ASSOC, BATTERY PROD DIV, UNION CARBIDE CORP, 76- Mem: AAAS; Am Chem Soc; Electrochem Soc. Res: Infrared spectra; theory of liquids; electron spin resonance; quantum chemistry; electrolyte transport; electrode processes. Mailing Add: Battery Prod Div Union Carbide Corp PO Box 6116 Cleveland OH 44101

BLOMMERS, ELIZABETH ANN, b Pittsburgh, Pa, Nov 21, 23. ORGANIC POLYMER CHEMISTRY. Educ: Bryn Mawr Col, AB, 45, MA, 46, PhD(chem), 50. Prof Exp: Instr chem, Wheaton Col, Mass, 50-51; instr, Wells Col, 51-52; chemist org res, Signal Corps, US Army, 52-53; chemist textile res, Quaker Chem Prods, 53-56; res chemist struct adhesives, Borden Chem Co, 56-62; sr chemist, Thiokol Chem Corp, 62-65; sr chemist, Koppers Co, Inc, 65-74; SR RES SCIENTIST, ARCO POLYMERS, INC, 74- Mem: Am Chem Soc. Res: Mechanisms of organic reactions; adhesives; analysis of thermoplastic polymers. Mailing Add: 218 Fairlawn Dr Monroeville PA 15146

BLOMQUIST, ALFRED THEODORE, b Chicago, Ill, Nov 16, 06; m 31; c 3. ORGANIC CHEMISTRY. Educ: Univ Ill, AB, 28, MS, 29, PhD(org chem), 32. Prof Exp: Nat Res Coun fel, 32-33, assoc, 41-42, actg asst prof chem, 42-43, asst prof, Nat Defense Res Comn Proj, 43-48, from assoc prof to prof, Univ, 48-75, EMER PROF CHEM, CORNELL UNIV, 75- Concurrent Pos: Consult, B F Goodrich Co, Akron, Ohio, 46-; Sloan vis prof, Harvard Univ, 62; Acad Press Naval Ord Develop award. Mem: Nat Acad Sci; fel AAAS; Am Chem Soc; fel NY Acad Sci; The Chem Soc. Res: Chemistry of small and medium size carbon rings; ketenes; organic chemistry of high polymers; amino acid and peptide synthesis. Mailing Add: Dept of Chem Cornell Univ Ithaca NY 14850

BLOMQUIST, CHARLES HOWARD, b Rockford, Ill, Oct 27, 33; m 62. BIOCHEMISTRY. Educ: Univ Ill, BS, 55; Univ Minn, PhD(biochem), 64. Prof Exp: Asst biochem, Univ Wash, 55-56; Olson Mem Found fel, Karolinska Inst, Sweden, 64-65; mem staff H-4 div, Los Alamos Sci Lab, 65-69; MEM STAFF BIOCHEM, UNIV MINN, MINNEAPOLIS, 69-; RES DIR, DEPT OBSTET-GYNEC, ST PAUL-RAMSEY HOSP, 69- Mem: AAAS; Am Chem Soc. Res: Heart cell culture; kinetics and mechanisms of dehydrogenase enzymes; fluorometric methods as applied to enzyme and protein chemistry. Mailing Add: Dept of Obstet & Gynec St Paul-Ramsey Hosp St Paul MN 55101

BLOMQUIST, CONRAD ALVIN, b Rockford, Ill, Dec 9, 13; m 41; c 2. ZOOLOGY. Educ: Univ Ill, BS, 40, MS, 47, PhD, 52. Prof Exp: From asst prof to assoc prof zool, 52-72, ASSOC PROF PHARM, COL OF PHARM, UNIV ILL, CHICAGO CIRCLE, 72-, ASST DEAN, 59- Res: Golgi material with reference to secretory activity and neural lesion as a result of oxygen lack. Mailing Add: Dept of Zool Col of Pharm Univ of Ill 833 S Wood St Chicago IL 60612

BLOMQUIST, GARY JAMES, b Iron Mountain, Mich, Apr 1, 47; m 70; c 1. BIOCHEMISTRY. Educ: Univ Wis, La Crosse, BS, 69; Mont State Univ, PhD(chem), 73. Prof Exp: ASST PROF CHEM, UNIV SOUTHERN MISS, 73- Mem: AAAS; Am Oil Chemists Soc; Am Chem Soc; Entom Soc Am. Res: Lipid biochemistry; insect biochemistry; hydrocarbon biosynthesis; insect hormones and pheromones; marine biochemistry. Mailing Add: Southern St Box 5347 Univ Southern Miss Hattiesburg MS 39401

BLOMQUIST, MARY M OSBORN, b Dexter, Iowa, Dec 14, 36; m 60; c 2. STATISTICS, MATHEMATICS. Educ: Iowa State Univ, BS, 58; Univ Minn, MS, 60, PhD(nonparametric statist), 66. Prof Exp: Res asst, Dept Textile Chem, Univ Minn, 58-60, res assoc, Dept Biomet, 61-62; mathematician & opers & systs analyst, US Dept Defense, Air Force, 63-65; statistician, NCent Regional Textiles Proj, Univ Minn, 65-67; pres, Biocentric, Inc, 68-75, SR STATISTICIAN, NAT BIOCENTRIC, INC, 68- Concurrent Pos: Pvt consult statist, math anal & exp design, Govt Acad & Indust Insts, 66-68; statist consult, Okla State Univ, 68; guest prof, Univ Minn, 69, lectr statist, Dept Forest Resources Develop, Univ Minn, 75. Honors & Awards: Patricia Kayes Glass Award, 65. Mem: Am Statist Asn; Biomet Soc; NY Acad Sci. Res: Design and analysis of experiments; engineering and biological field models; mathematical and statistical models; computer programming and analysis; biometrics; nonparametric statistics; coordination of central research functions and regional analytical programs; textile chemistry and analysis. Mailing Add: 3521 NSnelling St Paul MN 55112

BLOMQUIST, RICHARD FREDERICK, b Mankato, Minn, Nov 25, 12; m 37; c 3. ORGANIC CHEMISTRY. Educ: Coe Col, AB, 34; Univ Iowa, MS, 36, PhD(org chem), 37. Prof Exp: Prof chem & chmn dept, Doane Col, 37-42; from asst chemist to assoc chemist, Div Wood Preserv, Forest Prod Lab, 42-45, proj leader glues & gluing res, 47-60 & glues & glued prod res, 60-66, PROJ LEADER HOUSING RES, FORESTRY SCI LAB, US FOREST SERV, 66- Concurrent Pos: Marburg lectr, Am Soc Testing & Mat, 63; sr res fel, NZ Forest Serv, 65; prof, Sch Forestry Resources, Univ Ga, 68-; chmn ad hoc comt on aerospace struct adhesives, Nat Adv Bd, Nat Acad Sci, 72-74. Honors & Awards: Adhesives Award, Am Soc Testing & Mat, 62. Mem: Am Chem Soc; Am Soc Testing & Mat; Forest Prod Res Soc (pres, 74-75); fel Am Inst Chemists; fel Int Acad Wood Sci. Res: Better utilization of wood products in building construction; adhesives for construction; development of composite lumber products for building construction. Mailing Add: Forestry Sci Lab US Forest Serv Carlton St Athens GA 30602

BLOMQVIST, CARL GUNNAR, b Bararyd, Sweden, Dec 31, 31; m 61; c 2. CARDIOLOGY, CARDIOVASCULAR PHYSIOLOGY. Educ: Univ Lund, BM, 54, MD, 60; Karolinska Inst, Sweden, PhD(med), 67. Prof Exp: Res fel cardiovasc epidemiol, Physiol Hyg Lab, Col Med Sci, Univ Minn, Minneapolis, 60-61; resident dept med, Karolinska Inst, Sweden, 62-65; from instr to assoc prof med, 66-72, ASSOC PROF MED & PHYSIOL, UNIV TEX HEALTH SCI CTR DALLAS, 72- Concurrent Pos: Sr attend physician & dir EKG & exercise labs, Parkland Mem Hosp, 68-; estab investr, Am Heart Asn, 68-72, mem coun epidemiol; adj assoc prof med & physiol, Inst Technol, Southern Methodist Univ, 69-; consult, Dept Med, Vet Admin Hosp, Dallas, Tex, 72-; mem study sect appl physiol & bioeng, NIH, 74- Mem: Am Fedn Clin Res; fel Am Col Cardiol; Int Soc Cardiol; Swed Soc Cardiol. Res: Coronary heart disease; cardiovascular response to stress; exercise electrocardiography; quantitative non-invasive techniques in the study of circulatory physiology and pathophysiology. Mailing Add: Health Sci Ctr Univ of Tex Dallas TX 75235

BLOMSTER, RALPH N, b Lynn, Mass, May 18, 31; m 62; c 2. PHARMACOGNOSY. Educ: Mass Col Pharm, BS, 53; Univ Pittsburgh, MS, 57; Univ Conn, PhD(pharmacog), 63. Prof Exp: Asst prof pharmacog, Univ Pittsburgh, 63-68; PROF PHARMACOG & CHMN DEPT, SCH PHARM, UNIV MD, BALTIMORE, 68- Concurrent Pos: Am Found Pharmaceut Educ fel, 59-61; Lederle Labs res fel, 62-63. Mem: AAAS; Am Soc Pharmacog; Am Pharmaceut Asn. Res: Phytochemical investigations; isolation, purification and identification of plant constituents with biological activity, mainly the genera Heimia and Catharanthus. Mailing Add: Sch of Pharmacy Univ of Md Baltimore MD 21201

BLOMSTROM, DALE CLIFTON, b Lincoln, Nebr, Dec 13, 27; m 56; c 3. ORGANIC BIOCHEMISTRY. Educ: Univ Nebr, BS, 48, MS, 50; Univ Ill, PhD(chem), 53. Prof Exp: RES ORG CHEMIST, EXP STA, E I DU PONT DE NEMOURS & CO, 53- Mem: Am Chem Soc. Res: Heterocyclic compounds; restricted rotation; quinone diimides; polymers; organic reaction mechanisms; protein structures; hormone controls in adipose tissue; mechanism of herbicide action. Mailing Add: 3333 Rockfield Dr S Wilmington DE 19810

BLONDEAU, RENE, plant physiology, see 12th edition

BLONDIN, GEORGE ARTHUR, organic chemistry, biochemistry, see 12th edition

BLOOD, ALDEN EDWARD, organic chemistry, polymer chemistry, see 12th edition

BLOOD, BENJAMIN DONALD, b Wabash, Ind, June 11, 14; m 39; c 5. PRIMATOLOGY, SCIENCE ADMINISTRATION. Educ: Colo State Univ, DVM, 39; Harvard Univ, MPH, 49; Am Bd Vet Pub Health, dipl. Prof Exp: Chief adv vet affairs, Ministry Pub Health, Govt Korea, 46-48; chief vet pub health, Pan Am Sanit Bur, Regional Off WHO, 49-53, consult, Zones V & VI, Buenos Aires, 53-57; founder & dir, Pan Am Zoonoses Ctr, Pan Am Health Orgn, Azul, Arg, 57-64; int health rep, Off Int Health, USPHS, DC, 64-66, assoc dir int orgn affairs, 66-69, int health attache, US Mission to UN & Other Int Orgn in Geneva, 69-74; EXEC DIR, FED INTERAGENCY PRIMATE STEERING COMT, NIH, 75- Concurrent Pos: Mem WHO-Food & Agr Orgn expert comt on zoonoses, Geneva, Switz, 50 & Stockholm, Sweden, 58; secy directing coun, Pan Am Cong Vet Med, Lima, Peru, 51, Sao Paulo, Brazil, 54 & Kansas City, Mo, 59; mem US del to directing coun, Pan Am Health Orgn, 65, 67 & 68, US rep, Exec Comt, 66-69, vchmn, 69; mem US del, Pan Am Sanit Conf, 66; mem US del to World Health Assemblies, 66-71; alt mem exec bd, WHO, 69-72; secy, Am Bd Vet Pub Health, 51-53. Honors & Awards: Knight Comdr, Orden de Mayo, Govt Arg, 68. Mem: Fel Am Pub Health Asn; Am Vet Med Asn; Conf Pub Health Vets; Wildlife Dis Asn; NY Acad Sci. Res: International aspects of public health; epidemiology of zoonoses and other diseases common to man and animals; comparative medicine. Mailing Add: 1210 Potomac School Rd McLean VA 22101

BLOOD, CHARLES ALLEN, JR, b White River Junction, Vt, Sept 6, 23; m 48; c 1. ORGANIC CHEMISTRY. Educ: Univ Vt, BS, 50, MS, 52; Lehigh Univ, PhD(chem), 56. Prof Exp: Asst inorg & org chem, Univ Vt, 50-52; asst inorg chem, Lehigh Univ, 52-53, res fel org chem, 53-56; res chemist, E I du Pont de Nemours & Co, Inc, 56-62; assoc prof, 62-69, chmn dept, 68-74, PROF CHEM, STATE UNIV NY COL PLATTSBURGH, 69- Mem: Am Chem Soc. Res: Mechanisms and kinetics of organic reactions; cellophane development chemistry. Mailing Add: Dept of Chem State Univ of NY Col Plattsburgh NY 12901

BLOOD, FRANKLIN HARVEY, b Caldwell, Idaho, Nov 26, 36; m 62; c 2. INORGANIC CHEMISTRY. Educ: Col Idaho, BS, 58; Ind Univ, PhD(chem), 63. Prof Exp: Instr chem, Univ Puget Sound, 63-65; from asst prof to assoc prof, Northern Ariz Univ, 65-68; assoc prof, 68-73, chmn dept, 72-76, PROF CHEM, COL IDAHO, 73- Concurrent Pos: Res Corp grant, 70-72. Mem: Am Chem Soc. Res: Chemistry of the phosphates of lead, tin and germanium. Mailing Add: Dept of Chem Col of Idaho Caldwell ID 83605

BLOODGOOD, ROBERT ALAN, b Holyoke, Mass, Sept 24, 48; m 70; c 1. CELL BIOLOGY. Educ: Brown Univ, BS, 70; Univ Colo, Boulder, PhD(biol), 74. Prof Exp: FEL & RES ASSOC BIOL, YALE UNIV, 74- Concurrent Pos: Am Cancer Soc res fel, Yale Univ, 75-76, NIH fel, 76-78. Mem: Am Soc Cell Biol. Res: Mechanisms of cell motility, cilia, flagella, axostyles, mitosis, axoplasmic transport; structure and function of microtubules and microfilaments. Mailing Add: Dept of Biol Kline Biol Tower Yale Univ New Haven CT 06520

BLOODSTEIN, OLIVER, b New York, NY, Dec 2, 20; m 41; c 2. SPEECH PATHOLOGY. Educ: City Col New York, BA, 41; Univ Iowa, MA, 42, PhD(speech path), 48. Prof Exp: PROF SPEECH, BROOKLYN COL, 48- Mem: AAAS; fel Am Speech & Hearing Asn. Res: Stuttering. Mailing Add: Dept of Speech Brooklyn Col Brooklyn NY 11210

BLOODWORTH, JAMES MORGAN BARTOW, JR, b Atlanta, Ga, Feb 21, 25; m 47; c 3. PATHOLOGY. Educ: Emory Univ, MD, 48. Prof Exp: From intern to asst resident path, Presby Hosp, NY, 48-50; asst resident internal med, Univ Iowa Hosps, 50-51; from instr to prof path, Col Med, Ohio State Univ, 51-62, chief div path anat, 54-61, attend staff, Univ Hosp, 51-62; PROF PATH, SCH MED, UNIV WIS-MADISON, 62- Concurrent Pos: Instr, Col Physicians & Surgeons, Columbia Univ, 49-50; attend path, Columbus State Hosp, 54-57; chief labs, Madison Vet Admin Hosp, 62- Honors & Awards: Lilly Award, Am Diabetes Asn, 63. Mem: Am Soc Exp Path; Am Asn Path & Bact; Soc Exp Biol & Med; Am Soc Cell Biol; Endocrine Soc. Res: Degenerative vascular disease, especially diabetic microangiopathy; related aspects of diabetes mellitus, renal disease electron microscopy, enzyme histochemistry and radioautography. Mailing Add: Dept of Path Univ of Wis Sch of Med Madison WI 53706

BLOOM, ALLEN, b New York, NY, Sept 9, 43; m 67; c 2. ORGANIC CHEMISTRY, PHOTOCHEMISTRY. Educ: Brooklyn Col, BS, 65; Iowa State Univ, PhD(org chem), 71. Prof Exp: Res assoc x-ray crystallog, Cornell Univ, 69-72; MEM TECH STAFF ORG CHEM & PHOTOCHEM, RCA LABS, DAVID SARNOFF RES CTR, 72- Mem: AAAS; Am Chem Soc; Am Crystallog Asn. Res: Organic materials for optically recording and storing information; liquid crystal phenomena; electrical and photochemical properties of organic molecules. Mailing Add: RCA Labs David Sarnoff Res Ctr Princeton NJ 08540

BLOOM, ARNOLD LAPIN, b Chicago, Ill, Mar 7, 23; m 51; c 1. PHYSICS. Educ: Univ Calif, PhD(physics), 51. Prof Exp: Asst physics, Univ Calif, 48-51; physicist, Varian Assocs, 51-61; dir theoret res, Spectra-Physics, Inc, 61-68, chief scientist, 68-71; PHYSICIST, COHERENT RADIATION, INC, 72- Concurrent Pos: Lectr, Stanford Univ, 69-70, res assoc physics, 71-74. Mem: Fel AAAS; fel Am Phys Soc; fel Optical Soc Am. Res: Optical pumping; lasers; physical optics; gas lasers; organic dye lasers; optical thin films design and computation. Mailing Add: 1955 Oak Ave Menlo Park CA 94025

BLOOM, ARTHUR DAVID, b Boston, Mass, Oct 4, 34; m 57; c 3. GENETICS, PEDIATRICS. Educ: Harvard Univ, AB, 56; NY Univ, MD, 62. Prof Exp: Intern pediat, Bellevue Hosp, New York, 60-61; asst resident, Johns Hopkins Hosp, Baltimore, Md, 61; surgeon, USPHS, NIH, 62-64; sr asst resident pediat, Johns Hopkins Hosp, 64-65; cytogeneticist, Atomic Bomb Casualty Comn, Nat Res Coun, Hiroshima, Japan, 65-68; asst prof human genetics & res assoc pediat, Med Sch, Univ Mich, Ann Arbor, 68-70, assoc prof human genetics, 70-74; PROF PEDIAT, HUMAN GENETICS & DEVELOP, COL PHYSICIANS & SURGEONS, COLUMBIA UNIV, 74- Concurrent Pos: Consult, Bur Radiol Health, USPHS, 68-70; Atomic Bomb Casualty Comn, Nat Res Coun, 68-74. Mem: AAAS; Am Soc Human Genetics; Soc Pediat Res; Radiation Res Soc; Environ Mutagen Soc. Res: Cytogenetics and cell culture as applied to study of genetics and human disease. Mailing Add: Col of Physicians & Surgeons Columbia Univ New York NY 10032

BLOOM, ARTHUR LEROY, b Genesee, Wis, Sept 2, 28; m 53; c 3. GEOLOGY. Educ: Miami Univ, AB, 50; Victoria Univ, NZ, MA, 52; Yale Univ, PhD(geol), 59. Prof Exp: Instr geol, Yale Univ, 59-60; asst prof, 60-65, ASSOC PROF GEOL, CORNELL UNIV, 65- Concurrent Pos: Fulbright sr scholar, James Cook Univ NQueensland, Townsville, Australia, 73. Mem: AAAS; Geol Soc Am. Res: Geomorphology; Pleistocene geology. Mailing Add: Dept of Geol Sci Cornell Univ Ithaca NY 14853

BLOOM, BARRY MALCOLM, b Roxbury, Mass, Aug 12, 28; m 56; c 3. ORGANIC CHEMISTRY, MEDICINAL CHEMISTRY. Educ: Mass Inst Technol, SB, 48, PhD(org chem), 51. Prof Exp: Nat Res Coun fel, Univ Wis, 51-52; res chemist, 52-58, res supvr, 58-59, mgr, 59-63, dir med chem res, 63-68, dir med prod res & develop div, 68-71, vpres res & develop, Pfizer Pharmaceut, 69-71, PRES CENT RES, PFIZER, INC, 71-, MEM BD DIRS, 73- Concurrent Pos: Mem sr exec prog, Alfred B Sloan Sch Mgt, Mass Inst Technol, 67, mem corp vis comt, Dept Chem, 71-74 & Dept Biol, 73-; consult med chem, Walter Reed Army Inst Res, 69-; mem sci adv comt, Pharmaceut Mfg Asn Found, 75- Mem: Am Chem Soc; NY Acad Sci. Res: Synthetic medicinal agents; steroids; central nervous system drugs; structure-biological activity relationships; administration of research and development of human therapeutics; agricultural and fine/specialty chemical products. Mailing Add: Cent Res Pfizer Inc Groton CT 06340

BLOOM, BARRY R, b Philadelphia, Pa, Apr 13, 37; m 63. IMMUNOLOGY. Educ: Amherst Col, AB, 58; Rockefeller Univ, PhD(immunol), 63. Prof Exp: NSF fel, Wright-Flaming Inst, London, 63-64; from asst prof to assoc prof microbiol & immunol, 64-73, PROF MICROBIOL, IMMUNOL & CELL BIOL, ALBERT EINSTEIN COL MED, 73- Res: Cell-mediated immunity; virus-lymphocyte interactions. Mailing Add: Dept of Microbiol Albert Einstein Col of Med Bronx NY 10461

BLOOM, BEN, physiology, see 12th edition

BLOOM, EDA TERRI, b Los Angeles, Calif, May 11, 45. IMMUNOLOGY, CANCER. Educ: Univ Calif, Los Angeles, AB, 66, PhD(microbiol & immunol), 70. Prof Exp: Fel immunol, Sloan Kettering Inst Cancer Res, 70-71; lectr microbiol & immunol, 71-72, ASST RES SCIENTIST IMMUNOL, UNIV CALIF, LOS ANGELES, 72- Concurrent Pos: Dir, Cell Cult Facil, Cancer Ctr, Univ Calif, Los Angeles, 75- Mem: Am Asn Immunologists; Am Asn Cancer Res; Transplantation Soc; AAAS; Sigma Xi. Res: Tumor immunology with emphasis on definition of human tumor associated antigens and investigation of the immune response of cancer patients. Mailing Add: Dept of Microbiol & Immunol Sch of Med Univ of Calif Los Angeles CA 90024

BLOOM, EDWARD STANTON, organic chemistry, see 12th edition

BLOOM, ELLIOTT D, b New York, NY, June 11, 40; m 62; c 1. PARTICLE PHYSICS. Educ: Pomona Col, BA, 62; Calif Inst Technol, PhD(physics), 67. Prof Exp: Res assoc physics, 67-70, asst prof, 70-74, ASSOC PROF PHYSICS, LINEAR ACCELERATOR CTR, STANFORD UNIV, 74- Concurrent Pos: Consult, Sci Appln, Inc, 75- Mem: AAAS; Am Phys Soc. Res: Study of the nature of the structure of the nucleon, using high energy inelastic lepton scattering and electron positron colliding beams. Mailing Add: Dept of Physics Stanford Linear Accelerator Ctr Menlo Park CA 94305

BLOOM, HERBERT JEROME, b Chicago, Ill, Feb 15, 12; m 35; c 3. ORAL SURGERY, MAXILLOFACIAL SURGERY. Educ: Univ Mich, DDS, 35, MS, 38 & 43, PhD(path), 40; Am Bd Oral Surg, dipl, 47. Prof Exp: Lectr surg, Univ Mich, 43-52; ASSOC PROF ANAT, SCH MED, WAYNE STATE UNIV, 52- Concurrent Pos: Assoc ed, J Oral Surg, 47-66; ed, Sinai Hosp Bull; chmn dept oral surg, Mt Carmel Mercy Hosp, 44-73 & Sinai Hosp Detroit, 59-73; dir, Sinai Clift Palate Ctr, 55-73; mem bd dirs, Proj Hope, 64-; mem bd dirs, Detroit Area Coun World Affairs, 66-71; consult, Coun Int Rels, Am Dent Asn, 66-72, chmn rev comn advan educ, 68-71, consult, Coun Dent Educ, 68-71; trustee & chmn comt head & neck cancer, Mich Cancer Found, 69-; pres, Detroit Bd Health, 70-73; trustee, United Health Orgn, 70-; ed, Detroit Dent Bull, 70-72; consult, Am Bd Oral Surg, 71-74. Honors & Awards: Merit Award, Peru-NAm Cult Asn, 62; Medal of Honor, Peru, 63; Alpha Omega Recognition Awards, 65, 69, Nat Achievement Award, 70; Knights of Charity Award, Maryglade Col, 73; Alfred C Fones Award, State of Conn, 75. Mem: Am Soc Oral Surg; Am Acad Oral Path; Am Med Writer's Asn; fel Am Col Dent; fel Int Asn Oral Surg. Res: Comprehensive cancer control in area of head and neck. Mailing Add: 3066 Middlebelt Rd Orchard Lake MI 48033

BLOOM, JACK, b Centerville, Iowa, May 4, 19; m 44; c 3. FOOD CHEMISTRY. Educ: Univ Chicago, BS, 40, PhD(chem), 45. Prof Exp: Asst chemist, Continental Foods, Inc, 40; res dir, 40-60, V PRES RES & QUAL CONTROL, CONTINENTAL COFFEE CO, 60- Concurrent Pos: Mem, Off Sci Res & Develop. Mem: AAAS; Am Chem Soc; Inst Food Technol. Res: Rancidity of edible oils; stability of oil in water emulsions; roasting of coffee; food technology. Mailing Add: Continental Coffee Co 2550 N Clybourne Ave Chicago IL 60614

BLOOM, JAMES R, b Clearfield, Pa, Feb 20, 24; m 47; c 4. PLANT PATHOLOGY. Educ: Pa State Univ, BS, 50; Univ Wis, PhD, 53. Prof Exp: Asst plant path, Univ Wis, 50-53; from asst prof to assoc prof, 53-66, PROF PLANT PATH, PA STATE UNIV, 66- Mem: Am Phytopath Soc; Soc Nematol. Res: Phytonematology and disease of vegetable crops; interactions between plant pathogens. Mailing Add: Dept of Plant Path Buckhout Lab Pa State Univ University Park PA 16802

BLOOM, JOSEPH DUITCH, b Pittsburgh, Pa, July 6, 33; m 56; c 7. MEDICINE, INTERNAL MEDICINE. Educ: Univ Pittsburgh, BS, 54, MD, 58; Univ Rochester, MS, 60. Prof Exp: US Navy, 57-, intern, Naval Hosp, Bethesda, Md, 58-59, med officer, Naval Reactor Facil, Idaho Falls, 62, submarine med officer, Mare Island Naval Shipyard, Vallejo, Calif, 61-62, submarine squadron med officer, Submarine Squadron Five, San Diego, Calif, 63-64, resident internal med, Naval Hosp, Oakland, 64-67, fel chest dis, Naval Hosp, San Diego, 67-68, dir submarine med res lab, Submarine Med Ctr, 68-72, CHMN DEPT INTERNAL MED, NAVAL REGIONAL MED CTR, SAN DIEGO, 72- Mem: Fel Am Col Chest Physicians; AAAS; fel Am Col Physicians; NY Acad Sci; AMA. Res: Diving, hyperbaric and submarine medicine and physiology; pulmonary diseases; radiation biology. Mailing Add: Dept of Internal Med US Naval Regional Med Ctr San Diego CA 92134

BLOOM, MELVIN SIGMUND, chemistry, see 12th edition

BLOOM, MYER, b Montreal, Que, Dec 7, 28; m 54. PHYSICS. Educ: McGill Univ, BSc, 49, MSc, 50; Univ Ill, PhD(physics), 54. Prof Exp: Nat Res Coun Can traveling fel, Kamerlingh Onnes Lab, Univ Leiden, Holland, 54-56; res assoc physics, 56-57, from asst prof to assoc prof, 57-63, PROF PHYSICS, UNIV BC, 63- Concurrent Pos: Alfred P Sloan fel, 61-65; Guggenheim fel, 64-65. Honors & Awards: Steacie Prize, 67. Mem: Am Phys Soc; Can Asn Physicists; Royal Soc Can. Res: Nuclear magnetic resonance; spin relaxation. Mailing Add: Dept of Physics Univ of BC Vancouver BC Can

BLOOM, SHERMAN, b Brooklyn, NY, Jan 26, 34; m 60; c 2. EXPERIMENTAL PATHOLOGY, CELL BIOLOGY. Educ: NY Univ, AB, 55, MD, 60. Prof Exp: Res trainee exp path, NY Univ, 61-65, instr path, 65-66; from asst prof to assoc prof, Univ Utah, 66-73; ASSOC PROF PATH, UNIV S FLA, 73- Mem: AAAS; NY Acad Sci; Am Soc Exp Path; Am Asn Path & Bact; Int Study /roup Res Cardiac Metab. Res: Biochemical control of differentiation; differentiation of heart muscle; control of myocardial contractility. Mailing Add: Dept of Path Univ of SFla Col of Med Tampa FL 33620

BLOOM, STANLEY, b Plainfield, NJ, Oct 18, 24; m 48; c 4. ELECTRONICS. Educ: Rutgers Univ, BS, 48; Yale Univ, MS, 51, PhD(physics), 52. Prof Exp: MEM STAFF, RCA LABS, 52- Concurrent Pos: RCA Corp res fel, Cambridge Univ, 66-67. Honors & Awards: RCA Corp Awards, 53-60, 64 & 70. Mem: Inst Elec & Electronics Engrs; Am Phys Soc. Res: Microwave electronics; plasma physics; noise; quantum electronics; solid state. Mailing Add: RCA Labs Princeton NJ 08540

BLOOM, STANLEY MORTON, organic chemistry, see 12th edition

BLOOM, STEPHEN EARL, b Brooklyn, NY, Oct 18, 41; m 63; c 1. CYTOLOGY, CYTOGENETICS. Educ: Long Island Univ, BS, 63; Pa State Univ, MS, 65, PhD(genetics), 68. Prof Exp: Asst genetics, Pa State Univ, 63-68; asst prof, 68-74, ASSOC PROF CYTOGENETICS, CORNELL UNIV, 74- Concurrent Pos: Vis prof biol, Univ Tex M D Anderson Hosp & Tumor Inst Houston, 75. Mem: AAAS; Am Genetic Asn; Genetics Soc Am; Poultry Sci Asn; Am Inst Biol Sci; NY Acad Sci.

Res: Cytological, cytochemical and genetic studies of mitosis; avian cytology and cytogenetics; chromosome aberrations and neoplasia. Mailing Add: Dept of Poultry Sci Cornell Univ Ithaca NY 14850

BLOOM, STEWART DAVE, b Chicago, Ill, Aug 22, 23; m 49; c 3. NUCLEAR PHYSICS. Educ: Univ Chicago, PhD(physics, 52. Prof Exp: Assoc physicist, Brookhaven Nat Labs, 52-56; Guggenheim res fel, Cambridge Univ, 56-57; staff physicist, 57-69, PROF APPL SCI, LAWRENCE RADIATION LAB, 69- Concurrent Pos: FUlbright fel, France, 63. Mem: Fel Am Phys Soc. Res: Beta ray; nuclear reactions. Mailing Add: Lawrence Radiation Lab Bldg 216 Rm 104 Livermore CA 94551

BLOOM, VICTOR, b New York, NY, Aug 17, 31; m 54; c 2. PSYCHIATRY. Educ: Univ Mich, BS, MD, 57; Am Bd Psychiat & Neurol, dipl & cert psychiat, 63. Prof Exp: Intern med, Sinai Hosp Detroit, 57-58; resident psychiat, Lafayette Clin, 58-61, staff psychiatrist, 61-67, chief adult inpatient serv, 68-72; ASSOC PROF PSYCHIAT, SCH MED, WAYNE STATE UNIV, 72- Concurrent Pos: Instr, Sch Med, Wayne State Univ, 61-63, asst prof, 64-72; teacher, Lafayette Clin, 61-69; lectr psychiat, Mich State Dept Ment Health, 61-; consult group psychother, Neuropsychiat Inst, Med Sch, Univ Mich, 68-69 & Hemodialysis Unit, Mt Carmel Mercy Hosp, 71-; mem, Inst Bioenergetic Anal; consult group ther, Harper Hosp Sch Nursing, 72- Mem: Fel Am Psychiat Asn; fel Am Group Psychother Asn. Res: Psychoanalysis; psychotherapy; group psychotherapy; suicide prevention; schizophrenia; milieu therapy; bioenergetic analysis; bioenergetics in group therapy. Mailing Add: 1007 Three Mile Dr Grosse Pointe Park MI 48230

BLOOM, WALTER LYON, b Hamilton, Ont, Dec 14, 15; US citizen; m 42; c 2. CLINICAL MEDICINE. Educ: Yale Univ, MD, 40. Prof Exp: Intern med, New Haven Hosp, 40-41; resident med & res, Goldwater Mem Hosp, New York, NY, 42-43; physiol investr, Col Physicians & Surgeons, Columbia Univ, 43-44; from instr to assoc prof med, Sch Med, Emory Univ, 47-57; dir med educ & res, Piedmont Hosp, Atlanta, Ga, 57-67; asst to vpres acad affairs & prof appl biol, 67-68, actg vpres acad affairs, 68-69, ASSOC VPRES ACAD AFFAIRS, GA INST TECHNOL, 69- Concurrent Pos: Attend physician, Grady Mem Hosp, 47- & Lawson Vet Admin Hosp, 47-; assoc biochem, Emory Univ, 47-48, asst prof, 48-52, lectr, 52-57. Mem: Am Med Asn; Am Soc Clin Invest. Res: Endocrinology; metabolism; internal medicine; thermodynamics of human energy cycles; surface coating of blood vessels. Mailing Add: Ga Inst Technol Atlanta GA 30332

BLOOM, WILLIAM, anatomy, deceased

BLOOM, WILLIAM WHILEY, b Crystal City, Mo, Oct 7, 10; m 37, 60; c 4. BIOLOGY. Educ: Valparaiso Univ, AB, 39; Univ Chicago, MS, 49, PhD, 54. Prof Exp: From instr to prof, 43-66, chmn dept, 56-59, res prof, 66-68, PROF BIOL, VALPARAISO UNIV, 68- Mem: Bot Soc Am. Res: Botany; zoology; biology of the Marsileaceae. Mailing Add: Dept of Biol Valparaiso Univ Valparaiso IN 46383

BLOOMBERG, WILFRED, b Pittsburgh, Pa, Mar 25, 05; m 35. PSYCHIATRY, NEUROLOGY. Educ: Harvard Univ, SB, 24, MD, 28. Prof Exp: Chief neuro-psychiat serv, Cushing Vet Admin Hosp, Framingham, 46-52; chief neuro-psychiat serv, Vet Admin Hosp, Boston, 52-58; assoc prof psychiat & neurol, Sch Med, Boston Univ, 58; assoc clin prof psychiat & pub health, Sch Med, Yale Univ, 58-69; ment health adminr, Region VI, 69-71, DEP COMNR, DEPT MENT HEALTH, COMMONWEALTH OF MASS, 71- Concurrent Pos: Vis neurologist, Boston City Hosp, 34-49; hon consult, Surgeon Gen, US Dept Army; dir ment health, Southern Regional Educ Bd, Ga, 57-58; comnr ment health, State of Conn, 58-69; lectr, Harvard Med Sch, 70-; consult, Mass Gen Hosp, 70- Mem: Am Psychiat Asn; AMA; Am Neurol Asn; Am Acad Neurol. Mailing Add: 16 Farrar St Cambridge MA 02138

BLOOMER, HENRY HARLAN, b Roseville, Ill, Aug 3, 08; m 41; c 2. SPEECH PATHOLOGY. Educ: Univ Ill, AB, 30; Univ Mich, MA, 33, PhD(speech), 35. Prof Exp: Instr Eng, Lincoln Jr Col, 31-32; from instr to assoc prof speech, 35-47, prof otorhinolaryngol, Med Sch, 61-68, PROF SPEECH, UNIV MICH, ANN ARBOR, 48-, DIR SPEECH CLIN, 49-, PROF SPEECH PATH, DEPT PHYS MED & REHAB, MED SCH & HEAD SPEECH & HEARING SCI PROG, 69- Concurrent Pos: Mem, White House Conf Educ, 55; consult speech path, St Joseph Mercy Hosp, 58-, Vet Admin Hosp, 62- & Wayne County Gen Hosp, 68- Mem: Fel Am Speech & Hearing Asn (vpres, 51, pres, 55); Am Cleft Palate Asn; Am Cong Rehab Med; Int Asn Logopedics & Phoniatrics. Res: Respiration; voice disorders; cleft palate; diagnostic methods in speech pathology; physiology of palatopharyngeal function; oral diadokokinetics. Mailing Add: Univ of Mich Speech Clin 1111 E Catherine St Ann Arbor MI 48104

BLOOMER, JAMES L, b Knoxville, Tenn, Jan 6, 39. ORGANIC CHEMISTRY. Educ: Univ Tenn, BS, 49, MS, 61; Univ London, PhD(org chem), 64. Prof Exp: NSF fel org chem, Harvard Univ, 64-65; asst prof, 65-70, ASSOC PROF ORG CHEM, TEMPLE UNIV, 70- Concurrent Pos: NSF res grant, 66-68; NIH res grant, 69-71. Mem: Am Chem Soc; Brit Chem Soc. Res: Biosynthetic organic chemistry. Mailing Add: Dept of Chem Temple Univ Philadelphia PA 19122

BLOOMER, ROBERT OLIVER, b New York, NY, May 3, 12; m 37; c 3. PETROLOGY. Educ: Univ Va, BS, 37, MS, 38; Univ NC, PhD(petrol), 41. Prof Exp: Instr geol, Univ NC, 39-41; instr, Univ Va, 41-44; asst prof, Syracuse Univ, 45-47; prof & head dept, 48-65, James Henry Chapin Prof, 65-72, EMER JAMES HENRY CHAPIN PROF GEOL & MINERAL, ST LAWRENCE UNIV, 72- Concurrent Pos: Va Geol Surv, 39-52; assoc, Harvard Univ, 47-48; mining consult. Mem: AAAS; Soc Econ Geol; Geol Soc Am; Am Geophys Union; Nat Asn Geol Teachers. Res: Areal geology of the Blue Ridge; petrology of granitic rocks; structural geology; areal geology of Northern New York; publications. Mailing Add: Dept of Geol St Lawrence Univ Canton NY 13617

BLOOMFIELD, DANIEL KERMIT, b Cleveland, Ohio, Dec 14, 26; m 55; c 3. MEDICINE, BIOCHEMISTRY. Educ: US Naval Acad, BS, 47; Western Reserve Univ, MS & MD, 54. Prof Exp: From intern to asst resident, Beth Israel Hosp, 54-56; resident, Mass Gen Hosp, 56-57; fel biochem, Harvard Univ, 57-59; hon asst registr, Inst Cardiol, 59-60; asst med, Western Reserve Univ, 60-64; dir cardiovasc res, Community Health Found, 64-66; assoc med, Mt Sinai Hosp, 66-69; PROF MED & DEAN SCH BASIC MED SCI, UNIV ILL, URBANA-CHAMPAIGN, 70-, DEAN SCH CLIN MED, 74- Mem: Fel Am Col Physicians; Am Fedn Clin Res. Res: Biosynthesis of unsaturated fatty acids; cardiac arrhythmias; gas chromatography of steroids; cholesterol metabolism; congenital heart disease; hypertension anticoagulant therapy; mass cardiac screening. Mailing Add: Sch of Basic Med Sci Univ of Ill Urbana IL 61801

BLOOMFIELD, DENNIS ALEXANDER, b Perth, Western Australia, May 4, 33; m 64; c 3. CARDIOLOGY, INTERNAL MEDICINE. Educ: Univ Adelaide, MB, BS, 56. Prof Exp: Australian Heart Found traveling fel cardiol, Vanderbilt Univ Hosp, 62-65; fel, Inst Cardiol, London, Eng, 65-66; DIR MED & CHIEF CARDIOL, ST VINCENTS MED CTR, STATEN ISLAND, 67-; ASST PROF MED, STATE UNIV NY DOWNSTATE MED CTR, 67- Concurrent Pos: Attend physician, Coney Island Hosp, Maimonides Med Ctr & Kings County Hosp, Brooklyn, 67-; mem coun basic sci, Am Heart Asn, 67- Mem: NY Acad Sci; fel Am Col Chest Physicians; fel Am Col Med; fel Am Col Cardiol; Australian Med Asn. Res: Application of indicator-dilution techniques to the measurement of regurgitant and regional blood flows. Mailing Add: 4802 Tenth Ave Brooklyn NY 11219

BLOOMFIELD, GERALD TAYLOR, b Leeds, Eng, Apr 12, 38; m 67; c 2. ECONOMIC GEOGRAPHY, HISTORY OF TECHNOLOGY. Educ: Univ Nottingham, BA, 60, PhD(geog), 64. Prof Exp: Demonstr geog, Univ Nottingham, 61-63; lectr, Univ Auckland, 64-68; ASSOC PROF GEOG, UNIV GUELPH, 69- Concurrent Pos: Univ Auckland SPac Prog Comt, 65-67; NZ Univs Grants Comt, 66-68; fels, Univ Guelph, 69-72 & 72-73; Can Coun grant, 71-72; Nuffield Found travel fel, 75. Mem: Inst Brit Geogr; Can Asn Geogr; Asn Am Geogr; Soc Hist Archaeol; Soc Indust Archaeol. Res: World automotive industry evolution, distribution and structure; geography of administrative areas; geography of British Isles and of New Zealand; energy problems of pacific islands. Mailing Add: Dept of Geog Univ of Guelph Guelph ON Can

BLOOMFIELD, JORDAN JAY, b South Bend, Ind, Feb 25, 30; m 60; c 3. ORGANIC CHEMISTRY. Educ: Univ Calif, Los Angeles, BS, 52; Mass Inst Technol, PhD(org chem), 58. Prof Exp: Asst chem, Mass Inst Technol, 52-53, res, 55-57; instr, Univ Tex, 57-60; res assoc, Univ Ill, 60-61; res assoc, Univ Ariz, 61-62; asst prof, Univ Okla, 62-66 & assoc prof on leave, 66-67; SR RES SPECIALIST, MONSANTO CO, 67- Concurrent Pos: Vis prof & sr res group leader, 66-67; vis assoc prof, Univ Mo-St Louis, 75. Mem: AAAS; Am Chem Soc; The Chem Soc; Am Soc Photobiol. Res: Reaction mechanisms; transannular reactions; molecular rearrangements; small rings; bicyclic and polycyclic systems of unusual structure; synthetic photochemistry. Mailing Add: Monsanto Co 800 N Lindbergh Blvd St Louis MO 63166

BLOOMFIELD, PHILIP EARL, b Erie, Pa, Aug 28, 34; m 58; c 2. SOLID STATE PHYSICS, THEORETICAL PHYSICS. Educ: Univ Chicago, BA, 56, BS, 57, MS, 59, PhD(physics), 65. Prof Exp: Technician, Nat Bur Standards, Washington, DC, 56-57; physicist, Lab Appl Sci, Univ Chicago, 58-60; res asst solid state physics, Inst Study Metals, 60-64; asst prof physics, Univ Pa, 64-69; assoc prof, Drexel Univ, 69-70; assoc prof, City Col New York, 70-73; PHYSICIST, NAT BUR STANDARDS, GAITHERSBURG, MD, 74- Concurrent Pos: Instr, Univ Chicago Col, 58-59; res assoc, John Crerar Libr, Ill, 57; lectr, Ill Inst Technol, 62; ed, Accelerated Instr Methods, Ill, 62-63; consult, Pitman Dunn Labs, Frankford Arsenal, Pa, 67-73. Mem: Am Phys Soc. Res: Thermodynamics; statistical mechanics; electromagnetic phenomena in solids; ferroelectrics; magnetic materials; general relativity; optical properties of metals. Mailing Add: 21 W Dartmouth Rd Bala Cynwyd PA 19004

BLOOMFIELD, RICHARD ALLEN, comparative biochemistry, deceased

BLOOMFIELD, VICTOR ALFRED, b Newark, NJ, June 10, 38; m 62. BIOPHYSICAL CHEMISTRY. Educ: Univ Calif, Berkeley, BS, 59; Univ Wis, PhD(phys chem), 62. Prof Exp: NSF fel phys chem, Univ Calif, San Diego, 62-64; from asst prof to assoc prof, Univ Ill, 64-70; assoc prof, 70-74, PROF BIOCHEM & CHEM, UNIV MINN, ST PAUL, 74- Concurrent Pos: Alfred P Sloan Found fel, 69-71; NIH career develop award, 71-76; cochmn, Gordon Res Conf Physics & Phys Chem of Biopolymers, 74; mem NIH study sect BBCB, 75-78. Mem: AAAS; Am Chem Soc; Am Soc Biol Chem; Biophys Soc. Res: Physical chemistry of biological macromolecules and other high polymers; biological hydrodynamics; quasi-elastic laser light scattering. Mailing Add: Dept of Biochem Col of Biol Sci Univ of Minn St Paul MN 55108

BLOOMQUIST, EUNICE, b Worcester, Mass, Sept 9, 40; m 70; c 1. PHYSIOLOGY. Educ: Simmons Col, BS, 63; Boston Univ, PhD(physiol), 68. Prof Exp: From instr physiol to asst prof, 69-76, ASST PROF MED, SCH MED, TUFTS UNIV, 74-, ASSOC PROF PHYSIOL, 76- Mem: Am Physiol Soc; Soc Gen Physiologists. Res: Vascular smooth muscle and hypertension. Mailing Add: Dept of Physiol Tufts Univ Med Sch Boston MA 02111

BLOOR, BYRON MICHEL, b Moscow, Idaho, June 19, 21; m 42; c 4. MEDICINE, NEUROSURGERY. Educ: Univ Idaho, BS, 42; Duke Univ, MD, 45. Prof Exp: Damon Runyan res fel, Sch Med, Duke Univ, 50-52; assoc neurosurg, 53-57; chief neurosurg sect, US Vet Admin Hosp, 53-57; chief neurol surg, Cleveland Metrop Gen Hosp, 57-60; asst prof neurosurg, Sch Med, Western Reserve Univ, 57-60; asst, Highland View Cuyahoga County Hosp, 57-60; attend neurosurgeon, Crile Vet Admin Hosp, 57-60; consult neurosurg, US Vet Admin Hosp, Clarksburg, WVa, 60-70; PROF SURG & CHMN DIV NEUROL SURG, STRITCH SCH MED, LOYOLA UNIV, CHICAGO, 70- Concurrent Pos: Prof & chmn div neurosurg, Med Ctr, WVa Univ; consult, Hines Vet Admin Hosp. Mem: AAAS; AMA; Am Asn Neurol Surgeons; Cong Neurol Surgeons; fel Am Col Surgeons. Res: Cerebral hemodynamics and metabolism; cerebral trauma. Mailing Add: Div of Neurol Surg Loyola Univ Stritch Sch of Med Maywood IL 61053

BLOOR, COLIN MERCER, b Sandusky, Ohio, May 26, 33; m 59; c 3. PATHOLOGY, PHYSIOLOGY. Educ: Denison Univ, BS, 55; Yale Univ, MD, 60. Prof Exp: From intern to resident path, Yale Univ, 60-62; from asst prof to assoc prof, 68-74, PROF PATH, UNIV CALIF, SAN DIEGO, 74- Concurrent Pos: USPHS trainee path, 60-62; Life Ins Med Res Fund fel physiol, Nuffield Inst Med Res, 62-63 & fel path, Yale Univ, 63-64. Mem: Am Physiol Soc; Am Asn Pathologists & Bacteriologists; Am Soc Exp Path; Am Col Cardiol; Int Acad Pathologists. Res: Cardiovascular physiology and pathology, particularly the coronary circulation and coronary artery disease; effects of exercise on the development of coronary collateral circulation; cardiovascular genetics; exercise physiology. Mailing Add: Dept of Path Sch Med Univ of Calif at San Diego La Jolla CA 92093

BLOOR, JOHN E, b Stoke on Trent, Eng, Aug 31, 29; m 61; c 2. CHEMISTRY. Educ: Oxford Univ, BA, 52; Univ Manchester, PhD(chem), 59. Prof Exp: Res chemist, Clayton Aniline Co, Ciba Corp, 52-55; instr org chem, Univ Manchester, 55-57; assoc res chemist, BC Res Coun, Can, 60-63; sr res chemist, Franklin Inst Labs, 64-67; prof chem, Univ Va, 67-69; ASSOC PROF CHEM, UNIV TENN, KNOXVILLE, 69- Mem: Am Chem Soc. Res: Molecular and far infrared spectroscopy; organic metallic compounds; organic reactivity; quantum chemistry; application of computers to chemical problems; quantum biochemistry and theory of drug action. Mailing Add: Dept of Chem Univ of Tenn Knoxville TN 37916

BLOOR, ROBERT JOHN, b Boston, Mass, Mar 25, 16; m 42; c 3. RADIOTHERAPY. Educ: Univ Rochester, BA, 37; Harvard Univ, MD, 41. Prof Exp: Intern med clins, Univ Chicago, 41-42; asst resident radiol, Strong Mem Hosp, 42, resident, 46-47; from instr to asst prof, Univ Rochester, 47-56; radiation therapist, Henry Ford Hosp, Detroit, Mich, 56-69; CHIEF DEPT THERAPEUT RADIOL, WILLIAM BEAUMONT HOSP, 70- Concurrent Pos: Consult, Mich State Health Comnr Radiation Adv Comt; Mich rep, Med Liason Off Network, US Environ

Protection Agency. Mem: AMA; Radiol Soc NAm; Am Radium Soc; fel Am Col Radiol; Am Soc Therapeut Radiol. Res: Radiation damage to normal tissues in reference to radiation therapy; modification of radiation effects by oxygen or other agents. Mailing Add: Dept of Therapeut Radiol William Beaumont Hosp Royal Oak MI 48072

BLOSS, FRED DONALD, b Chicago, Ill, May 30, 20; m 46; c 3. MINERALOGY. Educ: Univ Chicago, BS, 47, MS, 49, PhD(geol), 51. Prof Exp: Asst, Univ Chicago, 47-50; instr phys sci, Univ Ill, 50-51; from asst prof to assoc prof geol & mineral, Univ Tenn, 51-57; assoc prof geol, Southern Ill Univ, 57-61, prof, 61-67; prof, 67-72, ALUMNI DISTINGUISHED PROF MINERAL, VA POLYTECH INST & STATE UNIV, 72- Concurrent Pos: Consult, Tenn Valley Authority, 55-; Geol Soc Am Penrose Fund grant-in-aid, 57-58; NSF sr fel, Cavendish Lab, Cambridge & Swiss Fed Inst Technol, 62-63; ed, The Am Mineralogist, 72-76. Mem: Fel Mineral Soc Am (vpres, 75-76, pres, 76-77); Mineral Asn Can; Mineral Soc Gt Brit & Ireland. Res: Crystallography. Mailing Add: Dept of Geol Sci Va Polytech Inst & State Univ Blacksburg VA 24061

BLOSS, HOMER EARL, b Cumberland, Md, Dec 3, 33. PLANT PATHOLOGY, PLANT BIOCHEMISTRY. Educ: Univ Md, BS, 59; Univ Del, MS, 61; Univ Ariz, PhD(plant path), 65. Prof Exp: Res technician plant physiol, USDA, Md, 57-59; plant pathologist, 63-65, ASST PROF PLANT PATH, UNIV ARIZ, 66- Mem: Am Phytopath Soc. Res: Physiology of fungi and fungus diseases of plants; biochemistry of disease resistance in plants. Mailing Add: Dept of Plant Path Univ of Ariz Tucson AZ 85721

BLOSS, RONALD EDWARD, b Capreol, Ont, Jan 20, 22; US citizen; m 43; c 3. ANIMAL NUTRITION. Educ: Purdue Univ, BS, 48, MS, 50; Mich State Univ, PhD(animal nutrit), 61. Prof Exp: Animal nutrit res, Gen Mills, Inc, 52-58; res scientist, 59-69, RES HEAD, UPJOHN CO, 70- Mem: Am Soc Animal Sci. Res: Use of drugs as means of improving growth, feed utilization or production of farm animals. Mailing Add: 9660-190-1 Upjohn Co Kalamazoo MI 49001

BLOSSER, HENRY GABRIEL, b Harrisonburg, Va, March 16, 28; m 51, 73; c 4. NUCLEAR PHYSICS. Educ: Univ Va, BA, 51, MS, 52, PhD(physics), 54. Prof Exp: Physicist, 86" Cyclotron Nuclear Res Group, Oak Ridge Nat Lab, 54-56, group leader, Cyclotron Analogue Group, 56-58; assoc prof, 58-61, PROF PHYSICS, MICH STATE UNIV, 61-, DIR CYCLOTRON PROJ, 58- Concurrent Pos: NSF sr fel, 66-67; Guggenheim fel, John Simon Guggenheim Found, 73-74. Mem: Fel Am Phys Soc. Res: Design and construction of sector focused cyclotrons of medium and high energy; nuclear reaction studies at medium energies. Mailing Add: Cyclotron Lab Mich State Univ East Lansing MI 48824

BLOSSER, TIMOTHY HOBERT, b Nappanee, Ind, Jan 7, 20; m 45; c 3. DAIRY SCIENCE, ANIMAL SCIENCE. Educ: Purdue Univ, BSA, 41; Univ Wis, MS, 47, PhD, 49. Prof Exp: Asst, Univ Wis, 41-42, 46-48; asst prof, Wash State Univ, 48-52, assoc prof dairy sci & assoc dairy scientist, 52-58, prof dairy sci & dairy scientist, 58-74, chmn dept dairy sci, 61-64, chmn dept animal sci, 64-74; STAFF SCIENTIST, NAT PROG STAFF, AGR RES SERV, USDA, 74- Mem: Am Dairy Sci Asn; Am Soc Animal Sci; Am Inst Nutrit; Poultry Sci Asn. Res: Parturient paresis and ketosis in dairy cows; nutritive value of dehydrated forages for ruminants; physiological changes associated with parturition in dairy cows; factors affecting forage quality. Mailing Add: Room 308A Bldg 005 USDA Agr Res Ctr W Beltsville MD 20705

BLOSTEIN, RHODA, b Montreal, Que, Nov 4, 36; m 57; c 1. BIOCHEMISTRY. Educ: McGill Univ, BS, 56, MS, 57, PhD(biochem), 60. Prof Exp: Res assoc biochem, Univ Ill, 60-63; ASST PROF BIOCHEM, McGILL UNIV, 63-, RES ASSOC, UNIV CLIN, ROYAL VICTORIA HOSP, 64- Concurrent Pos: Res fel, Univ Clin & hon lectr, McGill Univ, 63-64. Mem: AAAS; Can Biochem Soc. Res: Enzymology in relation to cellular functions. Mailing Add: Dept of Biochem McGill Univ Fac of Med Montreal PQ Can

BLOT, WILLIAM JAMES, b New York, NY, Aug 30, 43; m 65; c 2. BIOSTATISTICS, EPIDEMIOLOGY. Educ: Univ Fla, BS, 64, MS, 66; Fla State Univ, PhD(statist), 70. Prof Exp: Assoc sci researcher comput prog, Oak Ridge Nat Lab, 66-67; statistician med res, Atomic Bomb Casualty Comn, 70-72; asst prof, Johns Hopkins Sch Hyg & Pub Health, 72-74; BIOSTATISTICIAN, RES EPIDEMIOL BR, NAT CANCER INST, 74- Mem: Am Pub Health Asn; Am Statist Asn. Res: Study of the patterns and determinants, particularly environmental, of cancer in the United States and abroad. Mailing Add: Epidemiol Br Nat Cancer Inst Bethesda MD 20014

BLOUCH, RALPH IRVING, b Ann Arbor, Mich, June 29, 19; m; c 3. WILDLIFE MANAGEMENT. Educ: Univ Mich, BS, 40, PhD(wildlife mgt), 63; Pa State Univ, MS, 42. Prof Exp: Asst zool, Pa State Univ, 40-42; asst wildlife mgt, Univ Mich, 47-49; biologist, Game Div, Mich Dept Conserv, 49-56, biologist in charge, Houghton Lake Wildlife Exp Sta, 56-65, supvr wildlife res sect, Res & Develop Div, Mich Dept Natural Resources, 65-73, SR WILDLIFE EXEC, WILDLIFE DIV, MICH DEPT NATURAL RESOURCES, 73- Mem: Ornithologist Union; Wildlife Soc; Am Soc Mammalogists. Res: Game-bird ecology; bobwhite quail life history; relation of soils to pheasant ecology; deer management. Mailing Add: Mich Dept of Natural Resources Lansing MI 48926

BLOUET, BRIAN WALTER, b Darlington, Eng, Jan 1, 36; m 70. GEOGRAPHY. Educ: Univ Hull, BA, 60, PhD(geog), 64. Prof Exp: From asst lectr to lectr geog, Univ Sheffield, 64-69; assoc prof, 69-75, PROF GEOG, UNIV NEBR-LINCOLN, 75- Mem: Asn Am Geogr; Brit Geog Asn; Brit Econ Hist Soc; Inst Brit Geogr; fel Royal Geog Soc. Res: Historical geography with particular reference to the study of settlement patterns. Mailing Add: Dept of Geog Univ of Nebr Lincoln NE 68508

BLOUGH, HERBERT ALLEN, b Philadelphia, Pa, Dec 18, 29; m 54; c 4. VIROLOGY. Educ: Pa State Col, BSc, 51; Chicago Med Sch, MD, 55. Prof Exp: Intern, Cincinnati Gen Hosp, 55;56; univ fel, Univ Minn, 58-60, USPHS fel, 60-61; USPHS res fel virol, Univ Cambridge, 61-63; asst prof microbiol, 63-69, assoc prof ophthal, 70-75, PROF OPHTHAL, UNIV PA, 75-, HEAD DIV BIOCHEM VIROL & MEMBRANE RES, 70- Concurrent Pos: Consult, US Naval Hosp, Philadelphia, Pa; symp mem & lectr, Ciba Found, 64-; assoc mem, Comn Influenza, Armed Forces Epidemiol Bd; Nat Multiple Sclerosis Soc sr fel, Oxford Univ, 71-72; vis prof, Univ Helsinki, 72; mem virol study sect, NIH, 75- Mem: AAAS; NY Acad Sci; Am Asn Immunol; Am Soc Microbiol; Brit Biochem Soc. Res: Structure and assembly of enveloped viruses; attachment of viruses to artificial membranes; lipid biochemistry. Mailing Add: Scheie Eye Inst Univ of Pa Sch of Med Philadelphia PA 19104

BLOUIN, ANDRE, b Montreal, Que, Dec 18, 45; m 68. PHARMACOLOGY. Educ: Univ Montreal, DVM, 68, MSc, 70, PhD(pharmacol), 73. Prof Exp: Fel anat, Univ Bern, 73-74; ASST PROF PHARMACOL, FAC VET MED, UNIV MONTREAL, 75- Mem: Fr-Can Asn Advan Sci; Can Asn Anatomists; Can Asn Vet Anatomists. Res: Quantitation by stereological methods of the ultrastructural modifications following the use of different drugs. Mailing Add: Fac of Vet Med PO Box 5000 Univ of Montreal St Hyacinthe PQ Can

BLOUIN, FLORINE ALICE, b New Orleans, La, July 3, 31. ORGANIC CHEMISTRY. Educ: La State Univ, BS, 53, MS, 56; Univ Akron, PhD(polymer chem), 75. Prof Exp: CHEMIST, CELLULOSE CHEM RES, SOUTHERN REGIONAL RES LAB, 53- Mem: Am Chem Soc; Sigma Xi. Res: Textile, cellulose and polymer chemical research. Mailing Add: Southern Regional Res Lab PO Box 19687 New Orleans LA 70179

BLOUIN, LEONARD THOMAS, b Detroit, Mich, June 15, 30; m 67; c 3. CARDIOVASCULAR PHYSIOLOGY, ANALYTICAL CHEMISTRY. Educ: Mich State Univ, BS, 51, MS, 56; Univ Tenn, PhD(physiol), 59. Prof Exp: Anal develop chemist, Gen Elec Co, Wash, 51-55; lab asst, Mich State Univ, 55; lab asst, Univ Tenn, 56-59, instr clin physiol, 59; from assoc res physiologist to res physiologist, 59-65, sect leader cardiovasc-renal physiol, 65-66, sr res physiologist & head cardiovasc-renal sect, 66-67, assoc lab dir, 67-70, dir cardiovasc-renal sect, 70-74, SR CARDIOVASC PHYSIOLOGIST, CLIN RES DEPT, PARKE, DAVIS & CO, 74- Mem: AAAS; Am Soc Nephrol; NY Acad Sci; Asn Advan Med Instrumentation. Res: Renal physiology; hypertensive cardiovascular disease; diuretics; radioprotective compounds; electrocardiography; radio and clinical chemistry; congentive heart disease and edema; peripheral vascular disease; angina pectoris; cardiac arrhythmias. Mailing Add: Cardiovasc Dis Parke Davis & Co Ann Arbor MI 48106

BLOUKE, MORLEY MATTHEWS, solid state physics, semiconductor physics, see 12th edition

BLOUNT, CHARLES E, b Hobbs, NMex, Apr 8, 31; m 53; c 2. PHYSICS. Educ: Tex Western Col, BS, 54; Tex A&M Univ, MS, 60, PhD(physics), 63. Prof Exp: Instr math, Tex Western Col, 56-58; asst prof, 63-68, assoc prof, 68-72, PROF PHYSICS, TEX CHRISTIAN UNIV, 72- Concurrent Pos: Consult, El Paso Nat Gas Co, 56-58 & Gen Dynamics Ft Worth, 64- Mem: Am Phys Soc; Am Asn Physics Teachers; Am Soc Metals; Am Inst Mining, Metall & Petrol Eng. Res: Electronic spectra of molecular solids; imperfections in molecular solids; color centers in alkali halides. Mailing Add: Dept of Physics Tex Christian Univ Ft Worth TX 76129

BLOUNT, DON HOUSTON, b Cape Girardeau, Mo, Mar 25, 29; m 52; c 3. PHYSIOLOGY. Educ: Univ Mo, BA, 50, MA, 56, PhD(physiol), 58. Prof Exp: Asst instr med physiol, Col Med, Univ Mo, 56-58; from instr to asst prof physiol & biophys, Col Med, Univ Vt, 58-61; asst prof, Dept Physiol, Med Ctr, WVa Univ, 61-67; head physiol br, Life Sci Labs, Melpar, Inc, 67-69; scientist adminr, Spec Progs Br, 69-72, CHIEF CARDIAC FUNCTIONS BR, DIV HEART & VASCULAR DIS, NAT HEART & LUNG INST, 72- Mem: Am Physiol Asn. Res: Cardiovascular physiology; relationship between the biochemical and mechanical events of the heart. Mailing Add: Landow Bldg 7910 Woodmont Ave Bethesda MD 20014

BLOUNT, EUGENE IRVING, b Boston, Mass, May 26, 27; m 55; c 2. THEORETICAL SOLID STATE PHYSICS. Educ: Harvard Univ, AB, 47; Univ Chicago, MS, 51, PhD, 59. Prof Exp: Res physicist, Int Harvester Co, 50-54; res physicist, Westinghouse Elec Corp, 57-60; RES PHYSICIST, BELL LABS, 60- Concurrent Pos: Guggenheim fel, 63-64. Mem: Fel Am Phys Soc. Res: Behavior of electrons in crystals. Mailing Add: Bell Labs 1D-263 PO Box 216 Murray Hill NJ 07974

BLOUNT, FLOYD EUGENE, b Dallas, Tex, Dec 11, 22; m 48; c 5. INORGANIC CHEMISTRY, PHYSICAL CHEMISTRY. Educ: ETex State Univ, BS, 43. Prof Exp: Res technologist, Res & Develop Lab, Socony Mobil Oil Co, Inc, NJ, 47-52, sr res technologist, Field Res Lab, Tex, 52-63, eng assoc, 63-66, ENG ASSOC, FIELD RES LAB, MOBIL RES & DEVELOP CORP, 66- Mem: Am Chem Soc; Nat Asn Corrosion Eng; Sci Res Soc Am; Am Inst Chem Eng. Res: Analytical procedures, particularly spectrophotometric techniques; electro and surface chemistry related to corrosion of metals, primarily iron in many environments; water chemistry, especially solubilities and scaling tendencies in high dissolved solids brine. Mailing Add: Field Res Lab Mobil Res & Develop Corp Dallas TX 75221

BLOUNT, RAYMOND FRANK, b LaGrange, Ill, Oct 24, 00; m 33. ANATOMY. Educ: Univ Ariz, BSA, 24, MS, 26; Yale Univ, PhD(zool), 31. Prof Exp: Instr zool, Univ Ariz, 26-27; asst, Yale Univ, 27-29, asst anat, Sch Med, 29-30; from instr to asst prof, Med Sch, Univ Minn, 31-41; from assoc prof to prof, 42-71, EMER PROF, UNIV TEX MED BR GALVESTON, 71- Mem: AAAS; Am Asn Anat; Soc Exp Biol & Med; Am Soc Zoologists; Endocrine Soc. Res: Interstitial tissue of reptilian testis; transplantation of pituitary rudiment in urodele embryo; pituitary and size relationships; vascular and glomerular change; endocrines and development; control of pituitary differentation; endocrine interrelationships in mice; kidney structure. Mailing Add: Univ of Tex Med Br Galveston TX 77550

BLOUNT, STANLEY FREEMAN, b Detroit, Mich, June 12, 29; m 57; c 2. GEOGRAPHY, RESOURCE MANAGEMENT. Educ: Wayne State Univ, BA, 52, MA, 58; Northwestern Univ, PhD(geog), 62. Prof Exp: Mkt dir, Chrysler Corp, 52-58; asst prof geog, Univ Ill, 62-63; assoc prof, Kent State Univ, 63-67; chmn dept, 67-74, PROF GEOG, STATE UNIV NY ALBANY, 67- Concurrent Pos: Mem, Columbia Univ Sem Ecol, 70-; consult, NY State Legis Inst, 75- Mem: AAAS; Asn Am Geog; Am Geog Soc. Res: Economic development and planning in emerging countries, especially Latin America; environment perception and digitized land use mapping; geometric changes in high energy stream systems. Mailing Add: Dept of Geog & Environ Studies 111 Soc Sci Bldg State Univ NY Albany NY 12222

BLOUSE, LOUIS E, JR, b San Antonio, Tex, Nov 29, 31; m 58; c 3. BACTERIOLOGY, MICROBIOLOGY. Educ: Univ Tex, BA, 53, MA, 55; La State Univ, New Orleans, PhD(microbiol), 71. Prof Exp: Asst immunologist, Tex State Dept Health, 54-55; dir lab, El Paso City-County Health Dept, 57-59; chief bact br, US Air Force Epidemiol Lab, 59-68; res microbiologist & chief exp methods function, 71-72, MICROBIOLOGIST & CHIEF DIS SURVEILLANCE BR, EPIDEMIOL DIV, US AIR FORCE SCH AEROSPACE MED, 73- Honors & Awards: Superior Performance Award, US Air Force, 64, Cert Merit, Systs Command, 65; Fisher Sci Award, 72. Mem: AAAS; Am Soc Microbiol; Electron Micros Soc Am; Sigma Xi. Res: Isolation and characterization of bacteriophages; epizootiology of staphylococcal infections in canines; comparative studies on staphylococcal typing phages; combined viral-bacterial infection. Mailing Add: Epidem Div US Air Force Sch Aerospace Med Brooks AFB TX 78235

BLOUT, ELKIN ROGERS, b New York, NY, July 2, 19; m 39; c 2. BIOCHEMISTRY. Educ: Princeton Univ, AB, 39; Columbia Univ, PhD(chem), 42. Hon Degrees: AM, Harvard Univ, 62; DSc, Loyola Univ, 76. Prof Exp: Nat Res fel, 42-43, res assoc, 50-52 & 56-60, lectr biophys, 60-62, PROF BIOL CHEM, HARVARD MED SCH, 62-, EDWARD S HARKNESS PROF, 64- Concurrent Pos: Mem staff, Polaroid Corp, 43-48, assoc dir res, 48-58, vpres & gen mgr res, 58-62, consult, 62-; consult, Children's Med Ctr, 52- & NSF, 63-64; mem, Sci Adv Comt,

Ctr Blood Res, Inc, 67- & mem, Bd Dirs, 72-; mem, Dept Chem Vis Comt, Carnegie-Mellon Univ, 68-; mem, Sci Adv Comt, Mass Gen Hosp, Boston, 68-71; mem, Bd Visitors, Fac Health Sci, State Univ NY Buffalo, 68-; trustee, Boston Biomed Res Inst, 72-; mem, Exec Comt, Div Chem & Chem Technol, Nat Res Coun, 72-; mem, Corp Mus of Sci, Boston, 74-; mem, Adv Coun, Dept Biochem Sci, Princeton Univ, 74-; trustee, Bay Biochem Res, Inc, 73-; mem, Bd Dirs, CHON Corp, 74-; mem, Conseil de Surveillance, Compagnie Financiere du Scribe, 75- Mem: Nat Acad Sci; fel AAAS; fel Am Acad Arts & Sci; fel NY Acad Sci; fel Optical Soc Am. Res: Polypeptides and proteins; spectroscopy and rotatory phenomena; synthetic organic chemistry. Mailing Add: Dept of Biol Chem Harvard Med Sch Boston MA 02115

BLOZIS, GEORGE G, b Chicago, Ill, Nov 3, 29; m 56; c 3. PATHOLOGY. Educ: Ohio State Univ, DDS, 55; Univ Chicago, SM, 61; Am Bd Oral Path, dipl, 66. Prof Exp: Intern, Billing's Hosp, Univ Chicago, 55-56, instr oral path, Walter G Zoller Mem Dent Clin, 61-64, asst prof, 64-66; PROF ORAL DIAG & CHMN DEPT, COL DENT, OHIO STATE UNIV, 66- Concurrent Pos: Proj dir, Oral Cancer Demonstration grant, 63-66; spec consult, Chicago Bd Health, 63-64. Mem: Fel AAAS; Am Dent Asn; fel Am Acad Oral Path; Am Soc Cytol; Int Asn Dent Res. Res: Nephrotic syndrome in rats; application of exfoliative cytology to salivary gland disease; production of salivary gland lesions; oral exfoliative cytology. Mailing Add: Ohio State Univ Col of Dent 305 W 12th Ave Columbus OH 43210

BLU, KAREN ISOBELL, b Oak Park, Ill, Aug 20, 41. ANTHROPOLOGY. Educ: Bryn Mawr Col, AB, 63; Univ Chicago, MA, 65, PhD(social anthrop), 72. Prof Exp: Lectr anthrop, City Col New York, 68-71; ASST PROF ANTHROP, BARUCH COL, 72- Mem: Am Anthrop Asn; Am Ethnol Soc; Am Indian Hist Soc. Res: Ethnicity; cultural theory; symbolic analysis; kinship; North American Indians; American South. Mailing Add: Dept of Sociol & Anthrop Baruch Col 17 Lexington Ave New York NY 10010

BLUBAUGH, LOUIS VINCENT, pharmaceuticals, see 12th edition

BLUDMAN, SIDNEY ARNOLD, b New York, NY, May 13, 27; m 49; c 3. THEORETICAL PHYSICS. Educ: Cornell Univ, AB, 45; Yale Univ, MS, 48, PhD(physics), 51. Prof Exp: Instr physics, Lehigh Univ, 50-52; mem staff, Lawrence Radiation Lab, Univ Calif, Berkeley, 52-61; PROF PHYSICS, UNIV PA, 61- Concurrent Pos: Assoc, Oceanog Inst, Woods Hole, 51; vis mem, Inst Advan Studies, 56-57; lectr, Univ Calif, 59-60; vis res fel, Imp Col, Univ London, 67-68; vis prof, Tel-Aviv Univ, 71; Lady Davis vis prof, Hebrew Univ, 76. Mem: Am Phys Soc. Res: Elementary particle physics; theoretical astrophysics. Mailing Add: Dept of Physics Univ of Pa Philadelphia PA 19104

BLUE, JAMES LAWRENCE, b Pekin, Ill, Sept 20, 40; m 66; c 2. APPLIED MATHEMATICS, MATHEMATICAL PHYSICS. Educ: Occidental Col, AB, 61; Calif Inst Technol, PhD(physics), 66. Prof Exp: MEM TECH STAFF, BELL TEL LABS, 66- Res: Numerical analysis. Mailing Add: Bell Tel Labs Mountain Ave Murray Hill NJ 07974

BLUE, JOSEPH EDWARD, b Quitman, Miss, Sept 29, 36; m 62; c 2. ACOUSTICS. Educ: Miss State Univ, BS, 61; Fla State Univ, MS, 66; Univ Tex, Austin, PhD(acoust), 71. Prof Exp: Res physicist acoust, Mine Defense Lab, US Navy, 61-68; res physicist, Appl Res Labs, Univ Tex, 68-71; HEAD MEASUREMENTS BR, UNDERWATER SOUND REFERENCE DIV, NAVAL RES LAB, 71- Honors & Awards: Publ Award, Naval Res Lab, 72. Mem: Acoust Soc Am. Res: Methods and systems for characterizing transducers and other devices for the generation and reception of underwater acoustic signals and noise. Mailing Add: 5127 Leeward Way Orlando FL 32809

BLUE, MARTS DONALD, b Des Moines, Iowa, Feb 4, 32; m 60; c 4. SOLID STATE PHYSICS. Educ: Iowa State Univ, BS, 52, MS, 54, PhD(physics), 56. Prof Exp: Asst physics, Iowa State Univ, 53, asst, Inst Atomic Res, 55-56; res scientist, Minneapolis-Honeywell Res Ctr, 56-62, head res sect, Res Ctr, Honeywell Inc, 63-69, sci attache, Honeywell Europe, Inc, 69-72, mem opers staff, Honeywell Radiation Ctr, 72-74; PRIN RES SCIENTIST, ENG EXP STA, GA INST TECHNOL, 74- Concurrent Pos: Fulbright fel, Univ Paris, 56-57; consult, US Army Missile Res, Develop & Eng Lab, Huntsville, Ala, 75. Mem: Am Phys Soc; Sigma Xi. Res: Photographic emulsions; electron microscopy; transport properties of metals; semiconductors; thin films; infrared and submillimeter physics; radiation detection; semiconductors. Mailing Add: Eng Exp Sta Ga Inst of Technol Atlanta GA 30332

BLUE, MICHAEL HENRY, b Ovid, Colo, Nov 25, 29; div; c 2. COSMIC RAY PHYSICS. Educ: Colo State Univ, BS, 50; Univ Wash, PhD(physics), 60. Prof Exp: Fel physics, Colo State Univ, 53-54; engr, Boeing Co, Wash, 54-56; res asst physics, Univ Wash, 56-59, res instr, 60; physicist, Lawrence Radiation lab, Univ Calif, 60-64; ASSOC PROF PHYSICS, UNIV TEX, EL PASO, 64- Mem: AAAS; Am Phys Soc; Am Asn Physics Teachers; Italian Phys Soc; Am Mgt Asn. Res: Proton-proton scattering scattering in the Bev energy region; low and medium energy nuclear physics; Monte Carlo simulation of nucleon-nucleon collisions. Mailing Add: Campus Box 36 Univ of Tex El Paso TX 79968

BLUE, RICHARD ARTHUR, b Le Mars, Iowa, Nov 12, 36. EXPERIMENTAL NUCLEAR PHYSICS. Educ: Cornell Col, BA, 58; Univ Wis, MS, 60, PhD(physics), 63. Prof Exp: Fel physics, Ohio State Univ, 63-66; asst prof, 66-73, ASSOC PROF PHYSICS, UNIV FLA, 73- Mem: Am Phys Soc. Res: Low energy experimental nuclear physics; giant resonances observed via radiative capture reactions. Mailing Add: Dept of Physics Univ of Fla Gainesville FL 32611

BLUE, ROBERT DARWIN, electrochemistry, see 12th edition

BLUE, WILLIAM GUARD, b Poplar Bluff, Mo, Nov 7, 23; m 47; c 3. SOILS. Educ: Univ Mo, PhD(soils), 50. Prof Exp: Asst, Univ Mo, 47-50; from asst biochemist to assoc biochemist, 50-63, BIOCHEMIST, UNIV FLA, 63-, PROF BIOCHEM, 73- Concurrent Pos: Pasture agronomist, Univ Fla-Costa Rican Contract, San Jose, 58-60; consult, Guyana, 67; lectr trop soils, Orgn Trop Studies, Costa Rica, 69. Mem: Soil Sci Soc Am; Am Soc Agron; Am Forage & Grassland Coun; Int Grassland Asn. Res: Chemistry, fertility and biological studies of soils, especially chemistry and fertility of tropical soils in British Honduras, Costa Rica, Panama, Venezuela and Guyana. Mailing Add: 106 Newell Hall Dept of Soils Agr Exp Sta Univ of Fla Gainesville FL 32611

BLUEBOND-LANGNER, MYRA HONORE, b Philadelphia, Pa, June 29, 48; m 72. ANTHROPOLOGY. Educ: Temple Univ, BA, 69; Univ Ill, MA, 71, PhD(anthrop), 75. Prof Exp: ASST PROF ANTHROP, RUTGERS UNIV, CAMDEN, 74- Concurrent Pos: Mem, Adv Bd, Term-Care, Inc, 75-; vpres, Ars Moriendi, 76- Mem: Fel Am Anthrop Asn; Med Anthrop Soc; Sigma Xi; AAAS. Res: Childhood socialization and culture; death and dying; psychology. Mailing Add: Dept of Sociol & Anthrop Rutgers Univ Camden NJ 08102

BLUEFARB, SAMUEL M, b St Louis, Mo, Oct 15, 12; m 45; c 1. DERMATOLOGY. Educ: Univ Ill, BS, 35, MD, 37. Prof Exp: From instr to assoc prof, 42-57, PROF DERMAT & CHMN DEPT, MED SCH, NORTHWESTERN UNIV, CHICAGO, 57- Concurrent Pos: Consult, Res Vet Hosp, 64- Mem: AMA; Soc Invest Dermat; Am Acad Dermat; Am Dermat Asn; Am Col Physicians. Res: Syphilology; reticuloendothelial system in cutaneous disorders. Mailing Add: Dept of Dermat Northwestern Univ Med Sch Chicago IL 60611

BLUEMEL, VAN (FONKEN WILFORD), b Freeport, Ill, Feb 8, 34; m 63; c 2. QUANTUM OPTICS. Educ: Univ Mich, BS, 56; Univ Ill, MS, 60, PhD(nuclear physics), 67. Prof Exp: Asst prof, 66-73, ASSOC PROF PHYSICS, WORCESTER POLYTECH INST, 73- Mem: Am Phys Soc; Am Asn Physics Teachers. Res: Statistics of scattered light; applications of atomic coherent states. Mailing Add: Dept of Physics Worcester Polytech Inst Worcester MA 01609

BLUEMLE, LEWIS W, JR, b Williamsport, Pa, Mar 9, 21; m 53; c 4. INTERNAL MEDICINE. Educ: Johns Hopkins Univ, AB, 43, MD, 46. Prof Exp: From asst instr to instr med, Univ Pa, 50-52, assoc, 52-55, from asst prof to assoc prof, 55-68, dir clin res ctr, Univ Hosp, 61-68; prof med & pres, State Univ NY Upstate Med Ctr, 68-74; PROF MED & PRES, UNIV ORE HEALTH SCI CTR, 74- Concurrent Pos: Markle scholar, 55-60; vis consult, Vet Hosp, Philadelphia, 55-68; consult, Nat Inst Arthritis & Metab Dis, 67-74. Mem: Am Col Physicians; Am Fedn Clin Res; Am Clin & Climat Asn. Res: Renal disease. Mailing Add: Univ of Ore Health Sci Ctr Portland OR 97201

BLUESTEIN, ALLEN CHANNING, b Lynn, Mass, Aug 25, 26; m 47; c 5. ORGANIC CHEMISTRY. Educ: Univ Mass, BS, 49. Prof Exp: Jr chemist, Advan Res & Develop Measurements Lab, Gen Elec Co, 52-55; rubber chemist, Anaconda Wire & Cable Co, 55-56, sr develop chemist, 56-61; res dir, Aerovox Corp, 61-63; TECH DIR, COOKE COLOR & CHEM CO, INC, REICHHOLD CHEM INC, HACKETTSTOWN, 63-, VPRES, 68- Mem: Am Chem Soc; sr mem Inst Elec & Electronics Eng; fel Am Inst Chem. Res: Rubber compounding technology; cross-linking of thermoplastics; reactions between ozone and polymers; electrical properties of polymers; plastics technology. Mailing Add: 6 Highland Ave Succasunna NJ 07876

BLUESTEIN, BEN ALFRED, b Chicago, Ill, July 5, 18; m 47; c 3. ORGANOMETALLIC CHEMISTRY, POLYMER CHEMISTRY. Educ: Univ Chicago, BS, 38, PhD(org chem), 41. Prof Exp: Res assoc chem, Univ Chicago, 41-43, asst, Metall Lab, 43-44; asst, Los Alamos Sci Lab, 44-46; asst, Res Lab, 46-50, ASST, SILICONE PROD DEPT, GEN ELEC CO, 50- Concurrent Pos: Pittsburgh Plate Glass Co fel, Univ Chicago, 41-43. Mem: AAAS; Am Chem Soc; NY Acad Sci; fel Am Inst Chem; Sigma Xi. Res: Reactions of 9-formylfluorene; synthesis of organo-metallic resins; plastics; radiochemistry; organo-silicon chemistry. Mailing Add: Silicone Prod Dept Gen Elec Co Waterford NY 12188

BLUESTEIN, BERNARD RICHARD, b Philadelphia, Pa, Oct 7, 25; m 47; c 4. ORGANIC CHEMISTRY. Educ: Univ Pa, BS, 46; Univ Ill, MS, 47, PhD(org chem), 49; Fairleigh Dickinson Univ, MBA, 67. Prof Exp: Assoc chem, Rutgers Univ, 49-51; fel, Purdue Univ, 51-52; asst prof, Coe Col, 52-55; sr res chemist, Sonneborn Chem & Refining Co, 55-59, asst supt, 59-62; supvr chem res, Cent Res Labs, 62-66, ASST DIR CORP RES & DEVELOP, WITCO CHEM CO, INC, OAKLAND, 66- Concurrent Pos: Dir proj res, Sonneborn Chem & Refining Co, 60-62. Mem: Am Chem Soc; The Chem Soc. Res: Process development; synthetic organic chemistry; vapor phase catalysis; detergents; surface active agents; polymers; organophosphorus chemistry; petroleum waxes. Mailing Add: 358 Dunham Pl Glen Rock NJ 07452

BLUESTEIN, CLAIRE, b Philadelphia, Pa, May 3, 26; m 47; c 4. INDUSTRIAL ORGANIC CHEMISTRY. Educ: Univ Pa, AB, 47; Univ Ill, MS, 48, PhD(org chem), 50. Prof Exp: Fel, Purdue Univ, 52-53; instr, Ext Serv, Pa State Univ, 58-59; sr res chemist, Sonneborn Chem & Refining Co Div, 59-63, group leader, Cent Res Labs, 63-75, MGR NEW TECHNOL, WITCO CHEM CORP, 75- Mem: Am Chem Soc; Am Inst Chemists; Soc Mfg Engrs; Am Asn Textile Chemists & Colorists. Res: Polyurethane coatings, particularly aqueous emulsions, photocurables, solids, powders; polyurethane elastomers; polyurethane chemistry; fabric coating technology; polymer analytical and physical characterization; plastics processing aids; petroleum sulfonates and waxes; alkaline earth salt dispersions. Mailing Add: Cent Res Labs Witco Chem Corp 100 Bauer Dr Oakland NJ 07436

BLUESTONE, E MICHAEL, b New York, NY, Dec 26, 91; m 22. MEDICAL ADMINISTRATION, PUBLIC HEALTH. Educ: Columbia Univ, BS, 13, MD, 16. Prof Exp: Pvt pract, New York, 19-20; asst dir, Mt Sinai Hosp, 20-26; dir, Hadassah Med Orgn, Palestine, 26-28; dir, 28-51, CONSULT, MONTEFIORE HOSP, NEW YORK, 51- Concurrent Pos: Chmn med reference bd, Hadassah & Hebrew Univ, 40-50; consult, Prof Adv Comt, Off Voc Rehab, Fed Security Agency, 44-50; asst prof, Columbia Univ, 45-71, chmn int class, Sch Pub Health, 62-68; mem med comt, Rosenstock Med Found, 46-48; mem subcomt chronic care, Nat Health Assembly, Washington, DC, 48 & Nat Conf Aging, 50; mem comn chronic dis & old age, Int Hosp Fedn, 51-61; Int Hosp Fedn rep, UN Orgns, 53-67, former mem exec comt, Comn World Orgns Interested in Handicapped, UN; spec consult, Fed Hosp Coun, USPHS, 54; Rene Sand Mem lectr, Lucern, 55; mem expert panel on med care, UN, 55-67; James M Anders lectr, Col Physicians, Philadelphia, 57; Phizer lectr, Col Gen Practitioners, Eng, 58; prof pub admin, NY Univ, 58-65; Franklin I Harris lectr, San Francisco, 61; guest lectr numerous cols & univs, US, PR, Eng & Belg; consult numerous hosp & community projs; mem adv comt med social serv, United Hosp Fund; mem comt hosp treatment of alcoholism & exec comt, Res Coun Probs Alcohol; mem bd dirs, Straus Health Ctr, Jerusalem; mem res adv comt, Res Coun Econ Security; mem adv comt, Nat Comt Alcoholism; mem res coun, Geront Res Found; tech adv comn chronic illness, AMA-Am Hosp Asn-Am Pub Health Asn-Am Pub Works Asn; mem exec comt, Am Comt Ment Health in Israel; prof path, Hebrew Univ, Jerusalem, 75- Honors & Awards: Citations, Hosp Mgr, 56; Hadassah, 57; Bronze Medal of Charleroi, Belg, 58; Distinguished Serv Award, Am Hosp Asn, 61; Distinguished Serv Award, Am Asn Hosp Planning, 73. Dr E M Bluestone fel in hosp admin estab 1950; Dr E M Bluestone scholar in hosp admin estab Columbia Univ, 72. Mem: Fel AAAS; Am Asn Hosp Consult (pres, 51-53); fel Am Col Hosp Adminr; fel Am Col Prev Med; fel Am Pub Health Asn. Res: Social medicine. Mailing Add: 140 West End Ave 29 H New York NY 10023

BLUESTONE, HENRY, b Chicago, Ill, May 17, 14; m 38; c 2. ORGANIC CHEMISTRY. Educ: Univ Chicago, SB, 38, SM, 39. Prof Exp: Instr chem, Chicago City Jr Col, 39-42; res chemist, Velsicol Corp, 42-47; asst res dir, Julius Hyman & Co, 47-51; from res group leader to sr res group leader, 51-73, STAFF CHEMIST, DIAMOND SHAMROCK CORP, 73- Mem: AAAS; Am Chem Soc; Am Inst Chem. Res: Organic synthesis; diene reactions; heterocyclic compounds; polycyclic compounds; stereochemistry; bactericides; fungicides; herbicides; insecticides; plant growth regulators; nematocides. Mailing Add: Diamond Shamrock Corp PO Box 348 Painesville OH 44077

BLUHM, AARON LEO, b Boston, Mass, May 26, 24; m 48; c 1. ORGANIC

CHEMISTRY. Educ: Northeastern Univ, BS, 51; Boston Univ, PhD(chem), 56. Prof Exp: RES CHEMIST ORG CHEM, US ARMY NATICK LABS, 56- Mem: Am Chem Soc; Sigma Xi. Res: Synthesis; reaction mechanisms and kinetics; polymer structural research; infrared spectroscopy of organic and polymeric materials; photochromic materials; electron spin resonance studies of free radicals; photochemistry. Mailing Add: 5 Wentworth Rd Canton MA 02021

BLUHM, HAROLD FREDERICK, b Williamsport, Pa, Mar 12, 27. ORGANIC CHEMISTRY, EXPLOSIVES. Educ: Bucknell Univ, BS, 52; Pa State Univ, MS, 54. Prof Exp: Asst chem, Pa State Univ, 52-54; res chemist, Hercules Powder Co, 54-57; res chemist, Atlas Chem Industs, Inc, 57-66, supvr chem develop lab, 66-74, SUPVR CHEM RES & DEVELOP DEPT, ATLAS RES & DEVELOP LAB, ATLAS POWDER CO, 74- Mem: AAAS; Am Chem Soc. Res: Organometallic compounds; plasticizers; synthetic resins; polymers; new explosive compounds and propellants; development of new commercial explosives. Mailing Add: Atlas Res & Develop Lab Atlas Powder Co PO Box 271 Tamaqua PA 18252

BLUHM, LESLIE, b Winthrop, Mass, May 15, 40; m 64; c 2. FOOD SCIENCE, FOOD TECHNOLOGY. Educ: Univ Mass, BSc, 63; Univ Ill, MS, 67, PhD(food sci), 68. Prof Exp: SECT HEAD & RES MICROBIOLOGIST, MILES LABS, INC, 68- Mem: Am Soc Microbiol; Inst Food Technologists; Am Dairy Sci Asn; Soc Indust Microbiol; Am Cult Dairy Prod Inst. Res: Sublethal heat injury of vegetative cells of staphylococci and streptococci; dairy and food microbiology. Mailing Add: 22815 Londonderry Ct Elkhart IN 46514

BLUM, ALVIN SEYMOUR, b Boston, Mass, Jan 23, 26; m 51; c 1. BIOPHYSICS. Educ: Brown Univ, ScB, 48. Prof Exp: Radioisotope scientist, Vet Admin Hosp, Calif, 50-52, biochemist, La, 52-53, Fla, 53-54, prin scientist, 55-61; CHIEF SCIENTIST, BROWARD GEN HOSP, FT LAUDERDALE, 61- Concurrent Pos: Instr, Sch Med, Univ Miami, 56-; consult, Broward Gen Hosp, 58-61, Mt Sinai Hosp, 60-, Vet Admin Hosp, Miami, Miami Heart Inst, 71- & Doctor's Gen Hosp Plantation, 71- Mem: Soc Nuclear Med. Res: Circulation; kidney function; biochemistry of thyroid; iron metabolism; pulmonary function; radio pharmaceuticals; instrumentation development. Mailing Add: 2350 Del Mar Pl Ft Lauderdale FL 33301

BLUM, BARTON MORRILL, b Newark, NJ, May 17, 32; m 58; c 3. SILVICULTURE. Educ: Rutgers Univ, BS, 54; Yale Univ, MF, 57; State Univ NY Col Forestry, PhD, 71. Prof Exp: Res forester, Northeastern Forest Exp Sta, US Forest Serv, 57-64, from assoc silviculturist to silviculturist, 64-72, PROJ LEADER, NORTHEASTERN FOREST EXP STA, US FOREST SERV, 72- Mem: Soc Am Foresters. Res: Silviculture and ecology of northern hardwood forests; individual tree growth; physiology of maple sap flow and attendent problems; silviculture of northern conifers; northern conifer (spruce and fir) silviculture and ecology. Mailing Add: US Forest Serv NEFES USDA Bldg Univ of Maine Orono ME 04470

BLUM, EDWARD K, b New York, NY, Dec 1, 23; m 53. MATHEMATICS. Educ: Cooper Union, BME, 43; Columbia Univ, AM, 46, PhD(math), 52. Prof Exp: Engr heat transfer, Pratt & Whitney Aircraft, 43-44; lectr math, Columbia Univ, 47-51, instr, 51-52; mathematician, Naval Ord Lab, 53-56; staff mathematician, Space Tech Labs, Calif, 56-62, dir comput & data reduction ctr, 60-62; prof math & dir comput lab, Wesleyan Univ, 62-70; PROF MATH, UNIV SOUTHERN CALIF, 70- Mem: Am Math Soc. Res: Banach algebras; partial differential equations; numerical analysis; digital computers; celestial mechanics. Mailing Add: Dept of Math Univ of Southern Calif Univ Park Los Angeles CA 90007

BLUM, FRED A, b Austin, Tex, Nov 30, 39; m 61; c 2. SOLID STATE ELECTRONICS, LASERS. Educ: Univ Tex, BS, 62; Calif Inst Technol, MS, 64, PhD(physics), 68. Prof Exp: Res scientist, Appl Sci Lab, Gen Dynamics/Ft Worth, 63-64; staff mem solid state div, Lincoln Lab, Mass Inst Technol, 68-73; prog mgr, Cent Res Labs, Tex Instruments, 73-75; ASST DIR SOLID STATE ELECTRONICS DEPT, SCI CTR, ROCKWELL INT, 75- Mem: Am Phys Soc; Inst Elec & Electronic Engrs; AAAS; Optical Soc Am; Soc Photo-Optical Instrumentation Engrs. Res: Semiconductor lasers and optical devices; integrated optics; fiber optics. Mailing Add: Sci Ctr Rockwell Int Thousand Oaks CA 91360

BLUM, HAROLD FRANCIS, b Escondido, Calif, Feb 12, 99; m 30; c 1. PHYSIOLOGY, BIOPHYSICS. Educ: Univ Calif, AB, 22, PhD(physiol), 27. Prof Exp: Asst physiol, Univ Calif, 24-27; asst prof, Univ Ore, 27-28; instr, Harvard Med Sch, 28-30; from asst prof to assoc prof, Med Sch, Univ Calif, 30-38; res fel, Nat Cancer Inst, 38-41, sr pharmacologist, 41-43; prin biophysicist, Naval Med Res Inst, 43-45; Guggenheim fel, 46; physiologist, Nat Cancer Inst, 47-67; prof biol, State Univ NY Albany, 67-72; VIS PROF PHOTOBIOL, SKIN & CANCER HOSP, HEALTH SCI CTR, TEMPLE UNIV, PA, 72- Concurrent Pos: Guggenheim fels, 36, 46, 53; biophysicist, Wash Biophys Inst, 38; vis prof, Princeton Univ, 47-67; secy-gen, Int Photobiol Comt, 51-54; USPHS spec fel, 57. Mem: AAAS; Soc Gen Physiol (vpres, 47, pres, 48); Fr Asn Physiol; Soc Philomatique Paris. Res: General and medical physiology; biological effects of light; diseases caused by light; cancer; biological and cultural evolution. Mailing Add: 612 E Durham St Philadelphia PA 16119

BLUM, HAYWOOD, b New York, NY, Jan 31, 35; m 55; c 3. BIOPHYSICS, SOLID STATE PHYSICS. Educ: City Col New York, BS, 55; Univ Ill, MS, 57, PhD(physics), 62. Prof Exp: Asst physics, Univ Ill, 55-62; mem staff, Mass Inst Technol, 62-64; mem fac, Brookhaven Nat Lab, 64-67; ASSOC PROF PHYSICS, DREXEL UNIV, 67- Concurrent Pos: NIH spec res fel, Sch Med, Univ Pa, 72-73. Mem: Am Phys Soc; AAAS; NY Acad Sci. Res: Double resonance and high pressure studies of the hyperfine interaction in solids; molecular biophysics of cell processes. Mailing Add: 619 Loves Lane Wynnewood PA 19096

BLUM, JACOB JOSEPH, b New York, NY, Oct 3, 26; m 50, 60; c 4. PHYSIOLOGY, BIOPHYSICS. Educ: NY Univ, BA, 47; Univ Chicago, MS, 50, PhD(physiol), 52. Prof Exp: Merck fel, Calif Inst Technol, 52-53; asst prof biochem, Univ Mich, 56-58; chief biophys sect, Geront Br, Nat Heart Inst, 58-62; assoc prof physiol, 62-66, PROF PHYSIOL, DUKE UNIV, 66- Concurrent Pos: Guggenheim fel, Weizmann Inst Sci, Israel, 69-70. Mem: Am Physiol Soc; Biophys Soc; Soc Cell Biologists; Soc Gen Physiol; Protozool Soc. Res: Control and structural organization of intermediary metabolism; mechanochemistry of cilia and flagella. Mailing Add: Dept of Physiol & Pharmacol Duke Univ Durham NC 27710

BLUM, JOHN LEO, b Madison, Wis, May 2, 17; m 47; c 3. PHYCOLOGY. Educ: Univ Wis, BS, 37, MS, 39; Univ Mich, PhD(bot), 53. Prof Exp: Instr biol, Canisius Col, 41-43, asst prof, 46-51, from assoc prof to prof, 53-63; assoc prof, 63-66, assoc dean col letters & sci, 67-69, PROF BOT, UNIV WIS-MILWAUKEE, 66- Mem: Phycol Soc Am; Int Asn Theoret & Appl Limnol; Bot Soc Am; Int Phycol Soc; Brit Phycol Soc. Res: Ecology of algae, especially of fresh and brackish water; aquatic plants. Mailing Add: Dept of Bot Univ of Wis Milwaukee WI 53201

BLUM, JULIUS RUBIN, b Nuremberg, Ger, Feb 1, 22; nat US; m 44; c 2. MATHEMATICS. Educ: Univ Calif, AB, 49, PhD(math), 53. Prof Exp: From instr to assoc prof math, Ind Univ, 53-59; mem staff, Sandia Corp, NMex, 59-62; prof math, Univ NMex, 62-74; PROF MATH, UNIV WIS-MILWAUKEE, 74- Concurrent Pos: Chmn dept math, Univ NMex, 63-69. Mem: Am Math Soc; Math Asn Am; fel Inst Math Statist. Res: Probability theory; mathematical statistics; ergodic theory. Mailing Add: Dept of Math Univ of Wis Milwaukee WI 53201

BLUM, LENORE CAROL, b New York, NY, Dec 18, 42; m 61; c 1. MATHEMATICAL LOGIC, APPLIED MATHEMATICS. Educ: Simmons Col, BS, 63; Mass Inst Technol, PhD(math), 68. Prof Exp: Air Force Office Sci Res fel, 68-69; lectr math, 69-71, RES ASSOC MATH, UNIV CALIF, BERKELEY, 71-; HEAD DEPT MATH & COMPUT SCI, MILLS COL, 74- Concurrent Pos: Coun mem, Conf Bd Math Sci, 76- Mem: Asn Women in Math (pres, 75-77); Asn Women in Sci; Am Math Soc; Math Asn Am; Sigma Xi. Res: Applications of mathematical logic to mathematics, in particular, applications to differential algebra; mathematical theory of inductive inference. Mailing Add: 700 Euclid Ave Berkeley CA 94708

BLUM, LEON LEIB, b Telsiai, Lithuania, May 4, 08; nat US; m 36; c 2. MEDICINE, PATHOLOGY. Educ: Univ Berlin, MD, 33. Prof Exp: Res pathologist, Mt Sinai Hosp, Ill, 34-36; instr path, bact & trop dis, Sch Nursing, 37-57, PATHOLOGIST & DIR LABS, UNION HOSP, IND STATE UNIV, 37-, DIR SCH MED TECHNOL, 54- Concurrent Pos: Pathologist & dir labs, Assoc Physicians & Surgeons' Clin, 36-, St Anthony Hosp, 44-54, Clay County Hosp, Brazil, 50-73, Putnam County Hosp, Greencastle, 53-, Greene County Gen Hosp, Linton, 55- & Vermillion County Hosp, Clinton, 56-; sr partner, Terre Haute Med Lab, 47-; consult, Charles Pfizer & Co, 58-65; adj prof life sci & clin dir, Ctr Med Technol, Ind State Univ, 67-; med dir, Cert Lab Asst Prog, Ind Voc Tech Col, 73- Honors & Awards: Am Path Found Achievement Award, 61; Am Cancer Soc Cert Merit, 74. Mem: Am Soc Cytol; fel Col Am Path; Am Soc Clin Path; AMA; Pan-Am Med Asn. Res: Hematology. Mailing Add: 3200 Ohio Blvd Terre Haute IN 47803

BLUM, MANUEL, b Caracas, Venezuela, Apr 26, 38; m 61; c 1. MATHEMATICS. Educ: Mass Inst Technol, BS & MS, 61, PhD(math), 64. Prof Exp: Res assoc elec eng, Mass Inst Technol, 64-65, asst prof math, 65-69; assoc prof, 69-72, PROF ELEC ENG, UNIV CALIF, BERKELEY, 72- Concurrent Pos: Sloan Found fel, 71-72. Mem: Am Math Soc. Res: Theory of algorithms and its application to programming; minimization of computation time and of program size; mathematical theory of inductive inference. Mailing Add: Dept of Elec Eng & Comput Sci Univ of Calif Berkeley CA 94720

BLUM, MARVIN, b Manhattan, NY, June 18, 28; m 49; c 1. APPLIED MATHEMATICS, SYSTEM ANALYSIS. Educ: Brooklyn Col, BS, 48. Prof Exp: Physicist, Naval Proving Grounds, Va, 48-49; elec engr, Ord Div, Nat Bur Standards, 49-53; proximity fuze studies, Dept Army, 53-54; design specialist, Gen Dynamics, 54-59; sr opers res scientist, Syst Develop Corp, 59-62; sr res mathematician, Conductron Corp, 62-64; res mathematician, Rand Corp, 64-68; sr res scientist, Syst Develop Corp, 68-71; mem tech staff, Inst Defense Anal, 71-72; mem sr tech staff, Technol Serv Corp, 72-73; MEM RES & TECH STAFF, NORTHROP RES & TECHNOL CTR, 73- Concurrent Pos: Consult, Air Force Satellite Tracking Network. Mem: Soc Indust & Appl Math; Inst Elec & Electronics Engrs. Res: Detection; parameter estimation; statistical filter theory; real-time and recursive smoothing techniques; high energy laser applications; system analysis; laser propagation; scenario simulation; vulnerability and system effectiveness evaluation. Mailing Add: 1629 S Bentley Ave Los Angeles CA 90025

BLUM, MURRAY SHELDON, b Philadelphia, Pa, July 19, 29; m 53; c 4. ENTOMOLOGY, PHYSIOLOGY. Educ: Univ Ill, BS, 52, MS, 53, PhD(entom), 55. Prof Exp: Entomologist, USDA, 57-58; assoc prof entom, La State Univ, Baton Rouge, 58-67; PROF ENTOM, UNIV GA, 67- Concurrent Pos: Ed, Insect Biochemistry Entomologia Experimentalis et Applicata, J Chem Ecol, 1972-; consult, NSF, 74- Mem: AAAS; Entom Soc Am. Res: Chemistry of insect pheromones and defensive secretions; biochemistry of the insect reproductive system; regulation of insect behavior by chemical releasers; chemistry and functions of arthropod natural products. Mailing Add: Dept of Entom Univ of Ga Athens GA 30601

BLUM, NORMAN ALLEN, b Boston, Mass, Dec 29, 32; m 57; c 3. SOLID STATE SCIENCE. Educ: Harvard Univ, AB, 54; Brandeis Univ, MA, 59, PhD(physics), 64. Prof Exp: Sr res scientist, Res & Advan Develop Div, Avco Corp, 58-60; res staff mem, Nat Magnet Lab, Mass Inst Technol, 60-66 & NASA Electronics Res Ctr, 66-70; SR PHYSICIST, APPL PHYSICS LAB, JOHNS HOPKINS UNIV, 70- Mem: AAAS; Am Asn Physics Teachers; Am Phys Soc. Res: Solid state physics; magnetism; Mössbauer effect; instrumentation; amorphous films; optical and electronic properties of solids. Mailing Add: Appl Physics Lab Johns Hopkins Univ Laurel MD 20810

BLUM, PETER, b New York, NY, Jan 25, 39; m 64; c 1. ALGEBRA, GEOMETRY. Educ: Oberlin Col, BA, 60; Univ Calif, Berkeley, MA, 63, PhD(math), 66. Prof Exp: Instr math, Univ Calif, Berkeley, 66; J F Ritt instr, Columbia Univ, 66-69; ASST PROF MATH, UNIV ROCHESTER, 69- Concurrent Pos: NSF grant, 70-72. Res: Global differential algebra; complete models of differential fields; extensions of differential specializations; rational functions on Ritt manifolds and Ritt's conjecture concerning the values of such functions. Mailing Add: Dept of Math Univ of Rochester Rochester NY 14627

BLUM, RICHARD, b Cernauti, Romania, Sept 26, 13; Can citizen; m 36; c 2. MATHEMATICS. Educ: Univ Cernauti, Romania, Lic, 34; Univ Bucharest, PhD(math), 46. Prof Exp: Asst math, Univ Bucharest, 45-49; lectr, Univ Cluj, 49-50; asst prof, Acadia Univ, 53-54; lectr, 54-56, assoc prof, 56-63, PROF MATH, UNIV SASK, 63- Mem: Am Math Soc; Can Math Cong. Res: Differential geometry, particularly imbedding properties of Riemannian spaces in Euclidean spaces. Mailing Add: Dept of Math Univ of Sask Saskatoon SK Can

BLUM, ROBERT ALLAN, b Philadelphia, Pa, May 16, 38; m 59; c 3. PSYCHIATRY, PSYCHOANALYSIS. Educ: Mass Inst Technol, BS, 59, MS, 60; Univ Pa, MD, 64. Prof Exp: Engr, Gen Atronics Corp, 59-61; intern med, Mem Hosp Long Beach, Calif, 64-65; resident psychiat, Mass Ment Health Ctr, 65-66, chief resident, Clin Res Ctr, 66-67; staff assoc clin psychopharm, Div Spec Ment Health Res Progs, NIMH, 67-69, med officer, Nat Inst Alcoholism & Alcohol Abuse, 69-74; ASST PROF PSYCHIAT, GEORGETOWN UNIV, 72- Concurrent Pos: Teaching fel, Harvard Med Sch, 65-67; instr, Sch Med, George Washington Univ, 69-74; instr, Sch Med, Johns Hopkins Univ, 69- & Baltimore-DC Inst Psychoanal, 69-; consult, US Govt. Mem: AAAS; Am Psychiat Asn; Am Psychoanal Asn; Soc Gen Systs Res. Res: Information theory; error correction codes; cognitive and attentional mechanisms in schizophrenia; electrophysiologic correlates of behavior; mathematical models of behavior using automata theory; psychobiographical analysis; behavioral aspects of terrorism. Mailing Add: 9819 Hill St Kensington MD 20795

BLUM, SAMUEL EMIL, b New York, NY, Aug 28, 20; c 3. PHYSICAL CHEMISTRY. Educ: Rutgers Univ, BS, 42, PhD(phys chem), 50. Prof Exp: Prod chemist, US Rubber Co, 42-43; asst, Rutgers Univ, 46-50; prin res chemist, Battelle

Mem Inst, 52-59; MEM RES STAFF, THOMAS J WATSON RES LAB, IBM CORP, 59- Mem: AAAS; Sigma Xi; Electrochem Soc; fel Am Inst Chem. Res: Semiconductors; material science. Mailing Add: Thomas J Watson Res Lab IBM Corp PO Box 218 Yorktown Heights NY 10598

BLUM, STANLEY WALTER, b Sheboygan, Wis, July 3, 33; m 57; c 2. ORGANIC CHEMISTRY, BIOCHEMISTRY. Educ: Univ Wis, BS, 55; Univ Ill, PhD(org chem), 60. Prof Exp: Res fel org chem, Univ Minn, 60-61; res assoc chem microbial prod, Rutgers Univ, 61-62; scientist, 62-66, SR SCIENTIST, WARNER-LAMBERT RES INST, 66- Mem: Am Chem Soc. Res: Biochemical pharmacology; drug metabolism; myocardial metabolism and cardiovascular biochemistry; mitochondrial bioenergetics and membrane permeability; metabolic control mechanisms. Mailing Add: Drug Metab Warner-Lambert Res Inst 170 Tabor Rd Morris Plains NJ 07950

BLUM, UDO, b Lüdenscheid, Ger, Nov 29, 39; US citizen; m 68. BOTANY, ECOLOGY. Educ: Franklin Col, BA, 62; Ind Univ, Bloomington, MA, 65; Univ Okla, PhD(bot), 68. Prof Exp: Vis asst prof bot, Univ Okla, 68-69; asst prof, 69-74, fac res grant, 70, agr exp sta grant, 70-72, ASSOC PROF BOT, NC STATE UNIV, 74- Mem: Brit Ecol Soc; Am Forestry Asn; Ecol Soc Am; Am Soc Plant Physiologists. Res: Plant/plant interactions, such as competition and allelopathy; effects of natural and man-made stresses on the physiology of plants; old-field succession. Mailing Add: Dept of Bot NC State Univ Raleigh NC 27607

BLUM, VICTOR JOSEPH, b Defiance, Iowa, Mar 30, 07. GEOPHYSICS. Educ: Xavier Univ, Ohio, AB, 31; St Louis Univ, MA, 33, MS, 36, STL, 40, PhD(geophys), 44. Prof Exp: From instr to prof geophys, St Louis Univ, 44-68, prof earth & atmospheric sci, 68-75, from asst dean to dean, Inst Technol, 56-75, PROF EMER GEOPHYS, ST LOUIS UNIV, 75- Concurrent Pos: Trustee, Univ Corp for Atmospheric Res. Mem: AAAS; Am Soc Testing & Mat; Seismol Soc Am; Soc Explor Geophys; Am Meteorol Soc. Res: Magnetic field and geological formations and structures. Mailing Add: Dept of Earth & Atmospheric Sciences St Louis Univ 221 N Grand St Louis MO 63103

BLUMBERG, AVROM AARON, b Albany, NY, Mar 3, 28; m 55, 69; c 3. PHYSICAL CHEMISTRY. Educ: Rensselaer Polytech Inst, BS, 49; Yale Univ, PhD(phys chem), 53. Prof Exp: Lab instr, Rensselaer Polytech Inst, 49; Plate Glass fel, Mellon Inst, 53-59, res chemist polymer studies, 59-63; from asst prof to assoc prof, 63-75, PROF CHEM, DePAUL UNIV, 75-, HEAD DIV NATURAL SCI & MATH, 66- Concurrent Pos: Lectr, Univ Pittsburgh, 57-58. Mem: Am Chem Soc; The Chem Soc; Sigma Xi. Res: Reaction kinetics; solid-liquid reactions; glassy state; differential thermal analysis; x-ray diffraction structure calculations; polymer configurations; transport phenomena in polymers and through films; kinetics of organic growth; arms control; polymer degradation. Mailing Add: Dept of Chem DePaul Univ Chicago IL 60614

BLUMBERG, BARUCH SAMUEL, b New York, NY, July 28, 25; m 54; c 4. MEDICINE, MEDICAL ANTHROPOLOGY. Educ: Union Univ, NY, BS, 45; Columbia Univ, MD, 51; Oxford Univ, PhD, 57. Prof Exp: Intern & asst resident, First Med Div, Bellevue Hosp, 51-53; asst physician, Dept Med, Presby Hosp, NY, 53-55; vis scientist, Dept Biochem, Oxford Univ, 55-57; sr investr, Nat Inst Arthritis & Metab Dis, 57-59, chief geog med & genetics sect, 60-64; assoc prof med, 64-70, prof human genetics, 64-74, PROF MED, UNIV PA, 70-, PROF ANTHROP, 74-; ASSOC DIR CLIN RES & SR MEM, INST CANCER RES, 64- Concurrent Pos: Consult, Goldwater Mem Hosp, NY, 53-55; asst ed, Arthritis & Rheumatism, 59-63; res collabr, Brookhaven Nat Lab, 59-; ed, Progress in Rheumatol, 63-; consult, WHO, 69-70; vis prof, Pa Hosp, 70- Honors & Awards: Albion O Bernstein Award, Med Soc NY, 69; Grand Sci Award, Phi Lambda Kappa, 72; Eppinger Prize, Univ Feiburg, 73; Passano Award, 74; Gairdner Award, 75. Mem: Nat Acad Sci; Asn Am Physicians; Am Soc Human Genetics; Am Fedn Clin Res; Am Soc Clin Invest. Res: Genetics; clinical and cancer research. Mailing Add: Inst for Cancer Res 7701 Burholme Ave Fox Chase Philadelphia PA 19111

BLUMBERG, HAROLD, b Fairmont, WVa, June 19, 09; m 34; c 1. PHARMACOLOGY. Educ: Johns Hopkins Univ, ScD(biochem), 33. Prof Exp: Asst pediat, Sch Med, Johns Hopkins Univ, 33-35, instr, 35-36; assoc biochemist, Off Child Hyg, USPHS, 36-38; res biochemist, Dept Biochem, Sch Hyg & Pub Health, Johns Hopkins Univ, 38-42; assoc toxicologist, Army Indust Hyg, US War Dept, 42-44; sr biologist, Sterling-Winthrop Res Inst, NY, 44-47; dir biol labs, Endo Labs, Inc, 47-60, assoc dir res, 60-74; RES PROF PHARMACOL, NEW YORK MED COL, 74- Concurrent Pos: Nat Res Coun fel, Sch Hyg & Pub Health, Johns Hopkins Univ, 34-35. Mem: AAAS; Soc Toxicol; Am Soc Pharmacol & Exp Therapeut; Soc Exp Biol & Med; NY Acad Sci. Res: Vitamin E; lead poisoning; argyremia; manganese rickets; choline and liver cirrhosis; mercurial diuretics; theophylline derivatives; antispasmodics; naloxone and naltrexone narcotic antagonists; molindone tranquilizer; nalbuphine analgesic. Mailing Add: Dept of Pharmacol New York Med Col Valhalla NY 10595

BLUMBERG, JOE MORRIS, b Baltimore, Md, June 27, 09; m 35. PATHOLOGY. Educ: Emory Univ, BS, 30, MD, 33. Hon Degrees: MD, Cath Med Col, Seoul, 66. Prof Exp: Formerly dir, Armed Forces Inst Path & commanding gen, US Army Med Res & Develop Command; PATHOLOGIST, OSCAR B HUNTER MEM LAB, 69- Concurrent Pos: Clin prof, Georgetown Univ Sch Med, 63 & George Washington Univ Sch Med, 64- Honors & Awards: Hektoen Bronze Medal, AMA, 61; Stitt Award, Asn Mil Surg, 61, Founders Medal, 64; Seale Harris Award, Southern Med Asn, 63; Assoc Clin Scientist of Year, Emory Univ, 68; Ward Burdick Award, Am Soc Clin Path, 69. Mem: Fel Am Soc Clin Path; fel Col Am Path; AMA; Am Asn Path & Bact. Res: Infectious diseases; hematology; forensic pathology. Mailing Add: Oscar B Hunter Mem Lab 915 19th St NW Washington DC 20006

BLUMBERG, JOHN OTTO, b Elizabeth, NJ, Aug 21, 07; m 40; c 1. MATHEMATICS. Educ: Univ Pittsburgh, AB, 29, MA, 31, PhD(math), 40. Prof Exp: Asst, 29-37, from instr to asst prof, 37-47, ASSOC PROF MATH, UNIV PITTSBURGH, 47- Mem: Math Asn Am. Mailing Add: Dept of Math Univ of Pittsburgh Pittsburgh PA 15213

BLUMBERG, LEROY NORMAN, b Atlantic City, NJ, June 22, 29; m 57; c 3. PHYSICS. Educ: Mass Inst Technol, BSc, 51; Columbia Univ, MA, 55, PhD(physics), 62. Prof Exp: Physicist hydrodyn, Los Alamos Sci Lab, Univ Calif, 55-57, nuclear physics, 57-60; physicist, Union Carbide Nuclear Div, Oak Ridge Nat Lab, 62-65; physicist, Cambridge Electron Accelerator, Harvard Univ, 65-66; PHYSICIST, BROOKHAVEN NAT LAB, 66- Mem: AAAS; Am Phys Soc. Res: Shock wave hydrodynamics; fission fragment angular distributions; polarized proton scattering. Mailing Add: Accelerator Dept Brookhaven Nat Lab Upton NY 11973

BLUMBERG, RICHARD WINSTON, b Winston-Salem, NC, Nov 10, 14; m 69. PEDIATRICS. Educ: Emory Univ, BS, 35, MD, 38. Prof Exp: Res assoc, Children's Hosp, Cincinnati, 47-48; assoc, 48-51, from asst prof to assoc prof, 51-59, actg chmn dept, 51-59, PROF PEDIAT & CHMN DEPT, SCH MED, EMORY UNIV, 59- Mem: Am Acad Pediat; AMA; Am Pediat Soc. Res: Infectious diseases. Mailing Add: Emory Univ Sch of Med 69 Butler St SE Atlanta GA 30303

BLUMBERG, WILLIAM EMIL, b College Station, Tex, Dec 23, 30. PHYSICS, BIOPHYSICS. Educ: Univ Tex, BS, 52; Univ Calif, Berkeley, PhD(physics), 59. Prof Exp: Phys scientist, US Army Biol Labs, Md, 53-55; MEM TECH STAFF BIOPHYSICS, BELL LABS, INC, 59- Concurrent Pos: Assoc prof, Albert Einstein Col Med, 70-; adj assoc prof, Rockefeller Univ, 71-72. Mem: AAAS; fel Am Phys Soc; Biophys Soc; Am Soc Biol Chem; Am Soc Photobiol. Res: Magnetic interactions and magnetic resonance of electrons and nuclei; role of paramagnetic ions in biologically important reactions. Mailing Add: Bell Labs Inc Murray Hill NJ 07974

BLUMBERGS, PETER, organic chemistry, see 12th edition

BLUME, ARTHUR JOEL, b New York, NY, May 19, 41; m 68; c 2. NEUROCHEMISTRY. Educ: Univ Rochester, BA, 63; Syracuse Univ, MS, 66, PhD(molecular biol), 68. Prof Exp: USPHS fel neurochem, Lab Biochem Genetics, NIH, 68-71; ASST MEM NEUROCHEM, DEPT PHYSIOL CHEM, ROCHE INST MOLECULAR BIOL, 71- Mem: Sigma Xi. Res: Storage and transfer of information in the nervous system; emphasis on properties of neuronal membranes. Mailing Add: Dept of Physiol Chem Roche Inst of Molecular Biol Nutley NJ 07110

BLUME, FREDERICK DUANE, b Mishawaka, Ind, Aug 27, 33; m 55; c 8. ENVIRONMENTAL PHYSIOLOGY. Educ: Wabash Col, AB, 55; Univ Calif, Berkeley, PhD(altitude metab), 65. Prof Exp: Asst res physiologist, White Mountain Res Sta, Univ Calif, 64-69, asst dir, 69-72; PROF BIOL & CHMN DEPT, CALIF STATE COL, BAKERSFIELD, 72- Res: Environmental physiology; carbohydrate metabolism at altitude; growth and development and cardiovascular changes at altitude. Mailing Add: Dept of Biol Calif State Col Bakersfield CA 93309

BLUME, MARTIN, b New York, NY, Jan 13, 32; m 55; c 2. PHYSICS. Educ: Princeton Univ, AB, 54, Harvard Univ, AM, 56, PhD(physics), 59. Prof Exp: Fulbright res fel physics, Univ Tokyo, 59-60; res assoc theoret physics, Atomic Energy Res Estab, Eng, 60-62; assoc physicist, 62-65, physicist, 65-70, SR PHYSICIST, BROOKHAVEN NAT LAB, 70-; PROF PHYSICS, STATE UNIV NY STONY BROOK, 72- Concurrent Pos: Consult, Bell Tel Labs, NJ, 63-70 & Lawrence Radiation Lab, Univ Calif, 66-70; vis lectr, Yale Univ, 67-69. Mem: Fel Am Phys Soc. Res: Theoretical solid state and atomic physics; neutron scattering; magnetism; magnetic resonance. Mailing Add: Dept of Physics Brookhaven Nat Lab Upton NY 11973

BLUME, SHEILA BIERMAN, b Brooklyn, NY, June 21, 34; m 55; c 2. PSYCHIATRY. Educ: Harvard Med Sch, MD, 58. Prof Exp: Intern pediat, Med Ctr, Children's Hosp, 58-59; Fulbright fel psychiat biochem, Sch Med, Tokyo Univ, 59-60; resident psychiat, 62-65, CHIEF ALCOHOLISM UNIT, CENT ISLIP PSYCHIAT CTR, 64- Concurrent Pos: Clin asst prof psychiat, Sch Med, State Univ NY Stony Brook, 71- Mem: Fel Am Psychiat Asn; Am Med Soc Alcoholism (mem bd dirs, 73-); Am Asn Group Psychother & Psychodrama; Nat Coun Compulsive Gambling. Res: Clinical aspects, treatment, public policy and confidentiality of records of alcoholism. Mailing Add: Cent Islip Psychiat Ctr Central Islip NY 11722

BLUMEN, ISADORE, b Newark, NJ. STATISTICS. Educ: Univ Minn, BA, 41; Univ NC, MA, 48, PhD(statist), 54. Prof Exp: PROF STATIST, CORNELL UNIV, 49- Mailing Add: 122 Warren Rd Ithaca NY 14850

BLUMEN, WILLIAM, b Hartford, Conn, Aug 7, 31; m 60; c 2. FLUID DYNAMICS, METEOROLOGY. Educ: Fla State Univ, BS, 57, MS, 58; Mass Inst Technol, PhD(mountain-wave drag), 63. Prof Exp: Res assoc, Univ Oslo, 63-64; fel meteorol, Nat Ctr Atmospheric Res, 64-66; from asst prof to assoc prof, 66-74, PROF ASTRO-GEOPHYS, UNIV COLO, BOULDER, 74- Concurrent Pos: Lectr, Univ Colo, 56-66; res grants, NATO, 63-65, NSF, 68-70; NATO sr vis fel, Cambridge Univ, 70-71. Mem: Am Meteorol Soc. Res: Theoretical aspects of geophysical fluid dynamics, with application to the dynamics of atmospheric and oceanic circulations; wave motion and hydrodynamic instability. Mailing Add: Dept of Astro-Geophys Univ of Colo Boulder CO 80302

BLUMENBERG, KARL EDWARD, b Bellaire, Ohio, Oct 1, 14; m 37; c 3. INDUSTRIAL CHEMISTRY. Educ: Ohio State Univ, AB, 36, MS, 39, PhD(inorg chem), 40. Prof Exp: Asst chem, Ohio State Univ, 36-40; RES CHEMIST, PIGMENTS DEPT, E I DU PONT DE NEMOURS & CO, 40- Mem: Am Chem Soc. Res: Reaction of various elements with molten sodium amide; titanium dioxide manufactured. Mailing Add: Pigments Dept E I du Pont de Nemours & Co Edge Moor DE 19809

BLUMENFELD, HENRY A, b Amsterdam, Neth, May 31, 25; US citizen; m 56; c 2. PHYSICS. Educ: Harvard Univ, AB, 48; Columbia Univ, PhD(physics), 56. Prof Exp: Res assoc physics, Mass Inst Technol, 55-56 & Duke Univ, 56-57; mem physics staff, Princeton Univ, 58-67 & Europ Orgn Nuclear Res, Geneva, Switz, 67-70; MEM PHYSICS STAFF, SACLAY NUCLEAR RES CTR, FRANCE, 70- Mem: Am Phys Soc. Res: High energy nuclear physics; experimental, elementary particles; large and rapid cycling bubble chambers. Mailing Add: Saclay Nuclear Res Ctr BP 2 91 Gif-sur-Yvette France

BLUMENFELD, MARTIN, b Baltimore, Md, Sept 4, 41; m 66; c 2. CELL BIOLOGY, MOLECULAR BIOLOGY. Educ: Johns Hopkins Univ, BA, 63; Case Western Reserve Univ, PhD(biol), 68. Prof Exp: Fel zool, Univ Mich, 67-68; res assoc, Univ Tex, 68-72; fel genetics, Univ Wis, 72-73, res assoc, 73-74; ASST PROF ZOOL, UNIV MINN, MINNEAPOLIS, 74- Concurrent Pos: USPHS res grant, Inst Gen Med, 74. Res: Satellite DNA and molecular organization of chromosomes of drosophila. Mailing Add: Dept of Zool Univ of Minn Minneapolis MN 55414

BLUMENFELD, OLGA O, b Lodz, Poland, Apr 6, 23; nat US; m 44; c 1. BIOCHEMISTRY. Educ: City Col New York, BS, 46; Univ Colo, MS, 48; NY Univ, PhD(biochem), 57. Prof Exp: Fel, Nat Res Coun & Rockefeller Inst, 57-61; from asst prof to assoc prof, 61-72, PROF BIOCHEM, ALBERT EINSTEIN COL MED, 72- Concurrent Pos: Sr investr, Arthritis Found, 61-67; NIH career develop award, 67. Mem: Am Chem Soc; Fedn Am Soc Exp Biol. Res: Protein structure. Mailing Add: Dept of Biochem Albert Einstein Col of Med Bronx NY 10461

BLUMENFIELD, DAVID, b Philadelphia, Pa, Mar 7, 28; m 59; c 2. HORTICULTURE, PLANT PHYSIOLOGY. Educ: Del Valley Col, BS, 50; Rutgers Univ, MS, 56, PhD(hort, plant physiol), 59. Prof Exp: Spec rep tech sales, Wyeth Labs, Pa, 53-55; res asst hort, Rutgers Univ, 55-59; from asst prof to assoc prof, 59-71, PROF HORT, DEL VALLEY COL, 71- Mem: AAAS; Am Soc Hort Sci. Res: Plant nutrition; agricultural climatology. Mailing Add: Dept of Hort Del Valley Col Doylestown PA 18901

BLUMENSON, LESLIE ELI, b New York, NY, Mar 20, 34; m 57; c 2.

MATHEMATICAL BIOLOGY, CANCER. Educ: NY Univ, AB, 55, MS, 56, PhD(math), 62. Prof Exp: Sr res mathematician, Electronics Res Lab, Columbia Univ, 56-63; USPHS trainee, Univ Chicago, 63-65; from sr cancer res scientist to assoc cancer res scientist, 65-71, PRIN CANCER RES SCIENTIST, ROSWELL PARK MEM INST, 72- Concurrent Pos: Assoc res prof biostat, State Univ NY Buffalo, 69- Mem: AAAS; Am Asn Cancer Res; Am Math Soc; Math Asn Am; Soc Math Biol. Res: Mathematical models in biological sciences; cancer research; biostatistics. Mailing Add: Dept of Biostat Roswell Park Mem Inst Buffalo NY 14263

BLUMENSTEIN, MICHAEL, b New York, NY, Nov 16, 47. BIOCHEMISTRY. Educ: City Col New York, BS, 68; Calif Inst Technol, PhD(chem), 72. Prof Exp: Mem tech staff biochem, Bell Labs, 73-74; res assoc, Univ Ariz, 74-75; ASST PROF BIOCHEM, SCH MED, TUFTS UNIV, 75- Concurrent Pos: NIH fel, Univ Ariz, 75. Mem: AAAS; Am Chem Soc. Res: Nuclear magnetic resonance studies of biochemical systems; protein-hormone interactions; mechanism of action of adenine nucleotide coenzymes. Mailing Add: Sch of Med Tufts Univ 136 Harrison Ave Boston MA 02111

BLUMENSTOCK, DAVID A, b Newark, NJ, Feb 14, 27; m 52; c 3. MEDICINE. Educ: Union Col, NY, BS, 49; Cornell Univ, MD, 53; Am Bd Surg, dipl, 59; Am Bd Thoracic Surg, dipl, 60. Prof Exp: Resident surgeon, 54-60, NIH res career develop award, 62-63, SURGEON-IN-CHIEF, MARY IMOGENE BASSETT HOSP, 63-; PROF CLIN SURG, COL PHYSICIANS & SURGEONS, COLUMBIA UNIV, 74- Mem: Soc Univ Surgeons; Am Asn Thoracic Surg; Soc Thoracic Surgeons; Fedn Am Soc Exp Biol; Am Fedn Clin Res. Res: Transplantation of tissues and organs; preservation of living tissues; metabolic response to surgery. Mailing Add: Mary Imogene Bassett Hosp Atwell Rd Cooperstown NY 13326

BLUMENTHAL, GEORGE R, astrophysics, see 12th edition

BLUMENTHAL, HAROLD JAY, b New York, NY, Jan 21, 26; m 50; c 3. MICROBIOLOGY, MICROBIAL BIOCHEMISTRY. Educ: Ind Univ, BS, 47; Purdue Univ, MS, 49, PhD(bact), 53. Prof Exp: Teaching asst chem, Purdue Univ, 47-49; Am Cancer Soc fel, Inst Cancer Res, 53-54; from instr to assoc prof microbiol, Univ Mich, 54-65; PROF MICROBIOL & CHMN DEPT, STRITCH SCH MED, LOYOLA UNIV CHICAGO, 65- Concurrent Pos: Res assoc, Rackham Arthritis Res Unit, 54-56; NIH spec res fel, 63-64; mem adv comt study, NIH Res Training Grant Progs, Nat Acad Sci, 66-75; consult, Argonne Univs Asn-Argonne Nat Lab, 71-74; Am Acad Microbiol vis prof, Cath Univ Chile, 72. Mem: AAAS; Am Soc Microbiol; Asn Am Med Cols; Am Chem Soc; Brit Soc Gen Microbiol. Res: Pathways of carbohydrate catabolism in microorganisms; metabolism of hexaric acids; staphylococcal virulence. Mailing Add: Dept of Microbiol Loyola Univ Stritch Sch of Med Maywood IL 60153

BLUMENTHAL, HERBERT, b New York, NY, May 1, 25; m 50; c 2. BIOCHEMISTRY, TOXICOLOGY. Educ: City Col New York, BS, 48; Univ Southern Calif, PhD(biochem), 55. Prof Exp: Actg chief, Pharmaco-Dynamics Br, Div Pharmacol, Food & Drug Admin, 55-58; biochemist, Nat Inst Dent Res, NIH, 58-61; chief petitions res br, 61-71, dep dir, Div Toxicol, 70-74, ACTG DIR, DIV TOXICOL, FOOD & DRUG ADMIN, 74- Mem: Fel AAAS; NY Acad Sci; Soc Toxicol; Environ Mutagen Soc. Res: Toxicology of food additives, pesticides, colors, drugs. Mailing Add: Food & Drug Admin 200 C St NW Washington DC 20204

BLUMENTHAL, HERMAN T, b New York, NY, Apr 8, 13; m 40; c 2. PATHOLOGY. Educ: Rutgers Univ, BS, 34; Univ Pa, MS, 36; Washington Univ, PhD(path), 38, MD, 42; Am Bd Path, dipl. Prof Exp: Asst path, Sch Med, Washington Univ, 39-42; resident, Jewish Hosp, St Louis, 42-43; lab dir, Jewish Hosp, Louisville, 46; dir div labs, Jewish Hosp, St Louis, 50-57, inst exp path, 57-62; gerontol res assoc, 62-68, RES PROF GERONT, BIOPSYCHOL RES LAB, DEPT PSYCHOL, WASHINGTON UNIV, 68- Concurrent Pos: Attend pathologist, Vet Admin Hosp, Jefferson Barracks, Mo, 47-50, dir clin res prog aging, 60-66; co-chmn endocrinol sect, Int Cong Gerontol, London, 54, prog biol & clin med. San Francisco, 60; consult pathologist, Cochran Vet Admin Hosp, 64-; mem coun arteriosclerosis, Am Heart Asn. Honors & Awards: Berg Prize, 41. Mem: AAAS; Soc Exp Biol & Med; Am Asn Cancer Res; Am Asn Path & Bact; Col Am Path. Res: Cancer; endocrinology; aging; transplantation; virus diseases; amebiasis; arteriosclerosis. Mailing Add: Biopsychol Res Lab Washington Univ St Louis MO 63131

BLUMENTHAL, IRVING JACK, b New York, NY, Dec 14, 10; m 37; c 2. ELECTROENCEPHALOGRAPHY, PSYCHIATRY. Educ: City Col New York, BS, 31; Graz Univ, MD, 37; Am Bd Psychiat & Neurol, dipl, 49. Prof Exp: Assoc med officer, Vet Admin Hosp, Augusta, Ga, 41; staff physician, Vet Admin Hosp, Bedford, Mass, 41-50; assoc chief staff, Vet Admin Hosp, Northport, NY, 50-70; psychiatrist II, Hoch Psychiat Ctr, West Brentwood, 70-73; PSYCHIATRIST II, NORTHEAST NASSAU PSYCHIAT CTR, 73- Concurrent Pos: Clin asst prof, State Univ NY Downstate Med Ctr, 64- Mem: AMA; fel Am Psychiat Asn; Am Electroencephalog Soc; Am Epilepsy Soc; Am Acad Neurol. Res: Neurology and psychiatry, especially electroencephalographic survey of mentally ill epileptic veterans; spontaneous seizures following shock therapy; reading epilepsy; neurological disorders in a psychiatric hospital; management of schizophrenia in the Veterans Administration. Mailing Add: Northeast Nassau Psychiat Ctr Kings Park NY 11754

BLUMENTHAL, LEONARD MASCOT, b Athens, Ga, Feb 27, 02; m 26. MATHEMATICS. Educ: Ga Inst Technol, BS, 23; Univ Chicago, MS, 24; Johns Hopkins Univ, PhD(math), 27. Prof Exp: Instr math, Univ Mich, 24-25; from asst to instr, Johns Hopkins Univ, 26-29; instr, Rice Inst, 29-33; Nat Res fel, Inst Advan Study, 33-34 & Univ Vienna, 34-35; asst, Inst Advan Study, 35-36; from asst prof to prof, 36-68, Defoe distinguished prof, 68-71, EMER DEFOE DISTINGUISHED PROF, UNIV MO-COLUMBIA, 71- Concurrent Pos: Collabr, Math Rev, 40-; assoc ed, Am Math Monthly, 41-; mathematician, Inst Numerical Anal, Nat Bur Standards, 51-52; mem fel bd, Math Div, Nat Res Coun, 53-; Fulbright prof, Univ Leiden, 54-55, Univ Madrid, 62-63 & Univ Buenos Aires, 67. Mem: Am Math Soc; Math Asn Am; foreign mem Royal Spanish Acad. Res: Geometry; distance geometry; abstract spaces; theory of linear inequalities. Mailing Add: Dept of Math Univ of Mo-Columbia Columbia MO 65202

BLUMENTHAL, MONICA DAVID, b Tübingen, Ger, Sept 1, 30; US citizen; m 54; c 2. PSYCHIATRY. Educ: Univ Mich, BA, 52, MS, 53, MD, 57; Univ Calif, Berkeley, PhD(physiol), 62. Prof Exp: Intern, San Francisco Gen Hosp, 57-58; assoc res clin biochemist, Schizophrenia & Psychopharmacol Proj, 62-63, resident, Med Ctr, 63-65, dir psychopath in phenylketonuria heterozygotes proj, Surv Res Ctr, Inst Social Res, 63-69, res assoc, 68-71, from instr to asst prof, Univ, 65-72, PROG DIR, SURV RES CTR, INST SOCIAL RES & ASSOC PROF PSYCHIAT, UNIV MICH, ANN ARBOR, 72- Concurrent Pos: NIMH career investr develop award, 62-67; staff consult, Mass Media Task Force, President's Comn on Violence, 69; sr staff psychiatrist adult psychiat, Neuropsychiat Inst, Univ Mich, 72-; mem, Nat Res Coun, 73- Honors & Awards: Prize for Outstanding Res on Aggression & Violence, Am Psychiat Asn, 72. Res: Attitudes toward violence; violence, psychiatric epidemiology; biological psychiatry. Mailing Add: Inst for Social Res Univ of Mich Ann Arbor MI 48106

BLUMENTHAL, RALPH HERBERT, b New York, NY, Feb 24, 25; m 48; c 3. PHYSICS. Educ: Brooklyn Col, BA, 45, MA, 49; NY Univ, PhD(sci educ), 56. Prof Exp: Radio engr, Hamilton Radio Corp, 45; lectr physics, Brooklyn Col, 46-48; physicist & proj leader, US Naval Supply Activ, 48-52; physicist & supvr test group, Picatinny Arsenal, 52; lectr physics, Brooklyn Col, 52-54 & Queens Col, NY, 54; tutor, City Col New York, 54-56; sr engr & group leader, Sperry Gyroscope Corp, NY, 58-62, assoc mem tech & mgt staff, 62-63; staff engr, Grumman Aerospace Corp, 63-70; ADJ ASST PROF PHYSICS, QUEENSBOROUGH COMMUNITY COL & PHYSICS TEACHER, SEWANHAKA CENT HIGH SCH DIST, FRANKLIN SQ, 70- Mem: Fel AAAS; Am Phys Soc; Am Asn Physics Teachers; Nat Asn Res Sci Teaching. Res: Laser communication systems; electrooptical light modulation; space systems; microwave electronics; transistor physics; pyrotechnic radiation; heat transfer; programmed physics instruction. Mailing Add: 15 Bonnie Dr Westbury NY 11590

BLUMENTHAL, REUBEN R, b Philadelphia, Pa, Nov 5, 04; m 29; c 2. MARINE BIOLOGY. Educ: Univ Pa, BS, 26, MS, 27, PhD(physiol), 30. Prof Exp: Physiologist, Res & Biol Labs, E R Squibb & Sons, 30-40; asst to exec vpres, S B Penick & Co, 40-50, asst to vpres res, 50-70; RES ASSOC, MONELL CHEM SENSES CTR, UNIV PA, 70- Mem: Am Chem Soc. Res: Chemoreception in aquatic invertebrates. Mailing Add: Monell Chem Senses Ctr Univ of Pa Philadelphia PA 19104

BLUMENTHAL, ROBERT GEORGE, b Brooklyn, NY, July 6, 42. ALGEBRA. Educ: Brooklyn Col, BS, 63; Yale Univ, MA, 65, PhD(math), 68. Prof Exp: Instr math, Brown Univ, 68-69; asst prof, Univ Miami, 69-71 & Wellesley Col, 71-74; MEM STAFF, DELTA RES CORP, 74- Mem: Am Math Soc. Res: Commutative Banach algebras; uniform algebras of continuous functions; measure algebras; measure algebras; harmonic analysis. Mailing Add: Delta Res Corp 1515 Wilson Blvd Arlington VA 22209

BLUMENTHAL, ROBERT MCCALLUM, b Chicago, Ill, Feb 7, 31; m 52; c 2. MATHEMATICS. Educ: Oberlin Col, BA, 52; Cornell Univ, PhD, 56. Prof Exp: Instr, 56-57, from asst prof to assoc prof, 57-65, PROF MATH, UNIV WASH, 65- Mem: Inst Math Statist; Am Math Soc. Res: Probability theory; mathematical statistics. Mailing Add: 4734 NE 178th St Seattle WA 98155

BLUMENTHAL, SIDNEY, b New York, NY, June 24, 09; m 53; c 2. CARDIOLOGY. Educ: Univ Iowa, BS & MD, 33. Prof Exp: Prof clin pediat, Col Physicians & Surgeons, Columbia Univ, 60-70; assoc dean continuing educ, 70-74, PROF PEDIAT CARDIOL, SCH MED, UNIV MIAMI, 70- Mem: Am Pediat Soc; Am Acad Pediat; Am Heart Asn. Res: Pediatric cardiology. Mailing Add: Dept of Pediat Univ of Miami Sch of Med Miami FL 33152

BLUMENTHAL, WARREN BARNETT, b New Orleans, La, Nov 9, 12; m 39; c 1. INDUSTRIAL CHEMISTRY. Educ: Cornell Univ, BA, 33. Prof Exp: Anal chemist, Metrop Ref Co, 33-34; res chemist, Ansbacher-Siegle Corp, 34-45; from res chemist to chief chem res, Titanium Alloy Mfg Div, NL Industs, 45-74; INNOVATOR INDUST CHEM, BLUMENTHAL-ZIRCONIUM, 74- Concurrent Pos: Consult, NL Indust, Inc, 74- Honors & Awards: Jacob F Schoellkopf Medal, 71. Mem: Am Chem Soc; Am Inst Chem. Res: Synthetic pigments; zirconium and titanium chemistry; chemical behavior of zirconium; applications of inorganic chemical, particularly zirconium compounds, to a variety of chemical manufacturing fields, particularly paints, plastics and pharmaceuticals. Mailing Add: 747 Ohio Ave North Tonawanda NY 14120

BLUMER, MAX, b Basel, Switz, Aug 3, 23; nat US. GEOCHEMISTRY. Educ: Univ Basel, PhD(chem), 49. Prof Exp: Asst, Univ Basel, 48-50; Janggen-Poehn fel, Univ Minn, 50-51; geochemist, Scripps Inst Oceanog, 51-52, Royal Dutch Shell Co, 52-53 & Shell Develop Co, 53-58; res chemist, Rare Metals Div, CIBA, Ltd, 58-59; SR SCIENTIST ORG GEOCHEM, WOODS HOLE OCEANOG INST, 59- Mem: AAAS; Am Chem Soc. Res: Organic geochemistry; comparative biochemistry; origin, fate and effect of organic compounds; marine organic chemistry. Mailing Add: Woods Hole Oceanog Inst Woods Hole MA 02543

BLUMGART, HERRMAN LUDWIG, b Newark, NJ, July 19, 95; m 31; c 1. MEDICINE. Educ: Harvard Univ, SB, 17, MD, 21. Hon Degrees: SD, Harvard Univ, 62. Prof Exp: Intern, Peter Bent Brigham Hosp, 21-22; Mosely traveling fel from Harvard, Nat Health Inst & Univ Col Hosp, London, 23; asst med, Harvard Univ Hosp, 24-27, from instr to prof, 27-64, EMER PROF MED, HARVARD MED SCH, 64- Concurrent Pos: Asst, Thorndike Mem Lab, Boston City Hosp, 24-28; chief consult, China-Burma-India Theatre, 45; physician-in-chief & dir med res, Beth Israel Hosp, 46-64; consult, US Vet Admin, 46-66 & Harvard Univ Med Area Health Serv, 66-74; ed-in-chief circulation, AMA. Honors & Awards: James B Herrick Award, Am Heart Asn. Mem: Master Am Col Physicians; Am Soc Clin Invest (vpres); fel AMA; hon fel Am Heart Asn; hon fel Am Col Cardiol. Res: Heart and circulation; radioactive isotopes; cardiovascular disease; velocity of blood in health and disease. Mailing Add: 987 Memorial Dr Cambridge MA 02138

BLUMSTEIN, ALEXANDRE, b Grodno, Poland, Jan 13, 30; m 59; c 2. POLYMER CHEMISTRY. Educ: Univ Paris, BS, 51; Univ Toulouse, Chem Engr, 52; Univ Strasbourg, PhD(polymer chem), 60. Prof Exp: Res trainee polymer chem, Nat Sci Res Ctr, France, 52-54, researcher, 54-57 & 59-60, res asst, 60; sr proj engr, Instruments Div, Budd Co, 61, eng supvr, 61-62; NSF res assoc, Univ Del, 62-64; from asst prof to assoc prof chem, Lowell Technol Inst, 64-72; PROF CHEM, UNIV LOWELL, 72- Mem: Am Chem Soc. Res: Polymer structure; polymerization in special conditions; photoelasticity; matrix polymerization; polymerization in liquid-crystalline media; liquid crystalline order in polymers. Mailing Add: Dept of Chem Univ of Lowell Lowell MA 01854

BLUMSTEIN, ALFRED, b New York, NY, June 3, 30; m 58; c 3. OPERATIONS RESEARCH. Educ: Cornell Univ, BEP, 51, PhD(opers res), 60; Univ Buffalo, MA, 54. Prof Exp: Prin opers analyst, Aeronaut Lab, Cornell Univ, 51-61; mem staff systs anal, Inst Defense Anal, 61-69; PROF URBAN SYSTS, SCH URBAN & PUB AFFAIRS & GRAD SCH INDUST ADMIN & DIR URBAN SYSTS INST, CARNEGIE-MELLON UNIV, 69- Concurrent Pos: Mem staff, Off Naval Res, 54-55 & 57; vis assoc prof, Cornell Univ, 63-64; dir sci & technol task force, President's Comn Law Enforcement & Admin of Justice, 66-67; dir, Off Urban Res, 68-69; mem exec comt, Coun on Res, Nat Coun Crime & Delinquency, 68-70; panel chmn, Nat Acad Sci, 75- Mem: Opers Res Soc Am; Am Statist Asn; Inst Mgt Sci. Res: Law enforcement and criminal justice; urban transportation; family planning. Mailing Add: Urban Systs Inst Carnegie-Mellon Univ Pittsburgh PA 15213

BLUMSTEIN, GEORGE I, b Philadelphia, Pa, Nov 26, 04; m 35; c 1. MEDICINE. Educ: Temple Univ, MD, 29. Prof Exp: Instr, 32-42, assoc, 43-49, asst prof, 50-56, assoc prof, 59-63, CLIN PROF MED, SCH MED, TEMPLE UNIV, 64- Concurrent Pos: Clin asst med, Albert Einstein Med Ctr, 30-32, adj, 33-52, attend allergy, 52-

Mem: Am Acad Allergy (treas, 56-59, pres, 61-62). Res: Allergy. Mailing Add: 2039 Delancey St Philadelphia PA 19103

BLUNDELL, GEORGE PHELAN, b Yazoo City, Miss, July 29, 14; m 41; c 6. PATHOLOGY. Educ: Univ Miss, AB, 36, MA, 37; Vanderbilt Univ, MS, 38; Yale Univ, PhD(bact immunol), 41; McGill Univ, MD, CM, 48; Am Bd Path, dipl, 52. Prof Exp: Asst biol, Univ Miss, 36-37; asst bact & immunol, Sch Med, Yale Univ, 39-41; assoc, Jefferson Med Col, 41-45; instr bact & path, Univ Miss, 45; intern, Royal Victoria Hosp, Montreal, Que, 48-49; asst prof path, Dent Col Ala, 49-50; asst prof, Med Col Ala, 50-53, actg chmn dept, 52-53; pathologist, Third Army Area Med Lab, Ft McPherson, Ga, 53-54; asst chief infectious dis sect, Armed Forces Inst Path, 54-55; chief path div, Biol Labs, US Army, Ft Detrick, 55-61; ASSOC PATHOLOGIST, OSCAR B HUNTER MEM LAB, 61- Concurrent Pos: Assoc pathologist, Baptist Mem Hosp, Memphis, 53. Res: Pathology of infectious diseases and immunology of malignant tumors. Mailing Add: Oscar B Hunter Mem Lab 915 19th St NW Washington DC 20006

BLUNN, CECIL THOMAS, b Wichita, Kans, Mar 22, 07; m 30; c 4. GENETICS. Educ: Univ Calif, BS, 28, PhD(genetics), 34; Kans State Col, MS, 29. Prof Exp: Asst, Kans State Col, 28-29; technician, Div Animal Husb, Univ Calif, 29-31, asst, 34-37, instr, 37-38; from asst animal husbandman to assoc animal husbandman, Southwest Range & Sheep Breeding Lab, Bur Animal Indust, USDA, 38-45; from asst prof to prof, 45-72, EMER PROF ANIMAL SCI, UNIV NEBR, LINCOLN, 72- Concurrent Pos: Animal breeding specialist, Univ Nebr Group, Univ Ankara, 56-58; res assoc, Univ Agr, Rijadh, Saudi Arabia, 72-73; agr consult, Near East Found, Amman, Jordan, 73-74. Mem: AAAS; Am Genetic Asn; fel Am Soc Animal Sci. Res: Genetic improvement of domestic animals. Mailing Add: Dept of Animal Sci Univ of Nebr Lincoln NE 68503

BLUNT, HARRY WILLIAM, b Wilmington, Del, Aug 13, 37; m 58; c 4. POLYMER CHEMISTRY. Educ: Univ Del, BS, 60; Univ Pa, PhD(chem), 65. Prof Exp: Chemist, Hercules Powder Co, 60-62, res chemist, 65-67, SR RES CHEMIST, HERCULES, INC, 70- Mem: Am Chem Soc. Res: Homogeneous and heterogeneous Ziegler catalysis of olefin polymerizations; polymer structure-property relationships. Mailing Add: 500 Hemingway Dr Hickory Hills Hockessin DE 19707

BLUNT, ROBERT F, b Greeley, Kans, Aug 24, 20; m 53; c 2. SOLID STATE PHYSICS. Educ: Rice Univ, BA, 43, MA, 47, PhD(physics), 49. Prof Exp: PHYSICIST, NAT BUR STANDARDS, 49- Mem: Am Phys Soc. Res: Optical properties. Mailing Add: Solid State Physics Sect Nat Bur of Standards Washington DC 20234

BLURTON, KEITH F, b Grays, Eng, Apr 11, 40; m 66; c 1. PHYSICAL CHEMISTRY. Educ: Southampton Univ, BSc, 61, PhD(chem), 66. Prof Exp: Lectr, Portsmouth Polytech, 61-66; sr chemist, Leesona Corp, 66-68; RES DIR, ENERGETICS SCI INC, ELMSFORD, 68- Mem: Am Chem Soc; Electrochem Soc; AAAS; Brit Chem Soc; Royal Inst Chem. Res: Air pollution instrumentation; electrochemical power sources; catalysis and analytical chemistry. Mailing Add: 1248 McKeel St Yorktown Heights NY 10598

BLY, DONALD DAVID, b Bryan, Ohio, Sept 24, 36; m 59; c 3. ANALYTICAL CHEMISTRY. Educ: Kenyon Col, BA, 58; Purdue Univ, MS, 61, PhD(anal chem), 62. Prof Exp: Eli Lilly fel, Purdue Univ, 62-63; res chemist, Carothers Res Lab, 63-67, SECT SUPVR, CENT RES DEPT, E I DU PONT DE NEMOURS & CO, 67- Mem: Am Chem Soc; Am Soc Testing & Mat; Sigma Xi; Am Soc Mass Spectros; Soc Appl Spectros. Res: Polymer physical chemistry and polymer characterization; gel permeation chromatography; thermal methods; viscosity; mass spectrometry; gas and liquid chromatography. Mailing Add: DuPont Experimental Sta Wilmington DE 19898

BLY, ROBERT MALCOLM, analytical chemistry, see 12th edition

BLY, ROBERT STEWART, b Lakeland, Fla, Aug 10, 29; m 56. PHYSICAL ORGANIC CHEMISTRY. Educ: Fla Southern Col, BS, 51; Northwestern Univ, MS, 56; Univ Colo, PhD(org chem), 58. Prof Exp: Res chemist, Nylon Res Lab, E I du Pont de Nemours & Co, 57-59; NIH fel, Mass Inst Technol, 59-61; from asst prof to assoc prof chem, 61-70, head dept, 70-73, PROF CHEM, UNIV SC, 70- Mem: Am Chem Soc; The Chem Soc. Res: Mechanisms of reactions; carbonium ions; pi-complexes. Mailing Add: Dept of Chem Univ of SC Columbia SC 29208

BLYDENSTEIN, JOHN, ecology, see 12th edition

BLYE, RICHARD PERRY, b West Chester, Pa, Nov 11, 32; m 56; c 2. REPRODUCTIVE BIOLOGY. Educ: Trinity Coll, Conn, BS, 55; Rutgers Univ, PhD, 60. Prof Exp: Asst zool & gen biol, Rutgers Univ, 55-59; assoc scientist, Div Endocrinol, Ortho Pharmaceut Corp, 59-62, sr scientist, 62-63, Ortho res fel, 64-70; HEALTH SCIENTIST ADMINR, CDB, CPR, NAT INST CHILD HEALTH & HUMAN DEVELOP, 71- Concurrent Pos: Consult, Bur Drugs, Food & Drug Admin, 74. Mem: AAAS; Soc Sci Study Sex; NY Acad Sci; Soc Study Reprod; Am Chem Soc. Res: Endocrinology of reproduction. Mailing Add: Nat Insts of Health Landow Bldg Rm A-706 Bethesda MD 20014

BLYHOLDER, GEORGE DONALD, b Elizabeth, NJ, Jan 10, 31; m 55; c 3. PHYSICAL CHEMISTRY. Educ: Valparaiso Univ, BA, 52; Purdue Univ, BS, 53; Univ Utah, PhD(chem), 56. Prof Exp: Fel, Univ Minn, 56-57; assoc Johns Hopkins Univ, 57-59; from asst prof to assoc prof chem, 59-67, vchm dept, 68-72, PROF CHEM, UNIV ARK, FAYETTEVILLE, 67- Mem: Am Chem Soc. Res: Kinetics; catalysis; surface chemistry; molecular spectroscopy; molecular orbital theory. Mailing Add: Dept of Chem Univ of Ark Fayetteville AR 72701

BLYLER, LEE LANDIS, JR, b New Brunswick, NJ, Oct 22, 38; m 62; c 2. RHEOLOGY, POLYMER ENGINEERING. Educ: Princeton Univ, BSE, 61, MSE, 62, PhD(mech eng), 66. Prof Exp: Mem tech staff, 65-71, GROUP SUPVR, BELL LABS, 71- Mem: AAAS; Am Chem Soc; Soc Rheol; Soc Plastics Eng. Res: Polymer mechanics; flow and processing behavior of polymer melts; relationships between polymer properties and molecular structure; electrical charge storage in polymer dielectrics; polymeric materials for optical fiber waveguides. Mailing Add: Bell Labs Murray Hill NJ 07974

BLYSTONE, ROBERT VERNON, b El Paso, Tex, July 4, 43; m 64; c 1. CELL BIOLOGY, ELECTRON MICROSCOPY. Educ: Univ Tex, El Paso, BS, 65; Univ Tex, Austin, MA, 68, PhD(zool), 71. Prof Exp: ASST PROF BIOL & DIR ELECTRON MICROS LABS, TRINITY UNIV, 71- Mem: AAAS; Electron Micros Soc Am; Am Inst Biol Sci. Res: Fine structure of visual receptors; fine structure and cytochemistry of lung alveoli; fine structure of cultured cells. Mailing Add: Dept of Biol Trinity Univ San Antonio TX 78284

BLYTAS, GEORGE CONSTANTIN, b Cairo, Egypt, Dec 20, 30; US citizen; m 63; c 2. PHYSICAL CHEMISTRY. Educ: Am Univ Cairo, BSc, 56; Univ Wis, PhD(phys chem), 61. Prof Exp: Chemist, 61-72, SR STAFF RES CHEMIST, SHELL DEVELOP CO, 72- Concurrent Pos: Lectr electrochem, Univ Calif, Berkeley, 66-67 & 69. Mem: Sigma Xi; Am Chem Soc; Am Inst Chem Eng. Res: Research and development in novel separation systems and processes; adsorbents, zeolites, solvent extraction, hydrometallurgy, chemistry in nonaqueous solvents, energy systems, environmental chemistry and engineering. Mailing Add: 14323 Apple Tree Houston TX 77079

BLYTH, COLIN ROSS, b Guelph, Ont, Oct 24, 22; m 55; c 6. MATHEMATICAL STATISTICS. Educ: Queen's Univ, Ont, BA, 44; Univ Toronto, MA, 46; Univ Calif, PhD(statist), 50. Prof Exp: From asst prof to prof math, Univ Ill, Urbana, 50-74; PROF MATH, QUEEN'S UNIV, KINGSTON, 72- Mem: Am Math Soc; fel Inst Math Statist; fel Am Statist Asn; Math Asn Am; Can Math Cong. Res: Statistical inference. Mailing Add: Dept of Math Queen's Univ Kingston ON Can

BLYTH, MARY ISOBEL, b Ottawa, Ont, Mar 17, 15; nat US. MATHEMATICS. Educ: Mich State Col, BA, 36, MA, 42; Univ Mich, PhD(educ), 50. Prof Exp: Teacher, Grand Ledge High Sch, 36-41; asst, 41-42, from instr to assoc prof, 42-72, PROF MATH, MICH STATE UNIV, 72- Mem: Math Asn Am; Nat Coun Teachers Math. Res: Teaching of mathematics and preparation of mathematics teachers. Mailing Add: Dept of Math Mich State Univ East Lansing MI 48824

BLYTHE, JACK GORDON, b Kansas City, Mo, July 15, 22; m 48; c 4. GEOLOGY. Educ: Wichita State Univ, BA, 47; Northwestern Univ, MS, 50; Univ Okla, PhD(geol), 57. Prof Exp: Instr geol, Wichita State Univ, 49-51; assoc prof, Okla City Univ, 53-57; assoc prof, 57-63, chmn dept, 65-70, PROF GEOL, WICHITA STATE UNIV, 63- Mem: AAAS; Geol Soc Am; Am Asn Petrol Geol; Am Paleont Soc; Soc Econ Paleont & Mineral. Res: Stratigraphy; structural geology. Mailing Add: Dept of Geol Wichita State Univ Wichita KS 67208

BLYTHE, PHILIP ANTHONY, b Dewsbury, Eng, Mar 30, 37; m 63; c 2. APPLIED MATHEMATICS, FLUID MECHANICS. Educ: Univ Manchester, BSc, 58, PhD(fluid mech), 61. Prof Exp: Res fel aerodyn, Nat Phys Lab, Eng, 61-63, sr sci officer, 63-64; lectr, Imp Col, Univ London, 64-68; assoc prof, 68-70, PROF APPL MATH, LEHIGH UNIV, 70- Concurrent Pos: Vis prof, Univ Newcastle, Eng, 74-75. Res: Nonlinear wave propagation; nonequilibrium flows; perturbation techniques; asymptotic expansions. Mailing Add: Ctr for Appln Math Lehigh Univ 4 W 4th St Bethlehem PA 18015

BLYTHE, RUDOLPH HAMMA, b Roxbury, NY, Mar 23, 10; m 35; c 2. PHARMACEUTICAL CHEMISTRY. Educ: Union Univ, NY, PhC, 31; Columbia Univ, BS, 32, PharmD(pharmaceut chem), 34. Prof Exp: Res pharmacist, Smith Kline & French Labs, 34-40, develop, coordr, 40-47, head pharmaceut res sect, 47-51, dir pharmaceut chem, 51-57, dir pharmaceut res, 57-67; assoc prof pharm, Univ Fla, 67-75, asst dean col pharm, 70-75; RETIRED. Honors & Awards: Elbert Award, 50; Am Pharmaceut Asn Found Award, 63. Mem: AAAS; Am Chem Soc; Am Soc Hosp Pharmacists; Am Pharmaceut Asn. Res: Research administration; formulation of medicinal products; availability of medicinal agent to body; stability of product and consumer acceptance; medication processed to provide time absorption. Mailing Add: 7008 SW 30th Way Gainesville FL 32601

BLYTHE, WILLIAM BREVARD, b Huntersville, NC, Sept 23, 28; m 56; c 4. INTERNAL MEDICINE. Educ: Univ NC, AB, 48; Washington Univ, MD, 53. Prof Exp: Life Ins Med Res Fund fel, 58-60; from instr to assoc prof med, 60-70, assoc dir clin res unit, 65-66, DIR CLIN RES UNIT, SCH MED, UNIV NC, CHAPEL HILL, 66-, PROF MED, 70-, HEAD DIV NEPHROLOGY, 72- Mem: Am Fedn Clin Res; AMA; Am Physiol Soc; Am Soc Artificial Internal Organs; Am Soc Nephrology. Res: Renal physiology and disease; hemodialysis. Mailing Add: Dept of Med Univ of NC Sch of Med Chapel Hill NC 27514

BO, WALTER JOHN, b Chisholm, Minn, Aug 12, 23; m 48; c 3. ANATOMY, ENDOCRINOLOGY. Educ: Marquette Univ, BS, 46, MS, 47; Univ Cincinnati, PhD(anat), 53. Prof Exp: Instr zool, Xavier Univ, 47-49; from asst prof to assoc prof anat, Sch Med, Univ NDak, 53-60; assoc prof, 60-63, PROF ANAT, BOWMAN GRAY SCH MED, 63- Mem: Am Asn Anat; Asn Cancer Res; Histochem Soc; Soc Exp Biol & Med; Soc Study Reproduction. Res: Gross histochemistry and abnormal growth of the uterus, cervix and vagina; endocrinoloby; female reproductive system; intra uterine device effect on uterus mode of action. Mailing Add: Dept of Anat Bowman Gray Sch of Med Winston-Salem NC 27104

BOADE, RODNEY RUSSETT, b Armstrong, Iowa, Aug 17, 35; m 65; c 2. PHYSICS, ACOUSTICS. Educ: Augustana Col, SDak, BS, 57; Iowa State Univ, PhD(physics), 64. Prof Exp: STAFF MEM PHYSICS, SANDIA CORP, 64- Mem: Am Phys Soc. Res: Energy transfer processes in gases and their influence on the propagation of sound; propagation of shock waves in solids. Mailing Add: Sandia Corp Org 5167 Albuquerque NM 87115

BOAG, DAVID ARCHIBALD, b Edmonton, Alta, Can, Jan 24, 34; m 63; c 2. ANIMAL ECOLOGY. Educ: Univ Alta, BSc, 57, MSc, 58; Wash State Univ, PhD(zool), 64. Prof Exp: Lectr zool, Univ Alta, 58-59, asst prof, 59-60; researcher, Zool Sta, Naples, Italy, 61; asst prof, 63-69, ASSOC PROF ZOOL, UNIV ALTA, 69- ; DIR, R B MILLER BIOL STA, 63- Mem: Am Ornith Union; Cooper Ornith Soc; Ecol Soc Am; Can Soc Zool; Can Soc Wildlife & Fishery Biol. Res: Population ecology, especially waterfowl and grouse. Mailing Add: Dept of Zool Univ of Alta Edmonton AB Can

BOAG, THOMAS JOHNSON, b Liverpool, Eng, Apr 11, 22; Can citizen; m 50; c 4. PSYCHIATRY, PSYCHOANALYSIS. Educ: Univ Liverpool, MB, ChB, 44; McGill Univ, dipl, 53. Prof Exp: Clin asst psychiat, Royal Victoria Hosp, 52-53; lectr, McGill Univ, 53-55; from asst psychiatrist to assoc psychiatrist, Royal Victoria Hosp, 55-61; prof psychiat & chmn dept, Col Med, Univ Vt, 61-67; chief serv, DeGoesbriand Mem & Mary Fletcher Hosps, 61-67; prof psychiat & head dept, 67-75, DEAN FAC MED, QUEEN'S UNIV, ONT, 75- Concurrent Pos: Asst prof, McGill Univ, 57-61; asst dir psychiat, Allan Mem Hosp, 59-61; psychiatrist-in-chief, Kingston Gen Hosp, Ont, 67-75; consult, Hotel Dieu, Kingston Psychiat & St Mary's of the Lake Hosps & Dept Vet Affairs. Mem: Acad Psychoanal; AMA; Am Psychiat Asn; Int Psychoanal Asn; Can Psychiat Asn. Res: Human adaptation in Arctic; interaction recording and study of interview situation; psychiatric treatment milieu; history of psychiatry. Mailing Add: Fac of Med Queen's Univ Kingston ON Can

BOAK, RUTH ALICE, b Auburn, NY, May 25, 06; m 42; c 2. MEDICINE. Educ: Cornell Univ, BS & MS, 27, PhD(bact), 29; Univ Rochester, MD, 40. Prof Exp: Asst bact, Cornell Univ, 27-28; asst bacteriologist, Albany Hosp & instr bact, Albany Med Col, Union Col & Univ, NY, 28-30; assoc, Sch Med & Dent, Univ Rochester, 33-40; intern, Johns Hopkins Hosp, 40-41; assoc, Sch Med & Dent, Univ Rochester, 41-44; asst pediatrician, Greenwich Hosp, 44-45; assoc bact, Univ Rochester, 45-47; assoc prof, 48-57, PROF INFECTIOUS DIS, PEDIAT & PUB HEALTH, MED SCH,

UNIV CALIF, LOS ANGELES, 57- Concurrent Pos: Fulbright award, Univ Tokyo, 54-55 & Sch Med, Univ Teheran, 59-60; vis prof, Calif Proj, Sch Med, Airlangga Univ, Indonesia, 63-66. Mem: Fel Am Pub Health Asn; Am Venereal Dis Asn. Res: Syphilis; undulant fever. Mailing Add: Dept of Microbiol & Immunol Univ of Calif Sch of Med Los Angeles CA 90024

BOAKE, WILLIAM CHARLES, b Melbourne, Australia; US citizen; m 49; c 2. CARDIOLOGY, INTERNAL MEDICINE. Educ: Univ Melbourne, BSc, 42, MSc, 46, MD, 48. Prof Exp: Intern med, Royal Melbourne Hosp, 48-49; Nuffield Dom fel path, Oxford Univ, 50-52; resident, St Mary's Hosp, London, 53-55; instr, Western Reserve Univ, 55-56; asst, Univ Melbourne, 58-61; USPHS spec fel, Cardiovasc Lab, Univ Wis, 61-63; from asst prof to assoc prof, 63-73, PROF MED, CARDIOL SECT, UNIV WIS-MADISON, 73- Concurrent Pos: UNICEF fel, Int Children's Ctr, Paris, 55-; consult, Vet Admin Hosp, Madison, 67- Mem: Royal Col Physicians; Royal Australasian Col Physicians; Am Col Physicians; Am Col Cardiologists; Transplantation Soc. Res: Clinical cardiology; hypertension; cardiac transplantation. Mailing Add: Dept of Med Univ of Wis Madison WI 53706

BOAL, JAN LIST, b Canton, Ohio, Oct 20, 30; m 53; c 3. APPLIED MATHEMATICS. Educ: Ga Inst Technol, BME & MS, 54; Mass Inst Technol, PhD(math), 59. Prof Exp: Instr math, Mass Inst Technol, 59-60; from asst prof to assoc prof, Univ SC, 60-69; PROF MATH & CHMN DEPT, GA STATE UNIV, 69- Concurrent Pos: Vis lectr cols, Math Asn Am, 64 & 68. Mem: Am Math Soc; Math Asn Am; Am Soc Indust & Appl Math; Am Sci Affil. Res: Numerical analysis; differential equations; function approximation. Mailing Add: Dept of Math Ga State Univ Univ Plaza Atlanta GA 30303

BOARD, JOHN ARNOLD, b Altavista, Va, June 30, 31; m 59; c 3. OBSTETRICS & GYNECOLOGY. Educ: Randolph-Macon Col, BS, 53; Med Col Va, MD, 55. Prof Exp: Intern, Louisville Gen Hosp, Ky, 55-56; from jr asst resident to resident obstet & gynec, Med Col Va Hosps, 56-59; USPHS fel reproductive physiol & infertil, Sch Med, Yale Univ, 61-62; from instr to assoc prof obstet & gynec, 62-69, PROF OBSTET & GYNEC, MED COL VA, VA COMMONWEALTH UNIV, 69- Mem: Am Col Obstet & Gynec; Am Fertil Soc; Soc Study Reproduction; Endocrine Soc. Res: Gynecologic endocrinology and infertility. Mailing Add: Dept of Obstet & Gynec Med Col of Va Richmond VA 23298

BOARD, ROBERT DENNIS, b Princeton, Ind, Aug 31, 42; m 62; c 3. ANALYTICAL CHEMISTRY, COMPUTER SCIENCE. Educ: Univ Akron, BS, 66; Purdue Univ, PhD(chem), 69. Prof Exp: Chemist, Goodyear Tire & Rubber Co, 63-65; STAFF SCIENTIST, HEWLETT-PACKARD CO, 68- Mem: Am Chem Soc; Am Soc Mass Spectrometry. Res: Organic structural determination using high-resolution mass spectrometry; methods of data acquisition and reduction for mass spectrometry and other analytical instruments. Mailing Add: 3002 Fenwick Way San Jose CA 95122

BOARDMAN, DONALD CHAPIN, b Adna, Wash, Nov 18, 13; m 38; c 3. GEOLOGY. Educ: Wheaton Col, BS, 38; Univ Iowa, MS, 42; Univ Wis, PhD(geol), 52. Prof Exp: Dir visual educ, 38-40, from instr to assoc prof geol, 40-57, PROF GEOL, WHEATON COL, 57-, CHMN DEPT, 58-, DIR WHEATON COL SCI STA, SDAK, 52- Concurrent Pos: Seato Prof, Univ Peshawar, WPakistan, 59-60; Fulbright lectr, Univ Peshawar, Pakistan, 74-75. Mem: Soc Econ Paleont & Mineral; Nat Asn Geol Teachers; Am Asn Petrol Geol; fel Geol Soc Am; Soc Vert Paleont. Res: Stratigraphy and sedimentation; origin of dolomite; calcium-magnesium rations in sedimentary rocks. Mailing Add: 311 EFranklin St Wheaton IL 60187

BOARDMAN, HAROLD, b Eng, May 11, 17; nat US; m 43; c 3. INDUSTRIAL CHEMISTRY. Educ: Univ BC, BA, 40, MA, 42; Northwestern Univ, PhD(chem), 48. Prof Exp: Res chemist, 48-54, sr res chemist, 54-72, RES SCIENTIST, HERCULES INC, 72- Mem: Am Chem Soc; Sigma Xi. Res: Reaction mechanisms; polymerization; oxidation and hydroperoxide chemistry; applications and materials. Mailing Add: Box 523 RR 1 Chadds Ford PA 19317

BOARDMAN, JOHN, b Turlock, Calif, Sept 8, 32; m 63; c 2. THEORETICAL PHYSICS. Educ: Univ Chicago, AB, 52; Iowa State Univ, MS, 56; Syracuse Univ, PhD(physics), 62. Prof Exp: Lectr physics, Queens Col, NY, 61-62; instr, 62-65, ASST PROF PHYSICS, BROOKLYN COL, 65- Mem: Am Phys Soc; Fedn Am Sci. Res: General theory of relativity. Mailing Add: Dept of Physics Brooklyn Col Brooklyn NY 11210

BOARDMAN, JOHN MICHAEL, b Manchester, Eng, Feb 13, 38; m 67; c 2. MATHEMATICS. Educ: Cambridge Univ, BA, 61, PhD(math), 65. Prof Exp: Fel, Dept Sci & Indust Res, Eng, 64-66; instr math, Univ Chicago, 66-67; lectr, Univ Warwick, 67-68; visitor, Haverford Col, 69; assoc prof, 69-72, PROF MATH, JOHNS HOPKINS UNIV, 72- Res: Algebraic topology, particularly homotopy theory, stable homotopy theory and singularities of differentiable maps. Mailing Add: Dept of Math Johns Hopkins Univ Baltimore MD 21218

BOARDMAN, RICHARD STANTON, b Oak Park, Ill, July 16, 23; m 46; c 2. INVERTEBRATE PALEONTOLOGY. Educ: Univ Ill, BS, 48, MS, 52, PhD, 55. Prof Exp: Geologist, invert paleont, US Geol Surv, 52-57; assoc cur, 57-60, CUR, DEPT PALEOBIOL, NAT MUS NATURAL HIST, 60- Mem: AAAS; Paleont Soc; Am Asn Petrol Geol; Soc Econ Paleontologists & Mineralogists; Soc Syst Zool. Res: Fossil and recent Bryozoa, their mode of growth, functional morphology, classification and evolution. Mailing Add: Dept of Paleobiol Nat Museum of Natural Hist Washington DC 20560

BOARDMAN, ROBERT LELAND, b Denver, Colo, Aug 9, 23; m 45; c 6. ECONOMIC GEOLOGY. Educ: Yale Univ, BA, 44. Prof Exp: Subsurface explor geologist, Dominican Seaboard Oil Co, 44-45; subsurface explor geologist, Standard Oil Co, Cuba, 45-46; subsurface explor geologist, Dominican Seaboard Oil Co, 46-47; geologist, Trace Elements Off, 48-51, explor geologist, Mineral Deposits Br, 51-57, geologist, Prog Off, Geol Div, 57-61, prog off, 61-66, STAFF GEOLOGIST, US GEOL SURV, 66- Mem: Mineral Soc Am. Res: Exploration for mineral deposits including metals, non-metals, oil and gas. Mailing Add: US Geol Surv 345 Middlefield Rd Menlo Park CA 94025

BOARDMAN, SHELBY JETT, b Akron, Ohio, Nov 7, 44; m 66; c 2. GEOLOGY. Educ: Miami Univ, BA, 66; Univ Mich, MS, 69, PhD(geol), 71. Prof Exp: Petrol geologist, Mobil Oil Corp, 66; explor geologist, Bear Creek Mining Co, 67; res asst stratig, Mobil Res & Develop Corp, 68; ASST PROF GEOL, CARLETON COL, 71- Mem: Geol Soc Am; Sigma Xi; Nat Asn Geol Teachers; Nat Geog Soc. Res: Detailed petrology and geochemistry of the Tertiary igneous rocks of the eastern Bearpaw Mountains, Montana; Precambrian petrology of the Salida area, Chaffee County, Colorado. Mailing Add: Dept of Geol Carleton Col Northfield MN 55057

BOARDMAN, WILLIAM JARVIS, b Akron, Ohio, Aug 19, 39; m 65; c 3. PHYSICS, ASTRONOMY. Educ: Miami Univ, AB, 61, MS, 63; Univ Colo, PhD(astrogeophys), 68. Prof Exp: Asst prof physics, 68-74, ASSOC PROF MATH & PHYSICS, BIRMINGHAM-SOUTHERN COL, 74- Mem: Am Astron Soc; Am Phys Soc. Res: Radiative processes in the solar corona; temperature and density of coronal enhancement over regions of solar activity. Mailing Add: Dept of Physics Birmingham-Southern Col Birmingham AL 35204

BOARDMAN, WILLIAM WALTER, JR, b Rugby, NDak, June 2, 16; m 41; c 6. PHYSICAL INORGANIC CHEMISTRY. Educ: Coe Col, BA, 38; Univ Iowa, MS, 40, PhD(phys chem), 42. Prof Exp: Res chemist, Gaseous Diffusion Isotope Separation Plant, Carbide & Carbon Chem Corp, 45-49; supvr, Control Lab, Twin Cities Arsenal, 50-55; res chemist, Lithium Corp Am, Inc, 55-67; head dept chem, Biola Col, 69-70; res assoc chem, Creation-Sci Res Ctr, 70-71; chief chemist, Gen Monitors, Inc, 71-74; STAFF ANALYST, OCCIDENTAL RES CORP, 74- Mem: Am Chem Soc; Sigma Xi; Creation Res Soc. Res: Synthesis of inorganic compounds; development of catalytic and semiconductor devices for gas detection; analytical methods for oil shale and coal. Mailing Add: 11632 Grovedale Dr Whittier CA 90604

BOARDWAY, NANCY LOUISE, b Springfield, Mass, Apr 17, 34. ORGANIC CHEMISTRY. Educ: Mt Holyoke Col, AB, 55; Mass Inst Technol, PhD(org chem), 70. Prof Exp: Res chemist, Am Cyanamid Co Inc, 55-58; res chemist, Polaroid Corp, 58-61; teacher chem, Marian High Sch, Framingham, Mass, 61-66; teacher, Regis Col, Mass, 70-73; SR RES CHEMIST, DOW CHEM CO, 73- Mem: Am Chem Soc; Sigma Xi. Res: Organic synthesis; organic reaction mechanisms. Mailing Add: Dow Chem Co Box 400 Wayland MA 01778

BOAS, CHARLES WILLIAM, b Harrisburg, Pa, Sept 5, 26; m 54; c 4. GEOGRAPHY. Educ: Lafayette Col, AB, 49; Univ Va, MA, 50; Univ Mich, PhD, 56. Prof Exp: From instr to asst prof geog, Mich State Univ, 55-61; pvt pract, 61-66; prof soc sci & chmn dept, Harrisburg Area Community Col, 66-68; prof geog, York Col, Pa, 68-70, PROF SOC SCI, YORK COL, PA, 70- Mem: Asn Am Geog; Am Geog Soc; Nat Coun Geog Educ. Res: Geographic field techniques; air photo interpretation; industrial location analysis; Anglo-American geography; history of American circus industry. Mailing Add: Dept of Soc Sci York Col of Pa York PA 17405

BOAS, MARY LAYNE, b Prosser, Wash, Mar 10, 17; m 41; c 3. PHYSICS. Educ: Univ Wash, BS, 38, MS, 40; Mass Inst Technol, PhD(physics), 48. Prof Exp: Fel math, Univ Wash, 38-40; instr, Duke Univ, 40-43 & Tufts Col, 43-48; lectr, Wellesley Col, 49-50; consult, Northwestern Nuclear Res Lab, 50-52; vis lectr physics, Northwestern Univ, 52-53; lectr, 55-56, from asst prof to assoc prof, 57-75, PROF PHYSICS, DEPAUL UNIV, 75- Mem: Am Math Soc; Am Phys Soc; Math Asn Am. Res: Theoretical nuclear physics; photo-disintegration of H-3; mathematical physics; special relativity. Mailing Add: 2440 Simpson St Evanston IL 60201

BOAS, NORMAN FRANCIS, b New York, NY, Aug 4, 22; m 45; c 3. INTERNAL MEDICINE, RHEUMATOLOGY. Educ: Harvard Univ, MD, 45. Prof Exp: Intern & med resident, Michael Reese Hosp, 45-47; fel path, Mt Sinai Hosp, NY, 47-48, fel med, 48-51; sr asst surgeon, NIH, 51-54; RES DIR, NORWALK HOSP, 54-, SR ATTEND PHYSICIAN, 67- Mem: AAAS; AMA; Am Rheumatism Asn. Res: Clinical medicine. Mailing Add: 25 Partrick Lane Wilton CT 06897

BOAS, RALPH PHILIP, JR, b Walla Walla, Wash, Aug 8, 12; m 41; c 3. MATHEMATICS. Educ: Harvard Univ, AB, 33, PhD(math), 37. Prof Exp: Instr math, Harvard Univ, 36-37; Nat Res fel, Princeton & Cambridge Univs, 37-39; instr math, Duke Univ, 39-42; asst instr, US Navy Pre-Flight Sch, NC, 42-43; vis lectr, Harvard Univ, 43-45; res assoc, Brown Univ, 45-50; PROF MATH, NORTHWESTERN UNIV, 50- Concurrent Pos: Exec ed, Math Revs, 45-50; lectr, Mass Inst Technol, 48-49, Guggenheim fel, 51-52. Mem: AAAS; Am Math Soc (vpres, 59-60); Math Asn Am (pres, 73-75); London Math Soc; Soc Indust & Appl Math. Res: Laplace integrals; moment problems; Fourier series and integrals; entire functions; power series; approximation of functions. Mailing Add: 2440 Simpson St Evanston IL 60201

BOATMAN, EDWIN S, b London, Eng, July 24, 21; m 47. MICROBIOLOGY, ELECTRON MICROSCOPY. Educ: Univ London, BSc, 53; Univ Wash, MSc, 61, PhD(marine microbiol), 67. Prof Exp: Res asst bact, Royal Postgrad Med Sch, 51-52; res asst, St George's Hosp Med Sch, London, 52-55; head microbiologist, Royal Columbian Hosp, BC, 55-56; res asst microbiol, 56-58, res assoc prev med, 58-61, res instr, 61-67, ASSOC PROF, SCH PUB HEALTH, UNIV WASH, 67- Concurrent Pos: Fel, Inst Med Lab Sci, London, 53-; Josiah Macy Found fac award, Univ Bern, 74. Mem: Am Soc Microbiol; Electron Micros Soc Am; fel Royal Soc Health. Res: Relationship of ultra-structure to function in Mycoplasmatales and the anatomy of bacteria and viruses; growth and development of the lungs and the effects of air pollutants on the lungs. Mailing Add: Dept of Pathobiol & Environ Health Univ of Wash Sch of Pub Health Seattle WA 98195

BOATMAN, JAMES CLAUDE, inorganic chemistry, deceased

BOATMAN, JOSEPH BRASHER, b Missoorie, India, May 14, 20; US citizen; m 42; c 4. PHYSIOLOGY. Educ: Univ Ky, BS, 44; Univ Pittsburgh, MS, 51, PhD(physiol), 55. Prof Exp: Res assoc, Gibson Lab, Sch Med, Univ Pittsburgh, 49-54, asst dir, Gibson Lab, 54-56; head dept res physiol, Singer Res Lab, Allegheny Gen Hosp, Pittsburgh, 56-62; head lab environ physiol, Midwest Med Res Found, 62-64, dir res, 64-65; CHIEF DIV PHYSIOL & BIOPHYS, BATTELLE MEM INST, 65- Concurrent Pos: Lectr, Dept Biol Sci, Univ Pittsburgh; asst attend staff, Endocrinol Dept, Allegheny Gen Hosp. Mem: AAAS; Am Soc Zoologists; Am Soc Cell Biol; Ecol Soc Am; NY Acad Sci. Res: Comparative physiology and biochemistry of thyroid; water, electrolytes and endocrine metabolism in the cold; dynamics of neuro-endocrine regulation of metabolism. Mailing Add: Battelle Mem Inst 505 King Columbus OH 43201

BOATMAN, RALPH HENRY, JR, b Carlinville, Ill, Apr 20, 21; m 43; c 2. PUBLIC HEALTH. Educ: Southern Ill Univ, BS, 43; Univ NC, MPH, 47, PhD(pub health), 54. Prof Exp: Instr health educ & dir sch-community health proj, Southern Ill Univ, 47-54, chmn dept health educ, 48-54; dir community health serv, Cook County Tuberc Inst, Ill, 54-60; chmn dept pub health educ, 61-70, PROF PUB HEALTH EDUC, SCH PUB HEALTH, UNIV NC, CHAPEL HILL, 60-, DIR CONTINUING EDUC & FIELD SERV, 67-, DEAN ALLIED HEALTH SCI, 70-, DIR OFF CONTINUING EDUC HEALTH SCI, 74- Concurrent Pos: Lectr, Prof Schs, Univ Ill, 55-56, asst prof, 56-60; consult, Radiol Health Conf, WHO, 62; mem nat adv coun allied health prof, Dept Health, Educ & Welfare, 68-; chmn coun on baccalaureate & higher degrees, Asn Schs Allied Health Proj, 69-70, secy, 73-75. Mem: Fel Am Pub Health Asn; Am Soc Allied Health Prof (secy, 73-75, pres-elect, 75-76). Res: Public, school and allied health education. Mailing Add: Allied Health Sci Univ of NC Chapel Hill NC 27514

BOATMAN, SANDRA, b Tampa, Fla, Nov 1, 39. ORGANIC CHEMISTRY, BIOCHEMISTRY. Educ: Rice Univ, BA, 61; Duke Univ, PhD(chem), 65. Prof Exp: Fel, Univ NC, 65-66, NIH fel, 65-67; asst prof, 67-74, ASSOC PROF CHEM,

HOLLINS COL, 74- Mem: AAAS; Am Chem Soc; The Chem Soc. Res: Synthetic studies on polyanions of B-ketoaldehydes, pyridones; B-ketoenamines and related compounds; condensations of polyketo compounds; chemical modification of proteins and enzymes, especially at the active sites. Mailing Add: Dept of Chem Hollins College VA 24020

BOATNER, LYNN ALLEN, b Clarksville, Tex, Aug 3, 38; m 61; c 3. MAGNETIC RESONANCE. Educ: Tex Technol Col, BS, 60, MS, 61; Vanderbilt Univ, PhD(physics), 66. Prof Exp: Asst physics, Tex Technol Col, 60-61; asst, Vanderbilt Univ, 61-66; res scientist, LTV Res Ctr, 66-70, prin scientist, Advan Technol Ctr, Inc, 71-74; SR RES SCIENTIST, LAB EXP PHYSICS, FED POLYTECH SCH, SWITZ, 75- Concurrent Pos: Adj fac mem, Tex Tech Univ, 68- Mem: AAAS; Am Phys Soc; NY Acad Sci; assoc fel Am Inst Aeronaut & Astronaut. Res: Electron paramagnetic resonance investigations of transition metal ions in single crystals; studies of rare-earth and actinide ions; single crystal growth; superconductivity; investigations of the Jahn-Teller effect. Mailing Add: Fed Polytech Sch Lab of Exp Physics 33 av de Cour Lausanne Switzerland

BOATWRIGHT, GLENNIS O, soil science, see 12th edition

BOAZ, PATRICIA ANNE, b Chester, Pa, May 13, 22; div; c 5. PHYSICAL CHEMISTRY, ANALYTICAL CHEMSITRY. Educ: Vassar Col, AB, 44; Univ Iowa, PhD(phys chem), 51. Prof Exp: Chemist, Nat Aniline Div, Allied Chem & Dye Corp, 44-45; asst chem, Univ Iowa, 45-48; asst phys chem, Northwestern Univ, 48; instr chem, Smith Col, 49-50; asst prof, 67-70, ASSOC PROF CHEM, IND UNIV-PURDUE UNIV, INDIANAPOLIS, 70- Concurrent Pos: Consult stream monitoring, Indianapolis Ctr Advan Res, 75- Mem: Geol Soc Am; Am Chem Soc. Res: Raman, infrared, microwave spectroscopy; molecular structure; chemical evolution; water chemistry; environmental chemistry and geology. Mailing Add: Ind Univ-Purdue Univ 1201 E 38th St Indianapolis IN 46205

BOAZ, THURMOND DEWITTE, JR, b Shreveport, La, Aug 22, 13; m 40; c 2. PREVENTIVE MEDICINE. Educ: La State Univ, BS, 34, MB, 39, MD, 40; Harvard Univ, MPH, 50; Am Bd Prev Med, cert, 53. Prof Exp: Intern med, Charity Hosp of La, New Orleans, 39-40; resident, St Vincent's Hosp, Birmingham, Ala, 40-41; dir pub health, Winn-Jackson Parish Health Unit, La State Dept Health, 46-48, dir pub health, Ouachita Parish Health Univ, 48-49, assoc dir northern region, Div Local Health Serv, La State Bd Health, 52-56; asst chief prev med, Med Field Serv Sch, Tex, 57; chief dept health pract, Walter Reed Army Inst Res, Washington, DC, 58-59; chief prev med officer, N Area Command, Frankfurt, Ger, 60-62, chief Ninth Hosp Ctr, Heidelberg, 62-63, chief spec proj br, Life Sci Div, US Army Res Off, 63-64, chief nutrit br & prev med serv consult br, Prev Med Div, Off Surgeon Gen, 64-65, chief environ med br, 65-69, chief prev med div, 69-70; PROF ASSOC, DIV MED SCI, NAT RES COUN, 70- Mem: Am Col Prev Med; Soc Environ Geochem & Health; Am Asn Hist Med; Med Libr Asn; Royal Soc Health. Res: Public health; military preventive medicine; biological toxicants; environmental pollution. Mailing Add: Div of Med Sci Nat Res Coun 2101 Constitution Ave NW Washington DC 20418

BOAZ, WILLARD DENTON, b Roanoke, Va, Dec 29, 20; m 45; c 4. MEDICINE. Educ: Bridgewater Col, BA, 42; Univ Pa, MD, 45. Prof Exp: From instr to sr instr, Univ, 53-57, dir child psychiat sect, Univ Hosps, 56-74, ASST PROF CHILD PSYCHIAT, CASE WESTERN RESERVE UNIV, 57- Concurrent Pos: Consult, Family Serv Asn, 54-70; consult, Guid Dept, Orange Schs, Cuyahoga County, 58-67 & F Crittenton Serv, 73- Mem: Civil Aviation Med Asn; Am Psychiat Asn; Am Psychoanal Asn; Am Acad Child Psychiat. Res: Etiology of pica in children; training in child psychiatry; psychological factors in aircraft accidents. Mailing Add: 11328 Euclid Ave Cleveland OH 44106

BOBB, MARVIN LESTER, b Columbia, SC, Oct 3, 11; m 39; c 3. ENTOMOLOGY. Educ: Clemson Col, BS, 33; Va Polytech Inst, MS, 35; Univ Va, PhD(biol), 50. Prof Exp: Instr & asst econ entom, 33-36, asst entomologist, Agr Exp Sta, 35-47, assoc entomologist in charge entom res, Piedmont Fruit Res Lab, 47-60, prof entom & entomologist in charge entom res, 60-73, EMER PROF ENTOM, VA POLYTECH INST & STATE UNIV, 73- Concurrent Pos: Mem, Int Cong Entom; consult, Ag-Tec Corp, 74- Mem: Entom Soc Am. Res: Control of fruit insects; aquatic and semi-aquatic Hemiptera of southeastern United States. Mailing Add: 2400 Angus Rd Charlottesville VA 22901

BOBB, YVONNE DOLORES, b Seattle, Wash, June 21, 20. MICROBIOLOGY, BIOCHEMISTRY. Educ: Univ Wash, BA, 47, BS, 48, MA, 55; Univ Kans, PhD(microbiol), 59. Prof Exp: SR RES ASSOC, PALO ALTO MED RES FOUND, 59- Mem: AAAS; Am Soc Microbiol; Am Chem Soc. Res: Protein-DNA complexes; protein structure-function relationships; natural protein complexes; protein denaturation. Mailing Add: Palo Alto Med Res Found 860 Bryant St Palo Alto CA 94301

BOBBER, ROBERT JOHN, b Milwaukee,Wis, Mar 9, 18; m 48; c 3. ACOUSTICS. Educ: Univ Wis, MS, 47. Prof Exp: Physicist, US Navy Underwater Sound Ref Lab, 47-67, SUPT, UNDERWATER SOUND REF DIV, NAVAL RES LAB, 67- Concurrent Pos: Instr, Univ Wis, 47. Mem: Fel Acoustical Sco Am; Inst Elec & Electronics Engrs. Res: Underwater sound measurements; electroacoustic transducers. Mailing Add: Naval Res Lab PO Box 8337 Orlando FL 32806

BOBBIT, JESSE LEROY, b Yellow Springs, Ohio, Sept 18, 33; m 56. BIOCHEMISTRY. Educ: Berea Col, BA, 55; Univ Louisville, PhD(biochem), 64. Prof Exp: Teacher pub schs, Ky, 58-60; biochemist, US Army Med Res Labs, Ft Knox, Ky, 63-66; RES BIOCHEMIST, ELI LILLY & CO, 66- Mem: Am Chem Soc. Res: Protein isolation and characterization; mechanism of enzyme and hormone action. Mailing Add: 8101 Rosemead Lane Indianapolis IN 46240

BOBBITT, JAMES MCCUE, b Charlestown, WVa, Jan 18, 30; m 52; c 3. CHEMISTRY. Educ: WVa Univ, BS, 51; Ohio State Univ, PhD(chem), 55. Prof Exp: Fel, Wayne State Univ, 55-56; from instr to assoc prof, 56-68, PROF CHEM, UNIV CONN, 68- Concurrent Pos: NSF fel, Univ Zurich, 59-60; vis fel, Univ EAnglia, 64-65; guest prof, La Trobe Univ, Melbourne, Australia, 71-72. Mem: Am Chem Soc. Res: Insoquinoline alkaloid syntheses; alkaloid and glucoside structure proof, nitrogen heterocyclics; organic electrochemistry. Mailing Add: Dept of Chem Univ of Conn Storrs CT 06268

BOBBITT, JEFFREY L, b Houston, Tex, Dec 12, 42; m 66; c 1. ORGANIC CHEMISTRY, PHYSICAL CHEMISTRY. Educ: Univ St Thomas, Tex, BA, 65; Tex A&M Univ, MS, 67, PhD(chem), 70. Prof Exp: Teaching asst, Tex A&M Univ, 65-68; INSTR CHEM, S TEX COL, 70- Mem: AAAS; Am Chem Soc. Res: Education. Mailing Add: Dept of Chem STex Col Houston TX 77002

BOBBITT, OLIVER BIERNE, b Charleston, WVa, Jan 10, 17; m 43; c 3. CLINICAL PATHOLOGY. Educ: Univ Ga, BS, 39; Univ Va, MD, 43. Prof Exp: From instr to assoc prof, Med Sch, 47-57, chmn dept clin path, Med Sch & dir clin lab, Hosp, 52-71, PROF CLIN PATH, MED SCH, UNIV VA, 57- Mem: AMA; Col Am Path; Am Soc Clin Path. Res: Blood groups. Mailing Add: Univ of Va Hosp Charlottesville VA 22902

BOBE, ERNEST CHRISTOPH, b Brooklyn, NY, Nov 15, 20; m 50; c 3. ORGANIC CHEMISTRY. Educ: Polytech Inst Brooklyn, BSc, 50. Prof Exp: Chemist, Permatex Co, Inc, NY, 50-51; res chemist polymer chem, Thiokol Chem Co, NJ, 51-56; develop chemist, Gen Elec Co, NY, 56-66; group leader specialty tape, Mystik Tape Div, Borden Chem Co, 66-71; proj mgr, Chase-Foster Div, Keene Corp, East Providence, 71-72; CHIEF CHEMIST, WASHBURN WIRE CO, EAST PROVIDENCE, 73- Mem: Am Chem Soc; Am Soc Testing & Mat; fel Am Inst Chem. Res: Silicone resin chemistry; electrical insulation. Mailing Add: 151 Promenade St West Barrington RI 02890

BOBEAR, JEAN B, b Schenectady, NY. SYSTEMATIC BOTANY. Educ: Col St Rose, BS, 48; Cornell Univ, MS, 54; Trinity Col, Dublin, PhD(taxon bot), 64. Prof Exp: Eng asst, Control Dept, Knolls Atomic Power Lab, Gen Elec Co, 48-53; herbarium asst, Cornell Univ, 54-56; from asst prof to assoc prof, 56-65, PROF BIOL, STATE UNIV NY COL BROCKPORT, 65- Mem: AAAS; Bot Soc Am; Bot Soc Brit Isles; Int Asn Plant Taxon. Res: Experimental taxonomic treatment of critical genera; genus Euphrasia. Mailing Add: Dept of Biol Sci State Univ of NY Col Brockport NY 14420

BOBEAR, WILLIAM JOSEPH, b Schenectady, NY, Mar 19, 20; m 49; c 8. POLYMER CHEMISTRY. Educ: Harvard Univ, BS, 42. Prof Exp: Chemist, Hercules Powder Co, 42-44; chemist, 46-59, SPECIALIST SILICONE RUBBER, GEN ELEC CO, 59- Mem: AAAS; Am Chem Soc. Res: Silicone rubber, vulcanization, elasticity, high temperature stability, ablative thermal insulative properties. Mailing Add: Silicone Prod Dept Gen Elec Co Waterford NY 12188

BOBISUD, LARRY EUGENE, b Midvale, Idaho, Mar 16, 40; m 63. MATHEMATICS. Educ: Col Idaho, BS, 61; Univ NMex, MS, 63, PhD(math), 66. Prof Exp: Vis mem, Courant Inst Math Sci, NY Univ, 66-67; from asst prof to assoc prof, 67-74, PROF MATH, UNIV IDAHO, 74- Mem: Am Math Soc; Soc Indust & Appl Math. Res: Singular perturbation problems for partial differential equations; nonlinear oscillations; mathematical biology. Mailing Add: Dept of Math Univ of Idaho Moscow ID 83843

BOBKA, RUDOLPH J, b Akron, Ohio, July 27, 28; m 55; c 1. PHYSICAL CHEMISTRY. Educ: Marietta Col, AB, 50; Miami Univ, MS, 51; Western Reserve Univ, PhD(phys chem), 60. Prof Exp: Anal chemist, Army Chem Ctr, Md, 51-52; asst, Western Reserve Univ, 54-58; chemist, Res Lab, Union Carbide Corp, 58-68; assoc prof, 68-72, PROF CHEM, STATE UNIV NY COL PLATTSBURGH, 72- Mem: Am Chem Soc. Res: Low temperature calorimetry; surface chemistry; carbon technology. Mailing Add: Dept of Chem State Univ NY Col Plattsburgh NY 12901

BOBKO, EDWARD, b Cleveland, Ohio, Feb 15, 25; m; c 3. ORGANIC CHEMISTRY. Educ: Western Reserve Univ, BS, 49; Northwestern Univ, PhD(org chem), 52. Prof Exp: Instr chem, Northwestern Univ, 52-53; res chemist, Olin Industs, Inc, 53-54; asst prof chem, Washington & Jefferson Col, 54-55, instr, 55-56; from asst prof to assoc prof, 56-69, PROF CHEM, TRINITY COL, CONN, 69- Mem: AAAS; Am Chem Soc. Res: Organometallic compounds; pyrimidines; physical organic chemistry. Mailing Add: Dept of Chem Trinity Col Hartford CT 06106

BOBLITT, ROBERT LEROY, b Springfield, Ohio, Nov 1, 25; m 44; c 4. MEDICINAL CHEMISTRY. Educ: Ohio Northern Univ, BS, 48; Univ Minn, PhD(pharmaceut chem), 53. Prof Exp: Asst, 53-55, assoc prof, 56-70, PROF PHARMACEUT CHEM, COL PHARM, UNIV HOUSTON, 70-, ASST DEAN, 72- Mem: Am Chem Soc; Am Pharmaceut Asn. Res: Synthesis of amidines; physical and chemical properties of amidines; phytochemistry; synthesis of beta-adrenergic blocking agents. Mailing Add: Col of Pharm Univ of Houston Houston TX 77004

BOBO, EDWIN RAY, b Oklahoma City, Okla, May 17, 38. MATHEMATICS. Educ: Pa State Univ, BA, 60; Univ Va, MA, 62, PhD(math), 65. Prof Exp: Asst prof, 65-70, ASSOC PROF MATH, GEORGETOWN UNIV, 70- Mem: Am Math Soc; Math Asn Am. Res: Algebra; finite groups; automorphism groups of Jordan algebras; group representations. Mailing Add: Dept of Math Georgetown Univ Washington DC 20007

BOBONICH, HARRY MICHAEL, b Ashland, Pa, Sept 13, 24; m 49; c 2. INORGANIC CHEMISTRY. Educ: Susquehanna Univ, BA, 50; Bucknell Univ, MS, 56; Syracuse Univ, PhD(chem), 65. Prof Exp: Chemist, Barrett Div, Allied Div, Allied Chem Corp, 52-55; prof chem, Pa State Univ, 55-61 & 64-65; chmn dept, 69-74, PROF CHEM, SHIPPENSBURG STATE COL, 65-, ACTG DEAN GRAD STUDIES, 74- Mem: Am Chem Soc. Res: Synthesis of coordination compounds; use of magnetic measurements; infrared and ultraviolet spectra; conductance and x-rays to determine structure; development of new experiments for freshman chemistry programs. Mailing Add: Dept of Chem Shippensburg State Col Shippensburg PA 17257

BOBONIS, AUGUSTO, b Humacao, PR, June 20, 07; m 36; c 4. MATHEMATICS. Educ: Univ PR, AB, 27, MS, 34; Univ Chicago, PhD(math), 39. Hon Degrees: LLD, Inter-Am Univ PR, 75. Prof Exp: Instr math & physics, St Augustine Mil Acad, PR, 27-30; critic teacher physics, 30-34, from instr to prof, 34-48, chmn dept, 45-47, dir city col, 47-52, tech adv to secy educ, 52-53, dir sec educ, 53-57, prof math, 57-58, from actg dean to dean col educ, 58-70, prof math, 70-75, EMER PROF MATH & EMER DEAN, UNIV PR, RIO PIEDRAS, 75- Concurrent Pos: Vpres acad affairs, Inter-Am Univ PR, 70-75. Mem: Am Math Soc. Res: Boundary value problems. Mailing Add: Dept of Math Univ of PR Rio Piedras PR 00931

BOBROW, DANIEL G, b New York, NY, Nov 29, 35; m 63; c 2. COMPUTER SCIENCE. Educ: Rensselaer Polytech Inst, BS, 57; Harvard Univ, SM, 58; Mass Inst Technol, PhD(math), 64. Prof Exp: Mem staff, Bolt, Beranek & Newman Inc, 62-64; asst prof elec eng, Mass Inst Technol, 64-65; vpres, Info Sci Div, Bolt, Beranek & Newman, Inc, 65-69, dir comput sci div, 69-72; PRIN SCIENTIST, XEROX PALO ALTO RES CTR, 72- Concurrent Pos: Fulbright fel, 72-73; lectr, Stanford Univ, 73- Honors & Awards: Prog Lang Award, Asn Comput Mach, 74. Mem: AAAS; Asn Comput Mach; Asn Comput Ling; Inst Elec & Electronics Eng. Res: Development of computer languages and programming systems; computational linguistics, including development of a computer program that understands natural language; artificial intelligence. Mailing Add: Xerox Palo Alto Res Ctr Palo Alto CA 94304

BOBST, ALBERT M, b Zurich, Switz, Sept 10, 39; nat US; m 70; c 2. BIOPHYSICAL CHEMISTRY, ORGANIC BIOCHEMISTRY. Educ: Univ Zurich, PhD(chem), 65. Prof Exp: Res fel, Inst Biophys & Biochem, Paris, 66; Swiss Nat Found fel, Univ Calif, Berkeley, 67-68; res assoc biochem, Princeton Univ, 68-69; asst prof, 69-74, ASSOC PROF CHEM, UNIV CINCINNATI, 74- Concurrent Pos: Vis scientist, Nat Inst Arthritis, Metab & Digestive Dis, 74-75. Mem: AAAS; Am Chem Soc. Res: Application of electron spin resonance and circular dichroism spectroscopy to biological systems; biological redox mechanisms; synthesis of nucleic acid analogs. Mailing Add: Dept of Chem 172 Univ of Cincinnati Cincinnati OH 45221

BOCAGE, ALBERT J, b New Orleans, La, Apr 8, 30. PHYSIOLOGY, BIOCHEMISTRY. Educ: Xavier Univ La, BS, 57; La State Univ, MS, 62, PhD(physiol), 64. Prof Exp: Res assoc, 64-68, ASST PROF PHYSIOL, LA STATE UNIV SCH MED, NEW ORLEANS, 68- Res: Circulatory and biochemical effects of occlusion of a coronary artery with the object of reducing the incidence of mortality and arrythmias. Mailing Add: Dept of Physiol La State Univ Sch of Med New Orleans LA 70112

BOCCABELLA, ANTHONY VINCENT, b Brooklyn, NY, June 12, 29; m 51; c 2. ANATOMY. Educ: Seton Hall Univ, BS, 50; Univ Iowa, MS, 56, PhD(anat), 58. Prof Exp: Instr anat, Univ Iowa, 58-59; from instr to assoc prof, 59-67, PROF ANAT, COL MED & DENT NJ, 67-, CHMN DEPT, 71- Mem: Endocrine Soc; Am Asn Anat; NY Acad Sci. Res: Thyroid physiology; male and female reproductive systems; muscular dystrophy. Mailing Add: Dept of Anat Col of Med & Dent NJ Newark NJ 07103

BOCCHIERI, SAMUEL FRANCIS, b Chicago, Ill, Oct 5, 31; m 67; c 3. MICROBIOLOGY, BIOCHEMISTRY. Educ: Loyola Univ, Ill, BS, 54; Cornell Univ, MS, 56; Rutgers Univ, PhD(microbiol), 65. Prof Exp: Res chemist, Abbott Labs, Ill, 56-61; assoc microbiol, Sch Dent Med, Univ Pa, 65-68; ASST PROF MICROBIOL, ALBANY MED COL, 68- Concurrent Pos: Nat Inst Dent Res grants, 67-69. Mem: AAAS; Am Soc Microbiol; Am Chem Soc. Res: Folic acid biosynthesis and mechanism of sulfonilamide inhibition; microbial metabolism and control mechanisms; biochemical and genetic mechanisms of resistance to antimicrobial drugs. Mailing Add: Dept of Microbiol Albany Med Col Albany NY 12208

BOCEK, ROSE MARY, b Portland, Ore, Sept 2, 24. BIOCHEMISTRY, CLINICAL CHEMISTRY. Educ: Marylhurst Col, BS, 46; Univ Ore, MS, 54, PhD(biochem), 64. Prof Exp: Med technologist, Clin Lab, Univ Hosp & Clins, 47-50, chief technologist, Clin Chem Sect, 50-52, res assoc biochem, Med Sch, 54-57, instr, 57-64, ASST PROF BIOCHEM, MED SCH, UNIV ORE, 64-; ASSOC SCIENTIST, ORE REGIONAL PRIMATE RES CTR, 67- Concurrent Pos: Res assoc, Ore Regional Primate Res Ctr, 61-64, asst scientist, 64-67. Mem: AAAS; Am Chem Soc. Res: Intermediary metabolism of skeletal muscle, including mechanisms of energy production from metabolic fuels as influenced by hormones and other physiological factors. Mailing Add: Dept of Perinatal Physiol Ore Regional Primate Res Ctr Beaverton OR 97005

BOCH, ROLF, b Stiefenhofen, Ger, Nov 9, 28; Can citizen; m 57; c 3. ANIMAL BEHAVIOR, ANIMAL PHYSIOLOGY. Educ: Univ Munich, Dr rer nat, 55. Prof Exp: Res assoc, Zool Inst, Munich, 55-56; RES SCIENTIST, CAN DEPT AGR, 56- Res: Insect physiology and behavior; communications in social insects; apiculture. Mailing Add: Ottawa Res Sta Entom Sect Can Dept of Agr Ottawa ON Can

BOCHE, ROBERT DEVORE, b Prairie City, Ore, Feb 11, 12; m 38; c 3. GENETICS, VIROLOGY. Educ: Calif Inst Technol, BS, 34, PhD(genetics), 38. Prof Exp: Asst cytogenetics, Carnegie Inst, 37-39; asst prof zool, Univ NC, 39-40; instr, Univ Pa, 40-43, res assoc, Harrison Dept Surg Res, Sch Med, 42-44; chief radiol div, Manhattan Dept, Univ Rochester, 44-46; asst prof zool, Univ Chicago, 46-52; biologist, US Air Force Radiation Lab, Chicago, 52-54, assoc, Virus Lab, 56-58; from instr to asst prof, 58-63, ASSOC PROF PEDIAT, LOMA LINDA UNIV, 63- Mem: Radiation Res Soc; Am Soc Microbiol. Res: Biological effects of radiation, viruses, and toxic substances; mechanism of carcinogenesis and anticancer agents; electron microscopy; genetic diseases. Mailing Add: Dept of Pediat Virus Lab Loma Linda Univ Med Ctr Loma Linda CA 92354

BOCHNER, SALOMON, b Cracow, Austria-Hungary, Aug 20, 99. MATHEMATICS. Educ: Univ Berlin, PhD, 21. Prof Exp: Lectr, Univ Munich, 27-32; assoc, Princeton Univ, 33-34, from asst prof to prof, 34-59, Henry Burchard Fine prof, 59-68, EDGAR ODELL LOVETT PROF MATH, RICE UNIV, 68- Concurrent Pos: Vis prof, Univ Calif, 53; Int Educ Bd fel, Copenhagen Univ, Oxford Univ & Cambridge Univ, 25-27; consult, Los Alamos Proj, Princeton Univ, 51-52, Nat Sci Found & Air Res & Develop Command, 52- Mem: Nat Acad Sci; Am Math Soc. Res: Almost periodic functions; probability theory; complex functions and manifolds. Mailing Add: Dept of Math Rice Univ Houston TX 77001

BOCK, CARL E, b Berkeley, Calif, Mar 29, 42; m 64. ZOOLOGY, ECOLOGY. Educ: Univ Calif, Berkeley, AB, 64, PhD(zool), 68. Prof Exp: ASST PROF BIOL, UNIV COLO, 68-, ASSOC CHMN DEPT, 74-, ASSOC PROF ENVIRON, POP & ORGANISMIC BIOL, 74- Mem: Ecol Soc Am; Soc Study Evolution; Cooper Ornith Soc; Am Ornith Union; Brit Ecol Soc. Res: Animal ecology; evolution and ecology of birds; fire ecology. Mailing Add: Dept of Biol Univ of Colo Boulder CO 80302

BOCK, CHARLES WALTER, b Philadelphia, Pa, Jan 19, 45; m 68. QUANTUM CHEMISTRY, NUCLEAR PHYSICS. Educ: Drexel Univ, BS, 68, MS, 70, PhD(physics), 72. Prof Exp: ASST PROF MATH & PHYSICS, PHILADELPHIA COL TEXTILES & SCI, 72- Mem: Am Phys Soc. Res: Investigation of the most appropriate definitions of stabilization and destabilization energies; determination of field-theoretic nucleon-nucleon scattering models. Mailing Add: Philadelphia Col Textiles & Sci Henry Ave Philadelphia PA 19144

BOCK, ERNST, b Charkow, USSR, Feb 6, 29; nat Can; m 55; c 4. PHYSICAL CHEMISTRY. Educ: Univ Man, BSc, 56, MSc, 57, PhD, 60. Prof Exp: Lectr, 58-61, asst prof, 61-66, ASSOC PROF CHEM, UNIV MAN, 66- Res: Nuclear magnetic resonance spectroscopy. Mailing Add: Dept of Chem Univ of Man Winnipeg MB Can

BOCK, FRED G, b St Louis, Mo, Aug 28, 23; m 44, 65; c 4. BIOCHEMISTRY. Educ: Univ Minn, BA, 47, MS, 48; Univ Buffalo, PhD, 58. Prof Exp: Assoc scientist, Univ Minn, 52-53; from sr to assoc cancer res scientist, 53-66, PRIN CANCER RES SCIENTIST, ROSWELL PARK MEM INST, 66-, DIR, ORCHARD PARK LABS, 64-; ASSOC RES PROF, STATE UNIV NY BUFFALO, 70- Concurrent Pos: Asst res prof, State Univ NY Buffalo, 63-68. Mem: AAAS; Am Chem Soc; Am Asn Cancer Res; Soc Exp Biol & Med; Am Genetic Asn. Res: Carcinogenesis. Mailing Add: 321 North St East Aurora NY 14052

BOCK, JANE HASKETT, b Rochester, Ind, Oct 3, 36; m 64. BOTANY. Educ: Duke Univ, BA, 58; Indiana Univ Pa, MA, 60; Univ Calif, Berkeley, PhD(bot), 66. Prof Exp: Asst prof biol, Calif State Col, Hayward, 67-68; asst prof, 68-73, ASSOC PROF BIOL, UNIV COLO, 73- Concurrent Pos: Trustee, Res Ranch, Elgin, Ariz, 73- Mem: Bot Soc Am; Soc Study Evolution; Brit Ecol Soc; Ecol Soc Am; Int Asn Plant Taxon. Res: Reproductive systems in terrestrial flowering plants and determination of the evolutionary significance of these systems; reproductive biology of flowering plants; ecology of fire; conservation of native flora of the western US. Mailing Add: EPO Dept of Biol Univ of Colo Boulder CO 80302

BOCK, PHILIP KARL, b New York, NY, Aug 26, 34; m 57; c 3. ANTHROPOLOGY. Educ: Fresno State Col, AB, 55; Univ Chicago, MA, 56; Harvard Univ, PhD(social anthrop), 62. Prof Exp: PROF ANTHROP, UNIV NMEX, 62- Concurrent Pos: Vis prof, Columbia Univ, 67, Iberoamerican Univ, Mex, 70, Stanford Univ, 72 & Univ Autonoma Guadalajara, 75; co-ed, J Anthrop Res, 72- Mem: Am Anthrop Asn; Soc Appl Anthrop; Soc Ethnomusicol. Res: North American Indians; language and culture; research on peasant communities and formal models for the description of social structure. Mailing Add: Dept of Anthrop Univ of NMex Albuquerque NM 87131

BOCK, ROBERT MANLEY, b Preston, Minn, July 26, 23; m 47; c 2. BIOPHYSICAL CHEMISTRY. Educ: Univ Wis, BS, 49, PhD(chem), 52. Prof Exp: Res assoc, NSF, 52, from asst prof to assoc prof, 52-61, PROF BIOCHEM & MOLECULAR BIOL, UNIV WIS, 61-, DEAN GRAD SCH, 67- Mem: Am Chem Soc. Res: Physical chemistry of proteins and nucleic acids; metabolic regulation; mechanism of enzyme action; structure and function of transfer RNA. Mailing Add: Grad Sch Univ of Wis Madison WI 53706

BOCK, ROBERT OLIVER, b Kalamazoo, Mich, Jan 12, 12; m 34; c 3. PHYSICS, NAVIGATION. Educ: Kalamazoo Col, AB, 34; Univ Iowa, MS, 36; Cornell Univ, PhD(physics), 42. Prof Exp: Instr physics & appl math, Westminster Col, Utah, 36-38; asst physics, Cornell Univ, 38-42; develop engr, Arma Corp, 42-46; assoc prof physics, Colo State Univ, 46-51; sr res engr, Arma Corp, 51-56, mgr eng, 56-69; dir aerospace systs, AMBAC Industs, 67-69; assoc dir, 69-70, DIR, GUID & CONTROL SUBDIV, AEROSPACE CORP, 70- Concurrent Pos: Assoc mem, Defense Sci Bd, 75-76. Mem: Sigma Xi; Am Inst Aeronaut & Astronaut; Inst Navig. Res: Infrared absorption; naval gun-fire control and associated problems; magnetic compasses; inelastic electron collisions; optical constants of the divalent metals; bomber defense; inertial guidance. Mailing Add: Aerospace Corp PO Box 92957 Los Angeles CA 90009

BOCK, WALTER JOSEPH, b New York, NY, Nov 20, 33; m 57; c 3. VERTEBRATE MORPHOLOGY, EVOLUTION. Educ: Cornell Univ, BS, 55; Harvard Univ, MA, 57, PhD(biol), 59. Prof Exp: From asst prof to assoc prof zool, Univ Ill, 61-65; from asst prof to assoc prof, 65-73, PROF ZOOL, COLUMBIA UNIV, 73- Concurrent Pos: Nat Sci Found fel, Univ Frankfurt, 59-61; res assoc ornith, Am Mus Natural Hist, 65- Honors & Awards: Coues Award, Am Ornithologists Union, 75. Mem: Fel AAAS; fel Am Ornithologists Union; Soc Study Evolution; Am Soc Naturalists; Am Soc Zool. Res: Functi- onal and evolutionary morphology; morphology and classification of birds; general theories of major evolutionary changes. Mailing Add: Dept of Biol Sci Columbia Univ New York NY 10027

BOCK, WAYNE DEAN, b Truman, Minn, Nov 15, 32; m 68; c 4. MICROPALEONTOLOGY. Educ: Univ Wis, BS, 58, MS, 61; Univ Miami, PhD(oceanog), 67. Prof Exp: Instr geol, 64-67, res scientist, 67-69, ASST PROF MICROPALEONT, SCH MARINE & ATMOSPHERIC SCI, UNIV MIAMI, 67- Concurrent Pos: Prin investr, NSF Res Grants, 67- Mem: AAAS; Paleont Soc; Paleont Res Inst. Res: Paleoecology of ocean sediment cores; use of Foraminifera as indicators of pollution. Mailing Add: Sch of Marine & Atmospheric Sci Univ of Miami Miami FL 33149

BOCKELMAN, CHARLES KINCAID, b San Francisco, Calif, Nov 29, 22; m 50; c 1. NUCLEAR PHYSICS. Educ: Univ Wis, PhD(physics), 51. Prof Exp: Res assoc physics, Mass Inst Technol, 51-55; from asst prof to assoc prof, 55-65, PROF PHYSICS, YALE UNIV, 65-, DEP PROVOST SCI, 69- Concurrent Pos: Fel, Univ Inst Theoret Physics, Copenhagen, 58-59; Guggenheim fel, Oxford Univ, 70. Mem: Fel Am Phys Soc. Res: Nuclear structure. Mailing Add: 252 JWG Yale Univ New Haven CT 06520

BOCKHOFF, FRANK JAMES, b Tiffin, Ohio, Mar 26, 28; m 51; c 4. PHYSICAL CHEMISTRY. Educ: Case Inst Technol, BS, 50; Case Western Reserve Univ, MS, 52, PhD(phys chem), 59. Prof Exp: Asst chem, Western Reserve Univ, 50-51; instr chem & chem eng, Fenn Col, 51-54, from asst prof to assoc prof chem, 54-64, chmn dept, 62-64; PROF CHEM & CHMN DEPT, CLEVELAND STATE UNIV, 64- Concurrent Pos: Consult, Am Agile Corp, 53-61, Apex Reinforced Plastics Div, White Sewing Mach Co, 61-63 & Signal Chem Mfg Co, 60-62; tech consult, Re-Ox Corp, 62-67; consult, Incar, Inc, Am Smelting and Refining Co, 64-67 & Cleveland Mus Natural Hist, 74- Mem: AAAS; Am Chem Soc; Soc Plastics Eng; Soc Plastics Indust; Am Inst Chem. Res: Plastics and resins development; complex-ion; thermodynamics and quantum theory; kinetics and mechanisms of anionic polymerizations; polymer development and applications. Mailing Add: Dept of Chem Cleveland State Univ Cleveland OH 44115

BOCKHOLT, ANTON JOHN, b Westphalia, Tex, Sept 23, 30; m 52; c 2. PLANT BREEDING, GENETICS. Educ: Tex A&M Univ, BS, 52, MS, 58, PhD(plant breeding), 67. Prof Exp: Res asst corn breeding, 54-55, instr, 55-66, asst prof corn breeding & leader, corn improv prog, 66-72, ASSOC PROF AGRON, TEX A&M UNIV, 72- Mem: Am Soc Agron; Crop Sci Soc Am. Res: Plant breeding, genetics, pathology, biology-environment, entomology and nutrition and metabolism. Mailing Add: Dept of Soil & Crop Sci Tex A&M Univ College Station TX 77843

BOCKIAN, ALBERT HAROLD, organic chemistry, air pollution, see 12th edition

BOCKMAN, DALE EDWARD, b Winona, Mo, Feb 4, 35; m 53; c 3. ANATOMY. Educ: Southwest Mo State Col, BS, 56; Los Angeles State Col, MA, 58; Univ Ill, PhD(anat), 63. Prof Exp: Instr anat, Univ Ill, 62-63; from instr to asst prof, Univ Tenn Med Units, 63-68; from assoc prof to prof, Med Col Ohio, 68-75; PROF ANAT & CHMN DEPT, MED COL GA, 75- Res: Fine structure; development of lymphoid system; immunology. Mailing Add: Dept of Anat Med Col of Ga Augusta GA 30902

BOCKRATH, RICHARD CHARLES, JR, b St Louis, Mo, May 7, 38; m 64; c 2. RADIATION BIOPHYSICS, MICROBIOLOGY. Educ: Yale Univ, BS, 61; Pa State Univ, MS, 63, PhD(biophys), 65. Prof Exp: Vis res fel microbial genetics, Univ Sussex, 65-67; ASST PROF MICROBIOL, SCH MED, IND UNIV, INDIANAPOLIS, 68- Concurrent Pos: NIH fel, Univ Sussex, 66-67. Res: Effects of low energy beta particles on cells; molecular biology and control; cellular accumulation of streptomycin; mutagenesis by ultraviolet radiation. Mailing Add: Dept of Microbiol Ind Univ Med Sch Indianapolis IN 46202

BOCKRIS, JOHN O'MARA, b Johannesburg, SAfrica, Jan 5, 23; nat US; div; c 2. PHYSICAL CHEMISTRY. Educ: Univ London, BSc, 43, PhD, 45, DSc(electrochem), 52. Prof Exp: Lectr phys chem, Univ London, 45-56; prof chem, Univ Pa, 56-62 & electrochem, 62-71; chmn dept phys chem, 71-74, PROF PHYS CHEM, SCH PHYS SCI, FLINDERS UNIV, 74- Concurrent Pos: Founder, Inst Solar & Electrochem Energy Conversion. Honors & Awards: Medaille d'honneur, Univ Louvain, 53. Mem: Fel Royal Inst Chem; Int Soc Hydrogen Energy; Int Soc Electrochem. Res: Hydrogen on and in metals; metal deposition and dissolution; adsorption at solid electrolyte surfaces; constitution of molten salts and silicates; electrode reactions in molten salts; quantum mechanics; electrode processes; hydrogen economy; energy science. Mailing Add: Sch of Phys Sci Flinders Univ Adelaide Australia

BOCKSCH, ROBERT DONALD, b Detroit, Mich, May 5, 31; m 58; c 3. ORGANIC CHEMISTRY. Educ: Wayne State Univ, BS, 54, Univ Wis, PhD, 60. Prof Exp: From asst prof to assoc prof, 58-68, chmn dept, 63-75, PROF CHEM, WHITWORTH COL, 68-, CHMN DEPT HEALTH SCI, 75- Mem: Am Chem Soc; Brit Chem Soc. Res: Synthetic organic chemistry; natural product chemistry; polynuclear hydrocarbons. Mailing Add: Dept of Chem Whitworth Col Spokane WA 99218

BOCKSTAHLER, LARRY EARL, b La Junta, Colo, May 13, 34; m 65; c 1. VIROLOGY, BIOPHYSICS. Educ: Mich State Univ, BS, 56; Univ Wis-Madison, MS, 60, PhD(biochem), 64. Prof Exp: NIH fel biochem, Max Planck Inst Biol, 64-67; biochemist & Humboldt fel biochem, Max Planck Inst Cell Biol, 67-69; BIOCHEMIST, BUR RADIOL HEALTH, USPHS, 69- Mem: Electron Micros Soc Am; Am Soc Photobiol; Biophys Soc; Radiation Res Soc Am. Res: Molecular biology of plant viruses and sea urchin egg development; radiobiology of animal viruses. Mailing Add: USPHS Bur of Radiol Health 12709 Twinbrook Pkwy Rockville MD 20852

BOCKSTAHLER, THEODORE EDWIN, b Monticello, Ill, Nov 22, 20; m 43; c 2. INDUSTRIAL ORGANIC CHEMISTRY. Educ: Ind Univ, AB, 42; Univ Ill, MS, 48, PhD(chem), 49. Prof Exp: Instr, Chem Warfare Sch, Edgewood Arsenal, 45; group leader org chem res, 49-66, HEAD DEVELOP LAB, ROHM AND HAAS CO, 66- Mem: Am Chem Soc. Res: Process development for plastic chemicals in laboratory and plant. Mailing Add: Rohm and Haas Co Box 672 Deer Park TX 77536

BOCZKOWSKI, RONALD JAMES, b Bridgeport, Conn, Nov 14, 43; m 70; c 2. ANALYTICAL CHEMISTRY. Educ: Canisius Col, BS, 65; Univ Notre Dame, PhD(chem), 71. Prof Exp: Res fel, Univ Cincinnati, 71-72; lab mgr textiles, Armtex, Inc, 72-73; RES SCIENTIST PAPER, UNION CAMP CORP, 73- Mem: Am Chem Soc; Anal Am Chem Soc; Am Asn Textile Chemists & Colorists. Res: Analytical application of electrochemical techniques; preconcentration techniques in trace analysis; analysis of minor componenets in paper, pulp and process liquors. Mailing Add: Box 412 Res & Develop Union Camp Corp Princeton NJ 08540

BODA, JAMES MARVIN, b Mt Victory, Ohio, July 3, 24; m 53; c 1. COMPARATIVE PHYSIOLOGY. Educ: Univ Calif, BS, 48, PhD, 53. Prof Exp: Actg instr animal husb, Univ & animal husbandman, Exp Sta, 52-54, asst prof animal husb, Univ, 54-59, assoc prof animal husb, Univ & assoc animal husbandman, Exp Sta, 59-65, prof animal husb & animal physiol, Univ, 65-68, chmn dept animal physiol, 68-73, PROF PHYSIOL, UNIV & PHYSIOLOGIST, EXP STA, UNIV CALIF, DAVIS, 68- Concurrent Pos: Fulbright res scholar, 60-61. Mem: Am Physiol Soc; Am Soc Zoologists; Endocrine Soc. Res: Comparative endocrinology; physiological adaptations. Mailing Add: Dept of Animal Physiol Univ of Calif Davis CA 95616

BODAMER, GEORGE WILLOUGHBY, b Philadelphia, Pa, Apr 12, 16; m 44; c 4. CHEMISTRY. Educ: Univ Pa, BS, 38; Cornell Univ, PhD(org chem), 41. Prof Exp: Chemist, Rohm and Haas, Philadelphia, 41-59; sr scientist, ESB, Inc, 59-61, mgr phys chem labs, 61-64, assoc dir res, 64-70; CHIEF CHEMIST, INT HYDRONICS CORP, PRINCETON, NJ, 70- Mem: Am Chem Soc; Electrochem Soc. Res: Electrochemistry; ion exchange; industrial waste treatment. Mailing Add: 311 Franklin Ave Cheltenham PA 19012

BODANSKY, DAVID, b New York, NY, Mar 10, 24; m 52; c 2. PHYSICS. Educ: Harvard Univ, BS, 43, MA, 48, PhD(physics), 50. Prof Exp: Instr physics, Columbia Univ, 50-52, assoc, 52-54; from asst prof to assoc prof, 54-63, PROF PHYSICS, UNIV WASH, 63- Concurrent Pos: Alfred P Sloan res fel, 59-63; Guggenheim fel, 66-67 & 74-75. Mem: Am Asn Physics Teachers; fel Am Phys Soc. Res: Nuclear physics and astrophysics. Mailing Add: Dept of Physics Univ of Wash Seattle WA 98195

BODANSKY, OSCAR, b Elizabethgrad, Russia, Aug 21, 01; nat US; m 29; c 1. MEDICINE, BIOCHEMISTRY. Educ: Columbia Univ, AB, 21, MA, 22, PhD(biochem), 25; Univ Chicago, MD, 38. Prof Exp: Asst chem, Columbia Univ, 22-26; res assoc biochem, Univ Calif, 26-27; from instr to assoc prof pharm, Univ Tex, 27-30; instr pediat, NY Univ & chemist, Children's Med Div, Bellevue Hosp, 30-37; asst adj pediatrician, Beth Israel Hosp, New York, 40-42, assoc pediatrician, 46-48; assoc prof clin pharmacol, Sloan-Kettering Div, Med Col, Cornell Univ, 48-52, prof biochem, 52-71, mem & chief div biochem, Inst Cancer Res, 56-71, vpres, 66-71, EMER MEM, MEM-SLOAN-KETTERING CANCER CTR, 71-; EMER, MEM HOSP CANCER, 67-; GUEST INVESTR, ROCKEFELLER UNIV, 72- Concurrent Pos: Chmn dept biochem, Mem Hosp, 48-67, dir resident training, 67-71. Honors & Awards: Alfred P Sloan Award, 62; Van Slyke Award, 65; Lucy Wortham James Clin Res Award, 73. Mem: AAAS; fel AMA; Am Soc Biol Chemists; Soc Exp Biol & Med; Am Soc Pharmacol. Res: Kinetics of enzyme action; tissue phosphatases; vitamin A and D status in infants and children; bromide excretion; methemoglobinemia; biochemical aspects of disease; blood and tissue enzymes; biochemistry of human cancer. Mailing Add: Mem-Sloan-Kettering Cancer Ctr 1275 York Ave New York NY 10021

BODANSZKY, AGNES ADRIENNE, b Budapest, Hungary, May 17, 25; US citizen; m 50; c 1. ORGANIC CHEMISTRY. Educ: Pazmany Peter Univ, Budapest, MSc, 48. Prof Exp: Res chemist, Chinoin Pharmaceut Co, 48-52; sr res chemist, Res Inst Pharmaceut Chem, 52-56; res asst, Sloan Kettering Inst, 57-59; res staff mem, Frick Chem Lab, Princeton, 60-66; SR RES ASSOC ORG CHEM, CASE WESTERN RESERVE UNIV, 66- Res: Chemistry of steroids, peptides and proteins. Mailing Add: 18035 Fernway Rd Shaker Heights OH 44122

BODANSZKY, MIKLOS, b Budapest, Hungary, May 21, 15; US citizen; m 50; c 1. BIO-ORGANIC CHEMISTRY. Educ: Budapest Tech Univ, dipl chem eng, 39, DSc(org chem), 49. Prof Exp: Lectr med chem, Budapest Tech Univ, 50-56; res assoc biochem, Med Col, Cornell Univ, 57-59; sr res assoc peptide chem, Squibb Inst Med Res, 59-66; PROF CHEM & BIOCHEM, CASE WESTERN RESERVE UNIV, 66- Concurrent Pos: USPHS grants, Case Western Reserve Univ, 66- Mem: Am Chem Soc; The Chem Soc; Swiss Chem Soc; Am Soc Biol Chemists. Res: Peptide synthesis, structure and activity relationships of biologically active peptides; conformation of peptide hormones and peptide antibiotics. Mailing Add: Dept of Chem Case Western Reserve Univ Cleveland OH 44106

BODDY, DENNIS WARREN, b Portland, Ore, Aug 26, 22. ENTOMOLOGY, ZOOLOGY. Educ: Univ Wash, BS, 47, PhD, 55. Prof Exp: Instr zool, Columbia Basin Col, 55-59, head sci dept, 59-60, dean instr, 60-62; asst prof, 64-69, ASSOC PROF BIOL, PORTLAND STATE UNIV, 69- Mem: AAAS. Res: Taxonomy of the Coleoptera Tenebrionidae. Mailing Add: Dept of Biol Portland State Univ Portland OR 97201

BODDY, PHILIP J, b Birmingham, Eng, Mar 2, 33; m 55; c 2. PHYSICAL CHEMISTRY. Educ: Univ Birmingham, BSc, 54, PhD(phys chem), 57. Prof Exp: Nat Res Coun Can fel, 57-59; mem tech staff, 59-69, HEAD CHEM PROCESS TECHNOL DEPT, BELL LABS, 69- Mem: Am Chem Soc; Electrochem Soc. Res: Semiconductor surface physics; semiconductor electrochemistry; gas phase kinetics; diffusion; oxidation; photolithography; electrochemistry. Mailing Add: 22 Silver Spring Rd Short Hills NJ 07078

BODE, DONALD EDWARD, b Merrill, Wis, Aug 26, 22; m 44; c 2. SOLID STATE PHYSICS. Educ: Syracuse Univ, AB, 48, MS, 49, PhD(physics), 53. Prof Exp: Res assoc, Syracuse Univ, 48-54; assoc physicist, Int Bus Mach Corp, 54-56, proj physicist, 56-57, physicist, 57; head physics sect, 57-72, mgr detector res & develop dept, 72-74, MGR COMPONENTS LAB, SANTA BARBARA RES CTR, HUGHES AIRCRAFT CO, 74- Concurrent Pos: Chmn, IRIS Detector Specialty Group, 67-72; vchmn sci & eng coun, Santa Barbara, Inc, 75- Mem: Inst Elec & Electronics Eng. Res: Photoconductivity; semiconductor properties; xerography; transistors, infrared detectors. Mailing Add: Santa Barbara Res Ctr 75 Coromar Dr Goleta CA 93017

BODE, JAMES DANIEL, b Pittsburgh, Pa, Sept 6, 23. PHYSICAL CHEMISTRY. Educ: Univ Pittsburgh, BS, 46, PhD(chem), 52. Prof Exp: Grad asst, Univ Pittsburgh, 46-51; proj engr, Fisher Sci Co, 51-52, head anal chem, 52-55; sr res chemist, Jones & Laughlin Steel Corp, 55-58, res supvr, 58-60; mem tech staff, Bell Tel Labs, 60-63; sr res chemist, FMC Corp, 63-66; PRIN SCIENTIST, BENDIX RES LABS, BENDIX CORP, 66- Mem: AAAS; Optical Soc Am; Am Chem Soc; Electrochem Soc; The Chem Soc. Res: Instrumental methods; physics and chemistry of surfaces. Mailing Add: Bendix Res Labs Bendix Corp 20800 10 1/2 Mile Rd Southfield MI 48076

BODE, VERNON CECIL, b Marion County, Mo, Feb 19, 33; m 56; c 4. BIOCHEMICAL GENETICS. Educ: Univ Mo, BS, 55; Univ Ill, PhD(biochem), 61. Prof Exp: USPHS fel biochem, Sch Med, Stanford Univ, 61-64; from asst prof to assoc prof biochem, Med Sch, Univ Md, 64-70; PROF BIOL, KANS STATE UNIV, 70- Mem: Am Soc Biol Chem; Am Soc Microbiol; Genetics Soc Am. Res: DNA structure; phage morphogenesis; biochemical genetics; molecular and cell biology; regulation of embryological development in the mouse. Mailing Add: Div of Biol Kans State Univ Manhattan KS 66502

BODE, WILLIAM MORRIS, b Wooster, Ohio, Mar 18, 43; m 66; c 2. ENTOMOLOGY. Educ: Col Wooster, BA, 65; Ohio State Univ, PhD(entom), 70. Prof Exp: ASST PROF ENTOM, PA STATE UNIV, 70- Mem: Entom Soc Am; Soc Invert Path; Am Inst Biol Sci; Int Orgn Biol Control. Res: Development of new techniques for the protection of apple orchards from insect pests, with emphasis on reduction of insecticide usage and utilization of biological control agents. Mailing Add: Fruit Res Lab Pa State Univ Biglerville PA 17307

BODEL, JOHN KNOX, b Salt Lake City, Utah, Aug 6, 07; m 31, 55; c 4. PHYSICAL ANTHROPOLOGY. Educ: Wesleyan Univ, AB, 29; Harvard Univ, AM, 40, PhD(phys anthrop), 51. Prof Exp: Master sci, Hotchkiss Sch, 29, dir studies, 46-61, chmn dept, 54-72; RETIRED. Concurrent Pos: Asst prof, Univ Hawaii, 63-64. Mailing Add: Lakeville CT 06039

BODEL, PHYLLIS T, b Hartford, Conn, Feb 24, 34; m; c 3. INTERNAL MEDICINE, INFECTIOUS DISEASES. Educ: Radcliffe Col, AB, 54; Harvard Med Sch, MD, 58. Prof Exp: Intern, Boston City Hosp, 58-59; asst residency, New York Hosp, 59-60; fel, McGill Univ, 60-61; from res assoc to sr res assoc med, 61-72, ASSOC PROF MED, SCH MED, YALE UNIV, 72- Mem: Am Fedn Clin Res; Am Soc Clin Invest; Infectious Dis Soc Am; AAAS; Sigma Xi. Res: Fever and the role of leucocytes in host defense. Mailing Add: Dept of Med Yale Univ Sch of Med New Haven CT 06510

BODEMER, CHARLES WILLIAM, b Denison, Iowa, Jan 4, 27; m 48; c 3. EMBRYOLOGY, HISTORY OF MEDICINE. Educ: Pomona Col, BA, 51; Claremont Col, MA, 52; Cornell Univ, PhD(zool), 56. Prof Exp: Asst zool, Cornell Univ, 53-56; from instr to assoc prof biol struct, 56-68, from asst dean to assoc dean, 63-68, head div biomed hist, 65-68, PROF BIOMED HIST & CHMN DEPT, SCH MED, UNIV WASH, 68- Mem: AAAS; Am Soc Zoologists; Int Soc Hist Med; Am Asn Hist Med; Am Hist Asn. Res: History of embryology and medicine; behavioral sciences and biology; experimental embryology; amphibian limb and nerve regeneration; developmental neurophysiology. Mailing Add: Dept of Biomed Hist Univ of Wash Sch of Med Seattle WA 98105

BODEN, BRIAN PETER, b Capetown, SAfrica, Apr 15, 21; div; c 2. MARINE BIOLOGY. Educ: Univ Capetown, BSc, 45; Univ Calif, PhD(oceanog), 50. Prof Exp: Browne res fel, The Lab, Plymouth, Eng, 50-51; oceanographer, Bermuda Biol Sta, 51-54; res fel, Scripps Inst Oceanog, Univ Calif, 54-57, asst res biologist, 57-67, assoc res biologist, 67-72. Mem: AAAS; Soc Syst Zool; Am Micros Soc; Marine Biol Asn Gt Brit; Brit Soc Vis Scientists. Res: Plankton ecology; euphausiacea. Mailing Add: 1339 Justin Rd Cardiff CA 92007

BODEN, HERBERT, b Staten Island, NY, Apr 5, 32; m 57. ORGANIC CHEMISTRY. Educ: Univ Vt, BS, 54, MS, 56; Ohio State Univ, PhD(org chem), 60. Prof Exp: Instr gen & org chem, Ohio State Univ, 56-58; res chemist, Org Chem Dept, 60-65, res chemist, Freon Prod Div, 65-70, tech assoc, 70, territory mgr, 70-72, prod & develop mgr, 72-74, NAT ACCT MGR, FREON PROD DIV, E I DU PONT DE NEMOURS & CO, INC, 74- Mem: Am Chem Soc. Res: Biguanide, benzo phenanthrene, fluorine and polymer chemistry; chemistry of tetra substituted amino ethylenes. Mailing Add: Freon Prod Div E I du Pont de Nemours & Co Inc Wilmington DE 19898

BODEN, ROBERT HECTOR, physics, chemistry, see 12th edition

BODENHEIMER, PETER HERMAN, b Seattle, Washington, June 29, 37; m 65; c 2. ASTROPHYSICS. Educ: Harvard Univ, AB, 59; Univ Calif, Berkeley, PhD(astron), 65. Prof Exp: NSF fel, Princeton Univ, 65-66, res assoc astrophys, 66-67; asst astronr & asst prof, 67-70, ASSOC ASTRONR & ASSOC PROF ASTRON, UNIV CALIF, SANTA CRUZ, 70- Mem: Am Astron Soc; Int Astron Union. Res: Theoretical calculations of stellar structure and evolution, including problems involving hydrodynamics and rotation. Mailing Add: Lick Observ Univ of Calif Santa Cruz CA 95064

BODENLOS, ALFRED JOHN, b New York, NY, June 4, 18; m 47. GEOLOGY. Educ: Columbia Univ, BA, 40, MA, 42, PhD(geol), 48. Prof Exp: Geologist, 42-45, geologist, Foreign Geol Br, Brazil, 45-46, Peru, 47-50, US, 50-53, Mex, 54-56 & Brazil, 57-67, chief field party, US Agency Int Develop, 61-66, SULFUR COMMODITY GEOLOGIST, US GEOL SURV, 67- Mem: Fel Geol Soc Am; Soc Econ Geologists; Peruvian Geol Soc; Geol Soc Brazil; Mex Asn Petrol Geologists. Res: Economic geology; nonmetallic and base-metals deposits of Peru and Brazil; regional geology of Mexico. Mailing Add: US Geol Surv Nat Ctr Reston VA 22091

BODENSTEIN, DIETRICH H F A, b Corwingen, Ger, Feb 1, 08; nat US; m 33; c 1. ZOOLOGY. Educ: Univ Freiburg, PhD(zool), 54. Prof Exp: Asst exp morphol, Kaiser Wilhelm Inst, Berlin, 28-33; res assoc, Inst Marine Biol, Italy, 33-34; res assoc biol, Stanford Univ, 34-41; Guggenheim fel, Columbia Univ, 41-43; asst entom, Conn Exp

Sta, 44; insect physiologist, Med Labs, Army Chem Ctr, 45-58; embryologist, Geront Br, Nat Heart Inst, 58-60; chmn dept, 60-73, PROF BIOL, UNIV VA, 60- Mem: Nat Acad Sci; Am Acad Arts & Sci; Am Soc Naturalists; Am Soc Zool; Genetics Soc Am. Res: Experimental morphology of amphibians; developmental physiology of amphibians and insects; endocrinology of insects; developmental genetics of Drosophila. Mailing Add: Dept of Biol Univ of Va Charlottesville VA 22903

BODEY, GERALD PAUL, SR, b Hazelton, Pa, May 22, 34; m 56; c 3. ONCOLOGY, INFECTIOUS DISEASES. Educ: Lafayette Col, AB, 56; Johns Hopkins Univ, MD, 60; Am Bd Internal Med, dipl, 72, cert oncol, 73 & cert infectious dis, 74. Prof Exp: Intern, Osler Med Serv, Johns Hopkins Hosp, 60-61, resident, 61-62; clin assoc, Nat Cancer Inst, 62-65; resident, Univ Wash, 65-66; asst internist & asst prof, 66-72, assoc internist & assoc prof med, 72-75, CHIEF SECT INFECTIOUS DIS, UNIV TEX M D ANDERSON HOSP & TUMOR INST, 72-, INTERNIST, PROF MED & CHIEF CHEMOTHER, 75- Concurrent Pos: From adj asst prof & clin asst prof to adj assoc prof & assoc clin prof, Baylor Col Med, 69-75, adj prof microbiol, immunol & med, 75-; Leukemia Soc scholar, 69-74; assoc prof, Univ Tex Grad Sch Biomed Sci, Houston; consult prev med div, Space Flight Biotechnol Specialty Team, Manned Spacecraft Ctr; consult, Brooke Gen Hosp, Ft Sam Houston, Tex, 71-; assoc prof med & assoc prof pharmacol, Univ Tex Med Sch, Houston, 74- Honors & Awards: Am Chem Soc, Am Inst Chemists, Merck Chem & Robert B Youngman Greek Prizes, 56. Mem: AAAS; AMA; Am Asn Cancer Res; Am Fedn Clin Res; Am Soc Clin Oncol. Res: Cancer research; hematology; infectious diseases. Mailing Add: Dept of Develop Therapeut M D Anderson Hosp & Tumor Inst Houston TX 77025

BODFISH, RALPH E, b Denver, Colo, Jan 24, 22; m 52; c 2. MEDICINE, NUCLEAR PHYSICS. Educ: Duke Univ, BA, 49; Emory Univ, MD, 54; Am Bd Internal Med, dipl, 61. Prof Exp: Chief radioisotope serv, Vet Admin Hosp, Long Beach, 56-70, dep dir res, 63-70, dep chief of staff, 68-70; head dept med, Rancho Los Amigos Hosp, 70-74; CHIEF OF STAFF, VET ADMIN HOSP, LONG BEACH, 74-; ASST DEAN, UNIV CALIF, IRVINE-CALIF COL MED, 74- Concurrent Pos: Asst clin prof, Sch Med, Univ Southern Calif, 60-70; assoc clin prof, Univ Calif, Irvine-Calif Col Med, 67-70, clin prof radiol sci, 74- Consult, Community Hosp, Long Beach, 66-74, Scripps Clin & Inst, 66-74 & Vet Admin Hosp, Long Beach, 70-74. Mem: Sr mem Am Fedn Clin Res; Soc Nuclear Med; AMA; fel Am Col Physicians. Res: Radiation physics; control of thyroid function; renal disease; relationship of metabolic state to cardiopulmonary and bone marrow function. Mailing Add: Chief of Staff Vet Admin Hosp 5901 E Seventh St Long Beach CA 90801

BODI, LEWIS JOSEPH, b Racine, Wis, Dec 31, 24; m 50; c 4. PHYSICAL CHEMISTRY. Educ: DePauw Univ, BS, 50; Univ Wis, PhD(chem), 55. Prof Exp: Asst prof chem, Brooklyn Col, 54-60; eng specialist, Gen Tel & Electronics Labs, 60-67; dean sci & math, 67-70, DEAN FAC, YORK COL, CITY UNIV NEW YORK, 70-, DEAN ACAD AFFAIRS, 74-, PROF CHEM, 67- Mem: Am Chem Soc. Res: Luminescence; solid state chemistry and physics; science education. Mailing Add: Off of Dean of Acad Affairs York Col City Univ of New York Jamaica NY 11432

BODIAN, DAVID, b St Louis, Mo, May 15, 10; m 44; c 5. NEUROBIOLOGY. Educ: Univ Chicago, BA, 31, PhD(anat), 34, MD, 37. Prof Exp: Asst anat, Univ Chicago, 34-35, sr asst, 35-38; Nat Res fel, Univ Mich, 38; res fel, Johns Hopkins Univ, 39-40; asst prof anat, Western Reserve Univ, 40-42; assoc epidemiol, 42-45, from asst prof to prof anat & dir dept, 57-75, PROF NEUROBIOL, DEPT OTOLARYNGOL, SCH MED, JOHNS HOPKINS UNIV, 74- Concurrent Pos: Lectr, Johns Hopkins Univ, 41; Harvey lectr, 56; consult to Surgeon Gen, USPHS, Polio Hall of Fame & Warm Springs Found, 58; mem tech comt poliomyelitis vaccine, USPHS, 57-64; mem bd sci counrs, Div Biol Standards, NIH, 57-59 & Nat Inst Neurol Dis & Stroke, 68-74. Honors & Awards: E Mead Johnson Award, 41; USPHS Award, 56. Mem: Nat Acad Sci; Am Acad Arts & Sci; AAAS; Am Soc Cell Biologists; Am Asn Anatomists. Res: Histology of nervous system; pathology and pathogenesis of poliomyelitis; development of nervous system; neurohypophysis; fine structure of synapses. Mailing Add: Dept of Otolaryngol Traylor Bldg Johns Hopkins Univ Sch of Med Baltimore MD 21205

BODIG, JOZSEF, b Gonc, Hungary, Jan 20, 34; US citizen; m 62; c 3. WOOD SCIENCE, WOOD TECHNOLOGY. Educ: Univ BC, BSF, 59; Univ Wash, MF, 61, PhD(wood technol), 63. Prof Exp: Proj leader, Fiber Res, Inc, Wash, 60-61; asst mech properties for wood, Univ Wash, 61-63; asst prof & asst forester, Agr Exp Sta, 63-66, ASSOC PROF WOOD SCI & TECHNOL, COLO STATE UNIV, 66- Concurrent Pos: Vis prof wood physics, Univ Wash, 69-70; mem tech adv comt, Am Inst Timber Construct, 70- Mem: Forest Prod Res Soc; Soc Wood Sci & Tech; Am Soc Testing & Mat. Res: Mechanical and rheological properties of wood and wood composites; nondestructive testing of wood; wood engineering. Mailing Add: Dept of Forest & Wood Sci Col of Forestry & Natural Resources Colo State Univ Ft Collins CO 80521

BODILY, DAVID MARTIN, b Logan, Utah, Dec 16, 33; m 58; c 4. PHYSICAL CHEMISTRY. Educ: Brigham Young Univ, BA, 59, MA, 60; Cornell Univ, PhD(phys chem), 64. Prof Exp: Res assoc, Northwestern Univ, 64-65; asst prof chem, Univ Ariz, 65-67; asst prof, 67-69, ASSOC PROF FUELS ENG IN MINING, METALL & FUELS ENG, UNIV UTAH, 69- Mem: Am Chem Soc. Res: Physical chemistry of high polymers; radiation chemistry; coal science. Mailing Add: Dept of Mining Metal & Fuels Eng Univ of Utah Salt Lake City UT 84112

BODILY, HOWARD LYNN, bacteriology, see 12th edition

BODIN, JEROME IRWIN, b New York, NY, July 2, 30; m 58; c 2. PHARMACEUTICAL CHEMISTRY. Educ: Columbia Univ, BS, 52, MS, 54; Univ Wis, PhD(pharmaceut chem), 58. Prof Exp: Res analyst pharmaceut chem, Chas Pfizer & Co, Inc, 58-61; dir pharmaceut anal, Drug Standards Lab, Am Pharmaceut Asn Found, 62-63; chemist, Div of New Drugs, US Food & Drug Admin, 63-64; dir qual control, Wallace Labs Div, Carter Prod Inc, 64-69, DIR ANAL RES & DEVELOP, CARTER-WALLACE, INC, 69- Concurrent Pos: Mem, Am Found Pharmaceut Educ. Mem: Am Chem Soc; Am Pharmaceut Asn; Asn Off Anal Chemists; Acad Pharmaceut Sci. Res: Chemical kinetics and drug stability; drug availability and absorption; pharmaceutical quality control; federal drug regulation. Mailing Add: Carter-Wallace Inc Half Acre Rd Cranbury NJ 08512

BODINE, CHARLES DAVID, b Joliet, Ill, May 12, 20; m 44; c 3. OBSTETRICS & GYNECOLOGY. Educ: Univ Okla, BA, 40, MD, 43. Prof Exp: Fel gynecol path, Sch Med, Univ Md, 48-49; ASSOC PROF OBSTET, UNIV OKLA, 50- Concurrent Pos: Mem Am Bd Obstet & Gynec. Mem: Am Col Obstetricians & Gynecologists. Res: Gynecological surgery. Mailing Add: Medical Tower 3141 NW Expressway Oklahoma City OK 73112

BODINE, JOHN JAMES, b Oklahoma City, Okla, Sept 15, 34. ANTHROPOLOGY. Educ: Univ Okla, BA, 56; Tulane Univ, MA, 61, PhD(anthrop), 67. Prof Exp: Assoc cur ed, Milwaukee Pub Mus, 61-63; asst prof anthrop, Marquette Univ, 63-67, chmn dept sociol & anthrop, 67-68; assoc prof, 68-72, PROF ANTHROP, AM UNIV, 74-, CHMN DEPT, 72- Concurrent Pos: Fac res grant, Marquette Univ, 66; NIMH grant, 70. Mem: Fel Am Anthrop Asn; Ethnol Soc; Am Asn Mus. Res: Cultural anthropology of the Southwestern United States and Middle America; culture change and acculturation. Mailing Add: Dept of Anthrop Am Univ Washington DC 20016

BODINE, MARC WILLIAMS, JR, b Sayre, Pa, Aug 7, 29; m 53; c 2. GEOCHEMISTRY Educ: Princeton Univ, AB, 51; Columbia Univ, MA, 53, PhD(geol), 56. Prof Exp: Instr geol, Polytech Inst Brooklyn, 55-56 & Union Col, NY, 56; asst prof geol, Bowdoin Col, 56-62; lectr, Princeton Univ, 62-63; from assoc prof to prof geol, State Univ NY Binghamton, 63-75, chmn dept, 65-69; PROF GEOL & CHMN DEPT GEOSCI, N MEX INST MINING & TECHNOL, 75-; GEOLOGIST, US GEOL SURV, 57- Concurrent Pos: NSF fac fel, Princeton Univ, 61-62; vis prof, Inst Mineral & Crystallog, Univ Vienna, 69-70. Mem: Mineral Soc Am; Geol Soc Am; Geochem Soc; AAAS; Am Geophys Union. Res: Low-temperature aqueous geochemistry; mineral-solution equilibria in natural waters; silicate mineralogy and geochemistry in marine evaporites; geochemistry and mineralogy of shales; environmental geochemistry of fresh water resources. Mailing Add: Geosci Dept NMex Inst of Mining & Technol Socorro NM 87801

BODLAENDER, PETER, organic chemistry, biochemistry, see 12th edition

BODLEY, HERBERT DANIEL, II, b Cleveland, Ohio, Nov 6, 39; m 61; c 2. ANATOMY. Educ: Hamline Univ, BS, 61; Univ Minn, MS, 68, PhD(anat), 70. Prof Exp: From instr to asst prof anat, Univ Ariz, 70-75; ASSOC PROF ANAT, CHICAGO COL OSTEOP MED, 75- Mem: Am Asn Anatomists; Electron Micros Soc Am; Am Soc Zoologists. Res: Ultrastructure. Mailing Add: Dept of Anat Chicago Col of Osteop Med Chicago IL 60615

BODLEY, JAMES WILLIAM, b Portland, Ore, Oct 5, 37; m 60; c 3. BIOCHEMISTRY. Educ: Walla Walla Col, BS, 60; Univ Hawaii, PhD(biochem), 64. Prof Exp: Prod control chemist, Dole Corp, 60; fel biochem, Sch Med, Univ, Wash, 64-66, actg asst prof, 66; asst prof, 67-71, ASSOC PROF BIOCHEM, MED SCH, UNIV MINN, MINNEAPOLIS, 71- Concurrent Pos: Am Cancer Soc fel, 65-66 & sr fel, 73-74. Mem: Am Soc Biol Chemists. Res: Mechanism and control of biological reactions involving nucleic acids. Mailing Add: Dept of Biochem Univ of Minn Med Sch Minneapolis MN 55455

BODMER, ARNOLD R, b Frankfurt, Ger, May 23, 29; m 56; c 4. PHYSICS. Educ: Univ Manchester, BSc, 49, PhD(physics), 53. Prof Exp: Sci officer, Royal Armament Res & Design Estab, Eng, 53-56; res fel theoret physics, Univ Manchester, 56-57; res fel, Theoret Nuclear Physics, Europ Orgn Nuclear Res, Switz, 57-58; asst lectr theoret physics, Univ Manchester, 58-59, lectr, 60-65; PROF PHYSICS, UNIV ILL, CHICAGO CIRCLE, 65-; SR PHYSICIST, ARGONNE NAT LAB, 72- Resident res assoc, Argonne Nat Lab, 63-65, assoc physicist, 65-72; vis prof, Nuclear Physics Lab, Oxford Univ, 70-71. Mem: Am Phys Soc; Brit Inst Physics & Phys Soc. Res: Theoretical nuclear physics; nuclear structure and many-body problems; hypernuclei and baryon-baryon interactions; Heay-ion collisions. Mailing Add: 239 55th Pl Downers Grove IL 60515

BODNER, STEPHEN E, b Rochester, NY, June 20, 39. THEORETICAL PHYSICS. Educ: Univ Rochester, BS, 61; Princeton Univ, MA, 62, PhD(physics), 65. Prof Exp: Physicist, Lawrence Radiation Lab, Univ Calif, 64-74; BR HEAD, NAVAL RES LAB, 74- Concurrent Pos: Instr appl sci, Univ Calif, Davis-Livermore, 66-68. Mem: Am Phys Soc. Res: Plasma physics; turbulence; laser fusion controlled thermonuclear reaction research. Mailing Add: Naval Res Lab Washington DC 20375

BODOLA, ANTHONY, b New York, NY, Oct 1, 14; m 36; c 1. HYDROBIOLOGY. Educ: Fairmont Col, AB, 42; WVa Univ, MS, 45; Ohio State Univ, PhD(hydrobiol), 55. Prof Exp: Teacher pub schs, WVa, 42-44 & Ohio, 45-51; sr conserv fel, Ohio Dept Natural Resources & Fisheries, US Fish & Wildlife Serv, 53-55; asst prof biol, Shepherd Col, 55-58, assoc prof, 58-64; asst unit leader, Pa Coop Fishery Unit, 64-67; mosquito-wildlife coordr, Bur Sport Fisheries & Wildlife, 67-70, PROF BIOL & NATURAL RESOURCES, DEL STATE COL, 69- Concurrent Pos: Res asst, Stone Inst Hydrobiol, 46-52; biologist, Cancer Res Unit, Vet Admin Ctr, 57; fisheries biologist, WVa Dept Conserv, 57-58; attend, NSF Genetics Inst, 59. Mem: AAAS; Am Fisheries Soc; Am Soc Limnol & Oceanog; Ecol Soc Am; Am Inst Biol Sci. Res: Life histories of fresh-water fishes. Mailing Add: Dept of Agr & Natural Resources Del State Col Dover DE 19901

BODOR, NICOLAE STEFAN, b Satu Mare, Romania, Feb 1, 39; m 71; c 3. MEDICINAL CHEMISTRY. Educ: Univ Cluj, BSc, 59, DrChem(org chem), 65. Prof Exp: Res chemist, I Septembrie Works, Romania, 59-61; sr res sci group leader org-pharm chem, Chem-Pharm Res Inst, Cluj, 61-70; ASSOC DIR MED-PHARM CHEM, INTERx RES CORP, 72- Concurrent Pos: R A Welch fel, Univ Tex, Austin, 68-69 & 70-72; adj prof med chem, Sch Pharm, Univ Kans, 73-, consult, Med Ctr, 75- Mem: Am Chem Soc. Res: Synthesis of new drug derivatives; delivery of drugs through biological membranes; pro-drugs; drug activity and metabolism; quantum organic chemistry; new synthetic methods; photoelectron spectrometry; cancer chemotherapy. Mailing Add: INTERx Res Corp Lawrence KS 66044

BODRE, ROBERT JOSEPH, b Philadelphia, Pa, Mar 6, 21; m 45. ANALYTICAL CHEMISTRY. Educ: St Josephs Col, Pa, BS, 43; Univ Louisville, PhD(chem), 61. Prof Exp: Chemist, Publicker Industs, Pa, 43-44; res chemist, Whitaker Wool Co, 44-47; res chemist, Girdler Div, Chemetron Corp, Ky, 47-55, group leader, 55-57; sr res chemist, 60-67, PROCESS SPECIALIST, MONSANTO CO, 67- Mem: Am Chem Soc. Res: Development of analytical methods for industrial processes; monomer syntheses and polymerization reactions; catalyst development and applications; production of ethylbenzene and styrene. Mailing Add: 2501 Meadow Lane La Marque TX 77568

BODSON, HERMAN, b Brussels, Belg, Dec 21, 12; US citizen; m 48; c 3. PHYSICAL CHEMISTRY, PHYSICS. Educ: Univ Brussels, MS, 35, PhD(phys chem), 37. Prof Exp: ASSOC PROF PHYSICS, LAKE ERIE COL, 63- Mem: Am Phys Soc; Mineral Soc Am; Geol Soc Am. Res: Mineralogy. Mailing Add: Dept of Phys Sci Lake Erie Col Painesville OH 44077

BODVARSSON, GUNNAR, b Aug 8, 16. APPLIED MATHEMATICS, GEOPHYSICS. Educ: Tech Univ, Berlin, Dipl Eng, 43; Calif Inst Technol, PhD, 57. Prof Exp: Head geothermal res dept, State Elec Auth, Reykjavik, Iceland, 46-64; vis prof math & oceanog, 64-66, PROF MATH & OCEANOG, ORE STATE UNIV, 66- Concurrent Pos: UN tech expert, Mex, BWI, Cent Am & Chile, 51, 54, 63, & 65-75. Res: Geothermal energy; theoretical geophysics; applied mathematics. Mailing Add: Sch of Oceanog Ore State Univ Corvallis OR 97331

BODWELL, CLARENCE EUGENE, b Woodward, Okla, July 30, 35. NUTRITION. Educ: Okla State Univ, BS, 57; King's Col, Cambridge Univ, MSc, 59; Mich State Univ, PhD(food sci), 64. Prof Exp: NIH staff fel, Lab Biophys Chem, Nat Inst Arthritis & Metab Dis, 64-65, USPHS res assoc, 65-67; res chemist, Human Nutrit Res Div, 67-72, RES CHEMIST NUTRIT INST, PROTEIN NUTRIT LAB, USDA,

72- Mem: AAAS; Am Chem Soc; Inst Food Technologists; Am Soc Testing & Mat. Res: Evaluation of protein nutritive value in humans; human amino acid and protein requirements; chemical and biochemical methods for estimating protein nutritional value and amino acid bioavailability. Mailing Add: Nutrit Inst Rm 313 Bldg 308 Agr Res Serv USDA Beltsville MD 20705

BOE, ARTHUR AMOS, b Westfield Twp, Minn, Apr 27, 33; m 65; c 1. HORTICULTURE. Educ: Utah State Univ, BS, 62, PhD(plant sci), 66. Prof Exp: Teaching asst hort, Utah State Univ, 62-65, agr specialist proj, Rural Indust Tech Assistance, Brazil, 66-67; asst prof plant sci, 67-72, ASSOC PROF PLANT SCI & ASSOC PLANT PHYSIOLOGIST, UNIV IDAHO, 72- Mem: AAAS; Am Soc Hort Sci; Am Inst Biol Sci. Res: Fruit development and maturation as affected by plant hormones, light, temperature and magnetic fields; dormancy of woody plants and seeds. Mailing Add: Dept of Plant Sci Univ of Idaho Moscow ID 83843

BOECKER, BRUCE BERNARD, b Aurora, Ill, July 9, 32; m 60; c 2. RADIOBIOLOGY. Educ: Grinnell Col, BA, 54; Univ Rochester, MS, 60, PhD(radiobiol), 62. Prof Exp: Radiobiologist, 61-64, asst head dept radiobiol, 64-75, ASST DIR INHALATION TOXICOL RES INST, LOVELACE FOUND MED EDUC & RES, 75- Concurrent Pos: Radiation biologist div biol & med, USAEC, 70-71. Mem: Health Physics Soc; Radiation Res Soc; AAAS. Res: Deposition and retention of inhaled materials; inhalation toxicology; effects of inhaled materials on the lung. Mailing Add: Inhalation Toxicol Res Inst Lovelace Found PO Box 5890 Albuquerque NM 87115

BOEDEKER, EDWARD RICHARD, chemistry, see 12th edition

BOEDEKER, RICHARD ROY, b St Louis, Mo, Apr 8, 33; m 61; c 3. PHYSICS. Educ: St Louis Univ, BS, 55, MS, 57, PhD(physics), 59. Prof Exp: Res assoc physics, St Louis Univ, 59-61; res scientist, McDonnell Aircraft Corp, 61-62; from asst prof to assoc prof, 62-74, PROF PHYSICS, SOUTHERN ILL UNIV, EDWARDSVILLE, 74- Mem: Am Phys Soc. Res: Neutron scattering in solids and liquids; temperature determination in gases by neutron scattering; lasers; nonlinear optics; scattering theory. Mailing Add: Dept Physics Southern Ill Univ Edwardsville Campus Edwardsville IL 62025

BOEDEKER, ROY VINCENT, b St Louis, Mo, Apr 22, 09; m 41; c 6. OBSTETRICS & GYNECOLOGY. Educ: St Louis Univ, AB, 31, MD, 35, MOG, 39. Prof Exp: Asst prof obstet & gynec, 45-70, assoc prof clin obstet & gynec, 70-74, EMER ASSOC PROF OBSTET & GYNEC, SCH MED, ST LOUIS UNIV, 74- Mem: AMA; Am Col Surgeons; Int Col Surgeons. Res: Flicker photometry as preclinical index of toxemia pregnancy; induction of labor. Mailing Add: 1035 Bellevue Ave St Louis MO 63108

BOEDTKER, OLAF A, b Colombo, Ceylon, Feb 10, 24; US citizen; m 52; c 4. SOLID STATE PHYSICS. Educ: Swiss Fed Inst Technol, BS, 49; Calif Inst Technol, MS, 58, PhD(eng physics), 61. Prof Exp: Res fel phys sci, Calif Inst Technol, 60-62; res scientist metals physics, Lawrence Radiation Lab, Univ Calif, 62-63; ASSOC PROF PHYSICS, ORE STATE UNIV, 63- Mem: Am Phys Soc; Am Soc Metals. Mailing Add: Col of Sci Ore State Univ Corvallis OR 97331

BOEHLE, JOHN, JR, b New York, NY, July 5, 32; m 55; c 2. AGRONOMY, WEED SCIENCE. Educ: Univ NH, BS, 54; Pa State Univ, MS, 61, PhD(agron), 64. Prof Exp: Agronomist, Chevron Chem Co, 62-64; mgr agron serv, Kerr McGee Chem Co, 64-68, dir field res & res farms, Agr Div, 68-75, DIR BIOL RES, CIBA-GEIGY, 75- Mem: Am Soc Agron; Crop Sci Soc Am; Soil Sci Soc Am; Am Soc Hort Sci. Res: Weed control; plant protectants; micronutrient nutrition of plants; growth regulators for plants. Mailing Add: 4110 Redwine Dr Greensboro NC 27410

BOEHM, FELIX H, b Basel, Switz, June 9, 24; nat US; m 56; c 2. NUCLEAR PHYSICS, ATOMIC PHYSICS. Educ: Swiss Fed Inst Technol, Dipl physics, 48, Dr nat sci(physics), 51. Prof Exp: Res asst, Swiss Fed Inst Technol, 48-51; Boese fel, Columbia Univ, 52-53; res fel physics, Calif Inst Technol, 53-57, asst prof, 58; vis prof, Univ Heidelberg, 58-59; assoc prof, 59-61, PROF PHYSICS, CALIF INST TECHNOL, 61- Concurrent Pos: Sloan fel, 61-63; NSF sr fel, Univ Copenhagen, 65-66 & Europ Orgn Nuclear Res, Geneva, Switz, 71-72. Mem: Fel Am Phys Soc. Res: Nuclear structure; weak interaction; x-rays; meson physics. Mailing Add: 161-33 Calif Inst of Technol Pasadena CA 91125

BOEHM, JOHN JOSEPH, b Libertyville, Ill, May 27, 29; m 59; c 4. PEDIATRICS. Educ: Univ Notre Dame, BS, 51; Northwestern Univ, MD, 55. Prof Exp: Intern, St Luke's Hosp, Chicago, 55-56; resident pediat, Children's Mem Hosp, Chicago, 56-58; res fel newborn & premature pediat, Med Ctr, Univ Colo, 60-62; from instr to asst prof pediat, Med Sch, Univ Ky, 62-65; asst prof & vchmn dept, Univ Calif, Irvine, 65-66; asst prof, 66-69, asst dean med sch, 70-72, ASSOC PROF PEDIAT, MED SCH, NORTHWESTERN UNIV, CHICAGO, 69-, CHIEF PEDIAT, NORTHWESTERN MEM HOSP, 72- Mem: Am Acad Pediat; Am Soc Clin Invest. Res: Neonatology. Mailing Add: Northwestern Mem Hosp Superior St & Fairbanks Ct Chicago IL 60611

BOEHM, ROBERT LOUIS, b Buffalo, NY, Nov 17, 25; m 51; c 6. PAPER CHEMISTRY, PULP CHEMISTRY. Educ: Colgate Univ, AB, 48; Lawrence Col, MS, 50, PhD(chem), 53. Prof Exp: Res chemist, Mead Corp, 53-57, tech asst to mgr, Mass, 57, asst tech dir, Chillicothe Div, 57-60, mgr, Leominster Div, 60-67, bus planner, Gilbert Paper Div, 67-69; STAFF MGR PAPER & PAPERBOARD OPERS, AM CAN CO, GREENWICH, 69- Mem: Am Chem Soc; Tech Asn Pulp & Paper Indust. Res: Introduction of chlorine atoms into cellulose; research in pulping and bleaching of wood pulps; paper machines. Mailing Add: 6 Abbey Rd Darien CT 06820

BOEHME, DIETHELM HARTMUT, b Templin, Ger, Sept 23, 26; m 60; c 2. PATHOLOGY, NEUROPATHOLOGY. Educ: Univ Heidelberg, MD, 53. Prof Exp: Intern med, Behring Hosp, Berlin, Ger, 53-54; intern, Monmouth Med Ctr, Long Branch, NJ, 54-55; asst resident path, Victoria Hosp, Berlin, Ger, 55-56; res assoc exp path, Rockefeller Inst, 56-59; resident pathologist, New York Hosp, 59-61; pvt docent path, Univ Münster, 61-63; pvt docent & head physician, Free Univ Berlin, 63-64; asst prof & clin dir path, State Univ NY Buffalo, 64-66; assoc prof, Jefferson Med Col, 66-70; PROF PATH, COL MED & DENT NJ, NEWARK, 70-; CHIEF LAB SERV, VET ADMIN HOSP, EAST ORANGE, 70- Concurrent Pos: Chief lab serv, Vet Admin Hosp, Coatsville, Pa, 66-70. Mem: Am Soc Exp Path; Harvey Soc; Sigma Xi. Res: Nonspecific resistance against infection; allergic encephalomyelitis; autoimmune diseases; multiple sclerosis. Mailing Add: Vet Admin Hosp East Orange NJ 07019

BOEHME, HOLLIS CLYDE, b Liberty, Okla, Nov 28, 33; m 54; c 3. PHYSICS. Educ: Tex A&M Univ, BA, 60, MS, 61, PhD(physics), 67. Prof Exp: Engr, Boeing Airplane Co, 60; res asst, Tex A&M Univ, 60-61; engr, Tex Instruments Inc, 61-62; instr physics, 62-67; engr scientist, Tracor, Inc, 67-68, sr scientist, 68-70; PRES, SYNERGY, INC, 70- Concurrent Pos: Consult, Tex Instruments Inc, 62-63. Mem: Am Phys Soc. Res: Nuclear magnetic resonance; acoustics. Mailing Add: 320 King Arthur Court Austin TX 78746

BOEHME, WERNER RICHARD, b Englewood, NJ, Jan 15, 20; m 46; c 2. ORGANIC CHEMISTRY. Educ: Polytech Inst Brooklyn, BS, 42; Univ Md, PhD(chem), 48; Univ Chicago, MBA, 71. Prof Exp: Asst chemist, Gen Dyestuff Corp, 37-41; res chemist, Winthrop Chem Co, 42-44; res chemist, Naval Med Res Inst, 45-46; asst, Univ Md, 46; res chemist, Nat Drug Co, 48-52; sr chemist, Ethicon, Inc, 52-55, head med chem sect, 56-57, mgr dept org chem, 58-61; dir cent res, Shulton, Inc, 61-62; gen mgr & pres, A M Arnold Labs, Inc, 62-67; mgr chem opers & dir res develop, Dawe's Labs, Inc, 69-73; TECH DIR, FATS & PROTEINS RES FOUND, INC, 73- Mem: Am Chem Soc; NY Acad Sci; fel Am Inst Chem; Inst Food Technologists; Am Oil Chem Soc. Res: Pharmaceuticals and nutrition; relation of chemical structure to pharmacodynamic activity; animal by-products utilization; pollution control; triglycerides. Mailing Add: 2 S 403 Oaklawn Dr Glen Ellyn IL 60137

BOEHMER, HOWARD WILLIAM, b Duluth, Minn, Aug 21, 20; m 44; c 1. PHYSICS. Educ: Mont State Col, BS, 42; Cornell Univ, PhD(physics), 46. Prof Exp: Asst Off Sci Res & Develop contract, Cornell Univ, 43-45; res physicist, Strombert Carlson Co, NY, 46-47; from asst prof to assoc prof physics, Univ Colo, 47-53; staff mem, Mass Inst Technol, 51-55; chief engr, Hycon East, Inc, 55-58; from assoc mgr to mgr guidance & controls, 58-70, assoc mgr equip eng divs, 70-74, ASST GROUP EXEC, ELECTRO-OPTICAL & DATA SYSTS GROUP, HUGHES AIRCRAFT CO, 74- Mem: Am Phys Soc; Am Asn Physics Teachers; Inst Elec & Electronics Eng; Am Inst Navig; Am Ord Asn. Res: Electronics; cosmic rays; computers; data processing systems; guidance and control systems. Mailing Add: Electro-Optical Data Systs Group Hughes Aircraft Co Culver City CA 90230

BOEHMS, CHARLES NELSON, b Nashville, Tenn, Feb 1, 31; m 52; c 2. PHYSIOLOGICAL ECOLOGY, FRESHWATER BIOLOGY. Educ: George Peabody Col, BS, 56, MA, 57; Univ NC, Chapel Hill, PhD(zool), 71. Prof Exp: From instr to asst prof biol, Austin Peay State Univ, 57-61; instr zool, Univ NC, Chapel Hill, 62-64; assoc prof biol, 65-71, dean students, 68-71, PROF BIOL & VPRES STUDENT AFFAIRS, AUSTIN PEAY STATE UNIV, 71- Concurrent Pos: NSF fac fel, 64. Mem: Am Zool Soc; Soc Int Odonatologica. Res: Influence of temperature and photoperiod on seasonal regulation in fresh water invertebrates. Mailing Add: Dept of Biol Austin Peay State Univ Clarksville TN 37040

BOEHNE, JOHN WILLIAM, b Evansville, Ind, Feb 6, 21; m 46; c 2. NUTRITION. Educ: Ind Univ, BS, 42. Prof Exp: Chemist, Lederle Labs Div, Am Cyanamid Co, NY, 46-48; asst, Western Reserve Univ, 48-57; biochemist div nutrit, 57-59, asst to dir, 59-65, chief spec dietary foods br, 66-70, asst dir, 71-74, ASST ASSOC DIR NUTRIT & CONSUMER SCI, BUR FOODS, FOOD & DRUG ADMIN, 74- Honors & Awards: Award of Merit, Food & Drug Admin, 71 & 73. Mem: Am Inst Nutrit; Inst Food Technol; Animal Nutrit Res Coun; Asn Food & Drug Officials. Res: Bioanalytical nutrition; mineral metabolism; nutrient value of foods; food additives. Mailing Add: 10763 Kinloch Rd Silver Spring MD 20903

BOEHNING, ROCHELLE LLOYD, b Diamond, Mo, Nov 12, 32; m 52; c 3. GEOMETRY. Educ: Kans State Col Pittsburg, BS, 59, MS, 60. Prof Exp: Asst prof math, Col of the Sch of the Ozarks, 60-62; instr, Univ Wis-Green Bay, 62-65; asst prof, Wis State Univ, Whitewater, 65-67; ASSOC PROF MATH, MO SOUTHERN STATE COL, 67- Concurrent Pos: Vis lectr, Math Asn Am, 63-64; NSF res grant, Ill Inst Technol, 66-68. Mem: Am Math Soc; Am Math Asn. Res: Equilong geometry; dual and hyperbolic complex variables. Mailing Add: Dept of Math Mo Southern State Col Joplin MO 64801

BOEHNKE, DAVID NEAL, b Dayton, Ky, July 25, 39. PHYSICAL CHEMISTRY, THEORETICAL CHEMISTRY. Educ: Georgetown Col, BS, 61; Univ Ky, MS, 64; Univ Cincinnati, PhD(theoret chem), 68. Prof Exp: Instr chem & physics, Ky State Col, 64-65; res assoc, Univ NC, Chapel Hill, 67-69; asst prof, 69-74, ASSOC PROF CHEM, JACKSONVILLE UNIV, 74- Mem: Am Chem Soc. Res: Theory of critical viscosity; properties of critical fluids; kinetics of bimolecular reactions; synthesis and properties of detergents. Mailing Add: Dept of Chem Jacksonville Univ Jacksonville FL 32211

BOEK, WALTER ERWIN, b Holland Patent, NY, May 12, 23; m 46; c 2. ANTHROPOLOGY, ECONOMICS. Educ: Cornell Univ, BS, 46; Mich State Univ, MA, 48, PhD(sociol, anthrop), 53. Prof Exp: Dir res, Health Info Found, 50-52; asst to comnr, NY State Dept Health, 52-62; dir classified proj, Bur Soc Sci Res, 62-63; dir res, Montgomery County Ment Health Ctr, 63-64; res prof, Univ Md, 63-65; PRES, NAT GRAD UNIV, 65- Concurrent Pos: Assoc ed, Human Orgn, Soc Appl Anthrop, 48-52; vis lectr, Cornell Univ, 53-56; res assoc prof, Russell Sage Col, 54-59; lectr, Union Univ, 55-58 & Albany Med Ctr, 56-58; mem PhD comt, Harvard Univ, 57-58; chmn bd, Ctr Study Aging, 57-; lectr, Fla State Univ, 58; thesis adv, City Univ New York, 60-61; assoc dir, Inst Advan Med Commun, 60-62. Mem: Fel AAAS; fel Am Anthrop Asn; fel Soc Appl Anthrop; Human Serv Personnel Asn. Res: Acculturation of Indians in Manitoba; Eskimos and Indians of Canadian West Arctic; health, community decision making and human organization. Mailing Add: 5011 Lowell St NW Washington DC 20016

BOEKELHEIDE, VIRGIL CARL, b Chelsea, SDak, July 28, 19; m 45; c 3. ORGANIC CHEMISTRY. Educ: Univ Minn, AB, 39, PhD(org chem), 43. Prof Exp: Instr chem, Univ Ill, 43-46; asst prof org chem, Univ Rochester, 46-51, from assoc prof to prof, 51-60, PROF ORG CHEM, UNIV ORE, 60- Concurrent Pos: Guggenheim fel, Swiss Fed Inst Technol, Zurich, 53-54; Swiss-Am Found Lectr, 60; Frontiers of Sci & Welch Symposium lectr, 68; Karl Pfister lectr, Mass Inst Technol, 69; mem adv coun, Nat Inst Environ Health Sci; chmn div chem & chem technol, Nat Res Coun-Nat Acad Sci; mem, Coun Int Exchange Scholars, 75- Honors & Awards: Fulbright-Hays Distinguished Prof Award, Yugoslavia, 72. Mem: Nat Acad Sci; Am Chem Soc; Swiss Chem Soc; Ger Chem Soc; The Chem Soc. Res: Natural products, heterocycles and aromatic character in polycyclic molecules. Mailing Add: Dept of Chem Univ of Ore Eugene OR 97403

BOEKER, ELIZABETH ANNE, b Derby, Conn, Apr 30, 41. BIOCHEMISTRY. Educ: Radcliffe Col, AB, 62; Univ Calif, Berkeley, PhD(biochem), 67. Prof Exp: Am Cancer Soc fel biochem, Univ Wash, 67-70; staff fel biochem, NIMH, 70-72; res asst prof biochem, Univ Wash, 72-75; ASST PROF CHEM & BIOCHEM, UTAH STATE UNIV, 75- Concurrent Pos: Estab investr, Am Heart Asn, 72-77. Mem: AAAS; Am Chem Soc. Res: Structure and function of enzymes; neurochemistry. Mailing Add: Dept of Chem & Biochem Utah State Univ Logan UT 84321

BOELL, EDGAR JOHN, b Rudd, Iowa, Oct 30, 06; m 32; c 2. BIOLOGY. Educ: Univ Dubuque, AB, 29; Univ Iowa, PhD(zool), 35. Hon Degrees: MA, Yale Univ, 46; DSc, Univ Dubuque, 52. Prof Exp: Res assoc zool, Univ Iowa, 34-37; from inst to prof biol, 38-47, Ross G Harrison prof exp zool, 47-75, chmn dept, 55-62, actg dean Yale Col, 2nd term, 68-69, EMER ROSS G HARRISON PROF EXP ZOOL & SR

BIOLOGIST, YALE UNIV, 75- Concurrent Pos: Rockefeller Found fel, Cambridge Univ, 37-38; Fulbright award, Carlsberg Lab, Univ Copenhagen, 53-54; Guggenheim Found fel, Univ Rome, 63-64; actg master, Jonathan Edwards Col, 75-76. Mem: AAAS; Am Physiol Soc; Am Soc Zool; Soc Develop Biol; Am Soc Cell Biol. Res: Chemical embryology; developmental physiology; reproductive biology. Mailing Add: Kline Biol Tower Yale Univ New Haven CT 06520

BOELLSTORFF, JOHN DAVID, b Johnson, Nebr, May 14, 40; m 64; c 2. STRATIGRAPHY, GEOCHRONOLOGY. Educ: Univ Nebr, BS, 63, MS, 68; La State Univ, PhD(geol), 73. Prof Exp: Geologist, Western Labs, Inc, 64-65; res geologist stratig, 65-68, RES GEOLOGIST STRATIG, UNIV NEBR, 71- Mem: Geol Soc Am; Am Quarternary Asn. Res: Chronology and stratigraphy of Cenozoic deposits of the Great Plains; petrography of glacial deposits; fission track dating. Mailing Add: Conserv & Surv Div Univ of Nebr Lincoln NE 68508

BOELSCHE, ARRNELL, b New Ulm, Tex, May 18, 24; m 59. PEDIATRICS. Educ: Rice Univ, BA, 44; Univ Tex, MS, 49. Prof Exp: Intern, State Wis Gen Hosp, 49-50; res pediat, 51-52, from instr to asst prof, 53-63, ASSOC PROF PEDIAT, UNIV TEX MED BR GALVESTON, 63- Mem: AMA; Am Acad Pediat; Am Orthopsychiat Asn; Am Sch Health Asn. Res: Handicapped child; cerebral palsy; mental retardation; minimal brain injured. Mailing Add: Univ of Tex Med Br Galveston TX 77550

BOELTER, DON HOWARD, b Sanborn, Minn, Sept 1, 33; m 56; c 4. SOIL PHYSICS, PLANT PHYSIOLOGY. Educ: Iowa State Univ, BS, 55; Univ Minn, MS, 58, PhD(soil physics), 62. Prof Exp: RES SOIL PHYSICS, NCENT FOREST EXP STA, USDA, 59- Mem: Am Soc Agron; Soil Sci Soc Am; Int Soc Soil Sci; Soil Conserv Soc Am; Soc Am Foresters. Res: Physical and hydrologic properties of peats and organic soil and relationship to the watershed management of bog and swamp areas in north central United States. Mailing Add: N Cent Forest Exp Sta Grand Rapids MN 55744

BOEN, JAMES ROBERT, b Fergus Falls, Minn, July 21, 32; m 57; c 1. BIOSTATISTICS. Educ: Dartmouth Col, AB, 56; Univ Ill, MS, 57, PhD(math), 59. Prof Exp: Asst prof math, Southern Ill Univ, 59-60; res assoc, Univ Chicago, 60-61; lectr, Univ Mich, 61-62, fel biostatist, Stanford Univ, 62-64; from asst prof to assoc prof, 64-75, PROF BIOSTATIST, SCH PUB HEALTH, UNIV MINN, MINNEAPOLIS, 75- Mem: Am Statist Asn; Biomet Soc; Am Pub Health Asn. Mailing Add: Dept of Biomet Univ of Minn Sch of Pub Health Minneapolis MN 55455

BOENIG, HERMAN V, polymer chemistry, see 12th edition

BOENIG, ROBERT WILLIAM, b Brooklyn, NY, Jan 10, 17; m 44; c 1. BIOLOGY. Educ: Long Island Univ, BS, 38; Columbia Univ, MA, 40, EdD(sci educ), 63. Prof Exp: Instr chem & biol, Hartwick Col, 40-41 & high sch, NY, 41-43; prin, Round Lake High Sch, NY, 43-44, Richville High Sch, 44-45 & Oswegatchie High Sch, 45-46; instr, Schenectady Pub Schs, 46-54, prin, 54-56; assoc prof sci, 56-61, assoc prof biol & sci educ, 61-63, PROF BIOL, STATE UNIV NY COL FREDONIA, 63- Mem: AAAS; Asn Higher Educ; Nat Soc Study Educ. Res: Air pollution; thermophillic algae; teaching materials; educational administration; science education. Mailing Add: Dept of Biol State Univ Col at Fredonia Fredonia NY 14063

BOENIGK, JOHN WILLIAM, b Braddock, Pa, July 1, 18; m 47; c 4. PHARMACEUTICAL CHEMISTRY. Educ: Duquesne Univ, BS, 40; Western Reserve Univ, MS, 42; Purdue Univ, PhD(pharmaceut chem), 48. Prof Exp: Asst prof pharm, Univ Tex, 48-50; assoc prof, Med Col Va, 50-56; assoc prof, Univ Pittsburgh, 56-61; group leader pharm prod develop, 61-67, sect leader tablet develop, 67-69, sr res assoc, 69-70, PRIN INVESTR, MEAD JOHNSON & CO, 70- Mem: AAAS; Am Pharmaceut Asn. Res: Sustained release; stability; vitamins; antacids. Mailing Add: 2511 Bayard Park Dr Evansville IN 47714

BOER, FRANK PETER, chemical physics, see 12th edition

BÖER, KARL WOLFGANG, b Berlin, Ger, Mar 23, 26. SOLID STATE PHYSICS, ENERGY CONVERSION. Educ: Humboldt Univ, Berlin, dipl physics, 49, Dr rer nat(electronic noise), 52, Dr rer nat habil(photoconduction), 55. Prof Exp: Asst physics, Humboldt Univ, Berlin, 49-56, docent, 56-58, prof & head dept, 58-61; res prof, NY Univ, 61-62; from assoc prof to prof physics, 62-72, dir, Inst Energy Conversion, 72-75, PROF PHYSICS & ENG, UNIV DEL, 72-, CHIEF SCIENTIST, INST ENERGY CONVERSION, 73- Concurrent Pos: Ed, Fortschritte der Physik, 53-54; consult, Askania Corp Berlin Teltow, 53-55 & US Govt & industs, 62-; dir lab elec breakdown, Ger Acad Sci, Berlin, 56-61; ed-in-chief, Physica Status Solidi, 60-61, 72-; chmn bd, SES, Inc, 73- Mem: Fel Am Phys Soc; Ger Phys Soc; sr mem Inst Elec & Electronics Engrs; Int Solar Energy Soc (dir, US sect). Res: Solid state electronics; noise defect chemistry; radiation damage; semiconductivity; photoconductivity; photo-electromotive force; dielectric breakdown; field and current instabilities; surface and electrodes; reaction kinetics; electro-optics; semiconducting glasses; photovoltaic; CdS solar cell; solar energy conversion; Solar One house. Mailing Add: Buck Toe Hills Kennett Square PA 19348

BOERCKER, FRED D, b St Louis, Mo, July 16, 24; m 48; c 3. PHYSICS, SCIENCE EDUCATION. Educ: Univ Calif, Berkeley, BA, 45; Wash Univ, MA, 48, PhD(sci educ), 60. Prof Exp: Assoc sci, Educ Testing Serv, summer 59; asst prof physics, Western Ky State Univ, 60-62; dir manpower studies, Am Inst Physics, 62-64; assoc prof physics & math, Georgetown Col, 64-66; dir educ & employ studies, Nat Acad Sci-Nat Res Coun, 66-70; prof physics, 70-74, DIR PUB SERV, AUSTIN PEAY STATE UNIV, 74- Concurrent Pos: Off Sci Personnel observer, Sci Manpower Comn, 66-70; mem adv comt on manpower studies, Am Inst Physics, 68- Mem: Am Asn Physics Teachers. Res: Physics education; studies of education and utilization of scientific manpower. Mailing Add: Austin Peay State Univ Clarksville TN 37040

BOERSMA, P DEE, b Mt Pleasant, Mich, Nov 1, 46. ECOLOGY. Educ: Cent Mich Univ, BS, 69; Ohio State Univ, PhD(zool), 74. Prof Exp: Lectr zool, Ohio State Univ, 74; ASST PROF ENVIRON SCI, INST ENVIRON STUDIES, UNIV WASH, 74- Concurrent Pos: Adv, US Del, Status of Women Comn & World Pop Conf, UN, 74. Mem: AAAS; Ecol Soc Am; Am Ornith Union; Cooper Ornith Soc; Sigma Xi. Res: Ecology of the Galapagos penguin and marine iguana; adaptations of organisms to predictable and unpredictable environments, reproductive strategies, sexual dimorphism, temperature regulation; seabird growth rates in relationship to food supply and population dynamics. Mailing Add: 112 Sieg Hall FR-40 Environ Studies Univ Wash Seattle WA 98195

BOERTJE, STANLEY, b Pella, Iowa, Aug 1, 30; m 53; c 6. ZOOLOGY, PARASITOLOGY. Educ: Calvin Col, AB, 51; Univ Iowa, MS, 57; Iowa State Univ, PhD(zool), 66. Prof Exp: Teacher high sch, Iowa, 54-60; assoc prof biol, Dordt Col, 60-64, 65-67 & Midwestern Col, 67-70; PROF BIOL, SOUTHERN UNIV, NEW ORLEANS, 70- Mem: Am Soc Parasitol; Am Inst Biol Sci; Am Sci Affil. Res: Life cycle and host-parasite relationships of the cestode, Schistotaenia tenuicirrus. Mailing Add: Dept of Biol Southern Univ 6400 Press Dr New Orleans LA 70126

BOERWINKLE, FRED P, b Jacksonville, Fla, Nov 19, 41; m 61; c 3. ORGANIC CHEMISTRY. Educ: Univ Colo, Boulder, PhD(org chem), 68. Prof Exp: Res chemist, Celanese Chem Co, 68-69; sr chemist, 69-73, RES SUPVR, MINN MINING & MFG CO, 73- Res: Electrophilic addition reactions; solution polymerization. Mailing Add: Minn Mining & Mfg Co 3M Ctr Bldg 230-2S St Paul MN 55101

BOES, ARDEL J, b Wall Lake, Iowa, Sept 24, 37; m 63; c 2. MATHEMATICS. Educ: St Ambrose Col, BA, 59; Purdue Univ, MS, 61, PhD(math), 66. Prof Exp: Instr math, Marycrest Col, 63-64; asst prof, 66-69, ASSOC PROF MATH, COLO SCH MINES, 69- Res: Foundations of probability; measure theory. Mailing Add: Dept of Math Colo Sch of Mines Golden CO 80401

BOES, ELDON C, b Carroll, Iowa, July 29, 40; m 64; c 2. TOPOLOGY. Educ: Purdue Univ, MS, 64, PhD(math), 69. Prof Exp: Instr math, Roosevelt Univ, 66-68; ASST PROF MATH, NMEX STATE UNIV, 68- Mem: Am Math Soc; Math Asn Am. Res: K-rings of symmetric spaces; mathematics education. Mailing Add: Dept of Math Sci NMex State Univ Las Cruces NM 88001

BOESCH, DONALD FRIEDRICH, b New Orleans, La, Nov 14, 45; m 68; c 1. BIOLOGICAL OCEANOGRAPHY, SYSTEMATIC ZOOLOGY. Educ: Tulane Univ, BS, 67; Col William & Mary, PhD(marine sci), 71. Prof Exp: Fulbright-Hays fel marine ecol, Univ Queensland, Australia, 71-72; ASSOC MARINE SCIENTIST BIOL OCEANOG, VA INST MARINE SCI, 72-; ASST PROF MARINE SCI, COL WILLIAM & MARY & UNIV VA, 72- Mem: Am Soc Limnol & Oceanog; Ecol Soc Am; Estuarine Res Fedn; Australian Marine Sci Asn. Res: Ecology of coastal and estuarine benthos; mathematical ecology; systematics of marine invertebrates; effects of waste disposal, oil pollution and dredging on coastal ecosystems. Mailing Add: Va Inst Marine Sci Gloucester Pt VA 23062

BOESE, GILBERT KARYLE, b Chicago, Ill, June 24, 37; m 59; c 2. ANIMAL BEHAVIOR. Educ: Carthage Col, BA, 59; Northern Ill Univ, MS, 65; Johns Hopkins Univ, PhD(pathobiol), 73. Prof Exp: Teacher sci, Pub Sch, 60-62, teacher sci & head dept, 62-65; instr biol, Thornton Community Col, 65-67; asst prof biol, Elmhurst Col, 67-69; res assoc, 69-71, cur educ & res, 71-72, ASSOC DIR, BROOKFIELD ZOO, 72- Concurrent Pos: Primate adv comt mem, Baltimore Zool Soc, 71-72; adj prof, Northern Ill Univ, 72-73; vis lectr biol, Elmhurst Col, 71- Mem: Am Asn Zool Parks & Aquariums; Animal Behav Soc; EAfrican Wildlife Soc. Res: Social behavior and ecology of the Guinea baboon and comparison of social behavior with other species of the genus Papio. Mailing Add: Chicago Zool Park Brookfield IL 60513

BOESEL, MARION WATERMAN, b Columbus, Ohio, Mar 25, 01; m 33; c 1. ENTOMOLOGY, ZOOLOGY. Educ: Ohio State Univ, AB, 25, AM, 29, PhD(entom), 39. Prof Exp: Asst zool & entom, Ohio State Univ, 25-28; from instr to prof, 28-71, EMER PROF ZOOL & CUR ENTOM COLLECTIONS, MIAMI UNIV, 71- Concurrent Pos: Asst biologist, State Conserv Div, Ohio, 29-31; asst entomologist, State Nat Hist Surv Div, Ill, 38-39. Mem: AAAS; fel Entom Soc Am. Res: Biology and taxonomy of Ceratopogonidoe and Chironomidae. Mailing Add: 5141 Oxford-Milford Rd Oxford OH 45056

BOESGAARD, ANN MERCHANT, b Rochester, NY, Mar 21, 39; m 66. ASTRONOMY. Educ: Mt Holyoke Col, AB, 61; Univ Calif, Berkeley, PhD(astron), 66. Prof Exp: Res fel astrophys, Calif Inst Technol & Mt Wilson & Palomar Observ, 66-67; asst prof, 67-71, ASSOC PROF PHYSICS & ASTRON, INST ASTRON, UNIV HAWAII, 71- Concurrent Pos: Sabbatical leave, researcher, Nat Ctr Sci Res, France, 73-74; res assoc, Hale Observ, 74. Mem: AAAS; Am Astron Soc; Int Astron Union; Astron Soc Pac (vpres, 74-). Res: Stellar spectroscopy, nucleosynthesis and chemical composition of stars; stellar evolution. Mailing Add: Inst for Astron 2680 Woodlawn Dr Univ of Hawaii Honolulu HI 96822

BOESSENROTH, THEODORE, mathematics, see 12th edition

BOETTCHER, ARTHUR LEE, b Glasgow, Mont, Apr 27, 35; m 60; c 2. GEOCHEMISTRY, PETROLOGY. Educ: Mont Sch Mines, BS, 61; Pa State Univ, MS, 63, PhD(geol), 66. Prof Exp: Fel geochem, Univ Chicago, 66-67; from asst prof to assoc prof petrol, 67-72, chmn geochem & mineral, 71-75, PROF PETROL, PA STATE UNIV, 72- Concurrent Pos: Vis scientist, US Geol Surv, 75. Honors & Awards: Award, Mineral Soc Am, 72. Mem: Fel Mineral Soc Am. Res: High-pressure, high-temperature investigations of phase equilibria and mineral synthesis and stability; igneous and metamorphic petrology; planetary interiors. Mailing Add: Dept of Geosci Pa State Univ University Park PA 16802

BOETTCHER, F PETER, b Berlin, Ger, Aug 23, 32; US citizen; m 64; c 4. POLYMER CHEMISTRY. Educ: T H Darmstadt, DiplIng, 58, DrIng(org chem), 61. Prof Exp: Res assoc org chem, Univ Mich, Ann Arbor, 61-63; res chemist polymer chem, 64-68, sr res chemist textile technol, 68-74, RES SUPVR POLYMER CHEM, PIONEERING RES LAB, E I DU PONT DE NEMOURS & CO, INC, 74- Mem: Am Chem Soc. Res: Polymer synthesis; development of synthetic fibers. Mailing Add: E I du Pont de Nemours & Co Inc Exp Sta Wilmington DE 19898

BOETTIGER, EDWARD GEORGE, endocrinology, see 12th edition

BOETTLER, JAMES LEROY, b Summit, NJ, Dec 19, 35; m 63. SOLID STATE PHYSICS. Educ: Lafayette Col, BS, 58; Univ Ill, Urbana, MS, 59, PhD(physics), 66. Prof Exp: Asst physics, Univ Ill, Urbana, 59-65, asst sch sci curric proj, 65-66; asst prof physics, Ind Univ Pa, 66-69 & Western Ill Univ, 69-72; ASSOC PROF PHYSICS, TALLEDEGA COL, 72- Mem: AAAS; Am Asn Physics Teachers. Res: Optical properties, especially absorption, luminescence and polarization in near visible light, of crystals upon radiation damage or additive coloration; computational physics as applied to physics demonstrations and experiments. Mailing Add: Dept of Physics Talledega Col Talledega AL 35160

BOETTNER, EDWARD ALVIN, b Mich, Aug 28, 15; m 38; c 2. PHYSICS. Educ: Lawrence Inst Technol, BSc, 37; Univ Mich, MSc, 51. Prof Exp: Spectroscopist, Wyandotte Chem Corp, 37-46; physicist eng res inst, 46-59, res assoc, Inst Indust Health, 59-65, from asst prof to assoc prof indust health, 59-69, PROF INDUST HEALTH, UNIV MICH, 69- Mem: AAAS; Am Indust Hyg Asn; Optical Soc Am. Res: Applied spectroscopy; application of physics instrumentation to the assessment of the environment. Mailing Add: 3225 Farmbrook Court Ann Arbor MI 48104

BOETTNER, FRED EASTERDAY, b Murphysboro, Ill, Dec 10, 18; m 42; c 3. PHARMACEUTICAL CHEMISTRY. Educ: Carthage Col, AB, 40; Tulane Univ, MS, 42; Univ Ill, PhD(org chem), 47. Prof Exp: Asst chem, Tulane Univ, 40-42; res chemist, Monsanto Chem Co, Ala, 42-43, Higgins Indust, Inc, La, 43-44 & Armour Res Found, 44; asst org chem, Univ Ill, 44-47; SR SCIENTIST, ROHM AND HAAS CO, 47- Mem: Am Chem Soc; fel Am Inst Chemists; AAAS. Res: Lignin utilization; plastics and resins; synthesis of new anion exchange resins; structure of amino amide formaldehyde resins; insecticides; fungicides; herbicides; surface active compounds; pharmaceuticals. Mailing Add: 3872 Sidney Rd Huntingdon Valley PA 19006

BOEZI, JOHN A, b Binghamton, NY, Jan 21, 34; m 56; c 2. BIOCHEMISTRY, MOLECULAR BIOLOGY. Educ: St Bonaventure Univ, BS, 55; Univ Ill, PhD(microbiol), 61. Prof Exp: NSF fels, Biophys Group, Carnegie Inst, 60-61, Enzym Lab, Nat Ctr Sci Res, France, 61-62 & Dept Chem, Univ Ill, 62-63; from asst prof to assoc prof, 63-71, PROF BIOCHEM, MICH STATE UNIV, 71- Concurrent Pos: NIH grant, 64-; NSF grant, 69- Mem: Am Soc Microbiol; Am Soc Biol Chem. Res: Nucleic acids; protein biosynthesis; control mechanisms. Mailing Add: Dept of Biochem Mich State Univ East Lansing MI 48823

BOFFARDI, BENNETT PATSY, physical chemistry, see 12th edition

BOGACZ, JOHN, b Poland, May 12, 16. BACTERIOLOGY. Educ: Fordham Univ, MS, 52; Univ Paris, PhD(bact), 54. Prof Exp: Asst, Col Med, State Univ NY, 55-57; lectr, 57-59, asst prof, 59-68, ASSOC PROF BIOL, LA SALLE COL, 68- Concurrent Pos: Asst, Hahnemann Med Col, 57-60. Mem: Am Soc Microbiol; NY Acad Sci; Am Soc Cytol. Res: Cytologic study of carcinogenesis. Mailing Add: Dept of Biol La Salle Col Philadelphia PA 19141

BOGAN, DENIS JOHN, b Winchester, Mass, Aug 10, 41; m 69; c 2. CHEMICAL KINETICS, MOLECULAR SPECTROSCOPY. Educ: Northeastern Univ, AB, 65; Carnegie Mellon Univ, MS, 70, PhD(phys chem), 73. Prof Exp: Fel phys chem, Kans State Univ, 72-74, Nat Res Coun fel, 74-76, RES CHEMIST PHYS CHEM, NAVAL RES LAB, 76- Concurrent Pos: Assoc, Nat Res Coun-Nat Acad Sci, 74 & 75. Mem: Am Chem Soc; Sigma Xi. Res: Chemical kinetics in the bulk, dynamics of reactive collisions inferred from product quantum state populations obtained by spectroscopy; energy partitioning experimental and statistical theory thereof. Mailing Add: Chem Dynamics Br Code 6180 Naval Res Lab Washington DC 20375

BOGAN, ROBERT L, b Anderson, Ind, July 28, 26; m 48; c 3. DENTISTRY. Educ: Butler Univ, BS, 50; Ind Univ, DDS, 54, MSD, 67. Prof Exp: Asst prosthetics, 54-55, instr crown & bridge, 55-62, asst prof, 62-68, asst to dean, 64-67, asst dean, 67-73, ASSOC PROF CROWN & BRIDGE, SCH DENT, IND UNIV-PURDUE UNIV, INDIANAPOLIS, 68-, ASSOC DEAN, 67- Mem: Am Asn Dent Schs; Am Prosthodontic Soc; Int Col Dentists. Res: Clinical evaluation of factors contributing to stability in removable partial prosthodontics; laboratory study in soldering of high fusion gold alloys used in dental restorations. Mailing Add: Sch of Dent Ind Univ Indianapolis IN 46202

BOGAR, LOUIS CHARLES, b Bluefield, WVa, Aug 22, 32; m 54; c 3. RADIOCHEMISTRY, HEALTH PHYSICS. Educ: Mass Inst Technol, SB, 54. Prof Exp: Resident radiochemist, Bettis Atomic Power Lab Off, Chalk River Nuclear Labs, Can, 57-60; scientist, Bettis Atomic Power Lab, Westinghouse Elec Corp, 60-63, supvr mat activation, 63-64, supvr radiation sources, 65-67, mgr radiation eng, 68-69, mgr radiation control, 69-71, mgr radiation & safety, 71-74, SPEC ASSIGNMENT, DUQUESNE LIGHT CO, SHIPPINGPORT ATOMIC POWER STA, 74- Mem: AAAS; Am Nuclear Soc; Am Chem Soc; Am Ord Assn; fel Am Inst Chem. Res: Fission product chemistry; fuel element release; radioactivation and transport processes in reactor coolants; low level radioactivity measurements; high temperature aqueous corrosion mechanisms; radiological controls; in-vivo dosimetry; environmental radioactivities. Mailing Add: Duquesne Light Co PO Box 57 Shippingport Atomic Power Sta Shippingport PA 15077

BOGARD, ANDREW DALE, b Savannah, Mo, June 18, 15; m 46; c 2. RADIOCHEMISTRY. Educ: William Jewell Col, AB, 37; Univ Ark, MA, 39. Prof Exp: Asst chem, Univ Ark, 37-39; chemist coal prod & clay utilization, US Bur Mines, 39-42; chemist, Oak Ridge Nat Lab, 46-48; chemist liquid metals res, Naval Res Lab, 48-56; FEL ENGR, BETTIS ATOMIC POWER LAB, WESTINGHOUSE ELEC CORP, 56- Honors & Awards: Meritorious Civilian Serv Award, Naval Res Lab, 56. Mem: Am Chem Soc. Res: Coal utilization; liquid metals for nuclear reactors; fallout; fission product analysis; heavy element chemistry; gamma ray spectrometry; reactor coolant radiochemistry. Mailing Add: 4765 Baptist Rd Pittsburgh PA 15227

BOGARD, DONALD DALE, b Fayetteville, Ark, Feb 6, 40. GEOCHEMISTRY. Educ: Univ Ark, Fayetteville, BS, 62, MS, 64, PhD(isotope geochem). 66. Prof Exp: Res fel geol sci, Calif Inst Technol, 66-68; STAFF SCIENTIST, PLANETARY & EARTH SCI DIV, JOHNSON SPACE CTR, NASA, 68- Concurrent Pos: NSF fel, Calif Inst Technol, 66-67; assoc ed, J Geophys Res, Am Geophys Union, 75- Mem: AAAS; Am Geophys Union; Geochem Soc; Meteoritical Soc. Res: Age and origin of planetary objects in the solar system; geochronology; isotope geochemistry nuclear reactions in nature; meteorites. Mailing Add: NASA Johnson Space Ctr Code TN 7 Houston TX 77058

BOGARD, TERRY L, b Bicknell, Ind, Aug 8, 36; m 62; c 3. ORGANIC CHEMISTRY. Educ: Ohio State Univ, BSc, 58; Univ Calif, Los Angeles, PhD(org chem), 63. Prof Exp: Asst chem, Univ Calif, Los Angeles, 58-62; NIH fel, Swiss Fed Inst Technol, 62-64; fel, Brandeis Univ, 64-65; res chemist, Lederle Labs Div, Am Cyanamid Co, 65-70, MGR PROCESS DEVELOP, CYANAMID INT, LEDERLE, 71- Mem: Am Chem Soc; Brit Chem Soc. Res: Theoretical, synthetic and natural products organic chemistry; polycyclic bridged compounds, alkaloids, terpenes and reaction mechanisms. Mailing Add: 102 W Prospect St Nanuet NY 10954

BOGARDUS, CARL ROBERT, JR, b Hyden, Ky, June 26, 33; m 57; c 2. MEDICINE, RADIOLOGY. Educ: Hanover Col, BA, 55; Univ Louisville, MD, 59; Am Bd Radiol, cert therapeut radiol & nuclear med, 64; Am Bd Nuclear Med, dipl, 72. Prof Exp: Intern & resident therapeut radiol, Penrose Cancer Hosp, Colorado Springs, Colo, 63; from asst prof to assoc prof, 64-69, PROF RADIOL & VCHMN DEPT, MED CTR, UNIV OKLA, 69-, DIR DIV RADIATION THER, 66- Concurrent Pos: Res fel radiation physics, Mallinckrodt Inst Radiol, Sch Med, Wash Univ, 64. Mem: Am Soc Therapeut Radiol; Radiation Res Soc; Soc Nuclear Med; AMA; Am Col Radiol. Res: Cancer; radiation therapy, physics and biology; radioisotopes. Mailing Add: Dept of Radiol Univ of Okla Med Ctr Oklahoma City OK 73104

BOGARDUS, EGBERT HAL, b New York, NY, Feb 20, 31; m 53; c 3. SOLID STATE SCIENCE. Educ: Univ Tex, BS, 55; Pa State Univ, PhD(mat sci), 64. Prof Exp: Res assoc physics, Cornell Univ, 64-66; mem tech staff, Tex Instruments, Inc, Dallas, 66-68; staff physicist, Components Div, 69-73, ADV PHYSICIST, T J WATSON RES CTR, IBM CORP, 73- Mem: Am Phys Soc. Res: Optical characterization of III-V materials; ion implantation; magnetic properties of ferrofluids; surface properties of thin insulating films; electrical properties of ion-implanted devices. Mailing Add: T J Watson Res Ctr IBM Corp Yorktown Heights NY 10598

BOGART, BRUCE IAN, b New York, NY, Sept 11, 39; m 64; c 2. CELL BIOLOGY. Educ: Johns Hopkins Univ, BA, 61; NY Univ, PhD(basic med sci), 66. Prof Exp: From instr to asst prof, 66-74, ASSOC PROF ANAT, MED CTR, NY UNIV, 74- Mem: AAAS; Am Asn Anatomists; Am Soc Cell Biol. Res: Secretory dynamics of salivary glands; fine structural morphology, cytochemistry, pharmacology and cytopathology of exocrine secretion. Mailing Add: Dept of Cell Biol Div of Anat NY Univ Med Ctr New York NY 10016

BOGART, ELLIOT, b New York, NY, Apr 17, 36; m 65. BIOPHYSICS. Educ: Columbia Univ, BA, 57, MA, 60, PhD(physics), 66. Prof Exp: Res asst physics, Columbia Nevis Lab, 63-66; res assoc, Univ Pa, 66-72; BIOPHYSICIST, LAWRENCE LIVERMORE LAB, 72- Mem: Am Phys Soc. Res: Chromosome image analysis; pictorial pattern recognition. Mailing Add: Lawrence Livermore Lab L-523 Livermore CA 94566

BOGART, KENNETH PAUL, b Cincinnati, Ohio, Oct 6, 43; m 66; c 1. MATHEMATICS. Educ: Marietta Col, BS, 65; Calif Inst Technol, PhD(math), 68. Prof Exp: Asst prof, Marietta Col, 65-66; teaching asst, Calif Inst Technol, 66-68; asst prof, 68-74, ASSOC PROF MATH, DARTMOUTH COL, 74- Mem: Math Asn Am; Am Math Soc. Res: Abstract commutative ideal theory, noether lattices; lattice theory; ring theory; partially ordered sets; application of algebraic and combinatoric techniques to social and behavioral science problems. Mailing Add: Dept of Math Dartmouth Col Hanover NH 03755

BOGART, RALPH, b Lawson, Mo, Nov 30, 08; m 38; c 1. ANIMAL SCIENCE. Educ: Univ Mo, BS, 34; Kans State Col, MS, 36; Cornell Univ, PhD(genetics, physiol), 40. Prof Exp: Asst genetics, Kans State Col, 34-36; asst animal physiol & genetics, Cornell Univ, 36-38; agent, USDA & instr animal husb, Univ Mo, 38-42, asst prof, 42-44 & 45-47; prof animal genetics, 47-74, dir genetics inst, 64-74, EMER PROF ANIMAL GENETICS, ORE STATE UNIV, 74- Concurrent Pos: Technician, San Carlos Indian Agency, Ariz, 44. Mem: AAAS; Am Soc Zool; Am Asn Anat; Am Soc Animal Sci; Genetics Soc Am. Res: Inheritance and physiology of reproduction in farm animals. Mailing Add: Dept of Animal Sci Ore State Univ Corvallis OR 97331

BOGART, WILLIAM HAWKINS, JR, b Greensboro, NC, Mar 25, 27; m 60; c 2. ORGANIC CHEMISTRY. Educ: Davidson Col, BS, 49; NC State Col, MS, 51. Prof Exp: ASSOC PROF CHEM & HEAD DEPT, DANVILLE COMMUNITY COL, 68- Concurrent Pos: NSF fels, Philadelphia Col Pharm, 59, Ore State Univ, 61 & Univ Pac, 62; dir lab, Dibrell Bros, Leaf Tobacco. Res: Formation of amidines from Aliphatic Ortho Esters. Mailing Add: 137 London Bridge Dr Danville VA 24541

BOGATY, HERMAN, b New York, NY, Apr 9, 18; m 38; c 2. RESEARCH ADMINISTRATION, TECHNOLOGY. Educ: City Col NY, BS, 40. Prof Exp: From sci aide to textile technologist, Bur Standards, 37-47; res assoc, Harris Res Labs, Inc, 47-58; assoc res dir, Toni Div, Gillette Co, 58-62, lab dir & vpres res & develop, 62-68; vpres res & develop consumer prod group, Warner Lambert Co, 68-75; PRES, HERMAN BOGATY CONSULT CORP, 75- Mem: Am Chem Soc; Fiber Soc; Soc Cosmetic Chem; Acad Pharmaceut Sci. Res: Research and development in toiletries, proprietaries, chewing gum and confections. Mailing Add: Herman Bogaty Consult Co 22 Glen Brook Short Hills NJ 07078

BOGDAN, VICTOR MICHAEL, b Kiev, Ukrainia, Jan 4, 33; nat US; m 68; c 2. MATHEMATICS. Educ: Univ Warsaw, BS, 53, MS, 55; Polish Acad Sci, PhD(math), 60. Prof Exp: Asst math, Univ Warsaw, 52-53, asst, 53-55, from instr to asst prof, 55-61; res assoc, Univ Md, 61-62; from asst prof to assoc prof, Georgetown Univ, 62-64; assoc prof, 64-66, PROF MATH, CATH UNIV AM, 66- Concurrent Pos: Mem res staff math inst, Polish Acad Sci, 55-61; NSF res grant, 64-65; vis scientist, Univ Montreal, 70-71. Mem: Am Math Soc; Polish Math Soc; Soc Appl & Indust Math. Res: Functional analysis; linear methods of summability; integration theory; theory of distributions; differential equations in Banach spaces; almost periodic solutions of ordinary and partial differential equations; theory of random processes. Mailing Add: Dept of Math Cath Univ of Am Washington DC 20064

BOGDANOFF, DAVID ALLEN, mathematical statistics, applied statistics, see 12th edition

BOGDANOVE, EMANUEL MENDEL, b New York, NY, Feb 20, 25; m 55; c 6. ENDOCRINOLOGY. Educ: City Col New York, BS, 44; Univ Iowa, MS, 50, PhD(anat), 53. Prof Exp: Res asst pharmacol, Wayne State Univ, 46-47; from instr to asst prof anat, Albany Med Col, 53-60; from assoc prof to prof, Sch Med, Ind Univ, 60-71; PROF PHYSIOL, MED COL VA, 71- Concurrent Pos: USPHS spec trainee, Inst Psychiat, Univ London, 58-59. Mem: Endocrine Soc; Soc Exp Biol & Med; Am Asn Anatomists; Am Physiol Soc; Am Soc Zoologists. Res: Hypothalamic-pituitary relationships; kinetics of follicle-stimulating hormone and luteinizing hormone. Mailing Add: Dept of Physiol Med Col of Va Richmond VA 23298

BOGDANOVE, LASCA HOSPERS, b Hinsdale, Ill, May 24, 26; div; m 70; c 6. PHARMACOLOGY. Educ: Oberlin Col, AB, 46; Univ Iowa, MS, 50; Univ Buffalo, PhD(physiol), 58. Prof Exp: Res fel, Dept Pharmacol, Wayne Univ, 46-47; res asst, Iowa Pediat Lab, 48-51; supvr histol tech lab, Chicago Med Sch, 51-53; asst physiol, Univ Buffalo, 53-57; chemist, Arner Pharmaceut Labs, 57-58; pharmacologist, Miles-Ames Res Labs, Ind, 58-59; trainee pharmacol, Univ Ill, 59-60; assoc, 60-66, ASST PROF PHARMACOL, CHICAGO MED SCH RES INST, 66- Res: Neuropharmacology; psychopharmacology; electroencephelography. Mailing Add: Div of Behav Sci Rm 227 Chicago Med Sch Res Inst Chicago IL 60612

BOGDANOWICZ, MITCHELL JOSEPH, b New York, NY, Jan 18, 47; m 71. ORGANIC CHEMISTRY. Educ: Rochester Inst Technol, BS, 69; Univ Wis, PhD(org chem), 72. Prof Exp: Res chemist, E I du Pont de Nemours & Co, Inc, 72-74; RES CHEMIST, EASTMAN KODAK CO, 74- Res: New synthetically useful reactions. Mailing Add: Kodak Park Res Labs Bldg 82 Rochester NY 14650

BOGDANSKI, DONALD FRANK, b Port Chester, NY, Oct 31, 28; m 59; c 4. PHARMACOLOGY. Educ: Columbia Univ, BS, 50; Georgetown Univ, PhD, 54. Prof Exp: Lab asst pharmacol, Georgetown Univ, 52-53, lab asst biochem, 53; pharmacologist, Chem Pharmacol Lab, Nat Heart Inst, 54-58; spec res assoc from Med Sch, to Dept Neurophysiol, Walter Reed Army Inst Res, 58-64; PHARMACOLOGIST, LAB CHEM PHARMACOL, NAT HEART INST, 64- Mem: AAAS; Am Soc Pharmacol & Exp Therapeut. Res: Binding and release of chemical transmitters in the peripheral nervous system; phylogenetic distribution of amines; transport, storage and release of chemical transmitters in the peripheral and central nervous system. Mailing Add: 10596 Spotted Horse Lane Columbia MD 21043

BOGDEN, ARTHUR EUGENE, b DuBois, Pa, Oct 6, 21; m 48; c 2. IMMUNOLOGY. Educ: Univ Pa, BA, 52, MS, 53, PhD(med microbiol), 56. Prof Exp: Fel, Wistar Inst, 56-58, chief res serv, 58; chief div cancer immunol, Biochem Res Found, 58-63; DIR IMMUNOL, MASON RES INST, 63- Mem: Am Asn Cancer Res; Am Asn Immunol; NY Acad Sci. Res: Immunobiology; immunogenetics; rat and rhesus blood groups; tumor transplantation; cancer chemotherapy; radiobiology; anti-inflammatory and hormone immunoassay viral-hormonal cocarcinogenesis. Mailing Add: Mason Res Inst Harvard St Worcester MA 01608

BOGDONOFF, PHILIP DAVID, JR, b Lowell, Mass, June 11, 27; m 53; c 3. ANIMAL PHYSIOLOGY, ANIMAL NUTRITION. Educ: Univ Md, BS, 51, MS, 53, PhD(physiol), 55. Prof Exp: Res analyst animal & human sci, Nat Acad Sci, 55-

57; res nutritionist, Com Solvents Corp, 57-59, head animal nutrit res dept, 59-63; res assoc, Bio-Res Inst, Inc & Bio-Res Consult, Inc, 63-67; prof, N Shore Community Col, 70-72; VPRES, WHEELER ASSOCS, INC, 67- Res: Administration, research and educational development; consulting business and manufacturing systems rearranged work systems. Mailing Add: PO Box 255 Boxford MA 01921

BOGENSCHUTZ, ROBERT PARKS, b Ponca City, Okla, Mar 25, 33. PHYSIOLOGY, BIOLOGY. Educ: Okla State Univ, BS, 54, MS, 56; Cent State Col, Okla, BS, 61; Univ Okla, PhD(zool), 66. Prof Exp: Teacher, Am Sch London, 62-63; res assoc, Okla Univ Res Inst, 66; teacher, Am Int Sch, Vienna, 66-67; asst prof, 67-70, ASSOC PROF BIOL & PHYSIOL, CENT STATE UNIV, OKLA, 70- Mem: Am Inst Biol Sci; Am Soc Zool. Res: Effects of environmental factors on pituitary proteins and histology. Mailing Add: Dept of Biol Cent State Univ Edmond OK 73034

BOGER, ELIAHU, b Telaviv, Israel, Apr 27, 14; US citizen; m 63. ORGANIC CHEMISTRY, ENZYMOLOGY. Educ: Univ Nancy, Chem Eng, 35; Hebrew Univ, Israel, PhD(chem), 40. Prof Exp: Instr org chem, Hebrew Univ, Jerusalem, 40-48; chief chemist insecticides, Israel Dept Agr, 50-54; res assoc cancer res, Brandeis Univ, 54-61; res assoc org chem, Bio-res Inst, Cambridge, Mass, 62-74; ADJ PROF CHEM, BOSTON COL, 75- Mem: Am Chem Soc. Res: Active sites and mechanism of action of chemical carcinogens; inhibitors of carcinogenesis; fluorinated polycyclic hydrocarbons; alkylating agents with latent activity; enzymes related to malignancy; derivatives of pentaerythritol. Mailing Add: Dept of Chem Boston Col Chestnut Hill MA 02167

BOGER, WILLIAM PIERCE, b Johnstown, Pa, July 23, 13; m 43; c 2. MEDICINE. Educ: Bucknell Univ, BS, 34; Harvard Univ, MD, 38; Am Bd Internal Med, dipl, 45. Prof Exp: Intern, Philadelphia Gen Hosp, 38-40; asst resident med, Mass Gen Hosp, 41-42; chief resident & instr, Med Col Va, 42-43; dir dept internal med & assoc, St Luke's Hosp, WVa, 43-44; assoc med dir, Sharp & Dohme, Inc, 45-51, med dir, 51-56; instr, Grad Sch & Sch Med, Univ Pa, 46-50, assoc med, 50-62; DIR RES & CHIEF INFECTIOUS DIS, MONTGOMERY HOSP, NORRISTOWN, 62-; ASST CHIEF STAFF, MED SERV, VET ADMIN HOSP, COATESVILLE, 66- Concurrent Pos: Dir dept res therapeut, Norristown State Hosp, 52-62; dir internal med affairs, McNeil Labs, 62-64; corp med dir, Miles Labs, Inc, 64-65, vpres res & med affairs, Dome Labs Div, 65-66; ed, Chemotherapy. Mem: Am Col Physicians; fel Am Soc Pharmacol & Exp Therapeut; fel Endocrine Soc; fel Am Soc Clin Invest; fel Royal Soc Med. Res: Chemotherapy; antibiotics; internal medicine; metabolic disease.

BOGERT, BRUCE PLYMPTON, b Waltham, Mass, Sept 26, 23; m 49; c 2. PHYSICS. Educ: Mass Inst Technol, BS, 44, MS, 46, PhD(math), 48. Prof Exp: Staff mem, Radiation Lab, Mass Inst Technol, 44-45; MEM TECH STAFF, BELL LABS, 48- Honors & Awards: Biennial Award, Acoustical Soc Am, 58. Mem: Fel Acoustical Soc Am. Res: Digital transmission; seismic wave instrumentation and processing; underwater acoustics; digital computer analysis of time series. Mailing Add: Bell Labs Whippany NJ 07981

BOGGESS, ALBERT, III, b Dallas, Tex, Jan 30, 29; m 52; c 3. ASTROPHYSICS. Educ: Univ Tex, BA, 50; Univ Mich, MA, 52, PhD(astron), 54. Prof Exp: Fel, Johns Hopkins Univ, 54-55, consult, 55-56; physicist, Naval Res Lab, 55-58; astrophysicist interstellar medium sect, 59-69, head observational astron br, 69-72, HEAD ASTRON SYSTS BR, GODDARD SPACE FLIGHT CTR, NASA, 72- Mem: Am Astron Soc; Int Astron Union. Res: Galactic nebulae; interstellar medium; high altitude research; stellar atmospheres; airglow. Mailing Add: Code 673 Goddard Space Flight Ctr NASA Greenbelt MD 20771

BOGGESS, NANCY WEBER, b Philadelphia, Pa, Apr 25, 29; m 52; c 3. ASTRONOMY. Educ: Wheaton Col, Mass, BA, 47; Wellesley Col, MA, 49; Univ Mich, PhD, 67. Prof Exp: Res asst astrophys, Haverford Col, 49-50; ASTRONR, HQ, NASA, 67- Mem: Am Astron Soc. Res: Theoretical transition probabilities for elements of astrophysical interest; galactic structure of dwarf irregular galaxies. Mailing Add: Nat Aeronaut & Space Admin Washington DC 20546

BOGGESS, SAMUEL FOREST, b Auburn, Ala, Jan 1, 42. PLANT PHYSIOLOGY. Educ: Univ Ill, BS, 63; Duke Univ, PhD(bot), 70. Prof Exp: Fel plant physiol, Waite Inst, Univ Adelaide, 70-73; asst prof, Iowa State Univ, 74-75; RES ASSOC PLANT PHYSIOL, UNIV ILL, 75- Mem: Am Soc Plant Physiologists; Sigma Xi; Int Solar Energy Soc. Res: Changes in plant physiology and biochemistry induced by environmental factors. Mailing Add: Dept of Agron Univ of Ill Urbana IL 61801

BOGGESS, WILLIAM RANDOLPH, b Oakvale, WVa, Apr 9, 13; m 38; c 4. FORESTRY. Educ: Concord Col, AB, 33; Duke Univ, MF, 40. Prof Exp: Teacher pub sch, WVa, 33-35; res assoc forester, Ala Agr Exp Sta, Ala Polytech Inst, 39-48; prof & assoc forest res, Dixon Springs Exp Sta, 48-58, prof forestry, 58-73, head dept, 69-73, EMER PROF FORESTRY, UNIV ILL, URBANA, 73-; ENVIRON CONSULT, 73- Concurrent Pos: Ed, Am Water Resources Asn Bull, 65-69. Mem: Soc Am Foresters; Am Soc Agron; Soil Sci Soc Am; Ecol Soc Am; Am Water Resources Asn (pres, 70). Res: Forest ecology; soils; forest influences. Mailing Add: 1705 Parkside Lane Austin TX 78745

BOGGIO, JOSEPH E, b Holyoke, Mass, Mar 20, 36; m 58; c 2. PHYSICAL CHEMISTRY. Educ: Worcester Polytech Inst, BS, 58, MS, 60, PhD(phys chem), 63. Prof Exp: Res assoc surface chem, Brown Univ, 62-64; instr chem, 64-65, from asst prof to assoc prof, 65-75, PROF CHEM, FAIRFIELD UNIV, 75- Mem: Am Phys Soc; Am Chem Soc. Res: Theories of formation of very thin oxide films; contact potential studies of oxidizing metal surfaces; adsorption on the dropping mercury electrode; ellipsometry of metals and thin films. Mailing Add: Dept of Chem Fairfield Univ Fairfield CT 06430

BOGGS, DALLAS ERVIN, b Porter, WVa, Apr 20, 32; m 56; c 5. BIOCHEMISTRY. Educ: WVa Univ, BS, 54; Cornell Univ, MNS, 55, PhD(biochem), 59; Nat Registry Clin Chem, dipl. Prof Exp: Asst nutrit, Cornell Univ, 54-55, asst biochem, 55-58; asst prof nutrit & agr biochem, Univ Del, 58-61; res proj assoc pediat, Univ Wis, 61-63; res biochemist, Sonoma State Hosp, Eldridge, Calif, 63-65; res specialist biochem, Res Dept, Pac State Hosp, Pomona, Calif, 65-67; res biochemist, Clin Study Ctr, Children's Hosp, Columbus, Ohio, 67-70; pres, Metagen Labs, Inc, 70-74, dir biochem sect, 71-74; WITH BIO-MED DATA, INC, 74- Concurrent Pos: Adj asst prof, Ohio State Univ, 67- Mem: Am Chem Soc; Am Asn Clin Chemists. Res: Experimental phenylketonuria; nutrition and metabolism of amino acids; mental retardation; neurochemistry. Mailing Add: Bio-Med Data Inc PO Box 397 West Chicago IL 60185

BOGGS, DANE RUFFNER, b Orton, WVa, Apr 21, 31; m; c 5. HEMATOLOGY. Educ: Univ Va, BA, 52, MD, 56; Am Bd Internal Med, dipl, 63. Prof Exp: Intern, Univ Va Hosp, 57; clin assoc, Nat Cancer Inst, 57-59; asst resident med, Col Med, Univ Utah, 59-60, fel clin hemat, 60-61, from instr to asst prof med, 61-67; assoc prof, Med Sch, Rutgers Univ, 67-69; PROF MED, SCH MED, UNIV PITTSBURGH, 69- Concurrent Pos: Leukemia Soc scholar, 62-65; fac res assoc, Am Cancer Soc, 65-70. Mem: AAAS; Am Soc Clin Res; Am Soc Hemat; Reticuloendothelial Soc; Soc Exp Biol & Med. Mailing Add: Dept of Med Univ of Pittsburgh Sch of Med Pittsburgh PA 15213

BOGGS, JAMES ERNEST, b Cleveland, Ohio, June 9, 21; m 48; c 3. CHEMISTRY. Educ: Oberlin Col, BA, 43; Univ Mich, MS, 44, PhD(phys chem), 53. Prof Exp: Res chemist, Linde Air Prod Co, NY, 44-46; asst prof chem, State Norm Col, Mich, 48-52; instr, Univ Mich, 52-53; PROF CHEM, UNIV TEX, AUSTIN, 53- Mem: Am Chem Soc; Am Phys Soc. Res: Molecular structure and dynamics; microwave spectroscopy; collisional energy transfer; theoretical chemistry. Mailing Add: Dept of Chem Univ of Tex Austin TX 78712

BOGGS, JOSEPH D, b Bellefontaine, Ohio, Dec 31, 21. PATHOLOGY. Educ: Ohio Univ, AB, 41; Jefferson Med Col, MD, 45. Prof Exp: Instr & assoc path, Harvard Univ, 46-50; asst prof, 50-55, PROF PATH, NORTHWESTERN UNIV, CHICAGO, 55- Concurrent Pos: Dir labs, Children's Hosp, Chicago. Mem: Fel Am Soc Clin Pathologists; fel Col Am Pathologists. Res: Pediatric pathology. Mailing Add: 2300 Childrens Plaza Chicago IL 60614

BOGGS, LAWRENCE ALLEN, b Spokane, Wash, June 14, 10; m 47; c 1. CHEMISTRY. Educ: Univ Hawaii, BS, 38, MS, 39; Univ Minn, PhD(agr biochem), 51. Prof Exp: Asst chemist, Pac Chem & Fertilizer Co, Hawaii, 39-42; jr chemist, Pearl Harbor Testing Lab, US Dept Navy, 42-43; res assoc, Inst Paper Chem, 51-75; RETIRED. Mem: AAAS; Am Chem Soc. Res: Carbohydrates of spent sulfite liquor; constitution of polysaccharides. Mailing Add: 1705 Ravinia Pl Appleton WI 54911

BOGGS, NATHANIEL, JR, b Anniston, Ala, Dec 19, 26; m 58; c 4. PROTOZOOLOGY, CYTOLOGY. Educ: Howard Univ, BS, 51, MS, 55, PhD(zool), 63. Prof Exp: Physiologist, Walter Reed Army Inst Res, 56-59; prof biol, Va State Col, 63-75; PROF BIOL, FLA A&M UNIV, 75- Concurrent Pos: Reader, Advan Placement Exam in Biol, Educ Testing Serv, Princeton, 67 & 68; Danforth assoc, 68-70; Southern Fel Fund res fel, 69-70; vis prof, Univ Calif, Los Angeles, 69-70. Mem: AAAS; Soc Protozool; Nat Inst Sci; Am Soc Cell Biol; Am Micros Soc. Res: Systematics of ciliates; water flux through fresh water protozoans; motile behavior in ciliated protozoa; protozoan nutrition; biochemical and biophysical phenomena as related to cell biology. Mailing Add: 213 Perry-Paige Bldg Fla A&M Univ Tallahassee FL 32307

BOGGS, NORMAN TOWAR, III, b Kingston, NY, June 24, 40. ORGANIC CHEMISTRY. Educ: State Univ NY Buffalo, PhD(chem), 67. Prof Exp: Res asst org chem, Rutgers Univ, 66-68; ASST PROF CHEM, KIRKLAND COL, 68- Res: Synthesis, reactions and rearrangements in bicyclic organic molecules. Mailing Add: Dept of Phys Sci Kirkland Col Clinton NY 13323

BOGGS, SAM, JR, b King's Creek, Ky, July 26, 28; m 52; c 3. GEOLOGY. Educ: Univ Ky, BS, 56; Univ Colo, PhD(geol), 64. Prof Exp: Explor geologist, Phillips Petrol Co, 56-61; res geologist, Jersey Prod Res Co, Standard Oil Co, NJ, 64-65; ASSOC PROF GEOL SEDIMENTATION, UNIV ORE, 65-, HEAD DEPT GEOL, 74- Concurrent Pos: Vis prof, Inst Oceanog, Nat Taiwan Univ, 72-73; vis prof, US-China Coop Sci Prog, NSF, 72. Mem: Fel Geol Soc Am; Am Asn Petrol Geologists; Soc Econ Paleontologists & Mineralogists; Am Geol Inst. Res: Petrology of sedimentary rocks; sedimentology and modern sedimentation processes; marine geology. Mailing Add: Dept of Geol Univ of Ore Eugene OR 97403

BOGGS, STEPHEN TAYLOR, b Chicago, Ill, July 13, 24; m 48; c 3. ANTHROPOLOGY. Educ: Harvard Univ, AB, 47; Washington Univ, PhD(sociol, anthrop), 54. Prof Exp: Actg asst prof sociol, Stanford Univ, 53-56; soc anthropologist, NIMH, 57-60; exec secy, Am Anthrop Asn, 61-66; assoc prof, 66-74, PROF ANTHROP, UNIV HAWAII, 74- Mem: Fel Am Anthrop Asn. Res: Culture and personality; sociolinguistics; anthropology and education; applied anthropology. Mailing Add: Dept of Anthrop Univ of Hawaii Honolulu HI 96822

BOGGS, WILLIAM EMERSON, b Zanesville, Ohio, May 9, 24; m 53; c 2. PHYSICAL CHEMISTRY, ANALYTICAL CHEMISTRY. Educ: Denison Univ, AB, 48; Carnegie Inst Technol, BS, 49, MS, 62. Prof Exp: Anal chemist, State Health Dept, Ohio, 50; microchemist appl res lab, 51-58, sr technologist, 59-65, sr res engr, 65-66, ASSOC RES CONSULT, US STEEL CORP, 66- Mem: Am Chem Soc; Electrochem Soc; Electron Micros Soc Am; Am Microchem Soc. Res: Oxidation of metals and alloys, tin and iron; instrumental and micro methods of chemical analysis. Mailing Add: Res Lab US Steel Corp Monroeville PA 15146

BOGHOSIAN, CHARLES, b Concord, NH, Dec 22, 28; m 48; c 1. LOW TEMPERATURE PHYSICS, SOLID STATE PHYSICS. Educ: Univ NH, BS, 60; Duke Univ, PhD(physics), 66. Prof Exp: Res assoc physics, Duke Univ, 65-67; physicist, Solid State Prog, 67-75, ASSOC DIR PHYSICS DIV, US ARMY RES OFF, 75- Mailing Add: US Army Res Off PO Box 12211 Research Triangle Park NC 27709

BOGITSH, BURTON JEROME, b Brooklyn, NY, Feb 9, 29; m 51; c 4. PARASITOLOGY, CYTOCHEMISTRY. Educ: NY Univ, AB, 49; Baylor Univ, MA, 54; Univ Va, PhD(biol), 57. Prof Exp: From asst prof to prof biol, Ga Teachers Col, 57-63; asst prog dir, NSF, 63-64; assoc prof, 64-71, chmn dept, 72-75, PROF BIOL, VANDERBILT UNIV, 71- Concurrent Pos: Instr, Med Field Serv Sch, US Army, 52-54; NIH fel, Hammersmith, London, Eng, 73. Mem: Am Soc Parasitol; Am Micros Soc; Histochem Soc. Res: Histochemistry of host-parasite relationships; ultrastructural localization of enzymes in helminths. Mailing Add: Dept of Biol Vanderbilt Univ Box 1733 Sta B Nashville TN 37235

BOGLE, ROBERT WORTHINGTON, b Chicago, Ill, Dec 25, 18; m 42; c 3. EXPERIMENTAL PHYSICS. Educ: Univ Mich, BS, 41, MS, 42, PhD(physics), 48. Prof Exp: Asst, Off Sci Res & Develop, Nat Defense Res Comt, US Navy & Manhattan Dist contracts, Univ Mich, 42-43, assoc, 43-44, physicist, 44-48; physicist, Appl Physics Lab, Johns Hopkins Univ, 48-58; sr scientist, Santa Barbara Div, Curtiss-Wright Corp, 58-60; mgr Pac Range eng, Gen Elec Co, 60-64; mem, Appl Sci Staff, Gen Res Corp, 64-69; prog dir, Astro Res Corp, 69-71; MEM, RADAR ANAL STAFF, US NAVAL RES LAB, 71- Concurrent Pos: Parsons fel, Johns Hopkins Univ, 56-57. Mem: Am Phys Soc; Am Inst Aeronaut & Astronaut. Res: Experimental research in radar geophysics; applied research in physics; electronics; hypersonic fluid mechanics; system development and analysis. Mailing Add: Naval Res Lab Washington DC 20375

BOGLE, TOMMY EARL, b Logansport, La, Sept 4, 40; m 62; c 1. SOLID STATE PHYSICS. Educ: Va Polytech Inst, BS, 62; La State Univ, PhD(physics), 68. Prof Exp: Design engr, Tex Instruments, Inc, 62-63; asst prof, 68-70, ASSOC PROF PHYSICS, McNEESE STATE UNIV, 70- Mem: Am Phys Soc; Am Inst Physics. Res: Investigation of Fermi surface of metals at liquid helium temperatures using the magneto-acoustic effect. Mailing Add: Dept of Physics McNeese State Univ Lake Charles LA 70601

BOGNAR, KALMAN, physical geography, meteorology, see 12th edition

BOGNER, PHYLLIS HOLT, b Middleboro, Mass, Mar 26, 30; m 56 & 72; c 2. PHYSIOLOGY. Educ: Tufts Univ, BS, 52; Brown Univ, MSc, 54; Univ Md, PhD, 57. Prof Exp: Asst, Tufts Univ, 50, Brown Univ, 52-54, Columbia Univ, 54-55 & Univ Md, 55-57; instr physiol & pharmacol, Sch Med, Univ Pittsburgh, 57-61, from asst prof to assoc prof pharmacol, 61-70; ASSOC PROF PHARMACOL, SCH BASIC MED SCI, UNIV ILL MED CTR & ASSOC, EDUC SYSTS SECT, CTR FOR EDUC DEVELOP, 70- Mem: AAAS; Am Physiol Soc; Am Soc Zoologists; NY Acad Sci. Res: Development of mechanisms responsible for transport of monosaccharides across the intestinal wall; developmental physiology of the gastrointestinal tract; effectiveness of audiotutorials versus the lecture-mode of teaching at the medical school level. Mailing Add: 634 Downing Rd Libertyville IL 60048

BOGOCH, SAMUEL, b Saskatoon, Sask, Jan 13, 28; m 53. BIOCHEMISTRY, PSYCHIATRY. Educ: Univ Toronto, MD, 51; Harvard Univ, PhD(biochem), 56. Prof Exp: Intern, Toronto Gen Hosp, Ont, 51-52; resident psychiat, Crease Clin, Vancouver, BC & McLean Hosp, Mass, 52-53; resident psychiat & asst psychiatrist, Boston Psychopath Hosp Community Clin, 56-57, dir neurochem res lab, 56-61; asst res prof psychiat, 61-65, ASSOC RES PROF BIOCHEM & PSYCHIAT, SCH MED, BOSTON UNIV, 65-, MEM FAC GRAD SCH, 64-; DIR, FOUND RES NERV SYST, 61- Concurrent Pos: Fel, Harvard Med Sch, 56-57; sr psychiatrist, Mass Ment Health Ctr, 57-60; from asst to instr, Harvard Med Sch, 57-61; gen dir, Dreyfus Med Found, 68-, exec vpres. Mem: Am Psychiat Asn; Am Acad Neurol; Int Soc Neurochem. Res: Biochemistry, physiology and clinical disorders of central nervous system. Mailing Add: 36 Fenway Boston MA 02215

BOGORAD, LAWRENCE, b Tashkent, Russia, Aug 29, 21; nat US; m 43; c 2. PLANT PHYSIOLOGY. Educ: Univ Chicago, SB, 42, PhD(plant physiol), 49. Hon Degrees: MA, Harvard Univ, 67. Prof Exp: Instr bot, Univ Chicago, 48-51, from asst prof to assoc prof, 53-61, prof, 61-67; PROF BOT, HARVARD UNIV, 67-, CHMN DEPT BIOL, 74- Concurrent Pos: Nat Res Coun-Merck fel, Rockefeller Inst, 51-53; Fulbright res scholar & NSF sr fel div plant indust, Commonwealth Sci & Indust Res Orgn, Australia, 60; NSF sr fel biochem, Nobel Inst Med, Sweden, 61; NIH career award, 63-67; chmn sect molecular bot, Int Bot Cong, Seattle, 68; chmn, Gordon Conf Pyrrolic Compounds, 70; chmn bot sect, Nat Acad Sci, 74-77. Mem: Nat Acad Sci; Am Soc Plant Physiol (pres, 68); fel Am Acad Arts & Sci; Bot Soc Am; Am Soc Biol Chem. Res: Biosynthesis of porphyrins, plant growth and development; chloroplasts; eukaryotic organelle biology; molecular biology; porphyrin biosynthesis. Mailing Add: Biol Labs Harvard Univ 16 Divinity Ave Cambridge MA 02138

BOGUCKI, RAYMOND FRANCIS, b Wallingford, Conn, Mar 6, 28; m 54; c 5. INORGANIC CHEMISTRY. Educ: Col Holy Cross, BS, 53, MS, 54; Clark Univ, PhD(chem), 59. Prof Exp: Instr chem, Lafayette Col, 54-56; res fel, Clark Univ, 59; from asst prof to assoc prof, Boston Col, 59-68; PROF CHEM & CHMN DEPT, UNIV HARTFORD, 68- Concurrent Pos: Vis prof, Middle East Tech Inst, Ankara, 64-65, Purdue Univ, 66-68 & Tex A&M Univ, 74-75. Mem: AAAS; Am Chem Soc; Sigma Xi. Res: Reactions of metal chelate compounds in aqueous solution; polynuclear metal chelates; metal coordination polymers. Mailing Add: Dept of Chem Univ of Hartford West Hartford CT 06117

BOGUSCH, EDWIN ROBERT, b Mason, Tex, Jun 6, 05; m 29; c 2. ECOLOGY. Educ: Univ Tex, AB & AM, 28, PhD(plant physiol), 43. Prof Exp: Syst botanist, Univ Ill, 28-30; instr bot & cur herbarium, State Col Wash, 30-34; agent, USDA, 34-36; teacher schs, Tex, 36-41; prof biol, 41-70, sr prof, 70-74, chmn dept, 47-67, EMER PROF BIOL, TEX A&I UNIV, 74- Concurrent Pos: Assoc botanist, USDA with Dept Interior, 44; wildlife cinematophotographer and lectr conserv, 65-66; Nat Audubon wildlife film lectr. Res: Bioelectronics; symbiosis, growth and development of gonidia from species of Parmelia and Ramalina; bioecology, especially ecology of arid lands. Mailing Add: Dept of Biol Tex A&I Univ Kingsville TX 78363

BOGUSLASKI, ROBERT CHARLES, bioorganic chemistry, see 12th edition

BOGUSLAWSKI, GEORGE, b Kamyszlow, USSR, Sept 7, 41; US citizen; c 2. GENETICS, BIOCHEMISTRY. Educ: Warsaw Univ, MSc, 65; Cornell Univ, PhD(genetics), 73. Prof Exp: Fel, Sch Med, Washington Univ, 73-75; ASST PROF MICROBIOL, UNIV KANS, 75- Mem: Am Soc Microbiol. Res: Transcriptional processes in morphological differentiation of fungi. Mailing Add: Dept of Microbiol Univ of Kans Lawrence KS 66045

BOGYO, STEPHEN W, food technology, agronomy, see 12th edition

BOGYO, THOMAS P, b Budapest, Hungary, July 14, 18; US citizen; m 46; c 5. GENETICS, STATISTICS. Educ: NC State Univ, MES, 63; Jozsef Nador Univ, Budapest, DSc(agr bot), 41. Prof Exp: Plant breeder, Hungarian Plant Breeding Co, 41-45; asst prof genetics, Jozsef Nador Univ, Budapest, 45-46; asst dir, Hungarian Inst Genetics, 46-47; sr prof officer genetics, SAfrican Dept Agr, 56-58, chief prof officer genetics & agron, 56-60; asst statistician, 62-65, assoc statistician, 65-70, from asst prof to assoc prof bot & info sci, 64-66, assoc prof genetics, 66-70, PROF GENETICS & STATISTICIAN, WASH STATE UNIV, 70- Mem: Biomet Soc; Genetics Soc Am. Res: Genotype-environment interaction; simulation of genetic systems on electronic computers; experimental designs; selection. Mailing Add: 123 Ag-2 Bldg Wash State Univ Pullman WA 99163

BOHACHEVSKY, IHOR O, b Sokal, Ukraine, Sept 7, 28; US citizen. FLUID MECHANICS, APPLIED MATHEMATICS. Educ: NY Univ, BAE, 56, PhD(math), 61. Prof Exp: Assoc res scientist, NY Univ, 60-63, adj asst prof aeronaut, 61-63; aerodyn res engr, Cornell Aeronaut Lab, 63-67; prin res scientist, Avco-Everett Res Lab, 67-68; staff mem, Bell Tel Lab, Inc, 68-75; STAFF MEM, LOS ALAMOS SCI LAB, 75- Honors & Awards: Apollo Achievement Award, NASA, 69; Am Tel & Tel Recognition Cert, 69. Mem: Am Math Soc; Am Inst Aeronaut & Astronaut; Sigma Xi. Res: Magnetohydrodynamics; numerical methods in fluid mechanics; nonlinear wave propagation; geophysical fluid mechanics; mathematical economics; gas dynamics; problems related to laser applications and to laser initiated fusion reactor development. Mailing Add: 3210 A Walnut St Los Alamos NM 87544

BOHAN, THOMAS LYNCH, b Terre Haute, Ind, Feb 12, 38; m 60; c 3. BIOPHYSICS, SOLID STATE PHYSICS. Educ: Univ Chicago, BS, 60; Univ Ill, MS, 64, PhD(physics), 68. Prof Exp: Res physicist, Fansteel Metall Corp, 60-62; res assoc physics, Univ Ill, 68-69; ASST PROF PHYSICS, BOWDOIN COL, 69- Concurrent Pos: Vis prof physics & Fulbright lectr, Univ Nacional Mayor de San Marcos, Lima, Peru, 72-73. Mem: AAAS; Am Phys Soc. Res: Electron spin resonance, Mossbauer, and Raman spectroscopy of heme proteins; low temperature electron spin resonance and spin-lattice relaxation studies of rare earths; semiconductor device properties. Mailing Add: Dept of Physics Bowdoin Col Brunswick ME 04011

BOHANDY, JOSEPH, b Barnesville, Ohio, July 9, 38; m 59; c 2. SOLID STATE PHYSICS. Educ: Ohio State Univ, BS, 60, MS, 61, PhD(physics), 65. Prof Exp: SR PHYSICIST, APPL PHYSICS LAB, JOHNS HOPKINS UNIV, 65- Mem: Am Phys Soc. Res: Electron paramagnetic resonance used for study of magnetism and magnetic properties of crystalline solids. Mailing Add: Appl Physics Lab Johns Hopkins Rd Laurel MD 20810

BOHANNAN, LAURA, b New York, NY, July 20, 22; m 43; c 1. ANTHROPOLOGY. Educ: Univ Ariz, BA, 43, MA, 47; Oxford Univ, PhD, 51. Prof Exp: Instr Ger, Univ Ariz, 46-47; sr res fel anthrop, EAfrican Inst Soc Res, Uganda, 54-56; res assoc, Northwestern Univ, 60-65; assoc prof anthrop, 65-69, PROF ANTHROP, UNIV ILL, CHICAGO CIRCLE, 69- Concurrent Pos: Lectr, Univ Chicago, 62-63 & 64-65; res assoc, NIMH & NSF, 63-64; ed, Am Anthropologist, 70-73; Guggenheim fel, 74. Mem: AAAS; African Studies Asn; fel Am Anthrop Asn; fel Int African Inst; fel Royal Anthrop Inst Gt Brit & Ireland. Res: Family and household structure, especially in Africa; African witchcraft and magic. Mailing Add: Dept of Anthrop Univ of Ill at Chicago Circle Chicago IL 60680

BOHANNAN, PAUL JAMES, b Lincoln, Nebr, Mar 5, 20; m 43; c 1. ANTHROPOLOGY. Educ: Univ Ariz, BA, 47; Oxford Univ, BSc, 49, DPhil(anthrop), 51. Prof Exp: Univ lectr anthrop, Oxford Univ, 51-56; assoc prof, Princeton Univ, 56-59; prof anthrop & educ, Northwestern Univ, Evanston, 59-76, Stanley G Harris prof soc sci, 68-76; RES ASSOC, WESTERN BEHAV SCI INST, 74- Concurrent Pos: Field worker, Tiv of Cent Nigeria, 49-52, Wanga of Kenya, 55, divorcees in San Francisco, 63-64 & Downtown San Diego, 74-; exec officer, Human Environ in Mid Africa Proj, Nat Acad Sci, 58-60; NIMH fel, Inst Advan Study Behav Sci, 63-64; prof anthrop, Univ Calif, Santa Barbara, 76- Mem: AAAS; Am Anthrop Asn; Royal Anthrop Inst Gt Brit & Ireland; Brit Asn Social Anthrop. Res: General anthropology; urban anthropology; psychoanalysis. Mailing Add: Dept of Anthrop Univ of Calif Santa Barbara CA 93106

BOHANNON, GEORGE EDMOND, b Cushing, Okla, July 18, 47. NUCLEAR PHYSICS, PARTICLE PHYSICS. Educ: Okla City Univ, BA, 69; Mich State Univ, MS, 71, PhD(physics), 74. Prof Exp: APPOINTEE PHYSICS, LOS ALAMOS SCI LAB, 74- Mem: Am Phys Soc. Res: Theoretical low and medium energy nuclear physics, nucleon-nucleon scattering. Mailing Add: Group T5 Los Alamos Sci Lab Los Alamos NM 87545

BOHANNON, RANDOLPH F, b Ottawa, Ill, Mar 20, 43; m 63; c 1. PLANT PHYSIOLOGY, MOLECULAR BIOLOGY. Educ: St Olaf Col, BA, 65; Purdue Univ, PhD(plant physiol), 69. Prof Exp: ASSOC PROF BIOL, PAC LUTHERAN UNIV, 69- Mem: AAAS; Am Inst Biol Sci; Am Soc Plant Physiol. Res: Plant RNA metabolism; plant growth regulators. Mailing Add: Dept of Biol Pac Lutheran Univ Tacoma WA 98447

BOHANNON, ROBERT ARTHUR, b Holton, Kans, Feb 5, 22; m 46; c 3. SOIL SCIENCE. Educ: Mich State Univ, BS, 49; Kans State Univ, MS, 51; Univ Ill, PhD, 57. Prof Exp: County agent, Nemaha County, Kans, 51-52; exten specialist & agronomist crops & soils, 52-55, 56-61, asst to dean agr, 61-65, actg head dept agron, 65-66, dir int agr progs, 66-68, assoc prof, 51-61, PROF AGRON, KANS STATE UNIV, 61-, DIR EXTEN SERV, 68- Concurrent Pos: Fel, Univ Mich, 64-65; chmn exten comt, Great Plains Agr Coun, 71. Mem: Am Soc Agron; Soil Sci Soc Am. Res: Soil testing and fertility. Mailing Add: Div of Coop Exten Umberger Hall Kans State Univ Manhattan KS 66502

BOHANNON, ROBERT GARY, b Hollywood, Calif, Feb 17, 49; m 70. GEOLOGY. Educ: Glendale Col, AA, 69; Calif State Univ, Northridge, BS, 71; Univ Calif, Santa Barbara, PhD(geol), 76. Prof Exp: Teaching asst geol, Univ Calif, Santa Barbara, 71-74; GEOLOGIST, DEPT INTERIOR, US GEOL SURV, 74- Mem: Geol Soc Am. Res: Structural geology, geochemistry and sedimentology of the Muddy Mountains, southern Nevada; occurences of lithium in the western United States. Mailing Add: Dept of Interior US Geol Surv Box 25046 Fed Ctr Denver CO 80225

BOHART, GEORGE EDWARD, entomology, see 12th edition

BOHART, RICHARD MITCHELL, b Palo Alto, Calif, Sept 28, 13; m 39. ENTOMOLOGY. Educ: Univ Calif, BS, 34, MS, 35, PhD(entom), 38. Prof Exp: Asst, Univ Calif, 35-37; assoc entom, Univ Calif, Los Angeles, 38-39, instr & jr entomologist, Exp Sta, 39-46; from asst prof to assoc prof, 46-58, from vchmn to chmn dept, 57-67, PROF ENTOM, COL AGR, UNIV CALIF, DAVIS, 58- Mem: Am Entom Soc. Res: Taxonomy of Strepsiptera Vespidae and Sphecidae; biology; taxonomy and control of mosquitoes. Mailing Add: Entom Bldg Univ of Calif Davis CA 95616

BOHEN, JOSEPH MICHAEL, b Philadelphia, Pa, Feb 8, 46; m 69. ORGANIC CHEMISTRY. Educ: Temple Univ, AB, 68; Univ Pa, PhD(chem), 73. Prof Exp: SR RES CHEMIST, PENNWALT CORP, 73- Concurrent Pos: Instr, Univ Pa, 74-75. Mem: Am Chem Soc. Res: Organometallic compounds; heterocumulenes; cycloaddition reactions of heterocumulenes; photochemistry. Mailing Add: Pennwalt Corp 900 First Ave King of Prussia PA 19406

BOHINSKI, ROBERT CLEMENT, b Wilkes-Barre, Pa, Apr 11, 40; m 61; c 3. BIOCHEMISTRY. Educ: Scranton Univ, BS, 61; Pa State Univ, MS, 63, PhD(biochem), 65. Prof Exp: Res biochemist, Charles Pfizer & Co, Inc, 65-66; from asst prof to assoc prof chem, 71-74, PROF CHEM, JOHN CARROLL UNIV, 74- Honors & Awards: Distinguished Fac Award, John Carroll Univ, 73. Res: Growth and survival characteristics of bacteria under conditions of phosphate, sulfate and nitrogen deficiencies with specific emphasis on sulfur metabolism and turnover of ribosomes. Mailing Add: Dept of Chem John Carroll Univ Cleveland OH 44118

BOHL, EDWARD HOMER, b Georgetown, Ohio, Oct 16, 21; m 47; c 3. VETERINARY VIROLOGY. Educ: Ohio State Univ, DVM, 44, MSc, 48, PhD, 52. Prof Exp: Instr microbiol, Ohio State Univ, 48-52, from asst prof to prof, 52-63, PROF VET SCI, OHIO AGR RES & DEVELOP CTR, OHIO STATE UNIV, 63- Mem: Am Vet Med Asn; Am Soc Microbiol; Am Col Vet Microbiol; Conf Res Workers Animal Dis. Res: Animal virology; veterinary microbiology; leptospirosis; infectious diseases of animals; viral infection of the intestinal tract. Mailing Add: 564 McDonald St Wooster OH 44691

BOHLEN, HAROLD GLENN, b Raleigh, NC, Oct 11, 46; m 70; c 2. CARDIOVASCULAR PHYSIOLOGY. Educ: Appalachian State Univ, BA, 68; Bowman Gray Sch Med, PhD(physiol), 73. Prof Exp: Am Heart Asn fel physiol, Univ Ariz, 73-75; ASST PROF PHYSIOL, IND UNIV, INDIANAPOLIS, 75- Mem: Am Heart Asn. Res: Neural and local vascular control of the intestinal microcirculation. Mailing Add: Dept of Physiol Ind Univ Sch of Med Indianapolis IN 46202

BOHLEN, WALTER FRANKLIN, b Tarrytown, NY, June 21, 38; m 67; c 2. PHYSICAL OCEANOGRAPHY. Educ: Univ Notre Dame, BSEE, 60; Mass Inst Technol, PhD(oceanog), 69. Prof Exp: Res asst geophys, Woods Hole Oceanog Inst, 62-63; engr electronics, Robert Taggert Inc, 63-65; res asst oceanog, Woods Hole

Oceanog Inst, 65; res asst geol, Mass Inst Technol, 65-69; ASST PROF OCEANOG, UNIV CONN, 69- Concurrent Pos: Mem res planning adv comt, Long Island Sound Study, New Eng River Basins Comn, 72-75. Mem: Am Geophys Union; Marine Technol Soc; Sigma Xi. Res: Experimental investigations of sediment transport in coastal waters with particular emphasis on the factors governing variability, the relationship between laboratory and field data and the development of quantitative transport models. Mailing Add: Marine Sci Inst Univ of Conn Avery Point Groton CT 06340

BOHLIN, JOHN DAVID, b Valparaiso, Ind, Mar 19, 39; m 66; c 2. SOLAR PHYSICS. Educ: Wabash Col, AB, 61; Univ Colo, Boulder, PhD(solar physics), 68. Prof Exp: Res fel astron, Calif Inst Technol, 67-70; ASTROPHYSICIST SOLAR PHYSICS, NAVAL RES LAB, 70- Mem: Am Astron Soc; Int Astron Union. Res: Physics of the outer atmosphere of the sun, especially as inferred from ultraviolet wave lengths observed from space vehicles; structure and dynamics of the solar corona. Mailing Add: Code 7141B Naval Res Lab Washington DC 20375

BOHLIN, RALPH CHARLES, b Iowa City, Iowa, Nov 26, 43; m 71. ASTROPHYSICS. Educ: Univ Iowa, BA, 66; Princeton Univ, MA, 69, PhD(astrophysics), 70. Prof Exp: Nat Res Coun fel, 69; res assoc astrophysics, Lab Atmospheric & Space Physics, Univ Colo, 70-73; ASTRONR, GODDARD SPACE FLIGHT CTR, GREENBELT, 73- Concurrent Pos: Liaison astronr, Astron Spacelab Payloads Proj, 74- & Large Space Telescope Proj, 74- Mem: Am Astron Soc; Int Astron Union; Sigma Xi. Res: Spectroscopy and imaging of astronomical objects, including early-type stars, planetary nebulae, diffuse nebulae and Seyfert nuclei, using data from sounding rockets and orbiting satellites; emphasis on the gas and dust in the interstellar medium. Mailing Add: 9104 Horton Rd Laurel MD 20811

BÖHLKE, JAMES ERWIN, b Buffalo, Minn, Jan 11, 30; m 51; c 3. ICHTHYOLOGY. Educ: Stanford Univ, AB, 50, PhD(biol sci), 54. Prof Exp: Curatorial & res asst, Natural Hist Mus, Stanford Univ, 48-53, lab teaching asst vert, 50-51, spec asst zool collections, 51-54; fishery aide, US Fish & Wildlife Serv, 53-54, syst zoologist, 54; asst cur fishes, 54-57, assoc cur ichthyol & herpet, 58-62, CHAPLIN CHAIR ICHTHYOL, ACAD NATURAL SCI, 59-, CHMN, 60-, CUR, 63- Concurrent Pos: Exped, Mex, 48 & 70, Alaska, 51, Bahamas, 55-, Seychelles Islands, 64, Colombia, 69 & Lesser Antilles, 69; ichthyologist, Stanford Univ Sefton Found exped, Gulf of Calif, 52; ed sci publ & chmn publ comt, Acad Natural Sci, 70-; collabr, Fla State Mus, 58; mem adv panel vert, Smithsonian Oceanog Sorting Ctr, 64-66, chmn, 66-; res assoc fishes, Smithsonian Inst, 67-; adj prof marine sch, Sch Marine & Atmospheric Sci, Univ Miami, 67- Mem: AAAS; Am Soc Limnol & Oceanog; Soc Syst Zool; Am Soc Ichthyol & Herpet (treas, 58-60); Soc Study Evolution. Res: Systematics, including distribution, life history and ecology of fishes; fishery biology; comparative morphology. Mailing Add: Acad of Natural Sci 19th & Parkway Philadelphia PA 19103

BOHLMANN, EDWARD GUSTAV, b Milwaukee, Wis, July 19, 17; m 47; c 3. GEOCHEMISTRY, CORROSION. Educ: Univ Wis, BS, 39, MS, 41. Prof Exp: Asst sanit chem, Univ Wis, 39-40, asst phys chem, 40-41; chemist, Solvay Process Co, NY, 41-42; res chemist metal lab, Manhattan Proj, Univ Chicago, 43; res chemist, Oak Ridge Nat Lab, 43-48, group leader, 45-48; res engr, Battelle Mem Inst, 48-49; sr res chemist, 49-52, chief corrosion sect reactor exp eng div, 52-55, from asst dir to assoc dir, 55-61, assoc dir reactor chem div, 61-73, MEM SR RES STAFF CHEM DIV, OAK RIDGE NAT LAB, 73- Mem: AAAS; Am Chem Soc; Nat Asn Corrosion Eng; Sigma Xi. Res: Physical chemistry; inorganic chemistry; aqueous corrosion; materials and corrosion in nuclear power reactors; molten salt reactors; silica and scaling in geothermal brines. Mailing Add: Oak Ridge Nat Lab PO Box X Oak Ridge TN 37830

BOHM, ARNO, b Stettin, Ger, Apr 26, 36; m 65; c 2. HIGH ENERGY PHYSICS, THEORETICAL PHYSICS. Educ: Tech Univ, Berlin, DiplPh, 62; Univ Marburg, Dr rer nat, 66. Prof Exp: Sci asst theoret physics, Univ Karlsruhe, 63-64; Int Atomic Energy Agency res fel particle physics, Int Ctr Theoret Physics, Trieste, Italy, 64-66; res assoc, Syracuse Univ, 66-68; ASSOC PROF PARTICLE PHYSICS, UNIV TEX, AUSTIN, 68- Mem: Fel Am Phys Soc; Ger Phys Soc. Res: Theoretical and elementary particle physics; mathematical physics. Mailing Add: Dept of Physics Univ of Tex Austin TX 78712

BOHM, BRUCE ARTHUR, b Burlington, Wis, Apr 9, 35; Can citizen; m 61; c 2. BOTANY. Educ: Alfred Univ, BS, 56; Univ RI, MS, 58, PhD(chem), 60. Prof Exp: Fel bot, McGill Univ, 60-61; asst prof chem, Univ Sask, 61-63; asst res prof pharmacog, Univ RI, 63-66; from asst prof to assoc prof, 66-73, PROF BOT, UNIV BC, 73- Mem: Am Bot Soc; Am Soc Plant Taxonomists; Phytochem Soc NAm (pres, 69); AAAS; Res: Chemical plant taxonomy utilizing secondary metabolites and comparative protein structure. Mailing Add: Dept of Bot Univ of BC Vancouver BC Can

BOHM, HENRY, b Vienna, Austria, July 16, 29; US citizen; m 50; c 2. SOLID STATE PHYSICS. Educ: Harvard Univ, AB, 50; Univ Ill, MS, 51; Brown Univ, PhD(physics), 58. Prof Exp: Jr physicist, Gen Elec Co, 51 & 53-54; staff mem, Arthur D Little, Inc, 58-59; assoc prof physics, 59-64, actg chmn dept, 62-63, vpres grad studies & res, 68-71, vpres spec projs, 71-72, provost, 72-75, PROF PHYSICS, WAYNE STATE UNIV, 64- Concurrent Pos: Vis prof physics, Cornell Univ, 66-67; consult, Aerospace Syst Command, US Air Force, 60-62; mem bd dir, Ctr Res Libr, Chicago, Ill, 70-75, chmn, 73; comnr-examr, NCent Asn Schs & Cols, 70-; comnr, Comn Inst High Educ, 74- Mem: Am Phys Soc. Res: Study of Fermi surfaces of metals, using the magnetoacoustic technique at frequencies above 200 megacycles per second; cryogenics. Mailing Add: Dept of Physics Wayne State Univ Detroit MI 48202

BOHM, HOWARD ALLAN, b New York, NY, July 28, 43; m 67; c 1. MEDICINAL CHEMISTRY. Educ: NY Univ, BA, 66; Adelphi Univ, PhD(org chem), 71. Prof Exp: Res assoc, Israel Inst Technol, 71-72; res assoc, Princeton Univ, 72-73; SR RES CHEMIST MED CHEM, PENNWALT CORP, KING OF PRUSSIA, PA, 73- Res: Cardiovascular drugs; antihypertensives, antiarrhythmics, cardiotonics, antiangina drugs, diuretics; photochemistry of carbonyl compounds; organic synthesis of highly strained molecules; microencapsulation. Mailing Add: 750 Old Lancaster Rd Berwyn PA 19312

BOHMAN, VERLE RUDOLPH, b Peterson, Utah, Dec 29, 24; m 45; c 5. ANIMAL NUTRITION. Educ: Utah State Agr Col, BS, 49, MS, 51; Cornell Univ, PhD(animal nutrit), 52. Prof Exp: From asst prof & asst animal nutritionist to assoc prof animal husb & assoc animal nutritionist, 52-63, PROF ANIMAL HUSB & NUTRITIONIST, UNIV NEV, 63-, CHMN DIV ANIMAL SCI, 60- Concurrent Pos: Ed-in-chief, J Animal Sci, 70-72. Mem: AAAS; Am Inst Nutrit; Am Soc Animal Sci (pres-elect, 73-74, pres, 74-75); Am Dairy Asn. Res: Range livestock nutrition; antibiotics; fat, hormones, protein and mineral metabolism; radioactive fallout. Mailing Add: Animal Sci Div Univ of Nev Reno NV 89507

BÖHMER, HEINRICH EVERHARD, b u:nster, Ger, May 6, 31; m 62; c 1. PHYSICS. Educ: Univ Münster, dipl, 59, PhD(diffusion in metals), 61. Prof Exp: Asst solid state physics, Inst Metal Res, Univ Münster, 62; res assoc surface physics, 62-63 & plasma physics, 63-65, res asst prof, 65-69, ASST PROF PHYSICS, COORD SCI LAB, UNIV ILL, URBANA, 69- Concurrent Pos: Vis physicist, Dept Plasma Physics, Fontenay Nuclear Res Ctr, Fonteny, France, 72-73 & Dept Physics, Univ Calif, Irvine, 75- Mem: Am Phys Soc. Res: Plasma physics. Mailing Add: Dept of Physics Coord Sci Lab Univ of Ill Urbana IL 61801

BOHMFALK, ERWIN FREDERICK, JR, b Karnes City, Tex, Nov 2, 23; m 47. PHYSICAL CHEMISTRY. Educ: Tex Col Mines & Metal, BS, 48; Univ Colo, PhD(chem), 52. Prof Exp: Res chemist, E I du Pont de Nemours & Co, 51; asst, Univ Colo, 51-52; res chemist, 52-64, process supvr, 64-70, tech supvr, Tech Group-Fibers, Du Pont de Nemours (Nederland) NV, 70-73, PROCESS SUPVR, E I DU PONT DE NEMOURS & CO, INC, WAYNESBORO, VA, 73- Mem: Am Chem Soc. Res: Textile fibers, plastics; vapor phase calorimetry. Mailing Add: PO Box 986 Waynesboro VA 22980

BOHMONT, DALE W, b Wheatland, Wyo, June 7, 22; m 69; c 2. PLANT SCIENCE, ACADEMIC ADMINISTRATION. Educ: Univ Wyo, BS, 48, MS, 50; Univ Nebr, PhD, 52; Harvard Univ, MPA, 59. Prof Exp: Asst agronomist & asst prof agron, Univ Wyo, 48-53; assoc agronomist & head dept, 53-55, agronomist, prof & head dept, 55-60, head, Div Plant Sci, 59-61, assoc dir, Agr Exp Sta, Univ Colo, 61-63; DEAN & DIR, COL AGR, UNIV NEV, 63- Concurrent Pos: Asst agronomist, USDA, Univ Nebr, 50-51; conserv fel, Harvard Univ, 58-59; sr agr consult, Develop & Resources Corp NY, 68-74; pres, Enide Corp Nev, 74-; consult, TAMS Agr Develop Group, 75- Mem: AAAS; Am Soc Agron. Res: Crops, soils, weed control and public administration. Mailing Add: Col of Agr Univ of Nev Reno NV 89507

BOHN, CONRAD RATHMANN, b Austin, Tex, Oct 4, 26; m 48; c 4. PHYSICAL CHEMISTRY. Educ: Univ Tex, BA, 47, MA, 48; Univ Calif, PhD(chem), 52. Prof Exp: Asst, Univ Calif, 48-51; RES CHEMIST PHYS CHEM, E I DU PONT DE NEMOURS & CO, INC, 51- Res: Infrared spectroscopy; physical chemistry and properties of high polymers; electronic computer programming. Mailing Add: 111 Cambridge Dr Windsor Hills Wilmington DE 19803

BOHN, HINRICH LORENZ, b New York, NY, Aug 25, 34; m 62; c 3. SOIL CHEMISTRY. Educ: Univ Calif, Berkeley, BS, 55, MS, 57; Cornell Univ, PhD(soil chem), 63. Prof Exp: fel, Univ Calif, Riverside, 63-64; res chemist, Tenn Valley Authority, 64-66; ASSOC PROF AGR CHEM, UNIV ARIZ, 66-, PROF SOIL CHEM, 74- Mem: Am Chem Soc; Soil Sci Soc Am. Res: Soil-pollutant reactions. Mailing Add: Dept of Agr Chem & Soils Univ of Ariz Tucson AZ 85721

BOHN, RANDY G, b Toledo, Ohio, Feb 11, 41; m 65; c 1. LOW TEMPERATURE PHYSICS, CRYOGENICS. Educ: Univ Toledo, BS, 63; Ohio State Univ, PhD(physics), 69. Prof Exp: ASST PROF PHYSICS, UNIV TOLEDO, 69- Mem: Am Phys Soc. Res: Thermal conductivity in solid hydrogen; vitreous solids; thermal properties of solids at low temperatures. Mailing Add: 1611 Woodhurst Dr Toledo OH 43614

BOHN, ROBERT K, b New York, NY, Jan 24, 39; m 64; c 1. PHYSICAL CHEMISTRY. Educ: Univ Calif, Berkeley, BS, 59; Cornell Univ, PhD(phys chem), 64. Prof Exp: NATO fel electron diffraction, Univ Oslo, 64-65; asst prof, 65-71, ASSOC PROF CHEM, UNIV CONN, 71- Mem: Am Chem Soc; Am Phys Soc. Res: Molecular structure; electron diffraction; microwave spectroscopy. Mailing Add: Dept of Chem Univ of Conn Storrs CT 06268

BOHN, SHERMAN ELWOOD, b New England, NDak, Mar 11, 27; m 52; c 3. MATHEMATICS. Educ: Concordia Col, Moorhead, Minn, BA, 49; Univ Nebr, MA, 51, PhD(math, physics), 61. Prof Exp: Asst math, Univ Nebr, 49-51; res scientist, Nat Adv Comt Aeronaut, Va, 51-52; asst, Univ Minn, 52-53; from instr to asst prof, Concordia Col, Moorhead, Minn, 53-56; asst, Univ Nebr, 56-57, instr, 57-59; assoc prof, Wartburg Col, 59-61; from asst prof to assoc prof, Bowling Green State Univ, 61-64; assoc prof, 64-66, chmn dept, 65-71, PROF MATH, MIAMI UNIV, 66- Concurrent Pos: Chmn math sect, Ohio Acad Sci, 69-70. Mem: Math Asn Am; Am Math Soc; Soc Indust & Appl Math. Res: Mathematical analysis, function spaces and partial differential equations; topology and algebra; semi-topological groups. Mailing Add: Dept of Math Miami Univ Oxford OH 45056

BOHNENBLUST, FREDERIC, b Neuchatel, Switz, Mar 22, 06; nat US; m 37. MATHEMATICS. Educ: Polytech Sch, Zurich, licencie, 28; Princeton Univ, PhD(math), 31. Prof Exp: From instr to assoc prof math, Princeton Univ, 31-45; prof, Ind Univ, 45-46; prof, 46-74, dean grad studies, 56-71, EMER PROF MATH, CALIF INST TECHNOL, 74- Concurrent Pos: Mem, Nat Defense Res Comt. Mem: Am Math Soc; Math Asn Am. Res: Analysis; differential equations. Mailing Add: Dept of Math Calif Inst Technol Pasadena CA 91125

BOHNENBLUST, KENNETH E, b Bala, Kans, Dec 9, 23; m 45; c 3. PLANT BREEDING, PLANT PATHOLOGY. Educ: Kans State Univ, BS, 59; Univ Wyo, MS, 61; Univ Minn, St Paul, PhD(plant path), 66. Prof Exp: Instr agron, Univ Wyo, 59-61; asst prof biol, Wis State Univ, River Falls, 66-67; ASST PROF PLANT BREEDING, UNIV WYO, 67- Res: Potato diseases; Actinomycetes in plant disease; potato breeding; alfalfa breeding and pest resistance; virus diseases of plants; breeding cereals; morphology of cereal seedlings. Mailing Add: Plant Sci Div Univ of Wyo Laramie WY 82070

BOHNING, JAMES JOEL, b Cleveland, Ohio, Apr 11, 34; m. PHOTOCHEMISTRY. Educ: Valparaiso Univ, BS, 56; NY Univ, MS, 59; Northeastern Univ, PhD(phys chem), 65. Prof Exp: Instr chem, Wilkes Col, 59-62; res fel, Northeastern Univ, 62-64; from asst prof to assoc prof, 67-72, PROF CHEM, WILKES COL, 72-, CHMN DEPT, 70- Mem: Am Chem Soc. Mailing Add: Dept of Chem Wilkes Col Wilkes-Barre PA 18703

BOHNING, JOHN WILLIAM, b Cleveland, Ohio, Nov 14, 20; m 45; c 5. AGRONOMY. Educ: Univ Idaho, BS, 48. Prof Exp: Range conservationist, Range Surv, Umatilla Nat Forest, 48-50, 51; spec study, Rocky Mountain Sta, 50-51, range res, Southwestern Exp Sta, 51-54; range res, Rocky Mountain Exp Sta, 54-57, Forest Serv Regional Off, 57-59, Kaibab Nat Forest, 59-63 & Santa Fe Nat Forest, 63-68; RANGE & WILDLIFE STAFF MEM, PRESCOTT NAT FOREST, 68- Mem: Soc Range Mgt. Res: Grazing management of southwest native and reseeded ranges. Mailing Add: PO Box 441 Prescott AZ 86301

BOHNING, RICHARD HOWARD, b Hope Valley, RI, Sept 16, 19; m 38; c 1. BOTANY. Educ: RI State Col, BS, 40; Ohio State Univ, MSc, 41, PhD(bot), 48. Prof Exp: Asst bot, 40-42, from instr to assoc prof, 46-61, campus coordr, Agr Educ & Res Proj for India, 57-64, asst dean, Col Agr & Home Econ, 60-64, assoc dean & asst dir, Ohio Agr Res & Develop Ctr, 64-68, PROF BOT, OHIO STATE UNIV, 61-, DEAN COL BIOL SCI, 69- Mem: AAAS; Am Bot Soc; Am Soc Plant Physiol. Res: Photosynthesis; translocation; water relations. Mailing Add: Col of Biol Sci Ohio State Univ Columbus OH 43210

BOHNSACK, KURT K, b Stuttgart, Ger, Mar 23, 20; nat US; m 46; c 3. ECOLOGY, ACAROLOGY. Educ: Ohio Univ, BS, 46; Univ Mich, MS, 47, PhD(zool), 54. Prof Exp: Instr zool, Univ Mich, 50-51; from instr to asst prof biol, Swarthmore Col, 51-56; from asst prof to assoc prof, 56-63, chmn dept, 64-67, PROF ZOOL, SAN DIEGO STATE UNIV, 63- Concurrent Pos: Investr, Arctic Res Lab, 62-64; ed, Pedobiologia, 64-; mem bd dirs, San Diego Natural Hist Mus, 66- Mem: AAAS; Ecol Soc Am. Res: Entomology. Mailing Add: Dept of Zool San Diego State Univ San Diego CA 92182

BOHON, ROBERT LYNN, b Decatur, Ill, July 20, 25; m 47; c 3. ENVIRONMENTAL SCIENCES, CHEMISTRY. Educ: Univ Ill, BS, 46, PhD(phys chem), 50. Prof Exp: Asst phys chem, Univ Ill, 49-50; phys chemist & head infrared lab, Anderson Phys Lab, 50-56; sr chemist, Minn Mining & Mfg Co, 56-66, supvr, 66-69, mgr, 69-74, MGR, 3M ENVIRON LAB, 74- Mem: Am Chem Soc; Am Sci Affil. Res: General physical chemistry; thermodynamics of solutions; thermal analysis; solid state reaction kinetics; physical properties of polymers; material science; environmental chemistry and engineering. Mailing Add: 5960 Hobe Lane White Bear Lake MN 55110

BOHONOS, NESTOR, b Winnipeg, Man, Aug 22, 15; nat US; m 41; c 2. BIOCHEMISTRY. Educ: Univ Alta, BSc, 37; Univ Wis, MS, 39, PhD(biochem), 43. Prof Exp: Res chemist, Lederle Labs, NY, 41-44; res chemist, Upjohn Co, Mich, 44-46; assoc chemist & asst prof agr chem, Purdue Univ, 46-49; head fermentation biochem res dept & assoc dir chemother res sect, Lederle Labs, 49-67; head biochem labs, Children's Cancer Res Found, assoc path, Harvard Univ & res assoc, Children's Hosp, Boston, 67-73; DIR MICROBIOL, RES DEPT & STAFF SCIENTIST LIFE SCI, STANFORD RES INST, 73- Mem: Am Chem Soc; Am Soc Biol Chemists; Am Soc Microbiol; NY Acad Sci; Soc Indust Microbiol. Res: Biochemistry of microorganisms; mode of action of drugs; chemistry of vitamins, antibiotics, steroids, alkaloids, and natural products; chemotherapy; microbial biosynthesis; blood platelet physiology; mutagenesis and carcinogenesis; biodegradation of pesticides and pollutants. Mailing Add: Life Sci Div Stanford Res Inst Menlo Park CA 94025

BOHOR, BRUCE FORBES, b Chicago, Ill, May 4, 32; m 53; c 4. GEOLOGY, CLAY MINERALOGY. Educ: Beloit Col, BS, 53; Univ Ind, MS, 55; Univ Ill, PhD(geol), 59. Prof Exp: Res geologist, Prod Res Div, Res & Develop Dept, Continental Oil Co, Okla, 57-63, Explor Res Div, 63-65; assoc geologist, Clay Resources & Clay Mineral Technol Sect, Ill State Geol Surv, 65-74; GEOLOGIST, US GEOL SURV, BR COAL RESOURCES, 74- Mem: Soc Econ Paleont & Mineral; Am Geochem Soc; Clay Minerals Soc; Mineral Soc Am. Res: Clay minerals and geochemistry in coal exploration; industrial applications of clays and clay minerals; development of clay resources. Mailing Add: US Geol Surv Br Coal Resources Fed Ctr Box 25046 Mail Stop 972 Denver CO 80225

BOHR, DAVID FRANCIS, b Zurich, Switz, June 22, 15; US citizen; m 40; c 3. CARDIOVASCULAR PHYSIOLOGY. Educ: Univ Mich, MD, 42. Prof Exp: Life Inst Med Res fel, Univ Calif, San Francisco, 46-48, res fel path, 47-48; from asst prof to assoc prof physiol, Med Sch, Univ Mich, Ann Arbor, 48-55; vis prof pharmacol, Univ Calif, San Francisco, 55-56; PROF PHYSIOL, MED SCH, UNIV MICH, ANN ARBOR, 56- Concurrent Pos: Vis prof physiol, Univ Heidelberg, 61-62; mem coun high blood pressure & coun basic sci, Am Heart Asn; distinguished prof, Univ Mich, 73. Mem: Am Physiol Soc. Res: Vascular smooth muscle; hypertension. Mailing Add: Dept of Physiol Univ of Mich Ann Arbor MI 48104

BOHREN, BERNARD BENJAMIN, b Omaha, Nebr, Aug 15, 14; m 38; c 3. ANIMAL GENETICS. Educ: Univ Ill, BS, 37; State Col Wash, MS, 40; Kans State Col, PhD(genetics), 42. Prof Exp: Instr voc agr schs, Ill, 37-38; asst, State Col Wash, 38-39; asst, poultry husb, Kans State Col, 39-43; asst poultry husb & agr chem, 43-46, assoc prof genetics, 46-50, PROF GENETICS, PURDUE UNIV, 50- Honors & Awards: Res Award, Poultry Sci Asn, 44. Mem: Poultry Sci Asn; Am Genetic Asn; Genetics Soc Am. Res: Genetics of and selection for quantitative traits in fowl. Mailing Add: Dept of Animal Sci Purdue Univ Lafayette IN 47907

BOHREN, CRAIG FREDERICK, b San Francisco, Calif, Feb 24, 40; m 64. APPLIED PHYSICS. Educ: Calif State Univ, San Jose, BS, 63; Univ Ariz, MS, 55, PhD(physics), 75. Prof Exp: Res asst nuclear eng, Gulf Gen Atomic, 65-67; res assoc hydrol, Sch Renewable Natural Resources, Univ Ariz, 72-75, lectr, Inst Atmospheric Physics, 74-75; RES ASST, DEPT APPL MATH & ASTRON, UNIV COL CARDIFF, 75- Mem: AAAS; Am Geophys Union; Int Glaciological Soc; Am Asn Physics Teachers. Res: Light scattering by small particles; applications to atmospheric physics, biophysics and astronomy. Mailing Add: Univ Col Cardiff PO Box 78 Cardiff Wales

BOHRER, JAMES CALVIN, b Evansville, Ind, Oct 27, 23; m 49; c 5. ORGANIC CHEMISTRY. Educ: Evansville Col, AB, 44; Cornell Univ, PhD, 51. Prof Exp: SR PROJ CHEMIST, COLGATE-PALMOLIVE CO, PISCATAWAY, 50- Concurrent Pos: Instr, Univ Col, Rutgers Univ, 63- Mem: Am Chem Soc. Res: Many-membered ring compounds; detergents. Mailing Add: 19 Peach Orchard Dr East Brunswick NJ 08816

BOHRER, JOHN JUNIOR, b Wayland, NY, Nov 9, 19; m 43; c 1. CHEMISTRY. Educ: Alfred Univ, BS, 41; Polytech Inst Brooklyn, MS, 46, PhD(chem), 50. Prof Exp: Chemist phys testing of glass, Corning Glass Works, 41; chemist metallic & non-metallic protective coatings, Schori Process Corp, 41-46, res assoc copolymerization & free radical reactions, Polytech Inst Brooklyn, 46-49; instr physics, St Francis Col, 48-49; phys chemist struct & properties of waxes, Socony Vacuum Oil Co, 49-50; dir res Int Resistance Co, 50-62, vpres res & develop, 62-71, DIR GEN ADJOINT, TRW COMPOSANTS ELECTRONIQUES, 71- ASST DIR GEN, 73- Concurrent Pos: Lectr grad sch chem, Pa. Mem: Am Chem Soc; Am Inst Chem; Soc Plastics Eng; Electrochem Soc; Inst Elec & Electronics Eng. Res: Physical organic chemistry of polymeric reactions; semiconductors and thin metal films. Mailing Add: TRW Composants Electroniques SA Rue Rene Magne 33 Bordeaux France

BOHRER, ROBERT EDWARD, b St Paul, Minn, July 24, 39; m 60; c 1. MATHEMATICS, STATISTICS. Educ: SDak Sch Mines & Technol, BS, 61; Univ NC, Chapel Hill, PhD(statist), 66. Prof Exp: Statistician, Statist Res Div, Res Triangle Inst, 64-65; vis lectr statist, Univ Col Wales, 65-66; statistician, Statist Res Div, Res Triangle Inst, 66-68; asst prof, 68-72, ASSOC PROF MATH, UNIV ILL, URBANA-CHAMPAIGN, 72- Concurrent Pos: Res fel, Univ Kent at Canterbury, 72; vis assoc prof, Univ Calif, Berkeley, 74-75; vis scholar, Grad Theol Union, Berkeley, 74-75. Mem: Math Asn Am; Inst Math Statist; Royal Statist Soc. Res: Decision theory; confidence bounds; multiple inference; time series. Mailing Add: Dept of Math Univ of Ill Urbana IL 61801

BOHRER, VORSILA LAURENE, b Chicago, Ill, Jan 22, 31. ETHNOBOTANY. Educ: Univ Ariz, BA, 53, PhD(bot), 68; Univ Mich, MS, 54. Prof Exp: Lectr biol, Univ Mass, Boston, 69-71, asst prof, 71-73; RES ASSOC PROF, PALEOINDIAN INST, EASTERN N MEX UNIV, 73- Concurrent Pos: Consult, Sch Am Res, Sante Fe, 75, Univ Ariz Field Sch at Grasshopper, 75, San Juan Valley Archaeol Proj, 75 & Cent Ariz Ecotone Proj, 75. Mem: Am Soc Archaeol; Soc Econ Bot. Res: Inter-relationship of plants and pre-historic man in the American Southwest; ethnobotany, palynology, plant microfossils, plant macrofossils, paleoecology. Mailing Add: Dept of Anthrop Eastern NMex Univ Portales NM 88130

BOHROD, MILTON GEORGE, b Chicago, Ill, Sept 2, 04; m 26; c 2. PATHOLOGY, CLINICAL PATHOLOGY. Educ: Univ Ill, BS, 24, MD, 27. Prof Exp: Assoc path, Col Med, Univ Ill, 26-32; pathologist, Munic Tuberc Sanitarium, 29-30; dir labs, St Francis & Methodist Hosps, Peoria, Ill, 31-41; pathologist & dir labs, Rochester Gen Hosp, 42-70; ASST MED EXAMR, MONROE COUNTY, 70- Concurrent Pos: Rep, Int Cong Allergy, Zurich, Switz, 51; assoc clin prof, Col Med, Univ Rochester, 59- Mem: Am Soc Clin Pathologists; Am Col Pathologists; AMA; Philos Sci Asn. Res: Pathology of allergic and related disease; medical photography. Mailing Add: Off of Med Examr 435 E Henrietta Rd Rochester NY 14620

BOICE, LU BELLE, b Pasadena, Calif, May 8, 33. MOLECULAR GENETICS. Educ: Stanford Univ, BS, 55; Univ Ill, MS, 58, PhD(microbiol), 62. Prof Exp: Asst chem, Univ Ill, 55-57; NIH res fel bact viruses, Calif Inst Technol, 62-64; res fel microbial genetics, Univ Calif, Los Angeles, 64-66, asst res bacteriologist, 66-69; asst prof biol, Calif State Col, Fullerton, 69-71; asst res molecular biologist, Dept Molecular Biol & Biochem, Univ Calif, Irvine, 71-72; NIH res fel, Tumor Virology, Salk Inst Biol Studies, San Diego, Calif, 73-75; LECTR BIOL, LOYOLA MARYMOUNT UNIV, LOS ANGELES, 75- Concurrent Pos: Spec fel, USPHS & Nat Cancer Inst, 73. Mem: AAAS; Genetics Soc Am; Am Chem Soc; Am Soc Microbiol. Res: Genetic analysis of bacterial hybrids; genetic mapping of bacterial prophages; genetic and biochemical characterization of bacteriophages; microbial genetics and physiology; base sequence homology of DNA molecules by electron microscopy. Mailing Add: Dept of Biol Loyola Marymount Univ Loyola Blvd & W 80th St Los Angeles CA 90045

BOICOURT, WILLIAM CLOSSON, b Ambler, Pa, July 19, 44. PHYSICAL OCEANOGRAPHY. Educ: Amherst Col, BA, 66; Johns Hopkins Univ, MA, 69, PhD(phys oceanog), 73. Prof Exp: ASSOC RES SCIENTIST PHYS OCEANOG, CHESAPEAKE BAY INST, JOHNS HOPKINS UNIV, 73- Mem: Am Geophys Union. Res: Continental shelf dynamics; estuarine circulation. Mailing Add: Chesapeake Bay Inst Johns Hopkins Univ Baltimore MD 21218

BOIKESS, ROBERT S, b Brooklyn, NY, Feb 2, 37; m 69. ORGANIC CHEMISTRY. Educ: Columbia Univ, BA, 57, MA, 60, PhD(org chem), 61. Prof Exp: Res fel org chem, Univ Calif, Los Angeles, 61-62; asst prof chem, State Univ NY Stony Brook, 63-68; asst prof, 68-71, ASSOC PROF CHEM, DOUGLASS COL, RUTGERS UNIV, 71- Concurrent Pos: NSF fel, 62-63. Mem: Am Chem Soc; The Chem Soc. Res: Small rings; valency tautomerism; homoaromaticity; bicyclics; photochemistry. Mailing Add: Dept of Chem Douglass Col Rutgers The State Univ New Brunswick NJ 08903

BOIME, IRVING, b St Louis, Mo, Feb 5, 41; m 61; c 2. MOLECULAR PHARMACOLOGY. Educ: St Louis Col Pharm, BS, 64; Purdue Univ, MS, 66; Washington Univ, PhD(pharmacol), 70. Prof Exp: Fel molecular biol, NIH, 70-72; ASST PROF OBSTET, GYNEC & PHARMACOL, WASHINGTON UNIV, 72- Concurrent Pos: Am Cancer Soc fel, 70; NIH res grant, 73; Pop Coun res grant, 73. Res: Biosynthesis of peptide hormones; proinsulin and the protein hormones of human placenta. Mailing Add: Dept of Obstet & Gynec Washington Univ Sch of Med St Louis MO 63110

BOIS, PIERRE, b Oka, Que, Mar 22, 24; m 53; c 3. ANATOMY. Educ: Laval Univ, BA, 48; Univ Montreal, MD, 53, PhD(exp med), 57. Prof Exp: Res asst exp surg, Univ Montreal, 53-57, lectr endocrinol, 55-57, lectr advan endocrinol, 56-57, lectr pharmacol, 57-58; asst prof histol & embryol, Univ Ottawa, 58-60; from asst prof to assoc prof pharmacol, 60-64, PROF ANAT & HEAD DEPT, UNIV MONTREAL, 64-, DEAN FAC MED, 70- Concurrent Pos: Grants, Med Res Coun, Muscular Dystrophy, Nat Cancer Soc & Dept Pub Health. Mem: AAAS; Endocrine Soc; Histochem Soc; NY Acad Sci; Am Fedn Clin Res. Res: Lymphoid tissue; origin, distribution and function of mast cells, including the role of histamine, biosynthesis, physiology and pharmacology. Mailing Add: Fac of Med Univ of Montreal PO Box 6207 Montreal PQ Can

BOISSEVAIN, ETHEL (MRS ARTHUR LESSER, JR), b Rockville Center, NY, Feb 5, 13; m 38; c 4. ANTHROPOLOGY, ARCHAEOLOGY. Educ: Vassar Col, AB, 34; Prague Univ, PhD, 36. Prof Exp: Instr anthrop, Hunter Col, 37-45, lectr grad sch & sch gen studies, 59-67; asst prof, Drew Univ, 67-68; asst prof Sch Gen Studies, Lehman Col, 68-72, ASSOC PROF ANTHROP, SCH GEN STUDIES, LEHMAN COL, 73- Concurrent Pos: RI Hist Preserv Comn grants, 69-70; Lehman Col George N Shuster Fel Fund grants, 69 & 71; City Univ New York Res Ctr grant, 71. Mem: Fel Am Anthrop Asn; Am Asn Phys Anthrop; Soc Am Archaeol; Am Soc Human Genetics; Am Schs Oriental Res. Res: Ethnohistory of American Indians of the New England area; neolithic plastic art of Europe and its relationship with Near Eastern civilization. Mailing Add: Dept of Anthrop Sch of Gen Studies Herbert H Lehman Col Bronx NY 10468

BOISVENUE, RUDOLPH JOSEPH, b Windsor, Ont, Aug 30, 26; nat US; m 51; c 5. MICROBIOLOGY. Educ: Univ Western Ont, BSc, 49; Univ Detroit, MSc, 51; Mich State Univ, PhD(microbiol), 55. Prof Exp: Wildlife biologist, Kellogg Bird Sanctuary & Forest, 52-55; mgr parasitol, Ralston Purina Co, 55-62; RES SCIENTIST, LILLY RES LABS, 62- Concurrent Pos: Microtechnician, Sparrow Hosp, 53-55; asst, Mich State Univ, 54-55; mem fac, Dept Zool, Eve Div, Ind Cent Col, 62; mem, Livestock Insect Workshop Conf. Mem: Am Soc Parasitol; Am Asn Vet Parasitol; Wildlife Dis Asn; Am Heartworm Soc. Res: Veterinary parasitology and entomology; biological and chemical control of major helminthic species infecting domestic animals. Mailing Add: Dept of Parasitol Lilly Res Labs Box 708 Greenfield IN 46140

BOIVIN, ALBERIC, b Baie St Paul, Que, Feb 11, 19; m 45; c 4. PHYSICS. Educ: Laval Univ, BA, 40, BSc, 44, MSc, 47, PhD(diffraction optics), 60. Prof Exp: Asst, 44, lectr, 45, assoc prof, 50, dir Lab d'Optique & Hyperfrequences, 68-70, PROF PHYSICS, LAVAL UNIV, 55- Concurrent Pos: Guggenheim fel, Univ Rochester, 62-63. Honors & Awards: Laureate, Prix David, Prov Que, 65; Pariseau Medal, Fr-Can Asn Advan Sci, 67. Mem: Fel Optical Soc Am; Can Asn Physicists. Res: Scalar diffraction theory applied to image formation, circular gratings and optical resonators; electromagnetic diffraction theory applied to wide-angle antennas and optical systems; holography, with applications to interferometry and image deconvolution; microwave optics. Mailing Add: Dept of Physics Laval Univ Quebec PQ Can

BOJALIL, LUIS FELIPE, b Merida, Yucatan, Mar 12, 25; m 58; c 3. MEDICAL MICROBIOLOGY. Educ: Univ Campeche, BSc, 43; Nat Polytech Inst, Mex, PhD(microbiol), 49, DSc(biol), 64. Prof Exp: From assoc prof to prof gen microbiol, Nat Polytech Inst, Mex, 50-55; prof microbiol & chmn dept, Nat Univ Mex, 55-76, prof, Postgrad Med Sch, 60-62; MEM FAC, DIV BIOL & HEALTH SCI, FREE METROP UNIV, XOCHIMILCO, MEX, 76- Concurrent Pos: Consult, Nat

Campaign Against Tuberc, 57; WHO fel, 58; lectr, Univ PR, 62; mem permanent comt, Nat Coun Med Invest, 64; vis prof, Postgrad Sch Sci, Nat Polytech Inst, Mex, 64- Honors & Awards: Mex Acad Sci Invest Award, 64. Mem: Am Soc Microbiol; NY Acad Sci; Mex Acad Sci Invest; Mex Asn Microbiol (vpres, 58-60, pres, 61-62); Brit Soc Gen Microbiol. Res: Mycobacteria; epidemiology, physiology and immunology. Mailing Add: Div of Biol & Health Sci Univ Autonoma Metrop Xochimilco Mexico 23 DF Mexico

BOJANIC, RANKO, b Breza, Yugoslavia, Nov 12, 24; m 58; c 2. MATHEMATICAL ANALYSIS. Educ: Serbian Acad Sci, PhD(math), 53. Prof Exp: Docent math anal, Univ Skoplje, 54-56 & Sch Mech Eng, Univ Belgrade, 56-58; vis mem, Tata Inst Fundamental Res, India, 58-59; vis lectr math, Stanford Univ, 59-60; asst prof, Univ Notre Dame, 60-63; assoc prof, 63-66, PROF MATH, OHIO STATE UNIV, 66- Concurrent Pos: Fel, Univ Geneva, 63. Mem: Am Math Soc; Math Asn Am; Soc Indust & Appl Math. Res: Asymptotic analysis; Fourier series and the theory of approximation. Mailing Add: Dept of Math Ohio State Univ Columbus OH 43210

BOJAR, SAMUEL, b New York, NY, Jan 23, 15; m 47; c 2. PSYCHIATRY. Educ: Brown Univ, AB, 36; Univ Rochester, MA, 37; Johns Hopkins Univ, MD, 41. Prof Exp: Asst psychol, Univ Rochester, 36-37; intern, Jewish Hosp, Brooklyn, 41-42, intern & asst resident path, 43; asst, 47, res fel, 48-51, instr, 51-61, clin assoc, 61-67, asst clin prof, 67-73, ASSOC CLIN PROF PSYCHIAT, HARVARD MED SCH, 73-, PSYCHIATRIST, UNIV HEALTH SERV, 55- Concurrent Pos: Resident, Boston Psychopath Hosp, 47-48; asst, Peter Bent Brigham Hosp, 48-49, from jr assoc to assoc, 50-58, sr assoc, 58-; lectr, Harvard Divinity Sch, 63-, William James lectr, 71-72. Mem: AAAS; AMA; Am Psychiat Asn; Am Psychoanal Asn. Res: Psychoanalysis; psychosomatic and psychophysiological correlates of personality and behavior; motivational studies in higher education. Mailing Add: Harvard Med Area Health Serv 275 Longwood Ave Boston MA 02115

BOK, BART JAN, b Hoorn, Holland, Apr 28, 06; nat US; m 29; c 2. ASTRONOMY. Educ: State Univ Leiden, 26; State Univ Groningen, PhD(astron), 32. Prof Exp: Asst, Kapteyn Lab, State Univ Groningen, 27-29; from asst prof to prof astron, Harvard Univ, 33-57; prof, Australian Nat Univ & dir Mt Stromlo Observ, 57-66; prof, 66-74, dir Steward Observ, 66-70, EMER PROF ASTRON, UNIV ARIZ, 74- Concurrent Pos: Assoc dir observ, Harvard Univ, 46-53. Mem: Nat Acad Sci; AAAS; Am Acad Arts & Sci; Am Astron Soc; Int Astron Union (vpres). Res: Galactic structure and dynamics; interstellar matter. Mailing Add: Steward Observ Univ of Ariz Tucson AZ 85721

BOK, P DEAN, b Douglas Co, SDak, Nov 1, 39; m 64; c 3. ANATOMY, CELL BIOLOGY. Educ: Calvin Col, AB, 60; Calif State Univ, Long Beach, MA, 65; Univ Calif, Los Angeles, PhD(anat), 68. Prof Exp: Asst prof, 68-72, ASSOC PROF ANAT, UNIV CALIF, LOS ANGELES, 72-, ASSOC DIR, JULES STEIN EYE INST, 72- Concurrent Pos: Nat Coun Combat Blindness fel, Univ Calif, Los Angeles, 68-69, USPHS grant, 68-78. Honors & Awards: Gladys Shea Award, Univ Calif, Los Angeles, 68. Mem: AAAS; Asn Res Vision & Ophthal; Am Asn Anatomists. Res: Cell biology of retinal photoreceptors and pigment epithelium. Mailing Add: Jules Stein Eye Inst Univ of Calif Los Angeles CA 90024

BOKE, NORMAN HILL, b Mobridge, SDak, Mar 14, 13; m 48; c 1. PLANT ANATOMY, MORPHOLOGY. Educ: Univ SDak, AB, 34; Univ Okla, MS, 36; Univ Calif, PhD(bot), 39. Prof Exp: Asst bot, Univ SDak, 33-34 & Univ Okla, 34-36; asst bot, Univ Calif, 36-39, instr, 39-40, exp work in genetics & cytol, Bot Gardens, 40-41; inst biol, Univ NMex, 41-42; Johns Hopkins Univ, 42-44; opers analyst, US Army Air Force Pac Ocean Area & Seventh Air Force, 44-45; PROF BOT, UNIV OKLA, 45- Concurrent Pos: Guggenheim fel, 53-54; ed-in-chief, Am J Bot, Bot Soc Am, 70-75. Mem: Bot Soc Am; Torrey Bot Club; Bot Soc Mex; Mex Soc Cactology. Res: Developmental anatomy and morphology of vascular plants; development of foliar organs; floral histogenesis; developmental morphology of Cactaceae. Mailing Add: Dept of Bot & Microbiol Univ of Okla 770 Van Vleet Oval Norman OK 73069

BOKELMAN, DELWIN LEE, b Linn, Kans, Sept 13, 34; m 58; c 4. PATHOLOGY. Educ: Kans State Univ, BS & DVM, 58; Cornell Univ, PhD(path), 64. Prof Exp: Resident vet immunol, Plum Island Animal Dis Lab, USDA, NY, 58-59; instr path, NY State Vet Col, Cornell Univ, 60-61, NIH trainee, 61-64; pathologist, 64-67, dir path, 67-70, DIR PATH & TOXICOL, MERCK INST THERAPEUT RES, WEST POINT, 70- Mem: Fel Am Vet Med Asn. Res: Pathological changes associated with drug toxicity.

BOL, KEES, b Netherlands, June 16, 25; nat US; m 47; c 4. PLASMA PHYSICS. Educ: Stanford Univ, PhD(physics), 50. Prof Exp: Proj engr electron beams res, Sperry Gyroscope Co, 49-54; asst prof physics, Adelphi Univ, 54-57; NSF fel, Harvard Univ, 57-59; mem res staff, Proj Matterhorn, 59-64, MEM RES STAFF, PLASMA PHYSICS LAB, PRINCETON UNIV, 64- Mem: Am Phys Soc. Res: Measurement of velocity of light; electron beam problems. Mailing Add: Plasma Physics Lab Princeton Univ PO Box 451 Princeton NJ 08540

BOLAFFI, ANDRE, food science, biochemistry, see 12th edition

BOLAND, JAMES P, b Philadelphia, Pa, Mar 6, 31; m 63; c 5. CARDIOVASCULAR SURGERY, THORACIC SURGERY. Educ: St Joseph's Col, Pa, BS, 52; Jefferson Med Col, MD, 56. Prof Exp: Assoc prof, 64-73, PROF SURG, MED COL PA, 73- Mem: Soc Thoracic Surg; Am Col Surgeons. Mailing Add: Med Col of Pa 3300 Henry Ave Philadelphia PA 19129

BOLAND, WILLARD ROBERT, b Winter Haven, Fla, Sept 11, 37; m 60; c 2. NUMERICAL ANALYSIS. Educ: Davidson Col, BS, 59; Col William & Mary, MA, 63; Univ Colo, PhD(appl math), 68. Prof Exp: Mathematician, Langley Res Ctr, NASA, 59-63; instr math, Randolph-Macon Col, 63-64; NSF trainee, Univ Colo, 65-67; asst prof math, Drexel Univ, 67-70; asst prof, 70-72, ASSOC PROF MATH SCI, CLEMSON UNIV, 72- Mem: Am Math Soc; Math Asn Am; Soc Indust & Appl Math. Res: Integral equations; operations research. Mailing Add: Dept of Math Sci Clemson Univ Clemson SC 29631

BOLANDE, ROBERT PAUL, b Chicago, Ill, Apr 16, 26; m 54; c 4. PATHOLOGY. Educ: Northwestern Univ, BS, 48, MS & MD, 52; Am Bd Path, dipl, 58. Prof Exp: Intern, Chicago Wesley Mem Hosp, 52-53; resident pediat path, Children's Mem Hosp Chicago, 53-54; from resident to chief resident path, Inst Path, Case Western Reserve Univ, 54-56, from instr to assoc prof path, Sch Med, 56-72, sr instr pediat, 58-72, chg pediat path, Inst & Hosp, 56-72; PROF PATH, McGILL UNIV, 72-, PROF PEDIAT, 75-; DIR PATH, MONTREAL CHILDREN'S HOSP, 72- Concurrent Pos: USPHS grants, 58 & 64; consult, Inst Path, Case Western Reserve Univ, 66-; chief pathologist & dir labs, Akron Children's Hosp, 66-72; co-ed, Perspectives in Pediat Path, 70- Mem: Am Asn Path & Bact. Res: Application of electron microscopy, histochemistry and tissue culture to the study of disease in early life, especially neoplasia, cystic fibrosis, abnormalities of the terminal autonomic nervous system. Mailing Add: Montreal Children's Hosp 2300 Tupper St Montreal PQ Can

BOLANDER, RICHARD, b Parsons, Kans, Mar 26, 40; m 59; c 4. MATHEMATICS, PHYSICS. Educ: Mo Sch Mines, BS, 61; Tex Christian Univ, MS, 63; Univ Mo-Rolla, PhD(cloud physics), 69. Prof Exp: Assoc engr, Gen Dynamics, Ft Worth, 62; instr physics, Tex Woman's Univ, 63-65; asst prof, La Polytech Inst, 65-66; from instr to asst prof ceramic eng, Univ Mo-Rolla, 67-69, res assoc cloud physics, 69; ASSOC PROF MATH, GEN MOTORS INST, 69- Mem: Am Soc Eng Educ; Math Asn Am. Res: Engineering education. Mailing Add: Dept of Math & Eng Mech Gen Motors Inst 1700 W Third Ave Flint MI 48502

BOLAR, MARLIN L, b Lincoln, Nebr, Dec 10, 27; m 51; c 5. PLANT PHYSIOLOGY. Educ: Colo State Univ, BS, 49; Columbia Bible Col, BD, 52; Univ Nebr, PhD(bot), 63. Prof Exp: Instr, Ben Lippen High Sch, NC, 52-57, asst headmaster, 55-57; chmn dept, 68-74, PROF BIOL SCI, CALIF STATE UNIV, SACRAMENTO, 62- Mem: AAAS; Am Soc Plant Physiol. Res: Physiological relationship between plant hosts and obligate parasites, especially rust-producing organisms. Mailing Add: Dept of Biol Sci Calif State Univ Sacramento CA 95819

BOLD, HAROLD CHARLES, b New York, NY, June 16, 09; m 43. BOTANY. Educ: Columbia Univ, AB, 29, PhD(bot), 33; Univ Vt, MS, 31. Prof Exp: Instr bot & asst botanist, Exp Sta, Univ Vt, 29-31; instr bot, Vanderbilt Univ, 32-39, from assoc prof to prof, 45-57; vis lectr bot, Barnard Col, Columbia Univ, 39-40, asst prof, 40-43; chmn div biol sci, 69-71, PROF BOT, UNIV TEX, AUSTIN, 57- Concurrent Pos: Ed, Am J Bot, Bot Soc Am, 58-65. Mem: Nat Acad Sci; Am Acad Arts & Sci; Bot Soc Am (secy, 55-58, vpres, 65, pres, 66); Phycol Soc Am (vpres, 55, pres, 56); Torrey Bot Club (corresp secy, 41). Res: Plant morphology and cytology, especially of algae and bryophytes; cultivation of algae. Mailing Add: Dept of Bot Univ of Tex Austin TX 78712

BOLDEBUCK, EDITH MAUDE, b Downers Grove, Ill, Oct 18, 15. POLYMER CHEMISTRY. Educ: NCent Col, Ill, BA, 39; Univ Chicago, PhD(chem), 44. Prof Exp: Instr math, NCent Col, Ill, 39-40; RES CHEMIST, GEN ELEC CO, 45- Mem: AAAS; Am Chem Soc. Mailing Add: Res & Develop Ctr Gen Elec Co PO Box 8 Schenectady NY 12301

BOLDEN, THEODORE EDWARD, b Middleburg, Va, Apr 19, 20; wid. DENTISTRY, PATHOLOGY. Educ: Lincoln Univ, Pa, AB, 41; Meharry Med Col, DDS, 47; Univ Ill, Chicago Circle, MS, 51, PhD(path), 58; Am Bd Oral Med, dipl, 70; Am Bd Oral Path, dipl, 72. Prof Exp: Instr periodont, Meharry Med Col, 48-49; instr oral path, Sch Dent, Univ Ill, 55-57; asst prof oral diag & path, Seton Hall Col, 57-60, assoc prof, 60-62; prof oral path & oral med, 62-69, dir res, 62-73, assoc dean sch dent, 69-73, PROF ORAL PATH & CHMN DEPT, SCH DENT, MEHARRY MED COL, 69- Concurrent Pos: Lectr, Meharry Med Col & Univ Ill, 56; attend, Med Ctr, NJ, 58-62; chmn adv health comt, Town of Montclair, NJ, 59-60, comnr urban develop, 61-62; abstractor, NY State Dent J, 60 & J Oral Therapeut & Pharmacol, 64; consult, Vet Admin Hosp, Tuskegee, Ala, 62-, Nashville, Tenn, 68- & Murfreesboro, 70; consult, Inst Advan Educ Dent Res, Colgate-Palmolive Co, 67-; ed, Quart Nat Dent Asn, 75- Mem: Nat Dent Asn; Int Asn Dent Res; fel Am Acad Oral Path; NY Acad Sci. Res: Histology; periodontal disease; salivary gland pathology; lacrimal gland pathology; pigmentation effect of aging; experimental carcinogenesis; oral neoplasia. Mailing Add: Sch of Dent Meharry Med Col Nashville TN 37208

BOLDRIDGE, WILLIAM FRANKLIN, b Culpeper Co, Va, Sept 3, 17; m 42; c 2. PHYSICAL CHEMISTRY. Educ: Randolph-Macon Col, BS, 39; WVa Univ, MS, 41, PhD(chem), 53. Prof Exp: Asst prof physics, Randolph-Macon Col, 43-44; asst, WVa Univ, 39-41; chemist, Am Viscose Corp, Va, 41-43 & Clinton Labs, Oak Ridge, 44-45; from asst prof to assoc prof, 49-53, PROF CHEM, RANDOLPH-MACON COL, 53- Mem: Am Chem Soc. Res: Organic reagents as inorganic precipitants; radiochemistry; physics; radioactive tracer technique in the study of adsorption. Mailing Add: Dept of Chem Randolph-Macon Col Ashland VA 23005

BOLDT, CHARLES EUGENE, b Albuquerque, NMex, Nov 17, 28; m 50; c 4. FORESTRY. Educ: Colo State Univ, BS, 54, MS, 55. Prof Exp: Res forester, Lake States Forest Exp Sta, 55-58, RES FORESTER, ROCKY MOUNTAIN FOREST & RANGE EXP STA, 58- Mem: Soc Am Foresters. Res: Silviculture; mensuration. Mailing Add: 2219 Lockwood Dr Rapid City SD 57701

BOLDT, ELIHU (AARON), b New Brunswick, NJ, July 15, 31. PHYSICS. Educ: Mass Inst Technol, SB, 53, PhD(physics), 58. Prof Exp: Guest scientist, Brookhaven Nat Lab, 55-57; res scientist physics, Mass Inst Technol, 58; asst prof, Rutgers Univ, 58-64; Nat Acad Sci sr res assoc cosmic ray physics, 64-66, GROUP LEADER X-RAY ASTRON, GODDARD SPACE FLIGHT CTR, 66- Concurrent Pos: Res assoc physics lab, Polytech Sch, Paris, 60-62. Mem: Am Phys Soc; Am Astron Soc; Int Astron Union. Res: High energy physics; elementary particles. Mailing Add: Lab for High Energy Astrophys Goddard Space Flight Ctr Greenbelt MD 20771

BOLDT, ROGER EARL, b Milwaukee, Wis, Dec 23, 28; m 54; c 3. BIOCHEMISTRY. Educ: Univ Wis, BS, 51, MS, 53, PhD(biochem), 58. Prof Exp: Asst, Univ Wis, 51-58, proj assoc pediat, 58; USPHS cancer trainee, Univ Minn, 58-60; asst prof biochem, Univ Pittsburgh, 61; res biochemist, Directorate Med Res, Edgewood Arsenal, Md, 61-66; BIOCHEMIST, US ARMY ENVIRON HYG AGENCY, 66- Mem: Sigma Xi. Res: Changes in metabolic pathways induced by dietary, environmental or other means; kinetics of enzyme inhibitions. Mailing Add: Biochem Br US Army Environ Hyg Agency Aberdeen Proving Ground MD 21010

BOLDUC, REGINALD J, b Lac Drolet, Frontenac, Que, July 28, 39; m 64; c 1. PLANT PHYSIOLOGY, HISTOCHEMISTRY. Educ: Laval Univ, Bes A, 60, Bsc A, 64; Purdue Univ, PhD(plant physiol & biochem), 68. Prof Exp: Res asst plant physiol & biochem, Purdue Univ, 64-68; asst prof, Laval Univ, 68-72; MEM STAFF, RES STA, CAN AGR, 72- Concurrent Pos: Researcher for Minister of Nat Ctr Sci Res, 69. Mem: Can Soc Plant Physiol; Agr Inst Can. Res: Physiology and histochemistry of stresses, cold and pesticides in plants. Mailing Add: Res Sta Can Agr Ste-Foy PQ Can

BOLE, GILES G, JR, b Battle Creek, Mich, July 28, 28; m 51; c 2. INTERNAL MEDICINE, BIOCHEMISTRY. Educ: Univ Mich, BS, 49, MD, 53. Prof Exp: Resident physician, Univ Hosp, 54-56, instr med, Med Sch, 58-61, from asst prof to assoc prof, 61-70, PROF INTERNAL MED, RACKHAM ARTHRITIS RES UNIT, UNIV MICH, ANN ARBOR, 70-, PHYSICIAN-IN-CHG, 69- Concurrent Pos: Arthritis Found fel, 61-63, sr investr, 63-68; consult rheumatol, Wayne County Gen Hosp, Eloise, Mich & Vet Admin Hosp, Ann Arbor, 63- Res: Biochemistry and immunology of the inflammatory response using experimental animal model systems. Mailing Add: Univ of Mich Med Ctr 4633 Kresge Bldg Ann Arbor MI 48104

BOLEF, DAN ISADORE, b Philadelphia, Pa, June 10, 21; m 44; c 3. SOLID STATE PHYSICS. Educ: Columbia Univ, AM, 48, PhD(physics), 52. Prof Exp: Instr physics, Stevens Inst Technol, 48-49; from instr to asst prof math & physics, State Univ NY,

49-53; res physicist, Westinghouse Res Labs, 53-63; PROF PHYSICS, WASH UNIV, 63- Concurrent Pos: Vis prof, Empire State Col, 74-75. Mem: Fel Am Phys Soc. Res: Nuclear and electron spin-phonon interactions in solids; elastic properties of solids; acoustic magnetic resonance; acoustic absorption and dispersion in metals, ferromagnets; ultrasonic high frequency techniques. Mailing Add: Dept of Physics Wash Univ St Louis MO 63130

BOLEMON, JAY S, b Atlanta, Ga, Oct 14, 41; m; c 3. PHYSICS. Educ: Univ SC, BS, 64, PhD(physics), 69. Prof Exp: ASST PROF PHYSICS, FLA TECHNOL UNIV, 69- Mem: Am Phys Soc; Am Asn Physics Teachers. Res: Computer applications in physics; theoretical solid state physics. Mailing Add: Dept of Physics Fla Technol Univ PO Box 25000 Orlando FL 32816

BOLEN, DAVID WAYNE, b Whitby, WVa, May 23, 42; m 64; c 3. PHYSICAL BIOCHEMISTRY. Educ: Concord Col, BS, 64; Fla State Univ, PhD(biophys chem), 69. Prof Exp: NIH res fel biochem, Johns Hopkins Univ, 69-70 & Univ Minn, 70-71; ASST PROF BIOCHEM, SOUTHERN ILL UNIV, CARBONDALE, 71- Mem: Am Chem Soc. Res: Thermodynamics; kinetics; mechanisms of interacting systems. Mailing Add: Dept of Chem & Biochem Southern Ill Univ Carbondale IL 62901

BOLEN, ERIC GEORGE, b Plainfield, NJ, Nov 24, 37; m 67. WILDLIFE ECOLOGY. Educ: Univ Maine, BS, 59; Utah State Univ, MS, 62, PhD(wildlife), 67. Prof Exp: Instr biol, Tex A&I Univ, 65-66; from asst prof to assoc prof, 66-73, PROF RANGE & WILDLIFE MGT, TEX TECH UNIV, 73- Concurrent Pos: Asst dir, Welder Wildlife Found, 73-; adj prof, Tex A&I Univ, Corpus Christi, 74-; mem, Fish & Wildlife & Park Natural Sci Adv Comt for Secy Interior, 75-; ed, Wildlife Soc Bull, 75- Mem: Wildlife Soc; Ecol Soc Am; Am Ornith Union; Am Soc Mammal. Res: Waterfowl ecology and management; wetland ecology; non-game ecology and management; ornithology; plant ecology. Mailing Add: Welder Wildlife Found PO Drawer 1400 Sinton TX 78387

BOLEN, HOMER ROSCOE, b Shirley, Ind, July 30, 01; m 24; c 2. ZOOLOGY. Educ: Ind Univ, AB, 23, AM, 25; Univ Tex, PhD(physiol), 33. Prof Exp: Asst zool, Ind Univ, 23-25; instr biol, Southeast Mo State Teachers Col, 25-28; instr zool, Univ Tex, 29-32; prof biol, 32-52, prof zool & chmn div sci, 52-70, EMER PROF ZOOL, SOUTHEASTERN MO STATE COL, 70- Mem: AAAS; Nat Asn Biol Teachers; Nat Sci Teachers Asn. Res: Genetics; cellular respiration; specific dynamic action; growth rate of fishes; chromosomal translocations; genetics of Drosophila; oxidative metabolism of Planaria. Mailing Add: 1400 New Madrid St Cape Girardeau MO 63701

BOLEN, LEE NAPIER, JR, b Memphis, Tenn, June 11, 37; m 60; c 1. NUCLEAR PHYSICS, ACOUSTICS. Educ: Univ Miss, BS, 59; Univ Va, MS, 61, PhD(physics), 64. Prof Exp: Res assoc physics, Univ Va, 63-64; from asst prof to assoc prof, 64-70, PROF PHYSICS, UNIV MISS, 70- Mem: Am Phys Soc; Am Asn Physics Teachers; AAAS. Res: Nuclear spectroscopy; photonuclear physics; physical acoustics. Mailing Add: Dept of Physics Univ of Miss University MS 38677

BOLEN, MAX CARLTON, b Waynetown, Ind, Sept 23, 19; m 42; c 2. PHYSICS. Educ: Wabash Col, AB, 41; Univ Chicago, cert meteorol, 43; Purdue Univ, MS, 50; Tex A&M Univ, PhD, 61. Prof Exp: Teacher high sch, Ind, 41-42; asst physics, Purdue Univ, 46-48; from asst prof to assoc prof & chmn dept, Millikin Univ, 48-59; assoc prof & chmn dept, Trinity Univ, Tex, 59-63; assoc prof, Oklahoma City Univ, 63-65; head dept, 65-69, PROF PHYSICS, UNIV TEX, EL PASO, 65-, COORDR SCI EDUC, SCH SCI, 69- Concurrent Pos: Actg asst prof, Tex A&M Univ, 58-59, res fel, 59. Consult, Prog Develop Div, Tex Educ Agency, 68-71. Meteorologist, WTVP-TV Sta, Decatur, Ill, 53-55, KOCO, Oklahoma City, Okla, 64-65; sr res physicist, Southwest Res Inst, Tex, 63. Mem: Fel AAAS; Am Asn Physics Teachers; Am Phys Soc; Am Meteorol Soc; Nat Sci Teachers Asn. Res: Double lens beta spectrometry; physical properties of high polymers in phase changes under pressure; viscoelastic properties of both synthetic and natural biological high polymers; crystal growth, rate and morphology; biophysics in biopolymers; research in science education. Mailing Add: Dept of Physics Univ of Tex El Paso TX 79968

BOLENDER, CHARLES L, b Iowa City, Iowa, June 2, 32; m 55; c 3. DENTISTRY. Educ: Univ Iowa, DDS, 56, MS, 57. Prof Exp: From instr to assoc prof, 59-68, PROF PROSTHODONTICS, SCH DENT, UNIV WASH, 68-, CHMN DEPT, 63- Mem: Am Prosthodontic Soc; Am Dent Asn; fel Am Col Dent; fel Acad Denture Prosthetics. Res: Tissue response to artificial dentures. Mailing Add: Sch of Dent Univ of Wash Seattle WA 98195

BOLENDER, DAVID LESLIE, b Uniontown, Pa, June 11, 47; m 69; c 3. ANATOMY. Educ: Bethany Col, WVa, BS, 69; WVa Univ, PhD(anat), 75. Prof Exp: ASST PROF ANAT, SCH MED, TEX TECH UNIV, 74- Mem: Soc Study Reproduction; Soc Develop Biol. Res: Reproductive and developmental biology, specifically the effects of hormones on the morphology and metabolic activities of maturing oocytes and developing ovarian follicles. Mailing Add: Dept of Anat Sch of Med Tex Tech Univ PO Box 4569 Lubbock TX 79409

BOLENDER, ROBERT P, b Mt Kisco, NY, Sept 29, 38. CELL BIOLOGY. Educ: State Univ NY Albany, BS, 60; Columbia Univ, MA, 65; Harvard Univ, PhD(anat), 70. Prof Exp: Fel anat, Univ Bern, 70-71, from med asst to head med asst, 71-74; lectr, Harvard Univ, 74-75; ASST PROF BIOL STRUCT, UNIV WASH, 75- Concurrent Pos: Vis scholar biochem, Univ Louvain, 71-72. Mem: Am Soc Cell Biol; Union Swiss Socs Exp Biol; Int Soc Stereology. Res: Integration of stereological and biochemical information as a way of analyzing relationships of structure to function mathematically; biological membrane kinetics associated with drug inductions, secretory cells, and cellular autophagy. Mailing Add: Dept of Biol Struct Univ of Wash Sch of Med Seattle WA 98195

BOLER, REGINALD KEITH, b Salem, Ark, July 24, 37; m 59; c 1. ANATOMY, EXPERIMENTAL PATHOLOGY. Educ: Harding Col, BA, 59; Univ Ark, MS, 62; Univ Miss, PhD(anat), 66. Prof Exp: Asst prof anat & instr path, Sch Med, Univ Miss, 66-70; ASST PROF ANAT, COL MED, UNIV S FLA, 70- Mem: Am Asn Anatomists; Pan Am Asn Anatomists; Electron Micros Soc Am. Res: Ultrastructural morphology; experimental pathology of liver, kidney and gastric mucosa; endocrinology. Mailing Add: Dept of Anat Univ of S Fla Col of Med Tampa FL 33620

BOLES, JAMES RICHARD, b Richmond, Ind, Sept 3, 44; m 67; c 2. SEDIMENTARY PETROLOGY. Educ: Purdue Univ, BS, 66; Univ Wyo, MS, 68; Univ Otago, NZ, PhD(geol), 72. Prof Exp: Fel geol, Univ Wyo, 72-73; res geologist, Geosci Group, Atlantic Richfield Co, 73-75; ASST PROF GEOL, UNIV CALIF, SANTA BARBARA, 75- Res: Diagenesis of clastic sediments including sandstone cementation and clay diagenesis in Gulf Coast tertiary; zeolitization of volcanogenic sediments in New Zealand, Wyoming and Baja, California. Mailing Add: Dept of Geol Univ of Calif Santa Barbara CA 93106

BOLES, ROBERT JOE, b Brandsville, Mo, Nov 7, 16; m 49; c 2. AQUATIC BIOLOGY. Educ: Southwestern Col, Kans, AB, 38; Kans State Univ, MS, 49; Okla State Univ, PhD(zool), 60. Prof Exp: Teacher high schs, 39-42 & 45-58; from asst prof to assoc prof, 60-70, PROF BIOL, EMPORIA KANS STATE COL, 70- Concurrent Pos: Ed, Kans Sch Naturalist. Res: Farm pond dynamics and fish parasitology; productivity and management of small impoundments; parasites of freshwater fishes of Kansas. Mailing Add: Div of Biol Emporia Kans State Col Emporia KS 66801

BOLEY, FORREST IRVING, b Ft Madison, Iowa, Nov 27, 25; m 46; c 4. PLASMA PHYSICS, ASTROPHYSICS. Educ: Iowa State Univ, BS, 46, PhD(physics), 51. Prof Exp: Elec engr, Raytheon Mfg Co, 46-47; from asst prof to prof physics, Wesleyan Univ, 51-61; physicist, Lawrence Radiation Lab, Univ Calif, 61-64; PROF PHYSICS & ASTRON, DARTMOUTH COL, 64-, CHMN DEPT, 75- Concurrent Pos: Ed, Am J Physics, Am Asn Physics Teachers, 66-72. Mem: AAAS; Am Astron Soc; Am Phys Soc; Am Asn Physics Teachers. Res: Cosmic rays; plasma phenomena; astrophysics. Mailing Add: Dept of Physics & Astron Dartmouth Col Hanover NH 03755

BOLEY, LOYD EDWIN, b Topeka, Kans, Oct 21, 09; m 35; c 2. VETERINARY MEDICINE. Educ: Kans State Univ, DVM, 32; Univ Ill, MS, 42. Prof Exp: Pvt pract, Kans, 32-34; asst vet, Meat Inspection Div, USDA, 34-36; instr vet path & hyg, 36-38, assoc vet clin med, 38-41, from asst prof to assoc prof, Univ, 41-49, asst dean col vet med, 62-66, PROF VET CLIN MED & HEAD DEPT, UNIV ILL, URBANA-CHAMPAIGN, 49-, ASSOC DEAN COL VET MED, 66- Mem: Am Vet Med Asn. Res: Obstetrics; diseases of reproduction; clinical veterinary medicine; veterinary pathology. Mailing Add: Col of Vet Med Univ of Ill Urbana IL 61801

BOLEY, ROBERT B, b Big Spring, Tex, Aug 19, 28; m 60; c 1. BACTERIOLOGY, IMMUNOLOGY. Educ: Sam Houston State Univ, BS, 49; Tex A&M Univ, MS, 60; Ohio State Univ, PhD(microbiol), 63. Prof Exp: From instr to assoc prof biol, Tarleton State Col, 58-65; ASSOC PROF BIOL, UNIV TEX, ARLINGTON, 65- Mem: Am Soc Microbiol; Am Inst Biol Sci; NY Acad Sci; Brit Soc Gen Microbiol. Res: Intestinal flora and immune responses in small mammals. Mailing Add: Dept of Biol Univ of Tex Arlington TX 76010

BOLEY, ROBERT EUGENE, b Washington, DC, Nov 25, 25; m 50; c 1. GEOGRAPHY. Educ: George Washington Univ, AB, 53, MA, 58. Prof Exp: Geogr, Cent Intel Agency, 53-54; res assoc, US Army Environ Res Proj, George Washington Univ, 54-56; exec dir indust develop, Indust Develop Comt Prince George County, Md, 56-57; secy indust coun, ULI-the Urban Land Inst, 57-61, dir indust prog, 61-63, dir cent city & indust progs, 63-66, coun coord dir urban & econ planning & develop, 66-68, asst exec dir, Inst, 68-69, exec dir, Inst, Washington, DC, 69-73; EXEC VPRES, SOC INDUST REALTORS, 73- Mem: Asn Am Geog; Am Indust Develop Coun; Am Soc Asn Execs. Res: Economic and urban development; land planning; location of economic activities; environmental planning; land use policy. Mailing Add: 3430 N Randolph St Arlington VA 22207

BOLEY, SCOTT JASON, b Brooklyn, NY. PEDIATRIC SURGERY. Educ: Wesleyan Univ, AB, 46; Jefferson Med Col, MD, 49. Prof Exp: Assoc prof, 67-71, PROF SURG, ALBERT EINSTEIN COL MED, 71-; DIR PEDIAT SURG, MONTEFIORE HOSP & MED CTR, 67- Mem: Fel Am Acad Pediat; fel Am Col Surgeons; Am Pediat Surg Asn; Soc Surg Alimentary Tract; Am Med Writers' Asn. Res: Vascular diseases of the intestines; Hirschprung's disease. Mailing Add: Montefiore Hosp & Med Ctr 111 E 210th St New York NY 10467

BOLGER, EDWARD M, b Jersey City, NJ, Jan 9, 38; m 60; c 2. MATHEMATICS. Educ: St Peter's Col, NJ, BS, 59; Pa State Univ, MA, 61, PhD(probability, statist), 64. Prof Exp: Asst prof math, Bucknell Univ, 64-67; ASST PROF MATH, MIAMI UNIV, 67- Mem: Am Math Soc; Math Asn Am; Inst Math Statist. Res: Probability and statistics, particularly the exponential family of distributions. Mailing Add: Dept of Math Miami Univ Oxford OH 45056

BOLGIANO, NICHOLAS CHARLES, b Baltimore, Md, Oct 26, 23; m 52; c 4. ORGANIC POLYMER CHEMISTRY. Educ: Loyola Col, Md, BS, 44; Catholic Univ, MS, 49; Univ Md, PhD(chem), 54. Prof Exp: Res fel, Mellon Inst, 54-56; res chemist, 56-74, SR RES CHEMIST, ARMSTRONG CORK CO, 74- Mem: Am Chem Soc. Res: Monomer and polymer synthesis; foamed plastics; coatings; binders. Mailing Add: Res & Develop Ctr Armstrong Cork Co Lancaster PA 17604

BOLHOFER, WILLIAM ALFRED, b Jamaica, NY, May 18, 20; m 42; c 4. ORGANIC CHEMISTRY. Educ: Mass Inst Technol, BS, 42, PhD(chem), 50. Prof Exp: Chemist, Merck & Co, 42-47; asst, Mass Inst Technol, 47-49; sr res fel, 49-74, SR INVESTR, MERCK SHARP & DOHME RES LAB, 74- Mem: AAAS; Am Chem Soc; NY Acad Sci. Res: Medicinal chemistry; synthetic organic chemistry. Mailing Add: Merck Sharp & Dohme Res Lab West Point PA 19486

BOLIEK, IRENE, b Hickory, NC, Apr 18, 07. ZOOLOGY. Educ: NC Col for Women, AB, 29; Univ NC, AM, 33, PhD(zool), 36. Prof Exp: Teacher high sch, NC, 29-31; asst zool, Univ NC, 32-35; asst prof biol, Ala Col, 35-36; from instr to prof zool, Fla State Univ, 36-65; prof biol & chmn dept, Coker Col, 65-73; CUR CREATIVE MUS FOR YOUTH, 73- Res: Fresh water and marine invertebrates. Mailing Add: 223 Ninth Ave NW Hickory NC 28601

BOLIN, FONSOE M, b Niantic, Ill, Sept 22, 02; m 44. VETERINARY SCIENCE. Educ: Iowa State Col, DVM, 29. Prof Exp: Mem staff, Exp Sta, Ore State Col, 29-38; prof vet sci & veterinarian, Exp Sta, NDak State Univ, 38-71; RETIRED. Mem: Am Vet Med Asn; US Animal Health Asn; Wildlife Soc. Res: Bang's disease reinfection; enterohepatitis; pig anemia and its relation to hog cholera; diseases of the sheep; New Castle disease virus isolation; baby chick immunization; dwarfism in the bovine; ergotism of farm animals. Mailing Add: Dept of Vet Sci NDak State Univ Fargo ND 58102

BOLIN, HAROLD R, b Lefors, Tex, Jan 16, 30; m 45; c 2. FOOD CHEMISTRY, BIOCHEMISTRY. Educ: San Jose State Col, BS, 59; Utah State Univ, PhD(food sci & technol), 70. Prof Exp: Lab supvr canned fruits & veg, Dole Corp, 59-60; res chemist collabr dried fruits, 60-65, FOODS CHEMIST, WESTERN REGIONAL RES LAB, USDA, 65- Mem: Am Chem Soc; Inst Food Technol. Res: Dried fruits, antimicrobial protection, moisture detection, new products and nutrient. Mailing Add: Western Utilization R&D Div USDA 800 Buchanan St Albany CA 94710

BOLING, JOHN LANDRUM, b South Bend, Wash, June 19, 09; m 35; c 3. ZOOLOGY. Educ: Linfield Col, AB, 35; Brown Univ, PhD(zool), 39. Prof Exp: Asst anat, Sch Med, Yale Univ, 39-43; instr col med, NY Univ, 43-46; assoc prof, 46-47, dean men, 47-51 & 58-66, PROF BIOL, LINFIELD COL, 47- Concurrent Pos: Sr biologist, Linfield Res Inst, 58-59. Mem: Am Asn Anat; Am Soc Zool. Res: Rodent reproductive physiology and anatomy of rodent reproductive organs; electron microscopy of metals; pulse x-irradiation. Mailing Add: Dept of Biol Linfield Col McMinnville OR 97128

BOLING, JOHNNY C, entomology, see 12th edition

BOLINGER, EDGAR DARE, chemistry, see 12th edition

BOLINGER, ROBERT E, b Independence, Kans, May 31, 19. MEDICINE. Educ: Univ Kans, AB, 40, MD, 43. Prof Exp: From instr to assoc prof, 49-60, PROF MED & DIR CLIN RES CTR, MED CTR, UNIV KANS, 60- Res: Insulin metabolism; computerized health care. Mailing Add: Univ of Kans Med Ctr 39th & Rainbow Kansas City KS 66103

BOLKER, ETHAN D, b Brooklyn, NY, June 11, 38; m 60; c 2. MATHEMATICS. Educ: Harvard Univ, AB, 59, AM, 61, PhD(math), 65. Prof Exp: Instr math, Princeton Univ, 64-65; from asst prof to assoc prof, Bryn Mawr Col, 65-72; PROF MATH, UNIV MASS, BOSTON, 72- Concurrent Pos: Res fel, Univ Calif, Berkeley, 67-68. Mem: AAAS; Math Asn Am; Am Math Soc. Res: Combinatorics; geometry; convexity; number theory. Mailing Add: Dept of Math Univ of Mass Boston MA 02125

BOLKER, HENRY IRVING, b Montreal, Que, Feb 19, 26; m 53; c 1. WOOD CHEMISTRY. Educ: Queen's Univ, Ont, BA, 48, MA, 50; Yale Univ, PhD(chem), 52. Prof Exp: Fel biochem, McGill Univ, 51-54; chemist, Dominion Tar & Chem Co, 54 & DuPont of Can, Ltd, 54-60; chemist, 60-63, asst head wood chem div, 63-67, assoc div head chem, 67-69, head org chem sect, 69-72, HEAD CHEM PROCESSES SECT, PULP & PAPER RES INST CAN, 72-, RES ASSOC CHEM, McGILL UNIV, 63- Concurrent Pos: Mem exec comt, Youth Sci Found, 61-64, pres, 65-66, mem admin coun, 69-; chmn, Sci Affairs, 67-; mem admin coun, Int Coord Comt, UNESCO, 68- Mem: Chem Inst Can; Can Pulp & Paper Asn; Am Chem Soc; The Chem Soc. Res: Lignin; cellulose; chemical pulping and bleaching. Mailing Add: Pulp & Paper Bldg McGill Univ PO Box 6070 Montreal PQ Can

BOLL, THOMAS JEFFREY, b Milwaukee, Wis, Jan 29, 42; m 69; c 2. NEUROPSYCHOLOGY. Educ: Marquette Univ, BA, 63, PhD(psychol), 67; Univ Wis, MS, 66. Prof Exp: USPHS fel psychiat, Med Ctr, Ind Univ, 68-70; asst prof neurol surg, Med Sch, Univ Wash, 70-73; ASSOC PROF PSYCHIAT, PEDIAT & CLIN PSYCHOL, MED SCH, UNIV VA, 73- Concurrent Pos: Bd dir, Am Asn Psychiat Serv Children, 73-76. Mem: Sigma Xi; Int Neuropsychol Soc; Am Psychosom Soc; Am Psychol Soc; AAAS. Res: Influence of brain damage and neurodevelopmental abnormalities on human behavior; relationship of long-term physical disorders, personality style and life stress on coping and adaptation. Mailing Add: Depts of Psychiat & Pediat Univ of Va Med Sch Charlottesville VA 22901

BOLL, WILLIAM GEORGE, b Manchester, Eng, Dec 5, 21; m 54; c 3. PLANT PHYSIOLOGY. Educ: Univ Manchester, BSc, 48, PhD(plant physiol), 51. Prof Exp: Fel, Plant Res Inst, Univ Tex, 52-54, res scientist, 54-55; from asst prof to assoc prof, 55-70, PROF BOT, McGILL UNIV, 70- Mem: Can Soc Plant Physiol. Res: Physiological and biochemical study of growth and morphogenesis of plants; excised root, tissue and cell suspension culture. Mailing Add: Dept of Biol McGill Univ Montreal PQ Can

BOLLA, ROBERT IRVING, b Dansville, NY, Aug 18, 43; m 66; c 1. ZOOLOGY, MOLECULAR BIOLOGY. Educ: State Univ NY Buffalo, BS, 65; Univ Mass, Amherst, MSc, 67, PhD(zool), 70. Prof Exp: NIH res fel biol, Univ Notre Dame, 70-73; RES FEL, ROCHE INST MOLECULAR BIOL, 73- Concurrent Pos: Del, Int Cong Parasitologists, 70. Mem: AAAS; Am Soc Parasitol; NY Acad Sci. Res: Control of macromolecular synthesis in aging tissues. Mailing Add: 266 High St Passaic NJ 02055

BOLLAG, JEAN-MARC, b Basel, Switz, Feb 19, 35; m 59; c 4. SOIL MICROBIOLOGY. Educ: Univ Basel, PhD(plant physiol), 59. Prof Exp: Lectr plant physiol, Swiss Trop Inst, Basel, 58-59; Julius Baer fel, Switz, 63; fel plant genetics, Weizmann Inst Sci, Rehovot, Israel, 63-65; res assoc soil microbiol, Cornell Univ, 65-67; asst prof, 67-71, ASSOC PROF SOIL MICROBIOL, PA STATE UNIV, 71- Mem: Am Soc Microbiol; Am Soc Agron; Soil Sci Soc Am; Int Soc Soil Sci. Res: Microbial control of the environment; problems of soil pollution; degradation of pesticides by microbes; mode of action of growth regulators in plants. Mailing Add: Dept of Agron Pa State Univ University Park PA 16802

BOLLE, ARNOLD WILLIAM, b Watertown, Wis, Oct 5, 12; m 37; c 3. FORESTRY. Educ: Northwestern Col, BA, 34; Univ Mont, BS, 37; Harvard Univ, MPA, 55, DPA, 60. Prof Exp: Forester, US Forest Serv, 37-38; range conservationist, US Soil Conserv Serv, 38-40, work unit conservationist, Wyo, 40-44, area conservationist, 44-47 & Wash, 49-51, regional proj plans, Ore, 51-54, state soil conservationist, Md, 55; assoc prof forestry & conserv, 55-60, dean sch forestry & dir forest & conserv, Exp Sta, 62-73, PROF FORESTRY, UNIV MONT, 60- Concurrent Pos: Consult, US Forest Serv, US Soil Conserv Serv, US Forest Serv, Nat Park Serv & State agencies; resources prog staff, Off Secy, US Dept Interior. Mem: AAAS; Soc Am Foresters; Soil Conserv Soc Am. Res: Forest products and industry; marketing; economics of resource use; resource policy and administration; multiple use management of public lands in the Flathead Valley of Montana. Mailing Add: Sch of Forestry Univ of Montana Missoula MT 59801

BOLLEN, WALTER BENO, soil chemistry, bacteriology, see 12th edition

BOLLENBACHER, KATHARINA, botany, deceased

BOLLES, THEODORE FREDERICK, b LaCrosse, Wis, Nov 25, 40; m 62; c 2. INORGANIC CHEMISTRY, PHYSICAL CHEMISTRY. Educ: Univ Wis, BS, 62; Univ Ill, PhD(inorg chem), 66. Prof Exp: Res chemist, 66-71, SUPVR NUCLEAR MED RES LAB, NUCLEAR PROD PROJ, MINN MINING & MFG CO, 71- Mem: AAAS; Am Chem Soc. Res: Organometallic chemistry of group IV, carbon, silicon, germanium, tin and lead; thermodynamics; calorimetry; coordination chemistry; nuclear medicine; radiopharmaceuticals; biochemistry; drug metabolism. Mailing Add: 3234 Courtly Rd St Paul MN 55119

BOLLET, ALFRED JAY, b New York, NY, July 15, 26; m 54. MEDICINE. Educ: NY Univ, BS, 45, MD, 48; Am Bd Internal Med, dipl. Prof Exp: Resident physician, Chest Serv, Bellevue Hosp, New York, 50, asst resident physician, 52-53; clin res assoc, Nat Inst Arthritis & Metab Dis, 53-55; asst prof med, Col Med, Wayne State Univ, 55-59; assoc prof prev med & med, Sch Med, Univ Va, 59-65, prof internal & prev med, 65-66; prof med & chmn dept, Med Col Ga, 66-74; PROF MED & CHMN DEPT, STATE UNIV NY DOWNSTATE MED CTR, 74- Concurrent Pos: Asst, Sch Med, Johns Hopkins Univ, 54-55; consult physician, Baltimore City Hosp, 54-55; Markle scholar med sci, 56-61; mem, Am Bd Intern Med, 71- Mem: AAAS; Am Soc Clin Invest; Am Fedn Clin Res (pres, 63-64); Am Rheumatism Asn; Harvey Soc. Res: Clinical and laboratory studies in the rheumatic diseases. Mailing Add: Dept of Med State Univ of NY Downstate Med Ctr Brooklyn NY 11203

BOLLETER, WILLIAM THEODORE, b Lufkin, Tex, June 29, 27; m 52; c 3. ANALYTICAL CHEMISTRY. Educ: Tex Col Arts & Indust, BS, 48, MS, 50; Univ Tex, PhD(anal chem), 55. Prof Exp: Instr chem, San Antonio Col, 50-52; res chemist, Monsanto Chem Co, 55-62; group supvr, Utah, 62-64, asst supt, 64-66, sr tech specialist, Va, 66-68, asst tech dir, 68-74, TECH DIR, HERCULES INC, 74- Mem: Am Chem Soc. Res: Spectroscopy. Mailing Add: Hercules Inc Radford Army Ammunition Plant Radford VA 24141

BOLLEY, DON SHELDON, chemistry, see 12th edition

BOLLICH, CHARLES NELSON, agronomy, see 12th edition

BOLLINGER, FREDERICK WILLIAM, b Tyndall, SDak, Aug 5, 18; m 45; c 3. ORGANIC CHEMISTRY. Educ: Univ SDak, BS, 39; State Col Wash, MS, 41; Calif Inst Technol, MS, 43; Ill Inst Technol, PhD(chem), 51. Prof Exp: Asst chemist, State Chem Lab, Univ SDak, 39; asst chem, Ill Inst Technol, 48-50; SR CHEMIST, MERCK & CO, INC, 51- Honors & Awards: Thomas Emery McKinney Prize, 39. Mem: Am Chem Soc. Res: Stereochemistry; parkinsonism; natural products and related synthesis. Mailing Add: 607 Lawrence Ave Westfield NJ 07090

BOLLINGER, GILBERT A, b Edwardsville, Ill, Jan 4, 31; m 55; c 3. GEOPHYSICS, SEISMOLOGY. Educ: St Louis Univ, PhD(geophys), 67. Prof Exp: Geophysicist, Calif Oil Co, 56-61; res asst seismol, St Louis Univ, 61-67; ASSOC PROF GEOPHYS, VA POLYTECH INST & STATE UNIV, 67- Mem: Seismol Soc Am; Am Geophys Union; Soc Explor Geophys. Res: Exploration geophysics; earthquake and engineering seismology. Mailing Add: Dept of Geol Sci Va Polytech & State Univ Blacksburg VA 24061

BOLLINGER, JAMES NORMAN, b Stroudsburg, Pa, July 31, 34; m 63; c 2. BIOCHEMISTRY, NUTRITION. Educ: Lebanon Valley Col, BS, 56; Rutgers Univ, MS, 59; Tex A&M Univ, PhD(biochem), 63. Prof Exp: Asst, Tex A&M Univ, 60-63; res assoc, Oak Ridge Inst Nuclear Studies, 63-66; sr res biochemist, Dept Biol & Phys Sci, Southwest Res Inst, 66-73, ASSOC FOUND SCIENTIST, SOUTHWEST FOUND RES & EDUC, 73- Mem: Am Chem Soc; AAAS. Res: Interrelationship between inhaled oxidants, diet lipids, vitamin E, prostaglandin and the physiological responses such as platelet emboli and vasoconstriction; biological effects of electromagnetic radiation. Mailing Add: Southwest Found Res & Educ Dept of Environ Sci San Antonio TX 78284

BOLLINGER, JOSEPH MARTIN, physical organic chemistry, see 12th edition

BOLLINGER, LOWELL MOYER, b Greene Co, Va, Apr 28, 23; m 44; c 3. EXPERIMENTAL NUCLEAR PHYSICS. Educ: Oberlin Univ, AB, 43; Cornell Univ, PhD, 51. Prof Exp: Physicist, Nat Adv Comt Aeronaut, 43-46; from res asst to res assoc physics, Cornell Univ, 46-51; dir physics div, 63-72, PHYSICIST, ARGONNE NAT LAB, 51-, DIR SUPERCONDUCTING LINAC PROJ, 72- & PHYSICS DIV, 74- Mem: Am Phys Soc. Res: Nuclear physics; development of the technology and design of a superconducting heavy-ion linear accelerator. Mailing Add: Argonne Nat Lab Bldg 203 Argonne IL 60439

BOLLINGER, RICHARD COLEMAN, b Johnstown, Pa, Feb 1, 32; m 54; c 2. APPLIED MATHEMATICS. Educ: Univ Pittsburgh, BS, 54, MS, 57, PhD(math), 61. Prof Exp: Eng specialist, Westinghouse Res Labs, 54-59, res mathematician, 59-62; from asst prof to assoc prof math, 62-74, PROF-IN-CHARGE BEHREND COL GRAD CTR, PA STATE UNIV, 74- Mem: AAAS; Math Asn Am; Am Math Soc. Res: Numerical analysis; mathematical analysis and applied mathematics. Mailing Add: Dept of Math Behrend Col Grad Ctr Pa State Univ Station Rd Erie PA 16510

BOLLINGER, ROBERT OTTO, b Detroit, Mich, June 15, 39; m 62; c 2. CELL BIOLOGY. Educ: Wayne State Univ, BS, 62, MS, 65, PhD(biol), 72. Prof Exp: Asst prof zool, Eastern Ill Univ, 69-71; RES ASSOC CELL BIOL, CHILD RES CTR MICH, 71- Concurrent Pos: Adj asst prof, Col Pharm & Allied Health Prof, Wayne State Univ, 73- Mem: Am Soc Cell Biol; AAAS. Res: Electron microscopal studies in cell biology, pathology and pharmacology. Mailing Add: Child Res Ctr of Mich 3901 Beaubien Blvd Detroit MI 48201

BOLLMAN, HARRY THOMAS, chemistry, deceased

BOLLMAN, JESSE LOUIS, b Springfield, Ill, Feb 19, 96; m 22; c 3. PATHOLOGY. Educ: Univ Ill, AB, 17, BS, 22, MD, 23; Univ Mich, MS, 18. Prof Exp: Instr bact, Med Sch, Univ Mich, 17-18; asst physiol chem, Univ Ill, 19-22; first asst & instr exp path, Mayo Found, Univ Minn, 23-26, from asst prof to prof, 26-61; DIR RES, ROCHESTER STATE HOSP, 61- Mem: Am Chem Soc; Am Gastroenterol Soc; Am Soc Exp Path (vpres, 40, pres, 41); AMA; Soc Exp Biol & Med. Res: General physiology and pathology; cirrhosis of the liver; peptic ulcer; physiology of muscles; radioactive tracer in metabolic processes. Mailing Add: Rochester State Hosp Rochester MN 55901

BOLLMAN, VERNON LEROY, b Omaha, Nebr, April 26, 08; m 27; c 1. PHYSICS. Educ: Univ Nebr, BS, 31, MS, 33; Calif Inst Technol, PhD(x-ray, phys constants), 36. Prof Exp: Asst physics, Calif Inst Technol, 34-36; from instr to prof, 36-73, dean men, 40-43, chmn dept, 43-58, dean fac & vpres acad affairs, 58-67, EMER PROF PHYSICS, OCCIDENTAL COL, 73- Mem: Am Phys Soc; Am Asn Physics Teachers. Res: X-ray and crystal structure; tests of validity of x-ray crystal methods of determining e. Mailing Add: Occidental Col Los Angeles CA 90041

BOLLS, NATHAN J, JR, b De Witt, Mo, May 16, 31; m 62; c 1. VERTEBRATE PHYSIOLOGY. Educ: Kans State Univ, BS, 59, MS, 61, PhD(vert physiol), 63. Prof Exp: From instr to asst prof, 63-68, ASSOC PROF BIOL, WITTENBERG UNIV, 68- Concurrent Pos: Author & proj dir, NSF Matching Funds grant for undergrad physiol lab equip, 65-67. Mem: Am Inst Biol Sci; Am Soc Zool; Am Soc Mammal. Res: Role of glucocorticoids in energy utilization during bodily dehydration in mammals; ecological physiology of tree squirrels. Mailing Add: Dept of Biol Wittenberg Univ Springfield OH 45501

BOLLUM, FREDERICK JAMES, b Ellsworth, Wis, June 14, 27; m 48; c 4. BIOCHEMISTRY. Educ: Univ Minn, BA, 49, PhD(physiol chem), 56. Prof Exp: Asst physiol chem, Univ Minn, 53-55, instr, 55-56; USPHS fel, Univ Wis, 56-58; biochemist, Biol Div, Oak Ridge Nat Lab, 58-65; PROF BIOCHEM, MED SCH, UNIV KY, 65- Concurrent Pos: Consult radiation study sect, NIH, 67-68, biochem study sect, 69-73 & cancer treat rev panel, 75; vis prof, Univ Chile, 69; consult, Am Cancer Soc, 71-72; USPHS spec fel, Med Ctr, Univ Calif, 72-73. Honors & Awards: K A Forster Award & Lectureship, Mainz Acad Sci, Ger, 74. Mem: AAAS; Am Soc Biol Chem; Am Chem Soc. Res: Enzymes of nucleic acid synthesis; structure and function of nucleic acid; differentiation of lymphoid cells. Mailing Add: Dept of Biochem Univ of Ky Med Sch Lexington KY 40506

BOLMAN, WILLIAM MERTON, b Elyria, Ohio, Oct 8, 29; m 55; c 2. CHILD PSYCHIATRY. Educ: Harvard Univ, BA, 51, MD, 55; Am Bd Psychiat & Neurol, dipl psychiat, 63, dipl child psychiat, 69. Prof Exp: NIMH career develop award, Boston Univ & Boston City Hosp, 63-64; fel community psychiat, Harvard Lab Community Psychiat, 64-65; from asst prof to assoc prof psychiat, Med Sch, Univ Wis, 65-69; dir, Westside Community Ment Health Ctr, Inc, 69-70; PROF PSYCHIAT, SCH MED, UNIV HAWAII, 70-, PROF PEDIAT, 75- Concurrent Pos: Consult psychiat, Bur Testing & Guid, Univ Hawaii, 57-59 & Mass Div Legal Med, 60-63; consult child psychiat, Dane County Ment Health Ctr, 64-69, St Coletta's Sch, 67-69 & Joint Comn Ment Health Children, 67-68; consult prev psychiat, Wis State Comprehensive Ment Health Comt, 66-67; mem juv probs res rev comt, NIMH, 71-; actg dir, Child Guid Clin, Kauikeolani Children's Hosp, 71- Mem: Am Psychiat Asn; fel Am Orthopsychiat Asn; fel Am Pub Health Asn; fel Royal Soc Health. Res: Community mental health; child psychiatry; health delivery systems. Mailing Add: Dept of Psychiat Univ of Hawaii Sch of Med Honolulu HI 96816

BOLMARCICH, JOSEPH JOHN, b Philadelphia, Pa, Dec 22, 42; m 63; c 3. OPERATIONS RESEARCH, APPLIED MATHEMATICS. Educ: Univ Pa, PhD(appl math), 72. Prof Exp: Res physicist, Frankford Arsenal, 65-72; ASSOC, OPERS RES & MATH, DANIEL H WAGNER ASSOCS, 72- Honors & Awards: David Rist Prize, Mil Opers Res Soc, 75. Mem: Soc Indust & Appl Math. Res: Baysian information processing; combinatorial probability; bounds on operator norms; Perron-Frobenius theory. Mailing Add: Daniel H Wagner Assocs Sta Sq One Paoli PA 19301

BOLMER, PERCE W, b Springfield, Ohio, Nov 24, 28; m 56; c 1. ELECTROCHEMISTRY, SURFACE CHEMISTRY. Educ: Ohio Univ, BS, 50; Purdue Univ, MS, 52, PhD(anal chem), 54. Prof Exp: Sr res technologist, Field Res Lab, Socony Mobil Oil Co Inc, 54-66; sr res technologist, 66-73, HEAD FINISHING SECT, CTR FOR TECHNOL, KAISER ALUMINUM & CHEM CORP, 73- Mem: Am Chem Soc. Res: Electrochemistry of corrosion reactions; aluminum and other metal finishing; solid waste disposal; industrial pollution control; anodic oxides; electrodeposition; paint application technology. Mailing Add: Kaiser Aluminum & Chem Corp Ctr for Technol PO Box 870 Pleasanton CA 94566

BOLOGNESI, DANI PAUL, b Forgaria, Italy, Mar 19, 41; US citizen; m 64; c 2. VIROLOGY, IMMUNOLOGY. Educ: Rensselaer Polytech Inst, BS, 63, MS, 65; Duke Univ, PhD(virol), 67. Prof Exp: Res assoc virol, Med Ctr, Duke Univ, 67-68; NIH fel, Max-Planck Inst Virus Res, 68-71, Max-Planck res fel, 70-71; asst prof surg, 71-72, ASST PROF VIROL, MED CTR, DUKE UNIV, 71-, ASSOC PROF SURG, 72- Concurrent Pos: Mem tumor virus detection segment rev bd, Nat Cancer Inst, 71-; consult, M D Anderson Hosp & Tumor Inst, Tex & Frederick Cancer Res Ctr, Md, 73-; Nat Cancer Inst grant, Med Ctr, Duke Univ, 74-, USPHS grant, 74-75, Am Cancer Soc grant, 74-76 & Am Soc Fac Res award, 76-81. Mem: Am Soc Microbiol. Res: Tumor virology; tumor immunology; human cancer. Mailing Add: Dept of Surg Duke Univ Med Ctr PO Box 2926 Durham NC 27710

BOLON, DONALD A, b Cleveland, Ohio, Sept 7, 34; m 59; c 4. ORGANIC CHEMISTRY. Educ: Allegheny Col, BS, 56; Univ Minn, PhD(org chem), 60. Prof Exp: Fel, Harvard Univ, 60-62; RES CHEMIST, GEN ELEC CO, 62- Mem: Am Chem Soc. Res: Carbene chemistry; free radical rearrangements and reactions; autoxidation and catalytic oxidative coupling reactions; solvolitic reactions; photochemical crosslinking; acetylene chemistry; ultraviolet light cured coatings. Mailing Add: Gen Elec Co R&D Ctr PO Box 8 Schenectady NY 12345

BOLOTIN, HERBERT HOWARD, b New York, NY, Jan 11, 30; m 51; c 2. NUCLEAR PHYSICS. Educ: City Col New York, BS, 50; Ind Univ, MS, 52, PhD(exp nuclear physics), 55. Prof Exp: Physicist, US Naval Radiol Defense Lab, 55-58; res fel nuclear physics, Brookhaven Nat Lab, 58-60, asst physicist, 60-61; asst prof physics, Mich State Univ, 61-62; assoc physicist, Argonne Nat Lab, 62-71; CHAMBER MFRS PROF PHYSICS, UNIV MELBOURNE, 71- Concurrent Pos: Vis scientist, Univ Calif, Berkeley, 62 & Los Alamos Sci Lab, 66; chmn, Conf Slow-Neutron-Capture Gamma Rays, Argonne Nat Lab, 66; vis scientist, Brookhaven Nat Lab, 68; mem adv comt, Int Conf Neutron-Capture Gamma Rays Studies, Sweden, 69 & Int Conf Nuclear Structure Study Neutrons, Hungary, 72. Mem: Fel Am Phys Soc; fel Australian Inst Physics. Res: Experimental nuclear physics; beta and gamma ray, neutron and nuclear spectroscopy; lifetimes of excited nuclear states; neutron capture gamma rays; nuclear structure; interdisciplinary research in nuclear medicine investigations. Mailing Add: Sch of Physics Univ of Melbourne Parkville Victoria 3052 Australia

BOLSTAD, A N, organic chemistry, see 12th edition

BOLSTAD, LUTHER, b Floodwood, Minn, Feb 11, 18; m 54; c 2. PLASTICS CHEMISTRY. Educ: St Olaf Col, BA, 41; NDak State Univ, MS, 43; Purdue Univ, PhD(chem), 49. Prof Exp: Chemist, Plastics Res & Develop, Bakelite Corp, 43-45 & Honeywell, Inc, Minn, 51-65; CHEMIST, STACKPOLE CARBON CO, 65- Mem: AAAS; Am Chem Soc. Res: Aerospace materials; atmospheric contamination within manned spacecraft; carbon composition resistors; electrical properties of carbon dispersions in thermosetting resins and ceramics. Mailing Add: RD 1 Kersey PA 15846

BOLSTEIN, ARNOLD RICHARD, b New York, NY, Sept 2, 40; m 63; c 1. MATHEMATICAL ANALYSIS, OPERATIONS RESEARCH. Educ: Wagner Col, BA, 62; Purdue Univ, MS, 64, PhD(math), 67. Prof Exp: Grant, Univ NC, Chapel Hill, 68, asst prof math, 67-73; ASSOC PROF MATH, GEORGE MASON UNIV, 73- Mem: Math Asn Am; Opers Res Soc Am. Res: Linear and integer programming; network theory. Mailing Add: Dept of Math George Mason Univ Fairfax VA 22030

BOLSTERLI, MARK, b New Haven, Conn, Oct 3, 30. THEORETICAL PHYSICS. Educ: Wash Univ, AB, 51, PhD(physics), 55. Prof Exp: Res assoc, Argonne Nat Lab, 55; Fulbright student physics, Univ Birmingham, 55-56; res assoc, Wash Univ, 56-57; from asst prof to prof physics, Univ Minn, 57-68; STAFF MEM, LOS ALAMOS SCI LAB, 68- Concurrent Pos: Vis prof, Copenhagen Univ, 61-62; consult, Los Alamos Sci Lab, 63-64; Guggenheim fel, Oxford Univ, 64-65. Mem: Fel Am Phys Soc. Mailing Add: Los Alamos Scientific Lab T-9 PO Box 1663 Los Alamos NM 87544

BOLT, BRUCE A, b Largs, NSW, Australia, Feb 15, 30; m 56; c 4. SEISMOLOGY, APPLIED MATHEMATICS. Educ: Univ New Eng, Australia, BSc, 52; Univ Sydney, MSc, 54, PhD(math), 56, DSc(math), 72. Prof Exp: Lectr appl math, Univ Sydney, 54-61, sr lectr, 62; PROF SEISMOL & DIR SEISMOG STA, UNIV CALIF, BERKELEY, 63- Concurrent Pos: Fulbright res fel & res seismologist, Columbia Univ, 60; vis res scientist, Dept Geod & Geophys, Cambridge Univ, 61; consult seismologist, UK Atomic Energy Authority, 61; mem, Comt Seismol, Nat Acad Sci, 65-72 & Earthquake Eng Res Inst, 67-; consult, Vet Admin seismic adv comt, 71-; ed, Bull Seismol Soc Am, 65-; mem, Gov's Earthquake Coun, Calif, 72-74. Honors & Awards: H O Wood Award for Res in Seismol. Mem: Fel Am Geophys Union; Seismol Soc Am (pres, 74-75); fel Royal Astron Soc; Australian Math Soc. Res: Classical mechanics as applied to the propagation of waves and structure of the earth; statistical and computational methods of data reduction. Mailing Add: Seismog Sta Dept Geol & Geophys Univ of Calif Berkeley CA 94720

BOLT, DOUGLAS JOHN, b Isabel, Kans, Mar 2, 39; m 62. REPRODUCTIVE PHYSIOLOGY, ENDOCRINOLOGY. Educ: Kans State Univ, BS, 61, MS, 66, PhD(animal breeding), 68. Prof Exp: From res asst to res assoc animal husb, Kans State Univ, 62-67; fel obstet & gynec, Sch Med, Univ Mo, 67-68; RES PHYSIOLOGIST, DAIRY CATTLE RES BR, AGR RES SERV, USDA, 68- Mem: Am Asn Animal Sci; Soc Study Reproduction. Res: Endocrinology and physiology of reproduction, especially as related to corpus luteum function. Mailing Add: Agr Res Ctr USDA Beltsville MD 20705

BOLT, JOHN RYAN, b Grand Rapids, Mich, May 28, 40; m 64. VERTEBRATE ZOOLOGY, PALEOZOOLOGY. Educ: Mich State Univ, SB, 62; Univ Chicago, PhD(paleozool), 68. Prof Exp: Asst prof anat, Med Ctr, Univ Ill, 68-72; ASST CUR FOSSIL REPTILES & AMPHIBIANS, FIELD MUS NATURAL HIST, 72- Mem: Soc Vert Paleont; Soc Study Evolution; Am Soc Ichthyol & Herpet; Soc Study Amphibians & Reptiles; Paleontol Soc. Res: Evolution and origin of Paleozoic amphibia; evolution of Paleozoic vertebrate faunas. Mailing Add: Field Mus Natural Hist Roosevelt Rd at Lake Shore Dr Chicago IL 60605

BOLT, RICHARD HENRY, b Peking, China, Apr 22, 11; m 33; c 3. PHYSICS, ACOUSTICS. Educ: Univ Calif, AB, 33, MA, 37, PhD(physics), 39. Prof Exp: Nat Res Coun fel physics, Mass Inst Technol, 39-40; assoc, Univ Ill, 40; res assoc, Mass Inst Technol, 41-43; sci liaison officer, Off Sci Res & Develop, London, 43-44; chief tech aide, Div 6, Nat Defense Res Comt, New York, 44-45; assoc prof physics, Mass Inst Technol, 46-54, dir acoustics lab, 46-57, prof acoustics, 54-63, lectr polit sci, 63-70; CHMN BD, BOLT BERANEK & NEWMAN, INC, 53- Concurrent Pos: Pres, Int Comn Acoust, 51-57; prin consult, NIH, 57-59; assoc dir, NSF, 60-63; fel, Ctr Advan Study Behav Sci, 63-64; mem gov bd, Am Inst Physics. Honors & Awards: Biennial Award, Acoust Soc Am, 42. Mem: AAAS; fel Acoust Soc Am (pres, 48); Am Phys Soc; fel Inst Elec & Electronics Eng; fel Am Acad Arts & Sci. Res: Theoretical and experimental studies of sound distribution in enclosures; acoustics of rooms; bioacoustics; effects and control of noise; physics of sound; biophysics; science and public policy. Mailing Add: Tabor Hill Rd Lincoln MA 01773

BOLT, ROBERT JAMES, b Grand Rapids, Mich, Feb 23, 20; m; c 4. INTERNAL MEDICINE. Educ: Calvin Col, AB, 42; Univ Mich, MD, 45. Prof Exp: Instr internal med, Univ Mich, 51-52, instr & res assoc, Inst Indust Health, 52-54, from asst prof to prof internal med, 54-66; PROF MED & CHMN DEPT, UNIV CALIF, DAVIS, 66- Concurrent Pos: Am Cancer Soc fel, Univ Mich, 51-52; attend physician, Vet Admin Hosp, 53; lectr, Univ Medellin, Colombia, 56; lectr, Univ Louvain, 59-60; vis prof, St Francis Hosp, Honolulu; consult, Health Serv, Univ Mich, Vet Admin Hosp, Wayne County Hosp, 63- & David Grant Hosp, Travis AFB. Mem: AAAS; Am Soc Gastrointestinal Endoscopy; AMA; Am Gastroenterol Asn; Asn Am Med Cols. Res: Gastroenterology; small intestinal epithelial cell enzymes and histochemical aspects of the small bowel mucosa in normal and abnormal states. Mailing Add: Dept of Med UCD Prof Bldg Univ of Calif Davis 4301 X St Sacramento CA 95817

BOLT, ROBERT O'CONNOR, b Detroit, Mich, Aug 31, 17; m 43; c 3. ORGANIC CHEMISTRY. Educ: James Millikin Univ, BS, 39; Purdue Univ, MS, 42, PhD(chem), 44. Prof Exp: Res chemist, Calif Res Corp, 46-56, sr res chemist, 56-62, supvr res chemist, 62-66, sr res assoc, 66-69, MGR MKT SERV, CHEVRON RES CO, 69- Concurrent Pos: Mem Air Force Aircraft Nuclear Propulsion Radiation Damage Adv Group, 53-55, Mat Adv Group, 55-58 & AEC Org Reactor Working Group, 61-63. Mem: Am Chem Soc; Am Soc Lubrication Eng. Res: Radiolysis of materials; development of radiation resistant lubricants, fuels, nuclear reactor coolants, and other fluids; development of industrial lubricants; preparation and properties of synthetic oils; organic fluoride chemistry. Mailing Add: Chevron Res Co 576 Standard Ave Richmond CA 94802

BOLTE, EDOUARD, b Montreal, Que, July 23, 31; m 65; c 3. REPRODUCTIVE PHYSIOLOGY, MEDICINE. Educ: Univ Montreal, BA, 51, MD, 56. Prof Exp: Resident med, Univ Hosps, Columbia Univ, 58-60; Med Res Coun Can fel steroids, Columbia Univ, 60-62 & Univ Stockholm, 62-63; from asst prof to prof physiol, 64-75, PROF MED, UNIV MONTREAL, 75- Concurrent Pos: Mem endocrine serv, Hotel-Dieu Hosp, 64- Mem: Endocrine Soc; Can Soc Clin Invest; Int Soc Res Reproduction. Res: Various aspects of androgen and estrogen metabolism in human beings in vivo; placental and fetal steroidogenesis. Mailing Add: Endocrine Serv Hotel-Dieu Hosp 3840 St Urbain Montreal PQ Can

BOLTE, HENRY FREDERICK, b Rock Rapids, Iowa, Sept 9, 38. VETERINARY PATHOLOGY. Educ: Iowa State Univ, DVM, 63; Purdue Univ, MS, 66, PhD(vet path), 70. Prof Exp: Vet, pvt pract, 63-64; instr vet path, Purdue Univ, 64-67, Nat Inst Neurol Dis & Stroke spec fel, 67-70; RES GROUP LEADER, DEPT EXP PATH, LEDERLE LABS, 70- Mem: AAAS; NY Acad Sci; Int Acad Path; Am Vet Med Asn; Soc Pharmacol & Environ Path. Res: Ophthalmologic pathology; ultrastructural neuropathology dealing with isoniazid intoxication in ducklings and sodium ion intoxication in swine; diagnostic veterinary pathology. Mailing Add: Dept Exp Path Lederle Labs Div Am Cyanamid Co Pearl River NY 10965

BOLTE, JOHN R, b Waterloo, Iowa, Nov 19, 29; m 48; c 3. PHYSICS. Educ: Univ Northern Iowa, BA, 51, MA, 56; Okla State Univ, MS, 57; Univ Iowa, PhD(sci educ, physics), 62. Prof Exp: Instr pub schs, Iowa, 51-59; instr physics & sci educ, Univ Iowa, 59-62; from asst prof to assoc prof physics, San Diego State Col, 62-68; from asst dean to assoc dean, 68-72, PROF PHYSICS, FLA TECHNOL UNIV, 68-, ASSOC VPRES ACAD AFFAIRS, 72- Concurrent Pos: Consult, Educ Testing Serv, NJ, 66-68. Mem: Am Asn Physics Teachers; Optical Soc Am. Res: Carbon dioxide laser; use of lasers in communications applications. Mailing Add: Off Acad Affairs Box 25000 Fla Technol Univ Orlando FL 32816

BOLTER, ERNST A, b Mengen-Wuertt, Ger, Feb 16, 35. GEOCHEMISTRY, GEOLOGY. Educ: Univ Göttingen, PhD(mineral), 61. Prof Exp: Fel, Yale Univ, 61-63; fel, Rice Univ, 63-65; asst prof geol, 65-67, ASSOC PROF GEOCHEM, UNIV MO-ROLLA, 67- Mem: Mineral Soc Am; Geochem Soc; Am Geophys Union. Res: Geochemistry of hydrothermal ore deposits; water pollution; chemical oceanography; natural radioactivity. Mailing Add: Dept of Geol Univ of Mo Rolla MO 65401

BOLTJES, BEN HAROLD, b Wichita, Kans, Feb 18, 17; m 42; c 3. MEDICAL MICROBIOLOGY. Educ: Univ Wichita, AB, 40; Univ Kans, Md, 43; Univ Pa, PhD(microbiol), 49. Prof Exp: Physician, Armour & Co, 44-45; dir venereal dis control, Dept Pub Health, Kansas City, 45; res bacteriologist, Philadelphia Serum Exchange, Children's Hosp Philadelphia, 48-49, assoc immunol in pediat, 49; PROF MICROBIOL & CHMN DEPT, MED UNIV SC, 49- Concurrent Pos: Staff physician, St Mary's Hosp, Kans & Parkview Hosp, 45; bacteriologist, Roper Hosp, St Francis Hosp, Baker Hosp & Med Col Hosp; consult, WVa Pulp & Paper Co, 58-59. Mem: Am Soc Microbiol; AMA; Am Pub Health Asn; NY Acad Sci. Res: Physical antigenic fractionation of tubercle bacillus; nutritional requirements of hemophilus pertussis; effects of unidirectional pulsating magnetic fields on charged micro particles. Mailing Add: Dept of Microbiol Med Univ of SC Charleston SC 29401

BOLTON, ANTHONY PETER, b Glasgow, Scotland, Sept 20, 36; m 67. PHYSICAL CHEMISTRY. Educ: Univ London, BSc, 57, PhD(phys chem), 60. Prof Exp: Nat Res Coun Can fel, 61-63; chemist, Molecular Sieve Catalysis Group, Linde Div, 63-65, group leader, 65-69, supvr, 69-72, mgr prod develop, 72-75, MGR, CENTRAL SCIENTIFIC LABS, UNION CARBIDE CORP, 75- Mem: Am Chem Soc; The Chem Soc. Res: Heterogeneous catalysis; kinetics. Mailing Add: Tech Ctr Union Carbide Corp Tarrytown NY 10591

BOLTON, B A, b Wellsboro, Pa, Sept 5, 28; m 54; c 2. ORGANIC CHEMISTRY. Educ: Lafayette Col, BS, 50; Northwestern Univ, MBA, 58. Prof Exp: Chemist, Atlas Powder Co, 50-52; chemist, Standard Oil Co, 54-61; group leader org chem, 61-70, sect leader polymers & plastics, Res & Develop Dept, 70-72, RES SUPVR, AMOCO CHEM CORP, 72- Res: Organic chemicals; product applications; high temperature polymers; foamed polymers; vinyl thermoplastics; process and product development on polystyrenes. Mailing Add: Res & Develop Dept Amoco Chem Corp PO Box 400 Naperville IL 60540

BOLTON, CHARLES THOMAS, b Camp Forest, Tenn, Apr 15, 43. ASTRONOMY. Educ: Univ Ill, BS, 66; Univ Mich, MS, 68, PhD(astron), 70. Prof Exp: Fel astron, 70-73, ASST PROF ASTRON, DAVID DUNLAP OBSERV, UNIV TORONTO, 73- Concurrent Pos: Asst prof, David Dunlap Observ, Univ Toronto, 70-71; instr, Scarborough Col, Univ Toronto, 71-72; asst prof, Erindale Col, Univ Toronto, 72-73. Mem: Am Astron Soc; Can Astron Soc; AAAS; Astron Soc Pac; Illum Eng Soc. Res: Spectroscopy and photometry of x-ray and radio binaries, multiple and peculiar stars; spectroscopic studies of stellar atmospheres; light pollution control. Mailing Add: David Dunlap Observ Box 360 Richmond Hill ON Can

BOLTON, ELLIS TRUESDALE, b Linden, NJ, May 4, 22; m 43; c 2. BIOPHYSICS Educ: Rutgers Univ, BSc, 43, PhD, 50. Prof Exp: Res Coun fel, Rutgers Univ, 46, teaching asst genetics & biol, 46-47, instr, 47-50, dir res, Serol Mus, 47-50; mem staff, Dept Terrestrial Magnetism, Carnegie Inst, 50-64, assoc dir, 64-66, dir, 66-74; ADJ RES PROF, COL MARINE STUDIES, UNIV DEL, 74- Mem: AAAS; Biophys Soc. Res: Molecular biology; biosynthesis in microorganisms. Mailing Add: Col of Marine Studies Univ of Del Lewes DE 19958

BOLTON, JAMES R, b Swift Current, Sask, June 24, 37; m 59; c 2. BIOPHYSICAL CHEMISTRY. Educ: Univ Sask, BA, 58, MA, 60; Cambridge Univ, PhD, 63. Prof Exp: Boese fel & res assoc, Columbia Univ, 63-64; asst prof teaching & res, Univ Minn, 64-66, assoc prof, 66-69, prof assoc chmn, 69-70; PROF TEACHING & RES, UNIV WESTERN ONT, 70- Concurrent Pos: Alfred P Sloan Found fel, 66-68; grants, NSF, 64-70, Petrol Res Fund, 70-73, Nat Res Coun, 70-77. Mem: Fel Chem Inst Can; Brit Chem Soc; Am Soc Photobiol; Int Soc Magnetic Resonance; Solar Energy Soc Can. Res: Transient free radical intermediates in photochemistry and in photosynthesis using optical and electron spin resonance techniques; photochemical reactions with the potential to store solar energy. Mailing Add: Dept of Chem Univ of Western Ont London ON Can

BOLTON, LAURA LEE, b Royal Oak, Mich, Oct 13, 44; m 68; c 1. EXPERIMENTAL BIOLOGY. Educ: Univ Ill, BA, 66; Stanford Univ, MS, 69; Rutgers Univ, PhD(psychobiol), 75. Prof Exp: SR SCIENTIST SKIN BIOL, JOHNSON & JOHNSON RES, 72- Res: Exploration of variables affecting soft tissue repair. Mailing Add: Johnson & Johnson Res US Rte 1 North Brunswick NJ 08902

BOLTON, THOMAS ELWOOD, b Swastika, Ont, Mar 2, 24; m 51; c 1. INVERTEBRATE PALEONTOLOGY. Educ: Univ Toronto, BA, 47, MA, 49, PhD(geol), 55. Prof Exp: PALEONTOLOGIST, GEOL SURV CAN, 52- Mem: Am Paleont Soc; Geol Soc Am; Am Asn Petrol Geol; Geol Asn Can. Res: Ordovician and Silurian faunas of Canada, with emphasis on the corals, bryozoans and trilobites. Mailing Add: Geol Surv Can 601 Booth St Ottawa ON Can

BOLTON, WESSON DUDLEY, b Hardwick, Vt, Dec 30, 22; m 46; c 5. ANIMAL PATHOLOGY. Educ: Mich State Col, DVM, 44; Univ Vt, MS, 50. Prof Exp: Practicing vet, Vt, 44-47; asst animal pathologist, 47-50, PROF ANIMAL PATH & CHMN DEPT, UNIV VT, 51- Mem: Am Vet Med Asn. Res: Causes of infertility in dairy cattle. Mailing Add: Dept of Animal Path Univ of Vt Burlington VT 05401

BOLTRALIK, JOHN JOSEPH, b Mt Carmel, Pa, June 24, 26; m 51; c 2. BIOCHEMISTRY. Educ: Univ Pittsburgh, BS, 50, MS, 53. Prof Exp: Asst trop testing electronic equip, Univ Pittsburgh, 50-52; jr fel food technol, Mellon Inst Sci, 52-58; asst microbiol, Sch Med, Univ Pittsburgh, 58-60; jr biochemist, Ciba Pharmaceut Co, 60-67; res assoc, St Barnabas Med Ctr, 67-68; scientist, 68-70, sr res scientist biochem, 70-74, HEAD BIOCHEM, ALCON LABS INC, 74- Mem: Am Chem Soc; Acad Pharmaceut Sci. Res: Methods development for analysis of drugs from biological fluids; studies on absorption, distribution, metabolism and excretion of topically applied ophthalmic drugs; bioavailability of drugs and drug products. Mailing Add: Alcon Labs Inc 6201 S Freeway PO Box 1959 Ft Worth TX 76101

BOLTZ, DAVID FERDINAND, b Belfield, NDak, May 31, 16; m 42; c 2. ANALYTICAL CHEMISTRY. Educ: Univ Wis, BA, 38; Mo Sch Mines, MS, 40; Purdue Univ, PhD(chem), 46. Prof Exp: Instr chem eng, Mo Sch Mines, 41-44; from asst prof to assoc prof, 46-54, PROF CHEM, WAYNE STATE UNIV, 54- Mem: Am Chem Soc. Res: Spectrophotometry; electroanalytical methods; heteropoly chemistry; physical-chemical methods in analytical chemistry. Mailing Add: Dept of Chem Room 179 Chem Bldg Wayne State Univ Detroit MI 48202

BOLYN, ANTHONY EDWARD, b Jeddo, Pa, Aug 3, 15; m 42; c 4. MEDICAL MICROBIOLOGY. Educ: Lehigh Univ, BA, 36, MS, 38; Purdue Univ, PhD(bact), 55. Prof Exp: Instr bact, Purdue Univ, 39-41; bacteriologist, Pitman-Moore Co Div, Allied Labs, 41-43; sr res bacteriologist, Biol Labs Nat Drug Co, 43-55, assoc dir prod, 55-62, DIR ADMIN, BIOL LABS, NAT DRUG CO DIV, RICHARDSON-MERRELL, INC, 62- Mem: Am Chem Soc. Res: Preparation and purification of diphtheria and tetanus toxins and toxoids. Mailing Add: Woodland Rd Mt Pocono PA 18344

BOMBA, STEVEN JAMES, b Chicago, Ill, June 28, 37; m 59; c 2. APPLIED PHYSICS, SYSTEMS ENGINEERING. Educ: Univ Wis-Madison, BS, 59, MS, 61, PhD(physics), 68. Prof Exp: Res physicist, Mobil Res & Develop Corp, 68-70, sr res physicist, 70-72; mem tech staff, Tex Instruments, Inc, 72-75; SR SCIENTIST, COLLINS RADIO GROUP, ROCKWELL INT, 75- Concurrent Pos: Consult, S J Bomba & Assocs, 71- Mem: Am Asn Physics Teachers; Am Phys Soc; Inst Elec & Electronics Engrs; Sigma Xi; Am Radio Relay League Res: Electrodynamics; applied digital technology for communications, control and computation; miscellaneous applied classical physics; applied solid state physics in device and process development; systems science and engineering. Mailing Add: 7860 Querida Dallas TX 75248

BOMBARDIERI, CAURINO CESAR, b Calgary, Alta, Oct 21, 20; m 54; c 3. PETROLEUM CHEMISTRY. Educ: Univ Alta, BSc, 49, MSc, 52; McGill Univ, PhD(org chem), 54. Prof Exp: Qual control specialist, Ayerst McKenna & Harrison Pharmaceuts, 54-56; res specialist tire yarn textiles, Courtaulds Can Ltd, Courtaulds, Eng, 56-57; res petrol specialist, Imp Oil Ltd, Exxon Corp, 57-67, SR PETROL RES ENGR, IMP OIL LTD, 67- Res: Gas strengthened steam foam injection; improved thermal recovery processes. Mailing Add: 2436 Palisade Dr SW Calgary AB Can

BOMKE, HANS ALEXANDER, b Berlin, Ger, May 26, 10; nat US; m 32; c 1. PHYSICS. Educ: Univ Berlin, PhD(physics, math), 31; Tech Univ Berlin, Dr habil, 35. Prof Exp: Asst prof, Radiation Lab, Berlin, 31-33; mem sci staff, Ger Bur Stand, 33-37 & Kaiser Wilhelm Co, 38-41; mem, Aerial Navig Inst, 41-45; assoc prof, Med Sch, Univ Munich, 46-52; scientist, 52-58, dir explor res div, Surveillance Dept, 58-61, PRIN SCIENTIST, EXP RES INST, SIGNAL RES & DEVELOP LABS, US ARMY FT MONMOUTH, NJ, 61- Concurrent Pos: Adj prof, Fordham Univ, 61- Mem: Am Phys Soc; Optical Soc Am; Am Geophys Union; Am Inst Aeronaut & Astronaut; Ger Phys Soc. Res: Atomic and nuclear physics; geophysics; space. Mailing Add: 408 Central Ave Spring Lake NJ 07762

BOMPART, BILLY EARL, b Dallas, Tex, Dec 5, 33; m 57; c 3. MATHEMATICS. Educ: Univ Tex, Austin, BS, 56, PhD(math educ), 67; Southwestern Baptist Theol Sem, MRE, 60; NTex State Univ, MEd, 64. Prof Exp: Teacher high schs, Tex, 58-62; youth dir, Polytech Baptist Church, Ft Worth, 62-63; teacher high schs, Tex, 63-65; ASSOC PROF MATH, AUGUSTA COL, 67- Mem: Math Asn Am; Nat Coun Teachers Math; Sch Sci & Math Asn; Am Educ Res Asn; Am Asn Univ Prof. Res: Theory of numbers. Mailing Add: Dept of Math & Comput Sci Augusta Col Augusta GA 30904

BOMSE, FREDERICK, b New York, NY, Jan 24, 39; m 65; c 1. PHYSICS. Educ: Antioch Col, BS, 61; Johns Hopkins Univ, PhD(elem particle physics), 65. Prof Exp: Instr physics, Johns Hopkins Univ, 65-66; from instr to asst prof, Vanderbilt Univ, 66-71; MEM PROF STAFF, CTR NAVAL ANALYSES, 71- Mem: Am Phys Soc; Opers Res Soc Am. Res: Interactions of fundamental particles. Mailing Add: Ctr Naval Analyses 1401 Wilson Blvd Arlington VA 22209

BOMSTEIN, RALPH AARON, biochemistry, see 12th edition

BONAKDARPOUR, AKBAR, b Esfahan, Iran. RADIOLOGY. Educ: Univ Tehran, MD, 53; Temple Univ, MS, 58. Prof Exp: From instr to assoc prof, 64-72, PROF RADIOL, HEALTH SCI CTR, TEMPLE UNIV, 72- Mem: Asn Univ Radiologists; Am Col Radiol; Radiol Soc NAm; Am Roentgen Ray Soc. Res: Splanchnic circulation; contrast media reaction; skeletal and gastrointestinal systems. Mailing Add: Temple Univ Health Sci Ctr 3401 N Broad St Philadelphia PA 19140

BONAR, DANIEL DONALD, b Murraysville, WVa, July 7, 38; m 66. MATHEMATICS. Educ: WVa Univ, BS, 60, MS, 61; Ohio State Univ, PhD(math), 68. Prof Exp: From instr to asst prof math, Denison Univ, 65-67; teaching assoc, Ohio State Univ, 67-68; asst prof, Wayne State Univ, 68-69; asst prof, 69-71, ASSOC PROF MATH & CHMN DEPT, DENISON UNIV, 71- Mem: Am Math Soc. Res: Complex and real analysis; annular functions; topological idea of cluster sets. Mailing Add: Dept of Math Denison Univ Granville OH 43023

BONAR, LEE, b Belleville, WVa, Aug 26, 91; m 24; c 2. BOTANY. Educ: Univ Mich, AB, 18, AM, 20, PhD(bot), 22. Prof Exp: Asst Bot, Univ Mich, 19-20; from instr to prof, 22-58, chmn dept, 47-54, EMER PROF BOT, UNIV CALIF, BERKELEY, 58-, CUR, MYCOL COLLECTIONS, 40- Mem: AAAS; Bot Soc Am; Am Phytopath Soc; Mycol Soc Am. Res: Systematic mycology; California fungas flora; life history studies in fungi. Mailing Add: Dept of Bot Univ of Calif Berkeley CA 94720

BONAR, ROBERT ADDISON, b Kalamazoo, Mich, Aug 23, 25; m 51; c 4. BIOCHEMISTRY. Educ: Univ Calif, AB, 49, PhD(biochem), 53. Prof Exp: Res fel, Med Sch, Univ Calif, Los Angeles, 53-55; res assoc, Dept Surg, 55-59, asst prof, 59-63, ASSOC PROF BIOPHYS, DEPT SURG, SCH MED, DUKE UNIV, 63-; RES CHEMIST, VET ADMIN HOSP, 72- Concurrent Pos: Nat Cancer Inst fel, 53-55. Mem: AAAS; Electron Micros Soc Am; Am Soc Biol Chemists; Am Soc Microbiol; Am Asn Cancer Res. Res: Biochemical cytology; virology; tumor viruses. Mailing Add: Vet Admin Hosp 508 Fulton St Durham NC 27705

BONATTI, ENRICO, b Rome, Italy, Sept 28, 36; m 60; c 1. MARINE GEOLOGY. Educ: Univ Pisa, PhD(geol), 59. Prof Exp: Fulbright fel sedimentology, Yale Univ, 59-60; fel oceanog, Scripps Inst, Univ Calif, 60-64; from asst prof to assoc prof, 64-71, PROF MARINE SCI, UNIV MIAMI, 71- Res: Mineralogy and geochemistry of marine sediments; petrology of igneous rocks from the ocean floor. Mailing Add: Sch of Marine & Atmos Sci Univ of Miami Miami FL 33149

BONAVENTURA, CELIA JEAN, b Silver City, NMex, June 19, 41; m 60; c 2. PHOTOBIOLOGY. Educ: Calif State Univ, San Diego, BA, 64; Univ Tex, Austin, PhD(zool), 68. Prof Exp: Fel chem, Calif Inst Technol, 68-70; fel molecular biol, Regina Elena Inst Cancer Res, 71-72; RES ASST PROF BIOCHEM, DUKE UNIV MED CTR & MARINE LAB, 72- Mem: Soc Photochem & Photobiol; Am Chem Soc; AAAS; Sigma Xi. Res: Protein structure-function allosteric interactions in macromolecules; photochemical processes in proteins. Mailing Add: Dept of Biochem Duke Univ Med Ctr & Marine Lab Beaufort NC 28516

BONAVENTURA, JOSEPH, b Oakland, Calif, Feb 15, 42; m 60; c 2. MOLECULAR BIOLOGY. Educ: Calif State Univ, San Diego, BA, 64; Univ Tex, Austin, PhD(zool), 68. Prof Exp: Fel chem, Calif Inst Technol, 68-70; fel molecular biol, Regina Elena Inst Cancer Res, 71-72; RES ASST PROF BIOCHEM, DUKE UNIV MED SCH & MARINE LAB, 72- Concurrent Pos: Estab investr, Am Heart Asn, 75-80. Mem: Sigma Xi; Am Chem Soc; AAAS; Am Soc Zoologists. Res: Protein structure-function relationships; respiratory proteins as models for large multisubunit macromolecular complexes; biochemical basis for adaptation and acclimation. Mailing Add: Dept of Biochem Duke Univ Med Ctr & Marine Lab Beaufort NC 28516

BONAVENTURA, MARIA MIGLIORINI, b Somerville, Mass, June 29, 38; m 63. ORGANIC CHEMISTRY. Educ: Regis Col, BA, 60; Tufts Univ, PhD(chem), 65. Prof Exp: NSF res asst, Tufts Univ, 60-64; res assoc, Tufts Univ, 65-66; assoc prof, 65-71, PROF CHEM, SUFFOLK UNIV, 71-, CHMN DEPT, 72- Mem: Am Chem Soc; The Chem Soc; Sigma Xi. Res: Nonchair conformations of 2,5-dialkyl-1, 4-cyclohexanediols; conformational analysis. Mailing Add: Dept of Chem Suffolk Univ Boston MA 02114

BONAVITA, NINO LOUIS, b White Plains, NY, Sept 5, 31; m 65; c 2. PHYSICS, MAGNETISM. Educ: Fordham Univ, BS, 54; Cath Univ, MS, 63, PhD(physics), 71. Prof Exp: Physicist, Aberdeen Proving Ground, 54-55 & Naval Res Lab, 57-59; PHYSICIST, APPLN DIV, GODDARD SPACE FLIGHT CTR, NASA, 59- Mem: Am Phys Soc. Res: Statistical mechanics; theoretical solid state physics and theoretical physics; celestial mechanics; phase transitions and the physics of condensed matter. Mailing Add: Goddard Space Flight Ctr NASA Code 932 Greenbelt MD 20770

BOND, ALBERT F, b Clarksburg, WVa, July 27, 30; m 51; c 3. MATHEMATICS. Educ: WVa Univ, AB & MS, 56. Prof Exp: Develop engr, Goodyear Aircraft Corp, Ohio, 56-57; staff engr, Bendix Systs Div, 57-63; mgr systs eng dept, Brown Eng Co, 63-68; mgr Huntsville opers, Planning Res Corp, 68-70; gen mgr, Huntsville Div, Syst Sci, Inc, 70-72; DEPT MGR, COMPUT SCI CORP, HUNTSVILLE OPERS, 72- Concurrent Pos: Guest lectr, Univ Mich, 59. Mem: Math Asn Am; Opers Res Soc Am. Res: Systems engineering; operations research; management sciences; applied mathematics; software engineering. Mailing Add: Comput Sci Corp 515 Sparkman Dr Huntsville AL 35807

BOND, ALBERT HASKELL, JR, b Williamson, WVa, Sept 13, 40; m 68. NUCLEAR PHYSICS. Educ: Harvard Univ, AB, 61; Univ Wis, MS, 63, PhD(physics), 68. Prof Exp: Found for Aid to Res res assoc nuclear physics, Van Der Graaff Lab, Univ Sao Paulo, 68-74; ASSOC PROF PHYSICS, BROOKLYN COL, 74- Res: Fast neutron physics; fabrication techniques of lithium-germanium detectors; particle-gamma angular correlations. Mailing Add: Dept of Physics Brooklyn Col Brooklyn NY 11210

BOND, ANDREW, b Brownsville, Tenn, Aug 8, 27. BIOCHEMISTRY, AGRONOMY. Educ: Tenn Agr & Indust State Col, BS, 48; Tenn State Univ, MS, 52; Univ Minn, PhD(biochem), 65. Prof Exp: Instr agron, 54-58, assoc prof biochem, 65-69, ASSOC PROF ANIMAL SCI, TENN STATE UNIV, 69- Mem: Am Chem Soc. Res: Carbohydrate transformations in germinating cereal seeds. Mailing Add: Dept of Animal Sci Tenn State Univ Nashville TN 37203

BOND, ARTHUR CHALMER, b Salem, WVa, Feb 14, 17; m 45; c 1. PHYSICAL CHEMISTRY. Educ: Mich State Univ, BS, 39; Univ Mich, MS, 40, PhD(chem), 51. Prof Exp: Res chemist, Nat Defense Res Comt, Univ Chicago, 42-43, instr & asst, 43-46; from instr to asst prof phys chem, Univ Rochester, 51-57; from asst prof to assoc prof, 57-67, PROF CHEM, RUTGERS UNIV, 67- Concurrent Pos: Asst ed, J Phys Chem, 51-57. Mem: AAAS; Am Chem Soc. Res: Molecular structure; inorganic reaction mechanisms. Mailing Add: 27 Meadowbrook Lane Piscataway NJ 08854

BOND, ARTHUR DARROL, biochemistry, organic chemistry, see 12th edition

BOND, BERNARD BATSON, b Wiggins, Miss, Mar 28, 06; m 31, 53; c 1. CHEMISTRY. Educ: Miss Col, BS, 26. Prof Exp: Teacher pub sch, Miss, 26-28; architect draftsman, State Dept Educ, Miss, 28-30; clerk-bookkeeper, State Auditor's Off, 31-32; clerk-bookkeeper, State Dept Pub Welfare, 32, State Dept Hwys, 33-36, chemist, 36-42; mat eng supt, US Naval Air Sta, Pensacola, 42-67, mat eng supt, Naval Air Rework Facil, 67-70, DIR TECH SUPPORT CTR, DEPT DEFENSE EQUIP OIL ANAL PROG, US NAVAL AIR STA, PENSACOLA, 70- Honors & Awards: Naval Air Systs Command Award. Mem: Am Chem Soc; Am Inst Aeronaut & Astronaut; NY Acad Sci; fel Am Inst Chem; fel AAAS. Res: Aircraft materials; electroplated and organic coatings; preservative compounds; analysis of aircraft engine oils to determine engine condition. Mailing Add: 308 E Sunset Ave Warrington FL 32507

BOND, CARL ELDON, b Culdesac, Idaho, Sept 11, 20; m 42; c 2. ICHTHYOLOGY. Educ: Ore State Univ, MS, 48; Univ Mich, PhD(fisheries), 63. Prof Exp: Res asst zool, Univ Calif, 49; from asst prof to assoc prof fish & game mgt, 49-64, asst dean grad sch, 69-74, PROF FISHERIES, ORE STATE UNIV, 64- Concurrent Pos: NSF sci fac fel, 60 & 71; vis res prof, Tokyo Univ Fisheries, 72. Mem: Am Soc Ichthyol & Herpet; fel Am Inst Fishery Res Biol; Am Fisheries Soc. Res: Distribution, systematics and ecology of Northwest fishes. Mailing Add: Dept of Fish & Wildlife Ore State Univ Corvallis OR 97331

BOND, CLARENCE D, physics, mathematics, see 12th edition

BOND, CLIFFORD WALTER, b Buffalo, NY, Apr 7, 37. VIROLOGY. Educ: State Univ NY Buffalo, BA, 66; Univ Ky, PhD(microbiol), 73. Prof Exp: Technician virol, Case Western Reserve Univ, 66-67, res fel, 67-69; res fel, Univ Ky, 69-74; LEUKEMIA SOC AM FEL VIROL, UNIV CALIF, SAN DIEGO, 74- Mem: Am Soc Microbiol; AAAS. Res: Elucidation of interactions between neuropathogenic viruses and their host that result in diseases of the central nervous system. Mailing Add: Dept of Path Univ of Calif San Diego La Jolla CA 92093

BOND, DONALD C, b Jefferson City, Mo, Aug 19, 09; m 38; c 2. PHYSICAL CHEMISTRY. Educ: Univ Mo, PhD(chem), 36. Prof Exp: Asst dir res, Pure Oil Co, 37-65; HEAD OIL & GAS SECT, ILL STATE GEOL SURV & ASSOC PROF MINING, METALL & PETROL ENG, UNIV ILL, URBANA-CHAMPAIGN, 65- Mem: Am Chem Soc; Am Inst Mining, Metall & Petrol Engrs. Res: Exploration and production of petroleum. Mailing Add: Ill State Geol Surv Urbana IL 61801

BOND, DOUGLAS DANFORD, b Waltham, Mass, July 2, 11; m 37; c 4. PSYCHIATRY. Educ: Harvard Univ, BA, 34; Univ Pa, MD, 38; Am Bd Psychiat & Neurol, dipl. Hon Degrees: DSc, Heidelberg Col, 53. Prof Exp: Chief lab psychiat, Sch Aviation Med, Randolph Field, Tex, 42-43; psychiat consult, Eighth Air Force, Surgeon Gen Off, 44-46; dir dept psychiat, Univ & Univ Hosps, 46-69, dean sch med, 59-66, PROF PSYCHIAT, CASE WESTERN RESERVE UNIV, 46- Concurrent Pos: Consult army personnel policy, Dir Defense; med dir, Vet Admin, 46-48; med adv, Am Red Cross, 50-51; training analyst, Psychoanal Training Ctr, Cleveland, 54; mem, Res & Develop Bd, Dept Defense; mem, Nat Res Coun; mem ment health prog-proj comt, NIMH, 61-64, chmn bd sci counrs, 67-71, mem adv comt to dir, NIH, 66-70; trustee, Grant Found, 62-, pres, 66-; consult, AEC, 74- Mem: AAAS; Am Acad Arts & Sci; fel Am Psychiat Asn; Am Psychoanal Asn (secy, 55-57). Res: Neurophysiological aspects of human behavior. Mailing Add: Dept of Psychiat Univ Hosps 2040 Abington Rd Cleveland OH 44106

BOND, EDWIN JOSHUA, b Delta, Ont, June 9, 27, m 51; c 3. ZOOLOGY. Educ: Univ Toronto, BSA, 50; Univ Western Ont, MSc, 55; Univ London, DIC & PhD(insect toxicol), 59. Prof Exp: Tech officer forest entom, 50-51, RES SCIENTIST INSECT TOXICOL, CAN AGR, 51- Concurrent Pos: Mem toxicol res unit, Med Res Coun, London, Eng, 67-68; Food & Agr Orgn consult, Cent Food Technol Res Inst, Mysore, India, 71. Mem: Entom Soc Am. Res: Toxicology; application of fumigants for control of insects and other organisms; food storage and preservation. Mailing Add: Res Inst Can Agr Univ Sub PO London ON Can

BOND, ELIZABETH DUX, b Altoona, Pa, June 17, 23. ANALYTICAL CHEMISTRY. Educ: Pa State Univ, BS, 44, MS, 51; Rutgers Univ, PhD(chem), 73. Prof Exp: Instr chem, Orlando Col, 63-67; asst chem, Rutgers Univ, 68-73, asst prof, 73-74; RES CHEMIST, CITIES SERV CO, 74- Mem: Sigma Xi; Am Chem Soc; AAAS. Res: Chemistry of coal and its products; electron spin resonance spectroscopy; infrared and raman spectroscopy. Mailing Add: Cities Serv Co Drawer 4 Cranbury NJ 08512

BOND, FREDERICK THOMAS, b Brooklyn, NY, Oct 26, 36. ORGANIC CHEMISTRY. Educ: Mass Inst Technol, BS, 58; Univ Calif, Berkeley, PhD(chem), 62. Prof Exp: Asst prof chem, Ore State Univ, 62-67; asst prof, 67-70, ASSOC PROF CHEM, UNIV CALIF, SAN DIEGO, 70- Concurrent Pos: NIH spec fel, 72-73. Mem: Am Chem Soc; Sigma Xi; AAAS. Res: Chemistry of natural products; structure and reactivity relationships in simple and complex molecules; chemistry of divalent carbon; synthetic organic chemistry. Mailing Add: Dept of Chem Univ of Calif at San Diego La Jolla CA 92093

BOND, GARY CARL, b Lawrence, Kans, July 28, 42; m 65; c 1. MEDICAL PHYSIOLOGY. Educ: Univ Kans, BS, 65, PhD(physiol), 70. Prof Exp: Res physiologist, St Luke's Hosp, 70-73; ASST PROF PHYSIOL, MED CTR, UNIV ARK, LITTLE ROCK, 73- Concurrent Pos: Lectr physiol, Med Ctr, Univ Kans, 71-73; clin asst prof, Sch Med, Univ Mo-Kansas City, 72-73. Mem: AAAS; Am Heart Asn; Am Physiol Soc. Res: Cardiovascular physiology with emphasis on cardiovascular control mechanisms involved in blood pressure and blood volume regulation. Mailing Add: Dept of Physiol Univ of Ark Med Ctr Little Rock AR 72201

BOND, GEORGE CLEMENT, b Knoxville, Tenn, Nov 16, 36; m 68; c 1. ANTHROPOLOGY. Educ: Boston Univ, BA, 59; London Sch Econ, MA, 62, PhD(anthrop), 68. Prof Exp: Asst lectr sociol, Univ EAnglia, 66-68; asst prof, 68-74, ASSOC PROF ANTHROP, TEACHERS COL, COLUMBIA UNIV, 74- Mem: Am Anthrop Asn. Res: Religious and political organization of mainly non-western societies, specifically African; developing countries; United States southern communities. Mailing Add: Dept of Anthrop Columbia Univ New York NY 10027

BOND, HARLEY WILLIAM, b Pittsburgh, Pa, Feb 1, 35; m 62; c 5. NEUROPHARMACOLOGY, PHARMACY. Educ: Duquesne Univ, BS, 57; Univ Ill, MS, 60; Tulane Univ, PhD(pharmacol), 64. Prof Exp: From instr to asst prof pharmacol, Univ Ill, 64-66; pharmacologist, 66-70, mgr prod coord, 70-73, DIR TECH SERV, PARKE, DAVIS & CO, 73- Concurrent Pos: USPHS res support grant & Ill State Dept Ment Health res grant, 64-66; consult, Bioeng Consult, Inc, 66-67. Mem: AAAS; Am Pharmaceut Asn; Soc Neurosci. Res: Effect of chemicals on the direct current potential correlates of single unit activity in the neocortex; effect of chemicals on neuro-electrical correlates of behavior. Mailing Add: Parke Davis & Co Joseph Campau at the River Detroit MI 48232

BOND, HOWARD EDWARD, b Roseville, Calif, Mar 9, 30; m 59. BIOPHYSICS, CANCER. Educ: Univ Calif, BS, 53, DVM, 55; Cornell Univ, PhD(physiol), 59. Prof Exp: Asst physiol, State Univ NY Vet Col, Cornell Univ, 55-59; res assoc cellular physiol, Biol Div, Oak Ridge Nat Lab, 59-61; res biologist, US Naval Radiol Defense Lab, 61-65; res biologist, Nat Cancer Inst, Md, 65-66, head molecular separations unit, 66-69; DIR RES, ELECTRO-NUCLEONICS LABS, INC, 69- Concurrent Pos: Clin assoc prof, Sch Med, Georgetown Univ, 71- Mem: Am Soc Vet Physiol & Pharmacol; Biophys Soc; Am Soc Zoologists; NY Acad Sci. Res: Nutritional and metabolic diseases of ruminants; biochemistry and hormonal regulation of cellular proteins; cancer; ultracentrifugal analysis. Mailing Add: Electro-Nucleonics Labs Inc 4809 Auburn Ave Bethesda MD 20014

BOND, HOWARD EMERSON, b Danville, Ill, Oct 11, 42; m 67. ASTRONOMY. Educ: Univ Ill, BS, 64; Univ Mich, MS, 65, PhD(astron), 69. Prof Exp: Res assoc, 69-70, asst prof, 70-75, ASSOC PROF ASTRON, LA STATE UNIV, BATON ROUGE, 75- Concurrent Pos: Vis prof, Univ Wash, 75-76. Mem: AAAS; Am Astron Soc; Int Astron Union. Res: Astronomical spectroscopy and photometry. Mailing Add: Dept of Physics & Astron La State Univ Baton Rouge LA 70803

BOND, HOWARD WISSLER, b Hugo, Okla, Oct 31, 16; m 41; c 3. CHEMISTRY. Educ: Univ Ark, BS, 36; Univ Ill, MS, 38, PhD(chem), 41. Prof Exp: Asst chem, Univ Ill, 37-41; from jr chemist to assoc chemist, Nat Bur Standards, 41-46; from sr asst scientist to sr scientist, USPHS, 46-56, scientist dir, 56-66; chmn dept, 66-74, PROF MED CHEM, COL PHARM, UNIV RI, 66- Concurrent Pos: Lectr, George Washington Univ, 42-47. Mem: Am Chem Soc. Res: Mass spectroscopy; synthesis of isotopic compounds; synthesis of anthelmentic drugs; cancer chemotherapy; environmental health; molluscicides. Mailing Add: Dept of Med Chem Col of Pharm Univ of RI Kingston RI 02881

BOND, JACK J, soil science, see 12th edition

BOND, JAMES, b Denver, Colo, June 26, 23; m 58. ANIMAL SCIENCE. Educ: Univ Ill, BS, 50, PhD(animal sci), 55; Univ Wyo, MS, 51. Prof Exp: Animal husbandman, Animal & Poultry Husb Res Br, 55-58 & Animal Husb Res Div, 58-60, res animal husbandman, 60-70 & Animal Sci Res Div, 70-73, RES ANIMAL SCIENTIST, NUTRIT INST, RUMINANT NUTRIT LAB, AGR RES SERV, USDA, 73- Mem: Am Soc Animal Sci; Am Registry Cert Animal Scientists. Res: Effects of noise on farm animals; nutrition and reproduction interrelationships; behavior of domestic animals; environmental effects on farm animals; calf nutrition; grazing research with ruminants. Mailing Add: Ruminant Nutrit Lab Agr Res Ctr-East Beltsville MD 20705

BOND, JAMES, b Philadelphia, Pa, Jan 4, 00; m 53. TAXONOMIC ORNITHOLOGY. Educ: Cambridge Univ, BA, 22. Prof Exp: Res assoc, Acad Natural Sci, Philadelphia, 33-38, res fel Birds of Am, 38-39, assoc cur, 39-62, CUR BIRDS OF AM, ACAD NATURAL SCI, PHILADELPHIA, 62- Honors & Awards: Musgrave Medal, Inst Jamaica, 52; Brewster Medal, Am Ornith Union, 54. Mem: Fel Am Ornith Union; Cuban Soc Natural Hist. Res: West Indian avifauna; Peruvian ornithology. Mailing Add: Acad of Natural Sci 19th St & The Parkway Philadelphia PA 19103

BOND, JAMES ARTHUR, II, b Orangeburg, SC, July 11, 17; m 56; c 1. ECOLOGY. Educ: J C Smith Univ, BS, 38; Univ Kans, MA, 42; Univ Chicago, PhD, 61. Prof Exp: Teacher, Fla Pub Sch, 38-40; instr biol, Langston Univ, 46-48; John M Prather fel, Univ Chicago, 53-54; ASST PROF ZOOL, UNIV ILL, CHICAGO CIRCLE, 57-, ASST DEAN COL ARTS & SCI, 74- Mem: AAAS; Am Soc Zoologists; Ecol Soc Am; Biomet Soc; Soc Study Evolution. Res: Population ecology; problems of spacing in insect populations; aging phenomena. Mailing Add: 5134 S Dorchester Chicago IL 60615

BOND, JAMES OLIVER, b Wabash, Ind, Aug 16, 23; m 48; c 5. PUBLIC HEALTH. Educ: Earlham Col, AB, 48; Univ Chicago, MD, 50; Johns Hopkins Univ, MPH, 55; Univ Pittsburgh, DrPH, 67; Am Bd Prev Med, dipl. Prof Exp: Health officer, State Bd Health, Fla, 52-54, epidemiologist, 55-69; med officer, WHO, 69-71; MED OFFICER, PAN-AM HEALTH ORGN, 71- 43-46. Mem: Fel Am Pub Health Asn. Res: Epidemiology of arthropodborne diseases. Mailing Add: Pan-Am Health Orgn 525 23rd NW Washington DC 20037

BOND, JAMES WILLIAM, mathematics, see 12th edition

BOND, JENNY TAYLOR, b Kinston, NC, Sept 9, 39; m 76; c 2. NUTRITION. Educ: Meredith Col, AB, 61; Mich State Univ, PhD(nutrit, physiol), 72. Prof Exp: Exten agt youth, NC Agr Exten Serv, 61-64; from instr nutrit to instr pharmacol, 70-72, fel develop nephrology, 72-74, ASST PROF, DEPT FOOD SCI & HUMAN NUTRIT & DEPT HUMAN DEVELOP, MICH STATE UNIV, 74- Concurrent Pos: Specialist foods, NC Agr Exten Serv, 64. Mem: Am Soc Nephrology; Am Fedn Clin Res; Nutrit Soc Brit; Int Soc Nephrology. Res: Nutrition prenatally and postnatally; subsequent effects on development, specifically nutrition; drug interactions; hereditary hydronephrosis; nutritional status of migrant workers. Mailing Add: Dept of Food Sci & Human Nutrit Mich State Univ East Lansing MI 48824

BOND, JOHN GILBERT, b Missoula, Mont, Nov 29, 31; m 54; c 4. STRATIGRAPHY, REGIONAL GEOLOGY. Educ: Univ Idaho, BS, 54; Univ Wash, MS, 59, PhD(geol), 62. Prof Exp: Instr geol, Ore State Univ, 60-61; res geologist, Jersey Prod Res Co, 61-63; res geologist, Esso Explor, 64 & Esso Prod Res Co, 64-65, sr res geologist, 65-67; sr geologist, Humble Oil & Ref Co, 67-68; assoc prof geol, Univ Idaho, 68-69; sr geologist, Idaho Bur Mines & Geol, 69-74; prof geol, 70-74, ACTG DEAN COL MINES, UNIV IDAHO, 74-; CHIEF, IDAHO BUR MINES & GEOL, 74- Concurrent Pos: Secy, Idaho State Bd Regist Prof Geologists, 71-75. Mem: Geol Soc Am; Int Asn Volcanology & Chem Earth Interior. Res: Basalt stratigraphy and geologic evolution of the Columbia Plateau; origin of petroleum and diagnetic history; application of geologic and geochemical techniques for exploration in oil and ore deposits. Mailing Add: Idaho Bur Mines & Geol Moscow ID 83843

BOND, JOHN WALTER, JR, b Denver, Colo, Jan 13, 17; m 43; c 2. THEORETICAL PHYSICS. Educ: Univ Chicago, BS, 40; Univ NMex, PhD(theoret physics), 54. Prof Exp: Res physicist, Manhattan Dist, Columbia Univ, 43-45; staff scientist, Los Alamos Sci Lab, Univ Calif, 45-55 & Lockheed Aircraft Co, 55-56; head theoret physics group, Gen Elec Co, 56-57; chief physics, Convair, Calif, 57-59; sr scientist, Advan Res Proj Div, Inst Defense Anal, 59-60; dir res, Geophys Corp Am, 60-62; sr scientist, Kaman Nuclear Div, 62-63; head radiation physics, Aerospace Corp, 64-68; prin scientist, Heliodyne Corp, 68-69; vpres & dir res, Physics Technol Labs, Inc, 69-71; consult physicist, 71-73, RES PHYSICIST, US ARMY MOBILITY EQUIP RES & DEVELOP COMMAND, 73- Concurrent Pos: Consult, Kaman Nuclear Div, Naval Res Lab, Army Res Off & Aerospace Corp. Mem: AAAS; Am Inst Aeronaut & Astronaut; Am Phys Soc. Res: Hypervelocity impact and resultant spallation; solar photovoltaic technology; radiation and shock hydrodynamics; atomic physics; nuclear explosion physics. Mailing Add: 6621 Wakefield Dr 306/200 Alexandria VA 22307

BOND, JUDITH, b New York, NY, Sept 27, 40; m 74; c 5. BIOCHEMISTRY, PHYSIOLOGY. Educ: Bennington Col, BA, 61; Rutgers Univ, MS, 62, PhD(biochem & physiol), 66. Prof Exp: NIH fel, Vanderbilt Univ, 66-68; from instr to asst prof, 68-74, ASSOC PROF BIOCHEM, MED COL VA, VA COMMONWEALTH UNIV, 74- Mem: AAAS; Am Chem Soc; Am Soc Microbiol. Res: Intracellular protein degradation; regulation of enzyme activity and concentration; cancer biochemistry and chemotherapy; lysosomes. Mailing Add: Dept of Biochem Va Commonwealth Univ Richmond VA 23298

BOND, LORA, b Bryan, Tex, May 17, 17. BOTANY. Educ: Univ Tenn, BA, 38; Wellesley Col, MA, 41; Univ Wis, PhD(bot), 45. Prof Exp: Asst bot, Wellesley Col, 38-41 & Univ Wis, 41-43; actg instr, Univ Tenn, 43; instr biol & chem, Drury Col, 43-45 & bot, Wellesley Col, 45-48; assoc prof biol, 48-53, head dept, 49-74, chmn, Div Sci & Math, 52-55, 64-67, PROF BIOL, DRURY COL, 53- Concurrent Pos: Sci fac fel, NSF, 61. Mem: Fel AAAS; Bot Soc Am; Ecol Soc Am; Am Inst Biol Sci. Res: Morphology of angiosperms; colchicine induction of polyploidy in petunia; edaphic factors of Hempstead Plains, Long Island; plant morphology. Mailing Add: Dept of Biol Drury Col Springfield MO 65802

BOND, LYNDON HERRICK, b Bangor, Maine, Feb 2, 22; m 45; c 2. FISH BIOLOGY. Educ: Univ Maine, 40-43, 46-47, 48-49. Prof Exp: Fish hatchery asst, 41, asst fisheries, 42, fishery biologist, 47-53, asst chief fishery res & mgt div, 53-67, CHIEF FISHERY DIV, STATE DEPT INLAND FISHERIES & WILDLIFE, MAINE, 67- Concurrent Pos: Secy-treas, Atlantic Fisheries Biol, 51. Mem: Am Fisheries Soc. Res: Biology of fresh water fishes; growth control charts applied to Atlantic salmon; salmon restoration in Maine; man-made obstructions and logging practices in relation to certain salmonid fishes of northern Maine. Mailing Add: State Dept Inland Fish & Wildlife 284 State St Augusta ME 04330

BOND, PETER DANFORD, b Providence, RI, Jan 30, 40; m 68; c 2. EXPERIMENTAL NUCLEAR PHYSICS. Educ: Harvard Univ, BA, 62; Western Reserve Univ, MA, 63; Case Western Reserve Univ, PhD(physics), 69. Prof Exp: Res assoc physics, Stanford Univ, 69-72; from asst physicist to assoc physicist, 72-76, PHYSICIST, BROOKHAVEN NAT LAB, 76- Mem: Am Phys Soc; AAAS. Res: Study of the reaction mechanism in heavy ion induced transfer reactions. Mailing Add: Bldg 510A Brookhaven Nat Lab Upton NY 11973

BOND, RALPH HURD, b Columbus, Ohio, May 17, 09; wid; c 3. GEOLOGY. Educ: Ohio State Univ, BA, 31, MSc, 37. Prof Exp: Soil surveyor, Soil Conserv Serv, USDA, Washington, DC, 35-43; asst prof geol, Capital Univ, 43-49; assoc prof, Tex Tech Col, 49-55; prof, 55-72, EMER PROF GEOG & GEOL, CAPITAL UNIV, 72- Concurrent Pos: Jr geologist, US Geol Surv, Exp Sta, Ohio State Univ, 48-49; asst geologist, State Water Resources Bd, Ohio, 48-49. Mem: Geol Soc Am; Nat Asn Geol Teachers. Res: Micropaleontology; conodonts from Ohio shale. Mailing Add: Chambers Rd Rte 2 Amanda OH 43102

BOND, RICHARD GUY, b Beecher Falls, Vt, Dec 9, 16; m 53. ENVIRONMENTAL HEALTH. Educ: Univ NH, BS, 38; Univ Iowa, MS, 40; Univ Minn, MPH, 48; Environ Eng Intersoc Bd, dipl. Prof Exp: Asst, Univ NH, 38-39; asst, Univ Iowa, 39-40; pub health engr, State Dept Health, Iowa, 40-47; asst prof, Sch Civil Eng, Cornell Univ, 47-49; from asst prof to prof & pub health engr, Univ Health Serv, 49-62, PROF ENVIRON HEALTH, SCH PUB HEALTH, UNIV MINN, MINNEAPOLIS, 62- Concurrent Pos: Spec consult, NIH, USPHS & Off Surgeon Gen, US Army; tech staff, Task Force Environ Health, Nat Comn Community Health Serv; chmn planetary quarantine adv comt, Am Inst Biol Sci, 65-75; mem life sci adv comt, NASA, 71-74. Mem: Fel Am Pub Health Asn; Am Acad Environ Eng; fel Am Soc Civil Engrs; Nat Soc Prof Engrs; hon fel Royal Soc Health. Res: Public health engineering. Mailing Add: Sch of Pub Health Univ of Minn Minneapolis MN 55455

BOND, RICHARD MARSHALL, b New York, NY, Nov 5, 03; m 31, 56; c 4. ENVIROMENTAL BIOLOGY. Educ: Yale Univ, PhB, 26, PhD(zool, anat), 32. Prof Exp: Asst biol, Yale Univ, 28-30, instr, 32-33, Nat Res fel, 33-34; teacher, Santa Barbara Sch, 34-35; assoc wildlife technician, Nat Park Serv, 35-38; regional biologist, Soil Conserv Serv, USDA, 38-51, dir land & water develop prog, Virgin Islands Corp, 52, agriculturist in chg, Virgin Islands Agr Prog, Agr Res Serv, 53-67; independent consult biol, soils & trop agr, 67-70; res asst, Caribbean Res Inst, Col Virgin Islands, 70-72; SR SCIENTIST, ENVIRON CONSULTS, INC, 72- Concurrent Pos: Mem exped, Hispaniola, 33-34; consult, 70- Mem: Fel AAAS; Am Soc Mammal; Am Ornithologists Union; Cooper Ornith Soc; Wilson Ornith Soc. Res: Geographic distribution of plants and animals; taxonomy of Falco. Mailing Add: PO Box 377 Kingshill St Croix VI 00850

BOND, RICHARD RANDOLPH, b Lost Creek, WVa, Dec 1, 27; m 49; c 4. ACADEMIC ADMINISTRATION, ZOOLOGY. Educ: Salem Col, WVa, BS, 48; Univ WVa, MS, 49; Univ Wis, PhD(zool), 55. Prof Exp: Asst prof biol, Milton Col, 49-51; assoc prof, Salem Col, WVa, 55-58, dean men, 57-58; fel col admin, Univ Mich, 58-59; dean fac, Elmira Col, 59-63; prof acad admin, Cornell Univ & arts & sci consult, Univ Liberia, 63-64, chief party, Cornell Univ Proj in Lineria, 64-66; prof zool, vpres acad affairs & dean fac, Ill State Univ, 66-71; PRES, UNIV NORTHERN COLO, 71- Concurrent Pos: Mem, Bd Trustees, Salem Col; mem, Comn Fed Rels, Am Coun Educ, 75- & Comn Lib Learning, Asn Am Cols, 76- Mem: AAAS; Ecol Soc Am; Sigma Xi; Wilson Ornith Soc; Am Ornith Union. Res: Ecology and behavior of birds; autonomic responses; problems of higher education; international education. Mailing Add: Off of the Pres Univ of Northern Colo Greeley CO 80631

BOND, ROBERT FRANKLIN, b Pullman, Wash, Apr 9, 37; m 60; c 3. PHYSIOLOGY. Educ: Ursinus Col, BS, 59; Temple Univ, MS, 61, PhD(physiol), 64. Prof Exp: Assoc physiol & cardiovasc trainee, 64-65, from instr to assoc prof physiol, Bowman Gray Sch Med, 65-73; PROF PHYSIOL, CHMN DEPT & DIR CARDIOVASC RES LABS, KIRKSVILLE COL OSTEOP MED, 73- Concurrent Pos: Am Heart Asn advan res fel, 68-70; consult, Carolina Med Electronics & Electromagnetic Probe Co. Mem: Am Heart Asn; Am Physiol Soc; Microcirc Soc. Res: Autoregulation of blood flow; control of cardiovascular reflexes in shock and hypertension; control of regional vascular beds; myocardial metabolism; hyperbaric physiology. Mailing Add: Dept of Physiol Kirksville Col of Osteop Med Kirksville MO 63501

BOND, ROBERT SUMNER, b Cambridge, Mass, Oct 12, 25; m 50; c 2. FOREST ECONOMICS, RESOURCE ECONOMICS. Educ: Univ Mass, BS, 51; Yale Univ, MF, 52; State Univ NY Col Forestry, Syracuse Univ, PhD(forestry econ), 66. Prof Exp: Res forester, Fordyce Lumber Co, Ark, 52-54; dist forester, Div Forests & Parks, Mass Dept Natural Resources, 54-56; instr gen forestry, 56-63, asst prof forestry econ, 63-69, ASSOC PROF FORESTRY ECON, UNIV MASS, AMHERST, 69- Mem: Soc Am Foresters; Forest Hist Soc. Res: Forest oriented outdoor recreation; forest industry organization and labor; forest resource policy research. Mailing Add: Dept of Forestry & Wildlife Mgt Univ of Mass Amherst MA 01002

BOND, ROBERT WALLACE, b Weehawken, NJ, July 9, 02; m 44. CHEMISTRY. Educ: Princeton Univ, AB, 23, MA, 29, PhD(chem), 33; Columbia Univ, MA, 25. Prof Exp: Control chemist, Armstrong Cork Co, Pa, 23; asst instr chem, Princeton Univ, 25-27; instr, Emory Univ, 28-31; head dept, Morris Jr Col, NJ, 33-34, dean, 34-36; tech dir, Andrew Wilson, Inc, NJ, 35-38; chemist, Fries & Fries, Inc, NY & Ohio, 38-40 & Merck & Co, Inc, NJ, 40-51; tech dir, Chlorophyll, Inc, Kans, 51-52; prod chemist, S B Penick Co, 52-68; independent consult, 68-75; lectr chem, County Col Morris, 69-75; RETIRED. Mem: Am Chem Soc. Res: Insecticides; food flavors; fine chemicals; medicinal chemicals. Mailing Add: 28 Fairview Dr East Hanover NJ 07936

BOND, ROLSTON LYMAN, chemistry, see 12th edition

BOND, STEPHON THOMAS, b Lost Creek, WVa, Jan 25, 34; m 60; c 4. INORGANIC CHEMISTRY. Educ: Salem Col, BS, 60; WVa Univ, MA, 62; Kent State Univ, PhD(chem), 71. Prof Exp: Teacher chem, Belpre Pub Schs, 62-65; PROF CHEM, SALEM COL, 65-68 & 71- Mem: Am Chem Soc; Sigma Xi. Res: Organometallic complexes; computer control of mass spectroscopy. Mailing Add: Rte 1 Box 74 Jane Lew WV 26378

BOND, TED P, b Sweetwater, Tex, Apr 16, 26; m 53; c 1. BIOLOGY, PHYSIOLOGY. Educ: WTex State Univ, BS, 57; Univ Tex, MA, 63. Prof Exp: Contractor, 46-50; res assoc, 61-67, RES ASSOC PROF PHYSIOL, UNIV TEX MED BR, GALVESTON, 67- Concurrent Pos: Co-investr, NIH grant, 61-68 & Upjohn Co grant, 63-65; prin investr, Lloyd Bros Drug Co grant, 64-65. Mem: Am Physiol Soc; Microcirc Conf. Res: Microcirculation, especially normal and during extracorporeal circulation; hypothermia, after thermal and traumatic injury; blood coagulation and hemorrhagic diseases; fibrinolytic enzyme system and the clinical use of fibrinolytic activating agents. Mailing Add: Dept of Physiol Univ of Tex Med Br Galveston TX 77550

BOND, THOMAS JACKSON, b Ennis, Tex, Aug 16, 12; m 39; c 3. BIOCHEMISTRY. Educ: NTex State Col, BS, 38, MS, 39; Univ Tex, PhD(chem), 50. Prof Exp: Instr chem, Tex Col Arts & Industs, 42-43; from instr to assoc prof, 43-51, PROF CHEM, BAYLOR UNIV, 51-, CHMN DEPT, 64- Mem: Am Chem Soc; Am Soc Biol Chem; Sigma Xi. Res: Effects of carcinogenic chemicals on the fundamental chemistry of cellular metabolism. Mailing Add: Dept of Chem Baylor Univ Waco TX 76703

BOND, VICTOR POTTER, b Santa Clara, Calif, Nov 30, 19; m 46, 56; c 4. RADIATION BIOPHYSICS, EXPERIMENTAL MEDICINE. Educ: Univ Calif, AB, 43, MD, 45, PhD(med physics), 51. Hon Degrees: DSc, Long Island Univ, 65. Prof Exp: Asst, Med Sch, Univ Calif, 45; res group leader, Naval Radiol Defense Lab, 48-55, mem staff, 55-58, head, Div Microbiol, 58-62, chmn dept, 62-67, ASSOC DIR, MED DEPT, BROOKHAVEN NAT LAB, 67- Concurrent Pos: Prof Med, State Univ NY Stony Brook, 68-; prof radiol, Columbia Univ, 69-; pres, 5th Int Cong Radiation Res, 71-74. Mem: Am Phys Soc; Radiation Res Soc (pres, 73-74); Soc Exp Biol & Med; Soc Nuclear Med; Am Environ Mutagen Soc. Res: Biological effects of radiation; effects of radiation on mammalian systems; the use of particulate radiations in radiotheraphy. Mailing Add: Brookhaven Nat Lab Upton NY 11973

BOND, WALTER D, b Lebanon, Tenn, Mar 21, 32; m 55; c 3. PHYSICAL CHEMISTRY, CHEMICAL ENGINEERING. Educ: Mid Tenn State Univ, BS, 53; Vanderbilt Univ, PhD(phys chem), 57. Prof Exp: STAFF MEM RES & DEVELOP, OAK RIDGE NAT LAB, 57- Concurrent Pos: Guest scientist, Swiss Fed Inst Reactor Res, 71-72. Mem: Am Chem Soc; Am Nuclear Soc. Res: Chemistry of separations of transuranium elements; nuclear fuel reprocessing. Mailing Add: 6704 Stone Mill Rd Knoxville TN 37919

BOND, WILLIAM BRADFORD b Ithaca, NY, Oct 11, 29; m 52; c 4. PHYSICAL CHEMISTRY, ORGANIC CHEMISTRY. Educ: Cornell Univ, BA, 51; Fla State Univ, PhD(chem), 55. Prof Exp: Res chemist, 55-63, res supvr, 63-72, RES ASSOC, PLASTIC PROD DIV, PLASTICS DEPT, E I DU PONT DE NEMOURS & CO, 72- Mem: AAAS; Am Chem Soc; Sigma Xi. Res: Free radical chemistry; polyamidation kinetics; polymer chemistry; fibers. Mailing Add: Plastic Prod Div Plastics Dept E I du Pont de Nemours & Co Parkersburg WV 26105

BOND, WILLIAM H, b Toronto, Ont, Sept 20, 16; US citizen; m 49; c 4. INTERNAL MEDICINE, HEMATOLOGY. Educ: Univ Chicago, SB, 40, MD, 42. Prof Exp: From instr to assoc prof, 52-67, PROF INTERNAL MED, SCH MED, IND UNIV-PURDUE UNIV, INDIANAPOLIS, 67- Concurrent Pos: Consult, Vet Admin Hosp, 55- Mem: Am Fedn Clin Res; fel Am Col Physicians; Am Soc Hemat; Am Asn Cancer Res. Res: Cancer chemotherapy; clinical hematology. Mailing Add: Sch of Med Ind Univ-Purdue Univ Indianapolis IN 46202

BOND, WILLIAM PAYTON, b Franklinton, La, Nov 18, 41; m 69; c 2. PLANT PATHOLOGY. Educ: Southeastern La Col, BS, 63; La State Univ, MS, 66, PhD(plant path), 68. Prof Exp: Asst prof bot & plant path, 67-74, ASSOC PROF BIOL, SOUTHEASTERN LA UNIV, 74- Mem: AAAS; Am Phytopath Soc. Res: Plant virology; soil transmission of plant viruses; serology. Mailing Add: Dept of Biol Southeastern La Univ Hammond LA 70401

BONDAR, RICHARD JAY LAURENT, b New York, NY, Sept 4, 40; m 61. CLINICAL BIOCHEMISTRY, CLINICAL CHEMISTRY. Educ: McGill Univ, BSc, 62; Calif Inst Technol, MS, 65; Univ Calif, Riverside, PhD(biochem), 69. Prof Exp: Instr biol, La State Univ, New Orleans, 64-65; res assoc biochem, Mich State Univ, 69-71; DIR CLIN RES, WORTHINGTON BIOCHEM CORP, 71- Concurrent Pos: Mem subcomt on glucose ref methods, Standards Comt, Am Asn Clin Chem, 72-; mem subcomt on enzyme assay conditions, Standards Comt, Nat Comt on Clin Lab Standards, 73- Mem: AAAS; Am Chem Soc; Am Asn Clin Chemists; Sigma Xi. Res: Enzymology and protein chemistry; mechanisms of enzyme catalysis and mechanism of reactions of biological interest; analytical biochemistry; enzymology. Mailing Add: Worthington Biochem Corp Halls Mills Rd Freehold NJ 07728

BONDAREFF, WILLIAM, b Washington, DC, Apr 29, 30; m 58; c 2. ANATOMY. Educ: George Washington Univ, BS, 51, MS, 52; Univ Chicago, PhD(anat), 54; Georgetown Univ, MD, 62. Prof Exp: USPHS res fel, Nat Cancer Inst, 54-55; cytologist, Sect Aging, NIMH, 55-64; assoc prof, 64-70, PROF ANAT & CHMN DEPT, MED SCH, NORTHWESTERN UNIV, CHICAGO, 70- Mem: AAAS; Electron Micros Soc; Geront Soc; Am Asn Anatomists; Am Acad Neurologists. Res: Microscopic and ultrastructure of nervous system. Mailing Add: Dept of Anat Northwestern Univ Med Sch Chicago IL 60611

BONDE, ERIK KAUFFMANN, b Odense, Denmark, Oct 19, 22; nat US; m 53; c 3. PLANT PHYSIOLOGY. Educ: Univ Colo, BA, 47, MA, 48; Univ Chicago, PhD(bot), 51. Prof Exp: Res assoc bot, Univ Calif, Los Angeles, 51-53; NSF fel, Calif Inst Technol, 53-54; asst prof bot, Univ Mo, 54-55; from asst prof to assoc prof, 55-67, PROF BIOL, UNIV COLO, 67- Mem: AAAS; Ecol Soc Am; Am Soc Plant Physiol; Scandinavian Soc Plant Physiol. Res: Plant growth substances and inhibitors; photoperiodism; ecology. Mailing Add: Dept of Biol Univ of Colo Boulder CO 80302

BONDE, MORRIS REINER, b Presque Isle, Maine, Aug 7, 45; m·73. PLANT PATHOLOGY. Educ: Univ Maine, BS, 67; Cornell Univ, MS, 69, PhD(plant path), 75. Prof Exp: RES PLANT PATHOLOGIST, AGR RES SERV, USDA, 74- Mem: Am Phytopath Soc. Res: Epidemiology and control of diseases of major food crops. Mailing Add: Plant Dis Res Lab USDA Agr Res Serv PO Box 1209 Frederick MD 21701

BONDELID, ROLLON OSCAR, b Grand Forks, NDak, Jan 8, 23; m 73; c 3. NUCLEAR PHYSICS. Educ: Univ NDak, BS, 45; Washington Univ, MS, 48, PhD(physics), 50. Prof Exp: Mem staff, Los Alamos Sci Lab, 50-52; physicist, 52-63, HEAD CYCLOTRON BR, NAVAL RES LAB, 63- Mem: Am Phys Soc; Sigma Xi; Am Nuclear Soc. Res: Development of sector-focusing cyclotron research facility; correlation studies of few-nucleon systems; neutron cancer therapy; absolute energy measurements of nuclear reactions. Mailing Add: US Naval Res Lab Code 6610 Washington DC 20375

BONDERMAN, DEAN P, b Primghar, Iowa, July 6, 36; m 68; c 2. CLINICAL PATHOLOGY, TOXICOLOGY. Educ: Westmar Col, BA, 62; Univ Iowa, PhD(inorg chem), 68. Prof Exp: Sr chemist & instr pesticide toxicol, Dept Prev Med, Univ Iowa, 66-70; ASST PROF CLIN PATH, MED CTR, IND UNIV, INDIANAPOLIS, 70- Mem: Am Chem Soc; Am Soc Clin Path; Am Asn Clin Chem. Res: High temperature thermodynamics of inroganic compounds; enzyme response to environmental stimuli; pesticide detection; enzyme kinetics and thermodynamics. Mailing Add: Dept of Clin Path Ind Univ Med Ctr Indianapolis IN 46202

BONDI, AMEDEO, b Springfield, Mass, Dec 13, 12; m 39; c 4. MICROBIOLOGY. Educ: Univ Conn, BS, 35; Univ Mass, MS, 37; Univ Pa, PhD(bact), 42. Prof Exp: Instr bact, Sch Med, Temple Univ, 38-40, asst prof, 42-47; prof bact & chmn dept, 47-73, DEAN, GRAD SCH, HAHNEMANN MED COL, 73- Mem: AAAS; Am Soc Microbiol; Am Asn Immunol; Soc Exp Biol & Med; Am Pub Health Asn. Res: Immunology of Brucellosis and whooping cough; studies on the antibiotic agents; studies on staphylococcal infections. Mailing Add: Hahnemann Med Col 230 N 15th St Philadelphia PA 19102

BONDINELL, WILLIAM EDWARD, organic chemistry, see 12th edition

BONDURANT, CHARLES W, JR, b Abington, Va, Sept 4, 18. POLYMER CHEMISTRY, ANALYTICAL CHEMISTRY. Educ: Emory & Henry Col, BA, 39; Va Polytech Inst, MS, 51, PhD(polymer chem), 60. Prof Exp: Asst prof chem, Hampden-Sydney Col, 48-56; assoc prof, 56-65, PROF CHEM, ROANOKE COL, 65-, CHMN DEPT, 66- Mem: Am Chem Soc. Res: Characterization, fractionation and analysis of high polymers; mechanical properties of polymers; chemical education. Mailing Add: Dept of Chem Roanoke Col Salem VA 24153

BONDY, DONALD CLARENCE, b Harrow, Ont, June 5, 32; m 56; c 2. GASTROENTEROLOGY. Educ: Univ Western Ont, MD, 56; FRCPS(C), 61 & 67. Prof Exp: Instr med, 62-63, lectr, 63-65, lectr physiol, 65-67, assoc prof, 67-74, from asst prof med to assoc prof med & physiol, 68-73, PROF MED & PHYSIOL, UNIV WESTERN ONT, 73- Concurrent Pos: Consult gastroenterol, Westminster Hosp, 62-67, dir clin invest unit & chief of serv-gastroenterol, 67-72; consult gastroenterol, Victoria & St Joseph's Hosps, 62-67; chief of serv, Dept Gastroenterol, Univ Hosp, 73- Mem: Can Soc Clin Invest; Can Med Asn; Can Asn Gastroenterol. Res: Clinical gastroenterology and membrane biology. Mailing Add: Dept of Med Univ of Western Ont London ON Can

BONDY, STEPHEN CLAUDE, b Hilversum, Holland, Jan 10, 38; US citizen. BIOCHEMISTRY, NEUROBIOLOGY. Educ: Cambridge Univ, BA, 59, MA, 62; Univ Birmingham, MSc, 61, PhD(biochem), 62. Prof Exp: Res biochemist, Unilever Ltd, Bedford, Eng, 59; res asst, Univ Birmingham, 59-62; res scientist, NY State Psychiat Inst, 63-65; res biol chemist, Sch Med, Univ Calif, Los Angeles, 65-70; ASST PROF NEUROL, MED CTR, UNIV COLO, DENVER, 70- Concurrent Pos: NIH res career develop award, 71-75. Mem: AAAS; Am Soc Biol Chemists; Inst Soc Neurochem; Am Soc Neurochem. Res: Genetics expression in the nervous system; behavioral neurochemistry; neurobiology of cerebral development; axoplasmic transport; cerebral blood flow. Mailing Add: Dept of Neurol Univ of Colo Med Sch Denver CO 80220

BONE, DONALD ROBERT, b Minot, NDak, July 13, 44; m 66; c 1. VIROLOGY. Educ: Pasadena Col, BA, 66; Univ Calif, Davis, PhD(microbiol), 71. Prof Exp: Fel, Dept Virol & Epidemiol, Baylor Col Med, 71-74; RES VIROLOGIST VIRAL VACCINE, LEDERLE LAB, 74- Mem: Am Soc Microbiol. Res: Nucleic acid and protein syntheses in mammalian cells infected with animal viruses which can establish and maintain latent and/or persistent infections. Mailing Add: Lederle Lab Pearl River NY 10965

BONE, JACK NORMAN, b Montrose, Colo, Feb 10, 19; m 40; c 2. PHARMACY. Educ: Univ Colo, BS, 41, MS, 48; Univ Wash, PhD(pharm), 53. Prof Exp: Pharm res asst, Cutter Labs, Berkeley, Calif, 41-42, supvr blood fractionation, 42-45; instr pharm, Univ Colo, 45-48; from asst prof to assoc prof, 48-56, PROF PHARM, UNIV WYO, 56-, DEAN, COL PHARM, 66- & COL HEALTH SCI, 68- Concurrent Pos: Fulbright lectr, Univ Baghdad, 63-64. Res: Synthesis and biotesting of antibacterials and antifungals. Mailing Add: Col of Health Sci Univ of Wyo Laramie WY 82070

BONE, JESSE FRANKLIN, b Tacoma, Wash, June 15, 16; m 50; c 4. VETERINARY MEDICINE. Educ: State Col Wash, BA, 37, BS, 49, DVM, 50; Ore State Col, MS, 53. Prof Exp: Lab technician vet med, State Col Wash, 47-50; instr & res asst, 50-53, from asst prof to assoc prof, 53-67, PROF VET MED, ORE STATE UNIV, 67- Concurrent Pos: Fulbright lectr vet path, Univ Assiut, 64-65. Mem: Am Asn Vet Toxicologists; Am Asn Lab Animal Sci; Royal Soc Health. Mailing Add: Dept of Vet Med Ore State Univ Corvallis OR 97331

BONE, LARRY IRVIN, b Perry, Iowa, Aug 31, 35; m 57; c 2. CHEMICAL KINETICS, PHYSICAL CHEMISTRY. Educ: Coe Col, BA, 57; Ohio State Univ, MSc, 64, PhD, 66. Prof Exp: Res chemist, Aerospace Res Labs, Wright-Patterson Air Force Base, US Air Force, 64-67, nuclear scientist, US Dept Defense, 67-68; asst prof chem, 68-73, ASSOC PROF CHEM, EAST TEX STATE UNIV, 73-, ASST DEAN SCI & TECHNOL, 73- Mem: Am Chem Soc; Am Phys Soc; Am Soc Mass Spectrometry. Res: Radiation chemistry; mass spectrometric ion-molecule reactions; mass spectrometric ion-molecule studies. Mailing Add: Col of Sci & Technol East Tex State Univ Commerce TX 75428

BONE, ROBERT M, b Regina, Sask, June 20, 33; m 61; c 1. GEOGRAPHY. Educ: Univ BC, BA, 55; Univ Wash, MA, 57; Univ Nebr, PhD(geog), 62. Prof Exp: Geographer, Geog Br, Dept Mines & Tech Surv, Can, 59-63; from asst prof to assoc prof geog, Univ Sask, 63-73, actg head dept, 70-71, actg head, Inst Northern Studies, 72-73, PROF GEOG, UNIV SASK, 73- Concurrent Pos: Sabbatical leave, Scott Polar Res Inst, Eng, 71-72. Mem: Can Asn Geog. Res: Union of Soviet Socialist Republics; Polar lands; economic and regional geography. Mailing Add: Dept of Geog Univ of Sask Saskatoon SK Can

BONEHAM, ROGER FREDERICK, b Boston, Mass, July 26, 35; m 63; c 2. GEOLOGY. Educ: Mich Technol Univ, BS, 60; Wayne State Univ, MS, 65; Univ Mich, PhD(geol), 68. Prof Exp: Asst prof geol & bot, 66-71, ASSOC PROF GEOL, IND UNIV, KOKOMO, 71- Mem: AAAS; Am Asn Stratig Palynologists; Nat Asn Geol Teachers. Res: Tasmanites, chitinozoa and acritarchs; Paleozoic flora of the Eastern Interior Coal Basin. Mailing Add: Dept of Geol Ind Univ Kokomo IN 46901

BONEM, RENA MAE, b Tucumcari, NMex, May 6, 48. INVERTEBRATE PALEONTOLOGY, PALEOECOLOGY. Educ: NMex Inst Mining & Technol, BS, 70, MS, 71; Univ Okla, PhD(geol), 75. Prof Exp: Lectr geol, Univ Tex, Arlington, 73; ASST PROF GEOL, HOPE COL, 75- Concurrent Pos: Advan res award, Sigma Xi, 75. Mem: Paleont Soc; Int Paleont Asn; Geol Soc Am; Am Asn Petrol Geologists; Soc Econ Paleontologists & Mineralogists. Res: Development and succession of Paleozoic biohermal associations compared with modern patch reefs in Jamaica and the Florida Keys. Mailing Add: Dept of Geol Hope Col Holland MI 49423

BONETTI, GIOVANNI ALBERTO, organic chemistry, see 12th edition

BONEWITZ, ROBERT ALLEN, b Ottawa, Kans, Dec 18, 43; m 72. ELECTROCHEMISTRY, CORROSION. Educ: Lawrence Univ, BA, 65; Univ Fla, PhD(chem), 70. Prof Exp: Scientist corrosion, 70-73, sr scientist, 73-75, GROUP LEADER CORROSION, ALCOA LABS, 75- Mem: Electrochem Soc; Nat Asn Corrosion Engr. Res: Electrochemistry and mechanisms of corrosion of aluminum alloys. Mailing Add: Alcoa Labs Alcoa Center PA 15069

BONEY, WILLIAM ARTHUR, JR, b Iola, Tex, May 13, 16; m 41, 61; c 1. VETERINARY MEDICINE, IMMUNOLOGY. Educ: Agr & Mech Col Tex, BS, 40, DVM, 42, MS, 48. Prof Exp: Exten poultry vet, Exten Serv, Agr & Mech Col Tex, 42-44, poultry vet, Exp Sta, 44-46, assoc prof poultry dis & poultry pathologist, 46-54; pres & dir, Boney Labs, 54-59; mgr vet technol, Int Div, Schering Corp, 59-62; RES VET & VIROL PROJ LEADER, NAT ANIMAL DIS LAB, USDA, 63- Mem: Am Vet Med Asn; Am Asn Avian Path; US Animal Health Asn. Res: Microbiology; virology; bacteriology; avian diseases. Mailing Add: Nat Animal Dis Lab USDA Ames IA 50010

BONFIGLIO, MICHAEL, b Milwaukee, Wis, Apr 3, 17; m 43; c 3. BIOLOGY, ORTHOPEDIC SURGERY. Educ: Columbia Univ, AB, 40; Univ Chicago, MD, 43. Prof Exp: Instr orthrop surg, Univ Chicago, 49-50; assoc, 50-51, from asst prof to assoc prof, 51-62, PROF ORTHOP SURG, UNIV IOWA, 62- Mem: AMA; Am Acad Orthop Surg; Orthop Res Soc; Am Col Surg; Am Orthop Asn. Res: Pathology of irradiation fractures; aseptic necrosis of bone; bone transplant studies, histological and immunological; bone and skin transplant studies. Mailing Add: Children's Hosp Iowa City IA 52242

BONGA, JAN MAX, b Bern, Switz, Dec 5, 29; Can citizen; m 59; c 2. PLANT PHYSIOLOGY. Educ: State Univ Groningen, BSc, 54, MSc, 57; McMaster Univ, PhD(biol), 60. Prof Exp: Res officer forest path, Can Dept Forestry, 60-72, RES OFFICER FOREST PATH, MARITIMES FOREST RES CTR, CAN DEPT ENVIRON, 72- Mem: Can Soc Plant Physiologists. Res: Translocation of inorganic substances in plants; physiology of tree growth. Mailing Add: Can Dept of Environ Maritimes Forest Res Ctr Fredericton NB Can

BONGARD, STEVEN J, b New York, NY, July 1, 39; m 68. ANATOMY, EMBRYOLOGY. Educ: Hunter Col, AB, 62; Univ Vt, MS, 63; NY Univ, PhD(anat, embryol), 69. Prof Exp: Instr biol, Manhattan Community Col, 67-68; instr, 68-71, ASST PROF ANAT, CHICAGO MED SCH, 71-, ASST DEAN STUDENT AFFAIRS, 74- Mem: AAAS; Asn Am Med Cols; Am Asn Higher Educ; NY Acad Sci. Res: Development of the embryo nervous system. Mailing Add: Chicago Med Sch 2020 W Ogden Ave Chicago IL 60612

BONGIORNI, DOMENIC FRANK, b Ivoryton, Conn, June 19, 08. PHYSICAL ORGANIC CHEMISTRY. Educ: Conn State Col, BS, 33; Johns Hopkins Univ, PhD(chem), 36; Fordham Univ, LLB, 44. Prof Exp: Res chemist, Gen Chem Co, 36-39, spec consult, 40; patent dept, Union Carbide & Carbon Res Labs, Inc, 41-42; res chemist, Bakelite Corp, 42; chem res asst, Off Sci Res & Develop, War Res Div, Columbia Univ, 42-44; assoc patent lawyer, Campbell, Brumbaugh, Free & Graves, 44-51; patent lawyer, Gen Anil & Film Corp, 51-52; counr, Sci Design Co, Inc, 52-55; patent counr, Johnson & Johnson, 55-60; CONSULT, 60- Mem: AAAS; Am Chem Soc; fel Am Inst Chem; Am Bar Asn. Res: Reaction kinetics; methylene free radical and methyl, ethyl, acetonyl free radical studies; chain reactions; synthetic organic chemistry in drugs, insecticides; plastics; scientific-legal counsel. Mailing Add: PO Box 291 Ivoryton CT 06442

BONGIORNO, SALVATORE F, b Mt Vernon, NY, Aug 16, 39; m 68; c 1. VERTEBRATE ECOLOGY. Educ: Fordham Univ, BS, 61; Rutgers Univ, MS, 63, PhD(ecol), 67. Prof Exp: Asst prof ecol, La State Univ, New Orleans, 67-71; ASST PROF BIOL, FAIRFIELD UNIV, 71- Mem: AAAS; Ecol Soc Am; Am Ornith Union; Am Inst Biol Sci. Res: Nesting ecology of avifauna; pesticides environmental contamination. Mailing Add: Dept of Biol Fairfield Univ Fairfield CT 06430

BONGIOVANNI, ALFRED MARIUS, b Philadelphia, Pa, Sept 22, 21. BIOLOGY, MEDICAL SCIENCE. Educ: Villanova Col, BS, 40; Univ Pa, MD, 43. Prof Exp: Investr, Marine Biol Lab, Woods Hole, 39-40; resident physician path, Hosp, Univ Pa, 46; instr pharmacol, Philadelphia Col Pharm & Sci, 47-48; instr pediat, Sch Med, Univ Pa, 50-52, Nat Found Infantile Paralysis fel, 51-52, from asst prof to prof pediat, 54-67, chmn dept, 63-67; resident physician, 47-49, PHYSICIAN-IN-CHIEF, CHILDREN'S HOSP PHILADELPHIA, 67-, DIR ENDOCRINE LAB, 54- Concurrent Pos: Investr, Rockefeller Inst, NY, 49-50; asst prof, Johns Hopkins Univ, 52-54; mem study sect arthritis & metab, NIH, 63-67, child develop & ment retardation training comt, 66-, chmn, 67-69; US del to USSR on Physiol Develop Child, 65; mem joint coun, Nat Pediat Soc, 68; prof pediat, Univ Ife, Nigeria, 74-75; assoc adminr, Proj Hope, Cairo, Egypt, 75. Honors & Awards: Kolmer Medal, 40; Ciba Award, 57; Johnson Award, 58; Shaffrey Medal, St Joseph's Col, 65; Mendal Medal, Villanova Univ, 68. Mem: AAAS; Soc Pédiat Res; Endocrine Soc; Am Soc Clin Invest; fel NY Acad Sci. Res: Parthogenesis and fertilization of sea urchin eggs; liver metabolism in humans; fragility of erythrocytes; intermediary steroid metabolism; endocrine research. Mailing Add: Children's Hosp 34th & Civic Ctr Blvd Philadelphia PA 19104

BONHAM, CHARLES D, b Cleburne, Tex, Feb 4, 37; m 55; c 5. PLANT ECOLOGY, STATISTICS. Educ: Abilene Christian Col, BS, 59; Utah State Univ, MS, 65; Colo State Univ, PhD(plant ecol), 66. Prof Exp: Res conservationist, Colo State Univ, 62-64; plant ecologist, Cornell Aeronaut Labs, 66-67; from asst prof to assoc prof watershed mgt, Univ Ariz, 67-70; ASSOC PROF RANGE SCI, COLO STATE UNIV, 70- Mem: Am Soc Range Mgt; Ecol Soc Am; Brit Ecol Soc; Sigma Xi; Biomet Soc. Res: Quantitative plant ecology studies using statistical models and computer graphic techniques; vegetation description and plant community analyses using statistics and mathematical procedures; vegetation mapping by computer graphics. Mailing Add: Dept of Range Sci Colo State Univ Ft Collins CO 80523

BONHAM, HAROLD F, JR, b Los Angeles, Calif, Sept 1, 28; m 52; c 3. GEOLOGY. Educ: Univ Calif, Los Angeles, BA, 54; Univ Nev, MS, 62. Prof Exp: Field geologist, Southern Pac Co, 55-57, geologist, 57-61; from asst mining geologist to assoc mining geologist, Nev Bur Mines, Univ Nev, 63-74, MINING GEOLOGIST, NEV BUR MINES & GEOL, UNIV NEV, 74- & PROF GEOL, MACKAY SCH MINES, UNIV NEV, RENO, 74- Concurrent Pos: Co-prin investr, Off Water Resources Res grant, 67-68. Mem: Fel Geol Soc Am; Soc Econ Geol. Res: Geology and geochemistry of metaliferous ore deposits; petrology and petrography of volcanic rocks; environmental geology; geology of epithermal precious metal deposits. Mailing Add: Nev Bur Mines Univ of Nev Reno NV 89507

BONHAM, JAMES A, organic chemistry, see 12th edition

BONHAM, KELSHAW, b Seattle, Wash, Mar 19, 09; m 38; c 4. RADIOBIOLOGY. Educ: Univ Wash, BS, 31, MS, 35, PhD(zool), 37. Prof Exp: Asst prof fisheries, Agr & Mech Col Tex, 38-41; asst biologist, Fish & Wildlife Serv, US Dept Interior, Univ Wash, 41-42; biologist, State Fisheries Dept, Wash, 42-44; asst prof zool, Univ Hawaii, 44-45; res assoc prof lab radiation biol, 45-68, RES PROF, COL FISHERIES, UNIV WASH, 68- Mem: Am Fisheries Soc; Am Soc Ichthyol & Herpet; Am Soc Limnol & Oceanog; Marine Biol Asn; Radiation Res Soc. Res: Monogenea taxonomy; commercial biology of dogfish shark; low-level gamma-irradiation of salmon. Mailing Add: Col of Fisheries Univ of Wash Seattle WA 98105

BONHAM, LAWRENCE COOK, b Springfield, Mo, June 9, 20; m 46; c 4. GEOLOGY. Educ: Drury Col, BS, 42; Washington Univ, MS, 48, PhD, 50. Prof Exp: Photogram engr, US Coast & Geod Surv, 42-44; instr geol, Washington Univ, 46-48; supvry res geologist, Calif Res Corp, 50-64; explor geologist, Standard Oil Co, Calif, 64-66; SR RES ASSOC, CHEVRON OIL FIELD RES CO, 66- Concurrent Pos: Instr, Eve Div, East Los Angeles Col, 65-72. Mem: Soc Econ Paleont & Mineral; Am Asn Petrol Geologists. Res: Sedimentary petrology; organic geochemistry; computer applications in geology. Mailing Add: Chevron Oil Field Res Co Box 446 La Habra CA 90631

BONHAM, LAWRENCE DOUGLAS, b Corning, NY, Apr 30, 22; m 51; c 3. GEOLOGY. Educ: Iowa State Col, BS, 43; Univ Chicago, PhD(geol), 50. Prof Exp: GEOLOGIST, US GEOL SURV, 50- Mem: AAAS; Geol Soc Am. Res: Environmental geology. Mailing Add: 4856 Park Ave Washington DC 20016

BONHAM, RUSSELL AUBREY, b San Jose, Calif, Dec 10, 31; m 57; c 3. CHEMICAL PHYSICS. Educ: Whittier Col, BA, 54; Iowa State Univ, PhD(phys chem), 58. Prof Exp: Res assoc chem, Iowa State Col, 57-58; from instr to assoc prof, 58-65, PROF CHEM, IND UNIV, BLOOMINGTON, 65- Concurrent Pos: Res assoc, Nat Acad Sci-Nat Res Coun & US Naval Res Lab, 60; asst prof, Univ Col, Univ Md, 60; NSF grants, Conf Magnetism & Crystallog, Japan, 61 & Winter Inst Quantum Chem & Solid State Physics, Fla, 62; Alfred P Sloan Found fel, 64-66; Guggenheim Mem Found fel, 64-65; Fulbright Found res scholar, Univ Tokyo, 64-65; Royal Norweg Sci & Indust res grant, 65; vis scientist, NSF grant under US-Japan Coop Sci Prog, Univ Tokyo & Hokkaido Univ, Japan, 69-70; mem, Nat Acad Sci Joint Brazilian-US Comt Grad Teaching & Res in Chem, 69-76; assoc ed, J Chem Physics, 74-77. Mem: Am Chem Soc; Am Crystallog Asn; AAAS; Am Phys Soc. Res: High energy electron impact spectroscopy including scattering theory; Compton scattering by electrons; observation of extended x-ray absorption fine structure using electrons; time of flight Auger spectroscopy. Mailing Add: Dept of Chem Indiana Univ Bloomington IN 47401

BONHORST, CARL W, b Van Metre, SDak, Dec 31, 17; m 45; c 1. BIOCHEMISTRY. Educ: SDak State Univ, BS, 43; Pa State Univ, MS, 47, PhD(biol chem), 49. Prof Exp: Instr chem, Univ Portland, 49-51; res assoc pharmacol, Univ Va, 51-52; assoc biochemist, SDak State Univ, 52-56; assoc prof chem, 56-66, head dept, 68-73, PROF CHEM, UNIV PORTLAND, 66- Concurrent Pos: Mem teacher prep-cert study, Nat Asn State Dirs Teacher Educ, 61; vis prof & NSF fac fel, Univ Wis, 62-63. Mem: Am Chem Soc. Res: Physical properties of esters, peptide synthesis, selenium metabolism, active transport and Pasteur effect. Mailing Add: Dept of Chem Univ of Portland Portland OR 97203

BONI, KENNETH ARNOLD, polymer chemistry, see 12th edition

BONIC, ROBERT ALLEN, mathematics, see 12th edition

BONICA, JOHN JOSEPH, b Filicudi, Italy, Feb 16, 17; US citizen; m 42; c 4. ANESTHESIOLOGY. Educ: NY Univ, BS, 38; Marquette Univ, MD, 42; Am Bd Anesthesiol, dipl, 48. Prof Exp: Dir anesthesiol, Tacoma Gen & Pierce County Hosps, Tacoma, Wash, 47-63; PROF ANESTHESIOL & CHMN DEPT, SCH MED, UNIV WASH & ANESTHESIOLOGIST IN CHIEF, MED CTR, 60-, DIR ANESTHESIA RES CTR, 67- Concurrent Pos: Consult, US Army Med Corps, 47-, Vet Admin Hosp, 48-, Ministry of Health, Italy, 54-59, Brazil, 55-59, Arg, 59-63, Sweden, 69- & Ministry of Educ, Japan, 69; mem anesthesiol training comt, Nat Inst Gen Med Sci, 66-70, chmn gen med res prog-proj comt, 70-; sci adv comt, World Fedn Socs Anesthesiol, 68-; foreign ed, Minerva Anestesiologica, Italy. Honors & Awards: Order of Merit, Italy; Silver Medal, Swedish Med Soc, 70. Mem: Am Soc Anesthesiol (pres, 66); Acad Anesthesiol; Asn Am Med Cols; Int Anesthesia Res Soc; World Fedn Socs Anesthesiol. Res: Pharmacology of local anesthetics; neurophysiology; physiopathology and psychology of pain; obstetric anesthesia and perinatal biology; effects of regional and general anesthesia on human cardiovascular and respiratory functions. Mailing Add: Dept of Anesthesiol Univ of Wash Sch of Med Seattle WA 98195

BONIECE, WILLIAM SANFORD, bacteriology, see 12th edition

BONILLA, MANUEL GEORGE, b Sacramento, Calif, July 19, 20; m 49; c 3. GEOLOGY. Educ: Univ Calif, AB, 43; Stanford Univ, MS, 60. Prof Exp: Geologist, US Bur Reclamation, 46-47; GEOLOGIST, US GEOL SURV, 47- Honors & Awards: Claire P Holdredge Award, Asn Eng Geologists, 71; Meritorious Service Award, US Dept Interior, 75. Mem: Geol Soc Am; Nat Asn Geol Teachers; Seismol Soc Am; Asn Eng Geologists; Am Geophys Union. Res: Engineering geology; earthquake faults in relation to land use. Mailing Add: US Geol Surv 345 Middlefield Rd Menlo Park CA 94025

BONINI, WILLIAM EMORY, b Washington, DC, Aug 23, 26; m 54; c 4. GEOLOGY, GEOPHYSICS. Educ: Princeton Univ, BSE, 48, MSE, 49; Univ Wis, PhD(geol), 57. Prof Exp: From instr to assoc prof geol eng, 53-66, prof geophysics & geol, 66-70, MAGEE PROF GEOPHYSICS & GEOL ENG, PRINCETON UNIV, 70- Concurrent Pos: NSF sr fel, 63-64; vis scientist to Yugoslavia, Nat Acad Sci Eastern Europ Exchange Prog, 74. Mem: Am Asn Petrol Geol; Am Geophys Union; Geol Soc Am; Soc Explor Geophys; Asn Eng Geol. Res: Gravity anomalies and major tectonic features of the northwestern United States and Venezuela; geophysical prospecting as applied to groundwater development; engineering geophysics; environmental geology. Mailing Add: Dept of Geol & Geophys Sci Princeton Univ Princeton NJ 08540

BONK, JAMES F, b Menominee, Mich, Feb 6, 31. CHEMISTRY. Educ: Carroll Col, Wis, BS, 53; Ohio State Univ, PhD(chem), 58. Prof Exp: From asst prof to assoc prof, 59-74, PROF CHEM, DUKE UNIV, 74- Mem: Am Chem Soc. Res: Physical-inorganic chemistry. Mailing Add: Dept of Chem Duke Univ Durham NC 27706

BONKALO, ALEXANDER, b Hungary, Jan 24, 12; Can citizen; m 48; c 2. PSYCHIATRY, NEUROPHYSIOLOGY. Educ: Pazmany Peter Univ, Budapest, MD, 37. Prof Exp: Mem staff neurol & psychiat, Neuropsychiat Hosp, Univ Budapest, 37-47; res neuroradiol, Serafimer Hosp, Stockholm, Sweden, 48; res brain metab & neuropsychiat, Dept Metab Res, Wenner-Gren Inst, 49; mem staff, Dept Neuropath, Banting Inst, Ont, 50-51, mem staff, 51-52; clin teacher, 54-58, assoc, 58-62, from asst prof to assoc prof, 62-74, PROF PSYCHIAT, UNIV TORONTO, 74- Concurrent Pos: Resident neuropath & neurophysiol, Inst Brain Res, Berlin, Ger, 38-39; sr psychiatrist in-patient serv, Toronto Psychiat Hosp, Ont Dept Health, 53-66, med dir, Surrey Pl Ctr, 66-; consult, Clarke Inst Psychiat, 70- Mem: Can Psychiat Asn; Can Soc Electroencephalog. Res: Clinical psychiatry and neurology; pathological and neurophysiological aspects of psychiatry; electroencephalography; schizophrenia; biological psychiatry; psychopharmacology; epilepsy; sleep. Mailing Add: Dept of Psychiat Univ of Toronto Toronto ON Can

BONNELL, DAVID WILLIAM, b Wichita, Kans, Sept 29, 43; m 75. PHYSICAL INORGANIC CHEMISTRY, HIGH TEMPERATURE CHEMISTRY. Educ: Rice Univ, BA, 68, PhD(phys chem), 72. Prof Exp: Res chemist, Rice Univ, 64-68, mass spectrometrist, 68-69, res assoc chem, 72-73; dir res, MCR-Houston, Inc, 73-75; RES CHEMIST INORG CHEM, NAT BUR STAND, 75- Concurrent Pos: Consult, Rice Univ, 74-75. Res: Electron spectroscopic chem analysis instrumentation development; x-ray photoelectron and x-ray auger spectroscopy of inorganic and biologically interesting materials; levation calorimetry property measurements of high temperature materials; high temperature mass spectroscopy. Mailing Add: Nat Bur of Stand Inorg Materials Sect Washington DC 20234

BONNER, BILLY EDWARD, b Oak Grove, La, Dec 12, 39; m 67. NUCLEAR PHYSICS. Educ: La Polytech Inst, BS, 61; Rice Univ, MA, 63, PhD(physics), 65. Prof Exp: Fel physics, Rice Univ, 65-66; sr sci officer, Rutherford High Energy Lab, Sci Res Coun, Eng, 66-70; assoc res physicist, Crocker Nuclear Lab, Univ Calif, Davis, 71-72; MEM STAFF PHYSICS, LOS ALAMOS SCI LAB, UNIV CALIF, 72- Mem: Sigma Xi. Res: Medium energy research with neutrons, especially concerned with neutron proton interactions and few body problems. Mailing Add: Los Alamos Sci Lab Box 1663 Los Alamos NM 87545

BONNER, BRUCE ALBERT, b Jamestown, NDak, Apr 14, 29; m 53; c 4. PLANT PHYSIOLOGY. Educ: Univ Calif, AB, 52, PhD(bot) 57. Prof Exp: USPHS fel physiol genetics lab, Nat Cent Sci Res, France, 56-58; res assoc bot, Yale Univ, 58-61; instr, Harvard Univ, 61-63; asst prof, 63-71, ASSOC PROF BOT, UNIV CALIF, DAVIS, 71- Mem: Am Soc Plant Physiol; Bot Soc Am. Res: Mechanism of photomorphogenesis; growth and development. Mailing Add: Dept of Bot Univ of Calif Davis CA 95616

BONNER, DANIEL PATRICK, b Bayonne, NJ, Oct 9, 45; m 72. MICROBIOLOGY, CHEMOTHERAPY. Educ: Fairleigh Dickinson Univ, BS, 67; Rutgers Univ, MS, 69, PhD(microbiol), 72. Prof Exp: Biol sci asst microbiol, Dept Rickettsial Dis, Walter Reed Army Inst Res, Walter Reed Army Med Ctr, 69-71; RES ASSOC MICROBIOL, WAKSMAN INST MICROBIOL, RUTGERS UNIV, 72- Mem: AAAS; Am Soc Microbiol. Res: Chemotherapy of infectious diseases; biological properties of polyene macrolide antibiotics. Mailing Add: Waksman Inst of Microbiol Rutgers Univ New Brunswick NJ 08903

BONNER, FRANCIS TRUESDALE, b Salt Lake City, Utah, Dec 18, 21; m 46; c 3. CHEMISTRY. Educ: Univ Utah, BA, 42; Yale Univ, MS, 44, PhD(phys chem), 45. Prof Exp: Lab asst phys chem, Yale Univ, 42-44; assoc scientist, SAM Labs, Columbia Univ, 44-46; chemist, Clinton Nat Lab, Oak Ridge, 46-47; scientist, Brookhaven Nat Lab, NY, 47-48; asst prof chem, Brooklyn Col, 48-54; Carnegie Found fel, Harvard Univ, 54-55; phys chemist, Arthur D Little, Inc, 55-58; chmn dept chem, 58-70, PROF CHEM, STATE UNIV NY STONY BROOK, 58- Concurrent Pos: NSF sr fel, Saclay Nuclear Res Ctr, France, 64-65; consult ed, Addison-Wesley Publ Co, 56-; res collabr, Brookhaven Nat Lab, 58-; Rockefeller Found consult, Valle, Colombia, 61, 64, 68; Ford Found consult, Antioquia, Colombia, 62, 63, 68. Mem: Fel AAAS; Am Chem Soc; Am Geophys Union; Geochem Soc; NY Acad Sci. Res: Isotope chemistry, including exchange kinetics, reaction studies; inorganic solution chemistry of nitrogen. Mailing Add: Box 63 Setauket NY 11733

BONNER, HAZEL GARRISON, b Washington, DC, Oct 8, 28; m 57; c 3. BOTANY. Educ: Howard Univ, BS, 50; Univ Mich, MS, 51; Pa State Univ, PhD(bot), 57. Prof Exp: Instr biol, Southern Univ, 51-53 & Grambling Col, 53-54; asst prof bot, Howard Univ, 57-58; res assoc, 65-66, assoc prof, 67-72, PROF BIOL, HAMPTON INST, 72- Mem: Bot Soc Am; Nat Inst Sci. Res: Floral morphology of Fagus grandifolia; studies on Zostera marina, especially growth and development. Mailing Add: Dept of Biol Sci Hampton Inst Hampton VA 23668

BONNER, HUGH WARREN, b Chicago, Ill, Oct 27, 44; m 67. EXERCISE PHYSIOLOGY. Educ: Univ Minn, BS, 67; Calif State Col, Hayward, MS, 70; Univ Calif, Berkeley, PhD(phys educ), 72. Prof Exp: ASST PROF PHYS EDUC, UNIV TEX, AUSTIN, 72-, DIR EXER PHYSIOL LABS, 73- Mem: AAAS; Am Asn Health Phys Educ & Recreation; Am Col Sports Med. Res: Determination of the influence of exercise training upon calcium transport of mitochondria and sarcoplasmin reticulum in cardiac and skeletal muscle and morphometric changes of subcellular organelles. Mailing Add: Health Phys Educ & Recreation 222 Bellmont Hall Univ of Tex Austin TX 78712

BONNER, JAMES (FREDRICK), b Ansley, Nebr, Sept 1, 10; m 39. DEVELOPMENTAL BIOLOGY, BIOCHEMICAL GENETICS. Educ: Univ Utah, AB, 31; Calif Inst Technol, PhD(plant physiol, genetics), 34. Prof Exp: Nat Res Coun fel, State Univ Utrecht & Swiss Fed Inst Technol, 34-35; asst biol, 35-36, from instr to assoc prof plant physiol, 36-46, PROF BIOL, CALIF INST TECHNOL, 46- Concurrent Pos: Eastman prof, Oxford Univ, 63-64. Mem: Nat Acad Sci; Bot Soc Am; Am Chem Soc; Am Soc Plant Physiol; Am Soc Biol Chem. Res: Molecular biology of chromosomes; control of genetic activity. Mailing Add: Div of Biol Calif Inst of Technol Pasadena CA 91109

BONNER, JOHN FRANKLIN, JR, b Smullton, Pa, Sept 11, 17. BIOCHEMISTRY. Educ: Rensselaer Polytech Inst, BChE, 38; Univ Rochester, MS, 40, PhD(biophys), 48. Prof Exp: Res fel radiol, Univ Rochester, 39-47, from instr toxicol to asst prof radiation biol, 47-55; asst chief med res br, AEC, 55-61; PROF BIOCHEM, SCH MED, IND UNIV, INDIANAPOLIS, 61- Concurrent Pos: Fulbright lectr & guest prof, Univ Groningen, 54-55; vis prof molecular & genetic biol, Univ Utah, 67-68; USPHS spec fel, Univ Utah, 67-68. Mem: AAAS; Am Chem Soc; Biophys Soc. Res: Modification of biomolecules by ionizing radiation; radioprotective compounds. Mailing Add: Dept of Biochem Ind Univ Sch of Med Indianapolis IN 46202

BONNER, JOHN TYLER, b New York, NY, May 12, 20; m 42; c 4. DEVELOPMENTAL BIOLOGY. Educ: Harvard Univ, BS, 41, MA, 42, PhD(biol), 47. Hon Degrees: DSc, Middlebury Col, 70. Prof Exp: From asst prof to prof, 47-66, GEORGE M MOFFETT PROF BIOL, PRINCETON UNIV, 66-, CHMN DEPT, 65- Concurrent Pos: Instr embryol, Marine Biol Labs, Woods Hole, 51-52; Rockefeller traveling fel, France, 53; spec lectr, Univ Col, London, 57; Guggenheim fel, Scotland, 58 & 71-72; NSF sr fel, Cambridge Univ, 63; ed of two sect, Biol Abstr, 57-, trustee, 58-63; mem bd trustees, Princeton Univ Press, 64-68 & 71. Honors & Awards: Waksman Medal, 55. Mem: Nat Acad Sci; Am Philos Soc; Am Soc Nat; Soc Gen Physiol; Mycol Soc Am. Res: Development in lower organisms; cellular slime molds. Mailing Add: Dept of Biol Princeton Univ Princeton NJ 08540

BONNER, LYMAN GAYLORD, b Kingston, Ont, Sept 16, 12; m 37; c 3. CHEMISTRY. Educ: Univ Utah, AB, 32; Calif Inst Technol, PhD(chem), 35. Prof Exp: Nat Res fel chem, Princeton Univ, 35-37; from instr to asst prof physics, Duke Univ, 37-44; res assoc, George Washington Univ, 44-45; tech dir, Hercules Powder Co, 45-55, mgr, Explosives Res Div, 55-58, dir develop, Explosives & chem Propulsion Dept, 58-65; dir found rels, 65-67, asst to pres, 67-69, ASSOC CHEM, CALIF INST TECHNOL, 66-, DIR STUDENT RELS, 69- Res: Infrared and Raman spectroscopy; ballistics. Mailing Add: Calif Inst of Technol Pasadena CA 91125

BONNER, NORMAN ANDREW, b San Francisco, Calif, Aug 3, 20; m 45; c 1. CHEMISTRY. Educ: Univ Calif, BS, 42; Princeton Univ, MA, 44, PhD(chem), 45. Prof Exp: Res chemist, Manhattan Proj, Univ Calif, 41-42, Princeton Univ, 42-45, Los Alamos Sci Lab, Univ Calif, 45-46 & Gen Elec Co, NY, 46; instr chem, Wash Univ, 47-48, asst prof, 48-51; asst prof, Cornell Univ, 51-53; CHEMIST, LAWRENCE LIVERMORE LAB, UNIV CALIF, 53- Concurrent Pos: Mem adv coun, Bay Area Air Pollution Control Dist, 70- Mem: AAAS; Am Chem Soc; Air Pollution Control Asn. Res: Trace element analysis of environmental samples; radiochemistry; kinetics of reactions in solution; inorganic chemistry; chemistry of air pollution. Mailing Add: Lawrence Livermore Lab Univ of Calif PO Box 808 Livermore CA 94550

BONNER, OSCAR DAVIS, b Jackson, Miss, May 9, 17; m 40; c 3. PHYSICAL CHEMISTRY. Educ: Millsaps Col, BS, 39; Univ Miss, MS, 48; Univ Kans, PhD(chem), 51. Prof Exp: With Miss Testing Lab, 40-41 & Filtrol Corp, 46-47; res partic, Oak Ridge Nat Lab, 51; from asst prof to prof, 51-70, from actg head to head dept, 59-70, ROBERT L SUMWALT CHAIR CHEM, UNIV SC, 70- Concurrent Pos: Fulbright advan res award, 57-58; Russell fac res award, 60. Mem: Am Chem Soc. Res: Ion exchange; thermodynamics of solutions; polyelectrolytes; structure of water; interaction of water and salts with biological systems. Mailing Add: Dept of Chem Univ of SC Columbia SC 29208

BONNER, R ALAN, b Leominster, Mass, Dec 1, 39; m 64; c 3. MEDICAL RESEARCH. Educ: Mich State Univ, BS, 62, DVM, 64; Boston Univ, PhD(physiol), 72. Prof Exp: Vet intern, Rowley Mem Hosp, Springfield, Mass, 64-65; pvt pract, Dewitt Animal Hosp, North Attleboro, Mass, 65-67; fel, Boston Univ, 67-69; from instr to asst prof physiol, Hahnemann Med Col, 69-74; RES ASSOC CARDIOVASC PHYSIOL, BIOMED RES INST, CTR RES & ADVAN STUDY, UNIV MAINE, PORTLAND, 73-; VIS ASST PROF PHYSIOL, HAHNEMANN MED COL, 74- Concurrent Pos: NASA fel, Biospace Technol Training Prog, 68; fel, Pa Heart Asn, 71; consult, Maine Med Ctr, 71- Mem: Am Vet Med Asn; AAAS. Res: Development of animal model for sudden infant death syndrome studies; pathophysiologic mechanisms involved in sudden infant death syndrome; pharmacologic agents in an attempt to reverse irreversible hemorrhagic shock. Mailing Add: Biomed Res Inst Ctr Univ of Maine 246 Deering Ave Portland ME 04102

BONNER, ROBERT DUBOIS, b Camden, Ala, Apr 3, 26; m 57; c 3. BOTANY. Educ: Howard Univ, BS, 49, MS, 52; Pa State Univ, PhD, 58. Prof Exp: Asst Pa State Univ, 55-57; asst prof bot, Tex Southern Univ, 57-63; assoc prof, 63-69, PROF BIOL, HAMPTON INST, 69-, CHMN DEPT, 63-, DIR DIV PURE & APPL SCI, 74- Mem: AAAS; Nat Inst Sci; Bot Soc Am; Mycol Soc Am. Res: Taxonomy and ecology of animal parasitic fungi; marine microbiology. Mailing Add: Dept of Natural Sci Hampton Inst Hampton VA 23368

BONNER, TOM IVAN, b Boston, Mass, Mar 1, 42; m 66; c 1. MOLECULAR BIOLOGY. Educ: Rice Univ, BA, 63; Univ Wis, PhD(physics), 68. Prof Exp: Res assoc nuclear physics, Ctr Nuclear Studies, Univ Tex, Austin, 67-70; Carnegie fel, 70-73, RES ASSOC BIOPHYS, DEPT TERRESTRIAL MAGNETISM, CARNEGIE INST, 73- Mem: AAAS; Am Phys Soc; Biophys Soc. Res: Nucleic acid reassociation techniques; nucleic acid sequence organization and regulation; evolution of nucleic acids; RNA tumor viruses. Mailing Add: Dept of Terrestrial Magnetism Carnegie Inst 5241 Broadbranch Rd NW Washington DC 20015

BONNER, WALTER DANIEL, JR, b Salt Lake City, Utah, Oct 22, 19; m 44; c 2. CELL BIOLOGY. Educ: Univ Utah, BS, 40; Calif Inst Technol, PhD(bio-org chem), 46. Prof Exp: Asst, US Bur Mines, Utah, 41 & Calif Inst Technol, 41-46; res fel, Harvard Univ, 46-49; Am Cancer Soc fel, Molteno Inst, Cambridge Univ, 49-50, USPHS fel, 50-52; Smithsonian Inst, 52-53; asst prof bot, Cornell Univ, 53-58, assoc prof, 58-59; prof phys biochem & plant physiol, Johnson Res Found, 59-75, PROF BIOCHEM & BIOPHYS, SCH MED, UNIV PA, 75- Concurrent Pos: Guggenheim & overseas fels, Churchill Col, Cambridge Univ, 67-68; prog dir, Molecular Biol Sect, NSF, 74-76. Mem: Am Soc Plant Physiol; Am Soc Biol Chem; Am Chem Soc; Biophys Soc; Brit Biochem Soc. Res: Bioenergetics; cellular physiology; mechanisms of energy transfer and conservation in plant mitochondria and chloroplasts; metabolic control mechanism in plant and protist cells. Mailing Add: Dept of Biochem & Biophys Univ of Pa Philadelphia PA 19174

BONNER, WILLARD HALLAM, JR, b New Haven, Conn, May 22, 28; m 48; c 2. ORGANIC CHEMISTRY. Educ: Wesleyan Univ, BA, 47; Univ Buffalo, MA, 48; Purdue Univ, PhD(org chem), 52. Prof Exp: Res chemist, Humble Oil & Refining Co, 52-54; RES CHEMIST, E I DU PONT DE NEMOURS & CO, INC, 54- Mem: Am Chem Soc. Res: Reaction mechanisms; organic synthesis; petrochemicals; polymer chemistry; synthetic fibers. Mailing Add: E I du Pont de Nemours & Co Inc Du Pont & Brandywine Bldgs Wilmington DE 19801

BONNER, WILLIAM ANDREW, b Chicago, Ill, Dec 21, 19; m 44; c 4. CHEMISTRY. Educ: Harvard Univ, BS, 41; Northwestern Univ, PhD(org chem), 44. Prof Exp: Res chemist, Corn Prod Ref Co, Ill, 42; asst org chem, Northwestern Univ, 42-44, Nat Defense Res Comt res chemist & instr chem, univ col, 44-46; from instr to assoc prof, 46-59, PROF ORG CHEM, STANFORD UNIV, 59- Concurrent Pos: Guggenheim fel, 53; Nat Res Coun res assoc, 70. Mem: Am Chem Soc. Res: Carbohydrate and synthetic organic chemistry; application of radioactive carbon to elucidation of organic reaction mechanisms; isotope effects in organic reactions; mechanisms of heterogeneous catalytic reactions; natural products. Mailing Add: Dept of Chem Stanford Univ Stanford CA 94305

BONNER, WILLIAM D, b Rochester, Minn, Apr 28, 33; m 58; c 1. METEOROLOGY. Educ: Univ Chicago, BA, 52, MS, 60, PhD(meteorol), 65; Pa State Univ, BS, 54. Prof Exp: Res asst meteorol, Satellite & Meso-Meteorol Proj, Univ Chicago, 62-64, res assoc geophys sci, 65; res scientist, Meteorol Res, Inc, 65; asst prof meteorol, Univ Calif, Los Angeles, 65-70; res meteorologist, Tech Develop Lab, 70-72, CHIEF DATA ASSIMILATION BR, NAT METEOROL CTR, NAT WEATHER SERV, NAT OCEANIC & ATMOSPHERIC AGENCY, 72- Concurrent Pos: Vis lectr, Univ Md, 70- Mem: Am Meteorol Soc; Am Geophys Union. Res: Weather description and prediction; severe local storms; use and evaluation of satellite data in numerical weather prediction. Mailing Add: 610 C St SE Washington DC 20003

BONNET, DAVID DUDLEY, b Worcester, Mass, Oct 28, 14; m 38; c 3. MEDICAL ENTOMOLOGY. Educ: Harvard Univ, BS, 37, AM, 39, PhD(embryol), 41. Prof Exp: Fel, Radcliffe Col, 37-38; instr zool, Univ Hawaii, 41-43; jr biologist, Pub Health Serv, Honolulu, 43-44, asst sanitarian, 44-46, actg dir, Bur Mosquito Control, Dept Health, 46-48, med entomologist, 48-53; assoc res parasitologist, Pac Trop Dis Res Proj, Univ Calif, Los Angeles & Inst Med Res de L'Oceanie Francaise, 53-58; malaria adv, Agency Int Develop, Djakarta, Indonesia, 58-62; dept dean, Malaria Eradication Training Ctr, Manila, 62-64; training officer, Aedes Aegypti Eradication Br, Nat Commun Dis Ctr, US Pub Health Serv, 64-67, scientist-dir & training officer, Malaria Eradication Prog, 67-69, scientist dir, Foreign Quarantine, 69-73; RETIRED. Concurrent Pos: Res fel, Woods Hole Oceanog Inst, 40-41. Mem: Assoc Am Soc Zool; Am Soc Trop Med & Hyg. Res: Ecology of mosquitoes; medical entomology; filariasis. Mailing Add: 1108 Koohoo Pl Kailua HI 96734

BONNET, PHILIP D, b Worcester, Mass, May 28, 11; m 40; c 5. MEDICAL ADMINISTRATION. Educ: Wesleyan Univ, AB, 32; Harvard Univ, MD, 36. Prof Exp: Med dir, Lankenau Hosp, Philadelphia, 40-48; adminr, Univ Hosp, Boston, 48-66; PROF HEALTH CARE ORGN, SCH HYG & PUB HEALTH, JOHNS HOPKINS UNIV, 66- Mailing Add: Sch of Hyg & Pub Health Johns Hopkins Univ Baltimore MD 21205

BONNETT, RICHARD BRIAN, b Jamestown, NY, Feb 7, 39; m 64; c 2. GEOMORPHOLOGY, GEOLOGY. Educ: Allegheny Col, BS, 61; Univ Maine, MS, 63; Ohio State Univ, PhD(geol), 70. Prof Exp: Teaching assoc geol, Univ Maine, 61-63 & Ohio State Univ, 63-68; asst prof, 68-75, ASSOC PROF GEOL, MARSHALL UNIV, 75- Mem: AAAS; Geol Soc Am; Am Quaternary Asn. Res: Pleistocene-glacial geology; environmental geology; Pleistocene periglacial stratigraphy and geomorphology in tri-state area of Ohio, West Virginia and Kentucky. Mailing Add: Dept of Geol Marshall Univ Huntington WV 25701

BONNETT, ROBERT NELSON, chemistry, see 12th edition

BONNEVILLE, MARY AGNES, b Pittsfield, Mass, May 30, 31. CELL BIOLOGY. Educ: Smith Col, BA, 53; Amherst Col, MA, 55; Rockefeller Inst, PhD(cytol), 61. Prof Exp: Trainee anat, Col Physicians & Surgeons, Columbia Univ, 61-63; res assoc dermat, Sch Med, Tufts Univ, 63-64; asst prof biol, Brown Univ, 64-68 & Harvard Univ, 68-70; asst prof, 70-71, ASSOC PROF MOLECULAR, CELLULAR & DEVELOP BIOL, UNIV COLO, BOULDER, 71- Mem: Am Soc Cell Biol; Electron Micros Soc Am; Soc Develop Biol; Am Asn Anat. Res: Structure and formation of the cell surface. Mailing Add: Dept Molecular Cell & Develop Biol Univ of Colo Boulder CO 80302

BONNEY, ROBERT JOHN, b Rumford, Maine, Feb 6, 42; m 62; c 3. BIOCHEMISTRY. Educ: Univ Maine, BS, 64; State Univ NY Buffalo, MS, 67, PhD(biochem), 71. Prof Exp: Fel biochem, McArdle Lab Cancer Res, Univ Wis, 71-73; SR RES SCIENTIST BIOCHEM, DEPT HEALTH, NY STATE, 73- Concurrent Pos: Mem adj fac, W Alton Jones Cell Sci Ctr, 73- Mem: AAAS; Tissue Culture Asn. Res: The development of an in vitro system composed of adult liver parenchymal cells to serve as a control for studies on hepatoma cells in culture. Mailing Add: Three Victor Dr Albany NY 12203

BONNICHSEN, BILL, b Twin Falls, Idaho, Sept 25, 37; m 59; c 4. PETROLOGY, ECONOMIC GEOLOGY. Educ: Univ Idaho, BS, 60; Univ Minn, Minneapolis, PhD(geol), 68. Prof Exp: Instr econ geol, Univ Minn, Minneapolis, 66-67; res assoc geol, Minn Geol Surv, 67-69; ASST PROF PETROL, CORNELL UNIV, 69- Mem: Geol Soc Am; Am Geophys Union; Soc Econ Geologists. Res: Metamorphism of iron formations; pyroxene studies; geology and ore potential of Biwabik Formation and

Duluth Complex, Minnesota; petrology of and metal deposits in mafic igneous rocks. Mailing Add: Dept of Geol Sci Col of Eng Cornell Univ Ithaca NY 14853

BONO, VINCENT HORACE, JR, b Brooklyn, NY, July 23, 33; m 57; c 2. INTERNAL MEDICINE, BIOCHEMISTRY. Educ: Columbia Univ, AB, 53, MD, 57. Prof Exp: Intern internal med, Duke Univ Hosps, 57-58, univ res fel hemat, 58-59; resident internal med, Strong Mem Hosp, Univ Rochester, 59-60; clin assoc cancer chemother, Med Br, Nat Cancer Inst, Md, 60-63; trainee biochem, Dartmouth Med Sch, 63-64; sr investr cancer chemother, Chem Pharmacol Lab, 64-73, head molecular biol & methods develop sect, Drug Eval Br, 73-75, HEAD MOLECULAR BIOL & METHODS DEVELOP SECT, LAB MED CHEM & BIOL, NAT CANCER INST, 75- Res: Biochemical mechanisms of action of compounds used in the therapy of human neoplasia. Mailing Add: Nat Cancer Inst Bethesda MD 20014

BONSACK, WALTER KARL, b Cleveland, Ohio, Apr 14, 32; m 57. ASTROPHYSICS. Educ: Case Inst Technol, BS, 54; Calif Inst Technol, PhD(astron), 59. Prof Exp: Res fel astron, Calif Inst Technol, 58-60; from asst prof to assoc prof, Ohio State Univ, 60-66; assoc prof, 66-72, PROF PHYSICS & ASTRON, UNIV HAWAII, 72- Mem: Fel AAAS; Am Astron Soc; Int Astron Union; Astron Soc Pac. Res: Composition and physical conditions in stellar atmospheres; spectrum variations; spectroscopic instrumentation for astronomy. Mailing Add: Inst for Astron Univ of Hawaii Honolulu HI 96822

BONSIGNORE, PATRICK VINCENT, b New York, NY, May 10, 29; m 58; c 2. POLYMER CHEMISTRY. Educ: Queens Col, NY, BS, 51; Polytech Inst Brooklyn, PhD(org & polymer chem), 58. Prof Exp: Sr scientist plastics, Columbia Southern Div, PPG Industs, 58-59; sr scientist acrylic plastics, Rohm and Haas, 59-70; sr scientist hydrogels, Hydron Labs, 70; staff scientist reprographics, Memorex, 70-71; SR SCIENTIST FIRE RETARDANCY, ALUMINUM CO AM, 72- Concurrent Pos: Fel, Univ Ill, 57-58. Mem: Soc Plastics Engrs; Am Chem Soc; Sigma Xi; Fire Retardant Chem Asn. Res: Characterization and improvement of fire retardancy and smoke suppression in plastics; property improvement of plastics by microfiber reinforcement. Mailing Add: 3437 MacIntyre Dr Murrysville PA 15668

BONSNES, ROY WALTER, b Oslo, Norway, Mar 15, 09; m 41. BIOCHEMISTRY. Educ: Univ Conn, BS, 30; Yale Univ, PhD(physiol chem), 39. Prof Exp: Asst chem, Univ Conn, 30-31 & 32-34; asst biochem, Conn Exp Sta, 34-36; lab asst physiol chem, Med Sch, Yale Univ, 36-39, instr, 39-40; asst biochem, 40-41, from instr to asst prof, 41-50, from instr to assoc prof biochem in obstet & gynec, 41-72, CLIN PROF BIOCHEM IN OBSTET & GYNEC, MED COL, CORNELL UNIV, 72-, ASSOC PROF BIOCHEM, 50- Concurrent Pos: Chemist, Lying-In Hosp, New York, 43-, Cent Labs, NY Hosp, 58-60 & NY Hosp, 61- Mem: Am Chem Soc; Am Soc Biol Chemists; Soc Exp Biol & Med. Res: Biochemistry of anterior pituitary hormones; biochemistry of pregnancy with particular reference to eclampsia and kidney function in eclampsia; biochemical analytical methods. Mailing Add: Dept of Biochem Cornell Univ Med Col New York NY 10021

BONTE, FREDERICK JAMES, b Bethlehem, Pa, Jan 18, 22; m 52, 68; c 6. RADIOLOGY. Educ: Western Reserve Univ, BS, 42, MD, 45. Prof Exp: Fel radiobiol, Atomic Energy Med Res Proj, Western Reserve Univ, 48-49; resident physician, Univ Hosps, Cleveland, 49-52; instr radiol, Med Sch, Western Reserve Univ, 52-56; prof radiol & chmn dept, 56-73, DEAN, SOUTHWESTERN MED SCH, UNIV TEX HEALTH SCI CTR, 73- Concurrent Pos: Dir dept radiol, Parkland Mem Hosp, 56-73; consult, US Vet Admin Hosp Syst, 56-73 & Radiation & Med Res Found Southwest, Ft Worth, 58-73; mem bd, Nat Coun Radiation Protection & Measurements, 65-71; mem radiol training comt, Nat Inst Gen Med Sci, 65-70; mem med adv comt, Oak Ridge Assoc Univs, 68-70; trustee, Am Bd Radiol, 69-75 & Am Bd Nuclear Med, 71-73. Mem: AAAS; Am Col Radiol; Am Roentgen Ray Soc; Radiol Soc NAm; Sigma Xi. Res: Clinical and experimental nuclear medicine; radiobiology. Mailing Add: Southwestern Med Sch Univ of Tex Health Sci Ctr Dallas TX 75235

BONTEMPO, JOHN A, b Pesco, Italy, Sept 13, 30; US citizen; m 56; c 3. MICROBIOLOGY. Educ: Heidelberg Col, AB & BS, 56; Fairleigh Dickinson Univ, MS, 69; Rutgers Univ, PhD(microbiol), 75. Prof Exp: Asst scientist, Warner-Lambert Res Inst, NJ, 56-62; microbiologist, 62-70, sr microbiologist, 70-72, group leader, 72-75, TECH FEL, HOFFMANN-LA ROCHE, INC, 75- Mem: NY Acad Sci; Am Soc Microbiol; Sigma Xi. Res: Antimicrobial agents, analytical microbiology. Mailing Add: 9 Dalewood Rd Cedar Grove NJ 07009

BONTING, SJOERD LIEUWE, b Amsterdam, Holland, Oct 6, 24; nat US; m 51; c 4. BIOCHEMISTRY. Educ: Univ Amsterdam, BSc, 44, MSc, 50, PhD(biochem), 52. Hon Degrees: LTh, St Mark's Inst Theol, London, 75. Prof Exp: Instr anal chem, Univ Amsterdam, 47-49; biochemist, Neth Inst Nutrit, 50-52; res assoc physiol, Col Med, Univ Iowa, 52-55; asst prof physiol chem, Sch Med, Univ Minn, 55-56; asst prof biol chem, Col Med, Univ Ill & asst attend biochemist, Presby-St Luke's Hosp, Chicago, 56-60; head sect cell biol, Ophthal Br, Nat Inst Neurol Dis & Blindness, NIH, 60-65; PROF BIOCHEM & CHMN DEPT, UNIV NIJMEGEN, 65- Concurrent Pos: Instr, Univ Amsterdam, 51-52; fel, USPHS & Nat Cancer Inst, 52-54. Honors & Awards: Fight for Sight Award, Nat Coun Combat Blindness & Asn Res Ophthal, 61 & 62; Merit Award, Int Rescue Comt, 63; Arthur S Fleming Award, 64; Karger Mem Found Prize, 64. Mem: AAAS; Am Soc Biol Chem; Am Soc Cell Biol; Histochem Soc; Asn Res Vision & Ophthal. Res: Membrane and cellular biochemistry; quantitative histochemistry; visual mechanism. Mailing Add: Dept of Biochem Univ of Nijmegen Nijmegen Netherlands

BONTRAGER, MARION M, mathematics, see 12th edition

BONVENTRE, PETER FRANK, b New York, NY, Aug 18, 28; m 63. MICROBIOLOGY. Educ: City Col New York, BS, 49; Univ Tenn, MS, 55; Univ Mich, PhD, 57. Prof Exp: NSF fel, Inst Superiore di Sanita, Italy, 57-58; from asst prof to assoc prof, 58-68, PROF MICROBIOL, COL MED, UNIV CINCINNATI, 68- Concurrent Pos: Vis res assoc, Dept Pharmacol, Royal Vet Col Sweden, 60; USPHS career develop award, 63. Mem: AAAS; Am Soc Microbiol. Res: Host-parasite relationships in infectious processes; mechanisms of pathogenesis; in vivo action of bacterial toxins. Mailing Add: Dept of Microbiol Univ of Cincinnati Col of Med Cincinnati OH 45219

BONVICINO, GUIDO EROS, b Castrovillari, Italy, Feb 9, 21; nat US; m 52; c 1. ORGANIC CHEMISTRY. Educ: Fordham Univ, BS, 47, MS, 49, PhD(org chem), 52. Prof Exp: Instr org chem & anal, Fordham Univ, 47-52; res chemist org, Lederle Labs, 52-63; assoc prof org chem, 63-68, PROF ORG CHEM, C W POST COL LONG ISLAND, 68- Honors & Awards: Am Inst Chemists Medal, 47. Mem: AAAS; sr mem Am Chem Soc; NY Acad Sci. Res: Thiamine; chemical reactions; antibiotics and synthetic chemotherapeutic agents; chemopsychiatric agents; reaction mechanisms. Mailing Add: Dept of Chem C W Post Col Greenvale NY 11548

BOODMAN, DAVID MORRIS, b Pittsburgh, Pa, July 4, 23; m 48; c 2. SYSTEMS SCIENCE. Educ: Univ Pittsburgh, BS, 44, PhD(phys chem), 50. Prof Exp: Sr staff mem, Opers Eval Group, Mass Inst Technol, 50-60; sr staff mem, Opers Res Sect, 60-72, VICE PRES, ARTHUR D LITTLE, INC, 72- Concurrent Pos: Opers analyst, Staff, Comdr in Chief, Pac Fleet, US Navy, 51, Oper Develop Force, 53-54 & staff, Comdr First Fleet, 58. Mem: Fel AAAS; Math Asn Am; Opers Res Soc Am. Res: Systems science-operations research; military and industrial operations research. Mailing Add: Arthur D Little Inc 25 Acorn Park Cambridge MA 02140

BOODMAN, NORMAN S, b Pittsburgh, Pa, Feb 13, 27; m 63; c 3. FUEL SCIENCE. Educ: Univ Pittsburgh, BS, 49, MLitt, 57. Prof Exp: Asst chem, Univ Pittsburgh, 49-50; anal chemist, Vitro Mfg Co, Pa, 50-51; chemist, Koppers Co Res Ctr, 51-55; SECT SUPVR, US STEEL RES LAB, 56- Mem: Am Chem Soc. Res: Recovery and purification of chemicals from high-temperature carbonization coal tar; production of phenolics, alcohols and monomers via hydroperoxidation of alkylaromatic hydrocarbons; coal conversion via hydrogenation and fluidized-bed carbonization. Mailing Add: US Steel Res Lab Monroeville PA 15146

BOOE, JAMES MARVIN, b Austin, Ind, Nov 12, 06; m 38; c 3. ELECTROCHEMISTRY. Educ: Butler Univ, BS, 28. Prof Exp: Chemist, Indianapolis Plating Co; chief chemist, P R Mallory & Co, Inc, 29-45, charge electrochem res, 45-51; exec chem engr, Mallory Capacitor Co, 51-53, charge cent chem & metall res labs, 53-55, dir mat res labs, 55-57, cent chem res labs, 57-63, dir chem labs, 63-71; RETIRED. Concurrent Pos: Tech consult, P R Mallory & Co, Inc, 72- Mem: Am Chem Soc; Electrochem Soc; fel Am Inst Chem. Res: Aluminum and tantalum electrolytic capacitors; dry rectifiers; mercury batteries; electrostatic capacitors; electroplating; analysis; semiconductors and transistors; high temperature materials. Mailing Add: 548 N Audubon Rd Indianapolis IN 46219

BOOHAR, RICHARD KENNETH, b Philadelphia, Pa, Mar 7, 35; m 59; c 5. DEVELOPMENTAL BIOLOGY. Educ: Drew Univ, BA, 57; Univ Wis, MA, 60, PhD(zool), 66. Prof Exp: Asst zool, Univ Wis, 64; from instr to asst prof, Butler Univ, 64-67; asst prof, 67-71, ASSOC PROF ZOOL, UNIV NEBR, LINCOLN, 71- Mem: AAAS; Am Soc Zool; Am Inst Biol Sci; Inst Soc Ethics Life Sci. Res: Role of hormones in development of insect embryos; developmental biology of Volvox. Mailing Add: Sch of Life Sci Univ of Nebr Lincoln NE 68508

BOOK, DAVID LINCOLN, b Boston, Mass, Aug 4, 39. PLASMA PHYSICS. Educ: Yale Univ, BA, 59; Princeton Univ, MA, 61, PhD(physics), 64. Prof Exp: Staff assoc, Gen Atomic Div, Gen Dynamics Corp, 64-67; staff physicist, Lawrence Radiation Lab, 67-74; MEM STAFF, NAVAL RES LAB, 74- Mem: Am Phys Soc. Res: Theoretical plasma physics; controlled thermonuclear fusion; kinetic theory. Mailing Add: Naval Res Lab Code 7750 Washington DC 20375

BOOK, STEPHEN ALAN, b Newark, NJ, Dec 11, 41; m 68; c 3. STATISTICS. Educ: Georgetown Univ, AB, 63; Cornell Univ, MA, 66; Univ Ore, PhD(math), 70. Prof Exp: Asst prof math, 70-74, ASSOC PROF MATH, CALIF STATE COL, DOMINGUEZ HILLS, 74- Mem: Math Asn Am; Inst Math Statist; Am Math Soc; London Math Soc. Res: Limit theorems of probability theory; large deviation probabilities; laws of large numbers; measures of statistical dependence. Mailing Add: Dept of Math Calif State Col Dominguez Hills CA 90747

BOOK, STEVEN ARNOLD, b Albany, Calif, Dec 13, 45; m 68; c 2. PHYSIOLOGY. Educ: Univ Calif, Berkeley, AB, 57; Univ Calif, Davis, MA, 69, PhD(physiol), 73. Prof Exp: Res radioecologist, 69-73, ASST RES PHYSIOLOGIST, RADIOBIOL LAB, SCH VET MED, UNIV CALIF, DAVIS, 73-, LECTR RADIOL SCI, 74- Mem: AAAS; Health Physics Soc. Res: Metabolism of radioiodine and effects on thyroidal tissue associated with its assimilation particularly during prenatal and postnatal development. Mailing Add: Radiobiol Lab Univ of Calif Sch of Vet Med Davis CA 95616

BOOKE, HENRY EDWARD, b Brooklyn, NY, Jan 15, 32; m 59; c 1. ICHTHYOLOGY. Educ: Cornell Univ, BS, 60; Mich State Univ, MS, 62; Univ Mich, PhD(ichthyol), 68. Prof Exp: Conserv biologist, Cortland Exp Hatchery, NY State Conserv Dept, 61-63; US Bur Com Fisheries res fel, 63-65; res asst ichthyol, Mus Zool, Fish Div, Univ Mich, 65-66, teaching asst, Dept Wildlife & Fisheries, 66-67; instr biol, Yale Univ, 68-69; asst prof, Boston Univ, 69-72; ASSOC PROF BIOL, UNIV WIS, STEVENS POINT, 73-, ASST LEADER WIS COOP FISHERY UNIT, US DEPT INTERIOR FISH & WILDLIFE SERV, 73- Concurrent Pos: NSF res traineeship, Sport Fishing Inst Res grant & Theodore Roosevelt Mem fund grant, Am Mus Natural Hist, 66-67; Nat Oceanic & Atmospheric Admin sea grant, 75. Mem: AAAS; Am Fisheries Soc; Am Soc Ichthyologists & Herpetologists; Am Soc Zoologists. Res: Fish nutrition and hematology; biochemical systematics, cytotaxonomy and ecological physiology of fishes. Mailing Add: Col of Natural Resources Univ of Wis Stevens Point WI 54481

BOOKER, DOYLE RAY, meteorology, see 12th edition

BOOKER, HAROLD E, b Indianapolis, Ind, Sept 26, 32; m 57; c 2. NEUROLOGY. Educ: Ind Univ, MD, 57. Prof Exp: Intern, Marion County Gen Hosp, Ind, 57-58; resident neurol, Med Ctr, Ind Univ, 58-61; from asst prof to assoc prof, 64-72, PROF NEUROL, UNIV WIS-MADISON, 72- Concurrent Pos: Consult epilepsy to dir, Nat Inst Neurol Dis & Stroke, 67- Mem: Am Acad Neurol; Am Epilepsy Soc. Res: Clinical neurology, especially diagnosis and treatment of epilepsy. Mailing Add: Dept of Neurol Univ of Wis Madison WI 53706

BOOKER, HENRY GEORGE, b Barking, Eng, Dec 14, 10; nat US; m 38; c 4. RADIO PHYSICS. Educ: Univ Cambridge, BA, 33, PhD(radio physics), 36. Prof Exp: Fel, Christ's Col, Univ Cambridge, 35-48, lectr, Univ, 45-48; prof elec eng & eng physics, Cornell Univ, 48-65, dir sch elec eng & assoc dir ctr radio physics & space res, 59-63; PROF APPL PHYSICS, UNIV CALIF, SAN DIEGO, 65- Concurrent Pos: Vis scientist, Carnegie Inst, Univ Wash, 37-38; sci off, Ministry of Aircraft Prod of London, 40-45. Honors & Awards: Smith's Prize, Univ Cambridge, 35. Mem: Nat Acad Sci; Am Phys Soc; Inst Elec & Electronics Engrs; Am Meteorol Soc; Am Astron Soc. Res: Electromagnetic theory and radio wave propagation. Mailing Add: Dept Appl Physics & Info Sci Univ of Calif, San Diego La Jolla CA 92037

BOOKER, JOHN RATCLIFFE, b Swanage, Eng, Apr 28, 42; US citizen; m 64. GEOPHYSICS, HYDRODYNAMICS. Educ: Stanford Univ, BS, 63; Univ Calif, San Diego, MS, 65, PhD(earth sci), 68. Prof Exp: Fulbright-Hays fel, Meteorol Inst, Univ Stockholm, 68-69; res assoc & lectr geophys, Stanford Univ, 69-71; ASST PROF GEOPHYS, UNIV WASH, 71- Mem: AAAS; Am Geophys Union. Res: Propagation of waves in stratified and rotating fluids; motions in the earth's core; thermal instabilities in viscosity stratified fluids; role of pore pressure in earthquakes; earthquake prediction. Mailing Add: Geophys Prog Univ of Wash Seattle WA 98195

BOOKER, WALTER MONROE, b Little Rock, Ark, Nov 4, 07; m 32; c 2. PHARMACOLOGY. Educ: Morehouse Col, BA, 38; Univ Iowa, MS, 32; Univ Chicago, PhD(physiol), 42. Prof Exp: Instr biol & chem, Leland Col, 28-29; instr,

Prairie View Col, 29-31, head dept biol, 32-43; from instr to assoc prof, 43-53, actg chmn dept, 53-54, PROF PHARMACOL & CHMN DEPT, COL MED, HOWARD UNIV, 54- Concurrent Pos: Consult pharmacologist, Freedman's Hosp, 48- & Walter Reed Army Res Inst, 62-; sr Fulbright scholar, Heymans Inst, Ghent, Belg, 57-58; Am Pharmaceut Asn rep, Nat Res Coun, 59- Mem: Am Soc Pharmacol & Exp Therapeut; Am Physiol Soc; Endocrine Soc. Res: Carotid sinus response to various drugs; studies on alpha methyl dopa, nicotine and angiotensin. Mailing Add: Dept of Pharmacol Howard Univ Col of Med Washington DC 20001

BOOKHOUT, CAZLYN GREEN, b Gilboa, NY, Jan 28, 07; m 36; c 2. MARINE ZOOLOGY. Educ: St Stephens Col, AB, 28; Syracuse Univ, AM, 29; Duke Univ, PhD(zool), 34. Prof Exp: Instr biol, Women's Col, Univ NC, 29-31; asst zool, Duke Univ, 31-32; assoc prof biol, Elon Col, 34-35; from instr to assoc prof zool, 35-54, PROF ZOOL, DUKE UNIV, 54- Concurrent Pos: Dir marine lab, Duke Univ, 50-63 & 64-68, actg dir oceanog prog, 67-70. Mem: AAAS; Am Soc Zool; Am Soc Limnol & Oceanog. Res: Germ cells of mammals; embryology of invertebrates and polychaetes; growth in relation to molting in Crustacea; effect of pesticides on development of crabs. Mailing Add: Duke Univ Marine Lab Beaufort NC 28516

BOOKHOUT, THEODORE ARNOLD, b Salem, Ill, June 11, 31; m 52; c 2. WILDLIFE RESEARCH. Educ: Southern Ill Univ, BA, 52, MS, 54; Univ Mich, PhD(wildlife mgt), 63. Prof Exp: Asst wildlife mgt, Univ Mich, 56-59; game biologist, Mich Dept Conserv, 59-62; asst wildlife mgt, Univ Mich, 62, instr, 62-63; res asst prof vert ecol, Univ Md, 63-64; biologist, Ohio Coop Wildlife Res Unit, 64-70, assoc prof zool, 68-70, LEADER, OHIO COOP WILDLIFE RES UNIT, US FISH & WILDLIFE SERV, 64-, PROF ZOOL, OHIO STATE UNIV, 70- Concurrent Pos: ED, J Wildlife Mgt, Wildlife Soc, 75-76. Mem: Wildlife Soc; Am Soc Mammal. Res: Ecology and population dynamics of terrestrial game birds and mammals. Mailing Add: Ohio Coop Wildlife Res Unit Ohio State Univ 1735 Neil Ave Columbus OH 43210

BOOKMYER, BEVERLY BRANDON, b Glen Head, NY; c 2. ASTRONOMY. Educ: Chestnut Hill Col, AB, 46; Univ Pa, MS, 61, PhD(astron), 64. Prof Exp: Asst prof astron & math, Villanova Univ, 64-67; res asst prof, Univ Ariz, 67-71; ASSOC PROF PHYSICS & ASTRON, CLEMSON UNIV, 71- Concurrent Pos: NSF res grant, 66 & int travel grant, 73 & 75. Mem: Am Astron Soc; Am Phys Soc; fel AAAS. Res: Photoelectric photometry of variable stars; analysis of observations of eclipsing binary systems; applications of computerized synthetic light curve techniques to contact binary systems; mass exchange; stellar evolution. Mailing Add: Dept of Physics & Astron Clemson Univ Clemson SC 29631

BOOKSTEIN, ABRAHAM, b New York, NY, Mar 22, 40; m 67. INFORMATION SCIENCE. Educ: City Col New York, BS, 61; Univ Calif, Berkeley, MS, 66; Yeshiva Univ, PhD(physics), 69; Univ Chicago, MA, 70. Prof Exp: Asst prof libr sci, 71-75, asst prof behav sci, 74-75, ASSOC PROF LIBR & BEHAV SCI, UNIV CHICAGO, 75- Mem: Am Soc Info Sci; Asn Comput Mach. Res: Mathematical modelling in the information sciences; information storage and retrieval; research methods; operations research. Mailing Add: Grad Libr Sch Univ of Chicago Chicago IL 60637

BOOKSTEIN, JOSEPH JACOB, b Detroit, Mich, July 25, 29; m 54; c 3. RADIOLOGY. Educ: Wayne State Univ, BS, 50, MD, 54. Prof Exp: Am Cancer Soc fel, Stanford Univ, 59-60; from instr to prof radiol, Med Ctr, Univ Mich, Ann Arbor, 60-74; PROF RADIOL, HOSP, UNIV CALIF, SAN DIEGO, 74- Concurrent Pos: Consult, Vet Admin Hosp, Ann Arbor, Mich, 61-74; mem coun cardiovasc radiol, Am Heart Asn. Mem: Asn Univ Radiol; Radiol Soc NAm; NAm Soc Cardiac Radiol. Res: Cardiovascular radiology; pharmacoangiography; renovascular hypertension; magnification angiography. Mailing Add: Dept of Radiol Univ Calif at San Diego Hosp San Diego CA 92103

BOOLCHAND, PUNIT, b Varanasi, India. EXPERIMENTAL SOLID STATE PHYSICS, EXPERIMENTAL NUCLEAR PHYSICS. Educ: Punjab Univ, BS, 64, MS, 65; Case Western Reserve Univ, PhD(nuclear & solid state physics), 69. Prof Exp: Teaching asst physics, Case Western Reserve Univ, 65-67, grad fel, 67-69; from instr to asst prof, 69-75, ASSOC PROF PHYSICS, UNIV CINCINNATI, 75- Concurrent Pos: Cottrel res grant, Res Corp, 72; res assoc & vis scholar, Stanford Univ, 73-74; vis prof, Katholieke Univ, Leuven, Belg, 74-75; NSF res grant, 75; Am Philos Soc res grant, 75. Mem: AAAS; Am Phys Soc. Res: Hyperfine interactions studies by o:ssbauer effect, perturbed angular correlations, nuclear magnetic resonance and nuclear orientation; structure of amorphous solids; lattice location of implanted ions in semiconductors and metals. Mailing Add: Dept of Physics Univ of Cincinnati Cincinnati OH 45221

BOOLE, JOHN ALLEN, JR, b Exmore, Va, Dec 14, 21; m 49; c 3. PLANT MORPHOLOGY, PLANT PHYSIOLOGY. Educ: Univ Va, BA, 49; Va Polytech Inst, MS, 51; Univ NC, PhD(bot), 55. Prof Exp: Assoc prof, 55-58, head dept, 55-60, chmn div sci & math, 60-74, PROF BIOL, GA SOUTHERN COL, 58-, DIR ADVISEMENT, 74- Mem: AAAS; Bot Soc Am; Am Inst Biol Sci. Res: Systematic anatomy of the family Celastraceae; wood anatomy. Mailing Add: Div of Sci & Math Ga Southern Col PO Box 8042 Statesboro GA 30458

BOOLOOTIAN, RICHARD ANDREW, b Fresno, Calif, Oct 17, 27; m 67; c 3. INVERTEBRATE ZOOLOGY. Educ: Fresno State Col, BA, 51, MA, 53; Stanford Univ, PhD(biol), 57. Prof Exp: Asst comp physiol, Hopkins Marine Sta, Stanford Univ, 54-57, res assoc, 58; from asst prof to assoc prof zool, Univ Calif, Los Angeles, 57-66; staff consult biol sci curric study, Univ Colo, 67-68; pres, Visual Sci Prod, Los Angeles, 68-69; PRES, SCI SOFTWARE SYSTS, INC, 69- Concurrent Pos: NIH spec fel, 65; dir, Inst Visual Med, Los Angeles, 68-; pres, Instrnl Systs for Health Sci, 70- Honors & Awards: Lalor Found Award, 59, 61. Mem: Fel AAAS; Soc Gen Physiologists; NY Acad Sci; Am Soc Zoologists; Animal Behav Soc. Res: Comparative physiology of marine invertebrates. Mailing Add: Box 24787 Los Angeles CA 90024

BOOMAN, GLENN LAWRENCE, b Lynden, Wash, Sept 23, 29; m 53; c 2. ANALYTICAL CHEMISTRY. Educ: Western Wash Col, BA, 51; Univ Wash, PhD(chem), 54. Prof Exp: Chemist, Atomic Energy Div, Phillips Petrol Co, 54-66; supvr, Phys Chem Sect, Idaho Nuclear Corp, 67-71; supvr, Phys Chem Sect, Idaho Chem Prog, 71, SUPVR, MEASUREMENT SYSTS SECT, IDAHO NAT ENG LAB, ALLIED CHEM CORP, 71- Mem: Am Chem Soc; Am Nuclear Soc; Electrochem Soc; Am Vacuum Soc. Res: Electroanalytical chemistry; mechanisms of electrode reactions; reactions in solution and at electrodes by use of fast potentiostats; in-line instrumentation; real-time computer applications. Mailing Add: Allied Chem Corp CPP-637 550 Second St Idaho Falls ID 83401

BOOMAN, KEITH ALBERT, b Bellingham, Wash, Jan 15, 28; m 53; c 3. PHYSICAL ORGANIC CHEMISTRY. Educ: Univ Wash, BS, 50; Calif Inst Technol, PhD(chem), 56. Prof Exp: Res chemist, Res Labs, Rohm and Haas Co, Philadelphia, 56-59; res chemist, Redstone Arsenal Res Div, 59-63, head res lab, 63-70; TECH DIR, SOAP & DETERGENT ASN, 71- Mem: Am Chem Soc. Res: Surface chemistry; agricultural chemicals; detergents. Mailing Add: Soap & Detergent Asn 475 Park Ave New York NY 10016

BOOMS, ROBERT EDWARD, b Cleveland, Ohio, Mar 25, 45; m 68; c 1. PHOTOGRAPHIC CHEMISTRY. Educ: Ohio State Univ, BS, 67; Univ Calif, Los Angeles, PhD(org chem), 71. Prof Exp: CHEMIST, EASTMAN KODAK CO, 71- Mem: Am Chem Soc; Soc Photog Scientists & Engrs. Res: Color photographic systems. Mailing Add: Eastman Kodak Co Kodak Park Bldg 59 Rochester NY 14650

BOOMSLITER, PAUL COLGAN, b Urbana, Ill, Oct 24, 15; m 46; c 4. SPEECH PATHOLOGY, AUDIOLOGY. Educ: Univ WVa, AB, 35; Univ Iowa, MA, 38; Univ Wis, PhD(speech), 42. Prof Exp: Asst prof speech, Goucher Col, 40-46; asst prof, Cornell Univ, 46-48; chmn dept speech path & audiol, 69-73, PROF SPEECH, STATE UNIV NY ALBANY, 48- Concurrent Pos: Consult, Albany Vet Admin Hosp, 56-; prof dir, Northeastern NY Speech Ctr, Inc, 58-63; consult, Albany Child Guid Ctr & Study Ctr Learning Disabilities, 64-; res assoc prof, Dept Prev & Community Med, Albany Med Col, 68- Mem: AAAS; Acoust Soc Am; Am Speech & Hearing Asn; Mod Lang Asn Am; NY Acad Sci. Res: Organization as a factor in auditory perception as evidenced in language, musical forms, hearing testing, and hearing and speech therapy. Mailing Add: 8 Lawnridge Ave Albany NY 12208

BOON, DONALD ARTHUR, b Toledo, Ohio, Apr 2, 33; m 56; c 2. CLINICAL CHEMISTRY. Educ: Oberlin Col, AB, 55; Purdue Univ, MS, 58; Albany Med Col, PhD(biochem), 63. Prof Exp: Cancer res scientist, Roswell Park Mem Inst, 62-67; sr res scientist, Med Found Buffalo, 67-69; sr scientist, Children's Hosp Buffalo, 69-73; DIR CHEM, NIAGARA FRONTIER CLIN LAB, 73- Concurrent Pos: Res asst prof, Dept Pediat, State Univ NY Buffalo, 70-73; instr, Trocaire Col, 74- Res: Structure of estrogen conjugates; quantitative analysis of steroids; relationships among karyotype, behavior and endocrine function; circadian and developmental changes in sex hormone production. Mailing Add: Niagara Frontier Clin Lab 50 High St Buffalo NY 14203

BOON, JAMES ALEXANDER, b Sarasota, Fla, May 16, 46; m 68; c 1. ETHNOLOGY. Educ: Princeton Univ, BA, 68; Univ Chicago, MA, 69, PhD(anthrop), 73. Prof Exp: Anthrop researcher, Ctr Study Man, Smithsonian Inst, 69-71; asst anthrop, Inst Adv Study, Princeton, NJ, 73-74, vis mem soc sci, 74-75; ASST PROF ANTHROP, DUKE UNIV, 75- Mem: Am Anthrop Asn; Asn Asian Studies. Res: Balinese social structure and comparative Indonesian and Southeast Asian ethnology; history of cross-cultural theories of social institutions and belief systems; structuralism. Mailing Add: Dept of Anthrop Duke Univ Durham NC 27706

BOON, JOHN DANIEL, JR, b Ft Worth, Tex, Dec 29, 14; m 39; c 2. GEOLOGY. Educ: Southern Methodist Univ, BS, 37. Prof Exp: Instr geol, Southern Methodist Univ, 41; computer, Seismograph Party, Nat Geophys Co, 41-42; assoc prof geol & physics, 42-43, prof physics & actg head dept, 43-45, head dept geol, 45-71, PROF GEOL, UNIV TEX, ARLINGTON, 45- Mem: Fel Geol Soc Am; Am Asn Petrol Geologists. Res: Field geology of regions in trans-Pecos Texas; structural, general and economic geology; petroleum engineering; seismology. Mailing Add: Dept of Geol Univ of Tex Arlington TX 76019

BOONE, CHARLES J, physical chemistry, organic chemistry, see 12th edition

BOONE, CHARLES WALTER, b Berkeley, Calif, Dec 21, 25; m 65; c 2. PATHOLOGY, BIOCHEMISTRY. Educ: Univ Calif, San Francisco, MD, 51; Univ Calif, Los Angeles, PhD(biochem), 64; Am Bd Path, dipl, 60. Prof Exp: Gen pract med, Los Angeles, 52-56; resident path, 56-60; HEAD CELL BIOL SECT, VIRAL BIOL BR, NAT CANCER INST, NIH, 65- Concurrent Pos: Res fel, Albert Einstein Col Med, 64-65. Mem: Am Soc Exp Path; Am Soc Biol Chemists. Res: Isolation of pure tumor cells; production of tumor immunity; reaction between antibodies and cell surface antigens. Mailing Add: Nat Cancer Inst NIH Bldg 37 Rm 1008 Bethesda MD 20014

BOONE, DONALD JOE, b Spokane, Wash, May 12, 43; m 73. CLINICAL CHEMISTRY, CLINICAL PATHOLOGY. Educ: Cent State Univ, BS, 65; Iowa State Univ, MS, 67; Wash State Univ, PhD(chem), 70. Prof Exp: NSF fel biophys, Albert Einstein Col Med, 71-72; fel clin chem, Univ Iowa, 72-74; ASST PROF PATH, UNIV KY, 74- Concurrent Pos: Consult, Lexington Clin, 74-; clin chemist, Vet Admin Hosp, Lexington, 74- Mem: Am Chem Soc; Am Asn Clin Chemists. Res: Trace element pathology; kidney and liver disease. Mailing Add: Dept of Path Univ of Ky Lexington KY 40507

BOONE, DONALD MILFORD, b Phoenix, Ariz, Jan 7, 18; m 47; c 3. PLANT PATHOLOGY. Educ: Marion Col, Ind, BA & BS, 40; Univ Wis, PhD(plant path), 50. Prof Exp: Instr biol, Marion Col, Ind, 40-41; res asst, 46-49, proj assoc, 49-56, from asst prof to assoc prof, 56-69, PROF PLANT PATH, UNIV WIS, MADISON, 69- Mem: Am Phytopath Soc; Bot Soc Am; Mycol Soc Am. Res: Inheritance and nature of pathogenicity and disease resistance; genetics of microorganisms; diseases of small fruits. Mailing Add: Dept of Plant Path Univ of Wis Madison WI 53706

BOONE, GARY M, b Presque Isle, Maine, July 16, 29; m 52; c 2. PETROLOGY, MINERALOGY. Educ: Bowdoin Col, AB, 51; Brown Univ, MA, 54; Yale Univ, PhD(geol), 59. Prof Exp: Lectr geol, Univ Western Ont, 57-60, asst prof, 60-63; from asst prof to assoc prof, 63-71, PROF GEOL, SYRACUSE UNIV, 71- Concurrent Pos: Nat Res Coun Can res grants, 59-62; Off Naval res contract volcanic rocks, Azores, 65-67. Mem: Geol Soc Am; Mineral Soc Am; Mineral Soc Gt Brit & Ireland; Geol Asn Can. Res: Metamorphism and regional stratigraphy in northern Appalachians. Mailing Add: Dept of Geol Syracuse Univ Syracuse NY 13210

BOONE, JAMES LIGHTHOLDER, b San Antonio, Tex, Sept 11, 32; m 56; c 3. INORGANIC CHEMISTRY. Educ: St Louis Univ, BS, 54; Univ Southern Calif, PhD(chem), 60. Prof Exp: Res asst, Univ Southern Calif, 54-59; from res chemist to sr res chemist, 59-75, MGR CHEM ECON, US BORAX RES CORP, 75- Mem: Am Chem Soc. Res: Boron hydrides and their derivatives; organo-boron and organometallic compounds; high temperature boron compounds and refractories. Mailing Add: US Borax Res Corp 412 Crescent Way Anaheim CA 92803

BOONE, JAMES ROBERT, b Buffalo, NY, Aug 9, 39; m 61; c 2. MATHEMATICS. Educ: Tex A&M Univ, BA, 61, MS, 62; Tex Christian Univ, PhD(math), 68. Prof Exp: Opers res analyst, Gen Dynamics/Ft Worth, 63-64, aerodyn engr, 64-65; ASST PROF MATH, TEX A&M UNIV, 68- Mem: Am Math Soc; Math Asn Am. Res: General topology. Mailing Add: Dept of Math Tex A&M Univ College Station TX 77843

BOONE, JAMES RONALD, b Brownsville, Tenn, Nov 10, 46; m 70. ORGANIC CHEMISTRY. Educ: David Lipscomb Col, BA, 68; Ga Inst Technol, PhD(chem), 74. Prof Exp: Instr chem, Ga Inst Technol, 69-70, fel biochem, 74-75; ASST PROF CHEM, DAVID LIPSCOMB COL, 75- Mem: Am Chem Soc. Res: Stereoselective reductions of ketones by metal hydrides. Mailing Add: Dept of Chem David Lipscomb Col Nashville TN 37203

BOONE, MAX L M, b Peru, Ind, May 23, 31; c 5. RADIOTHERAPY. Educ: Ind Univ, Bloomington, AB, 52; Ind Univ, Indianapolis, MD, 56; Univ Tex, Houston, PhD(biophys, radiobiol), 68. Prof Exp: James Prof Exp: Assoc prof biophys & radiother, Grad Sch Biomed Sci, Univ Tex, 68-69; prof radiol & dir radiother ctr, Univ Wis Hosps, Madison, 69-72; DIR, ARIZ MED CTR RADIATION ONCOL, UNIV ARIZ, 72- Concurrent Pos: Assoc radiotherapist, Univ Tex M D Anderson Hosp & Tumor Inst, 68-69; consult, Vet Admin Hosp, Tucson, Ariz, 71- & Los Alamos Pi Meson Facil, NMex, 70-; mem comt radiation ther study, Nat Cancer Inst, 70- Honors & Awards: Bronze Medal, Am Roentgen Ray Soc, 66. Mem: Am Roentgen Ray Soc; Am Radium Soc; Am Inst Physics; fel Am Col Radiol; Am Asn Physicists in Med. Res: Radiation therapy; medical physics; radiobiology. Mailing Add: Ariz Med Ctr Radiation Oncol Univ of Ariz Tucson AZ 85724

BOONE, MERRITT ANDERSON, b Geneva, Nebr, Apr 27, 20; m 44; c 2. REPRODUCTIVE PHYSIOLOGY, POULTRY PHYSIOLOGY. Educ: Univ Nebr, BS, 41; Mich State Univ, MS, 47; Univ Ga, PhD(poultry), 62. Prof Exp: From asst to assoc poultry, 47-58, assoc prof, 58-66, PROF POULTRY, CLEMSON UNIV, 66-, POULTRY SCIENTIST, 58- Concurrent Pos: Animal Sci Advisor, USAID-Univ Fla, Nat Agr Inst, Saigon, Rep Viet Nam, 72-73. Honors & Awards: Educ Medal, Rep Viet Nam, 73. Mem: Poultry Sci Asn; World Poultry Sci Asn; Soc Study Reproduction. Res: Physiological effect of hormones, environmental temperatures and drugs on embryos, growing chickens and adult fowl. Mailing Add: Col of Agr Clemson Univ Clemson SC 29631

BOONE, PETER AUGUSTINE, b Corpus Christi, Tex, June 14, 40; m 65; c 3. STRATIGRAPHY, NATURAL RESOURCES. Educ: St Mary's Univ, Tex, BA, 63; Baylor Univ, MA, 66; Tex A&M Univ, PhD(geol), 72. Prof Exp: Geologist, Mobile Oil Corp, Corpus Christi, 65-67; res geologist, 71-74, CHIEF ENERGY RESOURCES DIV, GEOL SURV ALA, 74- Mem: Am Asn Petrol Geologists; Geol Soc Am; Soc Econ Paleontologists & Mineralogists; Sigma Xi. Res: Application of stratigraphic and sedimentologic principals in basin analysis for the assessment of energy resources and land use planning. Mailing Add: PO Drawer O University AL 35486

BOONE, WILLIAM WERNER, b Cincinnati, Ohio, Jan 16, 20; m 49; c 2. MATHEMATICAL LOGIC, ALGEBRA. Educ: Univ Cincinnati, AB, 45; Princeton Univ, MA, 48, PhD(math), 52. Prof Exp: Instr math, Princeton Univ, 45-47 & Rutgers Univ, 47-50; asst prof, Cath Univ Am, 50-54; mem, Inst Advan Study, Princeton, 54-56; Fulbright scholar, Univ Oslo, 56-57; Guggenheim fel, Oxford & Manchester Univs & Univ Münster, 57-58; assoc prof, 58-60, PROF MATH, UNIV ILL, URBANA-CHAMPAIGN, 60- Concurrent Pos: Consult mathematician, Marine Off Oper Res, 52-53, Am Car & Foundry Electronics Logic Proj, 54-56, Systs Develop Corp, 63-66 & Argonne Nat Labs, 64-; assoc mem, Ctr Advan Study, 60-62; mem, Inst Advan Study, Princeton, 64-66; vis prof, Math Inst & All Souls Col, Oxford Univ, 72-73. Mem: Am Math Soc; Asn Symbolic Logic. Res: Word problem for groups; related problems. Mailing Add: Dept of Math Univ of Ill Urbana-Champaign Urbana IL 61801

BOONSTRA, BRAM B, b Arnhem, Neth, June 4, 12; m 45; c 2. ELECTROCHEMISTRY. Educ: Univ Amsterdam, PhD(electrochem), 38. Prof Exp: Asst prof anal chem, Univ Amsterdam, 36-39; teacher high sch, Neth, 39-40; res chemist, Rubber Found, Neth, 40-43, mgr, Tech Rubber & Plastics Lab, 46-51; develop chemist, Royal Holland Cable Works, 43-46; sect head, 51-62, ASSOC DIR RES, RUBBER & PLASTICS LAB, GODFREY CABOT, INC, 62- Mem: Am Chem Soc. Res: High polymer physical chemistry and technology. Mailing Add: Cabot Corp Tech Ctr Concord Rd Bellvica MA 01821

BOOP, WARREN CLARK, JR, b Baltimore, Md, July 27, 33; c 3. NEUROSURGERY. Educ: Univ Tenn, BS & MD, 56; Univ Minn, MS, 64. Prof Exp: Intern, Baroness Erlanger Hosp, Chattanooga, Tenn, 56-57; resident, St Mary's Hosp, Knoxville, 57-58; resident neurosurg, Univ Minn, Minneapolis, 60-64; staff neurosurgeon, US Naval Hosp, Oakland, Calif, 64-66; head neurosurg serv, US Naval Hosp, Great Lakes, Ill, 66-69; ASSOC PROF NEUROSURG, MED CTR, UNIV ARK, LITTLE ROCK, 70- Concurrent Pos: US Navy rep, Nat Res Coun Spinal Cord Injuries, 67; vis prof, Wesley Mem Hosp, Chicago, 68; staff neurosurgeon, Vet Admin Hosp, Little Rock, 70-; consult neurosurg, Ark Children's Hosp & Ark Crippled Children's Div, Dept Pub Welfare Hot Springs Rehab Ctr, 70- Mem: Cong Neurosurg; Am Asn Neurol Surg; AMA; Neurosurg Soc Am; Am Col Surg. Res: Circulatory effects of increased intracranial pressure. Mailing Add: Dept of Neurosurg Univ of Ark Med Ctr Little Rock AR 72201

BOOR, JOHN, JR, organic chemistry, polymer chemistry, see 12th edition

BOORD, ROBERT LENNIS, b Masontown, Pa, July 29, 26; m 53; c 1. VERTEBRATE MORPHOLOGY, NEUROANATOMY. Educ: Washington & Jefferson Col, AB, 50; Univ Md, MS, 58, PhD(zool, comp anat), 60. Prof Exp: USPHS fel, 60-61; instr zool, Duke Univ, 61-62; asst prof biol sci, 62-69, ASSOC PROF BIOL SCI, UNIV DEL, 69- Mem: AAAS; Am Asn Anat; Am Soc Zoologists. Res: Comparative anatomy of vertebrate auditory system; comparative neuroanatomy. Mailing Add: Dept of Biol Sci Univ of Del Newark DE 19711

BOORMAN, EVELYN HUTTERER, b New York, NY, July 28, 45; m 67. ALGEBRA. Educ: Stanford Univ, BS, 67; Columbia Univ, MA, 70, PhD(math), 73. Prof Exp: ASST PROF MATH, UNIV MICH, ANN ARBOR, 73- Mem: Am Math Soc; Math Asn Am; Asn Women in Math. Res: Operations in representation theory; Lambda rings. Mailing Add: 576 Kellogg Ann Arbor MI 48105

BOORMAN, PHILIP MICHAEL, b Hitchin, Eng, May 11, 39; m 66; c 3. INORGANIC CHEMISTRY. Educ: Univ Nottingham, BSc, 61, PhD(inorg chem), 64. Prof Exp: Brit Titan res fel inorg chem, Univ Newcastle, 64-67; asst prof chem & asst dean fac arts & sci, 67-73, ASSOC PROF CHEM, UNIV CALGARY, 73- Concurrent Pos: Hon vis fel, Dept Chem, Univ Manchester, Eng, 73-74. Mem: Chem Inst Can; The Chem Soc. Res: Interaction of molybdenum and tungsten compounds with sulfur donor ligands; comparative study of phosphine, arsine and stibine chalcogenides. Mailing Add: Dept of Chem Univ of Calgary Calgary AB Can

BOORSE, HENRY ABRAHAM, b Norristown, Pa, Sept 18, 04; m 31; c 2. LOW TEMPERATURE PHYSICS. Educ: Columbia Univ, AM, 33, PhD(physics), 34. Prof Exp: Asst, Columbia Univ, 28-31, instr, 31-33; Lydig fel, Cambridge Univ, Eng, 34-35; instr physics, City Col New York, 35-37; asst prof & head dept, 37-43, from assoc prof to prof, 43-70, actg dean fac, 57, dean fac, 59-70, actg pres, 62 & 67, spec lectr & asst to pres, 70-74, EMER PROF PHYSICS & EMER DEAN, BARNARD COL, COLUMBIA UNIV, 70- Concurrent Pos: Adams fel, Columbia Univ, 38-40, res physicist, SAM Labs, Manhattan Dist, 41-45, div dir, 45-46; consult, US AEC, 46-58 & Brookhaven Nat Lab, 51-55; chmn, Calorimetry Conf, 56-57; mem, Comn I, Int Inst Refrig & vchmn subcomt basic sci, US Nat Comt, 58-72. Mem: Fel Am Phys Soc. Res: Low temperature physics; liquefaction of gases; superconductivity; properties of liquid helium. Mailing Add: 338 Summit Ave Leonia NJ 07605

BOOS, MARGARET (FULLER), b Beatrice, Nebr; m 27. MINERALOGY. Educ: Northwestern Univ, BS, 13; Univ Chicago, MS, 19, PhD(geol), 24. Prof Exp: Instr geol, Northwestern Univ, 21-27; actg park naturalist, Rocky Mt Nat Park, 28-30; geologist, Phillips Petrol Co, Okla, 32-33; prof geol & chmn dept, Univ Denver, 35-42; mineral technologist, US Bur Mines, Wash, 42-45; geologist, US Bur Reclamation, 45-47; CONSULT GEOL, 47- Concurrent Pos: Nat Res Coun awards, 32-34; Penrose grant-in-aid, Geol Soc Am, 34, 38, 40 & 54; mem, Int Geol Cong, London, 48, Algiers, 52; mineral technologist, US Bur Mines, Wash, 51-53. Mem: Fel AAAS; fel Geol Soc Am; fel Mineral Soc Am. Res: Engineering and mining geology; nonmetallics; petroleum geology; exploration petrography; petrology of igneous rocks; spectroscopy of pegmatite minerals. Mailing Add: 2036 S Columbine St Denver CO 80210

BOOS, RICHARD NEWTON, b Lancaster, Pa, Sept 5, 16; m 68; c 3. ANALYTICAL CHEMISTRY. Educ: Franklin & Marshall Col, BS, 38; NY Univ, MS, 40. Prof Exp: Sect leader in charge microanal lab, Merck & Co, Inc, 40-74; RETIRED. Mem: Am Chem Soc; Am Microchem Soc. Res: Organic and inorganic chemistry. Mailing Add: 661 Bryant St Rahway NJ 07065

BOOSALIS, MICHAEL GUS, b Faribault, Minn, Sept 20, 17; m 46; c 1. PLANT PATHOLOGY. Educ: Univ Minn, BS, 41, MS, 48, PhD(plant path), 51. Prof Exp: From asst prof to assoc prof, 51-57, PROF PLANT PATH, AGR EXP STA, UNIV NEBR, LINCOLN, 58-, CHMN DEPT, 64-, PROF BOT, 70- Mem: Am Phytopath Soc. Res: Ecology of fungi, causing root diseases of plants. Mailing Add: Dept of Plant Path Univ of Nebr Lincoln NE 68503

BOOSMAN, JAAP WIM, b Amsterdam, Holland, Mar 2, 35; US citizen; m 61. SCIENCE ADMINISTRATION, SEDIMENTOLOGY. Educ: Syracuse Univ, BS, 57, MS, 59; George Washington Univ, MPhil, 70, PhD(geol), 73. Prof Exp: Arctic proj officer, US Off Naval Res, 62-65; oceanogr, Off of Oceanogr, US Navy, 65-75; SCI STAFF ASST ENVIRON SCI, OFF CHIEF NAVAL OPERS, 75- Mem: Sigma Xi. Res: Clay mineral distribution and source, provenance and transport of sediments on the Argentine continental shelf; relict morphology of the Argentine continental shelf and evidence for sea level changes. Mailing Add: 6947 33rd St NW Washington DC 20015

BOOST, GERHARD, b Bitterfeld, Ger, Oct 14, 19; US citizen; m 46; c 4. CLINICAL MEDICINE. Educ: Univ Königsberg, MD, 43. Prof Exp: Pvt pract, 45-47; resident surg, Med Acad, Düsseldorf, Ger, 47-50; staff physician, Ore State Tuberc Hosp, Salem, 50-58; med dir int div, Parke, Davis & Co, 58-62; ASSOC MED DIR, SYNTEX RES, 62- Res: Chest diseases and chest surgery. Mailing Add: Syntex Res Stanford Indust Park Palo Alto CA 94304

BOOTE, KENNETH JAY, b Hull, Iowa, Sept 12, 45; m 69; c 1. CROP PHYSIOLOGY. Educ: Iowa State Univ, BS, 67; Purdue Univ, MS, 69, PhD(crop physiol), 74. Prof Exp: ASST PROF CROP PHYSIOL, UNIV FLA, 74- Mem: Am Soc Agron; Crop Sci Soc Am; Am Peanut Res & Educ Asn. Res: Relationships of physiological traits to productivity of peanut and soybean genotypes under varying environments; influence of reproductive sinks on photosynthesis; double cropping. Mailing Add: Dept of Agron Univ of Fla Gainesville FL 32611

BOOTH, ALFRED WHALEY, b Milwaukee, Wis, Oct 7, 08; m 37; c 3. GEOGRAPHY. Educ: Wis State Teachers Col, BEd, 29; Univ Wis, PhM, 32, PhD(geog), 36. Prof Exp: Asst, Univ Chicago, 33-34 & Univ Wis, 35-36; instr geog, Univ Ill, 36-43, asst prof, 46; assoc prof, State Col Wash, 46-47; from asst prof to assoc prof, prof & exec secy dept, 53-70, actg head dept, 65-66, EMER PROF GEOG, UNIV ILL, URBANA, 70- Mem: Asn Am Geog; fel Am Geog Soc; Regional Sci Asn; Nat Coun Geog Educ; Am Soc Photogram. Res: Transportation and recreation geography; geography of India; conservation of natural resources. Mailing Add: 404 W Nevada Urbana IL 61801

BOOTH, BRUCE L, b Philadelphia, Pa, Mar 22, 38; m 68; c 1. SOLID STATE PHYSICS, LASERS. Educ: Dartmouth Col, BA, 60; Northwestern Univ, PhD(physics), 67. Prof Exp: From res physicist to sr res physicist, 67-73, res supvr, 73-74, SR SUPVR, E I DU PONT DE NEMOURS & CO, 74- Mem: Am Phys Soc; Optical Soc Am. Res: Low temperature band structure studies using de Haas-Shubnikov effect in semiconductors, mainly gray tin; supervise development of techniques and applications of laser-optical measurements for industry. Mailing Add: Eng Dept Du Pont Exp Sta 1007 Market St Wilmington DE 19898

BOOTH, DAVID LAYTON, b Aurora, Ill, July 20, 39; m 62; c 2. PESTICIDE CHEMISTRY. Educ: Beloit Col, BS, 61; Univ Ore, PhD(org chem), 65. Prof Exp: Res chemist, Org Div, Morton Chem Co, 65-69, SUPVR ORG RES, MORTON INT, INC, 69- Mem: Am Chem Soc; Plant Growth Regulator Working Group; Coun Agr Sci & Technol. Res: Alkaloid chemistry; synthetic studies in pesticide research.

BOOTH, ERNEST SHELDON, b Lehman, Pa, Oct 8, 15; m 38; c 2. BIOLOGY. Educ: Pacific Union Col, BA, 38; Univ Wash, MS, 40; Wash State Univ, PhD(zool), 47. Prof Exp: From instr to assoc prof biol, Walla Walla Col, 38-47, prof biol & dir biol sta, 47-58; prof biol, Grad Sch, Loma Linda Univ, 62-74. Mem: Assoc Am Soc Mammalogists; assoc Am Ornith Union. Res: Distribution of mammals of Pacific Northwest; distribution of birds of West; mammalogy; ornithology; systematic review of land mammals of Washington. Mailing Add: PO Box 277 Anacortes WA 98221

BOOTH, EUGENE THEODORE, b Rome, Ga, Sept 28, 12; m; c 2. PHYSICS Educ: Univ Ga, BS, 32, MS, 34; Oxford Univ, PhD(physics), 37. Prof Exp: Lectr physics, Columbia Univ, 37-41, res scientist, 41-45; res scientist, Carbide & Carbon Chem Corp, 45-46; from asst prof to prof physics, Columbia Univ, 46-59; sci dir, Supreme Allied Comdr, Atlantic, Italy, 59-61; sci dir & vpres, Laser, Inc, Am Optical Soc, 61-66; dean grad studies, Stevens Inst Technol, 68-72; RETIRED. Concurrent Pos: Dir, Hudson Labs, 51-53; dir, Nevis Cyclotron Labs, 53-56; consult, Columbia Univ. Honors & Awards: AEC Award, 71. Mem: Fel Am Phys Soc; Optical Soc Am. Res: Optics; nuclear physics; plasma engineering. Mailing Add: 902 Pleasantville Rd Briarcliff Manor NY 10510

BOOTH, GARY EDWIN, b Campton, Ky, Dec 26, 40; m 62; c 2. ORGANIC CHEMISTRY. Educ: Eastern Ky State Col, BS, 62; Ohio State Univ, PhD(org chem), 65. Prof Exp: Res chemist, 65-69, SECT HEAD PHYS ORG CHEM, MIAMI VALLEY LABS, PROCTER & GAMBLE CO, 69- Mem: Am Chem Soc. Res: Conformational analysis; fluorescence of organic molecules; adsorption of dyestuffs onto cellulose; thin-layer chromatography. Mailing Add: 928 Springbrook Dr Cincinnati OH 45224

BOOTH, GARY MELVON, b Provo, Utah, Oct 9, 40; m 60; c 6. ENTOMOLOGY. Educ: Utah State Univ, BS, 63, MS, 66; Univ Calif, Riverside, PhD(entom), 69. Prof Exp: NIH fel, Univ Ill, 69-70, asst prof entom, 70-71, asst prof agr entom, 71-72; asst prof zool, 72-73, ASSOC PROF ZOOL, BRIGHAM YOUNG UNIV, 73- Concurrent Pos: Consult, Environ Protection Agency, 71; asst entomologist, Ill Nat Hist Surv, 71-72; consult, Thompson-Hayward Chem Co, 72-76; assoc ed environ entom, Entom Soc Am, 75. Honors & Awards: First Place Res Award, Sigma Xi, 73. Mem: Sigma Xi; Entom Soc Am; Weed Sci Soc Am. Res: Metabolism toxicity and environmental behavior of insecticides, herbicides, fungicides and rodenticides in model ecosystems; fish, insects, water and soil; development of pest management research programs. Mailing Add: Dept of Zool 697 WIDB Brigham Young Univ Provo UT 86402

BOOTH, JAMES SAMUEL, b LeFlore, Okla, Nov 10, 27; m 62; c 1. BACTERIAL PHYSIOLOGY. Educ: Calif State Col, BS, 59; Univ Southern Calif, MS, 62, PhD(staphylocoagulase), 68. Prof Exp: Lectr microbiol, Calif State Col, Los Angeles, 63-65; asst prof bact physiol, Univ NMex, 67-74; ASST PROF BIOL SCI, CALIF POLYTECH STATE UNIV, SAN LUIS OBISPO, 74- Mem: Am Soc Microbiol. Res: Degradation of ergothioneine by alcaligenes faecalis; factors influencing in vitro levels of staphylocoagulase. Mailing Add: Dept of Biol Sci Calif Polytech State Univ San Luis Obispo CA 93401

BOOTH, JOHN AUSTIN, b DuBois, Pa, Jan 27, 29; m 48; c 1. PLANT PATHOLOGY. Educ: Univ Ariz, BS, 51, MS, 58, PhD(plant path), 63. Prof Exp: Investr air pollution, Phelps Dodge Corp, Ariz, 51-57; asst prof, 63-69, ASSOC PROF PLANT PATH, NMEX STATE UNIV, 69- Mem: Am Phytopath Soc. Res: Disease physiology; antibiosis; air pollution. Mailing Add: Dept of Bot & Entom NMex State Univ Las Cruces NM 88003

BOOTH, KENNETH GORDON, b Victoria, BC, May 8, 18; m 45; c 2. CELLULOSE CHEMISTRY. Educ: Univ BC, BA, 40; McGill Univ, PhD(chem), 48. Prof Exp: Mem staff, Cent Res Dept, Crown Zellerbach Corp, 48-57; mgr res, 57-63, DIR RES, ABITIBI POWER & PAPER CO, LTD, CAN, 63- Mem: Tech Asn Pulp & Paper Indust; Can Pulp & Paper Asn. Res: Cellulose chemistry; pulp and paper research and development; hardboard. Mailing Add: Abitibi Paper Co Ltd Res Ctr Sheridan Park Mississauga ON Can

BOOTH, MAX HOWARD, b Longmont, Colo, Feb 11, 20; m 49; c 3. PHYSICAL CHEMISTRY. Educ: La State Univ, BS, 43, MS, 50; Univ Tex, PhD(chem), 53. Prof Exp: From chemist to sr res chemist, Benger Lab, 53-65, anal res supvr, Old Hickory Res & Develop Lab, 65-68, PROCESS SUPVR, E I DU PONT DE NEMOURS & CO, 68- Mem: Am Chem Soc. Res: Analytical chemistry of polymer systems. Mailing Add: 146 Cherokee Rd Hendersonville TN 37075

BOOTH, NEWELL ORMOND, b Pocatello, Idaho, Nov 23, 40; m 65; c 1. UNDERWATER ACOUSTICS, OCEAN ENGINEERING. Educ: Univ Calif, Berkeley, BA, 62; Univ Calif, Los Angeles, MS, 66, PhD(physics), 71. Prof Exp: Physicist optics, Naval Ord Test Sta, 63-65, RES PHYSICIST ACOUST HOLOGRAPHY & UNDERWATER ACOUST, NAVAL UNDERSEA CTR, 71- Mem: Inst Elec & Electronics Engrs. Res: Underwater imaging and acoustical holography; randomly distributed volumetric acoustic arrays. Mailing Add: Naval Undersea Ctr San Diego CA 92132

BOOTH, NICHOLAS HENRY, b Hannibal, Mo, Oct 22, 23; m 44; c 3. PHYSIOLOGY, PHARMACOLOGY. Educ: Mich State Univ, DVM, 47; Colo State Univ, MS, 51; Univ Colo, PhD, 59. Prof Exp: Pvt pract vet med, Hannibal, Mo, 47-48; from asst prof to prof physiol, Colo State Univ, 48-66, head dept, 56-66, dean col vet med & biomed sci, 66-71; dir div vet med res, Bur Vet Med, Food & Drug Admin, 71-74; PROF PHYSIOL & PHARMACOL, UNIV GA, 74- Concurrent Pos: Chmn vet drug panel, Nat Res Coun-Nat Acad Sci, 66-68; consult, Div Physician Manpower, Bur Health Manpower, Dept Health, Educ & Welfare, 67-68; mem vet med rev comt, Bur Health Physicians Educ & Manpower Training, Dept Health, Educ & Welfare, NIH, 69-; mem spec comt vet med educ, Southern Regional Educ Bd, 70. Mem: AAAS; Am Physiol Soc; Am Vet Med Asn; Am Heart Asn; Soc Exp Biol & Med. Res: Pharmacodynamic studies on d-tubo curarine chloride and succinylcholine chloride in the horse; meperidine studies in the cat; artificial respiration in large animals; electrical defibrillation of the dog heart; electro and phonocardiography studies in the lamb; cardiovascular studies in swine; veterinary pharmacology and toxicology. Mailing Add: 430 Sandstone Dr Athens GA 30601

BOOTH, NORMAN E, b Toronto, Ont, Mar 7, 30; m 58; c 2. ELEMENTARY PARTICLE PHYSICS. Educ: Univ Toronto, BA, 52; Queen's Univ, Ont, MA, 54; Univ Birmingham, PhD(physics), 57. Prof Exp: Res fel physics, Univ Birmingham, 57-58; res physicist, Lawrence Radiation Lab, Univ Calif, 59-62; from asst prof to assoc prof physics, Univ Chicago, 62-70; SR RES OFFICER, UNIV OXFORD, 70- Mem: Fel Am Phys Soc. Res: High energy physics. Mailing Add: Dept of Nuclear Physics Univ of Oxford Oxford England

BOOTH, RICHARD W, b Cincinnati, Ohio, Mar 17, 24; m 45; c 3. MEDICINE. Educ: Univ Cincinnati, MD, 52; Am Bd Internal Med, dipl. Prof Exp: Donald L Mahanna res fel, Ohio State Univ Hosp, 56-59; asst prof med & instr aviation & prev med, Ohio State Univ, 59-61; assoc prof, 61-64, PROF MED, SCH MED, CREIGHTON UNIV, 64-, ASSOC DEAN, 72- Concurrent Pos: Consult, Vet Admin Hosp, Dayton, Ohio, 56-61 & Wright Patterson AFB, 60-61; attend physician, Ohio State Univ, 59-61 & Vet Admin Hosp, Nebr, 63-64; fel coun clin cardiol, Am Heart Asn. Mem: Am Fedn Clin Res; Am Col Cardiol; Am Col Angiol; fel Am Col Physicians; Am Heart Asn. Res: Cardiovascular diseases. Mailing Add: Cardiac Ctr Creighton Univ Omaha NE 68108

BOOTH, ROBERT EDWIN, b New York, NY, Mar 28, 21; m 42; c 3. ORGANIC CHEMISTRY. Educ: NY State Col Forestry, Syracuse Univ, BS, 42, PhD, 50. Prof Exp: Asst instr org chem, Syracuse Univ, 49-50; res chemist, Durez Plastics & Chem Co, 50; res chemist chem & radiation labs, Chem Corps, US Army, 51-52; res chemist, Durez Plastics & Chem Co, 52-55; res chemist, Solvay Process Div, 55-68, SR RES CHEMIST, INDUST CHEM DIV, ALLIED CHEM CORP, 68- Mem: Am Chem Soc; The Chem Soc. Res: Plastics and polymers; vinyl ethers and chloride; polyesters; polyethers; phenolics; inorganic fluorine chemistry. Mailing Add: 174 Main St Hamburg NY 14075

BOOTH, ROGER ELWOOD, b Manton, Mich, Nov 12, 25; m 48; c 4. PHARMACEUTICAL CHEMISTRY. Educ: Univ Mich, PhD(pharmaceut chem), 53. Prof Exp: Pharmaceut chemist, Res Div, 52-54, anal chemist & asst head dept, Chem Control Div, 54-62, mgr prod control, 62-63 & control res & develop, 63-69, MGR, CONTROL RECORDS & SPEC PROJS, UPJOHN CO, 69- Mem: Am Pharmaceut Asn. Res: Instrumental analysis of crude drugs; pharmaceutical product development. Mailing Add: 429 Sun View Ave Kalamazoo MI 49001

BOOTH, SHELDON JAMES, b Marshalltown, Iowa, Feb 21, 45. MICROBIOLOGY. Educ: Univ Iowa, BA, 68; Univ Nebr, PhD(microbiol), 75. Prof Exp: Biol sci asst microbiol, US Army, Dugway, Utah, 68-70; RES ASSOC MICROBIOL, VPI ANAEROBE LAB, 75- Concurrent Pos: Fel, VPI Anaerobe Lab, 75. Mem: Am Soc Microbiol; AAAS. Res: Ultrastructure, genetics, and biochemistry of bacteriophage and bacteriocins of anaerobic bacteria; bacteriophage and bacteriocin typing of anaerobic bacteria; ultrastructure, genetics and biochemistry of anaerobic bacteria; biology of blue-green bacteria and their viruses. Mailing Add: VPI Anaerobe Lab Blacksburg VA 24060

BOOTH, WILLIAM THOMAS, JR, b Canton, Ohio, Apr 3, 19; m 41; c 1. ORGANIC CHEMISTRY. Educ: Mt Union Col, BSc, 41; Ohio State Univ, PhD(org chem), 47. Prof Exp: Res chemist, Am Cyanamid Co, Conn, 47-51, group leader, Develop Lab, 51-53, from asst chief chemist to chief chemist, 53-59, mgr, Resin Labs, Bridgeville Plant, 60-62; tech dir, Resins Dept, 63-66, asst mgr, Develop Dept, Tar & Chem Div, 66-69, MGR PROD DEVELOP, ORG MAT DIV, KOPPERS CO, 69- Mem: AAAS; Am Chem Soc; Am Soc Test & Mat; Am Inst Chemists. Res: Catalytic oxidations; esters; polyester and surface coating resins; coal tar chemicals. Mailing Add: Org Mat Div Koppers Co Inc 440 College Park Dr Monroeville PA 15146

BOOTHBY, WILLIAM MUNGER, b Detroit, Mich, Apr 1, 18; m 47; c 3. MATHEMATICS. Educ: Univ Mich, PhD, 49. Prof Exp: Swiss-Am Found for Sci Exchange fel, Swiss Fed Inst Technol, 50-51; from instr to asst prof math, Northwestern Univ, 48-59; assoc prof, 59-62, PROF MATH, WASH UNIV, 62- Concurrent Pos: NSF sr fel, Inst Advan Study, Princeton Univ, 61-62; vis mem, Inst Math, Univ Geneva, 65-66; prof associe, Univ Strasbourg, 71. Mem: AAAS; Am Math Soc. Res: Differential geometry and topology. Mailing Add: Dept of Math Wash Univ St Louis MO 63130

BOOTHE, JAMES HOWARD, b Okanogan, Wash, Nov 2, 16; c 2. ORGANIC CHEMISTRY. Educ: State Col Wash, BS, 39; Univ Minn, PhD(pharmaceut chem), 43. Prof Exp: RES CHEMIST, LEDERLE LABS, 43- Honors & Awards: Achievement Award, Univ Minn, 62. Mem: Am Chem Soc; NY Acad Sci. Res: Chemistry of pteridines; synthesis of folic acid and analogues; syntheses of thiobarbiturates; chemical modification of antibiotics to improve biological properties; chemistry and total synthesis of tetracyclines; chemistry of antibacterial substances. Mailing Add: Lederle Labs Pearl River NY 10965

BOOTHROYD, CARL WILLIAM, b Woodsville, NH, Jan 15, 15; m 41; c 2. PLANT PATHOLOGY. Educ: Dartmouth Col, AB, 38; State Col Wash, MS, 41; Cornell Univ, PhD, 50. Prof Exp: From asst prof to assoc prof, 49-57, assoc head dept, 63-69, PROF PLANT PATH, COL AGR & LIFE SCI, CORNELL UNIV, 57- Concurrent Pos: Guggenheim fel, 63. Mem: Am Inst Biol Sci; Am Phytopath Soc. Res: Diseases of corn. Mailing Add: Dept of Plant Path Cornell Univ Ithaca NY 14850

BOOTHROYD, ERIC ROGER, b Lennoxville, Que, Apr 2, 18; m 57; c 3. CYTOLOGY. Educ: Bishop's Univ, BSc, 38; McGill Univ, MSc, 40, PhD, 43. Prof Exp: Lectr chem & biol, Bishop's Univ, 45-46; from asst prof to assoc prof, 46-67, PROF GENETICS, McGILL UNIV, 67- Mem: AAAS; Genetic Soc Am; Genetics Soc Can; Am Soc Cell Biol. Res: Chromosome structure and behavior. Mailing Add: Dept of Genetics McGill Univ Montreal PQ Can

BOOTS, MARVIN ROBERT, b St Louis, Mo, Jan 29, 37; m 64. MEDICINAL CHEMISTRY. Educ: St Louis Col Pharm, BS, 58; Univ Wis, MS, 60; Univ Kans, PhD(pharmaceut chem), 63. Prof Exp: Asst prof pharmaceut chem, Univ Miss, 63-64; res org chemist, Gulf Res & Develop Co, 64-66; asst prof, 66-71, ASSOC PROF PHARMACEUT CHEM, MED COL VA, 71- Honors & Awards: Richardson Award, 63. Mem: Am Chem Soc. Res: Design and synthesis of potential therapeutic agents. Mailing Add: Dept of Pharmaceut Chem Va Commonwealth Univ Med Col Va Richmond VA 23219

BOOTS, SHARON G, b Grand Rapids, Mich, May 5, 39; m 64. ORGANIC CHEMISTRY. Educ: Univ Wis, BA, 60; Stanford Univ, PhD(org chem), 64. Prof Exp: NIH fel org chem, Mass Inst Technol, 64; assoc chemist, Midwest Res Inst, 65-66; res assoc, Am Tobacco Co, 66-67; RES ASSOC, MED COL VA, 67- Mem: Am Chem Soc. Res: Synthesis of biologically active compounds, steroids and heterocyclics; mechanisms of organic reactions. Mailing Add: 7801 Brown Rd Richmond VA 23235

BOOZER, CHARLES (EUGENE), b Nacogdoches, Tex, Oct 4, 27; m 53; c 6. RUBBER CHEMISTRY. Educ: Austin Col, BS, 49; Rice Inst, MA, 51, PhD(chem), 53. Prof Exp: Res assoc oxidation inhibitors, Iowa State Col, 53-54; assoc prof chem, La Polytech Inst, 54-59 & Emory Univ, 59-64; res supvr, 64-67, res mgr, 67-74, APPLN RES MGR, COPOLYMER RUBBER & CHEM CORP, 74- Concurrent Pos: Consult, Copolymer Rubber & Chem Corp, 56-64. Mem: Am Chem Soc. Res: Secondary isotope effects; mechanism of chlorosulfite decomposition; mechanism of action of oxidation inhibitors; nitric acid and air oxidations; free radical and Ziegler polymerization; rubber technology. Mailing Add: Copolymer Rubber & Chem Corp PO Box 2591 Baton Rouge LA 70821

BOOZER, REUBEN BRYAN, b Jacksonville, Ala, Aug 19, 25; m 54; c 4. ZOOLOGY, BIOLOGY. Educ: Jacksonville State Teachers Col, BS, 49; George Peabody Col Teachers, MA, 52; Auburn Univ, PhD(zool, bot), 69. Prof Exp: Teacher high sch, Ala, 50-54; asst prof biol, Jacksonville State Col, 54-63; instr zool, Auburn Univ, 63-68; PROF BIOL SCI, JACKSONVILLE STATE UNIV, 68-, DEAN COL ARTS & SCI, 74- Res: Histotechnique aging of mammals. Mailing Add: Dept of Biol Jacksonville State Univ Jacksonville AL 36265

BOPE, FRANK WILLIS, b Thornville, Ohio, Oct 30, 18; m 44; c 1. PHARMACEUTICAL CHEMISTRY. Educ: Ohio State Univ, BS, 41; Univ Minn, PhD(pharmaceut chem), 48. Prof Exp: Asst pharmaceut chem, Univ Minn, 41-42, jr scientist, 46-47; from asst prof to assoc prof pharm, 48-60, PROF PHARM, OHIO STATE UNIV, 60-, SECY, COL PHARM, 58-, ASST DEAN, COL PHARM, 70- Mem: Am Pharmaceut Asn; Acad Pharmaceut Sci; Am Asn Cols Pharm. Res: Chemistry and pharmacology of tannins; antioxidants; preliminary phytochemical and pharmacological investigation of the tannin obtained from Pinus Caribaea Morelet; synthesis of esters of gentisic acid; synthesis of urea analogs. Mailing Add: Ohio State Univ Col of Pharm 500 W 12th Ave Columbus OH 43210

BOPP, THOMAS THEODORE, b Glendale, Calif, Nov 29, 41; m 62, 73; c 4. PHYSICAL CHEMISTRY. Educ: Calif Inst Technol, BS, 63; Harvard Univ, PhD(chem), 68. Prof Exp: Asst prof, 67-75, ASSOC PROF CHEM, UNIV HAWAII, 75- Mem: Am Chem Soc. Res: Applications of magnetic resonance to liquid structure; molecular motion in liquids; computer simulation of spectra. Mailing Add: Dept of Chem 2545 The Mall Univ of Hawaii Honolulu HI 96822

BORAKER, DAVID KENNETH, b Los Angeles, Calif, Apr 21, 39; m 61; c 1. IMMUNOLOGY, IMMUNOGENETICS. Educ: Univ Calif, Santa Barbara, BA, 62; Univ Calif, Los Angeles, PhD(med microbiol), 67. Prof Exp: Immunologist, Dept Serol, Walter Reed Army Inst Res, DC, 67-69; asst prof, 69-72, ASSOC PROF MED

MICROBIOL, COL MED, UNIV VT, 72- Mem: AAAS. Res: Kinetics of induction of high and low dose immunological paralysis; comparative immunology of the double thymus of the South American degu. Mailing Add: Dept of Med Microbiol Univ of Vt Col of Med Burlington VT 05401

BORAX, EUGENE, b New York, NY, June 17, 19; m 50; c 2. GEOLOGY. Educ: Univ Calif, Los Angeles, MA, 52. Prof Exp: Sr geologist, Union Oil Co, 46-69, sr explor res geologist, Union Carbide Petrol Corp, 69-72; vpres, MOBWINC, Consult Geologists, 72-74; GEOLOGIST, PENNZOIL CO, 74- Mem: Geol Soc Am; Am Soc Photogram; Am Asn Petrol Geologists; Am Geophys Union; Geol Soc Australia. Res: Booming sand dunes; petroleum geology. Mailing Add: Pennzoil Co Pennzoil Pl Houston TX 77001

BORCH, RICHARD FREDERIC, b Cleveland, Ohio, May 22, 41; m 62. ORGANIC CHEMISTRY. Educ: Stanford Univ, BS, 62; Columbia Univ, MA, 63, PhD(chem), 65; Univ Minn, MD, 75. Prof Exp: Fel chem, Harvard Univ, 65-66; asst prof, 66-69, ASSOC PROF ORG CHEM, UNIV MINN, MINNEAPOLIS, 69- Mem: Am Chem Soc; The Chem Soc; AAAS. Res: Development of novel synthetic methods; the total synthesis of natural products. Mailing Add: Dept of Chem Univ of Minn Minneapolis MN 55455

BORCHARD, RONALD EUGENE, b Saginaw, Mich, Feb 16, 39; m 62; c 4. VETERINARY PHARMACOLOGY. Educ: Mich State Univ, BS, 66, DVM, 67; Univ Ill, PhD(vet pharm), 75. Prof Exp: Assoc vet, Northville Vet Clin, Mich, 67-69; instr vet pharm, Col Vet Med, Univ Ill, 69-73; ASST PROF VET PHARM, COL VET MED, WASH STATE UNIV, 73- Mem: Am Col Vet Toxicologists; Am Soc Vet Physiologists & Pharmacologists; Am Vet Med Asn. Res: Pharmacokinetics and tissue distribution of chloro-biphenyls in food producing animals. Mailing Add: Dept of Vet phys & Pharm Col of Vet Med Wash State Univ Pullman WA 99163

BORCHARDT, GLENN ARNOLD, b Watertown, Wis, July 28, 42; m 65; c 1. SOIL MINERALOGY. Educ: Univ Wis-Madison, BS, 64, MS, 66; Ore State Univ, PhD(soil mineral), 69. Prof Exp: Res asst soils, Int Minerals & Chem Corp, 64; Nat Res Coun res assoc, US Geol Surv, 69-71, GEOCHEMIST, CALIF DIV MINES & GEOL, 71- Concurrent Pos: Assoc ed, Geol Soc Am, 73-76. Mem: AAAS; Soil Sci Soc Am; Clay Minerals Soc; Geol Soc Am; Fedn Am Scientists. Res: Instrumental neutron activation analysis correlation of volcanic ash; multivariate similarity analysis and classification of samples using elemental ratios; quantitative soil clay mineralogy; chemical stabilization of landslides; soil smectites. Mailing Add: Div of Mines & Geol Ferry Bldg San Francisco CA 94111

BORCHARDT, HANS J, b Berlin, Ger, Jan 26, 30; US citizen; m 53; c 5. FLUORINE CHEMISTRY. Educ: Brooklyn Col, BS, 52, MA, 53; Univ Wis, PhD(phys chem), 56. Prof Exp: Inorg chemist, Gen Eng Lab, Gen Elec Co, 56-59; sr res chemist, Explosives Dept, 60-64, sr res chemist, Cent Res Dept, 65-69, sr res chemist org chem, 69-75, RES ASSOC, FREON PROD LAB, E I DU PONT DE NEMOURS & CO, 75- Mem: Am Chem Soc; Sigma Xi; Am Soc Heating, Refrig & Air Conditioning Engrs. Res: Differential thermal analysis; reactions in the solid state; rare earth solid state chemistry; phase equilibria; oxidation of metals; rare earth luminescence; lasers; crystal growth; fluorocarbon chemistry; chemical and physical properties of fluorocarbons. Mailing Add: E I du Pont de Nemours & Co Freon Prod Lab Chestnut Run Wilmington DE 19898

BORCHARDT, JOHN KEITH, b Evanston, Ill, June 2, 46. SYNTHETIC ORGANIC CHEMISTRY, PHYSICAL ORGANIC CHEMISTRY. Educ: Ill Inst Technol, BS, 68; Univ Rochester, PhD(chem), 73. Prof Exp: Lab technician chem, Ill Inst Technol, 68; fel, Univ Notre Dame, 72-74; RES CHEMIST, HERCULES, INC, 74- Mem: Am Chem Soc. Res: Mechanism of elimination reactions; transition metal promoted olefin formation; mechanism of cycloaddition reactions; synthesis of small and medium ring compounds. Mailing Add: Res Ctr Hercules Inc Wilmington DE 19899

BORCHARDT, KENNETH, b Chicago, Ill, Sept 20, 28; m 54; c 3. MICROBIOLOGY. Educ: Loyola Univ, Ill, BS, 50; Miami Univ, MS, 51; Tulane Univ, PhD(microbiol), 61. Prof Exp: US Army, 53-, asst chief vet dept, Med Lab, Ger, 54-56, asst chief bact, Fitzsimons Gen Hosp, 57-58, chief clin microbiol, Letterman Gen Hosp, 61-65, CHIEF CLIN MICROBIOL, US PUB HEALTH SERV HOSP, 65-, SCI DIR, 68- Mem: Fel Am Pub Health Asn. Res: Clinical research in infectious disease associated with problems in urology, dermatology, ophthalmology and internal medicine. Mailing Add: US Pub Health Serv Hosp San Francisco CA 94118

BORCHERS, CURTIS EDWARD, b Bellingham, Wash, Aug 17, 26; m 50; c 6. PHYSICAL CHEMISTRY. Educ: Wash State Univ, BS, 48; Ore State Univ, MS, 51; Iowa State Univ, MS, 53; Univ Ore, PhD(chem), 55. Prof Exp: Engr chemist, Phillips Petrol Co, 55-57; assoc prof chem & head dept, Ottawa Univ, 57-59; prof & head dept, Jamestown Col, 59-61; assoc prof electronics & instrumentation, Grad Inst Technol, Univ Ark, 61-66; dir chem labs, Dept Chem, 66-69, assoc chmn dept, 69-72, ASST DEAN COL ARTS & SCI, NORTHWESTERN UNIV, 72- Mem: AAAS; Instrument Soc Am. Res: Process instrumentation and control; electrode and solution kinetics. Mailing Add: Dean's Off Col of Arts & Sci Northwestern Univ Evanston IL 60201

BORCHERS, EDWARD ALAN, b Spring Lake, NJ, Feb 26, 25; m 53; c 2. PLANT BREEDING. Educ: Cornell Univ, BS, 51; Univ NC, MS, 53; Univ Calif, PhD(plant path), 57. Prof Exp: Asst hort, State Col Agr & Eng, Univ NC, 51-53; asst veg crops, Univ Calif, 53-56; plant breeder, 56-72, DIR, VA TRUCK & ORNAMENTALS RES STA, 73- Mem: Am Soc Hort Sci. Res: Breeding of vegetable crops. Mailing Add: Va Truck & Ornament Res Sta PO Box 2160 Norfolk VA 23501

BORCHERS, HAROLD ALLISON, b Chicago Heights, Ill, June 6, 35; m 60; c 3. ENTOMOLOGY. Educ: Iowa State Univ, BS, 61, MS, 64, PhD(entom, parasitol), 68. Prof Exp: Res assoc agr bee lab, Ohio State Univ, 63; instr zool & entom, Iowa State Univ, 65-68; ASSOC PROF BIOL, BEMIDJI STATE UNIV, 68- Mem: Entom Soc Am; Sigma Xi; Soc Agr Res. Res: Nest-cleaning behavior of honey bees; populations of mites symbiotic on nectrophilous beetles; diets of entomophagous fish. Mailing Add: Dept of Biol Bemidji State Univ Bemidji MN 56601

BORCHERS, RAYMOND (LESTER), b Juniata, Nebr, Apr 13, 16; m 42; c 3. BIOCHEMISTRY. Educ: Nebr State Teachers Col, BS, 38; Univ Iowa, MS, 40, PhD(biochem), 42. Prof Exp: Asst chem, Nebr State Teachers Col, 36-38; asst biochem, Univ Iowa, 38-42; instr sch med & actg head dept, Creighton Univ, 42-45; asst agr chemist, 45-49, assoc prof, 49-57, PROF BIOCHEM & NUTRIT, COL AGR, UNIV NEBR-LINCOLN, 57- Mem: Am Chem Soc; Poultry Sci Asn; Am Inst Nutrit; Am Soc Animal Sci; Soc Exp Biol & Med. Res: Amino acid metabolism and nutrition; nutrition value of soybean; unidentified nutrients; rumen metabolism. Mailing Add: 6200 Walker Ave Lincoln NE 68507

BORCHERS, ROBERT REECE, b Chicago, Ill, Apr 4, 36; m 60; c 3. NUCLEAR PHYSICS. Educ: Univ Notre Dame, BS, 58; Univ Wis, MS, 59, PhD(physics), 62. Prof Exp: From instr to assoc prof, 62-69, PROF PHYSICS, UNIV WIS-MADISON, 69-, ASSOC DEAN GRAD SCH, 70-, DIR PHYS SCI LAB, 72- Concurrent Pos: Sloan res fel, 64-66 & Inst Theoret Physics, Denmark, 64-65; Guggenheim fel, 71; consult, Laser Fusion, Lawrence Livermore Lab, 72- Honors & Awards: Sci Award, Univ Notre Dame; Kieckhofer Award, Univ Wis, 66. Mem: Fel Am Phys Soc. Res: Fast neutron spectroscopy; real time digital computation; electronics; nuclear instrumentation, especially cross sections of reactions induced by 14 mega electron volt neutrons. Mailing Add: Dept of Physics Univ of Wis Madison WI 53706

BORCHERT, JOHN ROBERT, b Chicago, Ill, Oct 24, 18; m 42; c 4. GEOGRAPHY. Educ: DePauw Univ, AB, 41; Univ Wis, MA, 46, PhD(geog), 49. Prof Exp: Instr geog, Univ Wis-Madison, 47-49; from asst prof to assoc prof, 49-56, PROF GEOG, UNIV MINN, MINNEAPOLIS, 56-, DIR CTR FOR URBAN & REGIONAL AFFAIRS, 68- Concurrent Pos: Urban res dir, Upper Midwest Econ Study, 61-64; consult, NSF, 67-69; chmn earth sci div, Nat Res Coun, 67-69; dir, Soc Sci - Res Coun; mem, Sci Adv Comt, Comt Pub Works & Transp, US House Rep, 74-75; consult, Minn State Planning Agency, 75- Honors & Awards: Van Cleef Gold Medal, Am Geog Soc, 70. Mem: Nat Acad Sci; Asn Am Geog (vpres, pres, 68-69); Am Geog Soc; Am Inst Planners. Res: Geography of land use and development in the United States. Mailing Add: Ctr for Urban & Regional Affairs Univ of Minn Minneapolis MN 55455

BORCHERT, PETER JOCHEN, b Allenstein, Ger, Feb 6, 23; m 58; c 1. ORGANIC CHEMISTRY. Educ: Univ Würzburg, BS, 48, MS, 51; Hannover Tech Univ, PhD(org chem), 53. Prof Exp: Chemist, Ind Farm Endoquimica, Sao Paulo, Brazil, 54, E F Drew, Sao Paulo, 55 & Fabrica de Pintura Montana, Caracas, Venezuela, 56-57; dir chem res lab, Marschall Div, 58-73, DIR RES & DEVELOP, SUMNER DIV, MILES LABS, INC, 73- Mem: Am Chem Soc. Res: Animal poisons; detergents; auxiliary textile products; polymers; paints; carbohydrates; organic synthesis. Mailing Add: Miles Labs Inc 1127 Myrtle St Elkhart IN 46514

BORCHERT, ROLF, b Frankfurt-Main, Ger, May 7, 33; m; c 3. PLANT PHYSIOLOGY. Educ: Univ Frankfurt, PhD(bot), 61. Prof Exp: Instr bot, Univ Frankfurt, 61-62; prof, Univ of the Andes, Colombia, 62-68; assoc prof, 68-74, PROF BOT, UNIV KANS, 74- Mem: AAAS; Am Soc Plant Physiol; Bot Soc Am; Ger Bot Soc; Scand Soc Plant Physiol. Res: Growth and development of tree tissues in vitro; tree development; control of cell division and differentiation during wound healing in tuber tissues. Mailing Add: Dept of Bot Univ of Kans Lawrence KS 66044

BORCHERTS, ROBERT H, b Mt Vernon, NY, Mar 22, 36; m 59; c 2. SOLID STATE PHYSICS. Educ: Gen Motors Inst, Eng dipl, 57; Univ Mich, MS, 59, PhD(electron spin resonance), 63. Prof Exp: Res asst radiation effects & electron spin resonance, Univ Mich, 61-63; res physicist, Inst Solid State Physics, Univ Tokyo, 63-64; RES SCIENTIST, SCI LAB, FORD MOTOR CO, 65- Mem: Am Phys Soc; AAAS; Inst Elec & Electronic Engrs. Res: Magnetic resonance; optical spectroscopy; Josephson junctions; magnetic suspension. Mailing Add: Ford Sci Lab PO Box 2053 Dearborn MI 48121

BORDA, ROBERT PAUL, b Akron, Ohio, Mar 13, 41; m 66. NEUROPHYSIOLOGY. Educ: Princeton Univ, AB, 63; Baylor Col Med, MS, 66, PhD(physiol), 69. Prof Exp: Instr, 68-71, ASST PROF PHYSIOL, BAYLOR COL MED, 71- Concurrent Pos: Consult neurophysiologist, Neurosurg Serv, Methodist Hosp, 70-, asst, Neurophysiol Serv, 71- Mem: Am EEG Soc; Am Inst Ultrasound in Med; Soc Neurosci; Int Soc Clin Electroretinography; Belg Soc Electromyog & Clin Neurophysiol. Res: Electrophysiology of vision; cerebral slow potentials; computer applications to diagnostic ultrasound. Mailing Add: Neurophysiol Dept Methodist Hosp Houston TX 77025

BORDELEAU, JEAN-MARC, b Montreal, Que, Sept 5, 24; m 49; c 2. PSYCHIATRY, PHARMACOLOGY. Educ: Col Ste Marie, BA, 45; Univ Montreal, MD, 51. Prof Exp: Psychiatrist, St Luc Hosp, Montreal, 55-58; psychiatrist, Inst Albert Prevost, Montreal, 58-61, res dir, 62-65; dir, Psychiat Ctr, Port-au-Prince, Haiti, 61-62; assoc prof, 62-73, PROF PSYCHIAT, UNIV MONTREAL, 73-; HEAD DEPT PSYCHIAT, NOTRE-DAME HOSP, 71- Concurrent Pos: Dir, Psychiat Res Inst, Joliette, 67-68; res dir psychiat, Hospital St Jean-de-Dieu, 65-71, med supt, 67-71. Mem: Fel Am Psychiat Asn; Can Psychiat Asn; Brit Med Asn. Res: Neuropsychopharmacology; clinical psychiatric research. Mailing Add: Dept of Psychiat Univ of Montreal Montreal PQ Can

BORDELON, DERRILL JOSEPH, b Cottonport, La, Aug 22, 21; m 47; c 3. MATHEMATICS, PHYSICS. Educ: La State Univ, BS, 42; Univ Md, MA, 56, PhD(math), 63. Prof Exp: Jr chem engr, Tenn Valley Auth, 42-43; instr physics, La State Univ, 43-44; sci staff asst, US Naval Ord Lab, 44-63; asst to asst secy gen sci affairs, NATO, 63-65; mem staff, US Naval Ord Lab, 65-67; HEAD ADVAN STUDY GROUP, US NAVAL UNDERWATER WEAPONS RES & ENG STA, 67- Mem: Acoustical Soc Am; Am Math Soc; Inst Math Statist. Res: Antisubmarine warfare; mathematical statistics; underwater acoustics; applied mathematics. Mailing Add: 12 Francis St Newport RI 02840

BORDEN, CRAIG W, b Springboro, Ohio, Aug 31, 15. INTERNAL MEDICINE. Educ: Oberlin Col, AB, 37; Harvard Univ, MD, 41; Am Bd Internal Med, dipl, 50. Prof Exp: Staff physician & chief blood bank sect, Vet Admin Hosp, Minneapolis, 47-51; asst prof med, Med Sch, Univ Minn, 50-53; assoc prof, 54-60, PROF MED, MED SCH, NORTHWESTERN UNIV, CHICAGO, 60- Concurrent Pos: Dir adult cardiac clin & cardiac catheterization lab, Variety Club Heart Hosp, Minn, 51-53; asst chief med serv & chief cardiovasc sect, Vet Admin Res Hosp, Chicago, 53-54, chief med serv, 54-74, sr physician, 71-; mem residency rev comt internal med, AMA Coun Med Educ, 66-71, vchmn, 70-71; secy-treas, Am Bd Internal Med, 67-68, vchmn, 68-70. Mem: AAAS; Soc Exp Biol & Med; fel Am Col Physicians; fel AMA; Am Fedn Clin Res. Res: Applied cardio-respiratory physiology. Mailing Add: Vet Admin Lakeside Hosp 333 E Huron St Chicago IL 60611

BORDEN, GEORGE WAYNE, b Shambaugh, Iowa, June 17, 37; m 59; c 2. ORGANIC CHEMISTRY, CHEMICAL ENGINEERING. Educ: Northwest Mo State Univ, BS, 59; Iowa State Univ, PhD(org chem), 63. Prof Exp: Proj scientist org chem res, Union Carbide Corp, 63-74; ASST DIR ADMIN PROCESS DEVELOP, PFIZER, INC, 74- Mem: Am Chem Soc. Res: Administration of an interdisciplinary process development group of chemists and chemical engineers involved in the development of new industrial specialty chemical products and upgrading of current manufacturing processes. Mailing Add: Cent Res Pfizer Inc Eastern Point Rd Groton CT 06340

BORDEN, KENNETH DUANE, b Floyd, NMex, May 4, 40; m 65; c 1. PHYSICAL CHEMISTRY, NUCLEAR CHEMISTRY. Educ: Eastern NMex Univ, BS, 62; Univ Ill, MS, 64; Univ Ark, PhD(nuclear chem), 68. Prof Exp: Asst prof chem, 68-71, ASSOC PROF CHEM, IND CENT COL, 71- Mem: Am Chem Soc; Sigma Xi. Res: Mass distribution from the fast neutron induced fission of heavy elements; use of radioactive tracers in reaction rate studies and reaction mechanisms. Mailing Add: Dept of Chem Ind Cent Col Indianapolis IN 46227

BORDEN, WESTON THATCHER, b New York, NY, Oct 13, 43; m 71. ORGANIC CHEMISTRY. Educ: Harvard Univ, BA, 64, MA, 66, PhD(chem), 68. Prof Exp: From instr to asst prof chem, Harvard Univ, 68-73; ASSOC PROF CHEM, UNIV WASH, 73- Mem: Am Chem Soc; The Chem Soc. Res: Theoretical organic chemistry; synthesis of molecules of theoretical interest; stereochemistry of organic reactions; reaction mechanisms. Mailing Add: Dept of Chem Univ of Wash Seattle WA 98195

BORDENCA, CARL, b Birmingham, Ala, Aug 13, 16; m 39; c 2. ORGANIC CHEMISTRY. Educ: Howard Col, BS, 36; Ga Inst Technol, MS, 38; Purdue Univ, PhD(org chem), 41. Prof Exp: Asst chem, Howard Col, 33-36, Ga Inst Technol, 36-38 & Purdue Univ, 38-41; instr org chem, Ala Polytech Inst, 41-43; res chemist, Visking Corp, 43-45 & Southern Res Inst, 45-56; asst to pres, Newport Industs, 56-57; dir res, Heyden Newport Chem Corp, 57-62; mgr res & develop, Org Chem Div, Glidden Co, 62-68, vpres, Biochem Dept, 68-73, MGR TECH LIAISON, GLIDDEN-DURKEE DIV, SCM CORP, 73- Concurrent Pos: Instr, Southern Col Pharm, 37-38. Mem: Am Chem Soc. Res: Chlorinolysis of hydrocarbons and their partially chlorinated derivatives; utilization of pine products; flavors and flavor components; coal tar by-products; pesticides. Mailing Add: Glidden-Durkee Div SCM Corp 900 Union Commerce Bldg Cleveland OH 44115

BORDER, JERRY R, animal nutrition, biochemistry, see 12th edition

BORDERS, ALVIN MARSHALL, b Mays, Ind, Mar 7, 14; m 37; c 3. CHEMISTRY. Educ: Ind Univ, AB, 35, PhD(chem), 37. Prof Exp: Res chemist, Goodyear Tire & Rubber Co, 37-41, group leader, 41-46; chief polymer res br, Off Rubber Reserve, Reconstruction Finance Corp, DC, 46; res coordr, Goodyear Tire & Rubber Co, 47; assoc, Inst Paper Chem, Lawrence Col, 48-50; group leader res, 50-52, assoc dir res, 52-58, res mgr, Reflective Prod Div, 58-62, dir, Appl Res Lab, Cent Res Labs, 62-65, dir, Chem Res Lab, 65-71 & Mat Sci & Process Res, 70-74, DIR, UNIV RELATED PROGS, CENT RES LABS, MINN MINING & MFG CO, 74- Mem: Am Chem Soc. Res: Chemistry of rubber and rubber-like materials; styrene-diene resins; paper; fluorochemicals and adhesives. Mailing Add: Minn Mining & Mfg Co Cent Res Labs PO Box 33221 St Paul MN 55133

BORDERS, CHARLES LAMONTE, JR, b Lebanon, Ky, June 13, 42; m 62; c 3. BIOCHEMISTRY. Educ: Bellarmine Col, Ky, BA, 64; Calif Inst Technol, PhD(chem), 68. Prof Exp: PROF CHEM, COL WOOSTER, 68- Mem: Am Chem Soc; Am Soc Biol Chemists. Res: Protein chemistry; chemical modification of enzymes. Mailing Add: Dept of Chem Col of Wooster Wooster OH 44691

BORDERS, DONALD B, b Logansport, Ind, Jan 12, 32; m 58; c 2. ORGANIC CHEMISTRY. Educ: Ind Univ, BS, 54, MA, 58; Univ Ill, PhD(chem), 63. Prof Exp: Res chemist, Rohm and Haas Co, 62-64; res chemist, 64-66, group leader fermentation & isolation dept, 66-75, GROUP LEADER MICROBIOL & CHEMOTHER RES DEPT, LEDERLE LABS, AM CYANAMID CO, 75- Mem: Am Chem Soc; The Chem Soc. Res: Syntheses of biologically active compounds; monomer and polymer syntheses; structure and isolation of natural products, especially antibiotics. Mailing Add: Microbiol & Chemother Res Dept Lederle Labs Am Cyanamid Co Pearl River NY 10965

BORDERS, JAMES ALAN, b Akron, Ohio, Aug 8, 41; m 69. SOLID STATE PHYSICS. Educ: Reed Col, BA, 63; Univ Ill, Urbana, MS, 65, PhD(physics), 68. Prof Exp: Res asst physics, Univ Ill, Urbana, 63-68; PHYSICIST, SANDIA LABS, 68- Mem: Am Phys Soc; Sigma Xi; Electrochem Soc. Res: Energetic ion backscattering analysis of solid surfaces and thin films; channeling location of ion-implanted impurities; formation of non-equilibrium alloys in metals by ion implantation. Mailing Add: Sandia Labs Div 5111 Albuquerque NM 87115

BORDINE, BURTON W, b Monroe, Mich, Nov 28, 34; m 61; c 5. MICROPALEONTOLOGY, PALEOBIOLOGY. Educ: Western Mich Univ, BS, 63; Brigham Young Univ, MS, 65; La State Univ, PhD(paleont), 74. Prof Exp: Geologist-micropaleontologist, Exxon Corp, 65-68; instr, Eastern Ariz Col, 68-70 & 74-75; ASST PROF, MIDDLE TENN STATE UNIV, 75- Mem: Am Asn Petrol Geologists; Soc Econ Paleontologists & Mineralogists. Res: Neogene foraminifera of northern South America; relationship of trace fossils to benthic foraminifera. Mailing Add: Dept of Geog & Earth Sci Middle Tenn State Univ Murfreesboro TN 37130

BORDLEY, JAMES, III, b Centreville, Md, Dec 7, 00; m 36; c 3. MEDICINE. Educ: Yale Univ, PhB, 23; Johns Hopkins Univ, MD, 27. Hon Degrees: ScD, Hartwick Col, 53 & Albany Med Col, 64. Prof Exp: Intern, Johns Hopkins Hosp, 27-28, resident physician, 32-34, from asst to assoc prof med, Sch Med, 28-47; clin prof, 47-75, EMER CLIN PROF, ALBANY MED COL, 75- Concurrent Pos: Nat Res Coun fel, Univ Pa, 30-32; from assoc clin prof to clin prof, Columbia Univ, 47-67; dir & physician-in-chief, Mary Imogene Bassett Hosp, NY, 47-66, emer dir, 66- Honors & Awards: Howell Award, 26. Mem: AAAS; Am Soc Clin Invest; Am Physiol Soc; Am Heart Asn; Am Clin & Climat Asn (secy, 41-51, pres, 56-57). Res: Physiology of kidney and blood vessels; arterial hypertension and nephritis. Mailing Add: 13 Main St Cooperstown NY 13326

BORDNE, ERICH FRED, b Mannheim, Ger, Apr 6, 23; nat US; m 50. GEOGRAPHY. Educ: Univ Cincinnati, BA, 48; Syracuse Univ, MA, 51, PhD(geog), 54. Prof Exp: From instr to asst prof geog, Yale Univ, 54-58; assoc prof, Univ Pittsburgh, 58-66; PROF GEOG, KENT STATE UNIV, 66- Mem: Asn Am Geog. Res: Regional water utilization; climatology; regional geography of Europe. Mailing Add: Dept of Geog Kent State Univ Kent OH 44242

BORDNER, CHARLES ALBERT, JR, b Salem, Mass, Nov 19, 37; m 60; c 2. PHYSICS. Educ: Colo Col, BS, 59; Harvard Univ, AM, 60, PhD(physics), 64. Prof Exp: Res assoc physics, Harvard Univ, 64-66; ASST PROF PHYSICS, COLO COL, 66- Concurrent Pos: Vis staff mem, Mass Inst Technol, 65-66. Mem: Am Phys Soc; AAAS; Am Asn Physics Teachers; Sigma Xi. Res: Experimental high energy physics. Mailing Add: Dept of Physics Colo Col Colorado Springs CO 80903

BORDNER, JON D B, b Massillon, Ohio, Jan 25, 40. ORGANIC CHEMISTRY, BIOCHEMISTRY. Educ: Case Inst Technol, BS, 62; Univ Calif, Berkeley, PhD(org chem), 66. Prof Exp: Fel x-ray crystallog, Calif Inst Technol, 56-69; asst prof, 69-73, ASSOC PROF CHEM, NC STATE UNIV, 73-, CHMN ORG DIV, 69- Concurrent Pos: Dreyfus teacher-scholar, 72; consult, Pfizer Inc, 74. Mem: Am Chem Soc; Am Crystallog Asn; Sigma Xi. Res: Structure and function of steroids; insect phermones; molecular neurochemistry; x-ray crystallographics. Mailing Add: Dept of Chem NC State Univ Raleigh NC 27607

BORDOLOI, KIRON, b Jorhat, India, Oct 1, 34; m 61; c 1. SOLID STATE PHYSICS. Educ: Gauhati Univ, India, BSc, 55; Univ Calcutta, MSc, 57; La State Univ, PhD(physics), 64. Prof Exp: Lectr physics, J B Col, India, 57-58; asst, La State Univ, 58-63; asst prof, Univ Southern Miss, 63-69; ASSOC PROF ELEC ENG & ENG PHYSICS, UNIV LOUISVILLE, 69- Res: Low temperature physics. Mailing Add: Dept Eng Physics Speed Sci Sch Univ of Louisville Louisville KY 40208

BORDT, DALE EMIL, b Mich, July 3, 27; m 52; c 4. MICROBIOLOGY. Educ: Albion Col, AB, 49; Mich State Univ, MS, 51, PhD(microbiol), 55. Prof Exp: Head biol prod develop, Pitman Moore Div, Dow Chem Co, 60-69; DIR BIOL RES, DIAMOND LABS, INC, 69- Mem: Am Soc Microbiol; Tissue Cult Asn; Soc Cryobiol. Res: Tissue culture methods; viral and bacterial vaccine research. Mailing Add: 1076 44th St Des Moines IA 50311

BORDWELL, FREDERICK GEORGE, b Marmarth, NDak, Jan 17, 16; m 39; c 2. ORGANIC CHEMISTRY. Educ: Univ Minn, BS, 37, PhD(org chem), 41. Prof Exp: Procter & Gamble fel, 41-42, from instr to prof, 42-74, CLARE HAMILTON HALL PROF CHEM, NORTHWESTERN UNIV, 74- Concurrent Pos: Humble lectr, 53 & 57; NSF sr fel, 57. Mem: AAAS; Am Chem Soc. Res: Mechanisms of organic reactions; chemistry of organic sulfur compounds. Mailing Add: Dept of Chem Northwestern Univ Evanston IL 60201

BOREHAM, MELVIN MURRAY, b Grass Valley, Calif, Jan 6, 37; m 59. MEDICAL ENTOMOLOGY. Educ: Calif State Univ, Sacramento, BA, 58; Calif State Univ, Fresno, MA, 65; Univ Utah, PhD(biol), 72. Prof Exp: Vector control specialist med entom, Bur Vector Control, Calif Dept Pub Health, 58-66; MED ENTOMOLOGIST, DIV SANITATION, CZ GOVT, 66- Mem: AAAS; Sigma Xi; Am Mosquito Control Asn; Entom Soc Am. Res: Mosquito taxonomy and bionomics studies; vector sampling and surveillance techniques; improved methods of applying insecticides for mosquito control; vector potentials of various mosquito species; surveys for mosquito pathogens with possible application in biological control of mosquitoes. Mailing Add: CZ Govt Med Entom Lab PO Box 92 Margarita CZ

BOREI, HANS GEORG, b Stockholm, Sweden, Feb 7, 14; m 38; c 3. INVERTEBRATE PHYSIOLOGY, ECOLOGY. Educ: Univ Stockholm, Fil Mag, 37, PhD(zool), 40, Fil Dr(biochem), 45. Prof Exp: Res asst biophys & exp biol, Univ Stockholm, 37-45, from asst prof to assoc prof, 45-52, head dept develop physiol & genetics, Wenner-Gren Inst, 47-50, from actg head dept to head dept biophys, 48-52, actg head inst, 48, 50; vis prof zool, 51, prog gen physiol, 55-60, PROF ZOOL, UNIV PA, 53- Concurrent Pos: Researcher, Carlsberg Lab, Copenhagen, Denmark, 46, Kristinebergs Zool Sta, Sweden, 46-52, 61, Millport Marine Sta, Scotland, 48, Molteno Inst, Cambridge, Eng, 48-49, Woods Hole, Mass, 51, 53, Mt Desert Island Biol Lab, Maine, 55-64 & Swans Island Marine Sta, Maine, 65- Honors & Awards: Aquist Award, Stockholm, 45, 46. Mem: Fel AAAS; Brit Biochem Soc; Am Chem Soc; Brit Soc Exp Biol; Int Soc Cell Biol. Res: Invertebrate embryology; chemical embryology; changes in cellular metabolism at fertilization; biochemistry; kinetics of enzymes; micromethods; respirometry; colorimetry; marine invertebrate ecology. Mailing Add: Joseph Leidy Lab of Biol Univ of Pa Philadelphia PA 19174

BOREK, BLANCHE ANN, b New York, NY, Oct 4, 21; m 51; c 1. BIOCHEMISTRY. Educ: Hunter Col, AB, 42; Polytech Inst Brooklyn, MS, 47; Columbia Univ, PhD(biochem), 53. Prof Exp: Instr chem, City Col New York, 47-48; res biochemist, Vet Admin Hosp, 53-74; ASST PROF SURG, STATE UNIV NY DOWNSTATE MED CTR, 62- Concurrent Pos: Asst, Sch Nursing, Columbia-Presby Med Ctr, 47-48; biochem consult. Mem: AAAS; Am Chem Soc. Res: Amino acid metabolism; brain metabolism; red cell metabolism and preservation; microbiology and microbiological determinations. Mailing Add: 9101 Shore Rd Brooklyn NY 11209

BOREK, CARMIA GANZ, b Tel Aviv, Israel, May 27, 37; US citizen; m 58; c 2. CELL BIOLOGY. Educ: Am Univ, BSc, 59; George Washington Univ, MSc, 61; Weizmann Inst Sci, PhD(cell biol, genetics), 67. Prof Exp: Res asst biochem, Sch Med, Georgetown Univ, 61-62; res asst, Sect Endocrinol, Nat Inst Arthritis & Metab Dis, 63; res assoc cell physiol, Col Physicians & Surgeons, Columbia Univ, 67-69; instr path, Sch Med, NY Univ, 69-71; ASST PROF RADIATION BIOL & PATH, COLUMBIA UNIV, 71- Concurrent Pos: Nat Cancer Inst fel physiol, Col Physicians & Surgeons, Columbia Univ, 67-69. Mem: Am Soc Cell Biol; Am Asn Cancer Res; Radiation Res Soc; Int Soc Develop Biol; Am Tissue Cult Asn. Res: Mammalian cell genetics and cancer. Mailing Add: Col of Physicians & Surgeons Columbia Univ New York NY 10032

BOREK, ERNEST, b Hungary, May 25, 11; nat US; m 38 & 51; c 2. BIOCHEMISTRY. Educ: City Col New York, BS, 33; Columbia Univ, AM, 34, PhD(biochem), 38. Prof Exp: Tutor, City Col New York, 34-38, from instr to prof chem, 38-69; PROF MICROBIOL, SCH MED, UNIV COLO, DENVER, 69- Concurrent Pos: Vis scholar, Pasteur Inst, Paris, 51; Guggenheim fel, 51-58; assoc, Columbia Univ, 55. Mem: Am Chem Soc; Am Soc Biol Chemists; Brit Biochem Soc. Res: Intermediate metabolism; bacterial metabolism; bacteriophage synthesis. Mailing Add: Dept of Microbiol Univ of Colo Sch of Med Denver CO 80220

BOREK, FELIX, b Cracow, Poland, May 5, 26; US citizen; m 57. IMMUNOCHEMISTRY, ORGANIC CHEMISTRY. Educ: Hobart Col, BSc, 50; Harvard Univ, MA, 52; Rutgers Univ, PhD(org chem), 56. Prof Exp: Res asst chem, Rutgers Univ, 52-55; res fel biochem, Col Physicians & Surgeons, Columbia Univ, 55-57; assoc biochemist, Armed Forces Inst Path, 58-63; vis scientist, Weizmann Inst, 63-65, sr scientist, 65-67; asst prof microbiol & immunol, Albert Einstein Col Med, 67-72; MANAGING ED, J IMMUNOL METHODS, 71- Concurrent Pos: Lectr, NIH, 60-63, spec res fel, 63-65. Mem: AAAS; Am Asn Immunol; Am Chem Soc; Brit Soc Immunol; Biochem Soc Israel. Res: Specificity of delayed hypersensitivity; immunosuppression and immune tolerance in adult hosts; relation between immunogenicity and the molecular size and structure of antigens; localization of labeled antigens in the host; steroid-protein conjugates as antigens; the function of the spleen in immune responses. Mailing Add: Dept of Microbiol Tel-Aviv Univ Tel-Aviv Israel

BOREL, ARMAND, b Switz, May 21, 23; m 52; c 2. MATHEMATICS. Educ: Swiss Fed Inst Technol, MS, 47; Univ Paris, PhD, 52. Prof Exp: Supplying prof algebra, Univ Geneva, 50-52; vis prof, Univ Chicago, 54-55; prof, Swiss Fed Inst Technol, 55-57; PROF ALGEBRA, INST ADVAN STUDY, 57- Concurrent Pos: Vis prof, Mass Inst Technol, 58 & Univ Paris, 64. Mem: Am Math Soc; Math Soc France; Swiss Math Soc. Res: Algebraic topology; lie groups; algebraic groups; differential geometry. Mailing Add: Inst for Advan Study Princeton NJ 08540

BORELLA, LUIS ENRIQUE, b Sept 29, 30; Can citizen; m 66; c 1. PHARMACOLOGY. Educ: Univ Buenos Aires, BSc, 50, MSc, 54; Univ Conn, MSc, 62, PhD(pharmacol), 64. Prof Exp: Pharmacist, 55-58; res asst clin chem, St Raphael's Hosp, New Haven, Conn, 58-60; Anna Bradbury Springer fel pharmacol, Univ Toronto, 64-65; SR RES ASSOC BIOCHEM & PHARMACOL, AYERST RES LABS, 65- Mem: Assoc AAAS; Can Pharmacol Soc. Res: Gastrointestinal physiology and pharmacology; drug effects of gastric acid secretion and experimental peptic ulcer; methodology in gastrointestinal experimental research, particularly gastric acid secretion; gastrointestinal drug absorption; pharmacological and biochemical effects of psychoactive drug interaction. Mailing Add: Ayerst Res Labs 1025 Laurentien Blvd St Laurent PQ Can

BOREN, ROGER BOATNER, b Potts Camp, Miss, June 1, 31; m 59; c 1. ENTOMOLOGY. Educ: Miss State Univ, BS, 53, MS, 57; Kans State Univ, PhD, 61. Prof Exp: Technologist, Shell Chem Co, 61-64, field rep, Pesticide Develop Dept, 66-69, technologist, Shell Develop Co, 64-66, supvr, Biol Sci Res Ctr, 69-73, TECH SUPPORT REP, SHELL CHEM CO, 73- Mem: Entom Soc Am; Weed Sci Soc Am. Res: Agricultural chemical screening and field development. Mailing Add: Shell Chem Co 500 First Bank & Trust B Bldg Richardson TX 75080

BORENFREUND, ELLEN, b Ger, Mar 15, 22; US citizen. BIOCHEMISTRY. Educ: Hunter Col, BS, 46; NY Univ, MS, 48, PhD(biol), 57. Prof Exp: Technician cancer res, Mt Sinai Hosp, 40-47; asst biochem, Col Physicians & Surgeons, Columbia Univ, 48-57; asst prof biochem, Sloan-Kettering Div, Grad Sch Med Sci, Cornell Univ, 61-68; from res assoc to assoc, 57-65, ASSOC MEM, SLOAN-KETTERING INST CANCER RES, 65-, ASSOC PROF BIOCHEM, SLOAN-KETTERING DIV, GRAD SCH MED SCI, CORNELL UNIV, 68- Concurrent Pos: NIH res career develop award, 63-66. Mem: Am Asn Cancer Res; Am Soc Cell Biol; Am Soc Biol Chem; Tissue Culture Asn. Res: Nucleic acid biochemistry; biochemical genetics; cellular physiology. Mailing Add: Mem Sloan-Kettering Cancer Ctr 1275 York Ave New York NY 10021

BORENSTEIN, BENJAMIN, b Jersey City, NJ, Nov 5, 28; m 52; c 2. FOOD CHEMISTRY, ORGANIC CHEMISTRY. Educ: Rutgers Univ, BS, 50, MS, 52, PhD(food sci), 54. Prof Exp: Res chemist, Fearn Foods, Inc, 54-59; res chemist, Nopco Chem Co, 59-62, head prod appl lab, 61-62; sr chemist, 62-66, MGR FOOD INDUST TECH SERV, HOFFMANN-LA ROCHE, INC, 66- Concurrent Pos: Adj prof food sci, Rutgers Univ, 73- Mem: AAAS; Am Chem Soc; Am Dairy Sci Asn; Inst Food Technol; fel Am Inst Chem. Res: Product development in food, feed and drug fields; meat and dairy chemistry; stabilization of enzymes and vitamins; nutrition; pectin chemistry and biochemistry. Mailing Add: Hoffmann-La Roche Inc Roche Chem Div Nutley NJ 07110

BORENSTEIN, SAMUEL R, b Montreal, Que, July 16, 36; m 62; c 1. PHYSICS, BIOLOGY. Educ: Imp Col, Univ London, PhD(high energy physics), 63. Prof Exp: Res assoc physics, Johns Hopkins Univ, 63-66; asst prof, Queens Col, NY, 67-68; ASST PROF PHYSICS, YORK COL, NY, 68- Mem: Am Phys Soc. Res: High energy physics; strong interactions in bubble chambers. Mailing Add: York Col 158-11 Jewell Ave Flushing NY 11365

BORENSZTAJN, DAVID ZELMAN, b Piaski, Poland, Aug 3, 14; m 51. MICROBIOLOGY. Educ: Univ Bordeaux, MD, 39; Univ Warsaw, cand, 54. Prof Exp: Adv epidemiol, Ministry Health, Poland, 47-51; sr res asst, NIH, 51-54, chief lab, 54-63; res assoc, Connaught Med Res Labs, Univ Toronto, 63-68, ASSOC PROF CLIN BIOCHEM, FAC MED, UNIV TORONTO, 68- Concurrent Pos: Consult, Nat Indust Pharmaceut Prod, 54-57. Mem: Can Soc Microbiol. Res: Biosynthesis of vitamin B12 and antibiotics, particularly tetracyclines and polypeptides; bacteriophages of bipolymyxa group. Mailing Add: 56 Alamosa Dr Willowdale ON Can

BORER, KATARINA TOMLJENOVIC, b Tuzla, Yugoslavia, Sept 17, 40; US citizen; m 64; c 3. NEUROENDOCRINOLOGY, COMPARATIVE PSYCHOLOGY. Educ: Univ Pa, BA, 62, PhD(zool), 66. Prof Exp: Lectr physiol & exp psychol, Alaska Methodist Univ, 68-69; lectr biol & comp vert anat, Anchorage Community Col, 68-69; NIH fel exp marine biol, Rosenstiel Sch Marine & Atmospheric Sci, Univ Miami, 69-70; res fel physiol psychol, 71-73, ASST RES SCIENTIST, NEUROSCI LAB, UNIV MICH, 73-, LECTR, DEPT PSYCHOL, 73- Mem: Am Soc Zoologists; Animal Behav Soc; Am Inst Biol Sci; Soc Neurosci. Res: Role of taste in control of feeding behavior; physiological mechanisms in control of octopus feeding behavior; neuroendocrine mechanisms in weight regulation and growth in rodents. Mailing Add: Neurosci Lab Univ Mich 1103 E Huron Ann Arbor MI 48104

BORG, ALFRED FRANCIS, b Pateros, Wash, Jan 13, 18; m 41; c 3. MICROBIOLOGY. Educ: Univ Wash, BS, 40, MS, 43, PhD(bact), 48. Prof Exp: Bacteriologist, Biochem Res Found, 48-49; asst prof bact, Univ Ill, 49-53; assoc prof bot, NC State Col, 53-57; prof bact & head dept, Kans State Univ, 57-67; PROG MGR, NAT SCI FOUND, 67- Concurrent Pos: Res intern, Northern Regional Lab, 52; consult, NSF, 58-66; NSF sci fac fel, 63-64. Mem: AAAS; fel Am Acad Microbiol; Am Soc Microbiol; Soc Gen Microbiol. Res: General microbiology. Mailing Add: Nat Sci Found Washington DC 20550

BORG, DONALD CECIL, b New York, NY, July 26, 26; m 49; c 3. MEDICAL RESEARCH. Educ: Harvard Univ, BS, 46, MD, 50. Prof Exp: Teaching fel med, Harvard Univ, 51-52; anal officer, Armed Forces Spec Weapons Proj, 52-54; teaching fel med, Washington Univ, 54-55; res assoc, 55-57, assoc scientist, 57-62, SCIENTIST, BROOKHAVEN NAT LAB, 62- Concurrent Pos: Consult, Armed Forces Spec Weapons Proj, 54-57; vis res assoc prof physiol, Mt Sinai Sch Med, 68-; biophysicist, Div Biol & Med, US Atomic Energy Comn, 70-72. Mem: AAAS; Soc Nuclear Med; NY Acad Sci; Am Physiol Soc; Radiation Res Soc. Res: Free radical mechanisms in biological function; electron paramagnetic resonance; health effects of energy production and use; radioactivation analysis; radiation effects; clinical research. Mailing Add: Med Res Ctr Brookhaven Nat Lab Upton NY 11973

BORG, IRIS Y P, b San Francisco, Calif, Oct 6, 28; m 50; c 1. EARTH SCIENCES, MINERALOGY. Educ: Univ Calif, Berkeley, BS, 51, PhD(geol), 54. Prof Exp: Asst geol, Princeton Univ, 54-56, vis fel, 56-59; res assoc, Univ Calif, Berkeley, 60-61; PETROLOGIST, LAWRENCE LIVERMORE LAB, UNIV CALIF, 61- Concurrent Pos: Consult, Shell Develop Co, Tex, 55-67; Geol Soc Am proj grant, 57; Guggenheim fel, 68-69; vis lectr, Am Univ of Cairo, Egypt, 71; lunar investr, NASA, 72; mem gov bd, Am Geol Inst, 75-77. Mem: Fel Geol Soc Am; Am Geophys Union; fel Mineral Soc Am. Res: Natural and artificial deformation of rocks and minerals; petrofabrics; high pressure x-ray technology; shock metamorphism; phenomenology of nuclear explosions; energy and natural resource issues. Mailing Add: Lawrence Livermore Lab Univ of Calif Livermore CA 94550

BORG, RICHARD JOHN, b Calif, Oct 18, 25; m 50; c 1. PHYSICAL CHEMISTRY. Educ: Univ Calif, BS, 50, MS, 52; Princeton Univ, PhD(chem), 57. Prof Exp: Microanalyst, Univ Calif, 51-52, res chemist, Radiation Lab, Livermore, 52-53; asst chem, Princeton Univ, 57-58, instr, 58-59; chemist, 59-75, DIV LEADER, LAWRENCE LIVERMORE LAB, UNIV CALIF, 75- Concurrent Pos: Lectr, Univ Calif, Davis, 59-65, prof, 65- Mem: Am Soc Metals; Am Phys Soc; Inst Mining, Metall & Petrol Eng; Soc Eng Educ; Am Nuclear Soc. Res: Diffusion in ionic and metallic solids; Mössbauer measurements in metals; nuclear technology and energy. Mailing Add: Lawrence Livermore Lab Univ of Calif Livermore CA 94550

BORG, ROBERT MUNSON, b Smithfield, Pa, May 19, 10; m 33; c 2. ECOLOGY. Educ: Univ Chicago, BS, 39, MS, 40. Prof Exp: Asst to staff, Harvard Univ, 29-31; asst forester, Dept Conserv, Mass, 33; tech forester, US Forest Serv, 33-35, supvry wildlife mgr, Mass, 35-36; asst dist agent, US Fish & Wildlife Serv, 40-45; fumigant engr, Dow Chem Co, Mich, 45; CHMN BD, BORG PESTICIDES, INC, 46- Mem: Soc Am Foresters. Res: Industrial and soil fumigation; agricultural chemicals; development of farm-forest ecosystems. Mailing Add: RFD Granite Ossipee NH 03864

BORGAONKAR, DIGAMBER SHANKARRAO, b Hyderabad, India, Sept 24, 32; m 63; c 2. GENETICS, CYTOLOGY. Educ: Osmania Univ, India, BScAgr, 53; Okla State Univ, PhD(genetics), 63. Prof Exp: Res asst, Indian Dept Agr, Hyderabad & Bombay, 55-59, lectr agr bot, Parbhani, 56-57; asst, Univ Minn, 59; res assoc bot, Okla State Univ, 59-63; asst prof biol, Univ NDak, 63-64; from instr to asst prof med & head chromosome lab, 64-71, ASSOC PROF MED, DIV MED GENETICS, SCH MED, JOHNS HOPKINS UNIV, 72- Concurrent Pos: Sigma Xi grant, 64; Am Philos Soc grant, 67; NSF grant, 66-68; NIMH grant, 69-73. Mem: AAAS; Am Inst Biol Sci; Am Soc Human Genetics; Am Fedn Clin Res. Res: Human cytogenetics and genetics; mammalian cytogenetics. Mailing Add: Div of Med Genetics Johns Hopkins Univ Sch of Med Baltimore MD 21205

BORGARDT, FRANK GOTTFRIED, organic chemistry, see 12th edition

BORGES, CARLOS REGO, b Sao Miguel, Port, Feb 17, 39; US citizen; m 58; c 3. MATHEMATICS. Educ: Humboldt State Col, BA, 60; Univ Wash, PhD(math), 64. Prof Exp: Asst prof math, Univ Nev, 64-65; asst prof, 65-69, ASSOC PROF MATH, UNIV CALIF, DAVIS, 69- Concurrent Pos: NSF grants, 65-68; Fulbright-Hays scholar, 71. Mem: Am Math Soc. Res: General topology. Mailing Add: Dept of Math Univ of Calif Davis CA 95616

BORGES, JOHN EDWARD, biochemistry, clinical chemistry, see 12th edition

BORGES, WAYNE HOWARD, b Cleveland, Ohio, Dec 8, 19; m 43; c 2. HEMATOLOGY. Educ: Kenyon Col, AB, 41; Western Reserve Univ, MD, 44. Prof Exp: Fel hemat, Harvard Med Sch, 51-52, instr pediat, 52; sr instr, Western Reserve Univ, 52-53, asst prof, 53-58; assoc prof, Univ Pittsburgh, 58-63; from assoc prof to prof, 63-71, THE GIVEN PROF PEDIAT, NORTHWESTERN UNIV, CHICAGO, 71-; MED DIR EDUC, CHILDREN'S MEM HOSP, 72- Concurrent Pos: Chief div hemat, Children's Mem Hosp, 63-72. Mem: Soc Pediat Res; Am Pediat Soc. Res: Pediatrics. Mailing Add: 2300 Children's Plaza Chicago IL 60614

BORGESE, THOMAS A, b New York, NY, Jan 21, 29; m 58; c 2. PHYSIOLOGY, BIOCHEMISTRY. Educ: NY Univ, BA, 50; Rutgers Univ, PhD(physiol, biochem), 59. Prof Exp: NIH fel, Harvard Med Sch, 59-61, res assoc biochem, 61-62; res assoc inst nutrit sci, Columbia Univ, 62-64, asst prof nutrit biochem, 64-69; asst prof, 69-74, ASSOC PROF BIOL SCI, HERBERT H LEHMAN COL, 74- Concurrent Pos: Assoc consult, St Luke's Hosp, New York, 62- Mem: AAAS; Am Chem Soc; Biophys Soc. Res: Cell physiology and biochemistry, especially control mechanisms in ion transport; cell metabolism; differential synthesis of hemoglobin. Mailing Add: Dept of Biol Sci Herbert H Lehman Col Bronx NY 10453

BORGESON, CARL, b Akeley, Minn, Jan 11, 07; m 34; c 2. AGRONOMY. Educ: Univ Minn, BS, 30, MS, 32. Prof Exp: From instr to assoc prof agron, Exp Sta, Univ Minn, St Paul, 35-74; RETIRED. Mem: Am Soc Agron. Res: Problems of recommendation, release, increase and distribution of foundation seedstocks of new varieties of field crops. Mailing Add: Inst of Agr Univ of Minn St Paul MN 55101

BORGESON, DAVID P, b Muskegon, Mich, Dec 6, 35; m 58; c 4. FISH BIOLOGY. Educ: Mich State Univ, BS, 58, MS, 59. Prof Exp: Fishery res biologist, Calif State Dept Fish & Game, 59-66; trout-salmon specialist, Mich Dept Conserv, 66, fisheries biologist, 66-70, IN-CHG INLAND FISHERIES SECT, MICH DEPT NATURAL RESOURCES, 70- Mem: Am Fisheries Soc; fel Am Inst Fishery Res Biologists. Res: Fisheries research, particularly trout and salmon. Mailing Add: Fish Div Mich Natural Resources S T Mason Bldg Lansing MI 48926

BORGLUM, GERALD BALTZER, b Penn Yan, NY, June 30, 33; m 54; c 4. BIOCHEMISTRY. Educ: Dana Col, BS, 53; Univ Nebr, PhD(dipeptidases), 66. Prof Exp: Teacher high sch, Nebr, 56-59; RES BIOCHEMIST, MILES LABS, 65- Mem: Am Chem Soc. Res: Characterization of industrial enzymes; development of enzyme products. Mailing Add: 60208 Robinhood Lane Elkhart IN 46514

BORGMAN, LEON EMRY, b Chickasha, Okla, Feb 16, 28; m 49; c 2. STATISTICS, GEOLOGY. Educ: Colo Sch Mines, 53; Univ Houston, MS, 59; Univ Calif, Berkeley, PhD, 63. Prof Exp: Oceanog engr, Shell Oil Co, 53-59; asst prof math, Univ Calif, Davis, 62-67, assoc prof eng geosci, Univ Calif, Berkeley, 67-70; PROF GEOL & STATIST, UNIV WYO, 70- Mem: Inst Math Statist; Am Geophys Union. Res: Statistics and mathematics applied to problems in geology, coastal engineering, hydrology, mining and various other earth science disciplines. Mailing Add: Dept of Geol Univ of Wyo Laramie WY 82070

BORGMAN, ROBERT JOHN, b Ft Madison, Iowa, Feb 4, 42; m 66; c 4. MEDICINAL CHEMISTRY. Educ: Univ Iowa, PhD(med chem), 70. Prof Exp: From asst prof to assoc prof med chem, WVa Univ, 70-75; GROUP LEADER MED CHEM, ARNAR-STONE LABS, INC, 75- Mem: Am Pharmaceut Asn; Am Chem Soc. Res: Design and synthesis of dopamine; receptor agonists in periphery and central nervous system. Mailing Add: Arnar-Stone Labs, Inc 601 E Kensington Rd Mt Prospect IL 60056

BORGMAN, ROBERT P, b Chicago, Ill, Apr 5, 30; m 57; c 3. PLANT PATHOLOGY, BOTANY. Educ: Cent Col, Iowa, BA, 57; Kans State Univ, MS, 59; Iowa State Univ, PhD(plant path), 62. Prof Exp: From asst prof to assoc prof biol, Univ Omaha, 62-69; PROF BIOL & CHMN DEPT, DIV NATURAL SCI, BUENA VISTA COL, 69- Mem: Am Phytopath Soc. Res: Fungal and vegetable crop diseases; virus diseases of cereal crop. Mailing Add: Div of Natural Sci Buena Vista Col Storm Lake IA 50588

BORGMAN, WILLIAM MARTIN, JR, b Detroit, Mich, Nov 20, 03; m 28; c 3. GEOMETRY. Educ: Univ Mich, BS, 24; Univ Chicago, MS, 31, PhD(math), 34. Prof Exp: Appraisal engr, Mich Bell Tel Co, Detroit, 24-25; instr math, Col City of Detroit, 25-33, from instr to assoc prof, 33-48, from asst dean admin to assoc dean admin, 48-74, EMER ASSOC DEAN ADMIN, WAYNE STATE UNIV, 74- Concurrent Pos: Supvr math comput, Jam Handy Orgn, 42-45. Mem: AAAS; Am Math Soc; Math Asn. Am; Am Soc Eng Educ. Res: Projective differential geometry. Mailing Add: 20114 Briarcliff Detroit MI 48221

BORGMANN, AUGUST RUSSELL, b Longmont, Colo, Dec 21, 17; m 42; c 1. TOXICOLOGY, PATHOLOGY. Educ: Colo State Univ, BS, 41; Kans State Col, MS, 42, DVM, 46, PhD(bact & path), 53. Prof Exp: Clinician, Dr Codd's Small Animal Hosp, 46, Dr Self's Small Animal Hosp, 46-47 & Dr Young's Small Animal Hosp, 47; clinician, teacher & pathologist, Kans State Col, 47-53; pathologist & toxicologist, Haskell Lab of Toxicol & Indust Hyg, 53-59; chief path & toxicol sect, Eaton Labs,

59-66; chief sect path & toxicol, 66, HEAD TOXICOL RES, ALCON LABS, INC, 66- Mem: Am Vet Med Asn. Res: Toxicity of fly sprays and base oils on cattle; studies on etiological agent of infectious atrophic rhinitis of swine; pathology and toxicology of chemicals. Mailing Add: Alcon Labs Inc 6201 S Freeway PO Box 1959 Ft Worth TX 76101

BORGNIS, FRITZ EDWARD, b Mannheim, Ger, Dec 24, 06; nat US; m 64. ELECTRONICS, BIOMEDICAL ENGINEERING. Educ: Inst Technol, Munich, MA, 29, PhD(elec physics), 36. Prof Exp: Sci asst, Inst Technol, Munich, 32-38; group leader, Res Labs, Telefunken Co, 38-40; prof physics & head dept appl physics, Graz Univ, 41-46; res assoc & lectr appl physics, Inst Technol, Zurich, 46-50; res assoc physics, Wesleyan Univ, 50, Calif Inst Technol, 51-54 & Harvard Univ, 55-57; dir res, Philips Lab, Hamburg & Aachen, Ger, 57-60; PROF HIGH FREQUENCY ELECTRONICS, SWISS FED INST TECHNOL, 60-, DIR INST HIGH FREQUENCY ELECTRONICS, 59- Concurrent Pos: Hon prof, Univ Hamburg; lectr, Inst Technol, Graz, 42-46; vis prof, Univ Innsbruck, 48-49 & Univ Calif, Berkeley, 55; chmn comn VII, Int Sci Radio Union. Mem: Fel Am Phys Soc; fel Inst Elec & Electronics Engrs; fel Acoust Soc Am; Swiss Soc Natural Sci; Ger Phys Soc. Res: Electromagnetic compatibility general electrodynamics; ultrasonics; plasma physics; medical electronics. Mailing Add: Swiss Fed Inst of Technol ETH 7 Sternwart Strasse Zurich 8006 Switzerland

BORGSTEDT, HAROLD HEINRICH, b Hamburg, Ger, Apr 21, 29; US citizen; m 57; c 2. PHARMACOLOGY, TOXICOLOGY. Educ: Univ Hamburg, MD, 56. Prof Exp: Asst, Dept Pharmacol, Univ Hamburg, 49-50; intern, Rochester Gen Hosp, 56-57; fel pharmacol & anat, 57-60, instr pharmacol, 60-63, sr instr pharmacol & res sr instr anesthesiol, 63-66, ASST PROF PHARMACOL & RES ASST PROF ANESTHESIOL, UNIV ROCHESTER, 66- Mem: AAAS; Am Soc Pharmacol & Exp Therapeut; Soc Toxicol; NY Acad Sci; Int Narcotic Enforcement Officers Asn. Res: Pharmacology and toxicology of anesthetics; toxicology of plastics; drug abuse. Mailing Add: Dept of Pharmacol Univ of Rochester Rochester NY 14642

BORGSTROM, GEORG ARNE, b Gustaf Adolf, Sweden, Apr 5, 12; m 37; c 3. FOOD SCIENCE. Educ: Univ Lund, ScD, 39. Prof Exp: Instr plant physiol, Univ Lund, 35-38, from asst prof to assoc prof bot & physiol, 38-43, head inst plant res & food storage, 42-48; head Swed Inst Food Preservation Res, 48-56; PROF FOOD SCI & HUMAN NUTRIT, MICH STATE UNIV, 56-, GEOG, 60- Concurrent Pos: Assoc prof, Chalmers Univ Technol, Sweden, 51-56; pres sci tech comn, Int Fedn Fruit Juice Prod, 51-56; mem sci bd, Int Union Nutrit Sci, 56-62; mem bd trustees, Pop Ref Bur, Washington, DC, 69-74. Honors & Awards: P Wahlberg Gold Medal, Royal Swedish Soc Anthrop & Geol, 74; Int Award, Inst Food Technologists, 75. Mem: AAAS; Inst Food Sci & Technol; World Acad Art & Sci; Am Geograph Soc; Am Inst Biol Sci. Res: Nutritional aspects of food processing; protein utilization; microbiology of frozen foods; fish processing; world food issues; food, water and energy relationships. Mailing Add: Dept Food Sci Mich State Univ East Lansing MI 48824

BORICK, PAUL MICHAEL, b Olyphant, Pa, Sept 27, 24; m; c 6. BACTERIOLOGY. Educ: Univ Scranton, BS, 47; Syracuse Univ, MS, 51, PhD(microbiol), 53. Prof Exp: Anal chemist dyes, Am Cyanamid Co, 47-48; teaching asst bact & microbiol, Syracuse Univ, 49-53, res adv dean men's staff, 51-53; head bio-lab & chg bact res, Wallace & Tiernan, Inc, 53-57, scientist in chg microbiol & bact, Bristol-Myers Co, 57-62; mgr microbiol res, Ethicon, Inc, NJ, 62-73; DIR RES & DEVELOP, MED SURG DIV, PARKE DAVIS & CO, 73- Concurrent Pos: Instr, Le Moyne Col, 52 & Columbia Univ, 71-73; consult, Environ Protection Agency, 71-74; mem, US Pharmacopoeia adv panel; indust rep, Pharm Mfrs Assoc & Health Indust Assoc. Mem: AAAS; Am Soc Microbiol; Am Chem Soc; NY Acad Sci; Soc Indust Microbiol. Res: End products of metabolism of wood-rotting Basidiomycetes; medical-surgical products; microbicides; pharmaceuticals; cosmetics, disinfection and sterilization. Mailing Add: Parke Davis & Co Greenwood SC 29646

BORIE, RICHARD D, JR, physical chemistry, see 12th edition

BORING, JOHN RUTLEDGE, III, b Gainesville, Fla, July 7, 30; m 59; c 2. MICROBIOLOGY. Educ: Univ Fla, BS, 53, MS, 55, PhD(microbiol), 61. Prof Exp: Nat Res Coun Can fel, 61-62; from asst chief to chief epidemic aid lab, Ctr Dis Control, 62-66; PROF PREV MED & ASSOC PROF MED, EMORY UNIV, 66- Concurrent Pos: Ctr for Dis Control grants, Emory Univ, 68-75. Mem: Am Soc Microbiol; Am Epidemiol Soc. Res: Host-parasite relationship between children with cystic fibrosis and the bacterium pseudomonas aeruginosa. Mailing Add: Dept of Prev Med Emory Univ Atlanta GA 30303

BORING, JOHN WAYNE, b Reidsville, NC, Oct 9, 29; m 57; c 3. PHYSICS. Educ: Univ Ky, BS, 52, MS, 54, PhD(nuclear physics), 61. Prof Exp: Sr scientist, Res Labs Eng Sci, 60-67, assoc prof eng physics, 67-71, asst dir ctr advan studies, 68-72, PROF ENG PHYSICS, UNIV VA, 71- Concurrent Pos: Consult, Int Comn Radiation Units & Measurement; vis prof, Inst of Physics, Univ Aarhus, Denmark, 74-75. Mem: Am Phys Soc; AAAS. Res: Atomic collision phenomena; atom-surface interactions. Mailing Add: Dept of Eng Sci & Systs Univ of Va Charlottesville VA 22901

BORISENOK, WALTER A, b Brooklyn, NY, Nov 10, 23; m 55; c 2. PHARMACY. Educ: Rensselaer Polytech Inst, BS, 50. Prof Exp: SR RES PHARMACIST & GROUP LEADER PARENTERAL RES & DEVELOP, STERLING-WINTHROP RES INST, 50- Mem: Am Pharmaceut Asn. Res: Parenteral pharmaceutical products. Mailing Add: Sterling-Winthrop Res Inst Rensselaer NY 12144

BORISH, IRVIN MAX, b Philadelphia, Pa, Jan 21, 13; m 36; c 1. OPTOMETRY. Educ: Northern Ill Col Optom, OD, 34, DOS, 35. Hon Degrees: LLD, Ind Univ, 68; DSc, Pa Col Optom, 75. Prof Exp: From instr to prof optom, Northern Ill Col Optom, 36-44, registr, 40-44; pvt pract, 44-72; PROF OPTOM, IND UNIV, BLOOMINGTON, 73- Concurrent Pos: Asst chief clins, Northern Ill Eye Clin, 36-41, dir clins, 41-44; lectr & vis prof optom, Ind Univ, 53-72; vpres, Ind Contact Lens Inc, 59-; vis lectr & prof var US univs & cols, 65-; consult, Nat Study Optom Educ, Nat Comn Educ, 72-73. Honors & Awards: Apollo Award, Am Optom Asn, 68. Mem: AAAS; Am Optom Asn; Am Acad Optom. Res: Contact lenses; refractive techniques, especially development of new instrumentation including use of vectographic design for binocular refractive methods of visual examination. Mailing Add: Ind Univ Sch of Optom 800 E Atwater Ave Bloomington IN 47401

BORISON, HERBERT LEON, b New York, NY, May 20, 22; m 44; c 2. PHARMACOLOGY. Educ: City Col New York, BS, 41; NY Univ, MS, 42; Columbia Univ, PhD(physiol), 48. Hon Degrees: MA, Dartmouth Col, 65. Prof Exp: Asst physiol, Col Physicians & Surgeons, Columbia Univ, 44-45, from asst instr to instr, 46-50; from instr to assoc prof pharm, Col Med, Univ Utah, 50-62; PROF PHARMACOL, DARTMOUTH MED SCH, 62- Concurrent Pos: Instr physiol, Hunter Col, 46-47; Guggenheim fel, 57-58; Rockefeller vis prof, Cali, Colombia, 67; Macy Found scholar, 74. Honors & Awards: Abel Award, 53. Mem: Am Physiol Soc; Am Soc Pharmacol & Exp Therapeut; Soc Exp Biol & Med; Am Acad Neurol; Soc Neurosci. Res: Neuropharmacology of brain stem functions. Mailing Add: Dept of Pharmacol Dartmouth Med Sch Hanover NH 03755

BORISON, SIDNEY LEE, physics, see 12th edition

BORISY, GARY GUY, b Chicago, Ill, Aug 18, 42; m 62; c 3. MOLECULAR BIOLOGY. Educ: Univ Chicago, BS, 62, PhD(biophys), 66. Prof Exp: NSF fel, 66-67; USPHS fel, 67-68; from asst prof to assoc prof, 68-75, PROF MOLECULAR BIOL & ZOOL, UNIV WIS-MADISON, 75- Concurrent Pos: NATO fel, 68. Mem: Am Soc Cell Biol. Res: Principles of biomolecular organization; assembly of macromolecules into functional structures at the cellular level, especially the mitotic spindle; cellular development. Mailing Add: Lab of Molecular Biol Univ of Wis Madison WI 53706

BORK, ALFRED MORTON, b Jacksonville, Fla, Sept, 18, 26; m 48; c 3. PHYSICS, HISTORY OF PHYSICS. Educ: Ga Inst Technol, BS, 47; Brown Univ, MS, 50, PhD(physics), 53. Prof Exp: From asst prof to prof physics, Univ Alaska, 52-63; from assoc prof to prof, Reed Col, 63-67; PROF PHYSICS, INFO & COMPUT SCI, UNIV CALIF, IRVINE, 68-, V CHMN PHYSICS DEPT, 73- Concurrent Pos: NSF fac fel, Harvard Univ, 62-63; chmn spec interest group for comput use & educ, 71, comt instrnl media, Am Asn Physics Teachers, 73- & conduit physics comt, 74- Mem: Am Asn Physics Teachers. Res: Philosophy and history of science; application of computers to teaching. Mailing Add: Dept of Physics Univ of Calif Irvine CA 92664

BORK, KENNARD BAKER, b Kalamazoo, Mich, Oct 13, 40; m 63; c 1. GEOLOGY, PALEONTOLOGY. Educ: DePauw Univ, BA, 62; Ind Univ, MA, 64, PhD(geol, paleont), 66. Prof Exp: Asst prof, 66-71, ASSOC PROF GEOL, DENISON UNIV, 71- Mem: AAAS; fel Geol Soc Am. Res: History of geology; invertebrate paleontology; biostratigraphy; sedimentary petrology. Mailing Add: 80 Beechwood Dr Granville OH 43023

BORKE, MITCHELL LOUIS, b Warsaw, Poland, Mar 23, 19; US citizen; m 49; c 2. PHARMACEUTICAL CHEMISTRY, NUCLEAR CHEMISTRY. Educ: Univ Ill, BS, 51, MS, 53, PhD, 57. Prof Exp: Asst chem, Univ Ill, 51-55, instr, 55-57; from asst prof to assoc prof, 57-64, PROF CHEM, DUQUESNE UNIV, 64- Concurrent Pos: Fulbright vis prof, Taiwan, 69-70. Honors & Awards: Elich Prize, 51. Mem: Am Pharmaceut Asn; Am Chem Soc. Res: Pharmaceutical and analytical chemistry; radiochemistry; neutron activation analysis using isotopic sources; gas-liquid chromatography; thin layer chromatography. Mailing Add: Duquesne Univ Mellon Hall Sch of Pharm Pittsburgh PA 15219

BORKON, ELI LEROY, b Chicago, Ill, Aug 11, 08; m 37; c 2. INTERNAL MEDICINE, NUCLEAR MEDICINE. Educ: Univ Chicago, BS, 31, PhD(physiol), 36, MD, 37; Am Bd Internal Med, dipl. Prof Exp: Asst physiol, Univ Chicago, 32-36; assoc prof anat, Chicago Med Sch, 37; assoc prof physiol & Health, Southern Ill Univ, 39-46; fel internal med, Wash Univ, 46; vis lectr physiol, Southern Ill Univ, 48, adj prof, 54; chmn dept med, Carbondale Clin, 54-71; clin prof med, 71-72, asst to dean, 71-74, ASST DEAN PROFESSIONAL DEVELOP, SCH MED, SOUTHERN ILL UNIV, CARBONDALE, 74-, PROF MED, 72- Concurrent Pos: Vpres, Ill State Med Soc, 75-76. Mem: AAMA; fel Am Col Physicians; Am Soc Internal Med; Soc Nuclear Med. Res: Motor activity of intestine; endocrines; effects of minerals on thyroidparathyroid; compensatory hypertrophy of remaining organs; hazards of anticoagulants. Mailing Add: 14 Pinewood Dr Carbondale IL 62901

BORKOVEC, ALEXEJ B, b Prague, Czech, Oct 17, 25; US citizen; m 51. ORGANIC CHEMISTRY, BIOCHEMISTRY. Educ: Prague Tech Univ, ChE, 49; Va Polytech Inst, MS, 54, PhD(org chem), 55. Prof Exp: Res org chem, Dow Chem Co, Tex, 55-56, sr res chemist, 56-58; asst prof org chem, Va Polytech Inst, 58-60; vis prof, Hollins Col, 60-61; res chemist, Agr Res Serv, 61-64, invests leader, 64-72, CHIEF, INSECT CHEMOSTERILANTS LAB, AGR RES SERV, USDA, 72- Honors & Awards: T L Pawlett Award, Univ Sydney, Australia, 71; Bronze Medal, 3rd Int Cong Pesticide Chem, 74. Mem: AAAS; Am Chem Soc; Entom Soc Am; Czech Soc Arts & Sci Am. Res: Carcinogenic hydrocarbons; reaction mechanism of epoxides and aziridines; synthesis and mode of action of insect chemosterilants; biochemistry of reproduction; pest control. Mailing Add: Agr Res Ctr US Dept of Agr Beltsville MD 20705

BORKOWSKI, RAYMOND P, b Kingston, Pa, Feb 5, 34; m 64; c 1. PHYSICAL CHEMISTRY. Educ: King's Col, BS, 55; Cath Univ, PhD(phys chem), 61. Prof Exp: Nat Acad Sci-Nat Res Coun res assoc photochem & radiation chem, Nat Bur Stand, 61-63; staff scientist, Aerospace Res Ctr, Gen Precision Inc, 63-67; asst prof, 67-71, ASSOC PROF CHEM, KING'S COL, PA, 71- Mem: AAAS; Am Chem Soc; Sigma Xi. Res: Chemical kinetics; photochemistry; radiation chemistry; phenomena of energy conversion and transfer. Mailing Add: 7 Salem Dr Laflin PA 18702

BORKOWSKI, WALTER LEONARD, b Camden, NJ, Sept 14, 21; m 48; c 3. ORGANIC CHEMISTRY. Educ: Univ Pa, BS, 43, MS, 48, PhD(chem), 50. Prof Exp: Chemist, Gen Aniline & Film Co, 43-44; instr inorg chem, Univ Pa, 48-50; sr chemist, Atlantic Refining Co, 50-55; asst mgr chem res, Foote Mineral Co, 55-60; proj mgr, Com Develop Dept, 60-73, SECT CHIEF ENG, SUN OIL CO, 73- Mem: Am Inst Chemists; Am Chem Soc. Res: Organometallic chemistry; high pressure polymerization of olefins; high pressure reactions; catalytic hydrogenation; lithium metal and lithium hydride chemistry; petroleum chemistry. Mailing Add: Sun Oil Co 1608 Walnut St Philadelphia PA 19103

BORLAND, JOHN RAYMOND, b Weehawken, NY, Nov 18, 14; m 37; c 3. COMMUNICATION SCIENCE. Educ: NY Univ, BS, 36, AM, 38, PhD(biol), 42. Prof Exp: Lab asst, NY Univ, 35-37, asst biol, Hofstra Col, 38-40, instr, 40-42; teacher high sch, 42-43; sr biologist, Naval Med Sch, 45-46; med ed, McGraw-Hill Book Co, 46-52, pres Physicians Publs, 52-71; DIR INFO SYSTS, IMS INT, INC, 71- Mem: Sigma Xi; Am Med Writers Asn; Nat Asn Sci Writers. Res: Scientific writing and communications. Mailing Add: IMS Int Inc 800 Third Ave New York NY 10022

BORLAUG, NORMAN ERNEST, b Cresco, Iowa, Mar 25, 14; m 40; c 2. MICROBIOLOGY. Educ: Univ Minn, BS, 37, MS, 41, PhD(plant path), 42. Hon Degrees: DSc, Punjab Agr Univ, Idia, 69; Royal Norweg Agr Col, 70, Luther Col, 70, Kanpur Univ, India, 70, Uttar Pradesh Agr Univ, India, 71; Mich State Univ, 71, Univ La Plata, Arg, 71, Univ Ariz, 72 & Univ Fla, 73; LHD, Gustavus Adolphus Col, 71; LLD, NMex State Univ, 73. Prof Exp: With US Forest Serv, 35-36, 37, 38; instr plant path, Univ Minn, 41; microbiologist, E I du Pont de Nemours & Co, Del, 42-44; res scientist in-chg wheat improv, Coop Mex Agr Prog, Mex Ministry Agr & Rockefeller Found, 44-60, assoc dir, Rockefeller Found assigned to Inter-Am Food Crop Prog, 60-63, DIR WHEAT RES & PROD PROG, INT MAIZE & WHEAT IMPROV CTR, MEX, 64- Concurrent Pos: Consult, Food & Agr Orgn UN, NAfrica & Asia, 60; consult & collabr, Nat Inst Invest Agr, Mex Ministry Agr, 60-64; mem, Citizen's Comn Sci, Law & Food Supply & Comn Critical Choices for Am, 73- Honors & Awards: Recognition Award, Agr Inst Can, 66; Int Serv Award, Am Soc Agron, 68; Meritorious Serv Award, Am Asn Cereal Chemists, 69; Nobel Prize for

Peace, 70; Recognition Award, Punjab Agr Res Inst, Pakistan, 71; Serv Award for Outstanding Contrib to Alleviation of World Hunger, Eight Latin Am Food Prod Conf, 72. Mem: Nat Acad Sci; Am Phytopath Soc; Soc Am Foresters; Am Chem Soc; hon mem Am Soc Agron. Res: Wheat breeding; agronomy; fungicides; weed killers; plant pathology; forestry. Mailing Add: CIMMYT Apartado Postal 6-641 Londres 40 Mexico 6 DF Mexico

BORLE, ANDRE BERNARD, b La Chaux-de-Fonds, Switz, May 27, 30; m 66; c 2. PHYSIOLOGY, ENDOCRINOLOGY. Educ: Univ Geneva, MD, 55. Prof Exp: Intern med, Mt Auburn Hosp, Cambridge, Mass, 56-57; res fel biochem, Harvard Med Sch, 57-59; resident, Cantonal Hosp, Geneva, Switz, 59-61; instr radiation biol, Atomic Energy Proj, Univ Rochester, 61-63; from asst prof to assoc prof, 63-74, PROF PHYSIOL, SCH MED, UNIV PITTSBURGH, 74- Concurrent Pos: Asst med, Peter Bent Brigham Hosp, Boston, 57-59; Lederle med fac award, 64-67. Honors & Awards: Andre Lichtwitz Prize, 70. Mem: AAAS; Am Physiol Soc; Endocrine Soc; Biophys Soc. Res: Cellular calcium metabolism; general physiology of calcium; mode of action of parathyroid hormone, calcitonin and vitamin D; membrane transport of calcium; calcium and phosphate metabolism. Mailing Add: Dept of Physiol Univ of Pittsburgh Sch of Med Pittsburgh PA 15213

BORM, ALFRED ERVIN, b Houston, Tex, Sept 2, 37; m 58; c 2. ALGEBRA, TOPOLOGY. Educ: Univ Tex, BSc, 58, PhD(math), 65; Univ Wash, MA, 61. Prof Exp: Radiation physicist, Col Med, Baylor Univ, 58; assoc res engr, Boeing Co, Wash, 59-62; teaching asst math, Univ Wash, 60-61 & Univ Calif, Berkeley, 61-62; spec instr, Univ Tex, 64-65; asst prof, 65-68, ASSOC PROF MATH, SOUTHWEST TEX STATE UNIV, 68- Mem: Am Math Soc. Res: Relationships between algebraic structure of set of continuous functions on a topological space and topological structure of the space; model theory. Mailing Add: Dept of Math Southwest Tex State Univ San Marcos TX 78666

BORMAN, ALECK, b Toledo, Ohio, July 13, 19; m 46; c 2. BIOCHEMISTRY. Educ: Univ Toledo, BS, 41; Univ Ill, PhD(biochem), 45. Prof Exp: Sr chemist & sect head biochem div, 45-52, head endocrinol res sect, Squibb Inst Med Res, 52-55, head endocrinol res dept, 55-62, dir physiol sect, 62-67, mgr sci personnel rels, 67, personnel mgr res & develop, 67-71, LABOR RELS MGR, E R SQUIBB & SONS, 71- Mem: AAAS; Soc Exp Biol & Med; Am Asn Cancer Res; NY Acad Sci; Endocrine Soc. Res: Protein and steroid hormones; endocrinology. Mailing Add: E R Squibb & Sons PO Box 4000 Princeton NJ 08540

BORMAN, WILLEM FREDERIK HENDRIK, organic chemistry, see 12th edition

BORMANN, FREDERICK HERBERT, b New York, NY, Mar 24, 22; m 52; c 4. ECOLOGY. Educ: Rutgers Univ, BS, 48; Duke Univ, MA, 50, PhD(bot), 52. Prof Exp: From instr to asst prof bot, Emory Univ, 52-56; from asst prof to prof, Dartmouth Col, 56-66; prof forest ecol, 66-69, OASTLER PROF FOREST ECOL & PROF BIOL, YALE UNIV, 69- Concurrent Pos: Vis scientist, Brookhaven Nat Lab, 63-64, fel Ezra Stiles Col, Yale Univ. Ecologist, Boston Univ exped, Alaska, 53. Honors & Awards: George Mercer Award, 54. Mem: Fel AAAS; Ecol Soc Am. Res: Ecology and physiology of Pinus; structure and function of root grafts; structure, function and development of forest ecosystems. Mailing Add: Sch of Forestry Yale Univ 205 Prospect St New Haven CT 06511

BORN, GORDON STUART, b Hammond, Ind, Apr 26, 33; m 57. BIONUCLEONICS, PHARMACY. Educ: Purdue Univ, BS, 55, MS, 64, PhD(bionucleonics), 66. Prof Exp: From teaching asst to teaching assoc, 62-64, from instr to assoc prof, 64-74, PROF BIONUCLEONICS, PURDUE UNIV, 74- Honors & Awards: Lederle Pharm Fac Award, Lederle Labs, 72. Mem: AAAS; Am Pharmaceut Asn; Health Physics Soc. Res: Drug and environmental toxicants and application of tracer techniques to analytical problems. Mailing Add: Bionucleonics Dept Purdue Univ West Lafayette IN 47907

BORN, HAROLD JOSEPH, b Evansville, Ind, Nov 22, 22; m 50; c 2. PHYSICS. Educ: Rose Polytech Inst, BS, 49; Iowa State Univ, MS, 58, PhD(physics), 60. Prof Exp: Equip engr, Phillips Petrol Co, Okla, 52-55; res asst, Ames Lab, AEC, 55-60; res physicist, Whirlpool Corp Res Labs, Mich, 60-61; assoc prof, 61-66, PROF PHYSICS & HEAD DEPT, ILL STATE UNIV, 66- Mem: Am Phys Soc; Am Asn Physics Teachers. Res: Low temperature thermoelectric effects; thermoelectric refrigeration; solid state physics. Mailing Add: Dept of Physics Ill State Univ Normal IL 61761

BORNE, RONALD FRANCIS, b New Orleans, La, Nov 17, 38; m 59; c 3. MEDICINAL CHEMISTRY, ORGANIC CHEMISTRY. Educ: Loyola Univ, La, BS, 60; Tulane Univ, MS, 62; Univ Kans, PhD(med chem), 67. Prof Exp: Res asst chem, Ochsner Med Found, 56-62; res chemist, C J Patterson Co, Mo, 62-63 & Mallinckrodt Chem Works, 67-68; asst prof pharmaceut chem, 68-70, assoc prof med chem, 70-72, PROF MED CHEM, UNIV MISS, 72- Concurrent Pos: NIH res grant, 75. Mem: Am Chem Soc; Am Acad Pharmaceut Sci; Int Soc Heterocyclic Chem; Am Asn Cols Pharm. Res: Medicinal chemistry, especially organic syntheses and conformational aspects of drug action. Mailing Add: Dept of Pharmaceut Chem Univ of Miss Sch of Pharm University MS 38677

BORNEMEIER, DWIGHT D, b Limon, Colo, Oct 29, 34; m 63; c 1. PHYSICS. Educ: NCent Col, Ill, BA, 56; Kans State Univ, MS, 60, PhD(physics), 65. Prof Exp: Res physicist, Naval Ord Test Sta, 56-59; instr physics, Kans State Univ, 64-65; assoc res physicist, Univ Mich, 65-69, asst prof elec eng, 68-69; VPRES, SENSORS INC, 69- Mem: Am Phys Soc; Optical Soc Am. Res: Nuclear spectroscopy; magnetism; coherent optics. Mailing Add: 1218 Van Dusen Dr Ann Arbor MI 48103

BORNMANN, JOHN ARTHUR, b Pittsburgh, Pa, May 1, 30; m 54; c 2. PHYSICAL CHEMISTRY. Educ: Carnegie Inst Technol, BS, 52; Ind Univ, PhD(phys chem), 58. Prof Exp: Res chemist, E I du Pont de Nemours & Co, 58-60; res assoc, Princeton Univ, 60-61; asst prof chem, Northern Ill Univ, 61-65; assoc prof, 65-68, PROF CHEM, LINDENWOOD COL, 68-, CHMN DEPT, 65-, CHMN DIV NATURAL SCI & MATH, 67-71, 74- Mem: AAAS; Am Chem Soc; fel Am Inst Chemists. Res: Interdisciplinary applications of physical chemistry. Mailing Add: Dept of Chem Lindenwood Col St Charles MO 63301

BORNMANN, ROBERT CLARE, b Pittsburgh, Pa, June 29, 31; m 63; c 4. MEDICINE, PHYSIOLOGY. Educ: Harvard Univ, AB, 52; Univ Pa, MD, 56, MS, 63. Prof Exp: Officer-in-chg, Cape Hallett Int Geophys Year Base, Antarctica, US Navy, 58-59, med officer, Underwater Swimmers Sch, Key West, Fla, 61-62, Deep Sea Divers Sch, Wash Navy Yard, Washington, DC, 63-65 & Exp Diving Unit, 65-68, dep asst med effects, Deep Submergence Systs Proj, 68-70, exchange officer underwater med, Royal Naval Physiol Lab, Alverstoke, Eng, 70-72, HEAD SUBMARINE & DIVING MED, NAVAL MED RES & DEVELOP COMMAND, BETHESDA, 72- Mem: AAAS; Aerospace Med Asn; Marine Technol Soc; Am Polar Soc; Undersea Med Soc. Res: Submarine, diving and industrial medicine; pathogenesis and treatment of decompression sickness and air embolism; development of decompression schedules for divers; design and development of diving equipment. Mailing Add: 11569 Woodhollow Ct Reston VA 22091

BORNONG, BERNARD JOHN, b Gilbertville, Iowa, Aug 24, 26; m 56; c 4. CHEMISTRY, MILITARY SYSTEMS. Educ: St Ambrose Col, BS, 47; Inst Textile Technol, MS, 49. Prof Exp: Res asst pediat lab, Children's Hosp, Iowa, 51-52; instr chem, St Ambrose Col, 52-56, from asst prof to assoc prof, 56-62; chemist, Rock Island Arsenal, 62-65; res chemist, US Army Weapons Command, 65-73; RES CHEMIST, GEN THOMAS J RODMAN LAB, 73- Mem: AAAS; Am Chem Soc; Am Soc Lubrication Engrs. Res: Lubrication and corrosion prevention; materials science. Mailing Add: Gen Thomas J Rodman Lab Rock Island Arsenal SARRI-LR-M Rock Island IL 61201

BORNS, HAROLD WILLIAM, JR, b Cambridge, Mass, Nov 28, 27; m 53; c 2. QUATERNARY GEOLOGY. Educ: Tufts Univ, BS, 51; Boston Univ, MA, 55, PhD(geol), 59. Prof Exp: From instr to assoc prof, 55-68, chmn dept, 71-74, PROF GEOL, UNIV MAINE, 68-, DIR, INST QUATERNARY STUDIES, 72- Concurrent Pos: Fel, Dept Geol, Yale Univ, 63-64; vis prof, Geol Inst, Univ Bergen, Norway, 75. Honors & Awards: Antarctic Serv Medal, Glaciol Soc, 61. Mem: Geol Soc Am; Glaciol Soc; Am Polar Soc. Res: Glacial geology; quaternary history of Northeast North America, Antarctica and Norway; quaternary climates; environments of early man. Mailing Add: Dept of Geol Sci Univ of Maine Orono ME 04473

BORNSIDE, GEORGE HARRY, b Wakefield, RI, Oct 8, 25; m 59; c 1. BACTERIOLOGY. Educ: Trinity Col, BS, 48; Univ Conn, MS, 50; Univ Iowa, PhD(bact), 55. Prof Exp: Instr, Univ Iowa, 55-56; assoc marine microbiol, Univ Ga, 56; microbiologist, Brooklyn Bot Garden, 56-57; from asst prof to assoc prof, 57-72, PROF SURG RES & MICROBIOL, SCH MED, LA STATE UNIV, NEW ORLEANS, 72- Mem: Am Soc Microbiol; Asn Gnotobiotics (pres, 74-75); Soc Exp Biol & Med. Res: Surgical bacteriology; bacterial virulence; antibacterial effects of hyperbaric oxygen; intestinal bacteria. Mailing Add: Dept of Surg La State Univ Sch of Med New Orleans LA 70112

BORNSTEIN, JOSEPH, b Boston, Mass, Feb 19, 25; m 54; c 3. SYNTHETIC ORGANIC CHEMISTRY. Educ: Boston Col, BS, 46; Mass Inst Technol, PhD(chem), 49. Prof Exp: Chemist, Tracerlab, Inc, 49-50; PROF CHEM, BOSTON COL, 50- Concurrent Pos: Consult, Qm, US Army. Mem: Am Chem Soc; Soc Chem Indust. Res: Synthesis of insecticides and synergists; organic fluorine compounds; heterocyclic compounds; carbon-14. Mailing Add: Dept of Chem Boston Col Chestnut Hill MA 02167

BORNSTEIN, LAWRENCE A, b New York, NY, Sept 15, 23; m 45; c 2. PHYSICS. Educ: City Col New York, BS, 44; NY Univ, MS, 51, PhD(physics), 57. Prof Exp: Tutor physics, City Col New York, 46-50; from instr to assoc prof, 50-68, assoc dean, 68-69, chmn dept, 69-73, PROF PHYSICS, NY UNIV, 68-, ASSOC CHMN DEPT, 73- Mem: AAAS; Am Phys Soc; Am Asn Physics Teachers. Res: Gaseous electronics and plasma physics. Mailing Add: Dept of Physics NY Univ Wash Sq New York NY 10003

BORNSTEIN, LEOPOLD FREY, b Krakow, Austria, Mar 27, 04; US citizen; m 27; c 1. PLASTICS CHEMISTRY. Educ: Univ Prague, PhD(chem), 27. Prof Exp: Tech dir plastics, Cable Factory, Ltd, Czech, 28-39; vpres, Synvar Corp, Del, 39-57; tech dir resins, Ga Pac Corp, 57-69; RETIRED. Concurrent Pos: Vpres & gen mgr, Nat Polytech Div, Fisons Ltd, 57-69; consult, Ga Pac Corp, 69- Mem: Am Chem Soc; Soc Plastics Indust; Tech Asn Paper & Pulp Indust; Forest Prod Res Soc; Am Inst Timber Construct. Res: Thermosetting resins and plastics for variety of industries. Mailing Add: 3324 B Northcrest Rd NE Atlanta GA 30340

BORNSTEIN, MICHAEL, b Zarki, Poland, May 2, 40; US citizen; m 67; c 2. PHYSICAL PHARMACY, ANALYTICAL CHEMISTRY. Educ: Fordham Univ, BS, 62; Univ Iowa, PhD(phys anal pharm), 66. Prof Exp: Sr pharmacist, Pitman-More Div, Dow Chem Co, 66-67; SR PHARMACEUT CHEMIST, ELI LILLY & CO, 67- Mem: Am Pharmaceut Asn; Acad Pharmaceut Sci. Res: Pilot plant operations in liquid and ointment development; color measurement; drug adjuvant interactions in solid dosage forms observed with diffuse reflectance spectroscopy; development of parenteral cephalosporids. Mailing Add: 138 Kenwood Circle Indianapolis IN 46260

BORNSTEIN, PAUL, b Antwerp, Belg, July 10, 34; US citizen; m 59; c 3. BIOCHEMISTRY, MEDICINE. Educ: Cornell Univ, BA, 54; NY Univ, MD, 58. Prof Exp: Intern surg, Yale-New Haven Med Ctr, 58-59, intern med, 59-60, resident, 60-62; Nat Arthritis & Rheumatism Found res fel immunol, Pasteur Inst, Paris, 62-63; res assoc biochem, NIH, 63-65, investr, 65-67; asst prof med, 67-69, asst prof biochem, 68-69, assoc prof, 69-73, PROF BIOCHEM & MED, SCH MED, UNIV WASH, 73- Mem: Am Soc Biol Chemists; Am Soc Clin Invest; Asn Am Physicians; Am Rheumatism Asn; Am Chem Soc. Res: Biochemistry of collagen; pathophysiology of the connective tissue diseases. Mailing Add: Dept of Biochem Univ of Wash Sch of Med Seattle WA 98195

BORNSTEIN, ROBERT D, b New York, NY, July 21, 42; m 64; c 2. METEOROLOGY. Educ: City Col New York, BS, 64; NY Univ, MS, 67, PhD(meteorol), 72. Prof Exp: Asst res scientist, NY Univ, 64-68; instr meteorol, State Univ NY Maritime Col, 68; ASST PROF METEOROL, SAN JOSE STATE UNIV, 69- Prof Exp: Consult, Ames Res Ctr, NASA, Calif, 69- Mem: Am Meteorol Soc. Res: Urban meteorology, especially temperature, wind and humidity distribution in an urban area. Mailing Add: Dept of Meteorol San Jose State Univ San Jose CA 95114

BOROFSKY, SAMUEL, b New York, NY, Mar 15, 07; m 34. MATHEMATICS. Educ: Columbia Univ, AB, 26, AM, 27, PhD(math), 31. Prof Exp: Asst math, Columbia Univ, 27-29, instr, 30-34; from instr to assoc prof, 34-53, chmn dept, 52-61, PROF MATH, BROOKLYN COL, 53- Concurrent Pos: Instr, Hunter Col, 34-36 Mem: Am Math Soc; Math Asn Am. Res: Dirichlet series; characterization of fields by a single operation. Mailing Add: Dept of Math Brooklyn Col Brooklyn NY 11210

BOROM, JOHN LEE, b Fairhope, Ala, Feb 5, 42; m 65; c 2. MARINE BIOLOGY, SCIENCE EDUCATION. Educ: William Carey Col, BS, 64; Univ Southern Miss, MS, 67, PhD(biol & sci educ), 75. Prof Exp: Instr biol, Fla Jr Col Jacksonville, 67-68, Enterprise State Jr Col, Ala, 68-69 & Mobile Col, Ala, 69-75; INSTR BIOL, JAMES H FAULKNER STATE JR COL, ALA, 75- Res: Seasonality of macroscopic fauna in estuaries. Mailing Add: PO Box 432 Fairhope AL 36532

BOROS, DOV LEWIS, b Budapest, Hungary, Mar 4, 32; US citizen; m 57; c 2. MICROBIOLOGY, IMMUNOLOGY. Educ: Hadassah Med Sch, Hebrew Univ, Jerusalem, MSc, 58, PhD(immunochem), 67. Prof Exp: Asst dir, Vaccine & Serum Inst, Jerusalem, 60-68; USPHS grant, Sch Med, Case Western Reserve Univ, 68-70, asst prof immunol, 70-74; ASSOC PROF IMMUNOL & MICROBIOL, SCH MED, WAYNE STATE UNIV, 74- Mem: Am Soc Trop Med & Hyg; Am Asn Immunologists. Res: Delayed hypersensitivity type granulomatous inflammation;

parasitic immunity. Mailing Add: Dept of Immunol & Microbiol Wayne State Univ Sch of Med Detroit MI 48202

BOROSH, ITSHAK, b Fes, Morocco, Oct 22, 38; Israel citizen; m 62; c 2. MATHEMATICS, NUMBER THEORY. Educ: Hebrew Univ, Israel, MSc, 61; Weizmann Inst Sci, Israel, PhD(math), 66. Prof Exp: Lectr, Bar-Ilan Univ, Israel, 66-70; vis lectr, Univ Ill, Urbana, 70-72; vis asst prof, 72-74, asst prof, 74-76, ASSOC PROF MATH, TEX A&M UNIV, 76- Mem: Israel Math Soc; Am Math Soc; Math Asn Am. Res: Diophantine approximations; relations between number theory and computing. Mailing Add: Dept of Math Tex A&M Univ College Station TX 77843

BOROUGHS, HOWARD, b New York, NY, Aug 21, 13; m 34. BIOCHEMISTRY. Educ: Univ Southern Calif, BA, 49; Calif Inst Technol, PhD(plant physiol), 53. Prof Exp: Nat Cancer Inst fel, Univ Paris, 52-53; fel, Univ Calif, Los Angeles, 53-54; res assoc, Hawaii Marine Lab, Univ Hawaii, 54-56, assoc prof, 56-58; chief nuclear energy prog, Inter-Am Inst Agr Sci, 58-62; staff assoc, NSF, Washington, DC, 62-66; asst dir div col support, Off Educ, 66-67; prof biol & dean facs, Portland State Univ, 67-70; assoc dean grad studies & res, Calif State Polytech Univ, San Luis Obispo, 70-74; RETIRED. Mem: Fel AAAS; Soc Gen Physiol. Res: Metabolism of radioisotopes in plants and marine animals. Mailing Add: Calif State Polytech Univ San Luis Obispo CA 93401

BOROVSKY, DOV, b Tel Aviv, Israel, Dec 4, 43; m 72. CHEMISTRY. Educ: Univ Calif, Los Angeles, BA, 67; Univ Miami, PhD(biochem), 72. Prof Exp: From res instr to res asst prof biochem, Univ Miami, 73-75; BIOCHEMIST, FLA MED ENTOM LAB, 75- Mem: AAAS; Am Chem Soc; Sigma Xi. Res: Starch and glycogen metabolism and structure; production of proteolytic enzymes and its regulation in insects; hormonal control of insects metabolism and egg development. Mailing Add: Fla Med Entom Lab Div of Health PO Box 520 Vero Beach FL 32960

BOROWIECKI, BARBARA ZAKRZEWSKA, b Warsaw, Poland, Nov 20, 24; US citizen; m 71; c 1. PHYSICAL GEOGRAPHY. Educ: Ind Univ, BS, 56, MS, 57; Univ Wis, PhD(geog), 62. Prof Exp: From instr to assoc prof, 60-69, PROF GEOG, UNIV WIS-MILWAUKEE, 69- Concurrent Pos: Vis prof geog, Ind Univ, 63-64. Mem: Asn Am Geog; Am Geog Soc; Am Asn Advan Slavic Studies. Res: Geography of Eastern Europe; geomorphology; spatial problems; valley evolution; loess terrain; numerical terrain analysis. Mailing Add: Dept of Geog Univ of Wis Milwaukee WI 53201

BOROWITZ, GRACE BURCHMAN, b New York, NY, Dec 7, 34; m 59; c 2. ORGANIC CHEMISTRY. Educ: City Col New York, BS, 56; Yale Univ, MS, 58, PhD(org chem), 60. Prof Exp: Asst, Yale Univ, 56-57; res chemist, Am Cyanamid Co, 60-62; lectr gen chem, Yeshiva Col, 67; US Dept Health, Educ & Welfare teaching fel org chem & biochem, Upsala Col, 67-68, asst prof, 68-73; asst prof, 73-75, ASSOC PROF CHEM, RAMAPO COL NJ, 75- Concurrent Pos: Sigma XI res grant, 68-71. Mem: Am Chem Soc; The Chem Soc; Sigma Xi; AAAS. Res: Mechanisms in organophosphorus chemistry, including nucleophilicities of tricovalent phosphorus compounds and lithium aluminum hydride reduction of phosphoranes; synthesis and studies of new ion-selective molecules. Mailing Add: Sch of Theoret & Appl Sci Ramapo Col of NJ Box 542 Mahwah NJ 07430

BOROWITZ, IRVING JULIUS, b Brooklyn, NY, May 15, 30; m 59; c 2. BIO-ORGANIC CHEMISTRY. Educ: City Col New York, BS, 51; Ind Univ, MA, 52; Columbia Univ, PhD(org chem), 56. Prof Exp: Fel biochem, Columbia Univ, 56-57; fel org chem, Yale Univ, 57-58; fel biochem, Columbia Univ, 58-59, res assoc, 59-60; instr, City Col New York, 60-62; from asst prof to assoc prof, Lehigh Univ, 62-66; assoc prof, 66-74, PROF CHEM, BELFER GRAD SCH SCI, YESHIVA UNIV, 74- Concurrent Pos: Fel org chem, Columbia Univ, 60-62; res grants, Sigma Xi, 61-62, Am Philos Soc, 61-64, NSF, 63-72, NIH, 64-68, 69-72 & 74, US Air Force, 65-69 & Petrol Res Fund, 66-69; grants, Res Corp, 73 & Health Res Coun, New York, 73-75; mem comt econ status, Am Chem Soc, 73-74. Mem: Am Chem Soc; NY Acad Sci. Res: Organic synthesis; selective ion-chelation; narcotic antagonists; antibiotic models. Mailing Add: Dept of Chem Yeshiva Univ New York NY 10033

BOROWITZ, JOSEPH LEO, b Columbus, Ohio, Dec 19, 32; m 63; c 2. PHARMACOLOGY. Educ: Ohio State Univ, BSc, 55; Purdue Univ, MS, 57; Northwestern Univ, PhD(pharmacol), 60. Prof Exp: Asst pharmacol, Purdue Univ, 55-57; pharmacologist, Sch Aerospace Med, Univ Tex, 60-62; fel pharmacol, Harvard Med Sch, 63-64; asst prof, Bowman Gray Sch Med, 64-69; assoc prof, 69-74, PROF PHARMACOL & TOXICOL, PURDUE UNIV, WEST LAFAYETTE, 74- Mem: Am Soc Pharmacol & Exp Therapeut. Res: Absorption of drugs from gastrointestinal tract; metabolism and release of catecholamines; release of granule bound substances; adenosine. Mailing Add: Dept of Pharmacol & Toxicol Purdue Univ West Lafayette IN 47906

BOROWITZ, SIDNEY, b New York, NY, June 12, 18; m 43; c 2. PHYSICS. Educ: City Col New York, BS, 37; NY Univ, MS, 40, PhD(physics), 48. Prof Exp: Sci aide, Nat Bur Standards, Washington, DC, 40-41; jr physicist, Navy Dept, 41-42; from asst engr to eng sect chief, Western Elec Co, NJ, 42-45; ed, Sci Publ Corp, 45-46; instr physics, NY Univ, 46-48; instr, Harvard Univ, 48-50; from asst prof to assoc prof, 50-59, PROF PHYSICS, NY UNIV, 59-, DEAN COL ARTS & SCI, 69-, ACTG PROVOST UNIV HEIGHTS, 71- Concurrent Pos: Assoc dir, Div Electromagnetics, Courant Inst Math Sci, 58-61, chmn dept physics, Univ, 61-69, actg head dept, 68-69. J F Kennedy Mem res fel, Weizmann Inst, 65-66; consult panel on optical masers, Nat Acad Sci. Mem: Fel Am Phys Soc. Res: Scattering theory; atomic and nuclear structure; many body theory. Mailing Add: University Col of Arts & Sci NY Univ New York NY 10453

BOROWSKA, ZOFIA KURYLO, b Lublin, Poland, May 13, 28. BIOCHEMISTRY. Educ: Gdansk Polytech Univ, MSc, 50, DSc(biochem), 58. Prof Exp: Asst biophys, Gdansk Polytech Univ, 50-52; asst biochem, Inst Marine Med, 52-54, adj, 54-61; res assoc, 62-67, asst prof, 67-72, ASSOC PROF BIOCHEM, ROCKEFELLER UNIV, 72- Concurrent Pos: Res fels, Inst Microbiol, Rutgers Univ, 61 & McArdle Inst Cancer Res, 62. Mem: NY Acad Sci; Am Soc Biol Chemists; Am Soc Microbiol. Res: Biosynthesis of amino acids and peptides antibiotics; biochemistry of neoplastic cells. Mailing Add: Dept of Biochem Genetics Rockefeller Univ New York NY 10021

BOROWSKY, HARRY HERBERT, b New York, NY, Apr 26, 14; m 40; c 2. CHEMISTRY. Educ: Brooklyn Col, BSc, 34. Prof Exp: Teacher pub schs, New York, 36-41 & 43-44; chemist, Onyx Oil & Chem Co, 44-57; tech dir, Intex Chem Co, 57-61 & Nuvite Chem Co, 61-69; vpres res & prod, Control Chem Corp, 69-73; CONSULT CHEMIST, 73- Mem: Am Chem Soc; Tech Asn Pulp & Paper Indust; assoc Am Dairy Sci Asn; assoc Am Pub Health Asn; Chem Specialties Mfrs Asn. Res: Detergents; paper chemistry; sanitary chemicals and germicides; paint removers; metal treating compounds; applied, pesticide and surface chemistry. Mailing Add: 1890 E 5th St Brooklyn NY 11223

BOROWSKY, RICHARD LEWIS, b New York, NY, Oct 21, 43. EVOLUTION, GENETICS. Educ: City Univ New York, BA, 64; Yale Univ, MPhil, 67, PhD(evolutionary biol), 69. Prof Exp: Asst prof biol, 70-75, ASSOC PROF BIOL, NY UNIV, 75- Res: Population genetics of fishes of the genus Xiphophorus; adaptive significance of morphological and biochemical variability in populations; relationships of behavior and evolution. Mailing Add: Dept of Biol NY Univ New York NY 10003

BORR, MITCHELL, b Poland, Feb 17, 24; Can citizen; m 49; c 3. CHEMISTRY. Educ: Univ Man, BSc, 46, MSc, 48; Purdue Univ, PhD(chem), 51. Prof Exp: Res scientist, Res Labs, Dom Rubber Co, 51-64, develop mgr, Gen Prod & Textiles Div, 64-65; SECT MGR RES & DEVELOP, UNIROYAL LTD, 66- Concurrent Pos: Chem lectr, Univ Man, 47. Mem: Am Chem Soc; Soc Automotive Eng; fel Chem Inst Can; Sigma Xi; Soc Plastics Engrs. Res: Organic and polymer chemistry; polymer physics and engineering. Mailing Add: 322 Gordon St Guelph ON Can

BORREGO, JOSEPH THOMAS, b Tampa, Fla, Sept 30, 39. MATHEMATICS. Educ: Univ Fla, BA, 61, MS, 62, PhD(math), 66. Prof Exp: Instr math, Univ Fla, 65-66; asst prof, 66-72, ASSOC PROF MATH, UNIV MASS, AMHERST, 73- Mem: Am Math Soc. Res: Topological semigroups. Mailing Add: Dept of Math Univ of Mass Amherst MA 01002

BORRELLI, NICHOLAS FRANCIS, b Philadelphia, Pa, Nov 30, 36; m 60; c 3. OPTICAL PHYSICS. Educ: Villanova Univ, BS, 58; Univ Rochester, MS, 60, PhD(chem eng), 62. Prof Exp: Sr res physicist, 62-74, RES ASSOC PHYSICS, CORNING GLASS WORKS, 74- Concurrent Pos: Lectr math, Elmira Col, NY, 69- Mem: AAAS; Am Phys Soc; Brit Soc Glass Technol. Res: Infrared and ultra-violet properties of glass; glass lasers; magneto-optics and electrooptic properties of glasses and glass-ceramics; optical properties of single crystals; photochromic materials. Mailing Add: Corning Glass Works Sullivan Res Lab Corning NY 14830

BORRELLI, ROBERT L, b Clarksburg, WVa, Mar 4, 32; m 56; c 4. APPLIED MATHEMATICS. Educ: Stanford Univ, BS, 53, MS, 54; Univ Calif, Berkeley, PhD(appl math), 63. Prof Exp: Mathematician, Gen Tel & Elec Labs, 59-62; asst prof math, US Naval Postgrad Sch, 62-63; mathematician, Philco Corp, 63-64; from asst prof to assoc prof, 64-73, PROF MATH, HARVEY MUDD COL, 73- Mem: Am Math Soc; Math Asn Am. Res: Partial differential equations; non-elliptic boundary problems. Mailing Add: Dept of Math Harvey Mudd Col Claremont CA 91711

BORRELLO, SEBASTIAN RONALD, b Syracuse, NY, Sept 9, 35; m 60; c 2. SEMICONDUCTORS. Educ: Syracuse Univ, BS, 57, MS, 62. Prof Exp: Geophysicist, US Coast & Geod Surv Antarctic Exped for 57-58 Int Geophys Year, 57-59; MEM TECH STAFF, CENT RES LAB, TEX INSTRUMENTS INC, 62- Mem: Am Phys Soc. Res: Semiconductor physics; photoconductive mercury doped germanium; indium-arsenic studies; mercury cadmium telluride detectors; surface barrier effects. Mailing Add: MS 202 Advan Technol Lab Tex Instruments Inc Box 6015 Dallas TX 75222

BORROR, ALAN L, b Ambridge, Pa, June 4, 34; m 59; c 2. ORGANIC CHEMISTRY. Educ: Drexel Inst Technol, BS, 57; Princeton Univ, PhD(org chem), 61. Prof Exp: Fel, Harvard Univ, 61-62; from asst prof to assoc prof chem, Drexel Inst Technol, 62-67; from scientist to sr scientist, 67-70, res group leader, 70-74, LAB MGR ORG CHEM, POLAROID CORP, 74- Mem: Am Chem Soc. Res: Synthesis and reactions of heterocyclic compounds; mechanisms of organic reactions. Mailing Add: Polaroid Corp 730 Main St Cambridge MA 02139

BORROR, ARTHUR CHARLES, b Columbus, Ohio, May 27, 35; m 57; c 2. PROTOZOOLOGY. Educ: Ohio State Univ, BSc, 56, MSc, 58; Fla State Univ, PhD(protozool), 61. Prof Exp: From asst prof to assoc prof, 61-74, PROF ZOOL, UNIV NH, 74- Mem: AAAS; Soc Protozool. Res: Morphology, ecology, systematics and distribution of marine ciliated Protozoa. Mailing Add: Dept of Zool Spaulding Bldg 203 Univ of New Hampshire Durham NH 03824

BORROR, DONALD JOYCE, b Columbus, Ohio, Aug 24, 07; m 31; c 1. ENTOMOLOGY, ORNITHOLOGY. Educ: Otterbein Col, BS, 28; Ohio State Univ, MS, 30, PhD(entom), 35. Prof Exp: Asst zool, 28-30, instr zool & entom, 30-42, asst prof, 42-44 & 46-47, assoc prof, 47-59, PROF ZOOL & ENTOM, OHIO STATE UNIV, 59- Concurrent Pos: Aide, Div Forest Insects, Bur Entom & Plant Quarantine, USDA, 35; instr, Audubon Camp, Maine, 38-41, 46, 48-53 & 55-62; res entomologist, Atlas Powder Co, 42. Honors & Awards: Award of Merit, N Cent Br, Entom Soc Am, 74. Mem: Cooper Ornith Soc; fel Entom Soc Am; Soc Syst Zool; Am Ornith Union; Wilson Ornith Soc. Res: Morphology, taxonomy and ecology of Odonata; recording and analysis of bird songs; insect morphology and systematics; acoustic behavior of birds. Mailing Add: Dept of Entom Ohio State Univ Columbus OH 43210

BORROWMAN, S RALPH, b Bedford, Wyo, Oct 29, 18; m 41; c 6. CHEMISTRY. Educ: Univ Idaho, BS, 41; Univ Utah, 44, PhD(chem), 50. Prof Exp: Chem engr, Deseret Chem Warfare Depot, 44; res chemist, 50-58, SUPVR RES CHEMIST, US BUR MINES, 58- Mem: Am Chem Soc. Res: Recovery of metals from low grade domestic ores by hydrometallurgical processes such as solvent extraction and ion exchange. Mailing Add: 158 W 2800 South Bountiful UT 84010

BORSA, JOSEPH, b Wakaw, Sask, Aug 5, 38; m 60; c 3. BIOPHYSICS, VIROLOGY. Educ: Univ Sask, BSc, 61, MSc, 63; Univ Toronto, PhD(biophys), 67. Prof Exp: Nat Cancer Inst Can fel, Wistar Inst Anat & Biol, Pa, 67-69; RES OFFICER, MED BIOPHYS BR, WHITESHELL NUCLEAR RES ESTAB, ATOMIC ENERGY CAN LTD, 69- Mem: Am Soc Microbiol; Can Soc Cell Biol. Res: Animal viruses; cell biology; antimetabolites; radiobiology. Mailing Add: Atomic Energy of Can Ltd Whiteshell Nuclear Res Estab Pinawa MB Can

BORSE, GAROLD JOSEPH, b Detroit, Mich, Dec 20, 40; m 63; c 1. THEORETICAL PHYSICS. Educ: Univ Detroit, BS, 62; Univ Va, MS, 64, PhD(physics), 66. Prof Exp: Asst prof, 66-71, ASSOC PROF PHYSICS, LEHIGH UNIV, 71- Mem: Am Phys Soc; Am Asn Physics Teachers. Res: Nuclear structure theory; meson-baryon bound states. Mailing Add: Dept of Physics Lehigh Univ Bethlehem PA 18015

BORSENBERGER, PAUL MICHAEL, b St Louis, Mo, Nov 2, 35; m 67; c 1. MATERIALS SCIENCE. Educ: Univ Mo, BS, 60; Stanford Univ, MS, 65, PhD(mat sci), 67. Prof Exp: Sr res chemist, 67-72, RES ASSOC, EASTMAN KODAK CO RES LABS, 72- Res: Photoelectronic processes in solids; electrography. Mailing Add: Eastman Kodak Res Lab 1669 Lake Ave Rochester NY 14650

BORSOOK, HENRY, b London, Eng, Nov 8, 97. BIOCHEMISTRY. Educ: Univ Toronto, BA, 21, MA, 22, PhD, 24, MB, 27, MD, 40. Prof Exp: Lectr biochem, Univ Toronto, 28-29; from asst prof to prof, 29-67, EMER PROF BIOCHEM, CALIF INST TECHNOL, 67-; VIS PROF MED PHYSICS, UNIV CALIF, BERKELEY, 67- Concurrent Pos: Mem food & nutrit bd & comt nutrit in indust, Nat Res Coun. Honors & Awards: Groedel Medal, Am Col Cardiol, 51. Mem: Am Soc Biol Chem; Soc Exp Biol & Med; assoc NY Acad Sci. Res: Application of thermodynamics to physiology and biochemistry; energy transfer in biological systems; nitrogen metabolism in animals and plants; nutrition; vitamins; erythropoiesis; protein synthesis; control mechanisms in erythropoiesis. Mailing Add: Space Sci Lab Univ Calif Berkeley CA 94720

BORSOS, TIBOR, b Budapest, Hungary, Mar 12, 27; US citizen; m 50; c 2. CANCER, IMMUNOLOGY. Educ: Cath Univ, BA, 54; Johns Hopkins Univ, ScD(hyg), 58. Prof Exp: Res fel microbiol, Johns Hopkins Univ, 58-60, asst prof, 60-62; res chemist, 62-66, HEAD IMMUNOCHEM SECT, NAT CANCER INST, 66-, ASSOC CHIEF BIOL BR, 71- Mem: Am Asn Immunol; Am Asn Cancer Res. Res: Tumor immunology; neoplasm of viral origin; immunochemistry; action, mechanism and characterization of complement components. Mailing Add: Immunochem Sect Biol Br Nat Cancer Inst Bethesda MD 20014

BORST, DARYLL C, b Pontiac, Mich, July 8, 40; m 65; c 1. LIMNOLOGY. Educ: Ferris State Col, BS, 62; Cent Mich Univ, MA, 64; Univ Ill, Urbana, PhD(zool), 68. Prof Exp: Chemist, Gt Lakes-Ill River Basins Proj, Fed Water Pollution Control Admin, 64; ASSOC PROF BIOL, QUINNIPIAC COL, 68- Mem: Am Soc Limnol & Oceanog; Am Micros Soc. Res: Limnological studies of reservoirs and streams in relationship to human useage; ecological studies of fresh water mussels in lakes. Mailing Add: Dept of Biol Quinnipiac Col Hamden CT 06518

BORST, LYLE BENJAMIN, b Chicago, Ill, Nov 24, 12; m 39; c 3. PHYSICS. Educ: Univ Ill, AB, 36, AM, 37; Univ Chicago, PhD(chem), 41. Prof Exp: Instr chem & gen sci, Univ Chicago, 40-41, res assoc, Metall Lab, 41-43; sr physicist, Clinton Lab, Oak Ridge, 43-46; asst prof chem, Mass Inst Technol, 46-51; prof physics, Univ Utah, 51-54; chmn dept, NY Univ, 54-61; master, Clifford Furnas Col, 68-73, PROF PHYSICS, STATE UNIV NY BUFFALO, 62- Concurrent Pos: Chmn dept reactor sci & eng, Brookhaven Nat Lab, 46-51. Mem: AAAS; fel Am Phys Soc. Res: Neutron and general nuclear physics; nuclear reactor design and development; infrared spectroscopy; liquid helium. Mailing Add: 17 Twin Bridge Lane Williamsville NY 14221

BORST, ROGER LEE, b Madison, Wis, Apr 14, 30; div; c 5. MINERALOGY. Educ: Univ Wis, BS, 56, MS, 58; Rensselaer Polytech Inst, PhD(geol), 65. Prof Exp: Assoc cur geol & mineral, NY State Mus & Sci Serv, 58-66; RES MINERALOGIST, PHILLIPS PETROL CO, 66- Mem: Fel Geol Soc Am; Mineral Soc Am; Clay Minerals Soc; Int Asn Study Clays; Sigma Xi. Res: Sedimentary petrology; clay mineralogy. Mailing Add: PO Box 793 Bartlesville OK 74003

BORST, WALTER LUDWIG, b Prague, Czech, Sept 12, 38; m 64; c 1. ATOMIC PHYSICS, ATMOSPHERIC PHYSICS. Educ: Univ Tubingen, BS, 60, MS, 64; Univ Calif, Berkeley, PhD(physics), 68. Prof Exp: Res assoc, Space Res Coord Ctr, Univ Pittsburgh, 68-69, res asst prof physics, 70; asst prof, 71-75, ASSOC PROF PHYSICS, SOUTHERN ILL UNIV, 75- Mem: Am Phys Soc; Am Geophys Union. Res: Atomic and molecular collision processes; reactions between electrons, ions and atmospheric gases; metastable spectroscopy; auroral phenomena; surface physics; solar energy research; solar heating and cooling of buildings; thermal design of buildings. Mailing Add: Dept of Physics Southern Ill Univ Carbondale IL 62901

BORSTING, JACK RAYMOND, b Portland, Ore, Jan 31, 29; m 53; c 2. MATHEMATICAL STATISTICS, OPERATIONS RESEARCH. Educ: Ore State Univ, BA, 51; Univ Ore, MA, 52, PhD(math statist), 59. Prof Exp: Instr math, Western Wash State Col, 53-54; from asst prof to assoc prof, 59-65, chmn dept opers anal, 64-71, chmn dept opers res & admin sci, 71-74, PROF MATH, US NAVAL POSTGRAD SCH, 65-, PROVOST & ACAD DEAN, 74- Concurrent Pos: Consult, Stanford Res Inst, Data Dynamics & Meteorol Int; IBM lectr, 66-69; mem Naval Res Adv Bd Personnel Labs, 71-; mem adv bd unified sci & math for elem schs, NSF Proj, 72- Mem: Inst Math Statist; Am Statist Asn; Math Asn Am; Opers Res Soc Am (pres, 75-76); Mil Opers Res Soc (pres, 71-72). Res: Statistical classification techniques; reliability. Mailing Add: Off of Provost Naval Postgrad Sch Monterey CA 93940

BORTH, RUDI, b Cologne, Ger, Oct 29, 14; m 45; c 2. ENDOCRINOLOGY, BIOMETRICS. Educ: Swiss Fed Inst Technol, dipl eng chem, 41, DSc(org chem), 46. Prof Exp: Head endocrinol lab, Univ Clin Gynec & Obstet, Univ Geneva, 47-67, lectr hormone assay, Fac Sci, 56-67, sr researcher endocrinol reprod, Dept Gynec & Obstet, 62-67; ASSOC PROF OBSTET & GYNEC, FAC MED, UNIV TORONTO, 67-; HEAD REPRODUCTIVE ENDOCRINOL RES UNIT, ST MICHAEL'S HOSP, TORONTO, 67- Concurrent Pos: Sci ed, Int Cong Gynaec & Obstet, Geneva, 54; adv bd mem, Acta Endocrinologica Cong, Geneva, 62; temporary adv, WHO, 63, 67 & staff scientist, Geneva, 72-74; corresp ed, Steroids, San Francisco, 63- Mem: Int Biomet Soc; Brit Soc Study Fertil; Can Soc Clin Invest; Swiss Soc Endocrinol; Swiss Soc Clin Chem. Res: Reproductive endocrinology; methodology and theory of hormone assay; bioengineering methods in ovulation detection. Mailing Add: Dept of Obstet & Gynec St Michael's Hosp Toronto ON Can

BORTHWICK, HARRY ALFRED, botany, deceased

BORTNER, CHARLES EUGENE, b McKeesport, Pa, Oct 21, 07; m 40; c 2. AGRONOMY. Educ: Univ Ky, BS, 30, MS, 33. Prof Exp: Asst agron, Univ Ky, 30-32 & 33-35, asst soils, 32-33, asst agronomist, 35-42, assoc agronomist, 46-48, agent & assoc agronomist, Exp Sta & Plant Sci Res Div, 48-50, AGRONOMIST, EXP STA, UNIV KY & PLANT SCI RES DIV, AGR RES SERV, USDA, 50- Mem: AAAS; Am Soc Agron; Soil Sci Soc Am. Res: Soil fertility as related to yield; quality and chemical composition of Burley tobacco. Mailing Add: Dept of Agron Univ of Ky Lexington KY 40506

BORTNICK, NEWMAN MAYER, b Minneapolis, Minn, May 14, 21; m 43; c 3. SYNTHETIC ORGANIC CHEMISTRY, ORGANIC POLYMER CHEMISTRY. Educ: Univ Minn, BA, 41, PhD(chem), 44. Prof Exp: Res chemist, Rohm and Haas Co, Bristol, 44-59, head high pressure res lab, 59-66, res supvr, 66-73, DIR PIONEERING PROCESS RES, ROHM AND HAAS CO, SPRING HOUSE, 73- Mem: Fel AAAS; Am Chem Soc; The Chem Soc. Res: Synthetic organic chemistry; polymerization. Mailing Add: 509 Oreland Mill Rd Oreland PA 19075

BORTOFF, ALEXANDER, b Cleveland, Ohio, Sept 13, 32; m 60; c 2. PHYSIOLOGY. Educ: Western Reserve Univ, BS, 53; WVa Univ, MS, 56; Univ Ill, PhD(physiol), 59. Prof Exp: Fel, Univ Ill, 59-60; instr ophthalmic res, Western Reserve Univ, 60-62, instr physiol, 61-62; from asst prof to assoc prof, 62-70, PROF PHYSIOL, STATE UNIV NY UPSTATE MED CTR, 70- Mem: AAAS; Am Physiol Soc. Res: Electrophysiology of smooth muscle. Mailing Add: Dept of Physiol State Univ of NY Upstate Med Ctr Syracuse NY 13210

BORTON, ANTHONY, b Bryn Mawr, Pa, June 6, 33; m 57; c 2. ANIMAL SCIENCE. Educ: Haverford Col, AB, 55; Mich State Univ, MS, 61, PhD(animal sci), 64. Prof Exp: Asst prof, 64-70, ASSOC PROF ANIMAL SCI, UNIV MASS, AMHERST, 70- Mem: Am Soc Animal Sci. Res: Reproduction in domestic animals; equine nutrition and physiology. Mailing Add: Dept of Vet & Animal Sci Univ of Mass Amherst MA 01002

BORTREE, ALFRED LEE, b Peterboro, NH, Oct 13, 16; m 43; c 2. VETERINARY SCIENCE. Educ: Pa State Univ, BS, 39; Mich State Univ, MS, 41, DVM, 44. Prof Exp: From instr to asst prof bact, Mich State Univ, 44-48; head dept, 53-74, PROF VET SCI, PA STATE UNIV, 48- Concurrent Pos: NSF sci fac fel, Univ Calif, Davis, 60-61. Mem: Am Vet Med Asn; Am Dairy Sci Asn; Conf Res Workers Animal Dis. Res: Veterinary physiology and diseases of cattle, especially mastitis. Mailing Add: Dept of Vet Sci Pa State Univ University Park PA 16802

BORTS, ROBERT BENJAMIN, physics, see 12th edition

BORUCKI, WILLIAM JOSEPH, b Chicago, Ill, Jan 26, 39; m 63; c 3. ATMOSPHERIC PHYSICS. Educ: Univ Wis-Madison, BS, 60, MS, 62. Prof Exp: Res scientist spectros, 62-72, RES SCIENTIST ATMOSPHERIC PHYSICS, NASA-AMES RES CTR, 72- Mem: Am Geophys Union. Res: Modeling of the earth's stratosphere and its perturbations caused by anthropogenic sources. Mailing Add: NASA-Ames Res Ctr Moffett Field CA 94035

BORUM, OLIN H, b Spencer, NC, Nov 3, 17; m 44; c 3. RESEARCH ADMINISTRATION, APPLIED CHEMISTRY. Educ: Univ NC, BS, 38, AM, 47, PhD(chem), 49. Prof Exp: Asst anal chem, Univ NC, 46-49; res chemist, Philadelphia Lab, E I du Pont de Nemours & Co, 49-50; interim res asst prof, Cancer Res Lab, Univ Fla, 50; res adminr chem div, Off Sci Res Hq, Air Res & Develop Command, 51-52; from instr to asst prof chem, US Mil Acad, 52-55; student, US Air Force Air Command & Staff Col, 55-56; res adminr res div, Chem Corps Res & Develop Command, DC, 56-60 & res & tech div, Wright-Patterson AFB, Ohio, 60-64; SCI ADMINR, HQ, ARMY MATERIEL COMMAND, RES, DEVELOP & ENG DIRECTORATE, 64- Honors & Awards: Dept of Army Cert of Achievement, Hq, Army Materiel Command, 71. Mem: Am Chem Soc; fel Am Inst Chemists. Res: Formation of acylamido ketones from amino acids; conversion of acylamido ketones to oxazoles and thiazoles; reactions of amino acids; diene synthesis with maleic anhydride and maleimide; synthesis of dioxaspiroheptane; chemistry of chemical warfare agents; materials science; explosives science. Mailing Add: 9002 Volunteer Dr Alexandria VA 22309

BORUN, THADDEUS W, b Milwaukee, Wis, June 10, 39; m 66; c 1. BIOCHEMISTRY. Educ: Univ Chicago, BS, 61, PhD(biochem), 65. Prof Exp: USPHS trainee, Albert Einstein Col Med, 65-67; asst prof biol sci, Columbia Univ, 67-69; asst prof zool, Univ Mysore, India, 69-70; asst prof biochem, Fels Res Inst, Med Sch, Temple Univ, 70-75; ASSOC MEM, WISTAR INST, 75- Mem: Am Chem Soc; Tissue Cult Asn; Am Soc Biol Chemists. Res: Histone messenger RNA; histone Fl phosphorylation, control of cell differentiation and proliferation; tissue culture. Mailing Add: Wistar Inst 36th & Spruce St Philadelphia PA 19104

BORWEIN, DAVID, b Kaunas, Lithuania, Mar 24, 24; m 46; c 3. MATHEMATICS. Educ: Univ Wiwatersrand, BSc, 45, BSc, 48; Univ London, PhD(math), 50, DSc(math), 60. Prof Exp: Lectr math, Univ St Andrews, 50-63; vis prof, 63-64, PROF MATH, UNIV WESTERN ONT, 64-, HEAD DEPT, 67- Concurrent Pos: Chmn res comt, Can Math Cong, 70-73. Mem: Am Math Soc; Can Math Cong (vpres, 73-75); London Math Soc; fel Royal Soc Edinburgh. Res: Theory of summability of series and integrals. Mailing Add: Dept of Math Univ of Western Ont London ON Can

BORYSENKO, MYRIN, b Berezhani, Ukraine, June 3, 42; US citizen; m 70; c 3. IMMUNOBIOLOGY. Educ: St Lawrence Univ, BS, 64; State Univ NY Upstate Med Ctr, PhD(anat), 68; Univ Calif, Los Angeles, cert immunol, 70. Prof Exp: NIH fel, Univ Calif, Los Angeles, 68-70; asst prof, 70-76, ASSOC PROF ANAT, SCH MED, TUFTS UNIV, 76- Mem: Am Asn Anatomists; Transplantation Soc; Am Soc Zoologists. Res: Phylogenetic and developmental aspects of the immune response; lymphoid organs, T and B cells, kinetics of cellular and humoral responses to antigenic stimulation. Mailing Add: Dept of Anat Sch of Med Tufts Univ Boston MA 02111

BORYSKO, EMIL, b Scranton, Pa, Sept 24, 18; m 54; c 3. BIOLOGY. Educ: Brooklyn Col, BA, 40; George Washington Univ, MA, 50; Johns Hopkins Univ, PhD(biol), 55. Prof Exp: Supvr, Optical Glass Plant, Nat Bur Stand, 41-46, fiber technologist, 47-50; instr, Johns Hopkins Univ, 55-56; res assoc, NY Univ, 56-58; PRIN RES SCIENTIST, ETHICON, INC, NJ, 58- Mem: Electron Micros Soc Am; NY Acad Sci; Am Soc Cell Biol. Res: Comparative optical and electron microscopic studies of the structure of cells and tissues; dynamic aspects of cellular growth and division; connective tissues and collagen; thermal microscopy of organic polymers; scanning electron microsocopy of biomedical materials. Mailing Add: 211 Love Rd Bridgewater NJ 08807

BORZELLECA, JOSEPH FRANCIS, b Norristown, Pa, Oct 3, 30; m 55; c 6. PHARMACOLOGY, TOXICOLOGY. Educ: St Joseph's Col, Pa, BS, 52; Jefferson Med Col, MS, 54, PhD(pharmacol), 56. Prof Exp: Res asst, Trudeau Found, Jefferson Med Col, 54-55; instr pharmacol, Woman's Med Col Pa, 56-57, assoc, 57-59; from asst prof to assoc prof, 59-67, PROF PHARMACOL, MED COL VA, VA COMMONWEALTH UNIV, 67- Mem: AAAS; Am Chem Soc; Am Soc Pharmacol & Exp Therapeut; Soc Exp Biol & Med; Soc Toxicol (pres). Res: Drug absorption, distribution and metabolism; toxicology of substances of economic importance. Mailing Add: Dept of Pharmacol Med Col of Va Richmond VA 23298

BOS, JANE, b Evanston, Ill, Aug 22, 50. NEUROANATOMY. Educ: Hope Col, BA, 72; Univ Ill Med Ctr, PhD(anat), 75. Prof Exp: FEL ULTRASTRUCTURE NEUROANAT, YERKES REGIONAL PRIMATE RES CTR, EMORY UNIV, 75- Mem: Soc Neurosci. Res: Anatomical investigation of the visual system of primates. Mailing Add: Yerkes Regional Primate Res Ctr Emory Univ Atlanta GA 30322

BOS, WILLIAM G, b Chicago, Ill, Jan 18, 37; m 58; c 4. INORGANIC CHEMISTRY, SOLID STATE CHEMISTRY. Educ: Calvin Col, AB, 58; Wayne State Univ, PhD(inorg chem), 63. Prof Exp: Res assoc phys chem, Univ Ill, 63-64; from asst prof to assoc prof, 64-72, PROF INORG CHEM, UNIV LOUISVILLE, 72-, ASSOC DEAN, COL ARTS & SCI, 74- Mem: Am Chem Soc. Res: Physical properties and structures of inorganic solids; transition metal hydrides; magnetic resonance of solids; rare earth chelate complexes. Mailing Add: Off of the Dean Univ of Louisville Louisville KY 40208

BOSARGE, W EDWIN, JR, control systems, numerical analysis, see 12th edition

BOSART, LANCE FRANK, b New York, NY, Aug 24, 42; m 69. DYNAMIC METEOROLOGY. Educ: Mass Inst Technol, BS, 64, MS, 66, PhD(meteorol), 69. Prof Exp: Res asst meteorol, Mass Inst Technol, 65-69; ASST PROF ATMOSPHERIC SCI, STATE UNIV NY ALBANY, 69- Concurrent Pos: Res Found NY fac res grant, 70-71; NSF grant, 70-72. Mem: Am Meteorol Soc; Royal Meteorol Soc. Res: Midtropospheric frontogenesis; mesoscale-synoptic scale interaction and winter interaction of midlatitude and tropical circulations; synoptic meteorology. Mailing Add: Dept of Meteorol State Univ of NY Albany NY 12203

BOSCH, ANTHONY, organic chemistry, see 12th edition

BOSCH, ARTHUR JAMES, b Luverne, Minn, Nov 15, 28; m 52; c 4. BIOCHEMISTRY. Educ: Cent Col, Iowa, BA, 51; Univ Wis, MS, 52, PhD(biochem), 58. Prof Exp: From instr to assoc prof, 58-69, PROF CHEM, CENT COL, IOWA, 69- Mem: AAAS; Am Chem Soc. Res: Organic, biological and general chemistry. Mailing Add: Dept of Chem Cent Col Pella IA 50219

BOSCH, WARREN LUTHER, b New York, NY, May 1, 40; m 62. ORGANIC CHEMISTRY, MATHEMATICS. Educ: Wittenberg Univ, BS, 61; Univ Kans, PhD(chem), 67. Prof Exp: Asst prof chem, Mary Baldwin Col, 67-71; asst prof chem & actg chmn dept, Col Steubenville, 71-74; ASST PROF CHEM, FLA INST TECHNOL, 74- Mem: Am Chem Soc. Res: Syntheses and reactions of cyclopropyl alcohols and cyclopropyl esters; metallo-organics. Mailing Add: Dept of Chem Fla Inst of Technol Melbourne FL 32901

BOSCHAN, ROBERT HERSCHEL, b Los Angeles, Calif, Oct 12, 25; m 60; c 2. ORGANIC CHEMISTRY, POLYMER CHEMISTRY. Educ: Univ Calif, Los Angeles, BS, 47, PhD(chem), 50. Prof Exp: Eli Lilly res fel, 50-51; res chemist synthetic med res, Merck & Co, Inc, 51-52; org chemist, Chem Div, Res Dept, US Naval Weapons Ctr, 52-59; res & develop specialist, Mat Methods Res & Eng Div, McDonnell Douglas Corp, 59-69; mgr chem instrumentation sect, Appl Sci Div, Analog Technol Corp, 69-71; HEAD CHEM SECT, HUGHES AIRCRAFT CO, 71- Concurrent Pos: Instr, Phys Sci Exten, Univ Calif, Los Angeles, 56- Mem: Am Chem Soc; The Chem Soc. Res: Organic reaction mechanisms; organic synthesis; kinetics; organic fluorine chemistry; chemistry based instrumentation; gas chromatography; mass spectroscopy. Mailing Add: 2012 Midvale Ave Los Angeles CA 90025

BOSCHERT, ULI (ULRICH), chemistry, see 12th edition

BOSCHMANN, ERWIN, b Fernheim, Paraguay, Jan 1, 39; m 62; c 2. BIOINORGANIC CHEMISTRY. Educ: Bethel Col, BS, 63; Univ Colo, MS, 65, PhD(metal chelates), 68. Prof Exp: Asst prof, 68-72, ASSOC PROF INORG CHEM, IND UNIV, INDIANAPOLIS, 72- Concurrent Pos: Vis prof, Agrarian Univ, Peru, 68-69, actg chief of party, Ford Found Proj, 69-70. Mem: Am Chem Soc. Res: Preparation and study of the chemistry of coordination compounds, especially metal chelates; preparation of new chelating agents; sulfoxide ligands. Mailing Add: Dept of Chem Ind Univ Indianapolis IN 46202

BOSCHUNG, HERBERT THEODORE, b Birmingham, Ala, July 16, 25; m 51; c 4. ICHTHYOLOGY. Educ: Univ Ala, BS, 48, MS, 49, PhD(zool, bot), 57. Prof Exp: Instr, 50-57, assoc prof, 57-66, PROF BIOL, UNIV ALA, 66-; DIR, ALA MUS NATURAL HIST, 66- Mem: Am Soc Ichthyologists & Herpetologists. Res: Fishes of Gulf of Mexico; freshwater fishes of southeastern United States. Mailing Add: Box 1927 Univ of Ala University AL 35486

BOSE, AJAY KUMAR, b Silchar, India, Feb 12, 25; m 50; c 6. ORGANIC CHEMISTRY. Educ: Univ Allahabad, India, BSc, 44, MSc, 46, Mass Inst Technol, ScD, 50. Hon Degrees: MEng, Stevens Inst Technol, 63. Prof Exp: Res fel chem, Harvard Univ, 50-51; lectr & asst prof, Indian Inst Technol, 51-56; res assoc, Univ Pa, 56-57; res chemist, Upjohn Co, 57-59; assoc prof, 59-61, PROF CHEM, STEVENS INST TECHNOL, 61- Concurrent Pos: Consult various companies. Honors & Awards: Jubilee Gold Medal, Univ Allahabad, 46; First Prize, Sci Essay Contest, Indian Sci News Asn, 56; Meghnad Saha Mem Prize, 57; Spec Commonwealth Prize, Sci Essay Contest, Eng, 57; Ottens Res Award, Stevens Inst Technol, 68. Mem: Am Chem Soc; Indian Chem Soc. Res: Stereochemistry; natural products; synthetic organic chemistry; biogenesis; popular science writing. Mailing Add: Dept of Chem & Chem Eng. Stevens Inst of Technol Hoboken NJ 07030

BOSE, ANIL KUMAR, b Calcutta, India, Apr 1, 29; m 59; c 1. PURE MATHEMATICS. Educ: Univ Calcutta, BS, 48, MS, 56; Univ NC, PhD(math), 64. Prof Exp: Lectr math, Goenka Col, Calcutta, 58-60; part-time instr, Univ NC, 60-63; vis prof, St Augustine's Col, 63-64; asst prof, Univ Ala, 64-68; ASSOC PROF MATH, CLEMSON UNIV, 68- Mem: Am Math Soc. Res: Differential equations on the properties of mean-value of the solutions of certain class of elliptic equations; mean-values are considered by taking various non-negative weight functions. Mailing Add: Dept of Math Clemson Univ Clemson SC 29631

BOSE, HENRY ROBERT, JR, b Chicago, Ill, Sept 20, 40. MICROBIOLOGY. Educ: Elmhurst Col, BS, 62; Univ Ind, Indianapolis, MS, 65, PhD(microbiol), 67. Prof Exp: NSF fel, 67-69, asst prof, 69-74, PROF MICROBIOL, UNIV TEX, AUSTIN, 74- Res: Replication of ribonucleic acid enveloped animal viruses; role of host cell membranes in the replication of animal viruses. Mailing Add: Dept of Microbiol Exp Sci 414 Univ of Tex Austin TX 78712

BOSE, RAJ CHANDRA, b Hoshangabad, India, June 19, 01; nat US; m 32; c 2. STATISTICS. Educ: Hindu Col, BA, 22, MS, 24; Univ Calcutta, DLitt, 47. Prof Exp: Lectr math, Asutosh Col, India, 30-34; statistician, Statist Inst, 34-40; lectr math, Postgrad Dept, Univ Calcutta, 40-41, lectr statist, 41-45, head dept, 45-49; prof, Univ NC, Chapel Hill, 49-66, Kenan prof, 66-71; PROF STATIST, COLO STATE UNIV, 71- Concurrent Pos: Lectr, Univ Calcutta, 38-40; statistician, Statist Inst, 42-49; vis prof, Columbia Univ, 47, Univ NC, 48, Case Inst Technol, 59 & Univ Geneva, 62; pres, Statist Sect, Indian Sci Cong, 47. Mem: Nat Acad Sci; Biomet Soc; fel Inst Math Statist (pres-elect, 70-71, pres, 71-); Int Statist Inst; fel Royal Statist Soc. Res: Design of experiments; multivariate analysis; mathematical statistics; combinatorics; non-Euclidean geometry; number theory; modern algebra; finite geometry; graph theory; Imgrossen problems of differential geometry; information theory; error correcting codes. Mailing Add: Dept of Math & Statist Colo State Univ Ft Collins CO 80521

BOSE, SAMIR K, b Decca, EPakistan, May 1, 34; m 61; c 3. THEORETICAL PHYSICS. Educ: Univ Delhi, BS, 56, MS, 58; Univ Rochester, PhD(physics), 62. Prof Exp: Mem, Sch Math, Inst Advan Study, 62-63; fel physics, Tata Inst Fundamental Res, India, 63-64; reader, Univ Delhi, 64-66; consult, Int Ctr Theoret Physics, UNESCO, Italy, 66; asst prof, 66-68, ASSOC PROF PHYSICS, UNIV NOTRE DAME, 68- Res: Theoretical research in elementary particles and their strong and weak interactions; application of group theory and algebraic techniques to the same. Mailing Add: Dept of Physics Univ of Notre Dame Notre Dame IN 46556

BOSE, SHYAMALENDU M, b Dacca, India, Aug 17, 39; m 63; c 2. SOLID STATE PHYSICS, METAL PHYSICS. Educ: Calcutta Univ, BS, 58, MS, 60; Univ Md, College Park, PhD(physics), 68. Prof Exp: Lectr physics, Midnapore Col, 60-61; fel, Cath Univ Am, 67-70; ASST PROF PHYSICS, DREXEL UNIV, 70- Concurrent Pos: Res assoc, Cath Univ Am, 70-71. Mem: Am Phys Soc. Res: Many-body problem; optical properties of metals; dilute magnetic alloys; chemisorption by metals; electronic properties of disordered solids; surface properties of solids. Mailing Add: Dept Phys & Atmos Sci Drexel Univ Philadelphia PA 19104

BOSE, SUBIR KUMAR, b Calcutta, India, Jan 1, 39; m 65; c 1. THEORETICAL PHYSICS, STATISTICAL MECHANICS. Educ: Bihar Univ, BSc, 58; Patna Univ, MSc, 60; Allahabad Univ, PhD(physics), 67. Prof Exp: Res fel physics, Univ Delhi, 66-67; res fel, St Louis Univ, 67-68; asst prof, 68-72, ASSOC PROF PHYSICS, SOUTHERN ILL UNIV, CARBONDALE, 72- Mem: Am Phys Soc; Sigma Xi. Mailing Add: Dept of Physics & Astron Southern Ill Univ Carbondale IL 62901

BOSE, SUBIR KUMAR, b Gaya, India, Sept 3, 31; US citizen; m 61. VIROLOGY, MOLECULAR Lucknow, BSc, 50, MSc, 52; Washington Univ, PhD(molecular biol), 63. Prof Exp: Lectr bot, Univ Lucknow, 52-58; Am Cancer Soc res assoc biochem, Med Sch, Univ Mich, 63-65; from asst prof to assoc prof, 65-75, PROF MICROBIOL, SCH MED, ST LOUIS UNIV, 76- Concurrent Pos: NIH res career develop award, 67-76; vis investr, Sch Med, Stanford Univ, 69. Mem: AAAS; Am Soc Biol Chemists; Am Soc Microbiol; Brit Soc Gen Microbiol; Biophys Soc. Res: Cell growth control; transformation by oncogenic viruses; regulation of gene activity; biochemistry of cancer. Mailing Add: Dept of Microbiol St Louis Univ Sch of Med St Louis MO 63104

BOSEE, ROLAND ANDREW, b Hagerstown, Md, Feb 12, 10; m 41; c 2. BIOCHEMISTRY. Educ: Temple Univ, AB, 32; Drexel Inst, BS, 34; Syracuse Univ, PhD(biochem), 37. Prof Exp: Chief chemist, Cheplin Biol Labs, NY, 34-37; res chemist, Endo Prod, Inc, 37-38, dir labs, 38-41, vpres & plant supt, 46-47; asst officer in charge physiol br, Air Test Ctr, US Navy, Md, 47-50, officer in charge, 50-52, test dir parachute unit, Auxiliary Air Sta, 53-56, dir air crew equip lab, Air Mat Ctr, Philadelphia, Pa, 56-64, head aviation med equip br, Bur Med & Surg & Human Factors Off, Bur Naval Weapons, 64-66, dir crew systs div, Naval Air Systs Command Hq, Washington, DC, 66-68; DIR RES, PHILADELPHIA GEN HOSP, 68- Concurrent Pos: Mem vision comt & hearing & bioacoustics comt, Nat Acad Sci-Nat Res Coun. Honors & Awards: Spec Aerospace Med Honor Citation, AMA; Cert of Merit, Bur Med & Surg, Navy Dept; Mosely Award, Aerospace Med Asn. Mem: AAAS; Am Chem Soc; fel Am Inst Chemists; Am Pharmaceut Asn; assoc Aerospace Med Asn. Res: Aviation physiology; carbon monoxide and its effects on aviation personnel; safety and survival equipment. Mailing Add: Philadelphia Gen Hosp Philadelphia PA 19104

BOSEN, SIDNEY FREDERICK, b Chicago, Ill, Feb 26, 41; m 61; c 3. FORENSIC SCIENCES, ANALYTICAL CHEMISTRY. Educ: George Washington Univ, BS, 63; Univ Ill, Chicago Circle, PhD(org chem), 69. Prof Exp: Asst chem, George Washington Univ, 63-65 & Univ Ill, Chicago Circle, 65-69; NIH fel, Univ Chicago, 69-71; ASST PROF ADMIN CRIMINAL JUSTICE, UNIV ILL, CHICAGO CIRCLE, 71- Concurrent Pos: Consult, Cent Regional Lab, US Environ Protection Agency, 75- Mem: AAAS; Am Chem Soc; NY Acad Sci. Mailing Add: 1246 W Pratt Chicago IL 60626

BÖSENBERG, WOLFRAM ARNULF (RÜDIGER), physics, see 12th edition

BOSHART, CHARLES RALPH, b Lowville, NY, June 9, 32; m 56; c 4. PHARMACOLOGY. Educ: Univ Buffalo, BS, 54, MS, 58; Purdue Univ, PhD(pharmacol), 59. Prof Exp: Instr pharmacol, Purdue Univ, 57-58; pharmacologist, 59-63, sr res scientist & group leader, 63-65, head dept endocrinol res, 65-69, HEAD DEPT TOXICOL EVAL, LEDERLE LABS, AM CYANAMID CO, 69- Mem: AAAS; Am Soc Pharmacol & Exp Therapeut. Res: Anticonvulsants; central nervous system barriers; biological use of tritium and other radioisotopes; metabolic effects of drugs. Mailing Add: Lederle Labs Am Cyanamid Co Pearl River NY 10965

BOSHART, GREGORY LEW, b Lowville, NY, Mar 21, 33. ORGANIC CHEMISTRY. Educ: St Lawrence Univ, BS, 55; Mass Inst Technol, PhD(org chem), 60. Prof Exp: Chemist, E I du Pont de Nemours & Co, 57, Esso Res & Eng Co, 60-61, 62-64 & Enjay Chem Co, 64-68; secy dept, 68-74, EXEC OFFICER & ACAD AIDE, DEPT CHEM, UNIV CHICAGO, 74-, PROF LECTR, 74-, ASST DEAN, PHYS SCI DIV, 74- Mem: AAAS; Am Chem Soc. Res: Polypeptide synthesis; carbodiimide chemistry; resolution of amino acids; resin plasticization. Mailing Add: Dept of Chem Univ of Chicago 5801 S Ellis Ave Chicago IL 60637

BOSHELL, BURIS RAYE, b Marion Co, Ala, Oct 9, 26; m 51; c 2. ENDOCRINOLOGY, INTERNAL MEDICINE. Educ: Ala Polytech Inst, BS, 47; Harvard Med Sch, MD, 53. Prof Exp: Intern med, Peter Bent Brigham Hosp, 53-54, from jr asst resident to sr asst resident, 54-56, asst, Thorn Lab, 56-57, jr assoc physician & asst dir, Diabetic Teaching Unit, 57-58, chief resident, 58-59; from asst prof to prof, 59-72, asst dir, Dept Med, 63-76, RUTH LAWSON HANSON PROF MED, SCH MED, UNIV ALA, BIRMINGHAM, 72-, CHIEF, DIV ENDOCRINOL & METAB, DIABETES RES & EDUC HOSP, 76- Concurrent Pos: Am Col Physicians res fel, Thorn Lab, Peter Bent Brigham Hosp, 56-57; asst, Harvard Med Sch, 58-59; clin investr, Vet Admin Hosp, 59-62, chief med serv, 62-; asst physician-in-chief, Univ Hosp & Hillman Clins, 63- Mem: Endocrine Soc; Am Diabetes Asn; Am Soc Clin Pharmacol & Therapeut; fel Am Col Physicians; AMA. Res: Diabetes mellitus; oral diagnostic test for diabetes; insulin in blood and etiological factors in insulin resistance. Mailing Add: Div of Endocrinol & Metab Diabetes Res & Educ Hosp Birmingham AL 35294

BOSHES, BENJAMIN, b Chicago, Ill, Feb 15, 07; m 31; c 2. NEUROLOGY. Educ: Northwestern Univ, BS, 29, BM, 30, MD, 31, MS, 34, PhD(neurol), 38. Prof Exp: Adj nerv & ment dis, Michael Reese Hosp, Chicago, 32-36; instr, 35-38, assoc, 38-41, from asst prof to assoc prof, 41-51, PROF NEUROL & PSYCHIAT & CHMN DEPT, MED SCH, NORTHWESTERN UNIV, CHICAGO, 51- Concurrent Pos: Psychiatrist, Inst Juv Res, 36-41; consult neuropsychiatrist, Ill Eye & Ear Infirmary; attend neuropsychiatrist, St Luke's Hosp & Passavant Mem Hosp, 39-53; sr consult, Vet Admin Hosp, 46-; chmn dept neurol, Northwestern Mem Hosp, 53-; consult neurologist & psychiatrist, Fed Aviation Agency. Mem: Am Psychiat Asn; Am Neurol Asn; fel AMA; fel Am Acad Neurol; Asn Res Nerv & Ment Dis. Res: Sympathetic nervous system; spinal cord injuries; Parkinsonism; role of the brain in learning disorders. Mailing Add: 251 E Chicago Ave Chicago IL 60611

BOSHES, LOUIS D, b Chicago, Ill, Oct 15, 08; m 42; c 2. NEUROPSYCHIATRY. Educ: Northwestern Univ, Evanston, BS, 31, Northwestern Univ, Chicago, MD, 36; Am Bd Psychiat & Neurol, dipl & cert psychiat, 47, cert neurol, 50 & cert child neurol, 69. Prof Exp: Fel psychiat, Ill Neuropsychiat Inst, Chicago, 40-42, 46-47; from instr to asst prof neurol & psychiat, Sch Med, Northwestern Univ, Chicago, 47-63; clin assoc prof, 63-70, CLIN PROF NEUROL, ABRAHAM LINCOLN SCH MED, UNIV ILL COL MED, 70-, DIR CONSULTATION CLIN FOR EPILEPSY, 63-; PROF NEUROL, COOK COUNTY GRAD SCH MED, 70- Concurrent Pos: Assoc & attend neurologist, Cook County Hosp, Chicago, 47-63, consult neurol, 70-; neuropsychiat consult, Ill State Psychiat Inst, Chicago, 48-70, Columbus Mem Hosp, 48-, Woodlawn Hosp, 50-68, Louis A Weiss Mem Hosp, 50-, Jackson Park Community Hosp, 60-68 & Micholas J Pritzker Ctr for Children, 64-70; sr consult neurol, Vet Admin Hosp, Downey, Ill, 54-60; attend physician, Res & Educ Hosps, Abrham Lincoln Sch Med, Univ Ill Col Med, 63-; chief neurol clins, Michael Reese Hosp & Med Ctr, 69-75, sr attend neurologist & psychiatrist, 65-; attend, Grant Hosp & St Joseph's Hosp, 74- Assoc examr, Am Bd Psychiat & Neurol, 57- & Am Bd Neurol Surg, 63-; assoc ed, Dis Nerv Syst, 60- & Int J Neuropsychiat, 66-68; consult ed, Current Med Dig, 65- & New Physician, 67-; mem adv bd & coun, Myasthenia Gravis Found, United Parkinson Found, Epilepsy Found Am & Nat Found-March of Dimes; ambassador, Int Bur Epilepsy & League Against Epilepsy; mem sci exhib adv comt, Mus Sci & Indust, Chicago. Mem: Fel Am Acad Neurol; fel Am Psychiat Asn; fel AMA; fel Am Col Physicians; fel Pan-Am Med Asn (pres, 73-74). Res: Neurology; child neurology; psychiatry; multiple sclerosis; parkinsonism; epilepsy; myasthenia gravis. Mailing Add: Abraham Lincoln Sch of Med 912 S Wood St Chicago IL 60680

BOSIN, TALMAGE R, b Fond du Lac, Wis, Mar 6, 41. ORGANIC CHEMISTRY. Educ: Wheaton Col, Ill, BS, 63; Ind Univ, PhD(org chem), 67. Prof Exp: NIH res fel org chem, Univ Calif, Berkeley, 67-69; asst prof, 69-73, ASSOC PROF PHARMACOL, MED SCI PROG, IND UNIV, BLOOMINGTON, 73- Mem: Am Chem Soc. Res: Synthetic organic and medicinal chemistry, particularly dealing with the central nervous system. Mailing Add: Med Sci Prog Myers Hall Ind Univ Bloomington IN 47401

BOSKEY, ADELE LUDIN, b New York, NY, Aug 30, 43; m 70; c 1. STRUCTURAL CHEMISTRY. Educ: Columbia Univ, BA, 64; Boston Univ, PhD(phys chem), 70. Prof Exp: Instr chem col lib arts, Boston Univ, 69-70; res fel crystallog, Imp Col Sci & Technol, 70-71; NIH fel & res scientist ultrastruct biochem, Hosp for Spec Surg, 71-74, RES ASSOC ULTRASTRUCT BIOCHEM MED COL, CORNELL UNIV, 74- Mem: Sigma Xi; Am Crystallog Asn; Am Chem Soc. Res: Mechanism of hard tissue mineralization; structure determination by x-ray crystallography and electron microscopy. Mailing Add: Hosp for Spec Surg Cornell Univ Med Col 535 E 70th New York NY 10021

BOSKIN, MARVIN JAY, organic chemistry, see 12th edition

BOSLEY, DAVID EMERSON, b Lundale, WVa, Dec 16, 27; m 52; c 4. PHYSICAL CHEMISTRY. Educ: Univ WVa, BS, 50; Mass Inst Technol, PhD(phys chem), 54. Prof Exp: RES CHEMIST, DACRON PLANT TECH SECT, E I DU PONT DE NEMOURS & CO, 54- Res: Physico-chemical properties of synthetic fibers. Mailing Add: Tech Sect E I du Pont de Nemours & Co Kinston NC 28501

BOSLEY, ELIZABETH CASWELL, b Wichita, Kans, July 4, 12; m 39; c 1. SPEECH PATHOLOGY. Educ: Friends Univ, AB, 33; Univ Kans, MA, 35. Prof Exp: Instr eng & math, Univ Kans, 36; teacher high sch, Kans, 36-38; instr clin logopedics, Inst Logopedics, Wichita State Univ, 39-40; supvr, 40-42, PRECEPTOR & SPEECH CLINICIAN, INST LOGOPEDICS, 42- Concurrent Pos: From instr to asst prof, Wichita State Univ, 40-67. Mem: Am Speech & Hearing Asn. Res: Teacher training and clinical work in correcting all types of speech defects; diagnosis of speech defects; physiology of speech structure; anatomy of speech structures; neurology. Mailing Add: Inst of Logopedics Speech Clin 2400 Jardine Wichita KS 67219

BOSLOW, HAROLD MEYER, b New York, NY, Apr 30, 15; m 43. PSYCHIATRY. Educ: Univ Va, MD, 39. Prof Exp: Staff psychiatrist, US Vet Admin, 48-50; instr, 50-66, ASST PROF PSYCHIAT, SCH MED, JOHNS HOPKINS UNIV, 66-, PSYCHIATRIST, OUTPATIENT DEPT, JOHNS HOPKINS HOSP, 50-; ASST PROF PSYCHIAT, SCH MED, UNIV MD, 66- Concurrent Pos: Med officer, Supreme Bench, Baltimore, Md, 50-54; consult psychiatrist, Surgeon Gen, US Dept Army, 52-53; dir, Patuxent Inst, 54-; WHO fel, 66; consult to White House on anti-social behav & delinq, 64, WHO & UN Social Defense Res Inst, 69-70, NIMH, 70, Govts PR & Que, 70 & Off Law Enforcement Assistance Agency. Mem: AAAS; AMA; Am Psychiat Asn; NY Acad Sci. Res: Personality structure and deviations; forensic psychiatry. Mailing Add: Jessup MD 20794

BOSMA, JAMES FREDERICK, b Grand Rapids, Mich, Apr 29, 16; m 42; c 8. PEDIATRICS. Educ: Calvin Col, AB, 37; Univ Mich, MD, 41. Prof Exp: Intern, Cleveland City Hosp & Western Reserve Hosp, 41-43, resident, 43-44; asst prof pediat, Med Sch, Univ Minn, 48-49; prof, Col Med, Univ Utah, 49-59; CHIEF ORAL & PHARYNGEAL DEVELOP SECT, NAT INST DENT RES, 61- Concurrent Pos: Nat Found Infantile Paralysis fel, Univ Minn, 44-46, Kellogg fel, 46-48; NIH fel, Karolinska Inst, Sweden & Wenner Gren Cardiovasc Res Lab, 59-61; pediatrician, Salt Lake Gen Hosp, 55-59. Mem: AAAS; Soc Pediat Res; Am Pediat Soc; Soc Res Child Develop; Int Soc Cranio-facial Biol. Res: Motor coordination of mouth, pharynx and larynx. Mailing Add: Nat Inst Dent Res Bethesda MD 20014

BOSMANN, HAROLD BRUCE, b Chicago, Ill, June 17, 42; m 66; c 2. BIOPHYSICS, PHARMACOLOGY. Educ: Knox Col, AB, 64; Univ Rochester, PhD(biophys), 66. Prof Exp: Asst prof, 68-74, ASSOC PROF PHARMACOL & TOXICOL, UNIV ROCHESTER, 74- Concurrent Pos: NSF fel biophys, Strangeways Lab, Cambridge Univ, 66-67; USPHS fel, Salk Inst Biol Studies, La Jolla, Calif, 67-68 & career develop award, Univ Rochester, 68-74; scholar award, Leukemia Soc Am, 74-79. Mem: Am Chem Soc; Am Soc Pharmacol & Exp Therapeut; Biophys Soc. Res: Membranes; glycoproteins; drug resistance; synaptosomes; oncogenic transformation; cell cycle events; mitochondria autonomy. Mailing Add: Dept of Pharmacol & Toxicol Sch Med & Dent Univ Rochester Rochester NY 14642

BOSNIACK, DAVID S, b New York, NY, July 10, 32; m 57; c 3. ORGANIC CHEMISTRY, PETROLEUM CHEMISTRY. Educ: City Col New York, BS, 54; NY Univ, PhD(org chem), 61. Prof Exp: Chemist, GAF Corp, 60-62; chemist, Exxon Govt Res Lab, 62-66, sr chemist, Exxon Prod Res Div, 66-70, RES ASSOC, PROD RES DIV, EXXON RES & ENG CO, 70-, PROJ HEAD, NEW PROJ DEVELOP, BAYTOWN PETROL RES LAB, 70- Concurrent Pos: Instr, Cooper Union, 61-63. Mem: Am Chem Soc. Res: Free radical reactions; petroleum chemistry; lubrication technology including synthetic lubricants, lubrication mechanisms and additive synthesis; acetylene chemistry; solid propellants; polymer synthesis. Mailing Add: Baytown Petrol Res Lab Exxon Res & Eng Co Box 4255 Baytown TX 77520

BOSNIAK, MORTON A, b New York, NY, Nov 13, 29; m; c 1. RADIOLOGY. Educ: Mass Inst Technol, BS, 51; State Univ NY, MD, 55. Prof Exp: Intern, Mt Sinai Hosp, NY, 55-56; resident radiol, NY Hosp, 56-57, 59-61; instr, Med Sch, Cornell Univ, 60-61; assoc prof, Sch Med, Boston Univ, 64-67; assoc prof, Albert Einstein Col Med, 67-69, sr attend radiologist, Hosp, 67-69, assoc dir dept, 68-69; PROF RADIOL, SCH MED, NY UNIV, 69- Concurrent Pos: Radiologist, Montefiore Hosp, 61-62, assoc attend, 63-; sr staff radiologist, Boston City Hosp, Mass, 64-67; assoc radiologist, Univ Hosp, Boston, 64-67; sr attend radiologist, Bronx Munic Hosp Ctr, NY, 67-; attend radiologist, NY Univ Hosp, 69-; vis attend radiologist, Bellevue Hosp; consult radiol, Vet Admin Hosp. Mem: Am Col Radiol; Radiol Soc NAm; Asn Univ Radiol; Am Roentgen Ray Soc. Mailing Add: Col of Med New York Univ New York NY 10016

BOSOMS, JOHN A, polymer chemistry, see 12th edition

BOSOMWORTH, DOUGLAS ROBERT, solid state physics, see 12th edition

BOSOMWORTH, PETER PALLISER, b Akron, Ohio, May 2, 30; m 56; c 4. MEDICINE, ANESTHESIOLOGY. Educ: Kent State Univ, BSc, 51; Univ Cincinnati, MD, 55; Ohio State Univ, MMedSc, 58. Prof Exp: Intern, Cincinnati Gen Hosp, Ohio, 55-56; asst resident anesthesiol, Ohio State Univ Hosp, 56-57, chief resident & asst instr, 57-58, instr, 58; chief anesthesia div, US Naval Hosp, Great Lakes, Ill, 58-60; dir anesthesia res, Ohio State Univ, 60-62; chmn dept anesthesiol, 62-70, assoc dean col med, 68-70, PROF ANESTHESIOL, UNIV KY, 62-, VPRES, MED CTR, 70- Concurrent Pos: Consult, Vet Admin Hosp & USPHS, Lexington, 62-Ireland Army Hosp, Ft Knox, 64-; chief staff, Univ Ky Hosp, 66-68; mem exec comt, Ohio Valley Regional Med Prog; mem bd dir, Ky Physicians Mutual, Inc; vpres, Health Develop Resources Inst. Mem: Am Soc Anesthesiol; Am Col Anesthesiol; Am Soc Clin Pharmacol & Therapeut; AMA; Int Anesthesia Res Soc. Res: Cardiovascular, uterine and renal physiology as influenced by anesthetic agents. Mailing Add: Univ of Ky Med Ctr Lexington KY 40506

BOSS, BRUCE DAVID, b Brooklyn, NY, Aug 17, 40; m 60; c 2. PHYSICAL CHEMISTRY. Educ: St Lawrence Univ, BS, 62; Univ Wis, PhD(phys chem), 66. Prof Exp: Res trainee phys chem, Cent Res Dept, Monsanto Co, 64-66; res chemist, Fuels Br, Naval Res Labs, Washington, DC, 66-74; FUELS MGR, FED ENERGY ADMIN, 74- Mem: Am Chem Soc; Sigma Xi. Res: Kinetics and mechanisms of oxidation reactions; pulsed nuclear magnetic resonance; fuel conversion technology. Mailing Add: 1 Devon Ct Rockville MD 20850

BOSS, KENNETH JAY, b Grand Rapids, Mich, Dec 5, 35. MALACOLOGY. Educ: Cent Mich Col, BA, 57; Mich State Univ, MSc, 59; Harvard Univ, PhD(biol), 63. Prof Exp: Res systematist mollusks, US Nat Mus, Dept Interior, 63-66; from asst cur to cur, 66-74, PROF BIOL, HARVARD UNIV, 70- Mem: Soc Syst Zool; Am Malacol Union; Marine Biol Asn UK. Res: Systematic and evolutionary studies of mollusks; comparative morphology of lamellibranchs; zoogeography. Mailing Add: Dept of Biol Harvard Univ Cambridge MA 02138

BOSS, MANLEY LEON, b Atlanta, Ga, Dec 24, 24; m 56; c 2. PLANT PHYSIOLOGY. Educ: Univ Miami, BS, 49; Inter-Am Inst Agr Sci, MAgr, 51; Iowa State Col, PhD, 55. Prof Exp: From instr to assoc prof bot, Univ Miami, 54-63; head macrobiol sect, 63-66, PROF BOT, FLA ATLANTIC UNIV, 63- Mem: NY Acad Sci; Am Soc Plant Physiol; Bot Soc Am; Am Geront Soc; Scand Soc Plant Physiol. Res: Environmental sciences. Mailing Add: Dept of Biol Sci Fla Atlantic Univ Boca Raton FL 33432

BOSS, WILLIS ROBERT, b Superior, Wis, May 10, 06; m 37. ZOOLOGY, PUBLIC HEALTH ADMINISTRATION. Educ: Wis State Col, BE, 30; Univ Iowa, MS, 33, PhD(zool), 43. Prof Exp: Head dept sci, Univ Exp Schs, Univ Iowa, 32-33; prof biol, Wentworth Mil Acad, 33-35; high sch teacher, Wis, 35-40; asst zool, Univ Iowa, 40-42, res assoc, 43; prof biol, Pa State Teachers Col, Edinboro, 46-47; from assoc prof to prof zool, Syracuse Univ, 47-59; sr sci attache, US Embassy, Tokyo, 59-61; chief training br, Nat Cancer Inst, 61-64; CHIEF CAREER DEVELOP REV BR, NIH, 64- Concurrent Pos: Physiologist, Div Biol & Med, US AEC, 53-55. Mem: AAAS; Am Soc Zoologists; Am Physiol Soc; NY Acad Sci. Res: Endocrinology; effects of steroid hormones upon birds; water metabolism; kidney physiology; radiobiology. Mailing Add: Career Develop Rev Br NIH Div Res Grants Bethesda MD 20014

BOSSARD, DAVID CHARLES, b Sellersville, Pa, Mar 23, 40; m 64; c 4. OPERATIONS RESEARCH. Educ: Drexel Univ, BSc, 62; Dartmouth Col, AM, 64 & 66, PhD(math), 67. Prof Exp: Assoc opers res, 67-73, VPRES OPERS RES, DANIEL H WAGNER ASSOCS, 73- Mem: Soc Indust & Appl Math. Mailing Add: Daniel H Wagner Assocs Sta Square One Paoli PA 19406

BOSSERT, ROY GARNER, b Monongahela, Pa, Feb 21, 08; m 33; c 2. ORGANIC CHEMISTRY. Educ: Col Wooster, BS, 30; Ohio State Univ, MS, 33, PhD(org chem), 36. Prof Exp: Chemist, Aluminum Res Labs, Aluminum Co Am, Pa, 30-31; asst chem, Ohio State Univ, 32-36; instr, Univ Ky, 36-37; from instr to prof chem, Ohio Wesleyan Univ, 37-72, chmn dept, 62-70; RETIRED. Concurrent Pos: Fel, Mellon Inst, 44-46. Mem: Am Chem Soc; fel Am Inst Chemists; Sigma Xi. Res: Synthetic organic chemistry; salts of acids; urethans; chemical education. Mailing Add: Dept of Chem Ohio Wesleyan Univ Delaware OH 43015

BOSSHARDT, DAVID KIRN, b Rochester, Minn, Apr 15, 16; m 43; c 1. ANALYTICAL BIOCHEMISTRY. Educ: Univ Minn, BS, 38; Rutgers Univ, MS, 40, PhD(dairy chem), 43. Prof Exp: Asst, NJ Exp Sta, 38-43; res assoc, Merck Sharp & Dohme Res Lab Div, Merck & Co, Inc, 43-71; res aide, Natural Resources Res Inst, Univ Wyo, 72-73; CHEMIST, WYO DEPT AGR, 73- Res: Silage preservation; carotinoid pigments in feed and butter; protein and fat metabolism in small animals; energy and vitamin requirements in small animals; nutritive value of end products of silage fermentation; effect of feed and storage on carotinoid pigments in butter; unidentified growth factors; cholesterol and bile acid metabolism. Mailing Add: 1707 Mitchell Laramie WY 82070

BOST, HOWARD WILLIAM, b Robstown, Tex, Sept 29, 24; m 50; c 3. ORGANIC CHEMISTRY. Educ: Southwest Tex State Teachers Col, BS, 48; Univ Tex, MA, 50, PhD(org chem), 55. Prof Exp: Res chemist, 50-51, res chemist & group leader liquid propellants, 54-57, res chemist & sr group leader, 57-60, group leader photochem reactions, 60-68, group leader oxidation processes, 68-70, group leader org sulfur chem, 70-74, TECH RECRUITMENT REP, PHILLIPS PETROL CO, 74- Concurrent Pos: Mem joint Army-Navy-Air Force comt on mono-propellant test methods, 59-60. Mem: Am Chem Soc. Res: Protective coatings; nitrogen chemicals; liquid rocket propellants; photo-chemical reactions; sulfur reactions; oxidation processes; organic sulfur chemistry; perfume ingredients; flame retardants; technical recruiting. Mailing Add: 1334 Quail Dr Bartlesville OK 74003

BOST, ROBERT ORION, b Elida, NMex, Jan 15, 43; m 66; c 1. TOXICOLOGY. Educ: Univ Tex, Austin, BS, 65, MA, 67; Univ Houston, PhD(chem), 70. Prof Exp: Lab asst, Southwestern Univ, 62-64; asst, Univ Tex, Austin, 65-67; teaching fel, Univ Houston, 67-69, res asst, 69-70; sci res specialist, La State Univ, New Orleans, 70-72; chemist, Vet Admin Hosp, New Orleans, 72-74; ASSOC TOXICOLOGIST, CUYAHOGA COUNTY CORONER'S OFF, 75- Mem: Am Chem Soc. Res: Forensic and clinical toxicology; gas chromatography; derivatives for chromatography; generation and reaction of carbenes; small ring compounds; arene-chromium complexes. Mailing Add: Cuyahoga County Coroner's Off 2121 Adelbert Rd Cleveland OH 44106

BOSTER, THOMAS ARTHUR, b Columbus, Ohio, Sept 28, 36; m 62; c 3. SOLID STATE PHYSICS. Educ: Capital Univ, BS, 58; Ohio Univ, MS, 60, PhD(physics), 66. Prof Exp: Res engr, Nortronics Div, Northrop Corp, 60-62; fel physics, Ohio Univ, 62-65; from asst prof to assoc prof, Okla City Univ, 65-69; sr physicist, 68-73, GROUP LEADER, LAWRENCE LIVERMORE LAB, UNIV CALIF, 73- Concurrent Pos: Consult, Okla Bur Invest, 67-68. Mem: Am Phys Soc; Am Asn

Physics Teachers; Am Soc Safety Engrs. Res: Absorption of x-rays; radiation damage; cryogenics; microwaves; x-ray diagnostics; x-ray spectrometers; solid state detectors. Mailing Add: 1035 Lynn St Livermore CA 94550

BOSTIAN, CAREY HOYT, b China Grove, NC, Mar 1, 07; m 29; c 3. GENETICS. Educ: Catawba Col, AB, 28; Univ Pittsburgh, MS, 30, PhD(genetics), 33. Hon Degrees: DSc, Catawba Col, 53, Wake Forest Col, 54 & Nat Univ Eng, Peru, 57. Prof Exp: Asst zool, Univ Pittsburgh, 28-30; asst prof zool, 30-36, assoc prof zool & assoc poultry genetics, 36-45, prof zool & res prof poltry genetics, 45-53, assoc dean agr, 48-53, chancellor, 53-59, prof genetics, 59-75, EMER PROF GENETICS, NC STATE UNIV, 75- Mem: Am Inst Biol Sci; Am Genetic Asn; Am Soc Human Genetics. Res: General and human genetics. Mailing Add: Dept of Genetics Box 5487 NC State Univ Raleigh NC 27607

BOSTIAN, LOGAN CHAPPELL, chemistry, see 12th edition

BOSTIC, CARLTON RAY, organic chemistry, see 12th edition

BOSTICK, EDGAR E, b Newville, Ala, Dec 10, 26; m 55; c 5. PHYSICAL ORGANIC CHEMISTRY. Educ: Ala Polytech Inst, BS, 50; Univ Akron, PhD(chem), 59. Prof Exp: Jr chem engr, Goodyear Tire & Rubber Co, 50-51, chem engr, 53-56; fel, Univ Akron, 59-60; res chemist, 60-69, mgr res & develop, Insulating Mat Dept, 69-76, WITH RES & DEVELOP, LEXON RESIN SECT, GEN ELEC CO, 76- Mem: Am Chem Soc; Am Inst Chem Engrs. Res: Ionic polymerization of vinyl, diene and heterocyclic compounds; reaction kinetics and mechanisms; characterization of polymers by chemical and physical techniques. Mailing Add: Lexon Resin Sect Gen Elec Co Hwy 69 S Mount Vernon IN 47620

BOSTICK, PETER EDWARD, biology, see 12th edition

BOSTICK, WARREN LITHGOW, b Dallas, Tex, July 28, 14; m 39; c 2. PATHOLOGY. Educ: Univ Calif, AB, 36, MD, 40. Prof Exp: From assoc prof to prof path, Sch Med, Univ Calif, San Francisco, 50-64, dir labs, Univ Hosps, 54-64; dean, Col Med, 64-74, PROF PATH, COL MED, UNIV CALIF, IRVINE, 64- Concurrent Pos: Trustee, Calif Physicians Serv, 62- Mem: AAAS; Col Am Path; Am Soc Exp Path (pres elect); Am Soc Clin Path. Res: Hodgkins disease; ameblasis; fibrocystic disease of pancreas; viral carcinogenesis. Mailing Add: Univ of Calif Col of Med Irvine CA 92664

BOSTICK, WINSTON HARPER, b Freeport, Ill, Mar 5, 16; m 42; c 3. PHYSICS. Educ: Univ Chicago, BS, 38, PhD(physics), 41. Prof Exp: Instr physics, Univ Chicago, 40; staff mem, Radiation Lab, Mass Inst Technol, 41-45, sect chief, 45-46, staff mem, Research Lab Electronics, 46-49; from asst prof to assoc prof physics, Tufts Col, 48-54; staff mem, Radiation Lab, Univ Calif, 54-56; prof, 56-69, head dept, 56-73, GEORGE MEADE BOND PROF PHYSICS, 69- Concurrent Pos: NSF sr fel, Nuclear Res Ctr, France & Culham Lab. Honors & Awards: Prizes, Gravity Res Found, 58 & 61. Mem: Fel Am Phys Soc. Res: Cosmic ray research with cloud chambers; design of pulse transformers; development of microwave linear accelerator; diffusion of gaseous ions across a magnetic field and plasma waves in presence of a magnetic field. Mailing Add: Dept of Physics Stevens Inst Technol Hoboken NJ 07030

BOSTOCK, JUDITH LOUISE, b Gardiner, Maine. THEORETICAL SOLID STATE PHYSICS. Educ: Trinity Col, AB, 60; Dickinson Col, BS, 63; Georgetown Univ, MS, 69, PhD(physics), 71. Prof Exp: Asst physics, Univ Saarland, 70-71; instr, 72-75, ASST PROF PHYSICS, MASS INST TECHNOL, 75- Concurrent Pos: Consult physicist crystal physics, Naval Res Lab, 72-75; consult sr physicist, BDM Corp, Va, 75- Mem: Am Phys Soc. Res: Lattice instabilities and superconductivity in metals at low temperatures; low temperature properties in theoretical solid state physics; non-equilibrium conditions in superconductors. Mailing Add: Dept of Physics Rm 6-204 Mass Inst of Technol Cambridge MA 02139

BOSTON, CHARLES RAY, b Bellaire, Ohio, Aug 4, 28; c 3. PHYSICAL CHEMISTRY. Educ: Ohio Univ, BS, 49; Northwestern Univ, PhD(chem), 53. Prof Exp: Chemist, 53-73, GROUP LEADER, OAK RIDGE NAT LAB, 73- Mem: AAAS; Sigma Xi; Am Chem Soc. Res: Fused salt chemistry; physical inorganic; absorption spectra; unusual oxidation states; management of interdisciplinary group doing environmental impact assessments. Mailing Add: Oak Ridge Nat Lab PO Box X Oak Ridge TN 37830

BOSTON, JAMES D, b Roscoe, Tex, July 7, 39; m 61; c 2. BIOCHEMISTRY, ENZYMOLOGY. Educ: Southwest Tex State Univ, BS, 61, MA, 62; Tex A&M Univ, PhD(biochem), 66. Prof Exp: Res & lab asst chem, Southwest Tex State Univ, 59-62; res asst biochem, Tex A&M Univ, 62-65, fel, 66; res assoc, Cardeza Found, Jefferson Med Col, 67-68; sr res biochemist, 68-69, res dir, 69-73, DIR BIOCHEM DIV, TRUETT LABS, 74-, VPRES, SOUTHWESTERN DRUG CORP, 74- Mem: AAAS; Am Chem Soc; NY Acad Sci; Sigma Xi. Res: Isolation and characterization of proteolytic enzymes of microbial origin; treatment of industrial wastewater by biological means. Mailing Add: 15515 Branchcrest Dallas TX 75240

BOSTON, JOHN ROBERT, b Evanston, Ill, Oct 16, 42; c 2. BIOMEDICAL ENGINEERING, AUDIOLOGY. Educ: Stanford Univ, BS, 64, MS, 66; Northwestern Univ, PhD(elec eng), 71. Prof Exp: Res assoc health care delivery, Hosp Res & Educ Trust, 71-72; asst prof elec eng, Univ Md, 72-75; ASST PROF BIOTECHNOL, CARNEGIE-MELLON UNIV, 75-; ASST PROF OTOLARYNGOL, SCH MED, UNIV PITTSBURGH, 75- Mem: Acoust Soc Am; AAAS; Inst Elec & Electronics Engrs; Soc Neurosci; Asn Advan Med Instrumentation. Res: Transduction processes in sensory hair cells; electrical potentials associated with hair cell transduction; computer applications in patient monitoring. Mailing Add: Biotechnol Prog Carnegie-Mellon Univ Pittsburgh PA 15213

BOSTON, NOEL EDWARD JAMES, b Vancouver, BC, May 28, 36; m 60; c 2. PHYSICAL OCEANOGRAPHY, MICROMETEOROLOGY. Educ: Univ BC, BASc, 59, PhD(phys oceanog), 71; Tex A&M Univ, MS, 63. Prof Exp: Asst scientist, Pac Oceanog Group, Fisheries Res Bd Can, 59-61; from asst prof to assoc prof phys oceanog, Naval Postgrad Sch, 68-75; PHYS OCEANOGR, BEAK CONSULTS LTD, 75- Concurrent Pos: Partner, Environ Res Assocs, 70-75; mem nat comt interaction atmosphere & ocean, Am Meteorol Soc, 75-78; proj supvr, First Global Atmospheric Res Prog global exp bouy array deployment plan, 75-76. Mem: Am Meteorol Soc; Am Geophys Union; AAAS; Sigma Xi. Res: Internal waves; air-sea interaction; turbulence; mixing of upper isothermal layer; machine processing of bathythermograph data; ocean thermal structure; surface circulation of southern oceans; ocean currents in Canadian Arctic archipelago. Mailing Add: Beak Consults Ltd 385-A Shell Rd Richmond BC Can

BOSTON, ROBERT WESLEY, b Athens, Ont, July 10, 32; m 58; c 2. PEDIATRICS. Educ: Queen's Univ, Ont, MD, 57; Royal Col Physicians & Surgeons Can, dipl pediat, 63. Prof Exp: Mead Johnson Educ Fund fel pediat, Harvard Med Sch & Boston Lying-In Hosp, 62-64; McLaughlin Found traveling fel, Univ Col Hosp, London, Eng, 64-65; asst prof, 65-74, ASSOC PROF PEDIAT, QUEEN'S UNIV, ONT, 74- Concurrent Pos: Attend physician, Kingston Gen Hosp, 65- Mem: Can Pediat Soc; Can Soc Clin Invest. Res: Neonatology; respiratory distress syndrome; assessment of the fetus in uterus and of immediate neonatal adaption. Mailing Add: Dept of Pediat Queen's Univ Kingston ON Can

BOSTRACK, JACK M, b Stoughton, Wis, Apr 30, 31; m 53; c 3. BOTANY. Educ: Wartburg Col, BA, 53; Univ Wis, MS, 59, PhD(bot hort), 62. Prof Exp: Asst bot, Univ Wis, 57-61; asst prof biol, Wayne State Col, 61-63; assoc prof, 63-70, PROF BIOL, WIS STATE UNIV, RIVER FALLS, 70- Mem: Bot Soc Am. Res: Morphogenesis of plant parts; determination of form in plants, as affected by exogenous and endogenous factors. Mailing Add: Dept of Biol Wis State Univ River Falls WI 54022

BOSTROM, CARL OTTO, b Port Jefferson, NY, Aug 18, 32; m 54; c 3. SPACE PHYSICS. Educ: Franklin & Marshall Col, BS, 56; Yale Univ, MS, 58, PhD(physics), 62. Prof Exp: Sr physicist, 60-64, supvr space physics group, 64-74, CHIEF SCIENTIST, SPACE DEPT, JOHNS HOPKINS UNIV, 74- Concurrent Pos: Instr, Evening Col, Johns Hopkins Univ, 75- Mem: AAAS; Am Phys Soc; Am Geophys Union. Res: Space particles and fields; satellite instrumentation; magnetospheric physics; solar physics; satellite systems design; research administration. Mailing Add: Appl Physics Lab Johns Hopkins Univ Laurel MD 20810

BOSTROM, KURT G V, b Stockholm, Sweden, Mar 25, 32; m 64. MINERALOGY, GEOCHEMISTRY. Educ: Univ Stockholm, BS, 58, PhD(mineral), 62. Prof Exp: Asst res oceanogr, Scripps Inst, Univ Calif, 63-67; from asst prof to assoc prof, 67-74, PROF MARINE SCI, UNIV MIAMI, 74- Mem: Mineral Soc Am; Geochem Soc; Swedish Geol Soc; Swedish Mineral Soc (secy, 60-62). Mem: Study of sediments and minerals and their formation and stabilities by physico-chemical calculations; hydrothermal synthesis; electron microprobe; electron and x-ray diffraction; mineral structures; thermodynamic data for minerals; geochemical balances. Mailing Add: Inst of Marine Sci Univ of Miami Coral Gables FL 33146

BOSTROM, NORMAN ALVIN, nuclear physics, see 12th edition

BOSTROM, ROBERT CHRISTIAN, b Edinburgh, Scotland, July 22, 20; m 52; c 3. GEOPHYSICS. Educ: Oxford Univ, BA, 48, MA, 52, DPhil(geophys), 61. Prof Exp: Geophysicist, Explor Consults Inc, 50-53 & Standard Oil Co Calif, 53-64; assoc prof, 64-69, PROF GEOL SCI & GEOPHYS, UNIV WASH, 69- Mem: Am Geophys Union; Soc Explor Geophys; Am Asn Petrol Geol; Brit Geol Soc. Res: Earth strains; tectono physics; marine geophysics. Mailing Add: Dept of Geol Sci Univ of Wash Seattle WA 98105

BOSTWICK, DAVID ARTHUR, b Helena, Mont, Nov 19, 17; m 42; c 2. INVERTEBRATE PALEONTOLOGY. Educ: Mont State Univ, BA, 42; Univ Wis, MA, 51, PhD(geol), 58. Prof Exp: From instr to asst prof, 53-62, ASSOC PROF GEOL, ORE STATE UNIV, 62- Mem: Soc Econ Paleont & Mineral; Geol Soc Am. Res: Micropaleontology. Mailing Add: Dept of Geol Ore State Univ Corvallis OR 97331

BOSWELL, CHARLES LELAND, b Buncome, Ill, Jan 31, 17; m 46; c 1. ORGANIC CHEMISTRY. Educ: SDak State Col, BS, 38; Purdue Univ, MS, 39, PhD(pharmaceut chem), 41. Prof Exp: Pharmaceut chemist, Eli Lilly & Co, Ind, 41-46; works mgr, Jamieson Pharmacal Co, Div Heyden Chem Corp, 46-48; dir pharmaceut develop, Commercial Solvents Corp, 48-51, supt, Pharmaceut Prods, 51-54; plant mgr & asst gen mgr, 54-66, vpres & gen mgr, 66-69, PRES & GEN MGR, DORSEY LABS DIV, SANDOZ-WANDER, INC, 69- Mem: Am Chem Soc; Am Pharmaceut Asn. Res: Vitamins; enzymes; antioxidants; development of stable combinations of vitamins for medical use; multiple compressed tablets. Mailing Add: Dorsey Labs NE US 6 & Interstate 80 Lincoln NE 68501

BOSWELL, DONALD EUGENE, b Durham, NC, Aug 12, 34; m 56; c 3. ORGANIC CHEMISTRY. Educ: Duke Univ, BS, 56; Va Polytech Inst & State Univ, MS, 61, PhD(chem), 63. Prof Exp: Res chemist, Cent Res Div Lab, Mobil Res & Develop Co, NJ, 63-65, asst supvr catalysis & math res, 65-67; mgr prod planning, 67-68, MGR RES & DEVELOP, PROD DIV, CINCINNATI MILACRON, 68- Mem: AAAS; Am Chem Soc; Am Soc Tool & Mfg Eng; Am Soc Lubrication Eng. Res: Chemistry of heterocyclic compounds; organometallic chemistry; metalworking synthetic lubricants; corrosion inhibitors; water purification techniques; organic analytical chemistry. Mailing Add: Cincinnati Milacron Prod Div PO Box 9013 Cincinnati OH 45209

BOSWELL, FRANK WILLIAM CHARLES, b Hamilton, Ont, July 11, 24; m 51; c 2. PHYSICS. Educ: Univ Toronto, BA, 46, MA, 47, PhD(physics), 50. Prof Exp: Sci officer, Dept Mines & Technol Surv, Can, 50-57, head metal physics sect, 57-60; assoc prof, 60-63, PROF PHYSICS, UNIV WATERLOO, 63-, ASSOC DEAN SCI FOR GRAD AFFAIRS, 67- Mem: Electron Micros Soc Am; Am Asn Physics Teachers; Can Asn Physicists. Res: Solid state physics; imperfections in crystals; electron microscopy and diffraction. Mailing Add: Dept of Physics Univ of Waterloo Waterloo ON Can

BOSWELL, FRED CARLEN, b Monterey, Tenn, Aug 20, 30; m 54; c 2. AGRONOMY, SOILS. Educ: Tenn Polytech Inst, BS, 54; Univ Tenn, MS, 56; Pa State Univ, PhD(agron, soils), 60. Prof Exp: Asst res agronomist, Agr Exp Sta, Univ Tenn, 55-56 & Exp Sta, Univ Ga, 56-57; asst, Pa State Univ, 58-60; asst soil chemist, Exp Sta, 60-64, ASSOC SOIL CHEMIST, EXP STA, UNIV GA, 64-, PROF AGRON & SOILS, 71- Mem: Am Soc Agron; Soil Sci Soc Am. Res: Soil chemistry, fertility, microbiology and closely related subjects; nitrogen transformations, management and micronutrients as related to plant growth and physiology. Mailing Add: Ga Sta Univ of Ga Col of Agr Experiment GA 30212

BOSWELL, GEORGE A, JR, b Hayward, Calif, Jan 26, 32; m 52; c 4. ORGANIC CHEMISTRY. Educ: Univ Calif, Berkeley, BS, 56, PhD(org chem), 59. Prof Exp: With Shell Develop Co, Calif, 59-61; res org chemist, 61-67, RES SUPVR, CENT RES DEPT, E I DU PONT DE NEMOURS & CO, INC, 67- Mem: Am Chem Soc. Res: Steroid chemistry; general organic synthesis; photochemical transformations of organic molecules; organic fluorine and medicinal chemistry. Mailing Add: Cent Res & Develop Dept E I du Pont de Nemours & Co Wilmington DE 19898

BOSWELL, JAMES LOUIS, b Wynnewood, Okla, Dec 10, 11; m 47; c 3. INVERTEBRATE ZOOLOGY, RADIATION BIOLOGY. Educ: ECent State Col, BS, 36; Univ Okla, MS, 38; Tex A&M Univ, PhD(zool), 66. Prof Exp: Assoc prof biol, ECent State Col, 42-46; assoc prof Oklahoma City Univ, 46-47; biologist, Res Found, Tex A&M Univ, 48-60, instr biol, 60-65; ASSOC PROF BIOL, MIDWESTERN UNIV, 65- Mem: AAAS; Am Soc Limnol & Oceanog. Res: Parasites of oysters; ecology of mollusca; ecology of crustacea, including effects of acute gamma radiation on Artemia; beginning electron microscope studies in invertebrate tissues. Mailing Add: Dept of Biol Midwestern Univ Wichita Falls TX 76308

BOSWELL, RUPERT DEAN, JR, b Marshall Co, Miss, Aug 11, 29; m 52; c 2. MATHEMATICS. Educ: Miss State Univ, BS, 50, MS, 51; Univ Ga, PhD(math), 57. Prof Exp: Asst math, Miss State, 50-51; instr Reinhardt Col, 51-53; asst, Univ Ga, 53-56; from assoc prof to prof, 57-62; chmn dept, 62-73, PROF MATH, MONMOUTH COL, ILL, 62- Mem: Am Math Soc; Math Asn Am. Res: General topology; abstract algebra. Mailing Add: Dept of Math Box 182 Monmouth Col Monmouth IL 61462

BOSWELL, THOMAS D, b Redlands, Calif, Apr 21, 41; m 66; c 1. GEOGRAPHY. Educ: San Diego State Univ, BA, 64, MA, 66; Columbia Univ, PhD(geog), 73. Prof Exp: Instr geog, Univ Northern Colo, 66-69; researcher, Res Inst Study of Man, 72-73; ASST PROF GEOG, UNIV FLA, 73- Mem: Asn Am Geogrs; Pop Asn Am; Pop Ref Bur; Conf Latin Americanist Geogrs; Am Geog Soc. Res: Population problems in Middle America and the Caribbean, particularly rural to urban migration and the various demographic consequences of modernization. Mailing Add: Dept of Geog 102 Bryan Hall Univ of Fla Gainesville FL 32611

BOSWICK, JOHN A, b Galatia, Ill, Aug 30, 26; m 49; c 3. SURGERY. Educ: Southern Ill Univ, Carbondale, BA, 51; Loyola Univ Chicago, MS, 52, MD, 56. Prof Exp: Assoc prof surg, Med Sch, Northwestern Univ, 61-70; prof surg, Univ Ill Med Ctr, 70-72; PROF SURG & CHMN SECT HAND SURG, MED CTR, UNIV COLO, DENVER, 72- Concurrent Pos: Dir burn unit & hand surg, Cook County Hosp, Chicago, 61-72; consult, Brooke Army Med Ctr, 70-, Great Lakes Naval Hosp, 70- & Vet Admin Hosp, Denver, 72- Mem: Am Asn Surg of Trauma (pres, 74-75); Am Burn Asn (pres, 74-75); Am Soc Surg of Hand (secy-treas, 72-76); Asn Am Med Cols; Int Soc Burn Injuries. Res: Hand surgery. Mailing Add: 4200 E Ninth Ave Denver CO 80220

BOTAN, EDWARD ALLAN, b Dorchester, Mass, Aug 18, 26; m 54; c 2. MICROBIOLOGY, BIOCHEMISTRY. Educ: NY Univ, BA, 48; Northeastern Univ, BS, 49; Univ Mass, MS, 50; McMaster Univ, PhD(microbiol), 57. Prof Exp: Res bacteriologist, Children's Hosp, Boston, 51; instr biol, Russell Sage Col, 52-54; demonstr microbiol, McMaster Univ, 54-56, lectr biol, 56-57; asst prof bact, Lowell Technol Inst, 57-62; sr staff scientist, Avco Res Ctr, Wilmington, Del, 62-67, group leader, Biol Res Group, 67-68, prin staff scientist, 68-69; pres, Hosp Serv Technol Corp, 69-73; DIR PROD DEVELOP, GIBCO DIAGNOSTICS, 73- Concurrent Pos: Vis instr, Lowell Gen Hosp, Mass, 57-62; researcher, USDA, 59; consult, Avco Res Ctr, 60-62. Mem: AAAS; Am Soc Microbiol; Aerospace Med Asn; Sigma Xi; Soc Indust Microbiol. Res: Exobiology; fluorescence microscopy; clinical bacteriology; sterilization. Mailing Add: 12 Marshall Ave Lowell MA 01851

BOTBOL, JOSEPH MOSES, b Cambridge, Mass, Oct 7, 37; m 66; c 1. GEOLOGY. Educ: St Lawrence Univ, BS, 59; Univ Utah, MS, 61, PhD(geol eng), 68. Prof Exp: Geologist explor, Hollinger N Shore Explor Co, Ltd, 58; geologist, Brit Nfld Explor Co, Ltd, 59 & Cummings-Roberts Fluorspar Co, Colo, 59-64; geologist & comput programmer mine & explor data, Kennecott Copper Corp, 64; mineral specialist mineral statist, US Bur Mines, 66; GEOLOGIST MINERAL RESOURCE APPRAISAL, US GEOL SURV, 67- Mem: Soc Explor Geochemists. Res: Computer and mathematical applications to geologic problems, including graphics, multi-variate exploration data, communications, and data bank design and implementation. Mailing Add: 2301 November Lane Reston VA 22091

BOTCH, WILLIAM DOMINIC, chemical physics, see 12th edition

BOTDORF, RUTH G, b Freeburg, Pa. ANALYTICAL CHEMISTRY, INORGANIC CHEMISTRY. Educ: Susquehanna Univ, BA, 45; Columbia Univ, MS, 48; Pa State Univ, PhD(chem), 69. Prof Exp: Instr chem, math & phys sci, DuBois Campus, 48-51, instr chem, Ogontz Campus, 51-61, instr, McKeesport Campus, 65-69, ASST PROF, BERKS CAMPUS, PA STATE UNIV, 69- Mem: AAAS; Am Chem Soc. Res: Solid state mercuric thiocyanates. Mailing Add: RD 5 Box 79 Harrisburg PA 17111

BOTELHO, STELLA YATES, b Japan, Jan 14, 19; US citizen. PHYSIOLOGY. Educ: Univ Pa, BA, 40; Med Col Pa, MD, 49. Prof Exp: From instr to assoc prof, 49-69, PROF PHYSIOL, SCH MED, UNIV PA, 69- Concurrent Pos: Vis prof, Cambridge Univ, 57-58; lectr, US Naval Hosp, Philadelphia, 64-70; sect chief, Philadelphia Gen Hosp, 63-74; consult, Children's Hosp Philadelphia, 63-; mem adv panel regulatory biol prog, NSF, 73- Mem: AAAS; Am Physiol Soc; Am Acad Neurol; Asn Res Nerv & Ment Dis; Asn Res Vision & Ophthal Res: Neuromuscular physiology; secretory mechanisms, especially of the orbital glands; diseases of neuromuscular system; ocular reflexes. Mailing Add: Rittenhouse Claridge Rittenhouse Sq Philadelphia PA 19103

BOTERO, J M, b Medellin, Colombia, Nov 12, 29; m 57; c 2. BIOCHEMISTRY. Educ: Univ Antioquia, BS, 48, MD, 55. Prof Exp: Head dept biochem, Univ Antioquia, 60-61; med & res dir western hemisphere, 61-67, med dir develop mkt, 64-67, med dir, Dome Labs, 67-69, VPRES MED AFFAIRS, AMES CO, MILES LABS, INC, 69- Concurrent Pos: Kellogg fel, 55-57; Rockefeller Found fel, 57-59. Honors & Awards: Cano Award, 48. Mem: NY Acad Sci; AMA; Am Chem Soc. Res: Synthesis of proteins and nucleic acids; chemical pathology. Mailing Add: 1127 Myrtle St Elkhart IN 46514

BOTH, EBERHARD, b Ahrweiler, Ger, Mar 21, 10; nat US; m 38; c 3. PHYSICAL CHEMISTRY. Educ: Univ Bonn, PhD(chem, physics, math), 34. Prof Exp: Sci asst mat res, Vacuumschmelze AG, Ger, 35-43, dir res, 43-47; consult magnetic mat, US Army Electronics Lab, Ft Monmouth, 47-57, chief mat br, 57-62, sr res scientist, 62-75; RETIRED. Mem: AAAS; sr mem Inst Elec & Electronics Engrs; Am Ceramic Soc. Res: Nuclear radiation effects; magnetic, dielectric and ferroelectric materials. Mailing Add: 1505 S Wanamassa Dr Ocean Township NJ 07712

BOTHAST, RODNEY JACOB, b Union City, Ind, Sept 18, 45; m 65; c 3. FOOD MICROBIOLOGY. Educ: Ohio State Univ, BS, 67; Va Polytech Inst & State Univ, MS, 70, PhD(food microbiol), 71. Prof Exp: Grad res asst food microbiol, Va Polytech Inst & State Univ, 67-71; MICROBIOLOGIST FOOD & GRAIN MICROBIOL, NORTHERN REGIONAL RES LAB, AGR RES SERV, USDA, 71- Concurrent Pos: Res assoc microbiol, Peoria Sch Med, Univ Ill, 71- Mem: Inst Food Technologists; Am Soc Microbiol; Sigma Xi. Res: Preservation of high moisture grains, solid substrate fermentations, fungal effects on flavor compounds in foods, food-borne infections and intoxications, mycotoxins and fungal interactions. Mailing Add: 1815 N University Peoria IL 61604

BOTHNER, RICHARD CHARLES, b New York, NY, May 16, 29; m 58; c 3. ECOLOGY, ANATOMY. Educ: Fordham Univ, BS, 51, MS, 57, PhD(biol), 59. Prof Exp: Asst biol, Fordham Univ, 55-57; mus asst herpet, Am Mus Natural Hist, New York, 57-58; from asst prof to assoc prof, 58-69, PROF BIOL, ST BONAVENTURE UNIV, 69- Mem: Am Soc Ichthyol & Herpet; Ecol Soc Am. Res: Herpetology; phylogenetic significance of ophidian circulatory systems. Mailing Add: Dept of Biol St Bonaventure Univ St Bonaventure NY 14778

BOTHNER, WALLACE ARTHUR, b Fitchburg, Mass, May 17, 41; m 63; c 3. PETROLOGY. Educ: Harpur Col, BA, 63; Univ Wyo, PhD(geol), 67. Prof Exp: Instr geol, Univ Wyo, 66-67; asst prof, 67-73, ASSO C PROF GEOL, UNIV NH, 73- Mem: AAAS; Geol Soc Am; Am Geophys Union; Mineral Soc Am. Res: Igneous and metamorphic petrology; metamorphic structural geology and the application of gravimetry to the study of subsurface igneous bodies. Mailing Add: Dept of Earth Sci Univ of NH Durham NH 03824

BOTHNER-BY, AKSEL ARNOLD, b Minneapolis, Minn, Apr 29, 21; m 49; c 2. BIOPHYSICAL CHEMISTRY. Educ: Univ Minn, BChem, 43; NY Univ, MS, 47; Harvard Univ, PhD(chem), 49. Prof Exp: Assoc scientist chem, Brookhaven Nat Lab, 49-53, scientist, 53; instr, Harvard Univ, 53-55, lectr, 55-58; staff fel, 58-59, from asst dir res to dir res, 59-62, mem adv comt, 62-67, chmn dept chem, 67-70, dean, Mellon Inst Sci, 71-75, PROF CHEM, CARNEGIE-MELLON UNIV, 70- Concurrent Pos: Am Cancer Soc fel, Univ Zurich, 52-53; consult, Retina Found, 57; Fulbright lectr, Univ Munich, 62-63; adj prof, Univ Pittsburgh, 64- Mem: Am Chem Soc. Res: Reaction mechanisms; nuclear magnetic resonance; structure of biopolymers. Mailing Add: Dept of Chem Carnegie-Mellon Univ Pittsburgh PA 15213

BOTHWELL, ALFRED LESTER MEADOR, b Springfield, Mo, Apr 29, 49; m 74. MOLECULAR BIOLOGY. Educ: Wash Univ, AB, 71; Yale Univ, MPh, 74, PhD(biol), 75. Prof Exp: Fel virol, Cold Spring Harbor Lab, 75-76; FEL CTR CANCER RES, MASS INST TECHNOL, 76- Concurrent Pos: Jane Coffin Childs Mem Fund Med Res fel, 75-77. Mem: Am Soc Microbiol; Sigma Xi. Res: RNA transcription and processing mechanisms of DNA tumor viruses. Mailing Add: Ctr Cancer Res Mass Inst Technol Cambridge MA 02139

BOTHWELL, MARVIN RALPH, physical chemistry, see 12th edition

BOTHWELL, MAX LEWIS, b Portsmouth, Va, Dec 5, 46; m 73. LIMNOLOGY. Educ: Univ Calif, Santa Barbara, BA, 68, MA, 72; Univ Wis-Madison, PhD(zool), 75. Prof Exp: Res assoc limnol, Ctr Gt Lakes Studies, 71-75; RES ASSOC STREAM ECOL, WEYERHAEUSER CO, 75- Mem: Am Soc Limnol & Oceanog; Int Soc Theoret & Appl Limnol. Res: Phytoplankton ecology; nutrient limitation and nutrient cycling; photosynthetic pigments and carbon fixation rates. Mailing Add: Res Lab A Weyerhaeuser Co Longview WA 98632

BOTIMER, LAURENCE WALLACE, b Jackson, Mich, Sept 29, 18; m 42; c 2. ORGANIC CHEMISTRY. Educ: Emmanuel Missionary Col, BA, 42; Univ Md, MS, 56, PhD(org chem), 59. Prof Exp: Assoc prof chem, Washington Missionary Col, 47-60; PROF CHEM, LOMA LINDA UNIV, 60- Mem: Am Chem Soc. Res: Reactions of heterocyclic compounds. Mailing Add: Dept of Chem Loma Linda Univ Loma Linda CA 92354

BOTKIN, DANIEL BENJAMIN, b Oklahoma City, Okla, Aug 19, 37; m 60; c 2. ECOLOGY. Educ: Univ Rochester, AB, 59; Univ Wis-Madison, MS, 62; Rutgers Univ, PhD(biol), 68. Prof Exp: Asst prof ecol, 68-74, ASSOC PROF ECOL, SCH FORESTRY & ENVIRON STUDIES, YALE UNIV, 74- Concurrent Pos: NSF res grants, 69-71 & 71-74; res grant, World Wildlife Fund, 74-76; mem planning & coord comt environ studies, Nat Acad Sci, 74-75; vis assoc prof, Ecosysts Ctr, Marine Biol Lab, Woods Hole, Mass, 75-76. Mem: AAAS; Ecol Soc Am; Am Inst Biol Sci; Sigma Xi; Am Soc Naturalists. Res: Mineral cycling and energy flow in ecosystems; ecosystem theory and models; interactions among plants and animals; population dynamics of long-lived and endangered species. Mailing Add: Ecosysts Ctr Marine Biol Lab Woods Hole MA 02543

BOTKIN, MERWIN P, b Fruita, Colo, Sept 19, 22; m 48; c 4. ANIMAL SCIENCE. Educ: Univ Wyo, BS, 48, MS, 49; Okla State Univ, PhD(animal breeding), 52. Prof Exp: Instr animal prod, Univ Wyo, 48-49; instr animal husb, Okla State Univ, 51-52; PROF ANIMAL BREEDING, UNIV WYO, 52- Mem: Am Soc Animal Sci. Res: Sheep breeding and physiology. Mailing Add: Animal Sci Div Univ of Wyo Laramie WY 82070

BOTROS, RAOUF, b Mit-Ghamr, Egypt, Aug 28, 32; m 67; c 1. CHEMISTRY. Educ: Alexandria Univ, Egypt, BSc, 55; Duquesne Univ, PhD(org chem), 62. Prof Exp: Res supvr dyestuff & intermediates, Egypt Dyes & Intermediates Corp, 62-67; SR RES CHEMIST & GROUP LEADER DYESTUFFS & INTERMEDIATES, AM COLOR & CHEM CORP, SUBSID N AM PHILIPS, 68- Mem: Am Chem Soc; Am Asn Textile Chemists & Colorists. Res: Synthesis of novel disperse dyes in the azo and anthraquinone classes for polyesters, acetates, nylon and polypropylene fibers; special interest in making intermediates and developing existing products. Mailing Add: Am Color & Chem Corp Lock Haven PA 17745

BOTSFORD, JAMES L, b Spokane, Wash, June 6, 42; m 64; c 1. MICROBIAL PHYSIOLOGY, BIOCHEMISTRY. Educ: Univ Idaho, BS, 64; Ore State Univ, PhD(microbial physiol), 68. Prof Exp: NIH fel, Univ Ill, 68-70; asst prof, 70-74, ASSOC PROF BIOL, N MEX STATE UNIV, 74- Mem: AAAS; Am Soc Microbiol. Res: Metabolism of methyl groups in yeast; regulation of enzymes synthesis in microorganisms; ecological significance of microbial regulatory mechanisms; genetics of tryptophanase in Escherichia coli. Mailing Add: Dept of Biol N Mex State Univ Las Cruces NM 88001

BOTSFORD, JAMES LAWRENCE, mathematics, deceased

BOTSTEIN, CHARLES, b Lodz, Poland, Jan 18, 11; m 38; c 3. RADIOLOGY. Educ: Univ Zurich, MD, 35; Am Bd Radiol, dipl, 55. Prof Exp: Intern, Lodz City Hosp, Poland, 35-36; asst path, Univ Zurich, 37-38, asst radiol, 38-46, sr asst, radiother dept, 46-49; adj radiotherapist, Mt Sinai Hosp, NY, 49-54; asst roentgenologist, Mem Hosp, 54-57; PROF RADIOTHER & HEAD DEPT, ALBERT EINSTEIN COL MED, 56-; HEAD DEPT, MONTEFIORE HOSP, 57- Mem: Am Radium Soc; AMA; fel Am Col Radiol; Am Roentgen Ray Soc; Radiol Soc NAm. Res: Radiotherapy. Mailing Add: Montefiore Hosp 210th St & Bainbridge Ave New York NY 10467

BOTSTEIN, DAVID, b Zurich, Switz, Sept 8, 42; US citizen; m 65. MOLECULAR GENETICS, MICROBIAL PHYSIOLOGY. Educ: Harvard Univ, AB, 63; Univ Mich, PhD(human genetics), 67. Prof Exp: Instr biol, 67-69, asst prof genetics, 69-74, ASSOC PROF GENETICS, MASS INST TECHNOL, 74- Mem: AAAS; Am Soc Microbiol; Genetics Soc Am. Res: Desoxyribonucleic acid replication; control of gene expression in temperate phages, bacteria and yeast. Mailing Add: Dept Biol Rm 56-721 Mass Inst of Technol Cambridge MA 02139

BOTT, KENNETH F, b Albany, NY, Dec 19, 36; div; c 2. MICROBIOLOGY, GENETICS. Educ: St Lawrence Univ, BS, 58; Syracuse Univ, MS, 60, PhD(microbiol), 63. Prof Exp: From instr to asst prof microbiol, Univ Chicago, 62-71; ASSOC PROF MICROBIOL, UNIV NC, CHAPEL HILL, 72- Concurrent Pos: Trainee microbial genetics & virol, Univ Chicago, 63-64; Merck & Co Found grant fac develop, 69; Fulbright grant & vis prof, Univ Paris, Orsay, 71. Mem: AAAS; Am Soc Microbiol; Genetics Soc Am; Am Chem Soc. Res: Regulation mechanisms for translation of genetic information into proteins; use of micro-organisms as tools for studying cellular differentiation; endospore formation in Bacillus species and those

Bacillus species exhibiting toxicity for insects. Mailing Add: Dept of Bact & Immunol Univ of NC Div of Health Affairs Chapel Hill NC 27514

BOTT, RAOUL H, b Budapest, Hungary, Sept 24, 23; m 47; c 2. MATHEMATICS. Educ: McGill Univ, MEng, 46; Carnegie Inst Technol, DrSc(math), 49. Prof Exp: Mem, Inst Adv Study, 49-51, 55-57; from instr to prof math, Univ Mich, 51-59; prof, 59-74, HIGGINS PROF MATH, HARVARD UNIV, 74- Honors & Awards: Veblen prize, Am Math Soc, 64. Mem: Nat Acad Sci; Am Math Soc. Res: Topology. Mailing Add: Dept of Math Harvard Univ Cambridge MA 02138

BOTT, THOMAS LEE, b Jersey City, NJ, June 15, 40; m 64; c 1. MICROBIOLOGY. Educ: Wheaton Col Ill, BS, 62; Univ Wis-Madison, MS, 65, PhD(bact, zool), 68. Prof Exp: NIH fel, Ind Univ, 68-69; ASST CUR AQUATIC MICROBIOL, LIMNOLOGY DEPT, ACAD NATURAL SCI PHILADELPHIA, 69- Concurrent Pos: Adj asst prof, Biol Dept, Univ Pa, 72- Mem: AAAS; Am Soc Microbiol; Am Soc Limnol & Oceanog. Res: Microbial ecology; aquatic microbiology; Clostridium botulinum type E in the Great Lakes; extreme thermophilic bacteria; bacterial and fungal decomposition activity; primary productivity. Mailing Add: Acad Natural Sci Philadelphia Stroud Water Res Ctr Box 286 Avondale PA 19311

BOTTA, JAMES ANTHONY, JR, physiology, pharmacology, see 12th edition

BOTTEI, RUDOLPH SANTO, b Old Forge, Pa, July 19, 29; m 60; c 4. ANALYTICAL CHEMISTRY, INORGANIC CHEMISTRY. Educ: Wilkes Col, AB, 50; Cornell Univ, MS, 52; Princeton Univ, PhD, 56. Prof Exp: Instr to assoc prof, 55-71, asst head dept, 66-74, PROF CHEM, UNIV NOTRE DAME, 74- Mem: Am Chem Soc. Res: Electroanalytical chemistry; complex ions; organometallic compounds; boron compounds; organic reduction. Mailing Add: Dept of Chem Univ of Notre Dame Notre Dame IN 46556

BOTTERELL, EDMUND HARRY, b Vancouver, BC, Feb 28, 06; m 35; c 2. NEUROSURGERY. Educ: Univ Man, MD, 30; Univ Toronto, MS, 36, FRCS(C), 37. Hon Degrees: DSc, McGill Univ, 71; LLD, Queen's Univ, Can, 73. Prof Exp: House surgeon, Winnipeg Gen Hosp, 29-30, res gen surg, 30-31; res physician, Montreal Gen Hosp, 31-32; tutor physiol & demonstr anat, Univ Toronto, 32-33; resident surgeon, Toronto Gen Hosp, 33-34; intern & extern neurol & neurosurg, Nat Hosp, London, Eng, 34-35; lectr neurophysiol & res neurosurgeon, Toronto Gen Hosp, 36-37, lectr & jr neurosurgeon, 37-39, consult neurosurgeon, 37-40, sr neurosurgeon, 45-62; assoc prof neurosurg, Univ Toronto, 52-62; dean fac med, 62-70, vprin health sci, 70-72, PROF CLIN NEUROANAT & SURG NEUROL, QUEEN'S UNIV, CAN, 72- Concurrent Pos: Consult neurosurgeon, Sunnybrook Hosp & Armed Forces Med Coun, 54-; mem comt health care syst, Ont Ministry Correctional Serv, 71-72; chmn, Nat Health Serv Adv Comt Can Penitentiary Serv, 73-; mem comt animal health care serv, Ont Ministry Agr & Food, 74- Mem: Am Acad Neurol Surg; Am Asn Neurol Surg; Can Med Asn; hon mem Brit Soc Neurosurg; hon mem Am Neurol Asn. Res: Brain injuries and complications; localized encephalitis, brain abscess and subdural empyema; paraplegia following war; experimental use of hypothermia in the neurosurgical management of ruptured aneurysms and arteriovenous cerebral malformations. Mailing Add: Queen's Univ Kingston ON Can

BOTTERON, DONALD GEORGE, b Columbus, Ohio, July 29, 16; m 41; c 2. ORGANIC CHEMISTRY. Educ: Univ Ill, AB, 36, MS, 37; Northwestern Univ, PhD(org chem), 42. Prof Exp: From instr to asst prof, 43-52, ASSOC PROF CHEM, SYRACUSE UNIV, 52- Mem: Am Chem Soc; Brit Chem Soc. Res: Amide pyrolyses; cyclopropanes; molecular structure; rearrangements. Mailing Add: Dept of Chem Syracuse Univ Syracuse NY 13201

BOTTGER, GARY LEE, b Los Angeles, Calif, Oct 27, 38; m 65. PHYSICAL CHEMISTRY. Educ: Univ Southern Calif, BS, 60; Univ Wash, PhD, 64. Prof Exp: Res asst infrared spectros, Univ Wash, 62-63, sr res chemist, 64-71, LAB HEAD, EASTMAN KODAK CO, 71- Mem: Sigma Xi; Soc Appl Spectros; AAAS; Am Chem Soc; Am Phys Soc. Res: Infrared and Raman spectroscopy; molecular structure; vibrational spectra of solids. Mailing Add: Eastman Kodak Co Res Labs Rochester NY 14650

BOTTGER, GILBERT TED, b Ollie, Iowa, Aug 6, 05; m 28; c 1. ECONOMIC ENTOMOLOGY. Educ: Iowa Wesleyan Col, BS, 27. Prof Exp: Asst entomol, Univ Ill, 27-28; biol aide, USDA, 28, entomologist res, 28-75; RETIRED. Concurrent Pos: Hqs entomologist, Foreign Agr Serv, Iran, 53-55. Mem: Am Entom Soc. Res: Toxicology; laboratory testing to evaluate insecticidal effectiveness of chemical compounds and plant materials; plant resistance to insect attack; the nutritive requirements of insects in relation to nutrients of the host; effects of varietal and seasonal differences; ecological investigations of Anthonomus weevils in Arizona and Sonora, Mexico. Mailing Add: 12205 San Tomas Ct San Diego CA 92128

BOTTICELLI, CHARLES ROBERT, physiology, see 12th edition

BOTTINI, ALBERT THOMAS, b San Rafael, Calif, June 13, 32; m 53; c 3. ORGANIC CHEMISTRY. Educ: Univ Calif, BS, 54; Calif Inst Technol, PhD(chem), 57. Prof Exp: From instr to assoc prof, 58-68, PROF CHEM, UNIV CALIF, DAVIS, 68- Mem: Am Chem Soc. Res: Small ring imines; reaction mechanisms. Mailing Add: Dept of Chem Univ of Calif Davis CA 95616

BOTTINO, MICHAEL LOUIS, b New York, NY, Dec 2, 35. GEOCHEMISTRY. Educ: Mass Inst Technol, BS, 59, PhD(geochem), 63. Prof Exp: Res asst, Mass Inst Technol, 59-60; asst prof chem, Old Dom Col, 63-64, assoc prof geol, 64-67; vpres, Ajax Tile & Marble Corp, 67-68; assoc prof geol, Marshall Univ, 68-70; vis assoc prof, Univ NC, 70-74; MEM FAC, DEPT GEOL, BROOKLYN COL, 74- Concurrent Pos: VPres, Ajax Co, Inc, 61-; res geochemist, Planetology Br, Goddard Flight Ctr, NASA, Md, 64- Mem: AAAS; Am Geophys Union; Geol Soc Am; Geochem Soc. Res: Geochemistry; geochronology; environmental chemistry. Mailing Add: Dept of Geol Brooklyn Col Brooklyn NY 11210

BOTTINO, NESTOR RODOLFO, b La Plata, Arg, Sept 3, 25; m 53; c 2. BIOCHEMISTRY. Educ: La Plata Nat Univ, DrChem, 54. Prof Exp: Asst prof biochem, La Plata Nat Univ, 56-59; Welch Found fel biochem & biophys, Tex A&M Univ, 59-61; assoc prof biochem, La Plata Nat Univ, 61-63; res assoc, 64-65, from asst prof to assoc prof, 65-74, PROF BIOCHEM & BIOPHYS, TEX A&M UNIV, 75- Mem: AAAS; Am Oil Chem Soc; Am Chem Soc; Soc Exp Biol & Med; Am Soc Biol Chemists. Res: Lipid metabolism and chemistry. Mailing Add: Dept of Biochem & Biophys Tex A&M Univ College Station TX 77843

BOTTINO, PAUL JAMES, b Price, Utah, Aug 3, 41; m 63; c 1. PLANT GENETICS. Educ: Utah State Univ, BS, 64, MS, 65; Wash State Univ, PhD(genetics), 69. Prof Exp: Res assoc, Brookhaven Nat Lab, 69-73; ASST PROF BOT, UNIV MD, COLLEGE PARK, 73- Mem: AAAS; Genetics Soc Am. Res: In vitro culture of plant cells; genetics of somatic plant cells; induced mutation in plant cells; application of tissue culture techniques to agricultural problems. Mailing Add: Dept of Bot Univ of Md College Park MD 20742

BOTTJER, WILLIAM GEORGE, b New York, NY, Feb 20, 31; m 58; c 2. INORGANIC CHEMISTRY. Educ: Hofstra Univ, BA, 53; Univ Conn, MS, 55; Univ NH, PhD(inorg chem), 64. Prof Exp: Res chemist, Gen Chem Res Lab, Allied Chem Corp, NJ, 55-59; res chemist org chem dept, Jackson Lab, 63-65, develop chemist, 65-68, sr res chemist, Photo Prod Dept, 68-72, SR RES CHEMIST, PHOTO PROD PLANT, E I DU PONT DE NEMOURS & CO, 72- Mem: Am Chem Soc; Soc Photog Scientists & Engrs; AAAS. Res: Solid state research; magnetic materials; transition metal and fluorine chemistry; metal oxide chemistry; photographic chemistry. Mailing Add: Photo Prod Plant E I du Pont de Nemours & Co Parlin NJ 08859

BOTTKA, NICHOLAS, b Huszt, Hungary, Dec 6, 39; US citizen; m 66; c 2. SOLID STATE PHYSICS, OPTICAL PHYSICS. Educ: Univ Calif, Los Angeles, BS, 63, MS, 66; Tech Univ, Berlin, PhD(physics), 70. Prof Exp: RES PHYSICIST, MICHELSON LAB, NAVAL WEAPONS CTR, US DEPT NAVY, 63- Mem: Am Phys Soc. Res: Optical properties of solids. Mailing Add: Code 6019 Michelson Lab Naval Weapons Ctr US Dept Navy China Lake CA 93555

BOTTOM, VIRGIL ELDON, b Douglas, Kans, Jan 6, 11; m 32; c 3. PHYSICS. Educ: Friends Univ, AB, 31; Univ Mich, MS, 38; Purdue Univ, PhD, 49. Prof Exp: Teacher high sch, Kans, 33-38; prof physics & head dept, Friends Univ, 38-42; asst prof, Colo State Univ, 42-43, assoc prof, 46-47; asst instr, Purdue Univ, 47-49; prof, Colo State Univ, 49-53, actg head dept, 52-53; res physicist, Motorola Res Lab, 53-58; prof, 58-73, EMER PROF PHYSICS, McMURRY COL, 73-; CONSULT, TYCO FILTERS DIV, TYCO LABS, INC, 73- Concurrent Pos: Chmn working group frequency control devices, US Dept Defense, 58-; Fulbright lectr, Univ Sao Paulo, 64-65. Mem: Am Phys Soc; Am Asn Physics Teachers; Inst Elec & Electronics Engrs; Phys Soc Brazil. Res: Semiconductors; piezoelectricity and applications to communication; ferromagnetism. Mailing Add: 3441 High Meadows Abilene TX 79605

BOTTOMLEY, FRANK, b Hatfield, Eng, May 14, 41. INORGANIC CHEMISTRY. Educ: Univ Hull, BS, 63, MS, 65; Univ Toronto, PhD(chem), 68. Prof Exp: Fel lectr chem, Univ Toronto, 67-69; asst prof, 69-74, ASSOC PROF CHEM, UNIV NB, 74- Mem: The Chem Soc; Can Inst Chem. Res: Reactions of coordinated ligands; transition metal chemistry. Mailing Add: Dept of Chem Univ of NB Fredericton NB Can

BOTTOMLEY, RICHARD H, b Arkansas City, Kans, Nov 24, 33; m 58; c 2. INTERNAL MEDICINE, ONCOLOGY. Educ: Univ Okla, BS, 54, MD, 58; Am Bd Internal Med, dipl. Prof Exp: Intern, Salt Lake County Gen Hosp, 58-59; resident internal med, Med Ctr, 59-61, clin asst, Sch Med, 61-64, instr internal med & sr investr oncol, 64-65, instr res med & asst prof res biol in biochem, 65-68, res med, 67-71, assoc prof med, 71-75, PROF MED, SCH MED, UNIV OKLA, 75-, ASSOC PROF RES MOLECULAR BIOL, 68-, HEAD ONCOL DIV, 72- Concurrent Pos: Nat Cancer Inst fels, McArdle Mem Lab, 61-62 & Okla Med Res Found, 61-63, asst head cancer sect, 65-72. Mem: Am Fedn Clin Res; AMA; Am Soc Clin Oncol; fel Am Col Physicians. Res: Biology, biochemistry and chemotherapy of human neoplastic disease. Mailing Add: Okla Med Res Found 825 NE 13th St Oklahoma City OK 73104

BOTTOMLEY, SYLVIA STAKLE, b Riga, Latvia, Mar 9, 34; US citizen; m 58; c 2. INTERNAL MEDICINE, HEMATOLOGY. Educ: Okla State Univ, BS, 54; Univ Okla, MD, 58; Am Bd Internal Med, dipl, 65, cert hemat, 72. Prof Exp: Intern med, Sch Med, Univ Utah, 59; resident, Med Ctr, 59-61, clin asst, Sch Med, 61-64, instr, 64-67, from asst prof to assoc prof med, 67-75, PROF MED & ASSOC PROF PATH, SCH MED, UNIV OKLA, 75-; ASST CHIEF HEMAT SECT, VET ADMIN HOSP, 69- Concurrent Pos: USPHS res fel hemat, 61-64; res assoc, Vet Admin Hosp, Oklahoma City, 64-65, clin investr, 65-68. Mem: Sigma Xi; Am Soc Hemat; fel Am Col Physicians. Res: Clinical hematology and research in erythropoiesis, especially heme synthesis and biochemical control mechanisms. Mailing Add: Hemat Sect Vet Admin Hosp Oklahoma City OK 73104

BOTTOMS, ALBERT MAITLAND, b Montclair, NJ, Sept 2, 25; m 57; c 2. OPERATIONS RESEARCH, SYSTEMS ANALYSIS. Educ: Univ Pa, BSChE, 49; Iowa State Univ, MS, 51; Mass Inst Technol, MS, 62. Prof Exp: Asst physics, Iowa State Univ, 49-51; res chemist, Am Cyanamid Co, 51-52; opers analyst, Opers Eval Group, Mass Inst Technol, 52-58; opers analyst, Weapons Systs Eval Group, Inst Defense Anal, US Dept Defense, 58-63, opers analyst, Res Anal Corp, 63-64, opers analyst, Weapons Systs Eval Group, Inst Defense Anal, 64-67; opers analyst, Off Chief Naval Opers, 67-68; dir opers res task force, Chicago Police Dept, 68-69; dir pub safety div, Urban Systs Res & Eng, Mass, 69-70; PRES, FUNDAMENTAL SYSTS, INC & RES ASSOC, MASS INST TECHNOL-HARVARD JOINT CTR URBAN STUDIES, 70-; WITH US NAVY UNDERWATER SYSTS CTR, 73- Concurrent Pos: Consult, US Mil Acad, 43-45, Nat Comn Causes & Prev Violence, 68-69 & Nat Acad Sci, 69- Mem: Fel AAAS; Opers Res Soc Am; Am Geophys Union; fel Geol Soc Am; fel Am Inst Chemists. Res: Public services systems analysis; military operations research; criminal justice system; police; undersea warfare. Mailing Add: Fundamental Systs Inc PO Box 432 Monument Beach MA 02553

BOTTOMS, GERALD DOYLE, b Holdenville, Okla, Apr 10, 30; m 52; c 3. PHYSIOLOGY, ENDOCRINOLOGY. Educ: ECent State Col, BS, 55; Okla State Univ, MS, 58, PhD(physiol), 66. Prof Exp: Sci teacher, Holdenville, Okla, 55-63; asst prof, 66-74, ASSOC PROF PHYSIOL, PURDUE UNIV, 74- Mem: AAAS; Am Physiol Soc. Res: Endocrinology as related to action sites of adrenal cortex steroids; binding sites of glucocorticoids within tissues; glucocorticoid effects on nucleic acid metabolism. Mailing Add: Dept of Vet Physiol Purdue Univ W Lafayette IN 47906

BOTTONE, EDWARD JOSEPH, b New York, NY, Feb 18, 34; m 62; c 2. MICROBIOLOGY. Educ: City Col New York, BS, 65; Wagner Col, MS, 68; St John's Univ, PhD(biol), 73. Prof Exp: Bacteriologist, Greenpoint Hosp, 62-64; sr bacteriologist, Elmhurst City Hosp, 64-68; from asst microbiologist to assoc microbiologist, 69-73, actg dir microbiol, 74-75, DIR MICROBIOL, MT SINAI HOSP, 75-; ASST PROF MICROBIOL, MT SINAI SCH MED, 73- Concurrent Pos: Consult, Elmhurst City Hosp of Mt Sinai Serv, 74- Mem: Am Soc Microbiol. Res: Clinical microbiology. Mailing Add: Dept of Microbiol Mt Sinai Hosp New York NY 10029

BOTTORFF, EDMOND MILTON, b Milroy, Ind, Sept 14, 16; m 48; c 2. ORGANIC CHEMISTRY. Educ: Hanover Col, AB, 37; Univ Ill, PhD(org chem), 41. Prof Exp: Res chemist, Rohm and Haas Co, Philadelphia, 41-47; RES CHEMIST, ELI LILLY & CO, 47- Mem: Am Chem Soc. Res: Action of the Grignard reagent on sterically hindered esters. Mailing Add: Eli Lilly & Co 231 E McCarty St Indianapolis IN 46206

BOTTS, JEAN, physical biochemistry, see 12th edition

BOTTS, TRUMAN ARTHUR, b De Land, Fla, Nov 26, 17; m 44, 64, 74; c 2. MATHEMATICS. Educ: Stetson Univ, BS, 38; Univ Va, MA, 40, PhD(math), 42. Prof Exp: Instr math, Univ Va, 41-42; asst prof, Univ Del, 46-48; from asst prof to assoc prof, Univ Va, 48-68; EXEC DIR, CONF BD MATH SCI, 68- Concurrent Pos: Ford Found fel, 53-54; fel comput sci, Nat Bur Stand, 59; vis prof, Univ PR, 63-64; exec dir comt support res math sci, Nat Acad Sci-Nat Res Coun, Columbia Univ, 65-66. Mem: AAAS; Am Math Soc; Math Asn Am. Res: Real function theory; calculus of variations; convex sets. Mailing Add: Conf Bd of Math Sci 2100 Pennsylvania Ave NW Washington DC 20037

BOUBOULIS, CONSTANTINE JOSEPH, b Piraeus, Greece, Feb 23, 28; US citizen; m 60; c 3. ORGANIC CHEMISTRY. Educ: Nat Univ Athens, dipl, 53; Columbia Univ, MA, 57; Univ Colo, PhD(org chem), 61. Prof Exp: Chemist, John Lafis & Co, Greece, 54-55; from res chemist to sr res chemist, 61-73, STAFF CHEMIST, EXXON RES & ENG CO, INC, 73- Mem: AAAS; Am Chem Soc; Am Inst Chemists; Sigma Xi. Res: Steroids; synthesis and rearrangements of bicyclic systems; condensation polymers; oxidation; amines; metal mining by solvent extraction. Mailing Add: 661 Golf Terr Union NJ 07083

BOUCHAL, ALEXANDER WAYNE, b Philadelphia, Pa, May 19, 24; m 46; c 2. COSMETIC CHEMISTRY. Educ: Pa State Univ, BS, 49, MS, 51, PhD(chem), 52. Prof Exp: Sr res chemist, 52-57, group leader skin prod, 57-60, sect head bioactive oral prod, 60-64, sect head oral prod, 64-67, assoc dir med res serv, 67-75, COORDR MED SERV, COLGATE-PALMOLIVE CO, 75- Mem: AAAS; Am Chem Soc. Res: Skin chemistry; oral biology; toxicology; industrial hygiene. Mailing Add: 671 Shadowlawn Dr Westfield NJ 07090

BOUCHARD, LOUIS-MARIE, b Chicoutimi, PQ, Aug 9, 38; m 64. URBAN GEOGRAPHY, CARTOGRAPHY. Educ: Univ Sherbrooke, BA, 59; Univ Montreal, lic geog, 62; Univ Strasbourg, Dr(geog), 71. Prof Exp: Prof geog, Ctr Teacher Develop, Chicoutimi, 63-69; PROF GEOG, UNIV QUEBEC, CHICOUTIMI, 69- Mem: Asn Can Geog; Fr-Can Asn Advan Sci. Res: Geography of small towns and Indian reserves; regional geography of the Saguenay Lake-St John area, Quebec. Mailing Add: Dept of Geog Univ of Quebec Chicoutimi PQ Can

BOUCHARD, RAYMOND WILLIAM, b Ft Bragg, Calif, May 28, 44; m 73; c 1. FRESH WATER BIOLOGY. Educ: Mass State Col, BSEd, 67; Univ Tenn, PhD(zool), 72. Prof Exp: Res assoc benthos & nekton, Ctr Wetland Resources, La State Univ, 73-74; RES BIOLOGIST CRUSTACEANS, OFF ENDANGERED SPECIES, US DEPT INTERIOR, 75- Concurrent Pos: Instr ichthyol, Reelfoot Lake Biol Sta, Tenn, 70; consult, Dept Anthrop, Univ Tenn & Tenn Valley Authority, Norris, Tenn, 72; instr zool, Univ Tenn, 73; fel, Nat Mus Natural Hist, Smithsonian Inst, 74-75. Mem: Am Fisheries Soc; Am soc Ichthyologists & Herpetologists; Int Asn Astacol; NAm Benthlog Soc; Soc Syst Zool. Res: Systematics; ecology, zoogeography and evolution of freshwater decapod crustaceans and fishes. Mailing Add: Dept of Invert Zool W-114 Mus Natural Hist Smithsonian Inst Washington DC 20560

BOUCHARD, RICHARD EMILE, b Colchester, Vt, Mar 31, 26; m 56; c 6. CARDIOLOGY, INTERNAL MEDICINE. Educ: Univ Vt, MD, 49, MS, 51. Prof Exp: Fel physiol, Nat Heart Inst, 50-51, fel cardiol, 54-55; from instr to asst prof, 55-65, ASSOC PROF MED, COL MED, UNIV VT, 65- Concurrent Pos: Fel coun clin cardiol, Am Heart Asn, 68. Mem: AMA; fel Am Col Physicians; fel Am Col Cardiol. Res: Clinical research in cardiology. Mailing Add: DeGoesbriand Unit Med Ctr Hosp of Vt Burlington VT 05401

BOUCHARD, ROBERT JOSEPH, inorganic chemistry, solid state chemistry, see 12th edition

BOUCHER, BERTRAND PHILLIP, b Fall River, Mass, Apr 3, 23; m 46; c 3. ECONOMIC GEOGRAPHY, GEOGRAPHY OF THE MIDDLE EAST. Educ: Univ Colo, BA, 50, MA, 51. Prof Exp: Chmn dept geog, 52-73, CHMN DEPT GEOG & URBAN STUDIES, MONTCLAIR STATE COL, 73- Concurrent Pos: Co-dir, NJ Land Use Surv, 59-61; vis prof, Eve Div, City Col New York, 68 & 69-70; consult & contribr, Gen Drafting Co & McGraw-Hill Bk Co, 68-69; consult, Indust Comn, Byram, NJ, 69. Mem: Am Geog Soc; Mid East Studies Asn; Asn Am Geogr; Nat Coun Geog Educ. Res: Problems of industrial location in New Jersey; petro dollar investments and economic development in the Arab oil producing states. Mailing Add: 19 Meadow Dr Little Falls NJ 07424

BOUCHER, FRANCIS R, water chemistry, see 12th edition

BOUCHER, GARY WYNN, b Colorado Springs, Colo, Nov 26, 41; m 64; c 2. GEOPHYSICS, SEISMOLOGY. Educ: Colo Col, BS, 63; Columbia Univ, PhD(geol), 69. Prof Exp: Res asst geophys, Lamont Geol Observ, 65-68; asst prof geophys & asst res geophysicist, Mackay Sch Mines, Univ Nev, Reno, 68-72; res assoc, Coop Inst Res in Environ Sci, Univ Colo, Boulder, 73-74; GEOLOGIST, US GEOL SURV, OFF MARINE GEOL & ENERGY RESOURCES, 74- Concurrent Pos: US Air Force Off Sci Res grant, 68-72; AEC res contract, 69-72; fel, Coop Inst Res in Environ Sci, Univ Colo, Boulder, 72-73. Mem: Soc Explor Geophys; Am Geophys Union; Seismol Soc Am. Res: Data analysis and modeling applied to seismic and geopotential field problems of continental margins, geotectonics, and petroleum exploration; collection of marine geophysical data. Mailing Add: US Geol Surv 345 Middlefield Rd Menlo Park CA 94025

BOUCHER, LAURENCE JAMES, b Yonkers, NY, Sept 16, 38; m 64; c 2. INORGANIC CHEMISTRY. Educ: Mich State Univ, BS, 61; Univ Ill, MS, 62, PhD(inorg chem), 64. Prof Exp: Resident res assoc, Chem Div, Argonne Nat Lab, 64-66; asst prof, 66-71, ASSOC PROF CHEM, CARNEGIE-MELLON UNIV, 71-, FEL, MELLON INST, 67- Mem: AAAS; Am Chem Soc. Res: Chemistry of coordination compounds; role of metal ions in biological systems; homogeneous and heterogeneous catalysis. Mailing Add: Dept of Chem Carnegie-Mellon Univ Pittsburgh PA 15213

BOUCHER, LOUIS JACK, b Ashland, Wis, May 24, 22; m 49; c 4. DENTISTRY, ANATOMY. Educ: Marquette Univ, DDS, 53, PhD(anat), 61. Prof Exp: Resident prosthodontics, Vet Admin Hosp, Wood, Wis, 53-55; instr, Sch Dent, Marquette Univ, 55-60, assoc prof, 60-65, dir grad dept, 63-65, dir postgrad dept, 64-65, assoc anat, Sch Med, 60-65; coordr grad studies & res, Col Dent, Univ Ky, 65-66; from asst dean to assoc dean sch dent, Med Col Ga, 66-71; PROF PROSTHODONTICS & DEAN SCH DENT, FAIRLEIGH DICKINSON UNIV, 71- Concurrent Pos: Spec fel grant, USPHS, 61-62, res career develop award, 62-65; consult, Vet Admin Hosp, Wood, Wis, 61-64, Nat Inst Dent Res, 64-66, Div Chronic Dis, USPHS, 65-67, Vet Admin Hosp, Augusta, Ga, 67-, US Army, Ft Jackson, SC, 67-, Ft Gordon, Ga, 68- & US Air Force Surgeon Gen, 69-; pres, Fedn Prosthodont Orgns, 67. Honors & Awards: Int Res Award, Int Asn Dent Res, 70. Mem: Am Acad Plastics Res in Dent; Am Equilibration Soc (pres, 67); Am Prosthodontic Soc; Int Asn Dent Res; Am Dent Asn. Res: Functional anatomy of temporomandibular joint; response of the stratum corneum to various dental materials; epithelial reaction to various implants; cineradiographic study of the temporomandibular joint. Mailing Add: Sch of Dent Fairleigh Dickinson Univ Hackensack NJ 07601

BOUCHER, RAYMOND EDWARD, b Brunswick, Maine, May 11, 21; m 47; c 2. ORGANIC CHEMISTRY. Educ: Bowdoin Col, BS, 45; Ind Univ, PhD(chem), 50. Prof Exp: Org chemist, Eli Lilly & Co, 50-56; chemist asst to tech dir res & develop, Polaks Frutal Works, 56-60; asst to pres, Gamma Chem Corp, 60-61; head org lab, Williamstown Plant, Anken Chem & Film Corp, 61-63; mgr plant lab, Geigy Chem Corp, 63-67; PRES, SEVEN UP BOTTLING CO, 67- Mem: AAAS; Am Chem Soc; Am Inst Chemists; NY Acad Sci. Res: Fine organics; pharmaceuticals; optical brighteners antioxidants; process development; laboratory management. Mailing Add: Seven Up Bottling Co 300 W Fourth St Plainfield NJ 07060

BOUCHER, ROGER, b Montreal, Que, Dec 17, 31; m 57; c 1. ORGANIC CHEMISTRY. Educ: Univ Montreal, BSc, 55, MSc, 57, PhD(chem), 59. Prof Exp: Sr res chemist, Clin Res Dept, Hotel-Dieu Hosp, 59-64; med res assoc, 64-66, assoc prof, 66-70, PROF MED, UNIV MONTREAL, 70-; MED RES ASSOC, CLIN RES DEPT, HOTEL-DIEU HOSP, 64-; DIR RENAL BIOCHEM LAB, CLIN RES INST, 67- Mem: Chem Inst Can. Res: Clinical research. Mailing Add: Clin Res Inst 110 Pine Ave W Montreal PQ Can

BOUCHEREAU, PAUL EMILE, plant pathology, see 12th edition

BOUCK, G BENJAMIN, b New York, NY, Oct 25, 33; m 61; c 3. BOTANY. Educ: Hofstra Univ, BA, 56; Columbia Univ, MA, 58, PhD(bot), 61. Prof Exp: Training fel, Harvard Univ, 61-62; from instr to assoc prof biol, Yale Univ, 62-71; PROF BIOL, UNIV ILL, CHICAGO, 71- Concurrent Pos: Vis assoc prof, Univ Calif, Berkeley, 68-69. Mem: AAAS; Bot Soc Am; Am Soc Cell Biol; Int Phycol Soc. Res: Fine structure of plant cells, especially the correlation and interpretation of physiological phenomena with cytological events at the ultrastructural level. Mailing Add: Dept of Biol Sci Univ of Ill Chicago IL 60680

BOUCK, GERALD R, b Owosso, Mich, Aug 22, 34; m 56; c 3. AQUATIC ECOLOGY, FISHERIES PHYSIOLOGY. Educ: Cent Mich Univ, BS, 60; Mich State Univ, MS, 63, PhD(fisheries physiol), 66. Prof Exp: Res asst fisheries physiol, Mich State Univ, 60-62, res instr, 65-66; chief biol effects br, Pac Northwest Water Lab, 66-70, DIR, WESTERN FISH TOXICOL STA, US ENVIRON PROTECTION AGENCY, 70- Mem: Ecol Soc Am; Am Fisheries Soc; Am Soc Limnol & Oceanog. Res: Stress physiology; pollution toxicology; serology; electrophoresis; enzymes; isozymes; histochemistry; Pacific salmon water quality requirements. Mailing Add: 1350 SE Goodnight Corvallis OR 97330

BOUCOT, ARTHUR JAMES, b Philadelphia, Pa, May 26, 24; m 48; c 4. GEOLOGY, PALEONTOLOGY. Educ: Harvard Univ, AB, 48, AM, 49, PhD(geol), 53. Prof Exp: Geologist, US Geol Surv, 49-56; Guggenheim fel, Europe, 56-57; from asst prof to assoc prof geol & geophys, Mass Inst Technol, 57-61; assoc prof, Calif Inst Technol, 61-68; prof geol, Univ Pa, 68-69; PROF GEOL, ORE STATE UNIV, 69- Concurrent Pos: Res assoc, Smithsonian Inst, 68-; US rep, Silurian Subcomt, Comt Stratig, IGC, 73-; chmn, Proj Ecostratig, IGCP-IUGS, 74- Mem: Geol Soc Am; Mineral Soc Am; Paleont Soc; Brit Palaeont Asn; hon cor mem Swed Geol Soc. Res: Silurian and Devonian stratigraphy and paleontology, especially brachiopods, zoogeography and paleoecology; rates of animal evolution and extinction and their controls. Mailing Add: Dept of Geol Ore State Univ Corvallis OR 97331

BOUDAKIAN, MAX MINAS, b New York, NY, Sept 8, 25; m 60; c 3. FLUORINE CHEMISTRY. Educ: City Col New York, BS, 49; Univ Mich, MS, 50; Purdue Univ, PhD(chem), 55. Prof Exp: Res chemist, Olin Mathieson Chem Corp, 50-52, group leader, 55-69, HEAD DEPTFLUORINE RES, OLIN CORP, 69-, LECTR, TECH SEM PROG, 69- Mem: Am Chem Soc. Res: Organic and aromatic fluorine chemistry; halopyridines, chemistry of heterocyclic compounds. Mailing Add: 30 Candlewood Dr Pittsford NY 14534

BOUDART, MICHEL, b Brussels, Belg, June 18, 24; m 48; c 4. PHYSICAL CHEMISTRY. Educ: Cath Univ Louvain, Cand Ing Civil, 44, Ing Civil Chim, 47; Princeton Univ, PhD(chem), 50. Prof Exp: Res asst, Princeton Univ, 50-53, res assoc, 53, asst to dir proj SQUID, 53-54, from asst prof to assoc prof chem eng, 54-61; prof, Univ Calif, Berkeley, 61-64; PROF CHEM ENG & CHEM, STANFORD UNIV, 64- Concurrent Pos: Consult, Exxon Res & Eng Co, 54- & Hoffmann-LaRoche, Inc, 69- Mem: Nat Acad Sci; Am Chem Soc; Am Inst Chem Eng. Res: Homogeneous and heterogeneous chemical kinetics; catalysis. Mailing Add: Dept of Chem Eng Stanford Univ Stanford CA 94305

BOUDREAU, ARMAND, food chemistry, biochemistry, see 12th edition

BOUDREAU, JAMES CHARLES, b Los Angeles, Calif, May 3, 36. NEUROPHYSIOLOGY. Educ: Univ Calif, Berkeley, BA, 57, PhD(psychol), 63. Prof Exp: Res psychologist, US Army Med Res Lab, 62-65; res physiologist, Vet Admin Hosp, Pittsburgh, 65-72; ASSOC PROF NEUROL SCI, SENSORY SCI CTR, UNIV TEX, HOUSTON, 73- Mem: Soc Neurosci; Am Physiol Soc; Acoust Soc Am; Europ Chemoreception Res Orgn. Res: Neurophysiological studies of sensory processes. Mailing Add: Univ of Tex Sensory Sci Ctr 6420 Lamar Flemming Houston TX 77032

BOUDREAU, PAUL E, mathematics, computer systems, see 12th edition

BOUDREAU, ROBERT DONALD, b North Adams, Mass, Mar 9, 31; m 51; c 3. METEOROLOGY. Educ: Tex A&M Univ, BS, 62, MS, 64, PhD(meteorol), 68. Prof Exp: Res meteorologist, Atmospheric Sci Lab, US Army, Ft Huachuca, Ariz, 65-68, Deseret Test Ctr, Ft Douglas, Utah, 68-70 & Meteorol Satellite Lab, Nat Environ Satellite Serv, Washington, DC, 70-71; ATMOSPHERIC SCIENTIST, EARTH RESOURCES LAB, MISS TEST FACIL, NASA, 71- Mem: Am Meteorol Soc; Am Geophys Union. Res: Radiative transfer of heat and remote sensing in the atmosphere; meteorological satellite studies; earth heat budget. Mailing Add: 2012 W Second St Apt 2F Long Beach MS 39560

BOUDREAULT, ARMAND, b Mont-Laurier, Que, Jan 24, 24; m 52; c 5. VIROLOGY, BIOLOGY. Educ: St Joseph Sem, BA, 45; Univ Montreal, BSc, 49, MSc, 51, cert, 50, PhD(biol), 55. Prof Exp: Probationer virol, 52-53, res asst, 53-63, lectr biol, Sch Hyg, 59-70, RES ASSOC VIROL, INST MICROBIOL & HYG, UNIV MONTREAL, 64-, CHIEF LAB, 67- Mem: Can Soc Microbiol. Res: Myxoviruses; interferon. Mailing Add: Inst of Microbiol & Hyg Univ of Montreal Laval des Rapides PQ Can

BOUDREAUX, EDWARD A, b New Orleans, La, Oct 30, 33; m 55; c 4. INORGANIC CHEMISTRY, CHEMICAL PHYSICS. Educ: Loyola Univ, La, BS, 56; Tulane Univ, MS, 59, PhD(chem), 62. Prof Exp: Res assoc chem, Tulane Univ, 56-62; asst prof, 62-65, ASSOC PROF CHEM, UNIV NEW ORLEANS, 65- Concurrent Pos: Consult, Kalvar Corp, 61-63; Petrol Res Fund grant, 64-66; Cancer Soc Greater New Orleans grant, 66-67; consult, USDA, New Orleans, 68-70; Fulbright teaching fel, Univ Zagreb, 70-71. Mem: Am Chem Soc; The Chem Soc;

Sigma Xi. Res: Inorganic solid compounds and complexes; ligand field theory; quantum chemical calculations; theoretical chemistry; molecular orbital calculations; diffuse reflectance electronic spectroscopy; magnetochemical research. Mailing Add: Dept of Chem Univ New Orleans New Orleans LA 70122

BOUDREAUX, HENRY BRUCE, b Scott, La, Nov 12, 14; m 41; c 1. ENTOMOLOGY. Educ: Southwest La Inst, BS, 36; La State Univ, MS, 39, PhD(entom), 46. Prof Exp: Asst biol, Southwest La Inst, 33-36, instr, 36-39, asst prof, 39-44; asst entom, La State Univ, Baton Rouge, 44-46; asst prof biol, Southwest La Inst, 46-47; from asst prof to assoc prof, 47-59, PROF ENTOM, LA STATE UNIV, BATON ROUGE, 60- Mem: Fel AAAS; Entom Soc Am; Soc Syst Zool; Acarological Soc Am. Res: Biology and taxonomy of aphids and spider mites; physiology of hemophilia. Mailing Add: Dept of Entom La State Univ Baton Rouge LA 70803

BOUGAS, JAMES ANDREW, b Bismarck, NDak, Jan 25, 24; m 53; c 2. THORACIC SURGERY, CARDIOPULMONARY PHYSIOLOGY. Educ: Harvard Med Sch, MD, 48. Prof Exp: Intern, Columbia Univ Serv, Bellevue Hosp, 48-49, resident surg, Presby Med Ctr & Bellevue Hosp, 49-53; dir cardiopulmonary lab, New Eng Deaconess Hosp, 55-65; ASSOC PROF SURG, SCH MED, BOSTON UNIV, 65- Concurrent Pos: Fel thoracic surg, Overholt Thoracic Clin, 53-55, assoc thoracic surgeon, 56-65; consult to numerous hosps, 56-; lectr, Tufts Univ & assoc staff, New Eng Med Ctr, 56-; sr active staff, New Eng Deaconess Hosp, 56-; sr active staff New Eng Baptist Hosp, Boston, 61-; vis staff Boston City Hosp, 65-; chief thoracic surg, Boston Univ Hosp, 67-70; chmn biomat, Gordon Res Conf, 67-; mem cardiac adv group, Regional Med Prog, NH, Mass & RI, 69- Mem: AAAS; Am Asn Thoracic Surg; Soc Thoracic Surg; Am Col Surg; Am Col Chest Physicians. Res: Physical biology; biomaterials; cardiovascular surgery. Mailing Add: Dept of Surg Boston Univ Sch of Med Boston MA 02115

BOUGH, WAYNE ARNOLD, b Stockton, Mo, Mar 21, 43; m 64; c 2. CHEMISTRY. Educ: Univ Mo, Columbia, BS, 65; Univ Minn, PhD(biochem), 69. Prof Exp: Res dir pollution control, Am Bact & Chem Res Corp, 69-72; ASST PROF POLLUTION CONTROL, DEPT FOOD SCI, GA EXP STA, UNIV GA, 72- Concurrent Pos: Consult, Environ Assoc Inc, 73-75 & SCS Eng, 74. Mem: Inst Food Technologists. Res: Recovery and utilization of by-products from food processing wastes and treatment systems; by-products evaluated for feed value; chitosan manufactured from shrimp wastes for coagulation of suspended solids in wastewaters. Mailing Add: Dept of Food Sci Ga Exp Sta Experiment GA 30212

BOUGHEY, ARTHUR STANLEY, b Stoke-on-Trent, Eng, July 15, 13; m 40; c 6. ECOLOGY. Educ: Univ Leicester, BSc, 35; Univ Edinburgh, PhD(microbiol), 38. Prof Exp: Asst lectr, Univ Edinburgh, 36-38; plant pathologist, Agr Res Serv, Sudan Govt, 38-45; lectr bot, Univ Exeter, 45-48; prof, Univ Ghana, 48-54 & Univ Rhodesia, 55-64; prof biol, 64-73, chmn dept pop & environ biol, 64-69, PROF PROG SOC ECOL, UNIV CALIF, IRVINE, 64- Mem: AAAS; fel Am Inst Biol Sci; Am Soc Naturalists; Brit Ecol Soc; Ecol Soc Am. Res: Effects of human occupation on ecosystems; human evolution and society succession; origin and distribution of flowering plants; automated storage, processing and retrieval of taxonomic information. Mailing Add: Dept of Pop & Environ Biol Univ of Calif Irvine CA 92664

BOUGHTON, JOHN HARLAND, b Niagara Falls, NY, May 7, 32; m 64; c 4. INORGANIC CHEMISTRY. Educ: Ariz State Univ, BS, 53, MS, 61; Univ Colo, PhD(inorg chem), 65. Prof Exp: Tech rep petrol ref, Tretolite Co, 55-58; teaching asst chem, Ariz State Univ, 58-60; teaching asst, Univ Colo, 60-61, res asst, 61-65; chemist, Spunbonded div, Textile Fibers Dept, 65-67, chemist, Pigments Dept, 67-72, SR RES CHEMIST, E I DU PONT DE NEMOURS & CO, 72- Mem: Am Chem Soc; Sigma Xi. Res: Pseudohalide complexes; solution chemistry; x-ray and light scattering. Mailing Add: Exp Sta E I du Pont de Nemours & Co Wilmington DE 19898

BOUHUYS, AREND, b Deventer, Neth, Oct 18, 25; US citizen; m 56; c 5. AIR POLLUTION, PNEUMOLOGY. Educ: State Univ Utrecht, Cand Med, 46, MD, 48; Univ Amsterdam, DrMedSc, 56. Honors & Awards: MA, Yale Univ, 68. Prof Exp: Chief asst physiol, Sch Med, State Univ Utrecht, 51-52, resident med, 52-55; resident chest dis, Beatrixoord Tuberc Sanitarium, 55-57; res physician, Neth Inst Prev Med, 57-58; asst prof lab clin physiol, Sch Med, State Univ Leiden, 59-62; asst prof physiol & med, Sch Med, Emory Univ, 62-64; assoc prof epidemiol, 64-68, PROF MED & EPIDEMIOL & DIR PULMONARY SECT, SCH MED, YALE UNIV, 68-, DIR LUNG RES CTR, 72- Concurrent Pos: NATO sci fel, 61; res grants, Nat Tuberc Asn, 61-64, Ga Heart Asn, 63-65 & USPHS, 63-74; John B Pierce Found fel, 64-; conf chmn & ed, Airway Dynamics, Int Union Physiol Sci, 68, mem comt respiration & environ health sci training comt, Nat Inst Environ Health Sci, 68-72; mem, Task Force Environ Health Sci Res Planning, 69; mem respiratory syst res eval comt, Vet Admin, 69-71; chmn subcomt byssinosis, Permanent Comn & Int Asn Occup Health, 69-; MacArthur Lect, Univ Edinburgh, 70; Josiah Macy, Jr fac scholar award, 75-76; managing ed, Pneumonologie, 76- Honors & Awards: Burger Prize, Neth Soc Occup & Indust Med, 63. Mem: Am Physiol Soc; fel Am Col Physicians; Am Soc Clin Invest; Am Soc Pharmacol & Exp Therapeut; Asn Physiologists. Res: Respiration physiology and lung diseases; including health hazards of dusts and air pollutants; respiratory pharmacology. Mailing Add: Yale Univ Sch of Med New Haven CT 06510

BOUIS, PAUL ANDRE, b Nice, France, Sept 21, 45; US citizen; m 67; c 2. ORGANIC CHEMISTRY. Educ: Va Mil Inst, BS, 67; Univ Tenn, PhD(org chem), 74. Prof Exp: Instr chem, Va Mil Inst, 67-68; chemist org chem, Ballistic Res Lab, 68-69; RES ASSOC ORG CHEM, RHODIA INC, 73- Mem: Am Chem Soc. Res: Novel synthesis of flavor and aroma chemicals; organic functional group transformations; structural elucidation of natural products; mechanism of organometalic reactions. Mailing Add: Rhodia Inc 297 Jersey Ave New Brunswick NJ 08903

BOUKNIGHT, JOSEPH WARD, b Irmo, SC, Dec 29, 07; m 47. INORGANIC CHEMISTRY. Educ: Univ SC, BS, 30; Northwestern Univ, MS, 33; Univ Wash, PhD(chem), 37. Prof Exp: Actg instr, Univ Wash, 38-39; instr chem, Agr & Mech Col Tex, 39-42, asst prof, 42; adj prof, Univ SC, 42-43, instr, Navy Flight Prep-Flight Sch, 43-44, from assoc prof to prof, 44-73, DISTINGUISHED EMER PROF CHEM, UNIV SC, 75- Mem: Am Chem Soc. Res: Investigation of complex compounds. Mailing Add: Dept of Chem Univ of SC Columbia SC 29208

BOULANGER, JEAN BAPTISTE, b Edmonton, Alta, Aug 24, 22; m 65; c 1. PSYCHOANALYSIS. Educ: Univ Montreal, BA, 41, MD, 48; McGill Univ, MA, 50; Univ Paris, dipl psychol, 51, lic psychol, 54; FRCP(C), 55. Prof Exp: Instr psychiat, 53, lectr, 54, from asst prof to assoc prof, 54-70, coordr behav sci course, 69-70, PROF PSYCHIAT, FAC MED, UNIV MONTREAL, 70- Concurrent Pos: Attend physician, neuropsychiat serv, Notre Dame Hosp, 53-61; training analyst, Can Inst Psychoanal, 57, med dir treatment serv, 59; consult & dir group psychother, Ste-Justine Hosp, 59-; consult, Inst Albert Prevost, 60-, Lakeshore Gen Hosp, 65- & Verdun Gen Hosp, 65-; pvt pract; dir, Congres des Psychanalystes de langues romanes, 60- Mem: Can Psychoanal Soc (pres, 58-60); fel Am Psychiat Asn; Int Psychoanal Asn; Can Psychiat Asn. Res: Medical undergraduate education; postgraduate psychiatric training; psychoanalytic education and training; group psychotherapy; forensic psychiatry; adult and child neuropsychiatry. Mailing Add: Univ of Montreal Fac of Med 2900 Blvd Edouard-Montpetitt Montreal PQ Can

BOULANGER, PAUL, b St Leon, Que, May 10, 18; m 46; c 3. SEROLOGY, VETERINARY MEDICINE. Educ: Univ Montreal, DVM, 41; Cornell Univ, MSc, 44. Prof Exp: Vet inspector, 42, agr asst brucellosis, 44-46, tech officer, 46-47, agr res officer, 47-64, RES SCIENTIST, ANIMAL PATH DIV, ANIMAL DIS RES INST, CAN DEPT AGR, 64-, HEAD SEROL, 60- Concurrent Pos: Mem & past coun mem, Am Leptospirosis Res Conf, 59- Mem: Am Asn Immunol; Conf Res Workers Animal Dis; Can Vet Med Asn; Can Soc Microbiol; Can Pub Health Asn. Res: Complement fixation test for avian and bovine viral infection; immunofluorescence in leptospirosis; taxoplasmosis; hog cholera, African swine fever, blue tongue and equine infectious anemia viruses. Mailing Add: Animal Dis Res Inst PO Box 11300 Sta H Ottawa ON Can

BOULDIN, DAVID RITCHEY, b Sedalia, Mo, Apr 6, 26; m 60; c 2. SOIL FERTILITY. Educ: Kans State Univ, BS, 52; Iowa State Univ, MS, 53, PhD(agron), 56. Prof Exp: Soil chemist, Tenn Valley Authority, 56-62; assoc prof, 62-69, PROF SOIL SCI, CORNELL UNIV, 69- Mem: AAAS; Am Soc Agron; Am Soc Plant Physiol. Res: Relationships between properties of the soil-fertilizer system and plant response. Mailing Add: Dept of Agron Cornell Univ Ithaca NY 14850

BOULDIN, RICHARD HINDMAN, b Florence, Ala, Feb 23, 42; m 70; c 1. MATHEMATICS. Educ: Univ Ala, BS, 64; Univ Chicago, MS, 66; Univ Va, PhD(math), 68. Prof Exp: Instr math, Univ Va, 68-69; asst prof, 69-73, ASSOC PROF MATH, UNIV GA, 73- Mem: Am Math Soc. Res: Perturbation theory and approximation theory bounded operators on a Hilbert space. Mailing Add: Dept of Math Univ of Ga Athens GA 30602

BOULET, MARCEL, b Montreal, Que, Dec 7, 19; m 47; c 4. FOOD SCIENCE, BIOCHEMISTRY. Educ: Univ Montreal, BSc, 43; McGill Univ, MSc, 45, PhD(agr chem), 48. Prof Exp: Res officer dairy chem, Nat Res Coun Can, 47-62; PROF FOOD SCI, LAVAL UNIV, 62- Mem: Am Dairy Sci Asn; Inst Food Technologists; Can Inst Food Sci & Technol; Int Inst Refrig; Fr-Can Asn Advan Sci. Res: Inorganic chemistry of milk; control of enzyme activity in foods of vegetable origin; storage of fresh fruits and vegetables; composition and properties of pectic substances in fruits. Mailing Add: Fac of Agr & Food Sci Laval Univ Quebec PQ Can

BOULLIN, DAVID JOHN, b London, Eng, Apr 21, 31; m 58; c 4. CLINICAL PHARMACOLOGY, PHYSIOLOGY. Educ: Univ London, BSc, 56 & 58, MSc, 60; St Andrews Univ, PhD(pharmacol), 63. Prof Exp: Demonstr pharmacol, St Bartholomew's Hosp Med Sch, London, 56-59; asst lectr, St Andrews Univ, 59-63; lectr, St Thomas Hosp Med Sch, London, 65-67; assoc prof, Univ Vt, 67-68; vis scientist, Nat Inst Ment Health, 68-73; SR BIOCHEMIST, MED RES COUN, RADCLIFFE INFIRMARY, OXFORD UNIV, 73- Concurrent Pos: USPHS fel, NIH, Md, 63-65; res grants, Ciba Ltd, Eng, 65-67 & Ciba Inc, 67-68; spec lectr pharmacol, George Washington Univ, 71- Mem: Am Soc Exp Pharmacol & Therapeut; Brit Nutrit Soc; Brit Pharmacol Soc; Brit Physiol Soc. Res: Physiology of intestinal peristalsis; mechanisms of uptake, binding and release of neuro-humours and anti-hypertensive drugs; significance of accumulation drugs by blood platelets; clinical pharmacology of phenothiazines and etiology of vasospasm in subarachnoid hemorrhage. Mailing Add: Dept of Clin Pharmacol Oxford Univ Radcliffe Infirmary Woodstock Rd Oxford England

BOULLION, THOMAS L, b Morse, La, Nov 4, 40; m 59; c 5. MATHEMATICS, STATISTICS. Educ: La State Univ, BS, 61; Univ Southwestern La, MS, 63; Univ Tex, PhD(math, statist), 66. Prof Exp: Asst prof math, Univ Southwestern La, 66-67; from asst prof to assoc prof, 67-74, PROF MATH & STATIST, TEX TECH UNIV, 74-, CHMN INTERDISCIPLINARY STATIST, 74- Concurrent Pos: Consult mathematician & statistician, Tex Ctr Res, 66- Mem: Am Statist Asn; Soc Indust & Appl Math. Res: Statistical inference; matrix theory. Mailing Add: Dept of Math Tex Tech Univ Lubbock TX 79407

BOULOS, BADI MANSOUR, b Alexandria, Egypt, July 3, 30; m 64; c 4. PHARMACOLOGY. Educ: Univ Alexandria, MB & BCh, 53, DPh, 58, DTM & H, 60; Univ Iowa, MS, 62; Univ Mo, PhD(pharmacol), 65. Prof Exp: Intern med, Univ Alexandria Hosps, 53-54; med officer, UN Relief & Works Agency, Gaza Strip, 54-56; instr pharmacol, Univ Alexandria, 56-60; asst radiation biol, Univ Iowa, 60-63; res assoc pharmacol, Univ Mo, 63-65; asst prof, Univ Alexandria, 65-66; asst prof, Univ Mo-Columbia & res assoc cancer res & chemother, Ellis Fischel State Cancer Hosp, 66-70; ASSOC PROF OCCUP & ENVIRON MED, UNIV ILL MED CTR, 72-; ASSOC DIR, ILL STATE ENVIRON HEALTH RESOURCE CTR, 72- Concurrent Pos: Vet Med Res Coun grants, 69-70; USPHS grant, 69-70; USPHS spec fel clin pharmacol & toxicol, Univ Kans Med Ctr, Kansas City, 70-72; Univ Ill grant, 73-74; NIHES contract, 73-76; scholarship indust toxicol, Wayne State Univ, 74; clin & indust toxicol consult; asst scientist, Cancer Res Ctr, 68-; mem, Int Cancer Cong. Mem: Soc Toxicol; AMA; Am Soc Vet Physiol & Pharmacol; NY Acad Sci; Egyptian Soc Pharmacol & Exp Therapeut. Res: Effect of antioxidant food additives on behavior and changes in biogenic amines; epidemiological studies to determine effect of barium in drinking water on hypertension; effect of polychlorinated-biphenyls in water on high risk population, for example, children, using feline species as a model. Mailing Add: Dept of Occup & Environ Med Sch of Pub Health Univ of Ill Med Ctr Chicago IL 60680

BOULTON, ALAN ARTHUR, b New Mills, Eng, Mar 14, 36; m 59; c 4. NEUROCHEMISTRY, PSYCHIATRY. Educ: Univ Manchester, BSc, 58, PhD(biochem), 62. Prof Exp: Res asst biochem, Univ Manchester, 60-62; mem med res coun unit res chem path ment dis, Birmingham Univ, 62-68; chief res biochemist & res assoc, 68-69, assoc prof, 69-75, PROF PSYCHIAT, UNIV SASK, 75-, DIR PSYCHIAT RES DIV, UNIV HOSP, 69- Concurrent Pos: Res fel biochem psychiat, Birmingham Univ, 62-68. Mem: Brit Biochem Soc; Can Biochem Soc; Soc Biol Psychiat; Int Neurochem Soc. Res: Neurobiology of anylalkyl amines and their role in psychiatry. Mailing Add: Div of Psychiat Res Univ Hosp Saskatoon SK Can

BOULTON, MARY, b Waltham, Mass, Nov 30, 07. ANALYTICAL CHEMISTRY. Educ: Col St Elizabeth, BA, 48; Fordham Univ, MS, 55. Prof Exp: From asst prof to assoc prof chem, Col St Elizabeth, 47-72, chmn dept, 58-70; RETIRED. Mem: AAAS; Am Chem Soc. Res: Cancer Research. Mailing Add: Dept of Chem Col of St Elizabeth Convent NJ 07961

BOULWARE, DAVID G, b Oakland, Calif, Nov 20, 37; m 65; c 2. THEORETICAL PHYSICS. Educ: Univ Calif, Berkeley, AB, 58; Harvard Univ, AM, 60, PhD(physics), 62. Prof Exp: Jr fel, Harvard Univ, 62-65; from asst prof to assoc prof, 65-73, PROF PHYSICS, UNIV WASH, 73- Mem: Am Phys Soc. Res: Quantum field theory;

quantum electrodynamics; elementary particles and relativity; general relativity and quantum gravity. Mailing Add: Dept of Physics Univ of Wash Seattle WA 98195

BOULWARE, RALPH FREDERICK, b Stamps, Ark, Feb 15, 17; m 41; c 1. GENETICS. Educ: Okla State Univ, BS, 49; Univ Nebr, MS, 50, PhD(animal genetics), 53. Prof Exp: Asst animal husb, Univ Nebr, 49-53; asst prof, Miss State Col, 53-55; mgr, Farrar Farms, La, 55-57; res assoc, 57-58, from instr to asst prof, 58-66, ASSOC PROF ANIMAL HUSB, LA STATE UNIV, BATON ROUGE, 66- Concurrent Pos: Vis prof, US AID-Nicaraguan Ministry Agr, 68-69. Mem: Fel AAAS; Am Soc Animal Sci; Am Meat Sci Asn; Inst Food Technol. Res: Animal breeding. Mailing Add: 1765 W Catalpa Dr Baton Rouge LA 70815

BOUMA, ARNOLD HEIKO, b Groningen, Netherlands, Sept 5, 32; m 60; c 3. MARINE GEOLOGY, SEDIMENTOLOGY. Educ: State Univ Groningen, BS, 56; State Univ Utrecht, MS, 59, PhD(sedimentol), 61. Prof Exp: Geol asst, State Univ Groningen, 54-56; sedimentol asst, State Univ Utrecht, 57-59, res fel, 60-62, lectr, 63-66; assoc prof, 66-70, PROF GEOL OCEANOG, TEX A&M UNIV, 70- Concurrent Pos: Fel Scripps Inst, Univ Calif, 62-63; ed-in-chief, Marine Geol, 63-66; mem pub & printing comt, Int Union Geol Sci, 65-; co-chmn panel sedimentary processes, Gulf Univ Res Corp, 67-70; mem Gulf of Mex panel, Joint Oceanog Insts Deep-Sea Drilling Proj. Mem: AAAS; Am Asn Petrol Geol; Geol Soc Am; Soc Econ Paleont & Mineral; Am Geophys Union. Res: Sediments, ancient and recent; sedimentary facies models; internal characteristics of sedimentary structures with regard to transport and sedimentation; turbidites; techniques used to study sedimentary structures; graphic presentations; pollution. Mailing Add: Dept of Oceanog Tex A&M Univ College Station TX 77843

BOUMAN, THOMAS DAVID, b Geneva, Ohio, Nov 23, 40; m 66; c 1. THEORETICAL CHEMISTRY. Educ: Wash Univ, AB, 62; Univ Minn, PhD(phys chem), 67. Prof Exp: Resident res assoc chem, Argonne Nat Lab, 67-69; asst prof, 69-72, ASSOC PROF CHEM, SOUTHERN ILL UNIV, 72- Concurrent Pos: Vis assoc prof, Dept Chem, Univ Va, 74-75; vis prof, Chem Lab IV, H C Oersted Inst, Copenhagen Denmark, 75-76; G C Marshall Mem Fund fel, 75; Fulbright-Hays travel fel, 75; Danish NATO Comt sr sci fel, 75. Honors & Awards: Eastman Kodak Award, Univ Minn, 65. Mem: AAAS; Am Chem Soc; Am Phys Soc. Res: Theory of optical activity; molecular orbital calculations on medium-sized molecules; group theory applications. Mailing Add: Dept of Chem Southern Ill Univ Edwardsville IL 62026

BOUNDS, HAROLD C, b Shreveport, La, Aug 13, 40; m 62; c 2. MICROBIAL ECOLOGY. Educ: Centenary Col, BS, 63; La State Univ, Baton Rouge, MS, 64, PhD(microbiol), 69. Prof Exp: From instr to asst prof, 69-74, ASSOC PROF BACT, NORTHEAST LA UNIV, 74- Mem: Am Soc Microbiol. Res: Herbicide influence on soil bacteria; microbial flora of ponds and sewage effluent; fungal contaminants in paper mills; antimicrobial properties of plant oils; sulfur oxidation by iron-sulfur autotrophs. Mailing Add: Dept of Biol Northeast La Univ 4001 De Siard Rd Monroe LA 71201

BOUNDS, JOHN HOWARD, b Oak Park, Ill, Mar 3, 31; m 58. GEOGRAPHY. Educ: Fla Southern Col, BS, 59; Univ Fla, MEd, 62; Univ Tenn, PhD(geog), 66. Prof Exp: Asst prof geog, Radford Col, 64-66 & Univ Southern Miss, 66-68; ASSOC PROF GEOG, SAM HOUSTON STATE UNIV, 68- Mem: Nat Coun Geog Educ; Asn Am Geog; Am Geog Soc. Res: Economic and classical geography; tropics; Latin America. Mailing Add: Dept of Geog Box 2042 Sam Houston State Univ Huntsville TX 77340

BOUNDY, RAY HAROLD, b Brave, Pa, Jan 10, 03; m 26; c 2. CHEMISTRY. Educ: Grove City Col, BS, 24; Case Inst Technol, BS, 26, MS, 30. Hon Degrees: ScD, Grove City Col, 47. Prof Exp: Phys chemist & mgr plastics div, Dow Chem Co, 30-50, vpres & dir res, 50-68; CONSULT MGT OF RES & DEVELOP, 68- Concurrent Pos: Mem insulation comt, Nat Res Coun; mem styrene technol comt, US Rubber Reserve; mem rubber adv comt, Army-Navy Munitions Bd; mem, Tech Intel Comt; dir, High Performance Technol Inc; consult to int exec, Govt Iran Serv Corps, 75-76. Honors & Awards: Indust Res Inst Medal, 64; Scroll Award NAm, 65. Mem: Nat Acad Eng; Am Chem Soc; Electrochem Soc; Am Inst Chem Engrs. Res: Production and use of metallic sodium; electrometric analysis; automatic control; production of bromine from seawater; utilization of olefine derivatives; plastic development. Mailing Add: 906 W Sugnet Rd Midland MI 48640

BOUNOUS, GUSTAVO, b Luserna-San Giovanni, Italy, July 10, 28; Can citizen. SURGERY, GASTROENTEROLOGY. Educ: Univ Turin, MD, 52. Prof Exp: Resident, Univ Parma, 52-56; res fel, Med Ctr, Ind Univ, 57-61; from asst prof to assoc prof, McGill Univ, 62-68; ASSOC PROF SURG, UNIV SHERBROOKE, 68- Concurrent Pos: Assoc, Med Res Coun Can, 68- Honors & Awards: Medal, Royal Col Surgeons of Can, 64. Mem: Can Soc Clin Invest. Res: Role of pancreatic proteolytic enzymes in the pathogenesis of ischemic enteropathy; use of an elemental diet in the protection of the intestine against the effects of chemotherapy and radiation. Mailing Add: Dept of Surg Univ of Sherbrooke Sherbrooke PQ Can

BOURASSA, RONALD RAY, b Oklahoma City, Okla, Sept 7, 40; m 62; c 3. SOLID STATE PHYSICS. Educ: Rice Univ, BA, 62; Univ Ill, MS, 64, PhD(physics), 67. Prof Exp: Sr res physicist, Pac Northwest Labs, Battelle Mem Inst, 67-68; asst prof, 68-72, ASSOC PROF PHYSICS, UNIV OKLA, 72- Concurrent Pos: Vis scientist, K F A Jülich, Ger, 74-75. Mem: Am Phys Soc. Res: Transport properties of solids; point defects in metals; anisotropic electron scattering. Mailing Add: Dept of Physics Univ of Okla 440 W Brooks Norman OK 73069

BOURBEAU, GERARD AUGUSTE, b Cold Lake, Alta, Nov 4, 16. SOIL MORPHOLOGY. Educ: St Francis Xavier Col, BA, 38; Laval Univ, BS, 43; Univ Wis, MS, 46, PhD(soils), 48; Univ Ottawa, BTh, 67, MTh, 69. Prof Exp: Instr chem, Laval Univ, 43-44; soil surv, Dept Agr, Que, 43-45; asst soil scientist, Conn Agr Exp Sta, 48-51; assoc prof agron, Univ Md, 52-61; dean fac arts & sci, Laurentian Univ, 61-65; PROF SOIL SCI, LAVAL UNIV, 69-, CHMN DEPT SOILS, 71- Concurrent Pos: Consult & collabr, Soil Serv Div, Soil Conserv Serv, USDA, 48-61; consult pedology, Mutual Security Agency, Belg Congo, 51-52; consult, Foreign Opers Admin, Univ Md & Brit Guiana, 54-55; Am secy comn V, Int Soil Sci Soc Cong, 60; res asst, Soil Res Inst, Univ Ottawa, 66-67; dir, Quebec Inst Pedology, 74. Mem: Am Soc Agron; Soil Sci Soc Am; Int Soil Sci Soc; Can Soc Soil Sci. Res: Soil mineralogy; morphology and classification. Mailing Add: Dept of Soils Laval Univ Fac of Agr Quebec PQ Can

BOURCHIER, ROBERT JAMES, b London, Ont, Dec 28, 27; m 52; c 3. FORESTRY. Educ: Univ Toronto, ScF, 51; Univ Alta, MSc, 55; State Univ NY Col Forestry, Syracuse, PhD, 60. Prof Exp: Mem staff, Can Dept Fisheries & Forestry, 71-73, dir prog opers, 73-75, DIR GEN, CAN FORESTRY SERV, CAN DEPT ENVIRON, 75- Mem: Mycol Soc Am; Can Bot Asn; Can Inst Forestry; Can Phytopath Soc. Mailing Add: 71 Ashgrove Crescent Ottawa ON Can

BOURDO, ERIC A, JR, b Muskegon, Mich, Jan 15, 17; m 42; c 3. FOREST MANAGEMENT. Educ: Mich Technol Univ, BS, 43; Univ Mich, MA, 51, PhD(forestry, bot), 55. Prof Exp: Compounder natural & synthetic rubber prods, Goodyear Tire & Rubber Co, Ohio, 43-46; consult forester, Pomeroy & McGowan Co, Ark, 46; from instr to assoc prof, 47-58, dir res, Ford Forestry Ctr, 55-68, PROF FORESTRY, MICH TECHNOL UNIV, 58-, DEAN SCH FORESTRY & WOOD PROD, FORD FORESTRY CTR, 68- Mem: Soc Am Foresters; Ecol Soc Am; Am Soc Agron; Wilson Ornith Soc. Res: Management and utilization of northern forest types. Mailing Add: Ford Forestry Ctr Mich Technol Univ L'Anse MI 49946

BOURGAULT, PRISCILLA C, b Winooski, Vt, Jan 1, 28. PHARMACOLOGY. Educ: Trinity Col, BS, 50; Loyola Univ, Ill, PhD(pharmacol), 66. Prof Exp: ASST PROF PHARMACOL, DENT SCH, LOYOLA UNIV, CHICAGO, 66- Res: Psychopharmacology; neuropharmacology. Mailing Add: 2160 S First Ave Maywood IL 60153

BOURGAUX, PIERRE, b Woluwe-St Pierre, Belg, Feb 25, 34; m 61; c 2. ANIMAL VIROLOGY, MOLECULAR BIOLOGY. Educ: Free Univ Brussels, MD, 59. Prof Exp: From lectr to asst prof microbiol, Med Sch, Free Univ Brussels, 60-68; assoc prof, 68-72, PROF MICROBIOL, MED SCH, UNIV SHERBROOKE, 72- Concurrent Pos: Med Res Coun res fel, Exp Virus Res Unit, Glasgow Univ, 63-64; Eleanor Roosevelt fel, Salk Inst Biol Studies, 68-69. Mem: Assoc mem Brit Soc Gen Microbiol. Res: Bacteriophage, particularly reactivation of biological properties after inactivation; oncogenic viruses, particularly biological properties and replication of the viral DNA in both lytic and transformation systems. Mailing Add: Dept of Microbiol Med Sch Univ of Sherbrooke Sherbrooke PQ Can

BOURGEOIS, LOUIS DOZAN, microbiology, see 12th edition

BOURGET, SYLVIO-J, b Quebec, Que, Jan 9, 30; m 53; c 4. SOIL PHYSICS, SOIL CONSERVATION. Educ: Laval Univ, BSc, 50; Univ Wis, MS, 51, PhD(soils), 54. Prof Exp: Res officer field husb div, Cent Exp Farm, Ottawa, Ont, 54-59, res officer, Soil Res Inst, 59-62; PROF SOILS, LAVAL UNIV, 62-; DIR RES STA, CAN DEPT AGR, 68- Mem: Soil Conserv Soc Am; Int Soc Soil Sci; Int Soc Soil Mech & Found Eng; Can Soc Soil Sci. Res: Movement and measurement of soil water; measurement and control of soil erosion; measurement of soil structure and temperature. Mailing Add: Dept of Soils Laval Univ Quebec PQ Can

BOURGIN, DAVID GORDON, b New York, NY, Nov 6, 00; m; c 2. MATHEMATICS. Educ: Harvard Univ, PhD(math, physics), 26. Prof Exp: Instr, Lehigh Univ, 25-27; assoc math, Univ Ill, 27-37, from asst prof to prof, 37-66; prof, 66-71, M D ANDERSON PROF MATH, UNIV HOUSTON, 71- Concurrent Pos: Mem, Inst Adv Study, 40-41, 48-49; lectr, NSF col math conf, Univ Ore, 54; Fulbright lectr, 54-55; Fulbright res grant, Rome, 55-56; ed, Ill J Math. Mem: Am Math Soc. Res: Applied mathematics; linear topological spaces; algebraic topology. Mailing Add: Dept of Math Univ of Houston Cullen Blvd Houston TX 77004

BOURGON, MARCEL, b Montreal, Que, Nov 9, 26; m 53. PHYSICAL CHEMISTRY, INORGANIC CHEMISTRY. Educ: Univ Montreal, BSc, 48, MSc, 50, PhD(phys chem), 53. Prof Exp: Res chemist, Metals Res Lab, Carnegie Inst Technol, 53-55; from asst prof to assoc prof, 55-69, PROF CHEM, UNIV MONTREAL, 69- Mem: Fel Chem Inst Can. Res: Physical properties of inorganic compounds in liquid state; electrical properties of molten metallic sulfides; semiconductivity. Mailing Add: Dept of Chem Univ of Montreal Montreal PQ Can

BOURGUIGNON, ERIKA EICHHORN, b Vienna, Austria, Feb 18, 24; US citizen; m 50. ANTHROPOLOGY. Educ: Queens Col, NY, BA, 45; Northwestern Univ, PhD(anthrop), 51. Prof Exp: From instr to assoc prof, 49-66, actg chmn dept, 71-72, PROF ANTHROP, OHIO STATE UNIV, 66-, CHMN DEPT, 72- Concurrent Pos: Lectr anthrop, Columbus State Hosp, Ohio, 62-65; Nat Insts Ment Health res grant, Ohio State Univ, 63-68; adv ed, Behav Sci Res, 75-78. Mem: Fel Am Anthrop Asn; fel Am Ethnol Soc; fel Ethnologia Europaea. Res: Cross-cultural studies of altered states of consciousness, as institutionalized in a religious framework, including ritual and psychological aspects of these states as well as their societal concomitants. Mailing Add: Dept of Anthrop Ohio State Univ Columbus OH 43210

BOURKE, ANNE ROSALEEN, b Ft Crockett, Tex, Oct 21, 14. PHARMACOLOGY. Educ: Univ Md, BS, 37; George Washington Univ, MS, 49, PhD(pharmacol), 51. Prof Exp: Pharmacologist, US Food & Drug Admin, 51-56; pharmacologist, Nat Cancer Inst, 56-63; PHARMACOLOGIST, DIV RES GRANTS, NIH, 63- Mem: Am Asn Cancer Res; Am Soc Exp Biol & Med; Am Soc Microbiol; Am Soc Pharmacol & Exp Therapeut. Res: Toxicology; drug evaluation; chemotherapy. Mailing Add: Div of Res Grants Nat Insts Health Bethesda MD 20014

BOURKE, JOHN BUTTS, b Tampa, Fla, Aug 29, 34; m 57; c 3. BIOCHEMISTRY. Educ: Colgate Univ, BA, 57; Ore State Univ, MS, 60, PhD(chem), 63. Prof Exp: Asst biochem, Ore State Univ, 62-63; res specialist, 63-65, asst prof chem, 65-71, ASSOC PROF CHEM & DIR ANAL DIV, FOOD SCI DEPT, CORNELL UNIV, 71- Mem: AAAS; Am Chem Soc; Am Soc Plant Physiol; Entom Soc Am; NY Acad Sci. Res: Metabolism and metabolic fate of pesticides in plants and animals; effect of ionizing radiation on the metabolism of plants; laboratory data acquisition systems using computers. Mailing Add: Anal Div Food Sci Dept Cornell Univ Geneva NY 14456

BOURLAND, CHARLES THOMAS, b Osceola, Mo, July 19, 37; m 63; c 1. FOOD SCIENCE. Educ: Univ Mo, Columbia, BS, 59, MS, 67, PhD(food sci & nutrit), 70. Prof Exp: Qual control dairy, Adams Dairy, 63-65; asst food sci, Univ Mo, Columbia, 65-69; RES SCIENTIST FOOD, TECHNOL INC, 69- Mem: Inst Food Technol; Dairy Sci Asn; Nutrit Today Soc; Asn Microbiol; Int Asn Milk, Food & Environ Sanit. Res: Food microbiology; food safety; aerospace food systems; food packaging. Mailing Add: 1111 Laurel Valley Houston TX 77058

BOURNE, EARL WHITFIELD, b Oklahoma City, Okla, July 6, 38. HISTOLOGY, CYTOLOGY. Educ: Westminster Col, Mo, AB, 60; Okla State Univ, MS, 62, PhD(zool), 68. Prof Exp: Asst prof biol, Bethany Col, WVa, 62-64; ASSOC PROF BIOL, UNIV NMEX, 68- Mem: Am Soc Cell Biol; Am Asn Cell Biol; Am Micros Soc. Res: Effects of carcinogenic hydrocarbons on cells in vitro; cytological effects of synthetic steroids. Mailing Add: Dept of Biol Univ of NMex Albuquerque NM 87131

BOURNE, FREDERICK MUNROE, b Victoria, BC, June 26, 10; m 43; c 3. INTERNAL MEDICINE. Educ: McGill Univ, BA, 31, MD, 37; Oxford Univ, BA, 34, MA, 42; FRCPS(C). Prof Exp: Assoc prof med, McGill Univ, 69-75; RETIRED. Concurrent Pos: Physician-in-chief, Reddy Mem Hosp, Montreql, Que, 67-75, consult, 75-; sr physician, Montreal Gen Hosp, 68-75, consult, 75- Mem: Fel Am Col Physicians. Mailing Add: 3550 Cote des Neiges Montreal PQ Can

BOURNE, GEOFFREY HOWARD, b Perth, Western Australia, Nov 17, 09; US citizen; m 64; c 2. EXPERIMENTAL PATHOLOGY, ANATOMY. Educ: Univ Western Australia, BSc, 30, MSc, 32, DSc, 35; Oxford Univ, DPhil, 43. Prof Exp: Biologist in chg exp work, Australian Inst Anat, Canberra, Australia, 34-36; tutor to extra-mural delegacy, Oxford Univ, 39-43; reader histol, Univ London, 47-57; chmn dept anat, 57-62, PROF ANAT, EMORY UNIV, 57-, DIR, YERKES REGIONAL PRIMATE RES CTR, 62- Concurrent Pos: Adj prof psychiat, Univ Ga, 70- & Ga State Univ, 74-; mem res adv comt biotechnol & human res div, NASA; vpres & founder & mem bd trustees, Zool Soc Atlanta; nutrit adv, Brit Mil Admin of Malaya. Mem: Fel Brit Inst Biol; fel Royal Soc Med; fel Am Geront Soc; fel Brit Zool Soc; Int Soc Cell Biol. Res: Nutrition; pathology of weightlessness. Mailing Add: Yerkes Regional Primate Res Ctr Emory Univ Atlanta GA 30322

BOURNE, JOHN ROSS, b Bryan, Tex, Aug 31, 44; m 68; c 1. BIOMEDICAL ENGINEERING, ELECTRICAL ENGINEERING. Educ: Vanderbilt Univ, BE, 66; Univ Fla, MSE, 67, PhD(elec eng), 69. Prof Exp: Asst prof, 69-73, ASSOC PROF ELEC & BIOMED ENG, VANDERBILT UNIV, 73- Mem: Inst Elec & Electronics Eng; Psychonomic Soc; Am Electroencephalog Soc; Asn Res Vision & Ophthal. Res: Visual electrophysiology; electroencephalogram analysis; biomedical computing. Mailing Add: Dept of Elec & Biomed Eng Vanderbilt Univ Nashville TN 37203

BOURNE, MALCOM CORNELIUS, b Adelaide, Australia, May 18, 26; m 53; c 5. FOOD SCIENCE. Educ: Univ Adelaide, BSc, 48; Univ Calif, MS, 61, PhD(agr chem), 62. Prof Exp: Chief chemist, Brookers Ltd, Australia, 48-58; from asst prof to assoc prof food sci, 62-73, PROF FOOD SCI & TECHNOL, CORNELL UNIV, 74- Mem: AAAS; Am Chem Soc; Inst Food Technol; Soc Rheol. Res: Physical measurement of quality of foods, food processing, kinetics and mechanisms of detergency; horticultural products; protein beverages. Mailing Add: NY State Agr Exp Sta Cornell Univ Geneva NY 14456

BOURNE, NEIL, b London, Ont, Aug 11, 29; m 55; c 4. INVERTEBRATE ZOOLOGY. Educ: McMaster Univ, BSc, 52, MSc, 53; Univ Toronto, PhD(limnol), 59. Prof Exp: Scientist, Offshore Group, 59-70, RES SCIENTIST, FISHERIES BIOL GROUP, FISHERIES RES BD CAN, 70- Mem: Nat Shellfisheries Asn; Can Soc Zoologists. Res: Estuarine ecology; ecology of marine bottom invertebrates; molluscs-bivalves; larval development of bivalves; molluscan aquaculture. Mailing Add: Pac Biol Sta Fisheries & Marine Serv Box 100 Nanaimo BC Can

BOURNE, SAMUEL G, b Liverpool, Eng, Apr 29, 16; US citizen. MATHEMATICS. Educ: Rutgers Univ, BS, 38; Johns Hopkins Univ, AM, 44, PhD(math), 50. Prof Exp: Instr math, Johns Hopkins Univ, 43-49, Univ Conn, 50-52 & Temple Univ, 52-54; asst prof, Univ Calif, 54-55 & Lehigh Univ, 56-58; res asst, Calif Inst Technol, 58-59; vis scholar, Univ Calif, Berkeley, 59-63; prof, Univ Fla, 63-64; RES MATHEMATICIAN, UNIV CALIF, BERKELEY, 64- Concurrent Pos: Mem, Inst Advan Study, 50; lectr, Math Inst, Hungarian Acad Sci, 75. Honors & Awards: Bogart Math Prize, 38. Mem: Fel AAAS; Am Math Soc; Math Asn Am; Indian Math Soc. Res: Structure of semirings and topological semirings; measure theory on locally compact semigroups. Mailing Add: PO Box 4583 Berkeley CA 94704

BOURNIQUE, RAYMOND AUGUST, b Arnold, Pa, Nov 5, 13; m 42; c 3. CHEMISTRY. Educ: Univ Toledo, BS, 35; Ohio State Univ, MS, 38, PhD(chem), 40. Prof Exp: Lab asst, Nat Supply Co, Toledo, 35-36; asst, Ohio State Univ, 36-39; res chemist, Tex Co, 40-43; res engr, Westinghouse Res Labs, 43-46; instr chem, Bowdoin Col, 46-48, asst prof, 48-54, assoc prof, 54-65, chmn dept, 62-67, PROF CHEM, MARQUETTE UNIV, 65- Mem: Am Chem Soc. Res: Spectrographic analysis of boron in steels; direct microelectrolysis of copper in steel; analysis of fluorine in highly fluorinated compounds; spectrophotometric methods of analysis; nonaqueous acid-base titrimetry; chelation in nonaqueous solvents; formal potentials of oxidation-reduction indicators. Mailing Add: Dept of Chem Marquette Univ Milwaukee WI 53233

BOURNS, ARTHUR NEWCOMBE, b Petitcodiac, NB, Dec 8, 19; m 43; c 4. ORGANIC CHEMISTRY. Educ: Acadia Univ, BSc, 41; McGill Univ, PhD(chem), 44. Hon Degrees: DSc, Acadia Univ, 68. Prof Exp: Res chemist, Dom Rubber Co, 44-45; lectr chem, Acadia Univ, 45-46; asst prof, Univ Sask, 46-47; from asst prof to assoc prof org chem, 47-52, dean fac grad studies, 57-61, chmn dept chem, 65-67, vpres div sci & eng, 67-72, actg pres, Univ, 70, PROF CHEM, McMASTER UNIV, 52-, PRES & V CHANCELLOR, 72- Concurrent Pos: Nuffield traveling fel sci, Univ Col, Univ London, 55-56; vchmn, Gordon Res Conf Chem & Physics of Isotopes, 59-60, chmn, 61-62; assoc ed, Can J Chem, 66-69; mem grant selection comt chem, Nat Res Coun Can, 66-69, chmn, 68-69, mem, Coun, 69-75; mem bd, Royal Bot Gardens, 72-; pres & chmn exec comt, Can Bur Int Educ, 73-; mem sci adv coun of Can Bd, Weizmann Inst Sci, 74-; mem bd dirs & exec comt, Asn Univs & Cols Can, 74-; mem bd dirs, Mohawk Col Appl Arts & Technol, 75- & Slater Steel Industs Ltd, 75- Mem: Royal Soc Can; fel Chem Inst Can; Am Chem Soc. Res: Gas phase reactions; hydrocarbon oxidations; kinetic isotope effects; organic reaction mechanisms. Mailing Add: Pres Off McMaster Univ Hamilton ON Can

BOURNS, THOMAS KENNETH RICHARD, b Vancouver, BC, Feb 9, 24; m 51. PARASITOLOGY. Educ: Univ BC, BA, 47, MA, 49; Rutgers Univ, PhD(zool), 55. Prof Exp: Tech officer, Med & Vet Entom Lab, Can Dept Agr, 49-52; asst zool, Rutgers Univ, 54-55 & bot, 55-56; lectr zool, Queen's Univ, Ont, 56-57; from lectr to assoc prof, 57-70, PROF ZOOL, UNIV WESTERN ONT, 70- Mem: Am Soc Parasitol; Wildlife Dis Asn; Can Soc Zool; fel Royal Soc Trop Med & Hyg; Can Pub Health Asn. Res: Serological study of schistosome trematode-host relationships; trematode life cycles; systematic serology. Mailing Add: Dept of Zool Univ of Western Ont London ON Can

BOURQUE, BRUCE JOSEPH, b Clinton, Mass, Dec 27, 43; m 65; c 2. ANTHROPOLOGY, ARCHAEOLOGY. Educ: Univ Mass, BA, 65; Univ Colo, MA, 67; Harvard Univ, PhD(anthrop), 71. Prof Exp: Asst prof anthrop, Skidmore Col, 70-71; ARCHAEOLOGIST, MAINE STATE MUS, 72-; PROF ANTHROP, BATES COL, 72- Concurrent Pos: Archaeol researcher, Nat Geog Soc, 74-75. Mem: Soc Am Archaeol; cor mem Am Anthrop Asn. Res: New World archaeology, especially eastern North America. Mailing Add: Maine State Mus Augusta ME 04330

BOURQUE, DON PHILIPPE, b St Louis, Mo, Nov 23, 42. MOLECULAR BIOLOGY. Educ: Johns Hopkins Univ, AB, 64; Duke Univ, MA, 67, PhD(bot), 69. Prof Exp: Res asst plant biochem, Dept Bot, Duke Univ, 64-69; fel molecular genetics, Dept Biol, Univ Calif, Los Angeles, 69-72, res assoc molecular biol, 72-73; ASST PROF AGR BIOCHEM, UNIV ARIZ, 73- Mem: AAAS; Am Soc Cell Biol; Am Soc Plant Physiologists. Res: Molecular biology of organelles, particularly chloroplasts of higher plants. Mailing Add: Dept of Nutrit & Food Sci Univ of Ariz Tucson AZ 85721

BOURQUIN, AL WILLIS J, b Castroville, Tex, Mar 2, 43. MARINE MICROBIOLOGY, MICROBIAL ECOLOGY. Educ: Univ Houston, BS, 65, MS, 68, PhD(biol), 71. Prof Exp: Res assoc microbiol, Lunar Receiving Lab, Brown & Root-Northrup Corp, 67-68; RES MICROBIOLOGIST, US ENVIRON PROTECTION AGENCY, 72- Concurrent Pos: Adj prof, Univ WFla, 73-75; co-ed, Develop Indust Microbiol, 74- Mem: Am Soc Microbiol; Soc Indust Microbiol; Sigma Xi. Res: Fate of pollutants in estuaries, particularly with relationship to hydrocarbons and pesticides in surface films. Mailing Add: US Environ Protection Agency Sabine Island Gulf Breeze FL 32561

BOUSFIELD, ALDRIDGE KNIGHT, b Boston, Mass, Apr 5, 41; m 68. TOPOLOGY. Educ: Mass Inst Technol, SB, 63, PhD(math), 66. Prof Exp: From instr to asst prof math, Brandeis Univ, 67-72; ASSOC PROF MATH, UNIV ILL CHICAGO CIRCLE, 72- Concurrent Pos: Off Naval Res assoc math, 66-67. Mem: Am Math Soc; Math Asn Am. Res: Homotopy theory; semisimplicial theory; algebraic topology. Mailing Add: Dept of Math Univ Ill Chicago Cir Box 4348 Chicago IL 60680

BOUSFIELD, EDWARD LLOYD, b Penticton, BC, June 19, 26; m 53; c 4. INVERTEBRATE ZOOLOGY. Educ: Univ Toronto, BA, 48, MA, 49; Harvard Univ, PhD(zool), 54. Prof Exp: Invertebrate zoologist, 50-64, CHIEF ZOOLOGIST, NAT MUS NATURAL SCI, 64-, ACTG CUR CRUSTACEANS, 74- Res: Marine zoology; Crustacea; Amphipoda; Cirripedia; Mysidacea; intertidal and estuarine ecology; marine fouling; stream biology; aquatic entomology; marine malacology. Mailing Add: Nat Mus of Natural Sci Ottawa ON Can

BOUSH, GEORGE MALLORY, b Norfolk, Va, June 5, 26; m 45; c 3. ENTOMOLOGY. Educ: Va Polytech Inst, BSc, 48; Ohio State Univ, MSc, 51, PhD, 55. Prof Exp: Asst entomologist, Va Agr Exp Sta, 49-50 & Rockefeller Found, 52-54; assoc prof entom, Univ Ky, 55-57 & Va Polytech Inst, 57-64; assoc prof, 64-68, PROF ENTOM, UNIV WIS-MADISON, 68- Concurrent Pos: Smith-Mundt fel, Iraq, 60-61; consult, USDA, 61. Mem: Entom Soc Am. Res: Microbial degradation of toxicants; insect microbial symbiotic associations; insect pathology; insect transmission of plant diseases. Mailing Add: 643 Russell Labs Dept of Entom Univ of Wis Madison WI 53706

BOUSQUET, WILLIAM F, b Milford, Mass, Sept 23, 33; m 57; c 2. PHARMACOLOGY, BIOCHEMISTRY. Educ: Mass Col Pharm, BS, 55; Purdue Univ, MS, 57, PhD(pharmacol), 59. Prof Exp: Asst prof bionucleonics, 59-61, from asst prof to assoc prof pharmacol, 61-68, PROF PHARMACOL, PURDUE UNIV, WEST LAFAYETTE, 68- Concurrent Pos: NIH res grant, 60; vis biologist, Am Inst Biol Sci, 61-63; Lederle Pharm fac res award, 63. Mem: AAAS; Am Chem Soc; Am Pharmaceut Asn; Am Soc Pharmacol & Exp Therapeut; NY Acad Sci. Res: Biochemical pharmacology interaction of drugs the cellular and subcellular structures; drug effects on carbohydrate and lipid metabolism; drug metabolism; cellular mechanisms in liver controlling enzyme synthesis. Mailing Add: Dept of Pharmacol Purdue Univ West Lafayette IN 47906

BOUSTANY, KAMEL, b Aleppo, Syria, Mar 22, 41; US citizen; m 64; c 4. RUBBER CHEMISTRY. Educ: Neuchatel Univ, Switz, chem eng, 64, PhD(organometallics), 67. Prof Exp: Sr res chemist rubber, 67-71, proj mgr, 71-74, SR PROJ MGR PROD DEVELOP, MONSANTO CO, 74- Mem: Am Chem Soc; Soc Plastic Engr. Res: The effect of short fibers on the reinforcement of elastomers and plastics to improve their physical properties, processibility and performances in tires and mechanical good products. Mailing Add: Monsanto Co 260 Springside Dr Akron OH 44313

BOUTHILET, ROBERT J, bacteriology, see 12th edition

BOUTHILLIER, LOUIS-PHILIPPE, b Montreal, Que, Oct 1, 12. BIOCHEMISTRY. Educ: Univ Montreal, BSc, 35, MSc, 36; Univ Ill, PhD(biochem), 45. Prof Exp: Asst chem, 36-41, from asst prof to assoc prof biochem, 44-59, PROF BIOCHEM, UNIV MONTREAL, 59- Mem: Am Chem Soc; Can Biochem Soc. Res: Problems of metabolism with special reference to amino acids. Mailing Add: Dept of Biochem Univ of Montreal PO Box 6128 Montreal PQ Can

BOUTILIER, ROBERT FRANCIS, b Lawrence, Mass, Nov 28, 37; m 63; c 1. GEOLOGY. Educ: Boston Univ, AB, 59, MA, 60, PhD(geol), 63. Prof Exp: Instr geol, Boston Univ, 62-65; from asst prof to assoc prof, 65-73, PROF EARTH SCI, BRIDGEWATER STATE COL, 73- Mem: AAAS; Geol Soc Am; Soc Econ Mineralogists & Paleontologists; Nat Asn Geol Teachers. Res: Origin of igneous and metamorphic rocks of eastern Massachusetts. Mailing Add: Dept of Earth Sci Bridgewater State Col Bridgewater MA 02324

BOUTON, EDWIN HARRY, chemistry, see 12th edition

BOUTON, THOMAS CHESTER, b Milwaukee, Wis, Dec 5, 39; m 62; c 3. POLYMER SCIENCE. Educ: Univ Wis, BS, 52; Univ Akron, MS, 67. Prof Exp: Jr engr, 62-64, supvr pilot plant, 65-67, MGR PROCESS DEVELOP, FIRESTONE RES, 67- Mem: Am Chem Soc; Am Inst Chem Engrs. Res: Structure-property relationships; process simulation; kinetics solution thermodynamics. Mailing Add: Firestone Res Wilbeth & S Main Akron OH 44317

BOUTROS, OSIRIS WAHBA, b Beni-Suef, U A R, Aug 16, 28; US citizen; m 65. c 4. PLANT PHYSIOLOGY. Educ: Cairo Univ, BSc, 52; Fla State Uniiv, MS, 62; Univ Pittsburgh, PhD(plant physiol), 68. Prof Exp: Instr, El-Mahallah El Kobra Sec Sch, Egypt, 52-53; instr chem, Ministry of Educ, Addis Abeba, Ethiopia, 53-59; tech asst biol & chem, Fla State Univ, 59-61; instr, Col Steubenville, 62-63; asst plant physiol, Univ Pittsburgh, 64-68, Am Cancer Soc fel, 68-69, ASSOC PROF BIOL, UNIV PITTSBURGH, BRADFORD, 69- Mem: Am Soc Plant Physiol; Am Inst Biol Sci; AAAS. Res: Physiological properties, modes of action and chemical nature of plant growth regulators. Mailing Add: Dept of Biol Univ of Pittsburgh Bradford PA 16701

BOUTROS, SUSAN NOBLIT, b Lock Haven, Pa, May 22, 42; m 65; c 4. CYTOGENETICS. Educ: Dickinson Col, BS, 64; Univ Pittsburgh, PhD(plant cytol), 67. Prof Exp: Instr biol, Univ Pittsburgh, 67-68, NIH traineeship human genetics, Grad Sch Pub Health, 68-69; auth gen biol lab book for col level, 69-71, ASST PROF BIOL, UNIV PITTSBURGH BRADFORD, 71- Mem: Am Soc Human Genetics; Genetics Soc Am; Soc Study Soc Biol. Res: Plant development and cytology; genetics and cytogenetics; improved identification techniques for chromosomes. Mailing Add: Dept of Biol Univ of Pittsburgh Bradford PA 16701

BOUTWELL, JOSEPH HASKELL, US citizen; m 43; c 3. CLINICAL BIOCHEMISTRY. Educ: Wheaton Col, BS, 39; Northwestern Univ, MS, 41, PhD(biochem), 47, MD, 49; Am Bd Clin Chemists, dipl, 54. Prof Exp: Asst prof biochem, Sch Med, Temple Univ, 49-66, prof, 66; ASST DIR BUR LABS, CTR DIS CONTROL, USPHS, 66- Mem: Am Chem Soc; Am Asn Clin Chem (pres); AMA; Am Soc Clin Path. Res: Analytical biochemical methods; enzyme systems; intracellular electrolytes; clinical medical education in biochemistry; quality control and proficiency testing in clinical laboratories. Mailing Add: 1600 Clifton Rd NE Atlanta GA 30333

BOUTWELL, ROSWELL KNIGHT, b Madison, Wis, Nov 24, 17; m 43; c 3. BIOCHEMISTRY. Educ: Beloit Col, BS, 39; Univ Wis, MS, 41, PhD(biochem), 44. Prof Exp: Asst, 39-44, instr cancer res, 45-49, from asst prof to assoc prof oncol, 49-67, PROF ONCOL, McARDLE LAB CANCER RES, UNIV WIS-MADISON, 67- Concurrent Pos: Fel, Univ Wis, 44-45; assoc ed, Cancer Res, Exp Therapeut Study

Sect, Nat Cancer Inst. Mem: Fel AAAS; Am Asn Cancer Res; Am Soc Biol Chem. Res: Interaction of carcinogens with tissue constituents and metabolic consequences. Mailing Add: McArdle Lab for Cancer Res Univ of Wis Madison WI 53706

BOUWKAMP, JOHN C, b Grant, Mich, Apr 20, 42; m 66. HORTICULTURE, PLANT BREEDING. Educ: Mich State Univ, BS, 64, MS, 66, PhD(hort), 69. Prof Exp: Asst prof, 69-74, ASSOC PROF HORT, UNIV MD, 74- Mem: AAAS; Am Genetic Asn; Am Soc Hort Sci; Am Soc Plant Physiol. Res: Genetics and physiology of vegetable crops. Mailing Add: Dept of Hort Univ of Md College Park MD 20742

BOUWMAN, FRED LUDWIG, b Grand Rapids, Mich, Jan 16, 16; m 47; c 3. ZOOLOGY. Educ: Calvin Col, AB, 40; Mich State Univ, MS, 47, PhD, 59. Prof Exp: Instr anat, Mich State Univ, 50-56; prof biol & chmn dept, Detroit Inst Technol, 56-62; chmn div natural sci, 70-72, PROF BIOL, AQUINAS COL, 63-, CHMN DEPT, 64- Mem: AAAS; Am Soc Zool. Res: Histology. Mailing Add: Dept of Biol Aquinas Col Grand Rapids MI 49506

BOUWSMA, WARD D, b Lansing, Mich, Jan 11, 35; m 60; c 2. MATHEMATICS. Educ: Calvin Col, AB, 56; Univ Mich, MA, 57, PhD(math), 62. Prof Exp: Asst prof math, Pa State Univ, 60-67; ASSOC PROF MATH, SOUTHERN ILL UNIV, 67- Mem: Am Math Soc; Math Asn Am. Res: Complex variables. Mailing Add: Dept of Math Southern Ill Univ Carbondale IL 62901

BOUYOUCOS, JOHN VINTON, b Lansing, Mich, Nov 9, 26; m 53; c 2. APPLIED PHYSICS. Educ: Harvard Univ, AB, 49, AM, 50, PhD(appl physics), 55. Prof Exp: Asst, Harvard Univ, 51-55, asst to dir acoustics res lab, 55-59; MGR, HYDROACOUSTICS, INC, 59- Mem: Fel Acoustical Soc Am; Inst Elec & Electronics E. Res: Hydrodynamic power conversion; physical acoustics; underwater sound; fluid mechanics; electric networks; electronics; servomechanisms. Mailing Add: Hydroacoustics Inc 321 Northland Ave PO Box 3818 Rochester NY 14610

BOVARD, FREEMAN CARROLL, b Eugene, Ore, July 18, 21; m 45; c 3. BIO-ORGANIC CHEMISTRY. Educ: Pomona Col, AB, 43; Iowa State Col, PhD(biochem), 52. Prof Exp: Jr chemist, Shell Develop Co, 43-45; res biochemist, Stine Lab, E I du Pont de Nemours & Co, 51-55; from asst prof to prof chem, Claremont Men's Col, 55-64, PROF CHEM, CLAREMONT MEN'S COL, PITZER COL & SCRIPPS COL, 64- Mem: AAAS; Am Chem Soc. Res: Amylolytic enzymes; carbohydrates; drug metabolism; peptide and protein synthesis; proteolytic enzymes. Mailing Add: 670 S College Claremont CA 91711

BOVARD, KENLY PAUL, b Pittsburgh, Pa, Mar 23, 28; m 50; c 3. ANIMAL HUSBANDRY. Educ: Cornell Univ, BS, 50; Iowa State Univ, MS, 54, PhD, 60. Prof Exp: Res assoc animal husb, Iowa State Col, 54-57; ASSOC PROF ANIMAL SCI, VA POLYTECH INST, 57- Mem: Fel AAAS; Am Soc Animal Sci; Biomet Soc; Am Genetic Asn. Res: Beef cattle breeding; problems dealing with heritabilities of economic characteristics of commercial production. Mailing Add: Beef Cattle Res Sta Agr Exp St Va Polytech Inst Front Royal VA 22630

BOVBJERG, RICHARD VIGGO, b Chicago, Ill, Sept 11, 19; m 42, 60; c 4. ANIMAL ECOLOGY. Educ: Univ Chicago, PhD(zool), 49. Prof Exp: Asst zool, Univ Chicago, 46-49; from instr to asst prof, Wash Univ, 49-55; from asst prof to assoc prof, 55-61, PROF ZOOL, UNIV IOWA, 61- Concurrent Pos: Ford Found fel, 51-52; dir, Iowa Lakeside Lab, 63- Mem: Ecol Soc Am; Soc Study Evolution; Am Soc Limnol & Oceanog; Am Soc Zoologists. Res: Ecology of aquatic invertebrates. Mailing Add: Dept of Zool Univ of Iowa Iowa City IA 52242

BOVE, JOSEPH RICHARD, b Orange, NJ, Apr 23, 26; m 52; c 3. HEMATOLOGY, LABORATORY MEDICINE. Educ: Univ Md, BS, 51, MD, 53. Hon Degrees: MA, Yale Univ, 70. Prof Exp: From instr to asst prof med, 59-66, assoc prof lab med, 66-70, PROF LAB MED, YALE UNIV, 70- Concurrent Pos: Consult, Vet Admin, 59- Mem: AAAS; Am Soc Hemat; Acad Clin Lab Physicians & Sci; Am Fedn Clin Res; Am Soc Clin Path. Res: Blood transfusion serology, especially alterations of red cell antigens and antibodies. Mailing Add: Yale-New Haven Hosp New Haven CT 06504

BOVEE, HARLEY HOWARD, b Edgecomb, Wash, Dec 12, 18; m 41; c 4. ANALYTICAL CHEMISTRY, OCCUPATIONAL HEALTH. Educ: Univ Wash, BS, 48, MS, 54, PhD(anal chem), 59. Prof Exp: Inorg chemist, US Nat Bur Standards, 48-50; anal chemist, US 13th Naval Dist, 50-53; res assoc, Univ Wash, 53-57; chief respiratory support systs group, Bioastronaut Sect, Aerospace Div, Boeing Airplane Co, 59-64; res asst prof, Dept Environ Health, Sch Pub Health & Community Med, Univ Wash, 64-72; OCCUP HEALTH OFFICER, FOOD & DRUG ADMIN, 73- Mem: Am Chem Soc; Am Indust Hyg Asn. Res: Air pollution and industrial hygiene analytical methods. Mailing Add: Food & Drug Admin HFA-400 Rockville MD 20852

BOVELL, CARLTON ROWLAND, b New York, NY, Nov 10, 24. MICROBIAL PHYSIOLOGY. Educ: Brooklyn Col, AB, 48, MA, 52; Univ Calif, PhD(microbiol), 57. Prof Exp: Instr biol & bact, Brooklyn Col, 50-52; actg instr bact, 54-56, from instr to assoc prof, 57-69, PROF MICROBIOL, UNIV CALIF, RIVERSIDE, 69- Mem: Am Soc Microbiologists; AAAS. Res: Comparative aspects of the autotrophic and heterotrophic physiology and metabolism of hydrogen bacteria. Mailing Add: Dept of Biol Univ of Calif Riverside CA 92502

BOVERMAN, HAROLD, b San Francisco, Calif, June 19, 27; m 56; c 3. PSYCHIATRY, PEDIATRICS. Educ: Univ Calif, AB, 50; Univ Chicago, MD, 56. Prof Exp: Resident, Stanford Univ, 57-60; asst prof, Univ Chicago, 62-68; assoc prof psychiat & dir div child psychiat, 68-70, prof psychiat & pediat, Sch Med, Univ Ore, 70-75; PROF PSYCHIAT & PEDIAT, SCH MED, UNIV CALIF, DAVIS, 75- Concurrent Pos: Fel child psychiat & pediat, Yale Univ, 60-62; Nat Inst Ment Health career teaching fel, 63-68; consult, US Peace Corps, 63- Mem: AAAS; AMA; Am Orthopsychiat Asn; Asn Am Med Cols; Am Psychoanal Asn. Res: Child psychiatry and pediatrics; clinical training and research; medical education; early ego development of infants; effects of foster care and adoption. Mailing Add: Dept of Psychiat Univ of Calif Davis CA 95616

BOVEY, FRANK ALDEN, b Minneapolis, Minn, June 4, 18; m 41; c 3. POLYMER CHEMISTRY. Educ: Harvard Univ, BS, 40; Univ Minn, PhD(phys chem), 48. Prof Exp: Asst chief chemist, Nat Synthetic Rubber Corp, 43-45; head polymer res dept, Minn Mining & Mfg Co, 48-55; res assoc, 55-62; MEM TECH STAFF, BELL LABS, 62-, HEAD POLYMER CHEM RES DEPT, 67- Honors & Awards: Polymer Chem Award, Am Chem Soc, 69; High Polymer Physics Award, Am Phys Soc, 74. Mem: Nat Acad Sci; Am Chem Soc; Am Phys Soc; NY Acad Sci. Res: Physical chemistry of polymers; nuclear magnetic resonance of polymers; emulsion polymerization; fluorescence; rates of conformational isomerization; optical rotary dispersion. Mailing Add: Polymer Chem Res Dept Bell Labs Murray Hill NJ 07974

BOVEY, RODNEY WILLIAM, b Craigmont, Idaho, July 17, 34; m 56; c 4. WEED SCIENCE. Educ: Univ Idaho, BS, 56, MS, 59; Univ Nebr, PhD(agron), 64. Prof Exp:

Instr & proj leader weed res, Univ Nebr, 59-64; res agronomist, Agr Res Serv, US Dept Agr, Tex A&M Univ, 64-67; proj leader defoliation res, Fed Exp Sta, Mayaguez, PR, 67; proj leader, 68-73, RES LEADER, BRUSH CONTROL RES, AGR RES SERV, US DEPT AGR, TEX A&M UNIV, 73- Concurrent Pos: Grants, Adv Res Projs Agency, US Dept Defense, 64-67, US Army, Ft Detrick, 67-68 & Dow Chem Co, Atlas Chem Co & Diamond Shamrock Chem Co, 69-70; spec adv to Secy Agr for 2,4,5-T hearing called by Environ Protection Agency, US Dept Agr, 72-74. Mem: AAAS; Soc Range Mgt; Scandinavian Soc Plant Physiol; Weed Sci Soc Am; Sigma Xi. Res: Field evaluation of herbicides for brush control; absorption and translocation of herbicides; effect of herbicides on ultrastructure, physiology, biochemistry, growth and anatomy of plants; herbicide residues in plants and soils. Mailing Add: Dept of Range Sci Tex A&M Univ College Station TX 77843

BOVILLE, BYRON WALTER, b Ottawa, Ont, Dec 14, 20; m 45; c 3. ATMOSPHERIC SCIENCE. Educ: Univ Toronto, BA, 42; McGill Univ, MSc, 58, PhD(meteorol), 61. Prof Exp: Meteorologist, Can Meteorol Serv, 42-58; from asst prof to prof meteorol, McGill Univ, 60-72, chmn dept, 68-70; DIR ATMOSPHERIC PROCESSES RES BR, ATMOSPHERIC ENVIRON SERV, 72- Concurrent Pos: Secy, Subcomt Meteorol & Atmospheric Sci, Nat Res Coun, 56-69, chmn, 69-; mem global atmospheric res prog, Int Ozone Comn, Working Group on Stratospheric Pollution; sabatical leave, Meteorol Nat France, 70-71. Mem: Am Meteorol Soc; Can Meteorol Soc; fel Royal Meteorol Soc. Res: Dynamic meteorology, especially on the general circulation and on the stratosphere and interlayer coupling. Mailing Add: Atmospheric Environ Serv 4905 Dufferin St Downsview ON Can

BOVING, BENT GIEDE, b Washington, DC, Feb 23, 20; m 44, 74; c 3. ANATOMY, REPRODUCTIVE BIOLOGY. Educ: Swarthmore Col, AB, 41; Jefferson Med Col, MD, 48. Prof Exp: Asst, Swarthmore Col, 41; asst zool, Univ Iowa, 41-42; intern, Wilmington Gen Hosp, Del, 48-49; mem staff dept embryol, Carnegie Inst, 51-70; PROF OBSTET, GYNEC & ANAT, SCH MED, WAYNE STATE UNIV, 70- Concurrent Pos: Nat Cancer Inst fel anat, Yale Univ, 49-51. Honors & Awards: Schering Award, 46. Mem: AAAS; Am Asn Anat; Soc Develop Biol; Int Soc Develop Biol; Soc Study Reproduction. Res: Rabbit blastocyst transport, spacing, orientation and attachment to the uterus; invasive growth chemistry and mechanics. Mailing Add: 41901 W Eight Mile Rd Northville MI 48167

BOWDEN, CHARLES MALCOM, b Richmond, Va, Dec 31, 33; m 60; c 3. QUANTUM OPTICS, STATISTICAL MECHANICS. Educ: Univ Richmond, BS, 56; Univ Va, MS, 59; Clemson Univ, PhD(physics), 67. Prof Exp: Physicist, US Naval Res Lab, 59-61; instr physics, Univ Richmond, 61-64; PHYSICIST, US ARMY MISSILE COMMAND, 67- Concurrent Pos: Mem part-time fac, Dept Physics, Univ Ala, Huntsville. Mem: Am Phys Soc; AAAS. Res: Quantum statistical mechanics of superradiance; cooperative processes in matter-radiation field interactions; laser physics; correlation phenomena in partially coherent radiation fields. Mailing Add: 716 Versailles Dr Huntsville AL 35803

BOWDEN, DAVID CLARK, b Tekamah, Nebr, Nov 23, 40; m 60; c 3. MATHEMATICAL STATISTICS, BIOMETRY. Educ: Colo State Univ, BS, 62, MS, 65, PhD(statist), 68. Prof Exp: From instr to asst prof, 65-74, ASSOC PROF STATIST, COLO STATE UNIV, 74- Concurrent Pos: Consult, Colo Game, Fish & Parks, 65- Res: Simultaneous confidence bands for linear regression models; discrimination and confidence bands on percentiles. Mailing Add: Dept of Statist Colo State Univ Ft Collins CO 80521

BOWDEN, DRUMMOND HYDE, b Wales, Mar 10, 24; Can citizen; m 48; c 2. PATHOLOGY. Educ: Bristol Univ, MB, ChB, 48, MD, 60. Prof Exp: Demonstr path, Bristol Univ, 51-52; asst pathologist, Hosp Sick Children, Toronto, 52-54; res assoc, 52-56; from asst prof to assoc prof, Sch Med, St Louis Univ, 56-64; assoc prof, 64-68, PROF PATH, UNIV MAN, 68- Concurrent Pos: Consult pathologist, Health Sci Ctr, Winnipeg & Deer Lodge Hosp. Mem: AAAS; Am Asn Path & Bact; Am Soc Exp Path; Can Soc Clin Invest. Res: Pulmonary pathology, human and experimental. Mailing Add: 770 Bannatyne Ave Winnipeg MB Can

BOWDEN, JOE ALLEN, b Dolores, Colo, July 13, 40; m 61. BIOCHEMISTRY, SOLAR ENERGY. Educ: Adams State Col, BA, 63; Univ NDak, MS, 65, PhD(biochem), 68. Prof Exp: Res assoc lipoproteins, Univ Fla, 68-69, NIH fel, 69-71; ASST PROF BIOCHEM, LA STATE UNIV, BATON ROUGE, 71- Concurrent Pos: Consult, Environ Protection Agency, 73-; consult energy problems, 73-; mem, Solar Energy Steering Comt, La State Univ, 74- Mem: AAAS; Sigma Xi; Am Sci Affil. Res: Enzymology of metabolic diseases; lipoprotein structure and function; biochemical aspects of mental retardation; solar energy and alternate energy sources. Mailing Add: Dept of Biochem La State Univ Baton Rouge LA 70803

BOWDEN, JOHN PHILIP, biochemistry, see 12th edition

BOWDEN, LEONARD WALTER, b Haxtun, Colo, Feb 28, 33; m 62; c 3. GEOGRAPHY. Educ: Univ Colo, BA, 64, MA, 61; Clark Univ, PhD(geog), 64. Prof Exp: Instr geog, RI Col, 62-63; asst prof, Univ Southern Calif, 63-64; asst prof, Univ Calif, Riverside, 64-66; head earth sci, Off Naval Res, 66-67; assoc prof, 68-75, PROF GEOG, UNIV CALIF, RIVERSIDE, 75- Concurrent Pos: NASA grant remote sensing, Univ Calif, Riverside, 62-72; prin investr, NASA Southern Calif Test Site, 66-75. Mem: AAAS; Asn Am Geog; Am Soc Photogram; Am Geog Soc. Res: Water resources and remote sensing of the environment. Mailing Add: Dept of Earth Sci Univ of Calif Riverside CA 92502

BOWDEN, MURRAE JOHN STANLEY, b Brisbane, Australia, Dec 15, 43; m 67; c 2. POLYMER CHEMISTRY. Educ: Univ Queensland, Australia, BSc, 64, 1st class hon, 65, PhD(polymer chem), 69. Prof Exp: Fel polymer chem, Univ Manchester, Eng, 69-71; MEM TECH STAFF POLYMER CHEM, BELL LABS, 71- Honors & Awards: Solid State Technol Young Authors Prize, Electrochem Soc, 74. Mem: Am Chem Soc; Royal Australian Chem Inst. Res: Effects of radiation on polymers and its application to resists for electron lithography; polymer coatings for optical fibers. Mailing Add: Bell Labs 600 Mountain Ave Murray Hill NJ 07974

BOWDEN, ROBERT LEE, JR, b Paris, Tenn, Apr 10, 33; m 68; c 2. MATHEMATICAL PHYSICS. Educ: Murray State Univ, AB, 55; Va Polytech Inst & State Univ, MS, 58, PhD(physics), 63. Prof Exp: Instr physics & math, Murray State Univ, 55-56; asst prof physics, Va Polytech Inst & State Univ, 63-68; vis assoc prof, Mid East Tech Univ, Turkey, 68-69; ASSOC PROF PHYSICS, VA POLYTECH INST & STATE UNIV, 69- Mem: Am Nuclear Soc; Am Phys Soc. Res: Linear and nonlinear operator theory analysis arising in neutral particle transport. Mailing Add: Dept of Physics Va Polytech Inst & State Univ Blacksburg VA 24061

BOWDLER, ANTHONY JOHN, b London, Eng, Oct 16, 28; m 55; c 2. HEMATOLOGY, EXPERIMENTAL MEDICINE. Educ: Univ London, BSc, 49, MB & BS, 52, MD, 63, PhD(hemat), 67; FRCP, 75. Prof Exp: Sr lectr med, Univ Col Hosp Med Sch, Univ London, 64-67; assoc prof, 67-71, PROF MED, MICH STATE UNIV, 71- Concurrent Pos: Pollard fel & Gould scholar, Univ Col Hosp Med Sch,

Univ London, 61-62; Buswell sr fel med, Sch Med & Dent, Univ Rochester, 62-64; mem med res cdun group study hemolytic dis, Univ Col Hosp Med Sch, Univ London, 64-67, hon consult, Univ Col Hosp, 67. Mem: Am Fedn Clin Res; NY Acad Sci; Brit Med Res Soc; Brit Med Asn; Royal Soc Med. Res: Myeloproliferative disorders; hemolysis; red cell membrane properties; hemodilutional anemias; hemagglutination reactions; pathophysiology of spleen. Mailing Add: Dept of Med Mich State Univ East Lansing MI 48823

BOWE, ARTHUR FREDERICK, b Somerville, Mass, Oct 25, 17; m 43; c 4. ANALYTICAL CHEMISTRY. Educ: Tufts Univ, BS, 39. Prof Exp: Chemist, Cities Serv Oil Co, Pa, 39-40, jr engr, 40-41, control chemist, 41-43 & 45-47, chief chemist, 43-45; anal supvr petrol chem div, 47-48, lab mgr, Gulf Coast Region, 48-49 & Eastern Region, 49-52, lab coordr, 52-56, off mgr petrol lab, 56-59; lab mgr, Western Region, 59-65 & El Monte Lab, 65-73, MGR POMONA PLANT, EI DU PONT DE NEMOURS & CO, INC, 73- Mem: Am Chem Soc. Res: Technical service; sales; administrative work. Mailing Add: 1000 Walnut Ave Pomona CA 91766

BOWE, JOSEPH CHARLES, b Chicago, Ill, Sept 17, 21; m 53; c 6. PHYSICS. Educ: St Procopius Col, AB, 43; DePaul Univ, MS, 46; Univ Ill, PhD(physics), 51. Prof Exp: Instr physics, DePaul Univ, 43-44; jr physicist, Metall Lab, Univ Chicago, 44-46; asst physics, Univ Ill, 46-51; assoc physicist, Argonne Nat Lab, 51-67; TEACHER PHYSICS, ILL BENEDICTINE COL, 67- Mem: Am Phys Soc; Am Asn Physics Teachers. Res: Electron mobility in gases; low energy electron-atom interactions; electronic instrumentation. Mailing Add: Dept of Physics Ill Benedictine Col Lisle IL 60532

BOWE, ROBERT LOOBY, b Worcester, Mass, Jan 25, 25; m 57; c 4. PSYCHOPHARMACOLOGY. Educ: Boston Col, BS, 50, MS, 57; Univ Tenn, PhD(clin physiol), 60. Prof Exp: Instr, New Prep Sch, Mass, 52-56; asst physiol, Boston Col, 56-57; assoc pharmacol, Med Col SC, 60-61, asst prof pharmacol & therapeut, 61-64; asst prof, 64-71, ASSOC PROF PHARMACOL, MED COL VA, 71- Concurrent Pos: Vis prof, Columbia State Hosp, 61-64. Mem: AAAS. Res: Cardiovascular and renal physiology and pharmacology; psychophysiology and pathophysiology; drug abuse. Mailing Add: Dept of Pharmacol Med Col of Va Richmond VA 23219

BOWEN, CHARLES CLARK, b Detroit, Mich, Mar 18, 17; m 47; c 3. CELL BIOLOGY, BOTANY. Educ: Mich State Col, BA, 49, MS, 50, PhD(bot), 53. Prof Exp: Teaching asst, Mich State Col, 49-52; Nat Cancer Inst fel, Brookhaven Nat Lab, 53-55; from asst prof to assoc prof, 55-63, asst dean col sci & humanities & chmn cell biol grad prog, 67-75, PROF BOT & PLANT PATH, IOWA STATE UNIV, 63- Concurrent Pos: Vis prof, Univ Tex Med Br, 75-76. Mem: Am Soc Cell Biol; Bot Soc Am. Res: Cell structure and function, especially the nucleus and mitosis; electron microscopy. Mailing Add: Room 3 Bessey Hall Iowa State Univ Ames IA 50011

BOWEN, CHARLES E, b St Louis, Mo, Apr 9, 36. BIOCHEMISTRY. Educ: San Jose State Col, AB, 59, BS, 63, MS, 65; Univ Calif, Davis, PhD(biochem), 69. Prof Exp: Asst prof, 69-74, ASSOC PROF CHEM, CALIF STATE POLYTECH COL, KELLOGG-VOORHIS, 74- Mem: Am Chem Soc. Res: Protein chemistry; enzyme mechanisms; marine biochemistry of invertebrates. Mailing Add: Dept of Chem Calif State Polytech Col Pomona CA 91766

BOWEN, CHARLES VERNE, b Gallipolis, Ohio, Feb 12, 98; m 24; c 1. CHEMISTRY. Educ: Denison Univ, BSc, 21; Washington & Jefferson Col, MS, 23. Hon Degrees: DSc, Denison Univ, 54. Prof Exp: Chemist, Clinchfield Prod Corp, Tenn, 23; jr chemist, Insecticide & Fungicide Bd, Washington, DC, 23-25; from instr to asst prof chem, Washington & Jefferson Col, 25-37; chemist, Div Insecticide Invest, Bur Entom & Plant Quarantine, US Dept Agr, 37-51; chemist, Insects Affecting Man & Animals, Entom Res Br, Orlando Sta, 51-56, in charge chem lab, Stored-prod Insect Sect, 56-59; chemist, Cancer Chemother Nat Serv Ctr, Nat Insts Health, US Pub Health Serv, 59-65; PRIVATE CONSULT, 65- Concurrent Pos: Lectr, US Dept Agr Grad Sch, 38-51. Mem: Am Chem Soc. Res: Tobacco alkaloids; synthetic organic insecticides; development of analytical methods; formulation of insecticides and insect repellants; insecticide residues; synthetic organic anticancer agents. Mailing Add: 1251 Lakeview Dr Winter Park FL 32789

BOWEN, CORNELIUS MONROE, b Jersey City, NJ, Nov 6, 12; m 66; c 2. PUBLIC HEALTH ADMINISTRATION. Educ: Rutgers Univ, CPHA, 37; NY Univ, BA, 56, MPA, 58. Prof Exp: Health officer, Closter & Paramus, NJ, 38-43, health officer, Wash Twp & Westwood, 39-43; regional health comn, Bergen County, 43-48; regional rep, Div Dent Pub Health & Resources, NY Regional Off, USPHS, 48-66, dep chief dis control br, Div Dent Health, 66-69, asst regional health dir planning & eval, Health Serv & Ment Health Admin, 69-73; dent health dir, State Del, 74; ADMINR, HEALTH CARE SYST, INC, 74- Concurrent Pos: Instr, Fairleigh Dickinson Univ, 50-62. Mailing Add: 319 Sunset Dr N St Petersburg FL 33710

BOWEN, DAVID HYWEL MICHAEL, b Gorseinon, Wales, July 1, 39; m 67. CHEMISTRY, CHEMICAL ENGINEERING. Educ: Univ Birmingham, BSc, 60, PhD(chem eng), 63. Prof Exp: Res engr, Eng Res Div, E I du Pont de Nemours & Co, Inc, 63-67; from asst ed to managing ed, 67-72, head journals dept, 73-75, DIR BOOKS & JOURNALS DIV, AM CHEM SOC, 75- Mem: Am Chem Soc; Am Inst Chem Eng. Res: Technical communication; information science. Mailing Add: Am Chem Soc 1155 16th St NW Washington DC 20036

BOWEN, DAVID VAUGHAN, biochemistry, spectroscopy, see 12th edition

BOWEN, DONALD EDGAR, b Brooklyn, NY, Apr 10, 39; m 65; c 1. SOLID STATE PHYSICS. Educ: Tex Christian Univ, BA, 61, MA, 63; Univ Tex, PhD(physics), 66. Prof Exp: Res assoc high pressure physics, Gen Dynamics/Ft Worth, 62-63; asst, 63-66, from asst prof to assoc prof, 66-71, PROF PHYSICS & CHMN DEPT, UNIV TEX, EL PASO, 71- Concurrent Pos: Res Corp Cottrell grant, 67; R A Welch Found grant, 68-71. Mem: AAAS; Am Asn Physics Teachers; Am Phys Soc. Res: Properties of metal-ammonia solutions and the investigations of the acoustical properties of these solutions. Mailing Add: Dept of Physics Univ of Tex El Paso TX 79999

BOWEN, DOUGLAS MALCOMSON, b Wellesley, Mass, Sept 24, 17; m 43; c 2. ORGANIC CHEMISTRY. Educ: Harvard Univ, AB, 37, AM, 39, PhD(org chem), 40. Prof Exp: Du Pont fel synthesis of tumor metabolites, Harvard Univ, 40-41, instr org chem, 41-45; asst prof, 45-53, chmn dept chem, 59-63, PROF CHEM, DARTMOUTH COL, 53-, REGISTR, 68- Concurrent Pos: NSF fac fel, Stanford Univ, 60-61. Mem: Am Chem Soc. Res: Synthesis and reactions of substituted quinolines; reactions of Grignard reagents with nitriles; qualitative organic analysis; entrainment in Grignard reactions; synthesis of compounds related to tremetone. Mailing Add: 8 Chase Rd Hanover NH 03755

BOWEN, EARL KENNETH, b Colonie, NY, Nov 13, 18; m 42; c 2. MATHEMATICAL STATISTICS. Educ: Univ Mass, BS, 40; Boston Univ, AM, 42. Prof Exp: Instr math, Northeastern Univ, 40-44; field serv consult, Off Sci Res & Develop, 44-45; PROF MATH & STATIST, BABSON COL, 62-, CHMN DEPT, 46- Concurrent Pos: Staff mem & consult opers res off, Johns Hopkins Univ, 51-52. Mem: Am Statist Asn; Inst Math Statist; Inst Mgt Sci; Asn Comput Mach; Am Inst Decision Sci. Res: Business research; operations research; applications of mathematics and statistics to problems of management. Mailing Add: Babson Col Babson Park MA 02157

BOWEN, EDWARD H, JR, b Fall River, Mass, Mar 7, 25; m 73; c 2. OCCUPATIONAL MEDICINE. Educ: Harvard Univ, MD, 49. Prof Exp: Intern med, Long Island Col Hosp, 49-50, from asst resident to chief resident, 50-52; resident fel cardiol, St Luke's Hosp, 52-53, attend staff, St Luke's Hosp & Long Island Col Hosp, 53-56; attend physician, Med Clin & Cardiac Clin, St Luke's Hosp, 53-56; mem clin staff, Smith Kline & French Labs, 56-59; dir clin res, Baxter Labs, 59-61; chief med electronics & space med, Melpar, Inc, 61-64; private practice, 64-68; sr physician, Cushing Hosp, Mass, 68-74; STAFF PHYSICIAN HEALTH CLIN, JOHN HANCOCK MUTUAL LIFE INS CO, 74- Concurrent Pos: Consult internal med, Int Ladies Garment Workers Union Health Ctrs, Fall River & Boston, Mass, 68- Mem: Am Therapeut Soc; Am Heart Asn; Am Fedn Clin Res; Am Occup Med Asn. Res: Etiology of hypertension and coronary heart disease. Mailing Add: Hancock Place Box 111 Boston MA 02117

BOWEN, FRANCIS JOHN, analytical chemistry, inorganic chemistry, see 12th edition

BOWEN, GEORGE HAMILTON, JR, b Tulsa, Okla, June 20, 25; m 48; c 5. PHYSICS. Educ: Calif Inst Technol, BS, 49, PhD(biophys), 53. Prof Exp: Assoc biologist, Oak Ridge Nat Lab, 52-54; from asst prof to assoc prof, 54-65, PROF PHYSICS, IOWA STATE UNIV, 65- Mem: AAAS; Am Asn Physics Teachers. Res: Physics teaching; optics. Mailing Add: Dept of Physics Iowa State Univ Ames IA 50011

BOWEN, HOLLIS HULON, b Doddridge, Ark, Dec 12, 31; m 53; c 2. PLANT BREEDING, GENETICS. Educ: Univ Ark, BS, 60, MS, 61; Rutgers Univ, PhD(hort), 64. Prof Exp: Res assoc plant breeding, Rutgers Univ, 61-65; asst prof, 65-71, ASSOC PROF PLANT BREEDING, TEX A&M UNIV, 71- Res: Breeding and culture of fruits. Mailing Add: Hort Sect Dept of Soil & Crop Sci Tex A&M Univ College Station TX 77843

BOWEN, IRA SPRAGUE, astronomy, deceased

BOWEN, JACOB VAN, JR, statistics, see 12th edition

BOWEN, JAMES MILTON, b Graham, Tex, Feb 1, 35; m 62; c 3. VIROLOGY, IMMUNOLOGY. Educ: Midwestern Univ, BS, 55; Ore State Univ, MS, 58, PhD(microbiol), 61. Prof Exp: Asst res biologist, Sterling-Winthrop Res Inst, 62-64; assoc prof, 64-73, PROF VIROL, UNIV TEX M D ANDERSON HOSP & TUMOR INST, 73- Concurrent Pos: USPHS trainee, Univ Tex, M D Anderson Hosp & Tumor Inst, 61-62. Mem: Am Asn Cancer Res; Am Soc Microbiol. Res: Tumor immunology; oncogenic viruses; biochemistry of oncogenic virus replication. Mailing Add: Univ of Tex M D Anderson Hosp & Tumor Inst Houston TX 77025

BOWEN, JOHN METCALF, b Quincy, Mass, Mar 23, 33; m 56; c 2. PHARMACOLOGY, BIOPHYSICS. Educ: Univ Ga, DVM, 57; Cornell Univ, PhD(physiol), 60. Prof Exp: From asst prof to assoc prof physiol, Kans State Univ, 60-63; fel pharmacol, Emory Univ, 63; assoc prof, 63-69, PROF PHARMACOL, COL VET MED, UNIV GA, 69- Mem: Am Soc Pharmacol & Exp Therapeut; Am Vet Med Asn; Am Soc Vet Physiol & Pharmacol. Res: Pharmacological and biophysical studies on neuromuscular transmission; electromyography. Mailing Add: Dept of Physiol & Pharmacol Univ of Ga Col of Vet Med Athens GA 30602

BOWEN, JULIUS ISADORE, physics, see 12th edition

BOWEN, KENNETH ALAN, b Boston, Mass, June 14, 42; m 63; c 1. MATHEMATICAL LOGIC. Educ: Univ Ill, Urbana-Champaign, BS, 63, MS, 65, PhD(math), 68. Prof Exp: ASST PROF MATH, SYRACUSE UNIV, 68- Mem: Am Math Soc; Math Asn Am; Asn Symbolic Logic. Res: Set theory and foundations of mathematics. Mailing Add: Syracuse Univ Dept of Math 15 Smith Hall Syracuse NY 13210

BOWEN, LAWRENCE HOFFMAN, b Lynchburg, Va, Dec 20, 34. PHYSICAL CHEMISTRY. Educ: Va Mil Inst, BS, 56; Mass Inst Technol, PhD(phys chem), 61. Prof Exp: Asst chem, Mass Inst Technol, 56-58 & nuclear chem, 60-61; from asst prof to assoc prof, 61-70, PROF CHEM, NC STATE UNIV, 70- Concurrent Pos: Fac res grant tracer chem, 62-63; US AEC nuclear teaching grant, 67; NSF res grant Mössbauer spectros, 68-75. Honors & Awards: Res Award, Sigma Xi, 70. Mem: Am Chem Soc; Am Phys Soc; Sigma Xi. Res: Mössbauer spectroscopy structure and bonding in Group V Compounds; nuclear and radiochemistry applied to physicochemical problems; thermodynamics in solutions and solids. Mailing Add: Dept of Chem NC State Univ Raleigh NC 27607

BOWEN, MARCIA ANN, b Rochester, NY, Dec 17, 50. AQUATIC ECOLOGY. Educ: Univ Rochester, BA, 72; Univ RI, MS, 75. Prof Exp: Biologist, New York Ocean Sci Lab, 75-76; MARINE SPECIALIST MARINE BIOL, VA INST MARINE SCI, 76- Mem: Sigma Xi. Res: Ecology of freshwater and marine ostracoda. Mailing Add: Dept of Invert Ecol Va Inst of Marine Sci Gloucester Point VA 23062

BOWEN, MARSHALL EVERETT, b Providence, RI, Jan 15, 38; m 69; c 4. GEOGRAPHY. Educ: Plymouth State Col, BEd, 60; Kent State Univ, MA, 63; Boston Univ, PhD(geog), 70. Prof Exp: Teacher geog, Streetsboro Pub Schs, Ohio, 60-63; lectr & instr, Boston Univ, 65; asst prof, 65-69, ASSOC PROF GEOG, MARY WASHINGTON COL, 69- Mem: Asn Am Geog; Am Geog Soc; Nat Coun Geog Educ. Res: Historical geography of the Great Plains and the dry Rocky Mountain basins, settlement patterns in the intermountain area; Anglo-American West. Mailing Add: Box 1306 College Station Fredericksburg VA 22401

BOWEN, MYLES FOSTER, b Spanish Fork, Utah, May 28, 07; m 54; c 1. ENTOMOLOGY. Educ: Utah State Univ, BS, 29, MS, 30. Prof Exp: Entomologist, US Dept Agr, 31-42 & hq, Sixth US Army, 44-66; RETIRED. Concurrent Pos: Asst, Ohio State Univ, 31-33. Mem: AAAS; Am Entom Soc. Res: Insect population; biometry; medical entomology. Mailing Add: 300 Castenada Dr Millbrae CA 94030

BOWEN, PAUL ROSS, b Catline, Ind, July 13, 02; m 44. BOTANY. Educ: DePauw Univ, AB, 25; Yale Univ, MA, 29, PhD(path), 31. Prof Exp: Teacher high sch, Wash, 25-27 & Mont, 27-28; asst bot, Yale Univ, 29-31, Sterling fel, 31-32; prof biol & head dept, High Point Col, 32-37; prof bot & zool, Beaver Col, 37-42; prof & head dept, Valley Forge Mil Jr Col, 42-46; PROF BIOL, DEL VALLEY COL, 46-, CUR HERBARIUM, 71- Res: Mycology; phytopathology of trees; taxonomy; plant materials for landscape design; gardening. Mailing Add: 1350 Fairy Hill Rd Jenkintown PA 19046

BOWEN, PETER, b Toronto, Ont, May 6, 32; m 65; c 1. INTERNAL MEDICINE, HUMAN GENETICS. Educ: Univ Toronto, MD, 56; FRCP(C), 61. Prof Exp: Instr med, Sch Med, Johns Hopkins Univ, 63-64; assoc prof, 64-71, PROF HUMAN GENETICS, UNIV ALTA, 71- Mem: Am Soc Human Genetics; Can Soc Clin Invest. Res: Human cytogenetics and somatic cell genetics; developmental defects, including chromosomal abnormalities; biochemical and population genetics. Mailing Add: Rm 4-101 Clin Sci Bldg Univ of Alta Edmonton AB Can

BOWEN, RAFAEL LEE, b Takoma Park, Md, Dec 27, 25; m 58; c 2. DENTISTRY, POLYMER CHEMISTRY. Educ: Univ Southern Calif, DDS, 53. Prof Exp: Pvt pract dent, San Diego, 53-55; dent res assoc, 56-70, ASSOC DIR, AM DENT ASN RES UNIT, NAT BUR STAND, 70- Mem: Am Dent Asn; Int Asn Dent Res; fel Am Col Dent; Sigma Xi. Res: Dental therapeutic materials and prevention of oral diseases; insoluble direct filling material approximating the properties and appearance of the anterior teeth; physical measurements. Mailing Add: 16631 Shea Lane Walnut Hill Gaithersburg MD 20760

BOWEN, RAYMOND COBB, parasitology, see 12th edition

BOWEN, RICHARD ELI, b Oskaloosa, Kans, Sept 27, 32; m 55; c 2. PARASITOLOGY. Educ: Univ Kans, BA, 54; Kans State Univ, MS, 60, DVM, 61; Univ Mass, PhD(microbiol), 65. Prof Exp: Instr vet sci, Univ Mass, 61-65; sr bacteriologist, Vet Res Div, 65-75, DEPT HEAD, PARASITOL RES, ELI LILLY & CO, 75- Mem: Am Vet Med Asn; Am Soc Microbiol; Am Asn Avian Path; Wildlife Dis Asn; Indust Vet Asn. Res: Avian leukosis; avian respiratory diseases; bacterial diseases of swine and wildlife; disinfectants; general diagnostic bacteriology. Mailing Add: Box 708 Dept of Parasitol Res Eli Lilly & Co Greenfield IN 46140

BOWEN, RICHARD LEE, b Bunn, NC, July 2, 29; m 57; c 3. PHYSICAL GEOLOGY. Educ: Univ NC, AB, 49; Ind Univ, MA, 51; Univ Melbourne, PhD(geol), 60. Prof Exp: Teaching fel, Ind Univ, 50; geologist, US Geol Surv, 51; geologist-geophysicist, Standard Oil Co Calif, 52-54; consult, Am Overseas Petrol Corp, 57 & petrol explor, Gulf Oil Corp, Africa, 59-61; US Agency Int Develop vis prof, Univ Rio Grande do Sul, Brazil, 62-64; assoc prof, 64-68, chmn dept, 68-74, PROF GEOL, UNIV SOUTHERN MISS, 68- Concurrent Pos: NATO sci conf grants, 68 & 70; instnl rep Gulf Univs Res Corp & mem sci panel. Mem: AAAS; Am Asn Petrol Geol; Am Geol Soc; Geophys Union; Am Inst Urban & Regional Affairs. Res: Petroleum and glacial geology; paleoclimatology; stratigraphy; sedimentology; Gondwana studies; oceanic geochemistry; continental margin tectonics; material balances in geology. Mailing Add: Univ of Southern Miss Southern Sta Hattiesburg MS 39401

BOWEN, RUTH JUSTICE, b Parkersburg, WVa, Jan 11, 42; m 72. INORGANIC CHEMISTRY, PHYSICAL CHEMISTRY. Educ: Glenville State Col, BA, 63; WVa Univ, PhD(chem), 68. Prof Exp: Asst prof inorg chem, 68-74, ASSOC PROF CHEM, CALIF STATE POLYTECH UNIV, POMONA, 74- Mem: Am Chem Soc. Res: Nuclear quadrupole resonance spectroscopy; magnetic susceptibility; stannic chloride adducts; magnetic and spectroscopic studies of N-bonded liquids to transition metals. Mailing Add: Dept of Chem Calif State Polytech Univ Pomona CA 91768

BOWEN, SAMUEL PHILIP, b Council Bluffs, Iowa, Oct 12, 39; m 62; c 2. THEORETICAL PHYSICS, LOW TEMPERATURE PHYSICS. Educ: Iowa State Univ, BS, 62; Cornell Univ, PhD(physics), 67. Prof Exp: Res physicist, Univ Calif, Berkeley, 67-69; asst prof physics, Univ Wis, Madison, 69-75; ASST PROF PHYSICS, VA POLYTECH INST & STATE UNIV, 75- Mem: AAAS; Am Phys Soc. Res: Dilute magnetic alloys; field theory; biophysics; solid state physics. Mailing Add: Dept of Physics Va Polytech Inst & State Univ Blacksburg VA 24060

BOWEN, SARANE THOMPSON, b Des Moines, Iowa, Dec 11, 27; m 54. GENETICS. Educ: Iowa State Univ, BS, 48, MS, 51, PhD(genetics, physiol), 52. Prof Exp: Instr biol, San Francisco Col Women, 53-54; from instr to asst prof, 56-62, assoc prof, 63-68, PROF BIOL, SAN FRANCISCO STATE UNIV, 69- Mem: Genetics Soc Am; Am Soc Human Genetics. Res: Developmental biology; physiological ecology; genetics of Artemia; hemoglobin characterization. Mailing Add: Dept of Biol San Francisco State Univ San Francisco CA 94132

BOWEN, THEODORE, b Evanston, Ill, Mar 19, 28; m 61; c 1. PHYSICS. Educ: Univ Chicago, PhB, 47, SM, 50, PhD, 54. Prof Exp: From asst to assoc cosmic ray res, Univ Chicago, 53-55; res fel, Brazilian Centre Physics Res, 54-55; from res asst to res assoc physics, Princeton Univ, 56-62, res physicist, 62; from assoc prof to prof physics, 62-75, PROF PHYSICS & RADIOLOGY, UNIV ARIZ, 75- Concurrent Pos: Nat Acad Sci sr res assoc, Goddard Space Flight Ctr, NASA, 68. Mem: Fel Am phys Soc; fel AAAS. Res: Cosmic ray search for rare or hypothetical new particles; properties of hypernuclei; radiotherapy and diagnostic radiography with protons and heavy ions; diagnostic ultrasound. Mailing Add: Dept of Physics Bldg 81 Univ of Ariz Tucson AZ 85721

BOWEN, THOMAS EARLE, JR, b Decatur, Ala, Oct 16, 38; m 61; c 2. PHYSIOLOGY, BIOPHYSICS. Educ: Birmingham Southern Col, BS, 61; Univ Ala, cert med technol, 61, PhD(physiol), 68. Prof Exp: Chief technologist adult cardiac path lab, Med Col, Univ Ala, Birmingham, 61-63, chief physiol monitoring technologist, 63-65, from instr to assoc prof physiol, 67-72; ASSOC PROF PHYSIOL, ASST VCHANCELLOR & COORDR EDUC RESOURCES, UNIV TENN CTR HEALTH SCI, 72- Concurrent Pos: Porter Found vis lectr, Am Physiol Asn, 70-71. Mem: Asn Am Med Cols; Am Asn Higher Educ. Res: Cardiovascular physiology; ventricular mechanics; cardiac muscle physiology; muscle mechanics. Mailing Add: Off Educ Resources Univ Tenn Ctr Health Sci Memphis TN 38163

BOWEN, VAUGHAN TABOR, b Buffalo, NY, Aug 23, 15; m 42; c 3. BIOCHEMISTRY. Educ: Yale Univ, BA, 37, PhD(zool), 48. Prof Exp: Chemist, Chevrolet Motor & Axle Div, Gen Motors Corp, 42-43; chemist, Electro Metall Co, Union Carbide & Carbon Corp, 43-45, chemist-in-chg anal labs, Area Plant, 45-46; asst, Yale Univ, 46-48; assoc biologist, Brookhaven Nat Lab, 48-52, biologist, 52-54; geochemist, 54-63, SR SCIENTIST, WOODS HOLE OCEANOG INST, 63- Concurrent Pos: Instr, Yale Univ, 48-52, lectr, 52-68; mem corp, Marine Biol Labs, Woods Hole, 57- & Bermuda Biol Sta, 59-; mem adv comt, Sears Found Marine Res, 61-, chmn, 62-63; mem panel on radioactivity in marine environ, Nat Acad Sci Comt Oceanog, 65-; mem panel on reference methods in marine radioactivity studies, Int Atomic Energy Agency, 68. Mem: Am Soc Nat. Res: Metabolism of heavy metals in insects; analytical studies of heavy metals in marine organisms; radioisotope tracer studies in geochemistry; use of radioisotope tracers in biology; oceanography; fallout studies. Mailing Add: Dept of Chem Woods Hole Oceanog Inst Woods Hole MA 02543

BOWEN, WILLIAM H, b Enniscorthy, Ireland, Dec 11, 33; m 58; c 5. ORAL MICROBIOLOGY, IMMUNOLOGY. Educ: Nat Univ Ireland, BDentSurg, 55; Univ Rochester, MSc, 59; Univ London, PhD(dent), 65; Univ Ireland, DSc(dent), 74; FFDRCSI, 65; FDSRCS, 74. Prof Exp: Pvt pract, 55-56; res fel dent, Eastman Dent Ctr, Rochester, NY, 56-59; Quinten Hogg fel, Royal Col Surgeons of Eng, 59-62, Nuffield Found fel, 62-65, sr res fel, 65-69, Sir Wilfred Fish fel, 69-73; ACTG CHIEF CARIES PREV & RES BR, NAT CARIES PROG, NAT INST DENT RES, 73- Concurrent Pos: Lectr, Univ London, 70; mem sci coun, Europ Orgn Caries Res, 70-73; C L Roberts Mem lectr, Univ Sheffield, 71; guest lectr, Fac Dent, Univ Oslo, 71. Honors & Awards: Colgate-Palmolive Prize, Brit Div, Int Asn Dent Res, 63; Int Dent Fedn Prize, 64; John Tomes Prize, Royal Col Surgeons Eng, 66. Mem: Europe Orgn Caries Res; Int Asn Dent Res; Am Asn Lab Animal Sci; Am Soc Zool; Int Dent Fedn. Res: Microbiology and immunology of dental caries; prevention of dental disease; oral biology of primates. Mailing Add: Rm 532 5333 Westbard Ave Bethesda MD 20014

BOWEN, WILLIAM R, b Iowa City, Iowa, Oct 15, 36; m 60; c 2. BOTANY, CELL BIOLOGY. Educ: Grinnell Col, BA, 60; Univ Iowa, MS & PhD(bot), 64. Prof Exp: Asst prof biol, Wis State Univ, Stevens Point, 64-65, Univ Ill, Chicago Circle, 65-66 & Western Ill Univ, 66-70; assoc prof, Ripon Col, 70-75; PROF BIOL, UNIV ARK, LITTLE ROCK, 75- Mem: AAAS; Am Inst Biol Sci; Nat Asn Biol Teachers; Bot Soc Am; Phycol Soc Am. Res: Developmental plant anatomy and ultrastructure of green algae; botanical teaching. Mailing Add: Dept of Biol Univ of Ark Little Rock AR 72204

BOWEN, ZEDDIE PAUL, b Rockmart, Ga, Mar 29, 37; m 55; c 2. GEOLOGY, PALEONTOLOGY. Educ: Johns Hopkins Univ, AB, 58; Harvard Univ, MA, 60, PhD(geol), 63. Prof Exp: From asst prof to assoc prof, 62-72, PROF GEOL, UNIV ROCHESTER, 72-, CHMN, DEPT GEOL SCI, UNIV ROCHESTER, 74- Concurrent Pos: NSF sci fac fel, 67-68. Mem: Geol Soc Am; Paleont Soc. Res: Invertebrate paleontology; evolution, paleoecology, paleobiogeography and taxonomy of the Brachiopoda; Silurian and Devonian biostratigraphy of eastern North America. Mailing Add: Dept of Geol Univ of Rochester Rochester NY 14627

BOWER, BARTON K, organic chemistry, see 12th edition

BOWER, CHARLES ARTHUR, b Shawnee, Okla, Feb 17, 16; m 41; c 4. SOIL CHEMISTRY. Educ: Okla State Univ, BS, 36; Univ Wis, PhD(soils), 41. Prof Exp: Soil scientist, US Soil Conserv Serv, 36-38; asst, Univ Wis, 38-41; chemist, NC State Dept Agr, 41-42; asst prof soils, Iowa State Col, 42-45; prin soil scientist, US Salinity Lab, 45-60, dir, 60-72; SOIL SCIENTIST, UNIV HAWAII, 72- Concurrent Pos: Prof soil sci, Univ Calif, Riverside, 65-72. Honors & Awards: Achievement Award, Am Soc Agron, 59. Mem: Soil Sci Soc Am; fel Am Soc Agron. Res: Chemistry of salt-affected and tropical soils; mineral nutrition of plants; cation exchange; water quality; agricultural development of arid and tropical lands. Mailing Add: PO Box 741 Captain Cook HI 96704

BOWER, DAVID ROY EUGENE, b Port Angeles, Wash, Sept 10, 34; m 55; c 4. FOREST BIOMETRY. Educ: Univ Idaho, BS, 58; Duke Univ, MF, 59. Prof Exp: Silviculturist, Southern Forest Exp Sta, US Forest Serv, 59-64, math statistician, 66-69; FOREST BIOMETRICIAN, WEYERHAEUSER CO, 69- Mem: Soc Am Foresters. Res: Biologic models to describe growth of trees and stands; experimental designs for testing effects of silvicultural treatments, effects of measurement error and biological variability on growth estimation. Mailing Add: 1016 Swanson Dr Centralia WA 98531

BOWER, FRANK ARNOLD, organic chemistry, see 12th edition

BOWER, GEORGE MYRON, b Arbuckle, Calif, May 26, 25; m 51; c 3. ORGANIC POLYMER CHEMISTRY. Educ: Ore State Univ, BS, 50; Univ Ore, MS, 54, PhD(chem), 57. Prof Exp: SR RES CHEMIST, WESTINGHOUSE RES & DEVELOP CTR, 56- Mem: AAAS; Am Chem Soc. Res: Polyimides and other thermally stable polymers; composite technology; phenol-formaldehyde polymers and melamine formaldehyde polymers. Mailing Add: 2031 Sonny St Pittsburgh PA 15221

BOWER, JOHN EDWIN, b Marietta, Ohio, Sept 1, 14; m 42; c 2. PHYSICAL CHEMISTRY. Educ: Miami Univ, AB, 36; Univ Mich, MS & PhD(chem), 51. Prof Exp: From asst prof to assoc prof, 51-58, PROF CHEM, NORTHERN ILL UNIV, 58- Concurrent Pos: NSF fel, Univ Colo, 57. Mem: AAAS; Am Chem Soc. Mailing Add: Dept of Chem Northern Ill Univ DeKalb IL 60115

BOWER, JOSEPH G, inorganic chemistry, see 12th edition

BOWER, OLIVER KENNETH, b Hindsboro, Ill, May 12, 02; m; c 2. MATHEMATICS. Educ: Univ Ill, AB, 24, AM, 27, PhD(math), 29. Prof Exp: Asst instr math, Univ Ill, 25-28, from instr to asst prof, 29-62; ASSOC PROF MATH, UNIV ARK, FAYETTEVILLE, 62- Mem: Am Math Soc. Res: Mathematics of the archery bow; applications of an abstract existence theorem to both differential and difference equations. Mailing Add: 20 Country Lane Fayetteville AR 72701

BOWER, RAYMOND KENNETH, b Kansas City, Mo, Oct 31, 27; m 59; c 2. MICROBIOLOGY, ANIMAL VIROLOGY. Educ: Univ Mo-Kansas City, BA, 49; Kans State Univ, MS, 51, PhD(microbiol), 54. Prof Exp: Res fel pediat, Sch Med, Univ Kans, 54-55; dir virus & rickettsia div, Ind State Dept Health, 55-57; USPHS fel microbiol, Sch Med, Univ Okla, 57-59; from instr to asst prof, Sch Med, Univ Ark, 59-63; asst prof, Grad Res Inst, Baylor Univ, 63-65, Col Dent & Grad Div, 65-68; asst prof, 68-74, ASSOC PROF, DEPT OF BOT & BACTERIOL, UNIV ARK, 74- Concurrent Pos: Consult microbiol, Div Pub Health Labs, Ark State Dept Pub Health,75 & Vet Admin Hosp, 75-76. Mem: Am Soc Microbiol; Am Asn Cancer Res. Res: Genetics of Newcastle disease virus and Rous sarcoma virus-host cell relationships; genetic and immunologic mechanisms underlying the relationship of Rous Sarcoma virus to the chicken. Mailing Add: 1764 Applebury Pl Fayetteville AR 72701

BOWERFIND, EDGAR SIHLER, JR, b Cleveland, Ohio, May 7, 24; m 56; c 4. INTERNAL MEDICINE. Educ: Western Reserve Univ, MD, 49; Am Bd Internal Med, dipl, 58. Prof Exp: From intern to resident, Univ Hosps Cleveland, 50-56; demonstr, 56-58, from instr to sr instr, 58-65, ASST PROF MED, SCH MED, CASE WESTERN RESERVE UNIV, 65- Concurrent Pos: Consult, Vet Admin Hosp, Cleveland, 56-; asst secy, Citizens Comn Grad Med Educ, 63-64, secy, 64-66. Mem: Am Soc Hemat; assoc Asn Am Med Cols. Res: Clinical hematology; lymphadenopathy; clinical medical education. Mailing Add: Univ Hosp of Cleveland 2065 Adelbert Rd Cleveland OH 44106

BOWERING, JEAN, b Yonkers, NY, Mar 16, 39. NUTRITION. Educ: Cornell Univ, BS, 60, MNS, 64; Univ Calif, Berkeley, PhD(nutrit), 69. Prof Exp: Res chemist, R T French Co, NY, 60-62; trainee, Inst of Nutrit for Cent Am & Panama, 64; res assoc nutrit, Cornell Univ, 64-66; instr dept pediat, Children's Hosp, DC & med sch, George Washington Univ,' 69-70; ASST PROF HUMAN NUTRIT & FOODS, CORNELL UNIV, 70- Res: Human metabolic studies—protein and energy utilization, purine metabolism; iron requirements of humans; applied clinical nutrition in low-income urban population. Mailing Add: Dept of Human Nutrit & Food Col of Human Ecol Cornell Univ Ithaca NY 14850

BOWERMAN, ERNEST WILLIAM, b Mitchell, SDak, Mar 11, 05; m 34; c 3. CHEMISTRY. Educ: Northern State Col, SDak, BS, 29; SDak Sch Mines & Technol, BS, 33; Ohio State Univ, PhD(chem), 37. Hon Degrees: LHD, SDak Sch Mines & Technol, 69. Prof Exp: Prin high sch, SDak, 29-30; asst chem, Ohio State Univ, 35-36, spec asst, 36-37; res chemist, Humble Oil & Ref Co, 37-42, sr res chemist, 42-43, asst sect head, Res & Develop Div, 43-46, sect head, 46-56, asst div head, 56-60; prod mgr, Raw Mat Div, Enjay Chem Co, 60-61; head dept, Esso Res & Eng Co, 61-65; prof & vchmn dept, 65-74, EMER PROF CHEM, OHIO STATE UNIV, 74- Mem: Fel Am Inst Chem; Am Chem Soc; Am Inst Chem Eng; Soc Res Adminstr. Res: Development work on catalytic processes including polymerization of olefins; alkylation of isoparaffins with olefins; hydrogenation of olefins; isomerization of normal paraffins; hydroforming naphthenes to aromatics; cracking of process gas oils. Mailing Add: 2761 Welsford Rd Columbus OH 43221

BOWERMAN, ROBERT FRANCIS, b Norfolk, Va, Oct 6, 44; m 73. NEUROPHYSIOLOGY, BEHAVIORAL BIOLOGY. Educ: Univ NC, Chapel Hill, AB, 66; Univ Miami, PhD(biol), 71. Prof Exp: Res asst oceanog, Marine Lab, Univ NC, 64-65; NIH fel neurophysiol, Dept Zool, Univ Tex, Austin, 71-73; ASST PROF ZOOL, UNIV WYO, 73- Mem: Soc Neurosci; Sigma Xi; Am Inst Biol Sci. Res: Investigation of the neural mechanism underlying behavior, particularly the control of posture and movement; analysis of the control of scorpion walking. Mailing Add: Dept of Zool & Physiol Univ of Wyo Laramie WY 82071

BOWERS, ALSTON GORDON, b Upper Tract, WVa, May 8, 09; m 31; c 2. PESTICIDE CHEMISTRY. Educ: Miami Univ, AB, 31; Ohio State Univ, MSc, 34. Prof Exp: Chemist, New Method Varnish Co, NY, 35-36; chief chemist, 36-40; chief chemist, Gerson Stewart Corp, 40-41, vpres, 41-51, pres, 52-57; pres, Pioneer Mfg Co, 57-75 & MacCarl Co, 65-75; RETIRED. Concurrent Pos: Chief chemist, Hunt Mfg Co, 41-57, pres, 52-57; instr, Fenn Col, 42-46 & Case Inst Technol, 48-49; mem bd dirs, Pioneer Mfg Co, 57- Honors & Awards: Cleveland Chem Prof Award, 54. Mem: AAAS; Am Chem Soc. Res: Preparation and stability of wax emulsions; composition and effectiveness of germicides, soaps and alkyl aryl sulfonates. Mailing Add: 5942 SW First Ave Cape Coral FL 33904

BOWERS, CLEMENT GRAY, plant breeding, deceased

BOWERS, DARL EUGENE, b Fresno, Calif, Sept 23, 21; m 44; c 3. ZOOLOGY, ECOLOGY. Educ: Univ Calif, Berkeley, AB, 48, MA, 52, PhD(zool), 54. Prof Exp: PROF BIOL, MILLS COL, 54- Concurrent Pos: NSF grant, 61-66. Mem: AAAS; Ecol Soc Am; Am Inst Biol Soc. Res: Correlation of the distribution and characteristics of animal forms with environmental gradients and physical factor constellations; ornithological ecology; correlation of the distribution of amphipods with physical factors of beaches. Mailing Add: Dept of Biol Mills Col Oakland CA 94613

BOWERS, DELORES MAUREEN, b Hicksville, Ohio, Aug 17, 43; div. ANALYTICAL CHEMISTRY, WATER CHEMISTRY. Educ: Adrian Col, BS, 65; Mich State Univ, MS, 68, PhD(anal chem), 72. Prof Exp: Anal technician biochem, Mich State Univ, 71; city chemist health & sanit testing, Water Pollution Control Ctr, 72-74; asst prof anal chem, Winona State Univ, 74-75; ASST PROF ANAL & PHYS CHEM, CARLETON COL, 75- Mem: Am Chem Soc. Res: Isolation and identification of trace materials in environmental and clinical samples; development of method of analysis of water and sediment samples for nutrients. Mailing Add: Dept of Chem Carleton Col Northfield MN 55057

BOWERS, FRANK DANA, b Fayetteville, Ark, Mar 21, 36; m 67. BRYOLOGY, PLANT TAXONOMY. Educ: Southwest Mo State Univ, BS, 66; Univ Tenn, Knoxville, MS, 68, PhD(bot), 72. Prof Exp: Asst cur herbarium, Dept Bot, Univ Tenn, 68-72, fel, 72-74, res assoc, 74-75; ASST PROF BOT, UNIV WIS-STEVENS PT, 75- Mem: Bryological & Lichenological Soc Am; Bot Soc Am; Soc Econ Bot; Am Soc Plant Taxonomists; Int Asn Plant Tax. Res: Moss flora of Mexico; mosses of Central America and Wisconsin; wildlife food plants; Heterotheca section Pityopsis, Compositae; edible and poisonous plants. Mailing Add: Dept of Biol Univ of Wis Stevens Point WI 54481

BOWERS, GEORGE HENRY, III, b Philadelphia, Pa, Oct 16, 23; m 49; c 4. ORGANIC CHEMISTRY. Educ: Washington & Lee Univ, BS, 44; Univ Pa, MS, 48, PhD(org chem), 51. Prof Exp: Asst instr chem, Univ Pa, 48-51; res chemist, 51-74, RES ASSOC, E I DU PONT DE NEMOURS & CO, INC, CHESTNUT RUN, 74- Mem: Am Chem Soc; Sigma Xi; Inst Elec & Electronics Eng. Res: Synthetic elastomer research; high temperature organic reactions; synthetic thermo plastic research; electrical insulation and systems research. Mailing Add: 612 Foulkstone Rd Wilmington DE 19803

BOWERS, JANE ANN (RAYMOND), b Fredonia, Kans, Aug 19, 40; m 63; c 2. FOOD SCIENCE. Educ: Kans State Univ, BS, 62, MS, 63, PhD(food, nutrit), 67. Prof Exp: Res assoc foods, Iowa State Univ, 63-64; from asst prof to assoc prof, 67-74, PROF FOODS, KANS STATE UNIV, 74-, ACTG HEAD, DEPT FOODS & NUTRIT, 75- Mem: Inst Food Technologists; Am Meat Sci Asn; Poultry Sci Asn; Am Home Econ Asn. Res: Chemical characteristics and sensory evaluation of meat and poultry. Mailing Add: Dept of Foods & Nutrit Kans State Univ Manhattan KS 66506

BOWERS, JOHN DALTON, b West Unity, Ohio, July 15, 15; m 41; c 2. AGRICULTURE. Educ: Ohio State Univ, BSc, 40. Prof Exp: Asst, Ohio State Univ, 40-42; lab dir, Borden, Morres & Ross, 42-43, supt milk dept, 46-47, lab dir, 47-51, lab dir qual control & res & develop, Mid-west Div, Borden Co, 51-70, dir prod develop, 70-72, DIR QUAL CONTROL & RES & DEVELOP, NORTHERN DIV, BORDEN, INC, 72- Concurrent Pos: Mem bd dirs, Am Cultured Prod Inst, 67-70. Mem: Am Dairy Sci Asn. Mailing Add: Northern Div Borden Inc 165 N Washington Ave Columbus OH 43216

BOWERS, JOHN LOWRY, horticulture, see 12th edition

BOWERS, KLAUS DIETER, b Stettin, Ger, Dec 27, 29; nat US; m 64; c 2. PHYSICS. Educ: Oxford Univ, BA, 50, MA, 53, PhD(physics), 53. Prof Exp: Res lectr, Christ Church, Oxford, 52-56; mem tech staff, Bell Labs, 56-59, supvr variable reactance amplifier group, 59-61, head microwave integrated device dept, 61-64, solid state optical device dept, 64-66, dir solid state device lab, 66-71; MANAGING DIR COMPONENT DEVELOP, SANDIA LABS, 71- Mem: Sr mem Inst Elec & Electronics Eng. Res: Quantum electronics; memory; piezoelectric and microwave devices. Mailing Add: Sandia Labs Albuquerque NM 87115

BOWERS, LANDON EMANUAL, b Mohawk, NY, Mar 16, 18; m 47; c 2. BACTERIOLOGY. Educ: Univ Ky, BS, 41, MS, 51; Univ Tex, PhD, 55. Prof Exp: Asst prof bact, Univ NMex, 55-57 & Syracuse Univ, 57-64; assoc prof, 64-67, PROF BIOL, ST LAWRENCE UNIV, 67- Mem: AAAS; Am Soc Microbiol; Am Soc Limnol & Oceanog. Res: Nutrition of anaerobic microorganisms; sporulation and lysis of anaerobes; bacteriology of water. Mailing Add: Dept of Biol St Lawrence Univ Canton NY 13617

BOWERS, MARY BLAIR, b Jackson, NC, Apr 2, 30. CELL BIOLOGY. Educ: Duke Univ, AB, 52, MA, 55; Harvard Univ, PhD(biol), 61. Prof Exp: Instr biol, Harvard Univ, 61-63; RES BIOLOGIST, NAT HEART INST, 69- Concurrent Pos: NIH fel, Harvard Univ, 61-63; staff fel, Nat Heart Inst, 66-69. Mem: AAAS; Electron Micros Soc Am; Am Soc Zool; Am Soc Cell Biol. Res: Correlation of cell structure with function; mechanisms of bulk transport. Mailing Add: Rm 322 Bldg 3 NIH Bethesda MD 20014

BOWERS, MAYNARD C, b Battle Creek, Mich, Nov 5, 30; m 52, 70; c 3. BOTANY. Educ: Albion Col, AB, 56; Univ Va, MEd, 60; Univ Colo, PhD(bot), 66. Prof Exp: Teacher pub sch, Mich, 56-57 & Fla, 57-59; asst prof biol, Towson State Col, 60-62; from asst prof to assoc prof, 66-75, actg head dept biol sci, 69-70, PROF BIOL, NORTHERN MICH UNIV, 75- Mem: Am Bryol Soc; Am Inst Biol Sci; Bot Soc Am; Int Asn Plant Taxon; Int Asn Bryologists. Res: Plant biosystematics; cytotaxonomy of bryophytes, especially the family Mniaceae; cytological effects of herbicides on non-target species. Mailing Add: Dept of Biol Sci Northern Mich Univ Marquette MI 49855

BOWERS, MICHAEL THOMAS, b Moscow, Idaho, June 6, 39; m 64; c 2. CHEMICAL KINETICS, CHEMICAL PHYSICS. Educ: Gonzaga Univ, BS, 62; Univ Ill, Urbana, PhD(phys chem), 66. Prof Exp: Asst prof, 68-73, ASSOC PROF PHYS CHEM, UNIV CALIF, SANTA BARBARA, 73- Mem: Am Chem Soc; Am Phys Soc; Am Soc Mass Spectros. Res: Gas phase ion chemistry; ion cyclotron resonance spectroscopy. Mailing Add: Dept of Chem Univ of Calif Santa Barbara CA 93106

BOWERS, NEAL MONROE, b Mishawaka, Ind, Oct 2, 06; m 47. GEOGRAPHY. Educ: Western Mich Univ, BS, 38; Univ Mich, MS, 39, PhD(geog), 51. Prof Exp: Asst prof geog, Mich State Norm Col, 45-46; instr, Amherst Col, 46-47; from assoc prof to prof, 49-73, chmn dept, 56-63, actg dir social sci res inst, 60-61, chmn SAsian studies prog, 67-70, EMER PROF GEOG, UNIV HAWAII, 73- Concurrent Pos: Chief ports & urban sect, Far East Div, Off Strategic Serv, 43-44, dep chief res br, 44-45; Fulbright prof, Pakistan, 52-53; vis prof, Inst Asian Studies, Philippines, 56-57; actg dir, Inst Advan Proj, East-West Ctr, 60-62; Fulbright res scholar, India, 63-64. Mem: AAAS; Asn Am Geogr; Asn Asian Studies; Nat Coun Geog Teachers. Res: Physical and social geography, especially Asia and the Pacific. Mailing Add: 4941 Maunalani Circle Honolulu HI 96816

BOWERS, PAUL APPLEGATE, b Big Run, Pa, Aug 4, 11; m 45; c 3. OBSTETRICS & GYNECOLOGY. Educ: Bucknell Univ, BS, 33; Jefferson Med Col, MD, 37. Prof Exp: Intern, Hosp, Jefferson Med Col, 37-39; resident, Chicago Lying-In-Hosp, Ill, 39-42; PROF OBSTET & GYNEC, JEFFERSON MED COL, 46- Concurrent Pos: Pvt pract, 46-; consult, Valley Forge Gen Hosp, Phoenixville, Pa, 49- Mem: AMA; Am Col Obstet & Gynec; Am Col Surgeons. Res: Physiologic obstetrics; disorder of the breast. Mailing Add: Dept of Obstet & Gynec Jefferson Med Col Philadelphia PA 19103

BOWERS, PETER GEORGE, b Sussex, Eng, May 14, 37. PHYSICAL CHEMISTRY. Educ: Cambridge Univ, BA, 61; Univ BC, PhD(chem), 64. Prof Exp: NATO res fel, Univ Sheffield, 64-66; vis asst prof chem, Boston Univ, 67-68; asst prof, 68-72, ASSOC PROF CHEM, SIMMONS COL, 72- Concurrent Pos: Vis assoc prof, Univ BC, 74-75. Mem: Can Inst Chem; Am Chem Soc; The Chem Soc. Res: Photochemistry; reaction kinetics. Mailing Add: Dept of Chem Simmons Col 300 Fenway Boston MA 02115

BOWERS, RAYMOND, b London, Eng, July 11, 27; nat US; m 54. PHYSICS. Educ: Univ London, BSc, 48; Oxford Univ, DPhil, 51. Prof Exp: Res fel, Univ Chicago, 51-53; res physicist, Westinghouse Res Labs, 53-60; dep dir prog sci, technol & Soc, 69-73, PROF PHYSICS, CORNELL UNIV, 60-, DIR PROG SCI, TECHNOL & SOC, 73- Concurrent Pos: With Off Sci & Technol, Exec Off of the President, 66-67; consult, NSF, State Dept & various indust orgns; chmn comt on NSF & sci & pub policy, AAAS; 72-76; mem comt on pub eng policy, Nat Acad Eng, 74- Mem: AAAS; fel Am Phys Soc; fel Am Acad Arts & Sci; fel Brit Inst Physics & Phys Soc. Res: Low temperature and solid state physics; science policy; interdisciplinary research on science, technology and society. Mailing Add: Clark Hall Cornell Univ Ithaca NY 14853

BOWERS, RAYMOND HAROLD, b Salem, Ore, Aug 14, 44; m 66; c 1. ANALYTICAL CHEMISTRY. Educ: SDak Sch Mines & Technol, BS, 66; Iowa State Univ, PhD, 71. Prof Exp: Chemist, 71-73, SECT HEAD INSTRUMENTAL ANAL, RES & DEVELOP CTR, SWIFT & CO, 73- Mem: Am Chem Soc; Am Oil Chemists Soc. Res: Instrumental analysis with special emphasis on development of methodology for the analysis of foods, adhesives and specialty chemicals using chromatographic, spectroscopic, thermal and electrochemical techniques. Mailing Add: Swift & Co Res & Develop Ctr 1919 Swift Dr Oak Brook IL 60521

BOWERS, RICHARD CHARLES, analytical chemistry, see 12th edition

BOWERS, ROBERT CHARLES, b Benton Harbor, Mich, Nov 24, 37; m 63. HORTICULTURE, PLANT SCIENCE. Educ: Mich State Univ, BS, 59, MS, 60; Univ Ariz, PhD(tree physiol), 66. Prof Exp: Res assoc hort, Univ Ariz, 64-66; assoc scientist, 66-68, SCIENTIST, UPJOHN CO, 68- Mem: Am Soc Hort Sci. Res: Research and development of agricultural chemicals. Mailing Add: 4116 Vista Way Davis CA 95616

BOWERS, ROY ANDERSON, b Racine, Wis, May 11, 13; m 40; c 2. PHARMACY. Educ: Univ Wis, BS, 36, PhD(pharmaceut chem), 41. Prof Exp: Asst instr pharm, Univ Wis, 37-40; asst prof, Univ Toledo, 40-41; from asst prof to assoc prof, Univ Kans, 41-45; prof pharm & dean col pharm, Univ NMex, 45-51; PROF PHARM & DEAN COL PHARM, RUTGERS UNIV, NEW BRUNSWICK, 51- Mem: Am Pharmaceut Asn (vpres, 51-52); Am Inst Hist Pharm; Am Asn Cols Pharm (pres, 63-64); NY Acad Sci. Res: Chemistry of fats and waxes; pharmaceutical formulation. Mailing Add: Col of Pharm Rutgers Univ New Brunswick NJ 08903

BOWERS, SPOTSWOOD D, JR, b New York, NY, Apr 2, 21; m 48; c 4. PHYSICAL CHEMISTRY, NUCLEAR CHEMISTRY. Educ: Yale Univ, BS, 42, PhD(phys chem), 51. Prof Exp: Res chemist luminescence, US Radium Corp, Pa, 51-54; fuel separations engr, Atomic Power Develop Asn, Mich, 55-59; asst secy corp affairs & classified advert mgr, NY Law Pub Co, 59-61; assoc prof, 62-74, chmn dept, 65-74, PROF CHEM, HARTWICK COL, 62- Mailing Add: Dept of Chem Hartwick Col Oneonta NY 13820

BOWERS, WAYNE ALEXANDER, b Bilbao, Spain, Mar 1, 19; m 44; c 4. PHYSICS. Educ: Oberlin Col, AB, 38; Cornell Univ, PhD(theoret physics), 43. Prof Exp: Asst physics, Cornell Univ, 38-42, instr, 42-44; physicist, Manhattan Proj, Los Alamos, NMex, 44-46; res assoc, Mass Inst Technol, 46-47; assoc prof, 47-55, PROF PHYSICS, UNIV N C, CHAPEL HILL, 55- Concurrent Pos: Vis prof, Mass Inst Technol, 68-69; NSF sci fac fel, 63-64. Mem: Am Phys Soc; Am Asn Physics Teachers. Res: Solid State physics. Mailing Add: Physics Dept Univ of N C Chapel Hill NC 27514

BOWERS, WILLIAM SIGMOND, b Chicago, Ill, Dec 24, 35; m 58; c 5. INVERTEBRATE PHYSIOLOGY, BIOCHEMISTRY. Educ: Ind Univ, AB, 57; Purdue Univ, MS, 59, PhD(entom), 62. Prof Exp: Sr insect physiologist, Agr Res Ctr, US Dept Agr, 62-72; PROF INSECT PHYSIOL, NY STATE AGR EXP STA, CORNELL UNIV, 72- Honors & Awards: Outstanding Young Scientist Agr Res Serv, US Dept Agr, 69, 70. Mem: AAAS; Entom Soc Am; Am Chem Soc. Res: Insect metabolism and biochemistry; hormonal regulation of invertebrate metamorphosis, reproduction and diapause; isolation identification and synthesis of insect hormones, anti-hormones and pheromones; insect-plant interactions. Mailing Add: NY State Agr Exp Sta Cornell Univ Geneva NY 14456

BOWERSOX, DAVID F, b Clinton, Iowa, Aug 14, 31; m 56; c 4. PHYSICAL CHEMISTRY, ANALYTICAL CHEMISTRY. Educ: Grinnell Col, BA, 53; Calif Inst Technol, PhD(chem), 58. Prof Exp: MEM STAFF RES & DEVELOP, LOS ALAMOS SCI LAB, UNIV CALIF, 57- Mem: Fel Am Inst Chem. Res: Inorganic chemistry; processing of reactor fuels; solid-gas reactions. Mailing Add: 204 Venado Los Alamos NM 87544

BOWERSOX, RALPH B, b Sandpoint, Idaho, Sept 4, 11; m 38; c 2. PHYSICS. Educ: Univ Chicago, BS, 33, MS, 34, PhD(physics), 38. Prof Exp: Asst, Univ Chicago, 34-38; asst prof physics, Univ Toledo, 38-42, head dept, 41-42; res assoc antisubmarine equip, Sound Lab, US Navy, Harvard Univ, 42-45, group leader systs, Res Lab, 45-46; assoc prof physics, Mich State Univ, 46-48; chief sect, Jet Propulsion Labs, Calif Inst Technol, 48-59; consult scientist, Lockheed Missile & Space Div, 59-60; mgr physics res, Martin Co, 60-64; PROF PHYSICS, COLO SCH MINES, 64-, CHMN DEPT, 68- Concurrent Pos: With Nat Defense Res Comt. Mem: Am Phys Soc; Am Asn Physics Teachers. Res: Line spectra; ultrasonic equipment; nuclear physics; electronics; solar collectors and detectors; noise control. Mailing Add: Dept of Physics Colo Sch of Mines Golden CO 80401

BOWERY, THOMAS GLENN, b Avalon, Pa, Dec 17, 21; m 45; c 2. HEALTH SCIENCES. Educ: Mich State Col, BSc, 43; Rutgers Univ, MSc, 48, PhD(entom), 51. Prof Exp: Res assoc insecticide screening, Rutgers Univ, 46-48 & chem & insecticide anal, Off State Chemist, NJ Agr Exp Sta, 48-51; asst entomologist, Everglades Exp Sta, Univ Fla, 51-53; from res assoc prof to res prof chem, NC Agr Exp Sta, 53-62; asst to assoc dir extramural progs, Off of Dir, 62-63 & extramural opers & procedures off, 63-65, asst chief div res facil & resources, 65-67, assoc dir opers, 67-69, DIR DIV RES RESOURCES, NIH, 69- Concurrent Pos: Mem toxicol study sect, Div Res Grants, NIH, 61; mem prof coun health sci, US Civil Serv Comn, 67- Mem: AAAS. Res: Pesticide metabolism; science and public policy. Mailing Add: Div of Res Resources NIH Bethesda MD 20014

BOWHILL, SIDNEY ALLAN, b Dover, Eng, Aug 6, 27; US citizen; m 59; c 2. AERONOMY. Educ: Cambridge Univ, BA, 48, MA, 50, PhD(physics), 54. Prof Exp: Res engr, Marconi's Wireless Tel Co, Eng, 53-55; assoc prof elec eng, Pa State Univ, 55-62; PROF ELEC ENG, UNIV ILL, URBANA-CHAMPAIGN, 62- Concurrent Pos: Fulbright grant, 55-56; ed, Antennas & Propagation Group Trans, 59-62 & Radio Sci, 67-73; assoc ed, Radio Propagation, 64-67; pres, Aeronomy Corp, 69- Mem radio stand panel, Nat Acad Sci, 61-64, comt poential contamination & interference from space exp, 63-, polar res comt, 67-70, chmn panel on upper atmospheric physics, 67-70, mem rocket res comt, 67-72, comt on data interchange and data centers, 67-, adv panel to World Data Ctr A, 69-, geophys data panel, 75-, earth sci panel for NSF postdoctoral fels, 68-69, comt on solar-terrestrial res, 69-, ad hoc panel on Jicamarca Radio Observ, 69- & ad hoc panel on New Upper Atmosphere Observ, 70-71; mem, US Nat Comt, Int Sci Radio Union, 64-75, chmn, US Comn 3, 64-67, mem, Int Coun Sci Unions comt on space res working group 4 on exp in upper atmsophere, 66-, co-chmn panel on interactions between neutral & ionized parts of the atmosphere, 66-, Int Coun Sci Unions spec comt on solar-terrestrial physics, 67-, chmn working group 11 on neutral & ion chem, 68-73, chmn atmospheric physics progs steering comt, 74-; mem, Int Union Radio Sci solar-terrestrial physics comt, 68-72, chmn working group on Int Ref Ionosphere, 67-68, comt on space res, 69-75, vchmn Int Comn 3, 69-72, chmn, 72-75. Mem: Nat Acad Eng; fel Inst Elec & Electronics Engrs; fel Am Astron Soc; Am Soc Eng Educ; Am Geophys Union. Res: Physics of ionosphere; radio propagation; rocket and satellite studies of upper atmosphere. Mailing Add: Dept of Elec Eng Univ of Ill Urbana-Champaign Urbana IL 61801

BOWIE, EDWARD JOHN WALTER, b Church Stretton, Eng, Mar 10, 25; m 48; c 4. HEMATOLOGY. Educ: Oxford Univ, MA, 50, BM & BCh, 52; Univ Minn, MS, 61. Prof Exp: From instr to assoc prof, 66-74, PROF MED & LAB MED, MAYO GRAD SCH MED, UNIV MINN, 74- Concurrent Pos: Consult internal med, Mayo Clin, 61- Honors & Awards: Trotter Medal, Univ Col Hosp, London, 51; Judson Daland Travel Award, 61. Mem: AAAS; Am Soc Hemat; Am Soc Exp Path; Int Soc Thrombosis & Haemostasis. Res: Blood coagulation and bleeding diseases, with particular interest in blood platelets and hemophilia. Mailing Add: Mayo Clin Rochester MN 55901

BOWIE, HAROLD E, b Durham, Maine, Jan 23, 01; m 21; c 2. MATHEMATICS. Educ: Univ Maine, BA, 28, MA, 32. Hon Degrees: DSc, Am Int Col, 71. Prof Exp: Prin high sch, Maine, 21-24 & 25-36, teacher, 24-25; instr math, Univ Maine, 36-38; prof, 38-69, chmn dept, 42-69, EMER PROF MATH, AM INT COL, 69- Concurrent Pos: Consult, G & C Merriam Co, 57-61; part-time prof, Continuing Educ Div, Univ Maine, 70-72. Mem: Am Math Soc; Math Asn Am. Res: Analysis. Mailing Add: RFD 2 Lisbon Falls ME 04252

BOWIE, OSCAR L, b Lisbon Falls, Maine, Nov 22, 21; m 44; c 3. APPLIED MATHEMATICS. Educ: Am Int Col, BA, 42. Prof Exp: Mathematician, Watertown Arsenal, 44-64, MATHEMATICIAN, ARMY MAT & MECH RES CTR, WATERTOWN, 64- Concurrent Pos: Secy of Army fel, 66-67. Mem: Soc Indust & Appl Math. Res: Mathematical theory of elasticity, including plane problems; axially symmetric three dimensional problems; theory of plates and shells; fracture mechanics with the application of conformal mapping techniques to plane crack problems. Mailing Add: 339 Valley Rd Concord MA 01742

BOWIE, ROBERT MCNEIL, b Table Rock, Nebr, Aug 24, 06; m 33; c 2. PHYSICS. Educ: Iowa State Col, BS, 29, MS, 31, PhD(physics), 33. Prof Exp: Asst, Iowa State Col, 29-33; physicist, Sylvania Elec Prod, Inc, 33-34, dir phys lab, 35-49, mgr physics, 42-49, dir eng, 49-55, dir res, 55-58, vpres, Sylvania Res Labs, 58-60 & Gen Tel & Electronics Labs, Inc, 60-65; tech mgt consult, 64-75; RETIRED. Concurrent Pos: Mem, Nat TV Syst Comt, 51-54; mem bd, Adv Coun Advan Indust Res & Develop, NY; mem & trustee, Palisades Inst Res Serv, Inc. Mem: Fel Am Phys Soc; fel Inst Elec & Electronics Engrs. Res: Thermionic emission from nickel; cathode ray tube research and design such as negative ion traps; metal bodied cathode ray tubes; microwave tube research and design; microelectronic circuits. Mailing Add: 101 Knickerbocker Rd Manhasset NY 11030

BOWIE, WALTER C, b Kansas City, Kans, June 29, 25; m 54; c 3. VETERINARY PHYSIOLOGY. Educ: Kans State Col, DVM, 47; Cornell Univ, MS, 55, PhD(physiol), 60. Prof Exp: From instr to assoc prof, 47-60, PROF PHYSIOL, SCH VET MED, TUSKEGEE INST, 60-, HEAD DEPT PHYSIOL & PHARMACOL, 50- Concurrent Pos: Prin investr, Mark L Morris Found Res Grant, 61-62; co-prin investr, NIH Res Grant, 61-; co-prin investr, US Army Res Grants, 61-62, prin investr, 64-71. Mem: Am Vet Med Asn; Am Soc Vet Physiologists & Pharmacologists. Res: Physiology, diagnosis and evaluation of cerebrospinal fluid of dogs; mechanisms of infection and immunity in listerosis; factors in virulence and immunogenicity in Listeria monocytogenes; cardiovascular and ruminant physiology; movements of the mitral valve. Mailing Add: 2011 Patterson St Tuskegee AL 36088

BOWIN, CARL OTTO, b Los Angeles, Calif, Jan 30, 34; m 56; c 4. GEOLOGY. Educ: Calif Inst Technol, BS, 55; Northwestern Univ, MS, 57; Princeton Univ, PhD(geol), 60. Prof Exp: Instr geol, Princeton Univ, 60-61; res assoc, 61-63, asst scientist, 63-65, ASSOC SCIENTIST GEOPHYS, WOODS HOLE OCEANOG INST, 65- Mem: AAAS; Geol Soc Am; Europ Asn Explor Geophys. Res: Development and use of automatic shipboard geophysical data processing system utilizing a digital computer; tectonics and gravity of earth features. Mailing Add: Woods Hole Oceanog Inst Woods Hole MA 02543

BOWKER, ALBERT HOSMER, b Winchendon, Mass, Sept 8, 19; m 42; c 3. MATHEMATICAL STATISTICS. Educ: Mass Inst Technol, BS, 41; Columbia Univ, PhD(statist), 49. Hon Degrees: LHD, City Univ, New York, 71. Prof Exp: Asst statistician, Mass Inst Technol, 41-43; asst dir statist res group, Columbia Univ, 43-45; from asst prof to prof statist, Stanford Univ, 47-59, head dept, 50-59, dean grad div, 58-63, asst to provost, 56-58; chancellor, City Univ New York, 63-71; CHANCELLOR, UNIV CALIF, BERKELEY, 71- Concurrent Pos: Mem-at-large, Div Math, Nat Res Coun-Nat Acad Sci, 62-65; mem adv comt, Off Statist Standards Statist Policy, 63; mem Nat Adv Coun Exten & Continuing Educ; mem, Nat Drug Abuse Coun, 72-; mem bd trustees, Univ of Haifa, 71- Honors & Awards: Frederick Douglass Award, NY Urban League, 69. Mem: Fel AAAS; fel Am Soc Qual Control; Opers Res Soc Am; Biomet Soc; fel Am Statist Asn (vpres, 62-64, pres elec, 63, pres, 64). Res: Industrial statistics; multivariate analysis. Mailing Add: Univ of Calif Berkeley CA 94720

BOWKER, JOHN, b Springfield, Mass, Apr 20, 28; m 69; c 1. ORTHOPEDIC SURGERY. Educ: State Univ NY, AB, 51; Albany Med Col, MD, 56. Prof Exp: From instr to assoc prof, 63-75, PROF ORTHOP, MED CTR, UNIV ARK, 75- Concurrent Pos: Consult, Vet Admin Hosp, Little Rock, 63-; attend staff, Ark Children's Hosp, 64-; consult, Ark Children's Colony, Conway, 64-; head sect rehab, Dept Orthop Surgery & dir educ & res, Ark Rehab Inst, 75. Res: Rehabilitation medicine. Mailing Add: Dept of Orthop Univ of Ark Med Ctr Sci Little Rock AR 72201

BOWKLEY, HERBERT LOUIS, b Pittston, Pa, July 9, 21; m 47; c 3. INORGANIC CHEMISTRY. Educ: Univ Mich, BSc, 50; Mo Sch Mines, MS, 51; Pa State Univ, PhD(inorg chem), 55. Prof Exp: Jr res chemist, Titanium Div, Nat Lead Co, 51-52; sr res chemist, Columbia-Southern Chem Co, 55-56 & High Energy Fuels Div, Olin Mathieson Chem Co, 56-57; assoc prof chem, Sch Mines & Metall, Univ Mo, 57-60; group leader, Explosives & Mining Chem Dept, Am Cyanamid Co, 60-63; sr res chemist, Elkton Div, Thiokol Chem Co, 63-64 & Armour Agr Chem Co, 64-65; assoc prof, 65-69, PROF CHEM, APPALACHIAN STATE UNIV, 69- Mem: Am Chem Soc. Res: Industrial inorganic chemicals; commercial explosives and rocket propellants research & development. Mailing Add: Dept of Chem Appalachian State Univ Boone NC 28607

BOWLAND, JOHN PATTERSON, b Man, Can, Feb 10, 24; m 46; c 2. ANIMAL NUTRITION. Educ: Univ Man, BSA, 45; Wash State Col, MS, 47; Univ Wis, PhD(biochem, animal husb), 49. Prof Exp: Agr supvr, Exp Farm, Brandon, Man, 45-46; from asst prof to prof animal sci, 49-75, DEAN AGR & FORESTRY, UNIV ALTA, 75- Concurrent Pos: Vis mem, Nat Inst Res Dairying, Eng, 59-60; guest prof, Animal Nutrit Inst, Swiss Fed Inst Technol, 68-69; mem comt animal nutrit, Nat Acad Sci-Nat Res Coun. Honors & Awards: Borden Award, Nutrit Soc Can, 66. Mem: Fel AAAS; Am Soc Animal Sci; Can Soc Animal Sci (secy-treas, 51-54); fel Agr Inst Can; Nutrit Soc Can. Res: Swine and rat nutrition, particularly energy, protein and amino acids. Mailing Add: Dean of Agr & Forestry Univ of Alta Edmonton AB Can

BOWLDEN, HENRY JAMES, b Hamilton, Ont, Apr 5, 25; nat US; m 48; c 2. COMPUTER SCIENCES. Educ: McMaster Univ, BA, 46; Univ Ill, MS, 47, PhD(physics), 51. Prof Exp: Asst physics, Univ Ill, 46-50; from asst prof to assoc prof physics & astron, Wayne State Univ, 50-57; res physicist, Parma Res Ctr, Union Carbide Corp, 57-61, group leader, 61-63; ADV PHYSICIST, COMPUT SCI DEPT, WESTINGHOUSE RES LABS, 63- Concurrent Pos: Res assoc, Detroit Inst Cancer Res, 52-57; consult, US Naval Res Labs, 56-60; mem working group WG 2-1, Int Fedn Info Processing, 70-, secy, 70-73. Mem: Asn Comput Mach. Res: Computer languages and programming systems; numerical methods; semiconductor theory; reaction kinetics. Mailing Add: Westinghouse Res Labs Pittsburgh PA 15235

BOWLER, PETER ALDRICH, b Boise, Idaho, Nov 27, 48. LICHENOLOGY. Educ: Bard Col, BA, 70; Univ Calif, Irvine, PhD(biol), 74. Prof Exp: NAT RES COUN FEL, MUS NATURAL SCI, NAT MUS CAN, 75- Mem: Am Bryol & Lichenological Soc; Brit Lichen Soc; Int Lichenological Asn. Res: Taxonomic, phytogeographic, chemotaxonomic and ecological studies in lichenology. Mailing Add: Mus of Natural Sci Nat Mus of Can Ottawa ON Can

BOWLER, WILLIAM WALLACE, rubber chemistry, see 12th edition

BOWLES, GORDON TOWNSEND, b Tokyo, Japan, June 25, 04; m 32; c 2. ANTHROPOLOGY. Educ: Earlham Col, AB, 25; Harvard Univ, PhD(anthrop), 35. Prof Exp: Asst anthrop, Harvard Univ, 32-35, res assoc, 35-38; asst prof, Univ Hawaii, 38-42; mem, Govt Mission to China, 42; res assoc phys anthrop, Bishop Mus, Honolulu, 38-42 & For Econ Admin & US Dept State, 42-47; exec secy, Conf Bd Assoc Res Couns, Comt Int Exchange of Persons, 47-51; vis prof anthrop, Tokyo Univ, 51-58 & Columbia Univ, 59-60; vis prof, Syracuse Univ, 60-62, prof, 62-72, EMER PROF ANTHROP, SYRACUSE UNIV, 72- Concurrent Pos: Assoc managing dir, Int House of Japan, 53-58; Fulbright-Hays lectr, Japan, 67-68. Mem: AAAS; fel Am Anthrop Asn; Am Asn Phys Anthrop; Asn Asian Studies; hon life mem Japanese Soc Ethnol. Res: Races and cultures of Asia; influences of environment on physique; relation of form to function in the human skeleton; interrelationships between biological and social factors. Mailing Add: 1023 Ackerman Ave Syracuse NY 13210

BOWLES, JOHN BEDELL, b Karuizawa, Japan, July 29, 33; US citizen; m 58; c 4.

ZOOLOGY. Educ: Earlham Col, AB, 56; Univ Wash, MS, 63; Univ Kans, PhD(biol), 71. Prof Exp: Teacher high sch, Hawaii, 59-61; asst dir, Waikiki Aquarium, Honolulu, 61-62; asst prof biol, William Penn Col, 63-67; from asst prof to assoc prof, 69-75, PROF BIOL, CENT COL, IOWA, 75- Mem: AAAS; Am Soc Mammalogists; Soc Syst Zool. Res: Acitivity and reproductive patterns of bats, especially Central American; effects of farming on small mammal distribution. Mailing Add: Cent Col Pella IA 50219

BOWLES, KENNETH LUDLAM, b Bronxville, NY, Feb 20, 29; m 54; c 3. PHYSICS. Educ: Cornell Univ, BEP, 51, MEE, 53, PhD(radio wave propagation), 55. Prof Exp: Physicist, Nat Bur Stand, Colo, 55-65; dir comput ctr, 67-72, PROF APPL PHYSICS & INFO SCI, UNIV CALIF, SAN DIEGO, 65- Concurrent Pos: Mem US nat comt, Int Sci Radio Union, 62-65. Honors & Awards: Gold Medal, US Dept Com, 62. Mem: Inst Elec & Electronics Engrs. Res: Plasma physics; physics and radar studies of the upper atmosphere, including measurements of ionization density, temperature, ionized constituents and nature of irregularities; radio and radar astronomy. Mailing Add: Dept of Appl Physics & Info Sci Univ of Calif at San Diego La Jolla CA 92037

BOWLES, WILLIAM HOWARD, b McCoy, Colo, Aug 1, 36; m 61; c 3. BIOCHEMISTRY. Educ: La Sierra Col, BA, 58; Univ Ariz, MS, 60, PhD(biochem), 64. Prof Exp: Asst prof chem, Southwestern Union Col, 63-65; from asst prof to assoc prof, 65-73, SPEC INSTR BIOCHEM, BAYLOR COL DENT, 73- Mem: AAAS; Am Chem Soc; Int Asn Dent Res. Res: Fat-soluble vitamins, lipids and hormones. Mailing Add: Dept of Biochem Baylor Col of Dent Dallas TX 75226

BOWLING, ARTHUR LEE, JR, b Roanoke, Va, May 14, 47; m 70. THEORETICAL HIGH ENERGY PHYSICS. Educ: Col of William & Mary, BS, 69; Univ Ill, MS, 70, PhD(physics), 74. Prof Exp: Asst vis prof physics, Swarthmore Col, 74-75; ASST PROF PHYSICS, OHIO STATE UNIV, MANSFIELD, 75- Res: Weak and electromagnetic interactions of leptons and hadrons. Mailing Add: Dept of Physics Ohio State Univ Mansfield OH 44903

BOWLING, CLARENCE C, b Salem, Ark, Nov 12, 26; m; c 3. ENTOMOLOGY, AGRICULTURE. Educ: Univ Ark, Fayetteville, BS, 54, MS, 55. Prof Exp: Asst entomologist, 55-56, assoc entomologist, 66-67, asst prof, 67-70, ASSOC PROF ENTOM, TEX A&M UNIV, 70- Concurrent Pos: Secy & prog chmn, Rice Tech Working Group, 68-70, chmn, 70-72. Mem: Entom Soc Am. Res: Biology, ecology, economic importance and methods of control of insect pests of the rice plant. Mailing Add: Tex Agr Exp Sta Rte 5 Box 784 Beaumont TX 77706

BOWLING, DAVID IVAN, b Los Angeles, Calif, May 26, 40; m 64; c 1. PHYSICS. Educ: Univ Calif, Los Angeles, AB, 62; San Diego State Col, MS, 64; Okla State Univ, PhD(physics), 68. Prof Exp: Physicist, Naval Electronics Labs, 64; asst prof, 68-74, ASSOC PROF PHYSICS, CENT MO STATE COL, 74- Mem: Soc Rheol; Am Asn Physics Teachers. Res: Acoustics; macromolecules; impedance of aperture in plates; effect of shipboard noise on job performance; electrooptic effects in solutions of rigid macromolecules. Mailing Add: Dept of Physics Cent Mo State Col Warrensburg MO 64093

BOWLING, FLOYD E, b Elizabethton, Tenn, Aug 28, 11; m 39; c 2. MATHEMATICS. Educ: Lincoln Mem Univ, AB, 34; Univ Iowa, MS, 38; Univ Tenn, EdD(math), 53. Prof Exp: Teacher & coach high sch, Tenn, 34-36; teacher math & physics, Lincoln Mem Univ, 37-42, prof math & head dept, 45-59; PROF MATH, HEAD DEPT & DEAN STUDENTS, TENN WESLEYAN COL, 59- Concurrent Pos: Lectr, NSF Insts. Mem: AAAS; Math Asn Am; Nat Coun Teachers Math; Am Asn Univ Prof. Res: Development of specific kinds of graph paper as a teaching aid in math; development of a technique for teaching college math, with more meaning and understanding on the part of the student. Mailing Add: Dept of Math Tenn Wesleyan Col Athens TN 37303

BOWLING, FRANKLIN LEE, b Guyman, Okla, Nov 2, 09; m 52; c 3. MEDICINE. Educ: Univ Colo, PhC & BS, 33, MD, 46; Univ Denver, MS, 47; US Air Force Sch Aerospace Med, cert, 47; Harvard Univ, MPH, 52; Mass Inst Technol, cert, 55; US Missile Test Ctr, Cape Kennedy, cert, 60; Univ Mich, cert, 67; Am Bd Prev Med, dipl & cert aerospace med. Prof Exp: Intern, Med Ctr, Univ Colo, 46-47; intern Med Corps, US Air Force, 48-67, dir base med serv, Hq, 15th Air Force, Strategic Air Command, Colorado Springs, 47-49, chief med output serv, March AFB, Calif, 49-50, hosp comdr, Saudi Arabia, 50-51, chief flight surgeon, Bolling AFB, DC, 51, chief prev med, Off Inspector Gen, Hq, 52-55, head dept prev med, US Air Force Sch Aviation Med, 55-58, chief prev med, Pac Air Forces Hq, 58-61, chief mil pub health & occup med, Off Surgeon Gen, 62-66, chief epidemiol br, Armed Forces Inst Path, 66-67; med dir, 67-71, MED SERV POLICY DIR, COLO STATE DEPT SOCIAL SERV, 71- Concurrent Pos: Rep, Armed Forces Epidemiol Bd, 61-66; consult, Am Bd Prev Med, 62-66; pres, Coun Fed Med Dir Occup Health, 65-66; consult to Surgeon Gen, Dept Air Force, 66-67; mem, Permanent Comn & Int Asn Occup Health, 66-; chmn comt toxicol, Nat Res Coun. Honors & Awards: Wisdom Award of Honor, 70. Mem: Fel Am Pub Health Asn; fel Am Col Prev Med; fel Indust Med Asn; fel Royal Soc Health; Int Health Soc US (pres, 67-68). Res: Anatomy, physiology and pharmacology of cystic innervation, Mus norvegicus; epidemiological factors in motor vehicle accidents for Army, Navy and US Air Force. Mailing Add: 1001 E Oxford Lane Cherry Hills Village Englewood CO 80110

BOWLING, LLOYD SPENCER, SR, b Newport, Md, Mar 29, 30; m 55; c 2. AUDIOLOGY, SPEECH PATHOLOGY. Educ: Univ Md, BS, 43, MA, 57, EdD(human develop), 64. Prof Exp: Clin audiologist, West Side Vet Admin Hosp, Chicago, 57-60; clin audiologist, Vet Admin Hosp, Washington, DC, 60-63, supvr audiol clin, 63-66, assoc chief audiol & speech path clin, 66-67; lectr, Sch Med, Georgetown Univ, 67; PROF SPEECH & CHMN DEPT SPEECH & DRAMA, GEORGE WASHINGTON UNIV, 67- Concurrent Pos: Consult, Parmly Hearing Inst, Ill, 60-61, Fairfax County Health Dept, Va & Vet Admin Hosp, Washington, DC, 67-; chmn, DC and Brasilia, Brazil Partners Rehab & Educ, 73- Mem: Am Speech & Hearing Asn. Res: Methods of aural rehabilitation; clinical auditory tests; intelligibility. Mailing Add: Dept of Speech George Washington Univ Washington DC 20052

BOWLING, ROBERT EDWARD, b Pauls Valley, Okla, Aug 9, 26; m 55; c 5. MICROBIOLOGY. Educ: Univ Okla, BS, 48, MS, 50, PhD, 57. Prof Exp: Asst bacteriologist, Okla State Health Dept, 50-51; asst, Med Ctr, Univ Okla, 51-52, 55-57; from instr to asst prof microbiol, 57-69, ASSOC PROF MICROBIOL & IMMUNOL, MED CTR, UNIV ARK, 69-, ASST DEAN COL MED, 73- Mem: Am Soc Microbiol; Brit Soc Gen Microbiol; Soc Cryobiol; Sigma Xi. Res: Antibiotic resistance; hypothermia; microbial antagonism; ecology; antibody production; pathogenesis and infection. Mailing Add: 4400 I St Little Rock AR 72205

BOWLING, SUE ANN, b Bridgeport, Conn, Feb 26, 41. METEOROLOGY. Educ: Harvard Univ, AB, 63; Univ Alaska, MS, 67, PhD(geophys), 70. Prof Exp: Partic, Advan Study Prog, Nat Ctr Atmospheric Res, 71-72; ASST PROF, GEOPHYS INST, UNIV ALASKA, 70-71 & 72- Mem: Am Meteorol Soc. Res: Local meteorological variations, climatic change, Alaskan climate, paleoclimatology. Mailing Add: Geophys Inst Fairbanks AK 99701

BOWLS, WOODFORD EUGENE, b Lewistown, Mo, Jan 9, 10; m 35; c 2. PHYSICS. Educ: Univ Calif, AB, 32, AM, 35, PhD(physics), 37. Prof Exp: Teaching asst, Univ Calif, 33-37; mem fac, 37-74, head dept physics, 66-71, dept phys sci, 43-66, EMER FAC MEM, CALIF POLYTECH COL, SAN LUIS OBISPO, 74- Mem: Am Asn Physics Teachers. Res: Electrical discharge through gases. Mailing Add: Dept of Physics Calif Polytech Col San Luis Obispo CA 93401

BOWMAN, ALLEN LEE, b Washington, DC, Jan 19, 31; m 52; c 4. PHYSICAL CHEMISTRY. Educ: Col William & Mary, BS, 51; Iowa State Univ, PhD(chem), 58. Prof Exp: Jr chemist, Electrodeposition Sect, Nat Bur Stand, 51; res asst phys chem, Iowa State Univ, 55-58; STAFF MEM PHYS CHEM, LOS ALAMOS SCI LAB, 58- Mem: Am Chem Soc; Am Crystallog Asn. Res: Gas-cooled reactor safety; crystal structures. Mailing Add: 10 Encino Los Alamos NM 87544

BOWMAN, BARBARA HYDE, b Mineral Wells, Tex, Aug 5, 30. GENETICS. Educ: Baylor Univ, BS, 51; Univ Tex, MA, 55, PhD(genetics), 59. Prof Exp: Bacteriologist, Tex State Dept Health, 54-55; technician human genetics lab, Dept Zool, Univ Tex, 55-59, res scientist, Genetics Found, 59-64; mem staff, Rockefeller Univ, 64-65, asst prof, 65-67; PROF HUMAN GENETICS & CHMN DEPT, MED BR, UNIV TEX, 67- Mem: Harvey Soc; Am Soc Human Genetics; Am Soc Biol Chem. Res: Biochemical genetics of humans; genetic control of protein structure and basic defects occurring in inherited diseases. Mailing Add: Dept of Human Biol Chem & Genetics Univ of Tex Med Br Galveston TX 77550

BOWMAN, BERNARD ULYSSES, JR, b Atlanta, Ga, Oct 1, 26; m 53; c 4. MEDICAL MICROBIOLOGY. Educ: Piedmont Col, BS, 50; Emory Univ, MS, 57; Univ Okla, PhD, 63. Prof Exp: Instr bact & biol, Ga State Col, 58-60; guest lectr virol, Okla State Univ, 62; instr microbiol, Med Ctr, Univ Okla, 63-64; asst prof path, 64-66, from asst prof to assoc prof med microbiol, 66-72, PROF MED MICROBIOL, COL MED, OHIO STATE UNIV, 72- Concurrent Pos: NIH trainee, Univ Okla, 63-64; Am Thoracic Soc grant, Ohio State Univ, 67-68, Nat Tuberc Asn grant, 67-68, NIH career develop award, 68-72 & grant, 69-72. Mem: Am Soc Microbiol; Am Thoracic Soc. Res: Mycobacteria phage-host relationships; etiology of sarcoidosis; lipids of mycobacteria and mycobacteriophages. Mailing Add: Dept of Med Microbiol Ohio State Univ Columbus OH 43210

BOWMAN, BRUCE TAMBLYN, b Oshawa, Ont, Jan 21, 42. SOIL CHEMISTRY, PHYSICAL CHEMISTRY. Educ: Ont Agr Col, Univ Toronto, BSA, 64; Univ Guelph, MSc, 66; Univ Minn, St Paul, PhD(soil sci), 69. Prof Exp: RES SCIENTIST SOIL CHEM, RES INST, CAN DEPT AGR, 69- Mem: Am Chem Soc; Am Soc Agron; Clay Minerals Soc. Res: Use of analytical tools such as gas-liquid chromatography, infrared spectroscopy and x-ray diffraction spectroscopy to study soil-insecticide interactions, elucidating bonding mechanisms and factors affecting insecticide adsorption onto soil colloids. Mailing Add: Res Inst Can Dept of Agr Univ Sub Post Off London ON Can

BOWMAN, CARLOS MORALES, b Mexico, DF, Mex, Mar 4, 35; nat US; m 55; c 6. INFORMATION SCIENCE, COMPUTER SCIENCE. Educ: Univ Utah, BA, 54, PhD(chem), 57. Prof Exp: Chemist, 57-61, info retrieval analyst, 61-64, group leader, 64-67, asst dir, 67-68, RES DIR, COMPUT RES LAB, DOW CHEM CO, 68- Concurrent Pos: Mem comt chem info, Nat Acad Sci-Nat Res Coun, 67-, chmn, 70-74; consult, NSF, 69-73; chmn comt div activities, Am Chem Soc, 75- Mem: AAAS; Am Chem Soc; Sigma Xi (pres, 64); Chem Notation Asn (pres, 71). Res: Information retrieval; chemical notation; computer applications. Mailing Add: 1414 Timber Dr Midland MI 48640

BOWMAN, CHARLES D, b Roanoke, Va, May 23, 35; m 56; c 2. NUCLEAR PHYSICS. Educ: Va Polytech Inst, BS, 56; Duke Univ, MA, 58, PhD(nuclear physics), 61. Prof Exp: Sr physicist, Lawrence Radiation Lab, Univ Calif, 61-68, res prog mgr, Livermore Electron Linac, 68-74; WITH NAT BUR STAND, 74- Concurrent Pos: Mem nuclear cross sect adv comt, US AEC, 69- Mem: Am Phys Soc; Am Nuclear Soc. Res: Neutron physics; development and application of techniques for partial neutron cross section studies on very heavy nuclei; resonance neutron spectrometry on medium weight nuclei; photodisintegration of heavy nuclei. Mailing Add: Nat Bur of Stand US Dept of Com Washington DC 20234

BOWMAN, CHARLES R, nuclear physics, molecular physics, see 12th edition

BOWMAN, CRAIG THOMAS, b Olney, Md, Sept 19, 39. CHEMICAL KINETICS, GAS DYNAMICS. Educ: Carnegie Inst Technol, BS, 61; Princeton Univ, MA, 64, PhD(aerospace & mech sci), 66. Prof Exp: Sr res scientist, United Aircraft Res Lab, 65-74, SR RES SCIENTIST, UNITED TECHNOL RES CTR, 74- Mem: Combustion Inst. Res: Kinetics of gas-phase reactions, especially in hydrocarbon oxidation; dissociation of hydrogen halides; air pollution, energy transfer processes; turbulent combustion. Mailing Add: United Technol Res Ctr East Hartford CT 06108

BOWMAN, DONALD EDWIN, b Orrville, Ohio, Nov 12, 08; m 34; c 2. BIOCHEMISTRY. Educ: Western Reserve Univ, AB, 33, AM, 35, PhD(biochem), 37. Prof Exp: Asst biol, Adelbert Col, Western Reserve Univ, 33-35, instr biochem, 37-41; from asst prof to assoc prof, 41-52, actg chmn dept biochem & pharmacol, 56-58, chmn dept biochem, 58-66, PROF BIOCHEM, SCH MED, IND UNIV, INDIANAPOLIS, 52- Mem: AAAS; Am Chem Soc; Am Soc Biol Chem; Soc Exp Biol & Med. Res: Enzymes; enzyme inhibitors. Mailing Add: Dept of Biochem Ind Univ Sch of Med Indianapolis IN 46202

BOWMAN, DONALD HOUTS, b Osage City, Kans, May 18, 11; m 35; c 2. AGRONOMY. Educ: Kans State Univ, BS, 33, MS, 35; Univ Wis, PhD(plant path, agron, 39. Prof Exp: Asst pathologist, Div Cereal Crops & Diseases & asst plant path, Ohio Exp Sta, 39-46, plant pathologist, Tex Agr Exp Sta, 46-48, AGRONOMIST, DELTA BR EXP STA, US DEPT AGR, 48- Mem: AAAS; Am Soc Agron. Res: Weed control, fertilization and general culture of rice; corn production. Mailing Add: Delta Br Exp Sta Stoneville MS 38776

BOWMAN, DONALD WHITNEY, b Aliance, Ohio, Feb 6, 07; m 31; c 2. PHYSICS. Educ: Mt Union Col, AB, 29; Ohio State Univ, AM, 30, PhD(physics), 40. Hon Degrees: DSc, Mt Union Col, 58. Prof Exp: Asst prof physics, Bluffton Col, 30-38; asst, Ohio State Univ, 38-40; from assoc prof to prof, Hendrix Col, 40-43; assoc prof, 43-49, PROF PHYSICS, BOWLING GREEN STATE UNIV, 49-, CHMN DEPT, 68- Concurrent Pos: Dir develop lab, Triplett Elec Instrument Co, 36-38; tech training consult, Frankford Arsenal, 52-54; consult, Detroit Arsenal, 55- Mem: AAAS; Am Asn Physics Teachers; Sigma Xi. Res: Zeeman effect spectroscopy; Zeeman and Paschen-Bach effect of krypton; industrial applications of x-rays. Mailing Add: Dept of Physics Bowling Green State Univ Bowling Green OH 43403

BOWMAN, DOUGLAS CLYDE, b St Louis, Mo, Oct 20, 25; m 55; c 3.

PHYSIOLOGY, PHARMACOLOGY. Educ: Col Puget Sound, BS, 48, BEd, 49; Univ Wash, MS, 57, PhD(zool), 58. Prof Exp: Instr physiol, Univ Wash, 58-59; from instr to asst prof, Dent Sch, Northwestern Univ, 59-64; asst prof, 64-68, ASSOC PROF PHYSIOL, SCH DENT, LOYOLA UNIV, CHICAGO, 68- Mem: NY Acad Sci; Int Asn Dent Res. Res: Inflammation; action of salicylates; functional innervation of peridontal ligament. Mailing Add: Dept of Physiol Loyola Univ Dent Sch Maywood IL 60153

BOWMAN, EDGAR CORNELL, b Ellensburg, Wash, Dec 23, 14; m 47. GEOLOGY. Educ: Univ Wash, BS, 37; Harvard Univ, PhD(geol), 50. Prof Exp: Geologist, Standard Oil Co, 39-41, 45-47; instr geol, Univ Calif, Los Angeles, 50-52; area geologist, Standard Oil Co, 52-54; geologist, Cuba Calif Oil Co, 54-55; supvr div geol, Standard Oil Co Calif, 55-61, geologist, West Div, Calif Oil Co, 61-63, geol sect supvr, Calif Res Corp, 63-64, chief geologist, West Div, 64-69, GEOL CONSULT, MINERALS STAFF, CHEVRON OIL CO, 69- Mem: Geol Soc Am; Am Asn Petrol Geol; Am Inst Mining, Metall & Petrol Engrs. Res: Regional tectonics; occurrence of petroleum; stratiform mineral deposits. Mailing Add: Minerals Staff Chevron Oil Co 225 Bush St San Francisco CA 94104

BOWMAN, EDWARD RANDOLPH, b Mercer Co, WVa, Feb 26, 27; m 54; c 1. PHARMACOLOGY, PHYSIOLOGY. Educ: Concord Col, BS, 52; Med Col Va, PhD, 63. Prof Exp: Bacteriologist, Va State Dept Health, 54-55; res asst pharmacol, 56-58, RES ASSOC PHARMACOL, MED COL VA, 61- Concurrent Pos: Vis investr, Royal Vet Col, Stockholm, Sweden, 61 & Inst Physiol, Santiago, Chile, 63. Mem: AAAS; Am Chem Soc; Am Soc Pharmacol & Exp Therapeut; Am Soc Exp Biol & Med; Sigma Xi. Res: Drug metabolism; nicotine metabolism; isolation and identification of urinary metabolites of nicotine; synthesis of pyridino compounds; whole body autoradiography. Mailing Add: Dept of Pharmacol Med Col of Va Richmond VA 23298

BOWMAN, EUGENE W, b North Powder, Ore, Mar 28, 10; m 38; c 2. MATHEMATICS. Educ: Univ Idaho, BS, 35, MS, 36, EdD(admin), 52. Prof Exp: Instr math & chem, Coeur d'Alene Jr Col, 36-38; supt pub schs, Wash, 38-43; PROF MATH & EDUC, SOUTHERN ORE COL, 47- Concurrent Pos: Educ adv, US Agency Int Develop, Ecuador, 60-62. Res: Personnel administration; comparative education. Mailing Add: Dept of Math Southern Ore Col Ashland OR 97520

BOWMAN, FAYE JOHNSON, b Carthage, Tenn, Aug 10, 39; m 62. PHARMACOLOGY, TOXICOLOGY. Educ: Tenn Polytech Inst, BS, 61, MA, 62; Vanderbilt Univ, PhD(pharmacol), 67. Prof Exp: Res assoc environ health, Univ Cincinnati, 67-69; RES ASSOC PHARMACOL, MED COL VA, 70- Mem: Soc Pharmacol & Exp Therapeut; Sigma Xi; Soc Toxicol. Res: Drug metabolism and excretion with identification of metabolites. Mailing Add: Dept of Pharmacol Med Col of Va Richmond VA 23298

BOWMAN, FERNE, b DeLeon, Tex, Dec 3, 02. FOODS, NUTRITION. Educ: West Tex State Univ, BS, 28; Univ Chicago, MS, 29; Univ Wis, PhD(foods), 49. Prof Exp: Instr foods & nutrit, Syracuse Univ, 29-30; asst prof, Elmira Col, 30-33; assoc prof, Univ Mo, 39-48; prof & head dept, 48-72, EMER PROF FOODS & NUTRIT, COLO STATE UNIV, 72- Concurrent Pos: Consult, Peshawar, 56; food scientist, Agr Res Serv, US Dept Agr, 58-59. Mem: Fel Am Home Econ Asn; Am Dietetic Asn; Inst Food Technol. Res: Food-culinary qualities of flours; frozen foods; preparation of food at high altitudes; baking quality of different varieties of potatoes; convenience foods; quality of fruits and vegetables. Mailing Add: 1005 N Wells St Pampa TX 79065

BOWMAN, JAMES ANTHONY, physical chemistry, see 12th edition

BOWMAN, JAMES DAVID, b White Plains, NY, Aug 24, 39; m 65; c 3. NUCLEAR PHYSICS, HIGH ENERGY PHYSICS. Educ: Calif Inst Technol, BS, 61, PhD(physics & math), 67. Prof Exp: Res fel nuclear & solid state physics, Calif Inst Technol, 67-68; prof nuclear physics, Univ Bonn, 68-70; res fel nuclear chem, Lawrence Berkeley Lab, Univ Calif, 70-73; STAFF MEM NUCLEAR PHYSICS, MESON PHYSICS FACIL, LOS ALAMOS SCI LAB, 73- Res: Medium energy physics; study of nuclear structure; nuclear stability; pion nucleus interaction and symmetry principles. Mailing Add: Medium Energy Physics Div Los Alamos Sci Lab Los Alamos NM 87545

BOWMAN, JAMES E, b Washington, DC, Feb 19, 23; m 50; c 1. PATHOLOGY, HUMAN GENETICS. Educ: Howard Univ, BS, 43, MD, 46. Prof Exp: Chmn dept path, Provident Hosp, Chicago, 50-53 & Nemazee Hosp, Shiraz Med Ctr, Iran, 55-61; vis assoc prof path, Fac Med, Pahlavi Univ, 57-59, vis prof & chmn dept, 59-61; hon res asst, Univ Col, Univ London, 61-62; asst prof med & path, 62-67, assoc prof path, med & biol, 64-71, PROF PATH, MED & GENETICS, UNIV CHICAGO, 71-, MED DIR BLOOD BANK, 62-, DIR LABS, 71-, DIR COMPREHENSIVE SICKLE CELL CTR, 73- Concurrent Pos: NIH spec res fel, Univ Col, Univ London, 61-62; USPHS res grant, 64-66 & 73-; chief path br, Med Nutrit Lab, Fitzsimmons Army Hosp, Colo. Mem: Fel Col Am Path; fel Am Soc Clin Path; Am Genetics Asn; Am Soc Human Genetics; NY Acad Sci. Res: Human and population genetics, particularly blood anthropology, serum and erythrocytic polymorphisms. Mailing Add: Dept of Path Univ of Chicago Chicago IL 60637

BOWMAN, JAMES FLOYD, II, b Orange, NJ, Feb 22, 32; m 68; c 4. SEDIMENTOLOGY, OCEANOGRAPHY. Educ: Rutgers Univ, AB, 56, PhD(geol), 66. Prof Exp: Lab technician ceramic eng, Rutgers Univ, 58-63; inspector eng geol, Yards Creek Hydroelec Proj, Jersey Cent Power & Light Co, 63-64; lab technician, Bur Mineral Res, Rutgers Univ, 64; lectr geol, Hunter Col, 65-66; instr, 66-69, ASST PROF GEOL, LEHMAN COL, 69-, PROJ DIR, X-RAY DIFFRACTION & FLUORESCENCE LAB, 70- Concurrent Pos: NSF matching funds grant USA, 52-54, Res, 54-60; col expert to media, Geol, Lehman Col, 72-; mem fac sedimentology, Univ Inst Oceanog, City Univ New York, 75- Mem: Geol Soc Am; Soc Econ Paleont & Mineral; Nat Asn Geol Teachers; Am Geophys Union; Int Asn Math Geol. Res: Pleistocene and Recent sedimentation; x-ray diffraction studies of weathering; statistical analysis of sedimentary facies; x-ray diffraction and x-ray fluorescence analysis of micrometeorites, clays; Hudson Estuary-Bight sediments; computer mapping. Mailing Add: Lehman Col City Univ of New York Bronx NY 10468

BOWMAN, JAMES SHEPPARD, b Orrville, Ohio, Oct 29, 28; m 56; c 3. ENTOMOLOGY. Educ: Ohio State Univ, BS, 51, MS, 54; Univ Wis, PhD(entom), 58. Prof Exp: Res asst, Ohio State Univ, 51, 53-54 & Univ Wis, 55-58; entomologist, Hazleton Labs, 58-61; res entomologist, Am Cyanamid Co, 61-70; tech rep, CIBA-Giegy Corp, 70-71; ASST PROF ENTOM, EXTEN ENTOMOLOGIST & PESTICIDE COORDR, UNIV N H, 71- Mem: Entom Soc Am. Res: Industrial development of agricultural pesticides; screening, metabolism and residue studies on insecticides, fungicides and herbicides. Mailing Add: Dept of Entom Nesmith Hall Univ of N H Durham NH 03824

BOWMAN, JAMES TALTON, b High Point, NC, Aug 2, 37; m 57; c 2. GENETICS. Educ: Duke Univ, BS, 61; Univ Calif, Davis, PhD(genetics), 65. Prof Exp: ASSOC PROF ZOOL, UTAH STATE UNIV, 65- Mem: AAAS; Genetics Soc Am. Res: Genetic finestructure and function in higher organisms. Mailing Add: Dept of Biol Utah State Univ Logan UT 84322

BOWMAN, JOEL MARK, b Boston, Mass, Jan 16, 48; m 72. THEORETICAL CHEMISTRY. Educ: Univ Calif, Berkeley, AB, 69; Calif Inst Technol, PhD(chem), 74. Prof Exp: Teaching asst chem, Calif Inst Technol, 69-74; ASST PROF CHEM, ILL INST TECHNOL, 74- Mem: Sigma Xi; Am Phys Soc. Res: Theoretical aspects of reaction dynamics of molecular systems involving exact and approximate quantum, semiclassical and quasiclassical techniques; theoretical studies of gas-surface interactions. Mailing Add: Dept of Chem Ill Inst of Technol Chicago IL 60616

BOWMAN, KIMIKO OSADA, b Tokyo, Japan, Aug 15, 27; US citizen; m; c 2. MATHEMATICAL STATISTICS. Educ: Radford Col, BS, 59; Va Polytech Inst, MS, 62, PhD(statist), 63. Prof Exp: From statistician to math statistician, Comput Technol Ctr, Nuclear Div, Union Carbide Corp, Tenn, 64-70; MEM STAFF, OAK RIDGE NAT LAB, 70- Mem: Fel AAAS; Biomet Soc; Int Asn Statist in Phys Sci; Am Statist Asn; Inst Math Statist. Res: Statistical research using high speed computing and estimating parameters. Mailing Add: Oak Ridge Nat Lab PO Box Y Oak Ridge TN 37830

BOWMAN, LEO HENRY, b Valeda, Kans, May 12, 34; m 52; c 3. ANALYTICAL CHEMISTRY. Educ: Univ Ottawa, Kans, BSc, 56; Mich State Univ, PhD(anal chem), 61. Prof Exp: Res chemist anal develop, Chem Div, Pittsburgh Plate Glass Co, 60-62; asst prof chem, Parsons Col, 62-65; chmn dept, Midwestern Col, 65-68; PROF CHEM, SOUTHWEST MINN STATE COL, 68- Concurrent Pos: NSF res participation summer fel, 63, 65, 67, 68. Mem: Am Chem Soc. Res: Nonaqueous solvents and complexametric reactions. Mailing Add: Dept of Chem Southwest Minn State Col Marshall MN 56258

BOWMAN, LEWIS WILMER, b Rothsville, Pa, Dec 8, 28; m 51; c 3. ORGANIC CHEMISTRY. Educ: Lebanon Valley Col, BS, 50; Univ Del, PhD(org chem), 54. Prof Exp: Instr chem, Lebanon Valley Col, 50-51 & Univ Del, 53-54; res chemist, Esso Res & Eng Co, 54-58, sr chemist, 58, sect head, 58-63, asst dir, 63-69, mgr chem planning & coord, 69-70, pres & managing dir, Esso Res SA, 70-73, gen mgr elastomers, 73, VPRES ELASTOMERS, ESSOCHEM EUROPE, INC, 73- Mem: Am Chem Soc. Res: Petrochemicals; synthetic rubber. Mailing Add: Essochem Europe Inc Nijverheidslaan 2 B-1920 Diegem Belgium

BOWMAN, MALCOLM JAMES, b Auckland, NZ, July 30, 42; m 67; c 3. PHYSICAL OCEANOGRAPHY, ELECTRONIC ENGINEERING. Educ: Univ Auckland, BSc, 65, MSc, 67; Univ Sask, PhD(elec eng), 70. Prof Exp: Res oceanogr, Defence oceanogr, Sci Estab, Auckland, NZ, 67; fel thin film magnetics, Univ Sask, 70-71; ASST PROF OCEANOG, MARINE SCI RES CTR, STATE UNIV NY STONY BROOK, 71- Mem: Am Geophys Union; Optical Soc Am; Inst Elec Electronics Engrs; Am Soc Limnol Oceanog. Res: Descriptive and dynamical oceanography of estuarine and coastal waters; water quality modelling; microstructure and turbulence. Mailing Add: Marine Sci Res Ctr State Univ NY Stony Brook NY 11794

BOWMAN, MARY LYNNE, b Greenville, Ohio, Aug 12, 36; c 2. ENVIRONMENTAL MANAGEMENT. Educ: Ohio State Univ, BS, 54, MA, 71, PhD(educ), 72. Prof Exp: ASST PROF SCH NAT RESOURCES, OHIO STATE UNIV, 72-, RES ASSOC EDUC RES INFO CTR, 74- Mem: Conserv Educ Asn. Res: Environmental attitudes assessments; status of education: a state by state report; bibliography of water resource materials. Mailing Add: Sch of Natural Resources Ohio State Univ 124 W 17th Ave Columbus OH 43210

BOWMAN, MAX I, b Weston, Ohio, Nov 4, 11; m 38; c 1. ORGANIC CHEMISTRY. Educ: DePauw Univ, AB, 32; Univ Ill, MS, 34; Ind Univ, PhD(org chem), 37. Prof Exp: Instr sci, N Ga Col, 36; prof chem & physics, Huron Col, 37-39; asst chemist, Mat Lab, Wright Field, Dayton, 39-43; instr physics, Naval Flight Prep Sch, DePauw Univ, 43; from asst prof to assoc prof org chem, Univ Louisville, 43-53; PROF CHEM, UNIV WIS-OSHKOSH, 53- Concurrent Pos: Smith-Mundt exchange prof, Royal Col Pharm & Chem, Iraq, 51-52. Mem: Am Chem Soc. Res: Synthetic organic chemistry; infrared spectroscopy. Mailing Add: Dept of Chem Univ of Wis Oshkosh WI 54901

BOWMAN, NEWELL STEDMAN, b Rocky Ford, Colo, Sept 4, 24; m 46; c 3. ORGANIC CHEMISTRY. Educ: US Naval Acad, BS, 46; Univ Md, BS, 51; Princeton Univ, AM, 54, PhD(chem), 55. Prof Exp: Fel, Purdue Univ, 54-56; from asst prof to assoc prof, 56-70, PROF CHEM, UNIV TENN, KNOXVILLE, 70- Concurrent Pos: Instr, Purdue Univ, 55-56; Fulbright lectr, Univ Karlsruhe, 69-70. Mem: Am Chem Soc. Res: Organic fluorine compounds; sterospecific reactions. Mailing Add: Dept of Chem Univ of Tenn Knoxville TN 37916

BOWMAN, NORMAN J, organic chemistry, see 12th edition

BOWMAN, PHIL BRYAN, analytical chemistry, see 12th edition

BOWMAN, PHILIP IRVIN, organic chemistry, see 12th edition

BOWMAN, RAY DOUGLAS, b Indianapolis, Ind, Mar 3, 42; m 69; c 2. PHYSICAL CHEMISTRY, MOLECULAR BIOLOGY. Educ: Ind Univ, BA, 64; Calif Inst Technol, PhD(phys chem), 71. Prof Exp: Res fel, Calif Inst Technol, 71-73; ASST PROF NATURAL SCI, UNIV N FLA, 73- Mem: AAAS; Biophys Soc. Res: Behavior of polymers in hydrodynamic flow fields; physical chemistry and electron microscopy of nucleic acids. Mailing Add: Dept of Natural Sci Univ of NFla Jacksonville FL 32216

BOWMAN, RAYMOND DONALD, biochemistry, botany, see 12th edition

BOWMAN, ROBERT CLARK, JR, b Dayton, Ohio, Oct 10, 45; m 65; c 2. CHEMICAL PHYSICS. Educ: Miami Univ, BS, 67; Mass Inst Technol, MS, 69. Prof Exp: From res chemist to sr res chemist, 69-75, RES SPECIALIST, MOUND LAB, MONSANTO RES CORP, 75- Mem: Am Phys Soc. Res: The impurity states, electronic structures and diffusion parameters of hydrides of metals and alloys as well as radiation damage behavior in metal tritides are being investigated by magnetic resonance techniques. Mailing Add: Monsanto Res Corp Mound Lab Miamisburg OH 45342

BOWMAN, ROBERT GOLDTHWAIT, b New Haven, Conn, Nov 29, 12; m 45; c 3. GEOGRAPHY. Educ: Dartmouth Col, AB, 35; Univ Calif, PhD(geog), 41. Prof Exp: Lectr geog, Univ NZ, 40; res assoc, Refugee Econ Corp, NY, 41; instr geog, Univ Calif, 42; from asst prof to assoc prof, Univ Iowa, 46-49; PROF GEOG, UNIV NEBR-LINCOLN, 49- Mem: AAAS; Asn Am Geogr; Am Geog Soc; Polar Soc; Nat Coun Geog Educ. Res: Prospects for land settlement; physiography; climatology;

tropical, regional and military geography. Mailing Add: 3440 Otoe St Lincoln NE 68506

BOWMAN, ROBERT HUNT, b Baytown, Tex, Sept 11, 38; m 63. MATHEMATICS. Educ: Univ Tex, BS, 61, MA, 63, PhD(math), 66. Prof Exp: From instr to asst prof math, Vanderbilt Univ, 65-70; ASSOC PROF MATH, ARK STATE UNIV, 70- Mem: Am Math Soc; Tensor Soc; Sigma Xi. Res: Higher order differential geometry and its applications to relativity. Mailing Add: Dept of Math Ark State Univ State University AR 72467

BOWMAN, ROBERT IRVIN, b Saskatoon, Sask, Can, Nov 19, 25; nat US; m 51; c 3. ORNITHOLOGY. Educ: Queen's Univ, Can, BA, 48; Univ Calif, PhD, 57. Prof Exp: Asst zool, Univ Calif, 48-51, res zoologist, 52-54; instr, 55-58, from asst prof to assoc prof, 58-72, chmn dept ecol & syst biol, 70-71, PROF BIOL, SAN FRANCISCO STATE COL, 72- Concurrent Pos: Mem, UNESCO Mission, Galapagos Islands, Ecuador, 57. Honors & Awards: Govt of Ecuador Medal of Honor, 64. Mem: AAAS; Am Ornith Union; Cooper Ornith Soc; Brit Ornith Union. Res: Functional anatomy; adaptive radiation in birds; evolution. Mailing Add: Dept of Ecol & Syst Biol San Francisco State Col San Francisco CA 94132

BOWMAN, ROBERT LEWIS, b New York, NY, May 1, 16; m 42; c 3. INSTRUMENTATION. Educ: Columbia Col, NY, AB, 38; NY Univ, MD, 42. Prof Exp: Intern, Montefiore Hosp, New York, 42-43; res fel, Goldwater Mem Hosp, 46-47; from res asst to instr med, Col Med, NY Univ, 47-50; CHIEF, LAB TECH DEVELOP, NAT HEART & LUNG INST, NIH, 50- Honors & Awards: Award, Am Chem Soc, 67; Meritorious Serv Award, USPHS, 70. Mem: Inst Elec & Electronics Engrs; Biomed Eng Soc; Am Soc Artificial Internal Organs. Res: Biomedical instrumentation; microanalysis; fluorescence; microbiology. Mailing Add: Lab of Tech Develop Bldg 10 5D18 Nat Heart & Lung Inst NIH Bethesda MD 20014

BOWMAN, ROBERT MATHEWS, b Belfast, Northern Ireland, July 18, 40; m 67; c 2. SYNTHETIC ORGANIC CHEMISTRY. Educ: Queen's Univ, Belfast, BS, 62, Hons, 63, PhD, 66. Prof Exp: Fel nat prod chem, Univ Minn, 66-68; fel photochem, McMaster Univ, 68-70; MEM STAFF, PHARMACEUT DIV, CIBA-GEIGY CORP, 70- Mem: Am Chem Soc. Res: Design and application of broad areas of synthetic organic chemistry leading to novel biologically significant agents, particularly those with antiallergic, anti-inflammatory and central nervous activity. Mailing Add: Pharmaceut Div CIBA-GEIGY Corp 556 Morris Ave Summit NJ 07901

BOWMAN, ROBERT SAMUEL, b Valley View, Pa, June 26, 17. PHYSICAL CHEMISTRY, ORGANIC CHEMISTRY. Educ: Pa State Col, BS, 40; Univ Pittsburgh, MS, 45, PhD(chem), 50. Prof Exp: Asst, Inst Animal Nutrit, Pa State Col, 40-41; res chemist, Jones & Laughlin Steel Corp, 41-45 & Gulf Res & Develop Co, 45-51; sr bone prod fel, 51-63, ST JOSEPH MINERAL CORP LABS SR FEL, MELLON INST, CARNEGIE-MELLON UNIV, 63- Mem: AAAS; Am Chem Soc; Am Inst Chemists. Res: Surface chemistry; solid phase reactions; coal tar Chemicals and technology; pyridine bases; petroleum hydrocarbons; polymerization; phenol and calcium phosphate chemistry; solid absorbents; catalysis, glass blowing; metal oxides. Mailing Add: Mellon Inst 4400 Fifth Ave Pittsburgh PA 15213

BOWMAN, ROGER HOLMES, b Ridgway, Pa, Apr 20, 24; m 47; c 3. BIOCHEMISTRY. Educ: Franklin & Marshall Col, BS, 47; Bucknell Univ, MS, 48; Cornell Univ, PhD(zool), 52. Prof Exp: Asst prof physiol, Philadelphia Col Osteop, 52-53; from asst prof to assoc prof, Bucknell Univ, 53-62; Am Heart Asn estab investr & asst prof physiol, Sch Med, Vanderbilt Univ, 62-66; ASSOC PROF PHARMACOL, STATE UNIV NY UPSTATE MED CTR & BASIC SCIENTIST, VET ADMIN HOSP, SYRACUSE, 66- Concurrent Pos: Am Heart Asn advan res fel biochem, Cambridge Univ, 60-62. Mem: AAAS; Am Zool Soc; Am Physiol Soc; Am Soc Nephrology; Brit Biochem Soc. Res: Metabolic control mechanisms; metabolism and function of the kidney. Mailing Add: 112 Pine Ridge Rd Fayetteville NY 13066

BOWMAN, THOMAS ELLIOT, b Brooklyn, NY, Oct 21, 18; m 43; c 3. ZOOLOGY. Educ: Harvard Univ, SB, 41; Univ Calif, MA, 48, PhD, 54. Prof Exp: Asst zool, Univ Calif, 45-48, res biologist, Scripps Inst, 48-53; asst prof marine biol, Narragansett Marine Lab, RI, 53-54; assoc cur div marine inverts, 54-65, CUR DIV CRUSTACEA, US NAT MUS, SMITHSONIAN INST, 65- Mem: Am Soc Limnol & Oceanog; Soc Syst Zool; Asn Trop Biol; Plankton Soc Japan. Res: Taxonomy and zoogeography of hyperiid amphipods, calanoid copepods and cymothoid isopods. Mailing Add: Div of Crustacea Smithsonian Inst Washington DC 20560

BOWMAN, WALLACE DEAL, b Jacksonville, Ill, Dec 10, 26; m 61; c 3. ENVIRONMENTAL MANAGEMENT, RESEARCH ADMINISTRATION. Educ: Wash Univ, AB, 49, MA, 52; Yale Univ, MS, 56. Prof Exp: Res asst psychoacoust, Wash Univ, 53-54; researcher, Agr Exp Sta, United Fruit Co, Honduras, 56-57; lectr, Interdisciplinary Land Use Sem, Univ Mich, 57-58; resource planner, Washtenaw County Planning Comn, Mich, 58-59; asst res dir, Conserv Found, NY, 60-61; resource planner, Off of Gov of Alaska, 61-62; exec officer, Conserv Found, NY, 63-64; sr reports officer, UN Develop Prog, NY, 65-67; SR SPECIALIST & CHIEF ENVIRON POLICY DIV, CONG RES SERV, LIBR CONG, 67- Concurrent Pos: Trustee, Conserv & Res Found, 69-; mem consult coun, Conserv Found DC, 70-; mem assembly, Inst Ecol. Mem: AAAS; Wildlife Soc; Foreign Policy Asn; Am Inst Biol Sci. Res: Natural resources; ecology; environmental pollution; legislation; public policy. Mailing Add: Cong Res Serv Libr of Cong Washington DC 20540

BOWMAN, WILFRED WILLIAM, b Piqua, Ohio, Dec 17, 41; m 59; c 2. CHEMISTRY. Educ: Wilmington Col, BA, 63; Univ Rochester, PhD(chem), 68. Prof Exp: Res assoc chem, Cyclotron Inst, Tex A&M Univ, 68-73; RES CHEMIST, SAVANNAH RIVER LAB, E I DU PONT DE NEMOURS & CO, INC, 73- Mem: Am Chem Soc; Am Phys Soc. Res: Gamma-ray spectroscopy; neutron activation analysis. Mailing Add: Savannah River Lab E I du Pont de Nemours & Co Inc Aiken SC 29801

BOWMAN, WILLIAM HENRY, b Reading, Pa, Apr 8, 38; m 60; c 2. CLINICAL BIOCHEMISTRY. Educ: State Col Pa, Kutztown, BS, 60; Pa State Univ, MS, 63, PhD(biochem), 65. Prof Exp: Asst prof chem, Ball State Univ, 65-69; res fel enzym, NIH, 69-71; res asst, 71-72, ASST DIR BIOCHEM, METROP LIFE INS CO, 72- Mem: Am Chem Soc; AAAS. Res: Assay of cardiovascular and atherosclerotic drugs in blood and urine; assay of enzymes and polyamines in urine as indicators of pathology. Mailing Add: Biochem Lab Metrop Life Ins Co One Madison Ave New York NY 10010

BOWMER, ERNEST JOHN, b Newcastle-upon-Tyne, Eng, Apr 10, 15; m 40; c 2. PATHOLOGY, MEDICAL MICROBIOLOGY. Educ: Univ Liverpool, MB & ChB, 38, MD, 62; Royal Col Physicians & Surgeons, Eng, DTM&H, 51 & Can, cert, 59; FRCP, 68. Prof Exp: Pvt pract, Eng, 38-40; DIR DIV LABS, HEALTH BR, BC DEPT PUB HEALTH, 56-; CLIN PROF MED MICROBIOL, UNIV BC, 69- Concurrent Pos: Sr bacteriologist, Microbiol Res Estab, Eng, 51-54 & War Off Enteric Invest Team, Egypt, 54-55; asst dir path, Mid East Land Forces, Egypt & Cyprus, 55-56; mem tech adv comt pub health lab serv, Dept Nat Health, 56-; mem hon consult staff, Vancouver Gen Hosp, BC, 62-; mem tech adv comt, Pub Health Lab Serv, Can Nat Minister of Health & Welfare, 62-, sr investr, Salmonella Proj, 62-66. Mem: AAAS; fel Am Pub Health Asn; Can Pub Health Asn (pres, 72-74); Defence Med Asn Can (pres, 68-69); Can Asn Med Microbiol (pres, 72-74). Res: Preparation and assay of the first international standards for Clostridium botulinum types A, B, C, D and E antitoxins; transmission and control of salmonellosis. Mailing Add: 828 W Tenth Ave Vancouver BC Can

BOWMER, RICHARD GLENN, b Spokane, Wash, Dec 4, 31; m 57; c 3. PLANT PHYSIOLOGY. Educ: Univ Idaho, BS, 53, MS, 57; Univ NC, PhD(bot), 60. Prof Exp: Asst prof biol, Lewis-Clark Norm Sch, 60-61; asst prof, 61-65, ASSOC PROF BOT, IDAHO STATE UNIV, 65- Mem: Bot Soc Am; Am Soc Plant Physiol. Res: Translocation of solutes in higher plants. Mailing Add: Dept of Biol Idaho State Univ Pocatello ID 83201

BOWN, DELOS EDWARD, b Provo, Utah, May 21, 23; m 49; c 4. ORGANIC CHEMISTRY. Educ: Brigham Young Univ, AB, 48, MA, 49; Mass Inst Technol, PhD(chem), 53. Prof Exp: Res specialist, Esso Res & Eng Co, 53-66; RES DIR, POLYCHROME CORP, 66- Mem: Am Inst Chem. Res: Polymer stabilization; polyolefins; offset lithography. Mailing Add: Polychrome Corp On the Hudson Yonkers NY 10702

BOWNDS, JOHN MARVIN, b Delta, Colo, Apr 21, 41; m 62; c 2. MATHEMATICS. Educ: Chico State Col, BA, 64; Univ Calif, Riverside, MA, 67, PhD(math), 68. Prof Exp: Mathematician, US Naval Weapons Ctr, Calif, 64-65; asst prof math, 68-74, ASSOC PROF MATH, UNIV ARIZ, 74- Mem: Am Math Soc. Res: Differential equations; Darboux continuity of functions of several variables. Mailing Add: Dept of Math Univ of Ariz Tucson AZ 85721

BOWNDS, M DERIC, b San Antonio, Tex, May 16, 42; m 68. NEUROBIOLOGY. Educ: Harvard Univ, BA, 63, PhD(biol), 67. Prof Exp: Instr neurobiol, Harvard Med Sch, 67-69; from asst prof to assoc prof, 69-75, PROF MOLECULAR BIOL & ZOOL, UNIV WIS-MADISON, 75- Concurrent Pos: NIH fel, 67-69 & res grant, 69- Mem: Soc Neurosci; Asn Res Vision & Ophthal. Res: Protein chemistry of visual pigments; chemistry and physiology of excitable nerve membranes. Mailing Add: Lab of Molecular Biol Univ of Wis Madison WI 53706

BOWNE, JOHN G, b Cedar Rapids, Iowa, Feb 1, 21; m 44; c 2. VIROLOGY, CYTOLOGY. Educ: Iowa State Univ, BS, 48, DVM, 53, MS, 56, PhD(neurol), 59. Prof Exp: Assoc prof anat, Iowa State Univ, 53-61; DIR VIROL, ANIMAL DIS RES LAB, USDA, 61- Concurrent Pos: Prof microbiol, Colo State Univ, 69-74. Mem: Am Vet Med Asn; Electron Micros Soc Am. Res: Cytology, histology and experimental neuropathology; virology and electron microscopy; cytopathology of arbor viruses in insect and mammalian tissues. Mailing Add: Agr Res Serv US Dept of Agr Bldg 45 Denver Fed Ctr Denver CO 80225

BOWNE, SAMUEL WINTER, JR, b New York, NY, Nov 19, 25; m 54; c 2. CHEMISTRY, GENETICS. Educ: Mass Inst Technol, BS, 48; Columbia Univ, MA, 52; Cornell Univ, MS, 53, PhD(genetics), 57. Prof Exp: Jr chemist, Merck & Co, Inc, 48-49; res assoc, Wash State Univ, 53-54; asst prof chem, Eastern Wash Col Educ, 57-59; assoc prof biol, Univ Wichita, 59-61; res assoc, Ohio State Univ, 61-62; from asst prof to assoc prof, 62-68, head dept sci, 64-69, PROF CHEM, EDINBORO STATE COL, 68- Mem: AAAS; Am Chem Soc. Res: Biochemical genetics; intermediate metabolism. Mailing Add: Dept of Chem Edinboro State Col Edinboro PA 16412

BOWNESS, COLIN, b London, Eng, Oct 26, 29; m 53; c 2. PHYSICS. Educ: Univ London, BSc, 50 & 51, PhD(physics), 56. Prof Exp: Microwave engr, Elec & Musical Industs, Eng, 51-56, lab mgr, 56-57; tech dir microwaves & lasers, 57-62, MGR SPEC MICROWAVE DEVICES OPER, RAYTHEON CO, 62- Mem: Sr mem Inst Elec & Electronics Engrs; Optical Soc Am; Brit Inst Physics & Phys Soc. Res: Lasers; microwave component design, especially ferrite devices. Mailing Add: Raytheon Co 130 Second Ave Waltham MA 02154

BOWNS, BEVERLY HENRY, b Ontario, Calif. COMMUNITY HEALTH, PUBLIC HEALTH. Educ: Columbia Univ, BA, 59; Univ Minn, MA, 60; Johns Hopkins Univ, DrPH, 68. Prof Exp: Adminr undergrad prog pub health nursing, Univ Calif, San Francisco, 61-63; asst prof, Univ Md, Baltimore, 68-70; assoc prof, Vanderbilt Univ, 70-72; PROF COMMUNITY HEALTH NURSING, MED UNITS, UNIV TENN, MEMPHIS, 72- Concurrent Pos: Consult primary care nurse clin prog, Tex Women's Univ, Pa State Univ, Yale Univ, Univ Kans, Meharry Med Col, Ariz State Univ & Ind Univ, 70- Mem: Am Pub Health Asn. Res: Adolescent pregnancy in girls age ten to fifteen and mother and daughter relationships; evaluative research of graduate programs and their graduates in family nursing in primary care. Mailing Add: Dept Commun & Family Nursing Univ of Tenn Ctr Health Sci Memphis TN 38107

BOWSER, CARL, b Compton, Calif, Apr 21, 37; m 60; c 1. GEOCHEMISTRY. Educ: Univ Calif, Riverside, BA, 59; Univ Calif, Los Angeles, PhD(geol), 65. Prof Exp: Asst geol, Inst Geophys, Univ Calif, Los Angeles, 60-64; from asst prof to assoc prof, 64-74, PROF GEOL, UNIV WIS, MADISON, 74- Mem: Geol Soc Am; Geochem Soc; Mineral Soc Am; Am Geophys Union. Res: Geochemistry of non-marine salt deposits; mineralogy and chemistry of fresh water ferromanganese nodules; chemical sedimentology of lakes; ore solution geochemistry. Mailing Add: Dept of Geol Univ of Wis Madison WI 53706

BOWSHER, ARTHUR LEROY, SR, b Wapakoneta, Ohio, Apr 29, 17; m 67; c 4. GEOLOGY. Educ: Univ Tulsa, BS, 41. Prof Exp: Mem staff, Tidewater Assoc Oil Co, 37-38; lab asst, Univ Tulsa, 38-41; lab asst, Univ Kans, 41-42, instr, 46-48; paleontologist, Smithsonian Inst, 48-52; chief lab, Navy Oil Unit, US Geol Surv, 52-57; staff geologist, Sinclair Oil & Gas Co, Okla, 57-69; sr geologist, Atlantic Richfield Oil Co, Tex, 69-70; SR GEOLOGIST, ARABIAN AM OIL CO, 70- Concurrent Pos: Asst geologist, Kans State Geol Surv, 41-42, geologist, 42. Mem: Am Asn Petrol Geol; Paleont Soc. Res: Petroleum geology; non-metallic resources; stratigraphy and paleontology. Mailing Add: Arabian Am Oil Co Box 2772 Dhahran Saudi Arabia

BOWSHER, HARRY FRED, b Lima, Ohio, Feb 26, 31; m 55; c 3. NUCLEAR PHYSICS. Educ: Ohio State Univ, BS, 55, MS, 56, PhD(nuclear physics), 60. Prof Exp: Asst prof nuclear physics, Univ Tenn, 60-66; assoc prof & chmn dept, 66-67, chmn dept, 71-74, PROF PHYSICS, AUGUSTA COL, 67- Concurrent Pos: Fulbright lectr, Dept Physics, Cheng Kung Univ, Taiwan, 69-70; consult, Oak Ridge Nat Lab, 60- Mem: Am Phys Soc. Res: Heavy-ion nuclear physics; reaction mechanisms. Mailing Add: Dept of Physics Augusta Col Augusta GA 30904

BOWYER, BEN, b Lakeland, Fla, Aug 16, 26; m 60; c 2. ECONOMIC GEOLOGY. Educ: La State Univ, BS, 49; Stanford Univ, MS, 52. Prof Exp: Geologist, US Geol Surv, 49-55 & Bell Explor & Develop Co, Bell Aircraft, 55-56; geol engr, AEC, 56-64, prog officer, 65-69, chief projs br, 69-73, dir, Resource Div, Grand Junction Off, 73-

75, ASST MGR, GRAND JUNCTION OFF, ENERGY RES & DEVELOP ADMIN, 75- Mem: Geol Soc Am; Soc Econ Geologists. Res: Radioactive and strategic minerals; structural geology; peaceful uses of atomic energy. Mailing Add: US Energy Resource & Develop Admin PO Box 2567 Grand Junction CO 81501

BOWYER, C STUART, b Toledo, Ohio, Aug 2, 34; m 57; c 3. ASTRONOMY, SPACE SCIENCE. Educ: Miami Univ, BS, 56; Cath Univ, PhD(physics), 65. Prof Exp: Physicist, Nat Bur Standards, 56-58 & Naval Res Lab, 58-66; asst prof space sci, Cath Univ, 66-67; assoc prof, 67-74, PROF ASTRON, UNIV CALIF, BERKELEY, 74- Concurrent Pos: Consult, NASA, 68-; sr vis fel, Sci Res Coun, Eng, 74; hon res fel, Univ Col, Univ London, 74. Mem: Am Geophys Union; Am Astron Soc; Int Astrophys Union; Astron Soc Pac. Res: Galactic and extra galactic x-rays; extreme ultraviolet radiation in earth's atmosphere and from astronomical objects. Mailing Add: Dept of Astron Univ of Calif Berkeley CA 94720

BOX, EDITH DARROW, b Glendale, Md, Jan 3, 22; m 51; c 2. PARASITOLOGY. Educ: Iowa State Col, BS, 43; Johns Hopkins Univ, DSc(parasitol), 48. Prof Exp: Student asst parasitol, Sch Hygiene, Johns Hopkins Univ, 43-48; res assoc, 48-49, asst prof, 49-53, ASSOC PROF MICROBIOL, MED BR, UNIV TEX, 53- Mem: AAAS; Am Soc Trop Med & Hyg; Am Soc Parasitol; Soc Protozool; Wildlife Dis Asn. Res: Chemotherapy and biology of avian and rodent malaria; arthropod transmission of blood and tissue protozoa; life cycles of intracellular protozoa; taxonomy and biology of avian coccidia. Mailing Add: Dept of Microbiol Univ of Tex Med Br Galveston TX 77550

BOX, GEORGE EDWARD PELHAM, b Gravesend, Eng, Oct 18, 19; m 59; c 2. STATISTICS. Educ: Univ London, BS, 48, PhD(statist), 52. Hon Degrees: DSc, Univ London, 61 & Univ Rochester, 75. Prof Exp: From statistician to head statist res sect, Dyestuffs Div, Imp Chem Indust Ltd, 48-56; dir res group, Dept Math, Princeton Univ, 56-59; RONALD AYLMER FISHER PROF STATIST, UNIV WISMADISON, 59- Concurrent Pos: Vis prof, NC State Col, 53-54 & Univ Essex, 70-71; Ford Found vis prof, Grad Sch Bus Admin, Harvard Univ, 65-66; consult, Pillsbury Co, Monsanto Co, Am Cyanamid Co & World Bank, 75- Honors & Awards: Brit Empire Medal, 46; Am Inst Chem Engrs Prof Progress Award, 63; Guy Medal, Royal Statist Soc, 64. Mem: Fel AAAS; fel Am Statist Asn (vpres, 60-63); Int Statist Inst; fel Inst Math Statist; fel Am Acad Arts & Sci. Res: Design and analysis of experiments; time series and forecasting; statistical inference. Mailing Add: Dept of Statist Univ of Wis Madison WI 53715

BOX, HAROLD C, b Clarence, NY, Aug 19, 25; m 51; c 5. PHYSICS. Educ: Canisius Col, BS, 48; Univ Buffalo, MA, 51, PhD(physics), 54. Prof Exp: Asst physicist, Cornell Aeronaut Lab, 51-54; sr cancer res scientist, 54-61, assoc cancer res scientist, 61-70, PRIN CANCER RES SCIENTIST, ROSWELL PARK MEM INST, 70- Concurrent Pos: Asst res prof, Roswell Park, Grad Div, State Univ NY Buffalo, 62-74, adj prof physics, 74- Mem: Am Phys Soc; Am Crystallog Asn. Res: Magnetic resonance studies of radiation effects in biological compounds; flying-spot television microscopy; gamma-ray spectroscopy. Mailing Add: Biophys Dept Roswell Park Mem Inst 666 Elm St Buffalo NY 14203

BOX, JAMES ELLIS, JR, b Georgetown, Tex, Sept 12, 31; m 56; c 4. SOIL PHYSICS. Educ: Tex A&M Univ, BS, 52, MS, 56; Utah State Univ, PhD(soil & irrig, 60. Prof Exp: Soil scientist, Soil Conserv Serv, US Dept Agr, 56-57; asst soil physics, Utah State Univ, 57-60; res soil scientist, Soil & Water Conserv Res Div, 60-61, supt, US Big Spring Field Sta, 62-65, dir, Southern Piedmont Soil & Water Conserv Res Ctr, 65, dir Southern Piedmont Conserv Res Ctr, 65-72, RES LEADER, SOUTHERN PIEDMONT CONSERV RES CTR, AGR RES SERV, US DEPT AGR, 72-, RES ADV, SOUTHEASTERN REGION, NAT ASN CONSERV DISTRICTS, 73- Concurrent Pos: US Dept Agr sponsoring scientist, Soil Physics Res, Israel Inst Technol, 66-; vis scientist, Am Soc Agron, 71-72; adv mem, Ga State Soil & Water Comt, Southeast Area Res Comt, Nat Asn Conserv Dist; mem grad fac, Univ Ga, 75-80. Mem: Am Soc Agron; Crop Sci Soc Am; Int Soil Sci Soc; Soil Conserv Soc Am; Soil Sci Soc Am. Res: Basic aspects of plant-soil-water relationships; thermodynamics of soil moisture; moisture conservation under dryland conditions. Mailing Add: Southern Piedmont Conserv Res Ctr Agr Res Serv PO Box 555 Watkinsville GA 30677

BOX, THADIS WAYNE, b Llano Co, Tex, May 9, 29; m 54; c 4. RANGE MANAGEMENT. Educ: Southwest Tex State Col, BS, 56; Agr & Mech Col, Tex, MS, 57, PhD, 59. Prof Exp: Asst range mgt, Agr & Mech Col, Tex, 57-59; asst prof, Utah State Univ, 59-62; from assoc prof to prof, Tex Technol Col, 62-68, dir, Int Ctr Arid & Semi-Arid Land Studies, 68-70; PROF RANGE SCI & DEAN COL NATURAL RESOURCES, UTAH STATE UNIV, 70- Concurrent Pos: Vis res scientist, Commonwealth Sci & Indust Res Orgn, Australia, 68-69; consult, Food & Agr Orgn, UN, EAfrica, 65, 69 & 70. Honors & Awards: E Harris Harbison Award Distinguished Teaching, 67. Mem: Am Soc Range Mgt; Wildlife Soc; Ecol Soc Am; Soil Conserv Soc Am. Res: Range ecology; interrelationships between wild and domestic animals; grazing management and applied ecology in range ecosystems. Mailing Add: Col of Natural Resources Utah State Univ Logan UT 84321

BOXER, ROBERT JACOB, b Brooklyn, NY, Apr 9, 35; m 63; c 2. ORGANIC CHEMISTRY. Educ: Brooklyn Col, BS, 56; Rutgers Univ, PhD(org chem), 61. Prof Exp: From asst prof to assoc prof chem, Oglethorpe Col, 61-64; assoc prof, 64-72, PROF ORG CHEM, GA SOUTHERN COL, 72- Mem: Am Chem Soc. Res: Organic halogenating agents; history of drugs of abuse. Mailing Add: Dept of Chem Ga Southern Col Statesboro GA 30458

BOXILL, GALE CLARK, b Indianapolis, Ind, Jan 28, 19; m 46; c 2. PHARMACOLOGY. Educ: Washington & Lee Univ, AB, 47; Univ Tenn, MS, 51, PhD(pharmacol), 54. Prof Exp: Jr pharmacologist, Wm S Merrell Co, 47-49; asst prof, Sch Pharm, Univ Ga, 52-53; pharmacologist, Res Lab, Mead Johnson & Co, 53-59; sr res assoc, Warner-Lambert Res Inst, 59-62, dir dept toxicol, 62-70; MGR TOXICOL, WYETH LABS, 70- Mem: Am Soc Pharmacol & Exp Therapeut; Sigma Xi; Soc Toxicol (secy, 75-77). Res: Cardiovascular effect of epinephrine on pressoreceptors; effects of various drugs on blood pressure to epinephrine; toxicological evaluation of new drugs. Mailing Add: Dept of Toxicol Wyeth Labs P O Box 861 Paoli PA 19301

BOYACK, GERALD (ARTHUR), b Montezuma, Iowa, Jan 6, 17; m 46; c 3. ORGANIC CHEMISTRY. Educ: Grinnell Col, BA, 39; Univ Minn, PhD(org chem), 47. Prof Exp: Jr chemist, Merck & Co, Ltd, NJ, 39-41; assoc, Squibb Inst Med Res, 45, 47-52; res chemist, 52-62, head agr formulations & anal, 62-67, head anal chem serv & rec, Agr Div, 67-69, ADMIN ASST TO DIR AGR RES, AGR DIV, UPJOHN CO, 69- Mem: Am Chem Soc. Res: Synthesis of vitamins; antibiotic extraction; agricultural formulations and analyses. Mailing Add: Agr Div Upjohn Co Kalamazoo MI 49001

BOYAR, ROBERT MARTIN, b New York, NY, Mar 10, 37; m 70; c 2. ENDOCRINOLOGY. Educ: Univ Ill, BS, 58; Albert Einstein Col Med, MD, 62. Prof Exp: Intern Parkland Mem Hosp, 62-63, asst resident, 63-64; asst resident, Med Ctr, Univ Calif, San Francisco, 64-66; fel endocrinol, Scripps Clin & Res Found, 67-68; staff physician nuclear med, US Naval Hosp, St Albans, NY, 68-70; INVESTR, INST STEROID RES, MONTEFIORE HOSP & MED CTR, 72-; ASST PROF MED, ALBERT EINSTEIN COL MED, 73- Concurrent Pos: Asst attend physician, 70-72, assoc attend physician, Montefiore Hosp & Med Ctr, 72-; Am Col Physicians traveling scholar, 72. Mem: Endocrine Soc; Am Soc Clin Invest; Am Fed Clin Res; Asn Psychophysiol Study Sleep; Lawson Wilkins Pediat Endocrine Soc. Res: Mechanism controlling the initiation of normal puberty; gonadotropin secretion in humans; prolactin and hypothalaminic-pituitary function in normal and disease states. Mailing Add: 111 E 210th St Bronx NY 10467

BOYARSKY, LILA HARRIET, b Brooklyn, NY, Apr 23, 21; m 41; c 2. BIOLOGY. Educ: Hunter Col, BA, 42; Univ Wis, MS, 43, PhD(genetics), 47. Prof Exp: Asst physiol reprod, Univ Wis, 42-43, animal husb, Purdue Univ, 43-45 & physiol, Univ Chicago, 45-47; instr biol, George Williams Col, 49-50; asst prof, 55-57, assoc prof, 58-61, PROF BIOL, TRANSYLVANIA COL, 61- Mem: AAAS; Am Zool Soc. Res: Genetics; endocrinology; physiology of reproduction; genetics of hamsters. Mailing Add: 1729 Traveller Rd Lexington KY 40504

BOYARSKY, LOUIS LESTER, b Jersey City, NJ, Sept 5, 19; m 41; c 2. NEUROPHYSIOLOGY. Educ: City Col New York, BS, 41; Purdue Univ, MS, 45; Univ Chicago, PhD(physiol), 48. Prof Exp: Psychophysiologist, Inst Juv Res, 49-50; from asst prof to assoc prof, 50-59, PROF PHYSIOL & BIOPHYS, COL MED, UNIV KY, 59- Concurrent Pos: Fulbright scholar, Univ Milan, 57-58; vis prof, Univ Hawaii, 68-69. Mem: Am Physiol Soc; Biophys Soc; Soc Exp Biol & Med. Res: Peripheral and central nervous system; biophysics. Mailing Add: Dept of Physiol & Biophys Univ of Ky Col of Med Lexington KY 40506

BOYARSKY, SAUL, b Burlington, Vt, July 22, 23; m 45; c 3. UROLOGY, PHYSIOLOGY. Educ: Univ Vt, BS, 43, MD, 46. Prof Exp: Instr urol, Sch Med, Duke Univ, 54; instr, NY Univ, 55-56; from asst prof to assoc prof, Albert Einstein Col Med, 56-63; prof urol & asst prof physiol, Sch Med, Duke Urologist, Duke Hosp, 63-70; chmn genitourinary surg, 70-73, PROF GENITOURINARY SURG, PROF BIOENG & ASSOC PROF PHARMACOL, SCH MED, WASH UNIV & UROLOGIST, BARNES HOSP, 70- Concurrent Pos: Fel surg, Col Med, Univ Vt, 47-48; USPHS fel physiol, Sch Med, NY Univ, 54-55; asst attend, Bellevue Hosp, New York, 54-56 & Univ Hosp, 56-63; asst chief urol, Bronx Vet Admin Hosp, 56-57, attend, 60-63; assoc attend, Bronx Munic Hosp, 56-62, vis urologist, 62-63; adj urologist, New Rochelle Hosp, 60-63; chief urol, Vet Admin Hosp, 63-70, dir rehab, 69-70; consult urol, St Louis City Hosp, 70-; consult, John Cochran Vet Admin Hosp, Jefferson Barracks, Jewish Hosp, Food & Drug Admin bur med, US Dept Health, Educ & Welfare, Vet Admin Spinal Cord Injury Serv & Mo Crippled Childrens Serv, Univ Mo, 70-; mem panel review gastroenterol-urol devices, Dept Health, Educ & Welfare, Food & Drug Admin. Honors & Awards: William P Burpeau Award. Mem: Am Physiol Soc; Am Asn Genitourinary Surgeons; Am Urol Asn; Am Col Surgeons; Soc Univ Urologists. Res: Renal, ureteral and bladder physiology and pharmacology; rehabilitation and bioengineering in urologic surgery; spinal cord injury care. Mailing Add: Div of Urol Wash Univ Sch of Med St Louis MO 63110

BOYCE, CHARLES, b Proctor, Vt, July 6, 28; m 51; c 4. OBSTETRICS & GYNECOLOGY. Educ: Marietta Col, BS, 49; Univ Vt, MD, 53; Am Bd Obstet & Gynec, dipl, 63, cert gynec oncol, 74. Prof Exp: Asst prof gynec oncol, Nat Naval Med Ctr, 60-62, asst chief obstet & gynec, US Naval Hosp, Chelsea, Mass, 62-63; assoc, Henry Ford Hosp, 63-66; from asst prof to assoc prof, Sch Med, 69-75, PROF & HEAD DIV GYNEC ONCOL, DEPT OBSTET & GYNEC, WAYNE STATE UNIV, 75- Concurrent Pos: Mich mortality consult, 63-68; consult, Vet Admin, 68- Mem: Am Col Obstet & Gynec; Soc Gynec Oncologists. Res: Gynecological oncology, especially trophoblast identification and disease research, immune identification and chemotherapeutic approach; study of human pelvic malignancies. Mailing Add: Div of Gynec Oncol Dept of Gynec & Obstet Wayne State Univ Detroit MI 48201

BOYCE, DONALD JOE, b Duncan, Okla, Dec 30, 31; m 52; c 6. MATHEMATICS. Educ: Cent State Col, BS, 56; Okla State Univ, MS, 57, EdD(math educ), 68. Prof Exp: Instr math, Cent State Col, 57-58; systs analyst, Tinker Air Force Base, 58-59; from asst prof to assoc prof, 59-71, PROF MATH, CENT STATE UNIV, 71- Concurrent Pos: NSF res grants for col teachers, 67-69. Mem: Math Asn Am. Res: Modern differential geometry, particularly the curvature of differentiable manifolds. Mailing Add: Dept of Math Cent State Univ Edmond OK 73034

BOYCE, FREDERICK FITZHERBERT, b Barbados, BWI, Sept 22, 03; nat US; m 41; c 3. SURGERY. Educ: Harvard Univ, BS, 27; Yale Univ, MD, 30. Prof Exp: From instr to asst prof surg, Sch Med, La State Univ, 32-39; from asst prof to assoc prof, 47-52, PROF CLIN SURG, SCH MED, TULANE UNIV, 52- Concurrent Pos: Pvt pract, 39- Honors & Awards: Gross Quinquennial Prize, 40. Mem: Fel AMA; fel Am Col Surgeons; Soc Exp Biol & Med; fel Am Col Chest Physicians; Int Soc Surg. Res: Liver and biliary tract; liver-kidney death; autolytic peritonitis; liver function tests and burns; acute appendicitis; regional enteritis. Mailing Add: 3801 Canal St New Orleans LA 70119

BOYCE, HENRY WORTH, JR, b Clinton, NC, Sept 21, 30; m 52; c 5. INTERNAL MEDICINE, GASTROENTEROLOGY. Educ: Wake Forest Col, BS, 52, MD, 55; Baylor Univ, MS, 61; Am Bd Internal Med, dipl, 62, cert gastroenterol, 65. Prof Exp: Intern, Tripler Gen Hosp, Honolulu, Hawaii, US Army, 55-56, resident internal med, Brooke Gen Hosp, San Antonio, Tex, 57-59, resident gastroenterol, 60-61; chief gastroenterol serv, Madigan Gen Hosp, Tacoma, Wash, 61-65 & Brooke Gen Hosp, San Antonio, Tex, 65-66, chief gastroenterol serv, Walter Reed Gen Hosp, 66-75; PROF MED & CHIEF SECT GASTROENTEROL, COL MED, UNIV S FLA, 75- Concurrent Pos: Consult, Surgeon Gen, US Army, 67-75. Mem: Fel Am Col Physicians; AMA; Am Soc Gastrointestinal Endoscopy; Asn Mil Surg US. Res: Clinical internal medicine and gastroenterology. Mailing Add: 13000 N 30th St Tampa FL 33612

BOYCE, JOHN SHAW, JR, b Portland, Ore, May 25, 21; m 46; c 4. FOREST PATHOLOGY. Educ: Yale Univ, BS, 42, MF, 48; Duke Univ, PhD(forest path), 51. Prof Exp: Pathologist, US Forest Serv, 50-61; prof plant path, Univ Ga, 61-66; prof & coordr math-sci div, Florence Regional Campus, Univ SC, 69-70; PROF BIOL & CHMN DEPT, FRANCIS MARION COL, 70- Mem: Am Phytopath Soc. Mailing Add: Dept of Biol Francis Marion Col Florence SC 29501

BOYCE, MOFFATT GRIER, mathematics, deceased

BOYCE, PETER BRADFORD, b New York, NY, Nov 30, 36; m 58; c 2. ASTRONOMY. Educ: Harvard Univ, AB, 58; Univ Mich, MA, 62, PhD(astron), 63. Prof Exp: Res astronomer, Lowell Observ, 63-75; adj prof, Ohio State Univ, 70-75; PROG DIR ASTRON DIV, NAT SCI FOUND, 73- Mem: AAAS; Am Astron Soc; Int Astron Union. Res: Photoelectric measurement of stellar, nebular and planetary spectra; astronomical instrumentation; planetary photometry. Mailing Add: Astron Div Nat Sci Found Washington DC 20550

BOYCE, RICHARD JOSEPH, b Schenectady, NY, Sept 15, 39; m 57; c 4. RUBBER CHEMISTRY. Educ: Rensselaer Polytech Inst, BChE, 61, PhD(phys chem), 66. Prof Exp: RES CHEMIST, E I DU PONT DE NEMOURS & CO, INC, 65- Mem: Am Chem Soc. Res: Polymer physical chemistry; rheology of elastomers. Mailing Add: 11 Harvest Lane Glen Farms Newark DE 19711

BOYCE, RICHARD P, b Pocatello, Idaho, Jan 27, 28; m 50; c 2. BIOPHYSICS. Educ: Univ Utah, BA, 55; Yale Univ, PhD(biophys), 61. Prof Exp: From res asst to res assoc radiobiol, Sch Med, Yale Univ, 61-64, from asst prof to assoc prof, 64-71; PROF BIOCHEM, SCH MED, UNIV FLA, 71- Concurrent Pos: Guggenheim fel, 68-69. Mem: Biophys Soc. Res: Enzymatic repair of irradiation damage in bacteria and bacterial viruses; molecular mechanisms of genetic recombination in bacteria. Mailing Add: Dept of Biochem Univ of Fla Sch of Med Gainesville FL 32601

BOYCE, RONALD REED, b Los Angeles, Calif, Jan 7, 31; m 55; c 2. URBAN GEOGRAPHY, HISTORICAL GEOGRAPHY. Educ: Univ Utah, BS, 56, MS, 57; Univ Wash, PhD(geog), 61. Prof Exp: Instr geog, Western Wash State Univ, 59; res assoc regional develop, Wash Univ, 60-61; asst prof planning, Univ Ill, Urbana, 61-62; assoc prof urban geog, Univ Iowa, 63-64; ASSOC PROF URBAN GEOG, UNIV WASH, 65- Concurrent Pos: S & W lectr econ develop, Highlands Univ, 65. Mem: Asn Am Geog. Res: Internal spatial structure of American cities with special emphasis on commercial and residential change; central business district studies, planned shopping centers and general business center decline; residential neighborhood change and mobility. Mailing Add: Dept of Geog DP10 Univ of Wash Seattle WA 98010

BOYCE, STEPHEN GADDY, b Anson Co, NC, Feb 5, 24; m 51. PLANT ECOLOGY. Educ: NC State Col, BS, 49, MS, 51, PhD(bot), 53. Prof Exp: Instr bot, NC State Col, 52-53; asst prof, Univ Ohio, 53-57; silviculturist, Cent States Exp Sta, 57-64, asst dir, 64-66, chief genetics res, 66-67, asst to dep chief res, 67-70, dir, 70-73, CHIEF FOREST ECOLOGIST, SOUTHEASTERN FOREST EXP STA, US FOREST SERV, 73- Mem: Ecol Soc Am; Soc Am Foresters. Res: Forest and coastal dunes ecology; tree improvement; wood growth relations. Mailing Add: Southeastern Forest Exp Sta US Forest Serv PO Box 2570 Asheville NC 28802

BOYCE, STEPHEN SCOTT, b Indianapolis, Ind, Feb 23, 42; m 65; c 1. MATHEMATICS, QUANTUM MECHANICS. Educ: Earlham Col, BA, 64; Univ Wis, MA, 65, PhD(math), 69. Prof Exp: ASST PROF MATH, BEREA COL, 69- Mem: Math Asn Am; Am Math Soc. Res: Comparison and development of mathematical models for quantum mechanics. Mailing Add: Dept of Math CPO-43 Berea Col Berea KY 40403

BOYCE, WILLIAM EDWARD, b Tampa, Fla, Dec 19, 30; m 55; c 3. APPLIED MATHEMATICS. Educ: Southwestern at Memphis, BA, 51; Carnegie Inst Technol, MS, 53, PhD(math), 55. Prof Exp: Universal Match Found fel appl math, Brown Univ, 55-56, res assoc, 56-57; from asst prof to assoc prof, 57-63, PROF MATH, RENSSELAER POLYTECH INST, 63- Concurrent Pos: Managing ed, SIAM Rev, 70- Mem: AAAS; Am Math Soc; Soc Indust & Appl Math; Math Asn Am; Am Asn Univ Prof. Res: Elasticity; plasticity; vibrations; differential equations; boundary value problems; stochastic problems in mechanics. Mailing Add: Dept of Math Rensselaer Polytech Inst Troy NY 12181

BOYCE, WILLIAM HENRY, b Ansonville, NC, Sept 22, 18; m 48; c 4. UROLOGY. Educ: Davidson Col, BS, 40; Vanderbilt Univ, MD, 44. Prof Exp: Asst res urologist, New York Hosp & Cornell Univ Med Ctr, 48-50; resident urologist, Univ Va, 50-52; from instr to assoc prof, 52-55, PROF UROL, BOWMAN GRAY SCH MED, 60- Honors & Awards: Am Urol Asn Prize, 51, 52 & 54. Mem: Soc Exp Biol & Med; Am Urol Asn; Am Col Surgeons; Soc Univ Surgeons; Int Soc Urol. Res: Plastic operations for congenital malformations of bladder and external genitalia; electromyography of urinary bladder; proteins, simple and conjugated, of urine in health and in various diseases; urinary calculous disease. Mailing Add: Dept of Urol Bowman Gray Sch of Med Winston-Salem NC 27103

BOYCE, WILLIAM MARTIN, b Florence, SC, Apr 21, 38; m 59; c 2. MATHEMATICS. Educ: Fla State Univ, BA, 59, MS, 60; Tulane Univ, PhD(math), 67. Prof Exp: Aerospace engr, Navig Anal Sect, Manned Spacecraft Ctr, NASA, 65-66, head, 66-67; mem tech staff, 67-70, HEAD MATH ANAL DEPT, MATH RES CTR, BELL TEL LABS, INC, 70- Mem: Am Math Soc; Soc Indust & Appl Math; Asn Comput Mach; Math Asn Am; Am Finance Asn. Res: Commuting functions; computational graph theory; mathematical economics; applied probability; operations research; computer program development. Mailing Add: 17 Rockage Rd Warren NJ 07060

BOYCKS, EDWARD CHARLES, analytical chemistry, physical chemistry, see 12th edition

BOYD, ALAN WILLIAM, b Calgary, Alta, May 25, 24; m 49; c 4. RADIATION CHEMISTRY. Educ: Univ BC, BA, 45, MA, 47; Univ Calif, PhD(chem), 51. Prof Exp: Instr chem, Univ Western Ont, 50-51; RES CHEMIST, ATOMIC ENERGY CAN, LTD, 51- Mem: Chem Inst Can. Res: Mechanisms in gas phase radiolysis; isotope effects in radiation chemistry. Mailing Add: Deep River ON Can

BOYD, ALFRED COLTON, JR, b Buffalo, NY, Dec 12, 29; m 59; c 3. INORGANIC CHEMISTRY. Educ: Canisius Col, BS, 51; Purdue Univ, MS, 53, PhD(inorg chem), 57. Prof Exp: Asst prof chem, 57-68, asst dean col arts & sci, 69-74, ASSOC PROF CHEM, UNIV MD, COLLEGE PARK, 69- Mem: Am Chem Soc. Res: Boron and organometallic chemistry. Mailing Add: Dept of Chem Univ of Md College Park MD 20742

BOYD, CARL EDMUND, b Kingston, Ont, Jan 7, 35; m 59; c 2. PHARMACOLOGY, TOXICOLOGY. Educ: Queen's Univ, Ont, MD & CM, 59. Prof Exp: Demonstr pharmacol, Queen's Univ, Ont, 59-60; intern, Hotel Dieu Hosp, 60-61; res scientist, Food & Drug Directorate, Ont, 61-63; resident obstet & surg, Ottawa Civic Hosp, 63-65; sci evaluator, 65-69, chief drug info, Bull Div, 69-75, EXEC SECY, SPEC ADV COMT REPRODUCTIVE PHYSIOL, HEALTH PROTECTION BR, DEPT NAT HEALTH & WELFARE, 75- Mem: Soc Toxicol; Can Med Asn. Res: Federal drug regulatory. Mailing Add: Bur of Drugs Health Protection Br Dept of Nat Health & Welfare Vanier ON Can

BOYD, CARL M, b Leavenworth, Kans, Mar 23, 33; m 55; c 4. BIOLOGICAL OCEANOGRAPHY. Educ: Univ Ind, AB, 55, MA, 56; Scripps Inst, Univ Calif, PhD(marine biol), 62. Prof Exp: From asst prof to assoc prof, 62-70, dir aquatron lab, 70-75, PROF MARINE BIOL, INST OCEANOG, DALHOUSIE UNIV, 70- Concurrent Pos: Nat Res Coun Can sr res fel, Sci Exchange, France, 68 & 69. Mem: Am Soc Limnol & Oceanog. Res: Ecology of marine zooplankton. Mailing Add: Inst of Oceanog Dalhousie Univ Halifax NS Can

BOYD, CHARLES ALEXANDER, b Snohomish, Wash, Mar 4, 17; m 42; c 3. CHEMISTRY. Educ: Univ Wash, BS, 39; Ore State Col, MS, 41; Univ Wis, PhD(phys chem), 48. Prof Exp: Res assoc, George Washington Univ, 42-45; chemist, Argonne Nat Lab, Chicago, 46-47; proj assoc, Univ Wis, 48-50, asst prof phys chem, 50-51; phys chemist, Camp Detrick, 51-53; sci warfare adv, Weapons Systs Eval Group, Off Secy Defense, 53-58, dir res, 58-63; chief scientist, Aeroprojects, Inc, 62-65; sr res assoc, Ord Res Lab, Pa State Univ, 65-70; TEACHER, SCI DEPT, STATE COL HIGH SCH, 70-, DIR ALTERNATIVE PROG, STATE COL AREA SCH SYST, 74- Concurrent Pos: Dir res, Weapons Systs Eval Div, Inst Defense Anal, 58-62. Mem: Fel AAAS; NY Acad Sci; Am Phys Soc; The Chem Soc. Res: Rocket weapons; electronic phenomena of solids; transport properties of gases; ultrasonics; aerosols; kinetics of biological processes; operational research; secondary science education. Mailing Add: Box 125 A RD 1 Port Matilda PA 16870

BOYD, CHARLES CURTIS, b Ottawa, Ill, Feb 18, 43; m 70; c 1. FORESTRY, AGRONOMY. Educ: Southern Ill Univ, BS, 66; Univ Hawaii, MS, 68; Wash State Univ, PhD(soils), 73. Prof Exp: PROJ LEADER NURSERY AGRONOMIST, WEYERHAEUSER CO, 74- Concurrent Pos: Res tech wheat res, US Agency Int Develop, 72. Mem: Agron Soc; Soil Sci Soc Am. Res: Develop and improve agronomic techniques, soil fertility, weed control, irrigation, seeding date and soil management that enhance the quantity and quality of conifer seedlings growing in a nursery. Mailing Add: Weyerhaeuser Forestry Res Ctr 505 N Pearl St Centralia WA 98531

BOYD, CLAUDE ELSON, b Hatley, Miss, Nov 14, 39; m 63; c 3. LIMNOLOGY, PLANT ECOLOGY. Educ: Miss State Univ, BS, 62, MS, 63; Auburn Univ, PhD(limnol), 66. Prof Exp: Aquatic biologist, Fed Water Pollution Control Admin, 66; asst prof res, Auburn Univ, 67-68; mem res staff, Savannah River ecol lab, 69-71; ASSOC PROF FISHERIES & ALLIED AQUACULT, AUBURN UNIV, 71- Mem: Am Soc Limnol & Oceanog; Ecol Soc Am; Soc Econ Bot. Res: Aquatic plant ecology and water chemistry in fish ponds; aquatic plant management. Mailing Add: Dept of Fisheries & Aquacult Auburn Univ Auburn AL 36830

BOYD, DAVID CHARLES, b Pittsburgh, Pa, Dec 31, 42; m 66; c 2. GLASS TECHNOLOGY. Educ: Lehigh Univ, BS, 64; Purdue Univ, MS, 66, PhD(metall engr), 69. Prof Exp: SR SCIENTIST GLASS COMPOSITION, CORNING GLASS WORKS, 69- Mem: Am Ceramic Soc. Res: Studies relating the properties of glasses to their compositions, particularly in the areas of television glasses and optical/ophthalmic glasses. Mailing Add: Glass Chem Res Dept Corning Glass Works Corning NY 14830

BOYD, DONALD MITCHELL, b Washington, DC, Mar 26, 19; m 43; c 3. BIOCHEMISTRY, MICROBIOLOGY. Educ: Univ Md, BS, 43; Georgetown Univ, MS, 50, PhD(biochem), 53. Prof Exp: Bacteriologist, Brucellosis Res, Animal Dis Sta, Beltsville Res Ctr, 46-47; bacteriologist, US Food & Drug Admin, 47-52; intel specialist, Medico-Biol Sci, Dir of Intel, US Air Force, 52-57; scientist, Sr Staff, Booz-Allen Appl Res, Inc, 57-58; medico-biol scientist, Res Anal Corp, 58-71; MGR ENVIRON AFFAIRS, SEVEN-UP CO, 71- Mem: Am Soc Microbiol; Inst Food Technol; Asn Mil Surg US. Res: Microbial physiology and chemistry; airborne routes of infection in bacterial, rickettsial and virus diseases; radiobiology and health hazards in nuclear research and development; air, water and solid waste pollution problems. Mailing Add: Seven-Up Co 121 S Meramec St Louis MO 63105

BOYD, DONALD WILKIN, b Newark, Ohio, Nov 1, 27; m 57. HISTORICAL GEOLOGY. Educ: Ohio State Univ, BS, 49; Columbia Univ, PhD(geol), 57. Prof Exp: Instr geol, Union Col, NY, 53-56; from asst prof to assoc prof, 56-66, PROF GEOL, UNIV WYO, 66- Concurrent Pos: Res assoc, Am Mus Natural Hist, 68- Mem: Geol Soc Am; Paleont Soc; Soc Econ Paleont & Mineral; Am Asn Petrol Geol. Res: Invertebrate paleontology; inorganic and organic sedimentary structures; Permian biostratigraphy and depositional environments. Mailing Add: Dept of Geol Univ of Wyo Laramie WY 82071

BOYD, EARL NEAL, b Trinity, Ky, Dec 20, 22; m 48; c 1. DAIRY SCIENCE. Educ: Eastern Ky State Col, BS, 48; Univ Ky, MS, 49; Ohio State Univ, PhD(dairy sci), 52. Prof Exp: Asst prof, Univ Ky, 52-55; res chemist, Swift & Co, 55-57; prin dairy technologist, US Dept Agr, 57-68; head dept food sci & technol, 68-70, ASST DEAN & DIR DIV BASIC SCI, COL AGR & LIFE SCI, VA POLYTECH INST & STATE UNIV, 70- Concurrent Pos: Vis res scientist, Dept Dairy Sci, Pa State Univ, 63-64. Mem: Am Dairy Sci Asn; Inst Food Technol. Res: Chemical and bacteriological research in relation to milk and milk products; effects of processing on milk constituents. Mailing Add: Col of Agr & Life Sci Hutcheson Hall Va Polytech Inst & State Univ Blacksburg VA 24061

BOYD, EDWIN TAYLOR, organic chemistry, see 12th edition

BOYD, ELDON MATHEWS, pharmacology, deceased

BOYD, ELEANOR H, b Philadelphia, Pa, Oct 7, 35; m 64. NEUROPHARMACOLOGY, NEUROPHYSIOLOGY. Educ: Wellesley Col, BA, 56; Univ Rochester, PhD(pharmacol), 68. Prof Exp: Assoc physiol, 71-72, ASST PROF PHARMACOL & TOXICOL, SCH MED, UNIV ROCHESTER, 72-, ASST PROF NURSING, SCH NURSING, 73- Concurrent Pos: USPHS fel physiol, Univ Rochester, 69-71. Mem: Soc Neurosci; AAAS. Res: Electrophysiologic effects of neuropharmacologic agents. Mailing Add: Dept of Pharmacol & Toxicol Univ of Rochester Med Ctr Rochester NY 14642

BOYD, ELIZABETH MARGARET, b Liverpool, Eng, July 8, 08; nat US. ZOOLOGY. Educ: Univ Edinburgh, BSc, 30; Mt Holyoke Col, MA, 33; Cornell Univ, PhD(zool), 46. Prof Exp: Asst zool, Univ Edinburgh, 30-31, Mt Holyoke Col, 31-33 & McGill Univ, 33-37; from instr to prof, 37-64, Alumnae Found prof, 64-74, chmn dept biol sci, 67-70, EMER PROF ZOOL, MT HOLYOKE COL, 74- Concurrent Pos: US Pub Health Serv fel, 63-64; fel trop med & parasitol, 64. Mem: Fel AAAS; Am Soc Zool; Nat Audubon Soc; Am Soc Parasitol; Am Ornith Union. Res: Parasitology; ornithology. Mailing Add: Dept of Biol Sci Mt Holyoke Col South Hadley MA 01075

BOYD, EUGENE STANLEY, b Waltham, Mass, May 11, 27; m 53; c 2. PHARMACOLOGY. Educ: Bowdoin Col, BA, 50; Univ Rochester, PhD(pharmacol), 53. Prof Exp: Jr scientist, Atomic Energy Proj, 53, from instr to asst prof, 53-63, ASSOC PROF PHARMACOL, UNIV ROCHESTER, 63- Mem: AAAS; Soc Pharmacol & Exp Therapeut; Soc Neurosci. Res: Drugs which affect behavior. Mailing Add: Dept of Pharmacol Univ of Rochester Rochester NY 14642

BOYD, FRANCIS R, b Boston, Mass, Jan 30, 26. GEOLOGY. Educ: Harvard Univ, AB, 49; Stanford Univ, MS, 50; Harvard Univ, PhD, 57. Prof Exp: Fel, Harvard Univ, 52-53; PHYSICAL CHEMIST, GEOPHYS LAB, CARNEGIE INST, 53- Mem: Mineral Soc Am; Geol Soc Am; Geochem Soc. Res: Phase equilibrium research at high pressures and temperatures. Mailing Add: 2801 Upton St NW Washington DC 20008

BOYD, FRANK MCCALLA, b Canton, Ga, Mar 16, 29. BACTERIOLOGY. Educ: NGa Col, BS, 48; Univ Tenn, MS, 50; Univ Wis, PhD(bact), 59. Prof Exp: Bacteriologist, Ga Poultry Lab, 50-51; communicable dis ctr, US Pub Health Serv, 51-54; res asst, Univ Wis, 54-58; asst poultry microbiologist, Univ Ga, 58-67; ASSOC PROF BIOL, NORTHEAST LA UNIV, 67- Mem: AAAS; Am Soc Microbiol; Poultry Sci Asn; Soc Gen Microbiol. Res: Agricultural and medical bacteriology. Mailing Add: Dept of Biol Northeast La Univ Monroe LA 71201

BOYD, FREDERICK TILGHMAN, b St Paul, Minn, Mar 9, 13; m 37; c 5. AGRONOMY. Educ: Univ Wis, BS, 34, PhD(soils, agron), 38. Prof Exp: Asst agron & soils, Univ Wis, 35-38; asst agronomist, Exp Sta, Fla, 38-41; agronomist, Raoul & Haney, Inc, Fla, 41-42; midwest agriculturist, Am Cyanamid Co, NY, 45-48; chief tech serv dept, Crow's Hybrid Corn Co, 48-50; supt plantation field lab & agronomist, Fla Exp Sta, 53-68; agronomist, Inst Food & Agr Sci, 68-74, EMER PROF AGRON, UNIV FLA, 74- Concurrent Pos: Consult, S Coast Sugar Corp, La, 48-52. Mem: AAAS; Am Soc Agron; Soil Sci Soc Am. Res: Fertilizer response on cultivated crops; effect of various cultural practices on growth of crops; chemicals as herbicides; corn, sorghum and grass breeding; nematodes in agronomic crops. Mailing Add: 5551 NW Fourth Pl Gainesville FL 32601

BOYD, GEORGE ADDISON, b Ambia, Tex, Nov 17, 07; m 44; c 4. RESEARCH ADMINISTRATION. Educ: Austin Col, AB, 29, AM, 30; Univ Iowa, MS, 35. Prof Exp: Res chemist, Phillips Petrol Co, Okla, 30-33; mgr info div, Develop & Patent Dept, Stand Oil Co, Ind, 35-37; asst to vpres in charge develop, Sharples Chem Co, 37-39; dir res, Ozone Processes Co, 40-41; res physicist, Biochem Res Found, Del, 41-44; res assoc, Sch Med & Dent, Univ Rochester, 44-50; prof biophys, Univ Tenn, 50-54; dir, Ariz Res Labs, 54-55; dir res grants & contracts, 55-65, ASSOC DIR CTR METEORITE STUDIES & MANAGING ED, METEORITICS, ARIZ STATE UNIV, 65- Concurrent Pos: Sr scientist, Oak Ridge Inst Nuclear Studies, 50-54; partic, NATO Int Inst Econ Forecasting on Sci Basis, Portugal, 67; develop officer, Gen Mercury Corp, 69-71; consult, Systs Mgt Corp, Mass, 70. Mem: AAAS; Am Soc Exp Biol. Res: Autoradiography; biophysics; meteoritics. Mailing Add: Ctr for Meteorite Studies Ariz State Univ Tempe AZ 85281

BOYD, GEORGE EDWARD, b Evansville, Ind, Sept 1, 11; m 42; c 1. PHYSICAL CHEMISTRY. Educ: Univ Chicago, BS, 33, PhD(phys chem), 37. Prof Exp: Rockefeller Found fel, Univ Chicago, 37-38, from instr to assoc prof, 38-48; assoc dir chem div, Oak Ridge Nat Lab, 49-54, asst lab dir, 54-70, sr sci adv, 70-73; PROF CHEM, UNIV GA, 73- Concurrent Pos: Sect chief metall lab, Univ Chicago & Clinton Lab, Manhattan Proj, 43-46; Fulbright scholar & Guggenheim fel, Univ Leiden, 52-53; Reilly lectr, Univ Notre Dame, 54; vis prof, Purdue Univ, 62; co-dir, NATO Advan Study Inst, 73; Welch Found lectr, 75. Assoc ed, J Phys Chem, 50-54 & Anal Chem, 53-55; mem adv comt nuclear data, Nat Res Coun, 55-60; mem, Atoms for Peace Mission, AEC, 57; chmn, Fulbright Comt Chem, Nat Acad Sci, 57-63; Am Nuclear Soc rep, Nat Res Coun, 66-69. Honors & Awards: Southern Chem Award, Am Chem Soc, 51, Nuclear Appln Award, 69. Mem: AAAS; Am Chem Soc; Am Phys Soc; fel Am Nuclear Soc; Coblentz Soc. Res: Physics and chemistry of surfaces; pure and applied nuclear chemistry; laser Raman and infrared spectra of molten salts and aqueous solutions; physical chemistry of polyelectrolyte solutions and gels. Mailing Add: Dept of Chem Univ of Ga Athens GA 30601

BOYD, GLENN D, b Mexico, NY, Jan 26, 19; m 44; c 4. FOOD SCIENCE. Educ: Univ Mass, BS, 40. Prof Exp: Control chemist, Campbell Soup Co, 40-42, sr chemist, 46-48, res technologist, 48-51, asst mgr prod res, 51-52, mgr prod res, 53-57, assoc dir res frozen prod, 57-62, dir frozen & dehydrated prod res, 62-66, VPRES PROD DEVELOP, CAMPBELL INST FOOD RES, CAMPBELL SOUP CO, 66- Mem: Inst Food Technol; Am Chem Soc. Res: Product development of canned, frozen, dehydrated and other food products. Mailing Add: 237 Redstone Ridge Cherry Hill NJ 08034

BOYD, HERMAN WAYNE, b Murfreesboro, Tenn, Aug 22, 35; m 66; c 4. NUCLEAR PHYSICS. Educ: Middle Tenn State Col, AB, 57; Vanderbilt Univ, MA, 59, PhD(physics), 64. Prof Exp: Instr physics, Middle Tenn State Col, 59-63; assoc prof, 64-68, PROF PHYSICS & HEAD DEPT, W GA COL, 68- Mem: AAAS; Am Phys Soc; Am Asn Physics Teachers. Res: Beta and gamma ray spectroscopy; decay schemes of intermediate even-even nuclei; internal conversion coefficients; photo electric cross sections; lithuim-drifted germanium gamma spectra. Mailing Add: Dept of Physics W Ga Col Carrollton GA 30117

BOYD, HOWARD ALLEN, JR, radiochemistry, see 12th edition

BOYD, IVAN LOUIS, b Spaulding, Iowa, Aug 4, 04; m 42; c 2. BIOLOGY. Educ: Simpson Col, AB, 31; Iowa State Col, MS, 39, PhD(bot), 44. Prof Exp: Instr jr high sch, Iowa, 28-37, high sch, 37-38; res asst, Soil Conserv Serv, USDA, 38-44; asst prof, 41-44, chmn sci div, 59-63, head dept, 41-74, EMER PROF BIOL, BAKER UNIV, 74- Mem: AAAS; Nat Audubon Soc; assoc Wilson Ornith Soc. Res: Plant ecology; zoology; establishment of native prairie grasses on eroded soils in southeastern Iowa; tannin content of native Iowa sumacs and their value in soil conservation; bird migration in Kansas. Mailing Add: Dept of Biol Baker Univ Baldwin KS 66006

BOYD, JAMES, b Kanowna, Western Australia, Dec 20, 04; nat US; m 32; c 4. GEOLOGY. Educ: Calif Inst Technol, BS, 27; Colo Sch Mines, MS, 32, DSc(geol), 34. Hon Degrees: DSc, Mont Sch Mines, 49, Mich Technol Univ, 62. Prof Exp: Field engr, Radiore Co, 27-29; from instr to assoc prof geol, Colo Sch Mines, 29-41, dean fac, 46-47; spec asst mineral matters, Secy Interior, Washington, DC, 47; dir bur mines, US Dept Interior, 47-51; vpres explor, Kennecott Copper Corp, 51-60; pres, 60-70, CHMN, COPPER RANGE CO, 70- Concurrent Pos: Defense minerals adminr, 50-51; chmn comt mineral res, NSF, 53-57. Honors & Awards: Rand Gold Medal, Am Inst Mining, Metall & Petrol Engrs, 63. Mem: Nat Acad Eng; fel Geol Soc Am; Am Inst Mining, Metall & Petrol Engrs (pres, 69); Soc Econ Geologists; Mining & Metall Soc Am (pres, 59-62). Res: Mining engineering; economics; geophysics; ore deposits. Mailing Add: Apt 1405 700 New Hampshire Ave NW Washington DC 20037

BOYD, JAMES BROWN, b Denver, Colo, June 25, 37; m 60; c 2. RADIATION GENETICS. Educ: Cornell Univ, BA, 59; Calif Inst Technol, PhD(biochem), 65. Prof Exp: Helen Hay Whitney fel, Beerman Div, Max Planck Inst Biol, 65-69; asst prof, 69-71, ASSOC PROF GENETICS, UNIV CALIF, DAVIS, 71- Concurrent Pos: Fels, NATO, 73-74 & Guggenheim, 74-75. Mem: Genetics Soc Am; Am Soc Cell Biol. Res: Biochemical characterization of gene action as revealed by polytene chromosomes; characterization of deoxyribonucleases; DNA repair; genetic and biochemical characterization of radiation sensitive mutants in Drosophila. Mailing Add: Dept of Genetics Univ of Calif Davis CA 95616

BOYD, JAMES C, b Washington, Oct 11, 16; m 46; c 3. DAIRY INDUSTRY. Educ: State Col Wash, BS, 39; Mont State Col, MS, 42; Mich State Col, PhD, 52. Prof Exp: Assoc prof dairy husb, Univ Idaho, 47-56; head dairy indust dept, 56-72, PROF AGR PROD UTILIZATION, MONT STATE UNIV, 56- Mem: Am Dairy Sci Asn. Res: Dairy manufacturing. Mailing Add: Dept of Animal Sci Mont State Univ Bozeman MT 59715

BOYD, JAMES EMORY, b Tignall, Ga, July 18, 06; m 34; c 2. NUCLEAR PHYSICS, OCEANOGRAPHY. Educ: Univ Ga, AB, 27; Duke Univ, MA, 28; Yale Univ, PhD(physics), 33. Prof Exp: Instr physics, Univ Ga, 28-30; head math-sci dept, WGa Col, 33-35; from asst prof to assoc prof physics, Ga Inst Technol, 35-42, prof, 46-61, actg pres, 71-72, res assoc, Eng Exp Sta, 46-50, head physics div, 50-54, from asst dir to dir, 54-61; pres, 61-71, EMER PRES, W GA COL, 74-; EMER PROF PHYSICS, GA INST TECHNOL, 74- Concurrent Pos: Vchancellor acad develop, Bd Regents, Univ Syst Ga, 72-74. Mem: Sigma Xi; Am Phys Soc; AAAS. Res: X-rays and microwave propagation. Mailing Add: 20 Ivy Ridge NE Atlanta GA 30342

BOYD, JAMES HAMILTON, high energy physics, electrophotography, see 12th edition

BOYD, JAMES ROBERT, b Bladen, Nebr, Oct 28, 36; m 53; c 2. SOLID STATE PHYSICS. Educ: Hastings Col, BA, 58; Univ Ill, MS, 60; Univ Tex, PhD(physics), 65. Prof Exp: Physicist, Int Bus Mach Components Lab, 60-62; res assoc physics, Case Western Reserve Univ, 65-67; ASST PROF PHYSICS, SOUTHERN METHODIST UNIV, 67- Mem: Am Phys Soc. Res: Ultrasonic studies of metals at liquid helium temperatures; De Haas-Van Alphen effect; fermi surface and electronic structure of metals and alloys. Mailing Add: Dept of Physics Southern Methodist Univ Dallas TX 75222

BOYD, JOHN EDWARD, b Albany, Ga, Jan 26, 32; m 54; c 2. AGRICULTURAL BIOCHEMISTRY, DRUG METABOLISM. Educ: State Teachers Col Pa, BS, 52; Lehigh Univ, MS, 54; Pa State Univ, PhD(agr chem), 57. Prof Exp: Asst monomer chem, Lehigh Univ, 52-54; asst pesticide chem, Pa State Univ, 54-57, instr, 55-56; res chemist, 57-62, group leader metab lab, 62-69, MGR METAB & ANAL CHEM SECT, AM CYANAMID CO, 69- Mem: Am Chem Soc; Entom Soc Am. Res: Metabolism and mode of action of pesticides; sub-cellular biology; protein structure and synthesis; structure and function of cell wall and plasma membrane. Mailing Add: Am Cyanamid Co PO Box 400 Princeton NJ 08540

BOYD, JOHN MANN, b Fayette, WVa, June 25, 03; m 32; c 2. CHEMISTRY. Educ: Univ Fla, BS, 26, MS, 28. Prof Exp: Instr physics, Univ Fla, 27-28 & chem, Univ Pittsburgh, 28-30; fel, Mellon Inst, 30-38; chief food processing res sect, Continental Can Co, 39-58, sr res engr, 58-65, adv scientist, 65-67, consult, 67-68; food technol consult, 68-75; RETIRED. Mem: Am Chem Soc; Inst Food Technol; fel Am Inst Chemists. Res: Food preservation; food processing equipment; food container linings; wire drawing lubricants; organic synthesis; means of eliminating air from food containers. Mailing Add: 166 Seminole St Clermont FL 32711

BOYD, JOHN PAUL, b Pontiac, Mich, May 26, 39; m 58; c 2. ANTHROPOLOGY, COMMUNICATIONS. Educ: Univ Mich, BA, 61, MS, 63, PhD(commun sci), 66. Prof Exp: Res asst soc sci, Ment Health Res Inst, Univ Mich, 60-65; asst prof anthrop & commun sci, Univ Calif, Irvine, 65-71, asst prof math anthrop, 71-74, ASSOC PROF MATH ANTHROP, UNIV CALIF, IRVINE, 74- Res: Mathematical models of kinship and other cognitive structures. Mailing Add: Sch of Social Sci Univ of Cal Irvine CA 92664

BOYD, JOHN ROBERT, b Coshocton, Ohio, July 28, 08; m 38; c 3. ANALYTICAL CHEMISTRY, SPECTROSCOPY. Educ: Denison Univ, BS, 30. Prof Exp: Engr, Westinghouse Elec Corp, 30-31; analyst, Youngstown Sheet & Tube, Ill, 34-35; chemist, Newcomerstown Plant, James B Clow & Sons, 35-36 & Coshocton Plant, 36-40, chief chemist, 40-63, chief chemist, Res Dept, Cast Iron Div, Clow Corp, 63-73; RETIRED. Mem: Am Chem Soc; Soc Appl Spectros; AAAS. Res: Analysis of cast iron; spectrophometric and colorimetric determinations; silicon and manganese analysis; emission spectroscopy; protective coatings; direct spectrometric analysis of magnesium in cast iron; pollution chemistry. Mailing Add: 687 Hill St Coshocton OH 43812

BOYD, JOHN WILLIAM, b Ft Worth, Tex, Aug 28, 31; m 57; c 3. PHYSICS. Educ: Mich State Univ, BS, 57, PhD(physics), 63. Prof Exp: Sr scientist, Systs Res Labs, Ohio, 63-66; asst prof, 66-73, ASSOC PROF PHYSICS, KNOX COL, ILL, 73- Mem: AAAS; Am Phys Soc; Am Asn Physics Teachers; Sigma Xi. Res: Atomic and molecular structure and spectroscopy; optical masers; optics. Mailing Add: Dept of Physics Knox Col Galesburg IL 61401

BOYD, JOSEPHINE WATSON, b St Paul, Minn, Feb 19, 27; m 56; c 3. WATER POLLUTION. Educ: Univ Minn, BChemE & BBusAdmin, 48, MS, 53, PhD(bact), 55. Prof Exp: Chemist & bacteriologist, Theo Hamm Brewing Co, 48-55; asst, Univ Minn, 51-55; res assoc bact, Arctic Inst NAm, 56-57, Ohio State Univ, 58-61 & Arctic Inst NAm, Alaska, 62-63 & Can, 64; fel microbiol, Colo State Univ, 68-74; ASST DIR WATER POLLUTION ASSESSMENT FACIL, CITY OF FORT COLLINS, 74- Concurrent Pos: Vis scientist, Norway, 65 & 70; fel, Colo State Univ, 68-73. Mem: AAAS; Am Soc Microbiol; Water Pollution Control Fedn. Res: Microbiology in the arctic environment; waste water and sewage sludge considered as raw materials rather than refuse. Mailing Add: 1313 Stover St Ft Collins CO 80521

BOYD, LEROY HOUSTON, b Arnett, Okla, May 29, 35; m 58; c 2. ANIMAL SCIENCE. Educ: Okla State Univ, BS, 57; Univ Ky, MS, 60, PhD(animal husb), 63. Prof Exp: Asst animal husbandman, Univ Ky, 57-60; asst prof, 63-66, ASSOC PROF ANIMAL HUSB, MISS STATE UNIV, 66- Mem: Am Soc Animal Sci. Res: Physiology, production and management of sheep and horses; ruminant nutrition. Mailing Add: PO Box 5228 Dept of Animal Sci Miss State Univ Mississippi State MS 39762

BOYD, LOUIS JEFFERSON, b Lynn Grove, Ky, Mar 14, 28; m 48; c 4. REPRODUCTIVE PHYSIOLOGY. Educ: Univ Ky, BS, 50, MS, 51; Univ Ill, PhD(dairy physiol), 56. Prof Exp: Field agent, Dairy Exten, Univ Ky, 51-53; assoc prof dairy, Univ Tenn, 56-62; from assoc prof to prof dairy physiol, Mich State Univ, 63-72; PROF ANIMAL SCI & CHMN DIV, UNIV GA, 72-, HEAD DEPT ANIMAL & DAIRY SCI, 74- Concurrent Pos: Mem staff, Brit Milk Mkt Bd, 70-71. Mem: AAAS; Am Dairy Sci Asn; Am Soc Animal Sci; Soc Study Reproduction; Brit Soc Study Fertil. Res: Factors affecting reproductive efficiency in dairy cattle, such as production, extension, preservation and use of bull sperm and conception and gestation in the dairy cow. Mailing Add: Div of Animal Sci Univ of Ga Athens GA 30602

BOYD, MARJORIE ELIZABETH, theoretical physics, see 12th edition

BOYD, RICHARD HAYS, b Columbus, Ohio, Aug 12, 29; m 51; c 2. PHYSICAL CHEMISTRY, POLYMER CHEMISTRY. Educ: Ohio State Univ, BSc, 51; Mass Inst Technol, PhD, 55. Prof Exp: Res chemist, E I du Pont de Nemours & Co, 55-62; assoc prof chem, Utah State Univ, 62-65, prof, 65-67; PROF CHEM ENG, MAT SCI & ENG & ADJ PROF CHEM, UNIV UTAH, 67- Mem: Am Chem Soc; Sigma Xi;

Am Phys Soc; Am Inst Chem Engrs. Res: Physical chemistry and physics of polymers; rheology, dielectric properties and kinetics; thermodynamics of solutions; thermochemistry. Mailing Add: Dept of Chem Eng Univ of Utah Salt Lake City UT 84112

BOYD, ROBERT EDWARD, SR, b Clarksville, Tenn, Nov 5, 27; m 60; c 3. MEDICINAL CHEMISTRY. Educ: Tenn State Univ, BA, 54; Fisk Univ, MA, 57; Univ RI, PhD(med chem), 74. Prof Exp: Res chemist, Nat Cancer Inst, NIH, 61-65, chemist, USPHS, 65-67; ASSOC PROF CHEM, LIVINGSTONE COL, 67- Mem: AAAS; Am Chem Soc. Res: Synthesis of potential antidepressant agents based on derivatives of biogenic amines; isolation and characterization of naturally occurring phospholipids. Mailing Add: Dept of Chem Livingstone Col Salisbury NC 28144

BOYD, ROBERT GERALD, solid state physics, see 12th edition

BOYD, ROBERT HENRY, b Norristown, Pa, Aug 27, 32; m 56; c 4. PHYSICAL CHEMISTRY. Educ: Lebanon Valley Col, BS, 54; Pa State Univ, PhD(chem), 59. Prof Exp: Res chemist, 58-72, RES SUPVR, E I DU PONT DE NEMOURS & CO, 72- Mem: Am Chem Soc; Soc Photog Sci & Eng. Res: Chemistry of photographic systems. Mailing Add: Photo Prods E I du Pont de Nemours & Co Towanda PA 18848

BOYD, ROBERT NEILSON, b Chicago, Ill, July 18, 14; m 61; c 2. ORGANIC CHEMISTRY. Educ: Univ Chicago, BS, 36; Univ Calif, Berkeley, PhD(chem), 39. Prof Exp: Asst chem, Univ Calif, Berkeley, 36-38; from instr to asst prof, Antioch Col, 39-47; from instr to prof, 47-74, EMER PROF CHEM, NY UNIV, 74- Concurrent Pos: NSF fel, Univ Ill, 58-59; res assoc, Vernay Labs, Ohio, 44-46. Mem: AAAS; Am Chem Soc. Res: Organic analysis; organic electrochemistry; x-ray diffraction; effect of temperature on structure of mercury. Mailing Add: Dept of Chem New York Univ New York NY 10003

BOYD, SAMUEL NEIL, JR, b Bristol, Tenn, Dec 25, 17; m 44; c 1. ORGANIC CHEMISTRY. Educ: Univ Va, BS, 42; Univ Ill, PhD(org chem), 45. Prof Exp: Lab asst org chem dept, E I du Pont de Nemours & Co, 36-39; res chemist, Comt Med Res & Off Sci Res & Develop, 44-45; res chem org chem dept, 45-53, head div dye res, 53-56, asst dir lab, 56-60, dir dyes & chem sect, 60-70, tech mgr dyes & chem div, MGR ENVIRON AFFAIRS, JACKSON LAB, E I DU PONT DE NEMOURS & CO, 74- Mem: Am Chem Soc; Am Asn Textile Chem & Colorists; Am Tech Asn Pulp & Paper Indust. Res: Chemistry of heterocyclic compounds; high polymers; dyes; textile chemicals; organic fluorine compounds. Mailing Add: Org Chem Dept E I du Pont de Nemours & Co Wilmington DE 19898

BOYD, THOMAS, organic chemistry, see 12th edition

BOYD, THOMAS CASEY, biochemistry, microbiology, see 12th edition

BOYD, VAUGHAN FRANK, biochemistry, see 12th edition

BOYD, VIRGINIA ANN LEWIS, b Shreveport, La, Nov 15, 44; m 64; c 2. VIROLOGY. Educ: Northwestern State Univ La, BS, 65, MS, 68; La State Univ, PhD(microbiol), 71. Prof Exp: Med technician path, Martin Army Hosp, Ft Benning, Ga, 65-66; instr biol, Jacksonville State Univ, 68-69; fel virol, Baylor Col Med, 71-73; SCIENTIST VIROL, FREDERICK CANCER RES CTR, 73- Mem: Am Soc Microbiol; Am Tissue Cult Asn; AAAS; Sigma Xi. Res: Transformation of Balb/c mouse cells with herpes simplex virus and fragments of herpes simplex virus DNA, including mechanism and consequence of herpes simplex virus induced transformation. Mailing Add: Frederick Cancer Res Ctr PO Box B Frederick MD 21701

BOYD, WILLIAM ADAM, b Frackville, Pa, Mar 25, 28; m 56; c 5. ANESTHESIOLOGY, PHYSIOLOGY. Educ: Pa State Univ, BS, 49, MS, 50; Univ Ill, PhD(physiol), 54; Boston Univ, MD, 58; Am Bd Anesthesiol, dipl, 64. Prof Exp: Intern med, Newton-Wellesley Hosp, 58-59; resident chief res anesthesiol, Boston City Hosp, 59-61; chief dept, Irwin US Army Hosp, 61-63; chief exp anesthesiol, Univ Tex M D Anderson Hosp & Tumor Inst, 63-65, assoc prof clin physiol, 65-68; ASST PROF ANESTHESIOL, UNIV UTAH MED CTR, 68- Concurrent Pos: Ed, Cancer Yearbk, 63- Mem: Am Soc Anesthesiol; AMA. Res: Cardiovascular and cardiopulmonary function; cellular metabolism; pharmacology of analgesics and anesthetics. Mailing Add: Dept of Anesthesiol Univ of Utah Med Ctr Salt Lake City UT 84112

BOYD, WILLIAM CLOUSER, b Dearborn, Mo, Mar 4, 03; m 31, 67; c 1. IMMUNOBIOLOGY. Educ: Harvard Univ, AB, 25, AM, 26; Boston Univ, PhD(chem), 30. Prof Exp: From asst prof to prof biochem, 35-69, univ lectr, 57, EMER PROF IMMUNOCHEM, SCH MED, BOSTON UNIV, 69-; RES ASSOC, UNIV CALIF, SAN DIEGO, 74- Concurrent Pos: Guggenheim fel, Europe, 35-36 & 37-38; Fulbright scholar, Pakistan, 52; res chemist, Evans Mem Hosp, 30-35, assoc mem, 35-45; spec assoc, Harvard Med Sch, 41-45; consult, USPHS, 46-49. Mem: AAAS; fel Am Acad Arts & Sci; Am Asn Phys Anthrop; Am Soc Human Genetics; Am Asn Immunol. Res: Blood grouping; immunochemistry; physical anthropology; human genetics. Mailing Add: 1241 Prospect St La Jolla CA 92037

BOYD, WILLIAM LEE, b Kingsport, Tenn, Apr 20, 26; m 56; c 3. BACTERIOLOGY. Educ: Univ Tenn, BA, 50; Univ Minn, MS, 52, PhD(bact), 54. Prof Exp: Asst bact, Univ Tenn, 49-50 & Univ Minn, 50-54; asst prof, Univ Ga, 54-55; prin investr, Arctic Res Lab, Point Barrow, Alaska, 55-57; res assoc, Col Dent, Ohio State Univ, 58-59, from asst prof to assoc prof, 59-64; assoc prof, 64-67, actg head dept, 66-67, PROF BACT, COLO STATE UNIV, 67- Concurrent Pos: Prin investr, NSF grant, US Antarctica Res Prog, 61-63 & 67-68; vis scientist, Troms Mus, Norway, 65, 70, Arctic Res Lab, Barrow, Alaska, 66, 71 & Mus Natural Hist, Reykjavik, Iceland, 71. Mem: AAAS; Am Soc Microbiol; Arctic Inst NAm; Ecol Soc Am; NY Acad Sci. Res: Ecology and physiology of Arctic and Antarctic microorganisms. Mailing Add: Dept of Microbiol Colo State Univ Ft Collins CO 80521

BOYD, WILLIAM WARREN, b Orange, Tex, Feb 28, 39; m 66; c 2. SOLID MECHANICS. Educ: Lamar Univ, BS, 62; Univ Wash, MS, 64; Tex A&M Univ, PhD(mech eng), 69. Prof Exp: Assoc engr, Boeing Co, 62-63; aerospace engr, Johnson Spacecraft Ctr, NASA, 64-67; MEM TECH STAFF SOLID MECH, TEX INSTRUMENTS INC, 69- Res: Bulk wave propagation in heterogeneous media and surface wave propagation in initially stressed piezoelectric material. Mailing Add: Tex Instruments Inc 13500 N Central Expressway Dallas TX 75222

BOYDEN, ALAN ARTHUR, b Milwaukee, Wis, June 16, 97; m 23; c 4. SYSTEMATIC ZOOLOGY. Educ: Univ Wis, AB, 21, PhD(zool), 25. Prof Exp: Asst zool, Univ Wis, 21-25; from instr to prof, 25-62, actg chmn dept, 47-48, chmn dept, 54-59, bur biol res, 36-39, dir serol mus, 48-62, EMER PROF ZOOL, RUTGERS UNIV, 62-, EMER DIR SEROLOGICAL MUS, 70- Concurrent Pos: Ed, Serological Mus, Rutgers Univ, 48-; lectr, Univ Calif, 29-30; grants, Rockefeller Found, 50-51, Johnson & Johnson Found, 51 & NSF, 57, 59; Rose Morgan vis prof, Univ Kans, 64, lectr, 66; Fulbright lectr, Queen Mary Col, Univ London, 60-61, 66-67. Mem: AAAS; Am Asn Immunol; NY Acad Sci; Am Soc Nat; Soc Study Evolution. Res: Systematic serology of vertebrates; invertebrates; precipitin reaction and photoelectric turbidimetry; comparative evolution; asexual reproduction; perspectives in zoology. Mailing Add: Dept of Zool Rutgers, The State Univ New Brunswick NJ 08903

BOYDEN, EDWARD ALLEN, b Bridgewater, Mass, Mar 20, 86; m 13; c 2. ANATOMY, EMBRYOLOGY. Educ: Harvard Univ AB, 09, AM, 11, PhD(anat), 16. Prof Exp: Asst zool, Harvard Univ, 08-11, from instr to asst prof comp anat, 16-26; assoc prof anat, Col Med, Univ Ill, 26-28, prof, 29; head dept, Univ Ala, 29-31; prof, 31-54, chmn dept, 40-49, head dept, 49-54, EMER PROF ANAT, UNIV MINN, 54-; RES PROF, UNIV WASH, 56- Concurrent Pos: Actg assoc prof, Stanford Univ, 21-22; managing ed, Anat Rec, 28-48; Am ed, Acta Anatomica, 45-; vis prof, Univ Wash, 54-56; redactor, 60-73. Honors & Awards: Beaumont Prize, Univ Ill, 28; Award, Mod Med, 60; Film Award, Am Col Chest Physicians, 62; Gray Award, Am Asn Anat, 70; 50 Year Award, Am Acad Arts & Sci, 74. Mem: AAAS; Am Asn Anat (vpres, 46-48, pres, 56); Soc Exp Biol & Med; Am Physiol Soc; hon fel Am Col Chest Physicians. Res: Experimental study of development of urogenital tract of avians and humans; anatomy, embryology and physiology of biliary tract of animals and humans; segmental anatomy and postnatal growth of human and development of monkey lung. Mailing Add: Dept of Biol Struct Univ of Wash Med Sch Seattle WA 98195

BOYDEN, MABEL (JOSEPHINE) GREGG, zoology, deceased

BOYE, CHARLES ANDREW, JR, b Appalachia, Va, Nov 15, 28; m 50; c 2. POLYMER PHYSICS. Educ: Emory & Henry Col, BS, 50; Univ Tenn, MS, 56. Prof Exp: Teacher pub sch, Va, 53-54; asst physics, Univ Tenn, 54-56; physicist, 56-70, RES ASSOC, TENN EASTMAN CO, 70- Concurrent Pos: Asst prof, Eastern Tenn State Col, 56-57. Mem: Am Phys Soc; Am Chem Soc; Sigma Xi. Res: X-ray diffraction of polymers; polymer morphology; macromolecular structure. Mailing Add: Res Labs Tenn Eastman Corp Kingsport TN 37660

BOYE, FREDERICK C, b Buffalo, NY, May 30, 23; m; c 3. ORGANIC CHEMISTRY. Educ: Univ Buffalo, BA, 49, MA, 51, PhD(org chem), 52. Prof Exp: Res chemist, Cornell Aeronaut Lab, 52-53 & Indust Chem Div, Allied Chem Corp, 53-67; sr res chemist, Cowles Chem Div, Stauffer Chem Co, 67-68, Eastern Res Ctr, 68-71; SR RES SCIENTIST, PAPER PROD RES DIV, AM CAN CO, 71- Mem: NY Acad Sci; fel Am Inst Chem. Res: Isocyanate resins; exploratory intermediates for colors; pharmaceuticals; polymers; curing agents; specialty chemicals; process variable studies; process development; plant demonstration; chelating chemicals; amino acids; detergent builders; organic acid anhydrides. Mailing Add: Am Can Co Paper Prod Res Rothschild WI 54474

BOYE, ROBERT JAMES, b Shattock, Okla, Jan 11, 34. NUCLEAR PHYSICS, THEORETICAL PHYSICS. Educ: Univ Okla, BS, 55; Johns Hopkins Univ, PhD(physics), 68. Prof Exp: ASSOC PROF PHYSICS, MORGAN STATE COL, 61- Mem: Am Phys Soc; Am Asn Physics Teachers. Res: Molecular structure; nuclear three-body problem. Mailing Add: Dept of Physics Morgan State Col Baltimore MD 21212

BOYER, ALVIN C, b Geneva, Ill, Apr 14, 34; m 56; c 3. BIOCHEMISTRY, ORGANIC CHEMISTRY. Educ: NCent Col, Ill, BA, 56; Univ Ill, MS, 58, PhD(enzyme kinetics), 60. Prof Exp: Res biochemist, USDA, 60-62; RES BIOCHEMIST, SHELL DEVELOP CO, 62- Mem: NY Acad Sci; Am Chem Soc. Res: Neurobiochemistry; enzymology of insect and mammalian nerve tissue. Mailing Add: Biol Sci Res Ctr Shell Develop Co PO Box 4248 Modesto CA 95352

BOYER, BARBARA CONTA, b Ithaca, NY, Jan 23, 42; m 68. DEVELOPMENTAL BIOLOGY. Educ: Univ Rochester, AB, 63; Univ Mich, MS, 64, PhD(zool), 69. Prof Exp: NIH trainee develop biol, Whitman Lab, Univ Chicago, 69-71; VIS ASST PROF BIOL, UNION COL, NY, 73- Mem: Sigma Xi; Am Soc Zool. Res: Experimental embryology and oogenesis of marine invertebrates, particularly turbellarian flatworms. Mailing Add: 1310 Garner Ave Schenectady NY 12309

BOYER, CARL BENJAMIN, b Hellertown, Pa, Nov 3, 06; m 35; c 4. MATHEMATICS. Educ: Columbia Univ, AB, 28, AM, 29, PhD(hist), 39. Prof Exp: Tutor, 28-33, instr, 34-40, asst prof, 41-47, assoc prof, 48-52, PROF MATH, BROOKLYN COL, 52- Concurrent Pos: Lectr, Rutgers Univ, 35-41; prof, Yeshiva Univ, 52-58; Guggenheim fel, 54-55. Mem: AAAS; Am Math Soc; Math Asn Am; Hist Sci Soc (vpres, 57-58); Int Acad Hist Sci. Res: History of science and mathematics, especially of calculus and analytic geometry. Mailing Add: 997 E 19th St Brooklyn NY 11230

BOYER, CHARLES CHESTER, b North Greece, NY, May 7, 17; m 41; c 4. ZOOLOGY. Educ: St Bonaventure Col, BS, 38; Duke Univ, MA, 40, PhD(zool, embryol), 48. Prof Exp: Asst, Duke Univ, 38-41, instr zool, 46; asst prof biol, Univ Louisville, 48-49; asst prof anat, Med Ctr, Univ Ala, 49-57; assoc prof, Med Ctr, WVa Univ, 57-62; from assoc prof to prof, 62-70, EMER PROF ANAT, IND UNIV, 70- Concurrent Pos: Vis prof, Med Col SC, 61 & Univ West Indies, 70; dir, Muncie Ctr Med Educ, Ball State Univ, 70-75. Mem: Am Asn Anatomists. Res: Vertebrate embryology; gross anatomy and neuroanatomy. Mailing Add: Dept of Anat Ind Univ Sch of Med Indianapolis IN 46202

BOYER, DELMAR LEE, b Salina, Kans, Oct 31, 26; m 61; c 2. MATHEMATICS. Educ: Kans Wesleyan Univ, AB, 49; Univ Kans, MA, 52, PhD(math), 55. Prof Exp: Asst prof math, Ariz State Col, 55-56, NMex State Col, 56-59 & Fresno State Col, 59-61; from asst prof to assoc prof, Univ Idaho, 61-65; PROF MATH, UNIV TEX, EL PASO, 65- Mem: Am Math Soc; Math Asn Am. Res: Abelian groups. Mailing Add: Dept of Math Univ of Tex El Paso TX 79999

BOYER, DON RAYMOND, b Lexington, Okla, Mar 31, 29; m 59; c 3. VERTEBRATE ZOOLOGY. Educ: Univ Okla, BS, 50, MS, 53; Tulane Univ, PhD(zool), 58. Prof Exp: Instr zool, Tulane Univ, 58; assoc prof, 58-66, PROF BIOL, WASHBURN UNIV, 66- Mem: AAAS; Am Soc Ichthyol & Herpet; Am Soc Zool; Am Inst Biol Sci. Res: Vertebrate zoology; physiology of temperature regulation and of respiratory and cardiovascular systemic responses in reptiles. Mailing Add: Dept of Biol Washburn Univ Topeka KS 66621

BOYER, DONALD D, biology, see 12th edition

BOYER, DONALD WAYNE, b Detroit, Mich, Jan 16, 29; m 56; c 4. PHYSICS, AEROPHYSICS. Educ: Univ Melbourne, BSc, 53; Univ Toronto, MASc, 54, PhD(aerophys), 60. Prof Exp: Asst, Inst Aerospace Studies, Univ Toronto, 53-59; res aerodynamicist, 59-63, PRIN AERODYNAMICIST, CALSPAN CORP, 63- Mem: AAAS. Res: Nonisentropic flows; viscous flows; spherical shock waves; experimental facility design; hypersonic aerodynamics; chemical and ionization non-equilibrium phenomena; magneto gas dynamics; plasma-microwave interactions; infrared radiation physics. Mailing Add: 92 Sherbrooke Ave Amherst NY 14221

BOYER, ERNEST WENDELL, b Oconto, Nebr, Mar 18, 37; m 58; c 2. MICROBIAL PHYSIOLOGY, MICROBIAL GENETICS. Educ: NCent Col, Ill, BA, 58; Iowa State Univ, PhD(bact), 69. Prof Exp: Res technician, Mayo Clin, Minn, 62-64; RES MICROBIOLOGIST, MILES LABS, INC, 68- Mem: AAAS; Am Soc Microbiol; Soc Indust Microbiol. Res: Regulation of the synthesis of extracellular enzymes by bacteria; genetic engineering of industrial microorganisms; extracellular enzymes and taxonomy of alkalophilic bacillus species. Mailing Add: Microbiol Res Marschall Div Miles Labs Inc Elkhart IN 46514

BOYER, HAROLD EDWIN, b St Lawrence, Pa, Aug 9, 25; m 48; c 2. ORAL SURGERY. Educ: Lebanon Valley Col, 46-48; Univ Pa, DDS, 52, MS, 58; Am Bd Oral Surg, dipl, 59. Prof Exp: From asst dean to dean, Sch Dent, 62-72, PROF ORAL SURG, UNIV LOUISVILLE, 59-, VPRES HEALTH AFFAIRS, 72- Concurrent Pos: Chief sect oral surg, Louisville Gen & Children's Hosps, 56-71; consult, Jewish & Vet Admin Hosps, 63-, Ireland Army Hosp, Ft Knox, Ky, 63- & Off Econ Opportunity, Proj Head Start, 65-66; pres, Am Bd Oral Surg. Mem: Am Soc Oral Surg; fel Am Col Dent; fel Int Col Dent. Res: Early recognition of oral cancer; oral exfoliative cytology; evaluation of emergency drugs and development of a practical office emergency kit. Mailing Add: Health Sci Ctr Univ of Louisville Louisville KY 40201

BOYER, HERBERT WAYNE, b Latrobe, Pa, July 10, 36; m 59; c 2. MICROBIOLOGY. Educ: St Vincent Col, AB, 58; Univ Pittsburgh, PhD(bact), 63. Prof Exp: From asst prof to assoc prof microbiol, 66-76, PROF DIV GENETICS, DEPT BIOCHEM & BIOPHYS, UNIV CALIF, SAN FRANCISCO, 76- Concurrent Pos: USPHS fel, Yale Univ, 63-66. Res: Molecular genetics. Mailing Add: Dept of Biochem & Biophys Univ of Calif San Francisco CA 94143

BOYER, JOHN FREDERICK, b Evanston, Ill, Oct 14, 41; m 68; c 2. POPULATION BIOLOGY. Educ: Amherst Col, BA, 64; Univ Chicago, PhD(biol), 71. Prof Exp: Fel genetics, Dept Zool, Univ Iowa, 71-73; ASST PROF BIOL, UNION COL, 73- Mem: Soc Study Evolution; Ecol Soc Am; Sigma Xi. Res: Tribolium populations; evolution of life histories; genetic variation. Mailing Add: Dept of Biol Sci Union Col Schenectady NY 12308

BOYER, JOHN STRICKLAND, b Cranford, NJ, May 1, 37; m 64; c 2. PLANT PHYSIOLOGY. Educ: Univ Wis, MS, 61; Duke Univ, PhD(bot), 64. Prof Exp: Vis asst prof bot, Duke Univ, 64-65; asst physiologist, Conn Agr Exp Sta, 65-66; from asst prof to assoc prof, 66-73, PROF BOT & AGRON, UNIV ILL, URBANA, 73- Mem: Am Soc Plant Physiol. Res: Water and salt relations of plants. Mailing Add: Dept of Bot Univ of Ill Urbana IL 61801

BOYER, JOSEPH HENRY, b Otto, Ind, Jan 4, 22. ORGANIC CHEMISTRY. Educ: Univ Ill, PhD(chem), 50. Prof Exp: Instr, Univ Mich, 50-52, res chemist, 51-52; asst prof chem, Tulane Univ, 53-60; asst prog adminr, Am Chem Soc Petrol Res Fund, 61-66; PROF CHEM, UNIV ILL, CHICAGO CIRCLE, 66- Mem: Am Chem Soc. Res: Organic azides; furoxanes; pyridine derivatives; imidazoles; saturated nitrogen heterocycles; tetrazoles. Mailing Add: Dept of Chem Univ of Ill Chicago IL 60680

BOYER, LEE EMERSON, b Gratz, Pa, July 27, 00; m 21; c 2. MATHEMATICS. Educ: Susquehanna Univ, AB, 26; Harvard Univ, EdM, 33; Pa State Univ, EdD, 39. Prof Exp: Teacher rural sch, Pa, 19-23, teacher & prin high sch, 26-27; teacher & supv prin, Tressler Orphans Home, 28-34; teacher math, Pa State Teachers Col, Millersville, 34-56; coordr, Div Math, State Dept Pub Instr, Pa, 56-60, Nat Defense Educ Act State In-Serv Prog, 60-62, dir, Bur State Cols, 63-64; lectr math, Harrisburg Area Community Col, 64-70; CONSULT. Concurrent Pos: Chmn dept math, Pa State Teachers Col, 34-56; prof, Franklin & Marshall Col, 45; head math dept, Lancaster Area Col, 45-47; sci fel, AAAS; 57. Mem: AAAS; Math Asn Am. Res: Curriculum construction in high school; college mathematics for prospective secondary school teachers. Mailing Add: Windsor Farms 1508 Pelham Rd Harrisburg PA 17110

BOYER, MICHAEL GEORGE, b Toronto, Ont, May 15, 25; m 52; c 3. PLANT PATHOLOGY. Educ: Ont Agr Col, BSA, 52, MS, 53; Iowa State Col, PhD(plant path), 58. Prof Exp: Lectr, Kemptville Agr Sch, Ont, 53-54; pathologist, Can Dept Forestry, 54-55, 58-65; asst prof, 65-68, ASSOC PROF BIOL, YORK UNIV, 68- Concurrent Pos: Mem, Can Comt Forest Tree Breeding. Res: Resistant poplar hybrids. Mailing Add: Dept of Biol York Univ Toronto ON Can

BOYER, NORMAN HOWARD, b Cleveland, Ohio, Nov 19, 08; m 34; c 2. MEDICINE. Educ: Tufts Univ, MD, 34. Prof Exp: Asst cardiol, Mass Gen Hosp, Boston, 39-41; from instr to assoc prof med, Sch Med, 41-75, FRAMINGHAM HEART STUDY, DEPT HEALTH, EDUC & WELFARE, BOSTON UNIV, 75- Concurrent Pos: Teaching fel physiol, Sch Med, Western Reserve Univ, 38-39. Mem: Fel Am Col Physicians; Am Heart Asn. Res: Clinical and physiological research on heart and circulation. Mailing Add: Dept of Med Boston Univ Boston MA 02118

BOYER, PAUL DELOS, b Provo, Utah, July 31, 18; m 39; c 3. BIOCHEMISTRY. Educ: Brigham Young Univ, BS, 39; Univ Wis, MS, 41, PhD(biochem), 43. Prof Exp: Asst biochem res, Univ Wis, 39-43; instr chem, Stanford Univ, 43-45, res assoc, 43-45; from asst prof to prof biochem, Univ Minn, 45-56, prof physiol chem, 56-63; PROF CHEM, UNIV CALIF, LOS ANGELES, 63- Concurrent Pos: Guggenheim fel, 55-56; mem, Biochem Study Sect, US Pub Health Serv, 58-62, chmn, 64- Honors & Awards: Paul Lewis award. Mem: Nat Acad Sci; AAAS; Am Soc Biol Chem; Am Chem Soc; Am Soc Exp Biol & Med. Res: Chemistry and mechanism of action of enzymes; isotopic studies of biological reaction mechanisms; oxidative phosphorylation; deoxyribonucleic acid structure in relation to function; protein structure and function relationships. Mailing Add: Dept of Chem Univ of Calif Los Angeles CA 90024

BOYER, PAUL S, geology, paleontology, see 12th edition

BOYER, PHILIP A, JR, b Philadelphia, Pa, July 9, 16; m 39; c 4. MEDICINE. Educ: Pa State Univ, BS, 36; Univ Pa, MD, 40. Prof Exp: Resident, Oakland County Tuberc Sanit, Mich, 41-42, Sunny Acres Sanit, Ohio, 42-44 & St Joseph's Mercy Hosp, Mich, 44-46; assoc med dir, Roosevelt Hosp, NJ, 48-49; assoc dir clin develop, Lederle Labs, 49-51; assoc med dir, Schenley Labs, 51-53; dir clin res, Pitman-Moore Co, 53-60; med dir, Plough, Inc, 60-61; dir clin res, Hoffmann-La Roche, Inc, 61-66; DIR DEPT MED, MARION LABS, INC, 67- Mem: AMA; Am Col Chest Physicians; Am Fedn Clin Res; Am Geriat Soc; Acad Psychosom Med. Res: Administration; new drug research. Mailing Add: Dept of Med Marion Labs, Inc Kansas City MO 64137

BOYER, RAYMOND FOSTER, b Denver, Colo, Feb 6, 10; m 36, 66; c 4. PHYSICS. Educ: Case Inst Technol, BS, 33, MS, 35. Hon Degrees: DSc, Case Inst Technol, 55. Prof Exp: Asst physics, Case Inst Technol, 33-35; mem staff, 35-45, asst dir phys res lab, 45-48, dir, 48-52, plastics res, 52-68, ASST DIR CORP RES & DEVELOP POLYMER SCI, DOW CHEM CO, 68- Concurrent Pos: Chmn adv panel to polymer div, Nat Bur Standards. Honors & Awards: Gold Medal & Int Award in Polymer Sci & Eng, Soc Plastics Eng, 68; Borden Award, Am Chem Soc, 70. Mem: AAAS; Soc Plastics Eng; Am Inst Chem; Am Chem Soc; Am Phys Soc. Res: Physics and physical chemistry of polystyrene family of high polymers; glass transition and related transitions or relaxations in polymers. Mailing Add: 415 W Main St Midland MI 48640

BOYER, ROBERT ALLEN, b Hummels Wharf, Pa, Aug 27, 16; m 39; c 2. PHYSICS. Educ: Susquehanna Univ, AB, 38; Syracuse Univ, MA, 40; Lehigh Univ, PhD(physics), 52. Prof Exp: Instr physics, Clarkson Col, 40-41; PROF PHYSICS & CHMN DEPT, MUHLENBERG COL, 41- Honors & Awards: Lindback Award, 61. Mem: Am Asn Physics Teachers. Res: Ultrasonics. Mailing Add: 20 Beverly Dr Allentown PA 18104

BOYER, ROBERT ERNST, b Palmerton, Pa, Aug 3, 29; m 51; c 3. GEOLOGY. Educ: Colgate Univ, BA, 51; Ind Univ, MA, 54; Univ Mich, PhD, 59. Prof Exp: From instr geol to assoc prof geol, 57-67, PROF GEOL SCI & EDUC, 67-, UNIV TEX, 67-, CHMN DEPT, 71- Concurrent Pos: Ed, Tex J Sci, 62-64 & J Geol Educ, 65-68. Mem: Fel Geol Soc Am; Am Asn Petrol Geol; Am Asn Geol Teachers; Am Soc Photogram. Res: Structural and field geology; mapping in Precambrian crystalline rocks of Wet Mountains, Colorado and Llano Region, Texas; heavy mineral studies of crystalline rocks; fracture pattern analysis; earth science teaching in secondary schools; space photography and remote sensing. Mailing Add: Dept of Geol Sci Univ of Tex Austin TX 78712

BOYER, RUTH MCDONALD, b Calls Fort, Utah, Oct 16, 18; m 39; c 3. ANTHROPOLOGY. Educ: Univ Calif, Berkeley, BA, 50, PhD(anthrop), 62. Prof Exp: Asst prof design, Univ Calif, Berkeley, 62-70, lectr, 71-72; PROF ANTHROP, HIST OF TEXTILES & COSTUME, CALIF COL ARTS & CRAFTS, 73- Concurrent Pos: Univ Calif, Berkeley fac res grants fieldwork in Peru, 66 & collection info for biog NMex Apache Woman, 69-70; res assoc, Lowie Mus Anthrop, Univ Calif, Berkeley, 72-76; Nat Endowment Arts grant study of Kashmir shawls, 74-75. Mem: Fel Am Anthrop Asn; fel Soc Appl Anthrop; Inst Andean Studies; Am Folklore Soc. Res: Socialization and personality studies among Athabascans; socialization and personality studies among Eskimos; worldwide folk arts, textiles, costumes, historical and contemporary. Mailing Add: Calif Col of Arts & Crafts Broadway at College Oakland CA 94618

BOYER, SAMUEL H, IV, b Duluth, Minn, Aug 16, 24; m 52; c 3. MEDICAL GENETICS. Educ: Stanford Univ, AB, 50, MD, 54. Prof Exp: From asst prof to assoc prof, 59-71, PROF MED & BIOL, SCH MED, JOHNS HOPKINS UNIV, 71- Concurrent Pos: USPHS fels, Johns Hopkins Univ, 56-58, Univ Mich, 58 & Univ Col, Univ London, 58-59; ed, Johns Hopkins Med J, 75- Mem: Am Soc Human Genetics (secy, 62-64); Am Soc Clin Invest. Res: Human biochemical genetics; genetic polymorphisms; comparative biochemistry of protein; regulation of protein synthesis; biology of single cells. Mailing Add: Traylor Bldg Johns Hopkins Univ Sch of Med 720 Rutland Ave Baltimore MD 21205

BOYER, TIMOTHY HOWARD, b New York, NY, Mar 20, 41. THEORETICAL PHYSICS. Educ: Yale Univ, BA, 62; Harvard Univ, MA, 63, PhD(physics), 68. Prof Exp: Fel physics, Univ Md, 68-70; asst prof, 70-73, ASSOC PROF PHYSICS, CITY COL NEW YORK, 74- Mem: AAAS; Am Phys Soc. Res: Relations between classical and quantum theories; classical electron theory; quantum zero-point energy; statistical thermodynamics; long-range van der Waals forces. Mailing Add: Dept of Physics City Col of New York New York NY 10031

BOYER, WILLIAM DAVIS, b Dayton, Ohio, Sept 27, 24; m 52; c 4. FOREST ECOLOGY. Educ: US Merchant Marine Acad, BS, 50; State Univ NY, BS, 51, MS, 54; Duke Univ, PhD, 70. Prof Exp: Res asst water resources & power, Comn Orgn of Exec Br of Govt, 54-55; RES FORESTER, USDA, 55- Mem: Ecol Soc Am; Soc Am Foresters; Am Geog Soc. Res: Silvicultural problems of southern pine. Mailing Add: 1223 Juniper Dr Auburn AL 36830

BOYER, WILLIAM MONTGOMERY, b Chicago, Ill, Aug 26, 26; m 48; c 2. ORGANIC CHEMISTRY. Educ: Ill Inst Technol, BS, 48, MS, 50. Prof Exp: Res supvr, W A Erickson & Co, 50-55; supvr pioneering res, Richardson Co, Ill, 55-62; sect mgr, Petrolite Corp, Mo, 62-64; dir res, Fiberfil Inc, 64-66, opers mgr, 66-67, exec vpres, Fiberfil Div, Dart Industs, Inc, 67-68, pres, 68-72; EXEC VPRES, VENTRON CORP, 72- Mem: Am Chem Soc. Res: Organic surfactant chemistry; acrylic, organosilicon and organophosphorus polymers; organosodium chemistry. Mailing Add: Ventron Corp Congress St Beverly MA 01915

BOYER, WILLIAM PAUL, b Escondido, Calif, Oct 28, 09; m 37; c 2. ENTOMOLOGY. Educ: Univ Ark, BSA, 37, MS, 57. Prof Exp: From asst county agent to county agent, Univ Ark, 37-51; farm mgr, Elms Planting Corp, 52-55; SURV ENTOMOLOGIST, UNIV ARK, FAYETTEVILLE, 57- Mem: Entom Soc Am. Res: Cooperative insect survey; survey methods. Mailing Add: A-312 Univ of Ark Fayetteville AR 72701

BOYER, WILLIAM PRESTON, organic chemistry, see 12th edition

BOYERS, HAROLD HENRY, dentistry, deceased

BOYES, JOHN WALLACE, b Sundridge, Ont Ont, Jan 27, 07; m 37; c 4. GENETICS. Educ: Univ Sask, BSc, 33, MSc, 36; Univ Wis, PhD(genetics, bot), 39. Prof Exp: Jr res cytogeneticist, Nat Res Coun Can, 39-40; asst prof genetics & plant breeding, Univ Alta, 40-42; mem, Munitions & Indust Alcohol Comn, 42-45; from asst prof to assoc prof genetics, 45-56, from actg chmn dept to chmn dept, 45-69, MOLSON PROF GENETICS, McGILL UNIV, 56- Concurrent Pos: Pres genetics sect, Int Union Biol Sci, 58-63. Honors & Awards: Mendel Mem Silver Medal, 65; Moravian Mus Medal, 70. Mem: Bot Soc Am; Genetics Soc Am; Am Soc Naturalists; Soc Study Evolution; Am Genetics Asn. Res: Interspecific hybrids in wheat; polyploidy; plant cytology; human heredity; cytology of higher flies. Mailing Add: Dept of Biol McGill Univ Montreal PQ Can

BOYES, ROBERT NICHOL, pharmaceutical chemistry, biochemistry, see 12th edition

BOYET, HOWARD, b New York, NY, Aug 2, 24. PHYSICS. Educ: City Col New York, BS, 44; NY Univ, PhD(physics), 53. Prof Exp: Physicist aerodyn, Nat Adv Comt Aeronaut, 44-47; instr physics, NY Univ, 48-53; mem tech staff, Bell Tel Labs, 53-57; sr proj mem tech staff, Radio Corp Am, 57-62; PROF ELEC ENG, PRATT INST, 62- Concurrent Pos: Res asst physics, Inst Math Sci, 52-53. Mem: Am Phys Soc. Res: Microwaves and solid state devices at microwave frequencies; quantum electronics. Mailing Add: Dept of Elec Eng Pratt Inst Brooklyn NY 11205

BOYETTE, JOSEPH GREENE, b Colerain, NC, May 10, 29; m 53; c 3. VERTEBRATE ECOLOGY, ACADEMIC ADMINISTRATION. Educ: E Carolina Col, BS, 56, MA, 57; NC State Univ, PhD, 66. Prof Exp: Chemist, NC State Dept Agr, 55-56; teacher, high sch, NC, 56-57; instr sci, 57-60, from asst prof to assoc prof, 60-69, asst dean grad sch, 70-73, PROF BIOL, E CAROLINA UNIV, 69-, DEAN

GRAD SCH, 73- Mem: Fel AAAS; Am Soc Mammalogists. Res: Behavior of the pine mouse. Mailing Add: Grad Sch E Carolina Univ Greenville NC 27834

BOYKIN, DAVID WITHERS, JR, b Montgomery, Ala, Jan 6, 39; m 59; c 1. ORGANIC CHEMISTRY. Educ: Univ Ala, BS, 61, MS, 63, PhD(org chem), 65. Prof Exp: Fel org chem, Univ Va, 66-67; asst prof, 65-66, 67-68, assoc prof, 68-72, PROF ORG CHEM, GA STATE UNIV, 72-, CHMN DEPT, 74- Concurrent Pos: Walter Reed Army Inst Res contract, 67- Mem: Am Chem Soc. Res: Reaction mechanisms of heterocycles, small rings and alpha, beta-unsaturated ketones; synthetic antimalarials. Mailing Add: Dept of Chem Ga State Univ Univerity Plaza Atlanta GA 30303

BOYKIN, LORRAINE STITH, b Crewe, Va, Feb 1, 31; m 53. NUTRITION. Educ: Va State Col, BS, 51, MS, 54; NY Univ, MA, 59; Long Island Univ, MS, 64; Columbia Univ, prof dipl nutrit, 67, EdD(nutrit), 70. Prof Exp: Dietitian, New York City Dept Hosps, 51-52, A&T Col NC, 52-53 & NY Univ Hosp, 54-55; nutritionist, New York City Dept Health, 55-64; teacher home econ, New York Bd Educ, 64-65; from instr to asst prof nutrit, Pratt Inst, 65-70; ASST PROF NUTRIT, HUNTER COL, 70- Concurrent Pos: Adj asst prof, NY Univ, 74- Mem: Fel AAAS; fel NY Acad Sci; fel Am Geriat Soc; fel Am Pub Health Asn; fel Royal Soc Health. Res: Geriatrics and nutrition; alcoholism and nutrition; cultural nutrition. Mailing Add: Bellevue Sch of Nursing Hunter Col 440 E 26th St New York NY 10010

BOYKINS, ERNEST ALOYSIUS, JR, b Vicksburg, Miss, Oct 5, 31; m 55; c 4. VERTEBRATE ZOOLOGY. Educ: Xavier Univ La, BS, 53; Tex Southern Univ, MS, 58, Mich State Univ, PhD(zool), 64. Prof Exp: Asst prof biol, Alcorn Agr & Mech Col, 54-61, prof, 64-71, dir div arts & sci, 70-71; PRES, MISS VALLEY STATE UNIV, 71- Mem: AAAS; Am Ornith Union. Res: Effects of dichlorodiphenyltrichloroethane on non-target animals. Mailing Add: Miss Valley State Univ PO Box 1178 Itta Bena MS 38941

BOYKO, EDWARD RAYMOND, b Passaic, NJ, May 16, 30. PHYSICAL CHEMISTRY. Educ: Univ Pa, AB, 51; Rutgers Univ, MS, 54, PhD(chem), 56. Prof Exp: Res assoc fuels res sect, Knolls Atomic Power Lab, Gen Elec Co, 55-57; sr engr solid state, Westinghouse Elec Corp, 57-59; suvpr solid state physics, Res Div, Am-Standard, 59-61; asst prof, 61-68, PROF PHYS CHEM, PROVIDENCE COL, 68- Mem: Am Crystallog Asn. Res: Crystal physics; relation of crystal structure to physical properties; solid state physics. Mailing Add: Dept of Chem Providence Col Providence RI 02910

BOYLAN, EDWARD, b New York, NY, Feb 8, 38. MATHEMATICS. Educ: Columbia Univ, AB, 59; Princeton Univ, MA, 61, PhD(math), 62. Prof Exp: Res assoc math, grad sch sci, Yeshiva Univ, 62-63; asst prof math, Rutgers Univ, 63-66; assoc prof math, Hunter Col, 66-68; ASSOC PROF MATH, RUTGERS UNIV, 68- Concurrent Pos: Consult, Hudson Inst, NY, 70- Mem: Am Math Soc. Res: Probability and inventory theory; possible connections between ergodic theory and martingale theory; strategic issues. Mailing Add: Dept of Math, Rutgers Univ Newark NJ 07102

BOYLAN, ELIZABETH SHIPPEE, b Shanghai, China; US citizen. VERTEBRATE EMBRYOLOGY, ONCOLOGY. Educ: Wellesley Col, AB, 68; Cornell Univ, PhD(zool), 72. Prof Exp: Teaching asst embryol, Cornell Univ, 68-72; trainee fertil & reprod physiol, Marine Biol Lab, Woods Hole, 70-71; res assoc biochem & oncol, Med Ctr, Univ Rochester, 72-73; ASST PROF BIOL, QUEENS COL, CITY UNIV NEW YORK, 73- Concurrent Pos: Nat Cancer Inst res grant, 75. Mem: Soc Develop Biol; Soc Study Fertil; Am Soc Zoologists; AAAS. Res: Effect of hormones on fetal tissue and its relation to carcinogenesis. Mailing Add: Dept of Biol Queens Col City Univ NY Flushing NY 11367

BOYLAN, FRANCIS JOSEPH, physical chemistry, see 12th edition

BOYLAN, JAMES CHARLES, b Elyria, Ohio, Jan 29, 37; m 59; c 2. PHARMACY. Educ: Ohio Northern Univ, BS, 59; Purdue Univ, MS, 61, PhD(indust pharm), 63. Prof Exp: Sr pharmaceut chemist, 63-69, head, liquid-ointment/parenteral prod pilot plants, 69-72, HEAD, PARENTERAL PROD DEVELOP, ELI LILLY & CO, 72- Concurrent Pos: Adj assoc prof, col of pharm, Univ Ky, 74- Mem: Am Pharmaceut Asn; Acad Pharmaceut Sci (vpres, 74-75). Res: Stability of pharmaceutical products; pharmaceutical product formulation, including injectables, topicals and tablets; rheology; oral sustained release drug product. Mailing Add: Eli Lilly & Co Indianapolis IN 46206

BOYLAN, JOHN W, b NY, Nov 25, 14; m 40; c 6. PHYSIOLOGY, MEDICINE. Educ: Georgetown Univ, BS, 37; Long Island Col Med, MD, 43; Am Bd Internal Med, dipl. Prof Exp: Asst prof, 55-60, PROF MED & PHYSIOL, SCH MED, STATE UNIV NY, BUFFALO, 60-, HEAD KIDNEY SECT, 62- Concurrent Pos: Spec res fel, Dept Exp Med, Cambridge Univ, 54-55 & Physiol Inst, Univ Göttingen, 60-61; attend physician, Edward J Meyer Mem Hosp, 59-; consult physician, Buffalo Gen Hosp & Vet Admin Hosp, 62-; vis scientist, Physiol Inst, Univ Munich, 67-68 & 73-74. Mem: Fel Am Col Physicians; Am Physiol Soc; Soc Exp Biol & Med; NY Acad Sci. Res: Body fluids; kidney function; kidney disease. Mailing Add: Dept of Med State Univ of NY Buffalo NY 14214

BOYLAN, WILLIAM J, b Bozeman, Mont, Dec 25, 29; m 58; c 4. ANIMAL BREEDING, QUANTITATIVE GENETICS. Educ: Mont State Univ, BS, 52; Univ Minn, MS, 59, PhD(genetics), 62. Prof Exp: From asst prof to assoc prof quant genetics, Univ Man, 62-66; assoc prof, 66-71, PROF ANIMAL SCI, UNIV MINN, ST PAUL, 71- Concurrent Pos: Sabbatical leave, Animal Breeding Res Orgn, Edinburgh, Scotland, 72-73. Mem: Am Soc Animal Sci; Genetics Soc Am; Am Genetic Asn. Res: Quantitative inheritance in animals; design and evaluation of systems of selection and crossbreeding with emphasis on genetic improvement of livestock. Mailing Add: Dept of Animal Sci Univ of Minn St Paul MN 55108

BOYLE, ANDREW JOSEPH, chemistry, deceased

BOYLE, DAVID JOSEPH, b Dubuque, Iowa, July 18, 41; m 69; c 1. EXPERIMENTAL SOLID STATE PHYSICS. Educ: Loras Col, BS, 63; Iowa State Univ, PhD(physics), 68. Prof Exp: Fel physics, Univ BC, 68-70; ASSOC PROF PHYSICS, THOMAS MORE COL, 70- Mem: Am Phys Soc; Am Asn Physics Teachers. Res: Mössbauer effect in monocrystalline ferrocene; educational research into means of teaching astronomy, electronics and physical science. Mailing Add: Thomas More Col Box 85 Covington KY 41017

BOYLE, EDWIN, b Sumter, SC, Sept 27, 23; m 51; c 4. MEDICINE. Educ: Univ NC, BA, 43; Jefferson Med Col, MD, 47. Prof Exp: Intern med, Philadelphia Gen Hosp, 47-48; asst resident, Watts Hosp, Durham, NC, 48-49; resident, Univ Hosp, Univ Va, 49-51; sr clin investr, NIH, 51-55; estab investr, Am Heart Asn, 56-61; assoc med, Med Col SC, 58-63, asst prof res med, 64-66; DIR RES, MIAMI HEART INST, 66- Concurrent Pos: Mem exec comt, Coun Arteriosclerosis, Am Heart Asn, 63-, mem, Coun Thrombosis & Coun Epidemiol; mem policy bd, Nat Coronary Drug Study. Mem: Am Heart Asn; Soc Exp Biol & Med; Am Diabetes Asn; Microcirc Soc; Am Col Cardiol. Res: Internal medicine and epidemiology as related to metabolism, atherosclerosis and occlusive vascular disease. Mailing Add: Miami Heart Inst 4701 N Meridian Ave Miami Beach FL 33140

BOYLE, JAMES MARTIN, b Edwardsville, Ill, June 21, 42; m 70. COMPUTER SCIENCES. Educ: Northwestern Univ, BS, 64, MS, 65, PhD(appl math), 70. Prof Exp: Res assoc, 67-69, asst comput scientist, 69-74, COMPUT SCIENTIST APPL MATH, ARGONNE NAT LAB, 74- Mem: Asn Comput Mach; Soc Indust & Appl Math; Sigma Xi. Res: Source-to-source transformations of computer programs, multiple realizations of computer programs; program correctness; mathematical software. Mailing Add: Appl Math Div Bldg 221 Argonne Nat Lab Argonne IL 60439

BOYLE, JAMES REID, soil science, forest ecology, see 12th edition

BOYLE, JOHN JOSEPH, b Middleport, Pa, Nov 23, 30; m 59. VIROLOGY. Educ: Pa State Univ, BS, 59, MS, 60, PhD(bact), 62. Prof Exp: Res biologist, Sterling-Winthrop Res Inst, 62; microbiologist, US Army Biolabs, 62-65; virologist, Smith Kline & French Labs, 65, sr virologist, 65-70; PROD SUPT VIRAL & BACT VACCINES, MERCK SHARP & DOHME, 70- Mem: AAAS; Am Soc Microbiol; Sigma Xi; NY Acad Sci. Res: Viral chemotherapy; arboviruses; viral and rickettsial diseases; virus vaccines. Mailing Add: Merck Sharp & Dohme West Point PA 19486

BOYLE, JOHN SAMUEL, b Laverne, Okla, June 7, 17; m 41; c 3. PLANT PATHOLOGY. Educ: Munic Univ Wichita, BA, 39; Univ Iowa, MS, 42; Univ Wis, PhD(plant path), 49. Prof Exp: Asst plant path, Univ Wis, 46-49; from asst prof to assoc prof, 49-64, PROF PLANT PATH, PA STATE UNIV, 64- Concurrent Pos: Fulbright lectr, Univ Assiut, 64-65. Mem: AAAS; Am Phytopath Soc; Brit Asn Appl Biol. Res: Diseases of deciduous fruit; virus diseases of plants; the influence of the host on virus variability. Mailing Add: Dept of Plant Path Buckhout Lab Pa State Univ University Park PA 16802

BOYLE, MARY MAURICE, b Phila, Pa, Dec 27, 22. PHYSICS, BIOPHYSICS. Educ: Villanova Univ, BA, 52, Cath Univ Am, MS, 59; St Thomas Inst, PhD(physics, biophys), 70. Prof Exp: Teacher elem parochial schs, Pa, 43-45, high sch, 45-59; mem fac math & sci, 59-63, actg registr & dir admissions, 61-63, mem fac physics & chem, 70-74, ADJ FAC EDUC COORDR ALLIED HEALTH PROGS, GWYNEDD-MERCY COL, 74- Mem: Am Physics Teachers Asn; Radiation Res Soc. Res: Effects of irradiation on enzymes; relationship of effects of chemotherapeutic agents and radiation on enzymes involved in the cancer problem. Mailing Add: Gwynedd-Mercy Col Gwynedd Valley PA 19437

BOYLE, PAUL EDMUND, b Somerville, Mass, Apr 20, 00; m 21, 37; c 3. PATHOLOGY. Educ: Harvard Univ, DMD, 23; Am Bd Oral Path, dipl. Prof Exp: Asst oper dent, Harvard Dent Sch, 26-28, instr oper dent, 28-41, instr path, 32-41, asst prof oral path & clin dent, 41-44; prof oral histol & path, Univ Pa, 44-56; prof, 56-70, dean sch dent, 56-69, EMER PROF DENT MED, SCH DENT & EMER PROF ORAL PATH, SCH MED, CASE WESTERN RESERVE UNIV, 70-, EMER DEAN SCH DENT, 69- Concurrent Pos: Oral pathologist, Children's Hosp, Boston, Mass, 43-44, stomatologist, 44; dentist, Peter Bent Brigham Hosp, 44; consult, Am Oncol Hosp, Philadelphia, Pa, 44-46; prof, grad sch med, Univ Pa, 45-56; consult, Crile Hosp & Vet Admin, Cleveland, Ohio, 58-71. Honors & Awards: Callahan Mem Award, Am Bd Oral Path, 66. Mem: Fel AAAS (vpres, 64); Am Dent Asn; fel Am Acad Oral Path (pres, 49); Am Asn Path & Bact; Int Asn Dent Res (pres, 55). Res: Histopathology of teeth and other oral tissues; effect of vitamin deficiencies on development and structure of teeth and jaws; experimental neoplastic disease. Mailing Add: 261 Landham Rd Sudbury MA 01776

BOYLE, RICHARD JAMES, b Westport, NY, Apr 19, 27; m 55; c 6. INDUSTRIAL ORGANIC CHEMISTRY. Educ: Univ Del, BS, 50; Univ Notre Dame, PhD(chem), 53. Prof Exp: Res chemist, Calco Chem Div, 53-59, group leader, Plant Tech Dept, 59-61, tech rep, Res Dept, 61-62, mgr sales develop, Textile Chem Dept, Org Chem Div, 62-67, prod mgr, Market Develop Dept, 67-69, prod mgr, Decision Making Systs Dept, 69, field sales supvr, 69-70, tech dir, Indust Prod Dept, Europe-African Region, Cyanamid Int, 70-73, prod mgr, Indust & Mining Chems, 73-74, PROD MGR, ORGANIC CHEMS, CYANAMID EUROPE-MIDEAST-AFRICA DIV, AM CYANAMID CO, 74- Mem: Am Chem Soc. Res: Analogs of podophyllotoxin; vat dyes; ultraviolet absorbers; textile chemicals; luminescent chemicals; elastomers. Mailing Add: Indust Prod Dept Am Cyanamid Co Wayne NJ 07470

BOYLE, ROBERT E, biochemistry, see 12th edition

BOYLE, ROBERT WILLIAM, b Wallaceburg, Ont, June 3, 20; m 45; c 2. ECONOMIC GEOLOGY, GEOCHEMISTRY. Educ: Univ Toronto, BASc, 49, MASc, 50, PhD(econ geol, geochem), 53. Prof Exp: Geologist, 53-67, prin res scientist, 67-70, RES SCIENTIST SPEC PROJS, GEOL SURV CAN, 70- Concurrent Pos: Spec lectr, Carleton Univ, 63- Honors & Awards: Barlow Medal, Can Inst Mining & Metal, 66; Miller Medal, Royal Soc Can, 71; Pub Serv Can Merit Award, Govt Can, 71. Mem: Soc Econ Geol; Am Geochem Soc; fel Royal Soc Can; Mineral Asn Can; Asn Explor Geochem (pres, 75-76). Res: Processes of ore deposition; geochemical prospecting; isotope geochemistry; metallogenesis. Mailing Add: 601 Booth St Ottawa ON Can

BOYLE, ROBERT WILLIAM, b St Paul, Minn, Feb 11, 08; m 31; c 3. PHYSIOLOGY. Educ: YMCA Col, BPE, 29, MPE, 31; Univ Chicago, PhB, 30; Col City Detroit, MS, 32; Univ Minn, PhD(physiol), 36; Univ Ark, MD, 40. Prof Exp: Instr phys educ, YMCA, Wis, 30-31; spec lectr, Col City Detroit, 31-32; from asst prof to prof physiol & pharm, Sch Med, Univ Ark, 36-44; chief phys med rehab serv, Vet Admin Hosp, 47-54; from asst prof to prof phys med & rehab, Sch Med, Marquette Univ, 54-65; chmn dept, 64-73, PROF PHYS MED & REHAB, MED COL WIS, 65- Concurrent Pos: Baruch fel, Mayo Clin & Bellevue Hosp, NY, 46-47; from actg head to head dept physiol & pharm, Sch Med, Univ Ark, 41-44; dir dept phys med & rehab, Milwaukee County Hosp. Mem: Am Acad Phys Med & Rehab (pres, 61-62, secy, 67-); fel Am Col Physicians; fel Am Col Chest Physicians; Am Cong Rehab Med. Res: Physiology of exercise; effects of toxins on the heart; physiological problems in physical medicine. Mailing Add: Dept of Phys Med & Rehab Med Col of Wis Wauwatosa WI 53226

BOYLE, WALTER GORDON, JR, b Seattle, Wash, June 27, 28; m 58; c 2. ANALYTICAL CHEMISTRY. Educ: Univ Wash, BS, 50, PhD(anal chem), 56. Prof Exp: RES CHEMIST, LAWRENCE LIVERMORE LAB, 56-, GROUP LEADER, 63- Mem: Am Chem Soc. Res: Determination of trace elements in metals and alloys; atomic absorption and flame emission spectroscopy; computer applications in analytical chemistry. Mailing Add: Lawrence Livermore Lab Chem Dept L-404 Livermore CA 94550

BOYLE, WILLARD STERLING, b NS, Aug 19, 24; m 46; c 4. PHYSICS. Educ:

McGill Univ, PhD(physics), 50. Prof Exp: Lectr physics, McGill Univ, 50-51; asst prof, Royal Mil Col, 51-53; mem tech staff, Bell Tel Labs, 53-62, dir explor studies, Bellcomm, Inc, 62-64, dir semiconductor device lab, Bell Tel Labs, 64-68, EXEC DIR, BELL TEL LABS, 68- Honors & Awards: Stuart Ballantine Medal Award, Franklin Inst, 73; Morris N Liebmann Award, Inst Elec & Electronics Engrs, 74. Mem: Nat Acad Eng; fel Inst Elec & Electronics Engrs; fel Am Phys Soc. Res: Study of electronic devices and systems relation to communications; solid state physics; semiconductors. Mailing Add: Bell Tel Labs Holmdel NJ 07733

BOYLE, WILLIAM SIDNEY, b Provo, Utah, May 27, 15; m 37; c 3. BOTANY. Educ: Brigham Young Univ, BS, 37; Univ Calif, MS, 39, PhD(bot), 43. Prof Exp: Jr chemist, Western Regional Res Lab, Bur Agr & Indust Chem, USDA, 43-45; from asst prof to assoc prof, 45-55, PROF BOT, UTAH STATE UNIV, 55- Mem: Am Genetic Asn; Radiation Res Soc. Res: Cytology, taxonomy and genetics of Gramineae; plant breeding; radiobiology. Mailing Add: Dept of Bot Utah State Univ Logan UT 84321

BOYLEN, JOYCE BEATRICE, b Weston, Ont, Aug 15, 26. BIOCHEMISTRY. Educ: Univ Toronto, BA, 47, MA, 49; McGill Univ, PhD(biochem), 60. Prof Exp: Jr res officer biol, Atomic Energy Can, Ltd, 49-53; Nuffield res asst genetics, Univ Leicester, 53-56; res asst biochem, McGill Univ, 57-60; sect head tissue culture, Frank W Horner, Ltd, Can, 60-63, virol, 63-64; res biochemist, 64-68, CLIN INFO SCIENTIST, MED RES DEPT, ICI UNITED STATES, INC, 68- Mem: Drug Info Asn; Can Biochem Soc. Res: Clinical studies of new drugs. Mailing Add: Med Res Dept ICI United States Wilmington DE 19899

BOYLES, JAMES GLENN, b Harrisburg, Pa, Mar 21, 37; m 60; c 2. PHYSICAL CHEMISTRY. Educ: Pa State Univ, BS, 59; Rutgers Univ, PhD(chem), 66. Prof Exp: Instr chem, Rutgers Univ, 65-66; asst prof, 66-73, ASSOC PROF CHEM, BATES COL, 73- Mem: AAAS; Am Chem Soc. Res: Chemical reaction kinetics and mechanisms, particularly coordination compounds; analysis of pesticides and herbicides. Mailing Add: Dept of Chem Bates Col Lewiston ME 04240

BOYLES, JAMES MCGREGOR, b Birmingham, Ala, Oct 1, 26; m 54; c 2. MEDICAL PARASITOLOGY. Educ: Univ Ala, BS, 51, MS, 52, PhD(biol), 66. Prof Exp: Instr biol, Mobile Ctr, Univ Ala, 56-64; from instr to asst prof, 64-71, ASSOC PROF BIOL, UNIV S ALA, 71- Mem: Am Soc Parasitol; Am Soc Trop Med & Hyg. Res: Amebic meningoencephalitis. Mailing Add: Dept of Biol Univ of SAla Mobile AL 36688

BOYLES, MARCIA V, physiology, see 12th edition

BOYNE, PHILIP JOHN, b Houlton, Maine, May 1, 24; m 46; c 2. ORAL SURGERY, ANATOMY. Educ: Tufts Univ, DMD, 47; Georgetown Univ, MS, 61. Prof Exp: Instr anat, Tufts Univ, 47-49; resident oral surg, US Naval Dent Sch, Nat Naval Med Ctr, 54-55, staff guest scientist, US Naval Med Res Inst, 55-57, dir dent res dept, 65-68; prof dent, Sch Dent, Univ Calif, Los Angeles, 68-69, prof oral surg & chmn div, 69-74, asst dean, Sch Dent, 71-74; DIR BONE RES LAB, UNIV TEX HEALTH SCI CTR SAN ANTONIO, DEAN DENT SCH & PROF ORAL SURG, DENT & MED SCHS, 74- Concurrent Pos: Guest lectr, Sch Dent, Georgetown Univ, 60- & US Naval Dent Sch, Nat Naval Med Ctr, 65-; mem adv comt, Am Bd Oral Surg, 66- Mem: Am Dent Asn; Am Soc Oral Surg; fel Am Col Dent; Int Asn Dent Res. Res: Oral surgical research; bone healing; response of osseous tissues to trauma bone growth and development; tissue transplantation in maxillo- facial surgery. Mailing Add: Univ of Tex Dent Sch 7703 Floyd Curl Dr San Antonio TX 78284

BOYNO, JOHN S, b Newark, NJ, Dec 8, 46; m 39; c 1. PHOTOGRAPHIC CHEMISTRY. Educ: Rutgers Univ, AB, 68; Univ Rochester, PhD(chem), 72. Prof Exp: Res assoc nuclear physics, Cyclotron Lab, Mich State Univ, 72-74; RES CHEMIST, EASTMAN KODAK RES LAB, 74- Mem: Sigma Xi. Res: Photographic technology. Mailing Add: Eastman Kodak Co B 59 Kodak Park Rochester NY 14650

BOYNTON, CHARLES F, physical chemistry, see 12th edition

BOYNTON, DAMON, b Chicago, Ill, Sept 27, 08; m 31; c 3. TROPICAL AGRICULTURE. Educ: Cornell Univ, BS, 31, PhD(pomol), 37. Prof Exp: From instr to prof pomol, Cornell Univ, 35-64, dean grad sch, 59-64; dean grad sch & sr adv, UN Spec Fund Proj, Inter-Am Inst Agr Sci, Costa Rica, 64-66; vis prof & acad adv, NC State Univ Mission, Agrarian Univ, Peru, 66-67, res adv, Ministry of Agr, 67-69, chief of party, 70-72; res adv, Ministry of Agr, El Salvador, 72-74; vis prof, Cornell Univ, 74-75; SR ADV, TROP CTR TRAINING & RES, COSTA RICA, 75- Concurrent Pos: Cornell traveling fel, Eng, 39; Guggenheim fel, Univ Calif, 45-46; dir res & teaching & head plant indust dept, Inter-Am Inst Agr Sci, Costa Rica, 54-55; consult, Rockefeller Bros Fund, Brazil, 55 & 58; ed, Proc Am Hort Soc, 56-59. Mem: Fel AAAS; Am Soc Plant Physiologists; Am Hort Soc; Am Soc Agron; Soil Sci Soc Am. Res: Plant physiology; soil science; fruit culture; farming systems; agricultural development research organization. Mailing Add: RFD 2 Ovid NY 14521

BOYNTON, JOHN E, b Duluth, Minn, June 3, 38. GENETICS. Educ: Univ Ariz, BS, 60; Univ Calif, Davis, PhD(genetics), 66. Prof Exp: NIH fel genetics, Univ Calif, Davis, 66 & Inst Genetics, Copenhagen Univ, 66-68; asst prof, 68-72, ASSOC PROF BOT, DUKE UNIV, 72- Concurrent Pos: Res career develop award, NIH, 72-77. Honors & Awards: Campbell Award, Am Inst Biol Sci, 67. Mem: Genetics Soc Am; Am Soc Cell Biol; Am Soc Nat. Res: Genetic control of organelle structure and function. Mailing Add: Dept of Bot Duke Univ Durham NC 27706

BOYNTON, ROBERT M, b Evanston, Ill, Oct 28, 24; m 47; c 4. PSYCHOLOGY, OPTICS. Educ: Amherst Col, AB, 48; Brown Univ, ScM, 50, PhD(psychol), 52. Prof Exp: From asst prof psychol to prof psychol & optics, Univ Rochester, 52-74, chmn dept psychol, 71-74; PROF PSYCHOL, UNIV CALIF, SAN DIEGO, 74- Concurrent Pos: Mem, Nat Res Coun-Armed Forces Comt Vision, 57-65, Exec Coun, 62-65; mem, US Nat Comt, Int Comn Illumination, 59-; NIH study fel, Eng, 60-61; dir, Ctr Visual Sci, Univ Rochester, 63-71; mem, Visual Sci Study Sect, NIH, 64-67; dir-at-large, Optical Soc Am, 66-69; vis prof, San Francisco Med Ctr, 69-70; chmn, Visual Sci B Study Sect, NIH, 73- Honors & Awards: Tillyer Medal, Optical Soc Am, 72. Mem: Fel AAAS; fel Optical Soc Am; fel Am Psychol Asn; Soc Exp Psychologists. Res: Psychology, physics and physiology of human vision. Mailing Add: Dept of Pyschol C-009 Univ of Calif at San Diego La Jolla CA 92093

BOYNTON, WALTER RAYMOND, b Lawrence, Mass, May 5, 47; m 73. MARINE ECOLOGY. Educ: Springfield Col, BS, 59; Univ NC, Chapel Hill, MS, 74; Univ Fla, PhD(environ eng), 75. Prof Exp: Res asst, 69-70, RES ASSOC MARINE ECOL, CHESAPEAKE BIOL LAB, UNIV MD, 75- Mem: Estuarine Res Fedn. Res: Population studies of estuarine fish populations; coastal zone management with special emphasis on power plant siting. Mailing Add: Hallowing Point Field Sta Rte 1 Prince Frederick MD 20678

BOYNTON, WILLIAM VANDEGRIFT, b Bridgeport, Conn, Oct 29, 44; m 74. GEOCHEMISTRY, COSMOCHEMISTRY. Educ: Wesleyan Univ, BA, 66; Carnegie-Mellon Univ, PhD(phys chem), 71. Prof Exp: Res assoc geochem, Ore State Univ, 71-74; ASST RES GEOCHEMIST, UNIV CALIF, 74- Mem: AAAS; Meteoritical Soc; Am Chem Soc. Res: Theromdynamics of trace element condensation from the solar nebula; trace element solid solution formation; neutron activation analysis. Mailing Add: Inst of Geophys & Planetary Physics Univ of Calif Los Angeles CA 90024

BOYS, DONALD W, b Atlanta, Ga, Apr 14, 40; m 64. SOLID STATE PHYSICS. Educ: Case Inst Technol, BS, 62; Iowa State Univ, PhD(thermal conductivity), 66. Prof Exp: Asst prof, 67-74, ASSOC PROF PHYSICS, UNIV MICH-FLINT, 74- Mem: Am Phys Soc; Am Asn Physics Teachers. Res: Thermal conductivity of rare earth single crystals. Mailing Add: Dept of Physics Univ of Mich Flint MI 48503

BOZAK, RICHARD EDWARD, b Aberdeen, Wash, Oct 13, 34; m 59; c 3. ORGANIC CHEMISTRY. Educ: Univ Wash, BS, 56; Univ Calif, Berkeley, PhD(org chem), 59. Prof Exp: Fel org synthesis, Moscow State Univ, 59-60; fel ferrocene chem, Univ Ill, 60-61; res chemist, Shell Oil Co, Calif, 61-64 & Shell Develop Co, 64; ASST PROF ORG CHEM, CALIF STATE UNIV, HAYWARD, 64-, COORDR RES, 66- Concurrent Pos: Consult, Lawrence Livermore Lab, 70-; Fulbright res fel, Inorg Inst, Munich Tech Univ, 72-73. Mem: Am Chem Soc; Sigma Xi. Res: Natural products; organic chemistry of ferrocene, the ferricenium ion. Mailing Add: Dept of Chem Calif State Univ 25800 Hillary St Hayward CA 94542

BOZARTH, GENE ALLEN, b Rumsey, Ky, Nov 10, 41; m 68. BOTANY, PLANT PATHOLOGY. Educ: Univ Ky, BS, 63; Auburn Univ, MS, 66, PhD(biochem), 69. Prof Exp: Fel fungus physiol, Univ Mo-Columbia, 69-71; sr res analyst, Space Sci Lab, 71-74, PRIN SCIENTIST BOT AREA, JOHNSON SPACEFLIGHT CTR, NORTHROP SERV, INC, 74- Mem: Am Phytopath Soc; Am Hort Soc. Res: Space biology; effect of zero gravity on growth and development of plants, also pathological aspects. Mailing Add: Northrop Serv Inc PO Box 34416 Houston TX 77034

BOZARTH, ROBERT F, b Herrin, Ill, Feb 23, 30; m 51; c 3. PLANT VIROLOGY. Educ: Univ Fla, BSc, 52, MSc, 57; Cornell Univ, PhD(plant path), 62. Prof Exp: Res assoc, Virol Pioneering Lab, USDA, 61-62; VIROLOGIST, BOYCE THOMPSON INST PLANT RES, 62- Res: Physiology of plants infected with virus diseases. Mailing Add: Boyce Thompson Inst for Plant Res Yonkers NY 10701

BOZEMAN, F MARILYN, b Washington, DC, Dec 3, 27; m 64. MICROBIOLOGY. Educ: Univ Md, BS, 48, MS, 50. Prof Exp: Bacteriologist, Am Type Cult Collection, 48-49; microbiologist, Dept Rickettsial Dis, Walter Reed Army Inst Res, 50-73; MICROBIOLOGIST, DIV VIROL, BUR BIOLOGICS, FOOD & DRUG ADMIN, 73- Mem: Am Soc Microbiol; Tissue Cult Asn; Am Asn Immunologists. Res: Rickettsiology and virology; cell and tissue culture. Mailing Add: Div Virol Bur Biologics Food & Drug Admin Bethesda MD 20014

BOZEMAN, JOHN RUSSELL, b Bleckley Co, Ga, Apr 2, 35; m 58; c 3. PLANT ECOLOGY, RESOURCE MANAGEMENT. Educ: Ga Southern Col, BS, 61; Univ NC, Chapel Hill, MA, 65, PhD(bot, ecol), 71. Prof Exp: Teaching asst bot, Univ NC, Chapel Hill, 61-64; instr biol, Ga Southern Col, 64-66; cur herbarium, Univ NC, Chapel Hill, 66-68, instr bot, 68; asst prof biol, 68-72, ASSOC PROF BIOL, GA SOUTHERN COL, 72- Concurrent Pos: Consult veg analyst, Inst Natural Resources, Univ Ga, 72-75; consult resource assessment & mgt, Dept Natural Resources, Off Planning & Res, State Ga, Atlanta, 74-75. Mem: Ecol Soc Am; Am Soc Photogram; Soil Conserv Soc Am. Res: Functional role of vegetaion in aquatic and terrestrial ecosystems; stabilizing and disruptive processes, mineral cycling and storage in ecosystems; vegetational surveys of the Coastal Plain, vegetational mapping and remote sensing. Mailing Add: Dept of Biol Ga Southern Col Statesboro GA 30458

BOZEMAN, SAMUEL RICHMOND, b Knoxville, Tenn, Nov 15, 15; m 44. BACTERIOLOGY. Educ: Univ Tenn, BA, 39, MS, 40; Va Polytech Inst, PhD(bact), 45. Prof Exp: From bacteriologist to asst dir biol prod div, Mich State Dept Health, 44-48; chief lab, Br Pub Health & Welfare, Gen Hq Supreme Comdr Allied Powers, 48-51, UN Command, 50-51; asst dir in chg qual control, Biol Labs, Pitman-Moore Co, 52, dir, 52-57, vpres, 57-63; asst mgr biol prod, Lederle Labs, 63-67, mgr clin lab aids dept, 67-69, gen mgr acquisitions, 69-70; assoc dean, Col Basic Med Sci, 70-72, PROF MICROBIOL, COL BASIC MED SCI, UNIV TENN CTR HEALTH SCI, MEMPHIS, 70-, EXEC ASST TO THE CHANCELLOR, 72- Concurrent Pos: Teaching fel bact, State Col Wash, 40-42. Mem: AAAS; Am Soc Microbiol; Am Chem Soc; Am Pub Health Asn; NY Acad Sci. Res: Iso and hetero-haemagglutinogens; autoantibodies; physiology of genus bacillus, antibiotics; methodology of vaccine, toxoid, antitoxin, antiserum production and control. Mailing Add: Univ of Tenn Ctr for the Health Sci Memphis TN 38163

BOZIAN, RICHARD C, b Springfield, Mass, Aug 12, 19; m 51; c 5. BIOCHEMISTRY, INTERNAL MEDICINE. Educ: Rutgers Univ, BS, 39; Albany Med Col, MD, 50; Am Bd Internal Med, dipl, 58; Am Bd Nutrit, dipl, 68. Prof Exp: Instr med, Sch Med, NY Univ-Bellevue Med Ctr, 53-55, USPHS trainee metab, 54-55, asst prof med & assoc vis physician & dir clins, 55-58; clin investr, Vet Admin Hosp, Nashville, Tenn, 61-63; assoc prof, 63-70, PROF MED, UNIV CINCINNATI, 70-, ASST PROF BIOCHEM & DIR DIV NUTRIT, COL MED, 63- Concurrent Pos: USPHS res grant, 58 & 64; mem coun arteriosclerosis, Am Heart Asn, 55-; assoc attend physician, Knickerbocker Hosp & consult, St Barnabas Hosp, New York, 56-58; consult, US Vet Admin Hosp, Jewish Hosp & Longview S Hosp & attend, Holmes Hosp, Cincinnati, 63-; Am Cancer Soc res grant, 63 & 64; lectr, Coun Food & Nutrit, 65- Mem: AAAS; Am Pharmaceut Asn; Harvey Soc; Am Soc Human Genetics; Am Fedn Clin Res. Res: Nutritional biochemistry. Mailing Add: Rm 6560 Med Sci Bldg Univ of Cincinnati Col of Med Cincinnati OH 45267

BOZICEVICH, JOHN, zoology, see 12th edition

BOZLER, EMIL, b Steingebronn, Ger, Apr 5, 01; nat US; m 33; c 3. PHYSIOLOGY. Educ: Univ Munich, PhD(zool), 23. Hon Degrees: DSc, Ohio State Univ, 75. Prof Exp: Asst & privat-docent, zool inst, Univ Munich, 24-31; Rockefeller Found fel, 28-29; vis fel, sch med, Univ Rochester, 29; fel med physics, Johnson Found, sch med, Univ Pa, 32-36; from asst prof to prof, 36-66, EMER PROF PHYSIOL, OHIO STATE UNIV, 66- Concurrent Pos: Fulbright award, 58. Mem: Am Physiol Soc; hon mem Ger Physiol Soc. Res: Physiology of primitive nervous systems; energetics; excitability; action potentials; mechanical properties of muscle. Mailing Add: Dept of Physiol Ohio State Univ 1645 Neil Ave Columbus OH 43210

BOZNIAK, EUGENE GEORGE, b Radway, Alta, Oct 15, 42; m 65. AQUATIC ECOLOGY, PHYCOLOGY. Educ: Univ Alta, BSc, 63, MSc, 66; Wash Univ, PhD(aquatic ecol), 69. Prof Exp: Asst prof, 69-74, ASSOC PROF BOT, WEBER STATE COL, 74- Mem: AAAS; Am Inst Biol Sci; Am Soc Limnol & Oceanog; Ecol Soc Am; Phycol Soc Am. Res: Natural and synthetic phytoplankton community cultivation and ecology; effects of synthetic organic compounds upon defined phytoplankton communities; algal ecology of the Ogden River drainage system. Mailing Add: Dept of Bot Weber State Col Ogden UT 84403

BRAASCH, NORMAN L, b Pierce, Nebr, July 29, 28; m 53; c 1. ENTOMOLOGY. Educ: Univ Nebr, BS, 50, MA, 55, PhD(entom), 65. Prof Exp: Instr, Nebr Pub Schs, 55-57; res asst entom, Univ Nebr, 58-63; ASST PROF ZOOL, SOUTHEAST MO STATE COL, 63- Mem: AAAS; Am Soc Zoologists; Entom Soc Am. Res: Invertebrate zoology. Mailing Add: Dept of Biol Southeast Mo State Col Cape Girardeau MO 63701

BRAATEN, MELVIN OLE, b Greenbush, Minn, Sept 6, 34; m 56; c 2. STATISTICS, ENGINEERING. Educ: NDak State Univ, BS, 56, MS, 61; NC State Univ, PhD(statist), 65. Prof Exp: Asst prof forest biomet, Duke Univ, 65-69; ASSOC PROF INDUST ENG & RADIOL SCI, UNIV MO-COLUMBIA, 69- Mem: Am Statist Asn; Biomet Soc; Opers Res Soc Am. Res: Computer applications in human engineering; operations research systems modeling in medical service; engineering statistical modeling. Mailing Add: Dept of Indust Eng Univ of Mo Columbia MO 65201

BRABANDER, HERBERT JOSEPH, b Brooklyn, NY, Apr 17, 32; m 63; c 3. ORGANIC CHEMISTRY. Educ: LI Univ, BS, 53; Stevens Inst Technol, MS, 58. Prof Exp: RES CHEMIST, LEDERLE LABS, AM CYANAMID CO, 53- Mem: Am Chem Soc. Res: Pharmaceutical chemical research for the development of new drugs, particularly agents which act upon the central nervous system. Mailing Add: Lederle Labs Am Cyanamid Co Bldg 65 Pearl River NY 10965

BRABB, EARL EDWARD, b Detroit, Mich, May 27, 29; m 57; c 2. GEOLOGY. Educ: Dartmouth Col, AB, 51; Univ Mich, MS, 52; Stanford Univ, PhD(geol), 60. Prof Exp: Actg instr geol, Stanford Univ, 58; GEOLOGIST, US GEOL SURV, 59- Mem: Am Asn Petrol Geol; Geol Soc Am. Res: Areal geology in Alaska and California; Tertiary and Paleozoic stratigraphy; earthquakes; air photo interpretation; environmental geology; San Andreas fault. Mailing Add: US Geol Surv 345 Middlefield Rd Menlo Park CA 94025

BRABSON, BENNET BRISTOL, b Washington, DC, July 29, 38; m 63; c 2. ELEMENTARY PARTICLE PHYSICS. Educ: Carleton Col, BA, 60; Mass Inst Technol, PhD(physics), 66. Prof Exp: Res assoc physics, Mass Inst Technol, 66-67, NSF fel, 67-68; asst prof, 68-72, ASSOC PROF PHYSICS, IND UNIV, BLOOMINGTON, 72- Mem: Sigma Xi. Res: Study of strong interactions, including meson spectroscopy, utilizing spark, streamer, bubble, proportioned chambers and counter systems. Mailing Add: Dept of Physics Swain Hall West Ind Univ Bloomington IN 47401

BRABSON, GEORGE DANA, JR, b Washington, DC, Feb 18, 35; m 59; c 3. PHYSICAL CHEMISTRY. Educ: Case Inst Technol, BS, 56; Univ Calif, Berkeley, MS, 62, PhD(chem), 65. Prof Exp: Chem engr, Dow Chem Co, Mich, 56; liaison officer, Defense Atomic Support Agency, Sandia Base, NMex, 56-61; from instr to assoc prof chem, US Air Force Acad, 62-70, chemist, F J Seiler Res Lab, 71-74; CHIEF, CHEM LASER BR, AIR FORCE WEAPONS LAB, 74- Mem: Am Phys Soc; Am Chem Soc. Res: Matrix isolation spectroscopy; IR, visible, UV and vacuum UV spectroscopy; molten salt electrochemistry; high energy gas lasers. Mailing Add: Air Force Weapons Lab/ALC Kirtland AFB NM 87117

BRABY, LESLIE ALAN, b Kelso, Wash, Jan 12, 41; m 66; c 2. RADIOLOGICAL PHYSICS, RADIATION BIOPHYSICS. Educ: Linfield Col, BA, 63; Ore State Univ, PhD(radiol physics), 72. Prof Exp: RES SCIENTIST RADIOL PHYSICS & BIOPHYS, PAC NORTHWEST LAB, BATTELLE MEM INST, 63- Mem: Sigma Xi; Radiation Res Soc; Am Phys Soc; Health Physics Soc; AAAS. Res: Investigation of the physical and early biological processes which determine the response of a biological system to ionizing radiation. Mailing Add: Battelle-Pac Northwest Lab 3746 Bldg 300 Area Richland WA 99352

BRACCO, DONATO JOHN, b Neresine, Lussino, Italy, Feb 16, 21; nat US; m 50; c 6. PHYSICAL CHEMISTRY, ANALYTICAL CHEMISTRY. Educ: City Col New York, BChE, 41. Prof Exp: Res chemist, Peter J Schweitzer, Inc, 41-42; chem engr, Titanium Div, Nat Lead Co, 42-45 & AEC, 47; sect head phosphor & cathode ray tube chem, Physics Labs, Sylvania Elec Prod, Inc, 47-55, mgr, Chem Lab, 55-57 & Planning Res Labs, 58-61, mgr mat anal, Gen Tel & Electronics Labs, 61-69, dir mat res lab, GTE Labs, 69-72, dir, Cent Serv Lab, 72-74, DIR, MAT ENG LAB, GTE LABS, WALTHAM, 75- Mem: Am Chem Soc; Electrochem Soc; Inst Elec & Electronics Engrs; Am Soc Qual Control; Solar Energy Soc. Res: Instrumental analysis; process monitoring; materials engineering; solar energy thermal conversion. Mailing Add: 348 Hayward Mill Rd Concord MA 01742

BRACE, C LORING, b Hanover, NH, Dec 19, 30; m 57; c 3. PHYSICAL ANTHROPOLOGY. Educ: Williams Col, BA, 52; Harvard Univ, MA, 58, PhD(anthrop), 62. Prof Exp: Instr anthrop, Univ Wis, Milwaukee, 60-61; from asst prof to assoc prof, Univ Calif, Santa Barbara, 61-67; assoc prof, 67-71, PROF ANTHROP, UNIV MICH, ANN ARBOR, 71-, CUR PHYS ANTHROP, MUS ANTHROP, 67- Concurrent Pos: Vis prof anthrop, Univ Auckland, 73. Mem: AAAS; Am Asn Phys Anthrop; fel Am Anthrop Asn; Soc Study Human Biol; Int Asn Human Biologists. Res: History of biological anthropology; evolutionary theory and study of human evolution as it is expressed in the human fossil record; analysis of the origin of contemporary human physical diversity. Mailing Add: Mus of Anthrop Univ of Mich Ann Arbor MI 48104

BRACE, JOHN WELLS, b Evanston, Ill, Jan 19, 26; m 50; c 5. PURE MATHEMATICS. Educ: Swarthmore Col, BA, 49; Cornell Univ, AM, 51, PhD(math), 53. Prof Exp: From instr to assoc prof, 53-63, PROF MATH, UNIV MD, COLLEGE PARK, 63- Mem: AAAS; Am Math Soc; Math Asn Am. Res: Applications of functional analysis, ranging through theoretical functional analysis. Mailing Add: Dept of Math Univ of Md College Park MD 20742

BRACE, KIRKLAND, b Sac City, Iowa, Apr 13, 21; m 50; c 3. RADIOBIOLOGY. Educ: Univ Iowa, BA, 42; Univ Ill, BS, 44, MD, 45; Am Bd Radiol, dipl, 60. Prof Exp: Fel radiobiol, Univ Chicago, 48-50; resident radiation ther, Univ Hosp, Univ Md, 57-58; radiation therapist, Nat Cancer Inst, 59-66, head radiother serv, USPHS, 67-71; CHMN RADIATION THER DEPT, WASHINGTON ADVENTIST HOSP, 71- Concurrent Pos: Assoc prof radiol, Georgetown Univ Med Sch, 70-; consult radiother, Nat Naval Med Ctr, 74- & USPHS Hosp, Baltimore, 74- Mem: Radiation Res Soc; Am Col Radiol; Am Soc Therapeut Radiol. Res: Radiation therapy. Mailing Add: Washington Adventist Hosp 7600 Carroll Ave Takoma Park MD 20012

BRACE, LARRY HAROLD, b Saginaw, Mich, Feb 19, 29; m 53; c 3. IONOSPHERIC PHYSICS. Educ: Univ Mich, BS, 58. Prof Exp: Res assoc, Space Physics Res Lab, Univ Mich, 58-60, dir, 60-62; RES PHYSICIST, GODDARD SPACE FLIGHT CTR, NASA, 62- Concurrent Pos: Mem ionosphere subcomt, NASA, 65-67. Honors & Awards: Except Sci Achievement Award, Goddard Space Flight Ctr, 73. Mem: Am Phys Soc; Am Geophys Union. Res: Rocket and satellite borne instruments in studies of the processes controlling the electron energy balance of the ionosphere. Mailing Add: Code 621 Goddard Space Flight Ctr Greenbelt MD 20771

BRACE, NEAL ORIN, b Osceola, Wis, Apr 12, 22; m 45; c 7. ORGANIC CHEMISTRY, FLUORINE CHEMISTRY. Educ: Univ Minn, BA, 46; Univ Ill, PhD(org chem), 48. Prof Exp: Res chemist, Tenn Eastman Corp, 48-49; res chemist, Org Chem Dept, E I du Pont de Nemours & Co, 49-63; assoc prof chem, N Park Col, 63-66, head dept, 66; assoc prof, 66-72, PROF CHEM, WHEATON COL, ILL, 72- Concurrent Pos: Consult, Ciba-Geigy Corp, 65-; sr US scientist, Alexander von Humboldt-Stiftung, Ger, 72. Mem: Am Chem Soc; Am Sci Affiliation. Res: Aromatic substitution; synthesis of vinyl monomers and dienes; organo-phosphorus compounds; fluorocarbon chemistry; surface active compounds; gas chromatography; free radicals; cyclization reactions. Mailing Add: Dept of Chem Wheaton Col Wheaton IL 60187

BRACE, ROBERT ALLEN, b Marlette, Mich, May 4, 48; m 69. CARDIOVASCULAR PHYSIOLOGY, CHEMICAL ENGINEERING. Educ: Mich State Univ, BS, 70, MS, 71, PhD(chem eng), 73. Prof Exp: Instr, 73-75, ASST PROF PHYSIOL & BIOPHYS, MED CTR, UNIV MISS, 75- Concurrent Pos: NIH cardiovasc traineeship, Dept of Physiol & Biophys, Med Ctr, Univ Miss, 73-75. Mem: Am Heart Asn; Am Inst Chem Eng; Am Physiol Soc. Res: Applied mathematics; electrophysiology; systems analysis. Mailing Add: Dept of Physiol & Biophys Univ of Miss Med Ctr Jackson MS 39216

BRACE, WILLIAM FRANCIS, b Littleton, NH, Aug 26, 26; m 55; c 3. STRUCTURAL GEOLOGY. Educ: Mass Inst Technol, BS, 49, PhD(geol), 53. Prof Exp: Fulbright fel, Austria, 53-54, from asst prof to assoc prof, 55-66, PROF GEOL, MASS INST TECHNOL, 66- Mem: Nat Acad Sci; Geol Soc Am; Am Soc Civil Eng. Res: Application of mechanics to problems of structural geology. Mailing Add: Dept of Earth & Planet Sci Mass Inst Technol Cambridge MA 02139

BRACELAND, FRANCIS JAMES, b Philadelphia, Pa, July 22, 00; m 38; c 2. PSYCHIATRY. Educ: La Salle Col, AB, 26; Jefferson Med Col, MD, 30. Hon Degrees: ScD, La Salle Col, 41, Col Holy Cross, 56, Cath Univ Am, 57, Northwestern Univ, 57, Trinity Col, 58, Fairfield Univ, 61; LHD, Canisius Col, 56 & St Joseph Col, Conn, 70; LLD, Manhattan Col, 56; DLitt, Univ Hartford, 64 & Jefferson Med Col, 65. Prof Exp: Intern, Jefferson Hosp, 30-32; fel, Pa Hosp Nerv & Ment Dis, 32-35, clin dir, 37-38; Rockefeller fel, Anstalt Burgholzli, Zurich, 35-36 & Nat Hosp, London, 36; practicing psychiatrist, Philadelphia, 38-41; prof psychiat & dean, Loyola Univ, Ill, 41-42; consult psychiatrist & head sect, Mayo Clin, 46-51, prof psychiat, Univ Minn, 46-51; psychiatrist-in-chief, 51-65, SR CONSULT, INST LIVING, 65- Concurrent Pos: Clin prof, Yale Univ, 51-68, emer clin prof, 68-; lectr, Harvard Univ, 60-66; ed, Am J Psychiat, 65-; ed dir, Yearbk Psychiat & Appl Ment Health, 68-, Med Insight, 70-74 & Psychiat Ann, 71-; mem vis comt bd overseers, Harvard Univ Health Serv, 70-74; vis prof psychiat, Univ Conn Health Ctr, 74- Chmn, Salmon Comt Psychiat & Ment Hyg, 59-; chmn adv comt alcoholism, NIMH, 67-69; chmn hist life sci study sect, USPHS, 68-70. Honors & Awards: Laetare Medal, Univ Notre Dame, 62; Stritch Medal, Loyola Univ, 65; Nat Human Rels Award, Nat Conf Christians & Jews, 69; E B Bowis Award, Am Col Psychiatrists, 71; Distinguished Serv Award, World Psychiat Asn, 71; Edward B Allen Gold Medal Award, Am Geriatrics Soc, 72; Salmon Medal for Distinguished Serv to Psychiat, Salmon Comt Psychiat & Ment Hyg, NY Acad Med, 72; William C Porter Award, Asn Mil Surgeons US, 74. Mem: Fel Am Col Psychiatrists; fel AMA; fel Am Col Physicians; fel Am Psychiat Asn (pres, 56-57); Asn Res Nerv & Ment Dis (pres, 57). Psychosomatic medicine; geriatrics; clinical psychiatry; mental hygiene. Mailing Add: 200 Retreat Ave Hartford CT 06106

BRACEWELL, RONALD NEWBOLD, b Sydney, Australia, July 22, 21; m 53; c 2. RADIO ASTRONOMY. Educ: Univ Sydney, BSc, 41, BE, 43, ME, 48; Cambridge Univ, PhD(physics), 50. Prof Exp: Sr res officer radiophys div, Commonwealth Sci & Indust Res Orgn, Australia, 42-46, 49-55; PROF ELEC ENG, STANFORD UNIV, 55- Concurrent Pos: Fulbright vis prof, Univ Calif, 54-55. Mem: Int Astron Union; fel Inst Elec & Electronics Eng; Int Sci Radio Union; Royal Astron Soc. Res: Solar physics; ionosphere; earth's atmosphere; interplanetary medium; radio wave propagation; antennas; radio telescopes; electromagnetic theory. Mailing Add: 836 Santa Fe Ave Stanford CA 94305

BRACHER, KATHERINE, b San Francisco, Calif, Oct 26, 38. ASTRONOMY. Educ: Mt Holyoke Col, AB, 60; Ind Univ, AM, 62, PhD(astron), 66. Prof Exp: From instr to asst prof astron, Univ Southern Calif, 65-67; ASST PROF ASTRON, WHITMAN COL, 67- Mem: Am Astron Soc. Res: Spectra of peculiar stars, particularly Wolf-Rayet stars; spectroscopic binary stars. Mailing Add: Dept of Astron Whitman Col Walla Walla WA 99362

BRACHFELD, BERNARD, food technology, see 12th edition

BRACHFELD, NORMAN, b New York, NY, Oct 16, 27. BIOCHEMISTRY, CARDIOVASCULAR PHYSIOLOGY. Educ: Columbia Univ, AB, 49; Wash Univ, MD, 53. Prof Exp: Intern med, Maimonides Hosp, Brooklyn, 53-54; intern surg, Peter Bent Brigham Hosp, Boston, 54-55, asst resident, 55-56, asst med, 57-59; asst resident med, NY Hosp, 59-60; from instr to asst prof, 60-68, ASSOC PROF MED, MED COL, CORNELL UNIV, 68- Concurrent Pos: Teaching & res fel, Peter Bent Brigham Hosp, Boston, 56-57; teaching fel, Harvard Univ Med Sch, 56-57, res fel med, 57-59, Samuel A Levine fel, 58-59; res fel physiol, Sloan Kettering Inst, 57; NY Heart Asn sr res fel, 61-62; City of New York Health Res Coun career scientist award, 62-68; NIH career develop award, 68-73; physician, Outpatient Dept, NY Hosp, 60-62, asst attend, 62-68, assoc attend, 68-; asst dir comprehensive care & teaching prog, Med Col, Cornell Univ, 60-65, assoc dir, 65-68; res assoc, Inst Muscle Dis, Inc, 60-66, head div myocardial metab & assoc mem, 66-73; med res collabr, Brookhaven Nat Lab, 61-63; asst vis physician, Sec Med Div, Bellevue Hosp, 63-68. Mem: Am Fedn Clin Res; Am Heart Asn; Harvey Soc; fel Am Col Physicians; fel Am Col Cardiol. Res: Metabolic and hemodynamic changes in the ischemic heart; cardiovascular research; clinical cardiology. Mailing Add: Dept of Med Cornell Univ Med Col New York NY 10021

BRACHMAN, ARMAND E, organic chemistry, see 12th edition

BRACHMAN, MALCOLM K, b Ft Worth, Tex, Dec 9, 26; m 51; c 3. THEORETICAL PHYSICS. Educ: Yale Univ, BA, 45; Harvard Univ, MA, 47, PhD(physics), 49. Prof Exp: Asst prof physics, Southern Methodist Univ, 49-50; assoc physicist, Argonne Nat Lab, 50-53; res physicist, Tex Instruments, Inc, 53-54; PRES, NORTHWEST OIL CO, 56- Concurrent Pos: Chmn and chief exec officer, Pioneer Am Ins Co, 54- Mem: Am Phys Soc; Inst Elec & Electronics Eng; Am Math Soc; Am Inst Mining; Soc Explor Geophys. Mailing Add: Northwest Oil Co 3217 Republic Bank Tower Dallas TX 75201

BRACHMAN, PHILIP SIGMUND, b Mulwaukee, Wis, July 28, 27; m 50; c 4. MEDICINE. Educ: Univ Wis, BS, 50, MD, 53. Prof Exp: Intern, Hosp, Univ Ill, 53-54; asst surgeon, Epidemiol Intel Serv, Commun Dis Ctr, Wistar Inst, Univ Pa, 54-58, med resident, Univ Pa Hosp, 58-60, sr surgeon & chief invest sect, Epidemiol Br, Commun Dis Ctr, 60-70, DIR BUR EPIDEMIOL, CTR DIS CONTROL, USPHS, 70- Mem: Am Pub Health Asn; Am Epidemiol Soc; Am Fedn Clin Res. Res:

Epidemiology, public health, preventive medicine. Mailing Add: USPHS Ctr for Dis Control 1600 Clifton Rd NE Atlanta GA 30333

BRACK, KARL, b Küttingen, AG, Switz, Nov 22, 23; US citizen; c 2. ORGANIC POLYMER CHEMISTRY, PHOTOCHEMISTRY. Educ: Swiss Fed Inst Technol, MS, 49, PhD(org chem), 51. Prof Exp: Res chemist pesticides, Hercules Inc, 52-57, res chemist polymers, 57-71; sr chemist inks & coatings, 71-75, RES ASSOC INKS & COATINGS, DENNISON MFG CO, 75- Mem: Am Chem Soc; Sigma Xi. Res: Synthesis of radiation-curable oligomers, prepolymers and polymers; research and development of radiation-curable inks, coatings and adhesives. Mailing Add: Res Div Dennison Mfg Co 300 Howard St Framingham MA 01701

BRACKELSBERG, PAUL O, b Mohall, NDak, Aug 27, 39; m 61; c 3. ANIMAL BREEDING. Educ: NDak State Univ, BS, 61; Univ Conn, 63; Okla State Univ, PhD(animal breeding), 66. Prof Exp: ASSOC PROF ANIMAL SCI, IOWA STATE UNIV, 66- Mem: Am Soc Animal Sci. Res: Application of genetic principles to applied animal breeding, with emphasis on growth and carcass traits; livestock production. Mailing Add: Dept of Animal Sci Iowa State Univ Ames IA 50010

BRACKEN, EVERETT CHRISTY, medical microbiology, deceased

BRACKEN, JEROME, b Pittsburgh, Pa, Oct 4, 35; m 57; c 4. OPERATIONS RESEARCH. Educ: Carnegie-Mellon Univ, BS, 56; George Washington Univ, MEA, 59; Harvard Univ, DBA, 63. Prof Exp: Res asst, Bus Sch, Harvard Univ, 60-61; res scientist, Logistics Res Proj, George Washington Univ, 62-64; mem tech staff, Res Anal Corp, 64-65, dep head, Advan Res Dept, 65-67; RES STAFF MEM, INST DEFENSE ANAL, 67- Concurrent Pos: Lectr, George Washington Univ, 61-71. Mem: Opers Res Soc Am; Inst Mgt Sci. Res: Defense analysis; optimization. Mailing Add: Inst for Defense Anal 400 Army-Navy Dr Arlington VA 22202

BRACKEN, MARILYN C, b Pittsburgh, Pa, Nov 5, 35; m 57; c 4. INFORMATION SCIENCE. Educ: Carnegie-Mellon Univ, BS, 57; Am Univ, MA, 69, PhD(pub admin), 71. Prof Exp: Chemist, Melpar, Inc, 57-58 & Nat Bur Standards, 62-64; consult, Nat Libr Med, 70-71; info systs analyst, Nat Agr Libr, 71-72; info systs analyst, Off Info Systs, USDA, 72-73; DIR, DIV SCI COORD, BUR BIOMED SCI, US CONSUMER PROD SAFETY COMN, 73- Concurrent Pos: Contribr ed, Bull Am Soc Info Sci, 74- Honors & Awards: Honor Award, US Consumer Prod Safety Comn, 74. Mem: AAAS; Am Soc Info Sci. Res: Scientific and technical information systems. Mailing Add: US Consumer Prod Safety Comn 5401 Westbard Ave Bethesda MD 20207

BRACKEN, RONALD CLAY, b Billings, Mont, Sept 22, 32; m 59; c 4. PHYSICAL CHEMISTRY. Educ: Rice Univ, BA, 56; Purdue Univ, PhD(chem), 61. Prof Exp: Res chemist, Physics Res Labs, Dallas, 61, res chemist gas solid reactions, Mat Res & Develop Labs, 61-69, sect head, Process Develop Lab, Components Group, Stafford, 69-71, mgr process eng, Circuits Design Pilot Line, 71-74, MGR PROCESS ENG, PHOTOMASK SERV DEPT, TEX INSTRUMENTS, INC, 74- Mem: Am Chem Soc; Electrochem Soc. Res: Catalytic activity of semiconductors; chemical vapor deposition of metals and ceramics; photoresist process development; aluminum anodization. Mailing Add: 7627 Meadowhaven St Dallas TX 75240

BRACKENBURY, ROBERT WILLIAM, b Long Beach, Calif, May 19, 48; m 69; c 1. DEVELOPMENTAL BIOLOGY. Educ: Calif Inst Technol, BS, 70; Brandeis Univ, PhD(biol), 76. Prof Exp: JANE COFFIN CHILDS FEL, ROCKEFELLER UNIV, 75- Res: Structure and function of cell-surface molecules and cell-cell interactions during development. Mailing Add: Rockefeller Univ 1230 York Ave New York NY 10021

BRACKENRIDGE, DAVID ROSS, b Buffalo, NY, Jan 31, 38; m 64; c 2. ORGANIC CHEMISTRY. Educ: Canisius Col, BS, 60; Ohio State Univ, MS, 65, PhD(org chem), 66. Prof Exp: CHEMIST RES & DEVELOP, ETHYL CORP, 66- Mem: Am Chem Soc. Res: Synthetic, organic chemistry; reaction mechanisms; bromo-organic compounds; flame retardants; lube oil additives. Mailing Add: 309 Amelia Ave Royal Oak MI 48073

BRACKENRIDGE, JOHN BRUCE, b Youngstown, Ohio, Apr 20, 27; m 54; c 4. CLASSICAL MECHANICS. Educ: Muskingum Col, BS, 51; Brown Univ, MS, 54, PhD(physics), 59; London Univ, MSc, 74. Prof Exp: Asst prof physics, Muskingum Col, 55-59; from asst prof to assoc prof, 59-66, PROF PHYSICS, LAWRENCE UNIV, 66- Concurrent Pos: Res grants, Res Corp, 55-60 and NSF, 60-65, 68; vis scholar, Imp Col, Univ London, 74-75. Honors & Awards: Chapman Chair of Physics, Lawrence Univ, 63. Mem: Am Acoust Soc; Am Asn Physics Teachers; Hist Sci Soc. Res: Subaqueous acoustics; fluid dynamics; application of schlieren interferometry systems to measurements of fluid flow. Mailing Add: Dept of Physics Lawrence Univ Appleton WI 54911

BRACKER, CHARLES E, JR, b Portchester, NY, Feb 3, 38; m 63; c 2. PLANT PATHOLOGY. Educ: Univ Calif, Davis, BS, 60, PhD(plant path), 64. Prof Exp: Res asst plant path, Univ Calif, Davis, 60-64; from asst prof to assoc prof, 64-73, PROF PLANT PATH, UNIV PURDUE, WEST LAFAYETTE, 73- Mem: AAAS; Mycol Soc Am; Bot Soc; Electron Micros Soc. Res: Fungal ultrastructure and the relation of fungus parasites to plant hosts; developmental cytology; subcellular symtomatology; endomembrane systems. Mailing Add: Dept of Bot & Plant Path Purdue Univ West Lafayette IN 47906

BRACKETT, BENJAMIN GAYLORD, b Athens, Ga, Nov 18, 38; m 59; c 3. VETERINARY MEDICINE, BIOCHEMISTRY. Educ: Univ Ga, BSA, 64, DVM, 62, MS, 64, PhD(biochem), 66; Am Col Theriogenologists, dipl. Hon Degrees: MA, Univ Pa, 71. Prof Exp: Mark L Morris Animal Found fel & Am Vet Med Asn fel, Univ Ga, 62-64, Nat Inst Child Health & Human Develop fel, 64-66; assoc reproductive biol, Dept Obstet & Gynec, Sch of Med, 66-68, from asst prof to assoc prof, 68-74, managing dir, Primate Colony, 66-74, PROF ANIMAL REPRODUCTION, DEPT CLIN STUDIES, SCH VET MED, UNIV PA, 74- Concurrent Pos: Consult, Nat Inst Child Health & Human Develop, 68-; mem sci adv bd, Mark L Morris Animal Found, 71-74, mem, Primate Res Ctrs Adv Comt, 74-78; consult, Ford Found & WHO. Mem: AAAS; Am Soc Theriogenology; NY Acad Sci; Am Asn Lab Animal Sci; Am Fertil Soc. Res: Preimplantation stages of mammalian reproduction, especially fertilization of mammalian ova in vitro. Mailing Add: Dept Clin Studies Sch of Vet Med Univ of Pa New Bolton Ctr Kennett Square PA 19348

BRACKETT, FREDERICK SUMNER, b Claremont, Calif, Aug 1, 96; m 18; c 2. PHYSICS. Educ: Pomona Col, AB, 18; Johns Hopkins Univ, PhD(physics), 22. Prof Exp: Lab asst, Bur Standards, 18-19; observer, Mt Wilson Observ, 19-20; instr physics, Johns Hopkins Univ, 20-22; asst physicist, Bur Standards, 21-22; from asst prof to assoc prof physics, Univ Calif, 22-29; sr physicist, fixed nitrogen res lab, USDA, 29; dir div radiation & organisms, Smithsonian Inst, 29-33, consult, 33-34; consult prin technologist, div cotton mkt, USDA, 34-36; sr physicist, div indust hyg, 36-38, prin physicist, 38-40 & Nat Cancer Inst, 40-41, div indust hyg, 41-47, biophysicist & chief sect photobiol, lab phys biol, 47-61, consult, Inst, 61-74, EMER SCIENTIST, NAT CANCER INST, NIH, 74- Concurrent Pos: Consult, Res Corp, 33-34 & Am Mach & Foundry Co, 61-66; secy & dir lab, Res Assocs, Inc, 35-36. Mem: AAAS; Am Phys Soc; Optical Soc Am. Res: Infrared spectroscopy; atomic and molecular structure; photochemistry; respiration; biophysics. Mailing Add: 6401 Tuckerman Lane Rockville MD 20852

BRACKETT, JOHN WASHBURN, b Brockton, Mass, May 26, 37; m 63. COMPUTER SCIENCE, SYSTEM ANALYSIS. Educ: Mass Inst Technol, BS, 59; Purdue Univ, PhD(phys chem), 63. Prof Exp: Mem comput sci div, Res Anal Corp, 63-64; res assoc comput sci, Mass Inst Technol, 64-69; dir graphic systs, 69-71, VPRES, SOFTECH, INC, 71- Concurrent Pos: Adj prof, Brown Univ, 74-75; nat lectr, Asn Comput Mach, 75- Mem: Asn Comput Mach; Sigma Xi. Res: Computer science and software engineering, particularly in the development of large systems; applications of computers in computer-based education; analysis and design techniques for computer-system specifications. Mailing Add: Softech Inc 460 Totten Pond Rd Waltham MA 02154

BRACKETT, ROBERT GILES, b Nyack, NY, Oct 8, 30; m 51; c 5. IMMUNOLOGY. Educ: Rutgers Univ, BS, 53, MS, 57, PhD(bact), 60. Prof Exp: Asst bact, Rutgers Univ, 56-59, instr, 59-60; asst res microbiologist cell cult, 60-62, res microbiologist cell cult & virol, 62-66, dir, Virol Sect, 66-70, DIR CLIN IMMUNOL, PARKE, DAVIS & CO, 70- Concurrent Pos: Jr microbiologist, Merck & Co, Inc, NJ, 56; bacteriologist, E R Squibb & Sons, 57-58. Mem: Am Soc Microbiol; Tissue Cult Asn; Soc Cryobiol. Res: Laboratory and/or clinical research and development of bacterial, plasmodial, and viral vaccines, skin test antigens, and blood products with emphasis on the immunology of these biological products. Mailing Add: Biol Res & Develop Dept Parke Davis & Co Detroit MI 48232

BRACKIN, EDDY JOE, b Town Creek, Ala, Feb 27, 45; m 66. ALGEBRA. Educ: Florence State Univ, BS, 67; Univ Ala, MA, 68, PhD(algebra), 70. Prof Exp: From instr to asst prof, 69-74, ASSOC PROF MATH, UNIV NORTH ALA, 74- Mem: Am Math Asn. Res: Theory of semirings. Mailing Add: Dept of Math Univ of North Ala Florence AL 35630

BRACKMANN, RICHARD THEODORE, b Kansas City, Mo, Nov 23, 30; m 54; c 3. MASS SPECTROMETRY, ATOMIC PHYSICS. Educ: Univ Kans, BS, 53. Prof Exp: Asst, Gen Atomic Div, Gen Dynamics Corp, 56-58, mem res staff, 58-63; assoc res prof phys & elec eng, Univ Pittsburgh, 63-72; pres, 66-72; MGR ENG, EXTRANUCLEAR LABS, INC, 73- Mem: Am Phys Soc; Am Soc Mass Spectrometry. Res: Atomic and upper atmosphere research using modulated crossed beam, mass spectrometric, cryogenic, counting and high vacuum techniques; development of state of the art quadrupole mass spectrometric equipment and techniques. Mailing Add: Extranuclear Labs 250 Alpha Dr Pittsburgh PA 15238

BRADBEER, CLIVE, b Tynemouth, Eng, Feb 20, 33; m 60; c 2. BIOCHEMISTRY. Educ: Univ Durham, BSc, 54, PhD(plant biochem), 57. Prof Exp: Jr res biochemist, Univ Calif, Berkeley, 57-59; jr res biochemist, Univ Calif, Davis, 59; proj assoc biochem, Univ Wis, 59-60; lectr microbiol, Queen Mary Col, Univ London, 60-62; vis scientist, Nat Heart Inst, 62-64; asst prof, 64-69, ASSOC PROF BIOCHEM, MED SCH, UNIV VA, 69- Mem: Am Soc Plant Physiol; Am Soc Microbiol; Am Soc Biol Chem. Res: Structure-function relationships in bacterial cell envelopes, with special attention to the transport of vitamin B12 and interactions of E colicins and bacteriophage BF23 with the cell envelope of Escherichia coli. Mailing Add: Dept of Biochem Univ of Va Med Sch Charlottesville VA 22903

BRADBURY, HELEN E, invertebrate zoology, morphology, see 12th edition

BRADBURY, JACK W, b Los Angeles, Calif, Sept 28, 41; m 69. ANIMAL BEHAVIOR. Educ: Reed Col, BS, 63; Rockefeller Univ, PhD(animal behav), 68. Prof Exp: NIH training grant to Rockefeller Univ taken at William Beebe Trop Res Sta, Simla, Trinidad, WI, 68-69; asst prof neurobiol & behav, Cornell Univ, 69-72; asst prof, Rockefeller Univ, 72-75; ASST PROF BIOL, UNIV CALIF, SAN DIEGO, 76- Concurrent Pos: Richard King Mellon fel, Rockefeller Univ, 73-75. Res: Field studies on resource utilization and social evolution in tropical bats. Mailing Add: Dept of Biol Univ of Calif San Diego La Jolla CA 92037

BRADBURY, JAMES CLIFFORD, b US, July 7, 18; c 3. GEOLOGY. Educ: Univ Ill, AB, 41; Harvard Univ, AM, 49, PhD, 58. Prof Exp: From asst geologist to assoc geologist, 49-59, GEOLOGIST, ILL STATE GEOL SURV, 59-, HEAD INDUST MINERALS SECT, 68- Mem: AAAS; Am Inst Mining, Metall & Petrol Eng; fel Geol Soc Am; Soc Econ Geologists. Res: Geology of mineral deposits of Illinois. Mailing Add: Apt 126 502 W Main Urbana IL 61801

BRADBURY, JAMES NORRIS, b Palo Alto, Calif, May 25, 35; m 61; c 2. EXPERIMENTAL PHYSICS. Educ: Pomona Col, BA, 56; Stanford Univ, PhD(physics), 65. Prof Exp: Res scientist, Lockheed Palo Alto Res Lab, 66-73; ALT GROUP LEADER, LOS ALAMOS SCI LAB, 73- Mem: Am Geophys Union; Am Phys Soc; Am Asn Physicists in Med. Res: Applications of LAMPF to biomedical and environmental research. Mailing Add: MS 844 Los Alamos Sci Lab Los Alamos NM 87544

BRADBURY, JAMES THOMAS, b Cody, Wyo, Apr 7, 06; m 29; c 2. ENDOCRINOLOGY. Educ: Mont State Col, BS, 28; Univ Mich, MS, 30, ScD(zool), 32. Prof Exp: Asst, Dept Obstet & Gynec, Univ Hosp, Univ Mich, 32-37, instr, 37-40; endocrinologist, Bur Dairy Indust, USDA, 40-44; from asst prof to assoc prof, 44-57, PROF OBSTET & GYNEC, UNIV HOSP, UNIV IOWA, 57- Concurrent Pos: Assoc prof, Univ Louisville, 49-52. Mem: Am Physiol Soc; Am Asn Anat; Endocrine Soc; Soc Exp Biol & Med. Res: Physiology, biochemistry and histology in study of the actions of hormones in processes of sex and reproduction. Mailing Add: Univ Hosp Univ of Iowa Iowa City IA 52241

BRADBURY, MARGARET G, b Chicago, Ill, July 15, 27. ICHTHYOLOGY. Educ: Roosevelt Univ, BS, 55; Stanford Univ, PhD(biol), 63. Prof Exp: Sci illusr zool, Field Mus Natural Hist, Chicago, 47-55; asst prof biol, MacMurray Col, 62-63; from asst prof to assoc prof, 63-71, chmn dept marine biol, 69-72, PROF BIOL, SAN FRANCISCO STATE UNIV, 71- Concurrent Pos: Asst prof, Stanford Univ, 64; res assoc, Calif Acad Sci, 68-; sr scientist, Stanford Univ oceanog expeds, cruises 5, 10 & 19. Mem: Am Soc Ichthyologists & Herpetologists; Soc Study Evolution; Soc Syst Zool. Res: Systematic ichthyology. Mailing Add: Dept Biol San Francisco State U 1600 Holloway San Francisco CA 94132

BRADBURY, NORRIS EDWIN, b Santa Barbara, Calif, May 30, 09; m 33; c 3. PHYSICS. Educ: Pomona Col, AB, 29; Univ Calif, PhD(physics, math), 32. Hon Degrees: ScD, Pomona Col, 51; LLD, Univ NMex, 53; DSc, Case Inst Technol, 56. Prof Exp: Nat Res fel physics, Mass Inst Technol, 32-34; from asst prof to prof, Stanford Univ, 34-51; prof physics, Univ Calif, 51-70, dir, Los Alamos Sci Lab, 45-70; RETIRED. Concurrent Pos: Mem sci adv bd nuclear weapons, US Air Force. Honors

& Awards: Enrico Fermi Award, Atomic Energy Comn, 70. Mem: Fel Nat Acad Sci; fel Am Phy Soc. Res: Conduction of electricity in gases; properties of ions; atmospheric electricity; nuclear physics. Mailing Add: 1451 47th St Los Alamos NM 87544

BRADBURY, PHYLLIS CLARKE, b Oakland, Calif. PROTOZOOLOGY. Educ: Univ Calif, Berkeley, AB, 58, MA, 61, PhD(zool), 65. Prof Exp: USPHS training fel, Rockefeller Univ, 65-67; asst prof, 67-72, ASSOC PROF ZOOL, NC STATE UNIV, 72- Concurrent Pos: Prin investr, USPHS grant, 68-71; NATO fel, Sta Biologique de Roscoff, 73-74. Mem: Soc Protozoologists (rep to Nat Res Coun, 66-72); Am Soc Zoologists. Res: Fine structure of morphogenesis in protozoa; taxonomy of ciliated protozoa; differentiation of sexual stages of malarial parasites; genesis of organelles in protozoa. Mailing Add: Dept of Zool NC State Univ Raleigh NC 27607

BRADBURY, TED CLAY, b Alameda, Calif, Jan 26, 32. THEORETICAL PHYSICS. Educ: Univ Nev, BS, 54; Cornell Univ, PhD(physics), 61. Prof Exp: From asst prof to assoc prof, 61-70, PROF PHYSICS, CALIF STATE UNIV LOS ANGELES, 70- Mem: Am Inst Physics; Am Asn Physics Teachers. Res: Classical mechanics, electricity and magnetism; relativity theory. Mailing Add: Dept of Physics Calif State Univ 5151 State College Los Angeles CA 90032

BRADBURY, WALTER CARLING, organic chemistry, analytical chemistry, see 12th edition

BRADDOCK, JAMES CONGER, b New York, NY, Sept 20, 13; m 34; c 1. ZOOLOGY. Educ: Williams Col, AB, 35; Univ Chicago, PhD(zool), 42. Prof Exp: Instr zool, Exten, Ind Univ, 38-42 & Kans State Col, 42-43; asst prof, Univ Miami, 43-44 & Univ Idaho, 44-47; assoc prof, 47-55, PROF ZOOL, MICH STATE UNIV, 55- Mem: Fel AAAS; assoc Am Soc Zool; Ecol Soc Am; fel Animal Behav Soc. Res: Social behavior of fishes; mass physiology of amphibian eggs; limnology of fresh water lakes. Mailing Add: Dept of Zool Mich State Univ East Lansing MI 48823

BRADDOCK, JOSEPH, b Hoboken, NJ, Dec 10, 29. PHYSICS. Educ: St Peters Col, BS, 51; Fordham Univ, MS, 52, PhD(physics), 59. Prof Exp: Instr physics, Fordham Univ, 53-58; asst prof, Iona Col, 58-60; dir syst studies, Braddock, Dunn & McDonald, Inc, 60-72; VPRES TECH PROGS, BDM CORP, 72- Mem: Am Phys Soc; Inst Elec & Electronics Engrs. Res: Weapon system design and analysis; neutron and solid state physics; atomic and molecular spectra. Mailing Add: 1920 Aline Ave Vienna VA 22180

BRADDOCK, WILLIAM A, b Rifle, Colo, Feb 3, 29; m 49; c 3. GEOLOGY. Educ: Univ Colo, BA, 51; Princeton Univ, PhD(geol), 59. Prof Exp: Geologist, US Geol Surv, 52-56; from instr to assoc prof, 56-70, PROF GEOL, UNIV COLO, 70- Concurrent Pos: Geologist, US Geol Surv, 56- Mem: Geol Soc Am; Am Geophys Union. Res: Structural geology; igneous and metamorphic petrology; Precambrian geology. Mailing Add: Dept of Geol Univ of Colo Boulder CO 80304

BRADDOCK-ROGERS, KENNETH, b Rochester, NY, Dec 15, 00; m 30; c 2. INORGANIC CHEMISTRY. Educ: Haverford Col, BS, 22; Univ Pa, MS, 25, PhD(chem), 28. Prof Exp: Instr chem, Univ Pa, 24-32, Drexel Inst Technol, 30-31 & sci dept, Haddonfield Mem High Sch, 35-38; prof chem, West Chester State Col, 38-64; prof phys sci, Peirce Jr Col, 64-72, actg chmn dept sci & math, 70-71; RETIRED. Concurrent Pos: Chmn chem & physics div, Curriculum Revision Unit, State Col, Pa, 47-50; critic, Chem Testing Prog, Nat League Nursing, 63. Mem: Franklin Inst. Res: Errors in chemistry, physics and astronomy which have appeared in print; effects of artificial lighting on roses. Mailing Add: 617 W Miner St West Chester PA 19380

BRADEN, CHARLES HOSEA, b Chicago, Ill, Mar 21, 26; m 52; c 2. NUCLEAR PHYSICS. Educ: Columbia Univ, BS, 46; Wash Univ, PhD(physics), 51. Prof Exp: From asst prof to assoc prof, 51-59, PROF PHYSICS, GA INST TECHNOL, 59- Concurrent Pos: Assoc dir physics prog, NSF, 59-60. Mem: Am Phys Soc. Res: Nuclear spectroscopy and structure. Mailing Add: Sch of Physics Ga Inst of Technol Atlanta GA 30332

BRADEN, CHARLES MCMURRAY, b Santiago, Chile, June 9, 18; US citizen; m 43; c 4. MATHEMATICS. Educ: Northwestern Univ, BS, 39; Univ Minn, MS, 50, PhD(math), 57. Prof Exp: Instr math, Inst Technol, Univ Minn, 46-56; from asst prof to assoc prof, 56-60, dean fac, 69-72, PROF MATH, MACALESTER COL, 60- Concurrent Pos: NSF fac fel, 59-60. Honors & Awards: Thomas Jefferson Award, 70. Mem: Am Math Soc; Math Asn Am. Res: Partial difference equations; logic; heat engines. Mailing Add: 80 Arthur Ave SE Minneapolis MN 55414

BRADEN, PATRICK O, b Houston, Tex, Feb 1, 24. PHYSICS, ENGINEERING. Educ: Rice Inst, BS, 44; Univ Tex, MS, 54, PhD(mech eng), 61. Prof Exp: Instr eng, Rice Inst, 46-47; instr math & physics, St Michael's Col, Ont, 49-53; asst prof physics, 54-58, chmn dept, 61-67, PROF PHYSICS & PRES, UNIV ST THOMAS, TEX, 67- Concurrent Pos: Partic, Denver Conf Physics Curric, 61 & Inst Mod Physics, Oak Ridge Inst Nuclear Studies, 65. Mem: AAAS; Am Asn Physics Teachers; Am Soc Mech Engrs. Res: Nuclear radiation studies; solar radiant energy and heat transfer. Mailing Add: Dept of Physics Univ of St Thomas 3812 Montrose Blvd Houston TX 77006

BRADEN, WILLIAM EDWIN, JR, organic chemistry, see 12th edition

BRADER, WALTER HOWE, JR, b Beaumont, Tex, Oct 30, 27; m 44; c 3. ORGANIC CHEMISTRY. Educ: Rice Univ, BA, 50; Ga Inst Technol, PhD(org chem), 54. Prof Exp: Res chemist, Standard Oil Co, Ind, 54-59; sr res chemist, 59-62, sr proj chemist, 62-64, supvr explor res, 64-67, coordr, 67-70, mgr commercial develop, 70-71, DIR RES & DEVELOP, JEFFERSON CHEM CO, 71- Mem: Am Chem Soc. Res: Petrochemicals synthesis; application of catalysis to organic synthesis; new process development; economic evaluations; development of new product sales. Mailing Add: Jefferson Chem Co PO Box 4128 NAS Austin TX 78765

BRADFIELD, RICHARD, b West Jefferson, Ohio, Apr 29, 96; m 23; c 6. AGRONOMY, SOIL FERTILITY. Educ: Otterbein Col, AB, 17; Ohio State Univ, PhD(soil chem), 22. Hon Degrees: ScD, Otterbein Col, 41; DSc, Ohio State Univ, 70. Prof Exp: From instr to assoc prof soils, Univ Mo, 20-29; prof soils & assoc agron, Exp Sta, Ohio State Univ, 30-37; prof, 37-62, head dept agron, 37-55, soil technologist, Exp Sta, 37-62, EMER PROF SOIL TECHNOL, CORNELL UNIV, 62-; SPEC CONSULT, ROCKEFELLER FOUND, FAR EAST, 62- Concurrent Pos: Soil scientist, Bur Plant Indust, Soils & Agr Eng, US Dept Agr; Guggenheim fel, Kaiser Wilhelm Inst, Univ Berlin, 27-28; regional dir agr, Rockefeller Found, Far East, 55-56, mem bd trustees, 57-61; proj leader, Cornell Grad Educ Prog, Philippines, 64; vis prof, Univ Philippines, 64-70; sr fel, East-West Ctr, Univ Hawaii, 70-71; vis prof, Univ Fla, 71-74. Mem: Hon mem Soil Sci Soc Am (pres, 36); Int Soc Soil Sci (pres, 56-60); Am Soc Agron (pres, 41); Am Chem Soc (vpres); Crop Sci Soc Am (pres). Res: Soil fertility, structure and colloids; physical chemistry of soils; techniques for increasing world food production. Mailing Add: 1715 NW 22nd Terrace Gainesville FL 32605

BRADFIELD, ROBERT B, b Columbia, Mo, Nov 15, 28; m 55; c 3. CLINICAL NUTRITION, BIOCHEMISTRY. Educ: Cornell Univ, BA, 51, MNS, 53, PhD(biochem, nutrit), 55. Prof Exp: For serv officer, Pub Health Div, US State Dept, Peru, 55-62; USPHS sr fel, Cornell Univ, 62-64; CLIN PROF HUMAN NUTRIT, UNIV CALIF, BERKELEY, 64- Concurrent Pos: Appt, Harvard Sch Pub Health, Harvard Univ; grants, William Waterman Fund, 57, 58, 60, 62, 64, 68; NIH, 62-67, 69-72, 73-75, Rockefeller Found, 58, 60, 62, 63-64, 67, 68, 69, 71, US Dept Agr, 69-71 & Brit Nutrit Found, 69, 71; fels, Rockefeller Found, 64, WHO, 65, 74 & Guggenheim, 69; consult, Inst Human Growth & Develop, NIH, 64, Pan Am Health Orgn, 66- & US Dept Agr, 70; consult & lectr, Peace Corps; ed, Am J Clin Nutrit, 68- Mem: Fel Am Inst Chem; fel Royal Soc Trop Med & Hyg; fel Royal Soc Health; fel Int Col Pediat; fel AAAS. Mailing Add: Dept of Nutrit Sci Univ of Calif Berkeley CA 94720

BRADFIELD, STILLMAN, b Westerly, RI, Aug 30, 30; m 56; c 2. ANTHROPOLOGY. Educ: Cornell Univ, BA, 56, MA, 57, PhD(anthrop), 63. Prof Exp: Asst prof anthrop, Pa State Univ, 63-65; assoc prof, 65-74, PROF SOCIOL & ANTHROP, KALAMAZOO COL, 74- Concurrent Pos: NSF res grant 64; Adv Res Projs Agency & Wenner-Gren Found Anthrop res grant, 66-68; expert, UN Develop Prog, Peru, 72; vis scientist, Int Ctr Trop Agr, Cali, Colombia, 73-74. Mem: Am Anthrop Asn; Soc Appl Anthrop; Asn Latin Am Studies; AAAS. Res: Migration; urban anthropology; industrial and economic development; cultural change; small farm agriculture in the lowland humid tropics. Mailing Add: Dept of Sociol & Anthrop Kalamazoo Col Kalamazoo MI 49001

BRADFORD, ARTHUR, bacteriology, deceased

BRADFORD, DAVID S, b Charlotte, NC, Oct 15, 36; c 3. ORTHOPEDIC SURGERY. Educ: Davidson Col, BS, 58; Univ Pa, MD, 62. Prof Exp: From resident gen surg to resident orthop surg, Columbia-Presby Med Ctr, 62-66, Annie C Kane jr fel, 68-69; asst prof, 70-73, ASSOC PROF ORTHOP SURG, UNIV MINN, MINNEAPOLIS, 73- Mem: Fel Am Col Surgeons; Orthop Res Soc; Am Acad Orthop Surgeons; Scoliosis Res Soc. Res: Connective tissue diseases; diseases of growth and development; idiopathic scoliosis; congenital malformation. Mailing Add: Box 133 412 Union St SE Minneapolis MN 55455

BRADFORD, EDWIN BERNARD, solid state physics, physical chemistry, see 12th edition

BRADFORD, GORDON ERIC, b Kingsey, Que, Nov 2, 29; m 54; c 4. GENETICS. Educ: McGill Univ, BS, 51; Univ Wis, MS, 52, PhD(genetics, animal husb), 56. Prof Exp: Res asst animal sci, Univ Wis, 51-55; asst prof, Macdonald Col, McGill Univ, 55-57; from asst prof to assoc prof, 57-69, PROF ANIMAL SCI, UNIV CALIF, DAVIS, 69- Mem: Am Soc Animal Sci; Genetics Soc Am; Soc Study Reproduction. Res: Genetics of growth and reproduction in livestock and laboratory animals. Mailing Add: Dept of Animal Sci Univ of Calif Davis CA 95616

BRADFORD, HENRY BERNARD, JR, b Baton Rouge, La, Mar 15, 42; m 64; c 2. VIROLOGY. Educ: La State Univ, BS, 64, MS, 67; Univ Ala, Birmingham, PhD(microbiol), 75. Prof Exp: Microbiologist, 67-69, chief virol unit, 70-71, ASST DIR BIRMINGHAM LAB, BUR LABS, STATE PUB HEALTH ALA, 71- Mem: AAAS; Am Soc Microbiol; Soc Gen Microbiol. Res: Molecular biology of molluscum contagiosum virus. Mailing Add: 2441 Jamestown Dr Birmingham AL 35226

BRADFORD, JAMES C, b Wichita Falls, Tex, Aug 28, 30; m 51; c 3. MATHEMATICS. Educ: NTex State Col, BS, 51, MS, 52; Univ Okla, PhD(math), 57. Prof Exp: Assoc prof math, Abilene Christian Col, 57-61; res mathematician, Teledyne Systs Co, 61-65; PROF MATH & CHMN DEPT, ABILENE CHRISTIAN COL, 65- Concurrent Pos: Consult, Chance-Vought Aircraft, 57-61; vis scientist, Tex Acad Sci, 60-61. Mem: Am Math Soc; Math Asn Am; Soc Indust & Appl Math; Inst Elec & Electronics Eng. Res: Functions of a real variable; abstract analysis; analysis stationary time series applied to geophysics. Mailing Add: Dept of Math Abilene Christian Col Sta ACC Abilene TX 79601

BRADFORD, JAMES CARROW, b Wilmington, Del, Dec 10, 30; m 53; c 3. TERATOLOGY, GASTROENTEROLOGY. Educ: Univ Del, BS, 54, MS, 60. Prof Exp: Sanitarian, Del Dept Pub Health, 54-55; from assoc res biologist to res biologist, 60-74, RES BIOLOGIST & GROUP LEADER, STERLING-WINTHROP RES INST, RENSSELAER, 74- Mem: NY Acad Sci; Teratology Soc. Res: Screen and evaluate drugs for teratogenic-embryotoxic potential; research and development of drugs for cure or relief of gastric ulcers. Mailing Add: Rensselaer NY

BRADFORD, JAMES MCCLELLAN, b Media, Ill, May 24, 04; m 31; c 2. PHYSICS. Educ: Monmouth Col, AB, 25; Univ Chicago, MS, 26. Prof Exp: Asst prof physics, Miami Univ, 26-27; actg prof & head dept, Muskingum Col, 29-31; asst, Univ Chicago, 31-33; from instr to asst prof, Cent YMCA Col, 33-45; assoc prof phys sci, George Williams Col, 45-47; assoc prof physics, Beloit Col, 47-59; prof, 59-72, chmn dept, 59-66, 69-70, EMER PROF PHYSICS, MUSKINGUM COL, 72- Concurrent Pos: Instr, Civilian Pilot Training Prog, 40-42; res physicist, Manhattan Proj, Univ Chicago, 43-44; mem, NSF Inst Teaching Physics, 54. Mem: Fel AAAS; Am Asn Physics Teachers. Res: Dielectric constant of Rochelle salt; crystal structure of inorganic salts; acoustics; solid state. Mailing Add: 170 Highland Dr New Concord OH 43762

BRADFORD, JOE MICHAEL, soil physics, see 12th edition

BRADFORD, JOHN NORMAN, b Carthage, Mo, May 28, 31; m 63; c 2. ATOMIC PHYSICS, NUCLEAR PHYSICS. Educ: Kans State Univ, BS, 54; Iowa State Univ, PhD(physics), 65. Prof Exp: Sr physicist, Gen Dynamics/Astronaut, 62-63; asst prof physics, Univ Mont, 65-67 & Kans State Univ, 67-69; RES PHYSICIST, AIR FORCE CAMBRIDGE RES LAB, BEDFORD, 69- Mem: Am Phys Soc. Res: Nuclear physics, particularly photonuclear reactions; chnanneling and energy loss; x-ray induced electron emission. Mailing Add: 39 Clarke St Lexington MA 02173

BRADFORD, REAGAN HOWARD, b Lawton, Okla, Dec 19, 32; m 53; c 2. BIOCHEMISTRY. Educ: Univ Okla, BS, 53, PhD(biochem), 57, MD, 61. Prof Exp: Res assoc, 59-62, from asst to actg head cardiovasc sect, 62-70, HEAD CARDIOVASC SECT, OKLA MED RES FOUND, 71-, SCI DIR, 74-; ASSOC PROF MED, SCH MED, UNIV OKLA, 72-, PROF BIOCHEM, 69- Concurrent Pos: Res fel, Okla Med Res Found, 57-59; instr res med, Univ Okla, 64-68, assoc prof res biochem, 64-69, asst prof med, 68-72. Mem: Am Physiol Soc; AMA; Am Fedn Clin Res; Am Chem Soc. Res: Lipid transport and metabolism. Mailing Add: Okla Med Res Found 825 NE 13th Oklahoma City OK 73104

BRADFORD, ROBERT R, agronomy, see 12th edition

BRADFORD, SPENCER GRAVES, b Philadelphia, Pa, Feb 25, 19; m 56; c 6. ACADEMIC ADMINISTRATION, PHYSIOLOGY. Educ: Philadelphia Col Osteop Med, DO, 42, MSc, 61. Prof Exp: Clin asst otolaryngol, 43-44, asst, 44-45, from instr to assoc prof, 45-63, PROF PHYSIOL & PHARMACOL, PHILADELPHIA COL OSTEOP MED, 63-, CHMN DEPT, 61-, ASST DEAN BASIC SCI, 72- Concurrent Pos: Vpres, Nat Bd Exam Osteop Physicians & Surgeons. Mem: AAAS; NY Acad Sci; Am Pub Health Asn; Am Heart Asn; Am Asn Lab Animal Sci. Res: Somaticovisceral relations. Mailing Add: Philadelphia Col of Osteop Med Philadelphia PA 19131

BRADFORD, WILLIAM DALTON, b Rochester, NY, Nov 2, 31; m 61; c 2. PATHOLOGY, PEDIATRICS. Educ: Amherst Col, AB, 54; Western Reserve Univ, MD, 58; Am Bd Pediat, dipl, 63; Am Bd Path, dipl anat path, 67. Prof Exp: Intern path, Children's Hosp Med Ctr, Boston, 58-59, asst resident pediat, 59-61; resident path, Boston Hosp for Women, 63-64; New Eng Deaconess Hosp, Boston, 64-65; asst prof path & assoc pediat, 66-70, ASSOC PROF PATH, ASST PROF PEDIAT & ASSOC DIR UNDERGRAD MED EDUC, DUKE UNIV, 70- Concurrent Pos: Teaching fel, Harvard Univ, 63-64; Mead Johnson res fel pediat, 63-64; fel path, Duke Univ, 65-66. Mem: Am Asn Path & Bact; Am Soc Exp Path. Res: Iron metabolism; subcellular pathology. Mailing Add: Dept of Path Duke Univ Durham NC 27710

BRADFORD, WILLIAM HENRY, b Newton, Miss, Aug 14, 12; m 39; c 1. APPLIED MATHEMATICS. Educ: Univ Southern Miss, BS, 36; La State Univ, MS, 39; Univ Tex, PhD(appl math), 53. Prof Exp: Teacher high sch, Miss, 36-37; from instr to prof math, McNeese State Col, 39-57, head dept math & sci, 42-45 & 47-50, dean div lib arts, 52-57; instr appl math, Univ Tex, 45-47; mem staff, Math Res Dept, Sandia Corp, 57-75; RETIRED. Mem: Math Asn Am. Res: Potential theory; polynomials with complex coefficients; operations research. Mailing Add: 2911 Arizona Pl NE Albuquerque NM 87110

BRADFORD, WILLIAM L, b Sedalia, Mo, June 8, 98; m 28; c 1. PEDIATRICS, BACTERIOLOGY. Educ: Univ Mo, AB, 20; Wash Univ, MD, 23. Prof Exp: Intern, St Louis City Hosp, 23-24; asst health comnr, Mo, 24-26; resident pediatrician, Strong Mem Hosp, 26-27, asst bact & path, 27-28; from instr to assoc prof pediat, 28-50, PROF PEDIAT, SCH MED & DENT, UNIV ROCHESTER, 50- Concurrent Pos: Asst dean sch med & dent, Univ Rochester, 47-54, chmn dept, 52-64, Richard Ham lectr, 69; Seyle lectr, Univ Wash, 58; distinguished prof, Univ Mo, 64, Dan Darrow lectr, 67. Mem: AAAS; Soc Pediat Res (vpres, 40); Am Pub Health Asn; Am Soc Microbiol; assoc Soc Exp Biol & Med. Res: Infectious diseases; nutrition and infection; immunity in pertussis; nasal culture method for diagnosis; immune rabbit serum in pertussis; para-pertussis. Mailing Add: Sch of Med & Dent Univ of Rochester Rochester NY 14620

BRADFORD, WILLIS WARREN, b Lincolnton, Ga, Mar 13, 22; m 41; c 2. AGRONOMY. Educ: Univ Ga, BSA, 47, MS, 48; Agr & Mech Col Tex, PhD(plant breeding), 54. Prof Exp: Assoc agronomist, Univ Ga, 48-50; geneticist, Inst de Fomento Algodonera, Colombia, SAm, 50; agronomist, Coastal Plain Exp Sta, Ga, 52-57; agronomist, Miss, 57-59, MEM STAFF ADMIN COTTON BREEDING, WESTERN DIV, DELTA & PINE LAND CO, 59- Res: Cotton breeding. Mailing Add: PO Box 1356 Brawley CA 92227

BRADHAM, GILBERT BOWMAN, b Sumter, SC, Aug 5, 31; m; c 2. SURGERY. Educ: Med Univ SC, MD, 56. Prof Exp: Assoc, 64-65, from asst prof to assoc prof, 65-72, PROF SURG, MED UNIV SC, 72- Mem: AMA; Soc Univ Surg. Mailing Add: Dept of Surg Med Univ of SC Charleston SC 29401

BRADHAM, LAURENCE STOBO, b Charleston, SC, Dec 3, 29; m 58; c 4. BIOCHEMISTRY. Educ: Univ of the South, BS, 51; Univ Tenn, MS, 53, PhD(biochem), 58. Prof Exp: Instr pharmacol, Vanderbilt Univ, 58-59; res assoc, Rockefeller Inst, 59-63; res biochemist, Vet Admin Hosp, Little Rock, 63-71; asst prof, 71-74, ASSOC PROF BIOCHEM, UNIV TENN, MEMPHIS, 74- Mem: Am Chem Soc; Am Soc Biol Chem. Res: Metabolism of sulfur compounds; biochemistry of hormone action; enzymology. Mailing Add: Dept of Biochem Univ of Tenn Memphis TN 38103

BRADLEY, ARTHUR, b New York, NY, Apr 6, 26; c 3. SURFACE CHEMISTRY. Educ: Columbia Univ, AB, 48, MA, 50, PhD(chem), 52. Prof Exp: Res assoc phys org chem, George Washington Univ, 52-53; org chemist, US Naval Ord Lab, 53-55; radiochemist, Associated Nucleonics, Inc, 55-58; dir res, Radiation Res Corp, 58-69; VPRES, SURFACE ACTIVATION CORP, 69- Mem: Am Chem Soc; Electrochem Soc. Res: Organic reaction mechanisms; nuclear batteries; plasma chemistry; thin organic polymer films; surface grafting; protective coatings; gas discharges. Mailing Add: 146 Beech St Floral Park NY 11001

BRADLEY, CHARLES CRANE, b Chicago, Ill, Jan 11, 11; m 42; c 2. EARTH SCIENCES. Educ: Univ Wis, PhB, 35, PhM, 47, PhD(geol), 50. Prof Exp: Actg instr geol, Univ Wis, 46-49; from instr to assoc prof, 50-57, dean div letters & sci, 57-68, PROF GEOL, MONT STATE UNIV, 57- Mem: AAAS; Geol Soc Am; Am Geophys Union. Res: Petrology; general geology; hydrogeology; snow mechanics and avalanches. Mailing Add: Dept of Earth Sci Mont State Univ Bozeman MT 59715

BRADLEY, DONALD BIGELOW, soil science, see 12th edition

BRADLEY, DORIS P, b Laurel, Miss, Oct 21, 32; m 57. SPEECH PATHOLOGY, AUDIOLOGY. Educ: Univ Southern Miss, BS, 53; Univ Fla, MA, 57; Univ Pittsburgh, PhD(speech path), 63. Prof Exp: Dir speech clin, Miss Soc Crippled Children, 53-54; speech therapist, Gulfport Schs, 54-56; speech clinician, Los Angeles Soc Crippled Children, 57-58; instr speech path, Med Sch, Univ Ore, 58-61; dir speech clin, Children's Hosp, Pittsburgh, Pa, 63-65; asst prof speech path & coordr grad prog speech & audiol, Med Sch, Univ Ore, 65-68; res assoc, Univ NC, Chapel Hill, 68-70, assoc prof oral biol & dir oral facial & commun dis prog, Dent Sch, 70-74; PROF DEPT OF LOGOPEDICS, WICHITA STATE UNIV & HEAD CLIN SERV, INST LOGOPEDICS, 74- Concurrent Pos: NIH grant, 65-; consult, United Cerebral Palsy, 54-56 & Ore State Dept Educ, 58-61; res assoc, Cleft Palate Res Ctr, Pa, 63-65. Mem: Am Cleft Palate Asn; Am Speech & Hearing Asn; Soc Res Child Develop. Res: Speech problems associated with cleft palate; development of normal human communication skills in children; modification of aerodynamics of speech production through speech therapy; development developmental aspects of oral port constriction and vowel length; multiple phonemic articulation therapy; development of models of program evaluation; effectiveness of language intervention systems. Mailing Add: Inst of Logopedics 2400 Jardine Dr Wichita KS 67219

BRADLEY, EDWARD CHARLES, b Philadelphia, Pa, July 18, 28. MEDICINE. Educ: St Joseph's Col, Pa, BS, 51; Jefferson Med Col, MD, 55. Prof Exp: Intern med, Lankenau Hosp, Philadelphia, 55-56; resident, Pa Hosp, 59-60; resident, Jefferson Med Col, 60-61; resident, Lankenau Hosp, Philadelphia, 61-62; fel Gothenburg Univ, 62-63; fel cardiovasc med, Sch Med, Univ Southern Calif, 63-64, instr med & res assoc cardiovasc med, 64-66, asst prof, 66-70, ASSOC PROF MED, SCH MED, UNIV SOUTHERN CALIF, 70- Mem: Am Heart Asn; Am Thoracic Soc; assoc Am Col Cardiol; fel Am Col Physicians. Res: Cardiovascular research; shock. Mailing Add: Dept of Med Univ Southern Calif Sch of Med Los Angeles CA 91103

BRADLEY, EDWIN LUTHER, JR, b Jacksonville, Fla, July 16, 43; m 63; c 2. STATISTICS. Educ: Univ Fla, BS, 64, MStat, 67, PhD(statist), 69. Prof Exp: Assoc sci programmer, Lockheed-Ga Co, 65; ASST PROF BIOSTATIST, UNIV ALA, BIRMINGHAM, 70- Concurrent Pos: Consult statistician, Patuxent Wildlife Res Ctr, Bur Sport Fisheries & Wildlife, Dept Interior, Fla, 68-69. Mem: Am Statist Asn; Biomet Soc. Res: Mathematical statistics; nonparametric statistics; application of statistical techniques to medical data. Mailing Add: Dept of Biostatist Univ of Ala University Sta Birmingham AL 35294

BRADLEY, FRANCIS J, b New York, NY, Jan 15, 26; m; c 5. HEALTH PHYSICS. Educ: Manhattan Col, BEE, 49; Univ Pittsburgh, MS, 53; Ohio State Univ, PhD(physics, chem), 61. Prof Exp: Engr, Westinghouse Elec Corp, 50-53; supt radiation safety, Ohio State Univ, 53-61; tech adv, Int Atomic Energy Agency, 61-62; asst prof health physics, Univ Pittsburgh, 62-65; head health physics sect, Isotopes, Inc, 65-66; PRIN RADIOPHYSICIST RADIOL HEALTH SECT, NY STATE DEPT LABOR, 66- Mem: AAAS; Am Phys Soc; Health Physics Soc. Res: Environmental factors effecting solid state dosimeters; high energy proton dosimeter and radiobiology studies; human radioisotope body burden assessment by bioassay and whole body counting; radiation protection administration and legislation. Mailing Add: Radiol Health Sect Off of Occup Health NY State Dept of Labor New York NY 10047

BRADLEY, GEORGE ALEXANDER, b Staunton, Va, Dec 4, 26; m 57; c 2. HORTICULTURE. Educ: Univ Del, BS, 51; Cornell Univ, MS, 53, PhD(veg crops), 55. Prof Exp: Res asst, Cornell Univ, 51-55; from asst prof to assoc prof, 55-63, dir sec sci training prog plant sci, 65, PROF HORT & FORESTRY, UNIV ARK, FAYETTEVILLE, 63-, HEAD DEPT HORT, 68- Mem: Fel AAAS; fel Am Soc Hort Sci. Res: Mineral nutrition; soils; physiology of vegetable crops. Mailing Add: 1700 Viewpoint Fayetteville AR 72701

BRADLEY, GEORGE EDGAR, b Indianapolis, Ind, Feb 21, 24; m 50; c 1. PHYSICS. Educ: Miami Univ, AB, 45; Univ Mich, MS, 47, PhD(physics), 52. Prof Exp: Assoc prof, 51-58, PROF PHYSICS & CHMN DEPT, WESTERN MICH UNIV, 58- Concurrent Pos: NSF fel, 59; mem staff, Ctr Res Nuclear Physics, Karlsruhe, Ger, 71-72. Mem: Am Phys Soc; Am Asn Physics Teachers. Res: Low energy nuclear physics. Mailing Add: 834 Ellensdale St Kalamazoo MI 49007

BRADLEY, HARRIS WALTON, b Charlotte, NC, Jan 13, 15; m 44; c 4. CHEMISTRY. Educ: Davidson Col, BS, 37; Univ Va, PhD(org chem), 42. Prof Exp: Res chemist, Jackson Lab, E I du Pont de Nemours & Co, 42-43, supvr plant opers, Chambers Works, 43-45, res chemist, Jackson Lab, 45-52, res supvr, Chambers Works, 52-57; tech dir, E R Carpenter Co, 57-60; consult, 60-61; mgr res & develop, Carolina Indust Plastics Div, Essex Wire Corp, 61-64; PRES, NC FOAM INDUSTS, INC, 64- Mem: AAAS; Am Chem Soc. Res: Textile chemicals; water-repellents for textiles; sulfoxidation of hydrocarbons; azo dyes; detergents; neoprene latex; rubber chemicals; blowing agents; polyurethanes. Mailing Add: NC Foam Industs, Inc Box 1112 Mt Airy NC 27030

BRADLEY, HUGH EDWARD, b Olean, NY, Nov 4, 34. OPERATIONS RESEARCH. Educ: Mass Inst Technol, SB, 57, SM, 57; Johns Hopkins Univ, PhD(opers res), 63. Prof Exp: Engr, Sperry Gyroscope Co, 57-60; asst prof indust eng, Univ Mich, 63-67; sr opers res analyst, 67-69, head math serv, 69-70, head mgt sci, 70-71, mgr sci mgt serv, 71-74, MGR MGT INFO SERV, UPJOHN CO, 74- Concurrent Pos: Adj assoc prof, Western Mich Univ, 68-74; ed, Int Abstr in Opers Res, 68-; adj prof, Grand Valley State Col, 75- Mem: Opers Res Soc Am; Inst Mgt Sci. Res: Sales forecasting; physical distribution; project management. Mailing Add: Upjohn Co Kalamazoo MI 49001

BRADLEY, JAMES HENRY STOBART, b London, Eng, Mar 26, 33; m 66. DYNAMIC METEOROLOGY, NUMERICAL ANALYSIS. Educ: Oxford Univ, BA, 54; Univ Toronto, MA, 60; Univ Mich, PhD(meteorol), 67. Prof Exp: Scientist isotope geol, Geol Serv Can, 57-59; scientist instrumentation, Can Meteorol Serv, 59-63; res asst meteorol, Univ Mich, 63-67; prof, Pa State Univ, 67-68; res assoc, McGill Univ, 68-70; PROF PHYSICS, DREXEL UNIV, 70- Mem: Soc Indust & Appl Math; Am Meteorol Soc; Instrument Soc Am; Royal Meteorol Soc; Int Asn Analog Comput. Res: Atmospheric planetary waves; computational partial differential equations; irreversible thermodynamics and non-equilibrium statistical mechanics; wind sensors. Mailing Add: 46 Rabbit Run Wallingford PA 19086

BRADLEY, JOHN A, geography, see 12th edition

BRADLEY, JOHN GOMERSOLL, chemistry, see 12th edition

BRADLEY, JOHN SAMUEL, b US, Feb 23, 23; m 51; c 3. GEOLOGY. Educ: Colo Sch Mines, GeolE, 48; Univ Wash, PhD(geol), 52. Prof Exp: Geologist, geol res sect, Humble Oil & Refining Co, 50-54; marine geologist, Inst Marine Sci, Tex, 54-56; geologist, res & develop dept, Atlantic Refining Co, 56-64; GEOLOGIST, GEOL RES DEPT, AMOCO PROD CO, 64- Mem: Geol Soc Am; Am Asn Petrol Geologists; Soc Econ Paleont & Mineral; Am Geophys Union. Res: Stratigraphy; organic geochemistry; photogeology. Mailing Add: Amoco Prod Res Ctr Box 591 Tulsa OK 74102

BRADLEY, JOHN SPURGEON, b Gulfport, Miss, Jan 28, 34; m 59; c 3. MATHEMATICS. Educ: Univ Southern Miss, BS, 55; George Peabody Col, MA, 56; Univ Iowa, PhD(math), 64. Prof Exp: Instr math, Univ Southern Miss, 56-57 & George Peabody Col, 59-60; from asst prof to assoc prof, 64-72, PROF MATH, UNIV TENN, KNOXVILLE, 72- Concurrent Pos: Vis prof, Univ Dundee, 71-72. Mem: Am Math Soc; Math Asn Am; Edinburgh Math Soc. Res: Boundary value problems; integral inequalities; oscillation theory. Mailing Add: Dept of Math Univ of Tenn Knoxville TN 37916

BRADLEY, JULIUS ROSCOE, JR, b Minden, La, Mar 25, 40; m 63; c 1. ENTOMOLOGY. Educ: La Polytech Inst, BS, 62; La State Univ, MS, 64, PhD(entom), 67. Prof Exp: Asst prof, 67-72, ASSOC PROF ENTOM, NC STATE UNIV, 72- Mem: Entom Soc Am. Res: Ecology and control of the boll weevil; pest management of Heliothis species; pesticide interactions. Mailing Add: Dept of Entom NC State Univ Raleigh NC 27607

BRADLEY, LEE CARRINGTON, III, b Birmingham, Ala, May 31, 26. OPTICAL PHYSICS. Educ: Princeton Univ, AB, 46; Oxford Univ, DPhil(physics), 50. Prof Exp: Instr physics, Princeton Univ, 50-54; asst prof, 54-64, STAFF PHYSICIST, LINCOLN LAB, MASS INST TECHNOL, 64- Mem: Am Phys Soc; Optical Soc Am. Res: Laser beam propagation; reentry physics. Mailing Add: Lincoln Lab Mass Inst Technol PO Box 73 Lexington MA 02173

BRADLEY, MARTIN PATRICK TIMOTHY, b Middlesbrough, Eng, Mar 25, 41; m 65; c 2. ANALYTICAL CHEMISTRY. Educ: Univ London, BSc, 63, ARCS, 63, PhD(anal chem), 67, Imp Col, dipl, 67. Prof Exp: Vis lectr, Sir John Cass Col, Univ London, 65-66; res analyst explor res & prod anal, Heavy Org Chem Div, Imp Chem Industs, 66-69; sr res chemist mass spectrometry & gas chromatog, 70-73, SUPVR ANAL & COMPUT TECH SERV, RES & DEVELOP DEPT, STANDARD OIL CO, OHIO, 73- Concurrent Pos: Short course prof, Am Chem Soc, 72- Mem: Am Chem Soc; Brit Soc Anal Chem. Res: Gas chromatography, on line instrumentation, automation and process control by use of chemical information; analytical chemistry of petrochemicals with particular emphasis on trace analysis and air pollution. Mailing Add: 18100 Scottsdale Blvd Cleveland OH 44122

BRADLEY, MURIEL VIRGINIA, b Colo, Aug 8, 08. BOTANY. Educ: Univ Calif, Los Angeles, AB, 31; Univ Calif, AM, 33, PhD(bot), 36. Prof Exp: Res asst bot, Univ Calif, Los Angeles, 31-32; res asst, Univ Calif, Berkeley, 33-34, teaching asst, 34-36; teaching asst, Univ Calif, Los Angeles, 37-38; res assoc cytogenetics, Univ Calif, Berkeley, 38-42, res assoc radiol, 41-44, Finney-Howell Found fel, 44-45, res assoc cytogenetics, 45-48, Guggenheim Found fel, 49-50; proj assoc, Univ Wis, 50-52, asst specialist pomol, 52-55; asst pomologist, 55-62, ASSOC POMOLOGIST, UNIV CALIF, DAVIS, 62-, LECTR POMOL, 55- Mem: AAAS; Bot Soc Am; Am Soc Hort Sci; Int Soc Plant Morphol. Res: Endopolyploidy; cytology, anatomy, morphology and physiology of fruits and fruit trees; effect of physiologically active substances on structure of cells and tissues in vitro and in vivo. Mailing Add: Dept of Pomology Univ of Calif Davis CA 95616

BRADLEY, RALPH ALLAN, b Can, Nov 28, 23; nat US; m 46; c 2. STATISTICS. Educ: Queen's Univ, Ont, BA, 44; Univ NC, PhD(math statist), 49. Prof Exp: Asst prof math, McGill Univ, 49-50; from assoc prof to prof statist, Va Polytech Inst, 50-59; prof, 59-70, DISTINGUISHED PROF STATIST, FLA STATE UNIV, 70-, HEAD DEPT, 59- Concurrent Pos: Ed, Biometrics, 57-62; chmn, Gordon Res Conf, 65; Ford Found Prog specialist, Cairo Univ, 66-67, Ford Found consult, 68-; mem, Comt Statist, Nat Acad Sci. Honors & Awards: Horsely Award, Va Acad, 57; Brumbaugh Award, Am Soc Qual Control, 56. Mem: Fel AAAS; Fel Inst Math Statist; fel Am Statist Asn (vpres, 76-79); Biomet Soc (pres, 65); Int Statist Inst. Res: Mathematical statistics; statistical methods for sensory difference testing. Mailing Add: 2522 Killarney Way Tallahassee FL 32303

BRADLEY, RICHARD CRANE, b Chicago, Ill, May 14, 22; m 47; c 4. PHYSICS. Educ: Dartmouth Col, AB, 43; Univ Calif, PhD(physics), 53. Prof Exp: Tech aid, Calif Inst Technol, 43; asst physics, Univ Calif, 47-53; res assoc, Cornell Univ, 53-56, from instr to assoc prof, 56-62; assoc prof, 62-66, chmn dept, 70-73, PROF PHYSICS, COLO COL, 66-, DEAN OF COL, 73- Concurrent Pos: Consult, Kaman Nuclear, Inc, 63 & Penrose Cancer Res Inst, 64- Mem: AAAS; fel Am Phys Soc; Am Asn Physics Teachers. Res: Sputtering; secondary ion emissions; surfaces of solids; mass spectrometry; field emission microscopy. Mailing Add: Colo Col Colorado Springs CO 80903

BRADLEY, RICHARD E, b Decatur, Ill, Oct 18, 27; m 48; c 4. PARASITOLOGY, VETERINARY MEDICINE. Educ: Fla State Univ, BS, 49; Univ Ga, MS, 53, DVM, 54; Univ Ga, PhD, 65. Prof Exp: Asst zool, Univ Ga, 49-50, lab asst bact & instr zool, 50-54, instr parasitol, 54-55; instr, Univ Ill, 55-56; pvt practr vet med, Ill, 56-62; instr zool, Danville Jr Col, 62 & vet med, Univ Ga, 63; asst parasitologist, 65-69, ASSOC PROF VET SCI, UNIV FLA, 69-, PROG DIR, FAC PARASITOL, 68- Mem: Am Vet Med Asn; Am Soc Parasitol; Am Soc Trop Med & Hyg; Am Asn Vet Parasitol. Res: Canine filariasis; ruminant nematode parasites; anthelmintics; parasitic protozoa; gnotobiology; pathology and pathogenesis of parasites; immunology of parasitism. Mailing Add: Dept of Vet Sci Univ of Fla Gainesville FL 32611

BRADLEY, RICHARD E, b Omaha, Nebr, Mar 9, 26; m 46; c 3. DENTISTRY, PERIODONTICS. Educ: Univ Nebr, BSD, 50, DDS, 52; Univ Iowa, MS, 58. Prof Exp: Instr periodont, Univ Iowa, 57-58; asst prof, Creighton Univ, 58-59; asst prof, 59-61, assoc prof, 61-65, PROF PERIODONT, COL DENT, UNIV NEBR-LINCOLN, 65-, CHMN DEPT, 61-, DEAN COL DENT, 68- Concurrent Pos: Consult, Vet Admin Hosp, Omaha, 60-, attend, Vet Admin Hosp, Lincoln, 60-; cent off consult, Vet Admin, 65- Mem: Am Soc Periodont (secy, 64); Am Acad Periodont; Am Dent Asn; Int Asn Dent Res. Res: Histochemistry of oral enzymology; clinical periodontics. Mailing Add: Col of Dent Univ of Nebr Lincoln NE 68503

BRADLEY, ROBERT LESTER, JR, b Beverly, Mass, Jan 14, 33; m 60; c 2. DAIRY TECHNOLOGY, FOOD SCIENCE. Educ: Univ Mass, BS, 58; Mich State Univ, MS, 60, PhD(food sci), 64. Prof Exp: ASSOC PROF FOOD SCI, UNIV WIS-MADISON, 64- Mem: Am Dairy Sci Asn; Am Chem Soc. Res: Fate of and antagonism of pesticides on humans and animals; functions of hydrocolloids in milk systems. Mailing Add: Dept of Food Sci Univ of Wis Madison WI 58706

BRADLEY, ROBERT LINCOLN, b Lima, Ohio, Feb 12, 15; m 53. INORGANIC CHEMISTRY. Educ: Bowling Green State Univ, BS, 41; Ohio Northern Univ, BS, 48; Ohio State Univ, MA, 53; Mich State Univ, PhD(higher educ), 62. Prof Exp: Foreman mach shop, Lima Locomotive Works, Inc, Ohio, 41-44, designer power shovels, 44-47; asst prof chem, Ohio Northern Univ, 48-55; from instr to assoc prof natural sci, 56-74, PROF NATURAL SCI, MICH STATE UNIV, 74- Mem: AAAS; Am Chem Soc. Res: General science education. Mailing Add: 121 Natural Sci Bldg Mich State Univ East Lansing MI 48823

BRADLEY, ROBERT MARTIN, b Bury, Eng, Oct 15, 39; m 68. SENSORY PHYSIOLOGY, DEVELOPMENTAL PHYSIOLOGY. Educ: Univ London, BDS, 63; Univ Wash, MSD, 66; Fla State Univ, PhD(biol sci), 70. Prof Exp: House surgeon dent, Royal Dent Hosp, London, 63; house officer, St Marys Hosp, London, 64; res assoc fetal physiol, Nuffield Inst Med Res, Eng, 70-72; asst prof, 72-75, ASSOC PROF DENT, SCH DENT, UNIV MICH, 75- Mem: Am Physiol Soc; Soc Neurosci; Int Asn Dent Res; Europ Chemoreceptor Res Orgn; Sigma Xi. Res: Functional and anatomical development of taste receptors and central taste pathways in fetal and neonatal mammals; development and significance of fetal swallowing function. Mailing Add: Dept of Oral Biol Sch of Dent Univ of Mich Ann Arbor MI 48109

BRADLEY, ROY HENRY EDWARD, b Perth, NB, Aug 10, 27; m 50; c 4. PLANT VIROLOGY. Educ: Univ NB, BSc, 48; Univ London, PhD(plant path), 50, DSc(plant path), 62. Prof Exp: MEM STAFF, RES OFF VIROL, CAN DEPT AGR, 50- Res: Insect vec- tors; aphids as vectors of plant viruses. Mailing Add: Res Sta PO Box 280 Fredericton NB Can

BRADLEY, STANLEY EDWARD, b Columbia, SC, Mar 24, 13; m 36; c 1. MEDICINE. Educ: Johns Hopkins Univ, AB, 34; Univ Md, MD, 38. Prof Exp: Intern, Univ Hosp, Univ Baltimore, 38-40; instr med, Sch Med, Boston Univ, 42-45, asst prof, 45-47; from asst prof to prof, 47-59, chmn dept, 59-70, BARD PROF MED, COL PHYSICIANS & SURGEONS, COLUMBIA UNIV, 60- Concurrent Pos: Commonwealth Fund fel, Col Med, NY Univ, 40-42; asst clin vis physician, Bellevue Hosp, New York, 40-42; asst vis physician, Evans Mem Hosp, Mass, 42-47 & Presby Hosp, New York, 47-51, assoc vis physician, 51-59, dir med serv, 59-70, attend physician, 59-; ed-in-chief, J Clin Invest, 52-57; mem bd sci consult, Sloan Kettering Inst, 62-72; trustee, Mt Desert Island Lab, 71- Honors & Awards: Gibbs Prize, NY Acad Med, 47. Mem: Asn Am Physicians; Am Soc Clin Invest (pres, 57); Am Physiol Soc; Am Acad Arts & Sci; Soc Exp Biol & Med. Res: Normal pathologic physiology of the kidney and liver; hemodynamic adjustments. Mailing Add: Presby Hosp 622 W 168th St New York NY 10032

BRADLEY, STERLING GAYLEN, b Springfield, Mo, Apr 2, 32; m 74; c 5. MICROBIOLOGY, GENETICS. Educ: Southwest Mo State Univ, BA & BS, 50; Northwestern Univ, MS, 52, PhD(biol), 54. Prof Exp: Instr biol, Northwestern Univ, 54; from instr to assoc prof microbiol, Univ Minn, 56-63, prof microbiol & genetics, 63-68; PROF MICROBIOL & CHMN DEPT, VA COMMONWEALTH UNIV, 68- Concurrent Pos: Eli Lilly fel, Univ Wis-Madison, 54-55, NSF fel, 55-56; consult, Upjohn Co, 60-68, Minneapolis Vet Admin Hosp, 61-68 & E R Squibb & Sons, 69-76; mem coun, Soc Exp Biol & Med, 64-66 & Am Soc Microbiol, 68-74; mem exec bd, Int Comt Syst Bacteria, 66-74; mem & chmn bd sci counr, Nat Inst Allergy & Infectious Dis, 68-72; ed, Am Soc Microbiol, 70- Mem: Soc Indust Microbiol (pres, 64-65); Am Asn Immunol; Am Inst Biol Sci; Am Acad Microbiol; Am Thoracic Soc. Res: Antibiotics with emphasis on chemotherapy and toxicology; biology of actinomycetes and actinophages; regulation of cellular phenotype. Mailing Add: Dept of Microbiol Va Commonwealth Univ Richmond VA 23298

BRADLEY, TED RAY, b Kansas City, Mo, Feb 1, 40. PLANT TAXONOMY, ECOLOGY. Educ: Rollins Col, BS, 62; Univ NC, MA, 66, PhD(plant taxon), 67. Prof Exp: Asst prof, 67-74, ASSOC PROF BIOL, GEORGE MASON UNIV, 74- Mem: Am Soc Plant Taxon. Res: Evolution of the genus Triodanis, especially hybridization between species. Mailing Add: Dept of Biol George Mason Univ Fairfax VA 22030

BRADLEY, THOMAS BERNARD, JR, b DuBois, Pa, Dec 2, 28; m 55; c 4. INTERNAL MEDICINE, HEMATOLOGY. Educ: Hamilton Col, AB, 50; Columbia Univ, MD, 54. Prof Exp: Intern med, Presby Hosp, New York, 54-55, asst resident, 55-56 & 58-59; sr instr, Seton Hall Col Med & Dent, 62-63, asst prof, 63-64; assoc, Albert Einstein Col Med, 64-65, asst prof, 65-69; assoc prof, 69-75, PROF MED, UNIV CALIF, SAN FRANCISCO, 75-; CHIEF HEMAT SECT, SAN FRANCISCO VET ADMIN HOSP, 69-, ACTG ASSOC CHIEF OF STAFF RES, 74- Concurrent Pos: Fel hemat, Hopkins Hosp, Baltimore, 59-60; Health Res Coun career scientist, 64-69. Mem: Am Fedn Clin Res; Am Soc Clin Invest; Am Soc Hemat (secy, 74-76). Res: Human hemoglobin variants, structure, function, genetics, pathophysiology; red cell metabolism. Mailing Add: 67 Reed Ranch Rd Tiburon CA 94920

BRADLEY, VIRGINIA, b San Marcos, Tex, July 30, 15. GEOGRAPHY. Educ: Southern Methodist Univ, BSc, 37; Univ Nebr, MA, 38; Univ Chicago, PhD(geog), 49. Prof Exp: Instr geog, 43-44 & 45-46, asst prof, 46-56, ASSOC PROF GEOG, SOUTHERN METHODIST UNIV, 56- Concurrent Pos: Mem, 19th & 22nd Int Geog Cong, Int Geog Union. Mem: Asn Am Geogrs; Am Geog Soc. Res: Regional geography of southwestern United States, Europe and Soviet Union. Mailing Add: Div of Geog Southern Methodist Univ Dallas TX 75275

BRADLEY, WILLIAM CRANE, b Madison, Wis, Feb 22, 25; m 58; c 3. GEOLOGY. Educ: Univ Wis, BS, 50; Stanford Univ, MS, 53, PhD(geol), 56. Prof Exp: From instr to assoc prof, 55-68, chmn dept, 68-72, PROF GEOL, UNIV COLO, BOULDER, 68- Concurrent Pos: Res scientist, Univ Tex, 65-66. Mem: AAAS; Geol Soc Am. Res: Geomorphology. Mailing Add: Dept of Geol Sci Univ of Colo Boulder CO 80302

BRADLEY, WILLIAM D, chemistry, see 12th edition

BRADLEY, WILLIAM FRANK, chemistry, deceased

BRADLEY, WILLIAM ROBINSON, b Minneapolis, Minn, Jan 31, 08; m 31; c 2. INDUSTRIAL HYGIENE, TOXICOLOGY. Educ: Cornell Col, BS, 31; Univ Iowa, MS, 40; Am Bd Indust Hyg, dipl. Prof Exp: Indust hygienist, Dept Health, Detroit, 37-43 & Fidelity & Casualty Co, NY, 43-44; asst dir environ health, Am Cyanamid Co, NY, 44-60; CONSULT, W R BRADLEY & ASSOC, 60- Concurrent Pos: Chmn, Air Pollution Control Comn, NJ, 55-64, expert witness. Mem: AAAS; Am Acad Indust Hyg; Air Pollution Control Asn; Am Indust Hyg Asn; Am Chem Soc. Res: Prevention of occupational diseases; industrial toxicology; radioactivity; air and stream pollution; workmen's compensation; industrial noise and safety. Mailing Add: 87 Homestead Rd Tenafly NJ 07670

BRADLEY, WILMOT HYDE, b New Haven, Conn, Apr 4, 99; m 22; c 2. GEOLOGY. Educ: Yale Univ, PhB, 20, PhD(geol), 27. Hon Degrees: DSc, Yale Univ, 47. Prof Exp: Field asst, US Geol Surv, 20, geol aid, 21-22, asst geologist, 22-25, assoc geologist, 25-29, geologist, 29-36, sr geologist, 36-43, prin geologist, 43-44, chief geologist, 44-59, res geologist, 59-69; RETIRED. Honors & Awards: Award of Merit, Wash Acad, 40. Mem: Nat Acad Sci; Geol Soc Am (pres, 65); Am Soc Limnol & Oceanog; Am Asn Petrol Geol; Int Asn Theoret & Appl Limnol. Res: Origin and stratigraphy of Tertiary sedimentary formations of the west; geomorphology; oceanic geology; paleoecology; military geology; paleolimnology; fossil freshwater algae. Mailing Add: Pigeon Hill Rd Milbridge ME 04658

BRADLOW, HERBERT LEON, b Philadelphia, Pa, Mar 21, 24; m 47; c 3. ORGANIC CHEMISTRY, BIOLOGICAL CHEMISTRY. Educ: Univ Pa, BS, 45; Univ Kans, MS, 47, PhD(chem), 49. Prof Exp: Fel, Univ Calif, 49-50; assoc, Sloan Kettering Inst, 51-63; ASSOC, INST STEROID RES, MONTEFIORE HOSP & MED CTR, 63-; PROF BIOCHEM, ALBERT EINSTEIN COL MED, 71- Concurrent Pos: Assoc prof biochem, Albert Einstein Col Med, 68-71; adj assoc prof, Rockefeller Univ, 71-74, adj prof, 74- Mem: Am Chem Soc; Am Soc Biol Chemists; Endocrine Soc; Brit Soc Endocrinol; The Chem Soc. Res: Fluoroaromatic compounds; cholesterol metabolism; steroid hormone metabolism. Mailing Add: Inst for Steroid Res Montefiore Hosp & Med Ctr Bronx NY 10467

BRADNER, HUGH, b Tonopah, Nev, Nov 5, 15; m 44; c 1. EXPERIMENTAL PHYSICS, GEOPHYSICS. Educ: Miami Univ, AB, 36; Calif Inst Technol, PhD(physics), 41. Hon Degrees: DSc, Miami Univ, 61. Prof Exp: Mem res dept, Champion Paper & Fiber Co, 36-37; asst, Calif Inst Technol, 39-41; mem res staff design & test magnetic mines, US Naval Ord Lab, Wash, 41-43; mem res staff, Manhattan Dist, Los Alamos, NMex, 43-46 & radiation lab, Univ Calif, Berkeley, 46-61; RES PHYSICIST, INST GEOPHYS & PLANETARY PHYSICS, UNIV CALIF, SAN DIEGO, 61-, PROF ENG PHYSICS & GEOPHYS, 64- Mem: Fel Am Phys Soc; Seismol Soc Am; Am Geophys Union. Mailing Add: Inst Geophys & Planetary Physics Univ of Calif at San Diego La Jolla CA 92093

BRADNER, NORMAN RICHARD, plant breeding, see 12th edition

BRADNER, WILLIAM TURNBULL, b NJ, Aug 16, 24; m 51; c 3. BACTERIOLOGY, ONCOLOGY. Educ: Lehigh Univ, BA, 48, MS, 49, PhD(bact),

52. Prof Exp: Assoc bact, Brown Univ, 53-55; assoc, 56-58, ASST PROF BACT, MED COL, CORNELL UNIV, 58-; ASST DIR PHARMACOL RES, BRISTOL LABS, SYRACUSE, 65- Concurrent Pos: Asst mem exp chemother, Sloan Kettering Inst, 55-58; sr res scientist cancer chemother screening, 58-65. Mem: AAAS; Am Soc Microbiol; Am Asn Cancer Res; NY Acad Sci. Res: Amebicidal agents; bacterial associates of Endamoeba histolytica; bacteriology of radiation infection; inhibition of anaerobic bacteria; experimental cancer chemotherapy; host defense against cancer. Mailing Add: 4903 Briarwood Circle Manlius NY 13104

BRADOW, RONALD L, organic chemistry, see 12th edition

BRADSHAW, AUBREY SWIFT, b West Sunbury, Pa, July 12, 10; m 34; c 1. FRESH WATER ECOLOGY. Educ: Univ Ky, AB, 34, MA, 44. Prof Exp: Asst zool, Univ Ky, 34-35; instr biol, Transylvania Col, 35-43, asst prof biol & geog, 44-51, prof biol, 51-53; from assoc prof to prof, 53-75, EMER PROF ZOOL, OHIO WESLEYAN UNIV, 75- Concurrent Pos: Asst, Cornell Univ, 49-51. Mem: Am Soc Syst Zool; Am Micros Soc; Am Soc Limnol & Oceanog; Int Asn Theoret & Appl Limnol; Int Asn Great Lakes Res. Res: Life histories, ecology and industrial bioassay of Cladocera and Copepoda. Mailing Add: 109 E Price Rd Oak Ridge TN 37830

BRADSHAW, BENJAMIN CRENSHAW, b Pembroke, Ky, Feb 21, 05. SURFACE PHYSICS. Educ: Transylvania Col, AB, 26; Vanderbilt Univ, AM, 27; Harvard Univ, PhD(phys chem), 32. Prof Exp: Instr, Univ Ky, 27-28 & Harvard Univ, 33-34; assoc prof chem, Maryville Col, 35-41; instr, NC State Col, 41 & Juniata Col, 41-42; prof math & physics, Lambuth Col, 43; res chemist, Manhattan Proj, Univ Chicago, 43-44; guest, Rockefeller Inst, 44-45; physicist & chemist, Squier Lab, Ft Monmouth, NJ, 45-75. Mem: AAAS; fel Am Inst Chemists. Res: Electrode potentials; conductivity; transference numbers; quantum yields; nuclear physics; equations relating time, adsorption, over voltage, catalytic decay and pressure. Mailing Add: Pembroke KY 42266

BRADSHAW, CLAIRE MARGARET, b Los Angeles, Calif, Dec 11, 39. IMMUNOLOGY, IMMUNOCHEMISTRY. Educ: Univ Calif, Los Angeles, BS, 61; Univ Miami, MS, 68, PhD(microbiol), 70. Prof Exp: Lab technician immunol, Scripps Clin & Res Found, 62-63, lab technician biochem, 63; lab technician immunol, Variety Childrens Res Found, Fla, 64; RES ASSOC MICROBIOL, UNIV MIAMI, 70- Concurrent Pos: Smith Kline & French Labs fel, 71. Mem: Am Soc Microbiol. Res: Cold agglutinins; secretory IgM; phylogeny of immunoglobulins; immunology of fish. Mailing Add: Lab of Virol Univ of Miami Miami FL 33152

BRADSHAW, GORDON VAN RENSSELAER, b Kansas City, Mo, Aug 8, 31; m 53; c 2. MAMMALOGY. Educ: Cent Mo State Col, BS, 53; Kans State Univ, MS, 56; Univ Ariz, PhD(zool), 61. Prof Exp: Asst prof biol, The Citadel, 60-61; TEACHER BIOL, PHOENIX COL, 61- Mem: AAAS; Am Soc Mammal. Res: Natural history of southwestern bats. Mailing Add: Dept of Biol Phoenix Col Phoenix AZ 85013

BRADSHAW, JERALD SHERWIN, b Cedar City, Utah, Nov 28, 32; m 54; c 2. ORGANIC CHEMISTRY. Educ: Univ Utah, BS, 55; Univ Calif, Los Angeles, PhD(chem), 62. Prof Exp: NFS fel, 62-63; chemist, Calif Res Corp, 63-66; assoc prof, 66-73, PROF CHEM, BRIGHAM YOUNG UNIV, 73- Concurrent Pos: Nat Acad Sci exchange prof, Univ Ljubljana, Yugoslavia, 72-73. Mem: Am Chem Soc; Int Soc Heterocyclic Chem. Res: Photochemistry of organic compounds in solution; benzyne chemistry; mechanisms of organic reactions; water pollution chemistry; synthesis of macrocyclic compounds. Mailing Add: 1616 Oaklane Provo UT 84601

BRADSHAW, JOHN ALDEN, b Ann Arbor, Mich, Dec 9, 19. ELECTROPHYSICS. Educ: Harvard Univ, BA, 41, MS, 48, PhD(electrophysics), 54. Prof Exp: Res assoc, res lab, Gen Elec Co, 54-62; ASSOC PROF ELEC ENG, RENSSELAER POLYTECH INST, 62- Mem: AAAS; Am Asn Physics Teachers; Inst Elec & Electronics Engrs; Int Solar Energy Soc. Res: Electron beam dynamics; wave propagation in overmoded guides and plasma; microwave scattering from waveguide obstacles. Mailing Add: 223 Green St Schenectady NY 12305

BRADSHAW, JOHN STRATLII, b High Wycombe, Eng, Oct 15, 27; nat US; m 48; c 4. BIOLOGICAL OCEANOGRAPHY. Educ: San Diego State Col, BS, 50; Univ Calif, MS & PhD(oceanog), 58. Prof Exp: Jr res biologist, Scripps Inst Oceanog, Univ Calif, San Diego, 58-67; ASSOC PROF BIOL, UNIV SAN DIEGO, 67- Mem: Am Soc Limnol & Oceanog; Ecol Soc Am. Res: Ecology of recent foraminifera; physiological ecology. Mailing Add: Dept of Biol Univ of San Diego San Diego CA 92110

BRADSHAW, LAWRENCE JACK, b Palo Alto, Calif, Dec 3, 24; m 49; c 4. MEDICAL MICROBIOLOGY. Educ: Stanford Univ, BA, 50, PhD(med microbiol), 56. Prof Exp: Instr microbiol, San Bernardino Valley Col, 54-63, assoc prof, 63-65; assoc prof, 65-68, PROF MICROBIOL, CALIF STATE UNIV, FULLERTON, 68- Mem: AAAS; NY Acad Sci. Res: Immunology; immunochemistry. Mailing Add: Dept of Biol Calif State Univ Fullerton CA 92631

BRADSHAW, RALPH ALDEN, b Boston, Mass, Feb 14, 41; m 61; c 2. BIOCHEMISTRY. Educ: Colby Col, BA, 62; Duke Univ, PhD(biochem), 66. Prof Exp: Res assoc biochem, Ind Univ, 66-67; sr fel, Univ Wash, 67-69; from asst prof to assoc prof, 69-74, PROF BIOCHEM, SCH MED, WASH UNIV, 74- Concurrent Pos: USPHS fel chem, Ind Univ, 66-67; USPHS fel, Univ Wash, 67-68; USPHS career res develop award, 70-74. Mem: AAAS; Am Chem Soc; NY Acad Sci; Am Soc Biol Chem; Am Soc Neurochem. Res: Structure-function relationships in proteins and enzymes; chemistry of peptides and related substances. Mailing Add: Dept of Biochem Wash Univ Sch of Med St Louis MO 63110

BRADSHAW, WILLARD HENRY, b Orem, Utah, Feb 11, 25. MICROBIOLOGY. Educ: Brigham Young Univ, BS, 52, MS, 53; Univ Calif, Berkeley, PhD(comp biochem), 57. Prof Exp: Asst biochem, Univ Calif, 56-57; res assoc, Univ Ill, 57-59; res asst, Brigham Young Univ, 59-60; sr biochemist, Wallace Labs, 60-61; asst prof, 61-65, ASSOC PROF BACT, BRIGHAM YOUNG UNIV, 65- Concurrent Pos: Hon lectr, Stanford Univ, 67-68. Mem: Am Soc Microbiol; Am Chem Soc; Brit Soc Gen Microbiol. Res: Microbial metabolism and physiology; microbial genetics; enzyme chemistry. Mailing Add: Dept of Microbiol Brigham Young Univ Provo UT 84601

BRADSHAW, WILLIAM EMMONS, b Orange, NJ, May 16, 42; m 71; c 1. ANIMAL PHYSIOLOGY, ANIMAL ECOLOGY. Educ: Princeton Univ, AB, 64; Univ Mich, MS, 65, PhD(zool), 69. Prof Exp: NIH res fel, Harvard Univ, 69-71; ASST PROF BIOL, UNIV ORE, 71- Mem: Am Soc Naturalists; Ecol Soc Am; Am Soc Zoologists; Entom Soc Am. Res: Population and community ecology of mosquitoes; photoperiodism and seasonal development; insect diapause. Mailing Add: Dept of Biol Univ of Ore Eugene OR 97403

BRADSHAW, WILLIAM NEWMAN, b Louisville, Ky, Nov 2, 28; m 56; c 2. ENVIRONMENTAL BIOLOGY. Educ: Austin Col, BS, 51; Univ Tex, MS, 56, PhD, 62. Prof Exp: Asst prof biol, McMurry Col, 57-61; lectr zool, Univ Tex, 61; from asst prof to assoc prof, 62-73, dir environ biol prof, 67-71, PROF BIOL, W VA UNIV, 73- Concurrent Pos: Vis assoc prof, M D Anderson Hosp & Tumor Inst, Univ Tex, 70-71; NIH spec res fel, 70-71; dir, Acad Assocs, Inc, 73- Mem: AAAS; Am Soc Mammal; Soc Study Evolution; Ecol Soc Am; Am Inst Biol Sci. Res: Speciation of small mammals; animal ecology; population biology; small mammal ecology; cytogenetics, speciation and behavior; terrestial and wildlife ecology. Mailing Add: Dept of Biol WVa Univ Morgantown WV 26506

BRADSHAW, WILLIAM S, b Salt Lake City, Utah, Oct 29, 37; m 61; c 5. BIOCHEMISTRY, DEVELOPMENTAL BIOLOGY. Educ: Harvard Univ, AB, 63; Univ Ill, PhD(biochem), 68. Prof Exp: Res fel biol div, US AEC, Oak Ridge Nat Lab, 68-70; ASST PROF ZOOL, BRIGHAM YOUNG UNIV, 70- Res: Biochemistry of development; synthesis of pancreatic enzymes during development; regulation of DNA synthesis during development of Rhynochosciara. Mailing Add: Dept of Zool Brigham Young Univ Provo UT 84601

BRADSHER, CHARLES KILGO, b Petersburg, Va, July 13, 12; m 38; c 3. ORGANIC CHEMISTRY. Educ: Duke Univ, AB, 33; Harvard Univ, AM, 35, PhD(org chem), 37. Prof Exp: Rohm & Haas fel, Univ Ill, 37-38, du Pont fel, 38; from instr to prof, 39-65, chmn dept, 65-70, JAMES B DUKE PROF CHEM, DUKE UNIV, 65- Concurrent Pos: Nat Res fel, 41-42; Fulbright lectr, State Univ Leiden, 51-52; NSF sr fel, Swiss Fed Inst Technol, 59- Mem: Am Chem Soc; Royal Netherlands Chem Soc. Res: Aromatic cyclodehydration; quinolizinium derivatives; cationic polar cycloaddition. Mailing Add: Dept of Chem Duke Univ Durham NC 27706

BRADSTREET, RAYMOND BRADFORD, b Salem, Mass, Aug 21, 01; m 30. ANALYTICAL CHEMISTRY, ORGANIC CHEMISTRY. Educ: Pratt Inst, dipl, 29; Polytech Inst Brooklyn, BS, 32, MS, 42. Prof Exp: Anal chemist, A C Lawrence Leather Co, 24-28; anal res chemist, Gen Labs, US Rubber Co, 29-36; anal res chemist, Standard Oil Develop Co, 36-42, group head anal res, 46; chief chemist, Creole Petrol Corp, Venezuela, 46-49; pres, Bradstreet Labs, Inc, 49-56; res dir, Carbon Solvents Lab, 56-59; supvr chem dept, US Testing Co, 59-61, petrol & leather div, 61-68; CONSULT, 68- Mem: Fel AAAS; Am Chem Soc; fel Am Inst Chemists; NY Acad Sci; fel The Chem Soc. Res: Kjedahl method for organic nitrogen; organic analysis methods; standard solutions. Mailing Add: PO Box 1 Cranford NJ 07016

BRADT, GLENN WARNER, mammalogy, deceased

BRADT, HALE VAN DORN, b Colfax, Wash, Dec 7, 30; m 58; c 2. PHYSICS. Educ: Princeton Univ, AB, 52; Mass Inst Technol, PhD(physics), 61. Prof Exp: From instr to assoc prof, 61-72, PROF PHYSICS, MASS INST TECHNOL, 72- Concurrent Pos: Assoc ed, Astrophys J Lett, 73- Mem: Am Phys Soc; Am Astron Soc. Res: Cosmic ray physics; x-ray astronomy. Mailing Add: 30 Clover St Belmont MA 02178

BRADT, OLIVER A, b Morrisburg, Ont, Sept 30, 13; m 42; c 2. HORTICULTURE. Educ: Univ Toronto, BSA, 38. Prof Exp: Asst, Hort Res Inst Ont, 38-42; lectr, Ont Agr Col, 46-48; asst res, 49-53, RES SCIENTIST, HORT RES INST ONT, 54- Honors & Awards: Wilder Medal, Am Pomol Soc, 73. Mem: Am Pomol Soc (pres, 69 & 70); Am Soc Hort Sci. Res: Breeding apricots, peaches and grapes. Mailing Add: Hort Res Inst Ont Vineland Station ON Can

BRADT, RUSSELL NEWTON, b Newton, Iowa, July 15, 22; m 49; c 4. MATHEMATICAL STATISTICS. Educ: Colo State Col Educ, AB, 46, MA, 48; Univ Kans, MA, 51; Stanford Univ, PhD, 54. Prof Exp: From asst prof to assoc prof, 54-69, PROF MATH, UNIV KANS, 70- Mem: Am Math Soc; Math Asn Am; Inst Math Statist; Am Statist Asn. Res: Comparison of experiments. Mailing Add: Dept of Math Univ of Kans Lawrence KS 60045

BRADWAY, KEITH E, b Indianapolis, Ind, Dec 31, 26; m 50; c 4. PULP & PAPER TECHNOLOGY, ORGANIC CHEMISTRY. Educ: Purdue Univ, BSChE, 48; Inst Paper Chem, MS, 50, PhD(org chem), 53 Prof Exp: Res chemist, Camp Mfg Co, 53-54, asst tech dir, 54-56, supt process eng, Union Bag Camp Paper Corp, 56-60, tech control, 60-62, RES SCIENTIST, UNION CAMP CORP, NJ, 62- Honors & Awards: Hugh D Camp Award, 68. Mem: Am Tech Asn Pulp & Paper Indust. Res: Technology of manufacture of white papers and unbleached paperboard, particularly with regard to printability and opacity. Mailing Add: 30 Pin Oak Dr Lawrenceville NJ 08648

BRADY, AL H, b Jonesboro, Ark, June 6, 38; m 67. MEDICAL BIOPHYSICS. Educ: Univ Dayton, BS, 60; Ariz State Univ, MS & PhD(physics), 64; Univ Miami, MD, 72. Prof Exp: Trainee, Univ Miami, 64-67; res assoc, Columbia Univ, 67-68; ASST PROF MED & BIOCHEM, SCH MED, UNIV MIAMI, 68- Concurrent Pos: NIH res career develop award, 75. Mem: Am Chem Soc; Am Soc Biol Chemists. Res: Structural relationships between proteins and their prosthetic groups; structure of myelin proteins; connective tissue diseases; protein optical activity. Mailing Add: Dept of Med Univ of Miami Sch of Med Miami FL 33152

BRADY, ALLAN JORDAN, b Fairview, Utah, May 23, 27; m 52; c 2. BIOPHYSICS. Educ: Univ Utah, BA, 51, MS, 52; Univ Wash, PhD(biophys), 56. Prof Exp: Asst physiol & biophys, Sch Med, Univ Wash, 55-56, Am Heart Asn fel, 56-57; Am Heart Asn fel, Physiol Lab, Univ Cambridge, 57-58; advan res fel, Los Angeles County Heart Asn Cardiovasc Res Lab, Med Ctr, 58-60, estab investr, 60-65, assoc prof in residence, Sch Med, 62-66, PROF PHYSIOL, SCH MED, UNIV CALIF, LOS ANGELES, 66-, CAREER DEVELOP AWARDS, LOS ANGELES COUNTY HEART ASN CARDIOVASC RES LAB, MED CTR, 65- Concurrent Pos: Mem basic sci exec comt, Am Heart Asn, 62- Mem: Biophys Soc; Am Physiol Soc. Res: Electrical and ionic basis of single fiber activity; excitation-contraction coupling; muscle mechanics. Mailing Add: Dept of Physiol Univ of Calif Sch of Med Los Angeles CA 90066

BRADY, ALLEN H, b Sewal, Iowa, Mar 17, 34; m 59; c 3. MATHEMATICS, COMPUTER SCIENCES. Educ: Univ Colo, BS, 56; Univ Wyo, MS, 59; Ore State Univ, PhD(math), 65. Prof Exp: Jr engr, Mat Res Lab, Collins Radio Co, Calif, 55; physicist, Cent Radio Propagation Lab, Nat Bur Standards, Colo, 56-64; asst prof comput sci, Univ Notre Dame, 65-66; dir, Comput Data Processing Ctr, Ball State Univ, 66-68; ASSOC RES PROF COMPUT SCI, DESERT RES INST, UNIV NEV, RENO, 68-, ASST DIR EDUC & RES COMPUT CTR, UNIV SYST, 69- Mem: Am Math Soc; Math Asn Am; Asn Comput Mach; Inst Elec & Electronics Engrs. Res: Artificial intelligence; turing computability; computer organization. Mailing Add: Comput Ctr Univ of Nev Syst Reno NV 89507

BRADY, ALLEN ROY, b Houston, Tex, Feb 7, 33; m 58; c 4. SYSTEMATIC ZOOLOGY. Educ: Univ Houston, BS, 55, MS, 59; Harvard Univ, PhD(biol), 64. Prof Exp: Res fel arachnology, Mus Comp Zool, Harvard Univ, 63-64; Kettering intern biol, Hope Col, 64-65; asst prof, Albion Col, 65-66; from asst prof to assoc prof, 66-72, PROF ZOOL, HOPE COL, 72- Concurrent Pos: Vis prof zool, Univ Fla, Gainesville, 72-73. Mem: Am Arachnol Soc; Am Soc Zoologists; Soc Syst Zool. Res: Systematics and zoogeography of spiders. Mailing Add: 97 W 14th Holland MI 49423

BRADY, BRIAN T, b Cleveland, Ohio, Sept 7, 38; m 65; c 3. MATERIALS SCIENCE, APPLIED MATHEMATICS. Educ: Univ Dayton, BSc, 61; Mass Inst Technol, MSc, 64; Colo Sch Mines, PhD(math, metall, mining), 69. Prof Exp: Geophysicist, Cities Serv Oil Co, Okla, 64-66; PHYSICIST, US BUR MINES, 67- Mem: Am Geophys Union; Am Inst Mining, Metall & Petrol Engrs. Res: Analysis of brittle rock failure and applications of results to earthquake seismology and mine failure prediction. Mailing Add: 1923 Sage Dr Golden CO 80225

BRADY, CARROLL PARKER, mathematics, see 12th edition

BRADY, EDWARD LEWIS, b Charleston, SC, Apr 21, 19; m 44. PHYSICAL CHEMISTRY. Educ: Univ Calif, Los Angeles, BA, 40, MA, 43; Mass Inst Technol, PhD(phys chem), 48. Prof Exp: Asst chem, Univ Calif, Los Angeles, 40-42 & metall lab, Univ Chicago, 42-43; res assoc, Clinton Labs, Oak Ridge, Tenn, 43-46, Mass Inst Technol, 46-48 & res lab, Gen Elec Co, 48-55, mgr coolant chem, Knolls Atomic Power Lab, 55-56, mgr exp equipment develop, 58-59; sr sci adv, US Mission to Int Atomic Energy Agency, Vienna, Austria, 59-61; asst chmn dept chem, Gen Atomic Div, Gen Dynamics Corp, 61-63; chief, off standard ref data, 63-69, ASSOC DIR INFO PROGS, NAT BUR STANDARDS, 69- Concurrent Pos: Lectr, Union Univ, NY, 52-56; sci rep, US Atomic Energy Comn, London, Eng, 56-58. Mem: AAAS; Am Nuclear Soc; Am Phys Soc; Am Chem Soc. Res: Chemistry and radiation characteristics of radioactive isotopes; physical chemistry and irradiation behavior of nuclear power reactor materials; information science and data center operation. Mailing Add: Nat Bur of Standards Washington DC 20234

BRADY, FRANKLIN PAUL, b Winnipeg, Man, Aug 14, 31; m 64. NUCLEAR PHYSICS. Educ: Univ Man, BS, 53; Princeton Univ, AM, 54, PhD(physics), 60. Prof Exp: Res assoc nuclear physics, Princeton Univ, 60-61; physicist, Schlumberger Ltd, 61-62; from asst prof to assoc prof, 62-74, PROF PHYSICS, UNIV CALIF, DAVIS, 74- Mem: Am Phys Soc. Res: Geophysics; experimental nuclear physics; nuclear reactions; cyclotron design. Mailing Add: Dept of Physics Univ of Calif Davis CA 95616

BRADY, FREDERICK JONATHON, b Canton, Pa, July 1, 08; m 32; c 3. TROPICAL MEDICINE. Educ: Univ Mich, MD, 31. Prof Exp: Intern, Hurley Hosp, Mich, 31-32; practicing physician, 32-33; physician officer, USPHS, 34-61, res, 37-50, int health rep, 50-57, chief prog officer, Bur State Serv, 57-59; dir, Pima County Health Dept, 61-69; dir, Pima County Health & Hosp Serv, 69-71; res assoc microbiol, Col Med, Univ Ariz, 62-75; RETIRED. Concurrent Pos: Asst chief div trop dis, NIH, 42-50; US deleg, WHO & Pan-Am Sanit Orgn. Mem: AMA; Am Soc Trop Med & Hyg (pres, 54); Am Pub Health Asn; Royal Soc Trop Med & Hyg. Res: Epidemiology of tropical diseases. Mailing Add: 3150 N Lodge Rd Tucson AZ 85715

BRADY, GEORGE W, b Quebec, Que, Jan 22, 21; m 49; c 2. PHYSICAL CHEMISTRY. Educ: Laval Univ, BSc, 42; McGill Univ, PhD, 49. Prof Exp: Asst, Univ Chicago, 49-51; res fel, Harvard Univ, 51-52; MEM TECH STAFF, BELL LABS, 52- Concurrent Pos: Adj prof, State Univ NY Albany, 76- Mem: Fel Am Phys Soc; fel NY Acad Sci. Res: Chemical kinetics; surface chemistry; structure of liquids; x-ray diffraction; critical state phenomena; structure of membranes and DNA. Mailing Add: 75 Elm Ave Delmar NY 12054

BRADY, HOWARD SHAUB, b Lancaster, Pa, May 31, 08; m 38; c 2. ORGANIC CHEMISTRY. Educ: Franklin & Marshall Col, BS, 30; Univ Pa, MS, 48; Univ Del, PhD(chem), 55. Prof Exp: Teacher, pub schs, Pa, 30-40; org res chemist, Barrett Div, Allied Chem Corp, 40-43; supvr org lab, Publicker Industs Inc, 43-47; asst, Univ Pa, 47-48; group leader, refrig res lab, Philco Corp, 49-51; sr res chemist, boron chem, Am Chem Prod Inc, 51-55; group leader, boron fuels res, Olin Mathieson Chem Corp, 55-59; res chemist, advan fuels synthesis, reaction motors div, Thiokol Chem Corp, 59-61; sr res chemist, heterocyclics synthesis, Hercules Powder Co, 61-63; res specialist, Apollo, NAm Aviation, Inc, 63-68; prof chem, Harford Jr Col, 68-73; RETIRED. Mem: AAAS; Am Chem Soc. Res: Organic synthesis; natural and synthetic fuels; high energy fuels; chemistry of boron and aluminum. Mailing Add: 1009 Winfield Dr Bel Air MD 21014

BRADY, JAMES EDWARD, b New York, NY, Jan 26, 38; m 60; c 1. INORGANIC CHEMISTRY. Educ: Hofstra Col, BA, 59; Pa State Univ, PhD(inorg chem), 64. Prof Exp: Staff scientist chem, Res Ctr, Aerospace Group, Gen Precision, Inc, 64-65; instr, 65-70, ASSOC PROF CHEM, ST JOHN'S UNIV, NY, 70- Mem: Am Chem Soc; NY Acad Sci. Res: Reaction mechanisms of coordination compounds by means of high pressure techniques; synthesis of inorganic luminescent materials at high temperatures. Mailing Add: Dept of Chem St John's Univ Jamaica NY 11432

BRADY, JAMES JOSEPH, b Oregon City, Ore, Nov 24, 04; m 32; c 2. SURFACE PHYSICS. Educ: Reed Col, AB, 27; Ind Univ, AM, 28; Univ Calif, PhD(physics), 31. Prof Exp: Asst physics, Ind Univ, 27-28; fel, St Louis Univ, 31-32, from instr to asst prof, 32-37; from asst prof to prof, 37-73, actg chmn dept, 65-66 & 69-71, EMER PROF PHYSICS, ORE STATE UNIV, 73- Concurrent Pos: Assoc group leader, Radiation Lab, Mass Inst Technol, 42-46; tech consult, Radiation Lab, Univ Calif, 48- Mem: Fel Am Phys Soc; AAAS; Am Asn Physics Teachers. Res: Photoelectric effect of thin films; cyclotron and nuclear physics; microwave optics. Mailing Add: 2015 SW Whiteside Corvallis OR 97330

BRADY, JOHN PAUL, b Boston, Mass, June 23, 28; m 63; c 4. PSYCHIATRY. Educ: Boston Univ, AB, 51, MD, 55. Prof Exp: Resident psychiat, Inst of Living, Hartford, Conn, 56-59; res psychiatrist, Sch Med, Ind Univ, 59-63; from assoc prof to prof, 63-74, KENNETH E APPEL PROF PSYCHIAT & CHMN DEPT, SCH MED, UNIV PA, 74- Concurrent Pos: Co-founder & assoc ed, Behav Ther, 69-; consult panel anal of impact of basic res on ment health, Comt Brain Sci, Nat Res Coun, 72- Mem: Asn Advan Behav Ther (pres, 70-71); Soc Biol Psychiat; Royal Col Psychiat; Pavlovian Soc NAm; Psychiat Res Soc. Res: Applications of principles of learning to disorders of behavior. Mailing Add: Dept of Psychiat Hosp Univ of Pa Philadelphia PA 19104

BRADY, JOSEPH VINCENT, b New York, NY, Mar 28, 22. BEHAVIORAL BIOLOGY. Educ: Fordham Univ, BS, 43; Univ Chicago, PhD, 51. Prof Exp: Chief clin psychologist, Neuropsychiat Ctr, Europ Command, Ger, 46-48; student officer psychol, Univ Chicago, 48-51; chief dept exp psychol, Walter Reed Army Inst Res, 51-64, dep dir, Div Neuropsychiat, 64-70; PROF BEHAV BIOL & DIR DIV, SCH MED, JOHNS HOPKINS UNIV, 67- Concurrent Pos: Prof, Univ Md, 55-69. Mem: Fel AAAS; fel Am Psychol Asn. Res: Experimental analysis of behavior; behavioral physiology; brain and behavior; comparative psychosomatics. Mailing Add: Div of Behav Biol Johns Hopkins Univ Sch Med Baltimore MD 21205

BRADY, LAWRENCE LEE, b Topeka, Kans, Nov 6, 36; m 60; c 2. GEOLOGY, ECONOMIC GEOLOGY. Educ: Kans State Univ, BS, 58; Univ Kans, MS, 67, PhD(geol), 71. Prof Exp: Eng geologist, US Army Corps Engr, Kansas City Dist, 58-63; asst prof geol, Okla State Univ, 71; RES ASSOC GEOL, KANS GEOL SURV, 71-, CHIEF MINERAL RESOURCES SECT, 75- Mem: Asn Eng Geologists; Am Asn Petrol Geologists; Soc Econ Paleontologists & Mineralogists. Res: Kansas coal deposits; quality and distribution of Kansas mineral resources; engineering geology related to mineral development. Mailing Add: Kans Geol Surv 1930 Ave A Campus West Univ of Kans Lawrence KS 66044

BRADY, LEONARD EVERETT, b Brooklyn, NY, Feb 24, 28; m 51; c 1. ORGANIC CHEMISTRY. Educ: Wagner Col, BS, 51; NC State Col, MS, 54; Mich State Univ, PhD(org chem), 58. Prof Exp: Asst, Mich State Univ, 53-57; res chemist, Abbott Labs, 57-61; res assoc chem, Univ Ill, 61-63, instr, 63-64; assoc prof chem, Ill State Univ, 64-66; assoc prof, 66-70, PROF CHEM, UNIV TOLEDO, 71-, CHMN DEPT, 75- Concurrent Pos: Vis scholar, Cambridge Univ, 73-74. Mem: Am Chem Soc; Am Soc Pharmacog. Res: Mechanisms; structure proof; stereochemistry; mass spectrometry; high-nitrogen compounds. Mailing Add: Dept of Chem Univ of Toledo Toledo OH 43606

BRADY, LUTHER WELDON, JR, b NC, 1925. MEDICINE, RADIOLOGY. Educ: George Washington Univ, AB, 46, MD, 48. Prof Exp: Asst prof med & radiol, 59-66, prof radiol, 66-70, PROF RADIATION THER & NUCLEAR MED & CHMN DEPT, HAHNEMANN MED COL & HOSP, 70-, DIR RADIATION THER, 59-, AM CANCER SOC CLIN ONCOL PROF, 75- Concurrent Pos: Consult, Mercy-Douglass Hosp, Philadelphia, 57-, Vet Admin Hosp, 60-, Crozer-Chester Med Ctr, Pa Hosp, Philadelphia, Lankenau Hosp, Philadelphia, West Jersey Hosp, Camden, NJ, Garden State Hosp, Cherry Hill, NJ, St Luke's Hosp, Bethlehem, Pa; asst prof, Harvard Med Sch, 62-63. Mem: Fel Am Col Radiol; Radiol Soc NAm; Am Radium Soc (pres); Am Soc Therapeut Radiol (past pres); Am Soc Clin Oncol. Res: Effects of radiation on pulmonary function on bone marrow using ferrochromokinetics in patients with established diagnosis of malignancy and on kidney function. Mailing Add: Radiation Ther & Nuclear Med Hahnemann Med Col & Hosp Philadelphia PA 19102

BRADY, LYNN R, b Shelton, Nebr, Nov 15, 33; m 57. PHARMACOGNOSY. Educ: Univ Nebr, BS, 55, MS, 57; Univ Wash, PhD(pharmacog), 59. Prof Exp: From asst prof to assoc prof, 59-66, PROF PHARMACOG, UNIV WASH, 66-, CHMN DEPT PHARMACEUT SCI, 72- Mem: Am Pharmaceut Asn; Am Soc Pharmacog; Am Asn Cols Pharm; fel Acad Pharmaceut Sci. Res: Constituents of higher fungi; poisonous plants; chemotaxonomy. Mailing Add: Sch of Pharm Univ of Wash Seattle WA 98195

BRADY, NYLE C, b Manassa, Colo, Oct 25, 20; m 36; c 4. AGRONOMY. Educ: Brigham Young Univ, BS, 41; NC State Col, PhD(agron), 47. Prof Exp: Jr agronomist, NC State Col, 42-44, res instr, 46, asst prof agron, 47; instr, NY State Col Agr, Cornell Univ, 48-49, assoc prof, 49-52, prof soil sci, 52-73, head dept agron, 55-63, dir agr exp sta, 65-73, assoc dean, 70-73; DIR, INT RICE RES INST, PHILIPPINES, 73- Concurrent Pos: Dir sci & educ, USDA, 63-65; chmn agr bd, Nat Res Coun, 67-70; ed, Proc Soil Sci Soc Am. Mem: AAAS; Soil Sci Soc Am; Am Soc Agron; Soil Conserv Soc Am. Res: Physiology of the peanut plant; fundamental effects of lime on soil; influence of fertilizer on the yield of corn, rice and coffee; influence of soil temperature on nutrient uptake. Mailing Add: Int Rice Res Inst Box 933 Manilla Philippines

BRADY, ROBERT FREDERICK, JR, b Washington, DC, July 20, 42; m 65; c 3. ORGANIC CHEMISTRY. Educ: Univ Va, BSChem, 64, PhD(org chem), 67. Prof Exp: Chemist, Nat Bur Stand, 67-72; chemist, US Customs Serv, 72-75; CHIEF, PAINTS BR, FED SUPPLY SERV, 75- Concurrent Pos: Lectr, Nat Bur Stand Grad Sch, 68. Mem: Am Chem Soc; Fed Socs Coatings Technol. Res: Synthetic, structural, and analytical organic chemistry of polymers and natural products; trace organic analysis by instrumental methods; coatings technology, testing and specifications. Mailing Add: 706 Hope Lane Gaithersburg MD 20760

BRADY, ROBERT JAMES, b Detroit, Mich, Apr 13, 27; m 55; c 4. MICROBIOLOGY. Educ: Univ Detroit, BS, 51, MS, 54; Univ Md, PhD(microbiol), 58. Prof Exp: Asst biol, Univ Detroit, 53-55; asst microbiol, Univ Md, 54-57; from asst prof to assoc prof microbiol, 57-69, PROF MICROBIOL, MIAMI UNIV, 69-, CHMN DEPT, 73- Mem: Am Soc Microbiol; Brit Soc Gen Microbiol. Res: Microbial physiology and genetics. Mailing Add: Dept of Microbiol Miami Univ Oxford OH 45056

BRADY, ROBERT NYLE, b Manassa, Colo, Apr 7, 37; m 60; c 3. BIOCHEMISTRY. Educ: Brigham Young Univ, BS, 62; Cornell Univ, PhD(biochem), 67. Prof Exp: NIH fel, Cornell Univ, 67 & Univ Calif, Berkeley, 67-69; ASST PROF BIOCHEM, VANDERBILT UNIV, 69- Concurrent Pos: Andrew Mellon Found fel & award, Vanderbilt Univ, 74. Mem: AAAS; Am Chem Soc. Res: Neurochemistry; membrane function and structure; receptors; lipid metabolism. Mailing Add: Dept of Biochem Vanderbilt Univ Sch of Med Nashville TN 37232

BRADY, ROBERT TOWNSEND, b Brooklyn, NY, Oct 23, 25; m 52; c 2. PETROLEUM GEOLOGY. Educ: Union Col, NY, 47; Univ Wyo, MA, 49. Prof Exp: Instr geol, Union Col, NY, 48-49; tutor, Queens Col, NY, 49-51; geologist, Gulf Oil Corp, 51-56; party chief, 56-58, subsurface geologist, 58-59, regional geologist, 59-64, head regional geologist, 64, supvr regional geol, 64-65, asst to explor mgr, 65-69, ASST EXPLOR CONSULT, AM OVERSEAS PETROL, LTD, 69- Mem: Am Asn Petrol Geologists; fel Geol Soc Am; Petrol Explor Soc Libya (vpres, 62, pres, 63). Res: Geology of Indonesia. Mailing Add: Am Overseas Petrol Ltd 380 Madison Ave New York NY 10017

BRADY, ROSCOE OWEN, b Philadelphia, Pa, Oct 11, 23. BIOCHEMISTRY. Educ: Harvard Univ, MD, 47. Prof Exp: Intern, Univ Hosp, Univ Pa, 47-48; officer-in-chg, Chem Labs, Naval Med Ctr, 52-54; chief lipid chem sect, 54-70, asst chief lab neurochem, 70-72, CHIEF DEVELOP & METAB NEUROL BR, NAT INST NEUROL & COMMUN DIS & STROKE, 72- Concurrent Pos: Nat Res Coun fel med sci, Univ Hosp, Univ Pa, 48-50, fel, 50-51, USPHS fel, 51-52, fel, Endocrine Sect, Med Clin, 52-54; prof lectr, Sch Med, George Washington Univ & Georgetown Univ. Honors & Awards: Superior Serv Award, US Dept Health, Educ & Welfare, 66, Distinguished Serv Award, 74; G Burroughs Mider Award, 70; Gairdner Found Int Award, 73; First Nat Lectr Award, Am Soc Biol Chem, 74. Mem: Nat Acad Sci; Am Acad Neurol; Am Soc Biol Chem; Am Chem Soc; Am Acad Ment Retarded. Res: Complex lipid metabolism; genetic diseases; viral carcinogenesis; neurochemistry; biosynthesis of fatty acids and complex lipids of the nervous system. Mailing Add: Rm 3DO3 Bldg 10 Nat Inst of Neurol & Commun Dis & Stroke Bethesda MD 20014

BRADY, RUTH MARY, b Bridgeport, Conn, Feb 18, 24. PHYSICAL CHEMISTRY. Educ: Albertus Magnus Col, AB, 44; Fordham Univ, MS, 55, PhD(phys chem), 65. Prof Exp: Asst prof chem, Col St Mary, Ohio, 63-65, Albertus Magnus Col, 65-68; sr scientist, York Res Corp, Conn, 68-69; PROF CHEM, ALCORN STATE UNIV, 69-, HEAD DEPT, 73- Mem: Am Chem Soc; AAAS. Res: Phase relations in isomeric systems; kinetics and mechanism of reactions of thiolacetic acid; effects of metal ions on growth of seedlings. Mailing Add: Dept of Chem Alcorn State Univ Lorman MS 39096

BRADY, STEPHEN FRANCIS, b New York, NY, Oct 17, 41; m 66. ORGANIC

CHEMISTRY. Educ: Columbia Univ, BA, 63; Stanford Univ, PhD(chem), 67. Prof Exp: Assoc chem, Dept Chem, Northwestern Univ, 67-70; SR RES CHEMIST, MERCK SHARP & DOHME RES LABS DIV, MERCK & CO, INC, 70- Mem: Am Chem Soc; The Chem Soc. Res: Synthesis of peptides; development of new methodology and protecting groups for peptide synthesis; chemistry and biology of hormonal peptides and proteins; synthesis of novel amino acids. Mailing Add: 1237 Mill Rd Meadowbrook PA 19046

BRADY, STEPHEN W, b Indianapolis, Ind, Jan 12, 41; m 65. MATHEMATICS. Educ: Ind Univ, BA, 63, MA, 65, PhD(math), 68. Prof Exp: ASST PROF MATH, WICHITA STATE UNIV, 67- Concurrent Pos: Consult, Math & Biol Corp, 67-68. Mem: AAAS; Am Math Soc; Math Asn Am; NY Acad Sci. Res: Functional and numerical analysis; differential equations; mathematics applied to physiology. Mailing Add: Dept of Math Wichita State Univ Wichita KS 67208

BRADY, THOMAS E, b Elizabeth, NJ, Apr 2, 41; m 70. SYNTHETIC ORGANIC CHEMISTRY, TEXTILE CHEMISTRY. Educ: St Vincent Col, BS, 63; Fordham Univ, PhD(org chem), 68. Prof Exp: Res chemist, Org Chem Div, 67-74, SR RES CHEMIST, CHEM RES DIV, AM CYANAMID CO, NJ, 74- Mem: Am Chem Soc; NY Acad Sci. Res: Synthesis of dyes and detergent additives; flame retardants. Mailing Add: 58 Evans Ave Piscataway NJ 08854

BRADY, ULLMAN EUGENE, JR, b Selma, Ala, July 18, 33; m 52; c 2. INSECT TOXICOLOGY, PHYSIOLOGY. Educ: Auburn Univ, BS, 60, MS, 61; Univ Ill, PhD(entom), 65. Prof Exp: Entomologist, Entom Res Div, Agr Res Serv, USDA, 65; asst prof, 65-69, ASSOC PROF ENTOM, UNIV GA, 69- Mem: Entom Soc Am. Res: Mode of action of insecticides; insect pheromones. Mailing Add: Dept of Entom Univ of Ga Athens GA 30601

BRADY, WILLIAM GORDON, b Zanesville, Ohio, May 31, 23; m 48; c 2. APPLIED MATHEMATICS. Educ: Univ Cincinnati, BAeroEng, 50; Brown Univ, MS, 53. Prof Exp: Asst res engr appl mech, Cornell Aeronaut Lab, Inc, NY, 52-55, assoc res engr, 55-58, res engr, 58-60, prin engr, 60-70; asst prof, 70-75, ASSOC PROF MATH, ERIE COMMUNITY COL, 75- Mem: Am Inst Aeronaut & Astronaut. Res: Aeroelasticity; low speed aerodynamics. Mailing Add: Dept of Math Erie Community Col Buffalo NY 14221

BRADY, WILLIAM THOMAS, b Ventura, Calif, Sept 25, 33; m 54; c 2. PHYSICAL ORGANIC CHEMISTRY. Educ: NTex State Univ, BA, 55, MS, 56; Univ Tex, PhD(chem), 60. Prof Exp: Res chemist, Tex Eastman Co, Eastman Kodak Co, 56-57; res scientist, Clayton Found Biochem Inst, 57-60; res chemist, Tex Eastman Co, Eastman Kodak Co, 60-62; from asst prof to assoc prof, 62-68, PROF CHEM, NTEX STATE UNIV, 68- Mem: Am Chem Soc. Res: Ketene chemistry. Mailing Add: Dept of Chem NTex State Univ Denton TX 78201

BRADY, WRAY GRAYSON, b Benton Harbor, Mich, July 20, 18; m 43; c 2. MATHEMATICS. Educ: Washington & Jefferson Col, BS, 40, MA, 42; Univ Pittsburgh, PhD(math), 53. Prof Exp: Instr math, Washington & Jefferson Col, 40-42; asst, Stanford Univ, 46-47; instr, Univ Wyo, 47-50; prof math, Washington & Jefferson Col, 51-66; Bernard prof math & chmn dept, Univ Bridgeport, 66-69; DEAN GRAD SCH, SLIPPERY ROCK STATE COL, 69-, PROF MATH, 72- Concurrent Pos: Consult, Atomic Energy Bd, Westinghouse Elec Bettis Plant. Mem: AAAS; Am Math Soc; Math Asn Am. Res: Infinite series; number theory. Mailing Add: Grad Sch Slippery Rock State Col Slippery Rock PA 16057

BRAEMER, ALLEN C, b Woodhaven, NY, May 8, 30; m 55; c 3. CHEMOTHERAPY, ANIMAL NUTRITION. Educ: State Univ NY Vet Col, Cornell Univ, DVM, 55. Prof Exp: Vet, Pine Tree Vet Hosp, Augusta, Maine, 55-56, Crawford Animal Hosp, Garden City, NY, 56-58 & Mindell Animal Hosp, Albany, 58-61; assoc res biologist, Sterling-Winthrop Res Inst, Rensselaer, NY, 61-62, group leader, 62-64, res biologist, 64-66, res vet & group leader, Exp Farm, 66-67; dir animal clin sect, Norwich Pharmacal Co, 67-69, chief vet bact sect, 69-73; HEAD DEPT ANIMAL SCI, SYNTEX RES, 73- Mem: AAAS; Am Vet Med Asn; Am Indust Vet Asn; Am Soc Microbiol; Am Soc Animal Sci. Res: Chemotherapy of microbial diseases. Mailing Add: 2375 Charleston Rd Mountain View CA 94043

BRAENDLE, DONALD HAROLD, b Flushing, NY, Nov 8, 27; m 57. MICROBIOLOGY. Educ: Rice Inst, BA, 52; Rutgers Univ, PhD(microbiol), 57. Prof Exp: Waksman-Merck fel, Rutgers Univ, 57-58; SR SCIENTIST, ABBOTT LABS, 58- Mem: AAAS; Am Soc Microbiol; Brit Soc Gen Microbiol. Res: Microbial genetics, especially that of filamentous microorganisms, their physiology, and their production of antibiotics and other microbial products. Mailing Add: Abbott Labs North Chicago IL 60064

BRAESTRUP, CARL BJORN, b Copenhagen, Denmark, Apr 13, 97; nat US; m 28; c 2. RADIOLOGICAL PHYSICS. Educ: Mass Inst Technol, BS, 22. Prof Exp: Develop engr, Bell Tel Labs, NY, 22-23; instr, NY Post-Grad Med Sch, Columbia Univ, 28-32, res assoc, Columbia Univ, 47-67; dir physics lab, Delafield Hosp, 50-67; CONSULT, LENOX HILL HOSP, NEW YORK, 64- Concurrent Pos: Dir physics serv, Dept Hosps, New York, 29-67; consult, Surgeon Gen Off, 45-69, US Vet Hosp, 46-67, AEC, NY, 47-63 & Oak Ridge Inst Nuclear Studies, 52-63; assoc prof, NY Univ, 48-49; mem, Expert Adv Panel Radiation, WHO, Int Comn Radiol Protection & chmn, Comn Shielding Design, Nat Coun Radiation Protection. Mem: Am Phys Soc; Radiol Soc NAm; fel Col Am Radiologists; Am Radium Soc; Radiation Res Soc. Res: Radiation protection. Mailing Add: Dept of Radiol Lenox Hill Hosp New York NY 10021

BRAGASSA, CHARLES B, data processing, statistics, see 12th edition

BRAGE, BURTON L, b Cannon Falls, Minn, May 18, 18; m 50; c 2. SOIL SCIENCE, GEOLOGY. Educ: Univ Minn, BS, 46, PhD(soils), 50. Prof Exp: Asst soils, Univ Minn, 48-50; from asst prof to assoc prof agron, 50-58, PROF AGRON, COL AGR & BIOL SCI, SDAK STATE UNIV, 58-, ASSOC DEAN AGR INSTR, 59- Mem: Am Soc Agron; Soil Sci Soc Am. Res: Soil management techniques, including use of fertilizers; nitrogen relations in soils, primarily the effect of different carriers of nitrogen upon germination. Mailing Add: Col of Agr & Biol Sci SDak State Univ Brookings SD 57006

BRAGG, DARRELL, b Sutton, WVa, May 24, 33; m; c 3. NUTRITION. Educ: WVa Univ, BS, 59, MS, 60; Univ Ark, Fayetteville, PhD(poultry nutrit), 66. Prof Exp: Res asst poultry nutrit, Univ Ark, Fayetteville, 60-64, res assoc, 64-67; from asst prof to assoc prof, Univ Man, 67-70; assoc prof, 70-74, PROF POULTRY NUTRIT, UNIV BC, 74-, ACTG CHMN DEPT, 75- Concurrent Pos: Mem, Subcomt Utilization Feed Grains, Tech Comt, Can Grain Coun, 70- & Can Comt Animal Nutrit, 74. Mem: Poultry Sci Asn; Agr Inst Can; Can Fedn Biol Sci; Nutrit Soc Can; World Poultry Sci Asn. Res: Poultry nutrition, particularly amino acid metabolism, mineral nutrition and utilization of cereal grain. Mailing Add: Dept of Poultry Sci Univ of BC Vancouver BC Can

BRAGG, DAVID GORDON, b Portland, Ore, May 1, 33; m 55; c 4. MEDICINE, RADIOLOGY. Educ: Stanford Univ, AB, 55; Univ Ore, MD, 59; Am Bd Radiol, dipl, 66. Prof Exp: Intern, Philadelphia Gen Hosp, 59-60; resident radiol, Col Physicians & Surgeons, Columbia Univ, 62-64, chief resident, 64-65, from instr to asst prof, Med Col, Cornell Univ, 65-70; PROF RADIOL & CHMN DEPT, COL MED, UNIV UTAH, 70- Concurrent Pos: Chmn dept diag radiol, Sloan-Kettering Cancer Ctr, New York, 67-70, consult, 70- Mem: AMA; Radiol Soc NAm; Am Roentgen Ray Soc; James Ewing Soc; Am Gastroenterol Asn. Res: Oncological and gastrointestinal radiology. Mailing Add: Dept of Radiol Univ of Utah Salt Lake City UT 84112

BRAGG, DENVER DAYTON, b Duffy, WVa, Apr 13, 15; m 41; c 2. POULTRY SCIENCE. Educ: WVa Univ, BS, 40; Va Polytech Inst & State Univ, MS, 53. Prof Exp: Poultry serviceman, Ralston Purina Co, 40; asst rural rehab supvr, Farm & Home Admin, 40-42; county 4-H Club agent, Agr Exten Serv, WVa, 46, from asst county agent to assoc county agent, 46-49; assoc prof poultry sci & exten specialist, 49-73, EMER ASSOC PROF POULTRY SCI, VA POLYTECH INST & STATE UNIV, 73- Mem: Poultry Sci Asn. Res: Egg production and processing. Mailing Add: 610 Alleghany St Blacksburg VA 24060

BRAGG, JOHN KENDAL, b Washington, DC, Nov 12, 19; m 43; c 4. CHEMICAL PHYSICS. Educ: Harvard Univ, BS, 41, PhD(chem physics), 48. Prof Exp: Asst prof chem, Cornell Univ, 48-50; res assoc, Gen Elec Co, 50-53; pres, Opers Res, Inc, 54-56; proj analyst, Gen Elec Co, 57-63, mgr res consult serv, 64-66, mgr res & develop consult serv, 66-70; asst vpres technol, 70-72, dir res, 72-75, CHIEF TECH OFFICER, SINGER CO, 75- Concurrent Pos: Consult, Weapons Syst Eval Group, 53-54 & Opers Res Off, 54; trustee, Textile Res Inst, 72- Mem: AAAS; Am Phys Soc; Opers Res Soc. Res: Applied physics. Mailing Add: 100 W 57th St New York NY 10019

BRAGG, LINCOLN ELLSWORTH, b Buffalo, NY, Jan 25, 36; m 61. MATHEMATICAL PHYSICS. Educ: Carnegie Inst Technol, BS, 59, MS, 60, PhD(continuum mech), 64. Prof Exp: Asst, US Steel Res Lab, Pa, 60; asst, Mellon Inst, 62 & 64; res asst prof continuum mech, Inst Fluid Dynamics & Appl Math, Univ Md, College Park, 63-66; Dunham Jackson instr math, Univ Minn, Minneapolis, 66-67; asst prof, Univ Ky, 67-71 & Fla Inst Technol, 71-75. Mem: Soc Natural Philos; Inst Elec & Electronics Engrs; Am Math Soc. Res: Electromagnetic behavior of deformable materials; theory of constitutive relations; concepts of electrodynamics and their historical development; relativistic gravitational theory; special solutions; application of exterior calculus in engineering sciences. Mailing Add: 9 East Woodland Dr Mechanicsburg PA 17055

BRAGG, LOUIS HAIRSTON, b Chicora, Miss, Sept 12, 28; m 54; c 3. BOTANY. Educ: NTex State Col, BS, 53, MS, 57; Univ Tex, PhD(bot), 64. Prof Exp: Tutor biol, NTex State Col, 53-54; teacher, Pub Schs, Tex, 54-60; instr, 60-62, asst prof, 64-69, ASSOC PROF BIOL, UNIV TEX, ARLINGTON, 69- Concurrent Pos: Res scientist, Univ Tex, 64. Res: Experimental plant ecology; palynology; biochemical systematics; plant cytology; morphology, taxonomy. Mailing Add: Dept of Biol Univ of Tex Arlington TX 76010

BRAGG, LOUIS RICHARD, b Weston, WVa, Aug 5, 31. MATHEMATICS. Educ: WVa Univ, AB, 52, MS, 53; Univ Wis, PhD(math), 55. Prof Exp: From instr to asst prof math, Duke Univ, 55-59; assoc prof, WVa Univ, 59-61; from asst prof to assoc prof, Case Inst Technol, 61-67; PROF MATH, OAKLAND UNIV, 67- Mem: Am Math Soc; Math Asn Am; Soc Indust & Appl Math. Res: Operational calculus as it relates to special functions and representation theory in partial differential equations; numerical integration. Mailing Add: Dept of Math Oakland Univ Rochester MI 48063

BRAGG, PHILIP DELL, b Gillingham, Eng, July 2, 32; m 58; c 3. BIOCHEMISTRY, ORGANIC CHEMISTRY. Educ: Bristol Univ, BSc, 54, PhD(org chem), 58. Prof Exp: Fel org chem, Queen's Univ, Ont, 57-59; res assoc biochem, sch med, La State Univ, 59-61; res assoc, 61-64, from asst prof to assoc prof, 64-74, PROF BIOCHEM, UNIV BC, 74- Concurrent Pos: Med Res Coun Can scholar, Univ BC, 64-69. Mem: AAAS; Am Soc Microbiol; Can Biochem Soc; The Chem Soc. Res: Biochemistry of microorganisms, especially carbohydrate metabolism; electron transport; oxidative phosphorylation; cellular membranes; control mechanisms; membrane transport. Mailing Add: Dept of Biochem Univ of BC Vancouver BC Can

BRAGIN, JOSEPH, b Brooklyn, NY, Jan 23, 39; m 68. PHYSICAL CHEMISTRY. Educ: Brooklyn Col, BS, 59; Univ Wis, PhD(chem), 67. Prof Exp: Fel chem, Univ SC, 67-70; asst prof, 70-74, ASSOC PROF CHEM, CALIF STATE UNIV, LOS ANGELES, 74- Res: Vibrational spectroscopy as a probe of chemical and biological activity. Mailing Add: Dept of Chem Calif State Univ Los Angeles CA 90032

BRAGOLE, ROBERT ANTHONY, b Somerville, Mass, Oct 17, 36; m 62; c 5. ORGANIC CHEMISTRY. Educ: Boston Univ, AB, 58; Northeastern Univ, MS, 60; Yale Univ, PhD(org chem), 65. Prof Exp: Teaching fel, Boston Univ, 57-58; teaching fel & res asst, Northeastern Univ, 58-60; teaching fel, Yale Univ, 60-62; res chemist, Carwin Res Labs, 65-66; sr res chemist, USM Chem Co, 66-69, mgr appl res, 70-73, LAB MGR, BOSTIK DIV, USM CORP, 73- Concurrent Pos: Instr, Southern Conn State Col, 65-66. Mem: Sr mem Am Chem Soc. Res: Adhesion theory and technology; photochemistry; polyolefins. Mailing Add: 5 Innis Dr Danvers MA 01923

BRAGONIER, JOHN ROBERT, b Cedar Falls, Iowa, July 4, 37; m 59; c 3. OBSTETRICS & GYNECOLOGY. Educ: Iowa State Univ, BS, 60; Univ Nebr, MS, 63, MD, 64, PhD(med sci, biochem), 67. Prof Exp: Intern, Univ Nebr Hosp, 64-65, resident obstet & gynec, 65; resident obstet & gynec & instr, Jefferson Med Col Hosp, 66-68; res assoc, Inst Child Health & Human Develop, NIH, 68-70; ASST PROF OBSTET & GYNEC, UNIV CALIF, LOS ANGELES, 70-; HEAD PHYSICIAN, DEPT OBSTET & GYNEC, HARBOR GEN HOSP, TORRANCE, 70- Mem: Fel Am Col Obstetricians & Gynecologyists; Am Fertil Soc. Res: Medical biochemistry; biochemical teratology, mechanisms by which purine antimetabolites produce experimental congenital malformations; social problems in obstetrics and gynecology, especially family planning, sex education, therapeutic abortion; use of paramedical personnel. Mailing Add: 1202 W 244th St Harbor City CA 90710

BRAGONIER, WENDELL HUGHELL, b Geneseo Twp, Iowa, Aug 5, 10; m 34; c 3. BOTANY. Educ: Iowa State Teachers Col, BA, 33; Iowa State Univ, MS, 41, PhD(plant path), 47. Prof Exp: High sch teacher, Iowa, 33-34, Tenn, 34-35 & Iowa, 35-39; instr bot, Iowa State Univ, 40-42, res assoc & asst to dir, Indust Sci Res Inst, 42-47, assoc prof bot, 47-49, prof & head dept bot & plant path, 50-63, assoc dir, Camp Dodge Br, 46-47; DEAN GRAD SCH, COLO STATE UNIV, 63- Concurrent Pos: Chmn sci adv coun, Am Seed Res Found, 59- Mem: Fel AAAS; Bot Soc Am; Am Inst Biol Sci. Res: Plant pathology and morphology; teaching of botany; umbrella disease of Rhus glabra L caused by Botryosphaeria ribis G & D. Mailing Add: Grad Sch Colo State Univ Ft Collins CO 80521

BRAHAM, ROSCOE RILEY, JR, b Yates City, Ill, Jan 3, 21; m 43; c 4. CLOUD

PHYSICS. Educ: Ohio Univ, BS, 42; Univ Chicago, SM, 48, PhD(meteorol), 51. Prof Exp: Sr analyst & officer in chg thunderstorm proj, US Weather Bur, 46-49; res asst meteorol, Univ Chicago, 49-50; res meteorologist cloud physics & weather modification, NMex Inst Mining & Technol, 50-51; sr meteorologist, 51-56, assoc prof, 56-65, PROF METEOROL, UNIV CHICAGO, 65- Concurrent Pos: Assoc ed, J Am Meteorol Soc, 53-69; dir, Inst Atmospheric Physics & assoc prof, Univ Ariz, 54-56; mem bd trustees, Univ Corp Atmospheric Res; prin investr, Proj Metromex, NSF, 70-; sci adv, Encycl Britannica Educ Films, 72-; mem adv comt, Proj Stormfury, Nat Oceanic & Atmospheric Admin, Nat Acad Sci, 74- Honors & Awards: Losey Award, Inst Aeronaut Sci, 50; Silver Medal, US Dept Com, 50. Mem: AAAS; fel Am Meteorol Soc; Am Geophys Union; Sigma Xi; Royal Meteorol Soc. Res: Cloud seeding and weather modification; urban meteorology; thunderstorms and severe weather. Mailing Add: 57 Longcommon Rd Riverside IL 60546

BRAHANA, THOMAS ROY, b Champaign, Ill, June 26, 26; m 51; c 4. MATHEMATICS. Educ: Univ Ill, AB, 47; Univ Mich, MA, 50, PhD(math), 55. Prof Exp: Instr math, Dartmouth Col, 53-54; from asst prof to assoc prof, 54-68, PROF MATH, UNIV GA, 68- Concurrent Pos: Mem, Inst Advan Study, 57-58; Fulbright lectr, Univ Zagreb, Yugoslavia, 71-72. Mem: Am Math Soc. Res: Algebraic topology and geometry. Mailing Add: Dept of Math Univ of Ga Athens GA 30602

BRAHEN, LEONARD S, b Philadelphia, Pa, Nov 28, 21; m 47; c 3. PHARMACOLOGY. Educ: Temple Univ, BS, 49, MS, 51; Univ Md, PhD(pharmacol), 54; Univ Louisville, MD, 58. Prof Exp: Asst pharmacol, Temple Univ, 49-51, res asst, 51; instr pharmacol, Sch Med, Univ Louisville, 54-56; chief investr, Ky Heart Asn grant, 56-58; asst surgeon, USPHS, NY, 58-59; assoc dir clin res, Chas Pfizer & Co, 59-60, dir clin res, 60-62, med dir, Leeming Pacquin Div, 62-64; med dir, Endo Labs, Inc, 64-68, corp med dir, 68-71; DIR MED RES & EDUC, NASSAU COUNTY DRUG ABUSE & ADDICTION COMN, 71- Concurrent Pos: NIMH fel, 57; Collins Found scholar, 57; mem drug adv comt, State Sen Dunne; chmn drug abuse subcomt, Nassau County Med Soc, NY; assoc clin prof, New York Med Col, 59-62. Mem: Am Pharmaceut Asn; fel Am Col Clin Pharmacol & Chemother; Am Therapeut Soc. Res: Mechanism of action of drugs on autonomic nervous system; developed quantitative method for determination of irritation; studies in vasodilation; drug action and metabolism in humans. Mailing Add: 22 Willow Rd Woodmere NY 11598

BRAID, MILTON, organic chemistry, see 12th edition

BRAID, THOMAS HAMILTON, b Heroit, Scotland, Dec 21, 25; m 51; c 2. NUCLEAR PHYSICS. Educ: Univ Edinburgh, BSc, 47, PhD(physics), 51. Prof Exp: Nat Res Coun Can res fel physics, Atomic Energy Can, Ltd, 50-52; res asst, Princeton Univ, 52-55, instr, 55-56, res assoc, 56; ASSOC PHYSICIST, ARGONNE NAT LAB, 56- Concurrent Pos: Vis physicist, Atomic Energy Res Establishment, Eng, 66-67. Mem: Am Phys Soc. Res: Nuclear reactions and structure; radiation detection and instrumentation. Mailing Add: Argonne Nat Lab Phys Div 9700 S Cass Ave Argonne IL 60439

BRAIDS, OLIN CAPRON, b Providence, RI, Apr 29, 38; m 62; c 1. SOIL CHEMISTRY, WATER CHEMISTRY. Educ: Univ NH, BA, 60, MS, 63; Ohio State Univ, PhD(agron, soil chem), 66. Prof Exp: Res assoc, Univ Ill, Urbana-Champaign, 66-67, asst prof soil org chem, 67-72; hydrologist, US Geol Surv, 72-75; SR SCIENTIST, GERAGHTY & MILLER, INC, 75- Concurrent Pos: Abstractor, Chem Abstr Serv, 67-; NSF guest lectr, Univ Calif, Riverside, 70. Mem: Am Chem Soc; Am Soc Agron; Soil Sci Soc Am; Int Soc Soil Sci. Res: Effect of land disposal of municipal sludges on soils and plants; influence of waste disposal practice on ground-water quality; chemistry of contaminants from waste in aquifers. Mailing Add: Geraghty & Miller Inc 44 Sintsink Dr E Port Washington NY 11050

BRAIDWOOD, CLINTON ALEXANDER, b Snover, Mich, Nov 14, 14; m 42; c 3. ORGANIC CHEMISTRY. Educ: Mich State Univ, BS, 40. Prof Exp: Mem staff motor prod develop, US Rubber Co, 40-42; asst dir res, Reichold Chem, Inc, 42-49; vpres mfg & res develop, 49-71, exec vpres, 71-73, PRES, SCHENECTADY CHEMS, INC, 73- Mem: AAAS; Am Inst Chemists; Am Chem Soc. Res: Alkylation of phenol and synthesis of phenolic resins from alkyl phenols; polymerization chemistry; terpene resins. Mailing Add: Schenectady Chems PO Box 1046 Schenectady NY 12301

BRAIDWOOD, ROBERT J, b Detroit, Mich, July 29, 07; m 37; c 2. ANTHROPOLOGY, ARCHAEOLOGY. Educ: Univ Mich, AB, 32, MA, 33; Univ Chicago, PhD(archaeol), 42. Hon Degrees: ScD, Ind Univ, 71; Dr, Univ Paris I-Sorbonne, 75. Prof Exp: Artist & draftsman, Univ Mich Exped, Iraq, 31-33; topog surveyor, Univ Chicago, 33, field asst, Oriental Inst, Syria, 33-38, publ asst, 38-44, from asst prof to assoc prof anthrop, 44-54, field archaeologist, Oriental Inst, 33-75 & prehist proj, 47-75, Oriental Inst Prof Old World Prehist, 54-75, EMER ORIENTAL INST PROF OLD WORLD PREHIST, UNIV CHICAGO, 75- Concurrent Pos: Wenner-Gren Found Anthrop Res fel field work in southwestern Asia, 54-64; NSF fels, 54-73; vis prof, Istanbul Univ, 63-64; co-field dir, Istanbul Univ-Univ Chicago. Honors & Awards: Medalist, Archaeol Inst Am, 71; Distinguished Lectr, Am Anthrop Asn, 71. Mem: Fel Nat Acad Sci; fel AAAS; fel Am Acad Arts & Sci; fel Am Anthrop Asn; fel Archaeol Inst Am. Res: Appearance of village-farming communities in southwestern Asia. Mailing Add: Oriental Inst Univ of Chicago Chicago IL 60637

BRAILOVSKY, CARLOS ALBERTO, b Buenos Aires, Arg, Oct 16, 39; Can citizen; m 65; c 3. CELL BIOLOGY, VIROLOGY. Educ: Buenos Aires Univ, MD, 61. Prof Exp: Buenos Aires Univ fel, Inst Sci Res Cancer, Vellejuif, France, 63-65, res assoc virol, 65-68; invited scientist path, Univ Montreal, 69; ASSOC PROF CELL BIOL, FAC MED, UNIV SHERBROOKE, 70- Honors & Awards: Squibb Award, Arg Med Asn, 57. Mem: French Soc Microbiol; Am Soc Cell Biol; Am Soc Microbiol; Am Tissue Cult Asn; Can Soc Cell Biol. Res: Oncogenic transformation and cell membrane modifications. Mailing Add: Dept of Cell Biol CHU Univ of Sherbrooke Sherbrooke PQ Can

BRAIN, DEVIN KING, b Mt Vernon, Ohio, Jan 7, 26; m 50; c 3. INDUSTRIAL ORGANIC CHEMISTRY. Educ: Univ Ariz, BS, 48, MS, 49; Ohio State Univ, PhD(org chem), 54. Prof Exp: RES CHEMIST, PROCTER & GAMBLE CO, 54- Mem: Am Chem Soc; Am Soc Mass Spectrometry. Res: Soap and detergent products; application of instrumental techniques, especially gas chromatography, gas chromatography/mass spectrometry, mass spectrometry, nuclear magnetic resonance, irridescent radiation and x-ray techniques, to product development. Mailing Add: Procter & Gamble Co Ivorydale Tech Ctr Cincinnati OH 45217

BRAIN, JAMES LEWTON, b Chigwell, Eng, Sept 12, 23; m 50; c 2. ANTHROPOLOGY. Educ: Univ London, LSE, 63; Syracuse Univ, PhD(anthrop), 68. Prof Exp: Community develop officer, Govt Tanganyika & Uganda, 51-63; lectr Swahili & anthrop, Syracuse Univ, 63-67; from asst prof to assoc prof anthrop, 67-72, PROF ANTHROP, STATE UNIV NY COL NEW PALTZ, 72- Concurrent Pos: Ford Found res grant, 65. Mem: Royal Anthrop Inst Gt Brit & Ireland; Soc Appl Anthrop; African Studies Asn; Am Anthrop Asn; African Studies Asn UK. Res: Belief systems; matrilineal descent. Mailing Add: Dept of Anthrop State Univ of NY Col New Paltz NY 12561

BRAINERD, BARRON, b New York, NY, Apr 13, 28; Can citizen. MATHEMATICS. Educ: Mass Inst Technol, SB, 49; Univ Mich, MS, 51, PhD(math), 54. Prof Exp: Instr math, Univ BC, 54-57; asst prof math, Univ Western Ont, 57-59; from asst prof to assoc prof, 59-67, PROF MATH, UNIV TORONTO, 67- Concurrent Pos: Vis fel, Australian Nat Univ, 62-63. Mem: AAAS; Ling Soc Am; Can Math Cong; Can Ling Asn; Australian, London & Mex Math Socs. Res: Theory of partially ordered rings; rings of continuous functions; operators on function spaces; mathematical linguistics; statistics of literary style. Mailing Add: 53 Wolfrey Ave Toronto ON Can

BRAINERD, JOHN WHITING, b Dover, Mass, Feb 14, 18; m 41; c 3. BIOLOGY. Educ: Harvard Univ, AB, 40, MA, 42, PhD, 49 Prof Exp: PROF BIOL, SPRINGFIELD COL, 49- Mem: Am Nature Study Soc (pres, 64); Am Ornith Union; Am Forestry Asn; Bot Soc Am; fel Asn Interpretive Naturalists. Res: Habitat analysis; conservation education; environmental planning for educational institutions. Mailing Add: Dept of Biol Springfield Col Springfield MA 01109

BRAINERD, WALTER SCOTT, b Des Moines, Iowa, May 27, 36; m 58; c 3. MATHEMATICS, COMPUTER SCIENCE. Educ: Univ Colo, BA, 58; Univ Md, MA, 61; Purdue Univ, PhD(comput sci), 67. Prof Exp: Instr math, Naval Postgrad Sch, 63-65, asst prof comput sci, 67-69; asst prof math statist, Columbia Univ, 69-72; prof math, Harvey Mudd Col, 72-73; SR SYSTS SPECIALIST, BURROUGHS CORP, 73- Concurrent Pos: Vis asst prof, Univ Calif, San Diego, 68. Mem: Asn Comput Mach; Am Nat Standards Inst. Res: Programming languages, theory of computation. Mailing Add: Burroughs Corp 460 Sierra Madre Villa Pasadena CA 91109

BRAITHWAITE, JOHN GEDEN NORTH, b Bishop Auckland, Eng, Oct 5, 20; US citizen; m 47; c 3. OPTICS. Educ: Cambridge Univ, BA, 47, MA, 51. Prof Exp: Prin sci officer, Royal Radar Estab, Eng, 48-57; sr scientist, Baird Atomic Inc, Mass, 58-61; sr proj leader, Block Assocs, 62; group leader, Res Inst, Ill Inst Technol, 62-63 & Bendix Systs Div, Mich, 64-65; res physicist, Univ Mich, Ann Arbor, 65-72; RES PHYSICIST, ENVIRON RES INST MICH, 73- Concurrent Pos: Consult, Manned Spacecraft Ctr, NASA & Denver Div, Martin Marietta Corp. Mem: AAAS; Optical Soc Am. Res: Infrared physics and technology; optical instrumentation; optical filters; spectrometry. Mailing Add: 1332 White St Ann Arbor MI 48104

BRAKE, JON MICHAEL, b Detroit, Mich, Apr 16, 35. BIOCHEMISTRY, ORGANIC CHEMISTRY. Educ: Univ Mich, BS, 57; Univ Ill, PhD(biochem), 62. Prof Exp: Res biochemist, res & develop div, Magna Corp, 61-63; fel, med res div, Vet Admin Hosp, Long Beach, Calif, 63-64; res biochemist, res & develop div, Magna Corp, 64-65 & TRW Systs, 65-66; res chemist, 66-73, SR RES SCIENTIST, McGAW LABS, DIV AM HOSP SUPPLY CORP, 73- Mem: AAAS; Am Chem Soc. Res: Mechanism of action of enzymes; biochemical fuel cells; detection of bacteria; aging of red cells; cryobiology; plasma substitutes. Mailing Add: McGaw Labs (Irvine) PO Box 11887 Santa Ana CA 92711

BRAKKE, KENNETH ALLEN, b Brooklyn, NY, Oct 1, 50. MATHEMATICAL ANALYSIS. Educ: Univ Nebr, BS, 72; Princeton Univ, PhD(math), 75. Prof Exp: ASST PROF MATH, PURDUE UNIV, 75- Mem: Math Asn Am; Am Math Soc. Res: Geometric measure theory; varifolds. Mailing Add: Dept of Math Purdue Univ West Lafayette IN 47907

BRAKKE, MYRON KENDALL, b Preston, Minn, Oct 23, 21; m 47; c 4. PLANT VIROLOGY. Educ: Univ Minn, BS, 43, PhD(biochem), 47. Prof Exp: Asst biochem, Univ Minn, 43-44, instr, 44-47; res assoc, Brooklyn Bot Garden, 47-52 & Dept Bot, Univ Ill, 52-55; CHEMIST, AGR RES SERV, USDA & PROF PLANT PATH, UNIV NEBR, LINCOLN, 55- Honors & Awards: Superior Serv Award, USDA & Ruth Allen Award, Am Phytopath Soc, 68. Mem: Nat Acad Sci; AAAS; Am Chem Soc; fel Am Phytopath Soc; Electron Micros Soc Am. Res: Purification and characterization of plant viruses and their nucleic acids; vectors of plant viruses and control of plant virus diseases. Mailing Add: 304 Plant Indust Bldg Col Agr Univ of Nebr Lincoln NE 68503

BRALEY, SILAS ALONZO, JR, b Sioux City, Iowa, July 15, 17; m 46; c 3. INDUSTRIAL CHEMISTRY. Educ: Univ Pittsburgh, BS, 47. Prof Exp: Mem staff metall, Pittsburgh Steel Co, 41-47; chemist, Mellon Inst, 47-51; supvr silastic develop lab, 51-59, exec secy, 59-63, DIR, CTR AID TO MED RES, DOW CORNING CORP, 63- Mem: Am Chem Soc; AAAS; Am Soc Artificial Internal Organs; Soc Biomat. Res: Medical use of silicones. Mailing Add: Ctr for Aid to Med Res Dow Corning Corp Midland MI 48640

BRALLEY, JAMES ALEXANDER, b Bath Co, Va, Aug 18, 16; m 46; c 3. ORGANIC CHEMISTRY. Educ: Univ Va, BS, 36, PhD(org chem), 41. Prof Exp: Res chemist, BF Goodrich Co, 41-46; develop chemist, Rohm and Haas Co, 46-56; dir res, AE Staley Mfg Co, 56-61, vpres res & develop, 61-70; VPRES RES & DEVELOP, PURITAN CHEM CO, 71- Mem: AAAS; Am Chem Soc; Am Oil Chemists Soc. Res: Morphine alkaloids; vinyl and acrylic resins; plasticizers; inhibitors; starch chemistry, processing and industrial uses; vegetable oils and proteins; epoxy resins; fermentation chemistry; amylose and amylopectin; amino acids; organic syntheses; detergents; disinfectants; insecticides; herbicides; waxes; synthetic polymer formulations. Mailing Add: Puritan Chem Co 916 Ashby St NW Atlanta GA 30318

BRALOW, S PHILIP, b Philadelphia, Pa, Aug 28, 21; m 49; c 2. GASTROENTEROLOGY, INTERNAL MEDICINE. Educ: Pa State Col, BS, 42; Temple Univ, MD, 45; Univ Ill, MS, 49. Prof Exp: From intern to resident gastrointestinal res, Michael Reese Hosp, 46-47; chief gastrointestinal clin & lab, Temple Univ Hosp, 54-63; from asst prof to assoc prof med, Temple Univ, 59-69; prof, Jefferson Med Col, 69-75; CLIN PROF MED, SCH MED, UNIV S FLA, 75- Concurrent Pos: NIH grant, Temple Univ Hosp; Am Cancer Soc & Nat Cancer Inst grants, Thomas Jefferson Univ Hosp; attend physician gastroenterol, Vet Admin Hosp, 54-63; consult, US Naval Hosp, 69; mem coun cancer, Am Gastroenterol Asn, 72; mem, Nat Sci Adv Bd, Nat Found Ileitis & Colitis, Inc & mem med & sci comt, Am Cancer Soc, 73. Honors & Awards: 25 Year Award, Colostomy Ileostomy Rehab Asn, 74. Mem: Am Cancer Soc; Am Gastroenterol Asn; Am Asn Cancer Res. Res: Experimental carcinogenesis in animal models; radiation injury to small intestines. Mailing Add: 1515 S Osprey Ave Sarasota FL 33579

BRAM, JOSEPH, b Chicago, Ill, June 19, 26; m 53; c 2. MATHEMATICS. Educ: Roosevelt Col, BS, 48; Univ Chicago, MS, 49, PhD(math), 53. Prof Exp: Mathematician, Ord Res Proj, Chicago, 48-50; sr mathematician, Eng Res Assoc, 53-54; mathematician, Nat Bur Standards, 54-56, Bur Ships, 56-61 & Taylor Model Basin, 61-62; mathematician, 62-74, SR MATHEMATICIAN, CTR NAVAL ANAL, 74- Mem: Am Math Soc. Res: Operator theory; Hilbert space; numerical analysis;

operations research. Mailing Add: Ctr for Naval Anal 1401 Wilson Blvd Arlington VA 22209

BRAM, LEILA DRAGONETTE, b Drexel Hill, Pa, July 22, 27; div; c 4. MATHEMATICS. Educ: Bryn Mawr Col, AB, 47; Univ Pa, MA, 49, PhD(math), 51. Prof Exp: Off Naval Res res assoc math, Univ Chicago, 51-53; mathematician, 53-56 & 60-65, head math br, 65-69, DIR MATH PROG, OFF NAVAL RES, 69- Mem: AAAS; Soc Indust & Appl Math; Am Math Soc; Math Asn Am. Res: Analytic number theory. Mailing Add: Math Prog Off of Naval Res Arlington VA 22217

BRAM, RALPH A, b Washington, DC, Dec 10, 32; m 56; c 4. ENTOMOLOGY. Educ: Univ Md, BS, 56, MS, 61, PhD(entom), 64. Prof Exp: Instr entom, Univ Md, 63-64; asst prof, Purdue Univ, 64-65; entomologist, Smithsonian Inst, 65-68 & animal parasite dis lab, entom res div, Agr Res Serv, 68-71, res entomologist, insects affecting man & animals res lab, 71-72, VET ENTOMOLOGIST, ANIMAL & PLANT HEALTH INSPECTION SERV, USDA, 72- Concurrent Pos: Consult, Food & Agr Orgn, Rome, Italy, 74- Mem: AAAS; Entom Soc Am. Res: Veterinary and medical entomology, especially mosquito systematics and zoogeography; general entomology and pathogen-vector relationships; tick and tick-borne disease control. Mailing Add: USDA-APHIS-VS Emergency Progs Fed Ctr Bldg Hyattsville MD 20782

BRAMAN, ROBERT STEVEN, b Lansing, Mich, Aug 31, 30; m 54; c 2. ANALYTICAL CHEMISTRY, INSTRUMENTATION. Educ: Mich State Univ, BS, 52; Northwestern Univ, PhD(chem), 56. Prof Exp: Group leader anal chem res & develop, Callery Chem Co, 56-59; res chemist, Res Inst, Ill Inst Technol, 59-64, sr chemist, 64-67; from asst prof to assoc prof, 67-73, PROF CHEM, UNIV S FLA, 73- Mem: Am Chem Soc; Marine Technol Soc. Res: Development of analytical instrumentation; environmental chemistry. Mailing Add: Dept of Chem Univ of S Fla Tampa FL 33620

BRAMANTE, PIETRO OTTAVIO, b Rome, Italy, May 21, 20; US citizen; m 57. PHYSIOLOGY, INTERNAL MEDICINE. Educ: Univ Rome, MD, 44, MS, 50; Drexel Univ, MS, 64. Prof Exp: Instr internal med, Sch Med, Univ Rome, 45-46, asst prof, 46-51; res fel, Sch Med, Univ Stockholm & King Gustaf's Res Inst, 51-52; from res assoc to assoc prof physiol, Sch Med, St Louis Univ, 52-65; assoc prof, 65-70, PROF PHYSIOL, UNIV ILL COL MED, 70- Concurrent Pos: Tech med consult, Court of Justice, Rome, 48-51; Fulbright fel, 52; Heart Inst fel, Drexel Univ, 62-64. Mem: AAAS; Am Physiol Soc; Am Thyroid Asn; Asn Am Med Cols; Soc Exp Biol & Med. Res: Pituitary in uremia; liver function; body electrolytes; hypotension; thyroid function; innervation of nasal mucosa; methodology of energy metabolism; allometry; experimental tumorigenesis; calorigenic drugs; exercise and metabolism; experimental myocardial necrosis; experimental dyslipoproteinemia. Mailing Add: Dept of Physiol Univ of Ill Col of Med Chicago IL 60680

BRAMBLE, JAMES H, b Annapolis, Md, Dec 1, 30; m 55; c 4. APPLIED MATHEMATICS, NUMERICAL ANALYSIS. Educ: Brown Univ, AB, 53; Univ Md, MA, 55, PhD(math), 58. Prof Exp: Mathematician, Gen Elec Co, 57-59 & US Naval Ord Lab, 59-60; res asst prof math, Univ Md, 60-63, from res assoc prof to res prof, 63-68; PROF MATH, CORNELL UNIV, 68- Concurrent Pos: Vis lectr, Soc Indust & Appl Math, 62-65; consult, Nat Bur Standards, 60-66; vis staff mem, Los Alamos Sci Lab, Univ Calif, 74-; chmn, Ed Comt, Math Comput, 75- Mem: Am Math Soc. Res: Partial differential equations; numerical methods. Mailing Add: 899 Cayuga Heights Rd Ithaca NY 14850

BRAMBLE, WILLIAM CLARK, b Baltimore, Md, Nov 7, 07; m 39; c 2. FORESTRY. Educ: Pa State Univ, BS, 29; Yale Univ, MF, 30, PhD(bot), 32. Prof Exp: Instr bot, Carleton Col, 32-37; Nat Res fel, Univ Zurich, 35-36; from asst prof to prof forestry, Pa State Univ, 37-58, head dept forest mgt, 54-58, actg dir sch forestry, 55-58; prof forestry & head dept forestry & conserv, 58-72, EMER PROF FORESTRY, PURDUE UNIV, 72- Concurrent Pos: Collabr, USDA, 34-72; mem div biol & agr, Nat Res Coun, 70-73. Mem: Soc Am Foresters; Ecol Soc Am; Swiss Bot Soc. Res: Forest ecology and management; silviculture. Mailing Add: Dept of Forestry & Conserv Purdue Univ West Lafayette IN 47906

BRAMBLETT, CLAUD ALLEN, b Crystal City, Tex, Oct 8, 39; m 61; c 2. PHYSICAL ANTHROPOLOGY, PRIMATOLOGY. Educ: Univ Tex, Austin, BA, 62, MA, 65; Univ Calif, Berkeley, PhD(anthrop), 67. Prof Exp: Mgr, Darajani Primate Res Ctr, Southwest Found Res, 63-64; asst prof, 67-73, ASSOC PROF ANTHROP, UNIV TEX, AUSTIN, 73- Concurrent Pos: Consult, Southwest Found Res & Educ, 71- & dept path, Univ Tex Med Sch San Antonio, 72-; mem proj comt, Arashiyama W Japanese Macaque Ranch, 72-; co-investr, Exp Atherosclerosis in Baboons, NIH, 73-78. Mem: Fel Am Anthrop Asn; Am Asn Phys Anthrop; Int Primatol Soc; Am Soc Mammalogists; Soc Study Human Biol. Res: Behavior as a risk factor in coronary heart disease; acquisition of social signals in developing vervet monkeys; behavioral regulators. Mailing Add: Dept of Anthrop Univ of Tex Austin TX 78712

BRAMBLETT, RICHARD LEE, b Dallas, Tex, Aug 15, 35; m 57; c 3. APPLIED PHYSICS, NUCLEAR PHYSICS. Educ: Rice Univ, BA, 56, MA, 57, PhD(physics), 60. Prof Exp: Physicist, Lawrence Radiation Lab, Univ Calif, 60-67; br mgr, Gulf Energy & Environ Systs, Calif, 68-73; SR PROG MGR, IRT CORP, 73- Mem: Am Phys Soc; Am Nuclear Soc; Inst Nuclear Mat Mgt. Res: Experimental photonuclear and neutron physics using electron accelerator technology; spectroscopy and cross-section measurements for photons and neutrons; application of accelerators, reactors and radioisotopes to non-destructive isotopic assays. Mailing Add: 5451 Sandburg Ave San Diego CA 92122

BRAME, EDWARD GRANT, JR, b Shiloh, NJ, Mar 20, 27; m 57. ANALYTICAL CHEMISTRY. Educ: Dickinson Col, BS, 48; Columbia Univ, MS, 50; Univ Wis, PhD(anal chem), 57. Prof Exp: Asst chem, Columbia Univ, 48-50; res anal chemist, Corn Prod Refining Co, 50-53; asst, Univ Wis, 53-56; res chemist, plastics dept, 57-64, RES CHEMIST, ELASTOMER CHEM DEPT, E I DU PONT DE NEMOURS & CO, 64- Concurrent Pos: Ed, Appl Spectros Rev; mem sci adv bd, Winterthur Mus, Del; mem postdoctoral res associateships eval panel, Nat Res Coun, 74-75; mem adv bd, Anal Chem, 74-76; chmn, Fedn Anal Chem & Spectros Socs, 76. Mem: Sigma Xi; Am Chem Soc; Soc Appl Spectros; NY Acad Sci. Res: Infrared spectroscopy; nuclear magnetic resonance spectroscopy; electron paramagnetic resonance spectroscopy; instrumental methods of analysis; nuclear quadrupole resonance spectroscopy. Mailing Add: E I du Pont de Nemours & Co Bldg 353 Rm 325 Exp Sta Wilmington DE 19898

BRAMLAGE, WILLIAM JOSEPH, b Dayton, Ohio, Mar 27, 37; m 67; c 3. HORTICULTURE. Educ: Ohio State Univ, BS, 59; Univ Md, MS, 61, PhD(hort), 63. Prof Exp: Horticulturist, agr mkt serv, USDA, 63-64; asst prof hort, 64-69, ASSOC PROF PLANT PHYSIOL, UNIV MASS, AMHERST, 69- Mem: AAAS; Am Soc Plant Physiol; Am Soc Hort Sci. Res: Post-harvest physiology; physiological disorders of fruit. Mailing Add: Dept of Plant & Soil Sci Univ of Mass Amherst MA 01002

BRAMLET, ROLAND C, b Wallowa, Ore, June 11, 21; m 61; c 2. RADIOLOGICAL PHYSICS. Educ: Ore State Univ, BS, 48; NY Univ, MS, 61; St John's Univ, NY, PhD(biol), 66. Prof Exp: Fel, Sloan-Kettering Inst Cancer Res, 58-59; jr physicist, Queens Gen Hosp, Jamaica, NY, 59-61, radioisotope physicist, 61-69; PHYSICIST, HIGHLAND HOSP, ROCHESTER, NY, 69- Concurrent Pos: Consult physicist, Rochester Gen Hosp, NY, 69- Mem: Soc Nuclear Med; Inst Elec & Electronics Engrs. Res: Clinical radioisotopic instrumentation; radiation biology. Mailing Add: Radiation Therapy Highland Hosp Rochester NY 14620

BRAMLETT, CHRISTOPHER L, b Canton, NC, Aug 31, 38; m 60; c 1. INORGANIC CHEMISTRY. Educ: Wake Forest Col, BS, 60, MA, 64; Univ Va, PhD(chem), 67. Prof Exp: Control engr, Union Carbide Corp, 60-62; asst prof, 67-71, asst dean col arts & sci, 70-73, ASSOC PROF CHEM, UNIV ALA, 71-, ASSOC DEAN GRAD SCH, 73- Mem: Am Chem Soc. Res: Carboranes; boron hydrides. Mailing Add: Off of Grad Sch Box W Univ of Ala University AL 35486

BRAMLETTE, WILLIAM (ALLEN), b Greenville, Tex, Aug 31, 11; m 32; c 3. GEOLOGY. Educ: Southern Methodist Univ, BS, 32; Univ Tex, MA, 34; Univ Kans, PhD(geol), 43. Prof Exp: Asst prof geol, Univ Tex, 43; div geologist, Carter Oil Co, 53, asst chief geologist, 54-55, chief geologist, 55-58, asst explor mgr, 58-61; opers mgr, Humble Oil & Refining Co, 61-68, exec vpres & dir, Esso Explor, Inc, 68-71, vpres, Esso Prod Res Co, 71-74, VPRES, EXXON PROD RES CO, 74- Mem: Am Asn Petrol Geol; Soc Econ Paleont & Mineral; fel Geol Soc Am. Res: Petroleum geology; stratigraphy; paleontology. Mailing Add: Exxon Prod Res Co PO Box 2189 Houston TX 77001

BRAMLITT, EDWARD TURNER, nuclear chemistry, see 12th edition

BRAMMER, JIMMIE DUANE, b Grand River, Iowa, Aug 30, 35; m 59; c 2. DEVELOPMENTAL BIOLOGY, NEUROBIOLOGY. Educ: Simpson Col, BA, 62; Purdue Univ, MS, 65, PhD(neurobiol), 68. Prof Exp: Nat Acad Sci-Nat Res Coun res fel neurobiol, US Air Force Sch Aerospace Med, 68-69; ASST PROF ZOOL, UNIV VT, 69- Mem: Am Soc Zoologists; Asn Res Vision & Ophthal. Res: Structure and function of the insect visual systems and the effects of vitamin A deficiency and light and light deprivation on such systems. Mailing Add: Dept of Zool Marsh Life Sci Bldg Univ of Vt Burlington VT 05401

BRAMS, STEWART L, b Greenville, Mich, July 10, 14; m 40; c 3. RUBBER CHEMISTRY. Educ: Univ Mich, BS, 35, MS, 36. Prof Exp: Jr chemist, Minn Mining & Mfg Co, Mich, 37-38; sect head & sr compounder, Inland Mfg Div, Gen Motors Corp, Ohio, 39-46; gen mgr, Dayton Chem Prod Labs, Inc, 46-50, pres, 50-68, gen mgr, Dayton Chem Prod Div, 68-70, sr staff consult, 70-72, gen Europ agent, 72-75, CONSULT, DAYTON CHEM PROD DIV, WHITTAKER CORP, 75- Mem: Am Chem Soc; Am Soc Testing & Mat; Fr Asn Eng Rubber & Plastics. Res: Adhesives; rubber adhesives and sealers; bonding of rubber to metal; synthetic rubber coatings; analytical methods for rubber compositions. Mailing Add: 3501 Meadow Lane Dayton OH 45419

BRAMWELL, FITZGERALD BURTON, b Brooklyn, NY, May 16, 45; m 73. PHYSICAL CHEMISTRY. Educ: Columbia Univ, BA, 66; Univ Mich, MS, 67, PhD(chem), 70. Prof Exp: Res chemist, Esso Res & Eng Co, 70-71; ASSOC PROF CHEM, BROOKLYN COL, 71- Concurrent Pos: Mem tech staff, Bell Tel Labs, 74-75. Mem: AAAS; Am Chem Soc; Am Phys Soc; NY Acad Sci. Res: Electron spin resonance of triplets; free radicals in solution; charge transfer complexes; John-Teller distortions. Mailing Add: Dept of Chem Brooklyn Col Brooklyn NY 11210

BRANCATO, FRANK PAUL, b New Haven, Conn, Sept 10, 15; m 55; c 2. MEDICAL MICROBIOLOGY. Educ: Long Island Univ, BS, 48; Boston Univ, AM, 49; State Col Wash, PhD(bact), 52. Prof Exp: Asst, State Col Wash, 49-52; microbiologist, 52-58, chief microbiol, 58-69, CHIEF DEPT MICROBIOL, USPHS HOSP, SEATTLE, WASH, 69- Concurrent Pos: Consult, Hall Health Ctr, Univ Wash, 67-, affil asst prof, 68- Mem: AAAS; Am Soc Microbiol; Am Acad Microbiol. Res: Staphylococcal pneumonias and infections among hospital patients and personnel; value of direct stained smear of clinical material as a diagnostic tool and in quality control microbiology; pyelonephritis in the male. Mailing Add: US Pub Health Serv Hosp Seattle WA 98114

BRANCH, CHARLES FRANKLIN, b Amherst, Mass, Aug 14, 97; m 24, 48; c 3. MEDICINE. Educ: Univ Vt, MD, 23. Prof Exp: Intern & asst path, Boston City Hosp, 23-25; instr & asst pathologist, Jefferson Med Col & asst pathologist, Philadelphia Gen Hosp, 25-26; from asst prof to prof, Sch Med, Boston Univ & pathologist, Mass Mem Hosp, 26-46; dir, Children's Hosp & Med Ctr for Children, 46-47; asst dir, Am Col Surg, Chicago, 47-50; dir labs & pathologist-in-chief, 50-68, CHIEF MED EXAMR, CENT MAINE GEN HOSP, STATE OF MAINE, 68- Concurrent Pos: Evans Mem traveling fel, Europe & Eng, 32; Asn Am Med Cols fel, Cent Am, 43; pathologist & consult pathologist, Hosps, 28-46; consult, Smith, Kline & French Industs, Inc, 37-; vchmn sect V, Int Cancer Cong, 39; assoc dir, Evans Mem & Mass Mem Hosp, 39-44; dean, Sch Med, Boston Univ, 44-46; trustee, Boston Med Libr & Univ Vt, 46-50; mem cancer control comt, Nat Cancer Inst, 48-60, chmn, 59-60; mem bd dir, Maine Law Enforcement Planning Asn, 68- Mem: Fel AMA; Am Asn Path & Bact; Am Soc Clin Path; Am Asn Cancer Res; fel Am Acad Forensic Sci. Res: General pathology; cancer research, particularly with carcinogenic or carcinolytic agents. Mailing Add: Cent Maine Gen Hosp Lewiston ME 04240

BRANCH, CHARLES HENRY HARDIN, b Hopkinsville, Ky, Feb 14, 08; m 37; c 2. PSYCHIATRY. Educ: Univ Fla, AB, 28; La State Univ, MD, 35; Am Bd Psychiat & Neurol, dipl. Prof Exp: Practicing physician, Calif, 37-42; exec med officer, Inst Pa Hosp & instr psychiat, Univ Pa, 48; prof & head dept, Col Med, Univ Utah, 48-70; dep dir, 70-71, PROG CHIEF, MENT HEALTH SERV, COUNTY OF SANTA BARBARA, 71- Concurrent Pos: Fel psychiat, Inst Pa Hosp, 46-48; dir, Am Bd Psychiat & Neurol, 54-62, secy-treas, 61, pres, 62; consult, US Army & US Air Force; mem, Nat Adv Ment Health Coun. Mem: Fel Am Psychiat Asn (secy, 58-61, pres, 62-63); Am Col Physicians; AMA; Am Geriat Soc; NY Acad Sci. Res: Psychotherapy; biochemical changes in mental illness. Mailing Add: Ment Health Serv 4444 Calle Real Santa Barbara CA 93110

BRANCH, DAVID REED, b Coaldale, Pa, Mar 12, 42; m 69; c 2. ASTROPHYSICS. Educ: Rensselaer Polytech Inst, BS, 64; Univ Md, PhD(astron), 69. Prof Exp: Res assoc solar physics, Goddard Space Flight Ctr, NASA, 69; res fel astrophys, Calif Inst Technol, 69-70; sr res fel, Royal Greenwich Observ, 70-73; ASST PROF ASTRON, UNIV OKLA, 73- Mem: Sigma Xi; Int Astron Union; Am Astron Soc; Royal Astron Soc. Res: Stellar spectroscopy and photometry; stellar and solar chemical composition; supernovae; extragalactic distance scale. Mailing Add: Dept of Physics & Astron Univ of Okla Norman OK 73069

BRANCH, GARLAND MARION, JR, b Plant City, Fla, Apr 16, 22; m 45; c 4. MEDICAL TECHNOLOGY. Educ: Stetson Univ, BS, 43; Cornell Univ, PhD(physics), 51. Prof Exp: Res asst nuclear reactor physics, Manhattan Proj, 43-46; asst physics, Cornell Univ, 46-48, res asst cosmic ray physics, 48-51; res assoc electron physics, Res Lab, 51-58, liaison scientist, 58-60, consult microwave physicist, Superpower Microwave Tube Lab, 60-63, mgr electron beam & circuit res, Tube Dept, 63-75, SR DEVELOP ENGR, GEN ELEC MED SYSTS DIV, MILWAUKEE, 75- Concurrent Pos: Adj prof, Union Col, NY, 67-72. Mem: Inst Elec & Electronics Engrs; Am Phys Soc; Am Asn Physics Teachers; NY Acad Acad Sci; Soc Photo-Optical Engrs. Res: Microwave electronics; electron optics; cosmic rays; satellite communication systems; high-energy particle accelerators. Mailing Add: 2468 N 88th St Wauwatosa WI 53226

BRANCH, JOHN CURTIS, b Buffalo, Okla, Oct 1, 34; m 60; c 3. PARASITOLOGY. Educ: Northwestern State Col, Okla, BS, 59; Univ Okla, MS, 62, PhD(prev med, pub health), 65. Prof Exp: Asst prof, 64-69, ASSOC PROF BIOL, OKLAHOMA CITY UNIV, 69-, CHMN DEPT, 66- Mem: AAAS; Am Soc Parasitol; Am Inst Biol Sci. Res: Parasite metabolism; transfer of learning via brain extracts. Mailing Add: Dept of Biol Oklahoma City Univ 2501 N Blackwelder Oklahoma City OK 73106

BRANCONE, LOUIS MARIA, b New York, NY, Nov 13, 15; m 46; c 3. CHEMISTRY. Educ: NY Univ, BA, 36, MS, 40. Prof Exp: Microchemist, Gen Aniline & Film Co, NJ, 41; chemist, Am Home Prod Co, 41; CHEMIST, LEDERLE LABS DIV, AM CYANAMID CO, 41- Mem: Am Chem Soc. Res: Microinorganic and organic analysis; organic synthesis. Mailing Add: 370 Wierimus Rd Hillsdale NJ 07642

BRAND, BENSON GLENN, b Morgantown, WVa, Apr 6, 10; m 46; c 1. CHEMISTRY. Educ: WVa Univ, AB, 32, MS, 34. Prof Exp: Res chemist, E I du Pont de Nemours & Co, Inc, Del, 35-41; res chemist, 46-66, sr res scientist, 66-67, SR CHEMIST, BATTELLE MEM INST, 67- Honors & Awards: Award, Roon Found, 60. Mem: Fel Am Chem Soc; fel Fedn Socs Paint Technol. Res: Paint and varnish; general organic coatings; white pigments; physical and organic chemistry; photography; radar; radio. Mailing Add: 4466 Olentangy Columbus OH 43214

BRAND, DONALD DILWORTH, b Chiclayo, Peru, Mar 6, 05; US citizen; m 32; c 2. GEOGRAPHY, ANTHROPOLOGY. Educ: Univ Calif, AB, 29, PhD(geog), 33. Prof Exp: Instr & lectr geog, Univ Calif, 34; from asst prof to prof anthropogeog, Univ NMex, 34-47, actg head dept anthrop, 35-36, head dept, 36-47; prof geog, Univ Mich, 47-49; chmn dept, 49-60, PROF GEOG, UNIV TEX, AUSTIN, 49- Concurrent Pos: Heller travelling fel from Univ Calif, Mexico, 30-31; cult geogr, Smithsonian Inst, Mex, 44-46. Mem: Fel AAAS; fel Am Geog Soc; Asn Am Geogr; fel Am Anthrop Asn; Soc Am Archaeol. Res: Latin America, especially Mexico; historical, plant and agricultural geography; coastal studies; history of scientific research in Mexico; hydrographic survey of the eastern Pacific. Mailing Add: Dept of Geog Univ of Tex Austin TX 78712

BRAND, EUGENE DEW, b CZ, Mar 10, 24; m 53, 72; c 6. PSYCHIATRY, PSYCHOPHARMACOLOGY. Educ: Univ Va, BA, 44; Harvard Univ, MD, 50. Prof Exp: Instr physiol, Sch Med, Univ Va, 48, intern med, Univ Hosp, 50-51; resident, Univ Utah, 51-52, asst pharm, 52-53; asst prof, Univ Va Hosp, 53-61, assoc prof pharmacol, Sch Med, 61-74, resident psychiat, Univ Hosp, 70-74; MED DIR & STAFF PSYCHIATRIST, MID PENINSULA NORTH NECK MENT HEALTH SERV, 74- Res: Physiology; traumatic shock; psychiatry; ethology. Mailing Add: Mid Peninsula North Neck Ment Health Serv Saluda VA 23149

BRAND, GLENN ELDON, physical chemistry, chemical engineering, see 12th edition

BRAND, JERRY JAY, b Waterloo, Ind, Sept 20, 41; m 67; c 2. PLANT PHYSIOLOGY, PHOTOBIOLOGY. Educ: Manchester Col, BS, 63; Purdue Univ, PhD(biol sci), 71. Prof Exp: Sec sch teacher chem & physics, Nigeria, 63-65; res assoc bot, Ind Univ, 71-74; ASST PROF BOT, UNIV TEX, AUSTIN, 74- Mem: AAAS; Am Soc Plant Physiologists; Am Chem Soc; Am Inst Biol Sci; Bot Soc Am. Res: Mechanism of photosynthesis; structure-function relationships in chloroplast membranes with particular emphasis on green and blue-green algae; role of galactolipids in photosynthetic membranes. Mailing Add: Dept of Bot Univ of Tex Austin TX 78712

BRAND, JOHN C, b Durban, S Africa, May 21, 21; m 43; c 1. PHYSICAL CHEMISTRY. Educ: Univ London, BSc, 41, MSc, 43, PhD, 46, DSc, 56. Prof Exp: Res chemist, May & Baker, Ltd, Eng, 42-43; asst lectr, chem, King's Col, Univ London, 43-46, lectr, 46-47; lectr, Glasgow Univ, 47-59, sr lectr, 59-64; prof, Vanderbilt Univ, 64-69; PROF CHEM & PHOTOCHEM, UNIV WESTERN ONT, 69- Concurrent Pos: Res assoc, Princeton Univ, 53-54. Mem: The Chem Soc. Res: Molecular electronic spectroscopy and molecular structure; rates and mechanisms of chemical reactions. Mailing Add: Dept of Chem Univ of Western Ont London ON Can

BRAND, JOHN ROBERT, b Onaga, Kans, Sept 26, 35; m 57; c 3. PHYSICAL INORGANIC CHEMISTRY. Educ: McPherson Col, BS, 61; Purdue Univ, PhD(chem), 67. Prof Exp: Investr chem, NJ Zinc Co, Gulf & Western Industs, Inc, 67-68; asst prof, Emporia Kans State Col, 68-72; SR RES SCIENTIST, NJ ZINC CO DIV, GULF & WESTERN, INC, 72- Mem: Am Chem Soc. Res: Thermodynamics and kinetics of inorganic reactions; thermochemistry; physical chemistry of surfaces; colloids. Mailing Add: Res Dept New Jersey Zinc Co Palmerton PA 18071

BRAND, JOHN S, b Buffalo, NY, May 2, 38; m 60; c 4. BIOPHYSICS. Educ: LeMoyne Col, BS, 59; Univ Rochester, MS, 64, PhD, 66. Prof Exp: Fel, 66-68, asst prof, 68-75, ASSOC PROF RADIATION BIOL & BIOPHYS, UNIV ROCHESTER, 75- Concurrent Pos: NIH career develop award, 75. Mem: AAAS. Res: Hormonal regulation of bone cell metabolism. Mailing Add: Dept of Radiat Biol & Biophys Univ Rochester Sch Med & Dent Rochester NY 14642

BRAND, KARL GERHARD, b Lübeck, Ger, June 10, 22; US citizen; m 49; c 2. MICROBIOLOGY. Educ: Univ Hamburg, MD, 49; Trop Inst Hamburg, cert, 54; Free Univ Berlin, DPH, 56. Prof Exp: Resident instr internal med, State Hosp Lübeck, 49-52; asst prof trop med, Trop Inst Hamburg, 52-54; assoc prof pub health & microbiol, Free Univ Berlin, 55-57; from asst prof to assoc prof microbiol, 57-72, PROF MICROBIOL, SCH MED, UNIV MINN, MINNEAPOLIS, 72- Mem: Am Asn Immunol; Soc Exp Biol & Med; Am Soc Microbiol. Res: Serological problems in influenza, mumps, psittacosis, viral pneumonia; foreign body carcinogenesis; antigenic structure of cells, tissues, tumors; cancer-aging relationships. Mailing Add: Dept of Microbiol Univ of Minn Sch of Med Minneapolis MN 55455

BRAND, LEONARD, b New York, NY, Dec 21, 23; m 51; c 3. MEDICINE, ANESTHESIOLOGY. Educ: Yale Univ, BS, 46; Columbia Univ, MD, 49; Am Bd Anesthesiol, dipl, 59. Prof Exp: Intern med, Long Island Col Hosp, 49-50; resident gen med, Leo N Levi Mem Hosp, Ark, 50-51; resident anesthesiol, Presby Hosp, 53-55; from instr to assoc prof, 55-72, PROF CLIN ANESTHESIOL, COLUMBIA UNIV, 72- Concurrent Pos: Asst anesthesiologist, Presby Hosp, New York, 55-57, asst attend anesthesiologist, 57-66, attend anesthesiologist, 66- Mem: Am Soc Anesthesiol. Res: Pharmacology and physiology of intravenous agents; fate and distribution of barbiturates in man; passage of substances into central nervous system; respiratory physiology; anesthesia for trauma and orthopedic cases. Mailing Add: Presby Hosp 622 W 168th St New York NY 10032

BRAND, LEONARD ROY, b Harvey, NDak, May 17, 41. VERTEBRATE ZOOLOGY. Educ: La Sierra Col, BA, 64; Loma Linda Univ, MA, 66; Cornell Univ, PhD(vertebrate zool), 70. Prof Exp: Asst prof, 69-74, ASSOC PROF BIOL, LOMA LINDA UNIV, 74-, CHMN DEPT, 71- Honors & Awards: A Brazier Howell Award, Am Soc Mammalogists, 67. Mem: AAAS; Am Soc Mammalogists; Animal Behav Soc; Ecol Soc Am. Res: Behavior and ecology of mammals; footprints of living and fossil animals. Mailing Add: Dept of Biol Loma Linda Univ Loma Linda CA 92354

BRAND, LUDWIG, b Vienna, Austria, Jan 3, 32; nat US; m 58. BIOCHEMISTRY. Educ: Harvard Univ, BA, 55; Ind Univ, PhD, 60. Prof Exp: Asst biochem, Ind Univ, 55-59; NSF fel, 59-62; from asst prof to assoc prof, 62-74, PROF BIOL, JOHNS HOPKINS UNIV, 74- Concurrent Pos: Fel, Weizmann Inst, 61-62. Res: Mechanism of enzyme action; fluorescence studies with proteins; fluorescence lifetimes. Mailing Add: Dept of Biol Johns Hopkins Univ Baltimore MD 21218

BRAND, PAUL JOHN, geography, see 12th edition

BRAND, PAUL W, b Oota Camund, India, July 17, 14; m 43; c 6. RECONSTRUCTIVE SURGERY, REHABILITATION. Educ: Univ London, MB & BS, 43; FRCS, 45. Hon Degrees: LLD, Wheaton Col, 71. Prof Exp: House surgeon, Univ Col Hosp, London, 42-43; resident surg officer, Hosp Sick Children, 43-44; 2nd asst surgeon, Univ Col Hosp, 44-46; lectr surg, Christian Med Col, India, 46-53, prof surg & lectr orthop, 53-65; CHIEF REHAB BR, USPHS HOSP, 65- Concurrent Pos: Hunterian prof, Royal Col Surgeons, Eng, 52 & 62; consult, Leprosy Mission & Am Leprosy Missions, 60-, consult, 67 & Pan Am Health Orgn, 65; mem expert panel on leprosy, WHO, 64-, clin prof, Med Sch, La State Univ, 66- Honors & Awards: Lasker Award, 60; Comdr, Order of the Brit Empire, 61; Medal, Am Asn Plastic Surgeons, 66; Founders Medal, Nat Rehab Asn, 67; Barclay Medal, Asiatic Soc, India, 67. Mem: Fel Royal Soc Med; fel Brit Orthop Asn; fel Am Col Surg; hon fel Am Surg Asn; hon mem Am Soc Surg Hand. Res: Reconstructive surgery of the hand in leprosy; reconstructive surgery of the foot; insensitivity and hypersensitivity; prevention of trophic ulceration; stress and strain in upper extremity orthotics. Mailing Add: USPHS Hosp Carville LA 70721

BRAND, RAYMOND HOWARD, b Highland Park, Mich, Sept 22, 28; m 51; c 2. ANIMAL ECOLOGY. Educ: Wheaton Col, Ill, BA, 50; Univ Mich, MS, 51, PhD(zool), 55. Prof Exp: Asst prof biol, Westmont Col, 55-57, actg chmn div sci, 57-59; assoc prof biol, 59-70, PROF BIOL & DIV CHMN, WHEATON COL, ILL, 70- Concurrent Pos: NSF basic res found grant, 62-64; grant, Inst Radiolbiol, Argonne Nat Lab, 65. Mem: Ecol Soc Am; Am Sci Affiliation. Res: Population dynamics; animal behavior; temperature adaptions; springtail insects. Mailing Add: Dept of Biol Wheaton Col Wheaton IL 60187

BRAND, RICHARD ROBERT, b New York, NY, Aug 30, 40; m 68; c 2. GEOGRAPHY. Educ: St John's Univ, NY, BA, 64; Columbia Univ, MA, 65, EdD(geog), 72. Prof Exp: Instr geog, McKenna Mem Jr High Sch, Massapequa, NY, 65-66; lectr, Univ Cape Coast, Ghana, 68-69; instr, DC Teachers Col, 69-70; from instr to asst prof, Univ RI, 70-73; res assoc, Pa State Univ, 73-74; ASSOC PROF GEOG, EDINBORO STATE COL, 74- Mem: Am Asn Geogrs; Am Geog Soc; Am Cong Surv & Mapping. Res: Application of cartographic and quantitative methods to land use planning questions. Mailing Add: Dept of Geog Edinboro State Col Edinboro PA 16444

BRANDALEONE, HAROLD, b New York, NY, Apr 27, 07; m 47; c 3. MEDICINE, CARDIOLOGY. Educ: NY Univ, BS, 28, MD, 31, ScD(med), 38; Am Bd Internal Med, dipl. Prof Exp: Intern, Third Med Div, Bellevue Hosp, 31-33; asst prof, 49-57, ASSOC PROF CLIN MED, NY UNIV, 57- Concurrent Pos: Assoc chief metab clin, NY Univ Hosp, 33-, asst attend physician, 51-57, assoc attend physician, 57-71, attend physician, 71-; assoc vis physician, Bellevue Hosp, 57-66, vis physician, 66-; med dir, Continental Tel Co. Mem: Fel Am Col Physicians; fel Am Col Cardiol; Am Diabetes Asn; AMA; Endocrine Soc. Res: Metabolic studies and accident prevention. Mailing Add: 116 E 63rd St New York NY 10021

BRANDAU, BETTY LEE, b Easton, Pa; m 62; c 1. INORGANIC CHEMISTRY, NUCLEAR CHEMISTRY. Educ: Ursinus Col, BS, 53; Carnegie Inst Technol, MS, 55; PhD(inorg chem), 60. Prof Exp: Chemist, Sun Oil Co, Pa, 54, Dow Chem Co, Mich, 56, Y-12 Plant, Union Carbide Nuclear Corp, Tenn, 60-64 & Oak Ridge Assoc Univs, 64-65 & 69; res assoc, Rosenstiel Inst Marine & Atmospheric Sci, Univ Miami, 65-68; asst dir, 69-71, ASSOC DIR, GEOCHRONOLOGY LAB, UNIV GA, 71- Mem: AAAS; Am Chem Soc; fel Am Inst Chemists; Sigma Xi. Res: Marine sediment and radiocarbon dating; geochemistry; natural radioactivity environmental problems; chemical archaeology. Mailing Add: Univ of Ga Geochronol Lab 110 Riverbend Rd Athens GA 30602

BRANDAU, ROBERT PAUL, b Chicago, Ill, Sept 25, 25; m 48; c 2. COSMETIC CHEMISTRY, COLLOID CHEMISTRY. Educ: Univ Chicago, PhB, 49, BS, 50. Prof Exp: Mgr tech serv raw mat, Stepan Chem Co, 53-59; proj leader soap pads, SOS Div, Gen Foods, 59-61; dir labs raw mat, Richardson Co, 61-64; dir tech serv cosmetic & toiletry raw mat, Henkel, Inc, 64-75; DIR RES & DEVELOP COSMETICS, YARDLEY OF LONDON, INC, TOTOWA, 75- Concurrent Pos: Instr shampoo technol, Educ Ctr, Upsala Col, 75- Mem: Soc Cosmetic Chemists; Am Chem Soc; Am Oil Chemists Soc; AAAS. Res: Investigation of mechanisms by which specific molecules and their derived cosmetic formulations may be used for the beneficial modification of normal and damaged human skin, scalp and hair.

BRANDE, EDWARD WOODROW, b Brooklyn, NY, Jan 3, 31. NUMBER THEORY, ALGEBRA. Educ: St Louis Univ, BA, 56, MS(R), 57, PhL, 57, PhD(math), 61; Woodstock Col Md, STL, 64. Prof Exp: Instr, high sch, NY, 57-58; chmn col philos & lett, 65-69, assoc chmn dept, 70-74, ASST PROF MATH, FORDHAM UNIV, 65-, DIR MST MATH PROG, 69-, PRES FAC SEN, 74- Mem: AAAS; Am Math Soc; Math Asn Am. Res: Number theory, especially quadratic forms and representation of integers; algebraic number theory. Mailing Add: Dept of Math Fordham Univ Bronx NY 10458

BRANDELL, BRUCE REEVES, b Detroit, Mich, Oct 4, 26; m 52; c 5. ANATOMY, ZOOLOGY. Educ: Univ Mich, BS, 49, MS, 50, PhD(zool), 63. Prof Exp: Teaching fel zool, Univ Mich, 54-57; instr biol, Univ Akron, 57-61; asst prof anat, Sch Med, Univ NDak, 62-65; asst prof, 65-69, ASSOC PROF ANAT, UNIV SASK, 69- Concurrent Pos: USPHS grant, 65-66; Dept Nat Health & Welfare, Can grants, 68-70. Mem: Am Soc Mammal; Am Asn Anatomists; Can Asn Anat; Int Soc Electromyography & Kinesiology. Res: Human gross anatomy; electromyographic and anatomical study of the finger moving muscles; comparative anatomy of nerve distribution patterns in mammalian forearm and hand muscles; fine distribution of blood vessels within skeletal muscles and tendons; investigations of locomotion by means of electromyography and cinematography. Mailing Add: Dept of Anat Univ of Sask Saskatoon SK Can

BRANDENBERGER, JERRY D, b Houston, Tex, Dec 20, 31; m 60; c 2. PHYSICS. Educ: Rice Univ, BA, 54; Univ Tex, MA, 57, PhD, 62. Prof Exp: Asst, Univ Tex, 55-60, res scientist, 60-63; asst prof, 63-69, ASSOC PROF PHYSICS, UNIV KY, 69- Mem: Am Phys Soc. Res: Nuclear structure physics; interaction of fast neutrons with matter; nuclear instrumentation. Mailing Add: Dept of Physics Univ of Ky Lexington KY 40506

BRANDENBERGER, JOHN RUSSELL, b Danville, Ill, May 13, 39; m 64; c 1. ATOMIC PHYSICS. Educ: Carleton Col, AB, 61; Brown Univ, ScM, 64, PhD(physics), 68. Prof Exp: Instr physics, Col Wooster, 64-66; asst prof, 68-75, chmn dept, 73-75, ASSOC PROF PHYSICS, LAWRENCE UNIV, 75- Concurrent Pos: Res fel, Harvard Univ, 75-76; consult, Los Alamos Sci Lab, Univ Calif, 75- Mem: Am Phys Soc. Res: Measurement of Sommerfeld fine structure constant by level-crossing in atomic hydrogen; fast atomic beam spectroscopy. Mailing Add: Dept of Physics Lawrence Univ Appleton WI 54911

BRANDENBERGER, STANLEY GEORGE, b Houston, Tex, Jan 18, 30; m 67; c 2. ORGANIC CHEMSTRY. Educ: Rice Univ, BA, 52; Univ Tex, PhD(org chem), 56. Prof Exp: Res chemist, Houston Res Lab, Shell Oil Co, 56-63, sr res chemist, 63-64, group leader, 64-68, sect head, Royal Dutch/Shell Lab, Holland, 68-69, staff res chemist, Houston Res Lab, Deerpark, 69-72, STAFF RES CHEMIST, SHELL DEVELOP CO, 72- Mem: Am Chem Soc. Res: Physical organic chemistry; heterogeneous catalysis. Mailing Add: 5726 Kuldell Houston TX 77035

BRANDENBURG, JAMES H, b Green Bay, Wis, July 17, 30; m 54; c 4. OTOLARYNGOLOGY. Educ: Univ Wis, BA, 52, MS, 56; Am Bd Otolaryngol, dipl, 63. Prof Exp: Intern, William Beaumont Gen Hosp, El Paso, Tex, 57, ear, nose & throat preceptorship, 58; ear, nose & throat resident, Brooke Gen Hosp, San Antonio, 61; from asst prof to assoc prof, 64-72, PROF & CHMN OTOLARYNGOL, MED SCH, UNIV WIS-MADISON, 72- Concurrent Pos: Attend consult otolaryngol, Vet Admin Hosp, Madison, 64-; mem fac, Home Study Course, Am Acad Otolaryngol & Ophthal, 67-; prin investr prototype comprehensive network demonstration proj head & neck cancer, NIH Contract, 74- Honors & Awards: Cert of Achievement, Armed Forces Inst Path, Walter Reed Hosp, 64. Mem: Fel Am Acad Ophthal & Otolaryngol; fel Am Laryngol, Rhinol & Otol Soc; Am Coun Otolaryngol; Am Soc Head & Neck Surg; AMA. Res: Combination therapy for epidermoid carcinoma of the head and neck; carcinoma of the larynx; study of the guinea pig cochlea by electron microscopy; study of the changes seen in the cochlea of the guinea pig following noise exposure; traumatic injuries to the larynx. Mailing Add: 1300 University Ave Madison WI 53706

BRANDENBURG, ROBERT O, b Minneapolis, Minn, Aug 5, 18; m 44; c 5. MEDICINE. Educ: Univ NDak, BS, 40; Univ Pa, MD, 43. Prof Exp: Intern, Presby Hosp, Pa, 44; intern, Nutrit Clin, Hillman Hosp, Ala, 47-48; fel, Mayo Clin, 48-51, from asst prof to assoc prof med, 58-69, PROF MED, MAYO MED SCH, 69-, CONSULT MED, MAYO CLIN, 51- Concurrent Pos: Chmn div cardiovasc dis, Mayo Clin, 69-74. Mem: AMA; Am Heart Asn; fel Am Col Physicians; fel Am Col Cardiol. Res: Congenital and rheumatic cardiovascular disease. Mailing Add: Mayo Clin Univ of Minn Rochester MN 55902

BRANDER, ROBERT BRUCE, ecology, wildlife management, see 12th edition

BRANDFONBRENER, MARTIN, b New York, NY, July 25, 27; m 56; c 4. INTERNAL MEDICINE, CARDIOLOGY. Educ: Albany Med Col, MD, 49. Prof Exp: Rotating intern, Med Ctr, Univ Ind, 49-50; resident internal med, Boston City Hosp, 50-51; resident, Dept Physiol, Western Reserve Univ, 51-52; res assoc cardiol, Sect Geront, Nat Heart Inst, 53-54; asst resident med, Univ Hosp, Univ Cleveland, 54-55; asst resident cardiol, Presby Hosp, New York, 55-56; fel cardiorespiratory med, Col Physicians & Surgeons, Columbia Univ, 56-58; chief cardiol sect, Vet Admin Res Hosp, Chicago, 58-63; assoc prof med, Univ NMex, 63-67; PROF MED, MED SCH, NORTHWESTERN UNIV, CHICAGO, 67- Concurrent Pos: USPHS fel, Sch Med, Western Reserve Univ, 51-52. Mem: Am Fedn Clin Res (secy, 62); fel Am Col Physicians; Am Physiol Soc. Res: Myocardial metabolism and coronary flow. Mailing Add: Dept of Med Northwestern Univ Med Sch Chicago IL 60611

BRANDHORST, CARL THEODORE, b Lincoln, Nebr, Aug 24, 98; m 22; c 8. ZOOLOGY, BOTANY. Educ: Ft Hays Kans State Col, BS, 39, MS, 42; Univ Nebr, PhD(zool), 61. Prof Exp: Teacher, pvt & pub schs, 17-38; from assoc prof to prof, 38-73, chmn dept, 57-62, EMER PROF BIOL, CONCORDIA TEACHERS COL, NEBR, 73- Concurrent Pos: Researcher & assoc prof, San Fernando Valley State Col, 62-63. Mem: AAAS; Entom Soc Am. Res: Plant-insect relation. Mailing Add: 55 Pearl St Seward NE 68434

BRANDMAN, HAROLD A, b Newark, NJ, Jan 29, 41; m 64; c 2. ORGANIC CHEMISTRY. Educ: Univ Pa, BA, 62; Seton Hall Univ, MS, 66, PhD(chem), 68. Prof Exp: Chemist, 68-74, GROUP LEADER, GIVAUDAN CORP, 75- Mem: Am Chem Soc; Sigma Xi. Res: Synthesis and development of antimicrobial agents used as industrial biocides. Mailing Add: Givaudan Corp 125 Delawanna Ave Clifton NJ 07014

BRANDNER, JOHN DAVID, b Bethlehem, Pa, Mar 23, 10; m 43; c 3. INDUSTRIAL CHEMISTRY. Educ: Lehigh Univ, BS, 32, MS, 34; Columbia Univ, PhD(chem), 43. Prof Exp: Chemist, Exp Lab, Atlas Powder Co, 34-37; asst, Columbia Univ, 38-40; asst dir control lab, Ravenna Ord Works, 41-42; chemist, 42-44, dir res lab, 44-47, asst dir cent res lab, 48-51, mgr, 51-55, mgr phys chem & anal sect, Res Dept, 55-61, assoc dir, Chem Res Dept, 61-70, DIR, CHEM RES DEPT, ATLAS CHEM INDUSTS, 70- Concurrent Pos: Mem comt specifications, Food Chem Codex Proj, Nat Res Coun & Food & Agr Orgn-WHO Expert Comt Food Additives. Mem: AAAS; Am Chem Soc; fel Am Inst Chem. Res: Calorimetry; kinetics; alkylene oxide reactions; food additives. Mailing Add: Chem Res & Develop Lab Atlas Chem Indust Wilmington DE 19899

BRANDOM, WILLIAM FRANKLIN, b Oklahoma City, Okla, Jan 2, 26; m 52; c 4. DEVELOPMENTAL BIOLOGY, CYTOGENETICS. Educ: Stanford Univ, AB, 51, MA, 54, PhD(biol), 59. Prof Exp: Asst biol, Stanford Univ, 53-58; Nat Cancer Inst fel, Princeton Univ, 58-59; from instr to asst prof zool, Newcomb Col, Tulane Univ, 59-65, asst prof path, Sch Med, 65-67; asst prof, Delta Primate Res Ctr, 67-68; assoc prof biol sci, 68-74, PROF BIOL SCI, UNIV DENVER, 74- Concurrent Pos: Nat Inst Child Health & Human Develop fel, Univ St Andrews, 66. Mem: AAAS; Am Soc Zool; Soc Develop Biol; Tissue Cult Asn. Res: Developmental biology of amphibians; developmental genetics; polyploidy; hybridization; pigment pattern formation in heteroploid hybrids; karyotyptes and idiograms of amphibian chromosomes; lampbrush chromosomes; somatic cell genetics; human radiation cytogenetics; chromosome biology. Mailing Add: Dept of Biol Sci Univ of Denver Denver CO 80210

BRANDON, FRANK BAYARD, b Indiana, Pa, Jan 2, 27; m 48; c 3. BIOPHYSICS, VIROLOGY. Educ: Carnegie Inst Technol, BS, 51; Univ Pittsburgh, MS, 54, PhD(biophys), 56. Prof Exp: Assoc res microbiologist, 56-58, res microbiologist, 58-63, sr res microbiologist, 63-67, SR SCIENTIST, PARKE, DAVIS & CO, 67- Mem: AAAS; Am Soc Microbiol; Biophys Soc. Res: Biophysical and immunologic properties of influenza virus; clinical immunology; biometrics. Mailing Add: Parke Davis & Co Res Div PO Box 118 Detroit MI 48232

BRANDON, RONALD ARTHUR, b Mt Pleasant, Mich, Dec 3, 33; m 54; c 3. VERTEBRATE ZOOLOGY. Educ: Ohio Univ, BS, 56, MS, 58; Univ Ill, PhD(zool), 62. Prof Exp: Asst prof biol, Univ Ala, 62-63; from asst prof to assoc prof, 63-74, PROF ZOOL, SOUTHERN ILL UNIV, CARBONDALE, 74- Mem: AAAS; Soc Study Amphibians & Reptiles; Am Soc Ichthyologists & Herpetologists; Soc Study Evolution. Res: Amphibian systematics; biology of salamanders. Mailing Add: Dept of Zool Southern Ill Univ Carbondale IL 62901

BRANDON, WALTER WILEY, JR, b Gainesville, Ga, Dec 1, 29; m 57; c 3. APPLIED PHYSICS. Educ: Emory Univ, BA, 52, MS, 53. Prof Exp: Scientist, Redstone Res Labs, Rohm and Haas Co, 53-65; res specialist, space div, Boeing Co, 65-67; scientist, Redstone Res Labs, Rohm and Haas Co, 67-71; PHYSICIST, US ARMY MISSILE COMMAND, 72- Mem: Assoc fel Am Inst Aeronaut & Astronaut. Res: Nuclear radiation spectroscopy; liquid and solid rocket propulsion; combustion instability of rockets; detonation of rocket propellants and explosives; instrumentation; microwaves; atmospheric optics. Mailing Add: 1902 Colice Rd SE Huntsville AL 35801

BRANDOU, JULIAN ROBERT, b Grand Rapids, Mich, Feb 22, 29; m 49; c 5. SCIENCE EDUCATION. Educ: Univ Mich, BS, 50, MA, 56; Mich State Univ, PhD(curriculum), 63. Prof Exp: Teacher, Pub Schs, Mich, 50-59; NSF traveling lectr sci, Mich State Univ, 59-60; consult sch sci, 60-63, from instr to assoc prof phys sci, 61-70, actg dir, Sci & Math Teaching Ctr, 66-68, PROF PHYS SCI, MICH STATE UNIV, 70-, DIR, SCI & MATH TEACHING CTR, 68- Mem: AAAS; Nat Asn Res Sci Teaching; Nat Sci Teachers Asn. Res: Pre-service and in-service preparation of elementary, secondary and collegiate teachers of science and mathematics, including the process of curriculum development and implementation. Mailing Add: Sci & Math Teaching Ctr Mich State Univ East Lansing MI 48823

BRANDOW, BAIRD H, b San Diego, Calif, June 9, 35; m 66; c 2. THEORETICAL PHYSICS. Educ: Calif Inst Technol, BS, 57; Cornell Univ, PhD(theoret physics), 64. Prof Exp: NSF fel, Niels Bohr Inst, Copenhagen Univ, 63-65; docent theoret nuclear physics, Nordic Inst Theoret Atomic Physics, Denmark, 65-66; instr physics, Cornell Univ, 66-69; mem staff, Battelle-Seattle Res Ctr, 69-71 & Battelle Mem Inst, Ohio, 71-74; MEM STAFF, LOS ALAMOS SCI LAB, 74- Mem: Am Phys Soc. Res: Nuclear many-body problem; linked-cluster expansions; theory of liquid and solid helium; theory of magnetic insulators and associated metal-insulator transitions. Mailing Add: Los Alamos Sci Lab Group T-9 PO Box 1663 Los Alamos NM 87544

BRANDRISS, MICHAEL W, b Brooklyn, NY, Oct 3, 31; m 55; c 4. IMMUNOLOGY, INFECTIOUS DISEASES. Educ: Kenyon Col, AB, 53; NY Univ, MD, 57. Prof Exp: Intern med, Johns Hopkins Hosp, 57-58, asst resident, 58-59; asst resident, Baltimore City Hosps, 59-60; clin assoc, Nat Inst Allergy & Infectious Dis, 60-62, clin investr, 62-63, sr investr, 63-65; asst prof, 65-69, ASSOC PROF MED, SCH MED & DENT, UNIV ROCHESTER, 69- Concurrent Pos: Nat Inst Allergy & Infectious Dis grants, 66-; consult med, Vet Admin Hosp, Batavia, NY, 65-; attend physician, Rochester Gen Hosp, NY; sr assoc physician, Strong Mem Hosp, Rochester. Mem: Am Fedn Clin Res; Am Asn Immunol; Infectious Dis Soc Am. Res: Delayed hypersensitivity in animals; antimetabolites in autoimmune states; immunologic studies on penicillin and related antigens; human immune response to haptens and to protein and polysaccharide antigens; human immune response to viral infections. Mailing Add: Dept of Med Univ of Rochester Sch of Med & Dent Rochester NY 14627

BRANDS, ALLEN J, b Kansas City, Mo, Sept 19, 14; m 34; c 1. PHARMACOLOGY. Educ: Univ Southern Calif, BS, 41. Hon Degrees: DSc, Philadelphia Col Pharm & Sci, 74. Prof Exp: Community pharmacist, 41-50; hosp pharmacist, USPHS, 50-51, personnel officer, 51-52, CHIEF PHARM BR, INDIAN HEALTH SERV, USPHS, 53-, PHARM LIAISON REP, 67- Concurrent Pos: Mem, Vis Comt, Col Pharm, Wayne State Univ; mem, Pub Health Serv Comt, Nat Asn Retarded Children, 69-; del, US Pharmacopiel Conv, 70 & 75. Honors & Awards: Andrew Craigie Award, Asn Mil Surgeons US, 73. Mem: Fel Am Col Clin Pharmacists; Fedn Int Pharmacists; Am Pharmaceut Asn; Am Soc Hosp Pharmacists. Res: Clinical pharmacy; rational drug therapy. Mailing Add: 3024 Tilden St NW Washington DC 20008

BRANDS, ALVIRA BERNICE, b Hader, Minn, July 9, 22; m 71; c 2. PSYCHIATRIC NURSING. Educ: Univ Minn, BS, 58, MNurse Admin, 60; Cath Univ Am, DSc, 75. Prof Exp: From head nurse to nursing instr, Anoka State Hosp, Minn, 54-63; chief nursing serv, Div Med Serv, Minn Dept Pub Welfare, St Paul, 63-69; psychiat nurse consult, 70-74, PROG ANALYST, MENT HEALTH CARE & SERV FINANCING BR, NIMH, ROCKVILLE, MD, 74- Concurrent Pos: Psychiat nurse consult, Jamestown State Hosp, NDak, 67, Woodward State Hosp & Sch, Iowa, 68, Sch Pharm, Univ Southern Calif, 71; mem pharm adv comt, Indian Health Serv, USPHS, 71-; mem liaison comt, Am Nurses Asn & Am Soc Hosp Pharmacists, 74- Mem: Am Asn Ment Deficiency; Am Nurses Asn; Am Pub Health Asn; Am Asn Univ Women; Nat League Nursing. Res: Drug use for long term psychiatric patients. Mailing Add: 3024 Tilden St NW Washington DC 20008

BRANDSBERG, JOHN WILBERT, mycology, see 12th edition

BRANDSTEIN, ALFRED GEORGE, b Brooklyn, NY, Oct 27, 38; m 62; c 3. MATHEMATICAL ANALYSIS. Educ: Brooklyn Col, BS, 59; Brown Univ, PhD(math), 72. Prof Exp: Asst prof math, Univ Conn, 65-72; MATHEMATICIAN, HARRY DIAMOND LABS, 72- Mem: Am Math Soc; Sigma Xi. Res: Invariant subspaces; numerical transforms; discrete stochastic modeling. Mailing Add: Harry Diamond Labs AMXDO-EM-2 Adelphi MD 20783

BRANDT, BRUCE LOSURE, b Seattle, Wash, Nov 3, 41; m 63; c 2. NEUROBIOLOGY, ELECTROPHYSIOLOGY. Educ: Pomona Col, BA, 63; Univ Rochester, PhD, 69. Prof Exp: RES ASSOC NEUROBIOL, SALK INST BIOL STUDIES, 73- Concurrent Pos: NIH fel genetics, Sch Med, Stanford Univ; Nat Multiple Sclerosis Soc fel. Mem: AAAS. Res: Properties of thin film superconductors; molecular genetics of bacteria and cultured cells; neuromuscular junction formation. Mailing Add: Dept of Neurobiol Salk Inst PO Box 1809 San Diego CA 92112

BRANDT, CARL DAVID, b Bridgeport, Conn, Jan 19, 28; m 64; c 2. VIROLOGY. Educ: Univ Conn, BS, 49; Univ Mass, MS, 51; Harvard Univ, PhD(bact), 58. Prof Exp: Instr vet sci, Univ Mass, 49-52 & 54; res virologist, Charles Pfizer & Co, 58-62; assoc, Dept Epidemiol, Pub Health Res Inst, New York, 62-66; instr, Med Sch, Georgetown Univ, 66-69; asst prof, 69-74, ASSOC PROF CHILD HEALTH,

DEVELOPMENT & MICROBIOLOGY, MED SCH, GEORGE WASHINGTON UNIV, 74-; RES ASSOC, VIROL SECT, CHILDREN'S HOSP RES FOUND, 66- Mem: AAAS; Am Soc Microbiol; fel Am Acad Microbiol; Soc Epidemiol Res. Res: Virus epidemiology; respiratory and enteric tract pathogens; virus vaccines. Mailing Add: Children's Hosp Res Found 2125 13th St NW Washington DC 20009

BRANDT, CARL STAFFORD, chemistry, see 12th edition

BRANDT, CHARLES LAWRENCE, b Prescott, Ariz, Dec 18, 25; m 54. PHYSIOLOGY. Educ: Stanford Univ, AB, 49, PhD(biol sci), 55. Prof Exp: Asst gen physiol, Stanford Univ, 50-54; instr zool, Univ Tex, 55-57; assoc prof, 57-66, chmn dept, 67-69, PROF BIOL, SAN DIEGO STATE UNIV, 66- Concurrent Pos: Res assoc, Hopkins Marine Sta, 56. Mem: AAAS; Soc Gen Physiol; Am Soc Photobiol. Res: Cellular and general physiology; photophysiology. Mailing Add: 6361 Rockhurst Dr San Diego CA 92120

BRANDT, DONALD PAUL, b New York, NY, Aug 10, 40; m. URBAN GEOGRAPHY, POPULATION GEOGRAPHY. Educ: Univ Okla, BA, 63, MA, 64; Ind Univ, PhD(geog), 69. Prof Exp: Instr geog, Mid Tenn State Univ, 65-67; assoc prof, 69-74, PROF GEOG, EDINBORO STATE COL, 74- Concurrent Pos: Consult, NW Pa Manpower Develop Corp, 71- Mem: Asn Am Geogr; Regional Sci Asn. Res: Urban land use expansion. Mailing Add: Dept of Geog Edinboro State Col Edinboro PA 16412

BRANDT, EDWARD NEWMAN, JR, b Oklahoma City, Okla, July 3, 33; m 53; c 3. BIOSTATISTICS, MEDICINE. Educ: Univ Okla, BS, 54; Okla State Univ, MS, 55, MD, 60, PhD(biostatist), 63. Prof Exp: From instr to prof prev med & pub health, Med Ctr, Univ Okla, 61-70, dir biostatist unit & med res comput ctr, 62-70, assoc prof, Dept Internal Med, 63-70, assoc dean sch med & assoc dir med ctr, 68-70; PROF PREV MED & FAMILY MED, UNIV TEX MED BR, GALVESTON, 70-, DEAN MED, 74- Mem: AAAS; Am Fedn Clin Res; AMA; Am Soc Clin Pharmacol & Chemother. Res: Educational administration. Mailing Add: Univ of Tex Med Br Galveston TX 77550

BRANDT, ELIZABETH ANNE, b Sanford, Fla, Oct 20, 45. ANTHROPOLOGY. Educ: Fla State Univ, BA, 67; Southern Methodist Univ, MA, 69, PhD(anthrop), 70. Prof Exp: Asst prof anthrop, Univ Ill, Chicago Circle, 70-74; ASST PROF ANTHROP, ARIZ STATE UNIV, 74- Concurrent Pos: Univ Res Bd fels, Univ Ill, Chicago Circle, 71-73. Mem: AAAS; fel Am Anthrop Asn; fel Soc Appl Anthrop; Ling Soc Am. Res: Sociolinguistics; application of computers to linguistics; cognitive anthropology; American Indian languages; Southwest ethnology; proxemics and social space. Mailing Add: Dept of Anthrop Ariz State Univ Tempe AZ 85281

BRANDT, GERALD BENNETT, b Pittsburgh, Pa, Apr 20, 38; m 61; c 4. OPTICAL PHYSICS. Educ: Harvard Univ, AB, 60; Carnegie Inst Technol, MS, 63, PhD(physics), 66. Prof Exp: Engr, 60-66, sr engr, 66-70, fel scientist, 70-74, MGR ELECTROOPTICS, OPTICAL PHYSICS DEPT, WESTINGHOUSE RES LABS, 74- Concurrent Pos: Lectr, Carnegie Mellon Univ, 69 & 70; assoc ed, J Optical Soc Am, 72- Mem: AAAS; Optical Soc Am; Am Phys Soc. Res: Integrated optics; optical design and instrumentation; coherent optics and holography; low temperature physics; electronic properties of metals; ultrasonics. Mailing Add: 208 W Swissvale Ave Pittsburgh PA 15218

BRANDT, GERALD H, b Lompoc, Calif, Feb 25, 33; m 58; c 3. SOIL PHYSICS. Educ: Ore State Univ, BS, 55; Mich State Univ, MS, 60, PhD(soil sci), 63. Prof Exp: Asst soil sci, Mich State Univ, 60-63; chemist, 63-72, GROUP LEADER, DOW CHEM CO, 73- Concurrent Pos: Mem, Comt Soil Physiochem Phenomena, Transp Res Bd, Nat Acad Sci, 71 & Soil Chem Stabilization Comt, 72- Mem: Am Soc Agron; Clay Minerals Soc; Soil Sci Soc Am; Int Soc Soil Sci; Sigma Xi. Res: Dissolved oxygen in soils; clay-chemical phenomena; modifying soil physical properties with chemicals; erosion and sediment control; soil stabilization; frost heave in soils; environmental effects of de-icing salts, general applications research. Mailing Add: Larkin Lab Dow Chem Co Midland MI 48640

BRANDT, IRA KIVE, b New York, NY, Nov 9, 23; m 47; c 4. PEDIATRICS. Educ: NY Univ, AB, 42; Columbia Univ, MD, 45. Prof Exp: Res fel pediat & biochem, Yale Univ, 55-57; from asst prof to assoc prof pediat, Yale Univ, 57-68; chmn dept, Children's Hosp, San Francisco, 68-70; clin prof, Univ Calif, San Francisco, 70; PROF, DEPTS PEDIAT & MED GENETICS, IND UNIV, INDIANAPOLIS, 70- Mem: Soc Pediat Res; Am Acad Pediatricians; Am Pediat Soc; Am Soc Human Genetics. Res: Metabolic disorders of children; developmental genetics. Mailing Add: Dept of Pediat Ind Univ Sch of Med Indianapolis IN 46202

BRANDT, J LEONARD, b New York, NY, Aug 3, 19; m 50; c 2. INTERNAL MEDICINE, NEPHROLOGY. Educ: Univ Mich, AB, 40; Long Island Col Med, MD, 43; Am Bd Internal Med, dipl. Prof Exp: Intern, Mt Sinai Hosp, Cleveland, 44; res assoc, Sch Trop Med, Columbia Univ, San Juan, PR, 45; asst res pathologist, Montefiore Hosp, New York, 48; from instr to assoc prof med, Col Med, State Univ NY, 51-59; lectr, 59-70, ASSOC PROF MED, SCH MED, McGILL UNIV, 70-; PHYSICIAN-IN-CHIEF, JEWISH GEN HOSP, MONTREAL, 59- Mem: AAAS; Am Fedn Clin Res; Soc Exp Biol & Med; Am Physiol Soc; Harvey Soc. Res: Hepatic and renal physiology. Mailing Add: 3755 Cote Ste Catherine Rd Montreal PQ Can

BRANDT, JAMES LEWIS, b Huntington, WVa, Nov 20, 25; m 48; c 3. PHYSICAL CHEMISTRY. Educ: St Francis Col, Pa, BS, 48; Purdue Univ, MS, 50, PhD(phys chem), 53. Prof Exp: Res chemist, Alcoa Res Labs, Aluminum Co Am, 52-57, asst chief phys chem div, 57-60, mgr tech info dept, 60-67, mgr chem metall div, 67-75, MGR ANAL CHEM DIV, ALCOA LABS, 75- Mem: Sigma Xi; Soc Naval Archit & Marine Eng; Am Inst Mech Eng; Soc Automotive Eng. Res: Analysis of gases in metals; physical properties of aluminum; information retrieval; corrosion and stress of aluminum and its alloys; analysis of aluminum and aluminum alloys. Mailing Add: Alcoa Res Labs Alcoa Tech Ctr PO Box 2970 Pittsburgh PA 15230

BRANDT, JOHN CONRAD, b St Louis, Mo, Aug 8, 34. ASTRONOMY, ASTROPHYSICS. Educ: Washington Univ, AB, 56; Univ Chicago, PhD, 60. Prof Exp: Asst prof astron, Univ Calif, Berkeley, 61-63; asst astronr, Kitt Peak Nat Observ, 63-65, assoc astronr, 65-66; sr res scientist, Inst Space Studies, NY, 66-67; head solar physics br, 67-70, chief lab solar physics, 70-74, CHIEF LAB SOLAR PHYSICS & ASTROPHYS, NASA GODDARD SPACE FLIGHT CTR, 74- Concurrent Pos: NSF fel, Mt Wilson & Mt Palomar Observs, 60-61. Mem: Fel AAAS; Am Astron Soc; Int Astron Union. Res: Solar system astrophysics; comets; interplanetary gas; planetary atmospheres; radiative transfer; structure of galaxies. Mailing Add: Lab for Solar Phys & Astrophys NASA Goddard Space Flight Ctr Greenbelt MD 20771

BRANDT, KARL GARET, b Galveston, Tex, Oct 15, 38; m 65; c 2. BIOCHEMISTRY. Educ: Rice Univ, BA, 60; Mass Inst Technol, PhD(org chem), 64. Prof Exp: NSF fel, Johnson Res Found, Univ Pa, 64-65; NIH fel, Cornell Univ, 65-66; from asst prof to assoc prof, 66-75, PROF BIOCHEM, PURDUE UNIV, 75- Mem: Am Chem Soc; Am Soc Biol Chemists. Res: Mechanism of enzyme action; structure-function relationships; kinetics of enzyme-catalyzed reactions. Mailing Add: Dept of Biochem Purdue Univ Lafayette IN 47907

BRANDT, LUTHER WARREN, b Gradan, Kans, Oct 1, 20; m 42; c 1. PHYSICAL CHEMISTRY. Educ: Ft Hays Kans State Col, AB, 41; Kans State Univ, PhD(phys chem), 50. Prof Exp: Chemist fission prod anal, Clinton Labs, E I du Pont de Nemours & Co, 44 & Hanford Eng Works, E I du Pont de Nemours & Co & Gen Elec Co, 44-47; res chemist high polymers, E I du Pont de Nemours & Co, 50-52; phys chemist, 52-59, chief res div, 59-61, chief helium res ctr, Tex, 61-62, res dir, 62-71, polymer chemist, Spokane Mining Res Ctr, Wash, 70-72, RES CHEMIST, TUSCALOOSA MINING RES LAB, US BUR MINES, 72- Mem: AAAS; Am Chem Soc; fel Am Inst Chemists; Sigma Xi. Res: Surface chemistry relating to mineral tailings; electrophoresis of serum proteins of certain animals; phase equilibria and thermodynamics of gases. Mailing Add: 22 Riverdale Tuscaloosa AL 35401

BRANDT, MANUEL, b Columbus, Ohio, Jan 7, 15; m 47; c 2. ANALYTICAL CHEMISTRY. Educ: Ohio State Univ, BA, 36, MA, 37. Prof Exp: Asst physiol chem, Ohio State Univ, 36-38; chemist, US Army Ord, 40-43 & Tenn Eastman Corp, 43-45; chemist, 45-50, group leader, 50-58, asst supvr, 58-66, SUPVR RES & DEVELOP DEPT, ETHYL CORP, 66- Mem: Am Chem Soc; Am Microchem Soc; Soc Appl Spectros. Res: Microchemical analyses; trace and ultratrace metal analyses. Mailing Add: Res & Develop Dept Ethyl Corp 1600 W Eight Mile Rd Detroit MI 48220

BRANDT, NEILL MATTESON, b Orange, NJ, May 12, 15; m 41; c 2. PHYSICS. Educ: St Lawrence Univ, BS, 37. Prof Exp: Instr, high sch, NY, 37-42; chemist, Bausch & Lomb Optical Co, 42-46; from fel to sr fel, Mellon Inst, 47-59; sr staff physicist, 59-68, CHIEF PHYSICIST, RES CTR, AM OPTICAL CO, 68- Mem: Am Chem Soc; fel Am Ceramic Soc; Brit Soc Glass Technol. Res: Variation of physical properties of optical glass with temperature and time; rheology of bituminous materials; lubrication of optical instruments; strength of glass. Mailing Add: Res Ctr Am Optical Co Southbridge MA 01550

BRANDT, PHILIP WILLIAMS, b Cleveland, Ohio, Sept 23, 30; m 54; c 2. ANATOMY, PHYSIOLOGY. Educ: Swarthmore Col, BA, 52; Univ Pa, MS, 57; Columbia Univ, PhD(anat), 60. Prof Exp: Asst instr anat, Univ Pa, 53-57; from asst to asst prof, 57-70, ASSOC PROF ANAT, COL PHYSICIANS & SURGEONS, COLUMBIA UNIV, 70- Concurrent Pos: Guggenheim fel, 68-69; NIH career develop award, 68- Mem: Am Soc Cell Biol; Am Asn Anat; Electron Micros Soc Am; Soc Gen Physiol. Res: Electron microscopy; muscle. Mailing Add: Col of Physicians & Surgeons Columbia Univ New York NY 10032

BRANDT, RICHARD BERNARD, b Brooklyn, NY, July 3, 34; m 56; c 3. BIOCHEMISTRY. Educ: Queens Col, NY, BS, 56; Brooklyn Col, MA, 61; NY Univ, PhD(biochem), 68. Prof Exp: Res asst, Sloan-Kettering Inst Cancer Res, NY, 56-58; substitute instr, Brooklyn Col, 59-61, chem & biol warfare grant proj leader, 61-62; res assoc, Inst Med Res & Studies, NY, 62-69; guest res assoc, Brookhaven Nat Labs, NY, 68-70; asst prof, 70-74, ASSOC PROF BIOCHEM, MED COL VA, VA COMMONWEALTH UNIV, 74- Concurrent Pos: Anal chemist, F D Snell, Inc, NY, 58. Mem: AAAS; Am Chem Soc; Soc Exp Biol & Med. Res: Fluorescent methods of analysis; melanoma detection and treatment; enzyme systems in cell growth, especially glyoxalase I and II. Mailing Add: Dept of Biochem Med Col of Va Richmond VA 23298

BRANDT, RICHARD CHARLES, b Philadelphia, Pa, Dec 18, 40; m 63; c 1. SOLID STATE PHYSICS. Educ: Calif Inst Technol, BS, 62; Univ Ill, MS, 64, PhD(physics), 67. Prof Exp: Staff mem solid state physics, Lincoln Lab, Mass Inst Technol, 67-71; RES ASST PROF PHYSICS, UNIV UTAH, 71- Mem: Am Phys Soc. Res: Infrared optical properties of solids. Mailing Add: Dept of Physics Univ of Utah Salt Lake City UT 84112

BRANDT, RICHARD GUSTAVE, b Albany, NY, Nov 2, 36; m 62; c 3. LOW TEMPERATURE PHYSICS. Educ: Yale Univ, BS, 58, MS, 59, PhD(physics), 65. Prof Exp: Res staff physicist, Yale Univ, 64-65; mem tech staff, Defense Res Corp, 65-68; PHYSICIST, OFF NAVAL RES, 68- Mem: Am Phys Soc; AAAS. Res: Nucleon-nucleon scattering; nuclear photodisintegration; gas-surface interactions; superconductivity; electrooptics; operations research. Mailing Add: 1030 E Green St Pasadena CA 91106

BRANDT, ROBERT WILLIAM, b Montrose, Mich, Aug 7, 25; m 50; c 3. FORESTRY. Educ: Mich State Univ, BS, 50, BSF, 51, MS, 55; Syracuse Univ, PhD(forest path), 58. Prof Exp: Forest pathologist, Northeastern Forest Exp Sta, 57-64, asst chief, Forest Dis Res Br, Div Forest Protection, Washington, DC, 64-68, br chief, Forest Dis Res, 68-71, asst dir, Eastern Regional Off, Int Progs Div, Agr Res Serv, 71-73, staff asst res, 73-74, DIR, INT FORESTRY STAFF, US FOREST SERV, WASHINGTON, DC, 74- Mem: Am Phytopath Soc; Soc Am Foresters; Am Forestry Asn. Res: Foliage disease; physiogenic diseases; research administration. Mailing Add: Int Forestry Staff US Forest Serv USDA Washington DC 20250

BRANDT, WALTER EDMUND, b Columbia, Pa, Apr 11, 35; m 57; c 2. VIROLOGY. Educ: Univ Md, BS, 60, MS, 64, PhD(microbiol), 67. Prof Exp: Microbiologist, 60-67, supvr, 67-70, ASST CHIEF VIROL, WALTER REED ARMY INST RES, 70- Concurrent Pos: Consult, NIH, 71-75; instr, eve div, Univ Md, 66-74, lectr microbiol, 70. Mem: Am Asn Immunologists; Am Soc Microbiol; Am Soc Trop Med & Hyg; AAAS; Sigma Xi. Res: Dengue fever virus infection of cultured vertebrate and invertebrate cells; isolation of particulate and soluble antigens for study of dengue in humans; attenuated live virus vaccines. Mailing Add: Dept of Virus Dis Walter Reed Army Inst Res Washington DC 20012

BRANDT, WERNER, b Kiel, Ger, May 19, 25; nat US. PHYSICS. Educ: Univ Heidelberg, BS, 48, MS, 50, DSc(physics), 51. Prof Exp: Fel, Inst Theoret Physics, Copenhagen, 51-52; res physicist, Exp Sta, E I du Pont de Nemours & Co, Inc, 52-57, res assoc, Radiation Physics Lab, 57-61; assoc prof physics & assoc dir radiation & solid state lab, 61-66, PROF PHYSICS & DIR RADIATION & SOLID STATE LAB, NY UNIV, 66- Concurrent Pos: Lectr, Univ Pa, 58-60; vis prof, Nordic Inst Theoret Atomic Physics, Denmark & Chalmers Univ Technol, Sweden, 63-64 & 66; vis prof, Grenoble Nuclear Res Ctr & Nat Inst Sci & Nuclear Technol, France, 68-69. Mem: Fel Am Phys Soc; Europ Phys Soc. Res: Radiation physics; solid state physics; positron physics. Mailing Add: Radiation & Solid State Lab NY Univ Dept of Physics New York NY 10003

BRANDT, WERNER WILFRIED, b Friedensdorf, Ger, Feb 13, 27; US citizen; m 52; c 4. PHYSICAL CHEMISTRY. Educ: Univ Heidelberg, BS, 49; Polytech Inst Brooklyn, PhD(phys chem), 56. Prof Exp: Anal chemist, Merck & Co, Inc, NJ, 52-53; res chemist, E I du Pont de Nemours & Co, Del, 55-59; asst prof chem, Ill Inst Technol, 59-65; ASSOC PROF CHEM, UNIV WIS-MILWAUKEE, 65- Mem: Am Chem Soc; Am Phys Soc; fel Am Inst Chemists. Res: Rate processes, especially diffusion; chemisorption. Mailing Add: Dept of Chem Univ of Wis-Milwaukee Milwaukee WI 53211

BRANDT, WILLIAM HENRY, b Great Falls, Mont, May 25, 27; m 53; c 2. PLANT PHYSIOLOGY. Educ: Univ Mont, BA, 50; Ohio State Univ, MSc, 51, PhD, 54. Prof Exp: Asst oak wilt, Ohio Agr Exp Sta, 52-54; res biologist, B F Goodrich Res Ctr, 54-56; from instr to asst prof, 56-65, ASSOC PROF BOT, ORE STATE UNIV, 65- Concurrent Pos: NIH spec res fel, 69-70; guest investr, Rockefeller Univ, 69-70. Mem: AAAS; Bot Soc Am; Mycol Soc Am; Am Soc Microbiol. Res: Physiology of fungi; morphogenesis in fungi. Mailing Add: Dept of Bot Ore State Univ Corvallis OR 97331

BRANDTS, JOHN FREDERICK, b Celina, Ohio, June 15, 34; m 52; c 6. PHYSICAL CHEMISTRY, BIOCHEMISTRY. Educ: Miami Univ, BA, 56; Univ Minn, PhD(phys chem), 61. Prof Exp: NIH fel, 61-62; from asst prof to assoc prof, 62-71, PROF CHEM, UNIV MASS, AMHERST, 71- Concurrent Pos: NIH res grant, 63-66. Mem: Am Chem Soc. Res: Physical biochemistry; conformation and stability of globular proteins. Mailing Add: Dept of Chem Univ of Mass Amherst MA 01002

BRANDVOLD, DONALD KEITH, b Maddock, NDak, Aug 12, 36; m 61; c 1. BIOCHEMISTRY, PHYSICAL CHEMISTRY. Educ: NDak State Univ, BS, 62, PhD(chem), 65. Prof Exp: Technician, NDak State Univ, 59-62; asst prof, 65-74, ASSOC PROF CHEM, NMEX INST MINING & TECHNOL, 74- Mem: AAAS; Am Chem Soc. Res: Submolecular and electronic biology; general molecular biology; vitamins and hormones mode of action; virus research; pesticide residues; ore leaching by bacteria; water chemistry and pollution. Mailing Add: Dept of Chem NMex Inst of Mining & Technol Socorro NM 87801

BRANDWEIN, BERNARD JAY, b Chicago, Ill, Apr 19, 27; m 51; c 3. BIOCHEMISTRY. Educ: Purdue Univ, BS, 48, MS, 51, PhD, 55. Prof Exp: Assoc prof, 55-69, PROF CHEM, SDAK STATE UNIV, 69- Mem: AAAS; Am Chem Soc. Res: Isolation, identification and proof of structure of naturally occurring carbohydrates, glycosides and vitamins; reversion products of carbohydrates. Mailing Add: Dept of Chem SDak State Univ Brookings SD 57006

BRANDWEIN, PAUL FRANZ, b Austria, Apr 2, 14; nat US; m 33. BOTANY. Educ: NY Univ, AB, 34, MS, 37, PhD(bot), 40; Colo Col, DSc, 52. Prof Exp: Asst, Littauer Pneumonia Lab, 29-31; asst pharmacol, NY Univ, 31-34, teaching fel biol, 34-36, lectr ecol, 36; instr, High Sch, New York, 36-40, chmn dept sci, 40-48; sci ed, Harcourt Brace & Co, 46-54, sr ed, 54-65, dir res, Harcourt Brace & World, Inc, Calif, 65, DIR RES & INSTR, HARCOURT BRACE JOVANOVICH, INC, NY, 65-, SR VPRES, 75- Concurrent Pos: Lectr, Columbia Univ, 42-54; consult, NY Pub Schs, 54-; sr ed, Science, 54-; assoc ed, Sci Teacher; dir educ, Conserv Found, 57-; dir, Pinchot Int Conserv Studies, 63-; adj prof conserv & educ, Univ Pittsburgh, 65- Mem: AAAS; Am Phytopath Soc; Nat Asn Res Sci Teaching; Nat Sci Teachers Asn. Res: Culture and physiology of spirogyra; culture and ecology of protozoa; smut infection in cereal grains; conservation. Mailing Add: Harcourt Brace Jovanovich Inc 757 Third Ave New York NY 10017

BRANEN, ALFRED LARRY, b Caldwell, Idaho, Jan 5, 45; m 72; c 3. FOOD SCIENCE. Educ: Univ Idaho, BS, 67; Purdue Univ, PhD(food sci), 70. Prof Exp: Asst prof food sci, Univ Wis, 70-73; ASST PROF FOOD SCI, WASH STATE UNIV, 74- Mem: Inst Food Technologists; Am Chem Soc; Am Oil Chemists Soc; Am Dairy Sci Asn; AAAS. Res: Toxicology of food additives; preservation of foods with naturally occurring biochemicals; autoxidation and antioxidant function. Mailing Add: Dept of Food Sci & Technol Wash State Univ Pullman WA 99163

BRANGES, LOUIS DE, b Paris, France, Aug 21, 32; US citizen; m 62. MATHEMATICS. Educ: Mass Inst Technol, BS, 53; Cornell Univ, PhD(math), 57. Prof Exp: Asst prof math, Lafayette Col, 58-59; mathematician, Inst Advan Study, 59-60; lectr, Bryn Mawr Col, 60-61; mem, Courant Inst Math Sci, 61-62; assoc prof, 62-63, PROF MATH, PURDUE UNIV, 63- Concurrent Pos: Fels, Sloan Found, 63-65 & Guggenheim Found, 67-68. Mem: Am Math Soc. Res: Number theory; quantum mechanics. Mailing Add: 331 Hollowood Dr West Lafayette IN 47906

BRANHAM, JOSEPH MORHART, b Washington, DC, Jan 31, 32; m 56; c 2. DEVELOPMENTAL BIOLOGY, ECOLOGY. Educ: Fla State Univ, BS, 56, MS, 58, PhD(exp biol), 63. Prof Exp: Asst prof biol, Oglethorpe Univ, 62-64, assoc prof, 63-65; NIH fel, Inst Animal Genetics, Univ Edinburgh, 65-67; asst prof zool, Univ Hawaii, 69-72; fel biol, Univ Utah, 72-73; SCI TEACHER, LEESBURG HIGH SCH, FLA, 74- Mem: AAAS; Am Soc Zoologists; Am Inst Biol Sci; NY Acad Sci; Sigma Xi. Res: Fertilization, sea urchin and rabbit gamete physiology; insect physiology and ecology, coral reef biology. Mailing Add: PO Box 38 Okahumpka FL 32762

BRANION, HUGH DOUGLAS, b Brownsville, Ont, June 18, 06; m 31; c 2. NUTRITION. Educ: Univ Toronto, BA, 28, MA, 29, PhD(biochem), 33. Prof Exp: Head nutrit lab, Ont Agr Col, 30-37, assoc prof animal nutrit, 37-38, prof & head dept nutrit, 38-64, chmn grad studies, 58-64, dean, 64-68, ASST TO PRES, UNIV GUELPH, 68- Concurrent Pos: Assoc ed, J Poultry Sci Asn, 35-38, ed, 49- Mem: AAAS; Am Inst Nutrit; fel Poultry Sci Asn (pres, 64-65); fel Agr Inst Can; fel Chem Inst Can. Mailing Add: Univ of Guelph Guelph ON Can

BRANLEY, FRANKLYN M, b New Rochelle, NY, June 5, 15; m 38; c 2. ASTRONOMY. Educ: NY Univ, BS, 42; Columbia Univ, MA, 46, EdD(sci educ), 57. Prof Exp: Teacher sci, NY Pub Schs, 36-54 & State Teachers Col NY, 54-56; assoc astronr, 56-63, asst chmn, 64-68, ASTRONR, AM MUS-HAYDEN PLANETARIUM, 63-, CHMN, 68- Concurrent Pos: Mem, Comn Teacher Educ & Cert, AAAS, 61-62 & Adv Comn, Nat Sci Exhibit, Worlds Fair, Wash, 62; proposals referee, NSF, 62. Mem: Fel AAAS; Am Astron Soc; Nat Sci Teachers Asn; fel Royal Astron Soc. Res: Science education, especially astronomy and full use of planetaria. Mailing Add: Am Mus-Hayden Planetarium 81st St at Central Park W New York NY 10024

BRANN, JAMES LEWIS, JR, b Norwood, Mass, June 24, 13; m 42; c 2. ENTOMOLOGY. Educ: Mass State Col, BS, 39; Cornell Univ, PhD(entom), 44. Prof Exp: Field asst, NY Exp Sta, Geneva, 40, investr entom, 42-44, asst prof, 44-48; assoc prof, 48-54, PROF ENTOM, CORNELL UNIV, 54-, EXTEN FRUIT SPECIALIST, 64- Concurrent Pos: Consult, United Fruit Co, 53-54; UN Food & Agr Orgn adv to Israel & Greece, 59; proj leader, NY State Apple Pest Mgt Prog, 73. Mem: Entom Soc Am. Res: Control of insect pests on vegetable and fruit corps; development of equipment for applying insecticides. Mailing Add: Dept of Entom Cornell Univ Ithaca NY 14850

BRANNEN, CECIL GRAY, b Seymour, Tex, Sept 18, 19; m 42; c 3. ORGANIC CHEMISTRY. Educ: Univ Ark, BSA, 40, MS, 41; Iowa State Col, PhD(chem), 51. Prof Exp: Chemist, Res Lab, Standard Oil Co, 51-64; CHEMIST, RES LAB, AMOCO CHEM CORP, 64- Mem: Am Chem Soc. Res: Organosilicon chemistry; lubricating grease; motor oil additives. Mailing Add: 29 W 478 Cape Ave West Chicago IL 60185

BRANNEN, ERIC, b Manchester, Eng, Sept 25, 21; m 46; c 3. NUCLEAR PHYSICS, QUANTUM ELECTRONICS. Educ: Univ Toronto, BA, 44, MA, 46; McGill Univ, PhD(nuclear physics), 48. Prof Exp: Demonstr, Univ Toronto, 45-46, McGill Univ, 46-48 & Univ Toronto, 48-49; instr, 49-50, lectr, 50-51, from asst prof to prof, 51-55, SR PROF PHYSICS, UNIV WESTERN ONT, 55- Mem: Am Phys Soc; Can Asn Physicists. Res: Racetrack microtron electron accelerator; submillimeter radiation; photon correlation; far infrared maser. Mailing Add: Dept of Physics Univ of Western Ont London ON Can

BRANNEN, JOSEPH P, b Conroe, Tex, Dec 27, 27; m 48; c 3. MATHEMATICS. Educ: Agr & Mech Col Tex, BS, 49; Univ Tex, PhD(math), 62. Prof Exp: Asst prof math, Sam Houston State Teachers Col, 56-57; spec instr, Univ Tex, 57-62; asst prof, La State Univ, 62-63; MEM STAFF, SANDIA CORP, 63- Mem: Am Math Soc; Soc Indust & Appl Math. Res: Teaching; summability theory; operations research. Mailing Add: 4011 Hannett Ave NE Albuquerque NM 87110

BRANNEN, WILLIAM THOMAS, JR, b Chicago, Ill, Sept 30, 36; m 57; c 5. ORGANIC CHEMISTRY. Educ: DePaul Univ, BS, 58; Northwestern Univ, PhD(org chem), 62. Prof Exp: Res asst org chem, Northwestern Univ, 59-62; proj chemist, Am Oil Co div, Standard Oil Co, Ind, 62-65, sr proj chemist, Amoco Chem Corp div, 65-67; tech dir, 67-71, V PRES RES & MFG, ELCO CORP, 71- Mem: AAAS; Am Chem Soc. Res: Mechanistic studies of organic and enzyme reactions; synthetic organic and inorganic chemistry. Mailing Add: 2050 Radcliffe Westlake OH 44145

BRANNOCK, KENT COMBS, organic chemistry, see 12th edition

BRANNON, DONALD RAY, b Fort Peck, Mont, May 30, 39; m 67. ORGANIC CHEMISTRY. Educ: Okla State Univ, BS, 61, PhD(org chem), 65. Prof Exp: Res asst, Okla State Univ, 61-65; sr scientist, 67-71, res scientist, 71-73, HEAD MICROBIOL & FERMENTATION PROD RES DIV, LILLY RES LABS, 73- Mem: Am Chem Soc. Res: Natural product chemistry; microbial transformations; biosynthetic pathways; organic synthesis with microbial enzymes; fermentation products. Mailing Add: Lilly Res Labs Ferment Prod Res 2511 E 46th St Indianapolis IN 46206

BRANNON, PAUL J, b Flint, Mich, Apr 12, 35; m 65; c 1. SPECTROSCOPY. Educ: Univ Mich, BS, 57, MS, 59, PhD(physics, infrared spectros), 65. Prof Exp: Mem staff upper atmospheric physics, Sandia Labs, 65-67; fel infrared spectros, Univ Tenn, 67-68; MEM STAFF UPPER ATMOSPHERIC PHYSICS, SANDIA LABS, 68- Res: Spectroscopy applied to upper atmospheric research, line shape and shifts in gases. Mailing Add: 207 Wells Dr NE Albuquerque NM 87123

BRANOVAN, LEO, b Rumania, Apr 17, 95; nat US; m 33; c 1. MATHEMATICS. Educ: Univ Wis, BS, 24; Univ Chicago, MS, 27. Prof Exp: Engr, Gen Elec Co, Ind, 24-25; instr math, Univ Minn, 27-31; appl mathematician, J P Goode Co, Ill, 32-34; consult mathematician, New York, NY, 35-38; instr math, Polytech Inst Brooklyn, 39-44; from instr to assoc prof, 49-69, EMER PROF MATH, MARQUETTE UNIV, 70- Mem: AAAS; Am Math Soc; Am Soc Eng Educ; Math Socs Belg, Italy, Denmark, Austria, Netherlands & Sweden. Res: Differential equations; differential geometry; umbilics in hyperspace. Mailing Add: 3201 N 48th St Milwaukee WI 53216

BRANS, CARL HENRY, b Dallas, Tex, Dec 13, 35; m 57; c 4. MATHEMATICAL PHYSICS. Educ: Loyola Univ, La, BS, 57; Princeton Univ, PhD(physics), 61. Prof Exp: From instr to assoc prof, 60-70, PROF PHYSICS, LOYOLA UNIV, LA, 70- Concurrent Pos: NSF grant, 62-66; vis prof physics, Princeton Univ, 73-74; res grant, Res Corp, 75. Mem: Am Phys Soc. Res: Varying gravitational constant in general relativity; mathematical methods in general relativity; differential geometry and topology; group theory. Mailing Add: Dept of Physics Loyola Univ New Orleans LA 70118

BRANSCOMB, ELBERT WARREN, b Yakima, Wash, Mar 31, 35; div; c 1. MOLECULAR BIOLOGY. Educ: Reed Col, BA, 57; Syracuse Univ, PhD(physics), 64. Prof Exp: Physicist theoret physics, 64-69, PHYSICIST MOLECULAR BIOL, LAWRENCE LIVERMORE LAB, 69- Res: Accuracy in the intracellular transfer of genetic information, particularly the consequences of inaccurate gene expression. Mailing Add: Lawrence Livermore Lab Biomed Div L-523 PO Box 808 Livermore CA 94550

BRANSCOMB, LEWIS MCADORY, b Asheville, NC, Aug 17, 26; m 51; c 2. ATOMIC PHYSICS, SCIENCE POLICY. Educ: Duke Univ, AB, 45; Harvard Univ, MS, 47, PhD(physics), 49. Hon Degrees: DSc, Western Mich Univ, 69, Univ Rochester, 71, Duke Univ, 71, 74 & Univ Colo, 73. Prof Exp: Instr physics, Harvard Univ, 50, jr fel, Soc Fels, 49-51; physicist, Nat Bur Stand, 51-54, chief atomic physics sect, 54-59, chief div, 59-62, chmn joint inst lab astrophys, 62-65, chief lab, Astrophys Div, 62-69, dir bur, 69-72; chief scientist, 72, VPRES & CHIEF SCIENTIST, IBM CORP, 72- Concurrent Pos: Rockefeller pub serv fel, Univ Col, Univ London, 57-58; prof-adjoint physics, Univ Colo, 62-69; mem, President's Sci Adv Comt, 65-68 & President's Comn for Medal of Sci, 70-74; mem, Int Comt Weights & Measures, 69-72; mem bd dirs, Am Nat Stand Inst, 69-72; ed, Rev Mod Physics, 69-72; chmn atomic & molecular physics & spectros comn, Int Union Pure & Appl Physics; mem adv comt sci & foreign affairs, US Dept State, 72-74; mem adv comt energy res & develop, Fed Energy Agency, 73-75; mem, Carnegie Inst Bd; mem bd dirs, Commonwealth Fund & Polytech Inst New York. Honors & Awards: Arthur Fleming Award, 58; Washington Acad Award, 59; Rockefeller Pub Serv Award; Samuel Wesley Stratton Award; Gold Medal Except Serv, US Dept Com; Proctor Prize, Sci Res Soc Am; Nat Civil Serv League Award. Mem: Nat Acad Sci; Inst of Med of Nat Acad Sci; Nat Acad Eng; fel Am Philos Soc; fel Am Acad Arts & Sci. Res: Gaseous electronics; spectra of diatomic molecules; physics of the upper atmosphere; physics of negative ions; science and technology policy. Mailing Add: IBM Corp Armonk NY 10504

BRANSOME, EDWIN D, JR, b New York, NY, Oct 27, 33; m 59; c 2. ENDOCRINOLOGY, BIOCHEMISTRY. Educ: Yale Univ, AB, 54; Columbia Univ, MD, 58. Prof Exp: House officer, Peter Brent Brigham Hosp, 58-59, asst med, 59-61, asst resident, 61-62; res assoc, Columbia Univ, 62-63; assoc endocrinol, Scripps Clin & Res Found, 64-66; from asst prof to assoc prof exp med, Mass Inst Technol, 66-70; instr med, Harvard Med Sch, 66-70; PROF MED & CHIEF DIV METAB & ENDOCRINE DIS, MED COL GA, 70- Concurrent Pos: NIH res fel, Harvard Med Sch, 59-61; vis fel biochem, Columbia Univ, 62-63; Am Cancer Soc fel, 62-64, fac res assoc award, 65-70; Am specialist, US Dept State, 61; asst physician, Vanderbilt Clin, 62-64. Mem: AAAS; Endocrine Soc; Am Physiol Soc; Am Fedn Clin Res; Am Chem Soc. Res: Internal medicine; mechanism of hormone and drug action on cellular metabolism and growth. Mailing Add: Dept of Med Med Col of Ga Augusta GA 30902

BRANSON, BRANLEY ALLAN, b San Angelo, Tex, Feb 11, 29; m 64; c 1. ICHTHYOLOGY, MALACOLOGY. Educ: Okla State Univ, BS, 56, MS, 57, PhD(zool), 60. Prof Exp: Spec instr zool, Okla State Univ, 56-57; asst prof, Kans State Col Pittsburg, 60-66; mem fac, 66-67, assoc prof, 67-70, PROF BIOL, EASTERN KY UNIV, 70- Concurrent Pos: Res grants, NSF, 60-61, Sigma Xi, 62-64 & NIH, 64-66. Mem: AAAS; Am Soc Ichthyologists & Herpetologists; Soc Study Evolution; Soc Syst Zool; Am Micros Soc. Res: Comparative morphology of the sensory systems of teleost fishes; ecological adaptations of fishes as regards their sensory system; zoogeography and taxonomy of fishes and mollusks. Mailing Add: Dept of Biol Eastern Ky Univ Richmond KY 40475

BRANSON, BRUCE WILLIAM, b Athol, Mass, Oct 20, 27; m 51; c 2. SURGERY. Educ: Loma Linda Univ, MD, 50; Am Bd Surg, dipl, 63. Prof Exp: Asst prof, 62-65, ASSOC PROF SURG, LOMA LINDA UNIV, 65- Mailing Add: Dept of Surg Loma Linda Univ Sch Med Loma Linda CA 92354

BRANSON, CARL COLTON, b Oberlin, Ohio, Sept 15, 06; m 48; c 3. STRATIGRAPHY. Educ: Univ Mo, AB, 26, AM, 27; Univ Chicago, PhD(geol), 29. Prof Exp: Instr paleont, State Col Wash, 29-30; from instr to asst prof geol, Brown Univ, 30-40; vis asst prof, Northwestern Univ, 40-41; assoc prof, Univ Ky, 41-44; res geologist, Shell Oil Co, Inc, 44-50; PROF GEOL, UNIV OKLA, 50- Concurrent Pos: Dir, Okla Geol Surv, 54-67; mem, Am Geol Inst House Soc Rep, 62-65. Mem: AAAS; fel Geol Soc Am; fel Paleont Soc; Am Asn Petrol Geologists; Soc Econ Paleontologists & Mineralogists. Res: Pennsylvanian and Permian stratigraphy and paleontology; petroleum geology. Mailing Add: 2310 Ashwood Lane Norman OK 73069

BRANSON, DOROTHY SWINGLE, b Modesto, Calif, June 17, 21; div; c 4. CLINICAL MICROBIOLOGY. Educ: Kans State Univ, BS, 42, MS, 44, PhD(bact), 64. Prof Exp: Field secy, Girl Scouts, S Oakland County, Mich, 44; instr zool, Univ Okla, 46-47; med lab technologist, Riverside County Gen Hosp, Calif, 59-60 & St Bernardine's Hosp, San Bernardino, Calif, 61; microbiologist & teaching supr, Columbia Hosp, Milwaukee, Wis, 64-68; microbiologist, Methodist Hosp & Med Ctr, St Joseph, Mo, 68-72; asst prof microbiol, Kirksville Col Osteopath Med, 73-74; MICROBIOLOGIST, GRANT HOSP, COLUMBUS, OHIO, 74- Mem: AAAS; Am Soc Microbiol; Am Soc Med Technol; Asn Practrs Infection Control. Res: Use by soil microorganisms of aromatic compounds; effects of host nutrition on parasitism. Mailing Add: Grant Hosp 309 E State Columbus OH 43215

BRANSON, FARREL ALLEN, b Coats, Kans, May 3, 19; m 47; c 2. PLANT ECOLOGY. Educ: Ft Hays Kans State Col, BS, 42, MS, 47; Univ Nebr, PhD(bot), 52. Prof Exp: Instr bot, Ft Hays Kans State Col, 47-48; asst prof range mgt, Mont State Univ, 51-57; BOTANIST, US GEOL SURV, 57- Mem: Fel AAAS; Ecol Soc Am; Am Inst Biol Sci; Soc Range Mgt. Res: Relationships of vegetation to hydrologic processes; vegetation indicators of quantities and qualities of soil water; effects of landtreatment practices on vegetation and hydrology. Mailing Add: 906 24th St Golden CO 80401

BRANSON, HERMAN RUSSELL, b Pocahontas, Va, Aug 14, 14; m 39; c 2. PHYSICS. Educ: Va State Col, BS, 36; Univ Cincinnati, PhD(physics), 39. Prof Exp: Instr math & physics, Dillard Univ, 39-41; prof & head physics dept, Howard Univ, 41-68; pres, Central State Univ, 68-70; PRES, LINCOLN UNIV PA, 70- Concurrent Pos: Chmn adv comt estab comn minorities in sci & eng, Nat Res Coun, 72-; mem adv comt major new children's prog in math, Educ Develop Ctr, 73-; mem adv coun status health sci in Black col, Inst Serv Educ, 74- Mem: Inst Med, Nat Acad Sci; AAAS; Soc Math Biol. Res: Mathematical biology; education. Mailing Add: Lincoln Univ of the Commonwealth Syst of Higher Educ Lincoln University PA 19352

BRANSON, ROY, b Lowell, Ariz, May 16, 21; m 52; c 3. SOIL CHEMISTRY. Educ: State Col Wash, BS, 50; Univ Wis, MS, 51, PhD(soils), 53. Prof Exp: Asst chemist, Citrus Exp Sta, 53-57, EXTEN SPECIALIST, AGR EXTEN SERV, UNIV CALIF, RIVERSIDE, 58- Concurrent Pos: Consult, UNESCO & Food & Agr Orgn, 68-71. Res: Soil salinity, water quality, soil fertility and plant nutrition. Mailing Add: 3455 Audubon Pl Riverside CA 92501

BRANSON, TERRY FRED, b Wichita, Kans, Feb 24, 35. ENTOMOLOGY. Educ: Colo State Univ, BS, 57, MS, 64; SDak State Univ, PhD(entom), 70. Prof Exp: Entomologist, 64-66, RES ENTOMOLOGIST, NORTHERN GRAIN INSECTS RES LAB, AGR RES SERV, USDA, 66-, PROJ LEADER, 70- Concurrent Pos: Tech adv, Morelos Inst of Super Agr Studies, Mex, 73. Mem: Entom Soc Am. Res: Host plant resistance to grain insects; mechanisms of host plant resistance; behavior of insect host selection, feeding, and reproduction. Mailing Add: Northern Grain Insects Res Lab Agr Res Serv USDA RR 3 Brookings SD 57006

BRANSTAD, WILLIAM, b Grantsburg, Wis, Aug 14, 07; m 52. DENTISTRY. Educ: Univ Minn, DDS, 30. Prof Exp: CLIN PROF ORAL PHYSIOL, SCH DENT, UNIV MINN, MINNEAPOLIS, 46- Mem: Am Acad Periodont; Am Acad Dent Radiol; Am Acad Restorative Dent (secy-treas, 61-64, pres elect, 64). Res: Restorative dentistry. Mailing Add: Sch of Dent Univ of Minn Minneapolis MN 55455

BRANT, ALBERT WADE, b Isabel, Kans, Mar 28, 19; m 40; c 2. FOOD TECHNOLOGY, POULTRY HUSBANDRY. Educ: Kans State Col, BS, 40; Mich State Col, MS, 42; Iowa State Col, PhD, 49. Prof Exp: Instr, State Col Wash, 42-44 & 46-47; poultry husbandman, Agr Res Serv, USDA, 49-57, chief, Poultry Res Br, Animal Husb Res Div, 57-59; food technologist, 59-73, AGRICULTURIST, COOP EXTEN, DEPT FOOD SCI & TECHNOL, UNIV CALIF, DAVIS, 73- Mem: AAAS; Poultry Sci Asn; Inst Food Technologists; Am Inst Biol Sci; World Poultry Sci Asn. Res: Methods of determining inferior quality of eggs; physical properties of eggs and poultry meat related to quality; microbiology of shell eggs; technology of animal products; processing of poultry and egg products; microbiology of poultry meat. Mailing Add: Dept of Food Sci & Technol OP-D Univ of Calif Davis CA 95616

BRANT, ARTHUR ALBERT, b Toronto, Ont, Oct 23, 10; m 40; c 4. GEOPHYSICS. Educ: Univ Toronto, BA, 32, MA, 33; Univ Berlin, PhD(geophys), 36. Prof Exp: Asst demonstr physics, Univ Toronto, 32-33, demonstr, 36-37, fel geophys, 37-40, from asst prof to assoc prof physics, 40-47; dir geophys explor, 48-74, DIR GEOPHYS DEPT, RES LAB, NEWMONT MINING CORP, 74- Concurrent Pos: Geol & geophys work, Ont Dept Mines, 37-42; in-chg geophys explor, Eldorado Uranium Mines, 44; practicing mining explorer, 45-47; prof eng, Ont & Ariz. Mem: Soc Explor Geophys; Am Inst Mining, Metall & Petrol Engrs; Am Geophys Union. Res: Mining geophysical exploration. Mailing Add: Geophys Dept Res Lab Newmont Mining Corp Danbury CT 06811

BRANT, CHARLES SANFORD, b Portland, Ore, Aug 11, 19; m 46; c 2. ANTHROPOLOGY. Educ: Reed Col, BA, 41; Yale Univ, MA, 43; Cornell Univ, PhD(anthrop), 51. Prof Exp: Asst, Yale Univ, 41-43; instr sociol & anthrop, Univ Buffalo, 46-47; asst, Cornell Univ, 48-49; vis asst prof, Colgate Univ, 51-52 & Univ Calif, 52-53; mem social sci fac, Sarah Lawrence Col, 54-56; social sci resident, Albert Einstein Col Med, 56-57; from asst prof to assoc prof anthrop, Portland State Col, 57-61; from assoc prof to prof, Univ Alta, 61-70, chmn dept, 66-70; PROF SOCIOL & ANTHROP, SIR GEORGE WILLIAMS UNIV, 70- Concurrent Pos: Can Coun sr fel, 67-68. Mem: Fel Am Anthrop Asn; Can Sociol & Anthrop Asn. Res: Applied anthropology; comparative cultures; social anthropology of developing areas; Arctic regions. Mailing Add: Dept of Sociol & Anthrop Sir George Williams Univ Montreal PQ Can

BRANT, DANIEL (HOSMER), b St Paul, Minn, Apr 10, 21; m 45, 65, 69; c 3. VERTEBRATE ECOLOGY. Educ: Univ Minn, BS, 43; Univ Wash, MS, 49; Univ Calif, PhD(zool), 53. Prof Exp: Asst, Univ Calif, 49-52; from asst prof to prof biol sci, 52-74, PROF ZOOL, HUMBOLT STATE UNIV, 74- Mem: AAAS; Ecol Soc Am; Am Soc Mammal. Res: Behavior of wild small rodents in the field and artificial runway systems. Mailing Add: Dept of Zool Humbolt State Col Arcata CA 95521

BRANT, RUSSELL ALAN, b Brooklyn, NY, Mar 18, 19; m 45; c 4. GEOLOGY. Educ: Univ Mich, BS, 48, MS, 49. Prof Exp: Asst chemist rubber chem, Monarch Rubber Co, 42-44; geologist, Fuels Br, US Geol Surv, 49-52; asst head coal sect, Ohio Geol Surv, Columbus, 52-55, head, 55-57, areal geol sect, 57-60, asst state geologist, 60-68, sr res geologist, 68; GEOLOGIST, OHIO RIVER VALLEY WATER SANIT COMN, 68- Concurrent Pos: Mine Drainage Specialist, Water Resources Ctr, Ohio State Univ, 68. Mem: Geol Soc Am; Am Inst Prof Geologists. Res: Coal; areal; geochemistry; water geology; acid mine water; mining; underground injection of waste water; aerial surveillance, air photo interpretation. Mailing Add: 8112 Bridgetown Rd Cleves OH 45002

BRANTIGAN, OTTO CHARLES, b Chattanooga, Tenn, Aug 31, 04; m; c 4. SURGERY. Educ: Northwestern Univ, BS, 31, BM, 33, MD, 34. Prof Exp: Prof, 56-74, EMER PROF CLIN ANAT & SURG, SCH MED, UNIV MD, BALTIMORE CITY, 74- Concurrent Pos: Chief thoracic surgeon, Church Home & St Josephs Hosp, 56-74; consult, Loch Raven Hosp. Mem: Am Surg Asn; Am Asn Thoracic Surgeons; Am Col Chest Physicians; AMA; Am Thoracic Soc. Res: Anatomy of the knee joint; methods of tracheal repair; surgical treatment of tuberculosis; cardiovascular subjects; pulmonary emphysema. Mailing Add: Univ of Md Sch of Med 104 W Madison St Baltimore MD 21201

BRANTLEY, BLAKE BRIDGES, JR, horticulture, see 12th edition

BRANTLEY, JOHN CALVIN, b Newtown, Mo, Aug 20, 21; m 47; c 2. RADIOLOGICAL HEALTH, RADIOCHEMISTRY. Educ: Northeastern Mo State Teachers Col, BA, 42; Univ Ill, PhD(chem), 49. Prof Exp: Chemist, Northern Regional Res Lab, Bur Agr & Indust Chem, USDA, 42-43 & 45-46; inorg chemist, Linde Air Prod Co, 49-56, res supvr, 56, asst mgr res, 56-59; asst dir res, Union Carbide Nuclear Co, NY, 59-60, dir, 60-66; pres, Nuclear Sci & Eng Corp, Pa, 66-68; mgr nuclides & sources div, 68-71, asst to exec vpres, 71-73, V PRES ADMIN, NEW ENG NUCLEAR CORP, 73- Mem: Am Chem Soc; Am Nuclear Soc; Soc Nuclear Med. Res: Organometallic compounds: chemistry of transition elements. Mailing Add: New Eng Nuclear Corp 549 Albany St Boston MA 02118

BRANTLEY, LEE REED, b Herrin, Ill, Sept 23, 06; m 30. PHYSICAL CHEMISTRY. Educ: Univ Calif, AB, 27; Calif Inst Technol, MS, 29, PhD(chem), 30. Prof Exp: Reader physics, Univ Calif, Los Angeles, 26-29; asst chem, Calif Inst Technol, 28-30; instr chem & physics, Occidental Col, 30-36, asst prof chem, 36-40, from assoc prof to prof, 40-67, head dept, 40-62; prof educ, Curric Res & Develop Group, Univ Hawaii, 67-72, consult, 72-73; CONSULT CHEM ENGR, 73- Concurrent Pos: Res fel, Calif Inst Technol, 36-43, asst & Nat Defense Res Coun consult, 43-44; teacher schs, Calif, 41-42; Petrol Res Fund award, 58-59; consult, Crescent Eng Co, 44-46, Pac Rocket Soc, 44-51, Nat Bur Standards, 51-54 & Corn Indust Res Found, 57-59; dir contracts, US Off Naval Res, 49-58 & Off Qm Gen, 52-58; vis prof, Lehigh Univ, 58-59; vpres, Alpha Chi Sigma Educ Found, 58-63, trustee, 63-; vis prof, Univ Hawaii, 62-63 & 65-66. Honors & Awards: John R Kuebler Award, Alpha Chi Sigma, 73. Mem: Fel AAAS; Am Chem Soc; Electrochem Soc; Nat Sci Teachers Asn; Am Inst Physics. Res: Preparation and properties of xenon trioxide; relation of air borne fungal spores and pollen to asthma in Hawaii; biofeedback instrumentation and applications. Mailing Add: 1025 Wilder Ave Penthouse B Honolulu HI 96822

BRANTLEY, WILLIAM HENRY, b Forsyth, Ga, Aug 23, 38; m 61; c 2. NUCLEAR PHYSICS. Educ: Mercer Univ, AB, 60; Vanderbilt Univ, MA, 63, PhD(nuclear physics), 66. Prof Exp: Sci cooperator low energy nuclear physics, Tech Univ Delft, 65-66; ASSOC PROF PHYSICS, FURMAN UNIV, 66-, CHMN DEPT, 67- Mem: Am Phys Soc. Res: Low energy nuclear physics, including beta and gamma ray spectroscopy. Mailing Add: Dept of Physics Furman Univ Greenville SC 29613

BRANTON, CECIL, b Shongaloo, La, July 3, 17; m 45; c 3. ANIMAL BREEDING. Educ: La State Univ, BS, 42; Cornell Univ, MS, 46, PhD(animal breeding), 50. Prof Exp: Teacher, High Sch, 42-43; res asst, Cornell Univ, 44-47; asst prof, 47-51, assoc prof dairy sci, 51-56, PROF DAIRY SCI, LA STATE UNIV, BATON ROUGE, 56- Mem: Am Dairy Sci Asn; Am Genetics Asn; Am Soc Animal Sci. Res: Dairy cattlebreeding and genetics; physiology of reproduction of cattle; environmental physiology of cattle. Mailing Add: Dept of Dairy Sci La State Univ Baton Rouge LA 70803

BRANTON, DANIEL, b Antwerp, Belgium, Jan 13, 32; US citizen; m 57; c 2. CELL BIOLOGY, BOTANY. Educ: Cornell Univ, AB, 54; Univ Calif, Davis, MS, 57; Univ Calif, Berkeley, PhD(plant physiol), 61. Prof Exp: NSF fel, Swiss Fed Inst Technol, 61-63; from asst prof to prof bot, Univ Calif, Berkeley, 63-72; PROF BIOL, HARVARD UNIV, 72- Concurrent Pos: Miller Res Prof, 67-68; Guggenheim fel, 70-71. Mem: AAAS; Am Soc Plant Physiologists; Am Soc Cell Biol; Biophys Soc. Res: Cell and membrane structure; cell biology; cytology. Mailing Add: Biol Labs Harvard Univ Cambridge MA 02138

BRANTON, PHILIP EDWARD, b Toronto, Ont, June 8, 43; m 70. VIROLOGY. Educ: Univ Toronto, BSc, 66, MSc, 68, PhD(med biophys), 72. Prof Exp: Fel biol, Mass Inst Technol, 72-74; asst prof cell biol, Univ Sherbrooke, 74-75; ASST PROF PATH, McMASTER UNIV, 75- Concurrent Pos: Nat Cancer Inst Can Hunter res fel, 72-74; Nat Cancer Inst Can res scholar, 74-; mem fel panel, Nat Cancer Inst Can, 76- Mem: Am Soc Microbiol; Can Soc Cell Biol. Res: Studies to elucidate the effect of oncogenic viruses on membrane metabolism at the molecular level; both Rous sarcoma virus and Herpes viruses are being studied. Mailing Add: Dept of Path McMaster Univ Sch of Med Hamilton ON Can

BRASCH, JAY, organic chemistry, see 12th edition

BRASCHO, DONN JOSEPH, b Syracuse, NY, Jan 9, 33; m 56; c 6. RADIOLOGY. Educ: Hobart Col, BS, 54; State Univ NY, MD, 58. Prof Exp: Asst chief radiol, Madigan Gen Hosp, Tacoma, Wash, 62-64, chief radiother, 62-64 & 65-66; radiologist, 121st Evacuation Hosp, Korea, 64-65; from asst prof to assoc prof, 66-74,

PROF RADIATION THER, UNIV ALA, BIRMINGHAM, 74- Mem: AMA; Radiol Soc NAm; Am Col Radiol; NY Acad Sci; Am Soc Therapeut Radiol. Res: Clinical radiation oncology. Mailing Add: Dept of Radiation Oncol Univ of Ala Birmingham AL 35233

BRASE, PETER CHARLES, b Corona, NY, June 19, 20; m 48; c 2. CHEMISTRY, ACADEMIC ADMINISTRATION. Educ: Tufts Univ, BS, 41; Columbia Univ, MA, 48, EdD, 64. Prof Exp: Instr chem, Univ Mass, 46-47; teacher pub sch, 48-50; prof chem, head dept & chmn div natural health sci, Jamestown Community Col, 51-65; dean, Fulton-Montgomery Community Col, 65-74; ACAD DEAN, ST FRANCIS COL, NY, 75- Mem: AAAS. Mailing Add: St Francis Col 180 Remsen St Brooklyn NY 11201

BRASEL, JO ANNE, b Salem, Ill, Feb 15, 34. PEDIATRIC ENDOCRINOLOGY, NUTRITION. Educ: Univ Colo, Boulder, BA, 56; Univ Colo, Denver, MD, 59. Prof Exp: From intern to resident pediat, NY Hosp-Cornell Univ Med Ctr, 59-62; fel pediat endocrinol, Sch Med, Johns Hopkins Univ, 62-65, asst prof, 65-68; from asst prof to assoc prof, Med Col, Cornell Univ, 69-72; ASSOC PROF PEDIAT, DIR DIV GROWTH & DEVELOP & ASST DIR INST HUMAN NUTRIT, COL PHYSICIANS & SURGEONS, COLUMBIA UNIV, 72- Concurrent Pos: Irma T Hirschl trust career scientist award, 73-77; Nat Inst Child Health & Human Develop res career develop award, 74-79; spec consult, Endocrine & Metab Adv Comt, Food & Drug Admin, 71-; mem nutrit study sect, NIH, 74- Mem: Soc Pediat Res (secy-treas, 73-); Lawson Wilkins Pediat Endocrine Soc; Endocrine Soc; Am Soc Clin Nutrit; Am Pediat Soc. Res: Problems of growth and development including endocrine disorders and altered nutrition; basic biochemical research, especially of factors related to cell division and DNA biosynthesis. Mailing Add: Col of Physicians & Surgeons Columbia Univ New York NY 10032

BRASELTON, WEBB EMMETT, JR, b Americus, Ga, July 17, 41; m 64; c 2. BIOCHEMISTRY, ENDOCRINOLOGY. Educ: Univ Wis, BS, 63, MS, 66, PhD(endocrinol), 69. Prof Exp: Assoc biol chem, Harvard Med Sch, 71-73; ASST RES PROF MED & ASST PROF ENDOCRINOL, MED COL GA, 73-, DIR, GAS CHROMATOGRAPHY-MASS SPECTROMETRY FACIL, 73- Concurrent Pos: Res fel biol chem, Harvard Med Sch, 69-71; fel biochem, Mass Gen Hosp, Boston, 69-73. Mem: AAAS; Sigma Xi; Am Soc Mass Spectrometry; Endocrine Soc. Res: Metabolism and pharmacokinetics of sulfonylurea drugs and oral contraceptives; extraction, purification and physiochemical characterization of the anterior pituitary gonadotropic hormones; enzymes and mechanisms involved in steroid biosynthesis. Mailing Add: Dept of Med BB R 650 Med Col of Ga Augusta GA 30902

BRASFIELD, TRAVIS WINFORD, b Midland, Ark, May 23, 11; m 36; c 1. MYCOLOGY, PLANT PHYSIOLOGY. Educ: Univ Ark, BA, 34; Univ Iowa, MS, 36, PhD(mycol), 38. Prof Exp: Head dept biol, Perkinston Jr Col, 38-40; sales mgr agr chem, US Rubber Co, 46-48, mgr, 48-56, dist mgr, Naugatuck Chem Div, 56-58, mgr, Kralastic Sales, 58-59; dir mkt, Velsicol Chem Corp, 59-66, corp mkt coordr, 66-67; chmn dept, 67-71, INSTR LIFE SCI, COL OF DU PAGE, 67- Mem: AAAS; Am Inst Biol Sci. Res: Entomology; growth regulants; herbicides; phytopathology. Mailing Add: Dept of Life Sci Col of Du Page Glen Ellyn IL 60137

BRASH, JOHN LAW, b Glasgow, Scotland, Mar 8, 37; m 64; c 3. PHYSICAL CHEMISTRY. Educ: Glasgow Univ, BSc, 58, PhD(chem), 61. Prof Exp: Nat Res Coun Can fel, Div Pure Chem, Ottawa, Ont, 61-63; res chemist, Yerkes Lab, E I du Pont de Nemours & Co, 63-64; polymer chemist, Stanford Res Inst, 64-69, sr polymer chemist, 69-72; ASSOC PROF CHEM ENG, McMASTER UNIV, 72- Mem: AAAS; Am Chem Soc; Am Soc Artificial Internal Organs. Res: Kinetics of polymerization mechanisms; photo and radiation chemistry; polymer degradation mechanisms; biomaterials; blood-surface interactions. Mailing Add: Dept of Chem McMaster Univ Hamilton ON Can

BRASHEARS, MAURICE LYMAN, b Washington, DC, Oct 4, 08; m 37; c 5. GEOLOGY. Educ: Mass Inst Technol, SB, 33. Prof Exp: Jr geologist, US Geol Surv, 36-40, asst geologist, 40-43, dist geologist, 43-52; PARTNER, LEGETTE, BRASHEARS & GRAHAM, CONSULT GROUNDWATER GEOLOGISTS, 52- Mem: Fel Geol Soc Am; Am Inst Prof Geologists; Am Water Works Asn; Am Geophys Union; Am Soc Econ Geol. Res: Ground water geology and hydrology. Mailing Add: 1211 N Westshore Blvd Suite 516 Tampa FL 33607

BRASHER, EUGENE PAUL, b Orrick, Mo, Dec 29, 08; m 35; c 1. HORTICULTURE. Educ: Univ Mo, BS, 30; Pa State Col, MS, 32; Mo Valley Col, DSc, 69. Prof Exp: Asst hort, Pa State Col, 30-32; dist supvr, State Food Conserv Prog, Mo, 33-35; county agr agent, Univ Mo, 35-36; instr hort, WVa Univ, 36-41; from asst prof to prof hort, 41-68, chmn dept, 43-68, prof plant sci, 68-74, EMER PROF PLANT SCI, UNIV DEL, 74- Concurrent Pos: Mem, Tomato Breeders Roundtable. Honors & Awards: Citation Award of Merit, Mo Valley Col, 54 & Mid-Atlantic Food Processors Asn, 67. Mem: Am Soc Hort Sci. Res: Vegetable breeding and nutrition. Mailing Add: Dept of Plant Sci Univ of Del Newark DE 19711

BRASHIER, CLYDE KENNETH, b Marion, La, May 21, 33; m 57; c 1. PLANT TAXONOMY, PLANT MORPHOLOGY. Educ: La Polytech Inst, BSc, 55; Univ Nebr, MSc, 57, PhD(bot), 61. Prof Exp: Prof bot & bact, Wis State Univ, Superior, 61-68; chmn dept sci & math, 67-74, ACAD DEAN, DAKOTA STATE COL, 74- Concurrent Pos: Grant, Wis State Univ, 63-64; NSF fel, 63 & 68. Mem: Bot Soc Am; Int Soc Plant Morphol. Res: Bacteriology; lake pollution research; environmental and career education; health service delivery systems. Mailing Add: Off of Acad Dean Dakota State Col Madison SD 57042

BRASHIER, GARY KERMIT, b Marion, La, Oct 6, 37; m 62; c 3. PHYSICAL CHEMISTRY. Educ: Northeast La State Col, BS, 60; La State Univ, PhD(chem), 64. Prof Exp: Asst prof chem, Northeast La State Univ, 64-67; asst prof, 67-70, asst dean acad affairs, 68-72, ASSOC PROF CHEM, LA STATE UNIV, SHREVEPORT, 70-, V CHANCELLOR ACAD AFFAIRS, 73- Mem: AAAS; Am Chem Soc. Res: Colloidal surfactants; nature of cationic detergents; kinetics of polymerization. Mailing Add: La State Univ 8515 Youree Dr Shreveport LA 71105

BRASLAU, NORMAN, b Galveston, Tex, July 21, 31; m 55; c 2. PHYSICS. Educ: Agr & Mech Col, Tex, BS, 51; Ohio State Univ, MS, 52; Univ Calif, Berkeley, PhD(physics), 60. Prof Exp: Physicist, US Air Force Cambridge Res Ctr, Mass, 52-56; staff mem, Lawrence Radiation Lab, Univ Calif, 60; RES STAFF MEM, IBM WATSON RES CTR, 61- Concurrent Pos: Vis fel, Ecole Normale Superieure, Paris, 60-61, vis prof, 75-76. Mem: Am Phys Soc; Sigma Xi. Res: Optical masers; non-linear optics; current instabilities in semiconductors; atmospheric physics. Mailing Add: IBM Watson Res Ctr Yorktown Heights NY 10598

BRASSARD, ANDRE, b Quebec City, Que, May 31, 33; m 59; c 2. COMPARATIVE PATHOLOGY, IMMUNOPATHOLOGY. Educ: Laval Univ, BA, 55; Univ Montreal, DVM, 60; Univ Pa, MSc, 62, PhD(comp path), 64. Prof Exp: Asst comp pathologist, Univ Pa, 60-62, assoc comp pathologist, 62-64; asst prof, 65-70, ASSOC PROF COMP PATH, LAVAL UNIV, 70- Mem: Can Soc Immunol. Res: Experimental autoimmune glomerulonephritis. Mailing Add: Biomed Ctr Fac of Med Laval Univ Quebec PQ Can

BRASTED, ROBERT CROCKER, b Lisbon, NDak, Aug 26, 15; m 42; c 4. INORGANIC CHEMISTRY. Educ: George Washington Univ, BS, 38, MA, 39; Univ Ill, PhD(inorg chem), 42. Prof Exp: Asst inorg & anal chem, George Washington Univ, 35-39 & Univ Ill, 39-42; res chemist, Celanese Corp Am, Md, 42-43; asst prof chem, Univ Hawaii, 43-47; from asst prof to assoc prof, 47-54, PROF CHEM, UNIV MINN, MINNEAPOLIS, 54- Concurrent Pos: Res chemist, Opers Res Off, Johns Hopkins Univ, 49; consult, Gould Nat Batteries, Inc, 54-57, Olin Mathieson Co, 56-59 & India educ, NSF, 66-69; Fulbright fel & NSF sr res scholar, Univ Heidelberg, 61; guest prof, Orgn Am States, Univ Costa Rica, 61; US AID guest prof, Jadavpur Univ, 64; lectr, Taiwan Nat Univ & Chinese Univ, Hong Kong, 70; Fulbright sr lectr & guest prof, Stuttgart Univ, 71. Honors & Awards: Am Inst Chemists Award, 39; Horace T Morse Award Contrib to Educ, Coun Lib Educ, 70; Award Excellence in Col Teaching, Mfg Chemists Asn, 71; Educ Award, Am Chem Soc, 73. Mem: Am Chem Soc; AAAS. Res: Optical studies of resolvable species in presence of resolved species; formation of large-ring chelate complexes; chemistry of sulfur-nitrogen ring systems; cyclic sulfur imides. Mailing Add: Dept of Chem Univ of Minn Minneapolis MN 55455

BRASWELL, EMERY HAROLD, b Brooklyn, NY, Jan 22, 32; m 52, 71; c 1. BIOPHYSICS, BIOPHYSICAL CHEMISTRY. Educ: Polytech Inst Brooklyn, BS, 52, MS, 55; NY Univ, PhD(phys chem), 61. Prof Exp: Res chemist, cent lab, Gen Foods Corp, 54-56; lectr chem, Hunter Col, 56-57; NIH fel, Univ Birmingham, 60-62; asst prof chem, 62-64, bact, 64-68, biol, 68-73, ASSOC PROF BIOL, UNIV CONN, 73- Mem: Am Chem Soc; Biophys Soc. Res: Physical biochemistry; physical chemistry of biomacromolecules; studies on dextran, gelatin and heparin, using techniques such as light scattering, ultracentrifugation, viscosity and dye binding; lactic dehydrogenase kinetics and subunit interaction. Mailing Add: U-125 Univ of Conn Storrs CT 06268

BRATENAHL, CHARLES GEORGE, b Cleveland, Ohio, Sept 27, 16; m 42; c 1. PATHOLOGY. Educ: Williams Col, BA, 40; Univ Pa, MD, 43; Am Bd Path, dipl, 50. Prof Exp: Resident path, US Naval Med Sch, 47-49 & sch med, Univ Mich, 49-50; pathologist, Naval Hosp, Key West, Fla, 50-53, Tripler Army Hosp, Honolulu, 53-56 & Naval Hosp, Camp Pendleton, Calif, 56-61; dep dir, Armed Forces Radiobiol Res Inst, Md, 61-66; head acad dept, US Naval Med Sch, Nat Naval Med Ctr, 66-67; PATHOLOGIST, NORTHERN VA PATH LABS, 67- Mem: Fel Am Soc Clin Path; Col Am Pathologists; Asn Mil Surgeons US. Res: Clinical pathology. Mailing Add: Northern Va Path Labs 11091 Main St Fairfax VA 22030

BRATHOVDE, JAMES ROBERT, b Glasgow, Mont, June 8, 26; m 49; c 4. ENVIRONMENTAL MANAGEMENT, ENVIRONMENTAL SCIENCES. Educ: Eastern Wash Col, BA, 50; Univ Wash, MS, 55, PhD(phys chem), 56. Prof Exp: Teacher, pub sch syst, Wash, 50-51; supvr, freshman labs, Univ Wash, 53-54, instr phys chem lab, 54; assoc prof chem, Whitworth Col, Wash, 56-57, chmn dept, 56-60; mem staff, phys sci res dept, Sandia Labs, NMex, 60-62 & aerospace nuclear safety prog, 62-63; assoc prog dir, undergrad educ sci, NSF, 63-64; prof chem & dir comput ctr, State Univ NY Binghamton, 64-67; prof chem & chmn dept, 67-70, dean col sci & humanistic studies, 70-72, PROF ENVIRON SCI & CHEM, NORTHERN ARIZ UNIV, 72- Mem: Am Crystallog Asn; Am Chem Soc. Res: X-ray and neutron diffraction; molecular and crystal structures; ambient air and water quality; waste water operations; small scale energy converters. Mailing Add: 519 N James Flagstaff AZ 86001

BRATINA, WOYMIR JOHN, b Sturie, Yugoslavia, Feb 21, 16; Can citizen; m 55. SOLID STATE PHYSICS. Educ: Univ Zagreb, Dipl mech eng, 40; Univ Ljubljana, dipl metall eng, 43; Univ Toronto, MASc, 52, PhD(metal physics), 54. Prof Exp: Res fel metal physics, 54-58, res scientist, 58-66, SR RES SCIENTIST METAL PHYSICS, ONT RES FOUND, 66- Mem: Am Phys Soc; Am Soc Metals. Res: Metal physics; mechanical properties of metals and alloys; elastic and plastic deformation; internal friction; ultrasonics; dislocation damping; cyclic stressing; fracture mechanics; biomaterials. Mailing Add: Ont Res Found Sheridan Park Mississauga ON Can

BRATT, ALBERTUS DIRK, b Bozeman, Mont, Apr 2, 33; m 55; c 4. ENTOMOLOGY. Educ: Calvin Col, BS, 55; Mich State Univ, MS, 57; Cornell Univ, PhD(entom), 64. Prof Exp: Instr biol, Calvin Col, 58-61; asst entom, Cornell Univ, 61-64; asst prof, 64-70, PROF BIOL, CALVIN COL, 70-, CHMN DEPT, 73- Mem: Am Inst Biol Sci. Res: Morphology and taxonomy of Acalyptrate Diptera; fresh-water biology; morphology of immature insects. Mailing Add: Dept of Biol Calvin Col Grand Rapids MI 49506

BRATT, PETER RAYMOND, b Syracuse, NY, Nov 11, 29; m 54; c 5. SOLID STATE PHYSICS. Educ: Syracuse Univ, BS, 54, PhD(physics), 65. Prof Exp: Sr res scientist, Spencer Labs, Raytheon Co, 60-63; MEM TECH STAFF, SANTA BARBARA RES CTR, 63- Mem: Am Phys Soc. Res: Photoconductivity in solids; semiconductor physics; infrared detection; solid state lasers. Mailing Add: 6505 Camino Venturoso Goleta CA 93017

BRATTAIN, WALTER HOUSER, b Amoy, China, Feb 10, 02; US citizen; m 35, 58; c 1. BIOPHYSICS. Educ: Whitman Col, BS, 24; Univ Ore, MA, 26; Univ Minn, PhD(physics), 29. Hon Degrees: ScD, Portland Univ, 52, Whitman Col, 55, Union Col, 55, Univ Minn, 63 & Gustavus Adolphus Col, 63; LHD, Hartwick Col, 64. Prof Exp: Asst physicist, radio sect, Nat Bur Standards, 28-29; res physicist, Bell Tel Labs, 29-67; VIS PROF PHYSICS, WHITMAN COL, 62- Concurrent Pos: Physicist, Nat Defense Res Comt, Columbia Univ, 41-43; vis lectr, Harvard Univ, 52; chmn comn semiconductors, Int Union Pure & Appl Physics, 57-69; mem, Defense Sci Bd, 62-65; mem adv bd, US Naval Ord Test Sta, 66-68; consult, Pac Northwest Labs, Battelle Mem Inst, 68- Honors & Awards: Nobel Prize for Physics, 56; Stuart Ballantine Medal, Franklin Inst, 52; John Scott Medal, 55. Mem: Nat Acad Sci; fel AAAS; fel Am Acad Arts & Sci; fel Am Phys Soc; Franklin Inst. Res: Electron collisions in mercury vapor; piezoelectric frequency standards, magnetometers; infrared detectors; thermionics; transistor physics; surface properties of semiconductors; phospholipid bilayers or membranes. Mailing Add: Dept of Physics Whitman Col Walla Walla WA 99362

BRATTON, GERALD ROY, b San Antonio, Tex, Sept 25, 42; m 65; c 3. VETERINARY ANATOMY. Educ: Tex A&M Univ, BS, 65, DVM, 66, MS, 70, PhD(vet anat), 76. Prof Exp: From instr to asst prof vet anat, Tex A&M Univ, 66-75; ASSOC PROF VET ANAT IN CHG ANAT PROGS, DEPT ANIMAL SCI & VET MED, UNIV TENN, KNOXVILLE, 75- Mem: Am Asn Vet Anatomists; Am Vet Med Asn. Res: Developing an immunological method for tracing anatomical connections of the basal ganglia, and the use of horseradish peroxidase tracer methods to localize the anatomical positions of the medullary respiratory centers. Mailing Add: Dept of Animal Sci Univ of Tenn PO Box 1091 Knoxville TN 37901

BRATTON, SUSAN POWER, b Wilmington, Del, Oct 11, 48. PLANT ECOLOGY.

Educ: Barnard Col, Columbia Univ, AB, 70; Cornell Univ, PhD(bot), 75. Prof Exp: RES BIOLOGIST & RES COORDR ECOL, FIELD RES LAB, GREAT SMOKY MOUNTAINS NAT PARK, US DEPT INTERIOR, 74- Mem: Ecol Soc Am; Brit Ecol Soc; Bot Soc Am; Wildlife Soc; Torrey Bot Club. Res: Impact of over-grazing and exotic species on native plant communities; ecology of the European wild boar; structure and diversity patterns of herbaceous plant communities; alpha and beta diversity in forest understories. Mailing Add: Great Smoky Mountains Nat Park Gatlinburg TN 37738

BRATTSTEN, LENA B, b Gothenburg, Sweden. BIOCHEMISTRY. Educ: Univ Lund, Sweden, Fil kand, 67; Univ Ill, Urbana, PhD(insecticide biochem & toxicol), 71. Prof Exp: Fel biochem, 71-72, RES ASSOC BIOCHEM, CORNELL UNIV, 72- Mem: Entom Soc Am. Res: Biochemistry and physiology of microsomal drug metabolizing enzymes in insects and of mitochondrial enzymes related to drug metabolism and detoxication. Mailing Add: Dept of Entom Comstock Hall Cornell Univ Ithaca NY 14850

BRATTSTROM, BAYARD HOLMES, b Chicago, Ill, July 3, 29; div; c 2. ZOOLOGY, ECOLOGY. Educ: San Diego State Col, BS, 51; Univ Calif, Los Angeles, MA, 53, PhD(zool), 59. Prof Exp: Asst cur herpet, Natural Hist Mus, San Diego, 49-51, dir educ, 50-51; asst zool, Univ Calif, Los Angeles, 52-55; fel paleoecol, Calif Inst Technol, 55-56; instr biol, Adelphi Univ, 56-60; from asst prof to assoc prof, 60-66, PROF ZOOL, CALIF STATE UNIV, FULLERTON, 66- Concurrent Pos: Res assoc, Los Angeles Co Mus, 61-; former pres & bd mem, Fullerton Youth Mus & Natural Sci Ctr; sr fel Monash Univ, Australia, 66-67. Mem: Fel AAAS; Am Soc Ichthyologists & Herpetologists; Ecol Soc Am; Am Soc Mammalogists; Am Ornith Union. Res: Ecology, behavior, paleontology, zoogeography and physiology of reptiles and amphibians; paleoecology and paleoclimates; social behavior of vertebrates; thermoregulation; ecology of tropics; population and repopulation problems. Mailing Add: Dept of Biol Calif State Univ Fullerton CA 92634

BRATZ, ROBERT DAVIS, b Sherman, Tex, Dec 14, 20; m 47; c 3. BIOLOGY, ECOLOGY. Educ: Sam Houston State Teachers Col, BS, 41; Ore State Univ, MS, 50, PhD(zool, bot), 52. Prof Exp: From asst prof to assoc prof, 53-63, PROF BIOL, COL IDAHO, 63- Mem: AAAS. Res: General biology and ecology; biogeography of Mexico and western United States. Mailing Add: Dept of Biol Col of Idaho Caldwell ID 83605

BRAUCHI, JOHN TONY, b Sayre, Okla, Dec 23, 27; m 48; c 3. PSYCHIATRY. Educ: Univ Okla, MD, 55; Am Bd Psychiat & Neurol, dipl, 62. Prof Exp: Intern, Hillcrest Med Ctr, 55-56; resident, Univ Hosps, Okla, 56-58; resident, Med Ctr, Univ Kans, 58-59, from instr to prof psychiat, 58-75, dir inpatient serv, 60-61, dir residency training, 64-69, dir prof serv psychiat, 64-75, assoc chmn dept psychiat, 68-75; PROF PSYCHIAT, FAMILY MED & CHMN DEPT PSYCHIAT, SCH MED, LA STATE UNIV, 75- Concurrent Pos: Dir, Atchison County Guid Clin, Kans, 59-60, consult, 60-61; psychiat consult, Vet Admin Consol Ctr, Wadsworth, 59-61; actg chief psychiat serv, Vet Admin Hosp, Kansas City, Mo, 61-63; psychiat consult, Vet Admin Hosp, Shreveport, La, 75-; chief psychiat, Confed Mem Med Ctr, Shreveport, 75- Mem: Fel Am Asn Social Psychiat; Am Asn Chmn Depts Psychiat; fel Am Psychiat Asn; fel Am Col Psychiat; AMA. Res: Scintillation measurement of gross locomotor behavior; electroconvulsive shock and retroactive inhibitions; sleep deprivation; concept identification in thinking disorders; psychopharmacology; government; biofeedback. Mailing Add: Dept of Psychiat La State Univ Med Sch Shreveport LA 71130

BRAUDE, ABRAHAM ISAAC, b Chicago, Ill, June 15, 17; m 42; c 2. MEDICINE. Educ: Univ Chicago, BS, 37, MD, 40; Univ Minn, PhD, 50. Prof Exp: Instr internal med, Univ Minn, 48-50; asst prof med, Univ Mich, 50-53; assoc prof, Med Sch, Univ Tex, 53-58; prof, Sch Med, Univ Pittsburgh, 58-69; PROF MED & PATH, UNIV CALIF, SAN DIEGO, 69- Mem: Am Soc Clin Invest; Am Soc Microbiol; Asn Am Physicians; Am Asn Immunologists. Res: Infectious diseases. Mailing Add: Dept of Med Univ Hosp of San Diego County San Diego CA 92103

BRAUDE, MONIQUE COLSENET, b Lisieux, France, Nov 13, 25; US citizen; m 49. PHARMACOLOGY, TOXICOLOGY. Educ: Univ Paris, dipl, 48; Ohio State Univ, MS, 54; Univ Md, PhD(pharmacol), 63. Prof Exp: Res pharmacist, Blaque Labs, Paris, 49-53; res assoc, Sch Med, Univ Md, Baltimore City, 62-65; pharmacologist, US Food & Drug Admin, 66-69; chief biomed sect, Ctr Studies Narcotics & Drug Abuse, NIMH, 70-74; CHIEF PRECLIN PHARMACOL, BIOMED RES BR, DIV RES, NAT INST DRUG ABUSE, 74- Mem: AAAS; Am Soc Pharmacol & Exp Therapeut; Soc Toxicol; Soc Exp Biol & Med. Res: Neuropharmacology, psychopharmacology and toxicology of drugs of abuse, marihuana, other central nervous system agents. Mailing Add: 2410 Parkway Cheverly MD 20785

BRAUER, FRED, b Königsberg, Ger, Feb 3, 32; nat Can; m 58; c 3. MATHEMATICS. Educ: Univ Toronto, BA, 52; Mass Inst Technol, SM, 53, PhD(math), 56. Prof Exp: From asst to instr, Mass Inst Technol, 53-56; instr, Univ Chicago, 56-58; from lectr to asst prof math, Univ BC, 58-60; from asst prof to assoc prof, 60-66, PROF MATH, UNIV WIS, MADISON, 66- Mem: Am Math Soc; Math Asn Am; Can Math Cong; Soc Indust & Appl Math. Res: Ordinary differential equations. Mailing Add: Dept of Math Univ of Wis Madison WI 53706

BRAUER, GEORGE ULRICH, b Ger, Mar 18, 27; m 68. MATHEMATICS. Educ: Univ Toronto, BA, 49; Univ Mich, MA, 50, PhD(math), 54. Prof Exp: Asst, Univ Mich, 50-53; from instr to asst prof, 53-66, ASSOC PROF MATH, UNIV MINN, MINNEAPOLIS, 66- Mem: Am & Calcutta Math Socs; Math Asn Am. Res: Summation of infinite series; Dirichlet series; functional analysis; real and complex variables. Mailing Add: Sch of Math Univ of Minn Minneapolis MN 55455

BRAUER, GERHARD MAX, b Berlin, Ger, Feb 5, 19; nat US; m 68; c 2. CHEMISTRY, DENTAL RESEARCH. Educ: Univ Minn, BS, 41; Univ NC, MS, 48, PhD(chem), 50. Prof Exp: Asst chem, Univ NC, 46-47; RES CHEMIST DENT & MED MAT, POLYMER DIV, NAT BUR STANDARDS, 50- Concurrent Pos: Sr res fel & lectr, Free Univ Berlin, 74-75; mem, Biomat Adv Comt, Nat Inst Dent Res & Comn Z-156 Dent Mat & Devices, Am Nat Standards Comt; mem, Comn F-4 Surgical Implants, Am Soc Testing & Mat. Honors & Awards: US Dept Com Meritorious Serv Medal, 64; Souder Award, Int Asn Dent Res, 75; US Sr Scientist Award, Humboldt Found, 74; US Dept Com Gold Medal, 75. Mem: Am Chem Soc. Res: Chemistry of dental materials; reaction mechanisms; polymerization; analysis; reactivity of tooth surfaces; adhesion. Mailing Add: Dent & Med Mat Br Nat Bur of Standards Washington DC 20234

BRAUER, RALPH WERNER, b Berlin, Ger, June 18, 21; nat US. PHARMACOLOGY. Educ: Columbia Univ, AB, 40; Univ Rochester, MSc, 41, PhD(biochem), 43. Prof Exp: Res chemist, Wyandotte Chem Co, 43 & Distillation Prod Co, NY, 43-44; instr pharmacol, Harvard Med Sch, 44-47; asst prof, Sch Med, La State Univ, 47-51; head pharmacol br, US Naval Radiol Defense Lab, Calif, 51-66; prof physiol & pharmacol, Sch Med, Duke Univ, 66-71; PROF MARINE PHYSIOL & DIR, INST MARINE BIO-MED RES, UNIV NC, WILMINGTON, 71- Concurrent Pos: Mary Scott Newbold lectr, Philadelphia Col Physicians, 62; dir, Wrightsville Marine Bio-Med Lab, 66-71. Mem: Marine Technol Soc; Am Soc Pharmacol & Exp Therapeut; Soc Exp Biol & Med; NY Acad Sci; Radiation Res Soc. Res: Liver physiology; bile secretion; enzyme and fat chemistry; neutral phosphate esters as cholinesterase inhibitors; bromsulphthalein excretion; liver and plasma proteins; autoxidation of betaeleostearic acid; delayed radiation effects; nutrition and aging; environmental physiology. Mailing Add: Wrightsville Marine Bio-Med Lab 7205 Wrightsville Ave Wilmington NC 28401

BRAUER, RICHARD DAGOBERT, b Berlin, Ger, Feb 10, 01; US citizen; m 25; c 2. MATHEMATICS. Educ: Univ Berlin, PhD, 25. Hon Degrees: DrMath, Univ Waterloo, 68; DSc, Univ Chicago, 69, Univ Notre Dame, 74 & Brandeis Univ, 75. Prof Exp: Asst, Univ Königsberg, 25-33, privat docent, 27-33; vis prof, Univ Ky, 33-34; asst, Sch Math, Inst Advan Study, 34-35; asst prof math, Univ Toronto, 35-43, assoc prof, 43-46, prof, 46-48; prof, Univ Mich, 48-52; prof, 52-67, Perkins Prof, 67-71, EMER PERKINS PROF MATH, HARVARD UNIV, 71- Concurrent Pos: Guggenheim fel, 41-42; assoc ed, Trans Can Math Cong, 43-49, ed, Proc, 54-57; assoc ed, Am J Math, 44, ed, 45-50; assoc ed, Can J Math, 49-59 & Duke Univ Math J, 51-56 & 63-69; assoc ed, Ann Math, 53-60. Honors & Awards: Nat Medal Sci, 70. Mem: Nat Acad Sci; Am Math Soc; Am Acad Arts & Sci; Royal Soc Can; Can Math Cong (vpres, 51-52, pres, 57-58). Res: Hypercomplex numbers; algebraic equations. Mailing Add: Dept of Math Sci Ctr Harvard Univ Cambridge MA 02138

BRAULT, ALBERT THOMAS, b Barton, Vt, May 24, 37; m 60; c 5. PHYSICAL INORGANIC CHEMISTRY. Educ: St Michael's Col, Vt, BA, 59; Northwestern Univ, PhD(inorg chem), 64. Prof Exp: Sr chemist, 63-69, res assoc, Res Labs, 69-74, LAB HEAD, RES LABS, EASTMAN KODAK CO, 74- Mem: Am Chem Soc. Res: Processing research in applied photography; chemistry of coordination compounds, particularly metal carbonyls; photographic science; chemistry of metal complexes; photochemistry. Mailing Add: 431 Thomas Ave Rochester NY 14617

BRAULT, MARY MARGARET, physical organic chemistry, see 12th edition

BRAULT, ROBERT GEORGE, b Hoquiam, Wash, Dec 4, 18; m 48; c 2. ORGANIC CHEMISTRY. Educ: Whitworth Col, Wash, BS, 42; Mich State Col, PhD(org chem), 48. Prof Exp: Asst, Armour & Co, 49-51; asst org chem, Sinclair Res Labs, Inc, 51-54; mkt res engr, Callery Chem Co, 54-59; RES CHEMIST, HUGHES AIRCRAFT CO, 59- Mem: Am Chem Soc; Sigma Xi; Soc Photog Scientists & Engrs. Res: Organic synthesis; fatty acid derivatives; organo-metallics; borane chemistry; organic photochromic compounds; photopolymerization; radiation resists. Mailing Add: 924 Princeton St Santa Monica CA 90403

BRAUMAN, JOHN I, b Pittsburgh, Pa, Sept 7, 37; m 64. CHEMISTRY. Educ: Mass Inst Technol, BS, 59; Univ Calif, Berkeley, PhD(chem), 63. Prof Exp: NSF fel chem, Univ Calif, Los Angeles, 62-63; from asst prof to assoc prof, 63-72, PROF CHEM, STANFORD UNIV, 72- Concurrent Pos: Mem, Chem Adv Panel, NSF, 74- Honors & Awards: Pure Chem Award, Am Chem Soc, 73. Mem: Nat Acad Sci; Am Chem Soc; The Chem Soc. Res: Physical organic chemistry; gas phase ionic reactions; photodetachment spectroscopy; reaction mechanisms. Mailing Add: Dept of Chem Stanford Univ Stanford CA 94305

BRAUMAN, SHARON KRUSE, b Elizabeth, NJ, Apr 14, 39; m 64. PHYSICAL ORGANIC CHEMISTRY. Educ: Mt Holyoke Col, BA, 61; Univ Calif, Berkeley, PhD(org chem), 65. Prof Exp: SR CHEMIST, STANFORD RES INST, 65-, ASSOC MGR, POLYMER CHEM GROUP, 75- Concurrent Pos: Lectr chem, Stanford Univ, 75-76. Mem: Am Chem Soc; AAAS; The Chem Soc. Res: Decarboxylation, dehydration and dehydrohalogenation reactions; reactions in aqueous acidic solutions; chemistry of difluoramino compounds; fire retardance in polymers; membrane transport. Mailing Add: Stanford Res Inst Menlo Park CA 94025

BRAUN, ALBERT E, botany, see 12th edition

BRAUN, ALVIN JOSEPH, b Chicago, Ill, July 10, 15; m 39; c 3. PLANT PATHOLOGY. Educ: Univ Chicago, BS, 37; Univ Wis, PhM, 38; Ore State Col, PhD(plant path), 47. Prof Exp: Asst, Univ Wis, 37-38 & Ore State Col, 38-42; anal chemist, Sherwin-Williams Co, Ill, 42-43; asst pathologist, Guayule Res Proj, Bur Plant Indust, Soils & Agr Eng, USDA, Calif, 43-44, pathologist, 44-45; from asst prof to assoc prof, 45-56, PROF PLANT PATH, NY STATE AGR EXP STA, GENEVA, 57- Concurrent Pos: Agr off, Food & Agr Orgn, UN, Res & Training Ctr for Rice Prod, Bangkok, Thailand, 66-67; vis prof, Res Inst Pomol, Poland, 75. Mem: Am Phytopath Soc; Soc Nematologists. Res: Fungus and virus diseases; nematode and physiological problems of small fruits and grapes; fungicides; fumigants; spray machinery. Mailing Add: Dept of Plant Path NY State Agr Exp Sta Geneva NY 14456

BRAUN, ANN BAKER, inorganic chemistry, analytical chemistry, see 12th edition

BRAUN, ARMIN CHARLES, b Milwaukee, Wis, Sept 5, 11. CELL BIOLOGY, PLANT BIOLOGY. Educ: Univ Wis, BS, 34, PhD(plant path), 38. Prof Exp: Visitor, Europ Labs, 36-37; agent, USDA & asst, Univ Wis, 37-38; fel, 38-40, asst, 40-46, assoc, 46-50, assoc mem, 50-57, assoc prof bact, 57-59, PROF BACT, ROCKEFELLER UNIV, 59-, HEAD LAB PLANT BIOL, 55- Concurrent Pos: Consult & mem, Adv Panel Develop Biol, Div Biol & Med Sci, NSF, 58-61; mem vis comt, Dept Biol, Brookhaven Nat Lab, 60-64; sci adv bd, Inst Cancer Res, Philadelphia, 60-65; mem, Adv Comt, Aspen Biol Inst, Colo, 65-68. Honors & Awards: Newcomb Cleveland Award, AAAS. Mem: Nat Acad Sci; AAAS; hon mem Harvey Soc; Am Acad Arts & Sci; Int Soc Differentiation. Res: Cellular mechanisms involved in the transformation of a normal cell to a tumor cell. Mailing Add: Rockefeller Univ 66th St & York Ave New York NY 10021

BRAUN, CHARLES LOUIS, b Webster, SDak, June 4, 37; m 58; c 2. CHEMICAL PHYSICS, EXPERIMENTAL SOLID STATE PHYSICS. Educ: SDak Sch Mines & Technol, BS, 59; Univ Minn, PhD(phys chem), 63. Prof Exp: Chief process control training sect, Eng Reactors Group, US Army, Ft Belvoir, Va, 64-65; res instr, 65-67, asst prof, 67-71, ASSOC PROF CHEM, DARTMOUTH COL, 71- Concurrent Pos: Dartmouth Col fac fel, Phys Inst, Univ Stuttgart, 69-70. Mem: Am Chem Soc; AAAS; Am Phys Soc. Res: Photoionization and photoconductivity in molecular liquids and solids; electronic energy transfer in molecular crystals; internal conversion of electronic excitation energy. Mailing Add: Dept of Chem Dartmouth Col Hanover NH 03755

BRAUN, CLAIT E, b Kansas City, Mo, Oct 4, 39; m 60; c 2. WILDLIFE BIOLOGY. Educ: Kans State Univ, BS, 62; Univ Mont, MS, 65; Colo State Univ, PhD(wildlife biol), 69. Prof Exp: Soil scientist, Soil Conserv Serv, 61-69; asst wildlife researcher, Colo Div Game, Fish & Parks, 69-73; WILDLIFE RESEARCHER, COLO DIV WILDLIFE, 73- Concurrent Pos: Assoc, Mont Forest Conserv Exp Sta, 63-65, Am Mus Natural Hist, 69- & Inst Arctic & Alpine Res, 70- Honors & Awards: USDA Soil Conserv Serv Merit Award, 65. Mem: AAAS; Wildlife Soc; Am Soc Mammal;

Cooper Ornith Soc; Wilson Ornith Soc. Res: Population ecology of grouse, native columbids and alpine ecology. Mailing Add: Wildlife Res Ctr PO Box 2287 Ft Collins CO 80522

BRAUN, DANIEL CARL, b San Diego, Calif, July 2, 05; m 29. INDUSTRIAL MEDICINE. Educ: Univ Pittsburgh, BS, 35, MD, 37; Am Bd Prev Med, dipl. Prof Exp: Chief med examr, Pittsburgh Coal Co, 38-44, med dir, 44-50; consult var indust, 50-52; med dir, Indust Hyg Found Am, Inc, 52-58; med dir, Homestead Dist Works, US Steel Corp, 58-61, asst med dir, 61-70; mgr occup med serv, 70-72, PRES, INDUST HEALTH FOUND, INC, 72- Concurrent Pos: Sr fel, Mellon Inst Indust Res, 52-58; lectr, Sch Med, Univ Pittsburgh, 45- & Grad Sch Pub Health, 50-; deleg, President's Conf Indust Safety, 49; mem, President's Comt Employ Handicapped. Mem: Fel Am Inst Chemists; AMA; Am Acad Occup Med; fel Am Col Chest Physicians. Res: Influence of toxic materials and hazardous conditions in work environment; development of effective medical programs in industry. Mailing Add: Indust Health Found Inc 5231 Centre Ave Pittsburgh PA 15232

BRAUN, DONALD E, b Dinuba, Calif, Dec 15, 30; m 56; c 2. ANALYTICAL CHEMISTRY. Educ: Fresno State Col, AB, 52, MA, 54; Univ of the Pac, PhD(chem), 65. Prof Exp: Lab technician, Biochem & Virus Lab, Univ Calif, Berkeley, 54-55; med technician, Bethel Deaconess Hosp, Newton, Kans, 56-57; PROF CHEM, PAC COL, CALIF, 57- Concurrent Pos: NSF lab asst, Univ of the Pac, 63, partic, NSF res participation chem teachers, Ore State Univ, 68-69; consult, Braun, Skaffs & Kevorkian, 66- Mem: Am Chem Soc. Res: Development and refinement of analytical procedures in water and biochemical samples. Mailing Add: Dept of Chem Pac Col Fresno CA 93702

BRAUN, ELDON JOHN, b Glen Ullin, NDak, Jan 14, 37. PHYSIOLOGY, COMPARATIVE PHYSIOLOGY. Educ: Concordia Col, Moorhead, Minn, BA, 60; Univ Ariz, PhD(zool, biochem), 69. Prof Exp: ASST PROF PHYSIOL, COL MED, UNIV ARIZ, 72- Concurrent Pos: NIH fel, Univ Ariz Col Med, 69-72. Mem: Am Ornith Union; Cooper Ornith Soc; Am Physiol Soc. Res: Comparative renal physiology; measurement of single nephron glomerular filtration rates; changes in intrarenal blood flow patterns. Mailing Add: Dept of Physiol Univ of Ariz Col of Med Tucson AZ 85724

BRAUN, GEORGE AUGUST, biochemistry, see 12th edition

BRAUN, JOACHIM WERNER, immunology, microbiology, deceased

BRAUN, JOHN CHARLES, organic chemistry, see 12th edition

BRAUN, JUERGEN HANS, b Hof, Ger, Nov 6, 27; US citizen; m 54; c 3. INORGANIC CHEMISTRY, PHYSICAL CHEMISTRY. Educ: Tech Univ, Berlin, Dipl Ing, 51; Univ Tex, PhD(inorg chem), 56. Prof Exp: Res chemist, 55-63, sr res chemist, 63-71, RES ASSOC, PIGMENTS DEPT, E I DU PONT DE NEMOURS & CO, 71- Mem: Am Chem Soc; Electrochem Soc. Res: Hydrothermal crystallization; crystal growth of semiconductors; chemistry of liquid ammonia; small particle technology; technology of white and colored pigments. Mailing Add: Pigments Dept E I du Pont de Nemours & Co Wilmington DE 19898

BRAUN, LEWIS TIMOTHY, b Sacramento, Calif, Oct 27, 23; m 50; c 3. ECONOMIC GEOLOGY. Educ: Univ Calif, Berkeley, BS, 48. Prof Exp: Jr geologist, Atlantic Refining Co, Wyo, 48-49; asst mining geologist, Calif Div Mines, 49-50; geologist, Geophoto Serv, Inc, Colo, 51-55, proj mgr, Cagayan Basin Geol Explor, Geophoto Explor, Ltd, Philippines, 55-57; admin geologist, Geophoto Serv, Inc, 57-59, chief geologist, Geophoto Serv, Ltd, Alta, 59-64, proj mgr, Geophoto Serv, Inc, 64-71, dep mgr geophoto resources, 71-72; MGR GEOPHOTO SERV, TEX INSTRUMENTS, INC, DALLAS, 72- Mem: Am Asn Petrol Geologists; Am Inst Prof Geologists; fel Geol Soc Am. Res: Regional stratigraphic and structural geology and its relation to petroleum, mineral deposits and geothermal energy. Mailing Add: Geophoto Serv Tex Instruments Box 5621 MS 956 Dallas TX 75222

BRAUN, LOREN L, b Waseca, Minn, June 12, 29; m 64; c 3. ORGANIC CHEMISTRY. Educ: Mankato State Col, BS, 51; Univ Nebr, MS, 53, PhD(chem), 56. Prof Exp: Chemist, Mead Johnson & Co, 56-57; from asst prof to assoc prof, 57-66, PROF CHEM, IDAHO STATE UNIV, 66- Mem: Am Chem Soc. Res: Aliphatic diazo compounds; N-Bromosuccinimide reactions; nitrogen bridgehead compounds. Mailing Add: Dept of Chem Idaho State Univ Pocatello ID 83201

BRAUN, MARTIN, b New York, NY, July 26, 41. APPLIED MATHEMATICS Educ: Yeshiva Univ, BA, 63; NY Univ, MS, 65, PhD(math), 68. Prof Exp: Instr math, NY Univ, 67-68, Courant Inst Math Sci, 68; asst res prof appl math, Brown Univ, 68-70, asst prof, 70-75; ASST PROF MATH, QUEENS COL, CITY UNIV NEW YORK, 75- Mem: Soc Indust & Appl Math; Am Math Soc. Res: Qualitative theory of ordinary differential equations. Mailing Add: Dept of Math Queens Col Flushing NY 11367

BRAUN, OTTO GODFREY, b St Louis, Mo, July 7, 13; m 46; c 2. INORGANIC CHEMISTRY. Educ: Wash Univ, BS, 34, MS, 37. Prof Exp: Chemist, Am Can Co, Maywood, Ill, 37-44, supvr, food inspection group, 44-51, sect mgr, cent lab, 51-68, asst to dir tech serv lab, Barrington, Ill, 68-75; RETIRED. Mem: Am Chem Soc; Inst Food Technologists. Res: Technology of canning and packaging materials. Mailing Add: 310 N Lincoln Lane Arlington Heights IL 60004

BRAUN, PETER ERICH, biochemistry, neurochemistry, see 12th edition

BRAUN, ROBERT A, organic chemistry, see 12th edition

BRAUN, ROBERT DENTON, b Santa Ana, Calif, June 28, 43; m 72. ANALYTICAL CHEMISTRY. Educ: Univ Colo, BA, 65; Univ Conn, MS, 70, PhD(anal chem), 72. Prof Exp: High sch teacher, Conn, 65-66; instr & fel anal chem, Univ Mich, 72-73; vis asst prof anal chem, Univ Ill, 73-74; ASST PROF CHEM, VASSAR COL, 74- Mem: Am Chem Soc; Sigma Xi. Res: Electrochemistry of biochemical compounds; electrochemistry and electroanalysis in nonaqueous solvents. Mailing Add: Dept of Chem Vassar Col Poughkeepsie NY 12601

BRAUN, ROBERT LEORE, b New England, NDak, Dec 25, 36; m 62; c 2. CRYSTALLOGRAPHY, PHYSICAL CHEMISTRY. Educ: Univ Wash, BS, 59, PhD(phys chem), 66. Prof Exp: Chemist, Hanford Labs, Gen Elec Co & Pac Northwest Labs, Battelle Mem Inst, 59-66; CHEMIST, LAWRENCE LIVERMORE LAB, 66- Honors & Awards: Robert Peele Mem Award, Am Inst Mining, Metall & Petrol Engrs, 75. Mem: Am Inst Mining, Metall & Petrol Engrs. Res: Computer modeling of physical and chemical processes related to in-situ recovery of natural resources. Mailing Add: Lawrence Livermore Lab PO Box 808 Livermore CA 94550

BRAUN, WILLI KARL, b Reutlingen, WGer, Sept 22, 31; m 58; c 3. MICROPALEONTOLOGY, STRATIGRAPHY. Educ: Univ Tübingen, Dr rer nat, 58. Prof Exp: From paleontologist to sr paleontologist, Shell Can Ltd, Alta, 58-64; from asst prof to assoc prof, 64-74, PROF PALEONT & STRATIG, UNIV SASK, 74- Mem: Geol Soc Am; Paleont Soc; Brit Palaeont Asn; fel Geol Asn Can. Res: Paleozoic microfaunas and biostratigraphy of Western Canada. Mailing Add: Dept of Geol Sci Univ of Sask Saskatoon SK Can

BRAUN, WINFRED QUENTIN, b Krem, NDak, Apr 26, 14; m 42. CHEMISTRY. Educ: St Olaf Col, AB, 35; Purdue Univ, PhD(chem), 41. Prof Exp: Res chemist, Wilson & Co, 41-60, tech dir, Refining Div, 60-64, head fats & oils res, 64-70, DIR BYPROD RES & TECH SERV, WILSON CERT FOODS, 70- Mem: Am Chem Soc; Am Oil Chemists' Soc. Res: Edible oil products; fat nutrition. Mailing Add: Wilson & Co Inc 4545 Lincoln Blvd Oklahoma City OK 73105

BRAUN, WOLFGANG G, b Vienna, Austria, Feb 20, 14; US citizen; m 45; c 1. PHYSICS. Educ: Vienna Tech Univ, MS, 37; Darmstadt Tech Univ, Dr Ing(physics), 57. Prof Exp: Asst, Elec Tube Factory, Ger, 38-40; from res assoc to group chief, Res Inst Aeronaut Electronics, 40-45; res assoc, Wright Air Develop Ctr, 47-56, group leader, Aerospace Res Labs, 58-69, RES PHYSICIST, AEROSPACE RES LABS, OFF AEROSPACE RES, US AIR FORCE, 69- Mem: Inst Physics; Am Inst Elec & Electronics Engrs. Res: Gaseous electronics; plasma physics and spectroscopy. Mailing Add: 1859 Trebein Rd RR 3 Xenia OH 45385

BRAUNDMEIER, ARTHUR JOHN, JR, b Granite City, Ill, Feb 28, 43; m 64; c 1. PHYSICS. Educ: Eastern Ill Univ, BSEd, 65; Univ Tenn, Knoxville, MS, 69; Oak Ridge Assoc Univ, PhD(physics), 70. Prof Exp: ASSOC PROF PHYSICS, SOUTHERN ILL UNIV, EDWARDSVILLE, 70- Concurrent Pos: Consult, Oak Ridge Nat Lab, 71- Mem: Am Phys Soc; Am Asn Physics Teachers; Optical Soc Am; Health Physics Soc; Am Vacuum Soc. Res: Vacuum and thin film technology; optical constants of solids in the vacuum ultraviolet; solid state plasma oscillations; spectroscopy as a tool for measuring air pollution; detectors for ultraviolet radiation. Mailing Add: Fac of Physics Southern Ill Univ Edwardsville IL 62025

BRAUNE, MAXIMILLIAN O, b Bethlehem, Pa, Nov 21, 32; m 61; c 1. VIROLOGY, IMMUNOCHEMISTRY. Educ: Moravian Col, BS, 55; Univ Maine, MS, 58; Pa State Univ, PhD(bact, biochem), 63. Prof Exp: Instr res vet sci, 58-63, ASST PROF VET SCI, PA STATE UNIV, UNIVERSITY PARK, 63- Mem: AAAS; Am Soc Microbiol. Res: Isolation and serological relationships of pleuropneumonia-like organisms; differential diagnosis of viral respiratory diseases by fluorescent antibody techniques. Mailing Add: Pa State Univ Wiley Lab University Park PA 16802

BRAUNER, KENNETH MARTIN, b San Francisco, Calif, Dec 29, 27; m 54; c 3. ANALYTICAL CHEMISTRY. Educ: Univ Calif, BS, 49; Univ Chicago, PhD, 59. Prof Exp: Chemist, A Schilling & Co, 49-50 & Tidewater Assoc Oil Co, 52-53; res chemist, E I du Pont de Nemours & Co, 57-63; CHIEF CHEM DIV, DUGWAY PROVING GROUND, 63- Mem: AAAS; Am Chem Soc; Soc Appl Spectros. Res: Electrochemistry; trace analyses; chemical separations; physical methods of analysis. Mailing Add: Chem Div Dugway Proving Ground Dugway UT 84022

BRAUNER, PHYLLIS AMBLER, b Natick, Mass, Oct 2, 16; m 43; c 2. ANALYTICAL CHEMISTRY, INORGANIC CHEMISTRY. Educ: Wheaton Col, BA, 38; Wellesley Col, MA, 40; Boston Univ, PhD, 59. Prof Exp: Head sci dept, Winnwood Sch, NY, 39; anal chemist, Bellevue Hosp, 41; asst, Purdue Univ, 41-43; anal chemist, Gen Elec Co, Mass, 44-45; asst, Northeastern Univ, 45-46; instr, Swarthmore Col, 46-48; asst chem, 49-51, from instr to assoc prof, 51-66, PROF CHEM, SIMMONS COL, 67- Concurrent Pos: NSF fel, Switz, 60-61; res assoc, Royal Inst Technol, Sweden, 64-65; vis res assoc, Swiss Fed Inst Water Res & Water Pollution Control, 70-71. Mem: AAAS; Am Chem Soc. Res: Physical methods of analysis; compleximetry and homogeneous precipitation; solution equilibria; ionic speciation in natural waters; preparation of complex inorganic compounds. Mailing Add: 15 Benton St Wellesley MA 02181

BRAUNFELD, PETER GEORGE, b Vienna, Austria, Dec 12, 30; US citizen; m 59; c 2. MATHEMATICS. Educ: Univ Chicago, AB, 49, BS, 51; Univ Ill, Urbana, MA, 52, PhD(math), 59. Prof Exp: Res asst prof math & coordr sci lab, 59-63, from asst prof to assoc prof math & educ, 63-68, PROF MATH & EDUC, UNIV ILL, URBANA-CHAMPAIGN, 68- Res: Mathematics education; mathematics curriculum development. Mailing Add: Dept of Math Univ of Ill Urbana-Champaign Urbana IL 61801

BRAUNGART, DALE CARL, b Syracuse, NY, June 7, 12. CYTOGENETICS. Educ: Syracuse Univ, AB, 33; Univ Ala, MA, 38; Univ Md, PhD(cytogenetics), 48. Prof Exp: Asst genetics, Carnegie Inst, 37; asst dean, grad sch arts & sci, 48-50, asst head dept biol, 65-67, chmn dept, 67-71, ASSOC PROF BIOL, CATH UNIV AM, 48- Mem: AAAS; NY Acad Sci; Am Inst Ultrasonics in Med. Res: Ultrasound in biology; effect of ultrasonics on permeability of living cells; physiology of krebs ascites tumor cells. Mailing Add: Dept of Biol Cath Univ of Am 620 Michigan Ave NW Washington DC 20017

BRÄUNLICH, PETER FRITZ, b Ger, Feb 25, 37; m 63; c 2. EXPERIMENTAL SOLID STATE PHYSICS, ATOMIC PHYSICS. Educ: Univ Marburg, pre-dipl, 58; Univ Giessen, dipl, 61, Dr rer nat(physics), 63. Prof Exp: Sci asst physics, Univ Giessen, 63-65; res assoc, mat res lab, Pa State Univ, 65-66; sr physicist, 66-70, SR PROJ PHYSICIST, BENDIX RES LABS, 70- Concurrent Pos: Adj prof eng sci, Wayne State Univ, 74- Mem: Am & Ger Phys Socs. Res: Luminescence; thermally stimulated processes; photoconductivity; electron kinetics; exolectron emission; laser spectroscopy; interaction of laser light with matter; surface physics. Mailing Add: Eng Physics Dept Bendix Res Labs Bendix Ctr Southfield MI 48075

BRAUNMULLER, ALBERT RICHARD, chemistry, deceased

BRAUNSCHWEIGER, CHRISTIAN CARL, b Wellsville, NY, Oct 18, 26; m 53; c 5. MATHEMATICS. Educ: Alfred Univ, BA, 50; Univ Wis, MS, 51, PhD(math), 55. Prof Exp: Teaching asst, Univ Wis, 52-55; instr math, Purdue Univ, 55-57; from asst prof to assoc prof, 57-67, chmn dept, 70-73, PROF MATH, MARQUETTE UNIV, 67- Concurrent Pos: NSF fac fel, Univ Heidelberg, 64-65. Mem: Am Math Soc; Math Asn Am. Res: Functional analysis and topology; geometric models of abstract linear topological spaces. Mailing Add: Dept of Math Marquette Univ Milwaukee WI 53233

BRAUNSTEIN, DAVID MICHAEL, b New York, NY, Dec 9, 42; m 65; c 2. ORGANIC CHEMISTRY, POLYMER CHEMISTRY. Educ: Polytech Inst Brooklyn, BS, 64; Purdue Univ, PhD(org chem), 68. Prof Exp: Assoc chemist, Gen Foods Corp, 64; res chemist, 68-74, group leader, 74-76, RES ASSOC, CELANESE CORP AM, 76- Concurrent Pos: Adj prof, Middlesex County Col, 71, Richmond Col, City Univ New York, 71. Mem: Am Chem Soc; Soc Plastics Engrs. Res: Polymer product and process development; thermoplastic resins, structural foams; stabilization, fire retardants; regulatory agency approvals; technical liaison with foreign affiliates. Mailing Add: Celanese Corp of Am 1000 Morris Ct Summit NJ 07901

BRAUNSTEIN, HELEN MENTCHER, b NY, Feb 5, 25; m 45; c 3. SCIENCE WRITING, INFORMATION SCIENCE. Educ: Univ Maine, BA, 64, MS, 65, PhD(phys chem), 71. Prof Exp: Res assoc solution chem, Reactor Chem Div, Oak Ridge Nat Lab, 72-73; DOCUMENT COORDR ENVIRON & HEALTH SCI, ENVIRON RESOURCE CTR, OAK RIDGE NAT LAB, 74- Mem: Am Chem Soc; Am Soc Info Sci; AAAS; Soc Tech Info. Res: Environmental and health literature research. Mailing Add: Environ Resource Ctr Bldg 2028 Oak Ridge Nat Lab Oak Ridge TN 37830

BRAUNSTEIN, HERBERT, b New York, NY, Jan 10, 26; m 54; c 4. PATHOLOGY. Educ: City Col New York, BS, 44; Hahnemann Med Col, MD, 50. Prof Exp: Intern, Montefiore Hosp, New York, 50-51; asst resident path, Univ Mich, 51-52; resident, Univ Cincinnati, 52-54, from instr to assoc prof, 54-64, dir, Inter-Dept Path Res Lab, 56-64; prof path, Chicago Med Sch & dir dept path, Michael Reese Hosp & Med Ctr, 64-65; from assoc prof to prof, Univ Ky, 65-70; CLIN PROF PATH, MED SCH, LOMA LINDA UNIV, 70-; DIR LABS, SAN BERNADINO COUNTY GEN HSOP, 70-; CLIN PROF PATH, SCH MED, UNIV CALIF, LOS ANGELES, 75- Concurrent Pos: Fel gastroenterol, Univ Cincinnati, 52-54; USPHS res career develop award, 58-64; asst & actg chief path, Vet Admin Hosp, 54-56, attend pathologist, 56-64; admin supvr, Cent Labs, Cincinnanti Gen Hosp, 56-64. Mem: Am Soc Exp Path; Am Asn Pathologists & Bacteriologists; Histochem Soc; Int Acad Pathologists. Res: Pathology of diseases of liver and cardiovascular systems; histochemistry of neoplasms and cardiovascular system; clinical microbiology and infectious disease. Mailing Add: San Bernadino County Gen Hosp San Bernadino CA 92404

BRAUNSTEIN, JERRY, b New York, NY, Dec 24, 22; m 45; c 3. PHYSICAL CHEMISTRY. Educ: City Col New York, BS, 42; Wesleyan Univ, MA, 47; Northwestern Univ, PhD(chem), 51. Prof Exp: Chemist, Manhattan Proj, Columbia Univ, 42-45; fel chem, Univ Wash, 50-52; physicist, Gen Elec Co, 52-54; from asst prof to prof chem, Univ Maine, 54-66; GROUP LEADER, OAK RIDGE NAT LAB, 66-, DIR, REACTOR CHEM DIV, 73- Concurrent Pos: Res assoc, Oak Ridge Nat Lab, 60-61; lectr, Univ Maine, 66- & Univ Tenn-Oak Ridge Grad Sch Biomed Sci, 71- Mem: AAAS; Am Chem Soc; fel Am Inst Chem. Res: Molten salts; thermodynamics; concentrated electrolytes; solution chemistry; theoretical chemistry. Mailing Add: Reactor Chem Div Oak Ridge Nat Lab PO Box X Oak Ridge TN 37830

BRAUNSTEIN, JULES, b Buffalo, NY, Nov 4, 13; m 34. STRATIGRAPHY. Educ: George Washington Univ, BS, 33; Columbia Univ, MA, 36. Prof Exp: Paleontologist, 37-41, stratigrapher, 41-52, spec probs, 52-55, sr stratigrapher, 55-61, area stratigrapher, 61-66, STAFF GEOLOGIST, SHELL OIL CO, 66- Honors & Awards: Distinguished Serv Award, Am Asn Petrol Geologists, 75. Mem: Fel AAAS; Am Asn Petrol Geologists; hon mem Soc Econ Paleontologists & Mineralogists; fel Geol Soc Am; Mex Asn Petrol Geologists. Res: Micropaleontology and stratigraphy of the Gulf Coast; lithology of carbonate rocks; petroleum geology of the southeastern Gulf Coast; regional stratigraphy, structure and petroleum geology of the US Gulf Coast. Mailing Add: Shell Oil Co PO Box 60775 New Orleans LA 70160

BRAUNSTEIN, RUBIN, b New York, NY, May 6, 22; m 48. PHYSICS. Educ: NY Univ, BS, 48; Syracuse Univ, MS, 51, PhD(physics), 54. Prof Exp: Asst physics, Syracuse Univ, 48-51, res asst molecular beams, 51-52; res assoc, Columbia Univ, 52-53; mem res staff, Solid State Physics, Radio Corp Am Labs, 53-64; PROF PHYSICS, UNIV CALIF, LOS ANGELES, 64- Concurrent Pos: Consult, RCA Labs, 64-; Sci Res Coun fel, Oxford Univ, 74-75. Mem: AAAS; fel Am Phys Soc. Res: Molecular beams; radiofrequency and microwave spectroscopy; solid state; quantum electronics. Mailing Add: Dept of Physics Univ of Calif Los Angeles CA 90024

BRAUNWALD, EUGENE, b Vienna, Austria, Aug 15, 29; nat US; m 52; c 3. INTERNAL MEDICINE, CARDIOLOGY. Educ: NY Univ, AB, 49, MD, 52. Prof Exp: Intern, Mt Sinai Hosp, New York, 52-53; clin assoc physiol, Nat Heart Inst, 55-56, resident med, 56-57; resident, Johns Hopkins Univ, 57-58, chief cardiol sect, Nat Heart Inst, 58-66, clin dir, 66-68; prof med & chmn dept, Sch Med, Univ Calif, San Diego, 68-72; HERSEY PROF MED, HARVARD MED SCH, 72- Concurrent Pos: Fel med, Mt Sinai Hosp, New York, 53-54 & Columbia Univ, 54-55; mem, US/USSR Res Proj Heart Dis, Nat Heart & Lung Inst, 72. Honors & Awards: Distinguished Achievement Award, Modern Medicine, 68; Res Achievement Award, Am Heart Asn; Hall Award; Jacobs Award; Selassie Award. Mem: Nat Acad Sci; Am Col Physicians; Am Physiol Soc; Am Fedn Clin Res (pres, 69-70); fel Am Acad Arts & Sci. Res: Cardiovascular hemodynamics and diagnostic techniques; clinical cardiology. Mailing Add: Harvard Med Sch Boston MA 02115

BRAUNWARTH, JOHN BERNARD, organic chemistry, see 12th edition

BRAUSE, ALLAN R, b New York, NY, July 27, 42. FOOD CHEMISTRY. Educ: Polytech Inst Brooklyn, BS, 63; Univ Wis, MS, 65; Univ Cincinnati, PhD(org chem), 67. Prof Exp: Teaching asst org chem, Univ Wis, 63-64; asst chem, Univ Cincinnati, 64-67; res chemist, Phys Res Lab, Dow Chem Co, 67-68; chemist, Coatings & Chems Lab, US Army, 68-70; chemist, Res Div, US Indust Chem Co, 70-72; SR FOOD CHEMIST, KROGER CO, 72- Mem: Am Chem Soc. Res: Platinum palladium and other transition metals complex chemistry; food and flavor chemistry; nutritional analysis; vitamins and fats by gas chromatography. Mailing Add: Kroger Co 1212 State Ave Cincinnati OH 45204

BRAUTH, STEVEN EARLE, b Trenton, NJ, Apr 12, 47. NEUROPSYCHOLOGY. Educ: Rensselaer Polytech Inst, BS, 67; NY Univ, PhD(psychol), 73. Prof Exp: Res fel biol, Calif Inst Technol, 72-74; res assoc anat, State Univ NY Stony Brook, 74-75; ASST PROF PSYCHOL, UNIV MD, COLLEGE PARK, 75- Res: Comparative neuroanatomy and neurophysiology of the vertebrate brain; study of the evolution of the brain-behavior relationships; comparative neurobiology of the basal ganglia. Mailing Add: Dept of Psychol Univ of Md College Park MD 20742

BRAVER, GERALD, b New York, NY, Mar 15, 24; m 49; c 1. GENETICS. Educ: Brooklyn Col, BS, 49; Univ Mo, MA, 50, PhD(zool), 55. Prof Exp: Jr res zoologist, Univ Calif, 54-55; actg instr zool, 55-56, res geneticist, 56-58; from asst prof to assoc prof, 58-67, PROF ZOOL, UNIV OKLA, 67- Mem: Genetics Soc Am. Res: Genetics of Drosophila; crossing over; genetic effects of microwave, electrostatic and electromagnetic fields. Mailing Add: Dept of Zool Univ of Okla Norman OK 73069

BRAVERMAN, IRWIN MERTON, b Boston, Mass, Apr 17, 29; m 55; c 3. MEDICINE, DERMATOLOGY. Educ: Harvard Univ, BA, 51; Yale Univ, MD, 55. Prof Exp: Intern, Med Serv, Yale Univ-New Haven Hosp, 55-56, asst resident med, 58-59; from asst prof to assoc prof, 62-73, PROF DERMAT, SCH MED, YALE UNIV, 73- Concurrent Pos: Helen Hay Whitney Found res fel dermat, Sch Med, Yale Univ, 59-62; vis scientist, Dept Biol Structure, Univ Wash, 69-70. Mem: Soc Invest Dermat; Am Fedn Clin Res; Am Acad Dermatologists; Am Dermat Asn. Res: Autoimmune disorders; lupus erythematosus; scleroderma and dermatomyositis, both from clinical and research aspects; microcirculation in skin, both from clinical and from electron microscopic aspects. Mailing Add: Sect of Dermat Yale Univ New Haven CT 06520

BRAVERMAN, JEROME DAVID, b Bethlehem, Pa, Nov 25, 26; m 53; c 2. APPLIED STATISTICS, STATISTICS. Educ: Muhlenberg Col, AB, 48; Univ Calif, Los Angeles, MBA, 63, PhD(statist), 66. Prof Exp: ASSOC PROF STATIST, TEMPLE UNIV, 66- Mem: Am Soc Qual Control; Am Inst Decision Sci; Am Statist Asn. Res: Quantification and utilization of prior information in decision making and in the design of economic acceptance sampling plans. Mailing Add: Dept of Math Temple Univ Philadelphia PA 19122

BRAVERMAN, MAXWELL HAROLD, developmental biology, see 12th edition

BRAVERMAN, SAMUEL WILLIAM, b Boston, Mass, Feb 17, 30; m 57; c 2. PLANT PATHOLOGY. Educ: Clark Univ, AB, 52; Pa State Univ, MS, 54, PhD(plant path), 57. Prof Exp: Agent biol aid, US Pasture Lab, Pa, 53-57; plant pathologist, 57-63, RES PLANT PATHOLOGIST, NEW CROPS RES BR, PLANT INTROD STA, NY STATE AGR EXP STA, 63-, ASSOC PROF, DEPT SEED INVESTS, 65- Concurrent Pos: Head, Div Quarantine Serv, EAfrica Agr & Forestry Res Orgn, Kenya, 72-74. Mem: Am Phytopath Soc. Res: Forage and vegetable diseases; seed-borne diseases; regulatory plant pathology. Mailing Add: Plant Introd Sta NY State Agr Exp Sta Geneva NY 14456

BRAVO, JUSTO BALADJAY, b Philippines, Dec 5, 17; nat US; m 42; c 3. INORGANIC CHEMISTRY. Educ: Adamson Univ, Manilla, BSChE, 40; Univ Kans, PhD(chem), 53. Prof Exp: Instr chem, Univ St Tomas, Philippines, 46-49; asst instr, Univ Kans, 49-50, asst, 50-53; res chemist, Wyandotte Chem Corp, Mich, 53 & Oldbury Electrochem Co, 54-56; res proj engr, Am Potash & Chem Corp, 56-57; sr res chemist, Foote Mineral Co, Pa, 57-59, group leader, 59-60; res supvr, Sun Oil Co, 60-63; sr res scientist, Glidden Co, Md, 63-64; chmn dept, 67-68, PROF CHEM, WEST CHESTER STATE COL, 64- Concurrent Pos: Mem, Bd Dirs & sr consult, Wastex Industs Co, Inc, 71- Mem: Am Chem Soc; fel Am Inst Chemists. Res: Uninegative rhenium; inorganic pigments and phosphorus compounds; lithium metal and inorganic lithium compounds; electrochemicals; gas-solid reactions; fused system electrolysis; powder metallurgy; electrode processes; fuel cell development; synthesis, structure and properties of inorganic complex compounds. Mailing Add: 10 Roberts Rd Malvern PA 19355

BRAWAND, HANS, soil science, see 12th edition

BRAWER, JAMES ROBIN, b Patterson, NJ, Dec 15, 44; m 68; c 3. NEUROCYTOLOGY, NEUROENDOCRINOLOGY. Educ: Tufts Univ, BS, 66; Harvard Univ, PhD(neurocytol), 71. Prof Exp: NIH fel, Harvard Med Sch, 71-72; from asst prof to assoc prof anat, Med Sch, Tufts Univ, 72-75; ASST PROF OBSTET & GYNEC & ANAT, MED SCH, McGILL UNIV, 75-, ASSOC SCIENTIST NEUROANAT, McGILL UNIV-ROYAL VICTORIA HOSP, 76- Concurrent Pos: Med Res Coun Can scholar, 76. Mem: Am Asn Anatomists; Sigma Xi. Res: Cytophysiology of neuroendocrine transducers in the medial basal hypothalamus. Mailing Add: Dept of OB/GYN Women's Pavilion Royal Victoria Hosp 687 Pine Ave W Montreal PQ Can

BRAWERMAN, GEORGE, b Biala Podlaska, Poland, June 12, 27; US citizen; m 53; c 2. BIOCHEMISTRY. Educ: Univ Brussels, BS, 48; Columbia Univ, PhD(biochem), 53. Prof Exp: Res assoc biochem, Col Physicians & Surgeons, Columbia Univ, 56-60, asst prof, 60-61; from asst prof to assoc prof, Yale Univ Sch Med, 61-70; PROF BIOCHEM, SCH MED, TUFTS UNIV, 70- Concurrent Pos: USPHS career develop award, 63-70. Mem: AAAS; Am Soc Biol Chemists. Res: Biochemistry of gene expression; structure and function of messenger RNA; protein biosynthesis. Mailing Add: Dept of Biochem & Pharmacol Tufts Univ Sch of Med Boston MA 02111

BRAWLEY, JOEL VINCENT, JR, b Mooresville, NC, Feb 2, 38; m 59; c 3. ALGEBRA. Educ: NC State Univ, BS, 60, MS, 62, PhD(math), 64. Prof Exp: Instr math, NC State Univ, 64-65; from asst prof to assoc prof, 65-72, PROF MATH, CLEMSON UNIV, 72- Concurrent Pos: Vis lectr statist, Southern Regional Educ Bd, 67-69; vis lectr, Math Asn Am, 68-; lectr, Nat High Sch & Jr Col Math Club, 70-; vis assoc prof, NC State Univ, 71-72; sect lectr, Math Asn Am, 75-76. Mem: Math Asn Am; Am Math Soc. Res: Linear and abstract algebra. Mailing Add: Dept of Math Clemson Univ Clemson SC 29631

BRAWN, ROBERT IRWIN, b Mifflinburg, Pa, Jan 13, 23; m 45; c 2. PLANT GENETICS, AGRONOMY. Educ: Pa State Univ, BSc, 43; Univ Wis, PhD(genetics), 56. Prof Exp: Res asst genetics, Univ Wis, 46-49; asst prof agron, Macdonald Col, McGill Univ, 49-56, from assoc prof to prog agron & genetics, 56-71, dean students, 66-71; CORN RESEARCHER, FUNK SEEDS, INT, 71- Concurrent Pos: Proj assoc, Univ Wis, 55-56. Mem: AAAS; Am Genetic Asn; Genetics Soc Am; Crop Sci Soc Am; Am Soc Agron. Res: Corn genetics and breeding; inheritance of earliness in corn; physiology of maturation in corn; plant breeding. Mailing Add: Funk Seeds Int 1300 W Washington St Bloomington IL 61701

BRAWNER, THOMAS ALLAN, b Cleveland, Ohio, Dec 8, 45; m 67; c 2. MICROBIOLOGY, VIROLOGY. Educ: Albion Col, BA, 67; Univ Tex, Austin, PhD(microbiol), 72. Prof Exp: Fel virol, Univ Tex, San Antonio, 71-73; ASST PROF MICROBIOL, MED SCH, UNIV MO-COLUMBIA, 73- Mem: AAAS; Am Soc Microbiol. Res: Process of viral RNA replication; RNA viral biochemistry and genetics. Mailing Add: Dept of Microbiol Univ Mo Sch of Med Columbia MO 65201

BRAXTON, WILBERT LEO, b Snow Camp, NC, Apr 22, 11; m 38; c 4. PHYSICS. Educ: Guilford Col, BS, 32; Haverford Col, MS, 33. Prof Exp: Teacher high sch, NC, 33-35; teacher, Friends Boys Sch, Palestine, 35-36 & Friends Boarding Sch, Ohio, 36-42; instr, Stanford Univ, 42-44; prin, Friends Boarding Sch, 44-47; head sci dept, 47-62, asst headmaster, 61-68, HEAD MASTER, WILLIAM PENN CHARTER SCH, 68- Concurrent Pos: Dir, Nat High Sch Prog, Am Friends Serv Asn, 57-58. Mem: Nat Sci Teachers Asn. Res: Absolute intensity of x-rays. Mailing Add: William Penn Charter Sch Philadelphia PA 19144

BRAY, DALE FRANK, b Paw Paw, Mich, Mar 2, 22; m 47; c 2. ENTOMOLOGY. Educ: Mich State Col, BS, 47, MS, 49; Rutgers Univ, PhD(entom), 54. Prof Exp: Asst entom, Mich State Col, 47-49; from instr to asst prof, Univ Del, 49-55; assoc entomologist, Bartlett Tree Res Labs, 55-58; PROF ENTOM & CHMN DEPT, UNIV DEL, 58- Mem: Entom Soc Am. Res: Economic entomology; shade tree and ornamental plant insects; social insects. Mailing Add: Dept of Entom & Appl Ecol Univ of Del Newark DE 19711

BRAY, DAVID FREDERICK, b St Catherines, Ont, Nov 24, 31; m 54; c 2. SCIENCE ADMINISTRATION, STATISTICS. Educ: Univ Toronto, BSA, 55; Purdue Univ, MSc, 57, PhD(genetics), 60. Prof Exp: Instr poultry genetics, Univ BC, 59-61; res officer, Animal Res Inst, Can Dept Agr, 61-63; assoc prof genetics, McGill Univ, 63-65; head biomet sect, Food & Drug Directorate, Dept Health & Welfare, Ottawa, 65-

67, head, Statist Serv Div, 67-70, chief, Statist & Info Sci Div, 70-72; DIR PLANNING & EVAL, HEALTH PROTECTION BR, HEALTH & WELFARE CAN, 73- Mem: Am Statist Asn; Can Statist Soc; Statist Sci Asn Can; Biomet Soc; Genetics Soc Am. Res: Food science policy; health science policy; quantitative biology; statistical consulting; sampling plans; survey designs; information systems. Mailing Add: Planning & Eval Directorate Health Protection Br Ottawa ON Can

BRAY, DONALD JAMES, b Anamosa, Iowa, Nov 8, 23; m 48; c 5. POULTRY NUTRITION. Educ: Iowa State Col, BS, 50; Kans State Col, MS, 52, PhD, 54. Prof Exp: From asst prof to assoc prof, 54-68, PROF ANIMAL SCI, UNIV ILL, URBANA-CHAMPAIGN, 68- Mem: AAAS; Poultry Sci Asn; World Poultry Sci Asn; Am Inst Nutrit. Res: Nutritional and environmental factors as they influence laying hens. Mailing Add: Dept of Animal Sci Univ of Ill Urbana-Champaign Urbana IL 61801

BRAY, GEORGE A, b Evanston, Ill, July 25, 31. MEDICINE. Educ: Brown Univ, AB, 53; Harvard Univ, MD, 57. Prof Exp: From asst prof to assoc prof med, Sch Med, Tufts Univ, 64-70; assoc prof, 70-72, PROF MED, SCH MED, UNIV CALIF, LOS ANGELES, 72-; DIR CLIN STUDY CTR, HARBOR GEN HOSP, TORRANCE, CALIF, 70- Concurrent Pos: NSF fel, Nat Inst Med Res, 61-62; NIH fel, New Eng Med Ctr Hosps, 62-64; NIH res grants, Harbor Gen Hosp, 67 & 70; assoc chief, Div Endocrinol & Metab, Harbor Gen Hosp, 70-; chmn, Fogarty Ctr Conf Obesity, 73; consult, Food & Drug Admin, 71 & Dept Health & Welfare, Can, 74; invited witness, Select Comt Nutrit & Human Needs, US Sen, 74. Res: Etiology, treatment and management of experimental animal and human obesity. Mailing Add: Clin Study Ctr Dept of Med Harbor Gen Hosp Torrance CA 90509

BRAY, JAMES WILLIAM, b Atlanta, Ga, Feb 6, 48. THEORETICAL SOLID STATE PHYSICS. Educ: Ga Inst Technol, BS, 70; Univ Ill, MS, 71, PhD(physics), 74. Prof Exp: PHYSICIST, GEN ELEC RES & DEVELOP CTR, 74- Mem: Am Phys Soc. Res: Theoretical study of magnetic and transport properties of quasi-one-dimensional system; superconductivity. Mailing Add: Gen Elec Res & Develop Ctr PO Box 8 Schenectady NY 12301

BRAY, JOHN ROGER, b Belleville, Ill, June 20, 29; m 61; c 3. ECOLOGY. Educ: Univ Ill, BA, 50; Univ Wis, PhD(bot), 55. Prof Exp: Asst, Univ Wis, 50-54; vis lectr, Univ Minn, 55-57; asst prof, Univ Toronto, 57-62; prin sci officer, Dept Sci & Indust Res, 63-66; CONSULT ECOLOGIST, 66- Mem: AAAS; Ecol Soc Am; Brit Ecol Soc; NZ Ecol Soc. Res: Quantitative techniques for sampling and classifying vegetation; savanna ecology; productivity and efficiency of terrestrial vegetation; ecologic theory; chronology of Neoglacial, historical climatology; influence of solar and volcanic activity on climate. Mailing Add: PO Box 494 Nelson New Zealand

BRAY, JOSEPH MOYER, b Freeland, Pa, June 21, 15; m 40; c 1. MINING GEOLOGY. Educ: Pa State Col, BS, 37; Mass Inst Technol, PhD(geol), 40. Prof Exp: Instr geol, Lafayette Col, 40-42; mining & geol consult, 46-47; head mining div, Osmose Wood Preserving Co, 47-50, vpres, 50-58; pres, Super Soils, Inc, 58-64; mgr spec prod, Welding & Steel Fabrication Co, 64-68, vpres, 68-73, VPRES, WSF INDUSTS, INC, 73- Mem: Am Inst Mining, Metall & Petrol Engrs; Am Meteorol Soc; Nat Soc Prof Engrs. Res: Spectroscopic studies of minor chemical elements in rocks and minerals; opaque minerals in igneous rocks; geology of metallic ores; mine-roof support; chemistry of mine-timber preservation and prevention of timber decay; mining methods; soils. Mailing Add: 200 Fayette Ave Kenmore NY 14223

BRAY, MALCOLM DAVONNE, b Noblesville, Ind, Feb 29, 16; m 41; c 2. PHARMACEUTICAL CHEMISTRY. Educ: Ind Univ, AB, 38; Mass Inst Technol, PhD(org chem), 41. Prof Exp: Res chemist, Tex Co, 41-45; res chemist, 45-64, dir prod develop div, 64-66, DIR, ANAL RES & DEVELOP DIV, ELI LILLY & CO, 66- Mem: AAAS; Am Pharmaceut Asn; Am Chem Soc. Res: Organic and physical chemistry; developing new and improving present petroleum products; pharmaceutical and medicinal products; organic dielectrics. Mailing Add: Anal Res & Develop Div Eli Lilly & Co Indianapolis IN 46206

BRAY, NORMAN FRANCIS, b Louisville, Ky, May 27, 38; m 66; c 2. PHYSICAL CHEMISTRY. Educ: Univ Louisville, BS, 60, PhD(phys chem), 63. Prof Exp: Res scientist, NASA, Ohio, 63-64; from instr to asst prof chem, Hunter Col, 64-69; ASST PROF CHEM, LEHMAN COL, 69- Mem: Am Chem Soc. Res: Structure determination by nuclear magnetic resonance; solution thermodynamics. Mailing Add: Herbert H Lehman Col Dept Chem Bedford Park Blvd W Bronx NY 10468

BRAY, PHILIP JAMES, b Kansas City, Mo, Aug 26, 25; m 51; c 3. SOLID STATE PHYSICS. Educ: Brown Univ, ScB, 48; Harvard Univ, MA, 49, PhD(physics), 53. Prof Exp: Asst prof physics, Rensselaer Polytech Inst, 52-55; assoc prof, 55-58, chmn dept, 63-68, PROF PHYSICS, BROWN UNIV, 58- Concurrent Pos: NSF sr fel, 61-62; John Simon Guggenheim fel, 68-69; vis prof, dept glass technol, Univ Sheffield, 61-62 & 68-69. Honors & Awards: George W Morey Award, Am Ceramic Soc, 70. Mem: Fel AAAS; fel Am Phys Soc; fel Am Acad Sci; fel Am Ceramic Soc. Res: Nuclear magnetic resonance and electron-spin paramagnetic resonance studies of the structure of glasses and crystalline materials; nuclear quadrupole resonance studies of electron distributions in compounds of biological significance. Mailing Add: Dept of Physics Brown Univ Providence RI 02912

BRAY, RALPH, b Russia, Sept 11, 21; nat US; m 48; c 3. SOLID STATE PHYSICS. Educ: Brooklyn Col, BA, 42; Purdue Univ, PhD(physics), 49. Prof Exp: From instr to assoc prof, 45-65, PROF PHYSICS, PURDUE UNIV, 65- Concurrent Pos: Nat Res Coun fel, Tech Univ Delft, 51-52; consult, Univ Reading, 52-53, Battelle Mem Inst & Nat Cash Register Co; vis scientist, Gen Atomic Div, Gen Dynamics Corp, 60-61; mem ed bd, J Appl Physics, 67-69; Guggenheim fel, 69-70; vis scientist, Clarendon Lab, Oxford Univ, 69-70 & Becton Ctr, Yale Univ, 70; fac adv, Tex Instruments, Gen Tel Res & Electronics Lab. Mem: Fel Am Phys Soc. Res: Semiconductors; nonequilibrium phenomena; acoustoelectric effects; instabilities; brillouin scattering. Mailing Add: Dept of Physics Purdue Univ West Lafayette IN 47907

BRAY, ROBERT WOODBURY, b Dodgeville, Wis, Oct 17, 18; m 43; c 2. MEAT SCIENCE. Educ: Univ Wis, BS, 40; Kans State Col, MS, 41, PhD(animal husb & biochem), 49. Prof Exp: Instr animal husb, 41-43, from instr to assoc prof animal husb, 46-53, prof meat & animal sci, 54-63, chmn dept, 54-65, asst dir agr exp sta & asst dean col agr, 66-67, ASSOC DIR AGR EXP STA & ASSOC DEAN COL AGR, UNIV WIS, MADISON, 67- Honors & Awards: Award for Meat Res, Am Soc Animal Sci, 62; Distinguished Meats Res Award, Am Meat Sci Asn, 69. Mem: Hon fel Am Soc Animal Sci; Inst Food Technologists; Am Meat Sci Asn. Res: Factors influencing carcass composition and meat quality; meat processing problems. Mailing Add: Col of Agr & Life Sci 136 Agr Hall Univ of Wis Madison WI 53706

BRAYER, FRANKLIN T, b Rochester, NY, Mar 7, 19; m 45; c 8. MEDICINE. Educ: Univ Rochester, AB, 42, MD, 44. Prof Exp: From jr scientist to sr scientist, Atomic Energy Proj, 55-59; head med radioisotopes, Sch Med, Georgetown Univ, 59-64, dir, Ctr Pop Res, 64-71; med dir, Northern Livingston Health Ctr, 71-76; ASST MED DIR, ST ANN'S HOME, ROCHESTER, NY, 76- Concurrent Pos: Cancer fel, Univ Tex M D Anderson Hosp Houston, 50-52 & Univ Rochester, 52-53; USPHS grants, 60-62 & Pop Coun Inc, 60-61; jr scientist, Atomic Energy Proj, Univ Rochester, 53; consult, Freedmen's Hosp, Washington, DC, 60-67, Radiation Tech, Inc, 62-68, Children's Hosp, DC, 60- & Columbia Res, Inc, 68- Mem: AAAS; Soc Nuclear Med. Res: Radiation biology; use of radioisotopes in research and chemical medicine; biomedical research to develop better methods for fertility control. Mailing Add: St Ann's Home Rochester NY 14621

BRAYMER, HUGH DOUGLAS, b Oklahoma City, Okla, Mar 28, 33; m 56; c 2. MOLECULAR BIOLOGY. Educ: Univ Okla, BS, 55, MS, 57, PhD(biochem), 60. Prof Exp: Biochemist, gene action res, radiobiol br, US Air Force Sch Aerospace Med, 60-63; USPHS fel biochem genetics, Stanford Univ, 63-66; from asst prof to assoc prof, 66-74, PROF MICROBIOL, LA STATE UNIV, BATON ROUGE, 74- Concurrent Pos: Mem fac, Trinity Univ, 60-63. Mem: Am Chem Soc; Genetics Soc Am; Am Soc Microbiol; Am Soc Biol Chemists; Brit Biochem Soc. Res: Biochemical genetics; gene action; protein structure; metabolic control mechanisms. Mailing Add: Dept of Microbiol La State Univ Baton Rouge LA 70803

BRAYSHAW, THOMAS CHRISTOPHER, botany, see 12th edition

BRAYTON, ROBERT KING, mathematics, see 12th edition

BRAZDA, FRED GEORGE, b Chicago, Ill, Feb 22, 09; m 40; c 1. BIOCHEMISTRY. Educ: Univ Chicago, SB, 31; Northwestern Univ, MS, 33, PhD(physiol chem), 37; Am Bd Clin Chem, dipl. Prof Exp: Instr biochem, Sch Med, Creighton Univ, 37-38; from instr to assoc prof, 38-49, PROF BIOCHEM, LA STATE UNIV MED CTR, 49-, HEAD DEPT, 46- Mem: AAAS; Am Soc Biol Chemists; Am Soc Exp Biol & Med. Res: Metabolism of glucuronic acid, antiscorbutic substances, amino acid derivatives and pyridine compounds and their effect on the liver. Mailing Add: Dept of Biochem La State Univ Med Ctr New Orleans LA 70112

BRAZEAU, PAUL, b Santa Barbara, Calif, Apr 22, 23. PHARMACOLOGY. Educ: Columbia Univ, AB, 49, PhD(pharmacol), 52. Prof Exp: Instr, Columbia Univ, 52-53, assoc, 53-55, asst prof, 55-57; from asst prof to assoc prof, 58-72, PROF PHARMACOL, ALBERT EINSTEIN COL MED, 72-, ACTG CHMN DEPT, 73- Mem: Am Soc Pharmacol; NY Acad Sci. Res: Renal physiology and pharmacology; mechanisms of electrolyte transport. Mailing Add: Dept of Pharmacol Albert Einstein Col of Med New York NY 10461

BRAZIER, MARY A B, US citizen; c 1. NEUROPHYSIOLOGY. Educ: Univ London, BSc, 26, PhD(biochem), 29, DSc(neurophysiol), 60. Prof Exp: Neurophysiologist, Mass Gen Hosp, Boston, 40-60; prof anat & physiol, 61-72, PROF ANAT, UNIV CALIF, LOS ANGELES, 72- Concurrent Pos: Res fel, Maudsley Hosp, London, 30-40; res assoc, Med Sch, Harvard Univ, 41-60 & Mass Inst Technol, 53-65; ed-in-chief, Electroencephalography & Clin Neurophysiol; assoc ed, Exp Neurol. Mem: Am Physiol Soc; Int Fedn Electroencephalog & Clin Neurophysiol (hon pres); hon mem Am Electroencephalog Soc; Am Neurol Asn; Am Acad Neurologists. Res: Brain research. Mailing Add: Brain Res Inst Univ of Calif Sch of Med Los Angeles CA 90024

BRAZZEL, JAMES ROLAND, b Hico, La, May 8, 21; m 42; c 1. ENTOMOLOGY. Educ: La State Univ, BS, 51, MS, 53; Tex A&M Univ, PhD(entom), 56. Prof Exp: Res assoc entom, La State Univ, 53, asst entomologist, 55-57; from assoc prof to prof, Tex A&M Univ, 57-63; head dept, Miss State Univ, 63-68; CHIEF OF STAFF, OFF METHODS IMPROV, PLANT PEST CONTROL, AGR RES SERV, USDA, 68- Mem: Entom Soc Am; Ecol Soc Am. Res: Biology and control of cotton insects. Mailing Add: 10801 Ashfield Rd Adelphi MD 20783

BREAKEY, DONALD RAY, b Snohomish, Wash, June 1, 27; m 48; c 2. BIOLOGY. Educ: Wilamette Univ, BS, 50; Mich State Univ, MS, 52; Univ Calif, Berkeley, PhD(zool), 61. Prof Exp: From instr to assoc prof, 54-67, PROF BIOL, WILLAMETTE UNIV, 67-, CHMN DEPT, 68- Mem: AAAS; Am Soc Mammal; Australian Mammal Soc. Res: Vertebrate mammalian populations and distribution; ecological habitats and their evaluation in relation to animal forms present. Mailing Add: Dept of Biol Wilamette Univ Salem OR 97301

BREAM, CHARLES ANTHONY, b Midland, Pa, Mar 15, 15; m 41; c 4. MEDICINE. Educ: Grove City Col, BS, 36; Temple Univ, MD, 40. Prof Exp: Intern, Mercy Hosp, Pa, 40-41; vis physician, St Vincent's Hosp, Pa, 46-48; instr radiol, Col Physicians & Surgeons, Columbia Univ, 51-52; from asst prof to assoc prof, 52-58, PROF RADIOL, SCH MED, UNIV NC, CHAPEL HILL, 58- Concurrent Pos: Mellon res fel, Mercy Hosp, Pa, 41-42; instr med, Sch Med, Univ Pittsburgh, 40-41; asst vis physician, Hamot Hosp, Pa, 46-48; attend radiologist, NC Mem Hosp, Univ NC, 52-; consult, Watts Hosp, 53- & Womack Army Hosp, 58-; vis attend radiologist, Vet Admin Hosp, Durham, 55- Mem: Roentgen Ray Soc; Radiol Soc NAm; Asn Am Med Cols; Asn Univ Radiologists; fel Am Col Radiol. Res: Improvement of techniques, methods and scope of investigation of the retioperitoneal space; extend the usefulness of excretory urography in various nephropathies. Mailing Add: Univ of NC Sch of Med Dept of Radiol Chapel Hill NC 27514

BREAULT, GEORGE OMER, b Providence, RI, Nov 30, 42; m 67; c 2. DRUG METABOLISM. Educ: Univ Notre Dame, AB, 64; Univ RI, MS, 67; Univ Mich, PhD(pharm chem), 70. Prof Exp: Sr res biochemist, 70-72, RES FEL DRUG METAB, MERCK INST THERAPEUT RES, 72- Mem: AAAS; Am Chem Soc. Res: Development of analytical methods for drugs in biological systems. Mailing Add: Merck Inst for Therapeut Res West Point PA 19486

BREAUX, RICHARD DIX, agronomy, plant breeding, see 12th edition

BREAZEALE, ALMUT FRERICHS, organic chemistry, polymer chemistry, see 12th edition

BREAZEALE, MACK ALFRED, b Leona Mines, Va, Aug 15, 30; m 52; c 3. PHYSICAL ACOUSTICS, SOLID STATE PHYSICS. Educ: Berea Col, BA, 53; Univ Mo-Rolla, MS, 54; Mich State Univ, PhD(physics), 57. Prof Exp: Asst physics, Univ Mo-Rolla, 53-54; asst, Mich State Univ, 54-57, asst prof, 57-58 & 59-62; Fulbright res grant, Phys Inst, Stuttgart Tech Inst, 58-59; assoc prof, 62-67, PROF PHYSICS, UNIV TENN, KNOXVILLE, 67- Concurrent Pos: Consult, Oak Ridge Nat Lab, 62- & Naval Res Lab, 72-74. Mem: AAAS; fel Acoust Soc Am; Am Phys Soc. Res: Ultrasonics; nonlinear solid state and liquid state phenomena. Mailing Add: Dept of Physics Univ of Tenn Knoxville TN 37916

BREAZEALE, ROBERT DAVID, b Ames, Iowa, Aug 30, 35. ORGANIC CHEMISTRY. Educ: SDak State Col, BS, 57; Univ Wash, PhD(org chem), 64. Prof Exp: Res chemist, Sinclair Res Labs, Inc, 57-59; res chemist, 64-71, RES SUPVR, FABRICS & FINISHES DEPT, RES DIV, E I DU PONT DE NEMOURS & CO, 71- Mem: Am Chem Soc. Res: Polymer and petroleum chemistry; non-benzenoid aromatic chemistry; finishes and adhesives technology. Mailing Add: Res Div Marshall Lab E I du Pont de Nemours & Co Philadelphia PA 19146

BREAZEALE, WILLIAM HORACE, b Greensboro, NC, Aug 30, 38; m 63; c 2. PHYSICAL CHEMISTRY. Educ: Univ SC, BS, 61; Univ Tenn, PhD(phys chem), 66. Prof Exp: Asst prof chem, Winthrop Col, 66-70; assoc prof, 70-71, assoc dean, 73-75, PROF CHEM & PHYSICS, FRANCIS MARION COL, 71-, CHMN DEPT, 70- Mem: Am Chem Soc. Res: Chemical education; thermodynamics of solutions; theory of ion exchange equilibria. Mailing Add: Dept of Chem & Physics Francis Marion Col Florence SC 29501

BREAZILE, JAMES E, b Rockport, Mo, Dec 31, 34; m 57; c 3. NEUROANATOMY, NEUROPHYSIOLOGY. Educ: Univ Mo, BS, 58, DVM, 58; Univ Minn, PhD(anat, physiol), 63. Prof Exp: Pvt pract vet med, 58-60; NIH fel, 61-63; asst prof physiol, Okla State Univ, 63-67; assoc prof vet anat, 67-68, PROF VET ANAT, UNIV MO-COLUMBIA, 68-, CHMN DEPT, 69-70 & 72-, DIR GRAD STUDIES VET ANAT, 72- Concurrent Pos: NIH res grant, 64- Mem: AAAS; Am Vet Med Asn; Am Asn Vet Anat; Am Soc Vet Physiol & Pharmacol; Am Asn Anatomists. Res: Studies of normal structure of the central nervous system of vertebrates correlated with physiological phenomena, especially pain and autonomic control mechanisms; gross anatomy. Mailing Add: Dept of Vet Anat Univ of Mo Columbia MO 65201

BREBNER, DONALD LEE, organic chemistry, see 12th edition

BREBRICK, ROBERT FRANK, JR, b Danvers, Mont, Oct 18, 25; m 57; c 4. SOLID STATE CHEMISTRY. Educ: Mont State Col, BS, 47; Cath Univ Am, PhD(chem), 52. Prof Exp: Asst, Cath Univ Am, 47-50; phys chemist, Naval Ord Lab, 52-61, group leader, mat group, solid state div, 57-61; phys chemist, solid state div, Lincoln Lab, Mass Inst Technol, 61-70; assoc prof, 70-73, PROF CHEM, DEPT MECH ENG MAT SCI PROG & DEPT CHEM, MARQUETTE UNIV, 73- Honors & Awards: Meritorious Civilian Serv Award, 56. Mem: Am Chem Soc; Am Phys Soc; Electrochem Soc. Res: Physical chemistry of semiconductors; theoretical and experimental studies of the homogeneity range and thermodynamic properties of semiconducting compounds. Mailing Add: Dept of Mech Eng Marquette Univ Milwaukee WI 53233

BRECHER, ARTHUR SEYMOUR, b New York, NY, Mar 30, 28; m 66; c 2. BIOCHEMISTRY. Educ: City Col New York, BS, 48; Univ Calif, Los Angeles, PhD, 56. Prof Exp: Jr res biochemist, NY State Psychiat Inst, 48-52; asst chem, Univ Calif, Los Angeles, 53-56; asst prof biochem, Purdue Univ, 56-58; biochemist pharmacol, Food & Drug Admin, Dept Health, Educ & Welfare, 58-60; assoc res scientist, NY State Dept Ment Hyg, 60-63; asst prof biochem, Sch Med, George Washington Univ, 63-69; ASSOC PROF BIOCHEM, BOWLING GREEN STATE UNIV, 69- Mem: AAAS; Am Chem Soc; Soc Exp Biol & Med. Res: Enzyme chemistry; intermediary metabolism; drug metabolism. Mailing Add: Dept of Chem Bowling Green State Univ Bowling Green OH 43402

BRECHER, AVIVA, b Bucarest, Romania, July 4, 45; m 65; c 1. SPACE PHYSICS, PALEOMAGNETISM. Educ: Mass Inst Technol, BSc & MSc, 68; Univ Calif, PhD(appl physics), 72. Prof Exp: RES ASSOC EARTH & PLANETARY SCI, MASS INST TECHNOL, 72- Mem: Am Phys Soc; Am Astron Soc; Am Geophys Union; Int Astron Union; Am Meteorit Soc. Res: Nature of lunar magnetism; planetary magnetic fields; terrestrial paleomagnetism and continental drift; physical properties of meteorites and models of planetary evolution. Mailing Add: Dept of Earth & Planetary Sci Mass Inst of Technol 54-1114 Cambridge MA 02139

BRECHER, CHARLES, b Brooklyn, NY, Nov 5, 32; m 69; c 1. PHYSICAL CHEMISTRY. Educ: Columbia Univ, BA, 54, MA, 55, PhD(phys chem), 59. Prof Exp: Res fel phys chem, Columbia Univ, 59-62; RES CHEMIST EXPLOR RES DEPT, GEN TEL & ELECTRONICS LABS, 62- Mem: Am Phys Soc; Optical Soc Am; Sigma Xi; Am Chem Soc. Res: Optical phenomena in condensed phases, especially spectra of solids and liquids as related to problems of molecular and crystal structure, molecular interactions; energy transfer, laser phenomena and light-generating chemical reactions. Mailing Add: Gen Tel & Electronics Lab 40 Sylvan Rd Waltham MA 02154

BRECHER, GEORGE, b Olmutz, Czech, Nov 5, 13; US citizen; m 59; c 5. CLINICAL PATHOLOGY, HEMATOLOGY. Educ: Univ Prague, MD, 38; Univ London, DTM&H, 40. Prof Exp: Sr asst surgeon, USPHS, 47, surgeon, 49-51, sr surgeon, 51-53, med dir, 54-56, chief hemat serv, Dept Clin Path, 53-66; PROF LAB MED & CHMN DEPT, UNIV CALIF, SAN FRANCISCO, 66- Concurrent Pos: Fel, Mayo Clin, 43; NIH fel, 46. Honors & Awards: Distinguished Serv Medal, USPHS, 68. Mem: Acad Clin Lab Physicians & Scientists (pres, 71); Am Soc Hemat (pres, 73); Am Asn Pathologists & Bacteriologists; Soc Exp Biol & Med; Int Soc Hemat. Res: Morphologic hematology; kinetics of bone marrow and blood cells; laboratory automation and quality control. Mailing Add: Dept of Lab Med Univ of Calif Sch of Med San Francisco CA 94143

BRECHER, GERHARD ADOLF, b Goldap, Ger, June 14, 09; nat US; m 41; c 3. PHYSIOLOGY. Educ: Duke Univ, AM, 30; Univ Hamburg, PhD, 32; Univ Kiel, MD, 37. Prof Exp: Instr physiol, Univ Kiel, 37-38; intern, San Francisco, Calif, 38-41; sr instr physiol, Tech Univ Prague, 41-45; from sr instr to assoc prof, Med Sch, Western Reserve Univ, 48-55; prof & dir inst res in vision, Col Med, Univ Ohio, 55-57; prof physiol & chmn dept, Sch Med, Emory Univ, 57-67; DISTINGUISHED PROF PHYSIOL, MED CTR, UNIV OKLA, 67- Concurrent Pos: Med consult res in vision, Gen Elec Co, Cleveland; asst resident, Orlando, Fla, 39-40; resident, Jacksonville, 40-41. Mem: Am Physiol Soc; fel Am Col Cardiol; Asn Res Vision & Ophthal; Am Heart Asn. Res: Eye physiology; cardiovascular physiology. Mailing Add: Dept of Physiol Univ of Okla Sch of Med Oklahoma City OK 73104

BRECHER, KENNETH, b New York, NY, Dec 7, 43; m 65; c 1. THEORETICAL ASTROPHYSICS. Educ: Mass Inst Technol, BS, 64, PhD(physics), 69. Prof Exp: Res physicist, Univ Calif, San Diego, 69-72; ASST PROF PHYSICS, MASS INST TECHNOL, 72- Mem: Am Phys Soc; Am Astron Soc; Int Astron Union; Sigma Xi. Res: Theoretical high energy astrophysics; x-ray astronomy; observational tests and consequences of gravitational theories; cosmology. Mailing Add: Dept of Physics Rm 6-201 Mass Inst of Technol Cambridge MA 02139

BRECHER, PETER I, b New York, NY, May 19, 40; m 64; c 2. BIOCHEMISTRY, ENDOCRINOLOGY. Educ: Ohio Univ, BS, 60; Boston Univ, PhD(biochem), 68. Prof Exp: Res assoc, Ben May Labs Cancer Res, Univ Chicago, 69-71; ASST PROF BIOCHEM, SCH MED, BOSTON UNIV, 71- Concurrent Pos: Res fel biochem, Sch Med, Boston Univ, 68-69; NIH trainee physiol, 68-69. Res: Mechanism of estrogen action, specifically the binding of estrogens to receptors present in the rat uterus and the fate of this steroid-receptor complex; angiotensin II binding to adrenal receptors; lipid metabolism in the arterial wall; cardiovascular diseases. Mailing Add: Boston Univ Sch of Med 80 E Concord St Boston MA 02215

BRECHNER, BEVERLY LORRAINE, b New York, NY, May 27, 36. MATHEMATICS. Educ: Univ Miami, BS, 57, MS, 59; La State Univ, Baton Rouge, PhD(math), 64. Prof Exp: Instr math, La State Univ, New Orleans, 62-64, asst prof, 64-68; asst prof, 68-71, ASSOC PROF MATH, UNIV FLA, 71- Mem: AAAS; Am Math Soc; Math Asn Am. Res: Topology; point set topology; spaces of homeomorphisms; locally setwise homogeneous continua; chainable continua; dimension theory. Mailing Add: Dept of Math Univ of Fla Gainesville FL 32611

BRECHT, EDWARD ARMOND, (JR), pharmaceutical chemistry, see 12th edition

BRECHT, PATRICK ERNEST, b Beaconsfield, Eng, Jan 8, 46; US citizen; m 67; c 1. PLANT PHYSIOLOGY, VEGETABLE CROPS. Educ: Whittier Col, BA, 68; Calif State Univ, Los Angeles, MS, 69; Univ Calif, Davis, PhD(plant physiol), 73. Prof Exp: ASST PROF POSTHARVEST PHYSIOL, CORNELL UNIV, 73- Mem: Am Soc Hort Sci; Am Soc Plant Physiologists; Inst Food Technologists; Sigma Xi. Res: Postharvest physiology of vegetables, especially carbohydrate and nitrogen metabolism as they relate to food quality; control and mechanism of fruit ripening. Mailing Add: Veg Crops Dept Cornell Univ Ithaca NY 14853

BRECK, DONALD WESLEY, b Wentworth, NH, Jan 5, 21; m 42; c 2. INORGANIC CHEMISTRY. Educ: Univ NH, BS, 42, MS, 48; Mass Inst Technol, PhD(chem), 51. Prof Exp: Instr chem, Univ NH, 46-48; asst, Mass Inst Technol, 49-51; res chemist, Linde Div, Union Carbide Corp, 51-56, res supvr, 56-59, sr res assoc, 59-64, staff mem, Union Carbide Res Inst, 64-67, sr scientist, Molecular Sieve Dept, 67-69, SR RES FEL, TARRYTOWN TECH CTR, LINDE DIV, UNION CARBIDE CORP, 69- Concurrent Pos: Mem, Petrol Res Fund Adv Bd, 62- Mem: Am Chem Soc; Am Ceramic Soc. Res: Fluorine chemistry; inorganic silicates; synthetic minerals; zeolites; hydrothermal synthesis; crystal growth; absorption; ceramics. Mailing Add: Union Carbide Corp Linde Div Tarrytown Tech Ctr Tarrytown NY 10591

BRECK, WALLACE GRAHAM, b Calgary, Alta, Can, Jan 1, 17; m 46; c 2. PHYSICAL CHEMISTRY. Educ: Queen's Univ, Ont, BSc, 50, MSc, 51; Cambridge Univ, PhD(phys chem), 54. Prof Exp: Lectr, 56-57, from asst prof to assoc prof, 57-67, PROF CHEM, QUEEN'S UNIV, ONT, 67- Mem: Faraday Soc. Res: Measurement of thermal diffusion and thermoelectric properties in aqueous solutions of electrolytes; aquatic chemistry of marine organisms and seston; natural redox levels. Mailing Add: Dept of Chem Queen's Univ Kingston ON Can

BRECKENRIDGE, BRUCE (MCLAIN), b Brooklyn, Iowa, Nov 7, 26; m 49; c 3. PHARMACOLOGY, PHYSIOLOGY. Educ: Iowa State Col, BS, 47; Univ Rochester, MS, 49, PhD(physiol), 56. Prof Exp: Instr physiol, Univ Rochester, 53-56; intern med, Barnes Hosp, St Louis, 56-57; from instr to assoc prof pharmacol, Washington Univ, 57-67; PROF PHARMACOL & CHMN DEPT, RUTGERS MED SCH, COL MED & DENT, NJ, 67- Concurrent Pos: Fel pharmacol, Washington Univ, 57-59; Markle scholar med sci, 59; mem, Pharmacol A Study Sect, NIH, 68-72. Mem: Am Soc Biol Chemists; Am Soc Pharmacol & Exp Therapeut; Histochem Soc. Res: Metabolism of the nervous system; cyclic nucleotides and drug action. Mailing Add: Dept of Pharmacol Rutgers Med Sch PO Box 101 Piscataway NJ 08854

BRECKENRIDGE, CARL, b Asphodel Twp, Ont, July 29, 42; m 65; c 2. CLINICAL BIOCHEMISTRY. Educ: Queen's Univ, Ont, BSc, 65; Univ Toronto, MSc, 66, PhD(biochem), 70. Prof Exp: Med Res Coun Can res fel, Ctr Neurochem, Strasbourg, France, 70-72 & Montreal Neurol Inst, 72; DIR CORE LAB & ASST PROF CLIN BIOCHEM, UNIV TORONTO, 72- Mem: Can Fedn Biol Socs; Am Oil Chemists Soc. Res: Studies of the structure of lipid and protein constituents of human lipoproteins and their role in hyperlipoproteinemia and the development of atherosclerosis. Mailing Add: Univ of Toronto Lipid Res Clin Rm 227 1 Spadina Crescent Toronto ON Can

BRECKENRIDGE, JOHN ROBERT, b Kingston, Pa, Oct 10, 20; m 45; c 1. MEDJCAL BIOPHYSICS. Educ: Univ Pa, BS, 42. Prof Exp: Student engr, Electrometall Co, Union Carbide Corp, Ohio, 42-45; physicist, Qm Res & Eng Ctr, Mass, 46-61; supvry physicist, 61-64, RES PHYSICIST, US ARMY RES INST ENVIRON MED, 64- Mem: Instrument Soc Am; NY Acad Sci; Am Soc Heat, Refrig & Air Conditioning Eng. Res: Heat and moisture transfer from clothed man to his environment in terms of clothing parameters and environmental factors. Mailing Add: US Army Res Inst Environ Med Natick MA 01760

BRECKENRIDGE, ROBERT GEORGE, b Jamestown, NY, Nov 14, 15; m 47. PHYSICS Educ: Cornell Univ, BA, 38, MA, 40; Mass Inst Technol, PhD(phys chem), 42. Prof Exp: Res assoc, Mass Inst Technol, 42-45, asst prof elec insulation, 45-48; chief solid state physics sect, Nat Bur Standards, 48-55; dir res, Nat Carbon Co, Union Carbide Corp, 55-59, Parma Res Lab, 59-63, dir res, Res Inst, 63-64; dir physics dept, Atomics Int Div, Rockwell Int, 64-71; CONSULT, 71- Concurrent Pos: Head physics br, Off Naval Res, 53-54; vchmn, Conf Elec Insulation, Nat Res Coun, 54, chmn, 55. Mem: Fel Am Phys Soc; Am Chem Soc; NY Acad Sci. Res: Solid state physics; semiconductors; dielectrics. Mailing Add: 19252 Kinzie St Northridge CA 91324

BRECKENRIDGE, ROBERT T, b Akron, Ohio, Sept 28, 29; m 59; c 3. INTERNAL MEDICINE, HEMATOLOGY. Educ: Amherst Col, BA, 51; Western Reserve Univ, MD, 55. Prof Exp: Demonstr med, Western Reserve Univ, 62-63, from instr to sr instr, 63-67; asst prof, 67-69, ASSOC PROF MED, UNIV ROCHESTER, 69- Concurrent Pos: Am Cancer Soc fel clin med, Sch Med, Western Reserve Univ, 61-62, Bishop fel, 63-67. Mem: AAAS; Am Soc Hemat; fel Am Col Physicians; Am Fedn Clin Res; Soc Exp Biol & Med. Res: Physiology of human hemostasis. Mailing Add: Rochester Gen Hosp Rochester NY 14621

BRECKER, LAWRENCE ROBERT, organic chemistry, see 12th edition

BRECKINRIDGE, CHARLES EDWARD, JR, b Louisville, Ky, June 21, 23; m 55; c 3. PHARMACY. Educ: Univ Ky, BS, 53; Purdue Univ, MS, 59, PhD(pharmaceut chem), 60. Prof Exp: Instr pharm, Univ Ky, 54-57; instr, Purdue Univ, 57-59; sr health physicist, Health Physics Div, Oak Ridge Nat Lab, 60-63; sr scientist, Pharmacol Opers, Hanford Labs, 63; assoc prof pharmaceut chem & head dept, Mercer Univ, 63-64; asst prof health physics, Purdue Univ, 64-65; prof radiation sci, chmn dept & dir radiol health training prog, Med Ctr, 66-70, prof environ health sci, 70-74, PROF CLIN PHARM, CHMN DEPT & DIR NUCLEAR PHARM EDUC, TRAINING & RES, COL PHARM, CTR MED SCI, UNIV ARK, LITTLE ROCK, 74- Mem: AAAS; Am Pharmaceut Asn; Sigma Xi. Res: Pharmaceutical education, training and research; clinical pharmacy; nuclear pharmacy. Mailing Add: Col of Pharm Univ of Ark for Med Sci Little Rock AR 72201

BRECKINRIDGE, JAMES BERNARD, b Cleveland, Ohio, May 27, 39; m 65; c 2. OPTICAL PHYSICS. Educ: Case Inst Technol, BSc, 61; Univ Ariz, MS, 70, PhD(optics), 76. Prof Exp: Asst astron, Lick Observ, Univ Calif, 61-64 & Kitt Peak Nat Observ, NSF, 64-66; engr physicist, Rauland Corp, Zenith Radio Corp, 66-67; ASSOC-IN-RES, KITT PEAK NAT OBSERV, NSF, 67- Mem: Optical Soc Am; Am

Astron Soc; Royal Astron Soc. Res: Optical interferometric techniques for research in astrophysics; image evaluation. Mailing Add: Kitt Peak Nat Observ PO Box 26732 Tucson AZ 85726

BRECKON, SYDNEY WILSON, b Georgetown, Ont, Sept 26, 18; m 45; c 5. PHYSICS. Educ: Queen's Univ, Ont, BSc, 41; McGill Univ, PhD(physics), 51. Prof Exp: Jr res physicist, Nuclear Physics Sect, Nat Res Coun, 46; demonstr physics, McGill Univ, 46-47, res asst, Radiation Lab, 47-50, res asst nuclear physics, 50-51, res assoc, 51-53; PROF PHYSICS & HEAD DEPT, MEM UNIV NFLD, 53- Mem: Am Phys Soc; Can Asn Physicists. Res: Nuclear physics; short half-life radioactivities; scintillation spectrometry; gamma rays; synchrocyclotron design. Mailing Add: Dept of Phys Mem Univ of Nfld St Johns NF Can

BREDAHL, EDWARD ARLAN, b Pelican Rapids, Minn, Dec 20, 37; div; c 2. PHYSIOLOGY, ECOLOGY. Educ: Concordia Col, Moorhead, Minn, BA, 59; Univ NDak, MST, 64, PhD(physiol ecol), 69. Prof Exp: Teacher high sch, Minn, 59-62; instr cell physiol, Univ NDak, 69; ASST PROF BIOL & ECOL, STOUT STATE UNIV, 69- Res: Physiological investigation of water balance in hibernating and nonhibernating Citellus franklinii and Citellus richardsonii to develop efficiency indices for water balance relative to the environment. Mailing Add: Dept of Biol Stout State Univ Menomonie WI 54751

BREDDERMAN, PAUL JOHN, b Cairo, NY, Dec 21, 36; m 64; c 1. CELL BIOLOGY, MOLECULAR BIOLOGY. Educ: Cornell Univ, BS, 54, MS, 67, PhD(phys biol), 71. Prof Exp: Fel biophys, Dept Radiation Biol & Biophys, Univ Rochester, 71-75; ASSOC PROF BIOPHYS, UNIV TENN-ENERGY RES & DEVELOP ADMIN COMP ANIMAL RES LAB, 75- Res: Membrane protein functional associations and protein mediation of metal ion transport with special reference to intestinal absorption. Mailing Add: Comp Animal Res Lab 1299 Bethel Valley Rd Oak Ridge TN 37830

BREDECK, HENRY E, b St Louis, Mo, Nov 5, 27; m 53; c 6. PHYSIOLOGY. Educ: St Louis Univ, BS, 50; Univ Mo, MS, 53, PhD(agr chem), 56. Prof Exp: Asst prof chem, Colo State Univ, 56-58, from asst prof to prof physiol, 58-66; res prog mgr, NASA, 66-67; prog dir, Gen Res Support Br, Div Res Resources, NIH, 67-71; ASSOC DIR OFF RES DEVELOP, MICH STATE UNIV, 71- Mem: Am Physiol Soc. Res: Endocrinology; physiology of reproduction; cardiovascular physiology. Mailing Add: Off of Res Develop Mich State Univ East Lansing MI 48823

BREDEHOEFT, JOHN DALLAS, b St Louis, Mo, Feb 28, 33; m 58; c 3. HYDROLOGY, GEOLOGY. Educ: Princeton Univ, BSE, 55; Univ Ill, MS, 57, PhD(geol), 62. Prof Exp: Geologist, Humble Oil & Refining Co Div, Standard Oil Co, NJ, 57-59; geologist, Desert Res Inst, Nev, 62; res geologist, 62-74, DEP ASST CHIEF RES HYDROLOGIST, WATER RESOURCES DIV, US GEOL SURV, 74- Concurrent Pos: Vis assoc prof, Univ Ill, 67-68; res assoc, Resources for Future, 68-70. Honors & Awards: R E Horton Award, Am Geophys Union & O E Meinzer Award, Geol Surv Am, 75. Mem: Am Asn Petrol Geologists; Geol Soc Am; Am Geophys Union. Res: Ground water hydrology; physical properties of ground water systems; physics of ground water motion; transport of chemical constituents in ground-water systems. Mailing Add: US Geol Surv Water Resources Div Reston VA 22092

BREDER, CHARLES MARCUS, JR, zoology, see 12th edition

BREDER, CHARLES VINCENT, b Atlantic City, NJ, Feb 15, 40; m 62; c 2. ORGANIC POLYMER CHEMISTRY, ANALYTICAL CHEMISTRY. Educ: Carson-Newman Col, BS, 62; Vanderbilt Univ, MS, 64, PhD(org chem), 68. Prof Exp: Res chemist, E I du Pont de Nemours & Co, 67-71; RES CHEMIST, FOOD & DRUG ADMIN, DEPT HEALTH, EDUC & WELFARE, 71- Mem: Am Chem Soc; Asn Off Anal Chemists. Res: Determination of the identity and quantity of chemicals that migrate from food packaging into foods. Mailing Add: Food & Drug Admin 200 C St SW Washington DC 20204

BREDEWEG, ROBERT ALLEN, b Forest Grove, Mich, Aug 15, 41; m 63; c 2. ANALYTICAL CHEMISTRY. Educ: Hope Col, BA, 63; Southern Ill Univ, MA, 65. Prof Exp: ANAL SPECIALIST II, DOW CHEM CO, 65- Honors & Awards: Vernon A Stenger Award, Dow Chem Anal Scientists, 75. Mem: Am Chem Soc. Res: Analytical methods development using gas and liquid chromatography. Mailing Add: Dow Chem Co 574 Bldg Midland MI 48640

BREDIG, MAX ALBERT, b Heidelberg, Ger, June 20, 02; nat US; m 44; c 1. PHYSICAL INORGANIC CHEMISTRY. Educ: Karlsruhe Technol Univ, Dipl Eng, 25; Berlin Technol Univ & Kaiser Wilhelm Inst, DEng, 26. Prof Exp: Fel, Univ Göttingen, 27 & Univ Munich, 27-28; res assoc, Kaiser Wilhelm Inst, 28-29; res chemist in charge x-ray & spectros lab, Bayer Stickstoffwerke, Berlin, 29-37; res fel, Univ Mich, 37-38; mgr appl chem div, Vanadium Corp Am, 38-46; proj leader chem div, 46-49, actg chief physics of solids sect, Physics Div, 49-50, assoc dir & res group leader, Chem Div, 50-67, CONSULT, OAK RIDGE NAT LAB, 67- Concurrent Pos: Consult, Glass Sci Inc, Pa, 44; chmn, Molten Salt Conf, Gordon Res Conf, 65. Mem: Fel AAAS; Am Chem Soc; Sigma Xi; Electrochem Soc; Am Crystallog Asn. Res: Crystal chemistry; calcium phosphates; calcium orthosilicate (bredigite); polymorphism of calcium, uranium carbides; high temperature x-ray diffraction; order-disorder transition in fluorite type crystals; radiation effects in solids; fused salt metal systems; molten salt nuclear reactor chemistry. Mailing Add: Chem Div Oak Ridge Nat Lab Oak Ridge TN 37830

BREDON, GLEN E, b Fresno, Calif, Aug 24, 32; m 63; c 2. MATHEMATICS. Educ: Stanford Univ, BS, 54; Harvard Univ, AM, 55, PhD(math), 58. Prof Exp: Mem, Inst Advan Study, 58-60; from asst prof to prof math, Univ Calif, Berkeley, 60-68; PROF MATH, RUTGERS UNIV, NEW BRUNSWICK, 68- Concurrent Pos: Sloan fel, 65-67; mem, Inst Advan Study, 66-67. Mem: AAAS; Am Math Soc; Math Asn Am. Res: Topological transformation groups; algebraic and differential topology. Mailing Add: Dept of Math Rutgers Univ New Brunswick NJ 08903

BREE, ALAN V, b Sydney, Australia, Nov 16, 32; m 61; c 2. PHYSICAL CHEMISTRY. Educ: Univ Sydney, BSc, 54, MSc, 55, PhD(phys chem), 57. Prof Exp: Nat Res Coun Can fel, 57-59; fel, Chem Dept, Univ Col, Univ London, 60-61; from asst prof to assoc prof, 61-71, PROF CHEM, UNIV BC, 71- Res: Electronic properties of aromatic molecules in the gas; solution and solid phases. Mailing Add: Dept of Chem Univ of BC Vancouver BC Can

BREE, MAX M, b Philadelphia, Pa, Dec 22, 15; m 40; c 4. VETERINARY MEDICINE. Educ: Univ Pa, VMD, 42; Am Col Lab Animal Med, dipl, 66. Prof Exp: Vet, Adrea Animal Hosp, Nutley, NJ, 46-63; res fel, 63-65, assoc prof, 65-75, PROF LAB ANIMAL MED, MED SCH, UNIV MICH, ANN ARBOR, 75- Concurrent Pos: Mem, Inst Lab Animal Resources, Nat Acad Sci, 65- & Am Col Lab Animal Med, 65; consult, Am Asn Accreditation Lab Animal Care, 65. Mem: Vet Med Asn; Asn Lab Animal Sci. Res: Laboratory animal medicine; clinical small animal medicine. Mailing Add: Animal Care Unit Univ of Mich Med Sch Ann Arbor MI 48104

BREED, BENNY RAY, b Hays Co, Tex, Dec 18, 36; m 64; c 2. ENVIRONMENTAL PHYSICS, SYSTEM THEORY. Educ: Univ Tex, BS, 58; Rice Univ, MA, 64, PhD, 65. Prof Exp: Mem staff, Los Alamos Sci Lab, Univ Calif, 65-67; eng scientist, Tracor, Inc, 67-69; sr scientist, Radian Corp, 69-75; INDEPENDENT CONSULT, 75- Concurrent Pos: Consult, IBM Corp. Mem: AAAS; Am Phys Soc; Sigma Xi. Res: Mathematical modeling; environmental modeling; statistical analysis. Mailing Add: 5918 Highland Hills Dr Austin TX 78731

BREED, ERNEST SPENCER, b Lyndonville, NY, May 27, 13; m 44; c 4. SURGERY, PHYSIOLOGY. Educ: Univ Mich, MD, 38; Am Bd Surg, dipl, 49. Prof Exp: Intern med & surg, NY Univ Div, Bellevue Hosp, 39-40, asst resident surg & path, 40-41, resident, 41-42; instr surg, 47-50, asst prof clin surg, 50-51 & surg, 51-60, ASSOC PROF SURG, SCH MED, NY UNIV, 60- Concurrent Pos: Res fel, Off Sci Res & Develop shock proj, Col Physicians & Surgeons, Columbia Univ, 42-44; Commonwealth Fund res fel physiol, 47-49; attend, Bellevue Hosp, 42-, Univ Hosp, 52- & Vet Admin Hosp, Manhattan, NY, 56-; consult, Glen Cove Community Hosp, Long Island, 58- & Misericordia Hosp, Bronx, 65- Mem: Soc Univ Surgeons. Res: Cardiac and renal vein catheterization in patients in shock; renal function in relation to surgical conditions in patients, including open heart surgery. Mailing Add: Dept of Surg NY Univ 550 First Ave New York NY 10016

BREED, HELEN ILLICK, b New Cumberland, Pa, Mar 12, 25; m 57; c 3. ZOOLOGY. Educ: Syracuse Univ, BS, 47, MS, 49; Cornell Univ, PhD(vert zool & biol), 53. Prof Exp: Asst, Dept Zool, Syracuse Univ, 47-49; teacher high sch, Lyons, NY, 49-50; instr zool & physiol, Dept Biol, Akron Univ, 54; Ford Found fel & instr physiol, Vassar Col, 54-55; asst prof biol, Russell Sage Col, 55-57; asst dir syst biol, NSF, DC, 57; res assoc prof conserv, Cornell Univ, 57-61; res assoc biol, Rensselaer Polytech Inst, 64-68; CONSULT, 68- Concurrent Pos: Am-Scand fel, 59-60; mem, Rensselaer Environ Mgt Coun, 71-77. Mem: AAAS; Am Soc Zoologists; Soc Syst Zool; Am Soc Ichthyol & Herpet; Am Inst Biol Sci. Res: Ichthyology; ecology; conservation; x-ray diffraction of vertebrate hard tissue. Mailing Add: RD 3 245 B Troy NY 12180

BREED, HENRY ELTINGE, b New York, NY, Dec 5, 15; m 57. PHYSICS. Educ: Colgate Univ, BA, 38; Rensselaer Polytech Inst, MS, 48, PhD, 55. Prof Exp: From instr to asst prof, 45-62, ASSOC PROF PHYSICS, RENSSELAER POLYTECH INST, 62- Concurrent Pos: Res asst, Gen Elec Res Lab, 46-48; consult, Ord Corps, US Army, 56-62; Fulbright fel, Inst Teorisk Kjemi, Trondheim, Norway, 59-60; Fulbright lectr optics, UNI, Lima, Peru, 73. Mem: Am Phys Soc; Am Asn Physics Teachers; fel Optical Soc Am. Res: Electron diffraction; optics. Mailing Add: Dept of Physics Rensselaer Polytech Inst Troy NY 12181

BREED, LAURENCE WOODS, b Decatur, Ill, Dec 16, 24; m 48; c 2. ORGANIC POLYMER CHEMISTRY. Educ: Park Col, BA, 48; Univ Kans, MA, 50. Prof Exp: Asst, Univ Kans, 48-50; res chemist, Battenfeld Grease & Oil Corp, 50-53; assoc chemist, 53-59, sr chemist, 59-62, prin chemist, 62-67, SR ADV CHEM, MIDWEST RES INST, 67- Mem: Am Chem Soc. Res: Organosilicon chemistry; organic and semiorganic polymers; organic synthesis; thermal analysis of materials. Mailing Add: Midwest Res Inst 425 Volker Blvd Kansas City MO 64110

BREED, WILLIAM JOSEPH, b Massillon, Ohio, Aug 3, 28; m 65; c 1. GEOLOGY. Educ: Denison Univ, BA, 52; Univ Ariz, BS, 55, MS, 60. Prof Exp: Stratigraphic aide, Shell Oil Co, Tex, 55; geol aide, Ground Water Br, US Geol Surv, Colo, 55-56; CUR GEOL, MUS NORTHERN ARIZ, 60- Mem: AAAS; Soc Econ Paleont & Mineral; Geol Soc Am; Soc Vert Paleont. Res: Lineation and morphology of cinder cones; Proterozoic rocks of Grand Canyon; Antarctic vertebrate fossils; general geology of northern Arizona. Mailing Add: Mus of Northern Ariz Rte 4 Box 720 Flagstaff AZ 86001

BREEDEN, JOHNNIE ELBERT, b Charlotte, Tenn, Nov 11, 31; m 53; c 3. PLANT ECOLOGY. Educ: Austin Peay State Univ, BS, 53, MA, 54; Vanderbilt Univ, PhD(plant ecol), 68. Prof Exp: PROF BIOL, DAVID LIPSCOMB COL, 68- Mailing Add: Dept of Biol David Lipscomb Col Nashville TN 37203

BREEDING, J ERNEST, JR, b Peoria, Ill, Mar 17, 38; m 70; c 1. PHYSICAL OCEANOGRAPHY. Educ: Drake Univ, BA, 60; Columbia Univ, PhD(geophys), 72. Prof Exp: Res asst physics, Univ Tenn, Knoxville, 60-61; RES PHYSICIST, NAVAL COASTAL SYSTS LAB, 62- Concurrent Pos: Res asst phys oceanog, Lamont-Doherty Geol Observ, Columbia Univ, 65-70. Mem: AAAS; Am Asn Physics Teachers; Am Geophys Union; Soc Explor Geophysicists; Sigma Xi. Res: Propagation of water waves in coastal waters; wave refraction; electromagnetic fields in sea water. Mailing Add: RR 4 Box 420 Panama City FL 32401

BREEDIS, CHARLES, b Morrisville, Pa, Aug 7, 11. PATHOLOGY. Educ: NY Univ, AB, 38; Cornell Univ, MD, 42. Prof Exp: Fel path, Grad Sch Med, 46, from asst instr to instr, 46-49, assoc, 49-51, from asst prof to assoc prof, 51-60, PROF PATH, SCH MED, UNIV PA, 60- Mem: Am Asn Path & Bact; Am Soc Exp Path; Am Asn Cancer Res. Res: Differentiation; neoplasia; regeneration. Mailing Add: Dept of Path Univ of Pa Sch of Med Philadelphia PA 19174

BREEDLOVE, CHARLES B, b Chico, Tex, Dec 5, 16; m 42; c 2. SCIENCE EDUCATION. Educ: McMurry Col, BS, 37; Southern Methodist Univ, MS, 38; Wayne State Univ, EdD, 63. Prof Exp: Teacher pub high schs, Tex, 38-42 & Mich, 43-58; head dept chem, Henry Ford High Sch, 58-64; PROF PHYSICS, EASTERN MICH UNIV, 64- Concurrent Pos: Mem, State Sci Curriculum Comt, Mich, 64-65; dir, Five Year NSF Proj, 70-75. Mem: Nat Asn Sci Teachers; Nat Asn Res Sci Teaching; Am Asn Physics Teachers. Res: Science teacher education and science curriculum projects. Mailing Add: 1007 Louise Ypsilanti MI 48197

BREEDLOVE, JAMES ROBBY, JR, b Tyler, Tex, Nov 10, 43; m 67; c 2. PHYSICS, COMPUTER SCIENCE. Educ: Mass Inst Technol, SB, 65; Wm Marsh Rice Univ, PhD(physics), 70. Prof Exp: Sr scientist, EG&G Inc, 70-73; staff mem, 73-75, ASST GROUP LEADER, LOS ALAMOS SCI LAB, 75- Mem: Optical Soc Am; Inst Elec & Electronics Engrs. Res: Processing of images by digital computer, atmospheric phenomena, digital filtering and image analysis. Mailing Add: Group C-8 Mail Stop 263 Los Alamos Sci Lab Los Alamos MN 87545

BREELAND, SAMUEL GLOVER, b Batesburg, SC, Aug 17, 26; m 52; c 1. ENTOMOLOGY. Educ: Univ Ga, BS, 50; NC State Univ, MS, 53; Univ Tenn, PhD(zool & entom), 57. Prof Exp: Entomologist, USPHS, 50-53; instr math & sci, Brewton-Parker Col, 53-54; asst entom, Univ Tenn Agr Exp Sta, 56-58; med entomologist, Canal Zone Govt, 58-60; res biol, Tenn Valley Authority, 60-64; prof & head dept biol, Hardin-Simmons Univ, 65-66; res biol, Tenn Valley Authority, 66-67; res entomologist, Cent Am Malaria Res Sta, 67-72 & Ctr Dis Control, Savannah, 72-73; CHIEF MED ENTOM BR, CTR DIS CONTROL, ATLANTA, 73- Mem:

Entom Soc Am; Am Mosquito Control Asn; Sigma Xi. Res: Bionomics and control of medically important arthropods, principally mosquitoes. Mailing Add: Med Entom Br Ctr for Dis Control Atlanta GA 30333

BREEN, GAIL ANNE MARIE, b Pembroke, Ont, Can. NEUROSCIENCES, MOLECULAR BIOLOGY. Educ: Univ Toronto, BSc, 70; Univ Calif, Los Angeles, PhD(neurosci), 74. Prof Exp: From res asst pharm to res asst immunol, Univ Toronto, 68-70; res asst neurosci, Univ Calif, Los Angeles, 70-74; MED RES COUN CAN FEL & RES ASSOC MOLECULAR BIOL, ROSWELL PARK MEM INST, 74- Concurrent Pos: Achievement Rewards for Col Scientists Found fel, 70-72. Mem: Soc Neurosci; Sigma Xi. Res: Genetics and biochemistry of various lysosomal enzymes, especially beta-galactosidase; regulation of gene expression during development and differentiation, especially of the nervous system. Mailing Add: Dept of Molecular Biol Roswell Park Mem Inst Buffalo NY 14263

BREEN, JAMES LANGHORNE, b Chicago, Ill, Sept 5, 26; m 51; c 5. OBSTETRICS & GYNECOLOGY. Educ: Northwestern Univ, BS, 48, MD, 52; Am Bd Obstet & Gynec, dipl. Prof Exp: Rotating intern, Walter Reed Army Hosp, 52-53, resident obstet & gynec, 54-57, asst chief, 57-58; from asst chief to chief, Second Gen Hosp, Landstuhl, Ger, 58-60; assoc prof, Seton Hall Col Med & Dent, 61-69; DIR DEPT OBSTET & GYNEC, ST BARNABAS MED CTR, 69-; ASSOC PROF OBSTET & GYNEC, MARTLAND HOSP UNIT, NJ COL MED & DENT, 69- Concurrent Pos: Res fel obstet, gynec & breast path, Armed Forces Inst Path, 60-61; attent obstetrician & gynecologist, Margaret Hague Maternity Hosp, Jersey City, 61-63; gynecologist & obstetrician-in-chief, Newark City Hosp, 63-; attend gynecologist, Med Ctr, Jersey City, 63-; consult, St Elizabeth Hosp & Monmouth Med Ctr, 63- & St Barnabas Med Ctr, 64- Mem: Fel Am Col Obstet & Gynec; fel Am Col Surgeons; Asn Mil Surgeons US; NY Acad Sci; Am Soc Cytol. Res: Oncology surgery; research cytology; biochemical assays of female genital malignancy; ovarian tumors. Mailing Add: St Barnabas Med Ctr Old Short Hills Rd Livingston NJ 07039

BREEN, MARILYN, b Anderson, SC, Nov 8, 44; m 75. GEOMETRY. Educ: Agnes Scott Col, BA, 66; Clemson Univ, MS, 68, PhD(math), 70. Prof Exp: Vis instr, 70-71, ASST PROF MATH, UNIV OKLA, 72- Mem: Math Asn Am; Am Math Soc. Res: Convexity and combinatorial geometry; convex polytopes; m-convex sets. Mailing Add: Dept of Math Univ of Okla Norman OK 73069

BREEN, MOIRA, b Madras, India, Dec 18, 23; US citizen. BIOCHEMISTRY. Educ: Univ Madras, BS, 44; Vasaar Col, MS, 51; Northwestern Univ, PhD(biochem), 60. Prof Exp: Asst biochem & physiol, Univ Madras, 44-49; asst physiol, Vassar Col, 49-52; instr, Sch Nursing, Johns Hopkins Univ, 52-53; asst nutrit, Harvard Univ, 53-54; instr biochem, Northwestern Univ, 60-62; res assoc med, Univ Va, 62-63; res assoc physiol, Univ Chicago, 64-65; SR RES BIOCHEMIST, RES-IN-AGING LAB, VET ADMIN HOSP, DOWNEY, ILL, 65- Concurrent Pos: Asst prof biochem, Sch Med, Northwestern Univ, 68- Mem: AAAS; Am Chem Soc; Am Soc Biol Chemists. Res: Complex carbohydrates in brain and connective tissue metabolism. Mailing Add: Res-in-Aging Lab Vet Admin Hosp Downey IL 60064

BREEN, ROBERT JAMES, physics, see 12th edition

BREENE, ROBERT GALE, JR, atomic physics, see 12th edition

BREENE, WILLIAM MICHAEL, b Adams, Wis, May 10, 30; m 63. FOOD SCIENCE, BIOCHEMISTRY. Educ: Univ Wis, BS, 57, MS, 61, PhD(dairy & food industs), 64. Prof Exp: Res asst dairy & food industs, Univ Wis, 58-64, proj asst, 64; res assoc dairy industs, Univ Minn, 64-66, asst prof food sci & industs, 66-67; asst prof, Mich State Univ, 67; asst prof, 68-72, ASSOC PROF FOOD SCI & NUTRIT, UNIV MINN, ST PAUL, 72- Mem: Inst Food Technologists; Am Soc Hort Sci. Res: Spray drying fruits and vegetables using skim milk as a carrier; carotenoid oxidation, fruit and vegetable texture. Mailing Add: Dept of Food Sci & Nutrit Univ of Minn St Paul MN 55101

BREESE, GEORGE RICHARD, b Richmond, Ind, Dec 27, 36; m 60; c 2. PHARMACOLOGY, NEUROBIOLOGY. Educ: Butler Univ, BS, 59, MS, 61; Univ Tenn, PhD(pharmacol), 65. Prof Exp: Instr pharmacol, Med Sch, Univ Ark, 65-66; res assoc pharmacol & toxicol, Nat Inst Gen Med Sci, 66-68; asst prof, 68-72, ASSOC PROF PSYCHIAT & PHARMACOL, SCH MED, UNIV NC, CHAPEL HILL, 72- Mem: AAAS; Am Soc Pharmacol & Exp Therapeut; Soc Neurosci; Am Col Neuropsychopharmacol. Res: Developmental neuropsychopharmacology; neuropharmacology; depression; hyperkinesis; neurocytotoxic compounds. Mailing Add: 226 Biol Sci Res Ctr Univ of NC Sch of Med Chapel Hill NC 27514

BREESE, SYDNEY SALISBURY, JR, b New York, NY, Apr 11, 22; m 53; c 3. VETERINARY VIROLOGY. Educ: Antioch Col, BS, 44; Univ Md, MS, 53. Prof Exp: Biophysicist virol, Walter Reed Army Inst Res, 48-53; physicist, Children's Cancer Res Found, 54-55; PHYSICIST, PLUM ISLAND ANIMAL DIS CTR, USDA, 55- Concurrent Pos: Prog chmn & ed, 4th Int Cong Electron Micros, 62; chmn, Gordon Res Conf Immuno-Electron Micros, 72. Mem: AAAS; Electron Micros Soc Am (pres, 64); Am Soc Cell Biol. Res: Physical properties and purification of viruses; electron microscopy and ultracentrifugation techniques for virology problems; development and identification of viruses; immunochemical techniques in electron microscopy. Mailing Add: Plum Island Animal Dis Ctr PO Box 848 Greenport NY 11971

BREG, WILLIAM ROY, b Arlington, Tex, Sept 6, 23; m 48; c 4. PEDIATRICS, CYTOGENETICS. Educ: Yale Univ, MD, 47. Prof Exp: Clin instr, 55-61, asst clin prof, 61-69, ASSOC CLIN PROF HUMAN GENETICS & PEDIAT, YALE UNIV, 69-; PHYSICIAN & DIR, RES LAB, SOUTHBURY TRAINING SCH, 54- Concurrent Pos: Consult, Nat Inst Child Health & Human Develop, 65-69. Mem: Am Acad Pediat; Am Asn Ment Deficiency; Soc Pediat Res; Am Soc Human Genetics; Am Pediat Soc. Res: Human chromosomal abnormalities. Mailing Add: Southbury Training Sch Southbury CT 06488

BREGER, IRVING A, b Boston, Mass, Jan 28, 20; m 43; c 2. ORGANIC GEOCHEMISTRY. Educ: Worcester Polytech Inst, BS, 41; Mass Inst Technol, SM, 47, PhD(geochem), 50. Prof Exp: Inspector powder & explosives, US War Dept, 41-42; chemist, Koppers United Co, 42-43 & Publicker Industs, 43-45; res asst physics, Mass Inst Technol, 45-47, res assoc geol, 47-52; CHEMIST, US GEOL SURV, 52- Concurrent Pos: Fulbright grants, Netherlands, 50-51; vis prof chem, Univ Md; adj prof, Am Univ, 57-; Guggenheim fel, Australia, 66-67. Mem: Fel AAAS; Am Chem Soc; Geochem Soc; fel Geol Soc Am; fel Am Inst Chemists. Res: Organic geochemistry; radiation chemistry; effects of alpha particles and deuterons on organic compounds; origin and nature of coal; origin of petroleum; geochemistry of uranium; other elements with naturally occurring carbonaceous substances. Mailing Add: US Geol Surv 923 Nat Ctr Reston VA 22092

BREGGER, JOHN TAYLOR, b Chicago, Ill, Jan 14, 96; m 32; c 3. HORTICULTURE. Educ: Mich State Univ, BS, 17; Cornell Univ, MS, 22. Prof Exp: Field asst white pine blister rust invests, USDA, 17-19, field asst cereal invests, 19; horticulturist, Exp Sta, Alaska, 21-22; head spec serv & orchard res dept, Stark Bros Nurseries, Mo, 23-27; supt in charge, Luther Burbank Exp Farms, Calif, 27-29; exten horticulturist, State Col Wash, 29-32; asst prof exten pomol, Cornell Univ, 32-33; ed in chief, Am Fruit Grower, 33-35; proj mgr, Soil Conserv Serv, USDA, 35-37, proj supvr in charge, 37-54, collabr, 54-57; ORCHARD CONSULT, 57- Mem: Am Soc Hort Sci; Am Pomol Soc (past pres). Res: Soil conservation in orchards; appraisal and nomenclature of fruit varieties. Mailing Add: 318 College Ave Clemson SC 29631

BREGMAN, ALLYN AARON, b Brooklyn, NY, Apr 29, 41; m 65; c 2. CYTOGENETICS. Educ: Brooklyn Col, BS, 62; Univ Rochester, MS, 64, PhD(cytogenetics), 68. Prof Exp: Asst prof, 67-71, ASSOC PROF BIOL, STATE UNIV NY COL NEW PALTZ, 71- Concurrent Pos: NIMH grant; vis scholar, Duke Univ, 75. Mem: AAAS; Environ Mutagen Soc; Sigma Xi. Res: Chromosome banding. Mailing Add: Dept of Biol State Univ of NY Col New Paltz NY 12561

BREGMAN, JACOB ISRAEL, b Hartford, Conn, Sept 17, 23; m 49; c 3. ENVIRONMENTAL MANAGEMENT. Educ: Providence Col, BS, 43; Polytech Inst Brooklyn, MS, 48, PhD(chem), 51. Prof Exp: Chemist, Fels & Co, 47-48; sr chemist, Nat Aluminate Corp, 50-52, head dept phys chem, 52-59; supvr phys chem, IIT Res Inst, 59-63, asst dir chem res, 63-65, dir chem sci res & mgr water res ctr, 65-67; dep asst secy, US Dept Interior, 67-69; PRES, WAPORA, INC, 69- Concurrent Pos: Chmn, Northeast Ill Metrop Area Air Pollution Control Bd, 62-63, Ill Air Pollution Control Bd, 63-67, Adv Comt Saline Water Conversion to Sci & Tech Comt, NATO Parliamentarians Coun, 63 & Water Resources Res Coun, 64-66; ed, Int Series Chem & Allied Sci, Spartan Book Co, 64-67; co-ed, Series on Water Pollution, Acad Press, 69- Mem: NY Acad Sci. Res: Ion exchange; corrosion; water treatment; saline water conversion; air and water pollution; fuel cells. Mailing Add: Wapora Inc 6900 Wisconsin Ave NW Washington DC 20015

BREGMAN, JUDITH, b New York, NY, July 13, 21. CHEMICAL PHYSICS. Educ: Bryn Mawr Col, AB, 42; Cornell Univ, PhD(phys chem), 50. Prof Exp: Instr chem, Barnard Col, Columbia Univ, 46-48; res assoc, Mass Inst Technol, 49-54; res assoc, Harvard Med Sch, 54-55; res assoc, 55-57, asst prof, 57-60, ASSOC PROF PHYSICS, POLYTECH INST NEW YORK, 62- Concurrent Pos: Weizmann Mem fel, Weizmann Inst Sci, 58-59, actg head dept x-ray crystallog, 59-61, sr scientist, 60-62. Mem: Am Chem Soc; Am Phys Soc; Am Crystallog Asn. Res: X-ray crystallography; organic structures; solid state chemistry; development of new instructional material, particularly films. Mailing Add: Dept of Physics Polytech Inst of New York Brooklyn NY 11201

BREHM, BERTRAM GEORGE, JR, b Cleveland, Ohio, Nov 26, 26; m 50; c 6. BOTONAY. Educ: Western Reserve Univ, BS, 50, MS, 52; Univ Tex, PhD(bot), 62. Prof Exp: Instr biol, Lee Col, Tex, 55-61; asst prof, 62-66, ASSOC PROF BIOL, REED COL, 66- Concurrent Pos: NSF grant, 63. Mem: AAAS; Bot Soc Am; Int Asn Plant Taxon. Res: Systematics and evolution of higher plants as determined by comparative chemistry. Mailing Add: Dept of Biol Reed Col Portland OR 97202

BREHM, JOHN JOSEPH, b Memphis, Tenn, Dec 6, 34; m 59; c 4. PHYSICS. Educ: Univ Md, BS, 56, PhD(physics), 63; Cornell Univ, MS, 59. Prof Exp: Physicist, US Naval Ord Lab, 58-62; NSF fel particle physics, Princeton Univ, 62-63; from asst prof to assoc prof high energy physics, Northwestern Univ, 63-67; assoc prof physics, 67-71, PROF PHYSICS, HASBROUCK LAB, UNIV MASS, AMHERST, 71- Mem: Am Phys Soc. Res: Theoretical, nuclear and elementary particle physics. Mailing Add: Dept of Physics Univ Mass Hasbrouck Lab Amherst MA 01002

BREHM, RICHARD KEITH, spectroscopy, see 12th edition

BREHM, SYLVIA PATIENCE, b Lancaster, Pa, Nov 13, 45. MICROBIOLOGY. Educ: Univ Kans, BS, 68; Northwestern Univ, PhD(biol sci), 72. Prof Exp: NIH fel microbiol, Scripps Clin & Res Found, 73-75; ASST PROF LIFE SCI, IND STATE UNIV, TERRE HAUTE, 75- Mem: AAAS; Sigma Xi; Am Soc Microbiol. Res: Concerted genetic and biochemical approach to the initiation of sporulation in Bacillus subtilis; role of DNA-binding proteins. Mailing Add: Dept of Life Sci Ind State Univ Terre Haute IN 47809

BREHM, WARREN JOHN, b New York, NY, Nov 2, 25; m 49; c 2. CHEMISTRY. Educ: NY Univ, AB, 44; Harvard Univ, AM, 46, PhD(org chem), 48. Prof Exp: Chemist, Carbide & Carbon Chem Corp, Tenn, 45-46; Jewett fel, Columbia Univ, 48-49; instr, NY Univ, 49-52; chemist, 52-56, from supvr to sr supvr, 56-68, RES LAB SUPT, PLASTICS DEPT, FLUOROCARBONS DIV, EXP STA, E I DU PONT DE NEMOURS & CO, INC, 68- Mem: Am Chem Soc; NY Acad Sci; The Chem Soc. Res: Structure of strychnine; antimalarials; reaction mechanisms; polymers and plastics. Mailing Add: Plastics Dept E I du Pont de Nemours & Co Inc Wilmington DE 19898

BREHME, ROBERT W, b Washington, DC, Mar 6, 30; m 54; c 4. THEORETICAL PHYSICS. Educ: Roanoke Col, BS, 51; Univ NC, MS, 54, PhD(physics), 59. Prof Exp: From asst prof to assoc prof, 59-68, PROF PHYSICS, WAKE FOREST UNIV, 68- Concurrent Pos: Res assoc, Univ NC, 64-65. Mem: Am Asn Physics Teachers; Am Phys Soc. Res: Field theory; quantum electrodynamics; relativity. Mailing Add: 1055 Peace Haven Rd Winston-Salem NC 27104

BREHMER, MORRIS LEROY, b Strasburg, Ill, Apr 10, 25; m 54. AQUATIC BIOLOGY. Educ: Eastern Ill Univ, BS, 50; Mich State Univ, MS, 56, PhD, 58. Prof Exp: Tech asst, Ill Nat Hist Surv, 51; fisheries biologist, Ill Dept Conserv, 51-55; tech asst, Inst Paper Chem, 55; student conserv aide, Mich Inst Fisheries Res, 55-58; assoc biologist, Va Inst Marine Sci, 58-63, sr marine scientist, 63-67, asst dir, 67-71, head div appl marine sci & ocean eng, 69-71; dir environ opers, 71-74, MGR ENVIRON OPERS, VA ELEC & POWER CO, 74- Mem: Am Soc Limnol & Oceanog; Int Asn Theoret & Appl Limnol; Water Pollution Control Fedn. Res: Aquatic ecology; primary production as related to nutrient levels, secondary productions as affected by pollution; pollution ecology; primary productivity; response of estuarine and freshwater organisms to elevated temperatures. Mailing Add: PO Box 26666 Richmond VA 23261

BREHOB, KENNETH RAYMOND, b Knox, Ind, July 15, 44; m 66; c 2. GEOGRAPHY. Educ: Ball State Univ, BS, 66, MA, 67; Univ Okla, PhD(geog), 74. Prof Exp: Instr geog, West Tex State Univ, 67-71; teaching asst, Univ Okla, 71-74; ASST PROF GEOG, UNIV NOTRE DAME, 74- Mem: Asn Am Geogrs. Res: Magnitude, extent and human perception of floods, tornadoes, hail and other climatological hazards; nutrient balances in urban lawn ecosystems. Mailing Add: Dept of Earth Sci Rm 121 Univ of Notre Dame Notre Dame IN 46556

BREIDENBACH, GEAROLD PETER, b Mt Victory, Ohio, Nov 14, 36. VIROLOGY. Educ: Ohio State Univ, BS, 61; Miami Univ, MS, 66; Tulane Univ, PhD(microbiol), 71. Prof Exp: Res asst microbiol, Med Ctr, Ind Univ, 61-64; asst, 64-66, instr, 71, ASST PROF MICROBIOL, MIAMI UNIV, 71- Mem: Am Soc Microbiol. Res:

Physiology of viral infected cells; herpes-type viruses. Mailing Add: Dept of Microbiol Miami Univ Oxford OH 45056

BREIDENBACH, LESTER, b New York, NY, Dec 21, 97; m 26; c 2. SURGERY. Educ: NY Univ, BS, 20, MD, 22; Am Bd Surg, dipl. Prof Exp: Intern, Bellevue Hosp, NY, 22-23, resident, 23-24; assoc prof clin surg, Col Med, 24-69, ASSOC PROF SURG, UNIV HOSP & DIR EMERGENCY SERV, MED CTR, NY UNIV, 69- Concurrent Pos: Surgeon & attend surgeon, Bellevue Hosp, Misericordia Hosp; attend surgeon, Doctor's Hosp & Westchester Square Hosp; dir surg, Grand Cent Hosp; consult surgeon, Mountain Side Hosp, NJ. Mem: AMA; fel Am Col Surg; Am Asn Surg Trauma; fel NY Acad Med. Res: Treatment of fractures; surgery of rectum and colon; blood vessel surgery; thyroid surgery. Mailing Add: 169 E 69th St New York NY 10022

BREIDENBACH, ROWLAND WILLIAM, b Dayton, Ohio, Feb 1, 35; m 57; c 2. PLANT PHYSIOLOGY, PLANT BIOCHEMISTRY. Educ: Univ Fla, BS, 59; Univ Calif, Davis, MS, 63, PhD(plant physiol), 66. Prof Exp: NIH fel biol sci, Purdue Univ, 66-67; asst prof, 70-71, ASST AGRONOMIST, EXP STA & LECTR AGRON & RANGE SCI, UNIV CALIF, DAVIS, 67- Concurrent Pos: NSF fel, 68-70; Du Pont Young fac award, Univ Calif, Davis, 70-71. Honors & Awards: NY Bot Garden Award, 68. Mem: AAAS; Am Soc Plant Physiol. Res: Cell physiology, particularly spacial organization of metabolic process in the cell; relationship between spacial organization and metabolic regulation; regulation of metabolism in developing/germinating seed tissues. Mailing Add: Dept of Agron & Range Sci Univ of Calif Davis CA 95616

BREIDENSTEIN, B C, animal science, see 12th edition

BREIG, MARVIN L, b St Marys, Mo, Oct 9, 34; m 61; c 1. SOLID STATE PHYSICS. Educ: Southeast Mo State Col, BS, 56; Univ Okla, MS, 59, PhD(physics), 63. Prof Exp: Asst prof, 63-64, ASSOC PROF PHYSICS, EASTERN ILL UNIV, 64- Res: Dislocation study in ionic crystals; pyroelectric effect in triglycene sulfate; electron spin resonance and nuclear magnetic resonance. Mailing Add: Dept of Physics Eastern Ill Univ Charleston IL 61920

BREIL, DAVID A, b Brockton, Mass, Mar 27, 38; m 63; c 2. BRYOLOGY, PLANT MORPHOLOGY. Educ: Univ Mass, Amherst, BS, 60, MA, 63; Fla State Univ, PhD(bot), 68. Prof Exp: Instr bot, Pa State Univ, 63-65; Tall Timbers Res Sta grant, Fla, 66-68; ASSOC PROF BIOL, LONGWOOD COL, 68- Mem: AAAS; Am Bryol & Lichenological Soc; Bot Soc Am; Am Inst Biol Sci; Am Soc Plant Taxon. Res: Liverwort flora of Florida; bryophytes of Virginia; ecology of bryophytes. Mailing Add: Dept of Natural Sci Longwood Col Farmville VA 23901

BREIL, SANDRA J, b Springfield, Mass, Apr 30, 37. CELL PHYSIOLOGY, PLANT HISTOCHEMISTRY. Educ: Univ Vt, BA, 58; Univ Mass, PhD(bot), 63. Prof Exp: Instr biol, Wilson Col, 63-65; fel, Inst Molecular Biophys, Fla State Univ, 66-68; ASSOC PROF BIOL, LONGWOOD COL, 69- Mailing Add: Dept of Biol Longwood Col Farmville VA 23901

BREILAND, JOHN GUSTAVSON, b Hjelmeland, Norway, Nov 21, 05; nat US; m 42; c 2. METEOROLOGY. Educ: Luther Col, Iowa, AB, 33; Univ Iowa, MS, 34; Univ Calif, Los Angeles, PhD(meteorol), 56. Prof Exp: Instr math, Luther Col, Iowa, 35-38 & Univ Pittsburgh, 38-39; observer meteorol, US Weather Bur, Iowa, 39-41; from instr to assoc prof physics & meteorol, 42-66, actg chmn dept physics, 57-58, 61-62, prof physics, 66-70, EMER PROF PHYSICS, UNIV N MEX, 70- Concurrent Pos: Asst dir, Air Force Meteorol Training Prog, 51-52, dir, 52-54. Mem: Am Meteorol Soc; Am Geophys Union. Res: Synoptic meteorology; thunderstorms; atmospheric ozone. Mailing Add: 908 Georgia S E Albuquerque NM 87108

BREINER, SHELDON, b Milwaukee, Wis, Oct 23, 36; m 62; c 2. GEOPHYSICS. Educ: Stanford Univ, BS, 59, MS, 62, PhD(geophys), 67. Prof Exp: Res geophysicist, Varian Assocs, 61-68; GEOPHYSICIST, GEOMETRICS, 69- Concurrent Pos: Res assoc, NSF grant, 65-67; res assoc, Dept Geophys, Stanford Univ, 67-, consult prof, Dept Appl Earth Sci, 74- Mem: AAAS; Soc Explor Geophys; Am Geophys Union; Europ Asn Explor Geophysicists; Sigma Xi. Res: Remote sensing, magnetic and gamma ray surveys for mineral and petroleum exploration; airborne and marine magnetic gradient studies; earthquake prediction; micropulsations; magnetic surveys for archaeological exploration. Mailing Add: Geometrics 395 Java Dr Sunnyvale CA 94086

BREININ, GOODWIN MILTON, b New York, NY, Dec 10, 18; m 47; c 2. OPHTHALMOLOGY. Educ: Univ Fla, BS, 39; Emory Univ, AM, 40, MD, 43; Am Bd Ophthal, dipl, 51. Prof Exp: Intern, US Marine Hosp, Stapleton, NY, 44; instr ophthal, 51-54, dir res, 54, from asst clin prof to assoc clin prof, 54-55, assoc prof, 56, KIRBY PROF OPHTHAL, SCH MED & POST-GRAD MED SCH, NY UNIV, 58-, CHMN DEPT, 59-, DIR EYE SERV, UNIV HOSP, 59-, UNIV HOSP & BELLEVUE HOSP, NEW YORK, 59- Concurrent Pos: Sr Heed fel, 54; chief consult, New York Vet Admin Hosp, 59-; consult ophthalmologist, NY Eye & Ear Infirmary, Manhattan Eye, Ear & Throat Hosp, Lenox Hill Hosp, Beth Israel Hosps, NY, St Vincent's Hosp, NY, St Clare's Hosp, US Marine Hosp, Stapleton & French Hosp, NY; consult, Surgeon Gen, USPHS; chmn, Vision Res Training Comt, Nat Inst Neurol Dis & Blindness; mem, Comt Vision, Armed Forces, Nat Res Coun; mem, Sci Adv Comt, Nat Coun Combat Blindness; mem, Am Orthop Coun; mem & chmn, Am Comt Optics & Visual Physiol; mem, Ophthal Adv Comt, Dept Health, NY; Gifford lectr, Univ Chicago, 70; Lloyd lectr, NY, 71 & May lectr, 74; dir & vchmn, Am Bd Ophthal, chmn, Residency Comn. Honors & Awards: Knapp Medal, AMA, 57; Holmes Award, 59; Heed Found Award, 68; Gifford Award, Univ Chicago, 70. Mem: AAAS; AMA; fel Am Acad Ophthal & Otolaryngol; Am Ophthal Soc; fel Am Col Surgeons. Res: Ophthalmic physiology, particularly the neurophysiology of ocular motility and clinical strabismus; neurophysiology of vision clinical strabismus and ocular disease; investigative ophthalmology. Mailing Add: NY Univ Sch of Med 550 First Ave New York NY 10016

BREISACHER, PETER, b Berlin, Ger, Oct 12, 30; nat US. PHYSICAL CHEMISTRY. Educ: State Univ NY, BA, 51; Cath Univ, MS, 53, PhD(chem), 57. Prof Exp: Chemist, Appl Physics Lab, Johns Hopkins Univ, 57-61; MEM TECH STAFF, AEROSPACE CORP, LOS ANGELES, 61- Mem: Am Chem Soc; Am Inst Aeronaut & Astronaut. Res: Epidemiology; health effects; air pollution; engine design; fuel modification; differential thermal analysis; chemical lasers. Mailing Add: 27731 Longhill Dr Palos Verdes Peninsula CA 90274

BREIT, GREGORY, b Nickolaev, Russia, July 14, 99; nat US; m 27. PHYSICS. Educ: Johns Hopkins Univ, AB, 18, AM, 20, PhD(physics), 21. Hon Degrees: DSc, Univ Wis, 54. Prof Exp: Nat Res fel, Univ Leiden, 21-22 & Harvard Univ, 22-23; asst prof physics, Univ Minn, 23-24; math physicist, Dept Terrestrial Magnetism, Carnegie Inst, 24-29; prof physics, NY Univ, 29-34 & Univ Wis, 34-47; prof, Yale Univ, 47-58, Donner prof, 58-68; distinguished prof, 68-74, EMER DISTINGUISHED PROF PHYSICS, STATE UNIV NY BUFFALO, 68- Concurrent Pos: Assoc ed, Phys Rev, 27-29, 39-41, 54-56 & 61-63; researcher, Swiss Fed Inst Technol, Zurich, 28; res assoc, Carnegie Inst Wash, 29-44; mem-at-large, Div Phys Sci, Nat Res Coun, 32-35, rep, Am Phys Soc, 38-41; vis mem, Inst Advan Study, 35-36; mem, Comt Uranium, Nat Defense Res Comt, 40-43, consult, Exec Comt, 42-44; mem staff, Naval Ord Lab, Washington, 40-41; info chief coordr fast neutron proj, Metrop Lab, Chicago, 42; group supvr, Johns Hopkins Univ, 43; head physicist, Ballistic Res Lab, Aberdeen Proving Ground, Md, 43-45; assoc ed, Il Nuovo Cimento, 64-67. Honors & Awards: Franklin Medal, 64; Nat Medal Sci, 67; Tom Bonner Prize, Am Phys Soc, 69. Mem: Nat Acad Sci; fel Inst Elec & Electronics Engrs; Am Acad Arts & Sci; Am Geophys Union; fel Am Phys Soc. Res: Quantum theory; nuclear physics; ionosphere studies; quantum electrodynamics; hyperfine structure. Mailing Add: 73 Allenhurst Rd Buffalo NY 14214

BREITBART, ANN, solid state physics, see 12th edition

BREITBEIL, FRED W, III, b Cincinnati, Ohio, Sept 25, 31; m 56; c 4. ORGANIC CHEMISTRY. Educ: Xavier Univ, Ohio, BS, 53, MS, 57; Univ Cincinnati, PhD(org chem), 60. Prof Exp: Res chemist, Procter & Gamble Co, 60-62; assoc, Iowa State Univ, 62-63; asst prof, 63-75, PROF ORG CHEM, DE PAUL UNIV, 75-, CHMN DEPT, 69- Mem: Am Chem Soc. Res: Enolate anions; synthesis and chemistry of cyclopropandiols; chemistry and physical properties of cyclic carbonates; nuclear magnetic resonance spectroscopy—force field shifts; halogen additions to heterocyclics. Mailing Add: Dept of Chem De Paul Univ Chicago IL 60614

BREITENBACH, ROBERT PETER, b Madison, Wis, Oct 10, 23; m 48; c 3. ZOOLOGY, ENDOCRINOLOGY. Educ: Univ Wis, BS, 49, MS, 50, PhD(zool, endocrinol), 58. Prof Exp: Instr zool, Univ Wis-Milwaukee, 56-59; from asst prof to assoc prof zool, 59-67, chmn dept, 64-68, PROF ZOOL, UNIV MO-COLUMBIA, 67-, ASSOC DIR DIV BIOL SCI, 75- Mem: AAAS; Am Physiol Soc; Soc Exp Biol & Med. Res: Avian physiology; development of immunological maturity, reproductive physiology, calcium metabolism, endocrines and development, and avian behavior. Mailing Add: Div of Biol Sci Univ of Mo Lefevre Hall Columbia MO 65201

BREITENBERGER, ERNST, b Graz, Austria, June 11, 24; c 4. PHYSICS. Educ: Univ Vienna, DrPhil(math), 50; Cambridge Univ, PhD(physics), 56. Prof Exp: Res assoc, Radiuminstitut, Univ Vienna, 50-51; lectr physics, Univ Malaya, 54-58; from assoc prof to prof, Univ SC, 58-63; PROF PHYSICS, OHIO UNIV, 63- Concurrent Pos: Guest prof, Univ Bonn, 69-70. Mem: Am Phys Soc; Brit Math Asn. Res: Theoretical physics; probability and statistics; mathematical methods. Mailing Add: Dept of Physics Ohio Univ Athens OH 45701

BREITER, JEROME JOHN, b Vandergrift, Pa, Sept 9, 36; m 58; c 3. ORGANIC CHEMISTRY. Educ: Pa State Univ, BS, 58; Purdue Univ, PhD(org chem), 62. Prof Exp: Res agr chemist, 62-68, supvr, Agr Chem Lab, 68-70, WASHINGTON REP, AGR CHEM LAB, HERCULES INC, 70- Mem: Am Chem Soc. Res: Organic synthesis and reaction mechanisms; agricultural chemicals. Mailing Add: Hercules Inc 1800 K St NW Washington DC 20006

BREITER, MANFRED WOLFGANG, b Langenbielau, Ger, Nov 5, 25; m 60; c 2. ELECTROCHEMISTRY, PHYSICAL CHEMISTRY. Educ: Munich Tech Univ, BS, 48, MS, 51, PhD(phys chem), 53. Prof Exp: Fel electrochem, Munich Tech Univ, 53-57; vis prof electrochem, La State Univ, 57-58; lectr phys chem, Munich Tech Univ, 58-60; RES ASSOC RES LAB, GEN ELEC CO, 61- Concurrent Pos: Adj prof chem, Rensselaer Polytech Inst; mem, Int Comt Electrochem Thermodyn & Kinetics, 59- Mem: Electrochem Soc; The Chem Soc. Res: Research and development in electrocatalysis and high energy density batteries. Mailing Add: Res & Develop Ctr Gen Elec Co Box 8 Schenectady NY 12301

BREITHAUPT, LEA JOSEPH, JR, b Natchez, Miss, Sept 16, 29; m 59; c 3. PAPER CHEMISTRY, CELLULOSE CHEMISTRY. Educ: St Louis Univ, BS, 51; La State Univ, MS, 56. Prof Exp: From res chemist to tech dir, Non-Wovens Div, 69-71, SR RES ASSOC, ERLING RIIS RES LAB, INT PAPER CO, 72- Mem: Am Chem Soc; Tech Asn Pulp & Paper Indust. Res: Pulp processing reactions, especially dissolving pulps, cellulose derivatives and reactions. Mailing Add: Int Paper Co PO Box 2328 Mobile AL 36601

BREITMAN, LEO, b Montreal, Que, Nov 18, 26; m 49; c 3. POLYMER CHEMISTRY, PHYSICAL CHEMISTRY. Educ: McGill Univ, BSc, 47, PhD(phys chem), 52. Prof Exp: Sr res chemist polymerization, Res & Develop Div, Polymer Corp, Ltd, 52-56, group leader emulsion polymers res, 56-59, proj supvr explor res, 59-61, proj supvr solution polymer res, 61-62, asst mgr, 62-67, mgr opers & develop planning, 67-69, prod mgr stereo, 69-71; bus planner, 71-72, TECH DEVELOP MGR SOLUTION RUBBERS, POLYSAR LTD, 72- Concurrent Pos: Secy, Can High Polymer Forum, 61-63, prog chmn, 63-64, chmn, 64-65; mem assoc comt high polymer res, Nat Res Coun Can, 63- Mem: Am Chem Soc; Chem Inst Can. Res: Physics and chemistry of high polymers; chemical kinetics. Mailing Add: Polysar Ltd Sarnia ON Can

BREITMAN, THEODORE RONALD, b Brooklyn, NY, Feb, 25, 31; m 58; c 2. BIOCHEMISTRY. Educ: City Col New York, BS, 53; Ohio State Univ, MS, 56, PhD(biochem), 58. Prof Exp: Nat Cancer Inst fel, USPHS, Brandeis Univ, 58-60; res biochemist, Lederle Labs, 60-62; RES BIOCHEMIST, NIH, 63- Mem: AAAS; Am Soc Microbiol; Am Soc Biol Chemists. Res: Nucleotide and nucleic acid metabolism with particular emphasis on control. Mailing Add: Lab of Physiol Nat Cancer Inst Bethesda MD 20014

BREIVIK, ORVILLE NEIL, b Vang, NDak, Jan 16, 15; m 42; c 1. FOOD TECHNOLOGY. Educ: St Olaf Col, AB, 35; Pa State Col, MS, 39, PhD(biochem), 41. Prof Exp: Chemist, Stand Brands, Inc, 40-42, res sect head, Fleischmann Labs, 44-70, SCI DIR, FLEISCHMANN LABS, STAND BRANDS, INC, 70- Mem: Am Oil Chem Soc; Am Chem Soc; Inst Food Technologists. Res: Fats and fat autoxidation; antioxidants; sterols; vitamin D; margarine, nut and candy technology; nutrition. Mailing Add: Fleischmann Labs Betts Ave Stamford CT 06904

BREIVOGEL, FRANCIS WILLIAM, JR, physical chemistry, see 12th edition

BRELAND, HERMAN LEROY, b Wiggins, Miss, Apr 21, 16; m 40; c 1. AGRONOMY. Educ: Miss State Col, BS, 42; Purdue Univ, MS, 48; Univ Ill, PhD(agron), 52. Prof Exp: Asst, Purdue Univ, 47-48; first asst soil chem, Univ Ill, 48-52; from asst soils chemist to assoc soils chemist, 52-74, ASSOC PROF SOIL SCI & SOIL CHEMIST, INST FOOD & AGR SCI, UNIV FLA, 74- Mem: Soil Sci Soc Am; Am Soc Agron. Res: Soil fertility and chemistry. Mailing Add: Dept Soil Sci 631 Wallace Bldg Inst Food & Agr Sci Univ of Fla Gainesville FL 32601

BRELAND, OSMOND PHILIP, b Decatur, Miss, Sept 17, 10; m 31; c 2. ZOOLOGY, ENTOMOLOGY. Educ: Miss State Univ, BS, 31; Ind Univ, PhD(zool), 36. Prof Exp: Asst zool, Ind Univ, 31-36; instr, NDak Col, 36-38; from asst prof to assoc prof, 38-50, PROF ZOOL, UNIV TEX, AUSTIN, 50- Concurrent Pos: Mem, Ind Univ Exped,

Mex & Guatemala, 35-36. Mem: Entom Soc Am; Am Mosquito Control Asn. Res: Cytology, biology and taxonomy of insects. Mailing Add: Dept of Zool Univ of Tex Austin TX 78712

BREM, THOMAS HAMILTON, b CZ, Oct 3, 10; m 38; c 3. MEDICINE. Prof Exp: Intern, Johns Hopkins Hosp, 37-38; asst path, Stanford Univ, 38-39; instr med, Univ Southern Calif, 39-48; from asst clin prof to clin prof, Univ Calif, Los Angeles, 49-54; prof, 54-72, head dept, 58-72, EMER PROF MED, UNIV SOUTHERN CALIF, 72- Concurrent Pos: Resident, Los Angeles County Hosp, 39-41, physician in chief, 59-; consult, Vet Admin, 54-, mem, Spec Med Adv Group, 60-, from vchmn to chmn, 63-65; mem, Am Bd Internal Med, 54-, chmn, 63-65. Mem: AMA; Am Col Physicians; Asn Am Physicians. Res: Cardiovascular and liver diseases. Mailing Add: Sch of Med Univ of Southern Calif Los Angeles CA 90007

BREMEL, ROBERT DUANE, b Eau Claire, Wis, Dec 19, 45; m 66; c 2. ANIMAL PHYSIOLOGY, BIOCHEMISTRY. Educ: Univ Wis-Madison, BS, 67; St Louis Univ, PhD(biochem), 72. Prof Exp: Fel anat-biophys, Med Ctr, Duke Univ, 72-74; ASST PROF DAIRY SCI, UNIV WIS-MADISON, 74- Mem: Am Dairy Sci Asn; Am Chem Soc; Sigma Xi. Res: Mechanistic studies of hormone action and the measurement and control of stress in food producing animals. Mailing Add: Dept of Dairy Sci Univ of Wis Col Agr & Life Sci Madison WI 53706

BREMER, HANS, b Hamburg, Ger, Aug 26, 27; m 57; c 4. MOLECULAR BIOLOGY. Educ: Univ Heidelberg, Vordiplom, 49; Univ Göttingen, PhD(biol), 57. Prof Exp: Res fel zool, Univ Göttingen, 57-59; asst genetics, Univ Cologne, 59-62; res microbiologist, Univ Calif, Berkeley, 62-65; asst genetics, Univ Cologne, 65-66; from asst prof to assoc prof biol, 66-73, PROF BIOL, SOUTHWEST CTR ADVAN STUDIES, UNIV TEX, DALLAS, 73- Concurrent Pos: Ger Res Soc fel, 58-59; Fulbright travel grant, 62. Res: Developmental physiology; synthesis of ribonucleic acid. Mailing Add: Univ of Tex Div of Biol PO Box 30365 Dallas TX 75230

BREMER, KEITH GEORGE, b Lorton, Nebr, Dec 20, 27; m 57; c 1. ORGANIC CHEMISTRY. Educ: Univ Nebr, BS, 48, MS, 50; Univ Iowa, PhD(chem), 52. Prof Exp: Res chemist, Ethyl Corp, 52-54; res chemist, 56-63, col rels rep, 63-66, coordr PhD recruitment, 66-70, COL RELS SUPVR, E I DU PONT DE NEMOURS & CO, INC, 70- Mem: Am Chem Soc. Res: Agricultural chemicals; fluorocarbon chemicals; chemistry of organo phosphorus and boron compounds; organic synthesis. Mailing Add: Employee Rels Dept E I du Pont de Nemours & Co Inc Wilmington DE 19898

BREMER, ROBERT F, b Bay City, Mich, Apr 5, 21; m 48; c 2. PHYSICAL CHEMISTRY. Educ: Mich State Col, BS, 48; Iowa State Col, PhD(chem), 51. Prof Exp: Chemist, Dow Chem Co, 51-58; chief res chemist, Leonard Refining Inc, 58-71, tech serv dir, 71-74, MGR TECH SERV, TOTAL LEONARD, INC, 74- Res: Chemical kinetics; plastic film and sheeting; petroleum refining and petrochemicals. Mailing Add: Total Leonard Inc E Superior St Alma MI 48801

BREMERMANN, HANS J, b Bremen, Ger, Sept 14, 26; nat US; m 54. MATHEMATICAL BIOLOGY, BIOPHYSICS. Educ: Univ Münster, MA & PhD(math), 51. Prof Exp: Instr math, Univ Münster, 52; res assoc, Stanford Univ, 52-53, vis asst prof, 53-54; asst prof, Univ Münster, 54-55; staff mem, Inst Advan Study, 55-57; asst prof, Univ Wash, 57-58; staff mem, Inst Advan Study, 58-59; assoc prof math, 59-64, assoc prof math & biophys, 64-66, PROF MATH & MED PHYSICS, UNIV CALIF, BERKELEY, 66- Concurrent Pos: Res fel, Harvard Univ, 53; indust consult; mem exec comt grad group biophys & med physics, Univ Calif, Berkeley, 64- Mem: Am Math Soc; Austrian Math Soc; Ger Soc Appl Math & Mech; Ger Math Asn. Res: Several complex variables; Schwartz distributions; physics; dispersion relations; renormalization; information theory; limitations of genetic control; evolution processes; self-organizing systems; biological algorithm; complexity theory; mathematical ethology; pattern recognition; model verification; optimization; control; medical applications of nonlinear control. Mailing Add: Dept of Math Univ of Calif Berkeley CA 94720

BREMMER, BART J, b Waddinxveen, Netherlands, Sept 4, 30; US citizen; m 53; c 3. ORGANIC CHEMISTRY, POLYMER CHEMISTRY. Educ: State Univ Leiden, BS, 50; Mich State Univ, MS, 62. Prof Exp: Anal chemist, Gouda Apollo, Netherlands, 50-51; res chemist, Grand Rapids Varnish Co, 54-55; chemist, Kelvinator Div, Am Motors Corp, 55-57; org res chemist, 57-64, proj leader org chem, 64-66, sect head, 66-69, develop mgr, 69-72, RES MGR ORG CHEM, DOW CHEM CO, 72- Mem: Am Chem Soc; Com Develop Asn; Soc Automotive Eng. Res: Thermoset resins, especially epoxy and phenolic resins with emphasis on fire-retardant resins; episulfide chemistry; chemistry of halogenated aromatic compounds, especially those containing amine groups; pyridine chemistry. Mailing Add: 214 W Meadowbrook Dr Midland MI 48640

BREMNER, JOHN MCCOLL, b Dumbarton, Scotland, Jan 18, 22; nat US; m 50; c 2. CHEMISTRY. Educ: Glasgow Univ, BSc, 44; Univ London, PhD(chem), 48, DSc, 59. Prof Exp: From sci officer to prin sci officer, Chem Dept, Rothamsted Exp Sta, Eng, 45-59; assoc prof soil biochem, 59-61, PROF AGRON & BIOCHEM, IOWA STATE UNIV, 61- Concurrent Pos: Rockefeller Found fel, 57-58; tech expert, Int Atomic Energy Agency, Yugoslavia, 64-65; Guggenheim Found fel, 68-69. Honors & Awards: Soil Sci Archievement Award, Am Soc Agron, 67. Mem: AAAS; Soil Sci Soc Am; Geol Soc Am; Brit Biochem Soc; Brit Soc Soil Sci. Res: Soil biochemistry and microbiology. Mailing Add: Dept of Agron Iowa State Univ Ames IA 50010

BREMNER, RAYMOND WILSON, b Bellingham, Wash, Dec 4, 04; m 35; c 3. ANALYTICAL CHEMISTRY. Educ: Univ Wash, BS, 28, MS, 32, PhD(chem), 37. Prof Exp: From instr to asst prof, Tex A&M Univ, 37-44; lab proj leader, Dow Chem Co, 44-47; from assoc prof to prof, 47-73, EMER PROF CHEM, CALIF STATE UNIV, FRESNO, 73- Concurrent Pos: Anal res chemist, Stanford Res Inst, 48; Fulbright prof & sr lectr, Univ Peshawar, 59-60; researcher, Pakistani Air Force, 60; mem policy bd, Moss Landing Marine Labs, Calif State Univs & Cols, 66-68; partic, Conf with Europ Prof Training & Placement of Anal Chemists, 71; Calif State Univ Fresno fac res fund grant, 72. Honors & Awards: Army-Navy Prod Award, 47; Cottrell Awards, Res Corp, 50-51; Smith-Mundt Award, US Dept State, 59; NSF Awards, 62, 64 & 66. Mem: Am Chem Soc. Res: Specific ion electrodes; spectrophotometry; water analysis; pollution; smog; dental caries; radioactive tracers; chemistry of the sea; instrumental and chemical analysis. Mailing Add: Dept of Chem Calif State Univ Fresno CA 93740

BRENCHLEY, JEAN ELNORA, b Towanda, Pa, Mar 6, 44. MOLECULAR BIOLOGY. Educ: Mansfield State Col, BS, 65; Univ Calif, San Diego, MS, 67, Univ Calif, Davis, PhD(microbiol), 70. Prof Exp: Res assoc biol, Mass Inst Technol, 70-71; asst prof, 71-75, ASSOC PROF MICROBIOL, PA STATE UNIV, 76- Concurrent Pos: Found lectr, Am Soc Microbiol, 75-76. Mem: Am Soc Microbiol; Am Soc Biol Chemists; AAAS; Am Inst Biol Sci; Genetics Soc Am. Res: Biochemical regulatory mechanisms; microbial genetics; regulation of ammonia assimilation and the synthesis and function of transfer RNA modified bases. Mailing Add: Dept of Microbiol S101 Frear Pa State Univ University Park PA 16802

BRENDEL, KLAUS, b Berlin, Ger, July 14, 33; m 57; c 3. PHARMACOLOGY. Educ: Free Univ Berlin, cand chem, 55, dipl chem, 59, Dr rer nat(chem), 62. Prof Exp: From asst prof to assoc prof pharmacol, Duke Univ, 67-70; ASSOC PROF PHARMACOL & TOXICOL, MED SCH, UNIV ARIZ, 70- Concurrent Pos: Fel Ger Chem Indust, Free Univ Berlin, 62-63; NATO fel, Univ of Pac, 63-64 & Duke Univ, 65-67; Am Heart Asn estab investr, Duke Univ & Univ Ariz, 65-70, fel, Coun Arteriosclerosis. Mem: Ger Chem Soc; Am Chem Soc; Am Soc Pharmacol & Exp Therapeut. Res: Bioorganic chemistry; drug metabolism; drugs and intermediary metabolism; toxicology; analytical organic and clinical chemistry. Mailing Add: Dept of Pharmacol Univ of Ariz Sch of Med Tucson AZ 85724

BRENDEMUEHL, RAYMOND HUGO, soils, silviculture, see 12th edition

BRENDER, ERNST VICTOR, b Wurttemberg, Ger, Feb 20, 05; nat US; m 34; c 2. FOREST MANAGEMENT. Educ: Univ Mich, BSF & MF, 37. Prof Exp: Forest supt, Toccoa Exp Forest, Appalachian Forest Exp Sta, US Forest Serv, 33-39, proj conservationist, Land Utilization Proj, Soil Conserv Serv, 40-46, proj leader, Southeastern Forest Exp Sta, Ga, 46-76; RETIRED. Concurrent Pos: Volunteer worker, Southeastern Forest Exp Sta, 76- Honors & Awards: Tech Award, Soc Am Foresters, 69, Ga Foresters Hall of Fame, 71. Mem: Sr mem Soc Am Foresters. Res: Silviculture of Piedmont Loblolly Pine. Mailing Add: 5070 Kathryn Dr Macon GA 31204

BRENDER, RONALD FRANKLIN, b Wyandotte, Mich, Sept 22, 43. INFORMATION SCIENCE. Educ: Univ Mich, BSE, 65, MS, 68, PhD(comput sci), 69. Prof Exp: SUPVR COMPUT LANG DEVELOP, SOFTWARE ENG DEPT, DIGITAL EQUIP CORP, 70- Concurrent Pos: Mem, Am Nat Stand Comt X3J3, Fortran Lang, 73- Mem: Asn Comput Mach. Res: Computer language design and development; implementation languages for system software; software development methodology. Mailing Add: Software Eng Dept Digital Equip Corp 146 Main St Maynard MA 01754

BRENDLEY, WILLIAM H, JR, b Philadelphia, Pa, Mar 26, 38; m 67. PHYSICAL CHEMISTRY, POLYMER CHEMISTRY. Educ: St Joseph's Col, Pa, BS, 60, MS, 62; Univ Pa, PhD(phys chem), 65. Prof Exp: Chemist, Rohm and Haas Co, 60-62; NSF fel, Univ Pa, 65-66; res chemist, 66-67, group leader thermoplastic polymer coatings, 67-72, lab head indust coatings, 72-74, DEPT MGR COATINGS TECH SERV, ROHM AND HAAS CO, 74- Mem: Am Chem Soc; Fedn Paint Socs; Am Inst Chemists. Res: Radiotracer and vacuum line techniques; volume expansion of alkali metals in liquid ammonia; physical chemistry of polymeric coatings and industrial application; polymeric coatings. Mailing Add: Rohm and Haas Co Springhouse PA 19477

BRENEMAN, EDWIN JAY, b Orrville, Ohio, Sept 29, 27; m 49; c 3. PSYCHOPHYSICS. Educ: Ohio Wesleyan Univ, BA, 50; Ohio State Univ, MSc, 52. Prof Exp: Res physicist visual perception & photog reproduction, Eastman Kodak Co Res Labs, 52-68; res assoc, Macbeth Corp, Kollmorgen Corp, 68-73; SR RES PHYSICIST VISUAL PERCEPTION & PHOTOG REPRODUCTION, EASTMAN KODAK CO RES LABS, 73- Mem: Optical Soc Am; Int Soc Color Coun; Soc Photog Scientists & Engrs. Res: Visual perception and photographic color reproduction. Mailing Add: Eastman Kodak Co Res Labs 1669 Lake Ave Rochester NY 14650

BRENEMAN, WILLIAM RAYMOND, b Indianapolis, Ind, June 3, 07; m 30; c 3. ZOOLOGY. Educ: Ind Cent Col, AB, 30; Ind Univ, PhD(zool), 34. Prof Exp: Asst zool, Ind Univ, 30-34; Nat Res Coun fel, Univ Wis, 34-35; instr, Miami Univ, 35-36; from asst prof to prof, 36-62, chmn dept, 66-69, LUTHER DANA WATERMAN PROF ZOOL, WATERMAN INST, IND UNIV, BLOOMINGTON, 62- Mem: AAAS; Am Soc Zool; NY Acad Sci; Poultry Sci Asn. Res: Endocrinology. Mailing Add: Dept of Zool Ind Univ Bloomington IN 47405

BRENGLE, KENNETH GORDON, b Perry, Okla, May 18, 26; m 44; c 3. SOIL SCIENCE, SOIL MANAGEMENT. Educ: Okla State Univ, BS, 53, MS, 54; Mich State Univ, PhD(soil sci), 65. Prof Exp: Asst prof, 55-65, scientist in-chg, San Juan Basin Res Ctr-Cortez Unit, 71-74, ASSOC PROF SOIL SCI, COLO STATE UNIV, 65- Mem: Am Soc Agron; Soil Conserv Soc Am; Soil Sci Soc Am. Res: Soil-water-plant relationships; physical properties of soils as affected by soil management; irrigation management. Mailing Add: Dept of Agron Colo State Univ Ft Collins CO 80521

BRENMAN, MANUEL, solid state physics, solid state chemistry, see 12th edition

BRENNAN, DANIEL JOSEPH, b South Bend, Ind, Sept 11, 29; m 55; c 7. STRATIGRAPHY, SEDIMENTOLOGY. Educ: Univ Notre Dame, BS, 51; SDak Sch Mines & Technol, MS, 53; Univ Ariz, PhD(geol), 57. Prof Exp: Geologist, Shell Oil Co, 57-60; from geologist to sr geologist, Sunray DX Oil Co, 60-64; from asst prof to assoc prof geol, Wichita State Univ, 64-68; assoc prof, 68-71, PROF GEOL, STATE UNIV NY COL CORTLAND, 71-, CHMN DEPT, 70- Concurrent Pos: Consult geologist, 68-; prin investr, NY State Sea Grant Prog, 71-74. Mem: Fel Geol Soc Am; Am Asn Petrol Geol; Soc Econ Paleont & Mineral; AAAS. Res: Petroleum and mineral exploration; stratigraphy; sedimentation and sediment-binding mechanism; higher education in marine science. Mailing Add: Dept of Geol State Univ of NY Col Cortland NY 13045

BRENNAN, DAVID MICHAEL, b Springfield, Mass, Jan 12, 29; m 58; c 2. ENDOCRINE PHYSIOLOGY. Educ: Tufts Univ, BS, 50; Purdue Univ, MS, 55, PhD(endocrinol), 57. Prof Exp: Asst prof physiol, Univ Notre Dame, 57-58; sr res endocrinologist, Squibb Inst, 58-60; sr endocrinologist, 60-66, head physiol res, 66-67, asst to vpres res & develop, 67-68, dir biochem & physiol res, 68-72, DIR BIOL, BIOCHEM & PHYSIOL RES, LILLY RES LABS, 72- Mem: AAAS; Endocrine Soc; NY Acad Sci. Res: Reproductive physiology; endocrine and central nervous system interrelationships; mechanisms of steroid antagonistic action; biochemistry of lipids. Mailing Add: Lilly Res Labs Indianapolis IN 46206

BRENNAN, DONALD GEORGE, b Waterbury, Conn, Apr 9, 26; m 69. MATHEMATICS, NATIONAL SECURITY. Educ: Mass Inst Technol, BS, 55, PhD(math), 59. Prof Exp: Mem staff, Crystal Res Labs, 49-51; mem staff & consult, Lincoln Lab, Mass Inst Technol, 53-62; pres, 62-64, DIR NAT SECURITY STUDIES, HUDSON INST, 64- Concurrent Pos: Instr, Ward Sch, Univ Hartford, 49-51; consult var govt agencies. Mem: Am Math Soc; Inst Elec & Electronics Engrs; Inst Strategic Studies; Coun Foreign Rels. Res: Radio engineering; probability theory; classical and abstract harmonic analysis; functional analysis; fluctuation phenomena; statistical theory of communication; military policy; implications of technology for military policy and foreign affairs; arms control; national security policy. Mailing Add: Hudson Inst Croton-on-Hudson NY 10520

BRENNAN, GERALD L, b Cape Girardeau, Mo, May 25, 33; m 54; c 2. INORGANIC CHEMISTRY. Educ: Southeastern Mo State Col, BS, 54; Iowa State

Univ, MS, 57, PhD(inorg chem), 60. Prof Exp: Res chemist, E I du Pont de Nemours & Co, 60-67, sr res chemist, Eastern Lab, 67-68, sr asst, Acct & Bus Anal Div, Explosives Dept, 68-70, SUPVR, ACCT & BUS ANAL DIV, EXPLOSIVES DEPT, E I DU PONT DE NEMOURS & CO, 70- Mem: Am Chem Soc. Res: Boron and fluorine chemistry. Mailing Add: E I du Pont de Nemours & Co Explosives Dept 1007 Market St Wilmington DE 19898

BRENNAN, HARRY MICHAEL, b Rifle, Colo, Oct 3, 23; m 52; c 2. PHYSICAL CHEMISTRY. Educ: Western State Col Colo, AB, 48; Wash Univ, PhD(chem), 52. Prof Exp: CHEMIST, STANDARD OIL CO (IND), 52- Mem: Am Chem Soc; Water Pollution Control Asn. Res: Catalysis; hydrogenation; petroleum technology; waste treating. Mailing Add: 1561 Swallow Naperville IL 60540

BRENNAN, J NORTON, physics, engineering mechanics, see 12th edition

BRENNAN, JAMES A, b Freeland, Pa, Nov 4, 20; m 43; c 1. ORGANIC CHEMISTRY, PETROLEUM CHEMISTRY. Educ: Temple Univ, BA, 50, MA, 52, PhD(org chem), 59. Prof Exp: Jr technologist, Mobil Oil Corp, 52-53, technologist, 53-56, sr res chemist, 56-68, res assoc, Appl Res & Develop Labs, Mobil Res & Develop Corp, 68-75, SR RES ASSOC, APPL RES & DEVELOP LABS, MOBIL RES & DEVELOP CORP, 75- Mem: AAAS; Am Chem Soc; Am Inst Chemists. Res: Synthetic lubricants; combustion; nitrogen heterocyclics; high temperature oxidation; organic synthesis; catalysis. Mailing Add: Mobil Res & Develop Corp Appl Res & Develop Labs Paulsboro NJ 08066

BRENNAN, JAMES GERARD, b Hazleton, Pa, Aug 30, 27; m 51; c 5. THEORETICAL PHYSICS. Educ: Univ Scranton, BS, 48; Univ Wis, MS, 50, PhD(physics), 52. Prof Exp: Asst, Univ Wis, 48-52; from instr to assoc prof, 52-61, chmn dept, 61-74, PROF PHYSICS, CATH UNIV AM, 61-, ASSOC DEAN GRAD PROG, ARTS & SCI, 74- Concurrent Pos: Consult, Naval Ord Lab, 55-; sci liaison, Off Naval Res, London, Eng, 65-66. Res: Theoretical nuclear physics; theoretical interior ballistics; atomic physics; ion implantation studies; electronic stopping power calculations. Mailing Add: Arts & Sci Cath Univ of Am Washington DC 20017

BRENNAN, JAMES MARKS, entomology, see 12th edition

BRENNAN, JAMES ROBERT, b Crawfordsville, Ind, Nov 14, 30; m 53; c 6. PLANT ANATOMY, CYTOLOGY. Educ: Va Polytech Inst, BS, 52, MS, 55; Univ Md, PhD(bot), 58. Prof Exp: From instr to asst prof biol, Norwich Univ, 58-61; from instr to assoc prof, 61-69, PROF BOT, BRIDGEWATER STATE COL, 69- Concurrent Pos: Dir, NSF Coop Col-Sch Proj, Bridgewater State Col, 66. Mem: AAAS; Bot Soc Am; Am Soc Plant Physiol. Res: Plant anatomical studies in relation to taxonomy and phylogeny; ultrastructural effects of chemical agents on meristematic plant cells. Mailing Add: 37 Maple Ave Bridgewater MA 02324

BRENNAN, JAMES THOMAS, b St Louis, Mo, Jan 12, 16; m 46; c 2. RADIOTHERAPY, RADIOBIOLOGY. Educ: Univ Ill, BA, 39; Univ Minn, MD, 43. Prof Exp: Investr radiobiol, Los Alamos Sci Lab, Univ Calif, 48-52; dir dept biophys, Walter Reed Army Inst Res, 52-54, resident radiol, Walter Reed Gen Hosp, 54-57, chief radiation ther, 60-61; consult to surgeon, US Army Europe, 57-60, dir, Armed Forces Radiobiol Res Inst, Md, 61-66; vis lectr radiobiol, 66-67, WILSON PROF RES RADIOL, UNIV PA, 67- Concurrent Pos: USPHS planning grant, 66-; mem, Main Comt, Nat Coun Radiation Protection, 64- Mem: AMA; Health Physics Soc; Radiol Soc NAm; Am Soc Therapeut Radiol; Radiation Res Soc. Res: Neutron radiobiology; experimental radiotherapy. Mailing Add: Dept of Radiol Univ of Pa Philadelphia PA 19104

BRENNAN, JOHN JOSEPH, b Boston, Mass, Sept 26, 38; m 61; c 3. PHYSICS. Educ: Boston Col, BS, 60; Worcester Polytech Inst, MS, 62; Ga Inst Technol, PhD(physics), 68. Prof Exp: Teaching asst physics, Ga Inst Technol, 62-66, instr, 66-68; asst prof, 68-71, ASSOC PROF PHYSICS, FLA TECHNOL UNIV, 71- Mem: Am Asn Physics Teachers. Res: Theoretical nuclear structure; Nilsson model; beta decay; nuclear magnetic resonance; molecular rotations in solids. Mailing Add: Fla Technol Univ Dept Physics PO Box 25000 Orlando FL 32816

BRENNAN, LAWRENCE EDWARD, b Oak Park, Ill, Jan 29, 27; m 47; c 4. ELECTRONICS. Educ: Univ Ill, BA, 48, MA, 49, PhD(elec eng), 51. Prof Exp: Res assoc electronic dynamics, Univ Ill, 49-51; specialist systs anal, NAm Aviation, Inc, 51-52; engr, Chicago Midway Labs, 52-55, Lockheed Missile Systs Div, 55, Aeronaut Systs Inc, 56, Off Naval Res, 57 & Rand Corp, 58-67; SR SCIENTIST, TECHNOL SERV CORP, 67- Mem: Inst Elec & Electronics Eng. Res: Radar systems; operations research. Mailing Add: Technol Serv Corp 2811 Wilshire Blvd Santa Monica CA 90403

BRENNAN, MICHAEL EDWARD, b Covington, Ky, July 8, 41; m 65; c 2. ORGANIC CHEMISTRY. Educ: Univ Dayton, BS, 62; Univ Fla, MS, 65, PhD(org chem), 67. Prof Exp: Sr res chemist, Monsanto Co, 67-69; res assoc chem, Univ Ga, 69-70; SR RES CHEMIST, JEFFERSON CHEM CO, INC, 70- Mem: Am Chem Soc. Res: Organic synthesis; homogeneous catalysis; reaction mechanisms; bridged polycyclic compounds; small-ring chemistry; photochemistry; nuclear magnetic resonance spectra. Mailing Add: Jefferson Chem Co Inc PO Box 4128 Austin TX 78765

BRENNAN, MICHAEL JAMES, b Mt Clemens, Mich, June 11, 21; m 45; c 8. MEDICINE. Educ: Univ Detroit, BS, 41; Stritch Sch Med, MD, 47. Prof Exp: Assoc physician, Hemat Dept, Henry Ford Hosp, Detroit, 50-52; chief med & lab sect, US Army Hosp, Ft Monmouth, NJ, 53-54; physician-in-chg oncol div, Henry Ford Hosp, 54-65, chief div, 65-66; PRES & DIR MED SCI, MICH CANCER FOUND, 66-; PROF MED, COL MED, WAYNE STATE UNIV, 66- Mem: AMA; Am Col Physicians; Am Asn Cancer Res; Am Soc Hematologists; Am Fedn Clin Res; Am Soc Clin Oncol (vpres, 64-65, pres, 65-66). Res: Clinical hematology and oncology; host tumor relationships; experimental and clinical chemotherapy. Mailing Add: Mich Cancer Found 110 E Warren Ave Detroit MI 48201

BRENNAN, PATRICIA CONLON, b Chicago, Ill, Nov 20, 32; m 58; c 3. MICROBIOLOGY, IMMUNOLOGY. Educ: Albion Col, AB, 54; Univ Wis, MS, 57; Loyola Univ, Ill, PhD(microbiol), 68. Prof Exp: Technician bact, G D Searle & Co, 54-55; bacteriologist, Animal Dis Diag Lab, Wis Dept Agr, 57-58; microbiologist, Univ Chicago Hosps & Clins, 60-63; sci asst microbiol, 63-68, asst microbiologist, 68-72, MICROBIOLOGIST, DIV BIOL & MED RES, ARGONNE NAT LAB, 72- Mem: AAAS; Am Soc Microbiol; NY Acad Sci; Radiation Res Soc; Am Asn Immunologists. Res: Mechanisms of pathogenicity; effect of low level gamma and neutron irradiation on cell-mediated immunity. Mailing Add: Argonne Nat Lab D202 BE 113 Argonne IL 60439

BRENNAN, ROBERT OWINGS, b Whiteboro, NY, Aug 17, 19. THEORETICAL PHYSICS. Educ: Boston Col, AB, 45, MA, 46; Cath Univ Am, PhD, 51; Woodstock Col, STL, 54. Prof Exp: Instr physics, Canisius Col, 46-47; instr, 55-58, chmn dept, 58-70, ASSOC PROF PHYSICS, LE MOYNE COL, 58- Mem: AAAS; Am Phys Soc; Am Asn Physics Teachers. Res: Molecular physics. Mailing Add: Dept of Physics Le Moyne Col Syracuse NY 13214

BRENNAN, THOMAS M, organic chemistry, see 12th edition

BRENNEMAN, FAITH NIELSEN, b New York, NY, Aug 24, 31; m 56; c 2. MICROBIOLOGY. Educ: Middlebury Col, AB, 53; Yale Univ, MS, 56, PhD(microbiol), 58. Prof Exp: Nat Found fel microbiol, Med Sch, Univ Va, 58-59; NIH fel biochem, 59-61, Nat Inst Arthritis & Metab Dis fel, 61-63, res microbiologist, 63-64, res assoc microbiol, 66-68, RES STAFF BIOLOGIST, YALE UNIV, 70- Res: Carbohydrate metabolism; nucleic acid synthesis and metabolism; plant development. Mailing Add: PO Box 5517 Hamden CT 06518

BRENNEMAN, JAMES ALDEN, b Elida, Ohio, Aug 26, 43; m 67; c 2. MYCOLOGY, PLANT PHYSIOLOGY. Educ: Goshen Col, BA, 65; WVa Univ, MS, 67; La State Univ, Baton Rouge, PhD(plant path), 70. Prof Exp: ASST PROF BIOL, UNIV EVANSVILLE, 70- Mem: AAAS; Am Inst Biol Sci; Inst Soc, Ethics & Life Sci; Nat Asn Biol Teachers ; Sigma Xi. Res: Use of fungi for converting agricultural waste products to human or animal food products. Mailing Add: Dept of Biol Univ of Evansville Evansville IN 47702

BRENNER, ABNER, b Kansas City, Mo, Aug 5, 08; m 36; c 4. ELECTROCHEMISTRY. Educ: Univ Mo, AB, 29; Univ Wis, MS, 30; Univ Md, PhD(chem), 39. Prof Exp: Asst to Dr Kahlenburg, Univ Wis, 29-30; jr chemist, Nat Bur Standards, 30-35, assoc chemist, 35-50, chief electrodeposition, 50-71; PRES, N-Q ELECTROCHEM RES CORP, 71- Honors & Awards: Gold Medal, US Dept Com, 63; Proctor Award, Am Electroplaters Soc, 49, Sci Achievement Award, 62; Hothersall Award, Brit Inst Metal Finishing, 61; Electrochem Eng & Technol Award, Electrochem Soc, 74. Mem: Am Electroplaters Soc; Am Chem Soc; Electrochem Soc; Brit Inst Metal Finishing. Res: Electrodeposition of metals and alloys; measurement of physical properties and of thickness and metal coatings; electrodeposition of metals from organic and fused salt baths; vapor deposition of tungsten; electroless plating; electrochemical calorimetry and galvanostalametry. Mailing Add: 7204 Pomander Lane Chevy Chase MD 20015

BRENNER, ALFRED EPHRAIM, b Brooklyn, NY, Sept 11, 31; m 58; c 3. PHYSICS. Educ: Mass Inst Technol, SB, 53, PhD(physics), 58. Prof Exp: Ford Found fel physics, Europ Orgn Nuclear Res, 58-59; instr & res fel, Harvard Univ, 59-62, asst prof, 62-66, sr res assoc, 66-70; PHYSICIST & DIR COMPUT, FERMI NAT ACCELERATOR LAB, 70- Mem: AAAS; Am Phys Soc; Asn Comput Mach. Res: Experimental high energy physics; computers and data processing. Mailing Add: Fermi Nat Accelerator Lab PO Box 500 Batavia IL 60510

BRENNER, BARRY MORTON, b Brooklyn, NY, Oct 4, 37; m 60; c 2. INTERNAL MEDICINE. Educ: Long Island Univ, BS, 58; Univ Pittsburgh, MD, 62. Prof Exp: From asst prof med to prof med & physiol, Univ Calif, San Francisco, 72-76; SAMUEL A LEVINE PROF MED, HARVARD MED SCH & DIR LAB KIDNEY & ELECTROLYTE PHYSIOL, PETER BENT BRIGHAM HOSP, BOSTON, 76- Concurrent Pos: Sr staff mem, Cardiovasc Res Inst, Univ Calif, San Francisco, 74-76; physician, Peter Bent Brigham Hosp, Boston, 76- Mem: Am Soc Clin Invest; Am Physiol Soc; Am Soc Nephrol; Asn Am Physicians. Res: Renal regulation of ion, water and macromolecule transport; regulation of renal glomerular filtration; pathophysiology of renal disorders. Mailing Add: Lab Kidney & Electrolyte Physiol Peter Bent Brigham Hosp Boston MA 02115

BRENNER, CHARLES, b Boston, Mass, Nov 18, 13; m 35; c 2. PSYCHOANALYSIS. Educ: Harvard Univ, AB, 31, MD, 35. Prof Exp: Asst neurol, Sch Med, Harvard Univ, 39-44; assoc, Col Physicians & Surgeons, Columbia Univ, 45-50; assoc clin prof, 50-67, LECTR PSYCHIAT, MED SCH, YALE UNIV, 67- Concurrent Pos: Freud fel, Boston Psychanal Soc & Inst, 39-44; assoc, Off Sci Res & Develop, 42-45; instr psychoanal, NY Psychoanal Inst, 55- Mem: Am Psychoanal Asn (pres, 67-68); Am Psychiat Asn; AMA; Int Psychoanal Asn. Res: Psychoanalytic research teaching and practice. Mailing Add: 1040 Park Ave New York NY 10028

BRENNER, DAEG SCOTT, b Reading, Pa, Aug 9, 39; m 64; c 2. NUCLEAR CHEMISTRY. Educ: Rensselaer Polytech, BS, 60; Mass Inst Technol, PhD(chem), 65. Prof Exp: Ford Fund fel nuclear chem, Neils Bohr Inst, Copenhagen, Denmark, 64, NATO fel, 65; res assoc, Brookhaven Nat Lab, 65-67; asst prof, 67-69, ASSOC PROF CHEM, CLARK UNIV, 69- Mem: Am Chem Soc; Am Phys Soc. Res: Nuclear fission; nuclear spectroscopy; reactions. Mailing Add: Dept of Chem Clark Univ Worcester MA 01610

BRENNER, DOUGLAS, b Washington, DC, Dec 31, 38; m 68; c 2. POLYMER SCIENCE. Educ: Johns Hopkins Univ, BA, 61; Harvard Univ, MA, 63, PhD(physics), 68. Prof Exp: Res fel atomic physics, Harvard Univ, 68-69, instr physics, 69; res physicist, 69-73, SR RES PHYSICIST, CORP RES LAB, EXXON RES & ENG CO, 73- Mem: AAAS; Am Phys Soc; Inst Elec & Electronics Engrs; Am Chem Soc; Am Vacuum Soc. Res: Physical aspects of elastomer and plastic multiphase systems and blends; polymers containing appended ionic groups; filled and reinforced polymers; photoelectron spectroscopy of polymer surfaces; radio frequency spectroscopy; adhesion studies. Mailing Add: Corp Res Lab Exxon Res & Eng Co PO Box 45 Linden NJ 07060

BRENNER, FIVEL CECIL, b Norfolk, Va, Sept 20, 18; m 47; c 2. CELLULOSE CHEMISTRY. Educ: Va Polytech Inst, BS, 40; Polytech Inst Brooklyn, MS, 42, PhD(chem), 50. Prof Exp: Asst, Polytech Inst Brooklyn, 40-42; teacher, Army Serv Col, Ger, 45-46; asst prof phys chem, Vanderbilt Univ, 49-51; sr res scientist, Johnson & Johnson, 51-60; sr group leader, Chemstrand Res Ctr, 60-66; chief textiles & apparel technol ctr, Inst Appl Technol, Nat Bur Standards, 66-67; CHIEF TIRE SYSTS DIV, SAFETY RES LAB, NAT HWY TRAFFIC SAFETY ADMIN, 67- Honors & Awards: Apparel Res Found Honor Award, 67. Mem: Am Chem Soc; Fiber Soc; Am Soc Testing & Mat; Brit Textile Inst. Res: Cellulose chemistry; fiber physics and mechanics; textile mechanics; statistical design; tire test method development; highway-vehicle interaction. Mailing Add: 2205 California St NW Washington DC 20008

BRENNER, FREDERIC J, b Warren, Ohio, Dec 25, 36. ECOLOGY, BEHAVIORAL BIOLOGY. Educ: Thiel Col, BS, 58; Pa State Univ, MS, 60, PhD(zool), 64. Prof Exp: Asst biol, Thiel Col, 57-58; asst zool, Pa State Univ, 58-64; first Kettering teaching intern, Denison Univ, 64-65; asst prof biol, Thiel Col, 65-69; asst prof, 69-71, ASSOC PROF BIOL, GROVE CITY COL, 71- Concurrent Pos: Grants, Am Philos Soc & Soc Sigma Xi, 65-66, 70, NSF, 66-67, Off Water Resources, 71-72, 75-76, Nat Wildlife Fedn, 73 & Nat Geog Soc, 73-75. Mem: AAAS; Am Soc Ichthyol & Herpet; Am Soc Mammal; Am Ornith Union; Am Soc Zool. Res: Environmental physiology; ecological and behavior aspects, especially hibernation and migration within animal populations; productivity and fish populations of strip mine lakes; limnology of strip mine lakes and mine drainage streams. Mailing Add: Dept of Biol Grove City Col Grove City PA 16127

BRENNER, GEORGE MARVIN, b Ottawa, Kans, Sept 19, 43; m 66; c 2. PHARMACOLOGY. Educ: Univ Kans, BS, 66, PhD(pharmacol), 71; Baylor Univ, MS, 68. Prof Exp: ASST PROF PHARMACOL, SDAK STATE UNIV, 71- Mem: Tissue Cult Asn; Sigma Xi. Res: Effect of prostaglandins on isolated myocardial cell contractility; effect of cytochalasin B on mammalian embryos maintained in vitro; motility of isolated blastomeres of early mammalian embryos. Mailing Add: Dept of Pharmacol SDak State Univ Brookings SD 57006

BRENNER, GERALD STANLEY, b Brooklyn, NY, July 8, 34; m 59; c 3. ORGANIC CHEMISTRY, PHARMACEUTICAL CHEMISTRY. Educ: City Col New York, BS, 56; Univ Wis, PhD(org chem), 61. Prof Exp: Sr chemist, 61-72, RES FEL, MERCK & CO, INC, 72- Mem: Am Chem Soc. Res: Structure determination; development of organic processes; synthetic organic chemistry; medicinal chemistry; physical pharmacy; preformulation research; biopharmaceutics; analytical chemistry. Mailing Add: Merck Sharp & Dohme Res Labs Div of Merck & Co Inc West Point PA 19486

BRENNER, GILBERT J, b New York, NY, Jan 18, 33; m 56; c 2. PALYNOLOGY. Educ: City Col New York, BS, 55; NY Univ, MS, 58; Pa State Univ, PhD(geol), 62. Prof Exp: Sci asst micropaleont, Am Mus Natural Hist, 56-58; teaching asst geol, Pa State Univ, 58-62; res paleontologist, Stand Oil Co Calif, 62-64; from asst prof to assoc prof geol, 64-70, res grant, 64-65, PROF GEOL, STATE UNIV NY COL NEW PALTZ, 70- Concurrent Pos: NSF res grant, 65-68; Am Philos Soc Penrose Fund grant, 70; vis res scientist, Geol Surv Israel, 70-71. Mem: AAAS; Am Paleont Soc; Nat Asn Geol Teachers; Am Asn Stratig Palynologists. Res: Paleogeography and ecology of Recent Foraminifera; Cretaceous angiosperm pollen and spores; origin of angiosperms; plant evolution. Mailing Add: Dept of Geol Sci State Univ of NY Col New Paltz NY 12561

BRENNER, HENRY CLIFTON, b Rochester, NY, Nov 21, 46. SOLID STATE CHEMISTRY. Educ: Mass Inst Technol, SB, 68; Univ Chicago, MS, 69, PhD(chem), 72. Prof Exp: Res assoc chem, Univ Calif, Berkeley, 72-75; ASST PROF CHEM, NY UNIV, 75- Concurrent Pos: NSF energy related fel, Univ Calif, Berkeley, 75. Mem: Am Phys Soc. Res: Electron spin resonance and optical spectroscopic studies of molecular crystals and biological systems; special interest in excited state energy transfer. Mailing Add: Dept of Chem NY Univ New York NY 10003

BRENNER, JOHN FRANCIS, b Charleston, SC, Sept 13, 41; m 64; c 1. MEDICAL RESEARCH. Educ: Mass Inst Technol, SB, 62, SM, 63; Tufts Univ, PhD(physics), 73. Prof Exp: Instr therapeut radiol, 72-75, ASST PROF THERAPEUT RADIOL, SCH MED, TUFTS UNIV, 75- Concurrent Pos: Physicist spec & sci staff radiother, New Eng Med Ctr Hosp, 72-; consult, Abbott Labs, 74-; mem task force automated differential cell counters, Ctr Dis Control, 76-; assoc dir image processing res, Dept Therapeut Radiol, New Eng Med Ctr Hosp, 76- Mem: Am Asn Physicists Med. Res: Computer assisted biomedical image processing. Mailing Add: Dept of Therapeut Radiol Tufts New Eng Med Ctr Boston MA 02111

BRENNER, JOSEPH, organic chemistry, see 12th edition

BRENNER, LORRY JACK, b Atlanta, Ga, May 16, 23; m 54; c 2. IMMUNOLOGY. Educ: Emory Univ, BA, 47; Univ Mich, MS, 49; Western Reserve Univ, PhD(immunol), 55. Prof Exp: Asst immunol, Dept Path, Western Reserve Univ, 54-57, instr microbiol, 57-65; from asst prof to assoc prof, 65-74, PROF BIOL, CLEVELAND STATE UNIV, 74-; IMMUNOLOGIST, HIGHLAND VIEW HOSP, 57- Concurrent Pos: Asst, St Luke's Hosp, Cleveland, 62; consult immunologist, Highland View Cuyahoga County Hosp, 62- Mem: Am Soc Zoologists; Soc Invert Path; Am Asn Immunologists; Soc Protozoologists. Res: Ciliates and antibodies; syphilis serology; immunohematology; leukemia. Mailing Add: Dept of Biol & Health Sci Cleveland State Univ Cleveland OH 44115

BRENNER, MARK, b Boston, Mass, June 19, 42; m 64. HORTICULTURE, PLANT PHYSIOLOGY. Educ: Univ Mass, BS, 64, MS, 65; Mich State Univ, PhD(hort), 70. Prof Exp: Asst prof, 69-75, ASSOC PROF HORT SCI & PLANT PHYSIOL, UNIV MINN, ST PAUL, 75- Mem: Am Soc Hort Sci; Am Soc Plant Physiol; Sigma Xi; Plant Growth Regulator Working Group. Res: Physiology of growth and development of plants; hormone metabolism in plant tissue; development and use of new methods for plant hormone extraction and identification. Mailing Add: Dept of Hort Sci Univ of Minn St Paul MN 55108

BRENNER, ROBERT MURRAY, b Lynn, Mass, Feb 6, 29; m 53; c 2. CYTOLOGY. Educ: Boston Univ, AB, 50, AM, 51, PhD(biol), 55. Prof Exp: From asst scientist to sr asst scientist, USPHS, 55-57; from instr to asst prof biol, Brown Univ, 57-64; ASSOC PROF & SCIENTIST, ORE REGIONAL PRIMATE RES CTR, MED SCH, UNIV ORE, 64- Concurrent Pos: Res grants, USPHS, 64-66, 67 & Pop Coun, 66-67. Mem: AAAS. Res: Histochemistry; autoradiography; electron microscopy endocrinology. Mailing Add: Ore Regional Primate Res Ctr 505 NY 185th Ave Beaverton OR 97005

BRENNER, RONALD JOHN, b Bethlehem, Pa, June 9, 33. PHARMACEUTICAL CHEMISTRY. Educ: Univ Cincinnati, BS, 55; Univ Fla, MS, 57, PhD(pharmaceut chem), 59. Prof Exp: Asst, Col Pharm, Univ Fla, 56-58; res pharmaceut develop chemist, McNeil Labs, Inc, 58-62, group leader pharm res dept, 62-64, mgr new prod, 64-66, dir develop res, 66-67, exec dir new prod, 67-70; asst to vchmn, Johnson & Johnson Int, NJ, 70-71, vpres, 71-74; EXEC VPRES, McNEIL LABS, INC, 74- Mem: AAAS; Am Pharmaceut Asn; Acad Pharmaceut Sci. Res: Pharmaceutical dosage forms; medicinal chemistry. Mailing Add: McNeil Labs Inc Camp Hill Rd Ft Washington PA 19034

BRENNER, STEPHEN LOUIS, b Brooklyn, NY, Apr 10, 48. BIOPHYSICS. Educ: State Univ NY Binghamton, BA, 69; Ind Univ, PhD(chem physics), 72. Prof Exp: Asst prof chem, Univ Ky, 72-74; STAFF FEL BIOPHYS, NIH, 74- Honors & Awards: Victor K LaMer Award, Colloid & Surface Chem Div, Am Chem Soc, 74. Mem: Am Chem Soc; Biophys Soc. Res: Interaction of charged macromolecules in solution; colloid stability, polymer interactions, biological assembly, liquid crystals; inelastic light scattering. Mailing Add: 12A/2007 Phys Sci Lab Comput Res NIH Bethesda MD 20014

BRENNIMAN, GARY RUSSELL, b Mt Clemens, Mich, Aug 4, 42; m 65; c 2. ENVIRONMENTAL HEALTH. Educ: Cent Mich Univ, BS, 64, MS, 66; Univ Mich, Ann Arbor, MPH, 69, PhD(environ health sci), 74. Prof Exp: Chief basic sci br, Brooke Army Med Ctr, US Army Med Field Serv Sch, 66-68; ASST PROF ENVIRON HEALTH SCI, SCH PUB HEALTH, UNIV ILL MED CTR, 74- Concurrent Pos: Consult, Argonne Nat Lab, 76. Mem: Sigma Xi. Res: Health effects of human exposure to metals, asbestos and chlorinated hydrocarbons in drinking water. Mailing Add: Sch of Pub Health PO Box 6998 Univ of Ill Med Ctr Chicago IL 60680

BRENOWITZ, A HARRY, b Brooklyn, NY, July 8, 18; m 48; c 2. BIOLOGY. Educ: Brooklyn Col, BA, 39; Columbia Univ, MA, 47, EdD, 58. Prof Exp: From asst prof to assoc prof, 46-64, PROF BIOL, ADELPHI UNIV, 64-, DIR, MARINE SCI INST, 68- Concurrent Pos: NSF fels, 59, 60 & 61; trustee, Affiliated Cols & Univs, NY Ocean Sci Lab. Mem: AAAS; Am Soc Zool; Am Soc Limnol & Oceanog; Am Inst Biol Sci. Res: Marine invertebrates; environmental studies of Great South Bay, Long Island, New York; biology of Chrsaora quinquecirrha. Mailing Add: Inst of Marine Sci Adelphi Univ Garden City NY 11530

BRENT, BENNY EARL, b Alton, Kans, July 3, 37; m 62. ANIMAL NUTRITION. Educ: Kans State Univ, BS, 59, MS, 60; Mich State Univ, PhD(animal husb, nutrit), 66. Prof Exp: Chemist, Mich State Univ, 64-66; asst prof animal husb, 66-69, ASSOC PROF ANIMAL SCI & INDUST, KANS STATE UNIV, 69- Mem: AAAS; Am Soc Animal Sci. Res: Trace mineral nutrition of swine; nonprotein nitrogen utilization by ruminants; control of feed intake in ruminants; enzymology of the rumen. Mailing Add: Dept of Animal Sci & Indust Kans State Univ Col of Agr Manhattan KS 66502

BRENT, BERNARD J, medicinal chemistry, endocrinology, see 12th edition

BRENT, CHARLES RAY, b Hattiesburg, Miss, June 12, 31; m 53; c 3. PHYSICAL CHEMISTRY, ORGANIC CHEMISTRY. Educ: Univ Southern Miss, BA, 53; Tulane Univ, MS, 60, PhD(phys chem), 63. Prof Exp: Asst chemist, Eagle Chem Co, Ala, 53; assoc prof phys chem, 60-70, asst dean, Col Arts & Sci, 66-70, PROF CHEM, UNIV SOUTHERN MISS, 70-, DIR, INST ENVIRON SCI, 74- Mem: AAAS; Am Chem Soc. Res: Thermochemical studies of nitrogen heterocyclic compounds; kinetic solvent effects. Mailing Add: Dept of Chem Univ of Southern Miss Hattiesburg MS 39401

BRENT, J ALLEN, b Carmi, Ill, Nov 21, 21; m 46; c 4. SCIENCE ADMINISTRATION. Educ: Eastern Ill State Col, BS, 43; Univ Fla, PhD(chem), 49. Prof Exp: Naval stores asst, Univ Fla, 46-49; head dept chem, Jacksonville Jr Col, 49-51; chief chem, Sect Res & Develop, Minute Maid Corp, 51-62, from assoc dir res to dir res, Minute Maid Co Div, Coca-Cola Co, 62-66, dir res & develop carbonated beverages, 66-68, VPRES, COCA-COLA USA DIV, COCA-COLA CO, 68- Concurrent Pos: With Manhattan Proj, 43-46. Mem: Am Chem Soc. Res: Vapor-liquid equilibrium; vacuum distillation; terpenes; citrus products and by-products. Mailing Add: PO Drawer 1734 Atlanta GA 30301

BRENT, MORGAN MCKENZIE, b Evanston, Ill, Jan 31, 23; m 55; c 3. MICROBIOLOGY. Educ: Northwestern Univ, BS, 48, MS, 49, PhD(biol sci), 53. Prof Exp: Jr res zoologist, Univ Calif, 53-54; instr microbiol, Jefferson Med Col, 54-57; assoc prof biol, 57-66, chmn dept, 64-67, PROF BIOL, BOWLING GREEN STATE UNIV, 66-, DIR APPL MICROBIOL, COL HEALTH & COMMUNITY SERV, 75- Mem: Am Micros Soc; Soc Protozool. Res: Effects of environmental factors on morphogenesis and transformation of amoebo-flagellates; nutrition of protozoa. Mailing Add: Dept of Biol Sci Bowling Green State Univ Bowling Green OH 43402

BRENT, ROBERT LEONARD, b Rochester, NY, Oct 6, 27; m 49; c 4. EMBRYOLOGY. Educ: Univ Rochester, AB, 48, MD, 53, PhD, 55. Prof Exp: Res assoc, AEC, Univ Rochester, 47-55, asst physics, 48-49; chief radiation biol, Div Nuclear Med, Walter Reed Army Inst Res, 55-57; assoc prof pediat, 57-60, PROF PEDIAT & RADIOL, JEFFERSON MED COL, 60-, PROF ANAT, 71-, CHMN DEPT PEDIAT, 66-; DIR, ELEANOR ROOSEVELT CANCER RES INST, 62- Concurrent Pos: Fel, Fitzwilliam Col, Cambridge Univ, 71; Royal Soc Med traveling fel, 71-72; intern, Mass Gen Hosp, 54-55; dir, Stein Res Ctr, Jefferson Med Col, 69-; mem, Embryol Study Sect, NIH, 70-74, chmn, Subcomt Prev Embryol, Fetal & Perinatal Dis, Fogarty Ctr. Honors & Awards: Ritchie Prize, 53. Mem: Teratology Soc (pres, 68-69); AAAS; Am Asn Anatomists; Am Soc Exp Path; Soc Pediat Res. Res: Experimental embryology; radiation biology; clinical pediatric research; immunology. Mailing Add: Dept of Pediat Jefferson Med Col Philadelphia PA 19107

BRENT, THOMAS PETER, b Leipzig, Ger, Nov 7, 37; Brit citizen; m 66. BIOCHEMISTRY. Educ: Cambridge Univ, Univ, BA, 62; Univ London, PhD(biochem), 66. Prof Exp: Res fel cell biol, Chester Beatty Res Inst, London, 66-68; asst prof cancer res & biochem, McGill Univ, 68-72; ASST MEM BIOCHEM, ST JUDE CHILDRENS RES HOSP, 72- Mem: Brit Biochem Soc; Biophys Soc. Res: Repair of DNA in response to radiation and alkylation; regulation of DNA replication and cell proliferation. Mailing Add: St Jude Childrens Res Hosp PO Box 318 Memphis TN 38101

BRENT, WILLIAM B, b Ky, June 28, 24. STRUCTURAL GEOLOGY, STRATIGRAPHY. Educ: Univ Va, BA, 49, JD, 66; Cornell Univ, MA, 52, PhD(geol), 55. Prof Exp: Asst prof geol, Okla State Univ, 55-61; assoc prof, La Tech Univ, 66-68; vis assoc prof, Univ Va, 69-70; GEOLOGIST, TENN DIV GEOL, 70- Mem: Geol Soc Am; Sigma Xi. Res: Field studies of mountain systems of the United States; chiefly structural and stratigraphic field studies in the complexly folded and faulted Southern Appalachians. Mailing Add: Tenn Div of Geol 4711 Old Kingston Pike Knoxville TN 37919

BRERETON, JOHN GROBE, dairy chemistry, biochemistry, see 12th edition

BRESCIA, FRANK, b New York, NY, May 19, 08; m 36. PHYSICAL CHEMISTRY. Educ: City Col New York, BS, 31; Columbia Univ, MA, 33, PhD(phys chem), 38. Prof Exp: Tutor chem, 33-37, from instr to prof, 37-70, dir undergrad res prog, 58-70, EMER PROF CHEM, CITY COL NEW YORK, 70- Concurrent Pos: Vis scholar, Columbia Univ, 39-41; vol investr, Nat Defense Res Comt, 41-43, res assoc, 43-45. Mem: Fel AAAS; Am Chem Soc; fel NY Acad Sci; fel Am Inst Chemists. Res: Reaction kinetics and dissociation constants in heavy water; properties of colloidal systems. Mailing Add: 637 Hillsdale Ave Hillsdale NJ 07642

BRESCIA, VINCENT THOMAS, b New York, NY, June 2, 30; m 65; c 4. MICROBIAL GENETICS. Educ: Cent Col, Iowa, BA, 55; Fla State Univ, MS, 65, PhD(genetics), 73. Prof Exp: Instr biol, Lees Jr Col, Ky, 60-63; ASST PROF BIOL, THE LINDENWOOD COL, 69- Mem: AAAS; Sigma Xi. Res: Isolation and characterization of Neurospora mutants with alpha-galactosidase activity. Mailing Add: Dept of Biol The Lindenwood Col St Charles MO 63301

BRESHEARS, WILBERT DALE, b Eugene, Ore, Apr 25, 39; m 60; c 4. PHYSICAL CHEMISTRY. Educ: Portland State Col, BS, 61; Ore State Univ, PhD(phys chem), 65. Prof Exp: NSF fel, Sch Chem Sci, Univ EAnglia, 65-66; staff mem, 66-74, GROUP LEADER PHYS CHEM, LOS ALAMOS SCI LAB, 74- Mem: AAAS; Am Chem Soc. Res: Chemical kinetics; molecular energy transfer; laser isotope separation. Mailing Add: Los Alamos Sci Lab Los Alamos NM 87544

BRESLER, EMANUEL H, b Pensacola, Fla, Mar 2, 19; m 50; c 3. MEDICAL RESEARCH, NEPHROLOGY. Educ: Univ Fla, BS, 41; Tulane Univ, MD, 47. Prof Exp: Intern, USPHS Hosp, Staten Island, NY, 47-48; instr physiol, Bellevue Med Sch, NY Univ, 50-51; pvt pract, 53-55; from asst prof to assoc prof, 55-62, CLIN PROF

MED, SCH MED, TULANE UNIV, 62-; MED INVESTR, VET ADMIN HOSP, NEW ORLEANS, 76- Concurrent Pos: NIH cardiovasc res fel, Sch Med, Tulane Univ, 48-50; assoc chief staff, Vet Admin Hosp, New Orleans, 57-75. Mem: AMA; Am Fedn Clin Res; Am Heart Asn; Am Physiol Soc. Res: Internal medicine; renal physiology; fluid balance; membrane transport. Mailing Add: Vet Admin Hosp 1601 Perdido St New Orleans LA 70114

BRESLER, JACK BARRY, b New York, NY, May 10, 23; m 52; c 4. RESEARCH ADMINISTRATION, HUMAN BIOLOGY. Educ: Univ Denver, BA, 48; Univ Okla, MS, 52; Univ Ill, PhD, 56. Prof Exp: Asst zool, Univ Okla, 50-52 & Univ Ill, 52-54; instr, Colgate Univ, 54-55; asst prof biol, Bard Col, 55-57 & Brown Univ, 57-62; assoc prof, Boston Univ, 62-66, dir res develop, 64-66; ASSOC PROF BIOL & ASST PROVOST, TUFTS UNIV, 66- Concurrent Pos: Consult human genetics, Nat Collab Study on Human Reproduction, NIH, 57-62; sem assoc, Columbia Univ, 57-62; consult biol educ, UNESCO, 70; co-chmn New Eng Region, Nat Coun Univ Res Adminr, 70-73; mem New Eng Adv Comt, Dept Health, Educ & Welfare, 70-73; mgt consult, NSF, 75-; genetics consult, Mus Sci, Boston, 75- Mem: Fel AAAS; Soc Study Social Biol; Behav Genetics Asn; Am Soc Human Genetics; Am Asn Phys Anthrop. Res: Genetic and social consequences of inter-ethnic matings; biological consequences of urban development; genetic and social consequences of human outcrossings; biological consequences of human intervention in environment. Mailing Add: 494 Ward St Newton MA 02159

BRESLIN, MAUREEN ELIZABETH, b Greenock, Scotland, May 6, 17; US citizen. ANALYTICAL CHEMISTRY. Educ: Col St Elizabeth, AB, 38; Cath Univ Am, MS, 55. Prof Exp: Teacher sci, Mt St Dominic Acad, 40-41 & St Mary High Sch, 41-48; from instr to prof chem & dean women, Caldwell Col, 48-69; prin, St Catherine Sch, 69-75; PROF CHEM, CALDWELL COL, 75- Mem: Am Chem Soc; Sigma Xi. Res: Chemical influence on cancer occurrence in mice. Mailing Add: Caldwell Col Caldwell NJ 07006

BRESLOW, ALEXANDER, b New York, NY, Mar 23, 28; m 50; c 3. PATHOLOGY. Educ: Univ Chicago, BS, 48, MS & MD, 53. Prof Exp: Instr path, Med Sch, Univ Wash, 59-61; from asst prof to assoc prof, 61-73, PROF PATH, SCH MED, GEORGE WASHINGTON UNIV, 74- Res: Surgical pathology and bacteriology. Mailing Add: Dept of Path Sch of Med Geo Washington Univ Washington DC 20037

BRESLOW, DAVID SAMUEL, b New York, NY, Aug 13, 16; m 46; c 3. ORGANIC CHEMISTRY, POLYMER CHEMISTRY. Educ: City Col New York, BS, 37; Duke Univ, PhD(org chem), 40. Prof Exp: Lab asst chem, Duke Univ, 37-40; asst org chem, Calif Inst Technol, 40-41; res fel, Radiation Lab, Univ Calif, 42; res assoc, Duke Univ, 42-44, instr, 44-45; from res chemist to sr res chemist, 46-63, res assoc, 63-71, SR RES ASSOC, HERCULES INC, 71- Concurrent Pos: Vis prof, Univ Munich, 64-65 & Univ Notre Dame, 71; adj prof chem, Univ Del, 72- Mem: Am Chem Soc. Res: Mechanisms of organic reactions; synthetic organic chemistry. Mailing Add: Res Ctr Hercules Inc Wilmington DE 19899

BRESLOW, ESTHER, M G, b New York, NY, Dec 23, 31; m 55; c 2. BIOPHYSICAL CHEMISTRY. Educ: Cornell Univ, BS, 53; NY Univ, MS, 55, PhD(biochem), 59. Prof Exp: Res assoc, 60-64, asst prof, 64-72, ASSOC PROF BIOCHEM, MED COL, CORNELL UNIV, 72- Concurrent Pos: NIH fel, Med Col, Cornell Univ, 59-61 & NIH res grants, 61-74; mem, Biophys & Biophys Chem Study Sect A, NIH, 73- Mem: AAAS; Am Chem Soc; Am Soc Biol Chemists; Harvey Soc. Res: Relationship between protein structure and biological and chemical reactivity. Mailing Add: Dept of Biochem Cornell Univ Med Col New York NY 10021

BRESLOW, LESTER, b Bismark, NDak, Mar 17, 15; m 39; c 3. PUBLIC HEALTH, PREVENTIVE MEDICINE. Educ: Univ Minn, BA, 35, MD, 38, MPH, 41. Prof Exp: Dist health officer, State Dept Health, Minn, 41-43; chief, Bur Chronic Dis, State Dept Pub Health, Calif, 46-60, chief, Div Prev Med Serv, 60-65, dir, 65-68; prof health serv admin, Sch Pub Health, 68-69, prev & social med & chmn dept, Sch Med, 69-73, DEAN, SCH PUB HEALTH, UNIV CALIF, LOS ANGELES, 72- Mem: Nat Acad Sci; Inst of Med of Nat Am Pub Health Asn (pres, 69); Pub Health Cancer Asn Am (pres, 53); Int Epidemiol Asn (pres, 67-68). Res: Chronic disease epidemiology; health services. Mailing Add: Sch of Med Univ of Calif Los Angeles CA 90024

BRESLOW, NORMAN EDWARD, b Minneapolis, Minn, Feb 21, 41; m 63; c 2. MEDICAL STATISTICS. Educ: Reed Col, BA, 62; Stanford Univ, PhD(statist), 67. Prof Exp: Vis res worker med statist, London Sch Hyg & Trop Med, 67-68; asst prof, 68-72, ASSOC PROF BIOSTATIST, UNIV WASH, 72- Concurrent Pos: Statistician, Children's Cancer Study Group, 68-72; mem, Nat Wilm's Tumor Study Comt, 69-; statistician, Inst Agency Res Cancer, WHO, 72-74; assoc mem, Fred Hutchinson Cancer Res Ctr, 74-; consult, NIH, 75- Mem: AAAS; Biomet Soc; Am Statist Asn; Inst Math Statist. Res: Statistics of childhood cancer, especially Wilm's tumor; cancer epidemiology, particularly in environmental carcinogenesis; statistical methodology for survival time studies and case-control studies. Mailing Add: Dept of Biostatist SC-32 Univ of Wash Seattle WA 98195

BRESLOW, RONALD, b Elizabeth, NJ, Mar 14, 31; m 55. ORGANIC CHEMISTRY. Educ: Harvard Univ, AB, 52, AM, 54, PhD(chem), 56. Prof Exp: Nat Res Coun fel, 55-56, from instr to prof chem, 56-67, S L MITCHILL PROF CHEM, COLUMBIA UNIV, 67- Concurrent Pos: Chmn div chem, Nat Acad Sci, 74-77. Honors & Awards: Fresenius Award, 66; Award Pure Chem, Am Chem Soc, 66, Harrison Howe Award, 75; Mark van Doren Medal, Columbia Univ, 69; Baekeland Medal, 69. Mem: Nat Acad Sci; Am Acad Arts & Sci; Am Chem Soc. Res: Aromaticity and small ring compounds; biochemical model systems; reaction mechanisms. Mailing Add: Box 566 Havemeyer Hall Columbia Univ New York NY 10027

BRESNICK, EDWARD, b Jersey City, NJ, Sept 7, 30; m 57. BIOCHEMISTRY. Educ: St Peters Col, BS, 52; Fordham Univ, MS, 54, PhD, 57. Prof Exp: Res assoc, Med Br, Univ Tex, 57-58; res biochemist, Burroughs, Wellcome & Co, 58-61; from asst prof biochem to prof pharmacol, Baylor Col Med, 70-72; PROF CELL & MOLECULAR BIOL & CHMN DEPT, MED COL GA, 72- Concurrent Pos: Consult, Nat Inst Gen Med Sci. Mem: AAAS; Am Chem Soc; Am Asn Cancer Res; Am Soc Cell Biol; Am Soc Biol Chemists. Res: Cancer research; enzymology; regulatory mechanisms; pyrimidine and nucleic acid metabolism. Mailing Add: Dept of Cell & Molecular Biol Med Col of Ga Augusta GA 30902

BRESNICK, GERALD IRWIN, b New York, NY, Dec 23, 49; m 71. LIMNOLOGY. Educ: State Univ NY Buffalo, BA, 71; Tulane Univ, PhD(biol), 75. Prof Exp: INSTR BIOL, TULANE UNIV, 75- Mem: AAAS; Am Fisheries Soc; Am Soc Ichthyologists & Herpetologists; Am Soc Limnol & Oceanog. Res: Effects of environmental stressors on aquatic organisms, particuarly the effects of domestic and industrial wastes on fish physiology. Mailing Add: Dept of Biol Tulane Univ New Orleans LA 70118

BRESSLER, BERNARD, b Milan, Mich, May 22, 17; m 48; c 2. MEDICINE. Educ: Washington Univ, AB, 38, MD, 42. Prof Exp: Assoc prof, 55-59, PROF PSYCHIAT, MED CTR, DUKE UNIV, 59- Concurrent Pos: Instr, Washington Psychoanal Inst, 58-60, training analyst, 60-; instr, Univ NC-Duke Univ Psychoanal Inst, 60-; consult, Ft Bragg, NC, 58. Mem: Am Psychoanal Asn; Am Psychiat Asn; AMA; Am Psychosomatic Soc; Am Col Psychiat. Res: Psychosomatic medicine; psychoanalysis. Mailing Add: Dept of Psychiat Duke Univ Med Ctr Durham NC 27706

BRESSLER, DAVID WILSON, b San Francisco, Calif, Sept 7, 23; m 49; c 2. MATHEMATICS. Educ: Univ Calif, Berkeley, AB, 49, MA, 51, PhD(math), 57. Prof Exp: Instr math, Univ BC, 57-58; instr, Tufts Univ, 58-59; from instr to asst prof, Univ BC, 59-68; assoc prof, 68-74, PROF MATH, CALIF STATE UNIV, SACRAMENTO, 74- Mem: Am Math Soc. Res: Set theory and measure theory. Mailing Add: Dept of Math Calif State Univ Sacramento CA 95819

BRESSLER, GLENN OTTO, b Hegins, Pa, Feb 13, 14; m 44; c 2. AGRICULTURAL ECONOMICS. Educ: Pa State Univ, BS, 35, MS, 40; Cornell Univ, PhD(agr econ), 49. Prof Exp: Teacher voc agr pub schs, Pa, 35-45; from instr to prof poultry sci, 45-75, EMER PROF POULTRY SCI, PA STATE UNIV, UNIVERSITY PARK, 75- Res: Poultry management; poultry housing and equipment. Mailing Add: 630 Franklin St State College PA 16801

BRESSLER, ROBERT S, b Brooklyn, NY, Dec 31, 39; m 62; c 3. CELL BIOLOGY, DEVELOPMENTAL ANATOMY. Educ: City Col New York, BS, 62; NY Univ, MS, 67, PhD(cell differentiation), 70. Prof Exp: Lectr biol, City Col New York, 62-63 & 64-67; teaching asst histol & ultrastruct, Med Ctr, NY Univ, 67-69; instr, 69-71, assoc, 71-72, ASST PROF ANAT, MT SINAI SCH MED, 72- Concurrent Pos: Lectr, Baruch Col, 69. Mem: Am Asn Anatomists; Am Soc Cell Biol; Am Soc Zoologists. Res: Cell differentiation, particularly role of hormones in gonadal development and function; testicular maturation; postnatal development of salivary glands. Mailing Add: Mt Sinai Sch of Med Fifth Ave & 100th St New York NY 10029

BRESSLER, WILBUR LEE, organic chemistry, see 12th edition

BRESSON, CLARENCE RICHARD, b Wooster, Ohio, July 30, 25; m 60; c 3. APPLIED CHEMISTRY. Educ: Col Wooster, BA, 50; Wash State Univ, MS, 55, PhD(chem), 58. Prof Exp: Baking technologist cereal chem, USDA, 51-54; chemist, 59-60, 61-65, SR RES CHEMIST, PHILLIPS PETROL CO, 66- Mem: Am Chem Soc. Res: Upgrading refinery by-product streams; oxo-alcohols, primary and secondary plasticizers, fluid coke, pyrolosis gasoline; applications for new products, sulfolanyl ethers; technical service; hydraulic barriers; asphaltic concrete flexure fatigue. Mailing Add: Rte 3 Box 396 Bartlesville OK 74003

BREST, ALBERT N, b Berwick, Pa, Dec 17, 28. MEDICINE. Educ: Temple Univ, MD, 53; Am Bd Internal Med, dipl, 61. Prof Exp: Intern, Philadelphia Gen Hosp, Pa, 53-54; researcher internal med, Temple Univ Hosp, 54; researcher, Albert Einstein Med Ctr, Pa, 56-58; instr med & dir hypertension unit, Hahnemann Med Col, 59-60, assoc, 60, from asst prof to assoc prof, 61-69, head sect hypertension & renology, 59-63, vasc dis & renology, 63-69, cardiol, 68-69; HEAD DIV CARDIOL, JEFFERSON MED COL & HOSP, PA, 69- Concurrent Pos: Am Heart Asn res fel, Hahnemann Hosp, 58-59. Mem: AMA; Am Heart Asn; NY Acad Sci; Am Fedn Clin Res; fel Am Col Cardiol. Res: Cardiovascular diseases. Mailing Add: Jefferson Med Col 1025 Walnut St Philadelphia PA 19107

BRETERNITZ, DAVID ALAN, b Fremont, Nebr, Nov 12, 29; m 52; c 3. ANTHROPOLOGY. Educ: Univ Denver, BA, 52; Univ Ariz, MA, 56, PhD, 63. Prof Exp: Asst anthrop, Univ Denver, 50-52; sr archaeologist, Southern Pac Co, 55; cur anthrop, Mus Northern Ariz, 56-59; asst, Lab Tree-Ring Res, Univ Ariz, 60-61; assoc prof anthrop, 66-70, PROF ANTHROP, UNIV COLO, BOULDER, 70- Concurrent Pos: Dir, Mesa Verde Res Ctr. Mem: Soc Am Archaeol. Res: Southwestern United States prehistory; archaeological interpretation of dated tree-ring specimens. Mailing Add: Dept of Anthrop Univ of Colo Boulder CO 80302

BRETHERTON, FRANCIS P, b Oxford, Eng, July 6, 35; m 59; c 3. METEOROLOGY. Educ: Univ Cambridge, BA, 58, MA & PhD(fluid dynamics), 62. Prof Exp: Instr math, Mass Inst Technol, 61-62; SAR dynamical meteorol, Univ Cambridge, 62-63, ADR, 63-64, univ lectr appl math, 66-69; prof meteorol & oceanog, Johns Hopkins Univ, 69-74; PRES, UNIV CORP ATMOSPHERIC RES, 73-; DIR, NAT CTR ATMOSPHERIC RES, 74- Concurrent Pos: Vis prof, Univ Miami, 68-69; mem, US Comt for Global Atmospheric Res Prog, 71-76; mem comt atmospheric sci, Nat Acad Sci, 71-77. Honors & Awards: Res Award, Area IV, World Meteorol Orgn, 71; Buchan Prize, Royal Meteorol Soc; Meisinger Award, Am Meteorol Soc, 72. Mem: Am Meteorol Soc; Royal Meteorol Soc. Res: Mesoscale and large scale dynamics of the atmosphere and ocean; applied mathematics and the general theory of wave propagation. Mailing Add: Univ Corp for Atmospheric Res Box 3000 Boulder CO 80303

BRETSCHER, MANUEL MARTIN, b River Forest, Ill, May 12, 28; m 55; c 5. NUCLEAR PHYSICS. Educ: Wash Univ, AB, 50, PhD(physics), 54. Prof Exp: Asst prof physics, Ala Polytech Inst, 54-56; from assoc prof to prof & co-chmn dept, Valparaiso Univ, 56-67; ASSOC PHYSICIST, APPL PHYSICS DIV, ARGONNE NAT LAB, 67- Concurrent Pos: Consult, Oak Ridge Nat Lab, 58-64. Mem: Am Phys Soc; Am Nuclear Soc; Am Asn Physics Teachers. Res: Particle accelerators; stripping reactions; coulomb excitation; scattering; reactor physics; plutonium capture-to-fission ratios; age measurements; mass spectrometry; inhomogeneous magnetic field spectrometers; breeding ratio measurements for liquid metal fast breeder reactors. Mailing Add: Appl Physics Div Bldg 316 Argonne Nat Lab Argonne IL 60439

BRETSCHNEIDER, CHARLES LEROY, b Red Owl, SDak, Nov 9, 20; m 48; c 2. PHYSICAL OCEANOGRAPHY, OCEAN ENGINEERING. Educ: Hillsdale Col, BS, 47; Univ Calif, Berkeley, MS, 50; Tex A&M Univ, PhD(phys oceanog), 59. Prof Exp: Engr, Waves & Wave Force Proj, 50-51; engr waves & wave forces on pilings, Res Found, Tex A&M Univ, 51-56; hydraul engr & chief oceanog br, Beach Erosion Bd, Corps Engr, US Army, 56-61; dir eastern opers, Nat Eng Sci Co, Washington, DC, 61-64, vpres, Vt, 64-65, geomarine technol, 66-67; PROF & CHMN DEPT OCEAN ENG, UNIV HAWAII, 67- Concurrent Pos: Lectr, Mass Inst Technol, 61-63. Honors & Awards: Outstanding Contrib Coastal Eng Res Prize, Am Soc Civil Eng, 59; Dept Army, Off Chief Engr, Outstanding Performance & Cash Award. Mem: Am Soc Oceanog; Am Soc Eng Educ; Soc Naval Archit & Marine Eng; Am Soc Civil Eng; Am Geophys Union. Res: Variability, spectra, forecasting, generation and decay of waves; wave forces on piles; hurricane surge and waves; rubble mound breakwater stability. Mailing Add: Dept of Ocean Eng Univ of Hawaii 2565 The Mall Honolulu HI 96822

BRETSKY, PETER WILLIAM, b Easton, Pa, Oct 26, 38; m 65; c 1. INVERTEBRATE PALEONTOLOGY. Educ: Lafayette Col, AB, 61; Southern Methodist Univ, MS, 63; Yale Univ, PhD, 68. Prof Exp: Instr geol, Northwestern Univ, 67-69, asst prof geol & biol sci, 69-70; ASSOC PROF EARTH & SPACE SCI & DIR ENVIRON PALEONT PROG, STATE UNIV NY STONY BROOK, 70- Mem: Soc Econ Paleont & Mineral; Paleont Soc; Marine Biol Asn UK. Res:

Paleoecology; biostratigraphy. Mailing Add: Dept of Earth & Space Sci State Univ of NY Stony Brook NY 11794

BRETSKY, SARA (SU) STEWART, b Clifton, Tex, Jan 16, 43; m 65; c 1. INVERTEBRATE PALEONTOLOGY. Educ: Southern Methodist Univ, BS, 64; Yale Univ, MS, 66, PhD(geol), 69. Prof Exp: Res asst geophys, Dallas Seismol Observ, 64; res assoc geol, Field Mus Natural Hist, 67-70 & Northwestern Univ, 69-70; res assoc, 71-73, ADJ ASST PROF EARTH & SPACE SCI, STATE UNIV NY STONY BROOK, 73- Concurrent Pos: Programmer, Harris Trust & Savings Bank, 68-69. Mem: Paleont Soc; Paleont Res Inst; Soc Syst Zool; Am Malacol Union. Res: Mollusca; systematics and evolution; taxonomy; ecology and paleoecology; statistics. Mailing Add: Dept of Earth & Space Sci State Univ of NY Stony Brook NY 11794

BRETT, CHARLES H, b Lincoln, Nebr, July 21, 09; m 31; c 2. ENTOMOLOGY. Educ: Univ Nebr, BS, 30, MS, 38; Kans State Univ, PhD(entom), 46. Prof Exp: Teacher, High Sch, Nebr, 31-43; from asst prof to assoc prof entom, Okla State Univ, 43-52; assoc prof, 52-60, PROF ENTOM, NC STATE UNIV, 60- Honors & Awards: Sigma Xi Res Award, Kans State Univ, 46. Mem: Entom Soc Am. Res: Resistance of vegetable varieties to insects. Mailing Add: Dept of Entom NC State Univ Raleigh NC 27607

BRETT, JOHN ROLAND, b Toronto, Ont, June 25, 18; m 48; c 2. FISHERIES. Educ: Univ Toronto, BA, 41, MA, 44, PhD, 51. Prof Exp: Asst, Ont Fisheries Res Lab, 40-41; from asst biologist to assoc biologist, 44-56, sr scientist, 56-58, PRIN SCIENTIST, BIOL STA, FISHERIES RES BD CAN, 58- Mem: Am Soc Limnol & Oceanog; Am Fisheries Soc; Can Soc Zool; Royal Soc Can; Prof Inst Pub Serv Can. Res: Limnology; experimental biology; temperature relations and metabolism of fish. Mailing Add: Biol Sta Fisheries Res Bd of Can Nanaimo BC Can

BRETT, ROBIN, b Adelaide, S Australia, Jan 30, 35; m 63; c 2. EARTH SCIENCES. Educ: Univ Adelaide, BSc, 56; Harvard Univ, Am, 60, PhD(geol), 63. Prof Exp: Geologist, US Geol Surv, Washington, DC, 64-69; chief, Geochem Br, Johnson Space Ctr, NASA, 69-74; GEOLOGIST, US GEOL SURV, RESTON, VA, 74- Concurrent Pos: Hon fel, Australian Nat Univ, 67-68; assoc ed, Geochimica et Cosmochimica Acta, 69- & J Geophys Res, 72-74. Honors & Awards: Lindgren Award, Soc Econ Geologists, 64; Exceptional Sci Achievement Medal, NASA, 73. Mem: Geochem Soc; Am Geophys Union; Mineral Soc Am; Meteorit Soc (pres, 73-74). Res: Mineralogy, petrology, geochemistry of lunar samples, meteorites, planetary geochemistry; meteorite impact structures, mineral deposits. Mailing Add: Stop 959 US Geol Surv Reston VA 22092

BRETT, THOMAS JOSEPH, JR, b Orange, NJ, May 11, 35. POLYMER CHEMISTRY. Educ: St Peters Col, BS, 57; Univ Notre Dame, PhD(chem), 63. Prof Exp: Res chemist, Res Ctr, US Rubber Co, NJ, 63-66, res scientist, 66-67; group leader solution polymers, 67-72, SECT MGR LATEX RES & DEVELOP, UNIROYAL CHEM, INC, 72- Mem: Am Chem Soc. Res: Conformational analysis of mobile cycloalkane systems; chemistry of vulcanization of elastomers; ethylene-propylene diene monomer rubber; paper coating and carpet latex development. Mailing Add: Uniroyal Chem Inc Naugatuck CT 06770

BRETT, WILLIAM JOHN, b Chicago, Ill, Mar 23, 23; m 48; c 1. PHYSIOLOGY. Educ: Northern Ill Univ, BS, 49; Miami Univ, MS, 50; Northwestern Univ, PhD(biol), 53. Prof Exp: Asst, Northwestern Univ, 50-53; asst prof biol, Millsaps Col, 53-56; assoc prof biol, 56-59, chmn dept life sci, 64-67, PROF BIOL, IND STATE UNIV, TERRE HAUTE, 59- Mem: Am Soc Zool; Am Inst Biol Sci; Animal Behav Soc. Res: Biological rhythms. Mailing Add: 2321 S Tenth St Terre Haute IN 47802

BRETTELL, HERBERT R, b Rock River, Wyo, Feb 1, 21; m 47; c 4. INTERNAL MEDICINE. Educ: Univ Wyo, BS, 42; Univ Rochester, MD, 50; Am Bd Family Pract, cert, 74. Prof Exp: Intern, Vanderbilt Univ, 50-51; resident, 51-55, asst, 54-55, from instr to assoc prof, 56-75, PROF FAMILY MED & MED & CHMN DEPT FAMILY MED, UNIV COLO MED CTR, DENVER, 75- Concurrent Pos: Head div family pract, Univ Colo Med Ctr, 71-75. Mem: Am Col Physicians; Soc Teachers Family Med; Am Col Clin Pharmacol & Chemother; Am Heart Asn; Am Acad Family Physicians. Res: Cardiovascular disease; diuretics. Mailing Add: Univ of Colo Med Ctr Dept of Family Med Denver CO 80220

BRETTHAUER, ROGER K, b Morris, Ill, Jan 4, 35; m 59; c 3. BIOCHEMISTRY. Educ: Univ Ill, BS, 56, MS, 58; Mich State Univ, PhD(biochem), 61. Prof Exp: NIH training grant bact, Univ Wis, 61-64; asst prof, 64-67, ASSOC PROF CHEM, UNIV NOTRE DAME, 67- Mem: AAAS; Am Chem Soc; Am Soc Biol Chemists. Res: Complex carbohydrates, biosynthesis, structure and function. Mailing Add: Dept of Chem Univ of Notre Dame Notre Dame IN 46556

BRETZ, HAROLD WALTER, b Indianapolis, Ind, Aug 17, 24; m 47. MICROBIOLOGY. Educ: Purdue Univ, BS, 45, PhD(bact), 58. Prof Exp: Bacteriologist, Microbiol Div, State Bd Health, Ind, 46-49, actg chief, 50, chief dairy microbiol div, 53-54; asst bact, Purdue Univ, 54-55; from instr to asst prof, 57-63, ASSOC PROF BACT, ILL INST TECHNOL, 63-, ASSOC DEAN GRAD SCH, 69- Concurrent Pos: NIH res grants, 58-66. Mem: AAAS; Am Soc Microbiol; Brit Soc Gen Microbiol; Am Soc Naturalists. Res: Bacterial genetics; cytology; single-cell studies; frozen storage and growth initiation; diagnostic methods; food bacteriology. Mailing Add: Dept of Biol Ill Inst of Technol Chicago IL 60616

BRETZ, MICHAEL, b Harvey, Ill, June 2, 38; m 64; c 2. LOW TEMPERATURE PHYSICS. Educ: Univ Calif, Los Angeles, 61; Univ Wash, PhD(physics), 71. Prof Exp: Asst scientist space physics, Lockheed Res Labs, 61-65; assoc physics, Univ Wash, 71-73; ASST PROF PHYSICS, UNIV MICH, ANN ARBOR, 73- Mem: Am Phys Soc; Fedn Am Scientists. Res: Properties of over-layer quantum films adsorbed on homogeneous substrates. Mailing Add: Dept of Physics Univ of Mich Ann Arbor MI 48105

BRETZLOFF, CARL WARREN, plant biochemistry, see 12th edition

BREUER, CHARLES B, b Brooklyn, NY, Aug 7, 31; m 62; c 1. CHEMISTRY. Educ: NY Univ, BA, 53; Rutgers Univ, MS, 59, PhD(biochem), 64. Prof Exp: Res scientist, Squibb Inst Med Res, Olin Mathieson Chem Corp, 59-61; res asst biochem, Rutgers Univ, 61-63; res scientist, 63-66, group leader, 66-67, sr res scientist, 67-70, head biotherapeut dept, 70-71, DIR DIAG RES & DEVELOP, LEDERLE LABS, AM CYANAMID CO, 71- Mem: AAAS; Am Chem Soc; Am Soc Biol Chem. Res: Diagnostics; immunochemistry; biology of serum proteins; research in new diagnostic agents and principles. Mailing Add: Lederle Labs N Middletown Rd Pearl River NY 10965

BREUER, FREDERICK WILLIAM, chemistry, see 12th edition

BREUHAUS, HERBERT CHARLES, b Elmhurst, Ill, Oct 2, 07; m 35; c 4. GASTROENTEROLOGY, INTERNAL MEDICINE. Educ: Univ Chicago, BS, 29, MD, 34. Prof Exp: ASST PROF MED, RUSH MED COL, 71- Concurrent Pos: Attend physician, Rush-Presby-St Luke's Hosp, Chicago, Ill. Mem: Am Gastroenterol Asn; fel Am Col Physicians. Mailing Add: 1023 N Elmwood Ave Oak Park IL 60302

BREUKELMAN, JOHN (WILLIAM), b Corsica, SDak, May 17, 01; m 25; c 2. ECOLOGY. Educ: Yankton Col, AB, 23; Univ Iowa, MS & PhD(zool), 29. Prof Exp: Teacher high schs, SDak, 23-25, supt schs, 25-27; asst zool, Univ Iowa, 27-29; head dept biol, 29-58, actg dean grad sch, 44-47, prof biol, 29-68, EMER PROF BIOL, EMPORIA KANS STATE COL, 68- Concurrent Pos: Ed, Am Biol Teacher, 42-53 & Kans Sch Naturalist, 54-68; Fulbright lectr, Netherlands, 64-65 & Colombian Asn of Univs, 68; educ consult, US AID, Caracas, Venezuela, 69. Mem: AAAS; Nat Asn Biol Teachers (pres, 57); Am Nature Study Soc; Nat Wildlife Fedn. Res: Science education. Mailing Add: 1715 E Wilman Emporia KS 66801

BREUNIG, HENRY LATHAM, b Indianapolis, Ind, Nov 19, 10; m 38; c 3. ENGINEERING STATISTICS. Educ: Wabash Col, AB, 34; Johns Hopkins Univ, PhD(phys chem), 38. Prof Exp: Chemist, 38-56, chem statistician, 56-59, SR STATISTICIAN, STATIST RES DEPT, ELI LILLY & CO, 59- Mem: Am Chem Soc; Am Statist Asn; fel Am Soc Qual Control; Sigma Xi. Res: Pharmaceutical quality control; introduction and evaluation of statistical quality control procedures; pharmaceutical sampling; homogeneity studies; design of experiments; evaluation of analytical methods; statistical research; interlaboratory collaborative studies. Mailing Add: Statist Res M 730 Eli Lilly & Co Indianapolis IN 46206

BREUSCH, ROBERT HERMANN, b Freiburg, Ger, Apr 2, 07; nat US; m 36. MATHEMATICS. Educ: Univ Freiburg, PhD(math), 32. Prof Exp: Teacher, Ger, 32-36; prof math, Santa Maria, Chile, 36-39; teacher, Shady Hill Sch, 40-43; from asst prof to prof, 43-72, EMER WALKER PROF MATH, AMHERST COL, 72- Mem: Am Math Soc; Math Asn Am. Res: Analytic functions; prime numbers. Mailing Add: 19 Dana Place Amherst MA 01002

BREW, DAVID ALAN, b Clifton Springs, NY, Nov 22, 30; m 58; c 4. GEOLOGY. Educ: Dartmouth Col, AB, 52; Stanford Univ, PhD(geol), 64. Prof Exp: GEOLOGIST, US GEOL SURV, 52- Concurrent Pos: Asst, Stanford Univ, 58-59; Fulbright scholar, Univ Vienna, 59-60. Honors & Awards: Career Serv Award, Nat Civil Serv League, 73; Meritorious Serv Award, US Dept Interior, 73. Mem: Geol Soc Am; Am Geophys Union; Am Asn Petrol Geologists; Geol Soc Vienna. Res: Relations of sedimentation and tectonics; structure and structural geometry; tectonic and plutonic history of northeastern Pacific rim; relations of mineral deposits to plutonic rocks. Mailing Add: US Geol Surv 345 Middlefield Rd Menlo Park CA 94025

BREW, DOUGLAS CROCKER, geology, see 12th edition

BREW, WILLIAM BARNARD, b Sibley, Ill, Apr 23, 13; m 41; c 5. CHEMISTRY. Educ: Wash Univ, BS, 35, MS, 36. Prof Exp: Res chemist, 36-38, chief anal chemist, 38-41, mgr org res lab, 41-55, mgr org & inorg res lab, 55-60, mgr spec prod res lab, 60-64, dir cent res labs, 64-73, DIV VPRES & DIR CENT RES LABS, RALSTON PURINA CO, 73- Mem: Am Chem Soc; Inst Food Technologists; Am Inst Chemists. Res: Vitamin chemistry; amino acid chemistry; chromatography; spectrophotometry. Mailing Add: Cent Res Labs Ralston Purina Co Checkerboard Sq St Louis MO 63188

BREWBAKER, JAMES LYNN, b St Paul, Minn, Oct 11, 26; m 54; c 4. PLANT GENETICS. Educ: Univ Colo, BA, 48; Cornell Univ, PhD(plant breeding), 52. Prof Exp: Asst genetics, Calif Inst Technol, 48, Univ Minn, 50 & Cornell Univ, 49-52; Nat Res Found fel, Univ Lund, Sweden, 52-53; asst prof plant breeding, Univ Philippines & Cornell Univ, 53-55; assoc geneticist, Brookhaven Nat Lab, 56-61; assoc prof, 61-64, PROF HORT, UNIV HAWAII, 64- Concurrent Pos: Geneticist, Rockefeller Found, Thailand, 67-68 & Int Atomic Energy Agency, Philippines, 70. Mem: Genetics Soc Am; Bot Soc Am; fel Am Soc Agron. Res: Genetics of incompatibility in plants, pollen cytology and physiology; maize genetics; breeding field and forage crops. Mailing Add: Dept of Hort Univ of Hawaii 3190 Maile Way Honolulu HI 96822

BREWER, ALAN WEST, b Montreal, Que, Mar 18, 15; m 39; c 3. ATMOSPHERIC PHYSICS. Educ: Univ London, BSc, 36, MSc, 37, PhD(physics), 49. Prof Exp: Sci officer, Meteorol Off, UK, 37-46; scientist, Elliott Bros, London, Ltd, 46-48; lectr & reader meteorol, Oxford Univ, 48-62; PROF PHYSICS, UNIV TORONTO, 62- Honors & Awards: Patterson Medal, Atmospheric Environ Serv Can. Mem: Am Meteorol Soc; Can Meteorol Soc (pres, 66-67); Royal Meteorol Soc (vpres); Brit Inst Physics & Phys Soc. Res: Meteorology. Mailing Add: Dept of Physics Univ of Toronto Toronto ON Can

BREWER, ALLEN A, b Chicago, Ill, May 21, 11; m 45; c 6. DENTISTRY. Educ: Loyola Univ, Ill, DDS, 34. Prof Exp: HEAD DEPT, EASTMAN DENT CTR & CLIN PROF PROSTHODONTICS, SCH DENT, STATE UNIV NY BUFFALO, 67- Concurrent Pos: Assoc clin prof, Univ Rochester, 71- Honors & Awards: Schweitzer Award, 68. Mem: Hon mem Brit Soc Study Prosthetic Dent. Res: Prosthodontic research; bio-electronics; radios in teeth, centric jaw relationship; comparative study of tooth forms. Mailing Add: Eastman Dent Ctr 800 Main St E Rochester NY 14603

BREWER, ARTHUR DAVID, b Evesham, Eng, Nov 14, 41; Can citizen; m 68; c 1. AGRICULTURAL CHEMISTRY. Educ: Cambridge Univ, BA, 63, MA, 67; Univ Strathclyde, Scotland, PhD(chem), 67. Prof Exp: Fel chem, Univ Ariz, 67-68; res chemist, 68-75, RES SCIENTIST CHEM, UNIROYAL RES LAB, 75- Mem: Fel The Chem Soc; Can Inst Chemists; Brit Ornithologists Union. Res: Basic synthesis research in heterocyclic chemistry with view to discovery of new agricultural and pharmaceutical chemicals. Mailing Add: Uniroyal Res Lab Huron St Guelph ON Can

BREWER, BURNAS W, mathematics, see 12th edition

BREWER, CARL ROBERT, b Indianola, Iowa, Sept 2, 12; m 42; c 2. MICROBIOLOGY. Educ: Simpson Col, AB, 34; Iowa State Univ, PhD(physiol bact), 39. Prof Exp: Lectr bact & immunity, McGill Univ, 39-42; asst prof bact, Univ Maine, 42-43; res chemist, Lederle Labs, NY, 43-44; chief nutrit br, Biol Labs, Chem Corps, US Army, 44-52, allied sci div, 52-56, res div, Res & Develop Command, 56-60; chief res grants br, Nat Inst Gen Med Sci, Md, 60-65; assoc dean grad sch biomed sci Houston, dep dir inst biomed sci & prof microbiol, Univ Tex, 65-66; chief prog rev br, Div Regional Med Progs, 66, chief gen res support br, Div Res Resources, 66-71, SPEC ASST INSTNL RELS, DIV RES RESOURCES, NIH, 71- Concurrent Pos: Dir, Found Advan Educ in Sci, Md, 66-, treas, 67-68, pres, 68-70. Honors & Awards: Except Civilian Serv Award, US War Dept, 45. Mem: AAAS; Am Soc Biol Chem; Am Soc Microbiol; Am Acad Microbiol. Res: Bacterial metabolism; nutritional requirements and intermediary metabolism of pathogenic microorganisms; biomedical research program development. Mailing Add: Div of Res Resources Nat Insts of Health Bethesda MD 20014

BREWER, DONALD, b Claygate, Eng, Dec 1, 27; m 55; c 2. MYCOLOGY. Educ:

Univ Durham, BSc, 48; Univ Toronto, MSA, 52, PhD(plant path), 55. Prof Exp: Asst res officer, 55-63, ASSOC RES OFFICER, ATLANTIC REGIONAL LAB, NAT RES COUN CAN, 63- Mem: Mycol Soc Am; Phytopath Soc Can. Res: Physiology of fungi. Mailing Add: 6302 Jubilee Rd Halifax NS Can

BREWER, DOUGLAS G, b Toronto, Ont, Dec 22, 35; m 61; c 1. INORGANIC CHEMISTRY, ANALYTICAL CHEMISTRY. Educ: Univ Toronto, BA, 58, PhD(chem), 61. Prof Exp: From asst prof to assoc prof anal & inorg chem, 61-72, PROF CHEM & CHMN DEPT, UNIV NB, 72- Concurrent Pos: Operating grants, Nat Res Coun Can, 61-71 & NB Res & Productivity Coun, 63-64. Mem: Chem Inst Can. Res: Stability constants of metal complexes as determined by differential potentiometry and related methods; infrared, optical rotary dispersion, ultraviolet-vision, nuclear magnetic resonance and x-ray studies of metal complexes. Mailing Add: Dept of Chem Univ of NB Fredericton NB Can

BREWER, FRANKLIN DOUGLAS, b Electric Mills, Miss, Sept 25, 38; m 66; c 3. ENTOMOLOGY. Educ: Miss Col, BS, 60; Univ Southern Miss, MS, 63; Miss State Univ, PhD(entom), 70. Prof Exp: Instr biol, Miss Woman's Univ, 64-66; from instr microbiol to res assoc biochem, Miss State Univ, 66-71; RES ENTOMOLOGIST, BIOL INSECT CONTROL LAB, AGR RES SERV, USDA, 71- Mem: Sigma Xi; Entom Soc Am; AAAS. Res: Insect nutrition and mass rearing techniques for producing a consistent yet economical number of insects for biological control purposes. Mailing Add: Biol Insect Control Lab Agr Res Serv USDA PO Box 225 Stoneville MS 38776

BREWER, GARY DAVID, b Los Angeles, Calif, Aug 26, 48; m 67; c 2. FISH BIOLOGY. Educ: Calif State Polytech Univ, BS, 70; Univ Southern Calif, MS, 72, PhD(biol), 75. Prof Exp: Biologist fisheries, Calif Dept Fish & Game, 70; ASSOC RES ZOOLOGIST MARINE ECOL, ALLAN HANCOCK FOUND, UNIV SOUTHERN CALIF, 73- Concurrent Pos: Sigma Xi grant-in-aid of res, 73; Nat Oceanic & Atmospheric Admin Sea Prog grant, 74-75. Mem: Am Soc Ichthyologists & Herpetologists; Am Fisheries Soc; Sigma Xi. Res: Zoogeography of marine fishes; ecology and development of larval fishes; physiological ecology of fishes including thermal tolerance and resistance; uptake of heavy metals and pesticides. Mailing Add: Allan Hancock Found Univ Southern Calif Los Angeles CA 90007

BREWER, GLENN A, JR, b New Haven, Conn, Nov 3, 27; m 57; c 4. CHEMISTRY. Educ: Hobart Col, BS, 49; Univ Wis, MS, 52, PhD(biochem), 54. Prof Exp: Sr res biochemist, Com Solvents Corp, 54-56; sr res microbiologist, 56-60, supvr anal res, 60-67, ASST DIR ANAL RES, E R SQUIBB & SONS, INC, 67- Mem: Am Chem Soc; NY Acad Sci; Acad Pharmaceut Sci. Res: Pharmaceutical analysis. Mailing Add: E R Squibb & Sons Inc New Brunswick NJ 08903

BREWER, HAROLD REID, b College Park, Ga, Aug 1, 24; m 56; c 3. THEORETICAL PHYSICS. Educ: Ga Inst Technol, BS, 49; Univ NC, PhD(physics), 56. Prof Exp: From asst prof to assoc prof, 53-66, PROF PHYSICS, GA INST TECHNOL, 66- Mem: Am Phys Soc. Res: Internal conversion; beta decay. Mailing Add: 3896 Parkcrest Dr Atlanta GA 30319

BREWER, HOWARD EUGENE, b Indianola, Iowa, Apr 30, 10. PLANT PHYSIOLOGY. Educ: Simpson Col, AB, 31, BS, 32; Iowa State Col, PhD(plant ecol), 42. Prof Exp: Asst bot, Iowa State Col, 37-41; instr plant physiol & ecol, State Col Wash, 41-42; from asst prof to assoc prof bot, Ala Polytech Inst, 46-49, assoc botanist, Exp Sta, 47-49; from asst prof to prof bot, 49-74, coordr gen studies sci, 68-74, EMER PROF BOT, WASH STATE UNIV, 74- Concurrent Pos: Coop agent, Soil Conserv Serv, USDA, 38-42. Mem: Ecol Soc Am; Bot Soc Am; Am Soc Plant Physiol; Brit Ecol Soc; Scand Soc Plant Physiol. Res: Seed physiology; nitrogen nutrition of cereals; water stress relations of plants. Mailing Add: 1828 Wheatland Dr NE Pullman WA 99163

BREWER, JAMES EDWARD, b Philadelphia, Pa, June 17, 20; m 46; c 4. ORNAMENTAL HORTICULTURE. Educ: Pa State Univ, BS, 48. MS, 54, PhD(hort), 62. Prof Exp: Instr hort, Univ RI, 55-58; from instr to asst prof ornamental hort, 58-74, ASSOC PROF ORNAMENTAL HORT, PA STATE UNIV, UNIVERSITY PARK, 74- Mem: Am Soc Hort Sci; Int Soc Hort Sci. Res: Nutrition of woody ornamental plants. Mailing Add: Dept of Hort Pa State Univ University Park PA 16802

BREWER, JAMES W, b West Palm Beach, Fla, May 29, 42; m 63; c 2. ALGEBRA. Educ: Fla State Univ, AB, 64, PhD(math), 68. Prof Exp: Asst prof math, Va Polytech Inst, 68-70; asst prof, 70-73, ASSOC PROF MATH, UNIV KANS, 73- Mem: Math Asn Am; Am Math Soc. Res: Commutative algebra. Mailing Add: Dept of Math Univ of Kans Lawrence KS 66044

BREWER, JESSE WAYNE, b Rives, Mo, Oct 10, 40; m 64; c 2. ENTOMOLOGY, PLANT PATHOLOGY. Educ: Cent Mich Univ, BS, 63, MA, 65; Purdue Univ, PhD(entom), 68. Prof Exp: Instr biol, Muskegon Community Col, 64-65; asst prof entom, 68-73, ASSOC PROF ENTOM, COLO STATE UNIV, 73- Mem: AAAS; Entom Soc Am. Res: Forest entomology; pollination biology. Mailing Add: Dept of Zool & Entom Colo State Univ Ft Collins CO 80521

BREWER, JOHN GILBERT, b Robinson, Ga, May 11, 37. ANALYTICAL CHEMISTRY. Educ: Univ Ga, BS, 58, MS, 63, PhD(chem), 66. Prof Exp: Anal develop chemist, Chemstrand Co, Monsanto Chem Co, Fla, 58-59; asst chem, Univ Ga, 60-66; asst prof, The Citadel, 66-68; assoc prof, 68-74, PROF CHEM, ARMSTRONG STATE COL, 74- Mem: Am Chem Soc. Res: Analysis of multicomponent mixtures through fitting of their excess functions via computer usage. Mailing Add: Dept Chem Armstrong State Col 11935 Abercorn St Savannah GA 31406

BREWER, JOHN HANNA, b Gorman, Tex, Nov 10, 09; m 31; c 1. MEDICAL BACTERIOLOGY. Educ: Simmons Univ, Tex, AB, 30, AM, 31; Johns Hopkins Univ, PhD(med bact), 38. Prof Exp: Lab technician, State Dept Health, Tex, 33-35; bacteriologist, Hynson, Westcott & Dunning, 38-40, dir biol res, 40-64; dir biol safety & control, Becton, Dickinson & Co, 64-69; PROF MICROBIOL, HARDIN-SIMMONS UNIV, 69- Mem: Am Soc Microbiol; fel Am Pub Health Asn; Am Pharmaceut Asn. Res: Antiseptics; anaerobes; laboratory equipment; biological production; biological tests for plastics. Mailing Add: Dept of Microbiol Hardin-Simmons Univ Abilene TX 79601

BREWER, JOHN ISAAC, b Milford, Ill, Oct 9, 30; m 28; c 1. OBSTETRICS & GYNECOLOGY. Educ: Univ Chicago, BS, 25, Rush Med Col, MD, 28, PhD(anat), 35. Prof Exp: From instr to prof, 35-74, chmn dept, 72-74, EMER PROF OBSTET & GYNEC, MED SCH, NORTHWESTERN UNIV, 74- Concurrent Pos: Emer chmn obstet & gynec, Passavant Hosp, Chicago. Mem: Am Gynec Soc (pres, 64-65); Am Asn Obstet & Gynec (pres, 68-69); Am Asn Anatomists; Am Col Obstet & Gynec (pres, 59-60). Res: Corpus luteum; human embryology; trophoblastic diseases. Mailing Add: Northwestern Univ Med Sch 303 E Chicago Ave Chicago IL 60611

BREWER, JOHN MICHAEL, b Garden City, Kans, May 13, 38; m 65; c 2. BIOCHEMISTRY. Educ: Johns Hopkins Univ, BA, 60, PhD(biochem), 63. Prof Exp: Res assoc chem, Univ Ill, Urbana-Champaign, 63-66; asst prof chem, 66-73, ASSOC PROF BIOCHEM, UNIV GA, 73- Concurrent Pos: NIH fel, 64-65. Mem: Am Chem Soc; Am Soc Biol Chemists; Biophys Soc. Res: Photosynthesis, protein chemistry and structure and the mechanism of enzyme action. Mailing Add: Dept of Biochem Univ of Ga Athens GA 30602

BREWER, LEO, b St Louis, Mo, June 13, 19; m 45; c 3. PHYSICAL CHEMISTRY. Educ: Calif Inst Technol, BS, 40; Univ Calif, PhD(chem), 43. Prof Exp: Assoc, Manhattan Dist Proj, 43-46, from asst prof to assoc prof, 46-55, PROF CHEM, UNIV CALIF, BERKELEY, 55-, HEAD INORG MAT RES DIV, LAWRENCE BERKELEY LAB, 61-, ASSOC DIR, 67- Concurrent Pos: Assoc, Lawrence Berkeley Lab, 47-60; Guggenheim Mem fel, 50; secy comn high temperature, Int Union Pure & Appl Chem, 57-61, assoc mem comn thermodyn & thermochem, 74- Robert S Williams lectr, Mass Inst Technol & Henry Werner lectr, Univ Kans, 63; O M Smith lectr, Okla State Univ, 64; G N Lewis lectr, Univ Calif, 64, fac lectureship, 66; assoc ed, J Chem Physics; mem exec comt, Off Critical Tables; mem rev comt for reactor chem div, Oak Ridge Nat Lab; mem, Calorimetry Conf. Honors & Awards: Baekeland Award, 53; E O Lawrence Award, 61; Palladium Medal, Electrochem Soc, 71. Mem: Nat Acad Sci; AAAS; Am Chem Soc; The Chem Soc; Coblenz Soc. Res: High temperature chemistry and thermodynamics; theory of bonding in metallic solutions. Mailing Add: Dept of Chem Univ of Calif Berkeley CA 94720

BREWER, LEROY EARL, JR, b Hagerstown, Md, June 1, 36; m 56; c 1. THERMAL PHYSICS. Educ: Univ Fla, BS, 60; Univ Tenn, MS, 65; Univ Brussels, PhD(appl sci), 75. Prof Exp: Res engr, Arnold Eng Develop Ctr, Aro, Inc, 60-66; physicist, Space Sci Lab, Gen Elec Co, Pa, 66-69; PRIN INVESTR, ARNOLD ENG DEVELOP CTR, ARO INC, 69- Mem: Am Phys Soc. Res: Fundamental and applied research in fields of plasma, combustion and infrared physics. Mailing Add: 1718 Country Club Dr Tullahoma TN 37388

BREWER, NATHAN RONALD, b Albany, NY, June 28, 04; m 36; c 3. VETERINARY PHYSIOLOGY. Educ: Mich State Univ, BS, 30, DVM, 37; Univ Chicago, PhD(physiol), 36. Prof Exp: Vet practitioner, 37-45; assoc prof physiol & dir animal quarters, 45-69, EMER PROF PHYSIOL & EMER DIR ANIMAL QUARTERS, UNIV CHICAGO, 69-; EXEC SECY, ILL SOC MED RES, 69- Concurrent Pos: Consult, Chicago Med Sch, Nat Res Coun, 53-60, Chicago Col Osteop Med, Mercy Hosp & Ill Inst Technol, Chicago, 69-; ed, Proc Asn Lab Animal Sci, 50-62, ed emer, Lab Animal Care, 63- Honors & Awards: Cert of Merit, Nat Soc Med Res, 59; Griffin Award, Asn Lab Animal Sci, 63- Mem: Hon mem Am Soc Vet Cardiol; Nat Soc Med Res; Asn Lab Animal Sci; Am Physiol Soc; Am Vet Med Asn. Res: Laboratory animal medicine. Mailing Add: 5526 Blackstone Ave Chicago IL 60637

BREWER, RICHARD (DEAN), b Murphysboro, Ill, June 17, 33; m 57; c 2. ECOLOGY, ORNITHOLOGY. Educ: Southern Ill Univ, BA, 55; Univ Ill, MS, 57, PhD(zool), 59. Prof Exp: Res asst zool, Univ Ill, 55-56; from instr to assoc prof, 59-71, PROF BIOL, WESTERN MICH UNIV, 71- Concurrent Pos: Asst dir, C C Adams Ctr Ecol Studies, 60-63, dir, 63-68. Mem: Fel AAAS; Ecol Soc Am; Am Ornith Union; Wilson Ornith Soc; Cooper Ornith Soc. Res: Competition, speciation and regulation of population size in birds; composition and structure of plant animal communities. Mailing Add: Dept of Biol Western Mich Univ Kalamazoo MI 49001

BREWER, RICHARD GEORGE, b Los Angeles, Calif, Dec 8, 28; m 54; c 3. QUANTUM PHYSICS, MOLECULAR PHYSICS. Educ: Calif Inst Technol, BS, 51; Univ Calif, Berkeley, PhD(chem), 58. Prof Exp: Researcher, Aerojet-Gen Corp, 51-53; instr, Harvard Univ, 58-60; asst prof, Univ Calif, Los Angeles, 60-63; res physicist, Res Lab, IBM Corp, 63-68; vis prof physics, Mass Inst Technol, 68-69; res physicist, 69-73, IBM FEL, RES LAB, IBM CORP, 73- Concurrent Pos: Consult, Space Technol Labs, Inc, 62-63; mem prog comt, Int Quantum Electronics Conf VII & VIII, 72 & 74; mem comt atomic & molecular physics, Nat Acad Sci-Nat Res Coun, 74-77; Japan Soc Prom Sci vis prof physics & appl physics, Univ Tokyo, 75. Mem: Fel Am Phys Soc. Res: Optical and radio frequency spectroscopy; molecular spectroscopy; optical pumping; nonlinear optic effects using lasers and nonlinear infrared spectroscopy of molecules, including coherent optical transients and molecular collision phenomena. Mailing Add: Res Lab IBM Corp San Jose CA 95193

BREWER, ROBERT FRANKLIN, b Woodbury, NJ, May 25, 27; m 47. HORTICULTURE. Educ: Rutgers Univ, BSc, 50, PhD(hort), 53. Prof Exp: Soil chemist, Dept Soils & Plant Nutrit, Citrus Exp Sta, 53-70, HORTICULTURIST, PLANT SCI DEPT, SAN JOAQUIN VALLEY RES CTR, UNIV CALIF, RIVERSIDE, 70- Mem: Am Soc Hort Sci; Int Soc Citricult; Air Pollution Control Asn. Res: Inorganic plant nutrition; effects of fluorine and ozone on plant growth; cold tolerance; frost protection; environment modification. Mailing Add: San Joaquin Valley Res Ctr 9240 S Riverbend Ave Parlier CA 93648

BREWER, ROBERT HYDE, b Richmond, Va, Dec 24, 31; m 60; c 1. ANIMAL ECOLOGY, INVERTEBRATE ZOOLOGY. Educ: Hanover Col, AB, 55; Univ Chicago, PhD(zool), 63. Prof Exp: Asst prof biol, Ill Col, 63-65; res fel & lectr entom, Waite Agr Res Inst, Univ Adelaide, 65-68; asst prof biol, 68-72, ASSOC PROF BIOL, TRINITY COL, CONN, 72- Mem: Fel AAAS; Ecol Soc Am; Am Soc Zool; Am Inst Biol Sci. Res: Physiological and population ecology of aquatic invertebrates; host-parasitoid relations; biological control. Mailing Add: Dept of Biol Trinity Col Hartford CT 06106

BREWER, ROBERT NELSON, b Philcampbell, Ala, Feb 24, 34; m 53; c 1. POULTRY PATHOLOGY. Educ: Auburn Univ, BS, 55, MS, 60; Univ Ga, PhD(poultry sci), 68. Prof Exp: Asst county agr agent, Exten Serv, Auburn Univ, 55-58; poultry specialist, Pillsbury Co, 60-65; asst prof poultry parasitol & path, 68-74, ASSOC PROF POULTRY SCI, AUBURN UNIV, 74- Mem: Poultry Sci Asn; Am Inst Biol Sci. Res: Poultry disease research, including immunity studies on roundworms, coccidiosis and epidemiological and etiological studies on Marek's disease; environment-disease interrelationships. Mailing Add: Dept of Poultry Sci Auburn Univ Auburn AL 36830

BREWER, STEPHEN WILEY, JR, b Spartanburg, SC, Jan 18, 41. ANALYTICAL CHEMISTRY. Educ: Univ Fla, BSCh, 62; Univ Wis-Madison, PhD(anal chem), 69. Prof Exp: Asst prof, 69-73, ASSOC PROF CHEM, EASTERN MICH UNIV, 73- Mem: Soc Appl Spectros; Spectros Soc Can. Res: Emission spectroscopy; electron probe microanalysis. Mailing Add: Dept of Chem Eastern Mich Univ Ypsilanti MI 48197

BREWER, STUART DEXTER, b Beverly, Mass, Aug 31, 15; m 42; c 4. ORGANOMETTALIC CHEMISTRY. Educ: Bowdoin Col, BS, 38; NY Univ, PhD(chem), 41. Prof Exp: Asst inorg & org chem, NY Univ, 38-41; res assoc, Explosives Res Lab, Nat Defense Res Comt, Off Sci Res & Develop, Carnegie Inst Technol, 41-45; res assoc, Res Lab, 45-49, appln engr, Silicone Prod Dept, 49-50, mgr anal serv, 50-56, mgr resin prod eng, 56-70, CONSULT PROCESS CHEM, SILICONE PROD DEPT, GEN ELEC CO, 70- Mem: Am Chem Soc. Res: Application technology and chemistry of silicones; mass spectrometry; explosives; amino acid chemistry. Mailing Add: 18 Parkwood Dr Burnt Hills NY 12027

BREWER, WILLIAM AUGUSTUS, b Oakland, Calif, May 27, 30. ENVIRONMENTAL MANAGEMENT, REMOTE SENSING. Educ: Univ Calif, Berkeley, BA, 54, MA, 55, PhD(eng), 63; Harvard Bus Sch, PMD, 67. Prof Exp: Sr geologist, Chile Explor Co, Anaconda Co, 55-60; assoc eng, Univ Calif, Berkeley, 60-63; dep div chief, Cent Intel Agency, 63-67; chief eng, Nat Photog Interpretation Ctr, 67-68; mgr preliminary design, Fed Systs, IBM Corp, 68-69; consult engr, 69-73; exec dir, Wash State Energy Policy Coun, 73-75; RES PROF & ACTG DIR ENERGY RES CTR, INST ENVIRON STUDIES, UNIV WASH, 75- Concurrent Pos: Prin investr geol, US Nat Aerospace Agency, 72; energy policy consult, Gov Off, State Wash & Nat Gov Conf, 73- Res: Energy and environmental policy studies; energy forecasting and planning; energy conversion; natural resources inventory and development; remote sensing and photogeologic mapping. Mailing Add: Inst for Environ Studies Univ of Wash Seattle WA 98195

BREWER, WILLIS RALPH, b Zell, SDak, June 8, 19; m 42; c 2. PHARMACOGNOSY. Educ: SDak State Col, BS, 42; Ohio State Univ, PhD(mat med), 48. Prof Exp: Asst, Ohio State Univ, 42-44, 46-48; asst prof pharm & med, Univ Utah, 48-49; from assoc prof to prof pharmacog, Col Pharm, 49-75, dean col, 52-75, ASST TO VPRES FOR HEALTH SCI, UNIV ARIZ, 75- Concurrent Pos: Dir, Univ Ariz Scholar & Awards & pres, Campus Christian Ctr, 67-69; pres, Pima Coun Alcoholism, 69-70. Mem: AAAS; Am Pharmaceut Asn. Res: Phytochemistry and phytophysiology of solanaceous drug plants; horticulture; medicinal plants of southwest United States and Mexico. Mailing Add: Off of Vpres for Health Sci Univ of Ariz Tucson AZ 85721

BREWER, WILMA DENELL, b Riley, Kans, Oct 18, 15. NUTRITION. Educ: Kans State Col, BS, 35; State Col Wash, MS, 39; Mich State Col, PhD, 50. Prof Exp: Instr home econ, Simpson Col, 39-40; instr foods & nutrit, Univ NH, 40-42, asst prof foods & nutrit & asst home economist, Exp Sta, 42-43; asst human nutrit, Mich State Col, 43-48, from asst prof to assoc prof foods & nutrit, 48-57; PROF FOODS & NUTRIT, IOWA STATE UNIV, 57-, HEAD DEPT, 61- Mem: AAAS; Am Home Econ Asn; Am Dietetic Asn; Am Chem Soc; Am Inst Nutrit. Res: Human nutrition; home economics; protein requirements of women. Mailing Add: Dept of Food & Nutrit Iowa State Univ Ames IA 50010

BREWSTER, JAMES HENRY, b Ft Collins, Colo, Aug 21, 22; m 54; c 3. ORGANIC CHEMISTRY. Educ: Cornell Univ, BA, 42; Univ Ill, PhD(chem), 48. Prof Exp: Chemist, Atlantic Refining Co, 42-43; asst, Univ Ill, 46-47; Fels Fund fel, Univ Chicago, 48-49; from instr to assoc prof, 49-60, PROF CHEM, PURDUE UNIV, 60- Mem: AAAS; Am Chem Soc; The Chem Soc. Res: Stereochemistry; displacement reaction and reductions. Mailing Add: Dept of Chem Purdue Univ West Lafayette IN 47907

BREWSTER, JOHN LA DUE, b Phoenix, Ariz, May 9, 30; m 54; c 2. PHYSICS. Educ: Pasadena Col, AB, 53; Univ Calif, Los Angeles, MA, 56, PhD(physics), 63. Prof Exp: Mem tech staff instrumentation, Hughes Aircraft Co, 56-57, mem tech staff electron optics, Res Labs, 57-60; res physicist, Linfield Res Inst, 60-62; res physicist, X-ray Systs, Field Emission Corp, 60-64, asst dir res & develop, 64-70, mgr res & develop & eng, 70-73; PROJ MGR, HEWLETT-PACKARD, 73- Concurrent Pos: Spec asst prof, Linfield Col, 61-63; prof physics, George Fox Col, 64-71. Mem: AAAS. Res: Acoustics; ultrasonics; superconductivity; electron optics; high power tube design; pulser design; x-ray system design; medical equipment design. Mailing Add: PO Box 149 Dundee OR 97115

BREWSTER, MARJORIE ANN, b Conway, Ark, Mar 7, 40. CLINICAL BIOCHEMISTRY. Educ: Univ Ark, Little Rock, BS, 64, MS, 66, PhD(biochem), 71; Am Bd Clin Chem, dipl, 74. Prof Exp: Clin biochemist, Clin Lab, 66-67, ASST PROF PATH, MED CTR, UNIV ARK, 69-, ASST PROF BIOCHEM, 73-; ASST DIR, ARK CHILDREN'S HOSP LAB, 76- Concurrent Pos: Mem, Bd Dirs, Nat Registry Clin Chemists. Mem: Am Asn Clin Chemists; Asn Clin Sci. Res: Birth defects; leukocyte chemistry; lipoprotein metabolism. Mailing Add: Clin Lab Univ of Ark Med Ctr Little Rock AR 72205

BREWTON, EDWARD S, pharmacy, see 12th edition

BREY, WALLACE SIEGFRIED, JR, b Schwenksville, Pa, June 6, 22; m 55; c 2. PHYSICAL CHEMISTRY. Educ: Ursinus Col, BS, 42; Univ Pa, MS, 46, PhD(chem), 48. Prof Exp: Asst chemist, Warner Co, 42-44; res chemist, Publicker Indust, 44; asst prof chem, De Pauw Univ, 48-49; from asst prof to assoc prof, St Joseph's Col, Pa, 49-52; from asst prof to assoc prof, 52-64, PROF CHEM, UNIV FLA, 64- Concurrent Pos: Ed, J Magnetic Resonance, 69- Mem: AAAS; Am Chem Soc; Am Phys Soc; The Chem Soc. Res: Magnetic resonance; heterogeneous catalysis and adsorption; molecular interactions in liquids; spectroscopy of biological molecules. Mailing Add: 800 NW 37th Dr Gainesville FL 32601

BREYER, ARTHUR CHARLES, b Brooklyn, NY, June 13, 25; m 48; c 3. PHYSICAL CHEMISTRY, INORGANIC CHEMISTRY. Educ: NY Univ, BA, 48; Columbia Univ, MA, 50; Rutgers Univ, PhD(chem), 58. Prof Exp: Instr chem, Columbia Univ, 48-49; from instr to asst prof, Upsala Col, 49-58; assoc prof, Ohio Wesleyan Univ, 59-63; prof, Harvey Mudd Col, 63-64; PROF CHEM & CHMN DEPT CHEM & PHYSICS, BEAVER COL, 64- Concurrent Pos: Chem study proj lectr, Gatlinburg Ion-Exchange Res Conf, 59; dir, NSF Undergrad Res Participation Prog, 60-64, dir, Inst High Sch Chem Teachers, 62-; consult, NSF, 62-, McGraw-Hill Book Co, Inc; Charles E Merrill Books, Inc & Wadsworth Publ Co. Mem: AAAS; Am Chem Soc; Am Sci Affil; Franklin Inst. Res: Chromatography and electrophoresis of plant pigments, dyes and surfactants; ion exchange catalysis; metal ion complexation. Mailing Add: Dept of Chem Beaver Col Glenside PA 19038

BREYERE, EDWARD JOSEPH, b Washington, DC, Apr 25, 27; m 51; c 1. IMMUNOGENETICS. Educ: Univ Md, BS, 51, MS, 54, PhD(zool), 57. Prof Exp: Asst, Univ Md, 54-55; assoc prof, 61-67, PROF BIOL, AM UNIV, 67-; DIR, IUMMOGENETICS LAB, SIBLEY MEM HOSP, 61- Concurrent Pos: Res fel, Nat Cancer Inst, 57-61. Mem: AAAS; Am Asn Cancer Res; Am Inst Biol Sci; Am Genetics Asn; Transplantation Soc. Res: Transplantation immunity; maternal-fetal relationships; immunology of oncogenic viruses. Mailing Add: Sibley Mem Hosp 5255 Loughboro Rd NW Washington DC 20016

BREZENOFF, HENRY EVANS, b New York, NY, July 9, 40; m 64; c 2. NEUROPHARMACOLOGY. Educ: Columbia Univ, BS, 62; NJ Col Med & Dent, PhD(pharmacol), 68. Prof Exp: From instr to asst prof, 69-74, ASSOC PROF PHARMACOL, COL MED & DENT NJ, 74- Concurrent Pos: Fel pharmacol, Sch Med, Univ Calif, Los Angeles, 68-69. Mem: AAAS; Am Soc Pharmacol & Exp Therapeut. Res: Pharmacology of the hypothalamus and hypothalamic control of autonomic functions; brain acetylcholine, tolerance to barbiturates. Mailing Add: Col of Med & Dent of NJ 100 Bergen St Newark NJ 07103

BREZENSKI, FRANCIS T, b Dudley, Mass, Apr 2, 32; m 62; c 1. MICROBIOLOGY, BIOCHEMISTRY. Educ: Boston Univ, AB, 60; Univ Mass, MS, 62. Prof Exp: Dir res & develop, 64-69, LAB DIR, ENVIRON PROTECTION AGENCY, 69- Concurrent Pos: Mem adv comt lab methodology, Fed Water Pollution Control Admin, 66, consult to chief microbiologist, 66-67. Mem: NY Acad Sci; Sigma Xi; Am Soc Microbiol; Am Soc Testing & Mat. Res: Food and water microbiology; marine viruses and bacteria. Mailing Add: Environ Protection Agency Raritan Arsenal Edison NJ 08817

BREZNER, JEROME, b New York, NY, July 18, 31; m 54; c 2. ENTOMOLOGY, BIOCHEMISTRY. Educ: Univ Rochester, AB, 52; Univ Mo, AM, 56, PhD(entom), 59. Prof Exp: Asst prof biol, Elmira Col, 59-61; res assoc, 61-62, from asst prof to assoc prof, 62-68, PROF ENTOM, STATE UNIV NY COL ENVIRON SCI & FORESTRY, 68- Concurrent Pos: Am Physiol Soc fel biol, Dartmouth Med Sch, 60; NIH res grant, 62-65; US Forest Serv grant, 66-69. Mem: AAAS; Entom Soc Am. Res: Insect enzyme proteins, the species specific nature of these molecules. Mailing Add: Dept of Entom State Univ of NY Col of Environ Sci & Forestry Syracuse NY 13210

BREZONIK, PATRICK LEE, b Sheboygan, Wis, July 17, 41; m 65; c 2. WATER CHEMISTRY, LIMNOLOGY. Educ: Marquette Univ, BS, 63; Univ Wis, MS, 65, PhD(water chem), 68. Prof Exp: Asst prof water chem & environ eng, 66-70, ASSOC PROF WATER CHEM, UNIV FLA, 70- Concurrent Pos: Res grants, Fed Water Pollution Control Admin, 68-71 & Off Water Resources Res, 69-71; NSF sci fac fel, Swiss Fed Inst Technol, Zurich, 71-72. Mem: AAAS; Am Chem Soc; Am Soc Limnol & Oceanog; Water Pollution Control Fedn. Res: Eutrophication of lakes; nitrogen dynamics in natural waters; nutrient chemistry; ionic equilibria; trace metals in natural waters; organic matter in water; biochemistry of waste water treatment. Mailing Add: Dept of Environ Eng Sci Univ of Fla Gainesville FL 32611

BRIAND, FREDERIC JEAN-PAUL, b Paris, France, Nov 1, 49. ECOLOGY. Educ: Univ Paris, BSc, 70; Univ Calif, Irvine, MS, 72, PhD(ecol), 74. Prof Exp: ASST PROF ECOL, DEPT BIOL, UNIV OTTAWA, 74- Mem: Ecol Soc Am; Fr Can Asn Advan Sci. Res: Role of competition and predation in the regulation of planktonic systems; selective grazing by freshwater and marine copepods; biological control of toxic algae; heavy metal binding by algae and algal byproducts. Mailing Add: Dept of Biol Univ of Ottawa Ottawa ON Can

BRICE, DAVID KENNETH, b Sulphur Springs, Tex, Apr 15, 33; m 64. SOLID STATE PHYSICS. Educ: ETex State Col, BS, 54; Univ Kans, MS, 56, PhD(physics), 63. Prof Exp: MEM TECH STAFF PHYSICS, SANDIA CORP, 63- Concurrent Pos: Assoc prof physics, NMex Inst Mining & Technol, 67-73. Mem: Am Phys Soc. Res: Ion implantation; radiation damage; lattice dynamics and electron spin resonance. Mailing Add: Sandia Corp Orgn 5111 Albuquerque NM 87115

BRICE, JAMES COBLE, b Union City, Tenn, Dec 21, 20. GEOMORPHOLOGY. Educ: Univ Ala, BS, 42; Univ Calif, PhD(geol), 50. Prof Exp: From asst prof to assoc prof, 50-64, PROF GEOL, WASH UNIV, 64- Concurrent Pos: Ford Found fel, 51-52; hydrologist, US Geol Surv, 53- Mem: Geol Soc Am; Asn Am Geographers. Res: Form and behavior of streams. Mailing Add: Dept of Earth Sci Wash Univ St Louis MO 63130

BRICE, LUTHER KENNEDY, b Spartanburg, SC, Jan 29, 28. PHYSICAL CHEMISTRY. Educ: Harvard Univ, BA, 49; Dartmouth Col, MA, 51; Duke Univ, PhD(chem), 55. Prof Exp: From asst prof to assoc prof, 54-65, PROF CHEM, VA POLYTECH INST & STATE UNIV, 65- Mem: Am Chem Soc. Res: Acid catalysis in aqueous and non-aqueous solvents. Mailing Add: Dept of Chem Va Polytech Inst & State Univ Blacksburg VA 24061

BRICE, THOMAS JACOB, fluorine chemistry, see 12th edition

BRICK, IRVING B, b Oakland, Calif, Apr 24, 14. INTERNAL MEDIICNE, GASTROENTEROLOGY. Educ: George Washington Univ, AB, 37, MD, 41; Am Bd Internal Med, dipl, 49; Am Bd Gastroenterol, dipl, 51. Prof Exp: From instr to assoc prof, 47-61, actg chmn dept, 68-69 & 70-72, PROF MED, SCH MED, GEORGETOWN UNIV, 61- Concurrent Pos: Sr med consult, Rehab & Vet Affairs Comn, Am Legion, 48-; consult, Clin Ctr, NIH, 57-; actg dir, Georgetown Univ Med Div, Washington, DC Gen Hosp, 64-65. Mem: Am Col Physicians; Am Gastroenterol Asn; Am Asn Study Liver Dis; Am Soc Gastroenterol Endocrinol; Am Fedn Clin Res. Res: Radiation effect on the gastrointestinal tract; liver disease; gastrointestinal endoscopy. Mailing Add: Georgetown Univ Hosp Washington DC 20007

BRICK, ROBERT WAYNE, b Dallas, Tex, May 24, 39; m 68; c 1. ANIMAL PHYSIOLOGY. Educ: Tex Technol Col, BS, 62; Univ Hawaii, MS, 70, PhD(zool), 75. Prof Exp: ASST PROF WILDLIFE & FISHERIES SCI, TEX A&M UNIV, 75- Mem: Am Soc Zoologists; Am Inst Biol Sci. Res: Physiology and culture of crustaceans, especially freshwater shrimps of the genus Macrobrachium and saltwater shrimps of the genus Penaeus. Mailing Add: Dept of Wildlife & Fisheries Sci Tex A&M Univ College Station TX 77843

BRICKBAUER, ELWOOD ARTHUR, b Elkhart Lake, Wis. AGRONOMY. Educ: Univ Wis, BS, 43, MS, 61. Prof Exp: Teacher, Pub Schs, 43-44; mgr, Peck Seed Farms, 44-46; county agent, 46-56, exten agronomist, 57-71, from asst prof to assoc prof, 57-70, PROF AGRON, UNIV WIS-MADISON, 70- Mem: Am Soc Agron. Res: Corn, soybeans, small grains and seed certification. Mailing Add: Dept of Agron Col of Agr Univ of Wis Madison WI 53706

BRICKER, CLARK EUGENE, b Shrewsbury, Pa, June 17, 18; m 42; c 3. CHEMISTRY. Educ: Gettysburg Col, BA, 39; Haverford Col, MS, 40; Princeton Univ, PhD(anal chem), 44. Hon Degrees: DSc, Pikeville Col, 70. Prof Exp: Res chemist, Heyden Chem Corp, NJ, 43-46; asst prof chem, Johns Hopkins Univ, 46-48; from asst prof to prof, Princeton Univ, 48-61; dean, Col Wooster, 61-63; PROF CHEM, UNIV KANS, 63- Mem: AAAS; Am Chem Soc. Res: Physicochemistry methods of analysis; electrochemistry; spectrophotometry; photochemistry. Mailing Add: Dept of Chem Univ of Kans Lawrence KS 66044

BRICKER, JEROME GOUGH, b Buffalo, NY, Jan 20, 28; m 49; c 2. MEDICAL PHYSIOLOGY. Educ: Univ Akron, BSc, 49; Ohio State Univ, MSc, 53, PhD(physiol), 63. Prof Exp: Res & develop adminr air force weapons, Air Res & Develop Command, US Air Force, 54-60, staff scientist, Air Force Systs Command, 63-66, chief life sci div, Off Aerospace Res, 66-69; SR RES SCIENTIST, GEOMET, INC, ROCKVILLE, 69- Mem: Sigma Xi; NY Acad Sci. Res: Endocrinology, especially intermediate metabolism; respiratory and bacterial physiology. Mailing Add: 6000 Cable Ave Camp Springs MD 20023

BRICKER, NEAL S, b Denver, Colo, Apr 18, 27; m 51; c 3. INTERNAL MEDICINE. Educ: Univ Colo, BA, 46, MD, 49; Am Bd Internal Med, dipl, 56. Prof Exp: Intern & resident, Post-Grad Med Sch, NY Univ-Bellevue Med Ctr, 49-52; clin asst, Sch Med, Univ Colo, 52-54; sr resident, Peter Bent Brigham Hosp, 54-55; instr, Harvard Med Sch, 55-56; from asst prof to prof med, Sch Med, Wash Univ, 56-72, dir renal div, 56-72; PROF MED & CHMN DEPT, ALBERT EINSTEIN COL MED, 72- Concurrent Pos: Fel, Howard Hughes Med Inst, 55-56; jr assoc & assoc dir, Cardiorenal Lab, Peter Bent Brigham Hosp, 55-56; estab investr, Am Heart Asn, 59-64; assoc ed, J Lab & Clin Med, 61-67; investr, Mt Desert Island Biol Labs, 62-66; mem sci adv bd, Nat Kidney Dis Found, 62-69, chmn res & fel grants comt, 64-65, mem exec comt, 68-71; USPHS career res award, 64-72; res & fel grants, Nat Kidney Dis Found, 64-; consult, NIH, 64-68, chmn gen med study sect, 66-68, chmn renal dis & urol teaching grants comt, 69-71; mem gen med study sect, US Dept Health, Educ & Welfare, 64-; mem drug efficacy comt, Nat Acad Sci, 66-68, mem comt space biol & med, 71-72, mem ad hoc panel renal & metab effects space flight, 71-72, chmn comt, 72-74, mem space sci bd comt space biol & med, 72; mem nephrol test comt, Am Bd Internal Med, 70, chmn, 74-, mem parent comt & mem bd, 72; chmn comt nat med policy, Am Soc Clin Invest, 73- Honors & Awards: Gold-Headed Cane Award, Univ Colo, 49. Mem: Am Soc Clin Invest (pres, 72-73); fel Am Col Physicians; Asn Am Physicians; Int Soc Nephrol (vpres, 66-69, treas, 69-); Am Soc Nephrol (past pres). Res: Nephrology and transport across isolated membranes. Mailing Add: Dept of Med Albert Einstein Col of Med Bronx NY 10461

BRICKER, OWEN P, III, b Lancaster, Pa, Mar 5, 36; m 58; c 4. GEOCHEMISTRY. Educ: Franklin & Marshall Col, BS, 58; Lehigh Univ, MS, 60; Harvard Univ, PhD(geol), 64. Prof Exp: Fel environ sci & eng, Harvard Univ, 64-65; from asst prof to assoc prof geol, Johns Hopkins Univ, 65-75; GEOLOGIST, MD GEOL SURV, 75- Mem: Geochem Soc; Am Chem Soc; Mineral Soc Am; Soc Econ Geol. Res: Mineral equilibria in the earth-surface environment; chemistry of weathering and supergene processes; chemistry of natural waters. Mailing Add: Md Geol Surv Baltimore MD 21218

BRICKER, VICTORIA REIFLER, b Hong Kong, China, June 15, 40; US citizen; m 64. ETHNOLOGY. Educ: Stanford Univ, AB, 62; Harvard Univ, AM, 63, PhD(anthrop), 68. Prof Exp: Vis lectr, 69-70, asst prof, 70-73, ASSOC PROF ANTHROP, TULANE UNIV, 73- Concurrent Pos: Bk rev ed, Am Anthropologist, Am Anthrop Asn, 71-73, ed, Am Ethnologist, 73-76; mem, Maya Cult Sem. Mem: Fel Am Anthrop Asn; fel Fr Soc Americanistes; fel Royal Anthrop Inst; Ling Soc Am; Am Ethnol Soc. Res: Indian rebellions in Latin America; relationship between oral tradition and written history. Mailing Add: Dept of Anthrop Tulane Univ New Orleans LA 70118

BRICKMAN, HARRY RUSSELL, b New York, NY, Feb 16, 24; m 48; c 2. PSYCHIATRY, PSYCHOANALYSIS. Educ: NY Univ, MD, 47. Prof Exp: Dir, Riverside State Ment Hyg Clin, Calif, 51-54; clin dir, Calif Youth Authority Reception Ctr & Clin, 54-56; asst prof psychiat, Sch Med & dir out-patient dept, Neuropsychiat Inst, Univ Calif, Los Angeles, 56-60; DIR, LOS ANGELES COUNTY DEPT MENT HEALTH, 60- Concurrent Pos: Consult, Calif Youth Authority, 56-57 & Calif Dept Corrections, 58-; clin prof, Med Sch, Univ Calif, Los Angeles, 69- Mem: Fel Am Psychiat Asn; AMA; fel Am Pub Health Asn. Res: Social psychiatry; preventive psychiatry; mental health. Mailing Add: L A County Dept Ment Health 1106 S Crenshaw Blvd Los Angeles CA 90019

BRICKMAN, LEO, b Can, Apr 8, 15; nat US; m 41; c 2. ORGANIC CHEMISTRY. Educ: Univ Man, BSc, 35, MSc, 36; McGill Univ, PhD(chem), 40. Prof Exp: Chief chemist, Mallinckrodt Chem Works, Can, 40-45 & Monsanto Ltd, 46-47; dir spec prod lab res & develop, 49-61, assoc dir surg adhesives, 62, dir adhesives & orthop lab, 63-70, nat dir tech serv, 70-74, NAT DIR QUAL ASSURANCE & TECH SERV, JOHNSON & JOHNSON, 74- Mem: AAAS; Am Chem Soc; Am Soc Qual Control. Res: Structure of lignin; synthetic organic chemistry; pharmaceutical chemistry; adhesive and orthopedic products; surgical dressings; plastics; gypsum; fabrics. Mailing Add: 95 Norris Ave Metuchen NJ 08840

BRICKS, BERNARD GERARD, b Pittsburgh, Pa, Nov 28, 41; m 63; c 2. LASERS. Educ: Memphis State Univ, BS, 63; Johns Hopkins Univ, MS, 69, PhD(physics), 71. Prof Exp: Fel, New Eng Inst, 71-72; LASER PHYSICIST, SPACE SCI LAB, GEN ELEC CO, 72- Mem: AAAS; Am Phys Soc; Am Asn Physics Teachers. Res: Research and developmental program in metal vapor lasers with principal emphasis on the copper vapor laser and its applications. Mailing Add: Space Sci Lab Gen Elec Corp King of Prussia PA 19406

BRICKWEDDE, FERDINAND GRAFT, b Baltimore, Md, Mar 26, 03; m 34; c 3. LOW TEMPERATURE PHYSICS. Educ: Johns Hopkins Univ, BA, 22, MA, 24, PhD(physics), 25. Prof Exp: Res physicist physiol optics & Munsell Color Co res assoc, Nat Bur Standards, 25-26, chief, Low Temperature Lab, 26-46, chief, Thermodynamics Sect, 46-52, chief, Heat & Power Div, 45-56; dean col chem & physics, 56-63, Evan Pugh res prof physics, 63-68, EMER EVAN PUGH RES PROF PHYSICS, PA STATE UNIV, 68- Concurrent Pos: Prof physics, Nat Bur Standards Grad Sch, Washington, DC, 31-42, USDA Grad Sch, 34-42 & Univ Md, College Park, 42-56; consult low temperature physics, Cryogenic Lab, Los Alamos Sci Lab, NMex, 48-72; physicist, Livermore Radiation Lab, Univ Calif, 52-53; mem adv comt thermometer, Int Comt Weights & Measures, Sevres, France, 58-, pres, 65-68. Honors & Awards: Hizzebrand Prize, Chem Soc Washington, 40; Award, Washington Acad Sci, 41; Meritorious Serv Gold Medal Award, US Dept Com, 53. Mem: Fel Am Phys Soc; fel AAAS; fel Acoustical Soc Am; Am Chem Soc; Am Asn Physics Teachers. Res: Experimental solid state physics at low temperatures, especially superconductivity—tunneling, superconducting energy-gap, pinning of magnetic flux; magnetoacoustic absorption in metals—quantum oscillations, geometric oscillations and Doppler shifted cyclotron resonances. Mailing Add: 104 Davey Lab Pa State Univ University Park PA 16802

BRIDEN, ROGER CLARENCE, b Waterloo, Iowa, Aug 20, 41; m 60; c 2. CLINICAL CHEMISTRY. Educ: Iowa State Univ, BS, 64; Univ Kans, PhD(org chem), 69. Prof Exp: Res chemist, 69-71, tech rep, 71-73, res chemist, 73-75, DEVELOP SUPVR, DU PONT PHOTO PROD, INSTRUMENT PROD DIV, E I DU PONT DE NEMOURS & CO, INC, 75- Mem: Am Chem Soc. Res: Development of clinical chemistry methods for automated analyzers; synthetic organic chemistry; photographic chemistry. Mailing Add: Photo Prod Dept E I du Pont de Nemours & Co Inc Wilmington DE 19898

BRIDGE, HERBERT SAGE, b Berkeley, Calif, May 23, 19; m 41; c 3. SPACE PHYSICS. Educ: Univ Md, BS, 41; Mass Inst Technol, PhD(physics), 50. Prof Exp: Mem res staff, Los Alamos Sci Lab, 43-46; res assoc cosmic ray res, 46-50, mem res staff, 50-55, res physicist, 55-65, PROF PHYSICS & ASSOC DIR CTR SPACE RES, MASS INST TECHNOL, 65- Concurrent Pos: Vis scientist, Europ Orgn Nuclear Res, Switz, 57-58. Honors & Awards: Except Sci Achievement Medal, NASA, 74. Mem: Am Geophys Union. Res: Cosmic rays; solar wind; high energy astrophysics. Mailing Add: Rm 37-241 Ctr for Space Res Mass Inst of Technol Cambridge MA 02139

BRIDGE, JOHN ROBERT, b Masontown, Pa, May 26, 25; m 57. STRUCTURAL CHEMISTRY, INFORMATION SCIENCE. Educ: St Vincent Col, BS, 44. Prof Exp: Metallogr & x-ray technician, Kennametal, Inc, 47-52; prin chemist, Battelle Mem Inst, 52-59; from asst ed to assoc ed, 59-67, group leader, 67-71, sr assoc ed, 71-74, SR ED, CHEM ABSTR SERV, 74- Mem: Am Chem Soc; Am Soc Metals. Res: Structure; solid state; metallurgy; metallography. Mailing Add: 3414 Clearview Ave Columbus OH 43221

BRIDGE, ROBERT RHEA, plant breeding, genetics, see 12th edition

BRIDGE, THOMAS E, b Apr 3, 25; m 47; c 4. GEOLOGY. Educ: Kans State Univ, BS, 50, MS, 53; Univ Tex, PhD(geochem), 66. Prof Exp: Teacher pub schs, 50-53, prin, 53-56; instr geol, Colo State Univ, 56-59; asst prof, Tex Technol Col, 63-66; ASSOC PROF GEOL, EMPORIA KANS STATE COL, 66- Mem: Am Geophys Union; Geol Soc Am; Geochem Soc. Res: Contact metamorphic rocks and ground water geology. Mailing Add: Dept of Geol Emporia Kans State Col Emporia KS 66801

BRIDGEO, WILLIAM ALPHONSUS, b St John, NB, Dec 15, 27; m 55; c 4. ORGANIC CHEMISTRY. Educ: St Francis Xavier Univ, BSc, 48; Univ Ottawa, PhD(chem), 52. Prof Exp: Instr gen chem, St Mary's Col, Ind, 52-53; assoc prof anal chem, 63-70, PROF ANAL CHEM, ST MARY'S UNIV, NS, 70-, DEAN SCI, 67- Concurrent Pos: Dir tech serv div, NS Res Found, Halifax, 53-65, dir chem div, 65-69; pres, Bridco Values Ltd, 71- Mem: Can Inst Chem. Res: Nitrogen and sulfur compounds; fuel cells and combustion of woods. Mailing Add: 5959 Spring Garden Rd Apt 1209 Halifax NS Can

BRIDGER, ROBERT FREDERICK, b Joplin, Mo, Oct 29, 34; m 57; c 3. ORGANIC CHEMISTRY. Educ: Univ Mo-Rolla, BS, 57, MS, 59; Iowa State Univ, PhD(chem), 63. Prof Exp: Instr chem, Univ Mo-Rolla, 57-59; res chemist, Cent Res Div, Socony Mobil Oil Co, 63-66, sr res chemist, Mobil Res & Develop Corp, 66-74, RES ASSOC, MOBIL RES & DEVELOP CORP, 74- Mem: Am Chem Soc. Res: Reactions of free radicals; autoxidation of hydrocarbons; chemistry of lubricants. Mailing Add: Cent Res Div PO Box 1025 Mobil Res & Develop Corp Princeton NJ 08540

BRIDGER, WILLIAM AITKEN, b Winnipeg, Man, May 31, 41; m 64; c 3. BIOCHEMISTRY. Educ: Univ Man, BSc, 62, MSc, 63, PhD(biochem), 66. Prof Exp: Med Res Coun Can res fel biochem, Univ Calif, Los Angeles, 66-67; asst prof, 67-71, ASSOC PROF BIOCHEM, UNIV ALTA, 71- Mem: Am Chem Soc; Can Biochem Soc(secy, 74-); Am Soc Biol Chemists. Res: Study of structure and mechanism of action of enzymes. Mailing Add: Dept of Biochem Univ of Alta Edmonton AB Can

BRIDGERS, BERNARD THOMAS, b Lasker, NC, Oct 6, 19. BOTANY, HORTICULTURE. Educ: Wake Forest Univ, BS, 40; NC State Univ, BS, 51; Univ Md, MS, 52. Prof Exp: Teacher high sch, Fla, 46-47; res analyst, Nat Acad Sci, 52-56; res assoc, George Washington Univ, 55-57; asst, Univ Md, 58-60; from asst prof to assoc prof, 60-70, PROF BIOL, MONTGOMERY COL, 70- Mem: Am Soc Plant Physiol; Bot Soc Am; Am Hort Soc. Res: Fern spore germination; development of fern prothalli; plant propagation; botany and horticulture for the architect and home gardener; ornamental plants. Mailing Add: 26800 Howard Chapel Dr Damascus MD 20750

BRIDGERS, WILLIAM FRANK, b Asheville, NC, July 26, 32. ACADEMIC ADMINISTRATION, NEUROBIOLOGY. Educ: Univ of the South, BA, 54; Washington Univ, MD, 59. Prof Exp: From instr to asst prof prev med & med, Sch Med, Washington Univ, 63-66; from asst prof to assoc prof med, Sch Med, Univ Miami, 66-68; assoc prof pediat, biochem & med, 68-70, prof psychiat & dir neurosci prog, Sch Med, 70-73, dir sponsored progs, 73-74, SPEC ASST TO VPRES HEALTH AFFAIRS, UNIV ALA, 75- Concurrent Pos: United Health Found fel, Dept Prev Med, Sch Med, Washington Univ, 63-65; assoc ed, Nutrit Rev, 63-67; mem staff, Nat Acad Sci/Nat Res Coun, 74-75. Mem: Am Soc Biol Chemists; Am Inst Nutrit; Am Soc Clin Nutrit; Am Soc Neurochem. Res: Neurochemistry and regulatory molecular biology of folic acid metabolism in developing, mammalian brain. Mailing Add: Off of VPres Health Affs Univ of Ala Birmingham AL 35294

BRIDGES, CHARLES HUBERT, b Shreveport, La, Feb 23, 21; m 45; c 2. VETERINARY PATHOLOGY. Educ: Agr & Mech Col, Tex, DVM, 45, MS, 54, PhD(vet path), 57; Armed Forces Inst Path, cert, 55; Am Col Vet Pathologists, dipl, 56. Prof Exp: Res assoc, Agr Exp Sta, La State Univ, 49-51; assoc prof, Univ, 55-59, prof, Agr Exp Sta, 59-60, PROF VET PATH & HEAD DEPT, TEX A&M UNIV, 60- Res: Experimental pathology as applied to natural diseases of animals. Mailing Add: Dept of Vet Path Tex A&M Univ College Station TX 77843

BRIDGES, KENT WENTWORTH, b Milwaukee, Wis, Aug 25, 41; m 65. ECOLOGY. Educ: Univ Hawaii, BA, 64, MS, 67; Univ Calif, Irvine, PhD(biol), 70. Prof Exp: Asst prof wildlife resources, Utah State Univ, 70-73; ASST PROF BOT, UNIV HAWAII, 73- Concurrent Pos: Chief modeler, Desert Biome, US Int Biol Prog, 70-73. Mem: Ecol Soc Am; Japanese Soc Pop Ecol; Asn Comput Mach. Res: Computer modeling of biological problems; systems ecology. Mailing Add: Dept of Bot Univ of Hawaii Honolulu HI 96822

BRIDGEWATER, ALBERT LOUIS, JR, b Houston, Tex, Nov 22, 41. ELEMENTARY PARTICLE PHYSICS. Educ: Univ Calif, Berkeley, BA, 63; Columbia Univ, PhD(physics), 72. Prof Exp: Fel physics, Lawrence Berkeley Lab, 70-73; STAFF ASST, PHYSICS SECT, NSF, 73- Concurrent Pos: Adjoint asst prof, Howard Univ, 75- Mem: AAAS; Am Phys Soc. Mailing Add: 11605 Vantage Hill Rd 11C Reston VA 22090

BRIDGFORTH, ROBERT MOORE, JR, b Lexington, Miss, Oct 21, 18; m 43; c 2. PHYSICAL CHEMISTRY. Educ: Iowa State Univ, BS, 40; Mass Inst Technol, SM, 48. Prof Exp: Instr chem, Mass Inst Technol, 40-43, staff mem, Div Indust Coop, 43-48; assoc prof physics & chem, Emory & Henry Col, 49-51; res specialist, Boeing Co, 51-58, chief propulsion syst sect, Syst Mgt Off, 58-59, chief propulsion res unit, 59-60; CHMN BD, ROCKET RES CORP, 60- Mem: Am Chem Soc; Am Astronaut Soc; Brit Interplanetary Soc; Am Inst Aeronaut & Astronaut; Am Rocket Soc (pres, Pac Northwest Sect, 55-56). Res: Thermodynamics of rocket propellants; thermodynamics of living systems. Mailing Add: 4325 87th Ave SE Mercer Island WA 98040

BRIDGHAM, CATHERINE MITCHELL, b Beaver Falls, Pa, June 30, 02; m 29. BIOCHEMISTRY. Educ: Univ Mich, BS, 25; Univ Pittsburgh, PhD(biochem), 32. Prof Exp: Apprentice engr, Nat Tube Co, Pa, 25-27; asst, Univ Pittsburgh, 27-32, asst prof, 32-33; from instr to assoc prof biol & chem, 34-52, prof chem, 52-69, EMER PROF CHEM, YOUNGSTOWN STATE UNIV, 69- Mem: Fel AAAS; Am Chem Soc; Sigma Xi. Res: Chemistry of pectin and other plant constituents; analysis of biochemical materials; ecology. Mailing Add: 11949 Ellsworth Rd North Jackson OH 44451

BRIDGMAN, ANNA JOSEPHINE, b Gainesville, Ga, Sept 26, 06. BIOLOGY. Educ: Agnes Scott Col, AB, 27; Univ Va, MS, 35; Univ NC, PhD(zool), 47. Prof Exp: Teacher schs, NC, 27-30 & Va, 30-35; assoc prof biol, Flora Macdonald Col, 35-40; prof, Limestone Col, 42-49; from assoc prof to prof biol, 49-74, chmn dept, 52-71, EMER PROF BIOL, AGNES SCOTT COL, 74- Concurrent Pos: With Nat Adv Comt Aeronaut, 31; corp mem, Marine Biol Lab, Woods Hole. Mem: AAAS; Am Soc Zool; Soc Protozool. Res: Excystment in freshwater ciliates; radiation studies on ciliates. Mailing Add: 715 Kirk Rd Decatur GA 30030

BRIDGMAN, CHARLES FLOYD, b Loma Linda, Calif, Oct 2, 23; m 49; c 5. ANATOMY. Educ: Univ Calif, Los Angeles, AB, 49, MS, 55, PhD(anat), 62; Univ Ill, cert med illustration, 51. Prof Exp: Asst prof anat & art, Sch Med, Univ Calif, Los Angeles, 63-65; asst prof anat & dir, Off Learning Resources, Univ Calif, San Diego, 65-70; dir, Nat Med AV Ctr, 70-73, ASSOC DIR FOR EDUC RESOURCES DEVELOP, NAT LIBR MED, 73- Concurrent Pos: Fel med commun, Sch Med, Univ Kans, 62-63; adj assoc prof anat, Sch Med, Emory Univ, 70-75. Mem: AAAS; Asn Med Illustrators; Sigma Xi; Am Asn Anatomists. Res: Medical education, programmed instruction, multi-media communication; morphology and physiology of proprioceptive sense organs. Mailing Add: Nat Libr of Med 8600 Rockville Pike Bethesda MD 20014

BRIDGMAN, GEORGE HENRY, b Minneapolis, Minn, 1940. MATHEMATICS. Educ: Hamline Univ, BA, 61; Univ Minn, Minneapolis, MA, 64, PhD(math), 69. Prof Exp: ASST PROF MATH, WARTBURG COL, 69- Mem: Math Asn Am. Res: Wiener integrals; mathematical analysis. Mailing Add: Dept of Math Wartburg Col Waverly IA 50677

BRIDGMAN, JOHN FRANCIS, b Kuling, China, Sept 6, 25; US citizen; m 52; c 4. PARASITOLOGY. Educ: Davidson Col, BS, 49; La State Univ, MS, 52; Tulane Univ, PhD(biol), 68. Prof Exp: Instr biol, Delta State Teachers Col, 52 & Univ Tenn, Martin Br, 52-54; lectr, Shikoku Christian Col, 56-67, from asst prof to prof, 57-72; ASSOC PROF BIOL, COL OZARKS, 72- Mem: AAAS; Am Soc Parasitologists; Japanese Soc Parasitologists; Am Soc Zoologists. Res: Invertebrate ecology; life cycles, bionomics and ecology of parasites together with their host-parasite relations. Mailing Add: Dept of Sci & Math Col of the Ozarks Clarksville AR 72830

BRIDGMAN, WILBUR BENJAMIN, b New Wilmington, Pa, Jan 28, 13; m 37; c 2. PHYSICAL CHEMISTRY. Educ: Wis State Teachers Col, BEd, 33; Univ Wis, PhD(chem), 37. Prof Exp: Instr chem, Univ Wis, 37-42, Rockefeller fel, 42-43; from asst prof to assoc prof, 43-59, PROF CHEM, WORCESTER POLYTECH INST, 59- Mem: Am Chem Soc. Res: Dipole moments; ultracentrifugal determination of particle size and shape; solution thermodynamics; surface chemistry. Mailing Add: Dept of Chem Worcester Polytech Inst Worcester MA 01609

BRIDGMON, GEORGE HARRISON, b Wheatland, Wyo, Mar 13, 21; m 46; c 3. PLANT PATHOLOGY. Educ: Univ Wyo, BS, 46, MS, 47; Univ Wis, PhD(plant path), 51. Prof Exp: Asst prof agron, 46-47, asst dir agr exp sta, 50-58, PROF PLANT PATH, UNIV WYO, 58-, HEAD SECT PLANT PATH & HORT, 67- Concurrent Pos: Rockefeller Found fel, Univ Minn, 56-57; dir res, Somali Repub Cent Agr Res Sta, Mogadiscio, 65-66; agr consult, World Bank & other orgns, 74. Mem: AAAS; Am Phytopath Soc; Am Soc Hort Sci. Res: Virus, rusts and diseases of beans; pollution studies. Mailing Add: Dept of Plant Path Univ of Wyo Laramie WY 82071

BRIDLE, ALAN HENRY, b Harrow, Eng, Sept 2, 42; m 68. ASTRONOMY. Educ: Cambridge Univ, BA, 63, MA, 67, PhD(radio astron), 67. Prof Exp: UK Sci Res Coun fel, 67; asst prof, 67-72, ASSOC PROF PHYSICS, QUEEN'S UNIV, ONT, 72- Concurrent Pos: Vis asst scientist, US Nat Radio Astron Observ, 68. Mem: Am Astron Soc; fel Royal Astron Soc; Can Astron Soc; Int Astron Union. Res: Extragalactic and galactic radio astronomy at decametre and centimetre wavelengths; observational cosmology; radio galaxies; galaxy clusters; interstellar radio communication. Mailing Add: Astrophys Lab Stirling Hall Queen's Univ Kingston ON Can

BRIEADDY, LAWRENCE EDWARD, b Syracuse, NY, May 13, 44; m 66; c 2. PHARMACEUTICAL CHEMISTRY, ORGANIC CHEMISTRY. Educ: LeMoyne Col, NY, BS, 66; NC State Univ, MS, 69. Prof Exp: RES SCIENTIST PHARMACEUT, BURROUGHS WELLCOME CO, 71- Mem: Am Chem Soc. Res: Prostaglandin synthetase inhibitors; central nervous system; monoamine oxidase inhibitors; anti-inflammatory analgesic fields. Mailing Add: Burroughs Wellcome Co 3030 Cornwallis Rd Research Triangle Park NC 27709

BRIEGER, GERT HENRY, b Hamburg, Ger, Jan 5, 32; US citizen; m 55; c 3. HISTORY MEDICINE. Educ: Univ Calif, Berkeley, AB, 53; Univ Calif, Los Angeles, MD, 57; Harvard Univ, MPH, 62; Johns Hopkins Univ, PhD, 68. Prof Exp: Asst prof hist of med, Johns Hopkins Univ, 66-70; assoc prof hist of med, Duke Univ, 70-75; PROF HIST OF HEALTH SCI & CHMN DEPT, UNIV CALIF, SAN FRANCISCO, 75- Mem: AAAS; Hist Sci Soc; Am Asn Hist Med. Res: History of American medicine and public health; history of medicine in the late 19th and 20th centuries. Mailing Add: Univ of Calif Dept of Health Sci Hist San Francisco CA 94143

BRIEGER, GOTTFRIED, b Berlin, Ger, Oct 27, 35; m 59; c 2. ORGANIC CHEMISTRY. Educ: Harvard Col, BA, 57; Univ Wis, PhD(chem), 61. Prof Exp: Asst prof chem, Univ Calif, Berkeley, 61-63; from asst prof to assoc prof, 63-72, PROF CHEM, OAKLAND UNIV, 72- Mem: AAAS; Am Chem Soc; Lepidopterists Soc; The Chem Soc. Res: Synthesis and structural determination of natural products, especially terpenes; chemistry and biochemistry of insect hormones. Mailing Add: Dept of Chem Oakland Univ Rochester MI 48063

BRIEGER, HEINRICH, medicine, see 12th edition

BRIEGLEB, PHILIP ANTHES, b St Clair, Mo, July 23, 06; m 35; c 2. FORESTRY. Educ: Syracuse Univ, BSF, 29, MF, 30. Prof Exp: Jr forester, Pac Northwest Forest Exp Sta, US Forest Serv, 29-34, asst forester, 35, from assoc forester to forester, 36-43, sr forester, Washington, DC, 44, Northeastern Forest Exp Sta, 45-46, chief div forest mgt res, Pac Northwest Forest Exp Sta, 46-51, dir, Cent States Forest Exp Sta, 51-53, Southern Forest Exp Sta, 54-63 & Pac Northwest Forest Exp Sta, 63-71; CONSULT FORESTER, 71- Concurrent Pos: Lectr, Univ Calif, 42; mem, US Forestry Mission, Chile, 43-44; guest, Brit Commonwealth Forestry Conf, Australia & NZ, 57; mem adv coun, Cascade Head Scenic Res Area, 75-78. Honors & Awards: USDA Super Serv Award, 60. Mem: Fel AAAS; fel Soc Am Foresters (pres, 64-65). Res: Forest survey; mensuration; management and research administration. Mailing Add: 4217 SW Agate Lane Portland OR 97201

BRIEHL, ROBIN WALT, b Vienna, Austria, June 21, 28; US citizen. MOLECULAR BIOLOGY, INTERNAL MEDICINE. Educ: Swarthmore Col, BA, 50; Harvard Med Sch, MD, 54. Prof Exp: Intern, Montefiore Hosp, New York, 54-55, asst resident med, 55-56; asst resident med & cardiol, Presby Hosp, 56-57, asst physician, 57-60; asst prof physiol, 62-66, assoc med, 65-69, assoc prof, 66-71, PROF PHYSIOL, ALBERT EINSTEIN COL MED, 71-, BIOCHEM, 73-, ASST PROF MED, 69- Concurrent Pos: NIH vis fel, Cardiopulmonary Lab, Columbia Univ, 57-60; NIH res fel biol, Harvard Univ, 60-62; career scientist, Health Res Coun, New York, 62-72; asst physician, Bronx Munic Hosp Ctr, 65- Mem: Am Soc Biol Chemists; AAAS; Am Fedn Clin Res; Am Soc Clin Invest; Am Physiol Soc. Res: Allostery; relations between the structure and function of hemoglobin; molecular bases of sickle cell disease. Mailing Add: Dept of Physiol Albert Einstein Col of Med Bronx NY 10461

BRIEN, FRANCIS STAPLES, b Windsor, Ont, Apr 9, 08; m 37; c 1. INTERNAL MEDICINE. Educ: Univ Toronto, BA, 30, MB, 33; FRCP(London) & FACP, 46; FRCP(C), 58. Prof Exp: Instr therapeut, Fac Med, Univ Toronto, 35-36; head dept, 45-70, prof, 45-73, HON PROF MED & CONSULT TO TEACHING HOSPS, UNIV WESTERN ONT, 73- Concurrent Pos: Med dir, Northern Life Assurance Co Can; mem, Exec Comt, Adv Med Bd, Ont Cancer Treatment & Res Found; consult, St Thomas Psychiat Hosp, Ont; hon mem, Col Family Med Can. Mem: Fel Am Geriatrics Soc; fel NY Acad Sci; Can Arthritis & Rheumat Soc. Res: Nutrition; absorption and malabsorption; aging; hypertensive states. Mailing Add: 144 Iroquois Ave London ON Can

BRIENT, CHARLES E, b El Paso, Tex, Oct 10, 34; m 59; c 4. NUCLEAR PHYSICS. Educ: Univ Tex, BS, 57, MA, 60, PhD(physics), 63. Prof Exp: Res assoc nuclear physics, Univ Tex, 63-64; asst prof, 64-68, ASSOC PROF PHYSICS, OHIO UNIV, 68- Concurrent Pos: NSF res grant, 65-66; Ohio Univ Fund res grant, 66-67; AEC grant, 68. Mem: Am Phys Soc. Res: Nuclear reaction mechanisms and structure. Mailing Add: Dept of Physics Ohio Univ Athens OH 45701

BRIENT, SAMUEL JOHN, JR, b Phoenix, Ariz, Mar 24, 30; m 56; c 2. SOLID STATE PHYSICS. Educ: Univ Tex, BS, 52, PhD(physics), 59. Prof Exp: Res scientist, Defense Res Lab, Univ Tex, Austin, 56, Spectros Lab, 59-60, instr physics, 60-62; assoc prof, 62-74, PROF PHYSICS, UNIV TEX, EL PASO, 74- Concurrent Pos: Consult, Air Force Off Sci Res, Holloman AFB, NMex, 63, Braddock, Dunn & McDonald Inc, Tex, 63-64, Los Alamos Sci Lab, 67-70 & Harry Diamond Labs, Army Res Off, 69. Mem: Fel AAAS; Am Vacuum Soc; Am Phys Soc; Am Asn Physics Teachers. Res: Conductive mechanisms in thin films; P-n junction theory; Green's function equation of motion technique. Mailing Add: 2011 N Kansas El Paso TX 79902

BRIENZA, MICHAEL JOSEPH, b Mt Vernon, NY, Aug 20, 39; m 62; c 3. ELECTROOPTICS, LASERS. Educ: Univ Notre Dame, BS, 60, PhD(physics), 64. Prof Exp: Res scientist, Res Labs, 64-67, sr res scientist, 67-68, chief appl laser technol, 68-71, CHIEF ELECTROOPTICS, NORDEN DIV, UNITED AIRCRAFT CORP, 71- Mem: Am Phys Soc. Res: Photoelectric and optical properties of metal films; hot electron ranges and devices; surface physics and thin film studies; ultra short laser pulse technology; acousto-optic signal processing; laser applications. Mailing Add: 1 Wisteria Lane Westport CT 06880

BRIER, GLENN WILSON, b Fairfield, Iowa, Apr 26, 13; m 40; c 3. STATISTICAL ANALYSIS, METEOROLOGY. Educ: Parsons Col, BS, 35; George Washington Univ, AM, 40. Prof Exp: Sci aide, USDA, Md, 36-38, jr statistician, 39-40, asst statistician, Ill, 41; assoc statistician, US Weather Bur, Washington, DC, 42-43, meteorologist, head verification sect, 44-47, chief meteorol statist sect, 48-61, head meteorol statist res proj, 62-65; meteorol statist br, Inst Atmospheric Sci, Environ Sci Serv Admin, 65-69; chief meteorol statist group, Res Labs, Nat Oceanog & Atmospheric Admin, 69-71; RES ASSOC ATMOSPHERIC SCI, COLO STATE UNIV, 72- Concurrent Pos: Vis staff mem, Mass Inst Technol, 62; vis scientist, Travelers Res Ctr, 66; exchange scientist, Brit Meteorol Off, 68. Mem: Fel Am Statist Asn; fel Am Meteorol Soc; assoc Am Geophys Union; Int Asn Statist in Phys Sci. Res: Weather forecasting; climatic trends; analysis of time series; verification of weather forecasts; atmospheric diffusion; extra-terrestrial effects on weather; weather modification. Mailing Add: Dept of Atmospheric Sci Colo State Univ Ft Collins CO 80523

BRIERLEY, GERALD PHILIP, b Ogallala, Nebr, Aug 14, 31; m 54; c 2. BIOCHEMISTRY. Educ: Univ Md, PhD(biochem), 61. Prof Exp: Fel, Inst Enzyme Res, Univ Wis, 60-62, asst prof biochem, 62-64; asst prof, 64-69, PROF PHYSIOL CHEM, OHIO STATE UNIV, 69- Concurrent Pos: Estab investr, Am Heart Asn, 64- Mem: Am Soc Biol Chemists; Am Chem Soc. Res: Active transport and oxidative phosphorylation in heart mitochondria. Mailing Add: Dept of Physiol Chem Ohio State Univ Col of Med Columbus OH 43210

BRIERLEY, JAMES ALAN, b Denver, Colo, Dec 22, 38; m 65. MICROBIOLOGY. Educ: Colo State Univ, BS, 61; Mont State Univ, MS, 63, PhD(microbiol), 66. Prof Exp: Fellow microbiol, Mont State Univ, 61-66; asst prof biol, NMex Inst Mining & Technol, 66-68; res scientist, Martin Marietta Corp, Colo, 68-69; ASST PROF BIOL, N MEX INST MINING & TECHNOL, 69- Mem: Am Soc Microbiol; Brit Soc Gen Microbiol; Ecol Soc Am. Res: Ecological aspects of chemolithotrophic bacteria in acid thermal waters; ecology of chemolithotrophic bacteria in dump-leaching; microbial leaching of metals; microbial ecology of river water. Mailing Add: Dept of Biol NMex Inst of Mining & Technol Socorro NM 87801

BRIERLEY, JEAN, b Dover, NH, Mar 17, 08. GENETICS. Educ: Univ NH, BS, 30; Univ Mich, MS, 31, PhD(genetics), 37. Prof Exp: Asst zool, Univ Mich, 30-36; asst prof biol, Tex State Col Women, 37-44; instr biol, Mich State Univ, 44-46, from asst prof to prof natural sci, 46-72; RETIRED. Mem: Genetics Soc Am; Soc Study Evolution; Am Soc Human Genetics. Res: Genetic aspect of evolution; Drosophila genetics. Mailing Add: 208 W Saginaw St East Lansing MI 48823

BRIÈRRE, ROLAND THEODORE, JR, b New Orleans, La, May 20, 32; m 56; c 1. POLYMER CHEMISTRY, POLYMER PHYSICS. Educ: Tulane Univ, BS, 54; La State Univ, MS, 60; Duke Univ, PhD(phys chem), 65. Prof Exp: Res chemist, Electrochem Dept, E I du Pont de Nemours & Co, Inc, 60-61; res chemist, Chemstrand Res Ctr, Monsanto Co, 64-66; res chemist, Beaunit Corp, 66-68; res chemist, Film Dept, 68-72, SR RES CHEMIST, TEXTILE FIBERS DEPT, E I DU PONT DE NEMOURS & CO, INC, 72- Mem: Sigma Xi. Res: Polymer solution characterization; fiber spinning research, rheology and studies of polymeric liquid crystals. Mailing Add: Textile Fibers Dept E I du Pont de Nemours & Co Inc Richmond VA 23261

BRIESE, FRANKLIN WAGNER, b St Paul, Minn, Dec 25, 37; m 60; c 3. BIOMETRY. Educ: Univ Minn, BA, 59, MS, 63, PhD(biostatist), 65. Prof Exp: Asst prof biomet & res assoc biomed data processing, Univ Minn, Minneapolis, 65-69; ASST PROF BIOMET, DIR SCI COMPUT CTR & ASSOC DIR COMPUT CTR, MED CTR, UNIV COLO, DENVER, 69- Mem: AAAS; Biomet Soc; Am Statist Asn; Math Asn Am; Asn Comput Mach. Res: Biostatistics; biomathematics; biomedical computer applications. Mailing Add: Dept of Prev Med & Comp Health Care Univ of Colo Med Ctr Denver CO 80220

BRIESKE, PHILIP RICHARD, nuclear physics, physics education, see 12th edition

BRIESKE, THOMAS JOHN, b Milwaukee, Wis, Sept 2, 39; m 63; c 3. MATHEMATICS EDUCATION. Educ: St Mary's Univ, Tex, BA, 61, MS, 63; Univ SC, PhD(math educ), 69. Prof Exp: Teacher, High Schs, Tex, 61-63; teaching asst math, Univ Tex, Austin, 63-64; sci programmer, A O Smith Corp, Wis, 64-65; instr math, Wis State Univ, Oshkosh, 65-66; teaching asst, Univ SC, 66-69; asst prof, 69-74, ASSOC PROF MATH, GA STATE UNIV, 74- Mem: Math Asn Am. Res: Learning of mathematics, in particular, the calculus. Mailing Add: Dept of Math Ga State Univ Atlanta GA 30303

BRIGGAMAN, ROBERT ALAN, b Hartford, Conn, Aug 14, 34; m 60; c 1. MEDICINE, DERMATOLOGY. Educ: Trinity Col, Conn, BS, 56; NY Univ, MD, 60. Prof Exp: Intern internal med, Univ Va Hosp, 60-61, asst resident, 61-62; resident, 64-66, from instr to assoc prof, 67-74, PROF DERMAT, SCH MED, UNIV NC, CHAPEL HILL, 74- Concurrent Pos: Res fel, Univ NC, 66-67; NIH spec fel, 67-70. Mem: AAAS; Am Acad Dermat; Soc Invest Dermat; Tissue Cult Asn; AMA. Res: Epidermal-dermal interactions; tissue culture; delayed hypersensitivity; electron microscopy. Mailing Add: NC Mem Hosp Chapel Hill NC 27514

BRIGGLE, LELAND WILSON, b Bismarck, NDak, Oct 6, 20; m 44; c 2. PLANT GENETICS. Educ: Jamestown Col, BS, 42; NDak State Univ, BS, 49, MS, 51; Iowa State Univ, PhD(plant breeding), 54. Prof Exp: Asst agron, NDak State Univ, 49-51; asst, Iowa State Univ, 52-53; geneticist, 54-55, res agronomist, 55-74, SCIENTIST, NAT PROG STAFF, AGR RES CTR, USDA, 75- Mem: AAAS; Am Soc Agron; Crop Sci Soc Am. Res: Wheat and oat breeding and genetics. Mailing Add: 701 Copley Lane Silver Spring MD 20904

BRIGGS, ARTHUR HAROLD, b East Orange, NJ, Nov 3, 30; m 53; c 2. PHARMACOLOGY, MEDICINE. Educ: Johns Hopkins Univ, BA, 52, MD, 56. Prof Exp: Intern internal med, Univ Hosp, Vanderbilt Univ, 56-57, asst resident, 57-58, res assoc pharmacol, Sch Med, 58-59; from asst prof to prof pharmacol, Sch Med, Univ Miss, 59-68, from instr to asst prof med, 60-68; PROF PHARMACOL & CHMN DEPT, MED SCH, UNIV TEX, SAN ANTONIO, 68- Concurrent Pos: US Govt fel, 58-60. Mem: Am Soc Pharmacol & Exp Therapeut; Soc Exp Biol & Med. Res: Smooth muscle and cardiac muscle pharmacology; clinical hypertension. Mailing Add: Dept of Pharmacol Univ of Tex Med Sch San Antonio TX 78229

BRIGGS, BEN THOBURN, b Portland, Ore, Dec 13, 11; m 37; c 2. CHEMISTRY. Educ: Willamette Univ, AB, 34; Univ Ill, PhD(phys chem), 38. Prof Exp: Res chemist, Rayonier, Inc, 37-55, res mgr, Olympic Res Div, 55-59; chem consult, 59-73; RETIRED. Mem: AAAS; Am Chem Soc; Am Tech Asn Pulp & Paper Indust. Res: Pulping and bleaching; wood cellulose; physical chemistry; wood, paper and pulp. Mailing Add: 825 Grant Ave Shelton WA 98584

BRIGGS, CHARLES FRANCIS, b Bay City, Mich, Apr 27, 21; m 52; c 2. MATHEMATICS. Educ: Wayne State Univ, BS, 45; Univ Mich, MA, 46, PhD(math), 54. Prof Exp: Teaching fel math, Univ Mich, 47-50; asst prof, Emory Univ, 50-53; jr instr, Univ Mich, 53-54; from instr to asst prof math, 54-65, asst dir, Comput Ctr, 59-62, res assoc mach trans, 58-60, ASSOC PROF MATH, WAYNE STATE UNIV, 65-, RES ASSOC RUSSIAN DICTIONARY PROJ, 63-, ASSOC DIR, COMPUT CTR, 62- Mem: AAAS; Am Math Soc; Math Asn Am; Am Asn Comput Mach. Res: Machine translation of Russian into English; Russian lexicons; numerical approximation; numerical solution of ordinary differential equations. Mailing Add: Dept of Math Wayne State Univ Detroit MI 48202

BRIGGS, DARINKA ZIGIC, b Belgrade, Yugoslavia, Sept 2, 32; US citizen; m 68. STRATIGRAPHY, INFORMATION SCIENCE. Educ: Univ Belgrade, Dipl Geol, 57. Prof Exp: Field geologist, Serbian Acad Sci, Belgrade, 56; res asst paleont, Royal Ont Mus, Toronto, 58; res geologist, Imp Oil Co, Ltd, Toronto, 59; assr geol, Univ Toronto, 58-60; teacher high sch, Tilbury, Can, 62; res assoc geol, 65-68, VIS SCIENTIST GEOL, SUBSURFACE LAB, UNIV MICH, ANN ARBOR, 68-; CONSULT GEOL, SYSTS ANAL RES INFO CO, 68- Mem: Am Asn Petrol Geologists; Geol Soc Am; Am Asn Univ Profs; Am Soc Info Sci; Int Asn Math Geol. Res: Theoretical and mathematical basis for stratigraphic analysis, and the application to historical reconstructions of paleogeographies, ancient environments, and their utilization in the search for fossil energy resources. Mailing Add: 3451 Burbank Dr Ann Arbor MI 48105

BRIGGS, DONALD K, b Northampton, Eng, Apr 10, 24; US citizen; m 58; c 2. MEDICINE, HEMATOLOGY. Educ: Cambridge Univ, MA, 44, MB & BCh, 46. Prof Exp: ASST PROF MED, NY UNIV, 57- Mem: Am Soc Hemat. Res: Genetics. Mailing Add: 170 E 79th St New York NY 10021

BRIGGS, EDGAR VAN, b Bruce, Wis, Nov 30, 07; m 33. PHYSICS. Educ: Lawrence Col, BA, 30; Univ Wis, PhD(physics educ), 54. Prof Exp: Teacher, Pus Sch, Wis, 30-42; civilian instr, Radiol Sch, US Air Force, 42-44; teacher, Pub Sch, Wis, 44-46; assoc prof physics & math, Richmond Prof Inst, Col William & Mary, 46-47; prof physics & phys sci methods, Univ Wis-Superior, 47-50; technologist, Forest Prod Lab, Mat & Containers Div, Univ Wis, 51-53; prof physics & phys sci methods, Univ Wis-Superior, 53-57; assoc prof physics, 57-64, PROF PHYSICS, CENT MICH UNIV, 64- Mem: AAAS; Am Asn Physics Teachers; Nat Sci Teachers Asn. Res: Physical science in secondary schools. Mailing Add: Dept of Physics Central Mich Univ Mt Pleasant MI 48858

BRIGGS, FRED NORMAN, b Oakland, Calif, Sept 12, 24; m 49; c 1. PHYSIOLOGY. Educ: Univ Calif, AB, 47, MA, 48, PhD(physiol), 53. Prof Exp: Radiologist, US Naval Radiol Defense Lab, 48-49; instr pharm, Dent Sch, Harvard Univ, 52-55, assoc pharmacol, Med Sch, 56-58; from asst prof to assoc prof, Med Sch, Tufts Univ, 58-61; prof physiol, Sch Med, Univ Pittsburgh, 61-71; PROF PHYSIOL & CHMN DEPT, MED COL VA, VA COMMONWEALTH UNIV, 71- Concurrent Pos: Res fel, Harvard Univ, 55-56; sect ed, Am J Physiol. Mem: Am Soc Pharmacol & Exp Therapeut; Am Physiol Soc. Res: Thyroid; adrenal; muscle; physiology. Mailing Add: Dept of Physiol Med Col of Va Va Commonwealth Univ Richmond VA 23219

BRIGGS, GARRETT, b Dallas, Tex, Dec 31, 34; m 57; c 3. SEDIMENTOLOGY. Educ: Southern Methodist Univ, BS, 58, MS, 59; Univ Wis, PhD(geol), 63. Prof Exp: Wis Alumni Res Found asst, Univ Wis, 59-60; geologist, Chevron Oil Co, La, 62-65; asst prof geol, Tulane Univ, 65-68; assoc prof, 68-74, interim head dept, 72-74, PROF GEOL & HEAD DEPT, UNIV TENN, KNOXVILLE, 74- Mem: Am Geol Soc; Am Asn Petrol Geologists; Am Soc Econ Paleont & Mineral. Res: General geology and sedimentation studies of late Paleozoic rocks in the Ouachita Mountains of Oklahoma; carboniferous sediments of Cumberland Plateau, Tennessee. Mailing Add: Dept of Geol Sci Univ of Tenn Knoxville TN 37916

BRIGGS, GEORGE MCSPADDEN, b Grantsburg, Wis, Feb 21, 19; m 41; c 3. NUTRITION. Educ: Univ Wis, BS, 40, MS, 41, PhD(biochem), 44. Prof Exp: From assoc prof to prof poultry nutrit, Univ Md, 45-47; from asst prof to assoc prof, Univ Minn, 47-51; chief nutrit unit, Lab Biochem & Nutrit, NIH, 51-58, exec secy biochem & pharmacol training comts, Div Gen Med Sci, 58-60; chmn dept nutrit sci, 60-70, PROF NUTRIT & BIOCHEMIST, EXP STA, UNIV CALIF, BERKELEY, 60- Concurrent Pos: Mem US nat comt, Int Union Nutrit Sci, 58-61 & 69-72, vchmn, 69-71, chmn, 71-72; exec ed, J Nutrit Educ. Honors & Awards: Borden Award, Poultry Sci Asn, 58. Mem: Am Chem Soc; Poultry Sci Asn; Am Inst Nutrit (secy, 57-60, pres, 67-68); fel Am Pub Health Asn; Soc Nutrit Educ (pres, 68-69). Res: Nutrition of laboratory animals; identification of new vitamins; fundamental vitamin studies; nutrition training and education. Mailing Add: Dept of Nutrit Sci Univ of Calif Berkeley CA 94720

BRIGGS, GEORGE ROLAND, b Ithaca, NY, May 21, 24; m 48; c 2. PHYSICS. Educ: Cornell Univ, AB, 47; Univ Ill, MS, 50, PhD(physics), 53. Prof Exp: Mem staff, Lincoln Lab, Mass Inst Technol, 52-53; mem tech staff, 54-69, MEM TECH STAFF, DIGITAL SYSTS RES LAB, DAVID SARNOFF RES CTR, RCA CORP, 69- Honors & Awards: Achievement Awards, RCA Corp, 54 & 60. Mem: Am Phys Soc; Inst Elec & Electronic Engrs; Asn Comput Mach. Res: Magnetic and ferroelectric computer memory and logic devices; electro-luminescent-magnetic computer and television displays; nuclear-radiation resistant computer memory and logic circuitry. Mailing Add: 77 Clearview Ave Princeton NJ 08540

BRIGGS, HILTON MARSHALL, b Cairo, Iowa, Jan 9, 13; m 35; c 2. ANIMAL HUSBANDRY. Educ: Iowa State Univ, BS, 33; NDak State Univ, MS, 35; Cornell Univ, PhD(nutrit), 38. Hon Degrees: DSc, NDak State Univ, 63; Dr Higher Educ & Admin, Univ SDak, 74. Prof Exp: Asst, NDak State Univ, 34-35 & Cornell Univ, 35-36; from asst prof to prof animal husb, Okla Agr & Mech Col, 36-50, assoc dean agr & assoc dir exp sta, 49-50; dean col agr & dir exp sta, Univ Wyo, 50-58; pres, 58-75, EMER PRES & DISTINGUISHED PROF AGR, S DAK STATE UNIV, 75- Honors & Awards: Distinguished Civilian Serv Medal, US Army, 73; Except Serv Medal, US Air Force, 75. Mem: Fel AAAS; fel Am Soc Animal Sci (secy, 47-50, vpres, 51, pres, 52). Mailing Add: SDak State Univ Brookings SD 57006

BRIGGS, JAMES EDWARD, animal nutrition, biochemistry, see 12th edition

BRIGGS, JEAN LOUISE, b Washington, DC, May 28, 29. ANTHROPOLOGY. Educ: Vassar Col, BA, 51; Boston Univ, MA, 60; Harvard Univ, PhD(cult anthrop), 67. Prof Exp: From asst prof to assoc prof, 67-75, PROF ANTHROP, MEM UNIV NFLD, 75- Concurrent Pos: Mem consult group on the individual, lang & society, Can Coun, 73-; coun mem, Soc Sci Res Coun Can, 75- Mem: Fel Am Anthrop Asn; Am Ethnol Soc; Arctic Inst NAm; Int Soc Res on Aggression; Can Ethnol Soc. Res: Eskimo emotional structure and ethnopsychology; Eskimo language. Mailing Add: Dept of Anthrop Mem Univ of Nfld St John's NF Can

BRIGGS, JOHN CARMON, b Portland, Ore, Apr 9, 20; m 48; c 9. ICHTHYOLOGY, ZOOGEOGRAPHY. Educ: Ore State Univ, BS, 43; Stanford Univ, MA, 47, PhD(biol), 52. Prof Exp: Aquatic biologist, US Fish & Wildlife Serv, Ore, 45-46; asst gen biol, Stanford Univ, 47-48; instr, Ore Inst Marine Biol, 48; asst gen biol, Hopkins Marine Sta, 49; biologist, Calif State Div Fish & Game, 50-51; res assoc, Natural Hist Mus, 52-54; from instr to asst prof biol, Univ Fla, 54-57, res assoc anat, 57-58; asst prof, Univ BC, 58-59, asst prof zool, 59-61; asst prof, Hopkins Marine Sta, 61; res scientist, Inst Marine Sci, Tex, 61-64; chmn dept biol, 69-71, PROF ZOOL, UNIV S FLA, 64-, DIR GRAD STUDIES, 71- Mem: Am Soc Ichthyol & Herpet (vpres, 59, 64); fel Am Inst Fishery Res Biologists; Am Soc Naturalists; Am Inst Biol Sci; Am Fisheries Soc. Res: Evolution, geographical distribution and behavior of fishes; marine zoogeography; evolutionary significance of zoogeographic patterns. Mailing Add: Dept of Biol Univ of SFla Tampa FL 33620

BRIGGS, JOHN DORIAN, b Santa Monica, Calif, Dec 20, 26; m 51; c 3. INSECT PATHOLOGY. Educ: Univ Calif, BS, 51, PhD(entom), 56. Prof Exp: Assoc entomologist, Ill Nat Hist Surv, 55-59; head entom dept, Bioferm Corp, 59-62; actg dean col biol sci, 68-69, PROF ENTOM, OHIO STATE UNIV, 62- Mem: AAAS; Entom Soc Am; Soc Invert Path (vpres, pres, 70-74). Res: Insect diseases; entomogenous bacteria, fungi, protozoa and viruses; invertebrate immunology; tissue culture; biological control; biology of aging. Mailing Add: Dept of Entom Ohio State Univ Columbus OH 43210

BRIGGS, JONATHAN, b White Plains, NY, Aug 21, 45; m 72; c 1. SOLID STATE PHYSICS. Educ: Dartmouth Col, AB, 67; Harvard Univ, MS, 68, PhD(solid state physics), 73. Prof Exp: Fel solid state physics, Naval Postgrad Sch, 73-74; ASST PROF PHYSICS, COLBY COL, 74- Mem: Am Phys Soc. Res: Elastic light scattering at visible frequencies from minute scatterers, involving the measurement of the intensity versus scattering angle and the time correlation of photon pulses. Mailing Add: Dept of Physics Colby Col Waterville ME 04901

BRIGGS, LLOYD CABOT, b Boston, Mass, June 27, 09; c 1. ANTHROPOLOGY. Educ: Harvard Univ, AB, 31, AM, 38, PhD(anthrop), 52; Univ Oxford, dipl anthrop, 32. Prof Exp: Chmn dept, 67-73, PROF ANTHROP, FRANKLIN PIERCE COL, 67- ; RES ASSOC, HARVARD UNIV, 52- Concurrent Pos: Tech consult, Lab Appl Anthrop, Univ Paris, 70- Honors & Awards: Chevallier, Order of Merit, Saharien, French Govt, 60 & Officier, Palmes Academiques, 71. Mem: Am Anthrop Asn; fel Am Geog Soc; Am Asn Phys Anthropologists; Royal Anthrop Inst Gt Brit & Ireland. Res: General anthropology, particularly of the Sahara and adjacent regions. Mailing Add: Dept of Anthrop Franklin Pierce Col Rindge NH 03461

BRIGGS, LOUIS ISAAC, JR, b Calif, Apr 24, 21; m 43; c 3. PETROLOGY. Educ: Fresno State Col, AB, 43; Univ Calif, PhD(geol), 50. Prof Exp: Asst geol, Univ Calif, 46-50; from instr to assoc prof, 50-60, PROF GEOL, UNIV MICH, ANN ARBOR, 60-, DIR SUBSURFACE LAB, 61- Mem: AAAS; Geol Soc Am; Soc Econ Paleont & Mineral; Geosci Info Soc. Res: Sedimentary petrology; physical stratigraphy; computer geology. Mailing Add: Dept of Geol & Mineral Univ of Mich Ann Arbor MI 48104

BRIGGS, NORMAN THEODORE, b Akron, Ohio, Feb 28, 29; m 53; c 4. PARASITOLOGY. Educ: Eastern Mich Univ, BS, 51; Wayne State Univ, MS, 55; Univ Tex, PhD(parasitol, prev med), 58. Prof Exp: Asst zool, Eastern Mich Univ, 50-51; asst, Wayne State Univ, 52-53, lab instr, 53-54; res assoc, Dept Prev Med & Microbiol, Univ Chicago, 60-62; asst prof biol, Wayne State Univ, 62-64; RES PARASITOLOGIST, DEPT MED ZOOL, INST RES, WALTER REED MED CTR, 64- Mem: Am Soc Parasitol; Am Soc Trop Med & Hyg. Res: Helminths of the domestic rat; immunology and host-parasite relationships in rodent filariasis, trichinosis, malaria, Leucocytozoon, ornithosis and histoplasmosis. Mailing Add: Dept of Med Zool Walter Reed Inst of Res Washington DC 20012

BRIGGS, PAUL CLAYTON, JR, b Schenectady, NY, Sept 1, 43. ORGANIC CHEMISTRY. Educ: State Univ NY Albany, BS, 65, State Univ NY Buffalo, PhD(org chem), 70. Prof Exp: RES CHEMIST, ELASTOMER CHEM DEPT, E I DU PONT DE NEMOURS & CO, 69- Res: Development of rubber-based adhesives. Mailing Add: Elastomers Lab E I du Pont de Nemours & Co Wilmington DE 19898

BRIGGS, REGINALD PETER, b Port Chester, NY, Mar 12, 29; m 56; c 3. GEOLOGY. Educ: Wesleyan Univ, BA, 51. Prof Exp: Geologist, 53-69, chief PR Coop Geol Mapping Proj, 65-70, PROJ DIR GREATER PITTSBURGH REGIONAL STUDIES, US GEOL SURV, 70- Mem: Fel Geol Soc Am; Am Geophys Union; Clay Minerals Soc; Am Inst Prof Geologists; Asn Eng Geologists. Res: Stratigraphy of marine and subaerial volcanic rocks; geomorphology; environmental geology applied to land-use planning and control. Mailing Add: US Geol Surv PO Box 420 Carnegie PA 15106

BRIGGS, RICHARD ALFRED, organic chemistry, physical chemistry, see 12th edition

BRIGGS, RICHARD JULIAN, b Shanghai, China, May 26, 37; US citizen; m 60; c 3. PLASMA PHYSICS. Educ: Mass Inst Technol, BS & MS, 61, PhD(elec eng), 64. Prof Exp: Instr elec eng, Mass Inst Technol, 61-64; plasma physicist, Lawrence Radiation Lab, Univ Calif, 64-66; from asst prof to assoc prof elec eng, Mass Inst Technol, 66-72; PHYSICIST & GROUP LEADER, LAWRENCE LIVERMORE LAB, UNIV CALIF, 72- Concurrent Pos: Consult, Lawrence Livermore Lab, 66-72, Microwave Assocs, 67-70 & Avco-Everett Res Lab, 67-72. Mem: Fel Am Phys Soc. Res: Controlled fusion; relativistic electron beams. Mailing Add: Lawrence Livermore Lab Univ of Calif Livermore CA 94550

BRIGGS, RICHARD M, b Auburn, Maine, Apr 13, 27. ORGANIC CHEMISTRY. Educ: Bates Col, BS, 49; Boston Univ, AM, 52, PhD(chem), 60. Prof Exp: From instr to asst prof, 56-66, ASSOC PROF CHEM, BATES COL, 66- Mem: AAAS; Am Chem Soc. Res: Reductions of N-nitro and N-nitroso compounds; synthesis and properties of organic derivatives of hydrazine; nature of the covalent bond; digital techniques in theoretical organic chemistry. Mailing Add: Dept of Chem Bates Col Lewiston ME 04240

BRIGGS, ROBERT EUGENE, b Madison, Wis, Apr 4, 27; m 48; c 4. AGRONOMY. Educ: Univ Wis, BS, 50, PhD(agron), 58; Mich State Univ, MS, 52. Prof Exp: Asst farm crops, Mich State Univ, 50-52; asst agron, Univ Wis, 53-56; asst plant breeder, 56-57, from asst prof to assoc prof agron, 57-70, PROF AGRON & PLANT GENETICS, UNIV ARIZ, 70-, AGRONOMIST, AGR EXP STA & EXTEN AGRONOMIST, COOP EXTEN SERV, 74- Concurrent Pos: Adv field crops, US AID-Brazil Prog, 64-66. Mem: Am Soc Agron. Res: Crop production; seed technology; cultural practices; fiber quality; cotton; physiology; management. Mailing Add: Dept of Agron Univ of Ariz Tucson AZ 85721

BRIGGS, ROBERT WILBUR, b Delavan, Ill, Mar 27, 34; m 62; c 1. GENETICS, BIOLOGY. Educ: Univ Ill, BS, 56, MS, 58; Univ Minn, PhD(genetics), 63. Prof Exp: Res assoc genetics, Brookhaven Nat Lab, 63-65, from asst geneticist to assoc geneticist, 65-69; GENETICIST, RES DEPT, FUNK BROS SEED CO, 69- Concurrent Pos: Fel biol, Brookhaven Nat Lab, 63-64. Mem: Genetics Soc Am; Crop Sci Soc Am. Res: Chemical and physical agents to affect inter- and intra-cistron genetic recombination in maize; chemical and physical mutagens in plants and corn breeding and genetics. Mailing Add: Funk Bros Seed Co Res Dept 1300 W Washington St Bloomington IL 61701

BRIGGS, ROBERT WILLIAM, b Watertown, Mass, Dec 10, 11; m 40; c 3. EMBRYOLOGY. Educ: Boston Univ, BS, 34; Harvard Univ, PhD(embryol), 38. Prof Exp: Asst, Harvard Univ, 36-38; asst, McGill Univ, 38-39, res fel, 39-41, res assoc zool, 41-42; biologist, Lankenau Hosp Res Inst & sr assoc mem, Inst Cancer Res; prof zool, Ind Univ, 56-63, res prof, 63-74, chmn dept, 69-74; WITH RES DEPT, FUNK SEEDS INT, INC, 74- Honors & Awards: Award, Fr Acad Sci, 72. Mem: Nat Acad Sci; AAAS; Am Soc Zoologists; Soc Study Develop & Growth; Genetics Soc Am. Res: Experimental embryology; development genetics and cytology. Mailing Add: Res Dept Funk Seeds Int Inc 1300 W Washington St Bloomington IL 61701

BRIGGS, RODNEY ARTHUR, agronomy, see 12th edition

BRIGGS, THOMAS, b New York, NY, May 24, 33; div; c 3. BIOCHEMISTRY. Educ: Yale Univ, BS, 54; Univ Pa, PhD(biochem), 60. Prof Exp: Res asst med, Yale Univ, 62-63, res assoc, 63-67; asst prof biochem, 67-70, ASSOC PROF BIOCHEM & MOLECULAR BIOL, COLS MED & DENT, UNIV OKLA, 70- Concurrent Pos: USPHS fel, Guy's Hosp Med Sch, London, 60-62. Mem: Sigma Xi; AAAS; Am Chem Soc; NY Acad Sci; Brit Biochem Soc. Res: Bile salt chemistry and metabolism; cholesterol metabolism; comparative biochemistry; biochemical evolution. Mailing Add: Dept of Biochem & Molec Biol Univ of Okla Health Sci Ctr Oklahoma City OK 73190

BRIGGS, WARREN STANLEY, physical chemistry, see 12th edition

BRIGGS, WILLIAM EGBERT, b Sioux City, Iowa, Mar 26, 25; m 47; c 4. NUMBER THEORY, ACADEMIC ADMINISTRATION. Educ: Morningside Col, BA, 48; Univ Colo, MA, 49, PhD(math), 53. Hon Degrees: DSc, Morningside Col, 68. Prof Exp: Instr math, Univ Colo, 48-53, res asst, 53-54; teacher pub sch, Colo, 54-55; from instr to assoc prof math, 55-64, actg dean col arts & sci, 63-64, PROF MATH & DEAN COL ARTS & SCI, UNIV COLO, BOULDER, 64- Concurrent Pos: Teacher high sch, Iowa, 47-48; dir acad yr inst high sch teachers, NSF, 56-60, actg chmn dept math, 59-60; NSF fel & res assoc, Univ Col, Univ London, 61-62; mem bd dirs, Educ Projs, Inc, 66-; mem regional selection comt, Woodrow Wilson Fel Found, 70-71; mem comn arts & sci, Nat Asn Land Grant Cols & Univs, 70-73. Mem: Coun Cols Arts & Sci (pres-elect, 74-75, pres, 75-76); Sigma Xi; Am Math Soc; Math Asn Am; London Math Soc. Res: Analytic number theory; prime number theory; dirichlet L-functions and Riemann zeta function; coef of power series; sequences generated by sieve processes. Mailing Add: 1440 Sierra Dr Boulder CO 80302

BRIGGS, WILLIAM PAUL, chemistry, pharmacy, see 12th edition

BRIGGS, WILLIAM SCOTT, b Shelton, Wash, Aug 15, 41; m 63. ORGANIC CHEMISTRY. Educ: Univ Wash, BS, 63; Stanford Univ, PhD(org chem), 68. Prof Exp: RES CHEMIST, BELLINGHAM DIV, GA-PAC CORP, 67- Mem: Am Chem Soc; Am Tech Asn Pulp & Paper Indust. Res: Wood chemistry and utilization of pulping wastes; application of chromatographic methods and instrumental techniques to the solution of organic structural problems; mass spectrometry. Mailing Add: 508 Chuckanut Dr Bellingham WA 98225

BRIGGS, WINSLOW RUSSELL, b St Paul, Minn, Apr 29, 28; m 55; c 2. PLANT PHYSIOLOGY. Educ: Harvard Univ, AB, 51, AM, 52, PhD(biol), 56. Prof Exp: From instr to assoc prof biol, Stanford Univ, 55-67; prof, Harvard Univ, 67-73; DIR DEPT PLANT BIOL, CARNEGIE INST WASH, 73- Mem: Nat Acad Sci; AAAS; Bot Soc Am; Am Soc Plant Physiol (pres, 75-76); Am Acad Arts & Sci. Res: Plant growth, development; physiology and biochemistry of photomorphogenesis. Mailing Add: Dept of Plant Biol Carnegie Inst of Wash 290 Panama St Stanford CA 94305

BRIGHAM, H IRVING, JR, ornamental horticulture, agronomy, see 12th edition

BRIGHAM, KENNETH LARRY, b Nashville, Tenn, Oct 29, 39; m 59; c 1. PULMONARY PHYSIOLOGY. Educ: David Lipscomb Col, BA, 62; Vanderbilt Univ, MD, 66. Prof Exp: From intern to asst resident internal med, Johns Hopkins Hosp, Baltimore, 66-68; med epidemiologist, Nat Commun Dis Ctr, USPHS Ecol Invest Prog, 68-70; instr internal med, Sch Med, Vanderbilt Univ, 70-71; NIH res fel pulmonary med, Cardiovasc Res Inst, Univ Calif, San Francisco, 71-73; asst prof internal med, 73-74, ASSOC PROF INTERNAL MED, SCH MED, VANDERBILT UNIV, 74-, DIR PULMONARY RES, 73- Concurrent Pos: Mem pulmonary study sect, Nat Heart & Lung Inst, 75-79; estab investr, Am Heart Asn, 75-80; mem lung res rev comt, Vet Admin, 76. Mem: AAAS; Am Fedn Clin Res; Am Thoracic Soc. Res: Humoral mechanisms in the pathogenesis of pulmonary edema resulting primarily from changes in the lung circulation. Mailing Add: Med Ctr Vanderbilt Univ Nashville TN 37232

BRIGHAM, M PRINCE, b Birmingham, Ala, Aug 6, 23; m 51; c 5. SURGERY, BIOCHEMISTRY. Educ: Univ Miami, BS, 44; Cornell Univ, MS, 46; Temple Univ, MD, 50, DSc(surg), 55. Prof Exp: Instr, 58-62, assoc, 62-66, asst dean student affairs, 67-73, ASST PROF SURG, MED SCH, TEMPLE UNIV, 66-, ASSOC DEAN STUDENT AFFAIRS, 73- Concurrent Pos: NIH fel, 58-59; guest investr biochem, Rockefeller Inst, 58-59. Mem: AAAS; AMA. Res: Amino acid metabolism, particularly sulfur amino acids; changes in tissues seen in burns. Mailing Add: Dept of Surg Temple Univ Sch of Med Philadelphia PA 19140

BRIGHAM, NELSON ALLEN, b Holyoke, Mass, Nov 6, 15. MATHEMATICS. Educ: Rutgers Univ, BS, 37, MS, 38; Univ Pa, PhD(math), 48. Prof Exp: Asst, Inst Math, Univ Pa, 38-41; sr res engr, Repub Aviation Corp, NY, 47; asst prof math, Swarthmore Ctr, Pa State Col, 47-48; asst prof, Univ Md, 48-51, mathematician, Appl Physics Lab, 51-56; sr scientist, Avco Mfg Corp, 56-59; prin mathematician, Baird-Atomic, Inc, 59-60; staff mem, Lincoln Lab, Mass Inst Technol, 60-62 & Mitre Corp, 62-64; assoc prof math, Southeastern Mass Tech Inst, 65; prof math & astron & head dept, Butler Univ, 65-70; prof math & chmn dept, Bradley Univ, 70-71; PVT TUTORING SERV, 71- Mem: Am Math Soc; Math Asn Am; Soc Indust & Appl Math. Res: Analytic additive theory of numbers; applied mathematics; digital and analog simulation and programming. Mailing Add: 753 W Wonderview Dr Dunlap IL 61525

BRIGHAM, RAYMOND DALE, b Stamford, Tex, Apr 1, 26; m 49; c 4. PLANT BREEDING. Educ: Tex Tech Col, BS, 50; Iowa State Col, MS, 52, PhD(crop breeding, plant path), 57. Prof Exp: Res agronomist, Agr Res Serv, USDA, 57-67; ASSOC PROF AGRON, TEX AGR EXP STA, TEX A&M UNIV AT LUBBOCK, 67- Concurrent Pos: Prog chmn, Int Sunflower Conf, Memphis, 70. Mem: AAAS; Am Phytopath Soc; Crop Sci Soc Am; Am Soc Agron; NY Acad Sci. Res: Breeding for disease resistance in oilseed crops; inheritance of quantitative and qualitative characteristics. Mailing Add: Tex Agr Exp Sta Rte 3 Lubbock TX 79401

BRIGHT, ARTHUR AARON, b Hanover, NH, Dec 31, 46; m 69; c 1. EXPERIMENTAL SOLID STATE PHYSICS. Educ: Dartmouth Col, AB, 69, MA, 70; Univ Pa, PhD(physics), 73. Prof Exp: Res investr physics, Univ Pa, 73-75; RES SCIENTIST PHYSICS, PARMA TECH CTR, CARBON PROD Res: Electronic properties of graphite and carbonaceous materials. Mailing Add: Union Carbide Corp Parma Tech Ctr 12900 Snow Rd Parma OH 44130

BRIGHT, DAVID BRUCE, organic chemistry, see 12th edition

BRIGHT, DONALD BOLTON, b Ventura, Calif, Nov 28, 30; m 55; c 2. MARINE ECOLOGY. Educ: Univ Southern Calif, AB, 52, MS, 55, PhD(biol), 67. Prof Exp: Res assoc biol, Univ Southern Calif, 57-60, lectr, 59-60; instr life sci, Fullerton Jr Col, 60-67; assoc prof biol, 67-70, PROF BIOL & CHMN DEPT, CALIF STATE UNIV, FULLERTON, 70- Concurrent Pos: Grants, US Dept Interior, Bur Com Fisheries, 57-70, Am Philos Soc, 67-69, NSF coop studies grant, 69 & Nat Geog Soc, 70; res assoc, Los Angeles County Mus Natural Hist, 66-; party chief, NSF US Antarctic Res Proj, Univ Southern Calif, 66; chmn, S Coast Regional Coastal Comn, Calif, 73-75; mem, Int Oceanog Found. Mem: AAAS; Am Inst Biol Sci; Ecol Soc Am; Soc Syst Zool. Res: Ecology of terrestrial crustaceans; diel patterns of midwater oceanic crustaceans. Mailing Add: PO Box 570 Long Beach CA 90808

BRIGHT, GORDON STANLEY, b Smethport, Pa, Apr 29, 15; m 39; c 3. CHEMISTRY. Educ: Tusculum Col, AB, 38; Univ Cincinnati, MS, 39. Prof Exp: Head dept sci, Pfeiffer Col, 39-40; res chemist, 40-54, supvr grease res dept, 54-62, RES ASSOC, TEXACO INC, 62- Mem: AAAS; fel Am Inst Chemists; Am Chem Soc; Am Oil Chem Soc; Am Soc Lubrication Eng. Res: Grease research. Mailing Add: 1010 31st St PO Box 1325 Nederland TX 77627

BRIGHT, HAROLD FREDERICK, b Smethport, Pa, Aug 6, 13; m 38; c 2. STATISTICS. Educ: Lake Forest Col, BA, 37; Univ Rochester, MS, 44; Univ Tex, PhD(educ psychol), 52. Prof Exp: Asst prof math, Denison Univ, 43-44 & Univ Rochester, 44-45; registr & dir guid, San Angelo Col, 45-49; assoc dir res, Am Asn Jr Cols, Univ Tex, 49-52; chief tech serv, Human Resources Res Off, George Washington Univ, 52-54, dept dir, 54-56; specialist opers res & synthesis, Gen Elec Co, 57-58; chmn dept statist, 58-64, assoc dean faculties, 64-66, PROF STATIST, GEORGE WASHINGTON UNIV, 58-, VPRES ACAD AFFAIRS, 66-, PROVOST, 69- Mem: AAAS; Am Psychol Asn; Inst Math Statist; Am Statist Asn; Math Asn Am. Res: Teaching of statistics; computing problems; operations research. Mailing Add: 314 Branch Circle SE Vienna VA 22180

BRIGHT, JOHN RUSSELL, chemistry, deceased

BRIGHT, NORMAN FRANCIS HENRY, physical chemistry, inorganic chemistry, see 12th edition

BRIGHT, PETER BOWMAN, b Gallipolis, Ohio, Dec 27, 37; div; c 3. MATHEMATICAL BIOLOGY. Educ: Antioch Col, BS, 60; Univ Chicago, PhD(math biol), 66. Prof Exp: Fel math biol, Univ Chicago, 66; res instr biophys, Ctr Theoret Biol, State Univ NY Buffalo, 66-69; asst prof bioinfo sci, Dept Surg, Southwestern Med Sch, Univ Tex, 69-70, asst prof biophys & surg, 70-73; MEM FAC, DEPT BIOMATH, SCH MED, UNIV CALIF, LOS ANGELES, 73- Concurrent Pos: Adj prof, Dept Statist, Southern Methodist Univ, 70- Mem: AAAS; Biophys Soc; NY Acad Sci. Res: Statistical mechanical foundations for the study of transients in biological transport in the presence of chemical reactions; information processing in the central nervous system and the physiological basis of memory. Mailing Add: Dept of Biomath Univ of Calif Sch of Med Los Angeles CA 90024

BRIGHT, ROBERT C, b Salt Lake City, Utah, Dec 27, 28. GEOLOGY. Educ: Univ Utah, BS, 52, MS, 60; Univ Minn, PhD(geol), 63. Prof Exp: Instr geol, Univ Minn, 59; NSF fel, Royal Univ Uppsala, 63-64; ASSOC PROF GEOL, UNIV MINN, MINNEAPOLIS, 64-, ASSOC PROF ECOL, 70-, CUR PALEONT, MUS NATURAL HIST, 64-, DIR, WASATCH-UINTA FIELD GEOL CAMP, 67- Mem: AAAS; Am Asn Petrol Geologists; fel Geol Soc Am; Paleont Soc; Am Soc Limnol &

Oceanog. Res: Pleistocene stratigraphy, paleontology; paleoecology and paleolimnology. Mailing Add: Mus of Natural Hist Univ of Minn Minneapolis MN 55455

BRIGHT, THOMAS J, b Millen, Ga, Sept 2, 37; m 60; c 2. BIOLOGICAL OCEANOGRAPHY. Educ: Univ Wyo, BS, 64; Tex A&M Univ, PhD(biol oceanog), 68. Prof Exp: Asst prof biol & invert zool, Jacksonville Univ, 68-69; asst prof oceanog, 69-74, ASSOC PROF OCEANOG, TEX A&M UNIV, 74- Mem: Am Soc Ichthyol & Herpet. Res: Ecology of coral reefs and hard-banks. Mailing Add: Dept of Oceanog Tex A&M Univ College Station TX 77843

BRIGHTMAN, I JAY, b New York, NY, Oct 28, 09; m 35; c 3. MEDICINE, PUBLIC HEALTH. Educ: NY Univ, BS, 30, MD, 34, MedScD(internal med), 40; Am Bd Internal Med, dipl, 41; Columbia Univ, MPH, 43; Am Bd Prev Med, dipl, 55. Prof Exp: Dir res proj, Div Syphilis Control, NY State Dept Health, 41-42, dir res proj, Veneral Dis Control, Buffalo, 43-45, asst dir div, 45-46, assoc physician, Div Med Serv, 46-48, asst dir div, 48-50, asst comnr, 50-51, asst comnr welfare med serv, 52-56, exec dir, Health Resources Bd, 56-60, asst comnr chronic dis control, 60-66, dep comnr med serv & res, 66-69; dir, New York Metrop Regional Med Prog, 69-72; CONSULT HEALTH PLANNING, 73- Concurrent Pos: Assoc prof, Albany Med Col, 51-; mem, Comt Welfare Serv, AMA, 54-60, consult, 60; mem, Nat Adv Community Health Comt, USPHS, 62- Mem: Fel AMA; Am Pub Health Asn; Am Col Physicians; Am Col Prev Med. Res: Chronic illness, especially natural history of disease, need for specialized resources and personnel, and medical economics. Mailing Add: 3 Rone Ct New York NY 10956

BRIGHTMAN, MILTON WILFRED, b Toronto, Ont, July 13, 23; nat US; m 53; c 2. NEUROANATOMY. Educ: Univ Toronto, AB, 45, AM, 48; Yale Univ, PhD(anat), 54. Prof Exp: NEUROANATOMIST, LAB NEUROANAT SCI, NAT INST NEUROL DIS & BLINDNESS, 54- Mem: Am Asn Anat. Res: Perivascular spaces in brain; neurosecretion; blood supply of spinal cord; neurocytology. Mailing Add: Lab of Neuroanat Sci Nat Inst Neurol Dis & Blindness Bethesda MD 20014

BRIGHTMAN, VERNON, b Brisbane, Australia, Dec 17, 30; m 62; c 3. DENTISTRY, ORAL MEDICINE. Educ: Univ Queensland, BDSc, Hons, 52, MDSc, 56; Univ Chicago, PhD(microbiol), 61; Univ Pa, DMD, 68. Prof Exp: Lectr dent, Univ Queensland, 52-56; asst, Zollar Dent Clin, Univ Chicago, 56-60; instr microbiol, Univ Pa, 60-62; sr lectr, Univ Queensland, 62-64; from assoc to assoc prof, 64-70, PROF ORAL MED, SCH DENT MED, UNIV PA, 70-, PROF OTORHINOLARYNGOLOGY, 73-, MEM STAFF, UNIV HOSP, 73- Concurrent Pos: Mem staff, Children's Hosp, Philadelphia, 60-62; chief dent res, Philadelphia Gen Hosp, 72. Honors & Awards: Univ Medal, Univ Queensland, 52. Mem: Fel AAAS; fel Am Acad Oral Path; fel Am Acad Oral Med; Int Asn Dent Res; Am Dent Asn. Res: Oral medicine and pathology. Mailing Add: Dept of Oral Med Univ of Pa Sch of Dent Med Philadelphia PA 19174

BRIGHTON, CARL T, b Pana, Ill, Aug 20, 31; m 54; c 4. ORTHOPEDIC SUGERY. Educ: Valparaiso Univ, BA, 53; Univ Pa, MD, 57; Univ Ill, Chicago Circle, PhD(anat), 69; Am Bd Orthop Surg, dipl, 65. Prof Exp: Staff orthopedist, Philadelphia Naval Hosp, US Navy, 62-63, Great Lakes Naval Hosp, Ill, 63-66 & USS Sanctuary, South China Sea, 66-67; from asst prof to assoc prof, 68-73, PROF ORTHOP SURG, UNIV PA, 73-, DIR ORTHOP RES, 68- Concurrent Pos: NIH res career develop award, 71-76. Mem: Orthop Res Soc; fel Am Acad Orhtop Surg; fel Am Col Surgeons. Res: Epiphyseal plate growth and development; fracture healing; arthritis and articular cartilage preservation and transplantation. Mailing Add: Dept of Orthop Univ of Pa Philadelphia PA 19104

BRIGHTON, KENNETH WILLIAM, organic chemistry, see 12th edition

BRIGHTWELL, WILLIAM THOMAS, horticulture, see 12th edition

BRIGLIA, DONALD DOMINICK, physics, see 12th edition

BRILES, CONNALLY ORAN, b Idabel, Okla, Dec 10, 19; m 45; c 2. IMMUNOGENETICS. Educ: Agr & Mech Col Tex, BS, 49, MS, 51; Univ Wis, PhD, 55. Prof Exp: Asst poultry husb, Agr Exp Sta, La State Univ, 55-57; asst to head poultry breeding sect, USDA, Md, 57-58; geneticist & dir blood grouping lab chickens, Arbor Acres Farm, Inc, 58-62; assoc prof animal sci & genetics, Macdonald Col, McGill Univ, 63-69; prof biol, 69-71, PROF ANIMAL SCI, TUSKEGEE INST, 72- Concurrent Pos: Breeding & incubation expert, Food & Agr Orgn, UN, Karachi, Pakistan, 71-72. Mem: Genetics Soc Am; Poultry Sci Asn. Res: Blood groups in domestic animals and humans; association of blood group genotypes with economic and morphological characteristics in chickens; general poultry breeding; basic and practical interaction of genetic traits and environment in poultry. Mailing Add: Dept of Agr Sci Tuskegee Inst Tuskegee Institute AL 36088

BRILES, GEORGE HERBERT, b Neodesha, Kans, Sept 15, 37. ORGANIC CHEMISTRY. Educ: Univ Kans, BS, 60; Whittier Col, MS, 61; Univ Mass, PhD(org chem), 66. Prof Exp: Chemist, US Borax Res Corp, 61; ASSOC PROF CHEM, SOUTHAMPTON COL, LONG ISLAND UNIV, 65- Mem: Am Chem Soc. Res: Properties and reactions of pentavalent organoantimony compounds. Mailing Add: Southampton Col Dept of Chem Long Island Univ Southampton NY 11968

BRILES, WORTHIE ELWOOD, b Italy, Tex, Jan 31, 18; m 41; c 3. IMMUNOGENETICS, POULTRY GENETICS. Educ: Univ Tex, BA, 41; Univ Wis, PhD(genetics, poultry breeding), 47. Prof Exp: Res asst genetics, Univ Wis, 41-42, 45-47, instr genetics & zool, 47; res asst genetics, Tex A&M Univ, 47-48, from asst prof to assoc prof poultry husb, 48-57; immunogeneticist, DeKalb AgRes, Inc, 57-70; PROF IMMUNOGENETICS, NORTHERN ILL UNIV, 70- Honors & Awards: Res Prize, Poultry Sci Asn, 51. Mem: AAAS; Genetics Soc Am; Am Genetic Asn; Poultry Sci Asn; Am Asn Immunol. Res: Blood groups of chickens; physiological effects of blood group genes; isoantigens and virus susceptibility. Mailing Add: Dept of Biol Sci Northern Ill Univ DeKalb IL 60115

BRILL, A BERTRAND, b New York, NY, Dec 19, 28; m 50; c 3. NUCLEAR MEDICINE. Educ: Grinnell Col, AB, 49; Univ Utah, MD, 56; Univ Calif, Berkeley, PhD(biophys), 61. Prof Exp: Med dir, Div Radiol Health, USPHS, 57-64; assoc prof med & radiol, 64-71, assoc prof biomed eng & biophys, 69-71, assoc prof physics, 70-71, PROF MED, RADIOL, BIOMED ENG, BIOPHYS & PHYSICS, SCH MED, VANDERBILT UNIV, 71- Mem: AAAS; Soc Nuclear Med; Radiation Res Soc; Am Thyroid Asn. Res: Radiation leukemogenesis; effects of radiation on thyroid function; diagnostic radioisotope studies. Mailing Add: Div of Nuclear Med & Biophys Vanderbilt Univ of Med Nashville TN 37232

BRILL, ARTHUR SYLVAN, b Philadelphia, Pa, June 11, 27; m 57; c 2. BIOPHYSICS. Educ: Univ Calif, Berkeley, AB, 49; Univ Pa, PhD(biophys), 56. Prof Exp: Res assoc, Electron Micros Lab, Cornell Univ, 50-61; asst prof biophys, Yale Univ, 61-64, assoc prof molecular biol & biophys, 64-68; mem ctr advan studies, 68-71, prof mat, 68-73, PROF PHYSICS, UNIV VA, 73- Concurrent Pos: Fel med physics, Univ Pa, 56-58; Donner fel, Med Sci Div, Nat Res Coun, Clarendon Lab, Oxford Univ, 58-59. Mem: AAAS; Am Phys Soc; Biophys Soc; Electron Micros Soc Am; Am Chem Soc. Res: Molecular biophysics; mechanisms of enzyme action; transition metal ions in biology; protein structure; fundamental limitations of measuring instruments. Mailing Add: Dept of Physics Univ of Va Charlottesville VA 22901

BRILL, DIETER RUDOLF, b Heidelberg, Ger, Aug 9, 33; m 71. THEORETICAL PHYSICS. Educ: Princeton Univ, AB, 54, MA, 56, PhD(physics), 59. Prof Exp: Instr physics, Princeton Univ, 59-60; Flick exchange res fel, Univ Hamburg, 60-61; from instr to assoc prof, Yale Univ, 61-70; PROF PHYSICS, UNIV MD, COLLEGE PARK, 70- Concurrent Pos: Vis prof, Univ Würzburg, 68. Mem: NY Acad Sci; fel Am Phys Soc. Res: General relativity and gravitation physics; foundation of quantum mechanics. Mailing Add: Dept of Physics & Astron Univ of Md College Park MD 20742

BRILL, EARL, b Hamtramck, Mich, Nov 1, 29; m 53; c 1. ORGANIC CHEMISTRY, PHARMACOLOGY. Educ: Univ Miami, BS, 57, MS, 58, PhD(org chem), 62. Prof Exp: Instr chem, 58-59, ASST PROF PHARMACOL, SCH MED, UNIV MIAMI, 62- Mem: Am Chem Soc; The Chem Soc. Res: Metabolism studies of the aromatic amines, especially carcinogenic amines and the heterocyclic nitrogen and oxygen quinoxaline and dibenzofuran. Mailing Add: Dept Pharmacol Sch Med Univ Miami Biscayne Annex Miami FL 33152

BRILL, HAROLD CLIFFORD, b Steubenville, Ohio, Mar 3, 09; m 40; c 1. CHEMISTRY. Educ: Muskingum Col, BS, 30; Ohio State Univ, MS, 33, PhD(chem), 35. Prof Exp: Asst, Ohio State Univ, 32-33; res chemist, E I du Pont de Nemours & Co, Inc, 35-50, res chemist, Tech Serv, 51-62, res assoc, 62-74; RETIRED. Mem: Am Chem Soc; Am Tech Asn Pulp & Paper Indust. Res: Boiler water; analytical chemistry; inorganic chemistry; colloid; paint and varnish; rare metals; paper. Mailing Add: 28 Riverside Dr Riverside Gardens Wilmington DE 19809

BRILL, KENNETH GRAY, JR, b St Paul, Minn, Nov 16, 10; m 39; c 2. GEOLOGY. Educ: Univ Minn, AB, 35; Univ Mich, MS, 38, PhD(geol), 39. Prof Exp: Asst geol, Yale Univ, 36-37; from instr to assoc prof geol & geog, Univ Chattanooga, 39-44; geologist, US Geol Surv, 44-45; from asst prof to assoc prof geol, 46-50, chmn dept geol & geol eng, 67-69, PROF GEOL, ST LOUIS UNIV, 50- Concurrent Pos: Spec coal consult, Econ Coop Admin, SKorea, 49; Fulbright lectr, Univ Tasmania, 53. Mem: Fel AAAS; fel Paleont Soc; fel Geol Soc Am; Am Asn Petrol Geol; Australian Geol Soc. Res: Permo-carboniferous stratigraphy and paleontology; paleozoic stratigraphy of Rocky Mountain region. Mailing Add: Dept of Earth & Atmospheric Sci St Louis Univ St Louis MO 63103

BRILL, NORMAN QUINTUS, b New York, NY, Aug 2, 11; m 37 & 70; c 3. PSYCHIATRY, NEUROLOGY. Educ: City Col New York, BS, 30; NY Univ, MD, 34. Prof Exp: Instr neurol, Col Physicians & Surgeons, Columbia Univ, 39-41; prof & head dept, Sch Med, Georgetown Univ, 46-49; chmn dept psychiat & med dir, Neuropsychiat Inst, 53-67, PROF PSYCHIAT, SCH MED, UNIV CALIF, LOS ANGELES, 53- Concurrent Pos: Chief neuropsychiat, State Hosp, Ft Bragg, NC, 41-44; chief psychiat br, Neuropsychiat Div, Off Surgeon Gen, DC, 44-45, dept dir, Neuropsychiat Div, 45-46; consult, Camarillo State Hosp, Calif, 67-; mem deans comt, Sepulveda Vet Admin Hosp, 67-68. Mem: Int Psychoanal Asn; Am Psychoanal Asn; fel Am Psychiat Asn; fel Am Col Psychiat; Int Asn Soc Psychiat. Res: Mechanism and effectiveness of various types of psychiatric treatment; psychotherapeutic process; socioeconomic factors in mental illness. Mailing Add: Dept of Psychiat Univ of Calif Sch of Med Los Angeles CA 90024

BRILL, ROBERT H, b Irvington, NJ, May 7, 29; m 57; c 1. PHYSICAL CHEMISTRY. Educ: Upsala Col, BS, 51; Rutgers Univ, PhD(phys chem), 55. Prof Exp: Asst prof chem, Upsala Col, 54-60; ADMINR SCI RES, CORNING MUS OF GLASS, 60- Concurrent Pos: Fel, Int Inst Conserv Hist & Artistic Works. Mem: Am Chem Soc; Am Ceramic Soc; fel Am Inst Chemists. Res: Archaeological chemistry; scientific examination of archaeological materials, particularly ancient glass; history of science and technology; glass chemistry; photochemistry. Mailing Add: Corning Mus of Glass Corning Glass Ctr Corning NY 14830

BRILL, THOMAS BARTON, b Chattanooga,Tenn, Feb 3, 44; m 66; c 2. INORGANIC CHEMISTRY, PHYSICAL CHEMISTRY. Educ: Univ Mont, BS, 66; Univ Minn, Minneapolis, PhD(chem), 70. Prof Exp: Asst prof, 70-74, ASSOC PROF CHEM, UNIV DEL, 74- Mem: Am Chem Soc. Res: Nuclear quadrupole resonance spectroscopy; infrared and Raman spectroscopy; organometallic chemistry; donor-acceptor complex chemistry; solid state effects. Mailing Add: Dept of Chem Univ of Del Newark DE 19711

BRILL, WILFRED G, b Albion, Ind, Aug 11, 30; m 54; c 2. PHYSICS. Educ: Manchester Col, BA, 52; Purdue Univ, MS, 55, PhD(physics), 64. Prof Exp: Res assoc physics, Purdue Univ, 64-67, asst prof, NCent Campus, 67-74, ASSOC PROF PHYSICS, PURDUE UNIV, NCENT CAMPUS, 74-Mem: Optical Soc Am; Am Asn Physics Teachers. Res: Atomic spectroscopy. Mailing Add: Dept of Physics NCent Campus Purdue Univ Westville IN 46391

BRILL, WILLIAM FRANKLIN, b Utica, NY, Feb 16, 23; m 51; c 3. ORGANIC CHEMISTRY. Educ: Univ Conn, BS, 45, MS, 48, PhD(chem), 50. Prof Exp: Res assoc org chem, Univ Calif, Los Angeles, 50-52; chemist, Polymers Sect, Olin Industs, 52-55; group leader oxidations, Petro-Tex, 55-60, res assoc, 60-66; dir chem res, Princeton Chem Res, 66-69; SR SCIENTIST, HALCON INT, 69- Res: Reaction mechanisms; oxidations; catalysis. Mailing Add: RD 1 Skillman NJ 08558

BRILL, WINSTON J, b London, Eng, June 16, 39; US citizen; m 65; c 1. MICROBIOLOGY, BIOCHEMICAL GENETICS. Educ: Rutgers Univ, BS, 61; Univ Ill, Urbana, PhD(microbiol), 65. Prof Exp: NIH fel biol, Mass Inst Technol, 65-67; PROF BACT, UNIV WIS-MADISON, 67- Concurrent Pos: Res grants, USPHS, 68- & NSF, 69-; panel mem, NSF Metab Biol Prog, 74- Mem: AAAS; Am Soc Microbiol. Res: Electron transport; ferredoxin; regulation of catabolism; nitrogen fixation; catabolite repression. Mailing Add: Dept of Bact Univ of Wis Madison WI 53706

BRILLHART, RUSSELL EDWARD, b York, Pa, Feb 9, 05; m 29; c 3. PHARMACOLOGY, BACTERIOLOGY. Educ: Univ Pa, AB, 27; Philadelphia Col Pharm, BSc, 32, MSc, 39, DSc(bact), 50. Prof Exp: Teacher high sch, Pa, 29-30; prof biol & head dept, Des Moines Col Pharm, 32-38; asst biol & pharmacog, Philadelphia Col Pharm, 38-40; asst prof pharmacog & pharmacol, Col Pharm, Rutgers Univ, 40-43; mem res staff & actg admin dir, Bact Div, William S Merrell Co, 44-45; instr pharmacog & zool, Philadelphia Col Pharm & prof bact, Pa State Col Optom, 45-46; prof microbiol & dean col pharm, Drake Univ, 46-50; prof pharmacol & allied sci, chmn dept & asst dean, RI Col Pharm, 50-57; prof pharmacog & microbiol & chmn dept, New Eng Col Pharm, 57-62; prof pharmacog & microbiol & asst dean col pharm, Northeastern Univ, 62-70; CONSULT, RI HEALTH DEPT, 70- Concurrent Pos: Lab bioassayist, H K Mulford Co, Pa, 27; mem staff res & prog penicillin,

Riechel Lab, John Wyeth & Co, Inc, 43; biol consult, 70- Mem: AAAS; Am Pharmaceut Asn; Am Soc Pharmacog; NY Acad Sci. Res: Disinfectants; antibiotics and oral vaccines. Mailing Add: 18 Merrick St Rumford RI 02916

BRILLINGER, DAVID ROSS, b Toronto, Ont, Oct 27, 37; m 60; c 2. STATISTICS. Educ: Univ Toronto, BA, 59; Princeton Univ, MA, 60, PhD(math), 61. Prof Exp: Soc Sci Res Coun fel, London Sch Econ, 61-62; mem tech staff statist, Bell Tel Labs, NJ, 62-64; lectr, London Sch Econ, 64-66, reader, 66-69; PROF STATIST, UNIV CALIF, BERKELEY, 69- Concurrent Pos: Lectr math, Princeton Univ, 62-64; mem anal adv comt, Nat Assessment of Educ Prog, 73-; Guggenheim fel, 75. Mem: Am Math Asn; fel Am Statist Asn; fel Inst Math Statist; Math Asn Am; Int Statist Inst. Res: Time series; applied probability; multivariate analysis. Mailing Add: Dept of Statist Univ of Calif Berkeley CA 94720

BRILLSON, LEONARD JACK, b New York, NY, Dec 15, 45; m 68. SURFACE PHYSICS. Educ: Princeton Univ, AB, 67; Univ Pa, MS, 69, PhD(physics), 72. Prof Exp: Assoc scientist, 72-74, SCIENTIST PHYSICS, XEROX CORP WEBSTER RES CTR, 74- Mem: Am Phys Soc; Am Vacuum Soc. Res: Surface electronic structure of semiconductors and insulators, their dependence on surface chemical composition and bonding, and their influence on charge transport across interfaces with metals. Mailing Add: Xerox Webster Res Ctr 800 Phillips Rd Webster NY 14580

BRIM, CHARLES A, b Spalding, Nebr, Apr 14, 24; m 48; c 2. PLANT BREEDING. Educ: Univ Nebr, BS, 48, MS, 50, PhD(agron), 53. Prof Exp: RES AGRONOMIST, USDA, NC STATE UNIV, 53-, PROF CROP SCI & GENETICS, UNIV, 69- Mailing Add: Dept of Crop Sci NC State Univ Raleigh NC 27607

BRIMHALL, JAMES ELMORE, b Burlington, Iowa, June 5, 36; m 58; c 2. PHYSICS. Educ: Hamline Univ, BS, 59; Univ Pittsburgh, MS, 65. Prof Exp: Teacher, 66-68, chmn dept, 68-74, ASSOC PROF PHYSICS, W VA STATE COL, 69- Mem: Am Phys Soc; Am Asn Physics Teachers; Nat Sci Teachers Asn; Soc Photog Technologists. Res: Positronium annihilation in gases; circular polarization analysis of x-rays via Compton scattering; plane polarization analysis of gamma rays via Compton scattering; momentum spectroscopy of charged particles. Mailing Add: Dept of Physics WVa State Col Institute WV 25112

BRIMIJOIN, WILLIAM STEPHEN, b Passaic, NJ, July 1, 42; m 64; c 2. PHARMACOLOGY, NEUROBIOLOGY. Educ: Harvard Col, BA, 64, Harvard Univ, PhD(pharmacol), 69. Prof Exp: Instr pharmacol, Mayo Grad Sch Med, 71-72; asst prof, 72-76, ASSOC PROF PHARMACOL, MAYO MED SCH, 76-, CONSULT, MAYO FOUND, 72- Concurrent Pos: NIH pharmacol res assoc training prog, 69-71; NIHM grant, 71-74; NIH grant & career develop award, 75-; assoc consult, Mayo Found, 71-72. Mem: Am Soc Pharmacol & Exp Therapeut; Am Soc Neurochem; Soc Neurosci. Res: Chemical biology of nerve cells; mechanisms and functions of exoplasmic transport. Mailing Add: Dept of Pharmacol Mayo Med Sch Rochester MN 55901

BRIMM, EUGENE OSKAR, b Sheboygan, Wis, July 7, 15; m 40; c 3. INORGANIC CHEMISTRY. Educ: Univ Wis, BS, 38; Univ Ill, MS, 38, PhD, 40. Prof Exp: Div head, Tonawanda Lab, Linde Div, Union Carbide Corp, 40-56, asst mgr, 56-63, mgr technol rels group, Union Carbide Europe S A, Switz, 63-68, mgr prod technol, Mat Systs Div, Union Carbide Corp, 68-70; mgr mkt develop, Stellite Div, Cabot Corp, Ind, 70-71, mgr int dept, 71-72; MGR RES & DEVELOP, FILTROL CORP, 72- Mem: Am Chem Soc. Res: Cracking and hydrotreating catalysts; adsorbents; metal carbonyls; organometallic compounds; molecular sieves. Mailing Add: 18121 Allegheny Dr Santa Ana CA 92705

BRIN, MYRON, b New York, NY, July 1, 23; m 44; c 3. BIOCHEMISTRY. Educ: NY Univ, BS, 45; Cornell Univ, BS, 47, MS, 48; Harvard Univ, PhD(med sci, biochem), 51. Prof Exp: Sr res scientist metab radiation, New Eng Deaconess Hosp, Boston, 51-53; instr biochem, Sch Med, Harvard Univ, 53-56; chief biologist, Food & Drug Res Labs, Inc, 56-58; assoc prof biochem & med, State Univ NY Upstate Med Ctr, 58-68; prof nutrit, Univ Calif, Davis, 68-69; asst dir biochem nutrit, 69-70, ASSOC DIR BIOCHEM NUTRIT, HOFFMANN-LA ROCHE INC, 70- Concurrent Pos: Instr biochem, Sch Med, Tufts Univ, 52-53; res fel & instr, Thorndike Mem Lab, Med Sch, Harvard Univ, 53-56, mem bd tutors, Harvard Univ, 54-56; NIH sr res fel, Hadassah Med Sch, Hebrew Univ Jerusalem, 67-68; adj prof, Columbia Univ, 69- Honors & Awards: A Cressy Morrison Award, NY Acad Sci, 62. Mem: AAAS; Am Soc Biol Chemists; Am Inst Nutrit; Am Soc Clin Nutrit; Am Chem Soc. Res: Nutrition; vitamin function in metabolism and health maintenance, particularly vitamin B1, B6 and E adequacy. Mailing Add: Biochem Nutrit Hoffmann-La Roche Inc 340 Kingsland St Nutley NJ 07110

BRINCKERHOFF, HAROLD GUION, b New Rochelle, NY, Aug 9, 18; m 41; c 2. CHEMISTRY. Educ: Univ Va, BSCh, 41. Prof Exp: Prod control chemist, Solvay Process Co, Va, 41-45; sr chemist, Texaco Exp, Inc, 45-55, tech serv mgr, 55-60, tech supvr, 60-69, SUPVR EMPLOYEE RELS, TEXACO RES LABS, 69- Res: Jet propulsion; chemical pyrolysis; combustion; technical personnel recruitment; laboratory personnel administration. Mailing Add: 2430 Pineway Dr Richmond VA 23225

BRINCK-JOHNSON, TRULS, b Stavanger, Norway, Oct 10, 26; m 52; c 3. ENDOCRINOLOGY, BIOCHEMISTRY. Educ: Univ Utah, MS, 56, PhD(anat, endocrinol), 59. Prof Exp: Res asst anal chem, Univ Oslo, 48-49; res asst, Inst Aviation Med, Norweg Air Force, 49-50; chief chemist, French & Norweg Trop & Antarctic Whaling Expeds, 50-52 & 53-54; res assoc med biochem, Univ Oslo, 52-53, investr steroid chem, Ullevaal Hosp, 59-63, chief biochemist, Endocrine Lab, Rikshosp, 63-64; res assoc steroid chem & asst prof path, 64-69, ASSOC PROF PATH, DARTMOUTH MED SCH, 69-, ASSOC PROF MED, 70- Concurrent Pos: Mem assoc staff, Mary Hitchcock Mem Hosp, Hanover, NH, 67- Mem: Endocrine Soc; Royal Soc Med. Res: Biochemistry and physiology of steroid hormones. Mailing Add: Dept of Path Dartmouth Med Sch Hanover NH 03755

BRINCKMAN, FREDERICK EDWARD, JR, b Oakland, Calif, June 24, 28; m 54; c 2. INORGANIC CHEMISTRY. Educ: Univ Redlands, BS, 54; Harvard Univ, AM, 58, PhD(chem), 60. Prof Exp: Res chemist & head propellant br, US Naval Ord Lab, Calif, 60-61; sci staff asst head anal chem div, Gen Res Div, US Naval Propellant Plant, Univ Md, 61-64; actg chief inorg chem sect, 74-75, RES CHEMIST, NAT BUR STAND, 64- Honors & Awards: Silver Medal, Dept Com, 74. Mem: AAAS; Am Chem Soc; The Chem Soc; NY Acad Sci; fel Am Inst Chemists. Res: Synthetic inorganic and organometallic chemistry; metal-nitrogen, metal-fluorine chemistry; coordination chemistry of main group elements; applications of magnetic and mass spectrometry to inorganic chemistry; dynamics of organometallic systems in aqueous phases; organometallics as intermediates in biotransformations of metals. Mailing Add: 5609 Granby Rd RR 1 Derwood MD 20855

BRINDELL, GORDON DWIGHT, organic chemistry, see 12th edition

BRINDLEY, CLYDE OWENS, b Temple, Tex, Feb 17, 17; m 52. MEDICINE. Educ: Univ Tex, AB, 38; Duke Univ, MD, 43; Univ Minn, MA, 54; Am Bd Internal Med, dipl, 55. Prof Exp: Intern, Duke Hosp, 43-44; house officer, Mass Gen Hosp, 49-50; clin investr, Nat Cancer Inst, 55-62; PHYSICIAN, VET ADMIN HOSP, 62- Res: Cancer chemotherapy. Mailing Add: Med Serv Vet Admin Ctr Mountain Home TN 37684

BRINDLEY, GEORGE W, b Stoke-on-Trent, Eng, June 19, 05; m 31; c 2. MINERALOGY, SOLID STATE PHYSICS. Educ: Univ Manchester, BSc, 26, dipl educ, 27, MSc, 28; Univ Leeds, PhD(physics), 33. Hon Degrees: DSc, Univ Louvain, 69. Prof Exp: Lectr physics, Univ Leeds, 29-48, reader x-ray physics, 48-53; res prof mineral sci, 53-55, head ceramic technol & prof solid state technol, 55-62, prof mineral sci, 62-73, EMER PROF MINERAL SCI, PA STATE UNIV, 73- Concurrent Pos: MacKinnon res student, Royal Soc London, 34-39; consult, Gulf Res & Develop Co, 54-59 & Jersey Prod Res Co, 60-62; vis prof, Tokyo Inst Technol, 61; consult, Ga Kaolin Co, NJ, 65- & W R Grace & Co, 72-; Edward Orton lectr, Am Ceramic Soc, 73. Honors & Awards: Roebling Medallist, Mineral Soc Am, 70. Mem: Fel Mineral Soc Am; Am Ceramic Soc; Am Crystallog Asn; fel Brit Inst Physics & Phys Sci; Clay Minerals Soc (pres, 69-70). Res: X-ray study of the arrangement of atoms in the solid state, especially the application of this method to minerals; x-ray crystallography; x-ray mineralogy; clay minerals; kinetics of solid state reactions. Mailing Add: 126 Mineral Sci Bldg Pa State Univ University Park PA 16802

BRINEN, JACOB SOLOMON, b Brooklyn, NY, Nov 16, 34; m 58; c 2. MOLECULAR SPECTROSCOPY, SURFACE CHEMISTRY. Educ: Brooklyn Col, BS, 56; Pa State Univ, PhD(chem & electronic spectros), 61. Prof Exp: From res chemist to sr res chemist, 61-74, PRIN RES CHEMIST, AM CYANAMID CO, 74- Mem: Am Chem Soc; Am Phys Soc. Res: Electronic absorption spectroscopy, fluorescence and phosphorescence; electron spin and photoelectron spectroscopy. Mailing Add: Am Cyanamid Co 1937 W Main St Stamford CT 06904

BRINER, ROBERT C, organic chemistry, see 12th edition

BRINER, WILLIAM WATSON, b Winchester, Ind, Dec 12, 28; m 54; c 4. MICROBIOLOGY. Educ: Ohio State Univ, BA, 53, MS, 57, PhD(microbiol), 59. Prof Exp: MICROBIOLOGIST, PROCTER & GAMBLE CO, 59- Concurrent Pos: Lectr, Eve Col, Univ Cincinnati, 63-70; assoc ed, J Dent Res, 73-75. Mem: AAAS; Am Soc Microbiol; Int Asn Dent Res; Am Asn Dent Res; Am Acad Microbiol. Res: Oral microbiology; dental diseases in animals. Mailing Add: Procter & Gamble Co Winton Hill Tech Ctr Cincinnati OH 45224

BRINEY, ROBERT EDWARD, b Benton Harbor, Mich, Dec 2, 33. MATHEMATICS. Educ: Northwestern Univ, AB, 55; Mass Inst Technol, PhD(math), 61. Prof Exp: Instr math, Mass Inst Technol, 61-62; asst prof, Purdue Univ, 62-68; assoc prof, 68-70, PROF MATH, SALEM STATE COL, 70-, CHMN DEPT, 71- Mem: Am Math Soc; Math Asn Am. Res: Algebraic geometry; commutative algebra. Mailing Add: Dept of Math Salem State Col Salem MA 01970

BRINGHURST, ROYCE S, b Murray, Utah, Dec 27, 18; m 45; c 6. PLANT BREEDING. Educ: Utah State Univ, BS, 47; Univ Wis, MS, 48, PhD(agron, genetics), 50. Prof Exp: PROF POMOL & CHMN DEPT, UNIV CALIF, DAVIS, 50- Mem: Genetics Soc Am; fel Am Soc Hort Sci; Am Phytopath Soc; Bot Soc Am. Res: Breeding and genetics of strawberries; evolution, genetics, cytogenetics and breeding of Fragaria polyploids. Mailing Add: Dept of Pomol Univ of Calif Davis CA 95616

BRINIGAR, WILLIAM SEYMOUR, JR, b Wichita, Kans, May 18, 30; m 56; c 3. BIOCHEMISTRY. Educ: Univ Kans, BS, 52, PhD(chem), 57. Prof Exp: Res assoc, Univ SC, 57-59 & Yale Univ, 59-66; asst prof chem, 66-67, ASSOC PROF CHEM, TEMPLE UNIV, 67- Mem: Am Chem Soc; The Chem Soc. Res: Mechanisms of enzyme reactions. Mailing Add: Dept of Chem Temple Univ Philadelphia PA 19122

BRINK, DAVID LIDDELL, b St Paul, Minn, July 7, 17; m 43; c 3. WOOD CHEMISTRY. Educ: Univ Minn, BS, 39, PhD(agr biochem, forestry), 54. Prof Exp: Asst, Univ Minn, 39-42; res chemist, Salvo Chem Corp, 43-45; asst, Univ Minn, 47-49; group leader lignin, Mead Corp, 49-53; chemist, Develop Ctr, Weyerhaeuser Timber Col, 54-55, chief chem sect, 55-57, chief org chem sect, Cent Res Dept, 57-58; assoc forest prod chemist & lectr forestry, 58-62, FOREST PROD CHEMIST & LECTR WOOD CHEM & PROF AGR CHEM, UNIV CALIF, BERKELEY, 62-, GRAD ADV, 72- Mem: Am Chem Soc; Forest Prod Res Soc; Tech Asn Pulp & Paper Indust. Res: Lignin, cellulose, pulping, bark, extractives. Mailing Add: Sch of Forestry & Conserv Univ of Calif Berkeley CA 94720

BRINK, FRANK, JR, b Easton, Pa, Nov 4, 10; m 39; c 2. BIOPHYSICS. Educ: Pa State Col, BS, 34; Calif Inst Technol, MS, 35; Univ Pa, PhD(biophys), 39. Prof Exp: Lalor fel, 39-40; instr physiol, Med Col, Cornell Univ, 40-41; Johnson Found fel & lectr biophys, Univ Pa, 41-47, asst prof, 47-48; assoc prof, Johns Hopkins Univ, 48-53; prof, 53-74, dean grad studies, 57-72, DETLEV W BRONK PROF, ROCKEFELLER UNIV, 74- Concurrent Pos: Chmn, President's Comt Nat Medal Sci, 63-65. Mem: Nat Acad Sci; Am Acad Arts & Sci; Am Physiol Soc; Soc Gen Physiol; Biophys Soc. Res: Physical chemistry of nerve cells. Mailing Add: Rockefeller Univ New York NY 10021

BRINK, GILBERT OSCAR, b Los Angeles, Calif, May 26, 29; m 57. EXPERIMENTAL ATOMIC PHYSICS. Educ: Col of the Pacific, BA, 53; Univ Calif, PhD(chem), 57. Prof Exp: Asst res physicist, Univ Calif, 57-59, physicist, Lawrence Radiation Lab, 59-63; asst res prof, Univ Pittsburgh, 62-63; prin physicist, Cornell Aeronaut Lab, 63-68; chmn dept physics & astron, 72-74, ASSOC PROF PHYSICS, STATE UNIV NY BUFFALO, 68- Mem: Am Phys Soc. Res: Atomic beam magnetic resonance; hyperfine structure anomaly measurements; nuclear spins and magnetic moments; atomic scattering by crossed beams; molecular structure; laser excitation of atomic states. Mailing Add: Dept of Physics & Astron State Univ of NY Buffalo NY 14214

BRINK, JOHN JEROME, b Secunderabad, India, Mar 18, 34; US citizen; m 60; c 2. NEUROSCIENCES. Educ: Univ Orange Free State, SAfrica, BSc, 55; Univ Witwatersrand, BSc(Hons), 56; Univ Vt, PhD(biochem), 62. Prof Exp: Biochemist, Stanford Res Inst, 62-64; NIH asst res biochemist, Ment Health Res Inst, Univ Mich, 64-66; ASSOC PROF BIOCHEM, CLARK UNIV, 66- Concurrent Pos: Vis lectr, Harvard Med Sch, 72. Mem: AAAS; Am Soc Neurochem; NY Acad Sci; Int Soc Neurochem. Res: Neurochemistry of brain proteins and nucleic acids; biochemical pharmacology of nucleic acid precursor antimetabolites. Mailing Add: Dept of Biol Clark Univ Worcester MA 01610

BRINK, KENNETH MAURICE, b Cleveland, Ohio, May 11, 32; m 56; c 2. HORTICULTURE. Educ: Purdue Univ, BS, 54, MS, 58, PhD, 65. Prof Exp: From instr to assoc prof hort, Purdue Univ, 58-68; PROF HORT & HEAD DEPT, COLO STATE UNIV, 68- Mem: AAAS; Am Soc Hort Sci. Res: Marketing horticultural crops; post-harvest physiology and economic factors. Mailing Add: Dept of Hort Colo State Univ Ft Collins CO 80523

BRINK, MARION FRANCIS, b Golden Eagle, Ill, Nov 20, 32. NUTRITION. Educ: Univ Ill, BS, 55, MS, 58; Univ Mo, PhD(nutrit), 61. Prof Exp: Res biologist, Biomed Div, US Naval Radiol Defense Lab, 61-62; from assoc dir to dir nutrit res, 62-71, PRES, NAT DAIRY COUN, 71- Mem: AAAS; Am Oil Chem Soc; Am Dairy Sci Asn; Sigma Xi. Res: Mineral requirements; toxicology; nutrition, biochemistry and physiology of humans and animals; effects of ionizing radiation upon nutrient requirements. Mailing Add: Nat Dairy Coun 111 N Canal St Chicago IL 60606

BRINK, NORMAN GEORGE, b Littleton, Colo, Aug 31, 20; m 47. HEALTH SCIENCES. Educ: Princeton Univ, AB, 42, MA, 43, PhD(chem), 44. Prof Exp: Res chemist, 44-50, asst dir org & biochem res, 50-56, dir bio-org chem, 56-66, DIR UNIV RELS, MERCK & CO, INC, 66- Mem: AAAS; Am Soc Biol Chem. Res: Isolation of biologically active natural products; intermediary metabolism in disease; enzyme inhibition; medicinal chemistry. Mailing Add: Merck & Co Inc Rahway NJ 07065

BRINK, RAYMOND WOODARD, mathematics, deceased

BRINK, ROBERT HAROLD, JR, b Cortland, NY, Dec 17, 29; m 54; c 3. BIOCHEMISTRY. Educ: Washington Col, Md, BS, 52; Univ Del, MS, 59, PhD(biochem), 63. Prof Exp: Res assoc biochem, Atlantic Richfield Co, 62-66, res assoc, Betz Labs, Inc, 66-74; LAB MGR, BETZ ENVIRON ENGRS, INC, 74- Mem: AAAS; Am Chem Soc; Water Pollution Control Fedn. Res: Treatability and recycling of waste materials, in particular water borne wastes subject to microbiological treatment. Mailing Add: Betz Environ Engrs Inc One Plymouth Meeting Hall Plymouth Meeting PA 19462

BRINK, ROYAL ALEXANDER, b Woodstock, Ont, Sept 16, 97; nat US; m 22, 63; c 2. GENETICS, BOTANY. Educ: Univ Toronto, BSA, 19; Univ Ill, MS, 21; Harvard Univ, ScD(genetics), 23. Prof Exp: Asst chemist, Western Can Flour Mills, 19-20; from asst prof to prof, 22-68, chmn dept, 39-51, EMER PROF GENETICS, UNIV WIS-MADISON, 68- Concurrent Pos: Nat Res fel, Berlin & Birmingham, 25-26; ed, Genetics, 52-57; Haight travelling fel, Univ London & Oxford Univ, 60-61; NSF sr fel, Commonwealth Sci & Indust Res Orgn, Canberra, Australia, 66-67. Mem: Nat Acad Sci; AAAS; Am Soc Naturalists (treas, 36-39, pres, 63); Genetics Soc Am (vpres, 53, pres, 57); Am Genetic Asn. Res: Pollen physiology; reciprocal translocations in maize; endosperm in seed development; gene action and mutation; genetics of maize; paramutation. Mailing Add: Lab of Genetics Univ of Wis Madison WI 53706

BRINK, VERNON CUTHBERT, agronomy, genetics, see 12th edition

BRINKER, KEITH CLARK, b Waterloo, Iowa, May 13, 21; m 42; c 3. ORGANIC CHEMISTRY. Educ: Iowa State Col, BS, 43; Univ Iowa, MS, 49, PhD(chem), 51. Prof Exp: Jr chemist, Shell Develop Co, 43-44; city chemist, Waterloo, Iowa, 46-48; CHEMIST, E I DU PONT DE NEMOURS & CO, INC, 51- Mem: Am Chem Soc. Res: General organic chemistry; high polymers. Mailing Add: 3214 Coachman Rd Surrey Park Wilmington DE 19803

BRINKER, WADE OBERLIN, b Fulton, Ohio, Oct 11, 12; m 39; c 3. VETERINARY SURGERY. Educ: Kans State Col, DVM, 39; Mich State Col, MS, 47. Prof Exp: Asst, 39-40, instr, 40-41, from asst prof to assoc prof, 46-67, head dept, 57-67, PROF VET SURG & MED, MICH STATE UNIV, 67- Honors & Awards: Centennial Award Distinguished Serv, Kans State Univ, 63. Mem: Am Vet Med Asn; Conf Res Workers Animal Dis. Res: Penicillin blood levels in dogs; bone and joint surgery in small animals. Mailing Add: 5009 N Okemos Rd East Lansing MI 48823

BRINKERHOFF, LLOYD ALLEN, b Provo, Utah, July 9, 15; m 38; c 3. PLANT PATHOLOGY. Educ: Univ Ariz, BS, 37, MS, 39; Univ Minn, PhD(plant path), 62. Prof Exp: Agt plant path, Univ Ariz, USDA, 37-39, Sacaton, Ariz, 39-40; asst apple dis, Cornell Univ, 40-42; plant pathologist, USDA, Ariz, 42-46 & Calif, 46-48; plant pathologist, USDA, Okla State Univ, 48-73, from assoc prof to prof bot, 48-73; RES PROF BOT, LANGSTON UNIV, 73- Concurrent Pos: Chmn comt cotton dis photog, US Cotton Dis Coun, 52-63, chmn comt bact blight of cotton, 57-58. Mem: Am Phytopath Soc. Res: Cotton diseases; characterization of stable plant disease resistance combining resistance to bacterial blight, Verticillium wilt, Fusarium wilt and root knot nematodes in upland cotton; variability of the blight bacterium; nature of the immune reaction effect of temperature on blight and Verticillum resistance; breeding for resistance; control of scab and black rot of apples; control of pecan root rot. Mailing Add: Langston Univ Res Prog Langston OK 73050

BRINKHOUS, KENNETH MERLE, b Clayton Co, Iowa, May 29, 08; m 36; c 2. PATHOLOGY. Educ: Univ Iowa, AB, 29, MD, 32. Hon Degrees: DSc, Univ Chicago, 67. Prof Exp: Asst path, Sch Med, Univ Iowa, 32-33, instr, 33-35, assoc, 35-37, from asst prof to assoc prof, 37-46; prof, 46-61, DISTINGUISHED ALUMNI PROF PATH, SCH MED, UNIV NC, CHAPEL HILL, 61- Concurrent Pos: Mem, Hemat Study Sect, USPHS, 48-52, chmn, 59-62, chmn path study sect, 57-59; mem, Subcomt Blood Coagulation, Nat Res Coun, 51-54, chmn, 54-62, mem, Comt Blood, 54-, mem, Thrombosis Task Force, 65-70; chmn, Med Adv Comt, Nat Hemophilia Found, 54-; mem, Sch Adv Comt, World Fedn Hemophilia, 64-; mem, Int Comt Haemostasis & Thrombosis, 54-, chmn, 64-66, secy gen, 66-; consult, Armed Forces Inst Path, 56-72; mem, Univs Assoc for Res & Educ in Path, 64-; pres, Fedn Am Socs Exp Biol, 66-67; mem, Arteriosclerosis Task Force & Coun, Nat Heart & Lung Inst, 70-71. Honors & Awards: Ward-Burdick Awards, Am Soc Clin Path, 41 & 63; O Max Gardner Award, Univ NC, 61; NC Award in Sci, 69; James F Mitchell Found Int Award for Heart & Vascular Res, 69; Modern Med Distinguished Achievement Award, 73. Mem: Nat Acad Sci; Am Soc Exp Path (secy-treas, 60-63, pres, 65-66); Am Asn Path & Bact (secy-treas, 68-71, pres, 73); Int Soc Thrombosis (pres, 72); AAAS. Res: Blood clotting; hemorrhagic diseases; vitamin K; vitamin E and muscular dystrophy; cold injury; platelet agglutination. Mailing Add: Dept of Path Univ of NC Sch of Med Chapel Hill NC 27514

BRINKLEY, AMIEL WORD, JR, b Memphis, Tenn, Sept 14, 27; m 53; c 3. PULP & PAPER TECHNOLOGY. Educ: Mass Inst Technol, BS, 50. Prof Exp: Res proj supvr, 50-53, res sect leader, 53-61, chief pulp res, 61-63, 2nd asst dir res, 63-66, asst dir res, 66-75, MGR RES SERV, MFG & ENG SERV, INT PAPER CO, 75- Mem: Tech Asn Pulp & Paper Indust. Res: Chemical and mechanical conversion of wood into unbleached and bleached pulps, papers, paperboards and into wood products; energy conservation and environmental improvements; extrusion coating and lamination of paper board. Mailing Add: Southern Kraft Div Int Paper Co Erling Riis Res Lab PO Box 2328 Mobile AL 36601

BRINKLEY, HOWARD J, b Bridgeport, WVa, Apr 27, 36; m 55; c 3. REPRODUCTIVE ENDOCRINOLOGY. Educ: WVa Univ, BS, 58; Univ Ill, MS, 60, PhD(endocrinol of reproduction), 63. Prof Exp: Trainee endocrinol, Univ Wis, 62-63; asst prof, 63-65, ASSOC PROF ENDOCRINOL, UNIV MD, COLLEGE PARK, 66- Concurrent Pos: Guest worker, Nat Inst Child Health & Human Develop, 70-71. Mem: Soc Study Reproduction; Am Soc Zool; Endocrine Soc; Am Inst Biol Sci; Am Soc Animal Sci. Res: Functioning of the ovary and its components, primarily the corpora lutea and their functional relationships to the hormonal secretions of the pituitary gland. Mailing Add: Dept of Zool Univ of Md College Park MD 20742

BRINKLEY, LINDA LEE, b Glendale, Calif, Aug 19, 43; m 73; c 1. DEVELOPMENTAL BIOLOGY. Educ: Calif State Univ, Northridge, BA, 65, MS, 67; Univ Calif, Irvine, PhD(biol), 71. Prof Exp: Fel cell biol, Dept Zool, 72, res assoc develop biol, Dept Oral Biol, Sch Dent, 72-74, RES INVESTR DEVELOP BIOL, DEPT ORAL BIOL, SCH DENT, UNIV MICH, ANN ARBOR, 74- Mem: AAAS; Soc Develop Biol; Tissue Cult Asn; Am Soc Cell Biologists. Res: Normal and abnormal development of secondary palate; in vitro techniques for studying craniofacial anomalies; epithelial cell behavior in vitro. Mailing Add: Dept of Oral Biol Univ of Mich Sch of Dent Ann Arbor MI 48104

BRINKMAN, GAIL LYNN, b Radisson, Wis, Sept 5, 35; m 61; c 2. NUTRITION, BIOCHEMISTRY. Educ: NCent Col, Ill, BA, 57; Va Polytech Inst, MS, 60, PhD(biochem, nutrit), 62. Prof Exp: Res scientist body compos in malnutrit, Dept Med, Med Sch, Univ Cape Town, 62-64; biochemist nutrit lab develop, Pan-Am Health Orgn & WHO, Trinidad, 64-68; asst prof nutrit sci, Col Human Develop, Pa State Univ, University Park, 68-75. Concurrent Pos: Biochemist, Inst Nutrit Cent Am & Panama & NIH Nutrit Surv Panama, 67. Mem: Am Inst Nutrit. Res: New protein foods including fish protein concentrate; utilization of amino acids from proteins; effects of processing on protein foods; iron utilization from food; molybdeum toxicity; body composition in malnutrition. Mailing Add: RD 1 Box 68 Warriors Mark PA 16877

BRINKMAN, JOHN ALLEN, b Fremont, Mich, June 30, 28; m 68. SOLID STATE PHYSICS. Educ: Mich State Univ, BS, 49, MS, 50. Prof Exp: Res engr, NAm Aviation, Inc, 51-53, sr res engr, 53-57, supvr radiation effects, Atomics Int Div, 57-58, res specialist, 58-61, assoc dir, Res Dept, 61-62, assoc dir, NAm Aviation Sci Ctr, 62-68, dir, Phys Sci Dept, Autonetics Div, 65-68, div dir res & develop, 68-69, exec dir tech advan, Aerospace & Systs Group, NAm Rockwell Corp, 69-70, dir, NAm Rockwell Sci Ctr, 70-71; DEP DIR, RES DEVELOP & ENG DIRECTORATE, US ARMY WEAPONS COMMAND, 71- Concurrent Pos: Adj prof mat eng div, Univ Iowa, 74- Mem: AAAS; fel Am Phys Soc; Am Defense Preparedness Asn. Res: Shock waves in solids; physics of metals; radiation effects; mechanical properties; lattice defects; diffusion. Mailing Add: 3215 E Locust Apt 63 Davenport IA 52803

BRINKMAN, LEONARD W, JR, b La Crosse, Wis, Jan 7, 33; m 61; c 2. GEOGRAPHY. Educ: Univ Wis, BSc, 54, MSc, 58, PhD(geog), 64. Prof Exp: Lectr geog, Melbourne Univ, 63-67; ASSOC PROF GEOG, UNIV TENN, KNOXVILLE, 67- Mem: Am Geog Soc; Asn Am Geogr; Agr Hist Soc. Res: Historical geography of the United States, especially the development of agriculture and agricultural regions in the nineteenth century; historical geography of southern Appalachia. Mailing Add: Dept of Geog Univ of Tenn Knoxville TN 37916

BRINKMAN, WILLIAM F, b Washington, Mo, July 20, 38; m 60; c 2. SOLID STATE PHYSICS. Educ: Univ Mo-Columbia, MS, 62, PhD(physics), 65. Prof Exp: NSF res fel physics, Oxford Univ, 65-66; mem tech staff, 66-74, DIR CHEM PHYSICS RES LAB, BELL LABS, 74- Mem: Am Phys Soc. Res: Solid state theory; itinerant magnetism and spin fluctuations; electron tunneling theory; theory of metal-insulator transitions. Mailing Add: Bell Labs 600 Mountain Ave Murray Hill NJ 07974

BRINKS, JAMES S, b Plymouth, Mich, Jan 2, 34; m 55; c 9. ANIMAL GENETICS. Educ: Mich State Univ, BS, 56, MS, 57; Iowa State Univ, PhD(statist, genetics), 60. Prof Exp: Res geneticist, Animal Husb Res Div, Agr Res Serv, USDA, 60-67; PROF ANIMAL SCI, COLO STATE UNIV, 67- Mem: Am Soc Animal Sci. Res: Estimation of genetic, environmental and phenotypic parameters; estimation of response to selection and inbreeding; comparisons of mating systems; performance testing. Mailing Add: Dept of Animal Sci Colo State Univ Ft Collins CO 80521

BRINLEY, FLOYD JOHN, JR, b Battle Creek, Mich, May 19, 30; m 55; c 3. BIOPHYSICS. Educ: Oberlin Col, AB, 51; Univ Mich, MD, 55; Johns Hopkins Univ, PhD, 61. Prof Exp: Intern, Los Angeles County Gen Hosp, 55-56; sr asst surgeon neurophysiol lab, NIH, 57-59; asst prof, 61-67, ASSOC PROF PHYSIOL, SCH MED, JOHNS HOPKINS UNIV, 67- Mem: Biophys Soc; Am Physiol Soc; Soc Neurosci. Res: Membrane phenomena; ionic transport. Mailing Add: 11106 Youngtree Ct Columbia MD 21044

BRINN, JACK ELLIOTT, JR, b Norfolk, Va, June 7, 42; m 65; c 3. ENDOCRINOLOGY, ELECTRON MICROSCOPY. Educ: ECarolina Col, BA, 64, MA, 66; Univ Wyo, PhD(zool), 71. Prof Exp: Instr biol, ECarolina Univ, 66-67; res assoc, Milton S Hershey Med Ctr, 70-72; ASST PROF ANAT, E CAROLINA UNIV, 72- Mem: Am Asn Anatomists; Am Soc Zoologists; Am Diabetes Asn; Sigma Xi. Res: Ultrastructure of polypeptide hormone synthesis and secretion; comparative endocrine cytology and ultracytochemistry. Mailing Add: Dept of Anat ECarolina Univ Sch of Med Greenville NC 27834

BRINSFIELD, TRUITT HICKS, reproductive physiology, deceased

BRINSON, MARK MCCLELLAN, b Shelby, Ohio, Oct 6, 43; m 71; c 1. PLANT ECOLOGY, LIMNOLOGY. Educ: Heidelberg Col, BS, 65; Univ Mich, Ann Arbor, MS, 67; Univ Fla, PhD(bot), 73. Prof Exp: Fisheries biologist, Peace Corps, Costa Rica, 67-69; ASST PROF BIOL, E CAROLINA UNIV, 73- Mem: Sigma Xi; Ecol Soc Am; Am Soc Limnol & Oceanog; Int Asn Theoret & Appl Limnol; Am Inst Biol Sci. Res: Nutrient cycling of nitrogen, phosphorus and sulfur in swamp forests, aquatic macrophyte communities and other wetland ecosystems. Mailing Add: Dept of Biol E Carolina Univ Greenville NC 27834

BRINSTER, RALPH LAWRENCE, b Montclair, NJ, Mar 10, 32. EMBRYOLOGY, REPRODUCTIVE PHYSIOLOGY. Educ: Rutgers Univ, BS, 53; Univ Pa, VMD, 60, PhD(physiol), 64. Prof Exp: Asst instr physiol, Sch Med, Univ Pa, 60-61, teaching fel, 61-64, instr, 64-65, from asst prof to assoc prof, Sch Vet Med, 65-70, PROF PHYSIOL, SCH VET MED, UNIV PA, 70- Mem: Am Vet Med Asn; Am Soc Cell Biol; Soc Study Reproduction; Brit Soc Study Fertil; Brit Biochem Soc. Res: Biochemistry and physiology of the cleavage stages of mammalian embryos; differentiation in the early mammalian embryo; fertilization in mammals and invertebrates; mechanism of implantation; regulation of ovulation. Mailing Add: Rm 530 Lippincott Bldg Univ Pa Sch of Vet Med Philadelphia PA 19103

BRINTON, CHARLES CHESTER, JR, b Pittsburgh, Pa, Aug 15, 26. MICROBIOLOGY, BIOPHYSICS. Educ: Carnegie Inst Technol, BS, 49; Univ Pittsburgh, MS, 52, PhD(biophys), 55. Prof Exp: Asst biophys, Univ Pittsburgh, 49-51; res fel, Geneva, Switz, 55-56; res assoc, Univ Pittsburgh, 56-59, from asst res prof to assoc prof, 59-69, PROF MICROBIOL & MEM GRAD FAC, MED SCH, UNIV

PITTSBURGH, 69- Mem: AAAS; Biophys Soc; Am Soc Microbiol. Res: Bacterial and viral genetics, biophysics, physiology and serology; electrophoresis; biochemistry and physical chemistry of proteins and nucleic acids. Mailing Add: Dept of Microbiol Univ of Pittsburgh Med Sch Pittsburgh PA 15213

BRINTON, EDWARD, b Richmond, Ind, Jan 12, 24; m 48; c 4. BIOLOGICAL OCEANOGRAPHY. Educ: Haverford Col, BA, 49; Bryn Mawr Col, MA, 50; Scripps Inst Oceanog, Univ Calif, PhD(oceanog), 58. Prof Exp: Res biologist, 53-59, from asst res biologist to assoc res biologist & lectr, 59-73, ASSOC RES PROF BIOL & LECTR MARINE LIFE RES GROUP, SCRIPPS INST OCEANOG, UNIV CALIF, 73- Concurrent Pos: Partic, AID-Univ Calif Proj, Southeast Asia, 60-61; UNESCO instr, Thailand, 62 & Pakistan, 64; UNESCO cur, Indian Ocean Biol Ctr, Indian Ocean Exped, Ernakulam, S India, 65-67. Mem: Am Soc Limnol & Oceanog; Challenger Soc. Res: Systematics, biology and zoogeography of Euphausiacea; distribution and ecology of zooplankton. Mailing Add: Scripps Inst of Oceanog Univ of Calif La Jolla CA 92037

BRINTON, ELIAS (LYLE) PATTERSON, b Salt Lake City, Utah, June 10, 31; m 54; c 5. DEVELOPMENTAL BIOLOGY. Educ: Brigham Young Univ, BS, 60, MS, 65; Univ Calif, Berkeley, PhD(entom), 69. Prof Exp: NIH SR SCIENTIST, HOST-PATHOGEN BIOL, ROCKY MOUNTAIN LAB, US DEPT HEALTH, EDUC & WELFARE, 68- Mem: Electron Micros Soc Am; Am Soc Zoologists. Res: Ultra-structure and developmental morphology of Ixodoidea; Acarina and host and/or vector-pathogen interaction between the arthropods and disease agents they transmit to man and animals. Mailing Add: Rocky Mountain Lab Hamilton MT 59840

BRINTON, ROBERT K, b Los Angeles, Calif, Jan 9, 15; m 46; c 3. PHYSICAL CHEMISTRY. Educ: Univ Calif, Los Angeles, AB, 36, MA, 38, PhD, 48. Prof Exp: Chemist, Gen Petrol Corp, 37-42; res chemist, Nat Defence Res Comt, 42-45; lectr chem, Univ Calif, Los Angeles, 48; from instr to assoc prof, 48-61, PROF CHEM, UNIV CALIF, DAVIS, 61- Concurrent Pos: Res Corp res grant, 49-55; NSF res grants, 58-; Nat Res Coun Can fel, 54-55; Guggenheim fel, Stuttgart Tech Univ, 61-62. Mem: Am Chem Soc. Res: Reaction kinetics in gas phase systems; photochemical and thermal study of elementary radical reactions. Mailing Add: 618 Oak Ave Davis CA 95616

BRIODY, BERNARD ALOYSIUS, b Bethlehem, Pa, Nov 12, 20; m 44; c 5. MEDICAL BACTERIOLOGY. Educ: Lehigh Univ, BA, 41; Yale Univ, PhD(med bact), 46. Prof Exp: From instr virol to asst prof microbiol, Sch Med, Yale Univ, 48-52; assoc prof microbiol & dir, Virus Lab, Hahnemann Med Col, 52-56, prof microbiol, 56-59; PROF MICROBIOL & CHMN DEPT, COL MED & DENT NJ, 59- Concurrent Pos: Nat Res Coun fel, Walter & Eliza Hall Inst, Australia, 46-47 & Univ Mich, 47-48. Mem: AAAS; Am Soc Microbiol; Soc Exp Biol & Med; Am Asn Immunologists; Am Pub Health Asn; Am Asn Cancer Res. Res: Viruses. Mailing Add: Dept of Microbiol Col of Med & Dent of NJ Newark NJ 07103

BRION, CHRISTOPHER EDWARD, b Eng, May 5, 37; m 61; c 3. PHYSICAL CHEMISTRY. Educ: Bristol Univ, BSc, 58, PhD(chem), 61. Prof Exp: Res fel chem, 61-63, teaching fel, 63-64, asst prof, 64-69, ASSOC PROF CHEM, UNIV BC, 69- Mem: Chem Inst Can. Res: Excited states of ions; penning ionization; energy loss electron spectroscopy; electron-electron coincidence techniques. Mailing Add: Dept of Chem Univ of BC Vancouver BC Can

BRISBANE, WILLIAM NEELY, b Los Angeles, Calif, Oct 12, 23; m 58; c 3. OPTOMETRY, PUBLIC HEALTH. Educ: Univ Calif, Los Angeles, BA, 49; Los Angeles Col Optom, OD, 52. Prof Exp: Clin instr, 52-59, Home: TOM & CLIN DIR, OPTOM CTR LOS ANGELES, SOUTHERN CALIF COL OPTOM, 59- Concurrent Pos: Consult optom, Med Care Div, Los Angeles Region, Calif State Off Health Care Serv & Life Systs Res Inst. Mem: Fel Am Acad Optom; Am Pub Health Asn. Res: Children's vision; school vision screening. Mailing Add: 3916 S Broadway Los Angeles CA 90037

BRISBIN, DOREEN A, b Edmonton, Alta, Dec 19, 26; m 49; c 3. BIOINORGANIC CHEMISTRY. Educ: Univ Alta, BSc, 46; Univ Toronto, PhD(chem), 60. Prof Exp: Res assoc chem, 60-61, lectr, 61-63, asst prof, 63-66, ASSOC PROF CHEM, UNIV WATERLOO, 66-, ASSOC DEAN SCI, 75- Mem: Chem Inst Can. Res: Coordination chemistry; stability of metal complexes; kinetics of metalloporphyrin formation; intermolecular complexes with porphyrins. Mailing Add: Dept of Chem Univ of Waterloo Waterloo ON Can

BRISBIN, I LEHR, JR, b Drexel Hill, Pa, Apr 2, 40. ZOOLOGY, ECOLOGY. Educ: Wesleyan Univ, AB, 62; Univ Ga, MS, 65, PhD(zool), 67. Prof Exp: Res assoc vert ecol, 67-68, AEC FEL, SAVANNAH RIVER ECOL LAB, 68-; ASST PROF POULTRY SCI, UNIV GA, 69- Concurrent Pos: Pop ecologist, US Energy Res & Develop Admin, 73-75. Mem: Am Soc Mammal; Am Ornithologists Union; Am Soc Ichthyologists & Herpetologists; Ecol Soc Am; Am Wildlife Soc. Res: Vertebrate ecology and bioenergetics; mammalogy, ornithology and herpetology; ecology of domestication; energy-related environmental impacts, radioecology. Mailing Add: Savannah River Ecol Lab PO Drawer E Aiken SC 29801

BRISBIN, WILLIAM CORBETT, geology, geophysics, see 12th edition

BRISCOE, ANNE M, b New York, NY, Dec 1, 18; m 55. BIOCHEMISTRY, PHYSIOLOGY. Educ: Adelphi Col, BA, 42; Vassar Col, AM, 45; Yale Univ, PhD(physiol chem), 49. Prof Exp: From res assoc to asst prof biochem, Med Col, Cornell Univ, 50-56; res assoc, Sch Med, Univ Pa, 56; assoc biochem, 56-72, ASST PROF MED, COL PHYSICIANS & SURGEONS, COLUMBIA UNIV, 72- Concurrent Pos: NIH fel, Univ Pa, 49-50; lectr, Sch Gen Studies, Hunter Col, 51-64, adj asst prof, Sch Health Sci, 73-; vis asst prof, Temple Univ, 56; consult, Vet Admin Hosp, Castle Point, NY, 57-67; career scientist, New York Health Res Coun, 60-66; lectr, Sch Gen Studies, Columbia Univ, 67-68, Sch Nursing, Harlem Hosp Ctr, 68- & Antioch Col at Harlem Hosp Ctr, 71-73. Mem: Fel Am Inst Chemists; Am Chem Soc; Am Soc Clin Nutrit; Asn Women Sci (pres, 74-76); Endocrine Soc. Res: Metabolism of calcium and magnesium in human subjects; metabolic acidosis; relation of plasma renin activity to hypertension. Mailing Add: Dept of Med Harlem Hosp Ctr New York NY 10037

BRISCOE, CHARLES VICTOR, b Abingdon, Va, May 5, 30; m 53; c 3. PHYSICS. Educ: King Col, BA, 52; Rice Inst, MA, 57, PhD(physics), 58. Prof Exp: Instr math, King Col, 52-53, instr physics, 53-54; res assoc, 58-60, from asst prof to assoc prof, 58-70, PROF PHYSICS, UNIV NC, CHAPEL HILL, 70- Concurrent Pos: Vis prof, Tech Univ Karlsruhe, 66 & 67. Mem: Am Phys Soc; Am Asn Physics Teachers. Res: Superconducting properties of metallic films; position annihilation in inert gas liquids. Mailing Add: Dept of Physics Univ of NC Chapel Hill NC 27514

BRISCOE, MADISON SPENCER, entomology, parasitology, see 12th edition

BRISCOE, WILLIAM ALEXANDER, b London, Eng, May 26, 18; m 55. PHYSIOLOGY, MEDICINE. Educ: Oxford Univ, BA, 39, MA, 40, BM, 42, DM, 51. Prof Exp: Registr, Med Sch, Univ London, 48-51; assoc physiol, Grad Sch Med, Univ Pa, 52-53, mem sci staff, Pneumoconiosis Res Unit, Cardiff, 54-55; from asst prof to assoc prof med, Col Physicians & Surgeons, Columbia Univ, 56-68; assoc prof, 68-71, PRO MED, MED COL, CORNELL UNIV, 71- Concurrent Pos: Brit Med Fedn traveling fel, Columbia Univ, 51-52. Mem: Am Physiol Soc; Am Soc Clin Invest; Brit Med Res Soc; Asn Am Physicians. Res: Respiration; clinical investigation of physiology of lungs and circulation; distribution of ventilation perfusion ratios. Mailing Add: 1300 York Ave New York NY 10021

BRISKEY, ERNEST J, b Hillsboro, Wis, Mar 30, 31. FOOD SCIENCE, BIOCHEMISTRY. Educ: Univ Wis, BS, 52, PhD(biochem, meat sci & muscle biol), 58; Ohio State Univ, MS, 55. Prof Exp: Asst prof, Univ Wis, 58-59; res fel, Danish Meat Res Inst & Low Temperature Res Sta, Cambridge, 59-60; from asst prof to prof muscle chem & physiol, Univ Wis, 60-70, H L Russell distinguished prof, 70, dir, Inst Muscle Biol, 70; vpres basic res, Campbell Inst Food Res, 70-73, vpres tech admin, Campbell Soup Co, 73-75, VPRES, CAMPBELL SOUP CO & PRES, TECHNOL RESOURCES, INC, 75- Concurrent Pos: Vibrans sr sci fel, Am Meat Inst Found, Univ Chicago, 62; vis scholar, Med Sch, Univ Calif, Los Angeles, 65-66. Mailing Add: Campbell Soup Co Campbell Pl Camden NJ 08101

BRISSEY, RUBEN MARION, b Auburn, WVa, July 12, 23; m 45; c 2. PHYSICAL CHEMISTRY. Educ: Salem Col, WVa, BS, 43; WVa Univ, MS, 48, PhD(phys chem), 50. Prof Exp: Instr chem, WVa Univ, 49; tech engr, Gen Elec Co, 50-52, group leader insulation develop, 52-53, mgr chem & insulation unit, 54-55, mgr lab 55-60, consult mat eng, 60-61, mgr mat lab, 61-66, consult bus planning & anal, 66-69, planning mgr, 69-73; dir res & engr, Lavion Div, Int Minerals & Chem Corp, 73-74; MGR RES & DEVELOP, NAT CAN CORP, 74- Mem: AAAS; Inst Elec & Electronics Engrs; Am Chem Soc. Res: Analysis of alloys by X-ray fluorescent spectrography; insulation problems in large rotating electrical machinery; materials selection in business systems; materials developments and applications for power transmission equipment; applied research and development to improve metal, glass and plastic packaging. Mailing Add: 3410 Hickory Trail Downers Grove IL 60515

BRISSON, GERMAIN J, b St Jacques, Que, Apr 12, 20; m 48; c 5. NUTRITION. Educ: Joliette Sem, BA, 42; Laval Univ, BSc, 46; McGill Univ, MSc, 48; Ohio State Univ, PhD(nutrit), 50. Prof Exp: Res officer nutrit, Can Dept Agr, 50-60, dir admin, Lennoxville Exp Sta, 60-62; PROF NUTRIT, LAVAL UNIV, 62-, DIR NUTRIT RES CTR, 68- Concurrent Pos: Sci ed, Can J Animal Sci, 63- Mem: Am Soc Animal Sci; Am Dairy Sci Asn; Nutrit Soc Can; Agr Inst Can; Can Soc Animal Production. Res: Nutritional requirements of young ruminants; protein and fat utilization by young animals. Mailing Add: Nutrit Res Ctr Laval Univ Quebec PQ Can

BRISTOL, BENTON KEITH, b Ansley, Nebr, Feb 21, 20; m 41; c 1. AGRICULTURE. Educ: Colo State Univ, BS, 46; Okla State Univ, MS, 50; Pa State Univ, DEd(agr educ), 59. Prof Exp: Teacher high sch, Colo, 46-51; from instr to asst prof agr educ, Pa State Univ, 55-63; Ohio State Univ-AID contract consult, Fac Educ, India, 63-65; PROF AGR, ILL STATE UNIV, 65- Concurrent Pos: Mem nat meeting comt, Am Inst Coop, 58-59; mem, Am Grassland Coun, 60-63; vpres, Am Asn Teacher Educators in Agr, 61-63; vchmn educ exhibits comt, Nat Grassland Conf & Field Days, 62-63; mem selection comt distinguised serv award, Am Asn Teacher Educators in Agr, 63-64; consult, Govt India Planning Comn, 64-65; fac develop grant, 70. Res: Agricultural mechanics shops in modern secondary schools; program planning guide in agricultural mechanics; creativity; evualuating creative ability; international education. Mailing Add: Dept of Agr Ill State Univ Normal IL 61761

BRISTOL, DOUGLAS WALTER, b Rochester, NY, Oct 31, 40; m 64; c 1. ORGANIC CHEMISTRY. Educ: St John Fisher Col, BS, 62; Syracuse Univ, PhD(chem), 69. Prof Exp: Res assoc chem, Columbia Univ, 69; instr, Univ Utah, 69-71; ASST PROF PESTICIDES CHEM, N DAK STATE UNIV, 71- Mem: AAAS; Am Chem Soc. Res: Free radical chemistry; reactions and properties of primary and secondary alkoxy-radicals; reactions of pyridinium salts with nucleophiles and dihydropyridine compounds with molecular oxygen; pesticides chemistry. Mailing Add: Dept of Biochem NDak State Univ Fargo ND 58102

BRISTOL, MELVIN LEE, b Hartford, Conn, Dec 3, 36; m 59; c 3. BOTANY. Educ: Harvard Univ, AB, 60, AM, 62, PhD(bot), 65. Prof Exp: USPHS fel, 65-66; asst prof bot & asst botanist, Lyon Arboretum, Univ Hawaii, 66-69; OWNER, LEE BRISTOL NURSERY, 69- Mem: Soc Econ Botanists; Asn Trop Biol. Res: Ethnobotany and floristics of Colombia and Samoa; South American narcotic plants; taxonomy and ethnobotany of Datura; plant domestication; horticulture. Mailing Add: Sherman CT 06784

BRISTOW, JOHN DAVID, b Pittsburgh, Pa, Dec 7, 28; m 50; c 3. MEDICINE. Educ: Williamette Univ, BA, 49; Univ Ore, MD, 53; Am Bd Cardiovasc Dis, dipl, 70. Prof Exp: Instr med, 60-61, dir cardiol lab, 62-68, from asst prof to assoc prof med, 62-70, chmn dept, 71-75, chief med serv, Univ Hosp, 69-75, PROF MED, MED SCH, UNIV ORE, 70-, CLIN RES ASSOC PHYSIOL, 65- Concurrent Pos: Nat Heart Inst spec fel, Cardiovasc Res Inst, Univ Calif, 61-62 & Radcliffe Infirmary, Eng, 67-68; Markle scholar, 64-69; Am Heart Asn Coun Clin Cardiol fel; consult, US Army Hosp, Ft Lewis, Wash, 66-68; mem, Heart & Lung Prog Proj Comt, NIH, 70-74. Mem: Fel Am Col Physicians; Asn Am Physicians; fel Am Col Cardiol; Am Soc Clin Invest; Am Fedn Clin Res. Res: Clinical cardiology; cardiovascular physiology; medical education. Mailing Add: Dept of Med Univ of Ore Med Sch Portland OR 97201

BRISTOWE, WILLIAM WARREN, b Philadelphia, Pa, Jan 14, 40; m 62; c 3. PHYSICAL CHEMISTRY. Educ: Temple Univ, BS, 61; Univ Del, PhD(phys chem), 65. Prof Exp: Res chemist, Textile Fibers Pioneering Res, E I du Pont de Nemours & Co, 65-67 & Polymer Sect, Atlas Chem Indust, 67-70; RES SUPVR POLYMER GROUP, CORP RES LAB, ICI UNITED STATES INC, 70- Concurrent Pos: AEC fel, 62-65. Mem: Am Chem Soc; Sigma Xi. Res: Radiation chemistry, effects of radiation on matter; characterization of fibers and foams; solution properties of water soluble polymers and uses; high temperature polymers synthesis and uses. Mailing Add: Corp Res Lab ICI United States Inc Wilmington DE 19897

BRITAIN, J W b Houston, Tex, May 17, 20; m 43; c 4. POLYMER CHEMISTRY. Educ: Rice Inst, AB, 42; Univ Minn, MS, 49, PhD(org chem), 50. Prof Exp: Res chemist, Manhattan Eng Proj, 42-46; asst, Univ Minn, 46-50; res chemist, Ansco Div, Gen Aniline & Film Corp, 50-53, res assoc, 54; sr res chemist, 55-60, res specialist, 61-63, sr group leader, 64-72, SR RES SPECIALIST, POLYURETHANE DIV, MOBAY CHEM CO, 72- Mem: Am Chem Soc. Res: Isocyanates; polyurethanes; catalysts; polyesters; elastomers; spandex; coatings; adhesives; thermoplastics; reaction injection molding; flexible foam. Mailing Add: 1228 Wren Dr New Martinsville WV 26155

BRITAN, NORMAN, b Chicago, Ill, July 22, 18; m 43; c 2. APPLIED ANTHROPOLOGY. Educ: Univ Chicago, BS, 39; Univ Chicago, MA, 52. Prof Exp: Coordr soc sci prog, Wright Jr Col, Ill, 53-59; chmn soc sci dept, Southeast Jr Col,

59-63; chmn div soc sci, 63-72, CHMN DEPT ANTHROP, NORTHEASTERN ILL UNIV, 72- Concurrent Pos: Mem, Ill State Coun Funding Fed Title I Proj, 66-69; consult, Moreland-Latchford Productions, Ltd, Toronto, Ont, 75- Mem: AAAS; Am Anthrop Asn; Am Ethnol Soc; fel Soc Appl Anthrop. Res: Minority problems; urban anthropology; American Indians. Mailing Add: Dept of Anthrop Northeastern Ill Univ Chicago IL 60625

BRITT, A D, b Childress, Tex, Nov 6, 34; m 63. PHYSICAL CHEMISTRY. Educ: WTex State Col, BS, 57; Wash Univ, MA, 61, PhD(chem), 64. Prof Exp: Asst chem, Wash Univ, 57-63; res assoc, Univ Chicago, 63-65; asst prof, 65-70, ASSOC PROF CHEM, GEORGE WASHINGTON UNIV, 70- Res: Chemical kinetics and magnetic resonance. Mailing Add: Dept of Chem George Washington Univ Washington DC 20006

BRITT, DANNY GILBERT, b Glasgow, Ky, Sept 19, 46; m 66; c 1. DAIRY NUTRITION. Educ: Western Ky Univ, BS, 68; Mich State Univ, MS, 72, PhD(dairy nutrit), 73. Prof Exp: Asst prof animal sci, Tex A&I Univ, 73-74; ASST PROF AGR, EASTERN KY UNIV, 74- Mem: AAAS; Am Dairy Sci Asn; Am Soc Animal Sci. Res: Use of high moisture corn in dairy rations; nutrient requirements of high producing cows. Mailing Add: Dept of Agr Eastern Ky Univ Richmond KY 40475

BRITT, EUGENE MAURICE, b Chicago, Ill, Mar 2, 24; m 47; c 3. MICROBIOLOGY. Educ: Valparaiso Univ, AB, 47; Univ Ill, MS, 49; Univ Mich, PhD, 57; Am Bd Med Microbiol, dipl. Prof Exp: Asst bacteriologist, Valparaiso Univ, 46-47; asst bacteriologist, Col Med, Ill, 47-48; res bacteriologist, Miles Labs, Inc, 48-53; asst med sch, Univ Mich, 53-56; microbiologist, Wayne County Gen Hosp, 56-59; MICROBIOLOGIST, ST JOSEPH MERCY HOSP, 59- Concurrent Pos: Adj asst prof, Univ Mich, 74. Mem: Am Soc Clin Path; Am Soc Microbiol; fel Am Acad Microbiol. Res: Clinical microbiology; infectious disease. Mailing Add: St Joseph Mercy Hosp Ann Arbor MI 48103

BRITT, HAROLD CURRAN, b Buffalo, NY, Sept 14, 34; m 56. NUCLEAR PHYSICS. Educ: Hobart Col, BS, 56; Dartmouth Col, MA, 58; Yale Univ, PhD(physics), 61. Prof Exp: Mem staff, Physics Div, Los Alamos Sci Lab, 61-72; vis prof physics, Univ Rochester, 72-73; GROUP LEADER, PHYSICS DIV, LOS ALAMOS SCI LAB, 74- Concurrent Pos: Mem staff, Nuclear Chem Div, Lawrence Radiation Lab, Univ Calif, 64-65; mem subpanel on instrumentation & tech, Nat Acad Sci-Nat Res Coun Nuclear Physics Surv, 70. Mem: AAAS; Am Phys Soc. Res: Experimental nuclear physics. Mailing Add: Los Alamos Sci Lab Box 1663 Los Alamos NM 84544

BRITT, HENRY GRADY, b Colerain, NC, Apr 24, 15. ZOOLOGY. Educ: Wake Forest Col, BS, 36, MA, 38; Univ Va, PhD(zool), 44. Prof Exp: Teaching fel biol, Wake Forest Col, 36-38, instr, 38-40; asst prof, Mary Washington Col, 44-47; from asst prof to prof, Wake Forest Col, 47-64; from asst prof to assoc prof, 64-72, PROF BIOL, LA STATE UNIV, ALEXANDRIA, 72- Res: Parasitology; cytotaxonomy; cytology of digenetic trematodes. Mailing Add: Dept of Biol La State Univ Alexandria LA 71301

BRITT, N WILSON, b Lucas, Ky, Jan 3, 13; m 41. LIMNOLOGY, ENTOMOLOGY. Educ: Western Ky State Col, BS, 39; Ohio State Univ, MS, 47, PhD(hydrobiol, entom), 50. Prof Exp: Pub sch teacher, Ky, 33-40; high sch prin, coach & teacher, 40-42; instr meteorol, Army Air Force Weather Sch, Ill, 42-43; asst ecol, Stone Inst Hydrobiol, 47-50, from asst prof to assoc prof, Univ, 50-68, PROF ENTOM, OHIO STATE UNIV, 68- Mem: Am Soc Limnol & Oceanog. Res: Ecology; limnology and aquatic entomology. Mailing Add: Dept of Entom Ohio State Univ Columbus OH 43210

BRITT, PATRICIA MARIE, b Los Angeles, Calif, May 2, 31. COMPUTER SCIENCE, BIOMATHEMATICS. Educ: Univ Chicago, BA, 49; Univ Calif, Los Angeles, BA, 51, PhD(philos), 59. Prof Exp: Mathematician, Bendix Comput Corp, Los Angeles, 58-63 & Control Data Corp, 63; sr systs engr, IBM Corp, 63-66; ASST DIR, HEALTH SCI COMPUT FACIL, UNIV CALIF, LOS ANGELES, 66- Concurrent Pos: Consult, Health Serv & Ment Health Admin, 71-73 & NIH, 72-; lectr, Dept Biomath, Univ Calif, Los Angeles, 72- Res: Application of computers to research in biomedicine. Mailing Add: Health Sci Comput Fac Univ of Calif Los Angeles CA 90024

BRITTAIN, JERE A, horticulture, plant physiology, see 12th edition

BRITTAN, MARTIN RALPH, b San Jose, Calif, Jan 28, 22; m 47; c 2. VERTEBRATE ZOOLOGY, ECOLOGY. Educ: San Jose State Col, AB, 46; Stanford Univ, PhD(biol sci), 51. Prof Exp: Teaching asst biol sci, Stanford Univ, 47-49; asst prof biol, SDak Sch Mines & Technol, 49-50; from instr to asst prof zool, San Diego State Col, 50-53; from asst prof to assoc prof biol sci, 53-62, PROF BIOL SCI, CALIF STATE UNIV, SACRAMENTO, 62- Mem: Soc Syst Zool; Am Soc Ichthyologists & Herpetologists; Wildlife Soc; Am Soc Limnol & Oceanog; Ecol Soc Am. Res: Ichthyology; hydrobiology; taxonomy; evolution. Mailing Add: Dept of Biol Sci Calif State Univ Sacramento CA 95819

BRITTELLI, DAVID ROSS, b Milwaukee, Wis, Oct 10, 44; m 73. ORGANIC CHEMISTRY, MEDICINAL CHEMISTRY. Educ: Univ Wis, BS, 66; Univ Ill, PhD(org chem), 69. Prof Exp: RES CHEMIST, CENT RES & DEVELOP DEPT, E I DU PONT DE NEMOURS & CO, 69- Mem: Am Chem Soc. Res: Chemistry of natural products; beta-lactams; alkaloids; heterocyclic chemistry; introduction of fluorine into natural products; new synthetic procedures; antibiotics; analgetics. Mailing Add: Cent Res & Develop Dept Exp Sta E I du Pont de Nemours & Co Wilmington DE 19898

BRITTEN, BRYAN TERRENCE, b Creston, Iowa, Dec 25, 33; m 58; c 3. BIOLOGY, ZOOLOGY. Educ: Merrimack Col, AB, 57; Villanova Univ, MS, 62; Univ Wyo, PhD(zool), 66. Prof Exp: ASSOC PROF BIOL, NIAGARA UNIV, 66- Mem: Wildlife Soc; Ecol Soc Am; Biomet Soc. Res: Biogeometry of bird eggs. Mailing Add: Dept of Biol Niagara University NY 14109

BRITTEN, EDWARD JAMES, b Rouleau, Sask, Mar 5, 15; m 40; c 3. PLANT CYTOGENETICS. Educ: Univ Sask, BSc, 40, MSc, 41; Univ Wis, PhD(genetics), 44. Prof Exp: Asst supt forage crops, Dominion Exp Farm, 44-47; asst prof bot, Univ Hawaii, 47-54, asst prof agron, 54-55, from assoc prof agr & assoc agronomist to prof agron, 55-64; head dept agr, 64-73, dean agr, 65-66 & 71, PROF AGR, UNIV QUEENSLAND, 64- Concurrent Pos: Sr res fel, Univ Melbourne, 61-62. Mem: Fel AAAS; Bot Soc Am; Am Genetic Asn; Am Soc Agron; Australian Inst Agr Sci. Res: Tropical agriculture; agricultural education; conservation; cytotaxonomy; legumes for semi-arid tropics. Mailing Add: Dept of Agr Univ of Queensland St Lucia Brisbane Australia

BRITTIN, WESLEY E, b Philadelphia, Pa, Apr 21, 17; m 41; c 4. THEORETICAL PHYSICS. Educ: Univ Colo, BS, 42, MS, 45; Princeton Univ, MA, 47; Univ Alaska, PhD, 57. Prof Exp: Asst physics, 42-44, instr, 44-45, from asst prof to assoc prof, 47-59, chmn dept, 57-60, chmn dept physics & astrophys, 60-74, PROF PHYSICS & ASTROPHYS, UNIV COLO, BOULDER, 59-, DIR, INST THEORET PHYSICS, 68- Concurrent Pos: Asst, Princeton Univ, 49-50. Mem: Am Phys Soc; Am Asn Physics Teachers. Res: Statistical mechanics. Mailing Add: Dept of Physics & Astrophys Univ of Colo Boulder CO 80302

BRITTON, DONALD MACPHAIL, b Toronto, Ont, Mar 6, 23; m 50; c 3. CYTOGENETICS. Educ: Univ Toronto, BA, 46; Univ Va, PhD(biol), 50. Prof Exp: Res scientist, Defense Res Bd Can, 51-52; Nat Res Coun Can fel biol, Univ Alta, 52-53; lectr genetics, 53; asst prof hort, Univ Md, 54-58; from asst prof to assoc prof bot, 58-71, PROF BOT, UNIV GUELPH, 71- Concurrent Pos: Sigma Xi res grant, 59-61. Honors & Awards: Gilchrist Prize, 44; Fleming Prize, 48. Mem: Bot Soc Am; Am Genetic Asn; Am Fern Soc; Genetics Soc Can. Res: Genetics; cytotaxonomy; plant breeding; biosystematics of ferns; floristics; pteridophytes of Canada. Mailing Add: Dept of Bot & Genetics Univ of Guelph Guelph ON Can

BRITTON, DOYLE, b Los Angeles, Calif, Mar 6, 30; m 62; c 3. STRUCTURAL CHEMISTRY. Educ: Univ Calif, Los Angeles, BS, 51; Calif Inst Technol, PhD(chem), 55. Prof Exp: From asst prof to assoc prof, 55-64, PROF INORG CHEM, UNIV MINN, MINNEAPOLIS, 65- Concurrent Pos: NSF sr fel, 63-64. Mem: Am Chem Soc; Am Crystallog Asn. Res: Molecular structure; x-ray crystallography. Mailing Add: Dept of Chem Univ of Minn Minneapolis MN 55455

BRITTON, JACK ROLF, b St Petersburg, Russia, Jan 27, 08; nat US; m 31; c 3. MATHEMATICS. Educ: Clark Univ, BA, 29; Univ Colo, PhD(math), 36. Prof Exp: High sch instr, Mass, 29; instr appl math, Univ Colo, 29-37, from asst prof to prof, 37-67, head dept, 62-65, chmn dept math, Univ WFla, 67; PROF MATH, UNIV S FLA, 67- Concurrent Pos: Mathematician, Res Dept, US Rubber Co, Mich, 42-43; instr, US Air Force Meteorol Prog, Univ Mich, 43, vis assoc prof, 48-49, vis prof, 59. Mem: Am Math Soc; Math Asn Am; Soc Indust & Appl Math. Res: Orthogonal polynomials; Laplace transformations. Mailing Add: Dept of Math Univ of S Fla Tampa FL 33620

BRITTON, MARVIN GALE, b Corning, NY, Jan 8, 22; m 48; c 3. MINERALOGY, CERAMICS. Educ: Alfred Univ, BS, 43; Ohio State Univ, MS, 49, PhD(mineral), 52. Prof Exp: Ceramics engr, Corning Glass Works, 43-46; res engr, Battelle Mem Inst, 48-49; res assoc mineral, Ohio State Univ, 48-52; res assoc ceramics, 52-56, develop mgr, 56-59, tech mgr govt serv, 59-61, staff res mgr, 61-63, MGR TECH LIAISON, CORNING GLASS WORKS, 63- Concurrent Pos: Mem comt aerospace & astronaut adv panel, Mat Adv Bd, Nat Acad Sci, 59-62, mem comt ceramic mat, 62-63, mem comt res-eng interaction, 65-66; instr, Fac Continuing Educ, Corning Community Col, 62- Mem: Fel Am Ceramic Soc. Res: Mineralogy of blast furnace slags and refractories; development of refractory ceramics. Mailing Add: Tech Staffs Div Corning Glass Works Corning NY 14830

BRITTON, MAXWELL EDWIN, b Hymera, Ind, Jan 26, 12; m 37, 57. BOTANY. Educ: Ind State Univ, AB, 34; Ohio State Univ, MS, 37; Northwestern Univ, PhD(bot), 41. Hon Degrees: DSc, Ind State Univ, 67. Prof Exp: Jr high sch teacher, Ind, 34-35; asst bot, Ohio State Univ, 35-37; from asst to assoc prof, Northwestern Univ, 37-55; sci officer arctic res, Geog Br, Off Naval Res, 55-66, dir arctic res prog, 66-70, actg dir, Earth Sci Div, 67-69; dir arctic develop & environ prog, Arctic Inst NAm, 70-73; consult environ sci, 73-74; BIOL SCIENTIST, US GEOL SURV, 74- Concurrent Pos: Mem exec comt geog & climatol, Nat Res Coun, 53-54; mem comt environ studies, Proj Chariot, AEC, 59-62; mem panel biol & med, Comt Polar Res, Nat Acad Sci, 71-74; mem, Environ Protection Bd, Winnipeg, Man, 72-74. Mem: AAAS; Bot Soc Am; Ecol Soc Am; Am Soc Limnol & Oceanog; Am Inst Biol Sci. Res: Taxonomy of freshwater algae; microclimatology; ecological relations of bog vegetation; soil temperatures; arctic ecology. Mailing Add: US Geol Surv MS 106 Nat Ctr Reston VA 22092

BRITTON, MICHAEL PAUL, b Wall, SDak, June 2, 25; m 45; c 2. PLANT PATHOLOGY. Educ: Mont State Univ, BS, 51, MS, 55; Purdue Univ, PhD(plant path), 58. Prof Exp: Soil conservationist, Soil Conserv Serv, USDA, Mont, 52-55; instr bot & plant path, Purdue Univ, 55-58; from asst prof to assoc prof plant path, Univ Ill, Urbana-Champaign, 58-69; BOTANIST, DIV NATURAL SCI, FLATHEAD VALLEY COMMUNITY COL, 69- Concurrent Pos: Consult air pollution, Anaconda Aluminum Co, 70- & Intalco Aluminum Co, 71- Mem: Am Phytopath Soc; AAAS; Bot Soc Am. Res: Disease of forest trees and agricultural crops; air pollution damage. Mailing Add: Flathead Valley Community Col PO Box 1174 Kalispell MT 59901

BRITTON, OTHA LEON, b Portales, NMex, Aug 6, 45; m 68; c 2. NUMERICAL ANALYSIS. Educ: Eastern NMex Univ, BS, 67; Drexel Univ, MS, 69, PhD(math), 72. Prof Exp: ASST PROF MATH, UNIV MISS, 72- Mem: Am Math Soc; Math Asn Am. Res: Efficient methods of finding approximate solutions to overdetermined systems of equations. Mailing Add: Dept of Math Univ of Miss University MS 38677

BRITTON, PHILIP STEPHEN, organic chemistry, see 12th edition

BRITTON, WALTER MARTIN, b Lasker, NC, Aug 10, 39; m 61; c 2. POULTRY NUTRITION. Educ: NC State Univ, BS, 61, MS, 63, PhD(nutrit), 67. Prof Exp: NIH trainee nutrit, Univ Calif, Berkeley, 67-69; ASST PROF POULTRY SCI, UNIV GA, 69- Mem: Poultry Sci Asn; World Poultry Sci Asn; Nutrit Today Soc. Res: Magnesium nutrition; egg quality. Mailing Add: 315 Ferncliff Dr Athens GA 30601

BRITTON, WILLIAM GIERING, b Wilkes-Barre, Pa, Sept 25, 21; m 43; c 4. PHYSICAL CHEMISTRY, INORGANIC CHEMISTRY. Educ: Millikin Univ, BS, 43; Univ Ill, MS, 47; Univ Colo, PhD(phys chem), 56. Prof Exp: PROF CHEM, BEMIDJI STATE UNIV, 47- Mem: Am Chem Soc. Res: Inclusion of nuclear science in undergraduate instruction. Mailing Add: Dept of Chem Bemidji State Univ Bemidji MN 56601

BRITZMAN, DARWIN GENE, b Watertown, SDak, May 13, 31; m 53; c 3. ANIMAL NUTRITION, POULTRY NUTRITION. Educ: SDak State Univ, BS, 53, PhD(animal sci), 64; Univ Minn, MS, 62. Prof Exp: Asst turkey mgr, Sunshine State Hatchery, Watertown, SDak, 53-54, hatchery mgr, Brookings, 54-55; territory mgr feed sales, McCabe Co, Minneapolis, 55-58; asst nutritionist, Farmer's Union Grain Terminal Asn, St Paul, 58-62; res scientist poultry, Swift & Co, Chicago, 63-65; DIR NUTRIT, FARMER'S UNION GRAIN TERMINAL ASN, 65- Mem: Am Soc Animal Sci; Am Dairy Sci Asn; Poultry Sci Asn. Res: Applied poultry and livestock nutrition and management. Mailing Add: GTA Feeds Box 1447 Sioux Falls SD 57101

BRIXEY, JOHN CLARK, b Mounds, Okla, June 28, 04; m 26; c 2. MATHEMATICS. Educ: Univ Okla, AB, 24, AM, 25; Univ Chicago, PhD(math), 36. Prof Exp: From instr to prof math, 25-74, consult prof, Sch Med, 60-74, EMER PROF MATH, UNIV OKLA, 74-, EMER ADJ PROF BIOSTATIST & EPIDEMIOL, 74- Mem: Am Math Soc; Math Asn Am. Res: Theory of numbers and algebra; mathematical statistics. Mailing Add: Dept of Math Univ of Okla Norman OK 73069

BRIXNER, LOTHAR HEINRICH, b Karlsruhe, Ger, Dec 30, 28; US citizen; m 55; c 1. INORGANIC CHEMISTRY. Educ: Univ Karlsruhe, BS, 51, MS, 53, PhD(inorg chem), 55. Prof Exp: Asst prof inorg chem, Univ Karlsruhe, 55-56; from res chemist to sr res chemist, 56-66, res assoc, Pigments Dept, 66-69, RES ASSOC INORG CHEM, CENT RES & DEVELOP DEPT, E I DU PONT DE NEMOURS & CO, 69- Concurrent Pos: Fel, Mass Inst Technol, 55; vis prof mat sci, Brown Univ, 68. Mem: Am Chem Soc. Res: Exploratory as well as crystal growth aspects of ferroelectric, ferroelastic and electrooptic materials; finding and developing novel phosphors. Mailing Add: Cent Res & Develop Dept Exp Sta E I du Pont de Nemours & Co Wilmington DE 19898

BRIZIARELLI, GIULIANO, b Citta di Castello, Italy, Dec 11, 24; m 57; c 3. PATHOLOGY. Educ: Univ Perugia, MD, 49; Ital Bd Path, cert, 56. Prof Exp: Resident path, Univ Perugia, 50-56; res assoc oncol, Ben May Lab Cancer Res, Univ Chicago, 57-59; pathologist, Warner-Vister Inst, Casatenovo, Italy, 59-69, PATHOLOGIST, WARNER-LAMBERT RES INST, 69- Concurrent Pos: Fels, Univ Graz, 52, Univ Marburg, 53 & Univ Bonn, 56; lectr path, Univ Milan, 59-68; Educ Coun Foreign Med Grad fel, 65. Honors & Awards: Recognition Award, AMA, 69. Mem: Environ Mutagen Soc; Europ Soc Study Drug Toxicity; Ital Cancer Soc; Ital Path Soc. Res: Teratology; general pathology; cancerology; endocrinology; genetics. Mailing Add: Warner-Lambert Res Inst 170 Tabor Rd Morris Plains NJ 07950

BRIZIO-MOLTENI, LOREDANA, b Savona, Italy, June 17, 27; US citizen; m 63; c 2. RECONSTRUCTIVE SURGERY. Educ: Univ Bologna, Italy, MD, 51, Bd Spec Surg, 56. Prof Exp: Asst prof surg, Univ Bologna, Italy, 52-59; instr, State Univ NY Buffalo, 71-73; ASST PROF SURG, UNIV MO-KANSAS CITY, 74- Mem: Fel Am Col Surgeons; AMA; Am Burn Asn; Am Soc Plastic & Reconstruct Surgeons. Res: Studies of the structure of the skin and connective tissue in normal and pathologic conditions such as hypertrophic scars, burns, aging metabolic diseases, both at light and at the electron microscope; pathogenesis of hypertensive disease in thermal burns. Mailing Add: 1441 Hillside Terr North Kansas City MO 64116

BRIZOLIS, DEMETRIOS, b Montreal, Que, July 1, 46; US citizen. ALGEBRA. Educ: Univ Calif, Los Angeles, BA, 68, MA, 69, PhD(math), 73. Prof Exp: Instr, 72-73, ASST PROF MATH, CALIF STATE POLYTECH UNIV, POMONA, 73- Concurrent Pos: Vis asst prof math, Univ Southern Calif, 74-75. Mem: Am Math Soc; Math Asn Am. Res: Algebraic, number-theoretic and geometric properties of rings of integral-valued polynomials. Mailing Add: Dept of Math Calif State Polytech Univ Pomona CA 91768

BRIZZEE, KENNETH RAYMOND, b Ogden, Utah, July 7, 16; m 44; c 3. NEUROANATOMY, NEUROPATHY. Educ: Univ Utah, BS, 39, MS, 41; St Louis Univ, PhD(neuroanat), 49; Univ Nebr, MD, 63. Prof Exp: From instr to assoc prof anat, Col Med, Univ Utah, 49-61; prof anat & asst res prof obstet & gynec, Col Med, Univ Nebr, 61-64; assoc prof neurol, Col Med, Univ Utah, 64-65, prof anat, 64-68, res prof neurol, 65-68; res assoc, Div Environ Health, Delta Regional Primate Res Ctr & adj prof anat, Tulane Univ, 68-71; prof anat & chmn dept, Sch Dent Med, Southern Ill Univ, Edwardsville, 71-72; RES SCIENTIST & HEAD DIV GEN BIOMED SCI, DELTA REGIONAL PRIMATE RES CTR, 72- Concurrent Pos: Prof anat, Tulane Univ, 72- Mem: AAAS; Am Asn Anatomists; Am Physiol Soc; fel Geront Soc; Radiation Res Soc. Res: Cerebral cortex; cell growth and aging; radiation effects; organization of central emetic appaRatus in nonhuman primates; motion sickness; experimental neuropathology. Mailing Add: Delta Regional Primate Res Ctr Covington LA 70433

BRO, MANVILLE I, chemistry, see 12th edition

BRO, PER, physical chemistry, chemical engineering, see 12th edition

BROACH, WILSON J, b Atkins, Ark, Aug 14, 15; m 42; c 4. PHYSICAL CHEMISTRY. Educ: Henderson State Teachers Col, BA, 37; Univ Ark, MS, 48, PhD(phys chem), 53. Prof Exp: Instr chem, Little Rock Jr Col, 46-48 & Univ Ark, 48-52; assoc prof, Southern State Col, 52-54; assoc prof, Northwestern State Col, La, 54-55, actg head div phys sci, 55-57; assoc prof, 57-59, PROF CHEM & DEAN DIV PHYS SCI, UNIV ARK, LITTLE ROCK, 59- Mem: Am Chem Soc; AAAS. Res: Reaction rates in aqueous and nonaqueous solutions. Mailing Add: Div of Phys Sci & Math Univ of Ark Little Rock AR 72204

BROAD, ALFRED CARTER, b Yonkers, NY, Apr 29, 22; m 50; c 1. INVERTEBRATE ZOOLOGY. Educ: Univ NC, AB, 43, MA, 51; Duke Univ, PhD, 56. Prof Exp: Chief shrimp invests, Inst Fisheries Res, Univ NC, 48-51; res investr, Marine Lab, Duke Univ, 53-57; from asst prof to assoc prof zool, Ohio State Univ, 57-64; chmn dept biol, 64-71, PROF BIOL, WESTERN WASH STATE COL, 64- Mem: Am Soc Zoologists; Am Micros Soc; Am Soc Protozool. Res: Larval development of Natantia; crustacean life histories; sand dollar biology; arctic littoral ecology. Mailing Add: Dept of Biol Western Wash State Col Bellingham WA 98225

BROADBENT, FRANCIS EVERETT, b Snowflake, Ariz, Mar 29, 22; m 44; c 6. SOIL MICROBIOLOGY. Educ: Brigham Young Univ, BS, 42; Iowa State Col, MS, 46, PhD(soil bact), 48. Prof Exp: Instr soils, Iowa State Col, 47-48; jr chemist, Citrus Exp Sta, Univ Calif, 48-50; assoc prof soil microbiol, Cornell Univ, 50-55; assoc prof, 55-61, PROF SOIL MICROBIOL, UNIV CALIF, DAVIS, 61- Concurrent Pos: Fulbright sr res scholar, New Zealand, 62-63. Mem: Fel Am Soc Agron; Soil Sci Soc Am. Res: Soil organic matter chemistry; metal organic complexes; nitrogen transformations; use of stable tracer isotopes in biological systems. Mailing Add: Dept of Soils & Plant Nutrit Univ of of Calif Davis CA 95616

BROADBENT, HYRUM SMITH, b Snowflake, Ariz, July 21, 20; m 42; c 8. CHEMISTRY. Educ: Brigham Young Univ, BS, 42; Iowa State Univ, PhD(org chem), 46. Prof Exp: Lab asst chem, Brigham Young Univ, 40-42; partic, Nat Defense Res Comt & Off Sci Res & Develop Projs, Iowa State Univ, 43-44; Milton Fund fel physico-org chem, Harvard Univ, 46-47; from asst prof to assoc prof, 47-52, PROF CHEM, BRIGHAM YOUNG UNIV, 52- Concurrent Pos: Group leader med chem, Schering Corp, 58-59; vis scientist, C F Kettering Labs, 62-63; res chemist, Eastman Kodak Res Labs, 70-71. Honors & Awards: Karl G Maeser Awards, Brigham Young Univ, 68 & 70. Mem: Am Chem Soc; The Chem Soc; Int Soc Heterocyclic Chem. Res: Heterocycles; organometallic compounds; physical and general synthetic organic chemistry; contact catalysis; medicinal chemistry. Mailing Add: Dept of Chem Brigham Young Univ Provo UT 84601

BROADBOOKS, HAROLD EUGENE, b Wilbur, Wash, Aug 29, 15; m 50; c 6. ZOOLOGY. Educ: Univ Wash, BA, 37; Univ Mich, MA, 40, PhD(zool), 50. Prof Exp: Asst prof biol, Stephen F Austin State Col, 49-50; asst prof zool, Univ Ariz, 50-52; res assoc surg, Sch Med, Univ Wash & Radioisotope Unit, Vet Admin Hosp, 53-54; biologist, State Dept Fisheries, Wash & Ore, 54-56; asst prof zool, Shurtleff Col, 56-57; from asst prof to assoc prof, 57-70, PROF ZOOL, SOUTHERN ILL UNIV, EDWARDSVILLE, 71- Mem: Am Soc Mammal; Soc Study Evolution; Ecol Soc Am; Cooper Ornith Soc; Am Behavior Soc. Res: Mammalogy; vertebrate ecology; behavior; distribution; evolution. Mailing Add: Dept of Biol Sci/Sci & Technol Div Southern Ill Univ Edwardsville IL 62025

BROADDUS, CHARLES D, b Irvine, Ky, Oct 17, 30; m 57; c 3. ORGANIC CHEMISTRY. Educ: Centre Col Ky, AB, 52; Auburn Univ, MS, 54; Univ Fla, PhD(org chem), 60. Prof Exp: Res chemist, 60-68, sect head res, 68-72, assoc dir food, paper & coffee technol div, 72-73, DIR FOOD, PAPER & COFFEE TECHNOL DIV, PROCTER & GAMBLE CO, 73- Mem: Am Chem Soc. Res: Metalation chemistry, base catalyzed exchange reactions. Mailing Add: Proctor & Gamble Co Miami Valley Labs PO Box 39175 Cincinnati OH 45247

BROADFOOT, ALBERT LYLE, b Milestone, Sask, Jan 8, 30; m 64; c 2. PHYSICS. Educ: Univ Sask, BE, 56, MSc, 60, PhD(physics), 63. Prof Exp: Engr, Defence Res Bd Can, 56-58; from asst physicist to assoc physicist, 63-71, PHYSICIST, PLANETARY SCI DIV, KITT PEAK NAT OBSERV, 71- Mem: Am Astron Soc; Am Geophys Union; Can Asn Physicists; Int Union Geod & Geophys; Int Asn Geomagnetism & Aeronomy. Res: Molecular spectroscopy; upper atmospheric physics; planetary atmosphere; airglow, aurora and associated phenomena. Mailing Add: Kitt Peak Nat Observ 950 N Cherry Ave Tucson AZ 85717

BROADFOOT, WALTER MARION, b Pulaski, Miss, Jan 13, 11; m 35; c 2. SOILS. Educ: Miss State Col, BS, 34; WVa Univ, MS, 38. Prof Exp: Asst agron, Exp Sta, Miss State Col, 34-35; asst, Exp Sta & instr, WVa Univ, 36-41, asst agronomist & asst prof, 46; soils scientist, Southern Forest Exp Sta, US Forest Se-v, 46-74; RETIRED. Mem: Am Soc Agron; Soil Sci Soc Am; Soil Conserv Soc Am. Res: Development of soil management systems and techniques for production of southern hardwood species while maintaining or improving site quality; determining forest soil requirements of commercial hardwood species. Mailing Add: 315 Deer Creek NW Leland MS 38756

BROADHEAD, GORDON CLIFFORD, b Chilliwack, BC, Mar 2, 24; m 48; c 2. FISH BIOLOGY, MARINE SCIENCE. Educ: Univ BC, BA, 49; Univ Miami, MS, 55. Prof Exp: Jr biologist, Fishery Res Bd, Can, 49-50; sr res asst marine lab, Univ Miami, 50-55; scientist, Inter-Am Trop Tuna Comn, 55-56, sr scientist, 56-63; sr analyst, Van Camp Seafood Co, 63-66, dir anal res, 66-68, dir res, 68-69; vpres res & proj develop, 69-70, PRES, LMR, INC, 70- Concurrent Pos: Consult, Harwell Knowles Assoc, 55. Res: Marine fishery management and population dynamics. Mailing Add: LMR Inc 11339 Sorrento Valley Rd San Diego CA 92121

BROADHURST, MARTIN GILBERT, b Washington, DC, Apr 28, 32; m 55; c 2. SOLID STATE PHYSICS. Educ: Western Md Col, BA, 55; Pa State Univ, MS, 57, PhD(physics), 59. Prof Exp: Res assoc physics, Pa State Univ, 59-60; physicist, 60-67, chief polymer dielectrics sect, Polymer Div, 67-70, chief, Dielectric & Thermal Properties Sect, 70-75, CHIEF BULK PROPERTIES SECT, NAT BUR STANDARDS, 75- Concurrent Pos: Chmn conf elec insulation, Nat Acad Sci, 75-77. Honors & Awards: Silver Medal, US Dept Com, 73. Mem: Nat Acad Sci; AAAS; Am Phys Soc. Res: Experimental and theoretical techniques for electrical, mechanical and thermal properties of organic and polymeric solids; relations between microscopic structure and bulk physical properties; thermodynamics, statistical mechanics, piezoelectric and pyroelectric properties. Mailing Add: Polymer Div Nat Bur of Standards Washington DC 20234

BROADIE, LARRY LEWIS, b Mangum, Okla, Feb 25, 40; m 62; c 1. NEUROPHARMACOLOGY. Educ: Southwestern State Col, BS, 62; Univ Kans, MS, 64; Univ Ariz, PhD(pharmacol), 70; Univ Okla, med, 74- Prof Exp: Assoc prof pharmacol, Southwestern State Col, 69-74. Res: Thermopharmacology, especially the effect of dextroamphetamine and other amine substances on body temperature controlling mechanisms in the central nervous system. Mailing Add: 3917 SE 14th Pl Del City OK 73115

BROADWATER, TOMMY L, physical chemistry, see 12th edition

BROBECK, JOHN RAYMOND, b Steamboat Springs, Colo, Apr 12, 14; m 40; c 4. PHYSIOLOGY. Educ: Wheaton Col, Ill, BS, 32; Northwestern Univ, MS, 37, PhD(neurol), 39; Yale Univ, MD, 43. Hon Degrees: LLD, Wheaton Col, Ill, 60; MA, Univ Pa, 71. Prof Exp: From instr to assoc prof physiol, Sch Med, Yale Univ, 43-52; prof & chmn dept, 52-70, HERBERT C RORER PROF MED SCI, SCH MED, UNIV PA, 70- Mem: Nat Acad Sci; fel Am Acad Arts & Sci; Am Physiol Soc (pres, 71-72); Am Soc Clin Invest; Am Inst Nutrit. Res: Physiological controls and regulations; control of energy balance; physiology of hypothalamus. Mailing Add: 224 Vassar Ave Swarthmore PA 19081

BROBERG, JOEL WILBUR, b Willmar, Minn, Aug 2, 10; m 37; c 2. INORGANIC CHEMISTRY. Educ: Macalester Col, BA, 32; Univ Minn, MA, 40, PhD(chem educ), 62. Prof Exp: From asst prof to assoc prof, 41-63, PROF CHEM, NDAK STATE UNIV, 63-, DIR, INST TEACHER EDUC, 69- Mem: Am Chem Soc. Res: Molar method of teaching chemistry; synthesis of coordination compounds. Mailing Add: Dept of Chem NDak State Univ Fargo ND 58102

BROBST, DONALD ALBERT, b Allentown, Pa, May 8, 25; m 50; c 1. ECONOMIC GEOLOGY. Educ: Muhlenberg Col, AB, 47; Univ Minn, PhD(geol), 53. Prof Exp: GEOLOGIST, US GEOL SURV, 48-, DEP CHIEF OFF MINERAL RESOURCES, 73- Mem: Fel Geol Soc Am; Mineral Soc Am; Soc Econ Geologists; Am Inst Mining, Metall & Petrol Eng. Res: Economic geology of pegmatites and barite deposits; assessment of mineral resources; petrology. Mailing Add: US Geol Surv Nat Ctr Stop 913 Reston VA 22092

BROBST, DUANE FRANKLIN, b Medicine Lake, Mont, Oct 8, 23; m 58; c 2. VETERINARY MEDICINE. Educ: Univ Southern Calif, AB, 49; Wash State Univ, DVM, 54; Univ Pittsburgh, MPH, 57; Univ Wis, PhD(vet sci), 62. Prof Exp: Pvt vet practice, Mont, 54-56; pub health veterinarian, Allegheny County Health Dept, Pa, 57-58; proj asst vet med, Univ Wis, 58-62; asst prof vet path, Sch Vet Sci & Med, Purdue Univ, 62-70; PROF VET MED, WASH STATE UNIV, 70- Concurrent Pos: Whitley County Cancer Asn, Ind & Delta Theta Tau res grant, 62-63; consult, Am Med Asn, 63- Mem: Am Vet Med Asn; Am Col Vet Path. Res: Pathology of animal disease; epidemiology. Mailing Add: Dept of Vet Clin Med Wash State Univ Pullman WA 99163

BROBST, KENNETH MARTIN, b Orangeville, Ill, Dec 18, 15; m 37; c 2. ANALYTICAL CHEMISTRY. Educ: Univ Ill, BS, 37. Prof Exp: Anal chemist, 37-46, res chemist, 46-61, head anal chem lab, 61-70, GROUP LEADER CHEM, A E STALEY MFG CO, 70- Mem: Am Chem Soc; Am Oil Chemists Soc. Res: Development of methods for the analysis of products from a corn and soybean processing industry, particularly the analysis of carbohydrate mixtures by chromatographic methods. Mailing Add: A E Staley Mfg Co Decatur IL 62525

BROCHMANN-HANSSEN, EINAR, b Hvitsten, Norway, June 18, 17; nat US; m 43; c 2. PHARMACEUTICAL CHEMISTRY. Educ: Univ Oslo, Cand Pharm, 41; Purdue Univ, PhD(pharmaceut chem), 49. Prof Exp: Pharmacist, Flekkefjord Apotek,

Norway, 42-44; res assoc, Univ Oslo, 44-46; from asst prof to assoc prof, 49-59, PROF PHARMACEUT CHEM, SCH PHARMACY, UNIV CALIF, SAN FRANCISCO, 59- Concurrent Pos: Instr, Univ Oslo, 45-46; partic scientist, UN Opium Res Prog, 58-; mem rev & exec comts, Nat Formulary, US Pharmacopeia, 60-70; vis prof, Robert Robinson Labs, Liverpool, 65-66; mem, US Pharmacopeial Conv, 70-75; vis prof, Nat Taiwan Univ, 74-75. Honors & Awards: Ebert Prize, 62; Powers Award, 63; Silver Medal, Univ Helsinki, 67. Mem: Am Soc Pharmacog; Am Chem Soc; Am Pharmaceut Asn; NY Acad Sci; Int Pharmaceut Fedn. Res: Isolation, structure and physiological activity of naturally occurring substances; analytical chemistry; chromatography; alkaloid chemistry and biosynthesis. Mailing Add: Sch of Pharmacy Univ of Calif San Francisco CA 94143

BROCHU, WILLIAM EUGENE, b Sumter, SC, May 29, 45; m 68; c 3. INDUSTRIAL PHARMACY, PHYSICAL PHARMACY. Educ: Mass Col Pharm, BS, 68, MS, 70; Purdue Univ, PhD(phys pharm), 74. Prof Exp: SR RES PHARMACIST, BAXTER LABS, INC, 74- Mem: Am Chem Soc; Am Pharmaceut Asn. Res: Physical-chemical characterization of compounds and drug delivery systems as it relates to drug products; devices for oral, topical and perenteral administration. Mailing Add: Baxter Labs Inc 6301 Lincoln Ave Morton Grove IL 60053

BROCK, CAROLYN PRATT, b Chicago, Ill, July 25, 46; m 72. STRUCTURAL CHEMISTRY. Educ: Wellesley Col, BA, 68; Northwestern Univ, PhD(chem), 72. Prof Exp: ASST PROF, DEPT CHEM, UNIV KY, 72- Mem: Am Chem Soc; Am Crystallog Asn. Res: Molecular structure as determined by x-ray diffraction; intra- and intermolecular forces; molecular packing in crystals. Mailing Add: Dept of Chem Univ of Ky Lexington KY 40506

BROCK, ERNEST GEORGE, b Detroit, Mich, Apr 7, 26; m 50; c 3. ELECTROPHYSICS. Educ: Univ Notre Dame, BS, 46, PhD(physics), 51. Prof Exp: Res assoc, Gen Elec Res Lab, 51-56; group leader, Linfield Res Inst, Ore, 56-58; sr physicist, Res Dept, Gen Dynamics/Electronics, 58-59, prin scientist, 59-61, mgr quantum physics lab, 61-66; head quantum electronics dept, Lab Opers, Aerospace Corp, Calif, 66-68, sr staff scientist, 68-69, sr staff engr, 69-71; sr engr, NAm Rockwell, 72; sr engr specialist, Garrett Corp, 73-74; DIV OFF TECH STAFF MEM, LASER DIV, LOS ALAMOS SCI LAB, UNIV CALIF, 74- Mem: AAAS; Am Phys Soc; sr mem Inst Elec & Electronics Engrs; Optical Soc Am; Sigma Xi. Res: Laser applications to fusion and isotope separation; applications of lasers and electronics technology to advanced strategic systems planning. Mailing Add: Los Alamos Sci Lab Univ of Calif Los Alamos NM 87544

BROCK, FRED VINCENT, b Chillicothe, Ohio, Nov 25, 32; m 60; c 2. METEOROLOGY. Educ: Ohio State Univ, BS, 54; Univ Mich, MSE, 60; Univ Okla, PhD(meteorol), 73. Prof Exp: Assoc res engr meteorol, Univ Mich, 60-69; spec instr, Univ Okla, 69-73; vis scientist, 73-75, STAFF SCIENTIST METEOROL, NAT CTR ATMOSPHERIC RES, 75- Concurrent Pos: Consult, Univ Corp Atmospheric Res & White Sands Missile Range, US Army, 70-73; Nat Ctr Atmospheric Res affil prof, Univ Okla, 75- Mem: Am Meteorol Soc. Res: Meteorological measurement systems including sensors, data loggers, system analysis and data processing. Mailing Add: Nat Ctr for Atmospheric Res PO Box 3000 Boulder CO 80303

BROCK, GEORGE WILLIAM, b Grant Co, Ind, Aug 27, 20; m 44; c 3. PHYSICS. Educ: Nebr State Teachers Col, AB, 39; NY Univ, cert, 43; Air Force Inst Technol, MS, 57; Purdue Univ, MS, 64. Prof Exp: Radio engr, Farnsworth TV & Radio Corp, Ind, 41-42; meteorologist, US Army Air Force, 42-46; physicist, Aeronaut Ice Res Lab, 46-48, chief res, 48-50; US Air Force, 50-, physicist, Wright Air Develop Ctr, Ohio, 50-59, mem staff, US Air Force Acad, 59-65, prof physics & actg head dept, 65-66, dir sci, Hq Air Force Systs Command, Andrews AFB, Md, 69-73, DIR, F J SEILER RES LAB, US AIR FORCE ACAD, 73- Concurrent Pos: Mem subcomt icing, Nat Adv Comt Aeronaut; US deleg, NATO DRG Panel Physics & Electronics, 70-75. Mem: AAAS; Am Phys Soc. Res: Solid state physics and electronic materials. Mailing Add: 3110 Nevermind Lane Colorado Springs CO 80917

BROCK, KATHERINE MIDDLETON, b Keokuk, Iowa, June 3, 38; m 71. MICROBIOLOGY, MICROBIAL ECOLOGY. Educ: Vassar Col, AB, 60; Univ Calif, Berkeley, MA, 63; Univ Mass, Amherst, PhD(microbiol), 67. Prof Exp: Res asst biochem, Tech Univ Norway, 63-64; asst prof microbiol, San Framcisco State Col, 67-70; res assoc, Ind Univ, Bloomington, 70-71; RES ASSOC BACT, UNIV WIS-MADISON, 71- Mem: AAAS; Am Soc Microbiol. Res: Microbial physiology; study of extreme environments, especially high temperature, high and low hydrogen-ion concentration and saline environments. Mailing Add: Dept of Bact Univ of Wis Madison WI 53706

BROCK, KENNETH JACK, b Pampa, Tex, Aug 22, 37; m 62; c 1. MINERALOGY. Educ: San Jose State Col, BS, 62; Stanford Univ, PhD(geol), 70. Prof Exp: Asst prof, 70-74, ASSOC PROF GEOL, IND UNIV NORTHWEST, 74- Mem: Mineral Soc Am; Mineral Soc Gt Brit & Ireland. Res: Mineralogy and genesis of skarns; mineralogy of zoned lithium micas. Mailing Add: Dept of Geol Ind Univ Northwest Gary IN 46408

BROCK, MARY ANNE, b Aurora, Ill, June 29, 32. BIOLOGY, PHYSIOLOGY. Educ: Grinnell Col, BA, 54; Radcliffe Col, MA, 56, PhD(biol), 59. Prof Exp: Res assoc med, Harvard Med Sch, 59-60; BIOLOGIST, GERONT RES CTR, NAT INST AGING, 60- Mem: Am Soc Cell Biol; NY Acad Sci; Am Soc Zoologists; Soc Cryobiol; Geront Soc. Res: Physiology of natural mammalian hibernation; erythrocyte longevity and physiology; ultrastructure of Cnidaria and ageing mammalian cell types; effect of temperature on cellular ageing; biorhythms. Mailing Add: Geront Res Ctr Baltimore City Hosps Baltimore MD 21224

BROCK, PAUL, b Brooklyn, NY, Mar 29, 23; m 46; c 3. APPLIED MATHEMATICS. Educ: Brooklyn Col, BA, 42; NY Univ, MS, 47, PhD(appl math), 51. Prof Exp: Instr math & physicist, Palmer Labs, Princeton, 42; computer, Kellex Corp, 43; instr math, Hunter Col, 46-48; head math group, Proj Cyclone, Reeves Instrument Corp, 48-52; mgr tech serv dept, Electrodata Corp, 52-56; assoc prof, Purdue Univ, 56-59; dir indust dynamics res, Hughes Aircraft Co, 59-60; head simulation & gaming dept, Stanford Res Inst, 60-63; sr scientist, Logistics Dept, Rand Corp, 64-65; mem prof staff, Tempo, Gen Elec Co, 65-69; PROF MATH, UNIV VT, 69- Concurrent Pos: Assoc prof, Univ Mich, 57-59; dir, Comput Instrument Corp & Inland Electronic Prod Corp, 62-; prof lectr, Univ Calif, Los Angeles, 64. Mem: Am Math Soc; Math Asn Am; Soc Indust & Appl Math; Soc Comput Simulation. Res: Differential equations; numerical methods; digital and analogue equipment; gaming and simulation. Mailing Add: Dept of Math Univ of Vt Burlington VT 05401

BROCK, THOMAS DALE, b Cleveland, Ohio, Sept 10, 26; m 52. MICROBIOLOGY. Educ: Ohio State Univ, BSc, 49, MSc, 50, PhD(microbiol), 52. Prof Exp: Res microbiologist, Upjohn Co, 52-57; asst prof biol, Western Reserve Univ, 57-59, fel, Sch Med, 59-60; from asst prof to prof bact, Ind Univ, Bloomington, 60-71; E B FRED PROF NATURAL SCI, UNIV WIS-MADISON, 71- Concurrent Pos: Vis biologist, Am Inst Biol Sci, 61-62; USPHS career develop award, 62-68. Mem: AAAS; Am Soc Microbiol; Brit Soc Gen Microbiol. Res: Microbial ecology; aquatic microbiology; biogeochemistry. Mailing Add: Dept of Bact Univ of Wis Madison WI 53706

BROCK, WILLIAM ELIHU, b Blackwell, Okla, Aug 10, 14; m 39; c 1. VETERINARY PATHOLOGY. Educ: Univ Wichita, AB, 37; Kans State Col, DVM, 44; Okla State Univ, MS, 55; Univ Okla, PhD(path), 58. Hon Degrees: DHC, San Carlos Univ, Guatemala, 67. Prof Exp: Vet pvt pract, 44-49; asst vet, 49-52, from asst prof to assoc prof vet med, 52-59, PROF VET MED, OKLA STATE UNIV, 59-, DEAN COL, 70- Mem: Am Vet Med Asn; US Animal Health Asn; Conf Res Workers Animal Dis. Res: Hematology; serology. Mailing Add: Col of Vet Med Okla State Univ Stillwater OK 74074

BROCKE, RAINER H, b Calcutta, India, Nov 24, 33; US citizen; m 57; c 3. MAMMALIAN ECOLOGY. Educ: Mich State Univ, BS, 55, MS, 57, PhD(mammalian bioenergetics), 70. Prof Exp: Res asst, Mich State Univ, 56-57, instr, 63-69, mus cur & naturalist, Nankin Mills Mus, Mich, 57-58; park naturalist, Huron-Clinton Metrop Authority, Mich, 58-63; SR RES ASSOC MAMMALIAN ECOL & BIOENERGETICS, STATE UNIV NY COL ENVIRON SCI & FORESTRY, 69- Concurrent Pos: Consult, Develop Sci Prog, Mich, 65-66; res grant, State Univ NY Res Found, 71-72. Mem: AAAS; Am Soc Mammal; Ecol Soc Am; Wildlife Soc. Res: Physiological ecology of mammals, with emphasis on bioenergetics. Mailing Add: Adirondack Ecol Ctr Newcomb NY 12852

BROCKELMAN, WARREN YALDING, ecology, see 12th edition

BROCKEMEYER, EUGENE WILLIAM, b Buckner, Mo, June 22, 29; m 51; c 2. PHARMACY. Educ: Univ Kans, BS, 51, MS, 52; Ohio State Univ, PhD(pharm), 54. Prof Exp: Res assoc, 57-67, asst dir prod develop, 67, mgr qual control & prod develop, 67-71, DIR RES DEVELOP & QUAL CONTROL, DORSEY LABS, 71- Mem: Am Pharmaceut Asn. Res: Rate of absorption of drugs from pharmaceutical vehicles and sustained release forms of medication. Mailing Add: Dorsey Lab Box 1113 Lincoln NE 68501

BROCKENBROUGH, EDWIN C, b Baltimore, Md, July 24, 30; m 68; c 5. SURGERY. Educ: Col William & Mary, BS, 52; Johns Hopkins Univ, MD, 56. Prof Exp: Asst surg, Johns Hopkins Univ, 57-58; clin assoc cardiac surg, NIH, 59-61; asst, 61-64, asst prof, 64-70, CLIN ASSOC PROF SURG, SCH MED, UNIV WASH, 70-; SR RES SCIENTIST, INST APPL PHYSIOL & MED, SEATTLE, 75- Concurrent Pos: Assoc surgeon-in-chief, Harborview Med Ctr, Seattle, 70-75. Res: Cardiovascular physiology; thoracic surgery; cerebrovascular diseases. Mailing Add: 1221 Madison St Seattle WA 98104

BROCKERHOFF, HANS, b Duisburg, Ger, July 8, 28; m 58; c 3. BIOCHEMISTRY, NEUROCHEMISTRY. Educ: Univ Cologne, Dr rer nat, 58. Prof Exp: Researcher, Univ Wash, 58-60 & Univ Calif, Berkeley, 60-61; assoc scientist, Fisheries Res Bd, Can, 61-63, sr scientist, 63-73; CHIEF SCIENTIST NEUROCHEM, NY STATE INST FOR BASIC RES IN MENT RETARDATION, 73- Mem: Am Soc Biol Chemists; Am Soc Neurochem. Res: Lipid chemistry; structure of phospholipids and triglycerides; marine lipids; lipolytic enzymes; myelin lipids; membrane structure. Mailing Add: NY State Inst Res Ment Retard 1050 Forest Hill Rd Staten Island NY 10314

BROCKETT, PATRICK LEE, b Monterey Park, Calif, Mar 29, 48; m 73. MATHEMATICAL STATISTICS. Educ: Calif State Univ, Long Beach, BA, 70; Univ Calif, Irvine, MA, 75, PhD(math), 75. Prof Exp: ASST PROF MATH, TULANE UNIV, 75- Mem: Am Math Soc; Inst Math Statist; Math Asn Am. Res: Probability theory and stochastic processes; specifically infinitely divisible distributions and infinitely divisible processes. Mailing Add: Dept of Math Tulane Univ New Orleans LA 70118

BROCKETT, ROGER WARE, b Wadsworth, Ohio, Oct 22, 38; m 60; c 2. APPLIED MATHEMATICS. Educ: Case Inst Technol, BS, 60, MS, 62, PhD(eng), 64. Prof Exp: From asst prof to assoc prof, Mass Inst Technol, 63-69; PROF APPL MATH, DIV ENG & APPL PHYSICS, AIKEN COMPUT LAB, HARVARD, 69- Concurrent Pos: Ford fel, 63-65. Mem: Inst Elec & Electronics Engrs; Am Math Soc. Res: Automatic control theory including optimal control and stability theory; nonlinear phenomena; mathematical system theory. Mailing Add: Div of Eng & Appl Physics Aiken Comput Lab Harvard Univ Cambridge MA 02138

BROCKETT, ROYCE MERRETT, b Chicago, Ill, Mar 17, 44; m 67; c 2. MEDICAL MICROBIOLOGY. Educ: Denison Univ, BS, 66; Univ NMex, MS, 68, PhD(microbiol), 71. Prof Exp: Microbiologist, Johnson Space Ctr, NASA, 71-74; CHIEF SPEC PATHOGEN SECT, SCH AEROSPACE MED, EPIDEMIOL DIV, BROOKS AFB, 74- Mem: Am Soc Microbiol; Soc Armed Forces Med Lab Scientists. Res: Epidemiology, surveillance and control of nosocomial infections; staphylococcus epidermidis and S aureus bacteriophage; epidemiology and control of staphylococcal infections; microbiology of hospital environments; microbiology of astronaut crews and the spacecraft environment. Mailing Add: 5942 Little Brandywine San Antonio TX 78233

BROCKHOUSE, BERTRAM NEVILLE, b Lethbridge, Alta, July 15, 18; m 48; c 6. SOLID STATE PHYSICS. Educ: Univ BC, BA, 47; Univ Toronto, MA, 48, PhD, 50. Hon Degrees: DSc, Univ Waterloo, 69. Prof Exp: Lectr physics, Univ Toronto, 49-50; res officer, Atomic Energy Can, Ltd, 50-60, head neutron physics br, 60-62; chmn dept physics, 67-70, PROF PHYSICS, McMASTER UNIV, 62- Concurrent Pos: Guggenheim fel, 70. Honors & Awards: Oliver E Buckley Prize, Am Phys Soc, 62; Duddell Medal & Prize, Brit Inst Physics & Phys Soc, 63. Mem: Am Phys Soc; Can Asn Physicists; fel Royal Soc Can; fel Royal Soc London. Res: Neutron physics as applied to physics of solid and liquids; philosophy of physics. Mailing Add: Dept of Physics SSC453 McMaster Univ Hamilton ON Can

BROCKINGTON, DONALD LESLIE, b Weslaco, Tex, Apr 28, 29; m 55; c 3. ANTHROPOLOGY. Educ: Univ NMex, BA, 54; Univ of the Americas, MA, 57; Univ Wis-Madison, PhD(anthrop), 65. Prof Exp: Admin asst, Univ of the Americas, 56, instr anthrop, 56-57; archaeologist, New World Archeol Found, 58-59; teaching asst anthrop, Univ Wis-Madison, 60-61, proj asst 62, res asst, 62-63; asst prof, San Diego State Col, 63-67; assoc prof anthrop, 67-73, assoc chmn dept, 72-73, PROF ANTHROP, UNIV NC, CHAPEL HILL, 73-, ASST DIR, RES LABS ANTHROP, 67- Concurrent Pos: Dir hwy salvage prog, State Hist Soc Wis, 60-62; fac res grant, Univ NC, 67-68; NSF res grants, Oaxaca, Mex, 69-70 & 72. Mem: AAAS; Soc Am Archaeol; fel Am Anthrop Asn. Res: Prehistory and archaeological inference with special reference to the development of urban civilization in Middle America. Mailing Add: Dept of Anthrop Univ of NC Chapel Hill NC 27514

BROCKINGTON, JAMES WALLACE, b Norfolk, Va, Apr 12, 43; m 63; c 2. ORGANIC CHEMISTRY. Educ: Univ Richmond, BS, 65; Univ Miami, MS, 67; Va Commonwealth Univ, PhD(chem), 70. Prof Exp: MEM RES STAFF, RICHMOND

RES LAB, TEXACO, INC, 70- Mem: Am Chem Soc. Mailing Add: Richmond Res Lab Texaco Inc Box 3407 Richmond VA 23234

BROCKMAN, ELLIS R, b St Louis, Mo, Feb 10, 34; m 59; c 2. MICROBIOLOGY. Educ: DePauw Univ, AB, 55; Univ Mo, AM, 60, PhD(bact), 64. Prof Exp: Instr bact, Univ Mo, 63; asst prof biol, Winthrop Col, 63-69; ASSOC PROF BIOL, CENT MICH UNIV, 69- Concurrent Pos: Consult, Warner Lambert Res Inst, 69- Mem: AAAS; Am Soc Microbiol; Soc Indust Microbiol. Res: Bacterial taxonomy; biology of the myxobacteria; bacteriology of aquatic ecosystems; screening anti-microbial compounds. Mailing Add: Biol Dept Cent Mich Univ Mt Pleasant MI 48858

BROCKMAN, HAROLD W, b Sidney, Ohio, Mar 31, 22; m 54; c 1. MATHEMATICS. Educ: Capital Univ, BSEd, 48; Ohio State Univ, MA, 50, PhD(math, math educ) 62. Prof Exp: From instr to assoc prof math, 49-64, PROF MATH, CAPITAL UNIV, 64- Mem: Am Math Soc; Math Asn Am. Res: Algebra; analysis. Mailing Add: Dept of Math Capital Univ Columbus OH 43209

BROCKMAN, HERMAN E, b Danforth, Ill, Dec 5, 34; m 56; c 6. GENETICS. Educ: Blackburn Col, BS, 56; Northwestern Univ, MA, 57; Fla State Univ, PhD(genetics), 60. Prof Exp: Res assoc, Biol Div, Oak Ridge Nat Lab, 60-61, Nat Acad Sci-Nat Res Found fel, 61-62, Coun fel, 61-62, USPHS fel, 62-63; assoc prof, 63-70, PROF GENETICS, ILL STATE UNIV, 70- Concurrent Pos: USPHS spec res fel, 69. Mem: AAAS; Genetics Soc Am. Res: Biochemical genetics; mutagenesis. Mailing Add: Dept of Biol Sci Ill State Univ Normal IL 61761

BROCKMAN, JOHN A, JR, b Kellogg, Idaho, Apr 4, 20; m 52; c 4. IMMUNOLOGY. Educ: Calif Inst Technol, BS, 42, PhD(org chem), 48. Prof Exp: Asst chem, Calif Inst Technol, 42-46, res fel, 48-49; RES CHEMIST, LEDERLE LABS, AM CYANAMID CO, 49- Mem: Am Chem Soc. Res: Analytical chemistry; isolation and structure determination of natural products; alkaloids of dichroa febrifuga; vitamins and growth factors; vitamin B12; leucovorin; thiotic acid; neurochemistry; immuno-chemistry; oncology. Mailing Add: Lederle Lab Div Am Cyanamid Co Pearl River NY 10965

BROCKMAN, ROBERT W, b Chester, SC, Dec 8, 24; m 48; c 2. BIOCHEMISTRY. Educ: Vanderbilt Univ, BS, 47, MS, 49, PhD(chem), 51. Prof Exp: Asst chem, Vanderbilt Univ, 50-51; sr chemist, 51-57, head drug resistance sect, 58-67, HEAD BIOL CHEM DIV, SOUTHERN RES INST, 67-, PRIN SCIENTIST, 64- Concurrent Pos: NIH spec fel, 64-66; mem adv comt, Am Cancer Soc, 66-69 & 75- Mem: Am Chem Soc; Am Soc Biol Chemists; Am Asn Cancer Res; Soc Exp Biol & Med. Res: Mechanisms of drug inhibition and resistance; cancer research. Mailing Add: Southern Res Inst 2000 Ninth Ave Birmingham AL 35205

BROCKMAN, WILLIAM WARNER, b Philadelphia, Pa, July 8, 42; m 67; c 2. MOLECULAR BIOLOGY, VIROLOGY. Educ: Cornell Univ, BS, 64, MD, 68. Prof Exp: Intern & resident med, Baltimore City Hosps, 68-70; fel microbiol & med, Johns Hopkins Univ, 70-74; res assoc mole cular biol, NIH, 74-76; ASST PROF MICROBIOL, MED SCH, UNIV MICH, ANN ARBOR, 76- Concurrent Pos: Sr surgeon, USPHS, 74-76. Mem: AAAS. Res: Molecular genetics of animal viruses; virus induced cellular transformation. Mailing Add: Dept Microbiol 6643 MS Bldg II Univ of Mich Med Sch Ann Arbor MI 48109

BROCKMANN, MAXWELL CURTIS, b Sioux Falls, SDak, Dec 2, 10; m 35; c 2. FOOD SCIENCE. Educ: Iowa State Univ, BS, 41, MS, 43; Ind Univ, PhD(physiol), 46. Prof Exp: Res chemist, Hiram Walker & Sons, Inc, Ill, 34-42; res supvr, Joseph E Seagram & Sons, Inc, Ky, 42-48, dir chem res, 48-51; sr scientist, Kingan-Hygrade Food Prod Corp, Ind, 51-53, dir res, 53-57; chief animal prod br, Qm Food & Container Inst, 57-63; CHIEF ANIMAL PROD BR, FOOD LAB, US ARMY NATICK LABS, 63- Concurrent Pos: Seagram res assoc, Ind Univ, 44-46. Mem: AAAS; Am Chem Soc; Am Soc Microbiol; Soc Indust Microbiol; Am Inst Chemists. Res: Technology of mammalian meats, poultry, fish and dairy products; concentrated and stabilized foods; food specifications. Mailing Add: Food Eng Lab US Army Res & Dev Command Natick MA 01760

BROCKMEIER, RICHARD TABER, b Grand Rapids, Mich, Apr 13, 37; m 64. NUCLEAR PHYSICS. Educ: Hope Col, AB, 59; Calif Inst Technol, MS, 61, PhD(physics), 65. Prof Exp: Res fel physics, Calif Inst Technol, 65-66; PROF PHYSICS, HOPE COL, 66- Mem: Am Phys Soc; Am Asn Physics Teachers; Sigma Xi. Res: Nuclear structure physics; x-ray isotope shifts. Mailing Add: Dept of Physics Hope Col Holland MI 49423

BROCKWAY, ALAN PRIEST, b Hanover, NH, Aug 21, 36; m; c 3. COMPARATIVE PHYSIOLOGY. Educ: St John's Col, Md, AB, 58; Western Reserve Univ, PhD(biol), 64. Prof Exp: Asst prof zool & entom, Ohio State Univ, 63-67; ASSOC PROF BIOL, UNIV COLO, DENVER, 67- Concurrent Pos: Vis assoc prof oral biol, Sch Dent, Univ Colo, 73- Mem: AAAS; Am Soc Zoologists; Am Inst Biol Sci; Nat Asn Biol Teachers; NY Acad Sci. Res: Insect respiration; control of water loss in insects; temperature effects on the metabolism of insects; seed development. Mailing Add: Dept of Biol Univ of Colo Denver CO 80202

BROCKWAY, BARBARA FINK, b Chicago, Ill, Feb 29, 36; m 59; c 2. VERTEBRATE ZOOLOGY, PHYSIOLOGY. Educ: Univ Calif, Santa Barbara, BS, 56; Cornell Univ, MS, 58, PhD, 62. Prof Exp: Instr biol, Western Reserve Univ, 61-63, NSF instnl grant, 62-63; asst prof zool, Ohio State Univ, 64-68; vis lectr biol, 68-69, lectr, 69-70, assoc prof, 70-73, PROF BIOL, SCH NURSING, UNIV COLO, DENVER, 73- Concurrent Pos: NSF res grant, 65- Mem: NY Acad Sci; AAAS. Res: Social behavior and reproductive physiology; nursing intervention and patient stress. Mailing Add: Univ of Colo Med Ctr 4200 E Ninth Ave Denver CO 80220

BROCKWAY, LAWRENCE OLIN, b Topeka, Kans, Sept 23, 07; m 32; c 1. PHYSICAL CHEMISTRY. Educ: Univ Nebr, BS, 29, MS, 30; Calif Inst Technol, PhD(chem), 33. Prof Exp: Sr res fel, Calif Inst Technol, 33-37; Guggenheim Mem Found fel, Oxford & Royal Inst, 37-38; from asst prof to assoc prof, 38-44, PROF CHEM, UNIV MICH, ANN ARBOR, 45- Concurrent Pos: Consult, Nat Adv Comt Aeronaut, 44-50, Nat Comt Crystallog, 52-53 & div phys sci, Nat Res Coun, 53-55. Honors & Awards: Prize, Am Chem Soc, 40. Mem: Am Chem Soc; Electron Micros Soc Am; Am Crystallog Asn. Res: Electron diffraction studies of gases and solid surfaces; mass spectrographic study of exchange reactions. Mailing Add: Dept of Chem Univ of Mich Ann Arbor MI 48104

BROCOUM, STEPHAN JOHN, b New York, NY, Feb 16, 41; m 69. STRUCTURAL GEOLOGY, TECTONICS. Educ: Brooklyn Col, BS, 63; Columbia Univ, PhD(geol), 71. Prof Exp: Res scientist geol, Lamont-Doherty Geol Observ, Columbia Univ, 71-73; asst prof, Tex Christian Univ, 73-75; PROJ GEOLOGIST, E D'APPOLONIA CONSULT ENGRS, INC, 75- Honors & Awards: Antarctic Serv Medal, NSF, 73. Mem: AAAS; Geol Soc Am; Am Geophys Union; Sigma Xi. Res: Improving geological criteria for siting nuclear power plants; relationships between mid-continent tectonics and earthquakes; tectonic and strain history of the Sudbury Basin; tectonic and metamorphic history of the Adirondack lowlands. Mailing Add: E D'Appolonia Consult Engrs Inc 10 Duff Rd Pittsburgh PA 15235

BROD, JOHN SYDNEY, b Dayton, Ohio, July 5, 13; m; c 4. ORGANIC CHEMISTRY. Educ: Swarthmore Col, AB, 34; Ohio State Univ, MS, 35, PhD(chem), 37. Prof Exp: Chemist, 37-55, assoc dir prod develop dept, 55-74, MGR FED TECH GOVT RELS, RES & DEVELOP DEPT, PROCTER & GAMBLE CO, 74- Mem: Am Chem Soc; AAAS. Res: Fats and oils; food products; soap; synthetic detergents.

BRODALE, GARY EDWARD, b Vanee Creek, Wis, July 15, 34. PHYSICAL CHEMISTRY. Educ: Univ Calif, Berkeley, BS, 55, PhD(chem), 60. Prof Exp: RES CHEMIST, UNIV CALIF, BERKELEY, 60- Res: Investigations of magneto thermodynamic properties of materials at low temperatures. Mailing Add: Dept of Chem Univ of Calif Berkeley CA 94720

BRODASKY, THOMAS FRANCIS, b New London, Conn, Oct 6, 30; m 60; c 2. PHYSICAL CHEMISTRY, ORGANIC CHEMISTRY. Educ: Northeastern Univ, BS, 53, MS, 55; Rensselaer Polytech Inst, PhD(chem), 61. Prof Exp: RES SCIENTIST, UPJOHN CO, 60- Mem: Am Chem Soc; NY Acad Sci. Res: Gas, high performance liquid, paper and thin-layer chromatographic separations; infrared, ultraviolet and visible absorption spectroscopy; data retrieval; antibiotic metabolism; kinetics of biological reactions; gas liquid chromatography-mass spectroscopy. Mailing Add: Upjohn Co Res Div 301 Henrietta St Kalamazoo MI 49001

BRODBECK, JOHN J, chemical engineering, petroleum chemistry, see 12th edition

BRODD, RALPH JAMES, b Moline, Ill, Sept 8, 28; m 50; c 3. PHYSICAL CHEMISTRY. Educ: Augustana Col, BA, 50; Univ Tex, MA, 53, PhD(chem), 55. Prof Exp: Electrochemist, Nat Bur Standards, 55-61; sr scientist, Res Ctr, Ling-Temco-Vought, Inc, 61-63; res chemist, Parma Res Ctr, 63-64, group leader, 64-65, TECH RES MGR, RECHARGEABLE BATTERIES, PARMA RES CTR, UNION CARBIDE CORP, 65- Concurrent Pos: Instr, USDA Grad Sch, 56-61, Am Univ, 58-59 & Georgetown Univ, 60-61. Mem: Fel Am Inst Chemists; Am Chem Soc; Electrochem Soc; Faraday Soc. Res: Metal-gas; metal-liquid interfaces; kinetics of electrochemical reactions. Mailing Add: Union Carbide Corp Parma Res Ctr Box 6116 Cleveland OH 44101

BRODE, GEORGE LEWIS, organic chemistry, polymer chemistry, see 12th edition

BRODE, HAROLD LEONARD, b Hamilton, Wash, Apr 18, 23; m 51; c 6. THEORETICAL PHYSICS. Educ: Univ Calif, Los Angeles, BA, 47; Cornell Univ, PhD(theoret physics), 52. Prof Exp: Physicist, Rand Corp, 51-71; PHYSICIST, R&D ASSOC, 71- Res: Theoretical nuclear physics; theoretical hydrodynamics; effects of nuclear weapons. Mailing Add: R&D Assocs PO Box 9695 Marina del Rey CA 90291

BRODE, ROBERT BIGHAM, b Walla Walla, Wash, June 12, 00; m 26; c 2. PHYSICS. Educ: Whitman Col, BSc, 21; Calif Inst Technol, PhD, 24. Hon Degrees: DSc, Whitman Col, 54; LLD, Univ Calif, Berkeley, 70. Prof Exp: Assoc physicist, Bur Standards, 24; Rhodes scholar, Oxford Univ, 24-25; Nat Res fel physics, Univ Göttingen, 25-26 & Princeton Univ, 26-27; from asst prof to prof physics, 27-67, acting dir space sci lab, 64-65, dir educ abroad prog, UK, 65-67, EMER PROF PHYSICS, UNIV CALIF, BERKELEY, 67- Concurrent Pos: Vis assoc prof, Mass Inst Technol, 32; deleg, Int Union Pure & Appl Physics, London, 34 & Warsaw, 63, vpres, 54-60; Guggenheim fel, Cambridge & London, 34-35; unit supvr, Appl Physics Lab, Johns Hopkins Univ, 41-43; group leader, Los Alamos Atomic Lab, 43-46; chmn adv bd, Naval Ord Test Sta, 48-55; Fulbright award, Univ Manchester, 51-52; mem nat comt codata, Nat Res Coun, 51-57, chmn 70-; mem, Comn High Altitude Res Sta, 54-58; deleg, Int Coun Sci Unions, Oslo, 55 & Wash 58; assoc dir res, NSF, 58-59; Bd Foreign scholar, 63-66. Mem: Nat Acad Sci; fel AAAS (vpres, 49); Am Acad Arts & Sci; fel Am Phys Soc; Am Asn Physics Teachers. Res: Proximity fuse; absorption coefficients of slow electrons in gases and metal vapors; cosmic rays; mass of the mu meson. Mailing Add: 1471 Greenwood Terr Berkeley CA 94708

BRODE, WALLACE REED, chemistry, deceased

BRODE, WILLIAM EDWARD, b McMinn Co, Tenn, Dec 17, 29; m 60; c 4. HERPETOLOGY, VERTEBRATE ECOLOGY. Educ: Univ Southern Miss, BS, 52, MA, 54, PhD(biol), 69. Prof Exp: Teacher biol, Copiah-Lincoln Jr Col, 55-57, Knox County Schs, Tenn, 59-60, Davidson County Schs, 60-62 & High Sch, Miss, 62-65; prof biol, Wesleyan Col, Ga, 69-74; CHIEF CURRENT PLANNING, TENN DEPT TRANSP, 74- Mem: AAAS; Am Soc Ichthyologists & Herpetologists; Soc Study Amphibians & Reptiles. Res: Amphibian and reptile taxonomy and ecology; hematological and serological studies on amphibians; impact of proposed highways on wildlife. Mailing Add: Tenn Dept of Transp Nashville TN 37219

BRODER, IRVIN, b Toronto, Ont, June 27, 30; m 54; c 3. IMMUNOLOGY. Educ: Univ Toronto, MD, 55; FRCPS(C), 60. Prof Exp: Intern, Toronto Gen Hosp, 55-56, asst resident med, 58-59, resident physician, 59-60; sr intern, Sunnybrook Hosp, Toronto, 56-57; clin instr allergy, Med Ctr, Univ Mich, 60-62; clin teacher med, 63-66, assoc, 66-68, asst prof, 68-71, ASST PROF PHARMACOL, UNIV TORONTO, 65-, PATH, 68-, ASSOC PROF MED, 71-, DIR, GAGE RES INST, 71- Concurrent Pos: Res fel endocrinol, Dept Path, Univ Toronto, 57-58; res fel immunol, Dept Pharmacol, Univ Col, Univ London, 62-63; res scholar, Med Res Coun Can, 63-66; res assoc, Med Res Coun Can, 66-; mem, Inst Med Sci, Univ Toronto, 68- Mem: Am Acad Allergy; Can Soc Immunol; Can Acad Allergy; Can Soc Clin Invest; Can Soc Immunol. Res: Mechanism of immunologic histamine release in the guinea pig lung; role of soluble antigen-antibody complexes in human and animal diseases; epidemiology of asthma and rhinitis. Mailing Add: Dept of Med Fac of Med Univ of Toronto Toronto ON Can

BRODER, SAMUEL B, b Hotin, Bessarabia, May 7, 01; nat US; m 31; c 2. NEUROLOGY, PSYCHIATRY. Educ: Univ Pa, BA, 26; Univ Chicago, MD, 32; Univ Ill, MS, 36. Prof Exp: Asst neuropsychiat, Col Med, Univ Ill, 33-36, instr, 36-39, assoc psychiat, 39-43, asst prof, 43-52; asst prof neurol & psychiat, 52-58, ASSOC PROF NEUROL & PSYCHIAT, CHICAGO MED SCH, 58- Concurrent Pos: Resident, Ill Neuropsychiat Inst, 33-36; fel, Inst Juvenile Res, 36-37, psychiatrist, Div Criminol, 37-40; instr, Silver Cross Hosp, Joliet, 37-47. Mem: Fel Am Psychiat Asn; fel AMA. Res: Therapy in mental illness; group psychotherapy with prisoners; psychological warfare; myelitis; teratology. Mailing Add: 111 N Wabash Ave Chicago IL 60602

BRODERICK, ALAN THOMAS, b Minneapolis, Minn, Oct 28, 17; m 40; c 2. ECONOMIC GEOLOGY. Educ: Univ Minn, BS, 39; Yale Univ, MS, 48, PhD(geol), 49. Prof Exp: Geologist, Cerro de Pasco Copper Corp, 41-44; chief geologist, Tex Mining & Smelting Co, 45-46; regional geologist, Bunker Hill & Sullivan Co, 49-50 & M A Hanna Co, 50-52; chief geologist, Inland Steel Co, 52-69, mgr ore develop, 69-75; VPRES, INLAND STEEL MINING CO & VPRES, INLAND STEEL COAL

CO, 75- Mem: AAAS; Soc Econ Geologists; Geol Soc Am; Am Inst Mining, Metall & Petrol Eng. Res: Geology of, exploration for, and evaluation of ore and coal deposits. Mailing Add: Inland Steel Mining Co PO Box 1001 Virginia MN 55792

BRODERICK, GRACE NOLAN, b Niagara Falls, NY; m 59; c 1. GEOLOGY. Educ: Univ Buffalo, BA, 47; Brigham Young Univ, MA, 50; Georgetown Univ, JD, 56. Prof Exp: Asst mineral, Pa State Univ, 50-52; geologist, US Geol Surv, 52-67; PHYS SCIENTIST, US BUR MINES, 67- VANADIUM SPECIALIST, DIV FERROUS METALS, 74- Mem: Am Asn Petrol Geologists; AAAS. Res: Geologic, economic and legal studies of the mineral industries, especially vanadium and other ferroalloy additives. Mailing Add: Div of Ferrous Metals US Bur of Mines Washington DC 20241

BRODERICK, LYNNE SECHRIST, b Milwaukee, Wis, Oct 26, 41; m 65; c 1. MICROBIOLOGY, GENETICS. Educ: Ohio Wesleyan Univ, BA, 63; Univ Wis-Madison, MS, 65; Ohio State Univ, PhD(bot), 69. Prof Exp: Asst prof biol, Ohio Dominican Col, 69-73; RES FEL, UNIV DAYTON, 74- Mem: AAAS; NY Acad Sci; Am Soc Microbiol; Phycol Soc Am. Res: Cellular slime molds; zygotic development, genetics and electron microscopy of Chlamydomonas; microbial metabolism and emulsification of hydrocarbons. Mailing Add: Dept of Biol Univ of Dayton Dayton OH 45469

BRODEUR, ARMAND EDWARD, b Penacook, NH, Jan 8, 22; m 47; c 6. RADIOBIOLOGY. Educ: St Louis Univ, MD, 47, MR, 52; Am Bd Radiol, dipl, 52. Hon Degrees: LLD, St Anselm's Col, 71. Prof Exp: Intern, St Mary's Group, 47-49; asst radiologist, City Hosp, St Louis, Mo, 52; instr radiol, 52-57, sr instr, 57-60, asst prof clin radiol, 60-62, assoc prof radiol, 62-70, ASSOC PROF PEDIAT, SCH MED, ST LOUIS UNIV, 66-, PROF RADIOL, 70-, CHMN DEPT, 75-, ASSOC DEAN SCH MED, 62- Concurrent Pos: Chief radiol sect, USPHS, 52-54, consult, Div Radiol Health, 63-; spec consult, Div Spec Health Serv, Firmin Desloge Hosp, St Louis, 54-, assoc radiologist, Hosp, 57; chief radiologist, Cardinal Glennon Mem Hosp Children, St Louis, 56; consult x-ray, Cath Hosp Asn, 56- Mem: Fel Am Col Radiol; AMA; Radiol Soc NAm; Soc Nuclear Med; Soc Pediat Radiol. Res: Pediatric radiology. Mailing Add: 400 Bambury Way Huntleigh Trails St Louis MO 63131

BRODEY, ROBERT S, b Toronto, Ont, Oct 30, 27; US citizen; m 52; c 2. VETERINARY SURGERY, ONCOLOGY. Educ: Ont Vet Col, DVM, 51; Univ Pa, MSc, 59. Prof Exp: From asst instr to instr vet surg, Univ Pa, 51-55, assoc, 55-59, asst prof, 59-63; assoc prof, Univ Calif, 63; assoc prof, 63-68, PROF VET SURG, UNIV PA, 68- Mem: Am Vet Med Asn; NY Acad Sci. Res: Canine surgery; canine and comparative oncology; pulmonary Osteoarthropathy. Mailing Add: Univ of Pa Sch of Vet Med Philadelphia PA 19104

BRODHAG, ALEX EDGAR, JR, b Charleston, WVa, Aug 23, 24; m 63; c 3. ORGANIC CHEMISTRY. Educ: Oberlin Col, AB, 48; Duke Univ, PhD(org chem), 54. Prof Exp: Asst, Duke Univ, 49-51; res assoc, Off Naval Res, 52; org res chemist, Union Carbide Chem Co, 53-60; asst ed, 60-66, assoc ed, 66-70, SR ED ORG ABSTR ED DEPT, CHEM ABSTR SERV, 70- Mem: Am Chem Soc. Res: Intramolecular rearrangements; acetylene derivatives; high polymers; dyes. Mailing Add: 1060 Woodmere Rd Columbus OH 43220

BRODIE, ANGELA (HARTLEY), b Oldham, Eng, Sept 28, 34; m 64; c 2. BIOCHEMISTRY. Educ: Univ Sheffield, BSc, 56, MSc, 58; Univ Manchester, PhD(steroid biochem), 61. Prof Exp: Jr sci officer serol, Nat Blood Transfusion Serv, Eng, 56-57; asst steroid biochem, Christie Hosp, Manchester, Eng, 57-59; NIH trainee, Steroid Training Prog, 61-62, SCIENTIST, WORCESTER FOUND EXP BIOL, 62- Res: Steroid biochemistry; aldosterone, particularly methodology, biosynthesis and control of this hormone; estrogen biosynthesis in reproduction and breast cancer. Mailing Add: Worcester Found for Exp Biol 222 Maple Ave Shrewsbury MA 01545

BRODIE, ARNOLD FRANK, b Boston, Mass, Dec 31, 23; m 48; c 2. BACTERIOLOGY, BIOCHEMISTRY. Educ: Northeastern Univ, BS, 46; Boston Univ, MA, 47; Univ Pa, PhD, 52. Prof Exp: Res assoc bact & immunol & biochemist, Leonard Wood Mem Lab, Harvard Med Sch, 54-57, assoc, 57-59, asst prof, 59-63; Hastings prof microbiol, 63-69, HASTINGS PROF BIOCHEM & CHMN DEPT, SCH MED, UNIV SOUTHERN CALIF, 69- Concurrent Pos: Nat Found Infantile Paralysis fel, Biochem Res Lab, Mass Gen Hosp, 52-54. Mem: AAAS; Am Soc Microbiol; Am Soc Biol Chemists. Res: Respiratory enzymes, quinines and oxidative phosphorylation; membrane structure and active transport of amino acids; physiology and metabolism of microorganisms. Mailing Add: Dept of Biochem Univ of Southern Calif Los Angeles CA 90033

BRODIE, BERNARD BERYL, b Liverpool, Eng, Aug 7, 09; nat US; m 50. PHARMACOLOGY. Educ: McGill Univ, BSc, 31; NY Univ, PhD(chem), 35. Hon Degrees: PhD, Univ Paris, 63 & Univ Barcelona, 66; DSc, Philadelphia Col Pharm & Sci, 65, NY Med Col, 70 & Univ Louvain, 71; DrMed, Karolinska Inst, Sweden, 68 & Univ Cagliari, Sardinia. Prof Exp: Asst chem, NY Univ, 31-35, instr pharmacol, Col Med, 35-41, res assoc biochem, 3rd Med Div, 41-50, asst prof pharmacol, 43-47, assoc prof biochem, 47-50; chief lab chem pharmacol, Nat Heart Inst, 50-70; VIS PROF PHARMACOL, COL MED, PA STATE UNIV, 71- & VIS PROF PHARMACOL, HERSHEY MED CTR, PA, 75- Concurrent Pos: Lectr, Shionogi Res Labs, Japan, 62; Beyer vis prof, Med Sch, Univ Wis, 62; Sturmer Mem lectr, Philadelphia Col Pharm & Sci, 62; res lectr & vis prof, Univ Calif, 62; Koch lectr, Univ Pittsburgh, 64; mem, Comt Environ Physiol & mem, Comt Appln Biochem Studies in Evaluating Drug Toxicity, Nat Acad Sci, 65-68, mem, Comt Probs Drug Safety, 68-71, mem, Drug Res Bd, 71-; Claude Bernard prof, Univ Montreal, 69; Paul Lamson lectr, Vanderbilt Univ, 71; Rosemary Cass Mem Lectr, Univ Dundee, 71; consult, Lab Chem Pharmacol, Nat Heart & Lung Inst, 71- & NIMH, 71-; sr consult, Hoffmann-La Roche Inc, 71-; ed & co-founder, Life Sci & Pharmacol; hon adv, Int Encycl Pharmacol & Therapeut; mem, Adv Bd, Pharmacol Res Commun; mem, Coun Neuropharmacol, Int Brain Res Orgn; vis prof pharmacol, Col Med, Univ Ariz, 72-75. Honors & Awards: Distinguished Serv Award, US Dept Health, Educ & Welfare, 58; Sollmann Award, 63; Mod Med Distinguished Achievement Award, 64; Distinguished Lectr Award, AAAS, 65; Albert Lasker Award, 67; Nat Medal Sci, 68; Schmeideberg-Plakette, Ger Pharmacol Soc, 69; Oscar B Hunter Mem Award, Am Therapeut Soc, 70; Golden Plate Award, Am Acad Achievement, 70; Inter-Sci Medal, 72; Pharmaceut Sci Award, 72. Mem: Nat Acad Sci; Nat Inst Med; fel AAAS; Am Chem Soc; Col Neuropsychopharmacol (pres, 64). Res: Body water; drug metabolism and enzymes; membrane permeability; biochemical evolution; biochemistry of function; mechanism of drug action; neurochemistry; biological control systems; biochemical mechanisms of drug-induced lesions. Mailing Add: Col of Med Pa State Univ Hershey PA 17033

BRODIE, BILL BURL, b Van Buren, Ark, Apr 25, 34; m 54; c 2. NEMATOLOGY. Educ: Okla State Univ, BS, 55, MS, 58; NC State Col, PhD(nematol), 62. Prof Exp: Plant pathologist, Cotton & Cordage Fibers Res Br, 56-64, NEMATOLOGIST, NORTHEASTERN REGION, AGR RES SERV, USDA, 64- Mem: Am Phytopath Soc; Soc Nematol. Res: Biological, cultural and chemical control of nematode parasites of plants; interaction of fungi, bacteria and nematodes in the plant disease syndrome. Mailing Add: Dept of Plant Path Cornell Univ Ithaca NY 14850

BRODIE, BRUCE ORR, b Allegan, Mich, Apr 19, 24; m 47, 63; c 6. VETERINARY MEDICINE. Educ: Mich State Univ, DVM, 51; Univ Ill, MS, 58. Prof Exp: Pvt pract, 51-54; from instr to assoc prof, 54-69, PROF VET MED, UNIV ILL, URBANA-CHAMPAIGN, 69- Mem: Am Vet Med Asn; Am Asn Vet Clinicians. Res: Infectious diseases of cattle, trichomoniasis and tuberculosis. Mailing Add: 405 W Elm St Urbana IL 61801

BRODIE, DAVID ALAN, b Albany, NY, June 2, 29; m 53; c 2. PHARMACOLOGY. Educ: Philadelphia Col Pharm & Sci, BSc, 51; Ohio State Univ, MSc, 53; Univ Utah, PhD(pharmacol), 56. Prof Exp: Instr pharmacol, Sch Med, Johns Hopkins Univ, 56-57; res assoc neuropharmacol, Merck Inst Therapeut Res, 57-62, sr investr gastroenterol, 62-64, dir gastroenterol, 66-70; mgr pharmacol, William H Rorer, Inc, Pa, 70-71; assoc dir, Smith Klein & French Labs, 71-72; mgr dept, 72-73, DIR DIV PHARMACOL & MED CHEM, ABBOTT LABS, 73- Concurrent Pos: Sr res fel, Merck Inst Therapeut Res, 64-66. Mem: Am Soc Pharmacol & Exp Therapeut; Am Physiol Soc; Am Gastroenterol Soc. Res: Drug effects on gastric secretion and experimental peptic ulcer. Mailing Add: Abbott Labs North Chicago IL 60064

BRODIE, DON E, b Bracebridge, Ont, Sept 8, 29; m 56; c 3. SOLID STATE PHYSICS. Educ: McMaster Univ, BSc, 55, MSc, 56, PhD(solid state physics), 61; Ont Col Educ, cert, 57. Prof Exp: Teacher physics, Humberside Collegiate Inst, 57-58; lectr, 58-59, from asst prof to assoc prof, 61-68, PROF PHYSICS, UNIV WATERLOO, 68- Concurrent Pos: Ed, J Can Asn Physicists, 64-68; consult, Atomic Energy Comn Can, 65-67. Mem: Can Asn Physicists. Res: Physics of thin films, amorphous and crystalline; electronic and optical properties. Mailing Add: Dept of Physics Univ of Waterloo Waterloo ON Can

BRODIE, DONALD CRUM, b Carroll, Iowa, Mar 29, 08; m 34. PHARMACY. Educ: Univ Southern Calif, BS, 34, MS, 38; Purdue Univ, PhD(pharmaceut chem), 44. Prof Exp: Lab asst, Col Dent, Univ Southern Calif, 36-38; asst, Col Pharm, Purdue Univ, 38-41, instr pharmaceut chem, 41-44; assoc pharm, Sch Med & Dent, Univ Rochester, 44-45; assoc prof pharmaceut chem, Col Pharm, Univ Kans, 45-47; assoc prof pharm, 47-53, lectr, Med Sch, 48-58, dir pharmaceut serv, 58-68, prof pharm & pharmaceut chem, Div Ambulatory & Community Med, Sch Med, 67-73, assoc dean prof affairs, Sch Pharm, 69-73, EMER PROF CHEM & PHARMACEUT CHEM, SCH PHARM, UNIV CALIF, SAN FRANCISCO, 73- Concurrent Pos: Res consult, Comn Outpatient Dispensing by Hosps & Related Facilities, 65; spec assignment, dir drug-related studies, Nat Ctr Health Serv Res & Develop, Health Serv & Ment Health Admin, Dept Health, Educ & Welfare, 70- Mem: AAAS; Am Chem Soc; Am Pharmaceut Asn. Res: Toxicity of drug agents; antihemorrhagic activity of the naphthoquinones; salicylate analgesics; physiology and pharmacology of vascular smooth muscle; delivery of health care. Mailing Add: Sch of Pharm Univ of Calif San Francisco CA 94122

BRODIE, EDMUND DARRELL, JR, b Portland, Ore, June 29, 41; m 62; c 2. HERPETOLOGY, BEHAVIORAL BIOLOGY. Educ: Ore Col Educ, BS, 63; Ore State Univ, MS, 67, PhD(zool), 69. Prof Exp: Asst prof zool, Clemson Univ, 69-74; ASSOC PROF BIOL, ADELPHI UNIV, 74- Mem: AAAS; Am Soc Ichthyologists & Herpetologists; Soc Study Amphibians & Reptiles; Am Soc Zoologists. Res: Amphibian skin toxins, related behavior and coloration; ecology and systematics of amphibians and reptiles. Mailing Add: Dept of Biol Adelphi Univ Garden City NY 11530

BRODIE, HARLOW KEITH HAMMOND, b Stamford, Conn, Aug 24, 39; m 67; c 3. PSYCHIATRY. Educ: Princeton Univ, AB, 61; Columbia Univ, MD, 65. Prof Exp: Intern med, Ochsner Found Hosp, New Orleans, 65-66; asst resident psychiat, Columbia-Presby Med Ctr, 66-68; clin assoc psychiat, Lab Clin Sci, NIMH, 68-70; asst prof, Sch Med, Stanford Univ, 70-74, prog dir, Gen Clin Res Ctr, 73-74; PROF PSYCHIAT & CHMN DEPT, DUKE UNIV MED CTR & CHIEF PSYCHIAT SERV, DUKE HOSP, 74- Concurrent Pos: Consult, Palo Alto Vet Admin Hosp, Calif, 70-72; assoc ed, Am J Psychiat, 73-; mem, Nat Adv Coun on Alcohol & Alcohol Abuse, 74-; examr, Am Bd Psychiat & Neurol, 74-; consult psychiat, Educ Br, NIMH, 73-75, Durham Vet Admin Hosp, NC, 74- & Asheville Vet Admin Hosp, NC, 75-; mem, President's Biomed Panel Interdisciplinary Cluster Pharmacol, Substance Abuse & Environ Toxicol, 75- Honors & Awards: Dean Echols Award, Ochsner Found Hosp, 65; A E Bennett Clin Res Award, Soc Biol Psychiat, 70. Mem: Fel Am Psychiat Asn; Soc Biol Psychiat; fel Am Col Psychiatrists; Am Psychopath Asn; Sigma Xi. Res: Psychobiology of mood disorders in man specifically as these are related to changes in endocrine function and neurotransmitter turnover in brain. Mailing Add: 63 Beverly Dr Durham NC 27707

BRODIE, HARRY JOSEPH, b New York, NY, Apr 25, 28; m 64; c 2. BIOCHEMISTRY, CHEMISTRY. Educ: Fordham Univ, BS, 50, MS, 52; NY Univ, PhD(org chem), 58. Prof Exp: Instr chem, Hunter Col, 53-56 & City Col New York, 57-58; chemist, Photo Prod Dept, E I du Pont de Nemours & Co, 58-60; fel, Clark & Worcester Found Exp Biol, 60-62; scientist, 62-67, SR SCIENTIST, WORCESTER FOUND EXP BIOL, 67- Mem: AAAS; Am Chem Soc; Endocrine Soc. Res: Mechanism of enzyme catalyzed reactions, especially estrogen biosynthesis and oxidation and reduction of steroids; role of steroid hormones in mammalian reproduction; steroid synthesis, including labeling. Mailing Add: 222 Maple Ave Shrewsbury MA 01545

BRODIE, JONATHAN, biological chemistry, see 12th edition

BRODIE, LAIRD CHARLES b Portland, Ore, Aug 30, 22; m 48; c 3. PHYSICS. Educ: Reed Col, BA, 44; Univ Chicago, MS, 49; Northwestern Univ, PhD(physics), 54. Prof Exp: Res engr, Lab Div, Radio Corp Am, 53-54; Ford Found intern, Reed Col, 54-55; from instr to assoc prof, 55-67, PROF PHYSICS, PORTLAND STATE UNIV, 67- Mem: Am Phys Soc. Res: Solid state physics. Mailing Add: Dept of Physics Portland State Univ Portland OR 97207

BRODISH, ALVIN, b Brooklyn, NY, June 11, 25; m 57; c 2. PHYSIOLOGY. Educ: Drake Univ, BA, 47; Univ Iowa, MS, 50; Yale Univ, PhD(physiol), 55. Prof Exp: Lab instr biol, Drake Univ, 47-48; asst physiol, Univ Iowa, 49-50; asst psychiat, Yale Univ, 51-53, asst physiol, 53-54, NSF fel, 55-57, from instr to assoc prof, 57-69; PROF PHYSIOL, COL MED, UNIV CINCINNATI, 69- Concurrent Pos: Investr, Howard Hughes Med Inst, 57-60. Mem: AAAS; Endocrine Soc; Am Physiol Soc. Res: Neuroendocrine systems; regulation of anterior pituitary secretions; nerve-muscle regeneration. Mailing Add: Dept of Physiol Univ Cincinnati Col of Med Cincinnati OH 45219

BRODKEY, JERALD STEVEN, b Omaha, Nebr, Jan 20, 34; m 62; c 2. NEUROSURGERY, BIOMEDICAL ENGINEERING. Educ: Harvard Univ, AB, 55; Univ Nebr, MS, 59, MD, 60; Am Bd Neurol Surg, dipl, 71. Prof Exp: Intern surg,

Barnes Hosp, St Louis, Mo, 60-61, asst resident, 61-62; asst resident neurosurg, Mass Gen Hosp, Boston, 63-67, clin asst, 67; dir sci comput sect & asst chmn dept bioeng, Presby-St Luke's Hosp, 67-69; asst prof bioeng, Univ Ill, Chicago, 65-69, clin instr neurol & neurosurg, Col Med, 67-69; asst prof, 69-73, PROF NEUROSURG & BIOMED ENG, CASE WESTERN RESERVE UNIV, 73-; CHIEF NEUROSURG, CLEVELAND VET ADMIN HOSP, 69- Concurrent Pos: Clin & res fel neurosurg, Mass Gen Hosp, Boston, 62-63; USPHS spec fel, 65-66; asst attend bioeng, Presby-St Luke's Hosp, 65-67, asst attend neurosurg, 67-69. Mem: AAAS; Inst Elec & Electronics Engrs; Soc Cybernet; Asn Comput Mach; NY Acad Sci. Res: Embedding of neurophysiological experiments in a neurological control system background; clinical neurosurgery, particularly sterotactic surgery. Mailing Add: Dept of Neurosurg Case Western Reserve Univ Cleveland OH 44106

BRODKORB, PIERCE, b Chicago, Ill, Sept 29, 08; m 31; c 1. ZOOLOGY. Educ: Univ Ill, AB, 33; Univ Mich, PhD(zool), 36. Prof Exp: Asst ornith, Field Mus, 30 & Cleveland Mus, Ohio, 31-32; asst mus zool, Univ Mich, 33-36, asst cur birds, 36-46; from asst prof to assoc prof, 46-55, PROF ZOOL, UNIV FLA, 55- Concurrent Pos: Mem exped, Idaho, 31-32, Black Hills, 35, Mex, 37, 39, 41, 53 & Bermuda, 60; consult, Fla Geol Surv, 57- & Govt of Bermuda, 60; mem Int Ornith Cong, 62- & Int Cong Zool, 63- Mem: Cooper Ornith Soc; fel Am Ornithologists Union; Wilson Ornith Soc; Paleont Soc; Soc Study Evolution. Res: Ornithology; zoogeography; avian paleontology and evolution. Mailing Add: Dept of Zool Univ of Fla Gainesville FL 32601

BRODMAN, ESTELLE, b New York, NY, June 1, 14. HISTORY OF MEDICINE, INFORMATION SCIENCE. Educ: Cornell Univ, AB, 35; Columbia Univ, BS, 36, MS, 43, PhD(med hist), 53. Hon Degrees: DSc, Univ Ill, 74. Prof Exp: Mem fac, Sch Med, Columbia Univ, 37-49; asst librn ref serv, Nat Libr Med, Wash, 49-61; assoc prof, 61-64, PROF MED HIST, SCH MED, WASHINGTON UNIV, 64-, LIBRN, 61- Concurrent Pos: Ed Bull, Med Libr Asn, 47-57; consult, NIH Libr, 50 & Am Hosp Asn, 59-63; vis prof, Keio Univ, Japan, 62; mem, Comn Libr Educ, Am Libr Asn, 62-; consult, Nat Clearinghouse Ment Health Info, NIMH, 64-68, Biomed Commun Study Sect, 71-, chmn, 73-; consult, USAID/Carolina Pop Ctr, 75; mem, President's Nat Adv Comn on Libr, 66-68; UN tech assistance expert, Cent Family Planning Inst, New Delhi, 67-68, Southeast Asia Regional Off, WHO, 70. Mem: Med Libr Asn (pres, 64-65); Spec Libr Asn; Am Libr Asn; Am Asn Hist Med. Res: Medical librarianship. Mailing Add: Sch of Med Washington Univ St Louis MO 63110

BRODMAN, KEEVE, b New York, NY, Aug 5, 06; m 28; c 1. CLINICAL MEDICINE. Educ: City Col New York, BS, 27; Cornell Univ, MD, 31. Prof Exp: Res asst, Biol Labs, Cold Spring Harbor, 27-30; instr physiol, 30-31, ASST PROF CLIN MED, MED COL, CORNELL UNIV, 34-; RES DIR, MEDATA FOUND, 66- Concurrent Pos: Res assoc, Off Strategic Serv, 44-46, Vet Admin, 47-51 & US Dept Army, 52-55; res dir, Med Data Corp, 62-66. Mem: Fel AAAS; fel NY Acad Sci; Asn Comput Mach; Am Psychosom Soc; AMA. Res: Psychiatry; medicine; automation in medicine and psychiatry. Mailing Add: 68 E 86th St New York NY 10028

BRODMANN, JOHN MILTON, b Savannah, Ga, Aug 20, 33. ORGANIC CHEMISTRY. Educ: Lynchburg Col, BS, 55; Emory Univ, MS, 59, PhD(natural prod synthesis), 67. Prof Exp: From instr to assoc prof, 57-67, PROF CHEM, CULVER-STOCKTON COL, 67-, CHMN DIV NATURAL SCI, 65- Mem: Am Chem Soc. Res: Grignard addition; reaction mechanisms. Mailing Add: Dept of Chem Culver-Stockton Col Canton MO 63435

BRODOWAY, NICOLAS, b Melfort, Sask, Dec 9, 22; US citizen; m 49; c 6. ORGANIC CHEMISTRY. Educ: Univ BC, BSc, 49; Univ Minn, PhD(org chem), 53. Prof Exp: RES CHEMIST, ELASTOMER CHEM DEPT, E I DU PONT DE NEMOURS & CO, INC, 53- Mem: Am Chem Soc; Sigma Xi. Res: Polymer synthesis, evaluation and process development. Mailing Add: 189 Honeywell Dr Claymont DE 19703

BRODRICK, HAROLD JAMES, JR, b Newport News, Va, July 18, 31; m 53; c 1. METEOROLOGY. Educ: Kans State Univ, BS, 53; Fla State Univ, MS, 58. Prof Exp: Res meteorologist, US Army Chem Corps, 58-60; RES METEOROLOGIST, NAT ENVIRON SATELLITE SERV, NAT OCEANIC & ATMOSPHERIC ADMIN, DEPT COM, 60- Mem: Meteorol Soc. Res: Development and evolution of satellite observed cloud patterns and their interpretations in terms of kinematic and dynamic parameters; dynamics of synoptic-scale atmospheric disturbances; atmospheric diffusion. Mailing Add: Nat Environ Satellite Serv Nat Oceanoc & Atmospheric Admin Washington DC 20233

BRODSKY, ALLEN, b Baltimore, Md, Nov 5, 28; m 51; c 3. HEALTH PHYSICS, BIOSTATISTICS. Educ: Johns Hopkins Univ, BE, 49, MA, 60; Univ Pittsburgh, ScD(biostatist, radiation health), 66; Am Bd Health Physics, dipl, 60; Am Bd Indust Hyg, dipl, 66; Am Bd Radiol, dipl, 75. Prof Exp: Head, Health Physics Unit, Naval Res Lab, Washington, DC, 50-52, physicist, Opers Ivy & Castle, Eniwetok & Bikini, 52-54; pres, Health Physics Servs, Inc, Baltimore, Md, 54-56; radiol defense officer, Fed Civil Defense Admin, Olney, Md, 56-57; health physicist, Health Protection Br, Div Biol & Med, US AEC, Washington, DC, 57-59, radiation physicist, Div Licensing & Regulation, Washington, DC, 59-61; res assoc health physics & epidemiol, Dept Occup Health, Grad Sch Pub Health, Univ Pittsburgh, 61-66, assoc prof, Dept Occup Health & Dept Radiation Health, 66-71; radiation physicist, Div Radiation Ther, Dept Radiol, Mercy Hosp, Pittsburgh, Pa, 71-75; SR HEALTH PHYSICIST, OCCUP HEALTH STAND BR, OFF STAND DEVELOP, US NUCLEAR REGULATORY COMN, WASHINGTON, DC, 75- Concurrent Pos: Mem, AEC Radiation Sci Fel Bd, Oak Ridge Assoc Univs, 68-71; adj res prof, Sch Pharm, Duquesne Univ, 71-; ed, Handbks Radiation Measurement & Protection, CRC Press, Inc, Cleveland, Ohio, 75- Honors & Awards: Distinguished Serv Award, Western Pa Chap, Health Physics Soc, 74. Mem: Am Asn Physicists in Med; Am Acad Indust Hyg; Health Physics Soc; Am Nuclear Soc; Soc Nuclear Med. Res: Radiation dose measurement and interpretation; radiation hazard evaluation and standards for radiation protection; mathematical models of carcinogenesis and risk estimation from environmental agents; epidemiologic studies of environmental exposures and effects. Mailing Add: Off of Stand Develop US Nuclear Regulatory Comn Washington DC 20555

BRODSKY, CARROLL M, b Lowell, Mass, Dec 23, 22; c 3. PSYCHIATRY, ANTHROPOLOGY. Educ: Cath Univ Am, AB, 49, MA, 50, PhD(anthrop), 54; Univ Calif, MD, 56. Prof Exp: Lectr anthrop, Cath Univ Am, 52; res assoc, Human Resources Res Off, 53; from instr to assoc prof, 60-70, PROF PSYCHIAT, SCH MED, UNIV CALIF, SAN FRANCISCO, 70- Concurrent Pos: NIMH grant, Langley-Porter Neuropsychiat Inst, 57-60 MEm: Acad Psychosom Med; Am Anthrop Asn; Am Col Psychiat. Res: Studies of human disability with emphasis on social and psychiatric factors delaying recovery from illness; problems of work. Mailing Add: Dept of Psychiat Univ of Calif Sch of Med San Francisco CA 94143

BRODSKY, MARC HERBERT, b Philadelphia, Pa, Aug 9, 38; m 66; c 1. SOLID STATE PHYSICS. Educ: Univ Pa, AB, 60, MS, 61, PhD(physics), 65. Prof Exp: Res assoc physics, Univ Pa, 65; res physicist, US Naval Ord Lab, 65-66; physicist, Night Vision Lab, US Army, 66-68; MEM TECH STAFF, T J WATSON RES CTR, INT BUS MACH CORP, 68- Concurrent Pos: Adj assoc prof, Columbia Univ, 72-74; exchange prof, Univ Paris VI, 74-75. Mem: Fedn Am Scientists; AAAS; Am Phys Soc. Res: Lattice vibrations; infrared spectroscopy; mixed crystals; semiconductor surfaces, amorphous semiconductors. Mailing Add: IBM Corp Watson Res Ctr PO Box 218 Yorktown Heights NY 10598

BRODSKY, MERWYN BERKLEY, b Chicago, Ill, Mar 4, 30; m 50; c 2. SOLID STATE PHYSICS. Educ: Roosevelt Univ, BS, 49; Ill Inst Technol, MS, 51, PhD(phys chem), 55. Prof Exp: Asst, Ill Inst Technol, 50-54; assoc chemist, Brookhaven Nat Lab, 54-58; assoc chemist, 58-66, group leader, 66-74, SR SCIENTIST, ARGONNE NAT LAB, 74- Concurrent Pos: Sr vis res fel, Imperial Col, Univ London, 71-73. Mem: Am Phys Soc; Am Inst Mining, Petrol & Metall Eng. Res: Electron transport and magnetism of actinides; fused salt electrorefining; physical chemistry of liquid metal solutions; high temperature thermodynamics; liquid metal reactor systems. Mailing Add: Mat Sci Div Argonne Nat Lab 9700 S Cass Ave Argonne IL 60439

BRODSKY, PHILIP HYMAN, b Philadelphia, Pa, July 7, 42; m 64; c 2. POLYMER CHEMISTRY. Educ: Cornell Univ, BChE, 65, PhD(chem eng), 69. Prof Exp: Sr res engr, 68-73, RES SPECIALIST, MONSANTO CO, 73- Mem: Am Chem Soc. Res: Compatibility in polymer blends; manufacturing processes and chemistry of condensation and addition polymers. Mailing Add: 190 Grochmal Ave Indian Orchard MA 01151

BRODSKY, STANLEY JEROME, b St Paul, Minn, Jan 9, 40; m 62; c 2. HIGH ENERGY PHYSICS, THEORETICAL PHYSICS. Educ: Univ Minn, BPhys, 61, PhD(physics), 64. Prof Exp: Res assoc theoret physics, Columbia Univ, 64-66; res assoc, 66-68, MEM RES STAFF, STANFORD LINEAR ACCELERATOR CTR, STANFORD UNIV, 68-, ASSOC PROF, 75- Concurrent Pos: Vis assoc prof, Dept Physics, Cornell Univ, 69; mem comt fundamental constants, Nat Res Coun-Nat Acad Sci, 74- & prog comt, Wilson Lab, Cornell Univ, 75- Mem: Fel Am Phys Soc. Res: Quantum electrodynamics; muonic x-rays; weak interactions; Zeeman structure; Lamb shift; lepton magnetic moments; elementary particles; electromagnetic interactions; colliding beam physics; large transverse momentum reactions; nuclear processes; photon-photon collisions; quark-model. Mailing Add: Stanford Linear Accelerator Ctr Stanford Univ Stanford CA 94305

BRODSKY, WILLIAM AARON, b Philadelphia, Pa, Jan 8, 18; m 50; c 2. PHYSIOLOGY. Educ: Temple Univ, BS, 38, MD, 41. Prof Exp: Intern, Philadelphia Gen Hosp, 41-43; instr pediat, Univ Pa, 43-44; instr, Univ Cincinnati, 48-49; from asst prof to assoc prof pediat, Univ Louisville, 51-60, prof exp med, 60-68; PROF BIOPHYS, MT SINAI SCH MED, 68- Concurrent Pos: Res fels pediat, Univ Cincinnati, 46-48 & 49-51; USPHS res fel, 62-; estab investr, Am Heart Asn, 55-60; physiol consult, Vet Admin Hosp, Louisville. Mem: Soc Pediat Res; Am Physiol Soc; Soc Exp Biol & Med; Am Soc Clin Invest; Biophys Soc (secy & exec coun, 67-72). Res: Osmotic properties of tissue; ion transport mechanisms; acid-base equilibria; renal transport mechanisms. Mailing Add: Mt Sinai Sch of Med 100th St & Fifth Ave New York NY 10029

BRODY, AARON LEO, b Boston, Mass, Aug 23, 30; m 53; c 3. FOOD TECHNOLOGY. Educ: Mass Inst Technol, BS, 51, PhD(food technol), 57; Northeastern Univ, MBA, 70. Prof Exp: Food technologist, Birdseye Fisheries Labs, Gen Foods Corp, 51-52 & Raytheon Mfg Co, 54-55; res food technologist, Whirlpool Corp, Mich, 57-61; mgr pkg & prod develop, M&M's Candies Div, Mars, Inc, 61-67; sr staff mem, Arthur D Little Inc, 67-73; NEW VENTURES MGR, MEAD PACKAGING, ATLANTA, 73- Concurrent Pos: Mem, US Navy Food Serv Adv Comt; sci lect, Inst Food Technologists, 72-75; consult, World Bank, 72-73. Honors & Awards: Willis H Carrier Award, Am Soc Heating, Refrig & Air Conditioning Eng, 60; Inst Food Technologists Indust Achievement Award, 64; Pkg Inst Leadership Award, 65, Excellence Award, 72. Mem: AAAS; Inst Food Technologists; Am Soc Heating, Refrig & Air Conditioning Eng; Pkg Inst US (vpres, 73-); NY Acad Sci. Res: Objective measurements of physical properties of foods; microwave heating of foods; refrigeration and freezing of food products; controlled atmosphere storage; packaging of foods. Mailing Add:

BRODY, ALFRED WALTER, b New York, NY, Feb 20, 20; m 43; c 4. PHYSIOLOGY, PHARMACOLOGY. Educ: Columbia Univ, BA, 40, MA, 41; Long Island Col Med, MD, 43; Univ Pa, DSc, 53. Prof Exp: Instr physiol & pharmacol, Grad Sch Med, Univ Pa, 51-54; Nebr State Heart Asn chair cardiovasc res, 58-67, from asst prof to assoc prof, 54-63, PROF PHYSIOL & PHARMACOL, SCH MED, CREIGHTON UNIV, 63-, PROF, MED, 60-, DIR PULMONARY LAB, 54-, CHIEF CHEST DIV, 62- Mem: AAAS; Am Physiol Soc; Am Thoracic Soc; NY Acad Sci. Res: Mechanics of respiration; respiratory physiology; lesser circulation; medicine. Mailing Add: Sch of Med Creighton Univ Omaha NE 68131

BRODY, BERNARD B, b New York, NY, June 24, 22; m 54; c 2. MEDICINE, CHEMISTRY. Educ: Univ Wis-Madison, BS, 43; Univ Rochester, MD, 51. Prof Exp: Res assoc chem, Univ Chicago, 43-45; res assoc, Monsanto Chem Corp, Ohio, 45-47; intern med, Strong Mem Hosp, Rochester, NY, 51-52, asst resident, 52-53; chief resident, Genesee Hosp, 55-56; from instr to asst prof, 56-67, ASSOC PROF MED, UNIV ROCHESTER, 67-; DIR CLIN LABS, GENESEE HOSP, 67-, DIR MED AFFAIRS, 74- Concurrent Pos: Pvt practr, 56-67. Mem: AAAS; AMA; Am Soc Internal Med; Am Asn Clin Chem; Am Col Physicians. Mailing Add: Genesee Hosp Clin Labs 224 Alexander St Rochester NY 14607

BRODY, DANIEL ANTHONY, cardiovascular diseases, deceased

BRODY, EUGENE B, b Columbia, Mo, June 17, 21; m 44; c 3. PSYCHIATRY, PSYCHOANALYSIS. Educ: Univ Mo, AB, 41, MA, 41; Harvard Med Sch, MD, 44; Am Bd Psychiat & Neurol, dipl, 50. Prof Exp: From intern to asst resident psychiat, Yale Univ, 44-46, chief resident, 48-49, from instr to assoc prof, 49-57; PROF PSYCHIAT, UNIV MD, BALTIMORE CITY, 57-, CHMN DEPT, 59-, DIR, INSTR PSYCHIAT & HUMAN BEHAV, 59- Concurrent Pos: Consult, WHO, 65-; dir, Interam Ment Health Studies Prog, Am Psychiat Asn, 66-68; ed-in-chief, J Nerv & Ment Dis, 67-; vis prof, Univ Brazil, 68 & Univ West Indies, 72. Mem: Fel Am Psychiat Asn. Res: Social psychiatry. Mailing Add: Dept Psychiat Univ Md Sch Med Inst of Psychiat & Human Behav Baltimore MD 21201

BRODY, FREDERICK, organic chemistry, see 12th edition

BRODY, GARRY SIDNEY, b Edmonton, Alta, Sept 21, 32; US citizen; m 57; c 3. PLASTIC SURGERY. Educ: Univ Alta, MD, 56; McGill Univ, MS, 59. Prof Exp: CHIEF PLASTIC SURG, SCH MED, UNIV SOUTHERN CALIF-RANCHO LOS AMIGOS HOSP, 69- Concurrent Pos: Abstract ed, J Plastic & Reconstruct Surg, 74-; chmn, Plastic Surg Res Coun, 76. Mem: Am Soc Plastic & Reconstruct Surg; Asn Surg of the Hand; Med Eng Soc. Res: Rheology of connective tissue and scar;

relationship of breast cancer and prosthetic insects; multiple clinical plastic surgery problems. Mailing Add: Suite 504 11411 Brookshire Ave Downey CA 90241

BRODY, GERALD, b Elizabeth, NJ, Aug 3, 26; m 48; c 3. RESEARCH ADMINISTRATION, CHEMOTHERAPY. Educ: NY Univ, AB, 48; Syracuse Univ, MA, 50; Mich State Col, PhD(parasitol), 53. Prof Exp: Asst, Syracuse Univ, 48-50; res asst, Sch Med, NY Univ, 50; teaching asst, Univ Ill, 50-51; res parasitologist, Moorman Mfg Co, 53-64; dir animal health & toxicol, Stauffer Chem Co, 64-67; sr parasitologist, Stanford Res Inst, 67-69, dir dept infectious dis, parasitol & toxicol, 69-72; dir res & develop, 72-75, VPRES, COOPER LABS, INC, 75- Mem: AAAS; Am Soc Parasitologists; Am Chem Soc. Res: Cardiovascular and bronchopulmonary pharmacology; oral health; chemotherapy of helminths and viruses. Mailing Add: Res & Develop Div Cooper Labs 110 E Hanover Ave Cedar Knolls NJ 07927

BRODY, HAROLD, b Cleveland, Ohio, May 15, 23; m 51; c 2. ANATOMY. Educ: Western Reserve Univ, BS, 47; Univ Minn, PhD(anat), 53; State Univ NY Buffalo, MD, 61. Prof Exp: Teaching asst anat, Univ Minn, 47-49, instr, 49-50; asst prof, Univ NDak, 50-54; from asst prof to assoc prof, 54-63, PROF ANAT, SCH MED, STATE UNIV NY BUFFALO, 63-, CHMN DEPT, 71- Fulbright sr res scholar, 63. Mem: AAAS; Am Asn Anatomists; Am Geront Soc (pres, 74-75); Int Asn Gerontol; Am Geriatric Soc. Res: Neuroanatomy; neuropathology; age changes in the human cerebral cortex; central nervous system changes due to increased body temperature. Mailing Add: Dept of Anat Sci State Univ of NY Sch of Med Buffalo NY 14214

BRODY, HOWARD, b Newark, NJ, July 11, 32; m 54; c 3. PHYSICS. Educ: Mass Inst Technol, SB, 54; Calif Inst Technol, MS, 56, PhD(physics), 59. Prof Exp: NSF fel, 59; from instr to asst prof, 59-64, ASSOC PROF PHYSICS, UNIV PA, 64- Mem: Am Phys Soc; Am Asn Physics Teachers. Res: Study of elementary particles and their interactions at high energy. Mailing Add: Dept of Physics Univ of Pa Philadelphia PA 19104

BRODY, JACOB A, b Brooklyn, NY, May 5, 31; m 69; c 1. EPIDEMIOLOGY, VIROLOGY. Educ: Williams Col, BA, 52; State Univ NY Downstate Med Ctr, MD, 56. Prof Exp: Intern, Roosevelt Hosp, NY, 56-57; mem staff surveillance of arthropod-borne dis, Surveillance Sect, Epidemiol Br, Commun Dis Ctr, USPHS, 57-59, chief poliomyelitis surveillance unit, 58-59, med officer, Virus Sect, Mid Am Res Unit, Nat Inst Allergy & Infectious Dis, CZ, Panama, 59-61, mem staff, Lab Trop Med, Md, 61-62, chief epidemiol sect, Arctic Health Res Ctr, Alaska, 62-65, CHIEF EPIDEMIOL BR, COLLAB & FIELD RES, NAT INST NEUROL DIS & STROKE, 65- Concurrent Pos: Exchange scientist to Inst Poliomyelitis & Virus Encephalitis, Moscow, 62; vpres, Muscular Dystrophy Asn Am, 68-, mem corp, 69-; mem comn geog neurol, World Fedn Neurol, 68-, mem res comt, 69-; assoc epidemiol, Sch Hyg & Pub Health, Johns Hopkins Univ, 70- Mem: Am Epidemiol Soc; Am Asn Immunol; Int Epidemiol Asn; fel Am Pub Health Asn; Soc Epidemiol Res. Res: Geographic influences on disease; arbo viruses; poliomyelitis; infectious disease of childhood; neurological diseases; possible etiological roles of autoimmunity and virus. Mailing Add: Epidemiol Br Collab & Field Res Nat Inst Neurol Dis & Stroke Bethesda MD 20014

BRODY, JEROME IRA, b New York, NY, Jan 24, 28; m 59; c 3. INTERNAL MEDICINE, HEMATOLOGY. Educ: NY Univ, AB, 47, AM, 48; Jefferson Med Col, MD, 52; Am Bd Internal Med, dipl. Prof Exp: Intern, Philadelphia Gen Hosp, 52-53, resident path, 53-54, resident cardiol, 54-55; resident internal med, Grad Hosp, Univ Pa, 57-58; chief hemat sect, Cet Admin Hosp, Coral Gables, Fla, 60-62; from asst prof to assoc prof internal med, Sch Med, Univ Pa, 62-75, dir hemat, Grad Hosp, 62-75; PROF MED, MED COL OF PA, 75-; CHIEF MED SERV, MED COL PA DIV, VET ADMIN HOSP, PHILADELPHIA, 75- Concurrent Pos: USPHS res fel hemat, Sch Med, Yale Univ, 58-60; asst instr path, Sch Med, Univ Pa, 53-54, asst instr internal med, 54-55, asst physician, Grad Hosp, 62-; asst attend physician, Grace-New Haven Community Hosp, Conn, 58-60 & Jackson Mem Hosp, Miami, Fla, 60-62; asst prof, Sch Med, Univ Miami, 60-62; consult, Variety Children's Hosp, Miami, 61-62; prin investr res grant, USPHS, 62-, prog dir training grant, 64-; consult, Walson Army Hosp & Naval Hosp, 63- Mem: AAAS; Am Soc Clin Invest; Am Fedn Clin Res; Am Soc Hemat; Am Soc Exp Path. Res: Immunologic alterations in leukemia and related disorders. Mailing Add: Med Col of Pa Philadelphia PA 19129

BRODY, JEROME SAUL, b Chicago, Ill, Dec 6, 34; m 55; c 3. PULMONARY DISEASES. Educ: Univ Ill, BS, 55, MD, 59. Prof Exp: NIH fel, Univ Pa, 65-67, from asst prof to assoc prof med & physiol, 67-73; CHIEF PULMONARY SECT, BOSTON CITY HOSP, 73-; ASSOC PROF MED, SCH MED, BOSTON UNIV, 73- Mem: Fel Am Col Physicians; Am Thoracic Soc; Am Fedn Clin Res; Am Physiol Soc; Am Soc Clin Invest. Res: Development, compensatory growth and repair of the mammalian lung viewed from morphologic, physiologic and biochemical points of view. Mailing Add: Pulmonary Sect Boston City Hosp 818 Harrison Ave Boston MA 02118

BRODY, MARCIA, b New York, NY, Dec 3, 29; m 50; c 3. PHOTOBIOLOGY, BIOCHEMISTRY. Educ: Hunter Col, AB, 51; Rutgers Univ, MS, 53; Univ Ill, PhD(biophys), 58. Prof Exp: Res asst photosynthesis proj, Univ Ill, 53-58, res assoc, 58-59; res assoc chem, Brandeis Univ, 59-61; from asst prof to assoc prof biol sci, 61-70, chmn dept, 63-69, PROF BIOL SCI, HUNTER COL, 70- Concurrent Pos: City Univ New York grant fac award, 61-62; NSF res grant, 62-65. Mem: Biophys Soc. Res: States of chlorophyll in vivo; light reactions in photosynthesis; low temperature fluorescence spectroscopy; physico-chemical properties of chromoproteins; role of accessory pigments in photosynthesis; organelle structure, enzymology and development; efficiencies of transfer of light energy; phototropism. Mailing Add: Dept of Biol Sci Hunter Col New York NY 10021

BRODY, MICHAEL J, b New York, NY, Aug 16, 34; m 56; c 2. PHARMACOLOGY. Educ: Columbia Univ, BS, 56; Univ Mich, PhD(pharmacol), 61. Prof Exp: From instr to assoc prof, 61-69, PROF PHARMACOL, UNIV IOWA, 69-, ASSOC DIR, CARDIOVASC CTR, 74- Concurrent Pos: USPHS spec fel, Univ Lund, 71; consult, Merck Inst Therapeut Res; mem, Pharmacol Test Comt, Nat Bd Med Examrs; fel, Am Heart Asn Coun Circulation, mem, Med Adv Bd, Coun High Blood Pressure & Exec Comt Basic Sci Coun. Mem: Am Soc Exp Biol & Med; Am Fedn Clin Res; Am Soc Pharmacol & Exp Therapeut; Am Heart Asn; Am Physiol Soc. Res: Cardiovascular and autonomic nervous system physiology, pathophysiology and pharmacology. Mailing Add: Dept of Pharmcol Univ of Iowa Iowa City IA 52242

BRODY, SELMA BLAZER, b Brooklyn, NY, May 2, 14; m 35. PHYSICS. Educ: NY Univ, AB, 34; Univ Va, MA, 35; Bryn Mawr Col, PhD(physics), 42. Prof Exp: Instr physics, Hunter Col, 41-42; res physicist, Camp Coles Sig Lab, Ft Monmouth, NJ, 42-44 & Bliley Elec Co, Pa, 44-46; instr physics & math, Sarah Lawrence Col, 46-47; instr physics, Brooklyn Col, 47-53; asst prof, Univ Conn, 53-55; assoc prof, Bard Col, 55-57; PROF PHYSICS, ST JOHN'S UNIV, NY, 57- Concurrent Pos: Res assoc, Molecular Beams Lab, Columbia Univ, 46-47; Huff Mem res fel, Bryn Mawr Col, 64-65. Mem: Am Phys Soc; Am Crystallog Asn. Res: X-ray analysis of crystal structure; surface structure; molecular beam determination of molecular moments; piezoelectricity. Mailing Add: Dept of Physics St John's Univ Jamaica NY 11439

BRODY, SEYMOUR STEVEN, b New York, NY, Nov 29, 27; m 49; c 2. BIOPHYSICS. Educ: City Col New York, BS, 51; NY Univ, MS, 53; Univ Ill, PhD(biophys), 56. Prof Exp: Res asst photosynthesis, Univ Ill, 53-56, res assoc, 56-59; sect chief photobiol, US Dept Air Force, 59-60; mgr biophys, Watson Lab, Int Bus Mach Corp, 60-65; assoc prof, 66-68, PROF BIOL, NY UNIV, 68- Concurrent Pos: Adj assoc prof, Dept Biol, Washington Square Col, NY Univ, 64-65; USPHS res career award, 66-70. Mem: AAAS; Biophys Soc. Res: Photobiology; fluorescence and absorption spectroscopy of molecules; heterogeneous photocatalysis; nonosecond phenomena in living systems and in molecules; monomolecular film models of membranes in vision, photosynthesis and respiration. Mailing Add: Dept of Biol NY Univ Washington Square New York NY 10003

BRODY, STUART, b Newark, NJ, June 25, 37; m 65; c 2. BIOCHEMICAL GENETICS. Educ: Mass Inst Technol, SB, 59; Stanford Univ, PhD(biol), 64. Prof Exp: Guest investr biol, Rockefeller Univ, 64-66, res assoc, 66, asst prof, 66-67; asst prof, 67-73, ASSOC PROF BIOL, UNIV CALIF, SAN DIEGO, 73- Concurrent Pos: Am Cancer Soc fel, 64-66; prin investr NSF grant, 66-75. Mem: AAAS; Am Soc Biol Chemists; Am Soc Microbiol; Genetics Soc Am. Res: Gene action; biochemical genetics; biochemistry and morphology of microorganisms. Mailing Add: Dept of Biol Univ of Calif at San Diego La Jolla CA 92037

BRODY, THEODORE MEYER, b Newark, NJ, May 10, 20; m 47; c 4. PHARMACOLOGY. Educ: Rutgers Univ, BS, 43; Univ Ill, MS, 49, PhD(pharmacol), 52. Prof Exp: From instr to prof pharmacol, Med Sch, Univ Mich, 52-66; PROF PHARMACOL & CHMN DEPT, MICH STATE UNIV, 66- Concurrent Pos: Mem, NIH Fel Rev Panel Pharmacol & Endocrinol, 64-68 & Nat Acad Sci-Nat Res Coun Pesticide Safety Adv Comt, 64-66; mem, Pharmacol & Exp Therapeut Study Sect, NIH, 69-73; mem, Bd Dirs, Fedn Socs Exp Biol, 73-76; US rep, Int Union Pharmacol, 73-; NSF distinguished scholar lectr, Univ Hawaii, 74; consult, Random House Dict Eng Lang; mem, Int Study Group Res Cardiac Metab; mem, Pharmacol Adv Bd, Pharmaceut Mfrs Asn Found; mem, Drug Abuse Rev Comt, Nat Inst Drug Abuse, 75- Mem: Am Soc Pharmacol & Exp Therapeut; Soc Neurosci; Soc Toxicol; Int Soc Biochem Pharamcol. Res: Mode of action of drugs and poisons; cardiac glycosides; ion transport; narcotics. Mailing Add: Dept of Pharmacol Life Sci Bldg Mich State Univ East Lansing MI 48824

BRODY, THOMAS PETER, b Budapest, Hungary, Apr 18, 20; m 52; c 4. SOLID STATE PHYSICS. Educ: Univ London, BS, 50, PhD(math physics), 53. Prof Exp: Lectr physics, Univ London, 53-55; res physicist, Brit Dielec Res Co, 55-57; lectr physics, Univ London, 57-59; sr physicist, 59-62, sect mgr dielec devices, 62-68, dept mgr, 68-72, DEPT MGR THIN FILM DEVICES, WESTINGHOUSE RES LABS, 72- Mem: Am Phys Soc; Inst Elec & Electronics Engrs; Soc Info Display. Res: Optoelectronics; pattern recognition. Mailing Add: Westinghouse Elec Corp Res Labs Churchill Borough Pittsburgh PA 15235

BRODZINSKI, RONALD LEE, b South Bend, Ind, Feb 14, 41; m 67; c 4. NUCLEAR SCIENCE, PLANETARY ATMOSPHERES. Educ: Purdue Univ, BS, 63, PhD(nuclear chem), 68. Prof Exp: SR RES SCIENTIST, PAC NORTHWEST LAB, BATTELLE MEM INST, 68- Concurrent Pos: Prin investr, NASA, 72- Res: High-energy charged particle reactions; cosmic-radiation activation of astronauts and spacecraft; nuclear waste management; environmental research lunar and space sciences; controlled thermonuclear reactor materials research. Mailing Add: Pac Northwest Labs PO Box 999 Richland WA 99352

BROECKER, WALLACE, b Chicago, Ill, Nov 29, 31; m 52; c 6. GEOCHEMISTRY. Educ: Columbia Col, AB, 53; Columbia Univ, MA, 56, PhD(geol), 58. Prof Exp: From instr to assoc prof, 56-64, Sloan fel, 64, PROF GEOL, COLUMBIA UNIV, 64- Mem: Geochem Soc; Am Geophys Union. Res: Pleistocene geochronology; carbon 14 and thorium 230 dating; chemical oceanography, including oceanic mixing based on radioisotope distribution. Mailing Add: Dept of Geol Columbia Univ New York NY 10027

BROEG, CHARLES BURTON, b Princeton, Ind, Mar 15, 16; m 46. CHEMISTRY. Educ: DePauw Univ, AB, 38; Okla Agr & Mech Col, MS, 40. Prof Exp: Asst, Okla Agr & Mech Col, 38-40; chemist, USDA, 41-43 & 46-58; head prod planning & develop, 58, dir tech serv, 58-64, VPRES & TECH DIR, SuCREST CORP, 64-, PRES, APPL SUGAR LABS, INC, 71- Concurrent Pos: Dir, US Regional Vchmn Res Prog Comt & mem exec comt, Int Sugar Res Found, Inc. Mem: Am Chem Soc; Am Soc Sugar Beet Technologists; Am Soc Sugar Cane Technologists; Am Inst Chemists; Inst Food Technologists. Res: Components of cane wax; sugar cane juice clarification; carbohydrate oxidation; addition compounds of benezene with polynitro aromatic compounds; standards for the color of sugar products; liquid sugar in production of confectionary; new product development. Mailing Add: Tech Serv Dept SuCrest Corp 120 Wall St New York NY 10005

BROEK, HOWARD WINDOLPH, physics, see 12th edition

BROEMELING, LYLE D, b Juneau, Alaska, Mar 17, 39; m 61; c 1. STATISTICS. Educ: Agr & Mech Col Tex, BA, 60, MS, 63; Tex A&M Univ, PhD(statist), 66. Prof Exp: Asst prof statist, Tex A&M Univ, 66-67 & Univ Tex, Houston, 67-68; asst prof, 68-70, ASSOC PROF STATIST, OKLA STATE UNIV, 70- Mem: AAAS; Am Statist Asn; Biomet Soc; Inst Math Statist; Royal Statist Soc. Res: Statistical inference. Mailing Add: Dept of Math & Statist Okla State Univ Stillwater OK 74074

BROENE, HERMAN HENRY, b Grand Rapids, Mich, Dec 28, 19; m 44; c 3. CHEMISTRY. Educ: Calvin Col, AB, 42; Purdue Univ, PhD(phys chem), 47. Prof Exp: Asst phys chem, Purdue Univ, 42-44; res chemist, Eastman Kodak Co, 46-56; assoc prof, 56-64, PROF CHEM, CALVIN COL, 64- Mem: Am Chem Soc. Res: Electrochemistry; surface chemistry of gelatin; thermodynamics of aqueous hydrofluoric acid solutions; mechanical properties of high polymers; chemical education. Mailing Add: Dept of Chem Calvin Col Grand Rapids MI 49506

BROERSMA, DELMAR B, b Lynden, Wash, July 2, 34; m 60; c 3. ENTOMOLOGY. Educ: Calvin Col, AB, 56; Syracuse Univ, MS, 63; Clemson Univ, PhD(entom), 65. Prof Exp: Teacher, Grand Rapids Christian High Sch, 56-57 & Western Mich Christian High Sch, 58-62; asst entom & zool, Clemson Univ, 63-65; res entomologist, Ill Natural Hist Surv, 65-69; asst prof, 69-71, ASSOC PROF ENTOM, PURDUE UNIV, WEST LAFAYETTE, 71- Mem: AAAS; Entom Soc Am. Res: Insect pathology; economic entomology. Mailing Add: Dept of Entom Purdue Univ West Lafayette IN 47906

BROERSMA, SYBRAND, b Harlingen, Netherlands, Sept 20, 19; nat US. MOLECULAR PHYSICS. Educ: Leiden Univ, BA, 39, MA, 41; Delft Inst Technol, PhD(physics), 47. Prof Exp: Asst, Leiden Univ & Delft Inst Technol, 39-46; int exchange fel, Northwestern Univ, 47; instr, Univ Toronto, 48; prof exp physics, Univ

Indonesia, 49-51; lectr, Northwestern Univ, 52, asst prof physics, 53-59; PROF PHYSICS, UNIV OKLA, 59- Mem: Am Phys Soc; Netherlands Phys Soc; Europ Phys Soc. Res: Magnetism; hydrodynamics; molecular spectroscopy. Mailing Add: Dept of Physics Univ of Okla Norman OK 73069

BROFAZI, FRED ROBERT, b Jersey City, NJ, Apr 21, 33; m 56; c 3. ANALYTICAL CHEMISTRY, PHARMACEUTICAL CHEMISTRY. Educ: Rutgers Univ, BS, 54, MS, 57, PhD(chem), 59. Prof Exp: Res anal chemist, Nat Cash Register Co, 59-60; res supvr, 60-64, group leader, 64-66, from asst dir to dir qual control, 66-71, EXEC DIR QUAL CONTROL, GEIGY PHARMACEUT DIV, CIBA-GEIGY CORP, 71- Concurrent Pos: Mem, Nat Formulary Comt Specif. Mem: Am Chem Soc; Am Pharmaceut Asn; Am Soc Hosp Pharmacists. Res: Development of analytical methods; pharmaceutical analysis; quality control. Mailing Add: Ciba-Geigy Pharmaceut Co 556 Morris Ave Summit NJ 07901

BROG, KENNETH CLAIR, b Grand Rapids, Mich, Sept 27, 35; m 55; c 4. PHYSICS. Educ: Albion Col, AB, 57; Case Inst Technol, BS, 59, PhD(physics), 63. Prof Exp: Sr physicist, 62-69, fel, 69-73, MGR RES SECT, COLUMBUS LABS, BATTELLE MEM INST, 73- Mem: Am Phys Soc. Res: Atomic spectroscopy; solid state physics, nuclear magnetic resonance in metals, transition metals and semiconductors; superconductivity; ferroelectricity; piezoelectricity. Mailing Add: Battelle Mem Inst 505 King Ave Columbus OH 43201

BROGAN, DONNA R, b Baltimore, Md, Aug 15, 39. APPLIED STATISTICS. Educ: Gettysburg Col, BA, 60; Purdue Univ, Lafayette, MS, 62; Iowa State Univ, PhD(statist), 67. Prof Exp: Instr statist, Univ Iowa, 66; asst prof biostatist, Sch Pub Health, Univ NC, 67-71; ASSOC PROF STATIST & BIOMET, EMORY UNIV, 71- Concurrent Pos: Mem statist coun, Am Pub Health Asn, 73-75 & health serv res study sect, Nat Ctr Health Serv Res, 73-77. Mem: Am Statist Asn; Am Pub Health Asn. Res: Program evaluation in mental health; delivery and evaluation of health services; sex role differences; women's studies. Mailing Add: Dept of Statist & Biomet Emory Univ Atlanta GA 30322

BROGAN, MARIANNE C, organic chemistry, see 12th edition

BROGDON, BYRON GILLIAM, b Ft Smith, Ark, Jan 22, 29; m 51; c 1. MEDICINE. Educ: Univ Ark, BS & BSM, 51, MD, 52. Prof Exp: Asst prof radiol, Col Med, Univ Fla, 60-63; assoc prof radiol & radiol sci, Sch Hyg & Pub Health, Sch Med, Johns Hopkins Univ, 63-67; asst dean hosp affairs, 70-72, PROF RADIOL & CHMN DEPT, SCH MED, UNIV N MEX, 67- Concurrent Pos: Radiologist-in-chg, Div Diag Roentgenol, Johns Hopkins Hosp, 63-; med dir, Bernalillo County Med Ctr, 69-72. Mem: AMA; Am Roentgen Ray Soc; Radiol Soc NAm; Soc Pediat Radiol; Asn Univ Radiologists. Res: Diagnostic roentgenology; clinical applications; determination of normal and abnormal variations. Mailing Add: Dept of Radiol Univ of NMex Sch of Med Albuquerque NM 87106

BROGE, ROBERT WALTER, b Cleveland, Ohio, Oct 27, 20; m 44; c 5. PHYSICAL CHEMISTRY. Educ: Harvard Univ, SB, 42; Cornell Univ, PhD(phys chem), 48. Prof Exp: Res assoc rocket res, Nat Defense Res Comt, 42-45; res chemist, 48-51, sect head basic res, 51-57 & prod res, 57-61, assoc dir, 61-67, dir prod develop, 67-71, DIR RES, PROCTER & GAMBLE CO, 71- Mem: AAAS; Am Chem Soc; Int Asn Dent Res. Res: X-ray crystallography; solutions of surface active agents; dental research; chemistry of keratin. Mailing Add: 221 Compton Ridge Dr Wyoming OH 45215

BROGLE, RICHARD CHARLES, b Boston, Mass, July 12, 27; m 59; c 3. FOOD SCIENCE. Educ: Mass Inst Technol, SB, 50, SM, 53, PhD(food sci), 60. Prof Exp: Sr chemist, Am Chicle Co, 60-62; group leader contract res, Am Chicle Co Div, Warner-Lambert Co, 63-66; dir clin res, 66-69, dir proprietary toiletries res, 69-72, dir clin regulatory affairs, 72-75, VPRES RES SERV, WARNER-LAMBERT CO, 75- Concurrent Pos: Mem indust liaison coun, Food & Nutrit Bd, Nat Acad Sci-Nat Res Coun, 73-; chmn sci affairs comt, Proprietary Asn, 75- Mem: AAAS; Inst Food Technologists; Acad Pharmaceut Sci; Soc Cosmetic Chemists; Am Soc Clin Pharmacol & Therapeut. Res: Antacid in vitro-in vira relationships; objective clinical testing methods; control of dental plaque. Mailing Add: Warner-Lambert Co 201 Tabor Rd Morris Plains NJ 07950

BROHN, FREDERICK HERMAN, b Flint, Mich, Mar 6, 40; m 69; c 2. BIOCHEMISTRY. Educ: Univ Mich, BS, 65; Wayne State Univ, PhD(biochem), 72. Prof Exp: Fel, 72-75, RES ASSOC PARASITOL, ROCKEFELLER UNIV, 75- Mem: Am Soc Protozoologists. Res: Biochemistry of host parasite interactions in intracellular protozoan infections. Mailing Add: Rockefeller Univ 1230 York Ave New York NY 10021

BROIDA, HERBERT PHILIP, b Aurora, Colo, Dec 25, 20; m 48; c 2. MOLECULAR SPECTROSCOPY. Educ: Univ Colo, BA, 44; Harvard Univ, MA, 45, PhD(physics), 49. Prof Exp: Instr physics, Wesleyan Univ, 44-45; asst, Harvard Univ, 45-49; physicist molecular spectros, Nat Bur Standards, 49-63; dir quantum inst, 73-76, PROF PHYSICS, UNIV CALIF, SANTA BARBARA, 63- Concurrent Pos: Guggenheim fel, 52-53; NSF sr fel, 59-60; mem & chmn eval panel optical physics div, Nat Bur Standards, 72-75; mem comt atomic & molecular physics, Nat Res Coun-Nat Acad Sci, 74-77. Honors & Awards: Flemming Award, 56. Mem: Fel AAAS; fel Am Phys Soc; fel Optical Soc Am; Am Asn Physics Teachers. Res: Planetary atmospheric physics; diatomic spectra; solar spectra; chemical kinetics; trapped radicals; properties of condensed gases; plasma and laser physics; isotope analysis; medical instrumentation. Mailing Add: Dept of Physics Univ of Calif Santa Barbara CA 93106

BROIDA, SAUL, b Pittsburgh, Pa, July 9, 20. PHYSICAL OCEANOGRAPHY. Educ: Univ Miami, BS, 57, MS, 59, PhD(marine sci), 66. Prof Exp: From instr to asst prof marine sci, Univ Miami, 61-71; CHIEF APPL PHYS OCEANOG BR, US COAST GUARD, 71- Mem: AAAS; Marine Technol Soc. Res: Kinetics and dynamics of the Florida current; circulation of the eastern Gulf of Mexico; Atlantic equatorial undercurrent measurements. Mailing Add: US Coast Guard Res & Develop Ctr Groton CT 06340

BROIDA, THEODORE RAY b Louisville, Ky, Dec 6, 28; m 52; c 2. RESEARCH ADMINISTRATION. Educ: Univ Calif, BS, 50, MS, 52. Prof Exp: Physicist, Nucleonics Div, US Naval Radiol Defense Lab, 50-53, Broadview Res Corp, 53-55 & Defense Atomic Support Agency, 55-57; sr opers analyst, Stanford Res Inst, 57-60; dir planning, Broadview Res Corp, 60-62; mgr techno-econ res div, Spindletop Res, Inc, 62-68, pres, 68-73; PRES, QRC RES CORP, 73- Concurrent Pos: Consult, US Off Educ, 68-; mem comt remote sensing agr, Nat Acad Sci-Nat Res Coun. Mem: Asn Comput Mach; Nat Soc Prof Engrs. Res: High intensity thermal radiation sources and instrumentation; measurement of air temperature; logistics; inventory management; electronic data processing; reconnaissance systems; applied industrial and regional economics; resource development; communications systems; research management, especially interdisciplinary teams of physical and social scientists. Mailing Add: 290 S Ashland Ave Lexington KY 40502

BROIDO, ABRAHAM, b Cherkassi, Russia, Sept 12, 24; nat US; m 44; c 2. PHYSICAL CHEMISTRY. Educ: Univ Chicago, SB, 43; Univ Calif, PhD(chem), 50. Prof Exp: Chemist, Metall Lab, Univ Chicago & Argonne Nat Lab, 43-46; res chemist, Clinton Labs & Oak Ridge Nat Lab, 46-48; chemist, Radiation Lab, Univ Calif, 48-50 & US Naval Radiol Defense Lab, 50-56; CHEMIST, PAC SOUTHWEST FOREST & RANGE EXP STA, US FOREST SERV, BERKELEY, 56- Concurrent Pos: Guest scientist, Hebrew Univ, Jerusalem, 64-65; res assoc, Statewide Air Pollution Res Ctr, Univ Calif, 66-, consult, Inst Eng Res, 56-; head thermal radiation br, US Naval Radiol Defense Lab, 51-53; mem, Calif Gov Radiol Defense Adv Comt, 62-66; Nat Acad Sci Adv Comt Civil Defense, 66-70. Mem: AAAS; Am Chem Soc; Opers Res Soc Am; Combustion Inst; Am Inst Chem. Res: Radiochemistry; ion exchange and solvent extraction; microchemistry; rare earth and transuranic elements; effects of thermal and nuclear radiation; forest fire and air pollution research.

BROIN, THAYNE LEO, b Kenyon, Minn, Sept 18, 22; m 49; c 3. GEOLOGY. Educ: St Cloud State Col, BS, 43; Univ Colo, MA, 52, PhD(geol), 57. Prof Exp: Asst physics, St Cloud State Col, 42, asst biol, 43; from instr to asst prof geol, Colo State Univ, 50-57; res geologist, Cities Serv Res & Develop Co, 57-65, supvr geol div, 61-65; res coordr, Explor Div, 65-72, CHIEF COMPUT GEOLOGIST, EXPLOR TECHNOL DEPT, CITIES SERV OIL CO, 72- Mem: AAAS; Soc Econ Paleontologists & Mineralogists; Geol Soc Am; Am Asn Petrol Geologists. Res: Sedimentation; stratigraphy; subsurface facies mapping; sedimentary petrology; petroleum geology; computer mapping. Mailing Add: Cities Serv Oil Co Box 300 Tulsa OK 74102

BROIS, STANLEY JAMES, organic chemistry, biochemistry, see 12th edition

BROITMAN, SELWYN ARTHUR, b Boston, Mass, Aug 30, 31; m 53; c 2. MICROBIOLOGY, MEDICAL EDUCATION. Educ: Univ Mass, BS, 52, MS, 53; Mich State Univ, PhD(microbiol), 56. Prof Exp: Res instr path, 63-64, from asst prof to assoc prof microbiol, 65-75, PROF MICROBIOL, SCH MED, BOSTON UNIV, 69- Concurrent Pos: Res assoc gastrointestinal res lab, Mallory Inst Path, Boston City Hosp, 56-, assoc med, Thorndike Mem Lab, 69-; assoc med, Harvard Med Sch, 69-; founding mem, Digestive Dis Found, 71; lectr, Div Allied Health Professions, Northeastern Univ, 71-73; lectr, Sargent Col Allied Health Professions, Boston Univ, 72-, assoc prof nutrit sci, Sch Grad Dent, 74. Mem: AAAS; Am Soc Microbiol; Brit Soc Appl Bact; Soc Exp Path; NY Acad Sci. Res: Experimental gastroenterology; malabsorptive diseases; colon cancer; intestinal flora; bacterial and viral enteric infections. Mailing Add: Dept of Microbiol Boston Univ Sch of Med Boston MA 02118

BROKAW, BRYAN EDWARD, b Pittsfield, Ill, Feb 27, 49; m 70; c 2. ANIMAL BREEDING. Educ: Abilene Christian Col, BS, 71; Ore State Univ, PhD(animal breeding & genetics), 75. Prof Exp: ASST PROF ANIMAL SCI, ABILENE CHRISTIAN COL, 75- Mem: Am Soc Animal Sci; Sigma Xi. Res: Relationships of physiological characteristics with performance traits in beef cattle and their value as selection criteria. Mailing Add: Abilene Christian Col Sta PO Box 8038 Abilene TX 79601

BROKAW, CHARLES JACOB, b Camden, NJ, Sept 12, 34; m 55; c 2. CELL BIOLOGY. Educ: Calif Inst Technol, BS, 55; Univ Cambridge, PhD(zool), 58. Prof Exp: Res assoc biol div, Oak Ridge Nat Lab, 58-59; asst prof zool, Univ Minn, 59-61; from asst prof to assoc prof biol, 61-68, PROF BIOL, CALIF INST TECHNOL, 68- Concurrent Pos: Guggenheim fel, Univ Cambridge, 70-71. Mem: AAAS; Soc Gen Physiol; Am Soc Cell Biol; Biophys Soc. Res: Motility and behavior of ciliated and flagellated cells. Mailing Add: Div of Biol Calif Inst of Technol Pasadena CA 91125

BROKAW, GEORGE YOUNG, b St Clairsville, Ohio, Sept 11, 21; m 44; c 4. APPLIED CHEMISTRY. Educ: Ohio Wesleyan Univ, BA, 41; Ohio State Univ, MS, 44, PhD(chem), 47. Prof Exp: Asst chem, Ohio State Univ, 41-43, asst, Univ Res Found, 43-44; chem engr, Naval Res Lab, Washington, DC, 44-45; res assoc, Distillation Prod Industs Div, Eastman Kodak Co, 47-63, head develop labs, 63-69; SR RES ASSOC, TENN EASTMAN CO, 69- Mem: AAAS; fel Am Chem Soc. Res: Fats and oils; food emulsifiers; fat-soluble vitamins. Mailing Add: Tenn Eastman Co Kingsport TN 37662

BROKAW, RICHARD SPOHN, b Orange, NJ, Mar 26, 23; m 47; c 4. PHYSICAL CHEMISTRY. Educ: Swarthmore Col, AB, 43; Princeton Univ, AM, 49, PhD(chem), 51. Prof Exp: Asst phys chem, Calco Chem Div, Am Cyanamid Co, 43-44 & Princeton Univ, 46-52; aeronaut res scientist phys chem & combustion, Lewis Res Ctr, NASA, 52-55, from assoc head to head, Combustion Fundamentals Sect, 55-57, chief, Phys Chem Br, 57-71, chief, Physics & Chem Div, 71-72, asst chief, Phys Sci Div, 72-73; LECTR CHEM, BALDWIN-WALLACE COL, 74- Honors & Awards: Medal for Except Sci Achievement, NASA, 72. Mem: AAAS; Am Chem Soc; Combustion Inst; Am Inst Aeronaut & Astronaut. Res: Combustion, especially burning velocity and spontaneous ignition; transport properties of gases, especially heat transport in chemically reacting gases; gas phase chemical kinetics. Mailing Add: 17403 Edgewater Dr Lakewood OH 44107

BROKER, THOMAS RICHARD, b Hackensack, NJ, Oct 22, 44; m 74. MOLECULAR GENETICS. Educ: Wesleyan Univ, BA, 66; Stanford Univ, PhD(biochem), 72. Prof Exp: Res fel chem, Calif Inst Technol, 72-75; SR STAFF SCIENTIST ELECTRON MICROS, COLD SPRING HARBOR LAB, 75- Concurrent Pos: Res fel, Helen Hay Whitney Found, 72-75. Mem: Am Soc Microbiol. Res: Development and application of electron microscope labeling techniques for identification of genes and protein binding sites on chromosomes; bacteriophage T4 DNA metabolism. Mailing Add: Cold Spring Harbor Lab PO Box 100 Cold Spring Harbor NY 11724

BROKKE, MERVIN EDWARD, b Silverton, Ore, Dec 13, 32; m 56; c 3. ORGANIC CHEMISTRY. Educ: Willamette Univ, BA, 54; Ore State Col, PhD(chem), 59. Prof Exp: Assoc res chemist, Parke-Davis & Co, 58-60; sr res chemist, 60-62, supvr, 62-67, mgr org synthesis, 67-72, MGR, SPECIALTIES DEPT, STAUFFER CHEM CO, 72- Res: Agricultural chemicals; fire retardants; functional fluids; catalysts. Mailing Add: Stauffer Chem Co Dobbs Ferry NY 10522

BROLLEY, JOHN EDWARD, JR, b Chicago, Ill, Jan 15, 19; m 46. ELEMENTARY PARTICLE PHYSICS. Educ: Univ Chicago, SB, 40; Ind Univ, MS, 48, PhD(physics), 49. Prof Exp: MEM STAFF, RES ELEM PARTICLE PHYSICS & ASTROPHYSICS, LOS ALAMOS SCI LAB, 49- Mem: Fel AAAS; fel Am Phys Soc; Archeol Inst Am; Am Astron Soc. Res: Nuclear and particle physics; laser physics. Mailing Add: Box 1663 Los Alamos NM 87544

BROLMANN, JOHN BERNARDUS, b Holland, Nov 20, 20; m 52; c 4. PLANT

BREEDING. Educ: State Agr Univ Wageningen, MS, 52; Univ Fla, PhD(agron), 68. Prof Exp: Plant breeder, Exp Sta Mas d'Auge, France, 52-60; ASST PROF AGRON & ASST LEGUME BREEDER, AGR EXP STA, UNIV FLA, 69- Mem: Am Soc Agron; Crop Sci Soc Am. Res: Selection and breeding of tropical and temperature climate legumes for pasture use. Mailing Add: Agr Res Ctr Box 248 Ft Pierce FL 33450

BROM, JOSEPH MARCH, JR, b Petersburg, Va, Oct 8, 42; m 67; c 2. PHYSICAL CHEMISTRY, MOLECULAR SPECTROSCOPY. Educ: Col St Thomas, BS, 64; Iowa State Univ, PhD(phys chem), 70. Prof Exp: Res assoc chem, Univ Fla, 70-73; asst res physicist, Univ Calif, Santa Barbara, 73-75; ASST PROF CHEM, BENEDICTINE COL, 75- Mem: AAAS; Am Chem Soc; Am Phys Soc. Res: Magnetic resonance and optical spectra of trapped radicals; laser chemistry; high temperature chemistry. Mailing Add: Dept of Chem Benedictine Col Atchison KS 66002

BROMAGE, PHILIP R, b London, Eng, Oct 28, 20; Can citizen; m 46; c 3. ANESTHESIOLOGY. Educ: Univ London, MB, BS, 44, dipl, 47; FRCS(C), 53. Prof Exp: Consult anesthetist, Portsmouth Hosps Group, Eng, 48-55; attend anesthetist, Royal Victoria Hosp, 55-68; from asst prof to assoc prof anesthesia, 62-70, PROF ANESTHESIA & CHMN DEPT, FAC MED, McGILL UNIV, 70-; ANESTHETIST-IN-CHIEF, ROYAL VICTORIA HOSP, 68- Mem: Am Soc Anesthesiol; Can Med Asn; Can Anaesthetists Soc; Brit Med Asn; NY Acad Sci. Res: Epidural analgesia for surgery, obstetrics and therapeutics; respiratory dynamics in anesthesia. Mailing Add: Dept of Anaesthesia Royal Victoria Hosp Montreal PQ Can

BROMAN, ROBERT FABEL, b Aitkin, Minn, Oct 19, 39; m 61; c 4. ELECTROANALYTICAL CHEMISTRY. Educ: Carleton Col, BA, 61; Northwestern Univ, PhD(anal chem), 65. Prof Exp: Res assoc chem, Univ NC, 64-65; asst prof, 65-70, coordr freshman chem, 68-72, ASSOC PROF CHEM, UNIV NEBR-LINCOLN, 70- Concurrent Pos: Res assoc chem, Ohio State Univ, 72-73. Mem: AAAS; Am Chem Soc. Res: Electroanalytical study of intermetallic compound formation in amalgams; electrochemistry of oxygen, oxygen-binding compounds, organometallics; spectroelectrochemistry; adsorption of electroactive species on electrodes. Mailing Add: Dept of Chem Univ of Nebr Lincoln NE 68588

BROMBERG, ELEAZER, b Toronto, Ont, Oct 7, 13; nat US; m 48; c 3. APPLIED MATHEMATICS. Educ: City Col New York, BS, 33; Columbia Univ, MA, 35; NY Univ, PhD(math), 50. Prof Exp: Asst demonstr physics, Univ Toronto, 36-37; instr, Worcester Polytech Inst, 41-42; res mathematician, Appl Math Panel, Nat Defense Res Comt, Columbia Univ, 43-44, NY Univ, 44-45; proj engr, Reeves Instrument Corp, 45-50 & 52; head mech br, Off Naval Res, 50-53; admin dir comput ctr & assoc prof math, 53-57, admin mgr, Courant Inst Math Sci, 57-59, asst dir, 59-66, vchancellor acad affairs, 70-72, PROF APPL MATH, NY UNIV, 57-, DEP CHANCELLOR, 72- Concurrent Pos: Res Scientist, Los Alamos Sci Lab, 56-57. Mem: AAAS; Soc Indust & Appl Math; Asn Comput Mach (secy, 53-56); Am Math Soc; Math Asn Am. Res: Nonlinear vibrations; elasticity; computing techniques; wave motion; simulation techniques. Mailing Add: NY Univ Washington Sq New York NY 10003

BROMBERG, J PHILIP, b Brooklyn, NY, Mar 25, 36; m 60; c 2. PHYSICAL CHEMISTRY. Educ: Mass Inst Technol, BS, 56; Calif Inst Technol, MS, 59; Univ Chicago, PhD(chem), 64. Prof Exp: Engr, Semiconductor Div, Westinghouse Elec Corp, Pa, 59-60; asst prof, 67-70, FEL PHYS CHEM, CARNEGIE-MELLON INST, 64-, ASSOC PROF CHEM, 70- Concurrent Pos: Lectr, Univ Pittsburgh, 65-68. Mem: Am Chem Soc; Am Phys Soc. Res: Nuclear magnetic resonance; paramagnetic susceptibility; electron impact spectra. Mailing Add: Mellon Inst 4400 Fifth Ave Pittsburgh PA 15213

BROMBERG, MILTON JAY b Brooklyn, NY, Sept 15, 23; div; c 1. INDUSTRIAL ORGANIC CHEMISTRY. Educ: City Col NY, BS, 44; Columbia Univ, MS, 48. Prof Exp: Asst res chemist, Columbia Univ, 43-44, asst, 46-48; chief chemist, H A Brassert Co, 48-51; chief chemist, Explosive Div, Olin Mathieson Chem Corp, 51-54, chief prod engr, 54-55, tech mgr, 55-57, proj mgr, High Energy Fuels Div, 57-58, mgr qual control, 58-60, staff consult org div, 60-62, mgr qual assurance, Chem Div, Doe Run Plant, 62-72, MGR QUAL ASSURANCE, CHEMICAL GROUP, LAKE CHARLES COMPLEX, OLIN CORP, 72- Mem: Am Soc Qual Control; Am Chem Soc; fel Am Inst Chemists. Res: Chemical management; quality control programs; fluoro-aromatics; petro chemicals; toluene diisocyanate; polypropylene glycols; explosives; high energy fuels; design of laboratories. Mailing Add: PO Box 5274 Lake Charles LA 70601

BROMBERGER, SAMUEL H, b New York, NY, Dec 21, 41; m 66; c 1. EXPLORATION GEOLOGY. Educ: City Col New York, BS, 62; Univ Iowa, MS, 65, PhD(geol), 68. Prof Exp: From instr to asst prof geol, Cornell Col, 67-69; geologist, Los Angeles Div, Texaco, Inc, 69-73, REGIONAL GEOLOGIST, TEXACO PROD SERV LTD, LONDON, 73- Mem: AAAS; Mineral Soc Am; Soc Econ Paleontologists & Mineralogists; Am Asn Petrol Geologists. Res: Stratigraphy, hydrocarbon potential analysis of sedimentary basins; regional geology. Mailing Add: Texaco Prod Serv Ltd 1 Knightsbridge Green London SW 1X England

BROMBERGER-BARNEA, BARUCH (BERTHOLD), b Adelnau, Ger, Aug 31, 18; nat US; m 55; c 2. ENVIRONMENTAL MEDICINE, PHYSIOLOGY. Educ: Tech Col Berlin, EE, 38; Univ Colo, PhD(physiol), 57. Prof Exp: Telecommun engr, Govt Israel, Tel Aviv, 46-51; asst, Med Sch, Univ Colo, 55-57, instr physiol res, 57-60, asst prof, 60-61; from asst prof to assoc prof, 62-74, PROF ENVIRON MED, SCH HYG & PUB HEALTH, JOHNS HOPKINS UNIV, 75- Concurrent Pos: Electrophysiologist, Nat Jewish Hosp, 57-62. Mem: Inst Elec & Electronics Engrs; Am Physiol Soc. Res: Electrophysiology; cardiac excitability; ventricular fibrillation; transmembrane potentials; circulation; coronary blood flow; environmental effects on coronary circulation. Mailing Add: Dept of Environ Med Johns Hopkins Univ Baltimore MD 21205

BROMEL, MARY COOK, b Colorado Springs, Colo, Mar 22, 17; m 41; c 2. BACTERIOLOGY, MICROBIAL ECOLOGY. Educ: Wayne State Univ, BS, 40, PhD(biol), 67. Prof Exp: Med technologist blood bank, Henry Ford Hosp, Detroit, Mich, 40-42; chief lab technologist, Kemp Clin, Birmingham, 43-48; prin prof asst bact & physiol, Wayne State Univ-US Army Res & Develop Command, 62-67, asst prof bact, Wayne State Univ, 67-68; assoc prof, 68-71, PROF BACT, N DAK STATE UNIV, 71- Mem: Am Soc Microbiol; Sigma Xi; AAAS. Res: Etiology of necroses in animals; water microbiology; microbial control of insect larvae; microbial degradation of toxic wastes; antibiotic resistance transfer in aquatic microorganisms. Mailing Add: Dept of Bact NDak State Univ Fargo ND 58102

BROMELS, EDWARD, b Hammond, Ind, Jan 19, 39; m. PHYSICAL CHEMISTRY, ENVIRONMENTAL CHEMISTRY. Educ: Purdue Univ, BS, 61; Univ Ill, MS, 63, PhD(phys chem), 68. Prof Exp: Chemist, E I du Pont de Nemours & Co, Inc, 67-73; SR PROJ RES CHEMIST, SCOTT PAPER CO, 73- Mem: Am Chem Soc; Tech Asn Pulp & Paper Indust. Res: Testing of tissue products; environmental considerations of new products and processes; development of new products and coatings. Mailing Add: Paper Res Scott Paper Co Scott Plaza III Philadelphia PA 19113

BROMER, WILLIAM WALLIS, b Racine, Wis, Oct 10, 27; m 52; c 3. BIOCHEMISTRY. Educ: DePauw Univ, AB, 49; Ind Univ, PhD(chem), 54. Prof Exp: Sr biochemist, 53-60, res assoc, 60-70, RES ADV, ELI LILLY & CO, 70- Mem: AAAS; Am Diabetes Asn; Am Chem Soc; Am Soc Biol Chemists. Res: Metabolism of essential fatty acids and cholesteryl esters; antibiotics isolation; isolation and structural analysis of proteins; glucagon; intrinsic factor and vitamin B12; insulin and proinsulin chemistry; glucagon structure and function; tubulin chemistry and vinca binding. Mailing Add: Lilly Res Lab Eli Lilly & Co Indianapolis IN 46206

BROMERY, RANDOLPH WILSON, b Cumberland, Md, Jan 18, 26; m 47; c 5. GEOLOGY, GEOPHYSICS. Educ: Howard Univ, BS, 56; Am Univ, MS, 62; Johns Hopkins Univ, PhD, 68. Hon Degrees: EdD, Western New Eng Col; DSc, Frostburg State Col. Prof Exp: Geophysicist, US Geol Surv, 48-67; assoc prof geophys, 67-69, head dept geol & geog, 69-70, vchancellor, 70-72, PROF GEOPHYS, UNIV MASS, AMHERST, 69-, CHANCELLOR, 72- Concurrent Pos: Prof lectr, Howard Univ, 61-65; consult, AID, US Dept State, 66-; consult, Kennecott Copper & Exxon; mem bd trustees, Fairfield Univ, Hampshire Col & Woods Hole Oceanog Inst. Mem: Nat Acad Sci; Am Geophys Union; Soc Explor Geophys; Geol Soc Am; Sigma Xi. Res: Gravity, magnetic, radioactivity and seismic surveying methods applied to geologic mapping programs. Mailing Add: Off of the Chancellor Univ of Mass Amherst MA 01002

BROMFIELD, CALVIN STANTON, b Freeport, NY, Feb 4, 23; m 43; c 4. GEOLOGY. Educ: Univ Ariz, BS, 48, MS, 50; Univ Ill, PhD, 62. Prof Exp: Geologist, Homestake Mining Co, 49-51; GEOLOGIST, US GEOL SURV, 51- Concurrent Pos: Asst, Univ Ill, 55-57. Mem: Geol Soc Am; Soc Econ Geologists. Res: Base metal ore deposits and regional geology, especially of Arizona, Colorado and Utah. Mailing Add: Geol Div US Geol Surv Bldg 25 Denver Fed Ctr Denver CO 80225

BROMFIELD, KENNETH RAYMOND, b Wilkes-Barre, Pa, June 22, 22; m 46; c 5. PLANT PATHOLOGY. Educ: Pa State Univ, BS, 49; Univ Minn, PhD(plant path), 57. Prof Exp: Plant pathologist epiphytology, US Army Biol Labs, 49-51; res asst, Dept Plant Path, Univ Minn, 51-52; plant pathologist epiphytology, US Army Biol Labs, 52-55; res asst, Dept Plant Path, Univ Minn, 55-56; plant pathologist forest dis, US Forest Serv, Northeastern Forest Exp Sta, NH, 57-59 & US Army Biol Defense Res Ctr, 59-71; RES PLANT PATHOLOGIST & RES LEADER, PLANT DIS RES LAB, USDA, 71- Concurrent Pos: Lectr, Frederick Community Col, 61- Mem: Am Phytopath Soc; Am Inst Biol Sci. Res: Etiology and epiphytology of plant rusts; evaluation of plant pathogens for damage potential; biological control of weeds by plant pathogens; ecology of plant pathogens; weather plant disease relationships; epiphytology; ecology. Mailing Add: USDA Plant Dis Res Lab PO Box 1209 Frederick MD 21701

BROMLEY, DAVID ALLAN, b Westmeath, Ont, May 4, 26; US citizen; m 49; c 2. NUCLEAR PHYSICS. Educ: Queen's Univ, Ont, BSc, 48, MSc, 50; Univ Rochester, PhD(physics), 52; Yale Univ, MA, 61. Prof Exp: Demonstr physics, Queen's Univ, Ont, 47; res officer, Nat Res Coun Can, 48; from instr to asst prof physics, Univ Rochester, 52-55; sr res officer, Atomic Energy Can Ltd, 55-60, sect head accelerators, 58-60; assoc prof physics & assoc dir heavy ion lab, 60-61, prof physics, 61-72, HENRY FORD II PROF, YALE UNIV, 72-, DIR, A W WRIGHT NUCLEAR STRUCT LAB, 61-, CHMN DEPT PHYSICS, 70- Concurrent Pos: Mem org comts, Int Conf Nuclear Struct, Int Union Pure & Appl Physics, Can, 60, Italy, 62, Tenn, 66, mem US deleg, Dubrovnick, 69, mem US nat comt, 70-, chmn, 74-; mem panel nuclear struct physics, NSF, 61, chmn nuclear sci comt, 66-, chmn nat physics surv comt, 69-; mem surv subcomt nuclear physics & intermediate energy physics, Nat Acad Sci, 64, mem naval studies bd, 74-; dir, United Nuclear Corp, 67-, Labcore Inc, 69-, Extrion Corp, 70- & United Illum Co, 73-; mem exec comt, div phys sci & mem-at-lg, Nat Res Coun, 68-; chmn off phys sci, 75-; consult, Oak Ridge Nat Lab, Brookhaven Nat Lab, Acad Press, Bell Tel Labs, NSF, McGraw-Hill, GT&E & Int Bus Mach Corp, 69-; mem high energy physics adv panel, Energy Resources Develop Admin, 73- Mem: Fel Am Phys Soc; Can Asn Physicists; fel Am Acad Arts & Sci; fel AAAS. Res: Nuclear structure and reaction mechanisms; heavy ion physics; accelerators. Mailing Add: Wright Nuclear Structure Lab Yale Univ New Haven CT 06520

BROMLEY, STEPHEN C, b Los Angeles, Calif, Aug 31, 38; m 67; c 2. DEVELOPMENTAL BIOLOGY. Educ: Brigham Young Univ, BS, 60; Princeton Univ, MA, 62, PhD(biol), 65. Prof Exp: Res asst ornith & mammal, Los Angeles County Mus, 57, 59-60; instr biol, Princeton Univ, 64-65; asst prof zool, Univ Vt, 65-69; res assoc, 69-70, ASSOC PROF ZOOL, ASSOC PROF BIOL SCI & DIR PROG, MICH STATE UNIV, 70- Mem: AAAS; Am Soc Zoologists. Res: Neural and endocrine factors in regeneration of amphibian limbs; response of amphibian limbs to culture conditions. Mailing Add: Dept of Zool Mich State Univ East Lansing MI 48823

BROMUND, RICHARD HAYDEN, b Oberlin, Ohio, Apr 28, 40; m 62; c 2. ANALYTICAL CHEMISTRY. Educ: Oberlin Col, AB, 62; Pa State Univ, PhD(chem), 68. Prof Exp: Asst prof, 67-73, ASSOC PROF CHEM, COL WOOSTER, 73- Mem: AAAS; Am Chem Soc. Res: Chemical instrumentation. Mailing Add: Dept of Chem Col of Wooster Wooster OH 44691

BROMUND, WERNER HERMANN, b Duluth, Minn, May 8, 09; m 35; c 2. ANALYTICAL CHEMISTRY, ORGANIC CHEMISTRY. Educ: Univ Chicago, BS, 32; Oberlin Col, AM, 35; NY Univ, PhD(chem), 42. Prof Exp: Asst chem, NY Univ, 34-37; from instr to assoc prof, 37-55, PROF CHEM, OBERLIN COL, 55- Mem: Am Chem Soc. Res: Amino sugar derivatives; adsorption characteristics of porous solids; macro and microchemical apparatus design; macro and microchemical methods of analysis; methods of analysis of pharmaceuticals; teaching methods in analytical chemistry. Mailing Add: Dept of Chem Oberlin Col Oberlin OH 44074

BRON, WALTER ERNEST, b Berlin, Ger, Jan 17, 30; m 52; c 2. SOLID STATE PHYSICS. Educ: NY Univ, BME, 52; Columbia Univ, MS, 53, PhD(metal physics), 58. Prof Exp: Res assoc, Eng Res & Develop Lab, 54-56; lectr, George Washington Univ, 55-56; res physicist, Watson Lab, IBM Corp, 57-58, staff mem physics group, Res Ctr, 58-66; vis prof, Phys Inst, Univ Stuttgart, 66-67; assoc prof, 67-69, PROF PHYSICS, IND UNIV, BLOOMINGTON, 69- Concurrent Pos: Res assoc & lectr, Columbia Univ, 52-58, 64; Guggenheim fel, 66-67; vis prof, Max Planck Inst for Solid State Res, 73-74. Honors & Awards: Award, Alexander Von Humboldt Found, 73. Mem: AAAS; Am Phys Soc; Sigma Xi. Res: Defects in solid crystals; optical properties of solids; lattice dynamics; phonon transport. Mailing Add: Dept of Physics Ind Univ Bloomington IN 47401

BRONCO, CHARLES JOHN, b Gary, Ind, Jan 25, 28. ATOMIC PHYSICS. Educ: Okla State Univ, BS, 49; Univ Tex, MA, 51; Univ Okla, PhD(physics), 59. Prof Exp: Sr systs design engr, Chance Vought Aircraft, Tex, 56-59, electronic systs engr, 59-60;

prin physicist, Phys Sci Div, Melpar, Inc, Va, 60-61, sr scientist gaseous laser res, 61-62; asst prof math & physics, Grad Inst Technol, Univ Ark, 62-68; ASSOC PROF PHYSICS, ARK POLYTECH COL, 68- Concurrent Pos: Consult, Melpar, Inc, 62. Mem: Am Phys Soc. Res: Resonance excitation cross-sections; high vacuum physics; electronic systems research; molecular dissociation by electron and photon impact; laser development; focal isolation methods; laboratory techniques. Mailing Add: Dept of Physics Ark Polytech Col Russellville AR 72801

BRONFIN, BARRY ROBERT, b Washington, DC, June 7, 39; m 63; c 2. RESEARCH ADMINISTRATION, PHYSICAL CHEMISTRY. Educ: Mass Inst Technol, SB, 60, SM, 61, ScD(chem eng), 63; Yale Univ, MA, 68. Prof Exp: Sr res engr, United Aircraft Res Labs, 65-68, supvr high energy molecular interactions group, 68-70, prin scientist, United Technol Res Ctr, 70-75; PRES, SCI LEASING INC, CONN, 75- Concurrent Pos: Lectr, SDak Sch Mines & Technol, 71. Mem: Am Inst Chem Eng; Am Chem Soc; Am Phys Soc; Combustion Inst. Res: Chemical reactions in plasmas; gas laser research; administration and direction of research. Mailing Add: Sci Leasing Inc 111 Founders Plaza East Hartford CT 06108

BRONIKOWSKI, THOMAS ANDREW, b Milwaukee, Wis, Oct 30, 32; m 61; c 3. APPLIED MATHEMATICS. Educ: Marquette Univ, BS, 59; Univ Wis, MS, 62, PhD(math), 65. Prof Exp: Asst prof 65-71, ASSOC PROF MATH, MARQUETTE UNIV, 71- Mem: Am Math Soc; Math Asn Am. Res: Stability theory of ordinary differential equations; asymptotic behavior of solutions of ordinary differential and intergrodifferential equations. Mailing Add: Dept of Math Marquette Univ Milwaukee WI 53233

BRONK, DETLEV WULF, physiology, biophysics, deceased

BRONK, JOHN RAMSEY, b Philadelphia, Pa, Dec 20, 29; m 55; c 2. BIOCHEMISTRY, CELL PHYSIOLOGY. Educ: Princeton Univ, AB, 52; Oxford Univ, PhD(biochem), 55. Prof Exp: From asst prof to prof zool, Columbia Univ, 58-66; PROF BIOCHEM, UNIV YORK, ENG, 66- Concurrent Pos: Guggenheim fel, 64-65. Mem: Am Soc Biol chemists; Brit Biochem Soc; The Physiol Soc; Am Chem Soc. Res: Control of metabolism, particularly as related to the mode of action of the thyroid hormones; oxidative phosphorylation and mitochondrial structure; amino acid transport; urea synthesis and gluconeogenesis in liver. Mailing Add: Dept of Biol Univ of York York England

BRONK, THEODORE TOBIAS, b Poland, Jan 12, 09; nat US; m 49; c 4. PATHOLOGY. Educ: Univ Pittsburgh, BS, 35; George Washington Univ, MD, 38. Prof Exp: Asst pathologist, Michael Reese Hosp, 49-50; from asst chief to actg chief lab serv, Vet Admin Hosp, 50-55; PATHOLOGIST & DIR LABS, MT ST MARY'S HOSP, 56- Concurrent Pos: From instr to clin assoc prof, Sch Med, State Univ NY Buffalo, 51- Mem: Fel Am Soc Clin Path; Am Med Asn; fel Am Col Physicians; fel Col Am Path; Int Acad Path. Mailing Add: Mt St Mary's Hosp Lewiston NY 14092

BRONNER, FELIX, b Vienna, Austria, Nov 7, 21; nat US; m 47; c 2. PHYSIOLOGY, BIOPHYSICS. Educ: Univ Calif, BS, 41; Mass Inst Technol, PhD(nutrit biochem), 52. Prof Exp: Asst, Mass Inst Technol, 51-52, assoc, 52-54; vis investr, Rockefeller Inst, 54-56, asst, 56; assoc, Hosp Spec Surg, Med Ctr, Cornell Univ, 57-58, Bicknell assoc mineral metab, 58-61, assoc scientist & asst prof biochem, Med Col, 61-63; assoc prof, Sch Med, Univ Louisville, 63-69; PROF ORAL BIOL, UNIV CONN, FARMINGTON, 69- Concurrent Pos: Helen Hay Whitney fel, Rockefeller Inst, 54-55 & Arthritis & Rheumatism fel, 55-56; vis, Pasteur Inst, 62 & Weizmann Inst Sci; consult, NIH & Acad Press; organizer, Gordon Res Conf Bones & Teeth, Sanger Symp, 24th Int Cong, 65; vis scientist, Nat Inst Health & Med Res, Paris, 72. Honors & Awards: Andre Lichtwitz Prize, Nat Inst Health & Med Res, Paris, 74. Mem: Fel AAAS; Am Fedn Clin Res; Am Physiol Soc; Biophys Soc; Orthop Res Soc. Res: Calcium and mineral metabolism; metabolic bone disease; biologic regulation; ion movement and transport. Mailing Add: Dept of Biol Univ of Conn Sch of Dent Med Farmington CT 06032

BRONOWSKI, JACOB, mathematics, philosophy of science, deceased

BRONSKILL, JOAN FRANCES, b Ottawa, Ont. INSECT HISTOLOGY & INSECT EMBRYOLOGY. Educ: Queen's Univ, Ont, BA, 48; Cornell Univ, PhD(insect embryol), 55. Prof Exp: Res technician insect histol & photog, 48-51, RES SCIENTIST INSECT HISTOL & EMBRYOL, ELECTRON MICROSCOPE CTR, CHEM & BIOL RES INST, CAN AGR, 55- Concurrent Pos: Consult histol, Can Dept Agr, London, Ont, 62-64 & Dept Entom, NY State Agr Exp Sta, 67. Mem: Fel AAAS; Entom Soc Am; Entom Soc Can; Can Soc Zoologists; Can Soc Cell Biol. Res: Defense reactions of insects to metazoan biotic agents; anti-tanning agent effects on immature Diptera; structure of insect pheromone glands and retrocerebral complex; embryology of Hymenoptera. Mailing Add: Electron Microscope Ctr Chem & Biol Res Inst Can Agr Ottawa ON Can

BRONSKILL, MICHAEL JOHN, radiation chemistry, biophysics, see 12th edition

BRONSKY, ALBERT J, b Waterbury, Conn, June 14, 28; m 55; c 2. BIOCHEMISTRY. Educ: Boston Col, BS, 54; Purdue Univ, MS, 57, PhD(biochem), 61. Prof Exp: Res biochemist, Res Dept, 59-71, head fermentation sect, 71-75, MGR RES & DEVELOP, JOSEPH E SEAGRAM & SONS, INC, 75- Mem: Am Chem Soc; Am Soc Brewing Chemists. Res: Enzymology; mechanism of action of esterases and phosphorylases; metabolism; formation of trace flavor components during fermentation. Mailing Add: Seagram & Sons Inc Res & Develop PO Box 240 Louisville KY 40201

BRONSKY, DAVID, b Chicago, Ill, Apr 18, 20. INTERNAL MEDICINE, ENDOCRINOLOGY. Educ: Univ Chicago, BS, 42; Chicago Med Sch, MD, 50. Prof Exp: Assoc dir endocrinol, Cook County Hosp, 54-58, dir, 68-71, assoc dir med, 64-70, actg dir, 70-71; assoc prof, 65-69, PROF MED, UNIV ILL COL MED, 69- Concurrent Pos: Assoc attend physician, Cook County Hosp, 54-59, attend physician, 59-; from clin instr to prof, Sch Med, Univ Ill, 56-; prof med, Cook County Grad Sch Med, 59-; attend physician, Res & Educ Hosps, Univ Ill, 64- Mem: Am Fedn Clin Res; AMA; Endocrine Soc. Res: Endocrinology and metabolism; parathyroid gland physiology; serum proteins. Mailing Add: 1355 Sandburg Chicago IL 60610

BRONSON, DAVID LEE, b Holland, Mich, Oct 29, 36; m 70. VIROLOGY. Educ: Hope Col, BA, 63; Iowa State Univ, MS, 66, PhD(microbiol), 69. Prof Exp: Fel virol & epidemiol, Baylor Col Med, '69-71; res assoc, Dept Microbiol, Sch Med, Univ Miami, 71-72; HEAD BIOCHEM VIROL, DEPT UROL SURG, MED SCH, UNIV MINN, MINNEAPOLIS, 72- Mem: Am Soc Microbiol; AAAS. Res: Molecular virology; oncogenesis. Mailing Add: Box 394 Univ of Minn Med Sch Minneapolis MN 55455

BRONSON, FRANKLIN HERBERT, b Pawnee City, Nebr; m 53; c 2. REPRODUCTIVE PHYSIOLOGY. Educ: Kans State Univ, BS, 57, MS, 58; Pa State Univ, PhD(zool), 61. Prof Exp: Staff scientist, Jackson Lab, 61-68; assoc prof, 68-72, PROF ZOOL, UNIV TEX, AUSTIN, 72- Mem: AAAS; Animal Behav Soc; Soc Study Reprod; Am Soc Zoologists; Ecol Soc Am. Res: Hormones and social behavior; mammalian pheromones. Mailing Add: Dept of Zool Univ of Tex Austin TX 78712

BRONSON, JEFF DONALDSON, b Dallas, Tex, Aug 12, 38. NUCLEAR PHYSICS. Educ: Rice Univ, BA, 59, MA, 61, PhD(physics), 64. Prof Exp: Res assoc physics, Rice Univ, 64; res fel, Univ Basel, 64-65 & Univ Wis, 65-67; asst prof, 67-74, MEM STAFF, CYCLOTRON INST, TEX A&M UNIV, 74- Mem: Am Phys Soc. Res: Nuclear forces using 3-body nuclear reactions; determination of atomic hyperfine fields and nuclear magnetic moments using perturbed angular correlations. Mailing Add: Cyclotron Inst Tex A&M Univ College Station TX 77843

BRONSON, ROY DEBOLT, b Reno, Nev, May 30, 20; m 47; c 3. SOIL SCIENCE. Educ: Mich State Univ, BS, 48, MS, 49; Purdue Univ, PhD, 59. Prof Exp: Soil technician, Agr Res Dept, Green Giant Co, Minn, 49-52; instr agron & supvr soil testing prog, Purdue Univ, 52-59, assoc prof, 59-60; head dept chem & soils, United Fruit Co, Honduras, 60-61; chief soils & fertil res br, Tenn Valley Authority, Wilson Dam, Ala, 61-63; prof agron, Purdue Univ, 63-68, soil scientist, Purdue-Brazil Proj, Agr Univ, Minas Gerais, 63-65, chief of party, 65-68, PROF AGRON & INT AGR, PURDUE UNIV, 68- Mem: Fel AAAS; Am Soc Agron; Am Chem Soc; Clay Minerals Soc; Int Soil Sci Soc. Res: Characterization and mineralogy of highly weathered soils, particularly of tropics, and relation to agricultural productivity potential; international agricultural education and development. Mailing Add: Dept of Agron Life Sci Bldg Purdue Univ West Lafayette IN 47906

BRONSTEIN, EUGENE L, b New York, NY, Aug 28, 24; m 61. MEDICINE. Educ: Univ Md, MD, 48; Am Bd Radiol, dipl, 53. Prof Exp: Intern, Queens Gen Hosp, 48-49, sr intern, 49-50, asst resident radiol, 50-51, sr resident, 51-52; resident radiation ther, Mem Hosp, 52-53, clin asst radiation therapist, 53-58; from instr to asst prof clin radiol, Med Col, Cornell Univ, 57-67; ASSOC PROF CLIN RADIOL, NY MED COL, 67- Concurrent Pos: Asst attend radiation therapist, Mem Hosp, 58-; asst attend radiologist, Long Island Col Hosp, 58-; asst vis radiation therapist, James Ewing Hosp, 58- Mem: Am Radium Soc; fel Am Col Radiol; AMA; Am Soc Therapeut Radiol; Radiol Soc NAm. Res: Clinical research in treatment of cancer and allied diseases. Mailing Add:

BRONZAN, JOHN BRAYTON, b Los Angeles, Calif, Apr 5, 37; m 63; c 1. HIGH ENERGY PHYSICS. Educ: Stanford Univ, BS, 59; Princeton Univ, PhD(physics), 63. Prof Exp: From instr to assoc prof physics, Mass Inst Technol, 63-71; ASSOC PROF PHYSICS, RUTGERS UNIV, NEW BRUNSWICK, 71- Mem: Am Phys Soc. Res: Theoretical high energy physics, especially S-matrix theory and higher symmetries. Mailing Add: Dept of Physics Rutgers Univ New Brunswick NJ 08903

BRONZO, JOSEPH ALEXANDER, b Philadelphia, Pa, Feb 18, 23; m 61; c 2. APPLIED MATHEMATICS, ELECTRICAL ENGINEERING. Educ: Columbia Univ, BS, 43; Brown Univ, PhD(appl math), 52. Prof Exp: Mem staff radar circuit design, Radiation Lab, Mass Inst Technol, 43-45 & Sperry Gyroscope Co, 45-47; res assoc ultrasonics, Brown Univ, 47-53; researcher aerosol dispersion, Aeroproj, Inc, 53-54; mem tech staff, 54-62, HEAD DEPT DATA PROCESSING SUPPORT, BELL TEL LABS, INC, 62- Mem: Math Asn Am; sr mem Inst Elec & Electronics Engrs. Res: Real-time data processing systems; computer utilization; software design; data analysis; probability and statistics. Mailing Add: Bell Tel Labs Inc Whippany Rd Whippany NJ 07981

BROODO, ARCHIE, b Wichita Falls, Tex, Feb 3, 25; m 51; c 4. PHYSICAL CHEMISTRY, CHEMICAL ENGINEERING. Educ: Agr & Mech Col Tex, BS, 48; Univ Tex, MS, 50, PhD(chem), 54. Prof Exp: Consult electronic mat & hybrid film microcircuits, 68-70; PRES & GEN MGR, ANAL INVEST DETERMINATION, INC, 70- Concurrent Pos: Ed, Southwest Retort, Am Chem Soc, 57-58. Mem: Am Chem Soc; Inst Elec & Electronics Engrs; Int Asn Arson Investr; Am Soc Metals. Res: Technical investigation of insurance losses; chemical and other product liability; arson determinations; explosion cause analyses; materials studies and analytical chemistry; material flammability studies; corrosion studies; tire defect analyses. Mailing Add: 2924 Blystone Lane Dallas TX 75220

BROOK, ADRIAN GIBBS, b Toronto, Ont, May 21, 24; m 54; c 3. ORGANIC CHEMISTRY. Educ: Univ Toronto, BA, 47, PhD(chem), 50. Prof Exp: Lectr chem, Univ Sask, 50-51; Nuffield res fel, Imp Col Sci & Technol, 51-52; res assoc, Iowa State Col, 52-53; lectr, 53-56, from asst prof to assoc prof, 56-62, from assoc chmn to chmn dept, 68-74, PROF CHEM, UNIV TORONTO, 62- Concurrent Pos: Vis prof, Sch Molecular Sci, Univ Sussex, 74-75. Honors & Awards: F S Kipping Award in Organosilicon Chem, Am Chem Soc, 73. Mem: Am Chem Soc; Chem Inst Can. Res: Organosilicon, organogermanium chemistry; organometallic compounds; stereochemistry. Mailing Add: Dept of Chem Univ of Toronto Toronto ON Can

BROOK, ALAN J, b Newcastle-upon-Tyne, Eng, Mar 5, 23; m 49; c 3. PHYCOLOGY. Educ: Univ Durham, BSc, 42, PhD(bot, phycol), 49; Univ Edinburgh, DSc(phycol), 60. Prof Exp: Lectr bot, Univ Khartoum, 49-52; sr sci officer, Freshwater Fisheries Lab, Scotland, 52-58; lectr bot, Univ Edinburgh, 58-64; PROF BOT, UNIV MINN, MINNEAPOLIS, 64-, HEAD DEPT ECOL & BEHAV BIOL, 67- Concurrent Pos: Joint secy to exec, Int Bot Cong, Univ Edinburgh, 61-64. Mem: Int Asn Theoret & Appl Limnol; fel Royal Soc Edinburgh; Brit Phycol Soc. Res: Ecology of algae in water supplies; algal-animal feeding relationships; effects of fertilizers on phytoplankton; use of algae as indicators of trophic status of lakes; desmid biology and taxonomy, especially genus Staurastrum; micro-stratification of lake phytoplankton. Mailing Add: Dept of Ecol & Behav Biol Univ of Minn Minneapolis MN 55455

BROOK, MARX, b New York, NY, July 12, 20; m 47; c 3. PHYSICS. Educ: Univ NMex, BS, 44; Univ Calif, Los Angeles, MA & PhD(physics), 53. Prof Exp: Asst, Univ NMex, 43-46; res physicist, Univ Calif, Los Angeles, 47-53; res physicist, 54-58, assoc prof, 58-60, PROF PHYSICS, NMEX INST MINING & TECHNOL, 60-, CHMN DEPT, 68- Mem: Fel Am Phys Soc; Am Asn Physics Teachers; Am Meteorol Soc; Am Geophys Union. Res: Upper atmosphere physics; cloud physics and lightning. Mailing Add: NMex Inst of Mining & Technol Socorro NM 87801

BROOK, ROBERT B, b Morristown, NJ, Jan 13, 44; m 67. TOPOLOGY. Educ: Trenton State Col, BA, 65; Wesleyan Univ, PhD(math), 69. Prof Exp: Asst prof math, Univ Hartford, 69-70; ASST PROF MATH, WINTHROP COL, 70- Mem: Am Math Soc; Math Asn Am. Res: Topological dynamics; transformation groups; topological groups. Mailing Add: Dept of Math Winthrop Col Rock Hill SC 29730

BROOK, TED STEPHENS, b San Antonio, Tex, Apr 19, 26; m 48; c 2. ENTOMOLOGY. Educ: Trinity Univ, Tex, BS, 48; Kans State Univ, MS, 50; Miss State Univ, PhD(entom), 66. Prof Exp: Asst entomologist, Tex Agr Exp Sta, 50-54; from asst entomologist to assoc entomologist, Agr Exp Sta, 56-70, from asst prof to assoc prof entom, Miss State Univ, 64-70; plant protection adv, Univ Mo-Columbia-US AID India Prog, Dept Agr, Bihar, 70-72; assoc prof entom, Miss Agr Exp Sta,

Miss State Univ, 72-75; PEST MGT SPECIALIST, MISS COOP EXTEN SERV, 75- Mem: Entom Soc Am. Res: Household and ornamental plant pests; termite biology and control; insect biology and ecology; general entomology. Mailing Add: 901 Howard Rd Starkville MS 39759

BROOKBANK, JOHN WARREN, b Seattle, Wash, Mar 3, 27; m 50; c 3. DEVELOPMENTAL BIOLOGY. Educ: Univ Wash, BA, 49, MS, 53; Calif Inst Technol, PhD, 55. Prof Exp: Asst biol, Calif Inst Technol, 52-55; from asst prof to prof zool, 55-72, PROF ZOOL & MICROBIOL, UNIV FLA, 72- Mem: Am Soc Zoologists; Soc Develop Biol. Res: Regulation of gene expression during development. Mailing Add: Dept of Zool Univ of Fla Gainesville FL 32611

BROOKE, CLARKE HARDING, b Evanston, Ill, June 11, 20; m 49; c 3. GEOGRAPHY. Educ: Univ Wash, BA, 42, MA, 50; Univ Nebr, PhD(geog), 56. Prof Exp: Instr geog, Harar Inst, Ethiopia, 52-54; from asst prof to assoc prof, 54-61, PROF GEOG, PORTLAND STATE UNIV, 61- Concurrent Pos: NSF grant, Tanzania, 60-61; Fulbright-Hayes res grant, MidE, 64-65; Fulbright-Hayes lectr, Kabul Univ, Afghanistan, 67-68; NSF res grant, 74-75. Mem: Asn Am Geogr; African Studies Asn; Am Geog Asn; MidE Studies Asn NAm; Afghan Studies Asn. Res: Grazing practices and sheep husbandry; survival of rare breeds domesticated animals; endangered breeds of sheep in Southern Europe and Western Turkey. Mailing Add: Dept of Geog Portland State Univ Portland OR 97203

BROOKE, CLEMENT EUSTACE, b Chicago, Ill, Aug 20, 17; m 41; c 4. PEDIATRICS. Educ: Univ Chicago, MD, 48. Prof Exp: Instr physiol, Univ Chicago, 47-48; lectr Roosevelt Col, 48; intern pediat, Univ Rochester, 48-49, resident, 49-50; chief clin investr biol warfare, US Army, 50-55; from instr to assoc prof, 55-65, PROF PEDIAT, UNIV MO-COLUMBIA, 65- Concurrent Pos: Consult, US Army, 55-56; Markle scholar, 56-61; trustee, Georgia Brown Blosser Convalescent Hosp, 62-; pediat consult, Pac State Hosp, Calif, 63-64; mem, Gov Comn Ment Retardation, Mo, 64-; med dir, Multiple Handicap Clin & Woodhaven Learning Ctr. Mem: AMA; fel Am Acad Pediat; Am Asn Ment Deficiency. Res: Child development; brain damage and retardation. Mailing Add: Dept of Pediat Univ of Mo Columbia MO 65201

BROOKE, JOHN PERCIVAL, b Chicago, Ill, Oct 20, 33; m 70. GEOPHYSICS, GEOCHEMISTRY. Educ: Mich Technol Univ, BS, 54; Univ Utah, MS, 59, PhD(geol eng), 64. Prof Exp: Geologist, Am Smelting & Refining Co, 60-61; asst prof geophys & struct geol, 61-70, ASSOC PROF GEOL, GEOPHYS & ENG GEOL, SAN JOSE STATE COL, 70- Mem: Soc Explor Geophys. Res: Instrumental and field analysis of geochemical alteration halos surrounding ore bodies; mineral exploration; operations research; geophysical and geochemical prospecting; engineering and environmental geology. Mailing Add: Dept of Geol San Jose State Col San Jose CA 95114

BROOKE, MARION MURPHY, b Atlanta, Ga, Dec 6, 13; m 40; c 3. PARASITOLOGY. Educ: Emory Univ, AB, 35, AM, 36; Johns Hopkins Univ, ScD(protozool), 42; Am Bd Microbiol, dipl. Prof Exp: Instr biol, Emory Jr Col, 36-38; asst protozool, Sch Hyg & Pub Health, Johns Hopkins Univ, 38-40, instr, 40-42, assoc parasitol, 42-44; assoc prof prev med, Col Med, Univ Tenn, 44-45; chief parasitol sect, Lab Br, Ctr Dis Control, USPHS, 45-51, chief parasitol & mycol sect, 51-57, chief microbiol sect, 57-62, chief lab consult & develop sect, 62-69, dep chief licensure & develop br, 69-72, ASSOC DIR HEALTH LAB MANPOWER DEVELOP, LAB TRAINING & CONSULT DIV, USPHS, 72- Concurrent Pos: Dir interstate malaria surv, Bd State Health Comn Upper Mississippi River Basin, 42; assoc prof, Med Sch, Emory Univ, 46- & Sch Pub Health, Univ NC, 63-; vis investr, Sch Trop Med, Univ PR, 46; mem joint dysentery unit, Armed Forces Epidemiol Bd Korea, 51, assoc mem comn enteric infections, 54-74; mem expert adv panel parasitic dis, WHO, 64- & Conf State & Prov Pub Health Lab Dirs. Honors & Awards: USPHS Meritorious Serv Award, 65. Mem: Am Soc Parasitol; Soc Trop Med & Hyg; Am Pub Health Asn; Am Acad Microbiol; Am Soc Allied Health Professions. Res: Proficiency examinations; health laboratory manpower; malaria; amoebiasis; intestinal parasites; toxoplasmosis; programmed instruction. Mailing Add: Health Lab Manpower Develop USPHS Atlanta GA 30333

BROOKEMAN, VALERIE ANN, b Dundee, Scotland, Aug 2, 43; m 65; c 2. RADIOLOGICAL PHYSICS. Educ: St Andrews Univ, BSc, 65, PhD(physics), 68. Prof Exp: From instr to asst prof, 68-72, ASSOC PROF RADIOL PHYSICS, UNIV FLA, 72- Concurrent Pos: Consult, Vet Admin Hosp, 71-; mem radiol device panel, Bur Med Devices & Diag Prod, Food & Drug Admin, 74-, res contract, 74- Mem: Am Asn Physicists in Med; Soc Nuclear Med; AAAS; Sigma Xi. Res: Nuclear medicine physics; internal radiation, absorbed dosimetry; gamma camera distortions and quality control; breath tests. Mailing Add: Dept of Radiol Box J-385 JHMHC Univ Fla Div Radiation Physics Gainesville FL 32610

BROOKER, FRANCIS MILTON, b Ill, May 7, 15; m 42; c 3. FOOD CHEMISTRY. Educ: Washington Univ, MS, 40; Univ Chicago, cert meteorol, 44. Prof Exp: Asst chem, Washington Univ, 38-40; chemist, Wonder Orange Co, 40-41; chief chemist, B T Fooks Mfg Co, 41-43; flavor chemist, 46-55, CHIEF CHEMIST, GRAPETTE CO, INC, 55- Mem: Am Chem Soc; Inst Food Technologists; Soc Soft Drink Technologists. Res: Improvement and stability of present soft drink flavors and development of new flavors. Mailing Add: 1033 Westwood Rd Camden AR 71701

BROOKER, GARY, b San Diego, Calif, Mar 24, 42; m 64; c 1. PHARMACOLOGY, BIOCHEMISTRY. Educ: Univ Southern Calif, BS, 66, PhD(cardiac pharmacol), 68. Prof Exp: Asst prof med, Sch Med, Univ Southern Calif, 68-72, asst prof biochem, 69-72; ASSOC PROF PHARMACOL, MED SCH, UNIV VA, 72- Concurrent Pos: Los Angeles County Heart Asn res grants, 68-70; Am Heart Asn res grant, 70-73; USPHS res grants, 71-74. Mem: AAAS; Am Fedn Clin Res; Am Soc Pharmacol & Exp Therapeut. Res: Mechanisms of hormone and drug action as it relates to cardiac metabolism and cardiac contraction; development of methods for rapid analysis of substances of biological interest. Mailing Add: Dept of Pharmacol Univ of Va Med Sch Charlottesville VA 22901

BROOKER, HAMPTON RALPH, b Columbia, SC, Sept 11, 34; m 62; c 2. NUCLEAR MAGNETIC RESONANCE. Educ: Univ Fla, BS, 56, MS, 58, PhD(physics), 62. Prof Exp: Asst prof physics, Abilene Christian Col, 61-63; fel, Univ Fla, 63-64; asst prof, 64-74, ASSOC PROF PHYSICS, UNIV S FLA, 74- Mem: Am Phys Soc; Am Asn Physics Teachers. Res: Quadrupole resonance. Mailing Add: Dept of Physics Univ of S Fla Tampa FL 33620

BROOKER, MURRAY H, physical chemistry, spectroscopy, see 12th edition

BROOKER, ROBERT MUNRO, b Troy Grove, Ill, Jan 3, 18; m 41; c 2. ORGANIC CHEMISTRY. Educ: Univ Mo, BS, 47, PhD(chem), 50. Prof Exp: Asst, Univ Mo, 46-50; head dept chem, 50-66, CHMN DIV SCI & MATH, IND CENT UNIV, 66- Concurrent Pos: Res Chemist, Pitman-Moore Co, 50-65; consult, Dow Chem Co, 65-; pres, Short Courses, Inc, 73- Mem: Am Chem Soc. Res: Acenaphthene; veratrum alkaloids; natural products. Mailing Add: Div of Sci & Math Ind Cent Univ Indianapolis IN 46227

BROOKES, JOHN A, protozoology, parasitology, see 12th edition

BROOKES, VICTOR JACK, b Batavia, Java, NEI, Mar 3, 26; nat US; m 56; c 4. ENTOMOLOGY. Educ: Univ Mich, BS, 50; Univ Ill, MS, 51, PhD(entom), 56. Prof Exp: Asst entom, Univ Ill, 51-56, res assoc, 56; res assoc entom & chem, 56, USPHS fel, 56-58, from asst prof to assoc prof entom, 58-70, PROF ENTOM, SCI RES INST, ORE STATE UNIV, 70- Concurrent Pos: NIH career develop award, 62; res fel, Biol Labs, Harvard Univ, 63-64. Mem: AAAS; NY Acad Sci; Am Entom Soc. Res: Insect physiology and biochemistry; insect metabolism of protein and lipids; biochemistry of insect development. Mailing Add: Dept of Entom Ore State Univ Corvallis OR 97331

BROOKHART, JOHN MILLS, b Cleveland, Ohio, Dec 1, 13; m 39; c 3. PHYSIOLOGY. Educ: Univ Mich, BS, 35, MS, 36, PhD(physiol), 39. Prof Exp: Asst physiol, Univ Mich, 35-39; instr, Sch Med, Loyola Univ, Ill, 40-42, assoc, 42-45, asst prof, 45-46; asst prof neurol, Med Sch, Northwestern Univ, 47-49; assoc prof, 49-52, PROF PHYSIOL & HEAD DEPT, MED SCH, UNIV ORE HEALTH SCI CTR, 52- Concurrent Pos: Fel, Inst Neurol, Med Sch, Northwestern Univ, 39-40; spec consult physiol study sect, USPHS, 51-55, neurol study sect, 57-60, gen clin res ctr comt, 61-63 & physiol training comt, 63-67; Fulbright res scholar, Univ Pisa, 56-67; mem physiol test comt, Nat Bd Med Exam, 59-62; mem bd sci counr, Nat Inst Neurol Dis & Blindness, 61-65; mem, Int Brain Res Orgn, 62-; ed-in-chief, J Neurophysiol, 64-74; mem cent coun, Int Brain Res Orgn, 64-68 & 74-; mem adv coun health res facil, NIH, 67-71; US del, Gen Assembly, Int Union Physiol Sci, 65, 68, 71 & 74, chmn, US Am Nat Comt, 69-75. Mem: AAAS; Am Physiol Soc (pres-elect, 64-65, pres, 65-66); Am Acad Arts & Sci. Res: Regional neurophysiology; systems analysis of postural control. Mailing Add: Dept Physiol Med Sch Univ Ore Health Sci Ctr Portland OR 97201

BROOKHART, MAURICE S, b Cumberland, Md, Nov 28, 42; m 65; c 2. ORGANIC CHEMISTRY. Educ: Johns Hopkins Univ, BA, 64; Univ Calif, Los Angeles, PhD(org chem), 68. Prof Exp: NATO fel, Univ Southampton, 68-69; ASSOC PROF ORG CHEM, UNIV NC, CHAPEL HILL, 69- Mem: Am Chem Soc; The Chem Soc. Res: Structure and rearrangements of carbonium ions; symmetry control of organic reactions; certain aspects of organometallic chemistry. Mailing Add: Dept of Chem Univ of NC Chapel Hill NC 27514

BROOKINS, DOUGLAS GRIDLEY, b Healdsburg, Calif, Sept 27, 36; m 61; c 2. GEOCHEMISTRY. Educ: Univ Calif, Berkeley, AB, 58; Mass Inst Technol, PhD(isotope geol), 63. Prof Exp: Asst emission spectrog, Mass Inst Technol, 58-59, asst geochronology, 59-63; asst prof geochem, Kans State Univ, 63-66, assoc prof geol, 66-71; PROF GEOL, UNIV NMEX, 71- Concurrent Pos: Consult, Mass Inst Technol, 63; with US Geol Surv, State Surv Maine, Conn & Kans; vis staff mem, Los Alamos Sci Lab, 74- Mem: AAAS; Am Geophys Union; Geol Soc Am; Geochem Soc; Meteoritical Soc. Res: Geochronological investigations on problems of petrogenesis and correlation in regionally metamorphosed areas by the rubidium-strontium methods phase equilibrium; kimberlites; carbonatites; uranium geochemistry. Mailing Add: Dept of Geol Univ of NMex Albuquerque NM 87131

BROOKMAN, DAVID JOSEPH, b Ft Collins, Colo, Oct 31, 43; m 65. ANALYTICAL CHEMISTRY. Educ: Colo State Univ, BS, 65; Univ Calif, Riverside, PhD(anal chem), 68. Prof Exp: Asst prof chem, Univ Wis-Madison, 68-70; from res chemist to sr res chemist, 70-73, GROUP SUPVR, STAUFFER CHEM CO, 74- Mem: AAAS; Am Chem Soc. Res: Environmental chemistry; analytical separations. Mailing Add: Stauffer Chem Co Western Res Ctr 1200 S 47th St Richmond CA 94804

BROOKS, ALBERT LAW, b Manchester, Conn, Jan 3, 29; div; c 1. BIOLOGICAL OCEANOGRAPHY. Educ: Univ RI, BS, 51, MS, 57, PhD(biol oceanog), 65. Prof Exp: Teacher, South Kingstown, RI, 57-58; asst to dir, Bermuda Biol Sta Res, Inc, 65-68; OCEANOGR, NAVY UNDERWATER SYSTS CTR, 68- Mem: Am Soc Limnol & Oceanog. Res: Deep ocean zooplankton. Mailing Add: Navy Underwater Systs Ctr New London Lab Ft Trumbull New London CT 06320

BROOKS, ALFRED AUSTIN, JR, b Swampscott, Mass, Aug 3, 21; m 43; c 2. PHYSICAL CHEMISTRY. Educ: Hobart Col, BA, 46; Ohio State Univ, PhD(chem), 50. Prof Exp: Chemist, Lake Ontario Ord Works, 43 & Manhattan Proj, 43-45; instr, Univ Chicago, 50-52; res chemist, Standard Oil Co, 52-54 & Upjohn Co, 54-56; phys chemist, Union Carbide Nuclear, 56-62 & Cent Data Processing Div, 62-70; head comput applns dept, Oak Ridge Nat Lab, 70-73; MGR COMPUT APPLN DEPT, COMPUT SCI DIV, NUCLEAR DIV, UNION CARBIDE CORP, 73- Mem: Am Chem Soc; Am Phys Soc; Asn Comput Mach; Am Soc Info Sci. Res: Application of high speed computers to research problems. Mailing Add: 100 Wiltshire Dr Oak Ridge TN 37830

BROOKS, ANTONE L, b St George, Utah, July 7, 38; m 63; c 3. CYTOGENETICS, RADIOBIOLOGY. Educ: Univ Utah, BS, 61, MS, 63; Cornell Univ, PhD(phys biol), 67. Prof Exp: MEM SR STAFF, LOVELACE FOUND, 67- Mem: Radiation Res Soc; Environ Mutagen Soc. Res: Distribution and uptake of radioactive fallout, effects of radiation on the chromosomes of somatic and reproductive tissue; repair of radiation-induced chromosomal damage; effects of internal emitters on chromosomes; hazards of alpha emitters. Mailing Add: Inhalation Toxicol Res Inst Lovelace Found PO Box 5890 Albuquerque NM 87115

BROOKS, ARTHUR M, physiology, see 12th edition

BROOKS, AUSTIN EDWARD, b Ft Wayne, Ind, Aug 10, 38; m 63; c 2. PHYCOLOGY. Educ: Wabash Col, AB, 61; Ind Univ, PhD(microbiol), 65. Prof Exp: Res assoc biol, Brown Univ, 65-66; asst prof, 66-72, ASSOC PROF BIOL, WABASH COL, 72- Concurrent Pos: Alexander von Humboldt Found fel, 74-75. Mem: AAAS; Soc Protozool; Phycol Soc Am; Int Phycol Soc. Res: Morphology, physiology and genetics of green flagellated algae; physiological effects of environmental chemicals on the green algae. Mailing Add: Dept of Biol Wabash Col Crawfordsville IN 47933

BROOKS, CHANDLER MCCUSKEY, b Waverly, WVa, Dec 18, 05; m 32. PHYSIOLOGY. Educ: Oberlin Col, AB, 28; Princeton Univ, MA, 29, PhD, 31; Berea Col, DSc, 69. Prof Exp: Asst physiol, Harvard Med Sch, 31-33; instr, Sch Med, Johns Hopkins Univ, 33-35, assoc, 35-42, assoc prof, 42-48; prof physiol & pharmacol, Long Island Col Med, 48-51; PROF PHYSIOL, STATE UNIV NY DOWNSTATE MED CTR, 51-, DEAN MED, 70- Concurrent Pos: Dir grad educ prog, State Univ NY Downstate Med Ctr, 56-, actg pres, 70-71; lectr, Post-Grad Comt Med, Australia; external examr, Sch Med, Univ Otago, NZ; vis prof, Med Schs, Kobe & Tokyo, 61-62; trustee, Int Found, 72, chmn grants comt, 74-; vis scholar, Marischal Col, Univ Aberdeen, 73; vis prof, Nat Defense Med Ctr, Taiwan, 74; mem neurol study sect & training grant comt, Nat Heart Inst; mem res career awards comt, Div Gen Med Sci, NIH. Mem: Am Physiol Soc; Soc Exp Biol & Med; Endocrine Soc; Am Soc Pharmacol & Exp Therapeut; Harvey Soc. Res: Functions and control of autonomic system; control of endocrine functions; cardiac physiology; neurophysiology; history of scientific thought. Mailing Add: Downstate Med Ctr State Univ NY Brooklyn NY 11203

BROOKS, CLYDE S, b Pittsburgh, Pa, May 1, 17; m 46; c 2. PHYSICAL CHEMISTRY. Educ: Duke Univ, BS, 40. Prof Exp: Koppers Co fel tar synthetics, Mellon Inst, 41-44, jr fel coal prod phys chem, 46-49; res chemist prod res, Shell Develop Co, 49-61; SR RES SCIENTIST, RES LABS, UNITED TECHNOL CORP, 61- Mem: AAAS; Sigma Xi (secy, 73-75); Am Chem Soc; Am Inst Chemists. Res: Surface chemistry; heterogeneous catalysis; chromatography; kinetics; colloids; application of physical chemistry to elimination of environmental pollution; applications of catalysis in fossil fuel and chemical conversion processes. Mailing Add: 41 Baldwin Lane Glastonbury CT 06033

BROOKS, COY CLIFTON, b Rose Bud, Ark, Sept 10, 18; m 40; c 1. ANIMAL NUTRITION. Educ: Ark State Col, BS, 42; Univ Mo, MS, 48, PhD(animal nutrit), 54. Prof Exp: Assoc prof animal husb, Ark State Col, 47-52; instr, Univ Mo, 52-54; prof, Va Polytech Inst, 54-66; PROF ANIMAL HUSB, UNIV HAWAII, 66- Mem: Am Soc Animal Scientists; Sigma Xi. Res: Protein quality; growth and development patterns as they affect muscle development; baby pig nutrition; use of tropical feeds in swine diets; vitamin K deficiency and heart lesions in swine. Mailing Add: Dept of Animal Sci Univ of Hawaii Honolulu HI 96822

BROOKS, DAVID W, physical organic chemistry, see 12th edition

BROOKS, DERL, b Headrick, Okla, July 25, 30; m 57; c 1. ENTOMOLOGY. Educ: Tex Tech Col, BS, 57, MS, 59; Iowa State Univ, PhD(entom), 62. Prof Exp: Assoc prof biol, Ark State Col, 62-65; assoc prof, 65-74, PROF BIOL, W TEX STATE UNIV, 74- Res: Acarology, chiefly nasal mites of birds. Mailing Add: Dept of Biol W Tex State Univ Canyon TX 79015

BROOKS, DOYLE, physics, see 12th edition

BROOKS, EDWARD MORGAN, b New Haven, Conn, Mar 19, 16; m 41; c 8. METEOROLOGY. Educ: Harvard Univ, AB, 37; Mass Inst Technol, SM, 39, ScD(meteorol), 45. Prof Exp: Asst observer & asst, Blue Hill Observ, Harvard Univ, 37-39; asst & map analyst, Radiosonde Sta, Mass Inst Technol, 39; map analyst & terminal forecaster, Pan Am Airways, Hawaii, 39-40, tutor meteorol, Calif, 40-41, instr, Fla, 41-42; instr, Mass Inst Technol, 42-46; from asst prof to prof geophys, Inst Technol, St Louis, 46-61; sr meteorologist, Allied Res Assocs, Mass, 61-63; staff scientist, Geophysics Corp Am, 63-65; lectr geophys, 65-67, PROF GEOL & GEOPHYS, BOSTON COL, 68- Concurrent Pos: Scholar, Milton Acad, 32-33; lectr, Boston Ctr Adult Educ, Mass, 43; ed, Mt Washington Observ News Bull, 65-; meteorologist, Wallace Howell Assoc, 67 & Edgerton, Germeshausen & Grier, 67; prof physics, State Univ NY Plattsburgh, 67-68; guest prof meteorol, McGill Univ, 67-68. Mem: AAAS; Am Meteorol Soc; Am Geophys Union; Am Astron Soc. Res: Physical oceanography; tropical cyclones; polar front in tropics; tropical cloudiness; solar radiation; microbarography; tornadoes; Coriolis force; pilot baloon accuracy; isentropic analysis; satellite meteorology. Mailing Add: Dept of Geol & Geophys Boston Col Chestnut Hill MA 02167

BROOKS, ELWOOD RALPH, b Charlevoix, Mich, Aug 10, 34; m 64; c 2. GEOLOGY, PETROLOGY. Educ: Mich Col Mining & Technol, BS, 56; Univ Calif, Berkeley, MS, 58; Univ Wis, PhD(geol), 64. Prof Exp: Mine geologist, White Pine Copper Co, Mich, 59-60; lectr geol, Univ Calif, Riverside, 64-65; PROF GEOL, CALIF STATE UNIV, HAYWARD, 65- Concurrent Pos: Res grants, 64, 66-67, 70-72 & 74. Mem: Geol Soc Am; Nat Asn Geol Teachers. Res: Igneous and metamorphic petrology, particularly volcanic petrology; structural analysis; optical mineralogy; field geology. Mailing Add: Dept of Earth Sci Calif State Univ Hayward CA 94542

BROOKS, FOSTER (LINDSEY), b Carrollton, Ohio, Sept 4, 08. MATHEMATICS. Educ: Mt Union Col, AB, 29; Ohio State Univ, PhD(math), 34. Prof Exp: Asst math, Ohio State Univ, 29-31; high sch instr, Ohio, 33-35; from instr to prof math, 35-73, EMER PROF MATH, KENT STATE UNIV, 73- Mem: Am Math Soc; Math Asn Am; Soc Indust & Appl Math; Opers Res Soc Am. Res: Theory of functions of a real variable. Mailing Add: Dept of Math Kent State Univ Kent OH 44242

BROOKS, FRANK PICKERING, b Portsmouth, NH, Jan 2, 20; m 42; c 3. PHYSIOLOGY. Educ: Dartmouth Col, AB, 41; Univ Pa, MD, 43, ScD(med), 51. Prof Exp: Intern & resident med, Hosp Univ Pa, 44-46 & 50-51, instr & assoc med, Sch Med, 52-54, from asst prof to assoc prof med & physiol, 54-70, PROF MED & PHYSIOL, SCH MED, UNIV PA, 62- Concurrent Pos: USPHS fel, Jefferson Med Col, 51-52; lectr, Univ Edinburgh, 55-56 & 74-75, Univ Calif, Los Angeles, 66-67; chief gastrointestinal sect, Hosp Univ Pa, 62-72, co-chief, 72-74; USPHS res career develop award, 63-70; res assoc, Vet Admin, Los Angeles, 66-67; mem training grants comt in gastroenterol & nutrit, Nat Inst Arthritis & Metab Dis, 66-70; mem gen med study sect A, Nat Inst Arthritis & Digestive Dis, 72-76; mem subspecialty bd gastroenterol, Am Bd Internal Med, 75-; consult, Food & Drug Admin, 75. Mem: Am Physiol Soc; Am Gastroenterol Asn; fel Am Col Physicians; Am Clin & Climat Asn; Asn Am Physicians. Res: Gastrointestinal physiology; nervous regulation of gastrointestinal function. Mailing Add: Gastrointestinal Sect Hosp of Univ of Pa Philadelphia PA 19104

BROOKS, FRANK TURNER, veterinary medicine, radiobiology, see 12th edition

BROOKS, FRANKLIN COOLIDGE, b Brooklyn, NY, Feb 16, 27. MATHEMATICAL PHYSICS. Educ: Yale Univ, BS, 47, MS, 48, PhD(physics), 50. Prof Exp: Sr scientist, Atomic Power Div, Westinghouse Elec Corp, 50-52; chmn armour group, Res Off, Johns Hopkins Univ, 52-54; dir combat oper res group, US Continental Army Command, 54-56; mgr west coast res off, 56-57, dir opers res, 57-61, corp fel, 62-65, SCI CONSULT, TECH OPERS, INC, 65- Concurrent Pos: Consult, US Army War Col, 58. Res: Theoretical physics; applied mathematics; military and commercial operations research. Mailing Add: 1832 La Coronilla Dr Santa Barbara CA 93109

BROOKS, GARNETT RYLAND, JR, b Richmond, Va, Nov 25, 36; m 59. ECOLOGY. Educ: Univ Richmond, BS, 57, MS, 59; Univ Fla, PhD(zool), 63. Prof Exp: From instr to asst prof, 62-68, ASSOC PROF BIOL, COL WILLIAM & MARY, 68- Mem: AAAS; Ecol Soc Am; Am Soc Ichthyologists & Herpetologists. Res: Physiological ecology of certain species of reptiles. Mailing Add: Dept of Biol Col of William & Mary Williamsburg VA 23185

BROOKS, GEORGE FRANK, JR, b Bellingham, Wash, Sept 17, 38; m 72; c 3. INTERNAL MEDICINE, INFECTIOUS DISEASES. Educ: Univ Wash, BA, 61, MS & MD, 66. Prof Exp: From intern to resident med, Cornell Med Div, Bellevue Hosp, NY, 66-68; Epidemic Intel Serv officer pub health, Ctr Dis Control, Atlanta, Ga, 68-70; fel infectious dis, Sch Med, Univ Wash, 70-72; asst prof med, 72-75, ASSOC PROF MED, SCH MED, IND UNIV, INDIANAPOLIS, 75- Concurrent Pos: Prin investr training grant, Sch Med, Ind Univ, 74-76; consult, NIH, 76. Mem: Fel Am Col Physicians; Am Fedn Clin Res; Am Soc Microbiol; Am Venereal Dis Asn. Res: Immunology and epidemiology of gonorrhea and other sexually transmitted diseases. Mailing Add: Emerson Hall Rm 302 Ind Univ Sch of Med Indianapolis IN 46202

BROOKS, GEORGE WILSON, b Warren, Vt, Feb 11, 20; m 43; c 4. PSYCHIATRY. Educ: Univ NH, BS, 41; Univ Vt, MD, 44; Am Bd Psychiat & Neurol, dipl psychiat, 55. Prof Exp: Intern, Mary Fletcher Hosp, Burlington, 44-45; resident psychiat, Vt State Hosp, Waterbury, 47-49, asst physician, 49-51; exchange resident, NH State Hosp, Concord, 52; sr psychiatrist, Vt State Hosp, Waterbury, 53-56, dir res & staff educ, 57-61, asst supt, 61-68, actg supt, 68; from instr to assoc prof, 53-70, PROF CLIN PSYCHIAT, COL MED, UNIV VT, 70-; SUPT, VT STATE HOSP, 68- Concurrent Pos: Smith, Kline & French Found fel, Mass Ment Health Ctr, Boston, 56-57. Honors & Awards: Citation Meritorious Serv, President's Comt Employ Physically Handicapped, 59; Citation, Nat Rehab Asn, 63. Mem: Fel Am Psychiat Asn; AMA; Asn Med Supt Ment Hosps; fel Am Col Physicians. Res: Biological and behavioral research; schizophrenia. Mailing Add: Vt State Hosp 103 S Main St Waterbury VT 05676

BROOKS, HAROLD KELLY, b Winfield, Kans, Nov 27, 24; m 62. GEOLOGY. Educ: Kans State Univ, BS, 47; Harvard Univ, AM, 50, PhD, 62. Prof Exp: Instr geol, Brown Univ, 50-51; instr, Oberlin Col, 51-52; asst prof, Univ Tenn, 52-54 & Univ Cincinnati, 54-56; from asst prof to assoc prof, 56-71, PROF GEOL, UNIV FLA, 71- Mem: AAAS; fel Geol Soc Am; Paleont Soc; Am Asn Petrol Geologists; fel Geol Soc London. Res: Marine geology; sedimentology; paleontology. Mailing Add: Dept of Geol Univ of Fla Gainesville FL 32601

BROOKS, HARVEY, b Cleveland, Ohio, Aug 5, 15; m 45; c 4. PHYSICS. Educ: Yale Univ, AB, 37; Harvard Univ, PhD, 40. Hon Degrees: DSc, Yale Univ, 62, Union Col, 62, Harvard Univ, 63, Kenyon Col, 63, Brown Univ, 64. Prof Exp: Mem staff, Underwater Sound Lab, Harvard Univ, 42-45; asst dir, Ord Res Lab, Pa State Univ, 45; res assoc, Res Lab & assoc lab head, Knolls Atomic Power Lab, Gen Elec Co, 46-50; dean div eng & appl physics, 57-74, PROF APPL PHYSICS, DIV ENG & APPL PHYSICS, HARVARD UNIV, 50- Concurrent Pos: Guggenheim fel, 56-57; ed-in-chief, J Physics & Chem Solids; mem adv comts reactor safeguards, progs & policies, AEC, 58; chmn solid state adv panel, Off Naval Res; chmn comt undersea warfare, Nat Res Coun; mem, President's Sci Adv Comt & Nat Sci Bd; chmn comt sci & pub policy, Nat Acad Sci. Mem: Nat Acad Sci; Nat Acad Eng; Am Acad Arts & Sci; Am Philos Soc; fel Am Phys Soc. Res: Solid state physics; underwater sound; nuclear reactors. Mailing Add: Div of Eng & Appl Physics Harvard Univ Cambridge MA 02138

BROOKS, HUGH CAMPBELL, b Seattle, Wash, June 19, 22; m 50; c 2. CULTURAL GEOGRAPHY. Educ: Univ Wash, BA, 47; Inst Int Rels, Geneva, MA, 49; Columbia Univ, MA, 51, EdD, 54. Prof Exp: Asst, Teachers Col, Columbia Univ, 52-54; asst prof, Ore State Univ, 54-55; assoc prof, NJ State Col, 58-61; ASSOC PROF & DIR CTR AFRICAN STUDIES, ST JOHN'S UNIV, NY, 61- Concurrent Pos: Lectr, Hunter Col, 52-54; Fulbright lectr, Univ witwatersrand, 55-58; NY State scholar incentive awards, Columbia Univ, 65 & Syracuse Univ, 68; consult ed, McGraw Hill Publ Co, 67-70, Nystrom Map Co, 67-73, Grolier Publ Co, 67-75, Sadlier Publ Co, 68-, Negro Univs Press, 70-73 & Hunter Develop Co, 71-75. Mem: AAAS; Am Geog Asn; Asn Am Geogr; African Studies Asn; Royal Geog Soc. Res: Industrial development and resource utilization in sub-Sahara Africa; human resources and cultural change. Mailing Add: Ctr for African Studies St John's Univ Jamaica NY 11439

BROOKS, JAMES ELWOOD, b Salem, Ind, May 31, 25; m 49; c 3. GEOLOGY. Educ: DePauw Univ, AB, 48; Northwestern Univ, MS, 50; Univ Wash, PhD(geol), 54. Prof Exp: From instr to assoc prof geol, 51-63, chmn dept, 62-70, assoc provost, 70-72, PROF GEOL, SOUTHERN METHODIST UNIV, 63-, VPRES & PROVOST, 72- Concurrent Pos: Consult geologist, Gulf Oil Corp, 52-53, DeGolyer & MacNaughton Inc, 54-59. Mem: Fel AAAS; fel Geol Soc Am. Res: Devonian and mid-Paleozoic stratigraphy and sedimentary petrology of north central and west Texas; stratigraphy and sedimentation; history of geologic concepts. Mailing Add: Southern Methodist Univ Dallas TX 75275

BROOKS, JAMES EUGENE, b Forest, Wash, Oct 10, 25; m 47; c 5. ECONOMIC GEOGRAPHY. Educ: Cent Wash State Col, BA, 49; Univ Wash, MA, 52, PhD(geog), 57. Prof Exp: Instr geog, Cent Wash State Col, 52; from instr to asst prof geog & geol, Eastern Wash State Col, 53-58; asst prof geog, Portland State Col, 58-59, asst to pres, 59-61; PROF GEOG & PRES, CENT WASH STATE COL, 61- Res: Pacific Northwest geography; water resources; conservation of natural resources. Mailing Add: 211 E Tenth Ellensburg WA 98926

BROOKS, JAMES KEITH, b Cleveland, Ohio, Sept 26, 38; m 61; c 1. MATHEMATICS. Educ: John Carroll Univ, BS, 59, MS, 61; Ohio State Univ, PhD(math), 64. Prof Exp: Asst prof math, Ohio State Univ, 64-66; vis lectr, Univ Southampton, 66-67; from asst prof to assoc prof, 67-74, PROF MATH, UNIV FLA, 74- Mem: Am Math Soc; Math Asn Am. Res: Functional analysis, measure and integration theory. Mailing Add: Dept of Math Univ of Fla Gainesville FL 32601

BROOKS, JAMES LEE, b Toledo, Ohio, Sept 11, 37; m 62; c 3. BIOCHEMISTRY. Educ: San Diego State Col, BS, 59, MS, 61; Univ Calif, Davis, PhD(biochem), 65. Prof Exp: Res assoc biochem, Univ Calif, Davis, 65, Scripps Inst, Univ Calif, 65-66 & Okla State Univ, 66-67; asst prof, 67-74, asst agr biochemist, 69-74, ASSOC PROF AGR BIOCHEM & ASSOC AGR BIOCHEMIST, W VA UNIV, 74- Mem: AAAS; Am Chem Soc. Res: Control of carbohydrate metabolism; fatty acid biosynthesis; sulfur metabolism in algae and bacteria; bacterial lipids; lipids in relation to membrane structure and cell permeability. Mailing Add: Plant Sci Div-Hort W Va Univ Morgantown WV 26506

BROOKS, JAMES O, b Evanston, Ill, July 7, 30; m 58; c 2. ALGEBRA. Educ: Oberlin Col, AB, 52; Univ Mich, MA, 53, PhD(math), 64. Prof Exp: Asst prof math, Haverford Col, 59-64; asst prof, 64-66, actg chmn dept, 68-69, ASSOC PROF MATH, VILLANOVA UNIV, 66-, CHMN DEPT, 69- Mem: Am Math Soc; Math Asn Am. Res: Algebraic number theory. Mailing Add: Dept of Math Villanova Univ Villanova PA 19085

BROOKS, JERRY R, b Barboursville, WVa, Sept 9, 30. REPRODUCTIVE ENDOCRINOLOGY. Educ: WVa Univ, BS, 53, MS, 58; Univ Mo, PhD(animal husb), 61. Prof Exp: Asst dairy husb, Univ Mo, 61-63; Ford Found res fel reproductive physiol, Worcester Found Exp Biol, Mass, 63-65; sr endocrinologist, 66-72, RES FEL, MERCK & CO, INC, 72- Mem: Am Soc Animal Sci; Soc Study Reproduction; Sigma Xi. Res: Causes of and therapy for benign prostatic hyperplasia; anti-acne agents; anti- progestational agents. Mailing Add: Merck Inst Rahway NJ 07065

BROOKS, JOHN BILL, bIsabella, Tenn, Aug 9, 29; m 51; c 2. BIOCHEMISTRY, MICROBIOLOGY. Educ: Western Carolina Univ, BS, 62; Va Polytech Inst,

PhD(biochem, microbiol), 69. Prof Exp: Chemist, Tenn Copper Co, 59; biol aide, Ctr Dis Control, 61-62; chemist, NIH, 62-63; biochemist, 63-69, RES CHEMIST, CTR DIS CONTROL, 69- Honors & Awards: Serv Award, Ctr Dis Control, 70. Mem: Am Soc Microbiol; Sigma Xi. Res: Develop methods for rapid identification of microorganisms by analysis of their metabolites or cellular constituents both in vivo and in vitro by advanced chemical procedures such as gas chromatography and spectrophotometry. Mailing Add: Bldg 4 Rm 112 Ctr for Dis Control Atlanta GA 30333

BROOKS, JOHN LANGDON, b New Haven, Conn, Feb 10, 20; m 53; c 1. ECOLOGY, SYSTEMATICS. Educ: Yale Univ, BS, 41, MS, 42, PhD(zool), 46. Prof Exp: Lab asst, Yale Univ, 42-45, asst instr, 45-46, from instr to asst prof zool, 46-56, assoc prof biol, 56-69; prog dir gen ecol, 69-75, sect head ecol & pop biol, 75, DEP DIV DIR, ENVIRON BIOL, NSF, 75- Concurrent Pos: Vis prof, Univ Rangoon, 48-49; ed, Syst Zool, 52-59. Mem: Am Soc Zoologists; Soc Syst Zool; Ecol Soc Am; Am Soc Limnol & Oceanog; Soc Study Evolution. Res: Ecology and evolution of fresh water organisms; history of evolutionary concepts. Mailing Add: Nat Sci Found Washington DC 20550

BROOKS, JOHN ROBINSON, b Cambridge, Mass, Nov 15, 18; m 44; c 4. SURGERY. Educ: Harvard Univ, AB, 40, MD, 43. Prof Exp: CHIEF SURG, HARVARD UNIV HEALTH SERV, 62-, PROF SURG, HARVARD MED SCH, 70- Mem: Am Col Surg; Soc Univ Surg; Am Surg Asn; Int Soc Surg; Soc Surg Alimentary Tract. Res: Gastrointestinal surgery; transplantation. Mailing Add: 721 Huntington Ave Boston MA 02115

BROOKS, KATHERINE, biology, see 12th edition

BROOKS, LESTER ALLEN, b Boston, Mass, May 20, 14; m 42; c 4. RUBBER CHEMISTRY. Educ: Mass Inst Technol, BS, 35; Univ Ill, PhD(org chem), 41. Prof Exp: Res chemist, Hooker Electrochem Co, Niagara Falls, 35-37; dir org res, Sprague Elec Co, Mass, 41-44; res chemist, 44-58, from asst res dir to res dir, Rubber Div, 58-73, PROD MGR, RUBBER, PETROL & PLASTIC DEPTS, R T VANDERBILT CO, INC, 73- Mem: Am Chem Soc. Res: Rubber chemicals; polymer stabilizers; oil additives; biocides; dielectrics; paint and paper chemicals. Mailing Add: 10 Pine Hill Rd East Norwalk CT 06855

BROOKS, MARGARET HOOVER, b Talas, Turkey, Feb 6, 13; m 38; c 4. CYTOGENETICS, BIOCHEMICAL GENETICS. Educ: Smith Col, AB, 33; Okla State Univ, PhD(genetics, plant breeding), 58. Prof Exp: Asst, Brown Univ, 33-34; asst genetics, Carnegie Inst, NY, 34-36, 37-38, Md, 36-37; asst cytogenetics, Okla State Univ, 54-55, from instr to assoc prof, 58-68; ASSOC PROF BIOL, OKLA CITY UNIV, 68- Mem: Am Soc Cell Biol; Soc Develop Biol; Bot Soc Am; Genetics Soc Am. Res: Genetics; botany; experimental teaching; biology for non-majors. Mailing Add: Dept of Biol Okla City Univ Oklahoma City OK 73106

BROOKS, MARION ALICE, b Mankato, Minn, Dec 17, 17; m 36, 75; c 1. ENTOMOLOGY. Educ: Univ Minn, BA, 47, MS, 50, PhD(zool), 54. Prof Exp: Res fel, 54-56, from instr to assoc prof, 56-71, PROF ENTOM, FISHERIES & WILDLIFE, UNIV MINN, ST PAUL, 71- Mem: NY Acad Sci; Entom Soc Am; Soc Invert Path; Am Soc Zoologists; Tissue Cult Asn. Res: Insect tissue culture; physiology of intracellular symbiotes; infectious diseases; nutrition and development of insects. Mailing Add: Dept of Entom Fish & Wildlife Univ of Minn St Paul MN 55108

BROOKS, MARVIN ALAN, b Trenton, NJ, Jan 28, 45; m 68. ANALYTICAL BIOCHEMISTRY. Educ: Lafayette Col, BS, 66; Univ Md, PhD(anal chem), 71. Prof Exp: SR BIOCHEMIST DRUG ANAL, HOFFMANN-LA ROCHE INC, 71- Mem: Am Chem Soc; Sigma Xi. Res: Analysis of drugs in biological fluids using chromatographic, spectrophotometric, spectrofluorometric and electrochemical methods; use of voltammetry to solve pharmaceutical problems. Mailing Add: Dept of Biochem & Drug Metab Hoffmann-La Roche Inc Nutley NJ 07110

BROOKS, MATILDA MOLDENHAUER, b Pittsburgh, Pa; m 17. PHYSIOLOGY. Educ: Univ Pittsburgh, AB & MS, 13; Harvard Univ, PhD(sci), 20. Prof Exp: Bacteriologist, Res Inst, Nat Dent Asn, 17; asst biologist, Hygienic Lab, USPHS, 20-24, assoc biologist, 24-26, lectr, 34 & 36, RES ASSOC PHYSIOL, UNIV CALIF, BERKELEY, 27- Concurrent Pos: Mem, Bermuda Biol Corp; mem, Marine Biol Lab Corp, Woods Hole; Bache grant, Nat Acad Sci, 27-32; grants, Nat Res Coun, Naples, 30-31, Permanent Sci Found, 49 & Am Philos Soc, 50-51; lectr under Coord of Inter-Am Affairs, SAm, 43-44. Mem: Am Physiol Soc. Res: Effects of solar light and ultraviolet light on sugar production and the four basic acids. Mailing Add: 630 Woodmont Ave Berkeley CA 94708

BROOKS, MERLE EUGENE, b Baldwin, Kans, Feb 8, 16; m 41; c 2. BIOLOGY, BOTANY. Educ: Kans State Teachers Col, BS, 46, MS, 47; Univ Colo, PhD(biol), 56. Prof Exp: From instr to assoc prof, Kans State Teachers Col, 47-59; assoc prof biol, 59-63, PROF BIOL, UNIV NEBR, OMAHA, 63- Mem: AAAS; Am Micros Soc; Nat Sci Teachers Asn; Nat Asn Biol Teachers; Bot Soc Am. Res: Limnology; Cladocera; plant morphology; science education; microbiology. Mailing Add: 8436 Loveland Dr Omaha NE 68124

BROOKS, NATHAN CYRUS, b Marietta, Okla, Feb 15, 16. GEOGRAPHY. Educ: Southeastern State Col, BA, 42; Univ Okla, MEd, 48, EdD, 56. Prof Exp: High sch teacher, Okla, 52-57; asst prof, 57-62, PROF GEOG, NORTHEASTERN OKLA STATE UNIV, TAHLEQUAH, 62- Mem: AAAS; Asn Am Geogr; Nat Coun Geog Educ. Res: Historical, political, economic and European geography. Mailing Add: Dept of Geog Northeastern State Col Tahlequah OK 74464

BROOKS, PHILIP RUSSELL, b Chicago, Ill, Dec 31, 38; m 60; c 4. PHYSICAL CHEMISTRY. Educ: Calif Inst Technol, BS, 60; Univ Calif, Berkeley, PhD(chem), 64. Prof Exp: From asst prof to assoc prof, 64-75, PROF CHEM, RICE UNIV, 75- Concurrent Pos: Alfred P Sloan fel, 70-72; Guggenheim Found fel, 74-75. Mem: Am Phys Soc; Am Chem Soc. Res: Molecular beam scattering. Mailing Add: Dept of Chem Rice Univ Houston TX 77001

BROOKS, ROBERT ALAN, b Gloversville, NY, Feb 23, 24; m 51; c 1. ORGANIC CHEMISTRY. Educ: Harvard Univ, BS, 44; Yale Univ, MS, 45, PhD(org chem), 49. Prof Exp: Instr chem, Yale Univ, 46-48; res chemist, Jackson Lab, 48-54, res supvr, 54-57, head div dyes, 57-59, head fluorine chem, 59-61, asst lab dir, 61-62, lab mgr, 62-69, tech mgr dyes & chem div, Org Chem Dept, 69-70, dir dyes & chem res, 70-75, TECH DIR FREON PROD DIV, E I DU PONT DE NEMOURS & CO, INC, DEEPWATER, 75- Mem: AAAS; Am Chem Soc; Am Asn Textile Chemists & Colorists; NY Acad Sci. Res: Synthetic and organic dyes; pigments and elastomers; chemistry of fluorine. Mailing Add: Seven Stars RD 2 Woodstown NJ 08098

BROOKS, ROBERT E, b Los Angeles, Calif, Aug 17, 21; m 50; c 2. EXPERIMENTAL PATHOLOGY. Educ: Univ Calif, Los Angeles, BS, 48; Univ Ore, MS, 64, PhD(path), 67. Prof Exp: Res asst biochem, Atomic Energy Proj, Univ Calif, Los Angeles, 48-50; res asst path, Sch Med, Univ Calif, San Francisco, 50-60; from instr to asst prof, 60-70, ASSOC PROF PATH, MED SCH, UNIV ORE, 70- Mem: Am Asn Cancer Res; Electron Micros Soc Am. Res: Ultrastructural analysis of spontaneous and experimental animal tumors; various normal and pathologic animal and human tissues. Mailing Add: Dept of Path Univ of Ore Med Sch Portland OR 97201

BROOKS, ROBERT FRANKLIN, b Columbus, Ohio, Mar 17, 28; m 59; c 1. ENTOMOLOGY. Educ: Ohio State Univ, BS, 54, MS, 55; Univ Wis, PhD(entom), 60. Prof Exp: Asst entom, Ohio State Univ, 54-55; asst, Univ Wis, 56-59, instr, 59-60; asst entomologist, Citrus Exp Sta, Univ Fla, 60-67, assoc prof entom & assoc entomologist, Inst Food & Agr Sci, 67-70, PROF ENTOM & ENTOMOLOGIST, INST FOOD & AGR SCI, UNIV FLA, 70- Concurrent Pos: Consult citrus, Standard Fruit Co Div, Castle & Cook, 66- Mem: Entom Soc Am; Sigma Xi; Int Orgn Biol Control. Res: Integrated control and management of citrus pests; improving and developing more efficient application equipment for use on citrus. Mailing Add: Inst Food & Agr Sci Agr Res & Educ Ctr Univ of Fla Lake Alfred FL 33850

BROOKS, ROBERT M, b Freeport, Tex, Jan 5, 38; m 60. MATHEMATICS. Educ: La State Univ, BS, 60, PhD(math), 63. Prof Exp: From instr to asst prof math, Univ Minn, 63-67; assoc prof, 67-72, PROF MATH, UNIV UTAH, 72- Mem: Am Math Soc. Res: Topological algebras; complex analysis. Mailing Add: Dept of Math Univ of Utah Salt Lake City UT 84112

BROOKS, RODNEY AVRAM, physics, see 12th edition

BROOKS, RONALD E, physical organic chemistry, see 12th edition

BROOKS, RONALD JAMES, b Toronto, Ont, Apr 16, 41; m 65; c 1. ETHOLOGY. Educ: Univ Toronto, BSc, 63, MSc, 66; Univ Ill, PhD(zool), 70. Prof Exp: ASST PROF ZOOL, UNIV GUELPH, 69- Mem: Am Soc Mammalogists; Animal Behav Soc; Can Soc Zoologists; Sigma Xi; Am Soc Naturalists. Res: Acoustic communication on Zonotrichia albicollis and Dicrostonyx groenlandicus; ontogeny of behavior and analysis of behavioral role in population changes in microtine rodents; behavior and population biology of Castor canadensis and Chelhydra sp. Mailing Add: Dept of Zool Univ of Guelph Guelph ON Can

BROOKS, SAM RAYMOND, b Austin, Tex, Apr 1, 40. MATHEMATICS. Educ: Univ Tex, BA, 62, MA, 64, PhD(math), 69. Prof Exp: ASST PROF MATH, MEMPHIS STATE UNIV, 66- Mem: Math Asn Am; Am Math Soc. Res: Group theory; semigroup theory and generalizations. Mailing Add: Dept of Math Memphis State Univ Memphis TN 38111

BROOKS, SAMUEL CARROLL, b Winchester, Va, May 12, 28; m 61; c 3. BIOCHEMISTRY, ENDOCRINOLOGY. Educ: Carnegie Inst Technol, BS, 51; Univ Wis-Madison, MS, 55, PhD(biochem), 57. Prof Exp: Res scientist, John L Smith Mem Lab Cancer Res, Charles Pfizer & Co, Inc, 57-59; from instr to assoc prof, 59-74, PROF BIOCHEM, SCH MED, WAYNE STATE UNIV, 74- Concurrent Pos: Res assoc, Detroit Inst Cancer Res, 59-70; Fulbright res scholar, Univ Louvain, 64-65; dir, Dept Chem, Mich Cancer Found, 70-74. Mem: Am Chem Soc; Soc Study Reprod; Endocrine Soc; Am Asn Cancer Res; Am Soc Biol Chem. Res: Steroid hormone activity and metabolism. Mailing Add: Dept of Biochem Wayne State Univ Sch Med Detroit MI 48201

BROOKS, SHEILAGH THOMPSON, b Tampico, Mex, Dec 10, 23; US citizen; m 51; c 2. PHYSICAL ANTHROPOLOGY. Educ: Univ Calif, Berkeley, BA, 44, MA, 47, PhD(phys anthrop), 51. Prof Exp: Assoc trop biogeog, Chihuahua & Durango, Mex, 55-58; res assoc physiol, Univ Calif, Berkeley, 58-62; lectr anthrop, Univ Southern Calif, 59-61; lectr, Pasadena City Col, 60-62; asst prof, Univ Colo, Denver & Boulder, 63-65; res asst, Mus, Southern Ill Univ, 65; asst prof, Univ Colo, Boulder, 65-66; assoc prof, 66-69, PROF ANTHROP, UNIV NEV, LAS VEGAS, 69-, CHMN DEPT, 73- Concurrent Pos: NSF fel arch & phys anthrop, Sarawak, Malaysia, 66; consult, Clark County Sheriff's Dept, 67-; mem coord coun, Nev Archaeol Surv, 68-; grants, Dept Health, Educ & Welfare, mus progs, Univ Nev, Las Vegas, 69-71, fac res comt, Photog Lab, 70 & Nat Endowment Humanities, Preserv Hist Sites, 72-73; mem bd, Gov Comn Nev Lost City Mus, 71-; consult, Nev Archaeol Surv, 73- Mem: Am Eugenics Soc; Inst Asn Human Biol; Soc Vert Paleont; Am Acad Forensic Sci; Am Asn Phys Anthrop. Res: Archaeologically recovered skeletal populations with emphasis on demography, paleopathology and morphological differences; analysis and identification of historical burials. Mailing Add: Dept of Anthrop Univ of Nev Las Vegas NV 89154

BROOKS, SIDNEY, physics, see 12th edition

BROOKS, STANLEY NELSON, b Laplata, NMex, Mar 1, 22; m 46; c 5. AGRONOMY, PLANT BREEDING. Educ: Colo Agr & Mech Col, BS, 48; Kans State Col Agr, MS, 49; Ore State Univ, PhD(genetics), 61. Prof Exp: Asst agron, Huntley Br, Mont Agr Exp Sta, 49-54; res agronomist, Crops Res Div, Ore, 55-68, invests leader, 61-68, asst chief oilseed & indust crops res br, 68-72, AREA RES DIR, ORE-WASH AREA, WESTERN REGION, AGR RES SERV, USDA, 72- Concurrent Pos: From asst prof to assoc prof, Ore State Univ, 55-68; collabr, Wash State Univ, 72- Mem: Fel AAAS; Am Soc Agron; Crop Sci Soc Am. Res: Administration of agricultural research programs in Oregon and Washington; breeding and production of hops. Mailing Add: Rm 219 Agr Sci Wash State Univ Pullman WA 99163

BROOKS, STUART MERRILL, b Cincinnati, Ohio, Apr 28, 36; c 3. PULMONARY DISEASE, OCCUPATIONAL MEDICINE. Educ: Univ Cincinnati, BS, 58, MD, 62; Am Bd Internal Med & Am Bd Pulmonary Dis, dipl, 69. Prof Exp: Resident internal med, Boston City Hosp, 63-64, 66-67, fel pulmonary dis, 67-69; res fel med, Sch Med, Tufts Univ, 68-69; from asst prof med to asst prof environ health, 69-73, ASSOC PROF ENVIRON HEALTH & MED & CHIEF DIV CLIN STUDIES, COL MED, UNIV CINCINNATI, 73- Concurrent Pos: Mem ad hoc comt case control studies on host factors as determinants of chronic obstructive pulmonary dis susceptibility, Nat Heart & Lung Inst, 70; attend physician, Cincinnati Gen Hosp & Vet Admin Hosp, Cincinnati, 73- Mem: Am Fedn Clin Res; Am Thoracic Soc; fel Am Col Physicians; fel Am Col Chest Physicians. Res: Pulmonary physiology; occupational lung diseases; pathogenetic mechanisms of bronchial asthma, non-respiratory functions of lung; corticosteroid metabolism. Mailing Add: Med Sci Bldg Rm 5251 Univ Cincinnati Col of Med Cincinnati OH 45267

BROOKS, THOMAS FURMAN, b Charlotte, NC, Oct 4, 43; m 70; c 1. ACOUSTICS. Educ: NC State Univ, BS, 68, PhD(acoust), 74. Prof Exp: Tool engr, Turbine Div, Westinghouse Elec Corp, 70; AEROSPACE TECHNOLOGIST ACOUST, LANGLEY RES CTR, NASA, 74- Mem: Acoust Soc Am. Res: Airflow-surface interaction noise; linear acoustics; unsteady aerodynamics. Mailing Add: Mail Stop 465 NASA Langley Res Ctr Hampton VA 23665

BROOKS, THOMAS JOSEPH, JR, b Starkville, Miss, May 23, 16; m 41; c 4. PREVENTIVE MEDICINE. Educ: Univ Fla, BS, 37; Univ Tenn, MS, 39; Univ NC, PhD(prev med), 52; Bowman Gray Sch Med, MD, 45; Am Bd Microbiol, dipl. Prof Exp: Instr bact & parasitol, Bowman Gray Sch Med, 42-45; intern, Bowman Gray Sch Med & NC Baptist Hosp, 45-46; assoc prof pharmacol, Sch Med, Univ Miss, 47-48; med dir, Fla State Univ Hosp, 48-52; Rockefeller Found travel grant, 52; asst dean in chg student affairs, 56-73, PROF PREV MED & CHMN DEPT, SCH MED, UNIV MISS, 52- Concurrent Pos: Officer-in-chg res unit, US Naval Training Ctr, Md, 53-55; consult, Vet Admin Hosp, Jackson, Miss, 56-; La State Univ Caribbean travel fel, 56 & 60; vis prof, Sch Med, Univ Costa Rica, 62-63, Keio Univ, Japan, 68 & Kyoto Univ, 68-69; NIH trainee, Univ Wis, 65; UN consult, India, 66; Alan Gregg fel, Japan & Southeast Asia, 68-69. Mem: Am Pub Health Asn; Am Soc Trop Med & Hyg; Asn Teachers Prev Med. Res: Treatment of filariasis; toxicology of antimony; treatment of canine filariasis with anthiomaline; tuberculosis in university students; epidemiology of streptococcal diseases; mitosis of Entamoeba histolytica; epidemiology of Echinococcus granulosus. Mailing Add: Dept of Prev Med Univ of Miss Med Ctr Jackson MS 39216

BROOKS, THOMAS WILLIAM, chemistry, see 12th edition

BROOKS, VERNON BERNARD, b Berlin, Ger, May 10, 23; nat Can; m 50; c 3. NEUROPHYSIOLOGY. Educ: Univ Toronto, BA, 46, PhD(physiol), 52; Univ Chicago, ScM, 48. Prof Exp: Lectr physiol, McGill Univ, 50-52, asst prof, 52-56; from asst prof to assoc prof neurophysiol, Rockefeller Inst, 56-64; prof physiol, New York Med Col, Flower & Fifth Ave Hosps, 64-71, chmn dept, 64-69; PROF PHYSIOL & CHMN DEPT, UNIV WESTERN ONT, 71- Concurrent Pos: Vis fel, Australian Nat Univ, 54-55. Mem: Am Physiol Soc; Can Physiol Soc. Res: Interaction of neurones; sensorimotor integration; motor control. Mailing Add: Dept of Physiol Univ of Western Ont London ON Can

BROOKS, WALTER LYDA, b Tazewell, Tenn, Mar 6, 23; m 49; c 1. PHYSICS, ASTRONOMY. Educ: Lincoln Mem Univ, BA, 43; NY Univ, MS, 50, PhD(physics), 53. Prof Exp: Instr physics, Lincoln Mem Univ, 46-47; asst, NY Univ, 47-53; consult scientist develop div, United Nuclear Corp, 53-71, Gulf United Nuclear Fuels Co, 71-74; NUCLEAR ENGR, US NUCLEAR REGULATORY COMN, 74- Mem: Am Nuclear Soc. Res: Reactor physics; electronics; computers; nuclear instrumentation. Mailing Add: 5925 Bradley Blvd Bethesda MD 20014

BROOKS, WAYNE MAURICE, b Lynchburg, Va, Mar 11, 39; m 61; c 2. INSECT PATHOLOGY, PROTOZOOLOGY. Educ: NC State Univ, BS, 61; Univ Calif, PhD(entom), 66. Prof Exp: Res asst entom, Univ Calif, 61-63, jr specialist, 63-66; asst prof, 66-71, ASSOC PROF ENTOM, NC STATE UNIV, 71- Mem: Entom Soc Am; Soc Invert Path. Res: Insect pathology with emphasis on entomophilic protozoa. Mailing Add: Dept of Entom NC State Univ Raleigh NC 27607

BROOKS, WENDELL V F, b Rockford Ill, Mar 7, 25; m 50; c 2. PHYSICAL CHEMISTRY, MOLECULAR SPECTROSCOPY. Educ: Swarthmore Col, BA, 48; Univ Minn, PhD, 54. Prof Exp: Instr chem, Univ Ariz, 52-54; NSF fel, Yale Univ, 54-55; from asst prof to prof chem, Ohio Univ, 55-67; PROF CHEM, UNIV NB, 67- Mem: Am Chem Soc; Am Phys Soc. Res: Infrared and microwave spectroscopy. Mailing Add: 218 Colonial Heights Fredericton NB Can

BROOKS, WENDELL WILKIE, organic chemistry, biochemistry, see 12th edition

BROOKSHEAR, JAMES GLENN, b Denton, Tex, Nov 27, 44; m 68; c 1. MATHEMATICS. Educ: N Tex State Univ, BS, 67; NMex State Univ, MS, 68, PhD(math), 75. Prof Exp: ASST PROF MATH, MARQUETTE UNIV, 75- Mem: Am Math Soc; Math Asn Am; Asn Comput Mach. Res: Rings of continuous functions and related areas. Mailing Add: Dept of Math & Statist Marquette Univ Milwaukee WI 53233

BROOM, ARTHUR DAVIS, b Panama, CZ, July 26, 37; m 60; c 3. BIO-ORGANIC CHEMISTRY, MEDICINAL CHEMISTRY. Educ: Univ Tex, Austin, BS, 59; Ariz State Univ, PhD(nucleoside methylation), 65. Prof Exp: Res assoc, Johns Hopkins Univ, 65-66; res assoc, 66-67, asst res prof, 67-69, from asst prof to assoc prof, 69-75, PROF MED CHEM, COL PHARM, UNIV UTAH, 75- Mem: AAAS; Am Chem Soc; Int Soc Heterocyclic Chem; Acad Pharmaceut Sci. Mailing Add: Dept of Med Chem Univ of Utah Col of Pharm Salt Lake City UT 84112

BROOM, KNOX MCLEOD, JR, b Jackson, Miss, Apr 23, 36; m 58; c 2. NUCLEAR CHEMISTRY. Educ: Southern Miss Univ, BA, 58; Univ Ark, MS, 61, PhD(nuclear chem), 63. Prof Exp: Sr chemist, Atomics Int Div, NAm Aviation, Inc, 63-66; sr chemist contract admin, Fast Breeder Reactor Fuels & Mat Develop, AEC, 66-67; mgr nuclear activities, Middle South Serv, Inc, 67-72; MGR POWER SERV, BROWN & ROOT, INC, 72- Mem: Am Soc Qual Control; Am Nuclear Soc; Am Chem Soc. Res: Nuclear and radiochemistry; cross-section and fission yield determinations; burnup of nuclear fuels; gamma-ray ray spectrometry; nuclear power engineering; nuclear fuel management. Mailing Add: Brown & Root Inc PO Box 3 Houston TX 77001

BROOMAN, ERIC WILLIAM, b London, Eng, Sept 15, 40; m 62; c 2. ELECTROCHEMISTRY. Educ: Univ Surrey, Dip Tech(metall), 63; Cambridge Univ, PhD(electrochem), 66. Prof Exp: Res electrochemist, 66-75, GROUP LEADER ELECTROCHEM ENERGY CONVERSION & STORAGE, COLUMBUS LABS, BATTELLE MEM INST, 75- Concurrent Pos: Res fel, Univ Salford, 72-73. Mem: Electrochem Soc; Brit Inst Metallurgists. Res: Electrochemical energy conversion and storage; secondary batteries; fuel cells; water electrolyzers; hydrogen production and storage; life support systems; electrocatalysis. Mailing Add: Battelle-Columbus Labs 505 King Ave Columbus OH 43201

BROOME, CARMEN ROSE, b Miami, Fla, June 19, 39. SYSTEMATIC BOTANY. Educ: Univ Miami, BS, 65; Univ SFla, MA, 68; Duke Univ, PhD(bot), 74. Prof Exp: ASST PROF BOT, UNIV MD, COLLEGE PARK, 73- Mem: Sigma Xi; Am Soc Plant Taxonomists; Int Asn Plant Taxon; Soc Study Evolution. Res: Systematics of angiosperms, particularly Gentianaceae, Orchidaceae; palynology, reproductive biology of angiosperms. Mailing Add: Dept of Bot Univ of Md College Park MD 20742

BROOME, ESTHER ROBERTS, b Kirbyville, Tex, Oct 9, 11; m 32; c 1. TEXTILES. Educ: Southwest Tex State Col, BS, 43; Tex Woman's Univ, MS, 52, PhD(textiles), 61. Prof Exp: High sch teacher, 43-59; instr textiles & clothing, 59-61, asst prof textile sci, 61-67, ASSOC PROF TEXTILE SCI, TEX WOMAN'S UNIV, 68- Concurrent Pos: Mem, Info Coun of Fabric Flammability. Mem: Fel Am Inst Chemists; Am Asn Textile Chemists & Colorists; Am Soc Testing & Mat; Am Home Econ Asn. Res: Utilization and chemical finishing of textile fabrics particularly those to which durable press, water repellency and nonflammable properties have been applied; performance of open-end spun yarns and the vapor phase durable press finish in relation to wearing apparel. Mailing Add: Box 23898 Univ Sta Tex Woman's Univ Denton TX 76204

BROOMFIELD, CLARENCE A, b Mt Morris, Mich, Sept 18, 30; m 56; c 3. BIOCHEMISTRY. Educ: Univ Mich, BS, 53; Mich State Univ, PhD(chem), 58. Prof Exp: Res assoc phys chem, Cornell Univ, 58-59; res biochemist, Chem Res & Develop Lab, 62-68, CHIEF PROTEIN CHEM, MED RES DIV, BIOMED LAB, EDGEWOOD ARSENAL, 68- Concurrent Pos: NIH fel, Cornell Univ, 60-62. Mem: Am Chem Soc; Sigma Xi; Am Soc Biol Chemists. Res: Secondary and tertiary protein structure; relationship of structure to biological activity; toxic proteins; cholinesterases; biochemistry of nerve transmission. Mailing Add: Protein Chem Br Med Res Div Biomed Lab Edgewood Arsenal MD 21010

BROPHY, GERALD PATRICK, b Kansas City, Mo, Sept 11, 26; m 51; c 3. GEOLOGY, MINERALOGY. Educ: Columbia Univ, AB, 51, MA, 53, PhD(geol), 54. Hon Degrees: MA, Amherst Col, 68. Prof Exp: Res asst geol, Columbia Univ, 51-54; from instr to assoc prof, 54-68, PROF GEOL, AMHERST COL, 68- Concurrent Pos: NIH res grants, 62-; Fulbright fel, Univ Baghdad, 65-66; cur, Pratt Mus Geol, 67-; NSF grants, 69-; consult, Bear Creek Mining Co, Cerro Corp, R T Vanderbilt & Co & Conyers Construct Co. Mem: Fel Geol Soc Am; Mineral Soc Am; Geochem Soc; Soc Econ Geologists; Yellowstone-Big Horn Res Asn (pres, 75-77). Res: Crystals chemistry of phosphates and sulphates; effects of pressure on crystallization of granitic magmas. Mailing Add: Dept of Geol Amherst Col Amherst MA 01002

BROPHY, JAMES JOHN, b Chicago, Ill, June 6, 26; m 49; c 3. SOLID STATE ELECTRONICS. Educ: Ill Inst Technol, BS, 47, MS, 49, PhD(physics), 51. Prof Exp: Res physicist, 51-53, supvr solid state physics, 53-56, asst mgr physics div, 56-61, dir tech develop, 61-63, vpres, 63-67, ACAD VPRES, RES INST, ILL INST TECHNOL, 67- Concurrent Pos: Mem, Bd Trustees, Underwriter Labs, Inc, 72. Mem: Fel Am Phys Soc; AAAS. Res: Solid state physics; semiconductors; fluctuation phenomena; secondary emission; magnetism; plasma physics. Mailing Add: Ill Inst of Technol Chicago IL 60616

BROPHY, JOHN ALLEN, b Rockford, Ill, Mar 30, 24; m 54; c 2. GEOLOGY. Educ: Univ Ill, AB, 48, MS, 49, PhD(geol), 58. Prof Exp: Asst instr geol, Univ Ill, 48-49; geologist, Magnolia Petrol Co, 49-51; asst geologist, Ill Geol Surv, 53-59; from asst prof to assoc prof geol, 59-67, chmn div natural sci, 67-70, PROF GEOL, NDAK STATE UNIV, 67-, CHMN DEPT, 70- Concurrent Pos: Leverhulme fel, Univ Birmingham, 67-68. Mem: AAAS; Geol Soc Am; Nat Asn Geol Teachers. Res: Pleistocene geology. Mailing Add: Dept of Geol ND State Univ Fargo ND 58102

BROQUIST, HARRY PEARSON, b Chicago, Ill, Jan 23, 19; m 42; c 2. BIOCHEMISTRY, NUTRITION. Educ: Beloit Col, BS, 40; Univ Wis, MS, 41, PhD(biochem), 49. Prof Exp: Group leader microbiol, Lederle Lab, Pearl River, NY, 41-46 & 49-58; from assoc prof to prof biol chem, Univ Ill, 58-69; PROF BIOCHEM, VANDERBILT UNIV, 69-, DIR DIV NUTRIT, 72- Concurrent Pos: Vis lectr biochem, Mich State Univ, 63; biochem consult interdept comt nutrit nat develop, US Nutrit Surv to Nigeria, 65; NSF sr fel, Karolinska Inst, Sweden, 65-66; mem nutrit study sect, NIH, 66-70; mem food & nutrit bd, Nat Res Coun, 74- Honors & Awards: Borden Award, Am Inst Nutrit, 68. Mem: Am Soc Biol Chemists; Am Inst Nutrit. Res: Amino acid metabolism in yeasts, molds and mammalian systems; nutritional biochemistry. Mailing Add: Dept of Biochem Vanderbilt Univ Sch of Med Nashville TN 37232

BROSBE, EDWIN ALLAN, b Burlington, NJ, Jan 8, 18; m 43; c 2. MEDICAL MICROBIOLOGY. Educ: Philadelphia Col Pharm, BS, 40; Univ Colo, MS, 47, PhD(bact), 51. Prof Exp: Bacteriologist, Vet Admin Hosp, Ft Logan, Colo, 47-50 & Denver, Colo, 50-53, bacteriologist & actg chief lab serv, NY, 53-55, chief tuberc res lab, Long Beach, Calif, 55-66, microbiologist, Little Rock, Ark, 66-68, clin microbiologist, Long Beach, 68-73; LECTR, CALIF STATE UNIV, LONG BEACH, 73- Mem: AAAS; Am Soc Microbiol; Soc Exp Biol & Med. Res: Bacteriology and chemotherapy of tuberculosis; biology and chemotherapy of coccidioidomycosis; host-parasite relationship. Mailing Add: Dept of Microbiol Calif State Univ Long Beach CA 90840

BROSEGHINI, ALBERT L, b Chicago, Ill, Sept 17, 32; m 58; c 2. ZOOLOGY. Educ: Northern Ill State Teachers Col, BS, 54; Iowa State Univ, MS, 56, PhD(zool), 59. Prof Exp: Asst prof biol, Fresno State Col, 59-62; assoc prof physiol, Col Osteop Med & Surg, 62-64; scientist adminr, Div Res Grants, NIH, 64-65, asst endocrinol prog dir, Nat Inst Arthritis & Metab Dis, 65-66, hemat prog dir, 66-69; DIR RES ADMIN, CHILDREN'S HOSP MED CTR, BOSTON, 69- Mem: AAAS. Res: Comparative physiology; histochemistry of insect tissues; hematology; erythropoiesis. Mailing Add: 64 Barber Rd Framingham MA 01701

BROSEMER, RONALD WEBSTER, b Oakland, Calif, Feb 17, 34; m 62; c 2. BIOCHEMISTRY. Educ: Univ Calif, BS, 55; Univ Ill, PhD(biochem), 60. Prof Exp: NSF res fel, Univ Marburg, 60-62; asst prof biochem, Univ Ill, 62-63; from asst prof to assoc prof, 63-72, PROF BIOCHEM, WASH STATE UNIV, 72-, ASSOC DEAN GRAD SCH, 75- Concurrent Pos: Fulbright scholar, Univ Konstanz, 70-71. Mem: AAAS; NY Acad Sci; Am Soc Biol Chemists. Res: Molecular basis of genetic diseases. Mailing Add: Prog in Biochem & Biophys Wash State Univ Pullman WA 99163

BROSHAR, WAYNE CECIL, b Boone Co, Ind, May 3, 33; m 54; c 4. SOLID STATE PHYSICS. Educ: Wabash Col, AB, 55; Univ Mich, MS, 56; Brown Univ, PhD(physics), 69. Prof Exp: Res engr, Convair San Diego, Gen Dynamics/Convair, 56-59; instr physics, Wabash Col, 59-61; from instr to asst prof, 66-70, ASSOC PROF PHYSICS, RIPON COL, 70- Mem: Am Asn Physics Teachers. Mailing Add: Dept of Physics Ripon Col Ripon WI 54971

BROSI, ALBERT RALPH, b Coatsburg, Ill, Nov 12, 07; m 37; c 2. CHEMISTRY. Educ: Univ Chicago, BS, 30, PhD(phys chem), 38. Prof Exp: Develop chemist, US Rubber Co, NJ, 30-33, res chemist, 37-43; res chemist, Metal Lab, Univ Chicago, 43; RES CHEMIST, OAK RIDGE NAT LAB, 43- Mem: AAAS; Am Chem Soc; Am Phys Soc. Res: Nuclear chemistry; surface and physical chemistry; decay schemes of radioactive isotopes. Mailing Add: 105 W Price Lane Oak Ridge TN 37830

BROSIN, HENRY WALTER, b Blackwood, Va, July 6, 04; m 49; c 1. MEDICINE. Educ: Univ Wis, BA, 27, MD, 33. Prof Exp: Intern, Cincinnati Gen Hosp, 33-34; Commonwealth fel psychiat, Sch Med, Univ Colo, 34-37; Rockefeller fel, Inst Psychoanal, Univ Chicago, 37, from instr to prof psychiat, Sch Med, 37-51; prof, 51-69, EMER PROF PSYCHIAT, SCH MED, UNIV PITTSBURGH, 69-; PROF PSYCHIAT, COL MED, UNIV ARIZ, 69- Concurrent Pos: Fel, Ctr Advan Study Behav Sci, Calif, 56 & 66; consult, Off Surgeon Gen; mem comt army med educ, Nat Res Coun; mem nat clearing house, NIMH. Mem: Fel Am Acad Arts & Sci; AAAS; fel AMA; Am Psychoanal Asn; Am Psychosom Soc. Res: Psychoanalysis; organic cerebral disease; psychosomatic medicine; communication theory. Mailing Add: Col of Med Univ of Ariz Tucson AZ 85724

BROSNAN, JOHN THOMAS, b Kenmare, Ireland, Feb 13, 43; m 70; c 1. METABOLISM. Educ: Nat Univ Ireland, BSc, 64, MSc, 66; Oxford Univ,

DPhil(biochem), 69. Prof Exp: Fel med res, Univ Toronto, 69-71; asst prof, 72-75, ASSOC PROF BIOCHEM, MEM UNIV NFLD, 75- Concurrent Pos: Mem grants comt metab, Med Res Coun Can, 75-76. Mem: Biochem Soc; Can Biochem Soc; Am Soc Biol Chemists. Res: Metabolic regulation, especially the regulation of glutamine metabolism and ammonia production by kidneys of normal and acidotic animals. Mailing Add: Dept of Biochem Mem Univ of Nfld St Johns NF Can

BROSS, HELEN H, mathematics, physics, see 12th edition

BROSSEAU, GEORGE EMILE, JR b Berkeley, Calif, July 24, 30; m 51; c 3. GENETICS. Educ: Univ Calif, Berkeley, BA, 52, PhD(genetics), 56. Prof Exp: Am Cancer Soc fel, Biol Div, Oak Ridge Nat Lab, 56-58, res assoc genetics, 58-59; from asst prof to prof zool, Univ Iowa, 59-70; PROG MGR, NAT SCI FOUND, 70- Mem: Am Soc Human Genetics; Genetics Soc Am; Asn Consumer Res; Geront Soc; Soc Social Biol. Res: Administration; applied science; social gerontology; consumer research. Mailing Add: Nat Sci Found Washington DC 20550

BROSTOFF, STEVEN WARREN, b Boston, Mass, Sept 10, 42; m 66. NEUROCHEMISTRY. Educ: Mass Inst Technol, BS, 64, PhD(biochem), 68. Prof Exp: Am Cancer Soc fel biochem, Eleanor Roosevelt Inst Cancer Res, Univ Colo Med Ctr, 68-70; res assoc neurochem, Salk Inst Biol Studies, 70; sr res biochemist, Dept Exp Biol, Marck Inst Therapeut Res, 71-72; asst prof, Dept Path, Albert Einstein Col Med, 72-73; ASSOC PROF NEUROCHEM, DEPTS NEUROL & BIOCHEM, MED UNIV SC, 73- Mem: AAAS; Am Chem Soc; Am Soc Neurochem; Int Soc Neurochem. Res: Chemistry of proteins and protein-lipid interactions in the nervous system; immunological properties of nervous system proteins. Mailing Add: Dept of Neurol Med Univ of SC Charleston SC 29401

BROSTROM, CHARLES OTTO, b Downsville, Wis, Nov 2, 42; m 65; c 1. PHARMACOLOGY, BIOCHEMISTRY. Educ: Wis State Univ, River Falls, BS, 64; Univ Ill, PhD(biochem), 69. Prof Exp: Asst prof biochem, 71-72, ASST PROF PHARMACOL, RUTGERS MED SCH, COL MED & DENT NJ, 72- Concurrent Pos: USPHS fel enzym, Univ Calif, Davis, 68-70, Health Sci Advan Award, 70-71. Res: Cyclic nucleotide metabolism in brain. Mailing Add: Rutgers Med Sch Dept Pharmacol Col of Med & Dent of NJ Piscataway NJ 08854

BROSTROM, MARGARET ANN, b Chicago, Ill, Aug 13, 41; m 65; c 1. BIOCHEMISTRY, PHARMACOLOGY. Educ: Clarke Col, BA, 63; Univ Ill, PhD(biochem), 68. Prof Exp: Sr teaching asst, 71-72, instr, 72-74, ASST PROF PHARMACOL, RUTGERS MED SCH, COL MED & DENT NJ, 74- Concurrent Pos: NIH fel, Univ Calif, 69-71. Mem: AAAS. Res: Neuropharmacology; protein biochemistry; enzymology. Mailing Add: Rutgers Med Sch Dept Pharmacol Col of Med & Dent of NJ Piscataway NJ 08854

BROT, FREDERICK ELLIOT, b Kalamazoo, Mich, May 28, 41; m 65; c 2. BIOCHEMICAL GENETICS. Educ: Univ Mich, BS, 62; Stanford Univ, PhD(org chem), 66. Prof Exp: NIH fel, Northwestern Univ, 66-68; sr res chemist, Monsanto Co, 68-71; RES INSTR & NIH TRAINEE, SCH MED, WASH UNIV, 71- Mem: AAAS; Am Chem Soc. Res: Protein purification and characterization; affinity chromatography; enzyme modification; application of enzymes for correction of genetic enzyme defects; synthesis and in vivo fate of site-directed drugs. Mailing Add: Div of Med Genetics Dept Pediat Wash Univ Sch of Med St Louis MO 63110

BROT, NATHAN, b New York, NY, July 27, 31; m 58; c 3. BIOCHEMISTRY. Educ: City Col New York, BS, 53; Univ Calif, PhD(biochem), 63. Prof Exp: Res chemist, Med Col, Cornell Univ, 53-58; chemist, Univ Calif, 62-63; USPHS res fel, NIH, 63-65, chemist, 65-67; CHEMIST, ROCHE INST MOLECULAR BIOL, HOFFMAN-LA ROCHE, INC, 67- Mem: AAAS; Am Chem Soc; Am Soc Microbiol; Am Soc Biol Chemists. Res: Vitamin B-12 metabolism; mechanism of protein synthesis in microorganisms. Mailing Add: Roche Inst of Molecular Biol Nutley NJ 07110

BROTHERS, ALFRED DOUGLAS, b Waukegan, Ill, Aug 6, 39; m 72. SOLID STATE PHYSICS. Educ: Northern Ill Univ, BS, 62; Iowa State Univ, MS, 65, PhD(physics & educ), 68. Prof Exp: Jr physicist, Ames Lab, US AEC, 65-66; asst prof physics, St Benedict's Col, 68-71; actg chmn, 71-72, CHMN DEPT PHYSICS, BENEDICTINE COL, 72- Mem: Am Asn Physics Teachers; Am Inst Physics; Sigma Xi. Res: Optical qualities of thin films and crystalline solids. Mailing Add: Dept of Physics Box N66 Benedictine Col Atchison KS 66002

BROTHERS, JAMES ALFRED, b North Sydney, NS, Feb 25, 19; m 52; c 4. INORGANIC CHEMISTRY. Educ: St Francis Xavier Univ, BSc, 42; Dalhousie Univ, MSc, 49. Prof Exp: Asst, NS Res Found, 49-51; chemist, Aluminum Co Can, 51-52 & NS Light & Power Co, 52-54; asst res officer, Atomic Energy Can, Ltd, 54-57; sect head glass res, Wyandotte Chem Corp, Mich, 57-67; assoc res officer, 67-69, DIR CHEM DIV, NS RES FOUND, 69- Mem: Am Chem Soc. Res: Analytical chemistry; radiochemistry; refractory metal recovery and reduction; actinide species identification; physical chemistry on molten glass refining stage. Mailing Add: Chem Div NS Res Found 100 Fenwick St PO Box 790 Dartmouth NS Can

BROTHERS, JOHN EDWIN, b Salt Lake City, Utah, July 6, 37. MATHEMATICS. Educ: Univ Utah, BA, 59, MS, 60; Brown Univ, PhD(math), 64. Prof Exp: Asst chmn dept, 71-73, ASSOC PROF MATH, IND UNIV, BLOOMINGTON, 66- Concurrent Pos: Co-recipient, NSF grant, 67-75; mem, Inst Advan Study, 73-74. Mem: Am Math Soc. Res: Geometric measure theory, branch of differential geometry. Mailing Add: Dept of Math Ind Univ Bloomington IN 47401

BROTHERSON, JACK DEVON, b Castle Dale, Utah, Sept 18, 38; m 64; c 4. PLANT ECOLOGY. Educ: Brigham Young Univ, BS, 64, MS, 67; Iowa State Univ, PhD(plant ecol), 69. Prof Exp: ASST PROF ECOL, BRIGHAM YOUNG UNIV, 69- Concurrent Pos: Consult, Indian Inst, Brigham Young Univ, 69- & Bur Reclamation, Cent Utah Proj, 72-73; mem, Am Inst Biol Comt Natural Areas State of Utah, 72-74; reviewer, Brown & Co, Publishers, 74. Mem: AAAS; Ecol Soc Am; Brit Ecol Soc; Soc Range Mgt; Nat Wildlife Fedn. Res: Ecological adaptation; niche metrics; evolutionary and environmental gradient accomodation of organisms, populations, and biotic communities in the Great Basin of North America. Mailing Add: Dept of Bot & Range Sci Brigham Young Univ Provo UT 84602

BROTHERTON, ROBERT JOHN, b Ypsilanti, Mich, Aug 4, 28; m 50; c 4. ORGANIC CHEMISTRY, RESEARCH ADMINISTRATION. Educ: Univ Ill, BS, 49; Wash State Univ, PhD, 54. Prof Exp: Du Pont fel & instr org chem, Univ Minn, 54-55; res chemist, Refinery Res Group, Union Oil Co, 55-57; res chemist, 57-64, MGR CHEM RES, US BORAX RES CORP, 64- Mem: Am Chem Soc. Res: Boron chemistry; agricultural chemicals. Mailing Add: US Borax Res Corp 412 Crescent Way Anaheim CA 92801

BROUGHTON, ROGER JAMES, b Montreal, Que, Sept 25, 36; m 59; c 3. NEUROPHYSIOLOGY, PSYCHOBIOLOGY. Educ: Queen's Univ, Ont, MD, CM, 60; Univ Aix-Marseille, dipl electroencephalog-neurophysiol, 64; McGill Univ, PhD(neurophysiol), 67. Prof Exp: Intern med, Univ Sask Hosp, 60-61; resident, Centre St Paul, Marseille, 62-64; asst prof clin neurophysiol, McGill Univ, 64-68; ASSOC PROF MED & PHARMACOL, UNIV OTTAWA, 68-; PHYSICIAN, OTTAWA GEN HOSP, 70- Concurrent Pos: Consult, WHO, 64-68; assoc, Med Res Coun Can, 68-; ed, Sleep Rev, 70- Mem: Am Electroencephalog Soc; Am Epilepsy Soc; Asn Psychophysiol Study Sleep; Int League Against Epilepsy; Can Soc Electroencephalog. Res: Mechanisms of precipitation of epileptic seizures; pharmacology of experimental and clinical epilepsy; clinical disorders of sleep and arousal; phylogeny of sleep; cerebral evoked potentials and their correlates in man. Mailing Add: Dept of Med Ottawa Gen Hosp Ottawa ON Can

BROUGHTON, WILLIAM ALBERT, b Ronan, Mont, Oct 12, 14; m 44; c 6. ECONOMIC GEOLOGY. Educ: Univ Wis, AB, 39. Prof Exp: Geologist, Wis Geol Surv, 36-41; instr petrol, State Col Wash, 41; geologist, State Div Geol, Wash, 41-45; geologist, Chicago, Milwaukee, St Paul & Pac Rwy, 45-46; consult geologist, 46; HEAD DEPT GEOL, UNIV WIS-PLATTEVILLE, 48- Concurrent Pos: With US Geol Surv, 51-54; geologist-in-chg, Mineral Develop Atlas, Wis Geol Surv, 61- Mem: Fel AAAS; Soc Econ Geologists; Am Inst Mining, Metall & Petrol Eng. Res: Sedimentation of Wisconsin lakes; tungsten and magnetic iron deposits of Washington; zinc-lead deposits; meteoritics. Mailing Add: 295 Bradford Platteville WI 53818

BROUILLETTE, WALTER, b Port Barre, La, Aug 13, 25; m 46. SOLID STATE PHYSICS. Educ: La State Univ, BSc, 48, MSc, 50; Univ Mo, PhD(physics), 55. Prof Exp: PHYSICIST, GEN ELEC CO, SYRACUSE, 55- Mem: Am Asn Physics Teachers; Am Phys Soc. Res: Physical electronics; electron tube circuitry; information theory; information storage and processing; electron beam lithography. Mailing Add: RD 1 Jamesville NY 13078

BROUMAND, HORMOZ, food chemistry, see 12th edition

BROUN, THOROWGOOD TAYLOR, JR, b Nashville, Tenn, Apr 4, 23; m 42; c 3. INDUSTRIAL CHEMISTRY. Educ: ETex State Univ, BS, 43; Purdue Univ, PhD(chem), 51. Prof Exp: Res analyst, Field Res Dept, Magnolia Petrol Co, 43-44; res chemist, Chem Div, PPG Industs, Inc, 51-56, res supvr, 56-67; sr res chemist, 67-75, RES COORDR, HOUSTON CHEMICAL CO DIV, PPG INDUSTS, INC, 75- Mem: Am Chem Soc. Res: Applied product and process research. Mailing Add: Houston Chemical Co PO Box 3785 Beaumont TX 77704

BROUNS, RICHARD JOHN, b Osakis, Minn, Oct 2, 17; m 52; c 6. CHEMISTRY. Educ: St John's Univ, Minn, BS, 42; Iowa State Col, MS, 44, PhD(anal chem), 48. Prof Exp: Plant supvr, Cardox Corp, Okla, 44-45; res chemist, Gen Elec Co, Wash, 48-52, supvr anal chem res, 52-61, opers res analyst, 61-64; RES ASSOC, PAC NORTHWEST LABS, BATTELLE MEM INST, 65- Mem: Am Chem Soc; Inst Nuclear Mat Mgt. Res: Analytical chemistry; electrochemistry; radiochemistry; instrumentation; development of nuclear safeguards systems through materials accounting measurements and statistics. Mailing Add: Battelle Mem Inst Pac Northwest Labs PO Box 999 Richland WA 99352

BROUS, JACK, b New York, NY, Nov 14, 26; m 51; c 2. INORGANIC CHEMISTRY, PHYSICAL CHEMISTRY. Educ: City Col New York, BS, 48; Univ Chicago, SM, 49; Polytech Inst Brooklyn, PhD(inorg chem), 53. Prof Exp: Develop engr, Gen Eng Labs, Gen Elec Co, 52-55 & Power Tube Dept, 55-57; develop engr, Radio Corp Am, 57-58, engr leader, Chem & Phys Lab, 58-68; dir process develop, Pyrofilm Corp, NJ, 69-70; mgr chem res & develop, 70-73, SR STAFF SCIENTIST, ALPHA METALS CORP, JERSEY CITY, 73- Mem: Am Chem Soc; Am Soc Testing & Mat; Sigma Xi. Res: Inorganic dielectrics; ferroelectrics; x-ray diffraction; thermionic electron emission studies; high vacuum technology; getter materials; thermal measurements; electronic materials; cleaning materials and processes; invented the ionograph for measurement of residual ionic contamination. Mailing Add: 35 Brandon Ave Livingston NJ 07039

BROUSSARD, LOIS MARY, b Houston, Tex, Apr 1, 47. TOPOLOGY. Educ: Univ Tex, Austin, BA, 67, MA, 69, PhD(math), 74. Prof Exp: Teaching asst math, Univ Tex, Austin, 67-74; ASST PROF MATH, NORTHERN ILL UNIV, 74- Mem: Am Math Soc; Math Asn Am; Asn Women in Math; Sigma Xi. Res: Study of geometric topology problems such as embeddings of Canter sets and manifolds in Euclidean space. Mailing Add: Dept of Math Sci Northern Ill Univ DeKalb IL 60115

BROUSSEAU, NICOLE, b Quebec, Can, Oct 5, 48. OPTICS. Educ: Laval Univ, BS, 70, MS, 72, PhD(optics), 75. Prof Exp: SCI RESEARCHER OPTICS, DEPT NAT DEFENCE, 75- Concurrent Pos: Res asst optics, Laval Univ, 75. Mem: Optical Soc Am; Can Asn Physicists; Fr-Can Asn Advan Sci. Res: Optical processing of synthetic aperture radar data. Mailing Add: DREO Shirley's Bay Ottawa ON Can

BROWDER, ELI JEFFERSON, b Sweetwater, Tenn, May 27, 94; m; c 2. SURGERY, NEUROSURGERY. Educ: Emory & Henry Col, BA, 15; Johns Hopkins Univ, MD, 20; Am Bd Surg, dipl, 37; Am Bd Neurol Surg, dipl, 40. Prof Exp: EMER PROF, STATE UNIV NY DOWNSTATE MED CTR, 59-; STAFF SURGEON, VET ADMIN HOSP, EAST ORANGE, 70-; PROF SURG, COL MED & DENT NJ, 70- Concurrent Pos: Mem, Am Bd Neurol Surg, 55-61, chmn, 61. Mem: Am Surg Asn; Asn Res Nerv & Ment Dis (pres, 43); Am Neurol Asn (vpres, 50-51); Soc Neurol Surg; Am Asn Neurol Surg. Res: Craniocerebral trauma; parkinsonism; myelodysplasia of distal spinal cord; spinal epidural abscess; dorsal, dorsolateral and anterolateral spinal cordotomies; anatomical study of cerebral dural sinuses and their tributaries. Mailing Add: Questover Farm Asbury Rd Hackettstown NJ 07840

BROWDER, FELIX EARL, b Moscow, Russia, July 31, 27; nat US; m 49; c 2. MATHEMATICAL ANALYSIS. Educ: Mass Inst Technol, SB, 46; Princeon Univ, PhD(math), 48. Prof Exp: Moore instr math, Mass Inst Technol, 48-51; instr, Boston Univ, 51-53; asst prof, Brandeis Univ, 55-56; from asst prof to prof, Yale Univ, 56-63; prof, 63-72, LOUIS BLOCK PROF MATH & CHMN DEPT, UNIV CHICAGO, 72- Concurrent Pos: Vis mem, Inst Advan Study, 53-54 & 63-64; Guggenheim fels, 53-54 & 66-67; NSF sr fel, 57-58; Sloan Found fel, 59-63; ed, Am Math Soc Bull, 60-67. Mem: Nat Acad Sci; Am Math Soc; fel Am Acad Arts & Sci. Res: Partial differential equations and nonlinear functional analysis. Mailing Add: Dept of Math Univ of Chicago Chicago IL 60637

BROWDER, HENRY POLK, JR, b Savannah, Ga, Oct 16, 28; m 52; c 3. BIOCHEMISTRY. Educ: Ga Inst Technol, BS, 52; Univ Tex, MA, 56, PhD(biochem), 59. Prof Exp: Res chemist, Electrochem Dept, E I du Pont de Nemours & Co, 59-61; sr scientist, 61-66, res assoc, 66-70, PRIN INVESTR, MEAD JOHNSON RES CTR, MEAD JOHNSON & CO, 70- Mem: Am Chem Soc; Am Soc Microbiol; Sigma Xi; NY Acad Sci. Res: Microbial fermentation; metabolism; mechanism of enzyme action. Mailing Add: Mead Johnson Res Ctr Mead Johnson & Co Evansville IN 47721

BROWDER, JAMES STEVE, b Goodwater, Ala, Aug 5, 39; m 62; c 3. SOLID STATE PHYSICS, OPTICS. Educ: Rollins Col, BS, 61; Univ Fla, MS, 63, PhD(physics), 67.

Prof Exp: Asst physics, Univ Fla, 64-67, res asst, 67-68, fel, 68; asst prof, Northwestern State Univ, 68-71; asst prof, 71-74, ASSOC PROF PHYSICS,JACKSONVILLE PHYSICS, JACKSONVILLE UNIV, 74- Concurrent Pos: NSF acad year exten grant, 69-71. Mem: Am Asn Physics Teachers; Optical Soc Am; Am Phys Soc. Res: Use of a three-terminal capacitance dilatometer to study the thermal expansion of solids—single crystals, polycrystals and glasses—in the temperature range from room temperature down to 16K. Mailing Add: Dept of Physics Jacksonville Univ Jacksonville FL 32211

BROWDER, LEON WILFRED, b Pueblo, Colo, Apr 19, 40; m 63; c 2. DEVELOPMENTAL BIOLOGY. Educ: Univ Colo, BA, 62; La State Univ, MS, 64; Univ Minn, PhD(zool), 67. Prof Exp: Res assoc, Univ Colo, 67-69; asst prof biol, 69-72, ASSOC PROF BIOL, UNIV CALGARY, 72- Concurrent Pos: NIH fels, 67-69; Nat Res Coun Can grant, 70- Mem: Can Soc Cell Biol; Soc Develop Biol; Genetics Soc Can; Am Soc Cell Biol. Res: Control of amphibian pigment cell differentiation; control of genic expression during oogenesis and early development in amphibians. Mailing Add: Dept of Biol Univ of Calgary Calgary AB Can

BROWDER, LEWIS EUGENE, b McQueen, Okla, Jan 29, 32; m 54; c 3. PLANT PATHOLOGY. Educ: Okla State Univ, BS, 54, MS, 56; Kans State Univ, PhD(plant path), 65. Prof Exp: Plant pathologist, Okla State Univ, 56-58; plant pathologist, USDA, 58-66, ASST PROF PLANT PATH, KANS STATE UNIV, 58-, RES PLANT PATHOLOGIST, USDA, 66- Mem: Am Phytopath Soc. Res: Physiologic specialization of cereal rusts. Mailing Add: Dept of Plant Path Kans State Univ Manhattan KS 66506

BROWDER, WILLIAM, bNew York, NY, Jan 6, 34; m 60; c 3. TOPOLOGY. Educ: Mass Inst Technol, BS, 54; Princeton Univ, PhD(math), 58. Prof Exp: Instr math, Univ Rochester, 57-58; from instr to assoc prof, Cornell Univ, 58-63; mem, Inst Advan Study, 63-64; PROF MATH, PRINCETON UNIV, 64- Concurrent Pos: NSF fel, 59-60; Guggenheim Found fel, 74. Mem: Am Math Soc. Res: Topology. Mailing Add: Dept of Math Fine Hall Princeton Univ Princeton NJ 08540

BROWE, JOHN HAROLD, b Burlington, Vt, Nov 17, 15; m 39; c 4. NUTRITION. Educ: Univ Vt, AB, 37, MD, 40; Columbia Univ, MPH, 50. Prof Exp: Res assoc med & clin dir nutrit study, Col Med, Univ Vt, 46-49; pub health physician in training, 49-50, actg dir bur nutrit, 50-51, DIR BUR NUTRIT, NY STATE DEPT HEALTH, 51- Concurrent Pos: Consult, USPHS, 47, 56-73 & Vt State Dept Health, 48-49; assoc, Albany Med Col, 52-55, instr, 58-66; mem, Am Bd Nutrit; mem panel nutrit, Life Sci Comt, Space Sci Bd, Nat Acad Sci, 62-70, coun epidemiol, Am Heart Asn, 66-, pub health, Am Bd Prev Med, 68-; mem consult group for dietitians, Albany Regional Med Prog, 69-73. Mem: Fel Am Pub Health Asn; Am Soc Clin Nutrit; Am Inst Nutrit; Asn State & Territorial Pub Health Nutrit Dirs (pres, 57-58); Latin Am Soc Nutrit. Res: Human nutrition. Mailing Add: Bur Nutrit NY State Dept Health Tower Bldg Empire State Plaza Albany NY 12237

BROWELL, EDWARD VERN, b Indiana, Pa, Feb 6, 47; m 69. OPTICS, ENVIRONMENTAL SYSTEMS & TECHNOLOGY. Educ: Univ Fla, BSAE, 68, MS, 71, PhD(appl optics), 74. Prof Exp: Res asst appl optics, Aerospace Eng Dept, Univ Fla, 69-74; SCIENTIST LASER APPL & ENVIRON SCI, NASA LANGLEY RES CTR, 74- Mem: Optical Soc Am; Am Inst Physics; Sigma Xi. Res: Remote sensing of pollutant gases using pulsed tunable laser systems; research into the chemical reactions of pollutant gases. Mailing Add: MS 401-A NASA Langley Res Ctr Hampton VA 23665

BROWER, FRANK M, b Burlington, Iowa, May 2, 21; m 49; c 2. ENVIRONMENTAL MANAGEMENT. Educ: Iowa Wesleyan Col, BA, 49; DePauw Univ, MA, 51; Univ Ky, PhD(org chem), 54. Prof Exp: Org chemist, 54-58; group leader, 58-65, asst dir sci proj lab, 65-72, DIR HYDROCARBONS & MONOMERS RES LAB & MGR ENVIRON CONTROL, MICH DIV, DOW CHEM CO, 72- Mem: Am Chem Soc; Am Inst Chem Engrs. Mailing Add: 628 Bldg Dow Chem Co Midland MI 48640

BROWER, JAMES CLINTON, b New Rochelle, NY, June 27, 34; m 38; c 2. INVERTEBRATE PALEONTOLOGY. Educ: Am Univ, BS, 59, MS, 61; Univ Wis, PhD(geol), 64. Prof Exp: From asst prof to assoc prof, 63-74, PROF GEOL, SYRACUSE UNIV, 74- Mem: Soc Syst Zool; Soc Econ Paleontologists & Mineralogists; Int Asn Math Geol; Paleont Soc; Int Paleont Asn. Res: Paleobiology; Paleozoic crinoids; Paleozoic arthropods; geostatistics. Mailing Add: Dept of Geol Syracuse Univ Syracuse NY 13210

BROWER, JAMES E, b Amsterdam, NY, Dec 11, 39; m 62; c 3. PHYSIOLOGICAL ECOLOGY. Educ: State Univ NY, Albany, BS, 61; Syracuse Univ, MS, 64, PhD(ecol), 69. Prof Exp: Instr, Cornell Univ, 67-69; ASST PROF BIOL SCI, NORTHERN ILL UNIV, 69- Mem: AAAS; Am Soc Mammalogists; Am Inst Biol Scientists. Res: Ecology and physiology of mammalian hibernation. Mailing Add: Dept of Biol Sci Northern Ill Univ De Kalb IL 60115

BROWER, JOHN HAROLD, b Augusta, Maine, June 8, 40. ENTOMOLOGY, RADIATION ECOLOGY. Educ: Univ Maine, BS, 62; Univ Mass, MS, 64, PhD(entom), 65. Prof Exp: Res assoc insect radiation ecol, Brookhaven Nat Lab, 61-65; RES ENTOMOLOGIST, STORED PROD INSECTS RES & DEVELOP LAB, AGR RES SERV, USDA, 65- Concurrent Pos: Consult, Int Atomic Energy Agency, 74- Mem: Entom Soc Am; Ecol Soc Am; Radiation Res Soc; Entom Soc Can. Res: Effects of increased levels of ionizing radiations on insect populations in nature; radiation effects on insect behavior, population dynamics, physiology, genetics and developmental success; possibilities of radiation control of insect pests. Mailing Add: 3 Althea Pkwy Savannah GA 31405

BROWER, KAY ROBERT, b Altus, Okla, June 7, 28; m 48; c 2. ORGANIC CHEMISTRY. Educ: Mass Inst Technol, BS, 48; Univ Maine, MS, 51; Lehigh Univ, PhD(chem), 53. Prof Exp: Instr chem, Lehigh Univ, 53-54; res assoc, Univ Ill, 54-56; assoc prof, 56-70, PROF CHEM & CHMN DEPT, NMEX INST MINING & TECHNOL, 70- Mem: Fel AAAS; Am Chem Soc. Res: Chlorination products of organic compounds of sulfur; coumarins and chromones; kinetics of aromatic nucleophilic substitution reactions; volumes of activation of organic reactions. Mailing Add: Dept of Chem NMex Inst of Mining & Technol Socorro NM 87801

BROWER, KEITH LAMAR, b South Haven, Mich, Oct 24, 36; m 60; c 3. PHYSICS. Educ: Hope Col, AB, 58; Univ Minn, MS, 61; Univ Ill, PhD(physics), 66. Prof Exp: Res assoc ion & plasma physics, Electronics Group, Gen Mills, 61-62; MEM TECH STAFF MAGNETIC RESONANCE, SANDIA LABS, 66- Mem: Am Phys Soc. Res: Radiation damage. Mailing Add: Sandia Labs Org 5111 Albuquerque NM 87115

BROWER, LINCOLN PIERSON, b Summit, NJ, Sept 10, 31; m 53; c 2. ECOLOGY. Educ: Princeton Univ, AB, 53; Yale Univ, PhD(zool), 57. Prof Exp: Fulbright scholar, Oxford Univ, 57-58; from instr to assoc prof, 58-68, PROF BIOL, AMHERST COL, 68- Concurrent Pos: NIH spec fel, Oxford Univ, 63-64. Mem: Fel AAAS; Am Soc Naturalists; Am Soc Zoologists; Am Inst Biol Sci; Am Sci Film Asn. Res: Experimental study of ecology; evolution; animal behavior; ecological chemistry; automimicry. Mailing Add: Dept of Biol Amherst Col Amherst MA 01002

BROWER, THOMAS DUDLEY, b Birch Tree, Mo, Mar 15, 24; m 48; c 2. ORTHOPEDIC SURGERY. Educ: Cent Col, BS, 45; Wash Univ, MD, 47. Prof Exp: Intern surg, Univ Chicago Clins, 47-48; fel orthoped surg, Albany Med Col, 48-49; resident orthoped surg, Univ Chicago Clins, 49-54, instr surg, 54-55; from asst prof to assoc prof orthoped, Univ Pittsburgh, 55-64; PROF ORTHOPED, MED CTR, UNIV KY, 64- Mem: Am Acad Orthoped Surg; Am Orthoped Soc. Res: Articular cartilage. Mailing Add: Div of Orthoped Surg Univ of Ky Med Ctr Lexington KY 40506

BROWMAN, LUDVIG GUSTAV, b DeKalb, Ill, Apr 23, 04; m 33; c 4. ZOOLOGY. Educ: Univ Chicago, BS, 28, PhD(zool), 35. Prof Exp: High sch teacher, Ill, 28-31; asst instr, Univ Chicago, 34-35; embryologist, Exp Sta, Agr & Mech Col Tex, 36-37; from instr to assoc prof zool, 37-46, prof zool & physiol, 46-72, EMER PROF ZOOL & PHYSIOL, UNIV MONT, 72- Mem: Fel AAAS; Genetics Soc Am; Soc Exp Biol & Med; Am Asn Anat. Res: Light and endocrines; neurosecretion in rats; embryology and inheritance of microphthalmus in rats; channels in ice. Mailing Add: 664 S Sixth East Missoula MT 59801

BROWN, ACTON RICHARD, b Ash Grove, Kans, Apr 3, 20; m 43; c 2. PLANT BREEDING. Educ: Kans State Col, BS, 42; Univ Wis, MS, 47, PhD(agron), 50. Prof Exp: Asst prof, 50-56, ASSOC PROF AGRON, UNIV GA, 56- Mem: Am Soc Agron; Sigma Xi. Res: Plant production; inheritance of leaf rust resistance in barley; barley loose smut; lysine content of barley; semidwarf wheats. Mailing Add: Dept of Agron Univ of Ga Col of Agr Athens GA 30602

BROWN, ALBERT, b Scranton, Pa, Feb 3, 19; m 56; c 2. PHYSICS. Educ: Wesleyan Univ, AB, 40; Columbia Univ, AM, 50. Prof Exp: Asst physics, Columbia Univ, 47-50, res assoc, 50-52; supvr & mem tech staff, Bell Tel Labs, 52-59; chief plans & systs eng div, Bur Res & Develop, Fed Aviation Agency, 59-61, dep dir systs res & develop serv, 61-65; consult, Dirs Off, US Bur Budget, 65-66; sr assoc, Transportation & Urban Affairs Dept, Planning Res Corp, McLean, Va, 66-68, prin, 68-71; vpres, 71-73, CORP DIR, TCR SERV INC, 71-; CORP DIR & TREAS, ADVAN TRANSP MGT INC, 72- Mem: AAAS; Am Phys Soc; Opers Res Soc Am. Res: Cryogenics; operations research; systems engineering. Mailing Add: 3023 Normanstone Tree NW Washington DC 20008

BROWN, ALBERT LOREN, b Rochester, NY, Aug 27, 23; m 49; c 3. VETERINARY MICROBIOLOGY. Educ: Cornell Univ, BS, 48, MS, 49, PhD(bact), 51. Prof Exp: Asst bact, Cornell Univ, 48-51; res assoc virol, Sharp & Dohme Div, Merck & Co, Inc, 51-55, tech asst to dir biol prod, 55-57; res biologist, 57-61, SR RES SCIENTIST, RES & DEVELOP, NORDEN LABS, 61- Concurrent Pos: NIH spec fel, Sloan-Kettering Inst Cancer Res, 64. Mem: Am Soc Microbiol; US Animal Health Asn; Conf Res Workers Animal Dis; Brit Soc Gen Microbiol; Int Asn Biol Stand. Res: Veterinary biological products; viruses; canine distemper; rabies; leptospira; tissue culture; immunology. Mailing Add: Norden Labs Res & Develop PO Box 80809 Lincoln NE 68501

BROWN, ALEXANDER, chemistry, see 12th edition

BROWN, ALEXANDER CYRIL, b Petrolia, Ont, July 24, 38; m 74. ECONOMIC GEOLOGY, GEOCHEMISTRY. Educ: Univ Western Ont, BSc, 62; Univ Mich, MS, 65, PhD(geol & mineral), 68. Prof Exp: Tech officer, Geol Surv Can, 62-63; ASSOC PROF ECON GEOL, POLYTECH SCH, UNIV MONTREAL, 70- Concurrent Pos: Nat Res Coun Can & NATO fel, Univ Liege, 68-70, Sigma Xi res grant-in-aid, 69; Nat Res Coun Can res grant, Polytech Sch, Univ Montreal, 70-76; Geol Surv Can res contracts, 74-76. Honors & Awards: Waldemar Lindgren Citation, Soc Econ Geologists, 71. Mem: Geol Asn Can; Geol Soc Belgium; Can Inst Mining & Metall; Soc Geol Appl Mineral Deposits. Res: Geochemistry of iron and manganese deposition in surface waters; mineralization of stratiform base metal deposits. Mailing Add: Univ of Montreal Polytech Sch CP 6079 Succursale A Montreal PQ Can

BROWN, ALFRED BRUCE, JR, b Oak Park, Ill, May 11, 20; m 56; c 3. ELECTRON PHYSICS. Educ: Lehigh Univ, BS, 42; Calif Inst Technol, MS, 47, PhD(physics), 50. Prof Exp: Res assoc, Gen Elec Res Lab, 50-56; mem tech staff, 56-62, DEPT HEAD, BELL TEL LABS, 62- Mem: AAAS; Am Phys Soc; Inst Elec & Electronics Engrs. Res: Information theory; data transmission. Mailing Add: Bell Tel Labs Holmdel NJ 07733

BROWN, ALFRED DANIEL, JR, organic chemistry, see 12th edition

BROWN, ALFRED EDWARD, b Elizabeth, NJ, Nov 22, 16; m 44; c 2. ORGANIC CHEMISTRY. Educ: Rutgers Univ, BSc, 38; Ohio State Univ, MSc, 40, PhD(chem), 42. Prof Exp: Org chemist, USDA, 42-45; res assoc, Off Sci Res & Develop, 45; res assoc & asst dir, Harris Res Labs, 45-53, from asst dir res to dir res, 53-61, from vpres to pres, 53-66; PRES, CELANESE RES CO, 66- Concurrent Pos: Mem adv bd mil personnel supplies & mem comt fire safety aspects polymeric mat, Nat Res Coun. Mem: Nat Acad Eng; Asn Res Dirs (pres, 72-73); Am Asn Textile Chemists & Colorists; Am Chem Soc; fel Am Inst Chemists. Res: Organic polymers; resins; fibers; coatings; films; carbohydrates and proteins; consumer products; structure-property relationships. Mailing Add: Celanese Res Co PO Box 1000 Summit NJ 07901

BROWN, ALLAN HARVEY, b Newark, NJ, Sept 14, 17; m 41; c 3. PLANT PHYSIOLOGY. Educ: Univ Md, BS, 39; Univ Rochester, MS, 40, PhD(bot), 43. Prof Exp: Teaching asst biol, Univ Rochester, 39-42, asst, Sch Med & Dent, 42-44; staff mem, Radiation Lab, Mass Inst Technol, 44-45; res assoc & instr chem, Univ Chicago, 45-47; from asst prof to prof bot, Univ Minn, 47-63; PROF BIOL, UNIV PA, 63- Mem: Bot Soc Am; Am Soc Plant Physiol; Soc Gen Physiol; Am Inst Biol Sci; cor mem Int Acad Astronaut. Res: Plant cell respiration and photosynthesis; morphological development and behavior of higher plants at uncommon g-levels. Mailing Add: Dept of Biol G-5 Univ of Pa Philadelphia PA 19174

BROWN, ANTHONY WILLIAM ALDRIDGE, b England, Nov 18, 11; m 38; c 2. INSECT PHYSIOLOGY, TOXICOLOGY. Educ: Univ Toronto, BSc, 33, MA, 35, PhD(biochem), 36. Prof Exp: Asst biol, Univ Toronto, 33-34; Royal Soc Can res fel, Univ London, 36-37; asst entomologist in charge Can forest insect surv, Dept Agr, Can, 37-42; assoc prof zool, Univ Western Ont, 47-49, prof & head dept, 49-68; vector biol & control, WHO, Switz, 69-73; JOHN A HANNAH DISTINGUISHED PROF INSECT TOXICOL & DIR PESTICIDE RES CTR, MICH STATE UNIV, 73- Concurrent Pos: Ed, Can Field-Naturalist, 41-42; biologist, WHO, Geneva, 56-58. Honors & Awards: Gold Medal Achievement, Entom Soc Can, 64. Mem: Entom Soc Am (pres, 67); fel Royal Soc Can; Entom Soc Can (pres, 62); Can Physiol Soc. Res: Insect biochemistry, especially nitrogenous metabolism; biology and epidemiology of forest insects; large-scale insect control with aircraft and aerosol generators; toxicity and repellency of organic compounds to insects; resistance of insects to insecticides. Mailing Add: Dept of Entom Mich State Univ East Lansing MI 48824

BROWN, ARLEN, b Goshen, Ind, Sept 5, 26; m 48, 64; c 3. MATHEMATICS. Educ: Univ Chicago, PhB, 48, BS, 49, MS, 50, PhD(math), 52. Prof Exp: From instr to assoc prof math, Rice Inst, 52-63; assoc prof, Univ Mich, 63-67; PROF MATH, IND UNIV, BLOOMINGTON, 67- Mem: Am Math Soc. Res: Operators on Hilbert spaces; theory of operators. Mailing Add: Dept of Math Ind Univ Bloomington IN 47401

BROWN, ARNOLD LANEHART, JR, b Wooster, Ohio, Jan 26, 26; m 49; c 5. PATHOLOGY. Educ: Med Col Va, MD, 49. Prof Exp: Resident path, Presby-St Lukes Hosp, 50-51; instr, Med Sch, Univ Ill, 53-59; from instr to assoc prof, Mayo Grad Sch Med, Univ Minn, 59-71, PROF PATH, MAYO MED SCH, 71-, CHMN DEPT PATH & ANAT, 68- Concurrent Pos: Resident, Presby-St Lukes Hosp, 53-56; NIH fel, 56-59; consult, Mayo Clin, 59-; mem, Nat Cancer Adv Coun, 70-72 & Nat Cancer Adv Bd, 72-74; consult, Nat Cancer Adv Bd Carcinogenesis & Nat Organ Site Progs, 74-, chmn comt determination carcinogenicity of cyclamates, 75-76. Mem: Am Asn Path & Bact; Am Soc Exp Path; Am Soc Clin Path; Am Gastroenterol Asn; Int Acad Path. Res: Fine structural and biochemical aspects of pathology. Mailing Add: 200 First St SW Rochester MN 55901

BROWN, ARTHUR, b New York, NY, Feb 12, 22; m 47; c 4. MICROBIOLOGY. Educ: Brooklyn Col, BA, 43; Univ Chicago, PhD(bact) 50. Prof Exp: Res assoc bact, Univ Chicago, 51; instr microbiol & immunol, Col Med, State Univ NY, 51-55; microbiologist & br chief, Virus & Rickettsia Div, Ft Detrick, Md, 55-68; PROF MICROBIOL & HEAD DEPT, UNIV TENN, KNOXVILLE, 68- Concurrent Pos: Sr fel, Inst Molecular Biol, Geneva, 63-64; consult, Virus Cancer Prog, Nat Cancer Inst, 69- & Oak Ridge Nat Lab, 69-; mem training grants comt, Nat Inst Allergy & Infectious Dis, 69-73; mem exec comt, ETenn Cancer Res Ctr, 74-; macebearer, Univ Tenn, 75-76. Mem: Soc Exp Biol & Med; Am Asn Immunol; AAAS; Am Soc Microbiol; Infectious Dis Soc. Res: Virology; molecular virology; viral pathogenesis and immunity; viral oncology; mechanism of FV-1 gene restriction of mouse leukemia viruses; mechanism of cross protection among arboviruses. Mailing Add: Dept of Microbiol Univ of Tenn Knoxville TN 37916

BROWN, ARTHUR BARTON, b Boston, Mass, Feb 10, 05; m 35; c 1. MATHEMATICAL ANALYSIS. Educ: Harvard Univ, AB, 25, Am, 26, PhD(math), 29. Prof Exp: Instr math, Harvard, 27-29; Nat Res Coun fel, Princeton, 29-30; from instr to asst prof math, Columbia Univ, 30-38; from assoc prof to prof, 38-75, EMER PROF MATH, QUEENS COL, NY, 75- Mem: Am Math Soc; Soc Indust & Appl Math; Math Asn Am. Mailing Add: Apt 4F 155-01 90th Ave Jamaica NY 11432

BROWN, ARTHUR CHARLES, b South Bend, Ind, Feb 7, 29; m 53; c 4. PHYSIOLOGY, BIOPHYSICS. Educ: Univ Chicago, AB, 48, MS, 54, Univ Wash, PhD(physiol, biophys), 60. Prof Exp: Asst physics, Univ Chicago, 51-52; asst physics, Univ Wash, 54-55, asst physiol, 55-56; instr, State Univ NY Upstate Med Ctr, 58-60; from instr to assoc prof, 60-70, PROF PHYSIOL & BIOPHYS, UNIV WASH, 70- Concurrent Pos: Vis lectr physiol & actg head dept, Fac Med, Univ Malaya, 69-70; chmn physiol sect, Am Asn Dent Schs. Mem: Biophys Soc; Am Physiol Soc; Int Asn Dent Res; Inst Elec & Electronics Engrs; Biomed Eng Soc. Res: Physiological control systems; temperature regulation; dental and oral physiology and biophysics. Mailing Add: Dept of Physiol & Biophys Univ of Wash Seattle WA 98195

BROWN, ARTHUR LLOYD, b Edmonton, Alta, Feb 16, 15; nat US; m 44; c 4. SOIL SCIENCE. Educ: Univ Alta, BSc, 39, MSc, 41; Univ Minn, PhD(soil sci), 46. Prof Exp: Agr asst, Can Dept Agr, 39-42; asst soil sci, Univ Minn, 43-45; analyst, Univ Alta, 46; soil surveyor, Res Coun Alta, 46-47; instr soils & jr soil technologist, 47-51, assoc specialist soils, 51-57, LECTR SOILS, EXP STA, UNIV CALIF, DAVIS, 51-, SPECIALIST, 57- Mem: AAAS; Am Soc Agron; Soil Sci Soc Am. Res: Soil chemistry, particularly heavy metals, especially zinc and cadmium, in the soil in relation to amounts aquired by food products; methods of diagnosing nutrient deficiencies and excesses. Mailing Add: Dept of Soils & Plant Nutrit Univ of Calif Davis CA 95616

BROWN, ARTHUR MORTON, b Winnipeg, Man, Mar 3, 32; div; c 4. PHYSIOLOGY, BIOPHYSICS. Educ: Univ Man, MD, 56; Univ London, PhD(physiol), 64. Prof Exp: Asst resident path, Columbia-Presby Med Ctr, 56-57; asst resident med, Col Med, Univ Utah, 57-59, trainee cardiol, 59-61; res fel, Cardiovasc Res Inst, Univ Calif, San Francisco Med Ctr, 61-66, asst prof med & phyisol, 66-69, from assoc prof to prof physiol & internal med, 69-73; PROF PHYSIOL & BIOPHYS & CHMN DEPT, UNIV TEX MED BR, GALVESTON, 73- Concurrent Pos: Chmn physiol study sect, NIH, 74-76; mem res comt B, Am Heart Asn, 76-77. Mem: AAAS; Am Physiol Soc; Soc Gen Physiol; Biophys Soc. Res: Membrane biophysics; phototransduction; neurocirculatory control; cardiology. Mailing Add: Dept of Physiol & Biophys Univ Tex Med Br Galveston TX 77550

BROWN, AUDREY KATHLEEN, b New York, NY, Feb 2, 23; m 54; c 1. PEDIATRICS, HEMATOLOGY. Educ: Columbia Univ, BA, 44, MA, 45, MD, 50; Am Bd Pediat, dipl, 56. Prof Exp: Intern med, Columbia Med Div, Bellevue Hosp, NY, 50-51, intern & asst resident, Children's Med Serv, 51-53; Holt fel pediat, Babies Hosp, Columbia Presby Med Ctr, 53-54; civilian pediatrician, Walter Reed Army Med Ctr, 54-55; from instr to asst prof pediat, Col Med, Wayne State Univ, 55-59; from asst prof to assoc prof, Sch Med, Univ Va, 59-67; prof, Med Col Ga, 67-73, actg chmn dept, 69-73; PROF PEDIAT, STATE UNIV NY DOWNSTATE MED CTR, 73- Concurrent Pos: Asst & asst pediatrician, Children's Hosp, Mich, 55-57, assoc pediatrician, 58, assoc hematologist, 58-59; sr res assoc, Child Res Ctr Mich, 55-59; assoc dir, Hemat Training Grant; pediatrician, Univ Va Hosp, 59-, dir pediat hemat clin, 60-; med coordr & dir res, Children's Rehab Ctr, Va, 59-; mem perinatal biol & infant mortality training grant rev comt, NIH. Honors & Awards: Columbia Univ Bicentennial Award. Mem: Fel Am Acad Pediat; Soc Pediat Res; NY Acad Sci; Am Fedn Clin Res; Am Pediat Soc. Res: Hyperbilirubinemia; neonatal development of the glucuronide conjugating system; infantile pyknocytosis; fetal hemoglobin development; nutrition and development. Mailing Add: 22 Plum Beach Point Dr Sands Point NY 11050

BROWN, AUSTIN ROBERT, JR, b Kansas City, Mo, Dec 13, 25; m 51; c 3. MATHEMATICS. Educ: Grinnell Col, BA, 49; Yale Univ, MA, 50, PhD(math) 52. Prof Exp: Mathematician, Ballistic Res Labs, 49-53, chief comput methods sect, 53-54; assoc prof, Drury Col, 54-56; chief ballistic comput br, Air Force Armament Ctr, 56-58 & Air Proving Ground Ctr, Eglin AFB, Fla, 58; opers analyst, Hq Air Defense Command, US Air Force, 58-65; assoc prof, 65-67, PROF MATH, COLO SCH MINES, 67-, DIR COMPUT CTR, 65- Concurrent Pos: Lectr, Univ Del, 52-54; adj prof, Colo Col, 62-63. Mem: Math Asn Am; Asn Comput Mach. Res: Computer science; numerical analysis; simulation; operations research. Mailing Add: 407 Peery Pkwy Golden CO 80401

BROWN, BARBARA BANKER, b Columbus, Ohio, Nov 16, 17. PSYCHOPHYSIOLOGY. Educ: Ohio State Univ, BA, 38; Univ Cincinnati, PhD(pharmacol), 50. Prof Exp: Head div pharmacol, William S Merrell Co, 53-57; res neuropharmacologist, Riker Labs, Inc, 57-62; consult neurophysiologist, Vet Admin Hosp, Sepulveda, 63-65; assoc prof pharmacol, Univ Calif, Irvine-Calif Col Med, 67-73; CHIEF EXPERIENTIAL PHYSIOL, VET ADMIN HOSP, 67- Concurrent Pos: NIH & NIMH var res grants, Vet Admin Hosp, Sepulveda, 57-; assoc clin prof pharmacol, Ctr Health Sci, Univ Calif, Los Angeles, 57-62, lectr, Dept Psychiat, Med Sch, 73 Mem: Biofeedback Res Soc (chmn, 69-70). Res: Behavior and brain electrical activity; biofeedback. Mailing Add: Vet Admin Hosp 16111 Plummer St Sepulveda CA 91343

BROWN, BARBARA ILLINGSWORTH, b Hartford, Conn, May 12, 24; m 51; c 3. PHYSIOLOGICAL CHEMISTRY. Educ: Smith Col, BA, 46; Yale Univ, PhD, 50. Prof Exp: Asst & res asst prof physiol chem, 50-64, res assoc prof biol chem, 64-74, PROF BIOL CHEM, SCH MED, WASH UNIV, 74- Concurrent Pos: Estab investr, Am Heart Asn, 66-71. Mem: Am Soc Biol Chemists; Brit Biochem Soc. Res: Carbohydrate metabolism; phosphorylase; glycogen storage disease. Mailing Add: Dept of Biochem Wash Univ Sch of Med St Louis MO 63110

BROWN, BARKER HASTINGS, biochemistry, see 12th edition

BROWN, BARREMORE BEVERLY, physics, see 12th edition

BROWN, BARRY LEE, b Kingman, Ariz, June 29, 44. ENDOCRINOLOGY, REPRODUCTIVE PHYSIOLOGY. Educ: Univ Ga, BSA, 66, MS, 68; Purdue Univ, PhD(physiol), 72. Prof Exp: Res asst, Dept Animal Sci, Univ Ga, 66-68, Purdue Univ, 68-72; res fel, Dept Pop Dynamics, Johns Hopkins Univ, 72-74; RES CHEMIST ENDOCRINOL, BIOL SCI RES UNIT, FED BUR INVEST LAB, 74- Concurrent Pos: Asst prof lectr, Dept Biol Sci, George Washington Univ, 75- Mem: Am Soc Animal Sci; Soc Study Reproduction; Am Acad Forensic Sci. Res: Biochemistry and physiology of the testis; development and application of techniques, relating principally to endocrinology and forensic science. Mailing Add: Biol Sci Res Fed Bur Invest Lab Washington DC 20535

BROWN, BARRY W, b Buffalo, NY, Dec 20, 39; m 63; c 1. MATHEMATICAL BIOLOGY. Educ: Univ Chicago, BS, 59; Univ Calif, Berkeley, MA, 61, PhD(math), 63. Prof Exp: Res assoc statist & asst prof biol sci, Comput Ctr, Univ Chicago, 63-65; ASSOC PROF, DEPT BIOMATH, UNIV TEX M D ANDERSON HOSP & TUMOR INST, 65- Concurrent Pos: Chief sect comput sci, M D Anderson Hosp, 72- Mem: Asn Comput Mach; Am Statist Asn. Res: Mathematical problems arising in biological sciences. Mailing Add: Dept of Biomath Univ of Tex M D Anderson Hosp & Tumor Inst Houston TX 77025

BROWN, BERNARD BEAU, b Philadelphia, Pa, Apr 15, 25; m 47; c 2. ORGANIC CHEMISTRY. Educ: Temple Univ, AB, 45; Univ Mich, MS, 47, PhD(chem), 49. Prof Exp: res chemist, Eng Res Inst, 49-51; sr res chemist, Olin Mathieson Chem Corp, 51-54, mgr chem res, Indust Chem Div, 54-58; dir synthetic chem res, 59-63, dir res labs, 63-69, dir res, 69-74, VPRES RES & APPLN DEVELOP, S B PENICK & CO, 74- Mem: Am Chem Soc; Weed Sci Soc Am; NY Acad Sci. Res: Organic azides; synthetic lubricants; organic phosphorus compounds; agricultural pesticides; pharmaceuticals. Mailing Add: S B Penick & Co Res Dept 215 Watchung Ave Orange NJ 07050

BROWN, BERT ELWOOD, b The Dalles, Ore, Sept 27, 26. PHYSICS. Educ: Wash State Univ, BS, 49; Calif Inst Technol, MS, 53; Ore State Univ, PhD(physics), 63. Prof Exp: Naval architect, Puget Sound Naval Shipyard, 53-54, physicist, 54-56; from instr to assoc prof, 60-73, PROF PHYSICS, UNIV PUGET SOUND, 73-, CHMN DEPT, 70- Mem: Am Phys Soc; Am Asn Physics Teachers. Res: Atomic physics; atmospheric physics; climatology. Mailing Add: Dept of Physics Univ of Puget Sound Tacoma WA 98416

BROWN, BERTRAM S, b Brooklyn, NY, Jan 28, 31; m 52; c 4. PSYCHIATRY. Educ: Brooklyn Col, BA, 52; Cornell Univ, MD, 56; Harvard Univ, MPH, 60; Am Bd Psychiat & Neurol, dipl, 63. Prof Exp: Intern pediat, Sch Med, Yale Univ, 56-57; resident psychiat, Harvard Med Sch & Mass Ment Health Ctr, 57-60; staff psychiatrist, ment Health Study Ctr, 60-67, dep dir, NIMH, 67-70, DIR, NIMH, 70-; DEP ADMINR, HEALTH SERV & MENT HEALTH ADMIN, DEPT HEALTH, EDUC & WELFARE, 71- Concurrent Pos: Sr psychiatrist, Mass Div Legal Med, 58-59; dir psychiat, Norfolk Prison Colony, 58-59; field work supvr, Dept Social Rels, Harvard Univ, 58-60; mem fac, Boston Univ Inst on Rehab Emotionally Handicapped, 58 & 59; consult, Children's Bur Demonstration Proj, Cambridge Health Dept, 59-60 & Outdoor Resources Recreation Rev Comn, Washington, DC, 60-61; staff dir, President's Panel on Ment Retardation, 61-62; consult, President's Comt Employ of Handicapped, 61-; dep spec asst to the President for ment retardation, 62-63; mem tech adv bd, Maurice Falk Med Found, Pa, 62-; assoc clin prof psychiat, Med Sch, George Washington Univ, 62-72, clin prof psychiat & behav sci, 71-; consult, Off of Spec Asst to President for Ment Retardation, 63-66; mem sr fac, Washington Sch Psychiat, 65-; consult, President's Comn on Admin Law Enforcement & Justice, 66-67, President's Comn on Causes & Prev of Violence, 68-69, Task Force Develop Policy Recommendations Health Aspects of Malnutrition, USPHS, 69- & US Comt on Int Cong Social Welfare, 69-; spec asst on drug abuse prev to secy, Dept Health & Welfare, 70-72; consult, Exec Secy, President's Task Force on Ment Handicapped, 70 & Hogg Found Nat Adv Coun, 70- Honors & Awards: Arthur S Flemming Award, 69. Mem: Fel Am Psychiat Asn; fel Am Pub Health Asn; fel Am Orthopsychiat Asn; Am Sociol Asn; Am Asn Ment Deficiency. Mailing Add: NIMH 5600 Fishers Lane Rockville MD 20852

BROWN, BILLINGS, b Seattle, Wash, June 23, 20; m 46; c 7. THERMODYNAMICS. Educ: Univ Wash, BS & MS, 51, PhD(phys chem), 53. Prof Exp: Assoc prof chem eng, Brigham Young Univ, 53-59; tech specialist thermodynamics, Hercules Inc, 59-63; supvr solid rockets, Boeing Co, 63-64; prof chem eng, Univ Wash, 64-65; sr tech specialist solid rockets, Hercules Inc, 65-66; mem tech staff, Inst Defense Analyses, 66-68; SR TECH SPECIALIST SOLID ROCKETS, HERCULES INC, 68- Concurrent Pos: Mem joint Army Navy Air Force Thermochem Panel, 58-71 & Interagency Working Group on Safety, 72- Mem: Am Inst Chem Engrs. Res: Explosion mechanisms of solid rocket fuels. Mailing Add: Hercules Inc PO Box 98 Magna UT 84044

BROWN, BRADFORD E b Worcester, Mass, Apr 1, 38; m 74; c 3. FISH BIOLOGY, ECOLOGY. Educ: Cornell Univ, BS, 60; Auburn Univ, MS, 62; Okla State Univ, PhD(statist zool), 69. Prof Exp: Asst fish cult, Auburn Univ, 61-62; fisheries biologist, Biol Lab, US Bur Com Fisheries, 62-65; asst leader, Okla Coop Fisheries Unit, US Bur Sport Fisheries & Wildlife, 65-70; LEADER FISHERY MGT BIOL INVEST, BIOL LAB, NAT MARINE FISHERIES SERV, US DEPT COM, 70- Concurrent Pos: Asst prof zool, Okla State Univ, 65-70; mem standing comt res & statist, Int Comn Northwest Atlantic Fisheries, 71- Mem: Biomet Soc; Am Fisheries Soc. Res: Population dynamics and ecology of fresh water environments; applications of statistical and computer technology to ecology; population dynamics of exploited marine species; ecosystem approach to marine fisheries management. Mailing Add: Northeast Fisheries Ctr Nat Marine Fisheries Serv NOAA Woods Hole MA 02543

BROWN, BRIAN ELLMAN, b Shoreham, Eng, Jan 11, 30; Can citizen; m 52, 68; c 3. NEUROPHYSIOLOGY. Educ: Univ Western Ont, BSc, 53, MSc, 55; Univ Ill, PhD(entom), 61. Prof Exp: Res officer nematol, Ont, 55-56, res officer insect toxicol, Sask, 56-58, RES OFFICER INSECT NEUROENDOCRINOL, RES INST, CAN DEPT AGR, ONT, 61- Concurrent Pos: Fel, Sittingbourne Res Ctr, Kent, Eng, 75-76. Mem: Soc Gen Physiologists; Can Soc Cell Biol; Am Soc Zoologists. Res: Insect homeostasis as regulated by neuroendocrines of the insect retrocerebral complex; chemistry and physiology of neuromuscular transmission in insect visceral muscle. Mailing Add: Res Inst Can Dept Agr Univ Sub PO London ON Can

BROWN, BRUCE ANTONE, b Adams, Wis, Jan 6, 28; m 51; c 5. FOREST ECOLOGY. Educ: Univ Minn, BS, 52, MF, 53, PhD(forest ecol), 58. Prof Exp: Res asst forestry, 52-53, from instr to assoc prof, 53-68, supt, Coloquet Forestry Ctr, 60-71, dir, 71-74, PROF FORESTRY, CLOQUET FORESTRY CTR, UNIV MINN, ST PAUL, 68- Mem: Soc Am Foresters; Ecol Soc Am; Am Inst Biol Sci. Res: Forest ecology and soils. Mailing Add: 175 University Rd Cloquet MN 55720

BROWN, BRUCE CLAIRE, b Burns, Ore, May 10, 44; m 75; c 2. HIGH ENERGY PHYSICS. Educ: Univ Rochester, BS, 66; Univ Calif, San Diego, MS, 69, PhD(physics), 73. Prof Exp: Res asst physics, Univ Calif, San Diego, 67-72; res assoc, Univ Mich, 72-73; RES ASSOC, FERMI NAT ACCELERATOR LAB, 73- Mailing Add: Fermi Nat Accelerator Lab PO Box 500 Batavia IL 60510

BROWN, BRUCE ELLIOT, b Chicago, Ill, May 15, 30; m 62. GEOLOGY, CRYSTALLOGRAPHY. Educ: Wheaton Col, Ill, BS, 52; Univ Wis, MS, 54, PhD(soil sci), 57, MS, 60, PhD(geol), 62. Prof Exp: Crystallogr, Allis-Chalmers Mfg Co, 61-65; assoc prof, 65-74, PROF GEOL SCI, UNIV WIS-MILWAUKEE, 74- Mem: Am Crystallog Asn. Res: Inorganic structure, especially clay minerals, layer silicates, feldspars and binary chalcogonides. Mailing Add: Dept of Geol Univ of Wis Milwaukee WI 53211

BROWN, BRUCE STILWELL, b Chicago, Ill, Mar 15, 45; m 67; c 1. SOLID STATE PHYSICS. Educ: Miami Univ, BA, 66; Univ Ill, MS, 68, PhD(physics), 72. Prof Exp: Fel solid state physics, 72-74, ASST PHYSICIST, MAT SCI DIV, ARGONNE NAT LAB, 74- Mem: Am Phys Soc. Res: Effect of neutron irradiation at liquid helium temperature on properties of superconductors. Mailing Add: Mat Sci Div Argonne Nat Lab Argonne IL 60439

BROWN, BRUCE WILLARD, b New York, NY, Nov 25, 27; m 50; c 2. ANALYTICAL CHEMISTRY, INORGANIC CHEMISTRY. Educ: Polytech Inst Brooklyn, BS, 49, MS, 52; Univ Wash, PhD(chem), 61. Prof Exp: Instr chem, Everett Jr Col, 58-63; asst prof chem, 63-67, asst dean fac, 68-70, actg dean undergrad studies, 70, asst dean acad affairs, 71, ASSOC PROF CHEM, PORTLAND STATE UNIV, 67- Mem: AAAS; Am Chem Soc; Am Crystallog Asn; The Chem Soc. Res: X-ray crystallography of transition metal coordination compounds; computer programing; solid state reactions; lattice defect compounds. Mailing Add: Dept of Chem Portland State Univ Portland OR 97207

BROWN, BRYCE CARDIGAN, b Harlingen, Tex, May 7, 20; m 45; c 5. VERTEBRATE ZOOLOGY. Educ: Univ Tex, BA, 42; Agr & Mech Col Tex, MS, 48; Univ Mich, PhD(zool), 54. Prof Exp: From instr to assoc prof biol, Univ & from asst cur to assoc cur, Strecker Mus, 47-70, PROF BIOL, UNIV & DIR, STRECKER MUS, BAYLOR UNIV, 70- Concurrent Pos: Baylor Univ res grant, 68- Mem: Am Soc Ichthyologists & Herpetologists; Soc Syst Zool; Soc Study Amphibians & Reptiles; Am Asn Mus. Res: Vertebrate field biology; herpetology. Mailing Add: Strecker Mus Baylor Univ Waco TX 76703

BROWN, BYRON WILLIAM, JR, b Chicago, Ill, Apr 21, 30; m 49; c 6. BIOSTATISTICS. Educ: Univ Minn, BA, 52, MA, 55, PhD, 59. Prof Exp: Instr biostatist, Univ Minn, 53-54; statistician, Fed Cartridge Corp, 54-55; instr biostatist, Univ Minn, 55-56; asst prof, La State Univ, 56-57; from lectr to prof & head div, Sch Pub Health, Univ Minn, Minneapolis, 57-68; PROF BIOSTATIST & HEAD DIV, STANFORD UNIV MED CTR, 68- Concurrent Pos: Consult, NIH, Nat Acad Sci, NASA & other govt agencies & pharmaceut firms; mem, Food & Drug Admin Biostatist & Epidemiol Adv Comt, 74- Mem: AAAS; Biomet Soc; fel Am Statist Asn; Am Heart Asn; Inst Math Statist. Res: Teaching, research and consulting in applied statistics for the biomedical sciences. Mailing Add: Div of Biostatist Stanford Univ Med Ctr Stanford CA 94305

BROWN, CALVIN HUGH, b Washington, DC, Apr 15, 43; m 67. SYSTEMATIC ENTOMOLOGY, INSECT ECOLOGY. Educ: Univ Ga, AB, 64; Univ Calif, Riverside, PhD(syst entom), 69. Prof Exp: ASST PROF BIOL, VALDOSTA STATE COL, 69- Res: Biosystematic revision of genus Sphinx Linnaeus; sphingid distribution in the Southeastern United States, especially Florida and Georgia. Mailing Add: Dept of Biol Valdosta State Col Valdosta GA 31601

BROWN, CARL DEE, b New Cambria, Kans, Oct 2, 19; m 44; c 2. ENTOMOLOGY. Educ: Okla Baptist Univ, BS, 47; La State Univ, MS, 47; Iowa State Col, PhD(entom), 51. Prof Exp: Asst entomologist, Dept Entom & Bot, Agr Exp Sta, Univ Ky, 50-52; from asst prof to assoc prof, 52-58, chmn dept, 62-69, PROF BIOL, MEMPHIS STATE UNIV, 58- Mem: AAAS; Am Entom Soc. Mailing Add: Dept of Biol Memphis State Univ Memphis TN 38111

BROWN, CHARLES A, b Pennington Gap, Va, Sept 23, 34; m 56. MATHEMATICS. Educ: Berea Col, BA, 56; Fla State Univ, MA, 58; Univ Miss, PhD(math), 75. Prof Exp: Instr math, Fla State Univ, 58-61; asst prof, 61-66, ASSOC PROF MATH, UNIV TENN, CHATTANOOGA, 66- Mem: Am Math Soc; Math Asn Am. Res: Complex analysis; univalent functions. Mailing Add: Dept of Math Univ of Tenn Chattanooga TN 37401

BROWN, CHARLES ALBERT, b St Clairsville, Ohio, Aug 18, 20; m 48; c 2. INORGANIC CHEMISTRY. Educ: Hiram Col, AB, 42; Cornell Univ, PhD(chem), 48. Prof Exp: Asst chem, Cornell Univ, 42-44 & 46-48; instr chem, Harvard Univ, 48-51; asst prof, Western Reserve Univ, 51-56; res chemist, Lamp Metals & Components Dept, 56-60, mgr anal res & serv, 60-67, mgr qual assurance opers, Refractory Metals Bus Sect, 67-70, mgr refractory metals lab, 70-75, MGR ENG, REFRACTORY METALS PROD DEPT, LAMP BUS DIV, GEN ELEC CO, 75- Res: Phosphors and luminescent processes; refractory metals; tungsten and molybdenum chemistry and metallurgy. Mailing Add: Refractory Metals Prod Dept Gen Elec Co 21800 Tungsten Rd Cleveland OH 44117

BROWN, CHARLES AUGUSTUS, cytology, deceased

BROWN, CHARLES JULIAN, JR, b Utica, Miss, Dec 31, 32; m 63. PHYSICAL CHEMISTRY. Educ: Miss Col, BS, 54; Univ Tex, PhD(chem), 59. Prof Exp: Fel, Univ Tex, 59-60; res chemist, Film Dept, Exp Sta, 60-65 & org chem dept, 65-73, SR RES CHEMIST, ORG CHEM DEPT, E I DU PONT DE NEMOURS & CO, INC, 73- Mem: Am Chem Soc; AAAS; Sigma Xi. Res: Polymers; reverse osmosis. Mailing Add: Org Chem Dept E I du Pont de Nemours & Co Wilmington DE 19898

BROWN, CHARLES MOSELEY, b Kansas City, Mo, July 15, 43. ATOMIC SPECTROSCOPY, MOLECULAR SPECTROSCOPY. Educ: Southern Ill Univ, BA, 65; Univ Md, PhD(physics), 71. Prof Exp: Nat Res Coun res assoc atomic & molecular spectros, 71-73; PHYSICIST, US NAVAL RES LAB, 73- Mem: Optical Soc Am. Res: Vacuum ultraviolet spectroscopy of atomic and molecular species. Mailing Add: Code 7146B US Naval Res Lab Washington DC 20375

BROWN, CHARLES MYERS, b Oswego, SC, Oct 16, 27; m 57; c 2. PLANT BREEDING. Educ: Clemson Col, BS, 50; Univ Wis, MS, 52, PhD(agron), 54. Prof Exp: Asst agron, Univ Wis, 50-54; first asst, 54-55, from asst prof to assoc prof, 55-68, PROF AGRON, UNIV ILL, URBANA-CHAMPAIGN, 68-, ASSOC HEAD DEPT, 71- Mem: Am Soc Agron. Res: Oat breeding; genetics; cytogenetics. Mailing Add: Dept of Agron Univ of Ill Urbana IL 61801

BROWN, CHARLES NELSON, b Victoria, BC, Can, June 3, 41; m 67; c 2. EXPERIMENTAL HIGH ENERGY PHYSICS. Educ: Univ BC, BSc, 63; Univ Rochester, AM, 66, PhD(physics), 68. Prof Exp: Res fel physics, Harvard Univ, 68-71, asst prof, 71-74; staff physicist, 74-75, HEAD MESON AREA DEPT, FERMI NAT ACCELERATOR LAB, 75- Mem: Am Phys Soc; AAAS. Res: Experimental study of lepton scattering and lepton production in very high energy collisions. Mailing Add: 128 Logan Ave Geneva IL 60134

BROWN, CHARLES QUENTIN, b Roanoke Rapids, NC, Sept 12, 28; m 50; c 2. GEOLOGY. Educ: Univ NC, BS, 51, MS, 53; Va Polytech Inst, PhD(geol), 59. Prof Exp: From instr to assoc prof geol, Clemson Univ, 54-67; chmn dept, 67-69, PROF GEOL, E CAROLINA UNIV, 68- DIR INSTNL DEVELOP, 69- Concurrent Pos: Mem, NC Marine Sci Coun, 71-75. Mem: Geol Soc Am; Soc Econ Paleont & Mineral. Res: Sedimentation; clay mineralogy of recent deposits. Mailing Add: Dept of Geol East Carolina Univ Greenville NC 27834

BROWN, CHARLES THOMAS, b Paterson, NJ, Oct 27, 28; m 53; c 2. PHYSICAL CHEMISTRY. Educ: Ohio Wesleyan Univ, BA, 53; Rensselaer Polytech Inst, MS, 55, PhD, 57. Prof Exp: Prin chemist, Battelle Mem Inst, 57-60; mgr chem sect, Nuclear Div, Combustion Eng Corp, 60-61; res chemist, United Aircraft Corp Res Labs, 61-75, SR RES CHEMIST, UNITED TECHNOL RES CTR, UNITED AIRCRAFT CORP, 75- Mem: Am Chem Soc; Electrochem Soc. Res: Electrochemistry of non-aqueous systems, including material compatibility. Mailing Add: United Technol Res Ctr East Hartford CT 06108

BROWN, CHRISTOPHER W, b Maysville, Ky, Apr 13, 38; m 61; c 3. PHYSICAL CHEMISTRY. Educ: Xavier Univ, BS, 60, MS, 62; Univ Minn, PhD(phys chem), 67. Prof Exp: Res fel chem, Univ Md, 66-68; asst prof, 68-72, ASSOC PROF CHEM, UNIV RI, 72- Honors & Awards: Award, Chicago Sect, Soc Appl Spectros, 66. Mem: Am Chem Soc; Am Phys Soc. Res: Infrared and Raman spectroscopy; molecular structure; intermolecular interactions; solid state infrared and Raman; vibrational-rotational interactions; pressure effects on optical properties; chemical applications of lasers. Mailing Add: Dept of Chem Univ of RI Kingston RI 02881

BROWN, CLAIR ALAN, b Port Allegany, Pa, Aug 16, 03; m 26, 63; c 1. TAXONOMY, PLANT ECOLOGY. Educ: Syracuse Univ, BS, 25; Univ Mich, AM, 26, PhD(mycol), 34. Prof Exp: Asst, Herbarium, Univ Mich, 25-26; instr bot, La State Univ, 26-29; asst, Univ Mich, 29-31; from asst prof to prof, 31-70, EMER PROF BOT, LA STATE UNIV, BATON ROUGE, 70-; BOT CONSULT, 70- Concurrent Pos: State surv mem, Isle Royale, Mich, 30; Guggenheim fel, Europe, 52; Edmund Niles Huyck fel, 58 & 59. Honors & Awards: Award, Conserv Educ Asn. Mem: Bot Soc Am; Am Soc Plant Taxon; Am Fern Soc (pres); Int Asn Plant Taxon; Ecol Soc Am. Res: Biological and ecological study of Taxodium distichum; wood rotting fungi; flora of Louisiana; chemical control of weeds; morphology and biology of some species of Odontia; Tertiary palynology. Mailing Add: 1180 Stanford Ave Baton Rouge LA 70808

BROWN, CLARENCE ERVIN, b Lancaster, Pa, Dec 26, 25; m 49; c 3. GEOLOGY. Educ: Franklin & Marshall Col, BS, 50. Prof Exp: GEOLOGIST, US GEOL SURV, 51-, CHIEF, EASTERN MINERAL RESOURCES BR, 74- Mem: Geol Soc Am; Soc Econ Geol. Res: Structure, stratigraphy and ore deposits in Saint Lawrence County, New York; industrial talc, zinc and lead. Mailing Add: Nat Ctr MS954 US Geol Surv Reston VA 22092

BROWN, CLAUD LAFAYETTE, genetics, see 12th edition

BROWN, CLAUDE HAROLD, b Galatia, Kans, Sept 25, 05; m 31; c 1. MATHEMATICS. Educ: Kans State Teachers Col, Pittsburg, BSEd, 31, MS, 34; Univ Kans, PhD(math), 40. Prof Exp: Pub sch teacher, Kans, 26-28, prin, 29-37, teacher, 37-38; asst instr math, Univ Kans, 38-40; supvr student teaching, Western Ill State Teachers Col, 40-43; prof, 43-69, head dept, 46-69, chmn div sci & math, 47-69, dean col arts & sci, 69-74, EMER DEAN COL ARTS & SCI, CENT MO STATE COL, 74- Mem: AAAS. Res: Teaching of mathematics; analysis. Mailing Add: Cent Mo State Col Warrensburg MO 64093

BROWN, CLAUDEOUS JETHRO DANIELS, b Farr West, Utah, Aug 15, 04; m 31; c 2. ZOOLOGY. Educ: Brigham Young Univ, BS, 27, MS, 28; Univ Mich, PhD(zool), 33. Prof Exp: Instr zool, Univ Mich, 31-33, res assoc, 33-34; asst aquatic biologist, Bur Fisheries, Utah, 34-35; instr zool, Mont State Col, 35-37; asst dir inst fisheries res, Mich State Dept Conserv, 37-44; tech dir, Wash State Pollution Comn, 44-45; aquatic biologist, US Fish & Wildlife Serv, Wash & Ore, 45-47; from assoc prof to prof, 47-72, EMER PROF ZOOL, MONT STATE UNIV, 73- Concurrent Pos: Fisheries adv, Food & Agr Orgn Paraguay, 57; fisheries consult, Ford Found, Egypt, 64 & 66. Mem: Am Soc Limnol & Oceanog; Am Micros Soc (secy-ed, 52-56, pres, 59); Am Fisheries Soc (pres, 70); Am Soc Ichthyol & Herpet; Wildlife Soc. Res: Limnology and freshwater fisheries management; fishes of Montana. Mailing Add: 1054 Copper Basin Rd Prescott AZ 86301

BROWN, CLINTON CARL, b Hamilton, Ohio, Mar 28, 21; m 69; c 1. PSYCHOPHYSIOLOGY. Educ: Univ Cincinnati, BA, 48, PhD(psychol), 53. Prof Exp: Res psychologist, Perry Point Vet Admin Hosp, 52-62; assoc prof med pyschol, Sch Med, Johns Hopkins Univ, 62-68; DIR BIOMED SCI, MD PSYCHIAT RES CTR, 68- Mem: Soc Psychophysiol Res (pres, 64-65); Neuroelec Soc; Am Psychol Asn. Res: Psychophysiological concomitants of emotion; somatic factors in mental disorder; biomedical electronics. Mailing Add: Md Psychiat Res Ctr Box 3235 Baltimore MD 21228

BROWN, CONNELL JEAN, b Everton, Ark, Mar 6, 24; m 46; c 2. ANIMAL BREEDING. Educ: Univ Ark, BSA, 48; Okla State Univ, MS, 50, PhD(animal sci), 56. Prof Exp: Asst animal husb, Okla State Univ, 48-50; asst prof, Univ Ark, Fayetteville, 50-53; asst, Okla State Univ, 53-54; from asst prof to assoc prof, 54-61,

PROF ANIMAL HUSB, UNIV ARK, FAYETTEVILLE, 61- Mem: Fel AAAS; NY Acad Sci; Am Soc Animal Sci; Genetics Soc Am. Res: Classification of beef type; performance records of beef cattle; swine and beef carcass quality; sheep breeding; beef cattle genetics and breeding. Mailing Add: Div of Animal Sci Univ of Ark Fayetteville AR 72701

BROWN, DAIL WOODWARD, b Columbus, Ohio, Aug 11, 42; m 65; c 1. BIOLOGICAL OCEANOGRAPHY. Educ: Denison Univ, BA, 64; Univ Calif, Santa Barbara, PhD(biol), 69. Prof Exp: Res asst biol oceanog, Lamont Geol Observ, 68-69; oceanogr, Smithsonian Inst, 69-72; oceanogr, 72-75, DIR DEEPWATER PORTS PROJ OFF, ENVIRON DATA SERV, NAT OCEANIC & ATMOSPHERIC ADMIN, 75- Concurrent Pos: US nat corresp, Coop Invests of the Mediter, 71. Mem: AAAS; Am Soc Ichthyol & Herpet; Marine Technol Soc; Am Soc Limnol & Oceanog. Res: Ichthyology; marine ecology. Mailing Add: Deepwater Ports Proj Off NOAA 3300 Whitehaven St NW Rm 162 Washington DC 20235

BROWN, DANIEL JOSEPH, b Elwood, Ind, June 3, 41; m 66; c 3. PHARMACOLOGY, TOXICOLOGY. Educ: Marian Col, BS, 63; Ind Univ, PhD(toxicol), 68. Prof Exp: ASST PROF PHARMACOL & TOXICOL, MED CTR, IND UNIV, INDIANAPOLIS, 69- Concurrent Pos: Fel pharmacol & toxicol, Univ Wash, 68-69. Mem: Am Acad Forensic Sci; Soc Toxicol; Int Asn Forensic Toxicol. Res: Pharmacology and toxicology of cannabinoids, alcohols and carbon monoxide. Mailing Add: Dept of Toxicol Ind Univ Med Ctr Indianapolis IN 46207

BROWN, DARRELL QUENTIN, b Manhattan, Kans, May 23, 32; m 56; c 2. RADIATION BIOPHYSICS. Educ: Univ Kans, AB, 54, MS, 59, PhD(biophys), 64. Prof Exp: Mathematician-comput programmer, Lewis Lab, NASA, Ohio, 57; lab instr radiation biophys, Univ Kans, 58; USPHS Trainee biophys, Univ Chicago, 59-64, res assoc, 64; asst prof radiation biol, Univ Tenn, Knoxville, 64-70; RADIOBIOLOGIST, AM ONCOL HOSP, 70- Concurrent Pos: Consult, Oak Ridge Nat Lab, 68-69. Mem: AAAS; Am Soc Cell Biol; Radiation Res Soc; Biphys Soc; NY Acad Sci. Res: Chemical and radiation carcinogenesis in vitro; effects of microbeam irradiation of parts of cells; quantitative microscopy; radiation dosimetry; radiobiological basis of radiotherapy. Mailing Add: Am Oncol Hosp Cent & Shelmire Ave Fox Chase Philadelphia PA 19111

BROWN, DAVID BASSET, b Camden, NJ, Apr 24, 43; m 66. INORGANIC CHEMISTRY. Educ: Wesleyan Univ, BA, 64; Northwestern Univ, MS, 65, PhD(chem), 68. Prof Exp: Mem tech staff, Bell Tel Labs, 68-69; ASST PROF CHEM, UNIV VT, 69- Mem: AAAS; Am Chem Soc. Res: Coordination chemistry, properties of cyanaide complexes; preparation and structure of organo-transition metal complexes. Mailing Add: Dept of Chem Univ of Vt Burlington VT 05401

BROWN, DAVID EDWARD, b Indianapolis, Ind, July 9, 09; m 39; c 2. OTOLARYNGOLOGY. Educ: Stanford Univ, AB, 32, MD, 36. Prof Exp: From asst to assoc, 44-52, from asst prof to assoc prof, 52-62, chmn dept, 62-71, PROF OTOLARYNGOL, MED CTR, IND UNIV, INDIANAPOLIS, 62- Mem: AMA; Am Acad Ophthal & Otolaryngol. Mailing Add: Dept of Otolaryngol Med Ctr Ind Univ Indianapolis IN 46202

BROWN, DAVID FREDERICK, b Bargoed, Wales, Sept 14, 28; US citizen; m 50; c 2. MEDICINE. Educ: Univ Wales, BSc, 47, MB, BCh, 50; Am Bd Internal Med, dipl. Prof Exp: House physician, Med Unit, Cardiff Royal Infirmary, 50-51 & Sully Hosp, Wales, 51; med registr, Morriston Hosp, 53-55; resident internal med, Cornell Univ Infirmary & Thompkins County Mem Hops, Ithaca, NY, 55-56; sr pub health officer, NY State Dept Health, 56-57; res assoc, Cardiovasc Health Ctr & asst med, 57-59, from instr to assoc prof med, 59-73, assoc dir, Cardiovasc Health Vtr, 59-69, PROF MED, ALBANY COL MED, 73- Concurrent Pos: Nat Heart Inst spec fel physiol chem, Univ Lund, 6364; asst dispensary physician, Clins, Albany Med Ctr, 57-62, assoc dispensary physician, 62-, clin asst, Hosp, 6062, asst attend physician, 62-, asst attend cardiologist, 62-66, attend cardiologist, 66-, dir adult cardiac clin; attend physician, Vet Admin Hosp, 66-; fel, Coun Arteriosclerosis & Coun Epidemiol, Am Heart Asn. Mem: Am Fedn Clin Res; Am Soc Clin Nutrit; fel Am Col Physicians. Res: Cardiovascular disease; lipid metabolism as related to atherosclerosis. Mailing Add: Dept of Med Albany Med Col Albany NY 12208

BROWN, DAVID HAZZARD, b Philadelphia, Pa, Apr 22, 25; m 55; c 2. CELL BIOLOGY. Educ: Rutgers Univ, BA, 51, PhD(plant path), 65; Univ Pa, MS, 54. Prof Exp: Res assoc, Rutgers Univ, 65; NIH fel, Biol Div, Oak Ridge Nat Lab, Union Carbide Nuclear Corp, 65-67, staff molecular anat prog, 67-70; SCIENTIST, OAK RIDGE ASSOC UNIVS, 70- Mem: AAAS; Am Inst Biol Sci; Am Soc Cell Biol. Res: Use of radiopharmaceuticals in cancer detection; comparative studies of the cellular and physiological uptake of scintiscanning agents in normal and neoplastic tissue. Mailing Add: Med & Health Sci Div Oak Ridge Assoc Univs Oak Ridge TN 37830

BROWN, DAVID HENRY, b Ely, Nev, June 17, 21; m 51; c 3. BIOLOGICAL CHEMISTRY. Educ: Calif Inst Technol, BS, 42, PhD(org chem), 48. Prof Exp: Asst, Nat Defense Res Comt, Off Sci Res & Develop & Comt Med Res Projs, Calif Inst Technol, 42-45; Merck fel, 48-50, from instr to assoc prof, 50-62, PROF BIOL CHEM, SCH MED, WASHINGTON UNIV, 62- Mem: Brit Biochem Soc; Am Soc Biol Chemists. Res: Biochemistry of amino sugars and mucopolysaccharides; mechanism of enzyme action; metabolism of glycogen; enzymes of glycogen metabolism. Mailing Add: Dept of Biol Chem Washington Univ Sch of Med St Louis MO 63110

BROWN, DAVID LYLE, b Victoria, BC, Mar 2, 43; m 62; c 1. CELL BIOLOGY. Educ: Univ BC, BSc, 66; Univ Calif, Davis, MSc, 68, PhD(biol), 69. Prof Exp: Fel biol, Yale Univ, 69-70; asst prof, 70-75, ASSOC PROF BIOL, UNIV OTTAWA, 75- Mem: Can Soc Cell Biol (secy, 71-75); Am Soc Cell Biol; AAAS; Electron Micros Soc Am. Res: Synthesis of microtubule protein and the regulation of microtubule assembly in unicellular flagellates. Mailing Add: Dept of Biol Univ of Ottawa Ottawa ON Can

BROWN, DAVID T, b Portland, Ore, May 20, 36; m 60; c 2. MATHEMATICS. Educ: Ottawa Univ, BA, 58; Syracuse Univ, MA, 61, PhD(math), 65. Prof Exp: Asst prof math, Hiram Col, 65-68 & Univ Pittsburgh, 68-74; ASST PROF MATH, BETHANY COL, WVA, 74- Mem: Am Math Soc; Math Asn Am. Res: Functional analysis; algebraic extensions of commutative Banach algebras. Mailing Add: Dept of Math Bethany Col Bethany WV 26032

BROWN, DAWN LARUE, b Frederick, Md, Nov 13, 48. NEUROSCIENCES. Educ: Boston Univ, BA, 70; Drake Univ, MA, 72; Univ NC, Greensboro, PhD(physiol psychol), 76. Prof Exp: Instr psychol, NC A&T State Univ, 74-75; FEL PSYCHOBIOL, FLA STATE UNIV, 75- Concurrent Pos: Asn Res Vision & Ophthal travel fel, 75. Mem: Soc Neurosci; Asn Res Vision & Ophthal; AAAS. Res: Neurophysiology of vision; development of mammalian visual system; correlation of behavioral and electrophysiological properties of visual system. Mailing Add: Dept of Psychol Fla State Univ Tallahassee FL 32306

BROWN, DELOS D, b Mansfield, La, Aug 2, 33; m 59; c 3. FOOD SCIENCE. Educ: La State Univ, BS, 56, MS, 63; Univ of Ga, PhD(food sci), 72. Prof Exp: Res assoc meat technol, La State Univ, 61-63; sr scientist & asst head sausage develop sect, Food Res Div, Armour & Co, Ill, 63-68; food scientist, Univ Ga, 68-73; DIR & FOUNDER, BAPTIST AGR PROG TRAINING IN SCI & TECHNOL, 73- Concurrent Pos: USPHS training grant; Armour Food Res Div grant. Mem: Am Meat Sci Asn. Res: Development of an agricultural experiment station; research with field crops, livestock and poultry, vegetable crops and food science. Mailing Add: BAPTIST Box 120 Petauke Zambia

BROWN, DENNISON ROBERT, b New Orleans, La, May 17, 34; m 56; c 2. MATHEMATICS. Educ: Duke Univ, BS, 55; La State Univ, MS, 60, PhD(math), 63. Prof Exp: Assoc math, La State Univ, 58-61; from asst prof to assoc prof, Univ Tenn, 63-67; assoc prof, 67-70, PROF MATH, UNIV HOUSTON, 70- Concurrent Pos: Vis lectr, Math Asn Am, 65-74; ed, Semigroup Forum, 70-; consult, Undergrad Prog Math, 73- Mem: Am Math Soc; Math Asn Am. Res: Topological and algebraic semigroups; semigroups of matrices. Mailing Add: Dept of Math Univ of Houston Houston TX 77004

BROWN, DILLON SIDNEY, b Genoa, Ill, Nov 3, 11; m. HORTICULTURE. Educ: Univ Ill, BS, 35; WVa Univ, MS, 39; Ohio State Univ, PhD(hort), 44. Prof Exp: Lab asst hort, Univ Ill, 35-37; asst, W Va Univ, 37-41; asst pomol, Univ Ill, 41-44, assoc, 44-45, asst prof & asst chief, 45-47; sr horticulturist, Hawaiian Pineapple Co, 47-49; asst prof pomol & asst pomologist, Exp Sta, 49-53, assoc prof pomol & assoc pomologist, 53-59, chmn dept pomol, 63-70, chmn div agr sci, 70-73, PROF POMOL & POMOLOGIST, UNIV CALIF, DAVIS, 59- Mem: Am Soc Hort Sci. Res: Effects of climate on deciduous fruits; soil management; plant nutrition. Mailing Add: Dept of Pomol Univ of Calif Davis CA 95616

BROWN, DONALD A, b Roswell, NMex, Dec 23, 16; m 41; c 3. SOIL CHEMISTRY. Educ: NMex Col Agr & Mech, BS, 39; Univ Mo, Columbia, MS, 47, PhD(soil chem), 50. Prof Exp: Asst prof agron, NMex Col Agr & Mech, 45-46; asst soil chem, Univ Mo, Columbia, 47-50; asst prof, 50-59, PROF AGRON, UNIV ARK, FAYETTEVILLE, 59- Mem: Fel AAAS; Soil Sci Soc Am; Clay Minerals Soc; fel Am Soc Agron. Res: Factors affecting the exchange diffusion of ions in soil systems and subsequent uptake by plants. Mailing Add: Dept of Agron Univ of Ark Fayetteville AR 72701

BROWN, DONALD D, b Cincinnati, Ohio, Dec 30, 31; m 57; c 3. DEVELOPMENTAL BIOLOGY. Educ: Univ Chicago, MS & MD, 56. Prof Exp: Intern, Charity Hosp, New Orleans, La, 56-57; res assoc biochem, NIMH, 57-59; spec fel, Pasteur Inst, Paris, 59-60; spec fel, 60-61, MEM STAFF BIOCHEM, CARNEGIE INST DEPT EMBRYOL, 61- Concurrent Pos: Prof, Johns Hopkins Univ, 69- Honors & Awards: US Steel Found Award Molecular Biol, Nat Acad Sci, 73; V D Mattia Lectureship Award, Roche Inst Molecular Biol, 75. Mem: Nat Acad Sci; Am Soc Biol Chem; Soc Develop Biol; Am Acad Arts & Sci; Am Soc Cell Biol. Res: Control of genes during development; isolation of genes. Mailing Add: Carnegie Inst Dept of Embryol 115 W University Pkwy Baltimore MD 21210

BROWN, DONALD FREDERICK MACKENZIE, b Chicago, Ill, Dec 9, 19; m 45. BIOLOGICAL SCIENCES. Educ: Univ Mich, BA, 49, MS, 51, PhD(bot), 58. Prof Exp: Asst prof natural sci, 56-60, assoc prof biol, 60-65, PROF BIOL, EASTERN MICH UNIV, 65- Mem: AAAS; Wilderness Soc; Am Inst Biol Sci; Nat Audubon Soc; Nat Parks Asn. Res: Taxonomy; morphology and ecology of fern genus Woodsia; zoology; botany; applied botany and zoology; pteridology. Mailing Add: Dept of Biol Eastern Mich Univ Ypsilanti MI 48197

BROWN, DONALD JOHN, b Utica, NY, June 1, 33; m 56; c 3. INORGANIC CHEMISTRY. Educ: Utica Col, AB, 55; Syracuse Univ, PhD(chem), 63. Prof Exp: From instr to asst prof, 60-70, ASSOC PROF CHEM, WESTERN MICH UNIV, 70- Concurrent Pos: NIH grant, 64; res fel with Prof J Lewis, Univ London, 68. Mem: AAAS; Am Chem Soc. Res: Magnetic and spectral investigations of transition metal complexes; determination of stability constants of complexes; synthesis and investigation of new chelating agents. Mailing Add: Dept of Chem Western Mich Univ Kalamazoo MI 49008

BROWN, DONALD MEEKER, b Evanston, Ill, Oct 28, 07; m 34; c 2. APPLIED MATHEMATICS. Educ: Univ Ill, BS, 29, AB, 30, MA, 32, PhD(math), 38. Prof Exp: Asst math, Univ Ill, 30-37; chmn dept math & physics, Sioux Falls Col, 37-42; vibrations engr, Boeing Aircraft Co, 42, flight test training engr, 43-44; head dept math & physics, MacMurray Col, 44-47; assoc prof math, Cent Mich Col, 47-48; res engr & head dept simulation & comput, Willow Run Res Ctr, Univ Mich, 48-57; ed staff consult, Remington Rand Univac, 57-62; prof math, head dept & supvr digital comput ctr, Norwich Univ, 62-64; actg assoc dir & head dept training, Comput Inst, US Dept Defense, Washington, DC, 64-65; PROF MATH, ALLEGANY COMMUNITY COL, 65- Mem: Am Math Soc; Math Asn Am; Asn Comput Mach. Res: Numerical analysis. Mailing Add: Dept of Math Allegany Community Col Cumberland MD 21502

BROWN, DONALD MURRAY, b Barrie, Ont, July 28, 28; m 53; c 3. AGRICULTURAL METEOROLOGY. Educ: Ont Agr Col, BSA, 51; Univ Toronto, MSA, 53; Iowa State Univ, PhD(agron), 58. Prof Exp: Res fel crop ecol, Dept Physiography, Ont Res Found, 53-65; ASSOC PROF SOIL SCI, ONT AGR COL, UNIV GUELPH, 66- Mem: Agr Inst Can; Can Soc Agron; Can Meteorol Soc. Res: Agricultural climatology; crop adaptation; resources planning. Mailing Add: Dept of Land Resource Sci Ont Agr Col Univ of Guelph Guelph ON Can

BROWN, DOUGLAS EDWARD, b Mt Morris, Mich, Mar 25, 26; m 50; c 2. OPTICAL PHYSICS. Educ: Univ Mich, BS, 49, MS, 51, PhD(physics), 57. Prof Exp: Physicist, Gen Elec Co, 51-52; res assoc, Willow Run Res Ctr, 52-53; res fel, Univ Mich, 53-56; physicist, Dow Chem Co, 56-58, Bendix Systs Div, 58-60 & Inst Sci & Technol, Univ Mich, 60-71; dir eng, Syscon Int, Ind, 71-74; PHYSICIST, DEPT OF DEFENSE, 74- Mem: Optical Soc Am. Res: Infrared; semiconductors; optics; thermal; optical data processing. Mailing Add: 5348 Lightningview Columbia MD 21045

BROWN, DOUGLAS FLETCHER, b Rouleau, Sask, Aug 13, 34. GENETICS. Educ: Univ Sask, BSA, 57, MSc, 58; Univ Wis, PhD(genetics), 63. Prof Exp: Res asst genetics, Univ Wis, 61-63; asst res officer biol, Atomic Energy Can, Ltd, 63-66; ASSOC PROF BIOL, BISHOP'S UNIV, 66- Mem: Genetics Soc Can; NY Acad Sci. Res: Mutagenesis in plants and microorganisms. Mailing Add: Dept of Biol Sci Bishop's Univ Lennoxville PQ Can

BROWN, DOUGLAS MARKHAM, b Plainfield, NJ, Jan 31, 23; m 45; c 3. BIOCHEMISTRY. Educ: Univ Utah, BS, 49. Prof Exp: Res assoc biochem, Metab Lab, Sch Med, Univ Utah, 47-64; ASSOC BIOCHEMIST, SCH MED, UNIV CALIF, LOS ANGELES, 64- Mem: AAAS; Biol Photog Asn. Res: Electrophoretic and ultracentrifugal investigation of proteins and related materials. Mailing Add: Dept of Biol Chem Univ of Calif Med Ctr Los Angeles CA 90024

BROWN, DOUGLAS RICHARD, b Sacramento, Calif, Mar 13, 42; m 66; c 2. Educ: Univ Calif, Berkeley, BS, 66, MA, 68; Univ Wash, PhD(physics), 74. Prof Exp: Res asst, Nuclear Physics Lab, Univ Wash, 66-74; RES ASST NUCLEAR PHYSICS, CYCLOTRON INST, TEX A&M UNIV, 74- Mem: Am Phys Soc. Res: Study of relatively simple modes of higher lying nuclear excitation; higher lying E22 and MI strength in nuclei. Mailing Add: Cyclotron Inst Tex A&M Univ College Station TX 77843

BROWN, DOUGLAS ROSS, b San Francisco, Calif, Feb 21, 46. SOLAR PHYSICS. Educ: Reed Col, BA, 68; Univ Colo, MS, 73, PhD(physics), 75. Prof Exp: SOLAR PHYSICIST, GODDARD SPACE FLIGHT CTR, NASA, 73- Mem: Am Astron Soc. Res: High resolution studies of solar chromosphere phenomena, based primarily on spectrographic data. Mailing Add: Code 683 NASA Goddard Space Flight Ctr Greenbelt MD 20771

BROWN, DUANE, b Fredonia, Ariz, Apr 1, 18; m 42; c 2. PHYSICAL CHEMISTRY. Educ: Brigham Young Univ, BS, 41; Princeton Univ & Mass Inst Technol, cert, 44; Cornell Univ, PhD(phys chem), 51. Prof Exp: Asst chem, Univ Nebr, 41-42 & Cornell Univ, 46-51; assoc prof, 51-58, PROF CHEM, ARIZ STATE UNIV, 58- Concurrent Pos: Res chemist, Nat Carbon Res Labs, Ohio, 57 & 58. Mem: AAAS; Am Chem Soc. Res: Chemical kinetics; thermodynamics; radiation chemistry; hydrometallurgy. Mailing Add: Dept of Chem Ariz State Univ Tempe AZ 85281

BROWN, EDGAR HENRY, JR, b Oak Park, Ill, Dec 27, 26; m 54; c 2. TOPOLOGY. Educ: Univ Wis, BS, 49; Wash State Univ, MA, 51; Mass Inst Technol, PhD, 54. Prof Exp: Instr math, Washington Univ, 54-55; instr, Univ Chicago, 55-57; res assoc, Brown Univ, 57-58; from asst prof to assoc prof, 58-63, PROF MATH, BRANDEIS UNIV, 63- Concurrent Pos: NSF sr fel, 61-62; Guggenheim fel, 65-66. Mem: Am Math Soc; Am Acad Arts & Sci. Res: Algebraic and differential topology. Mailing Add: Dept of Math Brandeis Univ Waltham MA 02154

BROWN, EDMOND, b Brooklyn, NY, July 22, 24; m 46; c 3. THEORETICAL SOLID STATE PHYSICS. Educ: Univ Ill, BS, 48; Cornell Univ, PhD(physics), 54. Prof Exp: From asst prof to assoc prof, 55-65, PROF PHYSICS, RENSSELAER POLYTECH INST, 65- Concurrent Pos: Vis prof, State Univ NY Albany, 67-68; consult, Phys Sci Div, Watervliet Arsenal, 73- Mem: Fel Am Phys Soc; Am Asn Physics Teachers. Res: Electron band theory of solids; quantum theory of electrons in perturbed periodic lattices. Mailing Add: Dept of Physics Rensselaer Polytech Inst Troy NY 12181

BROWN, EDWARD ALLAN, b Brooklyn, NY, Oct 4, 28; m 52; c 6. PHYSICS, MATHEMATICS. Educ: Fordham Univ, BS, 51; Univ NMex, MS, 55; Ohio State Univ, PhD(math), 59; Iona Col, MBA, 70. Prof Exp: Sr engr & exec mgr, Int Bus Mach Corp, 59-68; dir staff, Nat Alliance Businessmen, 68; founder, dir & vpres mkt, Greenwich Data Systs, Inc, 69-72; dep dir info systs, Corp Staff, Xerox Corp, 73-75; GEN MGR, US EASTERN REGION, I P SHARP CO, 75- Concurrent Pos: Assoc prof, MBA Grad Fac, Iona Col, 69-; prof bus admin, Long Island Univ/Mercy Col Joint MBA Prog, 75- Mem: Sigma Xi. Res: Operations research; systems engineering. Mailing Add: 28 Country Club Dr Larchmont NY 10538

BROWN, EDWARD MARTIN, b Philadelphia, Pa, Sept 24, 33; m 59; c 2. MATHEMATICS. Educ: Univ Pa, AB, 58, MA, 59; Mass Inst Technol, PhD(math), 63. Hon Degrees: AM, Dartmouth Col, 73. Prof Exp: Teaching asst math, Univ Pa, 57-59; teaching asst, Mass Inst Technol, 59-62, asst instr, 62-63, instr, 63-64; res instr, 64-65, from asst prof to assoc prof, 65-72, PROF MATH, DARTMOUTH COL, 72-, VCHMN DEPT, 74- Concurrent Pos: Staff mem, Math Res Inst, Univ Warwick, 70-71. Mem: Am Math Soc. Res: General knotting structure of complexes in manifolds; topology Mailing Add: Dept of Math Dartmouth Col Hanover NH 03755

BROWN, EDWIN AUGUSTUS, b Bethlehem, Pa, Dec 17, 19; m 45; c 2. ECONOMIC GEOLOGY. Educ: Lehigh Univ, BA, 41, MS, 42. Prof Exp: Jr geologist, US Geol Surv, 42-45, asst geologist, 46-52, assoc geologist, 52-54; geologist, 55-61, SR RES GEOLOGIST, SOUTHWEST RES INST, 62- Mem: Am Inst Mining, Metall & Petrol Engr; Geol Soc Am; Sigma Xi. Res: Mineral deposits in southeastern states; bauxite in Arkansas; phosphate in Wyoming; ground-water surveys in Indiana, Michigan and Texas; lignite, clay, cement and aggregate deposits and products in Texas and Arkansas; non-metallic mineral resources in Arizona. Mailing Add: Southwest Res Inst Bldg 128 PO Box 28510 San Antonio TX 78284

BROWN, EDWIN WILSON, JR, b Youngstown, Ohio, Mar 6, 26; m 52; c 3. COMMUNITY HEALTH, MEDICAL EDUCATION. Educ: Harvard Univ, MD, 53, MPH, 57. Prof Exp: Intern, E J Meyer Mem Hosp, 54-55; resident pub health, Arlington County Health Dept, Va, 55-56; asst prof prev med, Sch Med, Tufts Univ, 58-61; dep chief staff, Boston Dispensary, 61; vis prof social & prev med, Osmania Med Col, India, 61-63; asst dir, Div Int Med Educ, Asn Am Med Cols, 63-65; dir, Proj Vietnam, Washington, DC, 65-66; dir, Div Int Health, Med Ctr, Ind Univ, 66-69; DIR INT AFFAIRS, IND UNIV-PURDUE UNIV, INDIANAPOLIS, 69-, PROF COMMUNITY HEALTH SCI, 73-, ASSOC DIR INT AFFAIRS CTR, IND UNIV, BLOOMINGTON, 66- Concurrent Pos: Field dir, Harvard Epidemiol Proj, Greenland, 56-57; res assoc, Sch Pub Health, Harvard Univ, 59-60; consult, Boston City Health Dept, 58-60 & WHO, 73-; chmn bd dirs, Med Assistance Progs, Inc, Ill; chmn coun int educ in health prof, Midwestern Univs Consortium Int Activ, Inc; mem med adv comt, Iran Found; mem bd dirs, Paul Carlson Found & CARE-MEDICO, Int Students, Inc. Res: Problems of medical education in developing countries. Mailing Add: Regenstrief Inst 1001 W 10th St Indianapolis IN 46202

BROWN, ELEANOR MOORE, b East Liverpool, Ohio, Mar 19, 36; m 60; c 1. PHYSICAL BIOCHEMISTRY. Educ: Ohio Wesleyan Univ, BA, 58; Drexel Univ, MS, 67, PhD(chem), 71. Prof Exp: Chemist, Harshaw Chem Co, 58-61 & Calbiochem, 61-62; res asst biochem, Mich State Univ, 64; res asst phys chem, Drexel Univ, 67-71; RES CHEMIST, EASTERN REGIONAL RES CTR, USDA, 71- Concurrent Pos: Assoc, Nat Res Coun, 71-73; fel, Dairy Res Inc, 73-75. Mem: Am Chem Soc. Res: Metal-prot- ein and protein-protein interactions; porphyrins. Mailing Add: Eastern Regional Res Ctr USDA 600 E Mermaid Lane Philadelphia PA 19118

BROWN, ELISE ANN, b Jacksonville, Fla, Dec 5, 28; m 52; c 3. PHARMACOLOGY. Educ: George Washington Univ, BS, 49, MS, 50, PhD(pharmacol), 56. Prof Exp: Asst prof pharmacol, George Washington Univ, 56-57; PHARMACOLOGIST, NAT HEART & LUNG INST, 62- Concurrent Pos: Fel, McCollum-Pratt Inst, Johns Hopkins Univ & Sinai Hosp, Baltimore, Md, 57-59; lectr, Sch Med, Georgetown Univ, 64- Mem: Am Chem Soc; Am Soc Pharmacol & Exp Therapeut; Soc Gen Physiol; NY Acad Sci; Int Soc Biochem Pharmacol. Res: Pharmacology of lung tissue; effect of drugs on lipid metabolism; cancer chemotherapy; cystic fibrosis of the pancreas; para-amino-benzoic acid metabolism. Mailing Add: 6811 Nesbitt Pl McLean VA 22101

BROWN, ELLEN, b San Francisco, Calif, Apr 30, 12. MEDICINE. Educ: Univ Calif, BA, 34, MD, 39. Prof Exp: Intern, San Francisco Hosp, 38-39; from asst resident to resident med, Univ Calif Hosp, 39-43; clin instr, 43-44, from instr to assoc prof, 46-59, PROF MED, MED SCH, UNIV CALIF, SAN FRANCISCO, 59- Concurrent Pos: Asst physician, Cowell Mem Hosp, 40-42; Commonwealth Fund fel, Harvard Med Sch, 44-46; Guggenheim fel, Oxford Univ, 58. Mem: AAAS; Am Fedn Clin Res; Asn Am Med Cols; NY Acad Sci; Am Heart Asn. Res: Capillary pressure and permeability; blood volume and vascular capacity; cardiac failure; cardiac complications of pregnancy; peripheral circulation in relation to pain syndromes and vascular diseases. Mailing Add: Dept of Med Univ of Calif Sch Med San Francisco CA 94143

BROWN, ELLIS VINCENT, b Montreal, Que, Aug 6, 08; m; c 2. ORGANIC CHEMISTRY. Educ: Univ Ill, BS, 30; Iowa State Col, PhD(org chem), 36. Prof Exp: Asst, Iowa State Col, 30-36, instr chem, 36-37; res chemist, Chas Pfizer & Co, Inc, 37-43, group leader, 43-47; assoc prof org chem, Fordham Univ, 47-53; prof chem, Seton Hall Univ, 53-59, chmn dept, 56-59; PROF CHEM, UNIV KY, LEXINGTON, 59- Mem: Fel AAAS; Am Chem Soc; fel NY Acad Sci. Res: Chemistry and synthesis of vitamins, medicinals and antibiotics; reactions and synthesis in heterocyclics and carbohydrates; organic reaction mechanisms. Mailing Add: Dept of Chem Univ of Ky Lexington KY 40506

BROWN, ELMER BURRELL, b New York, NY, Apr 1, 26; m 54; c 4. INTERNAL MEDICINE, HEMATOLOGY. Educ: Oberlin Col, AB, 46; Wash Univ, MD, 50; Am Bd Internal Med, 57. Prof Exp: From intern to asst resident med, Presby Hosp, NY, 50-52; from instr to assoc prof, 55-71, chief hemat clin, 59-73, dir div hemat, 64-73, PROF MED, SCH MED, WASH UNIV, 71-, ASSOC DEAN CONTINUING MED EDUC, 73- Concurrent Pos: USPHS trainee hemat, Sch Med, Wash Univ, 54-55, Nat Res Coun fel, 55-57; USPHS spec res fel, Enzyme Sect, Lab Cellular Physiol, Nat Heart Inst, 57-59; consult, Wash Univ Clins, 55-57 & St Louis City Hosp, 55-57; Lederle Med Fac Award, 60-63; USPHS res career develop award, 63-; vis prof, Royal Postgrad Med Sch, London, 69-70. Mem: Am Soc Hemat; Am Fedn Clin Res; Soc Exp Biol & Med; Am Soc Clin Invest; Am Soc Exp Path. Res: Problems of iron metabolism at the clinical, physiological and biochemical levels; biochemistry. Mailing Add: Dept of Med Wash Univ St Louis MO 63110

BROWN, ELWYN S, b Belle Plaine, Iowa, Sept 1, 20; m 47; c 2. ANESTHESIOLOGY. Educ: Univ Iowa, BS, 42, MS, 47, MD, 50. Prof Exp: Jr engr, Shell Develop Co, 43-45; asst anesthesia, Wash Univ, 51-53; prin scientist cancer res, Roswell Park Mem Inst, 53-58, assoc chief, 58-63; ANESTHESIOLOGIST IN CHIEF, CHILDREN'S MERCY HOSP, 63- Concurrent Pos: Secy comt equip, Nat Found Infantile Paralysis, 51-53. Mem: AMA; Am Soc Anesthesiol. Res: Pulmonary physiology; resuscitation; endocrinology. Mailing Add: Children's Mercy Hosp 24th & Gillham Kansas City MO 64108

BROWN, ERIC REEDER, b Cortland, NY, Mar 16, 25; m 61; c 4. CHEMISTRY, VIROLOGY. Educ: Syracuse Univ, BA, 48, MS, 51; Univ Kans, PhD(virol, immunol, biochem), 57. Hon Degrees: DSc, Quincy Col, 66. Prof Exp: Instr, Col Med, Univ Ill, 57-59; asst prof hemat & immunol, Univ Ala, 59; asst prof, Univ Minn, 60-61; sr res assoc hemat, Hektoen Inst Med Res, 61-65; assoc prof microbiol, Med Sch, Northwestern Univ, 65-68; PROF MICROBIOL & CHMN DEPT, CHICAGO MED SCH, 67- Concurrent Pos: Assoc, Northwestern Univ, 63-; res fel, Am Cancer Soc, 60-63; scholar, Leukemia Soc, Inc, 65-70; mem med adv comt, Leukemia Res Found Inc, 70-; med adv, Ill State Dir Selective Serv Syst. Mem: NY Acad Sci; Am Asn Cancer Res; fel Am Inst Chemists; fel Am Acad Microbiol; fel Am Pub Health Asn. Res: Anthrax diagnosis; bacteriophage and transduction genetics; immunochemical tumor antigens; leukemia virus and antibodies; microbiology; epidemiology; pollution. Mailing Add: Dept of Microbiol Chicago Med Sch Chicago IL 60612

BROWN, ERIC RICHARD, b Ithaca, NY, Feb 12, 42; m 66; c 2. ANALYTICAL CHEMISTRY. Educ: Mich Technol Univ, BS, 63; Northwestern Univ, PhD(anal chem), 67. Prof Exp: RES CHEMIST, RES LABS, EASTMAN KODAK CO, 67- Mem: Am Chem Soc. Res: Electrochemistry, especially instrumentation and automatic data acquisition; investigation of mechanisms of electrodeposition and dissolution. Mailing Add: Eastman Kodak Co Kodak Park Bldg 59 Rochester NY 14650

BROWN, ERNEST BENTON, JR, b Mortons Gap, Ky, July 13, 14; m 34; c 2. MEDICAL ADMINISTRATION. Educ: Univ Ky, BS, 37, MS, 42; Univ Minn, PhD(physiol), 49. Prof Exp: Prin high sch, Ky, 38-41; asst, Univ Ky, 41-42, from instr to asst prof physiol, 42-46; asst, Univ Minn, 4648, instr physiol chem, 48-49, from asst prof to prof physiol, 49-61; prof physiol & chmn dept, 61-73, dean fac & acad affairs, 73-74, V CHANCELLOR FAC & ACAD AFFAIRS, UNIV KANS MED CTR, KANSAS CITY, 74- Concurrent Pos: Mem, NIH Physiol Fel Panel, 6265; ed, J Physiol & J Appl Physiol, 60-63; ed Proc, Soc Exp Biol & Med, 63-66. Mem: AAAS; Soc Exp Biol & Med; Am Physiol Soc; Am Heart Asn. Res: Anoxia and bends at high altitude; prolonged hyperventilation in man; mechanism of continued hyperpnoea following hyperventilation; effects of very high concentrations of carbon dioxide on cardiovascular system; acid-base balance; electrolyte metabolism. Mailing Add: Off of VChancellor Univ of Kans Med Ctr Kansas City KS 66103

BROWN, ESTHER MARIE, b Birmingham, Mich, May 12, 23; m 63. ANATOMY. Educ: Mich State Univ, BS, 46, MS, 51, PhD(path), 55. Prof Exp: Bacteriologist, Game Div, State Dept Conserv, Mich, 46-50; bacteriologist, Ft Detrick Proj, Mich State Univ, 50-54, asst vet path, 54-55, from asst prof to prof anat, 55-71, dir sch med technol, 66-71; PROF VET ANAT, SCH VET MED, UNIV MO-COLUMBIA, 71- Concurrent Pos: Mem vet med rev comt, NIH, 72-74. Mem: Am Soc Med Technol; Am Asn Anatomists; Tissue Cult Asn; Am Soc Cytol; NY Acad Sci. Res: Microscopic anatomy; cytology and tissue culture, particularly tissue responses of animal diseases; normal histology of the white rat. Mailing Add: Dept of Vet Anat Univ of Mo Sch of Vet Med Columbia MO 65201

BROWN, FARRELL BLENN, b Mt Ulla, NC, Nov 29, 34; m 58; c 2. STRUCTURAL CHEMISTRY. Educ: Lenoir-Rhyne Col, BS, 57; Univ Tenn, MS, 59, PhD(phys chem), 62. Prof Exp: Robert A Welch Found fel, Tex A&M Univ, 62-63; from asst prof to assoc prof, 63-73, PROF CHEM, CLEMSON UNIV, 73-, ASST DEAN GRAD STUDIES, 74- Mem: Am Chem Soc. Res: Molecular spectroscopy, structure and dynamics; far infrared spectral studies; molecular configurations; quantum mechanics of bonded systems. Mailing Add: Dept of Chem Clemson Univ Clemson SC 29631

BROWN, FERDINAND LOUIS, b Portsmouth, Ohio, Dec 9, 16. MATHEMATICS. Educ: Univ Notre Dame, AB, 38, MS, 45, PhD(math), 47. Prof Exp: From instr to asst prof math, 45-55, assoc vpres acad affairs, 68-70, ASSOC PROF MATH, UNIV NOTRE DAME, 55-, ASSOC PROVOST ACAD AFFAIRS, 70- Mem: Am Math Soc. Res: Modern algebra; algebra of analysis; tri-operational algebra; group theory. Mailing Add: 202 Admin Bldg Univ of Notre Dame Notre Dame IN 46556

BROWN, FIELDING, b Berlin, NH, Jan 2, 24; m 44; c 4. PHYSICS. Educ: Williams

Col, Mass, BA, 47, AM, 49; Princeton Univ, PhD, 53. Prof Exp: Res physicist, Sprague Elec Co, Mass, 52-59; from asst prof to assoc prof, 59-67, PROF PHYSICS, WILLIAMS COL, MASS, 67- Concurrent Pos: Vis scientist, Lincoln Lab, Mass Inst Technol, 61-62 & Francis Bitter Nat Magnet Lab, 68-; vis prof, Univ Tokyo, 65-66; consult, Mass Inst Technol, 73-74. Mem: Am Phys Soc; Sigma Xi. Res: Cosmic rays; ferrites; ferroelectricity; nonlinear optics; far infrared; lasers. Mailing Add: Dept of Physics Williams Col Williamstown MA 01267

BROWN, FOUNTAINE CHRISTINE, b Huffman, Ark, Oct 13, 23. BIOCHEMISTRY. Educ: Southeast Mo State Col, BA & BS, 47; Univ Mo, MA, 51; Univ Iowa, PhD(biochem), 55. Prof Exp: PROF BIOCHEM, COL MED, UNIV TENN, MEMPHIS, 58-, RES BIOCHEMIST, TENN PSYCHIAT HOSP, 58- Concurrent Pos: Fel, M D Anderson Hosp & Tumor Inst, Univ Tex, 55-58. Mem: AAAS; Am Soc Biol Chemists; Am Chem Soc; Am Soc Neurochem; Int Soc Neurochem. Res: Biochemistry of mental disease; isolation and purification of proteins; enzymes; metabolism. Mailing Add: Dept of Biochem Univ Tenn Col Med Memphis TN 38103

BROWN, FRANCES CAMPBELL, b Johnson City, Tenn, Dec 12, 06. ORGANIC CHEMISTRY. Educ: Agnes Scott Col, AB, 28; Johns Hopkins Univ, PhD(chem), 31. Prof Exp: From instr to prof, 31-73, EMER PROF CHEM, DUKE UNIV, 73- Mem: AAAS; Am Chem Soc. Res: Friedel-Crafts reaction; organic fungicides; heterocyclic sulfur compounds. Mailing Add: 1205 Dwire Pl Durham NC 27706

BROWN, FRANCIS EARL, organic chemistry, polymer chemistry, see 12th edition

BROWN, FRANCIS ROBERT, b Fairbury, Ill, Dec 19, 14; m 40; c 4. MATHEMATICS. Educ: Ill State Univ, BEd, 37; Columbia Univ, MA, 40; Univ Ill, EdD, 54. Prof Exp: Instr, Pub Sch, Ill, 37-42; asst prof math, Millikin Univ, 46-49; from instr to assoc prof, 49-59, dir div univ exten & field serv, 57-72, PROF MATH, ILL STATE UNIV, 59-, ASST DIR SUMMER SESSIONS, 68-, DIR DIV CONTINUING EDUC & PUB SERV, 72- Mem: Math Asn Am; Nat Coun Teachers Math; Adult Educ Asn US (pres, 62-63). Res: Effect of study of geometry on spatial visualization ability; mathematics education. Mailing Add: Div of Cont Educ & Pub Serv Ill State Univ Normal IL 61761

BROWN, FRANK ARTHUR, JR, b Beverly, Mass, Aug 30, 08; m 34; c 3. BIOLOGY. Educ: Bowdoin Col, AB, 29; Harvard Univ, AM & PhD(zool), 34. Prof Exp: Asst zool, Harvard Univ, 32-34; instr, Univ Ill, 34-37; from asst prof to assoc prof, 37-46, prof zool & chmn dept biol sci, 49-56, MORRISON PROF ZOOL, NORTHWESTERN UNIV, EVANSTON, 56- Concurrent Pos: Instr, Mt Desert Biol Lab, 40; vis prof, Univ Chicago, 41; assoc & rev ed, Physiol Zool, 42-; head invert zool, Marine Biol Lab, Woods Hole, 45-49, trustee, 46-, consult, 49-; trustee, John G Shedd Aquarium, Chicago; mem corp, Bermuda Biol Lab & Mt Desert Biol Lab; chmn adv panel biol, Off Naval Res, 53-58; Sigma Xi nat lectr, 68. Mem: AAAS; Am Soc Zool(treas, 48, vpres, 54); Am Soc Limnol & Oceanog; Am Physiol Soc; Soc Develop Biol. Res: Conditioned behavior in lower animals; phototaxis in Drosophila; color perception in fishes; swim-bladder physiology; plumage changes in birds; biological rhythms; chromatophores and color change; physiology of Crustacea; invertebrate endocrinology; physiology of compound eyes; biological clocks; biogeophysics. Mailing Add: Dept of Biol Sci Northwestern Univ Evanston IL 60201

BROWN, FRANK BURKHEAD, b Raleigh, NC, Mar 4, 19; m 43; c 3. PHYSICS. Educ: Univ SC, BS, 39; Ga Inst Technol, MS, 50; Univ NC, PhD(physics), 53. Prof Exp: Instr physics, Ga Inst Technol, 45-50; physicist, Melpar, Inc, 53-54, chief systs anal, Union Switch & Signal Div, 54-58; dir reliability & eval dept, Electronics & Ord Div, Avco Corp, 58-61; dir reliability progs, Opers Res, Inc, 61-64, assoc dir govt & indust systs div, 64-67, vpres-dir eng anal div, 67-70, vpres corp off, 70-72, PRES, COBRO CORP, 72- Mem: Opers Res Soc Am; Inst Mgt Sci. Res: Operations research; systems analysis; weapon systems; communication; transportation; systems analysis and integration of design, support and operational management. Mailing Add: 7100 Radnor Rd Bethesda MD 20034

BROWN, FRED, b Manchester, Eng, Feb 5, 23; Can citizen; m 45; c 3. SOLID STATE SCIENCE. Educ: Cambridge Univ, BA, 43, PhD(chem), 46. Prof Exp: Lectr chem, Univ NWales, 46-49; res officer, 49-76, BR HEAD SOLID STATE SCI, CHALK RIVER NUCLEAR LAB, ATOMIC ENERGY CAN, 76- Mem: Fel Chem Inst Can. Res: Atomic collisions in solids, nuclear methods of analysis, anodic oxidation. Mailing Add: Solid State Sci Br Atomic Energy of Can Ltd Chalk River ON Can

BROWN, FREDERICK CALVIN, b Seattle, Wash, July 6, 24; m 52; c 3. SOLID STATE PHYSICS. Educ: Harvard Univ, SB, 45, MA, 47, PhD(physics), 50. Prof Exp: Physicist, US Naval Res Lab, 50; assoc physicist, Appl Physics Lab, Univ Wash, 51; from instr to asst prof physics, Reed Col, 51-54; from asst prof to prof, Univ Ill, Urbana-Champaign, 55-73; prin scientist, Xerox Palo Alto Res Lab, 73-74; PROF PHYSICS, UNIV ILL, URBANA, 74- Concurrent Pos: NSF sr fel, Clarendon Lab, Oxford Univ, 64-65; assoc, Ctr Advan Study, Univ Ill, 69-70; consult prof appl physics, Stanford Univ, 73-74. Mem: Fel Am Phys Soc. Res: Photoconducting and optical properties of ionic crystals, especially silver and alkali halides; color center phenomena and cyclotron resonance; extreme ultraviolet spectroscopy and synchrotron radiation. Mailing Add: Dept of Physics Univ of Ill Urbana IL 61801

BROWN, FREDERICK LEROY, forest pathology, see 12th edition

BROWN, FREDERICK MARTIN, physiology, see 12th edition

BROWN, FREDERICK RONALD, b Pittsburgh, Pa, Oct 29, 45; m 67. FUEL SCIENCE. Educ: Lafayette Col, BS, 67; Univ Pittsburgh, PhD(phys chem), 71, Duquesne Univ, JD, 76. Prof Exp: Res chemist, US Bur Mines, 71-75; RES CHEMIST SPECTROS, US ENERGY RES & DEVELOP ADMIN, 75- Mem: Am Chem Soc. Res: Application of spectral techniques in determining the chemical structure of coal conversion catalysts. Mailing Add: Pittsburgh Energy Res Ctr 4800 Forbes Ave Pittsburgh PA 15213

BROWN, FREDERICK S, b Peoria, Ill, June 11, 40; m 70. BIOCHEMISTRY, ORGANIC GEOCHEMISTRY. Educ: Bradley Univ, BA, 62; Univ Ill, Urbana-Champaing, PhD(chem), 66. Prof Exp: Mem tech staff, 66-70, PROJ SCIENTIST, TRW SYSTS, 70- Concurrent Pos: Vis scientist, Univ Calif, Los Angeles, 67-69. Mem: Am Chem Soc. Res: Enzyme mechanism, structure, active site; bound enzymes; diagenesis of organic matter in recent marine sediment; planetary geochemistry; planetary biology and life detection; automated instrument systems. Mailing Add: TRW Systs R1/2094 1 Space Park Redondo Beach CA 90278

BROWN, G MALCOLM, b Can, July 16, 16; m 50; c 3. INTERNAL MEDICINE. Educ: Queen's Univ, Ont, MD, 38; Oxford Univ, DPhil, 40; FRCPS(C), 46; FRCP, 61. Hon Degrees: LLD, Univ Sask, 69; DSc, Mem Univ Nfld, 69; MD, Laval Univ, 69, Univ Montreal, 71; DSc, McGill Univ, 75. Prof Exp: From assoc prof to prof med, Queen's Univ, Ont, 46-65; PRES, MED RES COUN, 65- Concurrent Pos: Dir, Queen's Univ, Ont, Arctic exped, 47-49, 50 & 54. Mem: Am Fedn Clin Res; master Am Col Physicians; Am Soc Clin Invest; fel Royal Soc Can; Can Phys Soc. Res: Hematology and gastroenterology. Mailing Add: Med Res Coun Ottawa ON Can

BROWN, GENE MONTE, b Mo, Jan 21, 26; m 54; c 3. BIOCHEMISTRY. Educ: Colo State Univ, BS, 49; Univ Wis, MS, 50, PhD(biochem), 53. Prof Exp: Assoc, Univ Tex, 52-54; from instr to assoc prof, 54-67, PROF BIOCHEM, MASS INST TECHNOL, 67-, ASSOC HEAD DEPT BIOL, 72- Mem: Am Soc Biol Chem; Am Chem Soc. Res: Biosynthesis and function of vitamins and coenzymes. Mailing Add: Dept of Biol Mass Inst of Technol Cambridge MA 02139

BROWN, GEORGE BOSWORTH, b Birmingham, Ala, Apr 18, 14; m 40. BIO-ORGANIC CHEMISTRY. Educ: Ill Wesleyan Univ, BS, 34, DSc, 59; Univ Ill, MS, 36, PhD(org chem), 38. Prof Exp: Instr, Univ Ill, 35-38; asst, Med Col, Cornell Univ, 38-39, assoc biochem, 39-46; MEM, SLOAN-KETTERING INST CANCER RES, 46-; PROF BIOCHEM, SLOAN-KETTERING DIV GRAD SCH MED SCI, CORNELL UNIV, 51- Concurrent Pos: Panel mem, USPHS, Am Cancer Soc & NSF; asst prof, Med Col, Cornell Univ, 46-49, assoc prof, 49-51; trustee, Gordon Res Conf, 56-59 & 70-73; Fulbright scholar, Australia, 65; vpres, Sloan-Kettering Inst Cancer Res, 68-72. Mem: Am Chem Soc; Am Soc Biol Chem; Am Asn Cancer Res; Brit Biochem Soc. Res: Sulfur containing amino acids; biotin; penicillin; purines and nucleic acids; metabolism; pharmacology; oncogenicity of purine derivatives. Mailing Add: Sloan-Kettering Inst Cancer Res 145 Boston Post Rd Rye NY 10580

BROWN, GEORGE BRUCE, b Windsor, Ont, Sept 28, 12; m 46. METEOROLOGY. Educ: Univ Western Ont, BA, 34, Univ Toronto, MA, 46; Mass Inst Technol, MS, 59. Prof Exp: Teacher, Pub Schs, Can, 35-40; anal meteorologist, Meteorol Br, Dept Transport, Can, 40-64, actg chief prognostician, Cent Anal Off, Meteorol Serv Can, 64-66, supvr, Cent Anal Off, 66-72; RETIRED. Concurrent Pos: Asst meteorol, Mass Inst Technol, 56-58. Mem: Am Meteorol Soc; Can Asn Physicists; Royal Meteorol Soc. Res: Synoptic and dynamic meteorology-analysis and prognosis. Mailing Add: 256 Allard Ave Dorval PQ Can

BROWN, GEORGE D, JR, b Whiting, Ind, Sept 16, 31; m 59; c 2. GEOLOGY, PALEONTOLOGY. Educ: St Joseph's Col, Ind, BS, 53; Univ Ill, Urbana, MS, 55; Ind Univ, PhD(geol), 63. Prof Exp: Jr geologist, Pan-Am Petrol Corp, 55-57; sedimentary engr, Youngstown Sheet & Tube Co, 57-58; instr geol, Colgate Univ, 58-59; lectr, Ind Univ, 62; asst prof, 62-68, ASSOC PROF GEOL & CHMN DEPT GEOL & GEOPHYS, BOSTON COL, 68- Concurrent Pos: Sigma Xi grant, 66-67. Mem: Paleont Soc; Geol Soc Am; Int Bryozool Asn. Res: Paleozoic Trepostomatous Bryozoa; litho stratigraphy and biostratigraphy of Ordovician strata in the Central States area. Mailing Add: Dept of Geol & Geophys Boston Col Chestnut Hill MA 02167

BROWN, GEORGE EARL, b Weaubleau, Mo, Sept 4, 06; m 35; c 1. ORGANIC CHEMISTRY. Educ: Cent Col, Mo, AB, 31; Iowa State Col, PhD(org chem), 41. Prof Exp: Instr chem & math, Cent Col, Mo, 31-32; instr chem & physics, Culver-Stockton Col, 35-36, asst prof chem, 36-39, prof chem, 40-56, head dept phys sci, 40-46, head dept chem, 46-56; prof, Southeast Mo State Col, 56-62, chmn div natural sci, 56-60, chmn dept chem, 58-62; coordr phys sci gen studies, 62-65, PROF CHEM, SOUTHERN ILL UNIV, CARBONDALE, 62- Mem: Am Chem Soc. Res: Orientation studies with dibenzofuran organometallic studies; reactions of organolithium compounds with some organic phosphorus and nitrogen compounds. Mailing Add: Dept of Chem Southern Ill Univ Carbondale IL 62901

BROWN, GEORGE ELDON, b Mexico, Mo, Dec 28, 34; m 55; c 2. PLANT PATHOLOGY, PLANT GENETICS. Educ: Univ Mo, BS, 56, MS, 61; Univ Minn, PhD(plant path), 65. Prof Exp: Asst plant pathologist, 65-70, ASSOC PLANT PATHOLOGIST, CITRUS EXP STA, FLA CITRUS COMN, 70- Mem: Am Phytopath Soc. Res: Control of decay in fresh citrus fruit. Mailing Add: Citrus Exp Sta PO Box 1088 Lake Alfred FL 33850

BROWN, GEORGE L, b Athens, Greece, Jan 16, 31; US citizen; m 57; c 3. MICROBIOLOGY, IMMUNOLOGY. Educ: Univ RI, BS, 56, MS, 61, PhD(microbiol), 68; Nat Registry Microbiol, regist. Prof Exp: Med Dept, US Army, 57-, chief lab serv, 28th Gen Hosp, La Rochelle, France, 57-60, chief microbiol div, Brooke Gen Hosp, San Antonio, Tex, 62-65, chief microbiol dept, Fifth Army Med Lab, St Louis, 68-69, asst chief microbiol div, Army Med Res & Nutrit Lab, Fitzsimons Gen Hosp, 69-72, CHIEF IMMUNOL & MYCOBACT, CLIN INVEST SERV, FITZSIMONS ARMY MED CTR, 72- Concurrent Pos: Fac affil, Dept Microbiol, Colo State Univ. Honors & Awards: Army Commendation Medal, 65. Mem: Am Soc Microbiol; Am Acad Microbiol; Brit Soc Immunol. Res: Clinical and applied microbiology, especially diagnostic pathogenic bacteriology; serology and immunology applied in area of immunologic research in tuberculosis. Mailing Add: Clin Invest Serv Fitzsimons Army Med Ctr Denver CO 80240

BROWN, GEORGE LINCOLN, b Brookings, SDak, Dec 27, 21; m 43; c 4. CHEMISTRY. Educ: SDak State Col, BS, 41; Brown Univ, PhD(chem), 47. Prof Exp: Chemist, Standard Oil Co, La, 43-45; res chemist, Rohm and Haas Co, 47-51, head lab, 51-57, res supvr, 57-63; vpres res, Polyvinyl Chem, Inc, Peabody, 63-65; tech dir chem coatings div, Mobil Chem Co, 65-68, MGR COATINGS RES & DEVELOP, MOBIL OIL CORP, 69- Honors & Awards: Potter Prize, Brown Univ, 47. Mem: AAAS; Am Chem Soc. Res: Electrochemistry; properties of electrolytic solutions; surface chemistry; polymers; polymer dispersion coatings. Mailing Add: 23 Essex Rd Scotch Plains NJ 07076

BROWN, GEORGE MARSHALL, b Rochelle, Ga, Feb 14, 21; m 62; c 1. STRUCTURAL CHEMISTRY, X-RAY CRYSTALLOGRAPHY. Educ: Emory Univ, BA, 42, MS, 43; Princeton Univ, MA, 46, PhD(phys chem), 49. Prof Exp: From asst prof to assoc prof chem, Univ Md, 47-59; CHEMIST, OAK RIDGE NAT LAB, 59- Concurrent Pos: Res fel x-ray crystallog, Calif Inst Technol, 57-58; vis scholar protein crystallog, Univ Wash, 74-75. Mem: Am Chem Soc; Am Crystallog Asn. Res: X-ray and neutron crystal structure analysis; hydrogen atom location and hydrogen bonding; sugars; triarylamines; sandwich compounds; heteropoly acids; protein structure refinement. Mailing Add: Div of Chem Oak Ridge Nat Lab Oak Ridge TN 37830

BROWN, GEORGE RAYMOND, b Gainesville, Ga, Dec 13, 45; m 70. ENVIRONMENTAL PHYSICS, AIR POLLUTION. Educ: Ga Inst Technol, BS, 67; Duke Univ, PhD(physics), 72. Prof Exp: Res asst prof physics, Duke Univ, 70-72; ASST PROF PHYSICS, CLARK COL, 72- Mem: AAAS; Am Phys Soc. Res: Phase transitions in fluid systems, with particular emphasis on the gas to liquid transition of binary and multicomponent fluids which condense under atmospheric conditions to form aerosols. Mailing Add: Dept of Physics Box 167 Clark Col Atlanta GA 30314

BROWN, GEORGE WALLACE, b Warrensburg, Mo, Jan 31, 39; m 64; c 1. FOREST HYDROLOGY. Educ: Colo State Univ, BS, 60, MS, 62; Ore State Univ, PhD(forest hydrol), 67. Prof Exp: ASSOC PROF FOREST HYDROL, ORE STATE UNIV, 66-, HEAD DEPT FOREST ENG, 73- Mem: Soc Am Foresters; Am Geophys Union.

Res: Water quality prediction on small forested streams; temperature, sediment, dissolved oxygen. Mailing Add: Dept of Forest Eng Ore State Univ Corvallis OR 97331

BROWN, GEORGE WALLACE, b Chicago, Ill, May 28, 17; m 56; c 2. NEUROPHYSIOLOGY. Educ: Univ Iowa, BS, 40, MS, 42, PhD(physiol), 52. Prof Exp: Res physiologist neurophysiol, Aero Med Lab, 49-50; res assoc, Univ Iowa, 52-53; prin scientist, Gen Med Res Lab, Vet Admin Hosp, 53-67; asst prof civil eng, Univ Iowa, 67-71; TRANSPORTATION & TRAFFIC SAFETY CONSULT, 67- Concurrent Pos: Mem exec bd, Citizens for Environ Action; mem, Nat Air Conserv Comn & Scientist's Inst Pub Info. Mem: AAAS; Int Asn Accident & Traffic Med; Am Physiol Soc; Am Asn Automotive Med; NY Acad Sci. Res: Cerebral concussion; cerebral blood flow; convulsive disorders; electrical properties of cerebral structures; learning related to cerebral function and stimulation; motor vehicle accident analysis; vehicle and roadway safety design; mass transportation. Mailing Add: The Anchorage Rte 3 Solon IA 52333

BROWN, GEORGE WILLARD, JR, b Alameda, Calif, Oct 24, 24; m 48, 61; c 6. BIOCHEMISTRY, ENVIRONMENTAL CHEMISTRY. Educ: Univ Calif, Berkeley, BS, 50, MA, 51, PhD(comp biochem), 55. Prof Exp: Fel, NIH, 56; res fel & res assoc physiol chem, Univ Wis, 57, instr, 58, asst prof, 60-61; from asst prof to assoc prof biochem, Med Br, Univ Tex, 61-67; ASSOC PROF FISHERIES, COL FISHERIES, UNIV WASH, 67-, DIR WATER POLLUTION TRAINING, 67- Concurrent Pos: Consult, NIH, 58, Acad Press, 62-, NASA, 66-67, Highline Col, 67 & US Coast Guard, 70-73; vis biologist, Am Inst Biol Sci, 62-73. Honors & Awards: Lederle Med Faculty Award, 60; Belg Fourragere. Mem: Am Chem Soc; Am Soc Zool; Am Soc Biol Chem; fel Am Inst Chemists; Soc Protection Old Fishes (pres, 67). Res: Intermediary nitrogen metabolism; biochemistry of Amphibia and primitive fishes; enzymology; desert biology; biochemical ecology; water pollution. Mailing Add: Biochem Ecol Lab Univ Wash Col Fisheries Seattle WA 98195

BROWN, GEORGE WILLIAM, b Boston, Mass, June 3, 17; m 41; c 5. MATHEMATICS. Educ: Harvard Univ, AB, 37, AM, 38; Princeton Univ, PhD(math), 40. Prof Exp: Res statistician, R H Macy & Co, 40-42; res assoc, Princeton Univ, 42-44, res engr, 44-46; from assoc prof to prof math statist, Iowa State Col, 46-48; mathematician, Rand Corp, 48-52; sr staff engr, Int Telemeter Corp, 52-57; prof bus admin & eng, Univ Calif, Los Angeles, 57-67, dir, Western Data Processing Ctr, 57-64, chmn dept bus admin, 64-65; dean grad sch admin, 67-72, PROF ADMIN, UNIV CALIF, IRVINE, 67- Concurrent Pos: Dir & consult, Dataproducts Inc, 62-; dir, Comput Automation Inc, 69- Mem: Fel AAAS; fel Am Statist Asn; fel Inst Math Statist. Res: Dynamic decision processes; mathematical and applied statistics; management science; operations research. Mailing Add: Grad Sch of Admin Univ of Calif Irvine CA 92664

BROWN, GERALD EDWARD, b Brookings, SDak, July 22, 26; m 53; c 3. THEORETICAL PHYSICS. Educ: Univ Wis, PhB, 46; Yale Univ, MSc, 48, PhD, 50; Univ Birmingham, DSc, 57. Prof Exp: Lectr math physics, Univ Birmingham, 55-58, reader, 58-59, prof, 59-60; prof, Nordic Inst Theoret Atomic Physics, 60-64 & Princeton Univ, 64-68; PROF PHYSICS, STATE UNIV NY STONY BROOK, 68- Mem: Am Phys Soc. Res: Atomic and nuclear physics. Mailing Add: Dept of Physics State Univ of NY Stony Brook NY 11790

BROWN, GERALD LEONARD, b New York, NY, May 17, 36; m 67; c 1. APPLIED MATHEMATICS. Educ: Univ Miami, BS, 58; Mass Inst Technol, MS, 60; Univ Wis, PhD(math), 65. Prof Exp: Tech staff mem, Math Res Dept, Sandia Corp, 65-70; RES STAFF MEM, SYSTS EVAL DIV, INST DEFENSE ANAL, 71- Mem: Am Math Soc; Soc Indust & Appl Math; Opers Res Soc Am. Res: Partial differential equations; systems analysis. Mailing Add: Systs Eval Div Inst Defense Anal 400 Army-Navy Dr Arlington VA 22202

BROWN, GERALD RICHARD, b Poplar Bluff, Mo, Oct 22, 37; m 59; c 2. POMOLOGY, VITICULTURE. Educ: Univ Ark, BS, 59, MS, 63, PhD(bot), 74. Prof Exp: Res asst hort, Southwest Br Exp Sta, Univ Ark, Hope, 60-61; plant quarantine inspector, Agr Res Serv, USDA, 61-69; res asst genetics, Hort Dept, Univ Ark, 69-74; RES POMOLOGIST, STATE FRUIT EXP STA, SOUTHWEST MO STATE UNIV, 74- Honors & Awards: Krezdorn Award, Am Soc Hort Sci, 74. Mem: Am Soc Hort Sci; Sigma Xi. Res: Fruit production including the basic relationship of plant growth; especially the use of growth regulators, cultivar evaluation, nutrition, pruning, and irrigation. Mailing Add: Fruit Exp Sta Rte 3 Box 63 Southwest Mo State Univ Mountain Grove MO 65711

BROWN, GLEN FRANCIS, b Graysville, Ind, Dec 14, 11; m 75; c 1. GEOLOGY. Educ: NMex Sch Mines, BS, 35; Northwestern Univ, MS, 40, PhD(geol), 49. Prof Exp: Geologist, Philippine Bur Mines, 36-38; jr geologist, US Geol Surv, 38-41, from asst geologist to sr geologist, 41-44; sr geologist, For Econ Admin, 44-46; geologist, US Geol Surv, 46-48, actg chief, Mission to Thailand, 49, chief field party, Saudi Arabia, 50-54, chief, Saudi Arabian Proj, Washington, DC, 55-57; geol adv, Kingdom Saudi Arabia, 57-58; mem, World Bank Mission to Saudi Arabia, 60 & Kuwait, 62; chief field party, 63-69, SR STAFF GEOLOGIST FOR MIDEASTERN AFFAIRS, US GEOL SURV, 69- Concurrent Pos: US Geol Surv deleg, 22nd & 23rd World Geol Congs; mem, Joint Comn to Saudi Arabia, 74-76. Honors & Awards: Distinguished Serv Gold Medal, US Dept Interior, 64. Mem: Soc Econ Geologists; Geol Soc Am; Am Geophys Union; Am Asn Petrol Geologists; Soc Photogram. Res: Geology of metals and coal in the Philippine islands, Thailand and South China; geology of ground water, Mississippi, California, and Saudi Arabia; Precambrian geology in Saudi Arabia; tectonism of Arabian peninsula. Mailing Add: Apt 21C 2031 Royal Fern Court Reston VA 22091

BROWN, GLENN HALSTEAD, b Logan, Ohio, Sept 10, 15; m 43; c 4. CHEMISTRY. Educ: Ohio Univ, BS, 39; Ohio State Univ, MS, 41; Iowa State Col, PhD, 51. Hon Degrees: DSc, Bowling Green State Univ, 72. Prof Exp: Instr chem, Univ Miss, 41-42, asst prof, 43-50; asst prof, Univ Vt, 50-52; from asst prof to assoc prof, Univ Cincinnati, 52-60; prof, 60-68, chmn dept, 60-65, dean res, 63-69, REGENTS PROF CHEM, KENT STATE UNIV, 68-, DIR, LIQUID CRYSTAL INST, 65- Concurrent Pos: Instr, Iowa State Col, 45-48; Sigma Xi nat lectr, 70-71. Mem: Fel AAAS; fel Am Inst Chemists; Am Chem Soc; Am Crystallog Asn; NY Acad Sci. Res: Structures of solutions; x-ray diffraction; coordination compounds; structure and properties of liquid crystals; role of liquid crystals in life processes; photochromism; structure of concentrated salt solutions. Mailing Add: Liquid Crystal Inst Kent State Univ Kent OH 44242

BROWN, GLENN LAMAR, b Miami, Ariz, Oct 14, 23. PHYSICS. Educ: Univ Calif, Los Angeles, MA, 51, PhD, 55. Prof Exp: Asst physics, Univ Calif, Los Angeles, 49-50, res assoc geophys, 53-57; mem tech staff, Ramo-Wooldridge Corp, 57-59; mgr nuclear sci div, Space-Gen Corp, El Monte, 59-65; dir res, Am Nucleonics Corp, 65-72; PRES, RESOURCE ENG & PLANNING CO, 72- Mem: Am Phys Soc; Am Geophys Union. Res: Electromagnetic theory; nuclear weapons effects; communication systems; elastic wave theory. Mailing Add: 2604 Vista Dr SE Huntsville AL 35803

BROWN, GORDON CAMPBELL, epidemiology, deceased

BROWN, GORDON ELLIOTT, b Ellensburg, Wash, Sept 13, 36. MATHEMATICS. Educ: Calif Inst Technol, BS, 58; Cornell Univ, PhD(math), 63. Prof Exp: From instr to asst prof math, Univ Ill, Urbana, 62-66; asst prof, 66-69, ASSOC PROF MATH, UNIV COLO, BOULDER, 69- Mem: Am Math Soc. Res: Non-associative algebras; Lie algebras. Mailing Add: Dept of Math Univ of Colo Boulder CO 80302

BROWN, GORDON MANLEY, b Morpeth, Ont, Mar 17, 33; m 64; c 2. ORGANIC CHEMISTRY. Educ: Univ Western Ont, BSc, 54, MSc, 56; Laval Univ, DSc(chem), 59; Univ Montpellier, DSc(chem), 61. Prof Exp: Nat Res Coun overseas fel, Univ Montpellier, 59-61; lectr, 61-62, from asst prof to assoc prof, 62-75, PROF CHEM, UNIV SHERBROOKE, 75-, SECY FAC SCI, 70- Mem: Chem Inst Can. Res: Configurational and conformational studies of decaline and cyclohexane derivatives. Mailing Add: Dept of Chem Fac of Sci Univ of Sherbrooke Sherbrooke PQ Can

BROWN, GREGORY NEIL, b Detroit, Mich, Feb 10, 38; m 74; c 3. TREE PHYSIOLOGY, CRYOBIOLOGY. Educ: Iowa State Univ, BS, 59; Yale Univ, MF, 60; Duke Univ, DF, 63. Prof Exp: Plant physiologist, Oak Ridge Nat Lab, 63-66; from asst prof to assoc prof, 66-75, PROF FOREST PHYSIOL, UNIV MO-COLUMBIA, 75- Concurrent Pos: Mem, Nat Tree Physiol Comt, 68-72; NSF grant, 69-71. Mem: Soc Am Foresters; Am Soc Plant Physiol; Soc Cryobiol; Japanese Soc Plant Physiol; Scand Soc Plant Physiol. Res: Membrane structure and function and protein characterization during the induction of cold hardiness in woody plant species. Mailing Add: Sch of Forestry Univ of Mo Columbia MO 65201

BROWN, HARLEY PROCTER, b Uniontown, Ala, Jan 13, 21; m 42; c 1. INVERTEBRATE ZOOLOGY, ENTOMOLOGY. Educ: Miami Univ, AB & AM, 42; Ohio State Univ, PhD(zool), 45. Prof Exp: Asst zool, Ohio State Univ, 42-45; instr, Univ Idaho, 45-47 & Queen's Col, NY, 47-48; from asst prof to assoc prof, 48-62, PROF ZOOL, UNIV OKLA, 62- Concurrent Pos: Partic, NSF Inst Marine Biol & Trop Ecol, 64. Mem: Am Soc Zool; Soc Syst Zool; Am Micros Soc (pres, 75); Am Inst Biol Sci; Entom Soc Am. Res: Protozoology; ecology, life history and systematics of dryopoid beetles. Mailing Add: Dept of Zool Univ of Okla 730 Van Vleet Oval Norman OK 73069

BROWN, HARMON W, JR, b Mitchell, SDak, Jan 16, 32; m 61; c 3. CHEMISTRY. Educ: Northwestern Univ, BS, 54; Univ Calif, Berkeley, PhD(chem), 57. Prof Exp: Res chemist, Nat Bur Standards, 57-59; res assoc, Stanford Univ, 59-60; res chemist, Varian Assocs, 60-73; PROD MGR, SCI INSTR DIV, HEWLETT-PACKARD, 73- Mem: Am Chem Soc; Am Phys Soc. Res: Nuclear magnetic resonance; electron paramagnetic resonance spectroscopy; application of spectroscopy to molecular structure. Mailing Add: Hewlett-Packard 1601 California Ave Palo Alto CA 94304

BROWN, HAROLD, b New York, NY, Sept 19, 27; m 53; c 2. PHYSICS. Educ: Columbia Univ, AB, 45, MA, 46, PhD(physics), 49. Hon Degrees: DEng, Stevens Inst Technol, 64; LLD, Gettysburg Col, 67. Univ Calif, Los Angeles, 69 & Occidental Col, 69; ScD, Univ Rochester, 75. Prof Exp: Lectr & mem sci staff, Columbia Univ, 47-50; physicist, Lawrence Radiation Lab, Univ Calif, Berkeley, 50-52, mem staff, Livermore, 52-53, group leader, 53-55, div leader, 55-58, assoc dir, 58-59, from dep dir to dir, 59-61; dir defense res & eng, Off Secy Defense, 61-64, Secy Air Force, 65-69; PRES, CALIF INST TECHNOL, 69- Concurrent Pos: Lectr physics, Stevens Inst Technol, 49-50; mem, Polaris Steering Comt, 56-58; adv, US Deleg Conf Experts Detection Nuclear Weapons Tests, Geneva, 58; sr sci adv, US Deleg Conf Discontinuance Nuclear Weapons Tests, 58-59; mem sci adv comt ballistic missiles, Secy Defense, 58-61; mem, President's Sci Adv Comt, 61; deleg, Strategic Arms Limitation Talks, Helsinki, Vienna & Geneva, 69-75; mem gen adv comt, Arms Control & Disarmament Agency, 69-; mem exec comt, Trilateral Comn, 73-; chmn, Technol Assessment Adv Coun, 74-75. Honors & Awards: Distinguished Civilian Serv Award, US Navy, 61; Columbia Univ Medal, 63. Mem: Nat Acad Eng; Am Acad Arts & Sci; Am Phys Soc. Res: Nuclear and neutron physics; nuclear explosives and reactor design; weapons systems; management of research and development; technology and arms control. Mailing Add: Calif Inst of Technol 1201 E California Blvd Pasadena CA 91109

BROWN, HAROLD, b Lawrence, Mass, Sept 17, 17; m 50; c 3. MEDICINE. Educ: Harvard Univ, MD, 43; Am Bd Internal Med, dipl, 50. Prof Exp: Asst med, Harvard Med Sch, 46-47; from instr to assoc prof, Sch Med, Univ Utah, 47-61; actg chmn dept, 69-70, PROF MED, HEAD METAB SECT & DIR CLIN RES CTR, BAYLOR COL MED, 62-, VCHMN DEPT MED, 70- Concurrent Pos: House officer, intern & resident, Boston City Hosp, 43-47; from asst chief to chief med serv, Vet Admin Hosp, Salt Lake City, Utah, 48-61; vis prof, Univ Med Sci, Bangkok, 57-59; external exam med, Univ Hong Kong, 58; consult & dir metab res lab, Vet Admin Hosp, Houston, 62- Mem: Endocrine Soc; Am Fedn Clin Res; Am Col Physicians; AMA; NY Acad Sci. Res: Metabolism of corticosteroids; serotonin metabolism; liver disease. Mailing Add: Dept of Med Baylor Col of Med Houston TX 77030

BROWN, HAROLD DAVID, b Mishawaka, Ind, July 12, 34; m 58; c 2. APPLIED MATHEMATICS, COMPUTER SCIENCES. Educ: Univ Notre Dame, MSc, 63; Ohio State Univ, PhD(math), 66. Prof Exp: Asst to chmn dept math & asst dir, NSF Training Progs Math, Univ Notre Dame, 60-63; instr, Ohio State Univ, 63-66, asst prof, 67; vis mem, Courant Inst Math Sci, 67-68; from asst prof to assoc prof, Ohio State Univ, 68-73; ASSOC PROF COMPUT SCI, STANFORD UNIV, 73- Concurrent Pos: Dir, NSF Sec Sci Training Prog, Ohio State Univ, 64-71; vis prof, RWTH, Aachen, 71, 73 & 75. Mem: Am Math Soc; Math Asn Am; Asn Comput Mach. Res: Computational algebra and graph theory; biochemical structure determination; computer algorithms. Mailing Add: Dept of Comput Sci Stanford Univ Serra House Stanford CA 94305

BROWN, HAROLD HUBLEY, b Portsmouth, NH, Mar 14, 26; m 57; c 3. CLINICAL CHEMISTRY. Educ: Marietta Col, AB, 50; Wayne State Univ, PhD(biochem), 54. Prof Exp: Instr biochem, Albany Med Col, 54-56; biochemist, Pawtucket Mem Hosp, RI, 56-59 & Harrisburg Polyclin Hosp, 59-64; asst prof biochem & dir Core Lab Gen Clin Res Ctr, Albany Med Col, 64-65; biochemist, Harrisburg Polyclin Hosp, Pa, 65-70; clin chemist, Berkshire Med Ctr, Pittsfield, Mass, 70-73; ASSOC PROF BIOCHEM, ALBANY MED COL, 73-; DIR, DEPT CLIN CHEM, ALBANY MED CTR HOSP, 73- Mem: AAAS; Am Chem Soc; Am Asn Clin Chemists. Res: Methodology in clinical chemistry. Mailing Add: Albany Med Ctr Hosp Albany NY 12208

BROWN, HAROLD MACK, JR, b Salt Lake City, Utah, Mar 6, 36; m 59; c 3. VISUAL PHYSIOLOGY, BIOPHYSICS. Educ: Univ Utah, BS, 58, PhD(psychol, molecular biol), 64. Prof Exp: Res assoc, Scripps Inst Oceanog, 67-69; asst res prof physiol, Sch Med, Univ Calif, Los Angeles, 69-70; asst prof physiol, 70-72, ASST PROF MED, SCH MED, UNIV UTAH, 71-, ASSOC PROF PHYSIOL, 72- Concurrent Pos: Vet Admin fel neurol, Col Med, Univ Utah, 64-66; NIH spec fel,

Scripps Inst Oceanog, 67-69. Honors & Awards: Flanagan Award, Am Inst Res, 64. Mem: AAAS; Am Physiol Soc; Soc Gen Physiol; Biophys Soc; Am Soc Photobiol. Res: Visual biophysics and membrane biophysics. Mailing Add: Dept of Physiol Univ of Utah Med Ctr Salt Lake City UT 84132

BROWN, HAROLD PROBERT, b Granby, Mo, Aug 29, 08; m 34; c 1. ORGANIC POLYMER CHEMISTRY. Educ: Cent Mo State Univ, AB & BS, 28; Univ Mo, AM, 30; Univ Nebr, PhD(org chem), 33. Prof Exp: Mem dept chem, Univ Kansas City, 33-38, assoc prof, 38-42, chmn dept, 41-42; chemist, Polymerization Res Div, Res Labs, B F Goodrich Co, 42-43, res supvr, 43-48, res scientist, 48-57, sect leader specialty rubbers, 57-59 & adhesives, 59-62, sr res assoc, 62-64; ASST DEAN SCH ENG & APPL SCI, WASH UNIV, 65- Concurrent Pos: Consult chemist, 64- Mem: AAAS; Am Chem Soc. Res: Organic arsenicals; antimonials, organo-tin compounds and metallics; sulfonamides; naphthalene compounds; polymerization; synthetic rubber adhesives; specialty polymers; hydrogels. Mailing Add: 444 Woodlawn Estates Dr Kirkwood MO 63122

BROWN, HAROLD VICTOR, b Los Angeles, Calif, Nov 5, 18; m 46; c 2. ENVIRONMENTAL HEALTH, INDUSTRIAL HYGIENE. Educ: Univ Calif, Los Angeles, BA, 40, MPH, 65, DrPH, 70; Am Intersoc Acad Cert Sanit & Am Am Bd Indust Hyg, dipl. Prof Exp: Chemist, Eng Corps, US War Dept, 41 & 46; res chemist, Pioneer-Flintkote Co, 47; indust hygientist, State Dept Health, Calif, 47-52; indust hygientist, 52-65, ENVIRON HEALTH & SAFETY OFFICER, CTR HEALTH SCI, UNIV CALIF, LOS ANGELES, 65- Concurrent Pos: Mem, Calif Occup Safety & Health Stand Bd, 74- Mem: AAAS; Am Indust Hyg Asn; Health Physics Soc; fel Am Pub Health Asn; Conf Govt Indust Hygienists. Res: Chlinesterase inhibitors; laboratory safety. Mailing Add: 6008 Chariton Ave Los Angeles CA 90056

BROWN, HAROLD WILLIAM, b Muskegon, Mich, Jan 16, 02; m 33; c 2. PARASITOLOGY, PUBLIC HEALTH. Educ: Kalamazoo Col, AB, 24; Kans State Col, MS, 25; Johns Hopkins Univ, ScD(parasitol), 28; Vanderbilt Univ, MD, 33; Harvard Univ, DPH, 36. Hon Degrees: LHD, Kalamazoo Col, 46; LLD, Univ PR, 54. Prof Exp: Asst zool, Kans State Col, 24-25; asst helminth, Johns Hopkins Univ, 25-27; res assoc, Sch Med, Vanderbilt Univ, 27-34; Gen Educ Bd fel, London Sch Hyg & Trop Med, 34-35 & Harvard Sch Pub Health, 35-36; passed asst surgeon, USPHS, 36-37; prof pub health, Univ NC, 37-43, from asst dean to dean, 37-43; prof pub health & prev med, Duke Univ, 37-43; prof parasitol, Sch Pub Health, Columbia Univ, 43-71, from actg dir to dir, 47-55; CLIN PROF PATH, VANDERBILT UNIV, 71- Concurrent Pos: Vis prof, Sch Med, Nat Taiwan Univ, 55, 57, 58, 60, 62, 63 & 64 & Airlangga Univ, Indonesia, 62 & 64; mem, Panama Hookworm Exped, 26. Mem: Am Soc Parasitologists (secy, 48-51, vpres, 53, pres, 60); Am Soc Trop Med & Hyg (vpres, 44); Am Epidemiol Soc; Am Pub Health Asn; Royal Soc Trop Med & Hyg. Res: Anthelmintics; chemotherapy of amebiasis; malaria; medical education. Mailing Add: Dept of Path Vanderbilt Univ Nashville TN 37203

BROWN, HARRISON SCOTT, b Sheridan, Wyo, Sept 26, 17; m 38; c 1. GEOCHEMISTRY. Educ: Univ Calif, BS, 38; Johns Hopkins Univ, PhD(chem), 41. Hon Degrees: LLD, Univ Alta, 61, Johns Hopkins Univ, 69, Univ Calif, 70 & Univ Wyo, 71; ScD, Rutgers Univ, 64, Amherst Col, 66 & Cambridge Univ, 69. Prof Exp: Instr chem, Johns Hopkins Univ, 41-42; res assoc, Univ Chicago, 42-43; asst dir chem div, Clinton Labs, Oak Ridge, 43-46; from asst prof to assoc prof chem, Inst Nuclear Studies, Univ Chicago, 46-51; PROF GEOCHEM, CALIF INST TECHNOL, 51-, PROF SCI & GOVT, 69- Honors & Awards: Prize, AAAS, 47; Award, Am Chem Soc, 52; Lasker Found Award, 58; Mellon Inst Award, 71. Mem: Nat Acad Sci (foreign secy); AAAS; Am Chem Soc; Geol Soc Am. Res: Mass spectroscopy; thermal diffusion; fluorine and plutonium chemistry; meteoritics; planet structure; geochronology; planetary chemistry; population growth; resources; science, technology and economic development; environment. Mailing Add: Div of Humanities & Social Sci Calif Inst of Technol Pasadena CA 91125

BROWN, HARRY, physics, see 12th edition

BROWN, HARRY ALLEN, b New York, NY, Apr 26, 25. MAGNETISM. Educ: Univ Wis, PhD(physics), 54. Prof Exp: Instr physics, Oberlin Col, 54-55; asst prof, Miami Univ, 55-59; assoc prof, St John's Univ, 59-61; sr physicist, Lab Appl Sci, Univ Chicago, 61-63; Fulbright lectr, Univ Sao Paulo, 63-64; assoc prof, San Francisco State Col, 64-65; assoc prof, 65-70, PROF PHYSICS, UNIV MO-ROLLA, 72- Concurrent Pos: Consult, Nat Carbon Co, Ohio, 55-58 & Am Mach & Foundry Co, Conn, 60-61; Fulbright lectr, Univ Sao Paulo, 71-72. Mem: Am Phys Soc; Am Asn Physics Teachers; AAAS. Res: Solid state physics; magnetic properties of solids; statistical mechanics; phase transitions. Mailing Add: Dept of Physics Univ of Mo Rolla MO 65401

BROWN, HARRY DARROW, b Newark, NJ, July 21, 25; m 45; c 3. BIOCHEMISTRY. Educ: Long Island Univ, BS, 50; Columbia Univ, AM, 52, PhD(bot), 57. Prof Exp: Asst bot, Columbia Univ, 50-53; lectr biol, Hunter Col, 54-56; asst prof plant physiol, Loyola Univ, 57-61; assoc prof, Southern Ill Univ, 61-63; biochemist, Seed Protein Lab, USDA, 63-64; asst prof biochem, Univ Tex Med Br, 64-68; assoc prof, Sch Med, Univ Mo-Columbia, 68-74; PROF BIOCHEM & DEAN RES, COOK COL, RUTGERS UNIV, NEW BRUNSWICK, 74-; ASSOC DIR, NJ AGR EXP STA, 74- Concurrent Pos: Collabr, USDA, 63-64; chmn biochem sect & assoc dir, Cancer Res Ctr, 68-74. Mem: Am Asn Cancer Res; fel Am Inst Chemists; Biophys Soc; Biochem Soc. Res: Particulate enzymes; biochemical calorimetry; interfacial biochemical reactions. Mailing Add: Dept of Biochem Cook Col Rutgers Univ New Brunswick NJ 08903

BROWN, HARRY ESMOND, b Kuling, China, July 19, 24; nat US; m 45; c 3. FOREST HYDROLOGY. Educ: Univ Ill, BS, 47; Colo State Univ, MF, 50. Prof Exp: Forester, Rocky Mountain Forest & Range Exp Sta, Ariz, 50-74, PRIN RES HYDROLOGIST & GROUP LEADER, WATERSHED MGT RES STAFF, US FOREST SERV, USDA, DC, 74- Mem: Am Geophys Union. Res: Watershed management; studies of snow, streamflow, weather factors, infiltration, erosion, runoff, soils and vegetation; multiple-use evaluation of land treatments designed to increase streamflow. Mailing Add: S Agr Bldg Rm 808 RPE USDA Forest Serv Washington DC 20250

BROWN, HARVEY EARL, JR, b Norfolk, Va, Dec 25, 24; m 48; c 5. INTERNAL MEDICINE. Educ: Marquette Univ, MD, 48. Prof Exp: Intern, Mem Hosp, Charlotte, NC, 48-49; resident internal med, Vet Admin Hosps, Richmond, Va, 49-52; chief med serv, Coral Gables, Fla, 55-60; asst prof, 60-63, ASSOC PROF MED, SCH MED, UNIV MIAMI, 63- Mem: Am Rheumatism Asn; AMA; Asn Am Med Cols. Res: Rheumatology; new drug testing. Mailing Add: Dept of Med Univ of Miami Sch of Med Miami FL 33152

BROWN, HELEN BENNETT, b Greenwich, Conn, Oct 6, 02; m 28; c 2. NUTRITION, CARDIOVASCULAR DISEASES. Educ: Mt Holyoke Col, BA, 24; Yale Univ, PhD(biochem), 30. Hon Degrees: ScD, Mt Holyoke Col, 74. Prof Exp: Res chem technician, Dept Pediat, New Haven Hosp, 24-28; res asst biochem, Babies

& Children's Hosp, Western Reserve Univ, 28-31; instr bact, Sch Med, 42-44; res assoc biochem, Ben Venue Labs, Inc, 44-47; asst, Res Div, Cleveland Clin, 48-61, mem staff, 62-68, RES CONSULT BIOCHEM, CLEVELAND CLIN FOUND, 68- Concurrent Pos: Instr, Cleveland City Health Dept, 42-44; mem coun arteriosclerosis & coun epidemiol, Am Heart Asn; mem, Nat Diet-Heart Study Comt, 60-68; mem adv comt, Epidemiol & Biomet Sect, Nat Heart & Lung Inst, 69-71 & consult nutrit, Multiple Risk Factor Intervention Trial, 72-74; sci dir coop screening prog, Cleveland Chap, Am Heart & Diabetes Asn Greater Cleveland, 70-72. Mem: Am Heart Asn; Am Inst Nutrit; hon mem Am Dietetic Asn; NY Acad Sci. Res: Health problems in the general population and in the community; diet and heart disease. Mailing Add: Cleveland Clin Found 9500 Euclid Ave Cleveland OH 44106

BROWN, HENRY, b Erie, Pa, Feb 20, 20; m 45; c 5. SURGERY. Educ: Univ Mich, AB, 41; Univ Pa, MD, 44. Prof Exp: Clin assoc, 63-69, ASST CLIN PROF SURG, HARVARD MED SCH, 69- Concurrent Pos: Surgeon & mem clin res ctr, Mass Inst Technol; mem surg staff, New Eng Deaconess Hosp, Cambridge Hosp & Mt Auburn Hosp; dir hand surg, Vet Admin Hosp, Manchester, NH; mem Sears Surg Lab; hand surg consult, Ganta Hosp & Leprosarium Ganta, Liberia, WAfrica, 69, 71, 73 & 75; asst dir, Harvard Surg Unit, Boston City Hosp, 72-73. Mem: AMA; Am Soc Surg of Hand; Am Physiol Soc; Am Chem Soc; Am Col Surg. Res: Surgical physiology of the liver; surgery of the hand. Mailing Add: New Eng Deaconess Hosp 185 Pilgrim Rd Boston MA 02115

BROWN, HENRY, b Jersey City, NJ, Apr 5, 07; m 43; c 2. CHEMISTRY. Educ: Univ Kans, AB, 28; Univ Mich, MS, 29, PhD, 32. Prof Exp: Asst, Univ Mich, 28-32; res chemist, 34-50, RES DIR, UDYLITE CO, 50- Concurrent Pos: Staff mem, Manhattan Proj, Substitute Alloy Mat Labs, Columbia Univ, 43-44. Honors & Awards: Carl Heussner Mem Award, Am Electroplaters Soc, 53, George Hogaboom Mem Award, 63-64, Sci Achievement Award, 67; Westinghouse Award, Brit Inst Metal Finishing, 69; Thomas Midgley Award, Am Chem Soc, 71. Mem: AAAS; NY Acad Sci; Am Inst Chem; Nat Asn Corrosion Eng; Electrochem Soc. Res: Colloid chemistry; surface and interfacial tension; electrodeposition of metals. Mailing Add: 5270 Gulf of Mexico Dr Apt 502 Sarasota FL 33577

BROWN, HENRY CLAY, III, b Carbur, Fla, Mar 7, 19; m 44; c 3. ORGANIC CHEMISTRY. Educ: Duke Univ, BS, 41, MA, 42; Univ Fla, PhD(chem), 50. Prof Exp: Res chemist, Texas Co, 42-47; PROF CHEM & CHEM ENG, UNIV FLA, 52- Mem: Am Chem Soc; Am Inst Chem Engr; Am Inst Chemists; Soc Plastics Engr. Res: Fluorine chemistry and chemistry of fluorocarbon derivatives. Mailing Add: 215 NW 16th Ave Gainesville FL 32601

BROWN, HENRY SEAWELL, b Marion, NC, Mar 4, 30; m 51; c 5. ECONOMIC GEOLOGY. Educ: Berea Col, BA, 52; Univ Ill, MS, 54, PhD(geol), 58. Prof Exp: From asst prof to assoc prof geol, Berea Col, 55-58, head dept geol & geog, 55-58; from asst prof to assoc prof, 58-66, PROF GEOL, NC STATE UNIV, 66- Concurrent Pos: Consult, Tenn Copper Co, 60-61 & 66-67, Cranberry Magnetite Corp, 61- & Mineral Res & Develop Corp, 60-; pres, Geol Resources, Inc, 71- Mem: AAAS; Geol Soc Am; Am Inst Mining, Metall & Petrol Engrs; Am Petrol Inst; Am Inst Prof Geologists. Res: Origin and nature of mineral deposits. Mailing Add: Dept of Geosci NC State Univ Raleigh NC 27607

BROWN, HENRY TRUEHEART, b Galveston, Tex, Sept 7, 15; m 56. PHYSICAL CHEMISTRY. Educ: Johns Hopkins Univ, AB, 38, PhD(phys chem), 42. Prof Exp: Chemist, Standard Oil Co Ind, 42-43; chemist, Magnolia Petrol C, Tex, 43-44, group leader, 44-47; asst dir, Tex Res Found, 47-50; dir res, Geochem Surv, 50-67; PRES, KASHAN LABS, INC, 67- Mem: Am Chem Soc; Mineral Soc Am; Instrument Soc Am; NY Acad Sci. Res: Catalyst preparation and testing; surface area and particle size determination; petroleum exploration; crystal growth. Mailing Add: Kashan Labs Inc 2936 Blystone Lane Dallas TX 75220

BROWN, HERBERT, b South Irvine, Ky, Nov 10, 30; m 55; c 1. ANIMAL NUTRITION, ANIMAL HUSBANDRY. Educ: Univ Ky, BS, 52, MS, 56; Iowa State Univ, PhD(animal nutrit), 59. Prof Exp: Assoc prof animal sci, Ill State Univ, 59-61; area swine specialist, Univ Ky, 61-63; RES & DEVELOP SPECIALIST, LILLY RES LABS, ELI LILLY & CO, 63- Mem: Am Soc Animal Sci; Am Soc Cert Animal Scientists. Res: Development of products that will stimulate more efficient and economic production of livestock. Mailing Add: Lilly Res Labs Greenfield IN 46140

BROWN, HERBERT ALLEN, b Los Angeles, Calif, Jan 10, 40; m 65; c 2. VERTEBRATE ZOOLOGY. Educ: Univ Calif, Los Angeles, BA, 62; Univ Calif, Riverside, PhD(vert zool), 66. Prof Exp: Lectr biol, Univ Calif, Riverside, 66-67; ASST PROF BIOL, WESTERN WASH STATE COL, 67- Mem: Soc Study Evolution; Am Soc Ichthyologists & Herpetologists; Am Soc Zoologists; Soc Study Amphibians & Reptiles. Res: Physiological ecology of amphibians and reptiles; amphibian speciation; reproductive adaptations of amphibians and reptiles. Mailing Add: Dept of Biol Western Wash State Col Bellingham WA 98225

BROWN, HERBERT CHARLES, b London, Eng, May 22, 12; nat US; m 37; c 1. INORGANIC CHEMISTRY, ORGANIC CHEMISTRY. Educ: Univ Chicago, BS, 36, PhD(inorg chem), 38. Hon Degrees: DSc, Univ Chicago, 68. Prof Exp: Asst, Univ Chicago, 36-38, Eli Lilly fel, 38-39, instr chem, 39-43, res investr, 41-43; from asst prof to assoc prof, Wayne State Univ, 43-47; prof, 47-59, WETHERILL RES PROF CHEM, PURDUE UNIV, WEST LAFAYETTE, 59- Concurrent Pos: Clark lectr, WVa Univ, 53; Howe lectr, 53; Falk-Plaut lectr, 57; Stieglitz lectr, 58; Tishler lectr, 58; Franklin lectr, 60; Remsen lectr, 61; Edgar Rahs Smith lectr, 62. Working Group on Spec Fuels, Defense Dept, 55. Honors & Awards: Harrison Howe Award, 53; Nichols Medalist, 59; Award, Synthetic Org Chem Mfg Asn US, 60; Am Chem Soc Award, 60; Linus Pauling Medalist, 68; Nat Medal of Sci, 69; Roger Adams Medalist, 70. Mem: Nat Acad Sci; Am Acad Arts & Sci; AAAS; Am Chem Soc; The Chem Soc. Res: Hydrides of boron; reactions of atoms and free radicals; nature of Friedel-Craft catalysts; reaction mechanisms; chemistry of addition compounds; steric strains; selective reductions; hydroboration; chemistry of organoboranes. Mailing Add: Dept of Chem Purdue Univ West Lafayette IN 47907

BROWN, HERBERT ENSIGN, b Ogden, Utah, m 44; c 2. ANATOMY, PHYSIOLOGY. Educ: Univ Utah, BS, 49, MS, 51, PhD(anat), 55. Prof Exp: Instr, Westminster Col, 54-55; res instr, Univ Utah, 5556; asst prof, 56-61, ASSOC PROF ANAT, SCH MED, UNIV MO-COLUMBIA, 61- Mem: AAAS; Reticuloendothelial Soc; Am Asn Anatomists. Res: Cell biology; hematology; application of principles of endocrinology, biochemistry and immunology in understanding the morphology and function of cells of the blood and blood forming organs and connective tissues. Mailing Add: Dept of Anat Univ of Mo Sch of Med Columbia MO 65201

BROWN, HERBERT EUGENE, b Jackson, Ga, Dec 25, 15; m 40; c 5. MEDICINE. Educ: Emory Univ, AB, 36, AM, 37, MD, 46; Univ Calif, PhD(zool), 40. Prof Exp: Asst zool, Univ Calif, 37-40; instr, Univ Ga, 40-43; intern internal med, Grady Mem Hosp, Atlanta, 46-47; med officer, US Army, Brooklyn, 47 & Cristobal, CZ, 48-49; asst intermediate & sr resident internal med, Lawson Vet Admin Hosp, Ga, 49-52;

internist, Atomic Bomb Casualty Comn, Hiroshima, 52-53; from instr to assoc prof med, Sch Med, 54-73, ASSOC PROF MED, DEPT ORAL MED, SCH DENT, EMORY UNIV, 74-, PHYSICIAN UNIV STUDENT HEALTH, 73- Concurrent Pos: Physician, Fed Reserve Bank Atlanta, 73- Mem: AAAS; Am Fedn Clin Res. Res: Cytology of termite flagellates; human genetics; amebiasis; antibiotics. Mailing Add: 622 Park Lane Decatur GA 30033

BROWN, HORACE DEAN, b Gainesville, Ga, Feb 13, 19; m 40; c 3. MEDICINAL CHEMISTRY. Educ: Berry Col, BS, 39; Emory Univ, MS, 40; Iowa State Univ, PhD(chem), 47. Prof Exp: Chemist, Red Rock Cola Co, Atlanta, 40-41; instr chem, Iowa State Col, 41-43, res chemist, Manhattan Proj, 44-46, asst, Col, 46-47; mgr sci info, Merck & Co, Inc, 47-67, DIR PLANNING & SCI INFO, MERCK SHARP & DOHME RES LABS, 67- Concurrent Pos: Asst, Quaker Oats Co, 46-47. Mem: Am Chem Soc; NY Acad Sci. Res: Structure-activity relationships; insecticides; fungicides; antiparasitic agents; scientific information. Mailing Add: Merck Sharp & Dohme Res Labs Rahway NJ 07065

BROWN, HOWARD JUNIOR, public health, internal medicine, deceased

BROWN, HOWARD S, b Lakewood, Ohio, July 30, 21; m 49; c 4. BIOLOGY, GENETICS. Educ: Univ Calif, Los Angeles, BA, 43, MA, 49; Claremont Grad Sch, PhD, 60. Prof Exp: Assoc prof, 48-64, PROF BIOL, CALIF STATE POLYTECH COL, 64- Concurrent Pos: Vis prof, Chung Chi Col, Hong Kong, 61-62. Mem: AAAS; Bot Soc Am. Res: Cytogenetics; botanica genetics. Mailing Add: Dept of Biol Calif State Polytech Col Pomona CA 91766

BROWN, HUGH NEEDHAM, b Champaign, Ill, Sept 12, 28; m 59; c 2. NUCLEAR PHYSICS. Educ: Univ Ill, BS, 50, MS, 52, PhD(physics), 54. Prof Exp: PHYSICIST, ACCELERATOR DEPT, BROOKHAVEN NAT LAB, 56- Mem: Am Phys Soc. Res: Accelerator development and construction; high energy physics experimental facilities. Mailing Add: Accelerator Dept Brookhaven Nat Lab Upton NY 11973

BROWN, IAN DAVID, b Edgware, Eng, Apr 11, 32; Can citizen; m 61; c 3. CHEMICAL PHYSICS. Educ: Univ London, BSc, 55 & 56, PhD(crystallog), 59. Prof Exp: Fel crystallog, 59-63, from asst prof to assoc prof, 63-74, PROF PHYSICS, McMASTER UNIV, 74- Mem: Can Asn Physicists; Brit Inst Physics & Phys Soc; The Chem Soc; Chem Inst Can. Res: X-ray and neutron structure analysis. Mailing Add: Dept of Physics McMaster Univ Hamilton ON Can

BROWN, IAN MCLAREN, b St Andrews, Scotland, Apr 15, 35; m 62; c 2. MAGNETIC RESONANCE. Educ: St Andrews Univ, BSc, 57, PhD(physics), 62. Prof Exp: Res assoc chem, Wash Univ, 61-63 & physics, Argonne Nat Lab, 63-65; assoc scientist, 65-73, SCIENTIST, McDONNELL DOUGLAS CORP, 73- Mem: Am Phys Soc; Am Chem Soc. Res: Magnetic resonance of free radicals in polymers. Mailing Add: Dept 223 McDonnell Douglas Corp Box 516 St Louis MO 63166

BROWN, IRA CHARLES, b Victoria, BC, Sept 3, 16; m 42; c 3. HYDROLOGY. Educ: Queen's Univ, Ont, BS, 40, MSc, 42; Harvard Univ, PhD(geol), 49. Prof Exp: From geologist to head eng & groundwater geol sect, Geol Surv Can, 46-65, head groundwater sect, Water Res Br, 65-67; secy, Can Nat Comt Int Hydrol Decade, Can Dept Energy, Mines & Resources, 67-75; CHIEF SECY, INLAND WATERS DIRECTORATE, ENVIRON CAN, 75- Concurrent Pos: Exec secy, Assoc Comt Hydrol, 75- Honors & Awards: Barlow Mem Award, Can Inst Mining & Metall, 51. Mem: Can Inst Mining & Metall; Geol Asn Can; Int Water Resources Asn. Res: Water resources. Mailing Add: Inland Waters Directorate Environ Can Ottawa ON Can

BROWN, IRWIN, b New York, NY, May 24, 22; m 56; c 2. SPEECH PATHOLOGY. Educ: NY Univ, BA, 43; Univ Iowa, MA, 47; Univ Mich, PhD(speech path), 53. Prof Exp: Asst prof speech, Temple Univ, 47-51; asst prof, Univ Mich, 53-56; exec dir, Hearing & Speech Ctr Rochester, 56-62; exec dir, Cleveland Hearing & Speech Ctr & asst prof otolaryngol, Case Western Reserve Univ, 63-69; ASST CLIN PROF PREV MED, SCH MED & DENT, UNIV ROCHESTER & EXEC DIR, HEARING & SPEECH CTR ROCHESTER, 69- Concurrent Pos: Asst prof, Sch Med, Univ Rochester, 56-62; lectr, Syracuse Univ, 60-62. Mem: Fel Am Speech & Hearing Asn; Acad Aphasia. Res: Aphasic language disturbances. Mailing Add: Hearing & Speech Ctr 1000 Elmwood Ave Rochester NY 14620

BROWN, IRWIN FREDERICK, JR, plant pathology, see 12th edition

BROWN, IVAN WILLARD, JR, b Newfane, NY, July 6, 15; m 39; c 3. THORACIC SURGERY, CARDIOVASCULAR SURGERY. Educ: Duke Univ, BS & MD, 40; Am Bd Thoracic Surg, dipl; Am Bd Surg, dipl. Prof Exp: Instr path, 40-42, assoc surg, 45-51, from asst prof to prof, 51-66, JAMES B DUKE DISTINGUISHED PROF SURG, SCH MED, DUKE UNIV, 66-; THORACIC & CARDIOVASC SURGEON, WATSON CLIN, 68- Concurrent Pos: Markle scholar med sci, 48-53; asst med nat blood prog, Am Nat Red Cross, 48-49, consult, 49-55, consult, Hosps, 51-; consult & comt mem, Comn Plasma Fractionation, Harvard Univ, 51-64; mem comt blood & related probs, Nat Res Coun, 53-61; mem exec comt, Div Med, Nat Res Coun, 65- Honors & Awards: First Res Award, Glycerine Prod Asn, 52. Mem: Soc Exp Biol & Med; Am Surg Asn; Am Asn Thoracic Surg; Soc Vascular Surg; Int Cardiovasc Soc. Res: Problems related to thoracic surgery, extracorporeal circulation and perfusion hypothermia. Mailing Add: Watson Clin 1600 Lakeland Hills Blvd Lakeland FL 33802

BROWN, J MARTIN, b Doncaster, Eng, Oct 15, 41; m 67. RADIOBIOLOGY, ONCOLOGY. Educ: Univ Birmingham, BSc, 63; Univ London, MSc, 65; Oxford Univ, DPhil(radiation biol), 68. Prof Exp: NIH fel radiation biol, Med Ctr, 68-70, res assoc, 70-71, ASST PROF RADIATION BIOL, MED CTR, STANFORD UNIV, 71- Concurrent Pos: Am Cancer Soc Dernham sr fel, 71-74; mem adv comt biol effects of ionizing radiations, Nat Acad Sci, 71-; mem int comn radiol protection, 73-77. Mem: Brit Inst Radiol; Radiation Res Soc; AAAS. Res: Mammalian cellular radiobiology; tumor radiobiology; experimental chemotherapy; production and control of metastases; radiation carcinogenesis. Mailing Add: Dept of Radiol Stanford Univ Med Ctr Stanford CA 94305

BROWN, JACK BETHEL, mathematics, see 12th edition

BROWN, JACK HAROLD UPTON, b Nixon, Tex, Nov 16, 18; m 43. PHYSIOLOGY. Educ: Southwest Tex State Col, BS, 39; Rutgers Univ, PhD(biochem), 48. Prof Exp: Instr physiol, Univ Tex, 39-41; radio engr, US Air Force, Wis, 41-43; instr physics, Southwest Tex State Col, 43-44; instr phys chem, Rutgers Univ, 44-45, res assoc biochem & nutrit, 45-48; dir, Biol Labs, Mellon Inst, 48-50; asst prof physiol, Sch Med, Univ NC, 50-52; assoc prof, Emory Univ, 52-57, actg chmn dept, 57-59, prof, 58-60; exec secy training grant, Nat Inst Gen Med Sci, NIH, 60-62, chief spec res resources br, 62-63, chief gen clin ctr br, 63, asst chief div res facil & resources, 63-71, dir sci progs, Nat Inst Gen Med Sci, 70-71, from assoc dir to actg dir, 64-70; spec asst to adminr, Health Sci Ment Health Admin, 70-71, assoc dep adminr, 71-73; spec asst to admin, Health Resources Admin, 73-74; COORDR SOUTHWEST RES CONSORTIUM, 74-; PROF PHYSIOL, UNIV TEX HEALTH SCI CTR SAN ANTONIO, 74-, PROF ENVIRON SCI, UNIV TEX, SAN ANTONIO, 74- Concurrent Pos: Lectr, Univ Pittsburgh, 48-50, Med Schs, George Washington Univ, 61- & Georgetown Univ, 64-; consult, Vet Admin, Oak Ridge Inst Nuclear Studies & Lockheed Aircraft Co. Mem: Nat Acad Eng; Am Chem Soc; Soc Exp Biol & Med; Am Physiol Soc; Biomed Eng Soc (pres, 69). Res: Endocrines; adrenal steroid metabolism; biomedical engineering. Mailing Add: Southwest Res Consortium 8848 W Commerce St San Antonio TX 78284

BROWN, JACK STANLEY, b New Orleans, La, Mar 12, 29; m 48; c 2. FRESHWATER BIOLOGY. Educ: Tulane Univ, BS, 48; Univ Ala, MS, 49, PhD, 56. Prof Exp: Asst biol, Univ Ala, 48-49; asst prof, Jacksonville State Col, 49-52; instr zool, Auburn Univ, 54-55; med entomologist, US Air Force, 55-59; prof biol & chmn dept, Emory & Henry Col, 59-63; prof biol, Parsons Col, 63-65, dean, 65-68; PROF BIOL, UNIV N ALA, 68-, DIR INST FRESHWATER BIOL, 69-, HEAD DEPT BIOL, 74- Mem: Am Soc Trop Med & Hyg; Am Soc Ichthyol & Herpet; Am Inst Biol Sci; NAm Benthological Soc. Res: Herpetology; medical entomology and parasitology; natural history; freshwater benthos; water quality indicators. Mailing Add: Dept of Biol Univ of N Ala Florence AL 35630

BROWN, JAMES ALLEN, JR, b Akron, Ohio, May 23, 26; m 51. MEDICAL MICROBIOLOGY, PHARMACOLOGY. Educ: Univ Akron, BS, 49; Kent State Univ, MA, 52; Howard Univ, PhD(physiol, pharmacol), 68. Prof Exp: Bacteriologist, Cleveland Pub Health Dept, Ohio, 53-54; head dept bact, Vet Admin Hosp, Wilmington, Del, 54-56; div chief, Vet Admin Hosp, Indianapolis, Ind, 56-63; res microbiologist, Walter Reed Army Med Ctr, DC, 63-67; biol scientist, Nat Air Pollution Control Admin, DC, 67-69, supvry tech info specialist, Raleigh, NC, 69-70; RES SCIENTIST, UNIROYAL CHEM INC, 70- Mem: AAAS; Am Soc Microbiol; Am Inst Biol Sci; fel Am Pub Health Asn; affil Am Soc Clin Pathologists. Res: Identification, classification, propagation and isolation of various atypical mycobacteria, systemic fungus disease agents, etiological agent of plague and similar organisms; research and development pertaining to pollution abatement. Mailing Add: Uniroyal Chem Inc Elm St Res & Develop Bldg 122 Naugatuck CT 06770

BROWN, JAMES DOUGLAS, b Hamilton, Ont, July 17, 34; m 57; c 4. PHYSICAL CHEMISTRY. Educ: McMaster Univ, BSc, 57; Univ Md, PhD(phys chem), 66. Prof Exp: Asst solubility of plutonium compounds, Atomic Energy Can Ltd, Ont, 56; chemist, Res Dept, Imp Oil Ltd, Ont, 57-60; res chemist, Res Ctr, US Bur Mines, Md, 60-66; asst prof, 67-69, ASSOC PROF MAT SCI & CHMN GROUP, FAC ENG SCI, UNIV WESTERN ONT, 69- Concurrent Pos: Nat Res Coun Can fel, 67-; vis scientist, Metaalinstituut TNO Apeldoorn, Neth, 75-76. Mem: Am Chem Soc; Soc Appl Spectros; Electron Probe Anal Soc Am. Res: Structure and properties of vacuum deposited thin films; quantitative intensity; concentration relationships in electron probe microanalysis; x-ray spectroscopy; electrostatics. Mailing Add: 95 Cumberland Crescent London ON Can

BROWN, JAMES HAROLD, b Richwood, WVa, Nov 9, 31; m 53; c 1. FOREST ECOLOGY. Educ: WVa Univ, BSF, 53; Yale Univ, MF, 54; Mich State Univ, PhD(forestry), 67. Prof Exp: Res asst forestry, Northeastern Forest Exp Sta, US Forest Serv, 54-55; asst silviculturist, WVaUniv, 57-61, from asst prof to assoc prof forestry, 61-70; ASSOC PROF APPL SILVICULT, OHIO AGR RES & DEVELOP CTR, 70- Mem: Soc Am Foresters; Soil Sci Soc Am. Res: Problems of artifical establishment and growth of tree species, including ecological and genetic considerations. Mailing Add: Dept of Forestry Ohio Agr Res & Develop Ctr Wooster OH 44691

BROWN, JAMES HEMPHILL, b Ithaca, NY, Sept 25, 42; m 65; c 1. ECOLOGY. Educ: Cornell Univ, AB, 63; Univ Mich, PhD(zool), 67. Prof Exp: H H Rackham fel, Univ Calif, Los Angeles, 67-68, asst prof zool, 68-71; ASST PROF BIOL, UNIV UTAH, 71- Mem: AAAS; Ecol Soc Am; Am Soc Mammalogists. Res: Ecology and evolution, especially behavioral and physiological aspects of the ecology of small mammals. Mailing Add: Dept of Biol Univ of Utah Salt Lake City UT 84112

BROWN, JAMES HENRY, JR, b Oneco, Conn, July 27, 34; m 56; c 4. FOREST ECOLOGY, SOILS. Educ: Univ Conn, BS, 56; Univ RI, MS, 58; Duke Univ, DF(forest ecol), 65. Prof Exp: Res asst, 58-60, lectr, 60-62, asst prof, 64-69, actg chmn dept, 67-68, ASSOC PROF FORESTRY, UNIV RI, 69- Concurrent Pos: Vis lectr, Utah State Univ, 69-70. Mem: Soc Am Foresters; Sigma Xi. Res: Site-growth relationships; impact of fire on forests of southern New England; forest regeneration and forest water budget. Mailing Add: Dept of Forest & Wildlife Mgt Univ of RI Kingston RI 02881

BROWN, JAMES MELTON, b Clayton, Ala, Dec 29, 25; m 47; c 4. SOILS. Educ: Auburn Univ, BS, 49, MS, 51; NC State Univ, PhD(soils, plant nutrit), 60. Prof Exp: Asst prof soils, Auburn Univ, 51-53, asst agronomist, 53-55; res instr soils, NC State Univ, 55-58, exten specialist, Statewide Soils Educ Prog, 58-60; agronomist, Nat Plant Food Inst, DC, 60-61; agronomist, 61-68, mgr agron res, 68-69 & agron & pesticide residue res, 69-71, MGR PROD TECHNOL, NAT COTTON COUN AM, 71- Concurrent Pos: Alt rep mem pesticide study panel & nominating comt, Agr Res Inst, 74-75. Mem: Am Soc Agron; Soil Sci Soc Am; Weed Sci Soc Am; Int Soc Soil Sci; Entom Soc Am. Res: Soil fertility and chemistry; plant nutrition and physiology; weed control; air pollution; pesticide residues; plant diseases; insects; agricultural chemicals; legislation. Mailing Add: Nat Cotton Coun of Am 1918 North Pkwy PO Box 12285 Memphis TN 38112

BROWN, JAMES MILTON, b Niagara Falls, NY, Dec 16, 39; m 68. PLANT ECOLOGY, RESOURCE MANAGEMENT. Educ: Colgate Univ, AB, 61; NC State Univ, MS, 63; Univ Ariz, PhD(plant sci), 68. Prof Exp: Res scientist plant physiol, Forest Serv, USDA, 68-75; TEAM MEM PLANT ECOL, NAT STREAM ALTERATION TEAM OFF BIOL SERV, FISH & WILDLIFE SERV, US DEPT INTERIOR, 75- Mem: Ecol Soc Am; Am Bot Soc; Am Inst Biol Soc; Brit Ecol Soc. Res: Effect of management and land use upon streams and rivers; associated terrestrial ecosystems; impact of proposed stream alterations upon watershed ecosystems and associated fish and wildlife resources. Mailing Add: Nat Stream Alteration Team Rte 1 Columbia MO 65201

BROWN, JAMES RICHARD, b Charleston, Ill, Oct 6, 31; m 60; c 4. SOIL FERTILITY. Educ: Univ Ill, BS, 53, MS, 57; Iowa State Univ, PhD(soil fertil), 63. Prof Exp: Asst prof, 63-71, ASSOC PROF AGRON, UNIV MO-COLUMBIA, 71- Mem: AAAS; Am Soc Agron; Soil Sci Soc Am; Soil Conserv Soc Am. Res: Soil fertility and testing; micronutrient nutrition; plant analysis. Mailing Add: Dept of Agron Univ of Mo Columbia MO 65201

BROWN, JAMES ROY, b Wilkinsburg, Pa, June 14, 23; m 47; c 3. NUCLEAR PHYSICS. Educ: Allegheny Col, BA, 44; Calif Inst Technol, MS, 48, PhD(physics), 51. Prof Exp: Asst, Calif Inst Technol & Naval Ord Test Sta, 44-45; instr physics, Allegheny Col, 42-47; sr scientist nuclear physics, Reactors, Westinghouse Bettis

Atomic Power Div, 50-57, adv scientist exp reactor physics, 57-59; PRIN PHYSICIST, NEUTRON PHYSICS EXP, LINEAR ACCELERATOR & PHYSICIST CHG CRITICAL EXP, GULF GEN ATOMIC, INC, 59- Mem: AAAS; Am Nuclear Soc; Am Phys Soc. Res: Solid propellant rockets; elementary physics; precision spectroscopy of gamma rays using a curved crystal focusing spectrometer; present research on neutron and reactor physics. Mailing Add: Gulf Gen Atomic Div Gulf Energy & Environ Systs PO Box 608 San Diego CA 92112

BROWN, JAMES RUSSELL, b Portland, Ore, Mar 17, 32; m 67; c 5. MATHEMATICAL ANALYSIS. Educ: Ore State Univ, BA, 53, MS, 58; Yale Univ, PhD(math), 64. Prof Exp: Asst prof math, Univ Mass, 60-62; asst prof, 62-69, actg chmn dept, 71-73, ASSOC PROF MATH, ORE STATE UNIV, 69-, CHMN DEPT, 73- Concurrent Pos: Consult math, Boeing Airplane Co, 57-61 & Bell Tel Labs, 60-62; Fulbright lectr, Univ San Carlos, Philippines, 67-68. Mem: Am Math Soc. Res: Ergodic theory; topological dynamics; Markov processes; functional analysis. Mailing Add: Dept of Math Ore State Univ Corvallis OR 97331

BROWN, JAMES T, JR, b Jackson Heights, NY, Feb 23, 39; m 59; c 2. THEORETICAL PHYSICS. Educ: Univ Colo, BA, 61, PhD(physics), 68. Prof Exp: Physicist, Boulder Labs, Nat Bur Stand, Colo, 58-65; ASSOC PROF PHYSICS, COLO SCH MINES, 67- Mem: Am Phys Soc. Res: Hyperon-nucleon scattering theory; propagation of electromagnetic waves in the ionosphere; aerosol science and particulate control; condensation and nucleation properties of respirable dust. Mailing Add: Dept of Physics Colo Sch of Mines Golden CO 80401

BROWN, JAMES WALKER, JR, b Hampshire, Tenn, July 10, 30; m 53; c 2. BIOCHEMISTRY. Educ: Mid Tenn State Col, BS, 57; NC State Col, MS, 59, PhD(dairy chem), 62. Prof Exp: From asst prof to assoc prof, 62-69, PROF CHEM, MID TENN STATE UNIV, 69- Mailing Add: Dept of Chem Mid Tenn State Univ Murfreesboro TN 37132

BROWN, JAMES WALLACE, III, b Savannah, Ga, Feb 11, 38. PHOTOGRAPHIC CHEMISTRY. Educ: St Peters Col, BS, 59; Stevens Inst Technol, PhD(chem), 73. Prof Exp: Res chemist, Squibb Inst Med Res, E R Squibb & Sons, 60-63; res asst org chem, Ben May Labs Cancer Res, Univ Chicago, 63-67; SR RES CHEMIST, RES LABS, EASTMAN KODAK CO, 72- Mem: Am Chem Soc; Sigma Xi. Mailing Add: Eastman Kodak Res Labs 1669 Lake Ave Rochester NY 14650

BROWN, JAMES WARD, b Philadelphia, Pa, Jan 15, 34; m 57; c 2. APPLIED MATHEMATICS. Educ: Harvard Univ, AB, 57; Univ Mich, AM, 58, PhD(math), 64. Prof Exp: Asst prof math, Univ Mich, Dearborn, 64-66 & Oberlin Col, 66-68; assoc prof, 68-71, PROF MATH, UNIV MICH, DEARBORN, 71- Concurrent Pos: NSF res grant, 69; ed consult, Math Rev, 70- Mem: Sigma Xi; Math Asn Am; Am Math Soc. Res: Special functions of mathematical physics.

BROWN, JAMES WILSON, b Washington, DC, June 24, 13; m 42; c 3. PLANT PHYSIOLOGY. Educ: Univ Md, BS, 35; Duke Univ, AM, 37, PhD(forestry), 38. Prof Exp: Asst plant physiologist, USDA, Plant Indust Sta, Md, 44-47; plant physiologist, Crops Div, Ft Detrick, 47-54, prog coord officer, 54-59, dep chief crops div, 59-67, dep chief plant sci lab, 67-69; RETIRED. Concurrent Pos: Vegetation control adv & executor mission, Hq, First Army, 59, vegetation control adv, Third Army, 60; with, US Off Secy Defense, Advan Res Proj Agency, Repub Viet Nam & President Diem; adv, Gen Maxwell Taylor, Viet Nam & US, 61-62; adv vegetation control, US Army, Navy & Marine Forces, Can Armed Forces & Korean Armed Forces, 61-62; demonstator vegetation control, Viet Nam, 61-62, in charge supv & consult initial defoliation sprays, 61-62 & 65; in charge aerial spray systs, Eglin AFB, Fla, 62 & planning aerial spray modifications, 63; partic coop prog, US Naval Sta, Alameda, Calif & Jacksonville, Fla & US Air Force Langley Field, Va & Elgin AFB, Fla, 64-; with Res Analysis Corp, Viet Nam, 65; vegetation control adv, Camp David, Md, 67- 67-69. Mem: AAAS; emer Am Soc Plant Physiol. Res: Physiological effects and screening of plant growth regulating chemicals; information retrieval systems; world agriculture; vegetation control; aerial spray systems. Mailing Add: 468 Barrello Ln Cocoa Beach FL 32931

BROWN, JARVIS HOWARD, b Goshen, Ind, Aug 7, 37; m 59; c 3. CROP PHYSIOLOGY. Educ: Purdue Univ, BSA, 59; Iowa State Univ, MS, 62, PhD(crop physiol), 65. Prof Exp: Res assoc agron, Purdue Univ, 65-67; asst prof, 67-71, ASSOC PROF AGRON, MONT STATE UNIV, 71- Mem: Am Soc Agron; Crop Sci Soc Am. Res: Effects of leaf orientation, leaf width and chlorophyll concentration on the water use, growth, yield and photosynthesis of barley isotypes; cold hardiness of winter wheat. Mailing Add: Dept of Plant & Soil Sci Mont State Univ Bozeman MT 59715

BROWN, JEANETTE SNYDER, b Rochester, NY, Mar 6, 25; m 50; c 3. PHOTOBIOLOGY. Educ: Cornell Univ, BS, 45, MS, 48; Stanford Univ, PhD(biol), 52. Prof Exp: Asst chem, NY State Agr Exp Sta, Geneva, 46-47; bacteriologist, Am Inst Radiation, Calif, 48-49; STAFF BIOLOGIST, CARNEGIE INST WASH DEPT PLANT BIOL, 58- Mem: Am Soc Plant Physiol; Phycol Soc Am; Am Soc Photobiol. Res: Photosynthesis; photoreactive-pigment biochemistry; algal physiology. Mailing Add: Carnegie Inst of Wash 290 Panama St Stanford CA 94305

BROWN, JEROME ENGEL, b Buckhannon, WVa, Apr 29, 24; div; m 62; c 4. INDUSTRIAL CHEMISTRY. Educ: WVa Wesleyan Col, BS, 44; Univ Ill, PhD(chem), 49. Prof Exp: Asst res supvr, 49-60, commercial develop rep, 60-62, mgr mkt develop, 62-70, SR MKT RES ASSOC MKT DEVELOP, ETHYL CORP, 70- Res: Market development and research; gasoline additives; high-vacuum techniques. Mailing Add: PO Box 2332 Baton Rouge LA 70821

BROWN, JERRAM L, b Glen Ridge, NJ, July 19, 30; m 53; c 3. ETHOLOGY. Educ: Cornell Univ, AB, 52, MS, 54; Univ Calif, Berkeley, PhD(zool), 60. Prof Exp: NIH fel, 60-62; from asst prof to assoc prof, 62-72, PROF BIOL & BRAIN RES, UNIV ROCHESTER, 72- Mem: Soc Study Evolution; Am Ornith Union; Animal Behav Soc. Res: Neuro-ethology; neural bases of instinctive behavior; evolution of behavior; ecology of altruism in animals. Mailing Add: Dept of Biol Univ of Rochester Rochester NY 14627

BROWN, JERRY, b Boundbrook, NJ, Feb 22, 36; m; c 2. SOIL SCIENCE. Educ: Rutgers Univ, BS, 58, PhD(soils), 62. Prof Exp: SOIL SCIENTIST, US ARMY COLD REGIONS RES & ENG LAB, 61-, DIR TUNDRA BIOME, INT BIOL PROG, 70- Mem: AAAS; Soil Sci Soc Am; Arctic Inst NAm; Ecol Soc Am. Res: Arctic soil science; arctic ecology; permafrost. Mailing Add: PO Box 345 Hanover NH 03755

BROWN, JERRY L, b Malvern, Ark, Oct 3, 35; m 61; c 1. BIOCHEMISTRY. Educ: NTex State Univ, BS, 59, MS, 60; Univ Tex, PhD(biochem), 64. Prof Exp: Fel biol chem, Univ Calif, Los Angeles, 63-67; asst prof biochem, 67-73, ASSOC PROF BIOCHEM, MED SCH, UNIV COLO, DENVER, 73- Mem: Am Soc Biol Chemists. Res: Reactions resulting in structural modifications of proteins; reactions leading to blocked amino terminal amino acids are being investigated. Mailing Add: Dept of Biochem Univ of Colo Med Sch Denver CO 80220

BROWN, JERRY WILLIAM, b Wichita, Kans, July 4, 25; m 50; c 3. ANATOMY. Educ: Wichita State Univ, AB, 46; Univ Kans, MA, 49, PhD(anat), 51. Prof Exp: From asst to instr anat, Univ Kans, 47-51; instr, Sch Med, Univ Pittsburgh, 51-56; from asst prof to assoc prof, Sch Med, Univ Mo, 56-64; assoc prof, 64-70, PROF ANAT, MED CTR, UNIV ALA, BIRMINGHAM, 70- Concurrent Pos: Guest worker, Neth Cent Inst Brain Res, Univ Amsterdam, 65-66. Mem: AAAS; Am Acad Neurol; Soc Neurosci; Am Asn Anat. Res: Embryonic development of the brainstem and telencephalon; development of sensory nuclei of V in human embryos; development of telencephalon of bat embryos. Mailing Add: Dept of Anat Univ of Ala Med Ctr Birmingham AL 35294

BROWN, JIM MCCASLIN, b Minneapolis, Minn, Sept 29, 38; m 63; c 2. STRUCTURAL GEOLOGY, ROCK MECHANICS. Educ: Univ Alaska, BS, 60, MS, 63; Univ Wis-Madison, PhD(geol), 68. Prof Exp: Jr asst geologist, Pan Am Petrol Corp, 60; geologist, US Geol Surv, 62; sr asst field geol, Ont Dept Mines, 64; actg instr geol, Univ Wis-Madison, 67; asst prof geol & geol eng, St Louis Univ, 68-69; asst prof geol, Ind Univ-Purdue Univ, Indianapolis, 69-74; SR GEOLOGIST, R&M CONSULT, 74- Mem: Am Inst Mining, Metall & Petrol Eng; Geol Soc Am. Res: Structural analysis of multiply-deformed metamorphic tectonites; regional significance of metamorphic and orogenic events; rock mechanics and slope stability analysis with rock bolt design. Mailing Add: R&M Consult PO Box 2630 Fairbanks AK 99707

BROWN, JOE ROBERT, b Mt Pleasant, Iowa, Nov 24, 11; m 37; c 3. NEUROLOGY. Educ: Univ Iowa, BA, 33, MD, 37; Univ Minn, MS, 43. Prof Exp: Intern, Presby Hosp, Chicago, Ill, 37-38; first asst, Neurol Sect, Mayo Clin, 41-43; from asst prof to assoc prof, 46-63, PROF NEUROL, UNIV MINN, 63- Concurrent Pos: Fel neurol & psychiat, Mayo Found, Univ Minn, 39-43; chief neurol sect, Vet Admin Hosp, Minneapolis, 46-47, chief neuropsychiat, 47-49; consult, Vet Admin, 49-; consult to head neurol sect to sr consult, Mayo Clin, 49-; mem training grant comt, Nat Inst Neurol Dis & Blindness. Honors & Awards: Bronze Medal, AMA, 47. Mem: Fel Am Acad Neurol (secy-treas, 48-55, pres, 71); Am Neurol Asn; Asn Res Nerv & Ment Dis; Acad Aphasia. Res: Cerebellar function in man; cerebellar anomalies and diseases; psychologic changes in neurologic diseases; speech and language disorders; multiple sclerosis. Mailing Add: 102-110 Second Ave SW Rochester MN 55901

BROWN, JOEL EDWARD, b Middletown, NY, May 23, 37. NEUROPHYSIOLOGY. Educ: Mass Inst Technol, BS & MS, 60, PhD(physiol), 64. Prof Exp: From asst prof to assoc prof physiol, Mass Inst Technol, 64-71; prof anat, Vanderbilt Univ, 71-76, prof biochem, 74-76; PROF PHYSIOL, STATE UNIV NY STONY BROOK, 76- Concurrent Pos: NIH fel, Sch Med, Univ Chile, 64; corp mem, Woods Hole Marine Biol Lab, trustee, 73-77. Mem: Am Physiol Soc; Biophys Soc; Soc Gen Physiol. Mailing Add: Dept of Physiol & Biophys State Univ NY Stony Brook NY 11790

BROWN, JOHN A, economic geology, mineralogy, see 12th edition

BROWN, JOHN A, b Minneapolis, Minn, Mar 26, 30; m 52. MATHEMATICS. Educ: Univ Minn, BA, 51, MS, 59; Mont State Univ, PhD(math), 66. Prof Exp: Sr engr, Gen Mills, Inc, 56-59; instr math, Mont State Univ, 59-60; opers analyst, Opers Res Off, Johns Hopkins Univ, 60-61; asst prof math, Mont State Univ, 61-66 & Univ Mass, 66-67; Nat Acad Sci resident res assoc, NASA, 67-68; ASST PROF MATH, MONT STATE UNIV, 68- Res: Numerical analysis and computer sciences. Mailing Add: Dept of Math Mont State Univ Bozeman MT 59715

BROWN, JOHN A, JR, b Opelousas, La, Aug 28, 30; m 51; c 3. METEOROLOGY. Educ: Univ Southwestern La, BS, 51; Mass Inst Technol, SM, 57; Univ Colo, PhD(astrogeophys), 68. Prof Exp: Res meteorologist, US Weather Bur, 61-62 & Nat Ctr Atmospheric Res, 62-68; SUPVRY RES METEOROLOGIST, NAT METEOROL CTR, US WEATHER BUR, ROCKVILLE, 68- Mem: Am Meteorol Soc. Res: Dynamical meteorological research of large-scale atmospheric motions of the earth; numerical weather prediction; applied research. Mailing Add: 9306 Caldran Dr Clinton MD 20735

BROWN, JOHN ANGUS, b Lake Wales, Fla, Dec 13, 25; m 54; c 3. ORGANIC CHEMISTRY, PHYSICAL CHEMISTRY. Educ: Emory Univ, BA, 45; Ga Inst Technol, MS, 51, PhD(org chem), 54. Prof Exp: Engr, Law & Co, Ga, 45-48; teaching asst chem, Ga Inst Technol, 48-54; res chemist, Esso Res & Eng Co, NJ, 54-63, sr chemist, 63-68, head govt res mkt, 68-69, staff adv, 68-73; PRES, JOHN BROWN ASSOCS, INC, 73- Concurrent Pos: Mem var ad hoc study & planning panels, Dept of Defense. Mem: Am Chem Soc; Sigma Xi; Am Defense Preparedness Asn; Am Inst Chem Engrs. Res: Research and development of lubricating oil additives, contract research and development in safety and sensitivity of propellants and explosives; research and development marketing; advanced physical-chemical waste reclamation processes. Mailing Add: 15 York Pl Berkeley Heights NJ 07922

BROWN, JOHN BOYER, b Lexington, Ky, Nov 12, 24; m 49; c 3. PHYCIAL CHEMISTRY. Educ: Univ Ky, BS, 48; Northwestern Univ, PhD, 56. Prof Exp: From instr to assoc prof, 52-66, chmn dept, 61-64, 72-73 & 75-76, PROF CHEM, DENISON UNIV, 66- Concurrent Pos: Vis scientist, Swedish Inst Surface Chem, 73-74. Mem: Am Chem Soc; fel Am Inst Chemists; NY Acad Sci. Res: Cationic surfactants; physical properties of micelles; solubilization by detergents. Mailing Add: Dept of Chem Denison Univ Granville OH 43023

BROWN, JOHN CHARL, b Grantsville, Utah, Jan 31, 17; m 46; c 3. PLANT PHYSIOLOGY, SOIL CHEMISTRY. Educ: Brigham Young Univ, BS, 39; Mich State Univ, PhD, 49. Prof Exp: Soil scientist, Plant Indust Sta, 49-72, SOIL SCIENTIST, PLANT STRESS LAB, AGR RES CTR, USDA, 72- Honors & Awards: Serv Award, USDA, 62. Mem: Soil Sci Soc Am; Am Soc Plant Physiologists; Am Soc Agron. Res: Mineral nutrition of plants, including the physiological and biochemical aspects of plant growth as related to soil chemistry. Mailing Add: Plant Stress Lab Agr Res Ctr Beltsville MD 20705

BROWN, JOHN CLIFFORD, b Cullman, Ala, Feb 23, 43; m 66; c 2. BIOCHEMISTRY, IMMUNOLOGY. Educ: Auburn Univ, BS, 65, MS, 67; NC State Univ, PhD(biochem), 73. Prof Exp: FEL IMMUNOL, UNIV CALIF, BERKELEY, 73- Res: Study of conformational changes in hapten-specific antibody upon antigen binding and the possible effects of such changes on the in vitro immune response. Mailing Add: Dept of Bact & Immunol Univ of Calif Berkeley CA 94720

BROWN, JOHN FRANCIS, JR, b Providence, RI, Oct 11, 26; m 55; c 3. PHYSICAL ORGANIC CHEMISTRY. Educ: Brown Univ, BS, 47; Mass Inst Technol, PhD(chem), 50. Prof Exp: Res assoc, 50-56, mgr reaction studies unit, 56-61, org chemist, 61-65, MGR, LIFE SCI BR, RES & DEVELOP CTR, GEN ELEC CO, 65- Concurrent Pos: Fel clin path, State Univ NY Upstate Med Ctr, 68-69; mem, Study Critical Environ Problems, 70. Mem: AAAS; Am Chem Soc; NY Acad Sci. Res:

Biomedical surface chemistry. Mailing Add: Res & Develop Ctr Gen Elec Co PO Box 8 Schenectady NY 12301

BROWN, JOHN HAYNES, b Louisville, Ky, Sept 10, 34; m 55; c 4. PHARMACOLOGY, BIOCHEMISTRY. Educ: Bellarmine Col, Ky, BA, 59; Univ Louisville, PhD(biochem), 64. Prof Exp: Lectr chem, Catherine Spalding Col, 61-65; sr pharmacologist, Mead Johnson Co, 65-66, group leader, 66-68; asst prof pharmacol, Sch Pharm, Univ Miss, 68-69; asst prof, 69-71, ASSOC PROF PHARMACOL, LA STATE UNIV COL MED, NEW ORLEANS, 71- Concurrent Pos: Chief, Enzyme Sect, US Army Med Res Lab, Ky, 63-65; lectr chem, Univ Evansville, 65-68. Mem: AAAS; Soc Exp Biol & Med; Am Chem Soc; Sigma Xi; Am Soc Pharmacol & Exp Therapeut. Res: Arthritis and inflammation; anti-inflammatory drugs; adrenergic drugs; enzymology of snake venoms. Mailing Add: Dept of Pharmacol La State Univ Col of Med New Orleans LA 70112

BROWN, JOHN HENRY, b Kalamazoo, Mich, Nov 4, 24; m 46; c 3. WOOD SCIENCE, WOOD TECHNOLOGY. Educ: Mich State Univ, BS, 49, MS, 50; Syracuse Univ, PhD(wood prod eng), 62. Prof Exp: Trainee, Ga Pac Plywood Co, 50-51; buyer & salesman, Olympia Wood Preserving Co, 51-54; wood technologist, Res & Develop, Potlatch Forests, Inc, 54-56; instr & res asst, Syracuse Univ, 56-62, asst prof, 62-63; TECHNOLOGIST, WOOD PROD RES & DEVELOP DEPT, POTLATCH CORP, 63- Mem: Soc Wood Sci & Technol; Forest Prod Res Soc; Am Soc Testing & Mat. Res: Wood adhesion; electrical properties of wood; wood-moisture relations; wood combined with other materials—plastics, metals, paper, particleboard. Mailing Add: Wood Prod Res & Develop Dept Potlatch Corp PO Box 1016 Lewiston ID 83501

BROWN, JOHN KENNEDY, b Winnipeg, Man, Jan 12, 20; m 41; c 4. PHYTOCHEMISTRY, PHARMACOGNOSY. Educ: Univ Man, BSc, 59; Univ Wash, MSc, 62, PhD(pharmacog), 65. Prof Exp: Res asst pharmacog, Univ Wash, 60-61; asst prof, Col Pharm, Drake Univ, 65-67; from asst prof to assoc prof, 67-74, PROF PHARMACOG, SCH PHARM, UNIV OF THE PAC, 74- Mem: AAAS; Am Soc Pharmacog; Soc Econ Bot. Res: Investigation of the higher fungi for physiologically active constituents and some of the older medicinal plants as a source of new compounds. Mailing Add: Sch of Pharm Univ of the Pac Stockton CA 95211

BROWN, JOHN LAWRENCE, JR, b Ellenville, NY, Mar 6, 25. APPLIED MATHEMATICS, COMMUNICATION SCIENCE. Educ: Ohio Univ, BS, 48; Brown Univ, PhD(appl math), 53. Prof Exp: From asst prof to prof eng res, 52-69, PROF ELEC ENG, PA STATE UNIV, UNIVERSITY PARK, 69- Mem: Am Math Soc; fel Inst Elec & Electronics Eng; Math Asn Am; Soc Indust & Appl Math; Acoustical Soc Am. Res: Statistical communication theory; applied mathematics; underwater acoustics. Mailing Add: Dept of Elec Eng Pa State Univ University Park PA 16802

BROWN, JOHN LOTT, b Philadelphia, Pa, Dec 3, 24; m 48; c 4. PSYCHOPHYSIOLOGY. Educ: Worcester Polytech Inst, BSEE, 45; Temple Univ, MA, 49; Columbia Univ, PhD(psychol), 52. Prof Exp: Mem staff, Div Govt Res, Columbia Univ, 51-53, tech dir, US Air Force Res Proj, 53-54; head psychol div, Aviation Med Acceleration Lab, Naval Air Develop Ctr, 54-59; asst prof sensory physiol, Sch Med, Univ Pa, 57-62, assoc prof, 62-65, dir grad training prog physiol, 62-65; dean grad sch, Kans State Univ, 65-66, vpres acad affairs, 66-69; PROF PSYCHOL, UNIV ROCHESTER, 69-, DIR, CTR VISUAL SCI, 71- Concurrent Pos: USPHS sr res fel physiol, Sch Med, Univ Pa, 59-63; mem visual study sect, NIH, 67-71; staff adv, Comt Vision, Nat Acad Sci-Nat Res Coun, 68-; mem bd trustees, Worcester Polytech Inst, 71- Mem: Am Psychol Asn; Optical Soc Am; Am Physiol Soc; Soc Neurosci. Res: Psychophysiology and electrophysiology of vision; applications of psychology and physiology to man-machine systems; sensory physiology. Mailing Add: Ctr for Visual Sci Univ of Rochester Rochester NY 14627

BROWN, JOHN MAX, b Ft Wayne, Ind, Feb 15, 17; m 51. MEDICINAL CHEMISTRY, SCIENCE WRITING. Educ: Ind Univ, BS, 38, AM, 40. Prof Exp: Chemist, Ayerst, McKenna & Harrison, 40-42 & Parke, Davis & Co, 42-45; chemist, 45-62, SR PATENT ASSOC, G D SEARLE & CO, 62- Mem: Am Chem Soc. Mailing Add: 811 La Crosse Ave Wilmette IL 60091

BROWN, JOHN R, b London, Eng, Sept 18, 19; Can citizen; m 46; c 1. PHYSIOLOGY, TOXICOLOGY. Educ: Univ London, BSc, 50, MB, BS, 53, PhD(physiol), 59, MD, 66, DSc, 70; FRCP(C). Prof Exp: Lectr appl physiol, London Sch Hyg & Trop Med, 53-59; PROF PHYSIOL HYG & HEAD DEPT, UNIV TORONTO, 59-, PROF ENVIRON HEALTH, 69-, ASSOC PROF CIVIL ENG, 70- Concurrent Pos: Res grants appl physiol & environ health; dir, Health League Can, 60. Mem: Indust Med Asn; Can Physiol Soc; Brit Biochem Soc; Brit Ergonomics Res Soc; Brit Physiol Soc. Res: Problems of human ecology, physical and chemical medical history; physiological research into problems of physical fitness; pesticide study; water microbiology. Mailing Add: Dept Prev Med & Biostatist Univ Toronto 150 College St Toronto ON Can

BROWN, JOHN ROWLAND, JR, b Mansfield, Ohio, Mar 4, 12; m 38; c 3. RESEARCH ADMINISTRATION. Educ: Oberlin Col, AB, 33, MA, 35; Mass Inst Technol, ScD(chem eng), 39. Prof Exp: Instr chem eng, Mass Inst Technol, 38; res chemist, Esso Labs, Chem Div, Stand Oil Develop Co, 38-41, asst dir, 41-46; dir chem res, Pro-phy-lac-tic Brush Co, 46-48, tech dir, 48-49; dir res, Lambert Pharmacal Co, 49-52, vpres, 53; gen mgr res & develop, Spencer Chem Co, 53-56, vpres, 57; dir & vpres res & develop, Colgate-Palmolive Co, 57-73; RETIRED. Mem: AAAS; Am Chem Soc; Am Inst Chem Engrs. Res: Vulcanization of rubber; chemical derivatives of petroleum; production and evaluation of synthetic rubber; compression and injection molding of plastics; pharmaceuticals; soaps and detergents. Mailing Add: 19 Western Dr Short Hills NJ 07078

BROWN, JOHN STEWART, b Chicago, Ill, June 7, 18; m 43; c 3. CHEMISTRY. Educ: Univ Chicago, BS, 40. Prof Exp: Asst engr, Dept Bldgs & Grounds, Univ Chicago, 40-41; supvr res & prod explosives, E I du Pont de Nemours & Co, 41-45; plant supt, Plunkett Chem Co, Chicago, 45-47; res chemist, Standard Oil Co, Ind, 48-70, proj mgr packaging, Am Oil Co, Ind, 70-75, PROJ MGR, AMOCO OIL CO, STANDARD OIL CO, INC, 75- Mem: Fel AAAS; Am Chem Soc; Am Soc Lubrication Engr. Res: Industrial lubricants. Mailing Add: Amoco Oil Co Res & Develop Ctr Naperville IL 60540

BROWN, JOHN WELCH, b Fairfield, Iowa, Feb 15, 11; m 38; c 1. MEDICINE, PREVENTIVE MEDICINE. Educ: Univ Calif, AB, 31, MD, 35. Prof Exp: Intern, San Francisco Hosp, Calif, 34-35; asst resident physician, Hosp, Univ Calif, 35-36, resident physician, 38-39, instr, Sch Med, 39-42, asst prof, 42-45, asst vis physician, 39-46; prof prev med & dir dept student health, Med Sch, Univ Wis, 46-54; med officer, Bur Commun Dis, State Dept Pub Health, Calif, 56-68, med consult, Region II, Bur Health Facil, Licensing & Cert, 68-74. Concurrent Pos: Asst resident physician, Thorndike Mem Lab & Boston City Hosp, 36-38; asst vis physician, San Francisco Hosp, Calif, 39-46; dir clin labs, Hosp, Univ Calif, 41; consult, Med Res Unit No 4, US Naval Training Ctr, Ill, 50-54 & Unit No 1 & Student Health Serv, Univ Calif, 55-58; collabr, Sect Res Prog Microbiol, USPHS, 52-54; mem comt pub health & prev med, Nat Bd Med Examr, 53-57; lectr, Med Sch, Univ Calif, 57-; Fulbright Smith-Mundt vis prof, Med Sch, Univ Tehran, 59-60; adv, AID, Thailand, 61-63; prof & actg dean pub health col, Haile Selassie Univ, Gondar, Ethiopia, 67-68; res fel, Harvard Med Sch. Mem: Fel AAAS; fel Am Pub Health Asn; fel Am Col Physicians; Am Soc Clin Invest; Col Health Asn (pres, 55-56). Res: Hematology; bacterial infections of respiratory tract and of central nervous system; infectious hepatitis; epidemiology; chemotherapy and antibiotic therapy of infections; electroencephalogram in infectious diseases; animal diseases of man; industrial health; staphylococcal infection. Mailing Add: 5779 Gloria Dr Sacramento CA 95822

BROWN, JOHN WESLEY, b St Louis, Mo, Oct 13, 33; m 63; c 4. MATHEMATICS. Educ: Univ Mo, BA, 58; Univ Calif, Los Angeles, MA, 64, PhD(math), 66. Prof Exp: Mem tech staff, Douglas Aircraft Co, Inc, 58-59, Space Technol Labs, 59-61, Packard Bell Comput Corp, 61-63 & Hughes Aircraft Co, 63-65; specialist, Bell Tel Labs, 65-66; asst prof, 66-71, ASSOC PROF MATH, UNIV ILL, URBANA-CHAMPAIGN, 66- Mem: Am Math Soc; Math Asn Am. Res: Combinatorial mathematics. Mailing Add: Dept of Math Univ of Ill Urbana IL 61801

BROWN, JOHN WESLEY, b Chicago, Ill, Dec 2, 25; m 50; c 3. BIOLOGICAL CHEMISTRY. Educ: Elmhurst Col, BS, 50; Univ Ill, MS, 53, PhD, 56. Prof Exp: Tech rep nutrit, E I du Pont de Nemours & Co, Inc, 56-57; from asst prof to assoc prof biochem, 57-68, chmn dept chem, 68-71, PROF BIOCHEM, SCH MED, UNIV LOUISVILLE, 68-, ASSOC DEAN GRAD SCH & DIR SPONSORED PROG, 71- Mem: AAAS; Sigma Xi; Am Soc Biol Chem. Res: Bacterial nitrogen metabolism; mode of action of antibiotics; animal protein and amino acid nutrition. Mailing Add: Off Sponsored Prog Univ of Louisville Belknap Campus Louisville KY 40208

BROWN, JOSEPH ROSS, b Parkville, Mo, Sept 24, 20; m 48; c 4. MATHEMATICS. Educ: Park Col, AB, 41; Univ Kans, MA, 49, PhD(math), 53. Prof Exp: Instr math, Park Col, 46-49; PROF MATH, BRADLEY UNIV, 52- Mem: Math Asn Am. Res: Non-Euclidean geometry. Mailing Add: Dept of Math Bradley Univ Peoria IL 61606

BROWN, JOSHUA ROBERT CALLAWAY, physical chemistry, see 12th edition

BROWN, JOSHUA ROBERT CALLOWAY, b Switchback, WVa, Jan 6, 15; m 44; c 1. CELL BIOLOGY. Educ: Duke Univ, AB, 48, MA, 49, PhD(zool), 51. Prof Exp: Instr zool, Duke Univ, 51-52, Army Med Serv trainee biochem, Med Sch, 52-53; from asst prof to assoc prof, 53-68, PROF ZOOL, UNIV MD, 68- Concurrent Pos: Asst dir sci teaching improv prog, AAAS, 57-58; NSF fel, Walter Reed Army Inst Res, 63-64. Mem: AAAS; Am Soc Cell Biol; Am Soc Zool; Am Micros Soc; Tissue Cult Asn. Res: Cell physiology; isolated cell components; evolution and differentiation of cells in culture; effects of environmental factors on cells in culture. Mailing Add: 6513 40th Ave Univ Park Hyattsville MD 20782

BROWN, JOSIAH, b Centerfield, Utah, Dec 19, 23; m; c 3. MEDICINE. Educ: Univ Calif, Los Angeles, BA, 44; Univ San Francisco, MD, 47. Prof Exp: Intern, Univ Calif Serv, San Francisco Hosp, 47-48; jr asst resident path, Mallory Inst Path, Boston City Hosp, Mass, 48-49; resident med, Cincinnati Gen Hosp, Ohio, 49-50, sr asst, 50-51; chief int med clin, USPHS, Washington, DC, 51-53; res anatomist, 55-56, from instr to assoc prof, 55-67, PROF MED & CHIEF DIV ENDOCRINOL & METAB, UNIV CALIF, LOS ANGELES, 67- Concurrent Pos: Clin fel, Nat Cancer Inst, Md, 53-54; fel endocrinol, New Eng Ctr Hosp, Mass, 55-56; physician, Wadsworth Vet Admin Hosp, Los Angeles, 55- Mem: Endocrine Soc; Am Fedn Clin Res; Am Soc Clin Invest. Res: Diseases of thyroid, diabetes; effects of pituitary and adrenal hormones on intermediary metabolism. Mailing Add: Sch of Med Univ of Calif Los Angeles CA 90024

BROWN, JUDITH, b Teague, Tex, July 8, 36. QUANTUM OPTICS. Educ: Rice Univ, BA, 58; Univ Calif, Berkeley, PhD(chem), 62. Prof Exp: NATO fel, 62-63; NIH fel, 63-64; from instr to asst prof, 64-72, ASSOC PROF PHYSICS, WELLESLEY COL, 72- Mem: Am Phys Soc; Am Asn Physics Teachers; Sigma Xi. Res: Light scattering from macromolecules. Mailing Add: Dept of Physics Wellesley Col Wellesley MA 02181

BROWN, JUDITH ADELE, b Providence, RI, Dec 30, 44; m 73. ANIMAL BEHAVIOR, BIOLOGICAL RHYTHMS. Educ: Whittier Col, BA, 66; Northwestern Univ, MS, 69, PhD(biol), 73. Prof Exp: Asst prof, 69-75, ASSOC PROF BIOL, CALIF STATE COL, STANISLAUS, 75- Concurrent Pos: Res assoc, Inst Cult Resources, Calif State Col, Stanislaus, 73-; field res, Calif Fish & Wildlife Agency, 75- Mem: AAAS; Animal Behav Soc; Int Audio-Tutorial Cong. Res: Field research in animal behavior and laboratory research in patterns of biorhythmicity. Mailing Add: Dept of Biol Sci Calif State Col Stanislaus Turlock CA 95380

BROWN, KARL LESLIE, b Coalville, Utah, Sept 30, 25; m 48; c 5. PARTICLE PHYSICS. Educ: Stanford Univ, BS, 47, MS, 49, PhD(physics), 53. Prof Exp: Res assoc, Stanford Univ, 53-58; consult, Linear Accelerator Lab, Ecole Normale Superieure, France, 58-59; sr res assoc, 59-74, ADJ PROF, STANFORD LINEAR ACCELERATOR CTR, 74- Concurrent Pos: Consult, Varian Assocs, 66-; vis scientist, European Orgn Nuclear Res, 72-74. Mem: Am Phys Soc. Res: Experimental physics; charged particle accelerators and optics. Mailing Add: Stanford Linear Accelerator Ctr Stanford CA 94305

BROWN, KATHLEEN OLDER, b Charleston, WVa, Oct 21, 22; m 62. ANTHROPOLOGY. Educ: Morris Harvey Col, BS, 49; Marshall Univ, MS, 52; Am Univ, PhD(educ), 70. Prof Exp: Teacher social studies, Kanawha Co Bd Educ, Charleston, WVa, 49-62; teacher pilot prog cult anthrop, Lanier Intermediate Sch, Va, 62-74; MEM RESOURCE TEAM & LECTR ANTHROP, AREA IV SCHS, FAIRFAX, VA, 74- Mem: Fel Am Anthrop Asn; fel AAAS; Fedn Am Scientists; African Studies Asn; Coun Anthrop & Educ. Res: The bushmen hunters-gatherers of Botswana. Mailing Add: 13713 Pennsboro Dr Chantilly VA 22021

BROWN, KEITH BLANCHARD, b Montgomery, Iowa, Dec 18, 20; m 44; c 1. APPLIED CHEMISTRY. Educ: Univ SDak, BS, 43. Prof Exp: Actg asst prof chem, Univ SDak, 43; chemist, Tenn Eastman Corp, Oak Ridge, 44-46; sr chemist, Carbide & Carbon Chem Co, 46-50; sci prog dir & sect chief, 50-66, ASST DIR CHEM TECH DIV, OAK RIDGE NAT LAB, 66- Honors & Awards: Mining World Tech Achievement Award, 56; Cert of Merit, Am Nuclear Soc, 65. Mem: Am Chem Soc; Sigma Xi. Res: Uranium chemistry; heavy metal separations; solvent extraction; hydrometallurgy; processing in atomic energy. Mailing Add: Oak Ridge Nat Lab PO Box X Oak Ridge TN 37830

BROWN, KEITH CHARLES, b Beverley, Eng, Dec 2, 42; m 69; c 2. PHYSICAL ORGANIC CHEMISTRY. Educ: Univ Liverpool, BSc, 64, PhD(chem), 67. Prof Exp: Fel, Univ Rochester, 67-69; res officer, Elec Coun Res Ctr, Capenhurst, Eng, 69-73; scientist, Environ Impact Ctr Inc, Mass, 73-74; RES SCIENTIST, CLAIROL INC, DIV BRISTOL-MYERS CO, 74- Mem: The Chem Soc. Res: Kinetics and mechanisms of reactions of organic compounds, particularly compounds important in cosmetic chemistry. Mailing Add: Clairol Inc 2 Blachley Rd Stamford CT 06902

BROWN, KEITH H, b Salt Lake City, Utah, Sept 3, 39; m 63; c 3. PLASMA PHYSICS, HIGH PRESSURE PHYSICS. Educ: Brigham Young Univ, BA, 64, PhD(physics), 69; Univ Ill, Urbana-Champaign, MS, 65. Prof Exp: Asst prof, 68-72, ASSOC PROF PHYSICS, CALIF STATE POLYTECH UNIV, POMONA, 72- Concurrent Pos: Fel, Brigham Young Univ, 74-75. Mem: Am Phys Soc; Am Asn Physics Teachers. Res: Magnetic confinement of plasmas; x-ray diffraction of liquid metals at high pressures. Mailing Add: Dept of Physics Calif State Polytech Univ Pomona CA 91768

BROWN, KEITH IRWIN, b Hunter, Kans, Sept 28, 25; m; c 4. POULTRY SCIENCE, REPRODUCTIVE PHYSIOLOGY. Educ: Kans State Col, BS, 49, MS, 50; Univ Wis, PhD(zool), 56. Prof Exp: Asst zool, Kans State Col, 48-49 & Univ Wis, 50-56; from instr to asst prof physiol, Okla State Univ, 55-57; from asst prof to assoc prof poultry physiol, 57-66, PROF POULTRY PHYSIOL & ASSOC CHMN DEPT, OHIO AGR RES & DEVELOP CTR, 66- Mem: Poultry Sci Asn; Soc Study Reprod; AAAS. Res: Steroid hormones in bird and their role in egg production; physiology of stress and reproduction in domestic fowl; semen physiology. Mailing Add: Dept of Poultry Sci Ohio Agr & Res Develop Ctr Wooster OH 44691

BROWN, KENNETH E, b Guthrie, Okla, June 16, 04; m 49; c 1. MATHEMATICS. Educ: Okla State Col, BS, 35; Colo Col Educ, MA, 37; Columbia Univ, PhD(math), 43. Prof Exp: Head dept math & asst supt sch, Okla, 36-40; asst math, Columbia Univ, 40-41; instr, Adelphi Col, 41; head dept, NJ State Teachers Col, 41-42, E Carolina Teachers Col, 45-47 & Wagner Col, 47-48; assoc prof, Univ Tenn, 48-52; SPECIALIST MATH, US OFF EDUC, 52- Concurrent Pos: US del, Int Bur Educ Conf, UNESCO, Geneva, 57. Mem: AAAS; Math Asn Am; Nat Coun Teachers Math. Mailing Add: 9201 Davidson St College Park MD 20740

BROWN, KENNETH HENRY, physical chemistry, see 12th edition

BROWN, KENNETH HOWARD, b Chicago, Ill, Nov 21, 42; m 69; c 1. ORGANIC POLYMER CHEMISTRY. Educ: Univ Ill, BS, 65; Univ Minn, MS, 69; Loyola Univ Chicago, PhD(org chem), 71. Prof Exp: SR RES SCIENTIST ORG & POLYMER CHEM, CONTINENTAL CAN CO, 70- Concurrent Pos: Asst prof chem, eve div, Elmhurst Col, 75- Mem: Sigma Xi; Am Chem Soc. Res: Photopolymerization process of coatings; synthesis of photosensitizers and monomers for polymeric systems; scale-up of organic reactions. Mailing Add: Continental Can Co 1200 W 76th St Chicago IL 60620

BROWN, KENNETH TAYLOR, b Purcellville, Va, Apr 7, 22; m 48; c 2. PHYSIOLOGY. Educ: Swarthmore Col, BA, 47; Univ Chicago, MS, 49, PhD(psychol), 51. Prof Exp: Res psychologist, Aero-Med Lab, Wright Air Develop Ctr, 50-54; res assoc visual physiol, Brown Univ, 54-55; from instr to asst prof physiol optics, Wilmer Inst, Med Sch, Johns Hopkins Univ, 55-58; from asst prof to assoc prof, 58-66, PROF PHYSIOL, MED SCH, UNIV CALIF, SAN FRANCISCO, 66- Concurrent Pos: NSF fel, Brown Univ, 54-55; USPHS spec res fel, Wilmer Inst, Med Sch, Johns Hopkins Univ, 57-58; Nat Eye Inst res grant, Univ Calif, San Francisco, 59-; Commonwealth Fund fel, John Curtin Sch Med Res, Australian Nat Univ, 64-65; fac res lectr, Univ Calif, San Francisco, 69. Mem: Fel AAAS; Am Physiol Soc; Soc Neurosci; Asn Res Vision & Ophthal; Int Brain Res Orgn. Res: Neurophysiology of the vertebrate retina, with special reference to the photoreceptors; microelectrode techniques. Mailing Add: Dept of Physiol Univ Calif Sch of Med San Francisco CA 94143

BROWN, L CARLTON, b Mineral Springs, Ark, Mar 26, 15; m 38; c 1. PHYSICS. Educ: Henderson State Teachers Col, AB, 37; Fla State Univ, MA, 52; Ohio State Univ, PhD(physics), 55. Prof Exp: Instr math, Pulaski County Schs, Ark, 38-39; eng aide, Ark State Hwy Dept, 39-41 & US Army Corps Engrs, 41-43 & 46-47; training officer, US Vet Admin, 47-48; asst engr, Ark State Hwy Dept, 48-51; asst physics, Fla State Univ, 51-52; asst, 52-55, res assoc magnetic resonance, 55-57, from asst prof to assoc prof physics, 57-65, PROF PHYSICS, OHIO STATE UNIV, 65- Concurrent Pos: Co-supvr & chief investr, Ohio State Univ res grant, Off Naval Res & Air Force Off Sci Res, 55-62. Mem: Am Phys Soc; Am Asn Physics Teachers. Res: Nuclear magnetic resonance; electron paramagnetic resonance; x-ray and gamma ray scattering; electromagnetic theory. Mailing Add: Dept of Physics Ohio State Univ Columbus OH 43210

BROWN, LARRY CLYDE, radiochemistry, nuclear chemistry, see 12th edition

BROWN, LARRY NELSON, b Springfield, Mo, Dec 9, 37; m 58; c 3. MAMMALOGY, ECOLOGY. Educ: Southwest Mo State Col, BS, 58; Univ Mo, MA, 60, PhD(zool), 63. Prof Exp: NIH res fel, Univ BC & Ore State Univ, 63; asst prof zool, Univ Wyo, 64-67; ASSOC PROF ZOOL, DIR ECOL RES AREA & CUR MAMMALS, UNIV S FLA, 67- Concurrent Pos: Ecol consult, Fla State Agencies; vis prof biol, Calif Polytech State Univ, 75-76. Mem: Soc Study Reproduction; AAAS; Am Soc Mammal; Wildlife Soc; Am Soc Zool. Res: Studies of reproductive biology, population dynamics and behavior of mammals and insects. Mailing Add: Dept of Biol Univ of S Fla Tampa FL 33620

BROWN, LAUREN EVANS, b Waukesha, Wis, Sept 4, 39; m 68; c 2. ZOOLOGY. Educ: Carroll Col, Wis, BS, 61; Southern Ill Univ, MS, 63; Univ Tex, PhD(zool), 67. Prof Exp: Asst prof, 67-71, ASSOC PROF VERT ZOOL, ILL STATE UNIV, 71- Mem: Soc Study Evolution; Am Soc Ichthyologists & Herpetologists; Soc Study Amphibians & Reptiles; Australian Soc Herpetol. Res: Evolution; ecology; behavior; amphibian speciation; bioacoustics; herpetology; ecology of flood plains and prairies. Mailing Add: Dept of Biol Sci Ill State Univ Normal IL 61761

BROWN, LAURIE MARK, b Brooklyn, NY, Apr 10, 23; m 69; c 2. THEORETICAL PHYSICS. Educ: Cornell Univ, AB, 43, PhD(theoret physics), 51. Prof Exp: Res assoc, sam labs, Columbia Univ, 43-44; asst physics, Cornell Univ, 46-48, asst, Nuclear Studies Lab, 48-50; from instr to assoc prof, 50-61, PROF PHYSICS, NORTHWESTERN UNIV, EVANSTON, 61- Concurrent Pos: NSF fel, Inst Advan Study, 52-53; Fulbright res scholar, Italy, 58-60; consult, Argonne Nat Lab, 60-70; Int Atomic Energy Agency prof, Univ Vienna, 66; vis prof, Univ Rome, 67 & Univ Sao Paulo, 72-73; mem, NSF Panel Advan Sci Sem, 68, 69. Mem: Fel AAAS; fel Am Phys Soc. Res: Theoretical high energy physics; quantum electrodynamics. Mailing Add: Dept of Physics Northwestern Univ Evanston IL 60201

BROWN, LAWRENCE ALAN, b Erie, Pa, Nov 22, 35. URBAN GEOGRAPHY, POPULATION GEOGRAPHY. Educ: Univ Pa, BS, 58; Northwestern Univ, MA, 63, PhD(geog), 66. Prof Exp: Asst prof geog, Univ Iowa, 65-68; assoc prof, 68-71, PROF GEOG, OHIO STATE UNIV, 71- Mem: Asn Am Geog; Pop Asn Am; Am Sociol Asn; Regional Sci Asn. Res: Migration; diffusion and adoption of innovation; communication; regionalization. Mailing Add: Dept of Geog Ohio State Univ Columbus OH 43210

BROWN, LAWRENCE DAVID, mathematics, mathematical statistics, see 12th edition

BROWN, LAWRENCE ELDON, b Republic Co, Kans, June 19, 14; m 39, 56; c 1. MICROCHEMISTRY. Educ: Southwest Tex State Col, AB, 36. Prof Exp: Asst chemist in charge microanal lab, Southern Mkt & Nutrit Res Div, USDA, 42-50, chemist in charge, 50-75; RETIRED. Mem: AAAS; Am Chem Soc; Am Oil Chem Soc; fel Am Inst Chem; Sigma Xi. Res: Microchemical analysis; utilization research on oilseeds, meals and oils. Mailing Add: 805 Andrews Ave Metairie LA 70005

BROWN, LAWRENCE MILTON, b Boston, Mass, Oct 29, 27; m 51; c 2. PETROLEUM CHEMISTRY. Educ: Harvard Col, BS, 48; Howard Univ, MS, 50; Cath Univ Am, PhD(phys chem), 64. Prof Exp: Instr chem, Southern Univ, 50-51; phys chemist, Nat Bur Stand, 51-64; SR RES CHEMIST, CENT RES LAB, MOBIL RES & DEVELOP CO, INC, 64- Mem: AAAS; Am Phys Soc. Mailing Add: Cent Res Lab PO Box 1025 Mobil Res & Develop Co Inc Princeton NJ 08540

BROWN, LEE ROY, JR, b Warren, Ohio, Aug 20, 26; m 46; c 3. MEDICAL MICROBIOLOGY. Educ: Univ Ala, BS, 50; George Washington Univ, MS, 56, PhD(microbiol), 61. Prof Exp: Bacteriologist, Naval Med Field Res Lab, Camp Le Jeune, NC, 50-53; bacteriologist, Naval Dent Sch, Bethesda, Md, 53-61; asst prof microbiol, Sch Dent, Univ Mo-Kansas City, 61-62, assoc prof & chmn dept, 62-68; MEM & PROF, INST DENT SCI, UNIV TEX, HOUSTON, 68- Concurrent Pos: Consult, Res Div, Vick Chem Co, NY, 63-73. Honors & Awards: Civil Serv Superior Accomplishment Awards, US Dept Navy, 59 & 61. Mem: Am Soc Microbiol; Int Asn Dent Res. Res: Microbiological, immunological and chemical aspects of the etiology, prevention and treatment of diseases with manifestations in the oral cavity. Mailing Add: Inst of Dent Sci Univ of Tex Houston TX 77004

BROWN, LELAND ARTHUR, b Marion, Ohio, Nov 18, 97; m 23; c 2. ZOOLOGY. Educ: Denison Univ, BS, 22; Univ Pittsburgh, MA, 24; Harvard Univ, PhD(zool, gen physiol), 27. Hon Degrees: LLD, Transylvania Col, 56. Prof Exp: Asst genetics, Carnegie Inst Washington, 19-21; asst, Univ Pittsburgh, 21-23, instr zool, 23-25; Austin fel, Harvard Univ, 25-27; asst prof, Univ Iowa, 27-29; from assoc prof to prof biol, George Washington Univ, 29-72; dean col, 41-64, actg pres, 43-45, vpres acad affairs, 59-65, cur mus, 65-72, DISTINGUISHED SERV PROF BIOL, TRANSYLVANIA UNIV, 72- Mem: Am Soc Zoologists. Res: Sex control and distribution in Cladocera; early scientific apparatus. Mailing Add: Dept of Biol Transylvania Univ Lexington KY 40508

BROWN, LELAND RALPH, b Springdale, Ark, Mar 7, 22; m 49; c 3. ENTOMOLOGY. Educ: Univ Calif, BS, 46; Cornell Univ, 49. Prof Exp: PROF ENTOM & ENTOMOLOGIST, UNIV CALIF, RIVERSIDE, 69- Mem: AAAS; Entom Soc Am. Res: Insecticide applicators; insects affecting ornamental shrubs, shade trees and forests; insect photography; general entomology. Mailing Add: Dept of Entom Univ of Calif Riverside CA 92502

BROWN, LEON JOSEPH, b New York, NY, Nov 30, 19; m 47; c 4. CHEMISTRY. Educ: Ohio State Univ, BSc, 40; Ind Univ, MS, 42. Prof Exp: Asst chem, Ind Univ, 41-42, jr scientist, Dept Physics, Off Sci Res & Develop, 42-44; chemist, Metall Lab, Univ Chicago, 44; CHEMIST-PHYSICIST, LOS ALAMOS SCI LAB, UNIV CALIF, 44- Mem: AAAS; Am Chem Soc. Res: Radiochemistry; nuclear physics. Mailing Add: Los Alamos Sci Lab Box 1663 Los Alamos NM 87545

BROWN, LEONARD D, b Bremen, Ky, Oct 15, 30; m 50; c 2. ANIMAL NUTRITION, PHYSIOLOGY. Educ: Western Ky Univ, BS, 54; Univ Ky, MS, 55; Mich State Univ, PhD(nutrit, physiol), 61. Prof Exp: Instr animal nutrit, Univ Ky, 55-58; from instr to assoc prof, 58-66; PROF ANIMAL NUTRIT, WESTERN KY UNIV, 66- Mem: Am Dairy Sci Asn; Am Soc Animal Sci. Res: Silage fermentation; problems associated with pesticide residues in plant and animal products. Mailing Add: 904 Ridgecrest Dr Bowling Green KY 42101

BROWN, LEONARD FRANKLIN, JR, b Seminole, Okla, June 1, 28; m 56; c 1. GEOLOGY. Educ: Baylor Univ, BS, 51; Univ Wis, MS, 53, PhD(geol), 55. Prof Exp: Explor geologist, Stand Oil Co Tex, 55-57; res scientist, Bur Econ Geol, Univ Tex, 57-60; from asst prof to assoc prof geol, Baylor Univ, 60-66; res scientist, Bur Econ Geol & lectr geol, Univ, 66-71, ASSOC DIR RES, BUR ECON GEOL & PROF GEOL SCI, UNIV TEX, AUSTIN, 71- Concurrent Pos: Ed, Baylor Geol Studies Bull, 61-66. Mem: Am Asn Petrol Geol; Geol Soc Am. Res: Stratigraphic studies of the Pennsylvanian system; stratigraphy, environments and paleontology of rocks in Texas; terrigenous depositional systems; environmental geology in Texas; facies analysis; seismic-stratigraphy. Mailing Add: Univ of Tex Bur of Econ Geol Box X Univ Sta Austin TX 78712

BROWN, LEONARD KEITH, b Ames, Iowa, Feb 21, 33; m 55; c 3. ANTHROPOLOGY, ETHNOLOGY. Educ: Iowa State Univ, BS, 55; Univ Chicago, MA, 61, PhD(anthrop), 64. Prof Exp: Asst prof anthrop, Univ Nebr, 64-65; A W Mellon fel, 65-66, asst prof, 66-69, ASSOC PROF ANTHROP, UNIV PITTSBURGH, 69-, CHMN DEPT, 73- Mem: Am Anthrop Asn; Am Ethnol Soc; Soc Appl Anthrop; Asn Asian Studies; Royal Anthrop Inst. Res: Family, kinship and neighborhood adaptations to urbanization and modernization in Japan; family, kinship, community organization and demographic change in a Japanese town. Mailing Add: Dept of Anthrop Univ of Pittsburgh Pittsburgh PA 15260

BROWN, LEWIS RAYMOND, b Houston, Tex, Aug 11, 30; m 51; c 4. BACTERIOLOGY. Educ: La State Univ, BS, 51, MS, 53, PhD(bact), 58. Prof Exp: Asst br chief, US Civil Serv, Pine Bluff Arsenal, 53-55; instr bact, La State Univ, 55-58; operator, pvt consult & testing lab, 58-61; PROF MICROBIOL, MISS STATE UNIV, 61-, ASSOC DEAN RES, COL ARTS & SCI, 71- Mem: AAAS; Am Soc Microbiol; Am Acad Microbiologists; Int Water Resources Asn; Sigma Xi. Res: Biodeterioration of metals; microbial ecology; geomicrobiological prospecting; sewage and industrial waste disposal; petroleum degradation and water pollution control; isolation, characterization and metabolism of hydrocarbon oxidizing bacteria. Mailing Add: PO Drawer CU Mississippi State MS 39762

BROWN, LINDSAY DIETRICH, b Lynchburg, Va, Jan 14, 29; m 52; c 2. HORTICULTURE, PLANT PHYSIOLOGY. Educ: Lynchburg Col, BA, 51; Va Polytech Inst, MS, 58; Mich State Univ, PhD(hort), 62. Prof Exp: Jr instr biol, Johns Hopkins Univ, 55-56; instr hort, Va Polytech Inst, 57-58; instr, Mich State Univ, 61-63; asst prof, Univ Ky, 63-67; supvr agron serv, Southwest Potash Corp, 67-69; assoc prof hort, Va Polytech Inst & State Univ, 69-70; dir mkt serv, Amax Chem Corp, 70-72; DIR SPEC PROD MKT, INT MINERAL & CHEM CO, 72- Concurrent Pos: Consult, US AID, SE Asia, 66. Mem: Am Soc Hort Sci; Am Soc Agron. Res: Plant breeding; mineral nutrition of higher plants; physiology of greenhouse-grown plants. Mailing Add: Int Minerals & Chem Corp IMC Plaza Libertyville IL 60048

BROWN, LLOYD H, b Grants Pass, Ore, Mar 28, 12; m 50; c 5. INDUSTRIAL ORGANIC CHEMISTRY. Educ: Univ Ore, BS, 32. Prof Exp: Process engr adhesives, Boeing Aircraft Co, 41-45; res assoc resins, Arthur J Norton Labs, 45-49;

res chemist, Chem Div, Quaker Oats Co, 49-62, sect leader resins, Res Dept, 62-70, mgr mat appln, 70-71, asst dir chem res & develop, 71-75; RETIRED. Mem: Am Chem Soc. Res: Applications research on furans, chiefly in resin field. Mailing Add: 75 Victor Pkwy Crystal Lake IL 60014

BROWN, LLOYD LEONARD, b Topeka, Kans, Jan 20, 27; m 54; c 2. PHYSICAL CHEMISTRY. Educ: Wichita State Univ, BS, 50, MS, 52. Prof Exp: Chemist, Wichita Found Indust Res, 52; chemist, Oak Ridge Sch Reactor Technol, 52-53, CHEMIST, OAK RIDGE NAT LAB, 53- Mem: Am Chem Soc. Res: Chemistry of stable isotopes; chemical methods for enrichment of rare isotopes; kinetic effects; nuclear magnetic resonance spectroscopy. Mailing Add: Chem Div Oak Ridge Nat Lab PO Box X Oak Ridge TN 37830

BROWN, LORETTA ANN PORT, b Kingston, NY, July 30, 45; m 70. BIOCHEMICAL GENETICS, ENDOCRINOLOGY. Educ: State Univ NY Col New Paltz, BS, 67; Univ Mich, MS, 68. Prof Exp: NSF res partic, Albion Col, 66; USPHS trainee, Univ Mich, 67-69; lab technologist, Univ Mich, 69-70; res instr biol-med genetics, M D Anderson Hosp & Tumor Inst, 70; RES INSTR OBSTET & GYNEC, BAYLOR COL MED, 71- Mem: AAAS; Genetics Soc Am; Am Inst Chemists. Res: Suppressor mutations of the methionine-1 locus in Neurospora crassa; arginine and polyamine metabolism and control in Neurospora crassa; human protein hormones in endocrine problems. Mailing Add: Dept of Obstet & Gynec Baylor Col of Med Houston TX 77025

BROWN, LOUIS, b San Angelo, Tex, Jan 7, 29; m 52. NUCLEAR PHYSICS. Educ: St Mary's Univ, Tex, BS, 50; Univ Tex, PhD(physics), 58. Prof Exp: Res asst physics, Inst Physics, Univ Basel, 58-61; Carnegie fel, 61-64, MEM PHYSICS STAFF, CARNEGIE INST WASHINGTON, DEPT TERRESTRIAL MAGNETISM, 64- Honors & Awards: Amerbach Prize, Univ Basel, 64. Mem: Fel Am Phys Soc. Res: Experimental study of the interaction of polarized protons having energies up to three mega electron volts with the light nuclei; experimental study of x-rays produced in collisions of heavy ions with atoms. Mailing Add: Dept of Terrestrial Magnetism Carnegie Inst of Washington Washington DC 20015

BROWN, LOUIS MILTON, physical chemistry, see 12th edition

BROWN, LOWELL SEVERT, b Visalia, Calif, Feb 15, 34; m 56; c 1. PHYSICS. Educ: Univ Calif, Berkeley, AB, 56; Harvard Univ, PhD(physics), 61. Prof Exp: NSF fel physics, Univ Rome, 61-62 & Imp Col, Univ London, 62-63; res assoc, Yale Univ, 63-64, from asst prof to assoc prof, 64-68; assoc prof, 68-71, PROF PHYSICS, UNIV WASH, 71- Concurrent Pos: NSF sr fel, Imp Col, Univ London, 71-72. Mem: Fel Am Phys Soc; AAAS. Res: Theoretical physics; quantum field theory; elementary particle physics. Mailing Add: Dept of Physics Univ of Wash Seattle WA 98195

BROWN, LYNN RANNEY, b Waterloo, Iowa, Sept 26, 28; m 52; c 4. ANIMAL NUTRITION. Educ: Iowa State Univ, BS, 50, PhD(dairy nutrit), 59. Prof Exp: Teacher, Pub Schs, Iowa, 50-51 & 53-55; asst dairy nutrit, Iowa State Univ, 56-58, assoc, 58-59; from asst prof to assoc prof dairy exten, 60-74, PROF DAIRY EXTEN, UNIV CONN, 74- Mem: Am Dairy Sci Asn; Am Soc Animal Sci. Res: Physiological factors associated with bloat; nutrition of dairy calves and cows. Mailing Add: Dept of Animal Indust Univ of Conn Storrs CT 06268

BROWN, MARIANNE, b Philadelphia, Pa, Apr 25, 37; c 2. TOPOLOGY. Educ: Univ Pa, AB, 58, MA, 60; Mass Inst Technol, PhD(math), 71. Prof Exp: Instr math, Hanover High Sch, 68-69 & Canaan Col, 69-70; ASST PROF MATH, DARTMOUTH COL, 71- Mem: Am Math Soc; Math Asn Am. Res: Three-dimensional manifolds. Mailing Add: Dartmouth Col Hanover NH 03755

BROWN, MARK, mathematical statistics, operations research, see 12th edition

BROWN, MARK, b Miami, Fla, Dec 31, 25; m 53; c 3. RADIOLOGY. Educ: Univ Miami, BS, 46; Vanderbilt Univ, MD, 50. Prof Exp: Instr radiol, Sch Med, Wash Univ, 59-60; from asst prof to assoc prof radiol, 60-63, prof & chmn dept, 63-75, CHIEF NUCLEAR MED, MED COL GA, 75- Concurrent Pos: Consult, Augusta Vet Admin Hosp, Eisenhower Army Hosp & Ga Radiation Adv Coun. Mem: AMA; Radiol Soc NAm; fel Am Col Radiol; fel Am Col Nuclear Med. Res: Clinical radiology; medical education and research. Mailing Add: Dept Nuclear Med Med Col of Ga Augusta GA 30902

BROWN, MARVIN ROSS, b Douglas, Ariz, July 25, 47; m 63; c 1. MEDICAL PHYSIOLOGY, INTERNAL MEDICINE. Educ: Univ Ariz, BS, 69, MD, 73. Prof Exp: Fel, 74-75, RES ASSOC ENDOCRINOL, SALK INST BIOL STUDIES, 75- Mem: Am Diabetes Asn. Res: Study of the physiology and pharmacology of brain hormones. Mailing Add: Salk Inst for Biol Studies La Jolla CA 92037

BROWN, MAURICE VERTNER, b Durand, Mich, Jan 31, 08; m 34, 69; c 3. ACOUSTICS, GEOPHYSICS. Educ: Univ Mich, BS, 30, MS, 31; NY Univ, PhD(physics), 37. Prof Exp: Lab asst physics, Univ Mich, 30-31; from instr to prof, 31-74, EMER PROF PHYSICS, CITY COL NEW YORK, 74- Concurrent Pos: Instr, Pa State Col, 41; res physicist, Hudson Labs, Columbia Univ, 52-69; expert in acoust scattering, Naval Res Lab, Washington, DC, 69- Mem: Acoust Soc Am; Am Phys Soc. Res: Artificial radioactivity; biophysics of electric fish. Mailing Add: Dept of Physics City Col of New York New York NY 10031

BROWN, MERTON F, b Burlington, Vt, Sept 22, 35; m 57; c 3. MYCOLOGY, FOREST PATHOLOGY. Educ: Univ Maine, BS, 61, MS, 63; Univ Iowa, PhD(bot), 66. Prof Exp: Asst prof biol, Wis State Univ, Superior, 66-68; asst prof, 68-72, ASSOC PROF FORESTRY & PLANT PATH, UNIV MO-COLUMBIA, 72- Mem: Mycol Soc Am; Am Phytopath Soc; Bot Soc Am. Res: Ultrastructure and composition of fungal cell walls, plant pathogenic fungi and mycorrihizae. Mailing Add: Dept of Plant Path Univ of Mo Columbia MO 65201

BROWN, META (LOUISE) SUCHE, b San Antonio, Tex, Nov 21, 08; m 38; c 1. CYTOGENETICS. Educ: Univ Tex, AB, 31, A M, 33, PhD(genetics), 35. Prof Exp: Tutor zool, Univ Tex, 31-32, asst, 33-34, res assoc, 35-38; cytologist, Agr Exp Sta, 41-48, from assoc prof to prof, 48-74, EMER PROF AGRON, TEX A&M UNIV, 74- Honors & Awards: Fac Distinguished Achievement Award, 64; Cotton Genetics Award, 64. Mem: Genetics Soc Am; Bot Soc Am; Soc Study Evolution; Am Soc Naturalists; Genetics Soc Can. Res: Effects of radiation; cytogenetics of species hybrids of Gossypium; chromosome identification. Mailing Add: 700 Gilchrist College Station TX 77840

BROWN, MEYER, b Chicago, Ill, Oct 16, 10; m 34; c 3. NEUROLOGY, PSYCHIATRY. Educ: Univ Chicago, BS, 31, MD, 35; Northwestern Univ, MS, 37, PhD(nervous & ment dis), 39; Am Bd Psychiat & Neurol, dipl, 41. Prof Exp: Staff physician, Behav Clin, Chicago, 35-41; attend neuropsychiatrist, Student Health Serv, 39-61, assoc, 40-48, from asst prof to assoc prof neurol & psychiat, 48-73, PROF CLIN NEUROL & CLIN PSYCHIAT, MED SCH, NORTHWESTERN UNIV, CHICAGO, 73- Concurrent Pos: Attend physician, Evanston Hosp, 47-62, head div neurol & psychiat, 62-69, head div neurol, 69-75. Mem: Fel AMA; Am Psychiat Asn; Am Acad Neurol. Res: Epilepsy; sensation of vibration; constitutional differences between deteriorated and non-deteriorated epileptics; spinal cord injuries; stroke. Mailing Add: 1500 Sheridan Rd Apt 7J Wilmette IL 60091

BROWN, MICHAEL, b Philadelphia, Pa, Jan 8, 39; m 63; c 2. INDUSTRIAL PHARMACY. Educ: Temple Univ, BS, 60; Purdue Univ, MS, 62, PhD(phys pharm), 64. Prof Exp: Develop chemist pharm, Lederle Lab, Div Am Cyanamid, 64-67; group leader, Carter Prod, Div Carter-Wallace, Inc, 67-74; MGR PROPRIETARIES & TOILETRIES RES & DEVELOP, PLOUGH, INC, DIV SCHERING-PLOUGH, 74- Mem: Acad Pharmaceut Sci; Am Pharmaceut Asn; Sigma Xi. Res: Product research and development of proprietary, toiletries products; physical pharmacy. Mailing Add: Plough Inc 3030 Jackson Ave Memphis TN 38151

BROWN, MICHAEL MATHISON, b Palmyra, Pa, Dec 6, 40; m 63; c 2. PLANT PHYSIOLOGY, CYTOLOGY. Educ: Lebanon Valley Col, BS, 62; Univ Del, PhD(plant physiol), 69. Prof Exp: ASST PROF BIOL, WESTERN MD COL, 68- Mem: Am Soc Plant Physiol. Res: Chemistry and physiology of plant cell walls. Mailing Add: Dept of Biol Western Md Col Westminster MD 21157

BROWN, MILTON HERBERT, b Havelock, Ont, Sept 6, 98; m 50; c 2. MICROBIOLOGY. Educ: Univ Toronto, MB, 24, MD, 28, DPH, 39. Prof Exp: Asst dir, Connaught Med Res Lab, 52-69, prof pub health & head dept & assoc dir, Sch Hyg, 55-67, EMER PROF PUB HEALTH & SPEC LECTR, SCH HYG, UNIV TORONTO, 67-, EMER ASST DIR, CONNAUGHT MED RES LAB, 69- Concurrent Pos: Consult prev med, Can Forces Med Coun & to Dir Gen of Med Serv Order, Brit Empire, 43; med consult, Toronto Home Care Prog, 67- Mem: Am Asn Immunol; Am Pub Health Asn; fel Am Col Chest Physicians; Can Pub Health Asn; fel Royal Soc Med. Res: Immunological response to antigens used in human immunization. Mailing Add: 42 McRae Dr Toronto ON Can

BROWN, MORDEN GRANT, b Palo Alto, Calif, Dec 8, 06; m 34; c 2. BIOPHYSICS. Educ: Stanford Univ, AB, 29, AM, 36, PhD(biol), 38. Prof Exp: Asst zool, Stanford Univ, 34-37; res assoc, Wash Univ, 38-41, lectr, Univ Col, 39-41; res physicist instrument design, Spencer Lens Co, 41-45; chief develop physicist, Am Optical Co, 45-49, mgr develop & eng sect, 47-49, res supvr, 49-53, supvr res ctr, 53-72; RETIRED. Mem: Optical Soc Am; Am Soc Zool; Am Inst Biol Sci; Sigma Xi. Res: Scientific instrument design; experimental embryology; general physiology. Mailing Add: PO Box 2521 Laguna Hills CA 92653

BROWN, MORTON, b New York, NY, May 7, 31; m 57; c 3. POLYMER CHEMISTRY. Educ: Cornell Univ, AB, 52; Duke Univ, AM, 54; Mass Inst Technol, PhD(org chem), 57. Prof Exp: Fel, Mass Inst Technol, 57; res chemist, Cent Res Dept, 57-62, RES CHEMIST, ELASTOMER CHEM DEPT, E I DU PONT DE NEMOURS & CO, INC, 62- Mem: Am Chem Soc. Res: Fluorinated organo-sulfur compounds; cyano-carbons; polyurethanes; interfacial polymerizations; thermoplastic elastomers; effect of polymer melt rheology on processing characteristics. Mailing Add: E I du Pont de Nemours & Co Chestnut Run Wilmington DE 19898

BROWN, MORTON, b New York, NY, Aug 12, 31. MATHEMATICS. Educ: Univ Wis-Madison, BS, 53, MS, 55, PhD(math), 58. Prof Exp: Instr math, Univ Wis-Madison, 57 & Ohio State Univ, 57-58; Off Naval Res fel, 58-59, from asst prof to assoc prof, 59-64, PROF MATH, UNIV MICH, ANN ARBOR, 64- Concurrent Pos: NSF fel, Inst Advan Study, Princeton Univ, 60-62; Sloan Found fel, Cambridge Univ & Univ Mich, 63-65. Honors & Awards: Veblen Prize, Am Math Soc, 64. Mem: Am Math Soc; Math Asn Am. Res: Topology. Mailing Add: Dept of Math Univ of Mich Ann Arbor MI 48104

BROWN, MURRAY ALLISON, b Rochester, NY, Oct 29, 27; m 48; c 1. DAIRY SCIENCE. Educ: Mich State Univ, BS, 50; Tex A&M Univ, MS, 53, PhD(animal breeding), 56. Prof Exp: Instr dairy prod, Tex A&M Univ, 53 & 55-57, from asst prof to assoc prof dairy sci, 57-65; PROF AGR, SAM HOUSTON STATE UNIV, 65- Concurrent Pos: Animal sci consult. Mem: AAAS; Am Dairy Sci Asn; Nat Asn Col & Teachers Agr (pres, 69-70, secy-treas, 72-). Res: Bovine reproductive physiology; artificial insemination; dairy cattle breeding. Mailing Add: Sam Houston State Univ Dept Agr PO Box 2088 Huntsville TX 77340

BROWN, MYRTLE LAURESTINE, b Columbia, SC, June 1, 26. NUTRITION. Educ: Bennett Col, BS, 45; Pa State Univ, MS, 48, PhD(nutrit), 57. Prof Exp: Nutrit specialist, Human Nutrit Res Div, Agr Res Serv, USDA, 48-63; assoc prof home econ, Univ Hawaii, 63-65, nutrit, 65-66, pub health, 66-69; assoc prof nutrit, Rutgers Univ, 69-74; STAFF OFFICER, FOOD & NUTRIT BD, NAT ACAD SCI, 74- Mem: Am Dietetic Asn; Am Home Econ Asn; Am Pub Health Asn; Am Inst Nutrit. Res: Nutrition and reproduction. Mailing Add: Food & Nutrit Bd NAS 2101 Constitution Ave NW Washington DC 20418

BROWN, NANCY J, b Erie, Pa, Apr 19, 43; m 74. THEORETICAL CHEMISTRY. Educ: Va Polytech Inst, BS, 64; Univ Md, MS, 69, PhD(phys chem), 71. Prof Exp: Fel, Appl Phys Lab, Johns Hopkins Univ, 71-73; vis scholar combustion, 73, lectr chem, 73-75, RES ENGR COMBUSTION, UNIV CALIF, BERKELEY, 75- Concurrent Pos: Adv comt, Calif State Energy Resources Conserv & Develop Comn, 75- Mem: Am Chem Soc; Am Phys Soc; Combustion Inst. Res: Intermolecular forces; inelastic and reactive collisions; flame inhibition; fuel nitrogen chemistry and hydrocarbon combustion chemistry. Mailing Add: Dept of Mech Eng Univ of Calif Berkeley CA 94720

BROWN, NEAL CURTIS, b Vassalboro, Maine, Mar 15, 39; m 64; c 2. BIOCHEMISTRY, PHARMACOLOGY. Educ: Cornell Univ, DVM, 62; Yale Univ, PhD(biochem pharmacol), 66. Prof Exp: USPHS fel, Dept Chem, Karolinska Inst, Sweden, 66-68, Swed Cancer Soc fel, 68-69; asst prof, 69-73, ASSOC PROF CELL BIOL & PHARMACOL, SCH MED, UNIV MD, 73- Concurrent Pos: Assoc prof biochem, Med Sch, Univ Mass, 73-74, prof & chmn dept pharmacol, 74- Res: Metabolism of DNA and DNA precursors and its control in vitro and in vivo in prokaryotic and eukaryotic cells. Mailing Add: Dept of Cell Biol & Pharmacol Univ of Md Sch of Med Baltimore MD 21201

BROWN, NORMAN LOUIS, b Atlantic City, NJ, Dec 22, 23; m 57; c 3. PHYSICAL CHEMISTRY. Educ: Mass Inst Technol, SB, 47; Brown Univ, PhD(phys chem), 52. Prof Exp: Phys chemist, Electronics Lab, Gen Elec Co, 51-56; physicist, Nat Bur Standards, 57-63 & Nat Ctr Fish Protein Concentrate, Bur Commercial Fisheries, US Dept Interior, 63-67; prog leader, Fish Protein Concentrate Res Prog, 67-70; prof assoc, Bd Sci & Technol for Int Develop, Off Foreign Secy, Nat Acad Sci, 70-75; COUNTRY PROG SPECIALIST, OFF ASST ADMINR INT AFFAIRS, ENERGY RES & DEVELOP ADMIN, 75- Concurrent Pos: Consult, US Agency Int Develop, Pakistan, 70. Mem: AAAS; Inst Food Technol; Soc Int Develop. Res: Manufacture and use of fish and other protein concentrates; applications of science and technology to problems of development; small-scale energy technologies. Mailing Add: 1438 Iris St NW Washington DC 20012

BROWN, NORMAN RAE, b London, Ont, May 5, 19; m 43; c 1. FOREST ENTOMOLOGY. Educ: Univ Western Ont, BA, 41, MA, 43. Prof Exp: Agr scientist, Forest Biol Div, Res Br, Can Dept Agr, 43-46; PROF FOREST ENTOM, UNIV NB, 46- Mem: AAAS; Wilson Ornith Soc; Am Soc Mammal; Am Ornith Union; Can Inst Forestry. Res: Forest zoology; studies on siphonaptera; cone and seed insects; soil arthropods. Mailing Add: Dept of Forest Resources Univ of NB Fac of Forestry Fredericton NB Can

BROWN, OLEN RAY, b Hastings, Okla, Aug 18, 35; m 58; c 3. MICROBIOLOGY, BIOCHEMISTRY. Educ: Univ Okla, BS, 58, MS, 60, PhD(microbiol), 64. Prof Exp: Spec instr microbiol, Univ Okla, 63-64; from instr to asst prof, 64-70, ASSOC PROF MICROBIOL, UNIV MO-COLUMBIA, 70-, INVESTR, 68-, ASST DIR JOHN M DALTON RES CTR, 74- Mem: Fel Am Inst Chem; Undersea Med Soc; Sigma Xi. Res: Mechanisms of oxygen toxicity at the cellular level in bacteria and tissues; effects of hyperoxia on growth, respiration, permeability, oxidative phosphorylation and on enzymes and coenzymes of lipid metabolism. Mailing Add: John M Dalton Res Ctr Univ of Mo Columbia MO 65201

BROWN, OLIVER LEONARD INMAN, b Webster City, Iowa, Nov 15, 11; m 43; c 4. PHYSICAL CHEMISTRY. Educ: Univ Iowa, AB, 32, MS, 33; Univ Calif, PhD(phys chem), 36. Prof Exp: Instr gen & phys chem, Univ Mich, 36-41; from instr to asst prof chem, US Naval Acad, 41-46; asst prof, Syracuse Univ, 46-52; prof & chmn dept, 52-74, LUCRETIA L ALLYN PROF CHEM, CONN COL, 52- Concurrent Pos: Consult, Elec Boat Div, Gen Dynamics Corp, 56- Mem: AAAS; Am Chem Soc. Res: Low temperature heat capacities; heats of solution; free energies from equilibrium measurements; data of state of gases; vapor pressure relationships. Mailing Add: Dept of Chem Conn Col New London CT 06320

BROWN, PATRICIA LYNN, b Lafayette, La, Oct 1, 28. INFORMATION SCIENCE. Educ: Univ Southwestern La, BS, 47; Univ Tex, MA, 49. Prof Exp: Instr anal chem, Smith Col, 49-50; chemist, R & M Labs, 50; res assoc indust toxicol, Albany Med Col, 50-51; lit searcher info serv & res & eng dept, Ethyl Corp, 51-53, reference librn, 53-54, lit specialist, 54, ed asst, 54-55; sr tech writer & ed pub eng, Atomic Power Div, Westinghouse Elec Corp, 55-57, staff engr, 57; supvr info serv, Semiconductor-Components Div, Tex Instruments, Inc, 57-61, mgr info serv, Corp Res & Eng, 61-64, tech info consult, 64-66; SR RESEARCHER, BATTELLE MEM INST, 66- Mem: Am Chem Soc; Am Soc Info Sci. Res: Analysis and design of information systems; techniques of document preparation, acquisition, indexing, condensation, storage, retrieval, dissemination and use; design and development of methods for useful organization and presentation of information. Mailing Add: Battelle Mem Inst 505 King Ave Columbus OH 43201

BROWN, PATRICIA STOCKING, b Cadillac, Mich, Apr 25, 42; m 65. ZOOLOGY, ENDOCRINOLOGY. Educ: Univ Mich, BS, 63, MS, 66, PhD(zool), 68. Prof Exp: Res assoc develop biol, State Univ NY Albany, 68-69; asst prof, 69-75, ASSOC PROF BIOL, SIENA COL, NY, 75- Mem: Am Soc Zool; Am Soc Soc Icthyologists & Herpetologists; Am Soc Cell Biol. Res: Hormonal control of growth in amphibians and reptiles; salt and water balance in urodeles; function of prolactin in lower vertebrates. Mailing Add: Dept of Biol Siena Col Loudonville NY 12211

BROWN, PATRICK MICHAEL, b North Tonawanda, NY, 1938; m 59; c 4. PHYSICAL INORGANIC CHEMISTRY. Educ: Univ Miss, BS, 60; Syracuse Univ, PhD, 65. Prof Exp: Fel, State Univ NY Buffalo, 65-66; chemist, W R Grace & Co, 66-73, sr res chemist, 73; pres, Biolytic's Inc, 73-74; MGR CHEM RES, FOOTE MINERAL CO, 74- Mem: Sigma Xi; Am Chem Soc; Am Ceramic Soc. Res: Catalysis; materials science; lithium chemistry. Mailing Add: Foote Mineral Co Rte 100 Exton PA 19347

BROWN, PAUL BRUCE, b Greenwood, Miss, Aug 12, 09; m 37. ANIMAL SCIENCE. Educ: Miss State Col, BS, 32; Univ Ill, MS, 47, PhD(animal sci), 51. Prof Exp: Asst county agt, Boys 4-H, Miss State Col, 34-43; asst prof animal indust nutrit & meats, Univ Ark, 47-49; assoc prof animal sci nutrit, La State Univ, 51-66; dir res coord unit, Div Voc Educ, La Dept Educ, 66-72; RETIRED. Res: Beef cattle nutrition; utilization of low quality roughages and blackstrap molasses for beef cattle; utilization of cottonseed meal in swine rations; grazing trials for beef cattle and sheep; utilization of diammonium phosphate and urea as potential source of protein in steer fattening rations. Mailing Add: 432 Stanford Ave Baton Rouge LA 70808

BROWN, PAUL EDMUND, b Schuyler Co, Ill, Oct 17, 16; m 52; c 2. PHYSICAL CHEMISTRY. Educ: Western Ill Univ, BEd, 43; Purdue Univ, PhD(phys chem), 49. Prof Exp: Mem staff, 49-64, supvr coolant chem, Atomic Power Div, 64-69, supvr coolant technol, Bettis Atomic Power Div, 69-71, ADV SCIENTIST, BETTIS ATOMIC POWER DIV, WESTINGHOUSE ELEC CORP, 71- Concurrent Pos: Instr, Carnegie Inst Technol. Mem: AAAS; Am Chem Soc. Res: Radiation and steam system chemistry; corrosion; reactor development and decontamination; steam generator water chemistry. Mailing Add: 232 Gladstone Rd Pittsburgh PA 15217

BROWN, PAUL LAWSON, b Ash Grove, Kans, June 1, 18; m 46; c 2. SOIL FERTILITY. Educ: Kans State Univ, BS, 41, MS, 49; Iowa State Univ, PhD(soil fertil), 56. Prof Exp: Soil surveyor, Kans Agr Exp Sta, 46-48, soil scientist, Ft Hays Exp Sta, 48-56, res soil scientist, Agr Res Serv, Mont State Univ, 56-73, RES SOIL SCIENTIST, AGR RES SERV, NORTHERN PLAINS SOIL & WATER RES CTR, USDA, 73- Mem: Soil Sci Soc Am; Am Soc Agron; Can Soc Soil Sci; Int Soc Soil Sci. Res: Soil management for dryland crop production; soil, crop and land management for saline seep control. Mailing Add: PO Box 45 Ft Benton MT 59442

BROWN, PAUL LOPEZ, b Anderson, SC, Feb 18, 19; m 44; c 4. ZOOLOGY. Educ: Knoxville Col, BS, 41; Univ Ill, MS, 48, PhD(zool), 55. Prof Exp: From instr to assoc prof biol, Southern Univ, 48-58; prof, Fla Agr & Mech Univ, 58-59; PROF BIOL & HEAD DEPT, NORFOLK STATE COL, 59- Concurrent Pos: Mem staff, NSF, 63-64, consult, 64-; consult, Off Educ, US Dept Health, Educ & Welfare, 64- Mem: Am Soc Zoologists; Am Micros Soc. Res: Biology of crayfishes including taxonomy, life history, distribution, limnology of waters inhabited; ecology. Mailing Add: Dept of Biol Norfolk State Col Norfolk VA 23504

BROWN, PAUL WHEELER, b Teaneck, NJ, Mar 12, 36; m 57; c 2. VIROLOGY, INTERNAL MEDICINE. Educ: Harvard Univ, AB, 57; Johns Hopkins Univ, MD, 61; Am Bd Internal Med, cert, 69. Prof Exp: Intern, Johns Hopkins Hosp, 61-63; staff assoc virol, USPHS, NIH, 63-71; vis scientist virol, Nat Inst Health & Med Res, Paris, France, 71-72; MED DIR VIROL, NAT INST NEUROL & COMMUN DIS & STROKE, NIH, 72- Concurrent Pos: Resident, Univ Calif, San Francisco, 65-66 & Johns Hopkins Hosp, 66-67; fel, Nat Multiple Sclerosis Soc, 71-72; fel, Comt Control Huntington's Dis, 73-74. Mem: Am Col Physicians; Am Epidemiol Soc; Infectious Dis Soc Am. Res: Study of slow and latent viruses; epidemiology of diseases in Oceania. Mailing Add: Bldg 36 Rm 5B20 Nat Inst Neurol & Commun Dis & Stroke Bethesda MD 20014

BROWN, PAUL WOODROW, b New York, NY, Dec 28, 19; c 8. SURGERY, ORTHOPEDIC SURGERY. Educ: Univ Mich, BS, 42, MD, 50. Prof Exp: Resident orthop surg, Letterman Army Hosp, San Francisco, 51-55; fel hand surg, Walter Reed Army Hosp, Washington, DC, 62-63; div chmn orthop, Fitzsimons Gen Hosp, Denver, 66-69; prof surg, Med Ctr, Univ Colo, Denver, 69-72; prof, Sch Med, Univ Miami, 72-74; CHMN DEPT SURG, ST VINCENT'S HOSP, 74- Mem: Am Col Surg; Am Acad Orthop Surg; Am Soc Surg of the Hand; Int Soc Orthop Surg & Traumatol. Res: Reconstructive surgery of the hand; peripheral nerve lesions; mechanics of wound healing; fate of exposed bone. Mailing Add: Dept of Surg St Vincent's Hosp 2820 Main St Bridgeport CT 06606

BROWN, PAULA, b Chicago, Ill, Feb 24, 25; m 48, 66. ANTHROPOLOGY. Educ: Univ Chicago, BA, 43, MA, 48; Univ London, PhD(social anthrop), 50. Prof Exp: Asst lectr, Univ Col, Univ London, 48-51; res anthropologist, Univ Calif, Los Angeles, 52-55; lectr, Univ Wis, 55-56; res fel anthrop, Australian Nat Univ, 56-63, fel, 63-65, sr fel, 65-66; assoc prof, 66-68, PROF ANTHROP, STATE UNIV NY STONY BROOK, 68- Concurrent Pos: Vis lectr anthrop, Cambridge Univ, 65-66. Mem: Fel Am Anthrop Asn; Am Sociol Asn; Am Ethnol Soc. Res: Industrial relations; social structure of New Guinea; social and political change in developing countries; multi-ethnic communities. Mailing Add: Dept of Anthrop State Univ of NY Stony Brook NY 11790

BROWN, PETER, b Lincoln, Eng, Sept 12, 38. ORGANIC CHEMISTRY. Educ: Univ Southampton, BSc, 61, PhD, 64. Prof Exp: Res assoc mass spectrometry, Stanford Univ, 64-67; asst prof, 67-70, ASSOC PROF ORG CHEM, ARIZ STATE UNIV, 70- Mem: Am Chem Soc; The Chem Soc. Res: Cycloaddition reactions; photochemistry; mass spectrometry. Mailing Add: Dept of Chem Ariz State Univ Tempe AZ 85281

BROWN, PETER FRANK, b Buffalo, NY, May 28, 39; m 67; c 1. NUCLEAR PHYSICS. Educ: Canisius Col, BS, 61; Univ Pittsburgh, PhD(physics), 66. Prof Exp: Sr scientist, Bettis Atomic Power Lab, Westinghouse Elec Corp, 66-68; asst prof, 68-70, ASSOC PROF PHYSICS, DUQUESNE UNIV, 70- Concurrent Pos: Asst prof, Univ Pittsburgh, 67-68; prin investr, Mine Res Ctr, US Dept Interior, 69-70. Mem: Am Asn Physics Teachers; Am Phys Soc. Res: Experimental nuclear physics structure medium energy; physics education; coulomb excitation. Mailing Add: Dept of Physics Duquesne Univ Pittsburgh, PA 15219

BROWN, PHYLLIS R, b Providence, RI, Mar 16, 24; m 42; c 4. CHEMISTRY, BIOCHEMISTRY. Educ: George Washington Univ, BS, 44; Brown Univ, PhD(chem), 68. Prof Exp: Res asst chem, Harris Res Lab, 44-45; fel & res assoc, Dept Pharmacol, Brown Univ, 68-71, from instr to asst prof, 71-73; ASST PROF CHEM, UNIV RI, 73- Mem: Am Chem Soc; Am Inst Chemists; Am Asn Clin Chemists; Soc Appl Spectros. Res: Application of high pressure liquid chromatography to biomedical research with special emphasis on the analysis of nucleotides and other nucleic acid components. Mailing Add: Dept of Chem Pastore Lab Univ of RI Kingston RI 02881

BROWN, RALPH ANDRES, b Glenwood, Minn, Apr 19, 17; m 47; c 3. ANALYTICAL CHEMISTRY. Educ: Hamline Univ, BS, 39; Okla State Univ, MS, 41. Prof Exp: Sr chemist, Atlantic Res Co, 41-60; from res assoc to sr res assoc, Esso Res & Eng Co, 60-75, SCI ADV, EXXON RES & ENG CO, 75- Mem: Am Chem Soc; Am Inst Chem. Res: Application of mass and absorption spectroscopy; separations to composition of petroleum and other complex organics; new instrument development; methods for polycyclic aromatics; hydrocarbons in open ocean water. Mailing Add: 754 Crescent Pkwy Westfield NJ 07090

BROWN, RALPH CLARENCE, b Buffalo, NY, Nov 17, 22; m 44; c 1. GEOGRAPHY. Educ: Univ Buffalo, BA, 53, MA, 55; Syracuse Univ, PhD(geog), 64. Prof Exp: Asst prof geog, State Univ NY Col Buffalo, 60-61; from assoc prof to prof & chmn dept, California State Col, Pa, 61-66; prof & chmn dept, Univ Wis-Superior, 66-71; PROF GEOG, UNIV NDAK, 71- Mem: AAAS; fel Am Geog Soc; fel Asn Am Geogr; contrib mem Nat Coun Geog Educ; Can Asn Geog. Res: Changing rural settlement patterns; remote sensing; North America; Africa; water quality. Mailing Add: 1008 Darwin Dr Grand Forks ND 58201

BROWN, RANDALL EMORY, b Eugene, Ore, May 28, 17; m 50; c 2. GEOLOGY. Educ: Stanford Univ, AB, 38; Yale Univ, MS, 41. Prof Exp: Geologist, Glacier Peak Mine, M A Hanna Co, 41, State Dept Geol & Mineral Industs, Ore, 41-42, US Geol Surv, 42-45 & Corps Engrs, US Army, 45-47; sr geologist, Hanford Atomic Prod Opers, Gen Elec Co, 47-65; sr res scientist, Water & Land Resources Dept, Pac Northwest Lab, Battelle Mem Inst, 65-71; asst prof geol, Cent Wash State Col, 71-72; instr, Columbia Basin Col, 72-73; INSTR, JOINT CTR GRAD STUDY, RICHLAND, 73-; SR RES SCIENTIST, PAC NORTHWEST LABS, BATTELLE MEM INST, 73- Concurrent Pos: Adj assoc prof, Cent Wash State Col, 67-69. Mem: Fel AAAS; fel Geol Soc Am; Am Inst Mining, Metall & Petrol Engr; Sigma Xi. Res: Ground water geology, geology and hydrology of disposal of radioactive and other wastes; geology of eastern Washington; paleoecology. Mailing Add: 504 Road 49 N Pasco WA 99301

BROWN, RAY KENT, b Columbus, Ohio, Apr 7, 24; m 47; c 3. PROTEIN CHEMISTRY. Educ: Ohio State Univ, AB, 44, MD, 47, MS, 48; Harvard Univ, PhD(biochem), 51. Prof Exp: Intern, Boston City Hosp, 47-48; sr asst surgeon protein chem, NIH, 51-53; asst dir, Div Labs & Res, State Dept Health, NY, 53-59, assoc dir, 59-63; PROF BIOCHEM & CHMN DEPT, SCH MED, WAYNE STATE UNIV, 63- Concurrent Pos: From assoc prof to prof, Albany Med Sch, 56-63; mem, Coun Arteriosclerosis, Am Heart Asn. Mem: Am Chem Soc; Am Asn Immunol; Am Soc Biol Chem; Brit Biochem Soc. Res: Plasma proteins; erythrocyte membrane and transport; immunologic studies of ribonuclease, its derivatives and other proteins; ampholytes and isoelectric focusing. Mailing Add: Dept of Biochem Wayne State Univ Detroit MI 48201

BROWN, RAYMOND ARTHUR, b Sharon, Pa, Aug 25, 11; m 36; c 5. BIOCHEMISTRY. Educ: Pa State Col, BS, 33, MS, 34, PhD(agr biochem), 38. Prof Exp: Res biochemist, Parke, Davis & Co, Mich, 35-44, head div nutrit res, 44-51, dir bioassay div, 51-65, mgr biol control, 65-69, prod adminr, 69-73; RETIRED. Mem: Am Chem Soc; Soc Exp Biol & Med; Am Inst Nutrit; Am Soc Microbiol; Am Pharmaceut Asn. Res: Vitamin research; endocrinology; toxicology. Mailing Add: Sherwood Forest Rte 1 Box 125 Brevard NC 28712

BROWN, RAYMOND RUSSELL, b Calgary, Can, Dec 23, 26; nat US; m 52; c 3. BIOCHEMISTRY. Educ: Univ Alta, BS, 48, MS, 50; Univ Wis, PhD(physiol chem), 53. Prof Exp: Instr chem, Univ Alta, 48-50; fel cancer res, Univ Wis-Madison, 53-55, from asst prof to assoc prof surg, 55-65, PROF CLIN ONCOL, TUMOR CLIN, CANCER RES HOSP, UNIV WIS-MADISON, 65- Mem: Am Chem Soc; Am Asn Cancer Res. Res: Amino acid metabolism; chemical carcinogenesis. Mailing Add: Cancer Res Hosp Univ of Wis Madison WI 53706

BROWN, RELIS BASTIAN, b Albion, Mich, Aug 13, 13; m 41, 70; c 2. ZOOLOGY. Educ: Albion Col, AB, 33; Yale Univ, PhD(zool), 38. Prof Exp: Instr zool, DePauw Univ, 38-39 & biol sci, Cumberland Col, 40-43; asst prof biol, Wesleyan Col, 43-44; assoc prof, Univ Miss, 44-46 & Lawrence Col, 46-57; res assoc, Fla State Univ, 58-59, assoc prof biol sci, 59-69; PROF BIOL, W CHESTER STATE COL, 69- Concurrent Pos: Vis assoc prof biol, Vanderbilt Univ, 57-58. Mem: AAAS; NY Acad Sci. Res: Animal morphology and morphogenesis; biology curricula in college; history of biology. Mailing Add: 311 N Walnut St West Chester PA 19380

BROWN, RICHARD DEAN, b Mansfield, Ohio, Feb 6, 41; m 63; c 2. ANIMAL BEHAVIOR, BIOACOUSTICS. Educ: Columbia Union Col, BA, 64; Ohio State Univ, MS, 70, PhD(zool), 75. Prof Exp: Chmn dept sci, Garden State Acad, 66-70; ASST PROF BIOL, UNIV NC, CHARLOTTE, 75- Mem: AAAS; Am Ornith Union; Wilson Ornith Soc; Cooper Ornith Soc; Am Soc Zoologists. Res: Animal communication and associated behavior in social systems with emphasis on recording, playing back and analyzing bird songs; migratory bird studies involving banding. Mailing Add: Dept of Biol Univ of NC UNCC Sta Charlotte NC 28223

BROWN, RICHARD DON, b Alexandria, La, Mar 3, 40; m 63; c 1. PHARMACOLOGY, TOXICOLOGY. Educ: La Col, BS, 64; La State Univ, MS, 66, PhD(pharmacol), 68. Prof Exp: Instr pharmacol, Sch Med, La State Univ, New Orleans, 68-69; asst prof, 69-72, ASSOC PROF PHARMACOL, SCH MED, LA STATE UNIV, SHREVEPORT, 72- Concurrent Pos: USPHS gen res support grant, 68-69; Southern Med Asn res grant, 70-71; La State Univ Found res grant, 73-74; Nat Inst Gen Med Sci res grant, 75-78. Honors & Awards: Passano Award, La State Univ, Sch Med, Shreveport, 75; Frank R Blood Award, Soc Toxicol, 76. Mem: Am Soc Pharmacol & Exp Therapeut; Acoust Soc Am; Soc Neurosci; NY Acad Sci; Am Chem Soc. Res: Physiology and pharmacology of the cochlear efferents and of synaptic transmission; ototoxic action of aminoglycosides and the loop diuretics; mechanism of action of cholinergic drugs. Mailing Add: Dept of Pharmacol La State Univ Sch of Med Shreveport LA 71130

BROWN, RICHARD EDWARD, b Swarthmore, Pa, Oct 21, 23; m 43; c 4. NUTRITION, PHYSIOLOGY. Educ: Univ Md, BS, 48, MS, 51, PhD(diary sci), 54. Prof Exp: Instr, Univ Md, 53-54; from asst prof to prof nutrit, Univ Ill, Urbana-Champaign, 54-71; dir nutrit res, Animal Health Prod Div, Smith, Kline & French Labs, Pa, 71-73; VPRES & CONSULT, E S ERWIN & ASSOCS INC, 73- Concurrent Pos: Fulbright sr res grant, Australia, 62-63. Mem: Fel AAAS; Am Inst Nutrit; Am Soc Animal Sci; Am Dairy Sci Asn. Res: Ruminant nutrition; physiology and energy metabolism in ruminants. Mailing Add: E S Erwin & Assocs Inc Box 237 Tolleson AZ 85353

BROWN, RICHARD EDWIN, b Bloomington, Ill, May 28, 39; m 67; c 2. THEORETICAL CHEMISTRY. Educ: Univ Ill, BSc, 62; Ind Univ, PhD(chem), 67. Prof Exp: NIH fel, Quantum Chem Group, Univ Uppsala, 67-69; res fel chem, Queen's Univ, Ont, 69-71, asst prof, 71-73; prof chem & physics, Champlain Col, Que, 73-74; ASSOC PROF CHEM, INST CHEM, STATE UNIV CAMPINAS, BRAZIL, 74- Mem: Am Chem Soc. Res: Quantum mechanical determination of atomic and molecular properties; the effects of electronic correlation on molecular properties; the correct interpretation of Hunds' Rules; theoretical design of new drugs. Mailing Add: Inst Chem State Univ Campinas CP 1170 13100 Campinas Sao Paulo Brazil

BROWN, RICHARD EMERY, b Chatham, NJ, Apr 7, 29; m 51; c 4. ORGANIC CHEMISTRY. Educ: Moravian Col, BS, 51; Univ Md, MS, 52, PhD(org chem), 56. Prof Exp: Res chemist org chem, Allied Chem Corp, 55-60; ASSOC DEPT HEAD, WARNER LAMBERT RES INST, 60- Mem: Am Chem Soc. Res: Synthesis of new organic compounds for potential medicinal interest. Mailing Add: 16 Ridge Dr East Hanover NJ 07936

BROWN, RICHARD HARLAND, b Gloversville, NY, Nov 9, 21. MATHEMATICS. Educ: Columbia Univ, AB, 42, PhD(math), 51. Prof Exp: Mathematician, US Dept Navy, 42-46; lectr math, Columbia Univ, 46-49, instr, 50-54; asst prof, Boston Univ, 54-55, mem opers eval group, 55-60; assoc prof, 60-61, PROF MATH, WASHINGTON COL, 61- Concurrent Pos: Consult, Opers Res Off, 54-55 & Opers Eval Group, 60-63. Mem: Am Math Soc; Math Asn Am; Inst Math Statist; Asn Comput Mach; Soc Indust & Appl Math. Res: Finite differences; probability. Mailing Add: Box 25 Chestertown MD 21620

BROWN, RICHARD IRWIN, b Milwaukee, Wis, Sept 3, 34; m 63; c 2. ENVIRONMENTAL SCIENCES, PHYSICS. Educ: Kalamazoo Col, BA, 56; Univ Wis, MA, 58, PhD(nuclear physics), 61. Prof Exp: Res assoc physics, Univ Wis, 61-62 & Univ Sao Paulo, 62-64; res assoc & instr, Univ Wis-Madison, 64-65; from asst prof to assoc prof, State Univ NY Albany, 65-71, environ studies coordr, 71-72; ASSOC PROF, SCI DIV, EISENHOWER COL, 72- Mem: AAAS; Am Phys Soc. Res: Energy-society interaction; methods for improving energy education; design and development of integrated solar energy systems; studies of household energy usage. Mailing Add: Sci Div Eisenhower Col Seneca Falls NY 13148

BROWN, RICHARD JULIAN CHALLIS, b Sydney, Australia, Apr 10, 36; m 62; c 3. PHYSICAL CHEMISTRY. Educ: Univ Sydney, BSc, 57, MSc, 59; Univ Ill, PhD(phys chem), 62. Prof Exp: Asst prof chem, Queen's Univ, Ont, 62-66; mem staff, Australian AEC, 66-69; ASSOC PROF CHEM, QUEEN'S UNIV, ONT, 69- Res: Nuclear quadrupole resonance; nuclear spin relaxation and molecular motion in liquids; dielectric relaxation. Mailing Add: Dept of Chem Queen's Univ Kingston ON Can

BROWN, RICHARD KETTEL, b Long Branch, NJ, Feb 3, 28; m 55. MATHEMATICS. Educ: Muhlenberg Col, BS, 48; Rutgers Univ, MS, 50, PhD(math), 52. Prof Exp: Asst prof math, Rutgers Univ, 55-58; res mathematician, US Army Electronics Labs, 58-65; PROF MATH & CHMN DEPT, KENT STATE UNIV, 65- Mem: Am Math Soc; Math Asn Am. Res: Function of a complex variable; univalent and multivalent functions. Mailing Add: Dept of Math Kent State Univ Kent OH 44242

BROWN, RICHARD LELAND, b Bryn Mawr, Pa, Dec 2, 12; m 39; c 4. PHYSICS. Educ: Amherst Col, AB, 34; Wesleyan Univ, MA, 36; Mass Inst Technol, PhD(physics), 42. Prof Exp: Asst physics, Wesleyan Univ, 34-37; res assoc, Div Indust Coop, Mass Inst Technol, 42-47; PROF PHYSICS, ALLEGHENY COL, 47- Mem: AAAS; fel Am Acoust Soc; Inst Elec & Electronics Engrs; Am Phys Soc; Am Asn Physics Teachers. Res: Acoustics; sound-absorbing structures. Mailing Add: 211 Meadow St Meadville PA 16335

BROWN, RICHARD MALCOLM, JR, b Pampa, Tex, Jan 2, 39; m 61; c 1. BOTANY. Educ: Univ Tex, BS, 61, PhD(bot), 64. Prof Exp: Fel, Univ Hawaii & Univ Tex, 64-65; asst prof bot, Univ Tex, Austin, 65-68; assoc prof, 68-73, PROF BOT, UNIV NC, CHAPEL HILL, 73-, DIR ELECTRON MICROS LAB, 70- Concurrent Pos: NSF fel, Univ Freiburg, 68-69 & res grant, 70-72. Mem: AAAS; Am Soc Cell Biol. Res: Airborne algae; algal ecology; ultrastructure of algal viruses; algal ultrastructure; immunochemistry of algae; cytology; Golgi apparatus and cell wall formation; sexual reproduction amon algae; cellulose biogenesis. Mailing Add: Dept of Bot Univ of NC Chapel Hill NC 27514

BROWN, RICHARD MAURICE, b Cambridge, Mass, May 17, 24; m 46; c 4. COMPUTER SCIENCE. Educ: Harvard Univ, AB, 44, MA, 47, PhD(physics), 50. Prof Exp: Asst, Oceanog Inst, Woods Hole, 44-46; asst prof physics, State Col Wash, 49-54; res prof physics & elec eng, Coord Sci Lab, 55-67, PROF PHYSICS, UNIV ILL, URBANA-CHAMPAIGN, 67- Concurrent Pos: Res asst prof, Control Syst Lab, Univ Ill, 52-54. Mem: Fel Am Phys Soc; Inst Elec & Electronics Engr; Asn Comput Mach. Res: Electronic computer design and use. Mailing Add: Dept of Physics Univ of Ill Urbana IL 61801

BROWN, RICHARD MCPIKE, b San Diego, Calif, Feb 17, 26; m 60; c 3. PLANT TAXONOMY, PLANT ECOLOGY. Educ: Pomona Col, BA, 50; Harvard Univ, MA, 52. Prof Exp: Park naturalist, 57-66, RES BIOLOGIST, NAT PARK SERV, 66- Mem: Am Soc Plant Taxon; Am Bryol & Lichenological Soc. Res: Role of man in altering natural vegetation and fauna; restoration of natural biotic conditions and processes in national parks. Mailing Add: Point Reyes Nat Seashore Point Reyes CA 94956

BROWN, RICHARD W, solid state physics, see 12th edition

BROWN, RICHARD WALLACE, b Reading, Pa, Aug 21, 21; m 46; c 3. VETERINARY MEDICINE. Educ: Pa State Univ, BS, 43; Univ Pa, VMD, 45; Univ Minn, PhD, 56. Prof Exp: Gen vet pract, 45-47; RES VET, NAT ANIMAL DIS CTR, USDA, 51- Concurrent Pos: Vet, UNRRA, 46. Mem: NY Acad Sci; Sigma Xi. Res: Bovine mastitis; characterization and differentiation of pathogenic bacteria and factors in milk that affect growth of bacteria. Mailing Add: Nat Animal Dis Ctr Ames IA 50010

BROWN, ROBERT ALAN, b Los Angeles, Calif, June 11, 34; m 57; c 3. ATMOSPHERIC PHYSICS, GEOPHYSICS. Educ: Univ Calif, Berkeley, BS, 57, MS, 63; Univ Wash, PhD(geophys, atmospheric sci), 69. Prof Exp: Res engr fluid dynamics, Boeing Aircraft Co, Seattle, 64-66; res assoc boundary layer dynamics, Dept Atmospheric Sci, Univ Wash, 69-70; advan study prog fel atmospheric fluid dynamics, Nat Ctr Atmospheric Res, 70-71; PRIN SCIENTIST BOUNDARY LAYER DYNAMICS, ARCTIC ICE DYNAMICS JOINT EXP, DEPT ATMOSPHERIC SCI, UNIV WASH, 71- Mem: Am Geophys Union; Sigma Xi. Res: Planetary boundary layer modeling in connection with air-sea interaction, general circulation modeling, climate dynamics. Mailing Add: Dept of Atmospheric Sci-AIDJEX Univ of Wash Seattle WA 98105

BROWN, ROBERT BRUCE, b Portland, Ore, July 19, 38; m 62; c 3. MATHEMATICS. Educ: Harvard Univ, AB, 59; Univ Chicago, MS, 60, PhD(math), 64. Prof Exp: From instr to asst prof math, Univ Calif, Berkeley, 64-69; assoc prof, Univ Toronto, 69-70; ASSOC PROF MATH, OHIO STATE UNIV, 70- Mem: AAAS; Am Math Soc; Math Asn Am. Res: Algebra, combinatorics and their applications. Mailing Add: Dept of Math Ohio State Univ Columbus OH 43210

BROWN, ROBERT C, b Freeport, Ill, Feb 2, 21; m 43; c 3. MATHEMATICS. Educ: Ill Col, AB, 48; George Peabody Col, MA, 49, PhD(math), 56. Prof Exp: Instr math, George Peabody Col, 49-50; instr, 50-57, PROF MATH, SOUTHEASTERN LA UNIV, 57-, HEAD DEPT, 62- Mem: Math Asn Am. Res: Geometry; analysis. Mailing Add: Univ Sta Box 687 Hammond LA 70401

BROWN, ROBERT CALVIN, b Iredell Co, NC, Jan 7, 33; m 56; c 3. PATHOLOGY, MEDICINE. Educ: Erskine Col, AB, 55; Univ NC, MD, 59. Prof Exp: From intern to resident path, Sch Med, Univ NC & NC Mem Hosp, 5964, instr, 62-64; head cellular path group, Biol Div, Oak Ridge Nat Lab, 66-69; DIR DEPT PATH, RUTHERFORD HOSP, 71- Concurrent Pos: Assoc prof path, Sch Med, Univ NC, Chapel Hill, 69-74. Honors & Awards: Seard-Sanford Award, Am Soc Clin Path, 59. Mem: Am Soc Clin Path; Int Acad Path; Tissue Cult Asn; AMA; Soc Exp Biol & Med. Res: Megakaryocyte structure; platelet aggregation; radiation and viral leukemogenesis; ultrastructural pathology; mammalian cytogenetics; nuclear sex chromatin; malacoplakia; viruscell interactions. Mailing Add: Dept of Path Rutherford Hosp Rutherfordton NC 28139

BROWN, ROBERT CHARLES, b John Day, Ore, Feb 21, 38; m 60; c 3. ECONOMIC GEOGRAPHY, RESOURCE GEOGRAPHY. Educ: Ore State Univ, BSc, 60, MSc, 62; Mich State Univ, PhD(geog), 67. Prof Exp: From instr to asst prof geog, Okla State Univ, 65-67; from asst prof to assoc prof, 67-71, dean div gen studies, 71-72, DEAN FAC INTERDISCIPLINARY STUDIES, SIMON FRASER UNIV, 72- Concurrent Pos: Lectureships, Univ Calif, Davis, 68-69, Western Wash State Col, 69-70 & Humboldt State Col, 70-71. Mem: Asn Am Geog; Can Asn Geog. Res: Rural-urban fringe land re-allocation problems; primary resource development problems; spatial aspects of consumer price behavior. Mailing Add: Fac of Interdisciplinary Studies Simon Fraser Univ Burnaby BC Can

BROWN, ROBERT DILLON, b Paris, Ark, Nov 26, 33. MATHEMATICS. Educ: Univ Calif, Berkeley, AB, 55, PhD(math), 63. Prof Exp: Res mathematician, Univ Calif, Berkeley, 63; asst prof, 63-68, ASSOC PROF MATH, UNIV KANS, 68- Mem: Am Math Soc. Res: Partial differential equations. Mailing Add: Dept of Math Univ of Kans Lawrence KS 66044

BROWN, ROBERT DON, b Falls City, Nebr, May 30, 42; m 70; c 1. INDUSTRIAL HYGIENE. Educ: Northwest Mo State Univ, BS, 64; Univ Mich, MS, 69. Prof Exp: Health physicist, 69-72, indust hygienist, 72-74, SR INDUST HYGIENIST, KELSEY-SEYBOLD CLIN, 74- Mem: Health Physics Soc; Am Indust Hyg Asn; Am Acad Indust Hyg. Res: Environmental factors that may pose significant health hazards, particularly noise, chemical agents and electromagnetic radiation. Mailing Add: 8219 Lettie St Houston TX 77075

BROWN, ROBERT EUGENE, b Appleton City, Mo, July 7, 25. PHYSICAL CHEMISTRY. Educ: Univ Mo, AB, 49, PhD(phys chem), 57. Prof Exp: RES PHYSICIST, US NAVAL ORD LAB, 60- Mem: Am Chem Soc. Res: Measurements of thermal accommodation coefficients on clean metal surfaces; rare earth magnetism; development of magnetic materials; applications of superconductivity to devices. Mailing Add: 9144 Piney Branch Rd Silver Spring MD 20903

BROWN, ROBERT FRANCIS, b Plainfield, NJ, May 14, 21; m 63. BIOLOGY, ELECTRON MICROSCOPY. Educ: Rutgers Univ, BS, 50, MS, 52, PhD, 59. Prof Exp: Asst, Dept Entom, Rutgers Univ, 51-55; entomologist, Qm Res & Eng Ctr, US Army, 55-63; RES BIOLOGIST & ELECTRON MICROSCOPIST, DEPT PATH, US FOOD & DRUG ADMIN, 63- Mem: AAAS; Entom Soc Am; Sigma Xi; Electron Micros Soc Am. Res: Physiology; toxicology; human biology; medicine. Mailing Add: 13408 Kiama Ct Laurel MD 20810

BROWN, ROBERT FREEMAN, b Cambridge, Mass, Dec 13, 35; m 57; c 2. TOPOLOGY. Educ: Harvard Univ, AB, 57; Univ Wis, PhD(math), 63. Prof Exp: From asst prof to assoc prof, 63-73, PROF MATH, UNIV CALIF, LOS ANGELES, 73- Mem: Am Math Soc; Math Asn Am. Res: Algebraic topology. Mailing Add: Dept of Math Univ of Calif Los Angeles CA 90024

BROWN, ROBERT G, b Stockbridge, Mich, Oct 25, 31; m 55; c 3. PHYSICS. Educ: Yale Univ, BS, 53; Mich State Univ, MS, 57, PhD(physics), 59. Prof Exp: Physicist, Plastics Dept, 59-66, sr res physicist, 66-68, mkt rep, 68-73, res planner, Cent Res Dept, 73-75, MKT SPECIALIST, PLASTICS DEPT, E I DU PONT DE NEMOURS & CO, 75- Res: Infrared spectroscopy; Raman spectroscopy; electrical and physical properties of high polymers; optical properties of materials. Mailing Add: Plastics Dept E I du Pont de Nemours & Co Wilmington DE 19898

BROWN, ROBERT GEORGE, b Montreal, Que, Apr 7, 37; m 63; c 3. MICROBIAL BIOCHEMISTRY. Educ: Macdonald Col, McGill Univ, BSc, 59; McGill Univ, MSc, 61; Rutgers Univ, PhD(microbiol), 65. Prof Exp: Nat Res Coun Can overseas fel carbohydrate chem, Univ Stockholm, 65-67; asst prof biol, 67-71, ASSOC PROF BIOL, DALHOUSIE UNIV, 71- Concurrent Pos: Vis, Univ Stockholm, 74. Mem: Am Soc Microbiol; Can Soc Microbiol. Res: Microbial utilization of carbohydrates and the effect of chemical modification on subsequent enzymic breakdown; structure of microbial cell walls. Mailing Add: Dept of Biol Dalhousie Univ Halifax NS Can

BROWN, ROBERT GETMAN, b Gloversville, NY, Oct 25, 17; m 37; c 3. INDUSTRIAL CHEMISTRY. Educ: Cornell Univ, ChB, 39. Prof Exp: Chem microscopist, B F Goodrich Co, 39-44, chem purchasing agent, 44-46; chem salesman, Emergy Industs, 46-47; vpres, Perma Glaze Chem Corp, 47-53; asst to tech dir, Mohasco Industs, 53-57; vpres & tech dir, Perma Glaze Chem Corp, 57-69; PRES, KNIGHT OIL CORP & WELLS CHEM CO, INC, 69- Mem: Am Chem Soc. Res: Chemical specialties development; emulsion cleaners; polishes; sanitizers; industrial degreasing compounds; chrome tanning compounds; aerosol technology. Mailing Add: 251 N Comrie Ave Johnstown NY 12095

BROWN, ROBERT GLENN, b Norristown, Pa, Apr 16, 38; m 65; c 2. NUTRITION, PHYSIOLOGY. Educ: Drexel Inst, BSc, 60; Univ Tenn, Knoxville, PhD(nutrit & physiol), 64. Prof Exp: From asst prof to assoc prof nutrit, Drexel Inst, 64-69, actg chmn dept, 68-69; ASSOC PROF ANIMAL SCI, UNIV GUELPH, 69- Honors & Awards: Dr Francis S W Luken's Award, Del Valley Diabetes Asn, 69. Mem: AAAS; Am Chem Soc; Am Soc Animal Sci; Nutrit Soc Can; Am Inst Nutrit. Res: Genetic disorders of trace mineral metabolism; collagen and connective tissue metabolism; vitamin E and sulphur metabolic functions; membrane structure and metabolism. Mailing Add: Dept of Animal Sci Univ of Guelph Guelph ON Can

BROWN, ROBERT GOODELL, b Evanston, Ill, Apr 14, 23; m 48, 67; c 3. OPERATIONS RESEARCH. Educ: Yale Univ, BE, 44, MA, 48. Prof Exp: Mem staff, Opers Eval Group, US Navy, 48-50; head dept opers res, Willow Run Res Ctr, Univ Mich, 50-53; mem staff, Arthur D Little, Inc, 53-67; vpres & exec dir opers serv, Curtiss-Wright Corp, 67-68; indust consult, Distrib Industs Mkt, Int Bus Mach Corp, 68-71; consult, 71-75, PRES, MAT MGT SYSTS, INC, 75- Concurrent Pos: Vis lectr, Amos Tuck Sch Bus Admin, 64. Mem: Int Fedn Opers Res Soc (treas, 62-65); Opers Res Soc Am; Am Soc Mech Engr; Am Math Soc. Res: Statistical forecasting in discrete, nonstationary, time-series, inventory control, production planning and control systems; development, implementation, evaluation and education. Mailing Add: PO Box 332 Norwich VT 05055

BROWN, ROBERT HARRISON, b Edinburg, Tex, Feb 22, 38; m 60; c 2. VERTEBRATE ZOOLOGY. Educ: Univ Ariz, BS, 60, MS, 62, PhD(zool), 65. Prof Exp: Asst prof zool, Univ Idaho, 65-67; ASSOC PROF ZOOL, CENT WASH STATE COL, 67- Mem: Am Soc Zool. Res: Comparative osteology and myology of lizards and rodents. Mailing Add: Dept of Biol Sci Cent Wash State Col Ellensberg WA 98926

BROWN, ROBERT HENRY, b Sioux Falls, SDak, Aug 27, 15; m 42; c 2. GEOPHYSICS. Educ: Union Col, Nebr, BA, 40; Univ Nebr, MS, 42; Univ Wash, PhD(physics), 50. Prof Exp: Asst instr, Univ Nebr, 40-42; res engr, Sylvania Elec Co, NY, 42-45; head sci dept, Can Union Col, 45-47; from asst prof to prof physics, Walla Walla Col, 47-70, vpres student affairs, 61-70; prof physics & pres, Union Col, Nebr, 70-73; PROF GEOPHYS, ANDREWS UNIV, 73-; DIR, GEOSCI RES INST, 73- Concurrent Pos: Commun syst res specialist, Sylvania Elec Prod, NY, 51 & US Naval Radiol Defense Lab, 57. Mem: Am Phys Soc; Am Geophys Union; Geochem Soc. Res: Radioisotope dating; collision processes involving atoms and/or molecules in excited states; frequency-modulation system design; electronic bridge circuit for phase, frequency and impedance measurements; gamma ray shielding. Mailing Add: 225 Ridge Ave Berrien Springs MI 49103

BROWN, ROBERT JAMES SIDFORD, b Lawndale, Calif, Sept 7, 24; m 50, 61; c 5. PHYSICS. Educ: Calif Inst Technol, BS, 48; Univ Minn, MS, 51, PhD(physics), 53. Prof Exp: Electronic technician & asst, Los Alamos Sci Lab, 44-46; asst, Univ Minn, 48-53; SR RES ASSOC PHYSICS, CHEVRON OIL FIELD RES CO, STANDARD OIL CO CALIF, 53- Mem: Am Phys Soc; Soc Explor Geophys; Europ Asn Explor Geophys; Am Geophys Union. Res: Application of nuclear magnetic relaxation phenomena to study of liquids and their interactions with solid surfaces in porous media; application of electromagnetic and seismic techniques to oil exploration; study of elastic properties of porous materlas. Mailing Add: 916 W Fern Dr Fullerton CA 92633

BROWN, ROBERT KARL, organic chemistry, see 12th edition

BROWN, ROBERT LAMME, b Inglewood, Calif, Aug 17, 43; m 65; c 1. ASTROPHYSICS, ATOMIC PHYSICS. Educ: Univ Calif, Berkeley, BA, 65; Univ Calif, San Diego, MS, 67, PhD(physics), 69. Prof Exp: ASST SCIENTIST PHYSICS, NAT RADIO ASTRON OBSERV, 69- Mem: Am Astron Soc. Res: Interstellar absorption of cosmic x-rays; helium photoionization cross sections, x-ray production by suprathermal proton bremsstrahlung. Mailing Add: Nat Radio Astron Observ Edgemont Rd Charlottesville VA 22901

BROWN, ROBERT LAWRENCE, b Berlin, Ger, Dec 5, 12; US citizen; m 40; c 2. GEOPHYSICS, ASTRONOMY. Educ: Univ Colo, BA, 35, Columbia Univ, MA, 49. Prof Exp: Supt mining, Acme Cadillac Gold Mines, Ltd, 36-38 & Concordia Gold Mines, Ltd, 38-40; instr sci, 49-54, from asst prof to assoc prof earth sci, 54-75, dir planetarium & observ, 59-68, chmn dept earth sci, 66-68, EMER ASSOC PROF EARTH SCI, SOUTHERN CONN STATE COL, 75- Concurrent Pos: Consult radio & TV stas & Newspapers, Conn, 57-; sci consult, Newspaper Conf, Univ RI, 62- Mem: Fel AAAS; Am Geophys Soc; NY Acad Sci. Res: Entry paths of satellites and other extra-terrestrial bodies, especially angles of approach, apparent velocity of particle and terrestrial latitide; gravitational fields in space; mechanisms of capture of planets, satellites and other solar systems bodies; auto-tutorial instruction in earth sciences. Mailing Add: 1 Thompson Hill Rd Milford CT 06460

BROWN, ROBERT LEE, b Ranger, Tex, Nov 23, 32; m 56; c 2. BACTERIAL PHYSIOLOGY, BIOCHEMISTRY. Educ: Univ Houston, BS, 62, MS, 64; Univ Tex, PhD(microbiol), 67. Prof Exp: Asst biochemist & asst prof, Univ Tex M D Anderon Hosp & Tumor Inst, Houston, 68-69; asst prof microbiol, Dent Br, Univ Mo-Kansas City, 69-71; SR RES INVESTR, ANAL RES & DEVELOP DEPT, E R SQUIBB INST, 71- Concurrent Pos: NIH fel, Univ Tex Med Br Houston, 66-67; NIH fel biochem, Univ Tex M D Anderson Hosp & Tumor Inst, Houston, 6768; staff consult, Parkway Hosp, Houston, 69-; consult, BioControl, Inc, Houston, 67-69 & Bioassay, Inc, 68-69. Honors & Awards: O B Williams Res Award, Tex Soc Microbiol, 66. Mem: AAAS; Am Soc Microbiol; NY Acad Sci. Res: Clinical and industrial microbiology; clinical biochemistry and pathology; biochemistry of cancer. Mailing Add: Anal Res & Develop Dept E R Squibb & Sons New Brunswick NJ 08902

BROWN, ROBERT LEE, b Franklin, Pa, Feb 26, 08; m 40; c 3. SURGERY. Educ: Univ Mich, AB, 29; Harvard Univ, MD, 33. Prof Exp: Instr surg, Sch Med, Univ Rochester, 39-40, assoc, 4041; from assoc dir to dir, Robert Winship Clin, 45-66, instr, 4546, assoc, 46-54, from asst prof to assoc prof, 5466, PROF SURG, SCH MED, EMORY UNIV, 66-, DIR, EMORY UNIV CLIN, 66- Mem: Am Col Surgeons; AMA; Am Cancer Soc; Am Radium Soc (secy, 58-61, pres, 62); James Ewing Soc (pres, 52). Res: Cancer of the cervix, endometrium and melanoma; tumors of the neck; neoplastic diseases. Mailing Add: Emory Univ Clin 1365 Clifton Rd NE Atlanta GA 30322

BROWN, ROBERT LEONARD, physical chemistry, see 12th edition

BROWN, ROBERT MELBOURNE, b Richmond, Que, Sept 14, 24; m 50; c 3. ENVIRONMENTAL CHEMISTRY. Educ: Biship's Univ, Can, BSc, 44; McGill Univ, BSc, 47, PhD(phys chem), 51. Prof Exp: Jr res officer, Atomic Energy Can Ltd, 51-52; sci officer, Defence Res Bd, 52-62; assoc res officer, Environ Res Br, Atomic Energy Can Ltd, 62-70; sci officer, Sect Isotope Hydrol, Int Atomic Energy Agency, Vienna, 70-73; ASSOC RES OFFICER, ENVIRON RES BR, ATOMIC ENERGY CAN LTD, 73- Concurrent Pos: Mem sub-comt low level radioactivity in mat, Int Comn Radiol Units & Measurements, 60-62. Mem: Sr mem Chem Inst Can. Res: Low level radiochemistry, including early development of low background B counters; measurement of tritium and deuterium in natural waters and their use as hydrological and meteorological tracers. Mailing Add: Atomic Energy Can Ltd Chalk River ON Can

BROWN, ROBERT RAYMOND, b Akron, Ohio, June 30, 22; m 50; c 3. ORGANIC CHEMISTRY, POLYMER CHEMISTRY. Educ: Univ Akron, BS, 47, MS, 50; Ohio State Univ, PhD(org chem), 55. Prof Exp: Chemist, Firestone Tire & Rubber Co, Ohio, 47-50; sr res chemist, Int Latex Corp, 55-57, mgr polymerization res, 57-59, mgr polymer res, 59-65, DIR RES, STANDARD BRANDS CHEM INDUSTS, INC, 65-, DIR, 68- Mem: Am Chem Soc. Res: Polymer research, especially elastomers for both latex and dry polymer applications. Mailing Add: Standard Brands Chem Industs Inc PO Drawer K Dover DE 19901

BROWN, ROBERT REGINALD, b Alameda, Calif, Apr 4, 23; m 48; c 3. PHYSICS. Educ: Univ Calif, Berkeley, AB, 44, PhD(physics), 52. Prof Exp: Instr physics, Princeton Univ, 52-53; asst prof, Univ NMex, 53-56; asst prog dir, NSF, 56-57; lectr, 57-58, from asst prof to assoc prof, 58-65, PROF PHYSICS, UNIV CALIF, BERKELEY, 65- Concurrent Pos: Guggenheim fel & Fulbright scholar, 63-64. Mem: Am Geophys Union. Res: Cosmic ray time variations; auroral and ionospheric physics; geomagnetism. Mailing Add: Dept of Physics Univ of Calif Berkeley CA 94720

BROWN, ROBERT STANLEY, b High River, Alta, Can, Sept 16, 46; m 68; c 1. PHYSICAL ORGANIC CHEMISTRY. Educ: Univ Alta, BSc, 68; Univ Calif, San Diego, MSc, 70, PhD(chem), 72. Prof Exp: Nat Res Coun Can fel, Columbia Univ, 72-74; ASST PROF CHEM, UNIV ALTA, 74- Mem: Can Chem Soc; Am Chem Soc. Res: Photoelectron spectroscopy of theoretically interesting organic species; model enzyme studies. Mailing Add: Dept of Chem Univ of Alta Edmonton AB Can

BROWN, ROBERT STEPHEN, b New York, NY, May 21, 38; m 60; c 2. INTERNAL MEDICINE, NEPHROLOGY. Educ: Columbia Univ, MD, 63. Prof Exp: Intern & asst resident internal med, Bellevue Hosp, NY, 63-65; clin assoc med, Nat Cancer Inst, 65-67; sr asst resident internal med, Yale-New Haven Hosp, 67-68; fel med, Med Sch, Yale Univ, 68-69, asst prof, 69-72; ASST PROF INTERNAL MED, HARVARD MED SCH, 72- Concurrent Pos: Med dir, Kidney Transplantation & Dialysis Serv, Yale-New Haven Hosp, 69-72; clin chief renal unit, Beth Israel Hosp, Boston, 73- Mem: Am Fedn Clin Res; Am Soc Nephrology; fel Am Col Physicians; Int Soc Nephrology; Am Soc Artificial Internal Organs. Res: Potassium metabolism in chronic renal disease; kidney transplantation; metabolic aspects of uremia and nephrolithiasis. Mailing Add: Dept of Med Beth Israel Hosp Boston MA 02215

BROWN, ROBERT STEWART, physical chemistry, see 12th edition

BROWN, ROBERT THEODORE, b Bay City, Mich, Feb 16, 31; m 56; c 4. PHYSICS. Educ: Univ Calif, Riverside, BA, 58; Univ Mich, MS, 59, PhD(physics), 65. Prof Exp: Instr physics, Univ Mich, 65-66; asst prof space sci, Rice Univ, 66-68; sci specialist, EG & G, Inc, 68-72; STAFF MEM, LOS ALAMOS SCI LAB, 72- Concurrent Pos: Res Corp Cottrell grant, 67-68. Mem: AAAS; Am Phys Soc; Am Astron Soc. Res: Computational physics; atomic structure and spectra; x-ray scattering; Monte Carlo charged particle and photon transport; computer graphics. Mailing Add: Los Alamos Sci Lab PO Box 1663 Los Alamos NM 87545

BROWN, ROBERT THORSON, b Rochester, Minn, Sept 29, 23; m 53; c 4. PLANT ECOLOGY. Educ: Univ Wis, BS, 47 & 48, MS, 49, PhD(bot), 51. Prof Exp: Assoc prof, 51-66, PROF BIOL SCI, MICH TECHNOL UNIV, 66- Concurrent Pos: Mem staff radiation biol, Argonne Nat Lab, 66 & US AID, India, 68; Fulbright res fel, Univ Helsinki, 71-72; Sigma Xi lectr, 75. Mem: Fel AAAS; Am Inst Biol Sci; Ecol Soc Am; Sigma Xi; Nat Enviorn Educ Asn. Res: Air pollution; plant growth control; seed germination; northern forest ecology; radiation biology; exogenous growth substances; mycorrhizae on conifers. Mailing Add: Dept of Biol Sci Mich Technol Univ Houghton MI 49931

BROWN, ROBERT WALLACE, b Portland, Ore, May 20, 25; m 48; c 2. MATHEMATICS. Educ: Pac Univ, BS, 50; Ore State Univ, MS, 52, PhD(math), 58. Prof Exp: Asst math, Ore State Univ, 50-52; mathematician, Nat Bur Stand, Calif, 52-54; instr math, Ore State Univ, 56-58; mathematician, Boeing Airplane Co, 58-66; vis assoc prof, Ore State Univ, 66; PROF MATH & HEAD DEPT, UNIV ALASKA, FAIRBANKS, 67- Mem: Am Math Soc; Math Asn Am. Res: Nonlinear mechanics; integral equations; classical analysis. Mailing Add: Dept of Math Univ of Alaska Fairbanks AK 99701

BROWN, ROBERT WALTER, b Rockland, Maine, Feb 26, 22. CHEMISTRY. Educ: Bowdoin Col, BS, 43; Princeton Univ, MA, 46, PhD(org chem), 47. Prof Exp: Jr chem engr, Fed Tel & Radio Corp, 43; instr physics, Bowdoin Col, 43-44; sr res chemist, 47-56, group leader, 56, asst mgr colloidal res & develop, 56-60, MGR

RUBBER CHEM RES & DEVELOP, UNIROYAL INC, 60- Mem: AAAS; Am Chem Soc. Res: Catalysts; modifiers, monomers for synthetic rubber; antioxidants; organic sulfur compounds; polymethylol compounds. Mailing Add: Chem Div Uniroyal Inc Naugatuck CT 06770

BROWN, ROBERT WAYNE, b Atwood, Kans, June 27, 23; m 45; c 2. INTERNAL MEDICINE. Educ: Univ Colo, BA, 49; Univ Kans, MD, 55. Prof Exp: From asst prof to assoc prof, 60-69, PROF MED, UNIV KANS, 69- Concurrent Pos: Coordr, Kans Regional Med Prog, 69- Mem: Am Fedn Clin Res; Am Col Physicians; AMA. Res: Endocrinology and metabolism; diabetes. Mailing Add: Dept of Med Univ of Kans Med Ctr Kansas City KS 66103

BROWN, ROBERT WILLIAM, b St Paul, Minn, Oct 3, 41; m 60; c 2. PHYSICS. Educ: Univ Minn, BS, 63; Mass Inst Technol, PhD(physics), 68. Prof Exp: Res assoc physics, Brookhaven Nat Lab, 68-70; asst prof, 70-73, ASSOC PROF PHYSICS, CASE WESTERN RESERVE UNIV, 73- Res: High energy physics. Mailing Add: Dept of Physics Case Western Reserve Univ Cleveland OH 44106

BROWN, ROBERT ZANES, b Jackson, Mich, July 31, 26; m 48, 71; c 4. ECOLOGY, ANIMAL BEHAVIOR. Educ: Swarthmore Col, BA, 48; Johns Hopkins Univ, DSc(vert ecol), 52. Prof Exp: Res asst animal behavior, Am Mus Natural Hist, 48-53; sr asst scientist, Commun Dis Ctr, USPHS, 51-54; from assoc prof to prof zool, Colo Col, 54-63; NSF fac fel, Inst Marine Sci, Univ Tex, 63-64; PROF BIOL, DOWLING COL, 64- Concurrent Pos: Spec staff mem, Rockefeller Found, 68-70. Mem: Fel AAAS; Ecol Soc Am; Animal Behav Soc; Am Inst Biol Sci. Res: Role of animal behavior in population dynamics; bioenergetics; ecology of suburban areas. Mailing Add: Dept of Biol Dowling Col Oakdale NY 11769

BROWN, RODGER ALAN, b Ellicottville, NY, Mar 24, 37; m 59; c 3. METEOROLOGY. Educ: Antioch Col, BS, 60; Univ Chicago, MS, 62. Prof Exp: Res meteorologist, Univ Chicago, 60-65; assoc meteorologist, Cornell Aeronaut Lab, Inc, NY, 65-68, res meteorologist, 68-70; res meteorologist, Wave Propagation Lab, Environ Sci Serv Admin, Colo, 70; RES METEOROLOGIST, NAT SEVERE STORMS LAB, 70- Concurrent Pos: Consult meteorol, Soc Visual Educ, Inc, 64 & Instrnl Aids, Inc, 64-65. Honors & Awards: Antarctic Serv Medal, 65. Mem: Am Meteorol Soc; Am Geophys Union; Can Meteorol Soc; Royal Meteorol Soc; Sigma Xi. Res: Thunderstorm kinematics and dynamics; mesometeorology; Doppler weather radar. Mailing Add: Nat Severe Storms Lab 1313 Halley Circle Norman OK 73069

BROWN, RODNEY DUVALL, III, b Carbondale, Pa, Aug 28, 31; m 50; c 7. SOLID STATE PHYSICS, BIOPHYSICS. Educ: Univ Scranton, BS, 54; Columbia Univ, MA, 61; NY Univ, PhD(physics), 69. Prof Exp: Assoc engr xerography, IBM Develop Lab, 54-57, staff mem ferroelec, IBM Watson Lab, 57, staff mem hot electrons in germanium, 57-63, staff mem quantum effects in bismuth, 63-65, staff mem, Automated cytol, 65-67, mem interdiv comt med electronics, 66-70, staff mem quantum effects in bismuth, 67-68, mgr mach develop, 68-70, RES STAFF MEM, T J WATSON RES CTR, IBM CORP, 70- Mem: Am Phys Soc; NY Acad Sci. Res: NMRD investigation of metal and sugar binding properties of concanavalin A; nuclear magnetic relaxation in protein solutions. Mailing Add: Dept of Gen Sci IBM Corp PO Box 218 Yorktown Heights NY 10598

BROWN, ROGER E, b Marcellus, Mich, Feb 20, 20; m 63; c 3. ANATOMY, SURGERY. Educ: Mich State Univ, DVM, 50, MS, 60; Purdue Univ, PhD(anat), 64. Prof Exp: Instr surg, Mich State Univ, 50-53, instr anat, 60-62; pvt pract, 53-60; instr anat, Purdue Univ, 62-63; asst prof, Mich State Univ, 64-66, assoc prof anat & asst dir space utilization, 66-69; assoc vet med & surg & dir educ resources, 69-70, PROF VET MED & SURG & CHMN DEPT, UNIV MO-COLUMBIA, 70- Mem: Am Vet Med Asn. Res: Microcirculation of the myocardium. Mailing Add: Dept of Vet Med & Surg Univ of Mo Columbia MO 65201

BROWN, ROGER JAMES EVAN, b Toronto, Ont, Jan 17, 31; m 55; c 3. GEOGRAPHY, GEOLOGY. Educ: Univ Toronto, BA, 52, MA, 54; Clark Univ, PhD(geog), 61. Prof Exp: RES OFFICER, DIV BLDG RES, NAT RES COUN CAN, 53-, MEM PERMAFROST SUBCOMT, 60- Concurrent Pos: Mem bd gov & fel, Arctic Inst NAm; mem, Nat Res Coun Assoc Comt Quaternary Res, 69-72; mem comt permafrost, Nat Acad Sci, 75- Mem: Can Asn Geog (secy, 60-63). Res: Permafrost distribution in Canada and factors affecting its occurrence. Mailing Add: Div Bldg Res Nat Res Coun Montreal Rd Ottawa ON Can

BROWN, RONALD ALAN, b Cleveland, Ohio June 19, 36; m 63; c 3. THEORETICAL PHYSICS, SOLID STATE PHYSICS. Educ: Drexel Inst, BS, 59; Purdue Univ, MS, 61, PhD(physics), 64. Prof Exp: Asst prof metall & Ford Found fel eng, Mass Inst Technol, 64-67; asst prof physics, Kent State Univ, 67-71; ASSOC PROF PHYSICS, STATE UNIV NY COL OSWEGO, 71- Mem: Am Phys Soc; Am Asn Physics Teachers; Sigma Xi. Res: Recombination phenomena in semiconductors; electron-donor recombination in n-type germanium and silicon. Mailing Add: Dept of Physics State Univ of NY Col Oswego NY 13126

BROWN, RONALD DAVID, b Chicago, Ill, Sept 7, 45; m 69; c 2. DEVELOPMENTAL BIOLOGY. Educ: Mass Inst Technol, SB, 67; Univ Chicago, PhD(biochem), 71. Prof Exp: Am Cancer Soc fel, Lab Cell Biol, Nat Res Ctr, Rome, 71-73; fel embryol, Carnegie Inst Washington, 73-75; ASST PROF BIOL CHEM, MED CTR, UNIV CINCINNATI, 75- Res: The utilization of maternal messenger RNA during early development of Xenopus Laevis; molecular mechanisms by which development is controlled. Mailing Add: Dept of Biol Chem Univ of Cincinnati Med Ctr Cincinnati OH 45267

BROWN, RONALD FRANKLIN, b San Diego, Calif, May 22, 40; m 68. SOLID STATE PHYSICS. Educ: Univ Calif, Riverside, BA, 62, MA, 64, PhD(physics), 68. Prof Exp: Teaching fel physics, Harvey Mudd Col, 68-69, asst prof, 69-74; ASST PROF PHYSICS, CALIF POLYTECH STATE UNIV, 74- Concurrent Pos: Referee, Am J Physics, 75. Mem: Am Phys Soc; Am Asn Physics Teachers. Res: Low temperature solid state physics; properties of magnetic insulators; electron transport in semiconductors. Mailing Add: Dept of Physics Calif Polytech State Univ San Luis Obispo CA 93401

BROWN, RONALD FREDERICK, b Washington, DC, Apr 14, 10; m 35; c 2. ORGANIC CHEMISTRY. Educ: Univ Md, BS, 32; Harvard Univ, AM, 37, PhD(org chem), 39. Prof Exp: Chemist, Tolman Laundry, DC, 32-35; jr chemist, Qual Water Div, US Geol Surv, 35-36; asst org chem, Radcliffe Col, 37-39; instr, Harvard Univ, 39-40; instr gen chem, Purdue Univ, 40-42; from asst prof to prof chem, 42-75, head dept, 53-56 & 57-63, EMER PROF CHEM, UNIV SOUTHERN CALIF, 75- Concurrent Pos: Consult, Magnaflux Corp, Ill, 40-42; Fulbright sr res scholar, Imp Col, Univ London, 56-57; vis res assoc, Calif Inst Technol, 63-64. Mem: AAAS; Am Chem Soc; Brit Chem Soc. Res: Rearrangements; antimalarials; stability of ring compounds; free radicals; inductive effects. Mailing Add: 3398A Punta Alta Laguna Hills CA 92653

BROWN, RONALD HAROLD, b Dudley, Ga, July 25, 35; m 57; c 4. AGRONOMY, PLANT PHYSIOLOGY. Educ: Univ Ga, BS, 57, MS, 59; Va Polytech Inst, PhD(agron), 62. Prof Exp: Asst prof agron, Va Polytech Inst, 61-67; assoc prof, Tex A&M Univ, 67-68; assoc prof, 68-71, PROF AGRON, UNIV GA, 72- Mem: Am Soc Agron; Crop Sci Soc Am; Am Soc Plant Physiologists. Res: Physiology and microclimatology as related to growth of forage crops and peanuts. Mailing Add: Dept of Agron Univ of Ga Athens GA 30601

BROWN, ROSS DUNCAN, JR, b Fairmont, WVa, July 21, 35; m 58; c 2. BIOCHEMISTRY. Educ: WVa Univ, BS, 57; Univ Wis-Madison, MS, 65, PhD(biochem), 68. Prof Exp: Staff chemist, Bjorksten Res Lab, Wis, 60-62; ASST PROF BIOCHEM, VA POLYTECH INST & STATE UNIV, 67- Mem: AAAS; Am Chem Soc; Inst Food Technol. Res: Microbial biochemistry; cellulose degradation; mode of action of cellulases; denitrification. Mailing Add: Dept of Biochem & Nutrit Va Polytech Inst & State Univ Blacksburg VA 24061

BROWN, RUSSELL GUY, b Morgantown, WVa, Jan 10, 05; m 36. BOTANY. Educ: WVa Univ, BS, 29, MS, 30; Univ Md, PhD(plant physiol), 34. Prof Exp: Prof biol & head dept, New River State Col, 34-36; asst prof plant physiol & ecol, 36-45, from assoc prof to prof bot, 45-73, EMER PROF BOT, UNIV MD, COLLEGE PARK, 73- Mem: Bot Soc Am; Ecol Soc Am. Res: Taxonomy and ecology of native Maryland plants; environmental biology. Mailing Add: Dept of Bot Univ of Md College Park MD 20742

BROWN, RUSSELL HARDING, organic chemistry, see 12th edition

BROWN, RUSSELL VEDDER, b Tulsa, Okla, Mar 20, 25; m 53; c 2. GENETICS, IMMUNOLOGY. Educ: Univ Tulsa, BA, 48, MA, 50; Iowa State Univ, PhD(genetics), 62. Prof Exp: Geneticist, B-Bar-K Ranch, 53-59; res asst, Iowa State Univ, 59-62; asst prof biol, NTex State Univ, 62-68; assoc prof vet path, community health & med pract & biol sci, Sinclair Comp Med Res Farm, Univ Mo-Columbia, 68-74; PROF BIOL & CHMN DEPT, VA COMMONWEALTH UNIV, 74- Mem: AAAS; Genetics Soc Am; Am Genetic Asn; Poultry Sci Asn; World Poultry Sci Asn. Res: Immunogenetics, population and biochemical genetics; heritability in mice; blood groups in birds, animals and fish; molecular pathology. Mailing Add: Dept of Biol Va Commonwealth Univ Richmond VA 23220

BROWN, RUSSELL WILFRID, b Gray, La, Jan 17, 05; m 32. MEDICAL MICROBIOLOGY. Educ: Howard Univ, BS, 26; Iowa State Univ, MS, 32, PhD(physiol bact), 36. Hon Degrees: LLD, Tuskegee Inst, 71. Prof Exp: Assoc prof biol, Rust Col, 30-31 & Langston Univ, 32-33; head dept bact, Tuskegee Inst, 36-44, head div natural sci, 42-46, dir, George W Carver Found, 44-57, chmn grad prog, 46-62, prof bact, 46-70, dean res, 57-62, vpres & dean grad prog, 62-70; DISTINGUISHED PROF MICROBIOL, UNIV NEV, RENO, 70- Concurrent Pos: Res asst, Iowa State Univ, 42-43; mem, Tuskegee Inst Coun, Oak Ridge Inst Nuclear Studies, 51-62, bd dirs, 62-68; NSF sr fel, Yale Univ, 56-57; mem bd trustees, Stillman Col, 64-, chmn, 69-73; mem bd dirs, Southern Fel Fund, 65-; mem admin comt, Nat Fel Fund, 72- Mem: AAAS; fel Am Acad Microbiol; Am Soc Microbiol; Tissue Cult Asn; Am Soc Cell Biol. Res: Animal viruses; oncogenic viruses; medical virology. Mailing Add: PO Box 8097 Univ Sta Reno NV 89507

BROWN, SANBORN CONNER, b Beirut, Lebanon, Jan 19, 13; US citizen; m 40; c 3. PHYSICS. Educ: Dartmouth Col, AB, 35, MA, 37; Mass Inst Technol, PhD(physics), 44. Prof Exp: Asst physics, Dartmouth Col, 35-37; from instr to prof, 41-75, assoc dean grad sch, 63-75, EMER PROF PHYSICS, MASS INST TECHNOL, 75- Concurrent Pos: Tech adv, US Deleg, UN Int Conf Peaceful Uses Atomic Energy, Geneva, Switz, 58; mem, US Nat Comt, Int Union Pure & Appl Physics, 60-66, vchmn, 68-, pres, Comn Physics Educ, 60-66, secy plasma sub-comn, Comn Atomic & Molecular Physics & Spectros, 66-69, secy, Comn Plasma Physics, 69-; mem interunion comn teaching sci, Int Coun Sci Unions, 60-66; US deleg, Int Atomic Energy Agency, Conf Plasma Physics & Controlled Thermonuclear fusion, Salzburg, Austria, 61; mem, NASA Res Adv Comt Fluid Mech, 63-65; mem comt int exchange of persons, Conf Bd Assoc Res Coun, 65-67; mem comn I on radio measurement methods & standards, Int Sci Radio Union, 65-; chmn, Nat Acad Sci-Nat Res Coun Adv Comt Study Postdoctoral Educ US, 66-68; mem adv panel, Radio Standards Physics Div, Int Basic Standards, Nat Bur Standards, 67-70, chmn, 68-70. Mem: Fel AAAS; Am Phys Soc; Am Asn Physics Teachers (treas, 55-62); fel Am Acad Arts & Sci (secy, 64-67); Hist Sci Soc. Res: Plasma physics; history of science. Mailing Add: Hemlock Corner Henniker NH 03242

BROWN, SEVERN PARKER, b Chicago, Ill, Nov 18, 22; m 56; c 1. ECONOMIC GEOLOGY, STRUCTURAL GEOLOGY. Educ: Univ Rochester, AB, 43. Prof Exp: Jr geologist, US Geol Surv, 43-44; geologist, Am Smelting & Refining Co, 48-52 & St Joseph Lead Co, 53-57; asst prof geol, St Lawrence Univ, 57-61; consult geologist, 61-70; CONSULT GEOLOGIST, DUNN GEOSCI CORP, LATHAM, 71- Mem: Geol Soc Am; Am Inst Mining, Metall & Petrol Engr; Soc Econ Geol; Am Inst Prof Geol; Can Inst Mining & Metall. Res: Structural control of ore deposits; industrial mineral deposits. Mailing Add: 10 Hillside Rd Canton NY 13617

BROWN, SEWARD RALPH, b Glace Bay, NS, Mar 25, 20; m 52; c 4. GEOCHEMISTRY, ECOLOGY. Educ: Queen's Univ, Ont, BA, 50 & 51, MA, 52; Yale Univ, PhD(biogeochem), 62. Prof Exp: Lectr zool, Yale Univ, 56-58; from asst prof to assoc prof biol, 59-68, dir biol sta, 59-74, PROF BIOL, QUEEN'S UNIV, ONT, 68- Mem: Am Soc Limnol & Oceanog; Geochem Soc; Can Soc Zoologists. Res: Biogeochemistry; primary productivity of lakes, particularly the relationship between photosynthetic rates and absolute quantities of chlorophylls, carotenoids and their diagenetic derivatives; paleolimnology, using plant pigments from lake sediments as biochemical fossils. Mailing Add: Dept of Biol Queen's Univ Kingston ON Can

BROWN, SHELDON (JACK), b Los Angeles, Calif, Oct 13, 15; m 45; c 2. PHYSICS. Educ: Univ Calif, Los Angeles, AB, 39, PhD(physics), 51. Prof Exp: Assoc physics, Univ Calif, Los Angeles, 42-49 & 50-51; res fel, Univ Strasbourg, 52-54; asst prof physics, DePaul Univ, 54-56; from asst prof to assoc prof, 56-65, PROF PHYSICS, CALIF STATE UNIV, FRESNO, 65- Concurrent Pos: Guggenheim fel, 62. Mem: Am Phys Soc; Am Asn Physics Teachers. Res: Electron inertia effects; gryomagnetism; gravitation. Mailing Add: Dept of Physics Calif State Univ Fresno CA 93740

BROWN, SIDNEY OVERTON, physiology, see 12th edition

BROWN, SPENCER WHARTON, b Vermillion, SDak, Nov 26, 18; m 39. CYTOGENETICS. Educ: Univ Minn, AB, 38; Univ Calif, PhD(genetics), 42. Prof Exp: Asst prof bot, Univ Ga, 43-45; assoc genetics, 45-46, from instr & jr geneticist to assoc prof & assoc geneticist, 46-60, PROF GENETICS, UNIV CALIF, BERKELEY & GENETICIST, AGR EXP STA, 60- Concurrent Pos: Guggenheim fel, 56-57; secy gen, XIII Int Cong Genetics, Calif, 73. Mem: Genetics Soc Am; Bot Soc Am; Am Soc Naturalists; Int Genetics Fedn (pres, 73-78). Mailing Add: Dept of Genetics Univ of Calif Berkeley CA 94720

BROWN, STANLEY ALFRED, b Boston, Mass, Oct 9, 43; m 70. BIOMATERIALS. Educ: Mass Inst Technol, BS, 65; Dartmouth Col, DEng, 71. Prof Exp: Fel physiol, 70-72, res assoc, 72-74, ASST PROF SURG, DARTMOUTH MED SCH, 74- Concurrent Pos: Assoc staff, Mary Hitchcock Mem Hosp, 72-; adj asst prof eng, Dartmouth Col, 73- Mem: Asn Advan Med Instrumentation; Am Soc Artificial Internal Organs; Inst Elec & Electronics Engrs; Biomat Soc; Orthop Res Soc. Res: Biocompatibility of surgical implant materials; implant site infections; fracture healing, materials for fracture fixation and methods of assessment of healing; ultrasonics. Mailing Add: Surg Labs Dartmouth Med Sch Hanover NH 03755

BROWN, STANLEY D, organic chemistry, see 12th edition

BROWN, STANLEY GORDON, b Washington, DC, May 17, 39; m 70; c 1. ELEMENTARY PARTICLE PHYSICS. Educ: Harvard Univ, AB, 60; Univ Pa, MS, 63, PhD(physics), 66. Prof Exp: Instr & res assoc physics, Cornell Univ, 66-68; res assoc, Mass Inst Technol, 68-70; adj asst prof, Univ Calif, Los Angeles, 70-72; res assoc, State Univ NY Stony Brook, 72-74; ASST ED, PHYS REV, 74- Mem: Am Phys Soc. Mailing Add: Phys Rev Brookhaven Nat Lab Upton NY 11973

BROWN, STANLEY MONTY, b New York, NY, Apr 2, 43. PETROLEUM CHEMISTRY, SURFACE CHEMISTRY. Educ: Queens Col, NY, BA, 64; Northwestern Univ, PhD(org chem), 69. Prof Exp: Res chemist, Davison Chem Div, W R Grace & Co, Md, 69-73, res supvr, 73-74, res assoc, 74-75; GROUP LEADER, MINERALS & CHEM DIV, ENGELHARD MINERALS & CHEM CORP, 75- Mem: Am Chem Soc; Catalysis Soc; Sigma Xi. Res: Catalytic chemistry; organic reaction mechanisms; preparation, characterization and evaluation of heterogeneous catalysts and catalyst supports for industrial processes including fluid cracking, hydrotreating, hydrogenation, isomerization and polymerization. Mailing Add: Engelhard Minerals & Chem Corp Menlo Park Edison NJ 08817

BROWN, STEPHEN CLAWSON, b Caracas, Venezuela, June 16, 41; m 65. HISTOCHEMISTRY, INVERTEBRATE ZOOLOGY. Educ: George Washington Univ, BS, 63; Univ Mich, MS, 64, PhD(zool), 66. Prof Exp: Asst prof, 67-72, Res Found res grant, 68-70, ASSOC PROF BIOL, STATE UNIV NY ALBANY, 72- Mem: AAAS; Am Soc Zoologists. Res: Functional morphology and physiology of marine invertebrates; histochemistry and biochemistry of mucins. Mailing Add: Dept of Biol Sci State Univ of NY Albany NY 12203

BROWN, STEWART ANGLIN, b Peterborough, Ont, Apr 6, 25. PLANT BIOCHEMISTRY. Educ: Univ Toronto, BSA, 47; Mich State Univ, MS, 49, PhD(biochem), 51. Prof Exp: From asst res officer to assoc res officer, Prairie Regional Lab, Nat Res Coun Can, 51-62, sr res officer, 62-64; assoc prof chem, 64-68, PROF CHEM, TRENT UNIV, 68-, DEAN GRAD STUDIES, 73- Concurrent Pos: Vis res worker, Univ Chem Lab, Cambridge Univ, 55-56; vis prof, Col Pharm, Univ Minn, Minneapolis, 70-71. Mem: Phytochem Soc NAm (pres, Plant Phenolics Group, NAm, 63-64); Am Soc Pharmacog; Can Soc Plant Physiol (vpres, 60-61). Res: Biochemistry of higher plants and plant-parasite relations; chemistry of lignification, biosynthesis of coumarins and related compounds; methodology of biosynthetic investigation in plants. Mailing Add: Dept of Chem Trent Univ Peterborough ON Can

BROWN, STEWART CLIFF, b Philadelphia, Pa, Mar 15, 28; m 50; c 1. PLASTICS CHEMISTRY, POLYMER CHEMISTRY. Educ: Philadelphia Col Pharm & Sci, BSc, 50; Univ Del, MSc, 54, PhD(org chem), 57. Prof Exp: Res chemist, Tex-US Chem Co, 56-59 & E I du Pont de Nemours & Co, 59-61; res chemist & tech serv supvr polymer film develop, Avisun Corp, 61-65; MKT MGR & RES SUPVR POLYMER FILMS, HERCULES INC, 65- Res: Film forming plastics; effects of orientation on film properties, chemical structure and orientation versus film properties. Mailing Add: Hercules Res Ctr Hercules Rd Wilmington DE 19899

BROWN, STUART GRAEME, b Portland, Ore, May 9, 22; m 53; c 4. HYDROLOGY. Educ: Univ Portland, BS, 54. Prof Exp: Hydraul engr, 54-67, SUPVRY HYDROLOGIST, WATER RESOURCES DIV, US GEOL SURV, 67- Mem: AAAS. Res: Ground water hydrology; surface water-ground water relationships. Mailing Add: US Geol Surv Water Resources Div 301 W Congress Tucson AZ 85701

BROWN, STUART HOUSTON, b Bryn Mawr, Pa, Mar 25, 41; m 66; c 1. ORGANIC CHEMISTRY. Educ: Williams Col, BA, 63; Stanford Univ, PhD(org chem), 68. Prof Exp: NIH fel, Univ Liverpool, 68-69; res chemist, Lubricant Res Dept, 69-75, SR RES CHEMIST, GREASES & INDUST OILS DIV, CHEVRON RES CO, 75- Mem: Am Chem Soc. Res: Alkaloid structure and biosynthesis; x-ray crystallography; synthesis of lubricating oil additives; petroleum chemistry. Mailing Add: Chevron Res Co 576 Standard Ave Richmond CA 94802

BROWN, STUART IRWIN, b Chicago, Ill, Mar 1, 33. OPHTHALMOLOGY. Educ: Univ Ill, BMS, 55, MD, 57. Prof Exp: Intern, Jackson Mem Hosp, Miami, 57-58; resident ophthal, Harvard Med Sch, 58-59; resident, Eye, Ear, Nose & Throat Hosp, Med Sch, Tulane Univ, 59-61, Heed fel, 61-62; fel ophthal, Cornea Serv, Mass Eye & Ear Infirmary, 62-66; clin asst prof ophthal & asst attend surgeon, New York Hosp-Cornell Univ Med Ctr, 66-70, clin assoc prof & assoc attend surgeon, 70-74, dir, Cornea Serv & Cornea Res Lab, Med Ctr, 66-74; PROF OPHTHAL & CHMN DEPT, EYE & EAR HOSP PITTSBURGH & SCH MED, UNIV PITTSBURGH, 74- Mem: Am Acad Ophthal & Otolaryngol; Soc Contemporary Ophthal; Pan Am Asn Ophthal; Asn Res Vision & Ophthal. Res: Cornea. Mailing Add: Dept of Ophthal Eye & Ear Hosp Pittsburgh PA 15213

BROWN, TERENCE HORTON, organic chemistry, medicinal chemistry, see 12th edition

BROWN, TERRANCE R, chemistry, general science, see 12th edition

BROWN, THEODORE GATES, JR, b Boston, Mass, Sept 29, 20; m 45; c 5. PHARMACOLOGY. Educ: Univ Tenn, AB, 48, MS, 50; Med Col SC, PhD(pharmacol), 56. Prof Exp: Res asst pharmacol, Sterling-Winthrop Res Inst, 57-58, assoc mem, 58-61, group leader, 61-62, res biologist, 61-63, sr res biologist & sect head, 63-68; dir biomed res, Warren-Teed Pharmaceut, Inc, 68-74; MGR PRECLIN RES, ROHM & HAAS RES CTR, 74- Concurrent Pos: Lectr, Albany Med Col, 58-68. Mem: AAAS; Am Soc Pharmacol; Am Heart Asn. Res: Pharmacology and physiology of cardiovascular disease; gastrointestinal physiology and pharmacology; digestive diseases. Mailing Add: Rohm & Haas Co Res Labs Norristown & McKean Rds Spring House PA 19477

BROWN, THEODORE LAWRENCE, b Green Bay, Wis, Oct 15, 28; m 51; c 5. INORGANIC CHEMISTRY. Educ: Ill Inst Technol, BS, 50; Mich State Univ, PhD(chem), 56. Prof Exp: From instr to assoc prof, 56-65, PROF INORG CHEM, UNIV ILL, URBANA-CHAMPAIGN, 65- Concurrent Pos: Sloan fel, 62-66; NSF fel, 64-65; assoc ed, Inorg Chem, 68- Honors & Awards: Inorg Chem Award, Am Chem Soc, 72. Mem: Am Chem Soc; The Chem Soc. Res: Kinetics and mechanisms of organometallic reactions; nuclear quadrupole resonance spectroscopy. Mailing Add: Dept of Chem Univ of Ill Urbana-Champaign Champaign IL 61820

BROWN, THEODORE LLEWELLYN, b Banning, Calif, Feb 16, 23; m 48; c 2. ANALYTICAL CHEMISTRY, BIOCHEMISTRY. Educ: Univ Calif, Berkeley, BA, 47. Prof Exp: Jr chemist, Shell Develop Co, 46-48; res chemist, 48-58, sr anal chemist, 58-65, mgr anal serv dept, 65-74, MGR QUAL ASSURANCE OPERS, CUTTER LABS, 74- Mem: Am Chem Soc; NY Acad Sci. Res: Chemical and biological quality assurance testing of biological products; analytical methods research as applied to biochemical evaluation of drug action, purity of new synthetic drugs and stability of new synthetic drugs in pharmaceutical formulations. Mailing Add: Cutter Labs Fourth & Parker Sts Berkeley CA 94710

BROWN, THOMAS ANDREW, b Iowa City, Iowa, July 24, 32; m 57; c 3. MATHEMATICS. Educ: Univ Iowa, BA, 53; Oxford Univ, BA, 55; Harvard Univ, MA, 58, PhD(math), 62. Prof Exp: Mathematician, Rand Corp, 61-66, assoc head dept math, 66-74; RES DIR, PANHEURISTICS DIV, SCI APPLN,INC, 74- Mem: Am Math Soc; Soc Indust & Appl Math; Math Asn Am. Res: Graph theory; ordinary differential equations; statistics; systems analysis. Mailing Add: Panheuristics Suite 1221 1801 Ave of the Stars Los Angeles CA 90067

BROWN, THOMAS CRAIG, b Portland, Ore, May 30, 38; m 66. MATHEMATICS. Educ: Reed Col, BA, 60; Washington Univ, AM, 63, PhD(math), 64. Prof Exp: Instr math, Reed Col, 64-65; exchange student, Kiev Univ, 65-66; asst prof, 66-74, ASSOC PROF MATH, SIMON FRASER UNIV, 74- Mem: Am Math Soc; Math Asn Am. Res: Algebraic semigroups; ring theory; problems of local finiteness; scientology. Mailing Add: Dept of Math Simon Fraser Univ Burnaby BC Can

BROWN, THOMAS EDWARD, b Dayton, Ohio, May 19, 25; m 47; c 3. PLANT PHYSIOLOGY. Educ: Antioch Col, BA, 50; Ohio State Univ, MS, 51, PhD(plant physiol), 54. Prof Exp: Staff scientist, C F Kettering Res Lab, 55-69; assoc prof biol, 69-70-, PROF BIOL, ATLANTIC COMMUNITY COL, 70-, CHMN SCI DIV, 73- Mem: AAAS; Am Soc Plant Physiol; Phycol Soc Am; Brit Phycol Soc; Int Phycol Soc. Res: Photosynthesis and algal physiology; algal pollution control. Mailing Add: Dept of Biol Atlantic Community Col Mays Landing NJ 08330

BROWN, THOMAS TOWNSEND, b Zanesville, Ohio, Mar 18, 05; m 27; c 2. PHYSICS, BIOPHYSICS. Prof Exp: Mem staff electronics res, Denison Univ, 24-25; staff mem, Astrophys Res Lab, Swazey Observ, Ohio, 26-30; mem staff radiation & spectros, Naval Res Lab, Washington, DC, 30-33; state erosion engr, Fed Emergency Relief Admin, Ohio, 34; asst adminr relief for Ohio, dir fed student aid & dir selection, Civilian Conserv Corps, 34-35; staff mem cosmic radiation observations, Townsend Brown Found, Ohio, 36-37; mat & process engr, Glenn L Martin Co, Md, 39-40; officer in charge magnetic & acoust minesweeping res & develop bur ships, US Navy Dept, Washington, DC, 40-41; radar consult, Advan Design Sect, Lockheed Aircraft Corp, 44-45; staff mem cosmic radiation observations, Townsend Brown Found, Calif, 46-48; pvt res biophysics & plant growth, Island of Kauai, Hawaii, 48-52; staff mem radiation & field physics, Townsend Brown Found, Calif, 52-55; consult physicist, Soc Nat Construct Aeronaut, France, 55-56; chief consult res & develop, Whitehall-Rand Proj, Bahnson Co, NC, 57-58; pres, Rand Int, Ltd, 58-74; PRES, ENERGY RESOURCES GROUP, LTD, HONOLULU, 74- Concurrent Pos: Staff physicist, Int Gravity Exped to WI, US Navy Dept, 32; physicist, Johnson-Smithsonian Deep Sea Exped, Smithsonian Inst, 33; consult physicist, Pearl Harbor Navy Yard, 50 & Clevite-Brush Electronics Co, Ohio, 54. Mem: AAAS; Am Soc Naval Eng; Am Phys Soc; Am Geophys Union. Res: Electrohydrodynamics; magnetohydrodynamics; gravitation. Mailing Add: 1125-G Stewart Ct Sunnyvale CA 94086

BROWN, TORREY CARL, b Chicago, Ill, Feb 28, 37; m 57; c 2. MEDICINE. Educ: Wheaton Col, Ill, AB, 57; Johns Hopkins Univ, MD, 61; Am Bd Internal Med, dipl, 68. Prof Exp: From intern to asst resident, Johns Hopkins Hosp, 61-63; surgeon res assoc, Nat Heart Inst, 63-65; asst resident, Johns Hopkins Hosp, 65-66; instr, 66-68, ASSOC PROF MED & ASST PROF MED CARE & HOSPS & DIR ALCOHOLISM PROG, SCH MED, JOHNS HOPKINS UNIV, 68-, DEP DIR, OFF HEALTH CARE PROGS, 69-, DIR, OFF UNIV HEALTH SERV, 73- Concurrent Pos: Ed, Renal Dis Sect, Tice's Pract of Med; mem coun on circulation, Sect on Renal Dis, Am Heart Asn, 64; assoc dir drug abuse ctr, Johns Hopkins Univ, 70-74. Mem: Am Fedn Clin Res; Am Soc Nephrology; fel Am Col Physicians. Res: Hypertension; renal hypertension; reninaldosterone system; physiology of congestive heart failure; medical care delivery systems and research. Mailing Add: Johns Hopkins Hosp 601 N Broadway Baltimore MD 21205

BROWN, VIVIA JEAN, b Laclede, Mo, Dec 22, 02. PHARMACY. Educ: Univ Okla, BA, 28, MS, 47. Prof Exp: Pharmacist, Bethany Hosp, Kansas City, Kans, 30-33; community pharmacist, Brown's Pharm, Mo, 34-43; instr pharm, Univ Okla, 46-47; asst prof, Med Col SC, 47-48; from asst prof to prof, 49-73, EMER PROF PHARM, UNIV OKLA, 73- Mem: AAAS; Am Asn Cols Pharm; Am Pharmaceut Asn; Acad Pharmaceut Sic; Sigma Xi. Res: Dispensing pharmacy; pharmacy administration; compatibilities and incompatibilities of medications used in the treatment of various diseases; survey-prescription record systems. Mailing Add: 1357 Tarman Circle Norman OK 73069

BROWN, W RAY, b Lancaster, Pa, Sept 8, 39; m 63. VETERINARY PATHOLOGY. Educ: Pa State Univ, BS, 61; State Univ NY Vet Col, DVM, 64; Cornell Univ, PhD(vet path), 67; Am Col Vet Pathologists, dipl, 69. Prof Exp: Lab asst vet path, Cornell Univ, 64-67; res fel, Merck Inst Therapeut Res, 67-70, dir path, 70-71, dir path & toxicol, 71-73; PRES, RES PATH SERV INC, 73-; DIR, VET PATH LAB, 73- Concurrent Pos: Asst prof, Jefferson Med Col, 70-; consult exp & toxicol path. Mem: Am Soc Vet Clin Pathologists; Am Vet Med Asn; Int Acad Path. Mailing Add: 389 Creek Rd RR 1 Doylestown PA 18901

BROWN, WALTER CREIGHTON, b Butte, Mont, Aug 18, 13. BIOLOGY. Educ: Col Puget Sound, AB, 35, MA, 38; Stanford Univ, PhD, 50. Prof Exp: Head dept sci high , Wash, 38-42; actg instr biol, Stanford Univ, 49-50; instr, Northwestern Univ, 51-53; instr, 53-54, chmn div sci, 55-61, dean div sci & eng, 61-67, dean, col, 67-74, PROF BIOL, MENLO COL, 74- Concurrent Pos: Fulbright lectr zool, Silliman Univ, Philippines, 54-55; res assoc biol sci, Stanford Univ, 55- Mem: AAAS; assoc Am Soc Ichthyol & Herpet; assoc Am Soc Zool; assoc Soc Syst Zool. Res: Ecology, systematics and zoogeography of amphibians and reptiles in the islands of the Pacific. Mailing Add: Dept of Biol Menlo Col Menlo Park CA 94025

BROWN, WALTER ERIC, b Butte, Mont, Mar 17, 18; m 47; c 3. CHEMICAL PHYSICS. Educ: Univ Wash, BS, 40, MS, 42; Harvard Univ, PhD(chem physics), 49. Prof Exp: Physicist electron micros, B F Goodrich Co, 42-45; chemist phosphates, Tenn Valley Authority, 48-62; CHEMIST, AM DENT ASN RES UNIT, NAT BUR STANDARDS, WASHINGTON, DC, 62- Concurrent Pos: Rockefeller Found spec grant, Neth, 58-59. Mem: Am Chem Soc; Am Dent Asn; Int Asn Dent Res. Res:

Crystallography and physical chemistry of calcium phosphates in tooth and bone. Mailing Add: 500 Goldsborough Dr Rockville MD 20850

BROWN, WALTER HOWARD, b Heyworth, Ill, Oct 17, 09; m 33; c 3. PLANT ECOLOGY, PHYSIOLOGY. Educ: Ill State Norm Univ, BEd, 32; Univ Ill, AM, 37, PhD(bot), 39. Prof Exp: Asst bot, Univ Ill, 35-39; from asst prof to assoc prof biol, Oklahoma City Univ, 39-46; prof & head dept, Cent Col, Mo, 46-55; PROF BIOL, ILL STATE UNIV, 55- Mem: Bot Soc Am. Res: Pollen morphology; aerobiology; pollen analysis; aquatic ecology; limnology; marine biology. Mailing Add: Dept of Biol Ill State Univ Normal IL 61761

BROWN, WALTER JOHN, b North Providence, RI, Aug 25, 28; m 52; c 3. INTERNAL MEDICINE. Educ: Brown Univ, AB, 50; Univ RI, MS, 53; Med Col Ga, MD, 60. Prof Exp: Trainee, Clin Cardiovasc Res Prog, 62-63, asst prof med & assoc dir cardiovasc res training prog, 64-67, assoc prof, 67-71, PROF MED, MED COL GA, 71-; ASSOC DIR & CHIEF MED, UNIV HEALTH SERV, UNIV GA, 68-, PROF THERAPEUT MED, SCH PHARM & PROF MED & SURG, INST COMP MED, SCH VET MED, 69- Concurrent Pos: Grants, NIH coop study, 66-72, Ga Heart Asn, 66-68 & Cerebrovasc dis, Am Heart Asn, 68; attend physician, Vet Admin Hosp, Augusta, Ga, 6469; consult, Penitentiary Syst, State of Ga, 64-69; dir, Athens High Blood Pressure Ctr. Mem: Am Fedn Clin Res: Am Heart Asn; AMA; Am Col Clin Pharmacol & Chemother. Res: Epidemiology and hemodynamics of hypertension and peripheral vascular disease. Mailing Add: Univ Health Serv Univ of Ga Athens GA 30601

BROWN, WALTER LYONS, b Charlottesville, Va, Oct 11, 24; m 46; c 4. SOLID STATE PHYSICS. Educ: Duke Univ, BS, 45; Harvard Univ, AM, 48, PhD(physics), 51. Prof Exp: MEM TECH STAFF, BELL TEL LABS, 50- Mem: Fel Am Phys Soc. Res: Semiconductors; space physics; interaction of energetic particles with solids. Mailing Add: 138 Cambridge Dr Berkeley Heights NJ 07922

BROWN, WALTER REDVERS JOHN, physics, physiological optics, see 12th edition

BROWN, WALTER REED, science education, biology, see 12th edition

BROWN, WALTER VARIAN, b Leicester, Mass, Apr 3, 13; m 38; c 2. BOTANY. Educ: Brown Univ, AB, 37, ScM, 39; Duke Univ, PhD(bot), 43. Prof Exp: Instr biol, Greensboro Col, 42; instr bot, Brown Univ, 46-47; from asst prof to assoc prof, 47-60, PROF BOT, UNIV TEX, AUSTIN, 61- Mem: Bot Soc Am. Res: Cytogenetics of grasses; cytological studies in Gramineae; plant cytology; genetics; evolution and plant taxonomy. Mailing Add: Dept of Bot Univ of Tex Austin TX 78712

BROWN, WARREN SHELBURNE, JR, b Loma Linda, Calif, Sept 8, 44; m 66; c 2. PSYCHOPHYSIOLOGY. Educ: Pasadena Col, BA, 66; Univ Southern Calif, MA, 69, PhD(exp psychol), 71. Prof Exp: Trainee neurosci, Brain Res Inst, 71-73; asst res psychologist, Sch Med, 73-75, ASST PROF PSYCHOL, SCH MED, UNIV CALIF, LOS ANGELES, 75- Concurrent Pos: NIMH spec res fel, 75 & res scientist develop award, 75- Mem: AAAS; Am Psychol Asn; Soc Neurosci; NY Acad Sci. Res: Brain mechanisms of higher mental functions in humans, particularly language; electroencephalography as a means of studying cortical events. Mailing Add: 1419 Bresee Ave Pasadena CA 91104

BROWN, WELDON GRANT, b Saskatoon, Sask, Feb 4, 08; nat US; m. CHEMISTRY. Educ: Univ Sask, BSc, 27, MSc, 28; Univ Calif, PhD(chem), 31. Hon Degrees: LLD, Univ Sask, 59. Prof Exp: Nat Res fel, Univ Chicago, Univ Berlin & Univ Frankfurt, 31-33; res chemist, Titanium Pigment Co, NY, 33-34; assoc, Columbia Univ, 34-35; from asst prof to assoc prof, 35-70, PROF CHEM, UNIV CHICAGO, 70- Concurrent Pos: Sr chemist, Argonne Nat Lab, 64-73. Mem: AAAS; Am Chem Soc; The Chem Soc. Res: Thermochemistry; molecular spectra; organic tracer research; synthetic applications of metal hydrides; organic radiation chemistry; explosions and industrial accidents. Mailing Add: 5269 Meyer Dr Lisle IL 60532

BROWN, WENDELL STIMPSON, b Pompton Plains, NJ, Apr 4, 43; m 72. PHYSICAL OCEANOGRAPHY. Educ: Brown Univ, BS, 65, MS, 67; Mass Inst Technol, PhD(oceanog), 71. Prof Exp: Asst res oceanog, Inst Geophys & Planetary Physics, Univ Calif, San Diego, 70-74; ASST PROF OCEANOG, UNIV NH, 74- Mem: Am Geophys Union; Sigma Xi. Res: Seafloor pressure fluctuations on the continental shelf and in the deep ocean; estuarine circulation. Mailing Add: Dept of Earth Sci Univ of NH Durham NH 03824

BROWN, WILLARD ANDREW, b Seattle, Wash, Nov 13, 21; m 48; c 2. SCIENCE EDUCATION. Educ: Univ Wash, BS, 46; Wash State Col, MAT, 58; Univ Fla, EdD(phys sci), 63. Prof Exp: Data analyst, Boeing Co, 48-49; teacher high sch, 51-56; asst prof educ & supvr student teachers, Western Wash State Col, 56-60; asst prof physics, Cent Wash State Col, 60-61; asst prof phys sci & astron, San Francisco State Col, 63-66; coordr sci educ, 66-69, PROF PHYSICS & ASTRON, WESTERN WASH STATE COL & DIR PLANETARIUM, 66- Mem: Am Asn Physics Teachers; Nat Asn Res Sci Teaching. Res: Data processing; data analysis for implementation of educational objectives; air pollution control data analysis. Mailing Add: Dept of Physics & Astron Western Wash State Col Bellingham WA 98225

BROWN, WILLARD BRUCE, analytical chemistry, inorganic chemistry, see 12th edition

BROWN, WILLIAM ANDERSON, b Corpus Christi, Tex, Dec 21, 29; m 52; c 2. INORGANIC CHEMISTRY. Educ: Univ Tex, BS, 51, MA, 53, PhD(chem), 60. Prof Exp: Res scientist, Defense Res Labs, Univ Tex, 53-55; res engr, Alcoa Res Labs, Aluminum Co Am, 59-64; res chemist, Houston Res Lab, Indust Chem Div, Shell Chem Co, 64-68; res chemist, Ashland Chem Co, 68-71, SR RES CHEMIST, ASHLAND OIL, INC, 71- Mem: Am Chem Soc; Soc Appl Spectros; Fine Particle Soc. Res: Coordination compounds; catalysis; organic coatings; corrosion; fine particles; elastomers; surface chemistry. Mailing Add: 2694 McVey Blvd W Worthington OH 43085

BROWN, WILLIAM ARNOLD, b New York, NY, Feb 13, 33; m 57; c 2. ATOMIC PHYSICS. Educ: Cornell Univ, AB, 55; Univ Mich, MS, 60, PhD(physics), 65. Prof Exp: Res assoc, Inst Sci & Technol, Univ Mich, 57-61; RES SCIENTIST, PHYS SCI DEPT, LOCKHEED PALO ALTO RES LAB, LOCKHEED MISSILES & SPACE CO, 64- Mem: AAAS; Am Phys Soc; Optical Soc Am. Res: X-ray astronomy and solar physics; statistical analysis and design of experiments; shock tube spectroscopy; vacuum ultraviolet spectroscopy; atomic and molecular collisions. Mailing Add: 3525 Greer Rd Palo Alto CA 94303

BROWN, WILLIAM BERNARD, b Marinette, Wis, Apr 3, 36; m 58; c 3. POLYMER CHEMISTRY. Educ: Univ Wis, BS, 58; Univ Akron, PhD(polymer sci), 66. Prof Exp: Res chemist, Firestone Tire & Rubber Co, Ohio, 60-61 & Standard Oil Co (Ohio), 65-68; SR RES CHEMIST, GEN MOTORS RES LABS, 68- Mem: Am Chem Soc; Electrochem Soc; Am Inst Physics. Res: Chemical and physical aspects of providing protective coating to surfaces, including the failures of coatings due to both substrate deterioration and deterioration of the coatings. Mailing Add: 877 Hazelwood Birmingham MI 48009

BROWN, WILLIAM DUANE, b Hamlin, Tex, Dec 8, 29; m 54; c 1. BIOCHEMISTRY. Educ: Univ Tex, BS, 49, MA, 51, PhD(chem), 55. Prof Exp: Res chemist, Univ Calif, Davis, 55-59; asst res marine food scientist, Univ Calif, Berkeley, 59-63, from asst prof to assoc prof nutrit, 63-68, prof marine food sci, 68-70; PROF MARINE FOOD SCI, UNIV CALIF, DAVIS, 70- Mem: AAAS; Am Soc Biol Chem; Am Chem Soc; Inst Food Technol; Am Inst Nutrit. Res: Comparative biochemistry of hemoproteins and muscle proteins. Mailing Add: Inst of Marine Resources Univ of Calif Davis CA 95616

BROWN, WILLIAM E, b Oxford, Pa, May 26, 33; m 56; c 3. MICROBIOLOGY, FOOD SCIENCE. Educ: Univ Del, BS, 56; Univ Ga, MS, 60, PhD(food sci), 64. Prof Exp: Bacteriologist, Amerlab Inc, Ga, 56-58; res asst virol, Univ Ga, 58-60, res asst food sci, 60-64; microbiologist, Cornell Univ, 64-66; MICROBIOLOGIST, BIO-LAB INC, 67- Mem: Poultry Sci Asn; Inst Food Technol. Res: Environmental sanitation; enzymology. Mailing Add: Bio-Lab Inc PO Box 1489 Decatur GA 30032

BROWN, WILLIAM EMMETT, physical chemistry, see 12th edition

BROWN, WILLIAM ERNEST, b Benton Harbor, Mich, Aug 29, 22; m 44; c 3. DENTISTRY. Educ: Univ Mich, DDS, 45, MS, 47. Prof Exp: From instr to prof dent, Univ Mich, 47-69; actg provost, Health Sci Ctr, 73-75, DEAN, COL DENT, UNIV OKLA, 69- Concurrent Pos: Assoc dir, W K Kellogg Found Inst, 62-69; chmn, Am Bd Pedodontics, 64-65; mem dent educ rev comt, Bur Health Prof Educ & Manpower Training, 71-73; mem coun dent educ, Am Dent Asn, 74-; mem comn accreditation of dent & dent auxiliary educ progs, 75-; proj dir, Nat Dent Curric Study, 75-77. Mem: Am Soc Dent for Children (pres, 59-60); Am Acad Pedodontics (pres, 63-64); Am Dent Asn. Res: Vital partial pulpectomy technique; evaluation of the dental assistant utilization program at the University of Michigan. Mailing Add: Univ of Okla Sch of Dent Oklahoma City OK 73104

BROWN, WILLIAM EVERETT, b Kaslo, BC, Apr 13, 21; nat US; m 45; c 3. BIOCHEMISTRY, MICROBIOLOGY. Educ: Univ Alta, BSc, 43, MSc, 45; Univ Wis, PhD(biochem), 49. Prof Exp: Microbiologist, Nat Res Coun Can Labs, 45-46; microbiologist, Merck & Co, Ltd, Montreal, 49-50; head fermentation sect, Microbiol Pilot Plant, Com Solvents Corp, 50-51; res supvr, Dept Microbiol Develop, E R Squibb & Sons Div, Olin Mathieson Chem Corp, 51-61, dir, 6268, DIR MICROBIOL, SQUIBB INST MED RES, SQUIBB CORP, 68- Concurrent Pos: Consult, US Dept Health, Educ & Welfare. Honors & Awards: Can Soc Tech Agr Gold Medal, 43. Mem: AAAS; Brit Soc Gen Microbiol; NY Acad Sci; Am Chem Soc; Am Soc Microbiol. Res: Production of organic compounds biosynthetically; production of antibiotics, vitamins, dextran and therapeutic steroids; discovery of new anti-infectives and antibiotics. Mailing Add: Squibb Inst for Med Res Princeton NJ 08540

BROWN, WILLIAM FRANCIS, b Indianapolis, Ind, Mar 9, 29; m 50; c 2. INFORMATION SCIENCE. Educ: Purdue Univ, BS, 50; Butler Univ, MS, 62. Prof Exp: Assoc biochemist, 50-52, biochemist, 52-56, sr process engr, 56-59, statistician, 59-62, sr statistician, 62-66, res scientist, 66-75, RES ASSOC, ELI LILLY & CO, 75- Mem: Am Chem Soc; Am Soc Info Sci. Res: Applied statistics and biometrics as related to pharmaceutical and biological manufacture, development and control; development of scientific information systems; basic studies in information storage and retrieval. Mailing Add: Sci Info Serv Eli Lilly & Co Indianapolis IN 46201

BROWN, WILLIAM FRANCIS, b Cedar City, Utah, Feb 10, 35; m 61; c 3. VETERINARY MEDICINE. Educ: Wash State Univ, DVM, 60; Univ Minn, PhD(vet obstet), 67. Prof Exp: Instr vet med, Univ Minn, 60-61, instr vet obstet, 61-66; asst dir vet reproductive physiol, Syntex Res Ctr, Calif, 67-70; VPRES & GEN MGR, LIVESTOCK GENETICS DIV, SYNTEX AGRIBUSINESS, INC, 70- Res: Veterinary obstetrics and gynecology; animal reproductive physiology and pathology. Mailing Add: 1143 Par Rd Broomfield CO 80020

BROWN, WILLIAM FULLER, JR, b Lyon Mountain, NY, Sept 21, 04; m 36; c 1. PHYSICS. Educ: Cornell Univ, AB, 25; Columbia Univ, PhD(physics), 37. Prof Exp: Teacher, Pvt Sch, NC, 25-27; lectr physics, Columbia Univ, 28-38; asst prof, Princeton Univ, 38-43; sr physicist, Naval Ord Lab, 43-45; res physicist, Sun Oil Co, 46-55; sr res physicist, Minn Mining & Mfg Co, 55-57; prof elec eng, 57-73, EMER PROF ELEC ENG, UNIV MINN, MINNEAPOLIS, 73- Concurrent Pos: Instr, Brooklyn Col, 37-38; contract employee, Naval Ord Lab, 41-43; Fulbright scholar, Weizmann Inst Sci, Israel, 62; guest prof, Max Planck Inst Metall Res, 63-64. Honors & Awards: A Cressy Morrison Award, NY Acad Sci, 67. Mem: Fel AAAS; fel Am Phys Soc; Opers Res Soc Am; NY Acad Sci; fel Inst Elec & Electronics Engr. Res: Ferromagnetic domains and magnetomechanical effects; electromagnetics; elasticity and plasticity; applied theoretical physics; dielectrics; machine calculations; statistics; micromagnetics. Mailing Add: Dept of Elec Eng Univ of Minn Minneapolis MN 55455

BROWN, WILLIAM G, b Toronto, Ont, June 13, 38; m 69. MATHEMATICS. Educ: Univ Toronto, BA, 60, PhD(math), 63; Columbia Univ, MA, 61. Prof Exp: Asst prof math, Univ BC, 63-66; ASSOC PROF MATH, McGILL UNIV, 66- Mem: Am Math Soc; Math Asn Am; Can Math Cong; London Math Soc. Res: Combinatorial analysis; graph theory. Mailing Add: Dept of Math McGill Univ CP 6070 Montreal PQ Can

BROWN, WILLIAM HAMILTON, organic chemistry, see 12th edition

BROWN, WILLIAM HEDRICK, b Yakima, Wash, Mar 18, 33; m 57; c 2. ANIMAL NUTRITION. Educ: State Col Wash, BS, 55; Univ Md, MS, 57, PhD(animal nutrit), 59. Prof Exp: From asst prof to assoc prof dairy sci, 59-69, PROF ANIMAL SCI, UNIV ARIZ, 69- Concurrent Pos: Res grants, Dept HEW, 60-63, 64-67 & NSF, 62-64. Mem: Am Dairy Sci Asn; Sigma Xi. Res: Roughage utilization by ruminants; lipid and mineral metabolism; biochemistry of the bovine rumen; pesticide residues in milk; physiology of lactation. Mailing Add: Dept of Animal Sci Univ of Ariz Tucson AZ 85721

BROWN, WILLIAM HENRY, b Ogdensburg, NY, Oct 26, 32; m 60; c 4. ORGANIC CHEMISTRY, BIOCHEMISTRY. Educ: St Lawrence Univ, BS, 54; Harvard Univ, MA, 55; Columbia Univ, PhD(chem), 58. Prof Exp: From instr to asst prof org chem, Wesleyan Univ, 58-64; from asst prof to assoc prof, 64-72, PROF ORG CHEM, BELOIT COL, 73- Concurrent Pos: Res fel org chem, Calif Inst Technol, 63-64; res assoc, Univ Ariz, 70-71. Mem: Am Chem Soc. Res: Synthetic organic chemistry and chemistry of natural products. Mailing Add: Dept of Chem Beloit Col Beloit WI 53511

BROWN, WILLIAM JANN, b Camden, NJ, Nov 15, 19; m 47; c 1. NEUROPATHOLOGY. Educ: Univ Pa, AB, 43, MD, 46. Prof Exp: USPHS res fel,

Dept Path, Montefiore Hosp, New York, 50-52; from instr to assoc prof, 52-62, PROF PATH, SCH MED, UNIV CALIF, LOS ANGELES, 62-, PROF PSYCHIAT, NEUROPSYCHIAT INST, 72- Concurrent Pos: Attend pathologist, Los Angeles County Harbor Gen Hosp & Vet Admin Ctr, 52-58 & Vet Admin Hosp, Long Beach, 53-58; consult, Am Bd Anat, Path & Neuropath, Japan, 59-, Nat Path Training Grant Comt, 61-64, Wadsworth & Long Beach Vet Admin Hosps, Los Angeles County Harbor Gen Hosp, Camarillo State Hosp & Los Angeles County & Calif State Tumor Registry. Mem: AAAS; Am Acad Neurol; Am Acad Cerebral Palsy; NY Acad Sci. Res: Ultramicrochemical studies of peripheral nerves; enzyme changes in brain structures and nervous diseases, microchemically and histochemically; lysosomes in brain; heavy metal effects in brain; electron microscopy in central nervous system. Mailing Add: Dept of Path Univ of Calif Sch Med Los Angeles CA 90024

BROWN, WILLIAM JEFFREY, b Belper, Eng, Mar 30, 38. GEOGRAPHY. Educ: Univ London, BSc, 59, PhD(geomorphol), 65. Prof Exp: Lectr, 62-64, asst prof, 64-69, ASSOC PROF GEOG, UNIV MAN, 69-, HEAD DEPT, 74- Concurrent Pos: Res awards, Nat Res Coun Coun & Nat Adv Comt Geog Res, 67-68. Mem: Asn Am Geogr; Am Geog Soc; Can Asn Geogr; Inst Brit Geogr. Res: Drainage development in central southern England; slope development and cliff modification in western Manitoba. Mailing Add: Dept of Geog Univ of Man Winnipeg MB Can

BROWN, WILLIAM JOHN, b Ashland, Pa, June 7, 40; m 66; c 1. MICROBIOLOGY. Educ: Univ Scranton, BS, 63; Duquesne Univ, MS, 65; WVa Univ, PhD(microbiol), 69. Prof Exp: From instr to asst prof microbiol, 69-74, ASSOC PROF IMMUNOL & MICROBIOL, SCH MED, WAYNE STATE UNIV, 74- Concurrent Pos: Mich Kidney Found grant; consult microbiol, Detroit Gen Hosp. Mem: Am Soc Microbiol. Res: Basic research on lymphocytic choriomeningitis virus, particularly the virus-cell relationship; urinary tract infections from both the viral and bacteriological aspects. Mailing Add: Wayne State Univ Sch of Med 1400 Chrysler Freeway Detroit MI 48207

BROWN, WILLIAM LACY, b Arbovale, WVa, July 16, 13; m 42; c 2. GENETICS, PLANT BREEDING. Educ: Bridgewater Col, AB, 36; Wash Univ, MS, 40, PhD(cytogenetics), 42. Prof Exp: Teacher, High Sch, Va, 36; tech asst, Agr Exp Sta, Agr & Mech Col Tex, 36-37; geneticist, Green Sect, US Golf Asn, 41-42 & Rogers Bros Seed Co, 42-45; geneticist, Dept Plant Breeding, Pioneer Hi-Bred Corn Co, 45-58, asst dir res, 58-65, vpres & dir res, Pioneer Hi-Bred Int Inc, 65-73, exec vpres & dir res, 73-75; PRES, PIONEER HI-BRED INT, INC, 75- Concurrent Pos: Fulbright res scholar, Imp Col Trop Agr, Trinidad, 52-53; extramural prof, Wash Univ, 57-71; univ fel, Drake Univ, 70- Mem: AAAS; Genetics Soc Am; Bot Soc Am; Am Soc Plant Taxon; Am Soc Agron. Res: Cytogenetics of grasses; maize genetics and breeding. Mailing Add: Pioneer Hi-Bred Int Inc 1206 Mulberry St Des Moines IA 50308

BROWN, WILLIAM LEWIS, b Clover, SC, Nov 23, 28; m 51. FOOD MICROBIOLOGY. Educ: Clemson Col, BS, 49; NC State Col, MS, 51; Univ Ill, PhD(food microbiol), 56. Prof Exp: Asst, NC State Col, 49-51, instr meats, 51-54; asst, Univ Ill, 56; res food technologist, John Morrell & Co, 56-57, asst dir res, 57-64, vpres res, 64-67; PRES, AM BACT & CHEM RES CORP, 67- Mem: Fel Am Soc Microbiol; Sigma Xi. Res: Bacteriology; nutrition; biochemistry. Mailing Add: 3437 SW 24th Ave Gainesville FL 32608

BROWN, WILLIAM LOUIS, b Philadelphia, Pa, June 1, 22; m 46; c 3. BIOLOGY. Educ: Pa State Col, BS, 47; Harvard Univ, PhD, 50. Prof Exp: Asst, Pa State Col, 47; Parker fel from Harvard Univ, Australia, 50-51, Fulbright fel, 51-52; asst cur insects, Mus Comp Zool, Harvard Univ, 52-54, assoc cur, 54-60; from asst prof to assoc prof, 60-68, PROF ENTOM, CORNELL UNIV, 68- Concurrent Pos: Assoc, Mus Comp Zool, Harvard Univ, 60-; Guggenheim fel, 73-74. Honors & Awards: Donisthorpe Prize, 63. Mem: Am Entom Soc; Soc Syst Zool; Soc Study Evolution. Res: Systematics, ecology and behavior of ants; general evolutionary theory; zoogeography. Mailing Add: Dept of Entom Cornell Univ Ithaca NY 14850

BROWN, WILLIAM MALCOLM, JR, b Hastings, Nebr, June 20, 35; m 69; c 2. PLANT PATHOLOGY. Educ: Univ Calif, Davis, BSc, 58; Ore State Univ, PhD(plant path), 65. Prof Exp: Plant pathologist surv, Calif State Dept Agr, 58-59; asst bot, Ore State Univ, 59-62, instr, 62-65; plant pathologist, Colo State AID Nigeria Proj, 65-68; plant pathologist, Univ Ky AID Thailand Proj, 68-72; EXTEN SPECIALIST, FOOD AGR ORGN, KOREA PLANT PROTECTION PROJ, 73- Mem: Am Phytopath Soc; Int Soc Plant Pathologist; Korean Plant Protection Soc; Nigerian Soc Plant Protection; Agr Soc Thailand. Res: Characterization and control of diseases of legumes, especially peanut and soybean and tropical food and fiber crops. Mailing Add: FAO Plant Protection Serv Via delle Terme di Caracalla Rome Italy

BROWN, WILLIAM PAUL, b Muncie, Ind, Sept 26, 29; m 51; c 3. GENETICS, ANIMAL PHYSIOLOGY. Educ: Purdue Univ, BS, 51, MS, 56, PhD(genetics), 61. Prof Exp: Instr genetics, Purdue Univ, 59-60 & Univ Conn, 60-62; asst prof genetics & animal physiol, 62-68, ASSOC PROF BIOL, MARIETTA COL, 68- Mem: Am Inst Biol Sci. Res: Experimental population genetics; selection methods for fitness traits; plateaued populations. Mailing Add: Dept of Biol Marietta Col Marietta OH 45750

BROWN, WILLIAM RANDALL, b Staunton, Va, Oct 31, 13; m 42; c 3. GEOLOGY. Educ: Univ Va, BS, 38, MA, 39; Cornell Univ, PhD(mineral), 42. Prof Exp: From asst to assoc geologist, Va Geol Surv, 42-45; from asst prof to assoc prof, 45-52, PROF GEOL, UNIV KY, 52- Concurrent Pos: Consult, Calif Co, Shell Oil Co & US Geol Surv, 64- Mem: Fel AAAS; Soc Econ Paleont & Mineral; fel Geol Soc Am; Am Asn Petrol Geol. Res: Igneous, metamorphic and structural geology of Piedmont Province, Virginia; mica and feldspar deposits of Virginia; Piedmont zinc and lead deposits of Virginia; park basins of Colorado; Wind River Basin, Wyoming; Wyoming overthrust belt; Pennsylvanian of eastern Kentucky. Mailing Add: 253 Shady Lane Lexington KY 40503

BROWN, WILLIAM SAMUEL, JR, b Pottstown, Pa, Apr 25, 40; m 62; c 2. SPEECH PATHOLOGY. Educ: Edinboro State Col, BS, 62; State Univ NY Buffalo, MA, 67, PhD(speech sci), 69. Prof Exp: Speech & hearing therapist, Crawford County Operated Classes Handicapped, 62-65; res asst speech sci, State Univ NY Buffalo, 65-68; res assoc & fel, Commun Sci Lab, 68-70, ASST PROF SPEECH, UNIV FLA, 70-, AREA HEAD PHONETIC SCI, COMMUN SCI LAB, 74- Mem: Sigma Xi; Int Soc Phonetic Sci; Am Asn Phonetic Sci (treas, 73-76); Am Speech & Hearing Asn; Acoust Soc Am. Res: Experimental phonetics; physiological and aerodynamic study of speech articulatory behavior and laryngeal function utilizing both normal and pathological speakers. Mailing Add: Commun Sci Lab Dept of Speech Univ of Fla 35 ASB Gainesville FL 32611

BROWN, WILLIAM STANLEY, b Cleveland, Ohio, Feb 13, 35; m 56; c 2. COMPUTER SCIENCE. Educ: Yale Univ, BS, 56; Princeton Univ, PhD(physics), 61. Prof Exp: Instr physics, Princeton Univ, 60-61; mem tech staff, Math Physics Dept, 61-66, HEAD COMPUT MATH RES DEPT, BELL LABS, 66- Concurrent Pos: Mem, Chem Abstr Serv Adv Bd, 72-74; ed, Asn Comput Mach Trans on Math Software, 74. Mem: AAAS; Math Asn Am; Asn Comput Mach. Res: Computer algebra and languages; information transfer. Mailing Add: Bell Labs Murray Hill NJ 07974

BROWN, WILLIAM WARD, physics, see 12th edition

BROWNAWELL, WOODROW DALE, b Grundy Co, Mo, Apr 21, 42; m 65; c 1. NUMBER THEORY. Educ: Univ Kans, BA, 64; Cornell Univ, PhD(math), 70. Prof Exp: Asst prof math, Pa State Univ, 70-74; vis assoc prof, Univ Colo, 74-75; ASSOC PROF MATH, PA STATE UNIV, 75- Mem: Am Math Soc; Math Asn Am. Res: Independence properties of numbers and functions arising in classical analysis, including especially the exponential function and the Weierstrass elliptic functions. Mailing Add: Dept of Math Pa State Univ University Park PA 16802

BROWNE, CHARLES IDOL, b Atlanta, Ga, Feb 8, 22; m 42; c 1. RADIOCHEMISTRY. Educ: Drew Univ, AB, 41; Calif Inst Technol, MS, 48; Univ Calif, PhD(chem), 52. Prof Exp: Alt leader, Radiochem Group, 55-65, assoc div leader, Weapons Test Div, 65-70, alt div leader, 70-72, div leader, 72-74, ASST DIR, LOS ALAMOS SCI LAB, 74- Mem: Fel Am Phys Soc; fel Am Inst Chem. Res: Radiochemical diagnostics. Mailing Add: Los Alamos Sci Lab PO Box 1663 Los Alamos NM 87544

BROWNE, CHARLES MERMYN, solid state physics, see 12th edition

BROWNE, COLIN LANFEAR, b Buffalo, NY, Apr 11, 28; m 49; c 2. TEXTILE CHEMISTRY. Educ: Lafayette Col, BA, 49; Univ Va, MS, 51, PhD(chem), 53. Prof Exp: Textile res chemist, Rohm and Haas Co, 53-57; mgr indust prod develop, 57-66, mgr Dyeing & Finishing Lab, 66-70, mgr dyeing & finishing tech serv & develop, 70-74, RES & DEVELOP MGR, CELANESE FIBER CO, CELANESE CORP, CHARLOTTE, NC, 74- Mem: Am Chem Soc; Am Asn Textile Chemists & Colorists. Res: Development of smoking products and filters. Mailing Add: 5 Timberidge Dr Clover SC 29710

BROWNE, CORNELIUS PAYNE, b Madison, Wis, Oct 30, 23; m 57; c 2. NUCLEAR PHYSICS. Educ: Univ Wis, AB, 46, PhD(physics), 51. Prof Exp: Res assoc, Mass Inst Technol, 51-56; asstprof, 56-57, assoc prof, 58-64, PROF PHYSICS, UNIV NOTRE DAME, 64- Concurrent Pos: Vis prof, Univ Tex, Austin, 72-73. Mem: Am Asn Physics Teachers; fel Am Phys Soc; Sigma Xi. Res: Measurement of nuclear reaction energies and nuclear energy levels; nuclear reaction mechanisms. Mailing Add: Dept of Physics Univ of Notre Dame Notre Dame IN 46556

BROWNE, DOUGLAS TOWNSEND, organic chemistry, biochemistry, see 12th edition

BROWNE, EDMUND BROADUS, plant breeding, see 12th edition

BROWNE, EDWARD TANKARD, JR, b Chapel Hill, NC, Dec 17, 26; m 50; c 3. BOTANY. Educ: Univ NC, BA, 48, MA, 50, PhD(bot), 57. Prof Exp: Head bot slide dept, Carolina Biol Supply Co, Elon Col, 52-55; asst prof bot, Auburn Univ, 56-59 & Univ Ga, 59-60; from asst prof to assoc prof, Univ Ky, 60-67; assoc prof, 67-68, PROF BOT, MEMPHIS STATE UNIV, 68- Mem: Bot Soc Am; Am Soc Plant Taxon. Res: Contributions of plant embryology to systematic botany; flora of Kentucky; Liliaceae and Aletris systematics. Mailing Add: Dept of Biol Memphis State Univ Memphis TN 38152

BROWNE, JAMES CLAYTON, b Conway, Ark, Jan 16, 35; m 59; c 3. MOLECULAR PHYSICS, COMPUTER SCIENCE. Educ: Hendrix Col, BA, 56; Univ Tex, PhD(chem), 60. Prof Exp: Asst prof physics, Univ Tex, 60-64; NSF fel, 64-65; prof comput sci, Queen's Univ, Belfast, 65-68; PROF PHYSICS & COMPUT SCI, UNIV TEX, AUSTIN, 68-, RES SCIENTIST, 73- Concurrent Pos: Consult, NSF. Mem: Am Phys Soc; Asn Comput Mach; Soc Indust & Appl Math. Res: Atomic and molecular processes; operating systems; symbolic mathematics. Mailing Add: Dept of Physics Univ of Tex Austin TX 78712

BROWNE, JOHN C, b Pottstown, Pa, July 29, 42. EXPERIMENTAL NUCLEAR PHYSICS. Educ: Drexel Inst Technol, BS, 65; Duke Univ, PhD(nuclear physics), 69. Prof Exp: Instr basic physics, Duke Univ, 69-70; RES PHYSICIST, LAWRENCE LIVERMORE LAB, UNIV CALIF, 70- Concurrent Pos: Mem subcomt neutron data appln, US Nuclear Data Comt, 73-75. Mem: Am Phys Soc. Res: Fine structure of analog states; elastic proton scattering; neutron physics; nuclear astrophysics; nuclear fission Mailing Add: L-221 Lawrence Livermore Lab Univ of Calif Livermore CA 94550

BROWNE, JOHN MCDONALD, b Miami, Fla, Aug 13, 37; m 58; c 3. DEVELOPMENT BIOLOGY. Educ: Bethune-Cookman Col, BS, 62; Univ Miami, MS, 68, PhD(develop biol), 70. Prof Exp: Instr biol, Bethune-Cookman Col, 62-65; NIH res asst develop biol, Lab Quant Biol, Univ Miami, 65-68, teaching asst, Dept Biol, 68-69, res assoc, Lab Quant Biol, 70; assoc prof biol, Bethune-Cookman Col, 70-71; ASSOC PROF BIOL, ATLANTA UNIV, 71- Mem: Am Soc Zool; Soc Develop Biol; Am Soc Cell Biol; Teratol Soc. Res: Fluid and electrolyte imbalances role on development in avian and mammalian systems; effects of drugs on physiological activities in developing avian and mammalian embryos. Mailing Add: Dept of Biol Atlanta Univ Atlanta GA 30314

BROWNE, JOHN SYMONDS LYON, b London, Eng, Apr 13, 04. ENDOCRINOLOGY. Educ: McGill Univ, BA, 26, BSc, 27, MD, CM, 29, PhD(biochem), 32; FRCPS(C); FRCP(C). Prof Exp: Lectr med, 33-38, from asst prof to prof, 38-55, chmn dept, 47-55, prof invest med & chmn dept, 55-69, prof exp med, 69-71, EMER PROF MED, McGILL UNIV, 71 Concurrent Pos: Royal Soc Can traveling fel, 32-33; physician, Royal Victoria Hosp, 50-66, hon consult, 66-; dir, Univ Clin, 47-55. Honors & Awards: Distinguished Leadership Award, Endocrine Soc, 73. Mem: Can Soc Clin Invest; Asn Am Physicians; Am Soc Clin Invest (past pres); fel Am Col Physicians; Endocrine Soc (past pres). Res: Estrogenic substances; salt balances in Addison's Disease; sex hormone excretion in clinical cases; corpus luteum function; cortin in urine; histamine metabolism; shock; protein metabolism in illness; adaptation syndrome in man; adrenal cortical response to stress. Mailing Add: Apt 100 900 Sherbrooke St W Montreal PQ Can

BROWNE, MARJORIE LEE, mathematics, see 12th edition

BROWNE, MERVIN FOWLER, organic chemistry, see 12th edition

BROWNE, MICHAEL EDWIN, b Los Angeles, Calif, June 12, 30; m 51; c 7. PHYSICS. Educ: Univ Calif, Berkeley, BS, 52, PhD(physics), 55. Prof Exp: Assoc prof, San Jose State Col, 58-67; PROF PHYSICS & CHMN DEPT, UNIV IDAHO, 67- Concurrent Pos: Staff scientist, Lockheed Res Labs, 55-67; vis res physicist,

Physics Inst, Zurich, 63-64. Mem: Am Phys Soc; Am Asn Physics Teachers. Res: Solid state physics. Mailing Add: Dept of Physics Univ of Idaho Moscow ID 83843

BROWNE, PHILIP LINCOLN, b Calcutta, India, Feb 12, 18; m 48; c 2. HYDRODYNAMICS. Educ: Denison Univ, AB, 39; Mich State Univ, MA, 41; Ohio State Univ, PhD(physics), 51. Prof Exp: Instr math, Ill Inst Technol, 41-44; STAFF MEM GROUP TD5, LOS ALAMOS SCI LAB, 51- Mem: Am Phys Soc. Res: Numerical computations; high speed computers. Mailing Add: Group TD5 Los Alamos Sci Lab Los Alamos NM 87544

BROWNE, THEODORE CROWNINSHIELD, chemistry, deceased

BROWNE, WILLIAM RUTHERFORD, organic chemistry, see 12th edition

BROWNELL, ARNOLD S, b Idaho Falls, Idaho, May 30, 21; m 45; c 2. BIOPHYSICS. Educ: Univ Wyo, BS, 48; Univ Calif, Berkeley, PhD(biophys), 55. Prof Exp: Biophysicist, US Army Med Res Lab, Ft Knox, Ky, 55-73, head dept nonionizing radiation, 63-65; WITH LASER SAFETY TEAM, FRANKFORD ARSENAL, 73- Mem: AAAS; Biophys Soc. Res: Radiation biology; effects of electromagnetic radiation of cellular systems. Mailing Add: Laser Safety Team Frankford Arsenal Philadelphia PA 19137

BROWNELL, FRANK HERBERT, III, b New York, NY, Sept 20, 22; m 50; c 4. MATHEMATICS. Educ: Yale Univ, BA, 43, MS, 47; Princeton Univ, PhD(math), 49. Prof Exp: Fine instr, Princeton Univ, 49-50; from asst prof to assoc prof, 50-60, PROF MATH, UNIV WASH, 60- Concurrent Pos: Ford fel, Inst Advan Study, 53-54. Mem: Am Math Soc. Res: Nonlinear delay differential equations; operations and measures on Hilbert space and applications to quantum mechanics. Mailing Add: Dept of Math Univ of Wash Seattle WA 98105

BROWNELL, GEORGE H, b Minneapolis, Minn, Oct 23, 38; m 67; c 2. MICROBIOLOGY, BACTERIAL GENETICS. Educ: Univ Minn, BA, 61; Univ SDak, MA, 63, PhD(microbiol), 67. Prof Exp: Asst prof, 67-73, ASSOC PROF MICROBIOL, MED COL GA, 73- Concurrent Pos: Brown-Hazen study grant genophore homologies in nocardia. Mem: AAAS; Am Soc Microbiol; Genetics Soc Am. Res: Nocardial genetics; microbial genetics. Mailing Add: Dept of Cell & Molecular Biol Med Col of Ga Augusta GA 30902

BROWNELL, GEORGE L, b Hoosick Falls, NY, May 20, 23; m 56; c 3. ORGANIC CHEMISTRY. Educ: Rensselaer Polytech, BS, 47, MS, 48; Ohio State Univ, PhD(org chem), 53. Prof Exp: Res chemist, Aluminum Res Labs, 53-55; asst prof, Lehigh Univ, 55-58; res chemist, Miles Chem Co, 58-60; supvr indust chem res, Chem Div, 60-67, SECT HEAD RES LAB, US STEEL CORP, 67- Mem: Am Chem Soc; Sigma Xi; Am Inst Chemists; The Chem Soc. Res: Electrolytic reduction, oxidation and substitution of organic compounds; pharmaceutical chemistry; chemistry of coal; lignites; esterifications; liquid phase oxidations; maleic anhydride chemistry; polymerization studies; hydroformylation studies; polyester resins; plasticizers. Mailing Add: Res Lab MS 57 US Steel Corp Monroeville PA 15146

BROWNELL, GORDON LEE, b Duncan, Okla, Apr 8, 22; m 44; c 4. MEDICAL PHYSICS. Educ: Bucknell Univ, BS, 43; Mass Inst Technol, PhD(physics), 50. Prof Exp: Asst physics, 48-50, res assoc, 50-57, from asst prof to assoc prof, 57-70, PROF NUCLEAR ENG, MASS INST TECHNOL, 70-; LECTR, HARVARD MED SCH, 50- Concurrent Pos: From asst physicist to assoc physicist, Mass Gen Hosp, 49-61, physicist, 61-, head physics res lab; trustee, Neuro-Res Found; hon fac mem, Cuyo Univ, Arg, 54. Mem: AAAS; Am Asn Physicists in Med; Soc Nuclear Med; Am Phys Soc; Biophys Soc. Res: Imaging of positron emitting isotopes; computerized axiol tomography; reactor applications; analysis of tracer data; radiation effects and dosimetry. Mailing Add: Mass Gen Hosp Fruit St Boston MA 02114

BROWNELL, JAMES RICHARD, b Jeannette, Pa, Apr 20, 32; m 51; c 4. SOILS. Educ: Pa State Univ, BS, 54; Univ Minn, MS, 57; Univ Calif, Davis, PhD(geobiol), 70. Prof Exp: Jr voc instr soils & irrig, Fresno State Col, 58-59, asst prof chem, 59-60; agriculturist, Di Giorgio Fruit Corp, 60-64; technician soil fertil, Univ Calif, Davis, 64-69; PROF SOILS, CALIF STATE UNIV, FRESNO, 69- Mem: Soil Sci Soc Am; Soil Conserv Soc Am. Res: Land use planning; waste disposal; environmental impact studies; reclamation of saline and sodic soils; remote sensing; water measurement; irrigation scheduling; soil selection for adobe brick fabrication. Mailing Add: Dept of Plant Sci Calif State Univ Fresno CA 93740

BROWNELL, JOSEPH WILLIAM, b New Bremen, NY, May 25, 27; m 55; c 3. GEOGRAPHY. Educ: Syracuse Univ, AB, 50, MA, 51, DSS, 58. Prof Exp: From asst prof to assoc prof, 56-67, PROF EARTH SCI, STATE UNIV NY COL CORTLAND, 67-, CHMN DEPT GEOG, 71- Concurrent Pos: State Univ Res Found res grant, 67-69. Mem: Asn Am Geogr; Am Geog Soc; Nat Coun Geog Educ; Can Asn Geogr. Res: Land use; hamlets in areas of agricultural decline. Mailing Add: Dept of Geog State Univ of NY Cortland NY 13045

BROWNELL, KATHARINE ANNA, b Buffalo, NY, Apr 16, 02; m 67. PHYSIOLOGY. Educ: Univ Buffalo, AB, 25, AM, 30; Ohio State Univ, PhD(physiol), 40. Prof Exp: Asst physiol, 26-30; asst vital econ, Univ Rochester, 30; asst physiol, 31-34; asst, 34-40, res assoc, 40-46, from instr to assoc prof, 46-70, PROF PHYSIOL, OHIO STATE UNIV, 70- Concurrent Pos: Corp mem, Marine Biol Lab, Mass. Mem: Am Chem Soc; Am Physiol Soc; Soc Exp Biol & Med; Endocrine Soc. Res: Adrenal in relation to carbohydrate and fat metabolism; adrenal and energy metabolism; adrenal and hypertension. Mailing Add: Dept of Physiol Ohio State Univ Columbus OH 43210

BROWNELL, WILLIAM BOWLES, organic chemistry, analytical chemistry, see 12th edition

BROWNER, ROBERT HERMAN, b New York, NY, June 19, 43; m; c 1. NEUROBIOLOGY, COMPARATIVE NEUROLOGY. Educ: NY Univ, BA, 65, MA, 70, PhD(biol & neurobiol), 73. Prof Exp: Instr, 73-75, ASST PROF ANAT, NY MED COL, VALHALLA, NY, 75- Mem: AAAS; Soc Neurosci; Am Zoologists; Sigma Xi; Asn Biol Photogrs. Res: Comparative neurobiology of the central auditory pathway in vertebrates; light microscopy, electron microscopy, neurophysiology. Mailing Add: Dept of Anat NY Med Col Valhalla NY 10595

BROWNFIELD, ROBERT BRUCE, b Dillon, Mont, Oct 12, 27; m 51; c 3. ANALYTICAL CHEMISTRY, ORGANIC CHEMISTRY. Educ: Occidental Col, AB, 53; Univ Calif, Los Angeles, PhD(org chem), 58. Prof Exp: Teaching asst, Univ Calif, Los Angeles, 53-58; res chemist, Lederle Labs, 58-68; SR RES SCIENTIST, MILES LABS, 68- Mem: Am Chem Soc; fel Am Inst Chem. Res: Steroid chemistry; pharmaceutical analysis; qualitative and quantitative separation techniques. Mailing Add: Dome Labs Div Miles Labs 400 Morgan Lane West Haven CT 06516

BROWNIE, ALEXANDER C, b Bathgate, Scotland, Mar 6, 31; m 70; c 2. BIOCHEMISTRY. Educ: Univ Edinburgh, BSc, 52, PhD(biochem), 55. Prof Exp: Asst biochem, Univ Edinburgh, 53-55; Scottish Hosps Endowment Res Trust fel, Royal Infirmary, Edinburgh, 55-56; Organon res fel pharmacol, Univ St Andrews, 56-59, med res fel, 59-62, hon lectr, 60-62; USPHS fel, Univ Utah, 62-63; asst prof path & biochem, 63-67, assoc prof biochem, 67-70, PROF BIOCHEM, STATE UNIV NY BUFFALO, 70-, RES ASSOC PROF PATH, 67- Mem: Endocrine Soc. Res: Adrenal steroid biosynthesis; pituitary-adrenal function in disease; gas chromatography of steroids; hormones and hypertension. Mailing Add: Dept of Path Bell Facil State Univ of NY Buffalo NY 14207

BROWNING, CHARLES BENTON, b Houston, Tex, Sept 16, 31; m 56; c 6. ANIMAL NUTRITION. Educ: Tex Technol Col, BS, 55; Kans State Univ, MS, 56, PhD(animal nutrit), 58. Prof Exp: From asst prof to assoc prof dairy prod, Miss State Univ, 58-62, prof dairy sci, 62-66; chmn dept dairy sci, 66-69, DEAN COL AGR, UNIV FLA, 69- Mem: Am Soc Animal Sci; Am Dairy Sci Asn. Res: Dairy cattle nutrition and physiology. Mailing Add: Col of Agr Univ of Fla Gainesville FL 32601

BROWNING, DANIEL DWIGHT, b New Albany, Miss, Mar 24, 21; m 47; c 3. ORGANIC CHEMISTRY. Educ: Miss Col, BA, 41. Prof Exp: Control chemist, E I du Pont de Nemours & Co, 41; fel, Mellon Inst, 42-44 & 46-50; res chemist, 50-58, mgr paint res, 58-60, mgr plastic flooring res, 60-62, gen mgr bldg prod res, 62-68, ASST DIR RES, ARMSTRONG CORK CO, 68- Concurrent Pos: Mem, Bldg Res Inst. Mem: Am Chem Soc. Res: Building products. Mailing Add: 25 Eshelman Rd Lancaster PA 17601

BROWNING, EDWARD T, b Cleveland, Ohio, July 15, 39; m 68; c 2. PHARMACOLOGY, NEUROCHEMISTRY. Educ: Purdue Univ, BS, 61; Univ Ill Med Ctr, Chicago, MS, 64, PhD(pharmacol), 66. Prof Exp: Johnson Res Found fel phys biochem, Univ Pa, 66-68; asst prof, 68-74, ASSOC PROF PHARMACOL, RUTGERS MED SCH, 74- Res: Acetycholine synthesis; biochemistry of synaptic structures; choline metabolism. Mailing Add: Dept of Pharmacol Rutgers Med Sch Piscataway NJ 08854

BROWNING, GEORGE VERNON, physical chemistry, see 12th edition

BROWNING, HENRY (CHARLES), b Nottingham, Eng, Feb 25, 12; nat US; m 46; c 5. ANATOMY, ZOOLOGY. Educ: Bristol Univ, BSc, 37, EdDip, 38, PhD(zool), 42. Prof Exp: Demonstr zool, Bristol Univ, 39-41; lectr, Univ Sheffield, 41-43; res fel genetics, McGill Univ, 44-45, lectr zool, 46; sr res fel cancer, USPHS, 46-48; res assoc path, Sch Med, Yale Univ, 48-52, asst prof & res assoc anat & path, 50-52; assoc prof anat, Ind Univ, 52-53; assoc prof, Univ PR, 53; assoc prof, 54-57, PROF ANAT, UNIV TEX DENT BR HOUSTON, 57-, CHMN DEPT, 58-; PROF-IN-CHG ANAT, MED SCH, UNIV TEX HEALTH SCI CTR HOUSTON, 72- Concurrent Pos: Tech aide to chmn, Comt on Growth, Nat Res Coun, 48-51; vis prof anat, Med Col, Baylor Univ, 54-66; consult, Lackland AFB, 58-62; vis prof anat, Univ Houston, Univ St Thomas & Sacred Heart Col. Mem: AAAS; Am Soc Zoologists; Endocrine Soc; Soc Exp Biol & Med; Am Asn Anat. Res: Biology of neoplasia; comparative endocrinology. Mailing Add: Dept of Anat Univ of Tex Dent Br Houston TX 77025

BROWNING, HORACE LAWRENCE, JR, b Overton, Tex, Oct 8, 32; m 53; c 3. POLYMER CHEMISTRY, PHYSICAL CHEMISTRY. Educ: Stephen F Austin State Col, BA, 54; La State Univ, MS, 56, PhD(phys chem), 60. Prof Exp: Asst prof chem, Northeast La State Col, 60-63; res chemist, 63-65, SR RES CHEMIST, TENN EASTMAN CO, 65- Concurrent Pos: Spec lectr chem, King Col, 66-71. Mem: Am Chem Soc; fel Am Inst Chemists. Res: Gel permeation chromatography; ultracentrifugation; characterization of polymers; solution thermodynamics. Mailing Add: Tenn Eastman Co B 150A Kingsport TN 37662

BROWNING, IBEN, biophysics, see 12th edition

BROWNING, JOE LEON, b Huntington, WVa, June 24, 25; div; c 1. APPLIED CHEMISTRY, CHEMICAL ENGINEERING. Educ: Marshall Univ, BS, 47. Prof Exp: Chemist, C&O Rwy Co, 47-49; chemist, Allied Chem & Dye Corp, 49-50; chemist, Int Nickel Co, 50-51; head combustion studies, Naval Powder Factory, 51-53, head prod dept labs, 53-56, polaris rocket motor develop engr, Polaris Prog, Navy Spec Projs Off, 56-58, dir res & develop dept, Naval Propellant Plant, 58-62, TECH DIR, NAVAL ORD STA, NAVY DEPT, 62- Concurrent Pos: Mem, Md Govs Sci Adv Coun; mem, Navy Sr Scientists Coun; adj prof world bus, Thunderbird Grad Sch Int Mgt, 72-74. Honors & Awards: Pioneer in Space Award; Distinguished Civilian Serv Award & Distinguished Scientists Award, US Navy, 75. Mem: Am Inst Aeronaut & Astronaut; Am Defense Preparedness Asn; Am Chem Soc. Mailing Add: 7580 Bayside Lane Miami Beach FL 33141

BROWNING, JOHN ARTIE, b Kosse, Tex, Oct 3, 23; m 46; c 3. PLANT PATHOLOGY. Educ: Baylor Univ, BS, 47; Cornell Univ, PhD(plant path), 53. Prof Exp: Asst plant path, Cornell Univ, 49-53; from asst prof to assoc prof, 53-64, PROF PLANT PATH, IOWA STATE UNIV, 64- Concurrent Pos: Mem staff, Rockefeller Found agr prog, Colombia, 63-64. Mem: AAAS; Am Phytopath Soc; Am Soc Agron; Crop Sci Soc Am; Am Inst Biol Sci. Res: Diseases of small grains; epidemiology of cereal rusts; teaching epidemiology and plant pathology. Mailing Add: Dept of Bot & Plant Path Iowa State Univ Ames IA 50011

BROWNING, LUOLIN S, genetics, see 12th edition

BROWNING, ROBERT FRANCIS, invertebrate zoology, see 12th edition

BROWNING, ROBERT HAMILTON, b Oberlin, Ohio, Apr 7, 03; m 25, 58; c 6. MEDICINE. Educ: Oberlin Col, BA, 23; Western Reserve Univ, MD, 27; Am Bd Internal Med, dipl, 47. Prof Exp: From intern to asst resident med, Cleveland City Hosp, 27-29, asst resident surg, 29; resident, Trudeau Sanatorium, 30; supt & med dir, Cuyahoga County Tuberc Hosp, 31-48; dir col health serv, Oberlin Col, 48-51; prof, 51-73, EMER PROF MED, COL MED, OHIO STATE UNIV, 73- Concurrent Pos: Asst prof, Western Reserve Univ, 31-49; dir, Ohio Tuberc Hosp, 51-67; consult, US Vet Admin, 55-73 & USPHS, 56-72. Mem: Am Thoracic Soc (vpres, 62-63); fel, Am Col Physicians. Res: Pulmonary diseases. Mailing Add: 2138 Cheshire Rd Columbus OH 43221

BROWNLEE, JAMES LAWTON, JR, b Passaic, NJ, May 20, 32; m 58; c 3. NUCLEAR CHEMISTRY, RADIOCHEMISTRY. Educ: Univ Calif, Los Angeles, BS, 53; Univ Mich, MS, 59, PhD(chem), 60. Prof Exp: Eng asst, Union Oil Co Calif, 53; chemist, Wilmington-Dominguez Refining, Shell Oil Co, 56; chemist, Brea Res Ctr, Union Oil Co Calif, 57; asst nuclear chemist, Univ Mich, 58-60; from instr to asst prof chem, Univ Ill, 60-70; MEM STAFF RADIOCHEM DIV, LAWRENCE LIVERMORE LAB, UNIV CALIF, 70- Mem: Am Chem Soc. Res: Nuclear cross sections calculations measurement; gamma spectroscopy; nuclear instrumentation; radiochemical diagnostics; data analysis. Mailing Add: 3802 Pinot Ct Pleasanton CA 94566

BROWNLEE, PAULA PIMLOTT, b London, Eng, June 23, 34; m 61; c 3. ORGANIC CHEMISTRY. Educ: Oxford Univ, MA, 57, PhD(org chem), 59. Prof Exp: Res fel org chem, Univ Rochester, 59-61; res chemist, Stamford Res Labs, Am Cyanamid Co, 61-62; from instr to asst prof chem, 70-75, assoc dean, Douglass Col, 72-75, ACTG DEAN DOUGLASS COL & ASSOC PROF CHEM, RUTGERS UNIV, 75- Mem: The Chem Soc; Am Chem Soc. Res: Peptide synthesis; organic sulfur compounds; charge-transfer complexes incorporating polymers. Mailing Add: Douglass Col Rutgers Univ New Brunswick NJ 08903

BROWNLEE, ROBERT REX, b Zenith, Kans, Mar 4, 24; m 43; c 5. ASTROPHYSICS. Educ: Sterling Col, AB, 47; Univ Kans, MA, 51; Ind Univ, PhD(astron), 55. Hon Degrees: DSc, Sterling Col, 66. Prof Exp: Staff mem, 55-68, nuclear tech dir, 68-70, group leader, 71-74, assoc div leader, 74-75, ALT DIV LEADER, LOS ALAMOS SCI LAB, 75- Concurrent Pos: Mem, Joint Hazard Eval Group, 64-, chmn, 64-66, 70-; partic, Solar Eclipse Exped, 66 & 70; mem test eval panel, US Atomic Energy Comn, 66-, sci adv, Nev Test Site, 67- & Div Mil Appln, 70; sci dep comdr, Joint Task Force 8, 70-72. Mem: Am Astron Soc; Royal Astron Soc; Int Astron Union. Res: Stellar evolution; underground nuclear testing; effects of nuclear explosions; hazard evaluation. Mailing Add: 3007 Villa Los Alamos NM 87544

BROWNLEE, THOMAS HARLAND, organic chemistry, see 12th edition

BROWNLOW, ARTHUR HUME, b Helena, Mont, July 25, 33; m 59; c 4. GEOLOGY. Educ: Mass Inst Technol, SB, 55, PhD(geol), 60. Prof Exp: Asst prof geol, Univ Mo-Rolla, 60-65; asst prof, 65-67, ASSOC PROF GEOL, BOSTON UNIV, 67-, CHMN DEPT, 75- Concurrent Pos: NSF res grant, 70-72. Mem: AAAS; Geol Soc Am; Geochem Soc; Mineral Soc Am; Am Geophys Union. Res: Chemical analysis of rocks and minerals; igneous and metamorphic petrology; geochemistry of coal. Mailing Add: Dept of Geol Boston Univ Boston MA 02215

BROWNRIGG, LESLIE ANN, b Washington, DC, May 4, 42; m 71. APPLIED ANTHROPOLOGY. Educ: Columbia Univ, BA, 65, PhD(anthrop), 72. Prof Exp: Res assoc appl anthrop, Peru-Cornell Univ Proj, 65-66; asst prof anthrop, George Washington Univ, 70-72; ASST PROF ANTHROP, NORTHWESTERN UNIV, EVANSTON, 72- Concurrent Pos: Ed, Andean Highlands, Latin Am Studies Handbk, 72- Mem: Am Anthrop Asn; Nat Inst Hist & Anthrop Ecuador; Soc Appl Anthrop; Int Cong Americanists. Res: Andean highland ethnography; diachronic study of land tenure systems, especially hacienda system; landowning elites; community development; land reform; ethnographic cinema. Mailing Add: Dept of Anthrop Northwestern Univ Evanston IL 60201

BROWNSCHEIDLE, CAROL MARY, b Buffalo, NY, July 2, 46. HISTOCHEMISTRY. Educ: Niagara Univ, BS, 68; State Univ NY Buffalo, PhD(anat), 74. Prof Exp: ASST PROF ANAT, COL MED, UNIV CINCINNATI, 74- Res: Placental pathology and function in diabetes and pre-eclampsia; comparative morphology and histochemistry of orbital glands in mammals. Mailing Add: Dept of Anat Univ of Cincinnati Col of Med Cincinnati OH 45267

BROWNSCOMBE, EUGENE RUSSELL, b National City, Calif, July 9, 06; m 32; c 1. PHYSICAL CHEMISTRY. Educ: Yale Univ, BS, 28, PhD(phys chem), 32. Prof Exp: Res chemist, Biol Lab, Cold Spring Harbor, 32-34; res scientist, Atlantic Richfield Co, Pa & Tex, 34-69; DIR TECHNOL, SONICS INT, INC, 69-, DIR RES, 75- Concurrent Pos: Instr calculus, Bishop Col, 69-71. Mem: AAAS; Am Soc Petrol Engrs; Am Chem Soc; Soc Explor Geophys. Res: Band spectra; chemical effects of x-rays; solvent extraction; petroleum production and refining; flow of fluids through porous media; theoretical reservoir analysis; exploration methods. Mailing Add: Sonics Int Inc One Energy Sq Dallas TX 75206

BROWNSON, ROBERT HENRY, b Evanston, Ill, Mar 14, 25; m 57; c 4. NEUROANATOMY, NEUROPATHOLOGY. Educ: John Carroll Univ, BS, 48; George Washington Univ, MS, 50, PhD, 53. Prof Exp: Res analyst, Nat Res Coun, 49-51; instr, Univ Southern Calif, 52-54; from asst prof to prof, Med Col Va, 54-68, chmn dept, 67-68; actg chmn dept, 70-71, PROF HUMAN ANAT, SCH MED, UNIV CALIF, DAVIS, 68-, VCHMN DEPT, 71- Concurrent Pos: NIH grants, 56-64; Greenwell fel, NY Univ-Bellevue Med Ctr, 58; Am Cancer grant, 61-63; USAEC grant, 62-64; NIS spec fel, Univ C grant, 62-64; NIH spec fel, Univ Calif, Berkeley, 64-66; vis prof, Div Med Physics, Donner Lab, Univ Calif, Berkeley, 66-67; vis prof, Dept Anat, Univ of Helsinki, Finland, 75. Mem: Am Asn Neuropath; Radiation Res Soc; Am Asn Anat; Am Acad Neurol; Electron Micros Soc Am. Mailing Add: Dept of Human Anat Univ of Calif Sch Med Davis CA 95616

BROWNSTEIN, BARBARA L, b Philadelphia, Pa, Sept 8, 31. CELL BIOLOGY. Educ: Univ Pa, BA, 57, PhD(microbiol), 61. Prof Exp: Fel virol, Wistar Inst, 61-62; USPHS fel genetics, Karolinska Inst, Sweden, 62-64; assoc, Wistar Inst, 64-68; assoc prof, 68-75, PROF BIOL, TEMPLE UNIV, 75- Mem: AAAS; Am Soc Microbiol. Res: Genetic control of ribosome synthesis and function; mode of action of antibiotics; control of replication and motility in normal and malignant animal cells. Mailing Add: Dept of Biol Temple Univ Philadelphia PA 19122

BROWNSTEIN, KENNETH ROBERT, b New York, NY, May 6, 36. QUANTUM MECHANICS. Educ: Rensselaer Polytech Inst, BS, 57, PhD(physics), 66. Prof Exp: ASSOC PROF PHYSICS, UNIV MAINE, ORONO, 65- Mem: Am Phys Soc; Am Asn Physics Teachers; Am Soc Eng Educ; Sigma Xi. Res: Bounds for scattering phase shifts; approximation methods for bound and scattering states. Mailing Add: Dept of Physics Univ of Maine Orono ME 04473

BROWNSTEIN, SYDNEY KENNETH, b Regina, Sask, Mar 5, 31; m 54; c 3. ORGANIC CHEMISTRY. Educ: Univ Sask, BA, 52; Univ Chicago, PhD(chem), 55. Prof Exp: Chemist, Argonne Cancer Res Hosp, 55; fel & hon asst, Univ Col, Univ London, 55-56; instr chem, Cornell Univ, 56-59; RES OFFICER, NAT RES COUN CAN, 59- Concurrent Pos: Weizmann fel, 65-66. Mem: Chem Inst Can; The Chem Soc. Res: Application of nuclear magnetic resonance techniques to problems in chemistry. Mailing Add: Div of Chem Nat Res Coun Ottawa ON Can

BROWNSTONE, YEHOSHUA SHIEKY, b Winnipeg, Man, May 23, 29; m 50; c 3. CLINICAL BIOCHEMISTRY. Educ: Univ Man, BScA, 50, MSc, 56; McGill Univ, PhD(biochem), 58. Prof Exp: Chemist, Dominion Linseed Oil Ltd, 51-52; asst dept physiol & med res, Univ Man, 54-56; demonstr & asst dept biochem, McGill Univ, 56-58, res assoc, 58-59, lectr, 60; biochemist, Dept Biochem & Radioisotopes, Queen Mary Vet Hosp, 59-60; from asst prof to assoc prof, 61-68, PROF PATH CHEM, UNIV WESTERN ONT, 68-; ASST DIR, DEPT CLIN PATH, VICTORIA HOSP, 61- Mem: NY Acad Sci; Can Fedn Biol Soc; Can Biochem Soc; Can Soc Clin Chemists (pres, 73-74). Res: Clinical chemistry; enzymology of the normocyte; the pentose phosphate metabolic pathway; enzymes in neoplastic disease; blood preservation; effect of hormones on enzyme action; drugs; metabolism and detection. Mailing Add: Dept of Path Chem Univ of Western Ont London ON Can

BROYDE, BARRET, b New York, NY, Aug 24, 36; m 59; c 2. PHYSICAL CHEMISTRY. Educ: Yeshiva Univ, BA, 55; Polytech Inst Brooklyn, PhD(phys chem), 60. Prof Exp: Fel chem, NY Univ, 60-62; chemist, Esso Res & Eng Co, 62-67; sr chemist, 67-69, RES LEADER, MAT CHARACTERIZATION LAB, WESTERN ELEC RES & ENG CTR, 69- Mem: AAAS; Am Chem Soc; The Chem Soc. Res: Photochemistry and spectroscopy of dyes; molten salts; synthesis of high temperature refractories; colloid and surface phenomena; fuel cells; analytical chemistry; automation; water and air pollution. Mailing Add: 2000 N E Expressway Norcross GA 30071

BROYER, THEODORE CLARENCE, b Jersey City, NJ, Oct 13, 04; m 32, 54; c 5. PLANT NUTRITION. Educ: Univ Calif, BS, 27. Prof Exp: Assoc plant nutrit, 27-40, jr plant physiologist, 40-41, asst prof plant nutrit, 62-71, EMER PROF & EMER PLANT PHYSIOLOGIST, UNIV CALIF, BERKELEY, 71-; CONSULT, 71- Concurrent Pos: Muellhaupt fel, Ohio State Univ, 47-48; distinguished vis prof, Southern Ill Univ, 60-61; lectr, Univ Calif, Berkeley, 54-62, vchmn dept soils & plant nutrit, 61-64. Mem: Fel AAAS; fel Am Inst Chem; Am Inst Biol Sci; Am Chem Soc; Am Soc Plant Physiol (pres, 53, 54). Res: Movement of solutes and water into plants; translocation in plants; plant nutrition and biochemistry. Mailing Add: Dept of Soils & Plant Nutrit Univ of Calif Berkeley CA 94720

BROYLES, ARTHUR AUGUSTUS, b Atlanta, Ga, May 16, 23; m 43; c 4. PHYSICS. Educ: Univ Fla, BS, 42; Yale Univ, PhD(physics), 49. Prof Exp: Asst prof physics, Univ Fla, 49-50; mem staff, Los Alamos Sci Lab, 50-53; res physicist, Rand Corp, 53-59; assoc prof, 59-61, PROF PHYSICS & PHYS SCI, UNIV FLA, 61- Mem: Am Asn Physics Teachers; fel Am Phys Soc. Res: Quantum electrodynamics; equation of state. Mailing Add: Dept of Physics Univ of Fla Gainesville FL 32601

BROYLES, CARTER D, b Eckman, WVa, Nov 1, 24; m 55; c 2. PHYSICS, ENGINEERING SCIENCE. Educ: Univ Chattanooga, BS, 48; Vanderbilt Univ, PhD(physics), 52. Prof Exp: Res physicist, 52-56, supvr, Radiation Physics Div, 56-64, mgr, High Altitude Burst Physics Dept, 64-68, mgr test sci dept 9110, 68-72, dir effects exp orgn, 72-75, DIR FIELD ENG, SANDIA LAB, 75- Concurrent Pos: Mem staff, Vanderbilt Univ, 52. Mem: AAAS; fel Am Phys Soc; Am Asn Physics Teachers. Res: Magnetic spectrometer studies of radioactive isotopes; fluid dynamics; plasma physics; nuclear radiation measurements; radioactive waste disposal; in-situ fossil fuel technology. Mailing Add: 5310 Los Poblanos Lane NW Albuquerque NM 87107

BROYLES, RALPH EDWARD, chemistry, deceased

BROYLES, ROBERT HERMAN, b Kingsport, Tenn, Feb 16, 43; m 66; c 1. BIOCHEMISTRY, DEVELOPMENTAL BIOLOGY. Educ: Wake Forest Univ, BS, 65, Bowman Gray Sch Med, PhD(biochem), 70. Prof Exp: NIH fel, & res assoc biochem, Fla State Univ, 70-72; ASST PROF ZOOL, UNIV WIS-MILWAUKEE, 72- Concurrent Pos: Cottrell grant, Res-Cottrell, Inc, 73; res grant, US Sea Grant Prog, 75. Mem: AAAS; Am Soc Zoologists; Soc Develop Biologists. Res: Cellular and molecular interactions involved in red blood cell differentiation; effects of drugs and environmental contaminants on vertebrate embryos and on red blood cell differentiation. Mailing Add: Dept of Zool Univ of Wis Milwaukee WI 53201

BRUALDI, RICHARD ANTHONY, b Derby, Conn, Sept 2, 39; m 63; c 2. MATHEMATICS. Educ: Univ Conn, BA, 60; Syracuse Univ, MS, 62, PhD(math), 64. Prof Exp: Nat Acad Sci-Nat Res Coun fel, Nat Bur Stand, 64-65; from asst prof to assoc prof, 65-73, PROF MATH, UNIV WIS-MADISON, 73- Concurrent Pos: NATO fel, Univ Sheffield, 69-70. Mem: Am Math Soc; Math Asn Am. Res: Matrix theory and combinatorics. Mailing Add: Dept of Math Univ of Wis Madison WI 53705

BRUBAKER, CARL H, JR, b Passaic, NJ, July 13, 25; m 49; c 1. INORGANIC CHEMISTRY, ORGANOMETALLIC CHEMISTRY. Educ: Franklin & Marshall Col, BS, 49; Mass Inst Technol, PhD(chem), 52. Prof Exp: From asst prof to assoc prof, 52-61, PROF CHEM, MICH STATE UNIV, 61- Concurrent Pos: Fulbright prof radiochem, Univ Chile, 58; assoc ed, J Am Chem Soc, 64-; consult, Dow Chem Co, 70- Mem: Am Chem Soc; The Chem Soc; AAAS. Res: Oxidation-reduction reaction mechanisms; organometallic compounds of transition elements in lower oxidation states; organometallic compounds attached to polymers as catalysts for hydrogenation and nitrogen fixation. Mailing Add: 4466 Tacoma Blvd Okemos MI 48864

BRUBAKER, DAVID GORDON, physics, see 12th edition

BRUBAKER, GEORGE RANDELL, b New York, NY, July 17, 39; m 64; c 2. INORGANIC CHEMISTRY, BIOINORGANIC CHEMISTRY. Educ: Columbia Univ, BA, 60; Ohio State Univ, MSc, 63, PhD(chem), 65. Prof Exp: Res assoc inorg chem, Univ Pittsburgh, 65-66; asst prof, 66-72, ASSOC PROF CHEM, ILL INST TECHNOL, 72- Mem: Am Chem Soc; The Chem Soc. Res: Synthesis, characterization and reactions of transition metal complexes; biological significance of transition metal chemistry. Mailing Add: Dept of Chem Ill Inst of Technol Chicago IL 60616

BRUBAKER, INARA MENCIS, analytical chemistry, see 12th edition

BRUBAKER, KENTON KAYLOR, b Elizabethtown, Pa, Feb 17, 32; m 55; c 4. BIOLOGY, HORTICULTURE. Educ: Eastern Mennonite Col, BS, 54; Ohio State Univ, MSc, 57, PhD, 59. Prof Exp: Assoc prof biol, Eastern Mennonite Col, 59-62; horticulturist, Congo Polytech Inst, 62-64; chg de cours bot, Free Univ of the Congo, 64-65; chmn dept, 65-74, PROF BIOL, EASTERN MENNONITE COL, 65- Mem: Am Soc Hort Sci; Am Chem Soc; Am Soc Plant Physiologists; Am Sci Affil. Res: Chemical weed control in vegetable crops; carbohydrate translocation in greenhouse tomatoes; tropical horticulture; protein content of tropical vegetables. Mailing Add: Dept of Biol Eastern Mennonite Col Harrisonburg VA 22801

BRUBAKER, MERLIN L, b Live Oak, Calif, July 1, 22; m 43; c 4. INTERNAL MEDICINE, PUBLIC HEALTH. Educ: Calif Col Med, MD, 46; Univ London, dipl trop med & hyg, 52; La Verne Col, BA, 56; Univ Calif, Los Angeles, MA, 62. Prof Exp: Med supt, Garkida Leprosarium, Nigeria, 52-55; assoc prof med, Calif Col Med, 55-62, chmn curric, 56-62, dir clin pract, 59-62; chief prog planning & eval med div, Peace Corps, 62-64; asst dir, USPHS Hosp, Staten Island, NY, 64-65 & dir, Carville, La, 65-68, dir career develop global community health, USPHS, 68-70; REGIONAL ADV LEPROSY, VENEREAL DIS & TREPONEMATOSIS, WHO-PAN AM HEALTH ORGN, 70- Concurrent Pos: Col physician & dir student health servs, La Verne Col, 56-60; asst med officer in chg, USPHS Hosp, Staten Island, NY, 64-65 & dir prof training & res, Carville, La, 65, med officer in chg, 65-68; clin assoc prof, Sch Med, La State Univ, 65-68; assoc clin prof community med & int health, Sch Med, Georgetown Univ, 68-; La State Univ fel trop med & parasitol to Mid Am. Mem: Am Soc Trop Med & Hyg; Am Pub Health Asn; fel Royal Soc Med; Int Leprosy Asn; Soc Study Venereal Dis. Res: Preventive medicine; medical education. Mailing Add: 113 Southbrook Lane Bethesda MD 20014

BRUBAKER, PAUL EUGENE, cell biology, biochemistry, see 12th edition

BRUBAKER, ROBERT ROBINSON, b Wilmington, Del, Jan 15, 33; m 56; c 2. MICROBIOLOGY. Educ: Univ Del, BA, 57; George Washington Univ, MA, 61; Univ Chicago, PhD(microbiol), 66. Prof Exp: Microbiologist, US Army Biol Labs, 64-66; asst prof, 66-73, ASSOC PROF MICROBIOL, MICH STATE UNIV, 73- Concurrent Pos: Lectr, Found Microbiol, Am Soc Microbiol, 69-70. Mem: AAAS; Am Inst Biol Sci; Am Soc Microbiol. Res: Biochemical mechanisms of microbial virulence; genetic determinants of virulence; anti-tumor agents. Mailing Add: Dept of Microbiol & Pub Health Mich State Univ East Lansing MI 48824

BRUBAKER, WILSON MARCUS, b West Alexandria, Ohio, July 9, 06; m 34; c 3. PHYSICS. Educ: Miami Univ, AB, 32; Calif Inst Technol, PhD(physics), 36. Prof Exp: Asst physics, Calif Inst Technol, 32-36; instr, Ohio State Univ, 36-37; res physicist, Westinghouse Elec & Mfg Co, 37-40, sect mgr, Res Dept, 40-42; group leader, Manhattan Proj, Univ Calif, 42; sect mgr, Res Dept, Westinghouse Elec Corp, 43-52; staff physicist, Consol Electrodynamics Corp, 52-60; prin res physicist, Bell & Howell Res Ctr, 60-67; chief scientist, Teledyne/Earth Sciences, 67-69; sr staff scientist, Analog Technology Corp, 69-73; CONSULT, HEWLETT-PACKARD CORP, 72- Mem: Am Phys Soc; Am Vacuum Soc. Res: Nuclear physics; gaseous conduction phenomena; mass spectrometry; quadrupole mass filter; triode ion pumps; space charge; instrumentation for upper atmosphere analysis. Mailing Add: 1954 Highland Oaks Dr Arcadia CA 91006

BRUCE, ALAN KENNETH, b Nashua, NH, Aug 24, 27; m 48; c 4. RADIOBIOLOGY. Educ: Univ NH, BS, 51; Univ Rochester, MS, 54, PhD(radiation biol), 56. Prof Exp: Assoc Atomic Energy Proj, Univ Rochester, 53-56, jr scientist, 56; assoc, Biol Div, Oak Ridge Nat Lab, 56-57; asst prof, 57-62, ASSOC PROF BIOL, STATE UNIV NY BUFFALO, 62-, RADIATION SAFETY OFFICER, 57- Mem: AAAS; Am Soc Microbiol; Health Physics Soc; Radiation Res Soc. Res: Effects of radiation on cell permeability characteristics; applications of isotopic tracer techniques to biology; modification of radiation action on cells. Mailing Add: Dept of Biol State Univ of NY Buffalo NY 14214

BRUCE, CHARLES ROBERT, b Topeka, Kans, Jan 3, 25; m 49; c 5. PHYSICS. Educ: Wash Univ, AB, 50, PhD(physics), 56. Prof Exp: Res assoc, Wash Univ, 56-57; SR RES SCIENTIST, MARATHON OIL CO, 57- Mem: Am Phys Soc. Res: Nuclear magnetic resonance; geophysics and well logging; instrumentation; data acquisition; improving prospecting techniques. Mailing Add: Marathon Oil Co Denver Res Ctr PO Box 269 Littleton CO 80120

BRUCE, DAVID, b Berkeley, Calif, Jan 19, 16; m 43; c 4. FOREST MENSURATION. Educ: Yale Univ, BS, 36, MF, 37. Prof Exp: Forestry field asst, Northeastern Forest Exp Sta, Conn, 37 & Southern Forest Exp Sta, Fla, 38, forestry foreman, Black Hills Nat Forest, SDak, 38-39, jr forester, Southern Forest Exp Sta, New Orleans, 39-40, silviculturist, 45-52, res ctr leader, 52-54, chief, Div Forest Fire Res, 54-60, chief, Pac Northwest Forest & Range Exp Sta, 60-64, FOREST MENSURATIONIST, PAC NORTHWEST FOREST & RANGE EXP STA, US FOREST SERV, 64- Mem: Soc Am Foresters; Biomet Soc. Res: Forest measurement, management and fires. Mailing Add: Pac Northwest Forest & Range Exp Sta Box 3141 Portland OR 97208

BRUCE, DAVID LIONEL, b Champaign, Ill, Oct 27, 33; div; c 2. ANESTHESIOLOGY, MEDICINE. Educ: Univ Ill, MD, 60. Prof Exp: Resident anesthesia, Univ Pa, 61-62, USPHS fel anat, 62-63, resident anesthesia, 63-64; from instr to asst prof anesthesia, Univ Ky, 64-66; from asst prof to assoc prof, 66-74, PROF ANESTHESIA, MED SCH, NORTHWESTERN UNIV, CHICAGO, 74- Concurrent Pos: USPHS career develop award, Northwestern Univ, 67-72; consult, Food & Drug Admin, 74- Mem: Am Soc Anesthesiol; Soc Gen Physiol; Asn Univ Anesthetists; Int Anesthesia Res Soc. Res: Effects of volatile anesthetics on cell structure and function; toxicity of occupational exposure to anesthetics. Mailing Add: Dept of Anesthesia Northwestern Univ Med Sch Chicago IL 60611

BRUCE, DAVID STEWART, b Amherst, Ohio, Sept 17, 39; m 61; c 1. ANIMAL PHYSIOLOGY, ENVIRONMENTAL PHYSIOLOGY. Educ: Taylor Univ, BA & BS, 62; Purdue Univ, MS, 65, PhD(environ physiol), 68. Prof Exp: Instr physiol, DePauw Univ, 67; instr environ biol, Purdue Univ, 67, vis asst prof, 68; asst prof, 68-73, ASSOC PROF BIOL, SEATTLE PAC COL, 73- Mem: AAAS; Am Sci Affil. Res: Phenomenon of hibernation, especially in the bat, Myotis lucifungus and various biochemical and physiological functions of the species. Mailing Add: Sch of Natural & Math Sci Seattle Pac Col Seattle WA 98119

BRUCE, E IVAN, JR, b Center, Tex, Aug 20, 17; m 46; c 3. PSYCHIATRY. Educ: Univ Tex, BA, 39, MD, 42; Am Bd Psychiat & Neurol, dipl. Prof Exp: Resident psychiat, 46-49, from instr to assoc prof, 49-65, asst dir med br, 49-52, asst adminr, 52-59, actg chmn dept, 74-76, DIR, UNIV TEX MED BR GALVESTON, 59-, PROF PSYCHIAT, 65-, CHMN AD INTERIM DEPT, 76- Mem: Am Psychiat Asn; AMA; Am Col Psychiat. Res: Clinical and administrative psychiatry. Mailing Add: Dept of Psychiat Univ of Tex Med Br Galveston TX 77550

BRUCE, FRANCIS ROBERT, b Beverly, Mass, Feb 9, 19; m 44; c 3. CHEMISTRY. Educ: Tufts Univ, BS, 42. Prof Exp: Metallurgist, Remington Arms Co, NY, 42-43; chemist, 43-51, assoc div dir, Chem Tech Div, 51-59, dir safety & radiation control, 60-61, asst dep dir, 61-70, ASSOC DIR, OAK RIDGE NAT LAB, 70- Mem: Am Chem Soc; Am Nuclear Soc; Am Inst Chem Eng; Health Physics Soc. Res: Development of separation processes for the heavy metals; treatment and disposal of radioactive wastes; administration of radiation safety. Mailing Add: Oak Ridge Nat Lab PO Box X Oak Ridge TN 37830

BRUCE, HAROLD ASA, b Beaver Falls, Pa, Apr 17, 07; m 35; c 1. BIOLOGY. Educ: Geneva Col, BS, 29; Univ Pittsburgh, MS, 32, PhD(zool), 35. Prof Exp: Asst zool, Univ Pittsburgh, 30-34; from asst prof biol & chem to assoc prof biol, Beaver Col, 34-37; from assoc prof to prof, 37-74, dean freshmen, 43-57, dir student affairs, 57-63, dean students, 64-70, EMER PROF BIOL, GENEVA COL, 74- Concurrent Pos: Mem staff res lab, Mercy Hosp, Pittsburgh, 32; lectr, Beaver Valley Gen Hosp, 44-56 & Providence Hosp, 49-55. Mem: AAAS. Res: Ctyology; genetics; plant growth; conservation. Mailing Add: Dept of Biol Geneva Col Beaver Falls PA 15010

BRUCE, J P, meteorology, hydrology, see 12th edition

BRUCE, JOHN IRVIN, JR, b Ellicott City, Md, Aug 1, 29; m 67; c 1. PARASITOLOGY, PHYSIOLOGY. Educ: Morgan State Col, BS, 53; Howard Univ, MS, 65, PhD(physiol, biochem), 68. Prof Exp: Jr chemist, Metrop Hosp, New York, NY, 53-54; res asst, Columbia Univ, 54-55; med res technician, Walter Reed Army Inst Res, 55-59, parasitologist, 59-68, chief drug screening unit, 406th Med Lab, Camp Zama, Japan, 68-73; PROF BIOL SCI, LOWELL TECHNOL INST, 73- Mem: Am Soc Parasitologists; Wildlife Dis Asn; Am Inst Biol Sci; Am Soc Trop Med & Hyg; NY Acad Sci. Res: Chemotherapy of parasitic diseases, especially schistosomiasis; physiological and biochemical studies of parasites and host parasite relationship; aspects of parasitic immunology pertaining to host parasite relationship and epidemiology of parasitic diseases. Mailing Add: Col of Pure & Appl Sci Lowell Technol Inst Lowell MA 01854

BRUCE, JOHN MACMILLAN, JR, b Cleveland, Ohio, Aug 17, 22; m 44; c 3. ORGANIC CHEMISTRY. Educ: Hanover Col, BA, 47; Mont State Col, MS, 48; State Col Wash, PhD(org chem), 51. Prof Exp: RES ASSOC, EXP STA, E I DU PONT DE NEMOURS & CO, INC, 51- Res: Free radical polymerization; Cannizzaro reaction; antioxidants; fluorocarbons; low pressure polymerization. Mailing Add: 1422 Emory Rd Green Acres Wilmington DE 19803

BRUCE, RICHARD W, b Glasgow, Mont, Aug 22, 31; m 54; c 3. FOREST ECONOMICS. Educ: Wash State Univ, BS, 57, MA, 59, PhD(agr econ), 68. Prof Exp: Agr economist, USDA, 56-59; instr forest econ, Wash State Univ, 59-62; mkt analyst, Simpson Timber Co, 62-63; ASST PROF FOREST ECON, WASH STATE UNIV, 63- Concurrent Pos: Consult, Bright Logging Co, Ore, 59, Thurston County Resource Coun, Wash, 61, Western Wood Prod Asn, Ore, 67, US Coast Guard, Wash, 67-70, Potlatch Forests, Inc, Idaho, 70, US Coast Guard, Secy Transp, President, DC, 70, Can Econ Res Inst, 71 & Mid Columbia Econ Develop Dist, 72; Food & Agr Orgn-UN forest prod mkt officer, Brazil, 72-74. Mem: Soc Am Foresters (secy-treas, 68, 69); Forest Prod Res Soc; Am Mkt Asn (pres, 69, 70). Res: Forest products marketing; logging management; public policy related to forestry. Mailing Add: Dept of Forestry Wash State Univ Pullman WA 99163

BRUCE, ROBERT ARTHUR, b Somerville, Mass, Nov 20, 16; m 40; c 4. MEDICINE. Educ: Boston Univ, BS, 38; Univ Rochester, MS, 40, MD, 43. Prof Exp: Asst res physician, Strong Mem Hosp, Univ Rochester, 44-45, chief res physician, 45-46, instr med, 46-50; from asst prof to assoc prof, 50-59, PROF MED, SCH MED, UNIV WASH, 59-, CO-DIR, DIV CARDIOL, 73- Concurrent Pos: Buswell fel, Univ Rochester, 46-50; Commonwealth Fund fel, Sch Med, Univ Wash, 65-66. Mem: Asn Am Physicians; fel Am Col Physicians; Asn Univ Cardiol (secy-treas, 64-67, vpres, 67-68, pres, 68-69); Royal Soc Med. Res: Cardiology; exercise physiology; electrocardiography; cardiac physiology and rehabilitation; cardiovascular epidemiology. Mailing Add: Dept of Med Univ of Wash Sch of Med Seattle WA 98195

BRUCE, ROBERT BLACK, b Bamberg, SC, Feb 18, 18; m 47; c 3. DRUG METABOLISM. Educ: Col Charleston, BS, 41; Univ NC, PhD(biochem), 51. Prof Exp: Asst biochem, Univ NC, 49-51; res dir, Jr League Blood Ctr, Milwaukee, Wis, 51-52; biochemist, Hazelton Labs, Inc, 52-57, chief chemist, Western Div, Calif, 57-60; mgr drug metab, 60-71, DIR DRUG METAB, A H ROBINS CO, INC, 71- Mem: Am Chem Soc. Res: Metabolism and pharmacodynamics of drugs and pesticides. Mailing Add: 1211 Sherwood Ave Richmond VA 23220

BRUCE, ROBERT M, biochemistry, organic chemistry, see 12th edition

BRUCE, ROBERT RUSSELL, b Jasper, Ont, Mar 27, 26; m 51; c 1. SOIL PHYSICS. Educ: Ont Agr Col, BSA, 47; Cornell Univ, MS, 51; Univ Ill, PhD, 56. Prof Exp: Soil specialist, Ont Agr Col, 47-49; asst soil physics, Cornell Univ, 49-51; lectr, Ont Agr Col, 51-53; asst, Univ Ill, 53-55; from asst prof & asst agronomist to assoc prof & assoc agronomist, Miss State Univ, 55-65; RES SOIL SCIENTIST, SOUTHERN PIEDMONT CONSERV RES CTR, AGR RES SERV, USDA, 65- Concurrent Pos: Vis prof, Auburn Univ, 65- Mem: Fel AAAS; Soil Sci Soc Am; fel Am Soc Agron; Soil Conserv Soc Am; Agr Inst Can. Res: Soil moisture in relation to plant growth; soil moisture, air and heat movement in saturated and unsaturated media; factors affecting soil structure. Mailing Add: Southern Piedmont Conserv Res Ctr PO Box 555 Watkinsville GA 30677

BRUCE, RUFUS ELBRIDGE, JR, b New Orleans, La, Mar 20, 26; m 51; c 4. PHYSICS. Educ: La State Univ, BS, 49; Okla State Univ, MS, 62, PhD(physics, math), 66. Prof Exp: Engr, Liberty Mutual Ins Co, 50-54, sales rep, 54-55, resident mgr, 55-58; instr math & physics, Northeastern La State Col, 59-60; from res asst physics to res assoc, Res Found, Okla State Univ, 60-65; head dept math & physics, Ark State Col, 65-66; ASSOC PROF PHYSICS, UNIV TEX, EL PASO, 66- Concurrent Pos: Res physicist, Atmospheric Sci Off, Tex, 67-69. Mem: AAAS; Am Astron Soc; Am Astronaut Soc. Res: Radiative transfer; atmospheric and reentry vehicle radiative properties; atmospheric effects of electromagnetic wave properties. Mailing Add: Dept of Physics Univ of Tex El Paso TX 79902

BRUCE, THOMAS ALLEN, b Mountain Home, Ark, Dec 22, 30; m 60; c 2. INTERNAL MEDICINE, CARDIOVASCULAR DISEASE. Educ: Univ Ark, Fayetteville, BS, 51; Univ Ark, Little Rock, MD, 55; Am Bd Internal Med, dipl, 63; Am Bd Cardiovasc Dis, dipl, 70. Prof Exp: Cardiopulmonary trainee, Southwestern Med Sch, Univ Tex, 59-60; USPHS fel, Hammersmith Hosp, Postgrad Med Sch, Univ London, 60-61; from instr to prof med, Sch Med, Wayne State Univ, 61-68, asst dean, 66-68; prof med & head cardiovasc sect, Sch Med, Univ Okla, 68-74; DEAN COL MED, UNIV ARK, LITTLE ROCK, 74- Concurrent Pos: Mem coun arteriosclerosis & coun clin cardiol, Am Heart Asn, 68-, Okla & Ark rep, Coun Clin Cardiol, 70; chief cardiol sect, Vet Admin Hosp, Oklahoma City, 68-74. Mem: Fel Am Col Physicians; fel Am Col Cardiol; fel Am Heart Asn; Am Asn Med Cols; AMA. Res: Mechanism of cardiac control, using selective cardiac denervation techniques, plus studies on myocardial metabolism in man, especially lipids with coronary arteriovenous differences and flow. Mailing Add: Univ of Ark Med Sci Campus Little Rock AR 72201

BRUCE, VICTOR GARDINER, b Butte, Mont, Aug 10, 20; m 51; c 3. BIOLOGICAL RHYTHMS. Educ: Calif Inst Technol, BS, 42; Stanford Univ, PhD(eng mech), 50 Prof Exp: Res assoc appl mech, Radiation Lab, Mass Inst Technol, 42-45 & Polytech Inst Brooklyn, 50-52; Abbott fel, 53-54; USPHS fel, 54-56; res assoc, 56-60, LECTR BIOL, PRINCETON UNIV, 60- Mem: Am Soc Naturalists. Res: Biological clocks. Mailing Add: Dept of Biol Princeton Univ Princeton NJ 08540

BRUCE, WARREN, chemistry, chemical engineering, see 12th edition

BRUCE, WAYNE ROYAL, b Afton, Wyo, Feb 5, 44; m 64; c 2. ECONOMIC GEOLOGY. Educ: Ore State Univ, BS, 66, PhD(geol), 71. Prof Exp: EXPLOR GEOLOGIST, HANNA MINING CO, 70- Mem: Am Inst Mining Engrs; Soc Econ Geologists. Res: Age dating and its relationship to mineral deposits; stable isotope studies into the origin of certain ore deposits; basic geologic studies into the genesis of a specific type of gold deposit. Mailing Add: Hanna Mining Co PO Box 15787 Salt Lake City UT 84115

BRUCE, WILLIAM ALEXANDER, physics, see 12th edition

BRUCE, WILLIAM ROBERT, radiation biology, see 12th edition

BRUCH, CARL WILLIAM, b Kenosha, Wis, May 25, 29; m 53; c 2. MICROBIOLOGY. Educ: Mich State Univ, BS, 51; Univ Wis, MS, 53,

PhD(microbiol), 58. Prof Exp: Microbiologist, US Army Biol Labs, Ft Detrick, 55; res microbiologist, Process Develop Dept, Wilmot Castle Co, NY, 58-61, lab dir, 61-62; chief microbiologist, Schwarz Labs, Inc, 62-63; microbiologist, Off Space Sci & Appln, NASA, 63-66; chief bact br, 66-70, chief drug microbiol br, 71-73, DIR DIV RES & CLASSIFICATION, OFF MED DEVICES, FOOD & DRUG ADMIN, 73- Concurrent Pos: Fed exec fel, Brookings Inst, 68-69. Mem: AAAS; Am Soc Microbiol; Soc Indust Microbiol; NY Acad Sci; Soc Cosmetic Chemists. Res: Biocompatibility of medical plastics; toxicity of ethylene oxide; cosmetic and drug microbiology; yeast fermentations; sterilization resistance of microbial spores; gaseous sterilization; spray drying of microorganisms; rhizobiophages; soil microbiology. Mailing Add: Off of Med Devices HFK-400 FDA 5600 Fishers Lane Rockville MD 20852

BRUCH, HILDE, b Ger, Mar 11, 04; nat US; c 1. PSYCHIATRY, PSYCHOANALYSIS. Educ: Univ Freiburg, MD, 29. Prof Exp: Instr pediat, Columbia Univ, 34-43, assoc psychiat, Col Physicians & & Surgeons, 43-54, from assoc clin prof to clin prof, 55-64; PROF PSYCHIAT, BAYLOR COL MED, 64- Concurrent Pos: Rockefeller fel, 41-42; asst, Johns Hopkins Univ Hosp, 41-43; mem food comt, Nat Res Coun, 42-44; assoc psychoanalyst, Psychoanal Clin for Training & Res, 48-65; attend psychiatrist, NY State Psychiat Inst, 54-65. Mem: Am Psychiat Asn; fel Am Acad Child Psychiat; Am Psychoanal Asn. Res: Psychosomatic aspects of obesity and anorexia nervosa; elements and interaction in process of parent education; schizophrenia research. Mailing Add: Dept of Psychiat Baylor Col of Med Houston TX 77025

BRUCH, LUDWIG WALTER, b Rockford, Ill, Jan 23, 40; m 66; c 2. THEORETICAL PHYSICS. Educ: Univ Wis, BA, 59; Oxford Univ, BA, 61, MA, 65; Univ Calif, San Diego, PhD(physics), 64. Prof Exp: From asst prof to assoc prof, 66-75, PROF PHYSICS, UNIV WIS-MADISON, 75- Concurrent Pos: Partic NSF visiting scientist prog, Dept Physics, Tokyo Univ of Educ, Japan, 72-73. Mem: Am Phys Soc. Res: Chemical physics; theory of superfluid helium; statistical mechanics. Mailing Add: Dept of Physics Univ of Wis Madison WI 53706

BRUCHOVSKY, NICHOLAS, b Toronto, Ont, Sept 21, 36; m 68; c 1. ENDOCRINOLOGY. Educ: Univ Toronto, MD, 61, PhD(biophys), 66; FRCPS(C), 75. Prof Exp: Intern med, Toronto Gen Hosp, 61-62; fel med, Univ Tex Southwestern Med Sch Dallas, 66-68; from asst prof to assoc prof, 69-76, PROF MED & BIOCHEM, UNIV ALTA, 76- Concurrent Pos: Med Res Coun Can scholar, 70-75. Mem: Can Biochem Soc; Can Soc Clin Invest; Can Soc Endocrinol & Metab; Can Oncol Soc; Brit Biochem Soc. Res: Mechanism of action of androgens in prostate; control of cell proliferation in hormone-responsive and unresponsive tumors of the breast, endometrium and prostate. Mailing Add: Dept of Med Univ of Alta Edmonton AB Can

BRUCK, DAVID LEWIS, b New York, NY, July 17, 33. GENETICS, POPULATION BIOLOGY. Educ: Columbia Univ, BS, 55, AM, 57; NC State Univ, PhD(genetics), 65. Prof Exp: Res assoc genetics, NC State Univ, 65; res fel, Comt Math Biol, Univ Chicago, 65-66; asst prof biol, Queens Col, NY, 66-69; asst prof, 69-73, ASSOC PROF BIOL, UNIV PR, RIO PIEDRAS, 73- Mem: Genetics Soc Am; Soc Study Evolution; Animal Behav Soc; Asn Trop Biol; Statist Soc PR. Res: Ecological, including behavioral, and genetic studies of various local species; theoretical and experimental studies on the contributions of various components of behavior to speciation. Mailing Add: Dept of Biol Univ of PR Rio Piedras PR 00931

BRUCK, ERIKA, b Breslau, Ger, Apr 5, 08; nat US. PEDIATRICS. Educ: Friedrich-Wilhelms Univ, MD, 35. Prof Exp: Dir clin lab, 2nd Dept Internal Med, Istanbul Univ, 35-39; from instr to assoc prof, 46-69, PROF PEDIAT, SCH MED, STATE UNIV NY BUFFALO, 69-, ATTEND PHYSICIAN, DEPT PEDIAT, CHILDREN'S HOSP, BUFFALO, 46- Mem: Soc Pediat Res; Am Acad Pediat; Am Pediat Soc. Res: Renal and respiratory functions in children. Mailing Add: Children's Hosp 219 Bryant Ave Buffalo NY 14222

BRUCK, GEORGE, b Budapest, Hungary, Oct 20, 04; nat US; m 29. APPLIED PHYSICS. Educ: Tech Univ, Vienna, PhD(applied physics), 27. Prof Exp: Asst chief engr, Compagnie des Lamps, 29-36; resident engr, Fabbr Ital Magn Marelli, 36-41; res engr, Crosley Corp, 41-43; dir special studies, Nat Union, 43-44; chief engr, Hudson Am Corp, 44; res engr, Raytheon Mfg, 44-46; chief photo engr, Specialities Inc, 46-53; pres, Bruck Indust, Inc & consult, Radiation, Inc, 53-54; staff sci adv, Avco Corp, Ohio, 54-67, chief scientist, Electronic Div, 67-69; SCI ADV, 69- Mem: Fel Inst Elec & Electronics Eng; Am Phys Soc. Res: Electronic devices; solid state; military electronics. Mailing Add: 1632 RIAZ Switzerland

BRUCK, PETER, b Sheffield, Eng, Mar 13, 31; m 61; c 1. PHYSICAL ORGANIC CHEMISTRY. Educ: Univ Sheffield, BSc, 53, PhD(phys org chem), 56. Prof Exp: Res chemist, Benger Labs, UK, 56; staff demonstr org chem, Univ Leeds, 57-59; fel, Univ Calif, Los Angeles, 59-60, res scholar, 63-64; Imp Chem Indust fel, Univ Hull, 60-63; sr res chemist, Bell & Howell Res Labs, 64-71; sr res chemist, Calbiochem, 71-72; DIR RES, CYCLO CHEM DIV, TRAVENOL LABS, 73- Mem: Am Chem Soc; The Chem Soc. Res: Synthesis and behavior of non-classical carbonium ions; dehalogenation of halogenated polycyclic systems; carbazole chemistry; synthesis of steroids and heterocycles. Mailing Add: Cyclo Chem Div Travenol Labs 1922 E 64th St Los Angeles CA 90001

BRUCK, RICHARD HUBERT, b Pembroke, Ont, Dec 26, 14; nat US; m 40. MATHEMATICS. Educ: Univ Toronto, BA, 37, MA, 38, PhD(abstract algebra), 40. Prof Exp: Instr math, Univ Ala, 40-42; from instr to assoc prof, 42-52, PROF MATH, UNIV WIS-MADISON, 52- Concurrent Pos: Asst ed, Bulletin, Am Math Soc, 45-57 & Proceedings, 55-57; Guggenheim fel & Univ res fel, Univ Wis, 46-47; vis prof, Univ NC, Chapel Hill, 63-64. Honors & Awards: Chauvenet Prize, Math Asn Am, 56. Mem: Am Math Soc (assoc secy, 45-48); Math Asn Am. Res: Representation theory; tensor algebra; linear non-associative algebra; theory of loops; projective planes; theory of groups. Mailing Add: Dept of Math 213 Van Vleck Hall Univ of Wis Madison WI 53706

BRUCK, STEPHEN DESIDERIUS, b Budapest, Hungary, Feb 4, 27; nat US; m 54; c 1. POLYMER CHEMISTRY, BIOMEDICAL ENGINEERING. Educ: Boston Col, BS, 51; Johns Hopkins Univ, MA, 53, PhD(chem), 55. Prof Exp: Jr instr, Johns Hopkins Univ, 51-55; res chemist, Dacron Res Lab, E I du Pont de Nemours & Co, NC, 55-56, res chemist, Carothers Res Lab, Textile Fibers Dept, Exp Sta, Del, 56-60; proj leader, Nat Bur Standards, Washington, DC, 61-62; actg sect head coatings mat, NASA, 62; mem sr sci staff chem, Appl Physics Lab, Johns Hopkins Univ, 62-66; actg mgr polymer chem, Watson Res Ctr, Int Bus Mach Corp, 66-67; res prof chem eng, Cath Univ Am, 67-69; PROG MGR FOR BIOMAT, NAT HEART & LUNG INST, NIH, 69- Mem: Fel AAAS; sr mem Am Chem Soc; fel Am Inst Chemists; NY Acad Sci; fel The Chem Soc. Res: Research and development administration; structure and property correlation of high polymers; porphyrin chemistry; metal chelates and complexes; biopolymers; biocompatible materials. Mailing Add: Nat Heart & Lung Inst Nat Insts of Health Bethesda MD 20014

BRUCKENSTEIN, STANLEY, b Brooklyn, NY, Nov 1, 27; m 50; c 3. ANALYTICAL CHEMISTRY, ELECTROCHEMISTRY. Educ: Polytech Inst Brooklyn, BS, 50; Univ Minn, PhD(chem), 54. Prof Exp: From instr to assoc prof anal chem, Univ Minn, 54-62, prof anal chem & chief div, 62-68; PROF ANAL CHEM, STATE UNIV NY BUFFALO, 68- Mem: AAAS; Am Chem Soc; Electrochem Soc. Res: Aqueous and nonaqueous acid-base chemistry; electroanalytical methods. Mailing Add: Dept of Chem State Univ of NY Buffalo NY 14214

BRUCKER, EDWARD BYERLY, b Baltimore, Md, Mar 23, 31; m 63. PHYSICS. Educ: Johns Hopkins Univ, BE, 52, PhD(physics), 59. Prof Exp: Res assoc physics, Duke Univ, 58-61, vis asst prof, 59-60; asst prof, Fla State Univ, 61-63; vis lectr, Johns Hopkins Univ, 63-64; asst prof, Fla State Univ, 64-69; sr res assoc, Rutgers Univ, 69-70; ASSOC PROF PHYSICS, STEVENS INST TECHNOL, 71- Mem: Am Phys Soc; Sigma Xi. Res: High energy physics; interactions of fundamental particles. Mailing Add: Dept of Physics Stevens Inst of Technol Hoboken NJ 07030

BRUCKER, PAUL CHARLES, b Philadelphia, Pa, Nov 15, 31; m 57; c 3. FAMILY MEDICINE. Educ: Muhlenberg Col, BS, 53; Univ Pa, MD, 57. Prof Exp: Intern, Lankenau Hosp, Philadelphia, 57-58; resident, Hunterdon Med Ctr, Flemington, NJ, 58-59 & Lankenau Hosp, Philadelphia, 59-60; pvt family pract, Ambler, Pa, 60-73; ALUMNI PROF FAMILY MED & CHMN DEPT, JEFFERSON MED COL, 73- Concurrent Pos: Mem staff, Jefferson Med Col, Temple Univ & Univ Pa, 62-72; mem, Patient Care Systs Coun. Mem: Am Acad Family Physicians; AMA; Soc Teachers Family Med; Am Heart Asn. Mailing Add: Dept of Family Med Thomas Jefferson Univ Hosp Philadelphia PA 19107

BRUCKNER, ANDREW M, b Berlin, Ger, Dec 17, 32; US citizen; m 57; c 2. MATHEMATICS. Educ: Univ Calif, Los Angeles, BA, 55, PhD(math), 59. Prof Exp: From asst prof to assoc prof math, 59-68, actg dean grad div, 66-69, PROF MATH, UNIV CALIF, SANTA BARBARA, 68- Concurrent Pos: NSF res grant, 62-71. Mem: Am Math Soc; Math Asn Am. Res: Theory of functions of a real variable. Mailing Add: Dept of Math Univ of Calif Santa Barbara CA 93018

BRUCKNER, BENJAMIN HARRY, b New York, NY, Aug 5, 21; m 45; c 2. BIOCHEMISTRY. Educ: Columbia Univ, AB, 43; Georgetown Univ, PhD, 56. Prof Exp: Res chemist, Heyden Chem Corp, 46; biochemist, USDA, 48-51; org chemist, Nat Bur Standards, 51-54; res assoc chem, Georgetown Univ, 54-56; chief dept vet chem, Walter Reed Army Inst Res, 56-60 & US Food & Drug Admin, 60-63; chief environ radiochem prog, Res Br, Div Radiol Health, USPHS, 63-67, radiol sci adv, 67-69, res grants prog officer, Bur Radiol Health, Environ Health Serv, 69-72; DEP DIR OFF EXTRAMURAL ACTIVITIES, NAT INST OCCUP SAFETY & HEALTH, 72- Concurrent Pos: Lectr radiochem, Grad Sch, Georgetown Univ, 58-65; mem fac, USDA Grad Sch, 65-; consult, Pan-Am Health Orgn, 67 & 69; adv radiol technol prog, Montgomery Col, 69- Mem: Fel AAAS; fel Am Inst Chemists; Am Chem Soc; Sigma Xi; Health Physics Soc. Res: Radiobiochemistry; natural products; carbohydrates; biochemistry of microorganisms; radioisotopes in foods; environmental biochemistry; occupational safety and health. Mailing Add: Nat Inst Occup Safety & Health 5600 Fishers Lane Rockville MD 20852

BRUCKNER, DAVID ALAN, b Rhinelander, Wis, June 12, 41; m 65; c 2. MEDICAL PARASITOLOGY. Educ: Wis State Univ, Stevens Point, BS, 66; NDak State Univ, MS, 68; Johns Hopkins Univ, ScD(parasitol), 72. Prof Exp: Fel parasitol, 72-75, LECTR PARASITOL, UNIV CALIF, LOS ANGELES, 75- Mem: Am Soc Trop Med & Hyg; Am Soc Parasitologists. Res: Neurophysiological interactions of Schistosoma cercariae; in vitro cultivation of larval stages of cestodes and schistosomes. Mailing Add: Dept of Microbiol & Immunol Univ of Calif Sch of Med Los Angeles CA 90024

BRUCKNER, LAWRENCE ADAM, b Brooklyn, NY, Feb 16, 40; m 70. STATISTICAL ANALYSIS. Educ: Cath Univ, BA, 62, MA, 64, PhD(math), 68. Prof Exp: From instr to asst prof, Cath Univ, 66-68; mem tech staff, Sandia Corp, 68-71; assoc prof math, Col Santa Fe, 71-73; asst prof, Univ Maine, Portland, 73-74; MEM TECH STAFF, LOS ALAMOS SCI LAB, UNIV CALIF, 74- Mem: Am Statist Asn; Inst Math Statist; Am Math Asn. Res: Stochastic processes; statistical analysis. Mailing Add: Statist Grp C-5/M5254 Los Alamos Sci Lab Los Alamos NM 87545

BRUCKNER, ROBERT JOSEPH, b Jersey City, NJ, May 29, 21; m 54; c 3. DENTISTRY. Educ: Univ Md, DDS, 44; Western Reserve Univ, MS, 48; Am Bd Oral Path, dipl. Prof Exp: Teaching fel histol, embryol & anat, Sch Dent, Western Reserve Univ, 46-48, fel histol & embryol, Sch Med, 46-49, from instr to asst prof, 49-55, instr anat & oral diag, Sch Dent, 48-50, asst prof anat & path, 50-53, assoc prof path, 53-58, assoc prof path & periodont, 58-59, actg chmn dept oral path, 56-59; assoc prof 59-67, PROF DENT, DENT SCH, UNIV ORE, 67-, ASSOC DEAN ACAD AFFAIRS, 69- Mem: Am Dent Asn; fel Am Acad Oral Path; Int Asn Dent Res; fel Am Col Dent. Res: Calcification of dentin; aplasia of cementum; hypophosphatasia. Mailing Add: Dept of Path Univ of Ore Dent Sch Portland OR 97201

BRUDEVOLD, FINN, b Norway, June 12, 10; nat US; m 41; c 3. DENTISTRY. Educ: Norweg Dent Sch, LDS, 32; Univ Minn, DDS, 40; Univ Rochester, MS, 54. Hon Degrees: AM, Harvard Univ, 58; DOdontol, Univ Oslo, 65 & Royal Univ, Umea, Sweden, 68. Prof Exp: Asst prof operative dent, Dent Sch, Tufts Col, 45-47, asst prof prosthetic dent, 47-49; dir res, Eastman Dent Lab, 49-58; CHIEF PREV DENT, FORSYTH DENT CTR, 58- Concurrent Pos: Prof dent, Sch Dent Med, Harvard Univ, 58-70; mem dent study sect, USPHS, 65-69. Honors & Awards: Award Basic Res in Oral Therapeut, Int Asn Dent Res, 66, H Trendley Dean Award Fluoride Res, 69. Mem: Fel AAAS; Am Dent Asn; Int Asn Dent Res. Res: Chemistry of teeth and saliva. Mailing Add: Forsyth Dent Ctr 140 The Fenway Boston MA 02115

BRUDNER, HARVEY JEROME, b New York, NY, May 29, 31; m 63; c 2. BIOPHYSICS, ELECTRONICS. Educ: NY Univ, BS, 52, MS, 54, PhD(physics), 59. Prof Exp: Electronics engr, Pioneer Instrument Div, Bendix Corp, NJ, 52; instr atomic physics, NY Univ, 53-54; physicist, US Naval Ord Lab, 54; prin physicist, Emerson Res Labs, 54-57; res scientist, Courant Inst Math Sci, NY Univ, 57-62; prof math & physics, NY Inst Technol, 62-63, dean sci & technol, 63-64; sr res assoc, Princeton Lab, Am Can Co, 64-67; dir res & develop, 67, vpres res & develop, 67-71, PRES RES & DEVELOP, WESTINGHOUSE LEARNING CORP, 71- Concurrent Pos: Consult, Emerson Radio & Phonograph Corp, 57-59 & Emertron, Inc, 57-62; mem adv comt, Middlesex County Col; mem, Coun Latin Am; adv, William Patterson Col, NJ, 75- Mem: AAAS; Am Phys Soc; sr mem Inst Elec & Electronics Engrs; Int Fedn Med Electronics & Biol Eng; Nat Security Indust Asn. Res: Computers and multi-media as applied to advanced educational and training systems; Thomas-Fermi techniques for determining wave functions of excited states; electromagnetic wave propagation; biological effects of microwave radiation; medical instrumentation; operations research. Mailing Add: Westinghouse Learning Corp 100 Park Ave New York NY 10017

BRUDNER-WHITE, LILYAN A, social anthropology, anthropological linguistics, see 12th edition

BRUECKHEIMER, WILLIAM ROGERS, b Gary, Ind, Aug 19, 21; m 42; c 3. RESOURCE GEOGRAPHY. Educ: Univ Chicago, MA, 49; Univ Mich, MA, 52, PhD(geog), 54. Prof Exp: Instr geog, Fla State Univ, 49-51; asst prof, Southern State Col, 53-55; asst prof geog & geol, Western Mich Univ, 55-58, prof & chmn dept, 58-64; chmn dept geog, Fla State Univ, 64-71, dir, London Study Ctr, 71-72, PROF GEOG, FLA STATE UNIV, 64- Concurrent Pos: Vis scholar geog, Univ Mich, 74. Mem: Asn Am Geog; Royal Geog Soc. Res: Electoral geography of Florida; economic significance of tourist and recreation industry in Florida; environmental problems in Florida. Mailing Add: 1210 Waverly Rd Tallahassee FL 32303

BRUECKNER, HANNES KURT, b San Francisco, Calif, Apr 6, 40; m 66; c 1. GEOCHEMISTRY, STRUCTURAL GEOLOGY. Educ: Cornell Univ, BS, 62; Yale Univ, MS, 65, PhD(geol), 68. Prof Exp: Res assoc geosci, Univ Tex, Dallas, 67-70; asst prof, 70-75, ASSOC PROF, DEPT EARTH & ENVIRON SCI, QUEENS COL NY, 75- Concurrent Pos: Vis sr res assoc, Lamont-Doherty Geol Observ, 70-; NSF fels, 73-75 & 75-77. Mem: Am Geophys Union. Res: Structural, petrological and geochemical studies of relationships between mantle and crust in orogenic zones, particularly origins and derivations of eclogites, utramafic rocks and anorthosites. Mailing Add: Dept of Earth & Environ Sci Queens Col Flushing NY 11367

BRUECKNER, KEITH ALLAN, b Minneapolis, Minn, Mar 19, 24; m 60; c 3. THEORETICAL PHYSICS. Educ: Univ Minn, BA, 45, MA, 47; Univ Calif, PhD(physics), 50. Prof Exp: From asst prof to assoc prof physics, 51-55; physicist, Brookhaven Nat Lab, 55-56; prof physics, Univ Pa, 56-59; chmn dept physics, 59-61, dean letters & sci, 63-65, dean grad studies, 65; dir inst radiation physics & aerodyn, 65-67, dir inst pure & appl phys sci, 67-69, PROF PHYSICS, UNIV CALIF, SAN DIEGO, 59- Concurrent Pos: Consult, AEC, 53-70; vpres & dir res, Inst Defense Anal, Washington, DC, 61-62; vpres & tech dir, KMS Industs, Inc, 68-70, exec vpres, KMS Fusion, Inc, 71-74. Honors & Awards: Dannie Heinemann Prize, 63. Mem: Nat Acad Sci; fel Am Acad Arts & Sci; fel Am Phys Soc; fel Am Nuclear Soc. Res: Theoretical nuclear physics; statistical mechanics, plasma physics; interactions of lasers with matter; magnetohydrodynamics and theory of metals. Mailing Add: Dept of Physics B-019 Univ of Calif at San Diego La Jolla CA 92093

BRUECKNER, WERNER DIETRICH, b Königsberg, Ger, May 24, 09; m; c 4. GEOLOGY. Educ: Univ Basel, Dr phil (geol), 34. Prof Exp: Field geologist, Dr R Helbling's Surv Off, Switz, 37-38; asst geol dept, Nat Hist Mus Basel, 38-47; sr field geologist, Socony- Vacuum Oil Co, Colombia, 47-49; lectr & sr lectr, Univ Col Ghana, 51-58; prof geol & head dept, 58-68, J P HOWLEY PROF GEOL, MEM UNIV NFLD, 68- Concurrent Pos: Field geologist, Swiss Geol Comn, 38-. Mem: Geol Soc Am; Geol Asn Can. Res: Sedimentology; geomorphology; regional geology. Mailing Add: Dept of Geol Mem Univ of Nfld St Johns NF Can

BRUEHL, GEORGE WILLIAM, b Mentone, Ind, Sept 10, 19; m 49; c 2. PLANT PATHOLOGY. Educ: Univ Ark, BSA, 41; Univ Wis, PhD(plant path), 48. Prof Exp: Assoc pathologist, Div Cereal Crops & Dis, USDA, SDak State Col Exp Sta, 48-52, pathologist, Div Sugar Plant Invests, 52-54; PATHOLOGIST, DEPT PLANT PATH, WASH STATE UNIV, 54- Mem: Am Phytopath Soc. Res: Root rots of cereal crops; general pathology of cereals. Mailing Add: Dept of Path Wash State Univ Pullman WA 99163

BRUELS, MARK CHARLES, b Kansas City, Mo, Jan 2, 41; m 62; c 2. RADIOLOGICAL PHYSICS, MEDICAL PHYSICS. Educ: Mankato State Col, BA, 62; Univ Kans, MS, 65, PhD(theoret nuclear physics), 69. Prof Exp: Head dept physics, Pub Schs, 62-63; asst prof, Slippery Rock State Col, physicist, Naval Aerospace Med Res Labs, 72-73; fel, M D Anderson Hosp & Tumor Inst, 73-75; PHYSICIST RADIOTHER, VET ADMIN HOSP, 74- Concurrent Pos: NSF acad year exten grant, 70-72; col teacher res partic, La State Univ, Baton Rouge, 70; physicist, Baylor Col Med, 74. Mem: Am Phys Soc; Am Asn Physicists in Med. Res: Nuclear structure theory; electron molecule scattering; physics teaching devices; artificial bone implants; large irregular field linae dosimetry; automatic treatment planning optimization. Mailing Add: Vet Admin Hosp 54th St & 48th Ave S Minneapolis MN 55417

BRUEMMER, JOSEPH HENRY, biochemistry, see 12th edition

BRUENING, GEORGE EMIL, b Chicago, Ill, Aug 10, 38; m 60; c 3. BIOCHEMISTRY, PLANT VIROLOGY. Educ: Carroll Col, Wis, 60; Univ Wis, Madison, MS, 63, PhD(biochem), 65. Prof Exp: NSF fel, Virus Lab, Univ Calif, Berkeley, 65-66, asst prof, 67-70, ASSOC PROF BIOCHEM, UNIV CALIF, DAVIS, 71- Concurrent Pos: Vis scientist plant path, Cornell Univ, 74-75; Guggenheim Mem Found fel, 74-75. Mem: AAAS; Am Phytopath Soc; Am Soc Biol Chemists. Res: Biochemistry and chemistry of viruses, nucleic acids and proteins. Mailing Add: Dept of Biochem & Biophys Univ of Calif Davis CA 95616

BRUENNER, ROLF SYLVESTER, b Magdeburg, Ger, Dec 31, 21; US citizen; m 56; c 4. APPLIED CHEMISTRY. Educ: Munich Tech Univ, MS, 55, PhD(photochem), 57. Prof Exp: Sci asst photochem, Phys Chem Inst, Munich Tech Univ, 55-57; res chemist, US Army Signals Res & Develop Labs, Ft Monmouth, NJ, 57-59; from res chemist to sr res chemist, 59-75, SCIENTIST, AEROJET SOLID PROPULSION CO, GEN TIRE & RUBBER CO, 75- Mem: Am Chem Soc; Soc Ger Chem. Res: Mechanism of energy transfer from adsorbed dye molecules to crystal lattices; photopolymerization; photographic stabilization processing; catalysis of urethane formation; polymer chemistry; composite materials. Mailing Add: Dept 4400 Aerojet Solid Propulsion Co Sacramento CA 95813

BRUES, ALICE MOSSIE, b Boston, Mass, Oct 9, 13. PHYSICAL ANTHROPOLOGY. Educ: Bryn Mawr Col, AB, 33; Radcliffe Col, PhD(phys anthrop), 40. Prof Exp: Res assoc phys anthrop, Peabody Mus, Harvard Univ, 41-42; statistician, Anthropometric Unit, Wright Field, 42-44; anthropologist, Chem Warfare Serv, 45; from asst prof to prof anat, Sch Med, Univ Okla, 46-65; chmn dept anthrop, 68-71, PROF ANTHROP, UNIV COLO, BOULDER, 65- Mem: Am Asn Phys Anthrop (vpres, 66-68, pres, 71-); Am Soc Human Genetics; Soc Study Evolution; Am Soc Naturalists; fel Am Anthrop Asn. Res: Polymorphism; race formation and evolution of man; pigmentation; skeletal variation. Mailing Add: Dept of Anthrop Univ of Colo Boulder CO 80302

BRUES, AUSTIN M, b Milwaukee, Wis, Apr 25, 06. MEDICINE, RADIOBIOLOGY. Educ: Harvard Univ, AB, 26; Harvard Med Sch, MD, 30. Prof Exp: Intern neuropsychiat, Boston City Hosp, Mass, 29; med resident, Huntington Mem Hosp, Boston, 30-31; house officer, Mass Gen Hosp, Boston, 31-32; asst med, Harvard Med Sch, 32-34, res fel, 34-35, instr, 36-37, assoc, 37-42, asst prof, 42-45; assoc prof, Dept Med & Inst Radiobiol & Biophys, Univ Chicago, 45-52; sr biologist & dir, Biol Div, 46-50, sr biologist, Div Biol & Med Res, 50-71, dir, 50-62, actg assoc dir, 69-70, consult, Radiol Physics Div, 71-72, MED DIR, CTR HUMAN RADIOBIOL, RADIOL & ENVIRON RES DIV, ARGONNE NAT LAB, 72-; PROF MED, UNIV CHICAGO, 52- Concurrent Pos: From asst physician to assoc physician, Huntington Mem Hosp, Boston, 32-41; asst, Mass Gen Hosp, Boston, 32-34, asst physician, 42-45; instr & tutor biochem sci, Harvard Col, 34-42, freshman adv, 39-42; Moseley traveling fel, Harvard Univ & Royal Cancer Hosp, London, 35-36; asst, Peter Bent Brigham Hosp, Boston, 36-37, jr assoc, 40-41, assoc physician, 41-42; asst physician, Boston City Hosp, 38-39; internist, US Army Induction Ctr, Boston, 40-44; responsible investr, Off Sci Res & Develop, Boston, 41-44; sr biologist, Metall Lab, Univ Chicago, 44-46; partic, Russian Exchange Mission Radiobiol, 59; consult, Secy War, Japan, 46-47; chmn subcomt radiobiol, Comt Nuclear Sci, Nat Acad Sci, 47-50, mem adv comt, Atomic Bomb Casualty Comn, 47-; mem comt II internal emitters, Int Comn Radiol Protection, 57-65, mem comt I biol effects, 65-; expert consult, WHO Expert Comt Radiation, 57-; mem comt path effects of atomic radiation & chmn subcomt toxicity of internal emitters, Nat Res Coun, 58-65; mem ad hoc comt permissible somatic dose for gen pop, Nat Coun Radiation Protection & Measurements, 59, mem subcomt II internal emitters, 65-69; mem comn environ hyg, Armed Forces Epidemiol Bd, 62-71, mem adv comt metab, 62-63, mem comn radiation & infection, 65-69; mem comt space nuclear appln, US AEC, 70. Mem: Am Asn Anatomists; AAAS; Am Asn Cancer Res (pres, 54-55); Asn Am Physicians; Radiation Res Soc (pres, 55-56). Res: Epidemiology; war medicine; nuclear medicine; cell biology; cancer; physiology; biochemistry; radiation; theoretical biology; research techniques; philosophy of science; social implications. Mailing Add: Ctr for Human Radiobiol Argonne Nat Lab Argonne IL 60439

BRUESCH, JOHN F, b Detroit, Mich, Aug 31, 17; m 46; c 3. ORGANIC CHEMISTRY. Educ: Univ Mich, BS, 40, MS, 41. Prof Exp: Chemist, 41-42, res chemist, Conn, 42-51, develop chemist, 51-57, proj leader, 57-63, RES CHEMIST, AM CYANAMID CO, 63- Res: Vapor phase catalysis; physical and analytical chemistry; gas chromatography; emission spectroscopy. Mailing Add: Am Cyanamid Co Stamford CT 06904

BRUESCH, SIMON RULIN, b Norman, Okla, July 7, 14. ANATOMY. Educ: La Verne Col, AB, 35; Northwestern Univ, MS, 39, MB, 40, MD, 41. Hon Degrees: DSc, La Verne Col, 67. Prof Exp: From instr to prof, 41-60, GOODMAN PROF ANAT, CTR HEALTH SCI, UNIV TENN, MEMPHIS, 60- Mem: AAAS; Am Hist Sci Soc; Am Asn Hist Med; Am Asn Anat. Res: Visual system of vertebrates; afferent components of facial nerve; peripheral nerve endings; sweating by skin resistance methods. Mailing Add: Dept of Anat Univ of Tenn Ctr Health Sci Memphis TN 38163

BRUESKE, CHARLES H, b New Ulm, Minn, Dec 18, 37; m 61; c 2. PLANT PHYSIOLOGY. Educ: Elmhurst Col, BS, 59; Univ Southern Ill, MS, 61; Ariz State Univ, PhD(bot), 65. Prof Exp: Asst prof, 64-68, ASSOC PROF BIOL, MT UNION COL, 68- Mem: AAAS; Am Soc Plant Physiol; Am Inst Biol Sci. Res: Plant growth regulators; physiology of plant diseases. Mailing Add: Dept of Biol Mt Union Col Alliance OH 44601

BRUETMAN, MARTIN EDGARDO, b Buenos Aires, Arg, Aug 17, 32; US citizen; m 56; c 2. NEUROLOGY. Educ: Nat Col Buenos Aires, BS & BA, 49; Univ Buenos Aires, MD, 55. Prof Exp: Instr neurol, Baylor Col Med, 62-63, mem, Fac Comt Res Projs, 63-64; clin asst prof neurol & head sect, Inst Med Res, Univ Buenos Aires, 64-68; from assoc prof to prof, Chicago Med Sch, 68-74, chmn dept, 70-74; CHMN DEPT NEUROL & DIR RESIDENCY TRAINING PROG, MT SINAI HOSP MED CTR, CHICAGO, 74-; COORDR ILL REGIONAL MED PROG, 65- Concurrent Pos: Fel neurol, Baylor Col Med, 61-62; Nat Heart Inst grant, 64-68; coordr coop study cerebrovascular insufficiency, NIH, 61-64; assoc investr, Neurol Res Ctr, Inst Torcuato di Tella, Arg, 64-68; adv to secy pub health training progs, Govt Arg, 66-68; mem, Coun Cerebrovascular Dis, Am Heart Asn, 68- Mem: Am Epilepsy Soc; Am Acad Neurol; NY Acad Sci; Arg Soc Neurol; Arg Col Neurol. Res: Medical education; natural history and incidence of cerebrovascular diseases; epidemiology; community resources in relation to cerebrovascular diseases. Mailing Add: Dept of Neurol Mt Sinai Hosp Chicago IL 60608

BRUFFEY, JOSEPH ALAN, urban planning, transportation geography, see 12th edition

BRUGGE, JOHN F, b Brooklyn, NY, Apr 24, 37; m 60; c 3. PHYSIOLOGY. Educ: Luther Col, Iowa, BA, 59; Univ Ill, Urbana-Champaign, MS, 61, PhD(physiol), 63. Prof Exp: NIH fel, 63-66, asst prof, 66-71, ASSOC PROF NEUROPHYSIOL, SCH MED, UNIV WIS-MADISON, 71- Mem: Soc Neurosci; Am Physiol Soc. Res: Physiology and anatomy of auditory system. Mailing Add: Univ of Wis Sch of Med Madison WI 53706

BRUGGER, JOHN EDWARD, b Erie, Pa, Sept 3, 23; m 66. PHYSICAL CHEMISTRY. Educ: Gannon Col, BS, 45; Pa State Univ, MS, 46; Univ Chicago, PhD(phys chem), 54. Prof Exp: Asst prof chem, Gannon Col, 46-49; res & admin assoc, Res Insts, Univ Chicago, 52-57, chemist, Lab Appl Sci, 57-63; assoc chemist, Chem Eng Div, Argonne Nat Lab, 63-68; sr res chemist & mgr res, Microstatics Lab, Smith-Corona Marchant Div, SCM Corp, Ill, 68-69, res chemist & supvr phys systs group, Smith-Corona Marchant Res & Develop Lab, Palo Alto, 70-71; PHYS SCIENTIST & ADMINR, OIL & HAZARDOUS MATS SPILLS BR, INDUST ENVIRON RES LAB, US ENVIRON PROTECTION AGENCY, 71- Mem: Am Chem Soc; Am Phys Soc; Am Nuclear Soc; Soc Photog Sci & Eng. Res: Environmental science; chemical-physical methods for control of toxic spills; electrophotography; photophysics; high temperature calorimetry; refractory nonmetallics; infrared detectors; fluorescence and phosphorescence; photosynthesis; isotope applications; reaction kinetics. Mailing Add: PO Box 15 Metuchen NJ 08840

BRUGGER, ROBERT MELVIN, b Oklahoma City, Okla, Jan 13, 29; m 53; c 2. NUCLEAR PHYSICS. Educ: Colo Col, BA, 51; Rice Inst, MA, 53, PhD(physics), 55. Prof Exp: Asst, Rice Inst, 51-54; res physicist, Nuclear Physics Br, Atomic Energy Div, Phillips Petrol Co, 55-61, head solid state physics sect, 61-66; head solid state physics sect, Idaho Nuclear Corp, 66-68, mgr nuclear technol div, Aerojet Nuclear Co, 68-74; DIR RES REACTOR FACIL, UNIV MO, 74- Concurrent Pos: With UK Atomic Energy Res Estab, Harwell, 62-63; comr, Idaho Nuclear Energy Comn, 70-74. Mem: Fel Am Phys Soc; fel Am Nuclear Soc. Res: Direct research in neutron cross sections; neutron inelastic scattering; neutron diffraction; gamma detection; high pressure research; reactor conception; design and operation; metallurgy and materials testing. Mailing Add: Res Reactor Facil Res Park Univ of Mo Columbia MO 65201

BRUGGER, THOMAS C, b Fond du Lac, Wis, Jan 19, 27; m 54; c 5. CHILD PSYCHIATRY. Educ: Univ Wis, BNS, 48, BS, 50, MD, 53. Prof Exp: Staff psychiatrist, Winebago State Hosp, 53; resident, Cincinnati Dept Psychiat, Cincinnati Gen Hosp, 56-58; trainee child psychiat, Dept Child Psychiat, Child Guid Home, Ohio, 58-60; staff psychiatrist, instr & asst dir children's clins, Sch Med, Univ Cincinnati, 60-61; asst prof child psychiat & dir child guid & eval clin, Wash Univ, 61-75; COORDR CHILDREN'S SERV & ASST PROF CHILD PSYCHIAT, CASE WESTERN RESERVE UNIV, 75- Concurrent Pos: Consult, Cincinnati Speech &

Hearing Ctr, Ohio, 59-61, Diag Clin Ment Retarded Children, 59-61 & Miriam Sch, 63-75. Mem: AAAS; Am Psychiat Asn; Am Orthopsychiat Asn. Res: Child psychiatry. Mailing Add: Div Child Psychiat Case Western Reserve Univ Cleveland OH 44106

BRUHNS, KAREN OLSEN, b Santa Rosa, Calif, Apr 19, 41. ANTHROPOLOGY. Educ: Univ Calif, Berkeley, AB, 63, PhD(anthrop), 67. Prof Exp: Actg asst prof anthrop, Univ Calif, Los Angeles, 67-68; asst prof, Univ Calgary, 68-70; asst prof, Calif State Univ, San Jose, 70-72; ASST PROF ANTHROP, SAN FRANCISCO STATE UNIV, 72- Mem: AAAS; Am Anthrop Asn; Inst Andean Studies. Res: New World archaeology, especially the Northern Andes; Peru; Mesoamerica; primitive art; transcontinental culture contact; Central America. Mailing Add: Dept of Anthrop San Francisco State Univ San Francisco CA 94132

BRUICE, THOMAS CHARLES, b Los Angeles, Calif, Aug 25, 25; m; c 3. BIO-ORGANIC CHEMISTRY. Educ: Univ Southern Calif, BA, 50, PhD(biochem), 54. Prof Exp: Lilly fel, Univ Calif, Los Angeles, 54-55; from instr to asst prof, Yale Univ, 55-58; asst prof, Sch Med, Johns Hopkins Univ, 58-60; prof chem, Cornell Univ, 60-64; PROF BIOCHEM & CHEM, UNIV CALIF, SANTA BARBARA, 64- Concurrent Pos: USPHS sr fel, 57-58; career investr, NIH, 62- Mem: Am Chem Soc; The Chem Soc. Res: Organic compounds of sulfur; thryoxine analogues, activity and structure; spectral shifts occurring on substitution; mechanisms of organic reactions; model enzymes. Mailing Add: Dept of Chem Univ of Calif Santa Barbara CA 93106

BRUMAGE, WILLIAM HARRY, b Slick, Okla, Nov 18, 23; m 47; c 1. SOLID STATE PHYSICS. Educ: Okla State Univ, BS, 48, MS, 49; Univ Okla, PhD(physics), 64. Prof Exp: Instr physics, Ark Agr & Mech Col, 49-52; from asst prof to assoc prof, 52-64, PROF PHYSICS, LA TECH UNIV, 64-, CHMN DEPT, 65- Mem: Am Phys Soc; Am Asn Physics Teachers. Res: Paramagnetic susceptibilities of ions in host crystals. Mailing Add: Dept of Physics La Tech Univ Ruston LA 71270

BRUMAN, HENRY JOHN, b Berlin, Ger, Mar 25, 13; US citizen. CULTURAL GEOGRAPHY, GEOGRAPHY OF LATIN AMERICA. Educ: Univ Calif, Los Angeles, AB, 35; Univ Calif, Berkeley, PhD(geog), 40. Prof Exp: Asst prof geog, Pa State Col, 40-44; cult geogr, Inst Social Anthrop, Smithsonian Inst & M Proj, 44-45; from asst prof to assoc prof geog, 45-55, chmn dept, 57-61, coordr, Colombian & Brazilian Student Leader Sems, 58-66, actg dir, Latin Am Ctr, 62-63, assoc dir, 63-65, dir, NDEA Lang & Area Prog Latin Am, 62-65, dir, Univ Calif Study Ctr, Göttingen, Ger, 66-68, PROF GEOG, UNIV CALIF, LOS ANGELES, 55- Concurrent Pos: Mem, Nat Acad Sci-Nat Res Coun Adv Comt Geog, US Dept State, 61-65; Fulbright travel fel, Portugal, 63-64; vis prof, Univ Göttingen, 66-68. Honors & Awards: Alexander von Humboldt Gold Medal, Fed Repub Ger, 71. Mem: Fel AAAS; Asn Am Geogr; Int Geog Union. Res: Culture history of domesticated plants in America; agricultural colonization in Brazil; Alexander von Humboldt. Mailing Add: Dept of Geog Univ of Calif Los Angeles CA 90024

BRUMBAUGH, JOHN (ALBERT), b Buffalo, NY, May 3, 35; m 58; c 3. DEVELOPMENTAL GENETICS. Educ: Cedarville Col, BS, 58; Iowa State Univ, PhD(develop genetics), 63. Prof Exp: From instr to assoc prof biol, Cedarville Col, 59-64; from asst prof to assoc prof, 64-73, PROF ZOOL, SCH LIFE SCI, UNIV NEBR, LINCOLN, 73- Concurrent Pos: Attend, NSF Col Teachers Res Partic Prog, Purdue Univ, 64; Nat Inst Gen Med Sci res career develop award, 69-74. Mem: Am Genetic Asn; Am Soc Cell Biol; Genetics Soc Am; Soc Develop Biol. Res: Melanocyte differentiation in the fowl; fine structural variations in developing pigment granules of selected genotypes; genetic regulation in higher organisms; somatic cell genetics of melanocytes in culture. Mailing Add: Sch of Life Sci Univ of Nebr Lincoln NE 68588

BRUMBAUGH, RICHARD J, b Norristown, Pa, Dec 22, 41; m 69. MEDICINAL CHEMISTRY. Educ: Villanova Univ, BS, 63; Ohio Univ, Athens, PhD(chem), 70. Prof Exp: Asst prof, 69-74, ASSOC PROF CHEM, OHIO UNIV, ZANESVILLE, 74- Res: S-triazolo (4, 3-a) pyridine derivatives and pyrido-2, 3-furoxane derivatives; new compound synthesis and structure determination; synthesis of potential anti-cancer chemotherapeutic agents. Mailing Add: Dept of Chem Ohio Univ Zanesville OH 43701

BRUMBERGER, HARRY, b Vienna, Austria, Aug 28, 26; nat US; m 50; c 2. PHYSICAL CHEMISTRY. Educ: Polytech Inst Brooklyn, BS, 49, MS, 52, PhD(chem), 55. Prof Exp: Res assoc chem, Cornell Univ, 54-57; from asst prof to assoc prof, 57-69, PROF PHYS CHEM, SYRACUSE UNIV, 69- Concurrent Pos: Res leave, Graz Univ & Nat Bur Standards, 62-63, Weizmann Inst of Sci, 74. Mem: Am Chem Soc; Am Crystallog Asn. Res: Phase transitions and critical phenomena; small-angle x-ray scattering and light scattering of biopolymers in solution; crystallization and morphology of polymers. Mailing Add: Dept of Chem Syracuse Univ Syracuse NY 13210

BRUMFIELD, ROBERT THORNTON, botany, see 12th edition

BRUMLEVE, STANLEY JOHN, b Teutopolis, Ill, July 3, 24; m 55; c 3. PHYSIOLOGY. Educ: St Louis Univ, BS, 50, MS, 54, PhD(physiol, biochem), 57. Prof Exp: Asst physiol, St Louis Univ, 54; instr biol, Webster Col, 55-57; asst prof physiol & pharmacol, 57-64, actg chmn dept, 64-65, assoc prof, 64-73, PROF PHYSIOL & PHARMACOL, MED SCH, UNIV NDAK, 73-, CHMN DEPT, 72- Concurrent Pos: Asst, Univ Alaska, 54. Mem: Am Physiol Soc. Res: Environmental physiology; temperature regulation; high pressure and other stresses. Mailing Add: Dept of Physiol & Pharmacol Univ of NDak Med Sch Grand Forks ND 58201

BRUMLEY, CORWIN HOYT, b Pelham, NY, July 31, 23; m 48; c 3. ELECTRONICS, OPTICS. Educ: Mass Inst Technol, SB, 44. Prof Exp: Mem staff physics, Div Indust Coop, Mass Inst Technol, 47-48; head, Electronics Dept, 49-58, dir electronics measuring instruments, 58-64, vpres & dir res & develop, 64-70, vpres & gen mgr, Anal Systs Div, 70-74, PRES ANAL SYSTS, 74- DIV, BAUSCH & LOMB, INC, Mem: AAAS; Optical Soc Am; Inst Elec & Electronics Engrs; Am Soc Photogram. Res: Ophthalmic, optical and electronic instruments; ophthalmic lenses. Mailing Add: Anal Systs Div Bausch & Lomb Inc 820 Linden Ave Rochester NY 14625

BRUMLIK, JOSEPH V, b Rokycany, Czech, Jan 18, 97; nat US; m 25; c 1. MEDICINE. Educ: Charles Univ, Prague, MD, 21. Prof Exp: Asst med & cardiol, Charles Univ, Prague, 21-37, asst prof med, 32, dir dept cardiol, Univ Policlin, 37-39; res chief physician, Nat Inst Cardiol, Mex, 44; from instr to asst prof, 45-59, ASSOC PROF CLIN MED, SCH MED, NY UNIV, 59- Concurrent Pos: From asst vis physician to assoc vis physician, Bellevue Hosp, 45-67, vis physician, 67-, sr physician cardiac clin, 49-; asst attend physician, Clin, NY Univ, 48-50, from asst vis physician to assoc vis physician, Hosp, 50-67, vis physician, 67-; from asst attend physician to assoc attend physician, French Hosp, 49-58; consult cardiol, Vet Admin Regional Off; consult med, St Clare's Hosp. Mem: Harvey Soc; corresp mem Mex Nat Acad Med; corresp mem Mex Soc Cardiol; corresp mem Fr Soc Mil Med; Czech Cardiol Asn (gen secy, 30-39). Res: Clinical and experimental medicine and cardiology. Mailing Add: 11 Waverly Pl New York NY 10003

BRUMMER, JOHANNES J, b Graaff Reinet, SAfrica, Sept 2, 21; Can citizen; m 58; c 2. ECONOMIC GEOLOGY, EXPLORATION GEOLOGY. Educ: Univ Witwatersrand, c, 44 & 45, MSc, 51; McGill Univ, PhD(geol), 55. Prof Exp: Mine surveyor, E Geduld Mines Ltd, SAfrica, 45-47; mine geologist, Roan Antelope Copper Mines Ltd, Zambia, 47-49; chief geologist, Mufulira Copper Mines Ltd, 49-51 & Rhodesian Selection Trust Serv Ltd, 51-53; sr geologist, Kennco Explor Ltd, Que, 55-57, Vancouver, 57-58 & Toronto, 58-61; resident geologist, Cent Can, Falconbridge Nickel Mines Ltd, 61-68, explor mgr, Cent Div, 68-70; vpres explor, Occidental Minerals Corp Can, 70-72, EXPLOR MGR, MINERALS DIV, CAN OCCIDENTAL PETROL LTD, TORONTO, ONT, 72- Mem: Soc Econ Geologists; Geol Soc Am; Can Inst Mining & Metall; Geol Asn Can; Geol Soc SAfrica. Res: Exploration for economic deposits of nickel in ultramafic rocks; sedimentary copper and porphyry copper deposits; volcanic sulfide and pyrometasomatic copper deposits. Mailing Add: 11 Strath Humber Ct Islington ON Can

BRUMMER, S BARRY, b London, Eng, Oct 8, 35; m 61; c 3. PHYSICAL CHEMISTRY. Educ: Univ London, BSc, 57, DIC & PhD(chem), 60. Prof Exp: Royal Comn Exhib 1851 sr fel, Univ London, 60-62; staff scientist, Tyco Labs, Inc, 62-63, sr scientist, 63-67, head electrochem dept, 67-70, asst dir corp res div, 70-73; VPRES & DIR RES, EIC INC, 73- Mem: Am Chem Soc; Electrochem Soc; AAAS. Res: Theory of ionic migration in liquids; electrode processes; anodic oxidation and adsorption; surface oxidation studies; studies of anomalous water; non-aqueous electrochemistry. Mailing Add: EIC Inc 55 Chapel St Newton MA 02154

BRUMMETT, ANNA RUTH, b Ft Smith, Ark, Mar 25, 24. EMBRYOLOGY, CYTOLOGY. Educ: Univ Ark, BA, 48, MA, 49; Bryn Mawr Col, PhD(biol), 53. Prof Exp: Demonstr biol, Bryn Mawr Col, 50-52, instr, 52-53; from instr to asst prof, Carleton Col, 53-61; assoc dean col, 67-68, res status appt, 68-69, actg chmn dept, 69-70, ASSOC PROF BIOL, OBERLIN COL, 67-, CHMN DEPT, 73- Mem: AAAS; Tissue Cult Asn; Am Soc Zoologists; Am Soc Cell Biol; NY Acad Sci. Res: Morphogenetic movements and early differentiation in the teleost embryo. Mailing Add: Dept of Biol Oberlin Col Oberlin OH 44074

BRUMMETT, ROBERT E, b Concordia, Kans, Feb 11, 34; m 54; c 4. PHARMACOLOGY, OTOLARYNGOLOGY. Educ: Ore State Univ, BS, 59, MS, 60; Univ Ore, PhD, 64. Prof Exp: Asst prof pharmacog & pharmacol, Sch Pharm, Ore State Univ, 61-62; asst prof otolaryngol, 64-70, ASST PROF PHARMACOL, SCH MED, UNIV ORE, 64-, ASSOC PROF OTOLARYNGOL, 70- Mem: AAAS; Soc Exp Biol & Med; Am Acad Ophthal & Otolaryngol; Soc Toxicol; Soc Neurosci. Res: Effect of drugs on hearing; cochlear function; clinical pharmacology of drugs used in otolaryngology; electromyographic diagnosis of laryngeal paralysis. Mailing Add: Dept of Otolaryngol Univ of Ore Med Sch Portland OR 97201

BRUMMOND, DEWEY OTTO, b Towner, NDak, July 28, 25; m 54; c 2. BIOCHEMISTRY. Educ: NDak State Univ, BS, 50; Univ Wis, MS, 52, PhD(biochem), 54. Prof Exp: Fel, Nat Found Infantile Paralysis, 54-56; prin scientist, Radioisotope Serv, Vet Admin Hosp, Cleveland, Ohio, 56-64; assoc prof biochem, NDak State Univ, 65-66; PROF CHEM, MOORHEAD STATE UNIV, 66- Concurrent Pos: Sr instr, Western Reserve Univ, 56-64. Mem: Am Soc Biol Chemists. Res: Enzymatic synthesis of cellulose. Mailing Add: Dept of Chem Moorhead State Univ Moorhead MN 56560

BRUN, WILLIAM ALEXANDER, b NJ, Sept 10, 25; m 50; c 3. CROP PHYSIOLOGY. Educ: Univ Miami, BS, 50; Univ Ill, MS, 51, PhD(bot), 54. Prof Exp: Asst bot & plant physiol, Univ Ill, 50-54; asst prof bot, NC State Col, 54-57; plant physiologist, Agr Res Serv, USDA, 57-59 & United Fruit Co, 59-65; from asst prof to assoc prof, 65-75, PROF AGRON & PLANT GENETICS, UNIV MINN, ST PAUL, 76- Mem: Am Soc Plant Physiol; Scand Soc Plant Physiol. Res: Growth and reproduction; photosynthesis; water relations; nitrogen fixation. Mailing Add: Dept of Agron & Plant Genetics Univ of Minn St Paul MN 55108

BRUNAUER, STEPHEN, b Budapest, Hungary, Feb 12, 03; nat US; m 31, 61; c 2. COLLOID CHEMISTRY, SURFACE CHEMISTRY. Educ: Columbia Univ, AB, 25; George Washington Univ, MS, 29; Johns Hopkins Univ, PhD(chem), 33. Hon Degrees: DSc, Clarkson Col Technol, 69. Prof Exp: From jr chemist to chemist, Fixed Nitrogen Res Lab, Bur Plant Indust, USDA, 28-42; chief tech adminr explosives res & develop, Bur Ord, US Dept Navy, 46-50, chief chemist, 50-51; sr res chemist, Portland Cement Asn, 51-53, prin res chemist, 53-58, mgr, Basic Res Sect, 58-64; prof chem & chmn dept, 65-68, Clarkson prof, 68-73, EMER CLARKSON PROF CHEM, CLARKSON COL TECHNOL, 73- Concurrent Pos: Lectr, George Washington Univ, 37-45, prof lectr, 45-49; mem div chem & chem technol, Nat Res Coun, 48-51, vchmn, Gordon Conf Chem Interfaces, 56-57, chmn, 57-58; mem coun, Gordon Res Conf, 60-63; mem comt colloid & surface chem, Nat Acad Sci-Nat Res Coun, 61-; consult, Westvaco, 66-; vchmn comn colloid & surface chem, Int Union Pure & Appl Chem, 69-75. Honors & Awards: Order of Brit Empire, 46; Kendall Award, Am Chem Soc, 61; Publ Honor, J Colloid & Interface Sci, 72. Mem: Am Chem Soc. Res: Catalysis; heterogeneous kinetics and equilibria; adsorption; explosives; Portland cement; concrete; cement chemistry. Mailing Add: Dept of Chem Clarkson Col of Technol Potsdam NY 13676

BRUNBAUGH, JOE H, b Eldorado, Ohio, Sept 14, 30; m 61; c 3. MARINE BIOLOGY. Educ: Miami Univ, BSEd, 52; Purdue Univ, MS, 56; Stanford Univ, PhD(biol), 65. Prof Exp: High sch teacher, Ohio, 52-54; instr biol & bot, Wabash Col, 56-59; from asst prof to assoc prof biol, 64-71, PROF BIOL & CHMN DIV NATURAL SCI, 71- Mem: Am Soc Zoologists; Am Inst Biol Sci. Res: Anatomy, diet and feeding mechanisms in marine invertebrates, particularly holothurian echinoderms; functional morphology of marine invertebrates; natural history of invertebrates. Mailing Add: Div of Natural Sci Sonoma State Col Rohnert Park CA 94928

BRUNDAGE, ARTHUR LAIN, b Wallkill, NY, Dec 19, 27; m 51; c 4. DAIRY HUSBANDRY. Educ: Cornell Univ, BS, 50; Univ Minn, MS, 52, PhD, 55. Prof Exp: From dairy husbandman to res dairy husbandman, Alaska Agr Exp Sta, Univ Alaska-USDA, 52-68; PROF ANIMAL SCI, INST AGR SCI, UNIV ALASKA, 68- Mem: AAAS; Am Dairy Sci Asn; Am Soc Animal Sci; Brit Grassland Soc; Am Inst Biol Sci. Res: Dairy cattle management and production, especially ruminant nutrition and forage production and utilization. Mailing Add: Inst of Agr Sci Univ of Alaska Box AE Palmer AK 99645

BRUNDAGE, DONALD KEITH, b Traverse City, Mich, Oct 24, 13; m 38; c 5. ORGANIC CHEMISTRY. Educ: Eastern Mich Univ, AB, 33; Univ Mich, PhD(phys org chem), 40. Prof Exp: Asst chem, Univ Mich, 35-38; chemist, Bendix Aviation Corp, Mich, 41; instr chem, St Cloud State Col, 41-46; assoc prof, 46-58, PROF CHEM, UNIV TOLEDO, 58- Concurrent Pos: Res assoc, Univ Calif, Irvine, 67-68. Mem: Fel AAAS; Am Chem Soc. Res: Equilibrium; kinetics. Mailing Add: Dept of Chem Univ of Toledo Toledo OH 43606

BRUNDAGE, ROBERT EARL, b Tallahassee, Fla. COMPUTER SCIENCE. Educ: Fla State Univ, BA, 65; Ga Inst Technol, MS, 68; Univ Va, PhD(comput sci), 74. Prof Exp: Sci programmer, Lockheed-Ga Co, 65-67; systs analyst, Univ Va, 68-73; ASST PROF COMPUT SCI, FLA STATE UNIV, 74- Mem: Asn Comput Mach; Sigma Xi. Res: Computer system design and evaluation. Mailing Add: Dept of Math Fla State Univ Tallahassee FL 32306

BRUNDAGE, ROBERT SCOTT, b St Cloud, Minn, Aug 6, 42; m 68; c 2. BIOPHYSICAL CHEMISTRY. Educ: Univ Toledo, BS, 64; Brandeis Univ, PhD(biophys), 69. Prof Exp: RES BIOPHYSICIST, INSTRUMENTATION LAB, INC, 70- Mem: Am Chem Soc. Res: Cochlear energy transformations; biological and environmental kinetics and mechanisms; rapid and precise microsensor systems. Mailing Add: 31 Linden St Arlington MA 02174

BRUNDAGE, ROY CHARLES, b Otisville, NY, Apr 26, 02; m 30; c 3. FORESTRY. Educ: State Univ NY, BS, 25; Univ Mich, MS, 30. Prof Exp: Forest asst, US Forest Serv, 25-26; assoc prof, 30-45 & 46-70, EMER PROF FORESTRY & CONSERV, DEPT FORESTRY & CONSERV, AGR EXP STA, PURDUE UNIV, 70-; CONSULT, FOREST ECON & FOREST PROD MKT, 70- Concurrent Pos: Regional consult, Am Forestry Asn, 45. Mem: Fel AAAS; Soc Am Foresters; Sigma Xi. Res: Economic research in marketing and utilization of forest products. Mailing Add: 336 Park Lane West Lafayette IN 47906

BRUNDAGE, WILLIAM GREGORY, b Monroe, NY, Oct 18, 34; m 67; c 1. MEDICAL MICROBIOLOGY, IMMUNOCHEMISTRY. Educ: Northwestern State Col, La, BS, 61, MS, 62; La State Univ, PhD(microbiol), 67. Prof Exp: Asst prof, Dept Vet Sci, La State Univ, 67-68; ASSOC PROF MICROBIOL, UNIV SOUTHERN MISS, 68-, CHMN DEPT MED TECHNOL, 71-, ASST DEAN, COL SCI & TECHNOL, 74- Mem: NY Acad Sci; Am Soc Microbiol. Res: Host-parasite relationships; effects of pseudomonas pseudomallei and plasmodium lophurae upon their hosts. Mailing Add: Box 5165 Southern Sta Hattiesburg MS 39401

BRUNDIDGE, KENNETH CLOUD, b St Louis, Mo, May 31, 27; m 47; c 2. METEOROLOGY. Educ: Univ Chicago, BA, 52, MS, 53; Tex A&M Univ, PhD(meteorol), 61. Prof Exp: Lab instr & res asst cloud physics, Chicago Midway Labs, 52, asst meteorologist, 54; meteorologist jet stream res, 55, from instr to assoc prof, 55-68, asst dean student affairs, Col Geosci, 71-73, asst dean acad affairs, 73-75, PROF METEOROL, COL GEOSCI, TEX A&M UNIV, 68-, HEAD DEPT, 75- Mem: Am Meteorol Soc. Res: Dynamic and synoptic meteorology; mesoscale circulations; numerical simulation. Mailing Add: Dept of Meteorol Tex A&M Univ College Station TX 77843

BRUNDIN, ROBERT H, b Glenwood, Minn, Dec 2, 22; m 45; c 2. ANALYTICAL CHEMISTRY. Educ: US Mil Acad, BSc, 44; Ohio State Univ, MSc, 51. Prof Exp: Asst prof chem, US Mil Acad, US Air Force, 51-54, engr, Air Res & Develop Command, Wright Field, Ohio, 54-58, prog dir, Proj Mercury Booster, Space Systs Div, Calif, 59-62, assoc prof chem, US Air Force Acad, 62-65, prof & head dept, 65-66, dep head dept, 63-65; RES SCIENTIST & DIR ADMIN, KAMAN SCI CORP, 66- Res: Aircraft and missile development; manned space flight. Mailing Add: Kaman Sci Corp Garden of the Gods Rd Colorado Springs CO 80907

BRUNE, GUNNAR MAGNUS, b Pittsburgh, Pa, Apr 18, 14; m 39; c 1. GEOLOGY. Educ: Antioch Col, BS, 36. Prof Exp: Asst soil technologist, Soil Conserv Serv, USDA, Ohio, 35-39, sedimentation specialist, 39-42, geologist, Wis, 45-54 & Tex, 54-69; GEOLOGIST, TEX WATER DEVELOP BD, 69- Res: Engineering geology; dam site and ground water investigations. Mailing Add: Tex Water Develop Bd Box 13087 Austin TX 78711

BRUNE, JAMES N, b Modesto, Calif, Nov 23, 34; m 57; c 4. SEISMOLOGY, GEOPHYSICS. Educ: Univ Nev, BS, 56; Columbia Univ, PhD(seismol), 61. Prof Exp: Res scientist, Lamont Geol Observ, Columbia Univ, 58-64; geophysicist, US Coast & Geod Surv, 64; assoc prof geophys, Calif Inst Technol, 65-69; PROF GEOPHYS, SCRIPPS INST OCEANOG, UNIV CALIF, SAN DIEGO, 69-, ASSOC DIR, INST GEOPHYS & PLANETARY PHYSICS, 71-, CHMN, GEOL RES DIV, 74- Concurrent Pos: Adj assoc prof geol, Columbia Univ, 64. Honors & Awards: Macelwane Award, Am Geophys Union, 62, Karl Gilbert Award, 67. Mem: Am Geophys Union; Seismol Soc Am (pres, 70). Res: Seismology; earth structure; earth quake source mechanism; heat flow; geology. Mailing Add: Scripps Inst of Oceanog Univ Calif San Diego IGPP-A-025 La Jolla CA 92093

BRUNEAU, LESLIE HERBERT, b Cornwall, Ont, Nov 28, 28; nat US; m 53; c 2. GENETICS. Educ: McGill Univ, BSc, 50; Univ Tex, MA, 52, PhD(cytogenetics), 56. Prof Exp: From asst prof to assoc prof, 55-66, chmn biol sci, 60-74, PROF ZOOL, OKLA STATE UNIV, 66- Mailing Add: Sch of Biol Sci Okla State Univ Stillwater OK 74074

BRUNEL, PIERRE, b Montreal, Que, Mar 21, 31; m 55; c 4. MARINE ECOLOGY. Educ: Univ Montreal, BSc, 53; Univ Toronto, MA, 57; McGill Univ, PhD, 68. Prof Exp: Asst invertebrates, Royal Ont Mus Zool, 53-55; zoologist, Marine Biol Sta, 55-66; sr lectr, 66-68, asst prof, 68-70, ASSOC PROF, DEPT BIOL SCI, UNIV MONTREAL, 70- Concurrent Pos: Vpres & sci coordr, Interuniv Group Oceanog Res Que, 70- Mem: AAAS; Am Inst Biol Sci; Can Soc Zool; Soc Syst Zool; Am Soc Limnol & Oceanog. Res: Ecology of marine bottom invertebrates and communities of northern seas; carcinology; taxonomy; zoogeography, ecology of Peracarid Crustacea, Amphipoda. Mailing Add: 12224 James-Morrice St Montreal PQ Can

BRUNELL, GLORIA FLORETTE, b Meriden, Conn, Oct 6, 25. MATHEMATICS. Educ: Albertus Magnus Col, BA, 47; Fordham Univ, MA, 57; Yale Univ, PhD(math), 64. Prof Exp: Teacher, Holy Trinity High Sch, Ohio, 49-50, St Mary's Acad, 50-51, Holy Trinity High Sch, 51-55, Dominican Acad, NY, 55-57 & Watterson High Sch, Ohio, 57-58; from instr to assoc prof math, Albertus Magnus Col, 58-68; assoc prof, 68-71, PROF MATH, WESTERN CONN STATE COL, 71- Concurrent Pos: Vis lectr, Col St Mary of the Springs, 50-55. Mem: Am Math Soc; Math Asn Am. Res: Analysis; in service work with mathematics teachers. Mailing Add: Dept of Math Western Conn State Col Danbury CT 06810

BRUNELL, PHILIP ALFRED, b New York, NY, Feb 1, 31; m 52; c 3. PEDIATRICS, VIROLOGY. Educ: City Col New York, BS, 50; Univ Ill, MS, 52; Univ Buffalo, MD, 57. Prof Exp: Asst physiol, Univ Ill, 51-53; intern pediat, Meyer Mem Hosp, Buffalo, NY, 57-58; resident, Children's Hosp, Buffalo, 58-60; med off in chg virus reference unit, Nat Commun Dis Ctr, 63-64; from asst prof to assoc prof pediat, Sch Med, NY Univ, 69-75; PROF PEDIAT & CHMN DEPT, UNIV TEX HEALTH SCI CTR SAN ANTONIO, 75- Concurrent Pos: Mem adv comt, Human Resources Admin, Head Start, New York, 64-; res grants, NIH, 64-, Nat Commun Dis Ctr, 68- & WHO, 69-; consult, Am Acad Pediat, 66-, US Mil Acad, 69- & Northwick Pk Hosp, Harrow, Eng, 71-72; vis scientist, Clin Res Ctr, Harrow, 71-72; dir pediat, Bexar County Hosp, San Antonio, 75- Mem: Am Acad Pediat; Sigma Xi; Am Soc Microbiol; Soc Pediat Res; Infectious Dis Soc Am. Res: Clinical virology; infectious diseases of infants and children; varicellazoster virus; passive immunization. Mailing Add: Dept of Pediat Univ Tex Health Sci Ctr San Antonio TX 78284

BRUNELLE, RICHARD LEON, b Littleton, Mass, May 9, 37; m 58; c 2. FORENSIC SCIENCE. Educ: Clark Univ, BA, 60; George Washington Univ, MS, 72. Prof Exp: Res chemist, Worcester Found Exp Biol, 60-61; anal chemist, US Food & Drug Admin, 61-63; FORENSIC & REGULATORY CHEMIST, BUR ALCOHOL, TOBACCO & FIREARMS, US TREAS, 63-, CHIEF IDENTIFICATION LAB, 74- Concurrent Pos: Mem fac, George Washington Univ, 74- Honors & Awards: John A Dondero Mem Award, Int Asn Identification, 71. Mem: Int Asn Identification; Am Chem Soc; fel Asn Off Anal Chemists; Am Acad Forensic Sci. Res: Forensic science dealing with all areas, particularly new methods for the analysis of physical evidence such as documents, ink, paper, firearms, soils and paints. Mailing Add: 3820 Acosta Rd Fairfax VA 22030

BRUNELLE, THOMAS E, b Crookston, Minn, Feb 12, 35; m 58; c 3. BIOCHEMISTRY, ORGANIC CHEMISTRY. Educ: Col St Thomas, BS, 57; Univ Minn, MS, 62, PhD(biochem), 68. Prof Exp: Chemist, 57-60, group leader org chem, 60-64, asst mgr org & biol res, 67-69, MGR CORP RES SECT, RES & DEVELOP DEPT, ECON LAB, INC, 69-, VPRES CORP SCI & TECHNOL, 75- Mem: AAAS; Am Chem Soc; Am Oil Chemists' Soc. Res: Acid polysaccharide chemistry; surface active agents. Mailing Add: Res & Develop Dept Osborn Bldg Econ Lab Inc St Paul MN 55102

BRUNER, BUDDY LEROY, physical chemistry, chemical physics, see 12th edition

BRUNER, DORSEY WILLIAM, b Windber, Pa, Dec 25, 06; m 40. MICROBIOLOGY. Educ: Albright Col, BS, 29; Cornell Univ, PhD(bact), 33, DVM, 37. Prof Exp: Teacher high sch, Pa, 29-30; instr res, Dept Path & Bact, State Univ NY Vet Col, Cornell Univ, 31-37; asst bacteriologist, Dept Animal Path, Exp Sta, Univ Ky, 41-42, virologist, 46-48, bacteriologist, 48-49; vet bacteriologist, Dept Path & Bact, 49-65, vet microbiologist & chmn dept vet microbiol, 65-72, EMER PROF VET MICROBIOL, STATE UNIV NY VET COL, CORNELL UNIV, 72- Honors & Awards: 12th Int Vet Cong Prize, Am Vet Med Asn, 72. Mem: Soc Exp Biol & Med; Am Soc Microbiol; Am Vet Med Asn. Res: Genus Salmonella; equine virus abortion; neonatal isoerythrolysis; acidfast bacteria. Mailing Add: Dept of Vet Microbiol SUNY Vet Col Cornell Univ Ithaca NY 14850

BRUNER, EDWARD M, b New York, NY, Sept 28, 24; m 48; c 2. CULTURAL ANTHROPOLOGY. Educ: Ohio State Univ, BA, 48, MA, 50; Univ Chicago, PhD(anthrop), 54. Prof Exp: Instr anthrop, Univ Chicago, 53-54; asst prof, Yale Univ, 54-60; assoc prof, Univ Ill, 61-65, head dept, 66-70, PROF ANTHROP, UNIV ILL, URBANA, 65- Concurrent Pos: Ford Found grant-in-aid, Yale Univ, 57; Ford Found fel & NSF res grant, Sumatra, Indonesia, 57-58; NSF fel, Ctr Advan Study Behav Sci, 60-61; sr scholar, Inst Advan Proj, East West Ctr, Univ Hawaii, 63; NIMH res grant urbanization, Univ Ill, 65, Doris Duke Am Indian Oral Hist Proj, 67-; mem fac grants comt, Soc Sci Res Coun, 66; consult, Cult Anthrop Fel Rev Comt, NIMH, 66; chmn comt advan test anthrop, Educ Testing Serv, NJ, 67-69; Ford Found consult, Nat Assessment Educ Indonesia, 69-70; assoc mem, Ctr Advan Study, Univ Ill, 70-71; Asia Soc Southeast Asia Develop Adv Group res grant, Indonesian Acad Sci, 69-71. Mem: Am Anthrop Asn; Royal Anthrop Soc; Am Ethnol Soc; Soc Appl Anthrop; Asn Asian Studies. Res: Urbanization; processes of culture change; ethnic group relations. Mailing Add: Dept of Anthrop Univ of Ill Urbana IL 61801

BRUNER, ELMO CODY, JR, b Oklahoma City, Okla, July 27, 34; m 55; c 3. SPACE PHYSICS. Educ: Univ Ariz, BS, 57, MS, 59; Univ Colo, PhD(physics), 64. Prof Exp: Phys sci aide, Naval Ord Test Sta, China Lake, Calif, 55; teaching asst, Univ Ariz, 57-58; physicist, US Naval Ord Test Sta, China Lake, 58-60; teaching asst, Univ Colo, 60-62; res asst, 62-64, RES ASSOC, LAB ATMOSPHERIC & SPACE PHYSICS, UNIV COLO, BOULDER, 64- Concurrent Pos: Consult, Los Alamos Sci Lab, 75- Mem: Optical Soc Am; Am Astron Soc. Res: High resolution ultraviolet spectroscopy of the sun; structure and dynamics of the solar chromosphere and transition zone; design of space instrumentation; absolute ultraviolet spectroradiometry. Mailing Add: 3705 Emerson Ave Boulder CO 80303

BRUNER, FRANK HENRY, chemistry, see 12th edition

BRUNER, HARRY DAVIS, b Jeffersonville, Ind, July 18, 11; m 31; c 2. RADIOBIOLOGY, RESEARCH ADMINISTRATION. Educ: Univ Louisville, SB, 30, MD, 34, SM, 36; Univ Chicago, PhD(physiol), 39. Prof Exp: Asst, Univ Louisville, 33-34, asst physiol & pharmacol, 34-36; instr physiol, Med Col SC, 36-38; asst prof, Univ NC, 39-42; vis fel, Harrison Dept Surg Res, Univ Pa, 42-43, res assoc, 43-45, asst prof, 45-47; prof pharmacol, Univ NC, 47-49; chief scientist, Med Div, Oak Ridge Inst Nuclear Studies, 49-52; prof physiol, Emory Univ, 52-56; chief med res br, Div Biol & Med, AEC, 56-60, asst dir health & med res, 60-69, asst dir, 69-72, spec asst to chmn, 72-75; spec asst to asst secy for Oceans, Int Environ & Sci Affairs, Dept of State, 75; MED SCI ADV, US ENERGY RES & DEVELOP ADMIN, 75- Concurrent Pos: Tech adv, US deleg, UN Sci Comt Effects Atomic Radiations. Mem: Am Soc Pharmacol & Exp Therapeut; Soc Exp Biol & Med; Radiation Res Soc; Health Physics Soc. Res: Hematology; pulmonary physiology; isotopes. Mailing Add: 9404 Byeforde Rd Kensington MD 20795

BRUNER, LEON JAMES, b Ponca City, Okla, Dec 28, 31; m 60; c 3. BIOPHYSICS. Educ: Univ Chicago, AB, 52, MS, 55, PhD(physics), 59. Prof Exp: Staff mem physics, Watson Sci Comput Lab, Int Bus Mach Corp, 59-62; from asst prof to assoc prof, 62-73, PROF PHYSICS, UNIV CALIF, RIVERSIDE, 73- Mem: AAAS; Biophys Soc; Am Phys Soc. Res: Bioelectric phenomena; structure and transport properties of membranes. Mailing Add: Dept of Physics Univ of Calif Riverside CA 92502

BRUNER, LEONARD BRETZ, JR, b Wichita Falls, Tex, Oct 12, 21; m 47; c 1. CHEMISTRY. Educ: Univ Tulsa, ChB, 43; Univ Mich, MS, 49, PhD(chem), 57. Prof Exp: Jr chemist, Phillips Petrol Co, 43-44; instr chem, Univ Tulsa, 46-47; res chemist, Dow Corning Corp, 55-59; sr scientist, Trionics Corp, 59-61; assoc prof, Eureka Col, 61-66; TECH DIR, STAUFFER-WACKER SILICON CORP, 66- Concurrent Pos: Consult, Stauffer Chem Co, 63- Mem: Am Chem Soc. Res: Grignard reactions; organic azides; siloxane chemistry; polymerization; properties of siloxane polymers; electrical properties of ceramics. Mailing Add: SWS Silicones Div Stauffer Chem Co PO Box 428 Adrian MI 49221

BRUNETT, EMERY W, b Ovando, Mont, Dec 3, 27; m 60; c 4. MEDICINAL CHEMISTRY, PHARMACY. Educ: Univ Mont, BS, 53, MS, 56; Univ Wash, PhD(pharmaceut chem), 66. Prof Exp: Instr pharm, Univ Mont, 57-58, vis asst prof, 64-65; asst prof pharmaceut chem, Drake Univ, 66-69; asst prof, 69-75, ASSOC PROF PHARM, UNIV WYO, 75- Mem: AAAS; Am Asn Cols Pharm. Res: Aminothiophene chemistry; vitamin A analogs; hay fever pollens. Mailing Add: Univ of Wyo Sch of Pharm Box 3375 Univ Sta Laramie WY 82071

BRUNFIEL, CHARLES, b Matthews, Ind, Aug 13, 14; m 40; c 2. MATHEMATICS.

Educ: Ball State Teachers Col, BS, 39; Univ Chicago, MS, 44; Purdue Univ, PhD(math), 54. Prof Exp: Instr math, Ill Inst Technol, 45-46; from asst prof to prof, Ball State Teachers Col, 46-60; assoc prof, 60-62, PROF MATH, UNIV MICH, ANN ARBOR, 62- Concurrent Pos: Ford Found fel, 52-53. Mem: Math Asn Am; Am Math Soc. Res: Galois theory. Mailing Add: Dept of Math Univ of Mich Ann Arbor MI 48104

BRUNGARDT, VAL HILARY, meats, see 12th edition

BRUNGS, ROBERT ANTHONY, b Cincinnati, Ohio, July 7, 31. SOLID STATE PHYSICS. Educ: Bellarmine Col, NY, AB, 55; Loyola Sem, PhL, 56; St Louis Univ, PhD(physics), 62; Woodstock Col, Md, STL, 65. Prof Exp: Fel & res assoc physics, 66-70, ASST PROF PHYSICS, ST LOUIS UNIV, 70-, ASST PROF DOGMATIC & SYST THEOL, 72-; COUNR, MED CTR, 70- Mem: Am Phys Soc. Res: Experimental investigation of semiconductor properties of crystalline beta-rhombohedral boron. Mailing Add: St Louis Univ Dept Physics 221 N Grand Blvd St Louis MO 63103

BRUNGS, WILLIAM ALOYSIUS, JR, b Covington, Ky, Aug 10, 32; m 62; c 2. POLLUTION BIOLOGY. Educ: Ohio State Univ, BSc, 58, MSc, 59, PhD(wildlife mgt), 63. Prof Exp: Aquatic biologist, USPHS, 61-64; Cincinnati Water Res Lab, Fed Water Pollution Control Admin, 64-68, chief Newtown Fish Toxicol Lab, 68-71, ASST DIR WATER QUAL CRITERIA, ENVIRON RES LAB, DULUTH, US ENVIRON PROTECTION AGENCY, 71- Mem: Water Pollution Control Fedn; Int Asn Great Lakes Res; AAAS; Am Fisheries Soc. Res: Distribution of radionuclides in fresh-water environments; determination of acute and chronic effects of water pollution on fish and the relationship of environmental variables on these effects. Mailing Add: Environ Res Lab 6201 Congdon Blvd US Environ Protection Agency Duluth MN 55804

BRUNING, DONALD FRANCIS, b Boulder, Colo, Dec 18, 42; m 69; c 3. ORNITHOLOGY. Educ: Univ Colo, BA, 66, MA, 67, PhD(zool), 74. Prof Exp: Curatorial trainee ornith, 67-69, from asst cur to assoc cur, 69-74, CUR ORNITH, NEW YORK ZOOL SOC, 75- Concurrent Pos: Res assoc, Ctr Field Biol & Conserv, New York Zool Soc, 73-; consult, Time/Life Bks, 73-; adj assoc prof, Fordham Univ, 74- Mem: AAAS; Am Ornithologists Union; Wilson Ornith Soc; Am Asn Zool Parks & Aquaria; Nat Audubon Soc. Mailing Add: New York Zool Soc Bronx NY 10460

BRUNINGS, KARL JOHN, b Baltimore, Md, Dec 4, 13; m 40; c 2. ORGANIC CHEMISTRY, MEDICINAL CHEMISTRY. Educ: Johns Hopkins Univ, PhD(org chem), 39. Prof Exp: Res chemist, Eastman Kodak Co, 39-41; Graflin fel, Johns Hopkins Univ, 41-43, instr org chem, 43-44; asst prof, NY Univ, 44-46; from asst prof to assoc prof, 46-48; res chemist, Chem Res & Develop Div, Charles Pfizer & Co, Inc, 48-50, from asst dir to dir, 50-61, admin dir res, 61-62; pres res div, Geigy Chem Corp, 62-68, SR VPRES PHARM RES, CIBA-GEIGY CORP, 68- Concurrent Pos: Rennebohn lectr, 65; chmn, Int Pharmacol Mfg, Sweden. Mem: AAAS; corp assoc, Am Chem Soc; Asn Res Dirs; Am Inst Chem; NY Acad Sci. Res: Organic synthesis; pharmaceuticals; antibiotics; amino acids; proteins; industrial chemicals; color photography; pyrrole pigments; administration of and participation in medicinal and pharmaceutical research. Mailing Add: Ciba-Geigy Corp Saw Mill River Rd Ardsley NY 10502

BRUNK, CLIFFORD FRANKLIN, b Detroit, Mich, Feb 11, 40; m 62; c 3. BIOPHYSICS. Educ: Mich State Univ, BS, 61; Stanford Univ, MS, 62, PhD(biophys), 67. Prof Exp: Asst prof, 67-73, ASSOC PROF ZOOL, UNIV CALIF, LOS ANGELES, 73- Concurrent Pos: USPHS fel, Carlsberg Biol Inst, Copenhagen, Denmark, 67-68. Mem: Biophys Soc; Am Soc Cell Biol. Res: Research on repair of ultraviolet damaged DNA, in the absence of photoreactivating light; general systems for sequencing DNA. Mailing Add: Dept of Zool Univ of Calif Los Angeles CA 90024

BRUNK, HUGH DANIEL, b Manteca, Calif, Aug 22, 19; m 42; c 3. STATISTICS, MATHEMATICS. Educ: Univ Calif, AB, 40; Rice Inst, MA, 42, PhD(math), 44. Prof Exp: Asst math, Rice Inst, 40-44, from instr to asst prof, 46-51; mathematician, Sandia Corp, 51-52; from assoc prof to prof math, Univ Mo, 52-61; prof, Univ Calif, Riverside, 61-63; prof statist, Univ Mo, 63-69; PROF STATIST, ORE STATE UNIV, 69- Concurrent Pos: Fulbright lectr, Univ Copenhagen, 58-59; vis sr lectr, Univ Col Wales, 66-67; hon res assoc statist, Univ Col London, 71. Mem: Fel Am Statist Asn; Am Math Soc; Math Asn Am; fel Inst Math Statist; Int Statist Inst. Res: Analysis; mathematical statistics; probability. Mailing Add: Dept of Statist Ore State Univ Corvallis OR 97331

BRUNK, WILLIAM EDWARD, b Cleveland, Ohio, Nov 24, 28; m 57; div. ASTRONOMY. Educ: Case Inst Technol, BS, 52, MS, 54, PhD(astron), 63. Prof Exp: Res scientist, Lewis Flight Propulsion Lab, Nat Adv Comt, Aeronaut, 54-58; aerospace res engr, Lewis Res Ctr, NASA, 58-64, staff scientist, 64-65, actg chief, 65, PROG CHIEF, PLANETARY ASTRON, NASA HQS, 65- Mem: Am Astron Soc; Int Astron Union. Res: Stellar and planetary astronomy; aerodynamics; heat transfer; trajectory analysis. Mailing Add: Code SL NASA Hqs Washington DC 20546

BRUNKHORST, WILLA, endocrinology, see 12th edition

BRUNN, STANLEY DAVID, b Freeport, Ill, June 24, 39; m 62. GEOGRAPHY. Educ: Eastern Ill Univ, BS, 60; Univ Wis, MS, 62; Ohio State Univ, PhD(geog), 66. Prof Exp: Instr geog, Ohio State Univ, 65-66; asst prof, Univ Fla, 66-69; asst prof, 69-71, ASSOC PROF GEOG, MICH STATE UNIV, 71- Mem: Am Acad Arts & Sci; Asn Am Geogr; Am Geog Soc; Am Acad Polit & Social Sci; Am Sociol Asn. Res: Urban-economic geography; social geography; quantitative methods; dynamics of central places; geography of retail services; urban dynamics of developing countries; population changes in the United States; voting patterns. Mailing Add: Dept of Geog Mich State Univ East Lansing MI 48823

BRUNNER, EDWARD A, b Erie, Pa, July 18, 29; m 55; c 4. PHARMACOLOGY, ANESTHESIOLOGY. Educ: Villanova Univ, BS, 52; Hahnemann Med Col & Hosp, MD, 59, PhD(pharmacol), 62. Prof Exp: Instr pharmacol, Hahnemann Med Col & Hosp, 60-62, instr anesthesia, Sch Med, Univ Pa, 62-65, instr pharmacol, 64-65; from asst prof to assoc prof anesthesia, 66-71, PROF ANESTHESIA & CHMN DEPT, MED SCH, NORTHWESTERN UNIV, 71- Res: Carbohydrate metabolism in liver, brain and muscle; effects of anesthetic agents on metabolism; muscle relaxants; medical education. Mailing Add: Dept of Anesthesia Northwestern Univ Med Sch Chicago IL 60611

BRUNNER, JAY F, b Pendleton, Ore, May 2, 47; m 67; c 2. ECONOMIC ENTOMOLOGY. Educ: Williamette Univ, BA, 69; Wash State Univ, MS, 73, PhD(entom), 75. Prof Exp: ASST PROF ENTOM, MICH STATE UNIV, 75- Mem: Entom Soc Am; Sigma Xi. Res: Phenological development of tree fruit insect pests for purposes of determining optimum control strategies. Mailing Add: Dept of Entom Mich State Univ East Lansing MI 48823

BRUNNER, JAY ROBERT, b Royersford, Pa, Sept 17, 18; m 47; c 3. FOOD SCIENCE, DAIRY CHEMISTRY. Educ: Pa State Univ, BS, 40; Univ Calif, MS, 42; Mich State Univ, PhD(dairy & phys chem), 52. Prof Exp: Asst, Univ Calif, 40-42; from instr to assoc prof, 46-59, PROF FOOD SCI & HUMAN NUTRIT, MICH STATE UNIV, 59- Concurrent Pos: Mem EBC study sect, USPHS. Honors & Awards: Borden Award, Am Chem Soc, 64. Mem: Am Dairy Sci Asn; Inst Food Technol; Am Chem Soc; Am Asn Cereal Chemists. Res: Physical and chemical properties of milkfat and milk proteins; dairy products processing; chemistry of milk. Mailing Add: Dept of Food Sci & Human Nutrit Mich State Univ East Lansing MI 48824

BRUNNER, MICHAEL, b Brooklyn, NY, Apr 3, 43. MOLECULAR BIOLOGY. Educ: City Col New York, BS, 65; Pa State Univ, MS, 67, PhD(microbiol), 69. Prof Exp: Fel growth regulation, Sch Med, Washington Univ, 69-71; fel adenovirus, Inst Molecular Virol, 71-73; instr RNA tumor virus, Harvard Med Sch, 73-75; ASST PROF BIOL SCI, ST JOHN'S UNIV, NY, 75- Concurrent Pos: USPHS trainee grant, Sch Med, Washington Univ, 71-73; Leukemia Soc Am spec fel, 73. Res: Genetic interaction between leukemia-sarcoma viruses and the cells in which oncogenic changes are produced. Mailing Add: Dept of Biol Sci St John's Univ Jamaica NY 11439

BRUNNGRABER, ELINOR FLORA, b New York, NY, Aug 18, 30. BIOCHEMISTRY. Educ: Hunter Col, BA, 52, MA, 57; Cornell Univ, PhD(biochem), 62. Prof Exp: Chemist, 52-54; res worker, 62-63, res assoc, 63-71, ASST PROF BIOCHEM, COL PHYSICIANS & SURGEONS, COLUMBIA UNIV, 71- Mem: Am Chem Soc. Res: Enzymology; purification and characterization of nucleoside and nucleotide phosphotransferases. Mailing Add: Col of Physicians & Surgeons Columbia Univ New York NY 10032

BRUNNGRABER, ERIC GUSTAV, b New York, NY, Nov 4, 27; m 53; c 3. BIOCHEMISTRY. Educ: Columbia Univ, BA, 50, MA, 54; Univ Wis, PhD(physiol chem), 57. Prof Exp: Asst chem, Columbia Univ, 52-53; asst physiol chem, Univ Wis, 54-57; res assoc, Neuropsychiat Inst, Univ Ill, 57-59; admin res scientist, Ill State Psychiat Inst, 59-75; asst prof biochem, Univ Ill Col Med, 67-75; PROF PSYCHIAT, MO INST PSYCHIAT, ST LOUIS, 75-; PROF PSYCHIAT, SCH MED, UNIV MO-COLUMBIA, 75- Concurrent Pos: USPHS res grant, 64-72; prin investr, NIH, 64-72, NSF, 72-74. Mem: Fel AAAS; Am Chem Soc; Am Soc Biol Chemists; Am Soc Neurochem; Am Soc Neurosci. Res: Neurochemistry; glycoproteins; neuropathology. Mailing Add: Mo Inst Psychiat 5400 Arsenal St St Louis MO 63139

BRUNNING, RICHARD DALE, b Grand Forks, NDak, Mar 5, 32. HEMATOLOGY, PATHOLOGY. Educ: Univ NDak, BSc, 57; McGill Univ, MD, 59. Prof Exp: Intern, Ancker Hosp, St Paul, Minn, 59-60; fel path, Univ Minn, Minneapolis, 60-62, fel lab med, 63-64; from instr to assoc prof lab med & assoc dir hemat labs, 65-74, PROF LAB MED & CO-DIR HEMAT LABS, UNIV MINN, MINNEAPOLIS, 74- Concurrent Pos: Am Cancer Soc fel, 63-64. Mem: Am Soc Hemat; Am Asn Cancer Res. Res: Morphologic, ultrastructural and cytochemical characteristics of peripheral blood and bone marrow cells in inherited disorders; cytochemical and morphologic features of blast cells in leukemic disorders. Mailing Add: Dept of Lab Med Univ of Minn Health Sci Ctr Minneapolis MN 55455

BRUNNSCHWEILER, DIETER HEINZ, b Glarus, Switz, US citizen; m 49; c 1. GEOGRAPHY. Educ: Univ Zurich, MA, 49, PhD, 52. Prof Exp: Geographer, Land Use Surv, Dept Agr, PR, 50-51; res fel climatol, Harvard Univ, 53; asst prof geog, Clark Univ, 53-55; lectr, Univ Zurich, 55-57; assoc prof, 57-70, PROF GEOG, MICH STATE UNIV, 70- Concurrent Pos: Fulbright lectr, Asn Colombian Univs, 67 & 69 & San Marcos Univ, Lima, Peru, 73-74. Mem: Asn Am Geog; Am Soc Photogram; Am Meteorol Soc; Am Geog Soc; Latin Am Studies Asn. Res: Physical geography; climatic geomorphology, especially of Pleistocene; cold regions environment, Polar, Alpine and Andean; airphoto interpretation and remote sensing; natural hazards and colonization in tropical Andes. Mailing Add: Dept of Geog Mich State Univ East Lansing MI 48823

BRUNO, ARTHUR JOHN, b Whitinsville, Mass, Aug 12, 24; m 48; c 2. ORGANIC CHEMISTRY. Educ: Col of Holy Cross, BS, 45, MS, 47; Clark Univ, PhD(org chem), 50. Prof Exp: Res chemist, Finishes Div, E I du Pont de Nemours & Co, 47-48; sr res chemist, Johns-Manville Corp, 50-54, sect chief, 54-58, assoc tech mgr, Dutch Brand Div, 58-59, tech mgr, 59-62; dir res & eng, Audio Devices, Inc, 62-65, vpres, 65-73; TECH DIR, COLUMBIA MAGNETICS, CBS RECORDS, 73- Mem: Am Chem Soc. Res: Lacquer products; synthesis of amino acids and investigation of these materials as chelating agents; application of organic substances to building materials and asbestos containing products; pressure-sensitive tapes; adhesives; rubber products; magnetic tapes. Mailing Add: 97 Southgate La Southport CT 06490

BRUNO, CHARLES FRANK, b Westerly, RI, June 23, 36; m 58; c 6. BIOCHEMISTRY, BACTERIOLOGY. Educ: Roanoke Col, BS, 58; Va Polytech Inst, MS, 61, PhD(biochem), 63. Prof Exp: Sr res scientist microbiol & biochem, 63-71, process develop mgr fermentation, 71-75, DEPT HEAD FERMENTATION, SQUIBB INST MED RES, 75- Mem: Am Chem Soc; Am Soc Microbiol. Res: Fermentation, especially antibiotics from fungi and actinomyces; production and isolation of bacterial and fungal enzymes; production and purification of polysaccharide vaccine. Mailing Add: 11 Westwood Rd East Brunswick NJ 08816

BRUNO, GERALD A, b Red Bank, NJ, Sept 19, 35; m 58; c 4. BIONUCLEONICS, BIOCHEMISTRY. Educ: Purdue Univ, BS, 58, MS, 60, PhD(bionucleonics), 61. Prof Exp: Res chemist, Res & Develop Div, Nat Dairy Prod Corp, 61; asst prof pharm & dent res, SDak State Col, 63-64; sr res info scientist, 64-65, supvr radioisotopes res, 65-67, DIR DEPT DIAG RES & DEVELOP, E R SQUIBB & SONS, INC, 67- Mem: AAAS; Soc Nuclear Med; Am Pharmaceut Asn. Res: Development of liquid scintillation counting techniques; health physics; effects of protein anabolic agents on tooth decay; use of radio pharmaceuticals in clinical medicine; nuclear medicine. Mailing Add: Diag Res & Develop Dept E R Squibb & Sons Inc New Brunswick NJ 08903

BRUNO, GUENTER HUGO, organic chemistry, see 12th edition

BRUNO, JOHN JOSEPH, organic chemistry, see 12th edition

BRUNO, MERLE SANFORD, b Schenectady, NY, Jan 30, 39. SENSORY PHYSIOLOGY, SCIENCE EDUCATION. Educ: Syracuse Univ, BS, 60; Harvard Univ, MA, 63, PhD(biol), 71. Prof Exp: Staff developer, Educ Develop Ctr, 66-69; res assoc biol, Yale Univ, 71; ASST PROF BIOL, HAMPSHIRE COL, 71- Concurrent Pos: Consult, Workshop for Learning Things, 69-71; mem bd dirs, Yurt Found, 73-; NSF sch systs grant, 74-75. Mem: Sigma Xi; Am Soc Zoologists. Res: Comparative biochemistry of invertebrate visual systems; implementation of innovative science programs in elementary schools. Mailing Add: Sch of Natural Sci Hampshire Col Amherst MA 01002

BRUNS, CHARLES ALAN, b Baltimore, Md, Nov 1, 30; m 53; c 3. PHYSICS. Educ: Tufts Univ, BS, 52; Johns Hopkins Univ, PhD(physics), 61. Prof Exp: From instr to asst prof physics, Univ Mich, 61-64; asst prof, 64-69, ASSOC PROF PHYSICS, FRANKLIN & MARSHALL COL, 69- Mem: Am Phys Soc. Res: Low energy nuclear reactions; design and construction of cyclotrons. Mailing Add: Dept of Physics Franklin & Marshall Col Lancaster PA 17603

BRUNS, LESTER GEORGE, b Robertsville, Mo, Feb 11, 33; m 54; c 2. PHARMACY, BIOCHEMISTRY. Educ: St Louis Col Pharm, BS, 55, MS, 57. Prof Exp: From instr to asst prof pharmaceut chem, St Louis Col Pharm, 57-62; grad fel, 62-65, asst prof, 65-74, ASSOC PROF BIOCHEM, ST LOUIS COL PHARM, 74- Mem: AAAS; Am Chem Soc. Res: Bile acid metabolism and nutritional biochemistry. Mailing Add: St Louis Col of Pharm Dept Chem 4588 Parkview Pl St Louis MO 63110

BRUNS, PAUL DONALD, b Sioux City, Iowa, Oct 21, 14; m 43; c 7. OBSTETRICS & GYNECOLOGY. Educ: Trinity Col, Iowa, BS, 39; Univ Iowa, MS, 41. Prof Exp: Intern, Broadlawn Hosp, Des Moines, 41-42; resident, Univ Hosp, Johns Hopkins Univ, 45-49; from asst prof to prof, Med Ctr, Univ Colo, 49-67; PROF OBSTET & GYNEC & CHMN DEPT, GEORGETOWN UNIV HOSP, 67- Mem: AMA; Am Col Obstet & Gynec; Am Fedn Clin Res. Res: Reproductive physiology and pathology; gynecological physiology and pathology. Mailing Add: Georgetown Univ Hosp Washington DC 20007

BRUNS, PAUL ERIC, b New York, NY, July 30, 15; m 44; c 4. FORESTRY. Educ: NY Univ, AB, 37; Yale Univ, MF, 40; Univ Wash, PhD(forestry), 56. Prof Exp: Timber cruiser, Hamilton Veneer Co, SC, 41-42; instr aircraft engines & aerodyn, Southeastern Air Serv, 42-44, asst dir acads, 43 & dir, 44; assoc forester, Northeastern Forest Exp Sta, US Forest Serv, 44-45; woodlands mgr, North Troy Div, Blair Veneer Co, Vt, 45-47; assoc prof forestry, Univ Mont, 46-55; consult, 55-58; chmn dept, 59-68, PROF FOREST RESOURCES, UNIV NH, 59- Concurrent Pos: Mem adv comt, Northeastern Forest Exp Sta, US Forest Serv, 59-68, chmn, 65-66; mem, Nat Coun Forestry Sch Exec, 63-64; consult, Glastenbury Timberlands, Bennington, Vt, 69- Mem: Soc Am Foresters; Am Soc Photogram. Res: Forest management; remote sensing. Mailing Add: Inst of Natural & Environ Res Univ of NH Durham NH 03824

BRUNS, ROY EDWARD, b Breese, Ill, Sept 10, 41; m 74; c 1. THEORETICAL CHEMISTRY. Educ: Southern Ill Univ, BA, 63; Okla State Univ, PhD(phys chem), 68. Prof Exp: Asst prof chem, Univ Fla, 68-74; ASSOC PROF CHEM, UNIV CAMPINAS, BRAZIL, 71- Mem: Am Chem Soc; Brazilian Soc Advan Sci. Res: Classical and quantum mechanical investigations of the rotational vibrational spectral properties of molecules. Mailing Add: Chem Inst Univ of Campinas Campinas SP Brazil

BRUNSCHWIG, BRUCE SAMUEL, b Philadelphia, Pa, July 29, 44; m 66. PHYSICAL INORGANIC CHEMISTRY, CHEMICAL KINETICS. Educ: Univ Rochester, BA, 66; Polytech Inst NY, PhD(chem physics), 72. Prof Exp: Asst prof chem, 72-75, CHMN DEPT CHEM, HOFSTRA UNIV, 75- Concurrent Pos: Res collabr, Brookhaven Nat Labs, 74- Mem: Am Chem Soc; AAAS; Sigma Xi. Res: Study of electron-transfer reactions of transition metals in inorganic complexes and metalloproteins. Mailing Add: Dept of Chem Hofstra Univ Hempstead NY 11550

BRUNSKILL, GREGG J, b Kansas City, Kans, Sept 10, 41; m 70; c 2. LIMNOLOGY, GEOCHEMISTRY. Educ: Augustana Col, SDak, 63; Cornell Univ, PhD(biogeochem), 68. Prof Exp: RES SCIENTIST, FRESHWATER INST, ENVIRON CAN, 67- Concurrent Pos: Adj prof earth sci, Univ Man, 71- Mem: AAAS; Geochem Soc; Am Soc Limnol & Oceanog; Int Soc Limnol; Int Asn Gt Lakes Res. Res: Aqueous and recent sediment geochemistry of non-marine waters; lakes; aquatic ecology of arctic and subarctic ecosystems. Mailing Add: Freshwater Inst 501 University Crescent Winnipeg MB Can

BRUNSON, CLAYTON (CODY), b Colfax, La, Nov 3, 21; m 45; c 2. POULTRY HUSBANDRY. Educ: La State Univ, BS, 49; Okla Agr & Mech Col, MS, 51, PhD(animal breeding), 55. Prof Exp: Instr poultry husb, Okla Agr & Mech Col, 50-52; from asst prof to prof, La State Univ, 53-63; VPRES POULTRY RES, CAMPBELL INST AGR RES, CAMPBELL SOUP CO, 63- Mem: Poultry Sci Asn. Res: Poultry breeding. Mailing Add: Campbell Soup Co Fayetteville AR 72701

BRUNSON, GLENN SAMUEL, JR, b Midland, Tex, Oct 29, 22; m 50; c 6. NUCLEAR PHYSICS. Educ: US Mil Acad, BS, 45; Princeton Univ, MS, 50. Prof Exp: Tech off atomic weapons, Armed Forces Spec Weapons Proj, 50-54; PHYSICIST NUCLEAR REACTORS, ARGONNE NAT LAB, 54- Concurrent Pos: Consult, Off Spec Proj, USAEC, 60; Int Atomic Energy Agency tech asst expert, Nuclear Energy Group, Portugal, 63-64; sr officer, hq, Int Atomic Energy Agency, Vienna, 66-68; guest scientist, Fed Inst Reactor Res, Würenlingen, Switz, 75- Mem: AAAS; Am Nuclear Soc. Res: Statistics of chain related neutrons in a reactor and prompt neutron lifetimes inferred therefrom; neutron detection and spectrometry; fuel failure detection; promotion of science in developing countries; delayed neutrons; gamma spectrometry. Mailing Add: Argonne Nat Lab PO Box 2528 Idaho Falls ID 83401

BRUNSON, JOHN TAYLOR, b Kalamazoo, Mich, June 4, 40; m 71. PHYSIOLOGY. Educ: Hope Col, AB, 62; Univ Mich, Ann Arbor, MS, 64; Colo State Univ, PhD(zool), 73. Prof Exp: Instr biol & chem, State Univ NY Col Oneonta, 65-68; ASST PROF ZOOL, STATE UNIV NY COL OSWEGO, 71- Mem: Am Soc Zoologists; AAAS. Res: Physiological ecology, particularly respiratory physiology at low oxygen tension or high altitude; history of biology. Mailing Add: 211 Toad Hall Mexico NY 13114

BRUNSON, ROYAL BRUCE, b DeKalb, Ill, Feb 16, 14; m 36. ZOOLOGY. Educ: Western Mich Col Educ, BS, 38; Univ Mich, MS, 45, PhD(zool), 47. Prof Exp: Teaching fel zool, Univ Mich, 42-46; from instr to assoc prof, 46-57, PROF ZOOL, UNIV MONT, 57- Mem: Am Malacol Union; Am Soc Ichthyologists & Herpetologists; Am Soc Limnol & Oceanog; Am Micros Soc; Soc Syst Zool. Res: Taxonomy and natural history of North American Gastrotricha; taxonomy and distribution of Western Montana invertebrates; limnology of Western Montana lakes. Mailing Add: Dept of Zool Univ of Mont Missoula MT 59801

BRUNTON, GEORGE DELBERT, b McGill, Nev, Dec 20, 24. GEOLOGY, MINERALOGY. Educ: Univ Nev, BS, 50; Univ NMex, MS, 52; Ind Univ, PhD, 57. Prof Exp: Res asst, Pa State Univ, 52-53; geologist, Res Explor & Prod Res Lab, Shell Develop Co, 53-57, Pure Oil Res Ctr, 57-63 & US Gypsum Co, 63-64; GEOLOGIST, OAK RIDGE NAT LAB, 64- Mem: Mineral Soc Am. Res: Clay mineralogy; carbonate mineralogy; crystal structures; phase equilibria. Mailing Add: 112 Canterbury Rd Oak Ridge TN 37832

BRUNZIE, GERALD FRANKLIN, b Chicago, Ill, May 22, 36; m 60; c 3. ANALYTICAL CHEMISTRY. Educ: Loyola Univ, Ill, BS, 58; Univ Iowa, MS, 61, PhD(anal chem), 64. Prof Exp: Res chemist, Esso Res & Eng Co, Standard Oil Co, NJ, 62-64; res scientist, Marathon Oil Co, 64-68; SECT HEAD, WILLIAM S MERRELL CO, 68- Mem: Am Chem Soc. Res: Research and methods development in ethical pharmaceuticals instrumentation, polarography, mass spectrometry, dipole moments; chelates and coordination compounds. Mailing Add: 6278 Guinea Pike Loveland OH 45140

BRUS, LOUIS EUGENE, b Cleveland, Ohio, Aug 10, 43; m 70; c 1. CHEMICAL PHYSICS. Educ: Rice Univ, BA, 65; Columbia Univ, PhD(chem physics), 69. Prof Exp: Sci staff officer physics & chem, US Naval Res Lab, 69-73; MEM TECH STAFF PHYS CHEM, BELL LABS, 73- Mem: Am Phys Soc. Res: Molecular radiationless transitions of all sorts, including isolated molecule processes as well as condensed phase processes. Mailing Add: Bell Labs Murray Hill NJ 07974

BRUSCA, DONALD RICHARD, b Syracuse, NY, June 6, 39; m 61; c 2. CLINICAL CHEMISTRY. Educ: Rensselaer Polytech Inst, BS, 60; Univ Mass, PhD(chem), 64. Prof Exp: Sr investr male reproductive physiol, Pac Northwest Res Found, 64-66; assoc dir labs pharmaceut res, Enzomedic Labs, Inc Div, Eversharp, Inc, 66-69; biosci group leader, Corp Res Labs, Esso Res & Eng Co, 69-70; vpres, Nat Health Labs, Va, 70-75; VPRES, DAMON CORP, 75- Mem: Am Soc Clin Pathologists; Am Asn Clin Chemists; Am Asn Bioanalysts; Am Chem Soc; NY Acad Sci. Mailing Add: Damon Corp 115 Fourth Ave Needham Heights MA 02194

BRUSCA, GARY J, b Bell, Calif, Oct 10, 39; m 62; c 3. INVERTEBRATE ZOOLOGY, MARINE ECOLOGY. Educ: Calif State Polytech Col, BS, 60; Univ of the Pac, MA, 61; Univ Southern Calif, PhD(biol), 65. Prof Exp: Asst prof biol & asst dir Pac Marine Sta, Univ of the Pac, 64-67; asst prof biol, 67-70, ASSOC PROF BIOL, HUMBOLDT STATE UNIV, 70- Concurrent Pos: Dir, Humboldt State Marine Lab, Trinidad, Calif, 70- Res: Intertidal ecology; crustaceans. Mailing Add: Dept of Biol Humboldt State Univ Arcata CA 95521

BRUSCA, RICHARD CHARLES, b Los Angeles, Calif, Jan 25, 45; div; c 2. INVERTEBRATE ZOOLOGY, MARINE ECOLOGY. Educ: Calif State Polytech Univ, San Luis Obispo, BS, 67; Calif State Univ, Los Angeles, MS, 70; Univ Ariz, PhD(biol), 75. Prof Exp: Asst human physiol, Med Sch, Univ Calif, Irvine, 68; res biochemist, CalBioChem, Los Angeles, 69; teaching asst zool & cur insects & asst investr, Aquatic Insect Labs, Calif State Univ, Los Angeles, 69-70; resident dir, Univ Ariz-Univ Sonora Coop Marine Sta, Puerto Penasco, Mex, 70-72; teaching asst zool, Univ Ariz, 72-74; marine biologist, Biol Educ Exped, Vista, Calif, 74-75; ASST PROF BIOL & CUR CRUSTACEA, ALLAN HANCOCK FOUND, UNIV SOUTHERN CALIF, 75- Concurrent Pos: Vpres, Panamic Environ Consult, Ariz, 73-75; chief marine biologist, Scripps Alpha Helix Baja Exped, Scripps Inst Oceanog, La Jolla, Calif, 74. Mem: Sigma Xi; Soc Syst Zool. Res: Crustacean systematics and ecology; isopod systematics and biology; shallow water marine ecology; tropical marine ecosystems; marine symbiosis. Mailing Add: Allan Hancock Found Univ of Southern Calif Los Angeles CA 90007

BRUSCHTEIN, FABIO BENJAMIN, polymer chemistry, physical chemistry, see 12th edition

BRUSH, ALAN HOWARD, b Rochester, NY, Sept 29, 34; m 61; c 2. ZOOLOGY. Educ: Univ Southern Calif, BA, 56; Univ Calif, Los Angeles, MA, 57, PhD(zool), 64. Prof Exp: Res technician nuclear med & biophys, Univ Calif, Los Angeles, 56-61; fel zool, Cornell Univ, 64-65; asst prof zool, 65-69, ASSOC PROF BIOL SCI GROUP, UNIV CONN, 69- Concurrent Pos: NIH spec fel, Univ Calif, Berkeley, 71-72. Mem: AAAS; Cooper Ornith Soc; Am Soc Zool; Am Physiol Soc; fel Am Ornith Union. Res: Environmental physiology; biochemical systematics. Mailing Add: Biol Sci Group Univ of Conn Storrs CT 06268

BRUSH, GRACE SOMERS, b Antigonish, NS, Jan 18, 31; US citizen; m 53; c 3. FOREST ECOLOGY, PALYNOLOGY. Educ: St Francis Xavier Univ, BA, 49; Univ Ill, Urbana-Champaign, MS, 51; Radcliffe Col, PhD(biol), 56. Prof Exp: Technician, Geol Surv Can, 49-50, 51-52 & US Geol Surv, 56-57; lectr bot, George Washington Univ, 57-58; asst prof, Univ Iowa, 59-63; res assoc geol, Princeton Univ, 64-66, res staff mem, 66-70; RES SCIENTIST, DEPT GEOG & ENVIRON ENG, JOHNS HOPKINS UNIV, 70- Concurrent Pos: Asst prof, Rutgers Univ, 64-66. Mem: AAAS; Am Soc Limnol & Oceanog; Atlantic Estuarine Res Soc; Ecol Soc Am. Res: Relations between modern pollen distributions in water and surface sediments and vegetation; settling properties of pollen in water; effects of environmental changes on modern phytoplankton populations. Mailing Add: Dept of Geog & Environ Eng Johns Hopkins Univ Baltimore MD 21218

BRUSH, JAMES S, b Mankato, Minn, Mar 21, 29; m 58; c 4. BIOCHEMISTRY. Educ: Univ Chicago, PhD(biochem), 56. Prof Exp: Res assoc biochem, Univ Chicago, 56-57; clin chemist, Hurley Hosp, Flint, Mich, 58-60; fel biochem, Okla Med Res Inst, Oklahoma City, 60-63; biochemist, US Vet Admin, Wood, Wis, 63-69; from asst prof to assoc prof biochem, Col Basic Med Sci, Univ Tenn, Memphis, 69-74; ASSOC PROF BIOCHEM, FAC MED, UNIV PR, SAN JUAN, 75- Concurrent Pos: Instr biochem, Sch Med, Marquette Univ, 64-69; prin investr Vet Admin grants, 69-72 & NIH grant, 74-77; biochemist, US Vet Admin, Memphis, 69-74. Mem: AAAS; Am Chem Soc; Am Soc Biol Chemists; Endocrine Soc. Res: Protein sulfhydryl chemistry; enzyme kinetic mechanism of gulonolactone oxidase; insulin metabolism and mechanism of action; mechanisms of cyclic nucleotide action; cosmology of Swedenborg's Principia. Mailing Add: Dept of Biochem & Nutrit Univ of PR San Juan PR 00936

BRUSH, JOHN EDWIN, b Jefferson, Pa, Sept 2, 19; m 42; c 4. POPULATION GEOGRAPHY, URBAN GEOGRAPHY. Educ: Univ Chicago, AB, 42; Univ Wis-Madison, MA, 47, PhD(geog), 52. Prof Exp: Instr geog, St Louis Univ, 50-51; lectr, 51-52, from asst prof to assoc prof, 52-58, chmn dept, 67-73, PROF GEOG, RUTGERS UNIV, NEW BRUNSWICK, 58- Concurrent Pos: Lectr S Asia, Wharton Sch Finance & Com, Univ Pa, 56-57; John Simon Guggenheim Mem grant, Rutgers Univ, 57-58; Am Inst Indian Studies fac fel, Univ Poona & Univ Delhi, India, 66; sr res fel, Am Inst Indian Studies, New Delhi, India, 73. Mem: Asn Am Geog; Am Geog Soc; Asn Asian Studies. Res: Systematic geography of rural and urban settlement and population geography in the United States, especially New Jersey; foreign area concentration in above specialties in South Asia, especia especially India; spatial structure of population, density and other demographic variables in Indian cities. Mailing Add: Dept of Geog Rutgers Univ 185 College Ave New Brunswick NJ 08903

BRUSH, MIRIAM KELLY, b Boston, Mass, Nov 9, 15; m 42; c 4. NUTRITION. Educ: Mt Holyoke Col, AB, 37; Oberlin Col, AM, 39; Iowa State Col, PhD(nutrit), 46. Prof Exp: Assoc nutrit, Iowa State Col, 46-47; assoc cancer res, McArdle Mem Lab, Med Sch, Univ Wis, 47-50; assoc geront, Med Sch, Washington Univ, 51; lectr nutrit, Col Nursing, Rutgers Univ, 56-58; lectr, 57-69, PROF NUTRIT, DOUGLASS COL, RUTGERS UNIV, 69- Mem: AAAS; Am Inst Nutrit; Am Home Econ Asn; Am Dietetic Asn. Res: Protein and amino acid metabolism; nutrition and the aging process; protein-vitamin B6 inter-relationships; vitamin C nutrition in school children. Mailing Add: Dept of Nutrit Douglass Col Rutgers Univ New Brunswick NJ 08903

BRUSH, STEPHEN GEORGE, b Bangor, Maine; m 60; c 2. HISTORY OF SCIENCE. Educ: Harvard Univ, AB, 55; Osford Univ, DPhil(phys sci), 58. Prof Exp: NSF fel, 58-59; physicist, Lawrence Radiation Lab, Univ Calif, 59-65; res assoc physics & ed, Harvard Proj Physics, 65-68, lectr physics & hist of sci, Harvard Univ, 66-68; assoc prof hist & res assoc prof, Inst Fluid Dynamics & Appl Math, 68-71, PROF HIST OF SCI, UNIV MD, COLLEGE PARK, 71- Concurrent Pos: US nat rep, Comn Educ of Historians of Sci, Int Union Hist & Philos of Sci, 71- Mem: Am Phys Soc; Hist of Sci Soc. Res: History of physical science in 19th and 20th centuries, especially kinetic theory and statistical mechanics, geophysics, astrophysics; use of historical approach in teaching physics. Mailing Add: Inst for Fluid Dynamics Univ of Md College Park MD 20742

BRUSIE, JAMES POWERS, b North Egremont, Mass, July 3, 18; m 47; c 3. PHYSICAL CHEMISTRY. Educ: Lafayette Col, BS, 40; Yale Univ, PhD(phys chem), 43. Prof Exp: Res chemist, Manhattan Proj & Off Sci Res & Develop contract, Div War Res, Yale Univ, 42-43 & Columbia Univ, 43-44; sect supvr, Carbide & Carbon Chem Corp, Oak Ridge, 44-46; res chemist, Gen Aniline & Film Corp, 46-50, group leader, 50-52, res assoc, 52-55, prog mgr, 55-62, sr tech assoc, 62-64; tech dir, Girdler Catalysts Dept, Chemetron Corp, 64-68; prod mgr, Alrac Corp, 69-74; RES ASSOC, M W KELLOGG CO, 74- Mem: Am Chem Soc. Res: Catalysis, pressure acetylene reactions; polymerization of alpha-halo-acrylic esters; polymerization of 2-pyrrolidone to nylon-4; catalysis and coal conversion. Mailing Add: 10546 Idlebrook Dr Houston TX 77070

BRUSILOW, SAUL W, b Brooklyn, NY, June 7, 27; wid; c 3. PEDIATRICS, PHYSIOLOGY. Educ: Princeton Univ, AB, 50; Yale Univ, MD, 54. Prof Exp: Intern, Grace-New Haven Hosp, Conn, 54-55, asst resident, 55-56; asst resident, Hosp, 56-57, Nat Found Infantile Paralysis fel, Sch Med, 57-59, from instr to assoc prof, 59-74, PROF PEDIAT, SCH MED, JOHNS HOPKINS UNIV, 74- Mem: Am Pediat Soc; Am Physiol Soc; Soc Pediat Res. Res: Electrolyte physiology; nephrology. Mailing Add: 4804 Keswick Rd Baltimore MD 21210

BRUSON, HERMAN ALEXANDER, b Middletown, Ohio, July 20, 01; m 29; c 3. ORGANIC CHEMISTRY. Educ: Mass Inst Technol, BS, 23; Fed Polytech Inst, Zurich, DSc(org chem), 25. Prof Exp: Res chemist, Goodyear Tire & Rubber Co, Ohio, 25-28 & Rohm and Haas Co, 28-48; mgr polymer div, Indust Rayon Corp, 48-52, vpres org div, Olin Corp, 52-66; RETIRED. Concurrent Pos: Asst & lectr chem, Temple Univ, 38-48; Priestley lectr, Pa State Col, 44; mem plastics comt, Orgn Econ Coop & Develop, Paris, 69; lectr, Cheng Kung Univ, Taiwan, 70; consult US & foreign oil & chem co. Honors & Awards: Chem Pioneer Award, Am Chem Soc, 64; Award, Nat Asn Mfrs, 65; Wisdom Hall of Fame Award of Honor, 70. Mem: Am Chem Soc. Res: Polymers; pharmaceuticals; perfumes; adhesives; siccatives; pesticides; petrochemicals. Mailing Add: 98 Ansonia Rd Woodbridge CT 06525

BRUSSARD, PETER FRANS, b Reno, Nev, June 20, 38; m 62; c 1. POPULATION BIOLOGY. Educ: Stanford Univ, AB, 60, PhD(biol), 69; Univ Nev, Reno, MS, 66. Prof Exp: Asst prof, 69-75, ASSOC PROF ECOL, CORNELL UNIV, 75- Concurrent Pos: Ford Found fel, Stanford Univ, 69; partic, NATO adv study inst dynamics of numbers in pop, Oosterbeek, Netherlands, 70. Mem: AAAS; Ecol Soc Am; Cooper Ornith Soc; Am Soc Mammal; Soc Study Evolution. Res: Relations between phenetic variation, distribution and ecology in natural populations. Mailing Add: Sect of Ecol & Systs Cornell Univ 239 Langmuir Lab Ithaca NY 14853

BRUSSEL, MORTON KREMEN, b New Haven, Conn, Mar 31, 29; m 57; c 2 PHYSICS. Educ: Yale Univ, BS, 51; Univ Minn, PhD(physics), 57. Prof Exp: Res assoc fel physics, Brookhaven Nat Lab, 57-59, asst physicist, 59-60; res asst prof, 60-64, assoc prof, 64-68, PROF PHYSICS, UNIV ILL, URBANA-CHAMPAIGN, 68- Concurrent Pos: Assoc prog dir nuclear physics, Physics Sect, NSF, 72-73. Mem: Am Phys Soc; AAAS; Fedn Am Scientists; Am Asn Univ Profs; Sigma Xi. Res: Experimental nuclear physics; electron and photon scattering; scattering of light ions at medium energies; nuclear structure studies. Mailing Add: Dept of Physics Univ of Ill Urbana IL 61803

BRUST, DAVID, b Chicago, Ill, Aug 24, 35. SOLID STATE PHYSICS, SEMICONDUCTORS. Educ: Calif Inst Technol, BS, 57; Univ Chicago, MS, 58, PhD(physics), 64. Prof Exp: Res assoc physics, Purdue Univ, 63-64; res assoc, Northwestern Univ, 64-65, asst prof, 65-68; physicist, Lawrence Livermore Lab, Univ Calif, 68-73; PRES, MAT SYST ANALYSTS, 73- Concurrent Pos: NSF grant, 67-69. Mem: Am Phys Soc; Int Solar Energy Soc. Res: Electronic structure of crystalline and amorphous solids and application to their optical, photoelectric and transport properties; point defects in metals and semiconductors; dielectric properties of solids. Mailing Add: Mat Syst Analysts PO Box 13130 Oakland CA 94661

BRUST, DAVID PHILIP, b Albion, NY, Aug 13, 34; m 59; c 3. PHOTOGRAPHIC CHEMISTRY. Educ: State Univ NY Buffalo, BA, 55; Univ Rochester, PhD(org chem), 65. Prof Exp: Res chemist polymer chem, 58-61, sr res chemist, 65-72, RES ASSOC PHOTOG CHEM, EASTMAN KODAK CO RES LABS, 72- Mem: Am Chem Soc. Res: Polymers in photographic systems. Mailing Add: 239 Southridge Dr Rochester NY 14626

BRUST, HARRY FRANCIS, b Milwaukee, Wis, Jan 2, 14; m 39; c 3. ORGANIC CHEMISTRY. Educ: Univ Wis, BS, 38; Univ Pa, MS, 40, PhD(chem), 43. Prof Exp: Asst instr chem, Univ Pa, 38-41, 42-43; chemist & group leader, 43-68, sr res chemist, 68-73, SR RES SPECIALIST, DOW CHEM CO, 73- Mem: Am Chem Soc; Sigma Xi. Res: Chlorination; bromination of a variety of organic chemicals; agricultural chemicals, chiefly herbicides; new process development and reaction kinetics. Mailing Add: 601 Hillcrest St Midland MI 48640

BRUST, MANFRED, b Chemnitz, Ger, Oct 22, 23; nat US; m 57; c 3. PHYSIOLOGY. Educ: NY Univ, BA, 44, MSc, 46; Univ Ill, PhD(physiol), 54. Prof Exp: Asst physiol, Univ Chicago, 46-48; asst entom, Univ Ill, 50-53; res assoc zool, Syracuse Univ, 53-55; res assoc biol, Wash Sq Col, NY Univ, 55-59; res assoc, Div Physiol, Inst Muscle Dis, Inc, New York, 59-61, asst mem, 61-63; from asst prof to assoc prof rehab med, 63-70, ASSOC PROF PHYSIOL, STATE UNIV NY DOWNSTATE MED CTR, 70- Concurrent Pos: Asst, Babies Hosp, Columbia-Presby Med Ctr, 57. Mem: AAAS; Am Physiol Soc; NY Acad Sci; Biophys Soc. Res: Basic physiology of muscle contraction; excitation and muscle pharmacology in both normal and diseased conditions; muscular dystrophy; physiology of neuromuscular disease; fast and slow muscle; ultrasound. Mailing Add: Dept of Physiol State Univ NY Downstate Med Ctr Brooklyn NY 11203

BRUST, REINHART A, b Sibbald, Alta, Feb 7, 34; m 59; c 2. ENTOMOLOGY, INSECT ECOLOGY. Educ: Univ Man, BSc, 59, MSc, 60; Univ Ill, Urbana, PhD(entom), 64. Prof Exp: From asst prof to assoc prof, 64-72, PROF ENTOM, UNIV MAN, 73- Mem: Entom Soc Am; Entom Soc Can. Res: Ecology of North American species of aedine mosquitoes, disease transmission, biology and systematics of mosquitoes. Mailing Add: Dept of Entom Univ of Man Winnipeg MB Can

BRUST-CARMONA, HECTOR, b Colima, Mex, Mar 23, 35; m 63; c 4. PHYSIOLOGY, NEUROPHYSIOLOGY. Educ: Nat Univ Mex, BS, 51, MD, 58. Prof Exp: Assoc prof physiol, Med Sch, Nat Univ Mex, 54-63; vis assoc prof neurophysiol, NY State Vet Col, Cornell Univ, 63-65; chmn dept biol sci, 71-72, PROF PHYSIOL, MED SCH, NAT UNIV MEX, 65-, CHMN DEPT, 73-, SECY EXEC COUN, RES DIV, 73- Concurrent Pos: Lectr psychol, Cornell Univ, 64-65; Int Brain Orgn-UNESCO fel, Pharmacol Inst, Vienna, Austria, 68-69; Int Atomic Energy Agency expert neurophysiol, Inst Pharmacol, Univ Budapest, 69; adj prof physiol, Baylor Col Med, 69-; sci consult, Anahuac Univ, Mex, 70-74. Honors & Awards: Dr Miguel Galindo Award, State Cong Colima, Mex, 73. Mem: Mex Soc Physiol Sci (treas, 67-68); Latinam Soc Psychobiol. Res: Electrophysiological and neurochemical processes responsible for acquisition and maintenance of learned motor responses and also the autonomic responses. Mailing Add: Dept of Physiol Fac Med Nat Univ Mex Box 70250 Mexico City Mexico

BRUSVEN, MERLYN ARDEL, b Powers Lake, NDak, Mar 23, 37; m 59; c 3. ENTOMOLOGY. Educ: NDak State Univ, BS, 59, MS, 61; Kans State Univ, PhD(entom), 65. Prof Exp: Instr biol, Friends Univ, 61-63; ASST PROF ENTOM, UNIV IDAHO, 65- Mem: Am Soc Limnol & Oceanog; Entom Soc Am. Res: Insect ecology, especially range grasshopper ecology; grasshopper taxonomy; aquatic entomology, particularly population and community dynamics. Mailing Add: Dept of Entom Univ of Idaho Moscow ID 83843

BRUTCHER, FREDERICK VINCENT, JR, b Dorchester, Mass, Dec 5, 22. ORGANIC CHEMISTRY. Educ: Univ Mass, BS, 47; Yale Univ, MS, 49, PhD(org chem), 51. Prof Exp: Fel, Harvard Univ, 51-53; asst prof chem, 53-60, ASSOC PROF CHEM, UNIV PA, 60- Mem: Am Chem Soc. Res: Conformational analysis of substituted cyclopentanes; mathematics of organic chemistry; synthesis of indole and alicyclic compounds. Mailing Add: Harrison Lab Univ of Pa Dept of Chem Philadelphia PA 19104

BRUTLAG, DOUGLAS LEE, b Alexandria, Minn, Dec 19, 46; m 75. MOLECULAR BIOLOGY. Educ: Calif Inst Technol, BS, 68; Stanford Univ, PhD(biochem), 72. Prof Exp: Res scientist genetics, Commonwealth Sci & Indust Res Orgn, 72-74; ASST PROF BIOCHEM, STANFORD UNIV, 74- Res: The role of highly repeated DNA sequences in centromere function, particularly in chromosome disjunction at meiosis. Mailing Add: Dept of Biochem Stanford Med Ctr Stanford CA 94305

BRUTON, CHARLES WILLIAM, b New York, NY, May 31, 22; m 44; c 2. BIOLOGY, ANIMAL ECOLOGY. Educ: Okla State Univ, BS, 50, MS, 52; Univ NDak, PhD(biol, geol, climat), 69. Prof Exp: Field asst genetics, USDA, Stillwater, Okla, 52; pub sch instr, Okla, 52-57; from instr to assoc prof biol, 57-73, PROF BIOL SCI, ST CLOUD STATE COL, 73- Concurrent Pos: State Col Bd Res Div grant, St Paul, Minn, 69-71. Mem: Ecol Soc Am; Am Inst Biol Sci. Res: Vertebrate ecology; plant ecology. Mailing Add: Dept of Biol St Cloud State Col St Cloud MN 56301

BRUTON, JAMES DONALD, organic chemistry, see 12th edition

BRUTSAERT, WILFRIED, b Ghent, Belg, May 28, 34; nat US. HYDROLOGY. Educ: State Univ Ghent, BS, 58; Univ Calif, MSc, 60, PhD(eng), 62. Prof Exp: From asst prof to assoc prof, 62-74, PROF HYDROL, CORNELL UNIV, 74- Concurrent Pos: Vis scholar, Tohoku Univ, Japan, 69-70; Am Soc Civil Engrs Freeman fel, 75. Mem: Am Geophys Union; Am Soc Civil Engrs; Am Meteorol Soc. Res: Flow through porous media; permeability; infiltration and drainage; microclimatology; evaporation; surface water hydrology; hydrological systems. Mailing Add: Sch of Civil & Environ Eng Hollister Hall Cornell Univ Ithaca NY 14853

BRUUN, JOHANNES HADELN, b Ange, Sweden, Nov 8, 99; US citizen; m 30. CHEMISTRY, CHEMICAL ENGINEERING. Educ: Norweg Inst Technol, MS, 23; Johns Hopkins Univ, PhD(phys & org chem), 29. Prof Exp: Chemist, Kongsberg Arms Mfg Co, Norway, 23-24; prod mgr, Am Aniline Co, Pa, 24-27; sr res assoc, Nat Bur Standards, 27-32; mgr res, Sun Oil Co, Pa, 32-42; works mgr, Gen Aniline & Film Corp, 42-43, dir res & com develop, 43-52; dir res & develop, Hooker Chem Co, 52-58; STAFF SPECIALIST, OFF ADMINR, USDA, 59- Concurrent Pos: Trustee, Tome Sch, 39-42; mem adv comt, Norweg Indust Comn, 49-50. Mem: Am Chem Soc; Am Inst Chem; Asn Res Dirs (pres, 50-52); Com Chem Develop Asn. Res: Composition of petroleum; physical and organic chemistry; distillation; dyestuffs; oxidation; research administration; agricultural raw materials; thermal insulation. Mailing Add: 523 Paxinosa Rd E Easton PA 18042

BRUYERE, DONALD EUGENE, b Detroit, Mich, Apr 19, 29; div; c 4. GEOGRAPHY OF THE SOVIET UNION, POPULATION GEOGRAPHY. Educ: Wayne State Univ, BA, 51, MA, 52; Univ Mich, PhD(geog), 58. Prof Exp: Instr geog, Univ Ore, 55-56, Wayne State Univ, 57-58 & Modesto Jr Col, 58-61; res demogr, Metrop Water Dist Southern Calif, 61-64; PROF GEOG, UNIV WIS-OSHKOSH, 64- Concurrent Pos: Bd Regents grants, Univ Wis-Oshkosh, 66-67 & 70-71. Mem: Asn Am Geogr; Nat Coun Geog Educ; Nat Asn Foreign Student Affairs. Res: Population. Mailing Add: Dept of Geog Univ of Wis Oshkosh WI 54901

BRUYR, DONALD LEE, b West Mineral, Kans, Dec 3, 30; m 51; c 4. MATHEMATICS. Educ: Kans State Col, BS, 51, MA, 55; Okla State Univ, EdD(math), 64. Prof Exp: High sch teacher, Kans, 51-60; PROF MATH, KANS STATE TEACHERS COL, 60- Mem: Math Asn Am. Res: Topology and mathematics education. Mailing Add: Dept of Math Kans State Teachers Col Emporia KS 66801

BRYAN, ALAN LYLE, b Friday Harbor, Wash, June 21, 28; m 62; c 1. ANTHROPOLOGY. Educ: Univ Wash, BA, 52, MA, 54; Harvard Univ, PhD(anthrop), 62. Prof Exp: Archaeologist, Pac Northwest Gas Pipeline Co, 55-56; from asst prof to assoc prof, 63-75, PROF ANTHROP, UNIV ALTA, 75- Concurrent Pos: NSF fel, Univ Alta, 65-66 & Can Coun fel, 69-70; mem quaternary res comt, Nat Res Coun Can, 71-73. Mem: AAAS; Am Anthrop Asn; Soc Am Archaeol; Can Archaeol Asn; Am Soc Conserv Archaeol. Res: Early man in America. Mailing Add: Dept of Anthrop Univ of Alta Edmonton AB Can

BRYAN, ANDREW BONNELL, physics, see 12th edition

BRYAN, ASHLEY MONROE, b British West Indies, Apr 29, 17; m 48; c 4. PLANT BIOCHEMISTRY. Educ: Hampton Inst, BS, 48; Iowa State Univ, PhD(plant physiol), 53. Prof Exp: Prof biol, Talladega Col, 53-56; res assoc, Univ Pa, 56-59; assoc prof, Morgan State Col, 59-61; PROF BIOCHEM, STATE UNIV NY ALBANY, 61- Concurrent Pos: Mem comn on higher educ, Mid States Asn Cols & Sec Schs. Mem: Am Chem Soc; Sigma Xi. Res: Plant biochemistry; nucleic acids and protein synthesis in relation to growth, reproduction and cell differentiation. Mailing Add: Dept of Chem State Univ of NY Albany NY 12222

BRYAN, CARL EDDINGTON, b West Point, Miss, Jan 12, 17; m 42; c 2. CHEMISTRY. Educ: Univ Miss, BA, 37; Univ Minn, PhD(org chem), 42. Prof Exp:

Instr chem, Delta Jr Col, 37-38; asst chem, Univ Minn, 38-42; asst, Univ Ill, 42-44; res chemist, E I du Pont de Nemours & Co, 44-46, res chemist, Southern Res Inst, 46-50, head org sect, 48-50; res chemist, Res Ctr, US Rubber Co, 50-59, res scientist, 59-61; sr res chemist, Chemstrand Res Ctr, Monsanto Co, 61-67; res chemist, Mallinckrodt Chem Works, Mo, 67-68; RES ASSOC TEXTILES, NC STATE UNIV, 68- Concurrent Pos: With Off Sci Res & Develop; with off Rubber Reserve. Mem: AAAS; NY Acad Sci; Am Chem Soc; Am Inst Chem; Am Asn Textile Chemists & Colorists. Res: Organic synthesis and structure determinations; polymers and polymerization; textile chemistry and waste control. Mailing Add: 2631 St Mary's St Raleigh NC 27609

BRYAN, CHARLES A, b Livingston, Mont, May 19, 36; m 56; c 3. MATHEMATICS. Educ: Mont State Col, BS, 58; Univ Ariz, MS & PhD(math), 63. Prof Exp: Asst prof math, Ariz State Univ, 63-66; from asst prof to assoc prof, 66-73, PROF MATH, UNIV MONT, 73- Mem: Am Math Soc. Res: Numerical approximations to solutions of partial differential equations and ordinary differential equations. Mailing Add: Dept of Math Univ of Mont Missoula MT 59801

BRYAN, CHARLES F, b Louisville, Ky, Jan 17, 37; m 62. ICHTHYOLOGY, LIMNOLOGY. Educ: Bellarmine Col, Ky, BA, 60; Univ Louisville, PhD(zool), 64. Prof Exp: Asst cur fish, John G Shedd Aquarium, 64-66; fishery biologist, Marion Inserv Training Sch, US Bur Sport Fisheries, 66-67; asst leader, Calif Coop Fishery Unit, 67-70; LEADER, LA COOP FISHERY RES UNIT, 70- Mem: Am Soc Limnol & Oceanog; Soc Syst Zool; Am Soc Ichthyologists & Herpetologists; Am Fisheries Soc. Res: Stream and swamp limnology; water quality; ecology of fishes. Mailing Add: Rm 247 Agr Ctr La State Univ Baton Rouge LA 70803

BRYAN, CLIFFORD RANDALL, b Blackstonedge, Jamaica, Mar 19, 12; US citizen; m 43; c 3. GENETICS, ZOOLOGY. Educ: Howard Univ, BS, 49; Univ Wis, MS, 51, PhD(zool), 53. Prof Exp: Instr zool, Howard Univ, 53-56; assoc prof biol, Bethune-Cookman Col, 56-57, prof & chmn div natural sci, 57-58; assoc prof, 58-59, PROF BIOL & CHMN DIV NATURAL SCI, DILLARD UNIV, 59-, ASSOC DEAN ACAD AFFAIRS, 74- Mem: AAAS; Genetics Soc Am; Nat Inst Sci. Res: Immunogenetic studies of blood antigens in birds; screening of selected invertebrate fractions for growth factors. Mailing Add: Dept of Zool Dillard Univ New Orleans LA 70122

BRYAN, DOUGLAS EVERETT, b Auckland, NZ, Apr 8, 17; nat US; m 41; c 4. ENTOMOLOGY. Educ: Univ Calif, AB, 42, BS, 48, PhD(entom), 52. Prof Exp: Asst entom, Univ Calif, 48-52; from asst entomologist, Exp Sta & asst prof entom to assoc entomologist & assoc prof entom, Okla State Univ, 52-63, prof entom, 63-65; entomologist, Cotton Insects Br, Entom Res Div, 65-70, invests leader, Cotton Insects Biol Control Lab, 70-72, SUPVRY RES ENTOMOLOGIST, RES LEADER, COTTON INSECTS BIOL CONTROL LAB, AGR RES SERV, USDA, 72- Mem: AAAS; Entom Soc Am; Int Orgn Biol Control. Res: Forage and cotton insects; ecology; biological control of field crop insects. Mailing Add: Cotton Insects Biol Control Lab ARS USDA 2000 E Allen Rd Tucson AZ 85719

BRYAN, EDWIN HORACE, JR, b Philadelphia, Pa, Apr 13, 98; m 30; c 2. NATURAL HISTORY. Educ: Col Hawaii, BS, 20, Univ Hawaii, MS, 24; Yale Univ, PhB, 21. Prof Exp: From asst entomologist to entomologist, 19-27, cur collections, 27-41, 50-68, WILLIAM T BRIGHAM SR FEL, BISHOP MUS, 68-, MGR PAC SCI INFO CTR, 60- Concurrent Pos: Asst, Yale Univ, 20-21 & Kamehameha Schs, 21-22; instr, Univ Hawaii, 25-29, 39; partic, US Com Co Econ Surv, Micronesia, 46-47; consult, Pac Sci Bd, 47-70. Pac explor, Tanager Exped, 23, Whitney South Sea Exped, 24, Itasca Exped, 35, Guam Exped, 36 & Taney Exped, 38. Mem: AAAS. Res: Botany; entomology; Diptera of Hawaii; history, geography and bibliography of central Pacific; Polynesian and Micronesian ethnology and biogeography; popular astronomy; Hawaiian natural history. Mailing Add: Bernice P Bishop Mus PO Box 6037 Honolulu HI 96818

BRYAN, FRANK LEON, b Indianapolis, Ind, Aug 29, 30; m 52; c 2. BACTERIOLOGY, FOOD MICROBIOLOGY. Educ: Ind Univ, BS, 53; Univ Mich, MPH, 56; Iowa State Univ, PhD(bact), 65. Prof Exp: Pub health aide milk & environ sanit, USPHS, Durham & Chapel Hill Local Health Dept, NC, 53; sanitarian, Ind State Bd Health, 55; training off pub health, New Eng Field Training Sta, USPHS, 56-58, environ health, Nat Commun Dis Ctr, 58-63, SCIENTIST DIR & CHIEF FOODBORNE DIS ACTIV, CTR FOR DIS CONTROL, USPHS, 65- Concurrent Pos: Lectr, Univ Mass, 56-58. Mem: Am Soc Microbiol; Inst Food Technol; Sigma Xi; NY Acad Sci; Am Pub Health Asn. Res: Enteric bacteriology; Salmonella in turkeys, processing plants and turkey products; foodborne disease epidemiology and control; time-temperature factors in thawing, cooking, chilling and reheating turkey products and beef; Clostridium perfringens in beef; staphylococcal intoxication; miscellaneous foodborne diseases. Mailing Add: Bur of Training Ctr Dis Control USPHS Dept Health Educ & Welfare Atlanta GA 30333

BRYAN, FREDERICK ALLEN, JR, b Washington, DC, Oct 21, 33; m 56; c 4. SYSTEM ANALYSIS, APPLIED PHYSICS. Educ: Univ Mich, BS, 55, MS, 57; NC State Univ, PhD (physics, math), 65. Prof Exp: Res asst high temperature metall, Eng Res Inst, Univ Mich, 54-55, res asst nuclear systs, 55-56 & mach design, 56-57; proj physicist, Atomic Energy Div, Babcock & Wilcox Co, 57-58; sr physicist, Astra, Inc, 59-65; sr physicist & opers res anal, Res Triangle Inst, 65-71, HEAD HEALTH SYSTS RES, CTR HEALTH STUDIES, RES TRIANGLE INST, 71- Concurrent Pos: Lectr math, Univ Va, Lynchburg, 57-58; instr physics, NC State Univ, 58-61, lectr astron, 65. Mem: AAAS. Res: Health systems research and development; medical engineering and chronic diseases, information systems; systems analysis and operations research; environmental research; radiation transport and shielding; reactor physics; low energy nuclear physics. Mailing Add: Rte 4 Box 713 Raleigh NC 27606

BRYAN, GEORGE JENNINGS, physical chemistry, see 12th edition

BRYAN, GEORGE TERRELL, b Antigo, Wis, July 29, 32; m 54; c 5. MEDICINE, PHARMACOLOGY. Educ: Univ Wis, BS, 54, MD, 57, PhD(oncol, biochem), 63. Prof Exp: Intern med, NC Baptist Hosp, Winston-Salem, 57-58; instr cancer res, 61-63, from asst prof to prof clin oncol & surg, 63-75, PROF HUMAN ONCOL, SCH MED, UNIV WIS-MADISON, 75- Concurrent Pos: Consult, Food & Drug Admin & Nat Cancer Inst. Mem: fel Am Col Physicians; Am Asn Cancer Res; James Ewing Soc; Am Soc Exp Path; Am Soc Biol Chemists. Res: Chemical carcinogenesis; clinical and experimental cancer chemotherapy; metabolic disorders and normal human metabolism; nutrition. Mailing Add: C-713 Univ Hosp 1300 University Ave Madison WI 53706

BRYAN, GEORGE THOMAS, b Sewanee, Tenn, Nov 19, 30; m 52; c 2. MEDICAL EDUCATION. Educ: Univ Tenn, MD, 55; Am Bd Pediat, dipl, 69. Prof Exp: Intern, DC Gen Hosp, 55-56; resident pediat, State Univ Iowa Hosps, 56-58, fel pediat endocrinol, 58-59; clin assoc pediat, Nat Inst Allergy & Infectious Dis, 59-60, pediatrician, Clin Endocrinol Br, 61-63; from asst prof to assoc prof pediat, 63-73, asst dir clin study ctr & dir div endocrinol, Dept Pediat, 63-70, PROF PEDIAT, MED BR, UNIV TEX, GALVESTON, 73-, ASSOC DEAN CURRIC AFFAIRS, 74- Concurrent Pos: Markle Found scholar acad med, 67-72. Mem: AMA; Soc Pediat Res; Endocrine Soc; Am Fedn Clin Res. Res: Adrenal steroid biosynthesis; carcadian periodicity; hypertension; hypopituitarism. Mailing Add: Off of Dean of Med Univ of Tex Med Br Galveston TX 77550

BRYAN, GORDON HENRY, b Ashton, Idaho, Oct 22, 16; m 46; c 2. PHARMACOLOGY. Educ: Mont State Univ, BS, 42, MS, 47; Univ Md, PhD(pharmacol), 56. Prof Exp: Instr pharm, 4749, from asst prof to assoc prof pharmacol, 5056, PROF PHARMACOL, UNIV MONT, 56- Concurrent Pos: NIH fel, 64-65; vis lectr med, Univ Utah, 57. Mem: AAAS; Am Pharmaceut Asn. Res: Cardiology. Mailing Add: Sch of Pharm Univ of Mont Missoula MT 59801

BRYAN, HAROLD STEVER, b Bedminster, Pa, July 26, 20; m 45; c 3. VETERINARY MEDICINE. Educ: Mich State Univ, DVM, 44; Univ Ill, MS, 48, PhD, 53. Prof Exp: War food asst, Univ Ill, 44-45, asst vet path & hyg, 45-46, from instr to prof, 46-56; head dept vet res, Upjohn Co, 56-66; PROF VET MED EDUC, UNIV ILL, URBANA, 66- Mem: Am Vet Med Asn; US Animal Health Asn. Res: Bovine mastitis and brucellosis; leptospirosis. Mailing Add: Dept of Vet Path Univ of Ill Col of Vet Med Urbana IL 61801

BRYAN, HERBERT HARRIS, b Brunswick, Ga, Jan 25, 32; m 56; c 2. HORTICULTURE, PLANT PHYSIOLOGY. Educ: Univ Fla, BSA, 53; Cornell Univ, MS, 61, PhD(veg crops), 64. Prof Exp: Asst prof hort, NFla Exp Sta, 64-66, asst prof, Agr Res & Educ Ctr, 67-74, ASSOC PROF HORT & ASSOC HORTICULTURIST, AGR RES & EDUC CTR, UNIV FLA, 74- Concurrent Pos: Mem, Int Soc Hort Sci. Mem: Am Soc Hort Sci. Res: Nutrition and herbicides of vegetables and peaches; mechanical harvesting of tomatoes for fresh market; cultural practices; mulches; pole bean breeding; cover crops rotation with potatoes; bean and tomato growth regulators and plant populations. Mailing Add: Agr Res & Educ Ctr Univ of Fla Homestead FL 33030

BRYAN, HORACE ALDEN, b Bluff City, Tenn, Jan 1, 28. INORGANIC CHEMISTRY. Educ: King Col, BS, 50; Univ Tenn, PhD, 55. Prof Exp: Asst inorg chem, Univ Tenn, 51-55; from asst prof to assoc prof, 55-67, PROF CHEM, DAVIDSON COL, 67- Mem: Am Chem Soc. Res: Polarography of azo compounds; extraction of metal-containing anions; cation-ligand equilibria. Mailing Add: PO Box 503 Davidson NC 28036

BRYAN, HUGH D, b Hebron, Nebr, Aug 7, 21; m 43; c 3. PHARMACEUTICAL CHEMISTRY. Educ: Univ Nebr, BS, 48, MS, 51. Prof Exp: Instr pharm & pharmaceut chem, Univ Nebr, 49-53; res pharmacist, Smith-Dorsey Co, 53-54; res pharmacist, Mead Johnson & Co, 54-57, dir pharm, 57-59, dir pharmaceut prod develop, 59-69; asst prof, Sch Pharm & res projs coordr, Inst Pharmaceut Sci, Sch Pharm, Univ Miss, 69-70; head phys sci, Warrenteed Pharmaceut Inc, Rohm and Haas Co, 70-75; DIR COM DEVELOP, VIOBIN CORP, A H ROBINS CO, 75- Mem: AAAS; Am Chem Soc; NY Acad Sci; Am Pharmaceut Asn. Res: Pharmaceutical product development; protein supplements and enzymes. Mailing Add: 1 Walnut Lane Mahomet IL 61853

BRYAN, JACK T, b Birmingham, Ala, Apr 4, 25; m 45; c 2. PHARMACEUTICAL CHEMISTRY Educ: Howard Col, BS, 48; Univ Fla, MS, 50, PhD(pharmaceut chem), 52. Prof Exp: Pharmaceut chemist, Merck & Co, 52-53; assoc prof pharm, Howard Col, 53-57; res assoc, 57-71, mem staff sales mgt, 66-71, MGR FORMULATION & PHARM, UPJOHN CO, 71- Mem: Am Pharmaceut Asn; Am Asn Cols Pharm. Res: Synthesis of local anesthetics; crude drug extraction; new dosage forms of pharmaceutical products. Mailing Add: 6199 Horizon Heights Dr Kalamazoo MI 49001

BRYAN, JOHN HENRY DONALD, b London, Eng, Sept 18, 26; nat US; m 52; c 2. CELL BIOLOGY. Educ: Univ Sheffield, BSc, 47; Columbia Univ, AM, 49, PhD(zool), 52. Prof Exp: Lectr zool, Columbia Univ, 49-50; instr biol, Mass Inst Technol, 51-54; asst prof genetics, Iowa State Univ, 54-60, from asst prof to assoc prof zool, 60-67, chmn comt cell biol, 61-67; PROF ZOOL, UNIV GA, 67- Concurrent Pos: Consult, NSF-AID Indian Educ Prog, 68. Mem: Fel AAAS; Am Soc Cell Biol; Am Soc Zool; Genetics Soc Am; Soc Develop Biol. Res: Cell structure and function, especially the nucleus; cytochemistry; chromosome ultrastructure; genetic control of cell differentiation, especially gametogenesis. Mailing Add: Dept of Zool Univ of Ga Athens GA 30602

BRYAN, JOHN KENT, b Indianapolis, Ind, Mar 20, 36; m 58. PLANT PHYSIOLOGY, BIOCHEMISTRY. Educ: Butler Univ, BS, 58; Univ Tex, PhD(cellular physiol), 62. Prof Exp: Res asst biochem, Inst Psychiat Res, Sch Med, Ind Univ, 58-59; res asst electron micros, Plant Res Inst, Univ Tex, 62; res assoc molecular biol, Wash Univ, 62-64; asst prof bot, 64-69, ASSOC PROF BIOL, SYRACUSE UNIV, 70- Mem: AAAS; Am Soc Plant Physiol; Bot Soc Am. Res: Biochemistry of plant growth and development; control mechanisms in plant growth. Mailing Add: Dept of Biol Biol Res Labs Syracuse Univ 130 College Pl Syracuse NY 13210

BRYAN, JOSEPH GERARD, b Winnipeg, Man, Can, June 17, 16; US citizen; m 44; c 4. APPLIED MATHEMATICS. Educ: Mass Inst Technol, SB, 38; Harvard Univ, EdM, 42, EdD(educ measurement), 50. Prof Exp: Asst educ testing, Harvard Univ, 40-42; mem staff, Div Indust Coop, Mass Inst Technol, 42-45, dept dir, Statist Lab, 45-54, res mathematician, Div Sponsored Res, 54-57; sr opers res analyst, Cent Res Lab, Am Mach & Foundry Co, 57-60; chief statistician, Weather Res Dept, Travelers Ins Co, 60-61, chief statistician & consult scientist, Travelers Res Ctr, Inc, 61-70, SR RES SCIENTIST, TRAVELERS INS CO, 70- Concurrent Pos: Adj prof, Rensselaer Polytech Inst, 61- Mem: AAAS; Opers Res Soc Am; Am Statist Asn; Biomet Soc. Res: Probability models in operations research and acturial science. Mailing Add: 11 Newport Ave West Hartford CT 06107

BRYAN, KIRK, (JR), b Albuquerque, NMex, July 21, 29; m 56; c 2. METEOROLOGY. Educ: Yale Univ, BS, 51; Mass Inst Technol, PhD(meteorol), 57. Prof Exp: Res assoc meteorol, Woods Hole Oceanog Inst, 58-61; res meteorologist, Gen Circulation Res Lab, US Weather Bur, 61-68; OCEANOGR, GEOPHYS FLUID DYNAMICS LAB, NAT OCEANOG & ATMOSPHERIC ADMIN, PRINCETON UNIV, 68- Concurrent Pos: Vis lectr, Princeton Univ, 68-75; mem Panel Climatic Variation, Global Atmos Res Prog, Nat Acad Sci, 72-74. Honors & Awards: Distinguished Serv Medal, US Dept Commerce, 70; Sverdrup God M Gold Medal, Am Meteorol Soc, 70. Mem: Am Meteorol Soc; Am Geophys Union. Res: Dynamic meteorology; physical oceanography; general circulation of the atmosphere and the oceans. Mailing Add: Geophys Fluid Dynamics Lab NOAA Princeton Univ Princeton NJ 08540

BRYAN, LOREN ALDRO, b Emporia, Kans, Feb 4, 16; m 52; c 1. INDUSTRIAL CHEMISTRY. Educ: Kans State Teachers Col, BS & AB, 37; Kans State Univ, MS, 39; Northwestern Univ, PhD(org chem), 44. Prof Exp: Asst chem, Kans State Univ,

37-39; asst, Northwestern Univ, 41-43; res assoc, Miner Labs, 43-53; proj engr, Melpar, Inc, 53-54; sr mem chem res & develop sect, 54; chemist, FMC Corp, 54-60; sect head, Great Lakes Carbon Corp, 60-62, SR CHEMIST, GREAT LAKES RES CORP, 62- Mem: AAAS; Am Chem Soc; fel Am Inst Chem; NY Acad Sci. Res: Organic process development; alkaline earth chemicals; organic silicon compounds; sugar derivatives; fats and waxes; paint technology; carbon and graphite products. Mailing Add: Great Lakes Res Corp Box 1031 Elizabethton TN 37643

BRYAN, MARY LEO, b Philadelphia, Pa, Dec 31, 23. PHYSICAL ORGANIC CHEMISTRY, BIOCHEMISTRY. Educ: Rosemont Col, AB, 44; Cath Univ Am, MS, 46, PhD(chem), 53. Prof Exp: Teacher, Sch of the Holy Child, Sharon Hill, Pa, 45-46, 48-49 & St Walburga's Acad, NY, 49-50, 52-55; from instr to assoc prof chem, 55-70, PROF CHEM, ROSEMONT COL, 70-, CHMN DEPT, 67- Mem: AAAS; Am Chem Soc. Res: Production and reactions of gas-phase organic free radicals; kinetics and mechanism of enzyme reactions; spectroscopic studies of molecular structure. Mailing Add: Dept of Chem Rosemont Col Rosemont PA 19010

BRYAN, PHILIP STEVEN, b Lima, Ohio, Oct 5, 44; m 67; c 2. INORGANIC CHEMISTRY. Educ: Ohio State Univ, BS, 66; Univ Mich, PhD(chem), 70. Prof Exp: Asst prof chem, Macalester Col, 71-72; ASST PROF CHEM, COLGATE UNIV, 72- Mem: Am Chem Soc. Res: Synthesis and characterization of transition metal complexes containing multidentate ligands; macrocyclic complexes of manganese are of particular interest. Mailing Add: Dept of Chem Colgate Univ Hamilton NY 13346

BRYAN, ROBERT FINLAY, b Rhu, Scotland, May 15, 33. CRYSTALLOGRAPHY. Educ: Glasgow Univ, BSc, 54, PhD(chem), 57. Prof Exp: Res assoc chem, Fed Inst Technol, Zurich, 57-59; res assoc biol, Mass Inst Technol, 59-61; asst prof biophys, Sch Med, Johns Hopkins Univ, 61-67; ASSOC PROF CHEM, UNIV VA, 67- Concurrent Pos: Fels, Battelle, 57-59, Sloane, 59-60 & NIH, 60-61. Mem: Am Crystallog Asn; The Chem Soc. Res: X-ray analytical studies of molecular structure, especially of natural product tumor inhibitors. Mailing Add: Dept of Chem Univ of Va Charlottesville VA 22901

BRYAN, ROBERT NEFF, b Salt Lake City, Utah, Mar 18, 39; m 74. MATHEMATICS. Educ: Univ Utah, BA, 61, MA, 62, PhD(differential equations), 65. Prof Exp: Asst prof math, Ithaca Col, 65-69; asst prof, 69-70, ASSOC PROF MATH, UNIV WESTERN ONT, 70- Mem: Am Math Soc; Math Asn Am; Can Math Cong. Res: Linear differential systems with general boundary conditions. Mailing Add: Dept of Math Univ of Western Ont London ON Can

BRYAN, RONALD ARTHUR, b Portland, Ore, June 16, 32; m 53, 63; c 3 THEORETICAL PHYSICS. Educ: Yale Univ, BS, 54; Univ Rochester, PhD(theoret physics), 61. Prof Exp: Res assoc theoret physics, Univ Calif, Los Angeles, 60-63; physicist, Lab Nuclear Physics, Univ Paris, 63-64; physicist, Lawrence Radiation Lab, 64-68; vis lectr, 68-69, assoc prof, 69-73, PROF PHYSICS, TEX A&M UNIV, 73- Concurrent Pos: NATO fel, France, 63-64; long term vis staff mem, Los Alamos Sci Lab, 73-74; Nordic Inst Theoret Atomic Physics fel, 74-75; vis prof, Univ Helsinki, Finland, 74-75. Mem: Fel Am Phys Soc; fel Am Inst Physics. Res: Scattering theory; two-nucleon interaction, including time-reversal invariance violation. Mailing Add: Dept of Physics Tex A&M Univ College Station TX 77843

BRYAN, SARA E, b Yantley, Ala, Sept 22, 22. BIOCHEMISTRY. Educ: Auburn Univ, BS, 44; Baylor Univ, MS, 58, PhD(biochem), 64. Prof Exp: Med technologist, Hermann Hosp, Houston, 46-54; res assoc hemat, Southwestern Med Sch, Univ Tex, 54-56; instr chem, 58-65, asst prof biol sci, 66-68, ASSOC PROF BIOL SCI, UNIV NEW ORLEANS, 68- Concurrent Pos: Res assoc, Fla State Univ, 65-66; Brown-Hazen Fund Res Corp res grant, 67- Mem: Am Chem Soc. Res: Role of metals in biological processes, especially interactions of divalent metals with nucleic acids and proteins. Mailing Add: Dept of Biol Sci Univ of New Orleans New Orleans LA 70122

BRYAN, THOMAS ALAN, b Scott City, Kans, Apr 24, 43; m 70; c 2. POULTRY PATHOLOGY. Educ: Kans State Univ, BA, 65, DVM, 72, PhD(poultry nutrit), 72. Prof Exp: ASST PROF ANIMAL & VET SCI, UNIV MAINE, ORONO, 72- Mem: Poultry Sci Asn; Am Vet Med Asn. Res: Monitoring antibody response to infectious disease in chickens. Mailing Add: Dept of Animal & Vet Sci Hitchner Hall Univ of Maine Orono ME 04473

BRYAN, THORNTON EMRY, b Frankfort, Ky, Mar 16, 27; m 64. FAMILY MEDICINE. Educ: UNiv Ky, BS, 49; Univ Louisville, MD, 54. Prof Exp: Assoc prof family pract, Col Med, Univ Iowa, 71-74; PROF FAMILY MED & CHMN DEPT, CTR HEALTH SCI, UNIV TENN, 74 Concurrent Pos: AMA Physicians Recognition Award Continuing Med Educ, 70-73. Mem: AMA; Am Acad Family Physicians. Res: Clinical health care; ambulatory patient data, collection, storage and retrieval. Mailing Add: Dept of Family Med Univ of Tenn Health Sci Ctr Memphis TN 38104

BRYAN, VIRGINIA SCHMITT, b Ironwood, Mich, Jan 25, 22; m 41; c 2. BOTANY, ACADEMIC ADMINISTRATION. Educ: Univ Mich, BS, 48, MS, 50; Duke Univ, PhD(bot), 55. Prof Exp: Res assoc bot, 55-58 & 63-65, ASST PROF BOT & ACAD DEAN, 65- Mem: Am Bryol & Lichenological Soc (secy, 57-59); Bot Soc Am. Res: Chromosome studies in relation to the systematic positions of moss taxa. Mailing Add: Dept of Bot Duke Univ Durham NC 27706

BRYAN, WILBUR LOWELL, b South River, NJ, Feb 20, 21; m 47. PHARMACEUTICAL CHEMISTRY. Educ: Lafayette Col, BA, 42. Prof Exp: Chemist explosives, Hercules Powder Co, 42-43; chemist, 45-61, res investr, 61-69, SR RES INVESTR PHARMACEUT, E R SQUIBB & SONS, 69- Mem: Am Chem Soc; NY Acad Sci. Res: Isolation techniques of natural products, particularly antibiotics. Mailing Add: Squibb Inst for Med Res New Brunswick NJ 08903

BRYAN, WILFRED BOTTRILL, b Waterbury, Conn, Feb 18, 32; m 53; c 3. PETROLOGY, MARINE GEOLOGY. Educ: Dartmouth Col, BA, 54; Univ Wis-Madison, MA, 56, PhD(petrol), 59. Prof Exp: Geologist, Scripps Inst Oceanog, Revillagigedo Islands, Mex, 57, geol leader geophys exped, 57; geologist, M A Hanna Co, 59-61; sr lectr petrol, Univ Queensland, 61-67; sr fel, Carnegie Inst, Geophys Lab, 67-70; ASSOC SCIENTIST, DEPT GEOL & GEOPHYS, WOODS HOLE OCEANOG INST, 70- Concurrent Pos: Consult, Exoil Proprietary Ltd & Magellan Petrol Co, 63-65 & Tennant Mineral Develop Co Proprietary Ltd, 65-67; mem subcomt volcanology, Australian Acad Sci, 66-67. Mem: Australian Geol Soc; Am Geophys Union; Int Asn Volcanology; Mineral Soc Gt Brit & Ireland. Res: Igneous petrology; distribution and geological relationships of volcanic rocks in and around the ocean basins; compositional relationships and fractionation effects within genetically related rock suites; relations between geophysical phenomena and igneous activity. Mailing Add: Dept of Geol & Geophys Woods Hole Oceanog Inst Woods Hole MA 02543

BRYAN, WILLIAM PHELAN, b Chicago, Ill, June 4, 30; m 61. BIOCHEMISTRY. Educ: Univ Calif, Los Angeles, BS, 52, MS, 53; Univ Calif, Berkeley, PhD(chem), 57. Prof Exp: Grant, Calif Inst Technol, 57-59; fel, Carlsberg Lab, Copenhagen, Denmark, 59-60; chemist, Danish Atomic Energy Comn, Riso, 60-62; grant, Cornell Univ, 62-64; asst prof chem, Boston Univ, 64-69; ASSOC PROF BIOCHEM, MED CTR, IND UNIV, INDIANAPOLIS, 69- Mem: AAAS; Am Soc Biol Chemists; Am Chem Soc. Res: Physical chemistry of proteins, polypeptides and model systems; physical chemistry of membranes; membrane structure and function. Mailing Add: Dept of Biochem Ind Univ Med Ctr Indianapolis IN 46202

BRYAN, WILLIAM RAY, b Waco, Tex, Dec 25, 05; m 35; c 2. PHYSIOLOGY. Educ: Carson-Newman Col, BS, 28; Vanderbilt Univ, PhD(physiol), 31. Hon Degrees: ScD, Carson-Newman Col, 58. Prof Exp: Asst physiol, Sch Med, Vanderbilt Univ, 29-31, Denison fel, 31-32, from instr to asst prof, 32-36; asst prof path, Albany Med Col, 36-38; res fel, Nat Cancer Inst, 38-42, biologist, 42-58, head sect viral oncol, 58-61, chief lab, 61-64, assoc sci dir, 63-67, sci coordr viral oncol, 67-73; RETIRED. Mem: Am Physiol Soc; Am Asn Cancer Res. Res: Physiological mechanisms in parathyroid tetany; physiological variation in white blood cell counts; quantitative biological studies on the action of carcinogenic agents, particularly the Rous sarcoma and rabbit papilloma viruses; viruses in relation to human cancer. Mailing Add: Apt 118 18700 Walkers Choice Rd Gaithersburg MD 20760

BRYANS, ALEXANDER (MCKELVEY), b Toronto, Ont, Sept 16, 21; m 54; c 3. MEDICINE. Educ: Univ Toronto, MD, 44; Mich State Univ, MA, 71; FRCP, 52. Prof Exp: McLaughlin traveling fel, 56-57; assoc prof, 59-60, PROF PEDIAT, QUEEN'S UNIV, ONT, 60-, DIR HEALTH SCI OFF OF EDUC, 71- Mem: Fel Am Acad Pediat; Can Med Asn; Can Pediat Soc. Res: Medical education. Mailing Add: Dept of Pediat Queen's Univ Kingston ON Can

BRYANS, CHARLES IVERSON, JR, b Augusta, Ga, May 11, 19; m 46; c 5. OBSTETRICS & GYNECOLOGY. Educ: Univ Ga, BS, 40, MD, 43; Am Bd Obstet & Gynec, dipl. Prof Exp: Fel obstet & gynec, 48, PROF OBSTET & GYNEC, MED COL GA, 62- Concurrent Pos: Consult, US Army Hosp, Ft Gordon, Ga, 53-, Milledgeville State Hosp, Macon Hosp, Univ Hosp, Augusta & St Joseph's Hosp, 62-; consult, Greenville City Hosp & Mem Med Ctr, Savannah. Mem: Am Col Surg; Am Col Obstet & Gynecol; AMA; Int Col Surg. Res: Obstetrical analgesia and anesthesia; trophoblastic and other placental abnormalities; toxemia of pregnancy. Mailing Add: Dept of Obstet & Gynec Med Col of Ga Augusta GA 30902

BRYANS, JOHN THOMAS, b Paterson, NJ, June 1, 24; m 59; c 2. ANIMAL VIROLOGY. Educ: Fla Southern Col, BS, 49; Univ Ky, MS, 51; Cornell Univ, PhD(vet bact & path), 54. Prof Exp: Virologist, 54-60, PROF VET SCI, AGR EXP STA, UNIV KY, 60-, CHMN DEPT, 74- Mem: AAAS; Am Soc Microbiol; Conf Res Workers Animal Dis; NY Acad Sci. Res: Etiology, immunology and epizootiology of infectious diseases of domestic animals; pathogenesis and immunology of herpesviral infection. Mailing Add: Dept of Vet Sci Univ of Ky Lexington KY 40506

BRYANT, BEN S, b Seattle, Wash, Mar 28, 23; m 47; c 3. WOOD TECHNOLOGY. Educ: Univ Wash, BS, 47, MS, 48; Yale Univ, DFor(wood technol), 51. Prof Exp: From instr to assoc prof, 49-69, dir inst forest prod, 60-64, PROF WOOD SCI & TECHNOL, COL FOREST RESOURCES, UNIV WASH, 69- Concurrent Pos: Consult, forest prod industs, asn, chem & mach co, US & foreign govts in forest prod develop, adhesive develop, forest prod industs develop & surv, 51- Mem: Soc Am Foresters; Forest Prod Res Soc; Tech Asn Pulp & Paper Indust; Soc Wood Sci & Technol. Res: Structural utilization of wood; adhesion and gluing process technology; product development methods; forest utilization. Mailing Add: 4102 51st NE Seattle WA 98105

BRYANT, BERNARD, biophysics, see 12th edition

BRYANT, BILLY FINNEY, b McKenzie, Tenn, Nov 29, 22; m 46; c 3. MATHEMATICS. Educ: Univ SC, BS, 45; Peabody Col, BA, 48; Vanderbilt Univ, PhD(math), 54. Prof Exp: From instr to assoc prof, 48-66, PROF MATH, VANDERBILT UNIV, 66-, CHMN DEPT, 70- Concurrent Pos: Ford Found fac fel, Princeton Univ, 55-56; sci fac fel, Univ Calif, Berkeley, 67-68. Mem: Am Math Soc; Math Asn Am. Res: Point set topology. Mailing Add: Box 120 Vanderbilt Univ Nashville TN 37235

BRYANT, BRUCE HAZELTON, b New York, NY, Sept 25, 30; m 55; c 4. REGIONAL GEOLOGY. Educ: Dartmouth Col, AB, 51; Univ Wash, PhD(geol), 55. Prof Exp: Geologist, 55-63, res geologist, 63-69, GEOLOGIST, US GEOL SURV, 69- Mem: Mineral Soc Am; Geol Soc Am; AAAS. Res: Geologic mapping; structure; petrology of igneous and metamorphic rocks. Mailing Add: US Geol Surv Denver Fed Ctr Denver CO 80225

BRYANT, CARROLL WILLIAM, b Dousman, Wis, Sept 20, 07; m 30. PHYSICS, OPERATIONS RESEARCH. Educ: Beloit Col, AB, 28; Johns Hopkins Univ, PhD(physics), 35. Prof Exp: High sch instr, Wis, 28-30; jr instr physics, Johns Hopkins Univ, 30-35; assoc prof physics & math, Univ Tenn, Martin, 35-41; prof physics & head dept, Wichita State Univ, 41-43; opers analyst, US Air Force, Africa & Italy, 43-44; physicist, Nat Bur Standards, 44-45; prof physics & head dept, Wichita State Univ, 45-46; opers analyst, Hq Tactical Air Command, US Air Force, 46-51, chief opers anal, 51-62; PROF PHYSICS & HEAD DEPT, GA SOUTHERN COL, 62- Mem: AAAS; Am Asn Physics Teachers; Am Phys Soc; Optical Soc Am. Res: Optics; infrared reflection measurements; ultraviolet spectroscopy; applied physics. Mailing Add: Dept of Physics Ga Southern Col Statesboro GA 30458

BRYANT, ERNEST ATHERTON, b Brewster, Mass, Oct 13, 31; m 53; c 2. RADIOCHEMISTRY. Educ: Univ NMex, BS, 53; Wash Univ, PhD(radiochem), 56. Prof Exp: Instr chem, Wash Univ, 56-57; MEM STAFF RADIOCHEM, LOS ALAMOS SCI LAB, UNIV CALIF, 57- Mem: AAAS; Am Chem Soc. Res: High temperature chemistry; nuclear chemistry; radiochemical analysis; reactor technology. Mailing Add: 111 Beryl White Rock NM 87544

BRYANT, GEORGE MACON, b Anniston, Ala, Aug 3, 26; m 50; c 2. POLYMER CHEMISTRY. Educ: Auburn Univ, BS, 48; Inst Textile Technol, MS, 50; Princeton Univ, MA, 52, PhD(chem), 54. Prof Exp: Res chemist, Chem Div, 54-58, group leader, 58-63, res & develop mgr textile prod, 63-66, technol mgr, Fibers & Fabrics Div, 66-71, develop assoc res & develop, 71-75, CORP RES FEL, UNION CARBIDE CORP, 75- Honors & Awards: Award, Fiber Soc, 64. Mem: Am Chem Soc; Fiber Soc. Res: Physical chemistry of textile fibers; mechanical behavior of polymers; textile chemicals. Mailing Add: 1204 Williamsburg Way Charleston WV 25314

BRYANT, HAROLD HORN, b Alachua, Fla, Aug 19, 16; m 38; c 2. PHARMACOLOGY. Educ: Johns Hopkins Univ, BS, 53, Univ Md, MS, 54, PhD(pharmacol), 56. Prof Exp: Pharmacologist, Hynson, Westcott & Dunning, Inc, 44-62; pharmacologist & pres, Pharmacol Assocs, Inc, Towson, 62-65; pharmacologist, Huntingdon Res Ctr, Inc, 65-69, dir, 65-74, pres, 69-74; CONSULT, 74- Concurrent Pos: Assoc prof pharmacol, Sch Med, Univ Md, 62-65. Mem: Am Soc Pharmacol & Exp Therapeut. Res: Use of plastics in medicine; smooth muscle pharmacology;

gastrointestinal, reproduction and vascular or circulatory metabolic studies including isotopic. Mailing Add: Old Court Rd Brooklandville MD 21022

BRYANT, HARRY TALBOT, b Rumford, Maine, Oct 11, 21; m 46; c 2. AGRONOMY. Educ: Univ Maine, BS, 51, MS, 52; Univ Wis, PhD(agron), 55. Prof Exp: Asst, Univ Maine, 51-52 & Univ Wis, 52-55; ASSOC PROF AGRON, VA POLYTECH INST & STATE UNIV, 55- Concurrent Pos: Partic, Indust Reconstruct Inst, Brazil, 65-67. Mem: Am Soc Agron; Crop Sci Soc Am. Res: Evaluation of establishment, maintenance and nutritional value of grasses and legumes used as pasture; silage and hay. Mailing Add: Va Forage Res Sta Rte 2 Box 9 Middleburg VA 22117

BRYANT, HENRY CLAY, JR, pathology, see 12th edition

BRYANT, HERMAN GREY, JR, b Henry, Va, Oct 25, 39; m 59; c 2. PHYSICAL CHEMISTRY. Educ: Univ Va, BA, 61, PhD(chem), 66. Prof Exp: Res chemist, Celanese Corp, 66-68; SR CHEMIST, LIGGETT & MYERS TOBACCO CO, DURHAM, 68- Mem: Am Chem Soc; Am Phys Soc. Res: Kinetics; photochemistry; polymers; tobacco chemistry. Mailing Add: Rte 1 Box 91-5B Bahama NC 27503

BRYANT, HOWARD CARNES, b Fresno, Calif, July 9, 33; m 60; c 3. PHYSICS. Educ: Univ Calif, Berkeley, BA, 55; Univ Mich, MS, 57, PhD(physics), 61. Prof Exp: From asst prof to assoc prof, 60-71, PROF PHYSICS, UNIV N MEX, 71- Concurrent Pos: Vis scientist, Stanford Linear Accelerator Ctr, 67-69; consult, Los Alamos Sci Lab, 61-; sr vis fel, Queen Mary Col, Univ London, 75-76. Mem: Optical Soc Am; Am Phys Soc; Am Asn Physics Teachers. Res: Experimental high energy and nuclear physics; physical optics. Mailing Add: Dept of Physics Univ of NMex Albuquerque NM 87131

BRYANT, JAMES BERRY, JR, b New Waverly, Tex, Oct 22, 32; m 59. MICROBIOLOGY. Educ: Wiley Col, BS, 55; Tex Southern Univ, MS, 57; Pa State Univ, PhD(bact), 63. Prof Exp: Instr biol, Tex Southern Univ, 57-59; asst prof, 62-64, ASSOC PROF BACT, SOUTHERN UNIV, 64- Concurrent Pos: Acad vis, Med Sch, Tulane Univ, 72-73. Mem: AAAS; Am Soc Microbiol (secy-treas, 71-73). Res: Bacterial degradation of some aromatic and aliphatic herbicides, especially triazines, triazole, carbamates and pheny carboxylic acids; the genotypic and phenotypic effect of herbicid herbicides, Amitrole, Hyvar, Phenylureas and s-Triazines on soil microorganisms; synthesis of single cell protein from starchy foods. Mailing Add: Dept of Bact Southern Univ Southern Br PO Baton Rouge LA 70813

BRYANT, JAMES IRVIN, physical chemistry, see 12th edition

BRYANT, JAMES THOMAS, physical chemistry, see 12th edition

BRYANT, JAY CLARK, b Susquehanna, Pa, Feb 6, 05; m 40. BIOCHEMISTRY. Educ: Pa State Univ, BS, 32; Cornell Univ, MS, 55; Georgetown Univ, PhD, 63. Prof Exp: Asst, Cornell Univ, 33-37; soil technologist, USDA, NY, 38-40 & Md, 40-47; chemist, Nat Cancer Inst, NIH, Md, 47-57, res biochemist, 57-69; RETIRED. Concurrent Pos: Asst ed, In Vitro, Tissue Cult Asn, 70-74, rev ed, 75-76. Mem: AAAS; Am Chem Soc; NY Acad Sci; Tissue Cult Asn; Am Soc Cell Biol. Res: Tissue culture of mammalian cells; chemically defined culture media; massive agitated fluid suspension cultures; chemical analysis of run-off water; soil classification and morphology. Mailing Add: 1511 Glenallan Ave Silver Spring MD 20902

BRYANT, JOHN HARLAND, b Tucson, Ariz, Mar 8, 25; m 56; c 3. PUBLIC HEALTH. Educ: Univ Ariz, BA, 49; Columbia Univ, MD, 53. Prof Exp: Intern & asst resident, Presby Hosp, NY, 53-56; res fel biochem, Nat Found, NIH, Md, 56-57 & 58-59, Munich, 57-58; spec trainee hemat, Dept Med, Sch Med, Wash Univ, 59-60; asst prof med, Col Med, Univ Vt, 60-64, assoc prof & asst dean, 64-65; staff mem, Rockefeller Found & prof, Fac Med, Ramathibodi Hosp, Bangkok, Thailand, 65-71; DIR SCH PUB HEALTH, COL PHYSICIANS & SURGEONS, COLUMBIA UNIV, 71-, JOSEPH R DeLAMAR PROF PUB HEALTH, 73- Concurrent Pos: Asst attend physician, Mary Fletcher Hosp, 60-; asst attend physician, Degoesbriand Mem Hosp, 60-63, assoc attend physician, 63-; spec staff mem, Study Med Educ Develop Nations, Rockefeller Found, 64-65; chmn, Christian Med Comn, World Coun Churches, Geneva, 68- Mem: Nat Inst Med. Res: National and international approaches to evaluation and design of health care systems; education of health personnel; national health insurance and health manpower policies. Mailing Add: Sch Pub Health Columbia Univ New York NY 10032

BRYANT, JOHN LOGAN, b Corinth, Miss, Aug, 27, 40; m 58; c 2. MATHEMATICS Educ: Univ Miss, BS & MS, 62; Univ Ga, PhD(math), 65 Prof Exp: Asst prof math, Univ Miss, 65-66; from asst prof to assoc prof, 66-74, PROF MATH, FLA STATE UNIV, 74- Mem: Am Math Soc. Res: Geometric topology; properties of topological embeddings of polyhedra. Mailing Add: Dept of Math Fla State Univ Tallahassee FL 32306

BRYANT, JOHN PATRICK, soil fertility, soil genesis, see 12th edition

BRYANT, LESTER RICHARD, b Louisville, Ky, Sept 8, 30; m 51; c 2. MEDICINE, THORACIC SURGERY. Educ: Univ Ky, BS, 51; Univ Cincinnati, MD, 55, DSc(surg), 62; Am Bd Surg, dipl, 62; Am Bd Thoracic Surg, dipl, 63. Prof Exp: Fel physiol, Col Med, Baylor Univ, 61; instr, Univ Cincinnati, 61-62; from instr to prof surg, Med Ctr, Univ Ky, 70-73; PROF SURG & CHIEF THORACIC & CARDIOVASC SURG, LA STATE UNIV MED CTR, NEW ORLEANS, 73- Concurrent Pos: Resident coordr, Surg Adjuvant Breast Proj, Nat Coop Study, 61-62; responsible investr coop study on coronary artery dis & coop study of cirrhosis & esophago-gasteric varices, Vet Admin, 62-63, consult, 63-; consult, USPHS, 65-, Charity Hosp of La & Vet Admin Hosp, New Orleans, 73-; mem assoc staff, Southern Baptist Hosp, New Orleans & Hotel Dieu Hosp, New Orleans, 74- Mem: Am Asn Thoracic Surg; Am Col Surg; AMA; Asn Acad Surg; Int Soc Surg. Res: Cardiothoracic surgery; development of primate smoking model; evaluation of prophylactic antibiotics in thoracic trauma; comparison of incisions for tracheostomy; study of bacterial colonization profile in patients with tracheostomy. Mailing Add: Dept of Surg La State Univ Med Ctr New Orleans LA 70112

BRYANT, MARVIN PIERCE, b Boise, Idaho, July 4, 25; m 46; c 5. MICROBIAL ECOLOGY. Educ: State Col Wash, BS, 49, MS, 50; Univ Md, PhD(bact), 55. Prof Exp: Res asst, State Col Wash, 49-51; bacteriologist, Agr Res Serv, USDA, 51-64; assoc prof bact, 64-66, PROF MICROBIOL, UNIV ILL, URBANA-CHAMPAIGN, 66- Concurrent Pos: Ed, Appl Microbiol, 68-71, ed-in-chief, 71-; trustee, Bergey's Manual of Determinative Bact, 75- Honors & Awards: Superior Serv Award, USDA, 59. Mem: Fel AAAS; Am Soc Microbiol; Am Dairy Sci Asn; Brit Soc Gen Microbiol. Res: Ruminal bacteriology; systematics; nutrition; physiology; ecology; nonsporeforming anaerobic bacteria and ciliate protozoa; methanogenic bacteria. Mailing Add: 1003 S Orchard St Urbana IL 61801

BRYANT, MONROE DAVID, vertebrate zoology, see 12th edition

BRYANT, NEVIN ARTHUR, b Syracuse, NY, Sept 6, 42; m 65; c 2. GEOGRAPHY. Educ: McGill Univ, BA, 64; Univ Hawaii, MA, 67; Univ Mich, PhD(geog), 73. Prof Exp: Asst prof geog, Calif State Univ, Northridge, 72-75; MEM TECH STAFF, JET PROPULSION LAB, CALIF INST TECHNOL, 75- Mem: Asn Am Geographers; Am Geog Soc; Asn Asian Studies; Am Soc Photogram. Res: Design of two dimensional geographic information systems for urban land use monitoring via digital image processing of satellite imagery. Mailing Add: 5155 Stoneglen Rd La Canada CA 91011

BRYANT, PAUL JAMES, b Kansas City, Mo, May 11, 29; m 60. SOLID STATE PHYSICS. Educ: Rockhurst Col, BS, 51; St Louis Univ, MS, 53, PhD(physics), 57. Prof Exp: Prin physicist, Midwest Res Inst, 58-68; assoc prof, 68-74, PROF PHYSICS, UNIV MO-KANSAS CITY, 74- Mem: Am Phys Soc; Am Vacuum Soc. Res: Solid state structure studies; surface physics; field emission microscopy; friction of solids; ultra high vacuum technology. Mailing Add: Dept of Physics Univ of Mo 5100 Rockhill Rd Kansas City MO 64110

BRYANT, RALPH CLEMENT, b New Haven, Conn, Sept 27, 13; m 37; c 2. FOREST MANAGEMENT. Educ: Yale Univ, BS, 35, MF, 36; Duke Univ, PhD, 51. Prof Exp: From foreman to dist forest ranger, US Forest Serv, 36-46; assoc prof forest mgt & utilization & actg head dept, Colo Agr & Mech Col, 47-52; PROF FOREST MGT, NC STATE UNIV, 52- Concurrent Pos: Mem & vchmn, NC Forestry Coun, 75-79. Mem: AAAS; Soc Am Foresters; Can Pulp & Paper Asn. Res: Prescribed burning. Mailing Add: 1500 Lake Dam Rd Raleigh NC 27606

BRYANT, RHYS, b Swansea, Wales, Nov 28, 36; m 60; c 3. ORGANIC CHEMISTRY. Educ: Univ Wales, BSc, 57, PhD(org chem), 60. Prof Exp: Fulbright res scholar chem, Yale Univ, 60-61; fel org chem, Mellon Inst, 61; res chemist, Unilever Res Lab, Eng, 61-63; asst lectr chem, Univ Manchester, 63-65; group leader instrumentation & methods develop, Res Ctr, 65-67, sect leader, 67-68, DIR PHARMACEUT QUAL CONTROL, MEAD JOHNSON & CO, 68- Mem: Am Chem Soc; Sigma Xi; The Chem Soc; Royal Inst Chem. Res: Synthesis of oxygen heterocyclic compounds; sesquiterpenoids; analysis of experimental drug products. Mailing Add: Mead Johnson & Co Evansville IN 47721

BRYANT, ROBERT GEORGE, b Mineola, NY, Sept 13, 43; m 65; c 2. BIOPHYSICAL CHEMISTRY, INORGANIC CHEMISTRY. Educ: Colgate Univ, AB, 65; Stanford Univ, PhD(chem), 69. Prof Exp: Asst prof, 69-72, ASSOC PROF CHEM, UNIV MINN, MINNEAPOLIS, 72- Concurrent Pos: Teacher-scholar grant, Dreyfus Found, 74. Mem: AAAS; Am Chem Soc; Biophys Soc. Res: Investigation of metal ion protein interactions, water macromolecule interactions and metalloenzymes using NMR spectroscopy. Mailing Add: Dept of Chem 139 Smith Hall Univ of Minn Minneapolis MN 55455

BRYANT, ROBERT L, b New York, NY, Jan 3, 28; m 49; c 3. FORESTRY. Educ: La State Univ, BS, 53, MS, 54; Mich State Univ, PhD(forestry), 63. Prof Exp: Biologist wildlife invest, La Wildlife & Fisheries Comn, 52-53, res biologist refuge mgt, 53-54; asst prof forestry, McNeese State Col, 54-59; asst, Mich State Univ, 59-62; from asst prof to assoc prof, 62-69, PROF FORESTRY, McNEESE STATE UNIV, 69- Concurrent Pos: Collabr, Lake States Forest Exp Sta, 59-62. Mem: Soc Am Foresters; Am Forestry Asn; Ecol Soc Am. Res: Forest ecology; ecology of lowland hardwood forests, soils and ground-water; remote sensing of environment within forest structures. Mailing Add: Dept of Agr McNeese State Univ Lake Charles LA 70602

BRYANT, ROBERT WILLIAM, b Oxford, Ohio, June 8, 25; m 46; c 4. Educ: Miami Univ, AB, 47; Univ Ala, MA, 48. Prof Exp: Instr math, Miami Univ, 48-51; rating exam sci & tech personnel, Potomac River Naval Command, 51-52; mathematician, Underwater Sound Propagation, US Naval Res Lab, 52-56, supvry mathematician, 58-59; mathematician, Celestial Mech & Appl Orbit Anal, Goddard Space Flight Ctr, NASA, 59-66; mathematician, Systs Anal Staff, Anti-Submarine Warfare Spec Proj Off, 66-72, SYSTS ANALYST, ANTI-SUBMARINE WARFARE SYSTS PROJ OFF, HQS, NAVAL ORD LAB, WHITE OAK, 72- Mem: Fel AAAS; Am Math Soc. Res: Celestial mechanics; density of upper atmosphere; solar radiation pressure effects on echo type satellites; density interference from drag; signal processing. Mailing Add: 9706 Woodberry St Seabrook MD 20801

BRYANT, SHIRLEY HILLS, b Pittsfield, Mass, Dec 28, 24; m 46, 63; c 3. PHARMACOLOGY, PHYSIOLOGY. Educ: Aurora Col, BS, 48; Univ Chicago, PhD(physiol), 54. Prof Exp: Asst physiol, Univ Chicago, 50-54; from instr to assoc prof pharmacol, 55-69, PROF PHARMACOL, COL MED, UNIV CINCINNATI, 69- Concurrent Pos: USPHS career develop award, Univ Cincinnati, 59-69; vis lectr, Univ PR, 65-66 & Cambridge Univ, 72-73. Mem: AAAS; Soc Neurosci; Biophys Soc; Am Physiol Soc. Res: Physiology and pharmacology of excitation, conduction and synaptic transmission in excitable tissues; biophysics nerve; squid giant synapses; myotonia in goats; neuromuscular junction. Mailing Add: Dept of Pharmacol Univ Cincinnati Col of Med Cincinnati OH 45267

BRYANT, SUSAN VICTORIA, b Sheffield, Eng, May 24, 43; m 66. DEVELOPMENTAL BIOLOGY, CELL BIOLOGY. Educ: Univ London, BSc, 64, PhD(zool), 67. Prof Exp: Res fel zool, Case Western Reserve Univ, 67-69; lectr, 69-70, ASST PROF BIOL, UNIV CALIF, IRVINE, 70- Mem: Am Soc Zool. Res: Regeneration in vertebrates. Mailing Add: Dept of Develop & Cell Biol Univ of Calif Irvine CA 92664

BRYANT, TRUMAN RAI, botany, cell biology, see 12th edition

BRYANT, VAUGHN MOTLEY, b Dallas, Tex, Oct 5, 40; m 64; c 3. ANTHROPOLOGY, BOTANY. Educ: Univ Tex, Austin, BA, 64, MA, 66, PhD(bot), 69. Prof Exp: Asst prof anthrop, Wash State Univ, 69-71; asst prof, 71-74, ASSOC PROF ANTHROP & BIOL, TEX A&M UNIV, 74- Mem: AAAS; Am Asn Stratig Palynologists; Soc Am Archaeol; Bot Soc Am; Am Asn Quaternary Environ. Res: Pollen analytical studies of Quaternary paleoenvironments with special emphasis on areas in the American Southwest; pollen and macrofossil analysis of fossil human coprolites; pollen analytical studies of archaeological deposits. Mailing Add: Dept of Sociol & Anthrop Tex A&M Univ College Station TX 77843

BRYANT, WILLIAM RICHARDS, b Chicago, Ill, Feb 12, 30; m 65; c 3. MARINE GEOLOGY, OCEANOGRAPHY. Educ: Univ Chicago, MS, 60, PhD(geol), 66. Prof Exp: Oceanogr, Off Naval Res, 60-62; res scientist, 62-64, from asst prof to assoc prof oceanog, 64-71, PROF OCEANOG, TEX A&M UNIV, 71- Concurrent Pos: Proj supvr, Off Naval Res grant, 66-; NSF grant, 69- Mem: AAAS. Res: Marine geotechnique; geology and geophysics of the Gulf of Mexico and Caribbean; acoustic characteristics of marine sediments; sediment transport. Mailing Add: Dept of Oceanog Tex A&M Univ College Station TX 77843

BRYCE, DONALD HEWITT, b Rochester, NY, Oct 6, 35. RADIATION PHYSICS. Educ: Rensselaer Polytech Inst, BEE, 57, MEE, 60; Cornell Univ, PhD(nuclear eng), 64. Prof Exp: Instr elec eng, Rensselaer Polytech Inst, 57-60; asst prof, Univ Alaska,

60-61; res scientist, 64-71, PROG MGR, KAMAN SCI CORP, 73- Res: Nuclear weapon radiation effects on materials and systems; radiation effects on electronic components and systems. Mailing Add: Kaman Sci Corp PO Box 7463 Colorado Springs CO 80933

BRYCE, GALE REX, b Safford, Ariz, Nov 18, 39; m 63; c 7. STATISTICS. Educ: Ariz State Univ, BS, 67; Brigham Young Univ, MS, 70; Univ Ky, PhD(exp statist), 74. Prof Exp: Math programmer, Semiconductor Prod Div, Motorola Inc, 63-68; asst prof, 72-75, ASSOC PROF STATIST, BRIGHAM YOUNG UNIV, 75- Mem: Am Statist Asn; Int Biomet Soc; Sigma Xi. Res: Linear models with special emphasis in the analysis of unbalanced designs where a mixed model is appropriate; statistical problems in physiology. Mailing Add: Dept of Statist Brigham Young Univ Provo UT 84602

BRYCE, HUGH GLENDINNING, b Melville, Sask, Nov 30, 17; m 43; c 2. PHYSICAL CHEMISTRY. Educ: Univ Sask, BA, 38, MA, 40; Columbia Univ, PhD(chem), 43. Prof Exp: Res chemist, Off Sci Res & Develop contract, Columbia Univ, 42-43; res chemist, Los Alamos Lab, Univ Calif, 43-46; res chemist, Res Labs, Sharples Corp, Philadelphia, 46-49; head prod develop, Fluorochems Dept, 49-59, develop mgr, 59-61, asst tech dir, 61-66, tech dir, Chem Div, 66-73, exec dir, Cent Res Labs, 73-75, DIV VPRES, CENT RES LABS, MINN MINING & MFG CO, 75- Mem: AAAS; Am Chem Soc; Indust Res Inst. Res: Ultracentrifuge; low temperatures; ion exchange resins; ultracentrifugal behavior of starch and its triacetate and trimethyl derivatives; chemistry of carbon fluorine compounds. Mailing Add: Minn Mining & Mfg Co 3M Ctr St Paul MN 55101

BRYDEN, ELMER LOUIS, b Detroit, Mich, June 13, 27; m 48; c 3. GEOLOGY. Educ: Wayne State Univ, BA, 49, MS, 50. Prof Exp: SR EVAL ANALYST, CITIES SERV OIL CO, 50- Mem: Am Asn Petrol Geologists; Soc Petrol Engrs. Res: Petroleum geology; foreign exploration and petroleum economics. Mailing Add: 2163 N Waco Tulsa OK 74127

BRYDEN, HARRY LEONARD, b Providence, RI, July 9, 46; m 69; c 1. PHYSICAL OCEANOGRAPHY. Educ: Dartmouth Col, AB, 68; Mass Inst Technol, PhD(oceanog), 75. Prof Exp: Mathematician, US Naval Oceanog Off, 69-70 & US Naval Underwater Sound Lab, 70; res asst phys oceanog, Woods Hole Oceanog Inst, 70-75; RES ASSOC PHYS OCEANOG, ORE STATE UNIV, 75- Mem: Am Geophys Union; Am Meteorol Soc; Sigma Xi. Res: Dynamics of the Antarctic Circumpolar Current, coastal upwelling and mid-ocean low-frequency currents. Mailing Add: Sch of Oceanog Ore State Univ Corvallis OR 97331

BRYDEN, JOHN HEILNER, b Nampa, Idaho, Sept 9, 20; m 45; c 2. X-RAY CRYSTALLOGRAPHY. Educ: Col Idaho, BS, 42; Calif Inst Technol, MS, 45; Univ Calif, Los Angeles, PhD(chem), 51. Prof Exp: Chemist, Hercules Powder Co, 42-43; fel, Calif Inst Technol, 43-45; chemist, E I du Pont de Nemours & Co, 45-46; asst, Univ Calif, Los Angeles, 47-51; chemist, Res Dept, US Naval Ord Test Sta, 51-59; mem tech staff, Hughes Aircraft Co, 59-61; PROF CHEM, CALIF STATE UNIV, FULLERTON, 61- Mem: AAAS; Am Crystallog Asn; Am Chem Soc. Res: Crystal structures of organic and inorganic substances by x-ray diffraction. Mailing Add: 1416 Vista Del Mar Dr Fullerton CA 92631

BRYDEN, ROBERT RICHMOND, b Erie, Pa, July 19, 16; m 43; c 4. ANIMAL ECOLOGY. Educ: Mt Union Col, BS, 38; Vanderbilt Univ, PhD(zool), 51. Prof Exp: Instr biol, Univ Akron, 40-41; asst sci, Goodyear Aircraft Corp, 41-43; assoc prof biol, Mid Tenn State Univ, 46-51; biologist, AEC, 51-54; head biol dept & chmn sci div, Union Col, 54-58; prof biol, High Point Col, 58-61; PROF BIOL & HEAD DEPT, GUILFORD COL, 61- Concurrent Pos: AAAS & Tenn Acad res grant, 49; Am Acad Arts & Sci res grant, 58; Sigma Xi res grant, 59. Mem: AAAS; Am Soc Zoologists. Res: Limnology; physiology; invertebrate zoology. Mailing Add: Dept of Biol Guilford Col Greensboro NC 27410

BRYDON, JAMES EMERSON, b Portage la Prairie, Man, June 28, 28; m 51; c 3. MINERALOGY, SOIL SCIENCE. Educ: Univ Man, BSc, 51; Univ Mo, MSc, 54, PhD(soils), 56. Prof Exp: Res officer soil mineral, Soil Res Inst, Can Dept Agr, 51-71; RES MGR, ENVIRON PROTECTION SERV, ENVIRON CAN, 72- Mem: Mineral Soc Am; Am Clay Minerals Soc; Soil Sci Soc Am; Agr Inst Can; Can Soc Soil Sci (secy, 62-). Res: Soil and clay mineralogy, as related to weathering reactions, plant nutrient uptake and pedogenesis. Mailing Add: Environ Protection Serv Environ Can Ottawa ON Can

BRYNER, CHARLES LESLIE, b Dunbar, Pa, Oct 15, 14; m 47; c 2 BOTANY. Educ: Waynesburg Col, BS, 40; Univ WVa, MS, 48, PhD(bot), 57. Prof Exp: Teacher & supvry prin, Pub Schs, Pa, 40-42; assoc prof bot, 46-56, dean men, 53-58, chmn div sci, 58-71, asst acad vpres, 59, PROF BOT, WAYNESBURG COL, 57- Res: Genus Helianthus in West Virginia; taxonomic botany. Mailing Add: RD 3 Waynesburg PA 15370

BRYNER, CLARENCE SHELDON, b Juniata, Pa, Jan 1, 12; m 38; c 2. AGRONOMY. Educ: Pa State Univ, BS, 38, PhD(plant breeding), 57. Prof Exp: Jr soil survr, Soil Conserv Serv, USDA, 38-39, jr agronomist, Civilian Conserv Camp, 39-42, asst soil conservationist, 43-45; asst prof agron exten, 45-47, instr agron, 47-57, assoc prof agron exten, 57-64, prof, 64-74, EMER PROF AGRON EXTEN, PA STATE UNIV, 74-; CONSULT FOOD CROP PROD & SOIL MGT. Concurrent Pos: Adv seed improv, Pa State Univ-AID Proj, India, 67-69. Mem: Am Soc Agron; Crops Sci Soc; Potato Asn Am. Res: Potato and field crop production; field seed surveys; small grain breeding; conservation farm planning; soil survey. Mailing Add: 720 W Hamilton Ave State College PA 16801

BRYNER, JOHN C, b Salt Lake City, July 22, 31; m 53; c 4. APPLIED PHYSICS. Educ: Univ Utah, BA, 58, PhD(physics), 62. Prof Exp: Teaching asst physics, Univ Utah, 58-60; res specialist, NAm Rockwell, Inc, 62-70 & 72-76; res specialist, Hycon Mfg Co, 70-71; mem tech staff, Aerojet Electrosysts Co, Calif, 71-72; SR RES SCIENTIST, EYRING RES INST, 76- Concurrent Pos: Vis prof, Univ Southern Calif, 63-70. Mem: Am Asn Physics Teachers. Res: Plasma; lasers; infrared technology; electrostatics. Mailing Add: Eyring Res Inst 1455 West 820 North Provo UT 84601

BRYNER, JOHN HENRY, b Washington, Pa, Oct 28, 24; m 45; c 3. MICROBIOLOGY. Educ: Eastern Nazarene Col, AB, 49; Univ Wis, PhD, 68. Prof Exp: Bacteriologist, Animal Dis Sta, Bur Animal Indust, USDA, 50-61, SR RES MICROBIOLOGIST, NAT ANIMAL DIS LAB, AGR RES SERV, USDA, 61- Concurrent Pos: US AID consult, Vet Sch, Porto Alegre, Brazil, 72; managing ed, J Wildlife Dis, Wildlife Dis Asn, 73- Mem: Am Soc Microbiol; Wildlife Dis Asn; US Animal Health Asn; Am Asn Vet Lab Diagnosticians; Sigma Xi. Res: Infectious causes of sterility in livestock, specifically morphologic, serologic and biochemic characterization of the genus Vibrio; intensive study of wild animals. Mailing Add: Nat Animal Dis Lab PO Box 70 Ames IA 50010

BRYNER, LEONID, geology, see 12th edition

BRYNER, LOREN CONRAD, b Price, Utah, Jan 2, 04; m 30; c 3. BIOPHYSICAL CHEMISTRY. Educ: Brigham Young Univ, BS, 28, MS, 30; Iowa State Col, PhD(biophys chem), 34. Prof Exp: Asst chem, Brigham Young Univ, 28-30; asst, Iowa State Col, 30-35; from asst prof to prof, 35-75, EMER PROF CHEM, BRIGHAM YOUNG UNIV, 75- Mem: Am Chem Soc. Res: Physical chemistry; metabolism of sulfur by plants using radio sulfur as a tracer; catalysis; fermentations; biological oxidation of sulfur and sulfide minerals in the leaching of low grade copper ores. Mailing Add: 2194 Canyon Rd Provo UT 84601

BRYNES, PAUL JEFFREY, b Baltimore, Md, Nov 30, 47. MOLECULAR PHARMACOLOGY. Educ: Brandeis Univ, BA, 69; Cornell Univ, MS, 71, PhD(org chem), 75. Prof Exp: NIH fel, Rockefeller Univ, 74-76; ASST PROF PHARMACOL & CHEM, STATE UNIV NY STONY BROOK, 76- Mem: Am Chem Soc; AAAS. Res: Use of chemical probes as reporter groups for the study of drug receptors and membrane phenomena. Mailing Add: Dept of Pharmacol State Univ of NY Stony Brook NY 11790

BRYNILDSON, OSCAR MARIUS, b Wis, Mar 5, 18. ZOOLOGY. Educ: Univ Wis, BA, 47, MA, 50, PhD(zool), 58. Prof Exp: Asst fishery biol, Univ Wis, 48-52; fishery biologist, Wis Conserv Dept, 52-57, group leader cold water res, 57-69; LIMNOLOGIST, WIS DEPT NATURAL RESOURCES, 69- Mem: Am Soc Limnol & Oceanog; Am Fisheries Soc. Res: Ecology of trout and other cold water fishes in lakes and streams. Mailing Add: Wis Dept of Natural Resources Hartman Creek Waupaca WI 54981

BRYNJOLFSSON, ARI, b Akureyri, Iceland, Dec 7, 26; US citizen; m 50; c 5. NUCLEAR PHYSICS, GEOPHYSICS. Educ: Univ Copenhagen, Cand Phil, 49, Cand Mag & Mag Sci, 54, Dr Phil(nuclear physics), 73; Harvard Univ, AMP, 71. Prof Exp: Res physicist, Danish AEC Res Estab, Riso, Denmark, 57-58, head radiation res group, 57-59, dir radiation facil, 59-61; consult, US Army Natick Labs, 62-63; dir radiation facil, Danish AEC, Riso, 63-65; chief radiation sources div, 65-72, actg dir food irradiation, 72-74, CHIEF RADIATION PRESERV OF FOOD DIV & DIR RADIATION LAB, FOOD ENG LAB, US ARMY NATICK LABS, 74- Concurrent Pos: Consult, Int Atomic Energy Agency Panels, 70- & Iranian Govt, 72- Mem: Am Phys Soc; Radiation Res Soc; Am Nuclear Soc; 23 Am Soc Testing & Mat. Res: Theory of stopping of charged particles and degradation of radiation; basic radiation physics and effects of radiation on biological systems; design and operation of isotopes and electron accelerator irradiation facilities; irradiation of foods and medical products; research and industrial application. Mailing Add: Food Eng Lab US Army Natick Labs Natick MA 01760

BRYSK, HENRY, physics, see 12th edition

BRYSK, MIRIAM MASON, b Warsaw, Poland, Mar 10, 35; US citizen; m 55; c 2. BIOLOGICAL CHEMISTRY. Educ: NY Univ, BA, 55; Univ Mich, MS, 58; Columbia Univ, PhD(biol sci), 67. Prof Exp: Lectr biol, Queens Col, NY, 60-61; fel protein chem, Inst Muscle Dis, 67-69; res biologist, Dept Biol, Univ Calif, San Diego, 79-71; NIH FEL, DEPT CELLULAR CHEM, UNIV MICH, 74- Mem: Am Soc Microbiol; Brit Soc Gen Microbiol. Res: Amino acid and cyanide metabolism in microorganisms; synthesis of cell walls in plants; biochemistry and differentiation of the epidermis. Mailing Add: 3523 Larchmont Dr Ann Arbor MI 48105

BRYSON, GEORGE GARDNER, b Santa Barbara, Calif, Dec 16, 35. PHYSIOLOGY, PSYCHOPHYSIOLOGY. Educ: Univ Calif, Santa Barbara, BA, 57; San Francisco State Col, MA, 71. Prof Exp: Zoologist, 57-71, PHYSIOLOGIST, SANTA BARBARA COTTAGE HOSP RES INST, 71- Mem: Am Asn Cancer Res. Res: Psychophysiology of steroid hormones and catecholamines, pituitary-adrenal-gonadal function in aggression and reproductive behavior, carcinogenic mechanisms, solid state carcinogenesis; biologic transport of steroids. Mailing Add: Santa Barbara Cottage Hosp 320 W Pueblo Santa Barbara CA 93105

BRYSON, MARION RITCHIE, b Centralia, Mo, Aug 26, 27; m 47; c 4. OPERATIONS RESEARCH Educ: Univ Mo, BSEd, 49, MA, 50; Iowa State Univ, PhD(statist), 58. Prof Exp: Teacher math, Elgin High Sch, Ill, 50-52; instr, Univ Idaho, 52-53 & Drake Univ, 53-55; res assoc statist, Iowa State Univ, 55-58; dir spec res, Duke Univ, 58-68, from asst prof to assoc prof math & community health sci, 62-68; tech dir opers res, US Army Combat Develop Command Systs Anal Group, 68-72, SCI ADV, US ARMY COMBAT DEVELOP EXP COMMAND, 72- Concurrent Pos: Vis prof, Okla State Univ, 65; statist ed, J Parapsychol, 66-69; lectr, Dept Path, Vet Admin Hosp, 67-68. Mem: Inst Math Statist; Am Statist Asn; Opers Res Soc; Mil Opers Res Soc (pres, 75-76). Res: Military operations research; sampling theory; medical research. Mailing Add: US Army Combat Develop Exp Command Ft Ord CA 93941

BRYSON, MELVIN JOSEPH, b Providence, Utah, June 7, 16; m 42; c 2. CLINICAL BIOCHEMISTRY. Educ: Utah State Agr Col, BS, 46; Univ Utah, MS, 48; Agr & Mech Col, Tex, PhD(biochem & nutrit), 52. Prof Exp: Asst, Utah State Agr Col, 48-49; res biochemist & head drug enzym unit, Eaton Labs, 52-57; res assoc, 57-65, asst res prof, 65-71, ASSOC RES PROF, OBSTET & GYNEC, UNIV UTAH, 71- Concurrent Pos: Dir, Intermountain Perinatal Res & Serv Lab, 70- Mem: Endocrine Soc. Res: Synthesis and metabolism of steroid hormones and relation to intermediary metabolism; affect of drugs on enzymes systems; intestinal absorption; clinical endocrinology. Mailing Add: 5262 Woodcrest Dr Salt Lake City UT 84117

BRYSON, REID ALLEN, b Detroit, Mich, June 7, 20; m 42; c 4. METEOROLOGY. Educ: Denison Univ, BA, 41, Univ Chicago, PhD(meteorol), 48. Hon Degrees: DSc, Denison Univ, 71. Prof Exp: Asst prof meteorol & geol, Univ Wis, 46-48, from asst prof to assoc prof meteorol, 48-56, chmn dept, 48-50 & 52-54; prof, Univ Ariz, 56-57; prof, 57-68, PROF METEOROL & GEOG, UNIV WIS, chmn dept, 57-61, PROF METEOROL & GEOG, 68-, DIR INST ENVIRON STUDIES, 70- Concurrent Pos: Trustee, Univ Corp Atmospheric Res, 59-66; mem environ studies bd, Nat Acad Sci-Nat Acad Eng, 70-73; sci adv comt, Arctic & Alpine Res Inst; Wis Archaeol Surv. Mem: Soc Am Archaeol; Asn Am Geog; Am Meteorol Soc; Am Soc Limnol & Oceanog. Res: Physical limnology; paleoclimatology; dynamic climatology; interdisciplinary environmental studies. Mailing Add: Inst for Environ Studies Univ of Wis 1225 W Dayton St Madison WI 53706

BRYSON, ROBERT PEARNE, b Los Angeles, Calif, June 12, 11; m 44; c 2. GEOLOGY. Educ: Univ Calif, Los Angeles, AB, 34, AM, 37; Calif Inst Technol, MS, 37. Prof Exp: Geologist, US Geol Surv, 37-62, chief geologist's staff, 53-60; STAFF SCIENTIST LUNAR PHYSICS, LUNAR PROGS OFF, NASA, 62- Mem: AAAS; Am Geophys Union; fel Geol Soc Am; Am Asn Petrol Geologists. Res: Western coal fields; bauxite deposits of Arkansas; light metals and nonmetallic minerals; geologist staffing; geological problems related to the moon and its exploration. Mailing Add: Lunar Progs Off NASA Hq Code SM Washington DC 20546

BRYSON, THEODORE CORNELIUS, b Pittsburgh, Pa, Sept 29, 12; m 47; c 3. ANALYTICAL CHEMISTRY. Educ: Carnegie Inst Technol, BS, 44. Prof Exp: Anal chemist, Am Cyanamid Corp, 36-41; anal chemist, 41-43, lab supvr, 43-52, MGR ANAL CHEM, WESTINGHOUSE ELEC CORP, 52- Mem: Am Chem Soc; Am Soc Testing & Mat. Res: Analytical chemistry of metals; ferrous; non-ferrous; uranium; thorium; zirconium; hafnium; niobium; rare metals; general instrumental analysis and inorganic chemistry. Mailing Add: 1518 Maple Ave Verona PA 15147

BRYSON, THOMAS ALLAN, b Pittsburgh, Pa, July 24, 44; m 66; c 2. ORGANIC BIOCHEMISTRY. Educ: Washington & Jefferson Col, BA, 66; Univ Pittsburgh, PhD(org chem), 70. Prof Exp: NIH fel synthetic org chem, Stanford Univ, 70-71; asst prof, 71-75, ASSOC PROF ORG CHEM, UNIV SC, 75- Concurrent Pos: NSF trainee, 66-76. Mem: Am Chem Soc; Sigma Xi. Res: Synthetic organic chemistry; natural products chemistry; key biological compounds, particularly cortisone, juvenile hormone and nicotinamide-adenine dinucleotide. Mailing Add: Dept of Chem Univ of SC Columbia SC 29208

BRYSON, VERNON, b Detroit, Mich, Sept 17, 13; m 35; c 4. MICROBIOLOGY, GENETICS. Educ: Univ Calif, AB, 34; Columbia Univ, MA, 36, PhD(zool), 44. Prof Exp: Asst zool, Columbia Univ, 37-42; assoc, Carnegie Inst, 42-43, geneticist, Biol Lab, Cold Spring Harbor, NY, 43-55; prog dir genetic & develop biol, NSF, 55-56; assoc dir, Inst, 56-58, PROF MICROBIOL, INST MICROBIOL, RUTGERS UNIV, NEW BRUNSWICK, 56- Concurrent Pos: Lectr, Columbia Univ, 53-55 & 58-, State Univ NY & Adelphi Col, 55-56 & Douglass Col, Rutgers Univ, 58-; consult, NSF, 57-59; chmn bd, Am Type Cult Collection, 65- Mem: Genetics Soc Am; Radiation Res Soc; Am Soc Naturalists; Am Soc Microbiol; fel Am Acad Microbiol. Res: Genetic basis of drug and radiation resistance in bacteria; induced mutation; aerosols; cross resistance and indirect selection; turbidostat; microbial population genetics. Mailing Add: Inst of Microbiol Rutgers Univ New Brunswick NJ 08903

BRZENK, RONALD MICHAEL, b Jersey City, NJ, Jan 30, 49; m 71; c 2. NUMBER THEORY. Educ: St Peter's Col, NJ, BS, 69; Univ Notre Dame, PhD(math), 74. Prof Exp: ASST PROF MATH, HARTWICK COL, 74- Mem: Am Math Soc; Math Asn Am. Res: Axiomatic approaches to class field theory; computer applications in number theory and algebra. Mailing Add: Dept of Math Hartwick Col Oneonta NY 13820

BSHARAH, LEWIS, b Beckley, WVa, Feb 23, 24; m 50; c 2. CHEMISTRY. Educ: Iowa State Univ, BS, 49; Ohio State Univ, MS, 51, PhD(chem), 55. Prof Exp: Asst prof, Marine Lab, Univ Miami, 55-57; sr res chemist, Tex-US Chem Co, 57-63; group leader, Petrolite Corp, 63-65, mgr phys anal res, 65-68, dir phys anal lab, 68-72, dir corp res, 72-73; RES DIR, HODAG CHEM CORP, 73- Mem: AAAS; Am Chem Soc. Res: Physical and colloid chemistry of natural and synthetic polymers. Mailing Add: 538 Flanders Dr St Louis MO 63122

BUBAR, JOHN STEPHEN, b NB, Sept 13, 29; m 54; c 2. AGRONOMY. Educ: McGill Univ, BSc, 52, PhD(genetics), 57; Pa State Univ, MS, 54. Prof Exp: Asst agron, Pa State Univ, 52-53; asst, Macdonald Col, McGill Univ, 53-54, lectr, 54-59, asst prof, 59-67; assoc prof, 67-71, PROF AGRON, NS AGR COL, 71-, HEAD DEPT, 67- Concurrent Pos: Mem, Can Comt Forage Crops Breeding, 74- Mem: Agr Inst Can; Can Soc Agron; Am Soc Agron. Res: Crops production and breeding, specifically birdsfoot trefoil breeding and genetics; adaptation of cultivars to Atlantic provinces of Canada. Mailing Add: Dept of Plant Sci NS Agr Col Truro NS Can

BUBE, RICHARD HOWARD, b Providence, RI, Aug 10, 27; m 48; c 4. SOLID STATE PHYSICS, MATERIALS SCIENCE. Educ: Brown Univ, ScB, 46; Princeton Univ, MA, 48, PhD(physics), 50. Prof Exp: Res physicist, Labs Div, Radio Corp Am, 48-62; assoc prof, 62-64, PROF MAT SCI & ELEC ENG, STANFORD UNIV, 64-, CHMN DEPT, 75- Concurrent Pos: Ed, Jour Am Sci Affil; assoc ed, Ann Rev Mat Sci, 69- Mem: Fel AAAS; fel Am Phys Soc; fel Am Sci Affil; Am Soc Eng Educ. Res: Luminescence; photoconductivity; trapping in solids; phosphors; electronic crystal defect phenomena; semiconductors; photoelectronic properties of materials; photovoltaics; amorphous materials. Mailing Add: Dept of Mat Sci & Eng Stanford Univ Stanford CA 94305

BUBEL, HANS CURT, b Mannheim, Ger, Oct 3, 25; nat US; m 58; c 2. MEDICAL BACTERIOLOGY. Educ: Univ Utah, BS, 50, MS, 53, PhD(bact), 58. Prof Exp: Res assoc virol, Univ Utah, 53-58; from asst prof to assoc prof microbiol, 58-70, PROF MICROBIOL, UNIV CINCINNATI, 70- Mem: NY Acad Sci; Am Soc Microbiol. Res: Host-cell virus interactions; polio-virus; vaccinia virus; Newcastle disease virus. Mailing Add: Dept of Microbiol Univ Cincinnati Col of Med Cincinnati OH 45221

BUBEN, DAVID, organic chemistry, physical chemistry, see 12th edition

BUBLITZ, CLARK, b Merrill, Wis, Dec 8, 27; m 58; c 5. BIOCHEMISTRY. Educ: Univ Chicago, PhB, 49, PhD(biochem), 55. Prof Exp: Am Cancer Soc fel, 55-58; sr instr pharmacol, Sch Med, St Louis Univ, 59-60; asst prof biochem, 60-71, ASSOC PROF BIOCHEM, UNIV COLO MED CTR, 71- Mem: AAAS; Am Chem Soc; Am Soc Biol Chem. Res: Enzymology. Mailing Add: Dept of Biochem Univ of Colo Med Ctr Denver CO 80220

BUBLITZ, DONALD EDWARD, b Chicago, Ill, Dec 6, 35; m 57; c 2. ORGANIC CHEMISTRY. Educ: Univ Calif, Riverside, AB, 57; Univ Kans, PhD(chem), 61. Prof Exp: Res assoc chem, Univ Ill, 61-62; res chemist, 62-70, sr res chemist, 70-72, RES SPECIALIST, WESTERN DIV, DOW CHEM USA, 72- Mem: The Chem Soc; NY Acad Sci. Res: Synthesis of agricultural and pharmaceutical chemicals; organometallic compounds. Mailing Add: Dow Chem USA PO Box 1398 Pittsburg CA 94565

BUBLITZ, WALTER JOHN, JR, b Kansas City, Mo, Sept 26, 20. PULP CHEMISTRY, PAPER CHEMISTRY. Educ: Univ Ariz, BS, 41; Lawrence Col, PhD(paper chem), 49. Prof Exp: Res chemist, Lithographic Tech Found, 49-50; chemist, Munising Paper Co, 50-52; res chemist, Kimberly Clark Corp, 52-59; res chemist, Duplicating Prod Div, Minn Mining & Mfg Co, 59-60 & Paper Prod Div, 60-66; ASSOC PROF PULP & PAPER RES, ORE STATE UNIV, 66- Mem: Tech Asn Pulp & Paper Indust; Can Pulp & Paper Asn. Res: Research and development of specialty paper products; technical problems of lithography; 3M Brand Action paper. Mailing Add: Forest Res Lab Ore State Univ Corvallis OR 97331

BUCCI, ENRICO, b Este, Italy, May 27, 32; US citizen; m 59; c 4. PHYSICAL BIOCHEMISTRY. Educ: Univ Rome, MD, 56, PhD(biochem), 62. Prof Exp: From vol asst to asst biochem, Univ Rome, 56-65; asst prof, Med Sch, Ind Univ, 65-69; assoc prof, 69-74, PROF BIOCHEM, SCH MED, UNIV MD, 74- Mem: Am Soc Biol Chemists; Am Chem Soc. Res: Structure function relationships and protein folding of the hemoglobin system as seen in the physico-biochemical behavior of the entire molecule, its isolated subunits and large peptides obtained from them. Mailing Add: Dept of Biochem Univ of Md Sch of Med Baltimore MD 21201

BUCCI, THOMAS JOSEPH, b Smithfield, RI, July 21, 34; m 57; c 3. COMPARATIVE PATHOLOGY. Educ: Univ Pa, VMD, 59; Univ Rochester, MS, 62; Univ Colo Med Ctr, PhD(path), 74. Prof Exp: US Army, 59-, chief, Vet Div, 5th US Army Med Lab, 59-61, vet pathologist, Div Nuclear Med, Walter Reed Army Inst Res, 65-66, chief, Path Div, US Army Med Res & Nutrit Lab, 66-69, CHIEF, DEPT COMP MED, LETTERMAN ARMY INST RES, 70- Mem: Am Col Vet Pathologists; Am Soc Exp Path; Int Acad Pathologists; Am Asn Pathologists & Bacteriologists; Am Vet Med Asn. Res: Exercise physiology; comparative nutrition; vitamin metabolism; reproduction of nonhuman primates. Mailing Add: Letterman Army Inst of Res Presidio San Francisco CA 94129

BUCCINO, ALPHONSE, b New York, NY, Mar 14, 31; m 53; c 1. SCIENCE ADMINISTRATION, ALGEBRA. Educ: Univ Chicago, BS, 58, MS, 59, PhD(math), 67. Prof Exp: Asst prof math, Roosevelt Univ, 61-63; from asst prof to assoc prof, DePaul Univ, 63-70; SCI ADMINR, NSF, 70- Mem: Am Math Soc; Math Asn Am. Res: Matrices and linear algebra. Mailing Add: Off of Prog Integration Nat Sci Found Washington DC 20550

BUCCINO, RAYMOND, JR, b Bridgeport, Conn, Aug 10, 33; m 60; c 2. CLINICAL CHEMISTRY. Educ: Fairfield Univ, BSS, 55; Univ Conn, BS, 58, PhD(biochem), 68. Prof Exp: Am Cancer Soc instnl grant & res assoc pharmacol, Col Med, Baylor Univ, 68-69; asst prof chem & head dept, Defiance Col, 69-72, CLIN CHEMIST, DEFIANCE HOSP, 72- Mem: Fel Am Inst Chemists. Res: Clinical enzymology; endocrinology. Mailing Add: Dept of Path Defiance Hosp Defiance OH 43512

BUCCINO, SALVATORE GEORGE, b New Haven, Conn, Dec 23, 33; m 69; c 1. NUCLEAR PHYSICS. Educ: Yale Univ, BS, 56; Duke Univ, PhD(nuclear physics), 63. Prof Exp: Res assoc nuclear physics, Duke Univ, 62-63; asst physicist, Argonne Nat Lab, 63-65; from asst prof to assoc prof, 65-70, PROF PHYSICS, TULANE UNIV, 70- Concurrent Pos: Consult, Radiation Lab, Aberdeen Proving Grounds, 73 & 74; vis scholar physics, Duke Univ, 74-75. Mem: Am Phys Soc; Sigma Xi. Res: Coulomb excitation; fast neutron physics, including scattering; fast time-of-flight techniques; gamma ray spectroscopy, particularly (p,m) reactions. Mailing Add: Dept of Physics Tulane Univ New Orleans LA 70118

BUCHACEK, ROBERT JOSEPH, b Mosinee, Wis, Aug 19, 40; m 71; c 1. INORGANIC CHEMISTRY. Educ: Wis State Univ, Stevens Point, BS, 62; Univ Iowa, PhD(inorg chem), 72. Prof Exp: Chemist, Inst Paper Chem, A B Dick Co, US Army, 62-66; RES CHEM, E I DU PONT DE NEMOURS & CO, INC, 74- Concurrent Pos: Fel, State Univ NY Buffalo, 72-73. Mem: Am Chem Soc. Res: Titanium dioxide chemistry. Mailing Add: E I du Pont de Nemours & Co Inc New Johnsonville TN 37185

BUCHAL, ROBERT NORMAN, b Passaic, NJ, Apr 14, 30; m 57; c 3. APPLIED MATHEMATICS. Educ: St Lawrence Univ, BS, 51; Univ Conn, MA, 52; NY Univ, PhD(math), 59. Prof Exp: Anal engr, Bendix Aviation Corp, 52-54; res asst, Inst Math Sci, NY Univ, 54-58; asst prof, Math Res Ctr, US Army, Univ Wis, 58-60; assoc mathematician, Argonne Nat Lab, 60-70; mathematician, Off Naval Res, 70-75; MATHEMATICIAN, AIR FORCE OFF SCI RES, 75- Mem: Am Math Soc. Res: Asymptotic solutions and generalized solutions of differential equations; wave equations; mathematical biology. Mailing Add: Air Force Off of Sci Res/NM Bolling AFB Washington DC 20332

BUCHAN, DOUGLAS JOHN, b Saskatoon, Sask, Apr 2, 21; m 49; c 4. INTERNAL MEDICINE. Educ: Univ Sask, BA, 42; Univ Man, MD, 46; FRCP. Prof Exp: From asst prof to assoc prof, 58-70, PROF MED, UNIV SASK, 70- Res: Gastroenterology. Mailing Add: Dept of Med Univ of Sask Saskatoon SK Can

BUCHAN, GEORGE COLIN, b Seattle, Wash, Aug 30, 27; m 62; c 2. MEDICINE, NEUROPATHOLOGY. Educ: McGill Univ, MD, CM, 58. Prof Exp: From asst prof to assoc prof, 67-68, PROF PATH & HEAD DIV NEUROPATH, MED SCH, UNIV ORE, 68- Concurrent Pos: Fel neuropath, Univ Wash, 63-65. Mem: Am Asn Neuropath. Res: Factors influencing myelination in the nervous system. Mailing Add: Div of Neuropath Med Sch Univ of Ore Portland OR 97201

BUCHAN, RONALD FORBES, b Concord, NH, Sept 24, 15; m 40; c 3. MEDICINE. Educ: Univ NH, AB, 36; McGill Univ, MD & CM, 42. Prof Exp: Sanitarian, City of Concord & Eastern Health Dist, NH, 38; chief med unit, Bur Indust Hyg, State Dept Health, Conn, 43-46; asst prof indust med, Sch Med, Yale Univ, 46-48, clin dir, Inst Occup Med & Hyg, 47-48; assoc clin prof indust med, Postgrad Med Sch, NY Univ, 49-57; assoc clin prof prev med, Tufts New Eng Med Ctr, Boston Univ, 58-74; CHIEF MED CONSULT, MEDISCREEN DIV, BIOSCREEN, INC, BUFFALO, NY, 74- Concurrent Pos: Dir, Small Plant Coop Med Serv, Hartford, 46; assoc med dir, Prudential Ins Co, NJ, 48, dir employee health, 49-57, med dir, Northeastern Univ, 57-74; mem sci adv bd, Biomed Panel, USAF, 60-63; chmn res adv comt, Brattleboro Retreat, Vt, 64-; mem, Permanent Comn & Int Asn Occup Health, 65-; del, World Health Assembly, 69; consult to indust & community health serv, 75- Mem: AAAS; fel Indust Med Asn; AMA; fel Am Acad Occup Med (treas, 55-57, vpres, 57-58, pres, 58-59); fel Am Col Prev Med. Mailing Add: 1 Walton Alley Strawbery Banke Portsmouth NH 03801

BUCHANAN, BOB BRANCH, b Richmond, Va, Aug 7, 37; m 65; c 2. BIOCHEMISTRY, MICROBIOLOGY. Educ: Emory & Henry Col, AB, 58; Duke Univ, PhD(microbiol), 62. Prof Exp: NIH fel biochem, 62-63, PROF CELL PHYSIOL, UNIV CALIF, BERKELEY, 63- Concurrent Pos: Guggenheim Found fel, 74. Mem: AAAS; Am Soc Biol Chemists; Am Soc Microbiol; NY Acad Sci; Am Soc Plant Physiol. Res: Bacterial metabolism; enzymology; photosynthesis. Mailing Add: Dept of Cell Physiol Univ of Calif Berkeley CA 94720

BUCHANAN, BRUCE G, b St Louis, Mo, July 7, 40; m 61; c 2. COMPUTER SCIENCE. Educ: Ohio Wesleyan Univ, AB, 61; Mich State Univ, MA & PhD(philos), 66. Prof Exp: Res assoc comput sci, 66-73, RES COMPUT SCIENTIST, STANFORD UNIV, 73- Concurrent Pos: NIH career develop award, 71-76. Mem: Asn Comput Mach; Philos Sci Asn; AAAS. Res: Application of artificial intelligence to scientific inference. Mailing Add: Heuristic Prog Proj Stanford Univ Comput Sci Dept Stanford CA 94305

BUCHANAN, DAVID HAMILTON, b Indiana, Pa, July 4, 42; m 67; c 2. ORGANOMETALLIC CHEMISTRY. Educ: Case Inst Technol, BS, 64; Univ Wis-Madison, PhD(org chem), 69. Prof Exp: NIH fel organometallic chem, Univ Calif, Berkeley, 68-71; ASST PROF CHEM, EASTERN ILL UNIV, 71- Mem: Am Chem Soc; AAAS. Res: Mechanisms of organometallic reactions of transition metals, especially those of biological or synthetic importance. Mailing Add: Dept of Chem Eastern Ill Univ Charleston IL 61920

BUCHANAN, DAVID ROYAL, b Mansfield, Ohio, Mar 28, 34; m 57; c 4. TEXTILE PHYSICS. Educ: Capital Univ, BSc, 56; Ohio State Univ, PhD(chem), 62. Prof Exp: Res chemist, Chemstrand Res Ctr, Inc, NC, 62-68; sr res chemist, Phillips Petrol Co, Okla, 68-70, proj mgr, Phillips Fibers Corp, 70-72, mgr textile testing, 72-75; ASSOC PROF DESIGN & ENVIRON ANAL, CORNELL UNIV, 75- Mem: Am Phys Soc;

Am Crystallog Asn; Am Chem Soc; Fiber Soc; Am Soc Testing & Mat. Res: Structure and properties of polymers, fibers and textiles; flammability of textiles; x-ray diffraction from textile materials; polymer rheology. Mailing Add: Dept of Design & Environ Anal Cornell Univ Ithaca NY 14853

BUCHANAN, EDWARD BRACY, JR, b Detroit, Mich, Sept 12, 27. ANALYTICAL CHEMISTRY. Educ: Univ Detroit, BS, 50, MS, 54; Iowa State Univ, PhD, 59. Prof Exp: Instr chem, Iowa State Univ, 59-60; asst prof, 60-67, ASSOC PROF CHEM, UNIV IOWA, 67- Mem: Am Chem Soc; Electrochem Soc. Res: Electrochemistry; design of instrumentation. Mailing Add: Dept of Chem Univ of Iowa Iowa City IA 52240

BUCHANAN, GEORGE DALE, b Wichita Falls, Tex, Oct 1, 28. SCIENCE EDUCATION, ANATOMY. Educ: Rice Inst Technol, BA, 50, MA, 54, PhD(biol), 56. Prof Exp: From asst prof to assoc prof biol, Houston State Col, 56-58; res assoc, Rice Inst Technol, 58-59; from instr to assoc prof anat, Med Units, Univ Tenn, 59-68, assoc prof anat, Med Col Ohio, 68-75; ASSOC PROF NURSING & ANAT, FAC HEALTH SCI, McMASTER UNIV, 75- Concurrent Pos: Vis prof morphol & biol, Univ Valle, Columbia, 66-67. Mem: AAAS; Am Asn Anat; Endocrine Soc; Brit Soc Study Fertil; Soc Study Reproduction. Res: Reproductive biology; blastocyst implantation; reproduction in ferrets. Mailing Add: Sch of Nursing McMaster Univ 1200 Main St W Hamilton ON Can

BUCHANAN, GERALD WALLACE, b London, Ont, Mar 1, 43; m 67; c 1. ORGANIC CHEMISTRY, NUCLEAR MAGNETIC RESONANCE. Educ: Univ Western Ont, BSc, 65, PhD(org chem), 69. Prof Exp: Nat Res Coun Can fel, 69-70; vis asst prof chem, Univ Windsor, 70-71; ASST PROF CHEM, CARLETON UNIV, 71- Mem: Am Chem Soc; Chem Inst Can. Res: Conformational analysis of organic molecules; carbon-13 and proton nuclear magnetic resonance; organic reaction mechanisms; nuclear magnetic resonance spectroscopy. Mailing Add: Dept of Chem Carleton Univ Ottawa ON Can

BUCHANAN, HAYLE, b Teasdale, Utah, Apr 3, 25; m 53; c 4. PLANT ECOLOGY. Educ: Brigham Young Univ, BA, 51, MA, 53; Univ Utah, PhD(bot), 60. Prof Exp: Teacher & prin, Tabiona & Hurricane Latter-day Saints Sem, 53-56; teacher high sch, 58-60; asst prof bot, Church Col Hawaii, 60-62; instr life sci & bot, Am River Jr Col, 62-65; from asst prof to assoc prof, 65-71, PROF BOT, WEBER STATE COL, 71- Concurrent Pos: Partic range mgt res, US Forest Serv, 66- Mem: Ecol Soc Am; Nat Parks Asn. Res: Ecological study of an area for outdoor education; management of forests in national parks; ecological life history studies. Mailing Add: Dept of Bot 3730 Harrison Blvd Weber State Col Ogden UT 84403

BUCHANAN, HUGH, b New York, NY, Mar 20, 31; m 55; c 1. GEOLOGY, PALEONTOLOGY. Educ: Columbia Univ, BS, 61, PhD(geol), 70. Prof Exp: Res geologist, Explor Res Div, Continental Oil Co, 65-67; instr geol, Washington & Lee Univ, 67-68; from geologist to res geologist, Explor Div, Gulf Res & Develop Co, 68-74; MEM FAC, DEPT GEOL & GEOG, WVA UNIV, 74- Mem: Geol Soc Am; Paleont Soc; Soc Econ Paleontologists & Mineralogists. Res: Invertebrate paleontology; sedimentology; stratigraphy; paleoecology; paleobiogeography; carbonate geology. Mailing Add: Dept of Geol & Geog WVa Univ Morgantown WV 26506

BUCHANAN, JAMES BALFOUR, b Vancouver, BC, Mar 31, 21. ORGANIC CHEMISTRY Educ: 0Univ BC, BA, 44, MS, 46; Cornell Univ, PhD(chem), 50. Prof Exp: Res assoc, Cornell Univ, 50-52; from res chemist to sr res chemist, 52-69, RES ASSOC, E I DU PONT DE NEMOURS & CO, INC, 69- Mem: Am Chem Soc. Res: Structural studies; chemotherapeutic drugs; new agricultural chemicals. Mailing Add: E I du Pont de Nemours & Co Inc Wilmington DE 19898

BUCHANAN, JAMES WESLEY, b Union Mills, NC, May 5, 37; m 59; c 2. PHYSICAL CHEMISTRY, INORGANIC CHEMISTRY. Educ: Univ NC, AB, 59; Univ Fla, MS, 62, PhD(radiation chem), 68. Prof Exp: Instr chem, Ga Col Milledgeville, 62-63; head dept, Gaston Jr Col, 64-66; asst prof, 68-70, ASSOC PROF CHEM, SALEM COL, NC, 70-, CHMN DEPT CHEM & PHYSICS, 71- Mem: Am Chem Soc. Res: Radiation chemistry of both liquid and gas phase; radiolysis of ammonia-carbon tetrachloride solutions; radiolysis of phosphine and phosphine-containing gas mixtures; radiolysis of cyclohexane. Mailing Add: Dept of Chem & Physics Salem Col Winston-Salem NC 27108

BUCHANAN, JOHN DONALD, b Mesa, Ariz, Oct 1, 27; m 55; c 4. RADIOCHEMISTRY, HEALTH PHYSICS. Educ: Univ Ariz, BS, 49. Prof Exp: Radiochemist, Tracerlab, Inc, 50-56, sr chemist & sect head, 56-59; sr chemist, Gen Atomic, Inc, 59-62; sr radio- chemist, Hazelton/Nuclear Sci Corp, 62-67, mgr nuclear measurements dept, 65-67, mgr nuclear prod, appln & measurements, Palo Alto Labs, Teledyne-Isotopes, Inc, 67-71; mgr appl res, Int Nutronics, Inc, 71-73; supvr, Radiol Monitoring Progs, NUS Corp, 73-75; RADIOCHEMIST, US NUCLEAR REGULATORY COMN, 75- Mem: AAAS; Am Chem Soc; Am Nuclear Soc; Health Physics Soc; Am Inst Chemists. Res: Radiochemical analysis; environmental radioactivity, radioassay, radiological monitoring, nuclear regulatory standards.

BUCHANAN, JOHN MACHLIN, b Winamac, Ind, Sept 29, 17; m 48; c 4. BIOCHEMISTRY. Educ: DePauw Univ, BS, 38; Univ Mich, MS, 39; Harvard Univ, PhD(biochem), 43. Hon Degrees: DSc, Univ Mich, 61 & DePauw Univ, 75. Prof Exp: Instr biochem, Univ Pa, 43-46; Nat Res Coun fel med, Nobel Inst, Stockholm, 46-48; from asst to prof biochem, Univ Pa, 48-53; prof biochem & head div, 53-67, WILSON PROF BIOCHEM, MASS INST TECHNOL, 67- Concurrent Pos: Sabbatical leave, Salk Inst Biol Studies, 64-65. Mem fel comt, Div Med Sci, Nat Res Coun, 54-63, mem subcomt nomenclature of biochem, 60-66; mem biochem study sect, NIH, 59-63, chmn, 61-63; mem nat comt, Int Union Biochemists, 57-60, 63-69, secy-treas, 63-65, vchmn, 65-66; mem sci adv bd, St Jude Res Hosp, 64-65; mem bd, Fedn Am Socs Exp Biol, 64-67; mem, Biochem Training Grant Comt, 65-69. Honors & Awards: Eli Lilly Award Biol Chem, Am Acad Arts & Sci, 51. Mem: Nat Acad Sci; Am Soc Biol Chemists (secy, 69-72); Am Chem Soc; Am Acad Arts & Sci; Sigma Xi. Res: Synthesis of glycogen; oxidation of fatty acids and ketone bodies; synthesis of purines; isolation and purification of enzymes. Mailing Add: 56 Meriam St Lexington MA 02173

BUCHANAN, MARION ALEXANDER, b Toulon, Ill, Sept 14, 08; m 41; c 2. CHEMISTRY. Educ: Ill Col, AB, 30; Univ Iowa, MS, 31, PhD(org chem), 33. Prof Exp: Res assoc, Inst Paper Chem, 34-70, sr res assoc, 70-74; RETIRED. Mem: Am Chem Soc; Tech Asn Pulp & Paper Indust. Res: Lignin; tannin; coloring matters of wood; wood chemistry. Mailing Add: 501 E Grant St Appleton WI 54911

BUCHANAN, MARION LYNN, b Chelsa, Kans, Oct 17, 13; m 35; c 3. ANIMAL HUSBANDRY. Educ: Okla Agr & Mech Col, BS, 35; WVa Univ, MS, 37. Prof Exp: Asst animal husb, WVa Univ, 35-37, instr, 37-40, asst, 40; assoc prof, Univ Ga, 41-42; asst prof, Univ Idaho, 42-45; assoc prof, 45-46, PROF ANIMAL HUSB, NDAK STATE UNIV, 46- Concurrent Pos: Agr develop agent, Great North Rwy. Mem: AAAS; Am Soc Animal Sci; Genetics Soc Am. Res: Breeding and nutrition of farm livestock; genetics; application of statistical techniques to various fields; nutrition; general agriculture; animal breeding with emphasis on beef cattle improvement. Mailing Add: Dept of Animal Husb NDak State Univ Fargo ND 58102

BUCHANAN, MARY L, mathematics, deceased

BUCHANAN, O LEXTON, JR, b Atlanta, Ga, Jan, 31. MATHEMATICS EDUCATION. Educ: Ga State Col, BBA, 52; Univ Ga, MEd, 57; Univ Kans, MA, 62, PhD(math educ), 64. Prof Exp: Teacher high schs, Ga, 55-57; instr math, Univ Ga, 57-58; teacher high schs, Ga, 59-61; teaching asst, Univ Kans, 58-59 & math & educ, 61-64; asst prof math, Univ SC, 64-71; teacher high schs, Ga, 71-75, head, Dept Math, Fulton County Schs, 71-75, TEACHER GIFTED STUDENTS, FULTON COUNTY SCHS, 75- Concurrent Pos: Consult elem & high schs, SC & Ga. Mem: Math Asn Am; Nat Coun Teachers Math; Sigma Xi. Mailing Add: 1314 Hayes Dr Smyrna GA 30080

BUCHANAN, ROBERT ALEXANDER, b Detroit, Mich, Sept 8, 32; m 62; c 3. PEDIATRICS, PHARMACOLOGY. Educ: Univ Mich, MD, 57; Am Bd Pediat, dipl, 64. Prof Exp: Intern, Philadelphia Gen Hosp, Pa, 57-58; resident pediat, Med Ctr, Univ Mich, 60-62; pvt pract, 62-66; investr, 66-67, asst dir, 67-68, assoc dir, 69-74, DIR CLIN RES DEPT, PARKE, DAVIS & CO, 74- Concurrent Pos: Sr exam, Fed Aviation Agency, 63-; dir, Guthrie Clin Res Found, 68-; clin asst prof pediat, Med Sch, Univ Mich, 68- Mem: AMA; fel Am Acad Pediat; Am Therapeut Soc; Biomed Eng Soc. Res: Pediatric pharmacology, especially new drug development in anticonvulsants and antibiotics; medical electronic diagnostic instrument development. Mailing Add: Clin Res Dept Parke, Davis & Co Ann Arbor MI 48106

BUCHANAN, ROBERT AMBROSE, b Lodi, Calif, Jan 2, 31; m 52; c 2. PHYSICS. Educ: La Sierra Col, BA, 53; Univ Calif, Los Angeles, MA, 60, PhD(physics), 61. Prof Exp: Physicist, US Naval Ord Lab, 53-55; asst, Univ Calif, Los Angeles, 55-61; res scientist, Infrared Div, US Naval Ord Lab, 61-66; head magnetics group, Electronic Sci Lab, 66-70, head electro-optics group, Phys Sci Lab, 70-74, HEAD MAT EVAL GROUP, LOCKHEED RES LAB, 74- Mem: Am Phys Soc. Res: Solid state physics; solid state spectroscopy of lattice vibrations in crystals; electronic transitions of rare-earth ions in crystals; x-ray phosphors; x-ray imaging systems. Mailing Add: Lockheed Res Lab Dept 52-32 B201 3521 Hanover St Palo Alto CA 94304

BUCHANAN, ROBERT LESTER, JR, b Seattle, Wash, July 3, 46; m 74; c 1. FOOD SCIENCE. Educ: Rutgers Univ, BS, 69, MS, 71, MPhil, 72, PhD(food sci), 74. Prof Exp: Res assoc food microbiol, Dept Food Sci, Univ Ga, 74-75; ASST PROF FOOD SCI & NUTRIT, DEPT NUTRIT & FOOD, DREXEL UNIV, 75- Concurrent Pos: Dawson fel, Col Agr, Univ Ga, 74-75. Mem: Inst Food Technologists; Am Soc Microbiol; Sigma Xi. Res: Regulation of bacterial and fungal exotoxin production. Mailing Add: Dept of Nutrit & Food Drexel Univ Philadelphia PA 19104

BUCHANAN, RONALD JAMES, b Regina, Sask, Sept 15, 38; m 61; c 2. AQUATIC ECOLOGY, PHYCOLOGY. Educ: Univ BC, BS, 61, PhD(oceanog), 66. Prof Exp: Res assoc biol, Johns Hopkins Univ, 66-68; sr res assoc limnol, Univ Wash, 68-69; res officer aquatic ecol, 69-71, sr biologist, 71-74, DIV CHIEF ENVIRON STUDIES DIV, WATER INVESTS BR, WATER RESOURCES SERV, GOVT BC, 74- Mem: Phycol Soc Am; Am Soc Limnol & Oceanog; Ecol Soc Am; Am Fisheries Soc; AAAS. Res: Taxonomy and ecology of marine and freshwater planktonic algae and protozoa; lake eutrophication; planktonic microbiota as water quality indicators; watershed management for water quality control. Mailing Add: Water Invests Br Water Res Serv Parliament Bldgs Victoria BC Can

BUCHANAN, RONALD LESLIE, b Kirkland Lake, Ont, May 15, 37; m 61; c 3. MEDICINAL CHEMISTRY. Educ: Univ Western Ont, BSc, 59, PhD(org chem), 63. Prof Exp: Res fel, Univ BC, 63-64; SR RES CHEMIST, BRISTOL LABS, 64- Mem: Am Chem Soc. Res: Synthesis of biologically interesting monfluorinated compounds; biogenetic-type synthesis of natural products; synthesis of general medicinal agents. Mailing Add: Dept of Med Chem Bristol Labs Syracuse NY 13201

BUCHANAN, THOMAS STEWART, forest pathology, see 12th edition

BUCHANAN, W C, b Waverly, Tenn, Sept 18, 17; m 43; c 2. GEOGRAPHY. Educ: Austin Peay State Col, BS, 47; George Peabody Col, MA, 50. Prof Exp: From instr to assoc prof geog, Northeast La State Col, 48-67; ASSOC PROF GEOG & SOCIOL, COLUMBIA COL, SC, 67- Mem: Asn Am Geogr; Nat Coun Geog Educ. Res: Cultural anthropology and geography; geography of Anglo America. Mailing Add: Dept of Human Rels Columbia Col Columbia SC 29203

BUCHANAN-DAVIDSON, DOROTHY JEAN, b Monmouth, Ill, Dec 22, 25; m 57; c 3. BIOCHEMISTRY. Educ: Wash State Col, BS, 47; Univ Cincinnati, MS, 49, PhD(biochem), 51. Prof Exp: Asst inorg chem, Univ Cincinnati, 47-48; instr biochem, Sch Med, Vanderbilt Univ, 50-53; Nat Res Coun fel, Lister Inst, London, 53-55; USPHS fel immunochem, Pasteur Inst, Paris, 55-56; proj assoc biochem, Univ Wis, 56-60; ABSTRACTOR, CHEM ABSTR, 60-; INSTR, MADISON AREA TECH COL, 74- Concurrent Pos: Consult ed, Mother's Manual, 64-69; proj specialist, Water Resources Ctr, Univ Wis, 75- Mem: Sigma Xi. Res: Immunochemistry; chemistry of blood-group-specific substances and meconium; phosphorous poisoning; x-irradiation; immunochemical agar-gel diffusion techniques; immunochemistry and biological properties of synthetic polypeptides and modified proteins; immunochemistry of plant pro- teins. Mailing Add: 6278 Sun Valley Pkwy Oregon WI 53575

BUCHANAN-SMITH, JOCK GORDON, b Edinburgh, Scotland, Mar 9, 40; m 64. ANIMAL SCIENCE, ANIMAL NUTRITION. Educ: Aberdeen Univ, BSc, 62; Iowa State Univ, BS, 63; Tex Tech Col, MS, 65; Okla State Univ, PhD(animal sci), 69. Prof Exp: Res asst animal sci, Tex Tech Col, 64-65; res asst, Okla State Univ, 65-69; ASST PROF ANIMAL SCI, UNIV GUELPH, 69- Concurrent Pos: Nat Res Coun Can operating grants, 69-75; Agr Can operating grant, 75-76. Mem: AAAS; Am Soc Animal Sci; Am Dairy Sci Asn; Agr Inst Can. Res: Utilization of silages by ruminants; glucose metabolism and depression of milk fat in ruminants; dewatering of forages. Mailing Add: Dept of Animal & Poultry Sci Univ of Guelph Guelph ON Can

BUCHANNON, KENNETH WILLIAM, b Dauphin, Man, Sept 26, 21; m 42; c 1. PLANT BREEDING, GENETICS. Educ: Univ Man, BSA, 50, MSc, 52; Univ Sask, PhD, 61. Prof Exp: Barley breeding, 53-73, HEAD CEREAL BREEDING SECT, RES STA, CAN DEPT AGR, 73- Concurrent Pos: Adv wheat breeding, Can Int Develop Agency, Njoro, Kenya, E Africa, 69-71. Mem: Agr Inst Can; Genetics Soc Can; Can Soc Agron. Res: Breeding superior varieties of barley; genetics studies of disease reaction in barley. Mailing Add: Res Sta Can Dept of Agr 25 Dafoe Rd Winnipeg MB Can

BUCHDAHL, ROLF, b Frankfurt, Ger, Nov 2, 14; nat US; m 40; c 2. PHYSICS. Educ: Johns Hopkins Univ, PhD(physics), 40. Prof Exp: Res physicist, Sun Chem Corp, 41-

46; group leader chg physics & phys chem high polymers, Monsanto Chem Co, 46-54, sect leader chg physics, phys chem & explor res high polymers, 54-56, asst res dir fundamental & explor res polymers, 56-60; dir basic res, Chemstrand Res Ctr, Inc, NC, 60-68; DIR FUNDAMENTAL & EXPLOR RES, CORP RES DEPT, MONSANTO CO, 68- Concurrent Pos: Affil prof, Sch Appl Sci & Eng, Wash Univ; mem adv panel, Inst Mat Res, Nat Bur Standards, 71-74; mem adv panel eng mat, NSF, 74-75, metall & mat, 75-76. Mem: AAAS; Fiber Soc; fel Am Phys Soc; Am Chem Soc; The Chem Soc. Res: Solid state properties of high polymers; environmental problems. Mailing Add: Corp Res Dept Monsanto Co 800 N Lindbergh Blvd St Louis MO 63166

BUCHENAU, GEORGE WILLIAM, b Amarillo, Tex, May 23, 32; m 55; c 2. PLANT PATHOLOGY. Educ: NMex State Univ, BS, 54, MS, 55; Iowa State Univ, PhD(plant path), 60. Prof Exp: Asst prof, 59-63, ASSOC PROF PLANT PATH, SDAK STATE UNIV, 63- Mem: Am Phytopath Soc. Res: Wheat disease; resistance and chemical control. Mailing Add: Dept of Plant Sci SDak State Univ Brookings SD 57006

BUCHER, GORDON EDWARDS, b Toronto, Ont, Mar 28, 17; m 49; c 1. ENTOMOLOGY. Educ: Univ Toronto, BA, 37, MA, 39; Ohio State Univ, PhD(entom), 46. Prof Exp: RES SCIENTIST, CAN AGR, 46- Mem: Entom Soc Am; Soc Invert Path; Can Soc Microbiologists; Entom Soc Can. Res: Chalicid morphology and anatomy; insect physiology, particularly effects of temperature; parasitic insects and biological control methods; insect diseases and biological control; integrated control of insect pests of rape. Mailing Add: 159 Thatcher Dr Winnipeg MB Can

BUCHER, NANCY L R, b Baltimore, Md, May 4, 13. MEDICINE. Educ: Bryn Mawr Col, AB, 35; Johns Hopkins Univ, MD, 43. Prof Exp: Intern, Mass Mem Hosp, Boston, 43-44, clin fel, 44-45; res fel & assoc med, 45-59, clin assoc, 59-68, asst clin prof med, 68-72, ASSOC PROF MED, HARVARD MED SCH, 72-, ASSOC BIOLOGIST, MASS GEN HOSP, 52- Mem: AAAS; Soc Develop Biol; Am Asn Cancer Res; Am Physiol Soc; Am Soc Cell Biol. Res: Cancer; growth regulation; cholesterol biosynthesis; liver regeneration. Mailing Add: Mass Gen Hosp Boston MA 02114

BUCHHEIM, ARNO FRITZ GÜNTHER, b Döbeln, Ger, Oct 26, 24; m 53; c 2. SYSTEMATIC BOTANY. Educ: Free Univ Berlin, Dr rer nat(syst bot), 53. Prof Exp: Sci asst bot, Aachen Tech Univ, 54; taxon botanist, Bot Garden & Mus, Berlin-Dahlem, Ger, 55-63; BIBLIOGRAPHER, HUNT INST BOT DOC, CARNEGIE-MELLON UNIV, 63- Concurrent Pos: Secy subcomt for family names, Int Bot Cong, 59-, mem comt for Spermatophyta, 64-, standing comt on stabilization, 64- Mem: AAAS; Int Asn Plant Taxon; Int Soc Hort Sci. Res: Bibliography of botanical periodicals of all times and botanical books of the period 1730-1840; botanical nomenclature; taxonomy of higher plants. Mailing Add: Hunt Inst for Bot Doc Carnegie-Mellon Univ Pittsburgh PA 15213

BUCHHOLZ, ALLAN C, b Manchester, NH, July 20, 40; m 63; c 2. PHYSICAL ORGANIC CHEMISTRY, POLYMER CHEMISTRY. Educ: Univ Mass, BS, 62; Univ Ill, MSc, 64, PhD, 67. Prof Exp: Sr res chemist, 67-73, SUPVR INDUST TAPE DIV, 3M CO, 73- Mem: Am Chem Soc. Res: Preparation, nuclear magnetic resonance and mass spectral studies of carbon-13-labeled compounds; polymer synthesis and characterization, especially light-scattering of macromolecular systems; isocyanate and polyurethane chemistry; development and economic evaluation of new products and applications. Mailing Add: 3M Co 3M Ctr PO Box 33221 St Paul MN 55133

BUCHHOLZ, JEFFREY CARL, b Sheboygan, Wis, June 10, 47; m 70. SURFACE PHYSICS. Educ: Univ Wis, Eau Claire, BS, 69; Univ Wis, Madison, MS, 71 & 73, PhD(mat sci), 74. Prof Exp: Res assoc chem, Univ Calif, Berkeley, 74-76; ASSOC SR RES PHYSICIST, GEN MOTORS RES LABS, 76- Mem: Am Vacuum Soc; Am Phys Soc; Sigma Xi. Res: Study of the solid-gas and solid-liquid interface; adsorption on the solid surface using various electron and photon spectroscopies. Mailing Add: Physics Dept Gen Motors Res Labs Warren MI 48090

BUCHHOLZ, ROBERT E, b Milwaukee, Wis, July 5, 21; m 42; c 2. DENTAL RADIOLOGY, ORAL MEDICINE. Educ: Univ Minn, DDS, 50; State Univ Iowa, MS, 62. Prof Exp: Dentist pvt pract, 50-53; ASSOC PROF DENT RADIOL & DIAG, UNIV MICH, ANN ARBOR, 71- Mem: Am Dent Asn; Am Col Dentists; Am Acad Oral Radiol; Teachers Oral Diag. Mailing Add: Sch of Dent Univ of Mich Ann Arbor MI 48109

BUCHHOLZ, ROBERT HENRY, b Wakeeney, Kans, Aug 26, 24; m 47; c 3. PHYSIOLOGY. Educ: Ft Hays Kans State Col, BS, 49; Kans State Col, MS, 50; Univ Mo, PhD, 57. Prof Exp: Asst prof zool, 50-57, assoc prof biol, 57-63, PROF BIOL, MONMOUTH COL, ILL, 63- Concurrent Pos: Mem staff, Argonne Nat Lab, 62-63; NIH fel, Dept Physiol, Univ Edinburgh, 63-64. Mem: AAAS; Am Soc Zoologists; Am Soc Mammal; assoc mem Am Physiol Soc. Res: Gastrointestinal physiology; digestive enzymes in bat, rat and mole; pancreatic lipase in the mole and rat. Mailing Add: Dept of Biol Monmouth Col Monmouth IL 61462

BUCHI, GEORGE, b Baden, Switz, Aug 1, 21. ORGANIC CHEMISTRY. Educ: Swiss Fed Inst Technol, DSc, 47. Prof Exp: Firestone fel chem, Univ Chicago, 48-49, instr, 49-51; from asst prof to prof, 51-71, CAMILLE DREYFUS PROF CHEM, MASS INST TECHNOL, 71- Honors & Awards: Ruzicka Award, 57; Fritzche Award, 58; Am Chem Soc Award, 73. Mem: Nat Acad Sci. Res: Synthetic organic chemistry; natural products; free radical reactions. Mailing Add: Dept of Chem Mass Inst Technol 77 Massachusetts Ave Cambridge MA 02139

BUCHI, JULIUS RICHARD, b Porto Alegre, Brazil, Jan 31, 24; US citizen; m 58. MATHEMATICS. Educ: Swiss Fed Inst Technol, dipl, 48, DSc, 49. Prof Exp: Instr math, Ripon Col, 49-50; instr, Univ Mich, 50-55; asst prof math, Univ Ill, 55-57; guest prof, Univ Notre Dame, 57-58; assoc res mathematician, Univ Mich, 58-60, res mathematician, 60-61; prof appl math, Univ Mainz, 61-62; res mathematician, Univ Mich, 62-63; PROF MATH & COMPUT SCI, PURDUE UNIV, WEST LAFAYETTE, 63- Concurrent Pos: Swiss Rotary Club fel, Univ Chicago, 49; Fulbright lectr, Ger, Spain & Poland, 61-62; NSF grant, 64-; prof, Ohio State Univ, 64-65. Mem: Am Math Soc; Asn Symbolic Logic. Res: Lattice theory; set theory; mathematical logic; theory of automata. Mailing Add: Dept of Math Purdue Univ Lafayette IN 47907

BUCHIN, IRVING D, b New York, NY, Feb 17, 20; m 42; c 2. ORTHODONTICS. Educ: NY Univ, DDS, 43; Am Bd Orthod, dipl, 53. Prof Exp: Vis asst prof orthod, Col Med & asst prof, Grad Sch Dent, Boston Univ, 60-63, vis assoc prof & assoc prof, 63-68; assoc prof, 68-74, ADJ PROF ORTHOD, SCH DENT MED, UNIV PA, 74- Concurrent Pos: Instr, Exam & mem bd dir, Charles H Tweed Found Orthod Res, 53-; sect ed, Am J Orthod, 74-; mem orthod adv bd, Dept Health, New York. Mem: Fel Am Col Dent; NY Acad Sci; Int Asn Dent Res. Res: Clinical orthodontics; cephalometrics and the effects of clinical orthodontic procedures on growth and development on the dento-facial complex.

BÜCHLER, ALFRED, b Vienna, Austria, Jan 17, 27. PHYSICAL CHEMISTRY, HIGH TEMPERATURE CHEMISTRY. Educ: Univ Colo, BA, 49, MA, 51; Harvard Univ, PhD, 60. Prof Exp: Phys chemist, Arthur D Little, Inc, Mass, 55-63, group leader, 63-67; res assoc, Avco-Everett Res Lab, 67-69; vis lectr, Dept Mat Sci & Eng, Univ Calif, Berkeley, 70-71, res fel, Lawrence Berkeley Lab, Univ Calif, 72-74, RES EN 72-74, res engr, Space Sci Lab, Univ Calif, 73-75; NAT RES COUN SR RES ASSOC, NASA AMES RES CTR, 75- Mem: Am Phys Soc; Am Chem Soc; The Chem Soc. Res: High-temperature thermodynamics and molecular structure; mass spectrometry; infrared spectroscopy; combustion. Mailing Add: NASA Ames Research Ctr MS 223-6 Moffett Field CA 94035

BUCHLER, EDWARD RAYMOND, b Chicago, Ill, Sept 6, 42; m 65; c 3. ETHOLOGY. Educ: Calif State Polytech Col, BS, 64; Univ Calif, Santa Barbara, MA, 66; Univ Mont, PhD(zool), 72. Prof Exp: Res zoologist, US Army, Ft Detrick, MD, 66-68; res assoc ethology, Rockefeller Univ, 72-75; ASST PROF ETHOLOGY, UNIV MD, COLLEGE PARK, 75- Honors & Awards: Original Res Contribs Achievement Award, US Army, 68. Mem: AAAS; Animal Behav Soc; Am Soc Mammalogists; Am Asn Zool Parks & Aquaria. Res: Acoustic communication systems in animals; echolocation in small mammals; hunting strategies, prey selection and communication in insectivorous bats. Mailing Add: Dept of Zool Univ of Md College Park MD 20742

BUCHLER, IRA RICHARD, b New York, NY, Mar 6, 38; m 70. ANTHROPOLOGY. Educ: NY Univ, BA, 61; Univ Pittsburgh, PhD(anthrop), 64. Prof Exp: Instr anthrop, Boston Univ, 64-65; from asst prof to assoc prof, 65-73, PROF ANTHROP, UNIV TEX, AUSTIN, 73- Concurrent Pos: Off Naval Res grants, 66 & 68-69; NIH spec fel, Col France, 71-72. Mem: Fel Am Anthrop Asn. Res: Mathematical analysis of cultural systems; social structure; structural analysis of myth; ethnoscience. Mailing Add: Dept of Anthrop Univ of Tex Austin TX 78712

BUCHMAN, DAVID T, animal nutrition, see 12th edition

BUCHMAN, ELWOOD, medicine, see 12th edition

BUCHNEA, DMYTRO, organic chemistry, see 12th edition

BUCHSBAUM, DAVID ALVIN, b New York, NY, Nov 6, 29; m 49; c 3. MATHEMATICS. Educ: Columbia Col, NY, AB, 49; Columbia Univ, PhD, 54. Prof Exp: Instr math, Princeton, 53-54, NSF fel, 54-55; instr math, Univ Chicago, 55-56; asst prof, Brown Univ, 56-59, assoc prof, 59-60; assoc prof, 60-63, PROF MATH, BRANDEIS UNIV, 63- Concurrent Pos: NSF fel, 60-61. Mem: Am Math Soc. Res: Foundations and applications to commutative ring theory of homological algebra. Mailing Add: Dept of Math Brandeis Univ Waltham MA 02154

BUCHSBAUM, RALPH, b Chickasha, Okla, Jan 2, 07; m 33; c 2. INVERTEBRATE ZOOLOGY, ECOLOGY. Educ: Univ Chicago, BS, 28, PhD(zool), 32. Prof Exp: Asst zool, Univ Wis, 28-29; asst, Univ Chicago, 29-30, from instr biol to asst prof zool, 31-42, asst prof, 45-47, res assoc, Inst Radiobiol & Biophys, 47-50; prof, 50-71, EMER PROF BIOL, UNIV PITTSBURGH, 71- Concurrent Pos: Instr, Gary Col, 33-35; cur, Mus Sci & Indust, Univ Chicago, 45-47; Condon lectr, 54; hon res assoc, Carnegie Mus, 55-; chief adv biol, Encycl Britannica Films, 59-; Fulbright award, Thailand, 59-60; dir, Ctr Studies Learning, 62; mem staff, UNESCO educ mission, India, 64, chief consult biol teaching, Africa, 65-66; US AID-Univ Pittsburgh Ecuador Educ Proj, 65-67. Honors & Awards: Univ Chicago Prize, 40. Mem: Am Soc Zoologists; Ecol Soc Am; NY Acad Sci; AAAS; Biol Photog Asn. Res: Invertebrate biology; ecology of invertebrates; effects of irradiating living cells with various radiations in tissue cultures; cellular physiology; perfusion techniques in tissue culture; invertebrate physiology; ecology. Mailing Add: 183 Ocean View Blvd Pacific Grove CA 93950

BUCHSBAUM, SOLOMON JAN, b Stryj, Poland, Dec 4, 29; nat US; m 55; c 3. PHYSICS. Educ: McGill Univ, BS, 52, MSc, 53; Mass Inst Technol, PhD(physics), 57. Prof Exp: Asst physics, Mass Inst Technol, 53-55, staff mem, Res Lab Electronics, 57-58; mem tech staff, Bell Tel Labs, 58-65, head dept, 61-65, dir electronics res lab, 65-68; vpres res, Sandia Labs, 68-71; EXEC DIR RES COMMUN SCI DIV, BELL LABS, 71- Concurrent Pos: Assoc ed, Physics of Fluids, 63-64 & Rev of Mod Physics, 68-72; mem AEC standing comt controlled thermonuclear res, 65-72; chmn div plasma physics, Am Phys Soc, 68; mem, President's Sci Adv Comt, 70-73; consult, Nat Security Coun, 73-; mem fusion power coord comt, Energy Res & Develop Admin, 73-; chmn, Defense Sci Bd, 73-; mem exec comt, Assembly Eng, Nat Res Coun, 75- Mem: Nat Acad Sci; fel Am Phys Soc; fel Inst Elec & Electronic Engrs; fel Am Acad Arts & Sci; fel AAAS. Res: Plasma physics; gaseous electronics; plasmas in solids; controlled thermonuclear fusion; optical communications; communications systems. Mailing Add: Bell Labs Holmdel NJ 07733

BUCHTA, RAYMOND CHARLES, b Philadelphia, Pa, Dec 17, 42; m 68; c 1. ANALYTICAL CHEMISTRY. Educ: Pa State Univ, BS, 65; Univ Wis-Madison, PhD(anal chem), 69. Prof Exp: Technician, Rohm and Haas Co, 62-63, chemist, 65; res chemist, 69-75, SR RES CHEMIST, JACKSON LAB, E I DU PONT DE NEMOURS & CO, INC, 75- Mem: Am Chem Soc. Res: Mechanism of the electrochemical reduction of 1,3-diketones in aprotic solvents; electrolytic autocatalysis; trace moisture analysis; electrochemical detection in liquid chromatography; analysis of dyes and intermediates; auto exhaust analysis. Mailing Add: Jackson Lab E I du Pont de Nemours & Co Inc Wilmington DE 19898

BUCHWALD, CARYL EDWARD, b Medford, Mass, Oct 15, 37; m 59; c 3. ENVIRONMENTAL GEOLOGY. Educ: Union Col, BS, 60; Syracuse Univ, MS, 63; Univ Kans, PhD(geol), 66. Prof Exp: Teaching fel geol, McMaster Univ, 66-67; asst prof, 67-71, ASSOC PROF GEOL, CARLETON COL, 71- Mem: Geol Soc Am; Soc Econ Paleont & Mineral; Nat Asn Geol Teachers. Res: River hydrology; land use planning. Mailing Add: Dept of Geol Carleton Col Northfield MN 55057

BUCHWALD, HENRY, b Vienna, Austria, June 21, 32; US citizen; m 54; c 4. SURGERY. Educ: Columbia Univ, BA, 54, MD, 57; Univ Minn, Minneapolis, MS & PhD(surg), 66. Prof Exp: Intern, Columbia Presby Med Ctr, 57-58; from instr to asst prof, 65-70, ASSOC PROF SURG, UNIV MINN, MINNEAPOLIS, 70- Concurrent Pos: Helen Hay Whitney fel, 63-65; estab investr, Am Heart Asn, 65-70, mem coun arteriosclerosis, 66. Honors & Awards: Schering Award, 57; Sam D Gross Award, 67; Cine Clins Award, Am Col Surg, 69. Mem: AAAS; Am Col Surg; Soc Univ Surg; Asn Acad Surg; Am Therapeut Asn. Res: Atherosclerosis and the hyperlipidemias; development of surgical procedures for lipid reduction; cholesterol and bile acid metabolism; implantable infusion devices; hemodynamic rheology; dumping syndrome. Mailing Add: Box 290 Univ Of Minn Hosp Minneapolis MN 55455

BUCHWALD, JENNIFER S, b Okmulgee, Okla, Oct 20, 30; m 52; c 3. NEUROPHYSIOLOGY. Educ: Lindenwood Col, AB, 51; Tulane Univ, PhD(neuroanat), 59. Hon Degrees: LLD, Lindenwood Col, 70. Prof Exp: From asst prof to assoc prof physiol, 65-73, PROF PHYSIOL, SCH MED, UNIV CALIF, LOS ANGELES, 73- Concurrent Pos: Consult, NSF & NIMH; Ment Health Training Prog fel, 61; Parkinsonism Found sr fel, 63-66; USPHS res career develop grant, 64-69, spec fel award, 64-66. Mem: Neurosci Soc; Am Physiol Soc; Am Asn Anat. Res: Auditory physiology as related to models of learning and adaptive behavior; unit responses to single or interacting sensory stimuli; studies in simplified mammalian CNS preparations and in behaving animals. Mailing Add: Dept of Physiol Univ of Calif Los Angeles CA 90024

BUCHWALD, NATHANIEL AVROM, b Brooklyn, NY, July 19, 24; m 52; c 3. NEUROPHYSIOLOGY, NEUROANATOMY. Educ: Univ Miami, BS, 46; Univ Minn, Minneapolis, PhD(neuroanat, neurophysiol), 53. Prof Exp: Instr anat, Sch Med, Tulane Univ, 53-57; from asst res anatomist to assoc res anatomist, 57-61, from assoc prof to prof anat, 61-69, assoc dir res, Ment Retardation Res Ctr, 71-74, PROF ANAT & PSYCHIAT, UNIV CALIF, LOS ANGELES, 69-, DIR, MENT RETARDATION RES CTR, 74- Concurrent Pos: Consult, Fed and state agencies, 57-; USPHS career develop award, 58-67. Mem: AAAS; Am Physiol Soc; Soc Neurosci; Am Asn Anat. Res: Brain function with relation to behavior and mental retardation. Mailing Add: Mental Retardation Res Ctr Univ of Calif Los Angeles CA 90024

BUCHWALTER, FRANCIS, physical organic chemistry, see 12th edition

BUCK, ALAN CHARLES, environmental physiology, pulmonary physiology, see 12th edition

BUCK, ALFRED A, b Hamburg, WGer, Mar 9, 21; m 62; c 2. EPIDEMIOLOGY, TROPICAL MEDICINE. Educ: Univ Hamburg, MD, 45; Johns Hopkins Univ, MPH, 59, DrPH(epidemiol), 61. Prof Exp: Resident internal med, Hamburg City Hosp of Altona, Ger, 46-52; intern in chg internal & trop med, Gen Hosp, Makassar-Celebes, Repub Indonesia, 52-55; chief dept med, Ger Red Cross Hosp, Pusan, Korea, 56-58; asst prof epidemiol & pub health admin, Sch Hyg & Pub Health, Johns Hopkins Univ, 61-63, from assoc prof to prof epidemiol & int health, 63-74, res dir geog epidemiol, Univ, 64-74, dir div bact & mycol, 67-71; SR MED OFFICER & CHIEF EPIDEMIOL METHOD & CLIN PATH, DIV MALARIA & OTHER PARASITIC DIS, WHO, 74- Concurrent Pos: Consult, US Agency Int Develop, Ethiopia, 62 & Cent & WAfrica, 71; consult, WHO, Geneva, 71. Mem: Am Venereal Dis Asn; NY Acad Sci; Am Soc Trop Med & Hyg; fel Am Pub Health Asn. Res: International health; prevalence and interaction of diseases in Peru, Chad, Afghanistan and Cameroon by comprehensive epidemiologic investigations and laboratory studies; epidemiology and control of onchocerciasis and treponematoses. Mailing Add: Div Malaria & Parasitic Dis World Health Orgn Geneva Switzerland

BUCK, CARL J, organic chemistry, see 12th edition

BUCK, CAROL WHITLOW, b London, Ont, Apr 2, 25; m 46; c 2. PREVENTIVE MEDICINE. Educ: Univ Western Ont, MD, 47, PhD, 50; London Sch Trop Med, DPH, 51. Prof Exp: From asst prof to prof, Dept Psychiat & Prev Med, 52-67, PROF EPIDEMIOL & CHMN DEPT EPIDEMIOL & PREV MED, UNIV WESTERN ONT, 67- Concurrent Pos: Mem, Sci Coun Can, 70-73. Mem: Biomet Soc; Can Pub Health Asn. Res: Epidemiology; medical statistics. Mailing Add: Dept of Epidemiol & Prev Med Health Sci Ctr Univ Western Ont London ON Can

BUCK, CHARLES (CARPENTER), b Metamora, Ohio, June 28, 15; m 51; c 2. MATHEMATICS. Educ: Univ Mich, BS, 40, MS, 47, PhD, 54. Prof Exp: Instr math, Univ Nebr, 49-52 & Wayne Univ, 53; from asst prof to assoc prof, Univ Ala, 53-61; res assoc, Educ Res Coun, 61-67; ASSOC PROF MATH, CLEVELAND STATE UNIV, 67- Concurrent Pos: Consult, Educ Res Coun, 71- Mem: AAAS; Am Math Soc; Math Asn Am. Res: Foundations, non-Euclidean and differential geometry. Mailing Add: 1803 Wilton Rd Cleveland Heights OH 44118

BUCK, CHARLES ELON, b Linton, NDak, Jan 5, 19; m 48; c 3. BACTERIOLOGY. Educ: NDak State Col, BS, 42, MS, 47; Ohio State Univ, PhD(bact), 51. Prof Exp: From asst prof to assoc prof bact, 51-74, PROF MICROBIOL, UNIV MAINE, ORONO, 74- Mem: Am Soc Microbiol. Res: Virology and tissue culture. Mailing Add: Dept of Bact Univ of Maine Orono ME 04473

BUCK, CHARLES FRANK, b Grayson, Ky, June 19, 20; m 46; c 2. ANIMAL HUSBANDRY. Educ: Univ Ky, BS, 42, MS, 51; Cornell Univ, PhD(animal husb), 53. Prof Exp: Teacher high schs, 42-50; asst animal husb, Univ Ky, 50-51; asst, Cornell Univ, 51-53; from asst prof to assoc prof, 53-67, PROF ANIMAL HUSB, UNIV KY, 67- Mem: Am Soc Animal Sci. Res: Pasture utilization; protein study for ruminants; pasture grazing. Mailing Add: Parkers Mill Rd Rte 2 Lexington KY 40504

BUCK, CLAYTON ARTHUR, b Lyons, Kans, Mar 2, 37; m 61; c 2. BIOCHEMISTRY, CANCER. Educ: Kans State Univ, BS, 59; Mont State Univ, PhD(bact), 64. Prof Exp: NIH fel, Univ Calif, Irvine, 64-67; asst prof therapeut res, Univ Pa, 67-70; assoc prof biol, Kans State Univ, 70-75; PROF, WISTAR INST, 75- Mem: Am Soc Biol Chemists; Am Soc Microbiol. Res: Virology, protein biosynthesis and animal cell membranes; tRNA in mitochondria and glycoproteins from the surface of virus-transformed and normal animal cells. Mailing Add: Wistar Inst 36th at Spruce Philadelphia PA 19104

BUCK, DAVID HOMER, b Clifton, Ariz, Dec 31, 20; m 53; c 4. FISHERIES. Educ: Agr & Mech Col, Tex, BS, 43; Okla State Univ, PhD(zool), 51. Prof Exp: Aquatic biologist, State Game & Fish Comn, Tex, 46-48; dist biologist, Corps Engrs, Ft Worth Dist, 51-52; fisheries investr, State Game & Fish Dept, Okla, 54-56; assoc aquatic biologist, 56-68, AQUATIC BIOLOGIST, ILL NATURAL HIST SURV, 68- Mem: Am Fisheries Soc; Am Soc Limnol & Oceanog. Res: Productivity and population dynamics of freshwater fisheries; effects of turbidity on fish production; species interrelationships of pond fishes; culture of channel catfish; recycling of animal wastes for polyculture of fishes. Mailing Add: Ill Natural Hist Surv Rte 1 Box 104 Kinmundy IL 62854

BUCK, DOUGLAS L, b Frederic, Wis, May 9, 31; m 54; c 2. DENTISTRY. Educ: Univ Minn, BS, 54, DDS, 60, MSD, 62. Prof Exp: Asst orthod, Sch Dent, Univ Minn, 61-62; from asst prof to assoc prof, 62-71, PROF ORTHOD, DENT SCH, UNIV ORE, 71-, CHMN DEPT, 73- Concurrent Pos: Fulbright lectr, Ecuador, 68-69; vis prof, Sch Dent, Hokkaido Univ, 75. Mem: Am Asn Orthod; Int Asn Dent Res; Am Cleft Palate Asn. Res: Histology of tooth movement. Mailing Add: Dent Sch Univ of Ore 611 SW Campus Dr Portland OR 97201

BUCK, ERNEST MAURO, b Hartford, Conn, Apr 21, 30; m 54; c 4. MEAT SCIENCE. Educ: Univ Conn, BS, 55; NC State Col, MS, 57; Univ Mass, Amherst, PhD(food sci & technol), 66. Prof Exp: Assoc dean, Col Food & Natural Resources, 69-76, PROF FOOD SCI & NUTRIT, UNIV MASS, AMHERST, 76- Mem: Am Soc Animal Sci; Inst Food Technol. Res: Meats; physico-chemical changes in muscle after death. Mailing Add: Dept of Food Sci & Nutrit Univ of Mass Stockbridge Hall Amherst MA 01002

BUCK, GRIFFITH J, b Cincinnati, Iowa, Apr 19, 15; m 47; c 2. HORTICULTURE. Educ: Iowa State Col, PhD(hort, bot), 53. Prof Exp: From asst prof to assoc prof, 53-74, PROF HORT, IOWA STATE UNIV, 74- Mem: Am Soc Hort Sci; Sigma Xi. Res: Pelargonium and rose breeding; propagation and disease control. Mailing Add: 1108 Scott St Ames IA 50010

BUCK, JAMES GRAY, physics, deceased

BUCK, JOHN BONNER, b Hartford, Conn, Sept 26, 12; m 39; c 4. ZOOLOGY. Educ: Johns Hopkins Univ, AB, 33, PhD(zool), 36. Prof Exp: Asst zool, Johns Hopkins Univ, 33-36; Nat Res fel, Calif Inst Technol, 36-37; res assoc embryol, Carnegie Inst Technol, 37-39; from instr to asst prof zool, Univ Rochester, 39-45; cytologist, NIH, 45-47, sr biologist, 47-56, prin physiologist, 56-62, chief lab phys biol, 62-75, CHIEF SECT COMP PHYSIOL, NIH, 75- Concurrent Pos: Mem, Hopkins Exped, Jamaica, 36, 41 & 63; mem corp, Marine Biol Lab, Woods Hole, 37-, instr, 42-44 & 57-59, trustee, 59-; vis prof, Univ Wash, 51, Calif Inst Technol, 53 & Cambridge Univ, 63; chief scientist, Alpha Helix Exped, New Guinea, 69. Mem: AAAS; Am Soc Zool (vpres, 56); Soc Gen Physiol (secy-treas, 53-55, pres, 60). Res: Physiology and behavior of fireflies; chromosome and physiological cytology; biochemistry and physiology of insect blood; insect respiration; bioluminescence in invertebrates. Mailing Add: Nat Insts of Health Bethesda MD 20014

BUCK, JOHN DAVID, b Hartford, Conn, Oct 3, 35; m 60; c 3. BACTERIOLOGY, MARINE MICROBIOLOGY. Educ: Univ Conn, BA, 57, MS, 60; Univ Miami, PhD(marine sci), 65. Prof Exp: Asst instr bact, Univ Conn, 60-61; res instr marine microbiol, Univ Miami, 61-62 & 64-65; asst prof bact, 65-71, ASSOC PROF BIOL, UNIV CONN, 71- Mem: AAAS; Am Soc Microbiol; Am Soc Limnol & Oceanog; Am Soc Testing & Mat. Res: Enumeration and sampling methods in marine microbiology; nutrition and ecology of aquatic bacteria and yeasts. Mailing Add: Marine Res Lab Univ of Conn Noank CT 06340

BUCK, KEITH TAYLOR, organic chemistry, see 12th edition

BUCK, OTTO, b Ger, May 14, 33; m 61; c 2. METAL PHYSICS. Educ: Univ Stuttgart, BS, 56, MS, 59, PhD(physics), 61. Prof Exp: Asst metal physics, Univ Stuttgart, 61-64; mem tech staff, sci ctr, NAm Aviation Inc, 64-66 & Siemens Co, Ger, 66-68; mem tech staff, 68-75, GROUP LEADER, SCI CTR ROCKWELL INT, 75- Mem: Am Phys Soc; Am Soc Testing & Mat; Am Inst Mining, Metall & Petrol Engrs. Res: Plasticity of metals; dislocation theory; nonlinear theory of elasticity; internal friction; point-defect migration; fracture mechanics; surface properties; fatigue. Mailing Add: Rockwell Int Sci Ctr 1049 Camino Dos Rios Thousand Oaks CA 91360

BUCK, PAUL, b Highland Park, Mich, Sept 9, 27; m 50; c 2. PLANT ECOLOGY. Educ: Univ Tulsa, BS, 58, MS, 59; Univ Okla, PhD(bot), 62. Prof Exp: Asst prof, 62-70, ASSOC PROF BOT, UNIV TULSA, 70- Mem: AAAS; Ecol Soc Am. Res: Relationships of vegetational distribution, geological formations and soil types in Oklahoma. Mailing Add: Dept of Life Sci Univ of Tulsa Tulsa OK 74104

BUCK, PAUL ANDREWS, b Vancouver, BC, Oct 9, 22; nat US; m 51; c 6. PLANT PHYSIOLOGY, FOOD TECHNOLOGY. Educ: Univ BC, BSA, 44; Univ Calif, MS, 48, PhD(plant physiol), 54. Prof Exp: Chemist, Dominion Exp Sta, Can, 45-46; asst food technol, Univ Calif, 47-50; food technologist, Cent Exp Farm, Can, 50-51; rep west coast qual control, H J Heinz Co, Pa, 52-53; assoc prof dairy & food industs, Univ Wis, 54-59; ASSOC PROF FOOD SCI, CORNELL UNIV, 59- Concurrent Pos: Vis prof, US Army Natick Labs, 66-67; vis prof, Linus Pauling Inst Sci & Med, 76. Mem: Am Chem Soc; Inst Food Technol. Res: Biochemical and physiological changes of plant tissues; post-harvest physiological imbalances and freezing storage; imbalances causing abnormal flavors and textures changes; carotenoid development in carrots. Mailing Add: 293 Ellis Hollow Creek Rd Ithaca NY 14850

BUCK, RAYMOND WILBUR, JR, b Monticello, Maine, Apr 20, 19. CYTOGENETICS. Educ: Univ Maine, BS, 41; Univ Md, MS, 50, PhD(bot), 52. Prof Exp: Instr bot, Univ Maine, 46-48; geneticist, bur plant indust, soils & agr eng, USDA, 52-53 & tuber crops res br, Agr Res Serv, 53-68; assoc prof, 68-72, PROF BIOL & CHMN DEPT, MONTGOMERY COL, 72- Mem: AAAS; Bot Soc Am; Am Genetic Asn; Am Inst Biol Sci. Res: Cytogenetics of tuber-bearing Solanum species. Mailing Add: 4902 Laguna Rd College Park MD 20740

BUCK, RICHARD F, b Enterprise, Kans, Dec 8, 21; m 44; c 6. ATMOSPHERIC PHYSICS. Educ: Univ Kans, BS, 43; Okla State Univ, MS, 60. Prof Exp: Appl engr, radio tube dept, Tung Sol Lamp Works, Inc, 46-48; res physicist, 48-53, from asst proj dir to proj dir, electronics lab, 53-60, proj dir res found & instr physics, 60-67, DIR ELECTRONICS LAB, RES FOUND & ASST PROF PHYSICS, SCH ARTS & SCI, OKLA STATE UNIV, 67- Mem: Inst Elec & Electronics Engrs; Nat Soc Prof Engrs; AAAS; Sigma Xi. Res: Electronic instrumentation of rockets and satellites for measurement of physical parameters of space. Mailing Add: Res Found Okla State Univ Stillwater OK 74074

BUCK, RICHARD PIERSON, b Los Angeles, Calif, July 29, 29; m 59; c 3. PHYSICAL CHEMISTRY, ANALYTICAL CHEMISTRY. Educ: Calif Inst Technol, BS, 50, MS, 51; Mass Inst Technol, PhD(chem), 54. Prof Exp: Res chemist polarography & electrochem, Calif Res Corp, 54-56, asst to gen mgr, 56-58, res chemist combustion & high temperature inorg chem, 58-60, electrochem fuel cells, 60-61; prin res chemist, electrochem, electroanal chem & instrumentation, Bell & Howell Res Ctr, 61-65; sr scientist, Beckman Instruments, Inc, 65-67; assoc prof, 67-75, PROF CHEM, UNIV NC, CHAPEL HILL, 75- Concurrent Pos: VChmn, Gordon Res Conf Electrochem, 64, chmn, 65; vis prof, Bristol Univ, Eng, 76-77. Mem: Int Soc Electrochem; Am Soc Mass Spectro; Am Chem Soc; Electrochem Soc. Res: Electrochemistry; electrode processes; electroanalysis; chemical instrumentation; spark source mass spectroscopy; trace analysis; solid state transport; membrane electrochemistry. Mailing Add: Dept of Chem Univ of NC Chapel Hill NC 27514

BUCK, ROBERT CRAWFORTH, b London, Ont, Sept 23, 23; m 46; c 2. ANATOMY, HISTOLOGY. Educ: Univ Western Ont, MD, 47, MSc, 50; Univ London, PhD(path), 52. Prof Exp: From asst prof to assoc prof, 53-62, chmn dept, 67-73, PROF ANAT, UNIV WESTERN ONT, 62- Mem: Am Asn Anat; Can Asn Anat; Anat Soc Gt Brit & Ireland; Path Soc Gt Brit. Res: Cytology; electron microscopy in fields of blood vascular pathology and tumor pathology. Mailing Add: Dept of Anat Univ of Western Ont London ON Can

BUCK, ROBERT CREIGHTON, b Cincinnati, Ohio, Aug 30, 20; m 44; c 2. MATHEMATICAL ANALYSIS. Educ: Univ Cincinnati, BA, 41, MA, 42; Harvard Univ, PhD(math), 47. Prof Exp: Asst prof math, Brown Univ, 47-49; assoc prof math, 50-54, chmn dept, 64-66, dir math res ctr, 73-75, PROF MATH, UNIV WIS-MADISON, 54- Concurrent Pos: Ed, Proc, Am Math Soc, 53-55, 64-67; Guggenheim fel, 58-59; vis prof, Stanford Univ, 58-59; mem staff proj focus, Inst Defense Anal, 59-60; mem math div, Nat Res Coun, 61-64; mem, Nat Security Agency Adv Bd, 63-; mem, US Comn Math Instruct, 63-67; mem math adv panel, NSF, 65-70; adv bd, Sch Math Study Group, 66-69; Nat Adv Coun Educ Prof Develop, 70-71, 73-76; mem panel to evaluate, Nat Bur Stand Appl Math Prog, 74-77. Mem: AAAS; Am Math Soc (vpres, 72-74); Math Asn Am (vpres, 75-77); Soc Indust & Appl Math. Res: Complex variable theory; algebraic analysis; number theory; approximation theory; mathematics education; history of mathematics. Mailing Add: Dept of Math van Vleck Hall Univ of Wis Madison WI 53706

BUCK, ROBERT EDWARD, b Altmar, NY, Oct 16, 12; m 44; c 3. FOOD SCIENCE. Educ: Cornell Univ, AB, 33; Univ Mass, MS, 34, PhD(chem), 36. Prof Exp: Lab instr, Univ Mass, 36-42; res chemist, eastern regional res lab, Bur Agr Chem & Eng, USDA, 42-47; technologist, Quartermaster Food & Container Inst, Chicago, 47-48; chemist & head food technol, 48-73, SR MGR FOOD RES, H J HEINZ CO, 73- Mem: Am Chem Soc; Inst Food Technologists. Res: Process and product development in foods; pectin and pectic enzymes; technology and rheology ingredient materials as starch, flour, gums; flavoring materials. Mailing Add: H J Heinz Co PO Box 57 Pittsburgh PA 15230

BUCK, THOMAS M, b Elizabeth, Pa, June 29, 20; m 44; c 5. SURFACE PHYSICS. Educ: Muskingum Col, BS, 42; Univ Pittsburgh, MS, 48, PhD(chem), 50. Prof Exp: Asst, Univ Pittsburgh, 46-50; res chemist, Nat Lead Co, 50-52; mem tech staff, Bell Tel Labs, 52-63, supvr surface studies group, Semiconductor Device Technol Dept, 63-68, mem tech staff, Chem Electronics Res Dept, 68-71, MEM TECH STAFF, RADIATION PHYSICS RES DEPT, BELL LABS, 71- Mem: Am Phys Soc. Res: Surface analysis by ion scattering; ion neutralization behavior; chemical and electrical properties of semiconductor surfaces; semiconductor nuclear particle detectors; silicon diode array targets for television camera tubes. Mailing Add: Bell Labs Mountain Ave Murray Hill NJ 07974

BUCK, WARREN HOWARD, b Neptune, NJ, June 22, 42; m 68. PHYSICAL CHEMISTRY, POLYMER CHEMISTRY. Educ: Lehigh Univ, BA, 64; Univ Del, PhD(phys chem), 70. Prof Exp: RES CHEMIST, ELASTOMER CHEM DEPT, E I DU PONT DE NEMOURS & CO, INC, 69- Mem: Am Chem Soc. Res: Physical chemistry of macromolecules, especially elastomers. Mailing Add: Elastomer Chem Dept E I du Pont de Nemours & Co Inc Wilmington DE 19899

BUCK, WARREN LOUIS, b Normal, Ill, Jan 20, 21; m 44; c 2. RADIATION PHYSICS. Educ: Ill State Norm Univ, BS, 46; Univ Ill, MS, 47. Prof Exp: Asst physics, Univ Ill, 46-50; ASSOC PHYSICIST, ARGONNE NAT LAB, 50- Mem: Am Phys Soc. Res: Scintillation counters; luminescent materials; effects of radiation on materials; neutron flux monitors. Mailing Add: Components Technol Argonne Nat Lab 9700 S Cass Ave Argonne IL 60439

BUCK, WILLIAM BOYD, b Mexico, Mo, May 17, 33; m 52; c 6. VETERINARY TOXICOLOGY. Educ: Univ Mo, BS & DVM, 56; Iowa State Univ, MS, 63; Am Bd Vet Toxicol, dipl. Prof Exp: Area vet, animal dis eradication div, Agr Res Serv, USDA, 56-58, from vet to res vet, 58-64; assoc prof, 64-68, PROF VET TOXICOL, IOWA STATE UNIV, 68- Concurrent Pos: Mem panel on copper, Comt Med & Biol Effects Environ Pollutants, Nat Acad Sci, 73-, mem panel on arsenic, 73- Mem: Am Col Vet Toxicol; Soc Toxicol; Am Vet Med Asn; Conf Res Workers Animal Dis. Res: Toxicology of economic poisons and plants in animals; environmental toxicology; central nervous system physiopathology. Mailing Add: Toxicol Sect Vet Diag Lab Iowa State Univ Ames IA 50010

BUCKALEW, JOHN MCKINNEY, b Pa, Dec 25, 28. DAIRY PRODUCTION. Educ: Pa State Univ, BS, 49; Auburn Univ, MS, 51; Univ Mo, PhD(dairy prod), 55. Prof Exp: Asst dairy husb, Auburn Univ, 49-51; instr dairying, Univ Mo, 51-55; from asst prof to prof dairy prod, Miss State Univ, 55-69; ASSOC PROF DAIRY PROD, PA STATE UNIV, 69- Mem: Am Dairy Sci Asn. Res: Animal physiology of reproduction. Mailing Add: Dept of Dairy Sci Pa State Univ State College PA 16801

BUCKARDT, HENRY LLOYD, b Leland, Ill, Oct 24, 04; m 37. AGRONOMY. Educ: Univ Ill, BS, 26, MS, 29, PhD(agron), 32. Prof Exp: Asst prin & instr high sch, Ill, 26-27; asst crop prod, Dept Agron, Univ Ill, 28-33, Standard Oil Co fel, 33, erosion exten specialist coordr, 35-36; agronomist, Soil Conserv Serv, USDA, 34, Iowa, 37, DC, 38-40; head admin off, US Civil Serv Comn, 40-46; civil serv adminr & adv, Civil Affairs Div, US War Dept, Korea, 46; personnel policy specialist, Off Secy Defense, Nat Mil Estab, 48-50, staff mem personnel policy bd, US Dept Defense, 50-52, spec asst to Asst Secy Defense, Manpower & Personnel, 52-55; agr attache, Seoul, Korea, 56-57, Far Eastern are off, Foreign Agr Serv, USDA, Washington, DC, 58, agr attache, Montevideo, Uruguay, 58-69; PRES, AM WORK HORSE MUS, 70- Concurrent Pos: Hon mem, Asn Intensification Commerce Uruguay, Montevideo, 69. Mem: AAAS; Am Soc Agron; assoc Soc Personnel Admin. Res: Crop productions; physiological botany; soil conservation; weed eradication; personnel administration and training; civil service systems. Mailing Add: Paeonian Springs VA 22129

BUCKE, DAVID PERRY, JR, geology, sedimentary petrology, see 12th edition

BUCKELEW, ALBERT RHOADES, JR, b Washington, DC, Oct 22, 42; m 65; c 1. MICROBIOLOGY. Educ: Fairleigh Dickinson Univ, BS, 64; Univ NH, PhD(microbiol), 68. Prof Exp: ASST PROF BIOL, BETHANY COL, W VA, 69- Concurrent Pos: Am Lung Asn res grant, 74. Mem: AAAS; Am Soc Microbiol; Int Soc Toxinology. Res: Interaction of staphylococcal alpha toxin with monolayer films; properties of pulmonary surfactants. Mailing Add: Dept of Biol Bethany Col Bethany WV 26032

BUCKER, HOMER PARK, JR, b Ponca City, Okla, July 23, 33; m 55; c 2. ACOUSTICS, SPECTROSCOPY. Educ: Univ Okla, BS, 55, PhD(physics), 62. Prof Exp: Engr, Univ Okla, 55-60; physicist, US Navy Electronics Lab, 62-67; res physicist, Tracor, Inc, Tex, 67-69; head sound propagation br, Naval Res Lab, 69-70, RES PHYSICIST, NAVAL UNDERSEA CTR, 70- Mem: Acoustical Soc Am. Res: Propagation of sound in the ocean; infrared and Raman molecular spectroscopy. Mailing Add: Naval Undersea Ctr Code 503 San Diego CA 92132

BUCKEYE, DONALD ANDREW, b Lakewood, Ohio, Mar 12, 30; m 62; c 2. MATHEMATICS. Educ: Ashland Col, BS, 53; Ind Univ, MA, 61, EdD(math), 68. Prof Exp: Teacher, high sch, 53-54; instr math, Army Educ Ctr, Sendai, Japan, 54-56; teacher, high sch, 56-66; teaching assoc, Ind Univ, 66-68; PROF MATH, EASTERN MICH UNIV, 68- Concurrent Pos: Assoc instr, Ohio State Univ, Cleveland, 64-66. Res: Ways of increasing the creative ability of students in mathematics; use of a laboratory approach to the teaching of mathematics and its effects on attitude, creativity and achievement. Mailing Add: 1823 Witmire Blvd Ypsilanti MI 48197

BUCKHOLTZ, JAMES DONNELL, b Little Rock, Ark, Feb 25, 35; m 59; c 2. MATHEMATICAL ANALYSIS. Educ: Univ Tex, BA, 57, PhD(math), 60. Prof Exp: Instr math, Univ NC, 60-61, asst prof, 61-62; vis asst prof, Univ Wis, 62-63; asst prof math, Univ NC, 63-64; assoc prof, 64-69, PROF MATH, UNIV KY, 69- Mem: Am Math Soc; Math Asn Am. Res: Analytic functions; zeros of polynomials; polynomial expansions. Mailing Add: Dept of Math Univ of Ky Lexington KY 40506

BUCKINGHAM, FORREST MORGAN, b Hartford, Conn, Feb 12, 23; m 48; c 4. HYDROLOGY. Educ: Univ NB, BSc, 49; Harvard Univ, MF, 50; Duke Univ, DF, 66. Prof Exp: From asst prof to assoc prof forest mensuration, Univ NB, 51-57; from asst prof to assoc prof, 57-67, PROF FOREST MENSURATION, UNIV TORONTO, 67- Res: Growth and development of forest stands; effects of land-use practices, especially forest management, on water yields. Mailing Add: Fac of Forestry Univ of Toronto Toronto ON Can

BUCKINGHAM, JOHN HERBERT, b Caputa, SDak, Oct 5, 12; m 40; c 1. PHYSICAL CHEMISTRY. Educ: SDak Sch Mines, BS, 34; Ohio State Univ, PhD(phys chem), 40. Prof Exp: Actg instr gen chem, Hiram Col, 40; assoc prof, NDak Col, 40-41; instr, Colo Sch Mines, 41-43; from asst prof to assoc prof phys chem, 43-53, chmn dept, 59-74, PROF PHYS CHEM, MIAMI UNIV, 53-; PRES, MIAMI VALLEY ISOTOPE SERV, INC, 54- Mem: AAAS; Am Chem Soc. Res: Excitation potential of molecules; solubility studies; surface phenomena; radiochemistry; radioactive tracer studies; self-diffusion of ions. Mailing Add: Dept of Phys Chem Miami Univ Oxford OH 45056

BUCKINGHAM, WILLIAM THOMAS, b Lancaster, Ohio, Dec 21, 21; m 55; c 2. COMMUNICATIONS SCIENCE, PLANT BREEDING. Educ: Otterbein Col, BA, 45, BS, 46. Prof Exp: Prin chemist, Battelle Mem Inst, 42-46; asst prof metall chem, Ohio State Univ, 47-54; prin chemist, Battelle Mem Inst, 54-56; lab mgr & spec proj off, Foseco, Inc, 56-68; PRES, BUCKINGHAM ELECTRONIC DIV, BUCKINGHAM PROD, PRES & GEN MGR, BUCKINGHAM ORCHARDS, 68-; ASSOC SPONSOR, M L JONES & ASSOCS, 68- Concurrent Pos: Consult, Metall/Chem; consult & assoc sponsor, Dale Carnegie Courses, 68- Mem: Am Chem Soc. Res: Laboratory and general management; chemical treatment of metals; exothermics and explosives; development and growth of new fruit cultivars from origination to full scale commercial production. Mailing Add: M L Jones & Assocs 8803 Cheshire Rd Sunbury OH 43074

BUCKLAND, GOLDEN THADDEUS, b Crag, WVa, July 10, 13; m 39; c 2. MATHEMATICS. Educ: Appalachian State Univ, BS, 39, MA, 49; Pa State Univ, DEd, 54. Prof Exp: Admin & supv prin pub schs, Va, 39-48; teacher pub sch, NC, 48-49; from asst prof to assoc prof math, 49-59, PROF MATH SCI, GRAD FAC, APPALACHIAN STATE UNIV, 59- Concurrent Pos: Consult, Rowan County Schs, 58 & Montgomery County Sch, 59. Mem: Math Asn Am. Res: Foundations and algebraic theories. Mailing Add: Dept of Math Appalachian State Univ Boone NC 28607

BUCKLAND, ROGER BASIL, b Jemseg, NB, May 18, 42; m 65; c 1. POULTRY GENETICS, REPRODUCTIVE PHYSIOLOGY. Educ: McGill Univ, BSc, 63, MSc, 65; Univ Md, PhD(biochem genetics), 68. Prof Exp: Res scientist poultry physiol, Can Dept Agr, 67-71; asst prof, 71-73, ASSOC PROF ANIMAL SCI, MACDONALD COL, McGILL UNIV, 73- Honors & Awards: Poultry Sci Res Award, Poultry Sci Asn, 72. Mem: Am Soc Study Reproduction; Poultry Sci Asn; World Poultry Sci Asn. Res: Metabolism of chicken sperm; resistance to stress in poultry. Mailing Add: Dept of Animal Sci Macdonald Col Quebec PQ Can

BUCKLER, ERNEST JACK, b Birmingham, Eng, June 3, 14; m 42. CHEMISTRY. Educ: Cambridge Univ, MA, 35, PhD(phys chem), 38. Hon Degrees: LLD, Queens Univ, Ont, 59. Prof Exp: Res chemist, Trinidad Leaseholds, BWI, 38-41 & Imp Oil, Sarnia, 41-42; prod controller, St Clair Processing Corp, 42-45, tech supt, 45-48; mgr res, Polymer Corp, 48-53, vpres res & develop, 53-73; VPRES, POLYSAR LTD, 73- Mem: Fel Chem Inst Can; Am Chem Soc; Soc Chem Indust; Royal Inst Chemists; Indust Res Inst. Res: Chemical kinetics, gas phase; petroleum refining and petrochemicals; polymer synthesis and characterization; research management. Mailing Add: Polysar Ltd Sarnia ON Can

BUCKLER, ROBERT T, organic chemistry, medicinal chemistry, see 12th edition

BUCKLER, SHELDON A, b New York, NY, May 18, 31; m 52; c 3. ORGANIC CHEMISTRY. Educ: NY Univ, BA, 51; Columbia Univ, MA, 52, PhD(chem), 54. Prof Exp: Asst, Columbia Univ, 51-53; res fel, Univ Md, 55-56; res chemist & group leader, basic res dept, Am Cyanamid Co, 56-62; mgr org res, Am Mach & Foundry Co, 62-64; mem staff, 64-66, dir chem res & develop div, 66-69, asst vpres res, 69-72, vpres res, 72-75, GROUP VPRES, POLAROID CORP, 75- Mem: Am Chem Soc; Soc Photog Sci & Eng. Res: Research planning and evaluation; process research and development; synthetic and theoretical organophosphorus chemistry; photographic chemicals; carbohydrates; organometallics; organic peroxides. Mailing Add: Hiddenwood Lincoln MA 01773

BUCKLES, LAWRENCE CALVIN, b Helena, Mont, Aug 25, 15; m 53. PHYSICAL ORGANIC CHEMISTRY. Educ: Lehigh Univ, BS, 38. Prof Exp: Anal chemist, Edgewood Arsenal, 40-43, anal chemist, Rocky Mountain Arsenal, 43-44; phys chemist, US Army Res Labs, Edgewood Arsenal, 44-56, chief phys methods br, 56-63, chief agents properties br, Physical Chem Dept, 63-69, chief mat eng lab br, 69-72, chief process chem, Mfg Technol Directorate, Aberdeen Proving Grounds, Md, 72-74; RETIRED. Mem: AAAS; Sigma Xi; Am Chem Soc. Res: Surface chemistry; spectroscopy; purification; reaction kinetics; mechanisms; corrosion. Mailing Add: 11206 Sheradale Dr Kingsville MD 21087

BUCKLES, MARJORIE FOX, b Philadelphia, Pa, Sept 15, 22; m 53. MICROCHEMISTRY. Educ: Mt Holyoke Col, AB, 43. Prof Exp: Chemist, Beacon Res Labs, Tex Co, NY, 44-46, Victor Div, Radio Corp Am, Pa, 47-49, Ballistic Comput, US Dept Army, Tex, 49-51, Phys Br, 51-54, Org Br, 54-56, ANAL CHEM DEPT, MICROANAL BR, US ARMY RES LABS, EDGEWOOD ARSENAL, 56- Mem: Sigma Xi; Am; fel Am Inst Chem; Am Microchem Soc. Res: Micro and ultramicro methods of elemental and functional group analysis; chemical microscopy; physical chemical measurements; organic nomenclature, indexing and coding. Mailing Add: Anal Chem Dept Microanal Br US Army Res Labs Edgewood Arsenal MD 21010

BUCKLES, ROBERT EDWIN, b Fallon, Nev, Aug 11, 17; m 44; c 2. PHYSICAL ORGANIC CHEMISTRY. Educ: Univ Calif, BS, 39, MS, 40; Univ Calif, Los Angeles, PhD(phys-org chem), 42. Prof Exp: Asst, Ill Inst Technol, 40-41 & Univ Calif, Los Angeles, 41-42; du Pont fel & instr org chem, Univ Minn, 43-45; from instr to assoc prof, 45-59, PROF ORG CHEM, UNIV IOWA, 59- Mem: Am Chem Soc. Res: Molecular structure; addition reactions; replacement reactions; reaction mechanisms; halogen chemistry; chemistry education. Mailing Add: Dept of Chem Univ of Iowa Iowa City IA 52242

BUCKLEY, BERNARD PATRICK, b Sydney, NS, Feb 25, 15; m 46; c 4. PHYSICAL CHEMISTRY. Educ: St Francis Xavier Univ, BSc, 36; McGill Univ, PhD, 41. Prof Exp: Asst chem, St Francis Xavier Univ, 37-38; res chemist, Shawinigan Chem, Ltd,

41-56; CHEMIST, DOW CHEM OF CAN, 56- Mem: AAAS; Am Chem Soc; fel Chem Inst Can. Res: Effect of temperature on water sorption on cellulose; chemistry of acetylene based products; calcium carbide; applications chemistry of glycols and alkoxyalcohols; heat transfer media; corrosion inhibition. Mailing Add: 695 Kemsley Sarnia ON Can

BUCKLEY, CHARLES EDWARD, b Charleston, WVa, Sept 2, 29; m 55; c 4. INTERNAL MEDICINE, ALLERGY. Educ: Va Polytech Inst, BS, 50; Duke Univ, MD, 54. Prof Exp: Assoc med & immunol, 58-67, assoc immunol, 67-70, asst prof med, 67-70, asst prof microbiol & immunol, 70-74, ASSOC PROF MED, DUKE UNIV MED CTR, 70-, DIR ALLERGY & CLIN IMMUNOL LAB, 70- Concurrent Pos: USPHS res career develop award, 61-69; mem adv coun, Nat Inst Allergy & Infectious Dis, 75- Mem: Am Fedn Clin Res; Am Thoracic Soc; Am Acad Allergy; Am Col Physicians; Am Asn Immunol. Res: Allergy, clinical immunology; immunochemistry. Mailing Add: Duke Univ Med Ctr Box 3804 Durham NC 27710

BUCKLEY, DALE ELIOT, b Wolfville, NS, May 3, 36; m 64; c 3. GEOCHEMISTRY. Educ: Acadia Univ, BSc, 59; Univ Western Ont, MSc, 63; Univ Alaska, PhD(marine sci), 69. Prof Exp: Tech officer, Can Hydrographic Surv, Dept Mines & Tech Surv, 60-61 & Oceanog Res Div, 61-62; SCI OFFICER MARINE GEOL, ATLANTIC OCEANOG LAB, BEDFORD INST, 62- Concurrent Pos: Lectr, Calif State Col, Los Angeles, 65-66. Mem: Am Soc Limnol & Oceanog; Clay Minerals Soc; Prof Inst Pub Serv Can. Res: Marine inorganic geochemistry, particularly processes of chemical exchange between natural sediments and ionic metals in seawater, research to be applied to problems of pollution and waste disposal. Mailing Add: Atlantic Oceanog Lab Bedford Inst Dartmouth NS Can

BUCKLEY, EDWARD HARLAND, b Brandon, Man, Jan 30, 31; m 56; c 3. PLANT PHYSIOLOGY. Educ: Ont Agr Col, BSA, 54; Columbia Univ, MA, 56, PhD(plant physiol), 59. Prof Exp: Plant physiologist, Cent Res Labs, United Fruit Co, 58-60, plant biochemist, 61-65; PLANT BIOCHEMIST, BOYCE THOMPSON INST PLANT RES, INC, 65- Mem: NY Acad Sci; Am Soc Plant Physiol; Can Soc Plant Physiol; Ecol Soc Am; Am Soc Limnol & Oceanog. Res: Host-parasite relationships in plants; biosynthesis of flavor and aroma compounds and of phenolic compounds in banana; biochemical aspects of plant gametogenesis; estuarine ecology. Mailing Add: Boyce Thompson Inst for Plant Res, Inc 1086 North Broadway Yonkers NY 10701

BUCKLEY, JAMES THOMAS, b Ottawa, Ont, Can, Oct 3, 42; m 66; c 2. BIOCHEMISTRY. Educ: McGill Univ, BSc, 65, PhD(biochem), 69. Prof Exp: ASST PROF BIOCHEM, UNIV VICTORIA, 72- Mem: Can Biochem Soc. Res: Role of polyphosphoinositides in membrane structure and function regulation of their synthesis and breakdown influence of dietary lipid on growth and survival of salmonids. Mailing Add: Dept of Bact & Biochem Univ of Victoria Victoria BC Can

BUCKLEY, JAY SELLECK, JR, b Ansonia, Conn, Feb 16, 24; m 48; c 4. INFORMATION SCIENCE. Educ: Williams Col, BS, 44; Univ Minn, PhD(org chem), 49. Prof Exp: Res chemist, Winthrop Chem Co, 45-46; res chemist, 49-51, res supvr, 51-54, res mgr, 54-68, mgr, Comput Based Info Serv, 68-74; MGR TECH INFO, PFIZER, INC, 74- Mem: Am Chem Soc. Res: Organic chemistry; mechanisms of organic reactions and medicinal products; systems for encoding and retrieval of chemical and biological information. Mailing Add: 16 Birch Lane Groton CT 06340

BUCKLEY, JOHN LEO, b Binghamton, NY, Sept 22, 20; m 47; c 4. ECOLOGY. Educ: State Univ NY Col Forestry, Syracuse, BS, 42, MS, 47, PhD(wildlife mgt), 51. Prof Exp: Instr biol sci, Univ Alaska, 50-51, from assoc prof to prof wildlife mgt, 51-58; leader coop wildlife res unit, 51-58; asst chief br wildlife res, Bur Sport Fisheries & Wildlife, 58-59, chief off pesticides coord, 63-64, dir, Patuxent Wildlife Res Refuge, 59-63; tech specialist, Off Sci & Technol, Exec Off of President, 64-65; ecol res coordr, Off Sci Adv, US Dept Interior, 65-67, dir off ecol, 67-68; tech asst, Off Sci & Technol, Exec Off of President, 68-71; assoc dep asst adminr res, Agency, 71-73, dep dir off prog integration, 73-74, ACTG DEP ASST ADMINR, OFF PROG INTEGRATION, OFF RES & DEVELOP, ENVIRON PROTECTION AGENCY, 74- Concurrent Pos: Wildlife res biol supvr, US Fish & Wildlife Serv, 51-63; mem, Nat Res Coun. Mem: Fel AAAS; Wildlife Soc; Ecol Soc Am. Res: Wildlife ecology; population fluctuations; environmental contamination and quality; animal populations. Mailing Add: Off of Prog Integration EPA Off of Res & Develop Washington DC 20460

BUCKLEY, JOSEPH J, b Methuen, Mass, Sept 11, 22; m 45; c 4. ANESTHESIOLOGY. Educ: Dartmouth Col, AB, 44; NY Med Col, MD, 46; Univ Minn, MS, 58. Prof Exp: From instr to assoc prof, 54-61, PROF ANESTHESIOL, MED COL, UNIV MINN, MINNEAPOLIS, 61- Concurrent Pos: Consult, US Vet Admin, 58-; vis prof, Col Physicians & Surgeons, Columbia Univ, 62. Mem: AMA; Am Soc Anesthesiol; Acad Anesthesiol. Res: Cardiac and pulmonary physiology; mass spectrometry. Mailing Add: Rm C 596 Univ of Minn Hosp Minneapolis MN 55455

BUCKLEY, JOSEPH PAUL, b Bridgeport, Conn, Jan 12, 24; m 47. PHARMACOLOGY. Educ: Univ Conn, BS, 49; Purdue Univ, MS, 51, PhD(pharmacol), 52. Prof Exp: Instr, St Elizabeth Sch Nursing, 52; from asst prof to prof pharmacol & head dept, Sch Pharm, Univ Pittsburgh, 58-73, assoc dean, 69-73; PROF PHARMACOL, DIR CARDIOVASC RES & DEAN, COL PHARM, UNIV HOUSTON, 73- Mem: AAAS (secy, 62-66, vpres, 69); Am Pharmaceut Asn; Am Soc Pharmacol & Exp Therapeut; NY Acad Sci. Res: Mechanisms of action of antihypertensive drugs; effects of stress on blood pressure and animal behavior and how drugs affect this phenomenon. Mailing Add: Col of Pharm Univ of Houston Houston TX 77004

BUCKLEY, JOSEPH THADDEUS, b Boston, Mass, Apr 13, 37; m 61; c 4. MATHEMATICS. Educ: Boston Col, BS, 58; Ind Univ, PhD(math), 64. Prof Exp: Young res instr math, Dartmouth Col, 64-67; asst prof, Univ Mass, 67-69; asst prof, 69-74, ASSOC PROF MATH, WESTERN MICH UNIV, 74- Mem: Am Math Soc; Math Asn Am. Res: Group theory; group rings; cohomology of groups; extensions; nilpotent groups; homological algebra. Mailing Add: Dept of Math Western Mich Univ Kalamazoo MI 49001

BUCKLEY, NANCY MARGARET, b Philadelphia, Pa, Dec 2, 24. PHYSIOLOGY. Educ: Univ Pa, AB, 45, MD, 50. Prof Exp: Asst med dir, Columbus Blood Prog, Am Red Cross, 51-52; res assoc med, Ohio State Univ, 52, asst prof & res assoc physiol, 52-55; from asst prof to assoc prof, 55-74, PROF PHYSIOL, ALBERT EINSTEIN COL MED, 74- Mem: AAAS; Am Physiol Soc; NY Acad Sci. Res: Cardiodynamics; cardiac metabolism; cardiovascular development. Mailing Add: Dept of Physiol Albert Einstein Col of Med Bronx NY 10461

BUCKLEY, PATRICIA M, b Marion, Ohio, July 25, 23; m 41, 50; c 4. MICROBIOLOGY, BIOCHEMISTRY. Educ: Ohio Univ, BS, 48; Ore State Univ, MS, 58, PhD(microbiol), 63. Prof Exp: Med technologist, 48-55; chemist, Ore State Univ, 57-58; res asst virol, Stanford Univ, 62-63; asst res pomologist, Univ Calif, Davis, 64-68; asst prof biol, Fresno State Col, 68-71; asst res microbiologist, comp oncol lab, Univ Calif, Davis, 71-74; ASSOC PROF BIOL, LINFIELD COL, 74- Mem: Am Soc Microbiol. Res: Electron microscopy; cytology; virology. Mailing Add: Dept of Biol Linfield Col McMinnville OR 97128

BUCKLEY, PAUL ANTHONY, ethology, avian biology, see 12th edition

BUCKLEY, RAMON D, b Forest City, Iowa, July 12, 29; m 64; c 1. BIOCHEMISTRY, PHYSIOLOGY. Educ: Iowa State Univ, BS, 52; State Col Iowa, MEd, 58; Univ Iowa, PhD(physiol), 61. Prof Exp: From instr to asst prof biochem, Univ Southern Calif, 61-72; RES ASSOC, RANCHO LOS AMIGOS HOSP, 72- Res: Biochemical alterations in blood and lung tissues of animals exposed to air pollutants and humans exposed to smog. Mailing Add: Rancho Los Amigos Hosp 7601 E Imperial Hwy Downey CA 90242

BUCKLEY, REBECCA HATCHER, b Hamlet, NC, Apr 1, 33; m 55; c 4. PEDIATRIC ALLERGY, PEDIATRIC IMMUNOLOGY. Educ: Duke Univ, AB, 54; Univ NC, MD, 58. Prof Exp: Intern pediat, 58-59, from asst resident to resident, 59-61, from instr to assoc prof, 61-76, asst prof immunol, 69-72, ASSOC PROF IMMUNOL, SCH MED, DUKE UNIV, 72-, CHIEF DIV ALLERGY, IMMUNOL & PULMONARY DIS, 74-, PROF PEDIAT, 76- Concurrent Pos: Fel allergy, Sch Med, Duke Univ, 61-63, fel immunol, 63-65; Nat Inst Allergy & Infectious Dis acad award allergic dis, 74-79; mem immunol sci study sect, NIH, 76-80. Mem: Am Acad Allergy; Am Acad Pediat; Soc Pediat Res; Am Asn Immunologists; Sigma Xi. Res: Cellular basis of primary immunodeficiency; cell-mediated immune responsiveness in the atopic diseases; factors affecting the development of immunologic enhancement, and the characterization of enhancing antibodies. Mailing Add: Dept of Pediat Box 2898 Duke Univ Sch of Med Durham NC 27710

BUCKLEY, REGINALD R, b Jackson, Miss, May 7, 39; m 60; c 3. INORGANIC CHEMISTRY, PHYSICAL CHEMISTRY. Educ: Millsaps Col, BS, 61; Univ SC, PhD(inorg chem), 66. Prof Exp: TECH SUPVR, BELL TEL LABS, 65- Mem: Am Chem Soc; Am Soc Testing & Mat; Sigma Xi. Res: Research, development, and improvement of materials and processes of interest to the semiconductor and electronics industry, primarily solid state and electrochemical activities. Mailing Add: Bell Tel Labs Murray Hill NJ 07974

BUCKLEY, WILLIAM DEREK, solid state physics, see 12th edition

BUCKLIN, ROBERT VAN ZANDT, b Chicago, Ill, June 25, 16; m 67; c 5. PATHOLOGY, FORENSIC MEDICINE. Educ: Loyola Univ, Ill, BSM, 38, MD, 41; STex Col Law, JD, 69; Am Bd Path, dipl path anat, clin & forensic path. Prof Exp: Intern, St Josephs Hosp, Tacoma, Wash, 40-41; resident path, Tacoma Gen Hosp, 41-42; chief lab serv path, St Mary's Hosp & Saginaw Gen Hosp, Mich, 46-63; assoc med examr, Harris County, Tex, 64-68; prof path, Univ Tex Med Br, Galveston, 69-71; chief med examr, Galveston County, Tex & dir med legal affairs, Health Commun, Inc, 72-74; DEP MED EXAMR, LOS ANGELES COUNTY, 74- Mem: Am Col Physicians; Am Soc Clin Path; Am Col Legal Med; Am Acad Forensic Sci; AMA. Res: Medical and legal aspects of the Crucifixion; aging and dating of traumatic lesions. Mailing Add: 1480 Carla Ridge Beverley Hills CA 90210

BUCKMAN, ALFRED FLETCHER, b Philadelphia, Pa, Sept 30, 11; m 35; c 2. COLLOID CHEMISTRY. Educ: Muskingum Col, BS, 34. Prof Exp: Chief chemist, James Good, Inc, Pa, 34-37; res chemist, R M Hollingshead Corp, NJ, 37-40; chief chemist, J A Tumbler Labs, Md, 40-41 & A Penn Oil Co, Pa, 41-42; develop chemist, S C Johnson & Son, Inc, 46-50, chief control chemist, 50-51, prod control mgr, 51-56, asst dir res & develop, 56-58, dir, 58-63, vpres, 63-75; RETIRED. Mem: Am Chem Soc; fel Am Inst Chemists. Res: Self-polishing wax dispersions; paste and liquid waxes; soaps, emulsifiers, detergents, insecticides, disinfectants and automotive chemical specialties; research management and administration. Mailing Add: 1504 Westwood Circle Racine WI 53404

BUCKMAN, ALVIN BRUCE, b Omaha, Nebr, Dec 7, 41; m 66. ELECTROOPTICS, OPTICAL PHYSICS. Educ: Mass Inst Technol, BS, 64; Univ Nebr, MS, 66, PhD(elec eng), 68. Prof Exp: Instr elec eng, Univ Nebr, Lincoln, 66-68, from asst prof to assoc prof, 68-74, assoc prof elec eng, 73-74; ASSOC PROF ELEC ENG, UNIV TEX, AUSTIN, 74- Mem: Am Phys Soc; Optical Soc Am. Res: Optical properties and electronic structure of solids; especially surface and thin film effects; integrated optical devices. Mailing Add: Elec Eng Dept ENS103 Univ of Tex Austin TX 78712

BUCKMAN, ROBERT E, forestry, see 12th edition

BUCKMAN, WILLIAM GORDON, b Morganfield, Ky, Dec 1, 34; m 59; c 3. SOLID STATE PHYSICS. Educ: Western Ky Univ, BS, 60; Vanderbilt Univ, MS, 62; Univ NC, PhD(pub health, radiation physics), 67. Prof Exp: Instr radiation hyg, Univ NC, 62-64; chmn dept physics & math, Ky Wesleyan Col, 66-67; from asst prof to assoc prof, 67-73, PROF PHYSICS, WESTERN KY UNIV, 73- Mem: Am Phys Soc; Health Physics Soc. Res: Luminescence; solid state radiation dosimetry. Mailing Add: Dept of Physics Western Ky Univ Bowling Green KY 42101

BUCKMASTER, HARVEY ALLEN, b Calgary, Alta, Apr 8, 29; m 56, 68. MAGNETIC RESONANCE. Educ: Univ Alta, BSc, 50; Univ BC, MA, 52, PhD(physics), 56. Prof Exp: Nat Res Coun Can fel, Cavendish Lab, Cambridge Univ, 55-57; asst prof physics, Univ Alta, 57-60; from asst prof to assoc prof, 60-67, PROF PHYSICS, UNIV CALGARY, 67- Concurrent Pos: Leverhulme vis res fel, Univ Keele, 64-65; mem sci adv comt, Environ Conserv Authority, Prov Alta; mem bd gov, Univ Calgary, 75-78. Mem: Sr mem Inst Elec & Electronics Engrs; Can Asn Physicists; fel Brit Inst Physics; Am Asn Physics Teachers. Res: Electronic paramagnetic resonance and endor lanthanide elements in hydrated and anhydrous lattices; application determinations in biophysics; instrumentation development and sensitivity studies; electronic paramagnetic resonance; nuclear magnetic resonance normal/malignant human organ tissue. Mailing Add: Dept of Physics Univ of Calgary Calgary AB Can

BUCKMASTER, JOHN DAVID, b Belfast, Northern Ireland, Feb 2, 41; m 66; c 1. FLUID MECHANICS, APPLIED MATHEMATICS. Educ: Univ London, BSc, 62; Cornell Univ, PhD(appl math), 69. Prof Exp: Res engr fluids, Cornell Aeronaut Lab, Buffalo, NY, 62-65; lectr, Univ London, 68-69; asst prof, NY Univ, 69-72; asst prof eng & appl sci, Yale Univ, 72-74; ASSOC PROF MATH & MECH, UNIV ILL, URBANA, 74- Mem: Am Phys Soc; Soc Indust & Appl Math. Mailing Add: Dept of Math Univ of Ill Urbana IL 61801

BUCKMASTER, MARLIN DWIGHT, organic chemistry, polymer chemistry, see 12th edition

BUCKNAM, ROBERT CAMPBELL, b Lander, Wyo, Sept 29, 40; m 63; c 3. GEOLOGY. Educ: Colo Sch Mines, Geol Eng, 62; Univ Colo, PhD(geol), 69. Prof Exp: GEOLOGIST, US GEOL SURV, 68- Mem: Geol Soc Am; Seismol Soc Am.

Res: Structural geology; geology of earthquakes. Mailing Add: US Geol Surv Bldg 25 Fed Ctr Denver CO 80225

BUCKNER, CHARLES HENRY, b Toronto, Ont, Oct 8, 28; m 53; c 2. WILDLIFE ECOLOGY. Educ: Univ Toronto, BA, 52; Univ Man, MSc, 54; Univ Western Ont, PhD(zool), 59. Prof Exp: Res officer mammal, Can Dept Agr, 52-60 & Can Dept Fisheries & Forestry, 60-66, head vert biol, 66-70, HEAD ECOL IMPACT, CHEM CONTROL RES INST, 70-, DEP DIR, 75- Mem: Am Soc Mammal; Entom Soc Can. Res: Impact of pesticides on non-target organisms; vertebrate damage to forests; populations feeding and behavior; ornithology; population dynamics; vertebrate predation on forest insects. Mailing Add: Chem Control Res Inst Can Forestry Serv 25 Pickering Pl Ottawa ON Can

BUCKNER, DEAN A, mineralogy, see 12th edition

BUCKNER, EDWARD REAP, forestry, see 12th edition

BUCKNER, RALPH GUPTON, b Casper, Wyo, May 12, 21; m 49; c 2. VETERINARY MEDICINE. Educ: Westminster Col, AB, 47; Kans State Univ, BS, DVM, 56; Univ Okla, MS, 66. Prof Exp: Instr vet med & surg, 56-58, from asst prof to assoc prof vet path, 58-67, assoc prof vet med & surg, 67-69, prof vet path, 69-73, PROF VET MED & COORDR SMALL ANIMAL CLIN, OKLA STATE UNIV, 74- Mem: AAAS; Am Vet Med Asn; Am Soc Vet Clin Path; NY Acad Sci; Am Col Vet Int Med. Res: Clinical pathology, especially hematology and blood coagulation diseases; canine hemophilia. Mailing Add: Dept of Vet Med & Surgery Col of Vet Med Okla State Univ Stillwater OK 74074

BUCKNER, ROBERT CECIL, b Green County, Ky, July 20, 18; m 43; c 1. PLANT GENETICS. Educ: Univ Ky, BS, 47, MS, 48; Univ Minn, PhD(plant genetics), 55. Prof Exp: Asst agronomist, 48-55, assoc agronomist, 56, RES AGRONOMIST PLANT BREEDING, AGR RES SERV, USDA, UNIV KY, 56- Mem: Fel Am Soc Agron. Res: Grass breeding, palatability and disease resistance. Mailing Add: Agr Exp Sta Univ of Ky Lexington KY 40506

BUCKSER, STANLEY, physical chemistry, see 12th edition

BUCKWALTER, GARY LEE, b Kittanning, Pa, June 29, 34; m 57; c 2. PHYSICS. Educ: Pa State Univ, BS, 56; Cath Univ Am, MS, 61, PhD(physics), 66. Prof Exp: Instr, USNaval Acad, 59-62, asst prof, 62-65; fel, Cath Univ Am, 66; assoc prof, 66-69, PROF PHYSICS, INDIANA UNIV, PA, 69- Concurrent Pos: REs Corp Am res grant, 67- Mem: Am Asn Physics Teachers; Am Phys Soc. Res: High energy cosmic ray physics. Mailing Add: Dept of Physics Ind Univ of Pa Ind PA 15701

BUCKWALTER, GEOFFREY RIGGS, organic chemistry, see 12th edition

BUCKWALTER, HOWARD MCWILLIAMS, b Lancaster, Pa, June 1, 02; m 29; c 2. ORGANIC CHEMISTRY. Educ: Franklin & Marshall Col, BS, 24; Univ Pa, PhD(org chem), 29. Prof Exp: Asst chem, Franklin & Marshall Col, 24; instr org chem, Univ Pa, 26-29; res chemist, US Rubber Co, Passaic, NJ, 29-32, res chemist & mgr cellulose res lab, Detroit, 32-50, org res, 50-58; CONSULT CHEMIST, 58- Mem: AAAS; Am Chem Soc; fel Am Inst Chemists. Res: Organic analysis; cellulose; carbohydrates; natural resins. Mailing Add: 23401 SW 162nd Ave Homestead FL 33030

BUCKWALTER, JOSEPH ADDISON, b Royersford, Pa, Jan 4, 20; m 46; c 4. SURGERY. Educ: Colgate Univ, AB, 41; Univ Pa, MD, 44; Am Bd Surg, dipl. Prof Exp: Intern, Col Med, Univ Iowa, 44-45, resident gen surg, 45-46 & 49-52, assoc surg, 52-54, from asst prof to prof, 54-70; PROF SURG, SCH MED, UNIV NC, CHAPEL HILL, 70- Concurrent Pos: Fel anat, Sch Med, Univ NC, 48, fel path, 48-49; sr registr, Postgrad Med Sch, Univ London, 53-54; dir surg, Dorothea Dix Hosp, Raleigh, NC. Mem: Soc Exp Biol & Med; AMA; Am Col Surg; Soc Univ Surg; Int Soc Surg. Res: Clinical and basic studies of thyroid disorders, gastroenterology, blood groups, human genetics, parathyroid physiology and disease, blood volume dynamics. Mailing Add: Dept of Surg NC Mem Hosp Univ of NC Chapel Hill NC 27514

BUCKWALTER, TRACY VERE, JR, b Canton, Ohio, May 25, 18; m 42; c 2. GEOLOGY. Educ: Univ Mich, BS, 40, MS, 46, PhD(geol), 50. Prof Exp: Asst geol, Univ Mich, 39-41; jr geologist, US Engr Off, 42; geologist, Tex Co, 42-46; from asst prof to assoc prof geol, Univ Pittsburgh, 49-65, asst to pres, Titusville Campus, 63-65; PROF GEOL, CLARION STATE COL, 65-, CHMN DEPT GEOG & EARTH SCI, 72- Concurrent Pos: Coop geologist, State Geol Surv, Pa, 51-69. Mem: Fel Geol Soc Am; Am Asn Petrol Geologists; AAAS. Res: Igneous and metamorphic geology in eastern Pennsylvania; structural geology; geologic education; environmental geology. Mailing Add: Dept of Geog & Earth Sci Clarion State Col Clarion PA 16214

BUCKWOLD, SIDNEY JOSHUA, b Winnipeg, Can, Oct 30, 19; nat US; m 45; c 3. INORGANIC CHEMISTRY, ANALYTICAL CHEMISTRY. Educ: Univ Manitoba, BSA, 42, MSc, 44; Hebrew Univ, Israel, PhD(soil chem), 54. Prof Exp: Asst prof soils, NDak Agr Col, 45-46; sr asst irrig & soil chem, Univ Calif, Davis, 46-47; soil chemist & chief soil & water sect, Ministry Agr, Govt Israel, 48-50; fac assoc soil chem, Hebrew Univ, Israel, 50-54; res assoc clay mineral, Pa State Univ, 55-56; asst prof chem, NMex Inst Mining & Technol, 56-58; assoc prof & head dept, Col of St Joseph on the Rio Grande, 58-61; assoc prof, Am Int Col, 61-64, chmn dept, 63-69, prof, 65-69; ASSOC PROF CHEM, CENT CONN STATE COL, 69- Mem: Am Chem Soc; Am Geochem Soc. Res: Quantitative analysis; instrumental methods; coordination chemistry; ion exchange; physical chemistry of clays; soil-water relations. Mailing Add: Dept of Chem Cent Conn State Col New Britain CT 06050

BUCOLO, GIOVANNI, b Arezzo, Italy, Feb 2, 21; US citizen; m 55. ORGANIC CHEMISTRY, BIOCHEMISTRY. Educ: Univ Florence, DSc, 50. Prof Exp: Chief biochemist, Mayer Hosp & Pediat Clin, Florence, Italy, 50-53, New York Polyclin Med Sch & Hosp, 53-55, D N Sharp Mem Community Hosp, San Diego, Calif, 55-57 & Paterson Gen Hosp, NJ, 57-58; res assoc, Scripps Clin & Res Found, La Jolla, 58-61; dir clin lab, Calif Corp Biochem Res, Los Angeles, 61-62; indust hygienist, Gen Atomic Div, Gen Dynamics Corp, 62-69; dir diag res, Calbiochem, 69-75; DIR BIOCHEM RES & DEVELOP, COULTER DIAG, INC, 75- Mem: AAAS; Am Chem Soc; Am Asn Clin Chemists; Am Heart Asn; NY Acad Sci. Res: Clinical chemistry; metabolism of the red blood cell, particularly changes during storage; pentose phosphates pathway and nucleotides turnover. Mailing Add: 19600 W St Andrews Dr Hialeah FL 33015

BUCOVAZ, EDSEL TONY, b Eldorado, Ill, May 8, 28; m 53; c 2. BIOCHEMISTRY, ORGANIC CHEMISTRY. Educ: Southern Ill Univ, BA, 55, MA, 57; St Louis Univ, PhD(biochem), 62. Prof Exp: Control & res chemist, US Chem Co, 53-54; asst prof biochem, Ctr Health Sci, Univ Tenn, Memphis, 64-69; assoc prof, 69-75, prof biochem, 75, ASSOC DIR BASIC RES, MEMPHIS REGIONAL CANCER CTR, 74- Concurrent Pos: Res assoc fel biochem, Sch Med, Univ Tenn, 62-64. Mem: Sigma Xi; AAAS; Am Chem Soc; Am Asn Cancer Res; Am Soc Biol Chem. Res: Protein biosynthesis; cancer research; chemical carcinogenesis; human research related to fetal maturity and respiratory depression caused by analgysics; the comorosan effect on biological systems. Mailing Add: 4929 Mockingbird Ln Memphis TN 38117

BUCY, LAVERNE, b New Concord, Ky, Nov 21, 16; m 37; c 1. NUTRITION, BIOLOGY. Educ: Univ Ky, BS, 43, MA, 51; Univ Ill, PhD(nutrit), 54. Prof Exp: High sch teacher agr, Ky, 43-44 & 47-51; asst animal sci, Univ Ill, 51-54; PROF NUTRIT, CALIF POLYTECH STATE UNIV, SAN LUIS OBISPO, 55- Mem: Am Soc Animal Sci. Res: Animal nutrition, sheep, swine, beef cattle, poultry. Mailing Add: Dept of Nutrit Calif Polytech State Univ San Luis Obispo CA 93401

BUCY, PAUL CLANCY, b Hubbard, Iowa, Nov 13, 04; m 27; c 2. NEUROSURGERY, NEUROLOGY. Educ: Univ Iowa, BS, 25, MS & MD, 27. Hon Degrees: MD, Univ Thessaloniki, 70; Dr, State Univ Utrecht, 71. Prof Exp: Asst neuropath, Univ Iowa, 25-27; intern, Henry Ford Hosp, Detroit, 27-28; instr neurosurg, Univ Chicago, 28-33, from asst prof to assoc prof, 33-41, head div neurol & neurosurg, 39-41; from assoc prof to prof neurol & neurosurg, Univ Ill, 41-54; prof, 54-73, EMER PROF SURG, MED SCH, NORTHWESTERN UNIV, 73- Concurrent Pos: Gorgas lectr, Univ Ala, 44; John Black Johnston lectr, Univ Minn, 49; Commonwealth vis prof, Univ Louisville, 50; Max Minor Peet lectr, Univ Mich, 56; mem, World Fedn Neurosurg Socs, pres, 57-61, hon pres, 61-; mem adv coun, Nat Inst Neurol Dis & Blindness, 61-64, prog proj comt, 65-69; vis lectr, Free Univ Berlin, 63 & Sch Med, Johns Hopkins Univ, 64; vis prof, Southwest Med Sch, Univ Tex, 63, Univ Rochester, Montreal Neurol Inst & Harvard Univ, 69, Mass Gen Hosp & State Univ Utrecht, 69; mem, Ill Psychiat Training & Res Authority; W P Van Wagenen & J Hughlings Jackson lectrs, 65; chmn, Nat Comt Res in Neurol Dis, 69-75; ed, Surg Neurol. Honors & Awards: Hon dipl, Nat Cong Surgeons & Mex Conf Neurol & Phys Surgeons, 52. Mem: Soc Neurol Surg (pres, 59-60); Am Surg Asn; Am Physiol Soc; Am Asn Neurol Surg (pres, 51-52); Am Neurol Asn (vpres, 54-55, pres, 71-72). Res: Spinal cord injury; structure and function of the cerebral cortex; involuntary movements; intracranial neoplasms; clinical neurology and neurological surgery. Mailing Add: PO Box 1457 Tryon NC 28782

BUCY, RICHARD SNOWDEN, b Washington, DC, July 20, 35; m 61; c 2. APPLIED MATHEMATICS. Educ: Mass Inst Technol, BS, 57; Univ Calif, Berkeley, PhD(probability), 63. Prof Exp: Assoc mathematician, Appl Physics Lab, Johns Hopkins Univ, 57-60; mathematician, Res Inst Advan Studies, 60-61, 63-64; asst prof math, Univ Md, College Park, 64-65; assoc prof, Univ Colo, Boulder, 65-67; assoc prof aerospace & math, 67-70, PROF AEROSPACE ENG & MATH, UNIV SOUTHERN CALIF, 70- Concurrent Pos: Consult math dept, Rand Corp, 63-, Thompson Ramo Wooldridge Corp, Aerospace Corp & Electrac Inc; prof, Tech Univ Berlin, 75-76. Honors & Awards: US Sr Scientist Award, Alexander von Humboldt Found, 75. Mem: Am Math Soc; Soc Indust & Appl Math; Math Asn Am; fel Inst Elec & Electronics Eng. Res: Probability and control theory; optimum control and filtering theory; adaptive control; Markov processes; discrete potential theory. Mailing Add: 2035 Paseo del Sol Palos Verdes Estates CA 90274

BUDAY, PAUL VINCENT, b Jersey City, NJ, Aug 19, 31; m 56; c 5. PHARMACOLOGY. Educ: Fordham Univ, BS, 53; Temple Univ, MS, 55; Purdue Univ, PhD(pharmacol), 58. Prof Exp: Asst chem, Temple Univ, 53-55; asst pharmacol, Purdue Univ, 55-56; asst prof biol sci, Fordham Univ, 58-59; asst prof pharmacol, col pharm, Univ RI, 59-61; pharmacol ed, Lederle Labs, New York, 61-63; head sci info, Vick Div Res & Develop, Richardson-Merrell Inc, 63-65; med res assoc, Warner-Chilcott Labs, NJ, 65-66, asst to dir med serv, Warner-Lambert Res Inst, 66-67; head drug regulatory affairs, Sandoz Pharmaceut, 67-69; asst dir regulatory affairs, 69-75, DIR REGULATORY AFFAIRS, ETHICON, INC, 75- Concurrent Pos: Gustavus A Pfeiffer Mem res fel, Am Found Pharmaceut Ed, 58-59. Mem: Fel Am Med Writers' Asn; Am Chem Soc; AMA; Am Pharmaceut Asn; Int Soc Biochem Pharmacol. Res: Pharmacology of neurohormones; interrelationships between endocrine hormones and drug action; drug interactions and preclinical toxicity; medical writing; clinical testing of drugs; food and drug law. Mailing Add: Ethicon, Inc Rte 22 Somerville NJ 08876

BUDD, GEOFFREY COLIN, b London, Eng, Dec 2, 35; m 59; c 4. CELL BIOLOGY, PHYSIOLOGY. Educ: Chelsea Col Sci & Technol, Univ London, BSc, 58; Birkbeck Col, Univ London, BSc, 61; Royal Free Hosp Med, Univ London, PhD(zool), 66. Prof Exp: Tech officer, Med Res Coun Biophys Res Unit, Kings Col, Univ London, 58-61; asst lectr biol & radiation biol, Royal Free Hosp Sch Med, Univ London, 61-62, lectr, 6266; res assoc appl physics, neurobiol & behav, Cornell Univ, 6668, vis asst prof, 68-69; from asst prof to assoc prof, 6974, PROF PHYSIOL, MED COL OHIO, 74- Mem: Fel AAAS; Am Soc Cell Biol; Electron Micros Soc Am; Tissue Cult Asn; Histochem Soc. Res: Cytophysiology of liver; quantitative enzyme cytochemistry; development and application of quantitative electron microscope autoradiography. Mailing Add: Med Col of Ohio PO Box 6190 Toledo OH 43614

BUDD, THOMAS WAYNE, b Pittsburgh, Pa, Oct 1, 46. PLANT PHYSIOLOGY, MOLECULAR BIOLOGY. Educ: NDak State Univ, BS, 68, PhD(bot), 72. Prof Exp: ASST PROF BIOL, ST LAWRENCE UNIV, 72- Mem: AAAS; Am Soc Plant Physiologists; Sigma Xi. Res: Ultrastructure and cell type distribution of human adenoids; the role of the enzyme carbonic anhydrase in higher plants; microelectrode determination of plant cell potentials. Mailing Add: Dept of Biol St Lawrence Univ Canton NY 13617

BUDDE, MARY LAURENCE, b Covington, Ky, May 17, 29. ZOOLOGY, MICROBIOLOGY. Educ: Villa Madonna Col, AB, 53; Cath Univ Am, MS, 55, PhD(zool), 58. Prof Exp: Instr, 58-61, from asst prof to assoc prof, 61-75, PROF BIOL, THOMAS MORE COL, KY, 75- Mem: Am Soc Microbiol. Res: Endocrinology: regulation of calcium and phosphorus metabolism in bony fish; aquatic biology: power plant operations and microorganisms and fish population effects. Mailing Add: Thomas More Col PO Box 85 Fort Mitchell KY 41017

BUDDE, PAUL BERNARD, b Covington, Ky, June 24, 26; m 59; c 6. AGRICULTURAL CHEMISTRY. Educ: Xavier Univ, Ohio, BS, 50, MS, 51. Prof Exp: Chemist, 59-62, res chemist, 62-71, SR RES CHEMIST, EDGAR C BRITTON RES LAB, DOW CHEM CO, 71- Mem: Am Chem Soc; Am Inst Chem. Res: Synthesis of organic compounds for agricultural uses. Mailing Add: Agri-org Dept Bldg 9001 Dow Chem Co Midland MI 48640

BUDDE, WILLIAM L, b Cincinnati, Ohio, Dec 18, 34; m 59; c 5. ANALYTICAL CHEMISTRY. Educ: Xavier Univ, BS, 57, MS, 59; Univ Cincinnati, PhD(chem), 63. Prof Exp: Fel, Univ Calif, Riverside, 63-64; vis asst prof chem, Univ Kans, 64-65; assoc to sr chemist, Midwest Res Inst, 65-69; assoc dir, Mass Spectrometry Ctr, Purdue Univ, 69-71; GROUP LEADER ORG INSTRUMENTATION, OFF RES & DEVELOP, US ENVIRON PROTECTION AGENCY, 71- Mem: Am Chem Soc; Am Soc Mass Spectrometry. Res: Mass spectrometry: trace organic analysis by computerized gas chromatography; laboratory automation with digital computers. Mailing Add: Off Res & Develop US Environ Protection Agency Cincinnati OH 45268

BUDDEMEIER, ROBERT WORTH, b Chicago, Ill, Feb 24, 39; m 60; c 3. OCEANOGRAPHY, HYDROLOGY. Educ: Univ Ill, BS, 58; Univ Wash, Seattle, PhD(inorg chem), 69. Prof Exp: Chemist, Chem Dept, Univ Wash, Seattle, 65; asst prof inorg, radio- & geochem, Chem Dept & Hawaii Inst, 69-73, asst prof, 73-74, ASSOC PROF CHEM OCEANOG, DEPT OCEANOG & HAWAII INST GEOPHYS, UNIV HAWAII, 74- Mem: AAAS; Am Geophys Union. Res: Radiocarbon dating and geochronology; quaternary climatic changes; natural carbon cycle; coral, reef and atoll processes; groundwater hydrology; environmental radioactivity. Mailing Add: Dept of Oceanog Univ of Hawaii Honolulu HI 96822

BUDDEMEIER, WILBUR DAHL, b Sidney, Ill, Feb 26, 12; m 36; c 2. AGRICULTURAL ECONOMICS. Educ: Univ Ill, BS, 33, MS, 41, PhD(agr econ), 52. Prof Exp: Instr high sch, Ill, 33-39; fieldman, Blackhawk Farm Bur, Farm Mgt Serv, 43-47; asst agr econ, 39-43, asst prof, 47-50, asst prof agr econ & vocational agr, 50-54, assoc prof, 54-59, PROF FARM MGT, UNIV ILL, URBANA-CHAMPAIGN, 59-, ASST DIR INT AGR PROG, 69-, ASSOC DEAN, COL AGR, 70- Concurrent Pos: Adv & group leader, India, USAID, 59-63; contract admin adv & group leader, JN Agr Univ, Jabalpur, India, 64-67; consult, Off Technol Assessment, 74-75 & Food Adv Comt, Congress of US, 75-76. Mem: Agr Econ Asn; Int Conf Agr Econ; Soc Int Develop. Res: Farm management and international economy and agriculture. Mailing Add: 3004 Meadowbrook Ct Champaign IL 61820

BUDDENHAGEN, IVAN WILLIAM, b Ventura, Calif, Apr 26, 30; m 50; c 4. PLANT PATHOLOGY. Educ: Ore State Col, BS, 53, MS, 54, PhD, 57. Prof Exp: Plant pathologist, United Fruit Co, Honduras, 57-64; prof plant path, Univ Hawaii, 64-75; MEM STAFF CEREAL IMPROV PROG, INT INST TROP AGR, 75- Mem: Am Phytopath Soc. Res: Bacterial plant diseases; international plant path; host-parasite interaction; wilt diseases. Mailing Add: Int Inst of Trop Agr PMB 5320 Ibadan Nigeria

BUDDING, ANTONIUS JACOB, b Amsterdam, Netherlands, Jan 30, 22; m 53. GEOLOGY. Educ: Univ Amsterdam, BSc, 42, MSc, 48, PhD(geol), 51. Prof Exp: Instr geol & petrol, Geol Inst, Amsterdam, 46-51; geologist, Geol Surv, Sask, 51-52, prin geologist, 52-56; from asst prof to assoc prof, 56-72, PROF GEOL, N MEX INST MINING & TECHNOL, 72- Concurrent Pos: Consult, Los Alamos Sci Lab, 72- Mem: Fel Geol Soc Am; Ger Geol Soc; Am Geophys Union; Nat Asn Geol Teachers. Res: Metamorphic and igneous petrology; structural geology. Mailing Add: Dept of Geosci NMex Inst Mining & Technol Socorro NM 87801

BUDDINGH, GERRIT JOHN, b Oosterbeek, Neth, Aug 7, 04; US citizen; m 41. MICROBIOLOGY, PATHOLOGY. Educ: Calvin Col, AB, 29; Vanderbilt Univ, MD, 35. Prof Exp: Instr bact, Sch Med, Vanderbilt Univ, 36-38, instr path, 38-40, asst prof path, 40-42, from assoc prof to prof bact, 4248; prof microbiol, 48-74, EMER PROF MICROBIOL, MED CTR, LA STATE UNIV, NEW ORLEANS, 74- Concurrent Pos: Commonwealth travel fel, Europ Ctrs Virus Res, 38; assoc, Comt Influenza, Armed Forces Epidemiol Bd, 41-72; mem, Ekiri Comn Japan, 47; mem, Ad Hoc Comt Microbiol Probs Space Travel, NASA, 67-68. Mem: AAAS; Am Asn Pathologists & Bacteriologists; Soc Exp Biol & Med; Am Soc Infectious Dis. Res: Chick embryo technics for the study of the pathogenesis of infectious processes, viral, bacterial and mycological. Mailing Add: 1634 Foucher St New Orleans LA 70115

BUDDINGTON, ARTHUR FRANCIS, b Wilmington, Del, Nov 29, 90; m 24; c 1. GEOLOGY. Educ: Brown Univ, PhB, 12, MS, 13; Princeton PhD(geol), 16. Honors & Awards: Hon ScD, Brown Univ, 42; hon LLD, Franklin & Marshall Col, 58; hon degree, Univ Liege, 67. Prof Exp: Procter fel geol, Princeton Univ, 16-17; instr, Brown Univ, 17-19; petrologist, Geophys Lab, Carnegie Inst Technol, 19-20; from asst prof to assoc prof, 20-32, prof, 32-59, chmn dept, 36-50, sr res geologist, 62-64, EMER BLAIR PROF GEOL, PRINCETON UNIV, 59- Concurrent Pos: Geologist, US Geol Surv, 43-62; guest lectr, India Sci Cong & Geol Surv India, 57. Honors & Awards: Penrose Medal, Geol Soc Am, 54; Roebling Medal, Mineral Soc Am, 56; Dumont Medal, Geol Soc Belg, 60. Mem: Nat Acad Sci; Geol Soc Am (vpres, 43-47); Soc Econ Geol; fel Mineral Soc Am (pres, 42); fel Am Geophys Union. Res: Geology and mineral deposits of southeastern Alaska and the Adirondacks; geologic surveys in Newfoundland and Oregon Cascades; geology of magnetite iron ore deposits of New York, New Jersey and Pennsylvania; role of iron-titanium oxide minerals in paleomagnetism and petrogeny; origin of anorthosite. Mailing Add: 65 Pond St Cohasset MA 02025

BUDENSTEIN, PAUL PHILIP, b Philadelphia, Pa, June 27, 28; m 57; c 5. PHYSICS. Educ: Temple Univ, BA, 49; Lehigh Univ, MS, 51, PhD(physics), 57. Prof Exp: Instr physics, Lehigh Univ, 52-56; mem tech staff, Bell Tel Labs, 57-58; asst prof, 59-62, ASSOC PROF PHYSICS, AUBURN UNIV, 63- Concurrent Pos: Consult, US Army Missile Command, 69-71. Mem: Am Vacuum Soc; Electrochem Soc; Am Phys Soc; Am Asn Physics Teachers. Res: Solid state physics; thin films. Mailing Add: Dept of Physics Auburn Univ Auburn AL 36830

BUDERER, MELVIN CHARLES, b Sacramento, Calif, May 12, 41; m 64; c 2. PHYSIOLOGY. Educ: Univ Calif, Berkeley, AB, 63, PhD(physiol), 70. Prof Exp: Nat res coun resident res assoc, NASA L B Johnson Space Ctr, 70-72, RES PHYSIOLOGIST, TECHNOL INC, NASA L B JOHNSON SPACE CTR, 72- Mem: Aerospace Med Asn; Sigma Xi; Am Soc Mass Spectrometry. Res: Environmental physiology; space physiology. Mailing Add: 1707 Bowline Rd Houston TX 77058

BUDICK, BURTON, b Bronx, NY, May 22, 38; m 64; c 1. NUCLEAR PHYSICS, ATOMIC PHYSICS. Educ: Harvard Univ, AB, 59; Univ Calif, Berkeley, PhD(physics), 62. Prof Exp: Res assoc, Radiation Lab, Columbia Univ, 62-63, instr, 63-64; lectr, Hebrew Univ, Jerusalem, 64-67, sr lectr, 67-68; asst prof, 68-70, ASSOC PROF PHYSICS, NY UNIV, 70- Concurrent Pos: NSF fel Hebrew Univ, 65-67. Mem: Am Phys Soc. Res: Experimental nuclear physics; muonic and pionic atoms; atomic hyperfine structure. Mailing Add: Dept of Physics NY Univ Univ Heights Bronx NY 10453

BUDINGER, THOMAS FRANCIS, b Evanston, Ill, Oct 25, 32; m 65; c 3. MEDICAL PHYSICS, NUCLEAR MEDICINE. Educ: Regis Col, Colo, BS, 54; Univ Wash, MS, 57; Univ Colo, MD, 64; Univ Calif, Berkeley, PhD, 71. Prof Exp: Asst chem, Regis Col, Colo, 53-54; anal chemist, Indust Labs, 54; sr oceanogr, Univ Wash, 61-66; physicist, Lawrence Livermore Lab, Univ Calif, 6667, RES PHYSICIAN, DONNER LAB & LAWRENCE BERKELEY LAB, UNIV CALIF, 67-, DIR MED SERV, LAWRENCE BERKELEY LAB, 70- Concurrent Pos: Peter Bent Brigham Hosp, Boston, 64. Mem: AAAS; Am Geophys Union; Am Meterol Soc; NY Acad Sci; Soc Nuclear Med. Res: Imaging body functions; electrical, magnetic, sound and photon radiation fields; electron microscopy; trace elements in disease; mathematics and computers in biology; polar oceanography. Mailing Add: Lawrence Berkeley Lab Univ of Calif Berkeley CA 94720

BUDKE, CLIFFORD CHARLES, b Cincinnati, Ohio, Apr 3, 32; m 63; c 3. ANALYTICAL CHEMISTRY. Educ: Univ Cincinnati, BS, 54, MS, 67. Prof Exp: Anal chemist, 56-75, RES SUPVR, US INDUST CHEM CO, DIV NAT DISTILLERS & CHEM CORP, 72- Mem: Am Chem Soc; Am Soc Test & Mat. Res: Chemical and instrumental methods of analysis, especially trace analysis, functional group analysis, and electrometric methods of analysis. Mailing Add: US Indust Chem Co 1275 Sect Rd Cincinnati OH 45237

BUDNER, STANLEY, b New York, NY, Nov 12, 33; m 61; c 2. PUBLIC HEALTH. Educ: City Col New York, BA, 55; Columbia Univ, PhD(social psychol), 60. Prof Exp: Dir, Res & Demonstration Ctr, Sch Social Serv, Fordham Univ, 67-72; ASSOC PROF PUB HEALTH, FAC MED, COLUMBIA UNIV, 72- Concurrent Pos: Clin asst prof pediat, New York Med Col, 68-75; consult, Found Thanatology, 71-73 & United Neighborhood Houses of New York, 73-75; assoc ed, Man & Med, 72-; assoc dir, Joint Comn Neurol, 72-74. Mem: AAAS; Am Pub Health Asn; Am Psychol Asn; Am Sociol Asn; Am Asn Ment Deficiency. Res: The relation of patient management to therapeutic outcome; ethical and value issues in health care delivery and research. Mailing Add: Columbia Univ Fac of Med 630 W 168th St New York NY 10032

BUDNEY, MARY LILLIAN, b Worcester, Mass, Nov 10, 33. MICROBIOLOGY. Educ: Marywood Col, BS, 58; Univ Notre Dame, MS, 59; St Johns Univ, NY, PhD(microbiol), 65. Prof Exp: Teacher, St John Cantius Sch, 54-56; instr biol, 59-62, assoc prof, 64-71, PROF BIOL, HOLY FAMILY COL, PA, 72- Mem: AAAS; Nat Asn Biol Teachers. Res: Behavior of platelets and other clotting factors in artificial arterial prostheses; changes in the structure of DNA as a result of exposure to drugs; effects of cations, colloids and lipids on membrane strength. Mailing Add: Dept of Biol Grant & Frankford Aves Holy Family Col Philadelphia PA 19114

BUDNICK, JOSEPH IGNATIUS, b Jersey City, NJ, July 9, 29. PHYSICS. Educ: St Peters Col, BS, 51; Rutgers Univ, PhD(physics), 55. Prof Exp: Asst physics lab, Rutgers Univ, 51-52, asst atomic & nuclear lab, 52-53; proj physicist superconductivity, Res Ctr, Int Bus Mach, NY, 55; from assoc prof to prof, Fordham Univ, 55-74; PROF PHYSICS, UNIV CONN, 74- Concurrent Pos: Instr, Int Bus Mach Gen Educ, 57-58. Mem: AAAS; Am Phys Soc. Res: Superconductivity; low temperature solid state physics of metals. Mailing Add: Dept of Physics Univ of Conn Storrs CT 06268

BUDNITZ, ROBERT JAY, b Pittsfield, Mass, Oct 12, 40; m 61; c 2. ENVIRONMENTAL SYSTEMS AND TECHNOLOGY. Educ: Yale Univ, BA, 61; Harvard Univ, MA, 62, PhD(physics), 68. Prof Exp: Physicist, 67-71, coordr environ prog, 74-75, STAFF PHYSICIST ENVIRON SCI, LAWRENCE BERKELEY LAB, UNIV CALIF, 71-, ACTG HEAD ENERGY & ENVIRON DIV, 74- Concurrent Pos: Mem panel on sources & control techniques, Environ Res Assessment Comt, Nat Acad Sci, 75-76. Mem: AAAS; Am Phys Soc. Res: Instrumentation for measuring radiation in the environment; nuclear reactor environmental impact analysis; studies of indoor air pollution parameters and disease effects. Mailing Add: Lawrence Berkeley Lab Univ of Calif Berkeley CA 94720

BUDOWSKI, GERARDO, b Berlin, Ger, June 10, 25; m 58; c 2. FORESTRY, PLANT ECOLOGY. Educ: Univ Venezuela, BS, 48; Inter-Am Inst Agr Sci, MS, 54; Yale Univ, PhD(forest ecol), 62. Prof Exp: Forester, Ministry Agr, Venezuela, 48-50, head dept forest res, 50-52, forester, Northern Zone, Tech Coop Prog, Orgn Am States, 53-55 & Forestry Prog, Inter-Am Inst Agr Sci, 56-57, head dept forestry, 57-67; prog specialist ecol & conserv, Natural Resources Res Div, Paris, 67-70; DIR GEN, INT UNION CONSERV NATURE & NATURAL RESOURCES, UNESCO, SWITZ, 70- Res: Tropical forest ecology and dendrology; silviculture; conservation; national park and forest education planning. Mailing Add: Int Union for Conserv of Nature & Natural Resources 1110 Morges Switzerland

BUDRYS, RIMGAUDAS S, b Kaunas, Lithuania, Aug 2, 25; m 65; c 1. SURFACE CHEMISTRY. Educ: Univ Queensland, BSc, 59; Ill Inst Technol, PhD(phys chem), 64. Prof Exp: Chief chemist, Brisbane City Coun, Tennyson Power Sta, Queensland, Australia, 59-60; sr res chemist, res & develop ctr, Swift & Co, 63-65; sr res chemist, 65-67, ADV SCIENTIST, TECH CTR, CONTINENTAL CAN CO, INC, 68- Mem: AAAS; Am Chem Soc; Am Inst Physics. Res: Intermolecular bonding conformations and structural rearrangements in high polymers; characterization of metal surfaces; adhesion of polymers to metals; infrared spectroscopy of interfaces; physical chemistry of high polymers, coatings and surfaces. Mailing Add: 5833 S Sacramento Ave Chicago IL 60629

BUDY, ANN MARIE, b Waukegan, Ill, May 22, 13. PHYSIOLOGY, PHARMACOLOGY. Educ: Univ Chicago, BS, 46, PhD(pharmacol), 54. Prof Exp: Asst radiobiol, US Air Force Radiation Lab & Toxicity Lab, Univ Chicago, 48-54, physiologist, 47-54, res assoc, 54-55, instr, 55-58, asst prof, 58-66; res assoc histol, Col Dent, Univ Ill, 66-67, from asst prof to assoc prof med, Sch Med, 67-70, assoc researcher, Faculty Enrichment Proj, 70-73, assoc dean, Univ Hawaii Sch Nursing, 73; RESEARCHER, EPIDEMIOLOGIC STUDIES PROG, PBRC, UNIV HAWAII, 73-, ASST DIR, CLINICAL SCI UNIT, CANCER CTR, 75- Mem: AAAS; Orthop Res Soc; Soc Exp Biol & Med; Am Soc Pharmacol & Exp Therapeut; NY Acad Sci. Res: Physiology of bone; endocrinology; sex hormone effects on bone; radiobiology; genetics. Mailing Add: 4300 Waialae Ave B-1003 Honolulu HI 96816

BUDZILLOVICH, GLEB NICHOLAS, b Zlobin, Russia, Sept 7, 23; US citizen; m 50; c 1. PATHOLOGY, NEUROPATHOLOGY. Educ: Univ Munich, MD, 53. Prof Exp: Provisional asst pathologist, Med Ctr, Cornell Univ, 55-57; resident, Manhattan Vet Admin Hosp, 57-58, pathologist, 6061; fel, NY Univ Med Ctr, 61-64; asst prof, 6467, ASSOC PROF PATH & NEUROPATH, NY UNIV MED CTR, 67 Concurrent Pos: Am Cancer Soc fel, Francis Delafield Hosp, New York, 58-60, resident, 58-60; consult path, Manhattan Vet Admin Hosp, 63-; asst vis pathologist, Bellevue Hosp Med Ctr, New York, 65-; asst attend pathologist, Univ Hosp, NY Univ Med Ctr, 65-; neuropath consult, Off Med Examr, Suffolk County, NY, 67- Mem: Am Asn Neuropathologists. Res: General pathology of peripheral sensory and sympathetic ganglia; pathology of sympathetic nervous system in diabetes mellitus; role of diabetic neuropathy in pathogenesis of peripheral vascular disease. Mailing Add: Dept of Path NY Univ Med Ctr New York NY 10016

BUDZINSKI, WALTER VALERIAN, b Buffalo, NY, Aug 12, 37; m 65; c 3. EXPERIMENTAL SOLID STATE PHYSICS. Educ: Canisius Col, BS, 59; Univ Pittsburgh, MS, 64, PhD(physics, 67. Prof Exp: Asst prof, 67-70, ASSOC PROF PHYSICS, ST BONAVENTURE UNIV, 70- Concurrent Pos: Vis assoc prof physics, Univ Pittsburgh, 74-75. Mem: Am Phys Soc. Res: Tunneling into metals; microwave absorption by superconductors. Mailing Add: Dept of Physics St Bonaventure Univ St Bonaventure NY 14778

BUECHE, ARTHUR MAYNARD, b Flushing, Mich, Nov 14, 20; m 45; c 4. POLYMER CHEMISTRY, PHYSICAL CHEMISTRY. Educ: Univ Mich, BS, 43; Cornell Univ, PhD(phys chem), 47. Honors & Awards: Hon ScD, St Lawrence Univ, 69; ScD, Union Col, 73; LLD, Knox Col, 73; DEng, Rensselaer Polytech Inst, 75; ScD, Clarkson Col Technol, 75; ScD, Univ Akron, 75. Prof Exp: Res assoc, Cornell Univ, 47-50 & res lab, Gen Elec Co, 50-56, from unit mgr to sect mgr, 56-61, dept mgr, 61-65, VPRES RES & DEVELOP, 65- Concurrent Pos: Mem comt molecular chem, Nat Res Coun, 64-67, mem-at-large, Div of Chem & Chem Technol, 66-75; mem adv panel, Inst Mat Res, Nat Bur Standards, 66-72, mem vis comt, 71-75; chmn bd dirs, Gordon Res Conf, 67; adv & consult, Res & Technol Adv Coun, NASA, 69-74, mem of del, Conf of Ministers of Europe mem states responsible for sci policy, UNESCO, 70; mem sci & technol policy panel, President's Sci Adv Comt, 70-73; mem adv panel on energy conversion and utilization, NSF, 74; mem acad forum adv comt, Nat Acad Sci, 72-; mem, adv council, Nat Gov Coun Sci & Technol, 72-; mem standing comt res, Sci Technol Adv Group, US Air Force, 73-; mem indust panel on sci & technol, NSF, 74- Mem: Nat Acad Sci; fel AAAS; Am Chem Soc; fel Am Phys Soc; Indust Res Inst (pres, 75-). Res: Theory of management of research and development in electronics science and engineering, energy science and engineering, and materials science and engineering; physics and chemistry of polymers; effects of high-energy radiation on plastics. Mailing Add: Gen Elec Co PO Box 8 Schenectady NY 12301

BUECHE, FREDERICK JOSEPH, b Flushing, Mich, Aug 12, 23. PHYSICS. Educ: Univ Mich, BS, 44; Cornell Univ, PhD(physics), 48. Prof Exp: Asst eng physics, Cornell Univ, 44-48, res assoc phys chem, 48-52; asst prof, Univ Wyo, 52-53; physicist, Rohm and Haas Co, 53-54; prof physics, Univ Wyo, 54-59 & Univ Akron, 59-61; PROF PHYSICS, UNIV DAYTON, 61- Concurrent Pos: Peace Corps, 64-66. Mem: Am Phys Soc. Res: Physical properties of polymers; properties of liquids and gases. Mailing Add: Dept of Physics Univ of Dayton Dayton OH 45409

BUECHER, EDWARD JOSEPH, b Manchester, NH, Apr 2, 37. MICROBIOLOGY, NEMATOLOGY. Educ: St Anselm's Col, BA, 59; Ind Univ, MA, 62; Univ Calif, Davis, PhD(microbiol), 68. Prof Exp: Jr res scientist, Kaiser Res Labs, Calif, 63-64; res assoc, Clin Pharmacol Res Inst, Calif, 68-71; ASSOC PROF BIOL, COLO WOMEN'S COL, 71- Concurrent Pos: Consult, Oil Shale Corp, 75- Mem: AAAS; Soc Nematol; Am Soc Microbiol; Am Aging Asn; Am Soc Parasitol. Res: Assay and characteristics of beta galactosidase in Streptococcus faecium; physiology and cell wall biochemistry of yeast Saccharomycopsis guttulata; axenically cultured sereral free-living, insect, plant and animal parasitic nematodes; environmental effects on nematodes; axenic culture of sporocysts, Schistosoma mansoni. Mailing Add: Dept of Biol Colo Women's Col Montview Blvd & Quebec Denver CO 80220

BUECHLER, PETER ROBERT, b New York, NY, Oct 7, 19; m 46; c 5. CHEMISTRY. Educ: Fordham Univ, BS, 41; Univ Denver, MS, 46; Univ Cincinnati, PhD(appl sci), 49. Prof Exp: Anal chemist, NY Quinine & Chem Works, 41-43; asst prof appl sci, Univ Cincinnati, 46-48, res assoc & dir surg chem lab, 49-51; sr res chemist, Olin-Mathieson, 51-54; gen mgr & asst secy-treas, Tanners Res Corp, NJ, 55-57; group leader, Rohm & Haas, 57-66; sect head, Appl Res Sect, Mobil Chem Co, 66-69; DIR RES & DEVELOP, K J QUINN & CO, INC, 69- Concurrent Pos: Chem consult, Vet Admin, 49-51; mem fac, Eve Div, La Salle Col, 62-69. Mem: Am Chem Soc; Am Leather Chem Asn; Brit Soc Leather Trades Chem; Am Inst Chemists. Res: Analytical methods; structure of the fibrous proteins; chemical modification of skin and leather; treatment of burns; surgical sutures; condensation and addition polymers; leather chemistry and finishing; new polymers for new applications. Mailing Add: 5300 Richland Rd Gibsonia PA 15044

BUECHNER, HELMUT KARL, ecology, deceased

BUECHNER, HOWARD ALBERT, b New Orleans, La, Feb 1, 19; m 47. INTERNAL MEDICINE. Educ: Tulane Univ, BS, 39; La State Univ, MD, 43; Am Bd Internal Med, dipl, 53. Prof Exp: Chief pulmonary dis serv, Vet Admin Hosp, New Orleans, 50-56, chief med serv, 56-74; PROF MED, SCH MED, LA STATE UNIV, NEW ORLEANS, 73- Concurrent Pos: Prof med, Sch Med, Tulane Univ, 63-74; sr vis physician, Charity Hosp, La; Consult, USPHS Hosp, New Orleans; sr ed, J Am Col Chest Physicians. Mem: Fel Am Col Physicians; fel Am Col Chest Physicians; AMA; Am Thoracic Soc. Res: Pulmonary diseases especially tuberculosis, bagassosis, blastomycosis and lung cancer. Mailing Add: 1542 Tulane Ave New Orleans LA 70112

BUECHNER, WILLIAM WEBER, b Vallejo, Calif, May 12, 14; m 39. PHYSICS. Educ: Mass Inst Technol, BS, 35, PhD(physics), 39. Honors & Awards: Hon ScD, Nat Univ Mex, 54. Prof Exp: Res assoc, 39-42, from asst prof to assoc prof, 42-56, head dept, 61-67, PROF PHYSICS, MASS INST TECHNOL, 56- Concurrent Pos: Dir, High Voltage Eng Corp, 54-; vis prof, Nat Univ Mex, 52, 55, 56, 58; vis prof, US Dept State-India Wheat Loan Prog, Indian Univs, 57; Fulbright lectr, Cath Univ, Rio de Janeiro, 59; mem cmn nuclear masses, Int Union Pure & Appl Physics, 58-70. Honors & Awards: Naval Ord Develop Award, 46; Dudley Medal, Am Soc Test & Mat, 49. Mem: Fel Am Phys Soc; Am Asn Physics Teachers; fel Am Acad Arts & Sci; Mex Soc Physics. Res: Gas discharges; design of high voltage electrostatic accelerators; x-ray; nuclear reactions; nuclear spectroscopy. Mailing Add: Dept of Physics Mass Inst of Technol Cambridge MA 02139

BUEDING, ERNEST, b Frankfurt, Ger, Aug 19, 10; nat US; m 40; c 1. BIOCHEMISTRY. Educ: Univ Frankfurt, BA, 29; Univ Paris, MD, 36. Prof Exp: Asst biochem, Istanbul Univ, 36-38; asst prof pharmacol, Sch Med, Western Reserve Univ, 44-47, assoc prof, 47-54; prof & chmn dept, Med Sch, La State Univ, 54-60; PROF PATHOBIOL, SCH HYG & PUB HEALTH, JOHNS HOPKINS UNIV, 60-, PROF PHARMACOL & EXP THERAPEUT, SCH MED, 66- Concurrent Pos: Res fel, Col Med, NY Univ, 39-44; vis Fulbright prof, Oxford Univ, 59; consult, WHO, 61 & 63; mem sci adv bd, Nat Vitamin Found, 49-52; mem comn parasitol, US Armed Forces Epidemiol Bd, 53-72; mem study sect trop med & parasitol, Div Res Grants, NIH, 56-60; mem expert comn schistosomiasis, WHO, 59 & 62, mem expert adv panel parasitic dis, 63-75; mem panel metab biol, NSF, 62-65; mem parasitic dis panel, US-Japan Coop Med Sci Prog, 65-71; consult to surg gen, Dept Army, 73-; chmn Am Schistosomiasis Del, Nat Acad Sci, People's Repub China, 75; dir, Wellcome Labs Studies of Schistosomiasis, Sch Hyg, Johns Hopkins Univ, 69- Mem: Fel AAAS; Am Soc Biol Chem; Am Chem Soc; Am Soc Pharmacol & Exp Therapeut; Brit Biochem Soc. Res: Comparative biochemistry of parasites; chemotherapy of parasitic infections; glycogen metabolism; biochemistry of smooth muscle. Mailing Add: Dept of Pathobiol Johns Hopkins Univ Baltimore MD 21205

BUEHLER, CHARLES A, physical organic chemistry, see 12th edition

BUEHLER, EDWIN VERNON, b Alliance, Ohio, Sept 23, 29; m 50; c 4. TOXICOLOGY. Educ: Ohio State Univ, BA, 51, MSc, 56, PhD(bact), 58. Prof Exp: Bacteriologist, Res Div, 58-61, res toxicologist, 61-63, group leader, 64-70, HEAD TOXICOL, MIAMI VALLEY LABS, PROCTER & GAMBLE CO, 70- Mem: Soc Toxicol. Res: Immunology; mycology; antibacterials; contact sensitization; teratology. Mailing Add: Miami Valley Labs Procter & Gamble Co Cincinnati OH 45239

BUEHLER, FRITZ A, chemistry, see 12th edition

BUEHLER, JOHN A, b Philadelphia, Pa, Nov 17, 16; m 45; c 3. ORGANIC CHEMISTRY. Educ: Univ Pa, AB, 39; Anderson Col, BTh, 44; Ind Univ, PhD(chem), 49. Prof Exp: Chem operator, Barrett Coal Tar, 39-40 & Midvale Steel, 40-41; chemist, Empire Ord, 41-43; asst, Ind Univ, 45-47; asst prof, Anderson Col, Ind, 47-52, from assoc prof to prof, 52-60, chmn dept, 47-60 & sci div, 49-60; PROF CHEM & CHMN DEPT, LE MOYNE-OWEN COL, 60-, CHMN, DIV NAT SCI AND MATH, 73- Prof Exp: Instr, Adult Center, Ind Univ, 51-60; chemist, Anderson City Utilities, Ind, 47-60. Mem: Fel AAAS; Am Chem Soc. Res: Synthesis of amino acids; application of organic compounds to analytical determinations. Mailing Add: Dept of Chem Le Moyne-Owen Col Memphis TN 38106

BUEHLER, JOHN DAVID, b St Louis, Mo, Sept 14, 37; m 63; c 2. PHARMACEUTICAL CHEMISTRY. Educ: Philadelphia Col Pharm, BSc, 59, MSc, 61. Prof Exp: Pharmaceut chemist, Smith, Kline & French Labs, 63-64; process develop pharmacist, Merck, Sharp & Dohme Res Labs, 64-67; assoc scientist, Ortho Pharmaceut Corp, 68; sr pharmaceut chemist, 68-75, DEPT HEAD, WILLIAM H RORER, INC, 75- Mem: Am Pharmaceut Asn; Am Inst Chem Eng; Acad Pharmaceut Sci. Res: Application of physical pharmaceutical and biopharmaceutical principles in the development and processing of pharmaceutical products. Mailing Add: William S Rorer, Inc 500 Virginia Dr Ft Washington PA 19034

BUEHLER, MARTIN STOWELL, b Omaha, Nebr, Oct 23, 10; m 44; c 3. INTERNAL MEDICINE, CARDIOLOGY. Educ: Univ Minn, BS, 36, BM, 38, MD, 39, MS, 41; Am Bd Internal Med, dipl, 47. Prof Exp: Asst path, Univ Minn, 36, asst dermat & syphil, 37, teaching fel med, Univ Minn & Minneapolis Gen Hosp, 39-42, clin asst med & asst to prof med, Univ Minn, 41, chief med residents, Minneapolis Gen Hosp, 41-42; clin instr, 45-46, CLIN ASST PROF MED, UNIV TEX HEALTH SCI CTR DALLAS, 46- Concurrent Pos: Jr intern, Abbott Hosp, Minneapolis, 37; intern, Gallinger Munic Hosp, Washington, DC, 38-39; attend physician, Parkland City County Hosp, Dallas, 45-; pvt pract internal med, cardiol & diag, Dallas, 46-; courtesy staff, St Paul's Hosp, Methodist Hosp & Gaston Hosp, Dallas, 46-; chest & cardiac consult, Vet Admin, Dallas, 46-52, attend physician, Vet Admin Hosp, McKinney, Tex, 50-60; attend physician & mem staff, Baylor Med Ctr, Dallas, 46-74, consult category, 74-; attend physician, Vet Admin Hosp, Dallas, 50-; consult physician, Poliomyelitis Serv, Dallas City County Hosp, 50-53; attend physician & mem staff, Presby Hosp, Dallas, 66-; clin staff, Episcopal Hosp, Caruth Mem Rehab Ctr, Wadley Insts Molecular Med & Granville C Morton Cancer & Res Hosp, Dallas. Consult, Humble Oil Co, Am Airlines, Tex Instruments, Ling-Temco-Vought Co, Kaiser Industs & Schenley Distillers Co; lectr in med schs, London, Eng, Madrid, Spain, Lisbon, portugal, Panama, Rio de Janeiro, Brazil, Mexico City & Monterrey, Mex; mem bd dirs, Dallas Child Guid Clin, 52-55; mem, White House Comt on Aging, 62-65; mem courtesy staff, Med City Dallas Hosp, 75- Mem: Int Soc Internal Med; World Med Asn; Am Heart Asn; fel Am Col Physicians; fel Am Col Angiol. Res: Chemotherapy; heart. Mailing Add: 5616 Yolanda Circle Dallas TX 75229

BUEHLER, ROBERT JOSEPH, b Alma, Wis, May 1, 25; m 64; c 2. MATHEMATICAL STATISTICS. Educ: Univ Wis, BS, 48, MS, 49, PhD(math), 52. Prof Exp: Mem staff, Sandia Corp, NMex, 51-55; proj assoc chem & instr math, Univ Wis, 55-57; from asst prof to assoc prof, Iowa State Univ, 57-63; PROF STATIST, UNIV MINN, MINNEAPOLIS, 63- Concurrent Pos: Mem, Indian Statist Inst. Mem: Am Math Soc; Math Asn Am; fel Am Statist Asn; fel Inst Math Statist; Biomet Soc. Res: Statistical inference. Mailing Add: Sch of Statist Univ of Minn Minneapolis MN 55455

BUEHRING, GERTRUDE CASE, b Chicago, Ill, May 28, 40; m 62; c 2. CANCER. Educ: Stanford Univ, BA, 62; Univ Calif, Berkeley, PhD(genetics), 72. Prof Exp: Fel breast cancer, Nat Cancer Inst, 72-73; ASST PROF MED MICROBIOL, SCH PUB HEALTH, UNIV CALIF, BERKELEY, 73- Concurrent Pos: Co-prin investr, Nat Cancer Inst Res Contract, 74. Mem: Am Soc Cell Biol; Am Pub Health Asn; Am Asn Univ Prof. Res: Use of cell cultures of human mammary epithelium to study the cell biology, virology, biochemistry and genetics of human breast cancer. Mailing Add: Sch of Pub Health 208 Warren Hall Univ of Calif Berkeley CA 94720

BUEKER, ELMER DANIEL, b Hartsburg, Mo, July 30, 03; m 30; c 1. EMBRYOLOGY, NEUROANATOMY. Educ: State Teachers Col, Warrensburg, Mo, BS, 27; Univ Colo, AM, 29; Washington Univ, PhD(zool, anat), 42. Prof Exp: Asst instr, Univ Colo, 29-30, instr biol, Univ City Pub Schs, 3042; instr anat, Chicago Med Sch, 42-43; asst prof, Med Col SC, 43-47; assoc prof, Med & Dent Schs, Georgetown Univ, 47-50; assoc prof, Sch Med, Univ Mo, 50-55; prof, 55-71, EMER PROF ANAT, GRAD SCH & COL DENT, NY UNIV, 71- Concurrent Pos: USPHS res grants. Mem: Fel AAAS; Am Asn Anatomists; Soc Exp Biol & Med; fel NY Acad Sci; Harvey Soc. Res: Implantation and growth of tumors in the embryonic chick; nerve growth stimulating properties of mouse sarcomas; characterization of nerve growth, stimulating protein from salivary glands; regeneration of peripheral nerves; retrograde degeneration. Mailing Add: Col of Dent NY Univ New York NY 10010

BUELL, CARLETON EUGENE, b Grand Island, Nebr, Sept 2, 10; m 32; c 3. MATHEMATICS, METEOROLOGY. Educ: Oberlin Col, AB, 31; Ohio State Univ, AM, 32; Wash Univ, PhD(math), 35. Prof Exp: Instr me- teorol, math & physics, Parks Air Col, Inc, 35-38; meteorologist, Am Airlines, Inc, 38-39, chief meteorologist, 39-46; proj supvr, res & develop div, NMex Sch Mines, 46-48; div leader, Sandia Lab, 48-49; assoc prof math, Univ NMex, 49-57; sr scientist, Kaman Nuclear Div, Kaman Sci Corp, 57-75; RES ASSOC, UNIV COLO, COLORADO SPRINGS, 75- Concurrent Pos: Consult, Sandia Corp, 51-57. Mem: Am Math Soc; Math Asn Am; Am Meteorol Soc; Am Geophys Union; Nat Soc Prof Engrs. Res: Weather forecasting; statistics in meteorology. Mailing Add: 1601 Columbine Pl Colorado Springs CO 80907

BUELL, ELLIOTT LYNDON, b Syracuse, NY, Nov 5, 16; m 43; c 2. MATHEMATICS. Educ: Syracuse Univ, AB, 38; Mass Inst Technol, PhD(math), 41. Prof Exp: Asst math, Mass Inst Technol, 38-40; instr, Northwestern Univ, 41-44, math physicist, aerial measurements lab, 44-46, instr math, 46-47, mathematician, aerial measurements lab, 47-53, head analog div, 53-55, tech dir, 55-57; dir compt facil, 59-67, head dept, 60-68, PROF MATH, WORCESTER POLYTECH INST, 57- Mem: Am Soc Eng Educ; Math Asn Am. Res: Applied mathematics; mathematical theory of elasticity; digital and analog computing. Mailing Add: Dept of Math Worcester Polytechnic Inst Worcester MA 01609

BUELL, GEORGE CHRISTOPHER, b St Louis, Mo, Mar 24, 20; m 51; c 2. BIOCHEMISTRY, ORGANIC CHEMISTRY. Educ: Case Western Reserve Univ, BS, 47; Vanderbilt Univ, MS, 49; Tex A&M Univ, PhD(biochem), 58. Prof Exp: Chemist, Tilden Co, Mo, 34-40; dir prod develop, Vet-Line Labs, Tex, 58-60; RES CHEMIST, CALIF STATE DEPT PUB HEALTH, 60- Mem: Am Chem Soc; fel Am Inst Chemists; NY Acad Sci. Res: Enzymology; lipid metabolism; drug enzymology; blood lactate; steroid fractionation and relation to stress; bioenergetics; chemical changes in lung tissue following exposure to smog components; solubilization of lung collagen and elastin, and their ultraviolet absorption spectra. Mailing Add: 51 El Rancho Dr Pleasant Hill CA 94523

BUELL, GLEN R, b Lee's Summit, Mo, Jan 10, 31; m 61; c 2. ORGANIC CHEMISTRY, ANALYTICAL CHEMISTRY. Educ: Univ Mo, BS, 53, MA, 55;

Univ Kans, PhD(chem), 61. Prof Exp: Res chemist, Pittsburgh Plate Glass Co, 61-62; vis res assoc organo-silicon, Ohio State Res Found, 62-63; res chemist, Aerospace Res Labs, 63-75, RES CHEMIST, AIR FORCE MAT LAB, WRIGHT-PATTERSON AIR FORCE BASE, 75- Mem: Am Chem Soc. Res: Use of instrumental techniques for identification and characterization of organic and organo-metallic compounds with respect to structure and bonding. Mailing Add: 3548 Eastern Dr Dayton OH 45432

BUELL, KATHERINE MAYHEW, b Grand Island, Nebr, Nov 26, 12. ANATOMY, MORPHOLOGY. Educ: Oberlin Col, AB, 33; Univ Wash, MS, 35; Univ Wis, PhD(bot), 51. Prof Exp: From instr to prof, 36-72, DISTINGUISHED PROF BIOL, DOANE COL, 72- Mem: Am Bot Soc; Nat Asn Biol Teachers; Int Soc Plant Morphol. Res: Anatomy and morphology of seed plants; vertebrate morphology; embryology. Mailing Add: 730 E 12th Crete NE 68333

BUELL, MURRAY FIFE, botany, deceased

BUELL, WAYNE H, b Lewis, Ind, July 2, 13; m 39. PHYSICAL CHEMISTRY, CHEMICAL ENGINEERING. Educ: Lawrence Inst Technol, BChE, 36; Wayne State Univ, MS, 51. Hon Degrees: Hon DE, 58. Prof Exp: Instr chem, Lawrence Inst Technol, 33-44, prof, 46-48; res chemist, Aristo Corp, 44-46, dir res, 48-49, from vpres to exec vpres, Aristo Int, 49-64; PRES, LAWRENCE INST TECHNOL, 64- Honors & Awards: Award Sci Merit, Am Foundrymen's Soc, 70. Mem: Am Chem Soc; Am Foundrymen's Soc. Res: Administration in industry and education; organic and physical chemistry. Mailing Add: 21320 W Ten Mile Rd Southfield MI 48075

BUENING, GERALD MATTHEW, b Decatur Co, Ind, Sept 14, 40; m 64; c 2. VETERINARY IMMUNOLOGY, VETERINARY VIROLOGY. Educ: Iowa State Univ, MS, 66; Purdue Univ, DVM, 64, PhD(vet virol), 69; Am Col Vet Microbiologists, dipl, 70. Prof Exp: Instr vet microbiol, Purdue Univ, 66-69; asst prof, 69-72, ASSOC PROF VET MICROBIOL, UNIV MO-COLUMBIA, 72- Mem: Am Vet Med Asn; Conf Res Workers Animal Dis; Am Soc Microbiol; Asn Am Vet Cols; Am Animal Hosp Asn. Res: Cell-mediated immunity of domestic animals; in vitro methods are utilized to follow the cell-mediated immunity response during disease processes and recovery. Mailing Add: Dept of Vet Microbiol Univ of Mo Columbia MO 65201

BUENKER, ROBERT J, b Dubuque, Iowa, May 6, 42; m 69; c 1. PHYSICAL CHEMISTRY. Educ: Loras Col, BS, 63; Princeton Univ, PhD(chem), 66. Prof Exp: Asst, Princeton Univ, 63-64; res assoc phys chem, Mich State Univ, 66-67; from asst prof to assoc prof, 67-74, PROF CHEM, UNIV NEBR-LINCOLN, 74- Concurrent Pos: Vis prof, Univ Mainz, 70-71. Res: Ab initio self-consistent field and configuration interaction calculations; geometry and spectra of molecules; reaction surfaces. Mailing Add: Dept of Chem Univ of Nebr Lincoln NE 68508

BUERGER, ALFRED ARTHUR, b Buffalo, NY, Dec 22, 40; m 67; c 1. PHYSIOLOGY, PSYCHOBIOLOGY. Educ: Cornell Univ, AB, 62, MS, 64, PhD(animal physiol), 67. Prof Exp: Lectr physiol, Cornell Univ, 67-68; lectr psychol, Harvard Univ, 68-70; lectr psychobiol, 70-71, ASST PROF, UNIV CALIF, IRVINE, 71- Concurrent Pos: USPHS fel, Harvard Col, 68-70. Mem: AAAS; Am Physiol Soc; Am Psychol Asn: Soc Neurosci; Psychonomic Soc. Res: Physiology of learning and memory; some aspects of clinical neurophysiology. Mailing Add: Dept of Phys Med & Rehab Univ of Calif Irvine CA 92717

BUERGER, MARTIN JULIAN, b Detroit, Mich, Apr 8, 03; m 38; c 6. MINERALOGY, CRYSTALLOGRAPHY. Educ: Mass Inst Technol, SB, 25, SM, 27, PhD(mineral), 29. Hon Degrees: Dr, Univ Berne, 58. Prof Exp: From asst to instr geol, 25-29, from asst prof to assoc prof mineral & petrog, 29-44, from prof to inst prof mineral & crystallog, 44-68, chmn fac, 54-56, dir sch advan study, 56-63, EMER INST PROF MINERAL & CRYSTALLOG, MASS INST TECHNOL, 68-; HON RES ASSOC GEOL SCI, HARVARD UNIV, 73- Concurrent Pos: Mem, MacMillan Exped, Baffin Island, 37; co-ed, Int Tables X-ray Crystallog, 46- & Zeitschrift Krystallog, 53- Vis prof, Univ Rio de Janeiro, 48, Univ Minn, 71, Univ Ky, 71, Va Polytech Inst & State Univ, 74, 75; guest lectr, Inst Physics & Math, Univ Chile, 62; Univ prof, Univ Conn, 68-73. Mem solids comt, Nat Res Coun, 49-53, 56-62; Am del preliminary conv, Int Union Crystallog, 46, coun mem, 46-51, Nat Res Coun rep, gen assembly & cong, 51, mem apparatus comn, 51-56, US Nat Comt, 51-52, 56. Honors & Awards: Day Medal, Geol Soc Am, 51; Roebling Medal, Mineral Soc Am, 58; Isador Fankuchen Award, Am Crystallog Asn, 71. Mem: Nat Acad Sci; fel Mineral Soc Am (vpres, 42, pres, 47); fel Geol Soc Am (vpres, 48); Am Crystallog Asn (pres, 39-46, pres, Soc X-ray & Electron Diffraction. 48); hon mem Royal Span Soc Natural Hist. Res: Generalization of the Patterson function of crystallography to image sets and image functions; use of image algebra to explain some homometric sets. Mailing Add: Dept of Earth Sci Harvard Univ Hoffman Lab 20 Oxford St Cambridge MA 02138

BUERLE, DAVID E, b Jersey City, NJ. RESOURCE MANAGEMENT, WATER RESOURCES. Educ: Rensselaer Polytech Inst, BMgtE, 55; Columbia Univ, MS, 59; Clark Univ, PhD(geog), 65. Prof Exp: Jr gas engr, NY State Pub Serv Comn, 55; mgt analyst, Port of New York Authority, 59-60; assoc planner, Candeub, Fleissig & Assocs, Boston, Mass, 61-62; planner, Charles E Downe Co, Newton, 62; from instr to asst prof geog, Univ RI, 62-66; asst prof, State Univ NY Albany, 66-70, assoc res analyst & dir, NY State Dept Transp, 70-71; dir mgt study & anal, Southeast Water Supply Comn, NY, 71-74; water resources consult, NY State Catskill Study Comn, Rexmere Park, Stamford, NY, 74-75; COASTAL ZONE MGT COORDR, DIV STATE PLANNING, NY STATE DEPT OF STATE, 76- Concurrent Pos: Res & study fel, Univ Alta, 68-69; lectr polit geog, Boston Univ, 61-62; consult, NY State Educ Dept, Albany, 67-68 & Policy & Planning Br, Can Dept Energy, Mines & Resources, Ont, 69-70. Mem: AAAS; Asn Am Geog; Am Geog Soc; Int Oceanog Found; Am Inst Planners. Res: Examination of the uses to which resources of fresh and saline water bodies can be put while considering the environmental impact of such uses, particularly in and near urbanized regions. Mailing Add: Div of State Planning NY State Dept of State 162 Washington Ave Albany NY 12231

BUESCHER, BRENT J, b Galesburg, Ill, Sept 15, 40; m 65; c 1. SOLID STATE PHYSICS. Educ: Univ NC, BS, 62; Univ Ariz, PhD(physics), 69. Prof Exp: Asst, Univ NC, 61-62 & Univ Ariz, 62-69; res assoc physics, Rensselaer Polytech Inst, 69-71; assoc, Argonne Nat Lab, 71-73; PRIN ENGR, POWER GENERATION GROUP, BABCOCK & WILCOX CO, 73- Mem: Am Phys Soc; AAAS; Int Diffusion Researchers Asn. Res: Material properties of ceramics during irradiation; migration of defects in quenched metals; diffusion under pressure in the hexagonal metals. Mailing Add: Power Generation Group Babcock & Wilcox Co PO Box 1260 Lynchburg VA 24505

BUESCHER, EDWARD LOUIS, b Cincinnati, Ohio, July 24, 25; m 47; c 5. VIROLOGY. Educ: Univ Dayton, BS, 45; Univ Cincinnati, MD, 48; Am Bd Microbiol, dipl. Prof Exp: Virologist dept virus & rickettsial dis, Army Med Serv Grad Sch, 50-51, 54-56; virologist, Far E Med Res Unit & chief, Dept Virus & Rickettsial Dis, 406th Med Gen Lab, US Army, 51-54; chief dept virus dis, 56-58, dir div commun dis & immunol, 67-70, dep dir, Inst, 69-71, dir, 71-75; CHMN, ARMY VOLUNTEER PROG STUDY GROUP, WALTER REED ARMY INST RES, 75- Concurrent Pos: Consult, Surg Gen, US Army, 60-; mem study group arthropod-borne virus dis, WHO, 60 & USPHS study sect virol & rickettsiol, 60-69, 71-75; mem exec coun, Am Comt Arthropod-Brone Viruses, 62-68; consult, Off Sci & Technol, Exec Off & President, 64; mem comn on influenza, Armed Forces Epidemiol Bd, 65-72; mem vaccine develop comt, Nat Inst Allergy & Infectious Dis, 67-69, consult, Infectious Dis Br, 71-; clin prof pediat, Georgetown Univ, 70- Mem: Fel Am Acad Microbiol; AMA; Am Asn Immunol; Am Fedn Clin Res: Am Soc Microbiol. Res: Physiochemical, biological and immunological properties of neurotropic viruses; comparative pathogenesis of experimentally and naturally induced virus infections; ecology of virus infections. Mailing Add: Hqs US Army Med Res & Develop Com Forestal Bldg Washington DC 20314

BUESKING, CLARENCE W, b Ft Wayne, Ind, Apr 10, 19; m 43; c 4. MATHEMATICS. Educ: Ball State Univ, BA, 47; Purdue Univ, MS, 52. Prof Exp: Teacher math, Tri-State Col, 47, math & sci, Burris Lab Sch, Ball State Univ, 47-49 & pub schs, Ind, 50-57; PROF MATH & DIR COMPUT CTR, UNIV EVANSVILLE, 57-, HEAD COMPUT SCI DEPT, 70- Mem: Math Asn Am; Asn Comput Mach. Res: Undergraduate mathematics; computer programming. Mailing Add: Dept of Comput Sci Univ of Evansville Evansville IN 47701

BUESS, CHARLES MERLYN, b Forest, Ohio, Apr 22, 22; m 53; c 4. ORGANIC CHEMISTRY. Educ: Ohio State Univ, BA, 42; Western Reserve Univ, MS, 46; Univ Southern Calif, PhD(chem), 50. Prof Exp: Fel, Northwestern Univ, 49-51; asst prof, Univ Ga, 51-56; vis assoc prof, Univ Ky, 56-57; res assoc & vis prof, Univ Southern Calif, 57-61; assoc prof chem, 61-66, PROF CHEM, WICHITA STATE UNIV, 66- Mem: AAAS; Am Chem Soc; Brit Chem Soc. Res: Organic sulfur compounds; heterocyclic systems; steroids. Mailing Add: Dept of Chem Wichita State Univ Wichita KS 67208

BUESSELER, JOHN AURE, b Madison, Wis, Sept 30, 19; m 59. OPHTHALMOLOGY. Educ: Univ Wis-Madison, PhB, 41, MD, 44; Univ Mo, MS, 65. Prof Exp: Founding dean sch med, Tex Tech Univ, 70-73, vpres health affairs, Univ Complex, 70-75, vpres health sci, Sch Med, 72-74, PROF OPHTHAL, SCH MED, TEX TECH UNIV, 71-, PROF HEALTH ORGN MGT, GRAD SCH, 72-, UNIV PROF OPHTHAL & HEALTH ORGN MGT, UNIV COMPLEX, 73- Concurrent Pos: Assoc examr, Am Bd Ophthal, 59-74; mem, NASA Space Med Adv Group, 63-66; mem biol comt, Assoc Midwestern Univs, Argonne Nat Lab & AEC, 65-69; res consult ophthal, Argonne Nat Lab, Biol & Med Div & USPHS, US Dept Health, Educ & Welfare, Neurol & Sensory Dis Proj Rev Panel, 66-69; pres & mem bd dirs, Joint Comn on Allied Health Personnel in Ophthal, 72-76; mem adv panel on nat health ins, Subcomt Health, Ways & Means Comt, US House Rep, 75- Mem: Sigma Xi; AMA; AAAS; Asn Am Med Cols; Am Acad Ophthal & Otolaryngol. Res: Development of an administrative and health facilities organizational design model for state wide aeromedical evacuation systems; evaluation of hydroxyurea and interferon on cellular and viral DNA synthesis in the corneal epithelium of the eye. Mailing Add: Tex Tech Univ Complex PO Box 4250 Lubbock TX 79409

BUETOW, DENNIS EDWARD, b Chicago, Ill, June 20, 32; m 60; c 4. CELL BIOLOGY. Educ: Univ Calif, Los Angeles, AB, 54, MA, 57, PhD(zool), 59. Prof Exp: Biochemist, Baltimore City Hosps & biologist, biophys lab, NIH, Md, 59-65; assoc prof, 65-70, PROF PHYSIOL, UNIV ILL, URBANA-CHAMPAIGN, 70- Concurrent Pos: Mem biochem bd consult group, Off Naval Res; Nat Inst Gen Med Serv & NSF res grants, 73- Mem: AAAS; Am Inst Biol Sci; Am Soc Cell Biol; Am Physiol Soc; fel Gerontol Soc. Res: Mechanisms of RNA and protein synthesis; correlation of cellular structure and function; biology of aging; biochemistry of protozoa. Mailing Add: Dept of Physiol & Biophys Univ of Ill, Urbana-Champaign Urbana IL 61801

BUETTNER, ALBERT V, physical chemistry, see 12th edition

BUETTNER-JANUSCH, JOHN, b Chicago, Ill, Dec 7, 24; m 50. PHYSICAL ANTHROPOLOGY, PRIMATOLOGY. Educ: Univ Chicago, PhB, 48, SB, 49, AM, 53; Univ Mich, PhD, 57. Prof Exp: From instr anthrop to res asst prev med, Univ Utah, 53-55; tech asst, Heredity Clin, Univ Mich, 55-56, asst res, univ, 56-57, res asst, Lab Phys Anthrop, 57-58; from asst prof to assoc prof anthrop, Yale Univ, 58-65; assoc prof zool & anat, Duke Univ, 65-67, prof anthrop, 70-71, prof zool & anat, 67-73; PROF ANTHROP & CHMN DEPT, NY UNIV, 73- Concurrent Pos: Instr sociol & anthrop, 56; sem assoc, Columbia Univ, 59-; mem ed selection comt anthrop, Bobbs-Merrill, 61-; NSF sr fel anthrop, Yale Univ, 62-63; USPHS res career develop award, 63-73; assoc ed, Am J Phys Anthrop, 70-; mem adv panel anthrop, NSF, 71-74; chmn sci adv bd, Caribbean Primate Res Ctr, 71-; dir, Duke Univ Primate Facil, 65-73, assoc dir, 73- Mem: AAAS; Am Anthrop Asn; Am Asn Phys Anthrop; NY Acad Sci; Int Asn Human Biol. Res: Evolution, biochemical genetics and physical anthropology of the living primates, including Homo sapiens; history of anthropology and science. Mailing Add: Dept of Anthrop NY Univ 25 Waverly Pl New York NY 10003

BUFE, CHARLES GLENN, b Duluth, Minn, Jan 2, 38; m 67; c 2. SEISMOLOGY, TECTONICS. Educ: Mich Tech Univ, BS, 60, MS, 62; Univ Mich, PhD(geol), 69. Prof Exp: Res asst geophys, Willow Run Labs, Univ Mich, 64, res assoc, 67-69, assoc res geophysicist, 69; geophysicist, US Earthquake Mech Lab, Nat Oceanic & Atmospheric Admin, 69-73; GEOPHYSICIST, OFF EARTHQUAKE STUDIES, US GEOL SURV, 73- Concurrent Pos: Vis asst prof, Univ Wis, Milwaukee, 73. Mem: AAAS; Seismol Soc Am; Am Geophys Union; Soc Explor Geophys; Geol Soc Am. Res: Travel times and spectra of short-period body waves from explosions and earthquakes; micro-earthquake and fault-creep studies; earthquake prediction and control; induced seismicity; plate tectonics. Mailing Add: US Geol Surv 345 Middlefield Rd Menlo Park CA 94025

BUFF, FRANK PAUL, b Munich, Ger, Feb 13, 24; nat US; m 56; c 2. PHYSICAL CHEMISTRY. Educ: Univ Calif, BA, 44; Calif Inst Technol, PhD(chem), 49. Prof Exp: AEC fel, Calif Inst Technol, 49-50; from instr to assoc prof chem, 50-61, PROF CHEM, UNIV ROCHESTER, 61- Concurrent Pos: NSF sr fel, Inst Theoret Physics, Utrecht, Netherlands, 59-60; consult, Socony Mobil Co, Inc, Mobil Oil Corp, 62- Mem: AAAS; Am Chem Soc; fel Am Phys Soc. Res: Molecular theories of fluids; surface phenomena; solutions and chemical kinetics; nucleation processes. Mailing Add: Dept of Chem Univ of Rochester Rochester NY 14627

BUFFALOE, NEAL DOLLISON, b Leachville, Ark, Nov 15, 24; m 47; c 5. BIOLOGY. Educ: David Lipscomb Col, BS, 49; Vanderbilt Univ, MS, 52, PhD(biol), 57. Prof Exp: Instr biol, David Lipscomb Col, 49-54 & Vanderbilt Univ, 54-56; PROF BIOL, UNIV CENT ARK, 57- Concurrent Pos: Instr biol, George Peabody Col, 55. Mem: AAAS; Phycol Soc Am; Am Asn Biol Teachers. Res: Cytology and cytogenetics. Mailing Add: Box 1721 Univ of Cent Ark Conway AR 72032

BUFFETT, RITA FRANCES, b Beverly, Mass, Jan 18, 17. BIOLOGY,

PHYSIOLOGY. Educ: Boston Univ, AB, 47, AM, 48, PhD(biol), 54. Prof Exp: Instr biol, Boston Univ, 50-54; res asst exp path, Children's Cancer Res Found, Mass, 54-59; res scientist exp path, 59-61, res scientist viral oncol, 61-69, ASST RES PROF MICROBIOL, ROSWELL PARK MEM INST, 69- Mem: AAAS; Am Soc Microbiol; Am Asn Cancer Res. Res: Oncogenesis; radiation, hormonal and viral. Mailing Add: Roswell Park Mem Inst Buffalo NY 14203

BUFFINGTON, ANDREW, b Fall River, Mass, Dec 25, 38; m 68. ASTROPHYSICS. Educ: Mass Inst Technol, BS, 61, PhD(physics), 66. Prof Exp: Res asst physics, Mass Inst Technol, 66-68; RES PHYSICIST, SPACE SCI LAB & LAWRENCE BERKELEY LAB, UNIV CALIF, BERKELEY, 68- Mem: Sigma Xi; Am Phys Soc. Res: Measurement of cosmic-ray abundances; interpretation of these to deduce cosmic-ray history; search for cosmic ray antimatter; development of flexible telescope techniques to restore atmospherically perturbed images from astronomical telescopes. Mailing Add: Rm 50-232 Lawrence Berkeley Lab Univ of Calif Berkeley CA 94720

BUFFINGTON, EDWIN CONGER, b Ontario, Calif, Aug 21, 20; m 43; c 4. MARINE GEOLOGY. Educ: Carleton Col, BA, 41; Calif Inst Technol, MSc, 47. Prof Exp: Marine geologist, US Navy Electronics Lab, 48-69; head marine geol br, Navy Undersea Res & Develop Ctr, 69-74; MEM STAFF, MARINE GEOL BR, US GEOL SURV, 74- Concurrent Pos: Dir, Gen Oceanog, Inc. Mem: Fel Geol Soc Am; Soc Econ Paleontologists & Mineralogists; Am Asn Petrol Geologists; Asn Eng Geol; Am Geophys Union. Res: Bathymetry; sea floor geomorphology. Mailing Add: Marine Geol Br US Geol Surv 345 Middlefield Rd Menlo Park CA 94025

BUFFINGTON, JOHN DOUGLAS, b Jersey City, NJ, Nov 26, 41; m 65; c 2. ECOLOGY, ENTOMOLOGY. Educ: St Peter's Col, NJ, BS, 63; Univ Ill, Urbana-Champaign, MS, 65, PhD(zool), 67. Prof Exp: Asst prof, dept biol sci, Ill State Univ, 69-72; BIOLOGIST, ARGONNE NAT LAB, 72- Concurrent Pos: Adj assoc prof, Northern Ill Univ, 75- Mem: AAAS; Am Inst Biol Sci; Entom Soc Am; Ecol Soc Am. Res: Analysis of environmental effects of various energy sources. Mailing Add: Energy & Environ Systs Div Argonne Nat Lab Argonne IL 60439

BUFORD, WILLIAM HOLMES, JR, b Ruston, La, June 15, 34; m 53; c 2. NUCLEONICS. Educ: La State Univ, BS, 56. Prof Exp: Nuclear engr, Gen Dynamics/Convair, Tex, 56; proj engr, US Air Force Aeropropulsion Lab, Wright-Patterson AFB, 56-58; res engr, Marquardt Corp, 58-59, group supvr radiation shielding, 59-61, mgr physics res, 61-69; vpres, Satellite Positioning Corp, Calif, 69-71; vpres, Seismark Int, Reseda, 71-72; PRES, CODEVINTEC P PAC, INC, 72- Concurrent Pos: Lectr, Univ Calif, Los Angeles, 60-61. Mem: Am Nuclear Soc. Res: Development of marine navigation and ship automation systems and automatic nuclear activation analysis systems. Mailing Add: 19407 Shenango Dr Tarzana CA 91356

BUGBEE, ROBERT EARL, b Bemus Point, NY, Sept 12, 07; m 36; c 5. TAXONOMY. Educ: Allegheny Col, BS, 31; Univ Ind, PhD(zool), 36. Prof Exp: Head dept biol, Col Emporia, 36-39; asst prof zool, Ft Hays Kans State Col, 39-43; res assoc, Univ Ind, 42-45; asst prof zool, Univ Rochester, 45-47; PROF & HEAD DEPT BIOL, ALLEGHENY COL, 47- Mem: AAAS; Entom Soc Am; Soc Study Evolution; Soc Syst Zoology. Res: Taxonomy and phylogeny of Eurytoma parasites bred from galls. Mailing Add: Dept of Biol Allegheny Col Meadville PA 16335

BUGG, CHARLES EDWARD, b Durham, NC, June 5, 41; m 62. PHYSICAL CHEMISTRY. Educ: Duke Univ, AB, 62; Rice Univ, PhD(phys chem), 65. Prof Exp: Res fel, Calif Inst Technol, 65-66; res chemist, Dacron Res Lab, E I du Pont de Nemours & Co, 66-67; res fel, Calif Inst Technol, 67-68; asst prof biochem, 68-74, ASSOC PROF BIOCHEM, UNIV ALA MED CTR, 74- Mem: Am Crystallog Asn. Res: Crystal structures of compounds of biological interest. Mailing Add: Univ of Ala Med Ctr 1919 Seventh Ave S Birmingham AL 35233

BUGG, WILLIAM MAURICE, b St Louis, Mo, Jan 23, 31; m 54; c 4. PHYSICS. Educ: Wash Univ, AB, 52; Univ Tenn, PhD(physics), 59. Prof Exp: From asst prof to assoc prof, 59-69, PROF PHYSICS & HEAD DEPT, UNIV TENN, KNOXVILLE, 69- Concurrent Pos: Consult, Oak Ridge Nat Lab, 59-69, Danforth assoc, 69- Mem: Fel Am Phys Soc. Res: Elementary particle interactions with bubble chamber techniques. Mailing Add: Dept of Physics Univ of Tenn Knoxville TN 37919

BUGGS, CHARLES WESLEY, b Brunswick, Ga, Aug 6, 06; m 27; c 1. BACTERIOLOGY. Educ: Morehouse Col, AB, 28; Univ Minn, MS, 32, PhD(bact), 34. Prof Exp: Instr biol, Dover State Col, 28-29; prof chem, Bishop Col, 34-35; prof biol & chmn div sci, Dillard Univ, 35-43; from instr to assoc prof bact, Sch Med, Wayne Univ, 43-49; prof biol & chmn div sci, Dillard Univ, 49-56; prof microbiol, Col Med, Howard Univ, 56-71, head dept, 58-70; proj dir, Fac Allied Health Sci, Charles R Drew Postgrad Med Sch, Univ Calif, Los Angeles, 69-72, dean, 72; PROF MICROBIOL, CALIF STATE UNIV, LONG BEACH, 73- Concurrent Pos: Rosenwald fel, Woods Hole, 43; with Off Sci Res & Develop, wayne Univ, 43-47; vis prof, Sch Med, Univ Calif, Los Angeles, 69-72, vis prof, Univ Southern Calif, 69- Mem: Am Soc Microbiol; fel Am Acad Microbiol. Res: Resistance of bacteria to antibiotics. Mailing Add: 5600 Verdun Ave Los Angeles CA 90043

BUGL, PAUL G, mathematical physics, see 12th edition

BUGNOLO, DIMITRI SPARTACO, b Atlantic City, NJ, Feb 3, 29; m 58; c 3. PHYSICS, ELECTRICAL ENGINEERING. Educ: Univ Pa, BSEE, 52; Yale Univ, MEng, 55; Columbia Univ, ScD(elec eng), 60. Prof Exp: Computer components res, Burroughs Res Lab, 52-54; res asst, Yale Univ, 54-56; instr elec eng, Columbia Univ, 56-60; mem tech staff, Bell Tel Labs, 60-61, 63-66; asst prof elec eng, Columbia Univ, 61-63, sr res scientist, Hudson Labs, 66-67; prin engr adv res, Submarine Signal Div, Raytheon Corp, 67-69; CONSULT ENGR & PHYSICIST, 69- Concurrent Pos: Mem comn, Int Sci Radio Union, 64-68; consult & full prof, Nat Inst Sci Res, Univ Quebec, 71-73. Mem: Am Phys Soc; Inst Elec & Electronics Eng; Acoustical Soc Am; Am Phys Soc. Res: Applied plasma physics; wave propagation in random media; plasma turbulence; propagation theory; electromagnetic theory and practice; antenna theory and practice; radar and ocean engineering; anti-ballistic missile systems research. Mailing Add: 71 Canonchet Dr Portsmouth RI 02871

BUGOSH, JOHN, b Cleveland, Ohio, July 1, 24; m 52; c 3. PHYSICAL CHEMISTRY, COLLOID CHEMISTRY. Educ: Heidelberg Col & Adelbert Col, BS, 45; Western Reserve Univ, MS, 47, PhD(chem), 49; Escuela Interamericana de Verano, dipl, 48. Prof Exp: Res assoc ultrasonics, Western Reserve Univ, 48-49, lectr physics, 49-50; res supvr, Del, 50-68, planning consult, Pa, 68-72, DEVELOP MGR, WATER POLLUTION CONTROL FEDN, E I DU PONT DE NEMOURS & CO, INC, 72- Honors & Awards: Award, Am Chem Soc, 62. Mem: AAAS; Am Chem Soc. Res: Ultrasonic transducers; electrolytic solutions, inorganic colloids. Mailing Add: 1071 Squire Cheney Dr West Chester PA 19380

BUHAC, IVO, b Dubrovnik, Yugoslavia, Sept 4, 26; m 62; c 1. GASTROENTEROLOGY, INTERNAL MEDICINE. Educ: Univ Zagreb, MD, 52, ScD, 64; Univ Erlangen, MD, 62. Prof Exp: Intern med, Hosp Dr Stojanovic, Zagreb, Yugoslavia, 52-53; resident, Hosp Dr Novosel, 57-60, staff physician med & gastroenterol, 62-68; instr med, Med Col Va, 69-70; asst prof, 70-74, ASSOC PROF MED, ALBANY MED COL, 74-; CHIEF GASTROENTEROL, ALBANY VET ADMIN HOSP, 70- Concurrent Pos: Alexander von Humboldt fel, Med Sch, Univ Hamburg, 64-65; fel gastroenterol, Vet Admin Hosp, Richmond, Va, 68-70. Mem: Am Gastroenterol Asn; fel Am Col Physicians. Res: Peritoneal permeability; portal hypertension. Mailing Add: Div of Gastroenterol Albany Med Col Albany NY 12208

BUHL, DAVID, b Newark, NJ, Nov 20, 36; m 62; c 1. RADIO ASTRONOMY. Educ: Mass Inst Technol, BS & MS, 60; Univ Calif, Berkeley, PhD(elec eng), 67. Prof Exp: Electronics engr, Lawrence Radiation Lab, 61-64; from assoc scientist to scientist, Nat Radio Astron Observ, 67-74; SPACE SCIENTIST, NASA GODDARD SPACE FLIGHT CTR, 74- Mem: Inst Elec & Electronics Engrs; Am Geophys Union; Am Astron Soc; Int Astron Union. Res: Radio astronomy of the sun, moon, and planets; observations of complex molecules in interstellar clouds; chemical evolution of the interstellar medium; infrared molecular astronomy. Mailing Add: Code 691 Astrochem Br Lab for Extraterrestrial Physics NASA Goddard Space Flight Ctr Greenbelt MD 20771

BUHLE, EMMETT LOREN, b Moline, Ill, May 7, 18; m 47; c 4. INDUSTRIAL ORGANIC CHEMISTRY. Educ: Johns Hopkins Univ, AB, 41, MA, 42, PhD(chem), 48. Prof Exp: Res assoc, Mass Inst Technol, 48-50; res asst pharmacol, Johns Hopkins Univ, 50-54, instr, 52-54; chemist, Exp Sta, E I du Pont de Nemours & Co, 54-58; SR RES CHEMIST, WYETH LABS, INC, 58- Concurrent Pos: Mem surv antimalarial drugs, Nat Res Coun, 42-46. Mem: Am Chem Soc. Res: Antimalarial drugs; synthetic steroids; semisynthetic penicillins; prostaglandins; development of chemical processes. Mailing Add: Wyeth Labs, Inc West Chester PA 19380

BUHLER, DONALD RAYMOND, b San Francisco, Calif, Oct 11, 25; m 53; c 4. BIOCHEMICAL PHARMACOLOGY. Educ: Ore State Col, BA & BS, 50, Ore State Univ, PhD(biochem), 56. Prof Exp: Asst, Ore State Col, 51-55; staff biochemist, div exp med, Med Sch, Univ Ore, 55-56, USPHS fel, 56-58; biochemist, western fish nutrit lab, US Fish & Wildlife Serv, 58-59, Upjohn Co, Mich, 59-64, western fish nutrit lab, US Fish & Wildlife Serv, 64-66 & Pac Northwest Labs, Battelle Mem Inst, 66-68; assoc prof, 68-74, PROF, DEPT AGR CHEM, ORE STATE UNIV, 74- Mem: Am Chem Soc; Am Soc Biol Chemists; Soc Toxicol. Res: Biochemical mechanisms for toxicity; drug metabolism and pharmacokinetics; synthesis of radioactive compounds; fate of chemicals in the environment; heavy metal toxicity; biochemistry of fishes. Mailing Add: Dept Agr Chem & Env Health Sci Ctr Ore State Univ Corvallis OR 97331

BUHLER, JOHN EMBICH, b Marion, Ind, June 28, 08; m 40; c 2. DENTISTRY. Educ: Ind Univ, DDS, 35. Prof Exp: Intern, Med Ctr, Ind Univ, 35-36, instr clin dent, Sch Dent, 36, instr histopath & clin oral surg, 3642; assoc prof oral surg, Temple Univ, 42-47, prof oral diag, 47-48; dean sch dent, Emory Univ, 4861; exec vpres, Hanau Eng Co, NY, 61-64; dean col, 64-72, PROF ADMIN DENT, COL DENT MED, MED UNIV SC, 64- Concurrent Pos: Consult, Asst Surgeon Gen USPHS, 50, dent dir, 53-; consult, Surgeon Gen, US Dept Army, 54-61. Mem: Am Dent Asn; Am Asn Dent Schs (secy, 46-50, pres, 54); fel Am Col Dent. Res: Human growth and development; histopathology; periodontal disease and dental caries; problems in clinical oral surgery; dental public health. Mailing Add: Col of Dent Med Med Univ of SC Charleston SC 29401

BUHRER, CARL FREDERICK, inorganic chemistry, solid state physics, see 12th edition

BUHRKE, VICTOR E, analytical chemistry, physical chemistry, see 12th edition

BUHRMAN, ROBERT ALAN, b Waynesboro, Pa, Apr 24, 45; m 72. SOLID STATE PHYSICS. Educ: Johns Hopkins Univ, BES, 67; Cornell Univ, MS, 70, PhD(appl physics), 73. Prof Exp: ASST PROF APPL & ENG PHYSICS, CORNELL UNIV, 73- Mem: Am Phys Soc. Res: Superconducting quantum devices; low temperature properties of metals; size effects in metals; optical properties of composite materials. Mailing Add: Sch of Appl & Eng Physics Clark Hall Cornell Univ Ithaca NY 14853

BUHROW, CHARLES, JR, b Elliston, Ohio, Nov 20, 23; m 46; c 5. MEDICINE, RADIOLOGY. Educ: Univ Chicago, BS, 45, MD, 48; Am Bd Radiol, dipl, 57. Prof Exp: Intern, Hosp, USPHS, Ill, 48-49, med officer, Div Occup Health, 49-54, radiol resident, Hosp, Md, 54-57, staff radiologist, 57-58, chief radiologist, Hosp, Mich, 58-64, chief radiologist, Staten Island, 64-68, MED OFFICER IN CHG, OUTPATIENT CLIN, USPHS, 68- Mem: AMA; Am Col Radiol; Radiol Soc NAm. Res: Diagnostic and general radiology. Mailing Add: Outpatient Clin US Pub Health Serv San Pedro CA 90731

BUHS, RUDOLF PAUL, b Newark, NJ, Apr 3, 17; m 41; c 3. DRUG METABOLISM. Educ: Seton Hall Col, BS, 38; Rutgers Univ, MS, 44. Prof Exp: RES FEL, MERCK SHARP & DOHME RES LABS, 38- Mem: Am Chem Soc. Res: Development of analytical, isolation and identification procedures for drug metabolites in biological media; application of isotope dilution methods for assay and tracer purposes; separation techniques such as thin-layer and high performance liquid chromatography. Mailing Add: 2 Sylvan Way Short Hills NJ 07078

BUHSE, HOWAR EDWARD, JR, b Chicago, Ill, July 22, 34; m 57; c 3. ZOOLOGY. Educ: Grinnel Col, AB, 57; Univ Iowa, MS, 60, PhD(zool), 63. Prof Exp: Res fel microbiol, State Univ NY Upstate Med Ctr, 63-65; asst prof biol sci, 65-68, ASSOC PROF BIOL SCI, UNIV ILL, CHICAGO CIRCLE, 68- Concurrent Pos: NIH fel, 63-65; NSF develop biol grant, 67-69. Mem: Soc Protozool. Res: Morphogenesis in the ciliated protozoa. Mailing Add: Dept of Biol Sci Univ of Ill Box 4348 Chicago IL 60680

BUI, TIEN DAI, b Bac Ninh, Vietnam, Mar 22, 45; Can citizen; m 69; c 2. COMPUTER SCIENCE. Educ: Univ Saigon, BSc, 64; Univ Ottawa, BASc, 68; Carleton Univ, MEng, 68; York Univ, PhD(space sci), 71. Prof Exp: Res engr, Inst Aerospace Studies, Univ Toronto, 68-69; res assoc & asst prof lasers & comput, Dept Mech Eng, McGill Univ, 71-74; ASST PROF COMPUT, CONCORDIA UNIV, 74- Concurrent Pos: Prof Asn Res & Develop Fund assoc, McGill Univ, 71-74. Mem: Brit Inst Physics; Inst Elec & Electronics Engrs; Asn Comput Mach; Can Asn Physicists. Res: Computational methods for analysis of complex systems; numerical modeling; optimization and simulation of various physical phenomena; numerical analysis. Mailing Add: Concordia Univ 1455 de Maisonneuve St W Montreal PQ Can

BUIE, BENNETT FRANK, b Patrick, SC, Jan 9, 10; m 38; c 4. ECONOMIC GEOLOGY. Educ: Univ SC, BS, 30; Lehigh Univ, MS, 32; Harvard Univ, MA, 34, PhD(petrog, struct geol), 39. Prof Exp: Asst geol, Harvard Univ, 32-37; jr geologist, Seaboard Oil Co, NY & Amiranian Oil Co, Iran, 37-38; geologist, Indian Oil Concessions, Ltd, Standard Oil Co Calif, 39-41 & Standard Oil Co Tex, 41-42; prof, Univ SC & geologist, State Develop Bd, SC, 46-56; head dept, 56-61, chmn, 61-64,

PROF GEOL, FLA STATE UNIV, 56- Concurrent Pos: Fulbright res award, Iran, 51; chief geologist, Resources Develop Corp, Iran, 52; geologist, US Geol Surv, 53-57; ed, Soc Mining Engrs, Am Inst Mining Metall & Petrol Eng, 72-75; consult geol, J M Huber Corp, 58- Mem: Fel AAAS; fel Geol Soc Am; Am Asn Petrol Geol; Soc Mining Engrs, Am Inst Mining Metall & Petrol Eng. Res: Economic geology; petrography; industrial minerals; economic geology of mineral deposits, especially industrial minerals; world resources of kaolin clays and phospate rock; optical mineralogy; tritium in ground water; dewatering of Florida phospate slimes. Mailing Add: Dept of Geol Fla State Univ Tallahassee FL 32306

BUIKEMA, ARTHUR L, JR, b Evergreen Park, Ill, Feb 4, 41; m 62; c 6. AQUATIC BIOLOGY, ENVIRONMENTAL PHYSIOLOGY. Educ: Elmhurst Col, BS, 62, MA, 65, PhD(zool), 70. Prof Exp: Asst prof biol, St Olaf Col, 67-71; asst prof, 71-74, ASSOC PROF, VA POLYTECH INST & STATE UNIV, 74- Mem: AAAS; Am Inst Biol Sci; Am Soc Limnol & Oceanog; Am Soc Zool. Res: Pollution biology; ecology and physiology of invertebrates; bioassay technique development. Mailing Add: Dept of Biol Ctr Environ Studies Va Polytech Inst and State Univ Blacksburg VA 24061

BUIKSTRA, JANE ELLEN, b Evansville, Ind, Nov 2, 45. BIOLOGICAL ANTHROPOLOGY, ARCHAEOLOGY. Educ: DePauw Univ, BA, 67; Univ Chicago, MA, 69, PhD(biol anthrop), 72. Prof Exp: Instr anthrop, 70-71, ASST PROF ANTHROP, NORTHWESTERN UNIV, EVANSTON, 71- Concurrent Pos: NIH bio-med sci res grant, Northwestern Univ, 71. Mem: AAAS; Am Anthrop Asn; Am Asn Phys Anthrop; Soc Am Archaeol. Res: Intensive regional approach to the study of prehistoric skeletal populations emphasizing micro-evolutionary change and biological response to environmental stress. Mailing Add: Dept of Anthrop Northwestern Univ Evanston IL 60201

BUJAKE, JAHN EDWARD, JR, b New York, NY, May 23, 33; m 64; c 4. PHYSICAL CHEMISTRY, FOOD SCIENCE. Educ: Manhattan Col, BS, 54; Col Holy Cross, MS, 55; Columbia Univ, PhD(phys chem), 59; NY Univ, MBA, 63. Prof Exp: Res assoc phys chem & new prod develop, Lever Brothers Co, NY, 59-68; develop assoc, Foods Div, Coca-Cola Co, 68-69, asst to tech vpres, 69-70, mgr tech serv, 70-71, dir, Houston Res & Develop, 71-72; dir res & develop technol, 72-75, DIR RES & DEVELOP, FOODS DIV, QUAKER OATS CO, 75- Mem: Am Chem Soc; Inst Food Technol; Am Inst Phys. Res: Cereal, beverage, citrus, coffee, snack, protein food product development; processing and packaging; colloid chemistry; rheology, reaction kinetics. Mailing Add: 617 W Main St Barrington IL 60010

BUKANTZ, SAMUEL CHARLES, b New York, NY, Sept 12, 11; m 41; c 2. CLINICAL MEDICINE. Educ: NY Univ, BS, 30, MD, 34; Am Bd Int Med, dipl; Am Bd Allergy, dipl. Prof Exp: From instr to assoc prof med, Sch Med, Wash Univ, 47-54, assoc prof clin med, 54-58, asst dean, Sch Med, 48-54; med & res dir, Jewish Nat Home Asthmatic Children & Children's Asthma Res Inst & Hosp, Denver & assoc prof clin med, Med Ctr, Univ Colo, 58-63; assoc prof clin med, Sch Med, NY Univ, 64-72; PROF MED, UNIV S FLA, 72-; CHIEF SECT ALLERGY, MED SERV, VET ADMIN HOSP, TAMPA, 72- Concurrent Pos: Baruch fel, Harlem Hosp, New York, 38-40; fel allergy, Internal Med Dept, Sch Med, Wash Univ, 46-47; sr attend physician, Jefferson Barracks Vet Hosp, 46-48; chmn res comt, Vet Admin Hosp, St Louis, 52-54, secy dean's comt, 51-54; consult, Beth Israel Hosp, Newark, NJ, 63- & Clara Maass Hosp, Belleville, 63-; assoc dir med res clin invest, Schering Corp, 63-65; assoc vis physician, Bellevue Med Ctr, New York, 64-72; dir dept clin res, Hoffmann-La Roche Inc, NJ, 65-67; ed, Hosp Pract, 69- Mem: Am Soc Exp Biol; AMA; fel Am Col Physicians; Am Col Chest Physicians; Am Asn Immunol. Res: Clinical investigation with new drugs; experimental hyper- sensitive states; pathogenesis and therapy of bronchial asthma; immunochemistry. Mailing Add: Vet Admin Hosp 13000 N 30th St Tampa FL 33612

BUKATA, ROBERT PETER, physics, space science, see 12th edition

BUKER, ROBERT JOE, b Vancouver, Wash, June 2, 30; m 52; c 5. PLANT BREEDING. Educ: Wash State Univ, BS, 53; Purdue Univ, MS, 59, PhD(plant breeding, genetics), 63. Prof Exp: Instr agron, Purdue Univ, 56-61; dir res, 61-73, EXEC VPRES & GEN MGR, FARMERS FORAGE RES COOP, 73- Mem: Am Soc Agron; Am Forage & Grassland Coun; Nat Coun Commercial Plant Breeders (pres). Res: Breeding forage crop varieties; directing forage, turf, soybean and corn breeding programs; improving field plot techniques and equipment. Mailing Add: Farmers Forage Res Coop 4112 E State Rd 225 West Lafayette IN 47906

BUKHARI, AHMAD IQBAL, b Punjab, India, Jan 5, 43. MOLECULAR BIOLOGY, GENETICS. Educ: D J Col, Karachi, BSc, 61; Univ Karachi, MSc, 63; Brown Univ, MS, 66; Univ Colo, PhD(microbiol), 70. Prof Exp: Asst lectr microbiol, Univ Karachi, 63-64; staff investr, 72-75, SR STAFF INVESTR, COLD SPRING HARBOR LAB, 75-; ASST PROF MICROBIOL, STATE UNIV NY STONY BROOK, 74- Concurrent Pos: Fel, Cold Spring Harbor Lab, 70-71, Jan Coffin Childs fel, 71-72, Jane Coffin Childs grant, 72-74; grant, NSF, 74-, career develop award, 75-; Nat Cystic Fibrosis Found grant, 74- Mem: AAAS; Am Soc Microbiol; Genetics Soc Am. Res: Integration of viruses into the host genome; genetic engineering; intracellular protein turnover. Mailing Add: Cold Spring Harbor Lab Cold Spring Harbor NY 11724

BUKOVAC, MARTIN JOHN, b Johnston City, Ill, Nov 12, 29; m 36; c 1. HORTICULTURE, PLANT PHYSIOLOGY. Educ: Mich State Univ, BS, 51, MS, 54, PhD, 57. Prof Exp: Res asst, 54-56, res assoc, 56-57, from asst prof to assoc prof, 57-63, PROF HORT, MICH STATE UNIV, 63- Prof Exp: Lectr, Japan Atomic Energy Res Inst, Tokyo, 58; adv, Int AEC, Vienna, 61; NSF sr fel, Oxford Univ & Univ Bristol, 65-66; mem, Nat Acad Sci-Nat Res Coun subcomt on effect of pesticides on physiol of fruits & veg, 64-68; Nat Acad Sci exchange visitor, Coun Acads, Yugoslavia, 71; guest lectr, Polish Acad Sci, Warsaw; mem, US Nat Comn, Int Union Biol Sci, 71- Honors & Awards: Joseph Harvey Gourley award, Am Soc Hort Sci, 69; Citation for Meritorious Res, Am Hort Soc, 70; Distinguished Serv Award, Mich State Hort Soc, 74; M A Blake Award, Am Soc Hort Sci, 75; Marion W Meadows Award, Am Soc Hort Sci, 75. Mem: Fel AAAS; fel Am Soc Hort Sci; Am Hort Soc (pres, 74-75); Am Soc Plant Physiol; Bot Soc Am. Res: Plant growth and development; plant growth substances; mechanisms of foliar penetration. Mailing Add: Dept of Hort Mich State Univ East Lansing MI 48823

BUKOVSAN, LAURA A, b Norfolk, Va, Jan 2, 40; m 64; c 1. GENETICS. Educ: Univ Richmond, BS, 61; Ind Univ, MA, 63, PhD(genetics), 69. Prof Exp: Teaching asst zool, Ind Univ, 62-63; res assoc with Dr William Bukovsan, 67-74; lectr genetics, NY State Univ Col, 74- Mem: Genetics Soc Am. Res: Cytology of conjugation in Tokophrya lemnarum; genetics of mating type inheritance in the suctorian Tokophrya lemnarum. Mailing Add: Dept of Biol State Univ Col Oneonta NY 13820

BUKOVSAN, WILLIAM, b Chicago, Ill, June 4, 29; m 64; c 1. ENDOCRINOLOGY. Educ: Univ Ill, BS, 54, MS, 58; Ind Univ, PhD(zool), 67. Prof Exp: PROF BIOL, STATE UNIV NY COL ONEONTA, 67- Mem: AAAS; Am Soc Zool; NY Acad Sci. Res: Adrenal phosphorus metabolism of the newly hatched chick. Mailing Add: Dept of Biol State Univ of NY Oneonta NY 13820

BUKOWICK, PETER ANTHONY, b Westfield, Mass, Dec 15, 43; m 65; c 2. ORGANIC CHEMISTRY, BIOCHEMISTRY. Educ: Lafayette Col, BS, 65; Univ Va, PhD(chem), 68. Prof Exp: Res chemist, 68-74, res supvr, 74-75, TECH SUPT, HERCULES, INC, 75- Mem: Am Chem Soc; Sigma Xi. Res: Chemical process improvement; utilization of wastes; air and water pollution control. Mailing Add: Hercules, Inc Louisiana MO 63353

BUKRY, JOHN DAVID, b Baltimore, Md, May 17, 41. MICROPALEONTOLOGY. Educ: Johns Hopkins Univ, AB, 63; Princeton Univ, AM, 65, PhD(geol), 67. Prof Exp: Geologist, US Army Corps Engrs, 63; res asst micropaleont, Socony-Mobil Oil Co, 65; GEOLOGIST, US GEOL SURV, 67- Concurrent Pos: Biostratig consult, Deep Sea Drilling Proj, 68-; res assoc, Scripps Inst Oceanog, 70- Mem: AAAS; Geol Soc Am; Paleont Soc; Paleont Res Inst; Sigma Xi. Res: Phytoplankton micropaleontology and ocean stratigraphy; evolutionary trends, paleoecology, and taxonomy of calcareous nannoplankton and silicoflagellates. Mailing Add: US Geol Surv Box 271 La Jolla CA 92038

BULA, RAYMOND J, b Antigo, Wis, Aug 3, 27; m 52; c 8. AGRONOMY. Educ: Univ Wis, BS, 49, MS, 50, PhD(agron, bot), 52. Prof Exp: Asst prof, seed analyst & agronomist, NY Agr Exp Sta, Geneva, 52-53; agronomist, Agr Res Serv, Agr Exp Sta, USDA, Alaska, 53-56 & Crops Res Div, Purdue Univ, 56-74; AREA DIR, AGR RES SERV, USDA, 74- Mem: Fel AAAS; Crop Sci Soc Am; fel Am Soc Agron; Am Soc Plant Physiol. Res: Forage crop production; weed control; physiology of cold resistance; forage crop physiology; influence of environment on genetic stability and plant growth and development. Mailing Add: 2336 Northwestern Ave Agr Res Serv USDA West Lafayette IN 47906

BULAS, ROMUALD, b Chabno, Russia, Feb 2, 22; US citizen; m 50; c 3. PHYSICAL CHEMISTRY. Educ: Univ Heidelburg, BS, 49; Polytech Inst Brooklyn, MS, 63. Prof Exp: SR CHEMIST, CENT RES LABS, INMONT CORP, CLIFTON, 51- Mem: Am Chem Soc; Soc Rheology. Res: Colloid chemistry. Mailing Add: 388 Beech Spring Rd South Orange NJ 07079

BULAT, THOMAS JOSEPH, b Chicago, Ill, Oct 9, 26; m 49; c 6. BIOMEDICAL ENGINEERING, ACOUSTICS. Educ: St Ambrose Col, BS, 49; Univ Iowa, MS, 50, PhD(mycol), 53. Prof Exp: Mgr ultrasonic res, 53-60, chief engr, 60-62, mgr sonic prod, 62-70, dir indust & med prod, Instrument & Life Support Div, Bendix Corp, 70-73; MGR MAT APPLNS DEPT, MAT RES, DEERE & CO, 74- Concurrent Pos: Pres ultrasonic comt, Int Electrotech Comn, 68-; consult, US Army Sterilization Tech, 71- Mem: Acoust Soc Am; Soc Cryosurg; Am Soc Test & Mat; Soc Indust Microbiol. Res: Acoustics—applications of ultrasonics in industry, chemistry and biology; biomedical engineering—respiratory therapy, cryogenic surgery and sterilization techniques; metallurgy and chemistry. Mailing Add: 1114 Pine Acre Dr Bettendorf IA 52722

BULBENKO, GEORGE FEDIR, b Moshoryno, Ukraine, June 29, 27; nat US; m 53; c 2. ORGANIC CHEMISTRY. Educ: Hanover Col, AB, 52; Ind Univ, MA, 54, PhD(org chem), 58. Prof Exp: Sr res chemist, Org Sect, Thiokol Chem Corp, 58-59, supvr, Org & Polysulfides Sect, 60-68, head, Org Polysulfides & Anal Sect, 68-69; dir chem sci, 69-72, VPRES, PRINCETON BIOMEDIX INC, 72- Prof Exp: Asst prof, Trenton Jr Col, 60-66; assoc prof, Mercer County Commun Col, 67-71. Mem: AAAS; Am Chem Soc; Am Inst Chem; Am Asn Clin Chem; Sigma Xi. Res: Synthetic organic chemistry; sulfur and clinical chemistry; polymers; research on clinical chemical assays. Mailing Add: Princeton Biomedix Lab PO Box 2241 Princeton NJ 08540

BULBROOK, HARRY MARSHALL, b Greenville, Tex, Jan 10, 95; m 37; c 3. CHEMISTRY. Educ: Rice Inst, AB, 16, AM, 17. Prof Exp: Asst chemist, Naval Proving Ground, 17-20; chemist, Ft Worth Labs, 20 & Terrell Labs, 21-28; from chemist to owner & dir, 28-70, CONSULT, INDUST LABS, 70- & INDUST TESTS, 70- Concurrent Pos: Pres, Indust Tests, 50-70. Mem: Am Chem Soc; Am Oil Chemists Soc; Am Pub Health Asn. Res: Carbon dioxide recovery system; carbon dioxide cylinder. Mailing Add: 3001 Cullen St Ft Worth TX 76107

BULEN, WILLIAM ALFRED, biochemistry, deceased

BULGER, RUTH ELLEN, b Kansas City, Mo. ANATOMY, PATHOLOGY. Educ: Vassar Col, AB, 58; Radcliffe Col, AM, 59; Univ Wash, PhD(anat), 62. Prof Exp: Asst prof path, Med Sch, Univ Wash, 67-70; assoc prof anat, Univ NC, 70-72; from assoc prof to prof, Univ Md, Baltimore City, 72-76; PROF PATH, MED SCH, UNIV MASS, 76- Concurrent Pos: Fel anat, Harvard Med Sch, 63-64; fel path, Univ Wash, 64-67. Mem: Am Asn Anat; Am Soc Cell Biol; Am Soc Exp Path; Am Soc Nephrol; NY Acad Sci. Res: Kidney morphology and function. Mailing Add: Dept of Anat Univ of Mass Med Sch Worcester MA 01605

BULGRIN, VERNON CARL, b Cuyahoga Falls, Ohio, May 10, 23; m 54; c 2. PHYSICAL CHEMISTRY. Educ: Univ Akron, BS, 48; Iowa State Univ, PhD(chem), 53. Prof Exp: Asst, Iowa State Univ, 48-53; instr, 53-56, from asst prof to assoc prof, 56-62, head dept, 62-67, PROF CHEM, UNIV WYO, 62- Concurrent Pos: Res Corp grant, 56, 58; vis prof, Purdue Univ, 67-68. Mem: Fel AAAS; Am Chem Soc. Res: Kinetics of specific oxidation reactions; homogeneous equilibria. Mailing Add: Dept of Chem Univ of Wyo Laramie WY 82070

BULICH, ANTHONY ANDREW, b San Pedro, Calif, Feb 16, 44; m 67; c 2. MICROBIOLOGY. Educ: Univ Southern Calif, BA, 66; Iowa State Univ, MS, 68, PhD(microbiol), 72. Prof Exp: Res microbiologist, 72-75, SR RES MICROBIOLOGIST, FERMENTATION RES, A E STALEY MFG CO, 75- Mem: AAAS; Am Soc Microbiol; Soc Indust Microbiol; Am Chem Soc. Res: Isolation and characterization of carbohydrase producing microorganism; strain improvement thru genetic manipulation and fermentation optimization; characterization of thermophilic and acidophilic amylase producing microorganisms. Mailing Add: A E Staley Co Res Ctr 2200 Eldorado Ave Decatur IL 62525

BULKIN, BERNARD JOSEPH, b Trenton, NJ, Mar 9, 42; m 66; c 1. PHYSICAL CHEMISTRY, INORGANIC CHEMISTRY. Educ: Polytech Inst Brooklyn, BS, 62; Purdue Univ, PhD(phys chem), 66. Prof Exp: NSF fel, Swiss Fed Inst Technol, 66-67; asst prof chem, 67-69, ASSOC PROF CHEM, HUNTER COL, 70-, CHMN DEPT, 74- Concurrent Pos: Petrol Res Fund grant, 67-69; Am Cancer Soc res grant, 68-72; Army Res Off-Durham res grant, 69-72; partic, NSF High Sch Inst, 70-71. Mem: Am Chem Soc. Res: Infrared and Raman spectroscopy of liquid crystals; spectroscopy of organometallic compounds, particularly metal carbonyls. Mailing Add: Hunter Col Dept of Chem 695 Park Ave New York NY 10021

BULKLEY, GEORGE, b Pa, Apr 16, 17; m 41; c 4. MEDICINE, UROLOGY. Educ: Northwestern Univ, BS, 38, MD, 42. Prof Exp: Instr urol, Sch Med, Yale Univ, 49-51; assoc, 51-55, from asst prof to assoc prof, 56-73, PROF UROL, MED SCH,

NORTHWESTERN UNIV, 73- Concurrent Pos: From assoc attend to sr attend urologist, Chicago Wesley Mem Hosp, 53-; attend urologist, Vet Admin Res Hosp, 54-; consult urologist, Passavant Mem Hosp, 70- Mem: AMA; Am Urol Asn; Am Col Surg; Int Soc Urol. Res: Urologic clinical research; use of radioactive materials in treatment of carcinoma of the prostate; surgical and chemotherapy carcinoma of bladder. Mailing Add: 251 E Chicago Ave Chicago IL 60611

BULKLEY, ROSS VIVIAN, b Burley, Idaho, Nov 29, 27; m 49; c 3. FISHERIES. Educ: Utah State Agr Col, BS, 52; Utah State Univ, MS, 57; Iowa State Univ, PhD, 69. Prof Exp: Instr biol & agr, Liahona Col, Tonga, 52-56; aid fishery mgt, Utah Dept Fish & Game, 56-57; fishery res biologist, US Fish & Wildlife Serv, 57-62; proj leader, Ore Game Comn, 65-66; ASST LEADER, IOWA COOP FISHERY RES UNIT, US FISH & WILDLIFE SERV, IOWA STATE UNIV, 66-, PROF FISHERIES, 66- Concurrent Pos: Plantation mgr, Church Jesus Christ Latter-day Saints unified sch syst, Samoa & Tonga, 62-65. Mem: Am Fisheries Soc; Am Inst Fishery Res Biol. Res: Fish population dynamics; fish-pesticides relations; pollution; fish culture; effects of pollution and environmental factors on fish survival, reproduction and growth. Mailing Add: Iowa State Univ 3 Sci Hall II Ames IA 50010

BULL, ALICE LOUISE, b White Plains, NY, May 16, 24. DEVELOPMENTAL BIOLOGY. Educ: Middlebury Col, AB, 46; Mt Holyoke Col, MA, 48; Yale Univ, PhD(zool), 53. Prof Exp: Asst zool, Yale Univ, 48-50; asst, Mt Holyoke Col, 46-48, instr, 53-55; from instr to asst prof, Wellesley Col, 55-64; ASSOC PROF ZOOL, HOLLINS COL, 64- Concurrent Pos: Am Asn Univ Women fel, Univ Zurich, 52-53. Mem: AAAS; Genetics Soc Am; Am Soc Zoologists; Soc Develop Biol; Sigma Xi. Res: Oogenesis, fertility regulation and developmental genetics in Drosophila. Mailing Add: Dept of Biol Hollins Col Hollins College VA 24020

BULL, BRIAN S, b London, Eng, Sept 14, 37; US citizen; m 63; c 2. PATHOLOGY. Educ: Walla Walla Col, BS, 57; Loma Linda Univ, MD, 61. Prof Exp: Intern, Grace New Haven Med Ctr, 61-62, resident, 62-63; resident, NIH, 63-65, staff hematologist, 66-67; from asst prof to assoc prof, 68-72, PROF PATH & CHMN DEPT, LOMA LINDA UNIV, 73- Concurrent Pos: Fel lab med, NIH, 65-66, spec fel hemat, Nat Inst Arthritis & Metab Dis, 67-68. Mem: Am Soc Clin Path; AMA; Soc Acad Clin Lab Physicians & Scientists; Col Am Path; Am Soc Hemat. Res: Physiology of blood cells, particularly erythrocytes and platelets; clinical laboratory test development and automation. Mailing Add: Clin Lab Med Ctr Loma Linda Univ Loma Linda CA 92354

BULL, COLIN BRUCE BRADLEY, b Birmingham, Eng, June 13, 28; m 56; c 3. GEOPHYSICS. Educ: Univ Birmingham, BS, 48, PhD(physics), 51. Prof Exp: Geophysicist, Cambridge Univ, 52-55; Imp Chem Industs res fel geophys, Univ Birmingham, 55-56; sr lectr physics, Victoria Univ, NZ, 56-61; vis asst prof, 61-62, assoc prof, 62-65, dir inst polar studies, 65-69, chmn dept geol, 69-72, PROF GEOL, OHIO STATE UNIV, 65-, DEAN COL MATH & PHYS SCI, 72- Concurrent Pos: Geophysicist & chief scientist, Brit N Greenland Exped, 52-55; mem, Ross Dependency Res Comt, NZ Govt, 59-61; mem panels glaciol, geol & geophys & comt polar res, Nat Acad Sci. Honors & Awards: Polar Medal, 55; Antarctic Serv Award, US Cong, 74. Mem: Am Geophys Union; Arctic Inst NAm; Glaciol Soc; Geol Soc Am. Res: Geophysical and glaciological investigations in polar regions, particularly Greenland and Antarctica. Mailing Add: Col of Math & Phys Sci Ohio State Univ Columbus OH 43210

BULL, DON LEE, b Raymondville, Tex, Sept 3, 29; m 54; c 1. ENTOMOLOGY. Educ: Tex A&M Univ, BS, 53, MS, 60, PhD(entom), 62. Prof Exp: Analyst, Amoco Chem Corp, 56-57 & Shell Chem Co, 57-58; RES ENTOMOLOGIST TOXICOL, AGR RES SERV, USDA, 61- Mem: Entom Soc Am. Res: Insect toxicology and physiology. Mailing Add: Agr Res Serv USDA PO Drawer DG College Station TX 77840

BULL, HENRY BOLIVAR, b Stateburg, SC, June 16, 05; m 35; c 1. PHYSICAL BIOCHEMISTRY, PROTEIN CHEMISTRY. Educ: Univ SC, BS, 27; Univ Minn, MS, 28, PhD(biochem), 30. Prof Exp: Asst chem, Univ Minn, 27-28; asst biochem, Univ Rochester, 28-29; from instr to asst prof biochem, Univ Minn, 29-36; from asst prof to prof biochem, Med Sch, Northwestern Univ, 36-52; prof & head dept, 52-63, res prof, 63-73, EMER PROF BIOCHEM & RES SCIENTIST, COL MED, UNIV IOWA, 73- Concurrent Pos: Nat Res fel, Berlin-Dahlem, 31-32; prof, Calif Inst Technol, 43. Honors & Awards: Gold Medal, Iowa Sect, Am Chem Soc, 73. Mem: Am Chem Soc; Am Soc Biol Chemists; Biophys Soc; Soc Gen Physiologists. Res: Water and solute binding to proteins; protein denaturation. Mailing Add: Dept of Biochem Col of Med Univ of Iowa Iowa City IA 52242

BULL, JOHN, b New York, NY, Feb 28, 14; m 44; c 1. ORNITHOLOGY, ECOLOGY. Prof Exp: Customs examr wild bird plumage, US Customs, 42-59; spec investr & nature leader, Nat Audubon Soc, 60-61; res asst, 62-65, FIELD ASSOC ORNITH, AM MUS NATURAL HIST, 65- Concurrent Pos: Chief investr bird hazards to jet aircraft at Kennedy Int Airport, Fish & Wildlife Serv, US Govt, 65; Dan Caulkins & W Sanford writing grants NY State bird book, 66-; nature leader, Natural Hist Treasure World Tours, 67-68; del, Int Comt Bird Preservation, 66-70 & Int Ornith Cong, 66, 70; instr, New Sch Social Studies, 74- Honors & Awards: Outstanding title for gen libr collections, Library J, 75. Mem: Am Ornith Union; Cooper Ornith Soc; Wilson Ornith Soc; Brit Ornithologists' Union. Res: Birds of New York State; botany; entomology. Mailing Add: Am Mus Nat Hist Cen Park West and 79th St New York NY 10024

BULL, LEONARD SETH, b Westfield, Mass, Jan 31, 41; m 71. ANIMAL NUTRITION, PHYSIOLOGY. Educ: Okla State Univ, BSc, 63, MSc, 64; Cornell Univ, PhD(nutrit), 69. Prof Exp: NASA fel physiol, Med Sch Univ Va, 68-70; from asst prof to assoc prof, Univ Md, 70-75; ASSOC PROF NUTRIT, UNIV KY, 75- Mem: Am Dairy Sci Asn; Am Soc Animal Sci. Res: Animal nutrition with special interest in energy metabolism, digestive physiology and trace element interactions on metabolism; body compositon and animal calorimetry plus regulation of energy balance. Mailing Add: Dept of Animal Sci Univ of Ky Lexington KY 40506

BULL, RICHARD C, b Cedaredge, Colo, Nov 25, 34; m 59; c 2. ANIMAL NUTRITION, BIOCHEMISTRY. Educ: Colo State Univ, BS, 57, MS, 60; Ore State Univ, PhD(animal nutrit, biochem), 66. Prof Exp: Res assoc animal nutrit, Ore State Univ, 63-67; asst prof, 67-72, ASSOC PROF ANIMAL SCI & ASSOC ANIMAL SCIENTIST, UNIV IDAHO, 72- Mem: Am Inst Biol Sci; Am Soc Animal Sci. Res: Nutritional biochemistry, especially vitamin E and selenium. Mailing Add: Dept of Animal Sci Univ of Idaho Moscow ID 83843

BULL, WILLARD CLARE, chemistry, see 12th edition

BULL, WILLIAM BENHAM, b San Francisco, Calif, Apr 19, 30; m 53; c 2. GEOMORPHOLOGY. Educ: Univ Colo, BA, 53; Stanford Univ, MS, 57, PhD(geol), 60. Prof Exp: GEOLOGIST & RES HYDROLOGIST, WATER RESOURCES DIV, US GEOL SURV, 56-; PROF GEOL, UNIV ARIZ, 68- Mem: Am Geophys Union; Asn Eng Geologists; Geol Soc Am; Nat Asn Geol Teachers. Res: Tectonic and climatic geomorphology; fluvial geomorphology of arid regions; land subsidence due to ground water withdrawal and due to collapse of soils upon wetting. Mailing Add: Dept of Geosci Univ of Ariz Tucson AZ 85721

BULL, WILLIAM EARNEST, b Franklin Co, Mo, Jan 17, 33; m 55; c 3. INORGANIC CHEMISTRY. Educ: Univ Southern Ill, AB, 54; Univ Ill, AM, 55, PhD(chem), 57. Prof Exp: From asst prof to assoc prof, 57-70, PROF CHEM, UNIV TENN, KNOXVILLE, 70-, DIR GEN CHEM, 74- Concurrent Pos: Fulbright-Hayes award, Univ Col, Dublin, 70-71. Mem: AAAS; Am Chem Soc; BriThe Chem Soc. Res: Nonaqueous solvents; coordination compounds; photoelectron spectroscopy. Mailing Add: Dept of Chem Univ of Tenn Knoxville TN 37916

BULLA, LEE AUSTIN, JR, b Oklahoma City, Okla, May 1, 41; m 62; c 1. MICROBIOLOGY, BIOCHEMISTRY. Educ: Midwestern Univ, BS, 65; Ore State Univ, PhD(microbiol), 68. Prof Exp: Bacteriologist & chemist water purification, City of Wichita Falls, Tex, 65; microbiologist, Northern Regional Res Lab, USDA, 68-73, MICROBIOLOGIST, US GRAIN MKT RES CTR, USDA, KANS STATE UNIV, 73- Concurrent Pos: Affil instr chem, Grad Sch, Bradley Univ. Mem: AAAS; Am Soc Microbiol; Soc Invertebrate Path; NY Acad Sci. Res: Microbiology, especially physiology and biochemistry; physiology of bacteria associated with insects; electron microscopy of micro-organisms. Mailing Add: US Grain Mkt Res Ctr USDA Kans State Univ Manhattan KS 66502

BULLARD, EDWARD CRISP, b Norwich, Eng, Sept 21, 07; m 31; c 4. GEOPHYSICS. Educ: Cambridge Univ, BA, 29, MA, 30, PhD(natural sci), 32, ScD(natural sci), 48. Hon Degrees: Dsc, Mem Univ, Nfld, 70. Prof Exp: Demonstr geod, Cambridge Univ, 31-35; Smithsonian res fel geophys, Cambridge & Royal Soc, 36-43; reader, Cambridge Univ, 45-48; prof physics, Univ Toronto, 48-49; dir, Nat Phys Lab, 50-55; asst dir res geophys, Cambridge Univ, 56-60, reader, 60-64, prof, 64-74; PROF GEOPHYS, UNIV CALIF, SAN DIEGO, 63- Concurrent Pos: Exp officer, Admiralty, London, Eng, 39-45; dir, Bullard & Sons Ltd, 50-68; dir, IBM, UK, 64-75. Honors & Awards: Knight Bachelor, 53; Hughes Medal, Royal Soc, 53; Chree Medal, The Phys Soc, 56; Day Medal, Geol Soc Am, 59; Gold Medal, Royal Astron Soc, 65; Agassiz Medal of US, Nat Acad Sci, 65; Wollaston Medal, Geol Soc London, 67; Vetlesan Prize, 68; Bowie Medal, Am Geophys Union, 75. Mem: For assoc Nat Acad Sci; for hon mem Am Acad Arts & Sci; fel Royal Soc; fel Royal Astron Soc; fel Geol Soc London. Res: Geology of the oceans, the earth's magnetic field, the earth's upper mantle. Mailing Add: Inst Geophys & Planetary Physics Univ of Calif, San Diego La Jolla CA 92093

BULLARD, EDWIN ROSCOE, JR, b El Paso, Tex, Dec 15, 21; m 45; c 1. GEOPHYSICS, GEOLOGY. Educ: Univ Tex, El Paso, BS, 49 & 63. Prof Exp: Subsurface geologist, Kerr-McGee Oil Indust, Inc, 49-51, Tex Gulf Prod Co, 51-56, Ambassador Oil Corp, 56-60 & Ralph Lowe of Midland, Tex, 60-62; proj physicist, Schellenger Res Lab, Univ Tex, El Paso, 63-64; res geophysicist, water resources div, US Geol Surv, 64-66; res geophysicist, Globe Explor Co, 66-69, DIR, GLOBE UNIVERSAL SCI, ADVAN PROJ DIV, SHELL OIL CO, 69- Mem: Am Asn Petrol Geologists. Res: Bringing the state of the art to seismic and borehole geophysics for oil exploration. Mailing Add: 724 Kern Dr El Paso TX 79902

BULLARD, ERVIN TROWBRIDGE, b New York, NY, May 25, 20; m 48; c 3. HORTICULTURE. Educ: NC State Col, BS, 43; Cornell Univ, MS, 46; Purdue Univ, PhD(hort), 50. Prof Exp: Asst prof veg crops, Univ RI, 46-47; asst horticulturist, Purdue Univ, 50. Prof Exp: Asst prof veg crops, Univ RI, 46-47; asst horticulturist, Purdue Univ, 48-50; assoc prof veg res, Univ Idaho, 50-54; agr res adv, USAID, 54-64, agr prod adv, Brazil, 64-68, dep rural develop off, Repub Panama, 69-72, chief, Agr Div, India, 72-74, CHIEF AGR PROD, PAKISTAN, 74- Concurrent Pos: Fulbright res scholar, Univ Cairo, 51-52; vis prof, Dept Hort, Ohio State Univ, 68-69. Mem: Am Soc Hort Sci; Am Genetic Asn. Res: Cacao; coffee; vegetable crops. Mailing Add: Islamabad (ID), Dept of State Washington DC 20521

BULLARD, FRED MASON, b McLoud, Okla, July 20, 01; m; c 2. GEOLOGY, VOLCANOLOGY. Educ: Univ Okla, BS, 21, MS, 22; Univ Mich, PhD(geol), 28. Prof Exp: Field geologist, Okla Geol Surv, Norman, 21-23; consult geologist, 23-24; from instr to prof, 24-71, chmn dept, 29-37, EMER PROF VOLCANOLOGY, UNIV TEX, AUSTIN, 71- Concurrent Pos: Vis prof, Vassar Col, 49; Fulbright res scholar, Italy, 53; Fulbright lectr, Peru, 59; lectr on Paricutin Volcano, Am Asn Petrol Geologists, 43-45 & a volcanic cycle, 54; vis prof & chief party, US Tech Assistance Prog, Univ Baghdad, 62-64. Mem: Fel Geol Soc Am; fel Mineral Soc Am; Am Asn Petrol Geologists. Res: Igneous geology; meteorites; Paricutin Volcano, Mexico; volcanoes and geology of Latin America; active volcanoes of the world; Holocene volcanic activity of western United States. Mailing Add: Dept of Geol Sci Univ of Tex Austin TX 78712

BULLARD, REUBEN GEORGE, geology, archaeology, see 12th edition

BULLARD, TRUMAN ROBERT, b Miami, Fla, Jan 15, 39; m 60; c 2. PHYSIOLOGY. Educ: Ashbury Col, AM, 61; WVa Univ, MS, 63, PhD(physiol), 66. Prof Exp: Res assoc dept theoret & appl mech, WVa Univ, 64-65; asst prof, Christian Med Col, Ludhlana, Punjab, India, 68-69; asst prof, Rutgers Univ, 69-71; READER PHYSIOL, CHRISTIAN MED COL, VELLORE, INDIA, 71- Concurrent Pos: Fel physiol, Sch Med, Univ Va, 65-68; Sigma Xi grant, 68-69; Rutgers Res Coun grant, 69-70. Res: Effects of exercise on body composition and energy metabolism; interaction of endocrines and atrophy of disuse in bone. Mailing Add: Dept of Physiol Christian Med Col Vellore South India

BULLAS, LEONARD RAYMOND, b Lismore, New South Wales, Dec 8, 29; m 58; c 2. MICROBIAL GENETICS. Educ: Univ Adelaide, BSc, 53, MSc, 57; Mont State Col, PhD(genetics), 63. Prof Exp: Instr bact, Univ Adelaide, 53-58; asst genetics, Mont State Col, 59-62; from instr to asst prof microbiol, 62-70, ASSOC PROF MICROBIOL, SCH MED, LOMA LINDA UNIV, 70- Mem: Am Soc Microbiol; Am Genetics Soc. Res: Genetic basis of lysogeny in Salmonella; host-modification. Mailing Add: Sch of Med Loma Linda Univ Loma Linda CA 92354

BULLEN, ADELAIDE KENDALL, b Worcester, Mass, Jan 12, 08; m 29; c 2. ANTHROPOLOGY. Educ: Radcliffe Col, AB, 43. Prof Exp: Res anthropologist, Radcliffe Health Ctr, Radcliffe Col, 43-44; civilian consult anthrop, Dept Army, 46; anthropologist, Peabody Mus, Harvard Univ, 46-48; anthropologist, Fla State Mus, Univ Fla, 49-53; assoc anthrop, 54-69, RES ASSOC ANTHROP, FLA STATE MUS, UNIV FLA, 70- Concurrent Pos: Contrib ed, Handb Latin Am Studies, Hispanic Found, Library Cong, 69-71. Honors & Awards: Phipps Bird Award, Fla Acad Sci, 53. Mem: Fel AAAS; fel Am Anthrop Asn; Am Asn Phys Anthrop; Soc Res Child Develop; fel Royal Anthrop Inst Gt Brit & Ireland. Res: Comparative language behavior; variations in body build and temperament; nervous and mental fatigue; prehistoric and living Indian populations of Florida and the caribbean. Mailing Add: Fla State Mus Univ of Fla Gainesville FL 32611

BULLEN, MILES REX, b England, Mar 13, 26; Can citizen; m 57; c 4. GENETICS, CYTOGENETICS. Educ: Univ BC, BSA, 51; McGill Univ, MSc, 60; Univ Col Wales, PhD(genetics), 69. Prof Exp: Res officer agr genetics, 63-70, RES SCIENTIST, CAN DEPT AGR, 70- Mem: Genetics Soc Can; Brit Genetical Soc. Res: Genetic control of recombination of chiasma frequency; investigation of nuclear phenotype; inheritance of canopy shape and the outcome on yield in the grass timothy. Mailing Add: Can Dept of Agr Res Sta 2560 Chemin Gomin Quebec PQ Can

BULLEN, PETER SOUTHCOTT, b Portsmouth, Eng, Jan 19, 28; m 52; c 4. MATHEMATICS. Educ: Univ Natal, MSc, 49; Univ Cambridge, PhD(math), 55. Prof Exp: Lectr, Dept Pure & Appl Math, Univ Natal, 47-50, 53-56; from instr to assoc prof math, 56-70, PROF MATH, UNIV BC, 70- Res: Non-absolute integration. Mailing Add: Dept of Math Univ of BC Vancouver BC Can

BULLEN, THOMAS GERRARD, b Melling, Eng, Mar 5, 13. EXPERIMENTAL PHYSICS. Educ: Nat Univ Ireland, BS & MSc, 35, PhD(physics), 37, Higher Dipl Educ, 38. Prof Exp: Asst, Nat Univ Ireland, 35-37; res assoc, Bristol Univ, 38-39; head sci dept, St Brendan's Col, Eng, 39-41; head sci dept, St Joseph's Col, Eng, 41-46; head dept physics, 47-66, chmn sci div, 51-66, PROF PHYSICS, IONA COL, 47- Concurrent Pos: Consult, US Air Force Contract, 49-51; chief investr & coordr, US Army Ord, 59. Mem: AAAS; Electrochem Soc; Am Asn Physics Teachers; Brit Inst Physics & Phys Soc. Res: Teaching methods in physics; radioisotope techniques; thermal properties at high temperatures; neutron generators; soft x-ray spectroscopy. Mailing Add: 33 Beechmont Dr New Rochelle NY 10801

BULLER, CLARENCE S, b McPherson, Kans, June 21, 32; m 59; c 2. MICROBIOLOGY. Educ: Univ Kans, BA, 58, MA, 60, PhD(bact), 63. Prof Exp: Res fel microbiol, Western Reserve Univ, 63-66; res fel, 66-73, ASSOC PROF MICROBIOL, UNIV KANS, 73- Mem: Am Chem Soc; Am Soc Microbiol. Res: Physiology of bacteriophage infected coliforms; virus induced changes in cell surfaces. Mailing Add: Dept of Microbiol Univ of Kans Lawrence KS 66044

BULLER, ROBERT HENRY, b SDak, May 19, 19; m 56; c 3. PHARMACOLOGY. Educ: Ferris State Col, BS, 54; Purdue Univ, MS, 56, PhD(pharmacol), 58. Prof Exp: Group leader, Norwich Pharmacal Co, 57-63; dir pharmacol, Int Res & Develop Corp, 63-69; HEAD BIOCONTROL, UPJOHN CO, 69- Mem: AAAS; Am Soc Pharmacol & Exp Therapeut; Am Pharmaceut Asn; Soc Toxicol; NY Acad Sci. Res: Experimental pharmacology and pharmacodynamics; experimental toxicology; bioavailability of solid dosage forms; development of bioassay techniques. Mailing Add: Upjohn Co Dept 7832-41-2 7171 Portage Rd Portage MI 49081

BULLERMAN, LLOYD BERNARD, b Adrian, Minn, June 20, 39; m 60; c 4. FOOD MICROBIOLOGY, FOOD SCIENCE. Educ: SDak State Univ, BS, 61, MS, 65; Iowa State Univ, PhD(bact, food technol), 68. Prof Exp: Lab technician qual control, Dairy Prod, Inc, SDak, 61-63; food scientist prod develop, Green Giant Co, Minn, 68-70; asst prof, 70-75, ASSOC PROF FOOD SCI & TECHNOL, UNIV NEBR, LINCOLN, 75- Mem: Inst Food Technol; Am Soc Microbiol; Int Asn Milk, Food & Environ Sanit; Brit Soc Appl Bact. Res: Food microbiology and toxicology, particularly food-borne microorganisms that may be toxic or pathogenic to humans; food product development; waste disposal. Mailing Add: Dept of Food Sci & Technol Univ of Nebr Lincoln NE 68583

BULLINGTON, ROBERT ADRIAN, b Walnut, Ill, Feb 28, 08; m 35; c 2. SCIENCE EDUCATION, ECOLOGY. Educ: Eureka Col, BS, 31; Univ Ill, MS, 36; Northwestern Univ, PhD(sci educ, ecol), 49. Prof Exp: Instr biol, Eureka Col, 32; teacher pub schs, Ill, 32-45; asst & assoc prof sci, MacMurray Col, 45-47, prof, 48-50; instr physics, Northwestern Univ, 48; prof, 50-73, EMER PROF BIOL, NORTHERN ILL UNIV, 73- Concurrent Pos: Educ adv, US AID, US Dept State, Southeast Asia & Northern Rhodesia, 60-62; Fulbright-Hays lectr ecol & sci educ, Univ Philippines, 65-66. Mem: AAAS; Ecol Soc Am; Nat Asn Biol Teachers; Nat Sci Teachers Asn. Res: Training of biology teachers; science for general education; ecology and its applications in conservation; science teaching materials. Mailing Add: 1110 E Friendly Lane Horseshoe Bend AR 72512

BULLIS, GEORGE LEROY, b Eau Claire, Wis, Apr 26, 22; m 45; c 4. MATHEMATICS. Educ: Wis State Univ, Eau Claire, BS, 41; Univ Wis, MA, 42. Prof Exp: Instr math, Univ Wis, 43-45; assoc prof, 45-68, actg dean sch arts & sci, 68, PROF MATH & DEAN COL ARTS & SCI, UNIV WIS, PLATTEVILLE, 68- Mem: Math Asn Am. Mailing Add: Col of Arts & Sci Univ of Wis-Platteville Platteville WI 53818

BULLIS, HARVEY RAYMOND, JR, b Milwaukee, Wis, June 14, 24; m 44; c 4. MARINE BIOLOGY. Educ: Wis State Col, BS, 49; Univ Miami, MS, 51. Prof Exp: Asst zool, Univ Miami, 49-50; marine biologist, US Fish & Wildlife Serv, 50-70; assoc dir marine fisheries serv, Nat Oceanic & Atmospheric Admin, Dept Com, 70-71, DIR, SOUTHEAST FISHERIES CTR, NAT MARINE FISHERIES SERV, US DEPT COM, 71- Concurrent Pos: Dir, Bur Com Fisheries Explor Fishing Base, 55-56; consult, US AID, surv WAfrican fisheries, 61; mem, Nat Acad Sci comt technol & sci base Puerto Rican econ, 66-67, Sino-Am colloquium ocean resources, 71; US del, WCent Atlantic Fisheries Comn, 75; US asst nat coordr fisheries, Coop Invests Caribbean & Adjacent Regions, 71-76, asst int coordr, 75-76. Honors & Awards: Award, US Dept Interior, 54. Mem: Fel Am Inst Fishery Res Biologists; NY Acad Sci; Am Soc Ichthyol & Herpet; Soc Syst Zool; Am Fisheries Soc. Res: Zoogeography of tropical west Atlantic; fishery exploration and resource analysis. Mailing Add: 121 Island Dr Key Biscayne FL 33149

BULLIS, WILLIAM MURRAY, b Cincinnati, Ohio, Aug 29, 30; m 53; c 3. PHYSICS. Educ: Miami Univ, AB, 51; Mass Inst Technol, PhD(physics), 56. Prof Exp: Asst physics, Mass Inst Technol, 52-54; staff mem solid state physics, Lincoln Lab, 54 & Los Alamos Sci Lab, 56-57; physicist, Int Tel & Tel Labs, 57-59; mem tech staff, Tex Instruments Inc, Dallas, 59-65; res physicist, 65-68, CHIEF SEMICONDUCTOR CHARACTERIZATION SECT, NAT BUR STAND, 68-, ASST CHIEF SEMICONDUCTOR TECHNOL, ELECTRONIC TECHNOL DIV, 75- Concurrent Pos: Lectr elec eng, Univ Md, 66-70. Mem: Am Phys Soc; Electrochem Soc; Am Soc Testing & Mat. Res: Semiconductor physics; transport phenomena in solids and fluids. Mailing Add: Semiconductor Char Sect Nat Bur Stand Washington DC 20234

BULLITT, ORVILLE HORWITZ, JR, organic chemistry, see 12th edition

BULLOCK, AUSTIN LARNEL, b Tylertown, Miss, June 5, 17; m 40, 60; c 4. ORGANIC CHEMISTRY. Educ: Miss Col, BS, 49. Prof Exp: Teacher pub schs, Miss, 47-53; chemist, Southern Regional Res Lab, USDA, 53-62, chemist & proj leader, 62-71; res chemist, 71-75, ACTG DIR, TEXTILES & CLOTHING LAB, USDA, 74-75. Concurrent Pos: Lectr, Col Home Econ, Univ Tenn, 73-75. Mem: Am Asn Textile Chem & Colorists; Am Chem Soc; Sigma Xi; fel Am Inst Chem. Res: Chemical modification of cotton cellulose and carbohydrates; preparation of ion-exchange cellulose and starch; maintenance and care of consumer textiles. Mailing Add: USDA Textiles & Clothing Lab 2005 Lake Ave Knoxville TN 37916

BULLOCK, ERIC, b Hull, Eng, Dec 26, 28; m 57; c 2. ORGANIC CHEMISTRY. Educ: Univ Cambridge, BA, 53, MA & PhD(chem), 56. Prof Exp: Cancer Campaign fel, 56-57; Nat Res Coun Can fel, 57-59; Imp Chem Indust fel, 59-61; lectr chem, Univ Nottingham, 61-63; head dept, 63-74, PROF CHEM, MEM UNIV NFLD, 63- Mem: Brit Chem Soc. Res: Heterocyclic chemistry, especially pyrroles and porphyrins, application of nuclear resonance in this field. Mailing Add: Dept of Chem Mem Univ of Nfld St John's NF Can

BULLOCK, FRANCIS JEREMIAH, organic chemistry, see 12th edition

BULLOCK, GRAHAM LAMBERT, b Martinsburg, WVa, Mar 6, 35; m 55; c 5. BACTERIOLOGY, FISH PATHOLOGY. Educ: Shepherd Col, WVa, BS, 57; Univ Wis-Madison, MS, 59; Fordham Univ, PhD(biol sci), 70. Prof Exp: Bacteriologist qual control, Thomas J Lipton Inc, NJ, 59-60; RESEARCHER BACT FISH DIS, EASTERN FISH DIS LAB, FISH & WILDLIFE SERV, DEPT INTERIOR, 60- Concurrent Pos: Vis prof, Ore State Univ, 73-74. Mem: Am Soc Microbiol; Am Fisheries Soc; Sigma Xi. Res: Nature and control of bacterial infections of fish including the role of environmental factors, chemotherapy and serological methods for detection and identification of bacterial pathogens. Mailing Add: Rte 1 Box 17A Kearneysville WV 25430

BULLOCK, GREG A, organic chemistry, see 12th edition

BULLOCK, HOWARD R, b South Gate, Calif, Sept 17, 30; m 52, 68; c 2. ENTOMOLOGY, PHYSIOLOGY. Educ: Univ Utah, BS, 51, MS, 52; Univ Md, PhD(entom), 63. Prof Exp: Nat Acad Sci res assoc insect path, Pioneering Res Lab, Entom Res Div, USDA, 63-69, RES ENTOMOLOGIST, COTTON INSECTS PHYSIOL INVESTS, COTTON INSECTS RES BR, USDA, LA, 69- Mem: AAAS; Entom Soc Am; Am Micros Soc; Am Mosquito Control Asn. Res: Medical entomology; physiology; morphology and pathology of invertebrates. Mailing Add: Cotton Insects Res Br USDA 4115 Gourrier Ave Baton Rouge LA 70808

BULLOCK, JOHN, b Orange, NJ, Apr 24, 32; m 58; c 3. ENDOCRINOLOGY. Educ: Seton Hall Univ, BS, 57; Rutgers Univ, MS, 62; Colo State Univ, PhD(physiol), 65. Prof Exp: Res technician, Merck Inst Therapeut Res, 56-57; res asst, 57-61, staff mem & res assoc, 61-62, ASST PROF PHYSIOL, COL MED & DENT NJ, NEWARK, 65- Mem: AAAS; assoc mem Am Physiol Soc. Res: Androgenic control of connective tissue; biochemistry of tumors; mechanism of action of androgenic esters; connective tissue physiology; biochemistry of cancer; physiology of aging. Mailing Add: Col of Med & Dent of NJ 100 Bergen St Newark NJ 07103

BULLOCK, JONATHAN S, IV, b Mobile, Ala, Dec 8, 42; m 69; c 1. PHYSICAL CHEMISTRY. Educ: Ala Col, BS, 64; Tulane Univ, PhD(phys chem), 69. Prof Exp: Develop chemist, 68-76, MEM DEVELOP STAFF, NUCLEAR DIV, UNION CARBIDE CORP, 76- Mem: Am Chem Soc. Res: Solvent effect in kinetics; electrochemistry of fluid-electrode and solid-electrode systems; gas-solid interface reactions; corrosion; electrochemical energy systems; high-temperature thermochemistry. Mailing Add: 120 Euclid Circle Oak Ridge TN 37830

BULLOCK, KENNETH C, b Pleasant Grove, Utah, Sept 8, 18; m 38; c 4. GEOLOGY. Educ: Brigham Young Univ, BS, 40, MA, 42; Univ Wis, PhD(geol), 49. Prof Exp: From instr to assoc prof, 43-57, chmn dept, 56-62, PROF GEOL, BRIGHAM YOUNG UNIV, 57- Concurrent Pos: Geol engr, US Steel Corp, 53-54; geologist, Utah Geol & Mineral Surv, 67; res geologist, Columbia Iron Mining Co, 60. Mem: Fel Geol Soc Am; Am Inst Mining, Metall & Petrol Engrs; Nat Asn Geol Teachers; Mineral Soc Am. Res: Economic geology; mineralogy; petrology; iron and fluorite deposits of Utah. Mailing Add: Dept of Geol 129 ESC Brigham Young Univ Provo UT 84602

BULLOCK, LESLIE PATRICIA, b Los Angeles, Calif, May 20, 36. ENDOCRINOLOGY. Educ: Pomona Col, BA, 57; Univ Calif, Davis, DVM, 63. Prof Exp: Res fel endocrinol, Nat Cancer Inst, 68-70 & Med Col, Cornell Univ, 67-68; intern & mem staff, Angell Mem Hosp, Boston, 63-67; ASST PROF COMP MED & RES ASSOC ENDOCRINOL, MILTON S HERSHEY MED CTR, PA STATE UNIV, 71- Mem: AAAS; Endocrine Soc; Soc Study Reproductive Biol; Am Fedn Clin Res. Res: Interaction of androgens and other steroid and protein hormones at the biologic and molecular level; molecular mechanism of androgen action. Mailing Add: Div of Endocrinol Milton S Hershey Med Ctr Hershey PA 17033

BULLOCK, PAUL DAVID, b Waterloo, NY, Sept 18, 29; m 57; c 3. RESEARCH ADMINISTRATION, APPLIED STATISTICS. Educ: State Univ NY Albany, BA, 51; Univ Mich, MA, 56. Prof Exp: Res asst math, US Army Chem Ctr, 51-53; statist engr, Eastman Kodak Co, 54; res assoc statist sampling, Inst Social Res, Univ Mich, 54-56; statist engr, 56-75, DIV CHIEF, US STEEL RES LAB, 75- Mem: Am Statist Asn; Inst Mgt Sci. Res: Research and application of applied mathematics and computers to problems in raw materials, metals, chemicals and plastics concentrating in areas of physical sciences, finance, sales and production. Mailing Add: US Steel Res Lab MS 44 Monroeville PA 15146

BULLOCK, RICHARD MELVIN, b Glasco, Kans, July 8, 18; m 47; c 2. HORTICULTURE. Educ: Kans State Col, BS, 40; State Col Wash, MS, 42, PhD, 50. Prof Exp: Asst hort, State Col Wash, 40-42, asst & assoc horticulturist, Tree Fruit Exp Sta, 46-52; prof & head dept hort, Utah State Agr Col, 52-53; supt & horticulturist, Southwestern Wash Exp Sta, 53-58, North Willamette Exp Sta, Ore State Univ, 58-69; horticulturist, 69-72, ASST DIR HAWAII AGR EXP STA, UNIV HAWAII, MANOA, 69-, AGRONOMIST, 72-, PROF AGRON, 74- Mem: AAAS; Am Soc Plant Physiol; Am Soc Hort Sci. Res: Nutrition, production, storage and handling of fruit and vegetable crops; plant physiology; plant growth regulation; herbicides; research administration. Mailing Add: Hawaii Agr Exp Sta Univ of Hawaii Col of Trop Agr Honolulu HI 96822

BULLOCK, ROBERT CROSSLEY, b New York, NY, Oct 16, 24; m 52; c 2. ENTOMOLOGY. Educ: St Lawrence Univ, BS, 48; Univ Conn, MS, 50, PhD(entom), 54. Prof Exp: From asst entomologist to entomologist, trop res dept, Tela RR Co, 54-61; ASSOC ENTOMOLOGIST, AGR RES CTR, UNIV FLA, 61- Mem: Entom Soc Am. Res: Biology and control of insect pests of citrus. Mailing Add: Agr Res Ctr PO Box 248 Ft Pierce FL 33450

BULLOCK, ROBERT M, III, b Cincinnati, Ohio, June 30, 37; m 63; c 2. MATHEMATICS. Educ: Univ Cincinnati, BS, 61, MA, 62, PhD(math), 66. Prof Exp: Teacher pvt sch, Ohio, 62-63; instr math, Univ Cincinnati, 63-66; ASST PROF MATH & STATISTICS, MIAMI UNIV, 66- Mem: Math Asn Am; Am Math Soc; Soc Indust & Appl Math. Res: Theory of functional-differential equations. Mailing Add: Dept of Math Miami Univ Oxford OH 45056

BULLOCK, ROBERTS COZART, b Oxford, NC, Aug 23, 06; m 35; c 1. MATHEMATICS. Educ: Univ NC, AB, 26, MA, 28; Univ Chicago, PhD(math), 32. Prof Exp: Instr math, Univ NC, 27-30; head dept, Lambuth Col, 32-34 & Ark Tech,

34-35; from instr to assoc prof, 35-46, PROF MATH, NC STATE UNIV, 46- Mem: Am Math Soc; Am Soc Eng Educ; Math Asn Am. Res: Differential geometry; differential equations; non-conjugate osculating quadrics of a curve on a surface; theories of the motion of a spin-stabilized rocket during burning. Mailing Add: Dept of Math NC State Univ Raleigh NC 27607

BULLOCK, THEODORE HOLMES, b Nanking, China, May 16, 15; m 37; c 2. NEUROBIOLOGY. Educ: Univ Calif, AB, 36, PhD(zool), 40. Prof Exp: Sterling fel, Yale Univ, 40-41, Rockefeller fel, 41-42, res asst pharmacol & neuroanat, 42-43, instr neuroanat, 43-44; asst prof anat, Sch Med, Univ Mo, 44-46; from asst prof to prof zool, Univ Calif, Los Angeles, 46-66; PROF NEUROSCI, UNIV CALIF, SAN DIEGO, 66- Concurrent Pos: Mem corp, Marine Biol Lab, Woods Hole, 44-, trustee, 56-58; fel, Ctr Advan Study Behav Sci, 59-60; mem exec comt, Assembly of Life Sci, 73-76. Mem: Nat Acad Sci; Am Soc Zool (pres, 65); Am Physiol Soc; Am Philos Soc; Soc Neurosci. Res: Comparative neurophysiology. Mailing Add: Dept of Neurosci Univ of Calif at San Diego La Jolla CA 92093

BULLOCK, WILBUR LEWIS, b New York, NY, Mar 8, 22; m 44; c 4. PARASITOLOGY. Educ: Queen's Col, NY, BS, 42; Univ Ill, MS, 47, PhD(zool), 48. Prof Exp: Univ Va fel, Mountain Lake Biol Sta, 48; from instr to assoc prof, 48-61, actg chmn dept, 58-59, PROF ZOOL, UNIV NH, 61- Concurrent Pos: Asst instr, US Army Univ, France, 45; res fel biol, Rice Inst, 55-56; vis res prof, Fla Presby Col, 62-63; res affil, Harold W Manter Lab, Univ Nebr State Mus, 72- Mem: Fel AAAS; Wildlife Dis Asn; Am Soc Trop Med & Hyg; Am Soc Parasitol. Res: Morphology and taxonomy of Acanthocephala; fish intestinal histology and histopathology; blood protozoa of marine fish. Mailing Add: Dept of Zool Univ of NH Durham NH 03824

BULLOCK, WILLIAM HORACE, b Washington, DC, Nov 6, 19. INTERNAL MEDICINE, HEMATOLOGY. Educ: Howard Univ, BS, 40, MD, 44. Prof Exp: Fron instr to asst prof, 47-59, ASSOC PROF INTERNAL MED, HOWARD UNIV, 59- Concurrent Pos: Rockefeller fel hamat, Univ Mich, 50-51; hematologist & physician, Freedmen's Hosp, 51-; consult, Vet Hosp, DC, 58-59 & DC Gen Hosp, 58- Mem: Am Soc Hemat; Am Col Physicians; Am Soc Internal Med; AMA; Nat Med Asn. Res: Hemophilia; sickle cell anemia. Mailing Add: Howard Univ Col of Med Washington DC 20001

BULLOFF, JACK JOHN, b New York, NY, Dec 9, 14; m 42, 52; c 4. PHYSICAL INORGANIC CHEMISTRY. Educ: City Col New York, BS, 39; Rensselaer Polytech Inst, PhD, 53. Prof Exp: Jr chemist, Los Alamos Nat Lab, Calif, 46; asst prof chem, City Univ NY, Albany, 46-50; asst, Rensselaer Polytech Inst, 50-51; proj supvr, Commonwealth Eng Co, 53-56; prin chemist, Battelle Mem Inst, 56-68; PROF HIST SCI, STATE UNIV NY ALBANY, 68- Concurrent Pos: Consult, Rensselaer Polytech Inst, 73- Mem: AAAS; Am Chem Soc. Res: Actinides; metallic soaps; polyoxyanions; particle technology; graphic arts; photopolymerization; technological forecasting; science and technology assessment; research and development planning; oil and chemical spill amelioration; environmental chemistry; science policy; history of chemistry. Mailing Add: 399 Ridge Hill Rd Schenectady NY 12303

BULLOUGH, VAUGHN LYNN, b Salt Lake City, Utah, Sept 7, 24; m 47; c 4. ORGANIC CHEMISTRY. Educ: Univ Utah, BS, 52, PhD(chem), 55. Prof Exp: Develop chemist, 55-63, DIR APPL RES, REYNOLDS METALS CO, 63- Concurrent Pos: With Off Ord Res, 53, 54. Mem: Am Chem Soc; Am Inst Mining, Metall & Petrol Engrs. Res: Chemistry of coal, coke, pitches and carbon electrodes; electrochemical processes; high temperature processes. Mailing Add: Reynolds Metals Co PO Box 191 Sheffield AL 35660

BULMAN, WARREN EUGENE, b Woodville, Ala, Jan 3, 23; m 45; c 2. PHYSICS. Educ: Berea Col, AB, 48; Purdue Univ, MS, 51; Ohio State Univ, PhD, 58. Prof Exp: Lab asst physics & electronics, Berea Col, 41-43 & 46-48; instr physics, Purdue Univ, 48-51; electronics scientist, Nat Bur Stand, 51-52; prin physicist & group leader solid state physics, Battelle Mem Inst, 52-55; asst supvr electromagnetics, Antenna Lab, Ohio State Univ, 55-58; pres, Ohio Semiconductors, Inc, 56-60; vpres, Tecumseh Prod Co, 60-64; PRES, OHIO SEMITRONICS INC, 64- Concurrent Pos: Consult, Antenna Lab, Ohio State Univ, 58- Mem: Am Phys Soc; Electrochem Soc; Inst Elec & Electronics Engrs. Res: Solid state physics; electronics; new device design theory and various electromagnetic applications of materials. Mailing Add: Ohio Semitronics Inc 1205 Chesapeake Ave Columbus OH 43212

BULMER, GLENN STUART, b Windsor, Ont, May 3, 31; US citizen; m 53; c 4. MEDICAL MYCOLOGY, MICROBIOLOGY. Educ: Mich State Univ, BS, 53, PhD(mycol), 60. Prof Exp: From instr to assoc prof, 60-66, ASSOC PROF MICROBIOL, MED SCH, UNIV OKLA, 66- Concurrent Pos: NIH fel, Med Sch, Univ Okla, 60-62 & career develop award, 65-72; consult, Southeast Asian Ministers of Educ Orgn, 71-; instr, Sch Med, Univ Geneva, 72-73; assoc prof, Sch Med, Univ Saigon, 73- Honors & Awards: Serv Recognition Award, AMA, 73; Cult & Educ Medal, Govt South Vietnam, 74. Mem: Am Soc Microbiol; Med Mycol Soc of the Americas; Trop Med & Microbiol Soc SVietnam; Mycol Soc Am. Res: Fungus diseases; pathogenesis of cryptococcosis. Mailing Add: Univ of Okla Med Sch PO Box 26901 Oklahoma City OK 73190

BULOW, FRANK JOSEPH, b Oaklawn, Ill, July 11, 41; m 63; c 4. FISH BIOLOGY. Educ: Southern Ill Univ, BA, 64, MA, 66; Iowa State Univ, PhD(fisheries biol), 69. Prof Exp: ASSOC PROF BIOL, TENN TECHNOL UNIV, 69- Concurrent Pos: Res grants, Sport Fishing Inst, 68-69, Tenn Technol Univ, 69-75 & Bur Sport Fisheries & Wildlife Dingell-Johnson grant, 69-75. Mem: Am Fisheries Soc; Am Soc Ichthyologists & Herpetologists. Res: Propagation and rearing of channel catfish; relationship between nucleic acid concentrations and growth rate of fishes; water pollution biology; factors limiting fish production in reservoirs, streams and ponds. Mailing Add: Dept of Biol Tenn Technol Univ Cookeville TN 38501

BULS, ERWIN JULIUS, b Gresham, Nebr, July 5, 08; m 28; c 3. GEOGRAPHY. Educ: Valparaiso Univ, AB, 37; Univ Chicago, MS, 44. Prof Exp: Instr geog, 44-47, from asst prof geog to assoc prof geog & geol, 47-70, ASSOC PROF GEOG, VALPARAISO UNIV, 70- Mem: Asn Am Geogr; Nat Coun Geog Educ. Res: Regional geography of Anglo-America; American resources. Mailing Add: Dept of Geography Valparaiso University Valparaiso IN 46383

BULTMAN, JOHN D, b Chicago, Ill, Apr 11, 24; m 52; c 1. CHEMISTRY, MARINE BIOLOGY. Educ: George Washington Univ, BS, 50, MS, 54; Georgetown Univ, PhD(chem), 62. Prof Exp: Mem staff, E I du Pont de Nemours & Co, 42-43; chemist, Bur Mines, 50, RES CHEMIST, NAVAL RES LAB, US DEPT INTERIOR, 50- Mem: Am Chem Soc; Marine Technol Soc; Am Soc Limnol & Oceanog; Asn Trop Biol; Sigma Xi. Res: Biological deterioration of materials in the terrestrial and marine environments; surface chemistry of proteins. Mailing Add: US Naval Res Lab Washington DC 20390

BULUSU, SURYANARAYANA, b Ellore, India, Aug 22, 27; US citizen; ORGANIC CHEMISTRY, RADIATION CHEMISTRY. Educ: Andhra Univ, India, BSc, 48; Univ Bombay, BSc, 50, PhD(org chem), 54. Prof Exp: J N Tata Endowment Higher Educ Indians overseas fel chem, Yale Univ, 54, univ fel, 54-56; res assoc, Brookhaven Nat Lab, 56-59 & Univ Wis, Madison, 59-60; RES CHEMIST, EXPLOSIVES LAB, PICATINNY ARSENAL, US ARMY, 60- Honors & Awards: Outstanding Achievement Award, Army Sci Conf, US Army Off Chief Res & Develop, 68. Mem: AAAS; Am Chem Soc; Sigma Xi; Am Soc Mass Spectrometry; Int Soc Magnetic Resonance. Res: Organic mass spectrometry; applications to study of structural chemistry of explosives. Mailing Add: 22 Skyview Terr Morris Plains NJ 07950

BUMBY, RICHARD THOMAS, b Brooklyn, NY. NUMBER THEORY. Educ: Mass Inst Technol, SB, 57; Princeton Univ, MA, 59, PhD(math), 62. Prof Exp: From instr to asst prof, 60-68, ASSOC PROF MATH, RUTGERS UNIV, 68- Mem: Am Math Soc; Math Asn Am. Res: Number theory, especially diophantine problems; analogs of Littlewood's approximation problem; the Markoff spectrum; algebraic and combinatorial problems in number theory. Mailing Add: Dept of Math Rutgers Univ Brunswick NJ 08903

BUMCROT, ROBERT J, b Kansas City, Mo, Nov 24, 36; m 60; c 2. MATHEMATICS. Educ: Univ Chicago, BS, 59, MS, 60; Univ Mo, PhD(math), 62. Prof Exp: Asst prof math, Ohio State Univ, 62-66; lectr, Univ Sussex, 66-67; asst prof, Ohio State Univ, 67-68; assoc prof, 68-74, PROF MATH, HOFSTRA UNIV, 74-, CHMN DEPT, 69- Mem: Am Math Soc; Math Asn Am. Res: Geometry; abstract structures; lattice theory; algebraic geometry. Mailing Add: Dept of Mathematics Hofstra Univ Hempstead NY 11550

BUMGARDNER, CARL LEE, b Belmont, NC, Jan 8, 25; m 56; c 5. ORGANIC CHEMISTRY. Educ: Univ Toronto, BASc, 52; Mass Inst Technol, PhD(org chem), 56. Prof Exp: Res assoc, Mass Inst Technol, 56; res chemist, Redstone Res Div, Rohm and Haas Co, 56-64; assoc prof chem, 64-67, PROF CHEM, NC STATE UNIV, 67- Mem: Am Chem Soc. Res: Elimination reactions; photochemistry; nitrenes. Mailing Add: Dept of Chem NC State Univ Raleigh NC 27607

BUMGARDNER, JESS EDWARD, biochemistry, physiology, see 12th edition

BUMP, CHARLES KILBOURNE, b Pittsfield, Mass, June 1, 07; m 34; c 2. CHEMISTRY. Educ: Amherst Col, AB, 29; Cornell Univ, PhD(phys chem), 33. Prof Exp: Fel chem, Amherst Col, 34-35; teacher sci, Dorland-Bell Sch, NC, 35-36; inspector, Food & Drug Admin, USDA, 36-37; res chemist, plastics div, Monsanto Chem Co, 37-38, group leader, 38-46, asst dir res, 46-57, res personnel mgr, Monsanto Co, 58-70, res specialist, plastics prod & resins div, 70-72; RETIRED. Mem: Am Chem Soc. Res: Industrial problems dealing with plastics; resins for surface coatings. Mailing Add: 78 North Rd Hampden MA 01036

BUMP, DONALD DEAN, physical chemistry, see 12th edition

BUMPUS, DEAN FRANKLIN, b Newburyport, Mass, May 11, 12; m 39; c 1. PHYSICAL OCEANOGRAPHY. Educ: Oberlin Col, AB, 35. Prof Exp: Lab instr gen physiol & biochem, Brown Univ, 35-37; biol technician-oceanogr, 37-64, SR SCIENTIST, WOODS HOLE OCEANOG INST, 64- Concurrent Pos: Sr mem res comt, Int Panamaquoddy Fisheries Bd, 57-59; mem working group radioactive waste disposal into Atlantic and Gulf Coastal waters, Comt Oceanog, Nat Acad Sci-Nat Res Coun, 58-59. Mem: Am Soc Limnol & Oceanog. Res: Circulation on the continental shelf. Mailing Add: Dept of Phys Oceanog Woods Hole Oceanog Inst Woods Hole MA 02543

BUMPUS, FRANCIS MERLIN, b Rome, Ky, Dec 6, 22; m 47; c 2. ORGANIC CHEMISTRY, BIOCHEMISTRY. Educ: Purdue Univ, BS, 44; Univ Wis, MS, 47, PhD(biochem), 49. Prof Exp: Chemist, Allied Chem & Dye Co, 44-45; chemist, 49-61, asst dir res, 61-66, sci dir cardiovasc res, 66-67, CHMN RES DIV, CLEVELAND CLIN, 67-; PROF BIOL, CLEVELAND STATE UNIV, 70- Concurrent Pos: Mem, Coun High Blood Pressure Res, Am Heart Asn, mem, Coun Basic Sci & Study Sect; mem adv comt on hypertension, Nat Heart & Lung Inst, NIH, mem Gen Med Study Sect. Honors & Awards: Purdue Frederich Award, 67; Stouffer Prize, 68. Mem: AAAS; Am Chem Soc; Am Soc Biol Chemists; Am Heart Asn; Soc Exp Biol & Med. Res: Synthesis of vaccenic acid and peptides; structure of antimycin A; purification and structure determination of angiotonin; vapor phase oxidation, isolation and identification of vasopressor substances; mechanisms of action of angiotensin; alodsterone biosynthesis and release phenomena; peptide chemistry and pharmacology; etiology of cardiovascular diseases. Mailing Add: Cleveland Clin Found 9500 Euclid Ave Cleveland OH 44106

BUNAG, RUBEN DAVID, b Manila, Philippines, June 3, 31; m 56; c 2. CARDIOVASCULAR PHYSIOLOGY, PHARMACOLOGY. Educ: Univ Philippines, MD, 55; Univ Kans, MA, 62. Prof Exp: Instr physiol, Univ Philippines, 55-57; from asst prof to assoc prof pharmacol, Med Ctr, Univ of the East, Manila, 57-60; USPHS int res fel, 60-62; Life Ins Med Res Fund res fel, Med Sch, Case Western Reserve Univ, 62-63; res fel, Cleveland Clin Res Div, 63-69, assoc res staff, 69-70; assoc prof, 70-75, PROF PHARMACOL, MED CTR, UNIV KANS, 75- Concurrent Pos: Res pharmacologist, Vet Admin Hosp, Kansas City, Kans, 70-72. Mem: Sigma Xi; Am Soc Pharmacol & Exp Therapeut; Am Heart Asn; Am Physiol Soc. Res: Cardiovascular pharmacology; experimental hypertension; central neural mechanisms for cardiovascular regulation; autonomic pharmacology. Mailing Add: Dept of Pharmacol Univ of Kans Med Ctr Kansas City KS 66103

BUNBURY, DAVID LESLIE, b Georgetown, Brit Guiana, Feb 12, 33; m 57; c 4. PHYSICAL ORGANIC CHEMISTRY. Educ: Berea Col, BA, 52; Univ Notre Dame, PhD(phys chem), 56. Prof Exp: Fel chem, Univ Colo, 56-58; asst prof, 58-65, ASSOC PROF CHEM, ST FRANCIS XAVIER UNIV, 65- Concurrent Pos: Nat Res Coun Can grant, 59-71. Mem: Am Chem Soc; Chem Inst Can; The Chem Soc. Res: The photochemical decomposition of some ketones. Mailing Add: Dept of Chem St Francis Xavier Univ Antigonish NS Can

BUNCE, ELIZABETH THOMPSON, b Mineola, NY, Apr 25, 15. GEOPHYSICS. Educ: Smith Col, AB, 37, MA, 49. Hon Degrees: ScD, Smith Col, 71. Prof Exp: Instr physics, Smith Col, 49-51; res asst, 51-52, assoc scientist, 52-75, SR SCIENTIST, WOODS HOLE OCEANOG INST, 75- Mem: Fel Geol Soc Am; Soc Explor Geophys; Am Geophys Union; AAAS. Res: Marine seismology; underwater acoustics. Mailing Add: Oceanog Inst Woods Hole MA 02543

BUNCE, GEORGE EDWIN, b Nashville, Tenn, Dec 13, 32; m 55; c 2. NUTRITION, BIOCHEMISTRY. Educ: Va Polytech Inst, BS, 54, MS, 56; Univ Wis, PhD(biochem), 61. Prof Exp: Biochemist, US Army Med Res & Nutrit Lab, Fitzsimons Gen Hosp, Denver, Colo, 61-63; chief clin chem, Tripler Gen Hosps, Honolulu, Hawaii, 63-65; asst prof, 65-70, ASSOC PROF BIOCHEM, VA POLYTECH INST & STATE UNIV, 70- Concurrent Pos: Vis scientist, chem physiol lab, Cath Univ Louvain, 70-71. Mem: Am Inst Nutrit. Res: Mineral nutrition, particularly magnesium and its relation to soft tissue calcinosis and urolithiasis;

nutritional effects on cataract. Mailing Add: Dept of Biochem & Nutrit Va Polytech Inst & State Univ Blacksburg VA 24061

BUNCE, JAMES ARTHUR, b Troy, NY, June 19, 49; m 75. PHYSIOLOGICAL ECOLOGY. Educ: Bates Col, BS, 71; Cornell Univ, PhD(phys ecol), 75. Prof Exp: RES ASSOC CROP PHYSIOL, DUKE UNIV, 75- Mem: Sigma Xi; Am Soc Plant Physiologists. Res: Analysis of plant adaptations and responses to environmental factors, especially water, and consequences of these in determining distribution patterns and productivity. Mailing Add: Dept of Bot Duke Univ Durham NC 27706

BUNCE, NIGEL JAMES, b Sutton Coldfield, Eng, June 10, 43. ORGANIC CHEMISTRY. Educ: Oxford Univ, BA, 64, DPhil(chem), 69. Prof Exp: Asst prof, 69-73, ASSOC PROF CHEM, UNIV GUELPH, 73- Concurrent Pos: Killam Mem fel, Univ Alta, 67-69. Mem: Chem Inst Can; The Chem Soc. Res: Free radical substitution of alkanes; chemistry of acyl hypohalites; photochemistry of aromatic compounds. Mailing Add: Dept of Chem Univ of Guelph Guelph ON Can

BUNCE, PAUL LESLIE, b Fargo, NDak, Sept 24, 16; m 45; c 2. MEDICINE. Educ: Oberlin Col, AB, 38; Univ Chicago, MD, 42. Prof Exp: Asst instr pharmacol, Univ Pa, 46-48; house officer & resident urol, Johns Hopkins Univ, 48-51, instr, 50-52; from asst prof to assoc prof, 52-64, PROF SURG, UNIV NC, CHAPEL HILL, 64- Mem: AMA; Am Urol Asn. Res: Urology. Mailing Add: NC Mem Hosp Chapel Hill NC 27514

BUNCE, STANLEY CHALMERS, b Bayonne, NJ, Aug 21, 17; m 43; c 3. ORGANIC CHEMISTRY. Educ: Lehigh Univ, BS, 38, MA, 42; Rensselaer Polytech Inst, PhD(chem), 51. Prof Exp: Teacher, high sch, Pa, 39-41 & NJ, 41-43; res chemist, Johns-Manville Corp, 43-46; from instr to assoc prof, 46-58, PROF CHEM, RENSSELAER POLYTECH INST, 58- Mem: Fel AAAS; Am Chem Soc. Res: Organic reaction mechanisms; synthetic organic chemistry; small ring compounds; medicinal chemistry. Mailing Add: Dept of Chem Rennsselaer Polytech Inst Troy NY 12181

BUNCEL, ERWIN, b Presov, Czech, May 31, 31; m 56; c 2. PHYSICAL ORGANIC CHEMISTRY. Educ: Univ London, BSc, 54, PhD(chem), 57. Honors & Awards: DSc, Univ London, 70. Prof Exp: Res assoc chem & fel, Univ NC, 57-58; Nat Res Coun Can fel, McMaster Univ, 58-61; res chemist, Am Cyanamid Co, 61-62; from asst prof to assoc prof, 62-70, PROF CHEM, QUEEN'S UNIV, ONT, 70- Mem: Am Chem Soc; Chem Inst Can; Brit Chem Soc. Res: Isotope effects in organic reaction mechanisms; aromatic substitution; acid catalysis; solvolysis; model carbohydrates; reaction intermediates. Mailing Add: Dept of Chem Queen's Univ Kingston ON Can

BUNCH, HARRY DEAN, b Yukon, Okla, Oct 3, 15; m 41; c 3. AGRONOMY. Educ: Okla State Univ, BS, 41; Univ Tenn, MS, 42, PhD(agron, seed technol), 59. Prof Exp: Instr agron, Univ Ky, 42-43; asst agronomist, Agr Exp Sta, 46-55, assoc agronomist & supvr seed technol lab, 55-64, PROF AGRON, MISS STATE UNIV, 64-, DIR INT PROGS AGR & FORESTRY, 67- Mem: Am Soc Agron. Res: Seed storage; mechanical injury of seeds; electrostatic separation of seeds; magnetic seed cleaning. Mailing Add: PO Drawer NZ Miss State Univ Mississippi State MS 39762

BUNCH, JAMES R, b Globe, Ariz, Sept 23, 40. NUMERICAL ANALYSIS. Educ: Univ Ariz, BS, 62; Univ Calif, Berkeley, MA, 65, PhD(appl math), 69. Prof Exp: Asst mathematician, Agronne Nat Lab, 69-70; instr math, Univ Chicago, 70-71; asst prof comput sci, Cornell Univ, 71-74; ASSOC PROF MATH, UNIV CALIF, SAN DIEGO, 74- Concurrent Pos: Consult, Argonne Nat Lab, 74- Mem: Am Math Soc; Soc Indust & Appl Math; Asn Comput Mach. Res: Numerical solution of symmetric indefinite systems of linear equations and calculation of inertia. Mailing Add: Dept of Math Univ of Calif San Diego La Jolla CA 92037

BUNCH, PHILLIP CARTER, b Maryville, Tenn, Sept 14, 46; m 67. MEDICAL PHYSICS. Educ: Univ Chicago, AB, 69, MS, 71, PhD(med physics), 75. Prof Exp: RES PHYSICIST, RES LABS, EASTMAN KODAK CO, 75- Mem: Optical Soc Am; Am Asn Physicists in Med; Soc Photo-Optical Instrument Engrs; Soc Photog Scientists & Engrs. Res: Radiologic image analysis; Monte Carlo simulation of the optical characteristics of radiographic screen-film systems; objective measurement of optical and noise properties of radiologic imaging systems; application of visual psychophysics to radiography. Mailing Add: Eastman Kodak Co Res Labs Rochester NY 14650

BUNCH, THEODORE EUGENE, b Huntsville, Ohio, Mar 1, 36; m 63; c 2. ASTROGEOLOGY, GEOLOGY. Educ: Miami Univ, BA, 59, MS, 62; Univ Pittsburgh, PhD(geol), 66. Prof Exp: Res asst geosci, Mellon Inst Sci, 61-64; res asst impact craters & meteorites, Univ Pittsburgh, 64-66; NSF-Nat Res Coun fel, 66-69, RES SCIENTIST PLANETARY SCI & METEORITES, NASA-AMES RES CTR, 69- Concurrent Pos: Vis prof, Univ Cologne, 71 & Sonoma State Col, 74-; res assoc, Scripps Inst Oceanog, Univ Calif, San Diego, 74- Honors & Awards: Apollo Achievement Award, NASA, 70. Mem: Fel Mineral Soc Am; fel Meteoritical Soc (secy, 70-75). Res: Geological research on the moon, earth, Mars, asteroids, meteorites, interplanetary and interstellar dust. Mailing Add: 245-5 NASA-Ames Res Ctr Moffett Field CA 94035

BUNCH, WILBUR LYLE, b Pine Bluffs, Wyo, Apr 24, 25; m 46; c 5. PHYSICS. Educ: Univ Wyo, BS, 49, MS, 51. Prof Exp: Engr, Gen Elec Co, 51-62, sr physicist, 62-64; res assoc reactor shielding, Battelle Mem Inst, 65-66, mgr, 67-70; mgr radiation & shielding anal, Wadco, 70-72; MGR RADIATION & SHIELD ANAL, WESTINGHOUSE HANFORD CO, 72- Mem: Am Nuclear Soc. Res: Nuclear reactor shielding; nuclear instrumentation; reactor design. Mailing Add: 2403 Pullen St Richland WA 99352

BUNCH, WILTON HERBERT, b Walla Walla, Wash, Jan 12, 35; m 56; c 2. ORTHOPEDIC SURGERY, MUSCULAR PHYSIOLOGY. Educ: Walla Walla Col, BS, 56; Loma Linda Univ, MD, 60; Univ Minn, PhD(physiol), 67. Prof Exp: Mem attend staff orthop, Gillette State Hosp, 68-69; instr, Univ Minn, 68-69; assoc prof orthop, 69-74, prof orthop & pediat, Univ Va, 74-75; SCHOLL PROF ORTHOPED & PEDIAT, SCH MED, LOYOLA UNIV, 75- Concurrent Pos: Mem fac prosthetics & orthotics, Northwestern Univ, 60- Honors & Awards: Richards Award, Richards Co, Tenn, 68; Nicholas Andry Award, Asn Bone & Joint Surg, 70. Mem: Am Physiol Soc; Orthop Res Soc; Opers Res Soc Am; Scoliosis Res Soc. Res: The effect of denervation on skeletal muscle membrane. Mailing Add: 2160 S First Ave Maywood IL 60153

BUNCHER, CHARLES RALPH, b Dover, NJ, Jan 9, 38; c 64. BIOSTATISTICS, EPIDEMIOLOGY. Educ: Mass Inst Technol, BS, 60; Harvard Univ, MS, 64, ScD, 67. Prof Exp: Statistician, Atomic Bomb Casualty Comn, Nat Acad Sci, 67-70; chief biostatistician, Wm S Merrell Co-Nat Labs, 70-73; asst prof statist, Univ, 70-73, PROF STATIST & DIR DIV EPIDEMIOL & BIOSTATIST, MED COL, UNIV CINCINNATI, 73- Mem: AAAS; Am Pub Health Asn; Am Statist Asn; Biomet Soc; Soc Epidemiol Res. Res: Cancer epidemiology; screening, diagnosis, and treatment, as well as general and environmental epidemiology; statistical research, clinical trials, design of experiments, pharmaceutical research, teaching. Mailing Add: Univ of Cincinnati Med Col 3223 Eden Ave Cincinnati OH 45267

BUNDE, CARL ALBERT, b Ashland Co, Wis, Apr 22, 07; m 30. CLINICAL PHARMACOLOGY. Educ: Univ Wis, AB, 33, AM, 34, PhD(zool), 37; Southwestern Med Col, MD, 48. Prof Exp: Asst zool, Univ Wis, 34-37; instr, Sch Med, Univ Okla, 38-42; asst prof physiol & pharm, Col Med, Baylor Univ, 42-43; assoc prof, Southwestern Med Col, 43-44, assoc prof physiol, 44-49; vpres res, Pitman-Moore Co, 49-60; dir med res, Wm S Merrell Co, 60-72; RES CONSULT, 72- Concurrent Pos: Fel, Sch Med, Univ Okla, 37-38. Mem: Endocrine Soc; Am Physiol Soc; Soc Exp Biol & Med; AMA; Am Fedn Clin Res. Res: Endocrinology; circulation; metabolism; clinical pharmacology. Mailing Add: 3738 Donegal Dr Cincinnati OH 45236

BUNDE, DARYL E, b Sioux Falls, SDak, Oct 29, 37; m 59; c 3. ZOOLOGY, PHYSIOLOGY. Educ: Augustana Col, SDak, BA, 59; Univ Tex, MA, 62; Mont State Univ, PhD(zool), 65. Prof Exp: Assoc Rocky Mountain Univs fac orientation grant, Los Alamos Sci Lab, 65; asst prof zool, 65-74, ASSOC PROF ZOOL, IDAHO STATE UNIV, 74- Mem: Am Soc Zoologists. Res: Amino acid metabolism of an insect; metabolism of radioactive elements in mammals. Mailing Add: Dept of Zool Idaho State Univ Pocatello ID 83201

BUNDSCHUH, JAMES EDWARD, b St Louis, Mo, Nov 13, 41; m 65; c 4. PHYSICAL CHEMISTRY. Educ: St Louis Univ, BS, 63; Duquesne Univ, PhD(phys chem), 67. Prof Exp: Fel, Univ Ill, Chicago Circle, 67-68; asst prof, 68-75, ASSOC PROF CHEM & CHMN DEPT, WESTERN ILL UNIV, 75- Concurrent Pos: Guest lectr, Univ Stuttgart, 73-74. Mem: Am Chem Soc; Sigma Xi. Res: Surface chemistry; surface corrosion; nuclear magnetic resonance. Mailing Add: Dept of Chem Western Ill Univ Macomb IL 61455

BUNDY, FRANCIS PETTIT, b Columbus, Ohio, Sept 1, 10; m 36; c 4. PHYSICS. Educ: Otterbein Col, BS, 32; Ohio State Univ, MS, 32, PhD(physics), 37. Hon Degrees: DSc, Otterbein Col, 59. Prof Exp: Instr physics, Ohio Univ, 37-42; res physicist, underwater sound lab, Harvard Univ, 42-45; res physicist, Res & Develop Ctr, Gen Elec Co, 46-75; RETIRED. Concurrent Pos: Res physicist, Cooper-Bessemer Corp, Ohio, 39-42. Honors & Awards: Roozeboom Gold Medal, Netherlands Acad Sci, 69. Mem: Fel Am Phys Soc. Res: Physics of rocket gases; turbine shaft and bearing vibration; underwater sound transducers; physics of vacuum thermal insulation systems; ultra-high pressure apparatus and reactions. Mailing Add: 250 Alplaus Ave Alplaus NY 12008

BUNDY, GORDON LEONARD, b Akron, Ohio, Nov 27, 42; m 66; c 3. ORGANIC CHEMISTRY. Educ: Col Wooster, BA, 64; Northwestern Univ, PhD(org chem), 68. Prof Exp: RES CHEMIST, UPJOHN CO, 68- Mem: Am Chem Soc; The Chem Soc. Res: Chemistry and synthesis of naturally occurring materials, especially sesquiterpenes, steroids and prostaglandins; new synthetic methods. Mailing Add: Dept of Exp Chem Upjohn Co Kalamazoo MI 49001

BUNDY, HALLIE FLOWERS, b Los Angeles, Calif, Apr 2, 25. BIOCHEMISTRY. Educ: Mt St Mary's Col, BA, 47; Univ Southern Calif, MS, 55, PhD, 58. Prof Exp: Instr, Univ Southern Calif, 58-60; from asst prof to assoc prof, 60-67, PROF BIOCHEM, MT ST MARY'S COL, CALIF, 67- Concurrent Pos: Sci Fac Fel, Nat Sci Foundation, 69; Grant-in-Aid, Grad Women in Sci, 75. Mem: AAAS; Am Chem Soc. Res: Chemistry of proteins; proteolytic enzymes; zymogens and inhibitors. Mailing Add: Mount St Mary's Col 12001 Chaloxi Rd Los Angeles CA 90049

BUNDY, ROY ELTON, b Zanesville, Ohio, July 8, 24; m; c 2. PHYSIOLOGY. Educ: Ohio State Univ, BS, 46, MS, 48; Univ Wis, PhD(zool), 55. Prof Exp: Teacher pub sch & instr physics, Ohio State Univ, 46-47; instr biol, Transylvania Col, 48-51; asst, Univ Wis, 51-55; asst prof biol, Wagner Col, 55-57; asst prof biol & physiol, Sch Dent, Fairleigh Dickinson Univ, 57-61, assoc prof physiol & chmn dept, 61-68; CHMN DEPT BIOL, OKALOOSA-WALTON COMMUNITY COL, 68- Mem: AAAS; Am Physiol Soc. Mailing Add: Rte 1 Box 9 Mary Esther FL 32569

BUNDY, WAYNE MILEY, b Anderson, Ind, Jan 10, 24; m 45; c 3. CLAY MINERALOGY. Educ: Ind Univ, AB, 50, MA, 54, PhD, 57. Prof Exp: Geologist, NMex State Bur Mines, 51-53; petrogr, Ind Geol Surv, 53-57; dir labs, 57-67, dir res, 67-74, VPRES, DIR RES & COORDR CRUDE & FINISHED CLAY TECHNOL, GA KAOLIN CO, 74- Mem: Mineral Soc Am; Am Chem Soc; Mineral Soc Gt Brit & Ireland. Res: Geochemistry; petrology; surface chemistry. Mailing Add: Ga Kaolin Co 1185 Mary St Elizabeth NJ 07207

BUNGAY, HENRY ROBERT, III, b Cleveland, Ohio, Jan 22, 28; m 52; c 3. MICROBIAL BIOENGINEERING. Educ: Cornell Univ, BChE, 49; Syracuse Univ, PhD(biochem), 54. Prof Exp: Biochemist, Eli Lilly & Co, Ind, 54-62; prof sanit eng, Va Polytech Inst, 62-67, prof bioeng, Clemson Univ, 67-73; MEM STAFF, WORTHINGTON BIOCHEM CORP, 73- Mem: Am Chem Soc; Am Inst Chem Engrs; Am Soc Microbiol. Res: Bioengineering; fermentation processes; biological waste treatment. Mailing Add: Worthington Biochem Corp Halls Mill Rd Freehold NJ 07728

BUNGE, MARTA C, b Buenos Aires, Arg, Oct 31, 38; m 59; c 1. MATHEMATICS. Educ: Nat Inst Sec Prof, Buenos Aires, BA, 59; Univ Pa, MA, 64, PhD(math), 66. Prof Exp: Teaching asst, Univ Pa, 60-61; res assoc & fel, McGill Univ, 66-69, asst prof math, 69-74; MEM FAC, UNIV NAT MEX, 74- Mem: Am Math Soc. Res: Category theory, especially categories of set-valued functors—characterization, relative theory of triples and applications to relative functor categories, kan extensions and adjoints given by bifibrations, idempotent triples, tensor products of composable triples and closed categories. Mailing Add: Universidad Nacional de Mexico CIMAS Apart Postal 20-726 Mexico (20 DF) Mexico

BUNGE, MARY BARTLETT, b New Haven, Conn, Apr 3, 31; m 56; c 2. CYTOLOGY. Educ: Simmons Col, BS, 53; Univ Wis, MS, 55, PhD(zool, cytol), 60. Prof Exp: Asst med, Univ Wis, 53-56, asst zool, 56-60; res assoc anat, Col Physicians & Surgeons, Columbia Univ, 62-70; from res asst prof to res assoc prof, 70-74, ASSOC PROF ANAT, SCH MED, WASH UNIV, 74- Concurrent Pos: Nat Inst Neurol Dis & Blindness fel, Col Physicians & Surgeons, Columbia Univ, 60-62; res assoc, Harvard Med Sch, 68-69. Mem: Am Soc Cell Biol; Am Asn Anat; Electron Micros Soc Am. Res: Cell fine structure, particularly of the nervous system, normal and pathological; tissue culture. Mailing Add: Dept of Anat & Neurobiol Sch of Med Wash Univ St Louis MO 63110

BUNGE, RAYMOND GEORGE, b St Johns, Mich, Apr 8, 08; m 36; c 4. UROLOGY. Educ: Univ Mich, AB, 32, MD, 36; Am Bd Urol, dipl. Prof Exp: Mem fac, 38-53, PROF UROL, COL MED, UNIV IOWA, 53- Mem: AAAS; Am Urol Asn; Soc Exp Biol & Med; AMA; Am Col Surg. Res: Intersexuality; male infertility. Mailing Add: Univ Hosp Iowa City IA 55211

BUNGE, RICHARD PAUL, b Madison, SDak, Apr 15, 32; m 56; c 2. ANATOMY, CELL BIOLOGY. Educ: Univ Wis, BA, 54, MS, 56, MD, 60. Prof Exp: Asst anat, Univ Wis, 54-57, instr, 57-58; from asst prof to assoc prof anat, Col Physicians & Surgeons, Columbia Univ, 62-70; PROF ANAT, SCH MED, WASH UNIV, 70- Concurrent Pos: Vis prof, Harvard Med Sch, 68-69; Nat Multiple Sclerosis Soc fel surg, Col Physicians & Surgeons, Columbia Univ, 60-62; Lederle med fac award, 64-67. Mem: Am Asn Anat; Am Soc Cell Biol; Tissue Cult Asn; Am Asn Neutopath; Soc Neurosci. Res: Biology of cells of the nervous system in vivo and in vitro. Mailing Add: Dept of Anat & Neurobiol Sch of Med Wash Univ St Louis MO 63110

BUNGER, WILLIAM BOONE, b Alta Vista, Kans, Feb 14, 17; m 41; c 1. PHYSICAL CHEMISTRY. Educ: Washburn Col, BS, 40; Kans State Col, MS, 41, PhD(org chem), 49. Prof Exp: Chemist, Hill Packing Co, 39 & Hercules Powder Co, 41-45; instr, Kans State Col, 47-49; from asst prof to assoc prof, 49-57, assoc res prof, 57-65; PROF CHEM & CHMN DEPT, IND STATE UNIV, 65- Mem: Am Chem Soc. Res: Waxes; electrical conductivity of solutions; infrared spectrophotometry; chemical kinetics; organic solvents. Mailing Add: 1610 Rice Ave Terre Haute IN 47803

BUNGS, JANIS ARVIDS, polymer chemistry, see 12th edition

BUNKER, DON LOUIS, b San Fernando, Calif, July 12, 31; m 52, 61; c 3. PHYSICAL CHEMISTRY. Educ: Antioch Col, BS, 53; Calif Inst Technol, PhD(chem), 57. Prof Exp: Mem staff, Los Alamos Sci Lab, 57-65; assoc prof, 65-67, PROF CHEM, UNIV CALIF, IRVINE, 67- Mem: Am Chem Soc; Am Phys Soc. Res: Theoretical and experimental gas kinetics; computer simulation of reaction processes. Mailing Add: Dept of Chem Univ of Calif Irvine CA 92664

BUNKER, JOHN PHILLIP, b Boston, Mass, Feb 13, 20; m 44; c 4. MEDICINE. Educ: Harvard Univ, BA, 42, MD, 45; Am Bd Anesthesiol, dipl. Prof Exp: Instr anesthesia, Harvard Med Sch, 50-52, assoc, 52-55, asst clin prof, 55-60; PROF ANESTHESIA, SCH MED, STANFORD UNIV, 60- Concurrent Pos: Anesthetist, Mass Gen Hosp, Boston, 50-60; vchmn policy comt nat study instnl differences in death rates, Nat Res Coun. Mem: AMA; Am Soc Anesthesiol; Am Soc Pharmacol & Exp Therapeut. Res: Pharmacology of anesthesia; metabolic effects of blood transfusions. Mailing Add: Stanford Univ Sch of Med Stanford CA 94305

BUNKER, MERLE E, b Kansas City, Mo, Feb 8, 23; m 43; c 2. NUCLEAR PHYSICS. Educ: Purdue Univ, BS, 46; Ind Univ, PhD(physics), 50. Prof Exp: Mem staff physics div, 50-65, alt group leader, Group P-2, 65-74, GROUP LEADER, GROUP P-2, LOS ALAMOS SCI LAB, UNIV CALIF, 74- Concurrent Pos: NSF sr fel, 64-65; supvr res reactor, 54-; adv, Nuclear Data Group, Oak Ridge Nat Lab, 67- Mem: Fel Am Phys Soc. Res: Gamma-ray spectroscopy; nuclear structure; neutron activation analysis. Mailing Add: Reactor Group P-2 MS-776 Los Alamos Sci Lab Los Alamos NM 87545

BUNKFELDT, RUDOLF, b Milwaukee, Wis, June 6, 16; m 40; c 3. BIOCHEMISTRY. Educ: Univ Wis, BS, 37, MS, 39, PhD, 44. Prof Exp: Biochemist, Wander Co, 43, actg dir res, 44-46, dir nutrit res, 46-58; dir food res, Ovaltine Food Prod, 58-64; assoc dir res, Borden Spec Prod Co, Ill, 64-67, dir, Elgin Res Labs, 67, ASSOC DIR, SYRACUSE RES CTR, FOODS DIV, BORDEN, INC, 67- Res: Food technology. Mailing Add: Syracuse Res Ctr Food Div Borden, Inc 600 N Franklin St Syracuse NY 13204

BUNNELL, IVAN LEE, b Waterbury, Conn, Apr 17, 17; m 41; c 2. MEDICINE. Educ: Middlebury Col, AB, 38; Univ Buffalo, MS, 42, MD, 43. Prof Exp: ASSOC PROF MED, SCH MED, STATE UNIV NY BUFFALO, 54- Concurrent Pos: Attend physician, Buffalo Gen Hosp, dir prin clin, Angiol Lab, 59- Mem: AMA; Am Heart Asn; Am Col Cardiol; Am Col Physicians; NAm Soc Cardiac Radiol. Res: Cardiopulmonary field; hypertension; coronary angiography; pulmonary edema; coronary flow. Mailing Add: 100 High St Buffalo NY 14203

BUNNELL, RAYMOND HOWARD, biochemistry, animal nutrition, see 12th edition

BUNNETT, JOSEPH FREDERICK, b Portland, Ore, Nov 26, 21; m 42; c 3. ORGANIC CHEMISTRY. Educ: Reed Col, BA, 42; Univ Rochester, PhD(org chem), 45. Prof Exp: Res chemist, Western Pine Asn, Portland, 45-46; instr org chem,Reed Col, 46-48, asst prof chem, 48-52; from asst prof to assoc prof chem, Univ NC, 52-58; from assoc prof to prof, Brown Univ, 58-66, chmn dept, 61-64; PROF CHEM, UNIV CALIF, SANTA CRUZ, 66- Concurrent Pos: Fulbright fel, Univ London, 49-50; res grants, Res Corp, NSF petrol res fund, Am Chem Soc & NIH; Fulbright & Guggenheim fels, Univ Munich, 60-61; ed, Accounts of Chem Res, 67-; trustee, Reed Col. Mem: Am Acad Arts & Sci; Am Chem Soc; Brit Chem Soc. Res: Mechanisms of reactions of aromatic compounds with basic or nucleophilic reagents, of reactions of diazonium salts, of olefin-forming elimination reactions, of reactions in moderately concentrated mineral acids. Mailing Add: Natural Sci II Univ of Calif Santa Cruz CA 95064

BUNO, WASHINGTON HECTOR, b Montevideo, Uruguay, Nov 3, 08; m 34; c 1. HISTOLOGY. Educ: Univ Montevideo, BS, 26, MD, 36. Prof Exp: Chief lab histol, Inst Endocrinol, 37-46; PROF HISTOL & EMBRYOL & DIR DEPT, UNIV OF THE REPUB, 43-, DEAN SCH MED, 63- Concurrent Pos: Lectr, Univ Chile, 45; John Simon Guggenheim fel, 46-47; vis prof, Sch Med, Univ Ill, 61-62. Mem: AAAS; Am Asn Anat; Soc Exp Biol & Med; NY Acad Sci; corresp mem Int Soc Hist Med. Res: Histology of the nervous system and endocrine glands; experimental embryology; histochemistry. Mailing Add: Fac of Med Univ of the Repub Montevideo Uruguay

BUNT, LUCAS N H, b Edam, Netherlands, June 10, 05; m 32; c 2. MATHEMATICS, PHYSICS. Educ: Univ Amsterdam, Drs, 28; State Univ Groningen, PhD(math), 34. Prof Exp: Teacher high sch, 29-46; prof, State Univ Utrecht & State Univ Groningen, 46-68; PROF MATH, ARIZ STATE UNIV, 68- Concurrent Pos: Prof, State Univ Leiden, 52-53; head dept math & sci, Inst Training Sec Sch Teachers, State Univ Utrecht, 57-68; consult, Minister of Educ, Brazil, 56-57. Mem: Am Math Soc; Netherlands Math Soc. Res: Convex sets; training of teachers of mathematics. Mailing Add: Dept of Math Ariz State Univ Tempe AZ 85281

BUNTINAS, MARTIN GEORGE, b Klaipeda, Lithuania, Sept 1, 41; US citizen; m 66; c 3. MATHEMATICS. Educ: Univ Chicago, AB, 64; Ill Inst Technol, MS, 67, PhD(math), 70. Prof Exp: Instr math, Ill Inst Technol, 67-70; asst prof, 70-75, ASSOC PROF MATH, LOYOLA UNIV CHICAGO, 75- Concurrent Pos: Res fel, Alexander von Humboldt Found, 71-72. Mem: Am Math Soc; Math Asn Am. Res: Functional analysis and topological sequence spaces. Mailing Add: Dept of Math Loyola Univ of Chicago 6525 N Sheridan Rd Chicago IL 60626

BUNTING, ALBERT L, b Birmingham, Mich, Jan 15, 09; m 36; c 5. ORGANIC CHEMISTRY. Educ: Univ Mich, AB, 30, MS, 33; Detroit Col Pharm, ScD, 35. Prof Exp: Asst astron, Univ Mich, 28-31, lectr asst chem, 32-33; instr, Detroit Col Pharm, 35-38, prof, 38-39; pres, Union Res Corp, 39-40; pres, Chemicolor Corp, 40-42; vpres, 41-42, PRES, PLASTEEL CORP, 44- Concurrent Pos: Asst, Harvard Col Observ, 29; lectr, Henry Ford Community Col, 62-63. Mem: AAAS; Am Chem Soc. Res: Mechanism of the amino Grignards; production of direct colored photographic images; conversion of sea water to potable water; high pressure films for ferrous metals; continuous rotary molding of plastic foam plastics; electrostatic orientation of synthetic fibers. Mailing Add: 26970 Princeton Inkster MI 48141

BUNTING, BRIAN TALBOT, b Sheffield, Eng, Oct 15, 32; m 58; c 2. PHYSICAL GEOGRAPHY. Educ: Univ Sheffield, BA, 53, MA, 57; Univ London, PhD, 70. Prof Exp: Demonstr geog, Univ Col N Staffordshire, 54-55; lectr, Birbeck Col, London, 57-68; assoc prof, 68-74, mem Arctic Res Group, 69-73, PROF PEDOLOGY, McMASTER UNIV, 75- Concurrent Pos: Vis lectr, Univ Ibadan, 65-66; instr, Field Studies Coun, UK, 66-68; mem ed bd, Geoderma, 68- & Catena, 71-; mem sect biogeog, Int Geog Union, 70-72; guest prof, Aarhus Univ, 74-75. Mem: Fel Royal Geog Soc; Soil Sci Soc Am; Can Soc Soil Sci; Int Soc Soil Sci. Res: Studies of soil genesis and soil amelioration; geography of soils, especially of factorial influences on micromorphology. Mailing Add: Dept of Geog McMaster Univ Hamilton ON Can

BUNTING, DEWEY LEE, b Louisville, Ky, Sept 1, 32; m 57; c 3. ZOOLOGY. Educ: Univ Louisville, BA, 57, MS, 59; Okla State Univ, PhD(zool), 63. Prof Exp: From asst prof to assoc prof, 68-74, PROF ZOOL, UNIV TENN, 74- Concurrent Pos: USPHS res grant, 66-69; Fed Water Pollution Control Admin res grants, 67-69; res grants, Off Water Resources Res, 69-74. Mem: Am Soc Limnol & Oceanog; Int Asn Theoret & Appl Limnol. Res: Biological effects of water pollution; systematics of fresh-water invertebrates. Mailing Add: Dept of Zool Univ of Tenn Knoxville TN 37916

BUNTING, DONALD CHARLES, b Wrightstown, NJ, Dec 14, 12; m 38. METEOROLOGY. Educ: Dartmouth Col, AB, 34; Univ Fla, MS, 56. Prof Exp: Res lab technologist, Johns-Manville Corp, 35-36; meteorologist, United Air Lines, 37-38 & Pan Am World Airways, 38-48; assoc prof phys sci, Univ Fla, 48-75; RETIRED. Concurrent Pos: Grant, Colo State Univ, 64. Mem: Am Meteorol Soc; Am Geophys Union. Res: Ocean wave characteristics as related to meteorological factors; radar echoes from lightning discharges in thunderstorms; radar-rainfall correlations; radar observations of hurricanes. Mailing Add: 2038 NW 3rd Ave Gainesville FL 32603

BUNTING, GEORGE SYDNEY, JR, b Pocomoke City, Md, July 22, 27; m 64. PLANT TAXONOMY. Educ: Univ Md, BS, 50; Mich State Univ, MS, 51; Columbia Univ, PhD(bot), 58. Prof Exp: Tech asst, NY Bot Garden, 52-56; taxonomist, Mo Bot Garden, St Louis, 59-61; mem staff, L H Bailey Hortorium, Cornell Univ, 61-67; botanist, Univ Cent Venezuela, 67-73, PROF BOT, AGRO-INDUST REGION ANDES, UNIV INST TECHNOL, 73- Mem: Asn Trop Biol; Int Asn Plant Taxon. Res: Taxonomy of Araceae and cultivated plants; flora of tropical America. Mailing Add: Apartado Postal 40 San Cristobal Tachira Venezuela

BUNTING, JOHN WILLIAM, b Newcastle, Australia, Mar 30, 43; m 68; c 2. ORGANIC CHEMISTRY, BIOCHEMISTRY. Educ: Univ NSW, BSc, 63; Australian Nat Univ, PhD(med chem), 67. Prof Exp: NIH fel chem, Northwestern Univ, 67-68; asst prof, 68-74, ASSOC PROF CHEM & BIOCHEM, UNIV TORONTO, 74- Mem: Royal Australian Chem Inst; Chem Inst Can. Res: Physical organic chemistry; heterocyclic chemistry; enzymology. Mailing Add: Dept of Chem Univ of Toronto Toronto ON Can

BUNTING, MARY INGRAHAM, b Brooklyn, NY, July 10, 10; m 37; c 4. MICROBIOLOGY. Educ: Vassar Col, AB, 31; Univ Wis, AM, 32, PhD, 34. Prof Exp: Asst agr bact & chem, Univ Wis, 33-35; asst biol, Bennington Col, 36-37; instr physiol & hyg, Goucher Col, 37-38; asst bact, Yale Univ, 38-40; lectr bot, Wellesley Col, 46-47; lectr microbiol, Yale Univ, 48-55; prof bact & dean, Douglass Col, Rutgers Univ, 55-60; pres, Radcliffe Col, 60-72; asst to pres, Princeton Univ, 72-75; CONSULT, 75- Concurrent Pos: Comnr, Atomic Energy Comn, 64-65; mem nat sci bd, NSF, 65-70. Mem: Am Soc Microbiol. Res: Bacteriology; bacteriostatic action of dyes; role of carotenoids in bacteria and green plants; microbial genetics; color inheritance in Serratia. Mailing Add: Meadow Rd New Boston NH 03070

BUNTING, ROGER KENT, b Creston, Ill, Nov 27, 35; m 58; c 4. INORGANIC CHEMISTRY. Educ: Univ Ill, BS, 58, MS, 61; Pa State Univ, PhD(chem), 65. Prof Exp: Vis res assoc chem, Ohio State Univ, 65-66; asst prof, 66-69, ASSOC PROF CHEM, ILL STATE UNIV, 69- Concurrent Pos: Hon res fel, Birkbeck Col, Univ London, 74-75. Mem: Am Chem Soc; The Chem Soc. Res: Phosphorus-nitrogen and boron-nitrogen chemistry; inorganic heterocycles; Lewis acid-base interactions. Mailing Add: Dept of Chem Ill State Univ Normal IL 61761

BUNTING, THOMAS G, analytical chemistry, see 12th edition

BUNTING, WILLIAM, organic chemistry, organometallic chemistry, see 12th edition

BUNTLEY, GEORGE JULE, b Sheldon, Iowa, Feb 20, 24; m 47; c 2. SOIL MORPHOLOGY. Educ: SDak State Univ, BS, 49, MS, 50, PhD(soils), 62. Prof Exp: From asst prof to assoc prof agron, SDak State Univ, 50-68; ASSOC PROF AGRON, UNIV TENN, KNOXVILLE, 68- Concurrent Pos: Vis prof agron, Univ Tenn, 66-67. Mem: Am Soc Agron; Soil Sci Soc Am; Int Soil Sci Soc; Am Hort Soc. Res: Relationships between the physical and chemical properties of soils and specific land use alternatives. Mailing Add: Univ of Tenn PO Box 1071 Knoxville TN 37901

BUNTON, CLIFFORD A, b Chesterfield, Eng, Jan 4, 20; m 45; c 2. ORGANIC CHEMISTRY. Educ: Univ London, BSc, 42, PhD(org chem), 45. Prof Exp: Lectr, Univ London, 45-58, reader, 58-63; PROF CHEM, UNIV CALIF, SANTA BARBARA, 63-, CHMN DEPT, 67- Concurrent Pos: Commonwealth Fund fel, Columbia Univ, 48-49; Brit Coun vis lectr, Chile & Argentina, 60; vis lectr, Univ Calif, Los Angeles, 61 & Univ Toronto, 62; mem policy comt, Univ Chile-Univ Calif Coop Prog. Mem: Am Chem Soc; Brit Chem Soc; AAAS; Chilean Acad Sci. Res: Mechanisms of organic and inorganic reactions. Mailing Add: Dept of Chem Univ of Calif Santa Barbara CA 93106

BUNTROCK, ROBERT EDWARD, b Minneapolis, Minn, Nov 19, 40; m 61; c 2. ORGANIC CHEMISTRY, INFORMATION SCIENCE. Educ: Univ Minn, Minneapolis, BChem, 62, MA, 64, PhD(org chem), 67. Prof Exp: Asst org chem, Princeton Univ, 62-63; res chemist pesticide chem, Air Prod & Chem, Inc, 67-70; proj chemist, Am Oil Co, 70-71; RES INFO SCIENTIST, STANDARD OIL CO, IND, 71- Mem: AAAS; Am Chem Soc; Am Soc Info Sci; Sigma Xi. Res: Amide hemiaminals; 1,2,4-oxadiazetidines; cyclization reactions of 2,2-disubstituted biphenyls; syntheses of pesticides; heterocyclic chemistry; information retrieval; comparison of information systems; selective dissemination of information. Mailing Add: Amoco Res Ctr Standard Oil Co Ind PO Box 400 Naperville IL 60540

BUNYARD, GEORGE B, b Hollis, Okla, Jan 21, 29; m 51; c 4. NUCLEAR PHYSICS. Educ: Okla Baptist Univ, BS, 51; Vanderbilt Univ, MS, 56, PhD(physics), 60. Prof Exp: Physicist, Lawrence Radiation Lab, Univ Calif, 60-67; physicist & vpres res & develop, 67-75, PRES, UTI-SPECTROTHERM CORP, 75- Mem: AAAS; Am Phys

Soc; Am Vacuum Soc. Res: Mass spectrometry; vacuum technology; chemical-physical research and instrument development for process control in microelectronics, air pollution monitoring-control, medical clinical diagnostics via gas and liquid-gas analyses. Mailing Add: UTI-Spectrotherm Corp 325 North Mathilda Ave Sunnyvale CA 94086

BUOL, STANLEY WALTER, b Madison, Wis, June 14, 34; m 60; c 2. SOIL SCIENCE. Educ: Univ Wis, BS, 56, MS, 58, PhD(soils), 60. Prof Exp: Asst prof agr chem & soils, Univ Ariz, 60-64, assoc prof, 64-66; assoc prof soil sci, 66-69, PROF SOIL SCI, NC STATE UNIV, 69- Concurrent Pos: Consult, Sensory Systs Lab, 63-65. Mem: Am Soc Agron. Res: Micromorphology of soil profiles with related work in soil genesis and classification. Mailing Add: Dept of Soil Sci NC State Univ Raleigh NC 27607

BUONGIORNO, JOSEPH, b Golfech, France, Jan 15, 44; m 72; c 2. FOREST ECONOMICS. Educ: Advan Sch Forestry, Paris, Ingenieur, 67; State Univ NY Syracuse, MS, 69; Univ Calif, Berkeley, PhD(forest econ), 71. Prof Exp: Forestry officer, Food & Agr Orgn, UN, 71-75; ASST PROF FOREST ECON, UNIV WIS-MADISON, 75- Res: Investigation into the international patterns of growth of production, consumption and trade of forest products. Mailing Add: Dept of Forestry Univ of Wis Madison WI 53706

BUONI, JOHN J, b Philadelphia, Pa, Apr 25, 43; m 70; c 2. MATHEMATICAL ANALYSIS. Educ: St Joseph's Col, Pa, BS, 65; Univ Pittsburgh, MS, 68, PhD(math), 70. Prof Exp: Asst prof, 70-75, ASSOC PROF MATH, YOUNGSTOWN STATE UNIV, 75- Mem: Am Math Soc; Math Asn Am. Res: Functions of operators and essential spectra; behavior of essential spectra with regards to perturbation, products and mappings of operators. Mailing Add: Dept of Math Youngstown State Univ Youngstown OH 44555

BUONO, FREDERICK J, b Syracuse, NY, Apr 20, 39; m 64. MICROBIOLOGY, BIOCHEMISTRY. Educ: Syracuse Univ, BSc, 61, MSc, 64, PhD(microbiol), 67. Prof Exp: Res asst microbiol, Syracuse Univ, 61-64; res scientist, Rohm and Haas Co, Pa, 67-71; group leader, 71-72, SUPVR APPL DEVELOP, BIOCONTROL LAB, TENNECO CHEM, INC, 72- Mem: AAAS; Am Soc Microbiol. Res: Microbial physiology and enzyme control in microorganisms; applications development in biocides, paint additives, synthetic lubricants and fire retardants. Mailing Add: 18 Edgewood Rd Robbinsville NJ 08691

BUONOCORE, MICHAEL, b Brooklyn, NY, Dec 12, 18; m 47; c 2. ORGANIC CHEMISTRY. Educ: St John's Univ, NY, BS, 39; Tufts Univ, DMD, 45. Prof Exp: Pvt practice dent, 47-51; res assoc, 53-58, RES COODR, EASTMAN DENT CTR, 58- Concurrent Pos: Assoc clin prof dent, Univ Rochester, 54-; NIH career award dent res, 64-; spec consult, NIH, 69- Honors & Awards: Prev Dent Award, Am Dent Asn, 72; Sci Award, Int Asn Dent Res, 74. Mem: Am Dent Asn; Int Asn Dent Res; fel Am Col Dent. Res: Structure, composition and reaction of enamel and dentin; prevention of dental caries by chemical treatment; development of adhesive restorative materials in caries preventive agents to be applied to surfaces of teeth as coatings. Mailing Add: Eastman Dent Ctr 800 Main St East Rochester NY 14605

BUPP, LAMAR PAUL, b Grand Junction, Colo, July 25, 21; m 43; c 5. PHYSICAL CHEMISTRY. Educ: Univ Calif, BS, 43; Ore State Univ, PhD(chem), 51. Prof Exp: Asst, Ore State Univ; chemist, Pile Tech Div, Gen Elec Co, 50-53, mgr mat develop, Eng Dept, Hanford Atomic Prod Opers, 53-56, reactor eng develop, 56-57, chem res & develop, Hanford Labs, 57-61 & Vallicitos Atomic Lab, Pleasanton, 61-68; prof nuclear eng, Ore State Univ, 68-70; mgr res & eng, Jersey Nuclear Co, 70-74, VPRES RES & TECHNOL, EXXON NUCLEAR CO, 74- Mem: Am Chem Soc; Am Nuclear Soc; Atomic Indust Forum. Res: Surface chemistry; irradiation effects to materials; color centers in alkali halides; aqueous corrosion; nuclear fuels technology. Mailing Add: 8655 Juanita Dr NE Kirkland WA 98033

BUR, ANTHONY J, b Phildelphia, Pa, Dec 4, 35; m 63; c 4. MOLECULAR PHYSICS. Educ: St Joseph's Col, Pa, BS, 57; Pa State Univ, PhD(physics), 62. Prof Exp: Fel mech, Johns Hopkins Univ, 62-63; physicist polymer dielectrics, 63-71, PHYSICIST DENT RES, NAT BUR STANDARDS, 72- Mem: Am Chem Soc; Am Phys Soc. Res: Electrical properties of insulators; polymer dielectrics and physics; solution properties of polymers; piezoelectric effect; thermodynamics; physical properties of teeth, bone and skin. Mailing Add: Nat Bur Standards Washington DC 20234

BURAKEVICH, JOSEPH VINCENT, organic chemistry, see 12th edition

BURAS, EDMUND MAURICE, b New Orleans, La, Oct 24, 21; m 43; c 3. ORGANIC CHEMISTRY. Educ: Tulane Univ, BS, 41, MS, 47. Prof Exp: Chemist, southern regional res lab, Bur Agr & Indust Chem, USDA, 42-57; group leader, Harris Res Labs, 57-69; PRIN SCIENTIST, GILLETTE RES INST, 69- Mem: Am Chem Soc; Fiber Soc; Am Inst Chemists; Am Asn Textile Chemists & Colorists; Brit Textile Inst. Res: Physics and chemistry of cutting and shaving; absorbency; chemical modification of natural and synthetic fibers; textile literature; instrumentation. Mailing Add: Gillette Res Inst 1413 Research Blvd Rockville MD 20850

BURBA, JOHN VYTAUTAS, b Lithuania, Oct 1, 26; Can citizen; m 55; c 2. RADIOBIOLOGY, PHARMACOLOGY. Educ: Loyola Col Montreal, BSc, 51; Univ Ottawa, PhD(biochem), 63. Prof Exp: Control chemist, Merck & Co, Montreal, 51-52; tech sales rep, Can Lab Supplies, 52-54; tech sales rep, Mallinckrodt Chem Works, 54-57; med detail man, Frank W Horner, Ltd, 57-58; asst prof pharmacol, Univ Ottawa, 62-65; res scientist, 65-69, sci adv, Bur Drugs, 69-73, HEAD RADIOPHARMACOL SECT, RADIATION PROTECTION BUR, HEALTH PROTECTION BR, 73- Concurrent Pos: Fel, Dept Pharmacol, Yale Univ, 66-67. Mem: Soc Nuclear Med; Chem Inst Can; Pharmacol Soc Can. Res: Toxicity of drugs and their metabolites; drug metabolizing activity of human placenta; stereospecificity in enzymic reactions; effect of drugs on hepatic azo reductase activity. Mailing Add: Radiation Med Div Health Protection Br Ottawa ON Can

BURBAGE, JOSEPH JAMES, b Rural, Ohio, June 11, 14; m 41; c 2. INORGANIC CHEMISTRY. Educ: Miami Univ, BS, 35; Ohio State Univ, MS, 42, PhD(chem), 47. Prof Exp: Asst chem, Ohio State Univ, 39-40, asst physics, 43; res chemist, 43-44, group leader, 44-45, opers mgr, 45-46, asst lab dir, 46-50, dir Mound Lab, 50-55, dir develop, Inorg Div, 55-62, dir prod planning, 62-68, DIR FINANCIAL CONTROL, INORG DIV, MONSANTO CO, 68- Mem: AAAS; Am Chem Soc; Am Inst Chemists. Res: Phase rule; liquid ammonia; radioactivity; alpha-emitters; heavy chemicals; market research. Mailing Add: Monsanto Co 800 N Lindbergh Blvd St Louis MO 63166

BURBANCK, MADELINE PALMER, b Moorestown, NJ, Oct 27, 14; m 40; c 2. BOTANY, CYTOLOGY. Educ: Wellesley Col, AB, 35, AM, 38; Univ Chicago, PhD, 41. Prof Exp: Asst bot, Wellesley Col, 36-39 & Univ Chicago, 40; guest, Columbia Univ, 41-42; special instr, Drury Col, 42-46, asst prof, 46-50; herbarium asst, 56-65; RES ASSOC BIOL, EMORY UNIV, 50- Concurrent Pos: Mem corp, Marine Biol Lab, Woods Hole, Mass. Mem: AAAS; Ecol Soc Am. Res: Cytology and morphology of Cyathura polita. Mailing Add: 1164 Clifton Rd NE Atlanta GA 30307

BURBANCK, WILLIAM DUDLEY, b Indianapolis, Ind, Aug 20, 13; m 40; c 2. PROTOZOOLOGY, ANIMAL ECOLOGY. Educ: Earlham Col, AB, 35; Univ Chicago, MS, 36; PhD(zool), 41. Prof Exp: Instr biol, City Col NY, 41-42; guest, Dept Zool, Columbia Univ, 41-42; from asst prof to assoc prof, Drury Col, 44-45, prof, 45-50, chmn dept, 42-50; vis prof, 49-50, chmn dept, 52-57, PROF BIOL, EMORY UNIV, 50- Concurrent Pos: Mem corp, Marine Biol Lab, Woods Hole, Mass. Mem. Soc Protozool (vpres, 56-57); Int Asn Theoret & Appl Limnol; Ecol Soc Am; Marine Biol Asn UK; Estuarine Res Fedn. Res: Competition among protozoological populations; estuarine ecology, physiology and zoogeography of species of the isopod Cyathura. Mailing Add: 1164 Clifton Rd NE Atlanta GA 30307

BURBANK, BURR GAMALIEL, b Whiting, Ind, Aug 24, 12; m 40. PHYSICS. Educ: San Jose State Col, AB, 34; Stanford Univ, PhD(physics), 46. Prof Exp: Asst, Stanford Univ, 36-39; teacher high schs, Calif, 39-42; instr physics & eng, Stockton Jr Col, 42-47; asst prof physics, 47-51, chmn physics dept, 61-72, from asst prof to prof phys sci, 47-72, EMER PROF PHYS SCI, SAN FRANCISCO STATE COL, 72- Concurrent Pos: Asst prof, Col of the Pac, 45-47. Res: Methods of production of x-rays; direct and fluorescence excitation of the third level and the depths of production of continuous x-rays in thick targets of thorium. Mailing Add: 2400 Buchanan San Francisco CA 94115

BURBANK, ROBINSON DERRY, b Berlin, NH, Oct 3, 21; m 45; c 2. X-RAY CRYSTALLOGRAPHY. Educ: Colby Col, AB, 42; Mass Inst Technol, PhD(inorg chem), 50. Prof Exp: Asst spectros lab, Mass Inst Technol, 42-45, res asst, insulation res lab, 45-50; sr physicist, K-25 Labs, Carbide & Carbon Chem Co, 50-53; group leader crystallog, Olin Industs, Inc, 53-55; MEM TECH STAFF, BELL TEL LABS, 55- Concurrent Pos: Mem USA Nat Comt Crystallog, Nat Acad Sci, 68-70, 71-73, ex officio mem, 74-76. Mem: AAAS; Am Phys Soc; Am Crystallog Asn (treas, 65-68, vpres, 74, pres, 75). Res: X-ray crystallography of inorganic compounds; interhalogen compounds; noble gas compounds; phase transformations; thin films. Mailing Add: Bell Labs 600 Mountain Ave Murray Hill NJ 07974

BURBUTIS, PAUL PHILIP, b Haverhill, Mass, Nov 1, 27; m 52; c 2. ENTOMOLOGY. Educ: Univ Mass, BS, 50; Rutgers Univ, MS, 52, PhD, 54. Prof Exp: Asst entom, Rutgers Univ, 50-54, asst res specialist, 54-58; from asst prof to assoc prof, 58-72, PROF ENTOM, UNIV DEL, 72- Mem: Entom Soc Am; Int Orgn Biol Control; Am Entom Soc; Am Asn Univ Prof. Res: Economic entomology; biological insect control; teaching. Mailing Add: 4 E Rosemont Circle Newark DE 19711

BURCH, BENJAMIN CLAY, b Wilson, NC, Oct 23, 48. MATHEMATICS. Educ: NC State Univ, BS, 70; Tulane Univ, MS, 73, PhD(math), 75. Prof Exp: ASST PROF MATH, NORTHWESTERN UNIV, EVANSTON, 75- Mem: Am Math Soc. Res: Initial value, boundary value and free boundary problems for nonlinear partial differential equations. Mailing Add: Ddept of Math Northwestern Univ Evanston IL 60201

BURCH, CHARLES, b Erie, Pa, Jan 15, 19; m 46; c 3. PLANT MORPHOLOGY. Educ: Cornell Univ, PhD(bot), 51. Prof Exp: Asst, Cornell Univ, 47-51; prof, Calif State Col, Long Beach, 51-64; assoc prof, Hartwick Col, 64-67; PROF BIOL, WELLS COL, 67- Mem: AAAS; Bot Soc Am; Soc Study Evolution; Torrey Club; Int Soc Plant Morphol. Mailing Add: Dept of Biol Wells Col Aurora NY 13026

BURCH, CLARK WAYNE, b Harrisonville, Mo, Sept 8, 07; m 35. VETERINARY MEDICINE. Educ: Kans State Univ, BS, 32, DVM, 37. Prof Exp: With livestock sanit, State Dept Agr, Wis, 37-38; practicing vet, 38-42, 46-52; prof, 52-73, EMER PROF VET SCI, COL AGR, UNIV WIS-MADISON, 73- Mailing Add: 10419 High Dr Leawood KS 66206

BURCH, DAVID STEWART, b Sapulpa, Okla, Oct 9, 26; m 49. PHYSICS. Educ: Univ Wash, BS, 50, MS, 54, PhD(physics), 56. Prof Exp: Physicist, atomic physics lab, Nat Bur Standards, 56-58; from asst prof to assoc prof, 58-67, PROF PHYSICS, ORE STATE UNIV, 68- Mem: Am Phys Soc. Res: Electronic, ionic, atomic and molecular collisions; gaseous electronics; upper atmosphere physics; atomic and molecular structure; spectroscopy. Mailing Add: Dept of Physics Ore State Univ Corvallis OR 97331

BURCH, DEREK GEORGE, b Caerphilly, Gt Brit, June 26, 33; m 61; c 2. SYSTEMATIC BOTANY. Educ: Univ Wales, BSc, 54, MSc, 57; Univ Fla, PhD(bot), 65. Prof Exp: Plant pathologist, Cent Romana Corp, 57-58; nurseryman, Fairchild Trop Garden, 58-59, horticulturist, 59-61; asst prof, Wash Univ, 65-69; asst prof, 69-70, ASSOC PROF & DIR BOT GARDEN, UNIV S FLA, 70- Concurrent Pos: Botanist, Montreal Bot Garden, 59; asst botanist, Mo Bot Garden, 65-69, & chief horticulturist, 66-69. Mem: Soc Econ Bot; Asn Trop Biol; Int Asn Plant Taxon. Res: Revision of the New World members of the tribe Euphorbieae and systematic study of this group. Mailing Add: Dept of Biol Univ of SFla Tampa FL 33620

BURCH, ERNEST SUHR, JR, anthropology, see 12th edition

BURCH, GEORGE NELSON BLAIR, b Charlottetown, PEI, Can, Sept 22, 21; nat US; m 44. POLYMER CHEMISTRY. Educ: Mt Allison, BSc, 43; McGill Univ, MSc, 46; Ohio State Univ, PhD(chem), 49. Prof Exp: Asst anal chem, Ohio State Univ, 46-48; res chemist, Hercules Powder Co, 49-56, res supvr, 56-57, lab supvr, 57-62, sr res chemist, 62-68, mgr auxiliary group, 68-75, MGR UPHOLSTERY DEVELOP, HERCULES, INC, 75- Mem: Am Chem Soc; Sigma Xi. Res: Tall oil; organic fluorides; polyhydric alcohols; polymers; fibers, fiber design, prococessing and application in home furnishings. Mailing Add: Hercules Inc PO Box 12107 Res Triangle Park NC 27709

BURCH, HELEN BULBROOK, b Greenville, Tex, Nov 17, 06; m; c 1. BIOCHEMISTRY. Educ: Tex State Col Women, BS, 26; Iowa State Univ, MS, 28, PhD(org chem), 35. Prof Exp: Asst, Agr Exp Sta, Univ Wyo, 28-29; asst prof chem, Milwaukee-Downer Col, 29-36; consult, Pease Labs, NY, 36-39; head nutrit lab, Killian Res Labs, Inc, 39-42; anal res div, Standard Oil Co, 42-44; res assoc, Pub Health Res Inst New York, 44-48; lectr & res assoc nutrit biochem, Columbia Univ, 48-53; res assoc, 53-56, assoc prof, 57-74, PROF PHARMACOL, SCH MED, WASH UNIV, 74- Concurrent Pos: Nutrit survr, Bataan, Philippines, 48, 50; consult, Inst Nutrit, Cent Am & Panama, 52, 53, 56. Mem: AAAS; Am Chem Soc; fel Am Pub Health Asn; Am Soc Biol Chem; Am Inst Nutrit. Res: Assessing nutritional status; pyridine nucleotides, metabolism of carbon-14 labeled vitamin C; tissue enzymes and substrates during development; oxidative phosphorylation in riboflavin deficient mitochondria; quantitative histochemistry of kidney, enzymes and amino acid transport. Mailing Add: Dept of Pharmacol Sch of Med Wash Univ St Louis MO 63110

BURCH, JAMES LEO, b San Antonia, Tex, Nov 28, 42; m 65; c 3. MAGNETOSPHERIC PHYSICS. Educ: St Mary's Univ, Tex, BS, 64; Rice Univ, PhD(space sci), 68. Prof Exp: Physicist, Redstone Arsenal, Ala, 68-69; physicist, Sci Res Lab, US Mil Acad, 70-71; physicist, NASA-Goddard Space Flight Ctr, 71-74; PHYSICIST, NASA-MARSHALL SPACE FLIGHT CTR, 74- Concurrent Pos: Mem, US Panel Int Magnetospheric Study, 76- Mem: Am Geophys Union; Sigma Xi. Res: Experimental study of the geophysics and plasma physics of the earth's magnetosphere, particularly those processes that couple the magnetosphere, ionosphere and upper atmosphere. Mailing Add: Code ES 23 NASA-MSFC Huntsville AL 35812

BURCH, JOHN BAYARD, b Charlottesville, Va, Aug 12, 29; m 51; c 3. ZOOLOGY. Educ: Randolph-Macon Col, BS, 52; Univ Richmond, MS, 54; Univ Mich, PhD, 59. Prof Exp: Res assoc, 58-62, cur mollusks, Mus, 62-75, from asst prof to assoc prof, 62-70, PROF ZOOL, UNIV MICH, 70- Concurrent Pos: Res awards, Va Acad, 53 & NSF, 54; USPHS res career develop award, 64; ed-in-chief, Malacologia, ed, Malacol Rev, Inst Malacol. Mem: Inst Malacol (exec secy-treas, 61-62, treas, 63-67); Soc Syst Zool; Am Soc Zool; Ecol Soc Am; Am Micros Soc. Res: Cytology, comparative biochemistry and systematics of mollusks; medical malacology; Protozoa of mollusks. Mailing Add: Dept of Zool Univ of Mich Ann Arbor MI 48104

BURCH, JOSEPH EUGENE, b Winfield, Kans, June 3, 17; m 42. ORGANIC CHEMISTRY. Educ: Salem Col, WVa, BS, 42; Calif Inst Technol, MS, 44; Purdue Univ, MS, 48, PhD(org chem), 50. Prof Exp: Asst shift supvr inspection, Point Pleasant Ord Works, 42-43; res chemist vat dyes, Nat Aniline Div, 50-52; prin res chemist, 52-56, proj leader, Appl Res & Wood Preservation, 56-60, head chem econ group, 60-66, CHIEF CHEM & HEALTH ECON DIV, BATTELLE MEM INST, 66- Mem: Fel AAAS; Am Chem Soc; fel Am Inst Chemists. Res: Long-range planning; technological forecasting; chemical economics; chemical applications; international development. Mailing Add: Battelle Mem Inst 505 King Ave Columbus OH 43201

BURCH, ROBERT EMMETT, b St Louis, Mo, Oct 9, 33; m 56; c 2. INTERNAL MEDICINE, BIOCHEMISTRY. Educ: St Louis Univ, BS, 55, MD, 59. Prof Exp: Intern med & surg, Jewish Hosp, St Louis, 59-60; resident med, Firmin Desloge Hosp, St Louis Univ, 60-62; asst vis physician, Goldwater Mem Hosp, Columbia Univ, 65-68, assoc med, Univ, 65-69, asst prof, Col Physicians & Surgeons, 69-71; assoc prof, 71-73, PROF MED, SCH MED, CREIGHTON UNIV, 73- Concurrent Pos: Fel biochem, Western Reserve Univ, 62-65; asst vis physician, Francis Delafield Hosp, 68-70, assoc vis physician, 70-71; asst physician, Presby Hosp, 70; staff physician, Vet Admin Hosp, Omaha, 71-; mem coun arteriosclerosis, Am Heart Asn. Mem: NY Acad Sci; Am Fedn Clin Res; Am Soc Clin Nutrit; Am Inst Nutrit; Am Physiol Soc. Res: Acetoacetate production; cholesterol synthesis; Krebs cycle acid metabolism; trace elements. Mailing Add: Vet Admin Hosp 4101 Woolworth Ave Omaha NE 68105

BURCH, ROBERT J, physical chemistry, see 12th edition

BURCH, ROBERT RAY, b Edgard, La, May 28, 24; m 56; c 2. MEDICINE. Educ: Tulane Univ, BS, 48, MD, 51; Am Bd Internal Med, dipl, 58. Prof Exp: Intern med, Philadelphia Gen Hosp, 51-52; asst resident, Duke Univ Hosp, 52-53; from instr to asst prof, 54-66, ASSOC PROF CLIN & PREV MED, SCH MED, TULANE UNIV, 66- Concurrent Pos: Fel internal med, Sch Med, Tulane Univ, 53-54; Nat Heart Inst trainee, 54-55, fel internal med, 55-56; pvt pract, 56- Mem: AMA; Am Heart Asn; Am Geriat Soc; fel Am Col Physicians; NY Acad Sci. Res: Cardio-renal diseases. Mailing Add: 4303 Magnolia St New Orleans LA 70115

BURCH, STEPHEN H, geology, geophysics, see 12th edition

BURCH, THADDEUS JOSEPH, b Baltimore, Md, June 4, 30. SOLID STATE PHYSICS. Educ: Bellarmine Col, NY, AB, 54, PhL, 55; Fordham Univ, MA, 56, MS, 65, PhD(physics), 67; Woodstock Col, Md, STB, 60, STL, 62. Prof Exp: Asst Prof Physics, St Joseph's Col, Pa, 69-72; asst prof, Fordham Univ, New York, 72-74; ASSOC PROF, UNIV CONN, 74- Concurrent Pos: Res collab, Brookhaven Nat Lab, 68- Mem: Am Phys Soc; Am Asn Physics Teachers. Res: Study of magnetic and crystallographic phase transitions in ferromagnetic alloys and magnetic semiconductors by nuclear magnetic resonance, Mössbauer effect, specific heat, and resistivity measurements. Mailing Add: Physics Dept Univ Conn Storrs CT 06268

BURCH, THOMAS ADAMS, b El Paso, Tex, Dec 22, 18; m 42; c 2. EPIDEMIOLOGY. Educ: Univ Southern Calif, AB, 41, MS, 43, MD, 45; Johns Hopkins Univ, MPH, 58. Prof Exp: Physician, NIH, 46-71; CHIEF, OFF RES & STATIST, STATE HEALTH DEPT, HAWAII, 71- Concurrent Pos: Chief onchocerciasis res prof, Pan-Am Sanit Bur, Guatemala, 47-50; actg dir, Inst Am Found Trop Med, Liberia, 51-53; chief clin field studies unit, NIH, 65-69, chief southwestern field studies sect, 70-71; chmn, Int Symp Population Studies Rheumatic Dis, 66. Mem: Fel Am Pub Health Asn; Am Rheumatism Asn; Am Diabetes Asn; AMA; Int Epidemiol Asn. Res: Epidemiology of cancer, diabetes, arthritis, gall bladder disease, parasitic diseases. Mailing Add: Off of Res & Statist Hawaii State Health Dept Honolulu HI 96801

BURCH, WENDELL DALE, b McPherson, Kans, Oct 22, 41; m 60; c 2. INORGANIC CHEMISTRY. Educ: Bethany Col, Kans, BS, 63; Kans State Univ, PhD(inorg chem), 69. Prof Exp: Res scientist fertilizers, Plant Foods Div, Continental Oil Co, 67-70, res scientist surfactant appln, Petrochem Div, 70, tech serv rep, Conoco Chem, 71-72; supvr detergent develop, 72-73, TECH MGR DETERGENTS, CHEM PROD DIV, DE SOTO, INC, 73- Mem: Am Chem Soc; Am Oil Chemists Soc; Am Soc Testing & Mat; Am Asn Textile Chemists & Colorists. Res: Research, development and marketing of household laundry detergents, fabric softeners, bleaches, dishwashing detergents, all purpose cleaners, institutional laundry chemicals and personal care products. Mailing Add: Chem Prod Div De Soto Inc 1700 S Mt Prospect Rd Des Plaines IL 60018

BURCH, WILLIAM PAUL, b Knoxville, Iowa, Mar 5, 22; m 47; c 3. DENTISTRY. Educ: Iowa Wesleyan Col, AB, 51; Loyola Univ Ill, DDS, 55, MA, 66. Prof Exp: From asst prof to assoc prof pedodontics, Sch Dent, Loyola Univ, Chicago, 65-73, chmn dept, 65-73, dir clins, 68-73, chmn dept dent, Univ Hosp, 70-73; DIR CLINS, BAYLOR COL DENT, 73- Concurrent Pos: Mem, Am Asn Dent Schs. Mem: AAAS; Am Dent Asn; Am Soc Dent Children; Am Acad Gold Foil Opers; Am Pub Health Asn. Res: Pedodontics; dental anatomy; experimental immunoelectrophoresis. Mailing Add: Baylor Col of Dent 800 Hall St Dallas TX 75226

BURCHALL, JAMES J, b Brooklyn, NY, Feb 25, 32; m 60; c 3. BACTERIAL PHYSIOLOGY. Educ: St John's Univ, BS, 54; Brooklyn Col, MA, 59; Univ Ill, PhD(bact), 63. Prof Exp: Sr res microbiologist, Wellcome Res Labs, 62-68, HEAD, MICROBIOL DEPT, BURROUGHS WELLCOME & CO, 68- Concurrent Pos: Adj assoc prof, Duke Univ & Univ NC, 74- Mem: AAAS; Am Soc Microbiol. Res: Comparative biochemistry of single enzymes and metabolic pathways and their application to chemotherapy; folate biosynthesis and function. Mailing Add: Microbiol Dept Wellcome Res Lab Res Triangle Park NC 27709

BURCHAM, DONALD PRESTON, b Baker, Ore, Mar 11, 16; m 42; c 3. PHYSICS. Educ: Reed Col, BA, 37; Univ Wash, PhD(physics), 42. Prof Exp: Head physicist, Puget Sound Magnetic Surv Range, US Navy, 41-42, underwater ord, SW Pac Fleet, 43; physicist, appl physics lab, Univ Wash, 44-46; alt sect chief ord res, electronics div, Nat Bur Standards, 46-52, dep chief, ord electronics div, 52-53 & guided missile fuze lab, 53-54; dir, Emerson Res Labs, 54-57, div vpres, Emerson Radio & Phonograph Corp, 58-60, pres, Aerolab Develop Co, 61-62; proj mgr, Voyager, 63-67, mgr space sci div, 68-69, res & advan develop mgr space sci, 69-70, MGR SPACE SCI PROG OFF, JET PROPULSION LAB, CALIF INST TECHNOL, 70- Honors & Awards: Naval Ord Develop Award, 45. Mem: Am Astronaut Soc. Res: Applied research in the fields of electricity, optics and magnetism; radium analysis of submarine cores. Mailing Add: Jet Propulsion Lab 4800 Oak Grove Dr Pasadena CA 91103

BURCHAM, LEVI TURNER, b Ronda, NC, May 30, 12; m 42; c 2. ECOLOGY, GEOGRAPHY. Educ: Univ Calif, Berkeley, BSc, 41, PhD(ecol, geog), 56; Univ Nebr, MSc, 50. Prof Exp: Range exam, Gen Land Off Range Develop Serv, US Dept Interior, Nev, 41-42; forester, Calif Div Forestry, 47-57, asst dep state forester, 57-64; environ sci adv, Off Secy Defense, Adv Res Projs Agency, 64-66; ASST DEP STATE FORESTER, CALIF DIV FORESTRY, 66- Concurrent Pos: Consult ecol, forestry & geog, 48-64, 66-; Off Secy Defense, Adv Res Projs Agency, DC, 62-64; US Forest Serv, 66-67; US Army Res Off, 66-71; consult & mem bd dir, Cohron Indust, Inc, Ala, 68-73; lectr, Calif State Univ, Sacramento, 69-74; vpres & mem bd dirs, Natural Resources Mgt Corp, 70-74. Mem: AAAS; Am Geog Soc; Am Soc Range Mgt; Asn Am Geog; Ecol Soc Am. Res: Historical ecology and geography of range livestock industry in California and Mexico; ecology and geography of Pacific basin and Australia; quantitative description and interpretation of military significance of natural environment; plant succession and phenology of conifer forests of Sierra Nevada; soil and vegetation interrelationships. Mailing Add: 4701 Crestwood Way Sacramento CA 95822

BURCHAM, PAUL BAKER, b Fayette, Mo, Feb 22, 16; m 41; c 2. MATHEMATICS. Educ: Cent Col, Mo, BS, 35; Northwestern Univ, MA & PhD(math), 41. Prof Exp: Asst math, Northwestern Univ, 39-41; asst instr math, Cent Col, Mo, 41-42; asst prof, 46-54, PROF MATH, UNIV MO-COLUMBIA, 54- Mem: Am Math Soc; Math Asn Am. Res: Some inclusion relations in the domain of Hausdorff matrices; summability of series. Mailing Add: Dept of Math Univ of Mo-Columbia Columbia MO 65201

BURCHARD, HERMANN GEORG, b Würzburg, Ger, Dec 9, 34; m 61; c 4. MATHEMATICS. Educ: Univ Hamburg, Dipl-math, 63; Purdue Univ, PhD(comput sci), 68. Prof Exp: Res mathematician, Gen Motors Res Labs, Mich, 63-66; vis asst prof, Math Res Ctr, US Army, Univ Wis, Madison, 68-69; asst prof, Ind Univ, Bloomington, 69-72; ASSOC PROF MATH, OKLA STATE UNIV, 72- Concurrent Pos: Consult, Phillips Petroleum Co, 75- Mem: Am Math Soc; Soc Indust & Appl Math; Math Asn Am. Res: Approximation theory; numerical analysis; applied mathematics. Mailing Add: Dept Math Okla State Univ Stillwater OK 74074

BURCHARD, JEANETTE, b Wanesville, Mo, July 20, 17. MICROBIOLOGY, SEROLOGY. Educ: Southwest Mo State Col, BS, 39; Menorah Hosp Sch Med Technol, MT, 40; Univ Mich, MPH, 48. Prof Exp: Biologist, Southwest Br Lab, Mo Div Health, 41-42, biologist in charge, 42-45, sr biologist in charge, 45-58, dir, 48-63; microbiologist, Venereal Disease Res Lab, Commun Disease Ctr, USPHS, 63-65; chief microbiologist, 65-67, ASST DIR, BUR LAB SERV, MO DIV HEALTH, 67- Concurrent Pos: Guest lectr, AID course, Sch Pub Health, Univ WI. Mem: Fel Am Pub Health Asn; Am Soc Microbiol; Am Soc Med Technol; Asn State & Territorial Pub Health Lab Dirs. Res: Culture and serology of brucellosis; serology of poliomyelitis; microbiology of foods; syphilis serology. Mailing Add: Mo Div of Health Bur of Lab Serv Jefferson City MO 65101

BURCHARD, ROBERT P, b New York, NY, Nov 7, 38; m 67; c 2. BACTERIOLOGY. Educ: Brown Univ, BA, 60, MSc, 62; Univ Minn, Minneapolis, PhD(microbiol), 65. Prof Exp: Lectr microbiol, Univ Ife, Nigeria, 65-66; asst prof, 67-69, ASSOC PROF BIOL SCI, UNIV MD, BALTIMORE COUNTY, 69- Concurrent Pos: NSF res grants, 68-70, 71-73 & 75-; UK Sci Res Coun vis fel, dept of biochem, Univ Leeds, 73-74. Mem: Am Soc Microbiol; AAAS. Res: Developmental biology of the myxobacteria including mechanism of gliding motility, light-induced lysis and pigment synthesis; morphogenesis; microbial ecology of Chesapeake Bay. Mailing Add: Dept of Bio Univ of Md 5401 Wilkens Ave Catonsville MD 21228

BURCHELL, HOWARD BERTRAM, b Athens, Ont, Nov 28, 07; nat US; m 42; c 4. MEDICINE. Educ: Univ Toronto, MD, 32; Univ Minn, PhD(med), 40. Prof Exp: Instr med, Univ Pittsburgh, 36; from instr to prof, Mayo Found, 41-67, chief sect cardiol, 67-74, PROF MED, UNIV HOSP, UNIV MINN, MINNEAPOLIS, 67-; SR CARDIOLOGIST, NORTHWESTERN HOSP, MINNEAPOLIS, 74- Concurrent Pos: Ed., Circulation, 66-70. Mem: Am Physiol Soc; Am Fedn Clin Res; Asn Am Physicians; Am Heart Asn. Res: Clinical investigation; physiology of circulation. Mailing Add: Univ Unit Northwestern Hosp Chicago at 27th Minneapolis MN 55407

BURCHENAL, JOSEPH HOLLAND, b Milford, Del, Dec 21, 12; m 48; c 7. MEDICINE. Educ: Princeton Educ: Univ Pa, MD, 37. Prof Exp: Intern, Union Mem Hosp, Baltimore, 37-38; resident dept pediat, NY Hosp, Cornell Univ, 38-39; asst resident, Boston City Hosp, 40-42; assoc, Sloan-Kettering Inst Cancer Res, 48-52, vpres, 64-72; asst prof clin med, 49-50, asst prof med, 50-51, assoc prof, Sloan-Kettering Div, 51-52, prof, 52-55, PROF MED, MED COL, CORNELL UNIV, 55-; MEM, SLOAN-KETTERING INST CANCER RES, 52-, HEAD APPL THER LAB & FIELD COORDR HUMAN CANCER, 73- Concurrent Pos: Res fel med, Harvard Med Sch, 40-42; spec fel, Mem Hosp, New York, 46-49; intern, New York Hosp, 38-39; asst attend physician, Med Serv, Mem Hosp, New York, 49-52, attend physician, 52-, chief chemother serv, 52-64, med dir clin invest, 64-66, dir clin invest, 66-; consult, USPHS, Am Cancer Soc & Div Med Sci, Nat Res Coun, 54-56; mem, Nat Panel Consult Conquest of Cancer, US Senate Comt Labor & Pub Welfare, 70; chmn chemother adv comt, Nat Cancer Inst, 70-71; vpres med & sci affairs & chmn med & sci adv comt, Leukemia Soc Am, 70-75. Honors & Awards: Alfred P Sloan Award, 63; Albert Lasker Award Clin Cancer Chemother, 72; Prix Leopold Griffuel, 70; David A Kamoksky Mem Award, Am Soc Clin Oncol, 74; John Phillips Award, 74; James Ewing Soc Award, 75. Mem: Am Soc Clin Invest; Soc Exp Biol & Med; Asn Cancer Res (vpres, 64-65, pres, 65-66); Am Soc Trop Med & Hyg; Am Fedn Clin Res. Res: Chemotherapy of cancer and leukemia. Mailing Add: Mem Sloan-Kettering Cancer Ctr 1275 York Ave New York NY 10021

BURCHFIEL, BURRELL CLARK, b Stockton, Calif, Mar 21, 34; m 55; c 2. GEOLOGY. Educ: Stanford Univ, BS, 57, MS, 58; Yale Univ, PhD(geol), 61. Prof Exp: Geologist, US Geol Surv, 61; from asst prof to assoc prof, 61-70, prof, 70-75, CAREY CRONEIS PROF GEOL, RICE UNIV, 75- Concurrent Pos: Exchange prof, Geol Inst Belgrade, Yugoslavia, 68 & Geol Inst Bucharest, Romania, 70. Mem: Fel

Geol Soc Am; Am Asn Petrol Geol; Am Geophys Union. Res: Tectonics of the basin and range province; regional tectonics; orogenesis and geosynclinal development. Mailing Add: Dept of Geol Rice Univ Houston TX 77001

BURCHFIEL, JAMES LEE, b Los Angeles, Calif, Mar 16, 41; m 63; c 2. NEUROPHYSIOLOGY. Educ: Stanford Univ, BS, 63, PhD(pharmacol), 69. Prof Exp: Res assoc, 70-75, PRIN RES ASSOC NEUROL, SCH MED, HARVARD UNIV, 75- Res: Development and functional organization of visual and somatic sensory systems. Mailing Add: Seizure Unit Neurophysiol Childrens Hosp Med Ctr Boston MA 02115

BURCHFIELD, HARRY P, b Pittsburgh, Pa, Dec 22, 15; m 42, 63; c 5. BIOCHEMISTRY. Educ: Columbia Univ, AB & MA, 38, PhD, 56. Prof Exp: Analyst, Nat Oil Prod Co, NJ, 38-40; res chemist, Naugatuck Chem Div, US Rubber Co, 40-41, group leader, 41-50, dir plantations res dept, 59-61; inst scientist & mgr anal & biochem, Southwestern Res Inst, 61-65; officer-in-charge, pesticides res lab, USPHS, 65-67; SCI DIR & DIR BIOL SCI, GULF SOUTH RES INST, 67- Concurrent Pos: Adj prof, Univ Southwestern La, 67- Honors & Awards: Prize, Chicago Rubber Group, 46. Mem: AAAS; Am Inst Biol Sci; Soc Toxicol; Am Chem Soc; Am Phytopath Soc. Res: Biochemical applications of gas chromatography; metabolism and analysis of drugs, pesticides and natural products; mechanism of action of biocides. Mailing Add: Gulf South Res Inst Box 1177 New Iberia LA 70560

BURCHFIELD, PAUL EDWARD, chemistry, chemical engineering, see 12th edition

BURCHILL, BROWER RENE, b El Dorado, Kans, Dec 29, 38; m 60; c 2. CELL BIOLOGY. Educ: Phillips Univ, BA, 60; Fla State Univ, MS, 63; Western Reserve Univ, PhD(biol), 66. Prof Exp: NIH trainee, Fla State Univ, 63; postdoctoral appointee, Los Alamos Sci Lab, Univ Calif, 66-68; asst prof zool, 68-69 & physiol & cell biol, 69-71, ASSOC PROF PHYSIOL & CELL BIOL, UNIV KANS, 71-, CHMN DIV BIOL SCI, 73- Mem: AAAS; Am Soc Cell Biol; Biophys Soc; assoc Radiation Res Soc; Soc Protozool. Res: Nucleocytoplasmic interactions essential to oral differentiation in the ciliate protozoan Stentor coeruleus. Mailing Add: Dept of Physiol & Cell Biol Univ of Kans Lawrence KS 66045

BURCHILL, CHARLES EUGENE, b Makwa, Sask, Dec 12, 32; m 57; c 4. PHYSICAL CHEMISTRY. Educ: Univ Sask, BA, 55, MA, 61; Univ Leeds, PhD(phys chem), 68. Prof Exp: Instr chem, Victoria Col, 56-63, demonstr phys chem, Univ Leeds, 63-66; asst prof chem, 66-68, ASSOC PROF CHEM, UNIV MAN, 68- Res: Kinetics and mechanisms of free-radical reactions in solution; radiation chemistry and photochemistry of aqueous solutions. Mailing Add: Dept of Chem Univ of Man Winnipeg MB Can

BURCIK, EMIL JOSEPH, b Pittsburgh, Pa, Apr 19, 16; m 45; c 2. CHEMISTRY. Educ: Carnegie Inst Technol, BS, 37; Calif Inst Technol, PhD(chem), 41. Prof Exp: Chemist, US Bur Mines, 41-45; res chemist, Procter & Gamble Co, 45-49; assoc prof, Univ Okla, 49-52; ASSOC PROF PETROL ENG, PA STATE UNIV, UNIVERSITY PARK, 52- Concurrent Pos: Fel, Calif Inst Technol. Mem: Am Inst Mining, Metall & Petrol Eng. Res: Surface chemistry; petroleum reservoir engineering; polymers. Mailing Add: Col of Earth & Mineral Sci Pa State Univ University Park PA 16802

BURCK, LARRY HAROLD, b Detroit, Mich, Sept 25, 45; m 67; c 2. SOLID MECHANICS. Educ: Mich State Univ, BS, 67; Rensselaer Polytech Inst, MS, 71; Northwestern Univ, PhD(mat sci), 75. Prof Exp: Res engr, Advan Mat & Develop Lab, Pratt & Whitney Aircraft, 67-72; ASST PROF MAT SCI, UNIV WIS-MILWAUKEE, 75- Mem: Am Soc Metals; Am Inst Mining Metall & Petrol Engrs. Res: Fracture, fatigue and mechanical behavior of materials; analytical and experimental fracture mechanics. Mailing Add: 4131 Cherrywood Lane Brown Deer WI 53209

BURCK, PHILIP JOHN, b Milwaukee, Wis, Sept 21, 36; m 68; c 1. BIOCHEMISTRY. Educ: Lawrence Univ, AB, 58; Univ Ill, MS, 60, PhD(biochem), 62. Prof Exp: RES SCIENTIST BIOCHEM, LILLY RES LABS, ELI LILLY & CO, 62- Mem: AAAS; Am Chem Soc. Res: Enzyme and enzyme inhibitor purification and characterization; physiological role of proteases and protease inhibitors. Mailing Add: Lilly Res Labs Biochem Res Dept 307 E McCarty St Indianapolis IN 46206

BURCKBUCHLER, FREDERICK V, b New Haven, Conn, Aug 5, 35; m 56; c 3. SOLID STATE ELECTRONICS. Educ: Univ Conn, BA, 58, MS, 60, PhD(physics), 68. Prof Exp: Asst physics, Univ Conn, 58-60; mem tech staff, Bell Tel Labs, Pa, 60-61; asst prof physics, US Coast Guard Acad, 61-66; res asst solid state physics, Univ Conn, 66-68; MEM TECH STAFF, BELL TEL LABS, 68- Mem: Am Phys Soc. Res: Electrical charge transport and storage in metals; semiconductors and dielectric materials. Mailing Add: Bell Tel Labs Murray Hill NJ 07974

BURCKEL, ROBERT BRUCE, b Louisville, Ky, Dec 15, 39; m 67. MATHEMATICS. Educ: Univ Notre Dame, BS, 61; Yale Univ, MA, 63, PhD(math), 68. Prof Exp: From instr to asst prof math, Univ Ore, 66-70; actg chmn dept, 74-75, ASSOC PROF MATH, KANS STATE UNIV, 71- Mem: Am Math Soc. Res: Harmonic analysis, classical and abstract; function algebras; hardy inspaces, classical and abstract; classical complex analysis. Mailing Add: Dept of Math Kans State Univ Manhattan KS 66506

BURCKHALTER, JOSEPH HAROLD, b Columbia, SC, Oct 9, 12; m 43; c 3. MEDICINAL CHEMISTRY. Educ: Univ SC, BS, 34; Univ Ill, MS, 39; Univ Mich, PhD, 42. Prof Exp: Res assoc biol chem, Univ Mich, 39-40; sr res chemist, Parke, Davis & Co, 42-47; assoc prof med chem, Univ Kans, 47-50, prof & chmn dept, 50-60; PROF MED CHEM, UNIV MICH, 60-, CHMN DEPT, 67- Concurrent Pos: Fulbright prof, Inst Pharmaceut Chem, Tuebingen, Ger, 55-56; mem chemother adv comt, Nat Cancer Inst. Mem: AAAS; Am Chem Soc; Am Pharmaceut Asn; Am Soc Trop Med & Hyg; The Chem Soc. Res: Antimalarial, antiamebic and antischistosomal agents; fluorescent isothiocyanate labeling agents; steroids; Mannich reaction. Mailing Add: Dept of Med Chem Univ Mich Col Pharm Ann Arbor MI 48104

BURCSU, JAMES EDWARD, b Columbus, Ohio, Mar 11, 40; m 67. MEDICINAL CHEMISTRY. Educ: Ohio State Univ, BSc, 62; Univ Minn, Minneapolis, PhD(org chem), 66. Prof Exp: Res assoc chem, Univ Hawaii, 66-68; sr org chemist, 68-75, PLANNING & MFG DATA COORDR, WELLCOME RES LAB, BURROUGHS WELLCOME CO, 75- Mem: AAAS; Am Chem Soc. Res: Rearrangements of thietane dioxides; studies on the marine toxin palytoxin, stereochemistry and CNS agents. Mailing Add: 630 Ashe Ave Cary NC 27511

BURD, JOHN FREDERICK, b Ft Wayne, Ind, Aug 19, 46. BIOCHEMISTRY. Educ: Purdue Univ, BS, 68; Univ Wis, MS, 70, PhD(biochem), 75. Prof Exp: Assoc res scientist, 70-72, RES SCIENTIST BIOCHEM, AMES RES LABS, MILES LABS, INC, 75- Mem: Am Chem Soc. Res: Development of assays for therapeutic drugs in human serum. Mailing Add: Ames Res Labs Miles Labs Inc 1127 Myrtle St Elkhart IN 46514

BURD, JOHN W, organic chemistry, see 12th edition

BURDASH, NICHOLAS MICHAEL, b Hazleton, Pa, Sept 18, 41; m 65; c 2. IMMUNOLOGY, BACTERIOLOGY. Educ: Univ Scranton, BS, 63; Duquesne Univ, MS, 65; Ohio State Univ, PhD(microbiol), 69. Prof Exp: ASST PROF MICROBIOL & IMMUNOL, MED UNIV SC, 69- Mem: AAAS; Am Soc Microbiol; Electron Micros Soc Am; Am Inst Biol Sci. Res: Humoral and cellular aspects of autoimmune diseases, especially experimental allergic encephalomyelitis and experimental allergic neuritis. Mailing Add: Dept of Microbiol Med Univ of SC Charleston SC 29401

BURDEN, GEORGE STANLEY, b Lake Wales, Fla, Aug 11, 26; m 53; c 2. MEDICAL ENTOMOLOGY. Educ: Univ Fla, BSA, 50. Prof Exp: MED ENTOMOLOGIST, ENTOM RES DIV, US DEPT AGR, 53- Mem: Entom Soc Am. Res: Control methods; resistance and ecology pertaining to medical entomology; Blattidae, Cimicidae, Pediculidae, and Pulicidae. Mailing Add: Insects Affecting Man Res Lab USDA PO Box 14565 Gainesville FL 32604

BURDEN, HUBERT WHITE, b Elizabeth City, NC, Sept 12, 43; m 67; c 2. REPRODUCTIVE ENDOCRINOLOGY. Educ: Atlantic Christian Col, AB, 65; ECarolina Univ, MA, 67; Tulane Univ, PhD(anat), 71. Prof Exp: Instr biol, ECarolina Univ, 67-68; teaching asst anat, Med Sch, Tulane Univ, 68-71; asst prof, 71-75, ASSOC PROF ANAT, E CAROLINA UNIV, 75- Concurrent Pos: USPHS res grant, 75. Mem: Soc Study Reproduction; Soc Develop Biol; Am Asn Anatomists; Am Soc Zoologists; Pan Am Asn Anatomists. Res: Research on the role of the peripheral autonomic nervous system in reproductive function. Mailing Add: Dept of Anat ECarolina Univ Sch of Med Greenville NC 27834

BURDEN, STANLEY LEE, JR, b Aurora, Ill, Mar 9, 39; m 62; c 1. CHEMICAL INSTRUMENTATION, ANALYTICAL CHEMISTRY. Educ: Taylor Univ, BS, 61; Ind Univ, PhD(anal chem) 66. Prof Exp: Teacher, high sch, Ind, 61-62; from instr to assoc prof, 66-74, PROF CHEM, TAYLOR UNIV, 75- Concurrent Pos: NASA res fel, Manned Spacecraft Ctr, 69. Mem: Am Chem Soc; Am Sci Affiliation. Res: Instrumentation of analytical methods of analysis; environmental analysis and on-line computer applications in chemical instrumentation. Mailing Add: Sci Ctr Taylor Univ Upland IN 46989

BURDETT, LORENZO WORTH, b Pocatello, Idaho, Aug 9, 16; m 45; c 7. ANALYTICAL CHEMISTRY. Educ: Univ Idaho, BS, 43; Univ Ill, PhD(anal chem), 49. Prof Exp: Jr chemist, Shell Develop Co, 43-46; res chemist, Standard Oil Co Ind, 49-52; res chemist, 52-55, sect leader chem, 55-58, sr sect leader, 58-65, SUPVR RES DEPT, UNION OIL CO CALIF, 65- Mem: Am Chem Soc. Res: Method development and analytical research in electrochemical and chemical area. Mailing Add: Res Dept Union Oil Co of Calif PO Box 76 Brea CA 92621

BURDETTE, ALBERT CLARK, b Spurgeon, Ind, Feb 11, 05; m 27; c 2. MATHEMATICS. Educ: Oakland City Col, AB, 27; Ind Univ, AM, 31; Univ Ill, PhD(math), 36. Prof Exp: High sch teacher, Ind, 27-30; asst math, Univ Ill, 31-36; to instr, Col Agr, 36-42, from asst prof to prof, 42-74, EMER PROF MATH, UNIV CALIF, DAVIS, 74- Mem: Math Asn Am; Am Math Soc. Res: Functions of complex variables; difference equations; differential equations. Mailing Add: Dept of Math Univ of Calif Davis CA 95616

BURDETTE, ERNEST LINWOOD, b Birmingham, Ala, Apr 29, 45; m 66; c 2. OPTICS. Educ: Birmingham-Southern Col, BS, 66; Auburn Univ, MS, 68, PhD(physics), 75. Prof Exp: Asst prof physics, Ga Col, 71-73; INSTR PHYSICS, AUBURN UNIV, 75- Mem: Sigma Xi; Am Phys Soc. Res: Characteristics of visible Smith-Purcell radiation from small gratings; ruling of small gratings. Mailing Add: Dept of Physics Auburn Univ Auburn AL 36830

BURDETTE, WALTER JAMES, b Hillsboro, Tex, Feb 5, 15; m 47; c 2. SURGERY. Educ: Baylor Univ, AB, 35; Univ Tex, AM, 36, PhD(zool), 38; Yale Univ, MD, 42; Am Bd Surg, dipl, 50; Am Bd Thoracic Surg, dipl, 53. Prof Exp: Intern, Johns Hopkins Hosp, 42-43; Cushing fel surg, Yale Univ, 43-44; asst resident, New Haven Hosp, Conn, 44-46; from instr to assoc prof, Sch Med, La State Univ, 46-55, coordr cancer res & teaching, 48-55; prof surg & chmn dept, Sch Med, Univ Mo, 55-56; prof clin surg, Sch Med, St Louis Univ, 56-57; prof surg & head dept, Col Med, Univ Utah, 57-65; PROF SURG & ASSOC DIR, UNIV TEX M D ANDERSON HOSP & TUMOR INST, 65- Concurrent Pos: Asst vis surgeon, Charity Hosp of La, New Orleans, 46-47, vis surgeon, 47-55; vis surgeon, Southern Baptist Hosp, 53-55; vis investr, Chester Beatty Inst Cancer Res, London, Eng, 53; surgeon-in-chief, Univ Hosp, Univ Mo, 55-56; dir, St Louis Univ Surg Serv, Vet Admin Hosp, 56-57; dir, Lab Clin Biol, 57-; surgeon-in-chief, Salt Lake Gen Hosp, 57-65; Gibson lectr advan surg, Oxford Univ, 66; vis prof, Off Univ Congo, 68; lectr, Hosp Martinez, PR, 69, Univ Sendai, Japan, 70, Univ Melbourne, 70 & Univ Freiburg, 70. Consult, Oak Ridge Inst Nuclear Studies, 51 & Touro Infirmary, New Orleans, 53-55; chief surg consult, Vet Admin Hosp, Salt Lake City, 57-65. Mem morphol & genetics study sect, NIH, 55-58, chmn genetics study sect, 57-61, mem nat adv cancer coun, 61-65, mem nat adv heart coun, 65, dir working cadre on carcinoma of large intestine, Nat Cancer Inst; chmn res adv coun, Am Cancer Soc, 57-; mem adv comt smoking & health, Surgeon Gen US, 62-64; chmn Nat Acad Sci comt, Int Union Against Cancer, 62-; mem transplantation comt, Nat Acad Sci. Mem: Am Asn Cancer Res; Soc Exp Biol & Med; Am Col Surgeons; Soc Clin Surgeons (treas, 63-64); Am Surg Asn. Res: Genetics and cancer; metabolism of cardiac muscle; cardiovascular surgery; invertebrate hormones. Mailing Add: Dept of Surg Univ of Tex M D Anderson Hosp & Tumor Inst Houston TX 77025

BURDG, DONALD EUGENE, b Milford, Nebr, Aug 12, 29; m 52; c 2. MATHEMATICS. Educ: Colo State Univ, BS, 51; Univ Northern Colo, MA, 52; Ore State Univ, MS, 66. Prof Exp: Asst prof math, Cent Ore Community Col, 56-67; ASSOC PROF MATH, SOUTHWESTERN ORE COMMUNITY COL, 67- Mem: Math Asn Am. Mailing Add: Southwestern Ore Community Col Empire Lakes Coos Bay OR 97420

BURDGE, DAVID NEWMAN, b Sullivan, Ind, May 1, 31; m 53; c 2. PETROLEUM CHEMISTRY. Educ: Purdue Univ, BS, 52, MS, 56, PhD(org chem), 59. Prof Exp: Res chemist, Marathon Oil Co, 59-61 & Eastman Kodak Co, 61-62; res chemist, 62-67, mgr, Org Chem Dept, 67-70, mgr, Ref Sci & Eng Dept, 70-72, MGR, PETROLEUM CHEM DEPT, MARATHON OIL CO, 72- Mem: Am Inst Chem; Am Chem Soc; Sigma Xi; Soc Petroleum Engrs. Res: Organic sulfur chemistry; oxidation of alkyl aromatics; petroleum and refining chemistry; petroleum production chemistry; enhanced recovery chemicals and techniques. Mailing Add: Marathon Oil Co Denver Res Ctr PO Box 269 Littleton CO 80120

BURDI, ALPHONSE R, b Chicago, Ill, Aug 28, 35; m 69; c 1. DENTAL RESEARCH, CHILD GROWTH. Educ: Northern Ill Univ, BSEd, 57; Univ Ill, MS, 59; Univ Mich, MS, 61, PhD(anat), 63. Prof Exp: USPHS res trainee, 60-61, instr, 62-65, from asst prof to assoc prof, 65-74, PROF ANAT, UNIV MICH, 74- Concurrent Pos: Fel Inst Advan Educ Dent Res, 64. Mem: AAAS; Am Asn Anat; Am Asn Phys

Anthrop; Am Cleft Palate Asn; Tissue Cult Asn. Res: Human gross anatomy, embryology and prenatal craniofacial growth, identification of human remains in forensic medicine; biomechanics of the facial skeleton; developmental craniofacial biology; birth defects; mechanisms of tissue interactions; cephalometrics. Mailing Add: Dept of Anat Univ of Mich Ann Arbor MI 48104

BURDICK, ALLAN BERNARD, b Cincinnati, Ohio, Aug 16, 20; m 43; c 4. GENETICS, CYTOGENETICS. Educ: Iowa State Col, BS, 45, MS, 47; Univ Calif, Berkeley, PhD, 49. Prof Exp: Asst prof genetics & plant breeding, Univ Ark, 49-52; from asst prof to prof genetics, Purdue Univ, 52-63; prof & assoc dean sci, Am Univ Beirut, 63-66; prof biol & chmn dept, Adelphi Univ, 66-69, actg dir, Adelphi Inst Marine Sci, 67-68; chmn dept genetics, 69-70, group leader cytol & genetical sci, 70-75, PROF GENETICS, UNIV MO-COLUMBIA, 69- Concurrent Pos: Res collabr, Brookhaven Nat Lab, 55, 57; Fulbright res prof, Kyoto Univ, 59-60; Guggenheim fel, 59-60. Mem: Fel AAAS; Genetics Soc Am; Am Soc Naturalists; Am Soc Human Genetics; Biomet Soc. Res: Quantitative inheritance; heterosis; Drosophila gene structure; tomato genetics; mutation; mammalian meiotic cytogenetics. Mailing Add: Div of Biol Sci 205 Curtis Hall Univ of Mo-Columbia Columbia MO 65201

BURDICK, CHARLES LALOR, b Denver, Colo, Apr 14, 92; m 38; c 2. CHEMISTRY. Educ: Drake Univ, BS, 11; Mass Inst Technol, SB, 13, MS, 14; Univ Basel, PhD(chem), 15. Hon Degrees: DSc, Univ Del, 55; LLD, Drake Univ, 70. Prof Exp: Res assoc phys chem, Mass Inst Technol & Calif Inst Technol, 16-17; metall engr, Chile Copper Co & Guggenheim Bros, 19-24; vpres & consult engr, Anglo-Chilean Consol Nitrate Corp, 24-28; asst chem dir, ammonia dept, E I du Pont de Nemours & Co, 28-35 & develop dept, 36-39, asst to pres, 39-45, chmn bd, du Pont, SA & Cia Mexicana de Explosivos, Mexico City, 45-46, secy, high polymer comt, Wilmington, 46-50 & polyfibers comt, 50-57. Concurrent Pos: Mem vis comt biol & Bussey Inst, Harvard Univ, 49-64; dir, Planned Parenthood-World Pop, 61-67; mem exec comt, Int Planned Parenthood Fedn, 62-68, trustee & exec dir, Lalor Found; pres, Christiana Found, 60-73; emer trustee, Univ Del Res Found; trustee, Del Acad Med; life mem corp, Marine Biol Lab, Woods Hole. Mem: Fel AAAS; Am Chem Soc; NY Acad Sci; Am Inst Chem Engrs; Soc Study Fertil. Res: Administration of awards for research in mammalian reproductive physiology. Mailing Add: 4400 Lancaster Pike Wilmington DE 19805

BURDICK, DANIEL, b Syracuse, NY, Oct 21, 15; m 49; c 5. SURGERY. Educ: Syracuse Univ, AB, 37, MD, 50. Prof Exp: CLIN DIR TUMOR CLIN & ATTEND SURGEON, UNIV HOSP, STATE UNIV NY UPSTATE MED CTR, 54- Concurrent Pos: Attend, Syracuse Mem Hosp, 54-; surgeon, Community Gen Hosp. Mem: Am Col Surgeons. Res: Cancer research. Mailing Add: 713 E Genessee St Syracuse NY 13210

BURDICK, DAVID LEO, b Lemmon, SDak, Dec 26, 41; m 62; c 2. SOLID STATE PHYSICS. Educ: SDak Sch Mines & Technol, BS, 63, MS, 65; Mont State Univ, BA & PhD(physics), 70. Prof Exp: Res assoc physics, Mont State Univ, 70-71; physicist, Optical Design Br, 71-73, RES PHYSICIST, NAVAL WEAPONS CTR, 73- Mem: Am Phys Soc. Res: Optical properties of thin films and highly transparent bulk materials; UV filters; solar energy. Mailing Add: Code 6013 Michelson Lab Naval Weapons Ctr China Lake CA 93555

BURDICK, DONALD, b New York, NY, Nov 6, 34; m 59; c 4. BIOCHEMISTRY. Educ: Rutgers Univ, BS, 56, MS, 58; Pa State Univ, PhD(biochem), 62. Prof Exp: Res chemist, Eastern Utilization Res & Develop Div, Agr Res Serv, 62-67, head, Ky Coop Tobacco Invest, 67-69, CHIEF FORAGE & FEED LAB, RICHARD B RUSSELL AGR RES CTR, USDA, 69- Res: Chemical composition of tobacco and tobacco smoke; carbohydrates of forage crops. Mailing Add: R B Russell Agr Res Ctr USDA Box 5677 Athens GA 30604

BURDICK, DONALD LEE, organic chemistry, see 12th edition

BURDICK, DONALD SMILEY, b Newark, NJ, Feb 8, 37; m 58; c 3. MATHEMATICAL STATISTICS. Educ: Duke Univ, BS, 58; Princeton Univ, MA, 60, PhD(math statist), 61. Prof Exp: Instr, Princeton Univ, 61-62; asst prof, 62-68, ASSOC PROF MATH, DUKE UNIV, 68- Concurrent Pos: Consult, Army Res Off, 62-66; vis asst prof statist, Univ Wis, 65-66. Mem: Am Statist Asn; Royal Statist Soc. Res: Non-parametric techniques and the analysis of variance; linear statistical models; categorical data analysis. Mailing Add: Dept of Math Duke Univ Durham NC 27706

BURDICK, EVERETTE MARSHALL, b Champaign, Ill, Aug 9, 13; m 37. CHEMISTRY. Educ: Univ Miami, BS, 35; Purdue Univ, MS, 37, PhD(biochem), 43. Prof Exp: Chemist, northern regional res lab, Bur Agr & Indust Chem, USDA, 41-45; chemist, US Fruit & Veg Lab, Univ Tex, 45-46; dir res, Texsun Citrus Exchange, 46-52; dir lab, Am Chlorophyll Div, Strong, Cobb & Co, Inc, 52-53, vpres & dir res, 53-54; CHEM CONSULT, 54- Concurrent Pos: Tech consult, Rio Farms, Inc, 50-52, dir new prods div, Fla Citrus Mutual, 57-58; dir res & develop, True Taste Corp, 61-64; corp dir res & develop, Tex Western, Inc, 71- Mem: Fel AAAS; fel Am Inst Chemists; NY Acad Sci. Res: Enzymes; fermentations; analytical chemistry; citrus processing; product development; vitamins; private formulas; nutrition; submerged combustion; chlorophyll derivatives and chlorophyll; essential oils; humates; carotenoids; papaya products; vermiculites and perlite, processing and utilization. Mailing Add: 4821 Ronda St Coral Gables FL 33146

BURDICK, FRANK A, b Sapulpa, Okla, July 7, 21; m 42; c 4. PROSTHODONTICS. Educ: Sapulpa Jr Col, 39-42; Univ Mo-Kansas City, DDS, 50, MS, 65. Prof Exp: Pvt pract, 50-55; asst prof dent, Univ Tex, 55-62; asst prof, Univ Mo-Kansas City, 64-69; assoc prof, Sch Dent Med, Univ Pittsburgh, 69-72; PROF DENT, SCH DENT, UNIV TENN, MEMPHIS, 72-; CHIEF DENT SERV, VET ADMIN HOSP, 72- Concurrent Pos: Mem staff prosthodont, Vet Admin Hosp, Kansas City, Mo, 65-69; chief dent serv, Vet Admin Hosp, Pittsburgh, 69-72. Mem: AAAS; Am Dent Asn; Am Prosthodont Soc; Am Equilibration Soc; Am Col Dent. Res: Bone induction. Mailing Add: Vet Admin Hosp 1030 Jefferson Memphis TN 38108

BURDICK, GEORGE EDGAR, b Albion, Wis, Apr 29, 05; m 31; c 1. ZOOLOGY. Educ: Milton Col, BA, 27; Univ Wis, MA, 31. Prof Exp: Instr chem & biol, Wartburg Col, 28-34; teacher pub sch, NY, 35-37; prof sci, Chowan Jr Col, 37-41; from jr aquatic biologist to supvr aquatic biologist, NY Conserv Dept, 41-70, chief fish & wildlife ecologist, NY Dept Environ Cönserv, 70-72; RETIRED. Concurrent Pos: Spec consult, USPHS, 55-64, consult nat tech adv comt water qual criteria, Fed Water Pollution Control Admin, 67-68. Honors & Awards: Award, Am Motors, 64. Mem: Am Fisheries Soc. Res: Aquatic biology and the effect of pollutants on the aquatic environment; determination of toxicity of fish and other aquatic organisms; ecology of trout stream insects and their reaction to pollutants. Mailing Add: PO Box 381 Berlin NY 12022

BURDICK, HAROLD CHARLES, b North Loup, Nebr, Dec 20, 06; m 30; c 1. BIOLOGY, PHYSIOLOGY. Educ: Milton Col, BA, 29; Univ Iowa, MS, 32, PhD(zool), 36. Prof Exp: Prof biol chem & physiol, State Teachers Col, NDak, 36-45; PROF BIOL & PHYSIOL, UNIV MO-KANSAS CITY, 45- Res: Physiology of insect development; renal perfusion; temperomandibular joint and muscles of mastication. Mailing Add: Dept of Biol & Physiol Univ of Mo Kansas City MO 64110

BURDICK, MORTON LEON, biology, see 12th edition

BURDINE, HOWARD WILLIAM, b Big Stone Gap, Va, July 1, 09; m 34. PLANT NUTRITION, PHYSIOLOGY. Educ: Berea Col, BS, 35; Univ Ky, MS, 51; Cornell Univ, PhD, 56. Prof Exp: County agr agent, Ky, 37-44; soil conservationist, Soil Conserv Serv, US Dept Agr, 44-47, teacher vocational agr, 47-51, asst horticulturist, 56-59, asst & assoc soils chemist, 59-68, soils chemist, 68-70, PROF PLANT PHYSIOL, AGR RES & EDUC CTR, USDA, UNIV FLA, 70- Mem: Am Soc Plant Physiol; Am Soc Hort Sci; Soil Sci Soc Am; Am Soc Agron. Res: Nutrition and physiology of vegetable crops grown on organic soils of the everglades area of Florida. Mailing Add: Agr Res and Educ Ctr PO Drawer A Belle Glade FL 33430

BURDINE, JOHN ALTON, b Austin, Tex, Feb 7, 36; m 59; c 3. NUCLEAR MEDICINE. Educ: Univ Tex, Austin, BA, 59; Univ Tex Med Br Galveston, MD, 61. Prof Exp: Intern med, Med Ctr, Ind Univ, Indianapolis, 61-62; resident internal med, Univ Tex Med Br Galveston, 62-65; instr nuclear med, Dept Radiol, 65-66, from asst prof to assoc prof radiol, 66-74, actg chmn dept, 68-71, PROF RADIOL, BAYLOR COL MED, 74-, CHIEF NUCLEAR MED SECT, 65-; CHIEF NUCLEAR MED SERV, ST LUKE'S EPISCOPAL-TEX CHILDREN'S HOSPS, 69- Concurrent Pos: Consult, US Army, 66. Mem: Soc Nuclear Med. Res: Development and evaluation of new radiopharmaceuticals; design and implementation of a computer system for processing scintillation camera data, including development of radionuclide techniques for quantification of regional pulmonary function, bone densitometric studies. Mailing Add: Dept of Radiol Sect Nuclear Med Baylor Col of Med Houston TX 77025

BURDITT, ARTHUR KENDALL, JR, b Elizabeth, NJ, Feb 12, 28; m 52; c 4. ECONOMIC ENTOMOLOGY. Educ: Rutgers Univ, BS, 50; Univ Minn, MS, 53, PhD(entom), 55. Prof Exp: Asst entom, Univ Minn, 50-55; asst prof, Univ Mo, 55-57; entomologist, Entom Res Div, Agr Res Serv, USDA, 57-60, entomologist in chg citrus insects invest, Fla, 60-62; res entomologist & invest leader, Humid Areas Citrus Insects Invest, 62-64, asst to chief, Fruit & Veg Insects Res Br, Entom Res Div, 64-68, asst to dir, 68-71, staff asst plant sci & entom, 71-72, RES LEADER, SUBTROP HORT RES UNIT, AGR RES SERV, USDA, 72- Concurrent Pos: Adj prof, Univ Fla, Gainesville, 72- Mem: Fel AAAS; Entom Soc Am. Res: Subtropical fruit flies; agricultural entomology; chemical and biological control of insects; fruit insects and mites; insect ecology and populations; effects of radiation and use of radioisotopes on insects; commodity treatments. Mailing Add: Subtrop Hort Res Unit 13601 Old Cutler Rd Miami FL 33158

BURDSALL, HAROLD HUGH, JR, b Hamilton, Ohio, Nov 18, 40. MYCOLOGY. Educ: Miami Univ, BA, 62; Cornell Univ, PhD(mycol), 67. Prof Exp: Botanist, Forest Dis Lab, Laurel, Md, 67-71, BOTANIST, FOREST PROD LAB, CTR FOREST MYCOL RES, FOREST SERV, USDA, 71- Concurrent Pos: Lectr, Dept Forest Prod, Univ Wis-Madison, 71-, adj asst prof, Dept Plant Path, 74- Mem: Mycol Soc Am; Brit Mycol Soc; Am Inst Biol Sci; Int Asn Plant Taxon. Res: Taxonomic and systematic mycology of Aphyllophorales and other Basidiomycetous fungi, including culture studies, genetics, morphology and ecology with emphasis on forest decay organisms. Mailing Add: Ctr for Forest Mycol Res Forest Prod Lab PO Box 5130 Madison WI 53705

BUREK, ANTHONY JOHN, b Rockville Ctr, NY, June 14, 46. SOLAR PHYSICS, X-RAY ASTRONOMY. Educ: Princeton Univ, AB, 68; Univ Chicago, MS, 69, PhD(physics), 75. Prof Exp: FEL PHYSICS, LOS ALAMOS SCI LAB, 75- Mem: Am Phys Soc. Res: Solar x-ray spectroscopy; theory and evaluation of x-ray diffracting properties of crystals. Mailing Add: Los Alamos Sci Lab Group P-4 MS 436 Box 1663 Los Alamos NM 87545

BUREN, LAWRENCE LAMONT, b Glenmont, Ohio, Feb 21, 40; m 65; c 1. CROP PHYSIOLOGY, PLANT PHYSIOLOGY. Educ: Ohio State Univ, BS, 65; Iowa State Univ, MS, 68, PhD(crop prod, plant physiol), 70. Prof Exp: Asst agronomist, 70-73, ASSOC AGRONOMIST, HAWAIIAN SUGAR PLANTERS' ASN, 73- Mem: Am Soc Agron; Crop Sci Soc Am; Am Soc Plant Physiol. Res: Plant studies with drip irrigation; plant spacing patterns; efficiency of controlled release fertilizers and chemical ripeners for sugarcane. Mailing Add: Agron Dept Hawaiian Sugar Planters' Asn PO Box 1057 Aiea HI 96701

BURES, DONALD JOHN (CHARLES), b Winnipeg, Man, Jan 23, 38; m 61; c 4. MATHEMATICS. Educ: Queen's Univ, Ont, BA, 58; Princeton Univ, PhD(math), 61. Prof Exp: Asst prof math, Queen's Univ, Ont, 61-62; from asst prof to assoc prof, 62-71, PROF MATH, UNIV BC, 71-, HEAD DEPT, 73- Mem: Am Math Soc; Can Math Cong; Royal Soc Can. Res: Abstract analysis; von Neumann algebras. Mailing Add: Dept of Math Univ of BC Vancouver BC Can

BURES, MILAN F, b Prague, Czech, May 6, 32; m 59. EPIDEMIOLOGY, PUBLIC HEALTH. Educ: Charles Univ, Prague, MD, 58; Inst Postgrad Training Physicians, Prague, Czech, 64, MPH, 64. Prof Exp: Physician, County Hosp & Munic Med Ctrs, Czech, 58-61; res worker, Inst Radiation Hyg, Prague, 61-65; asst prof epidemiol & health, McGill Univ, 68-70; asst med dir, Northwestern Mutual Life Ins Co, 70-75; MED DIR, STATE MUTUAL LIFE ASSURANCE CO AM, 75- Concurrent Pos: Res fel radiation biol, Med Ctr, Univ Ala, 66-68. Res: Life insurance medicine; epidemiology of cardiovascular diseases; vital statistics; radiological health. Mailing Add: State Mutual Life Assurance Co Worcester MA 01605

BURFENING, PETER J, b Reno, Nev, Nov 8, 42; m 64; c 1. REPRODUCTIVE PHYSIOLOGY, GENETICS. Educ: Colo State Univ, BS, 64; NC State Univ, MS, 66, PhD(animal physiol), 68. Prof Exp: ASSOC PROF ANIMAL PHYSIOL, MONT STATE UNIV, 60- Mem: AAAS; Am Soc Animal Sci; Soc Study Reproduction. Res: Reproductive physiology of domestic animals with special emphasis on the relationship between genotype and the environment as it relates to reproductive events. Mailing Add: Dept of Animal & Range Sci Mont State Univ Bozeman MT 59715

BURFORD, ARTHUR EDGAR, b Olean, NY, Mar 5, 28; m 53; c 6. STRUCTURAL GEOLOGY. Educ: Cornell Univ, AB, 52; Univ Tulsa, MS, 54; Univ Mich, PhD(geol), 60. Prof Exp: Geologist, Pan Am Petrol Corp, Utah & Colo, 58-60; from asst prof to assoc prof, 60-68; actg head dept, 70-71, PROF GEOL, UNIV AKRON, 68-, HEAD DEPT, 71- Mem: Geol Soc Am; Am Asn Petrol Geol; Am Soc Photogram; Am Geophys Union. Res: Geology of Rocky Mountains, Appalachians and Alps; determining structural relations by fracture analysis; genesis and tectonics of mountains. Mailing Add: Dept of Geol Univ of Akron Akron OH 44325

BURFORD, HUGH JONATHAN, b Memphis, Tenn, Aug 5, 31; m 57; c 2. PHARMACOLOGY, MEDICAL EDUCATION. Educ: Millsaps Col, BS, 54; Univ Miss, MS, 56; Univ Kans, PhD(pharmacol), 62. Prof Exp: Asst chem, Univ Miss, 54-

56; asst pharmacol, Univ Kans, 57-60; asst prof, Bowman Gray Sch Med, 63-68; assoc prof pharmacol & dir MDL labs, Med Sch, Northwestern Univ, 68-71; assoc prof pharmacol & teaching assoc med sci, 71-73, ASSOC PROF PHARMACY, PHARMACY SCH, UNIV NC, CHAPEL HILL, 73- Concurrent Pos: USPHS fel, Tulane Univ, 62-63; Am Heart Asn, USPHS & Nat Fund Med Educ grants, 64-; fac fel, Kellogg Ctr Teaching Professions, 70-71. Mem: AAAS; NY Acad Sci; Am Soc Pharmacol & Exp Therapeut. Res: Structure function relationships in pharmacology; developmental pharmacology; optimizing education in pharmacology. Mailing Add: 201 Swing Bldg Univ of NC Sch of Med Chapel Hill NC 27514

BURFORD, MORTIMER GILBERT, b Brooklyn, NY, Oct 27, 10; m 34. PHYSICAL CHEMISTRY. Educ: Wesleyan Univ, AB, 32; Princeton Univ, AM, 33, PhD(chem), 35. Prof Exp: Asst chem, Princeton Univ, 33-34; instr, Cornell Univ, 35-36; from instr to assoc prof, 36-47, E B NYE PROF CHEM, WESLEYAN UNIV, 47-, ASSOC PROVOST, 69- Concurrent Pos: Consult & supvr, Conn State Dept Environ Protection, 42- & New Eng Interstate Water Pollution Control Comn, 49-; Fund Advan of Educ fac fel, 55-56; chmn bd dirs, Univ Res Inst Conn, 74- Mem: Fel AAAS; Am Chem Soc; Hist of Sci Soc; fel Am Inst Chemists. Res: Chemical reactions of hydriodic acid and ammonium iodide; analytical chemistry of difficulty soluble compounds; spectrophotometric methods of analysis; industrial trade waste disposal; chemical kinetics. Mailing Add: Hall Lab Wesleyan Univ Middletown CT 06457

BURFORD, THOMAS MAYNARD, b Independence, Kans, July 1, 29; m 52; c 1. MATHEMATICS. Educ: Univ Wis, PhD(elec eng), 55. Prof Exp: Res mathematician, Bell Tel Labs, Inc, 55-68; DIR SYSTS ANAL, SANDIA CORP, 68- Mem: Am Math Soc. Res: Communication and information theory. Mailing Add: Sandia Corp Albuquerque NM 87115

BURFORD, WILLIAM BERRYMAN, physical chemistry, deceased

BURG, ALAN WALTER, b Newark, NJ, Mar 11, 42; m 63. BIOCHEMISTRY. Educ: Cornell Univ, BS, 63; Mass Inst Technol, PhD(biochem), 67. Prof Exp: Sr scientist, 67-71, dir biochem lab, 71-72, SR CONSULT, ARTHUR D LITTLE, INC, 72- Mem: AAAS. Res: Enzymology applied to chemical, pharmaceutical, agricultural and environmental problems; clinical laboratory product development and evaluation; physiological disposition studies; cancer management; research applied to the improvement of diagnostic and therapeutic concepts. Mailing Add: Arthur D Little Inc Acorn Park Cambridge MA 02140

BURG, ANTON BEHME, b Dallas City, Ill, Oct 18, 04. INORGANIC CHEMISTRY. Educ: Univ Chicago, BS, 27, MS, 28, PhD(chem), 31. Prof Exp: Instr inorg chem, Univ Chicago, 31-39; from asst prof to assoc prof, 39-43, head dept, 40-50, PROF CHEM, UNIV SOUTHERN CALIF, 43- Concurrent Pos: Consult var govt agencies, 47- Honors & Awards: Tolman Medal, Am Chem Soc, 61, Award Distinguished Serv Advan Inorg Chem, 69. Mem: Fel AAAS (chmn, 54); Am Chem Soc. Res: Boron and silicon hydrides; fluorine compounds; vacuum technique; non-aqueous solvents and addition compounds; organo-phosphorus and inorganic polymers; metal-carbonyl analogues; fluorocarbon phosphines; study of new syntheses and compound-types in both fields. Mailing Add: Dept of Chem Univ of Southern Calif Los Angeles CA 90007

BURG, FREDRIC DAVID, b Chicago, Ill, May 23, 40; m 67; c 5. MEDICAL EDUCATION, PEDIATRICS. Educ: Miami Univ, BA, 61; Northwestern Univ, MD, 65. Prof Exp: Asst prof pediat, Med Sch, Northwestern Univ, 70-71; assoc dir bd, 71-75, DIR GRAD DEPT, NAT BD MED EXAMR, 75- Concurrent Pos: Mem, Am Bd Med Specialties, 70-, mem comt specialty bd eval procedures, 73-; consult, Am Bd Allergy & Immunol, 73-; secy staff res comt, Nat Bd Med Examr, 74-; proj dir comt to study role of prog dir & recert comt pediat, Am Bd Pediat, 74-, dir eval & res, 75-; clin asst prof pediat, Univ Pa, 74-76, adj assoc prof, 75- Mem: Ambulatory Pediat Asn; Am Educ Res Asn; Asn Am Med Cols. Res: Use of cognitive examinations to measure physician competency; use of medical audit, computers, simulators and rating scales to measure physician performance; development of methods to define competency in medical disciplines. Mailing Add: Nat Bd of Med Examr 3930 Chestnut St Philadelphia PA 19104

BURG, MARION, b Bridgeport, Conn, May 25, 21. ORGANIC CHEMISTRY. Educ: Queens Col, BS, 42; Cornell Univ, PhD(org chem), 47. Prof Exp: Asst instr chem,Queens Col, 42-44; res assoc org chem, Mass Inst Technol, 47-51; WITH E I DU PONT DE NEMOURS CO, 51- Mem: Am Chem Soc. Res: Synthetic organic chemistry; photochemistry. Mailing Add: 1911 Mt Vernon Lane Wilmington DE 19806

BURG, MAURICE B, b Boston, Mass, Apr 9, 31; m 66; c 4. NEPHROLOGY. Educ: Harvard Univ, BA, 52, MD, 55. Prof Exp: Res assoc, NIH, 57-75, CHIEF LAB OF KIDNEY & ELECTROLYTE METAB, NIH, 75- Mem: Am Physiol Soc; Am Soc Clin Invest; Am Fedn Clin Res; Biophys Soc; Soc Gen Physiol. Res: Renal and electrolyte physiology. Mailing Add: Lab Kidney & Electrolyte Metab Nat Insts of Health Bethesda MD 20014

BURG, RICHARD WILLIAM, b Ft Wayne, Ind, June 22, 32; m 61; c 2. MICROBIAL BIOCHEMISTRY. Educ: Wabash Col, AB, 54; Univ Ill, PhD(biochem), 58. Prof Exp: NSF res fel, Univ Calif, Berkeley, 58-60; res fel, 60-75, SR RES FEL, MERCK INST THERAPEUT RES, 75- Mem: AAAS; Am Chem Soc; Am Soc Microbiol; NY Acad Sci; Brit Soc Gen Microbiol. Res: Biochemistry of viruses; antibiotics and physiologically active products of microorganisms. Mailing Add: Merck Inst for Therapeut Res Rahway NJ 07065

BURG, STANLEY (PAUL), b Boston, Mass, Mar 17, 33; m 54; c 2. PLANT PHYSIOLOGY. Educ: Harvard Univ, BA, 54, PhD(biol), 59. Prof Exp: Instr biol, Harvard Univ, 58-60; asst prof physiol, 60-66, assoc prof med, biochem & biol, 66-70, adj assoc prof biol, 70-74, ADJ PROF BIOL, UNIV MIAMI, 74-; PLANT PHYSIOLOGIST, FAIRCHILD TROP GARDEN, 69- Mem: Am Soc Plant Physiol. Res: Growth hormones; post-harvest physiology; photobiology. Mailing Add: 4075 Malaga Ave Miami FL 33133

BURG, WILLIAM ROBERT, b Venango, Nebr, Aug 18, 29; m 61; c 2. INDUSTRIAL HYGIENE, ANALYTICAL CHEMISTRY. Educ: Nebr State Teachers Col, BS, 59; Univ Nebr, MS, 61; Kent State Univ, PhD(chem), 64. Prof Exp: Asst prof chem, Eastern NMex Univ, 64-67 & Nichols State Col, 67-69; ASST PROF ENVIRON HEALTH, UNIV CINCINNATI, 69- Mem: Am Chem Soc; Am Indust Hyg Asn. Res: Development of analytical methodology for the evaluation of the industrial environment. Mailing Add: Dept of Environ Health Univ of Cincinnati Cincinnati OH 45267

BURGAUER, PAUL DAVID, b St Gallen, Switz, May 21, 26; nat US; m 51; c 2. ORGANIC CHEMISTRY, PHARMACEUTICAL CHEMISTRY. Educ: Swiss Fed Inst Technol, Dipl Ing, PhD(org & pharmaceut chem), 52. Prof Exp: Res chemist, Hilton-Davis Chem Co, 53-54; patent liaison, E I du Pont de Nemours & Co, 54-59; WITH PATENT DEPT, ABBOTT LABS, 59- Concurrent Pos: Patent agent, 61- Res: Polymer chemistry; natural and synthetic drugs and polymers. Mailing Add: 1110 Woodview Dr Libertyville IL 60048

BURGE, BOYCE WILLIAM, virology, cell biology, see 12th edition

BURGE, DAVID E, physical chemistry, see 12th edition

BURGE, DENNIS KNIGHT, b Ogden, Utah, Dec 8, 35; m 65; c 2. PHYSICAL OPTICS. Educ: Univ Nev, BS, 56, MS, 61. Prof Exp: PHYSICIST, NAVAL WEAPONS CTR, 57- Mem: Optical Soc Am; Sigma Xi. Res: Optical properties of solids; thin films; ellipsometry. Mailing Add: Code 6018 Michelson Lab Naval Weapons Ctr China Lake CA 93555

BURGE, ROBERT ERNEST, JR, b Kansas City, Mo, Aug 21, 25; m 50. ORGANIC CHEMISTRY. Educ: Univ Ill, BS, 48; Cornell Univ, PhD(chem), 52; Bernard M Baruch Col, MBA, 69. Prof Exp: Asst org chem, Cornell Univ, 48-51; res chemist, Shell Chem Corp, 51-54, group leader, 54-58, sr chemist, 57-59, sr technologist, 59-60, sr chemist, 60-63, supvr, 63-68; asst prof, 68-72, ASSOC PROF CHEM, SUFFOLK COUNTY COMMUNITY COL, 72- Mem: Am Chem Soc; Nat Sci Teachers Asn. Res: Organic chemical synthesis; macrocyclic carbon compounds; epoxy resins. Mailing Add: 43 Huntington Rd Garden City NY 11530

BURGE, WYLIE D, b Denver, Colo, June 2, 25; m 58; c 2. MICROBIOLOGY, BIOCHEMISTRY. Educ: Colo State Univ, BS, 50, MS, 56; Univ Calif, Davis, PhD(soil microbiol), 60. Prof Exp: Jr soil scientist soil microbiol, Univ Calif, Berkeley, 56-60; assoc specialist soils res & sta supt, Antelope Valley Field Sta, Univ Calif, Riverside, 60-66; res soil scientist, Soils & Water Conserv Serv, 66-73, RES SOIL SCIENTIST, AGR ENVIRON QUAL INST, AGR RES SERV, USDA, 73- Mem: Soil Sci Soc Am; Am Soc Microbiol. Res: Ammonium fixation by soil organic matter; biochemistry of nitrification and chemistry of potassium fixation by soils; microbiology of pesticide degradation in soils. Mailing Add: Agr Environ Qual Inst Agr Res Serv USDA Beltsville MD 20705

BURGENER, FRANCIS ANDRE, b Visp, Switz, May 21, 42; m 70; c 1. RADIOLOGY. Educ: Univ Bern, MD, 67, Diss, 69; Am Bd Radiol, dipl & cert diag radiol, 72. Prof Exp: Resident radiol, Univ Bern, 67-68; fel exp med, Univ Zurich, 68-69; resident radiol, Univ Bern, 69-70; instr, Univ Mich, 70-71; asst prof, 71-76, ASSOC PROF RADIOL, UNIV ROCHESTER, 76- Concurrent Pos: Panelist, US Pharmacopeia Adv Panel, 75- Mem: Asn Univ Radiologists. Res: Radiological contrast media excretion and toxicity. Mailing Add: Dept of Radiol Univ of Rochester Med Ctr Rochester NY 14642

BURGER, ALFRED, b Vienna, Austria, Sept 6, 05; nat US; m 36; c 1. MEDICINAL CHEMISTRY. Educ: Univ Vienna, PhD(chem), 28. Honors & Awards: Hon DSci, Philadelphia Col Pharm & Sci, 71. Prof Exp: Chemist, Hoffmann-La Roche Co, 28-29; res assoc, Drug Addiction Lab, Nat Res Coun, 29-38, actg asst prof, 38-39, from asst prof to assoc prof, 39-52, prof, 52-70, chmn dept, 62-63, EMER PROF CHEM, UNIV VA, 70- Concurrent Pos: USPHS, 56-59; mem chem panel, Cancer Chemother Nat Serv Ctr, 56-59, med chem, 60-64; vchmn, Gordon Res Conf Med Chem, 58, chmn, 59; mem study sect on exp ther & pharmacol; vis lectr, Univ Calif, 63; NIH spec fel biochem, Univ Hawaii, 65; consult, Smith Kline & French Labs & Philip Morris Res Ctr; ed, J Med Chem; mem psychopharmacol study comt, Nat Inst Ment Health, 67-71. Mem: Am Chem Soc; Am Pharmacol Soc. Res: General organic chemistry; chemistry of opium alkaloids; syntheses of morphine substitutes; chemotherapy; antimalarials; antituberculous drugs; organic phosphorus compounds; antimetabolites; psychopharmacological drugs; synthesis and design of drugs. Mailing Add: Dept of Chem Univ of Va Charlottesville VA 22901

BURGER, AMBROSE WILLIAM, b Jasper, Ind, Nov 27, 23; m 46, 67; c 6. AGRONOMY, SCIENCE EDUCATION. Educ: Purdue Univ, BSA, 47; Univ Wis, MSA, 48, PhD(agron, plant physiol), 50. Prof Exp: Asst prof agron, Univ Md, 50-53; PROF AGRON, UNIV ILL, URBANA-CHAMPAIGN, 53- Honors & Awards: Agron Educ Award, Am Soc Agron, 65. Mem: Crop Sci Soc Am; AAAS; fel Am Soc Agron. Res: Forage crop production; pasture investigations; field crop science. Mailing Add: Dept of Agron Univ of Ill Urbana IL 61801

BURGER, CHARLES L, b Joliet, Mont, Feb 1, 27; m 52; c 4. RADIOBIOLOGY, VIROLOGY. Educ: Univ Ill, BS, MS, 51, PhD(zool), 58. Prof Exp: Teacher high sch, Univ Ill, 51-55; asst, Univ Ill, 55-58; asst prof biol, Wilson Col, 58-62; mem staff, Biol Div, Union Carbide Oak Ridge Nat Lab, 62-63 & virol sect, Chas Pfizer & Co, 63-64; asst prof radiol, 64-68, ASSOC PROF RADIOL & BIOCHEM, STATE UNIV NY UPSTATE MED CTR, 68- Mem: AAAS; Am Soc Cell Biol; Biophys Soc; Am Soc Microbiol; Am Asn Cancer Res. Res: Virology; virus isolation and characterization; subcellular fractionation; radiation effects. Mailing Add: Dept of Radiol Univ NY Upstate Med Ctr Syracuse NY 13210

BURGER, DIONYS, b Ambarawa, Indonesia, May 29, 23; Can citizen; m 52; c 2. FOREST ECOLOGY, SOILS. Educ: State Agr Univ Wageningen, MF, 52; Univ Toronto, PhD(bot), 65. Prof Exp: Forester, 53-61, RES SCIENTIST FOREST SOIL, RES BR, ONT MINISTRY NATURAL RESOURCES, 61- Concurrent Pos: Mem tech comt, Interprov Forest Fertil Prog, 68-; leader, Working Group Site Classification, Int Union Forest Res Orgn, 70- Mem: Can Inst Forestry; Can Soc Soil Sci; Int Soc Soil Sci; Ger Soc Forest Site Sci & Tree Breeding. Res: Physiographic site classification; mapping on aerial photographs; surface geology; forest nutrition and humus; soil weathering. Mailing Add: Ont Ministry of Nat Res Maple ON Can

BURGER, FRANCIS JOSEPH, analytical chemistry, see 12th edition

BURGER, GEORGE VANDERKARR, b Woodstock, Ill, Jan 22, 27; m 49; c 3. WILDLIFE CONSERVATION. Educ: Beloit Col, BS, 50; Univ Calif, MA, 52; Univ Wis, PhD(wildlife mgt), 59. Prof Exp: Asst zool & wildlife mgt, Univ Calif, 50-52; instr zool, bot & conserv, Contra Costa Jr Col, 52-54; asst wildlife mgt, Univ Wis, 54-58; field rep, Sportsmen's Serv Bur, NY, 58-62; mgr wildlife mgt, Remington Arms Co, 62-66; GEN MGR, McGRAW WILDLIFE FOUND, 66- Concurrent Pos: Ed, Wildlife Soc Bull, Wildlife Soc, 72-75. Honors & Awards: Leopold Award, Green Tree Club, 58. Mem: Wildlife Soc; Am Fisheries Soc. Res: Waterfowl and upland game bird management and ecology; shooting preserve and game farm management; wildlife and general conservation education. Mailing Add: PO Box 194 Dundee IL 60118

BURGER, HENRY G, b New York, NY, June 27, 23. CULTURAL ANTHROPOLOGY. Educ: Columbia Univ, BA, 47, MA, 65, PhD(cult anthrop), 67. Prof Exp: Pvt pract soc sci consult, NY, 56-67; anthropologist, Southwestern Coop Educ Lab, NMex, 67-69; assoc prof, 69-73, PROF ANTHROP & EDUC, UNIV MO-KANSAS CITY, 73- Concurrent Pos: Lectr innovational strategy, City Univ New York, 57-65; mem, Panel, var anthrop confs, 60-; adj prof educ anthrop, Univ NMex, 69; assoc, Coun Anthrop & Educ, 68-; NSF fac res grant, Univ Mo-Kansas City, 70-

71; consult, Vet Admin Hosp, Kansas City, Mo, 71-72; founding mem, Doc Fac, Univ Mo, 74- Mem: Fel Soc Appl Anthrop; fel Int Union Anthrop Ethnol Sci; fel World Acad Art & Sci; fel Am Anthrop Asn; fel Royal Anthrop Inst Gt Brit & Ireland. Res: Codification of the dynamics of monocultural and transcultural learning and of mental health; parsing language by cause and effect into a word tree; all with cultural-materialistic emphasis. Mailing Add: 7306 Brittany Shawnee Mission KS 66203

BURGER, HENRY ROBERT, III, b Pittsburgh, Pa, Aug 2, 40; m 63. STRUCTURAL GEOLOGY. Educ: Yale Univ, BS, 62; Ind Univ, AM, 64, PhD(geol), 66. Prof Exp: From asst prof to assoc prof, 66-75, PROF GEOL, SMITH COL, 75- Mem: AAAS; Am Geophys Union; Geol Soc Am; Nat Asn Geol Teachers. Res: Rock mechanics; structural analysis; petrofabrics. Mailing Add: Dept of Geol Smith Col Northampton MA 01060

BURGER, J C, analytical chemistry, see 12th edition

BURGER, JAMES WENDELL, b Philadelphia, Pa, Mar 1, 10; m 37; c 2. ZOOLOGY. Educ: Haverford Col, AB, 31; Lehigh Univ, AM, 33; Princeton Univ, PhD(zool), 36. Prof Exp: Asst biol, Lehigh Univ, 31-33; from instr to prof, 36-74, EMER PROF BIOL, TRINITY COL, CONN, 74- Concurrent Pos: Instr, Mt Desert Biol Lab, 38-41, trustee, 40-68, dir, 47-50, vpres, 63-68; chmn, Sch Nursing & dir, Hartford Hosp, 44-68; comnr, Conn State Geol & Natural Hist Surv, 61- Mem: AAAS; Am Soc Zoologists. Res: Comparative physiology; sex cycles of fish, amphibia, reptiles and birds, and factors which control same; hemodynamics; herpetology; sanitation. Mailing Add: 21 Glenbrook Rd West Hartford CT 06107

BURGER, JOANNA, b Schenectady, NY, Jan 18, 41. ETHOLOGY, ECOLOGY. Educ: State Univ NY Albany, BS, 63; Cornell Univ, MS, 64; Univ Minn, PhD(ecol & behav), 72. Prof Exp: Instr biol, State Univ NY Buffalo, 64-68; asst, Univ Minn, 68-72; res assoc ethology, Rutgers Inst Animal Behav, 72-73; ASST PROF BIOL, RUTGERS UNIV, 73- Mem: Am Ornithologists Union; Animal Behav Soc; Ecol Soc; AAAS. Res: Examination of the relationship between an animal's environment and its behavior, particularly in the ways animals partition the environment, such as food resources and nesting habitat. Mailing Add: Dept of Biol Livingston Col-Rutgers Univ New Brunswick NJ 08903

BURGER, JOHN ALLAN, b Price, Utah, Aug 19, 28; m 52; c 7. GEOLOGY. Educ: Univ Utah, BS, 52, MS, 55; Yale Univ, PhD(geol), 59. Prof Exp: Geologist, Texaco, Inc, 58-61; from asst prof to assoc prof, 61-72, PROF GEOL, BELOIT COL, 72- Mem: Soc Econ Paleont & Mineral; Geol Soc Am. Res: Stratigraphy and sedimentation. Mailing Add: 2658 E Collingswood Dr Beloit WI 53511

BURGER, JOHN MARTIN, b Niotaze, Kans, Nov 1, 15; m 43; c 2. EXPERIMENTAL NUCLEAR PHYSICS. Educ: Univ Kans, BS, 39, MS, 40, PhD, 54. Prof Exp: Instr math, Cent Mo State Col, 40-41; instr math & physics, Kemper Mil Acad, 41-42; asst prof physics, Sch Mines, Univ Mo, 46-51; asst physics, Univ Kans, 51-54; asst prof physics, Kans State Teachers Col, 54-57, from asst prof to assoc prof math, 57-59, head dept, 58-61, prof, 59-70; ASSOC CHMN MATH DEPT, EMPORIA KANS STATE COL, 70- Mem: Nat Coun Teachers Math; Math Asn Am; Am Asn Physics Teachers. Res: Nuclear magnetic resonance; teacher training. Mailing Add: Dept of Math Emporia Kans State Col Emporia KS 66801

BURGER, LELAND LEONARD, b Buffalo, Wyo, Nov 5, 17; m 42; c 3. PHYSICAL INORGANIC CHEMISTRY. Educ: Univ Wyo, BA, 39; Univ Wash, PhD(chem), 48. Prof Exp: Res chemist, Div War Res, Columbia Univ, 42-45 & Hanford Labs, Gen Elec Co, 48-64; RES ASSOC CHEM, PAC NORTHWEST LAB, BATTELLE MEM INST, 65- Mem: Fel AAAS; Am Chem Soc; Am Inst Physics; Am Nuclear Soc. Res: Solvent extraction mechanisms, absorption spectroscopy; actinide elements; nuclear fuel reprocessing; fluorine chemistry; radiation chemistry. Mailing Add: 1925 Howell Ave Richland WA 99352

BURGER, MAX M, biochemistry, see 12th edition

BURGER, OTHMAR JOSEPH, b Jasper, Ind, May 23, 21; m 43; c 3. AGRONOMY. Educ: Purdue Univ, BS, 43, MS, 47, PhD(agr chem), 50. Prof Exp: Asst agron, Purdue Univ, 46-47, asst agr chem, 47-50; from asst prof to prof agron, WVa Univ, 50-57; assoc prof, Iowa State Col, 57-59; prof, WVa Univ, 59-69, asst dean col Agr & forestry & dir resident instr, 59-68, asst to provost for instr, 68-69; PROF AGRON & DEAN SCH AGR SCI, CALIF STATE UNIV, FRESNO, 69- Concurrent Pos: From asst agronomist to agronomist, Agr Exp Sta, WVa Univ, 50-57. Mem: Fel Am Soc Agron; Crop Sci Soc Am. Res: Crop physiology; chemistry. Mailing Add: Sch of Agr Sci Calif State Univ Fresno CA 93740

BURGER, RICHARD MELTON, b New York, NY, Mar 23, 41. MOLECULAR BIOLOGY, BIOCHEMISTRY. Educ: Adelphi Col, BA, 62; Princeton Univ, PhD(biol), 69. Prof Exp: Spec trainee biochem, Brandeis Univ, 65-68; asst prof biol, MidE Tech Univ, Turkey, 71-72; ASSOC, SLOAN-KETTERING INST CANCER RES, 72-, ASST PROF BIOCHEM, SLOAN-KETTERING DIV, CORNELL UNIV, 73- Concurrent Pos: Nat Cancer Inst trainee, Univ Calif, Berkeley, 68-71; res assoc, Haskins Labs, New York, 59-. Mem: Soc Biol Chem; Harvey Soc. Res: Mechanism and regulation of DNA synthesis; inherited susceptibility to cancer. Mailing Add: Dept of Biochem Cornell Univ Grad Sch 1275 York Ave New York NY 10021

BURGER, ROBERT M, b Frederick, Md, Feb 14, 27; m 49; c 3. SOLID STATE PHYSICS, ELECTRONICS. Educ: Col William & Mary, BS, 49; Brown Univ, ScM, 52, PhD(physics), 56. Prof Exp: Physicist, US Dept Defense, 56-59; fel eng, Solid State Lab, Air Arm Div, Westinghouse Elec Corp, 59-62; dir solid state lab, 62-67, dir eng & environ sci div, 67-71, CHIEF SCIENTIST, RES TRIANGLE INST, 71- Concurrent Pos: Res affiliate, Univ Md, 56-61; adj assoc prof elec eng, Duke Univ, 62-69; chmn scientific review comt, Dental Res Ctr, Univ NC, 74- Mem: AAAS; Am Phys Soc; Inst Elec & Electronics Eng; Am Inst Aeronaut & Astronaut; Am Defense Preparedness Asn. Res: Solid state electronics and physics; electronic system; materials; air pollution; oceanography; instrumentation; signal and data processing; biomedical engineering; sensors. Mailing Add: PO Box 12194 Res Triangle Inst Res Triangle Park NC 27709

BURGER, ROBERT THORNTON, b Newark, NJ, Feb 25, 42; m 68; c 2. COMPUTER SCIENCE. Educ: Montclair State Col, BA, 64; Pa State Univ, MA, 66, PhD(comput sci), 69. Prof Exp: Sr analyst, 68-75, SR SCIENTIST, THE BDM CORP, 75- Mem: Asn Comput Mach. Res: Approximation theory-partitioned norms; development of automated simulation modeling systems. Mailing Add: The BDM Corp 1920 Aline Ave Vienna VA 22180

BURGER, WARREN CLARK, b Ripon, Wis, Feb 11, 23; m 44; c 2. PLANT SCIENCE. Educ: Univ Wis, BS, 48, MS, 50, PhD(biochem), 52. Prof Exp: Res chemist, 52-73, DIR, BARLEY & MALT LAB, AGR RES SERV, USDA, 73- Mem: Am Soc Plant Physiol; Am Chem Soc; Am Soc Brewing Chemists. Res: Proteolytic enzymes; seed germination; plant proteins; nutritional aspects of barley. Mailing Add: Barley & Malt Lab USDA 501 N Walnut St Madison WI 53705

BURGER, WILLIAM CARL, taxonomy, see 12th edition

BURGERS, JOHANNES MARTINUS, b Arnhem, Netherlands, Jan 13, 95; m 19, 41; c 3. PHYSICS, FLUID DYNAMICS. Educ: State Univ Leiden, Dr(math & phys sci), 18. Hon Degrees: Dr, Free Univ Brussels, 48, Univ Poitiers, 50. Prof Exp: Prof aerodyn & hydrodyn, Univ Delft, 18-55; res prof, 55-75, EMER RES PROF, INST FLUID DYNAMICS & APPL MATH, UNIV MD, COLLEGE PARK, 75- Honors & Awards: Panetti Medal, Acad Sci Turin, 61; Bingham Medal, Soc Rheol, 64; Gold Medal, Am Soc Mech Eng, 65. Mem: Royal Netherlands Acad Sci; Royal Inst Eng; for mem Acad Sci Turin; Am Inst Physics; Am Geophys Union. Res: Gas dynamics and kinetic theory of gases; magnetogasdynamics. Mailing Add: Inst Fluid Dynamics & Appl Math Univ of Md College Park MD 20742

BURGERT, BILL E, b Horton, Kans, Aug 9, 29. ORGANIC CHEMISTRY. Educ: Kans State Teachers Col, AB, 51; Kans State Col, MS, 52; Northwestern Univ, PhD(chem), 55. Prof Exp: Res chemist, 55-58, proj leader, 58-60, group leader, 60-64, div leader, 64-68, DIR PHYS RES LAB, DOW CHEM CO, 68- Mem: Am Chem Soc; Sigma Xi. Res: Polymer chemistry. Mailing Add: Dow Chem Co 1712 Bldg Midland MI 48640

BURGESON, ROBERT EUGENE, b Newhall, Calif, Aug 5, 45. MOLECULAR BIOLOGY, MEDICAL GENETICS. Educ: Univ Calif, Irvine, BS, 68; Univ Calif, Los Angeles, PhD(molecular biol), 74. Prof Exp: Res biochemist molecular biol, Sch Dent, 73-74, ASST RES BIOCHEMIST MED GENETICS, HARBOR GEN HOSP, UNIV CALIF, LOS ANGELES, 76- Concurrent Pos: NIH fel, 74-76. Mem: AAAS; Am Soc Cell Biol. Res: Investigations of the structure and function of connective tissue macromolecules with regard to human growth, development and inheritable disease. Mailing Add: Div of Med Genetics E-4 Harbor Gen Hosp Univ of Calif Torrance CA 90509

BURGESS, BENJAMIN FRANKLIN, JR, b Duluth, Ga, Dec 9, 21; m 45; c 3. PHYSIOLOGY, BIOCHEMISTRY. Educ: Mercer Univ, AB, 46; Univ Pa, MS, 55. Prof Exp: Head dept chem, Med Field Res Lab, Med Serv Corps, US Navy, 48-52, res biochemist, Aviation Med Acceleration Lab, Pa, 52-53, head aerospace med div, 54-59, liaison officer, Royal Air Force Inst Aviation Med, Eng, 59-61, dept dir, Aviation Med Acceleration Lab, 61-64, dir, 64; dir res, Philadelphia Gen Hosp, 64-68; DIR RES & PLANNING, PRESBY-UNIV PA MED CTR, 68- Concurrent Pos: Mem aerospace med panel, Adv Group Aeronaut Res & Develop, NATO, 59-; mem comt hearing, bioacoustics & biomech, Nat Res Coun, 64-; mem adv standardization comt, Aerospace & Life Support Systs for US, Can & Gt Brit, 64-; bd mem, West Philadelphia Corp, Pa Anvil Res Assocs, Can. Mem: AAAS; Aerospace Med Asn; Am Chem Soc; fel Am Inst Chemists; Soc Res Admin. Res: Aerospace medicine, particularly the cardiopulmonary effects of acceleration and problems of oxygen toxicity as related to space flight. Mailing Add: Presby-Univ of Pa Med Ctr 51 N 39th St Philadelphia PA 19104

BURGESS, CECIL EDMUND, b Happy, Tex, Jan 21, 20; m 48; c 2. MATHEMATICS. Educ: WTex State Univ, BS, 41; Univ Tex, PhD(math), 51. Prof Exp: Instr math, Univ Tex, 41-42; with Naval Ord Lab, 42-43; instr math, Univ Tex, 46-51; from instr to assoc prof, 51-61, PROF MATH, UNIV UTAH, 61-, CHMN DEPT, 67- Concurrent Pos: Vis lectr, Univ Wis, 56-57; vis mem, Inst Advan Study, 62-63. Mem: Am Math Soc; Math Asn Am. Res: Point set topology. Mailing Add: 2236 Logan Ave Salt Lake City UT 84108

BURGESS, CHARLES H, b Sheridan, Wyo, Apr 3, 10; m 34, 61; c 7. GEOLOGY. Educ: Harvard Univ, AB, 31, AM, 33, PhD(geol), 36. Prof Exp: Instr geol, Harvard Univ, 34-36; geologist, Anaconda Copper Mining Co, 36-38; self-employed mine lessee & consult, 38-41; analyst & dep chief, Admin Export Control Bd, Econ Warfare-Aluminum & Magnesium Sect, Off Price Admin, 41-42; chief wire, Rod & Bar Sect, Aluminum & Magnesium Div, War Prod Bd, 42-44; geologist, Hoover, Curtice & Ruby, NY, 44-46 & M A Hanna Co, 46-47; from dep dir to dir, Strategic Mat Div, Econ Coop Admin, 48-50; treas, United Elec Coal Co, Ill, 50-52; dist geologist, Bear Creek Mining Co, Minn, 52-56, pres & dir, NY & Utah, 56-60; vpres explor, Kennecott Copper Corp, 60-74; RETIRED. Mem: Am Inst Mining, Metall & Petrol Eng; Mining & Metall Soc Am; Soc Econ Geol. Mailing Add: 319 Church St Abbeville SC 29620

BURGESS, DAVID RAY, b Hobbs, NMex, Nov 21, 47. DEVELOPMENTAL BIOLOGY. Educ: Calif Polytech State Univ, BS, 69, MS, 71; Univ Calif, Davis, PhD(zool), 74. Prof Exp: RES ASSOC ZOOL, UNIV WASH, 74- Concurrent Pos: NIH fel, Univ Wash, 74- Mem: AAAS; Am Soc Zoologists; Sigma Xi; Soc Develop Biol. Res: Cellular and tissue morphogenesis; ultrastructure and biochemistry of motility in single cells and organ development; the role of calcium in the regulation of contractile proteins in non-muscle motile cells. Mailing Add: Friday Harbor Labs Friday Harbor WA 98250

BURGESS, EDWARD MEREDITH, b Birmingham, Ala, June 8, 34; m 57; c 2. ORGANIC CHEMISTRY. Educ: Auburn Univ, BS, 56; Mass Inst Technol, PhD(org chem), 62. Prof Exp: Instr chem, Yale Univ, 62-64; from asst prof to assoc prof, Ga Inst Technol, 64-74, PROF CHEM, GA INST TECHNOL, 74- Mem: Am Chem Soc; The Chem Soc. Res: Organic photochemistry; small ring heterocycle synthesis; new functional groups. Mailing Add: Dept of Chem Ga Inst of Technol Atlanta GA 30332

BURGESS, HOVEY MANN, b Stoneham, Mass, Oct 19, 16; m 39; c 1. FOOD SCIENCE, NUTRITION. Educ: Bowdoin Col, BS, 38; Columbia Univ, AM, 40. Prof Exp: From jr chemist to sect head, Cent Labs, Gen Foods Corp, 40-52, mgr res & develop, Gaines Div, 52-57, lab mgr, Post div, 57-61, mgr tech eval, 61-65, group res mgr technol, 65-69, dir tech appln, 69-70, fel basic sci, 70-74; CONSULT FOOD SCI, 74- Mem: AAAS. Res: Animal nutrition; dehydration; soluble coffee; dessert products; process for dehydrated potato product; cereals. Mailing Add: 555 Port Side Dr Naples FL 33940

BURGESS, JACK D, b Moline, Ill, July 16, 24; m 52; c 2. GEOLOGY, BOTANY. Educ: Univ Ill, BS, 49; Univ Mo, MA, 55. Prof Exp: Inspector qual control, Deere & Co, 50-52; mining geologist, Northern Pac RR, 55-56; geologist, sub-surface geol, 56, field geol, 56-59, paleontologist, 59-72, SR RES GEOLOGIST, GULF RES & DEVELOP CO, 72- Mem: Am Asn Petrol Geol; Soc Econ Paleont & Mineral; Brit Paleont Asn; Am Asn Stratig Palynol. Res: Fossil acid-insoluble palynomorphs from the entire geologic column within the Rocky Mountain Region; integrating microscopic kerogen description and thermal maturation with other hydrocarbon source rock studies in the Geochemistry section. Mailing Add: PO Drawer 2038 Pittsburgh PA 15230

BURGESS, JAMES HARLAND, b Portland, Ore, May 11, 29; m 51; c 3. PHYSICS. Educ: State Col Wash, BS, 49, MS, 51; Wash Univ, PhD(physics), 55. Prof Exp: Sr engr, Sylvania Elec Prod, Inc, 55-56; res assoc physics, Stanford Univ, 56-57, from instr to assoc prof, 57-73, PROF PHYSICS, WASH UNIV, 73- Mem: Am Phys Soc; Am Asn Physics Teachers. Res: Nuclear magnetic resonance; paramagnetic resonance; magnetic cooperative phenomena; biophysics. Mailing Add: Dept of Physics Wash Univ St Louis MO 63130

BURGESS, KENNETH ALEXANDER, b Stamford, Conn, June 27, 18; m 42; c 4. PHYSICAL CHEMISTRY. Educ: Princeton Univ, AB, 39; Pa State Univ, MS, 41. Prof Exp: Res chemist, Columbian Carbon Co, 47-52, chief chemist, 52-56, asst dir res, 56-61, dir carbon black res, 61-63, carbon & elastomers res, 63-68 & Columbian Div res, 68-71, DIR PETROCHEM RES, COLUMBIAN DIV, CITIES SERV CO, 71- Mem: Am Chem Soc. Res: Formation of carbon in flames; surface chemistry and physics of carbon black; applications of carbon black in polymers; interaction of carbon black with elastomers. Mailing Add: Cities Serv Co Columbian Div PO Box 4 Cranbury NJ 08512

BURGESS, PAUL RICHARDS, b Logan, Utah, May 2, 34; m 61. NEUROBIOLOGY. Educ: Reed Col, BA, 56; Oxford Univ, BA, 59; Rockefeller Inst, PhD(physiol), 65. Prof Exp: From instr to assoc prof, Univ Utah, 67-72; assoc prof, 71-75, PROF PHYSIOL, MED SCH, UNIV UTAH, 75- Concurrent Pos: Fel physiol, Univ Utah, 65-67. Mem: Soc Neurosci; Am Physiol Soc. Res: Somatic sensation; nerve growth. Mailing Add: Dept of Physiol Med Sch Univ of Utah Salt Lake City UT 84132

BURGESS, RICHARD ERNEST, b Detroit, Mich, Nov 15, 13; m 35, 66; c 3. BIOCHEMISTRY, BIOMEDICAL ENGINEERING. Educ: Pomona Col, BA, 35. Prof Exp: Res chemist, Dried Food Prod Co, 36-39; res group leader pharmaceut, Armour & Co, 39-46; res chemist, Don Baxter, Inc, 46-47, dir control, 47-57, tech serv, 57-65; DIR TECH SERV, PHARMASEAL DIV, AM HOSP SUPPLY CORP, 65- Mem: Nat Soc Med Res; Am Chem Soc; Am Med Writers Asn; Asn Advan Med Instrumentation. Res: Pharmaceutical products of animal origin; fluid electrolyte balance in human body; pyrogens; development and clinical applications of sterile disposable medical devices. Mailing Add: Pharmaseal Div of Am Hosp Supply 1015 Grandview Ave Glendale CA 91201

BURGESS, RICHARD RAY, b Mt Vernon, Wash, Sept 8, 42; m 67. MOLECULAR BIOLOGY, ONCOLOGY. Educ: Calif Inst Technol, BS, 64; Harvard Univ, PhD(molecular biol & biochem), 69. Prof Exp: ASST PROF ONCOL, McARDLE LAB CANCER RES, UNIV WIS-MADISON, 71- Concurrent Pos: Helen Hay Whitney Found fel, Inst Molecular Biol, Geneva, Switz, 69-71. Res: RNA polymerase and the regulation of gene expression. Mailing Add: McArdle Lab for Cancer Res Univ of Wis Madison WI 53706

BURGESS, ROBERT LEWIS, b Kalamazoo, Mich, Sept 12, 31; m 55; c 5. PLANT ECOLOGY, BOTANY. Educ: Univ Wis-Milwaukee, BS, 57; Univ Wis, Madison, MS, 59, PhD(plant ecol), 61. Prof Exp: Asst prof bot, Ariz State Univ, 60-63, dir desert inst, 63; from asst prof to assoc prof bot, NDak State Univ, 63-71; DEP DIR EASTERN DECIDUOUS FOREST BIOME, INT BIOL PROG, OAK RIDGE NAT LAB, 71-, PROG DIR ENVIRON SCI DIV, 72- Concurrent Pos: Res grants, Southwest Monuments Asn, 61-62 & 64, NSF, 62-63, 67, 71-75, Ariz Park Comn, 62-63, NDak Inst Regional Studies, 64 & Off Water Resources Res, 68-71; res collab, Nat Park Serv, Dept Interior, 64; vis prof bot, Pahlavi Univ, Iran, 65-66; consult, NSF, 74-75. Honors & Awards: NDak Conservationist of the Year Award, 69. Mem: Ecol Soc Am; Wilderness Soc; Sigma Xi. Res: Regional vegetation studies on composition, structure and dynamics of native plant communities in Wisconsin, Arizona and North Dakota; application of statistical methods to community analysis and interpretation; vegetation mapping. Mailing Add: Environ Sci Div Box X Oak Ridge Nat Lab Oak Ridge TN 37830

BURGESS, THOMAS EDWARD, b Pontiac, RI, Nov 19, 23; m 49. INORGANIC CHEMISTRY. Educ: RI State Col, BS, 49; Syracuse Univ, MS, 52; Univ Conn, PhD(chem), 59. Prof Exp: Chemist, Am Optical Co, 52-54; chemist, Sprague Elec Co, North Adams, 59-71; TECH DIR, BURGESS ANAL LAB, 71- Mem: Am Soc Test & Mat. Res: Analytical chemistry of semiconductor thin films; radiotracer studies of diffusion in solids; atomic absorption spectroscopy; optical emission spectroscopy. Mailing Add: 5 Grandview Dr Williamstown MA 01267

BURGESS, THOMAS EDWARD, b Rochester, NY, Aug 20, 46; m 73. CLINICAL CHEMISTRY. Educ: LeMoyne Col, BS, 68; Villanova Univ, MS, 71, PhD(biochem), 76. Prof Exp: NIH trainee clin chem, Hahnemann Med Col, 74-76; SUPVR SPEC CLIN CHEM, LAB PROCEDURES EAST, UPJOHN CO, 76- Mem: Am Asn Clin Chem; Am Soc Microbiol. Res: Biochemical and immunological investigations of the hepatitis B antigen and the hepatitis B virus. Mailing Add: Lab Procedures East Upjohn Co 1075 First Ave King of Prussia PA 19406

BURGESS, WILLIAM HOWARD, b Boston, Mass, Mar 13, 24; m 49; c 2. APPLIED CHEMISTRY. Educ: Cornell Univ, BChE, 49, MFS, 50, PhD(dairy chem), 54. Prof Exp: Asst food sci, Cornell Univ, 49-50, dairy chem, 50-53, nutrit & biochem, 53-541 from asst prof to assoc prof, 54-68, PROF CHEM ENG, UNIV TORONTO, 68- Mem: Am Chem Soc. Res: Reaction kinetics. Mailing Add: Dept of Chem Eng Univ of Toronto Toronto ON Can

BURGHARDT, ANDREW FRANK, b New York, NY, Apr 5, 24; Can citizen; m 56; c 5. GEOGRAPHY. Educ: Harvard Univ, BA, 49; Univ Wis, MS, 51, PhD(geog), 58. Prof Exp: Cartogr, Am Map Co, 49; cartogr, Univ Wis, 51; cartog compiler, Gousha Map Co, 51-52; cartog ed, Army Map Serv, 52-54; asst prof geog, Stanford Univ, 57-61; from asst prof to assoc prof, 61-71, PROF GEOG, McMASTER UNIV, 71- Concurrent Pos: Fulbright res grant, Vienna, Austria, 60-61; Can Coun res fel, McMaster Univ, 67-68; vis prof geog, Univ Minn, 70-71. Mem: Am Geog Soc; Asn Am Geogr; Can Asn Am Studies; Can Asn Geog. Res: The basis of claims to territory; the origins of cities less important, regional geography of United States and Central Europe. Mailing Add: Dept of Geog McMaster Univ Hamilton ON Can

BURGHARDT, GORDON MARTIN, b Milwaukee, Wis, Oct 11, 41; m 66. ETHOLOGY, BIOPSYCHOLOGY. Educ: Univ Chicago, BS, 63, PhD(biopsychol), 66. Prof Exp: Instr biol, Univ Chicago, 66-67; from asst prof to assoc prof, 68-74, PROF PSYCHOL, UNIV TENN, KNOXVILLE, 74- Concurrent Pos: NIMH res grant, 61-71 & 67-75; cur res, Knoxville Zool Park, 73-; NSF res grant, 75- Mem: Am Soc Ichthyol & Herpet; Animal Behav Soc; Am Psychol Asn; Psychonomic Soc; Int Soc Develop Psychobiol. Res: Chemical and visual perception; imprinting and early experience; evolution of behavior; behavior of lower vertebrates, especially reptiles; behavior of bears. Mailing Add: Dept of Psychol Univ of Tenn Knoxville TN 37916

BURGI, ERNEST JUNIOR, b Salt Lake City, Utah, Mar 22, 24; m 47; c 2. AUDIOLOGY, SPEECH PATHOLOGY. Educ: Ariz State Univ, BA, 50; Univ Denver, MA, 51; Univ Pittsburgh, PhD(audiol), 57. Prof Exp: Instr speech & speech path, Univ Nebr, 51-55; res assoc speech & hearing dis, Univ Pittsburgh, 55-57; asst prof audiol, Bowling Green State Univ, 57-58; assoc prof audiol & speech path, Univ Nebr, 59-62; assoc prof, 62-69, PROF SPEECH, UNIV PITTSBURGH, 69- Concurrent Pos: Consult, Cleft Palate Proj, Univ Pittsburgh, 58. Mem: Am Speech & Hearing Asn; Speech Asn Am. Mailing Add: Dept of Speech Univ of Pittsburgh Pittsburgh PA 15213

BURGIEL, J C, solid state physics, see 12th edition

BURGINYON, GARY ALFRED, b Spokane, Wash, June 29, 35; m 58; c 1. NUCLEAR PHYSICS. Educ: Wash State Univ, BS, 58; Yale Univ, MS, 61, PhD(physics), 66. Prof Exp: Physicist, Naval Res Lab, 61-62; res staff physicist, Yale Univ, 66-68; PHYSICIST, LAWRENCE LIVERMORE LAB, 68- Mem: Am Phys Soc; Am Astron Soc. Res: Collective nuclear structure studies among rare earth nuclei. Mailing Add: Lawrence Livermore Lab PO Box 808 Livermore CA 94550

BURGISON, RAYMOND MERRITT, b Baltimore, Md, Aug 17, 17. PHARMACOLOGY. Educ: Loyola Col, BS, 41; Univ Md, MS, 48, PhD(chem), 50; Johns Hopkins Univ, MLA, 68. Prof Exp: Chemist, Continental Oil Co, 41-42; chemist, Air Reduction Co, 42-45; chemist, US Indust Chem, 45-47; pharmacologist, 47-50, from asst prof to assoc prof, 50-63, PROF PHARMACOL, UNIV MD, BALTIMORE CITY, 63-, CHMN DEPT, 69- Concurrent Pos: Chmn chem panel, NIMH, 62-64. Mem: Am Chem Soc; Am Soc Pharmacol & Exp Therapeut. Res: Medicinal chemistry and pharmacology; hypotensive drugs; anticancer agents; cardiovascular drugs. Mailing Add: Dept of Pharmacol Sch of Dent Univ of Md Baltimore MD 21201

BURGMAIER, GEORGE JOHN, b Utica, NY, Jan 28, 44; m 68; c 1. ORGANIC CHEMISTRY, PHOTOGRAPHY. Educ: Univ Toronto, BSc, 66; Yale Univ, PhD(org chem), 70. Prof Exp: SR RES CHEMIST, EASTMAN KODAK CO, 70- Mem: Am Chem Soc. Res: Chemistry of highly strained small ring hydrocarbons; preparation of chemicals with photographic applications. Mailing Add: Eastman Kodak Co Res Labs 343 State St Rochester NY 14650

BURGNER, ROBERT LOUIS, b Yakima, Wash, Jan 16, 19; m 42; c 2. FISH BIOLOGY. Educ: Univ Wash, BS, 42, PhD(fisheries), 58. Prof Exp: Fishery biologist, 46-54, asst dir, 55, res assoc prof, 57-64, assoc prof, 64-69, asst dir, 57-67, PROF FISHERIES, FISHERIES RES INST, UNIV WASH, 69-, DIR INST, 67- Mem: AAAS; Am Fisheries Soc; Am Soc Limnol & Oceanog; Am Inst Fisheries Res Biol; Int Asn Theoret & Appl Limnol. Res: Life history and survival of salmonid; lake ecology and limnology. Mailing Add: Fisheries Res Inst Univ of Wash Seattle WA 98195

BURGOYNE, EDWARD EYNON, b Montpelier, Idaho, Sept 26, 18; m 50; c 4. ORGANIC CHEMISTRY. Educ: Utah State Univ, BS, 41; Univ Chicago, cert, 43; Univ Wis, MS, 47, PhD(chem), 49. Prof Exp: Alumni Res Found asst org chem, Univ Wis, 46-47, 48; res chemist, Phillips Petrol Co, 49-51; from asst prof to assoc prof, 51-59, PROF CHEM, ARIZ STATE UNIV, 59- Mem: AAAS; Am Chem Soc; fel Am Inst Chemists. Res: Catalytic hydrogenation; isomerization of hydrocarbons; condensation of alcohols; synthesis of some nitro-musks. Mailing Add: Dept of Chem Ariz State Univ Tempe AZ 85281

BURGOYNE, PETER NICHOLAS, b Bracknell, Eng, Oct 8, 32; Can citizen; m 62. MATHEMATICAL PHYSICS. Educ: McGill Univ, BSc, 55, MSc, 56; Princeton Univ, PhD(math physics), 61. Prof Exp: Instr math, PrincetonUniv, 59-61; asst prof, Univ Calif, Berkeley, 61-66; assoc prof, Univ Ill, Chicago, 66-68; assoc prof, 68-74, PROF MATH, UNIV CALIF, SANTA CRUZ, 74- Concurrent Pos: Sloan Found fel, 63-65. Res: Theoretical physics, particularly problems in the theory of elementary particles; pure mathematics, particularly finite group theory. Mailing Add: Dept of Math Univ of Calif Santa Cruz CA 95060

BURGSTAHLER, ALBERT WILLIAM, b Grand Rapids, Mich, July 10, 28; m 57; c 5. ORGANIC CHEMISTRY. Educ: Univ Notre Dame, BS, 49; Harvard Univ, MA, 50, PhD(chem), 53. Prof Exp: Instr org chem, Univ Notre Dame, 53-54; proj assoc, Univ Wis, 55-56; instr & res assoc, 56-57, from asst prof to assoc prof chem, 57-65, PROF CHEM, UNIV KANS, 65- Concurrent Pos: Sloan fel, 61-64. Honors & Awards: Notre Dame Centennial Sci Award, 65. Mem: Am Chem Soc; Int Soc Fluoride Res (pres, 70-73); Brit Chem Soc; Int Soc Res Civilization Dis & Vital Substances. Res: Structure determination, reactions and synthesis of natural products; newer synthetic methods and applications; stereochemistry and circular dichroism. Mailing Add: Dept of Chem Univ of Kans Lawrence KS 66045

BURGSTAHLER, SYLVAN, b Corvuso, Minn, Nov 7, 28; m 61; c 3. MATHEMATICS. Educ: Univ Minn, BS, 51, MS, 53, PhD(math), 63. Prof Exp: Asst prof, 61-70, head dept, 70-74, ASSOC PROF MATH, UNIV MINN, DULUTH, 74- Mem: Math Asn Am. Res: Flow of heat between materials having different physical constants. Mailing Add: Dept of Math Univ of Minn Duluth MN 55812

BURGUS, ROGER CECIL, b Osceola, Iowa, Sept 10, 34; m 55; c 3. BIOCHEMISTRY, NEUROENDOCRINOLOGY. Educ: Iowa State Univ, BS, 57, MS, 60, PhD(biochem), 62. Prof Exp: Asst chem, Iowa State Univ, 57-60, asst biochem, 60-62; res assoc physiol, Wayne State Univ, 62-65; from asst prof to assoc prof physiol & biochem, Col Med, Baylor Univ, 65-70; sr res assoc, 70-73, ASSOC RES PROF, SALK INST, 73- Mem: Am Soc Biol Chem; Int Soc Neuroendocrinol; AAAS; Am Chem Soc; Endocrine Soc. Res: Isolation and characterization of natural products; metabolism of pyrroles; vitamin B-twelve and related compounds in bacteria; chemistry of hormones originating in the hypothalmus and other areas of the brain. Mailing Add: Salk Inst PO Box 1809 San Diego CA 92112

BURGUS, WARREN HAROLD, b Burlington, Iowa, Oct 28, 19; m 51; c 2. RADIOCHEMISTRY. Educ: Univ Iowa, BS, 41; Wash Univ, PhD(chem), 49. Prof Exp: Asst metall lab, Univ Chicago, 42-43; res assoc, Clinton Labs, 43-46; asst, Wash Univ, 46-47; mem staff, Los Alamos Sci Lab, 48-54; head mat testing reactor chem sect, Atomic Energy Div, Phillips Petrol Co, 54-68, sr tech consult, Water Reactor Safety Prog Off, Atomic Energy Div, 68-69; SR SCIENTIST, AEROJET NUCLEAR CO, 71- Mem: Fel AAAS; Am Chem Soc. Res: Radiochemical procedures; hot atom chemistry; radioactive decay schemes; fission products; chemical processing of reactor fuels; cross sections. Mailing Add: 628 S Fanning Ave Idaho Falls ID 83401

BURGYAN, ALADAR, b Dicsoszentmarton, Transylvania, Dec 29, 17; nat US; m 44. INORGANIC CHEMISTRY. Educ: Tech Univ Budapest, Dipl, 39. Prof Exp: Lectr, Tech Univ Budapest, 39-41, asst prof, 41-45; color res chemist, 51-52, supvr, Color Res & Develop Lab, 52-59, DIR COLOR RES & DEVELOP LAB, FERRO CORP OF CLEVELAND, 59- Mem: Am Chem Soc; Am Ceramic Soc; Am Soc Testing & Mat; Soc Plastics Eng. Res: Heat stable inorganic pigments; applications in plastics, porcelain enamels, ceramics. Mailing Add: 3278 Rocky River Dr Cleveland OH 44111

BURHANS, ALLISON STILWELL, b Philadelphia, NY, June 29, 17; m 41; c 3. PLASTICS CHEMISTRY. Educ: Oberlin Col, AB, 39; Duke Univ, MS, 40. Prof Exp: Res chemist, Plastics & Protective Coatings, Bakelite Corp, 40-47; head polymer lab,

Plastics & Gum Res, Beech-Nut Packing Co, 47-52, asst dir res gum food & chems, 52-53, dir res, 53-56; group leader plastic develop, Union Carbide Plastics Co, 56-72, PROJ LEADER EPOXY MIL CONTRACTS PROG, UNION CARBIDE CORP, 66-, SR DEVELOP SCIENTIST, 72- Concurrent Pos: Consult, Arkell Safety Bag Co, 51 & 52; pres, Chem Specialties Co. Mem: Am Chem Soc; NY Acad Sci. Res: Thermoplastic resins; phenolic and epoxy resins; vinyl lattices; protective and decorative coatings; cycloaliphatic epoxide research and development for electrical and electronic applications. Mailing Add: 489 King George Rd Millington NJ 07946

BURHOLT, DENNIS ROBERT, b New York, NY, Sept 22, 42; m 64; c 2. CELL BIOLOGY, CANCER. Educ: Adelphi Col, BA, 63; Univ Fla, PhD(bot, zool), 68. Prof Exp: Asst prof biol, Southampton Col, 68-69; NIH fel, Brookhaven Nat Lab, 69-71; res asst, Inst Med Radiation Sci, Würzburg, WGer, 71-73; res asst cancer res, 73-75, RES ASSOC CANCER RES, ALLEGHENY GEN HOSP, 75- Mem: Radiation Res Soc. Res: Influence of radiation and chemotherapeutic agents on gastrointestinal epithelial cell proliferation. Mailing Add: Cancer Res Unit Allegheny Gen Hosp Pittsburgh PA 15212

BURI, PETER FREDERICK, b Minneapolis, Minn, Apr 20, 26; m 50; c 2. GENETICS. Educ: Univ Chicago, PhD, 55. Prof Exp: Asst prof zool, Univ Iowa, 55-58; asst prof biol, Kenyon Col, 58-60; assoc prof, San Francisco State Col, 60-65; COL, BIOL, NEW 65-, CHMN DIV NATURAL SCI, 65-70, 72-74 & 75- UNIV SOUTH FLA, Mem: Sigma Xi; AAAS. Res: Population genetics. Mailing Add: Div of Natural Sci New Col Univ South Fla Sarasota FL 33578

BURIKS, RUDOLF SIEGFRIED, organic chemistry, see 12th edition

BURINGTON, RICHARD STEVENS, mathematics, deceased

BURK, CARL JOHN, b Troy, Ohio, Dec 30, 35; m 66; c 1. PLANT TAXONOMY, PLANT ECOLOGY. Educ: Miami Univ, AB, 57; Univ NC, MA, 59, PhD(bot), 61. Prof Exp: Asst bot, Univ NC, 58-59, part-time instr, 60-61; from instr to assoc prof bot, 61-73, actg chmn dept biol sci, 70-71, chmn, 72-75, PROF BIOL SCI, SMITH COL, 73- Mem: AAAS; Bot Soc Am; Am Soc Plant Taxon; Ecol Soc Am. Res: Biosystematics of Quercus, Hepatica, Heterotheca; ecology, biogeography and floristics of coastal and freshwater ecosystems. Mailing Add: Dept of Biol Sci Smith Col Northampton MA 01060

BURK, CORNELIUS FRANKLIN, JR, b Sarnia, Ont, Jan 8, 33; m 59; c 2. GEOLOGY, INFORMATION SCIENCE. Educ: Univ Western Ont, BSc, 56; Northwestern Univ, PhD(geol), 59. Prof Exp: Explor geologist, Texaco, Inc, 59-60; geologist, 60-70, res scientist, 67-68, nat coordr, Secretariat Geosci Data, 68-70, NAT COORDR, CAN CTR GEOSCI DATA, GEOL SURV CAN, DEPT ENERGY, MINES & RESOURCES, 70- Concurrent Pos: Mem, Ad Hoc Comt Storage & Retrieval Geol Data, Nat Adv Comt Res Geol Sci, Can, 65-68, Subcomt Comput Appln, 68-73; ed, J Can Soc Petrol Geologists. Honors & Awards: Outstanding Serv Award, Can Soc Petrol Geologists, 68. Mem: Can Soc Petrol Geologists; Can Asn Info Sci; Geosci Info Soc (vpres, 69, pres, 70); Geol Asn Can; Geol Soc Am. Res: Regional stratigraphy, especially Silurian of eastern Canada and Upper Cretaceous of western Canada; computer-based storage and retrieval of geoscience data and national information systems for science generally; development and management of science information services. Mailing Add: Can Ctr Geosci Data EM&R Can 580 Booth St Ottawa ON Can

BURK, CREIGHTON, b Laramie, Wyo, Feb 1, 29; m 49; c 3. GEOLOGY. Educ: Univ Wyo, BSc, 52, MA, 53; Princeton Univ, PhD(geol), 64. Prof Exp: Field geologist, Stanolind Oil & Gas Co, 52-53; explor geologist, Richfield Oil Corp, 53-60; chief scientist, Am Miscellaneous Soc, Mohole Proj, Nat Acad Sci, 62-64; corp explor adv, Socony Mobil Oil Co, Inc, 64-69, chief geologist, & mgr regional geol, Mobil Oil Corp, 69-75; PROF GEOL SCI, CHMN DEPT MARINE STUDIES & DIR, MARINE SCI INST, UNIV TEX, AUSTIN, 75- Concurrent Pos: Instr, Princeton Univ, 60-61, vis prof geol & geophys sci, 67-75; NSF fel, 61-62; mem, US Geodynamics Comt; Pres Comn Marine Sci, Eng & Resources, Nat Oceanic Adv Comt, 68-, Circum-Pac Int Map Prog, 73- & Int Geol Correlation Proj, 73-; US del, World Petrol Cong, 75; mem & del, US-USSR Protocol on Oceanog, 74- Honors & Awards: Frank A Morgan Award, 56. Mem: Geol Soc Am; Am Geophys Union; Am Asn Petrol Geol; Marine Technol Soc (vpres, 74-); Geol Soc London. Res: Regional, structural and historical geology throughout the world; marine geology and geophysics; geology of continental margins. Mailing Add: Marine Sci Inst Univ of Tex Box 7999 Austin TX 78712

BURK, DAVID LAWRENCE, b Minn, July 20, 29; m 53; c 2. SOLID STATE PHYSICS, QUALITY ASSURANCE. Educ: Carnegie Inst Technol, BS, 51, MS, 55, PhD(physics), 57. Prof Exp: Asst, Carnegie Inst Technol, 51-54; res specialist physics, 57-67, mgr appl physics dept, 67-70, MGR QUALITY ASSURANCE, RES & DEVELOP LAB, ALLEGHENY LUDLUM STEEL CORP, 70- Mem: Am Phys Soc; Am Soc Metals. Res: Physics of metals. Mailing Add: Allegheny Ludlum Steel Corp Brackenridge PA 15014

BURK, DEAN, b Oakland, Calif, Mar 21, 04; m 29; c 3. BIOCHEMISTRY. Educ: Univ Calif, BS, 23, PhD, 27. Prof Exp: Assoc phys chemist fixed nitrogen res lab, Bur Chem & Soils, USDA, 29-37, chemist, 37-39; from sr chemist to chief chemist, Nat Cancer Inst, 39-74, head cytochem lab, 48-74; WITH DEAN BURK FOUND, 74- Concurrent Pos: Nat Res & Int Educ Bd fel, Univ Col, London, Kaiser Wilhelm Inst, Berlin & Harvard Univ, 27-29; USPHS fel, Kaiser Wilhelm Inst, Berlin, 50; guest lectr, Iowa State Col, 34; guest res worker, Biochem Inst, USSR Acad Sci, 35; assoc prof, Med Col, Cornell Univ, 39-41; res master, George Washington Univ, 47-; foreign mem, Max Planck Inst Cell Physiol, 53-, Max Planck Soc Advan Sci, Goettingen, 55- & Max Planck Inst Biochem, 73-; guest scientist, US Naval Med Res Inst, 74- Honors & Awards: Hillebrand Award, Am Chem Soc, 52; Gerhard Domagk Award Cancer Res, 65; Knight Comdr, Med Order Bethlehem, Rome, 70; Humanitarian Award, Nat Health Fedn, 71, Cancer Control Soc, 73 & Int Asn Cancer Victims & Friends, 74; Hon Pres, Ger Soc Med Tumorther, 73- Mem: Fel AAAS; Am Chem Soc; Am Soc Plant Physiol; Am Soc Biol Chem; Am Asn Cancer Res. Res: Biochemistry of cancer; biochemical systems; enzyme kinetics; vitamins; hormones; metal chelates. Mailing Add: Dean Burk Found 4719 44th St Washington DC 20016

BURK, LAWRENCE G, b New Orleans, La, June 1, 20; m 43. GENETICS. Educ: Univ Ga, BSA, 48, MS, 49. Prof Exp: Geneticist, Rubber Res Sta, 51-53, geneticist tobacco invest, Field Crops Res Br, Md, 53-67, RES GENETICIST, TOBACCO RES LAB, SOUTHERN REGION, AGR RES SERV, USDA, 67-; ASSOC PROF GENETICS, NC STATE UNIV, 67- Mem: Am Genetic Asn (treas, 56-62); Am Soc Agron; Bot Soc Am. Res: Crop improvement through interspecific hybridization; development of bridge-cross methods for overcoming sterility barriers; elucidation of tri-partite nature of tunica in Nicotiana tabacum; development of methods for the induction of haploid plantlets from aseptically cultured anthers of tobacco and converting haploids to diploids. Mailing Add: Oxford Tobacco Res Lab Oxford NC 27565

BURKA, EDWARD RICHARD, b Washington, DC, Dec 30, 30; m 58, 70; c 3. HEMATOLOGY, BIOCHEMISTRY. Educ: Princeton Univ, AB, 52; Columbia Univ, MD, 56; Am Bd Internal Med, dipl, 62. Prof Exp: Asst med, Sch Med, Georgetown Univ, 60-61; asst physician & assoc dir hemat lab, Presby Hosp, 64-66; from asst prof to assoc prof, 66-72, PROF MED, JEFFERSON MED COL, 72-, DIR BLOOD BANK, THOMAS JEFFERSON UNIV HOSP, 69- Concurrent Pos: Res fel, Presby Hosp, 61-64; NIH res fel hemat, Col Physicians & Surgeons, Columbia Univ, 61-64; attend physician, Thomas Jefferson Univ Hosp, 66- Mem: Harvey Soc; Am Soc Hemat; Am Fedn Clin Res; Am Soc Biol Chem; Am Soc Clin Invest. Res: Protein synthesis and nucleic acid metabolism in mammalian erythrocytes; thalassemia and other hemolytic anemias. Mailing Add: Cardeza Found 1015 Walnut St Philadelphia PA 19107

BURKARD, PERLE NELIUS, b Paxton, Ill, Dec 10, 11; m 41; c 4. INORGANIC CHEMISTRY, ANALYTICAL CHEMISTRY. Educ: Univ Ill, BS, 34; Syracuse Univ, MS, 36. Prof Exp: Chemist, Solvay Process Div, Wyandotte Chem Corp, 38-41, tech serv chemist, 41-43, asst to dir res, 43-45, dir tech serv, 45-52, dir indust & aircraft sales, 53-66, dir res, 66-70, TECH DIR CHEM SPECIALTIES DIV, BASF WYANDOTTE CORP, 70-, VPRES, WYANDOTTE CHEM OF CAN, LTD, 73- Mem: Am Chem Soc; fel Am Inst Chem Engrs; Am Electroplaters Soc; Am Soc Metals. Res: Cleaning and detergency related to metal cleaning, sanitation in food plants, commercial laundry work, dishwashing, floor waxes and finishes, zinc phosphating, paint stripping, commercial aircraft maintenance; chemical specialties in general. Mailing Add: BASF Wyandotte Corp Chem Spec Div 1609 Biddle Ave Wyandotte MI 48192

BURKART, BURKE, b Feb 23, 33; US citizen; m 66; c 1. GEOLOGY, GEOCHEMISTRY. Educ: Univ Tex, BS, 54, MA, 60; Rice Univ, PhD(geol), 65. Prof Exp: Fel geol, Univ Tex, 65; asst prof, Temple Univ, 65-70; asst prof, 70-74, ASSOC PROF GEOL, UNIV TEX, ARLINGTON, 74- Mem: Mineral Soc Am; Geochem Soc; Geol Soc Am. Res: Geology of Central America; geochemistry of phosphate minerals. Mailing Add: Dept of Geol Univ of Tex Arlington TX 76010

BURKART, LEONARD F, b Mosier, Ore, Oct 8, 20; m 49; c 3. WOOD CHEMISTRY, WOOD TECHNOLOGY. Educ: Univ Wash, BSF, 49, MF, 50; Univ Minn, MS, 61, PhD(wood chem), 63. Prof Exp: Technologist, Wash Veneer Corp, 50-52; staff missionary, Student Missionary Coun, Ore, 52-54; teacher sci & eng, Ore Tech Inst, 54-59; asst prof wood technol, 63-66, ASSOC PROF FORESTRY, STEPHEN F AUSTIN STATE COL, 66- Mem: Tech Asn Pulp & Paper Indust; Soc Am Foresters; fel Am Inst Chemists. Res: Utilization of wood residues as sources of organic chemicals, especially work with lignin from waste waters of the pulp and paper industry and lignin and carbohydrates from sawdust. Mailing Add: Dept of Forestry Stephen F Austin State Univ Nacogdoches TX 75961

BURKE, ANNA MAE WALSH, b New York, NY, July 20, 38; m 68; c 2. MOLECULAR PHYSICS. Educ: Manhattanville Col, BA, 60; Fordham Univ, MS, 62, PhD(physics), 65. Prof Exp: Asst prof physics, Merrimack Col, 64-67; staff scientist nuclear effects of electromagnetic theory, Avco Corp, 66-68; assoc prof physics & chem, Newton Col, 68-71; consult, 72-74, PROG DIR & PROF EDUC, NOVA UNIV, 74- Concurrent Pos: Adj prof physics, Fla Atlantic Univ, 73-; consult, Pine Crest Sch, Ft Lauderdale, 74- Mem: Am Phys Soc; Am Soc Physics Teachers. Res: Molecular structure studies including vibration-rotation interactions, solid state applications of molecular structure techniques. Mailing Add: Dept of Physics Nova Univ College Ave Ft Lauderdale FL 33314

BURKE, ARTHUR WADE, JR, b Richmond, Va, Jan 15, 27. CANCER, BIOPHYSICS. Educ: Univ Va, BA, 47, MA, 48; St Louis Univ, PhD, 57; Med Col Va, MD, 60. Prof Exp: Instr biol, Univ Va, 46-48; asst cytol & physiol, Harvard Univ, 48-49; asst & jr biologist radiation biol & microbiol, Oak Ridge Nat Lab, 49-52; res assoc low temperature biophys, Inst Biophys, St Louis Univ, 52-54, res assoc radiation biol & cancer, Dept Path, Dent Sch, 54-56; assoc cancer, Med Col Va, 56-60, med interm 60-61; res assoc, Dept Cancer Res, RI Hosp, Providence, 61-64; asst prof pharmacol, 64-72, MEM STAFF RADIOTHER, DIV THERAPEUT RADIOL & ONCOL, MED COL VA, 72- Concurrent Pos: Consult, Am Tobacco Co, 64-65, coordr biol res, 65-70, asst mgr basic mat res, 70-71. Mem: Radiation Res Soc; Am Col Radiol; Soc Toxicol. Res: Mechanisms of action of ionizing radiations, especially oxygen-nitrogen concentrations and chemical protective agents; action of low temperature on suvival of organisms; dhydration, low temperature and x-rays effects on ascites carcinoma; determination of leakage of chemotherapeutic agents during regional perfusion for cancer in humans; role of the centriole in cell division. Mailing Add: 2114 Shady Grove Rd Mechanicsville VA 23111

BURKE, BERNARD FLOOD, b Boston, Mass, June 7, 28; m 53; c 4. PHYSICS. Educ: Mass Inst Technol, SB, 50, PhD, 53. Prof Exp: Asst physics, Mass Inst Technol, 50-53; mem physics staff, Dept Terrestrial Magnetism, Carnegie Inst, 53-62, chmn radio astron sect, 62-65; PROF PHYSICS, MASS INST TECHNOL, 65- Concurrent Pos: Jr res assoc, Brookhaven Nat Lab, 52-53; mem, Nat Radio Astron Observ Adv Comt, 58-62; astron adv comt, NSF, 58-63, consult, Found, 63-, Int Sci Radio Union & NASA, 67- Honors & Awards: Warner Prize, Am Astron Soc, 63. Mem: Nat Acad Sci; fel AAAS; Am Acad Arts & Sci; Am Phys Soc; Am Astron Soc. Res: Radio astronomy; microwave spectroscopy; antenna arrays; low-noise electronic circuitry. Mailing Add: Room 26-459 Mass Inst of Technol Cambridge MA 02139

BURKE, DENNIS GARTH, b Bracken, Sask, June 4, 35; m 60; c 3. PHYSICS. Educ: Univ Sask, BE, 57, MS, 58; McMaster Univ, PhD(physics), 63. Prof Exp: Sci officer upper atmosphere physics, Defence Res Bd, 58-60; fel, Niels Bohr Inst, Copenhagen, 63-65; from asst prof to assoc prof physics, 65-74, PROF PHYSICS, McMASTER UNIV, 74- Concurrent Pos: Alfred P Sloan Found fel, 66-70. Mem: Can Asn Physicists. Res: Instrumentation of rocket nose cones for upper atmosphere studies; nuclear spectroscopy using decay scheme studies; nuclear structure studies using single nucleon transfer reactions. Mailing Add: Dept of Physics McMaster Univ Hamilton ON Can

BURKE, DENNIS KEITH, b Omaha, Nebr, June 29, 43; m 63; c 3. TOPOLOGY. Educ: Univ Wyo, BA, 64, MS, 66; Wash State Univ, PhD(math), 69. Prof Exp: Asst prof, 69-74, ASSOC PROF MATH, MIAMI UNIV, 74- Mem: Math Asn Am; Am Math Soc. Res: Study of covering properties; metrization theorems, continuous mappings, counter examples and base axioms for topological spaces. Mailing Add: Dept of Math & Statist Miami Univ Oxford OH 45056

BURKE, DOUGLAS WINSTON, plant pathology, see 12th edition

BURKE, EDMUND C, b Fargo, NDak, Nov 23, 19; m 45; c 7. PEDIATRICS. Educ: St Thomas Col, BS, 41; Univ Minn, MB, 44, MD, 45, MS, 51. Prof Exp: Consult, Mayo Clin, 52, from instr to asst prof, Mayo Found, 53-64, assoc prof clin pediat, Mayo Grad Sch Med, Univ Minn, 64-72, PROF PEDIAT, MAYO MED SCH, 72- Concurrent Pos: Fel pediat, Mayo Found, Univ Minn, 48-51. Res: Renal disease; salt and electrolyte disturbance; migraine in children. Mailing Add: Mayo Clin Rochester MN 55902

BURKE, EDWARD ALOYSIUS, b White Plains, NY, Oct 21, 29; m 55; c 5. THEORETICAL PHYSICS. Educ: NY Univ, BA, 54; Fordham Univ, MS, 55, PhD(physics), 59. Prof Exp: Electronics engr, Raytheon Mfg Co, 54; instr physics, US Maritime Acad, 55; asst, Fordham Univ, 55-58; asst prof phys sci, Montclair State Col, 58-59; from asst prof to assoc prof physics, St John's Univ, NY, 59-66, chmn dept, 62-65; assoc prof, 66-74, PROF PHYSICS, ADELPHI UNIV, 74-, DIR SPACE RELATED SCI, 66- Mem: Am Asn Physics Teachers; Am Phys Soc. Res: Theoretical atomic physics. Mailing Add: Dept of Physics Adelphi Univ Garden City NY 11530

BURKE, EDWARD WALTER, JR, b Macon, Ga, Sept 16, 24; m 46; c 2. PHYSICS. Educ: Presby Col, BS, 47; Univ Wis, MS, 49, PhD(physics), 54. Prof Exp: MARY REYNOLDS BABCOCK PROF PHYSICS, KING COL, 49-, CHMN DIV NATURAL SCI & MATH, 61- Concurrent Pos: Fulbright lectr, Univ Chile, 59. Mem: Am Phys Soc; Am Asn Physics Teachers. Res: Spectroscopy; isotope shift in the atomic spectra of boron; astrophysics; photoelectric study of variable stars. Mailing Add: Dept of Physics King Col Bristol TN 37622

BURKE, GAIL DE PLANQUE, b Orange, NJ, Jan 15, 45. RADIATION PHYSICS. Educ: Immaculata Col, AB, 67; Newark Col Eng, MS, 73. Prof Exp: PHYSICIST, HEALTH & SAFETY LAB, US ENERGY RES & DEVELOP ADMIN, 67- Concurrent Pos: Chmn health physics soc stand working group, Am Nat Stand Inst, 73-, co-chmn, Comt for Int Intercomparison of Environ Dosimeters, 74-, US expert deleg, Int Orgn Stand Comt for Develop of an Int Stand on Thermoluminescence Dosimetry. Mem: Am Nuclear Soc; Health Physics Soc. Res: Physics of radiation, radiation shielding and transport, radiation protection solid state dosimetry, reactor and personnel monitoring; design and testing of radiation instrumentation and calibration facilities. Mailing Add: Health & Safety Lab US Energy Res & Develop Admin 376 Hudson St New York NY 10014

BURKE, HANNA SUSS, b Karlsruhe, Ger, Oct 5, 26; nat US; m 65. ORGANIC CHEMISTRY. Educ: Goucher Col, BA, 48; Univ Pa, MS, 52, PhD(org chem), 56. Prof Exp: Chemist, Johns Hopkins Univ, 48-51 & Rohm & Haas Co, 55-63; TECH INFO SPECIALIST, FIBERS DIV, ALLIED CHEM CORP, 63- Mem: AAAS; Am Chem Soc; Am Soc Info Sci; Sigma Xi; Am Inst Chemists. Res: Separation of amino acids; isolation and synthesis of natural products; water soluble polymers; literature and patent searching; surveillance and indexing of literature and patents in fields of fiber technology and related subjects. Mailing Add: 4802 Cutshaw Rd Richmond VA 23230

BURKE, HOWARD JOSEPH, organic chemistry, see 12th edition

BURKE, J ANTHONY, b New York, NY, Jan 29, 37. ASTRONOMY, ASTROPHYSICS. Educ: Harvard Univ, AB, 58, AM, 59, PhD(astron), 65. Prof Exp: Res assoc astrophys, Brandeis Univ, 64-67; lectr & res assoc, Dept Astron, Boston Univ, 67-68; ASSOC PROF ASTRON, UNIV VICTORIA, 68- Mem: AAAS; Am Astron Soc; Royal Astron Soc; Int Astron Union. Res: Cosmogony; interstellar medium. Mailing Add: Dept of Physics Univ of Victoria Victoria BC Can

BURKE, JACK DENNING, b Clarksburg, WVa, Apr 24, 19; m 40; c 5. ANATOMY, PHYSIOLOGY. Educ: Univ Tenn, BA, 48; WVa Univ, MS, 49; Univ Fla, PhD(biol), 52. Prof Exp: Asst biol, Univ Fla, 50-52; asst prof, Longwood Col, 52-53; from asst prof to assoc prof, Univ Richmond, 53-63; assoc prof physiol, 63-64, PROF ANAT, MED COL VA, 64- Concurrent Pos: Adv, USAID-NSF Sci Improv Prog India, Univ Rajasthan, 68; Fulbright-Hays sr lectr, Univ Bogota, Colombia, 72. Mem: Am Physiol Soc; Am Asn Anat; Sigma Xi; Am Soc Cell Biol. Res: Cell biology; oxygen dissociation curves, blood oxygen capacity and blood volume studies in vertebrates; serum effects on carcinoma cells in vivo and in vitro. Mailing Add: Dept of Anat Med Col of Va Richmond VA 23298

BURKE, JAMES DAVID, b Darby, Pa, June 30, 37; m 68; c 2. ORGANIC CHEMISTRY. Educ: Spring Hill Col, BS, 61; Univ Calif, Berkeley, PhD(org chem), 65. Prof Exp: NIH fel, Columbia Univ, 65-66; sr res scientist, 66-72, PLACEMENT SUPVR, ROHM AND HAAS CO, 72- Mem: AAAS; Am Chem Soc. Res: Mass spectrometry and spectroscopic methods; molecular rearrangements; surface active agents for aqueous and non-aqueous systems; acrylate polymers; technology evaluation. Mailing Add: Rohm and Haas Co Independence Mall West Philadelphia PA 19105

BURKE, JAMES EDWARD, b Los Angeles, Calif, Oct 8, 31; m 54; c 3. APPLIED MATHEMATICS, MATHEMATICAL PHYSICS. Educ: San Jose State Col, BA, 53; Stanford Univ, MS, 55, PhD(math), 58. Prof Exp: Mathematician, Edwards AFB, 53-54; aeronaut res scientist syst anal, Ames Lab, NASA, 54-55; asst, Stanford Univ, 54-57; scattering specialist, Sylvania Electronics Defense Lab, Gen Tel & Electronics Co, 58-69; LAB MGR, ELECTROMAGNETIC SYST LABS, INC, 69- Mem: Sigma Xi; Am Math Soc; Soc Indust & Appl Math; Optical Soc Am; fel Acoust Soc Am. Res: Scattering of electromagnetic and acoustic waves by single and many objects. Mailing Add: 19960 Angus Ct Saratoga CA 95070

BURKE, JAMES JOSEPH, JR, b Chicago, Ill, July 26, 31; m 63; c 1. OPTICS. Educ: Univ Chicago, MS, 59; Univ Ariz, PhD(optics), 72. Prof Exp: Asst physicist optics, Ill Inst Technol Res Inst, 59-61; from res physicist to sr physicist, Optics Technol, Inc, 61-67; res assoc, 67-73, STAFF SCIENTIST & LECTR, OPTICAL SCI CTR, UNIV ARIZ, 73- Mem: Optical Soc Am. Res: Optical waveguides and integrated optics, image analysis and processing. Mailing Add: Optical Sci Ctr Univ of Ariz Tucson AZ 85721

BURKE, JAMES JOSEPH, JR, b Northampton, Mass, June 11, 37; m 65. POLYMER PHYSICS. Educ: Univ Mass, BS, 58; Mass Inst Technol, PhD(phys chem), 62. Prof Exp: Res assoc phys chem, Mass Inst Technol, 62-63; res chemist, Chemstrand Res Ctr, Inc, 63-66, sr res chemist, 66-67, res specialist, 67-68, res group leader, 68-72, sr res group leader, 72-74; SR RES GROUP LEADER, MONSANTO TRIANGLE PARK DEVELOP CTR, INC, 74- Mem: Am Chem Soc; Am Soc Test & Mat; NY Acad Sci; Fiber Soc. Res: Characterization and evaluation of materials; physical, mechanical, solution and thermal properties of polymers; testing of fibers, plastics, fabrics, foams, composites; dynamic mechanical properties of macromolecules; polymer and textile product development. Mailing Add: Monsanto Develop Ctr Inc PO Box 12274 Research Triangle Park NC 27709

BURKE, JAMES OTEY, b Richmond, Va, May 20, 12; m 42; c 4. MEDICINE. Educ: Va Mil Inst, BS, 33; Med Col Va, MD, 37. Prof Exp: Assoc prof med, 54-59, ASSOC PROF CLIN MED, MED COL VA, 59- Concurrent Pos: Consult, Surgeon Gen, US Dept Army, 50-54 & Vet Admin Hosp, 59-; med dir, A H Robins Co, Inc, 58-60, dir prev med, 60- Mem: Fel Am Col Physicians; AMA; Am Fedn Clin Res; Am Gastroenterol Asn; fel Am Occup Med Asn. Mailing Add: Dept of Clin Med Med Col of Va Richmond VA 23220

BURKE, JAMES RICHARD, nuclear physics, solid state physics, see 12th edition

BURKE, JERRY ALAN, b Elkins, WVa, June 30, 37; m 61; c 2. PESTICIDE CHEMISTRY. Educ: WVa Wesleyan Col, BS, 59. Prof Exp: Anal chemist, 59-65, sect head, 65-71, chief indust chem residues br, 71-72, chief residue & chem br, 72-74, CHIEF ANAL CHEM & PHYSICS BR, US FOOD & DRUG ADMIN, 74- Concurrent Pos: Mem working party, Comn on Pesticide Residues, Int Union Pure & Appl Chem, 67 & 68; mem panel anal methodology, Fed Working Group Pest Mgt, 70-75. Honors & Awards: Award of Merit, US Food & Drug Admin. Mem: Fel Asn Official Anal Chemists. Res: Development and standardization of broadly applicable analytical methodology for multiple pesticides and industrial chemicals residues in foods and related substrates. Mailing Add: 200 C St SW Washington DC 20204

BURKE, JOHN A, b Eastland, Tex, Dec 14, 36; m 62; c 1. INORGANIC CHEMISTRY. Educ: Tex Tech Col, BS, 59; Ohio State Univ, MSc, 61, PhD(inorg chem), 63. Prof Exp: PROF CHEM, TRINITY UNIV, 63-, CHMN DEPT, 69- Mem: Am Chem Soc. Res: Inorganic chemistry, especially coordination compounds. Mailing Add: Dept of Chem Trinity Univ 715 Stadium Dr San Antonio TX 78284

BURKE, JOHN FRANCIS, b Chicago, Ill, July 22, 22; m 50; c 4. SURGERY. Educ: Univ Ill, BS, 47; Harvard Med Sch, MD, 51. Prof Exp: Intern surg, Mass Gen Hosp, Boston, 51-52, resident, 52-56, resident surgeon, 57; instr surg, 58-60, tutor med sci, 60-65, clinassoc surg, 60-66, asst clin prof, 66-69, assoc prof surg, 69-75, PROF SURG, HARVARD MED SCH, 75-; CHIEF SURG, SHRINERS BURNS INST, BOSTON UNIT, 68-, CHIEF STAFF, 69- Concurrent Pos: Asst surg, Mass Gen Hosp, 58-60, asst surgeon, 61-63, assoc vis surgeon, 64-68, vis surgeon, 68- Mem: Am Thoracic Soc; Soc Univ Surg; Infectious Dis Soc Am; Am Surg Asn; Am Asn Surg Trauma. Res: Host defense against infection and the treatment of thermal injury. Mailing Add: Dept of Surg Mass Gen Hosp Boston MA 02114

BURKE, JOHN JAMES, b Wellsburg, WVa, Dec 25, 05; m 27; c 2. PALEONTOLOGY. Educ: Univ Pittsburgh, BS, 29, MSc, 31. Prof Exp: Asst paleont, Carnegie Mus, Pa, 31-39; metallurgist, US Steel Corp, 41-57, technologist & ed, Appl Res Lab, 57-62; geologist & ed, Ohio Div Geol Surv, 63-65; cur, Orton Mus, Ohio State Univ, 65-67; cur of collections & ed, 67-71, SR SCIENTIST, CLEVELAND MUS NATURAL HIST, 71- Concurrent Pos: Res assoc, WVa Geol Surv, 71- Mem: AAAS; fel Geol Soc Am; Paleont Soc; Soc Vert Paleont. Res: Eocene and Oligocene rodents and lagomorphs; Pennsylvanian crinoids, myriapods and gastropods; Pennsylvanian and Eocene stratigraphy. Mailing Add: Cleveland Mus of Natural Hist Wade Oval University Circle Cleveland OH 44106

BURKE, JOHN T, b Missouri Valley, Iowa, Mar 26, 29; m 57; c 2. INTERNAL MEDICINE. Educ: Univ Iowa, MD, 55. Prof Exp: Asst resident internal med, Vet Admin Hosp, San Francisco, 59-61; resident, Presby Med Ctr, 61-62; clin assoc med res, 62-65, DIR CLIN RES, MERCK SHARP & DOHME, 65-; DIR CLIN RES, MERRELL INT RES CTR, STRASBOURG, 73- Mem: AAAS; NY Acad Sci; Europ Soc Study Drug Toxicity; Royal Soc Health; AMA. Res: Clinical pharmacology. Mailing Add: 26 Avenue Montaigne Paris France

BURKE, JOSEPH ALOYSIUS, biology, see 12th edition

BURKE, KENNETH B S, b Willenhall, Eng, Dec 28, 35; m 61; c 2. GEOPHYSICS. Educ: Univ Leeds, BSc, 58, dipl, 59, PhD(appl geophys), 61. Prof Exp: Mining geophysicist, Henry Krumb Sch Mines, Columbia Univ, 61-62; asst prof appl geophys, Univ Sask, 62-68; vis prof, Univ Man, 68-69; ASSOC PROF GEOL, UNIV NB, 69- Mem: AAAS; Soc Explor Geophys; European Asn Explor Geophys. Res: Seismic experiments in real media; crustal seismology; physical properties of unconsolidated rocks, application of geophysics to regional geological mapping. Mailing Add: Dept of Geol Univ of NB Fredericton NB Can

BURKE, KEVIN CHARLES, b London, Eng, Nov 13, 29; m 61; c 3. GEOLOGY. Educ: Univ London, BSc, 51, PhD(geol), 53. Prof Exp: Lectr geol, Univ Ghana, 53-56; geologist nuclear raw mat, Geol Surv, Gt Brit, 56-60; adv, Govt Korea, Int Atomic Energy Agency, 60-61; sr lectr geol, Univ West Indies, 61-65; prof, Univ Ibadan, 65-71; vis prof, Univ Toronto, 71-72; PROF GEOL, STATE UNIV NY ALBANY, 72- Mem: AAAS; Geol Soc Am; Am Geophys Union; Nigerian Mining, Geol & Metall Soc. Res: Application of the findings of plate tectonics to interpretation of the geological history of the earth. Mailing Add: Dept of Geol Sci State Univ of NY Albany NY 12222

BURKE, LEONARDA, b Boston, Mass. MATHEMATICS, DATA PROCESSING. Educ: Emmanuel Col, AB, 26; Boston Col, MA, 28; Cath Univ Am, PhD(math), 31. Prof Exp: Elem teacher, Boston Pub Sch Syst, 21-23; sec teacher, Boston Diocesan Sch Syst, 26-28; prof math, 31-64, COORDR RES ACTIVITIES, RES CTR, REGIS COL, 64- Mem: Math Asn Am; Am Meteorol Soc; Sigma Xi; Am Math Soc. Res: Mathematical analysis as associated with climatology, meteorology, ionospheric studies and interplanetary studies. Mailing Add: Regis Col Res Ctr 235 Wellesley St Weston MA 02193

BURKE, MICHAEL FRANCIS, b Gallup, NMex, Jan 29, 39; m 60; c 4. ANALYTICAL CHEMISTRY. Educ: Regis Col, BS, 60; Va Polytech Inst, PhD, 65. Prof Exp: Teacher, New Orleans, La, 60-61; fel, Purdue Univ, 65-66, asst prof, 66-67; asst prof, 67-74, ASSOC PROF CHEM, UNIV ARIZ, 74- Res: Digital data handling and control of analytical instrumentation by means of small high speed computers; chromatographic separations; pyrolysis-gas chromatography; thermodynamics of gas-solid chromatography; high pressure chromatography. Mailing Add: Dept of Chem Univ of Ariz Tucson AZ 85721

BURKE, MICHAEL JOHN, b Westchester, Ill, July 7, 42; m 64; c 1. BIOPHYSICS. Educ: Blackburn Col, BA, 64; Iowa State Univ, PhD(biophys), 69. Prof Exp: Res asst biophys, Iowa State Univ, 64-69; fel chem, Univ Minn, 69-71, fel bot, 71-72, from res assoc to asst prof hort, Univ Minn, 72-76; ASSOC PROF HORT, COLO STATE UNIV, 76- Mem: Am Soc Hort Sci; AAAS; Am Chem Soc; Am Soc Plant Physiologists; Cryobiol Soc. Res: Plant stress physiology and physiological ecology; emphasis on biophysical properties of plants which allow them to survive low temperature and drought. Mailing Add: Dept of Hort Colo State Univ Ft Collins CO 80521

BURKE, MORRIS, b Hong Kong, Oct 25, 38; m 64; c 2. PROTEIN CHEMISTRY, ORGANIC CHEMISTRY. Educ: Univ Sydney, BSc, 60; Univ New South Wales, MSc, 63, PhD(wool chem), 66. Prof Exp: Res assoc radiation biochem, Mich State Univ, 66-67; res assoc phys biochem, Johns Hopkins Univ, 67-74, fel biol, 68-74; ASST PROF PHYSIOL, UNIV MD, SCH DENT, 74- Mem: Am Chem Soc; Biophys Soc. Res: Molecular basis of muscle contraction; mechanism of myosin and actomyosin Mg ATPase; selective chemical modification of proteins; peptide fractionation; associating macromolecules. Mailing Add: Dept of Physiol Sch of Dent Univ of Md Baltimore MD 21201

BURKE, NOEL IAN, organic chemistry, see 12th edition

BURKE, PAUL J, b New York, NY, Apr 21, 20. OPERATIONS RESEARCH. Educ: City Col New York, BS, 40; Harvard Univ, EdM, 50; Columbia Univ, PhD(math statist), 66. Prof Exp: MEM TECH STAFF PROBABILITY & STATIST, BELL LABS, 53- Mem: Opers Res Soc Am. Res: Telephone traffic theory; queuing theory. Mailing Add: Bell Labs Holmdel NJ 07733

BURKE, PHILIP WILLIAM, b Magdalene Islands, Que, Sept 22, 18; m 43; c 2. AGRICULTURE, APICULTURE. Educ: Ont Agr Col, BSA, 43. Prof Exp: ASSOC PROF APICULT, ONT AGR COL, UNIV GUELPH, 43- Concurrent Pos: Provincial apiarist, Ont Apicult. Res: Apiculture extension; apiary inspection; disease control; teaching. Mailing Add: Dept of Environ Biol Ont Agr Col Univ of Guelph Guelph ON Can

BURKE, RICHARD JAMES, physics, see 12th edition

BURKE, RICHARD LERDA, b San Francisco, Calif, Aug 30, 25; m 50; c 4. ORGANIC CHEMISTRY. Educ: Univ San Francisco, BS, 44, MS, 48; Mich State Univ, PhD(org chem), 52. Prof Exp: Asst chem, Univ San Francisco, 47-48 & Mich State Univ, 48-52; chemist, Calif Res Corp, 52-57; tech rep prod develop, Oronite Chem Co, 57-60, prod specialist indust chem, Oronite Div, Calif Chem Co, 60-63; sr res chemist, Res & Develop Dept, Colgate Palmolive Co, 64-69; TECH DIR, PAC SOAP CO, 69- Mem: Am Chem Soc. Res: Detergents and bleaches. Mailing Add: 5322 Soledad Rancho Ct San Diego CA 92109

BURKE, RICHARD MICHAEL, b Langdon, NDak, July 9, 03; wid. CLINICAL MEDICINE. Educ: Univ Minn, BS, 24, BMed, 28, MD, 30. Prof Exp: Staff physician, Glen Lake Santaorium, Minn, 30-33; staff physician & med dir, State Vet Hosp, Okla, 33-39; med dir, Western Okla Tuberc Sanatorium, 39-42; from instr to emer assoc clin prof med, Sch Med, Univ Okla, 42-75; dir, Div Tuberc & Chronic Respiratory Dis, 42-65, MED CONSULT, DIV TUBERC & RESPIRATORY DIS, OKLA STATE DEPT HEALTH, 65-; EMER ASSOC CLIN PROF MED, SCH MED, UNIV OKLA, 75- Concurrent Pos: Pvt pract, 42-65. Mem: AMA; fel Am Col Physicians; fel Am Col Chest Physicians; Am Soc Int Med; Am Thoracic Soc. Res: Pulmonary diseases. Mailing Add: Div of Tuberc & Respiratory Dis Okla State Dept of Health Oklahoma City OK 73105

BURKE, ROGER E, b Detroit, Mich, Aug 5, 36; m 70. INDUSTRIAL ORGANIC CHEMISTRY. Educ: Univ Chicago, BA, 59. Prof Exp: Sr chemist, 62-75, MGR RESIN DEVELOP, UNION CAMP CORP, 76- Mem: Am Paint & Varnish Asn. Res: Development of new resin products for printing inks and related technology; applications development of rosin derivatives for general coatings and binders. Mailing Add: Union Camp Corp PO Box 570 Savannah GA 47150

BURKE, SUSAN SCHILT, b Buffalo, NY, Aug 6, 47; m 69. SYNTHETIC ORGANIC CHEMISTRY, PESTICIDE CHEMISTRY. Educ: Univ Mich, BSChem, 69; Univ Colo, PhD(org chem), 74. Prof Exp: Fel, Univ Colo, 74; SR CHEMIST HERBICIDE RES, ROHM AND HAAS, 75- Mem: Am Chem Soc; AAAS. Res: Design and synthesis of small molecules useful for improving the growth of agronomic and horticultural crops, especially herbicides and plant growth regulators. Mailing Add: Rohm and Haas Res Labs Norristown & McKean Rds Spring House PA 19477

BURKE, TERENCE, b Leicestershire, Eng, 1931; m 55; c 2. GEOGRAPHY. Educ: Univ Birmingham, BA, 52, PhD(geog), 67. Prof Exp: Asst prof, 60-67, ASSOC PROF GEOG, UNIV MASS, AMHERST, 68- Mem: Asn Am Geogr; Inst Brit Geogr. Res: Political and economic geography, chiefly in Ireland. Mailing Add: Dept of Geog Morill Sci Ctr Univ of Mass Amherst MA 01002

BURKE, THOMAS JOSEPH, b Baltimore, Md, Mar 15, 38; m 65; c 2. PHYSIOLOGY. Educ: Niagara Univ, BS, 61; Adelphi Univ, MS, 66; Univ Houston, PhD(physiol), 70. Prof Exp: ASST PROF PHYSIOL, MED SCH, UNIV COLO MED CTR, DENVER, 73- Concurrent Pos: Nat Heat & Lung Inst fel, Med Sch, Duke Univ, 69-73. Mem: Am Soc Nephrol; Int Soc Nephrol; Am Fedn Clin Res. Res: Renal autoregulation of blood flow and filtration rate; micropuncture techniques; nephron function; sodium balance; renin-angiotension system. Mailing Add: Med Sch Dept of Physiol Univ of Colo Med Ctr Denver CO 80220

BURKE, WILLIAM HENRY, b Dallas, Tex, Sept 1, 24; m 53; c 3. NUCLEAR PHYSICS. Educ: Rice Inst, PhD(physics), 51. Prof Exp: Sr res physicist, 51-56, RES ASSOC, FIELD RES LABS, MOBIL RES & DEVELOP CORP, 56- Concurrent Pos: Consult, Grad Res Ctr, 64-65. Mem: Am Phys Soc. Res: Isotope geology; nuclear reactions induced by proton, deuteron and neutron bombardment; extremely low level radioactivity measurements; mass spectrometry; geochronometry. Mailing Add: Mobil Res & Develop Corp Field Res Lab Box 900 Dallas TX 75221

BURKE, WILLIAM JAMES, b Lowellville, Ohio, May 24, 12; m 40; c 4. POLYMER CHEMISTRY. Educ: Ohio Univ, AB, 34; Ohio State Univ, PhD(chem), 37. Prof Exp: Res chemist, Cent Chem Dept, E I du Pont de Nemours & Co, 37-46; assoc prof chem, Ohio Univ, 46-47; assoc prof, Univ Utah, 47-50, prof & head dept, 50-62; PROF CHEM & VPRES, ARIZ STATE UNIV, 62-, DEAN GRAD COL, 63- Concurrent Pos: Chmn, Midwest Conf Grad Study & Res, 69-70; generalist consult, Nat Coun Archit Registr Bd, 69-72; pres, Western Asn Grad Schs, 71-72; consult, US Army, 56-62, 72-; mem, Nat Archit Accrediting Bd, 72- Mem: Fel AAAS; Am Chem Soc. Res: Heterocyclic nitrogen compounds; organic sulfur compounds; carbohydrates; phenol-formaldehyde polymers; synthesis and characterization of polymers. Mailing Add: Dept of Chem Ariz State Univ Tempe AZ 85281

BURKE, WILLIAM L, b Bennington, Vt, July 5, 41; m. THEORETICAL PHYSICS. Educ: Calif Inst Technol, BS, 63, PhD(physics), 69. Prof Exp: Lectr astrophys, 70-71, ASST PROF PHYSICS & ASTROPHYS, LICK OBSERV, UNIV CALIF, SANTA CRUZ, 71- Res: General relativity; gravitational waves; singular perturbation theory; general astrophysics. Mailing Add: Lick Observ Univ of Calif Santa Cruz CA 95060

BURKE, WILLIAM THOMAS, JR, b Rochester, NY, July 30, 24; m 47; c 7. BIOLOGY, BIOCHEMISTRY. Educ: Univ Rochester, BA, 50, PhD(biol), 53. Prof Exp: Nat Cancer Inst fel, Atomic Energy Proj, Univ Rochester, 53-56; from instr to asst prof biol chem, Col Med & Dent, Seton Hall Univ, 56-60; assoc prof biochem, Sch Med, WVa Univ, 60-64; assoc prof exp path & biochem, NY Med Col, 64-67; dir div natural sci, Southampton Col, 67-73, PROF BIOL, SOUTHAMPTON COL, LONG ISLAND UNIV, 67-, DEAN COL, 73- Mem: AAAS; Am Soc Cell Biol; Am Asn Cancer Res; NY Acad Sci. Res: Protein and amino acid metabolism; metabolism of tumors; liver pathology; metabolic diseases; environmental action; social responsibility of scientists. Mailing Add: Off of the Dean Southampton Col Long Island Univ Southampton NY 11968

BURKEL, WILLIAM E, b Mankato, Minn, Oct 4, 38; m 66; c 2. ANATOMY. Educ: St John's Univ, Minn, BA, 60; Univ NDak, MS, 62, PhD(anat), 64. Prof Exp: NIH fel anat, Lab Electron Micros, Univ NDak, 64-66; asst prof, 66-71, ASSOC PROF ANAT, UNIV MICH, 71- Mem: AAAS; Electron Micros Soc Am; Am Soc Cell Biol; Am Asn Anat. Res: Electron microscopy; development of artificial blood vessels; organ culture of prostate gland. Mailing Add: Dept of Anat Univ of Mich Ann Arbor MI 48104

BURKELL, CHARLES CRAIG, b Yorkton, Sask, Jan 1, 13; m 42; c 4. RADIOLOGY. Educ: Univ Sask, BSc, 38; Univ Toronto, MD, 41; RCPS(E), DMR(T), 48; RCPS(C), cert specialist radiother, 49. Prof Exp: Demonstr anat, Sch Med, 43-45, from assoc to sr cancer assoc, Clin, 45-63, from asst prof to assoc prof med, Col Med, 52-63, PROF THERAPEUT RADIOL, COL MED, UNIV SASK, 63- Concurrent Pos: Dir, Saskatoon Cancer Clin, 63-71. Mem: Can Asn Radiol. Res: Clinical and statistical evaluation of various radiotherapeutic techniques in the treatment of cancer. Mailing Add: Dept of Radiol Col of Med Univ of Sask Saskatoon SK Can

BURKET, STANLEY CAMPBELL, physical chemistry, inorganic chemistry, see 12th edition

BURKETT, HOWARD (BENTON), b Putnam Co, Ind, Feb 26, 16; m 36; c 4. ORGANIC CHEMISTRY. Educ: DePauw Univ, BA, 38; Univ Wis, PhD(org chem), 42. Prof Exp: Asst org chem, DePauw Univ, 37-38; asst, Univ Wis, 38-40; res chemist, Eli Lilly & Co, Ind, 42-45; from asst prof to assoc prof, 45-54, PROF CHEM, DEPAUW UNIV, 54-, ROTATING HEAD DEPT, 64- Concurrent Pos: Fel, Univ Wash, 53-54, Petrol Res Fund grant, 62-63; sr eng assoc, Naka Works, Hitachi, Ltd, Japan, 70-71. Mem: Am Chem Soc. Res: Organic synthesis; barbituric acid derivatives; aliphatic nitro-compounds, basic condensations, physical-organic; acid catalysis. Mailing Add: 700 Shadowlawn Ave Greencastle IN 46135

BURKEY, BRUCE CURTISS, b Ravenna, Ohio, Nov 29, 38. SOLID STATE PHYSICS. Educ: Hiram Col, BA, 61; Mich State Univ, MS, 63, PhD(physics), 67. Prof Exp: PHYSICIST, EASTMAN KODAK CO, 67- Mem: Am Phys Soc; Sigma Xi. Res: Theory and analysis of both photoconductivity and solid state devices. Mailing Add: 37 Berkshire Dr Rochester NY 14626

BURKHALTER, ALAN, b Bloomington, Ind, Aug 11, 32; m 54; c 4. PHARMACOLOGY. Educ: DePauw Univ, BA, 54; Univ Iowa, MS, 56, PhD(pharmacol), 57. Prof Exp: Sr asst scientist, Lab Chem Pharmacol, Nat Heart Inst, 57-59; PROF PHARMACOL, MED CTR, UNIV CALIF, SAN FRANCISCO, 59- Concurrent Pos: USPHS grants, 60-63, 64-67; NIH spec fel, 63-64; consult, State of Calif, 60-62. Mem: AAAS; Am Soc Pharmacol & Exp Therapeut; Tissue Cult Asn. Res: Biochemical pharmacology; effect of drugs on enzyme regulation; histamine metabolism; effect of drugs during fetal and neonatal development; use of cell and organ culture in pharmacologic research. Mailing Add: Dept of Pharmacol Univ of Calif Med Ctr San Francisco CA 94122

BURKHALTER, JAMES HERBERT, b Rome, Ga, May 7, 22; m 46; c 2. MICROWAVE SPECTROSCOPY. Educ: Emory Univ, AB, 43; Ga Sch Technol, MS, 48; Duke Univ, PhD(physics), 50. Prof Exp: Instr physics, Ga Sch Technol, 46-47; asst prof, Univ Ga, 50-55; res physicist, Courtaulds Inc, 55-60, Sylvania Div, Gen Tel & Electronics Corp, 60-61 & Martin Co, 61-72; MEM STAFF, NAVAL RES LAB, 72- Concurrent Pos: Res physicist, Jacobs Instrument Co, 50. Mem: Am Phys Soc; sr mem Inst Elec & Electronics Engrs. Res: Lasers. Mailing Add: Naval Res Lab Code 7680 X-Ray Optics Washington DC 20390

BURKHARD, CHARLES (AUSTIN), b New York, NY, Feb 6, 16; m 40; c 7. ORGANIC CHEMISTRY. Educ: Ariz State Teachers Col, AB, 37; Purdue Univ, PhD(chem), 41. Prof Exp: Lab asst, Univ Ariz, 37-38; lab asst, Purdue Univ, 41, Nat Defense Res Coun fel, 41-42; chemist, 42-56, supvr insulation & non-metals lab, 56-57, MGR MAT & PROCESSES LAB, GEN ELEC CO, 57- Mem: Am Chem Soc; Inst Elec & Electronics Eng; Combustion Inst. Res: Organo-silicon research; certain esters of starch; synthetic resins and plastics; carbohydrates; mathematics; aliphatic nitrations. Mailing Add: Gen Elec Co 2901 E Lake Rd Erie PA 16531

BURKHARD, DONALD GEORGE, b NJ, Feb 10, 18; m 48; c 3. PHYSICS. Educ: Univ Calif, AB, 41; Univ Mich, MS, 47, PhD(physics), 50. Prof Exp: Eng aide, State Div Hwy, Calif, 36-39; asst eng aide, US Dept Eng, Los Angeles, 40-41; physicist, Radiation Lab, Univ Calif, 44; res physicist, Carter Oil Co, Okla, 44-45; res assoc, Ohio State Univ, 49-50; from asst prof to prof physics, Univ Colo, 50-64; PROF PHYSICS & ASTRON, UNIV GA, 65-; RES DIR, PHYSICS, ENG & CHEM RES ASSOCS, INC, 56- Mem: AAAS; Am Phys Soc; Optical Soc Am; Am Acoustical Soc; Am Asn Physics Teachers. Res: Theoretical physics; theoretical atomic and molecular structure; infrared and microwave spectroscopy; psychometrics; solid state physics; physical and geometrical optics. Mailing Add: Dept of Physics & Astron Univ of Ga Athens GA 30602

BURKHARD, MAHLON DANIEL, b Seward, Nebr, Jan 14, 23; m 45; c 4. ACOUSTICS. Educ: Nebr Wesleyan Univ, AB, 46; Pa State Col, MS, 50. Prof Exp: Teacher high sch, Nebr, 46-47; asst acoustics, Dept Physics, Pa State Univ, 47-50; physicist, Nat Bur Standards, 50-57; supvr, Acoustics Sect, IIT Res Inst, 57-60; MGR ACOUSTICAL RES, INDUST RES PROD, INC, 60- Concurrent Pos: Cert Superior Accomplishment, US Dept Com, 54. Mem: Sr mem Inst Elec & Electronics Eng; fel Acoustical Soc Am; Audio Eng Soc. Res: Atmospheric sound propagation; edge tone and whistle sound sources; calibration and standardization of audiometric instruments and microphones; acoustic impedance measurements; ultrasonic phenomena in gases, liquids, and solids; electro mechanical transducers; sound and noise control. Mailing Add: 1016 Raven Ct Palatine IL 60067

BURKHARD, RAYMOND KENNETH, b Tempe, Ariz, Aug 6, 24; m 48; c 4. BIOCHEMISTRY. Educ: Ariz State Univ, AB, 47; Northwestern Univ, PhD(chem), 50. Prof Exp: From instr to assoc prof, 50-65, PROF BIOCHEM, KANS STATE UNIV, 65- Concurrent Pos: Res fel, Inst Enzyme Res, Wis, 59-60. Mem: Am Soc Biol Chem; Am Chem Soc. Res: Protein interactions; microcalorimetry. Mailing Add: Dept of Biochem Kans State Univ Manhattan KS 66506

BURKHARDT, CHRISTIAN CARL, b Palmer, Nebr, Dec 26, 24; m 46; c 3. ENTOMOLOGY. Educ: Kans State Col, BS, 50, MS, 51; Univ Mo, PhD(entom), 67. Prof Exp: Field aide, European corn borer surv, Bur Entom & Plant Quarantine, USDA, 51; asst entomologist, Agr Exp Sta, Kans State Univ, 51-64, asst prof entom, 55-64; res entomologist, Univ Mo, 64-67; PROF ENTOM, UNIV WYO, 67- Mem: AAAS; Entom Soc Am; Am Soc Sugar Beet Technol. Res: Life history, biology and control of insects attacking cereal and forage crops; soil insects; biological control; ecological studies. Mailing Add: Dept of Entom Univ of Wyo Laramie WY 82070

BURKHARDT, HANS JOACHIM, biochemistry, see 12th edition

BURKHARDT, JAMES LEE, b Santa Monica, Calif, Oct 17, 29; m 51; c 2. EXPERIMENTAL PHYSICS. Educ: Mass Inst Technol, SB, 51, PhD(physics), 55. Prof Exp: Instr physics, Mass Inst Technol, 54-55; group leader, Hermes Electronics

Co, 55-60, dir res & develop div, 60; dir electronic res div, Itek Labs, Inc, 60-61; vpres & chief physicist, Elcon Lab, Inc, 61-63; PRES, APPL SCI ASSOCS, 63- Concurrent Pos: Consult, Sci Teaching Ctr & Exp Astron Lab, Mass Inst Technol, 63-68 & Oak Ridge Inst Nuclear Studies, 65-69; vpres, Adams-Russell Co Inc, 69-70. Mem: Am Phys Soc. Res: Educational apparatus; optical instruments and systems; electronic instrumentation. Mailing Add: 17 Centre St Watertown MA 02172

BURKHARDT, WALTER H, b Stuttgart, Ger, US citizen. COMPUTER SCIENCE. Educ: Univ Stuttgart, Vordiplom, 51, Dipl Phys, 54, Dr rer nat(physics), 59. Prof Exp: Sr asst programmer, World Trade Co & Data Syst Div, IBM Corp, 61-64; prin systs programmer, Univac Div, Sperry Rand Co, 64-66; sr staff res & develop, Comput Control Div, Honeywell, Inc, 65-66; staff prog mgr advan systs, Info Systs Div, RCA Corp, 66-69; assoc prof comput sci, Univ Pittsburgh, 69-74; CHMN & PROF HARDWARE, INST INFO, UNIV STUTTGART, 74- Concurrent Pos: Reviewer, Comput Rev, 66; dir, DPM Comput Leasing Co, 69; mem exec comt, Comput Ctr, Univ Pittsburgh, 72. Mem: Inst Mgt Sci; Asn Comput Mach; Am Mgt Asn. Res: Computer and operating systems and their languages; micro-programming and micro-implementation. Mailing Add: 12 Azenbergst D-7000 Stuttgart - 1 West Germany

BURKHART, BERNARD A, biochemistry, see 12th edition

BURKHART, HAROLD EUGENE, b Wellington, Kans, Feb 29, 44; m 71. FOREST BIOMETRY. Educ: Okla State Univ, BS, 65; Univ Ga, MS, 67, PhD(forest biomet), 69. Prof Exp: Asst prof, 69-73, ASSOC PROF FOREST BIOMET, VA POLYTECH INST & STATE UNIV, 73- Mem: AAAS; Soc Am Foresters; Biomet Soc. Res: Forest growth and yield. Mailing Add: Dept of Forestry & Forest Prod Va Polytech Inst & State Univ Blacksburg VA 24061

BURKHART, RICHARD DELMAR, b Kersey, Colo, June 26, 34; m 58; c 3. PHYSICAL CHEMISTRY. Educ: Dartmouth Col, AB, 56; Univ Colo, PhD(phys chem), 60. Prof Exp: Res chemist, Chem Div, Union Carbide Corp, 60-63; res assoc chem kinetics, Univ Ore, 63-65; from asst prof to assoc prof, 65-71, PROF PHYS & ANAL CHEM, UNIV NEV, RENO, 71- Mem: Am Chem Soc. Res: Chemical kinetics; diffusion of excited species; photochemistry. Mailing Add: Dept of Chem Univ of Nev Reno NV 89507

BURKHART, SARAH MAYBELLE, b Tulsa, Okla, Jan 16, 10. MATHEMATICS. Educ: Univ Tulsa, BS, 30, BM, 35; Univ Kans, MA, 31; Columbia Univ, PhD(math), 56. Prof Exp: Teacher pub schs, Okla, 32-40, asst to prin, 40-42; assoc prof math, Univ Tulsa, 46-65; SUPVR MATH, TULSA PUB SCHS, 65- Concurrent Pos: Asst supt in charge math curriculum, Tulsa County Schs, 54-59; vis assoc prof, Fla State Univ, 59-60; mem fac, NSF Inst Math Teachers, Southeastern State Col, 61-62 & Tulsa Conf Educ, 64; US del, Int Cong Math Educators, France, 69; math consult, Agency Int Develop Prog, India, 71; del, 2nd Int Cong Math Educators, Exeter, Eng, 72; founder Chi Alpha Mu, Nat Jr Math Club, 67, exec secy, 75-; regional rep, Nat Coun Teachers Math, 72-76; math consult, Educ Develop Co, 75-76. Mem: Math Asn Am; Nat Coun Teachers Math. Res: Generation of cubics in projective geometry; study on concepts of differential calculus; development of lessons in modern mathematics for junior high school; computer assisted instruction in mathematics; curriculum for grades 3-9. Mailing Add: 1208 E 21st St Tulsa OK 74114

BURKHEAD, MARTIN SAMUEL, b Ogden, Utah, May 23, 33; m 56; c 2. ASTROPHYSICS. Educ: Tex A&M Univ, BS, 55; Univ Calif, Los Angeles, MS, 57, PhD(astron), 64. Prof Exp: Sta chief, Smithsonian Satellite Tracking Prog, 57-60; res assoc astron, Univ Wis, 61-64; asst prof, 64-71, ASSOC PROF ASTRON, IND UNIV, BLOOMINGTON, 71-, ASSOC DIR, GOETHE LINK OBSERV, 74- Res: Photoelectric photometry. Mailing Add: Dept of Astron Ind Univ Bloomington IN 47401

BURKHOLDER, DAVID FREDERICK, b Grand Rapids, Mich, Mar 1, 31; m 54; c 1. PHARMACY. Educ: Ferris State Col, BS, 53; Univ Mich, MS, 61, PharD, 62. Prof Exp: Dir drug info ctr, asst dir pharm & res assoc, Med Ctr, Univ Ky, 62-67; assoc prof pharm & dir ctr pharmaceut pract, Sch Pharm, State Univ NY Buffalo, 67-70; ASSOC PROF PHARM, SCH PHARM, UNIV MO-KANSAS CITY, 70-, ASST DEAN SCH PHARM, 74- Concurrent Pos: Consult, Coun Dent Therapeut, Am Dent Asn, 62-; contrib ed, Int Pharmaceut Abstr, 64-; vis lectr, Coun Pharmaceut Educ, 64-; mem adv comt admis, Nat Formulary; comt revision, US Pharmacopoeia. Mem: Am Pharmaceut Asn; Am Soc Hosp Pharmacists. Res: Hospital pharmacy; rational drug therapy; statistical studies on drug selection and usage; drug literature organization and evaluation. Mailing Add: Sch of Pharm Univ of Mo Kansas City MO 64110

BURKHOLDER, DONALD LYMAN, b Octavia, Nebr, Jan 19, 27; m 50; c 3. MATHEMATICS. Educ: Earlham Col, BS, 50; Univ Wis, MS, 53; Univ NC, PhD(math statist), 55. Prof Exp: From asst prof to assoc prof, 55-64, PROF MATH, UNIV ILL, URBANA, 64- Mem: Am Math Soc; fel Inst Math Statist (pres, 75-76). Res: Probability and its applications to analysis. Mailing Add: Dept of Math Univ of Ill Urbana IL 61801

BURKHOLDER, JOHN HENRY, b Octavia, Nebr, July 11, 25; m 50; c 4. ZOOLOGY. Educ: McPherson Col, AB, 49; Univ Chicago, PhD(zool), 54. Prof Exp: Head dept, 54-70, PROF BIOL, McPHERSON COL, 53-, CHMN DIV NATURAL SCI, 67- Mem: AAAS; Genetics Soc Am; Am Inst Biol Sci. Res: Genetics; position effects in Drosophila melanogaster. Mailing Add: Div of Natural Sci McPherson Col McPherson KS 67460

BURKHOLDER, PAUL RUFUS, marine biology, deceased

BURKHOLDER, PETER M, b Cambridge, Mass, May 7, 33; m 56; c 2. PATHOLOGY, IMMUNOLOGY. Educ: Yale Univ, BS, 55; Cornell Univ, MD, 59; Am Bd Path, dipl, 64. Prof Exp: Intern path, New York Hosp-Cornell Med Ctr, 59-60; asst, Med Col, Cornell Univ, 60-62, from instr to asst prof, 62-65; asst prof, Med Ctr, Duke Univ, 65-70; assoc prof, 70-72, chmn dept, 72-74, PROF PATH, SCH MED, UNIV WIS-MADISON, 72- Concurrent Pos: USPHS trainee, 60-63, grant, 63-65; asst pathologist, New York Hosp, 60-63, asst attend pathologist, 63-65. Mem: Am Asn Path & Bact; Am Asn Immunol; Am Soc Exp Path; NY Acad Sci; Reticuloendothelial Soc. Res: Experimental pathology; immunochemistry of serum complement; immunohistochemistry of renal disease. Mailing Add: Dept of Path Med Sch Univ Wis Madison WI 53706

BURKHOLDER, WENDELL EUGENE, b Octavia, Nebr, June 24, 28; m 51; c 4. ENTOMOLOGY. Educ: McPherson Col, AB, 50; Univ Nebr, MSc, 56; Univ Wis, PhD(entom), 67. Prof Exp: From asst prof to assoc prof, 67-75, PROF ENTOM, UNIV WIS-MADISON, 75-; ENTOMOLOGIST, USDA, 56- Mem: Entom Soc Am. Res: Insect pheromones; reproductive biology; sex attractants; insect behavior; dermestid beetles and other stored product insects and mites. Mailing Add: Dept of Entom Univ of Wis Madison WI 53706

BURKI, HENRY JOHN, b Darby, Pa, Aug 8, 40; m 65. CELL BIOLOGY, RADIATION BIOLOGY. Educ: Rutgers Univ, BS, 62, MS, 63; Univ Rochester, PhD(biophys), 67. Prof Exp: Physics trainee & physicist, Aviation Med Lab, Pa, 58-62; AEC fel biophys, 67-68, actg asst prof med physics, 68-70, ASST PROF MED PHYSICS, UNIV CALIF, BERKELEY, 70- Concurrent Pos: Deleg, Int Atomic Energy Symp Biol Aspects Radiation Qual, Australia, 71. Mem: AAAS; Biophys Soc; Radiation Res Soc; Health Physics Soc. Res: Biophysical and radiobiological aspects of radiation quality in mammalian cells; organization of genetic information in mammalian cells; mammalian cellular physiology in cell culture. Mailing Add: 308 Donner Lab Univ of Calif Berkeley CA 94720

BURKIG, JACK WHIPPLE, b Weed, Calif, Mar 29, 22; m 52; c 2. NUCLEAR PHYSICS. Educ: Univ Redlands, AB, 43; Univ Calif, Los Angeles, MA, 48, PhD(physics), 51. Prof Exp: Res scientist solid state physics, Nat Adv Comt Aeronaut, 50-51; res engr, Hughes Aircraft Co, 51-53; instr physics, Reed Col, 53-54 & Univ Calif, Los Angeles, 54-55; sr engr, Magnavox Res Lab, 55-58 & Hughes Aircraft Co, 58-60; mem staff, AC Spark Plug Div, Gen Motors Corp, 60-64 & Space Tech Labs, Inc, Thompson-Ramo-Wooldridge, Inc, 64-75; RETIRED. Mem: Am Phys Soc. Res: Elastic scattering of protons; cold work in metals; proton-proton and proton-alpha scattering; thin film circuits; electro-optical systems design and analysis. Mailing Add: 2184 E Evans Creek Rd Rogue River OR 97537

BURKLE, JOSEPH S, b Philadelphia, Pa, July 28, 19; m 44; c 3. NUCLEAR MEDICINE. Educ: Univ Pa, AB, 40, MD, 43; Am Bd Internal Med, dipl, 55. Prof Exp: Intern, US Naval Hosp, Pa, US Navy, 44, resident med, Pa Hosp, 47-48, resident med, US Naval Hosp, Pa, 50-51, clin & lab instr, Radioisotope Lab, US Naval Hosp, St Albans, NY, 56-60, mem staff, Naval Hosp, Bethesda, Md, 60-63 & US Naval Sta Hosp, 63-66, dep dir, Navy Armed Forces Radiobiol Res Inst, Defense Atomic Support Agency, 66-67; CHMN DEPT NUCLEAR MED, YORK HOSP, 67-; ASST CLIN PROF MED, UNIV MD, 70- Mem: AMA; fel Am Col Physicians; Soc Nuclear Med; Radiation Res Soc; Am Inst Ultrasound in Med. Res: Internal medicine; isotopes in medicine; hematology. Mailing Add: Dept of Nuclear Med York Hosp York PA 17405

BURKMAN, ALLAN MAURICE, b Waterbury, Conn, Apr 23, 32; m 65; c 2. PHARMACOLOGY. Educ: Univ Conn, BS, 54; Ohio State Univ, MSc, 55, PhD(pharmacol), 58. Prof Exp: Teaching asst pharmacol, Ohio State Univ, 54-57; asst prof, Univ Ill, 58-63; assoc prof, Butler Univ, 63-66; assoc prof, 66-71, PROF PHARMACOL, OHIO STATE UNIV, 71- Honors & Awards: Mead Johnson Award for Undergrad Res Direction, 65. Mem: AAAS; Am Pharmaceut Asn; Soc Exp Biol & Med; Am Soc Pharmacol & Exp Therapeut. Res: Comparative avian pharmacology; pharmacodynamics of exotic aporphines; emetic and antiemetic mechanism. Mailing Add: Div of Pharmacol Col of Pharm Ohio State Univ Columbus OH 43210

BURKMAN, ERNEST, b Detroit, Mich, Oct 4, 29; m 54; c 4. SCIENCE EDUCATION. Educ: Eastern Mich Univ, BS, 52; Univ Mich, MS, 56, MA, 59, EdD(ed), 62. Prof Exp: Teacher high sch, Mich, 55-60; from asst prof to assoc prof, 60-67, head dept, 65-67, PROF SCI EDUC, FLA STATE UNIV, 67- Concurrent Pos: Consult sci curriculum develop, Various Sch Systs, 60-; estab nat high sch sci, Ford Found & Turkish Ministry Educ, 62-; regional consult biol sci curriculum study, 62-63; dir intermediate sci curriculum study, US Off Educ, 67-; dir, Intermediate Sci Curriculum Study, 67-71; dir, Educ Res Inst, Fla State Univ, 70-72, dir div instructional design & personnel develop, 72-74; dir, NSF Individualized Sci Instructional Syst, 72-75. Mem: AAAS; Nat Asn Biol Teachers; Nat Sci Teachers Asn; Nat Asn Res Sci Teaching. Res: Science curriculum development. Mailing Add: Col of Educ Fla State Univ Tallahassee FL 32306

BURKO, HENRY, b Detroit, Mich, Feb 24, 29; m 52; c 3. MEDICINE. Educ: Wayne State Univ, BS, 50, MD, 53. Prof Exp: From instr to asst prof radiol, State Univ NY Downstate Med Ctr, 59-64; assoc prof, 64-70, PROF RADIOL, VANDERBILT UNIV, 70- Concurrent Pos: Consult, Vet Admin Hosp, Brooklyn, 61-64, St Aubins Naval Hosp, 63-64 & Vet Admin Hosp, Nashville, 64- Mem: Soc Pediat Radiol. Res: Cardiac cineangiocardiography; cardiovascular and pediatric radiology. Mailing Add: Dept of Radiol Vanderbilt Univ Nashville TN 37203

BURKOTH, TERRY LEE, organic chemistry, see 12th edition

BURKS, BARNARD DE WITT, b East Las Vegas, NMex, Nov 12, 09. ENTOMOLOGY. Educ: Univ Ill, AB, 33, AM, 34, PhD(entom), 37. Prof Exp: Asst entomologist, State Natural Hist Surv, Ill, 35-42, assoc taxonomist, 46-49; assoc entomologist, 38, 41, entomologist, 49-64, COLLAB SCIENTIST, AGR RES SERV, USDA, 64- Mem: Fel Entom Soc Am. Res: Taxonomy of parasitic hymenoptera. Mailing Add: Syst Entom Lab USDA ARS-W 6B003 Agr Res Ctr Beltsville MD 20705

BURKS, ROBERT ELBERT, JR, b La Grange, Ga, Aug 24, 17; m 46; c 1. ORGANIC CHEMISTRY. Educ: Ga Inst Technol, BS, 38; Univ Wis, MS, 42, PhD(org chem), 48. Prof Exp: Chemist, 45 & 47-53, HEAD ORG CHEM SECT, SOUTHERN RES INST, 53- Mem: Am Chem Soc. Res: Catalytic hydrogenation; foods and flavors; radiation preservation of foods; butadiene dimerization; high temperature polymers; synthetic flocculants; depilatories; polyamides; fluorocarbon-siloxane polymers. Mailing Add: Southern Res Inst 2000 Ninth Ave S Birmingham AL 35205

BURKS, STERLING LEON, b Reydon, Okla, Mar 3, 38. ECOLOGY, LIMNOLOGY. Educ: Southwestern State Col, Okla, BS, 63; Okla State Univ, MS, 65, PhD(zool), 69. Prof Exp: Res assoc water pollution, 69-70, DIR ANAL LAB, RESERVOIR RES CTR, OKLA STATE UNIV, 70-, ASST PROF ZOOL, UNIV, 71- Concurrent Pos: Fel, Fed Water Pollution Control Admin, US Dept Interior, 68-70, proj dir, Off Water Resources Res, 70-72. Mem: Water Pollution Control Fedn. Res: Identification of fish toxins in surface waters with gas chromatography, mass spectrometry and atomic absorption spectrophotometry; determination of effects of chemical contaminants upon aquatic organisms. Mailing Add: Dept of Zool Okla State Univ Stillwater OK 74074

BURKS, THOMAS F, b Houston, Tex, Apr 3, 38; m 62. PHARMACOLOGY. Educ: Univ Tex, BS, 62, MS, 64; Univ Iowa, PhD(pharmacol), 67. Prof Exp: From instr to assoc prof pharmacol, Sch Med, Univ NMex, 67-71; assoc prof, 71-74, PROF PHARMACOL, UNIV TEX MED SCH, HOUSTON, 74- Concurrent Pos: USPHS fel, Nat Inst Med Res, Eng, 67-68; USPHS res grant, 69-; consult, US Air Force, 68- Mem: Am Soc Pharmacol & Exp Therapeut; Am Fedn Clin Res; Soc Exp Biol & Med; AAAS; Am Heart Asn. Res: Pharmacology of peripheral and central neurohumoral transmission; mode of action of narcotic analgesic agents. Mailing Add: Dept of Pharmacol Univ of Tex Med Sch Houston TX 77025

BURKSTRAND, JAMES MICHAEL, b Philadelphia, Pa, Jan 29, 46; m 67; c 2. SURFACE PHYSICS. Educ: Rensselaer Polytech Inst, BS, 67; Univ Ill, Urbana, MS, 69, PhD(physics), 72. Prof Exp: Res assoc surface physics, Univ Ill, 72; ASSOC SR RES PHYSICIST, GEN MOTORS RES LABS, 72- Mem: Am Vacuum Soc; Am Phys Soc; Sigma Xi. Res: Experimental investigation of geometrical and electronic properties of solid surfaces and the application of this information to the solution of technological problems. Mailing Add: Dept of Physics Gen Motors Res Labs Warren MI 48090

BURKWALL, MORRIS PATON JR, b Kansas City, Mo, May 16, 39; m 63; c 2. BIOCHEMISTRY, FOOD CHEMISTRY. Educ: Fla State Univ, BS, 61; Univ Minn, MS, 63, PhD(biochem), 66. Prof Exp: Group leader food chem, 66-70, sr group leader, 70-71, sect mgr semi-moist pet foods, 71-72, sect mgr dog food res, 72-75, MGR EXPLOR PET FOOD, JOHN STUART RES LAB, QUAKER OATS CO, 75- Honors & Awards: Sherwood Award, Am Asn Cereal Chem, 64. Mem: Am Chem Soc; Inst Food Technol; Am Asn Cereal Chem. Res: Proteins and enzymes of cereal grains; biochemical interactions in foods; water activity and semi-moist food concepts; carbohydrates and lipids of cereal grains. Mailing Add: Quaker Oats Res Div 617 W Main St Barrington IL 60010

BURKY, ALBERT JOHN, b Utica, NY, June 30, 42; m 70; c 2. PHYSIOLOGICAL ECOLOGY. Educ: Hartwick Col, BA, 64; Syracuse Univ, PhD(zool), 69. Prof Exp: Vis asst prof zool, Syracuse Univ, 69; fel, Cent Univ Venezuela, 69-70; instr biol, Case Western Reserve Univ, 71-73; ASST PROF BIOL, UNIV DAYTON, 73- Mem: AAAS; Am Soc Zoologists; Ecol Soc Am; Malacol Soc London. Res: Physiological ecology of freshwater clams, snails and terrestrial snails; energy budgets of natural populations, especially growth, reproduction and metabolism, and the physiological metabolic adaptations to environmental conditions. Mailing Add: Dept of Biol Univ of Dayton Dayton OH 45469

BURLAGA, LEONARD F, b Superior, Wis, Oct 1, 38. PHYSICS. Educ: Univ Chicago, BS, 60; Univ Minn, MS, 62, PhD(physics), 66. Prof Exp: Teaching asst physics, Univ Minn, 60-62, teaching assoc, 62-63, res assoc, 63-66; Nat Acad Sci-Nat Res Coun res assoc, 66-68, MEM STAFF AEROSPACE TECHNOL FIELDS & PARTICLES, GODDARD SPACE FLIGHT CTR, NASA, 68- Mem: Am Phys Soc; Geophys Union. Res: Interplanetary medium; propagation of cosmic rays. Mailing Add: Code 692 NASA Goddard Space Flight Ctr Greenbelt MD 20771

BURLAGE, HENRY MATTHEW, b Rensselaer, Ind, May 23, 97; m 25; c 1. PHARMACY, PHARMACEUTICAL CHEMISTRY. Educ: Ind Univ, AB, 19; Harvard Univ, AM, 21; Purdue Univ, BS, 24; Univ Wash, PhD(pharm), 27. Hon Degrees: DSc, Purdue Univ, 61. Prof Exp: Instr pharm, 22-27; assoc prof drug anal, Ore State Col, 27-29; from asst prof to assoc prof pharm & pharmaceut chem, Purdue Univ, 29-31; prof pharm, Univ NC, 31-47; dean, 47-62, prof pharm & pharmaceut chem, 47-71, EMER PROF PHARM, UNIV TEX, AUSTIN, 71- Concurrent Pos: Chemist, State Bd Pharm, Ore, 27-29; chmn, Nat Pharmaceut Syllabus Comt, 37-47. Mem: AAAS; Am Chem Soc; Am Pharmaceut Asn (hon pres); fel Am Col Apothecaries; Am Asn Cols Pharm (vpres, 59-60, pres, 60-61). Res: Volatile oils; plant chemistry; pharmaceutical products; phytochemical studies; chromatographic studies of plants; antibacterial activity of plant extracts. Mailing Add: 702 E 43rd St Austin TX 78751

BURLAGE, STANLEY R, b Rochester, NY, Dec 12, 32; m 59; c 3. APPLIED PHYSICS. Educ: John Carroll Univ, BS, 58, MS, 59. Prof Exp: Instr physics, John Carroll Univ, 59-60; physicist, Electronic Res Div, Clevite Corp, 60-65, res physicist, 67-69; mem res staff, Stand Oil Co, Ohio, 65-67; MGR ENG & APPL PHYSICS, RES DEPT, ADDRESSOGRAPH-MULTIGRAPH CORP, 69- Concurrent Pos: Lect physics, John Carroll Univ, 67- Mem: Acoust Soc; Inst Elec & Electronics Eng. Res: New product development; business equipment systems; electrophotography; lithography; micro film processing and retrieval; credit authorization systems; magnetic credit card terminals and systems development. Mailing Add: Res Dept Addressograph-Multigraph Corp Cleveland OH 44122

BURLANT, WILLIAM JACK, b Chicago, Ill, Oct 20, 28; m 55; c 2. ORGANIC CHEMISTRY. Educ: City Col New York, BS, 49; Brooklyn Col, MA, 51; Polytech Inst Brooklyn, PhD(chem), 55. Prof Exp: USPHS asst, Brooklyn Col, 50-52, instr org & gen chem, 53-54; Res Corp asst, Polytech Inst Brooklyn, 52-53; instr org & gen chem, Cooper Union, 54-55; chemist, Sci Lab, 55, mgr mat applns dept, Res Staff, 55-69, ASST DIR CHEM SCI, FORD MOTOR CO, 69- Mem: Am Chem Soc. Res: Radiation chemistry of organic systems; structure and properties of polymers; plastic applications; composites. Mailing Add: 29530 Woodhaven Lane Southfield MI 48075

BURLEIGH, BRUCE DANIEL, JR, b Augusta, Ga, June 23, 42; m 62. BIOCHEMISTRY, PROTEIN CHEMISTRY. Educ: Carnegie-Mellon Univ, BS, 64; Univ Mich, MS, 67, PhD(biochem), 70. Prof Exp: Vis scholar molecular biol, Med Res Coun Lab, Cambridge, Eng, 70-72, res staff, 72-73; ASST PROF BIOCHEM, M D ANDERSON HOSP & TUMOR INST, UNIV TEX SYSTS CANCER CTR & GRAD SCH BIOMED SCI, 73- Concurrent Pos: Am Cancer Soc fel, 70-72. Mem: AAAS; Am Chem Soc. Res: Structure and function of peptide hormones; hormonal control of cell states. Mailing Add: Univ of Tex Systs Cancer Ctr M D Anderson Hosp & Tumor Inst Houston TX 77025

BURLEIGH, JAMES REYNOLDS, b Fresno, Calif, Sept 6, 36; m 59; c 4. PLANT PATHOLOGY, GENETICS. Educ: Fresno State Col, BS, 58; Wash State Univ, MS, 62, PhD(plant path), 65. Prof Exp: Res asst agron, Wash State Univ, 59-61, res asst plant path, 61-64; RES PLANT PATHOLOGIST, AGR RES SERV, USDA, 64-; ASSOC PROF PLANT SCI, CALIF STATE UNIV, CHICO, 71- Concurrent Pos: Asst prof plant physiol, Kans State Univ, 64-71. Mem: Am Phytopath Soc; Am Mycol Soc. Res: Epidemiology of cereal rust diseases. Mailing Add: Dept of Plant Sci Calif State Univ Chico CA 95926

BURLEIGH, JOSEPH GAYNOR, b Crowley, La, Jan 20, 42; m 69; c 1. INSECT ECOLOGY. Educ: Univ Southwestern La, BS, 64; La State Univ, MS, 66, PhD(entom), 70. Prof Exp: Mem planning & recreation dept, La State Parks Comn, 66-68; res assoc entom, Okla State Univ, 70-72; ASSOC PROF ENTOM, UNIV ARK, PINE BLUFF, 72- Mem: Entom Soc Am. Res: Ecology and population dynamics of the bollworm and tobacco budworm; effects of cultural practices on soybean insects. Mailing Add: Dept of Agr Univ of Ark Pine Bluff AR 71601

BURLEIGH, MALCOLM BRUCE, organic chemistry, photochemistry, see 12th edition

BURLESON, CHARLES ALBERTIS, b Mart, Tex, Jan 21, 21; m 45; c 2. AGRONOMY, SOILS. Educ: Tex A&M Univ, BS, 48, MS, 50; Clemson Univ, PhD, 65. Prof Exp: Asst agronomist, Tex Agr Exp Sta, 50-52, supt, 52-54, assoc agronomist, 55-63; assoc prof, Clemson Univ, 65-66; mgr argichem develop, 66-71, mgr res & develop, 71-76, SR AGRONOMIST, ANSUL CO, 76- Mem: Am Soc Agron; Soil Sci Soc Am. Res: Land use; soil fertility; soil science; plant physiology; weed science. Mailing Add: 1105 W Fifth St Weslaco TX 78596

BURLESON, GEORGE ROBERT, b Baton Rouge, La, Oct 12, 33; m 60; c 2. EXPERIMENTAL NUCLEAR PHYSICS, EXPERIMENTAL HIGH ENERGY PHYSICS. Educ: La State Univ, BS, 55; Stanford Univ, MS, 57, PhD(physics), 60. Prof Exp: Res assoc high energy physics, Argonne Nat Lab, 60-62, asst scientist, 62-64; asst prof physics, Northwestern Univ, 64-71; ASSOC PROF PHYSICS, N MEX STATE UNIV, 71- Concurrent Pos: Vis staff mem, Los Alamos Sci Lab, 72, 73 & 74. Mem: Am Phys Soc; Sigma Xi. Res: Electron scattering at medium energies; hyperon polarization; mesonic x-rays; polarization in hadron-nucleon scattering; pion-nucleus scattering at medium energies. Mailing Add: Dept of Physics Box 3D NMex State Univ Las Cruces NM 88003

BURLESON, JAMES C, b San Antonio, Tex, Sept 10, 25; m 51; c 3. ORGANIC CHEMISTRY. Educ: Tex Col Arts & Indust, BS, 50. Prof Exp: Chemist, Delhi Oil Co, 50-53 & Monsanto Co, 53-58; teacher, Ft Davis Indust Sch Dist, 58-59; CHEMIST, MONSANTO CO, 59- Mem: AAAS; Am Chem Soc. Res: High temperature chemistry. Mailing Add: Monsanto Co 800 N Lindbergh St Louis MO 63166

BURLEW, JOHN SWALM, b Washington, DC, Sept 10, 10; m 34; c 2. PHYSICAL CHEMISTRY. Educ: Bucknell Univ, AB, 30; Johns Hopkins Univ, PhD(chem), 34. Hon Degrees: ScD, Bucknell Univ, 55; ScD, Drexel Univ, 56. Prof Exp: Sterling fel, Yale Univ, 34-36; from asst phys chemist to phys chemist, Carnegie Inst Geophys Lab, 36-52; tech dir, Cambridge Corp, 52-54; from asst dir to dir, Franklin Inst, 54-56, exec vpres, 56-60; dir res, Carrier Corp, 60-66; dir Conn Res Comn, 66-71; PRES, NEW DIRECTIONS INC, 71- Concurrent Pos: Consult, Nat Defense Res Comt, 41-42, secy div one, 42-44, tech aide, 44-46. Mem: AAAS; Am Chem Soc. Res: Application of science and technology to government. Mailing Add: PO Box 418 Glastonbury CT 06033

BURLEY, DAVID RICHARD, b Crooksville, Ohio, Nov 1, 42; m 66; c 2. PHYSICAL CHEMISTRY. Educ: Ohio State Univ, BS, 64; Univ Calif, PhD(phys chem), 69. Prof Exp: Fel photochem, Univ Col Swansea, Wales, 69-71; head chem res, Tile Coun Am, 71-75; SR CHEMIST, AM CYANAMID CO, 75- Mem: Am Chem Soc. Res: Photochemistry; surface chemistry; chemical technology; management. Mailing Add: 9 Meadowview Dr Cranbury NJ 08512

BURLEY, GORDON, b Giessen, Ger, Feb 15, 25; nat US; m 57, 72. PHYSICAL CHEMISTRY, SCIENCE ADMINISTRATION. Educ: Temple Univ, AB, 48; Univ Md, MS, 52; Georgetown Univ, PhD, 62. Prof Exp: Res assoc, Geophys Lab, Carnegie Inst Wash, 50-52; phys chemist, Nat Bur Stand, 52-67; phys chemist, Div Reactor Licensing, US AEC, 67-72; BR CHIEF, RADIATION PROG, US ENVIRON PROTECTION AGENCY, 72- Mem: Am Chem Soc; Am Crystallog Asn. Res: Solid state chemistry; environmental radiation standards; theoretical models for environmental transport and population impact of radionuclides. Mailing Add: Off of Radiation Progs US Environ Protection Agency Washington DC 20460

BURLEY, J WILLIAM ATKINSON, b Moundsville, WVa, Mar 31, 28; m 49; c 4. PLANT PHYSIOLOGY, RADIATION BIOLOGY. Educ: WVa Univ, BA & MS, 54; Ohio State Univ, PhD(bot), 60. Prof Exp: From instr to assoc prof bot, Ohio State Univ, 57-68; prof, Bowling Green State Univ, 68-72; PROF BIOL SCI & HEAD DEPT, DREXEL UNIV, 72- Concurrent Pos: Consult radioisotope assayist, 61-; NSF res grant, 63- Mem: AAAS; Am Soc Plant Physiol. Res: Synthesis and translocation of C14-labeled compounds in vascular plants; uptake and translocation of radioactive labeled ions in corn and pea roots; analysis of light intensity and quality in natural environments. Mailing Add: Dept of Biol Sci Drexel Univ Philadelphia PA 19104

BURLING, JAMES P, b Baltimore, Md, May 29, 30; m 61; c 2. MATHEMATICS. Educ: Grinnell Col, BA, 52; State Univ NY Albany, MA, 57; Univ Colo, PhD(math), 65. Prof Exp: High sch teacher, NY, 55-56; asst prof math, State Univ NY Col Oneonta, 57-60; PROF MATH, STATE UNIV NY COL OSWEGO, 65- Mem: AAAS; Math Asn Am; Am Math Soc. Res: Coloring problems of families of convex bodies. Mailing Add: Dept of Math State Univ of NY Col Oswego NY 13126

BURLING, RICHARD LANCASTER, b Ottawa, Ont, Sept 19, 16; nat US; div; c 2. PHYSICS. Educ: Univ Colo, BA, 37; Univ Wis, PhD(physics), 41. Prof Exp: Asst physics, Univ Wis, 37-41; instr chem, Johns Hopkins Univ, 41-42; Civilian Pub Serv assignee, 42-46; asst physicist, Sloan-Kettering Inst Cancer Res, New York, 46-47; asst prof physics, Wash Univ, 47-48; asst prof & actg chmn dept, Knox Col, 48-50; asst prof, Univ Hawaii, 50-53; computer, Western Geophys Co Am, 54-55; lectr physics, Univ PR, 55-56 & Univ Pa, 56-57; assoc prof, Antioch Col, 58-59; chmn dept, 60-65, PROF PHYSICS, CENT STATE UNIV, 60- Concurrent Pos: Fulbright lectr, Haile Selassie Univ, 65-67. Mem: AAAS; Am Asn Physics Teachers; Fedn Am Scientists. Res: Experimental nuclear physics; physics of soil conservation; geologic time; radiation and other environmental hazards; war prevention; population control. Mailing Add: 602 Robinwood Dr Yellow Springs OH 45387

BURLING, ROBBINS, b Minneapolis, Minn, Apr 18, 26; m 51; c 3. ANTHROPOLOGY. Educ: Yale Univ, BA, 50; Harvard Univ, PhD(anthrop), 58. Prof Exp: From instr to asst prof anthrop, Univ Pa, 57-63; assoc prof anthrop, 63-67, PROF ANTHROP & LING, UNIV MICH, ANN ARBOR, 67- Concurrent Pos: Fulbright lectr, Univ Rangoon, 59-60; fel, Ctr Advan Study Behav Sci, Stanford, Calif, 63-64; Guggen- heim Found fel, Toulouse, France, 71-72. Mem: Am Anthrop Asn; Asn Asian Studies; Ling Soc Am. Res: Social organization; sociolinguistics; Southeast Asia. Mailing Add: Dept of Anthrop Univ of Mich Ann Arbor MI 48104

BURLING, RONALD WILLIAM, b Masterton, NZ, Aug 4, 20; m 48; c 1. OCEANOGRAPHY. Educ: Univ NZ, BSc, 49, MSc, 50; Univ London, PhD(meteorol), 55. Prof Exp: Scientist, Oceanog Observ, Dept Sci & Indust Res, NZ, 49-51; mem staff, Nat Inst Oceanog, Eng, 53-55; scientist, Oceanog Observ, Dept Sci & Indust Res, NZ, 55-60; from asst prof to assoc prof phys & dynamical oceanog, 60-67, PROF OCEANOG, INST OCEANOG, UNIV BC, 67- Mem: Royal Meteorol Soc. Res: Physical and dynamical oceanography; wind generation of waves on water; wave observations; spectra of waves; correlations between wave properties and fluctuating properties of wind just above; air sea interaction; transfer of momentum, energy, heat and water vapor across the sea surface. Mailing Add: Inst of Oceanog Univ of BC Vancouver BC Can

BURLINGAME, ALMA L, b Cranston, RI, Apr 29, 37. PHYSICAL CHEMISTRY. Educ: Univ RI, BS, 59; Mass Inst Technol, PhD(chem), 62. Prof Exp: RES CHEMIST, SPACE SCI LAB, UNIV CALIF, BERKELEY, 63- Concurrent Pos: Mem, Lunar Sample Anal Planning Team & Lunar Sample Preliminary Exam Team, NASA Johnson Space Ctr, 69-73; Guggenheim fel, 70-71. Mem: AAAS; Am Chem Soc; Am Soc Mass Spectrometry; Am Geophys Union; Int Soc Study Origin of Life. Res: Development of real-time gas chromatography high resolution mass spectrometry; elucidation of organic molecular structures; applications to biomedical and environmental problems; field ionization mass spectrometry; origin and evolution of life on earth. Mailing Add: Space Sci Lab Univ of Calif Berkeley CA 94720

BURLINGTON, HAROLD, b Yonkers, NY, June 26, 25. PHYSIOLOGY. Educ: Franklin & Marshall Col, BS, 48; Syracuse Univ, MS, 49, PhD(physiol), 51. Prof Exp: Asst zool, Syracuse Univ, 48-51; instr physiol, NY State Col Med, Syracuse, 51-55; from asst prof to prof, Col Med, Univ Cincinnati, 55-68; PROF PHYSIOL, MT SINAI SCH MED, 68- Mem: AAAS; Am Physiol Soc; NY Acad Sci. Res: Cellular metabolism; regulation of growth and function. Mailing Add: Mt Sinai Sch of Med Fifth Ave at 100th St New York NY 10029

BURLINGTON, ROY FREDERICK, b South Bend, Ind, Oct 2, 36; m 59; c 3. ENVIRONMENTAL PHYSIOLOGY. Educ: Purdue Univ, BS, 59, MS, 61, PhD(environ physiol), 64. Prof Exp: Res physiologist, US Army Res Inst Environ Med, 66-70; PROF BIOL, CENT MICH UNIV, 70- Mem: Am Soc Zoologists; Am Physiol Soc. Res: Effects of pollution on aquatic invertebrate populations; hypoxic tolerance in and cellular physiology of hibernating mammals; effects of starvation on intermediary metabolism; myocardial metabolism and nutrition at high altitude. Mailing Add: Dept of Biol Cent Mich Univ Mt Pleasant MI 48858

BURLINSON, NICHOLAS EDWARD, b Bridgeport, Conn, Oct 14, 41; m 66; c 3. PHYSICAL ORGANIC CHEMISTRY. Educ: Fairfield Univ, BS, 63; Univ Md, MS, 66; Cath Univ Am, PhD(org chem), 72. Prof Exp: RES CHEMIST PHYS ORG CHEM, NAVAL ORD LAB, WHITE OAK, MD, 66- Mem: Am Chem Soc; Sigma Xi. Res: Chemistry of nitro aromatics and nitro aliphatics; photochemistry of nitro aromatics and structure-reactivity relationships of nitro compounds. Mailing Add: 15009 Butterchurn Lane Silver Spring MD 20904

BURMAN, HAROLD GARDNER, chemistry, see 12th edition

BURMAN, LOUIS ROBERT, b New York, NY, Aug 23, 06; m 32; c 2. PERIDONTOLOGY, ORAL MEDICINE. Educ: NY Univ, BS, 34, DDS, 40; George Washington Univ, AM, 35; Am Bd Periodont & Am Bd Oral Med, dipl. Prof Exp: Instr, 40-45, asst prof, 45-58, assoc prof, 58-71, CLIN PROF PERIODONT, COL DENT, NY UNIV, 71-, ASST PROF, POSTGRAD DIV, 51- Concurrent Pos: Instr anat, NY Univ, 41-51, asst prof, 51-58. Mem: Fel AAAS; fel Am Col Dent; fel Soc Geront; Sigma Xi; fel Am Acad Oral Med. Res: Spread of infections in head and neck; proliferation of gingival tissues; periodontol disease; application of special staining technics to diseases of the mouth; surgical anatomy of the mouth. Mailing Add: 200 Central Park S New York NY 10019

BURMAN, ROBERT L, b Chicago, Ill, June 13, 33; m 56; c 3. NUCLEAR PHYSICS. Educ: Mass Inst Technol, BS, 55; Univ Ill, MS, 57, PhD(physics), 61. Prof Exp: NSF fel, Bohr Inst Theoret Physics, Copenhagen, Denmark, 61-62, Ford Found fel, 62-63; asst prof physics, Univ Rochester, 63-68; STAFF MEM, LOS ALAMOS SCI LAB, 68- Mem: Am Phys Soc. Res: Beta decay; nucleon-nucleon forces; pion-nucleus interactions; neutrino physics. Mailing Add: Los Alamos Sci Lab MP-Div Los Alamos NM 87544

BURMASTER, CHESTER FREDERICK, organic chemistry, see 12th edition

BURMBLAY, RAY ULYSSES, b Azusa, Calif, Feb 8, 12; m 42; c 5. ANALYTICAL CHEMISTRY. Educ: Ind Univ, AB, 34; Univ Wis, PhD(inorg chem), 38. Prof Exp: Instr chem, Exten Div, Ind Univ, 38-43; instr tech subjects, Chem Warfare Sch, Edgewood Arsenal, Md, 45-46; from asst prof to prof chem, Univ Wis, Milwaukee, 46-69, chmn dept, 57-64, PROF CHEM, UNIV WIS, WAUSAU, 69- Mem: Am Chem Soc. Res: Single electrode potentials; electroplating of molybdenum; spectrophotometric analysis. Mailing Add: Dept of Chem Univ of Wis Wausau WI 54001

BURMEISTER, CHARLES W, b Pleasanton, Tex, Apr 4, 31; m 59. PHYSICS. Educ: Baylor Univ, BA & MA, 58; Univ Tex, PhD(physics), 63. Prof Exp: Asst prof physics, Trinity Univ, 58-63, assoc prof & chmn dept, 63-68, prof & dir res, 68-69; PRES, COMPUT KNOWLEDGE CORP, 69- Mem: AAAS; Am Phys Soc; Am Asn Physics Teachers. Res: Solid state physics; electronic properties of metals at low temperature; Fermi surface work. Mailing Add: Comput Knowledge Corp 4502 Centerview Dr San Antonio TX 78228

BURMEISTER, HARLAND RENO, b Seymour, Wis, Jan 10, 29; m 68; c 2. MICROBIOLOGY. Educ: Univ Wis, BS, 57; Iowa State Univ, PhD(bact), 64. Prof Exp: Jr chemist, A O Smith Corp, 58-60; RES MICROBIOLOGIST, NORTHERN UTILIZATION RES & DEVELOP DIV, AGR RES SERV, USDA, 64- Mem: Am Soc Microbiol; AAAS. Res: Mycotoxin investigations: Fusarium antibiotics and toxins. Mailing Add: Northern Util Res & Develop Div Agr Res Serv USDA Peoria IL 61604

BURMEISTER, JOHN LUTHER, b Fountain Springs, Pa, Feb 20, 38; m 60; c 2. INORGANIC CHEMISTRY. Educ: Franklin & Marshall Col, BS, 59; Northwestern Univ, PhD(coord chem), 64. Prof Exp: Instr inorg chem, Univ Ill, 63-64; from asst prof to assoc prof, 64-69, PROF CHEM, UNIV DEL, 69-, ASSOC CHMN DEPT, 74- Concurrent Pos: Mem ed bd, Inorganica Chimica Acta & Synthesis in Inorg & Metalorg Chem; exec secy, Intercollegiate Student Chemists, 68-71; Consult, Sun Oil Co, 69-74 & AMP, Inc, 71-73. Mem: Am Chem Soc; The Chem Soc. Res: Coordination chemistry of ambidentate ligands; inorganic linkage isomerism; activated bridge electron transfer reactions; oxidative addition reactions; polymer-bound transition metal catalyst systems. Mailing Add: Dept of Chem Univ of Del Newark DE 19711

BURMESTER, BEN ROY, b Petaluma, Calif, June 13, 10; m 33; c 3. VIROLOGY. Educ: Univ Calif, BS, 32, AM, 33, PhD(physiol), 36; Mich State Col, DVM, 51; Am Col Vet Microbiol, dipl, 64. Prof Exp: From asst to assoc, Univ Calif, 34-36; first asst & instr, Univ Ill, 36-40; sr biologist, Regional Poultry Res Lab, 40-63, dir lab, 64-75, COLLABR, REGIONAL POULTRY RES LAB, USDA, 75- Concurrent Pos: Mem res award comt, Poultry Sci Asn, 52-56; mem virol & rickettsiology study sect, Div Res Grants, NIH, 59-62; mem, Newman Mem Int Award Comt, 58- Honors & Awards: Award, Poultry Sci Asn, 41, Borden Award, 57; Jr Res Award, Sigma Xi, 48, Sr Res Award, 63; Certs Merit, USDA, 66, 71 & 72, Distinguished Serv Award, 75; Res Award, Am Poultry Hatchery Fedn, 71; Edward W Browning Award, Am Soc Agron, 73; Am Feed Mfrs Award, Am Vet Med Asn, 74. Mem: AAAS; fel Poultry Sci Asn; Am Asn Avian Path (pres, 61-62); fel Am Acad Microbiol; fel Leopoldina Ger Acad Scientists. Res: Oncogenic virology; avian diseases of poultry. Mailing Add: Regional Poultry Res Lab USDA 3606 E Mt Hope Rd East Lansing MI 48823

BURMESTER, MARY ALICE (HORSWILL), b Oakland, Calif, Sept 1, 09; m 33; c 3. PHYSIOLOGY. Educ: Univ Calif, AB, 30; Mich State Univ, MA, 48, EdD, 51. Prof Exp: Asst physiol, Univ Calif, 31-34; from asst prof to assoc prof natural sci, 45-65, PROF NATURAL SCI, MICH STATE UNIV, 65- Res: Methodology of science. Mailing Add: Dept of Natural Sci Mich State Univ East Lansing MI 48824

BURN, IAN, b Dewsbury, Eng, Nov 25, 37; US citizen; m 61; c 2. CERAMICS. Educ: Durham Univ, BSc, 60; Leeds Univ, PhD(ceramics), 66. Prof Exp: Physicist, Brit Oxygen Co Ltd, 60-62; res asst ceramics, Univ Leeds, 62-66; sr scientist, Pilkington Brothers Ltd, 66-67; SR TECH STAFF CERAMICS, SPRAGUE ELEC CO, 67-; ASSOC PROF, NORTH ADAMS STATE COL, 68- Mem: Brit Inst Physics. Res: Ceramic dielectrics, including interactions with base metals and glass-ceramic systems. Mailing Add: Res & Develop Ctr Sprague Elec Co North Adams MA 01247

BURNELL, EDWIN ELLIOTT, b St John's, Nfld, Dec 4, 43; m 71. CHEMICAL PHYSICS. Educ: Mem Univ Nfld, BSc, 65, MSc, 68; Bristol Univ, PhD(chem), 70. Prof Exp: Fel physics, Univ BC, 69-71 & Univ Basel, 71-72; ASST PROF CHEM, UNIV BC, 72- Mem: Am Phys Soc. Res: Nuclear magnetic resonance studies of molecules partially oriented in liquid crystal solvents and of soaps; model membranes and biomembranes; the physical basis of biological membrane function. Mailing Add: Dept of Chem Univ of BC Vancouver BC Can

BURNELL, JAMES MCINDOE, b Manila, Philippines, July 17, 21; US citizen; m 49; c 5. MEDICINE, PHYSIOLOGY. Educ: Stanford Univ, BA, 45, MD, 49. Prof Exp: Instr path, Columbia Univ, 49-50; asst med, 50-54, res assoc prof, 60-71, RES PROF MED, UNIV WASH, 71- Concurrent Pos: Res fel, Univ Wash & Pfizer Corp, 49-50. Res: Physiology of body fluid, potassium and acid-base; fluid and electrolytes. Mailing Add: Dept of Med Univ of Wash Seattle WA 98105

BURNELL, LOUIS A, b Belgium, Sept 15, 28; m 55; c 2. THEORETICAL CHEMISTRY. Educ: Univ Liege, BSc, 50, PhD(chem), 55. Prof Exp: Res molecular physics, Univ Liege, 55-60, lectr, 60-63; res physicist, Res Inst Advan Study, 63-66; assoc prof chem, NY Univ, 66-74; EXPERT, UNESCO, 74- Concurrent Pos: Belg govt travel grant, 55; Brit Coun scholar, Oxford Univ, 56-57; Nat Acad Sci grant, Ind Univ & Mass Inst Technol, 59-60. Honors & Awards: Stas-Spring Prize, Royal Academy Belg, 55. Mem: Am Chem Soc. Res: Quantum chemistry; interpretation of the electronic spectra of polyatomic molecules; theoretical problems concerned with the nature of the chemical bonds and molecular reactivity; history of science. Mailing Add: Apartado Aereo 3868 Bogota Columbia

BURNELL, ROBERT H, b Tondu, Wales, Nov 23, 29; m 50; c 3. ORGANIC CHEMISTRY. Educ: Sir George Williams Univ, BSc, 52; Univ NB, PhD(org chem), 55. Prof Exp: Lectr org chem, Univ West Indies, 55-62; investr nat prod, Venezuelan Inst Sci Res, 62-64; PROF CHEM, LAVAL UNIV, 64- Mem: Chem Inst Can; The Chem Soc. Res: Extraction and isolation of alkaloids from plants; structure elucidation of alkaloids. Mailing Add: Dept of Chem Laval Univ Quebec PQ Can

BURNESS, ALFRED THOMAS HENRY, b Birmingham, Eng, Feb 10, 34; m 59; c 2. VIROLOGY, MOLECULAR BIOLOGY. Educ: Univ Liverpool, BSc, 55, PhD(microbiol), 59. Prof Exp: Mem staff, Virus Res Unit, Brit Med Res Coun, 58-68; assoc, 68-73, ASSOC MEM, SLOAN-KETTERING INST CANCER RES, 74-; ASSOC PROF BIOL, SLOAN-KETTERING DIV, GRAD SCH MED SCI, CORNELL UNIV, 74- Concurrent Pos: USPHS int res fel, Virus Lab, Univ Calif, Berkeley, 62-63; asst prof biol, Grad Sch Med Sci, Cornell Univ, 71-74. Honors & Awards: Career Scientist Award, Health Res Coun of City of New York, 71. Mem: Brit Biochem Soc; NY Acad Sci; Am Soc Biol Chem; Am Soc Microbiol. Res: Relationship between structure and function in viruses; nature of cell receptors for viruses. Mailing Add: Sloan-Kettering Inst Cancer Res 145 Boston Post Rd Rye NY 10580

BURNESS, DONALD MAC ARTHUR, b Pittsfield, Mass, July 15, 17; m 41; c 2. ORGANIC CHEMISTRY. Educ: Worcester Polytech Inst, BS, 39; Univ Ill, PhD(org chem), 45. Prof Exp: Chemist, Eastman Kodak Co, 39-42; res chemist, Nat Defense Res Comt, Ill, 42-45; res chemist, 45-55, res assoc, 55-72, SR RES ASSOC, EASTMAN KODAK CO, 72- Mem: Am Chem Soc. Res: Organic compounds used in photographic emulsions; small ring compounds; synthetic Vitamin A; crosslinking agents; sulfur chemistry. Mailing Add: Eastman Kodak Co Res Labs Rochester NY 14650

BURNESS, JAMES HUBERT, b Philadelphia, Pa, Nov 20, 49; m 71. BIOINORGANIC CHEMISTRY. Educ: Rutgers Univ, BA, 71; Va Polytech Inst & State Univ, PhD(inorg chem), 75. Prof Exp: FEL BIOPHYS, MICH STATE UNIV, 75- Mem: Am Chem Soc; AAAS; Sigma Xi. Res: Model systems for vitamin B-12 and hemoglobin; binding of oxygen to metals; x-ray photoelectron spectroscopy of transition metal complexes; platinum complexes in cancer chemotherapy. Mailing Add: Dept of Biophys Mich State Univ East Lansing MI 48823

BURNETT, ALLISON L, b St Francis, Maine, Mar 3, 32; m 55; c 3. BIOLOGY. Educ: Bates Col, AB, 53; Cornell Univ, MS, 56, PhD(zool), 58. Prof Exp: Instr invert zool, Cornell Univ, 58-59; NSF fel zool, Univ Brussels, 59-60; asst prof biol, Univ Va, 60-61; from assoc prof to prof, Case Western Reserve Univ, 61-69; PROF BIOL, NORTHWESTERN UNIV, 69- Concurrent Pos: NSF grants, 60-; Am Cancer Soc grant, 61-62; NIH grants, 62- & NIH career develop award, 64-68; dir first nat teach-out on environ, Northwestern Univ, 70. Mem: AAAS; Soc Develop Biol; Am Asn Anat. Res: Developmental biology; growth and cell differentiation; chemical isolation of polarizing factors in coelenterates; problems of regeneration and senescence. Mailing Add: Dept of Biol Sci Northwestern Univ Evanston IL 60201

BURNETT, BRUCE BURTON, b Norristown, Pa, July 26, 27; m 50; c 2. ANALYTICAL CHEMISTRY, POLYMER PHYSICS. Educ: Lehigh Univ, BS, 50; Univ Ill, PhD(anal chem), 53. Prof Exp: Res chemist, Textile Fibers Dept, E I du Pont de Nemours & Co, 53-60, sr res chemist, 60-62; group leader anal chem, 62-69, SECT MGR ANAL & PHYS MEASUREMENTS, UNION-CAMP CO, 69- Mem: AAAS; Am Chem Soc; Tech Asn Pulp & Paper Indust. Res: Instrumental methods of analysis; spectroscopy; chromatography; x-ray physics of polymers and fibers; crystallization kinetics. Mailing Add: Res Div Union-Camp Co Box 412 Princeton NJ 08540

BURNETT, BRYAN REEDER, b Bethlehem, Pa, Aug 10, 45; m 75. INVERTEBRATE PHYSIOLOGY. Educ: San Diego State Univ, BS, 68, MS, 71; Scripps Inst Oceanog, Univ Calif, San Diego, MS, 75. Prof Exp: STAFF RES ASSOC, SCRIPPS INST OCEANOG, UNIV CALIF, SAN DIEGO, 75- Mem: AAAS; Am Soc Zoologists; Am Soc Limnol & Oceanog; Soc Syst Zool. Res: Morphology and physiology of blood circulation in the Crustacea. Mailing Add: Scripps Inst of Oceanog Univ of Calif at San Diego La Jolla CA 92093

BURNETT, CLYDE MARSHALL, b Cincinnati, Ohio, Jan 24, 23; m 44; c 5. TOXICOLOGY. Prof Exp: Mem res staff toxicol, Procter & Gamble Co, Inc, 50-63; mgr toxicol, Revlon, Inc, 63-73; ASSOC DIR RES TOXICOL, CLAIROL, INC, 73- Honors & Awards: Shaw Mudge Award, Soc Cosmetic Chemists, 72; CIBS Award, Cosmetic, Toiletry & Fragrance Asn, 74. Mem: Soc Toxicol. Res: Investigation of the toxicological properties of hair dyes and other chemicals used in the cosmetic industry. Mailing Add: Clairol Inc 2 Blachley Rd Stamford CT 06902

BURNETT, CLYDE RAY, b Nora Springs, Iowa, Dec 23, 23; m 47; c 5. PHYSICS. Educ: Univ Upper Iowa, BS, 46; Univ Wis, MS, 48, PhD(physics), 51. Prof Exp: Asst prof physics, SDak State Col, 50-53 & Pa State Univ, 53-57; physicist, Proj Matterhorn, Princeton Univ, 57-58; assoc prof physics, Pa State Univ, 58-63; res assoc & lectr, Univ Wis, 63-64; assoc prof & chmn dept, 64-70, PROF PHYSICS, FLA ATLANTIC UNIV, 70- Mem: Am Phys Soc; Optical Soc Am; Am Meteorol Soc; Am Asn Physics Teachers. Res: Atomic spectroscopy; plasma physics. Mailing Add: Dept of Physics Fla Atlantic Univ Boca Raton FL 33432

BURNETT, DONALD STACY, b Dayton, Ohio, June 25, 37; m 59; c 2. GEOCHEMISTRY. Educ: Univ Chicago, BS, 59; Univ Calif, Berkeley, PhD(chem), 63. Prof Exp: NSF res fel physics, 63-65, asst prof nuclear geochem, 65-68, ASSOC PROF NUCLEAR GEOCHEM, CALIF INST TECHNOL, 68- Res: Origin and abundances of the elements; nuclear chemistry. Mailing Add: Div of Geol Sci Calif Inst of Technol Pasadena CA 91109

BURNETT, EARL, b Brownfield, Tex, Sept 22, 22; m 44; c 2. SOIL SCIENCE. Educ: Tex Tech Univ, BS, 46, MS, 49; Ohio State Univ, PhD(soils), 52. Prof Exp: Instr agron, Cameron State Agr Col, 46-47; soil scientist, Soil Conserv Serv, USDA, 47-48; asst prof agron, Tex Tech Univ, 48-50; res fel soil sci, Ohio State Univ, 50-51; asst agronomist, Tex Agr Exp Sta, Tex A&M Univ, 51-57; res soil scientist, 57-69, scientist in charge, Blackland Res Ctr, 69-70, DIR BLACKLAND CONSERV RES CTR, SOIL & WATER CONSERV RES DIV, AGR RES SERV, USDA, 70- Concurrent Pos: Mem grad fac, Tex A&M Univ, 64-; dir, Am Soc Agron, 75-78. Mem: Fel AAAS; fel Am Soc Agron; fel Soil Sci Soc Am; fel Soil Conserv Soc Am; Soc Range Mgt. Res: Plant water use efficiency; soil compaction and soil structure; modification of soil environment for improved root development and water use efficiency. Mailing Add: Blackland Conserv Res Ctr Agr Res Serv USDA Box 748 Temple TX 76501

BURNETT, GEORGE WESLEY, b Graham, Tex, Nov 22, 14; m 43; c 2. MICROBIOLOGY, DENTISTRY. Educ: Tex Tech Col, BA, 37; Univ Tex, MA, 40; Wash Univ, DDS, 43; Univ Rochester, PhD, 50; Am Bd Med Microbiol, dipl. Prof Exp: Asst physiol, Wash Univ, 41-42; instr, Col Dent, Univ Tex, 46; chief dept dent res, Walter Reed Army Inst Res, DC, 50-60; chief dept oral biol, Army Inst Dent Res, 60-68, dir, Inst, 68; PROF ORAL BIOL, SCH DENT, MED COL GA, 68-, PROF CELL & MOLECULAR BIOL, SCH MED, 71-, ASSOC DEAN SCH DENT, 71- Concurrent Pos: Dent fel bact, Univ Rochester, 46-50; consult microbiol, Off Surgeon Gen, US Army, 60-68, mem dent res adv comt, 48-68; mem dent study sect, USPHS, 53-58; prof lectr, Georgetown Univ, 63-68. Mem: AAAS; Am Soc Microbiol; Int Asn Dent Res; Sigma Xi; fel Am Col Dent. Res: Oral disease; nutrition. Mailing Add: Sch of Dent Med Col of Ga Agusta GA 30902

BURNETT, JEAN BULLARD, b Flint, Mich, Feb 19, 24; m 47. BIOLOGICAL CHEMISTRY. Educ: Mich State Univ, 44, MS, 45, PhD(chem & math), 52. Prof Exp: Instr math, Mich State Univ, 46-49, 50-52, natural sci, 52-54, res assoc zool, 54-59, asst prof res, Biol Chem, 59-62; vis prof dermat, Harvard Univ, 62-64, assoc biol chem & dermat, 64-70, prin assoc biol chem & dermat, 70-73, mem fac med, 64-73; assoc biochemist, Mass Gen Hosp, 70-73; ASSOC PROF & ASST TO CHMN DEPT BIOMECH, COL OSTEOP MED, MICH STATE UNIV, 73- Mem: AAAS; Am Chem Soc; Genetics Soc Am; Am Inst Biol Sci; Soc Invest Dermat. Res: Biomechanical genetics; genetic control of protein synthesis; enzymes; biology of melanin; detection, diagnosis, prognosis and treatment of malignant melanoma; identification, physiological role, and mode of action of neutrophic factors. Mailing Add: Dept Biomech Col Osteop Med Mich State Univ East Lansing MI 48824

BURNETT, JERROLD J, b Mt Pleasant, Tex, May 31, 31; m 53; c 2. PHYSICS. Educ: Tex A&M Univ, BA, 53; Tex A&I Univ, MS, 59; Univ Okla, PhD(eng sci), 66. Prof Exp: Equip engr, Southwest Bell Tel Co, Tex, 55-57; instr physics & math, Univ Dallas, 58-59; res physicist, NMex Inst Mining & Tech, 59-61; assoc prof physics & head dept, Northwestern State Col, Okla, 61-64; asst prof, 66-69, ASSOC PROF PHYSICS, COLO SCH MINES, 69- Mem: Am Pub Health Asn; Am Asn Physics Teachers; Nat Sci Supvrs Asn; Nat Sci Teachers Asn. Res: Neutron radiography; activation analysis in bore-holes; x-ray fluorescence; uranium water analysis; equipment and system design. Mailing Add: Dept of Physics Colo Sch of Mines Golden CO 80401

BURNETT, JOHN LAMBE, b Bismarck, NDak, Dec 1, 34; m 57; c 3. INORGANIC CHEMISTRY, PHYSICAL CHEMISTRY. Educ: NDak State Univ, BS, 56; Univ Calif, PhD(chem), 64. Prof Exp: Chemist, Isotopes Div, Oak Ridge Nat Lab, 65-66 & Chem Div, 66-69; chemist, Div Phys Res, US AEC, 69-75, CHEMIST, DIV PHYS RES, US ENERGY RES & DEVELOP ADMIN, 75- Mem: AAAS. Res: Lanthanide and actinide thermodynamics and inorganic chemistry. Mailing Add: Div of Phys Res Energy Res & Develop Admin Washington DC 20545

BURNETT, JOHN LAURENCE, b Wichita, Kans, Aug 28, 32; m 54; c 2. GEOLOGY. Educ: Univ Calif, Berkeley, AB, 57, MS, 60. Prof Exp: Asst geologist, 58-64, ASSOC GEOLOGIST, CALIF DIV MINES & GEOL, 64- Concurrent Pos: Instr geol, Univ Calif Exten, Berkeley, Davis & Santa Cruz, 67-; courts expert, Superior Court Los Angeles, 67-70. Mem: Geol Soc Am; Asn Eng Geol. Res: Compilation of the geologic map of California; geologic mapping in the Sierra Nevada, California; mineral exploration for raw materials used in urban construction; geologic factors in urban development. Mailing Add: Calif Div of Mines & Geol Resources Bldg Sacramento CA 95814

BURNETT, JOHN NICHOLAS, b Atlanta, Ga, Aug 19, 39. ANALYTICAL CHEMISTRY. Educ: Emory Univ, BA, 61, MS, 63, PhD(chem), 65. Prof Exp: Res chemist, Org Chem Dept, E I du Pont de Nemours & Co, Del, 65-66; res assoc chem, Univ NC, Chapel Hill, 66-68; asst prof, 68-75, ASSOC PROF CHEM & CHMN DEPT, DAVIDSON COL, 75- Mem: AAAS; Am Chem Soc; The Chem Soc; Faraday Soc. Res: Oxidation reduction processes in biologically important compounds, organometallic complexes and electro-spectrochemical techniques. Mailing Add: Dept of Chem Davidson Col Davidson NC 28036

BURNETT, JOSEPH W, b Oil City, Pa, Mar 21, 33; m 60; c 3. DERMATOLOGY. Educ: Yale Univ, AB, 54; Harvard Med Sch, MD, 58. Prof Exp: Intern & asst resident, Johns Hopkins Univ, 58-61; resident dermat, Harvard Med Sch, 61-63; asst prof, 65-69, ASSOC PROF MED, SCH MED, UNIV MD, BALTIMORE CITY, 69- Concurrent Pos: Fel med, Johns Hopkins Univ, 58; fel dermat, Mass Gen Hosp, Boston, 61-63; fel trop pub health, Harvard Med Sch, 63-65. Mem: Am Soc Exp Path; Soc Invest Dermat; Am Fedn Clin Res; AMA; Am Acad Dermat. Res: Toxins and pathogens of skin. Mailing Add: Div of Dermat Univ of Md Sch of Med Baltimore MD 21201

BURNETT, LEO SETH, b New York, NY, July 26, 15; m 43. PHYSICAL ORGANIC CHEMISTRY. Educ: City Col New York, BS, 36; Columbia Univ, AM, 37, PhD(org chem), 42. Prof Exp: Asst org chem, Columbia Univ, 40, res chemist, Nat Defense Res Comt contract, 40-42; sr res chemist, Celanese Corp, 42-49 & Resin Div, Dexter Chem Corp, 49-50; head polymer group, Olin Mathieson Chem Corp, 50-54, chief plastics sect, 55-58, mgr plastics res, 58-59; tech dir, Dapon Dept, Chem Div, FMC Corp, 59-62, mgr res eval, 62-64, develop assoc, 64-65, mgr photosensitive prods, 65-68; asst to vpres tech opers & mgr res, 68-70, MGR TECH INFO, AZOPLATE CORP, 70- Concurrent Pos: Patent agent, Patent & Trademark Off, US Dept Com, 75- Mem: AAAS; Am Chem Soc; Soc Plastic Eng. Res: Vasopressor local anesthetics; boosters; structure, properties and applications of high polymers; plastic molding materials; coatings; polymers; graphic arts; chemistry and applications of photosensitive systems. Mailing Add: 100 Wildwood Lane Summit NJ 07901

BURNETT, MARVIN CLIFTON, b Marshall, Mo, Feb 10, 26; m 55; c 2. ORGANIC BIOCHEMISTRY. Educ: Mo Valley Col, BS, 49; Univ Kansas City, MA, 51; Univ Mo, PhD, 54. Prof Exp: Analyst & res chemist agr chem, Univ Mo, 51-54; chemist, Northern Regional Res Lab, USDA, 54-59; PROF CHEM, CALIF STATE UNIV, CHICO, 59- Mem: Am Chem Soc. Res: Cholic acid derivatives; volatile components of vacuum-stored dehydrated pork; plant lipides. Mailing Add: Dept of Chem Calif State Univ Chico CA 95926

BURNETT, ROBERT LAWRENCE, physical chemistry, see 12th edition

BURNETT, ROBERT WALTER, b Lima, Ohio, Aug 5, 44; m 65; c 2. CLINICAL CHEMISTRY, ANALYTICAL CHEMISTRY. Educ: Purdue Univ, BS, 66; Emory Univ, PhD(chem), 69. Prof Exp: DIR CLIN CHEM LAB, HARTFORD HOSP, 69-; ASST PROF LAB MED, SCH MED, UNIV CONN, 74- Mem: Am Asn Clin Chem; Am Chem Soc. Res: Analytical methods for clinical analyses; high accuracy spectrophotometry; blood pH and gas analysis; applied mathematics. Mailing Add: Clin Chem Lab Hartford Hosp Hartford CT 06115

BURNETT, THOMPSON HUMPHREY, b Bethlehem, Pa, July 25, 41; m 64; c 3. EXPERIMENTAL HIGH ENERGY PHYSICS. Educ: Univ Calif, Berkeley, AB, 63; Univ Calif, San Diego, PhD(physics), 68. Prof Exp: NSF fel, 69; res assoc physics, Princeton Univ, 68-70 & Univ Calif, San Diego, 70-76; ASST PROF PHYSICS, UNIV WASH, 76- Mailing Add: Dept of Physics FM 15 Univ of Wash Seattle WA 98195

BURNETT, WILLIAM CRAIG, b Lynn, Mass, Sept 20, 45; m 75. MARINE GEOCHEMISTRY. Educ: Upsala Col, BS, 68; Univ Hawaii, MS, 71, PhD(marine geochem), 74. Prof Exp: Res asst geol & geophys, Hawaii Inst Geophys, 69-73; teaching asst, Univ Hawaii, 73-74; lectr oceanog, Leeward Community Col, 74; ADJ ASST PROF OCEANOG, STATE UNIV NY STONY BROOK, 74- Concurrent Pos: Affil researcher, Hawaii Inst Geophys, 74-; vis chemist, Brookhaven Nat Lab, 75- Mem: AAAS; Geol Soc Am; Am Geophys Union. Res: Geochemical and radiochemical studies of sea-floor deposits; refinement of methods useful for Pleistocene and recent geochronologies; environmental aspects of the near shore and estuarine environments. Mailing Add: Dept of Earth & Space Sci State Univ NY Stony Brook NY 11794

BURNETT, WILLIAM THOMAS, JR, b Spartanburg, SC, Aug 13, 17; m 55; c 2. RADIOCHEMISTRY. Educ: Wofford Col, BS, 38; Pa State Col, PhD(biochem), 48. Prof Exp: Asst geol, Va Polytech Inst, 38-39; chemist, E I du Pont de Nemours & Co, Va, 41-45; chemist, Biol Div, Oak Ridge Nat Lab, 48-55; ASSOC PROF CHEM, LA STATE UNIV, 55- Concurrent Pos: Vis scientist, US Armed Forces Radiobiol Res Inst, 63-64. Mem: AAAS; Am Chem Soc; Am Nuclear Soc; Radiation Research Soc. Res: Radiation protection; radiobiology; mineral metabolism; activation analysis. Mailing Add: Dept of Chem Coates Lab La State Univ Baton Rouge LA 70803

BURNETTE, LLEWELLYN WILSON, b Davenport, Iowa, Nov 27, 18; m 42; c 4. ORGANIC CHEMISTRY. Educ: St Ambrose Col, BS, 40; Iowa State Col, PhD(chem), 43. Prof Exp: Res chemist, Cent Res Lab, 43-48, supt org prep lab, 48-53, sr res chemist, Process Develop Dept, 53-55, sect mgr, 55-58, asst tech mgr mkt dept, 58-62, MGR TECH INFO & SERV, GAF CORP, 62- Mem: AAAS; Am Chem Soc; Am Oil Chem Soc. Res: Vapor phase catalysis; surface active agents; dye intermediates; preparation of rubber from furfural; acetylene derivatives; toxicity evaluation. Mailing Add: GAF Corp 140 W 51st St New York NY 10020

BURNETTE, MAHLON ADMIRE, III, b Lynchburg, Va, May 18, 46. FOOD SCIENCE, NUTRITION. Educ: Va Polytech Inst, BS, 68; Rutgers Univ, MPhil, 73, PhD(nutrit), 74. Prof Exp: ASST DIR SCI AFFAIRS, GROCERY MFRS AM, INC, 74- Mem: Inst Food Technologists; AAAS; Asn Food & Drug Officials. Res: Population nutrition; food safety; federal food and drug regulations. Mailing Add: Grocery Mfrs of Am Inc 1425 K St NW Washington DC 20005

BURNEY, DONALD EUGENE, b Hartington, Nebr, Oct 17, 15; m 42; c 4. INDUSTRIAL ORGANIC CHEMISTRY. Educ: Univ SDak, BA, 37; Univ Ill, PhD(org chem), 41. Prof Exp: Res chemist, Amoco Chem Corp, 41-47, group leader, 47-57, sect leader, 57-65, div dir, 65-69, mgr org chem res & develop, 69-71, EXEC DIR, AMOCO FOUND, INC, 71- Mem: Am Chem Soc. Res: Chemicals from petroleum; propylene polymerization to make synthetic lubricating oils; synthetic detergents; reactions of 2, 8-dihydoxynaphtaldehyde; oxidation of hydrocarbons; polymerization of olefins to thermoplastics. Mailing Add: Amoco Found Inc 200 E Randolph Dr Chicago IL 60601

BURNEY, GLENN ADEEN, physical chemistry, see 12th edition

BURNEY, LEROY EDGAR, b Burney, Ind, Dec 31, 06; m 32; c 2. PUBLIC HEALTH. Educ: Ind Univ, BS, 28, MD, 30; Johns Hopkins Univ, MPH, 32. Hon Degrees: DSc, Jefferson Med Col, 57, DePauw Univ, 58 & Ind Univ, 59; LLD, Seton Hall Univ, 57. Prof Exp: Mem staff, USPHS, 32-45; secy & comnr health, State Bd Health, Ind, 45-54, asst surgeon gen & dep chief bur state servs, USPHS, 54-56, surgeon gen, 56-61; vpres health sci, Temple Univ, 61-70; PRES, MILBANK MEM FUND, 70- Concurrent Pos: Chief US deleg, WHO, 57-61. Mem: Fel Am Col Prev Med; fel Am Pub Health Asn; fel AMA; Am Col Physicians; fel NY Acad Med. Res: Public health administration. Mailing Add: Milbank Mem Fund 40 Wall St New York NY 10005

BURNHAM, BRUCE FRANKLIN, b New York, NY, Nov 12, 31; m 55; c 2. BIOCHEMISTRY. Educ: Univ Utah, BS, 53, MS, 54; Univ Calif, Berkeley, PhD(biochem), 60. Prof Exp: Nat Found fel, Nobel Med Inst, Sweden, 60-61; Jane Coffin Childs Mem Fund fel med res, Oxford Univ, 61-62, sr fel, C F Kettering Res Lab, Yellow Springs, Ohio, 62-63; vis asst prof chem, Cornell Univ, 63-64, asst prof, 64-66; asst prof med & biochem, Univ Minn, 66-68; from assoc prof to prof chem, Utah State Univ, 68-74, chmn div biochem, 69-74; PRES, PORPHYRIN PRODS, 74- Concurrent Pos: Vis prof chem, Univ Calif, Los Angeles, 72-73. Mem: Biochem Soc; Am Soc Biol Chem; Am Chem Soc. Res: Enzymology; microbiology; metal ion incorporation into poryphyrins and corrins; control mechanisms of tetraphrrole biosynthesis. Mailing Add: Porphyrin Prods Box 31 Logan UT 84321

BURNHAM, CHARLES WILSON, b Detroit, Mich, Apr 6, 33; m 58; c 2. MINERALOGY. Educ: Mass Inst Technol, SB, 54, PhD(mineral, petrol), 61. Hon Degrees: AM, Harvard Univ, 66. Prof Exp: Fel, Geophys Lab, Carnegie Inst, 61-63, petrologist, 63-66; assoc prof, 66-69, PROF MINERAL, HARVARD UNIV, 69- Concurrent Pos: Assoc ed, Am Mineralogist, 74-76. Mem: AAAS; fel Mineral Soc Am; Am Crystallog Asn; Am Geophys Union; Mineral Soc Gt Brit & Ireland. Res: Determination and refinement of the crystal structures of minerals; theoretical and experimental aspects of relationships between crystal structures, crystal chemistry and phase relations of minerals in natural systems. Mailing Add: Harvard Univ Hoffman Lab 20 Oxford St Cambridge MA 02138

BURNHAM, CLIFFORD WAYNE, b Murietta, Calif, Oct 24, 22; m 43; c 1.

GEOCHEMISTRY. Educ: Pomona Col, BA, 51; Calif Inst Technol, MS, 52, PhD(geochem), 55. Prof Exp: Asst prof econ geol, 55-59, assoc prof geochem, 59-65, PROF GEOCHEM, PA STATE UNIV, 65-, HEAD DEPT GEOSCI, 74- Mem: AAAS; fel Am Geophys Union; Geol Soc Am; Mineral Soc Am; Am Geochem Soc (pres, 73-74). Res: Experimental geochemistry; geochemistry of mineral deposits; igneous and metamorphic petrology. Mailing Add: 207 Deike Bldg Pa State Univ University Park PA 16802

BURNHAM, DONALD LOVE, b Lebanon, NH, Dec 6, 22; m 43; c 3. PSYCHIATRY. Educ: Dartmouth Col, BA, 43; Cornell Univ, MD, 46. Prof Exp: From resident psychiat to dir res, Chestnut Lodge, Md, 50-63, Ford Found grant, 57-63; RES SPYCHIATRIST, DIV CLIN & BEHAV RES, HIMH, 63- Concurrent Pos: From instr to pres, Wash Psychoanal Inst, 56-; trustee, William Alanson White Psychiat Found, 62-; ed, J Psychiat, 62-; mem ed adv bd, Schizophrenia Bull, 69- Mem: Am Psychoanal Asn; fel Am Psychiat Asn. Res: Personality development and mental illness, particularly schizophrenia, thought and language, and psychotherapy; psychobiographic study of August Strindberg. Mailing Add: 5003 Edgemoor Lane Bethesda MD 20014

BURNHAM, DWIGHT COMBER, b Macomb, Ill, Mar 17, 22; m 47; c 3. SOLID STATE PHYSICS, INFORMATION SCIENCE. Educ: Iowa State Univ, BS, 43; US Mil Acad, BS, 46; Univ Ill, MS, 56, PhD(physics), 59. Prof Exp: Teaching & res asst physics, Univ Ill, 54-59; physicist, Mil Photog Dept, Res Labs, NY, 59-61 & Solid State Phys Dept, 61-67, prod planning specialist, Bus Systs Mkt Div, 67-73, PROD PLANNING COORDR, BUS SYSTS MKT DIV, EASTMAN KODAK CO, 73- Mem: Am Phys Soc. Res: Ionic crystals; emphasis on magnetic resonance techniques and problems; computer-microfilm information systems. Mailing Add: Bus Systs Mkt Div Eastman Kodak Co 343 State St Rochester NY 14650

BURNHAM, GEORGE HYNDMAN, b Alamosa, Colo, Sept 4, 09; m 36; c 2. PHYSICS. Educ: Colo Col, AB, 30; NY Univ, MS, 33. Prof Exp: Asst physics, NY Univ, 30-35; from instr to assoc prof, Hofstra Col, 35-42, chmn dept, 39-42, coord civilian pilot training prog, 39-41; asst to dir, Am Inst Physics, 42-45; asst to pres, 45-50, from assoc prof to prof physics, 45-74, dir radiation facilities lab, 64-74, EMER PROF PHYSICS, NORWICH UNIV, 74- Concurrent Pos: Consult, E E Free Labs, 36-40. Mem: Am Soc Eng Educ; Am Asn Physics Teachers. Res: Effect of solvent on absorption spectrum of solutions; absorption spectrum in the near infrared of liquid hydrogen fluoride. Mailing Add: Dept of Physics Norwich Univ Northfield VT 05663

BURNHAM, JEFFREY C, b Beverly,M Mass, Aug 6, 42; m 64; c 2. MICROBIOLOGY. Educ: Dartmouth Col, AB, 64; Univ NH, PhD(microbiol), 67. Prof Exp: Asst prof, 69-73, ASSOC PROF MICROBIOL, MED COL OHIO, 73- Concurrent Pos: USPHS fel, Univ Ky, 67-69; adj prof biol, Bowling Green State Univ, 72- Mem: AAAS; Am Soc Microbiol; Electron Micros Soc Am; Phycol Soc Am; Can Soc Microbiol. Res: Relationship between microbial cell function and structure; bacterial control of algae populations; environmental effects on algae and yeasts; water pollution microbiology; electron cytochemistry. Mailing Add: Dept of Microbiol Med Col of Ohio PO Box 6190 Toledo OH 43614

BURNHAM, JOHN, b Mexico City, Mex, Aug 23, 11; nat US; m 38; c 1. PHYSICAL CHEMISTRY, ELECTRONICS. Educ: Pomona Col, BA, 32; Stanford Univ, MA, 33, PhD, 55. Prof Exp: Asst, Stanford Univ, 34-35; res chemist, Sprague Elec Co, 36-41, asst dir res, 41-46, actg head res & eng dept, 46-47, chief engr, 47-51; res & eng consult, 51-69; sr staff engr, 69-71, SR SCIENTIST, HUGHES AIRCRAFT CO, 71- Concurrent Pos: Pres, Ti-Tal, Inc, 56- Mem: AAAS; Am Chem Soc; sr mem Inst Elec & Electronics Eng. Res: Raman and absorption spectra; physical chemistry of dielectrics; ferromagnetics and semi-conductors; theory and design of capacitors; semiconducting devices. Mailing Add: Hughes Aircraft Co Centinela Ave & Teale St Culver City CA 90230

BURNHAM, KENNETH DONALD, b Chicago, Ill, Aug 10, 22; m 47; c 1. ZOOLOGY. Educ: Roosevelt Univ, BS, 48; Univ Iowa, MS, 51, PhD(zool), 57. Prof Exp: Teacher & asst prin pub schs, 48-49; asst zool, Univ Iowa, 50-53, zool & biol, 55-57; instr, Calif State Polytech Col, 53-55; from asst prof to assoc prof zool, Southeast Mo State Col, 57-63, inter-Am fel trop med & parasitol, Caribbean, 63; asst prof biol, Ball State Univ, 63-66; ASSOC PROF ZOOL, UNIV TENN, KNOXVILLE, 66- Concurrent Pos: NSF res grant, Inst Marine Biol, Univ Ore, 60; consult biol, Rensselaer Polytech Inst, 63; NSF res grant, Inst Desert Biol, Ariz State Univ, 68 & Inst Hist & Philos of Sci, Am Univ, 69; AAAS Chautaugua courses biol & human affairs, Clark Col, Atlanta, Ga, 71, genetics & societal probs, 74; instituted Univ Tenn trop & marine biol course, Jamaica. Mem: Am Inst Biol Sci; Nat Asn Biol Teachers. Res: Bioethics; improvement of teaching biology; marine biology; individualized instruction in biology. Mailing Add: Dept of Zool Univ of Tenn Knoxville TN 37916

BURNHAM, ROBERT DANNER, b Havre de Grace, Md, Mar 21, 44; m 65. SEMICONDUCTORS. Educ: Univ Ill, BS, 66, MS, 68, PhD(elec eng), 71. Prof Exp: MEM RES STAFF SEMICONDUCTORS, XEROX PALO ALTO RES CTR, 71- Mem: Electrochem Soc; Inst Elec & Electronics Engrs. Res: Growth and fabrication of light emitting and laser diodes for optical communication and integrated optics. Mailing Add: Xerox Palo Alto Res Ctr Gen Sci Lab 3333 Coyote Hill Rd Palo Alto CA 94304

BURNHAM, THOMAS K, b Berlin, Ger, June 6, 27; m 52; c 4. DERMATOLOGY, IMMUNOLOGY. Educ: Univ London, MB, BS, 52; Am Bd Dermat, dipl, 63. Prof Exp: Assoc physician, 60-62, STAFF PHYSICIAN DERMAT, HENRY FORD HOSP, 62-, DIR DERMAT RES LAB, 68- Honors & Awards: Gold Medal, Am Acad Dermat, 66. Mem: Soc Invest Dermat; Am Acad Dermat; AMA; Am Dermat Asn. Res: Auto-immunity; antinuclear factors; fluorescent antibody; evaluation of ten diagnostic and prognostic aspects of antinuclear antibodies detected by indirect immunofluorescence, being analyzed now immunologically by other immunologic techniques to determine responsible nuclear antigens. Mailing Add: Dept of Dermat Henry Ford Hosp Detroit MI 48202

BURNISON, BRYAN KENT, b Great Falls, Mont, Feb 26, 43; m 71. MICROBIAL ECOLOGY. Educ: Mont State Col, BS, 65; Ore State Univ, Corvallis, MS, 68, PhD(microbiol), 71. Prof Exp: RES SCIENTIST MICROBIAL ECOL, CAN CTR INLAND WATERS, 73- Concurrent Pos: Fel, Univ BC, Vancouver, 71-73. Mem: Am Soc Microbiol; Am Soc Limnol & Oceanog; Int Soc Limnol. Res: Role of bacteria in cycling carbon and phosphorus compounds in the lake environment; improving techniques for the measurement of microbial biomass and in situ metabolic activities. Mailing Add: Process Res Div Can Ctr Inland Waters Burlington ON Can

BURNISTON, ERNEST EDMUND, b Sheffield, Eng, Oct 26, 37; m 59; c 4. APPLIED MATHEMATICS. Educ: Univ London, BSc, 60, PhD(math), 62. Prof Exp: Lectr math, Univ London, 62-65; from asst prof to assoc prof, 65-72, PROF MATH, NC STATE UNIV, 72- Mem: Soc Indust & Appl Math; Math Asn Am. Res: Fracture mechanics; transport theory. Mailing Add: Dept of Math NC State Univ Raleigh NC 27607

BURNISTON, GEORGE KISSAM, b Summit, NJ, Sept 8, 18; m 62. ANALYTICAL CHEMISTRY. Educ: Yale Univ, BS, 40. Prof Exp: From jr chemist to chief chemist, Ciba Pharmaceut Co, 41-60, tech asst to dir contorl, 60-65, mgr pharmaceut inspection, 65-74, MGR QUAL ASSURANCE, AIRWICK INDUSTS, INC, DIV CIBA-GEIGY CORP, 75- Mem: Am Chem Soc; Am Pharmaceut Asn; Am Asn Chemists. Res: Chemical manufacturing; method development; quality control; contract liaison; editing and coding.

BURNS, ALLAN FIELDING, b Washington, DC, Nov 22, 36; m 58; c 4. RESEARCH ADMINISTRATION, SOIL CHEMISTRY. Educ: Cornell Univ, BS, 58; Purdue Univ, MS, 60, PhD(soil chem), 62. Prof Exp: Res chemist silicates, Johns-Manville Res & Eng Ctr, Johns-Manville Corp, 62-66, sr res chemist, 66-69, sect chief basic chem, 69-71; chemist pollution abatement, US Army Picatinny Arsenal, 72-73; TECH MGR SILICA & SILICATES, PHILADELPHIA QUARTZ CO, 73- Mem: Am Chem Soc; AAAS; Mineral Soc Am; Clay Minerals Soc; Am Ceramic Soc. Res: Clay chemistry; structural and surface modification of clays; chemistry of soluble alkali and organic ammonium silicates; amorphous synthetic and natural silicas; synthetic insoluble silicates; zeolites. Mailing Add: Philadelphia Quartz Co Res & Dev Ctr Box 258 Lafayette Hill PA 19444

BURNS, BERT E, b Preston, Iowa, May 2, 15; m 44; c 2. GEOGRAPHY. Educ: Hastings Col, BA, 47; Univ Nebr, MA, 49, PhD, 54. Prof Exp: PROF GEOG, MANKATO STATE COL, 50- Mem: Nat Coun Geog Educ; Asn Am Geogr. Mailing Add: Dept of Geog Mankato State Col Mankato MN 56001

BURNS, CHESTER RAY, b Nashville, Tenn, Dec 5, 37; m 62; c 2. HISTORY OF MEDICINE. Educ: Vanderbilt Univ, BA, 59, MD, 63; Johns Hopkins Univ, PhD(hist med), 69. Prof Exp: From asst prof to James Wade Rockwell asst prof hist med, Hist Med Div, 69-74, dir div, 69-74, James Wade Rockwell asst prof hist med, 74-75, JAMES WADE ROCKWELL ASSOC PROF HIST MED, DEPT PREV MED & COMMUNITY HEALTH, UNIV TEX MED BR GALVESTON, 75-, MEM & ASSOC DIR, INST MED HUMANITIES & ASSOC, GRAD SCH BIOMED SCI, 74- Honors & Awards: AMA Physician's Recognition Award, 69. Mem: Am Asn Hist Med; Soc Health & Human Values (pres-elect, 74-75); Int Soc Hist Med; Hist Sci Soc; Am Hist Asn. Res: History of medicine, especially medical education, medical science, medical ethics and medical jurisprudence. Mailing Add: Inst for Med Humanities Univ of Tex Med Br Galveston TX 77550

BURNS, DAVID JEROME, b Hobart, La, July 13, 22; m 46; c 2. AGRICULTURAL ECONOMICS, RESOURCE ECONOMICS. Educ: Univ Md, BS, 48 & 49, PhD(agr econ), 54. Prof Exp: From instr to asst prof agr econ, Univ Md, 51-56; assoc prof, 56-60, PROF AGR ECON, RUTGERS UNIV, NEW BRUNSWICK, 60- Mem: Am Agr Econ Asn. Res: Marketing of fruits and vegetables and land use planning. Mailing Add: Dept of Agr Econ Rutgers Univ New Brunswick NJ 08903

BURNS, DENVER PEEPER, b Bryan, Ohio, Oct 27, 40; m 65. RESEARCH ADMINISTRATION. Educ: Ohio State Univ, BSc, 62, MSc, 64, PhD(entom), 67. Prof Exp: From asst entomologist to assoc entomologist, 62-68, res entomologist, 68-72, asst dir, Southern Forest Exp Sta, 72-74, STAFF ASST TO DEP CHEIF FOR RES, US FOREST SERV, 74- Mem: AAAS; Entom Soc Am; Soc Am Foresters. Res: Sucking insect-host tree interactions; biology and bionomics of sucking insects; biology, ecology and behavior of borers attacking living hardwoods. Mailing Add: US Forest Serv 12th & Independence Washington DC 20250

BURNS, DONAL JOSEPH, b Belfast, Ireland, Mar 5, 41. ATOMIC PHYSICS. Educ: Queen's Univ, Belfast, BSc, 62, PhD(physics), 65. Prof Exp: Asst lectr physics, Queen's Univ, Belfast, 65-68; res assoc, 68, Univ Res Coun jr fac fel, 69, asst prof, 68-72, ASSOC PROF PHYSICS, UNIV NEBR, LINCOLN, 72- Mem: Am Phys Soc; Brit Inst Physics & Phys Soc. Res: Observations on forbidden transitions; measurement of atomic and molecular excitation cross sections; lifetimes of excited states and energy transfer effects; coherent beam foil spectroscopy. Mailing Add: Dept of Physics Univ of Nebr Lincoln NE 68508

BURNS, EDWARD COLUMBUS, b Mansfield, La, May 11, 20; m 43. ENTOMOLOGY. Educ: La State Univ, BS, 50, MS, 51; Iowa State Univ, PhD(entom), 57. Prof Exp: Asst prof entom, 54-55, 57-60, assoc prof, 60-65, PROF ENTOM, LA STATE UNIV, BATON ROUGE, 65- Mem: Entom Soc Am. Res: Medical and veterinary entomology. Mailing Add: Dept of Entom La State Univ Baton Rouge LA 70803

BURNS, EDWARD EUGENE, b Ft Wayne, Ind, Apr 13, 26; m 48; c 3. FOOD SCIENCE. Educ: Purdue Univ, BS, 50, MS, 52, PhD(food technol), 56. Prof Exp: Asst, Purdue Univ, 47-50, instr, 52-56; PROF HORT, TEX A&M UNIV, 56- Honors & Awards: Cruess Award, Inst Food Technol, 72. Mem: Am Hort Soc; Sigma Xi; Inst Food Technol; Am Soc Hort Sci. Res: Quality control techniques and methods; chemical and physiological indicators of quality; human foods of plant origin. Mailing Add: 1203 Park Pl College Station TX 77840

BURNS, EDWARD ROBERT, b Catskill, NY, Nov 6, 39; m 60; c 2. MICROSCOPIC ANATOMY, EXPERIMENTAL EMBRYOLOGY. Educ: Hartwick Col, BA, 61; Univ Maine, MS, 63; Tulane Univ, PhD(anat), 67. Prof Exp: From instr to asst prof, 68-73, ASSOC PROF ANAT, SCH MED, UNIV ARK, LITTLE ROCK, 73- Concurrent Pos: NIH fel exp path, George Washington Univ, 67-68; NIH res career develop award, 74-79. Mem: AAAS; Am Asn Anatomists; Am Soc Zoologists; Int Soc Chronobiol. Res: Experimental oncology; control of growth and differentiation in normal and neoplastic tissues; chronochemotherapy; chronobiology of neoplasia. Mailing Add: Dept of Anat Univ of Ark Med Ctr Little Rock AR 72201

BURNS, ERSKINE JOHN THOMAS, b Calgary, Alta, Oct 23, 44; m 75. PLASMA PHYSICS. Educ: Occidental Col, BA, 66; Calif State Univ, Los Angeles, MS, 68; Univ Calif, Davis, PhD(physics), 71. Prof Exp: Res physicist, 71-75, CHIEF X-RAY DIAG GROUP, AIR FORCE WEAPONS LAB, 75- Concurrent Pos: Consult mat surviveability & vulnerability br, Air Force Weapons Labs, 75- Mem: Am Phys Soc. Res: Radiation diagnostics of high temperature, high density plasmas. Mailing Add: AF Weapons Lab DYS Kirtland AFB NM 87117

BURNS, EUGENE ANTHONY, analytical chemistry, see 12th edition

BURNS, FRANCIS JOHN, b Ashtabula, Ohio, May 13, 14; m 48; c 6. SURGERY. Educ: John Carroll Univ, AB, 36; St Louis Univ, MD, 41, MS, 46; Am Bd Colon & Rectal Surg, dipl, 50. Prof Exp: Instr surg, 46-52, sr instr, 52-55, from asst prof to assoc prof clin surg, 55-73, CLIN PROF SURG, ST LOUIS UNIV, 73- Mem: AMA; Am Col Surg; Pan-Am Med Asn; Int Soc Univ Colon & Rectal Surg; Am Soc Colon & Rectal Surg. Mailing Add: Dept of Surg St Louis Univ Sch of Med St Louis MO 63103

BURNS, FRED PAUL, b New York, NY, Dec 27, 22; m 47; c 2. PHYSICS. Educ: City Col New York, BME, 47; Columbia Univ, PhD(physics), 54. Prof Exp: Asst prof mech eng, City Col New York, 47-54; mem tech staff, Bell Tel Labs, 54-57; mgr silicon rectifier develop, Tung Sol Elec, 57-59; mgr indust transistor design, Radio Corp Am, 59-60; mgr semiconductor devices, Solid State Radiations, Inc, 60-62; mgr opers, Korad Corp, 62-68; PRES, APOLLO LASERS, INC, 68- Mem: Am Phys Soc; Inst Elec & Electronics Engrs. Res: Semiconductor device development; germanium silicon transistors and rectifiers; solid state physics; lasers systems development and interaction of high intensity light with solids. Mailing Add: Apollo Lasers Inc 6357 Arizona Circle Los Angeles CA 90045

BURNS, FREDRIC JAY, b Wilmington, Del, Apr 20, 37; m 62; c 2. ONCOLOGY. Educ: Harvard Col, AB, 59; Columbia Univ, MA, 61; NY Univ, PhD(biol), 67. Prof Exp: ASST PROF ENVIRON MED, MED CTR, NY UNIV, 69- Concurrent Pos: NIH fel, Inst Cancer Res, Sutton, Eng, 67-69. Mem: Radiation Res Soc; Am Asn Cancer Res. Res: Cell population kinetics; radiation and chemical carcinogenesis in skin and liver; control of cell division; cell cycle models. Mailing Add: Inst Environ Med NY Univ Med Ctr New York NY 10016

BURNS, GEORGE, b Russia, June 4, 25; US citizen; m 60; c 1. CHEMISTRY. Educ: Columbia Univ, BS, 51; Princeton Univ, PhD(chem), 61. Prof Exp: Chemist, Gen Elec Co, 52-56; Nat Acad Sci-Nat Res Coun fel phys chem, Cambridge Univ, 61-62; from asst prof to assoc prof, 62-71, assoc chmn dept, 74-75, PROF CHEM, UNIV TORONTO, 71- Mem: Am Phys Soc; Am Chem Soc. Res: Lasers; shock waves; flash photolysis; theoretical and experimental gas phase chemical kinetics; recombination-dissociation reactions, theory and experiment. Mailing Add: Dept Chem Lash Miller Chem Lab Univ of Toronto Toronto ON Can

BURNS, GEORGE ROBERT, b Lineville, Ala, Nov 12, 31; m 54; c 3. SOIL FERTILITY. Educ: Auburn Univ, BS, 54; NC State Col, MS, 56; Iowa State Univ, PhD(soil fertil), 62. Prof Exp: Res soil scientist, US Soils Lab, 62-64, asst to dir agr res, Sulphur Inst, 64-67, BR CHIEF SOIL & WATER CONSERV RES DIV, AGR RES SERV, USDA, 67-, AREA DIR, SOUTHERN REGION, 72- Mem: Am Soc Agron; Soil Sci Soc Am; Soil Conserv Soc Am. Res: Chemistry of soil and plant interaction; chemistry of nitrogen and phosphorous in soil. Mailing Add: PO Box 5847 Raleigh NC 27607

BURNS, GEORGE W, b Cincinnati, Ohio, Nov 20, 13; m 42; c 3. GENETICS. Educ: Univ Cincinnati, AB, 37; Univ Minn, PhD(bot), 41. Prof Exp: Teaching fel bot, Univ Minn, 37-41, instr, 46; from asst prof to assoc prof, 46-54, actg vpres & dean, 57-59, actg pres, 58-59, vpres & dean, 59-61, PROF BOT & CHMN DEPT, OHIO WESLEYAN UNIV, 54- Concurrent Pos: Head insts sect, NSF, 61-62. Mem: Fel AAAS; Am Genetic Asn; Am Soc Human Genetics; Bot Soc Am. Res: Pleistocene flora. Mailing Add: Dept of Bot Ohio Wesleyan Univ Delaware OH 43015

BURNS, GERALD, b New York, NY, Oct 5, 32; m 54; c 2. PHYSICS. Educ: Rensselaer Polytech Inst, BS, 54; Columbia Univ, AM, 57, PhD, 62. Prof Exp: Res engr, Cornell Aerodyn Labs, 54; asst physics, Watson Labs, 54-57; RES PHYSICIST, RES LABS, IBM CORP, 57- Mem: Am Phys Soc; Inst Elec & Electronics Engrs. Res: Solid state physics. Mailing Add: IBM Corp Res Ctr Box 218 Yorktown Heights NY 10598

BURNS, GROVER PRESTON, b Putnam Co, WVa, Apr 25, 18; m 41; c 2. APPLIED MECHANICS, THEORETICAL PHYSICS. Educ: Marshall Col, AB, 37; WVa Univ, MS, 41. Hon Degrees: DSc, Colo State Christian Col, 73. Prof Exp: Teacher high sch, WVa, 37-40; instr physics, Univ Conn, 41-42; asst prof, Miss State Col, 42-44, actg head dept, 44-45; asst prof, Tex Tech Col, 46; assoc prof math, Marshall Col, 46-47; res physicist, Naval Res Lab, 47-48; from asst prof to assoc prof physics, Mary Washington Col, Univ Va, 48-69, chmn dept, 48-69; supvr statist anal sect, Am Viscose Div, FMC Corp, 50-67; MATHEMATICIAN, NAVAL SURFACE WEAPONS CTR, 74- Concurrent Pos: Pres, Burns Enterprises, Inc, 58- Mem: Am Phys Soc; Am Asn Physics Teachers; Am Defense Preparedness Asn; AAAS; Am Asn Univ Prof. Res: Superconductivity; electricity; mathematics; numerical integration; exterior ballistics; mathematical models for gunfire control systems; effects of Coriolis acceleration on projectile motion. Mailing Add: 600 Virginia Ave Fredericksburg VA 22401

BURNS, HUGH DONALD, b Scranton, Pa, Apr 17, 46; m 69; c 1. ORGANIC CHEMISTRY, NUCLEAR MEDICINE. Educ: Univ Scranton, BS, 68; Lehigh Univ, MS, 72, PhD(org chem), 74. Prof Exp: NIH fel, 74-75, ASST PROF NUCLEAR MED, DIV NUCLEAR MED, JOHNS HOPKINS MED INST, 75- Honors & Awards: Nat Res Serv Award, Nat Heart & Lung Inst, 75. Mem: Soc Nuclear Med; Am Chem Soc; Sigma Xi. Res: Design, synthesis and testing or radiopharmaceuticals; structure biodistribution relationships of technetium complexes and the design and synthesis of potential antitumor agents. Mailing Add: Johns Hopkins Med Inst 615 N Wolfe St Baltimore MD 21205

BURNS, JAY, III, b Lake Wales, Fla, Mar 22, 24; m 48; c 3. PHYSICS. Educ: Northwestern Univ, BS, 47; Univ Chicago, MS, 51, PhD, 59. Prof Exp: Physicist, Chicago Midway Labs, 51-57; sr physicist, Labs Appl Sci, Univ Chicago, 57-63, dir, 63-65; assoc prof astron, Northwestern Univ, 65-67; assoc dir res, Ranland Div, Zenith Radio Corp, 67-68; assoc prof astron & head lab exp astrophys, Northwestern Univ, 68-76, dir, Astro-Sci Workshops; PROF PHYSICS & HEAD DEPT PHYSICS & SPACE SCI, FLA INST TECHNOL, 76- Concurrent Pos: Vis prof physics, Univ Ill, Chicago Circle & sr res consult, Zenith Radio Corp, 71-76. Mem: AAAS; Am Phys Soc; Am Astron Soc; Am Phys Soc. Res: Secondary electron emission; photoelectric emission; solid state; surface physics; astronomical image tubes; atomic and molecular transition probabilities; electron emission from solids; soild surface physics. Mailing Add: Dept of Physics & Space Sci Fla Inst Technol Melbourne FL 32901

BURNS, JOHN ALLEN, b Little Rock, Ark, Aug 15, 45; m 65; c 1. APPLIED MATHEMATICS. Educ: Ark State Univ, BSE, 67, MSE, 68; Univ Okla, MA, 70, PhD(math), 73. Prof Exp: Spec instr math, Univ Okla, 72-73; asst prof, Lefschetz Ctr Dynamical Syst, Brown Univ, 73-74; ASST PROF MATH, VA POLYTECH INST & STATE UNIV, 74- Mem: Soc Indust & Appl Math; Math Asn Am. Res: Optimal theory; ordinary differential equations; functional differential equations. Mailing Add: Dept of Math Va Polytech Inst & State Univ Blacksburg VA 24061

BURNS, JOHN FRANCIS, b Minneapolis, Minn, Jan 10, 01; m 29; c 2. EXPERIMENTAL ATOMIC PHYSICS. Educ: Loras Col, BA, 22; Univ Wis, MA, 27; Univ Tenn, PhD(physics), 54. Prof Exp: Instr physics, Univ Wis, 25-27; instr physics & math, Gen Beadle State Teachers Col, 29-40; instr physics, Va Polytech Inst, 40-41; physicist, Radford Proving Ground, Va, 41-42, ballistics engr, Radford Proving Ground, Hercules Powder Co, 41-45; physicist, Oak Ridge Gaseous Diffusion Plant, 45-56; sr res physicist, Oak Ridge Nat Lab, Union Carbide Corp, Tenn, 56-66; assoc prof physics, Univ Tenn, Knoxville, 66-71; RETIRED. Concurrent Pos: Vis lectr, Univ Col, Dublin, 66; consult, Oak Ridge Nat Lab, 66-68. Mem: Am Phys Soc. Res: Gaseous ionization under electron impact; atomic and molecular structure; mass spectrometry; nondestructive testing; interior and external ballistics. Mailing Add: 131 Georgia Ave Oak Ridge TN 37830

BURNS, JOHN HOWARD, b Kilgore, Tex, Oct 8, 30; m 52; c 2. PHYSICAL CHEMISTRY. Educ: Rice Inst, BA, 51, MA, 53, PhD(chem), 55. Prof Exp: Res assoc chem, Oak Ridge Nat Lab, 55-56; asst prof, Univ Ky, 56-57; asst chemist, Argonne Nat Lab, 57-59, assoc chemist, 59-60; CHEMIST, OAK RIDGE NAT LAB, 60- Mem: Am Crystallog Asn. Res: X-ray and neutron diffraction; chemistry of transuranium elements. Mailing Add: Oak Ridge Nat Lab Chem Div PO Box X Oak Ridge TN 37831

BURNS, JOHN J, b Flushing, NY, Oct 8, 20. BIOCHEMISTRY. Educ: Queens Col, BS, 42; Columbia Univ, AM, 48, PhD(chem), 50. Prof Exp: Adj asst prof med, NY Univ, 50-60; dir res pharmacodynamics div, Wellcome Res Labs, 60-66; VPRES RES, HOFFMANN-LA ROCHE INC, 67- Concurrent Pos: Mem pharmacol & exp therapeut study sect, NIH, 58-62 & drug res bd, Nat Acad Sci-Nat Res Coun, 64-; vis prof, Albert Einstein Sch Med, 60-68 & Cornell Univ Med Col, 68-; mem, Inst Med, Nat Acad Sci. Mem: Am Chem Soc; Am Soc Pharmacol & Exp Therapeut; Am Soc Biol Chem; NY Acad Sci (vpres, 64); Nat Acad Sci. Res: Metabolism of vitamin C, pentoses, uronic acids; antirheumatic drugs; muscular relaxants; barbituates; anticoagulants; local anesthetics; mechanism of action of adrenergic blocking drugs. Mailing Add: Hoffman-La Roche Inc Nutley NJ 07110

BURNS, JOHN MCLAUREN, b Rochester, NY, June 6, 32; m 54; c 3. ZOOLOGY, EVOLUTION. Educ: Johns Hopkins Univ, AB, 54; Univ Calif, Berkeley, MA, 57, PhD(zool), 61. Prof Exp: Asst zool, Univ Calif, Berkeley, 54-57, asst entom, 57-58; asst prof biol, Wesleyan Univ, 61-69; ASSOC CUR LEPIDOPTERA, MUS COMP ZOOL, HARVARD UNIV, 69- Concurrent Pos: Mem, Lepidoptera Found. Mem: AAAS; Soc Study Evolution; Am Soc Naturalists; Soc Syst Zoologists; Genetics Soc Am. Res: Population differentiation and speciation in sexually reproducing animals; genetics and ecology of polymorphism, including electrophoretically detectable protein polymorphisms; systematics and behavior of Lepidoptera, especially Erynnis, other Hesperiidae and Colias. Mailing Add: Mus of Comp Zool Harvard Univ Cambridge MA 02138

BURNS, JOHN MITCHELL, b Hobbs, NMex, Dec 18, 40; m 63; c 2. PHYSIOLOGY, ENDOCRINOLOGY. Educ: NMex State Univ, BS, 63, MS, 66; Ind Univ, PhD(zool), 69. Prof Exp: Sci aide & comput programmer, Phys Sci Lab, NMex State Univ, 63-65, res asst microbial physiol, 65-66; teaching assoc zool, Ind Univ, 66-68; asst prof biol, 68-74, ASSOC PROF BIOL SCI, TEX TECH UNIV, 74- Mem: AAAS; Am Soc Zoologists; Am Chem Soc; Soc Develop Biol. Res: Steroid biosynthesis and metabolism. Mailing Add: Dept of Biol Tex Tech Univ Lubbock TX 74909

BURNS, JOSEPH A, b New York, NY, Mar 22, 41; m 67. PLANETARY SCIENCE, CELESTIAL MECHANICS. Educ: Webb Inst Naval Archit, BS, 62; Cornell Univ, PhD(space mechanics), 66. Prof Exp: Asst prof mech, 66-75, ASSOC PROF MECH & ASTRON, CORNELL UNIV, 75- Concurrent Pos: Nat Acad Sci-Nat Res Coun res assoc theoret div, Goddard Space Flight Ctr, NASA, 67-68; Nat Acad Sci exchange fel, Schmidt Inst, Moscow & Astron Inst, Prague, 73; sr scientist, Space Sci Div, NASA Ames Res Ctr, 75-76. Mem: AAAS; Am Geophys Union; Am Astron Soc; Int Astron Union; Sigma Xi. Res: Mechanics of solar system; celestial rotation; asteroid collisions; origin of solar system; orbital evolution; planetary satellites, dust dynamics; Saturn's rings and satellites celestial mechanics; teaching methods; art and science. Mailing Add: Dept of Theoret & Appl Mech 111 Thurston Hall Cornell Univ Ithaca NY 14853

BURNS, JOSEPH CHARLES, b Iowa City, Iowa, Feb 16, 37; m 56; c 6. PLANT PHYSIOLOGY, ANIMAL NUTRITION. Educ: Iowa State Univ, BS, 60, MS, 63; Purdue Univ, PhD(plant physiol, ecol), 66. Prof Exp: Dist sales mgr, Boeke Feed Co, Iowa, 61; res asst forage crop prod, Iowa State Univ, 61-63; res asst plant physiol & ruminant nutrit, Purdue Univ, 63-66; ASSOC PROF CROP SCI, N C STATE UNIV & RES PLANT PHYSIOLOGIST AGR RES SERV, USDA, 67- Mem: Am Soc Agron. Res: Initiation and accumulation of secondary plant products from primary plant metabolites induced through management, environment and cytogenetic control and relationship of these constituents to animal acceptance, intake and utilization. Mailing Add: Dept of Crop Sci NC State Univ Raleigh NC 27607

BURNS, KENNETH FRANKLIN, b Lebanon, Ind, June 6, 16; m 39. VIROLOGY. Educ: Ont Vet Col, DVM, 40; Univ Toronto, DVSc, 50; Univ Tokyo, PhD(microbiol), 52; Am Bd Vet Pub Health, dipl, 55; Am Col Lab Animal Med, dipl, 61. Prof Exp: Chief virol & vet br, Army Med Lab, Ft McPherson, Ga, 41-43, chief virol & vet br, 4th Army Med Lab, Ft Sam Houston, Tex, 46-47, chief dept virus & rickettsial dis, 406th Med Gen Lab, Japan, 47-51, dep chief vet microbiol div, Chem Corps Biol Warfare Labs, Ft Detrick, Md, 51-53, chief vet & virol br, 4th Army Area Med Lab, Brooke Army Med Ctr, Tex, 53-58, chief animal colonies div, Directorate Med Res, US Army Chem Warfare Labs, Army Chem Ctr, Md, 59-61, chief pub health officer, Civil Affairs Group & Lab & consult to High Comnr, Ryukyu Islands, 61-62; PROF COMP MED & CHMN DEPT VIVARIAL SCI & RES, TULANE UNIV, 62- Concurrent Pos: Consult, US Ord Command, 59-61. Mem: Am Vet Med Asn; Animal Care Panel. Res: Laboratory animal medicine. Mailing Add: Dept of Vivarial Sci & Res Tulane Univ New Orleans LA 70112

BURNS, MARY GRACE, b Monmouth Beach, NJ, Dec 28, 01. BIOCHEMISTRY. Educ: Georgian Court Col, AB, 23; Cath Univ Am, MA, 35, PhD, 39. Prof Exp: Teacher pub sch, 23-24; instr biol, Georgian Court Col, 27-30, 35-37, asst prof, 39-45, prof natural sci, 45-67; prof, Mt St Mary Jr Col, 67-73; RETIRED. Mem: AAAS; Am Chem Soc; Nat Sci Teachers Asn. Res: Muscle metabolism with reference to creatine content of muscle; tissue culture, especially Rous sarcoma. Mailing Add: Mt St Mary Motherhouse Rte 22 Terrill Rd North Plainfield NJ 07060

BURNS, MOORE J, b Wedowee, Ala, May 31, 17; m 39; c 3. BIOCHEMISTRY, PHYSIOLOGY. Educ: Auburn Univ, BS, 40, MS, 46; Purdue Univ, PhD(biochem), 50. Prof Exp: Instr animal sci, Auburn Univ, 46-47; res asst biochem, Purdue Univ, 47-50; assoc prof animal nutrit, 50-55, PROF PHYSIOL, AUBURN UNIV, 56- Mem: Ny Acad Sci; Am Chem Soc. Res: Vitamin A stability and utilization; nutrition in relation to cancer; parenteral nutrition; factors affecting cholesterol metabolism and atherosclerosis. Mailing Add: Dept of Physiol & Pharmacol Auburn Univ Basic Sci Bldg Auburn AL 36830

BURNS, NOEL M, physical chemistry, limnology, see 12th edition

BURNS, PAUL YODER, b Tulsa, Okla, July 4, 20; m 42; c 3. FORESTRY. Educ: Univ Tulsa, BS, 41; Yale Univ, MS, 46, PhD, 49. Prof Exp: From asst prof to assoc prof forestry, Univ Mo, 48-55; PROF FORESTRY & DIR SCH FORESTRY, LA STATE UNIV, BATON ROUGE, 55- Mem: Soc Am Foresters. Res: Forest soils;

silvics; meteorology; silviculture; economics; mensuration. Mailing Add: Sch of Forestry La State Univ Baton Rouge LA 70803

BURNS, RAYMOND EDWARD, b Muncie, Ind, Apr 24, 13; m 38; c 2. INORGANIC CHEMISTRY. Educ: Ball State Teachers Col, BS, 36; Purdue Univ, MS, 43, PhD(chem), 47. Prof Exp: Res chemist, Gen Elec Co, 47-65; mgr mat & process chem unit, 65-71, sect mgr chem technol dept, 71-73, STAFF SCIENTIST, PAC NORTHWEST LAB, BATTELLE MEM INST, 73- Mem: Am Chem Soc; Sigma Xi. Res: Radio chemistry; separations processes; materials of construction. Mailing Add: 1630 Davison Richland WA 99352

BURNS, RICHARD CHARLES, b Chicago, Ill, Oct 8, 30; m 55; c 4. BIOCHEMISTRY. Educ: Univ Wis, BS, 52, MS, 61, PhD(biochem), 63. Prof Exp: Fel biochem, C F Kettering Res Lab, 63-64, staff scientist, 64-67; biochemist, Exp Sta, 67-73, PROJ SUPVR, PHOTO PROD DEPT, E I DU PONT DE NEMOURS & CO, INC, 73- Res: Biological nitrogen fixation; hydrogen metabolism; intermediary metabolism of microorganisms; enzymology. Mailing Add: IPD Glasgow Site E I du Pont de Nemours & Co Inc Wilmington DE 19898

BURNS, RICHARD HENRY, b Battle Creek, Mich, Apr 11, 31; m 58; c 2. PHARMACOLOGY. Educ: Albion Col, AB, 52; Univ Mich, MS, 54. Prof Exp: Asst pharmacol, Univ Mich, 57-58; RES PHARMACOLOGIST, EATON LABS DIV, NORWICH PHARMACAL CO, 58- Res: Drug action on the central nervous system; narcotics; analgesics; narcotic antagonists; gross effect pharmacology. Mailing Add: Eaton Labs Div Norwich Pharmacal Co Norwich NY 13815

BURNS, RICHARD PRICE, b Bartlesville, Okla, Sept 19, 32; m 59. PHYSICAL CHEMISTRY, INORGANIC CHEMISTRY. Educ: Okla Baptist Univ, AB, 54; Univ Chicago, PhD(chem), 65. Prof Exp: Jr chemist, Lawrence Radiation Lab, Univ Calif, 55-58; res asst physics, Univ Chicago, 59-65; asst prof, 65-70, ASSOC PROF CHEM, UNIV ILL, CHICAGO CIRCLE, 70- Mem: Am Chem Soc; Am Phys Soc. Res: Applications of the mass spectrometer and thermal imaging techniques to high temperature chemistry, thermodynamics and kinetics. Mailing Add: Dept of Chem Univ of Ill at Chicago Circle Chicago IL 60680

BURNS, ROBERT B P, b Philadelphia, Pa, Sept 30, 33; m 59; c 3. MEDICINE, PHARMACOLOGY. Educ: St Joseph's Col, Pa, BS, 60; Jefferson Med Col, MD, 64. Prof Exp: Pharmacologist, Res & Develop Div, Menley & James Labs Div, Smith Kline & French Labs, Pa, 65-67, pharmacol sect head, 67-68; assoc dir clin res, Indust Biol Div, Food & Drug Res Labs, NY, 68-69; dir res & develop, Madison Labs, 69-70; DIR CLIN RES, SANDOZ PHARMACEUT, 70- Concurrent Pos: USPHS res fel pharmacol, Sch Med, Univ Pa & fel cardiol, Robinette Found, Univ Pa Hosp, 64-65. Mem: AAAS; Am Acad Clin Toxicol; NY Acad Sci. Res: Clinical pharmacology. Mailing Add: Sandoz Pharmaceut Rte 10 East Hanover NJ 07936

BURNS, ROBERT DAVID, b Detroit, Mich, Feb 28, 29; m 60; c 2. ZOOLOGY. Educ: Mich State Univ, BS, 51, MS, 54, PhD(zool), 58. Prof Exp: Asst prof zool, Univ Okla, 58-63; from asst prof to assoc prof 63-69, chmn dept, 69-72, PROF BIOL, KENYON COL, 69- Concurrent Pos: Vis prof, Biol Sta, Mich State Univ, 61; vpres, Ohio Acad Sci, 74-75. Mem: AAAS; Ecol Soc Am; Am Soc Mammal; Am Ornith Union; Wilson Ornith Soc. Res: Feeding and behavior in kangaroo rats; reproductive behavior in birds and mammals; physiology of the vestibular portion of the inner ear in kangaroo rats. Mailing Add: Dept of Biol Kenyon Col Gambier OH 43022

BURNS, ROBERT EARLE, b New York, NY, May 1, 25; m 51; c 2. OCEANOGRAPHY Educ: Col of Wooster, AB, 47; Lehigh Univ, MS, 50; Univ Wash, PhD(oceanog), 62. Prof Exp: Instr geol, Bucknell Univ, 49-51; oceanogr, div oceanog, US Navy Hydrographic Off, 51, head tides & currents unit, 51-52, dep head regional sect, 52-53, tech asst, 53-54, head geol oceanog unit, 54-56, supv oceanogr, 56-57; assoc oceanog, Univ Wash, 57-62; res oceanogr, Off Res & Develop, US Coast & Geod Surv, 62-65; chief environ sci serv admin complement, Joint Oceanog Res Group, Univ Wash, 65-69; res oceanogr, Pac Oceanog Labs, 70-73, ACTG DIR, PAC MARINE ENVIRON LAB, NAT OCEANIC & ATMOSPHERIC ADMIN, 73- Concurrent Pos: Chmn, Fed Interagency Subcomt Sedimentation, 63-64; vis geoscientist prog, Am Geol Inst, 64-65 & Am Geophys Union, 66-67, 70; chmn, Joint Oceanog Insts Deep Earth Sampling Pac Adv Panel, 68-73. Mem: Geophys Union; Marine Technol Soc. Res: Assessment of potential impact of man's activities on the marine environment. Mailing Add: Pac Marine Environ Lab 3711 15th Ave NE Seattle WA 98105

BURNS, ROBERT EMMETT, b Oxford, Iowa, May 9, 18; m 52; c 4. PLANT PHYSIOLOGY. Educ: Univ Iowa, BA, 40, MS, 47; Univ Calif, Santa Barbara, PhD(plant physiol), 50. Prof Exp: Plant physiologist, 50-64, ASSOC PLANT PHYSIOLOGIST, EXP STA, AGR RES SERV, USDA, UNIV GA, 64- Mem: AAAS; Am Soc Agron; Am Soc Plant Physiol. Res: Seed physiology; turf physiology. Mailing Add: Ga Exp Sta Univ of Ga Experiment GA 30212

BURNS, ROBERT HOMER, animal husbandry, deceased

BURNS, ROBERT KYLE, b Hillsboro, WVa, July 26, 96; m 24; c 3. ZOOLOGY, EMBRYOLOGY. Educ: Bridgewater Col, AB, 16; Yale Univ, PhD(zool), 24. Hon Degrees: ScD, Bridgewater Col, 53. Prof Exp: Asst zool & anat, Yale Univ, 20-24; instr zool, Univ Cincinnati, 24-25, asst prof, 25-28; from asst prof to assoc prof anat, Univ Rochester, 28-40; staff mem, Dept Embryol, Carnegie Inst, 40-62; interim prof zool, Bridgewater Col, 62-67; RETIRED. Concurrent Pos: Mem staff, Mt Lake Biol Sta, Univ Va, 40-45; hon prof, Johns Hopkins Univ, 45; Guggenheim fel, France, 55-56; exchange prof, Univ Paris, 56; vis lectr, Univ Calif, Santa Barbara, 65-66. Mem: Nat Acad Sci; AAAS; Am Asn Anat; Am Soc Nat; Soc Develop Biol. Res: Parabosis and experimental transformation of sex in amphibians; growth and development after heteroplastic grafting; interrelations of reproductive tract and hypophysis; pronephric duct and development of mesonephros; sex reversal in gonads and genital tract of opossum embryos induced experimentally by hormones; reproduction in natural populations of the opossum. Mailing Add: 303 N Second St Bridgewater VA 22812

BURNS, ROBERT L, organic chemistry, see 12th edition

BURNS, ROBERT OBED, b Emporia, Kans, Jan 16, 10; m 37; c 1. PHYSICS. Educ: Knox Col, BS, 31; Univ Ill, MS, 33, PhD(physics), 37. Prof Exp: Physicist, Celotex Corp, La, 37-41, NJ, 41 & Ohio, 41-42; asst dir eng res, Univ Chicago, 42-43; eng mgr, Div War Res, Univ Calif, 43-46; head develop div, Navy Electronic Lab, 46-50, head systs div, 50-52, chief scientist, Signal Corps, Electronic Warfare Ctr, 52-54; tech dir electronic proving ground, US Dept Army, 54-56; develop coordr, US Dept Navy, 56-59, dir develop planning group, Off Chief Naval Opers, 59-62, dir tech anal & adv group, 62-71, chief scientist, Res, Develop, 72-75; RETIRED. Concurrent Pos: Prof lectr, Am Univ, 67-75. Honors & Awards: Distinguished Civilian Serv Award, US Dept Navy, 74. Mem: AAAS; Inst Elec & Electronics Engrs; Opers Res Soc Am. Res: Sound absorbing materials; fiber and gypsum building materials; underwater sound and radar equipment; field information handling systems; science administration; military operations research. Mailing Add: 5007 Dodson Dr Annandale VA 22003

BURNS, ROGER GEORGE, b Wellington, NZ, Dec 28, 37; m 63; c 2. GEOCHEMISTRY, MINERALOGY. Educ: Victoria Univ, NZ, BSc, 59, MSc, 61; Univ Calif, Berkeley, PhD(geochem), 65. Hon Degrees: MA, Wadham Col, Oxford Univ, 68. Prof Exp: Demonstr chem, Victoria Univ, NZ, 58-60, sci officer, Dept Sci & Indust Res, 60-61; res assoc geochem, Dept Mineral Technol, Univ Calif, Berkeley, 64-65; sr res visitor, Cambridge Univ, 65-66; sr lectr, Victoria Univ, NZ, 66-67; lectr, Oxford Univ, 68-70; PROF GEOCHEM, DEPT EARTH & PLANETARY SCI, MASS INST TECHNOL, 70- Concurrent Pos: Vis lectr, Oxford & Cambridge Univs, 65-66; res visitor, Brit Coun, 65-66; sr res visitor, Natural Environ Coun, Eng, 65-66; ed, Chem Geol, 68- Honors & Awards: Min Soc Am Award, 75. Mem: Fel Geochem Soc; fel Mineral SocSoc Am; fel Mineral Soc Gt Brit & Ireland; fel The Chem Soc; fel Am Geophys Union. Res: Transition metal geochemistry and spectroscopic studies of minerals; bonding, distribution and properties of transition elements in silicate and ore minerals; Mössbauer, infrared and electronic absorption spectroscopy of minerals; geochemistry of deep-sea ferromanganese nodules; metallogenesis and plate tectonics. Mailing Add: 54-816 Dept of Earth & Planetary Sci Mass Inst Technol Cambridge MA 02139

BURNS, RUSSELL MACBAIN, b New York, NY, Aug 25, 26; m 48; c 3. SILVICULTURE, PLANT PHYSIOLOGY. Educ: Mich State Univ, BS, 50; Univ Miss, MS, 59; Univ Fla, PhD, 71. Prof Exp: Res forester, 51-67, SILVICULTURIST, US FOREST SERV, 67- Mem: Soc Am Foresters; Am Soc Plant Physiol; Soil Conserv Soc Am. Res: Regeneration problems; tree physiology and soil chemistry. Mailing Add: Crystal City 1600 S Eads St Apt 622 N Arlington VA 22202

BURNS, THOMAS WADE, b Dayton, Ohio, July 29, 24; m 52; c 4. INTERNAL MEDICINE, ENDOCRINOLOGY. Educ: Univ Utah, BA, 45, MD, 47, MS, 48; Am Bd Internal Med, dipl, 55. Prof Exp: Intern, Boston City Hosp, 48-49; asst resident, Harvard Med Sch, 49-50; clin investr, US Naval Hosp, Calif, 51-52; clin investr, US Naval Med Res Unit, Egypt, 52-54; resident physician, Med Ctr, Univ Calif, 55; asst prof, 55-57, assoc prof, 57-65, PROF MED, SCH MED, UNIV MO-COLUMBIA, 65-, PHYSICIAN, MED CTR, 55-, DIR DIV ENDOCRINOL & METAB, 75- Concurrent Pos: Fel, Harvard Med Sch, 49-50; fel med, Sch Med, Duke Univ, 50-51. Mem: AAAS; Endocrine Soc; Am Diabetes Asn; AMA; Am Col Physicians. Res: Endocrine and metabolic disorders. Mailing Add: Univ of Mo Columbia Med Ctr Columbia MO 65201

BURNS, VICTOR WILL, b Los Angeles, Calif, Nov 16, 25; m 50; c 3. BIOPHYSICS, CELL PHYSIOLOGY. Educ: Univ Calif, PhD(biophys), 55. Prof Exp: Assoc med physics, Univ Calif, 53-55; dir high energy proton biophys, Royal Univ Uppsala, 55-56; from res assoc to lectr, Stanford Univ, 56-64; assoc prof, 64-72, PROF BIOPHYS, UNIV CALIF, DAVIS, 72- Mem: Soc Gen Physiol; Radiation Res Soc; Biophys Soc. Res: Cell division and intracellular coordination; biological and physical effects of ionizing radiation and sonic energy; fluorescence analysis of cells and nucleic acids. Mailing Add: Dept of Physiol Sci Univ of Calif Davis CA 95616

BURNS, WILLIAM CHANDLER, b Fargo, NDak, Jan 5, 26; m 56; c 4. PARASITOLOGY. Educ: Ore State Col, BS, 50, MA, 52; Univ Wis-Madison, PhD(zool), 58. Prof Exp: Teaching asst zool, Ore State Col, 50-52; teaching asst, 52-57, actg instr, 57-58, from instr to assoc prof, 58-68, PROF ZOOL, UNIV WIS-MADISON, 68- Concurrent Pos: NSF fel, Ore State Col, 59-61. Mem: Am Soc Parasitol. Res: Ecology of helminths; coccidiosis; host-parasite relationships. Mailing Add: Dept of Zool Univ of Wis Madison WI 53706

BURNSIDE, CHARLES H, b Mitchell, SDak, Feb 17, 20. ORGANIC CHEMISTRY. Educ: Dakota Wesleyan Univ, BA, 47; Mont State Univ, MS, 52. Prof Exp: Res chemist, Phillips Petrol Co, 52-59; res chemist, Astrodyne Div, NAm Rockwell Corp, 59, rocketdyne div, 59-60, sr res chemist, 60-66, RES SPECIALIST, ROCKETDYNE DIV, ROCKWELL INT, 66- Mem: Am Chem Soc; Am Inst Aeronaut & Astronaut; Am Mgt Asn. Res: Development of rubber-based extrudable and castable composite solid propellants. Mailing Add: Rocketdyne Div Rockwell Int McGregor TX 76657

BURNSIDE, EDWARD BLAIR, b Madoc, Ont, Mar 16, 37; m 59; c 3. POPULATION GENETICS, ANIMAL BREEDING. Educ: Univ Toronto, BSA, 59, MSA, 60; NC State Univ, PhD(animal breeding, genetics, statist), 65. Prof Exp: Lectr animal breeding, Ont Agr Col, Univ Guelph, 60-61; res asst, NC State Univ, 61-64; asst prof, 64-66, ASSOC PROF ANIMAL BREEDING, UNIV GUELPH, 66- Mem: Am Dairy Sci Asn; Can Soc Animal Prod. Res: Estimation of genetic trend in populations; effect of environmental factors on milk yield; genotype by environment interaction; pedigree indexing of dairy cattle; sire evaluation methods. Mailing Add: Dept of Animal & Poultry Sci Univ of Guelph Guelph ON Can

BURNSIDE, MARY BETH, b San Antonio, Tex, Apr 23, 43. CELL BIOLOGY, ANATOMY. Educ: Univ Tex, Austin, BA, 65, MA, 67, PhD(zool), 68. Prof Exp: Instr anat, Harvard Med Sch, 71-72; asst prof anat, Univ Pa, 72-75; ASST PROF ANAT-PHYSIOL, UNIV CALIF, BERKELEY, 75- Concurrent Pos: Am Asn Univ Women fel, Hubrecht Lab, Utrecht, Holland, 68-69; NIH fels, Harvard Univ Biolabs, Cambridge, 69-70 & Harvard Med Sch, Boston, 70-71. Mem: AAAS; Am Soc Cell Biol; Am Asn Anat; Am Soc Zool; Asn Res Vision & Ophthal. Res: Mechanism of cell shape determination; elongation and apical constriction of neural epithelial cells during neurulation and photomechanical movements in teleost retinas; techniques are electron microscopy, biochemistry and tissue and cell culture. Mailing Add: Dept of Physiol-Anat Univ of Calif Berkeley CA 94720

BURNSIDE, ORVIN C, b Hawley, Minn, June 9, 32; m 54; c 2. WEED SCIENCE, AGRONOMY. Educ: NDak State Univ, BS, 54; Univ Minn, MS, 58, PhD(weed sci), 59. Prof Exp: From asst prof to assoc prof, 59-66, PROF AGRON, UNIV NEBR, LINCOLN, 66- Mem: Weed Sci Soc Am. Res: Weed control in agronomic crops; phenology and life history of weeds; herbicide dissipation. Mailing Add: Keim Hall 319 Dept of Agron Univ of Nebr Lincoln NE 68583

BURNSIDE, PHILLIPS BROOKS, b Columbus, Ohio, July 20, 27; m 54; c 3. PHYSICS. Educ: Ohio State Univ, BSc, 51, MSc, 54, PhD(physics), 58. Prof Exp: Instr physics, Ohio State Univ, 58-59; from asst prof to assoc prof, 59-69, PROF PHYSICS, OHIO WESLEYAN UNIV, 69- Mem: AAAS; Optical Soc Am; Am Asn Physics Teachers. Res: Physical biology; thermal physics; optical physics; philosophy of science. Mailing Add: Dept of Physics Ohio Wesleyan Univ Delaware OH 43015

BURNSTEIN, RAY A, b Harrisburg, Pa, Oct 5, 30; m 63. ELEMENTARY PARTICLE PHYSICS. Educ: Univ Chicago, BS, 52; Univ Wash, MS, 56; Univ Mich, PhD(physics), 60. Prof Exp: Asst prof physics, Univ Md, 60-66; assoc prof, 66-73, PROF PHYSICS, ILL INST TECHNOL, 73- Concurrent Pos: Vis assoc prof physics, Univ Md, 72-73. Mem: Am Phys Soc. Res: Experiments using visual detectors and counter techniques at high energy. Mailing Add: Dept of Physics Ill Inst of Technol Chicago IL 60616

BURNSTEIN, THEODORE, b Denver, Colo, Mar 18, 25; m 53; c 2. VIROLOGY. Educ: Colo Agr & Mech Col, DVM, 49; Cornell Univ, MS, 51, PhD(microbiol), 53. Prof Exp: Res fel biol, Johns Hopkins Univ, 53-55; res assoc, Sharp & Dohme, 55-59; res assoc prof microbiol, Med Sch, Univ Miami, 59-61; assoc prof, 61-65, PROF VIROL, SCH VET SCI & MED, PURDUE UNIV, 65- Mem: Am Asn Immunol; Am Soc Microbiol; Am Vet Med Asn; Tissue Cult Asn. Res: Measles encephalitis; tumor viruses; immunotherapy of canine neoplasia. Mailing Add: Dept of Vet Microbiol Purdue Univ Sch Vet Sci & Med Lafayette IN 47907

BUROW, DUANE FRUEH, b San Antonio, Tex, June 12, 40; m 69. STRUCTURAL CHEMISTRY, PHYSICAL CHEMISTRY. Educ: Univ Tex, Austin, BA, 61, PhD(chem), 66. Prof Exp: Instr chem, St Edward's Univ, 64-65; res phys chemist, Eng Physics Lab, E I du Pont de Nemours & Co, Inc, Del, 65-67; asst prof chem, Mich State Univ, 67-69; asst prof, 69-73, ASSOC PROF CHEM, UNIV TOLEDO, 73- Concurrent Pos: Petrol Res Fund grant, 68-70; Res Corp res grant, 70; NSF res grant, 74. Mem: Am Chem Soc; Optical Soc Am; The Chem Soc. Res: Optical activity; circular differential Raman spectroscopy; vibrational spectroscopy; non-aqueous solution chemistry. Mailing Add: Dept of Chem Univ of Toledo Toledo OH 43606

BUROW, KENNETH WAYNE, JR, b Humboldt, Nebr, Oct 24, 46; m 66; c 3. SYNTHETIC ORGANIC CHEMISTRY. Educ: Univ Nebr, BS, 68; Iowa State Univ, PhD(org chem), 73. Prof Exp: Res assoc, Univ Nebr, 72-73; SR ORG CHEMIST, ELI LILLY & CO, 73- Mem: Am Chem Soc; Chem Soc London. Res: Modification of ionosphoric compounds, including synthesis of crown ethers; ruminant nutrition; synthesis of sesquiterpenes. Mailing Add: Lilly Res Lab Box 708 Greenfield IN 46140

BURR, ALEXANDER FULLER, b Cambridge, Mass, July 18, 31; m 62; c 3. ATOMIC PHYSICS, SOLID STATE PHYSICS. Educ: Jamestown Col, BS, 53; Univ Edinburgh, MS, 58; Johns Hopkins Univ, PhD(physics), 66. Prof Exp: Res asst, Univ NDak, 50 & 53; physicist, Ballistics Res Lab, Aberdeen Proving Ground, Md, 54; res physicist, US Naval Res Lab, 65-66; asst prof, 66-69, ASSOC PROF PHYSICS, N MEX STATE UNIV, 69- Concurrent Pos: Nat Acad Sci-Nat Res Coun res assoc, 65-66; sr vis fel, Univ Strathclyde, Scotland, 73. Mem: Fel AAAS; Am Asn Physics Teachers; Am Phys Soc. Res: X-ray wavelengths and the x-ray wavelength scale; determination of electron binding energies; physics of soft x-rays; application of computers to physics education. Mailing Add: Dept of Physics NMex State Univ Las Cruces NM 88003

BURR, GEORGE OSWALD, b Conway, Ark, Oct 6, 96; m. BIOCHEMISTRY, PLANT PHYSIOLOGY. Educ: Hendrix Col, AB, 16, LLD, 36; Univ Ark, AM, 20; Univ Minn, PhD(biol chem), 22. Prof Exp: Prin high sch, Ark, 16-17; prof chem & physics, Ky Wesleyan Col, 17-18; asst agr biochem, Univ Minn, 20-22; Nat Res Fel chem, Univ Calif, 22-24, res assoc, 22-27; from assoc prof to prof plant physiol, Univ Minn, 27-40, head div physiol chem, 40-46; head dept physiol & biochem, Exp Sta, Hawaiian Sugar Planters Asn, 46-62; consult, 62-66; RES ADV, TAIWAN SUGAR CORP, 66- Concurrent Pos: Guggenheim Mem Found fel, 34; mem indust, Adv Comt, Sugar Res Found; del, First Int Conf on Peaceful Uses of Atomic Energy, Geneva, 55; ed, Arch Biochem. Honors & Awards: Outstanding Achievement Award, Univ Minn, 55. Mem: AAAS; Am Soc Naturalists; Am Soc Biol Chem; Am Chem Soc; Soc Exp Biol & Med. Res: Protein chemistry; nutrition; chemistry of reproduction of animals; chemistry and physiology of fats; photosynthesis. Mailing Add: 112 Niuiki Circle Honolulu HI 96821

BURR, HELEN GUNDERSON, b Iowa City, Iowa, Dec 30, 18; m 54; c 1. SPEECH PATHOLOGY, AUDIOLOGY. Educ: Stanford Univ, BA, 37; Univ Southern Calif, MA, 40; Columbia Univ, PhD(speech path & audiol), 49. Prof Exp: Speech pathologist, NY, 44-50; asst prof speech path & audiol & dir speech clin, State Univ NY, 50-53; from asst prof to assoc prof, 53-67, PROF SPEECH PATH & AUDIOL, 67-, CHMN DEPT, 62-, DIR SPEECH & HEARING CTR, 61- Concurrent Pos: Rehab Serv Admin training grant, 61-; US Off Educ training grant, 64-, res grant, 67-69; USPHS training grant, 67-70. Mem: AAAS; fel Am Speech & Hearing Asn; Ling Soc Am; NY Acad Sci. Res: Linguistics; verbal behavior. Mailing Add: Dept of Speech Path & Audiol Univ of Va Charlottesville VA 22903

BURR, HORACE KELSEY, b Manchester, Conn, Sept 9, 13; m 38; c 2. PHYSICAL CHEMISTRY, FOOD SCIENCE. Educ: Wesleyan Univ, AB, 35; Univ Wis, MS, 37, PhD(phys chem), 41. Prof Exp: Instr chem, Mich Col Mining & Technol, 37-38; from asst chemist to chemist, Forest Prod Lab, US Forest Serv, 41-46; from assoc chemist to prin chemist, Western Regional Res Lab, Bur Agr & Indust Chem, 46-61, chief veg lab, Western Utilization Res & Develop Div, Agr Res Serv, 61-69, RES CHEMIST, WESTERN REGIONAL RES CTR, AGR RES SERV, USDA, 69- Concurrent Pos: Adj assoc prof food sci, Univ Calif, Berkeley, 71- Mem: Fel AAAS; Am Chem Soc; Inst Food Technologists. Res: Diffusion in cellulosic materials; wood and paper base plastics; vegetable dehydration; chemistry and processing technology of fruits and vegetables. Mailing Add: Western Regional Res Ctr Agr Res Serv USDA Berkeley CA 94710

BURR, IRVING WINGATE, b Fallon, Nev, Apr 9, 08; m 30, 66; c 3. MATHEMATICS. Educ: Antioch Col, BS, 30; Univ Chicago, MS, 35; Univ Mich, PhD(math, statist), 41. Prof Exp: From instr to assoc prof math, Antioch Col, 30-41; asst prof math, 41-45, from assoc prof to prof statist & math, 45-74, EMER PROF STATIST, PURDUE UNIV, 74- Mem: Am Math Soc; Math Asn Am; Am Statist Asn; Inst Math Statist; fel Am Soc Qual Control. Res: Mathematical and engineering statistics. Mailing Add: PO Box 527 Ocean Park WA 98640

BURR, JOHN GREEN, b Ft Sill, Okla, Mar 12, 18; m 43; c 7. PHOTOCHEMISTRY, RADIATION CHEMISTRY. Educ: Mass Inst Technol, BS & MS, 40; Northwestern Univ, PhD(org chem), 48. Prof Exp: Chemist, Jackson Lab, E I du Pont de Nemours & Co, Inc, 40-41, 42; asst, Manhattan Dist, Chicago, 43-44; asst prof chem, Miami Univ, 47-48; sr chemist, Oak Ridge Nat Lab, 48-57; supvr, Atomics Inst, 57-62; group leader, NAm Sci Ctr, 62-69; PROF CHEM & RADIOL SCI, UNIV OKLA, 69- Concurrent Pos: USPHS spec fel, Univ London, Eng, 53-54; Guggenheim fel, Cambridge Univ, 65-66; Rosetta Brigel Bartin lectr, Univ Okla, 68; USPHS serv fel, 75-76; vis prof, Hebrew Univ & Max Planck Inst Radiation Chem, 75; NIH Serv fel, Nat Cancer Inst, 75-76. Mem: AAAS; Am Chem Soc; Radiation Res Soc; fel Am Inst Chem; Am Photobiol Asn. Res: Isotope tracers; organic radiation chemistry; organic reaction mechanisms; nucleic acid photochemistry; aqueous photochemistry. Mailing Add: Dept of Chem Univ of Okla 620 Parrington Oval Norman OK 73069

BURR, WILLIAM WESLEY, JR, b Lincoln, Nebr, Mar 12, 23; m 50; c 6. BIOCHEMISTRY, MEDICINE. Educ: Univ Nebr, AB, 47; Univ Ill, MS, 48, PhD(chem), 51; Univ Tex, MD, 60. Prof Exp: Asst prof biochem, Southwestern Med Sch, Tex, 51-53, assoc prof, 53-60, prof, 60-63; chief med res br, US AEC, 61-69, asst dir med & health res, 69-70, DEP DIR DIV BIOMED & ENVIRON RES, US ENERGY RES & DEVELOP ADMIN, 70- Concurrent Pos: Prof lectr, George Washington Univ, 63- Mem: AAAS; Am Inst Nutrit; Am Chem Soc; Am Soc Exp Biol & Med; Am Soc Biol Chem. Res: Protein and lipid metabolism. Mailing Add: 13008 Meadow View Dr Gaithersburg MD 20760

BURRAGE, R H, b Davidson, Sask, Mar 18, 20; m 54; c 5. ENTOMOLOGY. Educ: Ont Agr Col, BS, 49; Cornell Univ, PhD(entom), 53. Prof Exp: Entomologist field crop pests, 53-73, HEAD ENTOMOLOGIST, CAN DEPT AGR, 73- Mem: Entom Soc Am; Entom Soc Can. Res: Biology and control of wireworms in field crops in the prairie provinces. Mailing Add: Res Sta Res Br Can Dept of Agr Univ Campus Saskatoon SK Can

BURRELL, CRAIG DONALD, b Gravesend, Eng, July 5, 26; nat US; m 60; c 7. MEDICINE, ENDOCRINOLOGY. Educ: Univ Otago, NZ, MB, ChB, 51. Prof Exp: Intern, Wellington Hosp, NZ, 51-52; sr house officer, Children's Hosp, Nottingham, Eng, 53; house physician, Hammersmith Hosp & Post-grad Med Sch, London, 54, sr house officer, 55-56; registr & RMO, Welsh Nat Sch Med & Royal Infirmary, Cardiff, 56-60; asst prof med & med in psychiat, Med Col, Cornell Univ & New York Hosp, 60-61; dir med serv, 61-66, med dir, 66-69, vpres med affairs, Sandoz Pharmaceut, 69-72, VPRES, SANDOZ, INC, 73- Concurrent Pos: Brit Med Res Coun grant, 55; assoc clin prof, Col Med NJ. Mem: Endocrine Soc; hon fel Am Sch Health Asn; Am Col Clin Pharmacol & Therapeut; fel Royal Soc Med. Res: Thyroids; endocrine aspects of psychiatry; therapeutics. Mailing Add: Sandoz Inc East Hanover NJ 07936

BURRELL, DAVID COLIN, b England. CHEMICAL OCEANOGRAPHY. Educ: Univ Nottingham, BS, 61, PhD(geochem), 64. Prof Exp: Asst prof marine sci, 65-69, ASSOC PROF MARINE SCI, UNIV ALASKA, 69- Mem: Geochem Soc; Am Geophys Union; Arctic Inst NAm. Res: Marine trace metal chemistry. Mailing Add: Inst of Marine Sci Univ of Alaska Fairbanks AK 99701

BURRELL, ELLIOTT JOSEPH, JR, b Phoenix, Ariz, Feb 3, 29; m 50; c 4. PHYSICAL CHEMISTRY. Educ: Univ Notre Dame, BS, 50, MS, 51; Pa State Univ, PhD(phys chem), 54. Prof Exp: Res chemist, Fabrics & Finishes Div, E I du Pont de Nemours & Co, Inc, 54-58, from res chemist to sr res chemist, Radiation Physics Lab, 58-63; from asst dean to assoc dean sci, Col Arts & Sci, 70-74, asst prof, 63-65, ASSOC PROF CHEM, LOYOLA UNIV CHICAGO, 65- Concurrent Pos: Lectr, Temple Univ, 55-57, 58-63 & Univ Pa, 57-58. Mem: Am Chem Soc. Res: Nuclear and radiation chemistry; pulse radiolysis; photochemistry; kinetics; physical-organic chemistry; magnetic resonance spectroscopy; flash photolysis. Mailing Add: Dept of Chem Loyola Univ 6525 N Sheridan Rd Chicago IL 60626

BURRELL, HAROLD PAUL CHARLES, b Edmonton, Alta, May 16, 11; US citizen; m 52; c 1. ORGANIC CHEMISTRY. Educ: Yale Univ, BS, 34; Univ Conn, MS, 53. Prof Exp: Chemist rubber compounding, US Rubber Co, Conn, 34-37, anal control, Johns Manville, NJ, 37-38 & prod control develop orgn compounds & anal methods, Naugatuck Chem Co, Conn, 38-47; from instr to asst prof org chem & chem lit, Clarkson Tech Univ, 47-62; asst prof, 62-65, ASSOC PROF CHEM, ADIRONDACK COMMUNITY COL, 65-, ACTG CHMN DEPT, 62- Mem: Am Chem Soc. Res: Development of analytical methods for determination of polyamines; balancing organic redox equations; chemical literature; nomenclature; coding of organic compounds. Mailing Add: Dept of Chem Adirondack Community Col Glens Falls NY 12801

BURRELL, HARRY, b Dover, NJ, May 13, 13; m 33; c 3. POLYMER CHEMISTRY. Educ: Newark Col Eng, BS, 34, ChE, 38. Prof Exp: Res chemist, Ellis-Foster Co, 34-40 & Res Div, Heyden Chem Corp, 40-46; pres, Burrell & Neidig, Inc, 46-48; vehicle chemist, Finishes Div, Interchem Corp, NY, 47-57, mgr resin develop lab, 57-62, tech dir div, 62-68; TECH DIR BLDG & INDUST PROD, INMONT CORP, 68- Concurrent Pos: Asst prof, Stevens Inst Technol, 47-53; lectr, Xavier Univ, Ohio, 56-62; chmn, Gordon Conf, 56; trustee, Paint Res Inst, 58-, pres, 61-65; Mattiello lectr, 69. Honors & Awards: Roon Awards, 57, 61; Borden Award for Plastics & Coatings, Am Chem Soc, 68; George B Heckel Award, 70. Mem: Am Chem Soc. Res: Polymer chemistry as applied to coatings. Mailing Add: Inmont Corp 1255 Broad St Clifton NJ 07015

BURRELL, ROBERT GUTHRIE, b Springfield, Ohio, Aug 26, 33; m 55; c 2. IMMUNOLOGY, MICROBIOLOGY. Educ: Ohio State Univ, BS, 55, MS, 56, PhD(bact), 58. Prof Exp: Asst prof biol sci, Carnegie Inst Technol, 58-61; asst prof, 61-69, PROF MICROBIOL, SCH MED, W VA UNIV, 69- Mem: Am Asn Immunol. Res: Immune injury in pulmonary diseases; tissue antigens; serology of fungal antigens. Mailing Add: Dept of Microbiol Med Ctr W Va Univ Morgantown WV 26506

BURRELL, VICTOR GREGORY, JR, b Wilmington, NC, Sept 12, 25; m 56; c 4. MARINE SCIENCES. Educ: Col Charleston, BS, 49; Col William & Mary, MA, 68, PhD(marine sci), 72. Prof Exp: Res assoc, Va Inst Marine Sci, 66-68, from asst marine scientist to assoc marine scientist, 68-72; assoc marine scientist, 72-73, asst dir, 73-74, DIR, MARINE RESOURCES RES INST, SC, 74- Concurrent Pos: Adj prof biol, Col Charleston, 74-; mem, Tech Comt Fisheries, Atlantic States Marine Fisheries Comn, 74- Mem: Am Inst Fishery Res Biologists; Estuarine Res Fedn; Nat Shellfish Asn; Gulf & Caribbean Fisheries Inst. Res: Estuarine and marine zooplankton dynamics; copepod taxonomy; marine fisheries science. Mailing Add: Marine Resources Res Inst PO Box 12559 Charleston SC 29412

BURRESON, BURTON JAY, b Seattle, Wash, Aug 3, 42; m 68; c 2. NATURAL PRODUCTS CHEMISTRY. Educ: Ore State Univ, BA, 64; Univ Calif, Santa Barbara, PhD(chem), 69. Prof Exp: Sr chemist, Whittaker Corp Res & Develop, 68-72; RES ASSOC CHEM, UNIV HAWAII, 72- Concurrent Pos: NIH spec fel, 72-74. Mem: Am Chem Soc; AAAS. Res: Structure determination and synthesis of natural products of marine origin, especially those with physiological activity. Mailing Add: Dept of Chem 2545 The Mall Univ of Hawaii Honolulu HI 96822

BURRIDGE, KENELM OSWALD LANCELOT, b St Julians, Malta, Oct 31, 22; Brit citizen; c 1. HISTORY OF ANTHROPOLOGY, SOCIAL ANTHROPOLOGY. Educ: Oxford Univ, BA, 48, dipl anthrop, 49, MA, 50; Australian Nat Univ, PhD(social anthrop), 53. Prof Exp: Res fel, Univ Malaya, 53-56; prof anthrop, Univ Baghdad, 56-58; lectr ethnol, Pitt Rivers Mus, Univ Oxford, 56-68; PROF ANTHROP, UNIV BC, 68-, HEAD DEPT ANTHROP & SOCIOL, 73- Concurrent Pos: Vis lectr social anthrop, Univ Western Australia, 68; Guggenheim fel, Univ BC, 72-73. Mem: Royal Anthrop Inst Gt Brit & Ireland; Can Sociol & Anthrop Asn; Assoc Am Anthrop Asn. Res: Religious movements; mythology; social organization; history of anthropology; missiology. Mailing Add: Dept of Anthrop & Sociol Univ of BC Vancouver BC Can

BURRILL, CLAUDE WESLEY, b Akron, Iowa, Feb 19, 25; m 55; c 2. MATHEMATICS. Educ: Univ Iowa, BS, 48, MS, 50, PhD(math), 52. Prof Exp: Instr math, Univ Iowa, 48-52; asst prof, NY Univ, 53-56; mathematician, Int Bus Mach Corp, 56-57; assoc prof math, NY Univ, 57-67; sr staff mem, IBM Systs Res Inst, 67-72, SR STAFF MEM, IBM SYSTS SCI INST, 72- Concurrent Pos: Consult, Int Bus Mach Corp, 57-67; chmn, Bd Trustees, William Patterson Col NJ, 74- Mem: Am Math Soc; Math Asn Am; Opers Res Soc Am. Res: Mathematical analysis; financial model building. Mailing Add: IBM Systs Sci Inst 205 E 42 St New York NY 10017

BURRILL, DAN Y, b Chicago, Ill, May 9, 07; m 31; c 1. DENTISTRY. Educ: Univ Mich, BA, 29, LLB, 31; Northwestern Univ, DDS, 39, MS, 42. Prof Exp: From res assoc to asst prof oral path, Dent Sch, Northwestern Univ, 39-46; prof oral med & head dept, Sch Dent, Univ Louisville, 46-57; PROF ORAL DIAG & HEAD DEPT, DENT SCH, NORTHWESTERN UNIV, CHICAGO, 57- Mem: AAAS; Am Acad Oral Path; Int Asn Dent Res (asst secy-treas & secy-treas, 53-60, vpres, pres elect & pres, 60-63). Res: Oral diagnosis; analgesia; dental caries; periodontia; dental pulp; nutrition. Mailing Add: Northwestern Univ Dent Sch 311 E Chicago Ave Chicago IL 60611

BURRILL, ERNEST ALFRED, (JR), b Brockton, Mass, Apr 5, 17; m 42; c 2. NUCLEAR PHYSICS. Educ: Mass Inst Technol, BSc, 43. Prof Exp: Asst, High Voltage Lab, Mass Inst Technol, 39-43, 43-47; physicist, High Voltage Eng Corp, 47-52, mgr contract eng div, 52-57, vpres & sales mgr, 57-58, vpres & dir mkt, 58-67, vpres & mgr western region, 67-68; nuclear mkt consult, 68-70; exec vpres, Accelerators, Inc, 70-71; RADIATION APPLN CONSULT, 71- Honors & Awards: Naval Ord Award, 45; Dudley Medal, Am Soc Testing & Mat, 49. Mem: Am Phys Soc; Am Soc Testing & Mat; Am Nuclear Soc. Res: Van de Graaff, electron linear and low-voltage accelerator development and applications; radiation shielding. Mailing Add: 506 Cuesta Dr Aptos CA 95003

BURRILL, MEREDITH FREDERIC, b Houlton, Maine, Dec 23, 02; m 27; c 2. GEOGRAPHY. Educ: Bates Col, AB, 25; Clark Univ, MA, 26, PhD(geog), 30. Hon Degrees: ScD, Bates Col, 60. Prof Exp: Instr physiography, Lehigh Univ, 26-28; prof geog, Okla State Univ, 30-40; econ geogr, Gen Land Off, Dept Interior, 40-41, chief res & anal div, 41-43, dir Off Geog, 43-68, geogr, Defense Intel Agency, 68-72, geogr, Defense Mapping Agency, 72-73, exec secy, US Bd Geog Names, Dept Interior, 43-73, CONSULT, DEPTS STATE & DEFENSE, 73- Concurrent Pos: Indust consult, Montreal Bd Trade, 31; Soc Sci Res Coun southern regional grant in aid, Okla, 34-36; actg head land use planning, Region VIII, Resettlement Admin, 35-37; mem US nat comt, Int Geog Union, Nat Res Coun, 47-52, mem comt Am geog, 48-53 & vchmn, Comt Adv to Off Naval Res, 49-52; US mem, Int Comt Onomastic Sci, 49-; mem tech comt Antarctica, Dept State, 55-59; chmn, UN Group Experts Geog Names, 60, 66 & 67-; pres, UN Conf Standardization Geog Names, 67. Honors & Awards: El Sabio, Order of Alfonso X, Spain, 55; Citation, Cent Intel Agency, 58; Antarctic Medal, Dept Defense, 65; Distinguished Serv Medal, Dept Interior, 68; Meritorious Civilian Award, Defense Mapping Agency, 73; Distinguished Serv Award, Geog Soc Chicago, 73. Mem: AAAS (vpres sect geol & geog, 53); Asn Am Geogr (secy, 60-63, vpres, 65, pres, 66, parliamentarian, 70-); Am Name Soc (pres, 55); hon fel Am Geog Soc; Am Geophys Union. Res: International standardization of geographical names; geographic terminology; toponymy. Mailing Add: 5503 Grove St Chevy Chase MD 20015

BURRILL, ROBERT MEREDITH, b Oklahoma City, Okla, Oct 23, 33; m 73. BIOGEOGRAPHY, PHYSICAL GEOGRAPHY. Educ: Wesleyan Univ, BA, 56; Univ Chicago, MS, 61; Univ Kans, PhD(geog), 70. Prof Exp: Instr geog, Ohio Univ, 61-63; ASST PROF GEOG, UNIV GA, 66- Mem: Am Geog Soc; Asn Am Geogrs; Ecol Soc Am; AAAS. Res: Investigation of weather effects upon honeybee activity using an invention which makes measure of honeybee activity possible; geographic analysis of animal behavior. Mailing Add: Dept of Geog Univ of Ga Athens GA 30602

BURRIS, ALBERT, b Winchester, Ohio, May 15, 12; m 40. PHYSICS. Educ: Miami Univ, AB, 34; Iowa State Col, MS, 36; Mich State Univ, PhD(physics), 44. Prof Exp: Prof physics, Eastern NMex Col, 41-43; assoc prof, NMex State Univ, 43-44; sr physicist, Appl Physics Lab, Johns Hopkins Univ, 44-45; res physicist, Keystone Carbon Co, Pa, 46; physicist, Phys Sci Lab & prof physics, 46-74, EMER PROF PHYSICS, NMEX STATE UNIV, 74- Mem: Am Asn Physics Teachers. Res: Magnetic phenomena; powder metallurgy; electro-mechanical devices in rocket telemetry; solar energy. Mailing Add: Box 3 D Physics Dept NMex State Univ Univ Park Br Las Cruces NM 88001

BURRIS, B CULLEN, b Smithdale, Miss, June 17, 24; m 55; c 2. MEDICINE. Educ: Miss Col, BA, 43; Univ Tenn, MD, 46. Prof Exp: Pvt practr & asst attend staff mem, Chicago Wesley Mem Hosp, 53-58; clin dir psychiat, Milwaukee Sanitarium Found, Milwaukee Psychiat Hosp, 58-62, med dir, 62-67; ASSOC PROF PSYCHIAT, MED SCH, NORTHWESTERN UNIV, CHICAGO, 67- Concurrent Pos: Clin instr, Med Sch, Northwestern Univ, Chicago, 53-58; consult, Rehab Inst, Chicago, 5658 & Cent YMCA Counseling Ctr, Ill, 5758; assoc prof, Sch Med, Marquette Univ, 5867; consult, Peace Corps, 63-67; mem, Nat Asn Psychiat Hosps, 63-69; mem attend staff, Northwestern Mem Hosp, 67; pvt practr, 67- Mem: Fel Am Psychiat Asn. Res: Psychiatry. Mailing Add: 707 N Fairbanks Ct Chicago IL 60611

BURRIS, JOSEPH STEPHEN, b Cleveland, Ohio, Apr 18, 42; m 63; c 3. PLANT PHYSIOLOGY. Educ: Iowa State Univ, BSc, 64; Va Polytech Inst, MSc, 66, PhD(agron), 68. Prof Exp: Instr agron, Va Polytech Inst, 67; RES AGRONOMIST, SEED LAB, IOWA STATE UNIV, 68- Mem: Am Soc Agron; Crop Sci Soc Am; Asn Official Seed Analysts. Res: Carbohydrate and nitrogen metabolism relationships; corn and soybean seed and seedlings vigor. Mailing Add: Dept of Bot Iowa State Univ Ames IA 50011

BURRIS, MARTIN JOE, b Hebron, Nebr, Mar 30, 27; m 53; c 4. GENETICS, ANIMAL HUSBANDRY. Educ: Univ Nebr, BS, 49, MS, 51; Ore State Col, PhD(genetics), 53. Prof Exp: Asst animal breeding, Univ Nebr, 49-51; asst prof, Univ Ark, 53-54; assoc prof, Va Agr Exp Sta, 54-57; animal geneticist, Coop State Res Serv, 57-58, prin animal geneticist, 58-66, asst dir, Mont Agr Exp Sta, Mont State Univ, 66-67, ASSOC DIR MONT AGR EXP STA, MONT STATE UNIV, USDA, 67- Concurrent Pos: Vis prof animal sci, Purdue Univ, 64-65. Mem: AAAS; Am Genetic Asn; Am Soc Animal Sci; Genetics Soc Am. Res: Beef cattle breeding; hormone physiology; meat processing; growth in domestic animals; livestock breeding and genetics; research administration. Mailing Add: Mont Agr Exp Sta Mont State Univ Bozeman MT 59715

BURRIS, ROBERT HARZA, b Brookings, SDak, Apr 13, 14; m 45; c 3. BIOCHEMISTRY. Educ: SDak State Univ, BS, 36; Univ Wis, MS, 38, PhD(agr bact), 40. Hon Degrees: DSc, SDak State Univ, 66. Prof Exp: Asst agr bact, 36-40, instr, 41-44, from asst prof to assoc prof biochem, 44-51, chmn dept, 58-70, PROF BIOCHEM, UNIV WIS-MADISON, 51- Concurrent Pos: Nat Res Coun fel, Columbia Univ, 40-41; consult, NSF, 53-57 & NIH, 61-; Guggenheim fel, 54; chmn bot sect, Nat Acad Sci, 71. Mem: Nat Acad Sci; AAAS; Am Chem Soc; Soc Plant Physiol (pres, 60); Am Acad Arts & Sci. Res: Biological nitrogen fixation; respiration of plants; nitrogen metabolism of plants; photosynthesis; biological oxidations; cytochromes; hydrobiology. Mailing Add: Dept of Biochem Univ of Wis Madison WI 53706

BURRIS, WILLIAM EDMON, b Okla, Nov 14, 24; m 47; c 2. ENVIRONMENTAL SCIENCES. Educ: Okla State Univ, BS, 48, MS, 49, PhD(zool), 56. Prof Exp: Instr zool, Okla State Univ, 52; biologist, US Army Eng Dist, 52-59; from asst prof to prof biol, San Antonio Col, 59-67, from actg chmn to chmn dept, 60-67; biologist, 67-70, RECREATION RESOURCE SPECIALIST, SOUTHWESTERN DIV, CORPS ENGRS, 70- Mem: AAAS; Am Fisheries Soc; Am Inst Biol Sci; Am Soc Limnol & Oceanog; Sigma Xi. Res: Limnology; fisheries biology; environmental planning; planning, development and operation of multiple purpose reservoirs and other water resource projects. Mailing Add: Corps of Engrs Southwestern Div 1200 Main St Rm 525 Dallas TX 75202

BURRIS-MEYER, HAROLD, b Madison, NJ, Apr 6, 02; m 45; c 2. ACOUSTICS. Educ: City Col New York, BS, 23; Columbia Univ, AM, 26. Prof Exp: From asst prof to prof eng, Stevens Inst Technol, 29-54, dir theatre, 30-54, res in sound, 38-54; consult, 54-64; prof drama, 64-75, EMER PROF DRAMA, FLA ATLANTIC UNIV, 75-, DIR RES NON-VERBAL COMMUN, 68- Concurrent Pos: Mem Nat Defense Res Comt Prof, 41-43; vpres & dir, Muzak Corp, 43-47; dir, Assoc Prog Serv, NY, 45-46; vpres & dir, Magnetic Prog, Inc, 48-57; pres, Control Inc, 51-; consult to cities, art ctrs & architects, 34- Honors & Awards: Founders Award, US Inst Theatre Technol, 73. Mem: Fel Acoust Soc Am; fel Audio Eng Soc; Am Educ Theatre Asn; Am Nat Theatre & Acad; fel Am Coun Arts in Educ. Res: Control of sound, equipment and techniques; psychophysical reaction to auditory stimuli; functional applications of music; psychophysical measurements; control of sound in the theatre; theatre planning. Mailing Add: Fla Atlantic Univ Boca Raton FL 33432

BURROUGHS, ALBERT LAWRENCE, b Basin, Wyo, Feb 26, 16; m 45; c 2. VETERINARY VIROLOGY. Educ: Univ Calif, Berkeley, PhD(biol), 46; Tex A&M Univ, DVM, 58. ASSOC PROF PATH, KANS STATE UNIV, 60- Mem: Am Soc Mammalogists; Sigma Xi. Res: Sylvatic plague; tularemia; zoonoses; viral disease of cattle. Mailing Add: Dept of Infectious Dis Kans State Univ Manhattan KS 66502

BURROUGHS, JAMES EDWARD, b Washington, DC, Mar 4, 30; m 52; c 5. ANALYTICAL CHEMISTRY. Educ: Loyola Univ, Ill, BS, 59, MS, 67. Prof Exp: Technician, Crane Packing Co, 57-58; analyst, Cook Tech Ctr, 58-59; from asst res chemist to sr res chemist, Borg-Warner Corp, 59-69, group leader, 69-72, MGR ANAL CHEM, BORG-WARNER CORP RES CTR, 72- Mem: Soc Appl Spectros; Am Chem Soc; Sigma Xi (treas, Sci Res Soc Am, 60, secy, 67, vpres, 69, pres, 70); fel Am Inst Chemists; Am Soc Testing & Mat. Res: Gas chromatography of metal chelates; applications and extensions of the oxygen flask method; atomic absorption applied to organic analysis. Mailing Add: Borg-Warner Corp Res Ctr Wolf & Algonquin Rd Des Plaines IL 60018

BURROUGHS, RICHARD LEE, b Detroit, Mich, Aug 30, 32; m 61; c 2. PETROLEUM GEOLOGY. Educ: Okla State Univ, BS, 55; Univ Ariz, MS, 60; Univ NMex, PhD(geol), 72. Prof Exp: Explor geologist, Pure Oil Co, 56-58 & Monsanto Chem Co, 60-63; PROF GEOL & HEAD DEPT, ADAMS STATE COL, 63- Concurrent Pos: Consult hydrogeol, 73- Mem: Geol Soc Am; Am Asn Petrol Geologists; Nat Asn Geol Teachers. Res: Regional geology; volcanics; hydrogeology. Mailing Add: Dept of Geol Adams State Col Alamosa CO 81102

BURROUGHS, ROBERT ELI, b Everett, NC, Dec 26, 03; m 28; c 3. PHYSICS. Educ: NC State Col, BSc, 25; Duke Univ, AM, 26. Prof Exp: Physicist, Kodak Labs, 26-31; Kodak res fel, Purdue Univ, 31-33; physicist, Kodak Labs, 33-42; sect engr, Gen Eng Lab, Gen Elec Co, 46-47, mgr design eng, Plutonium Prod Areas, Hanford Works, 47-48, mgr eng, Aircraft Gas Turbine Dept, 48-50, mem, Gen Eng Staff, 50-51, mem, Switchgear Res Assignment, 51-53, mgr eng & labs, Rectifier Dept, 53-56; asst dir eng res inst, 56-58, dir res inst, 58-60, dir res admin, 60-74, EMER DIR RES ADMIN, UNIV MICH, ANN ARBOR, 74- Concurrent Pos: Instnl rep, Eng Col Res Coun, 58-69, vchmn, 65-66, chmn, 66-68; mem comt sponsored projs, Am Coun Educ, 64-67, chmn, 68. Mem: Inst Elec & Electronics Engrs; Am Soc Eng Educ (vpres, 66-68); Am Soc Naval Engrs. Res: Applied physics; frictional electrification; electrical circuit transients; instrumentation; solid state physics devices; spectroscopy. Mailing Add: 1120 N Wagner Rd Ann Arbor MI 48103

BURROUGHS, WISE, b Tipton, Iowa, Dec 19, 11; m 37; c 3. ANIMAL SCIENCE. Educ: Univ Ill, BS, 34, PhD, 39. Prof Exp: From asst to assoc animal sci, Ohio Agr Exp Sta, 39-46; assoc prof, Ohio State Univ, 46-51; assoc prof, 51-52, prof animal husb, 52-71, CHARLES F CURTISS DISTINGUISHED PROF AGR, IOWA STATE UNIV, 71- Honors & Awards: Am Feed Mfrs Award, 54; John Scott Medal Award, 58. Mem: Am Soc Animal Sci; Am Dairy Sci Asn; Am Inst Nutrit. Res: Animal nutrition; endocrinology; fermentation within digestive tract; beef cattle production. Mailing Add: Dept of Animal Sci 301 Kildee Hall Iowa State Univ Ames IA 50011

BURROUS, MERWYN LEE, b South Bend, Ind, Jan 2, 34; m 54; c 2. ORGANIC CHEMISTRY. Educ: Manchester Col, AB, 57; Purdue Univ, PhD(org chem), 61. Prof Exp: RES CHEMIST, CHEVRON RES CO, STAND OIL CO CALIF, 61- Mailing Add: Chevron Res Co 576 Standard Ave Richmond CA 94802

BURROUS, STANLEY EMERSON, b Elkhart, Ind, Mar 14, 28; m 56; c 3. CHEMOTHERAPY. Educ: Manchester Col, BA, 50; Ind Univ, MA, 53; Univ Ill, PhD(microbiol), 62. Prof Exp: Res asst virol, Upjohn Co, 58-60; asst prof bioeng, Okla State Univ, 62-63; sr res scientist, E R Squibb Div, Olin Mathieson Chem Corp, 63-65; sr biochemist, Smith Kline & French Labs, Pa, 65-67; unit leader, Eaton Labs Div, Norwich Pharmacal Co, 67-75; PRES, BURROUS ENTERPRISES, 75- Mem: AAAS; Am Soc Microbiol; Am Chem Soc; Soc Quantum Biol; Environ Mutagen Soc. Res: Microbiology. Mailing Add: RD 2 Box 415 Norwich NY 13815

BURROW, GERARD N, b Boston, Mass, Jan 9, 33; m 56; c 3. INTERNAL MEDICINE. Educ: Brown Univ, AB, 54; Yale Univ, MD, 58. Prof Exp: Intern internal med, Yale New Haven Hosp, 58-59, resident med, 61-63; instr, 65-66, from asst prof to assoc prof internal med, Sch Med, Yale Univ, 70-75; PROF MED, UNIV TORONTO, 75-; DIR ENDOCRINOL SECT & SR PHYSICIAN, TORONTO GEN HOSP, 76- Concurrent Pos: Fel metab, Yale New Haven Hosp, 63-65; res career develop award, 68-73; chief med res, Yale New Haven Hosp, 65-66; vis scientist, Fac Med, Univ Marseille, 72-73. Mem: Am Fedn Clin Res: Endocrine Soc; Am Thyroid Asn; fel Am Col Physicians. Res: Protein synthesis and hormone action; thyroid gland. Mailing Add: Dept of Internal Med Univ of Toronto Toronto ON Can

BURROWS, BENJAMIN, b New York, NY, Dec 16, 27; m 49; c 4. INTERNAL MEDICINE. Educ: Johns Hopkins Univ, MD, 49. Prof Exp: Instr med, Univ Chicago Clins, 55-56, asst prof, 56-61, assoc prof, 61-68; PROF MED & DIR DIV RESPIRATORY SCI, UNIV ARIZ, 68- Concurrent Pos: Consult, Tucson Vet Admin Hosp. Mem: Am Physiol Soc; Am Fedn Clin Res; Am Thoracic Soc; fel Am Col Chest Physicians; Am Soc Clin Invests. Res: Pulmonary disease; pulmonary physiology; diffusion. Mailing Add: Div Respiratory Sci Ariz Med Ctr Tucson AZ 85724

BURROWS, ELIZABETH PARKER, b Pittsburgh, Pa, Nov 5, 30; m 58. BIO-ORGANIC CHEMISTRY. Educ: Middlebury Col, AB, 52; Stanford Univ, MS, 54, PhD(chem), 57. Prof Exp: Res assoc pharmacol, Stanford Univ, 56-57; fel chem,

Wayne State Univ, 57-58; fel, Mass Inst Technol, 58-61, res assoc, 62-66; res assoc, Children's Cancer Res Found, Mass, 66-67; staff scientist, Worcester Found Exp Biol, 67-70; res assoc chem, Oakland Univ, 70-71; res assoc, 71-75, RES ASST PROF, VANDERBILT UNIV, 75- Res: Natural products; structure and biosynthesis of macrocyclic antibiotics; organic compounds of biological and physiological importance. Mailing Add: Dept of Chem Vanderbilt Univ Nashville TN 37235

BURROWS, GEORGE EDWARD, b Seattle, Wash, Feb 17, 35; m 55; c 6. VETERINARY PHARMACOLOGY, VETERINARY TOXICOLOGY. Educ: Univ Calif, Davis, BS, 64, DVM, 66; Wash State Univ, MS, 69, PhD(pharmacol, toxicol), 72. Prof Exp: From instr to asst prof vet med, Col Vet Med, Wash State Univ, 66-72; vis lectr, Univ Nairobi & Colo State Univ, 72-74; asst prof, Col Vet Med, Wash State Univ, 74-75; ASSOC PROF VET PHARMACOL & TOXICOL, UNIV IDAHO & NORTHWEST COL VET MED, 69-71. Mem: Am Vet Med Asn; Am Col Vet Toxicologists. Res: Cyanide intoxication and its therapy and prevention; initiation and perpetuation of endotoxic shock in domestic animals; antimethemoglobinemic therapeutic agents and their toxicity. Mailing Add: Dept of Vet Sci Univ of Idaho Moscow ID 83843

BURROWS, JOHN RONALD, b Huntington, WVa, July 3, 33; Can citizen; m 58; c 3. GEOPHYSICS, SPACE PHYSICS. Educ: Univ Toronto, BA, 55, MA, 56, PhD(physics), 61. Prof Exp: RES OFFICER COSMIC RAYS, NAT RES COUN CAN, 61- Mem: Am Geophys Union; Can Asn Physicists. Res: Space physics, magnetospheric physics and solar terrestrial problems; studies related to the charged particle environment of the earth. Mailing Add: Nat Res Coun of Can 100 Sussex Dr Ottawa ON Can

BURROWS, LESLIE RAYMOND, b La Junta, Colo, Nov 1, 29; m 51; c 4. ANATOMY, RESEARCH ADMINISTRATION. Educ: Univ Colo, BA, 53; Univ Kansas City, DDS, 57; Univ Rochester, PhD(anat), 65. Prof Exp: Nat Inst Dent Res trainee, 57-62; asst secy, Coun Dent Res, Am Dent Asn, 62-66; consult dent sch planning, 66-67, PROF DENT & DEAN SCH DENT, MED CTR, UNIV COLO, DENVER, 67- Concurrent Pos: Mem task force care of lab animals, AMA, 63-64; mem, Nat Coun Univ Res Admin, 65-67; secy-treas, Am Asn Accreditation Lab Animal Care, 65-66, chmn, 66-70; mem dent prog-proj comt, Nat Insts Dent Res, 68-70, chmn, 70; mem inst res progs eval comt, Vet Admin, 68-70. Mem: AAAS; Am Dent Asn; Int Asn Dent Res; Am Asn Lab Animal Sci. Res: Embryology and chemical composition of teeth; proteins of dental enamel. Mailing Add: Sch of Dent Univ of Colo Med Ctr Denver CO 80220

BURROWS, RAYMOND CLYDE, inorganic chemistry, industrial chemistry, deceased

BURROWS, ROBERT BECK, b Columbia, SC, Oct 29, 07; m 46; c 3. PARASITOLOGY. Educ: Emory Univ, AB, 29, MS, 30; Yale Univ, PhD(zool), 36. Prof Exp: Lab asst, Emory Univ, 29-30 & Yale Univ, 30-33; lab technician, State Bd Health, Ga, 33-34; biologist, US Bur Fisheries, WVa, 34-35, jr biologist, Conn, 35-36; assoc prof biol, Elon Col, 36-37; adj prof, Am Univ Beirut, 37-40, head dept, 39-40; parasitologist, State Hosp, SC, 40-43 & 46-49; actg chief dept parasitol, Army Med Ctr, 49-50; cmndg officer, Trop Res Lab, US Dept Army, PR, 50-51; chief parasitol br, Sixth Army Area Med Lab, 52-55; head, Parasitol Sect, Wellcome Res Labs, 55-72; RETIRED. Concurrent Pos: Med consult, 1st US Army Area, 63-68. Honors & Awards: Jefferson Award, SC Acad, 47. Mem: AAAS; Am Soc Parasitol; Am Soc Trop Med & Hyg. Res: Parathyroidism and bone; diagnosis, incidences, treatment and pathology of parasitic infections; morphology of amebae; experimental chemotherapy of helminth infection. Mailing Add: 8 Marlborough Dr Vincentown NJ 08088

BURROWS, VERNON DOUGLAS, b Winnipeg, Man, Jan 9, 30; m 54; c 2. PLANT BREEDING. Educ: Univ Man, BSA, 51, MSc, 53; Calif Inst Technol, PhD, 58. Prof Exp: Plant physiologist, 58-69, HEAD CEREAL SECT, RES STA, CAN DEPT AGR, 69- Honors & Awards: Grindley Medal, Agr Inst Can, 75. Res: Plant breeding in general and oats breeding in particular. Mailing Add: Cereal Sect Hq Res Sta Cent Exp Farm Can Dept of Agr Ottawa ON Can

BURROWS, WALTER HERBERT, b Columbia, SC, Oct 23, 11; m 35; c 2. INDUSTRIAL CHEMISTRY. Educ: Emory Univ, AB, 33, MS, 38. Prof Exp: Prof sci & math, Spartanburg Jr Col, SC, 38-41; from instr to asst prof chem, 41-56, res assoc prof & head indust prod br, Eng Exp Sta, 56-67, prin res chemist & head spec projs br, 67-73, HEAD INDUST CHEM LAB, ENG EXP STA, GA INST TECHNOL, 73- Concurrent Pos: Pres, W H Burrows, Consult, 72- Mem: AAAS; Am Chem Soc; Am Inst Chemists; Am Inst Chem Engrs; Nat Fire Protection Asn. Res: Surface chemistry, lubricant, coatings, adhesives sealants, detergents, corrosion; fire and combustion technology; nomography, graphical and mechanical computation; bioconversion of energy, anaerobic fermentation, waste utilization. Mailing Add: Eng Exp Sta Ga Inst of Technol Atlanta GA 30332

BURROWS, WILLIAM, b New Haven, Conn, Mar 6, 08; m 31; c 1. BACTERIOLOGY. Educ: Purdue Univ, BS, 28; Univ Ill, MS, 30; Univ Chicago, PhD(bact), 32. Prof Exp: Asst bact, Univ Ill, 28-30; asst, Univ Chicago, 32-35; Gen Educ Bd fel, NJ Exp Sta & Johns Hopkins Univ, 35-37; from asst prof to prof bact, 37-74, EMER PROF MICROBIOL, UNIV CHICAGO, 74- Concurrent Pos: Consult, Argonne Nat Labs; expert, Comt Cholera, WHO. Honors & Awards: Ricketts Prize, 32. Mem: AAAS; fel Am Acad Microbiol; Am Soc Microbiol; Soc Exp Biol & Med; Am Pub Health Asn. Res: Medical microbiology and immunity, especially enteric disease including Asiatic cholera. Mailing Add: Dept of Microbiol Univ of Chicago Chicago IL 60637

BURROWS, WILLIAM CHAPEL, b Albuquerque, NMex, Apr 28, 31; m 52; c 4. SOIL SCIENCE, AGRONOMY. Educ: Colo State Univ, BS, 52; Iowa State Col, MS, 57, PhD(soil physics), 59. Prof Exp: Asst soil physics, Iowa State Col, 52-56; soil scientist, Agr Res Serv, USDA, 56-65; soil physicist & climatologist, 65-73, SR STAFF SCIENTIST, DEERE & CO, 73- Mem: Am Soc Agron; Soil Sci Soc Am; Biomet Soc; Am Meteorol Soc. Res: Soil physics; statistics; agricultural climatology. Mailing Add: Dept of Eng Sci Tech Ctr Deere & Co Moline IL 61265

BURROWS, WILLIAM DICKINSON, b Saginaw, Mich, Dec 31, 30; m 58. ENVIRONMENTAL CHEMISTRY. Educ: Cornell Univ, BA, 53; Stanford Univ, PhD(chem), 56; Vanderbilt Univ, MS, 73. Prof Exp: Res assoc chem, Mass Inst Technol, 58-61; asst prof, Clarkson Col Technol, 61-62; res assoc, Mass Inst Technol, 62-66; sr vis scientist, US Army Natick Labs, 66-68; sr chemist, Garrett Res & Develop Co, Occidental Petrol Corp, 68-69; sr engr, Assoc Water & Air Resources Engrs Inc, 73-75; TECH COORDR, US ARMY MED BIOENG RES & DEVELOP LAB, 75- Mem: Am Chem Soc; The Chem Soc; Water Pollution Control Fedn; Sigma Xi. Res: Aquatic chemistry and toxicology of military and munitions production wastes. Mailing Add: Environ Protection Res Div US Army Med Bioeng R&D Lab Ft Detrick MD 21701

BURRUS, CHARLES ANDREW, JR, b Shelby, NC, July 16, 27; m 57; c 3. OPTICS. Educ: Davison Col, BS, 50; Emory Univ, MS, 51; Duke Univ, PhD, 55. Prof Exp: Res assoc physics, Duke Univ, 54-55; MEM TECH STAFF, BELL LABS, 55- Mem: AAAS; fel Inst Elec & Electronics Engrs; Am Phys Soc; Optical Soc Am. Res: Microwave spectroscopy in the shorter millimeter and sub-millimeter wave region; millimeter-wave diodes; electroluminescent diodes; optical-fiber devices; singe-crystal fibers. Mailing Add: Crawford Hill Lab Bell Labs Holmdel NJ 07733

BURRUS, HARRY OTTO, chemistry, see 12th edition

BURRUS, ROBERT TILDEN, b High Point, NC, July 15, 35; m 64; c 1. INORGANIC CHEMISTRY, PHYSICAL CHEMISTRY. Educ: Univ NC, BS, 57; Univ Tenn, PhD(Chem), 62. Prof Exp: Trainee, Westinghouse Elec Corp, 57-58; asst chem, Univ Tenn, 58-62; res chemist, 62-69, SR RES CHEMIST, E I DU PONT DE NEMOURS & CO, INC, 69- Mem: AAAS; Am Chem Soc. Res: Physical inorganic chemistry; electrochemistry in nonaqueous media; colloid and surface chemistry of polymeric materials; physical chemistry of synthetic fibers. Mailing Add: Dacron Res & Develop Lab Box 800 E I du Pont de Nemours & Co Inc Kinston NC 28501

BURRY, JOHN HENRY WILLIAM, b Nfld, June 10, 38; m 65; c 2. MATHEMATICS. Educ: Mem Univ Nfld, BAEd, 58; Dalhousie Univ, MSc, 61; Queen's Univ, Ont, PhD(math), 71. Prof Exp: Lectr, 61-63, from asst prof to assoc prof, 63-75, PROF MATH, MEM UNIV NFLD, 75- Mem: Can Math Soc; Can Asn Univ Teachers; Am Math Soc; Nat Coun Teachers Math. Res: Radon measures in Euclidian spaces; measure theory. Mailing Add: Dept of Math Mem Univ of Nfld St John's NF Can

BURSEY, CHARLES ROBERT, b Paris, Tenn, Feb 13, 40; m 64; c 2. INVERTEBRATE PHYSIOLOGY. Educ: Kalamazoo Col, BA, 62; Mich State Univ, MS, 65, PhD(zool), 69. Prof Exp: NIH fel, Inst Marine Sci, Univ Miami, 69-70; ASST PROF BIOL, PA STATE UNIV, 70- Mem: Am Soc Zoologists; Am Inst Biol Sci. Res: Microanatomy of Limulus heart ganglion and ventral ganglia; homeostasis in shrimp; structure and function of the cardiac ganglion of the spider Eurypelma marxi. Mailing Add: Dept of Biol Pa State Univ Shenango Valley Campus Sharon PA 16146

BURSEY, MAURICE MOYER, b Baltimore, Md, July 27, 39; m 70. ANALYTICAL CHEMISTRY. Educ: Johns Hopkins Univ, BA, 59, MA, 60, PhD(chem), 63. Prof Exp: Lectr chem, Johns Hopkins Univ, 63-64; asst prof, Purdue Univ, 64-66; from asst prof to assoc prof, 66-74, PROF CHEM, UNIV NC, CHAPEL HILL, 74- Concurrent Pos: Sloan fel, 69-71. Mem: Am Chem Soc; Am Soc Mass Spectrometry; The Chem Soc; Japanese Soc Mass Spectros. Res: Mass spectrometry; chemistry of gaseous ions by mass spectrometry and ion cyclotron resonance. Mailing Add: Dept of Chem Univ of NC Chapel Hill NC 27514

BURSH, TALMAGE POUTAU, b Leesville, La, Dec 24, 32; m 54; c 3. PHYSICAL CHEMISTRY. Educ: Southern Univ, BS, 56; Alfred Univ, PhD(chem), 63. Prof Exp: Asst instr phys sci, 57, assoc prof chem, 61-68, PROF CHEM, SOUTHERN UNIV, 68- Res: Interaction of water with silica, alumina and silica alumina co-oxide surfaces and the relationship of these interactions to catalytic activities of these oxides. Mailing Add: Dept of Chem Southern Univ PO Box 9208 Baton Rouge LA 70813

BURSKE, NORBERT WILLIAM, organic chemistry, see 12th edition

BURSNALL, JOHN TREHARNE, b Neath, UK, May 9, 40; m 66; c 2. STRUCTURAL GEOLOGY. Educ: Univ London, BSc, 68; Univ Cambridge, PhD(geol), 75. Prof Exp: Demonstr-tutor geol, Univ Cambridge, 68-72; tutor-organizer natural sci, Workers' Educ Asn, UK, 72-75; ASST PROF STRUCT GEOL, SYRACUSE UNIV, 75- Concurrent Pos: Vis lectr geol, Cambridgeshire Col Arts & Technol, 68-72. Mem: Fel Geol Soc London; Geol Soc Am. Res: Geological structure of parts of the northern Appalachians, central Maine, northwest Newfoundland; ophiolite emplacement and deformation history within orogenic belts, shear zones in foliated rocks. Mailing Add: Dept of Geol Heroy Geol Lab Syracuse Univ Syracuse NY 13210

BURSON, BYRON LYNN, b Hobart, Okla, Feb 24, 40; m 61; c 2. PLANT CYTOGENETICS, PLANT GENETICS. Educ: Okla State Univ, BS, 62; Tex A&M Univ, MS, 65, PhD(plant breeding, cytogenetics), 67. Prof Exp: Asst prof grass cytogenetics & asst agronomist, Miss State Univ, 67-71, assoc prof grass cytogenetics & assoc agronomist, 71-75; RES GENETICIST, GRASSLAND-FORAGE RES CTR, AGR RES SERV, USDA, 75- Mem: Am Soc Agron; Crop Sci Soc Am; Am Genetic Asn; Genetics Soc Can; Bot Soc Am. Res: Cytogenetics of grasses; reproductive systems, especially apomixis; species relationships within genus paspalum; interspecific hybridization. Mailing Add: Grassland Forage Res Ctr PO Box 748 Temple TX 76501

BURSON, SHERMAN LEROY, JR, b Pittsburgh, Pa, Dec 24, 23; m 44; c 4. ORGANIC CHEMISTRY. Educ: Univ Pittsburgh, BS, 47, PhD(org chem), 53. Prof Exp: Asst chem, Univ Pittsburgh, 47-48, 52; chemist, Lederle Lab Div, Am Cyanamid Co, 52-57; assoc prof chem, Pfeiffer Col, 57-60, prof & chmn dept, 60-63; prof chem, 63-67, chmn dept, 63-75, CHARLES H STONE PROF CHEM, UNIV NC, CHARLOTTE, 67-, DEAN COL SCI & MATH, 75- Mem: AAAS; Am Chem Soc; NY Acad Sci. Res: Biotin analogues; polysaccharides; natural products. Mailing Add: 7515 Shady Lane Charlotte NC 28215

BURST, JOHN FREDERICK, b St Louis, Mo, Oct 16, 23; m 56; c 4. CLAY MINERALOGY. Educ: Univ Mo-Rolla, MS, 47; Univ Mo-Columbia, PhD(geol), 50. Prof Exp: Mem staff geol, Shell Develop Co, Tex, 50-65; dir res & develop, Gen Refractories Co, 65, dir res, 65-70, tech dir, 70-73; tech adv, Certain-Teed Prods Co, 73-75; TECH DIR, DRESSER MINERALS, DIV DRESSER INDUST, 75- Mem: Fel Geol Soc Am; fel Mineral Soc Am; Clay Minerals Soc (pres, 71); Soc Petrol Geol; Am Ceramic Soc. Res: Chemistry and ceramics. Mailing Add: Dresser Minerals PO Box 6504 Houston TX 77005

BURSTAIN, ISRAEL G, physical organic chemistry, see 12th edition

BURSTEIN, ELIAS, b New York, NY, Sept 30, 17; m 43; c 3. SOLID STATE PHYSICS. Educ: Brooklyn Col, AB, 38; Univ Kans, AM, 41. Prof Exp: Assoc, Nat Defense Res Comt Proj, 42-44; res & develop proj engr, White Res Assocs Mass, 44-45; physicist, Crystal Br, Naval Res Lab, 20head semiconductor br, 58; PROF PHYSICS, UNIV PA, 58- Concurrent Pos: Mem solid state sci comt, Nat Res Coun-Nat Acad Sci; ed-in-chief, Solid State Commun; univ adv, Tex Instruments Cent Res Lab, 63-; consult, Thomas J Watson Res Ctr, IBM, 71-; co-ed, Comments on Solid State Physics, 72- Honors & Awards: Annual Award, Wash Acad Sci, 56; Civilian Meritorious Serv Award, Navy Dept, 57; Res Soc Am Pure Sci Award, 58. Mem: Fel Am Phys Soc; fel Optical Soc Am; Am Asn Physics Teachers. Res: Optical and acoustical spectroscopy of solids; dielectrics; photoconductors; crystal physics; electron tunneling; lattice dynamics; surface elastic and electromagnetic waves. Mailing Add: Dept of Physics Univ of Pa Philadelphia PA 19174

BURSTEIN, SAMUEL Z, b Brooklyn, NY, Oct 4, 35; m 59; c 1. APPLIED MATHEMATICS. Educ: Polytech Inst Brooklyn, BME, 57, MME, 58, PhD(mech eng), 62. Prof Exp: Instr mech eng, Polytech Inst Brooklyn, 57-62; ASSOC PROF APPL MATH, COURANT INST MATH SCI, NY UNIV, 62- Concurrent Pos: Assoc ed, J Comput Physics, 68-; vpres math res & dir, Artek Systs, Inc, 75- Mem: Soc Indust & Appl Math. Res: Combustion instability in rocket motors; numerical methids in hydrodynamics; computation of shock waves and hypersonic flow; transonic flow; reconstruction of brain blood flow; deformation of nonlinear elastic-plastic materials. Mailing Add: 1402 Beverly Rd Brooklyn NY 11226

BURSTEIN, SHLOMO, b Zasliai, Lithuania, June 10, 23; nat US; m 53; c 1. BIOCHEMISTRY. Educ: Hebrew Univ, Jerusalem, MSc, 46, PhD(hormones), 51. Prof Exp: Res asst hormones with Prof B Zondek, Hebrew Univ, Jerusalem, 49; res assoc biochem steroids, Worcester Found Exp Biol, 52-55; vis scientist, Nat Cancer Inst, USPHS, 55-56; res assoc, Columbia Univ, 56-58; sr scientist, Worcester Found Exp Biol, 59-68; head div steroid chem, Inst Muscle Dis, 68-73; prof cell biol, Baylor Col Med, 73-75; PROF BIOCHEM & OBSTET & GYNEC, MT SINAI SCH MED, NEW YORK, 75- Concurrent Pos: Mem study sect, NIH, 70-73. Mem: Am Chem Soc; Endocrine Soc; Am Soc Biol Chem. Res: Metabolism of steroid hormones; endocrinology. Mailing Add: Dept of Biochem Mt Sinai Sch of Med New York NY 10029

BURSTEIN, SUMNER HOWARD, organic chemistry, see 12th edition

BURSTONE, CHARLES JUSTIN, b Kansas City, Mo, Apr 4, 28. ORTHODONTICS. Educ: Wash Univ, DDS, 50; Ind Univ, MS, 55. Prof Exp: From asst prof to assoc prof orthod, Sch Dent, Ind Univ, 55-70; HEAD DEPT ORTHOD, SCH DENT MED, UNIV CONN, 70- Mem: AAAS; Am Asn Orthod; Am Dent Asn. Res: Application of biophysics to orthodontics; soft tissue morphology of the face; segmented arch therapy; growth and development. Mailing Add: Dept of Orthod Univ of Conn Sch Dent Med Farmington CT 06032

BURSTYN, HAROLD LEWIS, b Boston, Mass, Feb 26, 30; m 58; c 3. HISTORY OF SCIENCE, HISTORY OF TECHNOLOGY. Educ: Harvard Col, AB, 51; Univ Calif, MS, 57; Harvard Univ, PhD(hist sci), 64. Prof Exp: Instr hist sci, Brandeis Univ, 62-66; asst prof, Carnegie-Mellon Univ, 66-69, assoc prof hist sci & technol, 69-73; PROF MATH & NATURAL SCI, WILLIAM PATERSON COL, NJ, 73- Concurrent Pos: NSF fel, Imp Col, Univ London, 65-66; dean grad & res progs, William Paterson Col, NJ, 73-75; adv ed, Isis, 76-80. Honors & Awards: Henry Schuman Prize, Hist Sci Soc, 60. Mem: Hist Sci Soc; AAAS; Brit Soc Nautical Res; Int Ctr Hist Oceanog; NAm Soc Oceanic Historians. Res: History of marine science and technology, especially physical oceanography and its applications; history of marine exploration, especially the Challenger Expedition; environmental history, especially the fertilizer industry. Mailing Add: William Paterson Col of NJ 300 Pompton Rd Wayne NJ 07470

BURT, ALVIN MILLER, III, b Bridgeport, Conn, Aug 14, 35; m 61; c 2. NEUROCHEMISTRY, NEUROANATOMY. Educ: Amherst Col, BA, 57; Univ Kans, PhD(anat), 62. Prof Exp: Asst prof anat, Med Col Va, 62-63; instr, Sch Med, Yale Univ, 63-66; from asst prof to assoc prof, 66-74, PROF ANAT, SCH MED, VANDERBILT UNIV, 74- Concurrent Pos: USPHS res career develop award, Vanderbilt Univ, 68-73; vis scientist neurochem, Agr Res Coun, Inst Animal Physiol, Babraham, Cambridge, Eng, 72-73. Mem: AAAS; Am Soc Neurochem; Am Asn Anat; Soc Neurosci; Soc Develop Biol. Res: Developmental neurochemistry with particular reference to neurotransmitter systems and intermediary metabolism; mechanism of synthesis and action of acetylcholine. Mailing Add: Dept of Anat Vanderbilt Univ Nashville TN 37232

BURT, BRIAN AUBREY, b Melbourne, Australia, Jan 14, 39; m 65; c 2. DENTAL EPIDEMIOLOGY. Educ: Univ Western Australia, BDSc, 60; Univ Mich, Ann Arbor, MPH, 66; Univ London, PhD(dent epidemiol), 73. Prof Exp: Sr clin dentist, Perth Dent Hosp, Western Australia, 60-65; pub health adv dent, Div Dent Health, USPHS, 66-67; lectr, London Hosp Dent Sch, 68-74; ASSOC PROF DENT PUB HEALTH, SCH PUB HEALTH, UNIV MICH, ANN ARBOR, 74- Mem: Fel Royal Australian Col Dent Surgeons; Am Dent Asn; Int Asn Dent Res. Res: Epidemiology of dental disease; prevention of dental disease; delivery of dental care. Mailing Add: Prog in Dent Pub Health Univ of Mich Sch of Pub Health Ann Arbor MI 48109

BURT, DONALD MCLAIN, b East Orange, NJ, Oct 27, 43; m 72; c 1. MINERALOGY, PETROLOGY. Educ: Princeton Univ, AB, 65; Harvard Univ, AM, 68, PhD(geol), 72. Prof Exp: Lectr geochem, State Univ Utrecht, 72-73; Gibbs instr, Yale Univ, 73-75; ASST PROF MINERAL, ARIZ STATE UNIV, 75- Mem: Mineral Soc Am; Mineral Soc Can; Geol Soc Am; Am Geophys Union; Am Inst Mining, Metall & Petrol Engrs Res: Mineralogy and petrology of skarn, greisen and related ore deposits; phase equilibria; geochemistry of acid-base processes. Mailing Add: Dept of Geol Ariz State Univ Tempe AZ 85281

BURT, EVERT OAKLEY, b Londonderry, Ohio, Aug 24, 23; m 48. AGRONOMY. Educ: Ohio Univ, BS, 44, MS, 50; Ohio State Univ, PhD(agron), 54. Prof Exp: Instr high sch, 47-48; farm supt & instr agr, Ohio Univ, 48-51; asst agronomist, Univ Fla, 54-59; mem staff, Res Div, O M Scott & Sons, Ohio, 59-62; PROF ORNAMENTAL HORT, AGR RES CTR, UNIV FLA, 62- Mem: Am Soc Agron. Res: Chemical weed control in turf. Mailing Add: Univ of Fla Agr Res Ctr 3205 SW 70th Ave Ft Lauderdale FL 33314

BURT, GERALD DENNIS, b Dovray, Minn, Jan 3, 36; m 66. ORGANIC CHEMISTRY. Educ: Luther Col, BA, 57; Univ Iowa, MS, 60, PhD(org chem), 61. Prof Exp: Asst proj chemist, Am Oil Co, Ind, 61-64; res fel chem, Univ Tex, Austin, 64-65; SR RES CHEMIST, HARSHAW CHEM CO DIV, KEWANEE OIL CO, 66- Mem: AAAS; Am Chem Soc. Res: Synthetic organic chemistry; lube oil additives; organometallic chemistry; dyes pigments; metal plating additives. Mailing Add: 3290 Avalon Rd Shaker Heights OH 44120

BURT, GORDON WILLIS, b Bayshore, NY, Apr 9, 39; m 68; c 12. AGRONOMY. Educ: Tenn Technol Univ, BS, 61; Cornell Univ, MS, 64; Wash State Univ, PhD(agron), 67. Prof Exp: Res assoc, Herbicide Physiol, Wash State Univ, 67-68; ASST PROF AGRON, UNIV MD, COLLEGE PARK, 68- Mem: Weed Sci Soc Am; Am Soc Plant Physiol; Scand Soc Plant Physiol. Res: Influence of phosphorus and potassium on seedling development and soluble amino acids; action mechanism of 3-amino-1,2,4-triazole in Cirsiam arvense; influence of growth regulators in abscission; herbicide physiology: growth and physiology of Sorghum halepense as influenced by herbicides and the environment. Mailing Add: Dept of Agron Univ of Md College Park MD 20742

BURT, JAMES GORDON, organic chemistry, see 12th edition

BURT, JAMES KAY, b Des Moines, Mar 5, 34; m 54; c 5. VETERINARY RADIOLOGY. Educ: Iowa State Univ, DVM, 62, MS, 67; Am Col Vet Radiol, dipl. Prof Exp: Instr obstet & radiol, Iowa State Univ, 62-65, vet clin sci, 65-66, asst prof, 66-67; asst prof vet surg & radiol, 67-70, assoc prof vet clin sci, 70-74, PROF VET CLIN SCI, OHIO STATE UNIV, 74- Mem: Am Vet Med Asn; Am Vet Radiol Soc; Am Asn Vet Clinicians. Res: Teratologic effects of drugs on the embryo-fetus; dynamics of bone metabolism in eosinophilic panosteitis. Mailing Add: Col of Vet Med Ohio State Univ Columbus OH 43210

BURT, MICHAEL DAVID BRUNSKILL, b Colombo, Ceylon, Jan 19, 38; m 60; c 4. BIOLOGY, PARASITOLOGY. Educ: Univ St Andrews, BSc, 61, PhD(parasitol), 67. Prof Exp: Brit Coun scholar, Inst Zool, Univ Neuchatel, 63; asst prof, 61-62, 64-68, assoc prof, 68-75, PROF BIOL & CHMN DEPT, UNIV NB, 75- Mem: AAAS; Am Soc Parasitol; Am Inst Biol Sci; Electron Micros Soc Am; Can Soc Zool. Res: Cestode and Platyhelminth biology with reference to functional morphology and developmental changes using light and electron microscopy, life cycles, seasonal variation, and geographical distribution and systematics, host specificity, and evolution. Mailing Add: Dept of Biol Univ of NB Fredericton NB Can

BURT, PHILIP BARNES, b Memphis, Tenn, July 1, 34; m 55; c 4. THEORETICAL PHYSICS. Educ: Univ Tenn, AB, 56, MS, 58, PhD(physics), 61. Prof Exp: Sr scientist, Jet Propulsion Lab, Pasadena, Calif, 61-65; from asst prof to assoc prof, 65-73, PROF PHYSICS, CLEMSON UNIV, 73- Concurrent Pos: Vis asst prof, Univ Southern Calif, 62; consult, Oak Ridge Nat Lab, assoc physicist, 60, res partic, 66; vis, Stanford Linear Accelerator Ctr, 75-76. Mem: AAAS; Am Phys Soc. Res: Plasma quantum and field theories; elementary particles; scattering; theoretical high energy physics. Mailing Add: Dept of Physics Clemson Univ Clemson SC 29631

BURT, RICHARD LAFAYETTE, b Springfield, Mass, Dec 7, 15; m 42; c 4. OBSTETRICS & GYNECOLOGY. Educ: Springfield Col, BS, 38; Brown Univ, MS, 40, PhD(microbiol), 42; Harvard Univ, MD, 46. Prof Exp: Asst biol, Brown Univ, 38-41; intern, US Naval Hosp, Chelsea, Mass, 46-47; head nutrit facil, Naval Med Res Inst, 47-49; from asst to assoc prof, 49-60, PROF OBSTET & GYNEC, BOWMAN GRAY SCH MED, WAKE FOREST UNIV, 60- Concurrent Pos: Res fel dent med, Harvard Med Sch, 42-43. Mem: Am Gynec Soc; Soc Gyn Invest; Am Diabetes Asn; Soc Exp Biol & Med. Res: Cellular physiology; physiology of cell division; narcosis; respiratory enzymes; calcification; amino nitrogen metabolism in trauma; maternal nutrition; carbohydrate and lipid metabolism in pregnancy; amino nitrogen metabolism in pregnancy. Mailing Add: Bowman Gray Sch of Med Wake Forest Univ Winston-Salem NC 27103

BURT, ROBERT C, b Wilkinsburg, Pa, Sept 4, 16; m 42; c 2. MEDICINE, PATHOLOGY. Educ: Univ Pittsburgh, MD, 41. Prof Exp: Intern, 41-42, resident path & bacteriologist, 46-48, PATHOLOGIST, W H SINGER MEM RES INST, ALLEGHENY GEN HOSP, PITTSBURGH, 50- Mem: Am Asn Cancer Res: AMA; Am Soc Clin Path; NY Acad Sci; Int Acad Path. Res: Histochemistry; electron microscopy. Mailing Add: W H Singer Mem Res Inst 320 North Ave Pittsburgh PA 15212

BURT, WAYNE VINCENT, b South Shore, SDak, May 10, 17; m 41; c 4. OCEANOGRAPHY. Educ: Pac Col, BS, 39; Univ Calif, MS, 48, PhD(phys oceanog), 52; cert aerol, US Naval Acad Post Grad Sch, cert aerol, 44. Hon Degrees: ScD, George Fox Col, 63. Prof Exp: Mat engr, Kaiser Co, Inc, Wash, 42; instr math, Univ Ore, 46; asst oceanogr, Scripps Inst, Univ Calif, 46-48, assoc oceanogr, 48-49; asst prof oceanog & res oceanogr, Chesapeake Bay Inst, Johns Hopkins Univ, 49-53, asst dir, Inst, 53; res oceanogr, Univ Wash, 53-54; assoc prof, 54-59, chmn dept oceanog, 59-68, dir marine sci ctr, 64-72, PROF OCEANOG, ORE STATE UNIV, 59-, ASSOC DEAN RES, 68- Concurrent Pos: Mem, Nat Acad Sci Comt on Oceanog, 65-70; mem, Nat Comt on Oceans & the Atmosphere, 71-75. Honors & Awards: Centennial Award, 68; Gov Scientist Award, Ore Mus Sci & Indust, 69. Mem: Fel Am Meteorol Soc; Am Geophys Union; Am Soc Limnol & Oceanog. Res: Interaction between ocean and atmosphere; estuarine and inshore physical oceanography; light transmission in turbid water; physical limnology; reservoir temperature. Mailing Add: Sch of Oceanog Ore State Univ Corvallis OR 97331

BURT, WILLIAM ENOS, b Martin, Mich, Jan 12, 15; m 48; c 4. CHEMISTRY. Educ: Kalamazoo Col, AB, 38, MS, 39; Purdue Univ, PhD(org chem), 45. Prof Exp: Group leader, Off Sci Res & Develop, 42-43; org chemist, Emery Industs, Cincinnati, 43-45; proj leader, 45-58, asst res supvr new prods appln, 58-64, sr appln chemist, 64-70, SR DEVELOP ASSOC, ETHYL CORP, 70- Mem: Am Chem Soc. Res: Preparation of ephedrine type amines; halogenation of hydrocarbons; fats, fatty acids and derivatives; oxidation of hydrocarbons; fuels stability; organometallics; polymer applications. Mailing Add: Ethyl Corp Box 341 Baton Rouge LA 70821

BURT, WILLIAM HENRY, b Haddam, Kans, Jan 22, 03; m 28. MAMMALOGY. Educ: Univ Kans, AB, 26, AM, 27; Univ Calif, PhD(zool), 30. Prof Exp: Teaching fel zool, Univ Kans, 26-27; teaching fel, Univ Calif, 27-28, paleont, 28-29; res fel vert zool, Calif Inst Technol, 30-35; from instr to prof, 35-69, from asst cur to cur, Mammals Mus Zool, 35-69, EMER PROF ZOOL, UNIV MICH & ASSOC CUR, UNIV COLO MUS, 69- Concurrent Pos: Ed, J Am Soc Mammal, 47-52; mem, Nat Res Coun, 63-67. Mem: AAAS; Am Soc Mammal (vpres, 51-53, pres, 53-55); Soc Study Evolution; Soc Syst Zool; Soc Vert Paleont. Res: Vertebrate paleontology; taxonomy; ecology and populations of Mammalia; avian anatomy; Mammalian anatomy; effect of volcanic activity on vertebrates. Mailing Add: Univ of Colo Mus Boulder CO 80302

BURTI, UMBAY H, b Old Forge, Pa, June 19, 10; m 40; c 4. ORGANIC CHEMISTRY. Educ: Univ Scranton, BS, 35. Hon Degrees: LLD, Univ Scranton, 65. Prof Exp: From instr to assoc prof, 35-70, chmn dept 50-69, 69 & 72, PROF CHEM, UNIV SCRANTON, 70- Mem: Am Chem Soc. Mailing Add: Dept of Chem Col Arts & Sci Univ of Scranton Scranton PA 18510

BURTIS, CARL A, JR, b Flagstaff, Ariz, July 3, 37; m 59; c 3. BIOCHEMISTRY, ANALYTICAL CHEMISTRY. Educ: Colo State Univ, BS, 59; Purdue Univ, MS, 64, PhD(biochem), 67. Prof Exp: Chemist, Ind State Control Labs, 63-66; fel bioanal, Oak Ridge Nat Lab, 66-67, res assoc molecular anat, 67-69; sr chemist, Varian Aerograph, 69-70; GROUP LEADER GEMSAEC FAST ANALYZER PROJ, MOLECULAR ANAT PROG, OAK RIDGE NAT LAB, 70-, COORDR BIOTECHNOL PROG, 73- Mem: Am Asn Clin Chemists; Am Soc Biol Chemists; AAAS; Am Chem Soc. Res: Chemical inducement of ovine ketosis; separation and quantitation of body fluid constituents by liquid chromatography; separation of nucleic acid constituents by high efficiency liquid chromatography; rapid clinical analysis. Mailing Add: Oak Ridge Nat Lab Oak Ridge TN 37830

BURTNER, DALE CHARLES, b Portland, Ore, Oct 20, 26; m 50; c 2. ANALYTICAL CHEMISTRY. Educ: Reed Col, BA, 48; Univ Wash, Seattle, MS, 51, PhD(chem), 54. Prof Exp: Chemist, Shell Develop Co, 54-58; from asst prof to assoc prof, 58-67, chmn dept, 65-67, dean sch arts & sci, 67-69, PROF CHEM, FRESNO STATE COL, 67- Mem: Am Chem Soc. Res: Trace metal analysis; coordination compounds; spectrophotometry; flame photometry; microchemistry. Mailing Add: Dept of Chem Fresno State Col Fresno CA 93705

BURTNER, ROGER LEE, b Hershey, Pa, Mar 31, 36; m 65; c 1. GEOLOGY. Educ: Franklin & Marshall Col, BS, 58; Stanford Univ, MS, 59; Harvard Univ, PhD(geol), 65. Prof Exp: Assoc res geologist, Calif Res Corp Div, 63-64, res geologist, 64-68, explor geologist, Standard Oil Co Tex, 68-69, res geologist, 69-74, SR RES GEOLOGIST, CHEVRON OIL FIELD RES CO, STANDARD OIL CO CALIF, 74- Concurrent Pos: Chmn res proj adv comt, Am Petrol Inst, 72- Mem: Geol Soc Am; Electron Micros Soc Am; Soc Econ Paleont & Mineral; Am Asn Petrol Geol; Clay Minerals Soc. Res: Environmental studies of ancient sediments; sedimentary structures; paleocurrents; sedimentary petrography; stratigraphic correlation utilizing trace elements; quantative x-ray mineralogy; burial diagenesis of clay minerals; geochemistry formation waters; oxygen isotope geothermometry. Mailing Add: Chevron Oil Field Res Co PO Box 446 La Habra CA 90631

BURTON, ALAN CHADBURN, b London, Eng, Apr 18, 04; m 33; c 1. BIOPHYSICS. Educ: Univ London, BSc, 25; Univ Toronto, MA, 28, PhD(physics), 32. Hon Degrees: LLD, Univ Alta, 66; DSc, Univ Western Ont, 74. Prof Exp: Demonstr physics, Univ Col, Univ London, 25-26; sci master, Liverpool Collegiate Sch, 26-27; demonstr & asst physics, Univ Toronto, 27-28, lectr, 28-29; fel, Sch Med & Dent, Univ Rochester, 32-34; Gen Educ Bd fel, Sch Med, Univ Pa, 34-36; fel, Johnson Found Med, Pa, 36-40; Nat Res Coun Can fel, 41-45; from asst prof to assoc prof, 46-69, chmn dept biophys, 46-70, PROF BIOPHYS, FAC MED, UNIV WESTERN ONT, 69- Concurrent Pos: Mem, Order of Brit Empire, 48. Mem: Am Physiol Soc (pres, 57); Biophys Soc (pres, 67); Can Physiol Soc (secy, 49-53, pres, 63); Can Asn Physicists; fel Royal Soc Can. Res: Theory of control of cell division; rule of pH in control of cell division. Mailing Add: Dept of Biophys Univ of Western Ont London ON Can

BURTON, ALBERT FREDERICK, b Brandon, Man, Feb 7, 29; m 53; c 2. BIOCHEMISTRY, CANCER. Educ: Brandon Col, BSc, 53; Univ Western Ont, MSc, 56; Univ Sask, PhD, 58. Prof Exp: Asst prof, 62-67, ASSOC PROF BIOCHEM, UNIV BC, 67- Concurrent Pos: Nat Cancer Inst Can res fel biochem, Univ BC, 59-62. Mem: Soc Exp Biol & Med; Can Biochem Soc. Res: Metabolism of corticosteroids and their biological action, especially in fetal tissues and in relation to cancer. Mailing Add: Dept of Biochem Univ B C Vancouver BC Can

BURTON, ALEXIS LUCIEN, b Paris, France, Feb 4, 22; US citizen; m 52; c 3. HISTOLOGY, CYTOLOGY. Educ: Univ Geneva, BMSc, 43; Univ Strasbourg, MD, 46. Prof Exp: Surgeon, Dept Pub Health, Morocco, 51-56; asst prof microanat, Fac Med, Univ Montreal, 57-63; from asst prof to assoc prof, 64-72, PROF ANAT, UNIV TEX MED SCH SAN ANTONIO, 72- Concurrent Pos: Fel anat, Univ Tex Med Sch San Antonio, 63-64. Mem: Am Asn Anat; Tissue Cult Asn; Soc Motion Picture & TV Eng. Res: Scientific cinematography; cytology of the connective tissue; tissue culture and cinemicrography. Mailing Add: Dept of Anat Univ of Tex Med Sch San Antonio TX 78284

BURTON, ALICE JEAN, b Peiping, China, May 19, 34; US citizen. BIOCHEMISTRY, MICROBIOLOGY. Educ: Univ Mich, BS, 57; Univ Ill, PhD(bi . Prof Exp: NIH res fel, Calif Inst Technol, 61-63, res fel biol, 63-64; from asst biochemist to biochemist, Brookhaven Nat Lab, 64-70; ASSOC PROF BIOL, ST OLAF COL, 70- Concurrent Pos: Mem US nat comt, Int Union Pure & Appl Biophys, 75-78. Mem: Biophys Soc; Am Chem Soc; AAAS. Res: Biochemistry of nucleic acids, particularly the replication of bacteriophages and host modification of the replicative process. Mailing Add: Dept of Biol St Olaf Col Northfield MN 55057

BURTON, BENJAMIN THEODORE, b Wiesbaden, Ger, Aug 29, 19; nat US; m 52; c 2. NUTRITION. Educ: Univ Calif, BS, 41, MS, 43, PhD(microbiol, biochem), 47. Prof Exp: Res chemist, Mills Orchards Corp, 41-42; outside supvr & tech consult, Rosenberg Bros & Co, 42-48; vpres & tech dir, Pac States Labs, Inc, 52-55; staff consult nutrit, H J Heinz Co, 55-60; from nutritionist & spec asst to dir, 60-67, ASSOC DIR & CHIEF ARTIFICIAL KIDNEY PROG, NAT INST ARTHRITIS, METAB & DIGESTIVE DIS, 67- Honors & Awards: Superior Serv Award, US Dept Health, Educ & Welfare, 70. Mem: Am Inst Nutrit; Am Chem Soc; Inst Food Technologists; Am Soc Artificial Internal Organs; Am Soc Nephrology. Res: Human nutrition; intermediate metabolism; nutritional and metabolic diseases; malnutrition; chronic kidney failure and uremia; artificial kidney development; dialysis. Mailing Add: Nat Inst of Arthritis Metab & Digestive Dis Bethesda MD 20014

BURTON, BLENDIN L, b Island Falls, Maine, Mar 1, 18; div; c 1. PHYSICS. Educ: Univ Maine, BS, 41; Brown Univ, PhD(physics), 49. Prof Exp: Physicist, Naval Ord Lab, 41-46; STAFF MEM, LOS ALAMOS SCI LAB, UNIV CALIF, 49- Res: Ultrasonics; high pressure and shock wave phenomena in solids, especially metals; general wave propagation. Mailing Add: 939 Iris Los Alamos NM 87544

BURTON, CHARLES JEWELL, b Boston, Mass, July 27, 16; m 40; c 2. PHYSICS. Educ: Colgate Univ, AB, 37; NY Univ, MS, 43, PhD(physics), 47. Prof Exp: Mem staff, Am Cyanamid Co, 37-44, head electro-phys sect, Stamford Res Labs, 44-49; asst to vpres in chg res & develop, Am Optical Co, 49-51; vpres, Barnes Eng Co, 51-58; vpres, Res & Adv Develop Div, Avco Mfg Co, 58-61 & defense & indust prod group, Avco Corp, 61-65, vpres & gen mgr, Avco Appl Technol Div, 65-70; SR CONSULT, ARTHUR D LITTLE, INC, 70- Mem: Am Phys Soc; Am Inst Aeronaut & Astronaut. Res: Electron microscopy; electron optics; ultrasonics; industrial instrumentation; research administration. Mailing Add: Arthur D Little Inc Acorn Park Cambridge MA 02640

BURTON, CHARLES R, b Atchison, Kans, Apr 24, 18; m 46; c 2. MATHEMATICAL LOGIC. Educ: Univ Kans, BA, 40, MA, 47; Univ Calif, Berkeley, MA, 51, PhD(philos), 53. Prof Exp: From instr to asst prof philos, Wash Univ, St Louis, 52-59; from asst prof to assoc prof, 59-69, PROF MATH, SAN DIEGO STATE COL, 69- Mem: AAAS; Am Math Soc; Math Asn Am; Asn Symbolic Logic; Asn Comput Mach. Res: Computer science. Mailing Add: Dept of Math San Diego State Col San Diego CA 92115

BURTON, DANIEL FREDERICK, b Chicago, Ill, Oct 3, 15; m 50. BOTANY. Educ: Univ Chicago, MS, 40, PhD(bot), 47. Prof Exp: Instr bot, Miss State Col, 46-48; from asst prof to assoc prof, 48-62, PROF BOT, MANKATO STATE COL, 62- Concurrent Pos: Mem, Minn State Bd Educ, 71- Mem: AAAS; Bot Soc Am; Ecol Soc Am; Am Fern Soc; Asn Trop Biol. Res: Leaf morphogenesis; formative effects of certain substituted phenoxy compounds on bean leaves. Mailing Add: Dept of Biol Mankato State Col Mankato MN 56001

BURTON, DAVID LEE, b Norlina, NC, Oct 14, 21; m 54. HORTICULTURE, SOILS. Educ: Southeastern La Col, BS, 49; La State Univ, MS, 51, PhD(hort, bot), 57. Prof Exp: Asst horticulturist, La State Univ, 49-51; processed food inspector, fruits & veg, USDA, 51-55; asst horticulturist, La State Univ, 55-57; mkt specialist, Agr Mkt Serv & stand specialist, US Grade Stand, 57-67; FOOD TECHNOLOGIST, DIV FOOD TECHNOL, FOOD & DRUG ADMIN, 67- Concurrent Pos: Instr radio & electronic commun, US Coast Guard Res, 59- Res: Agronomy; botany. Mailing Add: Food & Drug Admin Div of Food Technol HFF-414 Washington DC 20204

BURTON, DAVID NORMAN, b Birkenhead, Eng, Feb 3, 41; m 65; c 2. BIOCHEMISTRY, MICROBIOLOGY. Educ: Univ Liverpool, BSc, 62, PhD(biochem), 65. Prof Exp: Fel biochem, Univ Calif, Davis, 65-66; chemist, Lipid Metab Lab, Vet Admin Hosp, Madison, Wis, 66-68; fel, 68-69, asst prof, 69-74, ASSOC PROF MICROBIOL, UNIV MAN, 74- Concurrent Pos: Nat Res Coun Can res operating grant, 69. Mem: Can Biochem Soc. Res: Mechanisms and control of lipid metabolism in microorganisms and animals; secondary metabolite production in fungi; biosynthesis of biologically active quinones in microorganisms. Mailing Add: Dept of Microbiol Univ of Man Winnipeg MB Can

BURTON, DONALD JOSEPH, b Baltimore, Md, July 16, 34; m 58; c 5. ORGANIC CHEMISTRY. Educ: Loyola Col, Md, BS, 56; Cornell Univ, PhD(org chem), 61. Prof Exp: Fel, Purdue Univ, 61-62; from asst prof to assoc prof, 62-70, PROF CHEM, UNIV IOWA, 70- Mem: Am Chem Soc; The Chem Soc. Res: Synthesis and chemistry of polyfluorinated compounds; chemistry of polyhalogenated organoboranes; metal halide catalysis of halogenated olefins; preparation and reactions of halogenated ylides. Mailing Add: Dept of Chem Univ of Iowa Iowa City IA 52240

BURTON, GILBERT W, b Covington, Va, Mar 2, 36; m 64; c 1. ORGANIC CHEMISTRY, POLYMER CHEMISTRY. Educ: Univ Calif, BS, 58; Univ Ill, PhD(org chem), 64. Prof Exp: Res Prof Exp: From res chemist to sr res chemist, Esso Res & Eng Co, 64-75; STAFF CHEMIST, EXXON CHEM CO, 75- Mem: AAAS; Am Chem Soc. Res: Organic photochemistry; chemistry of ozonization of organic compounds; anionic coordination polymerization; elastomer chemistry. Mailing Add: 1179 Puddingstone Rd Mountainside NJ 07092

BURTON, GLENN WILLARD, b Clatonia, Nebr, May 5, 10; m 34; c 5. AGRONOMY. Educ: Univ Nebr, BA, 32; Rutgers Univ, MS, 33, PhD(agron), 36. Hon Degrees: ScD, Rutgers Univ, 55; ScD, Univ Nebr, 62. Prof Exp: Asst agron, Rutgers Univ, 32-36; from agent to sr geneticist, Div Forage Crops & Dis, Bur Plant Indust, 36-53, PRIN GENETICIST, FORAGE & RANGE RES BR, CROPS RES DIV, AGR RES SERV, USDA, 53- Concurrent Pos: Chmn agron div, Univ Ga, 50-64. Honors & Awards: Stevenson Award & South Seedsman Asn Award, 49; Sears Roebuck Award, 53, 61; John Scott Award, 57; Superior Serv Award, USDA, 55; Life Mem, Ga Plant Food Educ Soc, Inc, Men's Garden Clubs Am Gold Medal Award & Am Grassland Coun Golden Medallion Award, 65; Am Agr Ed Asn Distinguished Serv to Agr Award & Ga Sci & Technol Comt First Citation for Distinguished Serv to Advan of Sci, 66; Nebr Centennial Notable Nebraskan Award, 67; Agr Inst Can Recognition Award & Distinguished & Meritorious Serv to Organized Agr Award, Am Farm Bur Fed, 68; Nat Coun Commercial Plant Breeders, 69; Edward W Browning Award, Am Soc Agron, 75. Mem: Nat Acad Sci; Am Soc Agron (pres, 62); Am Genetic Asn; Am Soc Range Mgt. Res: Grass breeding and genetics; grass cytology; grass seed production; grass fertilization and management; revegation of native ranges. Mailing Add: Ga Coastal Plain Exp Sta Tifton GA 31794

BURTON, HAROLD, b New York, NY, Jan 19, 43; m 63; c 1. NEUROPHYSIOLOGY. Educ: Univ Mich, BA, 64; Univ Wis, PhD(physiol), 68. Prof Exp: Nat Inst Neurol Dis & Blindness trainee neurophysiol, Med Sch, Univ Wis, 68-70; asst prof neurobiol & physiol, 70-76, ASSOC PROF NEUROBIOL & PHYSIOL, MED SCH, WASHINGTON UNIV, 76- Res: Central representation of pain and temperature; electrophysiological studies of synapses in tissue culture. Mailing Add: Dept of Neurobiol & Anat Washington Univ Sch of Med St Louis MD 63110

BURTON, JAMES SAMUEL, b Richmond, Ky, Aug 1, 36; m 59; c 1. PHYSICAL CHEMISTRY. Educ: Berea Col, AB, 58; Howard Univ, MS, 62, PhD(phys chem), 64. Prof Exp: Chemist res, Melpar, Inc, 63-66; sr chemist mat sci, Gen Technol Corp, 66-67, proj engr air pollution res, 67-69, dir environ sci res air pollution, 69; mem tech staff, Environ Systs Anal, The Mitre Corp, 69-73, dir, Solar Energy Lab, 73-74; dir, Water Resources Res Ctr, Washington Tech Inst, 74-75; ASST DIR RES, OFF WATER RES & TECHNOL, US DEPT INTERIOR, 75- Honors & Awards: Outstanding Young Mem of Am, Chicago, 69. Mem: Am Chem Soc; Water Pollution Control Fedn. Res: Solar energy research; water resources research; saline water conversion; molecular spectroscopy; materials research; infrared spectra of gas-solid interactions; air pollution control technology; detection of air pollutants. Mailing Add: 2446 Freetown Dr Reston VA 22091

BURTON, JOE COVINGTON, b Due West, SC, Aug 27, 14; m 43; c 5. AGRICULTURAL CHEMISTRY, MICROBIOLOGY. Educ: Clemson Col, BS, 35; Univ Wis, MS, 37, PhD(bact), 52. Prof Exp: Asst bacteriologist, 37-42, dir res bact, 46-47, VPRES RES & DEVELOP, NITRAGIN CO, INC, 67- Mem: Am Soc Microbiol; Soil Sci Soc Am; fel Am Soc Agron; Soil Conserv Soc Am; Can Soc Microbiol. Res: Agricultural bacteriology; cultural and symbiotic properties of species and strains of rhizobia and factors influencing these properties; practical and theoretical aspects of legume inoculation. Mailing Add: Nitragin Co Inc 3101 W Custer Ave Milwaukee WI 53209

BURTON, JOHN HESLOP, b Ottawa, Can, Aug 27, 38; m 69; c 2. ANIMAL NUTRITION. Educ: Univ Toronto, BSc, 62; Cornell Univ, MSc, 67, PhD(nutrit), 70. Prof Exp: Res assoc, 69-70, asst prof, 71-73, ASST PROF ANIMAL NUTRIT, UNIV GUELPH, 73- Mem: Can Nutrit Soc; Agr Inst Can; Am Dairy Sci Asn; Am Soc Animal Sci. Res: Role of metabolic hormones in controlling milk synthesis and production in dairy cattle and growth in the bovine in general. Mailing Add: Dept of Nutrit Univ of Guelph Guelph ON Can

BURTON, JOHN WILLIAMS, b Atlanta, Ga, Apr 15, 37; m 59; c 4. EXPERIMENTAL NUCLEAR PHYSICS. Educ: Carson-Newman Col, BS, 59; Univ Ill, MS, 61, PhD(physics), 65. Prof Exp: Asst physics, Univ Ill, 61-64; assoc prof & head dept, 64-66, PROF PHYSICS, CARSON-NEWMAN COL, 66- Concurrent Pos: Consult, Oak Ridge Nat Lab, 65-75. Mem: Am Phys Soc; Am Asn Physics Teachers. Res: Mössbauer effect on the surface of tungsten, involving ultra high vacuum techniques; Mössbauer spectroscopy; computer programming for data analysis. Mailing Add: Dept of Physics Carson-Newman Col Jefferson City TN 37760

BURTON, JOSEPH ASHBY, b Onley, Va, Aug 22, 14. SOLID STATE PHYSICS. Educ: Wash & Lee Univ, BS, 34; Johns Hopkins Univ, PhD, 38. Prof Exp: Mem tech staff, 38-58, dir semiconductor res, 58-64, dir chem physics res, 64-71, DIR PHYS RES, BELL TEL LABS, INC, 71- Mem: Fel Am Phys Soc (treas); Am Inst Phys; Am Chem Soc. Res: Neutron generator; bombardment induced conductivity; photoelectric and thermionic emission; physical chemistry of semiconductors; emission of hot electrons from crystals. Mailing Add: Bell Tel Labs Mountain Ave Murray Hill NJ 07974

BURTON, LEONARD PATTILLO, b Jasper, Ala, June 8, 18; m 42; c 3. MATHEMATICS. Educ: Univ Ala, AB, 39, MA, 40; Univ NC, PhD(math), 51. Prof Exp: Instr math, Univ Ala, 46-48; from instr to asst prof, Univ Calif, 51-54; from asst prof to prof, 54-65, HEAD PROF MATH, AUBURN UNIV, 65- Mem: Am Math Soc; Math Asn Am. Res: Ordinary differential equations. Mailing Add: Dept of Math Auburn Univ Auburn AL 36830

BURTON, LLOYD EDWARD, b Phoenix, Ariz, May 13, 22; m 68; c 3. PHARMACY, PUBLIC HEALTH. Educ: Univ Ariz, BS, 54, MS, 56, PhD(pharmacol), 64. Prof Exp: Instr pharm, 54-59, res assoc pharmacol, 59-60, instr pharm, 60-64, from asst prof to assoc prof pharmacol, 64-70, PROF PHARMACOL, COL PHARM, UNIV ARIZ, 70- Mem: Am Pharmaceut Asn; Am Pub Health Asn. Res: Brain neurohormone studies; development of effective antidote for oleander poisoning; investigation of toxicity of plant drugs; development of health care services. Mailing Add: Col of Pharm Univ of Ariz Tucson AZ 85721

BURTON, LOUIS LASSETER, b Cartersville, Ga, July 13, 40; m 62; c 3. PHYSICAL CHEMISTRY. Educ: Davidson Col, BS, 62; Univ Va, PhD(chem), 68. Prof Exp: Res chemist, Polyolefins Res Lab, Tex, 69-71 & Exp Sta, Del, 71-73, tech rep, 73-76, MKT REP, PLASTICS DEPT, E I DU PONT DE NEMOURS & CO, INC, 76- Concurrent Pos: Teaching fel, Univ Ala, Tuscaloosa, 68-69. Mem: Am Chem Soc; Sigma Xi; Soc Plastics Engrs. Res: Development of engineering plastics—high impact, high temperature and other aspects; physical organic chemistry; molten salt solution; proton magnetic resonance spectroscopy in molten salts; polymer science; polymer product development; polymer marketing. Mailing Add: E I du Pont de Nemours & Co Exp Sta PP & R Wilmington DE 19898

BURTON, MILTON, b Stapleton, NY, Mar 4, 02; m 34, 46; c 2. PHYSICAL CHEMISTRY. Educ: NY Univ, BS, 22, MS, 23, PhD(phys chem), 25. Prof Exp: Res investr, NY Univ, 35-36, asst, 36-37; vis fel, Univ Calif, 37-38; instr chem, NY Univ, 38-43; sect chief metall lab, Univ Chicago, 42-45; prof chem, 45-71, dir radiation lab, 46-71, EMER PROF CHEM, UNIV NOTRE DAME, 71- Concurrent Pos: Guggenheim fel; Fulbright lectr; sect chief, Clinton Labs, Tenn, 45-46; guest prof, Univ Göttingen, 55-56; mem adv comt isotopes & radiation develop, AEC, 67-71. Honors & Awards: S C Lind Lect Award, 69; AEC Citation & Medal, 71; Notre Dame Pres Award, 71. Mem: AAAS; Am Chem Soc; Radiation Res Soc (pres, 58-59); Am Phys Soc; Faraday Soc. Res: Photochemistry; free radicals; reaction mechanisms; radiation chemistry; chemistry of electric discharge; luminescence. Mailing Add: Radiation Lab Univ of Notre Dame Notre Dame IN 46556

BURTON, PAUL RAY, b Burnsville, NC, Dec 7, 31; m 59; c 2. ZOOLOGY, CELL BIOLOGY. Educ: Western Carolina Col, BS, 54; Univ Miami, MS, 56; Univ NC, PhD(zool), 60. Prof Exp: Asst prof biol, St Olaf Col, 60-63; NIH fel anat, Sch Med, Univ Wis, 63-64; from asst prof to assoc prof zool, 64-69, chmn dept physiol & cell biol, 73-76, PROF PHYSIOL & CELL BIOL, UNIV KANS, 69- Concurrent Pos: USPHS Career Develop Award, 68-73. Mem: Am Micros Soc; Electron Micros Soc Am; Am Soc Parasitol; Am Asn Anat; Am Soc Cell Biol. Res: Electron microscopy; invertebrate cytology and ultrastructure; differentiation; oogenesis and spermatogenesis; studies on the structure and function of microtubular elements, particularly in neuronal systems; studies of the ultrastructure and biology of the rickettsia, Coxiella burneti. Mailing Add: Dept of Physiol & Cell Biol Univ of Kans Lawrence KS 66045

BURTON, ROBERT CLYDE, b Borger, Tex, Feb 27, 29; m 51; c 4. GEOLOGY. Educ: Tex Tech Col, BA, 57, MSc, 59; Univ NMex, PhD(geol), 65. Prof Exp: Instr geol, Tex Tech Col, 58-59; from instr to asst prof, West Tex State Univ, 59-63; vis lectr, Univ NMex, 65; assoc prof, 65-66, PROF & HEAD DEPT GEOL, WEST TEX STATE UNIV, 66- Concurrent Pos: Petrol Res Fund grant, Am Chem Soc, 69-71. Mem: Nat Asn Geol Teachers (secy, 68, vpres, 69, pres, 70). Res: Biostratigraphy; conodont biostratigraphy in New Mexico. Mailing Add: Dept of Geol West Tex State Univ Canyon TX 79015

BURTON, ROBERT LOUIS, b Pueblo, Colo, Apr 18, 21; m 43; c 3. POLYMER CHEMISTRY. Educ: Colo Col, BA, 42; Rice Inst, MA, 43; Univ Ill, PhD(x-ray diffraction), 46. Prof Exp: Jr anal chemist, Shell Oil Co, Tex, 43-44; res chemist, 46-49, res supvr, 49-53, develop & process control supvr, 54-57, asst mgr, 57-59, lab dir, 59-66, mgr develop & qual control, Luxembourg, 66-70, TECH SUPT, E I DU PONT DE NEMOURS & CO, INC, OHIO, 70- Mem: Am Chem Soc; Am Soc Tes & Mat; Soc Plastics Indust. Res: Cellulose chemistry and technology; constitution of hydrous oxides; x-ray diffraction; investigation of the constitution of natural micro-crystalline waxes; chemistry and physics of polymers for films. Mailing Add: E I du Pont de Nemours & Co Inc Box 89 Circleville OH 43113

BURTON, ROBERT MAIN, b Oklahoma City, Okla, Mar 5, 27; m 47; c 5. BIOCHEMISTRY. Educ: Univ Md, BS, 50; Georgetown Univ, MS, 52; Johns Hopkins Univ, PhD(biol, biochem), 55. Prof Exp: Chemist, USDA, Md, 50; chemist, Nat Heart Inst, 50-52; chemist, Nat Inst Neurol Dis & Blindness, 55-57; asst prof, 57-64, ASSOC PROF PHARMACOL, SCH MED, WASHINGTON UNIV, 64- Concurrent Pos: Lectr, Georgetown Univ, 56-57; Am Chem Soc & Am Soc Biol Chemists travel award, Int Biochem Cong, Vienna, 58, Moscow, 61 & Tokyo, 67; NSF sr fel, Inst Phys Chem, Univ Cologne, 63; mem child health & human develop prog comt, NIH, 68-69, mem pop res & training comt, 69-70, chmn, 70-72; sci adv, Ctr Biochem Studies, Portugal, 65-; assoc ed, Lipids, 65-; mem med adv bd, Nat Tay-Sachs & Allied Dis, Inc, 75- Mem: Am Soc Pharmacol & Exp Therapeut; Am Chem Soc; Am Soc Biol Chemists; Am Soc Neurochem; fel Am Inst Chemists. Res: Biochemistry of nervous system; lipid and glycolipid metabolism; biochemical basis for neuropharmacological activity of drugs. Mailing Add: Dept of Pharmacol Washington Univ Sch of Med St Louis MO 63110

BURTON, ROBERT MCMAHON, b Madison, Wis, Nov 3, 40. BIOCHEMISTRY, IMMUNOLOGY. Educ: Univ Wis, BS, 63, PhD(physiol chem), 69. Prof Exp: INSTR OBSTET & GYNEC, SCH MED, UNIV LOUISVILLE, 72 Concurrent Pos: NIH fel, Univ Louisville, 69-72. Res: Chemistry and immunology of tumor antigens; immunodiagnosis; immunoglobulins; cell surface constituents. Mailing Add: Dept of Obstet & Gynec Univ of Louisville Sch of Med Louisville KY 40202

BURTON, RUSSELL ROHAN, b Chico, Calif, Jan 15, 32; m 58; c 2. PATHOLOGICAL PHYSIOLOGY. Educ: Univ Calif, Davis, BS, 54, DVM, 56, MS, 65, PhD, 70. Prof Exp: Pvt pract, 56-62; from assoc res specialist to res specialist, Univ Calif, Davis, 62-71; RES PHYSIOLOGIST, US AIR FORCE SCH AEROSPACE MED, 71- Concurrent Pos: Asst prof, Univ Tex Health Sci Ctr San Antonio, 73- Honors & Awards: Paul Bert Award, Aerospace Med Asn, 75. Mem: AAAS; Aerospace Med Asn; Am Physiol Soc. Res: Acceleration physiological pathology. Mailing Add: Biodynamics Br (VNB) US Air Force Sch Aerospace Med Brooks Air Force Base TX 78235

BURTON, SHERIL DALE, b Malad, Idaho, May 10, 35; m 59; c 3. MICROBIOLOGY, BIOCHEMISTRY. Educ: Brigham Young Univ, BS, 59, MS, 61; Ore State Univ, PhD(microbiol), 64. Prof Exp: Res asst bact, Brigham Young Univ, 59-61 & microbiol, Univ Nebr, 61-62; res asst microbiol, Ore State Univ, 62-64, asst prof & res grantee, 64-65; asst prof, Inst Marine Sci, Univ Alaska, 65-67; ASSOC PROF MICROBIOL, BRIGHAM YOUNG UNIV, 67- Mem: Am Soc Microbiol. Res: Thermodynamic constants of enzymes from psychrophiles; growth and metabolism of Beggiatoa and metabolism of ester producing yeasts. Mailing Add: Dept of Microbiol Brigham Young Univ Provo UT 84601

BURTON, STEPHEN DALE, pharmacognosy, see 12th edition

BURTON, THEODORE ALLEN, b Longton, Kans, Sept 7, 35; m 61. MATHEMATICS. Educ: Wash State Univ, BS, 59, MA, 62, PhD(math), 64. Prof Exp: Asst prof math, Univ Alta, 64-66; assoc prof, 66-71, PROF MATH, SOUTHERN ILL UNIV, CARBONDALE, 71- Concurrent Pos: Can Math Cong res fel, 65-66. Mem: Am Math Soc; Can Math Cong. Res: Stability theory of ordinary differential equations. Mailing Add: Dept of Math Southern Ill Univ Carbondale IL 62901

BURTON, THOMAS MAXIE, b West Carroll Parish, La, Nov 24, 41. AQUATIC ECOLOGY. Educ: Northeast La Univ, BS, 63, MS, 65; Cornell Univ, PhD(ecol), 73. Prof Exp: Teacher sci, Winnsboro High Sch, 65-66; instr biol, Univ NC, Charlotte, 66-67; res technician, US Army Res Inst Environ Med, Mass, 67-69; ecol consult, Eng-Sci, Cincinnati, 73; res assoc aquatic ecol, Fla State Univ, 74-75; ASST PROF AQUATIC ECOL, INST WATER RES, MICH STATE UNIV, 75- Mem: Ecol Soc Am; Am Soc Limnol & Oceanog; Am Inst Biol Sci; AAAS; Sigma Xi. Res: Watershed-stream studies of energy flow and nutrient cycling; effects of non-point source pollution on water quality. Mailing Add: Inst of Water Res Mich State Univ East Lansing MI 48824

BURTON, VERONA DEVINE, b Reading, Pa, Nov 23, 22; m 50; c 1. BOTANY. Educ: Hunter Col, AB, 44; State Univ Iowa, MS, 46, PhD(plant anat), 48. Prof Exp: Asst, State Univ Iowa, 44-48; from asst prof to assoc prof, 48-70, PROF BIOL SCI, MANKATO STATE UNIV, 70- Mem: AAAS; Bot Soc Am; Am Fern Soc; Int Soc Plant Morphol. Res: Floral abscission and plant embryology. Mailing Add: 512 Hickory St Mankato MN 56001

BURTON, WILLARD WHITE, b Richmond, Va, Feb 24, 22; m 48; c 3. ORGANIC CHEMISTRY. Educ: Univ Richmond, BS, 43. Prof Exp: From jr res assoc to res assoc, 46-68, supvr org chem sect, 68-70, supvr chem develop, Process Develop Lab, 70-71, SUPVR RES, DEPT RES & DEVELOP, AM TOBACCO CO, 71- Mem: Am Chem Soc; fel Am Inst Chem. Res: Tobacco chemical composition; browning reaction pigments; phenols; amino acids; carbohydrates and organic acids. Mailing Add: 6808 Greenvale Dr Richmond VA 23225

BURTON, WILLIAM BUTLER, b Richmond, Va, July 13, 40; m 72; c 2. ASTRONOMY Educ: Swarthmore Col, BA, 62; State Univ Leiden, DrS, 65, PhD(astron), 70. Prof Exp: Asst astron, State Univ Leiden, 64-65, sci officer, 66-70; from res assoc to asst scientist, 71-75, ASSOC SCIENTIST ASTRON, NAT RADIO OBSERV, 75- Mem: Am Astron Soc; Int Astron Union; Netherlands Astron Soc. Res: Galactic structure; interstellar medium. Mailing Add: Astron Observ Edgemont Rd Charlottesville VA 22901

BURTS, EVERETT C, b Mae, Wash, Sept 25, 31; m 54; c 2. ENTOMOLOGY. Educ: State Col Wash, BS, 54; Ore State Col, MS, 57, PhD(entom), 59. Prof Exp: Jr entomologist, Ore State Col & agent, Div Entom, Agr Res Serv, USDA, 54-58; ENTOMOLOGIST, WASH STATE UNIV, 58- Mem: AAAS; Entom Soc Am. Res: Biology and control of insects and mites of tree fruits; insect vectors of plant viruses; deciduous fruit insect pest management. Mailing Add: Tree Fruit Res Ctr 1100 N Western Ave Wenatchee WA 98801

BURTSELL, AUBISON TUTHILL, physical chemistry, analytical chemistry, see 12th edition

BURTT, BENJAMIN PICKERING, b Newburyport, Mass, June 7, 21; m 45; c 3. PHYSICAL CHEMISTRY, RADIATION CHEMISTRY. Educ: Ohio State Univ, BA, 42, PhD(chem), 46. Prof Exp: Asst chem & instr physics, Ohio State Univ, 42-45, instr chem, 45-46; instr, 46-48, from instr to assoc prof, 49-59, PROF CHEM, SYRACUSE UNIV, 59- Concurrent Pos: Assoc scientist, Brookhaven Nat Lab, 48-49. Mem: Am Chem Soc. Res: Mass spectrometric studies of electron impact phenomena; ion-molecule reactions. Mailing Add: Dept of Chem Syracuse Univ Syracuse NY 13210

BURWASH, RONALD ALLAN, b Edmonton, Alta, Sept 20, 25; m 52; c 3. GEOLOGY. Educ: Univ Alta, BSc, 45, BEd, 47, MSc, 51; Univ Minn, PhD(geol), 55. Prof Exp: Geologist, Sullivan Mines, BC, 51-52; petrogr, Shell Oil Co, Alta, 55-56; from asst prof to assoc prof, 56-65, PROF GEOL, UNIV ALTA, 65- Mem: Geochem Soc; Geol Asn Can. Res: Igneous and metamorphic petrology; Precambrian geology; geochemistry of Continental crust. Mailing Add: Dept of Geol Univ of Alta Edmonton AB Can

BURWASSER, HERMAN, b New York, NY, June 21, 27; m 54; c 3. PHYSICAL CHEMISTRY. Educ: Rutgers Univ, BS, 50; NY Univ, PhD(phys chem), 54. Prof Exp: Asst combustion chem, Princeton Univ, 54-56; res chemist, Atlantic Refining Co, 56-59; res scientist, AeroChem Res Labs, Inc, 59-60; proj chemist, Thiokol Chem Corp, 60-68; TECH ASSOC, GAF CORP, 66- Mem: Am Chem Soc. Res: Photo and radiation chemistry of organic materials, especially reversible photochemical reactions; organic photoconductors; electrophotography. Mailing Add: GAF Corp Wayne NJ 07470

BURWELL, JAMES ROBERT, b Anderson, Ind, Mar 28, 29; m 51; c 3. ELEMENTARY PARTICLE PHYSICS, HIGH ENERGY PHYSICS. Educ: Ind Univ, BS, 52, MS, 54, PhD(physics), 57. Prof Exp: Assoc high energy physics, Ind Univ, 57-58 & nuclear physics, 58-59; from asst prof to assoc prof physics, 59-68, chmn dept, 64-65, asst dean col arts & sci, 66-69, PROF PHYSICS, UNIV OKLA, 68-, ASSOC DEAN COL ARTS & SCI, 69- Mem: Am Phys Soc; Am Asn Physics Teachers. Res: High energy elementary particle physics. Mailing Add: Rm 221 Phys Sci Ctr Univ of Okla Norman OK 73069

BURWELL, ROBERT LEMMON, JR, b Baltimore, Md, May 6, 12; m 39; c 2. CATALYSIS. Educ: St John's Col, Md, AB, 32; Princeton Univ, AM, 34, PhD(phys chem), 36. Prof Exp: Instr chem, Trinity Col, Conn, 36-39; from instr to prof, 39-70, chmn dept, 52-57, IPATIEFF PROF CHEM, NORTHWESTERN UNIV, 70- Concurrent Pos: Chemist, Naval Res Lab, 43-45; mem bd dirs, Int Cong Catalysis, 56-64, mem coun, 64-, secy, 68-72, vpres, 72-76; chmn, Gordon Res Conf Catalysis, 57; mem subcomt heterogeneous catalysis, Int Union Pure & Appl Chem, 62-67, assoc mem comn colloid & surface chem, 67-69, mem, 69-; mem coun policy comt, Am Chem Soc, 68-72; mem chem adv panel, NSF, 68-71; mem bd dirs, Catalysis Soc, 64- Honors & Awards: Kendall Award, Am Chem Soc, 73. Mem: Am Chem Soc; Catalysis Soc (vpres, 68-73, pres, 73-77); The Chem Soc. Res: Heterogeneous catalysis and surface chemistry. Mailing Add: Dept of Chem Northwestern Univ Evanston IL 60201

BURZLAFF, DONALD FREDERICK, b Dodge Center, Minn, May 30, 23; m 48; c 2. RANGE MANAGEMENT, AGRONOMY. Educ: Univ Wyo, BS, 50, MS, 52; Utah State Univ, PhD(forestry), 60. Prof Exp: Instr supply, Univ Wyo, 52-53; exten agronomist, Univ Nebr, Lincoln, 53-57, assoc agronomist, 57-61, from assoc prof to prof agron, 61-72, vchmn dept, 71-72; PROF RANGE & WILDLIFE MGT & CHMN DEPT, TEX TECH UNIV, 73- Mem: Am Soc Agron; Soc Range Mgt; Soil Conserv Soc Am. Res: Soil vegetation relationships as influenced by agronomic practices imposed by modern agriculture; forage crop management; range livestock nutrition. Mailing Add: Dept of Range & Wildlife Mgt Tex Tech Univ Lubbock TX 79409

BURZYNSKI, NORBERT J, b South Bend, Ind, July 7, 29; m 61; c 2. EXPERIMENTAL PATHOLOGY. Educ: Ind Univ, AB, 52; St Louis Univ, DDS, 60, MS, 63. Prof Exp: USPHS fel oral path, St Louis Univ, 60-62; from asst prof to assoc prof, 65-73, PROF ORAL DIAG & ORAL MED & CHMN DEPT, SCH DENT, UNIV LOUISVILLE, 73- Concurrent Pos: Fel med genetics, Ind Univ, 70-71. Mem: Fel AAAS; Am Soc Human Genetics; Int Asn Dent Res; Am Dent Asn. Res: Chemistry of mucous membrane; neoplasia; salivary gland metabolism. Mailing Add: Univ of Louisville Sch of Dent Louisville KY 40202

BURZYNSKI, STANISLAW RAJMUND, b Lublin, Poland, Jan 23, 43. ONCOLOGY, PROTEIN CHEMISTRY. Educ: Med Acad, Lublin, Poland, MD, 67, PhD(biochem), 68. Prof Exp: Teaching asst chem, Med Acad, Lublin, Poland, 62-67, from intern to resident internal med, 67-70; res assoc biochem, 70-72, ASST PROF BIOCHEM, BAYLOR COL MED, 72- Concurrent Pos: Nat Cancer Inst grant, 74; West Found grant, 75. Mem: AAAS; Am Heart Asn; Fedn Am Scientists; Soc Neurosci; Sigma Xi. Res: Discovery of antineoplastons-components of biochemical defense system against cancer; elucidation of the structure of Ameletin-first substance known to be responsible for remembering sound in animal's brain. Mailing Add: 5 Concord Circle Houston TX 77024

BUSBICE, THADDEUS H, b Richland Parish, La, Feb 29, 32; m 64; c 3. GENETICS, PLANT BREEDING. Educ: Northeastern La State Col, BS, 56; Iowa State Univ, MS, 64, PhD(plant breeding), 65. Prof Exp: Asst prof crop sci, NC State Univ, 65-69; res agronomist, 65-70, RES GENETICIST, CROPS RES DIV, USDA, 70-; ASSOC PROF CROP SCI, NC STATE UNIV, 70- Mem: Am Soc Agron. Res: Genetics of medicago sativa and inbreeding and heterosis of this species. Mailing Add: Dept of Crop Sci NC State Univ Raleigh NC 27607

BUSBY, HUBBARD TAYLOR, JR, b Birmingham, Ala, May 20, 41; m 63; c 1. ORGANIC CHEMISTRY, INDUSTRIAL ORGANIC CHEMISTRY. Educ: Miss Col, BS, 63; Univ NC, PhD(org chem), 68. Prof Exp: Res chemist, Southern Dyestuff Co, 67-73; GROUP LEADER, DISPERSE DYE RES, MARTIN MARIETTA CHEM CO, 73- Mem: Am Chem Soc; Am Asn Textile Chem & Colorists. Res: Interannular substituent effects in ferrocene; synthesis and development of marketable dye stuff; organic chemical process development. Mailing Add: 3001 Wamath Dr Charlotte NC 28210

BUSBY, JOE NEIL, b Hartwell, Ga, Apr 9, 21; m 43; c 5. HORTICULTURE, BOTANY. Educ: Univ Fla, BSA, 47, PhD(fruit crops), 61. Prof Exp: Asst county agent, Agr Exten Serv, Univ Fla, 47-49, asst state 4-H agent, 49-52; asst chief plant inspector, State Plant Bd Fla, 53-55; from asst dir to assoc dir agr exten serv, 56-70, dean exten & continuing educ, 70-72, DEAN FLA COOP EXTEN SERV, UNIV FLA, 72- Res: Effects of air movement; external heat sources on plant and air temperatures in relation to injury of plants. Mailing Add: 1038 McCarty Hall Univ of Fla Gainesville FL 32601

BUSBY, ROBERT CLARK, b Darby, Pa, July 1, 40; m 64; c 1. MATHEMATICS. Educ: Drexel Inst, BS, 63; Univ Pa, AM, 64, PhD(math), 66. Prof Exp: Instr math, Drexel Inst, 65-67; asst prof, Oakland Univ, 67-69; asst prof, 69-70, ASSOC PROF MATH, DREXEL UNIV, 70- Concurrent Pos: Consult math, Off Emergency Preparedness, Exec Off Pres, 68- Mem: Am Math Soc; Soc Indust & Appl Math. Res: Modern analysis and applications; algebras of operators on a Hilbert space and group representations. Mailing Add: Dept of Math Drexel Univ Philadelphia PA 19104

BUSBY, WILLIAM FISHER, JR, b Pawtucket, RI, Oct 29, 39; m 62; c 1. BIOCHEMISTRY. Educ: Univ RI, BS, 61; Univ Calif, San Diego, PhD(marine biol), 66. Prof Exp: Res assoc biochem, Worcester Found Exp Biol, 66-71; RES ASSOC BIOCHEM, MASS INST TECHNOL, 71- Res: Interactions of carcinogens and other xenobiotic compounds with cellular macromolecules, and the characterization of the resultant alterations in biochemical function with emphasis on hormonal and other nuclear control mechanisms. Mailing Add: Dept of Nutrit & Food Sci Mass Inst Technol Cambridge MA 02139

BUSCEMI, PHILIP AUGUSTUS, b Mt Pleasant, Iowa, Mar 1, 26; m 50; c 3. LIMNOLOGY, ENVIRONMENTAL MANAGEMENT. Educ: Univ Colo, BS, 50, MA, 52, PhD(biol & zool), 57. Prof Exp: Asst gen biol & invert zool, Univ Colo, 50-52, instr, 52-53, gen biol, Denver Exten Cent, 55-56; from instr to asst prof zool, Univ Idaho, 56-65; asst prof, Okla State Univ, 65-66; assoc prof, 66-71, chmn dept biol sci, 66-75, PROF BIOL, EASTERN NMEX UNIV, 71-, PRES, INTERDISCIPLINARY ENVIRON INST, 72- Concurrent Pos: NSF fel, Ore Inst Marine Biol, 58; consult, Arctic Health Res Ctr, Anchorage, Alaska, 59; mem, Bio-Med Inst, Lawrence Radiation Lab, 70- Mem: Fel AAAS; Am Soc Zool; Am Soc Limnol & Oceanog; Ecol Soc Am; Sigma Xi. Res: Hydrobiology; animal ecology of mountain lakes; macroscopic bottom fauna; organic production in fresh water; zooplankton population dynamics; biochemical energy pathways in aquatic ecosystems. Mailing Add: Div of Natural Sci Eastern NMex Univ Portales NM 88130

BUSCH, DANIEL ADOLPH, b St Paul, Minn, May 31, 12; m 39; c 2. GEOLOGY. Educ: Capital Univ, BS, 34; Ohio State Univ, AM, 36, PhD(geol), 39. Hon Degrees: DSc, Capital Univ, 60. Prof Exp: Asst geol, State Geol Surv, Ohio, 35-36; instr, Univ Pittsburgh, 38-43; petrol geologist, State Topog & Geol Surv, Pa, 43-44; consult geologist, Huntley & Huntley, Pittsburgh, 44-46; sr res geologist, Carter Res Lab, 46-49; staff geologist, Carter Oil Co, 49-51; explor mgr, Zephyr Petrol Co, 51-54; CONSULT GEOLOGIST, 54- Concurrent Pos: Carnegie Mus field exped, 41; vis prof, Ohio State Univ, 63, Univ Tulsa, 63-64 & Univ Okla, 64-74; world-wide lectr, Oil & Gas Consults Int, Inc, 69- Honors & Awards: Orton Award, Ohio State Univ, 59, 60; Matson Award, Am Asn Petrol Geol, 59, Levorson Award, 71; President's Award, Am Asn Petrol Geol, 76. Mem: Am Asn Petrol Geol (vpres, 66-67, pres, 73-74); fel Geol Soc Am; Am Inst Prof Geol. Res: Silurian and Devonian stratigraphy of Appalachian Basin; Pennsylvania stratigraphy and sedimentology of eastern and north-central; subsurface stratigraphy of mid-continent; stratigraphy and structure of Mexican Gulf Coast. Mailing Add: 3757 S Wheeling Ave Tulsa OK 74103

BUSCH, DARYLE HADLEY, b Carterville, Ill, Mar 30, 28; m 51; c 5. INORGANIC CHEMISTRY. Educ: Univ Southern Ill, BA, 51; Univ Ill, MS, 52, PhD, 54. Prof Exp: From asst prof to assoc prof, 54-63, PROF CHEM, OHIO STATE UNIV, 63- Concurrent Pos: Consult, E I du Pont de Nemours & Co, 56-, Div Res Grants, NIH, 61-65, Div Res Grants, NSF, 65-68, Beaunit Fibers, 66-68 & Chem Abstr Serv, 67-; consult ed, Allyn & Bacon, Inc, 63- Honors & Awards: Inorg Chem Award, Am Chem Soc, 63, Morley Medal, 75 & Distinguished Serv in Advan of Inorg Chem, 76. Mem: Am Chem Soc; AAAS. Res: Coordination chemistry; stereo chemistry mechanisms of substitution reactions; magnetochemistry of transition metal ions; complexes of macrocyclic ligands; reactions of coordinated ligands; bioinorganic chemistry. Mailing Add: Dept of Chem Ohio State Univ Columbus OH 43210

BUSCH, HARRIS, b Chicago, Ill, May 23, 23; m 45; c 4. PHARMACOLOGY, BIOCHEMISTRY. Educ: Univ Ill, MD, 46; Univ Wis, PhD(biochem), 52. Prof Exp: Asst & sr asst surgeon, USPHS, 47-49; asst prof biochem, Yale Univ, 52-54, asst prof med & biochem, 54-55; assoc prof pharmacol, Univ Ill, 55-59, prof, 59-60; prof biochem & chmn dept, 60-62, PROF PHARMACOL & CHMN DEPT, BAYLOR COL MED, 60- Concurrent Pos: Baldwin scholar oncol, Sch Med, Yale Univ, 51-55, scholar cancer res, 54-55; vis prof, Univ Chicago, 68, Northwestern Univ, Ore State Univ, Ind Univ, Vet Hosp & Methodist Hosp, Houston; dir, Cancer Res Ctr; mem consult bd, Eli Lilly Co; mem pathogenesis panel, Am Cancer Soc; mem cancer chemother study sect, USPHS; consult, Uniformed Servs Univ, Bethesda, Md, 75; mem bd sci counrs, Div Cancer Treatment, Nat Cancer Inst, 75; vis prof, Univ Toronto, 75. Mem: AAAS; Biochem Soc; Am Chem Soc; Am Asn Cancer Res; Am Soc Biol Chemists. Res: Metabolism of nuclear proteins in tumor nuclei; isolation and metabolism of tumor nucleoli. Mailing Add: Dept of Pharmacol Baylor Col of Med Houston TX 77025

BUSCH, KARL HEINRICH DANIEL, b St Paul, Minn, Oct 23, 09; m 36; c 3. ZOOLOGY. Educ: Capital Univ, BS, 31; Ohio State Univ, MS, 37, PhD(zool), 40. Prof Exp: Lab asst, Capital Univ, 32-33; prin & teacher high sch, Ohio, 34-36; asst, Ohio State Univ, 36-40; head dept biol & physics, Corpus Christi Jr Col, 40-44; assoc prof biol, Houston Univ, 44-45; from assoc prof to prof, Augustana Col, SDak, 45-49, head dept biol, 45-49, chmn div natural sci, 47-49; from assoc prof to prof biol, Augustana Col, Ill, 49-52; prof & head dept, Millikin Univ, 52-56; prof biol, 56-74, head dept, 57-70, EMER PROF BIOL, UNIV NEBR, OMAHA, 74- Mem: AAAS; assoc Am Soc Zool; Am Inst Biol Sci. Res: Embryology of crayfish; freshwater bryozoa; teaching methods in biology. Mailing Add: Dept of Biol Univ of Nebr 60th & Dodge Sts Omaha NE 68101

BUSCH, KENNETH WALTER, b Mt Vernon, NY, Mar 29, 44; m 68. ANALYTICAL CHEMISTRY. Educ: Fla Atlantic Univ, BS, 66; Fla State Univ, PhD(anal chem), 71. Prof Exp: Teaching fel chem, Fla State Univ, 71-72; res assoc, Cornell Univ, 72-74; ASST PROF CHEM, BAYLOR UNIV, 74- Mem: Am Chem Soc; Chem Soc London; Soc Appl Spectros; Sigma Xi. Res: Development of chemical instrumentation for spectroscopic trace element analysis; simultaneous multi-element analysis and the theory of spectrochemical excitation in flames and plasmas. Mailing Add: Dept of Chem Baylor Univ Waco TX 76703

BUSCH, LLOYD VICTOR, b London, Ont, Nov 15, 18; m 43; c 4. PHYTOPATHOLOGY. Educ: Ont Agr Col, BSA, 42; Univ Toronto, MSA, 48; Univ Wis, PhD(plant path), 55. Prof Exp: From lectr to assoc prof, 46-71, PROF BOT, ONT AGR COL, UNIV GUELPH, 72- Mem: Potato Asn Am; Can Phytopath Soc (pres, 74-75); Sigma Xi. Res: Physiology and environmental influence on potato diseases and verticillium wilt; effects of mycorrhize on root rot; fusarium stock and cob rot of corn. Mailing Add: Dept of Environ Biol Univ of Guelph Guelph ON Can

BUSCH, PHILLIP MAXWELL, geography, geology, see 12th edition

BUSCH, ROBERT EDWARD, b Independence, Mo, May 13, 24; m 47; c 2. ANIMAL GENETICS. Educ: Univ Mo, BS, 45, MEd, 55; Ore State Col, PhD(genetics), 63. Prof Exp: Instr high schs, Mo, 45-59; res geneticist, US Fish & Wildlife Serv, 62-63; ASSOC PROF ANIMAL SCI, CALIF STATE UNIV, CHICO, 63- Mem: Am Soc Animal Sci. Res: Sheep genetics; genetics of fish; swine genetics and nutrition. Mailing Add: Dept of Agr Calif State Univ Chico CA 95926

BUSCH, ROBERT HENRY, b Jefferson, Iowa, Oct 22, 37; m 58; c 2. GENETICS, STATISTICS. Educ: Iowa State Univ, BS, 59, MS, 63; Purdue Univ, PhD(genetics, plant breeding), 67. Prof Exp: Technician, Iowa State Univ, 59-61, res assoc agron, 61-63; instr, Purdue Univ, 66-67; asst prof, 67-72, ASSOC PROF AGRON, NDAK STATE UNIV, 72- Mem: Am Soc Agron; Crop Sci Soc Am. Res: Quantitative inheritance of economic characteristics of spring wheat and more effective methods of effecting genetic control and selection. Mailing Add: Dept of Agron NDak State Univ Fargo ND 58102

BUSCHBACH, THOMAS CHARLES, b Cicero, Ill, May 12, 23; m 47; c 3. GEOLOGY. Educ: Univ Ill, BS, 50, MS, 51, PhD, 59. Prof Exp: From asst geologist to assoc geologist, 51-67, GEOLOGIST, ILL STATE GEOL SURV, 67- Concurrent Pos: Consult, US Nuclear Regulatory Comn, 75-76. Mem: Soc Petrol Engrs; fel Geol Soc Am; Am Asn Petrol Geol. Res: Cambrian and Ordovician stratigraphy; underground gas storage; siting of nuclear facilities. Mailing Add: 268 Natural Resources Bldg Ill State Geol Surv Urbana IL 61801

BUSCHER, HENRY NEIL, b Chanute, Kans, May 25, 37; m 63. ZOOLOGY, PARASITOLOGY. Educ: Kans State Col, Pittsburgh, BS, 61; Univ Okla, MS, 63, PhD(zool), 65. Prof Exp: Res assoc med zool, Sch Med, Univ PR, 65-66; res assoc parasitol, Univ Md Int Ctr Med Res & Training, Pakistan Med Res Ctr, Lahore, 66-68; from asst prof to assoc prof, 68-75, PROF BIOL, AUSTIN COL, 75-, CHMN DEPT, 70- Concurrent Pos: USPHS fel immunol & serol of parasitic infections, 65-66. Mem: Am Soc Parasitol. Res: Parasitic helminths. Mailing Add: Dept of Biol Austin Col Sherman TX 75090

BUSCHERT, ROBERT CECIL, b Preston, Ont, Nov 28, 24; m 48; c 3. SOLID STATE PHYSICS. Educ: Goshen Col, BA, 48; Purdue Univ, MS, 52, PhD(physics), 57. Prof Exp: Instr physics & math, Goshen Col, 48-50; mem tech staff, Bell Tel Labs, 56-58; asst prof physics, Purdue Univ, 58-65; PROF PHYSICS, GOSHEN COL, 65- Mem: Am Phys Soc. Res: Structure of semiconductors; germanium and silicon. Mailing Add: Dept of Physics Goshen Col Goshen IN 46526

BUSCHKE, HERMAN, b Berlin, Ger, Oct 15, 32; US citizen; m 57; c 2. NEUROLOGY, PSYCHOLOGY. Educ: Reed Col, BA, 54; Western Reserve Univ, MD, 58; Am Bd Psychiat & Neurol, dipl, 66. Prof Exp: Intern med, Bronx Munic Hosp Ctr, NY, 58-59, resident neurol, 59-62; from instr to asst prof, Sch Med, Stanford Univ, 62-69, res assoc, Inst Math Studies in Soc Sci, 68-69; assoc prof, 69-74, PROF NEUROL & NEUROSCI, ALBERT EINSTEIN COL MED, 74-, GLUCK DISTINGUISHED SCHOLAR NEUROL, 73- Concurrent Pos: USPHS fel, Albert Einstein Col Med, 59-62, res scientist develop award, 64-69; consult, Vet Admin Hosp, Palo Alto, Calif, 63-69; prin investr, USPHS Res Grant, 64-; sr investr, Rose F Kennedy Ctr Res Ment Retardation & Human Develop, 69- Mem: Psychonomic Soc; Acad Aphasia; Int Neuropsychol Soc; Am Acad Neurol; Am Psychol Asn. Res: Human memory; learning; psycholinguistics; perception; cognition; linguistic and cognitive dysfunction; neuropsychology. Mailing Add: Saul R Korey Dept of Neurol Albert Einstein Col of Med Bronx NY 10461

BUSCHMAN, WILLIAM OWEN, b Great Bend, Kans, Sept 13, 18; m 57; c 3. MATHEMATICS. Educ: Reed Col, AB, 41; Univ Ore, MEd, 47; Ore State Univ,

EdD, 53. Prof Exp: Engr, Kaiser Co, 41-43; marine engr, 43-45; instr high sch, Ore, 45-47; instr, Gen Exten Div, Ore State Syst Higher Educ, 46-53; asst prof, Portland State Col, 53-56; dir teacher training math, 56-66, dir comput ctr, 66-68, prof math & comput sci, 68-69, PROF COMPUT SCI & STATIST, CALIF POLYTECH STATE UNIV, SAN LUIS OBISPO, 69- Concurrent Pos: Instr math, Multonomah Col, 46-48; mem staff, Stanford Res Inst, 54; Inst Motivational Res, 54-56 & Med Sch, Univ Ore, 55-56; prof comput sci, Univ Calif, Santa Barbara Exten, 66-72. Res: Computer programming; statistics. Mailing Add: Dept of Comput Sci & Statist Calif Polytech State Univ San Luis Obispo CA 93407

BUSCHMANN, ROBERT JOHN, b Chicago, Ill, July 30, 42; m 70. PHYSIOLOGY, ELECTRON MICROSCOPY. Educ: Loyola Univ, Ill, BS, 64; Univ Ill, Urbana, MS, 66, PhD(physiol), 69. Prof Exp: PHYSIOLOGIST, VET ADMIN WEST SIDE HOSP, 69- Mem: AAAS; Am Soc Cell Biol; Electron Micros Soc Am. Res: Intestinal lipid absorption; cell physiology. Mailing Add: Dept of Path Vet Admin WSide Hosp Chicago IL 60612

BUSCOMBE, WILLIAM, b Hamilton, Ont, Feb 12, 18; m 42; c 8. ASTROPHYSICS. Educ: Univ Toronto, BA, 40, MA, 48; Princeton Univ, PhD(astron), 50. Prof Exp: Meteorologist, Dept Transp, Govt Can, 41-45; instr math & astron, Univ Sask, 45-48; res fel, Mt Wilson & Palomar Observ, 50-52; astronr, Mt Stromlo Observ, Australian Nat Univ, 52-68; PROF ASTRON, NORTHWESTERN UNIV, EVANSTON, 68- Concurrent Pos: Vis prof, Northern Ill Univ, 70-73. Mem: Am Astron Soc; Royal Astron Soc Can; Royal Astron Soc; Int Astron Union. Res: Stellar spectra; radial velocities; binary orbits; galactic kinematics; interstellar gas; variable stars. Mailing Add: Dearborn Observ Northwestern Univ Evanston IL 60201

BUSE, JOHN FREDERICK, b Charleston, SC, Oct 17, 21; m 56; c 1. INTERNAL MEDICINE. Educ: Univ SC, BS, 44, MD, 50. Prof Exp: Intern, Roper Hosp, 50-51, asst resident med, 51-52; asst resident, Univ Va, 52-53; from assoc to assoc prof, 5670, PROF MED, MED UNIV SC, 70 Concurrent Pos: Fel, Med Col SC, 53-54 & Cox Inst, Univ Pa, 54-56. Mem: Am Fedn Clin Res; Endocrine Soc; Am Diabetes Asn. Res: Diabetes mellitus; influence of muscular activity on insulin requirement. Mailing Add: Med Univ of SC Charleston SC 29401

BUSE, MARIA F GORDON, b Budapest, Hungary, July 17, 27; m 56; c 1. MEDICINE. Educ: Univ Buenos Aires, MD, 54. Prof Exp: Intern, Rivadavia Hosp, Arg, 53-54; res fel physiol, Inst Exp Biol & Med, Univ Buenos Aires, 54-55; fel med, Univ Pa, 55-56; res assoc med, 56-59, instr biochem, 59-61, assoc, 61-62, from asst prof to assoc prof res med, 62-72, PROF MED, MED UNIV SC, 72-, PROF BIOCHEM, 74- Concurrent Pos: NIH res career develop award, 63. Mem: AMA; Am Fedn Clin Res; Soc Nuclear Med; Endocrine Soc; Am Diabetes Asn. Res: Diabetes; action of insulin; intermediary metabolism. Mailing Add: Med Univ Hosp 80 Barre St Charleston SC 29401

BUSECK, PETER R, b Sept 30, 35; US citizen; m 60; c 4. GEOCHEMISTRY, ECONOMIC GEOLOGY. Educ: Antioch Col, AB, 57; Columbia Univ, MA, 59, PhD(geol), 62. Prof Exp: Fel, Geophys Lab, Carnegie Inst, 61-63; PROF GEOL & CHEM, ARIZ STATE UNIV, 63- Concurrent Pos: NSF Sci Fac fel, Oxford Univ, 70-71. Mem: Geol Soc Am; Mineral Soc Am; AAAS; Soc Econ Geol; Microbeam Anal Soc. Res: Electron microscopy and diffraction of minerals; electron microprobe analysis of air pollutants and minerals; geochemical exploration and ore deposition; meteoritics. Mailing Add: Depts of Geol & Chem Ariz State Univ Tempe AZ 85281

BUSEMANN, HERBERT, b Berlin, Ger, May 12, 05; nat US; m 39. MATHEMATICS. Educ: Humanistic Sch, Ger, AB; Univ Göttingen, PhD(math), 31. Hon Degrees: LLD, Univ Southern Calif, 71. Prof Exp: Asst, Univ Göttingen, 31-33; lectr, Copenhagen Univ, 33-36; asst math, Inst Advan Study, 36-39; instr, Swarthmore Col, 39 & Johns Hopkins Univ, 39-40; from instr to asst prof, Ill Inst Technol, 40-45; asst prof, Smith Col, 45-47; from prof to distinguished prof, 47-70, DISTINGUISHED EMER PROF MATH, UNIV SOUTHERN CALIF, 70- Concurrent Pos: Lectr, NY Univ, 36-38; mem, Royal Danish Acad, 64- Mem: Am Math Soc; Math Asn Am. Res: Differential geometry; convexity; foundations of geometry; area in general spaces. Mailing Add: 3839 Oak Trail Rd Santa Ynez CA 93460

BUSENBERG, EURYBIADES, b Jerusalem, Israel, Nov 13, 39; m 74. GEOCHEMISTRY. Educ: NY Univ, BA, 61, MS, 67; State Univ NY Buffalo, PhD(geochem), 75. Prof Exp: Chemist, Am Petrol Inst, 65-67; petrol geologist, Texaco Inc, 69-71; vis asst prof, 74-75, ASST PROF GEOCHEM, STATE UNIV NY BUFFALO, 75- Concurrent Pos: Hydrologist, US Geol Surv, 74- Mem: Sigma Xi; Clay Minerals Soc. Res: The dissolutions kinetics of aluminosilicate minerals; natural weathering reactions of aluminosilicates and their products; the aquatic chemistry of natural waters. Mailing Add: Dept of Geol Sci State Univ of NY Buffalo NY 14226

BUSENBERG, STAVROS NICHOLAS, b Jerusalem, Palestine, Oct 16, 41; nat US; m 69. MATHEMATICS. Educ: Cooper Union, BME, 62; Ill Inst Technol, MS, 64 & 65, PhD(math), 67. Prof Exp: Instr math, Loyola Univ, Ill, 66-67; fel, Sci Ctr, NAm Rockwell Corp, Calif, 67-68; asst prof, 68-74, ASSOC PROF MATH, HARVEY MUDD COL, 74- Mem: Am Math Soc. Res: Complex analysis; applied mathematics; fluid mechanics. Mailing Add: Dept of Math Harvey Mudd Col Claremont CA 91711

BUSER, KENNETH RENE, b Bloomington, Ind, Apr 13, 25; m 46; c 3. ORGANIC CHEMISTRY. Educ: Wabash Col, BA, 50; Purdue Univ, MS, 52, PhD(chem), 54. Prof Exp: Asst, Purdue Univ, 50-54; res chemist, Exp Sta, 54-69, STAFF CHEMIST, FABRICS & FINISHES DEPT, E I DU PONT DE NEMOURS & CO, INC, 69- Mem: Am Chem Soc. Res: Organic chemistry of sulfur, polymer chemistry. Mailing Add: 2813 Ambler Ct Skyline Crest Wilmington DE 19808

BUSER, MARY PAUL, b Wichita, Kans, Sept 7, 28. MATHEMATICS. Educ: Marymount Col, Kans, AB, 57; St Louis Univ, AM, 59, PhD(math), 61. Prof Exp: From instr to assoc prof math, 61-75, PRES, MARYMOUNT COL KANS, 75- Concurrent Pos: NSF, Inst Comput Sci, Univ Mo, Rolla, 64; partic, Asn Comput Mach, Inst Comput Sci, Purdue Univ, 71. Mem: Math Asn Am; Am Math Soc. Res: Approximating polynomial for the number of partitions of a positive integer into unit and prime summands; Waring's problem for congruences. Mailing Add: Off of Pres Marymount Col Salina KS 67401

BUSEY, RICHARD HOOVER, b Cairo, Ill, Apr 6, 19; m 42; c 4. PHYSICAL CHEMISTRY, THERMODYNAMICS. Educ: Southern Methodist Univ, BS, 41; Univ Calif, PhD(chem), 50. Prof Exp: Res anal chemist, Magnolia Petrol Co, 41-45; assoc, Univ Calif, 50-52; RES CHEMIST, OAK RIDGE NAT LAB, 52- Mem: AAAS; Am Chem Soc. Res: Low temperature calorimetry; spectroscopy; chemistry of Technetium and Rhenium; high temperature aqueous inorganic species equilibria. Mailing Add: Oak Ridge Nat Lab PO Box X Oak Ridge TN 37830

BUSH, ALFRED LERNER, b Rochester, NY, Dec 21, 19; m 42, 65; c 6. ECONOMIC GEOLOGY, ENVIRONMENTAL GEOLOGY. Educ: Univ Rochester, AB, 41, MS, 46. Prof Exp: GEOLOGIST, US GEOL SURV, 46- Concurrent Pos: Sci secy for

mineral resources, US deleg, UN Conf Sci & Technol, Switz, 63. Mem: Fel Geol Soc Am; Soc Econ Geol; Am Inst Mining Eng. Res: Uranium deposits of the Colorado Plateau; geology of the western San Juan Mountains, Colorado; economic geology of lightweight aggregates, United States and worldwide. Mailing Add: US Geol Surv MS 491 Box 25046 Denver Fed Ctr Denver CO 80225

BUSH, C ALLEN, b Rochester, NY, Aug 2, 38. BIOPHYSICAL CHEMISTRY. Educ: Cornell Univ, BA, 61; Univ Calif, Berkeley, PhD(chem), 65. Prof Exp: Res assoc chem, Cornell Univ, 66-68; asst prof, 68-73, ASSOC PROF CHEM, ILL INST TECHNOL, 73- Mem: Am Soc Biol Chemists. Res: Optical rotatory dispersion and circular dichroism of biopolymers; oligosaccharide structure in membrane glycoproteins. Mailing Add: Dept of Chem Ill Inst of Technol Chicago IL 60616

BUSH, DAVID CLAIR, b Malad City, Idaho, Nov 6, 22; m 44; c 4. ORGANIC CHEMISTRY. Educ: Univ Idaho, BS, 48; Ore State Col, MS, 50, PhD(org chem), 53. Prof Exp: Res chemist, Pittsburgh Plate Glass Co, 52-54; VPRES, IDAHO CHEM INDUSTS, 54- Mem: Am Chem Soc. Res: Metal pyridine salts; Mannich reaction; synthetic polymers; polyesters and epoxy resins. Mailing Add: 6820 McMullen Boise ID 83705

BUSH, DAVID GRAVES, b Westfield, Mass, Apr 21, 22; m 45; c 3. ANALYTICAL CHEMISTRY. Educ: Univ Mass, BS, 47; Univ Minn, MS, 50. Prof Exp: Asst synthetic rubber res, Univ Minn, 52; from res chemist to sr res chemist, 52-64, res assoc anal res & develop, 64-75, LAB HEAD, EASTMAN KODAK RES LABS, 75- Mem: Am Chem Soc. Res: Nonaqueous titrimetry; functional group analysis; separations chemistry; polymer identification and analysis; combustion analysis; atomic absorption spectroscopy. Mailing Add: Res Labs Kodak Park Bldg 82C Rochester NY 14650

BUSH, EVERETT HOMER b Westfield, Mass, Apr 18, 18; m 43. GEOGRAPHY. Educ: Mass State Teachers Col, Westfield, BSE, 42; Clark Univ, MA, 47. Prof Exp: From asst prof to assoc prof geog & chmn dept earth sci, 47-70, chmn dept geog, 64-70, PROF GEOG, WITTENBERG UNIV, 74- Concurrent Pos: Res sabbatical, associated with Univ Newcastle, 71. Mem: Nat Coun Geog Educ; Asn Am Geogr; Geog Soc NSW. Res: Economic geography; geography of Anglo-America. Mailing Add: Dept of Geog Wittenberg Univ Springfield OH 45501

BUSH, FRANCIS M, b Bloomfield, Ky, Sept 5, 33; m 59. ZOOLOGY. Educ: Univ Ky, BS, 55, MS, 57; Univ Ga, PhD(zool), 62. Prof Exp: From asst prof to assoc prof biol, Samford Univ, 60-64; asst prof, 64-69, ASSOC PROF ANAT, MED COL VA, VA COMMONWEALTH UNIV, 69- Concurrent Pos: Mary Glide Goethe travel awards, 59-60; Sigma Xi res grant, 63; Am Philos Soc grant, 64; USPHS grant, 65-68. Mem: Am Physiol Soc; Soc Gen Physiol; Am Asn Anat. Res: Relationship of pineal gland and antigonadotrophin in small laboratory mammals; homotransplantation; plasma protein biosynthesis and tissue enzymes of mammals and birds. Mailing Add: Health Sci Div Med Col of Va Va Commonwealth Univ Richmond VA 23219

BUSH, GEORGE CLARK, b St Catharines, Ont, Oct 18, 30; m 60; c 1. MATHEMATICS. Educ: McMaster Univ, BA, 54; Mass Inst Technol, SM, 56; Queen's Univ, Ont, PhD(math), 61. Prof Exp: Asst prof math, Queen's Univ, Ont, 61-67; assoc prof, Beirut Col Women, 67-71; spec lectr, Queen's Univ, Ont, 71-74; MEM FAC, COL ARTS & SCI, PAHLAVI UNIV, 74- Mem: Math Asn Am; Can Math Cong. Res: Algebra; numerical analysis. Mailing Add: Col of Arts & Sci Pahlavi Univ Shiraz Iran

BUSH, GUY L, b Greenfield, Iowa, July 9, 29; m 59; c 2. EVOLUTIONARY BIOLOGY. Educ: Iowa State Col, BS, 53; Va Polytech Inst, MS, 60; Harvard Univ, PhD(biol), 64. Prof Exp: Entomologist, USDA, 55-57; NIH fel, Univ Melbourne, 64-66; asst prof, 66-73, ASSOC PROF ZOOL, UNIV TEX, AUSTIN, 73- Mem: AAAS; Soc Study Evolution; Genetics Soc Am; Entom Soc Am; Soc Syst Zool. Res: Systematics of higher Diptera; evolutionary cytogenetics; mechanisms of sex determination; ecological and genetical aspects of phytophagous insect evolution. Mailing Add: Dept of Zool Univ of Tex Austin TX 78712

BUSH, IAN, b Briston, Eng, May 25, 28; m 67; c 5. PHYSIOLOGY, BIOCHEMISTRY. Educ: Cambridge Univ, BA, 49, PhD(endocrinol), 53, MB, BChir, 57. Prof Exp: Res asst, St Mary's Hosp, London, Eng, 53-56, asst endocrinol, External Sci Staff Med Res Coun, 58-60; prof physiol & chmn dept, Birmingham Univ, 60-64; sr scientist endocrinol, Worcester Found Exp Biol, 64-67; prof physiol & chmn dept, Med Col Va, 6770; PROF PHYSIOL, SCH MED, NY UNIV, 70- Concurrent Pos: Consult, Imp Chem Industs, Eng, 60-64; mem, Biol Res Awards Comt, Brit Med Res Coun, 62-64; vpres & dir res & develop, Cybertek, Inc, 70-73. Mem: AAAS; Am Physiol Soc; Endocrine Soc; fel Am Acad Arts & Sci; Brit Biochem Soc. Res: Analytical methods for steroids; secretion of adrenal cortex; metabolism of steroids; mechanism of action of steroid hormones; theory and practice of chromatography; automation of chromatographic estimation methods; neurochemistry. Mailing Add: Dept of Physiol NY Univ Sch of Med New York NY 10016

BUSH, JAMES, b New York, NY, Jan 10, 20; div; c 1. OCEANOGRAPHY. Educ: NY Univ, AB, 41; Ind Univ, AM, 49; Univ Wash, PhD, 58. Prof Exp: Meteorologist, Pan Am Airways, 42-43; geophysicist, United Geophys Co, 43-44; instr geol, Univ Miami, 47-48; asst prof, Univ Houston, 52-53; res geologist, Agr & Mech Col, Tex, 53-54; consult geologist, 54-60; consult geosci, Gen Elec Co, 60-63; sci adv, IIT Res Inst, 63-64; dir res, Ocean Systs, Inc, 64-65; pres, Marine Int, Inc, 65-69, PRES ACAD MARINE SCI, INC, 69- Mem: Fel AAAS; fel Geol Soc Am; Am Asn Petrol Geol; Paleont Soc; World Maricult Soc. Res: Marine geology and engineering; oceanographic systems; marine industries; aquaculture. Mailing Add: Box 2936 Palm Beach FL 33480

BUSH, KAREN JEAN, b Evansville, Ind, Oct 5, 43. BIOCHEMISTRY, ANALYTICAL CHEMISTRY. Educ: Monmouth Col, Ill, BA, 65; Ind Univ, Bloomington, PhD(biochem), 70. Prof Exp: Fel biol, Univ Calif, Santa Barbara, 70-71; instr biochem, Sch Med, Univ NC, Chapel Hill, 71-72; asst prof chem, Univ Del, 72-73; RES INVESTR, SQUIBB INST MED RES, 73- Mem: Am Chem Soc; AAAS. Res: Deuterium isotope effects with alcohol dehydrogenase; analytical biochemistry; high pressure liquid chromatography. Mailing Add: 651 Ridge Rd Monmouth Junction NJ 08852

BUSH, KENNETH ARTHUR, b Oneonta, NY, Feb 24, 14; m 42; c 3. MATHEMATICS. Educ: Columbia Univ, BA, 36, MA, 39; Univ NC, PhD(math statist), 50. Prof Exp: Instr math, US Naval Acad, 41-46; assoc prof, State Univ NY, 46-48 & 50-52; asst prof math statist, Univ Ill, 52-54; prof math & head dept, Univ Idaho, 54-61; PROF MATH, WASH STATE UNIV, 61- Concurrent Pos: Vis prof statist, Australian Nat Univ, 68; exchange prof, Univ Paris, 69. Mem: Am Math Asn; Am Inst Math Statist; Opers Res Soc Am; Economet Soc. Res: Mathematical statistics; combinatorial problems; matrices; mathematical economics. Mailing Add: Dept of Math Wash State Univ Pullman WA 99163

BUSH, LAURENS EARLE, b Martins, SC, Mar 24, 00; m 20; c 2. MATHEMATICS. Educ: The Citadel, BS, 19; Univ NC, SM, 26; Ohio State Univ, PhD(math), 31. Prof Exp: Teacher & prin pub schs, SC, 21-25; instr, Univ NC, 26-30 & Ohio State Univ, 31-33; prof math & chmn dept, Col St Thomas, 33-53; prof, 53-70, chmn dept, 53-64, EMER PROF MATH, KENT STATE UNIV, 70- Concurrent Pos: Dir, William L Putnam Math Competition, Math Asn Am, 48-65. Mem: AAAS; Am Math Soc; Math Asn Am. Res: Linear algebra. Mailing Add: 408 Burr Oak Dr Kent OH 44240

BUSH, LEON F, b Brooksville, Ky, Jan 28, 24; m 49; c 4. ANIMAL NUTRITION. Educ: Univ Ky, BS, 50, MS, 51; Cornell Univ, PhD(nutrit & physiol), 54. Prof Exp: ASSOC PROF ANIMAL SCI, S DAK STATE UNIV, 54- Mem: Am Soc Animal Sci. Res: Sheep management and nutrition; physiology of reproduction studies with sheep. Mailing Add: Dept of Animal Sci SDak State Univ Brookings SD 57006

BUSH, LINVILLE JOHN, b Winchester, Ky, May 7, 28; m 52; c 4. DAIRY NUTRITION. Educ: Univ Ky, BSc, 48; Ohio State Univ, MSc, 49; Iowa State Col, PhD(dairy nutrit), 58. Prof Exp: Field agent, Dairying, Ky, 53-55; ASSOC PROF DAIRY SCI, OKLA STATE UNIV, 58- Mem: Am Soc Animal Sci; Am Dairy Sci Asn. Res: Nutrition of young dairy calf; dietary requirements of lactating dairy cows; factors affecting rumen function. Mailing Add: Dept of Animal Sci & Indust Okla State Univ Stillwater OK 74074

BUSH, LOUISE FULTON, b Wichita, Kans, Feb 17, 07; m 36; c 2. MARINE BIOLOGY, TAXONOMY. Educ: Friends Univ, AB, 28; Univ Kans, MA, 31; Univ Minn, PhD(entomol), 38. Prof Exp: Asst, Univ Minn, 31-36, zool artist, 36-42; instr zool, Upsala Col, 55-57; from instr to prof, 56-72, EMER PROF ZOOL, DREW UNIV, 72- Concurrent Pos: Biol artist free lance, Chicago & New York publishers, 42-; sr vis invest, Syst-Ecol Prog, Marine Biol Lab, Woods Hole, 64-72. Mem: AAAS; Am Soc Zool; Am Micros Soc; Soc Syst Zool; Am Inst Biol Sci. Res: General invertebrate zoology; taxonomy of Turbellaria. Mailing Add: Dept of Zool Drew Univ Madison NJ 07940

BUSH, LOWELL PALMER, b Tyler, Minn, Mar 31, 39; m 62; c 3. PLANT PHYSIOLOGY. Educ: Macalester Col, BA, 61; Iowa State Univ, MS, 63, PhD(plant physiol), 64. Prof Exp: Fel plant path, Univ Minn, 64-66; assoc prof, 66-75, PROF AGRON, UNIV KY, 75- Mem: Am Soc Plant Physiol; Weed Sci Soc Am; Am Soc Agron. Res: Alkaloid metabolism in plants. Mailing Add: Dept of Agron Univ of Ky Lexington KY 40506

BUSH, MARTIN BRUCE, b Tex, Oct 9, 14; m 40. ANALYTICAL CHEMISTRY, CLINICAL CHEMISTRY. Educ: Wash Missionary Col, BA, 48; Univ Md, MS, 52. Prof Exp: Teacher pub schs, La, Ark & NC, 38-42; from instr to asst prof, Columbia Union Col, 48-65; clin chemist, Wash Sanitarium & Hosp, 65-69; DEVELOP COORD BIOCHEM, N ENG MEM HOSP, 69- Mem: AAAS; Am Chem Soc; Am Asn Clin Chem. Res: Photovoltaic effect; dairy foods; radiochemical analysis of liquid metals; insecticide research in dairy foods; analytical methods of clinical chemistry. Mailing Add: 146 Aldrich Rd Wilmington MA 01887

BUSH, MILTON TOMLINSON, b Watertown, Mass, May 16, 07; m 40; c 1. PHARMACOLOGY. Educ: Cornell Univ, BCh, 29, PhD(org chem), 38. Prof Exp: Asst phys chem, Cornell Univ, 29-35; from res assoc to assoc prof, 35-61, PROF PHARMACOL, SCH MED, VANDERBILT UNIV, 61- Mem: AAAS; Am Chem Soc; Am Soc Pharmacol & Exp Therapeut. Res: Chemotherapy; antibiotics; bacteriophage; antisepsis; anesthesia; drug metabolism; chemical toxicology. Mailing Add: Vanderbilt Univ Sch of Med Nashville TN 37203

BUSH, NORMAN, b New York, NY, Dec 10, 29; m 52; c 3. STATISTICS. Educ: City Col New York, BBA, 51, MBA, 52; NC State Col, PhD, 62. Prof Exp: Statistician, Army Chem Ctr, Md, 52-56, Patrick Air Force Base, Fla, 56-61 & Instrument Corp, Fla, 61-63; res dir statist, D Brown Assocs, Fla, 63-64; prin engr, Pan Am World Airways, 64-72; DIR DATA & COMPUT SCI DIV, ENSCO, INC, FLA, 72- Concurrent Pos: Statistician, Comput Ctr, Univ NC, 59-62; adj asst prof, Grad Eng Educ Syst, Univ Fla, Cape Kennedy, 67-70. Mem: Am Statist Asn; Opers Res Soc Am. Res: Technometrics; regression analysis; experimental design; computers. Mailing Add: ENSCO Inc 1127 S Patrick Dr Suite 21 PO Box 2578 Satellite Beach FL 32937

BUSH, OAKLEIGH ROSS, b Aliquippa, Pa, June 3, 23. GEOGRAPHY. Educ: Univ Pittsburgh, AB, 47, LittM, 48; Shakespeare Inst, Eng, dipl, 54; Ind Univ, MAT, 64. Prof Exp: Instr eng, Carnegie-Mellon Univ, 48-49; instr, Ind Univ, 49-51; asst prof, Samford Univ, 53-55; Fulbright teacher, Rotterdam, Netherlands, 55-56; lectr, Univ Md Overseas Prog, 56-57; prin, Am Sch Rotterdam, 60-61; assoc prof, Miss State Col Women, 61-64; ASST PROF GEOG, TEX CHRISTIAN UNIV, 64- Concurrent Pos: Addison H Gibson Found res grant, Univ Pittsburgh, Pa, 48. Mem: Asn Am Geog. Res: The effect of Common Market membership on the lives of individuals in the member nations. Mailing Add: Dept of Geog Tex Christian Univ Ft Worth TX 76129

BUSH, POWELL DANIEL, JR, b Athens, Ga, Oct 2, 24; m 48, 70; c 2. PHYSICS. Educ: Emory Univ, BS, 48, MA, 51. Prof Exp: Instr physics, Morehouse Col, 51-52; engr, Lockheed Aircraft Corp, 52-58; ASST PROF PHYSICS & ACTG CHMN DEPT, MERCER UNIV, 58- Mem: Exp Aircraft Asn; Am Asn Physics Teachers. Res: Infrared spectroscopy. Mailing Add: Dept of Physics Mercer Univ Macon GA 31207

BUSH, RICHARD WAYNE, b Cleveland, Ohio, Nov 5, 34; m 69. POLYMER CHEMISTRY. Educ: Mass Inst Technol, SB, 56; Univ Ill, PhD(org chem), 60. Prof Exp: Sr res chemist, 60-69, res supvr, 69-72, RES ASSOC, W R GRACE & CO, 72- Mem: Am Chem Soc. Res: Photopolymers; adhesives. Mailing Add: W R Grace & Co 7379 Rte 32 Columbia MD 21044

BUSH, STEWART FOWLER, b Charlotte, NC, Jan 8, 41; m 65; c 2. PHYSICAL CHEMISTRY. Educ: Erskine Col, AB, 63; Univ SC, PhD(phys chem), 67. Prof Exp: Asst prof chem, Erskine Col, 67-69; asst prof, 69-74, ASSOC PROF CHEM, UNIV NC, CHARLOTTE, 74- Mem: Am Chem Soc; Coblentz Soc. Res: Infrared, far-infrared and Raman spectroscopy as applied to vibrational analysis and molecular structure determination. Mailing Add: Dept of Chem Univ of NC UNCC Sta Charlotte NC 28213

BUSH, WALTER MONROE, b Huntingdon, Pa, Jan 31, 29; m 49; c 5. ORGANIC CHEMISTRY. Educ: Juniata Col, BS, 51; Rutgers Univ, PhD, 55. Prof Exp: Asst, Rutgers Univ, 51-53; res chemist, 55-63, RES ASSOC, EASTMAN KODAK CO, 63- Concurrent Pos: Asst lectr, Univ Rochester, 58-59. Res: Color photography. Mailing Add: Eastman Kodak Co Kodak Park B-59 Rochester NY 14650

BUSH, WARREN VAN NESS, b Montclair, NJ, June 10, 31; m 57; c 3. PETROLEUM CHEMISTRY, CHEMICAL ENGINEERING. Educ: Princeton Univ, BS, 53; Calif Inst Technol, PhD, 58. Prof Exp: Chemist, Shell Develop Co, Calif, 57-65, engr, Shell Oil Co, NY, 65-67, chemist, Shell Develop Co, 67-68, supvr, 68, group leader, Shell Oil, Martinez Refinery, 68-72, sr res eng, Shell Develop Co, 72-74, supvr staff, 74-76, STAFF RES ENGR, SHELL DEVELOP CO, 76- Mem: AAAS; Am Chem Soc; Am Inst Chem Eng. Res: Organic chemistry and chemical engineering, principally as applied to petroleum refining and air pollution control. Mailing Add: Shell Develop Co PO Box 1380 Houston TX 77001

BUSHAW, DONALD (WAYNE), b Anacortes, Wash, May 5, 26; m 46; c 5. MATHEMATICS. Educ: State Col Wash, BA, 49; Princeton Univ, PhD(math), 52. Prof Exp: Instr math, State Col Wash, 48-49; asst, Princeton Univ, 49-51; math consult, Stevens Inst Technol, 51-52; from instr to assoc prof, 52-62, actg chmn dept, 66-68, PROF MATH, WASH STATE UNIV, 62- Concurrent Pos: Vis scientist, Res Inst Advan Study, 62-63; NAS-PAN exchangee, Jagiellonian Univ, Cracow, Poland, 72-73. Mem: Am Math Soc; Math Asn Am; Soc Indust & Appl Math. Res: Ordinary differential equations. Mailing Add: PO Box 106 Pullman WA 99163

BUSHEY, ALBERT HENRY, b Mansfield, Ohio, Mar 18, 11; m 40; c 2. PHYSICAL CHEMISTRY, ANALYTICAL CHEMISTRY. Educ: Wittenberg Col, AB, 32; Univ Minn, PhD(phys chem), 40. Prof Exp: Asst chem, Univ Minn, 36-40; chemist, Res Lab, Aluminum Co Am, 40-44, asst chief anal div, 44-48; head anal res, Hanford Atomic Prod Oper, Gen Elec Co, 48-53, head chem res, 53-54, instr, Sch Nuclear Eng, 49-54, consult chemist, Gen Eng Lab, 54-62; head finishing res, 62-67, TECH SUPVR, ALUMINUM RES DIV, KAISER ALUMINUM & CHEM CORP, 67- Concurrent Pos: Instr, Pa State Univ, 42-43. Mem: Am Chem Soc. Res: Coprecipitation and aging of precipitates; chemistry and analyses of aluminum, uranium, transuranium and fission products; instrumental analyses; industrial separations processes for uranium, transurancis and fission products; electrochemistry; finishing of aluminum. Mailing Add: Kaiser Aluminum & Chem Corp Ctr for Technol PO Box 870 Pleasanton CA 94566

BUSHEY, GORDON LAKE, b Mission, Tex, Jan 13, 22; m 50; c 3. PHYSICAL CHEMISTRY. Educ: Rice Univ, BS, 43, MA, 44, PhD(chem), 48. Prof Exp: Instr chem, Univ Ill, 48-51; sci asst, Off Chief Chem Officer, 51-62, asst for sci affairs, US Defense Off for NAtlantic & Medit Areas, Paris, 62-64, asst chief scientist, US Army Materiel Command, 64-70, sci asst to dep foreign labs, 70-75, PHYS SCI ADMINR, OFF ASST DEP SCI & TECHNOL, US ARMY MATERIEL COMMAND, US DEPT ARMY, 75- Concurrent Pos: Exec officer, Div Chem & Chem Technol, Nat Res Coun, 51-53 & Nat War Col, 59-60. Mem: AAAS; Am Chem Soc. Res: Absorption by hydrous oxides; physical chemistry of soap crystals; x-ray diffraction studies of colloids and mixed crystals; arms control; international science organizations; technical information; physical sciences of military interest. Mailing Add: US Army Materiel Command 5001 Eisenhower Ave Alexandria VA 22333

BUSHEY, WILLIAM RAYMOND, b Lake Charles, La, Jan 9, 46; m 71. INORGANIC CHEMISTRY, CHEMICAL KINETICS. Educ: La State Univ, BS, 67; Iowa State Univ, PhD(inorg chem), 72. Prof Exp: Fel inorg chem, Cath Univ Am, 72-74; RES CHEMIST, E I DU PONT DE NEMOURS & CO, 74- Mem: Am Chem Soc. Res: Mechanisms of inorganic reactions, especially reactions involving carbon-metal bonds; relation between physical properties and effectiveness of refractory materials. Mailing Add: Exp Sta E I du Pont de Nemours & Co Wilmington DE 19898

BUSHKOVITCH, ALEXANDER VIATCHESLAV, b Leningrad, Russia, June 15, 06; nat US; m 41; c 1. THEORETICAL PHYSICS. Educ: Univ Pa, AB, 30, MS, 31, PhD(physics), 34. Prof Exp: Asst math, Univ Pa, 35-36; instr, Drexel Inst, 36-38; assoc prof physics & philos, Col Charleston, 38-47; assoc prof, 47-54, PROF PHYSICS, ST LOUIS UNIV, 54- Concurrent Pos: Physicist, Ballistic Res Lab, Aberdeen Proving Ground, Md, 43-45. Mem: Am Phys Soc; Am Asn Physics Teachers; Philos Sci Asn. Res: Quantum mechanics; mathematical analysis; philosophy of science; microwave spectroscopy; beta decay theory; group theory; elementary particle theory. Mailing Add: Dept of Physics St Louis Univ St Louis MO 63103

BUSHLAND, RAYMOND CECIL, b Browns Valley, Minn, Oct 5, 10; m 39; c 2. ENTOMOLOGY. Educ: SDak State Univ, BS, 32, MS, 34; Kans State Univ, PhD, 53. Prof Exp: Entomologist, USDA, 35-74; RETIRED. Concurrent Pos: Consult, Agr Res Serv, USDA, 74- & Nat Acad Sci, 75- Honors & Awards: US Typhus Comn Medal, 45; Nat Hide Asn Medal, 58; Distinguished Serv Award, Tex & Southwest Cattle Raisers Asn, 59; Hoblitzelle Nat Award, 60; John W Scott Medal, 61; Acad Achievement, Golden Plate Award, 62; Plaque, Southwest Animal Health Res Found, 63; Am Farm Bur Fedn Award, 64; USDA Award for Distinguished Serv, 67; Founders Mem Award, Entom Soc Am, 74; Distinguished Serv in Agr Award, Kans State Univ, 74. Mem: Fel AAAS; Entom Soc Am. Res: Biology and control of insects affecting man and animals; sterile male technique for insect control. Mailing Add: 1514 Xanthisma McAllen TX 78501

BUSHMAN, DONALD OTTO, geography, see 12th edition

BUSHMAN, JESS RICHARD, b American Fork, Utah, May 12, 21; div; c 5. GEOLOGY. Educ: Brigham Young Univ, BA, 49; Princeton Univ, PhD, 58. Prof Exp: From asst prof to assoc prof, 55-70, PROF GEOL, BRIGHAM YOUNG UNIV, 70- Concurrent Pos: Consult-geologist, Ministry Mines & Hydrocarbons, Venezuela, 53-55, 58 & 59-60. Mem: Geol Soc Am; Am Asn Petrol Geol; Asn Mining & Petrol Geol Venezuela; Nat Asn Geol Teachers; Am Asn Stratig Palynologists. Res: Sedimentation; polymology. Mailing Add: Dept of Geol Brigham Young Univ Provo UT 84601

BUSHMAN, JOHN BRANSON, b Farmington, NMex, Aug 26, 26; m 51; c 5. VERTEBRATE ZOOLOGY. Educ: Univ Utah, BS, 52, MS, 55. Prof Exp: Asst cur birds, Univ Utah, 46 & 49-52; from asst ecologist to ecologist, Univ Utah, 54-62, zoologist, 62-72; ZOOLOGIST, US ARMY CORPS ENGRS, WASHINGTON, 72- Mem: Wilson Ornith Soc; Cooper Ornith Soc; Am Ornith Union; Am Soc Mammal; Wildlife Soc. Res: Endangered species; avian and mammalian ecology; life histories; distribution and population dynamics. Mailing Add: US Army Corps of Engrs Civil Works Washington DC 20314

BUSHNELL, DAVID L, b Platteville, Wis, Apr 14, 29; m; c 3. PHYSICS. Educ: Univ Wis, BS, 51, MS, 53; Va Polytech Inst, PhD(physics), 61. Prof Exp: Instr physics, Mankato State Col, 53-55; from instr to asst prof, Va Polytech Inst, 56-61; ASSOC PROF PHYSICS, NORTHERN ILL UNIV, 61- Mem: Am Phys Soc; Am Nuclear Soc; Am Asn Physics Teachers. Res: Neutron capture gamma-ray studies; gamma-ray spectroscopy and nuclear level schemes. Mailing Add: Dept of Physics Northern Ill Univ DeKalb IL 60115

BUSHNELL, GORDON WILLIAM, b Oxford, Eng, June 17, 36; m 59; c 2. STRUCTURAL CHEMISTRY, CRYSTALLOGRAPHY. Educ: Oxford Univ, MA, 59, BSc, 63; Univ WI, PhD(chem), 66. Prof Exp: Sci officer, UK Atomic Energy Auth, 59-63; lectr chem, Univ WI, 63-67; ASSOC PROF CHEM, UNIV VICTORIA, 67- Mem: Brit Chem Soc; Am Crystallog Asn. Res: X-ray crystallography of coordination compounds; structure determination by single crystal x-ray diffraction. Mailing Add: Dept of Chem Univ of Victoria Victoria BC Can

BUSHNELL, JOHN HEMPSTEAD, b Minneapolis, Minn, May 3, 22; m 42; c 2. ANTHROPOLOGY. Educ: Univ Calif, AB, 48, PhD(anthrop), 55. Prof Exp: Instr anthrop, Vassar Col, 54-55, res assoc, Mary Conover Mellon Found, 55-59; dir res, State Comn for Human Rights, NY, 60-70; PROF ANTHROP, BROOKLYN COL, 70- Mem: Fel Am Anthrop Asn; Ethnol Soc; Soc Appl Anthrop; NY Acad Sci. Res: Ethnology of mesoamerica; culture and personality; US culture. Mailing Add: Dept of Anthrop Brooklyn Col Brooklyn NY 11210

BUSHNELL, JOHN HORACE, b Grand Rapids, Mich, Mar 17, 25; m 51; c 4. INVERTEBRATE ZOOLOGY. Educ: Vanderbilt Univ, BA, 48; Mich State Univ, MS, 56, PhD(zool), 61. Prof Exp: US State Dept spec lectr biol & English, Habibia Col, Kabul, Afghanistan, 48-50; asst secy & sales mgr, Wagemaker Boat Co & US Molded Shapes, Inc, 50-55; instr zool, Mich State Univ, 59-60; asst prof biol, Washington & Jefferson Col, 60-64; from asst prof to assoc prof, 64-71, PROF ENVIRON, POP & ORGANISMIC BIOL, UNIV COLO, BOULDER, 71- Concurrent Pos: AEC summer study & continuing isotopes res grant, 63-; consult ed, Barnes & Noble Publ House, 70- Mem: Ecol Soc Am; Am Soc Zool; Am Micros Soc; Animal Behav Soc; Marine Biol Asn UK. Res: Ecology, zoogeography and systematics of Ectoprocta; phenotypic plasticity, adaptation and physiological ecology of aquatic invertebrates; nutritional studies and sexual cycle studies on Ectoprocta and other invertebrates; Ectoproct population studies. Mailing Add: Dept of Biol Univ of Colo Boulder CO 80302

BUSHNELL, KENT O, b Tolland, Conn, Dec 12, 29; m 55. ENVIRONMENTAL GEOLOGY, GEOPHYSICS. Educ: Univ Conn, BA, 51; Yale Univ, MS, 52, PhD(geol), 55. Prof Exp: Geologist, Stand Oil Co Calif, 55-68; assoc prof, 68-74, PROF GEOL, SLIPPERY ROCK STATE COL, 74- Concurrent Pos: Dir, NSF seismic equip grant, 70-72. Mem: Am Asn Petrol Geol; Geol Soc Am; Soc Explor Geophys. Res: Gravity surveys for buried valleys; Pleistocene history of western Pennsylvania. Mailing Add: Dept of Geol Slippery Rock State Col Slippery Rock PA 16057

BUSHNELL, ROBERT HEMPSTEAD, b Wooster, Ohio, May 11, 24; m 65; c 2. ATMOSPHERIC PHYSICS. Educ: Ohio State Univ, BSc, 47, MSc, 48; Univ Wis, PhD(meteorol), 62. Prof Exp: Physicist, Hoover Co, 48-50, Goodyear Aircraft Co, 50-56, Radio Corp Am, 57-59, Univ Wis, 58-62 & Nat Ctr Atmospheric Res, Boulder, Colo, 62-74; CONSULT, 74- Concurrent Pos: Univ Corp Atmospheric Res fel, 61. Mem: AAAS; Inst Elec & Electronics Engrs; Am Meteorol Soc; Am Geophys Union; Int Solar Energy Soc. Res: Climatology and statistics of sunshine and temperature related to solar heating; thunderstorm structure; weather radar. Mailing Add: 502 Ord Dr Boulder CO 80303

BUSHNELL, VERNON CLIFFORD, b Salem, Ore, Aug 8, 12; m 34; c 2. SOIL CHEMISTRY. Educ: Willamette Univ, AB, 34; Ore State Col, MS, 36, PhD(soils, agr chem), 39. Prof Exp: Asst, Ore State Col, 34-38, anal chemist, Exp Sta, 35-39; chemist, Charlton Labs, 39-41; soil scientist, Soil Conserv Serv, USDA, Univ Idaho, 41-46; soil scientist, US Bur Reclamation, 46-59, chemist & chief regional soil & water lab, 59-67, phys scientist & regional water qual & pollution control coordr, 67-73; RETIRED. Concurrent Pos: Consult, Trail Smelter Co, Can, 37; asst, Exten Serv, Ore State Col, 39. Mem: Am Chem Soc; Nat Asn Corrosion Engrs; AAAS. Res: Physical and analytical chemistry; heats of solutions; fertility and irrigability of soils. Mailing Add: 3631 W Clement Rd Boise ID 83704

BUSHNELL, WILLIAM RODGERS, b Wooster, Ohio, Aug 19, 31; m 52; c 3. PLANT PHYSIOLOGY, PHYTOPATHOLOGY. Educ: Univ Chicago, AB, 51; Ohio State Univ, BS, 53, MS, 55; Univ Wis, PhD(bot), 60. Prof Exp: PLANT PHYSIOLOGIST, USDA, 60- Mem: Am Soc Plant Physiol; Am Phytopath Soc. Res: Physiology of parasitism, especially rust and mildew diseases of cereals. Mailing Add: Cereal Rust Lab Univ of Minn St Paul MN 55101

BUSHONG, ALLEN DAVID, b West Palm Beach, Fla, Mar 25, 31; m 71. GEOGRAPHY. Educ: Univ Miami, BA, 52; Univ Fla, MA, 54, PhD(geog), 61. Prof Exp: Map researcher, Libr Cong, 54-56; from instr to asst prof geog, Bowling Green State Univ, 61-66; ASSOC PROF GEOG, UNIV SC, 66- Mem: Asn Am Geogr. Res: History of geography; information sources in geography; geography of Middle America. Mailing Add: Dept of Geog Univ of SC Columbia SC 29208

BUSHONG, JEROLD WARD, b Dayton, Ohio, Sept 22, 35; m 55; c 2. PLANT PATHOLOGY. Educ: Miami Univ, BA, 57; Univ Ill, MS, 59, PhD(plant path), 61. Prof Exp: Sr plant pathologist, Fungicide Res, Niagara Chem Div, FMC Corp, 61-66; asst dir pesticide res, W R Grace & Co, 66-67; sr agrichem scientist, 67-69, supvr agr eval, 69-70, mgr agrichem, 70-73, mgr biol & field develop, 73-75, MGR INT MKT & PROD DEVELOP, MINN MINING & MFG CO, 75- Concurrent Pos: Ed staff, Plant Dis Reporter, 74- Mem: Am Chem Soc; Am Phytopath Soc; Weed Sci Soc Am. Res: Fungicide chemistry; legume and fruit pathology; mycology; entomology. Mailing Add: Minn Mining & Mfg Co 3M Ctr St Paul MN 55101

BUSHONG, STEWART CARLYLE, b Washington, DC, Nov 25, 36; m 58; c 3. RADIOLOGICAL HEALTH. Educ: Univ Md, BS, 59; Univ Pittsburgh, MS, 63, ScD(radiol health), 67. Prof Exp: Health serv officer, USPHS, Washington, DC, 59-61; health physicist, Univ Pittsburgh, 62-64; asst prof, 66-71, ASSOC PROF RADIOL SCI, BAYLOR COL MED, 71- Concurrent Pos: Assoc prof radiol, Univ Tex Dent Br, 70-; assoc prof radiol sci, Houston Community Col, 71- Mem: Am Asn Physicists Med; Health Physics Soc; Soc Nuclear Med; Am Indust Hyg Asn; Radiation Res Soc. Res: Investigations of radiation dose and dose distribution in patients receiving radiologic examination for medical diagnosis. Mailing Add: Dept of Radiol Baylor Col of Med Houston TX 77030

BUSHUK, WALTER, b Pruzana, Poland, Jan 2, 29; Can citizen; m 55; c 2. CEREAL CHEMISTRY. Educ: Univ Man, BSc, 52, MSc, 53; McGill Univ, PhD(phys chem), 56. Prof Exp: Chemist, Grain Res Lab, Man, Can, 53-61, sect head, 61-62; dir, Ogilvie Flour Mills Co, Ltd, 62-64; res chemist, Grain Res Lab, 64, head wheat sect, 64-66; PROF PLANT SCI, UNIV MAN, 66- Concurrent Pos: Nat Res Coun Can overseas fel, Macromolecule Res Ctr, France, 57-58. Mem: Am Asn Cereal Chem; fel Chem Inst Can; Can Inst Food Technol; Can Res Mgt Asn. Res: Enzyme kinetics; surface properties of flour; physicochemical properties of polymers; mechanism of flour quality improvement; wheat proteins. Mailing Add: Dept of Plant Sci Univ of Man Winnipeg MB Can

BUSHWELLER, CHARLES HACKETT, b Port Jervis, NY, June 21, 39; m 61; c 3. PHYSICAL ORGANIC CHEMISTRY, STRUCTURAL CHEMISTRY. Educ: Hamilton Col, AB, 61; Middlebury Col, MS, 63; Univ Calif, Berkeley, PhD(org chem), 66. Prof Exp: Sr chemist, Mobil Chem Co, 66-68; asst prof, 68-72, ASSOC PROF CHEM, WORCESTER POLYTECH INST, 72- Prof Exp: Alfred P Sloan res fel, 71-73; Camille & Henry Dreyfus teacher scholar, 72-75. Mem: AAAS; Am Chem Soc; NY Acad Sci; Am Inst Chem. Res: Nuclear magnetic resonance spectroscopy; stereodynamics of organic and inorganic systems; mechanisms of carcinostatic and carcinogenic activity. Mailing Add: Dept of Chem Worcester Polytech Inst Worcester MA 01609

BUSING, WILLIAM RICHARD, b Brooklyn, NY, June 21, 23; m 51; c 3. PHYSICAL CHEMISTRY. Educ: Swarthmore Col, BA, 43; Princeton Univ, MA, 48, PhD(chem), 49. Prof Exp: Res assoc chem, Brown Univ, 49-51; instr, Yale Univ, 51-54; CHEMIST, OAK RIDGE NAT LAB, 54- Mem: Am Crystallog Asn (pres, 71). Res: Molecular structure and spectroscopy; neutron diffractions; crystallographic computation; interionic and intermolecular forces in crystals. Mailing Add: Chem Div Oak Ridge Nat Lab Oak Ridge TN 37830

BUSINGER, JOOST ALOIS, b Haarlem, Neth, Mar 29, 24; m 49; c 3. METEOROLOGY. Educ: State Univ Utrecht, BS, 47, MSc, 50, PhD(meteorol), 54. Prof Exp: Technician, Inst Heating Eng, Neth, 48-51 & Inst Hort Eng, 51-56; assoc & proj dir, Dept Meteorol, Univ Wis, 56-58; from asst prof to assoc prof meteorol, 58-65, chmn geophys exec comt, 63-65, PROF METEOROL, UNIV WASH, 65- Mem: Fel Am Meteorol Soc; Am Geophys Union; AAAS. Res: Energy transfer in atmosphere; turbulence; radiation; micro and physical meteorology. Mailing Add: Dept of Atmospheric Sci Univ of Wash Seattle WA 98105

BUSINGER, PETER ARTHUR, mathematics, computer science, see 12th edition

BUSKIRK, ELSWORTH ROBERT, b Beloit, Wis, Aug 11, 25; m 48; c 2. PHYSIOLOGY. Educ: St Olaf Col, BA, 50; Univ Minn, MA, 51, PhD(physiol hyg), 54. Prof Exp: Assoc physiol, Lab Physiol Hyg, Univ Minn, 54; physiologist, Environ Protection Res Div, Q Res & Educ Ctr, Mass, 54-56, chief environ physiol sect, 56-57; physiologist, Metab Dis Br, Inst Arthritis & Metab Dis, 57-63; DIR LAB HUMAN PERFORMANCE RES INST SCI & ENG, PA STATE UNIV, UNIVERSITY PARK, 63- Concurrent Pos: Mem thermal factors subcomt, Nat Acad Sci, 63-64 & human adaptability subcomt, Int Biol Prog, 68-; mem comt on interaction of eng with biol & med, Nat Acad Eng-Nat Acad Sci, 68-; mem subcomt on calories food & nutrit bd, Nat Acad Sci-Nat Res Coun, 68-70; mem epidemiol sect, Am Heart Asn; sect ed, J Appl Physiol, Am Physiol Soc, 73- Honors & Awards: Citation, Am Col Sports Med, 73. Mem: AAAS; Am Physiol Soc; Aerospace Med Soc; Am Inst Nutrit; Am Soc Heating, Refrigeration & Air-Conditioning Eng. Res: Metabolism; physiology of exercise; environmental physiology; growth and development and aging; epidemiology of coronary heart disease and obesity. Mailing Add: 119 Noll Lab Pa State Univ University Park PA 16802

BUSKIRK, FRED RAMON, b Indianapolis, Ind, Dec 29, 28; m 58. THEORETICAL PHYSICS. Educ: Western Reserve Univ, BS, 51, MS, 55, PhD(physics), 58. Prof Exp: Proj assoc physics, Univ Wis, 58-59; instr, Case Western Reserve Univ, 59-60; asst prof, 60-65, ASSOC PROF PHYSICS, US NAVAL POSTGRAD SCH, 65- Res: Nuclear and field theory. Mailing Add: US Naval Postgrad Sch Monterey CA 93940

BUSKIRK, RUTH ELIZABETH, b Indianapolis, Ind, Mar 9, 44; m 75; c 1. ANIMAL BEHAVIOR, ANIMAL ECOLOGY. Educ: Earlham Col, AB, 65; Harvard Univ, MAT, 66; Univ Calif, Davis, PhD(zool), 72. Prof Exp: Instr biol, Southern Univ, 66-68; fel, Univ Calif, Davis, 72-74; ASST PROF BEHAV, CORNELL UNIV, 74- Mem: Animal Behav Soc; Ecol Soc Am; Am Soc Zoologists; Am Soc Arachnologists; Brit Arachnologists Soc. Res: Animal social behavior; behavioral ecology of predators. Mailing Add: Neurobiol & Behav Langmuir Lab Cornell Univ Ithaca NY 14853

BUSLIK, ARTHUR J, b Philadelphia, Pa, Mar 7, 33. REACTOR PHYSICS, SOLID STATE PHYSICS. Educ: Univ Pa, BA, 54; Columbia Univ, MA, 56; Univ Pittsburgh, PhD(physics), 62. Prof Exp: From assoc scientist to sr scientist, Bettis Atomic Power Lab, Westinghouse Elec Corp, 58-72; adv nuclear engr, Gulf United Nuclear Fuels Corp, 72-74; ASSOC PHYSICIST, BROOKHAVEN NAT LAB, 74- Mem: Am Phys Soc; Am Nuclear Soc. Res: Theoretical solid state physics; magnetic resonance line width; kinetics and stability; neutron resonance absorption; variational principles in reactor physics. Mailing Add: Brookhaven Nat Lab Bldg 130 Upton NY 11973

BUSS, CHARLES DELEVAN, b Boise, Idaho, Jan 21, 10; m 44; c 2. FOOD SCIENCE, FOOD TECHNOLOGY. Educ: Univ Southern Calif, BA, 31; Univ Calif, MS, 52. Prof Exp: Asst res dir, Anabolic Food Prod, Inc, 35-42; chemist, V R Smith Olive Co, 47-49; tech supvr, Tillie Lewis Foods, Inc, 52-75; CONSULT, 75- Mem: Am Chem Soc; Inst Food Technologists; Am Soc Qual Control. Res: Chemistry and engineering and food processing; design and production of special dietary foods. Mailing Add: 35 W Ellis St Stockton CA 95204

BUSS, DENNIS DARCY, b Gainesville, Fla, July 11, 42; m 64; c 2. SOLID STATE ELECTRONICS, SEMICONDUCTORS. Educ: Mass Inst Technol, SB, 63, SM, 64, PhD(solid state physics), 68. Prof Exp: Asst prof elec eng, Mass Inst Technol, 68-69; res staff theoret physics, Tex Instruments Inc, 69-74; vis assoc prof elec eng, Mass Inst Technol, 75; BR MGR SOLID STATE RES, TEX INSTRUMENTS INC, 75- Concurrent Pos: Mem prog comts, Int Electron Device Meeting & Int Solid State Circuits Conf, 75-; assoc ed, Trans Electron Devices, Inst Elec & Electronics Engrs, 76. Res: Development of advanced, charge-coupled solid state structures for digital memory and analog signal processing. Mailing Add: MS 134 Tex Instruments Inc Dallas TX 75222

BUSS, EDWARD GEORGE, b Concordia, Kans, Aug 28, 21; m 49; c 2. POULTRY GENETICS. Educ: Kans State Col, BS, 43; Purdue Univ, MS, 49, PhD(poultry genetics), 56. Prof Exp: Asst prof poultry husb, Colo State Univ, 49-55, actg head dept, 50-55; instr, Purdue Univ, 55-56; assoc prof, 56-65, chmn grad sch interdept prog genetics, 69-74, PROF POULTRY SCI, PA STATE UNIV, 65- Honors & Awards: Corrispondente Award, Ital Soc Advan Zootechnol, 72. Mem: Am Soc Zool; Poultry Sci Asn; Am Genetic Asn; Genetics Soc Am; Am Inst Biol Sci. Res: Poultry genetics and physiology of reproduction; pathogenesis in turkeys; biochemical nature of gene action (riboflavin, chondrodystrophy, diabetes insipidus), infertility and obesity. Mailing Add: 1420 S Garner St State College PA 16801

BUSS, GLENN RICHARD, b Easton, Pa, Apr 12, 40; m 61; c 2. PLANT BREEDING, PLANT GENETICS. Educ: Pa State Univ, BS, 62, MS, 64, PhD(genetics), 67. Prof Exp: ASST PROF AGRON, VA POLYTECH INST & STATE UNIV, 67- Mem: Am Soc Agron; Crop Sci Soc Am. Res: Cytogenetics; breeding and genetics of alfalfa and soybeans. Mailing Add: Dept of Agron Va Polytech Inst & State Univ Blacksburg VA 24061

BUSS, JACK THEODORE, b Benton Harbor, Mich, Oct 2, 43; m 65; c 2. VERTEBRATE PHYSIOLOGY. Educ: Bethel Col, BA, 65; Wayne State Univ, MS, 68; Univ Minn, PhD(zool), 71. Prof Exp: Instr biol, Bethel Col, 66-68; asst prof biol, William Jewel Col, 71-75; ASST PROF BIOL, FERRIS STATE COL, 75- Concurrent Pos: Dir res, Antigen Labs, Inc, 72-75. Mem: AAAS; Am Inst Biol Sci; Am Soc Zoologists; Am Sci Affil. Res: Interaction of leukocytes and allergens; in vitro analysis. Mailing Add: Dept of Biol Sci Ferris State Col Big Rapids MI 49307

BUSS, KEEN, b Easton, Pa, Oct 20, 18; m 49; c 3. ZOOLOGY. Educ: Pa State Univ,

BA, 52. Prof Exp: Fishery biologist in chg fish cult res, Pa Fish Comn, 51-66, chief aquatic biologist, 66-67, chief div fisheries, 67-70; group vpres res & prod, 70-75, VPRES, MARINE PROTEIN CORP, 75- Mem: Am Fisheries Soc. Res: Fish culture, genetics and management with emphasis on Salmonidae and Esocidae. Mailing Add: Marine Protein Corp 33 Nashua Rd Londonderry NH 03053

BUSS, WALTER RICHARD, b Provo, Utah, Nov 1, 05; m 28; c 4. GEOLOGY, GEOGRAPHY. Educ: Brigham Young Univ, AB, 30, MA, 33; Stanford Univ, PhD(geol), 64. Prof Exp: Asst geol, Brigham Young Univ, 29-33, libr asst, 29-30 & 32-33; prof, 33-74, EMER PROF GEOL & GEOG, WEBER STATE COL, 74- Concurrent Pos: Asst, Stanford Univ, 41-42; dir night sch, Weber State Col, 43-45; geologist, US Forest Serv, 55- Mem: Fel AAAS; fel Nat Asn Geol Teachers; Asn Am Geogrs; Sigma Xi. Res: Physical geography of Utah; creep of mantle material. Mailing Add: 2820 Liberty Ave Ogden UT 84403

BUSSARD, ROBERT W, b Washington, DC, Aug 11, 28; m 49, 66, 73; c 4. ENGINEERING PHYSICS. Educ: Univ Calif, Los Angeles, BS, 50, MS, 52; Princeton Univ, AM, 59, PhD(physics), 61. Prof Exp: Design engr, Falcon Prog, Hughes Aircraft Co, 49-51; engr, Aircraft Nuclear Propulsion Prog, Oak Ridge Nat Lab, 52-55; alternate group leader, Nuclear Reactor Design & dir res & develop nuclear propulsion, Los Alamos Sci Lab, 55-62; dir nuclear systs staff, Space Tech Labs, Thompson-Ramo-Wooldridge, Inc, 62-64, consult, 59-60; assoc mgr res & eng & corp chief scientist, Electro-Optical Systs Div, Xerox Corp, 64-69; corp chief scientist, CSI Corp, Calif, 69-70; electro-optical consult, 70-71; alternate div leader, Laser Div, Los Alamos Sci Lab, 71-73; asst dir, Div Controlled Thermonuclear Res, USAEC, 73-74; MGR, ENERGY RESOURCES GROUP, 74- Concurrent Pos: Consult, Aerospace Corp, 60-62, Inst Defense Anal, 60-67 & Adv Group Aeronaut Res & Develop NATO, France, 60-65; vpres tech activities, Am Inst Aeronaut & Astronaut, 65-66, mem bd dirs, 62-68; consult, Los Alamos Sci Lab, 62-71 & 74-, US Energy Res & Develop Admin, 74- & Lawrence Livermore Lab, 75- Mem: Fel Am Inst Aeronaut & Astronaut; Am Phys Soc. Res: Electrooptical systems; nuclear space power and propulsion; nuclear weapon effects; high density plasma physics and application; rocket propulsion systems; fusion plasmas; laser physics; energy conversion; technical economic parametric analysis of advanced energy conversion systems. Mailing Add: Energy Resources Group 1200 N Nash St Arlington VA 22209

BUSSE, EWALD WILLIAM, b St Louis, Mo, Aug 18, 17; m 41; c 4. PSYCHIATRY. Educ: Westminster Col, AB, 38, ScD, 60; Washington Univ, MD, 42; Am Bd Psychiat & Neurol, dipl. Prof Exp: From instr to prof psychiat, Med Ctr, Univ Colo, 46-53; prof psychiat & chmn dept, Sch Med, 53-74, ASSOC PROVOST & DIR MED & ALLIED HEALTH EDUC, DUKE UNIV MED CTR, 74- Concurrent Pos: Dir EEG lab, Colo Psychopath Hosp, 46-53; lectr, Sch Grad Educ, Univ Denver, 47-53; head div psychosom med, Colo Gen Hosp, 50-53; actg head ment hyg & child guid clin, Univ Colo, 52-53; dir ctr study aging, Duke Univ, 57-70; dir, Am Bd Psychiat & Neurol, 61-69; consult, US Navy, US Army, Vet Admin & NIH; mem, President's Biomed Res Panel, 75-76. Honors & Awards: Strecker Award, Pa Hosp Inst, 67; Edward B Allen Award, Am Geriat Soc, 67; Cert Commendation, Am Psychiat Asn, 67-70; Robert W Kleemeier Award, Geront Soc, 68, Cert Commendation, 68; William C Menninger Award, 71; Mod Med Award, 72; Seltzer Award, 75. Mem: Nat Inst Med; life fel Am Psychiat Asn (pres, 71-72); fel Am Col Physicians; fel Am Geriat Soc (pres-elect, 74-75, pres, 75-76); fel Geront Soc (pres, 67-68). Res: Electroencephalography; gerontology. Mailing Add: Med & Allied Health Educ Duke Univ Med Ctr Durham NC 27710

BUSSE, ROBERT FRANKLYN, b Plainfield, NJ, Apr 30, 37; m 59; c 2. BIOCHEMISTRY, ORGANIC CHEMISTRY. Educ: Univ Notre Dame, BS, 59; Rutgers Univ, PhD(org chem), 63. Prof Exp: Res chemist, Celanese Corp Am, 63-67, mgr tech progs, Celanese Fibers Mkt Co, 67-69, tech develop mgr, 69-72; CHMN SCI DEPT, PROVIDENCE DAY SCH, CHARLOTTE, NC, 73- Mem: Am Chem Soc. Res: Inhibition of enzymes and metabolism; separation scheme for the isolation of plant sterols; new product and process development and evaluation. Mailing Add: Providence Day Sch 5800 Sardis Rd Charlotte NC 28211

BUSSELL, ROBERT HARRY, b Capac, Mich, May 20, 28; m 50; c 4. VIROLOGY. Educ: Mich State Univ, BS, 50, MS, 52; Univ Tex, PhD(virol), 56. Prof Exp: Res assoc prev med & pub health, Univ Tex Med Br, 52-55, asst, 55-57; fel virol, Dept Pediat, Sch Med, State Univ NY Buffalo, 57-61, instr, Dept Bact & Immunol, 62-63; from asst prof to assoc prof, 63-71, PROF MICROBIOL, UNIV KANS, 71- Concurrent Pos: Res assoc, Dept Pediat, State Univ NY Buffalo, 61-63; NIH Res fel, Div Infectious Dis, Stanford Univ Med Ctr, 71-72. Mem: Am Soc Microbiol; Soc Gen Microbiol; AAAS; Am Inst Biol Sci; Sigma Xi. Res: Animal virology; tissue culture; paramyxoviruses, measles and canine distemper viruses; RNA tumor viruses; cell fusion induced by viruses. Mailing Add: Dept of Microbiol Univ of Kans Lawrence KS 66045

BUSSERT, JACK FRANCIS, b Chicago, Ill, Dec 13, 22; m 54; c 2. ORGANIC CHEMISTRY. Educ: DePaul Univ, BSc, 47; Purdue Univ, MSc, 50; Ohio State Univ, PhD(org chem), 55; Univ Chicago, MBA, 65. Prof Exp: Res chemist, Universal Oil Prod Co, 48 & Nat Aluminate Corp, 50-51; RES CHEMIST, AMOCO CHEM CORP, STANDARD OIL CO, IND, 55- Mem: Am Chem Soc. Res: Hydrocarbon synthesis; anti-oxidants-petroleum products; agricultural pesticides; hydrocarbon oxidation; chemicals marketing research; chemicals commercial development. Mailing Add: Amoco Chem Corp 200 E Randolph Dr Chicago IL 60601

BUSSEY, ARTHUR HOWARD, Can citizen. BIOLOGY. Educ: Univ Bristol, BSc, 63, PhD(microbiol), 66. Prof Exp: Fulbright fel microbial regulation & travel scholar, Dept Biol Sci, Purdue Univ, Lafayette, 66-69; asst prof, 69-73, ASSOC PROF BIOL, McGILL UNIV, 73- Mem: Can Genetics Soc; Am Soc Microbiol; AAAS. Res: Eukaryotic cell surface; energy coupling; yeast killer factor; cell-cell recognition; regulatory switching. Mailing Add: Dept of Biol McGill Univ PO Box 6070 Sta A Montreal PQ Can

BUSSEY, HOWARD EMERSON, b Yankton, SDak, Sept 14, 17; m 46; c 4. ELECTROMAGNETISM. Educ: George Wash Univ, BA, 43, MS, 51; Univ Colo, PhD(physics), 64. Prof Exp: Sci aide physics, 41-42, phys scientist, 46-49, PHYSICIST, NAT BUR STAND, 50- Mem: Am Meteorol Soc; Am Phys Soc; Inst Elec & Electronics Eng; Sigma Xi; Int Sci Radio Union. Res: Tropospheric propagation; dielectric and ferrimagnetic measurements; antenna theory; electromagnetic wave scattering theory and experiment; direct and inverse scattering problems applied to electromagnetic wave interactions with animals and biomaterials. Mailing Add: Electromag Div 276 07 Nat Bur Stand Boulder CO 80302

BUSSEY, ROBERT JAMES, physical organic chemistry, see 12th edition

BUSSIAN, ALFRED ERICH, b Milwaukee, Wis, Sept 9, 33; m 58; c 2. PARTICLE PHYSICS. Educ: Ripon Col, BA, 55; Univ Colo, Boulder, PhD(physics), 64. Prof Exp: Engr, A C Spark Plug Div, GMC, Wis, 57-58; res asst physics, Univ Colo, Boulder, 58-64; res assoc, Max Planck Inst Physics & Astrophys, 64-66; vis scientist, Nat Ctr Atmospheric Res, Colo, 66-67; res assoc physics, Univ Mich, 67-73; PROF PHYSICS, COMMUNITY COL DENVER, RED ROCKS CAMPUS, 73- Mem: Am Phys Soc. Res: Study of the role of general relativity in particle physics. Mailing Add: Jamestown CO 80455

BUSTA, FRANCIS FREDERICK, b Faribault, Minn, Sept 25, 35; m 57. FOOD MICROBIOLOGY. Educ: Univ Minn, Minneapolis, BA, 57; Univ Minn, St Paul, MS, 61; Univ Ill, PhD(food sci), 63. Prof Exp: From asst prof to assoc prof food sci, NC State Univ, 63-67; assoc prof, 67-71, PROF FOOD SCI & NUTRIT, UNIV MINN, ST PAUL, 72- Concurrent Pos: Vis scientist, Commonwealth Sci & Indust Res Orgn, Sydney, Australia, 74-75; educ res grant, Campbell Soup Co Res Inst, 74. Mem: Am Meat Sci Asn; Int Asn Milk, Food & Environ Sanit; Am Soc Microbiol; Inst Food Technol; Brit Soc Appl Bact. Res: Microbiological aspects of food processing; environmental stress on microorganisms; microbiological quality of food; food substances inhibitory or stimulative to microorganisms; bacterial spore physiology; microbial based food from wastes. Mailing Add: Dept of Food Sci & Nutrit Univ of Minn St Paul MN 55108

BUSTAD, LEO KENNETH, b Stanwood, Wash, Jan 10, 20; m 42; c 3. RADIOBIOLOGY. Educ: Wash State Univ, BS, 41, MS, 48, DVM, 49; Univ Wash, PhD, 60. Prof Exp: Mgr exp farm, Biol Labs, Hanford Labs, Gen Elec Co, 49-64; prof radiation biol & dir radiobiol & comp oncol labs, Univ Calif, Davis, 65-73; PROF PHYSIOL & DEAN COL VET MED, WASH STATE UNIV, 73- Concurrent Pos: Guest lectr, Wash State Univ, 55-65; NSF fel, 58; guest fac, Univ Wash, 60-; consult, US Air Force-AEC, 58-63; mem adv comt interdisciplinary conf prog, Am Inst Biol Sci, 63-; mem nat adv comt, Regional Primate Res Ctr, Univ Wash, 65-; consult scientist, Biol Dept, Battelle Mem Inst, 65-; assoc ed, Lab Animal Care, 67-; mem, Nat Coun Radiation Protection & Measurements, 69-75; mem panel consult, Nev Opers Off, AEC, 69-; mem bd regents, Calif Lutheran Col, 70-73; mem bd gov, Found Human Ecol, 70-; mem subcomt, Adv Comt Civil Defense, Nat Acad Sci-Nat Res Coun, 60-, comt prof educ, Inst Lab Animal Resources, 66-68, subcomt stand large animals, 67-70, adv coun, 69-74, subcomt radioactivity in food, 67- & food protection comt, 67-75; chmn, Nat Acad Sci-Nat Res Coun Comt Vet Med Sci, 75; mem NIH, Adv Res Resources Coun, 75; nat consult vet med, Surgeon General, US Air Force, 75. Mem: AAAS; Am Asn Lab Animal Sci; Am Vet Med Asn; Radiation Res Soc; Soc Exp Biol & Med; Am Physiol Soc. Res: Thyroid physiology; metabolism and toxicity of radionuclides; physiological response to irradiation; laboratory animal biology and medicine; veterinary medical education; cooperative regional curriculum; carcinogenesis and aging; health sciences education. Mailing Add: Col of Vet Med 305 Col Hall Wash State Univ Pullman WA 99163

BUSTEAD, RONALD LORIMA, JR, b Woburn, Mass, Mar 29, 30; m 61. FOOD TECHNOLOGY, FOOD CHEMISTRY. Educ: Mass Inst Technol, BS, 52; Northeastern Univ, MS, 66. Prof Exp: Dairy technologist, H P Hood & Sons, 56-57; chemist, Nat Lead Co, Mass, 57-58; food technologist, Schroeder Industs, Inc, 58-59, vending machine engr, 59-62; food technologist, H A Johnson Co, 62; qual control specialist, Armed Forces Food & Container Inst, 62-64, food technologist, 64-70, OPERS RES ANALYST, US ARMY NATICK LABS, 70- Mem: Inst Food Technol; Am Chem Soc; Am Soc Qual Control; Armed Forces Mgt Asn. Res: Feeding systems; military rations; space foods; new convenience foods; quality assurance for foods; application of operations research, systems analysis and computer sciences to feeding situations. Mailing Add: 124 Winter St Framingham MA 01701

BUSTEED, ROBERT CHARLES, b Milan, Ind, Sept 4, 07; m 32; c 4. BIOLOGY. Educ: Ind Univ, AB, 30, AM, 32, PhD(bot, zool), 36. Prof Exp: Instr bot & mycol, Ind Univ, 34-36; prof & head dept, Appalachian State Teachers Col, 37-46; prof bot & chmn dept biol, Univ Ga, 46-48; prof biol & head dept, 48-75, chmn div sci, 50-75, EMER PROF BIOL, W TEX STATE UNIV, 75- Concurrent Pos: Mem admin comt, Killgore Res Ctr, 68- Res: Grain sorghum genetics; canine genetics; plant pathology. Mailing Add: Rte 2 Box 20 Canyon TX 79015

BUSTOS-VALDES, SERGIO ENRIQUE, b Constitucion, Chile, July 6, 32; m 63; c 3. BIOCHEMISTRY. Educ: Univ Concepcion, Chile, DDS, 58; Univ Rochester, PhD(biochem), 68. Prof Exp: From instr to asst prof physiol sch med, Univ Concepcion, Chile, 58-67, from assoc prof to prof, Inst Biomed Sci, 67-71; assoc prof, 71-74, PROF BIOCHEM DEPT ORAL BIOL, MED COL GA, 74-, ASST PROF CELL & MOLECULAR BIOL, 71- Concurrent Pos: Fulbright scholar & US State Dept grant, Univ Rochester, 62-63, Orgn Am States grant, 63-64. Mem: Sigma Xi; Int Asn Dent Res. Res: Chromosomal protein metabolism, cell differentiation and regulation of gene transcription. Mailing Add: Dept of Oral Biol Med Col of Ga 1459 Gwinnett St Augusta GA 30902

BUSWELL, ROBERT JAMES, b New York, NY, Oct 18, 14; m 44; c 2. ANALYTICAL CHEMISTRY. Educ: Univ Ill, BA, 36, MA, 37. Prof Exp: Res chemist, Gen Mills, Inc, 37-46; res chemist, Armour & Co, Ill, 46-64, asst head anal sect, 64-65, HEAD ANAL CHEM SECT, ARMOUR FOOD CO, 65- Mem: Am Chem Soc; Am Oil Chem Soc. Res: Triglyceride chemistry; food analysis; edible fats and shortenings. Mailing Add: Anal Chem Sect Armour Food Co 15101 N Scottsdale Rd Scottsdale AZ 85260

BUSZA, WIT, b Ploesti, Roumania, Jan 14, 40; m 64. ELEMENTARY PARTICLE PHYSICS. Educ: Univ London, BSc, 60, PhD(nuclear physics), 64. Prof Exp: Res assoc physics, Univ Col, Univ London, 63-66 & Stanford Linear Accelerator Ctr, 66-69; asst prof, 69-73, ASSOC PROF PHYSICS, MASS INST TECHNOL, 73- Mem: Am Inst Physics; Brit Inst Physics & Phys Soc. Res: Study of elementary particles using counter techniques. Mailing Add: Rm 24-518 Mass Inst of Technol Cambridge MA 02139

BUTA, JOSEPH GEORGE, organic chemistry, see 12th edition

BUTAS, CONSTANDINA, b Brasov, Romania, Nov 12, 16; Can citizen. MEDICAL MICROBIOLOGY. Educ: Univ Cluj, MD, 39. Prof Exp: Asst prof bact, Hyg Inst, Univ Cluj, 39-47; Nat Res Coun res fel, Pasteur Inst, Paris, 47-51; resident commun dis, Hotel Dieu, Chicoutimi, PQ, 51-54; lectr, 54-56, asst prof, 56-62, ASSOC PROF MICROBIOL, McGILL UNIV, 62- Concurrent Pos: Sr bacteriologist, Royal Victoria Hosp, 70, consult infectious dis serv, 69. Mem: Am Soc Microbiol; Can Med Asn; Can Soc Microbiol; Can Asn Med Microbiol; Can Pub Health Asn. Res: Mycoplasma. Mailing Add: Dept of Microbiol & Immunol McGill Univ Montreal PQ Can

BUTCHART, JOHN HARVEY, b Hofei, China, May 10, 07; nat US; m 29; c 2. MATHEMATICS. Educ: Eureka Col, BS, 28; Univ Ill, AM, 29, PhD(math), 32. Prof Exp: Asst, Univ Ill, 29-32; instr math, Butler Univ, 33-36; prof, Phillips Univ, 36-39 & William Woods Col, 39-42; asst prof, Grinnell Col, 42-45; PROF MATH, NORTHERN ARIZ UNIV, 45- Concurrent Pos: Fel, Univ Iowa, 44. Mem: AAAS; Am Math Soc; Math Asn Am. Res: Geometry of the triangle and circle: differential geometry; pure solid geometry. Mailing Add: Dept of Math Northern Ariz Univ Flagstaff AZ 86003

BUTCHER, EARL ORLO, b Burlington, Ind, Sept 20, 03; m 29; c 3. BIOLOGY. Educ: DePauw Univ, AB, 25; Cornell Univ, AM, 26, PhD(histol & embryol), 28. Prof Exp: Instr histol & embryol, Cornell Univ, 26-28; from asst prof to prof biol, Hamilton Col, 28-43; from asst prof to assoc prof anat, Col Dent, NY Univ, 43-47, prof anat & chmn dept anat, Grad Sch, 47-72, from asst dean to assoc dean, Col Dent, 54-72, EMER PROF ANAT, GRAD SCH, NY UNIV, 72- Mem: AAAS; Am Col Dent; Am Asn Anat; Soc Exp Biol & Med; Harvey Soc. Res: Origin of germ cells; development of mesoderm in mammals; transplantation of eyes and skin; melanophoric responses in fish; endocrines and hair growth; innervation and growth of teeth. Mailing Add: 113 Laurel Rd Princeton NJ 08540

BUTCHER, FRED RAY, b Roohester, Pa, Aug 11, 43; m 65; c 2. BIOCHEMISTRY. Educ: Ohio State Univ, BSc, 65, PhD(biochem), 69. Prof Exp: NIH fel oncol, Univ Wis, 69-71; ASST PROF BIOCHEM, BROWN UNIV, 71- Res: Role of cyclic nucleotides and calcium in the regulation of target tissue physiology and biochemistry by autonomic agents, the rat parotid gland is used as a model with particular emphasis on regulation of exocytosis. Mailing Add: Div of Biol & Med Sci Brown Univ Providence RI 02912

BUTCHER, HARVEY RAYMOND, JR, b Creighton, Mo, Jan 27, 20; m 44; c 3. SURGERY. Educ: Cent Col, AB, 41; Harvard Univ, MD, 44. Prof Exp: From instr to assoc prof, 5264, PROF SURG, SCH MED, WASHINGTON UNIV, 64, CANCER COORDR & DIR DIV TUMOR SERV, 57- Concurrent Pos: Markle fel, 55- Mem: Am Cancer Soc; Soc Univ Surgeons. Res: Vascular disease; cancer. Mailing Add: Dept of Surg Washington Univ Med Sch St Louis MO 63110

BUTCHER, HENRY CLAY, IV, b Newton, Mass, July 31, 33; m 58; c 3. PLANT PHYSIOLOGY, BIOCHEMISTRY. Educ: Tufts Univ, BSc, 55; Ohio State Univ, MS, 61, PhD(plant physiol), 64. Prof Exp: Fel, Res Inst Advan Studies, Md, 64-65; asst prof, 65-69, ASSOC PROF BIOL, LOYOLA COL, MD, 69-, CHMN DEPT, 71- Concurrent Pos: Guest plant physiologist, Brookhaven Nat Lab, 75-76. Mem: AAAS; Am Soc Plant Physiol; NY Acad Sci; Am Inst Biol Sci. Res: Biochemistry of plant cell vacuoles; translocation of organic solutes in plants; photosynthesis; tissue culture; differentiation. Mailing Add: 6621 Queens Ferry Rd Baltimore MD 21212

BUTCHER, JAMES WALTER, b Ramsaytown, Pa, Feb 14, 17; m 44; c 2. ENTOMOLOGY, ECOLOGY. Educ: Univ Pittsburgh, BS, 43; Univ Minn, MS, 48, PhD(entom), 51. Prof Exp: Entomologist, USDA, 49-52; res & asst state entomologist, Minn, 51-57; assoc prof entom, 57-66, assoc dean col natural sci, 72-74, actg dean, 73, PROF ENTOM, MICH STATE UNIV, 66-, DEAN RES & GRAD TRAINING, COL NATURAL SCI, 69-, CHMN DEPT ZOOL, 74- Concurrent Pos: Fulbright sr res scholar, Univ Vienna, 66-67. Mem: Entom Soc Am; Ecol Soc Am; Entom Soc Can. Res: Forest entomology; insect ecology; pesticide side effects; soil biology. Mailing Add: Dept of Zool Mich State Univ East Lansing MI 48823

BUTCHER, JOHN EDWARD, b Belle Fourche, SDak, Aug 4, 23; m 51; c 3. ANIMAL PRODUCTION. Educ: Mont State Col, BS, 50, MS, 52; Utah State Univ, PhD(animal prod), 56. Prof Exp: Instr animal indust, Mont State Col, 49-50, range mgt, 52-53; from asst prof to assoc prof, 55-67, PROF ANIMAL SCI, UTAH STATE UNIV, 67- Concurrent Pos: Mem subcomt sheep nutrit, Nat Res Coun. Mem: Fel AAAS; Am Soc Animal Sci; Am Soc Range Mgt; Am Soc Farm Mgr & Rural Appraisers; Animal Behav Soc. Res: Ruminant nutrition and environment. Mailing Add: Dept of Animal Sci Utah State Univ Logan UT 84322

BUTCHER, REGINALD WILLIAM, b Bayshore, NY, May 4, 30; m 53; c 3. BIOCHEMISTRY, PHYSIOLOGY. Educ: US Naval Acad, BS, 53; Western Reserve Univ, PhD(pharmacol), 63. Prof Exp: From instr to assoc prof physiol, Vanderbilt Univ, 63-69, investr, Howard Hughes Med Inst, 66-69; PROF BIOCHEM & CHMN DEPT, MED SCH, UNIV MASS, 69- Mem: Endocrine Soc; NY Acad Sci; Am Physiol Soc; Am Soc Biol Chemists. Res: Mechanism of hormone action at the molecular level; control of cyclic adenosinemonophosphate levels by adenyl cyclase and phosphodiesterase activities and effects of cyclic AMP on cellular processes. Mailing Add: Dept of Biochem Univ of Mass Med Sch Worcester MA 01605

BUTCHER, ROY LOVELL, b Reedy, WVa, Dec 15, 30; m 55; c 3. REPRODUCTIVE PHYSIOLOGY. Educ: WVa Univ, BS, 53, MS, 59; Iowa State Univ, PhD(animal reproduction), 62. Prof Exp: From instr to asst prof, 62-75, ASSOC PROF OBSTET & GYNEC, WVA UNIV, 75-, ASSOC PROF ANAT, 71- Mem: Endocrine Soc; Soc Study Reproduction; Am Soc Animal Sci; Am Fertil Soc; Brit Soc Study Fertil. Res: Control of luteal function and effects of delayed ovulation on congenital anomalies. Mailing Add: Dept of Obstet & Gynec WVa Univ Morgantown WV 26506

BUTCHER, SAMUEL SHIPP, b Gaylord, Mich, Nov 12, 36; m 61; c 2. PHYSICAL CHEMISTRY. Educ: Albion Col, AB, 58; Harvard Univ, AM, 61, PhD(chem), 63. Prof Exp: Fel spectros, Nat Res Coun Can, 62-64; from asst prof to assoc prof, 64-74, PROF CHEM, BOWDOIN COL, 74- Mem: AAAS; Am Chem Soc; Am Meteorol Soc. Res: Molecular structure and spectroscopy; atmospheric chemistry. Mailing Add: Dept of Chem Bowdoin Col Brunswick ME 04011

BUTCHKO, GREGORY MICHAEL, b Passaic, NJ, Oct 12, 42; m 72; c 1. IMMUNOLOGY. Educ: Univ Fla, BS, 66; Univ SFla, MS, 68; Univ Ga, PhD(microbiol), 71. Prof Exp: ASST PROF MICROBIOL, MED CTR, UNIV MISS, 73- Concurrent Pos: NIH fel, Med Ctr, Univ Ill, 71-73. Mem: Sigma Xi; AAAS; Am Soc Microbiol. Res: Studies on the nature and specificity of guinea pig T lymphocyte cell surface antigen receptors; studies on the etiology and immunology of juvenile laryngeal papillomatosis. Mailing Add: Dept of Microbiol Univ of Miss Med Ctr Jackson MS 39216

BUTEAU, GEORGE H, JR, zoology, parasitology, see 12th edition

BUTEL, JANET SUSAN, b Overbrook, Kans, May 24, 41; m 67; c 2. VIROLOGY. Educ: Kans State Univ, BS, 63; Baylor Univ, PhD(virol), 66. Prof Exp: Fel, 66-68, ASSOC PROF VIROL, BAYLOR COL MED, 68- Mem: AAAS; Am Soc Microbiol; Am Asn Immunologists. Res: Tumor viruses; defective viruses; virus genetics; transformation of cells by viruses. Mailing Add: Dept of Virol Baylor Col of Med Houston TX 77025

BUTENSKY, IRWIN, b New York, NY, Jan 27, 36; m 69. COSMETIC CHEMISTRY, PHARMACEUTICAL CHEMISTRY. Educ: Columbia Univ, BS, 56; Univ Mich, MS, 59, PhD(pharmaceut chem), 61. Prof Exp: Develop chemist, Lederle Labs, Am Cyanamid Co, NY, 61-66, group leader process improve lab, 66-67; develop chemist pharmaceut res & develop, Vicks Div Res & Develop, 67-69, sect head, Vicks Div Res, 69-71, asst dir pharmaceut develop, 71-72, DIR SKIN CARE & TOILETRIES, VICKS DIV RES, RICHARDSON-MERRELL, INC, 72- Mem: AAAS; Am Pharmaceut Asn; Am Chem Soc; Soc Cosmetic Chem. Res: Tablet coating technology; spraying techniques; fine particle technology; suspensions of natural product; proprietary drug development; aerosols, creams and all other dosage forms; plastics; development of new cosmetic and dermatological products. Mailing Add: 1355 Dickerson Rd Teaneck NJ 07666

BUTERA, RICHARD ANTHONY, b Carnegie, Pa, Nov 30, 34. SOLID STATE CHEMISTRY, PHYSICAL CHEMISTRY. Educ: Univ Pittsburgh, BS, 60; Univ Calif, Berkeley, PhD(phys chem), 63. Prof Exp: Res assoc phys chem, Univ Pittsburgh, 57-60; res asst low temperature chem, Univ Calif, Berkeley, 60-63; asst prof, 63-68, ASSOC PROF PHYS CHEM, UNIV PITTSBURGH, 68- Res: Magneto-thermodynamic studies of magnetic phase transitions; heat capacity studies of intermetallic compounds; computer controlled experimentation; solar energy converters and fuel cells. Mailing Add: Dept of Chem Univ of Pittsburgh Pittsburgh PA 15260

BUTKOV, EUGENE, b Pancevo, Yugoslavia, Sept 16, 28; US citizen. THEORETICAL PHYSICS. Educ: Univ BC, BASc, 54, MA, 56; McGill Univ, PhD(theoret physics), 60. Prof Exp: Asst prof physics, St John's Univ, NY, 59-62 & Hunter Col, 62-65; asst prof, 65-68, ASSOC PROF PHYSICS, ST JOHN'S UNIV, NY, 68-, CHMN DEPT PHYSICS, 74- Mem: Am Phys Soc; Math Asn Am. Res: Theory of non-linear differential equations; quantum field theory. Mailing Add: Dept of Physics St John's Univ Jamaica NY 11432

BUTKUS, ANTANAS, b Lithuania, June 20, 18; US citizen; m 43; c 2. ANALYTICAL CHEMISTRY, CARDIOVASCULAR DISEASES. Educ: Univ Halle, BAgrSci, 43; Univ Bonn, DrAgrSci, 47. Prof Exp: Chief chemist, Processing Lab, Southland Frozen Meat, NZ, 54-62; chemist, Water Control Comn, Md, 62; asst staff mem lipids, Cleveland Clin, 62-67, assoc staff mem, 67-69, STAFF MEM, CLEVELAND CLIN FOUND, 69- Mem: AAAS; Am Chem Soc; Am Oil Chem Soc; Am Soc Exp Path. Res: Disorders of lipid metabolism and arteriosclerosis. Mailing Add: Res Div Cleveland Clin Found 9500 Euclid Ave Cleveland OH 44106

BUTLER, ALFRED BISBEE, b Spokane, Wash, Nov 27, 11; m 37; c 2. PHYSICS. Educ: State Col Wash, BS, 35, MA, 44. Prof Exp: Teacher pub sch, Wash, 36-43; supvr physics, Army Air Force Col Training Detachment, 43-44, from asst prof to assoc prof, 44-61, PROF PHYSICS, WASH STATE UNIV, 61- Concurrent Pos: Western Elec Fund award excellence in instr of eng stud, 68-69. Mem: AAAS; Am Asn Physics Teachers; Nat Sci Teachers Asn. Res: Teaching and visual aids in college physics. Mailing Add: Dept of Physics Wash State Univ Pullman WA 99163

BUTLER, ANN BENEDICT, b Wilmington, Del, Dec 2, 45; m 68. NEUROANATOMY. Educ: Oberlin Col, BA, 67; Case Western Reserve Univ, PhD(anat), 71. Prof Exp: NIH fel neuroanat, Brown Univ, 71-72 & Univ Va, 72-73; asst prof anat, George Washington Univ, 73-75; ASST PROF ANAT, GEORGETOWN UNIV, 75- Mem: Am Asn Anatomists; Soc Neurosci; Am Soc Zoologists. Res: Comparative neuroanatomy of the visual system; evolution of sensory system organization in the thalamus and telencephalon; evolution of neocortex and of limbic system. Mailing Add: Dept of Anat Sch Med & Dent Georgetown Univ Washington DC 20007

BUTLER, ARTHUR PIERCE, JR, b Morristown, NJ, June 23, 08; m 48; c 4. ECONOMIC GEOLOGY. Educ: Harvard Univ, AB, 30, MBA, 32, AM, 37, PhD(geol), 47. Prof Exp: From asst to instr geol, Harvard Univ, 38-40; recorder, US Geol Surv, 40-41, from jr geologist to geologist, 41-75; RETIRED. Mem: Geol Soc Am; Soc Econ Geol. Res: Stratigraphy and structure of volcanic rocks; geology of iron ore and of uranium resources. Mailing Add: 9625 W 36th Ave Wheat Ridge CO 80033

BUTLER, BYRON C, b Carroll, Iowa, Aug 10, 18; m 58; c 4. OBSTETRICS & GYNECOLOGY, BIOPHYSICS. Educ: Columbia Univ, MD, 43, DSc(med), 51. Prof Exp: Instr path, Col Physicians & Surgeons, Columbia Univ, 41-42, instr obstet & gynec, 50-53; DIR BIOPHYS, BUTLER RES FOUND, 63- Concurrent Pos: NIH grant, 52-53. Mem: Fel AAAS; AMA; Am Soc Fertil. Res: Hypnosis in relation to relief of pain in cancer; fibrinolysin enzyme system in relation to blood clotting defects in pregnancy; bacterial vaginal flora; magnetic field inhibition of cancer cells in tissue culture; infertility; diagnosis and treatment of cancer of cervix and early detection. Mailing Add: Med Sci Bldg 550 W Thomas Rd Phoenix AZ 85013

BUTLER, CALVIN CHARLES, b Fowler, Colo, Nov 8, 37; m 59; c 4. MATHEMATICAL STATISTICS. Educ: Colo State Univ, BS, 62, MS, 63, PhD(math statist), 66. Prof Exp: Asst prof math, Univ Colo, Boulder, 66-71; PROF MATH, COL SOUTHERN IDAHO, 71- Mem: Inst Math Statist. Res: Applied probability theory; stochastic processes. Mailing Add: Dept of Math Col of Southern Idaho Twin Falls ID 83301

BUTLER, CHARLES MORGAN, b Columbia, Tenn, Dec 16, 29; m 58; c 3. STATISTICS, OPERATIONS RESEARCH. Educ: US Mil Acad, BS, 53; Mass Inst Technol, MS, 62; Univ Ala, MA, 66, PhD(math), 70. Prof Exp: Asst prof indust eng, Univ Ala, 68-70; ASSOC PROF STATIST, MISS STATE UNIV, 70- Res: Application of statistical and other quantitative techniques for the solution of business and industrial problems. Mailing Add: Dept Bus Statist & Data Proc Miss State Univ Drawer DB Mississippi State MS 39762

BUTLER, CHARLES THOMAS, b Muskogee, Okla, Nov 30, 32; div; c 3. BIOPHYSICS. Educ: Iowa State Univ, BS, 54; Tex A&M Univ, MS, 57; Okla State Univ, PhD(physics), 72. Prof Exp: Sr scientist, Jet Propulsion Lab, Calif Inst Technol, 56-60; physicist, Oak Ridge Nat Lab, 60-70; MEM FAC, DEPT PHYSICS, OKLA STATE UNIV, 70- Mem: Am Phys Soc; Sigma Xi. Res: Physical properties of lipid monolayers and cell membranes. Mailing Add: Dept of Physics Okla State Univ Stillwater OK 74074

BUTLER, DANIEL KNOWLES, b New York, NY, Sept 12, 28; m 53; c 3. COMPUTER SCIENCE, REACTOR PHYSICS. Educ: Harvard Univ, AB, 50, MA, 52, PhD(nuclear physics), 55. Prof Exp: Physicist, Argonne Nat Lab, 55-73; MEM FAC, CLEVELAND STATE UNIV, 74- Mem: AAAS; Am Phys Soc; Am Nuclear Soc. Res: Nuclear reactions; experimental and theoretical reactor physics; data base management; reactor neutronics computations. Mailing Add: Cleveland State Univ Cleveland OH 44115

BUTLER, DONALD EUGENE, b Detroit, Mich, Dec 1, 33; m 62; c 3. ORGANIC CHEMISTRY. Educ: Wayne State Univ, BS, 55; Univ Fla, PhD(chem), 58. Prof Exp: From assoc res chemist to res chemist, 58-68, SR RES CHEMIST, PARKE, DAVIS & CO, 68- Res: Piperazine derivatives; synthetic organic medicinals; gastrointestinal and central nervous system pharmacology. Mailing Add: Res Div Parke Davis & Co Ann Arbor MI 48106

BUTLER, EDWARD BYRON, b Nashville, Tenn, Dec 9, 16; m 47; c 2. PHYSICAL CHEMISTRY. Educ: Univ Chicago, BS, 38, PhD(phys chem), 51. Prof Exp: Asst chemist, Tenn Valley Authority, 43; res chemist, Magnolia Petrol Co, 43-46; sr chemist, Socony Vacuum Oil Co, 46-47; res assoc, Nat Res Coun, 47-51; res chemist, Libbey Owens-Ford Glass Co, 51-56, Pan Am Petrol Co, 56-71 & Tulsa Surfchem,

71-74; SR RES SCIENTIST, AMOCO PROD CO, 74- Mem: Fel AAAS; fel Am Inst Chem; Am Chem Soc. Res: Surface chemistry. Mailing Add: 4332 S Canton Ave Tulsa OK 74135

BUTLER, EDWARD EUGENE, b Wilmington, Del, Dec 8, 19; m 47; c 5. PHYTOPATHOLOGY, MYCOLOGY. Educ: Univ Del, BS, 43; Mich State Univ, MS, 48; Univ Minn, PhD(plant path), 54. Prof Exp: Asst bot, Mich State Univ, 46-49; instr plant path, Univ Minn, 51-54; jr plant pathologist, 55-56, lectr plant path & asst plant pathologist, 57-61, from asst prof to assoc prof, 61-68, PROF PLANT PATH, UNIV CALIF, DAVIS, 68- Concurrent Pos: Vis prof, Univ PR, 66-67; assoc ed, Phytopath, 73-76. Mem: AAAS; Am Phytopath Soc; Bot Soc Am; Mycol Soc Am; Brit Mycol Soc. Res: Ecology, genetics, and taxonomy of plant pathogenic fungi. Mailing Add: Dept of Plant Path Univ of Calif Davis CA 95616

BUTLER, ELIOT ANDREW, b Snowflake, Ariz, Feb 13, 26; m 49; c 4. ANALYTICAL CHEMISTRY. Educ: Calif Inst Technol, BS, 52, PhD, 55. Prof Exp: From asst prof to assoc prof, 55-65, PROF CHEM, BRIGHAM YOUNG UNIV, 65- Mem: Am Chem Soc. Res: Precipitation from homogeneous solution; equilibria of ionic and molecular species in aqueous solutions; electrode reactions. Mailing Add: Dept of Chem Brigham Young Univ Provo UT 84601

BUTLER, FRANK ANDREW, b New York, NY, Jan 12, 40; m 61; c 1. SOLID STATE PHYSICS. Educ: Univ Miami, BSES, 61; Rensselaer Polytech Inst, PhD, 66. Prof Exp: ASST PROF PHYSICS & ASTRON, UNIV KY, 66- Mem: Am Phys Soc. Res: Electron states in metals, band structures with and without magnetic fields. Mailing Add: Dept of Physics & Astron Univ of Ky Lexington KY 40506

BUTLER, GEORGE BERGEN, b Liberty, Miss, Apr 15, 16; m 44; c 2. ORGANIC CHEMISTRY. Educ: Miss Col, BA, 38; Univ NC, PhD(org chem), 42. Prof Exp: Res chemist, Rohm & Haas Co, Philadelphia, 42-46; from instr to assoc prof, 46-57, dir ctr macromolecular sci, 70, RES PROF CHEM, UNIV FLA, 57- Concurrent Pos: Consult & past vpres, Peninsular Chem Res, Inc; consult, Calgon Corp, Chemstrand Res Ctr & Atlantic Ref Co; co-ed, J Macromolecular Sci Rev & Rev Macromolecular Chem. Honors & Awards: Annual Res Award, Fla Chap, Sigma Xi, 61. Mem: AAAS; Am Chem Soc. Res: Explosive resins; resins and plastics; pharmaceutical chemicals; relationship between tanning properties and chemical structure; plastics useful as explosive binders; ion exchange resins; reaction mechanisms; radioactive carbon syntheses, quaternary ammonium compounds and polymers; polymerization mechanisms. Mailing Add: 420 Space Sci Res Bldg Univ of Fla Gainesville FL 32601

BUTLER, GEORGE DANIEL, JR, b Newark, NJ, Apr 13, 23; m 49; c 3. ENTOMOLOGY. Educ: Univ Mass, BS, 48; Cornell Univ, PhD(entom), 51. Prof Exp: Asst prof entom, Univ Ariz, 51-59, assoc prof, 59-66; res entomologist, Cotton Insects Br, 66-72, RES ENTOMOLOGIST, WESTERN COTTON RES LAB, USDA, 72- Mem: Am Entom Soc; Ecol Soc Am. Res: Biological control of insects. Mailing Add: 4135 Broadway Phoenix AZ 85040

BUTLER, GILBERT W, b Ft Collins, Colo, Feb 22, 41. NUCLEAR CHEMISTRY. Educ: Ore State Univ, BS, 63; Univ Calif, Berkeley, PhD(nuclear chem), 67. Prof Exp: Res nuclear chemist, Lawrence Radiation Lab, 67-70 & Argonne Nat Lab, 70-72; STAFF MEM & CHEMIST, LOS ALAMOS SCI LAB, 72- Mem: Am Phys Soc; Am Chem Soc. Res: Study of high energy nuclear reactions using solid state detectors; high resolution gamma ray spectroscopy. Mailing Add: Los Alamos Sci Lab Group CNC-11 MS-514 Los Alamos NM 87545

BUTLER, GORDON CECIL, b Ingersoll, Ont, Sept 4, 13; m 37; c 4. BIOCHEMISTRY. Educ: Univ Toronto, BA, 35, PhD(biochem), 38. Prof Exp: 1851 Exhib scholar, Univ London, 38-40; res chemist, Chas E Frosst & Co, 40-42; mem, Nat Res Coun Can, 45-47; from assoc prof to prof biochem, Univ Toronto, 47-57; dir biol & health physics div, Atomic Energy of Can Ltd, 57-65; dir div radiation biol, 65-68, DIR DIV BIOL SCI, NAT RES COUN CAN, 68- Mem: AAAS; Health Phys Soc; Can Physiol Soc; Soc Biol Chem; Royal Soc Can. Res: Steroids; immunochemistry; radiation chemistry; nucleic acids and nucleoproteins; scientific criteria for environmental quality; radiation protection. Mailing Add: Div of Biol Sci Nat Res Coun of Can Ottawa ON Can

BUTLER, HAROLD S, b Ardmore, Okla, Sept 14, 31; m 54; c 1. COMPUTER SCIENCE. Educ: Phillips Univ, BA, 53; Kans State Univ, MS, 56; Stanford Univ, PhD(physics), 61. Prof Exp: Scientist, Lockheed Missile & Space Co, 56-61; staff mem, Stanford Linear Accelerator Ctr, 61-63; STAFF MEM, LOS ALAMOS SCI LAB, 63- Res: Computer control of accelerators; beam dynamics and magnet system design; application of computers to solution of problems in physics. Mailing Add: 10 Escondido Los Alamos NM 87544

BUTLER, HARRY, b Birmingham, Eng, Dec 29, 16; m 42; c 1. EMBRYOLOGY. Educ: Univ Cambridge, BA, 38, MB, BCh, 41, MA, 46, MD, 51. Prof Exp: Demonstr anat, Univ Cambridge, 46-50, lectr, 50-51; sr lectr, St Bartholomew's Hosp Med Col, London, 51; reader, Univ London, 51-55; prof, Univ Khartoum, 55-64; assoc prof, 64-68, PROF ANAT, COL MED, UNIV SASK, 68- Mem: Can Asn Anat; Int Primatol Soc. Res: Primatology; early development; implantation; placentation and reproductive cycle of the Prosimii. Mailing Add: Dept of Anat Univ of Sask Saskatoon SK Can

BUTLER, HUGH C, b Helena, Mont, Jan 7, 25; m 48; c 3. SURGERY, PHYSIOLOGY. Educ: Mont State Col, BS, 50; Wash State Univ, DVM, 54; Am Col Vet Surgeons, dipl. Prof Exp: Asst state veterinarian, Livestock Sanit Bd, Mont, 54-55; from instr to assoc prof clin med & surg, Wash State Univ, 55-64, assoc prof physiol & pharmacol, 64-66; head dept surg, Animal Med Ctr, New York, 66-68; PROF SURG, KANS STATE UNIV, 68- Concurrent Pos: Vis scientist exp surg, Mayo Clin, 63. Mem: Am Vet Med Asn; Conf Res Workers Animal Dis. Res: Experimental surgery with emphasis on cardiovascular orthopedic and tissue transplantation. Mailing Add: Dept of Surg Kans State Univ Col of Vet Med Manhattan KS 66502

BUTLER, IAN SYDNEY, b Newhaven, Eng, Aug 22, 39; m 66. INORGANIC CHEMISTRY. Educ: Bristol Univ, BSc, 61, PhD(inorg chem), 64. Prof Exp: Res assoc, Ind Univ, 64-65 & Northwestern Univ, 65-66; asst prof, 66-71, ASSOC PROF CHEM, McGILL UNIV, 71- Concurrent Pos: Co-ed, Can J Spectros, 72- Mem: Am Chem Soc; Chem Inst Can; Spectros Soc Can; The Chem Soc. Res: Preparation and studies of the physical properties of transition metal organometallic compounds, particularly metal carbonyls; infrared and laser Raman spectroscopy. Mailing Add: Dept of Chem McGill Univ Montreal PQ Can

BUTLER, JACKIE DEAN, b Raleigh, Ill, Mar 1, 31; m 57; c 2. HORTICULTURE. Educ: Univ Ill, Urbana, BS, 57, MS, 59, PhD, 66. Prof Exp: Asst county farm adv agr, Univ Ill, Urbana, 57-58, from res asst to res assoc hort, 58-63, from instr to assoc prof, 63-71; ASSOC PROF HORT, COLO STATE UNIV, 71- Mem: Am Soc Hort Sci; Am Soc Agron. Res: Principles and practices associated with turfgrass establishment and maintenance. Mailing Add: Dept of Hort Colo State Univ Ft Collins CO 80521

BUTLER, JAMES EHRICH, b Tenafly, NJ, Nov 29, 44; m 69. CHEMICAL PHYSICS. Educ: Mass Inst Technol, BS, 66; Univ Chicago, PhD(chem physics), 72. Prof Exp: Fel chem physics, NIH, Univ Chicago, 72-74; res assoc, James Franck Inst, Univ Chicago, 74-75; RES CHEMIST, CHEM DIV, US NAVAL RES LAB, 75- Mem: Am Phys Soc; Optical Soc Am. Res: Photochemistry; chemical kinetics; energy transfer; chemiluminescence. Mailing Add: Chem Div Code 6110 Naval Res Lab Washington DC 20375

BUTLER, JAMES JOHNSON, b Jackson, Tenn, May 5, 26; m 52; c 2. PATHOLOGY. Educ: Univ Mich, MD, 52. Prof Exp: Intern, Cincinnati Gen Hosp, Ohio, 52-53; resident path, Univ Iowa & Univ Cincinnati, 53-57; from jr pathologist to assoc pathologist, Armed Forces Inst Path, 57-61; asst pathologist, 59-60 & 62-64, assoc pathologist, 64-74, PATHOLOGIST, UNIV TEX M D ANDERSON HOSP & TUMOR INST HOUSTON, 74-, PROF PATH, 75- Concurrent Pos: Mem path panel, Lymphoma Clin Trials. Mem: AAAS; AMA; Am Soc Hemat; Int Acad Path. Res: Path-physiology of the reticuloendothelial system; cancer research. Mailing Add: Univ Tex M D Anderson Hosp Houston TX 77025

BUTLER, JAMES KEITH, b Temple, Tex, Sept 26, 26; m 49; c 1. CYTOLOGY. Educ: Univ Tex, BS, 50, MA, 52, PhD(zool), 61. Prof Exp: Asst anat, Med Br, Univ Tex, 52-53, res assoc ortop surg, 53-55; ASSOC PROF CELL & DEVELOP BIOL, UNIV TEX, ARLINGTON, 60- Concurrent Pos: Genetics Found fel, Univ Tex, Austin, 65; vis assoc prof anat, Univ Tex Southwestern Med Sch, Dallas, 67. Mem: AAAS. Res: Biological ultrastructure and cytochemistry, spermatogenesis and cytology of the reproductive duct systems of selected crustaceans. Mailing Add: Dept of Biol Univ of Tex Arlington TX 76019

BUTLER, JAMES NEWTON, b Cleveland, Ohio, Mar 27, 34; m 57, 66; c 3. PHYSICAL CHEMISTRY, ENVIRONMENTAL SCIENCES. Educ: Rensselaer Polytech Inst, BSc, 55; Harvard Univ, PhD(chem physics), 59. Prof Exp: Fel chem, Harvard Univ, 59; from instr to asst prof, Univ BC, 59-63; sr scientist chem physics, Tyco Labs, Inc, 63-66, head phys chem dept, 66-71; GORDON McKAY PROF APPL CHEM, HARVARD UNIV, 71-, MEM FAC GEOL SCI, 72- Concurrent Pos: Consult, Tyco Labs, Inc, 62-63, 71-; lectr, Harvard Univ, 70-71; trustee, Bermuda Biol Sta, 72- Mem: Am Soc Limnol & Oceanog; The Chem Soc; Electrochem Soc; AAAS; Am Chem Soc. Res: Chemical kinetics; ionic equilibria; electrochemistry; surface chemistry. Mailing Add: Div of Eng & Appl Physics Harvard Univ Pierce Hall Cambridge MA 02138

BUTLER, JAMES ROBERT, b Macon, Ga, Apr 17, 30; m 60; c 3. GEOLOGY, PETROLOGY. Educ: Univ Ga, BS, 52; Univ Colo, MS, 55; Columbia Univ, PhD(geol), 62. Prof Exp: Lectr geol, Columbia Univ, 59-60; from asst prof to assoc prof, 60-72, PROF GEOL, UNIV NC, CHAPEL HILL, 72- Mem: AAAS; fel Geol Soc Am; Mineral Soc Am; Nat Asn Geol Teachers. Res: Igneous, structural and metamorphic evolution of the crystalline southern Appalachians; petrology of Precambrian metamorphic rocks, Beartooth Mountains, Montana; differentiation and textural development of igneous rocks. Mailing Add: Dept of Geol Univ of NC Chapel Hill NC 27514

BUTLER, JERRY FRANK, b Lingle, Wyo, Apr 17, 38; m 61; c 2. VETERINARY ENTOMOLOGY, MEDICAL ENTOMOLOGY. Educ: Univ Wyo, BS, 62, MS, 64; Cornell Univ, PhD(entom), 68. Prof Exp: Res asst entom, Univ Wyo, 62-64; res asst, Cornell Univ, 64-68, res technician, 68; asst prof, 68-74, ASSOC PROF ENTOM, UNIV FLA, 74- Mem: Entom Soc Am; Acarology Soc Am. Res: Bionomics and control of arthropods of veterinary importance; special interest in population dynamics of ectoparasites. Mailing Add: Dept of Entom & Nematol Univ of Fla Gainesville FL 32611

BUTLER, JOHN, b Grantham, Eng, Nov 28, 23; nat US; m 60; c 5. MEDICINE. Educ: Birmingham Univ, MB, ChB, 46, MD, 57; FRCP(E), 69. Prof Exp: Lectr med, Birmingham Univ, 56-58; sr lectr & consult physician, Manchester Univ, 58-60; assoc clin prof & lectr, Cardiovasc Res Inst, San Francisco, 60-65; PROF & HEAD DIV RESPIRATORY DIS, UNIV WASH, 65- Concurrent Pos: Dir pulmonary SCOR, Univ Hosp. Mem: Am Physiol Soc; Am Fedn Clin Res; Brit Med Res Soc. Res: Mechanical forces in the lungs; pulmonary circulation. Mailing Add: Dept of Med Univ of Wash Seattle WA 98195

BUTLER, JOHN BEN, JR, b New York, NY, Aug 26, 23; m 46; c 2. MATHEMATICS. Educ: Swarthmore Col, BS, 45; NY Univ, MS, 47; Univ Calif, PhD, 54. Prof Exp: Instr math, Univ Calif, 55-56; asst prof, Univ Wash, 57-59 & Univ Ariz, 59-60; ASST PROF MATH, PORTLAND STATE UNIV, 61- Mem: Soc Indust & Appl Math; Am Math Soc. Res: Abstract analysis; differential equations. Mailing Add: Dept of Math Portland State Univ Portland OR 97207

BUTLER, JOHN C, b Port Clinton, Ohio, Oct 31, 41; m 65; c 2. GEOLOGY, MINERALOGY. Educ: Miami Univ, BA, 63, MS, 65; Ohio State Univ, PhD(mineral), 68. Prof Exp: Instr geol, Miami Univ, 66-68; asst prof, 68-71, ASSOC PROF GEOL, UNIV HOUSTON, 71- Mem: Mineral Soc Am. Res: Structural states of alkali feldspars; mineralogy of contact metamorphosed marbles; size analysis of lunar regolith. Mailing Add: Dept of Geol Univ of Houston Houston TX 77004

BUTLER, JOHN EARL, b Stockton, Kans, Sept 7, 18; m 47; div; c 2. BOTANY. Educ: Ft Hayes Kans State Col, AB, 40; Univ Wis, MS, 49, PhD(bot), 54. Prof Exp: With AEC, Univ Calif, 48-51; teacher bot & agron, Wis State Col, 54-55; asst prof, Fresno State Col, 55-58; assoc prof bot, 58-71, PROF BIOL, HUMBOLDT STATE UNIV, 71- Mem: Ecol Soc Am; Phycol Soc Am; Nat Asn Biol Teachers. Res: Ecology of life histories; secondary science education; freshwate algae. Mailing Add: Dept of Biol Humboldt State Univ Arcata CA 95521

BUTLER, JOHN EDWARD, b Rice Lake, Wis, Jan 10, 38; div; c 1. IMMUNOLOGY. Educ: Univ Wis-River Falls, BS, 61; Univ Kans, PhD(zool), 66. Prof Exp: Actg asst prof zool, Univ Kans, 65-66, res assoc immunobiol, 66-67; res biologist, Agr Res Serv, USDA, 67-71; asst prof, 71-74, ASSOC PROF MICROBIOL, UNIV IOWA, 74- Concurrent Pos: USPHS fel & NSF fel, 64-66; vis prof, Max Planck Inst, WGer, 73-74. Mem: Am Asn Immunologists; NY Acad Sci; Soc Exp Biol & Med. Res: Ungulate immunoglobulins; secretory immune systems; milk protein intolerance; cancer immunochemistry. Mailing Add: Dept of Microbiol Univ of Iowa Sch of Med Iowa City IA 52241

BUTLER, JOHN JOSEPH, b Rochester, NY, Oct 18, 20; m 48; c 5. HEMATOLOGY. Educ: Univ Toronto, BA, 42; Univ Rochester, MD, 44; Am Bd Internal Med, dipl, 51. Prof Exp: From instr to asst prof med, Georgetown Univ, 52-59; assoc prof, Seton Hall Col Med & Dent, 59-66; CHIEF HEMAT, CATH MED CTR BROOKLYN & QUEENS, 66- Mem: Am Col Physicians; Am Fedn Clin Res;

Am Soc Hemat; Int Soc Hemat. Res: Red and white cell metabolism. Mailing Add: 88-25 153rd St Jamaica NY 11432

BUTLER, JOHN L, zoology, limnology, see 12th edition

BUTLER, JOHN LOUIS, b Brockton, Mass, Aug 23, 34; m 59; c 4. PHYSICS. Educ: Northeastern Univ, BS, 57, PhD(acoust), 67; Brown Univ, ScM, 62. Prof Exp: Res engr, Melpar Inc, 61-62; sr engr, Harris ASW Div, Gen Instrument Corp, 62-66; res assoc physics & acoust, Parke Math Labs, Inc, 66-70; develop engr acoust, Massa Div Dynamics Corp of Am, 70-72; sr engr, Raytheon Co, Submarine Signal Div, 72-75; PRES, IMAGE ACOUST, INC, 75- Concurrent Pos: Lectr, Northeastern Univ, 68, 70, 73 & 75. Mem: Acoust Soc Am. Res: Acoustical radiation from underwater sound sources and arrays; acoustic transducers and arrays. Mailing Add: 205 Carolyn Circle Marshfield MA 02050

BUTLER, JOHN MANN, b Richmond, Va, Mar 23, 17; m 42; c 2. ORGANIC CHEMISTRY, POLYMER CHEMISTRY. Educ: Richmond Univ, BS, 37; Ohio State Univ, PhD(org chem), 40. Prof Exp: Res chemist, Bakelite Corp, NJ, 40-41; res chemist, Monsanto Co, 41-45, group leader, 45-53, sect leader, 53-54, asst dir res, 54-59, res specialist, 59-61, MGR ORG & POLYMER RES, MONSANTO RES CORP, 61- Res: Polymer synthesis and application; organic synthesis. Mailing Add: Monsanto Res Corp 1515 Nicholas Rd Dayton OH 45418

BUTLER, JOHN PARKMAN, b St Louis, Mo, Feb 15, 29; m 55; c 3. RESEARCH ADMINISTRATION. Educ: Amherst Col, AB, 50; Iowa State Univ, MS, 52, PhD(anal chem), 54. Prof Exp: Res asst anal chem, Inst Paper Chem, 54-57; mem tech serv staff, 57-76, supvr flexible pkg, 67-76, ASSOC DIR RES & DEVELOP, AM CAN CO, 76- Mem: Am Chem Soc. Res: Development of materials for use in food packaging. Mailing Add: 333 N Commercial St Neenah WI 54956

BUTLER, JOSEPH HERBERT, b Oyster Bay, NY, Feb 20, 26; m 49; c 4. ECONOMIC GEOGRAPHY. Educ: Cornell Univ, BCE, 47; Columbia Univ, PhD(geog), 60. Prof Exp: Engr, Westinghouse Elec Int Co, NY, 47-49, proj engr, 49-51; lectr geog, Columbia Univ, 54-56, asst to dean eng, 57-59; assoc prof geog, Mich Col Mining & Technol, 59-63; PROF GEOG, STATE UNIV NY BINGHAMTON, 63- Concurrent Pos: Lectr, Columbia Univ, 57-59. Mem: AAAS; Asn Am Geog; Am Soc Eng Educ. Res: Water resources; land use and conservation; theory of production; industrialization and resource utilization in developing areas, principally Latin America. Mailing Add: Dept of Geog State Univ NY Binghamton NY 13901

BUTLER, JOSEPH MILES, b Paducah, Ky, Dec 4, 23; m 46; c 3. MEDICAL ENTOMOLOGY, PARASITOLOGY. Educ: Univ Utah, BS, 48, MS, 50, PhD, 52. Prof Exp: Asst zool, Univ Utah, 48-52; proj officer, Parasitol Sect, Dugway Proving Ground, Dept Army, Utah, 52-53; malariologist, Tech Coop Admin, US Opers Mission, Saudi Arabia, USPHS, 53-54, malaria adv, Int Coop Admin, Thailand, 54-60 & Malaria Eradication Training Ctr, Kingston, Jamaica, 60, nat malaria training adv, US Int Coop Admin, Indonesia, 60-62, nat parasitol lab serv adv, 63, chief parasitol unit, Trop Dis Sect, Ecol Invest Prog, Nat Commun Dis Ctr, PR, 63-70, parasitologist, Dis Invests Sect, Ecol Invest Prog, Phoenix Labs, Ctr Dis Control, 70-71; PRES, ARIZ BUTLER CORP, 71- Concurrent Pos: Lectr, Univ PR, 69-70. Mem: AAAS; Am Soc Parasitologists; Am Soc Trop Med & Hyg; Royal Soc Trop Med & Hyg. Res: Helminthology; schistosomiasis; malaria; biological control; chemical control; insect pest control. Mailing Add: 5103 E Calle del Norte Phoenix AZ 85018

BUTLER, KARL DOUGLAS, SR, b Douglas, Ariz, Feb 4, 10; m; c 5. AGRICULTURE. Educ: Univ Ariz, BS, 31, MS, 33; Cornell Univ, PhD(plant sci), 40. Prof Exp: Asst plant path, Iowa State Col, 33-34; jr forest pathologist, Div Forest Path, USDA, 34; res asst & instr plant path, Univ Ariz, 34-36, asst plant pathologist, 37-40; agent rubber dis invests, Bur Plant Indust, USDA, 40-43, dir res agr chem, Coop Grange League Fedn Exchange, Inc, 43-45, dir educ & res, 45-47; pres, Am Inst Coop, Washington, DC, 47-50; farm counr, 50-67; BUS & AGR CONSULT, 67- Concurrent Pos: Mem, Am Rubber Surv, Bolivia & Brazil, 40-41; pres, Comn Increased Industs Use Agr Prods, 56-57; USDA farm mech exchange deleg to USSR, 58; agr specialist, Peace Corps, Latin Am Countries, 63-64, US AID, Taiwan & India, 64 & WAfrican Studies for Afro-Am Labor Ctr, New York, 66; chmn exec comt & mem res comt, Chemurgic Coun. Honors & Awards: Am Meat Inst Award, 63. Mem: Fel AAAS; Am Inst Animal Agr (secy); Am Soc Agr Engrs. Res: Farmer cooperatives; diseases of cotton, dates, lettuce, trees, cereal crops and watermelons. Mailing Add: 1583 E Shore Dr Ithaca NY 14850

BUTLER, KEITH HUESTIS, b Halifax, NS, July 5, 05; nat US; m 29; c 2. CHEMISTRY. Educ: Dalhousie Univ, BA, 25, MA, 26; McGill Univ, PhD(phys chem), 28. Prof Exp: Asst, Mass Inst Technol, 28-29; res chemist, E I du Pont de Nemours & Co, Inc, 29-44; eng specialist, Sylvania Lighting Prod Group, 44-46, mgr eng labs, 56-66, mgr phosphor lab, 66-69, mgr res lab chem, 69-70; RETIRED. Mem: Optical Soc Am; Electrochem Soc. Res: Fluorescent powders; pigments; lamps. Mailing Add: 89 Clifton Ave Marblehead MA 01945

BUTLER, KEITH WINSTON, b St Boniface, Man, Apr 12, 41; m 66; c 2. BIOPHYSICS. Educ: Univ Toronto, BSA, 63; Duke Univ, PhD(physiol), 68. Prof Exp: Univ fel, Aarhus Univ, 68-69; Nat Res Coun Can fel, 69-71, vis scientist, 71-72, ASST RES OFFICER BIOPHYS, NAT RES COUN CAN, 72- Mem: Biophys Soc. Res: Electron spin resonance studies of the structure and kinetics of real and model membrane systems. Mailing Add: Biol Sci Div Nat Res Coun of Can Ottawa ON Can

BUTLER, KENNETH, organic chemistry, see 12th edition

BUTLER, LARRY G, b Elkhart, Kans, Dec 14, 33; m 53; c 4. BIOCHEMISTRY. Educ: Okla State Univ, BS, 60; Univ Calif, Los Angeles, PhD(biochem), 64. Prof Exp: Asst prof natural sci & chmn dept, Los Angeles Baptist Col, 64-65; res assoc biochem, Univ Ariz, 65-66; ASSOC PROF BIOCHEM, PURDUE UNIV, WEST LAFAYETTE, 66- Concurrent Pos: NIH res career develop award, 70-75; vis prof chem, Univ Ore, 74-75. Mem: Am Chem Soc; Am Soc Biol Chem; Am Soc Plant Path. Res: Elucidation of the detailed mechanism of enzyme action, particularly those enzymes which hydrolyze high energy phosphate bonds; enzyme technology. Mailing Add: Dept of Biochem Purdue Univ West Lafayette IN 47906

BUTLER, LEONARD, b Finedon, Eng, Apr 13, 12; m 39; c 3. GENETICS. Educ: Ont Agr Col, BSA, 35; Univ Toronto, MSA, 37, PhD(genetics), 39. Prof Exp: Demonstr zool, Univ Toronto, 35-39; biologist, Hudson's Bay Co, 39-46; asst prof genetics, McGill Univ, 47-48; from asst prof to assoc prof, 48-63, PROF GENETICS, UNIV TORONTO, 63- Mem: Am Genetic Asn; Genetics Soc Am; Genetics Soc Can. Res: Quantitative inheritance in tomatoes, mice and Drosophila; linkage map of the tomato; population genetics of the color phases of red fox. Mailing Add: Dept of Zool Univ of Toronto Toronto ON Can

BUTLER, LEWIS CLARK, b Hornell, NY, July 11, 23; m 48; c 4. MATHEMATICS. Educ: Alfred Univ, BA, 44; Rutgers Univ, MS, 48; Univ Ill, PhD(math), 57. Prof Exp: Instr math, Alfred Univ, 47-49; asst, Univ Ill, 49-54; instr, Pa State Univ, 54-57; from asst prof to assoc prof, State Univ NY Col Ceramics, Alfred Univ, 57-62; DEAN GRAD SCH, ALFRED UNIV, 63- Concurrent Pos: Dir cent inst math, Concepcion Univ, 61-63. Mem: Am Math Soc. Res: Algebraic topology; fiber spaces. Mailing Add: Grad Sch Alfred Univ Alfred NY 14802

BUTLER, LILLIAN IDA, b Saginaw, Mich, Jan 31, 10. ANALYTICAL CHEMISTRY, BIOLOGICAL CHEMISTRY. Educ: Univ Mich, BS, 29, MS, 30. Prof Exp: Asst chem, Mich State Col, 30-42; assoc human nutrit & home econ, 42-43, chemist, 43-47, div bee cult, 47-51, pesticides chem res, 51-53, asst head chemist, 53-58, actg in charge, 57-68, RES CHEMIST & ASST HEAD, USDA, 58- Concurrent Pos: Mem, Gov Status of Women Comn, Wash, 63-65. Honors & Awards: Cert of Merit, Off Sci Res & Develop, 44; Unit Award, USDA, 64. Mem: Fel AAAS; Am Chem Soc; fel Am Inst Chem; Entom Soc Am. Res: Analytical methods, primarily in foods and crops; inorganic minerals; vitamins; insecticides; antibiotics and virus; development of methods; isolation of sex attractants from insects. Mailing Add: 2804 Arlington Yakima WA 98902

BUTLER, LINDA, b Martin, Tenn, Nov 11, 43. ENTOMOLOGY. Educ: Univ Ga, BS, 65, MS, 66, PhD(entom), 68. Prof Exp: From asst prof to assoc prof, 68-75, PROF ENTOM, WVA UNIV, 75- Honors & Awards: J Everett Bussart Mem Award, Entom Soc Am, 74. Mem: AAAS; Entom Soc Am; Electron Micros Soc Am. Res: Insect aging; insect ultrastructure and biochemistry; insect behavior, attractants and control. Mailing Add: Dept of Entom WVa Univ Morgantown WV 26506

BUTLER, MARGARET K, b Evansville, Ind, Mar 7, 24; m 51; c 1. MATHEMATICS. Educ: Ind Univ, AB, 44. Prof Exp: Statistician, US Bur Labor Statist, 45-46 & US Air Forces in Europe, 46-48; mathematician, Argonne Nat Lab, 48-49; statistician, US Bur Labor Statist, 49-51; MATHEMATICIAN, ARGONNE NAT LAB, 51-, DIR ARGONNE CODE CTR, 71- Mem: Fel Am Nuclear Soc; Asn Comput Mach. Res: Information systems; computer program interchange; computer system performance; documentation standards; scientific and engineering applications. Mailing Add: Argonne Code Ctr 9700 S Cass Ave Argonne IL 60439

BUTLER, MICHAEL ALFRED, b Chesterfield, Eng, Nov 24, 43; m 66. SOLID STATE PHYSICS. Educ: Rensselaer Polytech Inst, BS, 64; Univ Calif, Santa Barbara, MA, 66, PhD(physics), 69. Prof Exp: Res fel, Univ Calif, Santa Barbara, 69-70; mem tech staff, Bell Tel Labs, Inc, NJ, 70-75; MEM TECH STAFF, SANDIA LABS, 75- Mem: Am Phys Soc. Res: Use of solid state devices for conversion of solar energy to more useful forms. Mailing Add: Sandia Labs Albuquerque NM 87115

BUTLER, OGBOURNE DUKE, b Orange, Tex, Sept 29, 18; m 43; c 3. ANIMAL SCIENCE. Educ: Agr & Mech Col Tex, BS, 39, MS, 47; Mich State Univ, PhD(animal husb), 53. Prof Exp: From instr to assoc prof, 47-57, PROF ANIMAL SCI & HEAD DEPT, TEX A&M UNIV, 57- Mem: Am Soc Animal Sci; Inst Food Technol. Res: Meats; microbiology; biochemistry; statistics. Mailing Add: Dept of Animal Sci Col of Ag Tex A&M Univ College Station TX 77843

BUTLER, ORTON CARMICHAEL, b Millersburg, Ohio, June 9, 23; m 51; c 2. EARTH SCIENCE. Educ: Oberlin Col, AB, 48; Clark Univ, MA, 51; Ohio State Univ, PhD(geog), 69. Prof Exp: Mil intel res specialist geog, US Army Corps Engrs, 51-60; ASSOC PROF GEOG, MEMPHIS STATE UNIV, 60- Mem: Asn Am Geogr; Nat Coun Geog Educ; Int Geog Union; Am Meteorol Soc. Res: Geography of China; microclimatology and teaching methods in geography; classification of microclimates; measures of continentality. Mailing Add: Dept of Geog Memphis State Univ Memphis TN 38152

BUTLER, PATRICK COLIN, b Kingston, Jamaica, June 1, 32; m 57. SOIL FERTILITY. Educ: Cornell Univ, BS, 53; Univ Ill, MS, 55; NC State Col, PhD(soil chem), 57. Prof Exp: Assoc soil chemist, United Fruit Co, Honduras, 57-59, soil chemist, Cent Res Labs, Mass, 59-60, spec duties, Gen Off, 60-63, tech asst to dir res, 63-66, sr econ analyst, 66-68, mgr com info, 68-70, dir agr develop, 70-74, DIR TECH SERV, AGRIMARK GROUP, UNITED BRANDS CO, 74- Mem: Soil Sci Soc Am. Res: Fertility as related to the banana plant; efficiency of utilization of nitrogen fertilizers by the banana plant. Mailing Add: Agrimark Group United Brands Co Prudential Ctr Boston MA 02199

BUTLER, PHILIP ALAN, b Upper Montclair, NJ, May 15, 14; m 41; c 2. MARINE BIOLOGY. Educ: Northwestern Univ, BS, 35, PhD(zool), 40. Prof Exp: Fishery res biologist, US Fish & Wildlife Serv, 46-58, dir biol lab, 58-68, res consult, 68-70; RES CONSULT, ENVIRON PROTECTION AGENCY, 71- Mem: AAAS; Am Soc Limnol & Oceanog; Nat Shellfisheries Asn (secy-treas, 57-59, vpres, 59-61, pres, 61-63); Am Fisheries Soc. Res: Biology and economics of shellfish; effects of pesticides in estuaries. Mailing Add: Environ Protection Agency 106 Matamoros Dr Gulf Breeze FL 32561

BUTLER, ROBERT ALLAN, b Pittsfield, Mass, Mar 29, 23; m 52; c 4. PSYCHOACOUSTICS. Educ: Univ Fla, BA, 47; Univ Chicago, PhD(psychol), 51. Prof Exp: Instr psychol, Univ Wis, 51-53; res psychologist, Walter Reed Army Hosp, 53-57; res assoc biopsychol, 57-65, assoc prof surg & psychol, 65-72, PROF SURG & BEHAV SCI, UNIV CHICAGO, 72- Mem: Acoustical Soc Am; Asn Res Otolaryngol (pres, 75-76). Res: Psychophysics of hearing and the electrophysiology of the auditory system. Mailing Add: Dept of Surg Univ of Chicago Chicago IL 60637

BUTLER, ROBERT FRANKLIN, b Eugene, Ore, July 6, 46; m 67. PALEOMAGNETISM. Educ: Ore State Univ, BS, 68; Stanford Univ, MS, 70, PhD(geophys), 72. Prof Exp: Fel geophys, Univ Minn, Minneapolis, 72-74; ASST PROF GEOPHYS, UNIV ARIZ, 74- Mem: Am Geophys Union; AAAS. Res: Paleomagnetic polarity stratigraphy of cretaceous-lower tertiary deposits of the San Juan Basin, New Mexico; paleomagnetic recordings of transitional behavior of the geomagnetic field. Mailing Add: Dept of Geosci Univ of Ariz Tucson AZ 85721

BUTLER, ROBERT LEE, b Kansas City, Kans, Feb 12, 18; m 43; c 3. FISHERIES. Educ: Park Col, BA, 40; Univ Minn, BS, 47, MS, 50, PhD, 62. Prof Exp: Asst, Red Lake Com Fishery, Univ Minn, 49-52, res assoc insect & rodent, Univ, 52-54; fisheries biologist, Ill, State Dept Fish & Game, Calif, 54-62; res zoologist, Univ Calif, 62-63; asst prof & leader coop fishery unit, 63-69, assoc prof, 69-72, PROF ZOOL, PA STATE UNIV, 72- Mem: Am Fisheries Soc; Am Soc Limnol & Oceanog; Ecol Soc Am; Animal Behav Soc. Res: Life history studies of the freshwater drum, Aplodinotus grunniens; utilization of hatchery trout, Salmo gairdneri; stream and lake ecology; fish behavior. Mailing Add: 328 Life Sci Pa State Univ University Park PA 16802

BUTLER, ROBERT N, b New York, NY, Jan 21, 27; m 50; c 3. PSYCHIATRY. Educ: Columbia Univ, BA, 49, MD, 53. Prof Exp: Intern med, St Luke's Hosp, New York, 53-54; resident psychiat, Univ Calif, 54-55; res psychiatrist, NIMH, 55-62; ASSOC PROF PSYCHIAT, SCH MED, GEORGE WASHINGTON UNIV, 62-; ASSOC PROF PSYCHIAT, COL MED, HOWARD UNIV, 72- Concurrent Pos: Resident psychiat, NIMH, Chestnut Lodge, Md, 57-58, clin adminr psychother aged, 58-59, consult, 59-68; res psychiatrist, Wash Sch Psychiat, 62-; lectr, Wash Psychoanal Inst, 62-; consult, St Elizabeth's Hosp, Washington, DC, 64-, US Senate Comt Aging, 71- & NIMH; chmn, Washington, DC Adv Comt on Aging; mem bd, Nat Coun Aging; chmn comt aging, Group Advan Psychiat. Mem: Fel Am Psychiat Asn; fel Am Geriat Soc; Geront Soc. Res: psychotherapy theory and technique; nature of health and disorder in the aged; aging; schizophrenia; personality theory; psychoanalysis; methods in psychiatry; creativity. Mailing Add: 3815 Huntington St NW Washington DC 20015

BUTLER, ROBERT WESTBROOK, b Jamaica, Brit WI, June 23, 21; US citizen; m 54; c 2. POLYMER CHEMISTRY, ORGANIC CHEMISTRY. Educ: Imp Col Trop Agr, Brit WI, dipl, 43; McGill Univ, BSc, 46, PhD(cellulose chem), 50. Prof Exp: Res chemist cellulose mixed ester, Indust Cellulose Res Ltd, Ont, 50-52; res chemist, 52-58, res chemist, Mkt & New Prod Develop, 59-68, sales supvr, 69-70, mgr mkt serv, 71-73, MGR MKT DEVELOP, HERCULES INC, 73- Mem: Am Chem Soc. Res: Cellulose ethers and esters; starch derivatives; properties and applications for water soluble polymers. Mailing Add: 203 Lister Dr Hyde Park Wilmington DE 19808

BUTLER, RONALD G, b Fairbury, Ill, Sept 22, 40; m 63; c 2. MATHEMATICS. Educ: Ill State Univ, BS & MS, 63; Okla State Univ, EdD(math), 70. Prof Exp: Asst prof math, Radford Col, 63-65; assoc prof, 69-74, PROF MATH, EDINBORO STATE COL, 74- Mem: Math Asn Am; Am Math Soc. Res: Metrization of Moore spaces. Mailing Add: RD 1 105 Valley View Dr Edinboro PA 16412

BUTLER, TERENCE, b Cincinnati, Ohio, Nov 21, 29; m 56; c 5. MATHEMATICS. Educ: Mass Inst Technol, SB, 51; Harvard Univ, AM, 52; Univ Ind, PhD(math), 59. Prof Exp: From instr to assoc prof, 58-71, PROF MATH, RUTGERS UNIV, 71- Concurrent Pos: Fac res fel, Rutgers Univ, 60 & 63; consult, Off Res Anal, Holloman Air Force Base, NMex, 59, 60, 61 & 62. Res: Differential equations, similar to the Navier Stokes equations of fluid flow, for properties resembling turbulent flow phenomena; calculus of variations of optimization and control problems. Mailing Add: Dept of Math Rutgers The State Univ New Brunswick NJ 08903

BUTLER, THOMAS ARTHUR, b Farmington, Ark, May 14, 19; m 45, 53; c 3. CHEMISTRY. Educ: Southwest Mo State Col, AB, 41; Iowa State Col, MS, 51. Prof Exp: Asst chem, Iowa State Col, 41-42, jr res chemist, 42-51; SUPT, OAK RIDGE NAT LAB, 51- Mem: Am Chem Soc. Res: Preparation of uranium compounds; separation of rare earth elements; radioisotope research and development. Mailing Add: 119 Dana Dr Oak Ridge TN 37830

BUTLER, THOMAS CULLOM, b Phoenix, Ariz, Apr 2, 10; m 36; c 3. PHARMACOLOGY. Educ: Vanderbilt Univ, AB, 30, MD, 34. Prof Exp: From asst to asst prof pharmacol, Sch Med, Vanderbilt Univ, 34-44, assoc prof, 46; assoc prof pharmacol & exp therapeut, Sch Med, Johns Hopkins Univ, 46-50; PROF PHARMACOL, SCH MED, UNIV NC, CHAPEL HILL, 50- Mem: Am Soc Pharmacol & Exp Therapeut; Soc Exp Biol & Med. Res: Pharmacology of hypnotic and anesthetic drugs; metabolic fate of drugs; intracellular pH. Mailing Add: Univ of NC Sch of Med Chapel Hill NC 27514

BUTLER, THOMAS DANIEL, b Oklahoma City, Okla, Apr 19, 38; m 58; c 3. FLUID DYNAMICS. Educ: NMex Inst Mining & Technol, BS, 60; Univ NMex, MS, 64. Prof Exp: Staff mem group T-3, 60-72, assoc group leader T-3, 72-73, ALT GROUP LEADER T-3 PHYSICS, LOS ALAMOS SCI LAB, 73- Concurrent Pos: Consult, Battelle Columbus Labs, Durham Opers Off, 75- Res: Development of numerical methods to solve a wide variety of multidimensional, transient fluid dynamics problems. Mailing Add: Theoret Div Los Alamos Sci Lab Mail Stop 216 PO Box 1663 Los Alamos NM 87545

BUTLER, VINCENT JOHN, psychiatry, psychology, see 12th edition

BUTLER, VINCENT PAUL, JR, b Jersey City, NJ, Feb 16, 29. INTERNAL MEDICINE, IMMUNOCHEMISTRY. Educ: St Peter's Col, AB, 49; Columbia Univ, MD, 54. Prof Exp: Intern med, Presby Hosp, New York, 54-55, asst resident, 55-56 & 58-59; asst med & microbiol, Sch Med & Dent, Univ Rochester, 59-60, instr med, 60-61; from asst prof to assoc prof, 63-74, PROF MED, COL PHYSICIANS & SURGEONS, COLUMBIA UNIV, 74-; ATTEND PHYSICIAN, PRESBY HOSP, NEW YORK, 74- Concurrent Pos: Helen Hay Whitney Found fel, 60-63; vis res fel microbiol, Col Physicians & Surgeons, Columbia Univ, 61-63; asst attend physician, First Med Div, Bellevue Hosp, 63-68; sr investr, Arthritis Found, 63-68; from asst attend physician to assoc attend physician, Presby Hosp, New York, 68-74; asst vis physician, Harlem Hosp, 68-; career scientist, Irma T Hirschl Found, 73- Mem: Am Soc Exp Pharmacol & Therapeut; Am Asn Immunol; Am Heart Asn; Am Soc Clin Invest; Am Rheumatism Asn. Res: Immunochemical studies of rheumatoid factor; purine specific antibodies and their cross reactions with DNA; digoxin-specific antibodies; immunochemical studies of fibrinogen. Mailing Add: Dept of Med Columbia Univ Col Phys & Surg New York NY 10032

BUTLER, WALTER CASSIUS, b Rifle, Colo, Feb 10, 10; m 34; c 2. MATHEMATICS. Educ: Colo Agr & Mech Col, BS, 32; Colo State Col, MA, 42. Prof Exp: Teacher high sch, Colo, 32-33 & 34-35; miner, 33-34; prin high sch, Ill, 35-38; supt sch, Wyo, 38-46; prof math, Ft Lewis Agr & Mech Col, 46-50; from asst prof to assoc prof, 50-74, EMER ASSOC PROF MATH, COLO STATE UNIV, 74- Concurrent Pos: Dir, Jr Engr Scientists' Summer Inst, 60- Mem: Am Soc Mech Eng; Math Asn Am. Res: Training of junior and senior high school mathematics teachers. Mailing Add: Dept of Math Colo State Univ Ft Collins CO 80521

BUTLER, WARREN LEE, b Yakima, Wash, Jan 28, 25; m 51; c 3. BIOPHYSICS. Educ: Reed Col, BA, 49; Univ Chicago, PhD(biophys), 55. Prof Exp: Res assoc, Univ Chicago, 55-56; biophysicist, USDA, 56-64; PROF BIOL, UNIV CALIF, SAN DIEGO, 64- Concurrent Pos: Charles F Kettering res award, 63; spec fel, NIH, 64-65. Mem: Nat Acad Sci; Am Soc Plant Physiol; Biophys Soc; Am Inst Biol Sci. Res: Photobiology of plants, especially photosynthesis and photomorphogenesis. Mailing Add: Univ of Calif at San Diego PO Box 109 La Jolla CA 92037

BUTLER, WILLIAM ALBERT, b Independence, Kans, Jan 4, 22; m 44; c 3. PHYSICS. Educ: Univ Kansas City, BA, 42; Univ Ill, MS, 47, PhD(physics), 52. Prof Exp: Asst physics, Univ Ill, 46-52, & Carleton Col, 52-70; PROF PHYSICS & HEAD DEPT, EASTERN ILL UNIV, 70- Concurrent Pos: Vis assoc prof, Univ Calif, Berkeley, 60-61; vis scientist, Univ Ill, 67-68. Mem: Am Phys Soc; Am Asn Physics Teachers; Sigma Xi. Res: Nuclear and solid state physics. Mailing Add: Dept of Physics Eastern Ill Univ Charleston IL 61920

BUTLER, WILLIAM H, b Tallassee, Ala, Aug 13, 43; m 67; c 2. SOLID STATE PHYSICS. Educ: Univ Calif, San Diego, PhD(physics), 69. Prof Exp: Asst prof physics, Auburn Univ, 69-72; MEM RES STAFF, OAK RIDGE NAT LAB, 72- Mem: AAAS; Am Phys Soc. Res: Theory of electronic states in metals and alloys; theory of disordered alloys; electron phonon interaction in transition metals. Mailing Add: Metals & Ceramic Div Oak Ridge Nat Lab Oak Ridge TN 37830

BUTLER, WILLIAM THOMAS, b Ranger, Tex, Dec 10, 35; m 59; c 2. BIOCHEMISTRY. Educ: Baylor Univ, BS, 58; Vanderbilt Univ, PhD(biochem), 66. Prof Exp: Pub sch teacher, Tex, 58-60; res asst biochem, Vanderbilt Univ, 61-63; staff fel, Nat Inst Dent Res, 66-67, asst prof, Univ, 67-72, assoc prof biochem & asst prof physiol & biophys, 72-75, PROF BIOCHEM, MED CTR, UNIV ALA, BIRMINGHAM, 75-, SR SCIENTIST, INST DENT RES, 75- Res: Primary structure of collagen; special aspects of collagen chemistry; proteins of periodontal tissues; phosphoprotein of dentin. Mailing Add: Inst of Dent Res Univ of Ala Med Ctr Birmingham AL 35233

BUTOW, RONALD A, b Aug 1, 36; US citizen; m 58; c 1. BIOCHEMISTRY. Educ: Hobart Col, BS, 58; Cornell Univ, MNS, 60, PhD(biochem), 63. Prof Exp: NSF res fel biochem, Pub Health Res Inst New York, 63-65; asst prof, Princeton Univ, 65-71; ASSOC PROF BIOCHEM, UNIV TEX HEALTH SCI CTR DALLAS, 71- Concurrent Pos: USPHS res grant, 66. Mem: AAAS; Am Soc Biol Chemists. Res: Biological electron transport systems; oxidative phosphorylation bioenergetic reactions; structure; function and biogenesis of membranes. Mailing Add: Dept of Biochem Univ of Tex Health Sci Ctr Dallas TX 72535

BUTSCH, ROBERT STEARNS, b Owatonna, Minn, July 10, 14; m 41; c 1. MAMMALOGY. Educ: Univ Iowa, BA, 36, MA, 41; Univ Mich, PhD(zool), 54. Prof Exp: Cur, Arrowhead Mus, 36-37; chief preparator marine zool, Barbados Mus & Hist Soc, Barbados, BWI, 38-39; asst to dir, 51-55, assoc cur, 56-64, CUR EXHIBITS, EXHIBIT MUS, UNIV MICH, 64- Concurrent Pos: Lectr, Wildlife Films, Nat Audubon Soc. Mem: Am Soc Mammal; Am Asn Mus. Res: Ecology of vertebrates; design and construction of natural history exhibits; teaching of museum methods. Mailing Add: Univ of Mich Exhibit Mus Ann Arbor MI 48104

BUTSON, ALTON THOMAS, b Pa, Feb 18, 26; m 48; c 4. MATHEMATICS. Educ: Franklin & Marshall Col, BS, 50; Mich State Univ, MS, 51, PhD(math), 55. Prof Exp: Asst math, Mich State Univ, 50-54, instr, 54-55; from asst prof to assoc prof, Univ Fla, 55-59; res specialist, Boeing Co, Wash, 59-61; PROF MATH, UNIV MIAMI, 61- Mem: Am Math Soc; Math Asn Am. Res: Lattice, group and number theory; combinatorial analysis. Mailing Add: Dept of Math Univ of Miami Box 9085 Coral Gables FL 33124

BUTSON, KEITH D, b Iowa, Aug 15, 20; m 50; c 2. CLIMATOLOGY. Educ: Upper Iowa Univ, BS, 41; Iowa State Univ, MS, 47. Prof Exp: Climatologist, US Weather Bur, Washington, DC, 47-48 & 55-56, Ore, 48-50, River Forecast Ctr, 50-55, Fla, 56-66; PROJ COORDR, NAT CLIMATIC CTR, NAT OCEANIC & ATMOSPHERIC ADMIN, 66- Mem: Am Meteorol Soc; Am Geophys Union. Res: Climatology related to agriculture and general economy of Florida. Mailing Add: Nat Climatic Ctr Fed Bldg Asheville NC 28801

BUTT, ELIZABETH, b Atlanta, Ga, Mar 25, 12. MEDICAL MICROBIOLOGY. Educ: Wesleyan Col, Ga, AB, 34; Univ Mich, Ann Arbor, MPH, 55. Prof Exp: Technician microbiol, Ga Dept Pub Health, 34-37, bacteriologist, 37-42, actg dir, Albany Br Lab, 42-45; bacteriologist, Ill Dept Pub Health, 46-48; dir, Waycross Regional Lab, Ga Dept Pub Health, 48-57, chief tuberc & parasitol sect, 57-64, chief microbiol sect, 64-66; CHIEF BACT SECT, GA DEPT HUMAN RESOURCES, 66- Mem: Am Soc Microbiol; Med Mycol Asn Americas; Conf Pub Health Lab Dirs. Mailing Add: Ga Dept of Human Resources 47 Trinity Ave SW Atlanta GA 30334

BUTT, HUGH ROLAND, b Belhaven, NC, Jan 8, 10; m 39; c 4. INTERNAL MEDICINE. Educ: Univ Va, MD, 33; Univ Minn, MS, 37; Am Bd Internal Med, dipl. Prof Exp: First asst, Div Med, 37-38, from instr to assoc prof, 38-52, PROF MED, MAYO GRAD SCH MED, UNIV MINN, 52- Concurrent Pos: Attend physician, St Mary's, Methodist & Assoc Hosp, 38- Honors & Awards: Horsley Mem Prize, 38. Mem: AMA; master Am Col Physicians; Am Soc Clin Invest; Am Gastroenterol Asn; Asn Am Physicians. Res: Medicine; clinical research gastroenterology. Mailing Add: 200 First St SW Rochester MN 55901

BUTTAR, HARPAL SINGH, b Nanonkee, India, Nov 2, 39; Can citizen; m 62; c 3. PHARMACOLOGY, TOXICOLOGY. Educ: Col Vet Med, India, BVSc, 61; Univ Alta, MSc, 66, PhD(pharmacol), 71. Prof Exp: Lectr pharmacol, Col Vet Med, India, 61-64; teaching asst, Univ Alta, 64-70; res fel med, Sch Med, Wayne State Univ, 70-71; RES SCIENTIST TOXICOL, DRUG RES LABS, HEALTH PROTECTION BR, HEALTH & WELFARE CAN, 71- Mem: Chem Inst Can; Can Asn Res Toxicol. Res: Toxicological assessment of drugs, drug metabolites and cosmetics with reference to their absorption, distribution, biotransformation and excretion as well as drug interactions and embryotoxicity. Mailing Add: Drug Res Labs Health Protection Br Health & Welfare Can Ottawa ON Can

BUTTE, WALTER ALBERT, JR, organic chemistry, see 12th edition

BUTTER, FRANKLIN ALFRED, JR, mathematics, see 12th edition

BUTTER, STEPHEN ALLAN, b New York, NY, May 15, 37; m 63; c 1. INORGANIC CHEMISTRY. Educ: Brooklyn Col, BS, 59; Univ Del, PhD(inorg chem), 65. Prof Exp: Teaching asst, Univ Md, 59-60; chemist, Thiokol Chem Corp, 63; res chemist, Indust & Biochem Dept, E I du Pont de Nemours & Co, 64-66; res fel, Univ Sussex, 66-67; SR RES CHEMIST, MOBIL CHEM CO, 67- Mem: AAAS; Am Inst Chem; NY Acad Sci; Am Chem Soc. Res: Transition metals; organometallics; metallocenes; boron chemistry; coordination complexes; zeolites; homogeneous and heterogeneous catalysis. Mailing Add: Res & Develop Labs Mobil Chem Co PO Box 240 Edison NJ 08817

BUTTERBAUGH, DARREL J, chemistry, see 12th edition

BUTTERFIELD, DAVID ALLAN, b Milo, Maine, Jan 14, 46; m 68; c 1 BIOPHYSICAL CHEMISTRY. Educ: Univ Maine, BA, 68; Duke Univ, PhD(phys chem), 74. Prof Exp: Teacher chem high sch, United Methodist Church, Rhodesia, 68-71; NIH fel neurosci, Med Sch, Duke Univ, 74-75; ASST PROF CHEM, UNIV KY, 75- Mem: Am Chem Soc; Sigma Xi. Res: Biological applications of electron spin resonance, with particular emphasis on membrane systems; genetic, viral and carcinogenic effects on membrane structure and function; muscular dystrophy. Mailing Add: Dept of Chem Univ of Ky Lexington KY 40506

BUTTERFIELD, VELOY HANSEN, JR, b Fillmore, Utah, Sept 24, 42; m 66; c 2. COMPUTER SCIENCES. Educ: Univ Utah, BSEE, 67, MEA, 73. Prof Exp: Electronic engr, Design Div, Mare Island Naval Shipyard, 68-70; ASSOC SCIENTIST PROCESS INSTRUMENTATION & CONTROL, RES CTR, KENNECOTT COPPER CORP, 70- Honors & Awards: Res Award, Kennecott Copper Corp, 74. Mem: Inst Elec & Electronics Engrs. Res: Computer applications;

cybernetics; microelectronics applications. Mailing Add: 4876 Colony Dr Salt Lake City UT 84117

BUTTERFIELD, WALTER K, b Concord, NH, Mar 16, 32; m 57; c 1. VETERINARY VIROLOGY. Educ: Univ NH, BS, 57, MS, 61; Univ Conn, PhD(microbiol), 67. Prof Exp: Chemist, NH Dept Pub Health, 57-58; res asst virol, Dept Animal Dis, Univ Conn, 60-67; SR MICROBIOLOGIST, PLUM ISLAND ANIMAL DIS LAB, AGR RES SERV, USDA, 67- Mem: US Animal Health Asn; Am Asn Avian Pathologists. Res: Vaccine production and response to viral vaccines. Mailing Add: Plum Island Animal Dis Lab PO Box 848 Greenport NY 11944

BUTTERMAN, WILLIAM CHARLES, b Detroit, Mich, Aug 23, 28; m 62; c 2. MINERALOGY, GEOLOGY. Educ: Ohio State Univ, BSc, 58, MSc, 61, PhD(mineral), 65. Prof Exp: Res chemist, Explosives Dept, 65-69, PLANNING SPECIALIST MINERALS INDUSTS, PURCHASING DEPT, E I DU PONT DE NEMOURS & CO, INC, 69- Mem: Mineral Soc Am. Res: High temperature phase equilibria studies in inorganic oxide systems; use of explosively generated shock waves for phase conversion and synthesis of compounds; mineral economics. Mailing Add: Purchasing Dept E I du Pont de Nemours & Co Wilmington DE 19898

BUTTERWORTH, ALLEN VIRGIL, b Denison, Iowa, Mar 23, 19; m 48; c 2. MATHEMATICS. Educ: Univ Iowa, BS, 41; Univ Chicago, MS, 49. Prof Exp: Chem engr petrol ref, Universal Oil Prod Co, 41-46; asst to dir, Inst Air Weapons Res, Univ Chicago, 48-55; asst dir opers anal forecasting, AEC, 55-59; head math & eval studies dept, Defense Res Labs, 59-67, HEAD MATH DEPT, RES LAB, GEN MOTORS CORP, 67- Mem: Am Math Soc; Opers Res Soc Am. Res: Mathematics applied to study of operations; probability theory and inductive reasoning. Mailing Add: 1265 Robson Lane Bloomfield Hills MI 48013

BUTTERWORTH, BERNARD BERT, b Woodbine, Iowa, May 2, 23. ZOOLOGY. Educ: Univ Mich, BS, 52, MS, 54; Univ Southern Calif, PhD(zool), 60. Prof Exp: Asst biol, Univ Southern Calif, 55-58; lectr, Los Angeles State Col, 58-59; asst prof, Univ Wichita, 60-62, Sch Dent, Univ Southern Calif, 62-64 & Univ Ariz, 64; assoc prof, 64-70, PROF ANAT, SCH DENT, UNIV MO-KANSAS CITY, 70-, CHMN DEPT, 75-, COORDR GRAD STUDIES, 73- Concurrent Pos: Res partic, Inst Trop Biol, Costa Rica, 62. Mem: Soc Study Evolution; Ecol Soc Am; Soc Syst Zool; Am Soc Mammal; Am Soc Ichthyol & Herpet. Res: Mammalian behavior and reproduction; Heteromyidae. Mailing Add: 6501 E 55th Terr Kansas City MO 64129

BUTTERWORTH, CHARLES E, JR, b Lynchburg, Va, Mar 11, 23; m 46; c 3. INTERNAL MEDICINE. Educ: Univ Va, BA, 44, MD, 48. Prof Exp: Intern, Med Col Ala, 48-49, resident hemat, 49-50, resident internal med, 51-53; mem staff, Trop Res Med Lab, San Juan, PR, 55-57; mem staff, Walter Reed Army Inst Res, Washington, DC, 57-58; from instr to assoc prof, 58-66, PROF INTERNAL MED, SCH MED, UNIV ALA, 66- Concurrent Pos: Consult, NIH, 69- & WHO, 71. Mem: Am Fedn Clin Res; Soc Exp Biol & Med; Am Soc Hemat; Am Col Physicians; Am Soc Clin Nutrit (pres, 74-75). Res: Nutrition; hematology. Mailing Add: Dept of Med Univ of Ala Sch of Med Birmingham AL 35233

BUTTERWORTH, DOUGLAS STANLEY, b North Bergen, NJ, Sept 3, 30; m 59. CULTURAL ANTHROPOLOGY. Educ: Univ Heidelberg, dipl, 56; Univ of the Americas, BA, 60, MA, 62; Univ Ill, PhD(anthrop), 69. Prof Exp: Instr anthrop, Univ of the Americas, 62-63; from instr to asst prof, 67-72, ASSOC PROF ANTHROP, UNIV ILL, URBANA, 72- Concurrent Pos: Ctr Latin Am Studies fel Mex, Univ Ill, 69-70, 71 & 72. Mem: Am Anthrop Asn. Res: Urbanization in Latin America; peasant and Indian cultures of Mexico and Guatemala; Spanish-speaking peoples in the United States; social change in Cuba. Mailing Add: Dept Anthrop 109 Davenport Hall Univ of Ill Urbana IL 61801

BUTTERWORTH, FRANCIS M, b Philadelphia, Pa, Mar 29, 35; m 61; c 2. GENETICS. Educ: Columbia Univ, BA, 57; Northwestern Univ, PhD(genetics), 65. Prof Exp: USPHS fel, Univ Va, 65-66; asst prof, 66-69, ASSOC PROF GENETICS, OAKLAND UNIV, 69- Concurrent Pos: Vis prof, Univ Nijmegen, Holland, 73-74. Mem: AAAS; Am Soc Zool; Genetics Soc Am; Sigma Xi. Res: Genetic and endocrine control of development. Mailing Add: Dept of Biol Oakland Univ Rochester MI 48063

BUTTERWORTH, JULIAN SCOTT, b Iowa City, Iowa, Sept 24, 10; m 42; c 2. MEDICINE. Educ: Cornell Univ, BS, 32, MS, 33, MD, 37; Columbia Univ, MedSciD, 41. Prof Exp: Pvt practr med, 41-42; assoc med, NY Postgrad Med Sch, Columbia Univ, 45-48; asst prof, 48-53, ASSOC PROF MED, POSTGRAD SCH MED, NY UNIV-BELLEVUE MED CTR, 53- Concurrent Pos: Melville fel, NY Postgrad Med Sch, Columbia Univ, 3941. Mem: Fel Am Col Physicians; fel Am Heart Asn (pres, 6162). Res: Cardiology; heart disease; cardiovascular diseases. Mailing Add: Postgrad Med Sch NY Univ Bellevue Med Ctr New York NY 10016

BUTTERWORTH, THERON HERVEY, b New York, NY, Mar 31, 04; m 31; c 4. PUBLIC HEALTH. Educ: Princeton Univ, AB, 27; Univ Wis, MS, 29, PhD, 31; Univ NC, MSPH, 45. Prof Exp: Bacteriologist in chg acidophilus milk prod, Walker-Gordon Lab Co, Inc, NJ, 31-33; mem staff, Tech Dept, Borden's Cheese Co, Inc & res staff, Borden Co, NY, 34-37; res staff, Dept Health Lab, San Antonio, Tex, 37-38; consult dairy technologist, 38-39; chief bur dairy & milk inspection, Health Dept, 39-41; supvr milk sanit, Houston, 41; assoc milk specialist, Dist 4, USPHS, La, 42-44, milk specialist, DC, 45-46, health educ consult, 46-48, asst chief div pub health educ, 49-51; dep chief health educ sect, WHO, 51-55; health educ consult, Nat Soc Crippled Children & Adults, Inc, NY, 56-62; health educr, Off Health Educ, Div Med Care Admin, US Dept Health, Educ & Welfare, 68-69, educ consult, Div Health Care Serv, 69-70, health educr, Off Training & Staff Develop, Off Dir Community Health Serv, Health Serv & Ment Health Admin, 70-72; RETIRED. Concurrent Pos: Lectr, Univ Calif, 48, Johns Hopkins Univ, 49 & Univ Calcutta, 54; trustee, Am Nat Coun Health Educ of Pub, 59-61. Mem: Fel Am Pub Health Asn; fel Soc Pub Health Educr (pres, 68-69); Adult Educ Asn US; Int Union Health Educ. Res: Health education; small group discussion techniques. Mailing Add: 521 N Pollard St Apt 29 Arlington VA 22203

BUTTERY, BRIAN RICHARD, b London, Eng, Jan 10, 31; m 59; c 2. PLANT PHYSIOLOGY. Educ: Univ London, BSc, 54; Univ Southampton, PhD(bot), 59. Prof Exp: Botanist, Rubber Res Inst Malaya, 59-64; RES SCIENTIST, RES STA, CAN DEPT AGR, 65- Mem: Can Soc Plant Physiol; Brit Soc Exp Biol. Res: Physiological ecology; water relations; growth analysis; biochemical genetics. Mailing Add: Res Br Can Dept of Agr Harrow ON Can

BUTTIMER, ANNE, b Cork, Ireland, Oct 31, 39; US citizen. URBAN GEOGRAPHY, URBAN SOCIOLOGY. Educ: Nat Univ Ireland, BA, 57, MA, 58; Seattle Univ, teaching cert, 61; Univ Wash, PhD(geog), 65. Prof Exp: Belg-Am Educ Found Comn for Relief in Belg fel, Cath Univ Louvain, 65-66; Seattle Univ res & travel grant, Sorbonne & other French Univs, 66; lectr social sci, Seattle Univ, 66-68; lectr urban studies& Social Sci Res Coun fel, Glasgow Univ, 68-70; univ fel, 70-71, ASST PROF GEOG, CLARK UNIV, 71- Concurrent Pos: Assoc mem, Comn Hist Geog Thought, Int Geog Union, 59-76 & mem US Comt; mem, Le Havre Inst Sociol & Human Psychol. Mem: Asn Am Geogr; Am Geog Soc; Am Inst Planners; fel Royal Soc Health. Res: Residential area design; public participation in planning; other aspects of social space, particularly in urban areas; history and philosophy of social science, existentialism and phenomenology; time-space rhythms. Mailing Add: Grad Sch of Geog Clark Univ Worcester MA 01610

BUTTLAR, RUDOLPH O, b Chicago, Ill, Dec 31, 34; m 55; c 3. INORGANIC CHEMISTRY. Educ: Wheaton Col, BS, 56; Ind Univ, PhD(inorg chem), 62. Prof Exp: Asst prof chem, 62-71, asst dean col arts & sci, 67-70, assoc dean, 70-74, actg dean, 74-75, ASSOC PROF CHEM, KENT STATE UNIV, 71-, DEAN COL ARTS & SCI, 75- Mem: Am Chem Soc. Res: Boron hydride and boron-nitrogen chemistry, especially cyclic boron-nitrogen compounds. Mailing Add: Col of Arts & Sci Kent State Univ Kent OH 44242

BUTTOLPH, LEROY JAMES, physics, physical chemistry, deceased

BUTTON, ALLAN CLIFFORD, b Lake Geneva, Wis, July 11, 39; m 61; c 2. ORGANIC CHEMISTRY. Educ: Univ Wis-Madison, BS, 61; Univ Ill, Urbana-Champaign, PhD(org chem), 67. Prof Exp: SR RES CHEMIST, MINN MINING & MFG CO, 67- Mem: Am Chem Soc. Res: Structure and stereochemistry of natural products; dye chemistry, coatings and formulations. Mailing Add: Minn Mining & Mfg Co 235-F340 3M Ctr St Paul MN 55101

BUTTON, DON K, b Kenosha, Wis, Oct 2, 33; m 61; c 5. BIOCHEMISTRY, MARINE MICROBIOLOGY. Educ: Wis State Col, Superior, BS, 55; Univ Wis, MS, 61, PhD(biochem), 64. Prof Exp: From asst prof marine biochem to assoc prof marine sci, 64-74, PROF MARINE SCI, INST MARINE SCI, UNIV ALASKA, 74- Concurrent Pos: Res assoc, Dept Biophys, Univ Colo Med Ctr, 69-70; consult, continuous cult oper & design, oil slick biodegradation. Mem: Am Soc Microbiol; Am Soc Limnol & Oceanog; Am Chem Soc. Res: Microbial biochemistry; metabolism kinetics of marine heterotrophs. Mailing Add: Inst of Marine Sci Univ of Alaska College AK 99701

BUTTON, KENNETH J, b Rochester, NY, Oct 11, 22; m 50. PHYSICS. Educ: Univ Rochester, BS, 50, MS, 52. Prof Exp: Asst, Univ Rochester, 48-52; physicist, Lincoln Lab, 52-62, asst to dir, Nat Magnet Lab, 62-64, GROUP LEADER MAGNETO-OPTICS, NAT MAGNET LAB, MASS INST TECHNOL, 64- Mem: Fel Am Phys Soc; Optical Soc Am; sr mem Inst Elec & Electronics Engrs. Res: Ferromagnetic resonance; electromagnetic theory of propagation in tensor media; solid state microwave devices; crystal structure; optical and submillimeter spectroscopy; infrared lasers; nonlinear optical effects; electron-phonon interactions. Mailing Add: Nat Magnet Lab Mass Inst of Technol Cambridge MA 02139

BUTTON-SHAFER, JANICE, b Cincinnati, Ohio, Sept 13, 31; m 62; c 3. EXPERIMENTAL HIGH ENERGY PHYSICS. Educ: Cornell Univ, BEngPhys, 54; Univ Calif, Berkeley, PhD(physics), 59. Prof Exp: Physicist, Lawrence Radiation Lab, Univ Calif, 59-66, lectr physics, Univ Calif, Berkeley, 62-64; assoc prof, 66-70, PROF PHYSICS, UNIV MASS, 70- Mem: Am Phys Soc. Res: High-energy experimental physics with counters and bubble chambers. Mailing Add: Dept of Physics Univ of Mass Amherst MA 01002

BUTTREY, BENTON WILSON, b Craigmont, Idaho, Mar 25, 19; m 61. PROTOZOOLOGY. Educ: Univ Idaho, BS, 47, MS, 49; Univ Pa, PhD(zool), 53. Prof Exp: Asst zool, Univ Idaho, 47-49; asst instr, Univ Pa, 49-52; from asst prof to prof, Univ SDak, 53-61; PROF ZOOL & ENTOM, IOWA STATE UNIV, 61- Mem: AAAS; Soc Protozool; Am Soc Zool; Am Soc Parasitol. Res: Morphology; taxonomy and host-parasite relationships of the protozoa from amphibia and swine; morphological variations in trichomonad protozoa. Mailing Add: Dept of Zool & Entom Iowa State Univ Ames IA 50010

BUTTRILL, SIDNEY EUGENE, JR, b Corpus Christi, Tex, July 21, 44; m 65; c 1. ANALYTICAL CHEMISTRY. Educ: Mass Inst Technol, BS, 66; Stanford Univ, PhD(chem), 70. Prof Exp: ASST PROF CHEM, UNIV MINN, MINNEAPOLIS, 69- Mem: AAAS; Am Phys Soc; Am Chem Soc. Res: Kinetics and mechanisms of ion-molecule reactions; gas phase ion thermochemistry; calculations of ion-molecule reaction product distributions; chemical ionization mass spectrometry. Mailing Add: 139 Chem Bldg Univ of Minn Minneapolis MN 55455

BUTTS, DAVID, b Rochester, NY, May 9, 32; m 58; c 2. BIOLOGY, SCIENCE EDUCATION. Educ: Butler Univ, BS, 54; Univ Ill, MS, 60, PhD(bot, sci educ), 62. Prof Exp: Asst prof sci educ, Olivet Nazarene Col, 61-62; from asst prof to assoc prof, Univ Tex, Austin, 62-71, prof curric & instr, 71-74; PROF SCI EDUC & CHMN DEPT, UNIV GA, 74- Concurrent Pos: Consult, Robert E Lee Sci Curric Proj, 62-63; consult, Eval Ctr & mem, Stanford Writing Conf, Am Asn Advan Sci Elem Sci Proj, 63 & 64; dir, Sci In-Serv Proj, 65-66, TAB Sci Test Proj, 65- & Elem Process-Oriented Curric Proj; proj dir, Personalized Teacher Educ Prog, 66- Mem: AAAS; Nat Asn Res Sci Teaching. Res: Curriculum materials in life sciences for elementary and junior high students. Mailing Add: Dept of Sci Educ Univ of Ga Athens GA 30602

BUTTS, HUBERT S, b Burkburnett, Tex, Nov 7, 23; m 47; c 3. MATHEMATICS. Educ: NTex State Univ, BS, 47, MS, 48; Ohio State Univ, PhD(math), 53. Prof Exp: Instr math, NTex State Univ, 47-48; from asst prof to assoc prof, 53-62, PROF MATH, LA STATE UNIV, 62- Mem: Am Math Soc. Res: Algebraic number theory; ideal theory in commutative rings. Mailing Add: Dept of Math La State Univ Baton Rouge LA 70803

BUTTS, WILLIAM CUNNINGHAM, b Evanston, Ill, Jan 13, 42; m 63; c 2. LABORATORY MEDICINE. Educ: Purdue Univ, BS, 63; Iowa State Univ, PhD(anal chem), 68. Prof Exp: Res chemist, Oak Ridge Nat Lab, 68-71; fel clin chem, Univ Wash, 71-73; sr clin chemist, Insts Med Sci, San Francisco, 73-75; DIR CLIN CHEM, GROUP HEALTH OF PUGET SOUND, 75- Mem: Am Asn Clin Chem; Am Chem Soc; AAAS. Res: Identification of constituents isolated from physiologic fluids; development of separation techniques for biochemical systems; gas chromatography of inorganic species; individual biochemical profiles. Mailing Add: Chem Lab Group Health 200 15th Ave E Seattle WA 98112

BUTTS, WILLIAM LESTER, b Reynoldsburg, Ohio, Dec 7, 31; m 54; c 3. ENTOMOLOGY. Educ: Wilmington Col, Ohio, BSc, 53; Ohio State Univ, MSc, 54, PhD(entom), 64. Prof Exp: From instr to assoc prof entom, Purdue Univ, 57-66; assoc prof, 66-69, PROF BIOL, STATE UNIV NY COL ONEONTA, 69- Mem: Am Mosquito Control Asn; Entom Soc Am. Res: Medical and public health entomology; biology and control Reticulitermes species; classification and biology of immature stages of insects. Mailing Add: Dept of Biol State Univ of NY Col Oneonta NY 13820

BUTZ, ANDREW, b Perth Amboy, NJ, Feb 15, 31. INSECT PHYSIOLOGY. Educ: St Peter's Col, BS, 52; Fordham Univ, MS, 54, PhD, 56. Prof Exp: Asst physiol, Fordham Univ, 52-56; asst prof, 58-64, ASSOC PROF ZOOL, UNIV CINCINNATI, 64- Mem: AAAS; Am Soc Zoologists; Entom Soc Am. Res: Circadian rhythm; RNA studies; protein studies; pheromones; metabolism; insect ecology; soil arthropods; insecticidal relationships. Mailing Add: Dept of Biol Sci Univ of Cincinnati Cincinnati OH 45221

BUTZEL, HENRY M, JR, b Detroit, Mich, Nov 7, 22; m 43; c 2. GENETICS. Educ: Williams Col, AB, 43; Ind Univ, PhD(ctyogenetics), 53. Prof Exp: From asst prof to assoc prof biol, 53-63, PROF BIOCHEM GENETICS, UNION COL, NY, 63- Concurrent Pos: NIH fel, Inst Allergy & Infectious Dis, Col Med, Univ Miami & Res Div, Vet Admin Hosp, 61-62. Mem: Genetics Soc Am; Am Soc Human Genetics; Soc Protozool; Brit Soc Gen Microbiol; NY Acad Sci. Res: Cytogenetics and biochemistry of Paramecium aurelia. Mailing Add: Dept of Biol Sci Union Col Schenectady NY 12308

BUTZER, KARL WILHELM, b Ger. PHYSICAL GEOGRAPHY, ARCHAEOLOGY. Educ: McGill Univ, BSc, 54, MSc, 55; Univ Bonn, Dr rer nat(geog), 57. Prof Exp: Ger Acad grant, Univ Bonn, 57-59; from asst prof to assoc prof geog, Univ Wis, 59-66; PROF ANTHROP & GEOG, UNIV CHICAGO, 66- Concurrent Pos: NSF grants, 64-; Wenner-Gren grants, 65- Honors & Awards: Meritorious Contrib Award, Asn Am Geogr, 68. Mem: Am Quaternary Asn; Asn Am Geogr; Am Geog Soc. Res: Environmental archaeology; geomorphology; prehistoric environmental reconstructions; dating of archaeological sites. Mailing Add: Dept of Anthrop & Geog Univ of Chicago Chicago IL 60637

BUTZOW, JAMES J, b Chicago, Ill, Oct 17, 35. BIOCHEMISTRY. Educ: St Bonaventure Univ, BS, 57; Stanford Univ, PhD(biochem), 63. Prof Exp: Res chemist, Geront Br, Nat Heart Inst, 62-65, res chemist, Geront Res Ctr, Nat Inst Child Health & Human Develop, 67-75, RES CHEMIST, GERONT RES CTR, NAT INST AGING, 75- Concurrent Pos: Fel, Dept Biochem, Univ Göteborg, Sweden, 65-67. Mem: AAAS. Res: Physical chemistry of nucleic acids and proteins; interaction of metals with nucleic acids. Mailing Add: Geront Res Ctr Nat Inst on Aging NIH Baltimore City Hosps Baltimore MD 21224

BUURMAN, CLARENCE HAROLD, b Orange City, Iowa, July 31, 15; m 43; c 4. INDUSTRIAL ORGANIC CHEMISTRY. Educ: Univ Iowa, AB, 36, MS, 38, PhD(org chem), 41. Prof Exp: Semi works chemist, Iowa State Dept Health, 39-41; chief chemist, GAF Corp, 41-52, chief chemist surfactants sulfur color & chlorine area, 52-56, prod mgr vats-intermediates & org chem dept, 56-65, plant mgr, 67-70; V PRES & GEN MGR, TRYLON CHEM INC, EMERY INDUST, INC, 70- Mem: Am Chem Soc; Am Inst Chemists. Res: Organic research chemistry; photographic products; dye and pigment chemistry; surface active agents; agricultural chemistry; textile chemistry. Mailing Add: 19 Red Fox Trail Greenville SC 29607

BUXBAUM, EDWIN CLARENCE, b Milwaukee, Wis, Mar 19, 03; m 26, 54; c 2. ANTHROPOLOGY. Educ: Univ Wis, BA, 25; Univ Pa, MA, 64, PhD(anthrop), 67. Prof Exp: Lectr anthrop, Univ Pa, 64; from asst prof to prof, 67-75, EMER PROF ANTHROP, UNIV DEL, 75- Mem: Fel Am Anthrop Asn; Am Ethnol Soc; assoc Royal Photog Soc Gt Brit. Res: Cultural change; ethnic and minority groups; Greeks and other Mediterranean peoples. Mailing Add: Dept of Anthrop Univ of Del Newark DE 19711

BUXTON, DWAYNE REVERE, b Tremonton, Utah, Apr 14, 39; m 64; c 5. CROP PHYSIOLOGY. Educ: Utah State Univ, BS, 64, MS, 66; Iowa State Univ, PhD(agron), 69. Prof Exp: ASSOC PROF AGRON & PLANT GENETICS, UNIV ARIZ, 69-, ASSOC AGRONOMIST, AGR EXP STA, 74- Mem: AAAS; Am Soc Agron; Crop Sci Soc Am. Res: Cotton production and physiology. Mailing Add: Dept of Agron & Plant Genetics Univ of Ariz Tucson AZ 85721

BUXTON, JAY A, b Carrara, Italy, May 29, 19; m 46; c 3. ENTOMOLOGY. Educ: Southwest Tex State Teachers Col, BS, 48; Univ Tex, MA, 50; Ohio State Univ, PhD(entom), 57. Prof Exp: Tech res asst entom, Ohio State Res Found, 50-52; entomologist, Battelle Mem Inst, 52-57; assoc prof biol, Tex Col Art & Indust, 57-61; asst prof entom, Clemson Univ & asst entomologist, Exp Sta, 61-67; PROF BIOL & CHMN DEPT, CATAWBA COL, 67- Mem: AAAS; Entom Soc Am; Am Inst Biol Sci. Res: Taxonomy of immature Orthoptera and genus Melanoplus; fruit insects. Mailing Add: Dept of Biol Catawba Col Salisbury NC 28144

BUYERS, ARCHIE GIRARD, b Va, Feb 24, 23; m 47. ENVIRONMENTAL CHEMISTRY. Educ: Haverford Col, BS, 43; Univ Ill, MS, 47, PhD(inorg chem), 50. Prof Exp: Sr proj chemist, Colgate Palmolive-Peet Co, 50-52; from sr res engr to res specialist, NAm Aviation, Inc, 53-59; mem staff aeronaut, Ford Motor Co, 59-61; staff physicist, Hughes Res Lab, 61-67; consult solid state chem, 67-71; res assoc, Univ NMex, 71-74; HEALTH SCIENTIST, SCI LAB SYST, STATE OF NMEX, 74- Res: Inorganic phosphates and polyphosphates; high temperature-high vacuum classified subjects; environmental and health sciences. Mailing Add: 426 Richmond Dr NE Albuquerque NM 87106

BUYERS, WILLIAM JAMES LESLIE, b Aboyne, Scotland, Apr 10, 37; Can citizen; m 66; c 3. SOLID STATE PHYSICS. Educ: Aberdeen Univ, BSc, 59, PhD(physics), 63. Prof Exp: From asst lectr to lectr physics, Aberdeen Univ, 62-65; fel, Atomic Energy Can, Ltd, 65-66, assoc res officer, 66-75, SR RES OFFICER PHYSICS, ATOMIC ENERGY CAN, LTD, 75- Concurrent Pos: Sr res fel, Oxford Univ, 71-72. Mem: Can Asn Physicists; Am Phys Soc; Inst Physics. Res: Neutron scattering from spin waves, phonons in solids and liquids; study of crystal fields, exchange, soft modes phase transitions, defects and random alloys; positron studies of defects, angular correlation. Mailing Add: Atomic Energy of Can Ltd Chalk River ON Can

BUYNISKI, JOSEPH P, b Worcester, Mass, July 18, 41; m 73; c 1. PHARMACOLOGY. Educ: Univ Cincinnati, BS, 63, PhD(pharmacol), 67. Prof Exp: NIH cardiovasc res fel, Bowman Gray Sch Med, 67-69, instr, 68-69; head cardiovasc res, 69-72, asst dir, 72-74, DIR DEPT PHARMACOL, BRISTOL LABS, 75- Concurrent Pos: NC Heart Asn grants, 68 & 69. Mem: Am Soc Pharmacol & Exp Therapeut; assoc Am Physiol Soc; Am Heart Asn; NY Acad Sci. Res: Antidysrhythmic, antithrombotic and antihypertensive drug research; myocardial infarction and lethal arrhythmias; tissue catecholamine levels and cardiovascular hemodynamics; endotoxemia and platelet function; control mechanisms affecting cerebral blood flow. Mailing Add: Pharmacol Dept Bristol Labs PO Box 657 Syracuse NY 13201

BUYSKE, DONALD ALBERT, b Milwaukee, Wis, Aug 30, 27; m 53; c 4. BIOCHEMISTRY, ORGANIC CHEMISTRY. Educ: Drury Col, BS, 49; Univ Wis, MS, 52, PhD(biochem), 54. Prof Exp: Asst biochem, Univ Wis, 50-54; instr & res assoc chem, Duke Univ, 54-56; group leader pharmacol res, Lederle Labs, Am Cyanamid Co, head dept chem pharmacol, 61-65; dir res, Ayerst Labs, Que, 65-69; VPRES RES & DEVELOP ETHICAL DRUGS, WARNER LAMBERT CO, 69- Concurrent Pos: Mem, NSF Comt Sci & Technol, 74- Mem: Am Soc Microbiol; AAAS; Am Chem Soc; Am Soc Pharmacol & Exp Therapeut; Biochem Soc. Res: Rational drug design; drugs from natural products; biochemical pharmacology and chemotherapy; hydrophilic polymers. Mailing Add: 12 Glen Rd Mountain Lakes NJ 07046

BUZARD, JAMES ALBERT, b Warren, Ohio, Nov 2, 27; m 51; c 2. BIOCHEMISTRY. Educ: Kent State Univ, BS, 49; Univ Buffalo, MA, 51, PhD(biochem), 54. Prof Exp: Sr res biochemist, Sect Biol, Eaton Labs, Norwich Pharmacal Co, 54-55, unit head, 55-58, group leader, 58-59, chief biochem sect, 59-62, from asst dir to dir biol res, 62-65, dir res, 65-68; dir develop, 68-69, dir res, 69-70, vpres res & develop, 70-72, pres, Searle Labs, 72-75, vpres, 72-75, EXEC VPRES OPERS, G D SEARLE & CO, 75- Mem: Am Chem Soc; Am Soc Biol Chem; NY Acad Sci. Res: Intermediary metabolism; drug metabolism and mode of action; chemotherapy; analytical biochemistry; research management. Mailing Add: G D Searle & Co Box 1045 Skokie IL 60076

BUZAS, MARTIN A, b Bridgeport, Conn, Jan 30, 34; m 58; c 3. PALEOECOLOGY. Educ: Univ Conn, BA, 58; Brown Univ, MSc, 60; Yale Univ, PhD(geol), 63. Prof Exp: CUR INVERT PALEONT, SMITHSONIAN INST, 63- Concurrent Pos: Mem, Cushman Found. Mem: AAAS; Am Soc Limnol & Oceanog; Paleont Soc. Res: Quantitative ecology, paleoecology; Benthonic Forminifera. Mailing Add: Dept of Paleobiol Nat Mus of Natural Hist Smithsonian Inst Washington DC 20560

BUZZEE, DAVID H, b Pittsfield, Mass, Mar 11, 42; m 66; c 1. BIOCHEMISTRY. Educ: Univ Louisville, BS, 63; Univ Tenn, Knoxville, PhD(biochem), 68. Prof Exp: Fel bact & immunol, Harvard Med Sch, 68-69; assoc prof biochem, ETenn State Univ, 69-71; CHEMIST, ST ELIZABETH HOSP, 71- Mem: AAAS; Am Chem Soc; Am Soc Microbiol. Res: Purine nucleotide regulation in Enterobacteriaceae; clinical chemistry methodology. Mailing Add: c/o Lab St Elizabeth Hosp Covington KY 41014

BUZZELL, ANNE, b Wilmington, Mass, Apr 27, 22. BIOPHYSICS. Educ: Vassar Col, AB, 44; Cornell Univ, PhD(phys chem), 50. Prof Exp: Res assoc biophys, Univ Pittsburgh, 50-57, asst res prof, 57-60; MEM STAFF PHYS SCI DIV, US ARMY MED RES INST INFECTIOUS DIS, 60- Mem: AAAS; Am Chem Soc; Biophys Soc; Electron Micros Soc Am; NY Acad Sci. Res: Heats of protein inactivation; irradiation and thermal inactivation of viruses; physical properties of microorganisms. Mailing Add: US Army Med Res Inst Ft Detrick Frederick MD 21701

BUZZELL, JOHN GIBSON, b Delavan, Wis, Apr 29, 22; m 50; c 3. POLYMER CHEMISTRY. Educ: Univ Wis, BS, 48; Univ Iowa, PhD(chem), 55. Prof Exp: Chemist, Sherwin-Williams, 48-51; res chemist, 55-67, SR RES CHEMIST, E I DU PONT DE NEMOURS & CO, INC, 67- Mem: Am Chem Soc. Res: Photo polymer process research. Mailing Add: Photo Prod Dept E I du Pont de Nemours & Co Towanda PA 18848

BUZZELL, RICHARD IRVING, b Lincolnville, Maine, May 1, 29; m 56; c 4. PLANT BREEDING, PLANT GENETICS. Educ: Univ Maine, BS, 58; Iowa State Univ, MS, 60, PhD(plant breeding), 62. Prof Exp: RES SCIENTIST SOYBEAN BREEDING & GENETICS, CAN DEPT AGR, 62- Mem: Crop Sci Soc Am; Genetics Soc Can. Res: Tetraploid genetics of birdsfoot trefoil; biochemical genetics of soybeans; variety development. Mailing Add: Can Agr Res Sta Can Dept of Agr Harrow ON Can

BUZZELLI, DONALD EDWARD, b Detroit, Mich, July 24, 36. SCIENCE POLICY, PHILOSOPHY OF SCIENCE. Educ: Cornell Univ, BChE, 59; Mass Inst Technol, ScD, 64; Fordham Univ, PhD(philos sci), 74. Prof Exp: Sr res engr, Standard Oil Co Calif, 63-68; PROF ASST SCI POLICY, NAT SCI FOUND, 75- Res: Analysis of policy issues relating to public confidence in and support for science and technology; condition of research in universities under current economic and political conditions. Mailing Add: Nat Sci Found 1800 G St NW Rm 548 Washington DC 20550

BUZZELLI, EDWARD S, b Cleveland, Ohio, July 31, 39; m 60; c 4. PHYSICS, ELECTROCHEMISTRY. Educ: John Carroll Univ, BS, 62, MS, 65. Prof Exp: From jr physicist to physicist, Res & Develop Lab, Standard Oil Co Ohio, 62-65, sr physicist & proj leader, 65-67, sr res physicist, 67-70; sr engr, 70-73, PROJ MGR, RES & DEVELOP CTR, WESTINGHOUSE ELEC CORP, 73- Mem: Electrochem Soc. Res: Fused salt electrochemistry; secondary batteries; reactions in fused salt media; solid state reactions and devices; air electrode and metal-air battery development. Mailing Add: 39 Morris St Export PA 15632

BYALL, ELLIOTT BRUCE, b Fall River, Mass, May 31, 40; m 62; c 4. FORENSIC SCIENCE. Educ: Sterling Col, BS, 62; Ore State Univ, PhD(org chem), 67. Prof Exp: Res chemist, E I du Pont de Nemours & Co, 67-71; FORENSIC CHEMIST, BUR ALCOHOL, TOBACCO & FIREARMS, 71- Mem: Int Asn Identification. Res: New methods for forensic analysis; explosive and gunshot residue examinations. Mailing Add: Bur of Alc Tobac & Firearms Treasure Island CA 94130

BYARD, JAMES LEONARD, b Hartwick, NY, Dec 1, 41; m 64; c 2. TOXICOLOGY, BIOCHEMISTRY. Educ: Cornell Univ, BS, 64; Univ Wis, MS, 66, PhD(biochem), 68. Prof Exp: Arthritis fel, Harvard Med Sch, 68-70; asst prof toxicol & biochem, Albany Med Col, 70-74; ASST PROF ENVIRON TOXICOL, UNIV CALIF, DAVIS, 74- Mem: Soc Toxicol. Res: Metabolism and molecular mechanism of toxic action of environmental chemicals. Mailing Add: Dept of Environ Toxicol Univ of Calif Davis CA 95616

BYARD, PAUL L, b London, Eng, Feb 17, 32; m 56, 69; c 3. ASTRONOMY, ASTROPHYSICS. Educ: Univ Col, London, BSc, 56; Univ London, PhD(sci), 66. Prof Exp: Mem res & develop staff, Hilger & Watts, Ltd, 58-60; RES ASSOC ASTROPHYS, OHIO STATE UNIV, 61-, ASST PROF ASTRON, 70- Concurrent Pos: Lectr, Ohio State Univ, 66-70. Res: Laboratory astrophysics; measurement of atomic transition probabilities; spectroscopy of solar corona. Mailing Add: Dept of Astron Ohio State Univ W 18th Ave Columbus OH 43210

BYATT, PAMELA HILDA, b Cape Town, SAfrica, Nov 28, 16; nat US. VIROLOGY. Educ: Univ Calif, Los Angeles, BA, 43, MA, 47, PhD(microbiol), 50. Prof Exp: Asst microbiol, Univ Wash, 50-52; virologist, Dept Med Microbiol & Immunol & Clin Labs, 53-73, CHIEF VIROL LAB, MICROBIOL SECT, CLIN LABS, CTR HEALTH SCI, UNIV CALIF, LOS ANGELES, 73- Mem: AAAS; Am Soc Microbiol; Tissue Cult Asn; NY Acad Sci. Res: Staphylococci; antibiotics; viruses; microbiology. Mailing Add: Clin Labs Virol Univ of Calif Ctr for Health Sci Los Angeles CA 90024

BYCK, JOSEPH SYLVAN, organic chemistry, see 12th edition

BYCK, ROBERT, b Newark, NJ, Apr 26, 33; c 3. PHARMACOLOGY, PSYCHIATRY. Educ: Univ Pa, AB, 54, MD, 59. Prof Exp: Res assoc, NIH, 60-62; sr fel pharm, Yeshiva Univ, 62-64, asst prof rehab med & pharmacol, 64-69; fel psychiat

& lectr pharmacol, 69-72, ASSOC PROF PHARMACOL & PSYCHIAT, YALE UNIV, 72- Concurrent Pos: Vis scientist & vis asst prof physiol, Med Ctr, Univ Calif, 63; asst attend physician, Bronx Munic Hosp Ctr, 64-; NIMH sr fel & res career develop award, Dept Pharmacol, Albert Einstein Col Med, 67-69; Burroughs Wellcome Fund scholar clin pharmacol, 72- Mem: Am Soc Pharmacol & Exp Therapeut; Am Psychiat Asn. Res: Chemoreceptors control of respiration; neuropharmacology; cryobiology; psychopharmacology; clinical pharmacology. Mailing Add: Dept of Pharmacol Yale Univ Sch of Med New Haven CT 06510

BYCROFT, GEORGE NOEL, applied mechanics, applied mathematics, see 12th edition

BYDALEK, THOMAS JOSEPH, b Grand Rapids, Mich, Apr 22, 35; m 57; c 4. ANALYTICAL CHEMISTRY, INORGANIC CHEMISTRY. Educ: Aquinas Col, BS, 57; Purdue Univ, PhD(anal chem), 61. Prof Exp: From instr to asst prof chem, Univ Wis, 61-65; assoc prof, 65-68, PROF CHEM, UNIV MINN, DULUTH, 68- Mem: Am Chem Soc. Res: Kinetics of multidentate ligand complexes of the transition metals and their application in analysis; method development for analysis of metal ions at trace levels in natural waters. Mailing Add: Dept of Chem Univ of Minn Duluth MN 55812

BYER, NORMAN ELLIS, b Brooklyn, NY, July 17, 40; m 64; c 2. SOLID STATE PHYSICS. Educ: Cooper Union, BEE, 62; Princeton Univ, MSE, 63; Cornell Univ, PhD(appl physics), 67. Prof Exp: Res assoc appl physics, Cornell Univ, 66-67; mem sci staff injection laser studies, RCA Labs, 67-69; mem sci staff luminescence studies, Res Inst Advan Studies, 69-74, SR RES SCIENTIST, MARTIN MARIETTA LABS, 74- Mem: AAAS; Am Phys Soc; Inst Elec & Electronics Engrs. Res: Acoustic studies of rotational impurities in alkali halides; laser and electroluminescent diodes; optically active defects in infrared detection materials; studies of pryoelectric and other infrared detectors. Mailing Add: Martin Marietta Labs 1450 S Rolling Rd Baltimore MD 21227

BYERLEE, JAMES DOUGLAS, b Cairns, Australia, Aug 12, 27; m; c 3. GEOPHYSICS. Educ: Univ Queensland, BSc, 63; Mass Inst Technol, PhD(geol), 66. Prof Exp: Geophysicist, Md, 66-69, GEOPHYSICIST, NAT CTR EARTHQUAKE RES, US GEOL SURV, 69- Res: Mechanical properties of rocks and rock forming minerals. Mailing Add: Nat Ctr for Earthquake Res US Geol Surv 345 Middlefield Rd Menlo Park CA 94025

BYERLY, DON WAYNE, b Wilmington, Del, Mar 8, 33; m 55; c 4. GEOLOGY. Educ: Col Wooster, AB, 55; Univ Tenn, MS, 57, PhD(geol), 66. Prof Exp: Instr geol, Univ Tenn, 56-66; asst prof, Murray State Univ, 66-67; ASST PROF GEOL, UNIV TENN, KNOXVILLE, 67- Concurrent Pos: Consult, Tenn Div Geol, 63-65. Mem: AAAS; Geol Soc Am; Nat Asn Geol Teachers. Res: Geology of the northern portion of Dutch Valley in Anderson County, Tennessee; Greenville quadrangle in Greene County and Baileytown quadrangle in Greene and Hawkins Counties, Tennessee; the Pulaski fault near Greenville. Mailing Add: Dept of Geol Univ of Tenn Knoxville TN 37916

BYERLY, PAUL ROBERTSON, JR, b Lancaster, Pa, Sept 4, 22; m 47; c 5. PHYSICS. Educ: Washington & Jefferson Col, AB, 43; Univ Pa, PhD, 51. Prof Exp: Instr physics, Washington & Jefferson Col, 43-44; jr physicist, Tenn Eastman Corp, 44-45; instr physics, Univ Pa, 45-50; physicist, Radiation Lab, Univ Calif, 50-58; educ adv physics, Govt Philippines, 58-63; ASSOC PROF PHYSICS, UNIV NEBR-LINCOLN, 63- Concurrent Pos: Regional counr, physics, Nebr, 65-71. Mem: Am Phys Soc; Am Asn Physics Teachers. Res: Mass spectroscopy; nuclear physics; Mössbauer effect. Mailing Add: Dept of Physics Univ of Nebr Lincoln NE 68508

BYERLY, PERRY, b Clarinda, Iowa, May 28, 97; m 41; c 3. SEISMOLOGY. Educ: Univ Calif, Berkeley, AB, 21, MA, 22, PhD(seismol), 24. Hon Degrees: LLD, Univ Calif, Berkeley, 66. Prof Exp: Instr physics, Univ Nev, 24-25; in chg seismographic sta, 25-50, dir, 50-62, from asst prof to prof, 27-64, EMER PROF SEISMOL, UNIV CALIF, BERKELEY, 64- Concurrent Pos: Guggenheim Mem Found fel, Cambridge Univ, 28-29 & 52-53, Fulbright sr fel, 60-61; Condon lectr, Ore univs, 51; Smith-Mundt lectr, Univ Mex, 54. Mem: Nat Acad Sci; Seismol Soc Am (secy, 30-56, pres, 57-58); fel Geol Soc Am; Soc Explor Geophys; Int Asn Seismol & Physics Earth's Interior (pres, 60-63). Res: California seismology; earth structure; roots of mountains; energy in earthquake waves; nature of forces at source of earthquakes. Mailing Add: 5340 Broadway Terr Number 401 Oakland CA 94618

BYERLY, PERRY EDWARD, b Berkeley, Calif, Feb 2, 26; m 59; c 1. GEOPHYSICS. Educ: Univ Calif, AB, 49; Harvard Univ, AM, 51, PhD(geophys), 54. Prof Exp: From geophysicist to sr res geophysicist, US Geol Surv, 52-63; geophys rep, Calif Res Corp, Houston, 64-66; sr geophysicist, Chevron Explor Co, 67-69, sr res geophysicist, Chevron Oilfield Res Co, 69-70; SR RES GEOPHYSICIST, CONTINENTAL OIL CO, 70- Concurrent Pos: Asst prof, Johns Hopkins Univ, 54-55. Mem: Geol Soc Am; Am Geophys Union; Seismol Soc Am; Soc Explor Geophys. Res: Explosion seismology; gravity and magnetics; geothermal problems; general geophysics. Mailing Add: Continental Oil Co Drawer 1267 Ponca City OK 74601

BYERLY, THEODORE CARROLL, physiology, see 12th edition

BYERRUM, RICHARD UGLOW, b Aurora, Ill, Sept 22, 20; m 45; c 4. BIOCHEMISTRY. Educ: Wabash Col, AB, 42; Univ Ill, Urbana, PhD(biochem, org chem), 47. Hon Degrees: DSc, Wabash Col, 67. Prof Exp: From instr to prof chem, 47-59, asst provost, 59-62, actg dir, Inst Biol & Med, 61-62, PROF BIOCHEM, MICH STATE UNIV, 57-, DEAN COL NATURAL SCI, 62- Concurrent Pos: Sabbatical leave, Calif Polytech Inst, 56; travel awards, Int Cong Biochem, Vienna, 58 & Montreal, 59; sabbatical leave, Scripps Inst Oceanog, 73; consult, Am Chem Soc, 58-62; vis scientist, 62-64; mem bd, Mich Health Coun, 61-; consult, NCent Asn Sec Schs & Cols, 62-; consult, Nat Acad Sci, 72-74. Mem: AAAS; Am Chem Soc; Am Soc Plant Physiol; Soc Exp Biol & Med; Am Soc Biol Chemists. Res: Biosynthesis of alkaloids and other plant substances of medical interest. Mailing Add: Col of Natural Sci Mich State Univ East Lansing MI 48824

BYERS, ALFRED RODDICK, economic geology, see 12th edition

BYERS, BENJAMIN ROWE, b Austin, Tex, Nov 7, 36; m 57; c 2. MICROBIOLOGY. Educ: Univ Tex, Austin, BA, 58, MA, 60, PhD(microbiol), 65. Prof Exp: Bacteriologist, Standard Brands, Inc, 60-62; NIH fel, Univ Tex, Austin, 65-66, asst prof microbiol, 66; from asst prof to assoc prof, 67-73, PROF MICROBIOL, MED CTR, UNIV MISS, 73- Concurrent Pos: NIH res career develop award, Med Ctr, Univ Miss, 71-75, NIH res grant, 71-, NIH contract, 74- Mem: AAAS; Am Soc Microbiol. Res: Iron metabolism; mechanisms of active iron transport; iron chelating agents in treatment of iron storage disease. Mailing Add: Dept of Microbiol Univ of Miss Med Ctr Jackson MS 39211

BYERS, DOHRMAN HAROLD, b East Liverpool, Ohio, July 18, 14; m 39; c 2. INDUSTRIAL HYGIENE. Educ: Mt Union Col, BS, 36; Purdue Univ, MS, 38; Am Bd Indust Hyg, dipl. Prof Exp: Asst chem, Purdue Univ, 36-38; instr, Mich State Univ, 38-41; from chemist to dir, Div Occup Injury & Dis Control, USPHS, 41-69; assoc prof, 69-74, PROF INDUST HEALTH, SCH PUB HEALTH, UNIV MICH, ANN ARBOR, 74- Concurrent Pos: Ed jour, Am Indust Hyg Asn, 58-75. Mem: Am Indust Hyg Asn; fel Am Pub Health Asn; Am Conf Govt Indust Hygienists. Res: Analysis of air and tissues for industrial contaminants related to industrial toxicology; field and laboratory studies of industrial hygiene; information storage and retrieval; short course training. Mailing Add: Dept of Environ & Indust Health Univ of Mich Sch of Pub Health Ann Arbor MI 48104

BYERS, DON HARRISON, nuclear physics, see 12th edition

BYERS, DONALD JAMES, b Manchester, Iowa, Aug 28, 13; m 39; c 2. ORGANIC CHEMISTRY. Educ: Iowa State Col, BS, 36; Univ Minn, PhD(chem), 40. Prof Exp: Du Pont fel, Univ Ill, 40-41; res chemist, E I du Pont de Nemours & Co, 41-45; asst supvr graphic arts res, Battelle Mem Inst, 45-55; res chemist, Time, Inc, 55-62; ASSOC PROF GRAPHIC ARTS & CHEM, CARNEGIE INST TECHNOL, 62- Mem: Am Chem Soc. Res: Printing processes and materials. Mailing Add: 5120 Fifth Ave Pittsburgh PA 15232

BYERS, FLOYD MICHAEL, b Rush City, Minn, Feb 8, 47; m 67; c 2. ANIMAL NUTRITION. Educ: Univ Minn, BS, 69; SDak State Univ, MS, 72; Colo State Univ, PhD(ruminant nutrit), 74. Prof Exp: Lab technician plant res, Univ Minn, 65-67, nutrit res, 67-69; feedlot mgr, SDak State Univ, 70-71, dir, Ruminant Nutrit Lab, 71-72; res asst energy metab, Colo State Univ, 72-73, NDEA res fel, 73-74; ASST PROF BEEF CATTLE NUTRIT, OHIO AGR RES & DEVELOP CTR, 75- Mem: Am Soc Animal Sci. Res: Investigations concerning protein and energy requirements and efficiency of protein and energy utilization for maintenance and production of cattle, varying in mature size, fed rations varying in available energy. Mailing Add: Animal Sci Dept Ohio Agr Res & Develop Ctr Wooster OH 44691

BYERS, FRANK MILTON, JR, b Moline, Ill, Mar 5, 16; m 45; c 4. GEOLOGY. Educ: Augustana Col, AB, 38; Univ Chicago, PhD, 55. Prof Exp: GEOLOGIST, US GEOL SURV, 41- Mem: AAAS; fel Geol Soc Am; Geochem Soc; Am Geophys Union. Res: General and engineering geology; petrology; tectonics and earthquake faults of western Nevada. Mailing Add: 125 Everett St Lakewood CO 80226

BYERS, GEORGE WILLIAM, b Washington, DC, May 16, 23; m 55. ENTOMOLOGY. Educ: Purdue Univ, BS, 47; Univ Mich, MS, 48, PhD, 52. Prof Exp: Rackham fel zool, Univ Mich, 52-53; from asst prof to assoc prof entom, 56-65, asst cur, Snow Entom Mus, 56-73, chmn dept entom, 69-72, PROF ENTOM, UNIV KANS, 65-, CUR SNOW ENTOM MUS, 74- Concurrent Pos: Ed, Syst Zool, Soc Systs Zool, 63-66. Mem: Soc Study Evolution; Soc Syst Zool; Entom Soc Am; Am Soc Naturalists; Entom Soc Can. Res: Biology and classification of Tipulidae and Mecoptera. Mailing Add: Dept of Entom Univ of Kans Lawrence KS 66045

BYERS, GORDON CLEAVES, b Hancock, Mich, June 18, 18; m 49; c 1. MATHEMATICS. Educ: Univ Mich, BA, 41; Duke Univ, PhD(math), 53. Prof Exp: Instr math, Duke Univ, 51-52; from asst prof to assoc prof, 53-63, PROF MATH & HEAD DEPT, MICH TECHNOL UNIV, 63- Mem: Am Math Soc; Am Soc Eng Educ; Math Asn Am. Res: Algebraic numbers; analysis; numerical analysis. Mailing Add: Dept of Math Mich Technol Univ Houghton MI 49931

BYERS, HORACE ROBERT, b Seattle, Wash, Mar 12, 06; m 27; c 1. METEOROLOGY, CLOUD PHYSICS. Educ: Univ Calif, AB, 29; Mass Inst Technol, MS, 32, ScD(meteorol), 35. Prof Exp: Asst, Mass Inst Technol, 30-32 & Scripps Inst, Univ Calif, 32-33; instr meteorol, Transcontinental & Western Air, 33-35; meteorologist, US Weather Bur, 35-40; from assoc prof to prof meteorol, Univ Chicago, 40-65, chmn dept, 48-60; dean col geosci, 65-68, acad vpres, 68-71, distinguished prof, 71-74, EMER PROF METEOROL, TEX A&M UNIV, 74- Concurrent Pos: Dir, US Interdept Thunderstorm Proj, 46-50; vpres, Int Asn Meteorol & Atmospheric Physics, 54-60, pres, 60-63; chmn bd, Univ Corp Atmospheric Res, 63-65; chmn bd, Gulf Univ Res Corp, 66-69; vis prof, Univ Clermont-Ferrand, France, 75. Honors & Awards: Losey Award, Am Inst Aeronaut & Astronaut, 41; Award of Merit, Chicago Tech Sci Coun, 59; Brooks Award, Am Meteorol Soc, 60. Mem: Nat Acad Sci; Am Meteorol Soc (pres, 52-53). Res: Thunderstorms; physical and dynamic meteorology. Mailing Add: 1036 The Fairway Santa Barbara CA 93108

BYERS, JOHN ROBERT, b Stoughton, Sask, July 27, 37; m 63; c 2. ENTOMOLOGY. Educ: Univ Sask, BSA, 62, MSc, 63, PhD(entom, physiol), 66. Prof Exp: Nat Res Coun Can overseas fel, 66-67; asst prof biol, Univ Sask, 67-68; RES SCIENTIST, CAN DEPT AGR, 68- Mem: Can Soc Zoologists; Entom Soc Can. Res: Biochemistry and ultrastructure of insect tissues in relation to function and changes occurring as a response to environmental changes; ultrastructure of insects and nematodes in relation to function and systematics. Mailing Add: Exp Taxon Sect Biosysts Res Inst Cent Exp Farm Ottawa ON Can

BYERS, LARRY DOUGLAS, b Los Angeles, Calif, Feb 18, 47. BIOCHEMISTRY. Educ: Univ Calif, Los Angeles, BS, 68; Princeton Univ, MS, 70, PhD(biochem), 72. Prof Exp: Fel enzym, Univ Calif, Berkeley, 72-75; ASST PROF CHEM, TULANE UNIV, 75- Concurrent Pos: NIH fel, 73-75. Mem: Am Chem Soc. Res: Mechanism of action of enzymes; design of enzyme inhibitors. Mailing Add: Dept of Chem Tulane Univ New Orleans LA 70118

BYERS, LAWRENCE WALLACE, b Pulaski, Pa, Nov 12, 16; m 56; c 2. BIOCHEMISTRY. Educ: Westminster Col, Pa, BS, 38; Oberlin Col, MA, 40; Univ Ill, PhD(biochem), 48. Prof Exp: Teaching asst chem, Oberlin Col, 38-40; teaching asst, Exp Sta, Mich State Univ, 40-42; asst org res, Dow Chem Co, Mich, 42-45; res biochemist, Bristol Labs, 48-52; res biochemist, Ft Detrick, Md, 52-56; res biochemist, Dept Psychiat & Neurol, Sch Med, Tulane Univ, 56-66, asst prof biochem, 60-66; asst prof, 66-72, ASSOC PROF BIOCHEM, UNIV TENN MED UNITS, MEMPHIS, 72-; RES BIOCHEMIST, DEPT PATH, BAPTIST MEM HOSP, MEMPHIS, 66- Concurrent Pos: Biochem consult, Vet Hosp, Gulfport, Miss, 64-66. Mem: AAAS; NY Acad Sci; Am Asn Clin Chem; Soc Neurosci. Res: Carotenoids; fatty acids; silicones; phospholipids; antibiotics; nutrition; intermediary metabolism biochemical aspects of mental illness; antihypertensive factors of kidney medulla. Mailing Add: 3138 Dumbarton Rd Memphis TN 38128

BYERS, NINA, b Los Angeles, Calif, Jan 19, 30. PARTICLE PHYSICS, THEORETICAL PHYSICS. Educ: Univ Calif, BA, 50; Univ Chicago, MS, 53, PhD(physics), 56. Hon Degrees: MA, Oxford Univ, 67. Prof Exp: Res fel math physics, Univ Birmingham, 56-58; res assoc, Stanford Univ, 58-59, actg asst prof, 59-61; from asst prof to assoc prof, 61-67, PROF PHYSICS, UNIV CALIF, LOS ANGELES, 67- Concurrent Pos: Guggenheim fel & vis mem, Inst Advan Study, 63-64; fel, Somerville Col, 67-68, Janet Watson vis fel, 68-; fac lectr, Oxford Univ, 67-68, vis scientist, 69-74. Mem: Am Phys Soc; Fedn Am Scientists. Mailing Add: Dept of Physics Univ of Calif Los Angeles CA 90024

BYERS, ROBERT ALLAN, b Latrobe, Pa, Dec 6, 36; m 60; c 2. ENTOMOLOGY. Educ: Pa State Univ, BSc, 60; Ohio State Univ, MSc, 61; Purdue Univ, PhD(entom), 71. Prof Exp: Res entomologist, Entom Res Div, Tifton, Ga, 61-66, Lafayette, Ind, 66-70, RES ENTOMOLOGIST, ENTOM RES DIV, AGR RES SERV, USDA, UNIVERSITY PARK, PA, 70- Mem: Entom Soc Am. Res: Biology and control of the two-lined spittlebug on Coastal bermuda grass; effect of a complex of insects on the yield of Coastal bermuda grass; ability of the Hessian fly to stunt winter wheat; host plant resistance to meadow spittlebug, and clover root curculio in alfalfa. Mailing Add: US Regional Pasture Res Lab Agr Res Serv USDA University Park PA 16802

BYERS, RONALD ELNER, b Everett, Wash, Nov 8, 36; m 60; c 2. THEORETICAL PHYSICS. Educ: Wash State Univ, BS, 66, PhD(physics), 74. Prof Exp: Mathematician, Strategy & Tactics-Anal Group, US Army, 60-62; technologist, Battelle Northwest Lab, 62-65; TEACHER PHYSICS, CENT COL, 70- Res: Quantum hidden variables theory unifying essentials of quantum mechanics and general relativity; evolution processes in scientific theory making; philosophical grounding of scientific theories. Mailing Add: Dept of Physics Cent Col Pella IA 50219

BYERS, SANFORD OSCAR, b New York, NY, May 1, 18; m 42; c 5. BIOCHEMISTRY. Educ: Rensselaer Polytech Inst, BSc, 39; Univ Cincinnati, PhD(biochem), 44. Prof Exp: Instr biochem & physiol, Washington Univ, 44-45; res biochemist, Nat Drug Co, Philadelphia, 45-46; RES BIOCHEMIST, HAROLD BRUNN INST, MT ZION HOSP, 46- Concurrent Pos: Fel coun arteriosclerosis, Am Heart Asn. Mem: AAAS; Reticuloendothelial Soc; Am Chem Soc; Am Soc Exp Biol & Med; Am Physiol Soc. Res: Atherosclerosis; hyaluronic acid; nephrosis; metabolism of digitalis, cholesterol and lipids; catecholamines. Mailing Add: Harold Brunn Inst Mt Zion Hosp 1600 Divisadero St San Francisco CA 94115

BYERS, STANLEY A, b Ashland, Ohio, Jan 9, 31; m 66; c 2. CERAMICS. Educ: Antioch Col, BA, 55; Mich State Univ, BS, 57; Case Western Reserve Univ, PhD(mat sci), 74. Prof Exp: Metall & ceramics engr, Ohio Brass Co, 58-63; ceramic engr, Am Standard, Inc, 63-69 & Carborundum Co, 69-70; res scientist mat, Res Exp Sta, Ga Inst Technol, 72-75; MEM STAFF MAT RES, BALL CORP, 75- Mem: Am Ceramic Soc; Am Soc Metals. Res: Glass processing and formulation; glass coatings. Mailing Add: 2804 W Purdue Muncie IN 47304

BYERS, THOMAS JONES, b Philadelphia, Pa, Oct 12, 35; m 60; c 2. BIOLOGY. Educ: Cornell Univ, AB, 58; Univ Pa, PhD(zool), 62. Prof Exp: USPHS fel biophys, Carnegie Inst Dept Terrestrial Magnetism, 62-64; asst prof zool & entom, 64-68, dir develop biol prog, 72-75, ASSOC PROF MICROBIOL & CELL BIOL, OHIO STATE UNIV, 68- Mem: Am Soc Cell Biol; Soc Protozool. Res: Protozoan growth and reproduction; molecular biology of amoebic encystment. Mailing Add: Dept of Microbiol Ohio State Univ Columbus OH 43210

BYERS, VERA STEINBERGER, b Houston, Tex, Nov 18, 42; m 73. IMMUNOBIOLOGY. Educ: Univ Calif, Los Angeles, BA, 65, MS, 67, PhD(immunobiol), 69. Prof Exp: Bench chemist, Res Div, Abbott Labs, 69-71; res immunologist, Dept Med, 71-74, ADJ ASST PROF DERMAT, UNIV CALIF, SAN FRANCISCO, 75- Concurrent Pos: Nat Inst Allergy & Infectious Dis fels, 68 & 72; Arthritis Found fel, 72. Res: Human cellular immunology; lymphocyte regulation and maturation; mechanism of action and specificity of transfer factor. Mailing Add: Dept of Dermat Univ of Calif San Francisco CA 94143

BYERS, VIRGINIA PRATT, b Canora, Sask, Oct 1, 20; US citizen; m 45; c 4. GEOLOGY. Educ: Univ Denver, BA, 41. Prof Exp: GEOLOGIST, US GEOL SURV, 43-56, 61- Mem: Am Asn Petrol Geol; Geol Soc Am. Res: Stratigraphy; uranium. Mailing Add: Br of Uranium & Thorium Resources US Geol Surv Fed Ctr Bldg 25 Denver CO 80225

BYERS, WALTER HAYDEN, b Johnstown, Pa, July 6, 14; m 45; c 3. PHYSICS. Educ: Univ Fla, BS, 36, MS, 38; Pa State Col, PhD(physics), 42. Prof Exp: Physicist, US Air Force, 43-46; asst prof physics, Univ Okla, 46-47; asst prof elec eng, Univ Ill, 46-51; physicist, US Air Force, 51-54; weapon syst analyst, Sandia Corp, 54-56; PHYSICIST, US AIR FORCE, 56- Concurrent Pos: Asst prof eng, Clark Tech Col, 71-73, prof eng & natural sci, 73- Mem: Am Phys Soc. Res: Ultrasonics; astronautics; space system engineering. Mailing Add: Dept Eng & Natural Sci Clark Tech Col Springfield OH 45505

BYFIELD, JOHN ERIC, b Toronto, Ont, Nov 26, 36; US citizen; m 73; c 2. RADIOTHERAPY, ONCOLOGY. Educ: Univ Calif, Los Angeles, BA, 60, MD, 65, PhD(physiol), 70. Prof Exp: Intern path, 66-67, radiation therapist, 67-70, instr radiol, 69-70, asst prof & asst res physician, 70-74, assoc prof radiol & assoc res physician, Lab Nuclear Med & Radiation Biol, Sch Med, Univ Calif, Los Angeles, 74-75; ASSOC PROF RADIOL & CHIEF RADIATION THER, UNIV SAN DIEGO, 76- Concurrent Pos: Chief radiation ther, Harbor Gen Hosp, Torrance, Calif, 70-75. Mem: Am Soc Clin Oncol; fel Am Col Radiol; Am Soc Therapeut Radiol; Am Soc Cell Biologists; Am Asn Cancer Res. Res: Biochemistry of anticancer drugs; clinical chemotherapy; radiation biology; clinical radiation therapy. Mailing Add: Dept of Radiol Univ of San Diego San Diego CA 92103

BYFIELD, PATRICIA E, b Santa Monica, Calif, July 21, 43; div; c 1. IMMUNOLOGY. Educ: Univ Calif, Los Angeles, AB, 65, PhD(microbiol), 69. Prof Exp: Teaching asst bact, Univ Calif, Los Angeles, 66-67, asst prof microbiol, Calif State Univ, Los Angeles, 69-70; asst res biologist, Univ Calif, Los Angeles, 70; staff scientist, Wellcome Res Labs, 70-72; ASST PROF PEDIAT, MICROBIOL & IMMUNOL, UNIV CALIF, LOS ANGELES & HARBOR GEN HOSP, 72- Concurrent Pos: Nat Res Coun Italy grant, 74. Mem: AAAS; Tissue Cult Asn; Am Asn Immunol; Am Soc Microbiol; Brit Soc Immunol. Mailing Add: E-6 Lab Dept of Pediat Harbor Gen Hosp Torrance CA 90509

BYINGTON, KEITH H, b Plymouth, Iowa, Mar 14, 35; m 55; c 4. BIOCHEMICAL PHARMACOLOGY, PHARMACOLOGY. Educ: Univ Iowa, BS, 58; Univ SDak, PhD(pharmacol), 64. Prof Exp: Org chemist, Dr Salsbury's Labs, Iowa, 58-60; instr pharmacol, Univ SDak, 63-64; fel, Univ Fla, 64-65; fel biochem, Inst Enzyme Res, Univ Wis, 65-68; asst prof pharmacol, 68-74, ASSOC PROF PHARMACOL, SCH MED, UNIV MO-COLUMBIA, 74- Mem: Am Soc Pharmacol & Exp Therapeut; Am Chem Soc. Res: Drug metabolism; drug enzyme interaction; drug-membrane interactions; oxidative phosphorylation; toxicology. Mailing Add: Dept of Pharmacol Univ of Mo Sch of Med Columbia MO 65201

BYLER, DAVID MICHAEL, b Mishawaka, Ind, Dec 23, 45; m 71; c 1. PHYSICAL INORGANIC CHEMISTRY. Educ: Univ NC, Chapel Hill, AB, 68; Northwestern Univ, Evanston, MS, 69, PhD(inorg chem), 74. Prof Exp: Instr chem, Wilbur Wright Col & Kennedy-King Col, Chicago City Cols, 70-73; instr, Drexel Univ, 73-75; VIS ASST PROF CHEM, TEMPLE UNIV, 75- Mem: Am Chem Soc; The Chem Soc; Sigma Xi. Res: Infrared and Raman spectroscopy group Va pentahalide adducts; sulfur dioxide adducts; transition metal complexes involving trichlorogermane ligands; absorption spectroscopy of diatomic transition metal oxides, sulfides and halides in vapor phase. Mailing Add: Dept of Chem Temple Univ Philadelphia PA 19122

BYLES, PETER HENRY, b Exeter, Eng, 1931; c 2. ANESTHESIOLOGY. Educ: Univ London, MB & BS, 55; Am Bd Anesthesiol, dipl, 65. Prof Exp: Intern, King's Col Hosp, London, 55-56, sr house officer anesthesiol, 57-58; sr house officer, Plymouth Hosps Group, 56-57; res asst, Western Hosp, London, 58-61; resident, State Univ Hosp, Syracuse, NY, 62-63; from asst prof to assoc prof, 63-72, PROF ANESTHESIOL, STATE UNIV NY UPSTATE MED CTR, 72-, ASST DIR SCH ALLIED HEALTH PROF, 68- Concurrent Pos: Asst prof, State Univ Hosp, Syracuse, 63-67, assoc prof, 67-; attend anesthesiologist, Vet Admin Hosp, Syracuse, 64- Mem: AMA; Am Soc Anesthesiol; Am Asn Advan Med Instrumentation; fel Am Col Anesthesiol; Int Anesthesia Res Soc. Res: Medical instrumentation and electronics; new anesthetic drugs. Mailing Add: Dept of Anesthesiol State Univ of NY Upstate Med Ctr Syracuse NY 13210

BYNUM, WILLIAM LEE, b Carlsbad, NMex, June 28, 36; m 65. MATHEMATICS. Educ: Tex Technol Col, BS, 57; Univ NC, MA, 64, PhD(math), 66. Prof Exp: Engr, Bell Helicopter Corp, Tex, 57-61; instr math, Univ NC, 65-66; asst prof, La State Univ, 66-69; asst prof, 69-74, ASSOC PROF MATH, COL WILLIAM & MARY, 74- Mem: Am Math Soc; Math Asn Am. Mailing Add: Dept of Math Col of William & Mary Williamsburg VA 23185

BYRD, DANIEL MADISON, III, b Detroit, Mich, Dec 30, 40; m 63; c 3. PHARMACOLOGY. Educ: Yale Univ, BA, 63, PhD(pharmacol), 71. Prof Exp: Res assoc cell biol, Sch Med, Univ Md, 70; vis scientist pharmacol, Nat Cancer Inst, 71; res assoc oncol, Med Sch, Johns Hopkins Univ, 71-72; from asst cancer res scientist to cancer res scientist, Roswell Park Mem Inst, 72-75; ASST PROF PHARMACOL, COL MED, UNIV OKLA, 75- Mem: Am Chem Soc; Am Soc Microbiol; Sigma Xi. Res: Chemotherapy; mammalian defense and disease mechanisms, carcinogenesis, chemical and genetic alterations in viral host range, pharmacogenetics. Mailing Add: Dept of Pharmacol Univ of Okla Health Sci Ctr PO Box 26901 Oklahoma City OK 73190

BYRD, DAVID LAMAR, b Houston, Tex, June 3, 22; m 47; c 1. DENTISTRY. Educ: Univ Tex, DDS, 46; Northwestern Univ, MSD, 49. Prof Exp: Resident oral surg, Jackson Mem Hosp, Fla, 49-50; intern, Charity Hosp La, 50-51; PROF ORAL SURG & CHMN DEPT, BAYLOR COL DENT, 51-, CHIEF DENT, BAYLOR MED CTR, 64- Concurrent Pos: Consult, Attend Staff, Baylor Med Ctr, Parkland Mem Hosp & St Paul's Hosp; ed, Current Ther Dent, Vols I-III; mem, Am Bd Oral Surg. Mem: Am Soc Oral Surg; Sigma Xi; fel Am Col Dent. Res: Clinical oral pathology; diseases of soft tissue of bone of head and neck; research in facial and oral pain. Mailing Add: Dept Oral & Maxillofacial Surg Baylor Col of Dent Dallas TX 75226

BYRD, DAVID SHELTON, b Stephens, Ark, May 28, 30; m 53; c 2. ORGANIC CHEMISTRY, PHYSICAL CHEMISTRY. Educ: Southern State Col, BS, 52; Univ Ky, MS, 55; Univ Louisville, PhD(org chem), 60. Prof Exp: Instr chem, Univ Louisville, 57-60; from asst prof to assoc prof, 60-69, PROF CHEM, NORTHEAST LA UNIV, 69- Mem: Am Chem Soc. Res: Trace metal analysis of soil and water. Mailing Add: Dept of Chem Northeast La Univ Monroe LA 71201

BYRD, EARL WILLIAM, JR, b Pomona, Calif, Mar 17, 46. DEVELOPMENTAL BIOLOGY. Educ: San Francisco State Univ, BA, 68, MA, 70; Univ BC, PhD(zool), 73. Prof Exp: FEL DEVELOP BIOL, SCRIPPS INST OCEANOG, 73- Mem: Soc Develop Biol; Am Soc Zoologists; AAAS. Res: Analysis of development in vertebrates and invertebrates, particularly control and synthesis of macromolecules and membrane related changes at fertilization. Mailing Add: Scripps Inst of Oceanog La Jolla CA 92093

BYRD, ISAAC BURLIN, b Canoe, Ala, Mar 14, 25; m 49; c 3. FISHERIES MANAGEMENT, MARINE BIOLOGY. Educ: Auburn Univ, BS, 48, MS, 50. Prof Exp: Fisheries res asst, Auburn Univ, 49-51; chief fisheries sect, Ala Dept Conserv, 51-65; chief div fed aid, Fisheries Res & Develop, 65-74, CHIEF FISHERIES MGT DIV, NAT MARINE FISHERIES SERV, 74- Honors & Awards: Ala Gov Conserv Award, 64; Ala Fisheries Asn Freshwater Div Award, 65. Mem: AAAS; Am Fisheries Soc (pres, 65); Gulf & Caribbean Fisheries Inst; World Maricult Soc; fel Am Inst Fishery Res Biol. Res: Marine and freshwater fisheries; estuarine waters; fish population dynamics; reservoir management; aquaculture; aquatic botany and biology; ecology; water pollution; oceanography; public fishing lakes; hydrology. Mailing Add: Nat Marine Fisheries Serv Duval Bldg 9450 Gandy Blvd St Petersburg FL 33702

BYRD, J ROGERS, b Henderson, NC, Apr 30, 31; m 63. CYTOGENETICS, ZOOLOGY. Educ: Wake Forest Col, BS, 53; Univ Mich, MS, 57, PhD(zool), 60. Prof Exp: Asst prof biol, Col William & Mary, 59-63; NIH res fel endocrinol, 63-65, asst res prof, 65-69, ASSOC PROF ENDOCRINOL, MED COL GA, 69- Res: Human chromosome studies related to the inheritance of chromosomal aberrations as a causative factor in habitual spontaneous abortion. Mailing Add: Dept of Endocrinol Med Col of Ga Augusta GA 30902

BYRD, JAMES DOTSON, b Jackson, Miss, July 9, 32; m 54; c 3. ORGANIC POLYMER CHEMISTRY. Educ: Miss Col, BS, 54; Univ Ala, MS, 74. Prof Exp: Res asst chem, Purdue Univ, 54-55; chemist, Ethyl Corp, 55-61; sr chemist, Geigy Chem Corp, 61-63; unit chief polymer chem, George C Marshall Space Flight Ctr, NASA, 63-67; sr res chemist, 67-74, GROUP SUPVR, HUNTSVILLE DIV, THIOKOL CHEM CORP, 74- Honors & Awards: NASA Awards, 54-67. Mem: Am Chem Soc; fel Am Inst Chem; Soc Aerospace Mat & Process Eng. Res: New adhesive and insulation materials used in space environment, such as high and very low temperature, vacuum and radiation, including synthesis of new monomers and polymers. Mailing Add: 9032 Craigmont Rd SW Huntsville AL 35802

BYRD, JAMES WILLIAM, b Mt Olive, NC, Dec 4, 36; m 59; c 3. PHYSICS. Educ: NC State Univ, BS, 59, MS, 61; Pa State Univ, PhD(physics), 63. Prof Exp: Assoc prof, 62-64, PROF PHYSICS, E CAROLINA UNIV, 64-, CHMN DEPT, 65- Mem: Am Phys Soc; Am Asn Physics Teachers. Res: Plasma and fluid physics; mathematical methods. Mailing Add: Dept of Physics E Carolina Univ Box 2791 ECU Sta Greenville NC 27834

BYRD, KENNETH ALFRED, b Erwin, NC, Sept 17, 40; m 64; c 2. ALGEBRA. Educ: Duke Univ, BS, 62; NC State Univ, PhD(math), 69. Prof Exp: ASST PROF MATH, UNIV NC, GREENSBORO, 69- Mem: Am Math Soc; Math Asn Am. Res: Non-commutative ring theory; quotient rings; homological algebra as it applies to the module category of a ring. Mailing Add: Dept of Math Univ of NC Greensboro NC 27412

BYRD, MITCHELL AGEE, b Franklin, Va, Aug 16, 28; m 54. ZOOLOGY. Educ: Va Polytech Inst, BS, 49, MS, 51, PhD(biol), 54. Prof Exp: From asst prof to assoc prof, 56-63, PROF BIOL, COL WILLIAM & MARY, 63-, HEAD DEPT, 62- Mem: Am Soc Parasitol; Am Soc Mammal; Am Ornith Union. Res: Mammalogy; ecology;

taxonomy; wildlife diseases and parasites; avian ecology. Mailing Add: Dept of Biol Col of William & Mary Williamsburg VA 23185

BYRD, NORMAN ROBERT, b New York, NY, Mar 10, 21; m 46; c 2. ORGANIC POLYMER CHEMISTRY. Educ: Polytech Inst Brooklyn, BS, 49, PhD(org chem), 55. Prof Exp: Res chemist, Nat Starch Prod, NY, 48-51; asst, Polytech Inst Brooklyn, 51-52 & US Navy Proj, 52-54; res chemist, E I du Pont de Nemours & Co, Inc, 54-55; sr res chemist, Goodyear Tire & Rubber Co, 55-58; head fundamental sect, Rayonier, Inc, 58-60; res scientist, Aeronutronic, Calif, 61-62; head org-polymer sect, Astropower Labs, Douglas Aircraft Co, 62-69, MGR MAT RES, DOUGLAS AIRCRAFT DIV, McDONNEL DOUGLAS CORP, 69- Mem: AAAS; Am Chem Soc; Am Inst Chem; The Chem Soc. Res: Starch and cellulose chemistry; synthetic resins; elastomers; graft and block polymers; organic reactions macromolecules; electrical properties of organic polymers; semiconductor; adhesion studies; free-resistant materials. Mailing Add: 17991 Athens Ave Villa Park CA 92667

BYRD, RICHARD DOWELL, b Newport, Ark, Mar 14, 33; m 64; c 1. MATHEMATICS. Educ: Hendrix Col, BA, 58; Univ Ark, MS, 59; Tulane Univ, PhD(math), 66. Prof Exp: Instr math, La State Univ, New Orleans, 60-64; asst prof, Lehigh Univ, 66-67; from asst prof to assoc prof, 67-72, PROF MATH, UNIV HOUSTON, 72- Mem: Am Math Soc; Math Asn Am. Res: Lattice-ordered groups. Mailing Add: Dept of Math Col Arts & Sci Univ of Houston Houston TX 77004

BYRD, WILBERT PRESTON, b Burlington, NC, July 7, 26; m 47; c 1. STATISTICS, GENETICS. Educ: NC State Univ, BS, 49, MS, 52; Iowa State Univ, PhD(crop breeding), 55. Prof Exp: Res assoc, NC State Univ, 49-52; asst, Iowa State Univ, 52-55; asst prof, Ohio State Univ, 55-56; assoc prof & statistician, 56-66, PROF EXP STATIST & CHMN DEPT, CLEMSON UNIV, 66- Concurrent Pos: Agent, USDA, 52-54. Mem: Am Soc Agron; Am Statist Asn; Biomet Soc. Res: Experimental design; data processing; statistical genetics. Mailing Add: Dept of Exp Statist Clemson Univ Clemson SC 29631

BYRD, WILLIAM JOSEPH, b Dothan, Ala, Dec 30, 37. IMMUNOLOGY, PHARMACOLOGY. Educ: Samford Univ, BS, 65; Univ Ala Med Col, MS, 69; Univ Melbourne, PhD(immunol), 73. Prof Exp: ASST RES PROF IMMUNOPHARMACOL, SALK INST BIOL STUDIES, 72- Mem: AAAS; Am Asn Immunologists. Res: Delineation of mechanisms of action of immunoregulatory substances. Mailing Add: Salk Inst for Biol Studies PO Box 1809 San Diego CA 92112

BYRD, WILLIS EDWARD, b Athens, Ga, Dec 27, 22. PHYSICAL CHEMISTRY. Educ: Talladega Col, AB, 41; Univ Iowa, PhD(phys chem), 49. Prof Exp: Assoc prof, 49-54, PROF CHEM, LINCOLN UNIV, MO, 54-, HEAD DEPT, 64-, CHMN DIV NAT SCI & MATH, 74- Mem: AAAS; Am Chem Soc; fel Am Inst Chem. Res: Donor-acceptor complexes, including charge-transfer complexes; molecular spectra and structure. Mailing Add: Dept of Chem Lincoln Univ Jefferson City MO 65101

BYRN, ERNEST EDWARD, b Frederick, Okla, Apr 4, 24; m 52; c 2. PHYSICAL CHEMISTRY, ANALYTICAL CHEMISTRY. Educ: Univ Tenn, BS, 50, PhD, 54. Prof Exp: Res chemist, Buckeye Cellulose Corp, 54-55; asst prof chem, Univ Okla, 55-61; asst res dir, MacDermid, Inc, 61-62, res dir, 62-64; prof, George Peabody Col, 64-66; chmn dept, 66-70, PROF CHEM, EASTERN KY UNIV, 66- Mem: AAAS; Am Chem Soc. Res: Organic reagents for inorganic analysis; titrations in nonaqueous solvents; liquid-liquid extractions; surface activity and detergency; chemical education. Mailing Add: Dept of Chem Eastern Ky Univ Richmond KY 40475

BYRN, STEPHEN ROBERT, b New Albany, Ind, Oct 7, 44; m 69; c 2. BIOPHYSICAL CHEMISTRY, SOLID STATE CHEMISTRY. Educ: DePauw Univ, BA, 66; Univ Ill, Urbana, PhD(chem), 71. Prof Exp: Scholar, Univ Calif, Los Angeles, 71-72; ASST PROF MED CHEM, PURDUE UNIV, 72- Mem: Am Chem Soc; Am Crystallog Asn; The Chem Soc; Am Soc Pharmacog; Am Asn Cols Pharm. Res: Structural studies of physiologically active compounds in the solid state and in solution; X-ray crystallography; solid state reactions of organic crystals, particularly drugs. Mailing Add: Dept of Med Chem & Pharmacog Purdue Univ Sch of Pharm West Lafayette IN 47906

BYRNE, BARBARA JEAN MCMANAMY, b Baraboo, Wis, Aug 9, 41; m 68; c 2. ANIMAL GENETICS. Educ: Blackburn Col, BA, 62; Ind Univ, Bloomington, MA, 63, PhD(genetics), 69. Prof Exp: Lectr biol, Ind Univ, 71-72; res specialist genetics, Cornell Univ, 72-74; ASST PROF BIOL, WELLS COL, 74- Mem: Sigma Xi; Genetics Soc Am. Res: Behavioral genetics of Paramecium aurelia; examination of wild type and mutant stocks attempting to correlate behavioral abnormalities with changes in ultrastructure of membranes as revealed by freeze-etching. Mailing Add: Dept of Biol Wells Col Aurora NY 13026

BYRNE, BRUCE CAMPBELL, b Hammond, Ind, May 13, 45; m 68; c 2. ANIMAL GENETICS. Educ: Ind Univ, Bloomington, AB, 67, PhD(genetics), 72. Prof Exp: USPHS trainee genetics, Cornell Univ, 72-74; ASST PROF BIOL, WELLS COL, 74- Mem: AAAS; Genetics Soc Am. Res: Genetic analyses of behavioral responses of electrically active membrane by investigation on the swimming behavior among the ciliated protozoans, particularly Paramecium aurelia and Tetrahymena pyriformis. Mailing Add: Dept of Biol Wells Col Aurora NY 13026

BYRNE, FRANCIS PATRICK, b Kansas City, Mo, July 25, 13; m 42; c 6. ANALYTICAL CHEMISTRY. Educ: Rockhurst Col, BS, 35; Creighton Univ, MS, 38; Univ Tenn, PhD(chem), 49. Prof Exp: Instr chem, St Joseph Col, 40-42; res chemist, Westvaco Chlorine Prods, 42-46; instr chem, Univ Tenn, 46-49; MGR ANAL CHEM DEPT, WESTINGHOUSE RES & DEVELOP LABS, 49- Concurrent Pos: Gen chmn, Pittsburgh Conf Anal Chem & Appl Spectros, 64; mem NSF adv panel to Anal Div, Nat Bur Standards, 72-75. Honors & Awards: Lundel-Bright Award, Am Soc Testing & Mat, 74. Mem: Am Chem Soc; Am Soc Testing & Mat. Res: Detection and determination of pertinent materials in environmental situations. Mailing Add: Anal Chem Dept Westinghouse Res & Develop Labs Pittsburgh PA 15235

BYRNE, FRANK EDWARD, b Chicago, Ill, July 27, 07; m 47; c 2. GEOLOGY. Educ: Univ Chicago, BS, 27, PhD(geol), 40. Prof Exp: Asst geol, Univ Chicago, 26-30; from instr to prof geol, Kans State Col, 30-54; mem staff, Am Stratig Co, 54-55; res geologist, Tidewater Oil Co, 55-59; explor mgr, Amaden Petrol Co, 59-62; prof geol, Idaho State Univ, 62-65; geol consult, Colo, 65-66; prof geol, Wis State Univ-Whitewater, 66-68; chmn dept, 68-73, PROF REGIONAL ANAL, UNIV WIS-GREEN BAY, 68- Concurrent Pos: Geologist, US Geol Surv, Colo & Kans. Mem: Fel AAAS; fel Geol Soc Am; Am Inst Prof Geol; Am Asn Petrol Geol. Res: Stratigraphy; petroleum, ground water and engineering geology. Mailing Add: Regional Anal Concentration Univ of Wis Green Bay WI 54305

BYRNE, GEORGE D, b Earlham, Iowa, June 15, 33; m 60; c 5. NUMERICAL ANALYSIS. Educ: Creighton Univ, BS, 55; Iowa State Univ, MS, 61, PhD(appl math), 63. Prof Exp: Mathematician, White Sands Proving Grounds, NMex, 55-56; programmer, Sandia Corp, NMex, 56-58; asst prof math & comput sci, 63-67, ASSOC PROF MATH, UNIV PITTSBURGH, 67- Concurrent Pos: Adj assoc prof, Dept Chem & Petrol Eng, Univ Pittsburgh, 67-; consult, Lawrence Livermore Lab, Univ Calif, 73-; vis scientist, Appl Math Div, Argonne Nat Lab, 74-75. Mem: Am Math Soc; Soc Indust & Appl Math; Asn Comput Mach. Res: Numerical solution of ordinary differential equations and related software. Mailing Add: Dept of Math 822 Schenley Hall Univ of Pittsburgh Pittsburgh PA 15260

BYRNE, HUGH DESMOND, b Dublin, Ireland, Sept 14, 36; m 61; c 6. ENTOMOLOGY. Educ: Univ Col Dublin, BAgrSc, 60; Univ Md, MS, 65, PhD(entom), 68. Prof Exp: Instr high sch, ENigeria, 60-63; teacher pub sch, Eng, 63; res asst entom, Univ Md, 63-65, fac res asst, 65-68; res assoc, Cornell Univ, 68-69; entomologist, 69-73, STAFF SPECIALIST PROD DEVELOP, AGR RES LAB, ORTHO DIV, CHEVRON CHEM CO, 73- Mem: AAAS; Entom Soc Am. Res: Factors determining host plant selection by insects; forage crop entomology; teaching; laboratory rearing of insects; screening of candidate pesticides; monitoring of registration programs. Mailing Add: Agr Res Lab Ortho Div Chevron Chem Co 940 Hensley St Richmond CA 94809

BYRNE, HUGH MICHAEL, US citizen. OCEANOGRAPHY. Educ: Boston Col, BS, 69; Mass Inst Technol, MS, 74. Prof Exp: OCEANOGR, NAT OCEANIC & ATMOSPHERIC ADMIN, 73- Concurrent Pos: Mem, Seasat User Adv Group, NASA, 72-74 & Seasat Altimeter Team, 74-; consult environ data, Environ Systs Div, EG&G, 73. Honors & Awards: Spec Achievement Award, Nat Oceanic & Atmospheric Admin, 76. Mem: Am Geophys Union; Asn Comput Mach. Res: Satellite oceanography, use of orbiting sensors in visible, IR, and microwave regions to physical oceanography and marine geodesy; use is made of data from National Oceanic and Atmospheric Administration polar orbiters, Nimbus series, and GEOS-III and Seasat altimetric satellites. Mailing Add: Atlantic Oceanog & Meteorol Labs NOAA 15 Rickenbacker Causeway Miami FL 33149

BYRNE, JEFFREY EDWARD, b Minneapolis, Minn, July 15, 39; m 60; c 2. PHYSIOLOGY, PHARMACOLOGY. Educ: Univ NDak, BA, 62; Univ SDak, MA, 64, PhD(physiol & pharmacol), 66. Prof Exp: Fel & lectr pharmacol & exp therapeut, Univ Man, 66-69; sr scientist cardiovasc pharmacol, 69-73, SR INVESTR, MEAD JOHNSON & CO, 73- Concurrent Pos: Adj fac mem, Sch Nursing, Evansville Univ & Sch Med, Ind Univ, 72- Mem: AAAS; NY Acad Sci; Sigma Xi. Res: Cardiovascular physiology and pharmacology; cardiac arrhythmia and antiarrhythmic compounds. Mailing Add: 5120 New Harmony Rd Evansville IN 47712

BYRNE, JOHN E, physiology, animal behavior, see 12th edition

BYRNE, JOHN JOSEPH, b Morristown, NJ, Aug 30, 16. SURGERY. Educ: Princeton Univ, AB, 37; Harvard Univ, MD, 41. Prof Exp: Mem staff, 47-57, PROF SURG, SCH MED, BOSTON UNIV, 57-, PROF SOCIO-MED SCI & DIR STUDENT TEACHING, DEPT SURG, 73- Concurrent Pos: Asst chief, Neurosurg Sect, May Gen Hosp, Ill; dir hand serv & vis surgeon, Boston City Hosp; chief surg, Framingham Union Hosp; mem staff, Boston Univ Hosp. Mem: AAAS; Am Surg Asn; Am Soc Surg of Hand; Am Asn Surg Trauma; fel Am Col Surgeons. Res: Physiology of pulmonary embolism; serum amylase activity in intestinal obstruction; diseases of the hand; phlebitis; medical history; shock; pancreatitis. Mailing Add: Dept of Surg Boston Univ Sch of Med Boston MA 02118

BYRNE, JOHN MAXWELL, b Gassaway, WVa, May 7, 33; m 60; c 1. PLANT ANATOMY. Educ: Glenville State Col, BA, 60; Univ Miami, MA, 64, PhD(bot), 69. Prof Exp: Instr biol, Univ Miami, 64-66; asst prof, Va Polytech Inst & State Univ, 69-75; ASST PROF BIOL, KENT STATE UNIV, 75- Mem: AAAS; Bot Soc Am; Am Inst Biol Sci. Res: Developmental plant anatomy, root development, effects of pathogens and symbionts on root structure, structural responses of plant organs to exogenous growth regulators. Mailing Add: Dept of Biol Sci Kent State Univ Kent OH 44240

BYRNE, JOHN RICHARD, b Portland, Ore, Mar 23, 26; m 49; c 2. MATHEMATICS. Educ: Reed Col, BA, 47; Univ Wash, MSc, 51, PhD, 53. Prof Exp: Asst prof math, San Jose State Col, 53-54; instr, Portland State Univ, 54-56; asst prof, San Jose State Col, 56-57; from asst prof to assoc prof, 57-63, PROF MATH, PORTLAND STATE UNIV, 63-, HEAD DEPT, 67- Concurrent Pos: Asst consult curric study, Portland High Sch, 59; dir, NSF Insts Math Teachers, 61-66. Mem: Am Math Soc; Math Asn Am. Res: Modern algebra. Mailing Add: Dept of Math Portland State Univ Portland OR 97207

BYRNE, JOHN THOMAS, analytical chemistry, see 12th edition

BYRNE, JOHN VINCENT, b Hempstead, NY, May 9, 28; m 54; c 4. OCEANOGRAPHY. Educ: Hamilton Col, AB, 51; Columbia Univ, MA, 53; Univ Southern Calif, PhD(geol), 57. Prof Exp: Field asst, Newell Reef Study, Am Mus, 51 & Raroia Exped, Pac Sci Bd, 52; lab assoc, Univ Southern Calif, 53-55; geologist, Res Sect, Humble Oil & Ref Co, Tex, 57-60; assoc prof, 60-65, PROF OCEANOG, ORE STATE UNIV, 65-, CHMN DEPT, 68-, DEAN SCH OCEANOG, 72- Concurrent Pos: Dir oceanog prog, NSF, 66-67. Mem: AAAS; Am Geophys Union; Geol Soc Am; Am Asn Petrol Geologists; Soc Econ Paleontologists & Mineralogists. Res: Marine geology. Mailing Add: Sch of Oceanog Ore State Univ Corvallis OR 97331

BYRNE, RICHARD N, theoretical physics, see 12th edition

BYRNE, ROBERT HOWARD, b Omaha, Nebr, Apr 15, 41; m 67; c 1. OCEANOGRAPHY. Educ: Univ Chicago, BS, 64; DePaul Univ, MS, 67; Boston Univ, MA, 70; Univ RI, PhD(oceanog), 74. Prof Exp: RES ASSOC OCEANOG, UNIV RI, 74- Res: Investigation of the physical chemistry of seawater; and trace metal speciation in seawater with special emphasis on the speciation of ferric ions. Mailing Add: Grad Sch of Oceanog Univ of RI Kingston RI 02881

BYRNE, ROBERT JOHN, b Chicago, Ill, Sept 17, 32; m 56; c 2. OCEANOGRAPHY. Educ: Univ Chicago, MS, 61, PhD(geophys sci), 64. Prof Exp: Ford Found fel, Woods Hole Oceanog Inst, 64-65, asst scientist, dept geophys & geol, 65-67; res oceanogr, Land & Sea Interaction Lab, Environ Sci Serv Admin, 67-69; assoc prof marine sci, 69-72, SR MARINE SCIENTIST, VA INST MARINE SCI, 72- Concurrent Pos: Assoc prof, Sch Marine Sci, Col William & Mary, 72- Mem: Am Geophys Union. Res: Geological oceanography; coastal engineering; mechanics of sediment transport; coastal fluid and sediment processes; coastal zone planning. Mailing Add: Dept of Geol Oceanog Va Inst of Marine Sci Gloucester Point VA 23062

BYRNE, ROBERT JOSEPH, b Irvington, NJ, Dec 24, 22; m 46; c 3. VETERINARY SCIENCE. Educ: Cornell Univ, DVM, 44; George Washington Univ, MS, 58. Prof Exp: Med bacteriologist, Walter Reed Army Med Ctr, 53-54; assoc prof vet sci, Univ Md, 54-67; chief, Res Ref Reagents Br, 67-70, ASST SCI DIR COLLAB RES & CHIEF, RES RESOURCES BR, NAT INST ALLERGY & INFECTIOUS DIS, 70-

Concurrent Pos: Mem, WHO Bd Comp Virol. Mem: Am Vet Med Asn; Conf Res Workers Animal Dis. Res: Infectious diseases of domestic animals; leptospirosis; equine encephalomyelitis and bovine respiratory diseases; infectious diseases of horses; epidemiology. Mailing Add: Nat Inst Allergy & Infectious Dis Bethesda MD 20014

BYRNE, WILLIAM EDWARD, organic chemistry, see 12th edition

BYRNE, WILLIAM LAWRENCE, b Santa Fe, NMex, Feb 18, 27; m 47; c 4. BIOCHEMISTRY. Educ: Stanford Univ, BS, 48, MS, 50; Univ Wis, PhD(biochem), 53. Prof Exp: Assoc biochem, Duke Univ, 54-56, from asst prof to assoc prof, 56-68; PROF BIOCHEM & CHMN DEPT, UNIV TENN CTR HEALTH SCI, 68 Concurrent Pos: NIH fel biochem, 53-54; NIH fel, Univ Calif, Berkeley, 6566; dir, Brain Res Inst, 70- Mem: AAAS; Am Chem Soc; Am Soc Biol Chem; Soc Neurosci. Res: Enzymes; membranes; chemistry of learning and memory. Mailing Add: Dept of Biochem Univ of Tenn Ctr for Health Sci Memphis TN 38163

BYRNES, BERNARD CHRISTOPHER, b Barnesboro, Pa, Mar 6, 23; m 52; c 4. OCEANOGRAPHY. Educ: St Francis Col, Pa, BS, 48. Prof Exp: Geophysicist, US Coast & Geod Surv, 48-51; from supvr geophysicist to head geomagnetics br, Hydrographic Off, 51-61, dept dir, Marine Surv Div, 61-62, coordr, AGOR Task Force, Oceanog Off, 62-65, actg dir, Oceanog Surv Dept, 65-66, dir develop surv div, 66-74, DIR, INSHORE PROJS DIV, US NAVAL OCEANOG OFF, 74- Honors & Awards: Super Achievement Award, US Navy, 55, 60 & 64. Mem: Am Geophys Union; Marine Technol Soc; Sigma Xi. Res: Marine geophysics; coordination of design requirements for scientific outfitting of and operations aboard Navy oceanographic research ships and manned and unmanned submersible vehicles. Mailing Add: 2503 Lorring Dr SE Washington DC 20028

BYRNES, EUGENE WILLIAM, b Roselle, NJ, July 3, 33; m 62; c 2. ORGANIC CHEMISTRY. Educ: Rensselaer Polytech Inst, BS, 56; Univ NH, PhD(org chem), 64. Prof Exp: Prod chemist, Schering Corp, 56; instr chem, Ohio Northern Univ, 62-63; NIH fel biochem, Mich State Univ, 63-65; asst prof chem, West Liberty State Col, 65-66 & Alliance Col, 66-68; asst prof, 68-69, ASSOC PROF CHEM, ASSUMPTION COL, MASS, 69-, CHMN DIV NATURAL SCI & MATH, 72- Concurrent Pos: CP Snow lectr, Ithaca Col, 64; consult, Astra Pharmaceut Prod, Inc. Mem: Am Chem Soc. Res: Synthesis of antiarrhythmic and local anesthetic drugs. Mailing Add: 191 Nola Dr Holden MA 01520

BYRNES, FRANCIS CLAIR, b Vail, Iowa, July 24, 17; m 42; c 3. COMMUNICATION SCIENCE, ACADEMIC ADMINISTRATION. Educ: Iowa State Col, BS, 38; Mich State Univ, PhD(commun arts), 63. Prof Exp: News/Farm ed, Denison Rev, 38-41; info dir, Iowa Agr Conserv Comt, 41-42; chief tech info, US Air Force Res & Develop, 46-48; agr ed & prof, Agr Exp Sta, Ohio State Univ, 48-53; assoc dir, Nat Proj Agr Commun, 53-60; lectr & consult, Int Prog, Mich State Univ, 61-63; head, Off Commun, Int Rice Res Inst, Philippines, 63-67; dir training & confs, Int Ctr Trop Agr, Colombia, 68-75; HEAD TRAINING, RES & COMMUN, INT AGR DEVELOP SERV, 76- Concurrent Pos: Consult pub rels, Ohio Bell Tel Co, 50-53; mem field staff agr sci, Rockefeller Found, 63- Mem: Rural Sociol Soc; Soc Appl Anthrop; Int Commun Asn. Res: Factors influencing acceptance and use of modern technology by small and subsistence level farmers of developing countries; design, implementation and evaluation of integrated programs of agricultural and rural development. Mailing Add: Apt 2K 405 E 56th St New York NY 10022

BYRNES, WILLIAM RICHARD, b Barnesboro, Pa, Oct 12, 24; m 47; c 5. FORESTRY. Educ: Pa State Univ, BS, 50, MF, 51, PhD(agron, soils), 61. Prof Exp: Asst forestry, Pa State Univ, 50-51; soil scientist, Soil Conserv Serv, USDA, 51-52; instr forestry, Pa State Univ, 52-61, assoc prof, 61-62; assoc prof, 62-65, PROF FORESTRY, PURDUE UNIV, 65- Mem: Soc Am Foresters; Am Soc Agron; Weed Sci Soc Am. Res: Forest soils; watershed management; ecology; silviculture; physiology. Mailing Add: Dept of Forestry & Natural Resources Purdue Univ West Lafayette IN 47907

BYRNES, WILLIAM WINFIELD, b Green Bay, Wis, Mar 8, 20; m 48. ENDOCRINOLOGY. Educ: Univ Wis, BA, 42, MS, 49, PhD(zool, physiol), 53. Prof Exp: Asst, Univ Wis, 45-50; res scientist, 51-57, head dept nutrit & metab dis, 57-67, head dept metab dis, 67-68, PROD RES DIR, UPJOHN CO, 68- Mailing Add: 9566 W AB Otsego MI 49078

BYRON, JOSEPH WINSTON, b New York, NY, Apr 23, 30; m 61; c 1. PHARMACOLOGY. Educ: Fordham Univ, BSc, 52; Philadelphia Col Pharm, MSc, 55; Univ Buffalo, PhD(pharmacol), 59. Prof Exp: Asst pharmacol, Philadelphia Col Pharm, 53-55; assoc, Univ Buffalo, 56-59; NSF fel, Oxford Univ, 59-60, Am Cancer Soc Brit-Am Exchange fel, 61-62; from sr res scientist to prin res scientist, Christie Hosp & Holt Radium Inst, Eng, 62-73; ASSOC PROF PHARMACOL, SCH MED, UNIV MD, BALTIMORE CITY, 73- Concurrent Pos: Hon lectr, Univ Manchester, 68-73. Mem: Brit Asn Radiation Res; Europ Soc Radiation Res. Res: Hematology; radiation biology. Mailing Add: Dept of Pharmacol Univ of Md Sch of Med Baltimore MD 21201

BYRON, WILLIAM GLENN, b Denver, Colo, Jan 28, 23; m 49; c 4. GEOGRAPHY. Educ: Univ Calif, Los Angeles, AB, 48, MA, 51; Syracuse Univ, PhD(geog), 54. Prof Exp: From asst prof to assoc prof geog, 54-67, PROF GEOG, CALIF STATE UNIV, LOS ANGELES, 67- Concurrent Pos: NSF grant statist geog, Northwestern Univ, 61; grants cartog, Univ Wash, 63 & archaeol-geog res, Nayarit, Mex, 66-68; grant spec res, Chancellor's Off, Calif State Cols, 67. Mem: Asn Am Geog; Am Geog Soc. Res: Latin America; population geography; cartography. Mailing Add: Dept of Geog Calif State Univ 5151 State College Dr Los Angeles CA 90032

BYRUM, WOODROW ROBERT, b Phoebus, Va, Jan 24, 14; m 39; c 3. PHARMACOLOGY. Educ: Med Col Va, BS, 37; Ohio State Univ, PhD(pharmacol), 47. Prof Exp: From asst instr to instr pharm & pharmacol, Ohio State Univ, 42-44; asst prof, Univ Ga, 4546; from instr pharm to asst prof gen pharmacol, Ohio State Univ, 46-48; assoc prof, Univ Ariz, 48-50; prof pharmacol & head dept, Univ Ga, 50-52; dir div pharm, 52-67, dean sch, 67-72, EMER DEAN, SCH PHARM, SAMFORD UNIV, 72- Mem: Fel Am Found Pharmaceut Educ; Nat Pharmaceut Asn. Res: Fungicidal of thiomalic acid derivatives; pharmacology and bioassay of the viburnums; pharmacology of phlorizin; toxicology of gold compounds; sapote gum. Mailing Add: Sch of Pharm Samford Univ Birmingham AL 35209

BYSTRICKY, KARL M, b Vienna, Austria, Oct 7, 19; m 53; c 2. OPTICS. Educ: Tech Univ Vienna, BS, 47, MS, 49, PhD(optics), 68. Prof Exp: Mgr optical design, Eumig, Vienna, 54-68; sr optical designer, Zoomar, 68-69; SR OPTICAL DESIGNER, PERKIN-ELMER CORP, 69- Mem: Optical Soc Am; Ger Soc Appl Optics; Austrian Phys Soc. Res: Complex optical systems design, especially zoom lenses. Mailing Add: 75 Wilton Rd E Ridgefield CT 06877

BYSTROFF, ROMAN IVAN, b San Francisco, Calif, Nov 6, 31; m 56; c 4. ANALYTICAL CHEMISTRY. Educ: Univ Calif, Berkeley, BS, 53; Iowa State Univ, PhD(chem), 59. Prof Exp: SR CHEMIST, LAWRENCE LIVERMORE LAB, 58- Mem: Am Chem Soc. Res: Stability of coordination compounds in solution; inorganic analysis by solution chemistry; ultraviolet and visible spectrophotometry; flame spectroscopy, both emission and atomic absorption. Mailing Add: Lawrence Livermore Lab Box 808 Livermore CA 94551

BYSTROM, BARBARA GILLOOLY, b Los Angeles, Calif, Sept 9, 21; m 58; c 1. NEUROBIOLOGY. Educ: Univ Calif, Los Angeles, BA, 45, MA, 50. Prof Exp: Atomic Energy proj res lab technician, 50-58, res instr & supvr grad studies electron micros, 58-62, lab technician bot & plant biochem, 62-63, lab technician air pollution res, 63-65, ELECTRON MICROSCOPIST, ENVIRON NEUROBIOL LAB, BRAIN RES INST, UNIV CALIF, LOS ANGELES, 65- Mem: Am Inst Biol Sci; Electron Micros Soc Am; Ecol Soc Am; Bot Soc Am. Res: Study of brain structures at the subcellular level under various physiological and environmental conditions; observation of ultrastructural details of interest to many related disciplines. Mailing Add: Environ Neurobiol Lab Brain Res Inst Univ Calif Los Angeles CA 90024

BYTHER, RALPH SUMNER, plant pathology, botany, see 12th edition

BYVIK, CHARLES EDWARD, b Ladd, Ill, Mar 26, 40; m 64; c 3. SOLID STATE PHYSICS. Educ: Ill Inst Technol, BS, 63; Univ Mo-Rolla, MS, 64; Va Polytech Inst & State Univ, PhD(physics), 72. Prof Exp: PHYSICIST, NASA, HAMPTON, VA, 64- Mem: Sigma Xi; AAAS. Res: Characterization of semiconductor surface states, their interactions with the semiconductor bulk, and their interactions with electrolytes both electrochemically and photoelectrochemically. Mailing Add: 17 Garrett Dr Hampton VA 23669

BYWATERS, JAMES HUMPHREYS, b Washington Co, Tenn, Sept 17, 07; m 37. GENETICS. Educ: Ohio State Univ, BS, 29; Univ Ky, MS, 32; Iowa State Col, PhD(animal breeding & genetics), 36. Prof Exp: Asst animal husb, Cornell Univ, 29-30; prof agr, Eastern Ky State Teachers Col, 31; instr animal husb, Univ Ky, 36-39; from assoc geneticist to geneticist, regional poultry res lab, Bur Animal Indust, USDA, 39-47; prof in charge poultry husb res, Va Agr Exp Sta, Va Polytech Inst, 47-58; dir res, William H Miner Agr Res Inst, NY, 58-60; turkey specialist, Yoder Feeds, Iowa, 60-62; teacher, high sch, Iowa, 62-64; prof chem & biol, Abraham Baldwin Agr Col, 64-67; prof sci, Northeast Mo State Col, 67-74; prof biol, Coe Col, 74-75; RETIRED. Concurrent Pos: Mem NSF Inst, Bradley Univ, 63. Mem: AAAS; Poultry Sci Asn; Am Genetic Asn. Res: Poultry and animal breeding. Mailing Add: 400 Larick Dr Marion IA 52302

BZOCH, KENNETH R, b Chicago, Ill, Nov 6, 27; m 50; c 2. SPEECH PATHOLOGY, AUDIOLOGY. Educ: DePaul Univ, BA, 50; Northwestern Univ, MA & PhD(speech path), 56. Prof Exp: Instr, DePaul Acad, Ill, 50; lectr speech, Univ Col, Northwestern Univ, 51; asst prof, Grad Sch, Loyola Univ Ill, 53-57; prof grad fac & coordr, Cleft Lip & Palate Inst, Northwestern Univ, Chicago, 57-59 & 60-64; ASSOC PROF COMMUN DIS & CHMN DEPT, COL HEALTH RELATED SERV, UNIV FLA, 64- Concurrent Pos: Consult, St Francis Hosp, Evanston, Ill, 53-58, Vet Res Hosp, Chicago, 57-59, Nat Inst Dent Res, 64, 66-70 & Vet Hosp, Gainesville, 70- Mem: Fel Am Speech & Hearing Asn; Speech Asn Am; Am Cleft Palate Asn. Res: Cineflurographic studies of co-articulation in normal and abnormal speech; efficacy of cleft palate habilitation procedures; normal and abnormal language development in infancy; physiological phonetics. Mailing Add: Col of Health Related Prof Univ of Fla Gainesville FL 32601

BZOCH, RONALD CHARLES, b Chicago, Ill, Mar 16, 30; m 51; c 2. MATHEMATICS. Educ: DePaul Univ, AB, 53, MS, 54; Ill Inst Technol, PhD(math), 57. Prof Exp: Asst math, Ill Inst Technol, 54-57; asst prof math, Univ Minn, 57-60 & Univ Utah, 60-61; assoc prof math, La State Univ, 61-66, assoc chmn dept, 64-66; PROF MATH & CHMN DEPT, UNIV N DAK, 66- Mem: Am Math Soc; Math Asn Am. Res: Real variables; integration theory. Mailing Add: Dept of Math Univ of N Dak Grand Forks ND 58201

C

CABANA, ALDEE, b Beloeil, Que, July 20, 35; m 58; c 4. PHYSICAL CHEMISTRY, MOLECULAR SPECTROSCOPY. Educ: Univ Montreal, BSc, 58, MSc, 59, PhD(chem), 62. Prof Exp: Fel chem, Princeton Univ, 61-63; from asst prof to assoc prof, 63-71, PROF CHEM, UNIV SHERBROOKE, 71- Concurrent Pos: Pres, Syndicate Prof Univ Sherbrooke, 75- Mem: Chem Inst Can; Spectros Soc Can. Res: Infrared and Raman spectra of molecular crystals; high resolution infrared spectroscopy. Mailing Add: Dept of Chem Univ of Sherbrooke Sherbrooke PQ Can

CABANES, WILLIAM RALPH, JR, b Memphis, Tex, Nov 13, 32; m 62; c 1. ORGANIC CHEMISTRY, POLYMER CHEMISTRY. Educ: Univ Tex, Austin, BA, 53, MA, 55, PhD(org chem), 57. Prof Exp: Res fel chem, Univ Ariz, 62-64; lectr org chem, Univ Minn, Duluth, 64-65; ASSOC PROF ORG CHEM, UNIV TEX, EL PASO, 65- Concurrent Pos: R A Welch grant, 67-70. Mem: Fel AAAS; NY Acad Sci. Res: Organic polymers; polyampholytes. Mailing Add: Dept of Chem Univ of Tex El Paso TX 79999

CABANISS, GERRY HENDERSON, b Winter Haven, Fla, Apr 22, 35; m 62; c 3. GEOPHYSICS, TECTONICS. Educ: Dartmouth Col, AB, 57; Boston Col, MS, 68; Boston Univ, PhD(geol), 75. Prof Exp: GEOPHYSICIST, AIR FORCE CAMBRIDGE RES LABS, 58- Mem: AAAS; Am Geophys Union. Res: The temporal and spatial characteristics of earth crustal deformations at periods longer than one hour. Mailing Add: AF Cambridge Res Labs L G Hanscom AFB Bedford MA 01731

CABASSO, ISRAEL, b Jerusalem, Palestine, Nov 17, 42. POLYMER CHEMISTRY. Educ: Hebrew Univ Jerusalem, BS, 66, MS, 68; Weizmann Inst Sci, PhD(polymer chem), 73. Prof Exp: Asst res assoc & teacher chem, Hebrew Univ Jerusalem, 66-68; res asst, Dept Plastic, Weizmann Inst Sci, 68-73; fel, 73-74, sr investr polymer chem, 74-75, GROUP LEADER POLYMER CHEM, GULF SOUTH RES INST, 76- Concurrent Pos: Prof sci, Technicon High Sch, 68-72; sr res assoc, Weizmann Inst Sci, 73- Mem: Am Chem Soc; Israel Chem Soc. Res: Hollow fiber and synthetic polymeric membrane; synthesis and transport phenomena, for hemodialysis, reverse osmosis, pervaporation, and polymer materials for artificial organs, self extinguish material, and development of ligand polymers. Mailing Add: Gulf South Res Inst PO Box 26500 New Orleans LA 70186

CABASSO, VICTOR JACK, b Port Said, Egypt, June 21, 15; US citizen; m 48; c 2. VIROLOGY. Educ: Lycee Francais, Egypt, BA, 33; Hebrew Univ, Israel, MS, 38; Sorbonne & Univ Algiers, ScD(bact), 41; Am Bd Microbiol, dipl, 61. Prof Exp: Spec investr pub health & hyg, Off Foreign Relief & Rehab Opers, US Dept State, Tunis, 43; chief bacteriologist, Greece & Mid East, UNRRA, 44-46; res virologist, Lederle Labs, 46-58; head virus immunol res dept, Lederle Labs, Am Cyanamid Co, 58-67; dir microbiol res dept, 67-69, assoc dir res-microbiol, 69-74, DIR RES & DEVELOP, CUTTER LABS, 74- Concurrent Pos: Assoc mem, Pasteur Inst, Tunis; mem comt on stand methods for vet microbiol, Nat Acad Sci; mem working comt microbiol, Nat Comt Clin Lab Stand; mem subcomt rabies, Agr Bd, Nat Res Coun. Mem: AAAS; fel NY Acad Sci; fel Am Acad Microbiol; Am Pub Health Asn; Am Soc Microbiol. Res: Myxoviruses; picornaviruses; adenoviruses; viral oncolysis; rabies; hepatitis; measles; rubella; canine hepatitis; bluetongue; smallpox. Mailing Add: Cutter Labs Microbiol Res Dept Fourth & Parker St Berkeley CA 94710

CABAT, GEORGE ALAN, organic chemistry, polymer chemistry, see 12th edition

CABBINESS, DALE KEITH, b Binger, Okla, May 22, 37; m 58; c 2. ANALYTICAL CHEMISTRY. Educ: Southwestern State Col, BS, 60; Univ Ark, MS, 65; Purdue Univ, PhD(chem), 70. Prof Exp: Chemist, Dow Chem Co, 60-61; RES SCIENTIST, CONTINENTAL OIL CO, 69- Mem: Am Chem Soc; Sigma Xi. Res: Chemical methods of analysis using kinetic techniques, continuous flow procedures and automation of same. Mailing Add: PO Box 1267 Ponca City OK 74601

CABEEN, SAMUEL KIRKLAND, b Easton, Pa, Jan 22, 31. CHEMISTRY, INFORMATION SCIENCE. Educ: Lafayette Col, BA, 52; Syracuse Univ, MS, 54. Prof Exp: Asst librn, Am Metal Climax Inc, 56-58; librn, Ford Instrument Co Div, Sperry Rand Corp, 58-64; asst dir, 64-68, DIR, ENG SOCS LIBR, 68- Mem: Spec Libr Asn; Am Soc Info Sci; Am Libr Asn. Mailing Add: Eng Socs Libr 345 E 47th St New York NY 10017

CABELL, PAMELA WHITING, b Boston, Mass, Nov 12, 46; m 70. PLASTICS CHEMISTRY. Educ: Stonehill Col, BS, 67; Brown Univ, PhD(chem), 71. Prof Exp: Fel, State Univ Groningen, 71-73; PRES & CHMN, GAMMA PLASTICS CORP, 73- Mem: Soc Plastics Engrs; Am Soc Qual Control; Am Soc Testing & Mat; Royal Soc Encour Arts, Mfg & Com. Res: Investigation of high heat thermoplastic resins. Mailing Add: 370 Lenox St Norwood MA 02062

CABELLI, VICTOR JACK, b New York, NY, Dec 9, 26; m 49; c 4. MICROBIOLOGY, ENVIRONMENTAL HEALTH. Educ: Univ Calif, Los Angeles, AB, 48, PhD, 51. Prof Exp: Asst bact, Univ Calif, Los Angeles, 49-51; asst prof microbiol, Univ Mo, 51-56; chief, Agent Biol Sect, Dugway Proving Ground, Utah, 56-59, Biol Labs Br, 59-61 & Biol Div, 61-64; prin microbiologist, Northeastern Water Hyg Lab, 64-75, CHIEF, MARINE FIELD STA, HEALTH EFFECTS RES LAB-CIN, ENVIRON PROTECTION AGENCY, 75- Concurrent Pos: Adj prof microbiol, Univ RI, 65- Mem: Am Soc Microbiol. Res: Nutrition and physiology of Pasteurella tularensis; shellfish microbiology; host-parasite relationship; metabolism and taxonomy of coliforms; aerobiology; ecology of C botulinum; water pollution microbiology, epidemiology and public health. Mailing Add: Marine Field Sta HERL-Cin Environ Protection Agncy Box 277 West Kingston RI 02892

CABELLO, JULIO, biological chemistry, see 12th edition

CABIB, ENRICO, b Genoa, Italy, Jan 11, 25; m 55; c 3. BIOCHEMISTRY, ENZYMOLOGY. Educ: Univ Buenos Aires, PhD(chem), 51. Prof Exp: Investr, Inst Biochem Invest, 49-53; vis investr biochem, Col Physicians & Surgeons, Columbia Univ, 53-54; investr, Inst Biochem Invest, 55-58; investr, Sch Sci, Univ Buenos Aires, 58-67; vis scientist, 67-69, RES BIOCHEMIST, NAT INST ARTHRITIS, METAB & DIGESTIVE DIS, 69- Concurrent Pos: Instr, Sch Sci, Univ Buenos Aires, 50-53; career investr, Nat Coun Sci & Tech Invests, 60-67; ed, Arch Biochem & Biophys, 70-72 & J Biol Chem, 75- Mem: Am Soc Microbiol; Soc Complex Carbohydrates; Fedn Am Sci; Am Soc Biol Chemists. Res: Metabolism of carbohydrates, particularly studies on the biosynthesis of di- and poly-saccharides from sugar nucleotides; biosynthesis of yeast cell wall; regulation of glycogen synthesis in yeast and muscle. Mailing Add: Nat Inst of Arthritis Bldg 10 Rm 9N-111 Bethesda MD 20014

CABLE, CHARLES ALLEN, b Akeley, Pa, Jan 15, 32; m 55; c 2. ALGEBRA, NUMBER THEORY. Educ: Edinboro State Col, BS, 54; Univ NC, Chapel Hill, MEd, 59; Pa State Univ, PhD(math), 69. Prof Exp: Teacher, high sch, NY, 54-55 & Pa, 57-58; from instr to asst prof math, Juniata Col, 59-67; assoc prof, 69-75, PROF MATH, ALLEGHENY COL, 75-, CHMN DEPT, 70- Mem: Am Math Soc; Math Asn Am. Res: Group rings where ring is a finite field and group is Abelian. Mailing Add: Dept of Math Allegheny Col Meadville PA 16335

CABLE, DWIGHT RAYMOND, b Chicago, Ill, Dec 16, 16; m 48; c 3. FORESTRY. Educ: Univ Idaho, BS, 38; Univ Ariz, S, 59. Prof Exp: Jr range examr, Soil Conserv Serv, 38-41, range conservationist, 46-50; RANGE SCIENTIST, ROCKY MOUNTAIN FOREST & RANGE EXP STA, US FOREST SERV, 50- Mem: Am Soc Range Mgt. Res: Plant ecology; grazing management research. Mailing Add: Rocky Mt Forest & Range Exp Sta Box 4460 Tucson AZ 85717

CABLE, JOE WOOD, b Murray, Ky, Feb 17, 31; m 50; c 3. SOLID STATE PHYSICS. Educ: Murray State Univ, AB, 52; Fla State Univ, PhD(chem), 55. Prof Exp: PHYSICIST, OAK RIDGE NAT LAB, 55- Mem: Am Phys Soc. Res: Neutron scattering studies of magnetic materials. Mailing Add: Oak Ridge Nat Lab Oak Ridge TN 37830

CABLE, RAYMOND MILLARD, b Campton, Ky, Apr 22, 09; m 36; c 4. ZOOLOGY. Educ: Berea Col, AB, 29; NY Univ, MS, 30, PhD, 33. Hon Degrees: ScD, Berea Col, 55. Prof Exp: Assoc prof biol, Berea Col, 33-35; from asst prof to prof, 35-75, EMER PROF PARASITOL, PURDUE UNIV, 75- Concurrent Pos: Guggenheim fel, Univ PR, 51-52. Mem: AAAS; Am Soc Zool; Am Micros Soc; Am Soc Parasitol(vpres, 58, pres, 64); NY Acad Sci. Res: Parasitology. Mailing Add: Dept of Biol Sci Purdue Univ West Lafayette IN 47907

CABRERA, ANGEL LULIO, b Madrid, Spain, Oct 19, 08; m 38; c 3. SYSTEMATIC BOTANY, PHYTOGEOGRAPHY. Educ: La Plata Nat Univ, DNatSci, 31. Prof Exp: Botanist, Admin Agr Buenos Aires, 38-45; head div phytogeog, Inst Ministry Agr, 47-49; asst dean fac natural sci, 57, TITULAR PROF BOT, FAC NATURAL SCI & MUS, LA PLATA NAT UNIV, BIOGEOG, FAC HUMANITIES, HEAD DIV VASCULAR PLANTS & INTERIOR HEAD DEPT BOT, MUS, 49- Concurrent Pos: Hon head, Inst Spegazzini; mem Int Coun Study Arid Regions, UNESCO, 58; hon dir, J Soc Bot Arg. Mem: Soc Plant Taxon; Torrey Bot Club; Soc Natural Sci Arg; Soc Bot Arg (pres); Arg Soc Agron. Res: Composites and plant geography of South America. Mailing Add: Museo de La Plata La Plata Argentina

CABRERA, BLAS, b Paris, France, Sept 21, 46; US citizen; m 72; c 2. LOW TEMPERATURE PHYSICS. Educ: Univ Va, BS, 68; Stanford Univ, PhD(physics), 75. Prof Exp: RES ASSOC PHYSICS, STANFORD UNIV, 75- Res: Testing general relativity with an orbiting precise superconducting gyroscope referenced to a fixed star, using a London moment readout in an ultra-low magnetic field region. Mailing Add: Dept of Physics Stanford Univ Stanford CA 94305

CABRERA, EDELBERTO JOSE, b Pinar del Rio, Cuba, Nov 5, 44; US citizen; m 64; c 2. IMMUNOLOGY, MALARIOLOGY. Educ: Univ Ill, Urbana-Champaign, BS, 67, MS, 68, PhD(zool), 72. Prof Exp: Res asst malaria, Univ Ill, Urbana-Champaign, 69-72; RES SCIENTIST IMMUNOL, UNIV NMEX, 72- Mem: AAAS; Am Soc Parasitologists; Soc Protozoologists; Sigma Xi. Res: Investigation of the immune response of primates and rodents to plasmodial infections; the overall aim of the research is directed toward the preparation of a vaccine against human malaria. Mailing Add: Dept of Biol Univ of NMex Albuquerque NM 87131

CABRERA, NICOLAS, b Madrid, Spain, Feb 12, 13; US citizen; m 43; c 3. SOLID STATE PHYSICS. Educ: Univ Madrid, BSc, 35; Univ Paris, PhD(physics), 44. Prof Exp: Asst prof physics, Univ Madrid, 35-37; sci asst, Int Bur Weights & Measures, Paris, 38-52; res assoc, Bristol Univ, 47-50; from assoc prof to prof physics & chem, Univ Va, 52-74, chmn dept, 62-74; MEM FAC, UNIV MADRID, FAC SCI, DEPT PHYSICS, 74- Concurrent Pos: Consult, Oak Ridge Nat Lab; Radio Corp Am. Mem: Am Phys Soc. Res: Chemical physics. Mailing Add: Univ of Madrid Fac Sci Dept Physics Madrid 11 Spain

CABRI, LOUIS J, b Cairo, Egypt, Feb 23, 34; Can citizen; m 59; c 2. MINERALOGY, GEOCHEMISTRY. Educ: Univ Witwatersrand, BSc, 54, Hons, 55; McGill Univ, MSc, 61, PhD(geol, geochem), 65. Prof Exp: Geologist, Josan, SA, WAfrica, 56, Sierra Leone Mineral Syndicate, 57-58 & New Amianthus Mines, SAfrica, 59; sci officer mineral, Dept Mines & Tech Surv, Can, 64-65, RES SCIENTIST, DEPT ENERGY, MINES & RES CAN, 66- Honors & Awards: Waldemar Lindgren Citation, Soc Econ Geol, 66. Mem: Mineral Soc Am; Mineral Asn Can; Geol Asn Can; Can Inst Mining & Metall; Prof Inst Pub Serv Can. Res: Phase equilibrium research in sulfide and sulfide-type systems and applications to mineralogical problems and to genesis of ore deposits. Mailing Add: Mines Br 555 Booth St Ottawa ON Can

CACELLA, ARTHUR FERREIRA, b New Bedford, Mass, Nov 26, 20; m 50; c 5. POLYMER CHEMISTRY. Educ: NC State Col, BChE, 47; Inst Textile Technol, MSc, 49; Princeton Univ, MA, 52. Prof Exp: Res chemist, Benger Lab, El du Pont de Nemours & Co, Va, 54-58 & Inst Textile Technol, 59-60; dir res & Deveop, Globe Mfg Co, 60-62; VPRES & LAB DIR, AMELIOTEX, INC, ROCKY HILL, 62- Mem: Am Chem Soc; Am Inst Chemists; Am Asn Textile Chemists & Colorists; NY Acad Sci. Res: Stress relaxation of nylon; interaction of water vapor with natural and syntheic fibers; solubility of polymers; chemical modification of cotton; polyurethane chemistry of coatings and adhesives; hydrolysis; catalysis; spandex fibers. Mailing Add: 16 Hillwood Rd East Brunswick NJ 08816

CACERES, CESAR A, b Puerto Cortes, Honduras, Apr 9, 27. MEDICINE, COMPUTER SCIENCE. Educ: Georgetown Univ, BS, 49, MD, 53. Prof Exp: Intern med, Boston City Hosp, Mass, 53-54, resident, 54-55; resident, New Eng Med Ctr, 55-56; fel cardiol, George Washington Univ, 56-60, from asst clin prof to assoc prof med, 60-69, prof & chmn dept, 69-71; PROF ELEC ENG, UNIV MD, COLLEGE PARK, 71-; PRES, CLIN SYSTS ASSOC, 71- Concurrent Pos: Chief instrumentation field sta, Heart Dis Control Prog, USPHS, 60-69. Honors & Awards: Dept Health, hduc & Welfare Super Serv Awards, 63 & 66. Mem: Soc Advan Med Systs (past pres); Int Health Eval Asn (past pres); Asn Advan Med Instrumentation (past pres). Res: Electrocardiography; cholesterol; cardiology; medical diagnosis; computers in medicine. Mailing Add: 1759 Q St NW Washington DC 20009

CACERES, EDUARDO, b Lima, Peru, Jan 3, 15; m 44; c 4. ONCOLOGY. Educ: San Marcos Univ, Lima, BS, 36, MD, 42. Prof Exp: Clin asst, Mem Hosp Cancer & Allied Dis, New York, 49-51; DIR, NAT INST NEOPLASTIC DIS, 52- Concurrent Pos: Clin asst, James Ewing Hosp, New York, 50-51; consult, Mem Hosp, Wilmington, Del, 50-51; mem fel comt, Eleanor Roosevelt Cancer Found, 62-66; expert adv panel cancer, WHO, 63-68; vpres, Latin Am Int Union Against Cancer, 62-66, Int Cong Smoking & Health, 63 & Pan-Pac Surg Asn, 6466. Mem: AAAS; Am Asn Cancer Res; fel Am Col Surgeons; James Ewing Soc; Am Soc Cytol. Res: Cancer surgey. Mailing Add: Nat Inst of Neoplastic Dis 825 Avenida Alfonsa Ugarte Lima Peru

CADBURY, WILLIAM EDWARD, JR, b Germantown, Pa, Apr 19, 09; m 33; c 2. PHYSICAL CHEMISTRY. Educ: Haverford Col, BS, 31, AM, 32; Univ Pa, PhD(chem), 40. Hon Degrees: LLD, Haverford Col, 74. Prof Exp: From instr to asst prof chem, Haverford Col, 32-43; actg assoc prof, Univ NC, 43-44; from assoc prof to prof, 44-70, dean col, 51-66, dir post-baccalaureate fel prog, 66-70, EMER PROF CHEM, HAVERFORD COL, 70-; EXEC DIR, NAT MED FELS, INC, 70- Concurrent Pos: Assoc ed, J Chem Educ, 50-55. Mem: Fel AAAS. Res: Phase relations in the system sodium sulfate-sodium chromate-water; phase rule studies on aqueous solutions. Mailing Add: 404 Riverside Dr New York NY 10025

CADE, JAMES ROBERT, b San Antonio, Tex, Sept 26, 27; m 53; c 6. INTERNAL MEDICINE. Educ: Univ Tex, MD, 54. Prof Exp: Fel, Cornell Univ, 58-61; assoc prof med, 61-72, PROF RENAL MED & CHIEF MED SCH, UNIV FLA, 72- Mem: Am Fedn Clin Res; Am Soc Clin Invest; Am Physiol Soc. Res: Renal, electrolyte and exercise physiology. Mailing Add: Dept of Med Univ of Fla Gainesville FL 32601

CADE, ROGER, theoretical physics, see 12th edition

CADE, RUTH ANN, b Yazoo City, Miss, Nov 17, 37; m 74; c 2. APPLIED MATHEMATICS. Educ: Miss State Univ, BS, 63; Univ Ala, MA, 68, PhD(eng), 69. Prof Exp: Chem engr, Southern Res Inst, 63-64; instr eng mech, Univ Ala, 67-68; asst prof math, 68-75, ASSOC PROF COMPUT SCI, UNIV SOUTHERN MISS, 75- Mem: Am Soc Eng Educ; Math Asn Am. Mailing Add: Box 5257 Southern Sta Hattiesburg MS 39401

CADE, STEPHEN C, b Bloomington, Ill, Aug 22, 39; m 62; c 2. FORESTRY. Educ: Duke Univ, BS, 61, MF, 62; Univ Wash, PhD(forestry), 70. Prof Exp: Forest entomologist, Southeast Forest Exp Sta, US Forest Serv, 62-63; forest entomologist, 70-72, seed orchard scientist, 73-75, FORESTRY RES MGR, FORESTRY RES CTR, WEYERHAEUSER CO, 75- Mem: Soc Am Foresters; Entom Soc Am. Res: Forestry herbicides and pesticides; forest insect biology and control; forest tree seed production and protection. Mailing Add: Weyerhaeuser Co PO Box 9 Klamath Falls OR 97601

CADE, THOMAS JOSEPH, b San Angelo, Tex, Jan 10, 28; m 52; c 5. ORNITHOLOGY. Educ: Univ Alaska, AB, 51; Univ Calif, Los Angeles, MA, 55, PhD(zool), 58. Prof Exp: Assoc zool, Univ Calif, Los Angeles, 55-58; NSF fel, Mus Vert Zool, Univ Calif, Berkeley, 58-59; from asst prof to prof zool, Syracuse Univ, 59-67; PROF ORNITH, STATE UNIV NY COL AGR, CORNELL UNIV & RES DIR LAB ORNITH, 67-, MEM STAFF INSTR, RES & EXTEN AT ITHACA, 72- Concurrent Pos: NSF grants, 60-65 & sr fel, Transvaal Mus, SAfrica, 65-66; USPHS grant, 63-66. Mem: Am Soc Mammal; fel Am Ornith Union; Cooper Ornith Soc; Wilson Ornith Soc. Res: Population and physiological ecology; behavior of birds and mammals. Mailing Add: Cornell Univ Lab Ornith 159 Sapsucker Woods Rd Ithaca NY 14850

CADENHEAD, DAVID ALLAN, b Tillicoultry, Scotland, Aug 5, 30. PHYSICAL CHEMISTRY. Educ: Univ St Andrews, BSc, 53; Bristol Univ, PhD(chem), 57. Prof Exp: Can Nat Defence fel, Royal Mil Col, Can, 57-59; fel, Alfred Univ, 59-60; asst prof, 60-64, ASSOC PROF CHEM, STATE UNIV NY BUFFALO, 64-, RES ASSOC PROF BIOCHEM, 70- Concurrent Pos: State Univ NY Buffalo rep, Univs Space Res Asn, 74-; vis prof, San Francisco State Univ, 76. Mem: Am Chem Soc; The Chem Soc. Res: Adsorption and porosity; transition metal catalysis and hydrogenation; monomolecular film studies; surface characterization of phospholipids; molecular interactions in mixed monolayers; cell membrane structure; gas interactions with lunar materials. Mailing Add: Dept of Chem Acheson Hall State Univ NY Chemistry Rd Buffalo NY 14214

CADIEN, JAMES DAVID, b Hollywood, Calif, June 22, 41; m 66; c 2. PHYSICAL ANTHROPOLOGY. Educ: Univ Calif, Berkeley, BA, 65, PhD(anthrop), 70. Prof Exp: Asst prof anthrop, Case Western Reserve Univ, 69-73; ASST PROF ANTHROP, ARIZ STATE UNIV, 73- Mem: Am Anthrop Asn; Am Asn Phys Anthrop; Am Eugenics Soc. Res: Relationship between cusp number and size of human molar teeth; inheritance of dental characteristics. Mailing Add: Dept of Anthrop Ariz State Univ Tempe AZ 85281

CADIGAN, FRANCIS C, JR, b Boston, Mass, June 17, 30; m 56; c 1. INFECTIOUS DISEASES, TROPICAL MEDICINE. Educ: Boston Col, AB, 51, MS, 56; Tufts Univ, MD, 56; Johns Hopkins Univ, MPH, 68; Am Bd Pediat & Am Bd Prev Med, dipl. Prof Exp: Intern, Carney Hosp, 56-57; resident physician pediat, Boston Floating Hosp, Mass, 57-59; Med Corps, US Army, 59-, virologist, Walter Reed Army Inst Res, 59-64, dep dir, SEATO Med Res Lab, 64-67, cmndg officer, US Army Med Res Unit, Malaysia, 69-72, DIR MED RES HQ, US ARMY MED RES & DEVELOP COMMAND, WASHINGTON, DC, 72- Concurrent Pos: Clin asst prof pediat, Sch Med, Georgetown Univ, 64-67, clin assoc prof, 67-74, clin prof, 74-; mem bd studies & bd examr, Malaysian Nat Centre Trop Med & Pub Health, 69-72. Mem: Am Soc Trop Med & Hyg; AMA; fel Royal Soc Trop Med & Hyg; Soc Pediat Res; fel Royal Soc Health. Res: Malaria; scrub typhus; primate ethology. Mailing Add: Hq US Army Med Res & Develop Command Washington DC 20314

CADIGAN, ROBERT ALLEN, b Glen Falls, NY, June 10, 18; m 44; c 4. GEOLOGY. Educ: Col of Puget Sound, AB, 47; Pa State Univ, MS, 52. Prof Exp: Geologist, 48-74, RES TEAM LEADER, US GEOL SURV, 74- Concurrent Pos: Adv tech coord comt, Nat Hydrogeochem Surv, US Energy Resources Develop Admin, 75- Mem: Geol Soc Am; Am Asn Petrol Geologists; Am Soc Econ Paleont & Mineral. Res: Textural and geochemical properties of sedimentary rocks; geochemical exploration techniques applied to detecting areas in the United States containing economically important uranium ore deposits. Mailing Add: 9125 W Second Ave Lakewood CO 80226

CADLE, RICHARD DUNBAR, b Cleveland, Ohio, Sept 23, 14; m 40; c 3. PHYSICAL CHEMISTRY. Educ: Western Reserve Univ, AB, 36; Univ Wash, PhD(chem), 40. Prof Exp: Teaching fel, Univ Wash, 36-38; res chemist, Procter & Gamble Co, 40-47; unit head, Naval Ord Test Sta, Calif, 47-48; sr phys chemist, Stanford Res Inst, 48-54, mgr atmospheric chem physics, 54-63; prog scientist, 63-66, head chem dept, 66-73, PROJ SCIENTIST, NAT CTR ATMOSPHERIC RES, 73- Concurrent Pos: Asst lectr, Univ Cincinnati, 41-43; consult, Army Chem Corps, 55; vis prof, Ripon Col, 66; mem tech adv comt, Denver Regional Air Pollution Control Agency, 67-72; consult, Gulf South Res Inst, 68-70; mem adv comt air pollution, chem & physics, USPHS, 69-73; panel mem polar meteorol, Nat Acad Sci-Nat Res Coun, 69-73, panel mem atmospheric chem, 70-75; adj prof, Pa State Univ, 72-75. Mem: Sigma Xi; Am Chem Soc; Am Geophys Union; Am Meteorol Soc. Res: Properties of dilute solutions of colloidal electrolytes; preparation and identification of aerosols; chemical kinetics; atmospheric chemistry. Mailing Add: 4415 Chippewa Dr Boulder CO 80303

CADLE, STEPHEN HOWARD, b Cincinnati, Ohio, Feb 19, 46; m 68; c 2. ANALYTICAL CHEMISTRY, ENVIRONMENTAL CHEMISTRY. Educ: Univ Colo, BA, 67; State Univ NY Buffalo, PhD(chem), 72. Prof Exp: Asst prof chem, Vassar Col, 72-73; ASSOC SR RES CHEMIST, GEN MOTORS RES LABS, 74- Honors & Awards: Young Author Award, Electrochem Soc, 75. Mem: Am Chem Soc. Res: Analytical methods of analysis, especially those pertaining to atmospheric pollution problems. Mailing Add: Res Labs Gen Motors Tech Ctr Warren MI 48090

CADMUS, EUGENE L, b Newark, NJ, Aug 28, 22; m 46; c 3. MICROBIOLOGY. Educ: Rutgers Univ, BS, 49; Brooklyn Polytech, MS, 51. Prof Exp: Res chemist, F W Berk & Co, Inc, 51-56; tech dir chem, Woodridge Chem Corp, 56-68; TECH MGR CHEM, VENTRON CORP, 68- Mem: Am Soc Testing & Mat; Am Chem Soc; Soc Indust Microbiol. Res: Mechanisms of brocide activity in industrial applications, plastics; structure activity relationships for compounds showing microbiological activity. Mailing Add: Ventron Corp 154 Andover St Danvers MA 01923

CADMUS, ROBERT R, b Little Falls, NJ, June 16, 14; m 41; c 2. PREVENTIVE MEDICINE, MEDICAL ADMINISTRATION. Educ: Col Wooster, AB, 36; Columbia Univ, MD, 40. Hon Degrees: DSc, Col Med & Dent NJ, Newark, 71. Prof Exp: Dir prof serv patients, Vanderbilt Clin-Columbia-Presby Hosp, NY, 45-48; asst dir, Univ Hosps, Cleveland, Ohio, 48-50; dir, NC Mem Hosp, 50-62; prof hosp admin, Univ NC, 52-66, chmn dept, 62-66; consult dir, NC Mem Hosp & res dir, Educ & Res Found, 62-66; prof prev med & pres, Col Med & Dent, NJ, Newark, 66-71; dir, Med Ctr Southeastern Wis, 71-75; EXEC DIR, FOUND FOR MED CARE EVAL SOUTHEASTERN WIS, 75- Concurrent Pos: Consult, Nat Inst Arthritis & Metab Dis, 60-64 & Div Hosp & Med Facilities, USPHS, 63; mem comt, Clin Res Ctr, Div Res Facilities & Resources, 64-67; consult, Vet Admin Hosp, East Orange, NJ & USPHS Hosp, Staten Island, NY, 66-71; mem nat adv coun, Gen Med Sci, NIH; mem regional adv group, Health Serv & Ment Health Admin-Regional Med Prog; mem comprehensive health planning coun, Health Serv & Ment Health Admin. Mem: AMA; Asn Teachers Prev Med; Am Hosp Asn; Am Asn Hosp Consult. Res: Administrative medicine. Mailing Add: Found Med Care Eval of SE Wis 756 N Milwaukee St Milwaukee WI 53202

CADOFF, BARRY CARL, organic chemistry, see 12th edition

CADOGAN, KEVIN DENIS, b New York, NY, Feb 17, 39; m 61; c 4. PHYSICAL CHEMISTRY. Educ: Manhattan Col, BS, 60; Cornell Univ, PhD(phys chem), 66. Prof Exp: Res assoc chem, Cornell Univ, 66-67; res assoc chem, Lab Chem Biodynamics, Univ Calif, Berkeley, 67-69; asst prof, 69-74, ASSOC PROF CHEM, CALIF STATE COL, HAYWARD, 74- Concurrent Pos: NIH fel, 67-69; Fulbright-Hays lectr, US-UK Educ Comn, 70-71; Fulbright lectr, Inst Educ Technol, Univ Surrey, 70-71. Mem: Am Chem Soc. Res: Electronic spectroscopy and solid state photochemistry in aqueous and hydrocarbon solutions at cryogenic temperatures; environmental chemistry; induced and natural chemical cycles in the biosphere; scarce chemical research utilization; scientific utopia. Mailing Add: Dept of Chem Calif State Col 25800 Hillary St Hayward CA 94542

CADORET, REMI JERE, b Scranton, Pa, Mar 28, 28; m 50, 69; c 4. PSYCHIATRY. Educ: Harvard Univ, AB, 49; Yale Univ, MD, 53. Prof Exp: Intern rotating, Robert Packer Hosp, 53-54; res assoc parapsychol, Parapsychol Lab, Duke Univ, 56-58; from asst prof to assoc prof physiol, Med Sch, Univ Man, 58-65; resident psychiat, Sch Med, Wash Univ, 65-68, from asst prof to assoc prof, 68-73; PROF PSYCHIAT, MED SCH, UNIV IOWA, 73- Mem: Psychiat Res Soc. Res: Genetics of psychiatric illness; nosology of psychiatric conditions; personality; psychiatric epidemiology. Mailing Add: Dept of Psychiat State Psychopathic Hosp 500 Newton Rd Iowa City IA 52240

CADOTTE, JOHN EDWARD, b Minn, Jan 28, 25; m 48; c 5. ORGANIC CHEMISTRY. Educ: Univ Minn, MS, 51. Prof Exp: Res chemist, Merck & Co, Inc, 51-52, Clinton Foods, Inc, 52-56 & Wood Conversion Co, 56-65; polymer researcher, 65-71, chemist, 66-75, SR CHEMIST, N STAR DIV, MIDWEST RES INST, 75- Mem: Am Chem Soc. Res: Adhesives; starch; thermosetting resins; Fourdrinier formed board products and coatings; reverse osmosis membranes. Mailing Add: 2427 Western Ave N St Paul MN 55113

CADWALLADER, DONALD ELTON, b Buffalo, NY, June 14, 31; m 60; c 3. PHARMACY, MEDICINAL CHEMISTRY. Educ: Univ Buffalo, BS, 53; Univ Ga, MS, 55; Univ Fla, PhD(pharm), 57. Prof Exp: Interim res prof, mil med supply agency contract, Univ Fla, 57-58; res assoc pharm, Sterling-Winthrop Res Inst, 58-60; group leader, White Labs, Inc, 60-61; from asst prof to assoc prof, 61-68, PROF PHARM, UNIV GA, 68- Honors & Awards: Lunsford-Richardson Pharm Award, 57. Mem: Fel AAAS; Am Pharmaceut Asn; fel Acad Pharmaceut Sci. Res: Behavior of erythrocytes in various solvent systems; preparation and evaluation of parenteral products; stability of drugs; drug manufacturing procedures and processes; biopharmaceutics. Mailing Add: Dept of Pharm Univ of Ga Athens GA 30601

CADY, BLAKE, b Washington, DC, Dec 27, 30; m 60; c 3. SURGERY, ONCOLOGY. Educ: Amherst Col, AB, 53; Cornell Univ, MD, 57. Prof Exp: Fel surg, Sloan-Kettering Inst & Med Col, Cornell Univ, 65-67; ASST CLIN PROF SURG, MED SCH, HARVARD UNIV, 67- Mem: AMA; Am Col Surgeons; James Ewing Soc; Soc Head & Neck Surgeons; Soc Surg Alimentary Tract. Res: Clinical interest in human tumors and cancers. Mailing Add: Dept of Surg Lahey Clin Found 605 Commonwealth Ave Boston MA 02215

CADY, FOSTER BERNARD, b Middletown, NY, Aug 5, 31; m 53; c 3. STATISTICS. Educ: Cornell Univ, BS, 53; Univ Ill, MS, 56; NC State Univ, PhD, 60. Prof Exp: Soil scientist, Agr Res Serv, USDA, NC State Univ, 57-60; assoc prof statist, Iowa State Univ, 60-68; prof, Univ Ky, 68-71; PROF BIOL STATIST, CORNELL UNIV, 71- Mem: Am Statist Asn; Am Soc Agron; Biomet Soc. Res: Statistical methods and experimental design in the agricultural and biological sciences. Mailing Add: Biomet Unit Cornell Univ Ithaca NY 14850

CADY, GEORGE HAMILTON, b Lawrence, Kans, Jan 10, 06; m 29; c 2. INORGANIC CHEMISTRY, FLUORINE CHEMISTRY. Educ: Univ Kans, AB, 27, AM, 28; Univ Calif, PhD(chem), 30. Prof Exp: Asst prof chem, Univ S Dak, 30-31; instr inorg chem, Mass Inst Technol, 31-34; res chemist, gen labs, US Rubber Prod, Inc, 34-35 & Columbia Chem Div, Pittsburgh Plate Glass Co, 35-38; from asst prof to prof, 38-72, chmn dept, 61-65, EMER PROF CHEM, UNIV WASH, 72- Concurrent Pos: Sect leader, Manhattan Dist Proj, Columbia Univ, 42-43; GN Lewis Mem lectr, 67. Honors & Awards: Distinguished Serv Award, Am Chem Soc, 66, Award for Res Fluorine Chem, 72; US Navy Meritorious Pub Serv Citation, 70; Distinguished Serv Citation, Univ Kans, 72. Mem: AAAS; Am Chem Soc; Leopoldine Ger Acad Researchers Natural Sci. Res: Preparation and properties of fluorides; hypochlorous acid and hypochlorites; determination of rare gases. Mailing Add: Dept of Chem Univ of Wash Seattle WA 98105

CADY, HOWARD HAMILTON, b Sioux City, Iowa, Jan 15, 31; m 57; c 4. PHYSICAL CHEMISTRY. Educ: Univ Wash, BS, 53; Univ Calif, PhD(chem), 57. Prof Exp: Asst, Radiation Lab, Univ Calif, 54-57; staff mem, 57-74, SECT LEADER, LOS ALAMOS SCI LAB, UNIV CALIF, 74- Mem: AAAS; Am Chem Soc; Am Crystallog Asn; fel Am Inst Chemists. Res: General physical chemistry; crystallography. Mailing Add: 1373 40th St Los Alamos NM 87544

CADY, JOHN GILBERT, b Seneca Falls, NY, Jan 30, 14; m 40. SOILS. Educ: Syracuse Univ, BS, 36; Univ Wis, MS, 38; Cornell Univ, PhD(soils), 41. Prof Exp: Asst soils, Univ Wis, 36-38 & Cornell Univ, 38-42; instr agron, Univ Idaho, 42-43; soil scientist, Mil Geol Unit, US Geol Surv, 44-46; soil scientist, Bur Plant Indust, Soils & Agr Eng, USDA, 46-51 & Soil Surv Lab, Soil Conserv Serv, 52-69; LECTR SOIL SCI, JOHNS HOPKINS UNIV, 69- Concurrent Pos: Soil mineralogist, Soil Classification Mission, Econ Coop Admin-Belgian Congo Nat Inst Agron Studies, 51-52. Mem: AAAS; fel Geol Soc Am; fel Am Soc Agron. Res: Soil genesis; weathering; soil mineralogy; relation of soil characteristics to land use and natural vegetation. Mailing Add: Dept of Geog & Environ Eng Johns Hopkins Univ Baltimore MD 21218

CADY, LEE (DE), JR, b St Louis, Mo, Nov 22, 27; m 65; c 1. CARDIOLOGY, INDUSTRIAL MEDICINE. Educ: Washington Univ, AB, 47, MA, 48, MD, 51; Yale Univ, MPH, 57, DrPH, 59. Prof Exp: Intern, Vet Admin Hosp & Baylor Univ, 51-52; epidemic intel officer, USPHS, 52-54; health officer, Int Coop Admin, 54-56; from asst to assoc prof phys med & rehab, NY Univ, 59-64; prof biomath & chmn dept, Univ Tex, Houston, 64-67; staff dir, Comt Emergency Med Care for Los Angeles County, 70-72; CHIEF CARDIOPULMONARY LAB, LOS ANGELES COUNTY OCCUP HEALTH SERV, 72-; ADJ PROF MED, UNIV SOUTHERN CALIF, 67- Concurrent Pos: USPHS spec res fel, 58-61. Mem: Fel AAAS; fel Am Col Cardiol; fel Am Pub Health Asn; AMA; Asn Comput Mach. Res: Cardiovascular disease; epidemiology; preventive medicine; public health; computing machines; statistics. Mailing Add: 1718 Warwick Rd San Marino CA 91108

CADY, WALLACE MARTIN, b Middlebury, Vt, Jan 29, 12; m 42; c 3. GEOLOGY. Educ: Middlebury Col, BS, 34; Northwestern Univ, MS, 36; Columbia Univ, PhD(geol), 44. Prof Exp: Asst geol, Northwestern Univ, 34-36; asst, Columbia Univ, 38-40; substitute tutor, Brooklyn Col, 40-41; from jr geologist to prin geologist, 39-63, RES GEOLOGIST, US GEOL SURV, 63- Concurrent Pos: Coun Int Exchange Scholars lectr probs mod tectonics, Voronezh State Univ, USSR, 75. Mem: Fel AAAS; Soc Econ Geol; fel Geol Soc Am; Geochem Soc; fel Am Geophys Union. Res: Geology of northwestern Washington; regional geology of New England and adjacent Quebec; geotectonics; structural geology; metamorphic stratigraphy. Mailing Add: US Geol Surv Box 25046 Fed Ctr Denver CO 80225

CAESAR, CAMERON HULL, b Guelph, Ont, May 21, 10; m 37; c 4. PHYSICAL CHEMISTRY. Educ: Univ Toronto, BA, 33; Univ London, PhD(chem), 36. Prof Exp: Chief res chemist, 52-54, asst mgr res dept, 54-69, MGR RES DEPT, IMP OIL ENTERPRISES LTD, 69- Mem: Chem Inst Can; Royal Can Inst. Res: Petroleum processing and product quality; physical petroleum chemistry. Mailing Add: 1682 Lakeshore Rd Sarnia ON Can

CAESAR, PHILIP D, b New Haven, Conn, Mar 13, 17; m 40; c 2. PETROLEUM CHEMISTRY. Educ: Yale Univ, BS, 39; Univ Ill, MS, 48, PhD(chem), 50. Prof Exp: Chemist, res dept, Paulsboro Lab, Socony Mobil Oil Co, NJ, 39-45, res assoc, 45-56, supvr chem res, 56-60 & catalysis res, 60-64, mgr prod res sect, cent res div, 64-70, MGR EXPLOR PROCESS RES GROUP, PROCESS RES & TECH SERV DIV, PAULSBORO LAB, MOBIL RES & DEVELOP CORP, 70- Mem: AAAS; Am Chem Soc. Res: Thiophene chemistry; composition and properties of petroleum; catalysis research; petroleum processing; fuel cell applications. Mailing Add: Mobil Res & Develop Corp Paulsboro Lab Paulsboro NJ 08066

CAFARELLI, ENZO DONALD, b White Plains, NY, Feb 15, 42; m 67. EXERCISE PHYSIOLOGY. Educ: East Stroudsburg State Col, BS, 69, MEd, 70; Univ Pittsburgh, PhD(exercise physiol), 74. Prof Exp: VIS ASST FEL, JOHN B PIERCE FOUND LAB & FEL ENVIRON PHYSIOL, YALE UNIV, 74- Res: Physiological mechanisms that subserve the sensation of effort during muscular exercise. Mailing Add: John B Pierce Found Lab 290 Congress Ave New Haven CT 06519

CAFFEY, HORACE ROUSE, b Grenada, Miss, Mar 24, 29; m 54; c 4. AGRONOMY, PLANT BREEDING. Educ: Miss State Univ, BS, 51, MS, 55; La State Univ, PhD(agron), 59. Prof Exp: Asst, Miss State Univ, 54-55; agronomist, Miss Rice Grower Asn, 55-57; res assoc, Agron Dept, La Exp Sta, 57-58; assoc agronomist, Delta Br Exp Sta, 58-62, prof & supt, Rice Exp Sta, La State Univ, 62-70; ASSOC DIR, LA AGR EXP STA, 70- Mem: Am Soc Agron. Res: Rice breeding; fertilization and cultural practices. Mailing Add: La Agr Exp Sta PO Drawer E Univ Sta Baton Rouge LA 70803

CAFLISCH, EDWARD GEORGE, b Union City, Pa, Dec 6, 25; m 49; c 3. ORGANIC CHEMISTRY. Educ: Alleghany Col, BS, 48; Ohio State Univ, PhD(org chem), 54. Prof Exp: Chemist, Plaskon Div, Libbey-Owens-Ford Glass Co, 48-49; asst & instr org chem, Ohio State Univ, 49-52; CHEMIST, CHEM & PLASTICS DIV, UNION CARBIDE CORP, 54- Mem: Am Chem Soc. Res: Synthesis of aldehydes; chlorocarbons; fluorocarbons; polymerization of butadiene, pentadienes, and styrene; development of gas chromatography equipment; petrochemicals; organic analytical research. Mailing Add: Union Carbide Corp Tech Ctr 740-2109 South Charleston WV 25303

CAFRUNY, EDWARD JOSEPH, b New Castle, Pa, Dec 17, 24; m 48; c 3. PHARMACOLOGY. Educ: Ind Univ, AB, 50; Syracuse Univ, PhD, 55; Univ Mich, MD, 59. Prof Exp: From instr to assoc prof pharmacol, Med Sch, Univ Mich, 55-65; prof pharmacol, Sch Med, Univ Minn, 65-68; prof pharmacol & exp therapeut, Med Col Ohio, 68-73; PRES, STERLING-WINTHROP RES INST, 73- Concurrent Pos: Consult, Coun on Drugs, Am Med Asn, 63, 64; mem pharmacol & exp therapeut study sect, NIH, 64-68 & anesthesiol training grant comt, 68-71; mem sci adv bd, Pharmaceut Mfg Asn Found, Inc, 68-, pharmacol test comt, Nat Bd Med Examrs, 69-73 & comt on probs of drug safety, Nat Acad Sci-Nat Res Coun, 70-; adj prof pharmacol, Med Col, Cornell Univ. Mem: AAAS; Am Soc Pharmacol & Exp Therapeut; NY Acad Sci; Am Soc Nephrol; Am Heart Asn. Res: Renal pharmacology; mechanism of action of diuretic agents. Mailing Add: Sterling-Winthrop Res Inst Rensselaer NY 12144

CAGAN, ROBERT HOWARD, b Brooklyn, NY, Apr 8, 38; m 68. BIOCHEMISTRY. Educ: Northeastern Univ, BS, 60; Harvard Univ, PhD(biochem), 66. Prof Exp: USPHS fel, Nobel Med Inst, Karolinska Inst, Sweden, 66-68; ASST PROF BIOCHEM, MONELL CHEM SENSES CTR & DEPT BIOCHEM, SCH MED, UNIV PA, 68-; RES CHEMIST, VET ADMIN HOSP, PHILADELPHIA, 68- Mem: AAAS; Am Chem Soc; Swed Biochem Soc. Res: Biochemical basis of physiological function; enzymology of phagocytic cells; biophysics of membranes and flavoenzymes; biochemical mechanisms of chemoreceptor function; interactions of receptors with taste stimuli; role of plasma membrane; mode of action of materials with special taste properties. Mailing Add: Monell Chem Senses Ctr Univ of Pa Philadelphia PA 19104

CAGLE, FREDRIC WILLIAM, JR, b Metropolis, Ill, Dec 17, 24. CHEMISTRY. Educ: Univ Ill, BS, 44, MS, 45, PhD(chem), 46. Prof Exp: Mem sch math, Inst Advan Study, 47-48; fel chem, 48-49, res asst prof, 49-53, from asst prof to assoc prof, 53-60, PROF CHEM, UNIV UTAH, 60- Concurrent Pos: Consult, Dow Chem Co, 63- & Pac Northwest Pipeline Corp, 58-61. Mem: Am Chem Soc; Am Phys Soc; Am Crystallog Asn. Res: Chemical kinetics; theoretical chemistry; x-ray crystallography and crystal structures; applications of computers to chemical problems. Mailing Add: Dept of Chem Univ of Utah Salt Lake City UT 84112

CAHILL, ALLEN ELY, inorganic chemistry, see 12th edition

CAHILL, CHARLES L, b El Reno, Okla, Feb 23, 33; m 54; c 3. PHYSICAL CHEMISTRY, BIOCHEMISTRY. Educ: Okla Baptist Univ, AB, 55; Univ Okla, MS, 57, PhD(biochem), 61. Prof Exp: From asst prof to prof chem, Oklahoma City Univ, 61-71, chmn dept, 63-71, assoc dean col arts & sci, 67-71; V CHANCELLOR ACAD AFFAIRS & PROF CHEM, UNIV NC, WILMINGTON, 71- Concurrent Pos: NIH res grant, 62-70. Mem: AAAS; Am Chem Soc; Endocrine Soc; NY Acad Sci. Res: Physicochemical characterization of synthetic polymers and proteins. Mailing Add: Univ of NC PO Box 3725 Wilmington NC 28401

CAHILL, DONALD R, b Glasgow, Mont, Sept 2, 36; m 59; c 2. ANATOMY. Educ: Mont State Univ, BS, 60; Univ Wash, MS, 64; Temple Univ, PhD(anat), 68. Prof Exp: Instr anat, Temple Univ, 64-68; asst prof, Northwestern Univ, 68-71; ASSOC PROF ANAT, UNIV MIAMI, 71- Mem: Am Asn Anatomists. Res: Experimental study of the process of tooth eruption; morphology and function of the lumbar spine and ankle joint. Mailing Add: Univ of Miami Sch Med PO Box 520875 Biscayne Annex Miami FL 33152

CAHILL, GEORGE FRANCIS, JR, b New York, NY, July 7, 27; m 49; c 6. METABOLISM. Educ: Yale Univ, BS, 49; Columbia Univ, MD, 53. Hon Degrees: MA, Harvard Univ, 66. Prof Exp: Intern med, Peter Bent Brigham Hosp, 53-54, from jr resident to sr resident, 54-58; from asst to assoc prof, 58-70, PROF MED, HARVARD MED SCH, 70-; DIR, ELLIOTT P JOSLIN RES LAB, 64- Concurrent Pos: Res fel biol chem & Nat Res Coun fel med sci, Harvard Univ, 55-57; sr assoc, Peter Bent Brigham Hosp, 59-64, physician, 68- Mem: AAAS; Am Soc Clin Invest; Am Physiol Asn; Am Clin & Climat Asn; Endocrine Soc. Res: Lipids and carbohydrates as related to atherosclerosis or diabetes. Mailing Add: Joslin Res Lab 170 Pilgrim Rd Boston MA 02215

CAHILL, JERRY EDWARD, b Philadelphia, Pa, Oct 15, 42; m 64; c 4. CHEMICAL PHYSICS. Educ: St Joseph's Col, Pa, BS, 64; Princeton Univ, AM, 67, PhD(chem), 68. Prof Exp: Res assoc chem, Univ Southern Calif, 68-70; ASST PROF CHEM, DREXEL UNIV, 70- Concurrent Pos: Merck grant fac develop, Merck Co Found, 70. Res: Infrared and Raman spectra of pure and mixed molecular crystals and liquids; intermolecular interactions and motions in condensed matter; vibrational energy transfer and relaxation in condensed matter. Mailing Add: Dept of Chem Drexel Univ Philadelphia PA 19104

CAHILL, JOHN PHILIP, optical physics, optics, deceased

CAHILL, JOSEPH JAMES, JR, organic chemistry, see 12th edition

CAHILL, KEVIN M, b New York, NY, May 6, 36; m 61; c 5. MEDICINE, TROPICAL MEDICINE. Educ: Fordham Univ, AB, 57; Cornell Univ, MD, 61; Univ London & Royal Col Physicians, dipl trop med & hyg, 63; Am Bd Med Microbiol, dipl, 67; Am Bd Prev Med, dipl, 70. Prof Exp: CLIN ASSOC PROF MED, NEW YORK MED COL, 65-, DIR TROP DIS CTR, 65- Concurrent Pos: Lectr, Univ Cairo & Univ Alexandria, 63-65; consult, UN Health Serv & USPHS, 65-; mem sci adv bd, Am Found Trop Med, 66-; prof trop med & chmn dept, Royal Col Surgeons, Ireland, 70-; prof pub health & prev med, NJ Col Med, 74-; chmn, Health Planning Comn & Health Res Coun, NY State, 75-, asst to Governor for Health Affairs, 75- Mem: Fel Am Col Chest Physicians; Am Pub Health Asn; Am Soc Trop Med & Hyg; Royal Soc Trop Med & Hyg. Res: Chest medicine. Mailing Add: 850 Fifth Ave New York NY 10021

CAHILL, LAURENCE JAMES, JR, b Frankfort, Maine, Sept 21, 24; m 49; c 3. SPACE PHYSICS. Educ: US Mil Acad, BS, 46; Univ Chicago, SB, 50; Univ Iowa, MS, 56, PhD(physics), 59. Prof Exp: Asst prof physics, Univ NH, 59-62; chief physics, Hqs NASA, 62-63; from assoc prof to prof physics, Univ NH, 63-68, dir space sci ctr, 67-68; PROF PHYSICS & DIR SPACE SCI CTR, UNIV MINN, MINNEAPOLIS, 68- Concurrent Pos: Chmn, US-USSR Comt Co-op Space Res Geomagnetism, 63; mem working group, Int Year Quiet Sun, Int Comt Space Res, 63-64; consult, NASA, 63- Mem: Fel Am Geophys Union; Am Phys Soc. Res: Investigation of earth's magnetic field by rocket and satellite experiment. Mailing Add: Space Sci Ctr Univ of Minn Minneapolis MN 55455

CAHILL, THOMAS A, b Paterson, NJ, Mar 4, 37; m 65; c 2. NUCLEAR PHYSICS. Educ: Holy Cross Col, BA, 59; Univ Calif, Los Angeles, MA, 61, PhD(physics), 65. Prof Exp: Asst prof in residence physics, Univ Calif, Los Angeles, 65-66; NATO fel, Ctr Nuclear Studies, Saclay, France, 66-67; asst prof, 67-71, actg dir, Crocker Nuclear Lab, 72, dir, Inst Ecol, 72-75, ASSOC PROF PHYSICS, UNIV CALIF, DAVIS, 71- Mem: Am Phys Soc. Res: Investigation of very light nuclei through the techniques of nuclear scattering with emphasis on spin effects; application of nuclear and radiation techniques to ecological problems; air pollution. Mailing Add: Dept of Physics Univ of Calif Davis CA 95616

CAHILL, VERN RICHARD, b Tiro, Ohio, May 5, 18; m 46; c 3. MEAT SCIENCE. Educ: Ohio State Univ, BSc, 41, MSc, 42, PhD(meat), 55. Prof Exp: From instr to PROF ANIMAL SCI, OHIO STATE UNIV, 46- Mem: Am Soc Animal Sci; Am Meat Sci Asn; Inst Food Technologists. Res: Carcass yield of edible portion; meat tenderness and preservation; comminuted meat products. Mailing Add: 133 Aldrich Rd Columbus OH 43214

CAHILLY, GLENN MOYLAN, biochemistry, see 12th edition

CAHN, ARNO, b Cologne, Ger, Sept 6, 23; nat; m 50; c 3. ORGANIC CHEMISTRY. Educ: Queen's Univ, Ont, BSc, 46; Purdue Univ, PhD(chem), 50. Prof Exp: Asst, Purdue Univ, 46-48; sr res chemist, 50-53, prin res chemist, 53-56, res assoc, 56-58, chief org sgct, 58-63, chief detergent liquids prod develop sect, 63-65, res mgr chem & physics dept, 65-66, develop mgr household prod, 66-73, DIR DEVELOP, HOUSEHOLD PROD DIV, LEVER BROS CO, 73- Mem: Am Chem Soc; Am Oil Chemists Soc; fel Am Inst Chemists; Indust Res Inst. Res: Kinetics of pyridinium salt formation; synthesis of labelled compounds; structure and physical properties of surface-active agents. Mailing Add: Lever Bros Co Res Ctr 45 River Rd Edgewater NJ 07020

CAHN, HAROLD ARCHAMBO, physiology, parapsychology, see 12th edition

CAHN, JULIUS HOFELLER, b Chicago, Ill, Oct 29, 19; m 43, 51; c 2. ASTROPHYSICS. Educ: Yale Univ, BA, 42, MS, 47, PhD(physics), 48. Prof Exp: Jr engr, Signal Corps Labs, NJ, 42-43; lab asst, Yale Univ, 46-47; asst, gas discharge proj, US Navy, 47-48; asst prof physics, Univ Nebr, 48-50; prin physicist, Battelle Mem Inst, 50-51, asst div consult, 51-59; assoc prof physics & elec eng, 59-70, ASSOC PROF ELEC ENG, PHYSICS & ASTRON, UNIV ILL, URBANA-CHAMPAIGN, 70- Concurrent Pos: Consult, Boulder Labs, Nat Bur Standards. Mem: Am Phys Soc; Am Astron Soc; Royal Astron Soc; Int Astron Union. Res: High current electrical discharges in gases; effects of electrostatic interaction on the electronic velocity distribution function; continuous x-ray absorption; antenna impedance; friction; nondestructive testing; solid state physics; plasma physics; stellar evolution; planetary nebulae; galactic absorption distribution; stellar distances. Mailing Add: Univ of Ill Observ Urbana IL 61801

CAHN, PHYLLIS HOFSTEIN, b New York, NY, Sept 22, 28; m 47; c 2. FISH BIOLOGY, ANIMAL BEHAVIOR. Educ: NY Univ, BA, 48, MS, 51, PhD(biol), 57. Prof Exp: Teaching asst biol, Newark Col Arts & Sci, Rutgers Univ, 48-50; from instr to prof, Stern Col, Yeshiva Univ, 58-69; assoc prof, 68-69, PROF MARINE SCI, C W POST COL, LONG ISLAND UNIV, 69-, CHMN DEPT, 74- Concurrent Pos: Res fel, Ichthyol Dept, Am Mus Natural Hist, 57-64. Mem: AAAS; Am Soc Zoologists; Am Soc Ichthyologists & Herpetologists; Am Fisheries Soc; Animal Behav Soc. Res: Fish sensory systems, particularly behavior and physiology. Mailing Add: Dept of Marine Sci C W Post Col Long Island Univ Brookville NY 11548

CAHN, ROBERT DAVID, zoology, see 12th edition

CAHN, ROBERT NATHAN, b New York, NY, Dec 20, 44; m 65; c 2. THEORETICAL HIGH ENERGY PHYSICS. Educ: Harvard Univ, BA, 66; Univ Calif, Berkeley, PhD(physics), 72. Prof Exp: Res assoc physics, Stanford Linear Accelerator Ctr, 72-73; RES ASST PROF PHYSICS, UNIV WASH, 73- Mem: Am Phys Soc. Res: Phenomenology with an emphasis on current-induced reactions. Mailing Add: Dept of Physics Univ of Wash Seattle WA 98195

CAHN, ROBERT STERN, mathematics, see 12th edition

CAHNMANN, HANS JULIUS, b Munich, Ger, Jan 27, 06; nat US; m 45; c 3. ORGANIC CHEMISTRY. Educ: Univ Munich, Lic Pharm, 30, PhD(chem), 32. Prof Exp: Instr chem, Univ Munich, 31-33 & Univ Paris, 33-36; dir res, Labs, Crinex, 36-39; res assoc org chem, French Pub Health Serv, 39-40; res assoc biochem, Univ Aix Marseille, 40-41; dir res & develop, Biochem Prod Corp, NY, 41-42; res assoc, Mt Sinai Hosp, 42-44; sr chemist, Givaudan-Delawanna, 44-47 & William R Warner, 47-49; SR BIOCHEMIST, NIH, 50- Mem: Am Chem Soc; Am Thyroid Asn. Res: Biochemistry. Mailing Add: NIH Bethesda MD 20014

CAHOON, GARTH ARTHUR, b Vernal, Utah, Dec 23, 24; m 48; c 4. HORTICULTURE. Educ: Utah State Agr Col, BS, 50; Univ Calif, Los Angeles, PhD(hort sci), 54. Prof Exp: Res asst, Dept Floricult & Ornamental Hort, Univ Calif, Los Angeles, 50-54; asst horticulturist, Univ Calif, Riverside, 54-63; assoc prof, 63-67,

PROF HORT, OHIO STATE UNIV & OHIO AGR RES & DEVELOP CTR, 67- Concurrent Pos: US AID consult, India, 68 & 70; mem, Coun Soil Test & Plant Anal, chmn. Honors & Awards: Laurie Award, Am Soc Hort Sci, 56. Mem: Am Soc Hort Sci; Int Soc Hort Sci; Soil Sci Soc Am. Res: Floricultural plant nutrition; citrus physiology and nutrition; pomology and viticulture; physiology and nutrition. Mailing Add: Dept of Hort Ohio State Univ Columbus OH 43210

CAHOON, MARY ODILE, b Houghton, Mich, July 21, 29. CELL PHYSIOLOGY. Educ: DePaul Univ, BS, 54, MS, 58; Univ Toronto, PhD(cell physiol), 61. Prof Exp: From instr to assoc prof, 54-67, chmn dept, 61-73, acad dean, 63-67, PROF BIOL, COL ST SCHOLASTICA, 67-, CHMN DIV NATURAL SCI, 71- Concurrent Pos: Res assoc, McMurdo Sta, Antarctica, 74. Mem: Am Soc Zoologists. Res: Radiation effects on growth of bacterial cells; calcium variations during crustacean intermolt cycle; ultraviolet resistance of bacterial spores; cold adaptation in metabolism of Antarctic fauna. Mailing Add: Dept of Biol Col of St Scholastica Duluth MN 55811

CAHOON, NELSON COREY, b Bloomfield, Ont, Oct 12, 04; nat US; m 29; c 2. CHEMISTRY. Educ: Univ Toronto, BA, 25, MA, 26. Prof Exp: Control & develop chemist, Nat Carbon Co, Can & Ohio, 26-34, res chemist, 34-53, res group leader, 53-58, tech asst to mgr develop lab, 58-63, sr scientist, Consumer Prods Div, Union Carbide Corp, 63-70; RETIRED. Honors & Awards: George W Heise Medal, Electrochem Soc, 62. Mem: Am Chem Soc; Electrochem Soc (pres, 69-70); Nat Asn Corrosion Engrs. Res: Electrochemistry of primary batteries, dry batteries and alkaline battery systems. Mailing Add: 4660 W 210th St Fairview Park OH 44126

CAHOY, ROGER PAUL, b Colome, SDak, Feb 19, 27; m 50; c 4. ORGANIC CHEMISTRY. Educ: Dakota Wesleyan Univ, BA, 50; Univ SDak, MA, 52; Univ Nebr, PhD(chem), 56. Prof Exp: Develop chemist, silicone prod dept, Gen Elec Co, 56-60; res chemist, Spencer Chem Co, 60-63, SR RES CHEMIST, GULF OIL CHEM CO, 63- Mem: Am Chem Soc. Res: Organometallic compounds; catalysis; biological active organic compounds. Mailing Add: Gulf Oil Chem Co 9009 W 67th St Merriam KS 66202

CAILLE, GILLES, b Montreal, Que, May 5, 35; m 57; c 4. MEDICINAL CHEMISTRY. Educ: Univ Montreal, BSc, 59, MSc, 65, PhD(med chem), 69. Prof Exp: Monitor instrumental anal & pharmacog, 63-67, prof instrumental anal, 67-69, RES PROF PHARMACOL, UNIV MONTREAL, 69- Concurrent Pos: Res fel metab psychotrope, Joliette Res Inst, 69-71, co-dir, 71-74; mem, comt pharm, St-Charles Hosp, Joliette, 69-; consult, Burroughs Wellcome & Co, 70- & Santa Cabrini Hosp, 73-; pres, Biopharm Inc, 71-73. Mem: Am Acad Clin Toxicol; Fr Soc Toxicol. Res: Metabolism of psychotrope; clinical pharmacology and analytical toxicology. Mailing Add: Dept of Pharmacol Fac of Med Univ of Montreal Montreal PQ Can

CAILLEAU, RELDA, b San Francisco, Calif, Feb 1, 09. CELL PHYSIOLOGY. Educ: Univ Calif, AB, 30, MA, 31; Univ Paris, DSc, 37. Prof Exp: Fr Nat Res fel, Pasteur Inst, Paris, 37-40; asst microbiol vitamin assays, Univ Calif, 41-43; assoc nutritionist & home economist, USDA, 43-46; in charge res, Nat Ctr Sci Res, Ministry Educ, France, 47-49; jr asst res biochemist, Univ Calif, Berkeley, 50-53, asst res biochemist tissue cult, 53-55, res assoc oncol & assit res biochemist, cancer res inst, Med Sch, Univ Calif, San Francisco, 55-59, assoc res biochemist, 59-70; RES ASSOC MED, UNIV TEX MD ANDERSON HOSP & TUMOR INST, 70- Mem: Tissue Cult Asn; Soc Exp Biol & Med; Soc Cell Biol. Res: Effects of various growth media and chemotherapeutic agents upon normal and malignant mouse and human cells in vitro and their chromosomes; establishment of human breast carcinoma cell lines from pleural effusions. Mailing Add: Dept of Med Univ of Tex MD Anderson Hosp & Tumor Inst Houston TX 77025

CAILLEUX, ANDRE PAUL, b Paris, France, Dec 24, 07; m 31; c 12. GEOGRAPHY, ARCHAEOLOGY. Educ: Univ Paris, BA, 26, MS, 32 & 33, Agrege, 36, Dr(geol), 42; Univ Strasbourg, MS, 31. Hon Degrees: Dr, Univ Lodz, 61. Prof Exp: Prof biol, French Col Warsaw, 34-36; prof, Col Brest, 36-38 & Berthelot Col, 38-56; assoc prof geog, Univ Paris, 56-58, prof geol, 58-69; PROF EARTH SCI & RES DIR CTR NORDIC STUDIES, LAVAL UNIV, 69- Concurrent Pos: Assoc prof geog, Sch Advan Studies, France, 48; consult, Geol Surv France, 56; mem, French Nat Comt Geog, 57- & French Nat Comt Antarctic Res, 59-; lectr numerous US, Can & foreign univs, 58-; pres, Comt Int Union Quaternary Res, 61-65; expert, French Tribunals, 64; consult, Esso Oil Co & other co, 65- Honors & Awards: Prix Malouet, Acad Ethics & Pop Sci France, 46; Prix Cayeux, Acad Sci, France, 48; Medal, Dept Agr, France, 53; Medal, Liege Univ, Belg, 62; Albrecht Penck Medal, Ger Asn Quaternary Res, 72. Mem: Geol Soc France (vpres, 46-47); corresp mem Geol Soc Swed; Geol Soc Belg; Arg Acad Geog; corresp mem Göttingen Acad Sci. Res: Comparative morphology of earth and other planets and satellites; elementary physicochemical processes; acceleration in biological, psychological, technological and demographical progress from prehistory to present day; acceleration in monetary devaluation; epistemology. Mailing Add: Ctr Nordic Studies Laval Univ Quebec PQ Can

CAILLOUET, CHARLES W, b Baton Rouge, La, Dec 15, 37; m 59; c 4. FISH BIOLOGY. Educ: La State Univ, BS, 59, MS, 60; Iowa State Univ, PhD(fishery biol), 64. Prof Exp: Asst prof biol, Univ Southwestern La, 64-67; assoc prof, Sch Marine & Atmospheric Sci, Univ Miami, 67-72; supvry fisher res biologist, 72-74, FISHERY BIOLOGIST, RES ADMIN, NAT MARINE FISHERIES SERV, GULF COASTAL FISHERIES CTR, GALVESTON FACIL, 74- Concurrent Pos: Assoc ed marine invertebrates, Trans Am Fisheries Soc, 74-76; adj grad fac, Univ Houston, 75- Mem: AAAS; Am Fisheries Soc; Am Inst Fishery Res Biol. Res: Modeling and simulation of population dynamics of commercial shrimps. Mailing Add: Nat Marine Fisheries Serv Gulf Coastal Fisheries Ctr Galveston Facil 4700 Ave U Galveston TX 77550

CAILLOUX, MARCEL LOUIS, b Ottawa, Ont, Dec 8, 14; m 44; c 3. PLANT PHYSIOLOGY. Educ: Mt St Louis Col, BA, 35; Univ Montreal, LSc, 40; Univ Paris, DSc, 67. Prof Exp: Lab asst & illusr sci publ, Bot Inst, 36, from asst prof to assoc prof, 40-70, PROF PLANT PHYSIOL, UNIV MONTREAL, 70- Concurrent Pos: Vchmn res conf plant physiol, Can Nat Res Coun, 54. Mem: AAAS; Am Soc Plant Physiologists; Fr-Can Asn Advan Sci; Can Fedn Biol Socs; Can Soc Natural Hist (pres, 47). Res: Mechanism of the absorption of water by cells; culture of protoplasts; mechanisms of entry of water in plant protoplasts. Mailing Add: Dept of Biol Sci Univ of Montreal Montreal PQ Can

CAIN, ARTHUR SAMUEL, JR, b Leavenworth, Kans, Apr 13, 13; m 41. PHYSIOLOGY, SURGERY. Educ: Univ Kans, BA, 36, MD, 40. Prof Exp: Intern, Albany Hosp, NY, 40-41; fel path, Sch Med, Dartmouth Univ, 41-42; resident surg, Presby Hosp, Med Ctr, Columbia Univ, 42-45, asst surg, Col Physicians & Surgeons, Columbia Univ, 4445; asst, Sch Med, Univ Kans, 46-48, dir surg servs, Kansas City Gen Hosp, 48-51; instr physiol, Col Physicians & Surgeons, Columbia Univ, 52-55; dir res fels & grants, Am Heart Asn, 55-56; staff officer, Nat Acad Sci-Nat Res Coun, 56-60; spec asst to dir, Asn Am Med Cols, 60-61; CHIEF EDUC RES, US VET ADMIN, 61- Res: Medical and biological science education and research administration; cardiovascular physiology; tissue transplantations. Mailing Add: Cent Off US Vet Admin Washington DC 20420

CAIN, CARL, JR, b Chattanooga, Tenn, Jan 6, 31; m 61; c 2. ANALYTICAL CHEMISTRY, CHEMICAL ENGINEERING. Educ: Univ Chattanooga, BS, 52; Univ Tenn, MS, 54, PhD(chem), 56. Prof Exp: Proj engr, Cramet, Inc, 56-58; res engr, Develop Lab, Combustion Eng, Inc, 58-61; anal res chemist, Rock Hill Lab, Chemetron Corp, 61-62; staff chem engr, 62-70, SUPVR CHEM SECT, TENN VALLEY AUTHORITY, 70- Mem: Am Chem Soc; Am Soc Mech Eng; Am Soc Testing & Mat. Res: Power plant analytical chemistry, especially automatic monitors and controllers; corrosion of metals in aqueous environment; industrial water conditioning. Mailing Add: 617 Edney Bldg Chattanooga TN 37401

CAIN, CHARLES ALAN, b Tampa, Fla, Mar 3, 43. BIOENGINEERING. Educ: Univ Fla, BEE, 65; Mass Inst Technol, MSEE, 67; Univ Mich, PhD(elec eng), 72. Prof Exp: Mem tech staff, Bell Labs, Inc, 65-68; ASST PROF ELEC ENG, UNIV ILL, URBANA, 72- Mem: Inst Elec & Electronics Engrs. Res: Biological effects and medical applications of microwave energy; mathematical modeling of physiological systems. Mailing Add: Dept of Elec Eng Univ of Ill Urbana IL 61801

CAIN, CHARLES COLUMBUS, b Fields, La, Oct 29, 15; m 39; c 2. SOIL SCIENCE. Educ: La State Univ, BS, 37; Iowa State Univ, MS, 38, PhD, 56. Prof Exp: Jr soil survr, Soil Conserv Serv, USDA, 38-41; asst soil technologist, 41; instr agron & asst agronomist, La State Univ, 41-46; assoc prof, 46-53, head dept gen agr, 55-65, PROF AGRON, UNIV SOUTHWESTERN LA, 53-, DIR TECH STUDIES, EDUC CTR, NEW IBERIA, 65-, DIR FRESHMAN DIV, UNIV, 67- Mem: Am Soc Agron; Soil Sci Soc Am. Res: Soil microscopy; soil morphology and genesis. Mailing Add: 720 Girard Park Dr Lafayette LA 70501

CAIN, CHARLES EUGENE, b Ellerbe, NC, June 22, 32; m 58. ORGANIC CHEMISTRY. Educ: Univ NC, BS, 54; Duke Univ, MA, 57, PhD(chem), 59. Prof Exp: Jr res chemist, Shell Chem Corp, 54; res chemist, E I du Pont de Nemours & Co, 58-60; PROF CHEM & CHMN DEPT, MILLSAPS COL, 60- Mem: Am Chem Soc. Res: Organic reactions of organometallic compounds; reaction mechanisms by relative rates and stereochemical effects; electronic versus stereochemical effects in monomers and polymers. Mailing Add: Dept of Chem Millsaps Col Jackson MS 39210

CAIN, CORNELIUS KENNADY, b Owensboro, Ky, Sept 10, 10; m 43; c 2. MEDICINAL CHEMISTRY. Educ: Univ Ky, BS, 32; Univ Mass, MS, 37; Johns Hopkins Univ, PhD(org chem), 39. Prof Exp: Asst chem, Univ Mass, 35-37 & Johns Hopkins Univ, 37-38; spec asst org chem, Univ Ill, 39-41; instr biochem, St Louis Univ, 41-44; from instr to asst prof chem, Cornell Univ, 44-50; head org chem dept, 50-60, asst dir res & dir chem res, 60-68, DIR RES INFO, McNEIL LABS, 68- Mem: AAAS; Am Chem Soc. Res: Chemistry of natural products; heterocyclic nitrogen compounds; synthetic organic medicinals. Mailing Add: McNeil Labs Camp Hill Rd Ft Washington PA 19034

CAIN, DENNIS FRANCIS, b Republican City, Nebr, June 4, 30; m 61; c 5. BIOCHEMISTRY. Educ: Creighton Univ, BS, 52; Georgetown Univ, MS, 56; Univ Pa, PhD(biochem), 60. Prof Exp: NIH fel, 60-61; assoc biochem, Sch Vet Med, Univ Pa, 61-63, from asst prof to assoc prof, 63-68; biochemist, Surg Neurol Br, 68-74, SCIENTIST ADMINR, SPEC PROGS, SCI REV BR, NAT INST NEUROL DIS & STROKE, 74- Concurrent Pos: Res grant, Nat Inst Neurol Dis & Stroke. Mem: AAAS; Am Chem Soc. Res: Mechanochemistry of muscular contraction; metabolism of phosphorus, nucleic acids and protein in brain tissues. Mailing Add: Div of Res Grants NIH Westwood Bldg Rm 2A-15 Bethesda MD 20014

CAIN, GEORGE D, b Pittsburgh, Pa, June 1, 40; m 66; c 2. PARASITOLOGY. Educ: Sterling Col, BS, 62; Purdue Univ, MS, 64, PhD(biol), 68. Prof Exp: Fel, Univ Mass, 68-70; asst prof, 70-75, ASSOC PROF ZOOL, UNIV IOWA, 76- Mem: AAAS; Am Soc Parasitol; Am Soc Zoologists; Nat Speleol Soc. Res: Comparative biochemistry of invertebrate hemoglobins and collagens; oxygen-dependent biosynthetic mechanisms in parasitic helminths; lipid metabolism in lower invertebrates. Mailing Add: Dept of Zool Univ of Iowa Iowa City IA 52240

CAIN, GEORGE LEE, JR, b Wilmington, NC, Jan 13, 34; m 56; c 1. MATHEMATICS. Educ: Mass Inst Technol, BS, 56; Ga Inst Technol, MS, 62, PhD(math), 65. Prof Exp: Mathematician, Lockheed-Ga Co, 56-60; from instr to asst prof, 60-68, ASSOC PROF MATH, GA INST TECHNOL, 68-, ASST DIR SCH MATH, 73- Mem: Math Asn Am; Am Math Soc. Res: Topology; analysis. Mailing Add: Sch of Math Ga Inst of Technol Atlanta GA 30332

CAIN, H THOMAS, b Burlington, Wash, May 18, 13; m 41; c 3. ANTHROPOLOGY. Educ: Univ Wash, Seattle, AB, 38; Univ Ariz, MA, 45. Prof Exp: Cur anthrop, San Diego Mus of Man, 45-50; archaeologist, River Basin Survs, Smithsonian Inst, 52; dir, Heard Mus Anthrop, 52-68, CUR ANTHROP, HEARD MUS, 68- Concurrent Pos: Lectr, Ariz State Univ, 62- Mem: Fel Am Anthrop Asn; Am Asn Mus. Res: Museology as it applies to anthropological materials; cultural anthropology. Mailing Add: Heard Mus 22 E Monte Vista Rd Phoenix AZ 85004

CAIN, JAMES ALLAN, b Isle of Man, Eng, July 23, 35; m 59; c 2. GEOLOGY. Educ: Univ Durham, BSc, 58; Northwestern Univ, MS, 60, PhD(geol), 62. Prof Exp: From instr to asst prof geol, Western Reserve Univ, 61-66; assoc prof, 66-71, PROF GEOL, UNIV RI, 71-, CHMN DEPT, 67- Concurrent Pos: Res Corp res grant, 62-65; NSF res grant, 64-66; US Navy Underwater Sound Lab grant, 68; hon res assoc, dept geol sci, Harvard Univ, 73; US Geol Surv res grant, 75. Mem: Geol Soc Am; Brit Geol Soc. Res: Quantitative igneous and metamorphic petrology; mineral resources. Mailing Add: Dept of Geol Univ of RI Kingston RI 02881

CAIN, JAMES CLARENCE, b Kosse, Tex, Mar 19, 13; m 38; c 4. INTERNAL MEDICINE, GASTROENTEROLOGY. Educ: Univ Tex, BA, 33, MD, 37; Univ Minn, MMS, 47; Am Bd Internal Med, Am Bd Gastroenterol, dipl. Prof Exp: Intern, Protestant Episcopal Hosp, Philadelphia, Pa, 37-39; instr path, Med Sch, Univ Tex, Galveston, 39-40; instr path, 40-41 & 46-48, CONSULT MED, MAYO CLIN, 48-, PROF, MAYO GRAD SCH MED, UNIV MINN, MINNEAPOLIS, 72- Concurrent Pos: From asst prof to clin prof med, Mayo Grad Sch Med, Univ Minn, Minneapolis, 54-72; chmn, Combined Comt of Nat Adv Comt to Selective Serv for Selection of Doctors, Dentists, Vet & Allied Med Personnel & Nat Health Resources Adv Comt, 68-69, adv to dir, Selective Serv Syst, 69-70; mem, Nat Adv Heart Coun, Minn State Bd Med Exam & Nat Adv Comt to Health Manpower Comn; chmn, Fed Use of Health Manpower Panel; civilian consult to Surgeon Gen. Honors & Awards: Billings Gold Medal Award, AMA, 63. Mem: AAAS; AMA; Am Col Physicians; Am Asn Hist Med; Pan-Am Med Asn. Res: Lymphatic system; liver and intestinal tract; gastric ulcers. Mailing Add: Mayo Clin Rochester MN 55902

CAIN, JEROME RICHARD, b Piqua, Ohio, Apr 26, 47; m 69; c 2. ALGOLOGY. Educ: Miami Univ, BA, 69, MS, 72; Univ Conn, PhD(algol), 75. Prof Exp: Scientist, Tech Serv Div, Am Can Co, 69-70; teaching asst bot, Miami Univ, 70-72; res asst, Inst Water Resources, Univ Conn, 72-75; ASST PROF ALGOL, ILL STATE UNIV, 75- Mem: Phycol Soc Am; Bot Soc Am; Sigma Xi. Res: Algal ecology, nutrition and morphology; environmental regulation of sexual and asexual reproduction in microalgae; algal bioassays for nutrient potential in freshwater ecosystems; environmental influences on structure of benthic diatom communities. Mailing Add: Dept of Biol Sci Ill State Univ Normal IL 61761

CAIN, JOHN CARLTON, b Blakely, Ga, Oct 14, 11; m 35; c 2. PLANT PHYSIOLOGY. Educ: Univ Fla, BSA, 35; Cornell Univ, PhD(pomol), 47. Prof Exp: Lab asst, Exp Sta, Univ Fla, 30-37, asst horticulturist, 37-40; instr pomol, Cornell Univ, 40-42; from assoc prof to prof, 46-73, EMER PROF POMOL, NY AGR EXP STA, GENEVA, 73- Honors & Awards: Stark Award, Am Soc Hort Sci, 74. Mem: Fel AAAS; fel Am Soc Hort Sci. Res: Storage and preservation of citrus fruit; nutrition of fruit trees; interrelationships between calcium, magnesium and potassium with respect to the foliar diagnosis of the nutrient status of fruit trees; mineral nutrition of fruit plants; mechanical pruning and harvesting. Mailing Add: NY Agr Exp Sta Geneva NY 14456

CAIN, JOHN MANFORD, b Morrison, Ill, June 10, 32; m 55; c 2. RESOURCE MANAGEMENT. Educ: Univ Wis, BS, 55, MS, 56, PhD(soil sci), 67. Prof Exp: Soil scientist, Soil Conserv Serv, USDA, 58-60; teacher, Mich Jr High Sch, 60-61; soil scientist, Soil Conserv Serv, USDA, 61-65; asst prof soil sci, Wis State Univ, River Falls, 66-67; WATER RESOURCE PLANNER, DEPT NATURAL RESOURCES, STATE WIS, 67- Mem: Soil Conserv Soc Am. Res: Use of soil survey materials and techniques for nonagricultural interpretations. Mailing Add: Water Planning Sect Wis Dept of Natural Resources Madison WI 53702

CAIN, JOSEPH, biology, see 12th edition

CAIN, JOSEPH CARTER, b Georgetown, Ky, Oct 31, 30; m 68. GEOPHYSICS. Educ: Univ Alaska, BS, 52, PhD, 57. Prof Exp: Physicist, Geophys Inst, Univ Alaska, 52-57, asst prof geophys res, 57-59; physicist, Goddard Space Flight Ctr, NASA, 59-74; GEOPHYSICIST, ELECTROMAGNETICS & GEOMAGNETISM BR, US GEOL SURV, 74- Mem: Am Geophys Union; Int Asn Geomag & Aeronomy; Sigma Xi. Res: Magnetic field experiments from spacecraft; geomagnetism; magnetospheric and earth physics; numerical modelling of fields. Mailing Add: Stop 964 Box 25046 Lakewood CO 80225

CAIN, PAUL SYLVESTER, agronomy, see 12th edition

CAIN, ROBERT FARMER, b Winona, Idaho, Mar 12, 17; m 39; c 3. FOOD SCIENCE. Educ: Tex Tech Col, BS, 39; Tex A&M Univ, MS, 41; Ore State Univ, PhD(food technol), 52. Prof Exp: From instr to assoc prof hort processing, Tex A&M Univ, 41-52; from asst prof to assoc prof food technol, 52-60, PROF FOOD SCI & TECHNOL, ORE STATE UNIV, 60- Concurrent Pos: Consult, Chilean Canning Indust Asn, 61. Honors & Awards: Basic Res Award, Ore Agr Exp Sta, 59. Mem: Inst Food Technologists. Res: Processing of fruits and vegetables; irradiation of foods; effects of storage on foods. Mailing Add: Dept of Food Sci Ore State Univ Corvallis OR 97331

CAIN, ROLENE B, applied statistics, mathematics, see 12th edition

CAIN, ROY FRANKLIN, b Paris, Ont, Sept 23, 06; m 33; c 5. BOTANY. Educ: Univ Toronto, BA, 30, MA, 31, PhD(mycol), 33. Prof Exp: Fro lectr to assoc prof, 46-61, head dept bot, 59-61, PROF BOT, UNIV TORONTO, 61-, CUR, CRYPT HERBARIUM, 33-, ASSOC CHMN DEPT BOT, UNIV, 72- Mem: Mycol Soc Am (vpres, 67-68, pres, 69-70); Asn Trop Biol; Brit Mycol Soc; Can Phytopath Soc; Int Soc Plant Taxon. Res: Mycology; bryology; forest pathology. Mailing Add: Dept of Bot Univ of Toronto Toronto ON Can

CAIN, STANLEY ADAIR, b Jefferson Co, Ind, June 19, 02; m 40; c 1. BOTANY. Educ: Butler Univ, BS, 24; Univ Chicago, MS, 27, PhD(plant ecol), 30. Hon Degrees: DSc, Univ Montreal, 59, Williams Col, 67, Butler Univ, 69 & Drury Col, 70. Prof Exp: From instr to assoc prof bot, Butler Univ, 25-31; asst prof, Ind Univ, 31-33; res assoc, Waterman Inst, 33-35; from asst prof to prof, Univ Tenn, 35-46; botanist, Cranbrook Inst Sci, 46-50; chmn dept conserv, 50-61, prof bot & Charles Lathrop Pack prof conserv & dir, Inst Environ Qual, 50-74, CHARLES LATHROP PACK EMER PROF NATURAL RESOURCES & BIOL SCI, UNIV MICH, ANN ARBOR, 74- Concurrent Pos: Guggenheim fel, 40-41; chief sci sect, US Army Univ, France, 45-46; distinguished vis prof, Cent Wash State Col, 69. Consult ed, Biol Conserv, Eng, 68- & Harper & Row; consult, Pan Am Develop Found, Rockefeller Brothers Fund, 69. Mem adv bd, Conserv Found, 54-; ecol expert, Tech Asst Mission To Brazil, UNESCO, 55-56; chmn panel environ biol, NSF, 56-59; vpres, Int Bot Cong, Can, 59; mem, Mich Conserv Comn, 59-, chmn, 63-64; mem adv bd, Nat Park Serv, 58-, chmn, 64-, nat sci adv comt, 65-, coun, 68-, publ comt, 69-; mem adv bd wildlife, US Dept Interior, 62-, asst secy for fish, wildlife, parks & marine resources, 65-68; chmn ad hoc comt int biol prog, Nat Acad Sci, 63-64; mem, Task Force on Natural Resources, 68; mem ecol panel, US Plywood-Champion Papers, Inc, 69; chmn org comt, 18th Gen Assembly, Int Union Biol Sci, 70; pres first nat biol cong, Am Inst Biol Sci, 70. Mem, Int Phytogeog Excursion, Ireland, 49, Finland & Norway, 61. Rapporteur, UNESCO Biosphere Conf, Paris, 68, off, working group III, 69; vis prof environ studies, Univ Calif, Santa Cruz, 73- Honors & Awards: Cert of Merit & Golden Jubilee Award, Bot Soc Am, 56; Distinguished Achievement Award, Univ Mich, 59; Conservationist of the Year, Mich, 65; Mary Soper Pope Medal, Cranbrook Inst Sci, 69. Mem: Nat Acad Sci; AAAS (secy, 48-54, vpres, 54); Ecol Soc Am (treas, 38-40, vpres, 53, pres, 58); Soc Study Evolution (secy, 46-48, vpres, 54); Benjamin Franklin fel, Royal Soc Arts. Res: Plant ecology; conservation of natural resources. Mailing Add: 2019 Devonshire Rd Ann Arbor MI 48104

CAIN, STEPHEN MALCOLM, b Lynn, Mass, Oct 4, 28; m 51; c 2. PHYSIOLOGY. Educ: Tufts Col, BS, 49; Univ Fla, PhD(physiol), 59. Prof Exp: Physiologist, Chem Corps Med Lab, 53-56; asst physiol, Johns Hopkins Univ, 56; asst, Univ Fla, 56-59, trainee, 59; physiologist, US Air Force Sch Aerospace Med, 59-71, head sect, 60-71, chief respiratory sect, 68-71; PROF PHYSIOL & ASSOC PROF MED, UNIV ALA, BIRMINGHAM, 71- Mem: Am Physiol Soc; Aerospace Med Asn. Res: Pulmonary physiology; blood gas transport; tissue hypoxia. Mailing Add: Pulmonary Div Dept of Med Univ of Ala Med Ctr Birmingham AL 35294

CAIN, WILLIAM AARON, b Leesville, La, Apr 30, 39; m 59; c 2. IMMUNOLOGY, MICROBIOLOGY. Educ: Northwestern State Univ La, BS, 61, MS, 64; La State Univ, PhD(immunol), 66. Prof Exp: Res fel, Univ Minn, 66-67, USPHS res fel pediat, 67-69; ASSOC PROF MICROBIOL & IMMUNOL & ADJ ASSOC PROF MED & PEDIAT, HEALTH SCI CTR, UNIV OKLA, 69- Concurrent Pos: Immunol consult, Okla Med Res Found, 69-71; allergy immunol consult. Mem: Am Soc Microbiol. Res: Allergy and immediate hypersensitivity; cell mediated immunity and anergy. Mailing Add: Dept of Microbiol & Immunol Univ of Okla Health Sci Ctr Oklahoma City OK 73104

CAINE, DRURY SULLIVAN, III, b Selma, Ala, June 9, 32; m 61; c 2. ORGANIC CHEMISTRY. Educ: Vanderbilt Univ, BA, 54, MS, 56; Emory Univ, PhD(chem), 61. Prof Exp: Fel org chem, Columbia Univ, 61-62; from asst prof to assoc prof, 62-74, PROF CHEM, GA INST TECHNOL, 74- Mem: Am Chem Soc; Brit Chem Soc. Res: Synthesis of natural products; chemistry of enolate anions. Mailing Add: Dept of Chem Ga Inst of Technol Atlanta GA 30332

CAINE, T NELSON, b Barrow, Eng, Dec 17, 39; m 66. GEOMORPHOLOGY, HYDROLOGY. Educ: Univ Leeds, BA, 61, MA, 62; Australian Nat Univ, PhD(geomorphol), 66. Prof Exp: Lectr geog, Univ Canterbury, 66-67; ASSOC PROF GEOMORPHOL, INST ARCTIC & ALPINE RES, UNIV COLO, BOULDER, 68- Mem: Glaciol Soc; Am Geophys Union; Am Water Resources Asn; Australian & NZ Asn Advan Sci. Res: Evolution of hillslopes in the alpine environment; snow and glacier hydrology and water resources of the alpine area. Mailing Add: Inst of Arctic & Alpine Res Univ of Colo Boulder CO 80302

CAIRNCROSS, ALLAN, b Orange, NJ, May 27, 36; m 60; c 3. ORGANIC CHEMISTRY. Educ: Cornell Univ, BA, 58; Yale Univ, MS, 59, PhD(org chem), 63. ORG RES CHEMIST, CENT RES DEPT, E I DU PONT DE NEMOURS & CO, INC, 64- Mem: Am Chem Soc. Res: Stability of tropylium ions; thermal isomerization of tropilidenes; cycloaddition reactions of bicyclobutanes; substituent effects on tropilidene-norcaradiene equilibria, organocopper chemistry and structure, heterocycles, and polymers. Mailing Add: Cent Res Dept E I du Pont de Nemours & Co Wilmington DE 19898

CAIRNCROSS, STANLEY EVERETT, b Winnipeg, Can, Feb 3, 02; nat; m 26; c 2. CHEMISTRY. Educ: Univ Southern Calif, AB, 25, AM, 26; Columbia Univ, PhD(org chem), 32. Prof Exp: Asst chem, Columbia Univ, 28-30; mem staff, Bristol-Myers Co, 33-39; mem staff, 39-67, PART-TIME CONSULT, ARTHUR D LITTLE, INC, 67- Concurrent Pos: With Nat Res Coun; mem, Agr Res Inst. Mem: Am Chem Soc; Am Pharmaceut Asn; Inst Food Technologists. Res: Organic and pharmaceutical research; food and drug flavor problems. Mailing Add: Arthur D Little Inc 15 Acorn Park Cambridge MA 02140

CAIRNIE, ALAN B, b Ayr, Scotland, Nov 18, 32; m 58; c 4. RADIOBIOLOGY, DEVELOPMENTAL BIOLOGY. Educ: Glasgow Univ, BSc, 54; Aberdeen Univ, PhD(physiol), 58. Prof Exp: Sci officer physiol, Rowett Res Inst, Aberdeen, 56-58; lectr, Aberdeen Univ, 58-61; scientist biophys, Inst Cancer Res, London, 61-67; PROF BIOL, QUEEN'S UNIV, ONT, 67- Concurrent Pos: Vis asst prof, Univ Calif, San Francisco, 65-66; vis assoc prof, Univ Colo, 74. Mem: Radiation Res Soc; Soc Develop Biol. Res: Cell proliferation in renewal tissues; mammalian radiobiology; teratogenesis. Mailing Add: Dept of Biol Queen's Univ Kingston ON Can

CAIRNS, ELDON JAMES, plant nematology, see 12th edition

CAIRNS, ELTON JAMES, b Chicago, Ill, Nov 7, 32; m 59, 74; c 1. ELECTROCHEMISTRY, PHYSICAL CHEMISTRY. Educ: Mich Col Mining & Technol, BS(chem) & BS(chem eng), 55; Univ Calif, Berkeley, PhD(chem eng), 59. Prof Exp: Chemist, US Steel Corp, 53; soils res chemist, Corps Eng, US Army, 54-55; asst thermodyn, Univ Calif, 57-58; phys chemist electrochem, Gen Elec Res Lab, 59-66; phys chemist & group leader, Argonne Nat Lab, 66-70, sect head, 70-73; ASST HEAD, ELECTROCHEM DEPT, GEN MOTORS RES LABS, WARREN, MICH, 73- Concurrent Pos: Consult, US Dept Interior & US Dept Defense, 69-70, US Dept Defense, 73- & NASA, 71; consult, Mat Res Coun, 73; mem subpanel, US Dept Com Panel on Elec Powered Vehicles, 67; mem org comt, Intersoc Energy Conversion Eng Conf, 69, steering comt, 70-, gen chmn, Conf, 76; mem adv bd, Advances Electrochem & Electrochem Eng, 75-; chmn, Energy Conversion, Am Inst Chem Engrs; mem panel, Nat Acad Sci Comt Motor Vehicle Emissions, 72-73; sci deleg, NATO Conf Air Pollution, Eindhoven, 71; div ed, J Electrochem Soc, 69- Honors & Awards: Francis Mills Turner Award, Electrochem Soc, 63. Mem: AAAS; Electrochem Soc; Am Chem Soc; Am Inst Chem Eng; fel Am Inst Chem. Res: Electrochemical energy conversion; thermodynamics; transport phenomena; molten salts; liquid metals; surface chemistry. Mailing Add: Gen Motors Res Labs 12 Mile & Mound Rds Warren MI 48090

CAIRNS, GORDON MANN, b Brooklyn, NY, July 14, 11; m 38; c 2. DAIRY HUSBANDRY. Educ: Cornell Univ, BS, 36, MS, 38, PhD(animal husb), 40. Prof Exp: Asst animal husb, Cornell Univ, 36-39; from assoc prof to prof animal indust & head dept, Univ Maine, 39-45; head dept, 45-51, PROF DAIRY HUSB, UNIV MD, 45-, DEAN COL AGR, 50- Mem: Fel AAAS. Res: Dairy nutrition and dairy cattle breeding. Mailing Add: Univ of Md Col of Agr College Park MD 20742

CAIRNS, JOHN, JR, b Conshohocken, Pa, May 8, 23; m 44; c 4. LIMNOLOGY. Educ: Swarthmore Col, AB, 47; Univ Pa, MS, 49, PhD(protozool), 53. Prof Exp: From asst cur to cur limnol, Acad Natural Sci, Philadelphia, 48-67, res assoc, 68; prof zool, Univ Kans, 67-68; res prof, 68-70, UNIV PROF BIOL & DIR CTR FOR ENVIRON STUDIES, VA POLYTECH INST & STATE UNIV, 70- Concurrent Pos: Lectr, dept educ, Temple Univ, 62-63; mem fac, biol sta, Univ Mich, 64-70; trustee, Rocky Mountain Biol Lab, 62-; mem panel fresh-water aquatic life, Nat Acad Sci. Honors & Awards: Presidential Commendation, 71. Mem: Fel AAAS; Am Soc Limnol & Oceanog; Soc Protozool; Am Fisheries Soc; Am Micros Soc. Res: Ecology of fresh-water Protozoa; response of aquatic organisms to toxic substances; water management; rapid biological information systems; ecology of polluted waters; regional environmental analysis. Mailing Add: Ctr for Environ Studies Va Polytech Inst & State Univ Blacksburg VA 24061

CAIRNS, JOHN MACKAY, b Scranton, Pa, Dec 3, 12; m 48; c 4. EMBRYOLOGY. Educ: Hamilton Col, BA, 35; Univ Rochester, MS, 37; Washington Univ, PhD(embryol), 41. Prof Exp: Instr, Washington Univ, 41-42; instr, Univ Tex, 46-52, asst prof biol & embryol, 48-52; asst prof histol & embryol, Sch Med, Univ Okla, 52-56; engr, Bell Aircraft Corp, 56-57; sr cancer res scientist, 57-67, ASSOC CANCER RES SCIENTIST, ROSWELL PARK MEM INST, 67- Mem: Am Soc Zoologists; Am Asn Anatomists; Soc Develop Biol; Int Soc Develop Biol. Res: Experimental embryology; regional differentiation in mesoderm; application of control system concepts. Mailing Add: Springville Labs Roswell Park Mem Inst Springville NY 14141

CAIRNS, ROBERT EDWARD, b Rock Island, Que, Jan 2, 16; m 41; c 4. ORGANIC CHEMISTRY. Educ: Middlebury Col, BS, 38. Prof Exp: From works chemist to res chemist, Shawinigan Resins Corp, 38-47; res chemist, vinyl polymers, 47-48, res group leader, vinyl polymers & ethylene, 49-56, res sect leader, polyolefins, 57-59, asst res div, 59-64, mgr Saflex res, safety glass plastic, 64-70, ASSOC DIR RES & DEVELOP, PLASTIC PROD & RESINS DIV, MONSANTO CO, 70- Mem: Am Chem Soc; Soc Plastics Engrs; Soc Automotive Eng; Am Soc Testing & Mat. Res: Polymerization of vinyls and olefins; free radical and ionic initiation of polymerization; application of thermoplastics. Mailing Add: Monsanto Co 190 Grochmal Ave Indian Orchard MA 01151

CAIRNS, ROBERT WILLIAM, b Oberlin, Ohio, Dec 23, 09; m 32; c 4. PHYSICAL CHEMISTRY. Educ: Oberlin Col, AB, 30; Ohio State Univ, PhD(chem), 32; Harvard

Univ, AMP, 51; Univ Del, DSc, 69. Prof Exp: Bartol Res Found fel, Johns Hopkins Univ, 33-34; res chemist, Res Ctr, Hercules Inc, 34-40, asst to dir res, 40-41, dir Res Ctr, 41-45, from asst dir res to dir res, 45-65, dir, Co, 60-65, vpres, 65-71; dep asst secy sci & technol, US Dept Com, 71-72; EXEC DIR, AM CHEM SOC, 72- Concurrent Pos: Vchmn res & develop bd & dep asst secy defense, Dept Defense, 53-54; consult, Off Secy Defense, 55-69 & Defense Sci Bd, 56-61, mem exec comt, 65-69; mem nat comt, Int Union Pure & Appl Chem, 58-64; div chem & chem technol, Nat Res Coun, 58; dir, Indust Res, Indust Res Inst, 55-, pres, 59-60; mem bd trustees, Gordon Res Conf, 58-61, chmn, 60-61; chmn, Joint Comt Sci & Technol Commun, Nat Acad Sci-Nat Acad Eng, 66-70; mem, Nat Acad Eng, 69-, exec comt, 70-74; sci adv, Gov of Del, 69-71; mem panel on govt labs, President's Sci Adv Comt; chmn bd, Am Chem Soc, 72. Honors & Awards: Perkin Medal, Soc Indust Chem, 69; Patent Award, Freedman Found, 72; IRI Medal, Indust Res Inst, 74. Mem: Am Chem Soc (pres, 68); Int Union Pure & Appl Chem (pres, 76-77); hon mem Soc Indust Chem; Am Phys Soc; Com Develop Asn. Res: Technology of high explosives and propellants; chemical engineering; cellulose, rosin and terpene chemistry; industrial applications of chemicals; polymer chemistry; petrochemicals; research management. Mailing Add: Am Chem Soc 1155 16th St NW Washington DC 20036

CAIRNS, STEWART SCOTT, b Franklin, NH, May 8, 04; m 28; c 2. MATHEMATICS. Educ: Harvard, AB, 26, AM, 27, PhD(math), 31. Prof Exp: Instr math, Harvard Univ, 27-28 & Yale Univ, 29-31; from instr to asst prof math, Lehigh Univ, 31-38; asst prof, Queens Col, NY, 38-46; prof math & head dept, Syracuse Univ, 46-48; prof, 48-72, head dept, 48-58, EMER PROF MATH, UNIV ILL, URBANA, 72- Concurrent Pos: Mem, Inst Advan Study, 36-37, 59-60, 62-63 & 74-75; consult, Nat Defense Res Comt, 44-46, Rand Corp, 50-70, Res & Develop Bd, 50-52, Dept Defense, 51-70 & NSF, 51-53; mem, Nat Res Coun, 49-54; Fulbright prof, France, 54-55; vis prof, Nat Taiwan Univ, 73. Mem: AAAS; Am Math Soc; Math Asn Am. Res: Topology and analysis of manifolds. Mailing Add: 607 W Michigan Ave Urbana IL 61801

CAIRNS, THEODORE L, b Edmonton, Alta, July 20, 14; nat; m 40; c 4. ORGANIC CHEMISTRY. Educ: Univ Alta, BSc, 36; Univ Ill, PhD(org chem), 39. Hon Degrees: LLD, Univ Alta, 70. Prof Exp: Lab asst, Univ Ill, 37-39; instr org chem, Univ Rochester, 39-41; res chemist, 41-45, res supvr, 45-51, lab dir, 51-61, dir basic sci, 62-66, dir res, 66-67, asst dir cent res dept, 67-71, DIR CENT RES & DEVELOP DEPT, E I DU PONT DE NEMOURS & CO, 71- Concurrent Pos: Fuson lectr, Univ Nev, 68; mem, President Nixon's Sci Policy Task Force, 69, Gov Coun Sci & Technol, 69-72, President's Sci Adv Comt, 70-73 & President's Comt on Nat Medal Sci, 74-75. Honors & Awards: Am Chem Soc Award, 68; Synthetic Org Chem Mfrs Asn US Medal, 68; Perkin Medal, Am Sect, Soc Chem Indust, 73; Cresson Medal, Franklin Inst, 74. Res: Stereochemistry of biphenyls; chemistry of polyamides; reactions of acetylene and carbon monoxide; cyanocarbon chemistry. Mailing Add: Cent Res & Develop Dept E I du Pont de Nemours & Co 6032 Du Pont Bldg Wilmington DE 19898

CAIRNS, THOMAS W, b Hutchinson, Kans, Nov 13, 31; m 59; c 2. MATHEMATICS. Educ: Okla State Univ, BS, 53, MS, 55, PhD(math), 60. Prof Exp: From asst prof to assoc prof, 59-69, PROF MATH & HEAD DEPT, UNIV TULSA, 69-, CHMN DIV MATH SCI, 72- Mem: Math Asn Am; Am Math Soc; Soc Indust & Appl Math. Res: Computing; biomathematics; time series analysis. Mailing Add: Dept of Math Sci Univ of Tulsa Tulsa OK 74104

CAIRNS, WILLIAM LOUIS, b St Catharines, Ont, Oct 18, 42. BIOCHEMISTRY, PHOTOBIOLOGY. Educ: Univ Guelph, BSA, 65; Iowa State Univ, PhD(biochem), 70. Prof Exp: ASST PROF CHEM, UNIV ARK, FAYETTEVILLE, 70- Mem: Am Chem Soc; AAAS; Am Soc Photobiol. Res: Biochemical basis of circadian rhythms; mechanism of phase-shifting of circadian rhythms. Mailing Add: Dept of Chem Univ of Ark Fayetteville AR 72701

CALA, JOSEPH ANTHONY, polymer chemistry, see 12th edition

CALABI, EUGENIO, b Milano, Italy, May 11, 23; nat US; m 52; c 1. MATHEMATICS. Educ: Mass Inst Technol, BS, 46; Univ Ill, AM, 47; Princeton Univ, PhD(math), 50. Prof Exp: Asst & instr, Princeton Univ, 47-51; asst prof math, La State Univ, 51-55; from asst prof to prof, Univ Minn, 55-64; prof, 64-69, THOMAS A SCOTT PROF MATH, UNIV PA, 69-, CHMN DEPT, 73- Concurrent Pos: Mem, Inst Advan Study, 58-59; Guggenheim fel, 62-63. Mem: Am Math Soc. Res: Differential geometry of complex manifolds. Mailing Add: Dept of Math Grad Sch Univ of Pa Philadelphia PA 19104

CALABI, LORENZO, b Milan, Italy, Apr 11, 22; nat; m 49; c 4. APPLIED MATHEMATICS. Educ: Swiss Fed Inst Technol, dipl, 46; Univ Milan, PhD(math), 47; Univ Strasbourg, PhD(math), 51. Prof Exp: Res fel, Inst Advan Math, Italy, 51-52; asst prof, Boston Col, 52-56; res assoc, Parke Math Labs, Inc, 55-56, assoc dir res, 56-74; TECH DIR, SOLO CORP, 74- Concurrent Pos: Fulbright travel award, 52; lectr, Holy Cross Col, 53-54; assoc prof, Boston Col, 56-59, lectr, 59-62. Mem: Soc Gen Systs Res; Asn Comput Mach; Am Math Soc; Soc Indust & Appl Math; Ital Math Union. Res: Computer sciences applications. Mailing Add: Solo Corp One River Rd Carlisle MA 01741

CALABI, ORNELLA, US citizen. MICROBIOLOGY. Educ: Hebrew Univ, Jerusalem, MSc, 45; Univ Chicago, MSc, 51; Harvard Univ, DSc(microbiol), 57. Prof Exp: Res asst trop med, Hebrew Univ, Jerusalem, 40-45; res asst med, Univ Chicago, 48-49, res asst bact & parasitol, 49-51; instr bact, Brandeis Univ, 53-54; instr microbiol & in-chg trop med, Sch Med, Yale Univ, 56-57; res assoc microbiol, Dept Metall, Mass Inst Technol, 57-58; res assoc path, Children's Hosp, Harvard Med Ctr, Boston, 59-60; instr microbiol, Seton Hall Col Med, NJ, 60-61; res assoc path, New York Med Col, 61-63; microbiologist, Walter Reed Army Inst Res, Washington, DC, 64-70, US CIVIL SERV CAREER SCIENTIST, 67-, LEAVE OF ABSENCE, 70- Concurrent Pos: Vis scientist, Queen Elizabeth Col, Univ London, 71-72; adj dir, Cantonal Inst Bact & Serol, Lugano, Switz, 73-74. Mem: Am Soc Microbiol; AAAS; NY Acad Sci. Res: Pathogenesis of disease; blood and enteric bacterial infections; relapsing fever, shigellosis, melioidosis; humoral and cellular immunity, phagocytosis; chemotherapy, synergism of drugs; mathematical models of drug interaction and their application in biological systems. Mailing Add: 5100 Dorset Ave Kenwood MD 20015

CALABRESE, ANTHONY, b Providence, RI, Feb 25, 37; m 63; c 2. MARINE ECOLOGY. Educ: Univ RI, BS, 59; Auburn Univ, MS, 62; Univ Conn, PhD(zool, ecol), 69. Prof Exp: FISHERY BIOLOGIST, NAT OCEANIC & ATMOSPHERIC ADMIN, NAT MARINE FISHERIES SERV, 62- Mem: Am Fisheries Soc; Nat Shellfisheries Asn (secy-treas, 74-76); Atlantic Estuarine Res Soc. Res: Development of biological information concerning the effect of pollutants on marine organisms, including shellfish, fin fish and crustaceans, to provide a basis for environmental management. Mailing Add: Nat Marine Fisheries Serv Mid Atlantic Coastal Fisheries Ctr Milford CT 06460

CALABRESE, FRANCIS ANTHONY, b Waterbury, Conn, Oct 1, 42; m 70. ANTHROPOLOGY, ARCHAEOLOGY. Educ: Univ Colo, Boulder, BA, 65; Univ Mo-Columbia, MA, 69, PhD(anthrop), 71. Prof Exp: Asst state archaeologist, Kans State Hist Soc, 65-67; asst prof anthrop, Univ Tenn, Chattanooga, 71-73; supvry archaeologist, 73-75, CHIEF, MIDWEST ARCHEOL CTR, NAT PARK SERV, 75- Mem: Soc Am Archaeol; fel Am Anthrop Asn. Res: North American prehistory; plains; theory and method; material culture change; statistical methods applied to anthropology and archaeology. Mailing Add: Midwest Archeol Ctr Nat Park Serv 100 Centennial Mall N Lincoln NE 68508

CALABRESE, PHILIP G, b Chicago, Ill, Feb 21, 41; div. MATHEMATICS, PHYSICS. Educ: Ill Inst Technol, BS, 63, MS, 66, PhD(math), 68. Prof Exp: Instr math, Ill Inst Technol, 67-68; asst prof, Naval Postgrad Sch, 68-70; ASST PROF MATH, CALIF STATE COL, BAKERSFIELD, 70- Concurrent Pos: Instr, Rosary Col, 66-67. Mem: AAAS; Am Math Soc; Math Asn Am. Res: Mathematical philosophy, logic, probability and statistics; physics, especially quantum theory; probabilistic metric spaces; philosophy. Mailing Add: Calif State Col 9001 Stockdale Hwy Bakersfield CA 93309

CALABRESI, MASSIMO, b Ferrara, Italy, June 2, 03; nat US; m 29; c 2. PATHOLOGY. Educ: Univ Florence, MD, 26; Yale Univ, DrPH, 44. Prof Exp: Instr anat, Univ Florence, 26-27; from instr to assoc prof internal med, Univ Milan, 27-38, fac prof ther, 35-38; from asst clin prof to assoc clin prof, 59-71, PROF INTERNAL MED, YALE UNIV, 71- Mem: Am Heart Asn; AMA; Am Pub Health Asn; NY Acad Sci; Am Col Physicians. Res: Physiopathology of cardiac failure; physiology of circulation; cardiac hypertrophy. Mailing Add: 300 Ogden St New Haven CT 06511

CALABRESI, PAUL, b Milan, Italy, Apr 5, 30; US citizen; m 54; c 3. PHARMACOLOGY, CHEMOTHERAPY. Educ: Yale Univ, BA, 51, MD, 55; Am Bd Internal Med, dipl, 64. Prof Exp: Intern, Harvard Med Serv, Boston City Hosp, Mass, 55-56, asst resident, 58-59; proj assoc, Univ Wis, 5658; from instr to assoc prof med & pharmacol, Yale Univ, 60-68; PROF MED & CHMN DEPT, BROWN UNIV, 68- Concurrent Pos: Clin fel, Sch Med, Yale Univ, 59-60; field investr, Nat Cancer Inst, 56-60; attend physician, Vet Admin Hosp, West Haven & asst dir, Clin Res Ctr, Yale-New Haven Med Ctr, 60-; assoc physician, Univ Serv, Grace-New Haven Hosp, 60-64, asst attend physician, 64-, assoc coordr cancer teaching, 61- Mem: Am Soc Clin Invest; Am Soc Hemat; Am Soc Pharmacol & Exp Therapeut; Am Soc Oncol; Am Fedn Clin Res. Res: Medical oncology; hematology; virology; immunology; metabolism of nucleic acids; autoimmune disease; bone marrow function and physiology; antimetabolites in viral and cancer chemotherapy. Mailing Add: Roger Williams Gen Hosp 825 Chalkstone Ave Providence RI 02908

CALABRETTA, PETER JOSEPH, b Brooklyn, NY, Feb 10, 34; m 57; c 5. ORGANOMETALLIC CHEMISTRY. Educ: St John's Univ, NY, BS, 55, MS, 58, PhD(org-organometallic chem), 65. Prof Exp: Group leader polymers, Stauffer Chem Co, 65-67; asst dir res, Universal Oil Prod Fragrances, 67-69; DIR RES FRAGRANCES & FLAVOR CHEM, FELTON INT INC, 69- Concurrent Pos: Lectr, City Univ New York, 58-; fel chem, Columbia Univ, 65; consult, NRA, Inc, 64-65. Mem: AAAS; Am Chem Soc; fel Am Inst Chem; NY Acad Sci. Res: Synthesis of chemicals used in fragrances and flavor formulas. Mailing Add: Res Dept Felton Int Inc 599 Johnson Ave Brooklyn NY 11237

CALABRISI, PAUL, b Naples, Italy, Jan 27, 07; nat; m 40. ANATOMY. Educ: Cath Univ Am, BA, 31; George Washington Univ, MA, 40; Cambridge Univ, PhD(anat), 55. Prof Exp: From instr to prof anat, Sch Med, George Washington Univ, 39-74; VIS PROF ANAT, UNIV PR MED CTR, 74- Concurrent Pos: Vis lectr, Nat Naval Med Ctr, 47-48, Am Univ, 47-48, Washinton Hosp Ctr, 58, Armed Forces Inst Path, 58, Casualty Hosp, 59 & Nat Univ Athens, 69; consult, Naval Med Res Inst, 47-51; examr, Am Bd Orthop Surg, 48-; lectr, Columbian Hosp for Women & George Washington Hosp, 48-49; lectr & supvr, Trinity Col, Cambridge Univ, 53-55; vis prof, Queen's Univ, Belfast, 69; guest lectr, Prince Georges Hosp, Md. Honors & Awards: Cross of Eloy Alfaro, Panama. Mem: AAAS; Soc Exp Biol & Med; Am Asn Anatomists; Asn Am Med Cols; Anat Soc Gt Brit & Ireland. Res: Embryology; pathology. Mailing Add: 3 Meadow St Binghamton NY 13905

CALAMARI, TIMOTHY A, JR, b New Orleans, La, Nov 12, 36; m 64; c 2. TEXTILE CHEMISTRY. Educ: Loyola Univ, La, BS, 58; La State Univ, MS, 61, PhD(chem), 63. Prof Exp: RES CHEMIST, SOUTHERN UTILIZATION RES & DEVELOP LAB, USDA, NEW ORLEANS, 63- Mem: Am Asn Textile Chemists & Colorists; Sigma Xi. Res: Additive finishing techniques for the development of dimensionally stable cotton fabric; liquid ammonia based treatments for cotton textiles; new fire retardant finishes for cotton and cotton blends. Mailing Add: 1016 Rosa Ave Metairie LA 70005

CALAME, GERALD PAUL, b Lelocle, Switz, Nov 27, 30; US citizen; m 59; c 1. ASTROPHYSICS, PHYSICS. Educ: Col of Wooster, BA, 53; Harvard Univ, AM, 55, PhD(physics), 60. Prof Exp: Theoret physicist, Knolls Atomic Power Lab, Gen Elec Co, 59-61; from asst prof to assoc prof nuclear eng, Rensselaer Polytech Inst, 61-66; assoc prof, 66-69, PROF PHYSICS, US NAVAL ACAD, 69- Concurrent Pos: Consult, Knolls Atomic Power Lab, 61-66. Mem: AAAS; Am Phys Soc; Am Asn Physics Teachers. Res: Cosmic rays; nuclear physics. Mailing Add: Dept of Physics US Naval Acad Annapolis MD 21402

CALANDRA, ALEXANDER, b New York, NY, Jan 13, 11; m 47. PHYSICS, CHEMISTRY. Educ: Brooklyn Col, BS, 35; NY Univ, PhD(statist), 40. Prof Exp: Asst prof chem, Brooklyn Col, 38-45; asst prof physics, Univ Chicago, 45-47; from asst prof to assoc prof, 47-70, PROF PHYSICS, WASH UNIV, 70- Concurrent Pos: Consult, Am Coun Educ, 38-43; US Off Educ grant, 58; Stephens Col Found, TV Teaching, Sci, 58; sci consult, St Louis Pub Schs, 59; regional counselor, Am Inst Physics, Mo, 62-; chmn dept sci, Webster Col, 69-, assoc prof sci & actg chmn dept, 74-; mem staff, Educ TV, CBS, Magic People & KETC, Sci & Math; ed, Reporter, Am Chem Soc, 64- Mem: Am Chem Soc. Res: Statistical techniques in tests and measurements; integrated courses in elementary science and mathematics. Mailing Add: Dept of Physics Wash Univ St Louis MO 63130

CALANDRA, JOSEPH CARL, b Chicago, Ill, Mar 17, 17; m 44; c 4. TOXICOLOGY, BIOCHEMISTRY. Educ: Ill Inst Technol, BS, 38; Northwestern Univ, PhD(chem), 42, MD, 50; Am Bd Clin Chem, dipl, 55; Am Bd Indust Hyg, dipl, 62. Prof Exp: Instr chem, Ill Inst Technol, 38; from instr chem to assoc prof path, 38-53, chmn dept path, 49-53, PROF PATH, MED SCH, NORTHWESTERN UNIV, 53-; PRES, INDUST BIOTEST LABS, INC, 69- Concurrent Pos: NIH grant, 58-63; dir toxicol, Indust Bio-Test Labs, Inc, 53-59. Mem: Fel AAAS; fel Am Soc Clin Pharmacol & Chemother; AMA; Am Indust Hyg Asn. Res: Enzymology; pharmaceutical chemistry; experimental pathology. Mailing Add: Dept of Path Northwestern Univ Med Sch Chicago IL 60611

CALAPRICE, JOHN ROBERT, population biology, population genetics, see 12th edition

CALARCO, JOHN RICHARD, b Washington, DC, Mar 4, 40; m 61; c 3. NUCLEAR PHYSICS. Educ: George Washington Univ, BS, 63; Univ Ill, Champaign-Urbana, MS, 65, PhD(physics), 69. Prof Exp: Res assoc nuclear physics, Univ Wash, 69-71; asst prof physics, Stanford Univ, 71-75; RES PHYSICIST NUCLEAR PHYSICS, HIGH ENERGY PHYSICS LAB, STANFORD UNIV, 75- Mem: Am Phys Soc; Am Asn Univ Prof. Res: Photonuclear reactions, specifically in the structure of giant multipole resonances and the reaction mechanisms involved in their excitation and electron scattering and photopion reactions. Mailing Add: High Energy Physics Lab Stanford Univ Stanford CA 94305

CALARESU, FRANCO ROMANO, b Divaccia, Italy, Apr 12, 31; Can citizen; m 58; c 2. PHYSIOLOGY. Educ: Univ Milan, MD, 53; Univ Alta, PhD, 64. Prof Exp: Intern, Columbus Hosp, Chicago, 54-55, resident med, 55-56; resident radiother, Jefferson Med Col, 56-58; demonstr physiol, Univ Sask, 58-60; demonstr, Univ Alta, 60-64; scientist, Gatty Marine Lab, Scotland, 64-65; asst prof physiol, Univ Alta, 65-67; assoc prof, 67-72, PROF PHYSIOL, UNIV WESTERN ONT, 72- Concurrent Pos: Nat Res Coun Can fel, 58-60; Nat Heart Found Can fel, 60-65; Med Res Coun Can scholar, 65-67; Med Res Coun Can vis prof, Inst Cardiovasc Res, Univ Milan, 74-75. Mem: AAAS; Am Physiol Soc; Can Physiol Soc; Soc Neurosci. Res: Neural antonomic control; control systems analysis. Mailing Add: Dept of Physiol Univ of Western Ont London ON Can

CALAVAN, EDMOND, b Scio, Ore, Jan 13, 13; m 38; c 4. PLANT PATHOLOGY. Educ: Ore State Col, BS, 39, MS, 41; Univ Wis, PhD(plant path), 45. Prof Exp: Jr plant pathologist, 45-47, from asst plant pathologist to assoc plant pathologist, 47-59, PLANT PATHOLOGIST, CITRUS EXP STA, UNIV CALIF, RIVERSIDE, 59-, PROF PLANT PATH, UNIV, 62- Mem: AAAS; Am Phytopath Soc. Res: Diseases of citrus; fungal diseases; viral diseases; regulatory plant pathology; mycoplasma diseases of plants. Mailing Add: Dept of Plant Path Univ of Calif Riverside CA 92502

CALAWAY, PAUL KENNETH, b Bethesda, Ark, Mar 31, 10; m 45; c 3. ORGANIC CHEMISTRY. Educ: Ark Col, AB, 31; Ga Inst Technol, MS, 33; Univ Tex, PhD(chem), 38. Prof Exp: Instr chem, Ga Inst Technol, 33-35 & 38-42, from asst prof to prof, 42-54, head dept, 48-54, dir eng exp sta, 54-57; head dept, 57-64, prof, 64-75, EMER PROF CHEM, TEX A&M UNIV, 75- Concurrent Pos: Instr, Univ Tex, 35-38. Mem: Am Chem Soc. Res: Aryloxyketones; arylthicketones; substituted quinoline acids. Mailing Add: 1201 Ashburn College Station TX 77840

CALBERT, HAROLD EDWARD, b Edinburg, Ind, Mar 20, 18; m 42; c 3. BIOCHEMISTRY. Educ: Allegheny Col, BA, 39; Univ Wis, MS, 47, PhD(dairy & food technol, biochem), 48. Prof Exp: From instr to assoc prof dairy & food industs, 48-60, PROF FOOD SCI & CHMN DEPT, UNIV WIS-MADISON, 61- Concurrent Pos: Chmn environ sci training comt, USPHS. Mem: AAAS; Am Dairy Sci Asn; Inst Food Technologists; Am Fisheries Soc. Res: Dairy and food chemistry; food plant sanitation; product development; aquaculture. Mailing Add: Dept of Food Sci 105 Babcock Hall Univ of Wis Madison WI 53706

CALBO, LEONARD JOSEPH, b New York, NY, Feb 17, 41; m 64; c 1. INDUSTRIAL ORGANIC CHEMISTRY. Educ: Manhattan Col, BS, 62; Seton Hall Univ, MS, 64, PhD(org chem), 66; Univ Conn, MBA, 73. Prof Exp: RES CHEMIST, AM CYANAMID CO, 66- Mem: Am Chem Soc. Res: Surface coatings; cross-linking agents; thermoset amino and polyester resins; development of new resins and cross-linking agents for electrocoating application. Mailing Add: Am Cyanamid Co 1937 W Main St Stamford CT 06904

CALCAGNO, PHILIP LOUIS, b New York, NY, Feb 27, 18; m 51. PEDIATRICS. Educ: Univ Ga, BS, 40; Georgetown Univ, MD, 43. Prof Exp: Rotating intern, Morrisannia City Hosp, New York, 44, resident pediat & internal med, 44-45; chief resident contagious dis, Willard Packer Hosp, 45-46; pediatrician, Floating Hosp, 48; res asst, Children's Hosp, Buffalo, 48-52, assoc pediat, Sch Med, Univ Buffalo, 52-54, from asst prof to assoc prof, 54-62; PROF PEDIAT & CHMN DEPT, GEORGETOWN UNIV, 62 Mem: Soc Pediat Res; Am Acad Pediat; NY Acad Sci; Am Pediat Soc; Soc Exp Biol & Med. Res: Renal function in infants; physiological studies of the postsurgical state; neurosensory disorders in mental retardation. Mailing Add: Georgetown Univ Med Ctr 3800 Reservoir Rd NW Washington DC 20007

CALCOTE, HARTWELL FORREST, b Meadville, Pa, May 20, 20; m 43; c 4. PHYSICAL CHEMISTRY. Educ: Cath Univ Am, BACh, 43; Princeton Univ, PhD(phys chem), 48. Prof Exp: Instr electronics, Princeton Univ, 43-45, res asst combustion, 45-48; phys chemist, Exp Inc, 48-51, head thermokinetics dept, 52-54, dir res, 55-56; PRES & DIR RES, AEROCHEM RES LABS, INC, 57- Concurrent Pos: Mem papers comt, 8th-14th Int Symp on Combustion; mem adv panel, phys chem div, Nat Bur Standards, 66-69; mem, Joint Army-Navy-Air Force Tables Thermochem Working Group, 66-71. Mem: AAAS; Am Chem Soc; Am Phys Soc; Am Inst Aeronaut & Astronaut; Combustion Inst. Res: Physical chemistry of combustion processes and electrical discharges; electrical properties of flames; radar interference effects in rocket exhausts; instrumentation; air pollution. Mailing Add: AeroChem Res Labs Inc PO Box 12 Princeton NJ 08540

CALDECOTT, RICHARD S, b Vancouver, BC, Apr 15, 24; nat; m 47; c 3. GENETICS. Educ: Univ BC, BSA, 46; State Col Wash, MS, 48, PhD(radiation genetics), 51. Prof Exp: Asst, State Col Wash, 46-49, res assoc, 51; asst prof, Univ Nebr, 51-53; assoc radiobiologist, Brookhaven Nat Lab, 53-54; geneticist, res br, Agr Res Serv, USDA, 54-60 & US Atomic Energy Comn, Washington, DC, 60-63; assoc prof, grad sch, 55-65, PROF GENETICS & DEAN COL BIOL SCI, UNIV MINN, ST PAUL, 65-; GENETICIST, CROPS RES DIV, AGR RES SERV, USDA, 63- Concurrent Pos: Mem, Nat Acad Sci-Nat Res Coun subcomt on radiation biol; vpres & mem bd dirs, Fresh Water Biol Res Found; mem bd trustees, Argonne Univs Asn & St Paul Sci Mus. Mem: AAAS; Genetics Soc Am; Radiation Res Soc; Am Inst Biol Sci. Res: Radiobiology; biophysics; cytogenetics. Mailing Add: Col of Biol Sci Snyder Hall Univ of Minn St Paul MN 55101

CALDER, DALE RALPH, b St Stephen, NB, Apr 16, 41; m 65; c 2. MARINE SCIENCE. Educ: Acadia Univ, BSc, 64; Col William & Mary, MA, 66, PhD(marine Sci), 68. Prof Exp: Assoc marine scientist, Va Inst Marine Sci, 69-73; ASSOC MARINE SCIENTIST, MARINE RESOURCES RES INST, 73- Concurrent Pos: Nat Res Coun Can fel, 68-69; asst prof marine sci, Univ Va, 69-73; asst prof marine biol, Col Charleston, 74- Mem: Southeastern Estuarine Res Soc; Estuarine Res Fedn. Res: Ecology of marine fouling organisms; taxonomy and ecology of marine invertebrates, particularly the Hydrozoa and Scyphozoa; ecology of marine benthos. Mailing Add: Marine Resources Res Inst Charleston SC 29412

CALDER, JOHN ARCHER, b Baltimore, Md, Oct 18, 42; m 64. CHEMICAL OCEANOGRAPHY, ORGANIC GEOCHEMISTRY. Educ: Southern Methodist Univ, BS, 64; Univ Ill, Urbana, MS, 65; Univ Tex, Austin, PhD(chem), 69. Prof Exp: ASST PROF OCEANOG, FLA STATE UNIV, 69- Mem: Geochem Soc; Am Soc Limnol & Oceanog; Am Chem Soc. Res: Chemistry of the organic compounds in water and sediment; chemistry of the stable isotopes of carbon and oxygen; biological isotope effects. Mailing Add: Dept of Oceanog Fla State Univ Tallahassee FL 32306

CALDER, WILLIAM ALEXANDER, III, b Cambridge, Mass, Sept 2, 34; m 55; c 2. PHYSIOLOGICAL ECOLOGY. Educ: Univ Ga, BS, 55; Wash State Univ, MS, 63; Duke Univ, PhD(zool), 66. Prof Exp: Instr zool, Duke Univ, 66-67, res assoc, 67; asst prof, Va Polytech Inst, 67-69; assoc prof, 69-74, PROF, DEPT ECOL & EVOLUTIONARY BIOL, UNIV ARIZ, 74- Mem: AAAS; Am Ornith Union; Cooper Ornith Soc; Am Physiol Soc; Ecol Soc Am. Res: Physiology of temperature regulation and respiration; environmental physiology; microclimate selection and heat exchange of hummingbirds; energetic consequence of body size. Mailing Add: Dept of Ecol & Evolutionary Biol Univ of Ariz Tucson AZ 85721

CALDERON, NISSIM, b Jerusalem, Israel, Apr 1, 33; US citizen; m 61; c 2. POLYMER CHEMISTRY, ORGANOMETALLIC CHEMISTRY. Educ: Hebrew Univ, Jerusalem, MSc, 58; Univ Akron, PhD(polymer chem), 62. Prof Exp: Sr res chemist, 62-67, sect head, 67-75, MGR RES & ADMIN, GOODYEAR TIRE & RUBBER CO, 75- Mem: Am Chem Soc. Res: Developer of the olefin metathesis reaction of homogeneous catalysts; discoverer of the feasibility of cleavage of carbon-to-carbon double bonds by transition metal catalysts. Mailing Add: Goodyear Tire & Rubber Co Res Div 142 Goodyear Blvd Akron OH 44316

CALDERONE, JULIUS G, b Detroit, Mich, Aug 2, 28. MEDICAL BACTERIOLOGY, MEDICAL MYCOLOGY. Educ: San Jose State Univ, BA, 50; Univ Calif, Los Angeles, PhD(microbiol), 64. Prof Exp: Med microbiologist, Tulare County Health Dept, Calif, 51-53; clin lab technologist, US Navy, 53-55; med microbiologist, Orange County Health Dept, Calif, 55-58; asst med microbiol, Univ Calif, Los Angeles, 59-63, fel, 64-65; from asst prof to assoc prof, 65-73, PROF MED MICROBIOL, CALIF STATE UNIV, SACRAMENTO, 73- Mem: Am Soc Microbiol; Sigma Xi; Wildlife Dis Asn. Res: Bacterial and fungal diseases of animals which may be transmitted to humans. Mailing Add: Dept of Biol Sci Calif State Univ Sacramento CA 95619

CALDERWOOD, KEITH WRIGHT, b Henefer, Utah, July 28, 24; m 50; c 5. GEOLOGY. Educ: Brigham Young Univ, BS, 50, MS, 51. Prof Exp: Geologist, Phillips Petrol Co, 51-56, staff geologist, 57-59, dist geologist, 59-69; consult geologist, 69-71, PRES, CALDERWOOD & MANGUS, INC, 71- Mem: Am Inst Prof Geologists; Am Asn Petrol Geologists. Res: Petroleum and mineral exploration and development. Mailing Add: Calderwood & Mangus Inc 425 G St Anchorage AL 99501

CALDWELL, ALFRED CRAIG, b Ont, Sept 11, 10; nat US; m 38; c 3. SOIL CHEMISTRY. Educ: Univ Alta, BSc, 36, MSc, 38; Univ Minn, PhD(soils), 41. Prof Exp: Instr soils, Univ Sask, 41-42; prof soils, 42-68, PROF PLANT PHYSIOL, UNIV MINN, MINNEAPOLIS, 68- Mem: Am Soc Agron; Soil Sci Soc Am; Int Soc Soil Sci. Res: Fertility of soils; soil phosphorus; soil lime; soil potassium. Mailing Add: Dept of Soil Sci Univ of Minn Minneapolis MN 55455

CALDWELL, ARCHIE LEE, JR, physiology, biochemistry, see 12th edition

CALDWELL, AUGUSTUS GEORGE, b Belleville, Can, Apr 7, 23; US citizen; m 48; c 4. SOIL CHEMISTRY. Educ: Univ Toronto, BSA, 46, MSA, 48; Iowa State Univ, PhD(soil fertil), 55. Prof Exp: Soil surv scientist, Ont Agr Col, 46-52, lectr soil fertil res, 52-54; from asst prof to assoc prof soil chem, Tex A&M Univ, 54-62; PROF AGRON, LA STATE UNIV, BATON ROUGE, 62- Mem: Am Soc Agron; Soil Sci Soc Am; Am Chem Soc. Res: Soil fertility; soil organic phosphorus; soil tests for phosphorus; plant response to fertilizers; soil characterization; economics of fertilizer use; soil mineralogy. Mailing Add: Dept of Agron La State Univ Baton Rouge LA 70803

CALDWELL, BILLY EDWARD, agronomy, plant breeding, see 12th edition

CALDWELL, CARLYLE GORDON, b Little Rock, Ark, Mar 13, 14; m 40; c 1. PLANT CHEMISTRY. Educ: Iowa State Col, BS, 36, PhD, 40. Prof Exp: Mem staff, 40-47, res dir, 47-55, vpres res, 55-62, dir, 62-66, exec vpres, 66-68, PRES, NAT STARCH & CHEM CORP, 69-, CHIEF EXEC OFFICER, 75-; DIR, RES CORP, 73- Honors & Awards: Claude S Hudson Award, Am Chem Soc, 65. Mem: Am Chem Soc; Indust Res Inst; Am Inst Chemists; Asn Res Dirs. Res: Fractionation of starch; starch derivatives; polymers. Mailing Add: Nat Starch & Chem Corp 10 Finderne Ave Bridgewater NJ 08876

CALDWELL, DABNEY WITHERS, b Charlottesville, Va, Mar 26, 27; m 62; c 4. GEOLOGY. Educ: Bowdoin Col, AB, 49; Brown Univ, MA, 53; Harvard Univ, PhD(geol), 59. Prof Exp: Geologist, US Geol Surv, 52-54; from instr to asst prof geol, Wellesley Col, 55-67; ASSOC PROF GEOL, BOSTON UNIV, 67- Mem: AAAS; Am Geophys Union. Res: Glacial geology and fluvial processes in geomorphology; environmental geology. Mailing Add: 725 Commonwealth Ave Boston MA 02215

CALDWELL, DANIEL R, b Oakland, Calif, Feb 29, 36; m 61; c 2. MICROBIOLOGY, BIOCHEMISTRY. Educ: Reed Col, BA, 58; Univ Md, MS, 65, PhD(dairy sci), 69. Prof Exp: Microbiologist, Agr Res Serv, USDA, Md, 60-69; asst prof microbiol, 69-75, ASSOC PROF MICROBIAL PHYSIOL, UNIV WYO, 75- Mem: AAAS; Am Soc Microbiol; Sigma Xi. Res: Anaerobic bacterial nutrition and metabolism; rumen microbiology; porphyrin biosynthesis; bacterial carbohydrate fermentation; carbon dioxide fixation; mineral nutrition; anaerobic electron transport; cytochromes in ruminal bacteria. Mailing Add: Div of Microbiol & Vet Med Univ of Wyo Laramie WY 82070

CALDWELL, DAVID KELLER, b Louisville, Ky, Aug 6, 28. VERTEBRATE BIOLOGY. Educ: Washington & Lee Univ, AB, 49; Univ Mich, MS, 50; Univ Fla, PhD(biol), 57. Prof Exp: Asst zool, Univ Fla, 52-55, res asst, 55; curatorial asst fishes, State Mus, Fla, 55-57; fishery res biologist, US Fish & Wildlife Serv, 57-60; cur ichthol & marine mammals, Los Angeles County Mus Natural Hist, 60-67; cur & dir res, Marineland of Fla, 67-70; ASSOC PROF SPEECH, UNIV FLA, 70- Concurrent Pos: Res assoc, State Mus, Fla, 57-70, field assoc, 75-; res assoc, Ichthol Inst Jamaica, 58-70 & Los Angeles County Mus Natural Hist, 68- Mem: Am Fisheries Soc; Am Soc Mammalogists; Am Soc Ichthyologists & Herpetologists; fel Am Inst Fishery Res Biologists; fel AAAS. Res: Marine ichthyology, herpetology, mammalogy, and ecology; zoogeography of fishes; cetology; animal communication. Mailing Add: Rte 1 Box 121 St Augustine FL 32084

CALDWELL, DAVID ORVILLE, b Los Angeles, Calif, Jan 5, 25; m 50; c 2. EXPERIMENTAL HIGH ENERGY PHYSICS. Educ: Calif Inst Technol, BS, 47; Univ Calif, Los Angeles, MA, 49, PhD(physics), 53. Prof Exp: From instr to assoc prof physics, Mass Inst Technol, 54-63; vis assoc prof, Princeton Univ, 63-64; lectr, Univ Calif, Berkeley, 64-65, PROF PHYSICS, UNIV CALIF, SANTA BARBARA, 65- Concurrent Pos: Consult, Radiation Lab, Univ Calif, 50-51, 65-70, Am Sci & Eng, 59-60, Inst Defense Anal, 60-67 & Defense Commun Planning Group, 66-70; NSF fel,

Univ Calif, Los Angeles & Edigenossiche Tech Univ, 53-54; physicist, Radiation Lab, Univ Calif, 57, 58 & 64-67; NSF sr fel, 60-61 & Ford Found fel, 61-62; Guggenheim fel, Europ Orgn Nuclear Res, 71-72; mem high energy physics adv panel, Electronic Resources Develop Agency, 74- Mem: Fel Am Phys Soc. Mailing Add: Dept of Physics Univ of Calif Santa Barbara CA 93106

CALDWELL, DOUGLAS RAY, b Lansing, Mich, Feb 16, 36; m 61; c 2. OCEANOGRAPHY. Educ: Univ Chicago, AB, 55, BS, 57, MS, 58, PhD(physics), 64. Prof Exp: NSF fel, Cambridge Univ, 63-64; res geophysicist, Inst Geophys & Planetary Physics, Univ Calif, San Diego, 64-68; asst prof, 68-73, ASSOC PROF, DEPT OCEANOG, ORE STATE UNIV, 73- Res: Physics of fluids; hydrodynamics and magnetohydrodynamics; stability theory; measurements in the oceans. Mailing Add: Dept of Oceanog Ore State Univ Corvallis OR 97331

CALDWELL, ELWOOD F, b Gladstone, Man, Apr 3, 23; nat US; m 49; c 2. FOOD SCIENCE. Educ: Univ Man, BSc, 43; Univ Toronto, MA, 49, PhD(nutrit), 53; Univ Chicago, MBA, 56. Prof Exp: Chemist, Lake of Woods Milling Co, 43-48; res chemist, Can Breweries Ltd, 48-49; chief chemist, Christie, Brown & Co, 49-51; res assoc nutrit, Univ Toronto, 51-53; group leader processing & packaging res, Quaker Oats Co, 53-62, asst dir foods res, 62-69, dir res & develop, 70-72; PROF FOOD SCI & NUTRIT & HEAD DEPT, UNIV MINN, 72- Concurrent Pos: Chmn bd dirs, Dairy Qual Control Inst, Inc, St Paul, Minn, 73-. Honors & Awards: Cert Appreciation Patriotic Civilian Serv, Dept Defense, Army Mat Command, 70. Mem: Am Home Econ Asn; Can Inst Food Sci & Technol; Inst Food Technol; Am Asn Cereal Chem; Can Physiol Soc. Res: Food chemistry and technology; experimental, clinical and community nutrition; food service management; educational and research administration. Mailing Add: Dept of Food Sci & Nutrit Univ of Minn St Paul MN 55108

CALDWELL, FRED T, JR, b Hot Springs, Ark, May 12, 25; m 47; c 2. SURGERY. Educ: Baylor Univ, BS, 46; Washington Univ, MD, 50. Prof Exp: Assoc prof surg, Univ Hosp, State Univ NY Upstate Med Ctr, 58-67; PROF SURG, MED CTR, UNIV ARK, LITTLE ROCK, 67- Concurrent Pos: Res fel, Washington Univ, 57-58; Cancer Soc clin fel, 57-58; surg consult, Little Rock Vet Admin Hosp, 67- Mem: Am Col Surg; Am Physiol Soc; Soc Univ Surg; Am Surg Asn. Res: Energy balance following trauma, particularly after thermal burns; pathogenesis of cholesterol gall stones. Mailing Add: Univ of Ark Med Ctr 4301 W Markham St Little Rock AR 72204

CALDWELL, GEORGE CHARLES, mathematics, see 12th edition

CALDWELL, HENRY CECIL, JR, b Walnut Grove, Miss, Jan 21, 30; m 54; c 3. PHARMACEUTICAL CHEMISTRY, ORGANIC CHEMISTRY. Educ: Univ Miss, BS, 52, MS, 54; Univ Kans, PhD(pharmaceut org chem), 57. Prof Exp: Sr pharmaceut chemist, 57-62, group leader pharmaceut chem sect, 62-75, ASST DIR, SMITH KLINE & FRENCH LABS, 75- Mem: Am Chem Soc; Am Pharmaceut Asn. Res: Pharmaceutical product development; Mannich reaction; air suspension coating; drug latentiation; spasmolytics. Mailing Add: Smith Kline & French Labs 1500 Spring Garden St Philadelphia PA 19101

CALDWELL, JERRY, b Tallassee, Ala, May 1, 38; m 62; c 3. IMMUNOGENETICS, MOLECULAR GENETICS. Educ: Auburn Univ, BS, 60, MS, 62; Tex A&M Univ, PhD(animal breeding), 69. Prof Exp: Asst prof animal sci, Tex A&M Univ, 68-69; fel immunogenetics, Univ Calif, Davis, 69-70; ASST PROF ANIMAL SCI, TEX A&M UNIV, 70- Mem: Am Soc Animal Sci; Am Dairy Sci Asn. Res: Immunogenetics, especially antigenic components of red blood cells and spermatozoa molecular genetics, especially genetic variation of enzymes and other proteins of mammalian tissues. Mailing Add: Immunogenetics Lab Tex A&M Univ Dept Animal Sci College Station TX 77843

CALDWELL, JOHN R, b Middletown, Conn, Oct 11, 18; m 47; c 5. INTERNAL MEDICINE. Educ: Lafayette Col, BA, 40; Temple Univ, MD, 43; Am Bd Internal Med, dipl, 55. Prof Exp: Med practitioner, Del, 47-50; staff physician, Med Clins, 52-54, physician-in-charge, Div Hypertension, 54-67, CHIEF HYPERTENSION SECT, HENRY FORD HOSP, DETROIT, 67- Concurrent Pos: Mem coun arteriosclerosis & mem med adv bd coun high blood pressure res, Am Heart Asn; clin assoc prof internal med, Med Sch, Univ Mich. Mem: AAAS; fel Am Col Physicians; AMA; Am Soc Internal Med. Res: Causes and effects and treatment of high arterial pressure; renal artery stenosis; pheochromocytoma; primary aldosteronism; pyelonephritis; social and psychologic factors favoring development of hypertension. Mailing Add: Metab Dis Div Henry Ford Hosp Detroit MI 48202

CALDWELL, JOHN RICHARD, b Akron, Ohio, Sept 18, 10; m 46; c 2. POLYMER CHEMISTRY. Educ: Univ Akron, BS, 32; Ohio State Univ, MS, 34, PhD(colloid chem), 36. Prof Exp: Asst chem, Ohio State Univ, 32-36; res chemist, Tenn Eastman Co, 36-50, from res assoc to sr res assoc, 50-70, res fel, 70-75; RETIRED. Honors & Awards: Div Org Coatings & Plastics Chem Award, Am Chem Soc, 67. Mem: Am Chem Soc. Res: Polyesters; polyamides; vinyl polymers; synthetic fibers and plastics. Mailing Add: 238 Sixth St W Bonita Springs FL 33923

CALDWELL, JOHN THOMAS, b Mercedes, Tex, June 26, 37; m 61; c 2. EXPERIMENTAL NUCLEAR PHYSICS. Educ: Rice Univ, BA, 59; San Jose State Univ, MA, 61; Univ Calif, Davis, PhD(appl sci), 67. Prof Exp: Res scientist nuclear physics, Lawrence Livermore Lab, 61-69; RES SCIENTIST NUCLEAR PHYSICS, LOS ALAMOS SCI LAB, 69- Mem: Am Phys Soc. Res: Photonuclear physics, photofission, spontaneous fission, nuclear safeguards and nuclear radiation detector development. Mailing Add: Los Alamos Sci Lab Los Alamos NM 87545

CALDWELL, JOSEPH RALSTON, anthropology, see 12th edition

CALDWELL, LARRY D, b Manton, Mich, July 13, 32; m 59; c 4. ECOLOGY. Educ: Mich State Univ, BSc, 54, MS, 55; Univ Ga, PhD(zool), 60. Prof Exp: Asst prof biol, Southeastern La Col, 59-61; from asst prof to assoc prof, 61-69, PROF BIOL, CENT MICH UNIV, 69- Mem: AAAS; Am Soc Mammalogists; Am Ornith Union; Wildlife Soc. Res: Population and physiological ecology; mammalian population phenomena; competition in rodent populations; lipid levels and bird migration. Mailing Add: Dept of Biol Cent Mich Univ Mt Pleasant MI 48858

CALDWELL, LOREN THOMAS, b Fairmont, Ind, Aug 17, 02; m. GEOLOGY. Educ: Earlham Col, BS, 25; Univ Chicago, MS, 36; Ind Univ, EdD, 52. Prof Exp: Teacher, Ill Pub Sch, 28-29; prof geol, astron & geog, Northern Ill Univ, 29-70, head dept earth sci, 49-70; teacher adult educ courses, US Int Univ, 70-73; RETIRED. Concurrent Pos: Guest prof earth sci, US Int Univ, San Diego, 71-73; mem, Rancho Bernardo Educ Comn; educ consult, Reuben E Fleet Space Theater, San Diego, 72-75. Mem: Fel AAAS; Am Asn Petrol Geologists; Nat Asn Res Sci Teaching. Res: Glacial geology and regional physiography; sedimentary petrology and petroleum geology; science curriculum research; earth science in the public school science program. Mailing Add: 16463 Gabarda Rd Rancho Bernardo San Diego CA 92128

CALDWELL, MARTYN MATHEWS, b Denver, Colo, June 28, 41; m 67. PLANT ECOLOGY, PLANT PHYSIOLOGY. Educ: Colo State Univ, BS, 63; Duke Univ, PhD(bot), 67. Prof Exp: From asst prof to assoc prof, 67-75, PROF ECOL, UTAH STATE UNIV, 75- Concurrent Pos: NSF fel, Innsbruck, Austria, 68-69; chmn biol prog, Climatic Impact Assessment Prog, Dept Transp, 73-75; mem nat comt photobiol, Nat Acad Sci, 74-78. Mem: Ecol Soc Am; Am Soc Plant Physiologists; Brit Ecol Soc; Scand Soc Plant Physiologists; Soc Range Mgt. Res: Physiological ecology of plants under stress in arid and tundra environments; plant photosynthesis, transpiration, water relations, root growth and carbon balance; effects of solar ultraviolet radiation on plants. Mailing Add: Ecol Ctr Utah State Univ Logan UT 84322

CALDWELL, MARY ESTILL, b Columbus, Ohio, Feb 12, 96; m 25. BACTERIOLOGY. Educ: Univ Ariz, BS, 18, MS, 19; Univ Chicago, PhD(bact), 32. Prof Exp: From instr to asst prof biol, 19-32, from assoc prof to prof bact, 32-56, head dept bact, 35-56, PROF PHARMACOL & RES PHARMACOLOGIST, UNIV ARIZ, 57- Mem: Am Soc Microbiol; Am Pub Health Asn. Res: Viability mycobacterium tuberculosis; variation of certain Salmonellas; antibiotics; cancer chemotherpy; carcinogenesis; anti-inflammatory research. Mailing Add: Col of Pharm Univ of Ariz Tucson AZ 85721

CALDWELL, MELBA CARSTARPHEN, b Augusta, Ga, May 4, 21; m 60; c 2. VERTEBRATE ZOOLOGY, COMMUNICATIONS SCIENCE. Educ: Univ Ga, BS, 41; Univ Calif, Los Angeles, MA, 63. Prof Exp: Fishery aide marine zool, US Fish & Wildlife Serv, Ga, 56-59, fishery res biologist, 59-60; staff res assoc, Allan Hancock Found, Univ Southern Calif, 63-67; asst cur & assoc dir res, Marineland of Fla, 67-69; RES INSTR, INST ADVAN STUDY COMMUN PROCESSES, UNIV FLA, 70- Concurrent Pos: Res assoc, Los Angeles County Mus Natural Hist, 61-; field assoc, Fla State Mus, 75. Res: Odontocete cetaceans, especially communication, life history and other aspects of general biology; systematics and distribution of marine fishes and odontocete cetaceans. Mailing Add: Rte 1 Box 121 St Augustine FL 32084

CALDWELL, RALPH MERRILL, b Brookings, SDak, June 27, 03; m 31; c 1. PHYTOPATHOLOGY. Educ: SDak State Univ, BS, 25; Univ Wis, MS, 27, PhD(plant path, plant breeding), 29. Prof Exp: Asst bot, Univ Wis, 25-28; state leader barberry eradication, Div Cereal Crops & Dis, Bur Plant Indust, USDA, Wis, 28-30, assoc pathologist, 30-37; assoc botanist, Exp Sta, 30-37, prof, 37-71, head dept, 37-54, EMER PROF BOT & PLANT PATH, PURDUE UNIV, 71- Concurrent Pos: Wheat consult, DeKalb, AgRes, Inc, 72-; ed jour, Am Phytopath Soc, 54-58. Mem: Fel AAAS; Am Phytopath Soc (treas, 44-46); fel Am Soc Agron. Res: Breeding of nationally produced, disease and insect resistant wheat, oats and barley varieties; general resistance to cereal rusts, nature and value; physiology of stomatal penetration by Puccinia graminis and P recondita. Mailing Add: 1705B Trinity Pl College Station TX 77840

CALDWELL, RICHARD AVERILL, physical organic chemistry, see 12th edition

CALDWELL, RICHARD LOUIS, nuclear physics, see 12th edition

CALDWELL, RICHARD STANLEY, b Loma Linda, Calif, Nov 23, 40; m 60, 70; c 4. PHYSIOLOGICAL ECOLOGY. Educ: Calif State Polytech Col, BS, 62; Duke Univ, PhD(zool), 67. Prof Exp: US Atomic Energy Comn fel biol, Pac Northwest Lab, Battelle Mem Inst, 67-68; from fel to res assoc agr chem, 68-70, ASST PROF FISHERIES, ORE STATE UNIV, 70- Mem: AAAS; Am Soc Zoologists; Ecol Soc Am. Res: Physiological ecology of marine animals; temperature acclimation; marine pollution biology. Mailing Add: Marine Sci Ctr Ore State Univ Newport OR 97365

CALDWELL, ROBERT CRAIG, dentistry, deceased

CALDWELL, ROBERT EDWARD, b Lockhart, Fla, Aug 2, 15; m 40; c 1. SOILS. Educ: Univ Fla, BS, 37, MS, 41; Purdue Univ, PhD(agron), 53. Prof Exp: Sci asst, Subtrop Res Sta, USDA, 37-39; res asst, Fla Agr Exp Sta, 41; soil surveyor, State of Fla, 41-48; asst prof soils, Univ Fla, 48-51; Res Found fel, Purdue Univ, 51-53; from asst prof to assoc prof, Univ, 53-68, PROF SOILS, UNIV & SOIL CHEMIST, FLA AGR EXP STA, UNIV FLA, 68- Concurrent Pos: Fac mem, Sch Inter-Am Studies, 53-; agr consult, Pan-Am Sch Agr, Cent Am, 55 & Jamaican Sch Agr, 70; sr mem grad fac, Univ Fla, 56-, mem senate, 65-, chmn, Soil Sci Dept Characterization Comt, 65-73, assoc chief marshall, Convocation Marshall's Comt, 68-, chmn, Student Financial Aid Comt & mem, ROTC Army Scholarship Bd, 69- & chmn col agr awards & honors comt, 69-72; chmn procedures & correlation comt, Southern Regional Soil Surv Conf, 57-58, chmn org soils comt, 61-62; res consult, Econ Res Serv, USDA, 65; consult, Trop Soil Workshop, PR & VI, 69. Mem: AAAS; Am Soc Agron; Soil Sci Soc Am; Int Soc Soil Sci. Res: Physical, chemical and mineralogical study of soils as related to their genesis, morphology, classification and use as land surface units. Mailing Add: 2169 McCarty Hall Univ of Fla Gainesville FL 32611

CALDWELL, ROBERT GRANT, physical chemistry, see 12th edition

CALDWELL, ROBERT WILLIAM, b Brunswick, Ga, Dec 27, 42; m 65; c 2. PHARMACOLOGY, PHYSIOLOGY. Educ: Ga Inst Technol, BS, 65; Emory Univ, PhD(basic health sci), 69. Prof Exp: Pharmacologist, Div Med Chem, Walter Reed Army Inst Res, 70-72; asst prof, 72-75, ASSOC PROF PHARMACOL, CTR HEALTH SCI, UNIV TENN, MEMPHIS, 75-, CHMN CARDIOVASC MODULE, COL MED, 74- Mem: AAAS; Am Soc Pharmacol & Exp Therapeut; Soc Exp Biol & Med. Res: Endotoxin and hemorrhagic shock; central nervous system control of hypertension; search for less toxic cardiac glycosides. Mailing Add: Dept of Pharmacol Univ of Tenn Ctr Health Sci Memphis TN 38163

CALDWELL, ROGER LEE, b Los Angeles, Calif, May 12, 38; m 66; c 2. BIOCHEMISTRY, PLANT PATHOLOGY. Educ: Univ Calif, Los Angeles, BS, 61; Univ Ariz, PhD(chem), 66. Prof Exp: Nat Acad Sci res assoc, US Food & Drug Admin, Washington, DC, 66-67; asst prof, 67-72, ASSOC PROF PLANT PATH, UNIV ARIZ, 72-, ASSOC PLANT PATHOLOGIST, 74- Mem: Am Chem Soc; Am Phytopath Soc; Am Inst Biol Sci; Coun Agr Sci & Technol; Soil Conserv Soc Am. Res: Physiological studies of plant diseases; environmental interactions of agriculture and society; effects of air pollution on plants. Mailing Add: Dept of Plant Path Univ of Ariz Tucson AZ 85721

CALDWELL, WARREN W, b Davenport, Iowa, Dec 28, 25; m 52; c 1. ANTHROPOLOGY. Educ: Stanford Univ, BA, 48, MA, 49; Univ Wash, PhD, 56. Prof Exp: Assoc anthrop, Stanford Univ, 48-50; cur anthrop & hist, Seattle Mus Hist & Indust, 51-52; asst cur anthrop, State Mus, Univ Wash, 53-54; archaeologist, Mo Basin Proj, Smithsonian Inst, 56-63, chief, 63-65, dir, River Basin Surv, 65-68; PROF ANTHROP, UNIV NEBR, LINCOLN, 68-, CHMN DEPT, 70- Concurrent Pos: Ed, Plains Anthropologist & Pub Salvage Archaeol. Mem: Fel AAAS; Soc Am Archaeol; fel Am Anthrop Asn. Res: Pre-history of North America, specifically the cultural backgrounds of the Pacific Northwest, Great Basin and Plains; the fur trade and its relation to changing cultural patterns in those areas. Mailing Add: Dept of Anthrop Univ of Nebr Lincoln NE 68508

CALDWELL, WILLIAM ELMER, chemistry, deceased

CALDWELL, WILLIAM GLEN ELLIOT, b Millport, Scotland, July 25, 32; m 61; c 3. GEOLOGY. Educ: Glasgow Univ, BSc, 54, PhD, 57. Prof Exp: Asst lectr geol, Glasgow Univ, 56-57; spec lectr, 57-58, from asst prof to assoc prof, 58-70, PROF GEOL, UNIV SASK, 70-, HEAD DEPT GEOL SCI, 74- Mem: Paleont Soc; Brit Geol Asn; fel Geol Soc Am; Brit Paleont Asn; fel Geol Asn Can. Res: Stratigraphy and paleontology of Devonian, Carboniferous and Cretaceous systems. Mailing Add: Dept of Geol Sci Univ of Sask Saskatoon SK Can

CALDWELL, WILLIAM L, b Honolulu, Hawaii, Nov 12, 29; div; c 6. RADIOLOGY. Educ: Stanford Univ, AB, 51, MD, 55. Prof Exp: Nat Cancer Inst spec fel, Inst Cancer Res, Eng, 63-64; asst prof radiol, Stanford Univ Hosp, 64-66; from assoc prof to prof, Vanderbilt Univ, 66-71, dir radiother & radiation res, Univ Hosp, 66-71; prof radiol, 71-75, PROF HUMAN ONCOL, UNIV WIS CTR HEALTH SCI, MADISON, 75-, DIR RADIOTHER CTR, 71- Concurrent Pos: Consult radiotherapist, Palo Alto Vet Admin Hosp, 64-66, Nashville Vet Admin Hosp & Nashville Gen Hosp, 66-71 & Madison Vet Admin Hosp, 71- Mem: Am Col Radiol; Radiol Soc NAm; Radiation Res Soc; Am Soc Therapeut Radiol; Am Radium Soc. Res: Recovery kinetics in radiobiology; time-dose relationships in radiotherapy; radiotherapy of genitourinary tumors. Mailing Add: Radiother Ctr Univ of Wis Hosp 1300 University Ave Madison WI 53706

CALDWELL, WILLIAM P, plant breeding, plant pathology, see 12th edition

CALDWELL, WILLIAM V, b Boyd, Tex, Sept 3, 17; m 53; c 2. MATHEMATICS. Educ: Tex Christian Univ, BA, 51; Univ Mich, MA, 56, PhD(math), 60. Prof Exp: Anal engr, transonic lab, Mass Inst Technol, 51-54; asst prof math, Univ Del, 59-61; from asst prof to assoc prof, 61-68, PROF MATH, UNIV MICH, FLINT, 68-, CHMN DEPT, 66- Mem: AAAS; Am Math Soc; Math Asn Am. Res: Vector spaces and algebras of light interior functions; theory of quasiconformal functions. Mailing Add: Dept of Math Univ of Mich 1321 E Court St Flint MI 48503

CALE, WILLIAM GRAHAM, JR, b Philadelphia, Pa, Dec 10, 47; m 74. ECOLOGY. Educ: Pa State Univ, BS, 69; Univ Ga, PhD(ecol), 75. Prof Exp: Consult ecosyst modeling, Colo State Univ, 74-75; ASST PROF ENVIRON SCI, UNIV TEX, DALLAS, 75- Mem: AAAS. Res: Ecosystem modeling and systems analysis; theoretical ecology; terrestrial ecology. Mailing Add: Environ Sci Fac Box 688 Univ of Tex at Dallas Richardson TX 75080

CALE, WILLIAM ROBERT, b Paris, Ont, May 4, 13; m 46; c 2. INORGANIC CHEMISTRY, CHEMICAL ENGINEERING. Educ: Univ Toronto, BASc, 35, MASc, 36. Prof Exp: Asst to works chemist, Elec Reduction Co Can, Ltd, 36-38, works chemist, 38-53, tech serv mgr, 53, mgr res technol serv, 54-55, mgr planning & mkt res div, 56-58, mgr eng serv div, 58-62, mgr chlorate develop dept, 63-64, mgr patents & inventions dept, 64-68, MGR PATENTS & INFO, ERCO INDUSTS LTD, 68- Concurrent Pos: Mem exec bd, Can Comt Int Water Pollution Res, 75-76. Mem: Fel Chem Inst Can; Can Pulp & Paper Asn. Res: Production processes and applications of phosphorus compounds; chlorates and chlorine dioxide; long-range planning market research and patents. Mailing Add: Erco Industs Ltd 2 Gibbs Rd Islington ON Can

CALEF, WESLEY CARR, b Wis, June 22, 14. RESOURCE GEOGRAPHY. Educ: Univ Wis-Madison, BA, 36; Univ Calif, Los Angeles, MS, 44; Univ Chicago, PhD(geog), 48. Prof Exp: From asst prof to prof geog, Univ Chicago, 47-69; PROF GEOG, ILL STATE UNIV, 70- Concurrent Pos: Consult, President's Water Resources Policy Comn, 50, Area Develop Div, USDept Commerce, 51-63 & Pub Land Law Rev Comn, 69-70. Mem: Asn Am Geogr (secy, 67-70). Res: Resource management; environmental management; regional development. Mailing Add: Dept of Geog Ill State Univ Normal IL 61761

CALENDAR, RICHARD, b Hackensack, NJ, Aug 2, 40; m 69; c 1. MOLECULAR BIOLOGY. Educ: Duke Univ, BS, 62; Stanford Univ, PhD(biochem), 67. Prof Exp: Fel microbial genetics, Karolinska Inst, Sweden, 67-68; asst prof, 69-74, ASSOC PROF MOLECULAR BIOL, UNIV CALIF, BERKELEY, 74- Res: Gene control and morphogenesis of bacterial viruses. Mailing Add: Dept of Molecular Biol Univ of Calif Berkeley CA 94720

CALESNICK, BENJAMIN, b Philadelphia, Pa, Dec 27, 15; m 45; c 1. PHARMACOLOGY. Educ: St Joseph's Col, Pa, BS, 38; Temple Univ, AM, 41; Hahnemann Med Col, MD, 44. Prof Exp: Lab asst pharmacol, 40-41, assoc, 46-57, asst prof, 57-62, dir human pharmacol, 57-70, PROF PHARMACOL & CLIN PROF MED, HAHNEMANN MED COL, 62-, DIR DIV HUMAN PHARMACOL, 70- Concurrent Pos: Vis instr, Univ Pa, 50; lectr, Women's Med Col Pa, 51-52; chief hypertension clin, St Joseph Hosp, 58- Mem: Am Soc Pharmacol & Exp Therapeut; Am Diabetes Asn; NY Acad Sci; Soc Exp Biol & Med; Soc Toxicol. Res: Human and clinical pharmacology. Mailing Add: Div of Human Pharmacol Hahnemann Med Col & Hosp Philadelphia PA 19102

CALEY, DAVID WILLIAM, b Havertown, Pa, Oct 2, 32; m 59; c 2. FAMILY MEDICINE. Educ: Calif State Col, Long Beach, BA, 61; Univ Calif, Los Angeles, PhD(anat), 66; Univ Va, MD, 74. Prof Exp: Asst prof anat, Baylor Col Med, 66-69; ASST PROF ANAT, SCH MED, UNIV VA, 69- Concurrent Pos: Rotating intern, Riverside Hosp, Newport News, Va, 74-75. Mem: Am Asn Family Physicians; AMA. Res: Ultrastructure of the nervous system during development. Mailing Add: 50 Memorial Dr Luray VA 22835

CALEY, EARLE RADCLIFFE, b Cleveland, Ohio, May 14, 00; m 25; c 3. ANALYTICAL CHEMISTRY. Educ: Baldwin-Wallace Col, BS, 23; Ohio State Univ, MS, 25, PhD(anal chem), 28. Hon Degrees: DSc, Baldwin-Wallace Col, 67. Prof Exp: Asst chem, Baldwin-Wallace Col, 21-23; teacher, high sch, Ohio, 23-24, prin, 25-27; from instr to asst prof chem, Princeton Univ, 28-42; chief chemist, Wallace Labs, 42-46; from assoc prof to prof, 46-70, EMER PROF CHEM, OHIO STATE UNIV, 70- Prof Exp: Chemist, Agora Excavation Staff, Athens, Greece, 37; lectr & instr sci & mgt defense & war training, Princeton Univ, 42-45. Honors & Awards: Lewis Prize, Am Philos Soc, 40; Ohio J Sci Res Prize, 52; Dexter Award, Am Chem Soc, 66. Mem: AAAS; Am Chem Soc; Hist Sci Soc; fel Am Numis Soc. Res: Early history of chemistry; direct determination of sodium; analytical chemistry of the alkali and alkaline earth metals; new laboratory apparatus for analytical operations; reactions of hydriodic acid with detection and separation of difficultly soluble compounds; precipitations in nonaqueous solutions; application of chemistry to archaeology. Mailing Add: Dept of Chem Ohio State Univ 140 W 18th Ave Columbus OH 43210

CALEY, WENDELL J, JR, b Philadelphia, Pa, Jan 16, 28; m 52; c 5. PHYSICS. Educ: Houghton Col, BS, 50; Univ Rochester, MS, 59; Temple Univ, PhD(physics), 63. Prof Exp: Develop engr, Eastman Kodak Co, 53-59; from instr to asst prof physics & head dept, Gordon Col, 61-66; assoc prof, 66-72, head dept, 67-74, PROF PHYSICS, EASTERN NAZARENE COL, 72- Mem: Am Asn Physics Teachers; Am Sci Affil. Res: Hypervelocity projectiles; optics; solid state; laser light scattering in liquids. Mailing Add: Dept of Physics Eastern Nazarene Col Quincy MA 02170

CALFEE, ROBERT FRANCIS, b Norton, Kans, Jan 15, 16; m 38; c 3. ATMOSPHERIC PHYSICS. Educ: Ft Hays Kans State Col, BS, 43; Univ Ill, MS, 46. Prof Exp: Instr physics, Kans State Col, 43-44; physicist, Tenn Eastman Co, 44-45; instr physics, Pueblo Jr Col, 46-47; asst prof, Univ Denver, 47-54; res physicist, Denver Res Inst, 54-60; atmospheric physicist, Boulder Labs, Nat Bur Standards, 60-67 & Environ Sci Serv Admin, 67-70; PHYSICIST, NAT OCEANIC & ATMOSPHERIC ADMIN, 70- Mem: Am Phys Soc; Optical Soc Am; Sigma Xi. Res: Infrared spectroscopy. Mailing Add: 280 Devon Pl Boulder CO 80302

CALHOON, ROBERT ELLSWORTH, b Los Angeles, Calif, Dec 29, 38; m 69. QUANTITATIVE GENETICS. Educ: San Diego State Univ, AB, 61, MS, 67; Purdue Univ, PhD(genetics), 72. Prof Exp: Fel, Purdue Univ, 72-73; ASST PROF BIOL, QUEENS COL, NY, 73- Mem: Genetics Soc Am; Am Inst Biol Sci. Res: Genetic selection and correlated response to quantitative characters in Tribolium and Drosophila. Mailing Add: Dept of Biol Queens Col Flushing NY 11367

CALHOON, STEPHEN WALLACE, JR, b Morrow Co, Ohio, Oct 21, 30; m 52; c 3. ANALYTICAL CHEMISTRY. Educ: Houghton Col, BS, 53; Ohio State Univ, MSc, 58, PhD(chem), 63. Prof Exp: From instr to assoc prof, 56-64, PROF CHEM, HOUGHTON COL, 64-, HEAD DEPT, 71- Concurrent Pos: Asst instr, Ohio State Univ, 60-63. Mem: Am Chem Soc; Am Sci Affiliation. Res: Simultaneous analysis of carbon, hydrogen and nitrogen in organic compounds; effects of impurities on properties of ultrapure metals; cardiac-pacemaker electrode decomposition product analysis. Mailing Add: Dept of Chem Houghton Col Houghton NY 14744

CALHOON, THOMAS BRUCE, b Columbus, Ohio, Dec 15, 26; m 51; c 2. PHYSIOLOGY. Educ: Ohio State Univ, BA, 48, PhD(physiol), 52. Prof Exp: Asst physiol, Ohio State Univ, 48-51, asst instr, 51-52; from asst prof to assoc prof, Med Col SC, 52-63; from assoc prof to prof, Ohio State Univ, 63-67; asst dean basic sci affairs, 70-73, PROF PHYSIOL & BIOPHYS & CHMN DEPT, SCH MED, UNIV LOUISVILLE, 67-, ASSOC DEAN CURRICULAR AFFAIRS, 73- Mem: AAAS; Asn Am Med Cols; Am Heart Asn; Am Physiol Soc; NY Acad Sci. Res: Adrenal and thyroid related to total oxygen consumption; electrolyte shifts through connective tissues; influence of adrenal cortex; electrolytes and metabolism of cartilage; stress agents. Mailing Add: Dept of Physiol & Biophys Univ of Louisville Sch of Med Louisville KY 40202

CALHOUN, ALEXANDER JOHN, fish biology, see 12th edition

CALHOUN, BERTRAM ALLEN, b Petoskey, Mich, May 30, 25; m 48; c 5. MAGNETISM. Educ: Univ Man, BSc, 47; Wesleyan Univ, MA, 48; Mass Inst Technol, PhD(physics), 53. Prof Exp: Asst grad div appl math, Brown Univ, 48-49; asst, Lab Insulation Res, Mass Inst Technol, 49-53; res engr, Westinghouse Res Labs, 53-56; res staff mem, Res Ctr, 56-66, SR PHYSICIST, DEVELOP LAB, IBM CORP, 66- Mem: AAAS; fel Am Phys Soc. Res: Magnetic properties of ferrites and garnets. Mailing Add: IBM Corp 5600 Cottle Rd San Jose CA 95193

CALHOUN, CALVIN L, b Atlanta, Ga, Jan 7, 27; m 48; c 1. CLINICAL NEUROLOGY, ANATOMY. Educ: Morehouse Col, BS, 48; Atlanta Univ, MS, 50; Meharry Med Col, MD, 60. Prof Exp: Instr biol, Morehouse Col, 50-51; from instr to assoc prof anat & actg chmn dept, 51-72, PROF ANAT & CHMN DEPT, MEHARRY MED COL, 72-, ASSOC PROF MED & DIR DIV NEUROL, 66- Concurrent Pos: Resident fel neurol, Univ Minn, 62-65, Nat Inst Neurol Dis & Blindness res fel, 65-66; resource consult, Elem Curric, Minneapolis City Schs, 65-66. Mem: AAAS; Am Acad Neurol; Am Asn Anatomists; Nat Med Asn. Res: Microscopic anatomy; electron microscopic evaluation of the ultra structure of the evolution of experimental cerebral infarction. Mailing Add: Dept of Anat Meharry Col of Med Nashville TN 37208

CALHOUN, DAVID H, b Chattanooga, Tenn, Nov 9, 42. MICROBIOLOGY, BIOCHEMISTRY. Educ: Birmingham Southern Col, BA, 65; Univ Ala, Birmingham, PhD(microbiol), 69. Prof Exp: NIH fels, Baylor Col Med, 69-71 & Univ Calif, Irvine, 71-72, instr microbiol, 72-73; ASST PROF MICROBIOL, MT SINAI SCH MED, 73- Concurrent Pos: NSF res grant, Mt Sinai Sch Med, 74-76. Mem: Am Soc Microbiol. Res: Control of gene expression. Mailing Add: Dept of Microbiol Mt Sinai Sch of Med New York NY 10029

CALHOUN, GEORGE MILTON, b Tampa, Fla, Apr 24, 20; m 51; c 4. ORGANIC CHEMISTRY. Educ: Ga Inst Technol, BS, 42; Northwestern Univ, PhD(chem), 54. Prof Exp: Res chemist, Hercules Powder Co, 46-50 & Shell Develop Co, 54-60; tech dir, Elco Corp, 61-67; lab dir, Sadtler Res Labs, Inc, 68-69, vpres, 69-70; tech dir, 70-74, PRES, NEWPORT OF N AM, INC, PHILADELPHIA, 74- Mem: Am Chem Soc. Res: Oxidation of aromatic hydrocarbons; hydrolysis of sodium alkyl sulfates. Mailing Add: 641 Radnor Valley Dr Villanova PA 19085

CALHOUN, GORDON MAXWELL, b Ft Worth, Tex, Nov 24, 11; m 35; c 2. PHYSICAL CHEMISTRY. Educ: NTex State Univ, BA, 33; Columbia Univ, MA, 34, PhD, 37. Prof Exp: Asst chem, Columbia Univ, 34-37; res chemist, Remington Arms Co, 37-41, chief process engr, Denver Ord Plant, 41-44, prod eng supt, Lake City Ord Plant, Mo, 44-45; res mgr, Ammunition Div, 45-50, dir res & develop, 50-74, DIR SUPPORTING RES, AMMUNITION DIV, REMINGTON ARMS CO, 74- Mem: Am Chem Soc. Res: Ammunition components; explosives; paper; plastics; sporting firearms; traps and targets; abrasive products; custom powder metal parts. Mailing Add: Ammunition Div Remington Arms Co Bridgeport CT 06602

CALHOUN, JOHN BUMPASS, b Elkton, Tenn, May 11, 17; m 42; c 2. ECOLOGY. Educ: Univ Va, BS, 39; Northwestern Univ, MS, 42, PhD(zool), 43. Prof Exp: Instr biol, Emory Univ, 43-44; instr zool & ornith, Ohio State Univ, 44-46; res assoc parasitol, Johns Hopkins Univ, 46-49; NIH spec fel, Jackson Mem Lab, 49-51; psychologist, US Army Med Serv Grad Sch, 51-54; psychologist, Ment Health Intramural Res Prog, 54-73, CHIEF SECT, LAB BRAIN EVOLUTION & BEHAV, DIV BIOL & BIOCHEM RES, NIMH, 73- Mem: Ecol Soc Am; Wildlife Soc; Am Soc Mammal; Am Soc Naturalists; Soc Gen Systs Res. Res: Twenty-four hour activity rhythms; vertebrate ecology and social behavior; natural selection; ecology of the Norway rat; zoogeography; population and mental health; dialogue among scientists; theory of emotion and motivation; environmental design. Mailing Add: Lab Brain Evolution & Behav Div Biol & Biochem Res NIMH Bethesda MD 20014

CALHOUN, JOHN RAILEY, chemistry, see 12th edition

CALHOUN, MARY LOIS, b Lake City, Iowa, Mar 7, 04. HISTOLOGY, HEMATOLOGY. Educ: Iowa State Col, BS, 24, MS, 31, DVM, 39, PhD(histol, hemat), 46. Prof Exp: Technician vet anat, Iowa State Col, 28-33, from asst to instr, 33, 43; from instr to prof, 43-72, chmn dept, 48-67, EMER PROF ANAT, MICH STATE UNIV, 72- Mem: Am Vet Med Asn; Women's Vet Med Asn; Am Asn Vet Anat; Am Asn Anatomists. Res: Microscopic anatomy of the dog rat and of digestive tract of the chicken; bone marrow of the horse and cow; blood of mouse and swine; comparative histology of the integuments of domestic animals. Mailing Add: 421 Centerlawn East Lansing MI 48823

CALHOUN, MILLARD CLAYTON, b Philadelphia, Pa, Aug 25, 35; m 54; c 7. ANIMAL NUTRITION. Educ: Univ Del, BS, 58, MS, 60; Univ Conn, PhD(animal nutrit), 67. Prof Exp: Res asst animal sci, Univ Del, 58-60; res asst, Univ Conn, 61-66; asst prof animal nutrit, Univ Del, 67-68; asst prof, 68-71, ASSOC PROF ANIMAL SCI, TEX A&M UNIV, 71- Mem: Am Soc Animal Sci. Res: Endocrine factors in bovine ketosis; fat soluble vitamins and their interrelationships in the bovine; vitamin A and cerebrospinal fluid dynamics; primary acute hypovitaminosis A; effect of alkali hydroxides on adaptation of lambs to high concentrate rations. Mailing Add: Agr Res Ctr Tex A&M Univ McGregor TX 76657

CALHOUN, NOAH ROBERT, b Clarendon, Ark, Mar 23, 21; m 50; c 2. ORAL SURGERY, DENTAL RESEARCH. Educ: Howard Univ, DDS, 48; Tufts Univ, MSD(oral surg), 55. Prof Exp: Oral surgeon, Vet Admin Hosp, Tuskegee, Ala, 50-52 & Kesler AFB, Miss, 52-53; oral surgeon, Vet Admin Hosp, Tuskegee, Ala, 53-55, chief dent serv, 55-57, oral surgeon, 57-65; oral surgeon, 65-74, CHIEF DENTAL RES, VET ADMIN HOSP, WASHINGTON, DC, 69-, CHIEF DENT RES SERV, 74- Concurrent Pos: Mem, Adv Comt, Am Bd Oral Surgeons, 67-71; vis prof oral surg, Howard Univ, 68; prof lectr, Dent Col, Georgetown Univ; consult, DC Dent Examr Oral Surgeons, 70; mem, Nat Cancer Control Comt, 74 & Cancer Training Control Grant Comt, NIH, 72. Mem: Inst Med, Nat Acad Sci; Am Dent Asn; Am Soc Oral Surg; fel Am Col Dent; Int Asn Dent Res. Res: The effects of zinc on bone metabolism. Mailing Add: Dent Serv Vet Admin Hosp 50 Irving St Washington DC 20422

CALHOUN, WHEELER, JR, b Columbus, NMex, Nov 19, 16; m 42; c 3. AGRONOMY. Educ: Ore State Univ, BS, 46, MS, 53. Prof Exp: Res asst seed prod, 48-51, asst prof res farm oper, 55-65, ASSOC PROF NEW CROP RES, ORE STATE UNIV, 65, FARM SUPT, 51- Concurrent Pos: Mem sci adv comt, Nat Flax Seed Inst, 63-; oilseed crop specialist, Khuzestan Water & Power Authority, Iran, 72-74. Mem: Crop Sci Soc Am; Am Soc Agron; Am Soc Oil Chemists; Int Asn Mechanization Field Exp. Res: New crop adaptation; agronomic production; variety testing of grass and legumes for seed production potentials. Mailing Add: Dept of Crop Sci Ore State Univ Corvallis OR 97331

CALHOUN, WILLIAM KENNETH, b West Frankfurt, Ill, Oct 7, 22. NUTRITION. Educ: Univ Ill, BS, 47, PhD(animal nutrit), 51. Prof Exp: Nutritionist, Qm Food & Container Inst, 51-54; res nutritionist, Am Inst Baking, 54-60; supvry nutritionist, Armed Forces Food & Container Inst, 60-62; CHIEF, NUTRIT GROUP, FOOD SCI LAB, US ARMY NATICK DEVELOP CTR, 62- Mem: AAAS; Inst Food Technologists; Am Asn Cereal Chemists; Asn Vitamin Chemists (pres, 61); Am Inst Nutrit. Res: Relationships between nutrition and hormones; nutritional value of wheat and wheat products; chemical and physiological evaluation of nutritional adequacy of diets; nutritional requirements and dietary management. Mailing Add: Nutrit Group Food Sci Lab US Army Natick Devlop Ctr Natick MA 01760

CALI, C THERSA, biology, see 12th edition

CALIGUIRI, LAWRENCE ANTHONY, b McKees Rocks, Pa, Aug 10, 33; m 66; c 1. VIROLOGY, PEDIATRICS. Educ: Bethany Col, WVa, BS, 55; Loyola Univ, Ill, MD, 58; Am Bd Pediat, dipl, 64. Prof Exp: Intern, Med Ctr, Univ Pittsburgh, 58-59; from resident to chief resident pediat, Children's Hosp, Pittsburgh, 59-62; guest investr animal virol, 64-66, res assoc, 66-68, from asst prof to assoc prof, 68-73, ADJ PROF ANIMAL VIROL, ROCKEFELLER UNIV, 73-; PROF MICROBIOL & IMMUNOL & CHMN DEPT, ALBANY MED COL, 73- Concurrent Pos: Consult microbiologist, Albany Med Ctr Hosp, 73- Mem: Soc Gen Microbiol; Am Soc Cell Biologists; Am Soc Microbiol; Am Acad Pediat; Am Asn Immunol. Res: Viral biosynthesis and virus-cell interactions. Mailing Add: Dept of Microbiol & Immunol Albany Med Col Albany NY 12208

CALIO, ANTHONY JOHN, b Philadelphia, Pa, Oct 27, 29; m 71; c 4. SPACE PHYSICS. Educ: Univ Pa, BA, 53. Hon Degrees: DSc, Wash Univ, St Louis, 74. Prof Exp: Scientist, Bettis Atomic Power Div, Westinghouse Elec Corp, 56-59; mgr nuclear physics sect, Am Mach & Foundry Co, Va, 59-61; exec vpres & mgr opers, Mt Vernon Res Co, 61-63; mem electronic res task group, NASA, DC, 63-64, chief res eng, Electronics Res Ctr, Boston, 64-65 & Manned Space Sci Prog Off, 65-67, asst dir planetary progs, Off Space Sci & Applns, 67-68, dep dir projs, Sci & Applns Div, Johnson Spacecraft Ctr, 68-69, dir sci & applns, 69-75, DEP ASSOC ADMINR SPACE SCI, NASA HQ, WASHINGTON, DC, 75- Concurrent Pos: Sloan fel, Stanford Univ Grad Sch Bus, 74-75; mem bd dirs, Ctr Community Design & Res, Rice Univ, 75- Honors & Awards: Except Serv Medal, NASA, Except Achievement Medal, 71 & Distinguished Serv Medal, 73. Mem: NY Acad Sci; Am Phys Soc; Am Geophys Union; Am Inst Aeronaut & Astronaut. Mailing Add: 7701 Hamilton Spring Rd Bethesda MD 22034

CALISHER, CHARLES HENRY, b New York, NY, July 14, 36; m 65; c 3. MICROBIOLOGY. Educ: Philadelphia Col Pharm, BSc, 58; Univ Notre Dame, MS, 61; Georgetown Univ, PhD(microbiol), 64. Prof Exp: Chief cell develop unit, Microbiol Assoc, Inc, Md, 61-65; chief isolation & serol lab, Commun Dis Ctr, 65-69, res microbiologist, 69-74; RES MICROBIOLOGIST, USPHS, 74- Concurrent Pos: Mem subcomt interrelationships among catalogued arboviruses, Am Comt Arthropod-borne Viruses, 69-; mem dengue task force, Comn Viral Infections, Armed Forces Epidemiol Bd, 70-, bd mem, 71-; mem arboviruses comt, Res Reference Reagents Bd, Nat Inst Allergy & Infectious Dis, 70-; assoc prof, Univ NC & Ga State Univ, 71- Mem: AAAS; Am Soc Microbiol; Am Soc Trop Med & Hyg; Sigma Xi. Res: Murine viruses; in vitro growth of tumor cells; cancer viruses; arbovirus ecology and epidemiology. Mailing Add: US Pub Health Serv PO Box 2087 Ft Collins CO 80522

CALKIN, MELVIN GILBERT, b New Glasgow, NS, May 20, 36; m 61; c 4. PHYSICS. Educ: Dalhousie Univ, BSc, 57, MSc, 58; Univ BC, PhD(physics), 61. Prof Exp: With Naval Res Estab, Defence Res Bd Can, 61-62; from asst prof to assoc prof, 62-72, PROF PHYSICS, DALHOUSIE UNIV, 72- Res: Classical and quantum electromagnetic theory. Mailing Add: Dept of Physics Dalhousie Univ Halifax NS Can

CALKIN, PARKER, b Syracuse, NY, Apr 27, 33; m 55; c 2. GEOLOGY. Educ: Tufts Univ, BS, 55; Univ BC, MSc, 59; Ohio State Univ, PhD(geol), 63. Prof Exp: Res geologist, Tufts Univ, 60-61 & Inst Polar Studies, Ohio State Univ, 61-62; from instr to assoc prof, 63-75, PROF GEOL, STATE UNIV NY BUFFALO, 75- Concurrent Pos: Vis res scholar, Scott Polar Res Inst, Cambridge Univ, 70. Mem: Geol Soc Am; Glaciol Soc. Res: Geomorphology and glacial geology, particularly in northeastern United States and in polar areas. Mailing Add: Dept Geol Sci State Univ of NY 4240 Ridge Lea Rd Buffalo NY 14226

CALKINS, CARROL OTTO, b Sioux Falls, SDak, June 4, 37; m 59; c 2. ECOLOGY, ENTOMOLOGY. Educ: SDak State Univ, BS, 59, PhD(entom), 74; Univ Nebr, MS, 64. Prof Exp: Res entomologist, Forage Insect Lab, Lincoln, Nebr, 60-64 & Northern Grain Insects Res Lab, Brookings, SDak, 64-72, RES ENTOMOLOGIST, INSECT ATTRACTANTS, BEHAV & BASIC BIOL RES LAB, AGR RES SERV, USDA, 72- Concurrent Pos: Instr, Univ Nebr, 61-64 & SDak State Univ, 64-72. Mem: Ecol Soc Am; Entom Soc Am; Sigma Xi. Res: Ecological basis of insect behavior, population dynamics, distribution and abundance. Mailing Add: USDA Behav & Basic Biol Res Lab PO Box 14565 Gainesville FL 32604

CALKINS, CHARLES RICHARD, b Racine, Wis, May 30, 21; m 44; c 3. CHEMISTRY. Educ: Lawrence Col, BA, 42, MS, 47, PhD, 49. Prof Exp: Corp dir res & develop sect, Riegel Paper Corp, 51-65, vpres res & develop, 65-72, pres, Riegel Prod Corp, 72-74; EXEC V PRES, KERR GLASS MFG CORP, 74- Mem: AAAS; Am Chem Soc; fel Tech Asn Pulp & Paper Indust; Am Inst Chemists. Res: Development of specialty and packaging papers. Mailing Add: RD 1 Kintnersville PA 18930

CALKINS, HARMON ELDRED, b Ann Arbor, Mich, July 11, 12; m 38; c 3. BACTERIOLOGY. Educ: Transylvania Col, AB, 33; Univ Ky, MS, 37; Univ Pa, PhD(bact), 41. Prof Exp: Asst chem & biol, Transylvania Col, 30-33; asst bact, Univ Pa, 38-41 & Johnson Found, 39-41; res bacteriologist, Upjohn Co, 41-48; asst prof bact, Univ Ga, 48-54; assoc prof, SDak State Col, 54-64; assoc prof & assoc bacteriologist, Agr Exp Sta, Univ Idaho, 64-68; assoc prof biol, 68-74, PROF BIOL & CHMN DIV NATURAL SCI & MATH, PAUL QUINN COL, 74- Concurrent Pos: Technician, Bur Animal Indust, USDA, 34-35. Res: Serological reactivity of protein monolayers; virucidal activity of germicides; development of influenza virus vaccine; bovine liver abscess. Mailing Add: Div of Natural Sci & Math Paul Quinn Col Waco TX 76704

CALKINS, JAMES A, b Joliet, Ill, Jan 5, 23; m 52; c 1. GEOLOGY. Educ: Univ Calif, Berkeley, BA, 50; Univ Ore, MS, 54; Pa State Univ, PhD(geol), 66. Prof Exp: GEOLOGIST, US GEOL SURV, 52- Mem: Geol Soc Am. Res: Geology of the Northwest Himalaya; data storage and retrieval. Mailing Add: 2124 Powhatan St Falls Church VA 22043

CALKINS, MYRON EUGENE, b Syracuse, NY, Dec 8, 22; m 45; c 2. ANALYTICAL CHEMISTRY. Educ: Univ Ala, BS, 48, MS, 53, PhD(chem), 56. Prof Exp: Supvr control lab, Cellulose Prod Div, Dow Chem Co, 48-49, chemist use develop methyl cellulose, 49-51; chemist nuclear mat mgt, Savannah River Opers Off, AEC, 55-56, chemist chem processing, 56-57, chemist chem processing prod develop & qual assurance, 57-62, supvry nuclear engr qual assurance, Albuquerque Opers Off, 62-70, develop engr, Weapons Develop, 70-75; RETIRED. Mem: Am Chem Soc; Am Soc Qual Control. Res: Use of ion exchange and polarography in quantitative analytical chemistry. Mailing Add: 5816 E Ponderosa NE Albuquerque NM 87110

CALKINS, RUSSEL CROSBY, b Cedar Falls, Iowa, Dec 31, 21; m 43; c 3. ANALYTICAL CHEMISTRY. Educ: Univ Northern Iowa, BA, 48; Univ Wis, MS, 51, PhD(chem), 53. Prof Exp: Instr chem, Univ Northern Iowa, 48-49; from asst to res asst, Univ Wis, 49-53; chemist, Dow Chem Co, 53-59; sr res chemist, 59-69, STAFF RES CHEMIST, KAISER ALUMINUM & CHEM CORP, 69- Mem: AAAS; Soc Appl Spectros; Am Chem Soc; Sigma Xi. Res: Exchange properties of nickel complexes; polymer chemistry; analytical research. Mailing Add: PO Box 870 Pleasanton CA 94566

CALKINS, WILLIAM GRAHAM, b Chicago, Ill, May 29, 26; m 48; c 4. INTERNAL MEDICINE, GASTROENTEROLOGY. Educ: Univ Mich, BS, 46, MD, 50. Prof Exp: Intern med, US Naval Hosp, St Albans, NY, 50-51; resident, Gen Hosp, Kansas City, Mo, 51-52; resident, Med Ctr, 54-56, from instr to asst prof, Sch Med, 56-64, ASSOC PROF MED, SCH MED, UNIV KANS, 64- Concurrent Pos: Staff physician, Vet Admin Hosp, Kansas City, 61-64, chief med serv, 64-70, staff physician, 70- Mem: Fel Am Col Physicians; Am Gastroenterol Asn; AMA. Mailing Add: Vet Admin Hosp 4801 Linwood Blvd Kansas City MO 64128

CALKINS, WILLIAM HAROLD, b Toronto, Ont, May 28, 18; US citizen; m 43; c 2. INDUSTRIAL CHEMISTRY, POLYMER CHEMISTRY. Educ: Univ Calif, BS, 40, PhD(chem), 47. Prof Exp: Res chemist, 47-53, sr res supvr, 53-58, res mgr, 58-68, assoc lab dir, plastics dept, 68-75, MGR FEEDSTOCKS RES, ENERGY & MAT DEPT, E I DU PONT DE NEMOURS & CO, 75- Mem: Fel Am Chem Soc; Soc Plastics Engrs; fel AAAS; fel Sigma Xi. Res: Petrochemicals; coal liquefaction and gasification; heterogeneous catalysis. Mailing Add: Energy & Mat Dept E I du Pont de Nemours & Co Wilmington DE 19898

CALL, EDWARD PRIOR, b Kent, Ohio, Dec 24, 26; m 54; c 3. REPRODUCTIVE PHYSIOLOGY, AGRICULTURE. Educ: Ohio State Univ, BSc, 51; Kans State Univ, PhD(animal breeding), 67. Prof Exp: Asst herdsman artificial insemination, Cent Ohio Breeding Asn, 50-51, technician, 51-52; asst prof, 52-58, res asst, 58-63, ASSOC PROF DAIRY SCI, KANS STATE UNIV, 63- Mem: AAAS; Am Dairy Sci Asn; Am Soc Animal Sci; Soc Study Reproduction. Res: Morphology and histology of bovine genitalia; etiology of ovarian dysfunction and reproductive failure; cyclic changes of peripheral hypophyseal and gonadal hormones in the female bovine. Mailing Add: Dept of Dairy Sci Col of Agr Kans State Univ Manhattan KS 66502

CALL, JUSTIN DAVID, b Salt Lake City, Utah, Aug 7, 23; m 51; c 3. PSYCHIATRY. Educ: Univ Utah, BA, 44, MD, 46. Prof Exp: Pvt practr, 46-47; intern pediat, Albany Hosp, NY, 47-48; pediat path res, Children's Hosp, Boston, Mass, 48-49; resident pediat, NY Hosp, 49-51; child neuropsychiatrist, Emma Pendleton Bradley Home, 51; resident psychiat, Strong Mem Hosp, 52; from instr to assoc prof, Univ Calif, Los Angeles, 54-68, PROF PSYCHIAT & HUMAN BEHAV & CHIEF DIV CHILD PSYCHIAT, UNIV CALIF, IRVINE, 68- Concurrent Pos: Mem, Adv Bd, Little Village Nursery Sch, Los Angeles, Exceptional Children's Found; mem, Clin Res Proj Rev Comt, NIMH, 68-72; chmn, Ment Health Adv Bd, Los Angeles County Head 24 Proj, 69-70. Mem: Am Psychiat Asn; Am Orthopsychiat Asn; Am Acad Pediat; Am Psychoanal Asn. Res: Preventive psychiatry; personality development. Mailing Add: Dept of Psychiat Univ of Calif Irvine CA 92650

CALL, REGINALD LESSEY, b Rigby, Idaho, Apr 23, 26; m 50; c 3. PHYSICS. Educ: Brigham Young Univ, BS, 51; Univ Utah, PhD(physics), 58. Prof Exp: Mem tech staff, Bell Labs, 58-68; ASSOC PROF ELEC ENG, UNIV ARIZ, 68- Res: Solid state physics; photovoltaics; solar energy. Mailing Add: Dept of Elec Eng Univ of Ariz Tucson AZ 85721

CALL, RICHARD A, b Gooding, Idaho, July 16, 20; m 44; c 3. PATHOLOGY. Educ: Univ Utah, BA, 42, MD, 44; Am Bd Path, dipl. Prof Exp: From instr to asst prof, 49-53, ASSOC CLIN PROF PATH, COL MED, UNIV UTAH, 53- Concurrent Pos:

Asst dir labs, Salt Lake Gen Hosp, 51-52; chief lab serv, Vet Hosp, No 5252, Salt Lake City, 52-54; fel, Medici Publici, Col Med, Univ Utah, 65; med dir, Utah Valley Latter-Day Saint Hosp, 69-; mem adv bd, Health, Educ & Welfare, Health Ins Benefits Adv Coun Comt, 74-78. Mem: AMA; Col Am Pathologists; Am Soc Clin Pathologists; Am Acad Forensic Sci; Int Acad Path. Res: Penicillin blood levels; virulence studies in mice of M tuberculosis; causes of death as determined by autopsy in Utah;fluoridosis in man; forensic pathology. Mailing Add: 3201 N Shadowbrook Circle Provo UT 84601

CALL, TRACEY GILLETTE, b Afton, Wyo, May 31, 15; m 41; c 4. PHARMACOLOGY, HEMATOLOGY. Educ: Brigham Young Univ, AB, 47; Idaho State Col, BS, 40; Univ Md, MS, 44; Univ Minn, PhD(pharmacol), 56. Prof Exp: Asst prof pharmacog & pharmacol, Duquesne Univ, 45-46; lectr, Brigham Young Univ, 46-47; asst prof pharmacog & pharmacol, Univ Wyo, 47-49; assoc prof, Mont State Univ, 49-57; pharmacologist, Sunkist Growers, Inc, 57-60 & Indust Res, 60-62; MEM FAC PHYSIOL, CALIF POLYTECH STATE UNIV, SAN LUIS OBISPO, 62- Concurrent Pos: Asst, Univ Wash, 41-42 & Univ Md, 42-44; fel, Am Found Pharmacol Educ, 51-52. Mem: AAAS; Am Soc Pharmacog; Am Pharmaceut Asn; NY Acad Sci. Res: Sources, isolation and pharmacological activities of benzopyrone compounds from plants. Mailing Add: Calif Polytech State Univ San Luis Obispo CA 93407

CALLAGHAN, EUGENE, b Snohomish, Wash, Jan 10, 04; m 35; c 2. ECONOMIC GEOLOGY. Educ: Univ Ore, AB, 26, AM, 27; Columbia Univ, PhD(geol), 31. Prof Exp: Asst, Univ Ore, 26-27; geologist, Metrop Dist, Water Supply Comn, Boston, 29; from asst geologist to sr geologist, US Geol Surv, 30-45; prof econ geol, Ind Univ, 46-49; dir, NMex Bur Mines & Mineral Resources, 49-57; chief geol, Gold Prog, Mineracao Hannaco Ltd, Brazil, 58-60; sr resident geologist, Cyprus Mines Corp, Skouriotissa, 60-65; res prof mineral, 65-68, chmn dept geol & geophys, 68-70, prof geol, 70-72, EMER PROF GEOL & GEOPHYS SCI, UNIV UTAH, 72- Concurrent Pos: Geologist, Barton Exped, Persia, 28; econ geologist, Ind State Dept Conserv, 46-49; sr geologist, Utah Geol Surv, 65-67, asst dir, 67-68, assoc dir, 68- Mem: Soc Econ Geol; fel Geol Soc Am; Am Inst Mining, Metall & Petrol Engrs; Brit Inst Mining & Metall. Res: Economic geology of magnesite, brucite, gold, tungsten, tin, antimony, copper and clay; Cascade Mountains, Oregon; Eastern Cordillera of Bolivia, Great Basin, Nevada, high plateaus, Utah, Indiana, New Mexico and countries of the Mediterranean region. Mailing Add: 2500 Kensington Ave Salt Lake City UT 84108

CALLAHAM, MAC A, b Ft Payne, Ala, Aug 30, 36; m 60; c 3. WILDLIFE BIOLOGY, FISHERIES. Educ: Univ Ga, BS, 58, PhD(wildlife fisheries), 68; George Peabody Col, MA, 61, EdS, 63. Prof Exp: Assoc prof biol, Belmont Col, 61-63; assoc prof, 63-72, PROF BIOL, NGA COL, 72- Mem: AAAS; Am Fisheries Soc. Res: Fishery management; fish blood proteins. Mailing Add: Dept of Biol NGa Col Dahlonega GA 30533

CALLAHAM, ROBERT ZINA, b San Francisco, Calif, May 24, 27; m 49; c 2. FOREST ECOLOGY, GENETICS. Educ: Univ Calif, BS, 49, PhD(bot), 55. Prof Exp: Res forester ecol, Bur Entom & Plant Quarantine, Berkeley Forest Insect Lab, USDA, 49-54; geneticist, Calif Forest & Range Exp Sta, 54-58, leader, Inland Empire Res Ctr, Intermountain Forest & Range Exp Sta, 58-60 & Inst Forest Genetics, 60-63, asst dir, Pac Southwest Forest & Range Exp Sta, 63-64, chief, Br Forest Genetics Res, 64-66, asst to dep chief res, 66-70, dir, Div Forest Insects & Dis Res, 70-73, DIR, FOREST ENVIRON RES STAFF, US FOREST SERV, 73- Concurrent Pos: Mem exec bd, Int Union Forestry Res Orgns; ed, Silvae Genetica, 62-66; asst exec dir, Int Union Socs Foresters, 70-74, mem exec bd, 75-; Cong fel, 72-73. Mem: Soc Am Foresters. Res: Resistance of pine species and species hybrids to bark beetles and other forest insects; pine hybridization and improvement; geographic variation in forest trees; forestry research administration; renewable resources technical information systems. Mailing Add: Forest Environ Res US Forest Serv Washington DC 20250

CALLAHAN, FRANCIS P, JR, mathematics, see 12th edition

CALLAHAN, JAMES LOUIS, b Cleveland, Ohio, Sept 25, 26; m 49; c 2. INORGANIC CHEMISTRY. Educ: Baldwin-Wallace Col, BS, 50; Western Reserve Univ, MS, 54, PhD, 57. Prof Exp: From tech res specialist to supvr catalysis res, 50-70, MGR CHEM & CATALYSIS RES, STANDARD OIL CO OHIO, 70- Mem: Am Chem Soc. Res: Catalysis; catalytic conversion of hydrocarbons; solid state chemistry; petrochemicals. Mailing Add: Standard Oil Co 4440 Warrensville Center Rd Cleveland OH 44128

CALLAHAN, JEFFREY EDWIN, b Boston, Mass, Sept 24, 43; m 67; c 2. PHYSICAL OCEANOGRAPHY. Educ: US Naval Acad, BS, 65; Johns Hopkins Univ, MA, 69, PhD(phys oceanog), 71. Prof Exp: Weapons officer, 71-74, MIL ASST TO DIR SCI & ENG, US NAVAL OCEANOG OFF, 74- Mem: Am Geophys Union; Sigma Xi. Res: Large-scale oceanic circulation and its effect on the distributions of properties in the ocean, particularly polar oceans and their influence on abyssal circulation worldwide. Mailing Add: 6306 Davis Blvd Camp Springs MD 20023

CALLAHAN, JOHN JOSEPH, b Taylor, Pa, Apr 1, 25; m 53. PHYSICAL CHEMISTRY. Educ: Johns Hopkins Univ, BS, 54. Prof Exp: Phys chemist, Physicochem Res Div, Chem Res & Develop Lab, 54-60, asst chief chem, Phys Methods Br, 60-62, chief, Agents Reactions Br, 62-66, supvry chemist, Phys Chem Dept, 66-74, ACTG CHIEF, PHYS CHEM BR, CHEM RES DIV, CHEM LAB, EDGEWOOD ARSENAL, US DEPT ARMY, 74- Res: Measurement of physical properties; instrmental analysis; purification; reaction mechanisms. Mailing Add: Phys Chem Br Chem Res Div Chem Lab Edgewood Arsenal Aberdeen Proving Ground MD 21010

CALLAHAN, JOHN WILLIAM, b Welland, Ont, July 9, 42; m 69; c 2. BIOCHEMISTRY, NEUROCHEMISTRY. Educ: Univ Windsor, BSc, 65, MSc, 66; McGill Univ, PhD(biochem), 70. Prof Exp: Med Res Coun Can fel, Univ Calif, Los Angeles, 70-72; INVESTR NEUROSCI, RES INST, HOSP SICK CHILDREN, 72-; ASST PROF PEDIAT, FAC MED, UNIV TORONTO, 73-, ASST PROF BIOCHEM, 75- Concurrent Pos: Med Res Coun Can scholar, 73- Mem: AAAS; Can Biochem Soc; Am Soc Neurochem; Can Soc Clin Invest. Res: Isolation and characterization of lysosomal hydrolases, especially those enzymes involved in the lysosomal storage diseases; the study of the altered gene products in these diseases. Mailing Add: Neurosci Div Hosp Sick Children 555 University Ave Toronto ON Can

CALLAHAN, JOSEPH THOMAS, b Concord, NH, Mar 31, 22; m 43; c 3. GEOLOGY. Educ: Univ NH, BS, 49; Univ Ariz, MS, 51. Prof Exp: Geologist, Ground Water Br, US Geol Surv, 51-55, dist geologist, Ga, 55-61, from assoc chief to chief, Ground Water Br, 61-66; geologist-ground water tech adv, US AID, 66-71; chief underground waste disposal invests, Water Resources Div, 71, chief ground water br, 71-72, REGIONAL HYDROLOGIST, US GEOL SURV, 72- Honors & Awards: Meritorious Serv Award, US Dept Interior, 74. Mem: Geol Soc Am; Am Water Works Asn; Am Geophys Union. Res: Geologic and hydrologic investigations of humid and arid zones, especially in the lower and middle latitudes, and in Korea and Japan. Mailing Add: US Geol Surv Water Resources Div Reston VA 22092

CALLAHAN, KEMPER LEROY, b Mt Morris, Pa, July 14, 29; m 59; c 2. PLANT PATHOLOGY. Educ: Waynesburg Col, BS, 51; WVa Univ, MS, 53; Univ Wis, PhD(plant path), 57. Prof Exp: Cur mus, Waynesburg Col, 47-51; asst, WVa Univ, 51-53; asst, Univ Wis, 53-57; plant pathologist, Forest Serv, USDA, 57-58; plant pathologist, Entom & Path Div, Can Dept Forestry, 58-65; mem fac biol, 65-73, PROF BIOL, GREENSBORO COL, 73- Mem: Am Phytopath Soc; Can Phytopath Soc; Am Inst Biol Sci; Am Forestry Asn; Soc Am Foresters. Res: Virus diseases of forest trees; natural sciences. Mailing Add: Dept of Biol Greensboro Col Greensboro NC 27420

CALLAHAN, KENNETH PAUL, b Pasadena, Calif, Dec 15, 43. INORGANIC CHEMISTRY. Educ: Univ Santa Clara, BS, 65; Univ Calif, Riverside, PhD(inorg chem), 69. Prof Exp: Scholar inorg chem, Dept Chem, Univ Calif, Los Angeles, 70-74; ASST PROF CHEM, BROWN UNIV, 74- Mem: Am Chem Soc; The Chem Soc. Res: Metalloborane synthesis; synthesis, spectroscopy and optical activity of coordination compounds; design and synthesis of new ligands; organic chemistry of carboranes; metal-sulfur chemistry; early transition metal chemistry. Mailing Add: Dept of Chem Brown Univ Providence RI 02912

CALLAHAN, LLOYD MILTON, b Hobart, Okla, Mar 28, 34; m 59; c 4. AGRONOMY. Educ: Okla State Univ, BS, 59, MS, 61; Rutgers Univ, PhD(agron), 64. Prof Exp: From asst prof to assoc prof agron, 64-71, ASSOC PROF ORNAMENTAL HORT, UNIV TENN, KNOXVILLE, 71- Mem: Am Soc Agron; Weed Sci Soc Am; Crop Sci Soc Am; Sigma Xi. Res: Turfgrass management; cellular morphology and anatomy; plant physiology; herbicide phytotoxicity to turfgrasses; disease and sod webworm control. Mailing Add: Univ of Tenn Dept of Ornamental Hort & Landscape Design Knoxville TN 37916

CALLAHAN, LYNN THOMAS, III, b Washington, DC, May 11, 45. MICROBIOLOGY. Educ: Wake Forest Univ, BS, 67, MS, 70, PhD(microbiol), 72. Prof Exp: Bur Med & Surg Nat Res Coun fel, 72-74; MEM RES STAFF MICROBIOL, NAVAL MED RES INST, 74- Mem: Am Soc Microbiol. Res: Purification and characterization of virulence factors produced by pathogenic bacteria; development of vaccines for prophylactic treatment and assay systems for diagnostic purposes. Mailing Add: Dept of Microbiol Naval Med Res Inst Bethesda MD 20014

CALLAHAN, MARY VINCENT, b Bridgeport, Conn, July 2, 22. CHEMISTRY, INSTRUMENTATION. Educ: Col Notre Dame, Md, AB, 43; Cath Univ Am, MS, 45, PhD(chem), 66. Prof Exp: Teacher, Inst Notre Dame High Sch, 44-45; teacher, Prep Sch, Col Notre Dame, Md, 46-48, instr chem, Col, 48-50; teacher, NY High Sch, 50-54; from instr to assoc prof, 54-66, PROF CHEM & PHYSICS, COL NOTRE DAME, MD, 66-, CHMN DEPT CHEM, 64- Concurrent Pos: Vis prof, Summer Sch, Cath Univ Am, 65- Mem: AAAS; Am Chem Soc; Soc Appl Spectros. Res: Free radicals; pyrolysis of hexachloropropylene; instrumental analysis; infrared spectroscopy; analytical chemistry; molecular structure; amino acids; peptides; complexes. Mailing Add: Col of Notre Dame 4701 N Charles St Baltimore MD 21210

CALLAHAN, PHILIP SERNA, b Ft Benning, Ga, Aug 29, 23; m 49; c 4. ENTOMOLOGY. Educ: Univ Ark, BA & MS, 53; Kans State Col, PhD(entom), 56. Prof Exp: From asst prof to assoc prof entom, La State Univ, 56-63; prof entom, Univ Ga, 63-69; entomologist, Southern Grain Insects Res Lab, Agr Res Serv, 62-69, ENTOMOLOGIST, INSECT ATTRACTANTS, BEHAV & BASIC BIOL RES LAB, USDA, UNIV FLA, 69- Mem: Am Ornith Union; Entom Soc Am; NY Acad Sci. Res: Ecology and behavior of Lepidoptera; insect biophysics with special reference to theories of infrared and microwave electromagnetic attraction between insects and host plants; methods of insect biotelemetry. Mailing Add: Insect Attractants USDA Univ Fla Gainesville FL 32604

CALLAHAN, WILLIAM PAXTON, III, b Shreveport, La, Aug 9, 31; m 54; c 2. ANATOMY. Educ: Southern Methodist Univ, BS, 53; Univ, BS, 53, MS, 54; Univ Tex, PhD(anat), 58. Prof Exp: Lab asst zool, Southern Methodist Univ, 52-53, lab instr, 53-54; asst anat, Med Br, Univ Tex, 54-58; from instr to asst prof, Hahnemann Med Col, 58-61; asst prof, Col Med, Univ Fla, 61-67; vis prof, Univ Med Sci, Bangkok, 67-72; ASSOC PROF ANAT, COL MED, UNIV S ALA, 72- Concurrent Pos: Mem spec staff, Rockefeller Found, 67- Mem: AAAS; Am Asn Anatomists; Electron Micros Soc Am; Geront Soc. Res: Hormonal regulation of skin and hair; ultrastructure of endocrine glands and liver; application of stains in electron microscopy. Mailing Add: Dept of Anat Univ of SAla Col Med Mobile AL 36688

CALLAHAN, WILLIE RUSSELL, b Critz, Va, Mar 28, 10. APPLIED MATHEMATICS. Educ: Va Polytech Inst, BS, 34, MS, 35; Cath Univ, PhD, 62. Prof Exp: Instr math, Va Mil Inst, 40-42; instr elec eng, Harvard Univ, 42-44; lectr, Yale Univ, 44-46; asst prof mech eng, Northeastern Univ, 46-47; asst prof appl mech, NY Univ, 47-51; mathematician, Int Bus Mach Co, NY, 51-52; engr, Gen Elec Co, NY, 52-55; lectr math, Univ Mich, 55-57; asst prof, Cath Univ, 57-66; prof, St John's Univ, NY, 66-69; prof mech technol, NY Inst Technol, 69-74; PROF MATH, ST JOHN'S UNIV, NY, 74- Mem: Am Math Soc. Res: Mathematical applications to problems in physics and engineering. Mailing Add: Dept of Math St John's Univ Jamaica NY 11432

CALLAN, EDWIN JOSEPH, b Floral Park, NY, Nov 29, 22; m 46; c 2. ATOMIC PHYSICS. Educ: Manhattan Col, BS, 43; Ohio State Univ, MSc, 60. Prof Exp: Physicist & chief thermal res sect, Concrete Res Div, Corps Engrs, Miss, 46-53; res & develop adminstr, Directorate Res, Wright Air Develop Ctr, 53-56, chief plans & anal, Off Aerospace Res Labs, 56-68, SCI ADV, OFF AEROSPACE RES LABS, 68- Concurrent Pos: Scholar, Dublin Inst Advan Studies, 64-65; mem conf comt, Nat Conf Admin Res, 72-75. Mem: Am Phys Soc; Am Chem Soc; Am Inst Aeronaut & Astronaut. Res: Atomic transition rates; thermal and elastic properties of concreting materials; radiation shielding; research management; atomic and ionic energy level calculations and methods. Mailing Add: 541 Chaucer Rd Dayton OH 45431

CALLANAN, MARGARET JOAN, b Washington, DC, July 31, 26. PHYSICAL CHEMISTRY. Educ: Trinity Col, Washington, DC, AB, 48; Cath Univ Am, MS, 50. Prof Exp: Chemist, NIH, 50-58; res asst, Nat Res Coun, 58-61; assoc prog dir, 61-74, SPEC ASST, NSF, 74- Concurrent Pos: Fel educ for pub mgt, Civil Serv Comn, 72. Mem: Fel AAAS; Am Chem Soc. Res: Physico-chemical properties of proteins, nucleic acids and protamines; science education; public understanding of science; finding ways to increase the number of women with science careers. Mailing Add: NSF 1800 G St NW Washington DC 20550

CALLANTINE, MERRITT REECE, b Hammond, Ind, Jan 6, 36; m 56; c 2. PHYSIOLOGY, ENDOCRINOLOGY. Educ: Purdue Univ, BS, 56, MS, 58, PhD(endocrinol), 61. Prof Exp: Teaching asst biol sci, Purdue Univ, 58-60; head sect endocrinol, Parke-Davis Res Labs, Mich, 61-70; dir sci & regulatory affairs, 70-72, SR

CLIN INVESTR, MEAD JOHNSON RES CTR, 72- Concurrent Pos: Vis assoc prof, Hahnemann Med Col, 73- Mem: Endocrine Soc; Am Physiol Soc; Soc Exp Biol & Med; Soc Study Reproduction; AAAS. Res: Endocrinology of reproduction, especially control of gonadotrophin secretion, ovarian function and uterine contractility; mechanism of hormone action; nonsteroidal antifertility agents; hormone antagonists; clinical investigation, antineoplastic agents, antibiotics, estrogen replacement therapy and agents to control uterine contractility. Mailing Add: 12 Caranza Ave Evansville IN 47710

CALLARD, GLORIA VINCZ, b Perth Amboy, NJ, July 20, 38; m 62. COMPARATIVE ENDOCRINOLOGY. Educ: Tufts Univ, BS, 59; Rutgers Univ, MS, 63, PhD(zool), 64. Prof Exp: Endocrine biochemist, Personal Prod Co Div, Johnson & Johnson, 64-67; lectr, Col William & Mary, 67-72; asst prof, Tufts Univ, 72-73; res assoc & lectr, Boston Univ, 74-75; RES ASSOC OBSTET & GYNEC, HARVARD MED SCH, 75- Mem: AAAS; Brit Soc Endocrinol; Am Soc Zoologists. Res: Reproductive physiology; steroid synthesis. Mailing Add: Lab of Human Reproduction Harvard Med Sch Boston MA 02215

CALLARD, IAN PETER, endocrinology, reproductive physiology, see 12th edition

CALLAS, GERALD, b Beaumont, Tex, Oct 14, 32; m 67; c 2. ANATOMY, ELECTRON MICROSCOPY. Educ: Lamar State Col, BS, 59; Univ Tex, Galveston, MA, 62, PhD(anat), 66, MD, 67. Prof Exp: Intern, John Sealy Hosp, 67-68; asst prof anat, 68-75, ASSOC PROF ANAT, MED BR, UNIV TEX, 75- Mem: Am Asn Anatomists. Res: Electron microscopy of the lung secondary to altered thyroid function; fine structural changes occurring in the adrenal gland in relation to varying functions of the thyroid gland. Mailing Add: Dept of Anat Univ of Tex Med Br Galveston TX 77550

CALLAS, NICHOLAS P, mathematics, see 12th edition

CALLAWAY, ENOCH, III, b La Grange, Ga, July 12, 24; m 48; c 2. PSYCHIATRY. Educ: Columbia Univ, AB, 44, MD, 47. Prof Exp: Intern med, Grady Mem Hosp, Emory Univ, 47-48; resident psychiat, Worcester State Hosp, Mass, 48-49; from instr to assoc prof, Psychiat Inst, Univ Md, 52-58; CHIEF RES, LANGLEY PORTER NEUROPSYCHIAT INST, UNIV CALIF, SAN FRANCISCO, 58- Concurrent Pos: USPHS res career investr, 54-58; prof in residence, Sch Med, Univ Calif, San Francisco, 59-, chmn training comt, Interdisciplinary Training Prog, 65-; consult, US Vet Admin Hosp, Oakland, 59-63 & US Naval Hosp, 59-; assoc ed, Psychophysiology; mem psychopharmacol study sect, NIMH, 63-67, mem alcohol & alcohol probs study sect, 68-72; mem adv bd, Biol Psychiat, 70- Mem: Am Col Neuropsychopharmacol; Soc Biol Psychiat; Soc Psychophysiol Res: Psychol Res Soc. Res: Behavioral neurophysiology as related to psychiatric problems. Mailing Add: Dept of Psychiat Univ of Calif Med Ctr San Francisco CA 94143

CALLAWAY, JASPER LAMAR, b Cooper, Ala, Apr 5, 11; m 41; c 3. DERMATOLOGY. Educ: Univ Ala, BS, 31; Duke Univ, MD, 32. Prof Exp: Fel & instr dermat, Univ Pa, 35-37; PROF DERMAT, SCH MED, DUKE UNIV, 37- Concurrent Pos: Consult, Vet Admin Hosp, USPHS & Surgeon Gen, US Air Force; mem, Nat Adv Serol Coun; pres, Am Bd Dermat & Syphilol, 58-59; mem, Spec Adv Group, Vet Admin; mem, Cutaneous Dis Sect, Nat Res Coun. Mem: Am Dermat Asn (secy, pres, 5859); Soc Invest Dermat (pres, 55-56); fel Am Col Physicians; Am Asn Prof Dermatologists (pres, 65); Am Acad Dermat. Res: Syphilology; manual clinical mycology. Mailing Add: Dept of Dermat Duke Hosp Durham NC 27706

CALLAWAY, JOSEPH, b Hackensack, NJ, July 1, 31; m 49; c 3. THEORETICAL PHYSICS. Educ: Col William & Mary, BS, 51; Princeton Univ, MA, 53, PhD(physics), 56. Prof Exp: Asst prof physics, Univ Miami, 54-60; from assoc prof to prof physics, Univ Calif, Riverside, 60-67; chmn dept physics & astron, 70-73, PROF PHYSICS, LA STATE UNIV, BATON ROUGE, 67- Concurrent Pos: Consult, res labs, Philco Corp, 61-67; sr vis fel, Imp Col, Univ London, 74-75. Mem: Fel Am Phys Soc. Res: Electron-atom and atom-atom scattering; energy band theory; theory of ferromagnetism; imperfections in crystalline solids. Mailing Add: Dept of Physics & Astron La State Univ Baton Rouge LA 70803

CALLAWAY, RICHARD JOSEPH, b Washington, DC, Oct 31, 28; m 56; c 2. PHYSICAL OCEANOGRAPHY. Educ: Univ Wash, BS, 56; Ore State Univ, MS, 59. Prof Exp: Oceangr, US Fish & Wildlife Serv, Hawaii, 56-58; asst oceanog, Ore State Univ, 58-59; oceangr, US Fish & Wildlife Serv, Wash, 59-60 & US Pub Health Serv, 60-66; OCEANGR, PAC NORTHWEST WATER LAB, ENVIRON PROTECTION AGENCY, 66- Mem: AAAS; Am Soc Limnol & Oceanog; Am Geophys Union; Am Meteorol Soc. Res: Marine pollution; field and laboratory investigations of municipally and/or industrially caused pollution of marine waters. Mailing Add: Pac Northwest Water Lab Environ Protection Agency Corvallis OR 97330

CALLCOTT, THOMAS ANDERSON, b Columbia, SC, Jan 13, 37; m 61; c 3. SOLID STATE PHYSICS, SURFACE PHYSICS. Educ: Duke Univ, BS, 58; Purdue Univ, MS, 62, PhD(physics), 65. Prof Exp: Res assoc physics, Purdue Univ, 65; mem tech staff, res div, Bell Tel Labs, 65-68; asst prof, 68-72, ASSOC PROF PHYSICS, UNIV TENN, KNOXVILLE, 72- Concurrent Pos: Consult, Oak Ridge Nat Lab, 68-; NSF res grants, 69-77. Mem: AAAS; Am Phys Soc. Res: Radiation damage effects in crystalline solids; optical and photoemission measurements on solids; physics of solid surfaces. Mailing Add: Dept of Physics Univ of Tenn Knoxville TN 37916

CALLEN, EARL ROBERT, b Philadelphia, Pa, Aug 28, 25; m 50; c 4. SOLID STATE PHYSICS. Educ: Univ Pa, AB, 48, MA, 51; Mass Inst Technol, PhD(physics), 54. Prof Exp: Physicist, Nat Security Agency, 55-59; sr scientist physics, US Naval Ord Lab, 59-68; PROF PHYSICS, AM UNIV, 68- Mem: Fel Am Phys Soc. Res: Metal-semiconductor transitions; magnetostriction; amorphous magnetism; statistical mechanics of social systems; science and civil liberties; history of science, Greece; United States-Soviet scientific exchange and international scientific freedom. Mailing Add: Dept of Physics Am Univ Washington DC 20016

CALLEN, HERBERT BERNARD, b Philadelphia, Pa, July 1, 19; m 45; c 2. STATISTICAL MECHANICS. Educ: Temple Univ, BS, 41, AM, 42; Mass Inst Technol, PhD(physics), 47. Prof Exp: Physicist, Kellex Corp, New York, NY, 44-45; electronic res, Princeton Univ, 45; res assoc, Mass Inst Technol, 45-48; from asst prof to assoc prof, 48-56, PROF PHYSICS, UNIV PA, 56- Concurrent Pos: Consult, Sperry Rand Corp, 51-; mem adv panel physics, NSF, 66-69, chmn, 69; chmn fac sen, Univ Pa, 70-71; Guggenheim fel, 72; mem, Comn Statist Mech, Int Union Pure & Appl Physics, 72- Mem: Fel Am Phys Soc; Am Asn Physics Teachers. Res: Theoretical physics; theory of solid state; thermodynamics; irreversible statistical mechanics. Mailing Add: Dept of Physics Univ of Pa Philadelphia PA 19104

CALLEN, JAMES DONALD, b Wichita, Kans, Jan 31, 41; m 61; c 2. PLASMA PHYSICS. Educ: Kans State Univ, BS, 62, MS, 64; Mass Inst Technol, PhD(appl plasma physics), 68. Prof Exp: NSF fel, Inst Advan Study, Princeton Univ, 68-69; asst prof aeronaut & astronaut, Mass Inst Technol, 69-72; staff mem, 72-75, MGR, THEORY DEPT, THERMONUCLEAR DIV, OAK RIDGE NAT LAB, 75- Concurrent Pos: Consult, Oak Ridge Nat Lab, 69-72. Mem: Am Nuclear Soc; Am Phys Soc. Res: Plasma physics, mainly microinstabilities of confined plasmas; nuclear reactor engineering. Mailing Add: Thermonuclear Div PO Box Y Oak Ridge Nat Lab Oak Ridge TN 37830

CALLEN, JOSEPH EDWARD, b Moulton, Iowa, Mar 24, 20; m 44; c 2. ORGANIC CHEMISTRY, RESEARCH ADMINISTRATION. Educ: Univ Iowa, BS, 42, MS, 43, PhD(org chem), 46. Prof Exp: Asst chem, Univ Iowa, 42-43; res chemist, Nat Defense Res Comt, 43-45; res chemist, 46-60, ASSOC DIR RES, PROCTOR & GAMBLE CO, 60- Concurrent Pos: Instr, Univ Iowa, 44. Mem: Am Chem Soc. Res: Synthetic detergents; fats and oils; fluorescent dyes; infrared and ultraviolet spectroscopy; amines; gas chromatography. Mailing Add: 9465 Sherborn Dr Cincinnati OH 45231

CALLEN, ROBERT BRIEN, analytical chemistry, physical chemistry, see 12th edition

CALLENBACH, JOHN ANTON, b Merchantville, NJ, Apr 9, 08; m 33; c 3. ENTOMOLOGY. Educ: Univ Wis, BS, 30, MS, 31, PhD(econ entom), 39. Prof Exp: Asst econ entom, Univ Wis, 30-33; Va Smelting Co fel, Va Truck Exp Sta, 33-36; instr econ entom, Univ Wis & agent, Bur Entom & Plant Quarantine, USDA, 36-42; asst prof entom, Univ Idaho, 42-44; asst state entomologist, Mont, 44-46; asst prof zool & entom, Mont State Col, 46-48, assoc prof, 48-53; state entologist, NDak, & prof agr entom & chmn dept, 53-57, assoc dean agr & assoc dir agr exp sta, 57-69, prof entom, 69-74, chmn dept, 69-73, EMER PROF ENTOM, N DAK STATE UNIV, 74- Mem: Entom Soc Am. Res: Truck crop and cereal crop insects; mosquitoes. Mailing Add: 606 Ninth St S Fargo ND 58102

CALLENDER, E DAVID, b New York, NY, May 5, 34; m 54; c 3. APPLIED MATHEMATICS, COMPUTER SCIENCE. Educ: Drew Univ, AB, 55; Ind Univ, MA, 57; Stanford Univ, PhD, 59. Prof Exp: Teacher, Ind Univ, 55-57 & Stanford Univ, 57-58; proj engr, Philco Corp, 58-63; mem tech staff & dept head, 63-69, ASSOC DIR MATH & COMPUT CTR, AEROSPACE CORP, 69- Mem: Am Math Soc; Soc Indust & Appl Math. Res: Partial differential equations; satellite orbit determination and symbolic manipulation using large scale computers. Mailing Add: Aerospace Corp Box 92957 Los Angeles CA 90009

CALLENDER, JONATHAN FERRIS, b Los Angeles, Calif, Nov 7, 44; m 67; c 3. GEOLOGY, TECTONICS. Educ: Calif Inst Technol, BS, 66; Harvard Univ, AM, 68, PhD(geol), 75. Prof Exp: Teaching fel geol, Harvard Univ, 67-70, res asst mineral, 70; ASST PROF GEOL, UNIV N MEX, 72- Concurrent Pos: Consult, Pub Serv Co, NMex, 73-74 & Sandia Labs, 74-; vis staff scientist, Los Alamos Sci Lab, 74-; proj corresp, US Geodynamics Comt, 75- Mem: Geol Soc Am; Sigma Xi; Am Geophys Union; Mineral Soc Am; Am Asn Petrol Geologists. Res: Structural geology and petrology of the Precambrian of northern New Mexico; tectonics of the Rio Grande rift; structural petrology of cracks in rocks; fabric studies of salt; remote sensing; geotectonics. Mailing Add: Dept of Geol Univ of NMex Albuquerque NM 87131

CALLENDER, WADE LEE, b Kingsville, Ohio, May 27, 26; m 50; c 5. PHYSICAL CHEMISTRY. Educ: Col of Wooster, AB, 48; Univ Rochester, PhD(chem), 51. Prof Exp: Res chemist, Shell Oil Co, 51-66; staff res chemist, Koninklijke-Shell Lab, Amsterdam, 66-68 & Shell Oil Co, 68-73, SR STAFF RES CHEMIST, SHELL DEVELOP CO, 73- Mem: Catalysis Soc; Am Chem Soc; Am Soc Testing & Mat. Res: Mechanism and kinetics of catalytic reactions; analog and digital computing; catalytic reforming. Mailing Add: Shell Develop Co PO Box 1380 Houston TX 77001

CALLENDINE, GEORGE WEIRICH, JR, radiation physics, see 12th edition

CALLIGHAN, OWEN HUGH, b Johannesburg, SAfrica, Oct 2, 27; m 59; c 3. IMMUNOCHEMISTRY. Educ: Univ Witwatersrand, MSc, 56; Univ Sheffield, PhD(biochem), 58. Prof Exp: Vis scientist, Oxford Univ, 58; res assoc immunol, 59-60, from fac assoc to assoc prof, 61-70, PROF BIOCHEM, CHICAGO MED SCH, 70- Concurrent Pos: NIH res career develop award, 63. Mem: Brit Biochem Soc; Am Asn Immunologists. Res: Enzymology; immunology; histamine release; neurochemistry. Mailing Add: Dept of Biochem Chicago Med Sch 2020 W Ogden Ave Chicago IL 60612

CALLIHAN, ALFRED DIXON, b Scarbro, WVa, July 20, 08; m 36. PHYSICS. Educ: Marshall Col, AB, 28; Duke Univ, AM, 31; NY Univ, PhD(physics), 33. Hon Degrees: DSc, Marshall Col, 64. Prof Exp: From asst to instr physics, Marshall Col, 26-29; asst, Duke Univ, 29-30 & NY Univ, 30-34; tutor, City Col New York, 34-37, from instr to asst prof, 37-48; res physicist, Manhattan Dist Proj, Div War Res, Columbia Univ, 41-45; res physicist, 45-73, CONSULT, NUCLEAR DIV, UNION CARBIDE CORP, 73-; CONSULT, US ARMY BALLISTICS RES LAB, 73- Concurrent Pos: Ed, Nuclear Sci & Eng, 65-; vchmn nuclear standards mgt bd & exec comt, Am Nat Standards Inst; mem atomic safety & licensing bd panel, US Nuclear Regulatory Comn. Mem: Fel Am Phys Soc; fel Am Nuclear Soc; NY Acad Sci. Res: Reactor physics. Mailing Add: 102 Oak Lane Oak Ridge TN 37830

CALLIS, CLAYTON FOWLER, b Sedalia, Mo, Sept 25, 23; m 49; c 2. INORGANIC CHEMISTRY. Educ: Cent Col, Mo, AB, 44; Univ Ill, MS, 46, PhD(inorg chem), 48. Prof Exp: Instr physics & chem, Cent Col, Mo, 44-45; asst, Univ Ill, 45-47; res chemist, Gen Elec Co, 48-50; res chemist, Monsanto Co, Ala, 51-52 & Ohio, 52-54, res group leader, 54-57 & Mo, 57-59, sr res group leader, 59-60, asst dir res, 60-62, mgr res, 62-69, dir res & devel, Inorg Chem Div, 69-71, Detergents & Fine Chem Div, 71-73 & Detergent & Phosphate Div, 73-75, DIR TECHNOL PLANNING & EVAL, MONSANTO INDUST CHEM CO, 75- Concurrent Pos: Civilian with Atomic Energy Comn, 48-50; alt, Indust Res Inst, 75- Honors & Awards: St Louis Award, Am Chem Soc, 71. Mem: Sigma Xi; Am Inst Chemists; Am Chem Soc; AAAS. Res: Chemistry of phosphorus and its compounds; surfactants; detergents; antimicrobials; dentifrices. Mailing Add: 2 Holiday Lane St Louis MO 63131

CALLIS, JERRY JACKSON, b Parrott, Ga, July 28, 26. VETERINARY MEDICINE. Educ: Ala Polytech Inst, DVM, 47; Purdue Univ, MS, 49. Prof Exp: Vet, Brucellosis Test Lab, Purdue Univ, 48-49, State Vet Res Inst, Holland, 49-51 & Animal Dis Sta, Md, 52-53; in charge res opers, 53-56, asst dir, 56-62, DIR, PLUM ISLAND ANIMAL DIS CTR, 63- Honors & Awards: USDA Award, 57; XII Int Vet Cong Prize, Am Vet Med Asn, 74. Mem: AAAS; Am Vet Med Asn; Tissue Cult Asn; US Animal Health Asn. Res: Tissue culture; virology and immunology; foot-and-mouth disease. Mailing Add: Plum Island Animal Dis Ctr Box 848 Greenport NY 11944

CALLISON, GEORGE, b Blue Rapids, Kans, June 1, 40; m 62; c 2. EVOLUTIONARY BIOLOGY, VERTEBRATE MORPHOLOGY. Educ: Kans State Univ, BS, 62; Univ Kans, MA, 65, PhD(zool), 69. Prof Exp: Asst prof zool & asst cur vert paleont, SDak Sch Mines & Technol, 67; asst prof, 69-72, ASSOC PROF BIOL, CALIF STATE UNIV, LONG BEACH, 72- Concurrent Pos: Res assoc vert paleont & herpet, Los Angeles County Mus Natural Hist, 69-; consult, Badlands Nat Monument Master Plan, Nat Park Serv, 69; Long Beach Calif State Col Found res grant, 70. Mem: Am

Soc Ichthyologists & Herpetologists; Soc Vert Paleont; Soc Syst Zool; Soc Study Amphibians & Reptiles. Res: Evolutionary interrelationships of primitive snakes; fossil amphibians and reptiles from Mesozoic and Cenozoic rocks; comparative anatomical and functional morphological concepts. Mailing Add: Dept of Biol Calif State Univ Long Beach CA 90801

CALLOW, ALLAN DANA, b Somerville, Mass, Apr 9, 16; m 43; c 3. SURGERY. Educ: Tufts Col, BS, 38, MS, 48, PhD, 52; Harvard Univ, MD, 42; Am Bd Surg, dipl, 52. Prof Exp: From instr to assoc prof, 48-56, PROF SURG, MED SCH, TUFTS UNIV, 56- Concurrent Pos: Surgeon, New Eng Med Ctr, 54-62, assoc chief gen surg, 62-75, vchmn dept surg & chief gen & vascular surg, 75-; trustee, Tufts Univ, 71- Mem: Asn Am Med Cols; fel Am Col Surg; Am Heart Asn; Soc Vascular Surg; Int Cardiovasc Soc (secy-gen, 68-, pres, NAm chap, 74-75). Res: Clinical surgery; cardiovascular disease; general and vascular surgery. Mailing Add: New England Med Ctr Hosp 171 Harrison Ave Boston MA 02111

CALLOWAY, DORIS HOWES, b Canton, Ohio, Feb 14, 23; div; c 2. NUTRITION. Educ: Ohio State Univ, BS, 43; Univ Chicago, PhD(nutrit), 47; Am Bd Nutrit, dipl, 51. Prof Exp: Intern dietetics, hosp, Johns Hopkins Univ, 44; res dietician, dept med, Univ Ill, 45; consult nutrit, Med Assocs Chicago, 48-51; nutritionist, QM Food & Container Inst, 51-58, head metab lab, 58-59, chief nutrit br, 59-61; chmn dept food sci & nutrit, Stanford Res Inst, 61-64; PROF NUTRIT, UNIV CALIF, BERKELEY, 63- Concurrent Pos: Assoc ed, Nutrit Rev, 62-68; mem ed bd, J Nutrit, 67-72, Environ Biol & Med, 69-, J Am Dietetic Asn, 74- & Interdisciplinary Sci Rev, 75- Mem bd dirs, Am Bd Nutrit, 68-71; mem panel, White House Conf Food, Nutrit & Health, 69; mem bd trustees, Nat Coun Hunger & Malnutrit in US, 69-71; mem Nat Acad Sci-Nat Res Coun Food & Nutrit Bd, 72-75; mem vis comt, Mass Inst Technol, 72-74; mem expert adv panel on nutrit, WHO, 72-77; mem adv coun, Nat Inst Arthritis, Metab & Digestive Dis, NIH, 74-77; consult Nutrit Div, Food & Agr Orgn, UN, 74-75. Honors & Awards: Meritorious Civilian Serv Award, Dept of Army, 59. Mem: Am Inst Nutrit (secy); Am Dietetic Asn. Res: Human nutrition; protein and energy; gastrointestinal functions. Mailing Add: Dept of Nutrit Sci Univ of Calif Berkeley CA 94720

CALLOWAY, E DEAN, b Louisville, Miss, Sept 25, 24; m 52. PHYSICAL CHEMISTRY. Educ: Millsaps Col, BS, 48; Univ Ala, MS, 54, PhD(chem), 56. Prof Exp: Asst prof chem, Delta State Col, 54-56; assoc prof, Birmingham-Southern Col, 56-58 & Millsaps Col, 58-60; supvr, Chemstrand Co Div, Monsanto Co, Ala, 60-64; asst prof chem, Memphis State Univ, 64-66; assoc prof, 66-69, PROF CHEM, BIRMINGHAM-SOUTHERN COL, 69- Mem: Am Chem Soc. Res: Ultraviolet absorption and molecular spectroscopy. Mailing Add: Dept of Chem Birmingham-Southern Col Birmingham AL 35204

CALLOWAY, JEAN MITCHENER, b Indianola, Miss, Dec 18, 23; m 52; c 2. NUMBER THEORY. Educ: Millsaps Col, BA, 44; Univ Pa, MA, 49, PhD(math), 52. Prof Exp: Asst instr math, Millsaps Col, 44; instr, McCallie Sch, 44-47; asst instr, Univ Pa, 47-52; from asst prof to assoc prof, Carleton Col, 52-60, actg chmn dept, 53-54 & 58-59; chmn dept, 60-73, PROF MATH, KALAMAZOO COL, 60- Concurrent Pos: Mem, Inst Advan Study, 59; res assoc, Stanford Univ, 68-69. Mem: Am Math Soc; Math Asn Am. Mailing Add: Dept of Math Kalamazoo Col Kalamazoo MI 49001

CALMA, VICTOR CHARLES, b Houston, Tex, July 13, 17; m 62. MEDICAL EDUCATION, PEDIATRICS. Educ: Rice Univ, BA, 39; Univ Tex, MD, 42. Prof Exp: Assoc prof physiol, Univ Houston, 50-51; asst prof, Univ Tex Med Br, Galveston, 51-59, instr pediat, 57-59, asst prof, 59-66, dir adolescent med, 60-66; VPRES MED EDUC, MEM MED CTR, 66- Concurrent Pos: Adj prof health sci, Tex A&I Univ, Corpus Christi, 75- Mem: AMA; fel Am Acad Pediat; Asn Hosp Med Educ; Am Acad Family Physicians. Res: Pediatrics; adolescent medicine. Mailing Add: Mem Med Ctr 2606 Hosp Blvd Corpus Christi TX 78405

CALMAN, JACK, b New York, NY, July 16, 47. PHYSICAL OCEANOGRAPHY. Educ: City Col New York, BS, 69; Harvard Univ, SM, 70, PhD(appl physics), 75. Prof Exp: RES ASSOC PHYS OCEANOG, MASS INST TECHNOL, 75- Res: Understanding fluid flow in submarine canyons with application to oceanic tidal, continental shelf, and mixing phenomena; laboratory experiments on fluid dynamical instability with application to oceanic microstructure. Mailing Add: Dept of Earth & Planetary Sci Mass Inst of Technol Cambridge MA 02139

CALMON, CALVIN, chemistry, management, see 12th edition

CALNEK, BRUCE WIXSON, b Manchester, NY, Jan 29, 32; m 54; c 2. VETERINARY VIROLOGY. Educ: Cornell Univ, DVM, 55, MS, 56. Prof Exp: Actg asst prof poultry dis, Cornell Univ, 56-57; assoc prof vet sci, Univ Mass, 57-61; PROF AVIAN DIS, CORNELL UNIV, 61- Concurrent Pos: Nat Cancer Inst fel virol, Univ Calif, Berkeley, 67-68; vis scientist oncol, Houghton Poultry Res Sta, Eng, 74-75; cancer res fel, Int Union Against Cancer, 74-75; consult, Virol Study Sect, NIH, 75- Mem: Am Vet Med Asn; Am Asn Avian Pathologists; World Vet Poultry Asn. Res: Viral oncology, with particular emphasis on the pathogenesis of avian neoplastic diseases of chickens induced by DNA-containing viruses, known as Marek's disease, and RNA-containing viruses causing lymphoid leukosis. Mailing Add: Dept of Avian & Aquatic Med Col of Vet Med Cornell Univ Ithaca NY 14853

CALPOUZOS, LUCAS, b Detroit, Mich, Oct 20, 27; m 50; c 2. PLANT PATHOLOGY. Educ: Cornell Univ, BS, 50; Harvard Univ, AM, 52, PhD(biol), 55. Prof Exp: Plant pathologist, R T Vanderbilt Co, 50-51; instr bot, Harvard Univ, 52-53; plant pathologist, Fed Exp Sta, PR, USDA, 55-62; NSF sr fel, Bristol Univ, 62-63; res plant pathologist, Crops Res Div, Agr Res Serv, USDA, 63-67; from assoc prof to prof plant path, Univ Minn, St Paul, 67-71; PROF PLANT SCI & HEAD DEPT PLANT & SOIL SCI, UNIV IDAHO, 71- Mem: Am Phytopath Soc; Am Soc Agron; Am Soc Hort Sci. Res: Crop loss estimates; chemical control of plant diseases; fungus physiology. Mailing Add: Dept of Plant Soil Sci Univ of Idaho Moscow ID 83843

CALTAGIRONE, LEOPOLDO ENRIQUE, b Valparaiso, Chile, Mar 1, 27; nat; m 53; c 3. ENTOMOLOGY. Educ: Univ Chile, ErAgron, 51; Univ Calif, PhD(entom), 60. Prof Exp: Entomologist, Chilean Ministry Agr, 49-62; from jr entomologist to assoc entomologist, 62-75, PROF & ENTOMOLOGIST, UNIV CALIF, BERKELEY, 75- Concurrent Pos: John Simon Guggenheim fel, 57-59; lectr, Sch Agr, Cath Univ Chile, 60-62. Mem: AAAS; Entom Soc Am; Am Inst Biol Sci. Res: Biological control of agricultural insect pests; biology of parasitic Hymenoptera. Mailing Add: 1050 San Pablo Ave Albany CA 97406

CALTRIDER, PAUL GENE, b Mineral Wells, WVa, Jan 14, 35; m 56; c 2. MICROBIAL PHYSIOLOGY. Educ: Glenville State Col, BS, 56; WVa Univ, MS, 58; Univ Ill, PhD(plant path, microbiol), 62. Prof Exp: Sr microbiologist, 62-66, MGR ANTIBIOTIC QUAL CONTROL & TECH SERV, ELI LILLY & CO, 66- Mem: Am Soc Microbiol; Soc Indust Microbiol; Am Chem Soc. Res: Physiology and biochemistry of microorganisms, especially carbohydrate metabolism, antibiotic fermentations and biosynthesis of antibiotics. Mailing Add: PO Box 99 Clinton IN 47842

CALUSDIAN, RICHARD FRANK, b Watertown, Mass, Feb 6, 35; m 62; c 2. THEORETICAL PHYSICS. Educ: Harvard Univ, BA, 57; Univ NH, MS, 59; Boston Univ, PhD(physics), 65. Prof Exp: Physicist, Mat Res Agency, US Dept Army, 65, fel, Natick Labs, 65-66; assoc prof, 66-69, PROF PHYSICS, BRIDGEWATER STATE COL, 69-, CHMN DEPT, 66- Mem: Am Asn Physics Teachers. Res: Electron-phonon interaction; theory of electrical and thermal conductivity. Mailing Add: Dept of Physics Bridgewater State Col Bridgewater MA 02324

CALVANICO, NICKOLAS JOSEPH, b New York, NY, Aug 18, 36; m 58; c 2. BIOCHEMISTRY. Educ: City Col New York, BS, 58; DePaul Univ, MS, 60; Univ Vt, PhD(biochem), 64. Prof Exp: Trainee immunol, Div Exp Med, Col Med, Univ Vt, 64-65; res instr med, Sch Med, State Univ NY Buffalo, 68-73; ASST PROF IMMUNOL, MAYO FOUND SCH MED, UNIV MINN, 73- Concurrent Pos: Arthritis Found fel, Sch Med, State Univ NY Buffalo, 6568; consult, Erie County Labs, 68-73. Mem: NY Acad Sci. Res: Organic chemistry; immunochemistry; structure and function of immunoglobulins in normal and pathological conditions. Mailing Add: Dept of Immunol Mayo Found Sch of Med Univ Minn Rochester MN 55901

CALVER, JAMES LEWIS, b Pontiac, Mich, June 15, 13; m 45, 70; c 2. MINERALOGY, PHYSICAL GEOLOGY. Educ: Univ Mich, AB, 36, MS, 38, PhD(geol), 42. Prof Exp: Asst geol, Univ Mich, 35-39; instr geol, Univ Wichita, 40-42; asst prof, Univ Mo, 42-43; assoc geologist, Tenn Valley Authority, 43-47; geologist, Fla Geol Surv, 47-57; COMNR MINERAL RESOURCES & STATE GEOLOGIST, VA DIV, 57- Mem: Mineral Soc Am; Geol Soc Am; Am Inst Mining, Metall & Petrol Engrs; Am Asn Petrol Geologists; Am Geophys Union. Res: Economic geology. Mailing Add: 1614 Oxford Rd Charlottesville VA 22903

CALVERLEY, JOHN ROBERT, b Hot Springs, Ark, Jan 14, 32; m 53; c 2. NEUROLOGY. Educ: Univ Ore, BS, 53, MD, 55. Prof Exp: Resident neurol, State Univ Iowa Hosps, 56; fel med, Mayo Found, 57, resident neurol, 57-59; neurologist, Wilford Hall US Air Force Hosp, San Antonio, Tex, 60-62, chief neurol serv, 62-64; from asst prof to assoc prof, 64-70, chief div, 66-73, PROF NEUROL, MED BR, UNIV TEX, GALVESTON, 70-, CHMN DEPT, 73- Concurrent Pos: Asst examr, Am Bd Neurol & Psychiat, 65-; nat consult neurol, Surgeon Gen, US Air Force, 73- Mem: Am Neurol Asn; Asn Res Nerv & Ment Dis; Am Acad Neurol; Am Epilepsy Soc; Asn Univ Prof Neurol. Res: Medical education. Mailing Add: Dept of Neurol Univ of Tex Med Br Galveston TX 77550

CALVERT, ALLEN FISHER, b San Diego, Calif, Aug 1, 27; m 55; c 4. BIOCHEMICAL GENETICS. Educ: Univ San Francisco, BS, 52; Univ Calif, PhD(biochem), 62; Am Bd Clin Chem, dipl, 72. Prof Exp: Fel biochem, Univ Hawaii, 62-64; biochemist, Clin Invest Ctr, US Naval Hosp, Oakland, Calif, 64-65; biochemist, Med Genetics Unit, Col Med, 65-74, ASSOC PROF PEDIAT & CHILD HEALTH, DEPT PEDIAT, COL MED, HOWARD UNIV, 74- Mem: Am Soc Human Genetics; Am Asn Clin Chem. Res: Inborn errors of metabolism; screening for children's diseases of genetic origin; ultra micro methods; cyclic nucleotide metabolism in health and in disease. Mailing Add: Dept of Pediat Col of Med Howard Univ Washington DC 20059

CALVERT, DAVID VICTOR, b Chaplin, Ky, Feb 26, 34; m 57; c 2. SOIL FERTILITY. Educ: Univ Ky, BS, 56, MS, 58; Iowa State Univ, PhD(soil fertil), 62. Prof Exp: Asst soils chemist, Citrus Exp Sta, 62-67, ASSOC PROF SOIL FERTIL, AGR RES CTR, CITRUS DIV, UNIV FLA, 68- Concurrent Pos: Off collabr, Soil & Water Conserv Div, Agr Res Serv, USDA, 69-76; consult, Jamaican Sch Agr, 70. Mem: AAAS; Am Soc Agron. Res: Chemistry of soil phosphorus; soil-plant tracer studies with Nitrogen-fifteen isotope; soil chemistry-fertility, drainage and irrigation with citrus. Mailing Add: Agr Res Ctr Citrus Div Box 248 Univ of Fla Ft Pierce FL 33450

CALVERT, DEANE NESBITT, pharmacology, deceased

CALVERT, JACK GEORGE, b Inglewood, Calif, May 9, 23; m 46; c 2. PHOTOCHEMISTRY, PHYSICAL CHEMISTRY. Educ: Univ Calif, Los Angeles, BS, 44, PhD(chem), 49. Prof Exp: Nat Res Coun Can fel, 49-50; from asst prof to prof, 50-74, chmn dept, 64-68, KIMBERLY PROF CHEM, OHIO STATE UNIV, 74- Concurrent Pos: Chmn air pollution manpower develop comt, Nat Air Pollution Control Admin, 68; chmn, Conserv Found Air Qual Adv Comt, 69-; mem bd trustees, Gordon Res Cong, 69- Mem: AAAS; Am Chem Soc; fel Am Inst Chemists. Res: Atmospheric chemistry. Mailing Add: Dept of Chem Ohio State Univ Columbus OH 43210

CALVERT, JAMES BOWLES, b Columbus, Ohio, May 28, 35. OPTICAL PHYSICS. Educ: Mont State Col, BS, 56, MS, 59; Univ Colo, PhD(physics), 63. Prof Exp: Res physicist, 59-63, from lectr to asst prof, 62-66, ASSOC PROF PHYSICS, UNIV DENVER, 66- Mem: Am Inst Physics; Am Phys Soc; Am Asn Physics Teachers; Acoust Soc Am; Nat Soc Prof Engrs. Res: Energy transfer in molecular collisions; fluctuations in light beams; coherence theory. Mailing Add: Dept of Physics & Astron Univ of Denver Denver CO 80210

CALVERT, LAURISTON DERWENT, b Tasmania, Dec 10, 24; m 54; c 2. CHEMISTRY, CRYSTALLOGRAPHY. Educ: Univ NZ, BS, 46, BA, 48, MS, 49, PhD(crystal struct anal), 52. Prof Exp: Nat Res Coun fel, 52-54; from asst res officer to sr res officer, 54-73, PRIN RES OFFICER, NAT RES COUN CAN, 73- Concurrent Pos: Co-ed, Struct Reports; secy, Can Nat Comt Crystallog, 69- Res: Low and high temperature and high pressure x-ray diffraction; single-crystal structure analysis; inter-metallic compounds. Mailing Add: Nat Res Coun Can Ottawa ON Can

CALVERT, OSCAR HUGH, b Dallas, Tex, Oct 28, 18; m 44; c 4. PLANT PATHOLOGY. Educ: Okla State Univ, BS, 43; Univ Wis, MS, 45, PhD(plant path), 48. Prof Exp: Asst bot & plant path, Okla State Univ, 42-44; asst, Penicillin Proj, US Army & Dept Bot, Univ Wis, 44-45, asst, Dept Plant Path, 45-48; asst plant pathologist, Dept Plant Physiol & Path, Agr & Mech Col Tex, 48-52, plant pathologist, USDA, 52-54; consult plant path & nursery retail sales mgr, Lambert Landscape, 54-58; asst prof field crops, 58-67, from asst prof to assoc prof plant path, 67-75, PROF PLANT PATH, UNIV MO-COLUMBIA, 75- Concurrent Pos: Vis prof, Res Inst Plant Protection, Budapest, Hungary, 69-70. Mem: AAAS; Am Soc Agron; Mycol Soc Am; Am Phytopath Soc. Res: Diseases of field crops. Mailing Add: Dept of Plant Path Univ of Mo Columbia MO 65201

CALVERT, RALPH LOWELL, b Howard, Kans, Jan 10, 10; m 37; c 2. MATHEMATICS. Educ: Southwestern Col, Kans, AB, 37; Univ Ill, AM, 38, PhD(math, astron), 50. Prof Exp: From instr to assoc prof math & astron, Utah State Univ, 40-49; asst prof, Univ Wyo, 49-51; res mathematician, Sandia Corp, 51-73;

RETIRED. Res: Systems analysis. Mailing Add: 415 Adams St NE Albuquerque NM 87108

CALVERT, WYNNE, b Columbus, Ohio, Mar 4, 37; m 58; c 3. ATMOSPHERIC PHYSICS. Educ: Mont State Col, BS, 58; Univ Wis, MS, 59; Univ Colo, PhD(astrogeophys), 62. Prof Exp: Physicist, Nat Bur Stand, 60-62; asst prof physics, Mont State Col, 62-63; proj leader cent radio propagation lab, Nat Bur Stand, 63-65; chief rocket & satellite exp sect, Aeronomy Lab, 65-67, CONSULT TO AERONOMY LAB, ENVIRON RES LABS, NAT OCEANIC & ATMOSPHERIC ADMIN, 67- Mem: Int Sci Radio Union; Am Geophys Union; Sigma Xi. Res: Ionospheric physics and plasma physics. Mailing Add: 4686 Talbot Dr Boulder CO 80303

CALVIN, CLYDE LACEY, b Winlock, Wash, June 22, 34; m 60; c 2. PLANT ANATOMY. Educ: Wash State Univ, BS, 60; Purdue Univ, MS, 62; Univ Calif, Davis, PhD(bot), 66. Prof Exp: NIH fel, Univ Calif, Santa Barbara, 65-66; asst prof bot, Calif State Col, Long Beach, 66-67; assoc prof biol, Ore Col Educ, 67-68; ASSOC PROF BIOL, PORTLAND STATE UNIV, 68- Mem: Bot Soc Am; Int Soc Plant Morphol. Res: Plant anatomy and ultrastructure; tissue relations between dicotyledonous parasites and their hosts. Mailing Add: Dept of Biol Portland State Univ Portland OR 97207

CALVIN, LYLE D, b Nebr, Apr 12, 23; m 52; c 3. STATISTICS. Educ: Parsons Jr Col, AB, 43; NC State Univ, BS, 47, PhD, 53; Univ Chicago, BS, 48. Prof Exp: Asst, NC State Univ, 47-50, asst statistician, 52-53; biometrician, G D Searle & Co, 50-52; assoc prof, 53-57, PROF STATIST, ORE STATE UNIV, 57-, CHMN DEPT, 62-, STATISTICIAN, EXP STA, 53-, DIR, SURV RES CTR, 73- Concurrent Pos: Vis prof, Univ Edinburgh, 67 & Univ Cairo, 71-72; mem epidemiol & biomet training comt, Nat Inst Gen Med Sci, 68-72; mem Am Statist Asn comt, Bur of Census, 71-77. Mem: Fel AAAS; Biomet Soc (pres, NAm Region, 64-65); fel Am Statist Asn; Royal Statist Soc; Int Statist Asn. Res: Experimental design and analysis; sampling methods. Mailing Add: Dept of Statist Ore State Univ Corvallis OR 97331

CALVIN, MELVIN, b St Paul, Minn, Apr 8, 11; m 42; c 3. ORGANIC CHEMISTRY. Educ: Mich Col Mines, BS, 31; Univ Minn, PhD(chem), 35. Hon Degrees: DSc, Mich Col Mines, 55, Univ Nottingham, 58, Oxford Univ, 59, Northwestern Univ, 61, Wayne State Univ, 62, Gustavus Adolphus Col, 63, Univ Minn, 69 & State Agr Univ, Belg, 70; LLD, Univ Notre Dame, 65 & Whittier Col, 71. Prof Exp: Rockefeller grant, Univ Manchester, 35-37; instr chem, 37-40, from asst prof to prof, 41-71, UNIV PROF CHEM, UNIV CALIF, BERKELEY, 71-, PROF MOLECULAR BIOL, 63-, DIR BIO-ORG DIV, LAWRENCE RADIATION LAB, 46-, DIR LAB CHEM BIODYNAMICS, 60- Concurrent Pos: Off investr, Nat Defense Res Comt, Univ Calif, 41-44, Manhattan Dist Proj, 44-45; mem US deleg, Int Conf Peaceful Uses Atomic Energy, Geneva, Switz, 55; mem joint comt, Int Union Pure & Appl Chem, 56; chem adv comt, US Air Force Off Sci Res, 51-55; vchmn exec coun, Armed Forces-Nat Res Coun Comt Bio-Astronaut, 58-62; mem, President's Sci Adv Comt, 63-; bd sci counr, NIMH, 68-71, vchmn comt sci & pub policy, Nat Acad Sci, 71, chmn, 72-73. Honors & Awards: Sugar Res Found Prize, 50; Flintoff Medal & Prize, The Chem Soc, 55; Hales Award, Am Soc Plant Physiol, 56; Richards Medal, Am Chem Soc, 56, Award Nuclear Appln in Chem, 57, Nichols Medal, 58; Nobel Prize in Chem, 61; Davy Medal, Royal Soc, 64. Mem: Nat Acad Sci; AAAS; Am Chem Soc (pres-elect, 70, pres, 71); fel Am Phys Soc; Am Soc Biol Chem. Res: Physical chemistry; photosynthesis and chemical biodynamics; plant physiology; chemical evolution; molecular biology; chemical and viral carcinogenesis. Mailing Add: Lab of Chem Biodynamics Univ of Calif Berkeley CA 94720

CALVIN, WILLIAM HOWARD, b Kansas City, Mo, Apr 30, 39. NEUROPHYSIOLOGY, BIOPHYSICS. Educ: Univ Wash, PhD(physiol & biophys), 66. Prof Exp: Asst physics, Northwestern Univ, 61; from instr neurol surg & physiol-biophys to asst prof neurol surg, 67-73, ASSOC PROF NEUROL SURG, UNIV WASH, 74- Concurrent Pos: Nat Acad Sci travel grants to int cong, 66 & 71; NIH grant, 71. Mem: Biophys Soc; Am Physiol Soc; Soc Neurosci. Res: Neuronal processes involved in repetitive discharge, coding of information, integration at single cell level; neuronal mechanisms of epilepsy and central pain. Mailing Add: Dept of Neurol Univ of Wash Seattle WA 98195

CALVO, CRISPIN, b Paterson, NJ, Jan 24, 30; m 57; c 4. PHYSICAL CHEMISTRY. Educ: Rutgers Univ, BSc, 51, PhD(chem), 54. Prof Exp: Res assoc, Cornell Univ, 54-55; engr, David Sarnoff Labs, Radio Corp Am, NJ, 55-60; from asst prof to assoc prof, 60-68, PROF CHEM, McMASTER UNIV, 68- Mem: Am Phys Soc; Am Crystallog Asn; Can Asn Physicists. Res: X-ray crystallography and structure determination; electron spin resonance. Mailing Add: Dept of Chem McMaster Univ Hamilton ON Can

CAMACHO, ALVRO MANUEL, b Trinidad, WI, Mar 1, 27; US citizen; m 56; c 4. PEDIATRIC ENDOCRINOLOGY. Educ: Creighton Univ, 46-49, MD, 53. Prof Exp: Intern, Charity Hosp, New Orleans, La, 53-54; resident pediat, Med Br Hosps, Univ Tex, 55-58; fel pediat endocrinol, Children's Hosp, Ohio State Univ, 58-60; fel, Johns Hopkins Hosp, Baltimore, 60-61, instr pediat, 61-63; asst prof, Sch Med, Wayne State Univ, 63-64; from asst prof to assoc prof, 64-72, PROF PEDIAT, COL MED, UNIV TENN, MEMPHIS, 72-, CHIEF SECT PEDIAT ENDOCRINOL, 64- Honors & Awards: Lederle Med Fac Award, 66. Mem: Am Fedn Clin Res; Endocrine Soc; Soc Pediat Res; Am Pediat Soc. Mailing Add: Dept of Pediat Univ of Tenn Memphis TN 38103

CAMBEY, LESLIE ALAN, b Sydney Australia, May 15, 34; m 56; c 4. PHYSICS. Educ: Univ New SWales, BS, 55, PhD(physics), 60. Prof Exp: Fel physics, McMaster Univ, 60-62; sr res scientist, Nuclide Corp, Pa, 62-65, dir res, 65-67; mgr custom prod, CEC/Anal Instruments Div, Bell & Howell Co, Calif, 67-70; VPRES, MAT RES CORP, 70- Mem: Inst Elec & Electronic Engrs. Res: Atomic mass determinations; ion sputtering; ion optics; mass spectroscopy. Mailing Add: Mat Res Corp Orangeburg NY 10962

CAMBRAY, JOSEPH, b San Diego, Calif, Mar 11, 48. PHYSICAL ORGANIC CHEMISTRY, BIOORGANIC CHEMISTRY. Educ: Wayne State Univ, BS, 70; Univ Calif, Berkeley, PhD(chem), 75. Prof Exp: Anal technician chem, Uniroyal Inc, 68-70; FEL CHEM, STANFORD UNIV, 75- Mem: Am Chem Soc. Res: The study of synthetic model analogs of the prosthetic groups of metalloproteins, especially those involving non-heme bound iron. Mailing Add: Dept of Chem Stanford Univ Stanford CA 94305

CAME, PAUL E, b Dover, NH, Feb 25, 37; m 60; c 2. VIROLOGY, ONCOLOGY. Educ: St Anselm's Col, AB, 58; Univ NH, MS, 60; Hahnemann Med Col, PhD(microbiol), 64. Prof Exp: Instr microbiol, Med Units, Univ Tenn & res virologist, St Jude Hosp, 63-65; res assoc biophys cytol, Rockefeller Univ, 65-67; sr virologist, 67-70, mgr dept virol, 70-74, assoc dir microbiol sci & virol, Schering-Plough Corp, 74-75; HEAD DEPT MICROBIOL, STERLING-WINTHROP RES INST, 75- Mem: AAAS; NY Acad Sci; Am Soc Microbiol. Res: Mechanisms of cellular resistance to viruses; mechanisms of virus oncogenicity; chemotherapy of virus and microbial diseases. Mailing Add: Dept of Microbiol Sterling-Winthrop Res Inst Rensselaer NY 12144

CAMERINI-DAVALOS, RAFAEL, b Buenos Aires, Arg, July 3, 16; US citizen; m 43; c 4. MEDICINE. Educ: Univ Buenos Aires, MD, 43, DSc(nutrit), 45; Arg Dept Health, dipl, 47. Prof Exp: Physician, Nat Inst Nutrit, Arg, 42-51; chief dept endocrinol & metab, Nat Inst Allergy, Arg, 5055; resident physician, Joslin Serv, New Eng Deaconess Hosp, 55-57; chief diabetes sect, Inst Endocrinol, NIH, Arg, 58-59; res assoc, Baker Clin Lab & Dept Med, Harvard Med Sch, 59-64; assoc prof, 64-69, PROF MED, NY MED COL, 69-, DIR DIABETES RES UNIT & CHIEF DIABETES RES, MATERNAL-CHILD HEALTH CTR, 64- Concurrent Pos: Lilly foreign fel, 55-56; fel, Joslin Serv, New Eng Deaconess Hosp, Boston, 55-57; travel fel, Univ Heidelberg, 57; Health Res Coun career scientist award, 65; from instr to assoc prof, Univ Buenos Aires, 45-59. Mem: Hon mem Arg Med Asn & Span Diabetes Asn; Colombian Med Soc. Res: Early changes in endocrine-metabolic diseases, especially diabetes mellitus. Mailing Add: Dept of Med NY Col of Med New York NY 10029

CAMERINO, PAT WILLIAM, b Niles, Ohio, May 25, 35; m 58; c 3. BIOCHEMISTRY, ACADEMIC ADMINISTRATION. Educ: Kent State Univ, BS in Ed & BS(chem), 57; Cornell Univ, PhD(biochem), 61. Prof Exp: USPHS trainee biochem, Dartmouth Med Sch, 61-62, instr, 62-63; asst prof chem, Sci Res Inst, Ore State Univ, 63-65; grants assoc, Div Res Grants, NIH, 65-66, asst endocrinol prog dir, Nat Inst Arthritis & Metab Dis, 66-67, chief anal & eval br, 67-69, asst dir, Div Res Resources, NIH, 69-70; assoc grad dean for res, 70-75, DIR OFF GRANT & CONTRACT ADMIN, UNIV MASS, AMHERST, 75- Res: Respiratory enzymology and energy metabolism; science administration. Mailing Add: Grad Res Ctr Univ of Mass Amherst MA 01002

CAMERMAN, NORMAN, b Vancouver, BC, Apr 12, 39. MOLECULAR BIOLOGY, X-RAY CRYSTALLOGRAPHY. Educ: Univ BC, BSc, 61, PhD(chem), 64. Prof Exp: Nat Res Coun Can overseas fel, Royal Inst Gt Brit, 64-66; Med Res Coun Can prof asst, 67-69, ASST PROF BIOCHEM, UNIV TORONTO, 69- Concurrent Pos: Med Res Coun Can scholar, 69- Mem: AAAS; Am Crystallog Asn. Res: Molecular basis of biological activity; molecular structure determinations of drugs, hormones and other biologically-active substances; correlation of structure and actions. Mailing Add: Dept of Biochem Univ of Toronto Toronto ON Can

CAMERON, ALEXANDER MENZIES, b Yonkers, NY, Sept 5, 30; m 52; c 3. VETERINARY PATHOLOGY. Educ: Colo State Univ, BS, 61, DVM, 62; Univ Calif, Davis, PhD(comp path), 70. Prof Exp: Vet clinician small animals, pvt pract, NMex, 62-63 & Calif, 64-66; field vet dis eradication, USDA, Calif, 63-64; asst prof vet path, Col Vet Med, Univ Ill, 70-76; MEM STAFF PATH SERV PROJ, NAT CTR TOXICOL RES, 76- MED, UNIV ILL, 70- Mem: Am Col Vet Pathologists; Am Vet Med Asn; Int Acad Path; Soc Pharmacol & Environ Pathologists; Am Asn Lab Animal Sci. Res: Development of mammary neoplasms in animals. Mailing Add: Path Serv Proj Nat Ctr Toxicol Res Mail Code 16 Jefferson AR 72079

CAMERON, ANGUS EWAN, b Sylvania, Pa, Oct 14, 06; m 33; c 3. MASS SPECTROMETRY. Educ: Oberlin Col, AB, 28; Univ Minn, PhD(phys chem), 32. Prof Exp: Nat Res fel chem, Univ Rochester, 32-33; res chemist, Eastman Kodak Co, 33-38, Eisendrath Mem Lab, 38-41 & Indust Tape Corp, 41-42; res chemist, Eastman Kodak Co, 42-43, dept supt, Clinton Eng Works, Tenn Eastman Corp, 43-47; lab supvr, Distillation Prod, Inc, 47-48; res chemist, Carbide & Carbon Chem Co, 48; physicist, Oak Ridge Gaseous Diffusion Plant, 48-54; Fulbright res fel, Max-Planck Inst Chem, Ger, 54-55; dept head, stable isotopes div, Oak Ridge Nat Lab, 55-57, sect chief, Anal Chem Div, 57-60, asst dir, 60-71; RETIRED. Mem: Am Chem Soc. Res: Adsorption; photographic theory; electrometric studies of solutions; mass spectroscopy; high vacuum application; isotope separation. Mailing Add: 114 W Malta Rd Oak Ridge TN 37830

CAMERON, AUSTIN WEST, vertebrate zoology, see 12th edition

CAMERON, BARRY WINSTON, b NS, Feb 7, 40; US citizen; m 68. PALEOECOLOGY. Educ: Rutgers Univ, AB, 62; Columbia Univ, AM, 65, PhD(geol), 68. Prof Exp: Teaching asst geol, Columbia Univ, 62-64; from instr to asst prof geol, 67-73, asst prof biol, 69-73, ASSOC PROF BIOL, BOSTON UNIV, 73- Concurrent Pos: NSF grant, 70-72. Mem: AAAS; Paleont Soc; Geol Soc Am; Soc Econ Paleont & Mineral; Brit Paleont Asn. Res: Trace fossils, especially algal, sponge and worm borings; calcareous algae; paleoecology, sedimentation and stratigraphy of medial Ordovician limestones of New York and Ontario; evolution of benthonic marine invertebrate fossil communities; geologic computer applications. Mailing Add: Boston Univ Dept of Biol 725 Commonwealth Ave Boston MA 02215

CAMERON, BRUCE FRANCIS, b Damariscotta, Maine, Sept 10, 34; m 57; c 2. PHYSICAL BIOCHEMISTRY, HEMATOLOGY. Educ: Harvard Univ, BS, 56; Univ Pa, MD, 60, PhD(biochem), 62. Prof Exp: Vis res assoc chem, Am Univ Beirut, 62-63; res assoc, Univ Ibadan, 63-65; sr staff scientist, New Eng Inst Med Res, 66-69; ASSOC PROF MED, SCH MED, UNIV MIAMI, 69-; SR SCIENTIST, PAPANICOLAOU CANCER RES INST, 71- Concurrent Pos: NIH fel, 60-63; Am Heart Asn advan res fel, 67-69; vis scientist, Am Univ Beirut, 65; investr, Howard Hughes Med Inst, 69-71; adj prof life sci, New Eng Inst Grad Sch, 71-74; estab investr, Am Heart Asn, 72- Mem: AAAS; Am Chem Soc; fel Am Inst Chemists; NY Acad Sci; Am Soc Biol Chemists. Res: Physical chemistry of hemoglobin; thermodynamics of reactions of abnormal human hemoglobins; studies in sickle cell anemia; biophysical instrumentation with minicomputer systems. Mailing Add: Papanicolaou Cancer Res Inst PO Box 236188 Miami FL 33123

CAMERON, CHARLES W, animal husbandry, see 12th edition

CAMERON, COLIN ROBERT, b Toronto, Ont, Mar 31, 15; m 50; c 2. PHYSIOLOGY, BIOCHEMISTRY. Educ: Univ Toronto, BA, 49, DVM, 59; Ont Agr Col, MSA, 51. Prof Exp: Asst nutrit, Ont Agr Col, 51; from asst prof to assoc prof physiol, Ont Vet Col, 52-66; SCI ADV DIV VET MED, FOOD & DRUG DIRECTORATE, CAN DEPT HEALTH & WELFARE, 66- Concurrent Pos: Res asst, Hosp for Sick Children, Toronto, 52- Mem: Can Biochem Soc; Can Soc Clin Chem. Res: Nutrition cholesterol; protein and fat metabolism; immunology. Mailing Add: 175 Clearview Ave Ottawa ON Can

CAMERON, DALE CORBIN, b Hendley, Nebr, July 10, 12; m 36; c 2. PSYCHIATRY. Educ: Univ Nebr, AB, 33, MD, 36; Johns Hopkins Univ, MPH, 51. Prof Exp: Intern, USPHS Hosp, San Francisco, 36-37, med officer, Evansville, Ind, 37-38, med officer, New Orleans, 38-39; USPHS resident psychiat, Univ Colo, 39-40; clin dir, USPHS Hosp, Ft Worth, 40-42, exec officer, 42-43; med officer, US Coast Guard Acad, 43-45; chief psychiat serv, USPHS Hosp Ellis Island, NY, 45; asst chief, NIMH, 45-50; chief coop health serv-indust hyg, USPHS, Washington, DC, 51-53, med consult, Off Health Emergency Planning, 53-54; med dir, Minn State Dept Pub

Welfare, 54-60; med officer, NIMH, 60; asst supt med, St Elizabeths Hosp, Washington, DC, 60-62, supt, 62-67; chief, Drug Dependence, WHO, 67-74; RETIRED. Concurrent Pos: Mem comt probs drug dependence, Nat Acad Sci-Nat Res Coun, 47-48 & 60-61, chmn, 61-67; clin prof psychiat & neurol, Sch Med, Univ Minn, 54-60; clin prof psychiat, George Washington Univ, 60-74; mem expert adv panel ment health, WHO, 66-67. Honors & Awards: Outstanding Achievement Award, Alcohol & Drug Prob Asn NAm, 72. Mem: Fel Am Psychiat Asn (treas, 63-68); AMA. Res: Drug dependence. Mailing Add: 12546 Plaza Centrada San Diego CA 92128

CAMERON, DAVID GLEN, b Great Falls, Mont, Dec 6, 34; m 61; c 2. POPULATION GENETICS, EVOLUTIONARY BIOLOGY. Educ: Princeton Univ, AB, 57; Stanford Univ, PhD(biol), 66. Prof Exp: Res assoc environ med, US Army Med Res Labs, Ft Knox, Ky, 57-59; asst prof genetics, Univ Alta, 64-69; assoc prof zool & head dept zool & entom, 70-74, ASSOC PROF BIOL & GENETICS, MONT STATE UNIV, 74- Concurrent Pos: Secy & trustee, Rocky Mountain Biol Labs, Crested Butte, Colo, 68-74. Mem: AAAS; Am Inst Biol Sci; Genetics Soc Am; Soc Study Evolution; Am Soc Human Genetics. Res: Ecological, populational, genetic and evolutionary aspects of genetic polymorphisms in natural populations of small rodents, European rabbits, elk and trout; blood groups serum proteins isoenzymes; human genetics. Mailing Add: Dept of Biol Mont State Univ Bozeman MT 59715

CAMERON, DONALD EUGENE, b Sayre, Pa, Oct 1, 22; m 44; c 5. FOOD MICROBIOLOGY. Educ: Cornell Univ, BS, 43. Prof Exp: Proj leader microbiol, Cent Labs, Gen Foods Corp, 43-53, prod & process develop, 53-59, qual control mgr, Birds Eye Baby Foods, 59-62, technologist, Jell-O Plant, 62-64, bacteriologist, Tech Serv, 64-66, asst qual control supvr, 66-67, QUAL CONTROL SUPT BACT, JELL-O PLANT, GEN FOODS CORP, 68- Mem: Inst Food Technol; Am Soc Microbiol. Res: Processing of raw cacao and green coffee; microbiology of food plant processes; new product development. Mailing Add: Dover Oper Gen Foods Corp PO Box 600 Dover DE 19901

CAMERON, DONALD FORBES, b Edmonton, Alta, Aug 19, 20; m 44; c 4. ANESTHESIOLOGY. Educ: Univ Alta, BA, 47, MD, 49; FRCPS(C) cert. Prof Exp: Lectr pharmacol, 50-62, asst prof pharmacol & anesthesia & asst dean fac med, 62-65, assoc dean, 65-74, assoc prof, 65-73, PROF ANESTHESIA, UNIV ALTA, 73-, DEAN FAC MED, 74- Concurrent Pos: Mem, Med Coun Can, 62-, pres, 74-75; mem bd gov, Univ Hosp, 74-; mem gen coun, Can Med Asn, 75- Honors & Awards: Order of Brit Empire. Mem: Fel Am Col Anesthesiol. Mailing Add: Fac of Med Univ of Alta Edmonton AB Can

CAMERON, DONALD PETER, organic chemistry, see 12th edition

CAMERON, DOUGLAS EWAN, b New Brunswick, NJ, Apr 5, 41; div; c 2. TOPOLOGY. Educ: Miami Univ, BA, 63; Univ Akron, MS, 65; Va Polytech Inst & State Univ, PhD(topol), 70. Prof Exp: Instr math, Community & Tech Col, Univ Akron, 65-66 & Va Polytech Inst & State Univ, 66-69; from instr to asst prof, 69-74, ASSOC PROF MATH, UNIV AKRON, 74- Mem: Am Math Soc; Math Asn Am. Res: Properties of maximal and minimal topologies, specifically covering axioms; filters and their use in the study of point set topology. Mailing Add: Dept of Math & Statist Univ of Akron Akron OH 44325

CAMERON, DOUGLAS GEORGE, b Eng, Mar 11, 17; m 46; c 6. MEDICINE. Educ: Univ Sask, BSc, 37; McGill Univ, MD & CM, 40; Oxford Univ, Prof Exp: Res asst, Nuffield Dept Clin Med, Oxford Univ, 48; sr med res fel, Nat Res Coun Can, 49-52; assoc physician, 51-57, asst dir, 52-57, DIR & PHYSICIAN-IN-CHIEF, UNIV CLIN, MONTREAL GEN HOSP, 57-; PROF MED, FAC MED, McGILL UNIV, 57-, CHMN DEPT, 74- Concurrent Pos: Asst prof med, McGill Univ, 49-57, chmn dept, 64-69. Mem: Fel Am Col Physicians; Am Soc Clin Invest; Asn Am Physicians; Am Clin & Climat Asn. Mailing Add: 227 Portland Ave Montreal PQ Can

CAMERON, DUNCAN MACLEAN, JR, b St Albans, Vt, May 15, 31; m 55; c 3. ZOOLOGY, ECOLOGY. Educ: Univ Maine, BS, 54, MEd, 60; Univ Calif, Davis, PhD, 69. Prof Exp: Teacher, Pub Schs, Maine & Ohio, 56-61; instr bot, Univ Maine, 61-62; res zoologist, Univ Calif, Davis, 62-64, teaching asst zool, 64-66, assoc, 66-67, lectr, 67-68; fel, Univ BC, 68-69; asst prof biol, 69-73, ASSOC PROF BIOL, YORK UNIV, 73- Mem: AAAS; Am Inst Biol Sci; Ecol Soc Am; Am Soc Mammal; Can Soc Zool. Res: Population ecology of small mammals; ecological processes, especially competition; human ecological systems. Mailing Add: Dept of Biol York Univ Toronto ON Can

CAMERON, EDWARD ALAN, b Brockville, Ont, July 4, 38; m 64; c 2. FOREST ENTOMOLOGY. Educ: Ont Agr Col, BSA, 60; Univ Calif, Berkeley, MS, 67, PhD(entom), 74. Prof Exp: Entomologist, Commonwealth Inst Biol Control, 60-65; asst prof, 70-75, ASSOC PROF ENTOM, PA STATE UNIV, UNIVERSITY PARK, 75- Concurrent Pos: Collabr, USDA, 71- Mem: Entom Soc Can; Entom Soc Am; Am Inst Biol Sci; Int Orgn Biol Control. Res: Bionomics, ecology and control of forest insects; gypsy moth management, especially through disruption of adult chemical communication. Mailing Add: Dept of Entom Pa State Univ University Park PA 16802

CAMERON, EDWARD ALEXANDER b Manly, NC, Nov 10, 07; m 37; c 1. MATHEMATICS. Educ: Univ NC, AB, 28, AM, 29, PhD(math), 36. Prof Exp: From instr to prof, 29-72, EMER PROF MATH, UNIV NC, CHAPEL HILL, 72- Concurrent Pos: Ford Found fel, 51-52; NSF sci fac fel, 65-66; consult, NSF. Mem: Am Math Soc; Math Asn Am (treas, 68-72). Res: Algebra. Mailing Add: 404 Laurel Hill Rd Chapel Hill NC 27514

CAMERON, EUGENE NATHAN, b Atlanta, Ga, Aug 10, 10; m; c 3. ECONOMIC GEOLOGY. Educ: NY Univ, BS, 32; Columbia Univ, AM, 34, PhD(geol), 39. Prof Exp: Asst geol, NY Univ, 30-35; asst, Columbia Univ, 36-37, lectr, 37-39, instr, 39; assoc geologist, US Geol Surv, 42-44, geologist, 46, sr geologist, 46-50; from assoc prof to prof econ geol, 47-70, chmn dept geol, 55-60, VAN HISE DISTINGUISHED PROF ECON GEOL, UNIV WIS-MADISON, 70- Concurrent Pos: Geologist, US Geol Surv, 37-39; asst, State Geol & Natural Hist Surv, Conn, 41-42; consult, NASA, 62-63 & 64-67; deleg, Int Comn Ore Micros, 62-64; mem, Comt Mineral Resources & Environ, 73-75; mem, US Nat Comt Geol, 74-77. Mem: Fel Geol Soc Am; fel Mineral Soc Am; Soc Econ Geologists (secy, 61-69, vpres, 69, pres, 74); Am Inst Mining, Metall & Petrol Engrs; Geol Soc S Africa. Res: Mineralogy; economic geology of pegmatite mineral deposits; internal structure of granitic pegmatites; optical properties of ore minerals; chromite deposits; lunar samples; investigation of origin of chromite deposits in the Bushveld Complex; analysis of mineral resource problems. Mailing Add: Dept of Geol & Geophys Univ of Wis Madison WI 53706

CAMERON, GEORGE HARVEY, b Streetsville, Ont, Sept 8, 02; m 28; c 3. PHYSICS. Educ: Univ Sask, BSc, 22; Calif Inst Technol, PhD(physics), 26. Prof Exp: Res assoc, Calif Inst Technol, 26-28; physicist, E I du Pont de Nemours & Co, 28-32; from assoc prof to prof, 32-72, EMER PROF PHYSICS, HAMILTON COL, 72- Concurrent Pos: Instr, US Army Univ (France), 45-46; consult, Speech & Hearing Ctr, Utica, 50-60. Res: Cosmic rays; x-ray diffraction and particle size. Mailing Add: 299 College Hill Rd Clinton NY 13323

CAMERON, GUY NEIL, b San Francisco, Calif, May 1, 42. ECOLOGY, POPULATION BIOLOGY. Educ: Univ Calif, Berkeley, BA, 63; Calif State Col, Long Beach, MA, 65; Univ Calif, Davis, PhD(zool), 69. Prof Exp: Asst res zoologist, Univ Calif, Berkeley, 69-71; ASST PROF BIOL, UNIV HOUSTON, 71- Concurrent Pos: Consult, Gulf Univs Res Consortium, 73, Dames & Moore Environ Engrs, 74 & 75. Mem: Ecol Soc Am; Am Soc Naturalists; Am Soc Mammalogists; Brit Ecol Soc; Soc Study Evolution. Res: Investigation of role of interspecific interactions in demographic and genetic aspects of natural populations by experimental perturbation; contribution of decomposer organisms to community dynamics; stability of arthropod food chains. Mailing Add: Dept of Biol Univ of Houston Houston TX 77004

CAMERON, H RONALD, b Oakland, Calif, June 30, 29; m 57; c 2. PLANT PATHOLOGY. Educ: Univ Calif, BS, 51; Univ Wis, PhD(plant path), 55. Prof Exp: Agt, USDA, Univ Wis, 53-55; from asst plant pathologist to assoc prof pathologist, 55-70, PLANT PATHOLOGIST, ORE STATE UNIV, 70- Concurrent Pos: NSF fel, Stanford Univ, 62-63; NATO sr fel, East Malling Res Sta, Eng, 72. Res: Virus and bacterial diseases of fruit and nut trees; genetics of pathogenicity. Mailing Add: Dept of Bot & Plant Path Ore State Univ Corvallis OR 97331

CAMERON, IRVINE R, b Fredericton, NB, June 24, 19; m 45; c 2. POLYMER CHEMISTRY. Educ: Univ NB, BS, 41; Univ Toronto, MS, 42. Prof Exp: From sci officer to dir eng planning & coord, Chief Tech Serv Br, Can Forces Hq, 70-72; DIR GEN MAT DEVELOP PLANNING, DEPT NAT DEFENCE, CAN, 72- Mem: Fel Chem Inst Can; assoc fel Can Aeronaut & Space Inst; Am Inst Aeronaut & Astronaut; Prof Inst Pub Serv Can; Can Soc Chem Eng. Res: High explosives; propellants for guns and rockets; development of composite rocket propellants of polyurethane type resulting in the Black Brant series of sounding rockets; equipment development; military applications. Mailing Add: 4 Riopelle Ct Kanata ON Can

CAMERON, IVAN LEE, b Los Angeles, Calif, Jan 4, 34; m 55; c 3. ZOOLOGY, ANATOMY. Educ: Univ Redlands, BS, 56; Univ Southern Calif, MS, 59; Univ Calif, Los Angeles, PhD(anat), 63. Prof Exp: Biologist, Oak Ridge Nat Lab, 6365; asst prof anat, State Univ NY Upstate Med Ctr, 65-68; ASSOC PROF ANAT, UNIV TEX HEALTH SCI CTR, SAN ANTONIO, 68- Mem: Am Asn Anatomists; Am Soc Cell Biol. Res: Nuclear-cytoplasmic environmental interactions; chemical changes during the cell-growth-duplication cycle; feedback control of cellular proliferation. Mailing Add: Dept of Anat Univ of Tex Health Sci Ctr San Antonio TX 78284

CAMERON, J A, b Pittsburg, Okla, Oct 7, 29; m 53; c 2. MEDICAL MICROBIOLOGY. Educ: Maryville Col, BS, 51; Univ Tenn, MS, 55, PhD, 58. Prof Exp: Asst, Univ Tenn, 56-58; from instr to assoc prof bact, 58-73, PROF, BIOL SCI GROUP, UNIV CONN, 73- Concurrent Pos: Res assoc, Calif Inst Technol, 68-69. Mem: AAAS; Am Soc Microbiol; Soc Gen Microbiol. Res: Host parasite relationships in bacterial diseases. Mailing Add: Microbiol Sect Biol Sci Group Univ of Conn Storrs CT 06268

CAMERON, JAMES N, b Hanover, NH, May 28, 44; m 63; c 2. PHYSIOLOGY, ECOLOGY. Educ: Univ Wis, Madison, BS, 66; Univ Tex, Austin, PhD(zool), 69. Prof Exp: NIH fel, Univ BC, 69-73; ASSOC PROF ZOOPHYSIOL, UNIV ALASKA & INST ARCTIC BIOL, 73- Res: Ecology of warm water fishes; physiology of respiration; response to anemia; respiratory control. Mailing Add: Dept of Zool Univ of Alaska Fairbanks AK 99701

CAMERON, JAMES WAGNER, b Falls City, Nebr, Apr 23, 13; m 38; c 2. PLANT GENETICS. Educ: Univ Mo, AB, 38; Harvard Univ, AM, 43, PhD(genetics), 47. Prof Exp: Asst genetics, Harvard Univ, 41-42; res assoc, Off Sci Res & Develop Contract, 43-45; asst genetics, Harvard Univ, 45-46; from jr geneticist to assoc geneticist, Citrus Res Ctr, 47-60, lectr hort sci, Univ, 60-64, GENETICIST, CITRUS RES CTR, UNIV CALIF, RIVERSIDE, 60-, PROF HORT SCI, UNIV, 64- Concurrent Pos: Sr scholar, East-West Ctr, Univ Hawaii, 62-63; Fulbright lectr, Aegean Univ, Turkey, 70-71. Mem: Am Soc Agron; Am Soc Hort Sci; Sigma Xi; Am Genetic Asn. Res: Breeding and genetics of citrus and maize. Mailing Add: Dept of Plant Sci Univ of Calif Riverside CA 92502

CAMERON, JAMES (WILLIAM) MACBAIN, entomology, see 12th edition

CAMERON, JOHN ALEXANDER, b Toronto, Ont, May 19, 36; m 58; c 3. NUCLEAR PHYSICS. Educ: Univ Toronto, BA, 58; McMaster Univ, PhD(physics), 62. Prof Exp: NATO fel nuclear orientation, Clarendon Lab, Oxford Univ, 62-64; from asst prof to assoc prof, 64-74, PROF PHYSICS & ASSOC DEAN SCH GRAD STUDIES, McMASTER UNIV, 74- Concurrent Pos: Sloan Found fel, 67-68; vis scientist, Cent Res Inst Physics, Hungary, 70; mem fac sci, Univ Paris, 71. Mem: Can Asn Physicists. Res: Atomic beams; nuclear orientation; angular correlation studies; hyperfine structure in atoms and solids; nuclear moments. Mailing Add: Dept of Physics McMaster Univ Hamilton ON Can

CAMERON, JOHN RODERICK, b Wis, Apr 21, 22; m 47; c 2. MEDICAL PHYSICS. Educ: Univ Chicago, BS, 47; Univ Wis, MS, 49, PhD(nuclear physics), 52. Prof Exp: Asst prof, Univ Sao Paulo, 52-54; proj assoc, Univ Wis, 54-55; asst prof nuclear physics, Univ Pittsburgh, 55-58; from asst prof to assoc prof, 58-65, PROF RADIOL & PHYSICS, UNIV WIS-MADISON, 65- Concurrent Pos: Secy-gen, Int Orgn Med Physics, 69-76. Mem: Am Phys Soc; Am Asn Physicist in Med (pres, 68); Soc Nuclear Med; Health Physics Soc; Radiation Res Soc. Res: Radiation dosimetry; physics of diagnostic radiology; physics of bones; general applications of physics to medicine. Mailing Add: 118 N Breese Terr Madison WI 53705

CAMERON, JOSEPH MARION, b Edinboro, Pa, Apr 6, 22; m 46; c 4. MATHEMATICAL STATISTICS. Educ: Univ Akron, BS, 42; NC State Col, MS, 47. Prof Exp: Math statistician, 47-63, chief statist eng lab, 63-68, CHIEF OFF MEASUREMENT SERV, INST BASIC STANDARDS, NAT BUR STANDARDS, 68- Mem: AAAS; Math Asn Am; Inst Math Statist; fel Am Statist Asn; Biomet Soc. Res: Application of statistics in physical sciences. Mailing Add: Inst for Basic Standards Nat Bur of Standards Washington DC 20234

CAMERON, LOUIS MCDUFFY, b Richmond, Va, Dec 16, 35; m 59. APPLIED PHYSICS. Educ: Univ Richmond, BS, 57; George Washington Univ, MS, 62; Georgetown Univ, PhD(physics), 66. Prof Exp: Physicist, NIH, 58-61 & Naval Res Lab, Washington, DC, 61-66; physicist, 66-70, SUPVRY PHYSICIST, NIGHT VISION LAB, FT BELVOIR, VA, 70- Concurrent Pos: Edison Mem training scholar, Naval Res Lab, Washington, DC, 64-65. Honors & Awards: Ann Award, Night Vision Lab, 75; Res & Develop Achievement Award, US Army, 75. Mem: Am Phys Soc. Res: Liquid crystal physics; triton induced nuclear reactions employing a 2 MV Nan De Graaff accelerator; basic research including thermodynamics, light scattering, x-ray, microscope; development of a set of common modules utilized in infrared thermal imaging systems for all of the Department of Defense. Mailing Add: Night Vision Lab Ft Belvoir VA 22060

CAMERON, MALCOLM LAURENCE, b NS, Oct 23, 18; m 52; c 3. PHYSIOLOGY. Educ: Dalhousie Univ, BSc, 50, MSc, 51; Cambridge Univ, PhD(insect physiol), 53. Prof Exp: Nat Res Coun Can fel biol, Univ NB, 53-55; from asst prof to assoc prof, Univ Sask, 55-65; assoc prof, 65-67, PROF BIOL, DALHOUSIE UNIV, 67-, ASST DEAN ARTS & SCI, 69- Res: Insect hormones, enzymes and virology; excretory processes; invertebrate fine structure. Mailing Add: Dept of Biol Dalhousie Univ Halifax NS Can

CAMERON, MARGARET DAVIS, b Montgomery, Tex, Apr 25, 20. ORGANIC CHEMISTRY. Educ: Tex Women's Univ, BA, 42; Univ Houston, MS, 48; Tulane Univ, PhD(chem), 51. Prof Exp: Fulbright scholar & Ramsay mem fel, Univ Leeds, 51-52; USPHS fel, Ohio State Univ, 52-53; res chemist, Monsanto Chem Co, 53-56; from assoc prof to prof chem, 56-72, REGENTS PROF CHEM, LAMAR UNIV, 72-, HEAD DEPT, 74- Mem: Am Chem Soc; The Chem Soc. Res: Chemistry of acetylenic and heterocyclic compounds; chelation of metal ions by organic ligands. Mailing Add: 4060 Howard St Beaumont TX 77705

CAMERON, ROBERT ALAN, b Edmonton, Alta, Oct 19, 26; m 52; c 3. EXPLORATION GEOLOGY. Educ: Dalhousie Univ, BSc, 48; Univ Toronto, MASc, 53; McGill Univ, PhD(geol), 56. Prof Exp: Lectr geol, Univ NB, 52; geologist, Imperial Oil Ltd, Alta, 52-53 & Malartic Gold Fields Ltd, Que, 56-57; chief geologist, E Malartic Mines Ltd, 57-62; area supvr, McIntyre Porcupine Mines Ltd, 62; asst prof mining, NS Tech Col, 62-68; prin, Sudbury Campus, Cambrian Col, 68-69; ASSOC PROF GEOL, LAURENTIAN UNIV, 69- Mem: Fel Geol Asn Can; Can Inst Mining & Metall; Can Geophys Union. Res: Geological and geophysical exploration methods. Mailing Add: 1818 Hawthorne Dr E Sudbury ON Can

CAMERON, ROBERT HORTON, b Brooklyn, NY, May 17, 08; m 31; c 2. MATHEMATICS. Educ: Cornell Univ, AB, 29, AM, 30, PhD(math), 32. Prof Exp: Instr math, Cornell Univ, 29-33; Nat Res fel, Brown Univ, Princeton Univ & Inst Advan Study, 33-35; from instr to assoc prof, Mass Inst Technol, 35-45; prof, 45-74, chmn dept, 57-63, EMER PROF MATH, UNIV MINN, MINNEAPOLIS, 74- Concurrent Pos: Mem panel, Off Sci Res & Develop Proj, Appl Math Group, NY Univ, 44-45. Honors & Awards: Chauvenet Prize, Math Asn Am, 44. Mem: AAAS; Am Math Soc; Math Asn Am. Res: Mathematical analysis; almost periodic functions and transformations; partial differential equations; integration in function space; non-linear integral equations. Mailing Add: 3519 Stinson Blvd NE Minneapolis MN 55418

CAMERON, ROY (EUGENE), b Denver, Colo, July 16, 29; m 56; c 2. SOIL MICROBIOLOGY, ECOLOGY. Educ: Wash State Univ, BS, 53 & 54; Univ Ariz, MS, 58, PhD(plant sci), 61. Prof Exp: Res chemist, Hughes Aircraft Co, 55-56; sr scientist, Jet Propulsion Lab, Calif Inst Technol, 61-68, mem tech staff, 68-73; pres, Darwin Res Inst, Calif, 73-74; DEP DIR LAND RECLAMATION LAB, ENVIRON IMPACT STUDIES DIV, ARGONNE NAT LAB, 75- Concurrent Pos: Consult, Med Sch, Baylor Univ, 66-68; Jet Propulsion Lab team leader, Antarctic Exped, 66-74; vis scientist, Am Soc Agron, 67-69; prin investr, NSF res grant Antarctic microbial ecol, 69-74; Smithsonian Inst study grant Moroccan desert, 69-70; consult, Ecol Ctr, Utah State Univ, 70-72. Mem: AAAS; Am Soc Agron; Soil Sci Soc Am; Am Inst Biol Sci; Ecol Soc Am. Res: Desert soil microbiology; blue-green algae; microbial ecology; soil science; polar ecology; environmental monitoring; coal mine reclamation. Mailing Add: Argonne Nat Lab 9700 S Cass Ave Argonne IL 60439

CAMERON, SIDNEY HERBERT, b Hunmanby, Eng, Sept 18, 97; nat US; m 26; c 3. HORTICULTURE. Educ: Univ Calif, BS, 22, MS, 23, PhD(plant physiol), 27. Prof Exp: Instr subtrop hort, 23-27, from asst prof to prof, 29-65, chmn dept, 45-60, dean agr, 61-65, EMER PROF SUBTROP HORT & EMER DEAN AGR, UNIV CALIF, LOS ANGELES, 65- Concurrent Pos: Assoc dir, Calif Agr Exp Sta, 61-65; consult, Admin Lab Nuclear Med & Radiation Biol, Univ Calif, Los Angeles, 65- Mem: Am Soc Hort Sci; Am Soc Plant Physiol; Bot Soc Am. Res: Carbohydrate and nitrogen cycles in citrus trees; plant propagation; pruning of evergreen fruit trees. Mailing Add: 2404 Loring St San Diego CA 92109

CAMERON, THOMAS BROWN, inorganic chemistry, deceased

CAMERON, WILLIAM MAXWELL, b Battleford, Sask, July 24, 14; m 61; c 2. PHYSICAL OCEANOGRAPHY. Educ: Univ BC, BA & MA, 40; Univ Calif, PhD(oceanog), 51. Prof Exp: Sci asst biol, Pac Biol Sta, 38-41; meteorologist, Meteorol Serv Can, 41-44; assoc scientist, Pac Oceanog Group, 48-49; defense res serv officer, Defence Res Bd, 49-56, dir plans, 58-60, dir oceanog res, Dept Mines & Tech Surv, 60-62, dir marine sci, 62-67, dir, Dept Energy, Mines & Resources, 68-69, dir marine sci, Dept Environ, 69-71; PRES, CAMARINE CONSULT LTD, 74- Concurrent Pos: From hon assoc prof to hon prof, Inst Oceanog, Univ BC, 49-55; dir sci serv, Royal Can Navy, 56-58. Mem: Am Meteorol Soc; Am Soc Limnol & Oceanog; Am Geophys Union; Marine Biol Asn UK. Res: Estuarine dynamics; arctic oceanography. Mailing Add: 4448 Ross Crescent West Vancouver BC Can

CAMERON, WINIFRED SAWTELL, b Oak Park, Ill, Dec 3, 18; m 53; c 2. ASTRONOMY. Educ: Northern Ill Univ, BE, 40; Ind Univ, MA, 52. Prof Exp: Asst astron, Weather Forecasts, Inc, 42-46 & 49-50; instr, Mt Holyoke Col, 50-51; astron res asst, US Naval Observ, 51-58; aerospace technologist, Lab Space Physics, 59-71, AEROSPACE TECHNOLOGIST & ACQUISITION SCIENTIST, NAT SPACE SCI DATA CTR, GODDARD SPACE FLIGHT CTR, NASA, 71- Honors & Awards: Asteroid 1575 named Winifred, Minor Planets Ctr, 52; Spec Act Award, NASA, 66, Apollo Achievement Award, 70 & Exceptional Contrib to Educ Award, 71. Mem: Int Astron Union; Am Astron Soc; Am Geophys Union; Int Asn Planetology; Int Platform Asn. Res: Measurements of sunspots; lunar surface and scientific aspects research; scientific results of Mercury program of manned spaceflight; science data acquisition and analyses for Data Center. Mailing Add: Nat Space Sci Data Ctr Code 601 Goddard Space Flight Ctr Greenbelt MD 20771

CAMHI, JEFFREY MARTIN, b New York, NY, May 16, 41; m 65; c 1. NEUROPHYSIOLOGY, ANIMAL BEHAVIOR. Educ: Tufts Univ, BS, 63; Harvard Univ, PhD(biol), 67. Prof Exp: ASST PROF BIOL, CORNELL UNIV, 67- Mem: AAAS; Soc Neurosci. Res: Single-unit neurophysiology and behavior in arthopods; development of behavior and of the nervous system. Mailing Add: Sect of Neurobiol & Behav Cornell Univ Langmuir Lab Ithaca NY 14850

CAMIEN, MERRILL NELSON, b Redlands, Calif, Dec 10, 20; m 44; c 2. BIOLOGICAL CHEMISTRY. Educ: Univ Calif, Los Angeles, AB, 43, MA, 45, PhD(microbiol), 48. Prof Exp: Res assoc biochem, Univ Calif, Los Angeles, 48-50; Fulbright res scholar, Univ Liege, 50-51; jr res chemist, Univ Calif, Los Angeles, 51-53, from asst res chemist to assoc res chemist, 53-62; res chemist, Vet Admin Ctr, Los Angeles, 62-63; res assoc, Mt Sinai Hosp, 64-65, sr scientist, 65-68; res physiologist, Dept Med, Ctr Health Sci, Univ Calif, Los Angeles, 68-70; RES BIOCHEMIST & HEAD BIOCHEM & MICROBIOL, NUCLEIC ACID RES INST, INT CHEM & NUCLEAR CORP, 69- Mem: Fel AAAS; Am Inst Chemists; Am Soc Microbiol; Am Soc Biol Chem; Soc Exp Biol & Med. Res: Studies relating to biochemistry of memory and learning; evaluation of learning capability in experimental animals; specific antifungal, antibacterial and antiparasitic activities of unnatural nucleoside derivatives. Mailing Add: Nucleic Acid Res Inst Int Chem & Nuclear Corp Irvine CA 92664

CAMIENER, GERALD WALTER, b Detroit, Mich, Aug 15, 32; m 53; c 3. BIOCHEMISTRY, MICROBIOLOGY. Educ: Wayne State Univ, AB, 54; Mass Inst Technol, PhD(biochem), 59. Prof Exp: Teaching asst microbiol, Wayne State Univ, 53-54; res biochemist, Upjohn Co, 59-72; PRES, ICN LIFE SCI GROUP, ICN PHARMACEUT, 72- Mem: AAAS; Am Chem Soc. Res: Biosynthesis of thiamine; isolation and purification of enzymes; tissue culture and antitumor antibiotics; enzyme inhibitors; immunology; immunosuppression; delayed hypersensitivity. Mailing Add: ICN Life Sci Group 26201 Miles Rd Cleveland OH 44128

CAMILLI, CONCETTO THOMAS, organic chemistry, see 12th edition

CAMILLO, VICTOR PETER, b Bridgeport, Conn, Jan 23, 45; m 66; c 1. PURE MATHEMATICS. Educ: Univ Bridgeport, BA, 66; Rutgers Univ, PhD(math), 69. Prof Exp: Asst prof math, Univ Calif, Los Angeles, 69-70; asst prof, 70-74, ASSOC PROF MATH, UNIV IOWA, 74- Mem: Am Math Soc. Res: Theory of rings. Mailing Add: Dept of Math Univ of Iowa Iowa City IA 52242

CAMIN, JOSEPH HARVEY, b Cleveland, Ohio, June 6, 22; m 43; c 2. ZOOLOGY. Educ: Ohio State Univ, BSc, 46, MSc, 47, PhD(parasitol), 50. Prof Exp: Asst zool, Ohio State Univ, 47-50; asst prof, SDak State Col, 50-52; cur invert zool, Chicago Acad Sci, 52-58; assoc prof entom, 58-62, PROF ENTOM, UNIV KANS, 62- Concurrent Pos: Vis lectr, Inst Acarology, Univ Md, 54-61 & Ohio State Univ, 62-73. Mem: Am Soc Parasitol; Entom Soc Am; Soc Syst Zool (treas, 56-58); Soc Study Evolution; Am Soc Zool. Res: Sensory behavior and ecology of Acari; medical acarology; taxonomy of mites of the suborder Mesostigmata. Mailing Add: Dept of Entom Physiol & Cell Biol Univ of Kans Lawrence KS 66044

CAMINITA, BARBARA HOBSON, b Cranston, RI, Aug 18, 11; div; c 1. MICROBIOLOGY. Educ: Univ Md, BS, 38, MS, 48. Prof Exp: From jr bacteriologist to assoc bacteriologist, Div Indust Hyg, NIH, Md, 40-51; microbiologist, Off Naval Res, 51-62 & US Naval Weapons Lab, Va, 62-68; ASST PROF BIOL, MERCER UNIV, 70- Concurrent Pos: NIH fel, Univ Md, 48-50. Mem: AAAS; Am Soc Microbiol; Biophys Soc. Res: Air-borne infection; Entamoeba histolytica; Neisseria gonorrhoeae. Mailing Add: Dept of Biol Mercer Univ Macon GA 31204

CAMISHION, RUDOLPH C, surgery, see 12th edition

CAMMARATA, ARTHUR, b Montclair, NJ, Jan 6, 40; m 62; c 1. PHYSICAL ORGANIC CHEMISTRY. Educ: Upsala Col, BS, 60; Rutgers Univ, PhD(phys org chem), 65. Prof Exp: Asst prof chem, Med Col Va, 65-68; assoc prof, 68-71, chmn, Pharmaceut Sect, 69-72, PROF CHEM, SCH PHARM, TEMPLE UNIV, 71- Mem: AAAS; Am Chem Soc; Acad Pharmaceut Sci; Am Asn Cols Pharm. Res: Molecular pharmacology; stereochemistry of medicinals; computers in pharmaceutical sciences. Mailing Add: Dept of Med Chem Temple Univ Sch of Pharm Philadelphia PA 19140

CAMMARATA, PETER S, b Chicago, Ill, Dec 26, 20; m 52; c 3. BIOCHEMISTRY, PHYSICAL ORGANIC CHEMISTRY. Educ: Univ Chicago, BS, 43, MS, 47; Univ Wis, PhD(physiol chem), 51. Prof Exp: Res assoc cancer res, Univ Chicago, 47-49; asst prof biochem, Yale Univ, 52-54; sr chemist, 54-55, biochem supvr, 56-62, nat prod supvr, 62-64, asst dir biochem, 64-70, dir biochem res div, 70-74, RES ADV, G D SEARLE & CO LABS, 75- Honors & Awards: Am Inst Chemists Award, 39. Mem: Am Chem Soc. Res: Enzymology; sulfated polysaccharides; proteases; prostaglandins; biochemical pharmacology and endocrinology; medicinal chemistry. Mailing Add: G D Searle & Co Box 5110 Chicago IL 60680

CAMOUGIS, GEORGE, b Concord, Mass, May 10, 30; m 61; c 3. RESEARCH ADMINISTRATION. Educ: Tufts Univ, BS, 52; Harvard Univ, MA, 57, PhD(biol), 58. Prof Exp: Asst comp anat, Tufts Univ, 50-52; lab asst parasitol, Harvard Univ, 55; from asst prof to assoc prof physiol, Clark Univ, 58-64; sr neurophysiologist, Astra Pharmaceut Prod, Inc, 64-66, head neuropharmacol sect, 66-68; PRES & RES DIR, NEW ENG RES, INC, 68- Concurrent Pos: Res scientist, NY State Dept Health, 60; panel mem undergrad sci equip prog, NSF, 64 & 65; affil assoc prof, Clark Univ, 64-68, affil prof, 68-; mem corp, Bermuda Biol Sta Res, Inc, 68-; adj prof, Worcester Polytech Inst, 70- Mem: AAAS; Am Soc Zoologists; Biophys Soc; Am Physiol Soc; NY Acad Sci. Res: Management of research and development; animal physiology; pharmacology; neurobiology; pesticide research; environmental toxicology; aquatic ecosystems; environmental impact assessments; environmental management. Mailing Add: 10 Wheeler Ave Worcester MA 01609

CAMP, BENNIE JOE, b Greenville, Tex, Mar 19, 27; m 52; c 2. BIOCHEMISTRY. Educ: ETex State Univ, BS, 49; Tex A&M Univ, MS, 53, PhD(biochem), 56. Prof Exp: From asst prof to assoc prof biochem, 56-66, PROF BIOCHEM & BIOPHYS & VET PHYSIOL & PHARMACOL, TEX A&M UNIV, 66- Mem: Am Chem Soc; hon fel Am Col Vet Toxicol; Am Inst Chemists; Sigma Xi. Res: Chemistry of poisonous plants and toxicology of environmental pollutants. Mailing Add: Col of Vet Med Tex A&M Univ College Station TX 77843

CAMP, DAVID BENNETT, b Va, Jan 30, 10; m 48; c 2. ORGANIC CHEMISTRY. Educ: Col William & Mary, BS, 41; Univ Rochester, PhD(org chem), 50. Prof Exp: Instr chem, Col William & Mary & Va Polytech Inst, 41-46; assoc prof phys sci, Univ Idaho, 49-50; assoc prof chem, Oglethorpe Univ, 50-52 & Univ SDak, 52-54; assoc prof, 54-57, PROF CHEM, UNIV OF THE SOUTH, 57- Mem: AAAS; Am Chem Soc. Res: Synthesis of labeled amino acids; gamma ray dosimetry. Mailing Add: Dept of Chem Univ of the South Sewanee TN 37375

CAMP, DAVID CONRAD, b Atlanta, Ga, Aug 20, 34; m 61; c 2. EXPERIMENTAL NUCLEAR PHYSICS. Educ: Emory Univ, BA, 55; Ind Univ, MA, 58, MS, 59, PhD(physics), 63. Prof Exp: SR STAFF PHYSICIST, RADIOCHEM DIV, LAWRENCE LIVERMORE LAB, UNIV CALIF, 63- Concurrent Pos: Mem staff, Inter-Univ Reactor Inst, Delft Technol Univ, 71-72; consult, Mobil Res & Develop Corp, 74; adj asst prof radiol, Sch Med Univ Calif, 74- Mem: AAAS; Am Phys Soc. Res: Inorganic trace element concentrations in biomedical samples; multi-element multi-technique intercomparison studies; design of Si(Li) and Ge(Li) electron, x-ray fluorescence and gamma-ray spectrometers; x-ray, conversion electron, gamma-ray spectroscopy; nuclear medicine instrumentation. Mailing Add: Lawrence Livermore Lab Univ of Calif Livermore CA 94550

CAMP, EARL D, b Magazine, Ark, June 12, 18; m 48; c 1. PLANT PATHOLOGY. Educ: Tex Tech Univ, BS, 41; Univ NMex, MS, 43; Univ Iowa, PhD(bot), 52. Prof Exp: War res staff & investr med mycol, Columbia Univ, 43-44; from instr to assoc prof biol, 45-59, dept chmn, 59-71, PROF BIOL, TEX TECH UNIV, 59- Mem: Fel

AAAS. Res: Plant morphology and pathology; developmental anatomy of woody monocots. Mailing Add: Dept of Biol Sci Tex Tech Univ Lubbock TX 79409

CAMP, ELDRIDGE KIMBEL, b Tamaqua, Pa, Sept 6, 15; m 47; c 2. ELECTROCHEMISTRY. Educ: Pa State Univ, BS, 38. Prof Exp: Apprentice exec, Collins & Aikman Corp, 39-40, foreman dyer, 40-42; res chemist metallic corrosion, Westinghouse Res Labs, 46-56; staff engr elec contacts, Prod Develop Lab, Int Bus Machine Corp, 56-57, mgr finishes & corrosion lab, 57-59, mem tech staff, Mat & Processes Labs, 59-60; DIR RES & MEM TECH STAFF, AM CHEM & REFINING CO, INC, 60- Mem: Am Electroplaters Soc; Am Chem Soc; Electrochem Soc; Am Soc Test & Mat. Res: Electrode processes applied to electrodeposition; anodizing; electropolishing; galvanic corrosion; surface studies; the study of mechanical and physical properties of precious metal film deposits and processes. Mailing Add: Am Chem & Refining Co PO Box 4067 10 Sheffield Waterbury CT 06714

CAMP, FRANK RUDOLPH, JR, b North Adams, Mass, Oct 6, 19; m 44. IMMUNOLOGY. Educ: Emory Univ, AB, 48, MS, 54. Prof Exp: Chief lab serv, US Army Hosp, Bad Kreuznach, Ger, 52-54, instr immunohemat, Med Field Serv Sch, Ft Sam Houston, Tex, 54-57, asst dir, Europe Blood Bank, US Army, Landstuhl, Ger, 57-60, fel immunohemat, Walter Reed Army Inst Res, Washington, DC, 60-61, immunohematologist, Europe Blood Bank, 61-65, dir blood transfusion div, US Army Med Res Lab, Ky, 65-72, cmndg officer, Lab, 72-74; SCI DIR-DIR, LOUISVILLE REGIONAL RED CROSS BLOOD PROG, 74- Concurrent Pos: Mem comt blood & blood transfusion, Nat Res Coun, 65-66. Mem: AAAS; Am Acad Forensic Sci; Genetics Soc Am; Am Soc Human Genetics; Int Soc Hemat. Res: Immunohematology; quantitative hemagglutination; erythrocyte metabolism and preservation; blood group genetics; tissue compatibility. Mailing Add: Regional Red Cross Blood Ctr 510 E Chestnut St Louisville KY 40201

CAMP, LEON (W), b Justin, Tex, Sept 22, 04; m 30; c 3. PHYSICS. Educ: NTex State Teachers Col, BA, 24, BS, 26; Columbia Univ, AM, 28; Univ Tex, PhD(physics), 41. Prof Exp: Res chemist, Western Union Tel Co, 28-30; chief chemist, NJ Coal & Tar, 30-31; instr physics, Univ Tex, 39-40, instr math & physics, Col Mines & Metal, 40-43; res assoc underwater sound equip, Harvard Univ, 43-45; assoc prof eng res, Pa State Col, 45-48; dir transducer res & develop, Pac Div, Bendix Aviation Corp, 48-56, dir acoust res, Bendix Corp, 56-70, consult, Electrodynamics Div, 70-74; ELECTRODYNAMICS CONSULT, BENDIX CORP, 74-; RES ASSOC ENG & APPL SCI, UNIV CALIF, LOS ANGELES, 70- Mem: Am Phys Soc; fel Acoust Soc Am; Inst Elec & Electronics Eng. Res: Geophysical instruments; magnetostrictive properties of metals; transducer design; acoustic absorption; electroacoustic transducers; ultrasonic processing. Mailing Add: Bendix Corp 15825 Roxford Sylmar CA 91342

CAMP, LESLIE WILFORD, b San Luis Obispo, Calif, Apr 16, 26; m 46; c 7. GEOLOGY. Educ: Brigham Young Univ, BA, 50; Johns Hopkins Univ, AM, 51. Prof Exp: Geologist, Fuels Br, US Geol Surv, 51-54; sr explor geologist, Columbia Geneva Div, US Steel Corp, 54-63; opers supvr, Crest Explor, Ltd, Stand Oil Co Calif, 63-66; sr geologist, Bear Creek Mining Co, 66-68; proj mgr indust minerals div, Kennecott Explor Inc, 68-70, proj mgr finaicial eval, 70-71, prof mgr, Kennecott Explor Servs, 72-74, MGR ADMIN, KENNECOTT CORP EXPLOR GROUP, 75- Mem: Geol Soc Am; Am Inst Mining, Metall & Petrol Eng; Can Inst Mining & Metall; Soc Econ Geol. Res: Economic geology. Mailing Add: 2300 West 1700 South Salt Lake City UT 84104

CAMP, MARK JEFFREY, b Toledo, Ohio, Dec 19, 47. PALEOECOLOGY, MALACOLOGY. Educ: Univ Toledo, BSc, 70, MSc, 72; Ohio State Univ, PhD(geol), 74. Prof Exp: ASST PROF GEOL, EARLHAM COL, 74- Mem: Paleont Soc. Res: Comprehensive study of Pleistocene Lacustrine deposits in Ohio, Michigan and Indiana, with emphasis on ecologic relationships of fauna and flora, distribution of species, glacial history, stratigraphy and origin of sediments. Mailing Add: 856 McKinley Ave Toledo OH 43605

CAMP, PAUL R, b Middletown, Conn, Dec 29, 19; m 58; c 3. SOLID STATE PHYSICS. Educ: Wesleyan Univ, BA, 41; Harvard Univ, MA, 47; Pa State Col, PhD(physics), 51. Prof Exp: Physicist, US Naval Res Lab, 41-44; instr, Wesleyan Univ, 47-48; physicist, Res Lab, Radio Corp Am, 51-53; Ford intern physics, Reed Col, 53-54; from asst prof to assoc prof, Polytech Inst Brooklyn, 54-61; chief, Mat Res Br, Cold Regions Res & Eng Lab, 61-63, physicist at large, 63-65; staff physicist, Comn Col Physics, Univ Mich, 65-67; head dept physics, 67-73, PROF PHYSICS, UNIV MAINE, ORONO, 73- Concurrent Pos: Consult, Rand Corp, 56-61. Mem: AAAS; Am Phys Soc; Am Asn Physics Teachers. Res: Physics of ice. Mailing Add: Dept of Physics Bennett Hall Univ of Maine Orono ME 04473

CAMP, RUSSELL R, b Corning, NY, Mar 26, 41; m 65. PLANT PATHOLOGY, ELECTRON MICROSCOPY. Educ: Baldwin-Wallace Col, BS, 64; Miami Univ, MA, 66; Univ Wis, PhD(bot), 70. Prof Exp: Asst prof, 70-75, ASSOC PROF BIOL, GORDON COL, 75- Mem: AAAS; Am !nst Biol Sci; Mycol Soc Am; Bot Soc Am; Am Phytopath Soc. Res: Electron microscopy of host parasite relations of fungal pathogens and their respective plant host. Mailing Add: Gordon Col Dept of Biol 255 Grapevine Rd Wenham MA 01984

CAMPAGNA, ELZEAR (ALEXANDRE), b St-Paul-de-Chester, Oct 17, 98; m 26; c 4. PLANT PATHOLOGY. Educ: Col Nicolet, BA, 20; Laval Univ, BSA, 24, DSc(bot, plant path), 40. Prof Exp: Prof plant path, Fac Agr, Laval Univ, 24-68, dir res sta potato plants, Saguenay, 38-68; PROF BIOL, ACAD ST LOUIS, 68- Concurrent Pos: Vpres, Agron Corp, Que, 38; pres, Nat Weed Comt, 53. Mem: French-Can Asn Advan Sci; Agr Inst Can; Can Phytopath Soc. Res: Ragweed survey and eradication in Gaspecia and lower Saint Lawrence; pollen fungeous spores survey in eastern Canada; potato varieties resistant to late blight; Bordeaux mixture and other protectants for potatoes. Mailing Add: Acad of St Louis Dept of Biol 1445 Blvd de l'Entente Quebec PQ Can

CAMPAIGNE, ERNEST EDWIN, b Chicago, Ill, Feb 13, 14; m 41; c 3. MEDICINAL CHEMISTRY. Educ: Northwestern Univ, BS, 36, MS, 38, PhD(biochem), 40. Prof Exp: Lab asst, Northwestern Univ, 36-40, res fel chem, 41-42; instr, Bowdoin Col, 40-41; instr prev med & pub health, Sch Med, Univ Tex, 42-43; from instr to assoc prof chem, 43-53, PROF CHEM, IND UNIV, BLOOMINGTON, 53- Concurrent Pos: Assoc biochemist, M D Anderson Hosp Cancer Res, Univ Tex, 42-43; with Comt Med Res, 44; res partic, Oak Ridge Inst Nuclear Studies, 55; consult, Off Surgeon Gen, Walter Red Army Inst Res, 59-62; vis lectr, Univ Calif, San Francisco, 62; consult, NIH, 60-64, mem Coun Pub Comt, 60-62 & Policy Comt, 62-64; consult, Drug Dvelop Comt, Div Cancer Treatment, Dept Health Educ & Welfare, 75-79. Mem: AAAS; Am Chem Soc; Soc Exp Biol & Med; fel NY Acad Sci; The Chem Soc. Res: Thiophene chemistry; thiocarbonyls; antimetabolites; physiologically active and organic sulfur compounds. Mailing Add: Dept of Chem Ind Univ Bloomington IN 47401

CAMPAIGNE, HOWARD HERBERT, b Chicago, Ill, Apr 6, 10; m 34; c 1. MATHEMATICS, COMPUTER SCIENCES. Educ: Northwestern Univ, BSc, 33, MS, 34, PhD(math), 38. Prof Exp: From instr to asst prof math, Univ Minn, 38-46; mathematician, US Dept Navy, 46-49; asst chief res, Nat Security Agency, 50-57, chief res, 57-70; PROF MATH, SLIPPERY ROCK STATE COL, 70- Concurrent Pos: Lectr, Am Univ, 52-58; adj prof, 58-70; mem comput sci & eng bd, Nat Acad Sci, 70-74. Mem: Am Math Soc; Math Asn Am; Asn Comput Mach; fel Inst Elec & Electronic Engrs. Res: Extensions of the notion of group; statistics; chemistry; properties of hypergroups; learning machines; information storage and retrieval. Mailing Add: Dept of Math Slippery Rock State Col Slippery Rock PA 16057

CAMPANA, RICHARD JOHN, b Everett, Mass, Dec 5, 18; m 45; c 2. FOREST PATHOLOGY. Educ: Univ Idaho, BSF, 43; Yale Univ, MF, 47, PhD(forest path), 52. Prof Exp: Instr forestry, Pa State Univ, 47; asst prof bot, NC State Univ, 47-49; asst plant pathologist, USDA, 49-52; from asst plant pathologist to assoc plant pathologist, Ill Nat Hist Surv, 52-58; chmn plant sci PhD prog, 68-72, head dept bot & plant path, 58-68, PROF BOT & FOREST PATH, UNIV MAINE, 58- Concurrent Pos: Consult, Salt Prod Assocs, Ill, 55-57, State Chamber Com, Ill, 55-58, Morton Arboretum, 55-58, Tower Grove Park, Mo, 56-57 & Ill Munic League, 56-58; vis prof, NY State Col Forestry, Syracuse Univ, 67; guest botanist, Brookhaven Nat Lab, 67-71; ed, Phytopath News, 70-76; chmn, Orono Conserv Comn, 70-; pres, Maine Asn Conserv Comn, 72-73. Honors & Awards: Award of Merit, Orono Conserv Comn, 75; Award, Int Soc Arborists, 75. Mem: AAAS; Bot Soc Am; Am Phytopath Soc; Can Phytopath Soc; Int Soc Arborists (pres, 66-67). Res: Diseases of forest shade and ornamental trees and woody shrubs. Mailing Add: -5 Sunrise Terr Orono ME 04473

CAMPANELLA, PAUL JOSEPH, II, ECOLOGY, ENVIRONMENTAL MANAGEMENT. Educ: Brown Univ, BA, 67; Syracuse Univ, PhD(ecol), 72. Prof Exp: Fel behav ecol, Smithsonian Trop Res Inst, 72-74; res scientist pop ecol, Gorgas Mem Lab Trop Med, 74-76; INSTR ANIMAL BEHAV, POP BIOL & HUMAN ECOL, CANAL ZONE COL, 73-; ENVIRON & ENERGY CONTROL OFFICER ENVIRON MGT, PANAMA CANAL CO, 73- Mem: AAAS; Ecol Soc Am. Res: Territoriality and spacing systems; evolution of mating systems; community stability. Mailing Add: Box 298 Balboa CZ

CAMPAU, EDWARD JUNIOR, b Alto, Mich, July 5, 16; m 51; c 2. ENTOMOLOGY. Educ: Mich State Col, BS, 38; Stanford Univ, MA, 40; Univ Wis, PhD(econ entom), 42. Prof Exp: Asst entom, Mich State Col, 38-39; tech asst, Stanford Univ, 39-40; field inspector, State Dept Agr, Mich, 40; Cent Wis Canneries, 41; asst, Wis Alumni Res Found, 41-42; insecticides & fungicides group leader, Stand Oil Co (Ind), 46-58; entomologist, 58-70, STAFF ASST PLANT SCI RES, LILLY RES LABS, 70- Mem: AAAS; Entom Soc Am. Res: Field control of codling moth, grapeberry moth and pea aphid; morphology of beetles; fungicides; weed killers; aerosols; household and agricultural insecticides; influence of clay diluents on toxicity of rotenone in ground cube when used for pea aphid control. Mailing Add: Lilly Res Labs Box 708 Greenfield IN 46140

CAMPBELL, ADA MARIE, b Jewell, Iowa, Apr 1, 20. FOOD CHEMISTRY. Educ: Iowa State Univ, BS, 42, MA, 45; Cornell Univ, PhD(food sci, nutrit), 56. Prof Exp: Instr food & nutrit, NDak State Univ, 45-50; asst home economist exp sta, NMex State Univ, 50-52; actg asst prof food & nutrit, Univ Calif, Los Angeles, 52-53; asst foods, Cornell Univ, 53-56; asst prof food & nutrit, Univ Calif, Los Angeles, 56-61; assoc prof food sci, 61-67, PROF FOOD SCI, UNIV TENN, 67- Mem: AAAS; Am Asn Cereal Chem; Inst Food Technol; Am Oil Chem Soc. Res: Food lipids. Mailing Add: Col of Home Econ Univ of Tenn Knoxville TN 37916

CAMPBELL, ALAN, b Alexandria, Egypt, Aug 28, 44; Can & Brit citizen; m 71. ENTOMOLOGY, ECOLOGY. Educ: McGill Univ, BSc, 67; Univ Man, MSc, 69; Simon Fraser Univ, PhD(biol), 74. Prof Exp: Fel stored prod insects, Agr Can, 73-75; RES PROJ DIR TICK RES, DEPT BIOL, ACADIA UNIV, 75- Mem: Can Soc Zoologists; Entom Soc Am; Entom Soc Can; Sigma Xi. Res: Ecology of the American dog tick in Nova Scotia. Mailing Add: Dept of Biol Acadia Univ Wolfville NS Can

CAMPBELL, ALAN NEWTON, b Halifax, Eng, Oct 29, 99; m 31; c 1. PHYSICAL CHEMISTRY. Educ: Univ London, PhD(phys chem), 24; Aberdeen Univ, DSc, 29. Prof Exp: Asst chem, Aberdeen Univ, 25-30; from asst prof to prof phys chem, 30-69, head dept, 45-68, EMER PROF PHYS CHEM, UNIV MANITOBA, 69- Concurrent Pos: Mem, Nat Res Coun Can, 51-57. Honors & Awards: Chem Inst Can Medal, 71. Mem: Fel Royal Soc Can; fel Royal Inst Chem. Res: Electrochemistry; phase rule. Mailing Add: Dept of Chem Univ of Manitoba Winnipeg MB Can

CAMPBELL, ALFRED, b Glasgow, Scotland, Sept 29, 23; m 49; c 2. ORGANIC CHEMISTRY. Educ: Univ Glasgow, BSc, 44; Birkbeck Col, London, PhD(org chem), 52. Prof Exp: Res chemist, Stafford Allen & Son, Eng, 47-52; sr res chemist, Parke Davis & Co, Eng, 52-57 & Univ Mich, 57-66, COORDR VET PROG, PARKE DAVIS & CO, 66- Res: Synthesis of organic compounds with potential value as pharmacological or chemopharmaceutic agents. Mailing Add: 1256 Westport Rd Ann Arbor MI 48103

CAMPBELL, ALFRED DUNCAN, b Zion, Ill, Dec 25, 19; m 43; c 2. BIOCHEMISTY. Educ: Univ Ill, BS, 43; Purdue Univ, MS, 48, PhD(agr biochem), 50. Prof Exp: Asst, State Exp Sta, Ind, 46-50; asst div head, Fleischmann Labs, Stand Brands Inc, 50-57; asst res dir, Clinton Corn Processing Co, 57-62; chief food contaminants br, 62-71, DEP DIR DIV CHEM & PHYSICS, FOOD & DRUG ADMIN, 71- Concurrent Pos: Mem food contaminants comn, Int Union Pure & Appl Chemists, 69-; res coordr off sci, Food & Drug Admin, 73- Mem: AAAS; Am Chem Soc; Inst Food Tech. Res: Electrophoretic and amino acid analysis of corn proteins; gelatin; starches; flavors; milk products; food additives and contaminants; mycotoxins. Mailing Add: 3415 Rusticway Lane Falls Church VA 22044

CAMPBELL, ALICE DEL CAMPILLO, b Santurce, PR, May 30, 28; m 58; c 2. BIOCHEMISTRY. Educ: Columbia Univ, AB, 47; NY Univ, MS, 53; Univ Mich, PhD(biochem), 60. Prof Exp: Asst biochem, Pub Health Res Inst, NY, 47-48; dept pharmacol, NY Univ, 48-54; instr, Sch Med, Univ PR, 54-56; res assoc biol, Univ Rochester, 60-68; RES ASSOC BIOL SCI, STANFORD UNIV, 68- Mem: Am Chem Soc; fel Am Inst Chem. Res: Enzymology; intermediary metabolism; lipid metabolism; carboxylation reactions; bacterial and physiological genetics. Mailing Add: Dept of Biol Sci Stanford Univ Stanford CA 94305

CAMPBELL, ALLAN BARRIE, b Winnipeg, Man, Mar 28, 23; m 50; c 1. PLANT BREEDING. Educ: Univ Man, BSA, 44, MSc, 48; Univ Minn, PhD(genetics, plant breeding), 54. Prof Exp: RES SCIENTIST WHEAT BREEDING, CAN DEPT AGR, 49- Mem: Agr Inst Can; Can Soc Agron. Res: Breeding improved varieties of common wheat with special emphasis on resistance to stem rust and leaf rust. Mailing Add: Res Sta Agr Can 25 Dafoe Rd Winnipeg MB Can

CAMPBELL, ALLAN MCCULLOCH, b Berkeley, Calif, Apr 27, 29; m 58; c 2. MICROBIAL GENETICS. Educ: Univ Calif, BS, 50; Univ Ill, MS, 51, PhD(bact), 53. Prof Exp: Instr bact, Sch Med, Univ Mich, 53-57; res assoc genetics, Carnegie Inst,

57-58; from asst prof to prof biol, Univ Rochester, 58-68; PROF BIOL, STANFORD UNIV, 68- Concurrent Pos: Fel, Nat Found Inst Pasteur, 58-59; USPHS res career award, 62-; Found Microbiol lectr, 70-71; mem genetics study sect, NIH, 64-69. Mem: Nat Acad Sci; AAAS; Am Soc Microbiol; Am Soc Nat. Res: Genetics of bacteriophage; lysogeny; biochemical genetics. Mailing Add: Dept of Biol Sci Stanford Univ Stanford CA 94305

CAMPBELL, ALLEN JAMES, b Evansville, Ind, Nov 12, 41; c 4. ORGANIC CHEMISTRY, POLYMER CHEMISTRY. Educ: Univ Evansville, AB, 63; Purdue Univ, PhD(org chem), 69. Prof Exp: Chemist process develop, 69-72, advan res & develop, 72-75, SPECIALIST QUAL ASSURANCE, GEN ELEC CO, IND, 75- Res: Proprietary research. Mailing Add: 338 Lawrence Dr Mt Vernon IN 47620

CAMPBELL, ARDEN RAY, plant breeding, genetics, see 12th edition

CAMPBELL, ARTHUR WILLIAM, organic chemistry, see 12th edition

CAMPBELL, BARBARA KNAPP, b Denver, Colo, June 14, 08; m 33. MEDICINAL CHEMISTRY. Educ: Univ Chicago, BS, 29, MS, 31; Pa State Col, PhD(org chem), 37. Prof Exp: Teacher, Maret Sch, Washington, DC, 33; res assoc chem, Univ Notre Dame, 37-44; lectr, Univ Ind, 45-53; consult, Mead Johnson Res Labs, 53-73; CONSULT & SCI TRANSLATOR, CAMPBELL & ASSOCS, 73- Concurrent Pos: Instr, St Mary's Col (Ind), 39-40; with Off Sci Res & Develop; Nat Defense Res Comt, 42-45. Mem: Fel AAAS; Am Chem Soc; Sigma Xi. Res: Synthetic drugs, especially for malaria and cancer; amino alcohols and ethylenimines; aliphatic hydrocarbons; reduction of organic compounds; preparation and reactions of aliphatic tertiary aminoalcohols; infrared spectroscopy. Mailing Add: 8216 Petersburg Rd Evansville IN 47711

CAMPBELL, BENEDICT JAMES, b Philadelphia, Pa, Oct 17, 27; m 52; c 2. BIOCHEMISTRY. Educ: Franklin & Marshall Col, BS, 51; Northwestern Univ, PhD(biochem), 56. Prof Exp: Res biochemist, Glidden Co, 55-58 & Armour Co, 58-60; assoc prof, 60-67, PROF BIOCHEM, SCH MED, UNIV MO-COLUMBIA, 67-, CHMN DEPT, 73- Mem: AAAS; Am Chem Soc; NY Acad Sci; Am Soc Biol Chemists; Biophys Soc. Res: Biochemical research; graduate education. Mailing Add: Dept of Biochem Univ of Mo Med Ctr Columbia MO 65201

CAMPBELL, BERRY, b St Paul, Minn, Mar 21, 12; m 33; c 4. NEUROLOGY. Educ: Univ Calif, Los Angeles, AB, 32; Johns Hopkins Univ, PhD(anat), 35. Prof Exp: Asst zool, Univ Calif, 32; asst prof anat, Sch Med, Univ Okla, 37-42; from asst prof to prof anat, Med Sch, Univ Minn, 43-58; res prof neurosurg, Loma Linda Univ, 58-66; actg chmn dept, 66-72, PROF PHYSIOL, MED SCH, UNIV CALIF, IRVINE, 66- Concurrent Pos: Nat Res Coun fel med sci, Western Reserve Univ, 35-37; Guggenheim fel, Rockefeller Inst, 40-42; fel neurol, Col Physicians & Surgeons, Columbia Univ, 42-43; hon res fel, Univ Col, Univ London, 53-54; field naturalist, Roosevelt Wildlife Forest Exp Sta, Syracuse Univ, 32; vis asst prof, Sch Med, Univ Tenn, 42; vis prof, Columbia Univ, 53; res prof, Calif Col Med, 64-66; mem attend staff, Los Angeles County Hosp & Rancho Los Amigos, 64-68; consult physiol, Ctr Marital & Sexual Studies, 71- Mem: AAAS; Am Acad Neurol; Harvey Soc; Am Soc Ichthyologists & Herpetologists; Am Neurol Asn. Res: Distribution of vertebrates; evolution of mammals; anatomy and physiology of the nervous systems; encephalitis and multiple sclerosis; immunology; milk antibodies. Mailing Add: Dept of Physiol Univ of Calif Med Sch Irvine CA 92664

CAMPBELL, BONNALIE OETTING, b Springfield, Mo, Aug 21, 33; m 60; c 2. PHYSIOLOGY, ENDOCRINOLOGY. Educ: Southwest Mo State Univ, AB, 55; Northwestern Univ, MS, 58, PhD(physiol), 64. Prof Exp: Instr biochem, Univ Houston, 66-71, instr physiol, 71-73, ASST PROF PHYSIOL, BAYLOR COL MED, 73- Concurrent Pos: NASA grants, 70-; consult endocrinol, Vet Admin Hosp, Houston, 73- Mem: AAAS; Am Soc Zoologists; assoc Am Physiol Soc. Res: Adrenocorticotropic hormone; circadian rhythms. Mailing Add: Dept of Physiol Baylor Col of Med Houston TX 77025

CAMPBELL, BRUCE HENRY, b Madison, SDak, Oct 27, 40; m 63; c 3. ANALYTICAL CHEMISTRY. Educ: Univ Kans, BS, 62; Univ SDak, MA, 65; Univ Tex, PhD(chem), 68. Prof Exp: From asst prof to assoc prof chem, Univ Southern Miss, 67-72; fel, Clarkson Col Technol, 72-74; MGR RES ANAL SERV, J T BAKER CHEM CO, 74- Concurrent Pos: Ed, Critical Rev Anal Chem, 75- Mem: Am Chem Soc; Electrochem Soc; Chem Notation Asn. Res: Applied and theoretical electrochemisty. Mailing Add: 222 Red School Lane Phillipsburg NJ 08865

CAMPBELL, BRUCE (NELSON), JR, b Northampton, Mass, Apr 21, 31; m 56; c 4. BIOCHEMISTRY, ORGANIC CHEMISTRY. Educ: Williams Col, AB, 52; Univ Conn, PhD(chem), 58. Prof Exp: Asst instr chem, Univ Conn, 55-56; from asst prof to prof chem, MacMurray Col, 57-73, chmn dept chem, 72-73; PROF CHEM & CHMN DEPT, STATE UNIV NY POTSDAM, 73- Concurrent Pos: NSF sci fac fel, Mich State Univ, 65-66; vis prof, Dept Biochem, Univ Ill, Urbana, 73. Mem: Sigma Xi; AAAS; Am Chem Soc. Res: Enzyme isolation and reactions; micelle catalysis; theoretical and synthetic organic chemistry; innovation in chemical education. Mailing Add: Dept of Chem State Univ of NY Potsdam NY 13676

CAMPBELL, CARL WALTER, b Decatur, Ill, Jan 10, 29; m 51; c 5. PLANT PHYSIOLOGY, HORTICULTURE. Educ: Ill State Norm Univ, BSEd, 51; Kans State Col, MS, 52; Purdue Univ, PhD(plant physiol), 57. Prof Exp: Plant physiologist, Agr Mkt Serv, USDA, 57-60; asst horticulturist, 60-66, assoc horticulturist, 66-70, PROF HORT & HORTICULTURIST, AGR RES & EDUC CTR, UNIV FLA, 70- Mem: AAAS; Am Inst Biol Sci; Am Pomol Soc; Am Soc Hort Sci; Soc Econ Bot. Res: Plant growth regulators; post-harvest physiology of fruits; horticultural and physiological aspects of selection, propagation and production of tropical and subtropical fruit crops. Mailing Add: Agr Res & Educ Ctr Univ of Fla 18905 SW 280th St Homestead FL 33030

CAMPBELL, CARLOS BOYD GODFREY, b Chicago, Ill, July 27, 34; div; c 3. NEUROANATOMY, ZOOLOGY. Educ: Univ Ill, BS, 55, MS, 57, MD, 63, PhD(anat), 65. Prof Exp: Surgical intern, Presby-St Lukes Hosp, Chicago, Ill, 63-64; neuroanatomist, Walter Reed Army Inst Res, Washington, DC, 64-67; from asst prof to assoc prof anat & physiol, Ctr Neural Sci, Ind Univ, Bloomington, 67-74; resident physician, Dept Neurol, Los Angeles County-Univ Southern Calif Med Ctr, 73-74; ASSOC CLIN PROF ANAT, CALIF COL MED, UNIV CALIF, IRVINE, 75-, RESIDENT PHYSICIAN, DEPT RADIOL SCI, 75- Mem: Am Asn Anat; Soc Study Evolution; Soc Syst Zool; Am Soc Ichthyol & Herpet; Soc Neurosci. Res: Comparative neuroanatomy; systematic zoology; comparative neurology of motor and sensory systems; marine zoology; primate nervous systems; primate evolution. Mailing Add: Dept of Anat Calif Col of Med Univ of Calif Irvine CA 92664

CAMPBELL, CATHERINE CHASE, b New York, NY, July 1, 05; m 30; c 1. PALEONTOLOGY, GEOMORPHOLOGY. Educ: Oberlin Col, BA & MA, 27; Radcliffe Col, MA, 30, PhD(micropaleont), 32. Prof Exp: Instr geol, Mt Holyoke Col, 27-29; tech writer meteorol, Long Range Weather Forecasting Unit, US Army Air Force, 43-45; tech ed rocket proj, Calif Inst Technol, 45-46; underwater ord, US Naval Ord Test Sta, 47-51, supvr pub ed, 51-61; GEOLOGIST, US GEOL SURV, 61- Mem: Am Geol Soc; Geosci Info Soc; Asn Earth Sci Ed; Soc Tech Commun. Res: Environmental geology. Mailing Add: US Geol Surv 345 Middlefield Rd Menlo Park CA 94025

CAMPBELL, CHARLES DUNCAN, b Ann Arbor, Mich, Nov 12, 05; m 38. GEOLOGY. Educ: Univ Mich, BS, 30, MS, 31; Stanford Univ, PhD(petrol), 34. Prof Exp: From instr to prof geol, Wash State Univ, 34-71, chmn dept, 50-61; RETIRED. Concurrent Pos: Assoc geologist, US Geol Surv, 42-45. Mem: Fel Geol Soc Am; fel Mineral Soc Am; Nat Asn Geol Teachers (vpres, 60, pres, 61). Res: Petrology of alkaline igneous rocks and of metamorphic rocks; stratigraphy and lead-zinc deposits in Northeastern Washington; granitization in Northeastern Washington; petrology and magnetism of the Columbia River basalts; history and uses of geology. Mailing Add: Country Club Dr Carmel Valley CA 93924

CAMPBELL, CHARLES HAYWOOD, b Sanford, NC, Dec 8, 24; m 47; c 4. VETERINARY MICROBIOLOGY. Educ: Univ NC, BA, 49, MSPH, 51, PhD(parasitol), 54. Prof Exp: Instr parasitol, Univ NC, 50-54; instr microbiol & immunol, State Univ NY Downstate Med Ctr, 54-56; immunologist, Plum Island Animal Dis Lab, USDA, 56-58; sr cancer res scientist virol, Roswell Park Mem Inst, 58-60; RES MICROBIOLOGIST, PLUM ISLAND ANIMAL DIS CTR, USDA, 60- Mem: Am Soc Microbiol; Am Acad Microbiol. Res: Antigenic studies of Trichinella spiralis; natural resistance to foot-and-mouth disease; virus selection and recombination. Mailing Add: USDA Plum Island Animal Dis Ctr Box 848 Greenport NY 11944

CAMPBELL, CHARLES J, b Nanton, Alta, Can, Nov 25, 15; US citizen; m. FISH BIOLOGY, FISHERIES MANAGEMENT. Educ: Wash State Univ, BS, 38. Prof Exp: Jr fishery biologist, US Forest Serv, 39-40; from asst to chief opers, Fishery Div, Ore Wildlife Comn, 41-59, chief fisheries, 59-75, ASST CHIEF FISHERIES, ORE DEPT FISH & WILDLIFE, 75- Mem: Am Fisheries Soc (pres, 73); Am Inst Fisheries Res Biol; Wildlife Soc. Res: Fishery biology, management and adminstration; ecology of fish. Mailing Add: Ore Dept Fish & Wildlife Box 3503 Portland OR 97208

CAMPBELL, CHARLOTTE CATHERINE, b Winchester, Va, Dec 4, 14. MEDICAL MYCOLOGY. Educ: George Washington Univ, BS, 51; Am Bd Med Microbiol, dipl. Hon Degrees: DSc, Lowell Tech Inst, 72. Prof Exp: Technician, Dept Bact, Walter Reed Army Inst Res, 41-43, bacteriologist, 43-46, med mycologist, 46-49, chief mycol sect, 49-62; from assoc prof to prof med mycol, Sch Pub Health, Harvard Univ, 62-73; PROF MED MYCOL & CHMN DEPT MED SCI, SCH MED, SOUTHERN ILL UNIV, 73- Concurrent Pos: Consult, Vet Admin-US Armed Forces Comn Histoplasmosis & Coccidioidomycosis, 54-62 & Mid Am Res Unit, CZ, 57-62; assoc ed, Sabouraudia; assoc in med, Peter Bent Brigham Hosp, 63-73; mem sci adv bd, Gorgas Mem Inst; mem panel rev skin test antigens, Food & Drug Admin. Honors & Awards: Medals, US War Dept, 48 & US Dept Army, 61. Mem: Fel AAAS; Am Pub Health Asn; Am Thoracic Soc; Am Acad Microbiol; Med Mycol Soc of the Americas (pres, 70). Res: Antigenic analysis of systemic mycotic agents; epidemiology and ecology of histoplasmosis; serologic diagnosis and chemotherapy of systemic mycoses. Mailing Add: Dept of Med Sci Southern Ill Univ Sch of Med Springfield IL 62708

CAMPBELL, CLARENCE L, JR, b Indianapolis, Ind, Sept 24, 21; m. VETERINARY MEDICINE. Educ: Ohio State Univ, DVM, 45. Prof Exp: Vet pvt pract, Ill, 45; field vet, Fla Livestock Sanit Bd, Fla Dept Agr, 45-48, asst state vet, 48-52, state vet & secy, 53-61, STATE VET & DIR DIV ANIMAL INDUST, FLA DEPT AGR & CONSUMER SERV, 61- Concurrent Pos: Pres, Nat Assembly State Vet, 56-57. Honors & Awards: Meritorious Serv Award, USDA, 62; Cert Serv Award, Am Vet Med Asn, 68. Mem: US Animal Health Asn (pres, 65-66); Am Vet Med Asn; Am Asn Equine Practitioners. Res: Regulatory veterinary medicine. Mailing Add: Fla Dept of Agr & Consumer Serv Rm 328 Mayo Bldg Tallahassee FL 32304

CAMPBELL, CLEMENT, JR, b Milton, Pa, Oct 22, 30; m 66; c 1. CHEMISTRY. Educ: Bucknell Univ, BS, 51. Prof Exp: Res chemist, Pyrotechnics Lab, 51-74, RES CHEMIST, EXPLOSIVES DIV, PICATINNY ARSENAL, 74- Mem: AAAS; Am Chem Soc; Sigma Xi; NAm Thermal Anal Soc. Res: Pre-ignition and ignition reactions of explosive, propellant and pyrotechnic materials instrumentation; reactions of powdered metals; decomposition of inorganic oxidants; thermoanalysis. Mailing Add: 29 Cory Rd Flanders NJ 07836

CAMPBELL, CLYDE DEL, b Wheeling, WVa, Apr 1, 30; m 56; c 1. ORGANIC CHEMISTRY, BIOCHEMISTRY. Educ: WLiberty State Col, AB & BS, 53; NC State Col, MS, 55; WVa Univ, PhD(biochem, org chem), 58. Prof Exp: Asst biol chem, NC State Col, 53-55 & WVa Univ, 55-58; instr chem, WLiberty State Col, 58-61; sr res chemist, Mobay Chem Co, WVa, 61-68; chmn div sci & math, 68-70, assoc acad dean, 70-73, DEAN ADMIN, WLIBERTY STATE COL, 73- Mem: AAAS; Am Chem Soc. Res: Isolation and identification of chlorophyll and carotenoid pigments; nitrogen metabolism in ruminants; function of vitamin B12 in nitrogen metabolism; carbohydrate analysis of foodstuffs; polyurethanes; isocyanates; plastics and synthetic resins; intermediary metabolism. Mailing Add: WLiberty State Col West Liberty WV 26074

CAMPBELL, COLIN, b Washington, DC, June 24, 27; m 52; c 3. OBSTETRICS & GYNECOLOGY, MEDICAL ADMINISTRATION. Educ: Stanford Univ, AB, 49; Temple Univ, EdM, 67; McGill Univ, MD, CM, 53. Prof Exp: Instr obstet & gynec, Temple Univ, 61-64; from asst prof to assoc prof, 64-71, PROF OBSTET & GYNEC, UNIV MICH, ANN ARBOR, 71-, ASST DEAN STUDENT AFFAIRS, 72- Concurrent Pos: Consult, Wayne County Gen Hosp, 65-; examr, Am Bd Obstet & Gynec, 69- Mem: Fel Am Col Obstet & Gynec. Res: Erythroblastosis fetalis; medical education. Mailing Add: Med Sci I Bldg Univ of Mich Med Sch Ann Arbor MI 48104

CAMPBELL, COLIN ARTHUR, b Wildwood, Alta, Apr 16, 34; m 63; c 3. FOOD SCIENCE. Educ: Walla Walla Col, BA, 57, MA, 63; Loma Linda Univ, PhD(biol), 68. Prof Exp: Asst prof biol, Union Col, Nebr, 65-67; microbiologist, United Med Labs, 67-68, dir united anal labs div, 68-72, dir, Infectious Dis Lab, 72-74; TECH VPRES, COLUMBIA LABS, INC, 74- Mem: AAAS; Am Soc Microbiol; Mycol Soc Am; Int Asn Milk, Food & Environ Sanitarians. Res: Use of microbiological laboratory data to determine food quality and safety; microbial physiology; metabolism of fungal spores. Mailing Add: Columbia Labs Inc PO Box 40 Corbett OR 97019

CAMPBELL, CONSTANTINE ALBERGA, b Montego Bay, W Indies, Jan 18, 34; m 60. SOIL CHEMISTRY. Educ: Univ Toronto, BSA, 60, MSA, 61; Univ Sask, PhD(soil chem), 65. Prof Exp: Lectr soil chem & physics, Univ Sask, 63-64; RES SCIENTIST SOIL CHEM, RES STA, CAN FED DEPT AGR, 65- Mem: Can Soc Soil Sci; Am Soc Agron. Res: Influence of environmental conditions on changes in soil nitrogen status and soil nitrogen availability to plants. Mailing Add: Fed Dept Agr Res Sta Swift Current SK Can

CAMPBELL, DAN HAMPTON, immunochemistry, deceased

CAMPBELL, DAN NORVELL, b Mt Enterprise, Tex, Sept 10, 28; m 51; c 2. ANALYTICAL CHEMISTRY. Educ: Southern Methodist Univ, BS, 52, MS, 53. Prof Exp: From res chemist to sr res chemist, 55-70, PROCESS SPECIALIST, MONSANTO CO, 70- Mem: Am Chem Soc. Res: Chromatographic separations in analytical chemistry involving both gas and liquid chromatography; separation and determination of alkylbenzenes; thermal cracking of hydrocarbons to ethylene and propylene. Mailing Add: Process Technol Dept Monsanto Co Texas City TX 77590

CAMPBELL, DAVID KELLY, b Long Beach, Calif, July 23, 44; m 67; c 1. THEORETICAL HIGH ENERGY PHYSICS. Educ: Harvard Col, AB, 66; Cambridge Univ, PhD(theoret physics), 70. Prof Exp: Res assoc theoret high energy physics, Univ Ill, 70-72; mem, Inst Advan Study, Princeton, 72-74; STAFF MEM, THEORET DIV, LOS ALAMOS SCI LAB, 74- Concurrent Pos: Fel, Ctr Advan Study, Univ Ill, 70-72; Oppenheimer fel, Los Alamos Sci Lab, 74- Mem: Am Phys Soc. Res: Relativistic quantum field theory models of elementary particle physics; pion condensation in nuclear matter at high density; phenomenology of high-energy multi-particle production. Mailing Add: Theory Div T-8 Los Alamos Sci Lab Los Alamos NM 87545

CAMPBELL, DAVID OWEN, b Merriam, Kans, Nov 11, 27; m 54; c 3. RADIOCHEMISTRY. Educ: Univ Kansas City, BA, 47; Ill Inst Technol, PhD(chem), 53. Prof Exp: CHEMIST, OAK RIDGE NAT LAB, 53- Mem: Am Chem Soc. Res: Decontamination of nuclear equipment; molten salt and fluoride volatility fuel processing; transuranium element chemistry and chemistry of protactinium; nuclear reactor fuel reprocessing; radioactive waste treatment. Mailing Add: 102 Windham Rd Oak Ridge TN 37830

CAMPBELL, DAVID PAUL, b Seattle, Wash, May 12, 44; m 69; c 1. GENETICS. Educ: Western Wash State Col, BA, 67; Wash State Univ, PhD(genetics), 76. Prof Exp: ASST PROF GENETICS, CALIF STATE POLYTECH UNIV, POMONA, 75- Mem: AAAS; Am Soc Plant Physiologists; Int Plant Tissue Cult Asn. Res: Ultrastructural changes in germinating seed tissues of the Solanaceae. Mailing Add: Dept of Biol Sci Calif State Polytech Univ Pomona CA 91768

CAMPBELL, DEWAYNE E, b Ligonier, Pa, Aug 2, 23; m 46; c 4. FISHERIES. Educ: Univ Mich, BSF, 50. Prof Exp: Fisheries res technician, Inst Fisheries Res, Fisheries Div, Mich Dept Conserv, 47-49; fishery biologist, Benner Spring Fish Res Sta, Pa Fish Comn, 50-60, southwest regional fisheries mgr, 60-62; chief biol sect, Environ Planning Div, Neilan Engrs, Inc, Pa, 62-65; ZONE FISHERIES BIOLOGIST, NAT FORESTS MICH, US FOREST SERV, 65- Concurrent Pos: Consult mem, Environ Res Inst Inc, DC, 62-65. Mem: Am Fisheries Soc; Wildlife Soc; Am Soc Limnol & Oceanog. Res: Evaluation of aquatic resources. Mailing Add: Hiawatha Nat Forest US Forest Serv Escanaba MI 49829

CAMPBELL, DONALD BRUCE, b New South Wales, Australia. PLANETARY SCIENCES. Educ: Univ Sydney, BS, 63, MS, 65; Cornell Univ, PhD(astron), 71. Prof Exp: Res assoc astron, Cornell Univ, 71-73; mem staff, Haystack Observ, Northeast Radio Observ Corp, 73-74; RES ASSOC, ARECIBO OBSERV, CORNELL UNIV, 74- Mem: Am Astron Soc; Am Geophys Union. Res: Investigation of planetary surfaces and atmospheres by means of ground based radar. Mailing Add: Arecibo Observ Box 995 Arecibo PR 00612

CAMPBELL, DONALD EDWARD, b Brooklyn, NY, Sept 3, 28; m 52; c 4. ANALYTICAL CHEMISTRY, CERAMICS. Educ: Union Col, BS, 49; Rensselaer Polytech Inst, PhD(chem), 52. Prof Exp: Asst AEC contract, Rensselaer Polytech Inst, 50-52; from instr to asst prof chem, SDak State Col, 52-55; res chemist glass res & develop lab, 55-63, leader anal chem group tech serv res dept, 63-65, mgr chem anal res dept, 65-73, SR RES ASSOC, CORNING GLASS WORKS, 73-, MGR PHYS ANAL CHEM RES SERV, CORNING, CENTRE EUROP RECHERCHE, 75- Mem: Am Chem Soc; Am Phys Soc; Brit Soc Glass Technol; Am Ceramics Soc. Res: Chlorosilanes; aluminum soaps; solvent extraction of inorganic ions; chemical analysis of ceramic materials; materials characterization. Mailing Add: Corning Centre Europ de Recherche BP No 3 77210 Avon France

CAMPBELL, DONALD FERGUS, economic geology, mining engineering, see 12th edition

CAMPBELL, DONALD GRAY, b Valley Stream, NY, Aug 14, 43; m 64; c 3. REPRODUCTIVE PHYSIOLOGY. Educ: Cornell Univ, BS, 65; Univ Guelph, MS, 67; Rutgers Univ, PhD(reproductive physiol), 75. Prof Exp: Lab mgr semen processing, Select Sires, Inc, 67-70; SECT HEAD PHYSIOL, ANIMAL HEALTH RES DIV, SCHERING CORP, 70- Mem: Am Soc Animal Sci; Soc Study Reproduction. Res: Estrous synchronization in cattle, horses, dogs and cats. Mailing Add: Animal Health Res Div Schering Corp PO Box 608 Allentown NJ 08501

CAMPBELL, DONALD H, organic chemistry, see 12th edition

CAMPBELL, DONALD H, geology, see 12th edition

CAMPBELL, DONALD L, b Waverly, Iowa, July 16, 40. INORGANIC CHEMISTRY. Educ: Iowa State Univ, BS, 62; Univ Ill, Urbana, PhD(inorg chem), 69. Prof Exp: Chemist, Liquid Carbonic Div, Gen Dynamics Corp, 62-65; ASST PROF CHEM, UNIV WIS-EAU CLAIRE, 69- Mem: Am Chem Soc. Res: Coordination chemistry of the lanthanide ions; thermodynamic parameters for complex formation and factors effecting complex stability; heavy metals in the environment. Mailing Add: Dept of Chem Univ of Wis Eau Claire WI 54701

CAMPBELL, DONALD R, b Youngstown, Ohio, Oct 12, 30; m 56; c 2. ANALYTICAL CHEMISTRY, RADIOCHEMISTRY. Educ: Univ Akron, BS, 53. Prof Exp: Jr res chemist, 53-59; sr res chemist, 60-68, res scientist, 69-71, GROUP LEADER, GEN TIRE & RUBBER CO, 71- Mem: Am Chem Soc; Am Inst Chemists. Res: Analytical chemistry of high polymers; application of radiochemical techniques to elucidation of the structure and composition of polymers and to the determination of functional groups in polymers; synthesis of isotopically-substituted compounds. Mailing Add: Gen Tire & Rubber Co Akron OH 44329

CAMPBELL, DOUGLAS ARTHUR, b Duluth, Minn, Feb 17, 42. MOLECULAR GENETICS. Educ: Ind Univ, Bloomington, BA, 64; Univ Wash, PhD(genetics), 69. Prof Exp: Lectr microbiol, Dept Life Sci, Univ Calif, Riverside, 69-70; NIH fel biol, Univ Chicago, 70-72 & postdoctoral fel, Dept Genetics, Univ Calif, Berkeley, 73-75; ASST PROF BIOL, COL HOLY CROSS, WORCESTER, MASS, 75- Mem: Genetics Soc Am; AAAS. Res: Genetics and radiation genetics of bacteriophage T4; effects of x-rays on genetic recombination in yeast; chromosome behavior in yeast aneuploids; mechanism of genetic recombination in yeast. Mailing Add: Dept of Biol Col of the Holy Cross Worcester MA 01610

CAMPBELL, DOUGLAS MICHAEL, b San Pedro, Calif, May 4, 43; m 66; c 4. MATHEMATICAL ANALYSIS. Educ: Harvard Univ, BS, 67; Univ NC, Chapel Hill, PhD(math), 71. Prof Exp: Asst prof, 71-75, ASSOC PROF MATH, BRIGHAM YOUNG UNIV, 75- Concurrent Pos: Russian translr, Am Math Soc, 71-; reviewer, Math Rev, 74- Mem: Am Math Soc; Sigma Xi; Math Asn Am. Res: Geometric function theory. Mailing Add: Dept of Math Brigham Young Univ Provo UT 84602

CAMPBELL, EARL WILLIAM, b Bowling Green, Ohio. HEMATOLOGY. Educ: Harvard Col, BA, 58; Univ Rochester, MD, 62. Prof Exp: USPHS fel, Sch Med, Univ Utah, 67-68; instr med, WVa Univ, 68-69, asst prof, Med Ctr, 69-70; pvt pract, Md, 70-73; asst prof, 73-74, ASSOC PROF MED, MICH STATE UNIV, 74- Concurrent Pos: Investr, Acute Leukemia Group B, 68-72, Nat Polycythemia Rubra Vera Group, 73- & Mich Hemophilia Home Care Proj, 74-; consult hemat, Baltimore Cancer Res Inst, Nat Cancer Inst, 71-72; dir, Hemophilia Clin, Lansing Area, 74- Mem: AMA; fel Am Col Physicians; Am Soc Hemat. Res: Clinical investigation of polycythemia and related myeloproliferative diseases with the Polycythemia Rubra Vera Group; projects on antithrombin III, platelet factor IV. Mailing Add: Dept of Med Mich State Univ Rm B220 Life Sci East Lansing MI 48824

CAMPBELL, EARLE MALCOLM, radiation physics, see 12th edition

CAMPBELL, EDWARD CHARLES, b Brooklyn, NY, Dec 25, 13; m 36; c 3. PHYSICS. Educ: Univ Mich, BS, 34; Ohio State Univ, PhD(physics), 38. Prof Exp: Instr physics, Minn State Teachers Col, 38-42; vis asst prof, Princeton Univ, 42-46; sr physicist, Oak Ridge Nat Lab, 46-68; chmn dept physics, 68-73, PROF PHYSICS, NDAK STATE UNIV, 68- Concurrent Pos: Res physicist, Ctr Study Nuclear Energy, Belg, 57-58. Mem: Fel Am Phys Soc; Sigma Xi; Am Nuclear Soc. Res: Nuclear isomers; reactor physics; pulsed neutron techniques. Mailing Add: Dept of Physics NDak State Univ Fargo ND 58102

CAMPBELL, EDWIN STEWART, b Ada, Ohio, Aug 18, 26; m 49. CHEMICAL PHYSICS. Educ: Johns Hopkins Univ, AB, 45; Univ Mich, MSc, 48; Univ Calif, PhD, 51. Prof Exp: Instr quant & qual anal, St Martin's Col, 48; asst, Univ Calif, 48-51; lectr & fel, Univ Southern Calif, 51-52; proj assoc, Naval Res Lab, Univ Wis, 52-55; asst prof chem, 55-61, ASSOC PROF CHEM, NY UNIV, 62- Mem: AAAS; Am Chem Soc; Am Phys Soc; Philos Sci Asn; Fedn Am Scientists. Res: Hydrogen bonding and crystal structure of ice; theory of flame propagation; statistical mechanics. Mailing Add: Dept of Chem NY Univ Bronx NY 10453

CAMPBELL, EUGENE PAUL, b St Paul, Minn, July 22, 07; m 36; c 1. PUBLIC HEALTH, INTERNAL MEDICINE. Educ: Univ Calif, Los Angeles, BA, 29; Johns Hopkins Univ, MD, 33; Univ Pa, MPH, 42. Prof Exp: Asst physiol, Sch Med, Johns Hopkins Univ, 33; intern & asst resident med, Baltimore City Hosp, 33-35; ward officer, Commun Dis Hosp, Walter Reed Hosp, 35-39; asst prof epidemiol, Sch Pub Health, Univ Pa, 39-42; chief coop health prog, Inst Inter-Am Affairs, Guatemala, 42-43; field dir, Cent Am Coop Health Prog, 43-45; chief coop health prog, Inst Inter-Am Affairs & Foreign Opers, Brazil, 45-55; from dep chief to chief pub health div, Int Coop Admin, 55-59, dir, Off Pub Health, 59-63; chief health adv & attache int develop, AID, New Delhi, 63-66, chief health adv, Brazil, 66-70; CONSULT INT PUB HEALTH, 70- Concurrent Pos: Mem bd dirs, Am Sch, Rio de Janeiro, 53-55 & Gorgas Mem Inst, 61-64; adv to US deleg, WHO Gen Assembly, Geneva, 57 & Minneapolis, 58; consult, Am Pub Health Asn & AID, 70-76. Honors & Awards: Order Med Merit, Brazil, 55; Merit Citation, Int Coop Admin, 56 & Nat Civil Serv League, 58. Mem: Fel Am Pub Health Asn; fel Am Col Physicians; fel Royal Soc Trop Med & Hyg; Royal Soc Health; Indian Asn Advan Med Educ. Res: Communicable diseases; epidemiology. Mailing Add: 4701 Willard Ave Chevy Chase MD 20015

CAMPBELL, EVAN EDGAR, b Texarkana, Tex, Spet 13, 22; m 47; c 3. INDUSTRIAL HYGIENE. Educ: Univ Denver, BS, 47, MS, 53. Prof Exp: Indust hyg chemist, Med Sch, Univ Colo, 48-51; chemitoxicologist, Denver Gen Hosp, 51-53; indust hyg chemist, 53-62, sect leader indust hyg & radiobioassay, 62-73, asst group leader indust hyg group, 73-74, ALT GROUP LEADER, INDUST HYG GROUP, LOS ALAMOS SCI LAB, UNIV CALIF, 74- Mem: Am Indust Hyg Asn (pres, 76-77). Res: Air sampling and analysis of work room air; distribution of nuclides in tissue of workers exposed to nuclides; radio bioassay to evaluate worker exposure. Mailing Add: 423 Estante Way Los Alamos NM 87544

CAMPBELL, FERRELL RULON, b Afton, Wyo, Nov 14, 37; m 70; c 1. ANATOMY. Educ: Utah State Univ, BS, 60, MS, 63; Univ Chicago, PhD(anat), 66. Prof Exp: From instr to asst prof anat, Stanford Univ, 66-73; ASSOC PROF ANAT, UNIV LOUISVILLE, 73- Mem: Am Asn Anat. Res: Hematology. Mailing Add: Health Sci Ctr Dept of Anat Univ of Louisville Louisville KY 40201

CAMPBELL, FINLEY ALEXANDER, b Kenora, Ont, Jan 5, 27; m 53; c 3. GEOLOGY. Educ: Univ Man, BSc, 50; Queen's Univ, Ont, MA, 56; Princeton Univ, PhD(geol), 58. Prof Exp: Geologist, Prospectors Airways Co, Ltd, 50-56; explor geologist, Mining Corp Can, 56-58; asst prof geol, Univ Alta, 58-65; assoc prof & head dept, 65-70, PROF GEOL & VPRES CAPITAL RESOURCES, UNIV CALGARY, 70-, ACAD VPRES, 71- Mem: Royal Soc Can. Res: Economic geology; mineralogy; petrology; geochemistry. Mailing Add: 3408 Benton Dr NW Calgary AB Can

CAMPBELL, FRANCIS JAMES, b Toledo, Ohio, July 29, 24; m 48; c 7. PHYSICAL CHEMISTRY, PHYSICS. Educ: Univ Toledo, BS, 48. Prof Exp: Anal chemist, Dow Chem Co, 48-51, chem engr, 51-53, mat engr, Dow Corning Corp, 53-58; HEAD RADIATION APPLN SECT, US NAVAL RES LAB, 58- Concurrent Pos: US del, Int Electrotechnol Comn, 64- Mem: Am Chem Soc; Inst Elec & Electronics Eng; Am Soc Testing & Mat. Res: Radiation curing adhesives and composites; radiation and thermal effects on spacecraft materials, thermal control coatings, solar arrays and electrical insulation. Mailing Add: Div 6625 Naval Res Lab Washington DC 20375

CAMPBELL, FRANK LESLIE, b Philadelphia, Pa, Sept 5, 98. ENTOMOLOGY. Educ: Univ Pa, BS, 21; Rutgers Univ, MS, 24; Harvard Univ, ScD(entom), 26. Prof Exp: Jr chemist, Insecticide Invests, Japanese Beetle Lab, Riverton, NJ, 21-22; sewage invests, NJ Exp Sta, 22-24; asst prof biol, NY Univ, 26-27; assoc entomologist, Bur Entom, USDA, 27-30, entomologist & sr entomologist, Bur Entom & Plant Quarantine, 30-36, asst leader, Div Control Invests, 34-36; assoc prof entom, Ohio State Univ, 36-40, prof, 40-43, ed, Sci Monthly, 43-48; mem, Cent Intel Agency, 48-53; exec secy div biol & agr, Nat Res Coun, 53-64; vis prof, Zool Inst, Univ Vienna, 64-65; vis prof zool, Univ New Eng, 65-66; res assoc, Smithsonian Inst, 66-75; RETIRED. Concurrent Pos: Consult, Chem Div, Off Agr War Rel, USDA, 42; exec asst, Ohio State Univ Res Found, 43; Entom Soc Am rep, Nat Res Coun, 68-71; vis prof entom, Va Polytech Inst & State Univ, 70-74. Mem: AAAS; hon mem, Entom Soc Am; hon mem Nat Pest Control Asn; Am Inst Biol Sci. Res: Relative toxicity of stomach poison insecticides to insects; toxicity of rotenone-bearing insecticides; methods for determining effectiveness of household insecticides; chitin in insects;

synthetic organic insecticides; insect growth and development. Mailing Add: 2475 Virginia Ave NW Washington DC 20037

CAMPBELL, GARY THOMAS, b Granite City, Ill, Nov 11, 46; m 68; c 2. NEUROENDOCRINOLOGY. Educ: Wash Univ, BS, 68; Northwestern Univ, PhD(biol sci), 72. Prof Exp: Res biologist, Monsanto Chem Co, 68; res assoc biol sci, Northwestern Univ, 69, teaching asst, 70-72; instr physiol, Med Col Va, 72-73, asst prof, 73-75; ASST PROF PHYSIOL, UNIV NEBR MED CTR, OMAHA, 75- Mem: AAAS; Sigma Xi; Am Soc Zoologists. Res: Description of the cytoarchitecture in brain-pituitary control systems using immunohistochemical methods. Mailing Add: Dept of Physiol & Biophys Univ of Nebr Med Ctr Omaha NE 68105

CAMPBELL, GAYLON SANFORD, b Blackfoot, Idaho, Aug 20, 40; m 64; c 6. AGRICULTURAL METEOROLOGY, SOIL PHYSICS. Educ: Utah State Univ, BS, 65, MS, 66; Wash State Univ, PhD(soils), 68. Prof Exp: Captain, US Army Atmospheric Sci Lab, 68-70; asst prof, 70-75, ASSOC PROF BIOPHYS & SOILS, WASH STATE UNIV, 75- Concurrent Pos: Consult, Pac Northwest Labs, Battelle Mem Inst, 73- Mem: Am Meteorol Soc; Soil Sci Soc Am; Am Soc Agron. Res: Evapotranspiration, plant and soil water measurement, computer models of evaporation, transpiration and water uptake by plants; energy budgets of plants and animals. Mailing Add: Dept of Agron & Soils Wash State Univ Pullman WA 99163

CAMPBELL, GEORGE MELVIN, b Prospect, Pa, May 14, 29; m 59; c 2. PHYSICAL CHEMISTRY, ANALYTICAL CHEMISTRY. Educ: Hiram Col, BA, 54; Vanderbilt Univ, MS, 56, PhD(chem), 63. Prof Exp: Mem health physics staff, Argonne Nat Lab, 56-58; instr pub health, Univ Minn, 58-60; MEM CHEM STAFF, LOS ALAMOS SCI LAB, UNIV CALIF, 63- Mem: Am Chem Soc; fel Am Inst Chemists. Res: Electrochemical investigations in fused salts and other nonaqueous media; thermodynamic properties of plutonium compounds. Mailing Add: CMB-11 Los Alamos Sci Lab Los Alamos NM 87544

CAMPBELL, GEORGE WASHINGTON, JR, b Loma Linda, Calif, Sept 22, 19; m 42; c 4. INDUSTRIAL CHEMISTRY. Educ: Univ Calif, Los Angeles, AB, 42; Univ Southern Calif, MS, 47, PhD, 51. Prof Exp: Asst, Manhattan Proj, Univ Chicago, 42-43; rubber compounder, Goodyear Tire & Rubber Co, Calif, 43-45; lab assoc chem, Univ Southern Calif, 45-47, asst, 47-51, res assoc, Off Naval Research Proj, 51-52; assoc prof chem, Univ Houston, 53-57, chief investr, Off Ord Res Proj, 52-57; sr res chemist, US Borax Res Corp, 57-63, res supvr, 63-71, MGR PILOT PLANT RES DEPT, US BORAX RES CORP, 71- Concurrent Pos: From instr to prof, George Pepperdine Col, 45-53, head dept, 45-53. Mem: Am Chem Soc; fel Am Inst Chemists; Am Soc Testing & Mat; Nat Asn Corrosion Eng. Res: Boron chemistry; corrosion; process development. Mailing Add: US Borax Res Corp Pilot Plant Res Dept Boron CA 93516

CAMPBELL, GERALD ALLAN, b Cincinnati, Ohio, May 30, 46; m 67. ORGANIC POLYMER CHEMISTRY. Educ: Univ Cincinnati, BS, 67; Ohio State Univ, PhD(org chem), 71. Prof Exp: SR RES CHEMIST POLYMER CHEM, EASTMAN KODAK CO, 71- Mem: Am Chem Soc. Res: Synthesis of monomers and their subsequent polymerizations to materials of interest in photographic science. Mailing Add: Eastman Kodak Co Res Labs 1669 Lake Ave Rochester NY 14650

CAMPBELL, GILBERT SADLER, b Toronto, Ont, Jan 4, 24; US citizen; m 47, 61; c 6. SURGERY. Educ: Univ Va, BA, 43, MD, 46; Univ Minn, MS, 49, PhD(surg), 54. Prof Exp: Intern, Univ Minn Hosp, 46, asst physiol, Med Sch, 47-48, instr, 48-49, chief resident surg, Hosp, 53-54, from instr to asst prof, 54-58; prof & chief thoracic surg, Med Ctr Okla & chief surgeon, Vet Admin Hosp, 58-65; PROF SURG & HEAD DEPT MED CTR, UNIV ARK, LITTLE ROCK, 65- Concurrent Pos: Markle scholar, Univ Minn, 54-58. Honors & Awards: Horsley Prize, 54. Mem: Soc Exp Biol & Med; Soc Univ Surg; Am Asn Thoracic Surg; Am Col Surg; Am Physiol Soc. Res: Cardiovascular surgery; pulmonary physiology. Mailing Add: Dept of Surg Univ of Ark Med Ctr Little Rock AR 72205

CAMPBELL, GRAHAM HAYS, b Houston, Tex, Aug 17, 36; m 60; c 2. COMPUTER SCIENCES. Educ: Rice Univ, BA, 57; Yale Univ, MS, 58; Univ Calif, Berkeley, PhD(physics), 65. Prof Exp: From asst physicist to assoc physicist appl math, 66-69, COMPUTER SCIENTIST, BROOKHAVEN NAT LAB, 69- Mem: Asn Comput Mach. Res: Computer system modeling and simulation; symbolic algebraic manipulation by computers; operating systems theory. Mailing Add: Dept of Appl Math Brookhaven Nat Lab Upton NY 11973

CAMPBELL, HALLOCK COWLES, b Cortland, NY, June 4, 10; m 36, 63; c 2. CHEMISTRY. Educ: Hamilton Col, BS, 32; Harvard Univ, AM, 34, PhD(chem), 36. Prof Exp: Master in chg chem, Browne & Nichols Sch, Mass, 35-38; instr, Queens Col (NY), 38-43; res chemist, Arcos Corp, 43-45, assoc dir res & eng, 45-56, dir res, 56-68, dir res & technol, 68-73; MGR EDUC, AM WELDING SOC, 73- Concurrent Pos: Asst lectr, Eve Tech Sch, Temple Univ, 45-, co-dir metall, 55-61, dir, 61-73. Honors & Awards: Nat Meritorious Cert, Am Welding Soc, 73. Mem: AAAS; Am Chem Soc; Am Welding Soc; fel Am Soc Metals; fel Am Inst Chemists. Res: Kinetics of thermal explosions; arc welding electrodes; properties of low alloy and high alloy weld metal; thermal explosion of ethylazide gas. Mailing Add: Am Welding Soc 2501 NW Seventh St Miami FL 33125

CAMPBELL, HAROLD ALEXANDER, b Zion, Ill, June 27, 09; m 38; c 3. BIOCHEMISTRY. Educ: Univ Ill, BS, 35; Univ Wis, PhD(agr chem), 39. Prof Exp: Asst, Exp Sta, Univ Wis, 35-39; res chemist & mgr, Cent Labs, Gen Foods Corp, 39-61; vis investr, Walker Lab, Sloan-Kettering Inst Cancer Res, 62-69; RES ASSOC ONCOL, McARDLE LAB CANCER RES, UNIV WISMADISON, 69- Mem: Am Chem Soc. Res: Isolation of dicumarol hemorrhagic agent in sweet clover disease and its use for the treatment of humans as an anticoagulant; the relationship of L-asparaginase enzyme activity to the antileukemia activity of certain preparations including guinea pig serum; isolation and assay of biological materials of medical significance. Mailing Add: 5113 St Cyr Middleton WI 53562

CAMPBELL, HAYWARD, b Abbeville, La, Apr 30, 34; m 59; c 2. VIROLOGY, BIOCHEMISTRY. Educ: Southern Univ, BS, 54; Univ Iowa, MS, 61, PhD(bact), 62. Prof Exp: Biologist, NIH, 56-57; sr bacteriologist, 62-63, head dept biol assay develop, 63-65, mgr biol develop, 65-66, head, Ampoule Pilot Plant, 66-69, DIR, ELI LILLY & CO, 69- Mem: Am Soc Microbiol. Res: Development of purified and concentrated Coxsackie virus antigens; mechanisms of action of avian leucosis virus and Rous sarcoma virus on chick embryo tissue culture. Mailing Add: 2021 Brewster Indianapolis IN 46224

CAMPBELL, HERBERT NOEL, chemistry, see 12th edition

CAMPBELL, HOWARD, b Oklahoma City, Okla, Aug 29, 11; m 38; c 1. WILDLIFE BIOLOGY, HERPETOLOGY. Educ: Univ NMex, BS, 37. Prof Exp: Asst, Phys Testing Lab, Gates Rubber Co, Colo, 46; BIOLOGIST, STATE DEPT GAME & FISH, N MEX, 47- Mem: Wildlife Soc; Soc Study Amphibians & Reptiles. Res: Biology and management of upland game birds; economics of hunting and fishing; conservation of endangered species. Mailing Add: 7500 Leah Dr NE Albuquerque NM 87110

CAMPBELL, HOWARD ERNEST, b Detroit, Mich, Sept 20, 25; m 50; c 4. MATHEMATICS. Educ: Univ Wis, BS, 46, MS, 47, PhD(math), 49. Prof Exp: Asst, Univ Wis, 46-49; instr math, Univ Pa, 49-51; asst prof, Emory Univ, 51-56; from asst prof to assoc prof, Mich State Univ, 56-63; PROF MATH & CHMN DEPT, UNIV IDAHO, 63- Concurrent Pos: Chmn comt exam math test, Gen Exam, Col Level Exam Prog, Col Entrance Exam Bd & Educ Testing Serv. Mem: Am Math Soc; Math Asn Am. Res: Nonassociative algebras. Mailing Add: Dept of Math Univ of Idaho Moscow ID 83843

CAMPBELL, HOWARD WALLACE, b Baltimore, Md, Oct 23, 35; m 65; c 3. ENVIRONMENTAL SCIENCES, POPULATION BIOLOGY. Educ: Univ Fla, BA, 58; Univ Calif, Los Angeles, MA, 63, PhD(sensory physiol), 67. Prof Exp: Fel ctr biol natural systs, Wash Univ, 67-68 & ctr neurobiol sci, Med Sch, Univ Fla, 68-69; asst cur, Fla State Mus, 69-70; asst prof zool, Univ Fla, 70-72; consult ecologist, Jack McCormick & Assoc, 72-73; staff scientist, Off Endangered Species, Washington, DC, 73-74, CHIEF FIELD STA, NAT FISH & WILDLIFE LAB, FISH & WILDLIFE SERV, 74- Concurrent Pos: Mem, Crocodile Spec Group, Int Union Conserv Nature & Natural Resources Survival Serv Comn, 71- Mem: Am Soc Zool; Sigma Xi; Ecol Soc; Soc Study Amphibians & Reptiles. Res: Impact of land-use practices on ecosystem dynamics; population biology, ecology, behavior and physiology of crocodilians and Sirenians; behavior and ecology of amphibians and reptiles. Mailing Add: Nat Fish & Wildlife Lab 2820 E University Ave Gainesville FL 32601

CAMPBELL, HUGH JOHN, b Winnipeg, Man, July 29, 33. m 63; c 2. ORGANIC CHEMISTRY, PHYSICAL ORGANIC CHEMISTRY. Educ: Univ Man, BSc, 55, MSc, 57; McGill Univ, PhD(phys org chem), 61. Prof Exp: Res chemist, Charles E Frosst & Co, Que, 57-58; res chemist, E I du Pont de Nemours & Co, NY, 61-62; assoc res scientist, 62-68, RES SCIENTIST TEXTILES, ONT RES FOUND, 62- Res: Chemical modification of cellulosics via crosslinking; polymer deposition; protonation equilibria of ketones in strong acid solutions; flammability of textile materials; test methods research; toxic gas emissions. Mailing Add: Ont Res Found Sheridan Park Res Community Sheridan Park ON Can

CAMPBELL, IAN, b Bismarck, NDak, Oct 17, 99; m 30; c 1. ENVIRONMENTAL GEOLOGY. Educ: Univ Ore, AB, 22, AM, 24; Harvard Univ, PhD(econ geol), 31. Prof Exp: Asst prof geol, La State Univ, 25-28; instr mineral, Harvard Univ, 28-30, instr mineral & Petrol, 30-31; from asst prof to prof petrol, 31-59, res assoc, 59-70, assoc chmn div geol sci, 40-52, exec officer, 52-59, EMER PROF PETROL, CALIF INST TECHNOL, 70- Concurrent Pos: Sedimentologist, Vacuum Oil Co, 26-27; petrologist, Panama Corp, Ltd, 27-28; res assoc, Carnegie Inst, 34-39; leader, Carnegie Inst & Calif Inst Technol Exped, Grand Canyon, Ariz, 37; sr training engr & chief educ sect, Div War Res, Univ Calif & US Navy Radio & Sound Lab, San Diego, 44-46, field engr, Bur Ships, US Navy, Seattle, 45; mineralogist & chief, Calif State Div Mines, 59-61, geologist & chief, Calif State Div Mines & Geol, 61-69; mem exec comt, Am Geol Inst, 60-63, pres, 61; mem adv coun, Inst Marine Resources, Univ Calif, 62-68; US nat comt, Int Union Geol Sci, 61-65; dir, Calif State Dept Conserv, 66-67; res assoc, Calif Acad Sci, 69-, secy, 70-71, pres, 71-; mem, Calif State Bd Regist for Geologists, 69-72, pres, Calif State Bd Regist for Geologists & Geophysicists, 72-74. Honors & Awards: Hardinge Award, Am Inst Mining, Metall & Petrol Engrs, 62; Ben Parker Mem Award, Am Inst Prof Geologists, 70; Pub Serv Award, Am Asn Petrol Geologists, 73. Mem: Hon mem Asn Eng Geologists; distinguished mem Soc Mining Engrs; Asn Am State Geologists (pres, 65); fel Mineral Soc Am (vpres, 40, 61, pres, 62); fel Geol Soc Am (pres, 68). Res: Petrology; industrial minerals; geoscience and man. Mailing Add: 1333 Jones St Apt 906 San Francisco CA 94109

CAMPBELL, IAN MACLEAN, b Lloydminster, Sask, Feb 21, 27; m 58; c 3. ENTOMOLOGY, GENETICS. Educ: Univ Alta, BSc, 50; Univ Toronto, MA, 54, PhD(genetics), 58. Prof Exp: Tech officer, Forest Insect Lab, Can Dept Agr, 50-53, agr res officer, Sect Cytol & Genetics, Forest Biol Br, 53-63; assoc prof insect genetics, Iowa State Univ, 63-67; ASSOC PROF ZOOL, SCARBOROUGH COL, UNIV TORONTO, 67- Mem: Entom Soc Can; Genetics Soc Can. Res: Genetics of physiological variation within species of insects relative to their population dynamics and evolution. Mailing Add: Dept of Biol Univ of Toronto Toronto ON Can

CAMPBELL, IVOR EUGENE, inorganic chemistry, see 12th edition

CAMPBELL, JACK ALLEN, b Omaha, Nebr, Aug 25, 14; m 41; c 1. PHYSICAL CHEMISTRY. Educ: Munic Univ Omaha, AB, 37; Univ Iowa, MS, 39, PhD(chem), 41. Prof Exp: Instr phys chem, Univ Ky, 41-42; instr chem & physics, Utah State Univ, 43-44; asst prof chem, Univ Idaho, 44-47 & Univ Buffalo, 47-50; indust res, 50-53; PROF CHEM & PHYSICS, BLACKBURN COL, 53- Mem: AAAS; Am Chem Soc; Am Phys Soc. Mailing Add: Dept of Chem Blackburn Col Carlinville IL 62626

CAMPBELL, JACK JAMES RAMSAY, b Vancouver, BC, Mar 29, 18; m 42; c 4. BACTERIOLOGY. Educ: Univ BC, BSA, 39; Cornell Univ, PhD(bact), 44. Prof Exp: Asst bact, Cent Exp Farm, Ottawa, 39-40; asst, Queen's Univ (Ont), 44-46; assoc prof dairying, 46, prof, 47-65, PROF MICROBIOL & HEAD DEPT, UNIV BC, 65- Concurrent Pos: Vis lectr, Johns Hopkins Univ, 52; res assoc, Univ Ill, 54. Mem: AAAS; Am Soc Microbiol; Brit Soc Gen Microbiol; Can Soc Microbiol; Royal Soc Can. Res: Bacterial physiology. Mailing Add: Dept of Microbiol Univ of BC Vancouver BC Can

CAMPBELL, JAMES, b Glasgow, Scotland, July 18, 07; m 54; c 4. PHYSIOLOGY, BIOCHEMISTRY. Educ: Univ Toronto, BA, 30, MA, 32, PhD, 38. Prof Exp: Res assoc physiol, McGill Univ, 32-33; res assoc, Sch Hyg, 33-40, from asst prof to assoc prof, 40-58, PROF PHYSIOL, UNIV TORONTO, 58- Mem: Am Physiol Soc; Am Chem Soc; Am Diabetes Asn; Can Biochem Soc; Brit Biochem Soc. Res: Proteolytic enzymes; digestion; nutrition; hormones; metabolism. Mailing Add: Dept of Physiol Univ of Toronto Toronto ON Can

CAMPBELL, JAMES A, b Moweaque, Ill, Nov 29, 17; m 44; c 3. MEDICINE, CARDIOLOGY. Educ: Knox Col, Ill, AB, 39; Harvard Med Sch, MD, 43. Prof Exp: Resident path, Univ Chicago, 41-42; from intern to resident med, Boston City Hosp, 43-44; Harvey Cushing fel surg, Johns Hopkins Univ, 47-48; asst prof med, Presby Hosp, Univ Ill, 48-51; dean, Albany Med Col, 51-53; prof med & chmn dept, Presby-St Luke's Hosp, Univ Ill, 53-71; PROF MED, RUSH MED COL, 71-; PRES, RUSH-PRESBY-ST LUKE'S MED CTR, 69- Concurrent Pos: Pres, Presby-St Luke's Hosp, 64-69. Res: Cardiac physiology. Mailing Add: 1753 W Congress Pkwy Chicago IL 60612

CAMPBELL, JAMES A, b Chipley, Fla, Apr 12, 28; m 55; c 5. BIOLOGY, ECOLOGY. Educ: Fla Agr & Mech Univ, BS, 51, MEd, 56; Pa State Univ, DEd(biol sci), 62. Prof Exp: Sci teacher, Roulhac High Sch, 51-56; div chmn biol, Rosenwald Jr

Col, 58-64; assoc prof, Tenn State Univ, 64-74; MEM STAFF, DEPT BIOL, AM BAPTIST COL, 74- Mem: AAAS; Nat Asn Biol Teachers. Res: Fresh water ecology; influence of selected abiotic factors on population density. Mailing Add: Dept of Biol Am Baptist Col Nashville TN 37207

CAMPBELL, JAMES ALEXANDER, b Guelph, Ont, Oct 10, 13; m 39; c 2. FOOD SCIENCE, NUTRITION. Educ: Univ Toronto, BSA, 36; McGill Univ, MSc, 38, PhD(agr chem), 47. Prof Exp: Agr asst animal nutrit, Chem Div, Sci Serv, Can Dept Agr, Ottawa, 38-41, chemist, Vitamin & Physiol Res Lab, 41-48; chief vitamins & nutrit lab, 48-62, dir res labs, 63-67, sr sci adv foods, 67-71, actg dir nutrit bur, 71-73, DEP DIR CARIBBEAN FOOD & NUTRIT INST, FOOD & DRUG DIRECTORATE, DEPT NAT HEALTH & WELFARE, 73- Concurrent Pos: Past vpres & dir, Prof Inst Pub Serv Can, 52-56; consult, Protein Adv Group, WHO-Food & Agr Orgn-UNICEF, 61-; vis prof food technol & nutrit, Am Univ Beirut, 62-63; Nutrit Soc Can rep, Nat Comt, Int Union Nutrit Sci, 64, chmn comt food stand. Mem: Am Inst Nutrit; Animal Nutrit Res Coun; fel Chem Inst Can; Can Inst Food Technol; Nutrit Soc Can (past pres). Res: Chemical, microbiological and biological assays for vitamins; studies with rats; protein evaluation; physiological availability of drugs and vitamins; evaluation of drugs in oral prolonged action dosage forms; food chemistry; nutrition surveys; food and nutrition policy. Mailing Add: Caribbean Food & Nutrit Inst PO Box 140 MONA Kingston 7 Jamaica

CAMPBELL, JAMES ARTHUR, b Elyria, Ohio, Oct 1, 16; m 38; c 2. PHYSICAL CHEMISTRY. Educ: Oberlin Col, AB, 38; Purdue Univ, MS, 39; Univ Calif, PhD(chem), 42. Prof Exp: Instr chem, Manhattan Proj, Univ Calif, 42-45, instr plutonium res, 44-45; from asst prof to prof chem, Oberlin Col, 45-57; dir chem educ mat study, 60-63; PROF CHEM & CHMN DEPT, HARVEY MUDD COL, 57- Concurrent Pos: Advan educ fel, Cambridge Univ, 52-53; Guggenheim fel, Kyoto Univ & Cambridge Univ, 63-64; UNESCO sci teaching adv, Asia, 69-70; NSF fac fel, Ctr Pop Studies, Harvard, 70; resident scholar, Villa Serbelloni, 72; AAAS exchange lectr, 73; vis prof, Chinese Univ Hong Kong, 75-76. Honors & Awards: Mfg Chem Teaching Award, 62; James Flack Norris Award, Am Chem Soc, 64, Sci Apparatus Brokers' Award, 72. Mem: AAAS; Am Chem Soc; The Chem Soc; Int Union Pure & Appl Chem. Res: X-ray diffraction; vapor pressure measurement; absorption spectrometry; films for education in chemistry. Mailing Add: Dept of Chem Harvey Mudd Col Claremont CA 91711

CAMPBELL, JAMES ATLAS, pharmaceutical chemistry, see 12th edition

CAMPBELL, JAMES B, b Fraserburgh, Scotland, Sept 16, 39; m 66. MEDICAL MICROBIOLOGY. Educ: Aberdeen Univ, BSc, 62; Univ Alta, PhD(biochem), 65. Prof Exp: Asst biochem, Univ Alta, 62-65, fel, 65-66; trainee virol, Wistar Inst, 66-67, res asst, 67-68; asst prof, 68-70, ASSOC PROF VIROL, UNIV TORONTO, 70- Concurrent Pos: Spec lectr, Sch Hyg, Univ Toronto, 67-68; consult virol & biochem, Extendicare Diag Serv, Div Extendicare, Can Ltd, Willowdale, Ont, 71- Mem: Am Soc Microbiol; Tissue Cult Asn; Can Soc Microbiol. Res: Pathogenesis and biochemistry of animal viruses; tissue culture. Mailing Add: Dept of Microbiol & Parasitol Fac of Med Univ of Toronto Toronto ON Can

CAMPBELL, JAMES DOW, b Nashville, Tenn, Oct 26, 07; m 49. MATHEMATICS. Educ: Vanderbilt Univ, BA, 33; Univ Ill, MA, 34, PhD(math), 45. Prof Exp: From instr to prof, 38-73, EMER PROF MATH, RENSSELAER POLYTECH INST, 73- Mem: Am Math Soc; Math Asn Am. Res: Applied mathematics; differential equations; parametric theory of singular parabolic partial differential equations. Mailing Add: Dept of Math Rensselaer Polytech Inst Troy NY 12181

CAMPBELL, JAMES F, JR, solid state physics, see 12th edition

CAMPBELL, JAMES FULTON, b Philadelphia, Pa, Nov 24, 32; m 55, 69; c 3. PSYCHOPHYSIOLOGY, SCIENCE WRITING. Educ: Rutgers Univ, BSc, 55; Univ Va, MA, 60, PhD(psychol), 64. Prof Exp: Res assoc autonomic psychophysiol, Fels Res Inst, 62-64, USPHS fel psychophysiol, 64-65; asst prof, Antioch Col, 63-65; res assoc psychol, McGill Univ, 65-68, asst prof, 68-73; ASSOC PROF PSYCHOL, CARLETON UNIV, 73- Concurrent Pos: Proj officer, Ont Educ Commun Authorities, 75- Mem: AAAS; Am Psychol Asn. Res: Autonomic nervous system control of behavior; electrical brain stimulation and behavior. Mailing Add: Dept of Psychol Carleton Univ Ottawa ON Can

CAMPBELL, JAMES L, b Los Angeles, Calif, Sept 28, 24; m 60; c 3. INVERTEBRATE ZOOLOGY. Educ: Univ Calif, Berkeley, AB, 49, MA, 51; Univ Calif, Los Angeles, PhD(zool), 67. Prof Exp: ASSOC PROF BIOL, LOS ANGELES VALLEY COL, 55- Mem: AAAS. Res: Malacology, study of digestive system of Haliotis cracherodii; echinoderm biology, study of echinoid hoemol system; freshwater primary productivity; marine invertebrates. Mailing Add: Dept of Biol Los Angeles Valley Col Van Nuys CA 91401

CAMPBELL, JAMES NICOLL, b St Thomas, Ont, June 15, 30; m 54; c 4. MICROBIOLOGY, BIOCHEMISTRY. Educ: Univ Western Ont, BA, 51; Univ BC, BA, 55, MSc, 57; Univ Chicago, PhD(microbiol), 60. Prof Exp: From asst prof to assoc prof, 60-69, PROF MICROBIOL, UNIV ALTA, 69- Mem: Am Soc Microbiol; Can Soc Microbiol; Brit Soc Gen Microbiol. Res: Heterotrophic bacterial metabolism of inorganic nitrogen and sulfur compounds; isolation and characterization of bacterial enzymes; isotope fractionation by bacterial enzymes; characterization of bacterial pigments. Mailing Add: Dept of Microbiol Univ of Alta Edmonton AB Can

CAMPBELL, JAMES STEWART, b Bear River, NS, June 10, 23; m 47; c 5. PATHOLOGY. Educ: Dalhousie Univ, BSc, 43, MD, 47; Am Bd Path, dipl, 52; Royal Col Path, fel, 71; FRCPS(C). Prof Exp: Asst path, Med Sch, Tufts Univ, 49-52, instr, 52-53; from asst prof to prof path, Univ Ottawa, 53-74; PROF PATH & CHMN DEPT, MEM UNIV NFLD, 74- Concurrent Pos: From jr asst pathologist to asst pathologist, New Eng Med Ctr, 49-53; registr, Registry for Tissue Reactions to Drugs, Ottawa, 70-; consult pathologist, Can Tumor Ref Ctr, 58- & Nat Defence Med Ctr, Ottawa, 61-; pathologist, Ottawa Gen Hosp, 53-74; chmn dept lab med, St John Gen Hosp, 76- Mem: Int Acad Path; Can Med Asn; Can Asn Path; fel Am Col Path. Res: Endocrine and gynecologic pathology. Mailing Add: Dept of Path Mem Univ of Nfld St John's NF Can

CAMPBELL, JAMES WAYNE, b Highlandville, Mo, Mar 2, 32; m 60; c 2. COMPARATIVE BIOCHEMISTRY, COMPARATIVE PHYSIOLOGY. Educ: Southwest Mo State Col, BS, 53; Univ Ill, MS, 55; Univ Okla, PhD(zool), 58. Prof Exp: Asst zool, Univ Ill, 53-55 & Univ Okla, 55-56; Nat Acad Sci-Nat Res Coun fel, Johns Hopkins Univ, 58-59; from instr to assoc prof biol, 59-70, PROF BIOL, RICE UNIV, 70-, CHMN DEPT, 74- Concurrent Pos: NIH spec fel, Univ Wis, 64-65; USPHS career develop award, 66-70; consult prev med, NASA Manned Spacecraft Ctr, Houston; consult, BMS, NSF, 72-73 & 74-75, prog dir, Regulatory Biol, 73-74. Mem: Fel AAAS; fel Am Inst Chemists; Am Physiol Soc; Am Soc Zool; Am Soc Biol Chemists. Res: Comparative biochemistry of nitrogen metabolism. Mailing Add: Dept of Biol Rice Univ Houston TX 77001

CAMPBELL, JEPTHA EDWARD, JR, b Atlanta, Ga, Sept 16, 23; m 49; c 3. FOOD SCIENCE. Educ: Rollins Col, BS, 47; Univ Wis, MS, 49, PhD(biochem, physiol), 51. Prof Exp: Proj assoc biochem, Univ Wis, 51-52; group leader radiobiol, Mound Lab, Monsanto Chem Co, 52-55; chief food chem, USPHS, 55-70; chief microbiol biochem br, Div Microbiol, 70-71, ASST DIR BUR FOODS, FOOD & DRUG ADMIN, 71- Honors & Awards: Serv Award, US Dept Health, Educ & Welfare, 62. Mem: Am Chem Soc; Am Inst Nutrit; Am Pub Health Asn; Asn Off Agr Chem. Res: Nutrition; radiobiology; food technology; analytical food chemistry. Mailing Add: Food & Drug Admin 1090 Tusculum Ave Cincinnati OH 45226

CAMPBELL, JOHN ALEXANDER, b Detroit, Mich, July 7, 40; m 69. MEDICINE, CHEMICAL ENGINEERING. Educ: Univ Mich, BSE, 62, MSE, 64, PhD(bioeng), 67; Rush Med Col, MD, 74. Prof Exp: Chem engr, Process Develop Dept, Parke, Davis & Co, Mich, 64-65; asst attend biomed eng, Presby-St Luke's Hosp, 67-74; HOUSE STAFF, DEPT INTERNAL MED, UNIV ILL, CHICAGO, 74- Mem: Am Inst Chem Eng; assoc Am Col Physicians. Res: Mathematical models of physiological systems; patient monitoring, using computers; artificial organs. Mailing Add: 502 Stratford Elmhurst IL 60126

CAMPBELL, JOHN ARTHUR, b Muskogee, Okla, Nov 2, 30; m 53; c 3. PETROLOGY. Educ: Univ Tulsa, BGeol, 55; Univ Colo, MS, 57, PhD(geol), 66. Prof Exp: From instr to assoc prof geol, Colo State Univ, 57-74; GEOLOGIST, US GEOL SURV, 74- Mem: Fel AAAS; fel Geol Soc Am; Am Asn Petrol Geol; Soc Econ Paleont & Mineral. Res: Petrology, stratigraphy and depositional environ environment of uranium bearing Paleozoic Rocks of the Southwestern United States. Mailing Add: Uranium-Thorium Br US Geol Surv Denver Fed Ctr Denver CO 80225

CAMPBELL, JOHN B, organic chemistry, see 12th edition

CAMPBELL, JOHN BRYAN, b Fairmont, Nebr, Mar 30, 33; m 56; c 4. ENTOMOLOGY. Educ: Univ Wyo, BS, 61, MS, 63; Kans State Univ, PhD(entom), 66. Prof Exp: Entomologist, Agr Res Serv, USDA, Univ Nebr-Lincoln, 66-70; ASSOC PROF ENTOM, UNIV NEBR, NORTH PLATTE STA, 70- Mem: Entom Soc Am. Res: Biology and ecology of rangeland grasshoppers and livestock insects; biology and control of livestock insects in Central Plains. Mailing Add: Dept of Entom Univ of Nebr North Platte NE 69101

CAMPBELL, JOHN DUNCAN, b Hamilton, Ont, Apr 22, 23. PALEOBOTANY. Educ: McMaster Univ, BA, 44; Univ BC, MA, 49; McGill Univ, PhD, 52. RES OFFICER, ALTA RES COUN, 53- Res: Coal distribution and systemstics; palaeoecology and wood of Cretaceous and Tertiary continental deposits of Central Canada. Mailing Add: Alta Res Coun 11315 87th Ave Edmonton AB Can

CAMPBELL, JOHN HOWLAND, b Oklahoma City, Okla, Mar 9, 38; m 62; c 2. ANATOMY. Educ: Calif Inst Technol, BA, 60; Harvard Univ, PhD(biol), 64. Prof Exp: ASSOC PROF ANAT, SCH MED, UNIV CALIF, LOS ANGELES, 64- Concurrent Pos: NSF grants, Pasteur Inst, Paris, France, 64-65 & Commonwealth Sci & Indust Res Orgn, Canberra, Australia, 65-66. Mem: AAAS; Am Asn Anat. Res: Microbial genetics; evolution at the molecular level. Mailing Add: Dept of Anat Univ of Calif Sch of Med Los Angeles CA 90024

CAMPBELL, JOHN HYDE, b Ithaca, NY, Dec 2, 47; m 68. PHYSICAL CHEMISTRY. Educ: Rochester Inst Technol, BS, 70; Univ Ill, MS, 72, PhD(phys chem), 75. Prof Exp: Chemist, Eastman Kodak Co, 67-70; RES CHEMIST ENERGY SYSTS, LAWRENCE LIVERMORE LAB, UNIV CALIF, 75- Mem: Am Phys Soc. Res: Chemical kinetics related to in-situ coal gasification; ground water pollution from in-situ coal gasification. Mailing Add: Lawrence Livermore Lab Univ of Calif Livermore CA 94550

CAMPBELL, JOHN MARTIN, b Sedro Woolley, Wash, May 5, 27; m 49; c 3. ANTHROPOLOGY. Educ: Univ Wash, BA, 50; Yale Univ, PhD, 62. Prof Exp: From asst prof to assoc prof anthrop, George Washington Univ, 59-64; assoc prof, 64-68, chmn dept, 64-72, PROF ANTHROP, UNIV NMEX & CUR COLLECTIONS, MAXWELL MUS ANTHROP, 68- Mem: Fel AAAS; fel Am Anthrop Asn; Soc Am Archaeol; fel Arctic Inst NAm; Am Ornith Union. Res: Archaeology and ethnology of western North America, Pacific Basin and the northern regions. Mailing Add: Dept of Anthrop Univ of NMex Albuquerque NM 87106

CAMPBELL, JOHN RICHARD, b Pratt, Kans, Jan 16, 32; m 62; c 3. PEDIATRIC SURGERY. Educ: Univ Kans, BA, 54, MD, 58; Am Bd Surg, dipl, 64, cert pediat surg, 75. Prof Exp: Intern, Univ Pa Hosp, 58-59; surg resident, Med Ctr, Univ Kans, 59-63; asst instr pediat surg, Sch Med, Univ Pa, 65-67; from asst prof to assoc prof, 67-72, PROF SURG & PEDIAT, MED SCH, UNIV ORE, 72-, CHIEF PEDIAT SURG, 67- Concurrent Pos: Surg resident, Children's Hosp Philadelphia, 65-67. Mem: Am Pediat Surg Asn; Am Col Surg; Am Acad Pediat; AMA. Res: Pediatric surgery and oncology; shock in the newborn infant. Mailing Add: Dept of Surg Univ of Ore Med Sch Portland OR 97201

CAMPBELL, JOHN ROY, b Goodman, Mo, June 14, 33; m 54; c 3. DAIRY HUSBANDRY. Educ: Univ Mo, BS, 55, MS, 56, PhD(nutrit), 60. Prof Exp: From asst prof to assoc prof, 60-70, PROF DAIRY HUSB, UNIV MO-COLUMBIA, 70- Concurrent Pos: Southern Ice Cream Mfrs fel, 55-56; Danforth assoc, 70. Honors & Awards: Distinguished Teaching Award, Ralston-Purina Co, 74. Mem: AAAS; Am Dairy Sci Asn; Am Soc Animal Sci; NY Acad Sci. Res: Dairy cattle physiology, especially health, nutrition, production and management; recycling corrugated paper through ruminants in the production of meat and milk. Mailing Add: Dept of Dairy Husb 209 Eckles Hall Univ of Mo Columbia MO 65201

CAMPBELL, JOSEPH DEMPSEY, b Harris, Sask, June 21, 10; m 55; c 2. PLANT NUTRITION. Educ: Univ Sask, BSA, 41; Univ Western Ont, DBA, 47; McGill Univ, MSc, 50; Mich State Univ, PhD(hort), 53. Prof Exp: Elem teacher, Sask, 29-37; hort asst, Melfort Exp Farm, 37-41; asst, Can Industs Ltd, Que, 47-48; asst hort, MacDonald Col, McGill Univ, 48-50 & Mich State Univ, 50-52; instr veg crops, Univ Mass, 52-53; horticulturist, Olin Mathieson Chem Corp, 53-59; proj mgr, Monsanto Chem Co, 59-61; assoc prof veg crops, 61-73, prof plant sci, 73-75, RESEARCHER PLANT SCI, UNIV MAN, 75- Concurrent Pos: Colombo Plan adv, Khonkaen Univ, Thailand, 67-69; dir, Protected Growing Res Proj. Mem: AAAS; Am Soc Hort Sci; Int Soc Hort Sci. Res: Tissue testing for greenhouse production of vegetable crops and flowers. Mailing Add: Dept of Plant Sci Univ of Man Winnipeg MB Can

CAMPBELL, KENNETH BRUCE, b Gridley, Calif, Nov 30, 40; m 71; c 3. CARDIOVASCULAR PHYSIOLOGY. Educ: Univ Calif, Davis, BS, 63, DVM, 68, PhD(physiol), 73. Prof Exp: Assoc physiol, Backus Res Inst, 73-74, ASST PROF BIOENG & PHYSIOL, UNIV PA, 74- Res: Comparative cardiovascular physiology; arterial hemodynamics; circulatory system component interactions; neural control of circulatory system. Mailing Add: Dept of Bioeng 458 Moore Sch D2 Univ of Pa Philadelphia PA 19174

CAMPBELL, KENNETH NEILSEN, b Hillsdale, Mich, May 31, 05; m 33. MEDICINAL CHEMISTRY. Educ: Univ Chicago, BS, 28, PhD(org chem), 32. Prof Exp: Asst, Univ Chicago, 29-32; fel, Pa State Col, 33-34, asst chem, 34-35; Univ Ill, 35-36; instr org chem, Univ Notre Dame, 36-38, asst prof, 38-40, assoc prof, 40-45, prof, 45-54; dir med chem, Mead Johnson Res Ctr, 53-70, consult, 70-73; PHARMACEUT CONSULT, CAMPBELL & ASSOCS, 73- Concurrent Pos: Pharmaceut indust consult, 44-53; researcher cancer chemother, NIH, 45-54; consult, USPHS, 49-52; lectr, Univ Tex, 58. Honors & Awards: Ind Technol Col Citation of Merit, 58; Univ Notre Dame Alumni Sci Award, 65; Tri-State Coun Sci & Eng Tech Achievement Award, 73. Mem: Fel AAAS; fel Am Inst Chem; Am Chem Soc; Sigma Xi; NY Acad Sci. Res: Aliphatic hydrocarbons; reduction; synthetic drugs; antimalarials and growth inhibitors; amines; quinolines; ethyleneimines; acetylenes. Mailing Add: 8216 Petersburg Rd Evansville IN 47711

CAMPBELL, KENNETH WILFORD, b Ingersoll, Ont, Aug 21, 42; m 65; c 2. PLANT BREEDING, PLANT CYTOGENETICS. Educ: Univ Guelph, BSc, 71, PhD(plant breeding & genetics), 75. Prof Exp: RES SCIENTIST, AGR CAN RES BR, 74- Concurrent Pos: Mem, Barley Subcomt, Can Comt Grain Breeding, 75- Mem: Can Soc Agron; Genetics Soc Can; Crop Sci Soc Am. Res: Breeding superior six-row, and recently two-row barley for the Canadian prairies, with emphasis on malting quality; intercrosses of two-row and six-row types and haploidy. Mailing Add: Agr Can Res Br Res Sta PO Box 610 Brandon MB Can

CAMPBELL, KIRBY I, b San Jose, Calif, Jan 12, 33; m 54; c 1. ENVIRONMENTAL HEALTH. Educ: Univ Calif, Davis, BS, 55, DVM, 57; Harvard Univ, MPH, 64. Prof Exp: Vet, Butte Vet Hosp, Oroville, Calif, 59-61; proj vet officer, Air Pollution Res Ctr, Univ Calif, Riverside, USPHS, 61-63, actg chief chronic & explor toxicol unit, Health Effects Res Prog, Nat Air Pollution Control Admin, 64-68, dep chief, 68-70, chief vet med, 68-70; CHIEF COMP & REPROD TOXICITY SECT, TOXICOL DIV, US ENVIRON PROTECTION AGENCY, CINCINNATI, 72- Concurrent Pos: Mem proj group, Air Pollution Res Adv Comt, Coord Res Coun, 68-72. Mem: AAAS; Am Vet Med Asn; Am Asn Lab Animal Sci; Am Col Vet Toxicologists. Res: Experimental biology, environmental toxicology, including inhalation; experimental laboratory animal science; mammalian, submammalian, in vivo, in vitro bioassay systems; immune response and defense systems; reproductive funciton and developmental biology. Mailing Add: Toxicol Div Health Effects Res Lab US Environ Protection Agency Cincinnati OH 45268

CAMPBELL, LARRY EDWIN, b Gas City, Ind, Feb 21, 41; c 4. PHYSICAL INORGANIC CHEMISTRY. Educ: Purdue Univ, BSChE, 63, PhD(chem), 67. Prof Exp: Sr res chem, Corning Glass Works, 66-68, mgr silicate chem res, 68-71, mgr surface chem res, 71-75; MGR CHEM RES & DEVELOP, ENGLEHARD INDUST DIV, ENGLEHARD MINERALS & CHEM CO, 75- Mem: Am Chem Soc; Am Inst Chem Engrs; Am Soc Testing & Mat. Res: Catalyst preparation, characterization, evaluation; preparation of precious metal salts, compounds; refining electrochemistry; fuel cells; biological use of precious metals. Mailing Add: Englehard Minerals & Chem Menlo Park Edison NJ 08817

CAMPBELL, LARRY ENOCH, b Brookville, Pa, July 21, 38; m 65; c 3. SOLID STATE PHYSICS. Educ: Carnegie Inst Technol, BS, 60, MS, 62, PhD(physics), 66. Prof Exp: Appointee physics, Argonne Nat Lab, 66-68; ASST PROF PHYSICS, HOBART & WILLIAM SMITH COLS, 66- Concurrent Pos: Sci asst, Tech Univ Munich, 72-73. Mem: Am Phys Soc. Res: Application of the Mössbauer effect to solid state and nuclear physics problems. Mailing Add: Dept of Physics Hobart & William Smith Cols Geneva NY 14456

CAMPBELL, LINZY LEON, b Panhandle, Tex, Feb 10, 27; m 53. MICROBIOLOGY, BIOCHEMISTRY. Educ: Univ Tex, BA, 49, MS, 50, PhD(bact, biochem), 52. Prof Exp: Res scientist I marine microbiol, Dept Bact, Univ Tex, 47-50, res scientist II food bact, 50-51; Nat Microbiol Inst res fel plant biochem, Univ Calif, 52-54; from asst prof & asst bacteriologist to assoc prof hort & assoc bacteriologist, State Col Wash, 54-59; assoc prof microbiol, Western Reserve Univ, 59-62; prof microbiol, Univ Ill, Urbana-Champaign, 62-72, dir sch life sci, 71-72, head dept, 63-71; PROF MICROBIOL & PROVOST & VPRES ACAD AFFAIRS, UNIV DEL, 72- Concurrent Pos: Ed, J Bact, Am Soc Microbiol, 64-65, ed-in-chief, 65- Mem: AAAS; Am Chem Soc; Am Soc Biol Chem; Am Soc Microbiol (pres, 73-74); Brit Soc Gen Microbiol. Res: Microbial metabolism; enzymes; fermentations; food microbiology; thermophilic microorganisms. Mailing Add: 104 Hullihen Hall Univ of Del Newark DE 19711

CAMPBELL, LOIS JEANNETTE, b Toledo, Ohio, Nov 16, 23. GEOLOGY. Educ: Univ Mich, BS, 44; Ohio State Univ, PhD(glacial geol), 55. Prof Exp: Jr geologist statist & eval, Humble Oil & Refining Co, 45-47; instr geol, 54-58, ASST PROF GEOL, UNIV KY, 58- Mem: Am Asn Petrol Geologists; Geol Soc Am; Nat Asn Geol Teachers. Res: Pleistocene geology; late paleozoic invertebrates; geomorphology. Mailing Add: Dept of Geol Univ of Ky Lexington KY 40506

CAMPBELL, LORNE ARTHUR, b Saskatoon, Sask, Nov 10, 23; m 47. PHARMACOLOGY. Educ: Univ BC, BScAg, 53, MSc, 54; Univ Calif, Davis, PhD(comp pharmacol & toxicol), 62. Prof Exp: Fel, Univ Calif Med Sch, 62-63; dir path & toxicol, Smith Kline & French Lab, 63-67; asst prof pharmacol, Tex A&M Univ, 67-69; head toxicol, Abbott Lab, 69-71; staff officer, Comt Food Protection, Nat Res Coun, Nat Acad Sci, 71-73; FOOD REGULATIONS ADMINR, SUNKIST GROWERS, INC, 73- Concurrent Pos: Chmn, Nutrit Labeling Comt, United Fresh Fruit & Veg Asn, 74-; mem, Bd Dirs, Indust Comt Citrus Additives & Pesticides, 74- Mem: Inst Food Technologists; Asn Food & Drug Offs; Sigma Xi. Res: The role of dietary fiber in nutrition; role of pectin as a dietary fiber; the safety of low methoxyl pectin. Mailing Add: Sunkist Growers Inc PO Box 7888 Van Nuys CA 91409

CAMPBELL, LOUIS LORNE, b Winnipeg, Man, Oct 20, 28; m 54; c 3. MATHEMATICS. Educ: Univ Man, BSc, 50; Iowa State Univ, MS, 51; Univ Toronto, PhD(math), 55. Prof Exp: Defence sci serv officer, Defence Res Telecommunications Estab, Can, 54-58; from asst prof to assoc prof math, Univ Windsor, 58-63; assoc prof, 63-67, PROF MATH, QUEEN'S UNIV, ONT, 67- Mem: Am Math Soc; Soc Indust & Appl Math; Can Math Cong; Inst Elec & Electronics Engrs. Res: Information and communication theory. Mailing Add: Dept of Math Queen's Univ Kingston ON Can

CAMPBELL, MALCOLM JOHN, b Wollongong, Australia, Aug 16, 37; m 70. SPACE PHYSICS. Educ: Univ Sydney, BSc, 58, PhD(physics), 63. Prof Exp: Fel space res, Goddard Space Flight Ctr, NASA, Md, 63-65; assoc, Ctr Radiophysics & Space Res, Cornell Univ, 65-72; res assoc, State Univ NY at Syracuse Univ, 72-73; ASST PROF SPACE PHYSICS, WASH STATE UNIV, 73- Mem: Am Geophys Union. Res: Atmospheric physics and chemistry. Mailing Add: Col of Eng Wash State Univ Pullman WA 99163

CAMPBELL, MARY KATHRYN, b Philadelphia, Pa, Jan 20, 39. BIOPHYSICAL CHEMISTRY. Educ: Rosemont Col, BA, 60; Ind Univ, PhD(phys chem), 65. Prof Exp: Instr radiol sci, Johns Hopkins Univ, 65-68; asst prof chem, 68-74, ASSOC PROF CHEM, MT HOLYOKE COL, 74- Concurrent Pos: Vis scientist, Inst Molecular Biol, Univ Paris, 74-75. Mem: Am Chem Soc. Res: Specific interactions between proteins and nucleic acids; physical chemistry of nucleic acids and proteins. Mailing Add: Dept of Chem Mt Holyoke Col South Hadley MA 01075

CAMPBELL, MICHAEL DAVID, b Lancaster, Ohio, Aug 8, 41; m 65; c 3. GROUNDWATER GEOLOGY, EXPLORATION GEOLOGY. Educ: Ohio State Univ, BA, 66; Rice Univ, MA, 76. Prof Exp: Res asst, Ohio State Univ, 64-66; staff geologist, Continental Oil Co of Australia, Ltd, 66-69; dist geologist, United Nuclear Corp, Wyo, 69-70; dir res, Nat Water Well Asn Res Facil, 70-75, CONSULT RES GEOLOGIST, DEPT GEOL, RICE UNIV, 73- Concurrent Pos: Abstr ed, Ground Water, 66-70; tech consult, Water Well J, 71-73; res dir, Nat Demonstration Water Proj & Comn Rural Water, Washington, DC, 73-74. Mem: Am Asn Petrol Geologists; Nat Water Well Asn; Geol Soc Am; Am Inst Mining Engrs; Am Soc Testing & Mat. Res: Ground water and mineral development; exploration; pollution; geochemistry; rural water engineering; strata bound mineral exploration using groundwater geochemistry. Mailing Add: Dept of Geol Rice Univ Houston TX 77005

CAMPBELL, MICHAEL FLOYD, b Sparta, Ill, Nov 5, 42; m 64; c 2. FOOD SCIENCE Educ: Univ Ill, BS, 64, MS, 66, PhD(food sci), 68. Prof Exp: Microbiologist, R J Reynolds Indust, 68-70; sr food scientist, Libby McNeill & Libby, 70-71; sr food scientist, 71-73, GROUP LEADER VEG PROTEINS, A E STALEY MFG CO, 73- Mem: Inst Food Technologists; Am Soc Oil Chemists; Am Meat Sci Asn. Res: Vegetable protein products and process devélopment; corn sweeteners; food starches; food microbiology. Mailing Add: 1885 Ravina Park Rd Decatur IL 62526

CAMPBELL, MILTON HUGH, b Billings, Mont, Sept 2, 28; m 52; c 3. ANALYTICAL CHEMISTRY, NUCLEAR ENGINEERING. Educ: Mont State Col, BS, 51; Univ Wash, MS, 61. Prof Exp: Head analyst chem, Ideal Cement Co, 51-53 & 55; jr chemist, Hanford Atomic Prod Oper, Gen Elec Co, 55-57, anal chemist, 57-59, process chemist, 59-61, sr engr, 61-65; sr engr, Chem Processing Div, Isochem Inc, 65-67, staff engr, 67, mgr separations chem lab, Res & Develop, Atlantic Richfield Hanford Co, 67-73, mgr waste mgt & storage technol, 73-74; SR FUEL REPROCESSING ENGR, EXXON NUCLEAR CO, INC, 74-Mem: Am Chem Soc; Am Soc Testing & Mat; Am Nuclear Soc; Inst Nuclear Mat Mgt. Res: Inorganic ion exchange and liquid-liquid separations; nucleonics; gas chromatography; spectrophotometric analysis; thermochemistry; radioactive material measurements and waste management. Mailing Add: 2119 Beech Richland WA 99352

CAMPBELL, NEIL, b Medicine Hat, Alta, Apr 27, 14; m 44; c 3. GEOLOGY. Educ: Univ Alta, BSc, 37; Mass Inst Technol, PhD(econ geol), 43. Hon Degrees: LLD, Univ Alta, 70. Prof Exp: From dist geologist to explor supt, Consol Mining & Smelting Co, 50-65; chief geologist, Cominco Am Inc, Wash, 65-68; CONSULT GEOLOGIST, 68-; PRES, SILVERMONT INC, 69- Honors & Awards: Barlow Mem Gold Medal, Can Inst Mining & Metall, 47. Mem: Royal Soc Can; Geol Soc Am; Soc Econ Geologists; Can Asn Prof Engrs; Am Asn Petrol Geologists. Res: Economic and petroleum geology. Mailing Add: 607 Northtown Off Bldg Spokane WA 99207

CAMPBELL, NEIL JOHN, b Los Angeles, Calif, Aug 26, 25; Can citizen; m 54; c 2. PHYSICAL OCEANOGRAPHY. Educ: McMaster Univ, BSc, 50, MSc, 51; Univ BC, PhD(physics), 55. Prof Exp: Phys oceanogr Pac oceanog group, Fisheries Res Bd Can, 55, Atlantic oceanog group, 55-59, oceanogr in chg, 59-63; CHIEF OCEANOGR, MARINE SCI BR, DEPT ENERGY, MINES & RESOURCES, 63- Concurrent Pos: Lectr, Dalhousie Univ, 60-63. Res: Arctic and environmental oceanography. Mailing Add: Marine Sci Br Dept of Energy, Mines & Res Ottawa ON Can

CAMPBELL, NORMAN E ROSS, b Ft William, Ont, Oct 11, 20; m 47; c 2. MICROBIOLOGY. Educ: Ont Agr Col, BSA, 44; Univ Man, MSc, 49, PhD(microbiol), 60. Prof Exp: From asst prof to assoc prof, 49-68, PROF MICROBIOL, UNIV MAN, 69-, CHMN DIV BIOL SCI, 74- Mem: Can Soc Microbiol; Arctic Inst NAm. Res: Soil microbiology; nitrogen fixation in subarctic and arctic soils. Mailing Add: Div of Biol Sci Univ of Man Ft Garry Winnipeg MB Can

CAMPBELL, PAUL ANDREW, b Frankfort, Ind, Dec 25, 02; m 33; c 1. AEROSPACE MEDICINE. Educ: Univ Chicago, ScB, 24, MD, 28; Am Bd Otolaryngol, dipl, 38; Am Bd Prev Med, dipl, 54. Prof Exp: Asst prof otolaryngol, Rush Med Col, Chicago, 39-41; chief, Depts Ophthal & Otolaryngol, Sch Aviation Med, Dept Air Force, 41-42, dir res div, 42-45, 50-53, asst air attache, Am Embassy, London, 53-56, asst comdr, Air Force Off Sci Res, 56-58, chief space med div, Sch Aviation Med, 58-59, adv studies group, Aerospace Med Ctr, 59-61, comdr, Air Force Sch Aerospace Med, 61-62; consult aerospace med, 63-69; DIR SPACE SCI, TRINITY UNIV, 69- Concurrent Pos: Vpres, Am Med Soc, Vienna, 34-35 & 44-; chief consult, Mid-West Area, US Vet Admin, 46-50; nat consult to Surgeon Gen, US Dept Air Force & Surgeon Gen, US Dept Army, 48-50. Honors & Awards: Legion of Merit, 46; Oak Leaves, 63; Louis Bauer Founder's Award, 63; Bexar Library Award; Hubertus Strughold Award, Space Med Asn; Melbourne Boynton Award Flight Safety, Am Astronaut Soc. Mem: Fel Aerospace Med Asn; Am Otol Soc; Am Acad Ophthal & Otolaryngol; Am Col Surg; Am Inst Aeronaut & Astronaut. Res: Space medicine; aviation medicine; otolaryngology. Mailing Add: Apt 401 700 E Hildebrand San Antonio TX 78212

CAMPBELL, PAUL GILBERT, b Minneapolis, Minn, Sept 25, 25; m 51. ORGANIC CHEMISTRY. Educ: Univ Md, BS, 50; Univ SC, MS, 53; Pa State Univ, PhD(org chem), 57. Prof Exp: Chemist, NIH, 50-51; sr chemist, J T Baker Chem Co, 57-59 & Allied Chem Corp, 59-60; CHEMIST, NAT BUR STAND, 60- Honors & Awards: Silver Medal, Dept Com, 67. Mem: AAAS; Am Chem Soc; Am Soc Test & Mat. Res: Organic synthesis; bituminous building materials; organic coatings. Mailing Add: Nat Bur of Stand Bldg Res Div B-348 Washington DC 20234

CAMPBELL, PHIL M, JR, theoretical physics, astrophysics, see 12th edition

CAMPBELL, PRISCILLA ANN, b Mineola, NY, Aug 19, 40. IMMUNOLOGY. Educ: Colo Col, BA, 62; Univ Colo, MS, 65, PhD(cell biol), 68. Prof Exp: Fel immunol, Univ Calif, San Diego, 68-70 & Nat Jewish Hosp, Denver, 70-72; ASST PROF PATH, UNIV COLO MED CTR, DENVER, 72- Concurrent Pos: Mem staff, Div Allergy & Clin Immunol, Nat Jewish Hosp, Denver, 72-; NIH grant, 75-76. Mem: AAAS; Am Asn Immunologists; Am Soc Exp Path; Am Soc Microbiol. Res: Cell biology of immune response. Mailing Add: Div of Allergy & Clin Immunol Nat Jewish Hosp Denver CO 80206

CAMPBELL, RALPH EDMUND, b Providence, Utah, Sept 22, 27; m 52; c 5. FOREST SOILS. Educ: Utah State Univ, BS, 50, MS, 54. Prof Exp: Soil scientist, Exp Sta, SDak State Univ & USDA, 50-54 & NMex Agr Ext Sta, Tucumcari, 54-55, soil scientist, Southern Mountain Res Ctr, Agr Res Serv, Mont, 55-64, res soil scientist, 64; RES SOIL SCIENTIST, ROCKY MOUNTAIN FOREST & RANGE EXP STA, US FOREST SERV, 64- Mem: Soil Sci Soc Am; Am Soc Agron. Res: Soil fertility, moisture and management; forest soils problems related to nutrient cycling, sediment movement, water quality and forest climate. Mailing Add: Rocky Mountain Forest & Range Exp Sta US Forest Serv Flagstaff AZ 86001

CAMPBELL, RAYMOND EARL, b Ranger, Tex, Jan 4, 41; m 65; c 2. HORTICULTURE. Educ: Okla State Univ, BS, 63, MS, 66; Kans State Univ, PhD(hort), 72. Prof Exp: County exten agent, Coop Exten Serv, Okla State Univ, Coal County, 63-64 & Delaware County, 66-67; instr agr, Friends Univ, 67-68; county exten agent hort, Coop Exten Serv, Kans State Univ, 68-70; exten specialist voc hort, Va Polytech Inst & State Univ, 72-74; EXTEN SPECIALIST VEG CROPS, OKLA STATE UNIV, 74- Mem: Am Soc Hort Sci. Res: Investigation of environmental influence on vegetable crop production and yield with particular emphasis on temperature and humidity; investigation of applications of artificial soil mixes to greenhouse vegetable growing and vegetable crop alternatives. Mailing Add: 23 Preston Circle Stillwater OK 74074

CAMPBELL, RICHARD BRADFORD, geology, see 12th edition

CAMPBELL, RICHARD DANA, b Oklahoma City, Okla, June 12, 39; m 63; c 2. DEVELOPMENTAL BIOLOGY. Educ: Harvard Univ, BA, 61; Rockefeller Inst, PhD(biol), 65. Prof Exp: Asst prof organismic biol, 65-70, assoc prof develop & cell biol, 70-74, PROF DEVELOP & CELL BIOL, UNIV CALIF, IRVINE, 74- Mem: Am Soc Zool; Am Soc Cell Biol; Soc Develop Biol. Mailing Add: Dept of Develop & Cell Biol Univ of Calif Irvine CA 92664

CAMPBELL, ROBERT A, b Toledo, Ohio, Dec 21, 24; m 49; c 3. PEDIATRICS. Educ: Univ Calif, Berkeley, AB, 54; Univ San Francisco, MD, 58; Am Bd Pediat, dipl, 63, cert pediat nephrol, 74. Prof Exp: Res asst, Cancer Res Genetics Lab, Univ Calif, Berkeley, 54-58; USPHS fel pediat biochem, 61-63, assoc prof, 63-71, PROF PEDIAT, MED SCH, UNIV ORE, 71-, DIR PEDIAT RENAL-METAB LAB, 63- Mem: AAAS; Am Acad Pediat. Res: Metabolic disorders of infancy and childhood; cancer. Mailing Add: Dept of Pediat Univ of Ore Health Sci Ctr Portland OR 97201

CAMPBELL, ROBERT BENONI, b Providence, Utah, Feb 14, 21; m 42; c 5. AGRONOMY. Educ: Utah State Agr Col, BS, 43, MS, 51. Prof Exp: Asst soil scientist, USDA, 46-51; sr agronomist, Exp Sta, Hawaiian Sugar Planters Asn, 53-66; SOIL SCIENTIST, AGR RES SERV, USDA, 66- Mem: Am Soc Agron. Res: Agricultural physics related to the physical environment of plants. Mailing Add: USDA Agr Res Serv PO Box 3039 Florence SC 29501

CAMPBELL, ROBERT BRUCE, solid state physics, see 12th edition

CAMPBELL, ROBERT CALVIN, b Fort Worth, Tex, July 31, 48; m 69. MATHEMATICAL ANALYSIS. Educ: Trinity Univ, Tex, BA, 70; Univ Colo, MA, 72, PhD(math), 75. Prof Exp: VIS ASST PROF MATH, ROCKHURST COL, 75- Mem: Math Asn Am. Res: Characterization of potentially stable matrices. Mailing Add: Dept of Math Rockhurst Col Kansas City MO 64110

CAMPBELL, ROBERT DALE, b Omaha, Nebr, Dec 2, 14. GEOGRAPHY. Educ: Univ Colo, BA, 38, MA, 40; Clark Univ, PhD(geog), 49. Prof Exp: Prof geog, George Washington Univ, 47-66; regional planner, Ford Found, Calcutta, India, 64-66; vpres area systs, Matrix Corp, Arlington, Va, 66-70; PROF GEOG, UNIV NMEX, 70- Concurrent Pos: Prin investr var res projs, Off Naval Res, Off Qm Gen, Army Res Off, Outdoor Recreation Res Comn, Nat Park Serv, Bur Indian Affairs & local planning agencies, 47-; Fulbright prof, Univ Alexandria, 52-53 & Univ Peshawar, 57-58; vis lectr, Conf Am Studies, Oxford Univ, 55; mem Am studies adv comt, Am Coun Learned Socs, 62-63. Mem: Asn Am Geog. Res: Man-environment systems, including model building, synthesis and theory; psychological geography, including model personality and national character and cross-cultural communication. Mailing Add: Dept of Geog Univ of NMex Albuquerque NM 87106

CAMPBELL, ROBERT HENRY, b Williamson, WVa, Nov 5, 28; m 5 55; c 3. ANALYTICAL CHEMISTRY, ORGANIC CHEMISTRY. Educ: Marshall Univ, BS, 50, MS, 56; Purdue Univ, PhD(anal chem), 59. Prof Exp: Res anal chemist, Monsanto Res Corp, Mass, 59-61, sr res chemist, Monsanto Co, WVa, 61-66, res specialist, 66-69, res specialist, Rubber Chem Res Labs, 69-70, RES GROUP LEADER, MONSANTO CO, 70- Mem: AAAS; Am Chem Soc; Am Soc Testing & Mat. Res: Chemical analyses employing vapor phase chromatography; absorption spectroscopy; polarographic techniques. Mailing Add: Rubber Chem Res Labs Monsanto Co 260 Springside Dr Montrose OH 44313

CAMPBELL, ROBERT LOUIS, b Westerville, Ohio, Nov 16, 25; m 49; c 3. NEUROSURGERY. Educ: Baldwin-Wallace Col, BS, 45; Ohio State Univ, MD, 49; Am Bd Neurol Surg, dipl, 60. Prof Exp: From instr to asst prof, 57-64, PROF NEUROL SURG & DIR MED CTR, IND UNIV, INDIANAPOLIS, 65- Concurrent Pos: Consult, Vet Admin Hosp, Indianapolis, 57- & Naval Hosp, Great Lakes, Ill, 63-; mem, Subarachnoid Hemorrhage Study Sect, NIH, 62-; mem consult staff, Marion County Gen Hosp, 65- Mem: Cong Neurol Surg; Am Asn Neurol Surg; fel Am Col Surgeons; fel Int Col Angiol; AMA. Res: Diagnostic cerebral angiography; ventriculography and myelography with absorbable radiopaque media; treatment of intracranial aneurysms. Mailing Add: Dept of Neurol Surg Ind Univ Med Ctr Indianapolis IN 46202

CAMPBELL, ROBERT NOE, b Fairmont, Minn, Nov 16, 29; m 54; c 3. PLANT PATHOLOGY. Educ: Univ Minn, BS, 52, MS, 54, PhD(plant path), 57. Prof Exp: Asst plant path, Univ Minn, 52-54, instr, 54-56; plant pathologist, Forest Prod Lab, 57-59; from asst prof to assoc prof plant path, 59-69, PROF PLANT PATH & PLANT PATHOLOGIST EXP STA, UNIV CALIF, DAVIS, 69- Mem: Am Phytopath Soc; Bot Soc Am. Res: Cause and control of diseases of vegetable crops; fungal transmission of plant viruses. Mailing Add: Dept of Plant Path Univ of Calif Davis CA 95616

CAMPBELL, ROBERT SAMUEL, b La Harpe, Ill, Aug 15, 04; m 32, 54; c 4. BOTANY. Educ: Univ Chicago, BS, 25, MS, 29, PhD(plant ecol), 32. Prof Exp: Range examr, Jornada Exp Range, US Forest Serv, NMex, 25-34, asst chief div range res, Washington, DC, 34-43, in chg range utilization studies, 37-43, chief div range & watershed mgr res, Southern Forest Exp Sta, 43-63; CONSULT ECOLOGIST, 69- Concurrent Pos: Ed, J Range Mgt, 50-52 & 63-69. Honors & Awards: Superior Serv Award, USDA, 61. Mem: Ecol Soc Am; Am Inst Biol Sci; Soc Am Foresters; Am Soc Range Mgt (vpres, 57, pres, 58). Res: Range ecology; forest range management; research methods. Mailing Add: RR 7 S 36th St Quincy IL 62301

CAMPBELL, ROBERT SEYMOUR, b Saskatoon, Sask, Sept 9, 13; nat US; m; c 4. ZOOLOGY, LIMNOLOGY. Educ: Univ Sask, BA, 34, MA, 35; Univ Mich, PhD(zool), 39. Prof Exp: Teaching asst biol, Univ Sask, 32-35; teaching asst zool & asst limnol, Univ Mich, 35-39; asst prof biol, Cent Mich Univ, 39-44; assoc prof zool, 44-52, PROF ZOOL, COOP FISHERY UNIT SCH FORESTRY, FISHERIES & WILDLIFE, UNIV MO-COLUMBIA, 52- Mem: Assoc Am Soc Limnol & Oceanog; assoc Ecol Soc Am; assoc Am Fisheries Soc; Wildlife Soc; fel Am Inst Fishery Res Biol. Res: Limnology and fisheries biology; acid pollution in strip-mine lakes; thermal pollution. Mailing Add: Sch of Forestry Fisheries & Wildlife Univ of Mo-Columbia Columbia MO 65201

CAMPBELL, ROBERT TERRY, b Greenwood, Miss, July 27, 32; m 54; c 2. PULP & PAPER TECHNOLOGY. Educ: Miss Col, BS, 54; Univ NC, Chapel Hill, MA, 59. Prof Exp: From res chemist to chief pulp res, 59-75, MGR PULP RES, ERLING RIIS RES LAB, INT PAPER CO, 75- Mem: Tech Asn Pulp & Paper Indust. Res: Development and improvement of processes to convert wood and other fiber sources into bleached pulp for papermaking and regenerated cellulose manufacture. Mailing Add: 55 Hawthorne Pl N Mobile AL 36608

CAMPBELL, ROBERT W, forest entomology, see 12th edition

CAMPBELL, ROBERT WARREN, organic chemistry, see 12th edition

CAMPBELL, ROBERT WAYNE, b Concordia, Kans, Dec 19, 40; m 60; c 3. POLYMER CHEMISTRY. Educ: Kans State Univ, BS, 62; Purdue Univ, PhD(org chem), 66. Prof Exp: Res chemist, 66-68, group leader polymer res, 68-73, PROJ LEADER EXPLOR PLASTICS & FIBERS, PHILLIPS PETROL CO, 73- Mem: Am Chem Soc; Soc Plastics Engrs (secy, 75, treas, 76). Res: Synthetic and reaction mechanism aspects of organo-sulfur chemistry; mass spectral fragmentation of sulfonate esters and related materials; polymer synthesis and characterization, including structure-property correlation. Mailing Add: 1330 Cherokee Hills Ct Bartlesville OK 74003

CAMPBELL, RONALD JAMES, physical chemistry, see 12th edition

CAMPBELL, RONALD WAYNE, b Cherryvale, Kans, July 26, 19; m 43; c 3. HORTICULTURE. Educ: Kans State Univ, BS, 43, MS, 46; Mich State Univ, PhD(hort), 57. Prof Exp: From asst prof to assoc prof, 46-61, PROF HORT, KANS STATE UNIV, 61-, HEAD DEPT HORT & FORESTRY, 66- Mem: AAAS; Am Soc Hort Sci; Am Soc Plant Physiol; Weed Sci Soc Am; Am Pomol Soc. Res: Physiological studies of influences of agricultural chemicals on plants; winter hardiness and water relations studies; stock and scion relationships. Mailing Add: Dept of Hort & Forestry Kans State Univ Manhattan KS 66502

CAMPBELL, RUSSELL HARPER, b Bakersfield, Calif, Apr 20, 28; m 53; c 2. EARTH SCIENCE, ENGINEERING GEOLOGY. Educ: Univ Calif, BA, 51. Prof Exp: GEOLOGIST, US GEOL SURV, 51- Mem: Geol Soc Am; Asn Eng Geol; Soc Econ Paleontologists & Mineralogists. Res: Engineering geology, economic geology, structure and stratigraphy of the western Transverse Ranges, California; landslides and mudflows; stratigraphy and structure of northwestern Alaska; uranium deposits of southeastern Utah. Mailing Add: US Geol Surv Eng Geol Br 345 Middlefield Rd Menlo Park CA 94025

CAMPBELL, SAMUEL GORDON, b Oban, Scotland, Dec 10, 33; m 61; c 3. IMMUNOLOGY. Educ: Glasgow Univ, BVMS, 56; Toronto Univ, MVSc, 59; Cornell Univ, PhD(vet microbiol), 64. Prof Exp: House physician vet med, Vet Sch, Univ Glasgow, 56-57; asst vet microbiol, Ont Vet Col, 57-59; NY State Col Vet Med, Cornell Univ, 61-64; sr lectr, Sch Vet Sci, Melbourne Univ, 64-66; asst prof, NY State Col Vet Med, 67-70, dir int educ, 68-74, ASSOC PROF VET MICROBIOL, NY STATE COL VET MED, CORNELL UNIV, 70-, ACTG HEAD DEPT MICROBIOL, 74- Concurrent Pos: Consult, Vet Educ, Rockefeller Found, Brazil & Peru, 68 & Dept Microbiol, Tuskegee Inst, 73; external examr, Univ Guelph, 73, Melbourne Univ, 74; vis prof crib death res, Prof Coombs, Cambridge Univ, 74-75. Mem: Royal Col Vet Surg. Res: Immunological responses of neonatal cattle, sheep, pigs and humans. Mailing Add: Dept of Microbiol Cornell Univ NY State Col of Vet Med Ithaca NY 14853

CAMPBELL, SUZANN KAY, b New London, Wis, Apr 19, 43. PHYSICAL THERAPY. Educ: Univ Wis-Madison, BS, 65, MS, 68, PhD(neurophysiol), 73. Prof Exp: Phys therapist, Cent Wis Colony & Training Sch, 65-68; instr phys ther, Sch Med, Univ Wis-Madison, 68-70, consult, Univ Family Health Serv, 71-72; ASST PROF PHYS THER, SCH MED, UNIV NC, CHAPEL HILL, 72- Concurrent Pos: Consult, Div Dis of Develop & Learning, Biol Sci Res Ctr, Univ NC, Chapel Hill, 72- Mem: Am Phys Ther Asn; Soc Res Child Develop; Soc Behav Kinesiology. Res: Infants with central nervous system dysfunction; identification, psychoaffective and sensorimotor development, interaction with mother, and effectiveness of physical therapy in promoting development. Mailing Add: Div of Phys Ther Univ of NC Sch of Med Chapel Hill NC 27514

CAMPBELL, THOMAS COLIN, b Annandale, NJ, Mar 14, 34; m 62; c 4. BIOCHEMISTRY, NUTRITION. Educ: Pa State Univ, BS, 56; Cornell Univ, MS, 57, PhD(animal nutrit), 62. Prof Exp: Res biologist, Woodard Res Corp, 61-63; res assoc toxicol, Mass Inst Technol, 63-65; from asst prof to prof biochem, Va Polytech Inst & State Univ, 65-75; PROF NUTRIT BIOCHEM, DIV NUTRIT SCI, CORNELL UNIV, 75- Concurrent Pos: Campus coordr, Philippine Nat Nutrit Prog; NIH res career develop award, 74. Mem: Am Inst Nutrit; Am Soc Pharmacol & Exp Therapeut; Soc Toxicol; Sigma Xi. Res: Study of metabolism; mechanism of action and human consumption patterns of aflatoxin; nutrition-toxin interactions; environmental health; international nutrition. Mailing Add: Div of Nutrit Sci Cornell Univ Ithaca NY 14886

CAMPBELL, THOMAS COOPER, b Decatur, Ill, Feb 29, 32; m 57; c 2. PHYSICAL ORGANIC CHEMISTRY. Educ: Millikin Univ, BS, 57; Emory Univ, MS, 63; Ga Inst Technol, PhD(chem), 68. Prof Exp: Res chemist, Monsanto Co, 67-69, sr res chemist, 69-74; RES & DEVELOP ASSOC, PHILADELPHIA QUARTZ CO, 74- Mem: Am Chem Soc; Sigma Xi. Res: Surface chemistry; detergency research and theory; sodium silicate applications and technology; surfactant applications and formulations; hard surface detergency; metal cleaning. Mailing Add: Philadelphia Quartz Co Res & Develop Ctr PO Box 258 Lafayette Hill PA 19444

CAMPBELL, THOMAS HODGEN, b Toronto, Ohio, Dec 9, 24. MYCOLOGY. Educ: Ohio State Univ, BSc, 46, MSc, 49; Univ Wis, PhD(bot), 52. Prof Exp: Instr bot, Univ Tenn, 52-53; biologist, Am Cyanamid Co, NY, 53-54; from asst prof to assoc prof bot, Univ Tenn, 54-61; assoc prof biol, 61-65, head dept, 65-69, PROF BIOL, COL STEUBENVILLE, 65- Mem: AAAS; Mycol Soc Am; Bot Soc Am; Am Phytopath Soc; Mycol Soc France. Res: Morphology, cytology and variation of Penicillium chrysogenum; taxonomy and history of mycology. Mailing Add: Dept of Biol Col of Steubenville Steubenville OH 43952

CAMPBELL, THOMAS NOLAN, b Munday, Tex, July 3, 08; m 40; c 1. ANTHROPOLOGY, ARCHAEOLOGY. Educ: Univ Tex, Austin, BA, 30, MA, 36; Harvard Univ, MA, 40, PhD(anthrop), 47. Prof Exp: From instr to assoc prof anthrop, 38-52, dir res anthrop, 47-66, PROF ANTHROP, UNIV TEX, AUSTIN, 52- Concurrent Pos: Collabr, Nat Park Serv, 56-68; ed, Soc Am Archaeol, 62-66. Mem:

AAAS; Am Anthrop Asn; Soc Am Archaeol; Mex Anthrop Soc. Res: Archaeology of Texas and adjacent areas; ethnohistory; primitive technology; archaeological bibliography. Mailing Add: Dept of Anthrop Univ of Tex Austin TX 78712

CAMPBELL, W P, b Lloydminster, Sask, Aug 1, 22; m 50; c 3. PLANT PATHOLOGY. Educ: Univ Alta, BSc, 49; Univ Toronto, PhD(bot), 53. Prof Exp: Res officer plant path, 49-73, CHIEF PLANT INSPECTION & QUARANTINES, PLANT PROTECTION DIV, CAN DEPT AGR, 73- Mem: Am Phytopath Soc; Can Phytopath Soc; Agr Inst Can. Res: Ergot of cereals and grasses; pesticides. Mailing Add: Plant Protection Div Can Agr Cent Exp Farm Ottawa ON Can

CAMPBELL, WALLACE G, JR, b Lockport, NY, July 25, 30; m 57; c 4. PATHOLOGY. Educ: Harvard Univ, AB, 53; Cornell Univ, MD, 57. Prof Exp: Asst path, Cornell Univ, 58-61, instr, 61-62; from asst prof to assoc prof, 64-71, PROF PATH, EMORY UNIV, 71- Concurrent Pos: USPHS trainee path, 59-62. Mem: Am Soc Exp Path; Int Acad Path; Am Asn Path & Bact. Res: Experimental hypertension; cardiovascular and renal pathology; coagulation of blood; cellular pathology; arthritis; pneumocystis. Mailing Add: Dept of Path Emory Univ Atlanta GA 30322

CAMPBELL, WALLACE HALL, b New York, NY, Feb 6, 26; m 56; c 2. ATMOSPHERIC PHYSICS, GEOMAGNETISM. Educ: La State Univ, BS, 50; Vanderbilt Univ, MA, 53; Univ Calif, Los Angeles, PhD(physics), 59. Prof Exp: Grad res geophysicist, Inst Geophys, Univ Calif, Los Angeles, 55-57, jr res geophysicist, 57-59; asst prof geophys res, Geophys Inst, Univ Alaska, 59-60; group leader ultra low frequency res, Cent Radio Propagation Lab, Nat Bur Stand, 60-65, chief geomagnetism res, Aeronomy Lab, Inst Telecommun Sci & Aeronomy, Environ Sci Serv Admin, 65-67, dir geomagnetism lab, Environ Res Labs, Nat Oceanic & Atmospheric Agency, 67-71, head space magnetism res, 71-73; RESEARCHER EXTERNAL GEOMAGNETIC FIELD, US GEOL SURV, 73- Concurrent Pos: Mem various working groups, Int Asn Geomagnetism & Aeronomy, 64- Honors & Awards: Boulder Scientist Award, Sigma Xi, 63. Mem: Sigma Xi; Am Geophys Union; Int Sci Radio Union; Soc Terrestrial Magnetism & Elec Japan. Res: Upper atmospheric physics; geomagnetic phenomena; natural ultra low frequency field variations; auroral luminosity fluctuations. Mailing Add: US Geol Surv Fed Ctr Bldg 25 Denver CO 80225

CAMPBELL, WARREN ADAMS, b Berkeley, Calif, June 14, 36; m 58. ASTRONOMY. Educ: Willamette Univ, BA, 58; Univ Wis, MS, 60, PhD(astron), 65. Prof Exp: Asst prof math, Wash State Univ, 65-70; ASST PROF PHYSICS, WIS STATE UNIV, RIVER FALLS, 70- Mem: Am Astron Soc. Res: Forbidden oxygen and nitrogen lines in the spectra of planetary nebulae. Mailing Add: Dept of Physics Wis State Univ River Falls WI 54022

CAMPBELL, WILLIAM (ALOYSIUS), b Newcastle, NB, Apr 18, 06; US citizen; m 32; c 5. DEVELOPMENTAL ANATOMY, ELECTRON MICROSCOPY. Educ: St Francis Xavier Univ, BA, 27; St Francis Col (Pa), MS, 29. Prof Exp: Asst prof biol & physics, St Francis Col (Pa), 27-29, prof biol & head dept, 29-38; asst prof, Canisius Col, 38; from asst prof to assoc prof, 38-73, EMER ASSOC PROF BIOL, COL HOLY CROSS, 73- Mem: Nat Asn Biol Teachers. Res: Comparative anatomy, histology and cytology; developmental anatomy; electron microscopy; experimental embryology. Mailing Add: 73 Willow Hill Rd Cherry Valley MA 01611

CAMPBELL, WILLIAM ANDREW, b Paterson, NJ, Jan 29, 06; m 36; c 3. FOREST PATHOLOGY. Educ: Pa State Teachers Col, BS, 29; Univ Colo, AM, 31; Pa State Col, PhD(mycol, plant path), 35. Prof Exp: Asst forest pathologist, Div Forest Path, Bur Plant Indust, USDA, 36-42, forest pathologist, Guayule Res Proj, Salinas, Calif, 42-46, pathologist, Div Forest Path, Georgia, 46-53, sr pathologist, 53-54, plant pathologist, Southeastern Forest Exp Sta, US Forest Serv, 54-71; prof plant path, Univ Ga, 71-73; CONSULT FORESTRY, 73- Mem: Am Phytopath Soc. Res: Diseases of native tree species, particularly root diseases and wood decay. Mailing Add: 260 Milledge Heights Athens GA 30601

CAMPBELL, WILLIAM BRYSON, b Sulphur Springs, Tex, Mar 25, 47; m 75. PHARMACOLOGY. Educ: Univ Tex, Austin, BS, 70; Univ Tex Southwestern Med Sch, PhD(pharmacol), 74. Prof Exp: Fel pharmacol, Med Col Wis, 74-75, instr, 75-76; ASST PROF PHARMACOL, UNIV TEX SOUTHWESTERN MED SCH, 76- Mem: Am Fedn Clin Res; AAAS; Am Heart Asn. Res: Pharmacology of vasoactive substances and their relationship to the kidney, the adrenal gland and the peripheral vasculature in hypertension. Mailing Add: Dept of Pharmacol Univ of Tex Health Sci Ctr Dallas TX 75235

CAMPBELL, WILLIAM CECIL, b Londonderry, Ireland, June 28, 30; m 62; c 2. ZOOLOGY. Educ: Univ Dublin, BA, 52; Univ Wis, MS, 54, PhD(zool, vet sci), 57. Prof Exp: Asst parasitol, Univ Wis, 53-57; res assoc, 57-66, DIR PARASITOL, MERCK INST THERAPEUT RES, 66- Mem: Am Soc Parasitol; Am Soc Trop Med & Hyg. Res: Parasitology; helminthology; protozoology; chemotherapy. Mailing Add: Merck Inst for Therapeut Res Rahway NJ 07065

CAMPBELL, WILLIAM FRANK, b Mt Vernon, Ill, Sept 23, 28; m 54; c 4. RADIATION BOTANY, PLANT PHYSIOLOGY. Educ: Univ Ill, BS, 56, MS, 57; Mich State Univ, PhD(radiation bot), 64. Prof Exp: Collabr hort, Agr Res Serv, USDA, 64-68, ASSOC PROF PLANT PHYSIOL, UTAH STATE UNIV, 68- Mem: AAAS; Am Soc Hort Sci; Am Inst Biol Sci; Am Soc Agron; Bot Soc Am. Res: Effects of ionizing radiation in successive generations on developing and dormant plant embryos; plant growth, physiological responses histo and cyto-chemical and ultrastructural changes as influenced by environmental stresses; ultrastructural plant cytology. Mailing Add: Dept of Plant Sci Utah State Univ Logan UT 84321

CAMPBELL, WILLIAM H, b Fayette, Ala, July 20, 40. MATHEMATICS. Educ: Univ Ala, Tuscaloosa, BA, 62, MA, 63, PhD, 69. Prof Exp: Mathematician, Gen Dynamics Corp, Conn, 63-64; instr math, Univ Ala, Huntsville, 64-66, ASST PROF MATH, UNIV ALA, BIRMINGHAM, 69- Mem: Am Math Soc; Math Asn Am. Mailing Add: Dept of Math Univ of Ala Birmingham AL 35233

CAMPBELL, WILLIAM HOWARD, b Lakeview, Ore, Aug 13, 42; m 67; c 2. PHARMACOLOGY. Educ: Ore State Univ, BS, 65, MS, 68; Purdue Univ, PhD(pharm), 71. Prof Exp: Asst prof pharm, Ore State Univ, 71-75; CHMN DEPT PHARM PRACT, SCH PHARM, UNIV WASH, 75- Concurrent Pos: Sr investr, Health Serv Res Ctr, Ore Region, Kaiser Found, Portland, 71-; sr consult, Vet Admin Hosp, 75- Mem: Am Pub Health Asn; Am Pharmaceut Asn. Res: Assessment of drug systems operation within general medical care, specifically adverse drug reactions, drug use review and control, and technological applications. Mailing Add: Sch of Pharm Univ of Wash Seattle WA 98195

CAMPBELL, WILLIAM JACKSON, b Wichita Falls, Tex, Oct 23, 29; m 51; c 3. CLINICAL BIOCHEMISTRY. Educ: NTex State Univ, BA, 49; Univ Tex, BS, 52, MS, 53; Ohio State Univ, PhD(phys chem), 60. Prof Exp: Chief dept biochem, Walter Reed Army Inst Res, 60-64; prog adminr, clin chem, Nat Inst Gen Med Sci, 64-74; EXEC DIR CLIN CHEM, AM ASN CLIN CHEMISTS, 74- Concurrent Pos: Lectr, Am Univ, 61-73, prof lectr, 63-65, adj prof, 65-73. Mem: Sigma Xi; Am Chem Soc; Am Asn Clin Chemists. Mailing Add: Am Asn Clin Chemists 1725 K St NW Washington DC 20006

CAMPBELL, WILLIAM JOSEPH, b Washington, DC, July 31, 26; m 48; c 2. PHYSICAL CHEMISTRY. Educ: Univ Md, BS, 50, PhD(chem), 56. Prof Exp: Asst phys chem, Univ Md, 51; phys chemist, 51-56, supvry phys chemist, 56-62, SUPVRY RES CHEMIST, US BUR MINES, 62- Honors & Awards: Dept Interior Meritorious Serv Award, 62. Mem: AAAS; Am Chem Soc; Soc Appl Spectos; Am Inst Min, Metall & Petrol Eng; fel Am Inst Chem. Res: Fluorescent x-ray spectroscopy; x-ray diffraction; high temperature physical chemistry; auger spectroscopy; electron optics; chemical-instrumental methods of analysis. Mailing Add: 2507 Romona Dr District Heights MD 20028

CAMPBELL, WILLIAM JOSEPH, b Brooklyn, NY, May 1, 30. METEOROLOGY, OCEANOGRAPHY. Educ: Univ Alaska, BS, 55; Univ Wash, MS, 58, PhD(meteorol, oceanog), 64. Prof Exp: Meteorologist, Drifting Sta Alpha, Univ Wash, 57-58, micrometeorologist, Blue Glacier, 58-59; meteorologist, Ross Ice Shelf, Antarctica, Univ Mich, 62-63; METEOROLOGIST, US GEOL SURV, UNIV PUGET SOUND, 64-, CHIEF ICE DYNAMICS PROJ, 69- Concurrent Pos: Affil assoc prof oceanog, Univ Wash, 69; vis prof, Dartmouth Col, 69; Brown & Haley lectr, Univ Puget Sound, 71; prin investr sea ice, NASA Spacecraft Oceanog Ctr, 69; mem, Arctic Ice Dynamics Joint Exp Pilot Exped, Arctic Ocean. Mem: AAAS; Am Meteorol Soc; Glaciol Soc; Am Geophys Union. Res: Arctic meteorology; sea ice dynamics; polar oceanography; glacier and climate variations; glacier hydrology and meteorology; remote sensing of ice and lakes. Mailing Add: Ice Dynamics Proj US Geol Surv Univ of Puget Sound Tacoma WA 98416

CAMPBELL, WILLIAM ROBERT, b Roanoke, Va, Feb 7, 42. ENTOMOLOGY. Educ: Va Polytech Inst, BS, 63, MS, 65; Univ Ill, PhD(entom), 70. Prof Exp: Asst prof insect biochem, Purdue Univ, 69-75; RES SPECIALIST, CIBA-GEIGY CORP, 75- Concurrent Pos: Partic health sci advan award, NIH, 67-72, res grant viral pathogens, 70-73; fel biol receptors, Agr Exp Sta Proj, 69-75; consult, Cent Soya, Inc, 75 & Pest Mgt Assoc, Inc, 73- Mem: AAAS; Enotm Soc Am; Sigma Xi. Res: Insect biochemistry; neurochemical receptors, viral penetration and replication in insect tissues; effects of foreign compounds on insect development; development of new, selective insecticides. Mailing Add: Ciba-Geigy Corp PO Box 1090 Vero Beach FL 32960

CAMPBELL, WILLIAM VERNON, b Chester, SC, May 4, 24; m 47; c 2. ENTOMOLOGY. Educ: Mass State Col, BS, 51, MS, 52; NC State Col, PhD(entom), 58. Prof Exp: Entomologist, Insect Control, Entom Res Br, Agr Res Serv, USDA, Tidewater Field Sta, Va, 52-53 & wheat stem sawfly proj, Cereal & Forage Insects Sect, Entom Res Br, NDak, 53-55; asst entom, 55-58, from asst prof to assoc prof, 58-69, PROF ENTOM, NC STATE UNIV, 70- Mem: Am Entom Soc. Res: Chemical control of insects; plant resistance to insects attacking field and forage crops; plant and insect histology. Mailing Add: Dept of Entom NC State Univ Raleigh NC 27607

CAMPENOT, ROBERT BARRY, b East Orange, NJ, Mar 30, 46. NEUROBIOLOGY. Educ: Rutgers Univ, BA, 68; Univ Calif, Los Angeles, MS, 71; Mass Inst Technol-Woods Hole Oceanog Inst, PhD(biol oceanog), 76. Prof Exp: RES FEL NEUROBIOL, HARVARD MED SCH, 75- Res: Synaptic physiology of sympathetic neurons grown in tissue culture. Mailing Add: Dept of Neurobiol Harvard Med Sch 25 Shattuck St Boston MA 02115

CAMPER, NYAL DWIGHT, b Lynchburg, Va, May 12, 39. PLANT PHYSIOLOGY, PLANT BIOCHEMISTRY. Educ: NC State Univ, BS, 62, PhD(crop sci), 67. Prof Exp: Res asst herbicide physiol, NC State Univ, 62-66; asst plant physiol, 66-71, ASSOC PROF PHYSIOL, CLEMSON UNIV, 71- Concurrent Pos: Dir student sci training prog, NSF, 68-70. Mem: Am Chem Soc; Am Soc Plant Physiol; Weed Sci Soc. Res: Physiology and biochemistry of herbicide action, mechanisms of action and the biological systems involved. Mailing Add: Dept of Plant Path & Physiol Clemson Univ Clemson SC 29631

CAMPILLO, ANTHONY JOSEPH, b Newark, NJ, June 30, 42; m 68; c 3. LASERS, BIOPHYSICS. Educ: Newark Col Eng, BS, 64; Princeton Univ, MS, 66; Cornell Univ, PhD(appl physics), 73. Prof Exp: Staff mem, GT&E Labs Inc, 66-68; STAFF MEM, LOS ALAMOS SCI LAB, UNIV CALIF, 72- Mem: Am Phys Soc. Res: Use of picosecond laser techniques to probe very fast molecular events in physics, chemistry and biology; nonlinear optics. Mailing Add: Univ of Calif Los Alamos Sci Labs MS 552 Los Alamos NM 87545

CAMPION, JAMES J, b Philadelphia, Pa, July 14, 39; m 69. ANALYTICAL CHEMISTRY, PHYSICAL CHEMISTRY. Educ: La Salle Col, BA, 61; Univ Pittsburgh, PhD(nonaqueous solutions), 66. Prof Exp: Off Saline Water res assoc, Univ Pittsburgh & Mellon Inst, 67-68; asst prof, 68-73, ASSOC PROF ANAL CHEM, STATE UNIV NY COL NEW PALTZ, 73- 68-73. Mem: Am Chem Soc; Sigma Xi. Res: Solution chemistry of electrolytes and nonelectrolytes in aqueous and nonaqueous solvents. Mailing Add: PO Box 889 Tillson NY 12486 12486

CAMPION, JOHN EMMETT, (JR), organic chemistry, see 12th edition

CAMPION, RAYMOND JOSEPH, physical chemistry, inorganic chemistry, see 12th edition

CAMPISI, LOUIS SEBASTIAN, b New York, NY, Aug 9, 35; m 63; c 1. INORGANIC CHEMISTRY. Educ: City Col New York, BS, 56; Fordham Univ, MS, 60, PhD(inorg chem), 64. Prof Exp: Asst prof chem, 62-69, ASSOC PROF CHEM, IONA COL, 69- Mem: Am Chem Soc; Sigma Xi. Res: Chemistry of uranium and thorium; coordination chemistry. Mailing Add: Dept of Chem Iona Col New Rochelle NY 10804

CAMPO, ROBERT D, b New York, NY, Feb 18, 30; m 57; c 5. BIOCHEMISTRY, BIOLOGY. Educ: St John's Univ, NY, BS, 52, MS, 57; Rockefeller Inst, PhD(biochem), 63. Prof Exp: From instr to asst prof biochem, 63-70, ASSOC PROF ORTHOP SURG, SCH MED, HEALTH SCI CTR, TEMPLE UNIV, 70- Mem: Soc Exp Biol & Med; Am Rheumatism Asn; Orthop Res Soc. Res: Radiation biology; acid mucopolysaccharides; protein polysaccharides of cartilage; bone formation; calcification and degradation enzymes of cartilage; organic matrices of cartilage and bone; metabolism of selenium. Mailing Add: Dept of Orthop Surg Temple Univ Health Sci Ctr Philadelphia PA 19140

CAMPOLATTARO, ALFONSO, b Naples, Italy, Jan 4, 33; m 60; c 3. THEORETICAL PHYSICS. Educ: Univ Naples, PhD(theoret physics), 59. Prof Exp: Res physicist, Univ Naples, 60-65, assoc prof theoret physics, 63-65; NATO fel, Palmer Physics Lab, Princeton Univ, 65-66; vis asst prof, Univ Calif, Irvine, 66-67, actg assoc prof, 67-68; vis assoc prof, 68-70, assoc prof, 70-71, PROF PHYSICS,

UNIV MD, BALTIMORE COUNTY, 71- Honors & Awards: Young Physicist Prize, Ital Physics Soc, 62. Mem: AAAS; Am Phys Soc. Res: Mathematical aspects of quantum field theory; relativistic astrophysics; solid state physics; particle accelerators. Mailing Add: Dept of Physics Univ of Md Baltimore MD 21228

CAMRAS, MARVIN, b Chicago, Ill, Jan 1, 16; m 51; c 5. MAGNETISM, ELECTRONICS. Educ: Armour Inst Technol, BS, 40; Ill Inst Technol, MS, 42. Hon Degrees: LLD, Ill Inst Technol, 68. Prof Exp: Physicist, 40-45, sr physicist, 45-59, sr engr, 59-65, sci adv, 65-69, SR SCI ADV, IIT RES INST, 69- Concurrent Pos: Engr & draftsman, Delta Star Elec Co, Ill, 39; ed, Trans on Audio, Inst Elec & Electronics Eng, 58-64; chmn, Nat Comt II, Int Electrotech Comn; mem S4 comt, Am Nat Stand Inst. Honors & Awards: Distinguished Serv Award, Ill Inst Technol, 48; Scott Medal, 55; Citation, Ind Tech Col, 58; Achievement Award, Inst Elec & Electronics Eng, 58, Consumer Electronics Award, 64 & Broadcasting Papers Award, 64; US Camera Award, 59; Idust Res Prod Award, 66; John Potts Medal, Audio Eng Soc, 69. Mem: Fel AAAS; fel Inst Elec & Electronics Eng; Electronic Industs Asn; fel Acoust Soc Am. Res: Magnetic recording; stereophonic sound; electronics; magnetism; video recording. Mailing Add: Technol Ctr IIT Res Inst 10 W 35th St Chicago IL 60616

CAMU, PIERRE, economic geography, transportation, see 12th edition

CANADA, ROBERT, b El Dorado, Kans, Jan 26, 19; m 40; c 4. NUCLEAR PHYSICS. Educ: Ind Univ, PhD(physics), 51. Prof Exp: STAFF MEM WEAPONS DESIGN, LOS ALAMOS SCI LAB, 51- Mem: Am Phys Soc. Res: Nuclear spectroscopy. Mailing Add: Los Alamos Sci Lab PO Box 1663 Los Alamos NM 87544

CANADY, VAN LEONARD, analytic chemistry, organic chemistry, see 12th edition

CANADY, WILLIAM JAMES, b New York, NY, Dec 8, 24; m 55; c 1. BIOCHEMISTRY. Educ: Fordham Univ, BS, 46; George Washington Univ, MS, 50, PhD(biochem), 55. Prof Exp: From instr to assoc prof, 58-69, PROF BIOCHEM, MED CTR, W VA UNIV, 69- Concurrent Pos: Fel chem, Univ Ottawa, 55-57. Mem: Am Chem Soc; Soc Exp Biol & Med. Res: Vitamin K; methodology; mechanism of enzyme action; thermodynamics of ionization and solution processes. Mailing Add: Med Ctr WVa Univ Morgantown WV 26506

CANALE-PAROLA, ERCOLE, b Frosinone, Italy, Sept 13, 29; US citizen; m 54; c 2. MICROBIOLOGY. Educ: Univ Ill, BS, 56, MS, 57, PhD(microbiol), 61. Prof Exp: NIH fel, Hopkins Marine Sta, Stanford Univ, 61-63; from asst prof to assoc prof microbiol, 63-73, PROF MICROBIOL, UNIV MASS, 73- Mem: Am Soc Microbiol; Soc Gen Microbiol. Res: Bacterial physiology; microbial ecology; spirochetes; sarcinae; bacterial pigments; evolution of microorganisms. Mailing Add: Dept of Microbiol Univ of Mass Amherst MA 01002

CANARY, JOHN JOSEPH, b Mineola, NY, Jan 9, 25; m 51; c 7. METABOLISM, ENDOCRINOLOGY. Educ: St John's Col, NY, BS, 47; Georgetown Univ, MD, 51. Prof Exp: NIH trainee metab dis, Georgetown Univ, 54-55; from instr to assoc prof, 56-68, actg dir radioisotope lab, Univ Hosp, 58, PROF MED, SCH MED, GEORGETOWN UNIV, 68-, DIR GEN CLIN RES CTR, 73-, DIR DIV & CLINICS ENDOCRINOL & METAB, UNIV HOSP, 59- Concurrent Pos: Res fel, NIH, 55-57; clin instr biochem, Sch Med, Georgetown Univ, 57-60, spec lectr, 70-; consult & lectr endocrinol, Nat Naval Med Ctr, 58-; consult metab, Walter Reed Army Inst Res, 61- Mem: AAAS; Am Fedn Clin Res; Endocrine Soc; Am Diabetes Asn; Am Soc Clin Invest. Res: Study of the density and chemical composition of human bone in various disease states, particularly osteoporosis and the effects of therapy thereon; body composition and density. Mailing Add: Sch of Med Georgetown Univ Washington DC 20007

CANAVAN, FREDERICK LOUIS, b Ridgefield Park, NJ, Sept 18, 19. NUCLEAR PHYSICS. Educ: Boston Col, BS, 45, MA, 46; Cath Univ Am, PhD(physics), 52. Prof Exp: Res assoc, Radiation Lab, Univ Calif, 55-56; asst prof physics, Fordham Univ, 56-66, assoc prof & dean sch gen studies, 66-70; dean planning, Medgar Evers Col, 70-74; DIR ADVAN OPPORTUNITY PROG, WESTCHESTER COALITION, 74- Mem: Am Phys Soc. Res: Low energy nuclear physics. Mailing Add: 70 Church St White Plains NY 10601

CANAVAN, ROBERT I, b Ridgefield Park, NJ, July 31, 27. MATHEMATICS. Educ: Woodstock Col, Md, AB, 50; NY Univ, PhD(math), 57; Univ Innsbruck, lic theol, 61. Prof Exp: Instr math, LeMoyne Col, NY, 51-52; from asst prof to assoc prof, St Peter's Col, NJ, 62-69; assoc prof, 69-74, PROF MATH, MONMOUTH COL, NJ, 74- Mem: Am Math Soc; Math Asn Am; Soc Indust & Appl Math. Res: Ordinary and partial differential equations. Mailing Add: Dept of Math Monmouth Col West Long Beach NJ 07764

CANAVERA, DAVID STEPHEN, b Norway, Mich, Sept 4, 43; m 69; c 1. FOREST GENETICS. Educ: Mich Technol Univ, BS, 65; Mich State Univ, MS, 67, PhD(forest genetics), 69. Prof Exp: Peace Corps vol genetics, Nat Sch Forestry, Curitiba, Brazil, 69-72; asst prof forestry, Tuskegee Inst, 72-74; ASST PROF FORESTRY, UNIV MAINE, ORONO, 74- Mem: Am Soc Foresters; Am Forestry Asn. Res: Provenance testing, individual tree selection and breeding; species hybridization; accelerated growth techniques and containerized seedling production. Mailing Add: Sch of Forest Resources Univ of Maine Orono ME 04473

CANBY, HENRY FAWCETT, b Mt Pleasant, Iowa, June 21, 08; m 37; c 3. DENTISTRY. Educ: Univ Iowa, DDS, 32; Yale Univ, MPH, 42. Prof Exp: DENT DIR, USPHS, 32- Mem: Am Dent Asn; Am Col Dent. Mailing Add: 7013 Winslow St Bethesda MD 20034

CANBY, JOEL SHACKELFORD, b Denver, Colo, Aug 1, 19; m 61; c 3. CULTURAL ANTHROPOLOGY, ETHNOLOGY. Educ: Colo Col, BA, 41; Harvard Univ, MA, 48, PhD(anthrop), 50. Prof Exp: Prof anthrop & dir, Ctr Advan Anthrop Studies, 49-50; prof social rels & chmn col arts & sci, Univ Baghdad, 51-54; prof behav sci & coordr, State Univ NY, New Paltz, 55-57; supvr systs training, Syst Develop Corp, Calif, 58-62; supvr life sci, NAm Rockwell Corp, Calif, 62-70; PROF ANTHROP & CHMN DEPT, CALIF STATE COL, STANISLAUS, 70-; DIR, EMANUEL MENT HEALTH CTR DEMOG STUDIES, 72- Concurrent Pos: Wenner-Gren Viking Fund grant, Inst Anthrop & Hist, Guatemala, 49-50; res fel anthrop, Middle East Inst, Iraq, 50-52. Mem: AAAS; Am Anthrop Asn; Soc Appl Anthrop; Am Acad Polit & Social Sci; Sigma Xi. Res: Ethnohistory of the Middle East; dynamics of culture change; applied anthropology; pastoral nomadism. Mailing Add: Dept of Anthrop Calif State Col at Stanislaus Turlock CA 95380

CANCEL, CRUZ A, b Lajas, PR, Aug 6, 26; m 68; c 1. SPEECH PATHOLOGY, AUDIOLOGY. Educ: Univ PR, Bachellor, 49; Univ Mich, MS, 51; Ohio State Univ, PhD(speech & hearing), 57. Prof Exp: Asst prof speech path & audiol, Auburn Univ, 59-61; NIH fel, Bowman Gray Sch Med-NC Baptist Hosp, 61-62; speech pathologist, Winston-Salem Vet Admin, NC, 62 & Wiesbaden Air Force Hosp, Ger, 64-67; assoc prof speech audiol, Clarion State Col, 67-68; PROF SPEECH SCI & DIR SCH SPEECH PATH, UNIV PR, 68- Concurrent Pos: Pres, Ctr Invests & Resources in Speech, Hearing & Lang, PR, 75- Mem: Am Speech & Hearing Asn; Am Asn Sch Adminr; Am Mgt Asn; Int Asn Audiol; Latin Am Group Acoustitians. Res: Testing and development of diagnostic tools and resource materials for the bilingual communicatively handicapped; establishment of hearing levels for spoken Spanish and identification of language and culture barriers for the bilinguals. Mailing Add: Ctr Invest & Resources in Speech Hearing & Lang GPO Box 7133 San Juan PR 00937

CANCIO, MARTA, b San Sebastian, PR, Dec 8, 28; m 52, 64; c 3. BIOCHEMISTRY. Educ: Univ PR, BS, 49; Univ Mo, MS, 52, PhD(biol chem), 54. Prof Exp: Assoc biochem, Sch Med, Univ PR, 54-67, asst prof biochem & nutrit, 67-71; chief biochemist, 54-66, SUPVRY RES CHEMIST, VET ADMIN HOSP, 66-; ASSOC PROF BIOCHEM & NUTRIT, SCH MED, UNIV PR, RIO PIEDRAS, 71- Mem: NY Acad Sci; Am Fedn Clin Res; Am Inst Nutrit; Am Soc Clin Nutrit; Latin Am Nutrit Soc. Res: Lipid and protein chemistry; malabsorption; tropical sprue; immunochemistry; hyperlipidemias. Mailing Add: 39 Jazmin St San Francisco Rio Piedras PR 00927

CANCRO, ROBERT, b New York, NY, Feb 23, 32; m 56; c 2. PSYCHIATRY. Educ: Fordham Univ, 48-51; State Univ NY, MD, 55, DrMedSci, 62; Am Bd Psychiat & Neurol, cert psychiat, 62. Prof Exp: Dir alcohol res ward & instr psychiat, State Univ NY Downstate Med Ctr, 62-66; psychiatrist, Menninger Found, 66-69, mem fac, Menninger Sch Psychiat, 67-69; vis assoc, Ctr Advan Study & vis prof, Dept Comput Sci, Univ Ill, Urbana, 69-70; PROF PSYCHIAT, UNIV CONN, FARMINGTON, 70- Concurrent Pos: Consult, Dept Comput Sci, Univ Ill, 67 & Tokepa State & Vet Admin Hosps, Topeka, Kans, 67-69. Mem: AAAS; Am Psychiat Asn; Asn Am Med Cols; AMA; NY Acad Sci. Res: Prediction of outcome in schizophrenia; nature of pathology in schizophrenia and in addictions. Mailing Add: Dept of Psychiat Univ of Conn Farmington CT 06032

CANDELAS, GRACIELA C, b PR, 22; US citizen. CELL BIOLOGY, MOLECULAR BIOLOGY. Educ: Univ PR, BS, 44; Duke Univ, MS, 59; Univ Miami, PhD(molecular biol), 66. Prof Exp: Instr biol, 51-57, from asst prof to assoc prof, 61-71, PROF BIOL, UNIV PR, RIO PIEDRAS, 71- Concurrent Pos: Scientist, Isotopes Div, PR Nuclear Ctr, 61-; vis prof biol, Syracuse Univ, 69-71 & City Col New York, 74-75; prof cell & molecular biol, Med Col Ga, 72-74. Mem: Am Soc Cell Biol; Sigma Xi; Int Cell Res Orgn; AAAS. Res: Control of protein synthesis in early development; regulation and control of cell cycle in epidermis. Mailing Add: Dept of Biol Univ of PR Rio Piedras PR 00931

CANDER, LEON, b Philadelphia, Pa, Oct 7, 26; m 54; c 2. INTERNAL MEDICINE, CLINICAL PHYSIOLOGY. Educ: Temple Univ, MD, 51. Prof Exp: Intern, Southern Div, Einstein Med Ctr, 51-52; fel physiol, Dept Physiol & Pharmacol, Grad Sch Med, Univ Pa, 52-54, instr, 54-55, assoc, 55-56; from asst resident to resident med, Beth Israel Hosp, Boston, 56-58; sr instr, Med Sch, Tufts Univ, 58-60; from asst prof to assoc prof, Hahnemann Med Col & Hosp, 60-66, head, Sect Chest Dis, 60-66; prof physiol & internal med & chmn dept, Univ Tex Med Sch, San Antonio, 66-72; PROF MED, JEFFERSON MED COL, 72-; CHMN, DEPT MED, DAROFF DIV, ALBERT EINSTEIN MED CTR, PHILADELPHIA, 72- Concurrent Pos: Asst lab instr, Sch Auxiliary Med Sci, Univ Pa, 53-55, Am Col Physicians res fel, 54-55; asst, Harvard Med Sch, 57-58; fel, Nat Acad Sci-Nat Res Coun; Markle scholar acad med, Hahnemann Med Col & Hosp, 61-66. Mem: Am Fedn Clin Res; Am Physiol Soc; Am Col Physicians. Res: Clinical pulmonary physiology. Mailing Add: 317 Cherry Lane Wynnewood PA 19096

CANDIA, OSCAR A, b Buenos Aires, Arg, Apr 30, 35; m 60; c 3. PHYSIOLOGY, BIOPHYSICS. Educ: Univ Buenos Aires, MD, 59. Prof Exp: Instr basic physics, Univ Buenos Aires, 60-61, head lab & res assoc biophysics, 62-63; res assoc, Univ Louisville, 64-65, asst prof, 65-68; ASSOC PROF BIOPHYS, MT SINAI SCH MED, 68- Concurrent Pos: Res fel electrophysiol, Univ Buenos Aires, 60-62; NIH career develop award, 66-71; res assoc, Arg Nat Res Coun, 60-63. Mem: Biophys Soc; Asn Res Ophthal; Am Physiol Soc. Res: Ion transport in biological membranes, models; instrumentation. Mailing Add: Dept of Ophthal Mt Sinai Sch of Med New York NY 10029

CANE, DAVID EARL, b Sept 22, 44. BIO-ORGANIC CHEMISTRY, NATURAL PRODUCTS CHEMISTRY. Educ: Harvard Col, AB, 66; Harvard Univ, MA, 67, PhD(chem), 71. Prof Exp: NIH fel, 71; res assoc org chem, Swiss Fed Inst Technol, 71-73; ASST PROF CHEM, BROWN UNIV, 73- Mem: Am Chem Soc; The Chem Soc. Res: Biosynthesis of natural products; stereochemistry and synthetic methods. Mailing Add: Dept of Chem Brown Univ Providence RI 02912

CANELLAKIS, EVANGELO S, b Tientsin, China, June 20, 22; nat US; m 48; c 2. BIOCHEMISTRY. Educ: Nat Univ Athens, BS, 47; Univ Calif, PhD(biochem), 51. Prof Exp: Res asst, Univ Calif, 48-50, asst, 50-51; Nat Found Infantile Paralysis fel, Dept Physiol Chem, Univ Wis, 51-54; Squibb fel pharmacol, Yale Univ, 54-55, from instr to assoc prof, 55-64, PROF PHARMACOL, YALE UNIV, 64- Concurrent Pos: Travel grant, Int Biochem Cong, Austria, 58; scholar, USPHS, 59-, res career award, 64; ed, Biochimica Biophysica Acta, 69. Mem: Am Soc Biol Chem; Am Asn Cancer Res; Brit Biochem Soc; Hellenic Chem Soc. Res: Amino acid metabolism; pyrimidine metabolism; mechanisms in nucleic acid synthesis; interrrelationships of biological pathways. Mailing Add: Dept of Pharmacol Yale Univ New Haven CT 06510

CANELLAKIS, ZOE NAKOS, b Lowell, Mass, Sept 7, 27; m 48; c 2. BIOCHEMISTRY. Educ: Vassar Col, BA, 47; Univ Calif, MS, 51; Univ Wis, PhD(physiol chem), 54. Prof Exp: Asst biochem, 54-55, instr, 55-59, res assoc, 59-67, SR RES ASSOC, DEPT PHARMACOL, MED SCH, YALE UNIV, 67-, ASST DEAN GRAD SCH, 72- Res: Amino acid metabolism; protein and nucleic acid synthesis. Mailing Add: Dept of Pharmacol Yale Univ Med Sch New Haven CT 06520

CANELLOS, GEORGE P, b Boston, Mass, Nov 1, 34; m 58; c 3. INTERNAL MEDICINE, ONCOLOGY. Educ: Harvard Univ, AB, 56; Columbia Univ, MD, 60; Am Bd Internal Med, dipl, 68. Prof Exp: Intern, Mass Gen Hosp, Boston, 60-61; clin fel, Harvard Med Sch, 61-62; asst resident, Mass Gen Hosp, Boston, 62-63; clin assoc, Nat Cancer Inst, 63-65; sr resident, Mass Gen Hosp, Boston, 65-66; res asst hemat, Royal Postgrad Med Sch London, 66-67; sr investr, Nat Cancer Inst, 67-75, asst chief med br, 73-75, clin dir, 74-75; CHIEF DIV MED, SIDNEY FARBER CANCER CTR, 75-; ASSOC PROF MED, HARVARD MED SCH, 75-; SR ASSOC MED, PETER BENT BRIGHAM HOSP, 75- Concurrent Pos: Asst clin prof, Dept Med, Georgetown Univ, 68-74, assoc clin prof, 74- Mem: Am Soc Hemat; Am Fedn Clin Res; Am Asn Cancer Res; Am Soc Clin Oncol; fel Am Col Physicians. Res: Cell biology as related to disorders of hemopoiesis. Mailing Add: Harvard Med Sch Boston MA 02115

CANERDAY, THOMAS DONALD, b Florence, Ala, June 6, 39; m 60; c 2. INVERTEBRATE PATHOLOGY, ECONOMIC ENTOMOLOGY. Educ: Auburn Univ, BS, 61, MS, 63, PhD(microbiol, entom), 67. Prof Exp: From instr to asst prof entom, Auburn Univ, 63-68; assoc prof & head dept, 68-74, prof entom-fisheries & head dept, 74-75, PROF ENTOM & DIV CHMN, UNIV GA, ATHENS, 75- Mem: Entom Soc Am. Mailing Add: Div of Entom Hoke Smith Annex Univ of Ga Athens GA 30602

CANFIELD, CRAIG JENNINGS, b Pasadena, Calif, May 11, 32; m 54; c 7. CLINICAL PHARMACOLOGY, MALARIOLOGY. Educ: Univ Ore, BS, 55, MD, 57. Prof Exp: Resident med, Walter Reed Gen Hosp, 60-63; chief dept med clin res, SEATO Med Res Lab, 64-66; fel hemat, Walter Reed Gen Hosp, 68-69, asst dir malaria & chief clin pharm, Walter Reed Army Inst Res, 70-75, DIR DIV MED CHEM, WALTER REED ARMY INST RES, 75- Concurrent Pos: Prin investr antimalarial notices of claimed investigational exemption for a new drug, US Army Med Res & Develop Command, 70-; mem malarial chemother task force, WHO, 75- Mem: Am Col Physicians; Am Soc Trop Med & Hyg; Am Fedn Clin Res; Am Soc Hemat. Res: Antiparasitic drug development, especially malaria; hematology; clinical research and pharmacology; diagnosis and treatment of malaria. Mailing Add: Div of Med Chem Walter Reed Army Inst of Res Washington DC 20012

CANFIELD, EARLE LLOYD, b Des Moines, Iowa, Oct 24, 18; m 47; c 3. MATHEMATICS, STATISTICS. Educ: Drake Univ, BA, 40; Northwestern Univ, MA, 44; Iowa State Univ, PhD, 50. Prof Exp: Teacher high schs, Iowa, 40-46, prin, 43-46; from instr to assoc prof math, 46-58, PROF MATH, DRAKE UNIV, 58-, DEAN GRAD STUDIES, 57- Concurrent Pos: Consult, Des Moines Secondary Sch Math Teachers, 51-52; Des Moines consult sch math study group, Yale Univ, 60-61 & Stanford Univ, 61-62; consult, Coun Grad Schs. Mem: AAAS; Math Asn Am; Am Educ Res Asn. Res: Statistical analysis, experimental educational data; mathematical statistics; teaching mathematics, experimental materials. Mailing Add: Dept of Math Drake Univ 25th & University Ave Des Moines IA 50311

CANFIELD, ELMER RUSSELL, b Lorraine, Kans, Aug 9, 16; m 40; c 1. FOREST PATHOLOGY, MYCOLOGY. Educ: Univ Idaho, BS, 64, PhD(forest path), 69. Prof Exp: Vis asst res prof forest path, Col Agr, Univ Ariz, 70-72; forest consult, 72-73; res assoc, 73-75, ASSOC PROF FORESTRY, COL FORESTRY, WILDLIFE & RANGE SCI, UNIV IDAHO, 75- Mem: Mycol Soc Am; Soc Am Foresters. Res: Diseases of forest trees; taxonomy of wood-inhabiting fungi. Mailing Add: Col of Forestry Wildlife & Range Sci Univ of Idaho Moscow ID 83843

CANFIELD, EUGENE H, b San Antonio, Tex, Dec 5, 23; m 58; c 2. NUCLEAR PHYSICS. Educ: Tex A&M Univ, BS, 43; Univ Calif, Berkeley, MS, 54, PhD(nuclear theory), 62. Prof Exp: PHYSICIST, LAWRENCE LIVERMORE LAB, UNIV CALIF, 56- Concurrent Pos: Lectr, Rice Univ, 61. Mem: Am Nuclear Soc. Res: Theory and codes for reactor neutronics calculations; Monte Carlo applications in transport theory. Mailing Add: Lawrence Livermore Lab PO Box 808 L-71 Livermore CA 94550

CANFIELD, JAMES HOWARD, b Elmhurst, Ill, Dec 25, 30; m 50; c 2. ORGANIC CHEMISTRY, INFORMATION SCIENCE. Educ: Purdue Univ, BSc, 51; Univ Calif, PhD(chem), 54. Prof Exp: Res chemist, Elastomers Chem Dept, Jackson Lab, E I du Pont de Nemours & Co, Inc, 56-58; res proj chemist, Whittier Res Lab, Am Potash & Chem Corp, 58-59; res mgr, Magna Prods, Inc, 59-64; propellant specialist, Foreigh Tech Div, 65-67, chem specialist, 67-69, TECH ADV, INFO SYSTS DIV, AIR FORCE SYST COMMAND, WRIGHT-PATTERSON AFB, 70- Mem: Am Chem Soc; The Chem Soc; Am Soc Info Sci. Res: Organic synthesis; general polymer chemistry; organoboron derivatives. Mailing Add: 2298 Jacavanda Dr Dayton OH 45431

CANFIELD, JOHN M, theoretical physics, see 12th edition

CANFIELD, NORMAN L, b New London, Conn, Apr 26, 22; m 44; c 3. METEOROLOGY, CLIMATOLOGY. Educ: Univ NH, BS, 42; Univ Mich, MS, 64. Prof Exp: Meteorologist aviation forecasting, US Weather Bur, 46-51, marine climat, 51-54, appl climat, Nat Weather Records Ctr, 54-65, regional climat, Eastern Region, Nat Weather Serv, 65-73, chief data acquisition div, 73-75; PHYS SCIENTIST, NAT OCEANIC & ATMOSPHERIC ADMIN, 75- Concurrent Pos: Proj leader, Navy Marine Climatic Atlas World, Vols I-VII. Honors & Awards: US Dept Com Bronze Medal. Mem: Am Meteorol Soc; Air Pollution Control Asn. Res: Applied climatology, particularly marine climatology. Mailing Add: 6010 Executive Blvd Rm 717 Rockville MD 20852

CANFIELD, NORTON, b Montclair, NJ, Oct 4, 03; m 44; c 2. OTOLOGY, LARYNGOLOGY. Educ: Dartmouth Col, BS, 25; Univ Mich, MD, 29. Prof Exp: Intern, Univ Hosp, Univ Mich, 2930, resident otol, 30-31, instr, Sch Med, 30-33; assoc prof, Yale Univ, 36-62; DIR, HEARING & SPEECH CTR FOR VIRGIN ISLANDS, 62- Concurrent Pos: Consult, US Vet Admin & US Dept Army. Honors & Awards: Legion of Merit & Bronze Star. Mem: Fel AMA; fel Am Acad Ophthal & Otolaryngol; fel Am Otol Soc; fel Am Laryngol Asn; Am Speech & Hearing Asn. Res: Aero-otitis media; hearing testing in children and recording; professional and administrative implications of audiology to human hearing and speech; hearing and deafness. Mailing Add: Harwood Hosp Box 99 Christiansted St Croix VI 00820

CANFIELD, RICHARD CHARLES, b Detroit, Mich, Dec 9, 37; m 61; c 3. SOLAR PHYSICS. Educ: Univ Mich, BS, 59, MS, 61; Univ Colo, PhD(astrogeophysics), 68. Prof Exp: Vis scientist, High Altitude Observ, Nat Ctr Atmospheric Res, 68-69; fel, Neth Orgn Sci Res, 69-70; ASTROPHYSICIST, SACRAMENTO PEAK OBSERV, 70- Concurrent Pos: Mem adv comn, High Altitude Observ Nat Ctr Atmospheric Res, 74- Mem: Am Astron Soc; Int Astron Union. Res: Radiative transfer; application and development of techniques of spectral line formation to solar problems, including chromospheric heating, solar flares and solar velocity field. Mailing Add: Sacramento Peak Observ Sunspot NM 88349

CANFIELD, ROBERT E, b New York, NY, June 4, 31; m 54; c 3. MEDICINE, ENDOCRINOLOGY. Educ: Lehigh Univ, BS, 52; Univ Rochester, MD, 57. Prof Exp: From intern to asst resident med, Presby Hosp, 57-59; res assoc, NIH, 59-62; NIH spec res fel, Enzymol Lab, Nat Ctr Sci Res France, 62-63; from asst prof to assoc prof, 63-72, PROF MED, COL PHYSICIANS & SURGEONS, COLUMBIA UNIV, 72- Mem: Am Soc Biol Chemists; Am Soc Clin Invest; Endocrine Soc; fel Royal Soc Med; Harvey Soc. Res: Protein chemistry, especially related to endocrine and metabolism disorders in man; studies of lysozyme, human chorionic gonadotropin, fibrinogen and Pagets disease of bone. Mailing Add: 2 Edgewood Ave Hastings-on-Hudson NY 10706

CANFIELD, WILLIAM H, b Oklahoma City, Okla, May 24, 20. SPEECH PATHOLOGY, AUDIOLOGY. Educ: Northwestern Univ, BS, 42; Columbia Univ, MA, 50, EdD(speech), 59. Prof Exp: Instr speech, Hofstra Col, 50-57; instr speech path, Teachers Col, Columbia Univ, 57-58, res assoc, 58-60, asst prof, 60-63; assoc prof, 63-68, PROF SPEECH PATH & AUDIOL, ADELPHI UNIV, 68- Concurrent Pos: Consult speech ther, St Barnabas Hosp, Bronx, NY, 59- & St Luke's Hosp, New York, 59- Mem: Am Speech & Hearing Asn; Speech Commun Asn; Am Cleft Palate Asn. Res: Laryngectomy, cleft palate, Parkinson's disease. Mailing Add: Dept of Speech Path & Audiol Adelphi Univ Garden City NY 11530

CANHAM, JOHN EDWARD, b Buffalo, NY, Sept 10, 24; m 47; c 7. MEDICINE, NUTRITION. Educ: Columbia Univ, MD, 49. Prof Exp: US Army, 49-, intern, Letterman Gen Hosp, San Francisco, Calif, 49-50, resident internal med, 51-53, physician, 8th Sta Hosp, Kobe, Japan, 50-51, mem staff internal med, Us Army Hosp, Ft Belvoir, Va, 54-56, chief med serv, US Army Hosp, Wurzburg, Ger, 57-60, chief metab div, US Army Med Res & Nutrit Lab, 61-64, dir res lab, 64-66, comdr, 121st Evacuation Hosp, Ascom, Korea, 66-67, dir res lab, US Army Med Res & Nutrit Lab, 67-73, DIR RES LAB, LETTERMAN ARMY INST RES, 73- Concurrent Pos: Chief prev med, Wurzburg Med Serv Area, Ger, 57-60; clinician, Interdept Comt Nutrit for Nat Defense, Uruguay Nutrit Surv, 62; affil prof, Colo State Univ, 64-66 & 68-; US Army liaison rep, Nutrit Study Oect, NIH & Food & Nutrit Bd, Nat Acad Sci-Nat Res Coun, 64-; US deleg, Far East Conf Nutrit, Manila, Philippines, 67; consult nutrit to Surgeon Gen, US Army, 69-; mem food formulation res panel, US Dept Defense, 70- Honors & Awards: Joseph Goldberger Award Clin Nutrit, AMA, 71. Mem: Am Inst Nutrit; Am Soc Clin Nutrit; Am Col Clin Nutrit; Asn Mil Surg US; Am Bd Nutrit. Res: Nutrition research including applied, clinical and parenteral nutrition, nutrient requirements and nutritional biochemistry; acclimatization, metabolic and nutritional aspects of environment stress. Mailing Add: Letterman Army Inst of Res Presidio of San Francisco CA 94129

CANHAM, PETER BENNET, b Toronto, Ont, Apr 26, 41; m 64; c 2. BIOPHYSICS. Educ: Univ Toronto, BASc, 62; Univ Waterloo, MSc, 64; Univ Western Ont, PhD(biophys), 67. Prof Exp: Proj engr, Can Govt, 62-63; lectr biophys, 67-68, asst prof, 68-74, ASSOC PROF BIOPHYS, UNIV WESTERN ONT, 74- Res: Physics of the blood vessel wall, in particular at bifurcations. Mailing Add: Dept of Biophys Univ of Western Ont London ON Can

CANHAM, RICHARD GORDON, b Arlington, Va, Aug 30, 28; m 52; c 3. PHYSICAL CHEMISTRY. Educ: Col William & Mary, BS, 50; Johns Hopkins Univ, MA, 54, PhD, 59. Prof Exp: Chemist, Nat Bur Stand, 50-55; from asst prof to assoc prof chem, Col William & Mary, 56-62; assoc prof, Col Charleston, 62-64; assoc prof chem, 64-69, chmn dept phys sci, 71-73, PROF CHEM, OKLA BAPTIST UNIV, 69- Mem: Am Chem Soc. Res: Electromotive force of cells; pH in aqueous and nonaqueous media. Mailing Add: 4002 N Aydelotte Shawnee OK 74801

CANIS, WAYNE F, b Elmira, NY, Aug 30, 39; m 68. GEOLOGY. Educ: Colgate Univ, AB, 61; Univ Mo-Columbia, MA, 63, PhD(geol), 67. Prof Exp: Explor geologist, Shell Oil Co, 67-70; ASSOC PROF PHYS SCI, LIVINGSTON UNIV, 70- Mem: AAAS; Am Asn Petrol Geol; Brit Palaeont Asn; Paleont Soc; Geol Soc Am. Res: Biostratigraphy; conodonts; holothurian sclerites. Mailing Add: Dept of Phys Sci Livingston Univ Livingston AL 35470

CANIZARES, CLAUDE ROGER, b Tucson, Ariz, June 14, 45. X-RAY ASTRONOMY. Educ: Harvard Univ, AB, 67, MS, 68, PhD(physics), 72. Prof Exp: Mem res staff, Ctr Space Res, 71-74, ASST PROF PHYSICS, MASS INST TECHNOL, 74- Mem: Am Phys Soc; Am Astron Soc. Res: Optical and X-ray studies of X-ray sources; design and construction of X-ray satellite experiments; optical astronomy instrumentation. Mailing Add: Rm 37-501 Mass Inst of Technol Cambridge MA 02139

CANIZARES, ORLANDO, b Havana, Cuba, May 27, 10; nat US; m 38; c 3. DERMATOLOGY. Educ: Univ Paris, MD, 35. Prof Exp: PROF CLIN DERMAT & SYPHILOL, GRAD MED SCH, NY UNIV, 53-; VIS DERMATOLOGIST & SYPHILOLOGIST, BELLEVUE HOSP, 54- Concurrent Pos: Consult, Vet Admin, NY & USPHS Hosp, Staten Island; chief dermat serv, St Vincent's Hosp. Mem: AAAS; Am Dermat Asn; fel Am Col Physicians. Mailing Add: 3 E 69th St New York NY 10021

CANN, GORDON L, plasma physics, dynamics, see 12th edition

CANN, JOHN RUSWEILER, b Bethlehem, Pa, Dec 11, 20; m 46; c 3. BIOPHYSICAL CHEMISTRY, MOLECULAR BIOPHYSICS. Educ: Moravian Col, BS, 42; Lehigh Univ, MS, 43; Princeton Univ, MA, 45, PhD(phys chem), 46. Prof Exp: Res asst, Manhattan Proj, Princeton Univ, 43-46; res asst, SAM Lab, Carbon & Carbide Chem Corp, 46; res assoc & instr, Cornell Univ, 47; from asst prof to assoc prof, 51-63, PROF BIOPHYS, UNIV COLO MED CTR, DENVER, 63- Concurrent Pos: Res fel, Calif Inst Technol, 47-48, sr res fel, 48-50; NIH res grants biophys, Univ Colo Med Ctr, Denver, 51-; USPH spec res fel, Carlsberg Found Biol Inst, Denmark, 61-62; abstractor, Excerpta Medica, 52-; mem planning group biophys mat, Nat Inst Gen Med Students, 65, ad hoc mem biophys & biophys chem B study sect, 67; mem adv panel molecular biol, Div Biol & Med Sci, NSF, 67-70. Mem: Fel AAAS; Am Chem Soc; Am Asn Immunol; Am Asn Biol Chemists; Biophys Soc. Res: Separation, purification and characterization of proteins; electrophoresis; ultracentrifugation; interaction of proteins with each other and with small molecules; theory of electrophoresis and ultracentrifugation of reacting macromolecules; CD of proteins and peptides. Mailing Add: Dept of Biophys & Genetics Univ of Colo Med Ctr Denver CO 80220

CANN, MALCOLM CALVIN, b Yarmouth, NS, Feb 9, 24; m 49. ORGANIC CHEMISTRY, BIOCHEMISTRY. Educ: Sir George Williams Col, BSc, 53; McGill Univ, MSc, 55, PhD(biochem), 58. Prof Exp: Chemist, Food & Drug Directorate, Can Dept Nat Health & Welfare, 57-66; instr biochem sci, Eastern Ont Inst Technol, 66-67; MASTER SCH TECHNOL, ALGONQUIN COL, 67- Mem: Can Biochem Soc; NY Acad Sci; Can Inst Chem. Res: Bioassay of corticotropin by means of isolated adrenal tissue, adrenal ascorbic acid and plasma corticosteroids; absorption and toxicity of pesticides in rats of various age groups. Mailing Add: Algonquin Col 200 Lees Ave Ottawa ON Can

CANN, MICHAEL CHARLES, b Schenectady, NY, May 6, 47. ORGANIC CHEMISTRY. Educ: Marist Col, BA, 69; State Univ NY Stony Brook, MA, 72, PhD(org chem), 73. Prof Exp: NSF fel & assoc instr org chem, Univ Utah, 73-74; lectr org chem, UNiv Colo, Denver, 74-75; ASST PROF ORG CHEM, UNIV SCRANTON, 75- Mem: Am Chem Soc. Res: The synthesis of aromatic, heterocyclic cations; mechanism of the decomposition 3, 6-Dihydro-1,2-Oxazenes; mechanism of the formation of the Grignard reagent. Mailing Add: Dept of Chem Univ of Scranton Scranton PA 18510

CANNELL, GLEN H, b Abraham, Utah, Aug 5, 19; m 42, 66; c 6. SOIL PHYSICS. Educ: Utah State Univ, BS, 48, MS, 50; Wash State Univ, PhD(agron, soil physics), 55. Prof Exp: Soil scientist, Agr Res Serv, USDA, NDak, 54-56; soil physicist, 56-74, PROF SOIL PHYSICS, UNIV CALIF, RIVERSIDE, 74- Concurrent Pos: Assoc prog dir, Div Pre-Col Educ Sci, NSF, Washington, DC, 69-71. Mem: Am Soc Agron; Soil Sci Soc Am; Am Soc Hort Sci. Res: Soil physics, with emphasis on soil-water-plant relations, soil water movement, soil physical properties and instrumentation for measurement of soil water. Mailing Add: Dept of Soil Sci & Agr Eng Univ of Calif Riverside CA 92502

CANNELL, LAWRENCE GEORGE, physical chemistry, organic chemistry, see 12th edition

CANNEY, FRANK COGSWELL, b Ipswich, Mass, Oct 8, 20; m 44; c 1. GEOLOGY. Educ: Mass Inst Technol, SB, 42, PhD(geol), 52. Prof Exp: Asst geol, Mass Inst Technol, 49-51; geologist, 51-70, CHIEF EXPLOR RES BR, US GEOL SURV, 70- Mem: Soc Econ Geol; Am Inst Mining, Metall & Petrol Engrs; Geol Soc Am; Am Geochem Soc; Am Chem Soc. Res: Geochemical prospecting for mineral deposits; geochemistry of minor elements in the weathering cycle; remote sensing applied to exploration for mineral deposits. Mailing Add: US Geol Surv Bldg 25 Fed Ctr Denver CO 80225

CANNON, ALBERT, b Charleston, SC, Jan 15, 21; m 43; c 4. CLINICAL PATHOLOGY. Educ: Col Charleston, BS, 41; Med Col SC, MD, 49. Prof Exp: Intern surg, George Washington Univ Hosp, 49-50; mem surg staff, Charleston Naval Hosp, 50; resident path, US Naval Med Sch, 51-55, instr, 55-57, chief anat path, 57-58; asst prof path, Med Sch, Georgetown Univ, 58-59; asst prof, 59-70, PROF CLIN PATH, MED UNIV SC, 70- Concurrent Pos: Lectr lab med, US Naval Hosp, Charleston, SC; med dir, ARC Blood Ctr. Mem: AMA; Asn Clin Scientists. Res: Anatomic pathology of central nervous system; hematology; histochemistry; clinical chemistry; blood banking. Mailing Add: Dept of Clin Path Med Univ of SC Charleston SC 29401

CANNON, DONALD CHARLES, b Independence, Mo, Nov 14, 34; m 58; c 2. CLINICAL PATHOLOGY, IMMUNOLOGY. Educ: Harvard Univ, BA, 56; Univ Chicago, MD, 60, PhD(path), 64. Prof Exp: Intern path, Med Ctr, Univ Calif, Los Angeles, 6061; instr, Univ Chicago, 63-64; instr, Univ NC, 64-65; asst prof, State Univ NY Upstate Med Ctr & asst attend pathologist, State Univ Hosp, 65-68, chief diag reagents sect, Div Biol Stand, 68-69, ASST SECT CHIEF CLIN CHEM SERV, CLIN PATH DEPT, NIH, 69- Mem: AMA; Asn Am Med Cols; Col Am Path; Am Soc Clin Path; Int Acad Path. Res: Histology of immune response; spleen cell migrations; clinical serology; blood banking. Mailing Add: Clin Ctr Rm 4N-309 Nat Insts of Health Bethesda MD 20014

CANNON, DONALD JOSEPH, b Boston, Mass, Sept 28, 40; m 68; c 2. BIOCHEMISTRY. Educ: Harvard Univ, AB, 62; Boston Univ, MA, 65, PhD(med sci, biochem), 68. Prof Exp: STAFF SCIENTIST, BOSTON BIOMED RES INST, 72- Concurrent Pos: Res fel biochem, Univ Hawaii, 68-69; fel in aging, Boston Biomed Res Inst, 70-72. Mem: Am Chem Soc; Geront Soc. Res: Chemistry of connective tissue proteins; enzymology; gerontology. Mailing Add: Boston Biomed Res Inst 20 Staniford St Boston MA 02114

CANNON, EDWARD WHITNEY, b Cannon, Del, June 20, 07; m 43; c 2. MATHEMATICS. Educ: Univ Del, BS, 28, MS, 31, DSc, 65; Johns Hopkins Univ, PhD(math), 35. Prof Exp: Testman, Gen Elec Co, 28-29; instr math, Univ Del, 29-31; jr instr, Johns Hopkins Univ, 34-35; from instr to asst prof, Univ Del, 35-41; asst chief, Appl Math Div, Nat Bur Stand, 46-53; head naval logistics res proj, George Washington Univ, 53-55; chief, Appl Math Div, Nat Bur Stand, 55-71; CONSULT, 71- Mem: Am Math Soc; Soc Indust & Appl Math; Asn Comput Mach. Res: Design of computer systems; numerical analysis; data handling systems; logistics research; applied mathematics. Mailing Add: 5 Vassar Circle Glen Echo MD 20768

CANNON, GEORGE WESLEY, b Dickey, NDak, Jan 22, 17; m 44; c 3. ORGANIC CHEMISTRY. Educ: Dakota Wesleyan Univ, BA, 39; Univ Ill, MS, 41, PhD(org chem), 43. Prof Exp: Res chemist, Rohm and Haas Co, Pa, 43-48; assoc prof chem, 48-56, PROF CHEM, UNIV MASS, AMHERST, 56- Mem: Am Chem Soc. Res: Synthesis and structures of amino acids; synthesis of acrylic monomers; acylations of alicyclic ketones; condensations of lactones; reactions of nitriles and halides; cyclopropanes. Mailing Add: Dept of Chem Univ of Mass Amherst MA 01002

CANNON, GLENN ALBERT, b Easton, Md, Apr 11, 40; m 62; c 2. PHYSICAL OCEANOGRAPHY. Educ: Drexel Inst, BS, 63; Johns Hopkins Univ, MA, 65, PhD(oceanog), 69. Prof Exp: Res asst prof oceanog, Univ Wash, 69-70, res oceangr, Pac Marine Environ Lab, Nat Oceanic & Atmospheric Agency, 70-73; prog dir, NSF, 73-75; RES OCEANOGR, PAC MARINE ENVIRON LAB, NAT OCEANIC & ATMOSPHERIC ADMIN, UNIV WASH, 75- Concurrent Pos: Mem adv panel oceanog, NSF, 72-73, 75-76. Mem: Am Geophys Union. Res: Physical oceanography; estuarine and coastal circulation. Mailing Add: Pac Marine Environ Lab NOAA Dept of Oceanog Univ of Wash Seattle WA 98105

CANNON, HELEN LEIGHTON, b Wilkinsburg, Pa, Apr 30, 11; m 35; c 1. GEOCHEMISTRY. Educ: Cornell Univ, AB, 32; Univ Pittsburgh, MS, 34. Prof Exp: Asst geol, Northwestern Univ, 32-33 & Univ Okla, 34-35; geologist, Oil Geol, Gulf Oil Co, 35-36; geologist minor metal commodity admin, 42-46, geologist geochem prospecting methods, 46-62, GEOLOGIST GEOCHEM CENSUS, US GEOL SURV, 62- Concurrent Pos: Chmn subcomt geochem environ in health & dis, Nat Res Coun, 69-73. Honors & Awards: Meritorious Award & Nominee, Fed Woman of the Year, Dept Interior, 70, Distinguished Serv Award, 75. Mem: Fel AAAS (chmn geol sect, 76); Soc Econ Geologists; fel Geol Soc Am; Am Geochem Soc; Int Asn Geochem & Cosmochem. Res: Botanical methods of prospecting; trace element distribution in soils and plants as related to geology, health and disease. Mailing Add: US Geol Surv Fed Ctr Bldg 25 Denver CO 80225

CANNON, JERRY WAYNE, b Lambert, Miss, June 24, 42; m 63. BIOCHEMISTRY. Educ: Univ Miss, BSc, 64; Drexel Univ, PhD(chem), 69. Prof Exp: Instr biochem, Sch Med, Univ Miss, 68-70; ASST RES PROF CHEM, MISS COL, 70- Mem: Am Chem Soc. Res: Pathway of steroid biosynthesis; synthesis and biological testing of chemical analogs of digitalis. Mailing Add: Dept of Chem Miss Col Clinton MS 39056

CANNON, JOHN BURNS, b Spartanburg, SC, Mar 5, 48. ORGANOMETALLIC CHEMISTRY, BIOINORGANIC CHEMISTRY. Educ: Duke Univ, BS, 70; Princeton Univ, PhD(chem), 74. Prof Exp: RES CHEMIST, DEPT CHEM, UNIV CALIF, SAN DIEGO, 74- Mem: Am Chem Soc; Sigma Xi. Res: Kinetics and model studies of heme enzymes such as myoglobin, hemoglobin and cytochrome P-450. Mailing Add: Dept of Chem B-017 Univ of Calif at San Diego La Jolla CA 92093

CANNON, JOHN FRANCIS, b Monroe, Utah, Dec 14, 40; m 62; c 8. PHYSICAL CHEMISTRY, INORGANIC CHEMISTRY. Educ: Brigham Young Univ, BS, 65, PhD(phys chem), 69. Prof Exp: Fel, Georgetown Univ, 69-70; assoc dir high pressure data ctr, 70-72, ASSOC DIR CTR HIGH PRESSURE RES, BRIGHAM YOUNG UNIV, 72- Mem: Am Chem Soc; Am Crystallog Asn; Sigma Xi. Res: High pressure, high temperature synthesis of inorganic compounds, especially intermetallics and compounds containing lanthanide elements; x-ray crystallography. Mailing Add: Ctr for High Pressure Res 226 ESC Brigham Young Univ Provo UT 84602

CANNON, JOHN ROZIER, b McAlester, Okla, Feb 3, 38; m 57; c 3. MATHEMATICS. Educ: Lamar State Col, BA, 58; Rice Univ, MA, 60, PhD(math), 62. Prof Exp: Assoc mathematician, Brookhaven Nat Lab, 62-64, NATO fel, 64-65; mem fac math, Purdue Univ, 65-66, prof, 68-69; assoc prof, Univ Minn, Minneapolis, 66-68; PROF MATH, UNIV TEX, AUSTIN, 69- Concurrent Pos: Consult, Mobil Oil Corp, 67-69; vis prof, Tex Tech Univ, 73-74. Mem: Am Math Soc; Soc Indust & Appl Math; Sigma Xi; Math Asn Am. Res: Ordinary and partial differential equations; numerical analysis; function theory. Mailing Add: Dept of Math Univ of Tex Austin TX 78712

CANNON, JOSEPH G, b Decatur, Ill, Sept 30, 26; m 60; c 4. MEDICINAL CHEMISTRY. Educ: Univ Ill, BS, 51, MS, 53, PhD(pharmaceut chem), 57. Prof Exp: Asst prof pharmaceut, Univ Wis, 56-60, assoc prof pharmaceut chem, 60-62; assoc prof, 62-65, PROF MED CHEM, UNIV IOWA, 65- Mem: Am Chem Soc; fel Am Inst Chemists; fel Acad Pharmaceut Sci. Res: Organic synthesis; structure-activity relationships; nitrogen heterocycles. Mailing Add: Col of Pharmacy Univ of Iowa Iowa City IA 52242

CANNON, LAWRENCE ORSON, b Logan, Utah, June 11, 35; m 59; c 3. MATHEMATICS. Educ: Utah State Univ, BS, 58; Univ Wis, MS, 59; Univ Utah, PhD(math), 65. Prof Exp: From instr to asst prof math, Utah State Univ, 61-63; instr, Univ Utah, 63-64; asst prof, 65-68, ASSOC PROF MATH, UTAH STATE UNIV, 68-, HEAD DEPT, 69- Concurrent Pos: Vis prof, Rutgers Univ, 68-69. Mem: Am Math Soc. Res: Upper semicontinuous decompositions of 3-manifolds; wild and tame surfaces. Mailing Add: Dept of Math Utah State Univ Logan UT 84321

CANNON, MARVIN SAMUEL, b Toledo, Ohio, Feb 10, 40; m 73. HUMAN ANATOMY, CELL BIOLOGY. Educ: Univ Toledo, BS, 60, MS, 65; Ohio State Univ, PhD(human anat), 69. Prof Exp: Asst prof biol sci, Capital Univ, 71-73; ASST PROF ANAT, UNIV TEX MED BR GALVESTON, 73- Concurrent Pos: Bremer Found Fund fel, Ohio State Univ, 71-73. Mem: AAAS; Am Asn Anat; Pan Am Asn Anat; Am Inst Biol Sci; Soc Gnotobiotics. Res: Adrenal gland; venom glands. Mailing Add: Dept of Anat Univ of Tex Med Br Galveston TX 77550

CANNON, MELVIN CROXALL, b Salt Lake City, Utah, July 25, 13; m 37; c 4. CHEMISTRY. Educ: Univ Utah, BS, 33, MS, 38; Boston Univ, PhD(chem), 41. Prof Exp: Res chemist, Merck & Co, 41; indust fel, Mellon Inst, 42-44; dir res sapphire prod div, Elgin Nat Watch Co, Ill, 44-46; asst prof chem, Univ Denver, 46-47; assoc prof, 47-52, head dept, 55-68, PROF CHEM, UTAH STATE UNIV, 52- Mem: Am Chem Soc. Res: Inorganic elevated temperature reactions; analytical chemistry. Mailing Add: Dept of Chem Utah State Univ Logan UT 84321

CANNON, ORSON SILVER, b Salt Lake City, Utah, Nov 21, 08; m 34; c 5. PHYTOPATHOLOGY. Educ: Utah State Col, BS, 35, MS, 37; Cornell Univ, PhD(plant path), 42. Prof Exp: Asst plant path, Utah State Agr Col, 33-37; asst exten plant pathologist, Pa State Col, 42-43; head dept, Crop Res Lab, H J Heinz Co, Ohio, 43-48; plant pathologist, Bur Plant Indust, Soils & Agr Eng, USDA, 48-57, head dept bot & plant path, 57-74, EMER PROF BOT & PLANT PATH, UTAH STATE UNIV, 74- Mem: AAAS; Am Phytopath Soc; Am Soc Hort Sci; Am Soc Plant Physiol. Res: Bacterial wilt of alfalfa; vegetable diseases; mosaic resistance in cucumbers; anthracnose resistance in tomatoes; curly top and wilt resistance in tomatoes. Mailing Add: 1407 E 17th N Logan UT 84321

CANNON, PAUL ROBERTS, b Lexington, Ill, Aug 25, 92; m 17; c 1. PATHOLOGY. Educ: James Millikin Univ, AB; Univ Chicago, PhD(bact), 21; Rush Med Col, MD, 26. Prof Exp: Instr bact, Univ Chicago, 19-20; prof path & bact, Univ Miss, 20-23; from asst prof to prof, 25-57, chmn dept, 40-57, EMER PROF PATH, SCH MED, UNIV CHICAGO, 57- Concurrent Pos: Chief ed, Arch Path, AMA, 54-64; past ed, Am J Path & J Am Asn Immunologists. Honors & Awards: Burdick Award, Am Soc Clin Path; Gerhard Medal, Philadelphia Path Soc, 48; Groedel Medal, Col Cardiol, 58; Gold-Headed Cane Award, Am Asn Path & Bact, 65. Mem: Nat Acad Sci; AAAS; Am Soc Exp Path (pres, 47); Am Asn Path & Bact (pres, 42-46); Am Asn Immunologists (pres, 40). Res: Intestinal bacteriology; cellular and tissue immunity; spleen; anemia; experimental malaria; respiratory infections; lipid pneumonia; allergy; chemotherapy; experimental pathology; nutrition and protein metabolism; immunology. Mailing Add: RR 2 Box 79 Yorkville IL 60560

CANNON, PETER, b Chatham, Eng, Apr 20, 32; m 55; c 4. PHYSICAL CHEMISTRY. Educ: Univ London, BSc, 52, PhD(chem), 55. Prof Exp: Mem staff, Overseas Chem Dept, Procter & Gamble Co, 55-56; phys chemist, Gen Elec Res Lab, 56-65, mgr opers anal, Gen Elec Info Systs, 65-67; mgr sensors & mat, Gen Elec Mfg & Process Automation Bus Div, 67-69, mgr sensors & microelectronics, 69-72, strategy develop automation & machine tools bus, 72-73; dir new prod develop, 73-75, vpres, Bus Develop, Util & Indust Opers Div, 75-76; VPRES SCI CTR, ROCKWELL INT CORP, 76- Concurrent Pos: Adj prof, Polytech Inst Brooklyn, 64-69; lectr, Grad Sch Bus Admin, Univ Va, 65-67. Mem: Am Chem Soc; Am Phys Soc; Asn Comput Mach; Royal Inst Chemists. Res: Administrative product development programs and business strategy planning; physics and chemistry of surfaces and the solid state; catalysis; super-pressure reactions, especially diamonds; operations analysis and research, especially mathematical programming. Mailing Add: Rockwell Int Sci Ctr 1049 Camino Dos Rios Thousand Oaks CA 91360

CANNON, PHILIP JAN, b Washington, DC, Nov 15, 40; m 61; c 2. GEOMORPHOLOGY. Educ: Univ Okla, BS, 65, MS, 67; Univ Ariz, PhD(geol), 73. Prof Exp: Geologist, US Geol Surv, 67-72; res scientist geol, Tex Bur Econ Geol, 72-74; ASST PROF GEOL, UNIV ALASKA, 74- Mem: Am Soc Photogram. Res: Geomorphic investigations of Alaska and geologic mapping of coastal areas using side-looking airborne radar imagery. Mailing Add: Dept of Geol Univ of Alaska Fairbanks AK 99701 1663 Los Alamos NM 87544

CANNON, RALPH SMYSER, JR, geology, see 12th edition

CANNON, RAYMOND JOSEPH, JR, b Hartford, Conn. MATHEMATICS. Educ: Col Holy Cross, AB, 62; Tulane Univ, PhD(math), 67. Prof Exp: Asst prof math, Vanderbilt Univ, 67-69; asst prof, Univ NC, Chapel Hill, 69-74; ASST PROF MATH, STETSON UNIV, 74- Concurrent Pos: Off Naval Res fel, Univ Mich, 68-69. Mem: Am Math Soc; Math Asn Am. Res: Quasiconformal mappings; functions of a complex variable; topological analysis. Mailing Add: Dept of Math Stetson Univ De Land FL 32720

CANNON, ROBERT H, JR, dynamics, see 12th edition

CANNON, ROBERT YOUNG, b Boise, Idaho, Sept 11, 17; m 48; c 3. DAIRY SCIENCE. Educ: Iowa State Univ, BS, 39; Ohio State Univ, MS, 40; Univ Wis, PhD(dairy indust), 49. Prof Exp: PROF FOOD SCI, AUBURN UNIV, 48- Mem: Am Dairy Sci Asn; Inst Food Technol; Int Asn Milk, Food & Environ Sanit. Res: Dairy foods processing; qualtiy control of foods; food plant sanitation. Mailing Add: Dept of Animal & Dairy Sci Auburn Univ Auburn AL 36830

CANNON, THEODORE WILES, b Eugene, Ore, Aug 28, 33; m 69. CLOUD PHYSICS. Educ: Ore State Univ, BS, 56, PhD(nuclear physics), 66; Univ Ore, MS, 60. Prof Exp: Elec engr, Hanford Atomic Prod Oper, 56-59; teaching asst, Univ Ore, 59-60; instr, Westmont Col, 60-61; asst, Ore State Univ, 61-66; cloud physicist & fel, 66-69, PhD SCIENTIST, NAT CTR ATMOSPHERIC RES, 69- Honors & Awards: Soc Photographic Scientists & Engrs Serv Award, 75. Mem: Am Asn Physics Teachers; Soc Photog Scientists & Engrs; Am Meteorol Soc. Res: Coalescence process in rainclouds; photography of airborne atmopsheric particles. Mailing Add: Nat Ctr for Atmospheric Res Boulder CO 80303

CANNON, WILLIAM FRANCIS, III, b Troy, NY, Aug 11, 40; m 69. GEOLOGY. Educ: Syracuse Univ, AB, 62, PhD(geol), 68; Miami Univ, MS, 64. Prof Exp: GEOLOGIST, US GEOL SURV, 67- Concurrent Pos: Prof lectr, George Washington Univ, 70. Mem: Mineral Soc Am; Geol Soc Am. Res: Petrology and structural geology of Precambrian rocks in the Lake Superior-Lake Huron area. Mailing Add: US Geol Surv Nat Ctr Stop 920 Reston VA 22092

CANNON, WILLIAM NATHANIEL, b Atlanta, Ga, Oct 15, 27; m 51; c 4. ORGANIC CHEMISTRY. Educ: NGa Col, BS, 48; Univ Ga, MS, 50. Prof Exp: Chemist, Rohm & Haas Co, Redstone Arsenal, Ala, 50-52; org chemist, 52-57, appln res chemist, 57-66, process chemist, 66-67; sr org chemist, 67-74, RES SCIENTIST, ELI LILLY & CO, 74- Mem: Am Chem Soc; Soc Indust Microbiol (secy, 66-67). Res: Synthesis of organic compounds for testing and evaluation as agricultural chemicals.

CANNON, WILLIAM NELSON, JR, b Wilmington, Del, Jan 7, 32; m 62; c 3. ENTOMOLOGY. Educ: Univ Del, BS, 53, MS, 60; Ohio State Univ, PhD(entom), 63. Prof Exp: Sanitarian, Del State Bd Health, 56-58; res assoc entom, Ore State Univ, 63-65; RES ENTOMOLOGIST, FOREST INSECT & DIS LAB, NORTHEASTERN FOREST EXP STA, US FOREST SERV, 65- Mem: Entom Soc Am. Res: Insect-host plant relationships; biology and control of insects; insect sterility. Mailing Add: Forest Insect & Dis Lab USDA NE Forest Exp Sta Delaware OH 43015

CANNONITO, FRANK BENJAMIN, b New York, NY, Oct 19, 26; m 53; c 2. ALGEBRA, MATHEMATICAL LOGIC. Educ: Columbia Univ, BS, 59, MA, 61; Adelphi Univ, PhD(math), 65. Prof Exp: Res mathematician, Res Dept, Grumman Aircraft Eng Corp, 60-62; staff mathematician, Tech Anal Off, Hughes Aircraft Co, 62-64, mgr info sci progs, 64-66; asst prof math, 66-70, vchmn dept, 69-72, ASSOC PROF MATH, UNIV CALIF, IRVINE, 70- Concurrent Pos: Sr staff mathematician, Hughes Aircraft Co, 66-69; prin investr, Air Force Off Sci Res grant, 66-73; Army Res Off Conf & NSF Conf grants, 69; assoc ed, Info Sci, 68-. Mem: Am Math Soc; Asn Symbolic Logic. Res: Fine degrees of solvability of the word problem and related decision problems in group theory and in recursive function theory, particularly as applied to group theory. Mailing Add: Dept of Math Univ of Calif Irvine CA 92664

CANO, GILBERT LUCERO, b Mesilla, NMex, Jan 7, 32; m 52; c 4. EXPERIMENTAL ATOMIC PHYSICS. Educ: NMex State Univ, BS, 54, MS, 60, PhD(physics), 64. Prof Exp: MEM RES STAFF, SANDIA CORP, 64- Mem: Am Phys Soc. Res: Charge spectroscopy of laser-induced blow-off; atomic stopping power of thin metallic films; ion-atom interactions. Mailing Add: Sandia Corp Lab 5242 Sandia Base Albuquerque NM 87115

CANODE, CHESTER LANG, b Keystone, Okla, Oct 10, 20; m 42; c 2. AGRONOMY, PLANT BREEDING. Educ: Okla State Univ, BS & MS, 46; Univ Wis, PhD(agron), 56. Prof Exp: Instr agron, Murray State Sch Agr, 46-48; asst agronomist, Okla State Univ, 48-52 & Univ Idaho, 52-56; PROF AGRON, WASH STATE UNIV & RES AGRONOMIST, USDA, 56- Res: Grass seed production research and grass breeding. Mailing Add: Dept of Agron Wash State Univ Pullman WA 99163

CANOLTY, NANCY LEMMON, b Washington, Ind, Mar 1, 42; m 68; c 2. NUTRITION. Educ: Purdue Univ, BS, 63, MS, 68; Univ Calif, Berkeley, PhD(nutrit), 74. Prof Exp: ASST PROF NUTRIT, UNIV CALIF, DAVIS, 73- Mem: Inst Food Technologists; Sigma Xi. Res: Influence of nutrition upon human lactation, infant nutrition; energy metabolism and protein quality. Mailing Add: Dept of Nutrit Univ of Calif Davis CA 95616

CANONICO, PETER GUY, b Tunis, Tunisia, June 12, 42; US citizen; m 69; c 2. CELL BIOLOGY. Educ: Bucknell Univ, BS, 64; Univ SC, MS, 66; Rutgers Univ, PhD(physiol), 69. Prof Exp: RES SCIENTIST CELL PHYSIOL, US ARMY MED RES INST INFECTIOUS DIS, 69- Concurrent Pos: Adj prof, Hood Col, 75-; fac mem, NIH Found Advan Educ Sci, 76- Honors & Awards: Res & Develop Medal, Dept Army, 75. Mem: Am Soc Cell Biol; Am Physiol Soc; AAAS; Am Soc Microbiol; Soc Exp Biol & Med. Res: Effects of infections and inflammation on cellular functions and physiology and morphology of subcellular organelles; studies of host cellular defense mechanisms to microbial infection. Mailing Add: Bacteriol Div USAMRIID Frederick MD 21701

CANOSA, JOSE M, b Huelva, Spain, Oct 26, 32; US citizen; m 57; c 3. MATHEMATICAL PHYSICS, APPLIED MATHEMATICS. Educ: Univ Barcelona, BSc, 55; Harvard Univ, SM, 61, PhD(appl physics), 64. Prof Exp: Physicist, Vallecitos Atomic Lab, Gen Elec Co, Calif, 64-67; STAFF MEM, SCI CTR, IBM CORP, PALO ALTO, 67- Mem: Am Nuclear Soc; Am Phys Soc; Soc Indust & Appl Math. Res: Plasma physics; computational physics; nuclear reactor theory; nonlinear differential equations. Mailing Add: 120 Merritt Ct Los Altos CA 94022

CANRIGHT, JAMES EDWARD, b Delaware, Ohio, Mar 1, 20; m 43; c 4. PLANT MORPHOLOGY, PALEOBOTANY. Educ: Miami Univ, AB, 42; Harvard Univ, AM, 47, PhD(biol), 49. Prof Exp: Teaching fel biol, Harvard Univ, 46-49; from instr to prof bot, Ind Univ, 49-63; chmn dept, 64-72, PROF BOT, ARIZ STATE UNIV, 64- Concurrent Pos: Am Philos Soc res grant, 53-54; NSF travel grant, Int Bot Cong, Paris, 54, Edinburgh, 64 & Leningrad, 75; res grants, 54-; consult, Cent Univ Venezuela, 58; Guggenheim fel, Indonesia & Malaya, 60-61; vis scientist US-China coop sci prog, Nat Taiwan Univ, 71. Mem: Fel AAAS; Int Orgn Paleobot; Bot Soc Am; Am Inst Biol Sci; Int Soc Plant Morphol. Res: Floral morphology; wood anatomy; Tertiary and Cretaceous palynology; paleobotany of the Paleozoic. Mailing Add: Dept of Bot & Microbiol Ariz State Univ Tempe AZ 85281

CANTAROW, ABRAHAM, b Hartford, Conn, Jan 27, 01; m 32; c 1. BIOCHEMISTRY, MEDICINE. Educ: Jefferson Med Col, MD, 24, ScD, 69. Prof Exp: Res chemist, Jefferson Hosp, 24-25, res physician, 25-27, fel inorg metab, 27-29; asst demonstr med, 29-31, instr, 31-34, assoc, 34-37, from asst prof to assoc prof, 37-46, prof biochem & head dept, 45-66, EMER PROF BIOCHEM, JEFFERSON MED COL, 66-; RES PLANNING OFFICER, PROG ANAL & FORMULATION BR, NAT CANCER INST, 66- Concurrent Pos: Biochemist, Jefferson Hosp, 31-45; consult, Bur Med & Surg, US Navy. Mem: AAAS; Endocrine Soc; Am Asn Cancer Res (pres, 69-70); Am Physiol Soc; Am Soc Pharmacol & Exp Therapeut. Res: Clinical medicine; calcium metabolism; renal and hepatic functions; physiology of endocrine glands and of the liver; cancer research; steroid metabolism; nutrition. Mailing Add: Prog Anal & Formulation Br Nat Cancer Inst Bethesda MD 20014

CANTE, CHARLES JOHN, b Brooklyn, NY, Oct 31, 41; m 73. PHYSICAL CHEMISTRY, SURFACE CHEMISTRY. Educ: City Col New York, BS, 63, MA, 65, PhD(phys chem), 67; Iona Col, MBA, 72. Prof Exp: Lectr chem, City Col New York, 64-67; phys chemist, Phys Res Lab, Edgewood Arsenal, Dept Army, 67-68, res & develop coordr, 68-69; sr chemist, Tech Ctr, 69-71, group leader, 71-73, personnel assoc, 73-75, MGR NEW TECHNOL RES, PET FOODS DIV, GEN FOODS CORP, 75- Concurrent Pos: E I du Pont de Nemours teaching award, 66-67. Mem: AAAS; Am Chem Soc. Res: Colloid chemistry; food chemistry. Mailing Add: Tech Ctr Gen Foods Corp 250 North St White Plains NY 10602

CANTELO, WILLIAM WESLEY, b Medford, Mass, Sept 11, 26; m 58; c 3. ENTOMOLOGY. Educ: Boston Univ, AB, 48; Univ Mass, MS, 50, PhD(entom), 52. Prof Exp: Asst entomologist, Bartlett Tree Res Labs, 52-54, assoc entomologist, 54-55; staff entomologist, US Naval Forces, Marianas, 56-61; entom adv, US Opers Mission, Ministry Agr, Thailand, 61-66; SUPVRY RES ENTOMOLOGIST, USDA, 66- Mem: AAAS; Entom Soc Am. Res: Life history and control of ornamental and shade tree insects; tropical agriculture; economic entomology; insect population dynamics and ecology. Mailing Add: USDA BARC-E Bldg 470 Beltsville MD 20705

CANTER, JOSEPH M, b Springfield, Mass, Nov 15, 40; m 63; c 2. EXPERIMENTAL HIGH ENERGY PHYSICS. Educ: Yale Univ, BS, 62; Carnegie Inst Technol, MS, 64, PhD, 68. Prof Exp: Asst, Carnegie Inst Technol, 62-64, res asst, 64-65, proj physicist, 65-67; res assoc, State Univ NY Stony Brook, 67-70; ASST PROF PHYSICS, TUFTS UNIV, 70- Mem: Am Phys Soc. Mailing Add: Dept of Physics Tufts Univ Medford MA 02155

CANTER, NATHAN H, b Philadelphia, Pa, Nov 17, 42; m 64; c 1. POLYMER PHYSICS. Educ: Temple Univ, AB, 63; Princeton Univ, MS, 65, PhD(chem), 66. Prof Exp: RES ASSOC, CORP RES LABS, EXXON RES & ENG CO, 66- Mem: AAAS; Am Chem Soc; Am Phys Soc. Res: Statistical and solid state physics; physical chemistry; polymer physics and chemistry. Mailing Add: Exxon Res & Eng Corp Res Lab PO Box 45 Linden NJ 07036

CANTERINO, PETER J, organic chemistry, see 12th edition

CANTERO, ANTONIO, b Ofena, Italy, Nov 4, 02; Can citizen; m 39; c 2. BIOCHEMISTRY, PATHOLOGY. Educ: Univ Ottawa, BA, 22; McGill Univ, MD & CM, 27. Prof Exp: Gastroenterologist & intern, Radium Inst & Verdun Gen Hosp, Univ Montreal, 32-35; chief gastroenterologist, Reddy Mem Hosp, 36-40; asst exp cancer & endocrinol, Univ Montreal, 42-48; res dir, Montreal Cancer Inst, Notre Dame Hosp, 48-68, consult, 69-75; res prof med, Univ Montreal, 64-68, consult, 68-75. Concurrent Pos: Res assoc, Radium Inst, 36-37; res fel, Univ Montreal, 40; vis scientist, Inst Oswaldo Cruz, Brazil, 49; co-dir & sci adv, Nat Cancer Inst & Brazil Dept Health & Educ, 51; vis prof, Cent Univ Venezuela, 55; med dir, Can Broadcasting Corp, 69-75. Mem: Fel Am Col Gastroenterol; fel NY Acad Sci; fel Royal Soc Can; Fr-Can Asn Advan Sci; Can Asn Path. Res: Experimental bacteriology and pathology; hormonal dependant tumors; estrogen and mammary cancer; precancerous lesions in liver and its relation to vitamin deficiency; enzymatic changes in experimental neoplasia. Mailing Add: 98 Elmwood Place Outremont PQ Can

CANTIN, MARC, b Que, Aug 7, 33; m 59; c 1. MEDICINE. Educ: Laval Univ, MD, 58; Univ Montreal, PhD, 62. Prof Exp: Asst, Inst Exp Med & Surg, Univ Montreal, 58-62; instr path, Sch Med, Univ Chicago, 64-65; from asst prof to assoc prof, 65-74, PROF PATH, UNIV MONTREAL, 74- Concurrent Pos: USPHS fel, 62-63 & grant, 64-66; fel path, Sch Med, Univ Chicago, 62-64; Chicago Heart Asn fel, 63-65 & grant, 64-66; Ill Heart Asn fel, 63-65. Mem: Am Heart Asn; Can Med Asn. Res: Relationships between juxtaglomerular apparatus and adrenal cortex in various experimental situations; physiopathology of congestive heart failure; atrial specific granules structure and function; ultrastructural cytochemistry. Mailing Add: Dept of Path Univ of Montreal Montreal PQ Can

CANTINO, EDWARD CHARLES, b Berkeley, Calif, Oct 31, 21; m 45; c 2. MYCOLOGY. Educ: Univ Calif, AB, 43, PhD(plant physiol), 48. Prof Exp: Sr lab technician, Div Plant Nutrit, Univ Calif, 43-44; chemist, Am Cyanamid Co, Calif, 44-45; res chemist, Aerojet Eng Corp, 45; from asst prof to assoc prof bot, Univ Pa, 48-56; PROF BOT, MICH STATE UNIV, 56- Concurrent Pos: Guggenheim fel, 50-51; res collabr, Brookhaven Nat Lab, 55; vis prof, Univ Geneva, 60; consult, NIH, 63-66 & NSF, 63-; ed-in-chief, Experimental Mycol, 75- Honors & Awards: Distinguished Fac Award, Mich State Univ, 64. Mem: Am Soc Microbiol; Am Acad Microbiol; Bot Soc Am; Am Soc Cell Biol; Mycol Soc Am. Res: Physiological mycology; physiology of fungi; biology of aquatic phycomycetes; cell differentiation. Mailing Add: Dept of Bot & Plant Path Mich State Univ East Lansing MI 48824

CANTLIFFE, DANIEL JAMES, b New York, NY, Oct 31, 43; m 65; c 4. PLANT PHYSIOLOGY, VEGETABLE CROPS. Educ: Delaware Valley Col, BS, 65; Purdue Univ, MS, 67, PhD(plant physiol), 71. Prof Exp: Res asst, Purdue Univ, 65-69; res assoc, Cornell Univ, 69-70; res scientist, Hort Res Inst Ont, 70-74; ASST PROF SEED PHYSIOL, UNIV FLA, 74- Mem: Am Soc Hort Sci; Am Soc Plant Physiologists; Am Soc Agron; Crop Sci Soc Am; Sigma Xi. Res: Basic studies and applications with seeds as plant growth units, including the physiology of fruit development, seed formation, seed germination, seed dormancy, and seedling vigor. Mailing Add: Dept of Veg Crops Univ of Fla Gainesville FL 32611

CANTLON, JOHN EDWARD, b Sparks, Nev, Oct 6, 21; m 44; c 4. ECOLOGY. Educ: Univ Nev, BS, 47; Rutgers Univ, PhD(bot), 50. Prof Exp: Asst prof bot, George Washington Univ, 50-52, assoc prof, 52-53; sr ecologist, Phys Res Lab, Boston Univ, 53-54; assoc prof bot, 54-58, prof, 58-69, provost, 69-76, VPRES RES & GRAD STUD, MICH STATE UNIV, 76- Concurrent Pos: Mem adv panel environ biol, NSF, 61-64, prog dir environ biol, 65-66, mem adv comt, div environ sci, 66-69, adv comt instnl relations, 70-74; gov bd, Am Inst Biol Sci, 63-66; adv comt health physics, Oak Ridge Nat Lab, 66-69, adv coun, 71-75; exec comt, Div Biol & Agr, Nat Res Coun, 67-71, coun natural resources, 73- Mem: AAAS; Ecol Soc Am (secy, 58-63, vpres, 67-68, pres, 68-69); Am Inst Sci; Bot Soc Am; Am Soc Naturalists. Res: Pattern in communities; physiological ecology; Alaskan tundra vegetation; research administration. Mailing Add: Admin Bldg Mich State Univ East Lansing MI 48823

CANTOR, ABRAHAM, b Philadelphia, Pa, Jan 21, 11; m 33; c 2. BIOCHEMISTRY. Educ: Univ Pa, BA, 33, MA, 36, PhD(pub health, prev med), 40. Prof Exp: Instr bact, Univ Pa, 39-42, assoc pub health & prev med, 42-45; bacteriologist, Eastern Regional Res Br, Agr Res Serv, USDA, 45-46; res dir, Int Amino Corp, 46-48; bacteriologist, Hahnemann Med Col, 48-49; asst prof bact, Univ Akron, 49-53; pres, W Labs, Inc, 64-71, SR VPRES RES & DEVELOP, W CHEM PROD, INC, 71- Concurrent Pos: Res dir, Protein Prod Corp, 48-49. Mem: AAAS; Am Chem Soc; Am Soc Microbiol;

Inst Food Technologists; NY Acad Sci. Res: Microbiological chemistry. Mailing Add: W Chem Prod Inc 42-16 West St Long Island City NY 11101

CANTOR, CHARLES ROBERT, b Brooklyn, NY, Aug 26, 42; m 64. BIOPHYSICAL CHEMISTRY. Educ: Columbia Col, AB, 63; Univ Calif, Berkeley, PhD(chem), 66. Prof Exp: From asst prof to assoc prof, 66-72, PROF CHEM, COLUMBIA UNIV, 72- Concurrent Pos: Alfred P Sloan fel, 69-71; NIH study grant, 71-75; mem ed bd, J Molecular Biol, 72-; Guggenheim fel, 73-74; Sherman Fairchild Distinguished vis scholar, Calif Inst Technol, 75-76. Honors & Awards: Eastman Kodak Award, 65. Mem: Harvey Soc; Am Soc Biol Chem; Biophys Soc. Res: Optical properties and conformation of nucleic acids and proteins; structure of the ribosome; mechanism of protein synthesis; affinity labelling; association of multi-subunit proteins. Mailing Add: Box 608 Havenmeyer Columbia Univ Dept of Chem New York NY 10027

CANTOR, DAVID GEOFFREY, b London, Eng, Apr 12, 35; US citizen; m 58; c 2. MATHEMATICS, COMPUTER SCIENCES. Educ: Calif Inst Technol, BS, 56; Univ Calif, Los Angeles, PhD(math), 60. Prof Exp: Asst prof math, Univ Wash, 62-64; PROF MATH & COMPUT SCI, UNIV CALIF, LOS ANGELES, 64- Concurrent Pos: Sloan Found fel, 67; prin investr, NSF, 68- Mem: Am Math Soc; Math Asn Am; Soc Indust & Appl Math; Asn Comput Mach; Inst Elec & Electronics Engrs. Res: Number theory; combinatorics; algorithms. Mailing Add: Dept of Math Univ of Calif Los Angeles CA 90024

CANTOR, ENA D, b Montreal, Que, Mar 1, 20; m. MICROBIOLOGY, IMMUNOLOGY. Educ: McGill Univ, BA, 40, BSc, 42, PhD(microbiol, immunol), 68; Harvard Univ, MA, 46. Prof Exp: Technician, Royal Victoria Hosp, Can, 42-44; res technician, RI Hosp, Providence, 54-55; res technician, Mass Mem Hosp, 58-59; chief technician, Jewish Gen Hosp, Can, 60-62; techician, McGill Univ, 62-63; mem staff clin virol, Royal Victoria Hosp & McGill Univ, 68-69; chief sanit virol sect, 69-71, CHIEF GEN BACT SECT, MINISTRY SOCIAL AFFAIRS, MONTREAL, 71- Mem: Am Soc Microbiol; Can Pub Health Asn; Can Fedn Biol Sci; Can Soc Microbiol; Can Soc Immunol. Res: Enteroviruses; serum proteins; population studies of norms; enterococci, especially chemical and immunological studies; lysogeny of Lancefield group B streptococci. Mailing Add: Div of Microbiol Ministry of Social Affairs Labs Laval-des-Rapides PQ Can

CANTOR, MARVIN H, b Brooklyn, NY, Nov 17, 35. CELL PHYSIOLOGY. Educ: Boston Univ, AB, 57; Mass Inst Technol, SM, 59; Univ Calif, Los Angeles, PhD(zool), 64. Prof Exp: Fel zool, Syst-Ecol Prog, Marine Biol Lab, Woods Hole, 64-65; from asst prof to assoc prof biol, 65-71, PROF BIOL, CALIF STATE UNIV, NORTHRIDGE, 71- Mem: AAAS; Soc Protozool; Am Soc Zool; NY Acad Sci. Res: Cell growth and metabolism; cell synchrony; regulation of metabolism. Mailing Add: Dept of Biol Calif State Univ Northridge CA 91324

CANTOR, STANLEY, b Brooklyn, NY, Sept 23, 29; m 50; c 3. PHYSICAL CHEMISTRY. Educ: Tulane Univ, BS, 51, MS, 53, PhD(chem), 55. Prof Exp: MEM RES STAFF, OAK RIDGE NAT LAB, 55- Concurrent Pos: Exchange fel, Atomic Energy Res Estab, Eng, 63-64. Mem: AAAS; Am Chem Soc; Sigma Xi. Res: Properties of molten salts; thermodynamics; nuclear reactor chemistry; chemistry of energy systems. Mailing Add: PO Box X Oak Ridge Nat Lab Oak Ridge TN 37830

CANTOW, MANFRED JOSEF RICHARD, b Oberhausen, Ger, Mar 21, 26; m 61. PHYSICAL CHEMISTRY, POLYMER PHYSICS. Educ: Univ Mainz, BS, 52, MS, 55, PhD(phys chem), 59. Prof Exp: Sr res chemist, Calif Res Corp, 60-66, Chevron Res Co, 66-67; asst dir, Airco Cent Res Labs, 67-71; ADV CHEMIST, IBM CORP, 71- Concurrent Pos: Lectr exten, Univ Calif, Berkeley, 62- Mem: Am Chem Soc. Res: Characterization of polymers; polymer fractionation; physical properties of polymers; polymerization and modification of vinyl polymers. Mailing Add: IBM Corp Monterey & Cottle Rds San Jose CA 95193

CANTRALL, EDWARD WARREN, organic chemistry, see 12th edition

CANTRALL, IRVING JAMES, b Springfield, Ill, Oct 6, 09; m 32; c 2. ENTOMOLOGY. Educ: Univ Mich, AB, 35, PhD(entom), 40. Prof Exp: Asst Orthoptera, mus zool, Univ Mich, 34-37; tech asst, 37-42; jr aquatic biologist, Tenn Valley Authority, 42, asst aquatic biologist, 42-43; asst prof biol, Univ Fla, 46-49; from asst prof to assoc prof zool, 49-73, PROF ZOOL, UNIV MICH, ANN ARBOR, 73- Concurrent Pos: Mem Univ Mich exped, South & Southwest US, 35, Mex, 41, 53, 59, Guatemala, 56, Cent Am, 61; cur, Edwin S George Reserve, Mus Zool, 49-59, cur insects, 59- Mem: AAAS; Soc Syst Zool; Ecol Soc Am; Soc Study Evolution; Entom Soc Am. Res: Taxonomy of new world Acridoidea. Mailing Add: Mus of Zool Univ of Mich Ann Arbor MI 48104

CANTRELL, CYRUS D, III, b Bartlesville, Okla, Oct 4, 40; m 72. PHYSICS. Educ: Harvard Univ, AB, 62; Princeton Univ, MA, 64, PhD(physics), 68. Prof Exp: From asst prof to assoc prof physics, Swarthmore Col, 67-74; STAFF MEM LASER RES & TECHNOL DIV, LOS ALAMOS SCI LAB, 73- Concurrent Pos: Vis res fel, Princeton Univ, 70-71. Mem: Am Phys Soc; Am Asn Physics Teachers; Optical Soc Am; Inst Elec & Electronics Engrs. Res: Lasers; molecular spectroscopy; quantum optics; solid state physics. Mailing Add: Laser Res & Technol Div Los Alamos Sci Lab Los Alamos NM 87545

CANTRELL, ELROY TAYLOR, b Mobile, Ala, May 10, 43; m 67; c 1. PHARMACOLOGY, CANCER. Educ: Ark State Univ, BS, 65; Univ Tenn, MS, 68; Baylor Col Med, PhD(pharmacol, 71. Prof Exp: Fel pharmacol, Baylor Col Med, 71-72; res assoc med genetics, M D Anderson Hosp, 72-73; ASST PROF PHARMACOL, TEX COL OSTEOP MED, N TEX STATE UNIV, 73- Mem: Am Soc Pharmacol & Therapeut; Am Thoracic Soc. Res: The metabolism of chemical carcinogens by human tissues. Mailing Add: Dept of Pharmacol Tex Col Osteop NTex St Univ Denton TX 76203

CANTRELL, GRADY LEON, b Louisville, Ky, Feb 18, 36; m 54; c 1. MATHEMATICS. Educ: Univ Louisville, BA, 64; Univ Ky, MS, 66, PhD(math), 68. Prof Exp: Asst prof, 68-71, ASSOC PROF MATH, MURRAY STATE UNIV, 71- Mem: Math Asn Am. Res: Analytic function theory; continuous functions. Mailing Add: Dept of Math Murray State Univ Murray KY 42072

CANTRELL, JAMES CECIL, b Palmersville, Tenn, Sept 14, 31; m 54; c 3. TOPOLOGY. Educ: Bethel Col, BS, 53; Univ Miss, MS, 55; Univ Tenn, PhD(math), 61. Prof Exp: Instr, Dresden High Sch, 53-54; instr math, Univ Tenn, 61-62; from asst prof to assoc prof, 62-70, PROF MATH, UNIV GA, 70-, HEAD DEPT, 74- Concurrent Pos: Mem, Inst Advan Study, 66-67; Alfred P Sloan res fel, 66-68. Mem: Am Math Soc. Res: Topological embeddings of manifolds. Mailing Add: Dept of Math Univ of Ga Athens GA 30601

CANTRELL, JAMES R, b Norman, Okla, Aug 8, 22; m 53; c 5. SURGERY. Educ: Johns Hopkins Univ, AB, 44, MD, 46. Prof Exp: Intern, Johns Hopkins Univ Hosp, 46-47, asst resident surgeon, 47-48 & 50-52, from resident surgeon to surgeon, 52-60, dir tumor clin, 58-60, asst surg, Sch Med, Johns Hopkins Univ, 47-48 & 50-52, from instr to assoc prof, 52-60; PROF SURG, UNIV WASH, 60- Concurrent Pos: Consult, Vet Admin Hosp & Madigan Gen Hosp, Tacoma, 60- Mem: Asn Thoracic Surg; Am Col Surgeons; AMA; Soc Univ Surgeons; Am Surg Asn. Res: Neoplastic and cardiopulmonary disease. Mailing Add: Dept of Surg Univ of Wash Med Ctr Seattle WA 98105

CANTRELL, JOSEPH SIRES, b Parker, Kans, July 31, 32; m 58; c 3. PHYSICAL CHEMISTRY, SOLAR PHYSICS. Educ: Kans State Teachers Col, AB, 54; Kans State Univ, MS, 57, PhD(phys chem), 61. Prof Exp: Res chemist, Procter & Gamble Co, 61-66; mem fac chem, 65-69, ASSOC PROF CHEM, MIAMI UNIV, 69- Mem: AAAS; Am Chem Soc; Am Crystallog Asn; Int Solar Energy Soc. Res: Chemical kinetics; x-ray crystal structure; mesomorphic structure; thin film and interfacial structure; electron diffraction; chemical applications of solar energy; photogalvanic processes; electrochemistry. Mailing Add: Dept of Chem Miami Univ Oxford OH 45056

CANTRELL, THOMAS SAMUEL, b Spartanburg, SC, Aug 29, 38. ORGANIC CHEMISTRY. Educ: Univ SC, BS, 58, MS, 59; Ohio State Univ, PhD(chem), 64. Prof Exp: NSF fel org chem, Columbia Univ, 64-65; asst prof chem, Rice Univ, 65-70; res chemist, NIH, 70-71; asst prof chem, 71-74, ASSOC PROF CHEM, AM UNIV, 74- Mem: Am Chem Soc; The Chem Soc. Res: Organic photochemistry; nonbenzenoid aromatic compounds. Mailing Add: Dept of Chem Am Univ Washington DC 20016

CANTRELL, WILLIAM ALLEN, b Everton, Ark, Nov 6, 20; m 45; c 2. PSYCHIATRY. Educ: McMurry Col, BS, 40; Univ Tex Med Br Galveston, MD, 43; Am Bd Psychiat & Neurol, cert psychiat, 51. Prof Exp: Intern, US Naval Hosp, Corona, Calif, 43-44; resident neuropsychiat, Univ Tex Med Br Hosps, 49-50, asst prof, 50-51; pvt pract, Houston, 51-63; clin prof, 63-68, PROF PSYCHIAT, BAYLOR COL MED, 68- Concurrent Pos: Asst prof neuropsychiat, Univ Tex Med Br, 51-54; clin asst prof psychiat, Baylor Col Med, 51-63; from asst chief to sr attend, Psychiat Serv, Methodist Hosp, 52-; clin assoc prof psychiat, Univ Tex Grad Sch Biomed Sci Houston, 59-; attend staff, Psychiat Serv, Ben Taub Gen Hosp, 63-; asst examr, Am Bd Psychiat & Neurol, 64-; consult psychiat, Social Security Admin, Bur Hearings & Appeals, 64- & Vet Admin Hosp, Houston, 67-; adj prof, Inst Relig, Houston, 68- Mem: AAAS; Cent Neuropsychiat Asn (pres elect, 75-76); fel Am Psychiat Asn; Am Col Psychiat. Res: Undergraduate medical education, especially methodology and evaluation; doctor-patient relationship as a factor in the quality of health care delivery. Mailing Add: Dept of Psychiat Baylor Col of Med Houston TX 77030

CANTRELL, WILLIAM FLETCHER, b Young Harris, Ga, Oct 29, 16; m 47; c 2. CHEMOTHERAPY. Educ: Univ Ga, BS, 38, MS, 39; Univ Chicago, PhD, 49. Prof Exp: Instr zool, Univ Ga, 39-40; asst parasitol & pharmacol, Univ Chicago, 41-46; from asst prof to prof pharmacol, Univ Louisville, 49-59; mem staff, Lab Parasite Chemother, NIH, 59-62; ASSOC PROF PHARMACOL, UNIV TENN, MEMPHIS, 63- Mem: Am Soc Pharmacol; Am Soc Parasitol; Am Soc Trop Med & Hyg; Soc Exp Biol & Med. Res: Transmission of malaria by mosquitoes; chemotherapy of malaria; mode of action of chemotherapeutic drugs; antigenic variation in trypanosomes; drug resistance. Mailing Add: 1682 Estate Dr Memphis TN 38117

CANTRILL, JAMES EGBERT, b Frankfort, Ky, Sept 2, 33; m 56; c 6. ORGANIC CHEMISTRY. Educ: Univ Notre Dame, BS, 55; Mass Inst Technol, PhD(org chem), 59. Prof Exp: Res chemist, Eastman Kodak Co, NY, 59-60; instr chem, St Thomas Moore Col, 60-61, asst prof & chmn dept, 61-63; develop chemist, Plastics Dept, Gen Elec Co, 63-67, mgr res & advan develop, 67-69; mgr res lab, 69-74, BUS DEVELOP MGR, FOSTER GRANT CO, INC, 74- Mem: Am Chem Soc. Res: Polymer synthesis; reaction kinetics and mechanisms. Mailing Add: Foster Grant Co Inc 289 N Main St Leominster MA 01453

CANTU, ANTONIO ARNOLD, b Laredo, Tex, Jan 5, 41. FORENSIC SCIENCE, QUANTUM CHEMISTRY. Educ: Univ Tex, Austin, BS, 63, PhD(chem physics), 67. Prof Exp: Fel, Univ Alta, 68-71; Orgn Am States fel & vis prof, Inst Physics, Univ Mex, 70; res assoc, Res Coun Alta, 71-72; chemist, Law Enforcement Assistance Admin, US Dept Justice, 72-73; FORENSIC CHEMIST, BUR ALCOHOL, TOBACCO & FIREARMS, US DEPT TREAS, 73- Mem: Asn Appl Spectros; Am Chem Soc; Sigma Xi. Res: Development of techniques to improve conventional methods of ink and paper analysis in the examination of questioned documents; application of statistical pattern recognition techniques to the individualization of physical evidence. Mailing Add: Bur of ATF Lab US Treas Dept 1111 Constitution Ave Washington DC 20226

CANTWELL, GEORGE E, b Darlington, Pa, May 5, 29; m 50; c 10. INSECT PATHOLOGY. Educ: Kent State Univ, BS, 51, MA, 55; Univ Md, PhD(systematics), 60. Prof Exp: Histologist, 57-58, INSECT PATHOLOGIST, USDA, 58- Concurrent Pos: Lectr, Prince Georges Community Col, 60- Mem: Soc Invert Path. Mailing Add: Insect Path Lab USDA Beltsville MD 20705

CANTWELL, JOHN CHRISTOPHER, b St Louis, Mo, Aug 12, 36; m 64. MATHEMATICS. Educ: St Louis Univ, BS, 57; Univ Notre Dame, PhD(math), 62. Prof Exp: Asst math, Inst Advan Study, 62-64; asst prof, Univ Iowa, 64-67; from asst prof to assoc prof, 67-73, PROF MATH, ST LOUIS UNIV, 73- Mem: Am Math Soc; Math Asn Am; Nat Speleol Soc. Res: Convex geometry; foliations; morse theory. Mailing Add: 2351 Parkridge St Louis MO 63144

CANTWELL, ROBERT MURRAY, b St Louis, Mo, June 18, 31. THEORETICAL PHYSICS. Educ: St Louis Univ, BS, 52; Wash Univ, PhD(physics), 56. Prof Exp: NSF fel, Inst Theoret Physics, Copenhagen, Denmark, 56-57; sr physicist, Bettis Atomic Power Lab, Westinghouse Elec Corp, 58-63; STAFF MEM, LOS ALAMOS SCI LAB, 63- Mem: Am Phys Soc. Res: Radiation transport; nuclear reactor physics. Mailing Add: 3789 Gold St Apt 4 Los Alamos NM 87544

CANTWELL, THOMAS, b Buffalo, NY, June 25, 27; m 51; c 3. GEOPHYSICS. Educ: Mass Inst Technol, SB, 48, PhD(geophys), 60; Harvard Univ, MBA, 51. Prof Exp: Chem engr, Ionics Inc, 51-52; liaison officer, Mass Inst Technol, 52-54, res assoc nuclear eng, 54-58, asst prof geophys, 60-63; pres, Geosci, Inc, 63-70; PRES, PETROL HOLDINGS, INC, 70- Concurrent Pos: Lectr, Mass Inst Technol, 63-65; vpres data processing, Mandrel Industs, 66-69, pres, 69-70. Mem: Am Phys Soc; Am Asn Petrol Geol; Soc Explor Geophys; Am Inst Mining, Metall & Petrol Eng; fel Royal Geog Soc. Res: Seismic prospecting methods; sedimentary basin studies; energy source evaluation studies; petroleum exploration; mineral exploration. Mailing Add: 3914 Fairhill Dr Houston TX 77042

CANVIN, DAVID T, b Winnipeg, Man, Nov 8, 31; m 57; c 3. PLANT PHYSIOLOGY, PLANT BIOCHEMISTRY. Educ: Univ Man, BSA, 56, MSc, 57; Purdue Univ, PhD(plant physiol), 60. Prof Exp: Res assoc plant sci, Univ Man, 60-63, assoc prof, 63-65; PROF BIOL, QUEEN'S UNIV, ONT, 65- Concurrent Pos: Secy-treas, Biol Coun Can. Mem: AAAS; Am Soc Plant Physiol; Can Soc Cell Biol; Can Soc Plant

Physiol. Res: Intermediary metabolism in plants; environmental physiology. Mailing Add: Dept of Biol Queen's Univ Kingston ON Can

CAPALDI, EUGENE CARMEN, b Philadelphia, Pa, Apr 10, 37; m 63; c 2. ORGANIC CHEMISTRY. Educ: Univ Pa, BS, 59; Univ Del, MS, 61, PhD(org chem), 64. Prof Exp: Res chemist, 63-66, SR RES CHEMIST, ARCO CHEM CO, DIV ATLANTIC RICHFIELD CORP, 66- Mem: Am Chem Soc. Res: Synthesis of new monomers and chemical intermediates; polyester and polyurethane chemistry. Mailing Add: Arco Chem Co 500 S Ridgeway Ave Glenolden PA 19036

CAPE, JOHN ANTHONY, b Helena, Mont, Nov 2, 29; m 62; c 3. SOLID STATE PHYSICS. Educ: Carroll Col (Mont), AB, 51; Mont State Univ, MS, 53; Univ Notre Dame, PhD(physics), 58. Prof Exp: Prof math & physics, Carroll Col (Mont), 55-56; asst physics, Univ Notre Dame, 57, instr, 57-58; res assoc, Univ Ill, 58-60; res specialist, Atomics Int, 60-64; mem tech staff, 64-72, GROUP LEADER SOLID STATE SCI, SCI CTR, ROCKWELL INT CORP, 72-, PROG MGR ENERGY CONVERSION & STORAGE, 74- Concurrent Pos: Vis res assoc, Stanford Univ, 69. Mem: Fel Am Phys Soc. Res: Magnetic properties of materials; optical properties of magnetic materials; semiconductors; physical properties of surfaces and interfaces; superconductivity. Mailing Add: Sci Ctr Rockwell Int Corp Thousand Oaks CA 91360

CAPE, RONALD ELLIOT, b Montreal, Can, Oct 11, 32; m 56; c 2. BIOCHEMISTRY. Educ: Princeton Univ, AB, 53; Harvard Univ, MBA, 55; McGill Univ, PhD(biochem), 67. Prof Exp: Pres, Prof Pharmaceut Corp, 57-67, chmn, 67-73; managing partner, Cetus Sci Lab, 71-72, PRES, CETUS CORP, 72- Concurrent Pos: Med Res Coun Can Centennial fel, Univ Calif, Berkeley, 67-70; pres, Cape Farley Inc, 71-75. Mem: Am Soc Microbiol; Can Biochem Soc; Royal Soc Health; Sigma Xi; Soc Cosmetic Chemists. Res: Mutation and gene manipulation of industrial microorganisms. Mailing Add: Cetus Corp 600 Bancroft Way Berkeley CA 94710

CAPECCHI, MARIO RENATO, b Verona, Italy, Oct 6, 37; US citizen; m 63. CELL BIOLOGY. Educ: Antioch Col, BS, 61; Harvard Univ, PhD(biophys), 67. Prof Exp: Soc Fels jr fel biophys, Harvard Univ, 66-68, from asst prof to assoc prof biochem, Med Sch, 68-73; PROF BIOL, UNIV UTAH, 73- Concurrent Pos: Estab investr, Am Heart Asn, 69-72; NIH Career Develop Award, 72-74; Am Cancer Soc Fac Res Award, 74-79. Honors & Awards: Am Chem Soc Biochem Award, 69. Mem: Am Soc Biol Chem; Am Biochem Soc. Res: Gaining an understanding of how the information encoded in the gene is translated by the cell; expression in Eucaryotic and Procaryotic cells; somatic cell genetics. Mailing Add: Dept of Biol Univ of Utah Salt Lake City UT 84112

CAPECI, NICHOLAS ERNEST, b Port Chester, NY, Jan 17, 25; m 55; c 5. INTERNAL MEDICINE. Educ: Columbia Univ, MD, 47. Prof Exp: Asst dir res & develop, 69-70, VPRES MED RES, SCHERING CORP, 70- Mem: AMA. Res: Cardiovascular-renal disease. Mailing Add: Schering Corp 60 Orange St Bloomfield NJ 07003

CAPEL, CHARLES EDWARD, b Troy, NY, Dec 26, 22; m 45; c 2. MATHEMATICS. Educ: NY State Teachers Col, Albany, BA, 47; Univ Rochester, MA, 50; Tulane Univ, PhD(math), 53. Prof Exp: Instr math, Geneseo State Teachers Col, 47-49 & Tulane Univ, 50-51; asst prof, Miami Univ, 53-58; res mathematician, Westinghouse Res Lab, Pa, 58-60; PROF MATH, MIAMI UNIV, 60- Mem: Am Math Asn; Math Soc Am. Res: Inverse limit spaces; functions; fixed points. Mailing Add: Dept of Math Miami Univ Oxford OH 45056

CAPEN, CHARLES CHABERT, b Tacoma, Wash, Sept 3, 36; m 68. VETERINARY PATHOLOGY. Educ: Wash State Univ, DVM, 60; Ohio State Univ, MSc, 61, PhD(vet path), 65; Am Col Vet Pathologists, dipl. Prof Exp: Res assoc vet pathobiol, 60-62, from instr to prof, 62-72, PROF ENDOCRINOL, COL MED, OHIO STATE UNIV, 72- Concurrent Pos: Consult path, Food & Drug Admin, 73-; mem coun, Am Col Vet Pathologists. Honors & Awards: Borden Res Award, Am Vet Med Asn, 75. Mem: AAAS; Am Vet Med Asn; Endocrine Soc; Int Acad Path; Am Soc Exp Path. Res: Comparative and veterinary pathology; endocrine and metabolic diseases; calcium metabolism; ultrastructure of thyroid and parathyroid glands; metabolic bone disease. Mailing Add: Dept of Vet Pathobiol Ohio State Univ Columbus OH 43210

CAPEN, CHARLES FRANKLIN, JR, b Gilman, Ill, Jan 1, 26; m 56; c 3. PLANETARY SCIENCES. Educ: Spartan Col Aeronaut Eng, dipl, 46 & 49. Prof Exp: Dir astrophys, Smithsonian Astrophys Observ, Shiraz, Iran Sta, Int Geophys Year, 57-61; resident astronr planets, Table Mountain Observ, Jet Propulsion Lab, Calif Inst Technol, 62-70; ASTRONR PLANETS, PLANETARY RES CTR, LOWELL OBSERV, 70- Honors & Awards: Inst Environ Sci Award, Inst Environ Sci Calif, 69; Bruce Blair Gold Medal, 70. Mem: Am Astron Soc. Res: Observation and research on planets, comets, meteors; color photography and colorimetry of planets and comets; martian meteorology and surface variations; history of planetary science and its modern applications. Mailing Add: Lowell Observ Box 1269 Flagstaff AZ 86001

CAPERON, JOHN, b Milford, Utah, Apr 14, 29; m 64; c 3. ECOLOGY, OCEANOGRAPHY. Educ: Univ Utah, BS, 52; Scripps Inst Oceanog, Univ Calif, PhD(oceanog), 65. Prof Exp: Scientist, Int Bus Mach Corp, 53-59; oceanogr, Supreme Allied Comdr Atlantic Antisubmarine Warfare Res Ctr, Off Naval Res, London, 65-69; assoc prof, 69-74, PROF OCEANOG, UNIV HAWAII AT MANOA, 74- Mem: Ecol Asn Am. Res: Population dynamics; marine ecology. Mailing Add: Dept of Oceanog Univ of Hawaii Honolulu HI 96822

CAPERS, EVELYN LORRAINE, b Los Angeles, Calif, Dec 20, 25; m 58; c 1. CLINICAL MICROBIOLOGY. Educ: Univ Calif, Berkeley, BA, 46; Univ Southern Calif, PhD(bact), 71. Prof Exp: Med technologist, Childrens Hosp Los Angeles, 47-66; med microbiologist, Martin Luther King Jr Gen Hosp, Los Angeles County Dept Health Serv, 71-74; ASST DIR MICROBIOL, BIO-SCI LABS, 74- Mem: Am Soc Microbiol; Sigma Xi. Res: Improvement of existing diagnostic laboratory procedures and the development of new tests of value to the medical community. Mailing Add: Bio-Sci Lab 7600 Tyrone Ave Van Nuys CA 91405

CAPIZZI, ROBERT L, b Philadelphia, Pa, Nov 20, 38; m 65; c 4. ONCOLOGY, CLINICAL PHARMACOLOGY. Educ: Temple Univ, BS, 60; Hahnemann Med Col, MD, 64. Prof Exp: Fel clin pharmacol, Hahnemann Med Col, 65-66; fel med & pharmacol, 67-69, asst prof med & pharmacol, 72-75, ASSOC PROF MED & PHARMACOL, SCH MED, YALE UNIV, 75- Concurrent Pos: Consult, West Haven Vet Admin Hosp & Med Res Lab, Edgewood Arsenal, Md, 72-; fac develop award clin pharmacol, Pharmaceut Mfrs Asn, 73. Mem: Am Asn Cancer Res; Am Soc Clin Oncol; Environ Mutagen Soc; AAAS. Res: Cancer chemotherapy; chemical mutagenesis. Mailing Add: Sect of Med Oncol Sch of Med Yale Univ 333 Cedar St New Haven CT 06510

CAPLAN, ARNOLD I, b Chicago, Ill, Jan 5, 42; m 65; c 2. DEVELOPMENTAL BIOLOGY, BIOCHEMISTRY. Educ: Ill Inst Technol, BS, 63; Johns Hopkins Univ, PhD(biochem), 66. Prof Exp: Fel anat, Johns Hopkins Univ Med Sch, 66-67; fel biochem, Brandeis Univ, 67-68, fel biol, 68-69; asst prof biol, 69-74, ASSOC PROF BIOL & ANAT, CASE WESTERN RESERVE UNIV, 74- Mem: AAAS; Soc Develop Biol; Soc Cell Biol. Res: Biochemical control of phenotypic expression, especially in chick embryo limb mesodermal cells. Mailing Add: Dept of Biol Case Western Reserve Univ Cleveland OH 44106

CAPLAN, GERALD, b Eng, Mar 6, 17; nat US; m 42; c 1. PSYCHIATRY, PSYCHOANALYSIS. Educ: Univ Manchester, BSc, 37, MB & ChB, 40, MD, 45; Royal Col Physicians & Surgeons, DPM, 42. Hon Degrees: MA, Harvard Univ, 70. Prof Exp: Asst med officer, Birmingham City Ment Hosp, Eng, 40-43; dep med supt, Swansea Ment Hosp, Eng, 43-45; psychiatrist, Tavistock Clin, London, 45-48; adv psychiat, Ministry Health, Israel, 48-49; psychiat dir, Lasker Ment Hyg & Child Guid Ctr, Jerusalem, 49-52; lectr ment health, Sch Pub Health, Harvard Univ, 52-54, assoc prof, Sch Pub Health, dir community ment health prog & psychiat dir, Harvard Family Guid Ctr, 54-64, clin prof psychiat, Harvard Med Sch, 64-70, PROF PSYCHIAT, HARVARD MED SCH, 70-, DIR LAB COMMUNITY PSYCHIAT, 64- Concurrent Pos: Chmn, Mass Adv Coun Ment Health & Retardation, 68; sr psychiat consult, Peace Corps, 61-71 & Off Econ Opportunity, 67-70. Mem: Law & Soc Asn; fel Am Pub Health Asn; Am Psychiat Asn; Am Orthopsychiat Asn; Int Asn Child Psychiat (hon pres). Mailing Add: 93 Fresh Pond Pkwy Cambridge MA 02138

CAPLAN, MILTON, organic chemistry, see 12th edition

CAPLAN, PAUL E, b Far Rockaway, NY, Feb 29, 24; m 49; c 5. CHEMICAL ENGINEERING, INDUSTRIAL HYGIENE. Educ: Middlebury Col, AB, 44; Univ Colo, BSChE, 48; Univ Calif, Berkeley, MPH, 49; Am Acad Indust Hyg, dipl, 62; Bd Cert Safety Prof, cert, 71. Prof Exp: From asst indust hyg engr to sr indust hyg engr, Bur Occup Health, Calif State Dept Pub Health, 49-60 & adv, Bur Radiol Health, 60-71; DEP DIR DIV TECH SERVS, NAT INST OCCUP SAFETY & HEALTH, 71- Concurrent Pos: Mem threshold limits value comt, air sampling instruments comt, agr health comt & hyg guides comt, Am Conf Govt Indust Hyg. Mem: Am Conf Govt Indust Hyg; Am Indust Hyg Asn; Health Physics Soc; Am Pub Health Asn; Am Soc Testing & Mat. Res: Evaluation and control of hazards involved in use of agricultural chemicals and compressed air; development of techniques of air sampling for general environmental hazards evaluation. Mailing Add: Nat Inst Occup Safety & Health Div of Tech Serv Rm 506 PO Bldg Cincinnati OH 45230

CAPLAN, PAULA JOAN, b Springfield, Mo, July 7, 47; m 72; c 4. NEUROPSYCHOLOGY. Educ: Harvard Univ, AB, 69; Duke Univ, MA, 71, PhD(psychol), 73. Prof Exp: RES ASSOC NEUROPSYCHOL RES UNIT, HOSP SICK CHILDREN, 74- Mem: Soc Res Child Develop. Res: Effects of methylphenidate on the behavior of hyperactive children; sex differences in aggressive and achievement behavior; children's learning; hand preference development in infants. Mailing Add: Rm 9105 Neuropsychol Res Div Hosp for Sick Children 555 University Ave Toronto ON Can

CAPLAN, RICHARD MELVIN, b Des Moines, Iowa, July 16, 29; m 52; c 4. DERMATOLOGY. Educ: Iowa State Univ, BS, 49; Univ Iowa, MA, 51, MD, 55. Prof Exp: From asst prof to assoc prof, 61-69, PROF DERMAT, COL MED, UNIV IOWA, 69-, ASSOC DEAN CONTINUING MED EDUC, 70- Mem: Am Acad Dermat; Soc Invest Dermat; Am Dermat Asn. Mailing Add: Dept of Dermat Univ of Iowa Col of Med Iowa City IA 52240

CAPLAN, YALE HOWARD, b Baltimore, Md, Dec 27, 41; m 65; c 2. TOXICOLOGY. Educ: Univ Md, PhD(med chem), 68. Prof Exp: Asst pharmaceut chem, Sch Pharm, Univ Md, 64-65; res assoc toxicol & cancer chemother, Sinai Hosp, Baltimore, Md, 68-69, supvr surg res div, 69; asst toxicologist, 69-74, CHIEF TOXICOLOGIST, OFF CHIEF MED EXAMR, STATE OF MD, BALTIMORE, 74- Concurrent Pos: Res consult, Sinai Hosp, Baltimore, Md, 70-74; consult toxicologist, Cent Labs Assoc Md Pathologists, 71-72; instr toxicol, Sch Med, Univ Md, 72-73, clin asst prof path, 73-; lectr, Sch Hyg & Pub Health, Johns Hopkins Univ, 73-, consult appl physics lab, 73-; dir grad prog legal med (toxicol), Univ Md, Baltimore City, 74-; prof lectr, Dept Forensic Sci, George Washington Univ, 75- Mem: AAAS; Am Chem Soc; Am Acad Forensic Sci; Int Asn Forensic Toxicol; fel Am Inst Chemists. Res: Analytical forensic toxicology, particularly development of procedures for analysis and chemical diagnosis in drug related death; experimental toxicology, particularly relationship of toxic concentrations and effects of drugs. Mailing Add: 8100 Tapscott Ct Pikesville MD 21208

CAPLE, GERALD, b International Falls, Minn, Apr 3, 35; m 64. ORGANIC CHEMISTRY. Educ: St Olaf Col, BA, 57; Fla State Univ, PhD(org chem), 63. Prof Exp: Res assoc chem, Ore State Univ, 63-65, asst prof, 65-66; asst prof, 66-70, ASSOC PROF CHEM, NORTHERN ARIZ UNIV, 70- Mem: Am Chem Soc. Res: Organic synthesis and mechanisms. Mailing Add: Dept of Chem Northern Ariz Univ Flagstaff AZ 86001

CAPLE, RONALD, b International Falls, Minn, Dec 7, 37; m 59; c 4. ORGANIC CHEMISTRY. Educ: St Olaf Col, BA, 60; Univ Mich, MS, 62, PhD(org mechanisms), 64. Prof Exp: NSF fel, Univ Colo, 64-65; from asst prof to assoc prof, 65-74, PROF ORG CHEM, UNIV MINN, DULUTH, 74- Mem: Am Chem Soc. Res: Intermediates in bridged polycyclic compounds. Mailing Add: Dept of Chem Univ of Minn Duluth MN 55812

CAPLENOR, DONALD, b Lebanon, Tenn, Feb 10, 22; m 46; c 3. BOTANY. Educ: George Peabody Col, BS, 48, MA, 49; Vanderbilt Univ, PhD(biol), 54. Prof Exp: Instr biol, George Peabody Col, 49-51, asst prof, 55-57; instr, Vanderbilt Univ, 51-53; asst prof & chmn dept, Ga Southern Col, 53-55; prof & chmn dept, Millsaps Col, 57-63; prof & chmn dept, George Peabody Col, 63-67, chmn div sci & math, 66-67; prof biol & dean col arts & sci, Tenn Technol Univ, 67-70; prof biol & dean fac, Univ Ala, Huntsville, 70-72; dean lib arts, Univ Tenn, Martin, 72-73; dean undergrad studies, 73-76, ASSOC VPRES RES, TENN TECHNOL UNIV, 76- Mem: Ecol Soc Am. Res: Phytosociology; plant ecology. Mailing Add: Tenn Technol Univ Cookeville TN 38501

CAPLIN, SAMUEL MILTON, b Cleveland, Ohio, Oct 28, 17; m 42, 53; c 3. BOTANY. Educ: Univ Akron, BS, 39, MS, 41; Univ Chicago, PhD(plant physiol), 46. Prof Exp: NIH fel growth & metabolism plant tissue cult, 47-49; res assoc bot, Univ Rochester, 49-50, asst prof, 50-56; asst prof, ELos Angeles Col, 56-60; from asst prof to assoc prof bot, 60-67, chmn dept bot & coordr biol, 67-69, PROF BOT, CALIF STATE UNIV, LOS ANGELES, 67- Mem. AAAS; Am Soc Plant Physiol; Bot Soc Am; Tissue Cult Asn; Int Asn Plant Tissue Cult. Res: Growth of excised carrot root tissues. Mailing Add: Dept of Biol Calif State Univ Los Angeles CA 90032

CAPLIS, MICHAEL E, b Ypsilanti, Mich, July 25, 38; m 56; c 7. BIOCHEMISTRY, TOXICOLOGY. Educ: Eastern Mich Univ, BS, 62; Purdue Univ, MS, 64, PhD(biochem), 70. Prof Exp: Teacher physics, math & chem, St John High Sch, 58-62; asst biochem, Purdue Univ, 62-67, res asst, 67-69; instr, 69-70, ASST PROF

CLIN BIOCHEM, DIV ALLIED HEALTH, SCH MED, IND UNIV, 71-; DIR CLIN BIOCHEM & CONSULT PATH LABS, ST MARY MERCY HOSP, 69- Concurrent Pos: Anal biochemist, State of Ind, 62-67; toxicologist, Ind Criminal Justice Comn, Region I, 70; dir, Northwest Ind Criminal & Toxicol Lab, 70- Mem: AAAS; Am Asn Clin Chem; Am Chem Soc; Am Acad Clin Toxicol; Am Asn Crim Lab Dirs. Res: Analytical biochemistry-toxicology; mechanism of action of toxic chemicals and drugs; resistance to toxic action of chemicals and drugs; development of instrumental analytical procedures for study of chemicals, drugs and their metabolites in biological systems; applications of spectrochemical and electrochemical methods. Mailing Add: St Mary Mercy Hosp 540 Tyler St Gary IN 46402

CAPLOW, MICHAEL, b New York, NY, May 20, 35; m 59; c 2. BIOCHEMISTRY, ENZYMOLOGY. Educ: NY Univ, DDS, 59; Brandeis Univ, PhD(biochem), 63. Prof Exp: Asst biochem, Yale Univ, 63-70; assoc prof, 70-73, PROF BIOCHEM, UNIV NC, CHAPEL HILL, 73- Mem: AAAS; Am Chem Soc. Res: Enzyme kinetics and mechanisma mechanisms; chemistry of microtubules, biosynthesis of dextrans by cariogenic microorganisms. Mailing Add: Dept of Biochem Univ of NC Chapel Hill NC 27514

CAPO, BERNARDO GUILLERMO, b Guayama, Dec 28, 08; m 30; c 2. AGRICULTURE. Educ: Univ PR, BS, 29; Cornell Univ, MS, 41, PhD(agr), 42. Prof Exp: Anal chemist, Fertil & Feed Control, PR Dept Agr, 29-36; soils chemist, 36-42, head soils div, 42-43, biometrician, 43-48, head dept agron & hort, 44-52, asst dir res, 48-49, actg dir, 49-50, assoc dir, 52-67, TECH CONSULT, AGR EXP STA, UNIV PR, RIO PIEDRAS, 67- Concurrent Pos: Tech dir, Ancram Paper Mills, NY, 46-48; prof col soc sci, Univ PR, 48-49; proj mgr, UN Spec Fund Proj 101, Damascus Agr Res Sta, 64-65; statist consult, Urban Renewal & Housing Admin, 68-69; consult, Agr Coun PR, 69-71. Mem: Am Statist Asn; Inst Math Statist; Int Soc Sugar Cane Technol. Res: New fertilizer equation; leaf composition as index of availability of nutrients in the soil; optimum economic quantity of fertilizer based on leaf analysis; statistical analysis of research data. Mailing Add: 1749 Santa Praxedes St Rio Piedras PR 00926

CAPOBIANCO, MICHAEL F, b Brooklyn, NY, Oct 4, 31; m 65. MATHEMATICS, STATISTICS. Educ: Polytech Inst Brooklyn, BChE, 52, MChE, 54, PhD(math), 64; Columbia Univ, MA, 57. Prof Exp: Statistician, Am Cyanamid Co, 54-55; from instr to asst prof math, St John's Univ, 55-63; mathematician, Repub Aviation Corp, 63; lectr math, Polytech Inst Brooklyn, 63-64, asst prof, 64-66; assoc prof, 66-70, PROF MATH, ST JOHN'S UNIV, NY, 70- Concurrent Pos: Consult, Maimonides Hosp Brooklyn, 66- Mem: Am Math Soc; Math Asn Am; Biomet Soc; Soc Indust & Appl Math. Res: Statistical decision theory; statistical inference in digraphs; traffic analysis; combinatorics; statistics in the life sciences, especially psychology, economics, sociology, biophysics and cybernetics; tensor analysis and differential geometry. Mailing Add: Dept of Math St John's Univ Jamaica NY 11432

CAPON, BRIAN, b Wallasey, Eng, Dec 27, 31. BOTANY. Educ: La Sierra Col, BA, 58; Univ Chicago, MS, 60, PhD(bot), 61. Prof Exp: From asst prof to assoc prof, 61-71, chmn dept, 69-71, PROF BOT, CALIF STATE UNIV, LOS ANGELES, 71- Mem: AAAS. Res: Physiology of desert plants. Mailing Add: Dept of Biol Calif State Univ Los Angeles CA 90032

CAPONE, DONALD LEONARD, geography, cartography, see 12th edition

CAPONE, JAMES JOSEPH, b White Plains, NY, Jan 14, 45; m 68; c 1. IMMUNOCHEMISTRY, BACTERIOLOGY. Educ: NY Univ, BA, 67; Fla State Univ, MS, 69, PhD(biochem), 71. Prof Exp: Dir prod, Kallestad Labs, Inc, 71-74; DIR RES & DEVELOP, BIOMED DIV, INOLEX CORP, 74- Concurrent Pos: Hartford grant, Univ Ill, 71. Mem: Am Inst Chemists; Am Chem Soc. Res: Tumor associated antigens; cellular immunology and development of immunologically oriented diagnostics; biochemical differentiation for clinical microbiology. Mailing Add: Inolex Corp Biomed Div 3 Science Dr Glenwood IL 60430

CAPONETTI, JAMES DANTE, b Boston, Mass, Mar 15, 32; m 66. BOTANY. Educ: Mass Col Pharm, BS, 54, MS, 56; Harvard Univ, AM, 59, PhD(biol), 62. Prof Exp: Asst prof, 61-71, ASSOC PROF BOT, UNIV TENN, KNOXVILLE, 71- Mem: Fel AAAS; Bot Soc Am; Int Soc Plant Morphol; Am Fern Soc; Soc Develop Biol. Res: Plant morphogenesis; plant tissue cultural and growth promoting compounds. Mailing Add: Dept of Bot Univ of Tenn Knoxville TN 37916

CAPONIO, JOSEPH FRANCIS, b Canton, Mass, Mar 25, 26; m 57; c 2. BIOCHEMISTRY. Educ: St Anselm's Col, AB, 51; Georgetown Univ, PhD(biochem), 59. Prof Exp: Sr analyst, Libr Cong, 55-58; chief bibliog div, Off Tech Serv, US Dept Com, 58-61; dir tech info, Defense Document Ctr, Div Sci Area, US Dept Defense, 61-64; sci info officer, NIH, 64-70; assoc dir, Nat Agr Libr, 70-74; DIR ENVIRON SCI INFO CTR, NAT OCEANIC & ATMOSPHERIC ADMIN, DEPT COM, 74- Concurrent Pos: Mem off critical tables, Nat Acad Sci-Nat Res Coun, 64- Mem: Am Chem Soc; Am Soc Info Sci. Res: Protein and enzyme chemistry; information science. Mailing Add: 8417 Ft Hunt Rd Alexandria VA 22308

CAPOTOSTO, AUGUSTINE, JR, b Providence, RI, Mar 24, 38; m 60; c 3. ANALYTICAL CHEMISTRY. Educ: Providence Col, BS, 59; Univ RI, PhD(anal chem), 63. Prof Exp: Sr chemist & engr, Res Chem Group, 62-65, supvr, 65-68, prog mgr & res specialist, 68-71, CHIEF LAB SERV, RES CHEM GROUP, ELEC BOAT DIV, GEN DYNAMICS CORP, 71- Mem: Fel Am Inst Chem; Am Chem Soc. Res: Liquid crystals; atomic absorption spectroscopy; fluorescence; synthesis of organic peroxides, superoxides and ozonides; chelate chemistry. Mailing Add: Dept 460 Elec Boat Div Gen Dynamics Corp Groton CT 06340

CAPOZZA, RICHARD CARL, organic chemistry, polymer chemistry, see 12th edition

CAPP, GRAYSON L, b Seattle, Wash, Aug 11, 36; m 59; c 3. BIOCHEMISTRY. Educ: Seattle Pac Col, BS, 58; Univ Ore, MS, 62, PhD(biochem), 67. Prof Exp: Instr chem, Los Angeles Pac Col, 59-60; res assoc biochem, Ore Primate Ctr, 61-63; NIH fel protein struct, Duke Univ, 66-68; asst prof biochem, 68-70, asst prof chem, 70-73, ASSOC PROF CHEM & PRE-PROF HEALTH SCI ADV, SEATTLE PAC COL, 73- Concurrent Pos: NIH sr fel, Univ Wash, 69-70. Mem: AAAS; Am Chem Soc. Res: Protein subunit structure and function with studies of glutamic dehydrogenase and human hemoglobin. Mailing Add: Dept of Chem Seattle Pac Col Seattle WA 98119

CAPP, MICHAEL PAUL, b Yonkers, NY, July 1, 30; m 57; c 4. RADIOLOGY, PEDIATRICS. Educ: Roanoke Col, BS, 52; Univ NC, MD, 58. Prof Exp: Intern pediat, Med Ctr, Duke Univ, 58-59, resident radiol, 59-62, assoc, 6263, from asst prof to assoc prof, 63-70, dir diag radiol, 66-70, asst prof pediat, 68-70; PROF RADIOL & CHMN DEPT, COL MED, UNIV ARIZ, 70- Mem: Soc Pediat Radiol; AMA; Radiol Soc NAm; Am Roentgen Ray Asn; Am Col Radiol; Asn Univ Radiologists. Res: Congenital heart disease, particularly left ventricular function. Mailing Add: Dept of Radiol Univ of Ariz Col of Med Tucson AZ 85724

CAPPANNARI, STEPHEN CLEMENT, cultural anthropology, deceased

CAPPAS, C, b Cairo, Egypt, Mar 14, 26; US citizen; m 56. PHYSICAL CHEMISTRY, MATHEMATICS. Educ: Berea Col, BA, 56; Univ Fla, PhD(phys chem), 62. Prof Exp: Res assoc phys chem, Princeton Univ, 62, group leader dielectrics, 62-64; assoc prof chem, Oglethorpe Univ, 65-67; ASSOC PROF CHEM, UNIV S ALA, 67- Mem: AAAS; Am Chem Soc. Res: Chemical kinetics; determination of physical constants; polymer chemistry; solid state chemistry. Mailing Add: Dept of Chem Univ of S Ala Mobile AL 36608

CAPPEL, CARL ROBERT, b Connersville, Ind, Nov 17, 42. ORGANIC CHEMISTRY, PHOTOGRAPHIC CHEMISTRY. Educ: Ball State Univ, BS, 69; Univ Ill, MS, 72, PhD(chem), 73. Prof Exp: Asst chem, Univ Ill, 69-73; SR RES CHEMIST, EASTMAN KODAK CO, 73- Mem: Sigma Xi; Soc Photog Scientists & Engrs. Res: Investigation and application of organic chemistry to photographic science. Mailing Add: Eastman Kodak Co Res Labs 1669 Lake Ave Rochester NY 14650

CAPPELL, SYLVAIN EDWARD, b Brussels, Belg, Sept 10, 46; US citizen; m 66; c 3. TOPOLOGY. Educ: Columbia Univ, 66; Princeton Univ, PhD(math), 69. Prof Exp: Princeton nat fel, Princeton Univ, 66-69, from instr to asst prof math, 69-74; ASSOC PROF MATH, COURANT INST, NY UNIV, 74- Concurrent Pos: Woodrow Wilson Found fel, 66-67; NSF fel, 66-68; Danforth Found fel, 66-69; vis lectr, Harvard Univ, 70-71; Sloan Found fel, 71-73; vis prof, Weizmann Inst, Israel, 72 & Inst Advan Studies Sci, 73. Res: Manifolds and submanifolds. Mailing Add: Courant Inst of Math Sci 251 Mercer St New York NY 10012

CAPPELLETTI, RONALD LOUIS, b Waterbury, Conn, Sept 26, 40; m 65; c 3. EXPERIMENTAL SOLID STATE PHYSICS, METAL PHYSICS. Educ: Fairfield Univ, BS, 62; Univ Ill, Urbana, MS, 64, PhD(physics), 66. Prof Exp: Res assoc physics, Ames Lab, Iowa State Univ, 66-68; PROF PHYSICS, OHIO UNIV, 68- Concurrent Pos: Consult, Oak Ridge Assoc Univs, 74-76. Mem: Am Phys Soc. Res: Experimental physics. Mailing Add: Dept of Physics Ohio Univ Athens OH 45701

CAPPELLINI, RAYMOND ADOLPH, b Pittsburgh, Pa, Jan 30, 26; m 50; c 4. PLANT PATHOLOGY. Educ: Duquesne Univ, BEd, 50; Pa State Univ, MS, 52; Cornell Univ, PhD, 55. Prof Exp: Asst res specialist plant path, 55-61, assoc prof, 61-67, PROF PLANT PATH, RUTGERS UNIV, 67-, CHMN DEPT PLANT BIOL, 71- Mem: Am Phytopath Soc; Mycol Soc Am. Res: Market diseases; fungus physiology. Mailing Add: Dept of Plant Biol Cook Col Rutgers Univ New Brunswick NJ 08903

CAPPS, DAVID BRIDGMAN, b Jacksonville, Ill, Jan 25, 25; m 48; c 3. MEDICINAL CHEMISTRY. Educ: Ill Col, AB, 48; Univ Nebr, MS, 50, PhD(org chem), 52. Prof Exp: Res chemist, Chemstrand Corp, 52-56; from assoc res chemist to sr res chemist, 56-71, RES SCIENTIST ORG CHEM, RES LABS, PARKE, DAVIS & CO, 71- Mem: Am Chem Soc; AAAS. Res: Synthesis of experimental drugs for the control of infectious diseases, and the study of relationships between molecular structure and biological activities. Mailing Add: 1406 Brooklyn Ann Arbor MI 48104

CAPPS, JULIUS DANIEL, b Opelika, Ala, Nov 26, 13; m 38; c 2. ORGANIC CHEMISTRY. Educ: Ala Polytech Inst, BS, 34, MS, 35; Univ Nebr, PhD(org chem), 38. Prof Exp: Asst chem, Ala Polytech Inst, 34-35, instr, 35; asst, Univ Nebr, 35-36; instr, Ala Polytech Inst, 38-41; instr, St Louis Univ, 41-42; mfg chemist, Merck & Co, 42; assoc animal nutritionist, Exp Sta, 42-43, assoc prof org chem, 43-47, assoc res prof chem, 47-53, res prof, 53-70, prof, 70-75, EMER PROF CHEM, AUBURN UNIV, 75- Concurrent Pos: Res chemist, Tenn Eastman Corp, Tenn, 44 & Procter & Gamble Co, Ohio, 47; actg dir res & develop, Redstone Div, Thiokol Chem Corp, Ala, 51-52, adv & consult, 52-62; consult, Royal Crown Cola Co, Ga, 59- Mem: Am Chem Soc. Res: Synthesis in the quinoline series; arsonic acids; amine aldehyde condensations; vitamin C and carotene content of plant tissue; vitamin A and carotene content of dairy products; vegetable oils; separation of metallic elements; rocket and guided missile fuels; artificial sweeteners; essential oils. Mailing Add: Rte 5 Box 335-C Opelika AL 36801

CAPPS, MARY JAYNE, b Konawa, Okla, Mar 18, 41; m 70. PHYSIOLOGICAL PSYCHOLOGY, SENSORY PHYSIOLOGY. Educ: Univ Okla, BA, 62, MS, 64; Univ Ill, Urbana, PhD(psychol), 67. Prof Exp: NSF fel elec eng, Univ Ill, Urbana, 67-68, res assoc elec eng & asst prof psychol, 68-69; asst prof otolaryngol, 69-72, ASSOC PROF OTOLARYNGOL, MED SCH, UNIV MINN, MINNEAPOLIS, 72- Mem: AAAS; Am Asn Anat; Acoust Soc Am; Psychonomic Soc. Res: Auditory and vestibular system physiology; behavioral conditioning. Mailing Add: Dept of Otolaryngol Med Sch Univ of Minn Minneapolis MN 55455

CAPPS, RAYMOND HAUL, b Ky, Nov 11, 13; m 38; c 1. CHEMISTRY. Educ: Western Ky State Col, BS, 50. Prof Exp: Anal chemist, EI du Pont de Nemours & Co, 41, tech supvr & engr, 42-45; res chemist, Union Carbide Chem Co, 45-54; res engr, Union Carbide Metals Co, 55-56, process develop engr, 56-59, tech supvr, 59-60, supt refractory metals dept, 60-65; asst supt electrolytics, mining & metals div, 65-70, chem, Ferroalloys Div, 70-71, SUPT OPERS TANTALUM, COLUMBIUM, METALS DIV, UNION CARBIDE CORP, 71- Mem: Am Chem Soc. Res: Fluorine chemistry; liquid extraction; fused salt electrolysis; nitrocellulose manufacture; separation and purification of refractory metals. Mailing Add: 1226 Glendale Rd Rte 3 Marietta OH 45750

CAPPS, RICHARD BROOKS, b Chicago, Ill, Mar 14, 06; m 30; c 1. MEDICINE. Educ: Princeton Univ, BS, 27; Harvard Univ, MD, 31. Prof Exp: Med intern, Mass Gen Hosp, 31-33; asst med, Harvard Med Sch, 33-35; asst, Northwestern Univ, 36-37, instr, 37-39, assoc, 39-46, from asst prof to assoc prof, 46-57; from assoc clin prof to clin prof, Med Sch, Univ Ill, 57-71; PROF MED, RUSH MED COL, 71- Concurrent Pos: Asst resident physician, Thorndike Mem Lab, Boston City Hosp, 33-35; assoc, Cook County Hosp, 36-37; sr attend physician, St Luke's Hosp, Ill, 36-59 & Presby-St Luke's Hosp, 59- Mem: Fel Am Col Physicians; fel AMA; Am Clin & Climat Asn (vpres, 62-63); Am Heart Asn; Am Asn Study Liver Dis (pres, 58). Res: Cardiovascular system; liver disease. Mailing Add: 1725 W Harrison St Chicago IL 60612

CAPPS, RICHARD H, b Wichita, Kans, July 1, 28; m 55. THEORETICAL PHYSICS. Educ: Univ Kans, AB, 50; Univ Wis, MS, 52, PhD(physics), 55. Prof Exp: Fel, Univ Calif, Berkeley, 55-57; actg asst prof physics, Univ Wash, 57-59; fel, Cornell Univ, 58-60; from asst prof to prof, Northwestern Univ, 60-67; PROF PHYSICS, PURDUE UNIV, 67- Concurrent Pos: Fulbright & Guggenheim fel, Univ Rome, 62-63; consult, Argonne Nat Lab, Ill, 60- Mem: Am Phys Soc. Res: Theory of the interactions of fundamental particles. Mailing Add: Dept of Physics Purdue Univ West Lafayette IN 47907

CAPPUCCI, DARIO TED, JR, b Plains, Pa, Aug 19, 41. VETERINARY MEDICINE. Educ: Univ Calif, Davis, BS, 63, DVM, 65, MS, 66; Am Bd Vet Pub Health, dipl, 73; Am Registry Cert Animal Scientists, dipl, 75; Univ Calif, San Francisco, PhD(comp path), 76. Prof Exp: Head carnivore unit, NIH Animal Ctr, Md, 66-67; vet epidemiologist, Zoonoses Surveillance Unit, Nat Communicable Dis Ctr, Ga, 68; researcher vet med & sci, Independent Invest Studies, 69; vet, Vet Lab Serv, Calif Dept Agr, 69-70; PUB HEALTH VET, CALIF DEPT HEALTH, 70- Concurrent Pos: Mem subcomt pub health, Nat Brucellosis Comt, 68-69. Mem: Am Vet Med Asn; Am Soc Animal Sci; AAAS; Wildlife Dis Asn; NY Acad Sci. Res: Reproductive pathophysiology; veterinary public health; comparative medicine; supervision of laboratory animal breeding colony; animal science. Mailing Add: 1077 Sanchez St San Francisco CA 94114

CAPRA, J DONALD, b Burlington, Vt, July 20, 37; m 58; c 2. MEDICINE, IMMUNOLOGY. Educ: Univ Vt, BS, 59, MD, 63. Prof Exp: Intern, St Luke's Hosp, New York, 64, resident, 65; sr surgeon, NIH, 6567; guest investr, Rockefeller Univ, 6769; assoc prof microbiol, Mt Sinai Sch Med, 69-74; PROF MICROBIOL, UNIV TEX HEALTH SCI CTR, DALLAS, 74- Concurrent Pos: USPHS fel, 65-67. Mem: Am Asn Immunologists; Am Rheumatism Asn. Res: Immunogenetics; protein sequences; antibody combining site. Mailing Add: Dept of Microbiol Univ of Tex Health Sci Ctr Dallas TX 75235

CAPRANICA, ROBERT R, b Los Angeles, Calif, May 29, 31; m 58. NEUROBIOLOGY, ELECTRICAL ENGINEERING. Educ: Univ Calif, Berkeley, BS, 58; NY Univ, MEE, 60; Mass Inst Technol, ScD(elec eng), 64. Prof Exp: Mem tech staff commun res, Bell Tel Labs, NJ, 58-69; assoc prof neurobiol & elec eng, 69-75, PROF NEUROBIOL & BEHAVIOR & ELEC ENG, CORNELL UNIV, 75- Mem: AAAS; Am Soc Zool; fel Acoust Soc Am; Am Physiol Soc; Soc Neurosci. Res: Animal sound communication; auditory neurophysiology. Mailing Add: Langmuir Lab Sect of Neurobiol Cornell Univ Ithaca NY 14850

CAPRETTA, UMBERTO, b Nereto, Italy, Mar 5, 22; US citizen; m 50; c 2. ANALYTICAL CHEMISTRY. Educ: Univ Naples, Dr indust chem, 52. Prof Exp: Teacher high sch, Italy, 52-55; lab technician, 55-60, anal chemist, 60-68, SR ENGR, EASTMAN KODAK CO, 68- Mem: AAAS; fel Am Inst Chemists; Am Chem Soc; NY Acad Sci. Res: Instrumental analysis; x-ray emission; polarography; atomic absorption; development of new methods and techniques. Mailing Add: 187 River St Rochester NY 14612

CAPRI, ANTON ZIZI, b Czernowitz, Romania, Apr 20, 38; Can citizen; m 60; c 3. PARTICLE PHYSICS, MATHEMATICAL PHYSICS. Educ: Univ Toronto, BASc, 61; Princeton Univ, MA, 65, PhD(physics), 67. Prof Exp: Res physicist, Kimberly-Clark Corp, 61-63; fel, 67-68, vis asst prof theoret physics, 68-69, asst prof, 69-73, ASSOC PROF THEORET PHYSICS, UNIV ALTA, 73- Concurrent Pos: Sr res fel, Alexander von Humboldt Found, 75-76. Mem: Can Asn Physicists. Res: Higher spin field theories; existence of solutions in quantum field theory; nuclear-nucleon interaction. Mailing Add: Theoret Physics Inst Univ of Alta Edmonton AB Can

CAPRIO, JOSEPH MICHAEL, b New Brunswick, NJ, Nov 7, 23; m 51; c 3. AGRICULTURAL METEOROLOGY, MICROCLIMATOLOGY. Educ: Rutgers Univ, BS, 47, MS, 50; Calif Inst Technol, BS, 48; Utah State Univ, PhD(biometeorol), 70. Prof Exp: Agrometeorologist, Am Inst Aerological Res, 50-53; statistician & agronometeorologist, Citrus Exp Sta, Univ Calif, 53-55; from asst prof to assoc prof agr climat, 55-63, PROF AGR CLIMAT, MONT STATE UNIV, 63- Concurrent Pos: Mem & past chmn & secy, Tech Comt Western Regional Proj W-48, 57-72; mem, Tech Comt Great Plains Regional Proj GP-1, 59-61; climat expert, UN World Meteorol Orgn, Iran, 62-63; mem phenology panel, US Int Biol Prog, 67-70; co-invest, Earth Resources Technol Satellite-1, Phenology Satellite Exp, NASA, 72-74; vis scientist, Div Land Use Res, Commonwealth Indust & Sci Orgn, Canberra, Australia, 73-74; mem, UN World Meteorol Orgn Working Group on Methods of Forecasting Agr Crop Develop & Ripening, 75- Mem: Am Meteorol Soc; Int Soc Bioclimat & Biometeorol; Am Soc Agron; Soil Sci Soc Am. Res: Biometeorology; agrometeorology, including statistical climatic analysis; mapping of climatic elements in mountainous areas; study of weather effects on crop production; plant phenology; soil physics; hydrology. Mailing Add: Dept of Plant & Soil Sci Mont State Univ Bozeman MT 59715

CAPRIOGLIO, GIOVANNI, b Rome, Italy, Aug 9, 32; m 58; c 1. ELECTROCHEMISTRY. Educ: Univ Milan, DSc(indust chem), 56. Prof Exp: Asst prof electrochem, Univ Modena, 57-59 & Univ Milan, 59-62; consult, Gen Atomic Div, Gen Dynamics Corp, 61-62, staff mem, 62-67; proj mgr, Gulf Gen Atomic, Inc, 67-70; assoc dir lab, 70-73, MGR MAT & CHEM DEPT, GEN ATOMIC CO, 73- Mem: Int Soc Electrochem; Electrochem Soc; Am Nuclear Soc. Res: High temperature materials; corrosion; batteries; thermochemical hydrogen production. Mailing Add: Gen Atomic Co PO Box 81608 San Diego CA 92138

CAPRIOLI, RICHARD MICHAEL, b New York, NY, Apr 12, 43; m 71. BIOCHEMISTRY, MASS SPECTROMETRY. Educ: Columbia Univ, BS, 65, PhD(biochem), 69. Prof Exp: Res assoc chem, Purdue Univ, 69-70, from asst prof to assoc prof, 70-75; ASSOC PROF BIOCHEM, UNIV TEX MED SCH HOUSTON, 75- Mem: Am Soc Biol Chemists; Am Chem Soc; Am Soc Mass Spectrometry. Res: Peptide sequencing by mass spectrometry; intermediary metabolism using stable isotopes; mechanisms of enzyme action. Mailing Add: Dept of Biochem Univ of Tex Med Sch Houston TX 77025

CAPRIOTTI, EUGENE RAYMOND, b Brackenridge, Pa, June 20, 37; m 60; c 1. ASTRONOMY, PHYSICS. Educ: Pa State Univ, BS, 59; Univ Wis, PhD(astron), 62. Prof Exp: Res fel astron, Calif Inst Technol, 62-63; asst prof, Univ Calif, Berkeley, 63-64; sr engr, Westinghouse Elec Corp, 64; from asst prof to assoc prof astron, 64-73, PROF ASTRON, OHIO STATE UNIV, 73- Concurrent Pos: Vis assoc prof, Steward Observ, Univ Ariz, 69-70. Mem: Am Astron Soc. Res: Gaseous nebulae; electric arcs; radiation losses in plasmas. Mailing Add: Dept of Astron Ohio State Univ Columbus OH 43210

CAPSTACK, ERNEST, b Fall River, Mass, June 9, 30; m 58; c 2. ORGANIC CHEMISTRY, BIOCHEMISTRY. Educ: Mass Inst Technol, BS, 52; Univ RI, MS, 54; Brown Univ, PhD(chem), 59. Prof Exp: Res chemist, E I du Pont de Nemours & Co, 58-60; res fel, Steroid Training Prog, Worcester Found Exp Biol, 60-61; res assoc, Clark Univ, 61-62; scientist, Worcester Found Exp Biol, 62-64; vis asst prof, Clark Univ, 62-64; assoc prof chem, 64-66, PROF CHEM, W VA WESLEYAN COL, 66-, CHMN DEPT, 67- Mem: Am Chem Soc; Sigma Xi. Res: Organic and steroid synthesis; steroid and terpene biosynthesis and metabolism. Mailing Add: Dept of Chem WVa Wesleyan Col Buckhannon WV 26201

CAPURRO, LUIS R A, b Buenos Aires, Arg, Jan 11, 20; m 47; c 1. PHYSICAL OCEANOGRAPHY, MARINE METEOROLOGY. Educ: Univ Calif, MSc, 49; Univ Buenos Aires, DSc(oceanog), DSc, 51. Prof Exp: Cmndg officer res vessel, Arg Hydrographic Off, 52-54, head dept oceanog, 54-57, cmndg officer icebreaker, Arg Navy, 57-58; res scientist, Tex A&M Univ, 59-61; chief, Arg Hydrographic Off, 62-65; res scientist, Tex A&M Univ, 65-74; WITH OFF OCEANOG/UNESCO, PARIS, FRANCE, 74- Concurrent Pos: Pres, Sci Comt Oceanic Res, vpres, Intergovt Oceanog Comn & mem bur, Int Asn Phys Oceanog & Int Upper Mantle Bur. Mem: AAAS; Am Geophys Union. Res: Oceanic circulation; remote sensing of the ocean. Mailing Add: Off of Oceanog/UNESCO Pl de Fontenoy 75 Paris France 7

CAPUTI, ROGER WILLIAM, b Newark, NJ, Jan 9, 35; m 59; c 3. PHYSICAL CHEMISTRY. Educ: Calif Inst Technol, BS, 57, PhD(phys chem), 65. Prof Exp: Radiol chemist, US Naval Radiol Defense Lab, 57-60; teaching asst, Phys Chem Lab, Calif Inst Technol, 60-64; chemist, US Naval Radiol Defense Lab, 64-68; chemist, Vallecitos Nuclear Ctr, 68-70, mgr prod qual & applns lab, 70-72; SR SCIENTIST, TRACK ETCH DEVELOP LAB, VALLECITOS NUCLEAR CTR, GEN ELEC CO, 72- Mem: Sigma Xi; NY Acad Sci. Res: Etching of ion damage tracks in plastics; development of microbiological culturing techniques on membrane filters; modeling of membrane filter theory; instrumentation; analysis of nuclear weapons effects; oceanography. Mailing Add: Gen Elec Vallecitos Nuclear Ctr Vallecitos Rd Pleasanton CA 94566

CAPUTO, JOSEPH ANTHONY, b Jersey City, NJ, May 10, 40; m 65; c 2. PHYSICAL ORGANIC CHEMISTRY. Educ: Seton Hall Univ, BS, 62, MS, 64; Univ Houston, PhD(phys org chem), 67. Prof Exp: Fel org chem, Duke Univ, 67-68; from instr to asst prof, 68-71, ASSOC PROF ORG CHEM, STAT UNIV NY COL BUFFALO, 71-, CHMN DEPT CHEM, 74- Mem: The Chem Soc; Am Chem Soc. Res: Linear free energy relationships; electronic transmission; reaction mechanisms; diazoalkanes; phosphorus heterocycles; antimalarial compounds. Mailing Add: Dept of Chem State Univ of NY Col Buffalo NY 14222

CAPWELL, ROBERT J, b Binghampton, NY, July 6, 40; m 62; c 2. PHYSICAL CHEMISTRY. Educ: Ohio State Univ, BS, 63; Pa State Univ, MS, 65; Univ Pittsburgh, PhD(phys chem), 70. Prof Exp: Chemist, Gulf Res & Develop Co, 65-68; res scientist phys chem, 70-72, sect leader phys & anal chem 72-73, dept head anal & tech specialist, Cent Res Lab, 73-76, TECH SUPPORT MGR, INDUST CHEM DIV, N L INDUST, INC, 76- Mem: Am Chem Soc; Soc Appl Spectros. Res: Infrared and Raman spectroscopy; structure of matter; chemical instrumentation; molten salt chemistry; flame retardants; reaction mechanisms. Mailing Add: 139 Dorchester Dr East Windsor NJ 08520

CARABATEAS, PHILIP M, b New York, NY, Jan 20, 30; m 60; c 2. ORGANIC CHEMISTRY, MEDICINAL CHEMISTRY. Educ: Polytech Inst, Brooklyn, BS, 55; Rensselaer Polytech Inst, PhD(chem), 62. Prof Exp: From asst res chemist to res chemist, 55-70, SR RES CHEMIST, ORG CHEM, STERLING-WINTHROP RES INST, 70- Mem: Am Chem Soc. Res: Synthesis of nitrogen heterocycles, analgesics, antitussives and antifertility agents. Mailing Add: Sterling-Winthrop Res Inst Rensselaer NY 12144

CARADUS, SELWYN ROSS, b Auckland, NZ, Nov 10, 35; m 59; c 2. MATHEMATICAL ANALYSIS. Educ: Univ Auckland, BSc, 57, MSc, 58; Univ Southern Calif, MA, 62; Univ Calif, Los Angeles, PhD(math), 65. Prof Exp: Jr lectr math, Univ Auckland, 58-60; asst prof, 64-68, ASSOC PROF MATH, QUEEN'S UNIV, ONT, 69- Mem: Am Math Soc; Can Math Cong. Res: Theory of linear operators in Banach space. Mailing Add: Dept of Math Queen's Univ Kingston ON Can

CARAPELLUCCI, PATRICIA ANNE, physical chemistry, photochemistry, see 12th edition

CARASSO, ALFRED SAMUEL, b Alexandria, Egypt, Apr 9, 39; m 64; c 1. MATHEMATICS. Educ: Univ Adelaide, BS, 60; Univ Wis-Madison, MS, 64, MA, 65, PhD(math), 68. Prof Exp: Meteorologist, Bur Meteorol, Australia, 60-62; asst prof math, Mich State Univ, 68-69; asst prof, 69-74, ASSOC PROF MATH & STATIST, UNIV NMEX, 74- Mem: Am Math Soc; Soc Indust & Appl Math. Res: Partial differential equations; numerical analysis. Mailing Add: Dept of Math Univ of NMex Albuquerque NM 87106

CARAWAY, PRENTICE ALVIN, b Goldthwaite, Tex, Dec 22, 15; m 41; c 1. BIOLOGY. Educ: Tex A&M Univ, BS, 38; Mich State Univ, MS, 48, PhD(zool), 51. Prof Exp: Asst, Tex Agr Exp Sta, 38-40; asst zool, Mich State Univ, 47-49; asst fisheries biol, Mich Agr Exp Sta, 49-51; res assoc, Gulf Coast Res Lab, Miss, 52; prof zool & head dept biol, Tarleton State Col, 53-65; PROF BIOL, W TEX STATE UNIV, 65-, DIR ALLIED HEALTH SCI PROGS & HEAD DEPT ALLIED HEALTH SCI, 74- Concurrent Pos: NSF fels, Harvard Univ, 58 & Univ Wash, 59; scientist, Oak Ridge Inst Nuclear Studies, 65. Res: Aquatic and developmental biology; effects of drugs in embryonic development. Mailing Add: Dept of Allied Health Sci WTex State Univ Canyon TX 79015

CARAWAY, WENDELL THOMAS, b Xenia, Ohio, Nov 4, 20; m 44. BIOCHEMISTRY. Educ: Wilmington Col, BS, 42; Miami Univ, MA, 43; Johns Hopkins Univ, PhD(biochem), 50; Am Bd Clin Chem, dipl. Prof Exp: Asst chemist, Ohio River Div Labs, US Corps Engr, 43-45, mat engr & head chem lab, 45-46; biochemist, RI Hosp, 50-57; BIOCHEMIST, FLINT MED LAB, 57- Concurrent Pos: Mem bd dirs, Nat Registry Clin Chem, 67-73. Honors & Awards: Ames Award in Clin Chem, 70. Mem: AAAS; Am Chem Soc; Am Asn Clin Chem (pres, 65). Res: Kinetics and mechanisms of reactions; serum enzyme assays; ultramicrochemical methods of blood analysis. Mailing Add: 1102 Woodside Dr Flint MI 48503

CARBAJAL, BERNARD GONZALES, III, b New Orleans, La, Feb 15, 33; m 54; c 1. SOLID STATE CHEMISTRY, SEMICONDUCTORS. Educ: Univ Minn, PhD(chem), 58. Prof Exp: Asst chem, Univ Minn, 54-56; asst prof, Col St Thomas, 57-60; mem tech staff, 60-69, mgr multilevel tech prog, Components Group, 69-73, mgr advan circuit technol, 72-74, MGR DESIGN SUPPORT, TEX INSTRUMENTS, INC, 74- Mem: Am Chem Soc. Res: Organic semiconductors; thin film polymers; electrical conduction in organic materials; seminconductor processing; metal-oxide-semiconductor devices; thin films; interconnection technology. Mailing Add: Tex Instruments Inc PO Box 5012 Dallas TX 75222

CARBALLO-QUIROS, ALFREDO, b San Jose, Costa Rica, Nov 28, 19; m 45; c 3. PLANT BREEDING, QUANTITATIVE GENETICS. Educ: Iowa State Col, MS, 57; NC State Col, PhD(plant breeding), 62. Prof Exp: Head maize sect, Ministry Agr, Costa Rica, 48-51, head lowland crops, 51-54; coordr, Cent Am Coop Maize Proj, Rockefeller Found, 55-56 & 58-59; geneticist, Inter-Am Inst Agr Sci, Orgn Am States, Costa Rica, 62-63; dir maize prog, Nat Inst Agr Invest & Mendoza Found, Venezuela, 63-66; consult maize prog, Food & Agr Orgn, UAR, 66-67; geneticist maize prog, 69-70, GENETICIST, INT MAIZE TESTING PROG, INT MAIZE & WHEAT IMPROV CTR, MEX, 70- Concurrent Pos: Del, Inter-Am Conf Crop Sci, Peru, 63 & Colombia, 70; del annual meeting, Cent Am Coop Maize Improv Proj, 67. Mem: Am Soc Agron; Crop Sci Soc Am; Latin Am Soc Crop Sci; Mex Soc Plant Genetics; Agron Eng Soc Costa Rica. Res: Intra-population improvement of maize populations through a quantitative genetics approach. Mailing Add: Apartado 2591 San Jose Costa Rica

CARBERRY, EDWARD ANDREW, b Milwaukee, Wis, Nov 20, 41; m 67; c 1. INORGANIC CHEMISTRY, ORGANOMETALLIC CHEMISTRY. Educ: Marquette Univ, BS, 62; Univ Wis, PhD(inorg chem), 68. Prof Exp: PROF CHEM, SOUTHWEST MINN STATE UNIV, 68- Mem: Am Chem Soc; The Chem Soc. Res: Preparation, spectroscopic and bonding studies of inorganic and organometallic compounds, especially organosilicon compounds; catenated group IV chemistry emphasizing polysilanes and polygermanes linear and cyclic. Mailing Add: Dept of Chem Southwest Minn State Univ Marshall MN 56258

CARBON, JOHN ANTHONY, b Sharon, Pa, Jan 1, 31; m 50; c 2. BIOCHEMISTRY. Educ: Univ Ill, BS, 52; Northwestern Univ, PhD(biochem), 55. Prof Exp: Res assoc biochem, Med Sch, Northwestern Univ, 55-56; sr res chemist, Org Chem Dept, Abbott Labs, 56-63, res assoc, Biochem Dept, 63-68; assoc prof biochem, 68-70, PROF BIOCHEM, UNIV CALIF, SANTA BARBARA, 70- Concurrent Pos: Mem ed bd, J Biol Chem, 73-; consult, Abbott Labs, 74- Mem: Am Asn Biol Chemists. Res: Nucleic acids; transfer RNA; viral DNA; bacterial plasmids; molecular cloning. Mailing Add: Dept of Biol Sci Univ of Calif Santa Barbara CA 93106

CARBONE, GABRIEL, b New York, NY, Sept 4, 27; m 53; c 2. ENVIRONMENTAL CHEMISTRY. Educ: Brooklyn Col, BS, 49. Prof Exp: Chemist, Food & Water Chem Lab, 55-67, sr chemist, 67-70, PRIN CHEMIST & DIR FOOD & WATER CHEM LAB, BUR LABS, NEW YORK CITY HEALTH DEPT, 70- Concurrent Pos: Mem fumigant bd, New York City Health Dept, 70- Res: Differentiation of residual and reactivated phosphatase in dairy products; nitrates in meat products. Mailing Add: Food & Water Chem Lab Bur Labs New York City Health Dept 125 Worth St New York NY 10013

CARBONE, JOHN VITO, b Sacramento, Calif, Dec 13, 22; m 46; c 3. MEDICINE. Educ: Univ Calif, BA, 45, MD, 48. Prof Exp: From instr to assoc prof, 51-66, PROF MED, SCH MED, UNIV CALIF, SAN FRANCISCO, 66- Concurrent Pos: Giannini fel med, 54-55; consult to Surgeon Gen, US Army, 58 & Letterman Hosp, Travis AFB, San Francisco. Res: Gastroenterology; metabolic aspects of liver disease. Mailing Add: Sch of Med Univ of Calif San Francisco CA 94122

CARBONE, PAUL P, b White Plains, NY, May 2, 31; m 54; c 7. ONCOLOGY. Educ: Albany Med Col, MD, 56. Prof Exp: Intern med, USPHS Hosp, Baltimore, Md, 56-57, mem med staff, Savannah, Ga, 57-58, resident internal med, San Francisco, Calif, 58-60; sr investr, Nat Cancer Inst, 60-65, head solid tumor serv, 65-68, chief med br, 68-72, assoc dir med oncol, Div Cancer Treatment, 72-76; PROF HUMAN ONCOL & MED & CHIEF CLIN ONCOL, UNIV WIS, 76- Mem: Am Soc Hemat; Am Fedn Clin Res; Am Asn Cancer Res; Am Soc Clin Oncol; Am Soc Clin Invest. Res: Research in cancer chemotherapy, immunology and hematology. Mailing Add: Sch of Med Univ of Wis Madison WI 53706

CARBONE, ROBERT JAMES, b Hartford, Conn, Aug 17, 30; m 53; c 3. PHYSICS. Educ: Univ Conn, BA, 52, MA, 53, PhD, 56. Prof Exp: Asst, Univ Conn, 54-56; physicist, Lincoln Lab, Mass Inst Technol, 56-60 & 62-68 & Bomac Labs, Inc, 61-62; physicist, Philips Res Labs, Technol Univ Eindhoven, 68-69 & Lincoln Lab, Mass Inst Technol, 70-72; ALTERNATE LIFE TESTING GROUP LEADER, LOS ALAMOS SCI LAB, UNIV CALIF, 72- Mem: Am Phys Soc. Res: Radiofrequency gas discharges; heavy ion-atom interactions; chemical and gaseous lasers; direct current discharges; laser applications in fusion and isotope separation. Mailing Add: 2191 45th St Los Alamos NM 87544

CARBONELL, ROBERT JOSEPH, b El Salvador, Mar 20, 27; m 50; c 5. FOOD SCIENCE. Educ: Nat Univ El Salvador, BS, 48. Prof Exp: Res chemist, Sanit Eng Div, Inst Inter Am Affairs, El Salvador, 45-46, asst head dept chem, Ctr Nat Agron, 46-52, head green coffee processing div, 52-53; assoc prof biochem, Sch Agron, Nat Univ El Salvador, 51-52, assoc prof org anal, Sch Pharmaceut Chem, 51-53; res chemist, Res & Develop Food Prod, Fleischmann Labs, 53-54, group leader, 54-57, div head, 57-59, dir consumer prod res dept, 59-63, dir divisional res & develop, 63-65, mgr, 65-69, dir res, 69-73, corp dir prod & develop planning, 73-74, mgr Fleischmann Indust Mfg, 74-75, mgr grocery prod mfg, 75, VPRES RES & DEVELOP, STAND BRANDS INC, 75- Concurrent Pos: Mem, Coffee Processing Surv Comn, El Salvador Govt, 52; abstr, Chem Abstr, 53-54. Mem: Am Chem Soc; Inst Food Tecnol. Res: Composition of fats and oils form tropical species of Salvadorean Flora; nutritional value of coffee pulp; treatment of waste waters from coffee processing plants; effect of crossing upon chemical composition of sunflower oil; coffee bean mucilage. Mailing Add: Stand Brands Inc Betts Ave Stamford CT 06904

CARBONI, RUDOLPH A, b Yonkers, NY, Nov 10, 22; m 49; c 4. ORGANIC CHEMISTRY. Educ: Columbia Univ, AB, 47, AM, 48; Mass Inst Technol, PhD(org chem), 53. Prof Exp: Res chemist, Charles Pfizer & Co, 48-50; res chemist, Cent Res Dept, 53-61, res supvr, Org Chem Dept, 62-67, div head explor intermediates, 67-68, lab mgr, 68-70, res & develop mgr permasep prod, 70-73, TECH MGR ELECTRONIC PROD DIV, PHOTOPROD DEPT, E I DU PONT DE NEMOURS & CO, 73- Mem: Am Chem Soc; Am Asn Textile Chem & Colorists. Res: Cyanocarbon chemistry; new polynitrogen systems; organofluorine chemistry; antibiotic research; inductive and field effects; dye studies and oxidation catalysts; reverse osmosis; photopolymer systems for printed circuit and microelectronic device fabrication. Mailing Add: Photoprod Dept E I du Pont de Nemours & Co Wilmington DE 19803

CARBONNEAU, ROCH, b Ste Justine, Que, Aug 3, 31; m 57; c 2. PHYSIOLOGY, ENDOCRINOLOGY. Educ: Univ Montreal, BA, 54, BSc, 59, MSc, 61, PhD(physiol), 64. Prof Exp: Asst prof physiol, Univ Montreal, 64-65; instr anat, State Univ NY, 65-67; asst prof physiol, 67-71, ASSOC PROF PHYSIOL, UNIV MONTREAL, 71-, ASSOC DIR STUDENTS ACAD AFFAIRS SERV, FAC ARTS & SCI, 70- Concurrent Pos: NATO res fel, State Univ NY Downstate Med Ctr, 65-67; NIH fel instrumental physiol, Tex Med Ctr, Houston, summer 64. Mem: Nutrit Soc Can; Can Physiol Soc. Res: Fatty liver related to amino acids contents of the diet, protein level and neutral fats; different annual endocrine cycles in fishes, histochemical approach and chemical approach. Mailing Add: 768 Ave Marsolais Laval PQ Can

CARD, KENNETH D, b Ont, Jan 23, 37; m 59; c 2. GEOLOGY. Educ: Queen's Univ (Ont), BSc, 59; Princeton Univ, MA, 62, PhD(geol), 63. Prof Exp: Resident geologist, 63-66, field geologist, 66-74, SR GEOLOGIST, ONT DIV MINES, 74- Mem: Fel Geol Soc Am; fel Geol Asn Can. Res: Regional geology; precambrian stratigraphy, structure and metamorphism. Mailing Add: Ont Dept of Mines Geol Br Sudbury ON Can

CARDE, RING RICHARD TOMLINSON, b Hartford, Conn, Sept 18, 43; m 74. ENTOMOLOGY, BEHAVIORAL BIOLOGY. Educ: Tufts Univ, BS, 66; Cornell Univ, MS, 68, PhD(entom), 71. Prof Exp: Fel, NY State Agr Exp Sta, Cornell Univ, 71-75; ASST PROF ENTOM, MICH STATE UNIV, 75- Mem: Entom Soc Am; Entom Soc Can; AAAS; Sigma Xi; Soc Study Evolution. Res: Insect pheromones and behavior; pheromone identification; use of pheromones in pest management; biosystematics of the Lepidoptera. Mailing Add: Pesticide Res Ctr Mich State Univ East Lansing MI 48824

CARDELL, ROBERT RIDLEY, JR, b Atlanta, Ga, Nov 11, 31; m 59; c 3. CELL BIOLOGY. Educ: Ga Southern Col, BS, 56; Univ Va, MS, 59, PhD(biol), 62. Prof Exp: Asst biol, Univ Va, 56-59, instr, 59-60; asst biophys, Edsel B Ford Inst Med Res, 60-62, res assoc, 62-64; res assoc biol, Harvard Univ, 64-67; assoc prof, 67-70, PROF ANAT, SCH MED, UNIV VA, 70- Mem: Am Soc Cell Biol; Am Asn Anat; Endocrine Soc. Res: Ultrastructure of cells; morphological action of hormones; cell biology; cellular endocrinology. Mailing Add: Dept of Anat Univ of Va Sch of Med Charlottesville VA 22901

CARDENAS, CARLOS GUILLERMO, b Laredo, Tex, June 25, 41; m 65; c 2. ORGANIC CHEMISTRY. Educ: Univ Tex, BS, 62, PhD(org chem), 65. Prof Exp: Asst prof chem, US Naval Postgrad Sch, 65-67; res chemist, Phillips Petrol Co, 67-69, group leader, 69; sr chemist, 69-71, sect mgr, 71-74, MGR RES, GLIDDEN-DURKEE DIV, SCM CORP, 75- Concurrent Pos: Off Naval Res grant, 65-67. Mem: Am Chem Soc; The Chem Soc. Res: Olefin synthesis and reaction mechanisms; nuclear magnetic resonance; Diels-Alder reactions; base catalysis; organometallics; terpenes. Mailing Add: Glidden-Durkee Div SCM Corp PO Box 389 Jacksonville FL 32201

CARDENAS, MANUEL, b San Diego, Tex, Sept 18, 42; m 68; c 2. EXPERIMENTAL STATISTICS. Educ: Tex A&I Univ, BS, 68, MA, 70; Tex A&M Univ, PhD(statist), 74. Prof Exp: Comput programmer, Tex Real Estate Res Ctr, 72-73; ASST PROF STATIST, N MEX STATE UNIV, 74- Mem: Am Statist Soc; Biometric Soc. Res: Compartmental modeling; the derivation of the distribution of models with time-dependent transition probabilities, particularly as applied to bio-medical sciences. Mailing Add: Dept of Exp Statist Box 3130 NMex State Univ Las Cruces NM 88003

CARDENAS, MARY JANET M, b Miami, Okla, Jan 31, 42; m 60; c 4. PROTEIN CHEMISTRY, ENZYMOLOGY. Educ: Okla State Univ, BA, 63; Univ Ill, MS, 65, PhD(biochem), 67. Prof Exp: Res assoc protein synthesis, Ore State Univ, 67; instr chem, Col Nueva Granada, 69-71; fel, Ore State Univ, 71-73, ASST PROF BIOCHEM & BIOPHYSICS, ORE STATE UNIV, 73- Mem: AAAS; Am Chem Soc; Am Asn Biol Chemists; Acad Sci; Sigma Xi. Res: Studies on pyruvate kinase isozymes and isozyme hybrids; enzymes which phosphorylate or dephosphorylate proteins. Mailing Add: Dept of Biochem & Biophysics Ore State Univ Corvallis OR 97331

CARDER, DEAN SAMUEL, seismology, deceased

CARDER, KENDALL L, b Norfolk, Nebr, Sept 11, 42. PHYSICAL OCEANOGRAPHY. Educ: Ore State Univ, MS, 67, PhD(oceanog), 70. Prof Exp: Asst prof oceanog, 69-74, ASSOC PROF OCEANOG, UNIV S FLA, ST PETERSBURG, 74- Concurrent Pos: Del, Int Oceanog Cong, Tokyo, 70; NSF grant, 70-71; Off Naval Res grant, 72-; Off Water Resources res grant, 74-; del, Int Union Geod Geophys, Grenoble, 75. Mem: Am Geophys Union; Optical Soc Am. Res: Distribution and dynamics of marine particulates and their effects on the submarine light field; modeling of nearshore and estuarine environments. Mailing Add: Marine Sci Inst Univ of SFla 830 First St S St Petersburg FL 33701

CARDIFF, ROBERT DARRELL, b San Francisco, Calif, Dec 5, 35; m 62; c 3. PATHOLOGY, VIROLOGY. Educ: Univ Calif, Berkeley, BS, 58, PhD(zool), 68; Univ Calif, San Francisco, MD, 62; Am Bd Path, dipl, 69. Prof Exp: Rotating intern, Kings County Gen Hosp, Brooklyn, 62-64; instr path & resident anat path, Sch Med, Univ Ore, 64-66; staff pathologist, Dept Neuropsychiat, Walter Reed Army Inst Res, Washington, DC, 68-71; ASSOC PROF PATH, SCH MED, UNIV CALIF, DAVIS, 71- Mem: AAAS; Am Soc Microbiol; Am Asn Cancer Res; Int Acad Path; Am Asn Path & Bact. Res: Mammary tumor systems; control of viral expression; molecular biology of RNA tumor viruses; host-virus relationships; arbovirus structure and function; pathogenesis of arbovirus disease. Mailing Add: Dept of Path Univ of Calif Sch of Med Davis CA 95616

CARDILLO, FRANCES M, b Rome, NY, Apr 19, 32. BOTANY. Educ: Col St Rose, BS, 53, MA, 61; St Bonaventure Univ, PhD(biol), 67. Prof Exp: Instr bot, Immaculate Conception Jr Col, 66-67; asst prof, Ladycliff Col, 67-69; assoc prof, Quincy Col, 69-71; ASSOC PROF, LADYCLIFF COL, 71- Concurrent Pos: NSF Acad Year exten grant, 69-71. Mem: AAAS; Am Inst Biol Sci; Bot Soc Am; Am Fern Soc; NY Acad Sci. Res: Anatomy of Lycopodium species; biochemical studies of Crossosoma. Mailing Add: Ladycliff Col Highland Falls NY 10928

CARDILLO, MARK J, b Passaic, NJ, Aug 20, 43. CHEMICAL PHYSICS. Educ: Stevens Inst Technol, BS, 64; Cornell Univ, PhD(chem), 70. Prof Exp: Res assoc chem, Brown Univ, 69-71; MEM STAFF, BELL LABS, 71- Concurrent Pos: Nat Res Coun Italy fel, Inst Physics, Univ Genoa, 71-72. Mem: Am Chem Soc; Am Phys Soc. Res: Gas-surface interactions; molecular beams; scattering; reaction kinetics; chemical dynamics on surfaces; high temperature mass spectrometry. Mailing Add: Bell Labs Murray Hill NJ 07974

CARDILLO, THOMAS E, b Rochester, NY, Oct 19, 24; m 48. MEDICINE. Educ: Univ Rochester, BA, 47, MD, 51; Am Bd Internal Med, dipl, 60. Prof Exp: Intern asst med, Univ Rochester Strong Mem Hosp, 51-52; from asst resident to resident, State Univ NY Upstate Med Ctr, 52-54; intern asst med, Univ Rochester Strong Mem Hosp, 54-55; instr, Sch Med & Dent, Univ Rochester, 55-60, clin sr instr, 60-64, clin asst prof, 64-65; clin asst prof, Med Col, Univ Ala, 65-68; ASSOC PROF MED, SCH MED & DENT, UNIV ROCHESTER, 68- Concurrent Pos: Dir cardiol, Monroe Community Hosp, 68- Honors & Awards: Distinguished Serv Award, Am Heart Asn, 65. Mailing Add: Sch of Med & Dent Univ of Rochester Rochester NY 14642

CARDIN, DAVID WALTON, poultry nutrition, poultry biochemistry, see 12th edition

CARDINAL, ANDRE, b Valleyfield, Que, May 12, 35; m 59; c 2. PHYCOLOGY. Educ: Univ Montreal, MSc, 61; Univ Paris, DSc(phycol), 64. Prof Exp: Biologist, Marine Biol Sta Grande-Riviere, 64-68; PROF BIOL, UNIV LAVAL, 68- Mem: Phycol Soc Am; Brit Phycol Soc; Phycol Soc France; Int Phycol Soc. Res: Ecology, taxonomy and morphology of marine algae; marine phycology. Mailing Add: Dept of Biol Univ of Laval Quebec PQ Can

CARDINALE, GEORGE JOSEPH, b New York, NY, Mar 30, 36; m 64; c 2. BIOCHEMISTRY. Educ: Fordham Univ, BS, 57; Ohio State Univ, PhD(org chem), 65. Prof Exp: Fel biochem, Brandeis Univ, 65-67, res assoc, 67-70; ASST MEM PHYSIOL CHEM, ROCHE INST MOLECULAR BIOL, 70- Concurrent Pos: Adj assoc prof, Dept Pharmacol & Toxicol, Univ RI, 74- Mem: Int Inflammation Club. Res: Prolyl hydroxylase and its role in collagen biosynthesis; factors affecting collagen synthesis; role of collagen in atherosclerosis. Mailing Add: Roche Inst of Molecular Biol Nutley NJ 07110

CARDINET, GEORGE HUGH, III, b Oakland, Calif, Oct 28, 34; m 57; c 4. COMPARATIVE PATHOLOGY. Educ: Univ Calif, BS, 60, DVM, 63, PhD(comp path), 66. Prof Exp: From asst prof to assoc prof anat, Neuromuscular Res Lab, Kans State Univ, 66-74; ASSOC PROF ANAT, UNIV CALIF, DAVIS, 74- Concurrent Pos: NIH fel, 63-66 & 73-74. Mem: AAAS; Am Asn Anat; Am Soc Zool; Electron Micros Soc Am; Am Asn Vet Anat. Res: Investigations of myopathies in domestic animals including clinical enzymology; enzyme histochemistry and electromicroscopy. Mailing Add: Dept of Anat Univ of Calif Sch of Vet Med Davis CA 95616

CARDMAN, LAWRENCE SANTO, b Mt Vernon, NY, Oct 7, 44; m 68; c 2. NUCLEAR PHYSICS. Educ: Yale Univ, BA, 66, PhD(physics), 72. Prof Exp: Actg instr physics, Electron Accelerator Lab, Yale Univ, 71-72; Nat Acad Sci-Nat Res Coun res fel, Ctr Radiation Res, Nat Bur Standards, 72-73; ASST PROF PHYSICS, PHYSICS RES LAB, UNIV ILL, 73- Mem: Sigma Xi; Am Phys Soc. Res: Photonuclear physics; elastic and inelastic electron scattering studies of nuclear structure. Mailing Add: Dept of Physics Univ of Ill Urbana IL 61801

CARDON, BARTLEY PRATT, b Tucson, Ariz, Oct 1, 13; m 39; c 4. MICROBIOLOGY, NUTRITION. Educ: Univ Ariz, BS & MS, 40; Univ Calif, PhD(microbiol), 46. Prof Exp: From asst prof to assoc prof animal path & husb, Univ Ariz, 46-54; res dir, Ariz Flour Mills, 54-62; pres, Erly-Fat Livestock Feed Co, 62-68; pres, 68-74, SR EXEC OFFICER, ARIZ FEEDS, 74- Mem: AAAS; Am Soc Microbiol. Res: Digestion in ruminants; animal pathology; decomposition of proteins and amino acids. Mailing Add: Ariz Feeds PO Box 5526 Tucson AZ 85703

CARDON, PHILLIPPE VINCENT, JR, b Salt Lake City, Utah, June 17, 22; m 54; c 2. MEDICINE. Educ: Yale Univ, AB, 43; Columbia Univ, MD, 46. Prof Exp: Intern, Bellevue Hosp, NY Univ, 46-47, asst resident, 49-50, resident psychiat, Med Div, 50-51; instr, Med Col, Cornell Univ, 51-53; chief psychosom, Lab Clin Sci, NIMH, 53-74, ASSOC DIR CLIN CTR, NIH, 74- Concurrent Pos: Commonwealth fel, NY Hosp, 51-53. Mem: Am Psychosom Soc; Soc Psychophysiol Res; Soc Neurosci. Res: Interrelationships of the nervous and circulatory systems; clinical research in the general area of psychobiology. Mailing Add: 2209 Parker Ave Wheaton MD 20902

CARDON, SAMUEL ZELIG, chemistry, see 12th edition

CARDONA, EDWARD, physiology, biochemistry, see 12th edition

CARDONA-MORALES, NESTOR ANIBAL, biochemistry, see 12th edition

CARDONE, VINCENT J, b Brooklyn, NY, July 12, 41; m 66; c 2. METEOROLOGY. Educ: NY Univ, BS, 63, MS, 65, PhD(meteorol), 70. Prof Exp: Lectr meteorol, NY Univ, 63-68, res asst prof, 69-73; sr res scientist oceanog & ocean meteorol, City Col New York Res Found, 73-74, higher educ assoc, Inst Marine & Atmospheric Sci, 74-76, ASSOC PROF OCEANOG & OCEAN METEOROL, CITY UNIV NEW YORK INST MARINE & ATMOSPHERIC SCI, CITY COL NEW YORK, 76- Concurrent Pos: Chmn, Seasat Simulation Studies Steering Comt, NASA, 76-; consult, Ocean Routes, Inc, Calif & Shell Develop Co. Mem: Am Meteorol Soc; Am Geophys Union. Res: Air-sea interaction; remote sensing of the marine environment; numerical weather prediction; ocean wave forecasting. Mailing Add: Inst of Marine & Atmospheric Sci Wave Hill 675 W 252 St Bronx NY 10471

CARDOSO, SERGIO STEINER, b Belem-Para, Brazil, June 26, 27; m 54; c 4. PHARMACOLOGY, BIOLOGICAL RHYTHMS. Educ: Univ Brazil, MD, 52; Univ Sao Paulo, PhD(pharmacol), 64. Prof Exp: From asst prof to assoc prof pharmacol, Univ Sao Paulo, 61-68, asst prof, 67-69, ASSOC PROF PHARMACOL, SCH MED, UNIV TENN, MEMPHIS, 69- Mem: Am Soc Pharmacol & Exp Therapeut. Res: Possible role of circadian mitotic rhythms as related to cancer chemotherapy. Mailing Add: Dept of Pharmacol Univ of Tenn Med Sch Memphis TN 38103

CARDULLO, MARIA ANN, b Boston, Mass, Apr 11, 42. CELL PHYSIOLOGY, BACTERIOLOGY. Educ: Emmanuel Col, Mass, AB, 63 & 64; Boston Col, MS, 67, PhD(biol), 71. Prof Exp: RES FEL BIOL, BOSTON COL, 71- Mem: Am Soc Microbiol; Sigma Xi. Res: Effects of d-camphor and other terpenes on bacterial growth, physiology, metabolism and ultrastructure; membrane structure and function; bacterial physiology. Mailing Add: Dept of Biol Boston Col Chestnut Hill MA 02167

CARDUS, DAVID, b Barcelona, Spain, Aug 6, 22; m 51; c 4. CARDIOLOGY, BIOMATHEMATICS. Educ: Univ Montpellier, BA, 42; Univ Barcelona, MD, 49. Prof Exp: Intern, Hosp Clin, Univ Barcelona, 49-50; resident, Sanitarium of Puig de Olena, Barcelona, 50-53; res assoc physiol, Postgrad Sch Cardiol, Med Sch, Univ Barcelona, 54-55; from instr to asst prof physiol & rehab, 60-65, assoc prof rehab, 65-69, assoc prof physiol, 65-73, PROF PHYSIOL, COL MED, BAYLOR UNIV, 73-, PROF REHAB & HEAD CARDIOPULMONARY LAB, 69-, DIR DIV BIOMATH, 70- Concurrent Pos: French Govt res fel, 53-54; Brit Coun res fel, 57; Inst Int Educ fel, Lovelace Found, 57-60; head work tolerance eval unit, Tex Inst Rehab & Res, 60; adj prof math sci, Rice Univ, 70- Mem: AAAS; Am Col Cardiol; Am Col Chest Physicians; AMA; NY Acad Sci. Res: Experimental exercise and respiratory physiology; mathematical and computer applications to the study of physiological systems; body comcomposition of humans with extensive muscular paralysis; physiology of urinary bladder. Mailing Add: Tex Inst for Rehab & Res PO Box 20095 Houston TX 77025

CARDWELL, ALVIN BOYD, b Lenoir City, Tenn, Oct 16, 02; m 30; c 2. SOLID STATE PHYSICS. Educ: Univ Chattanooga, BS, 25; Univ Wis, MS, 27, PhD(physics), 30. Hon Degrees: DSc, Univ Chattanooga, 61. Prof Exp: Asst physics, Univ Wis, 26-29; from asst prof to assoc prof, Tulane Univ, 30-36; prof, 36-73, head dept physics & physicist in charge, Eng Exp Sta, 37-53, physicist in charge, Agr Exp Sta, 47-53, assoc dean sch arts & sci, 53-55, dir bur gen res, 54-67, head dept physics & physicist in charge, Eng Exp Sta, 57-67, Agr Exp Sta, 57-73, EMER PROF PHYSICS, KANS STATE UNIV, 73- Concurrent Pos: Res physicist, Clinton Eng Works, Tenn Eastman Corp, Oak Ridge, Tenn, 43-46. Mem: AAAS; fel Am Phys Soc; Am Asn Physics Teachers. Mailing Add: Cardwell Hall Dept of Physics Kans State Univ Manhattan KS 66502

CARDWELL, DUDLEY H, b Lawyers, Va, Oct 21, 01; m 35; c 1. PETROLEUM GEOLOGY, STRUCTURAL GEOLOGY. Educ: Univ Va, BS, 23, MS, 25. Prof Exp: Field geologist, Sun Oil Co, 29-44, dist geologist, 44-59, sr geologist, 59-66; PETROL GEOLOGIST, WVA GEOL SURVEY, 67- Concurrent Pos: Mem, Comt Statist on Drilling & Nat Petrol Coun, 68-; mem, Oil Reserves Subcomt for WVa, Am Petrol Inst, 69-, chmn, 74-75. Mem: Am Asn Petrol Geologists. Res: Petroleum and structural geology. Mailing Add: WVa Geol Surv PO Box 879 Morgantown WV 26505

CARDWELL, JOE THOMAS, b Vernon, Tex, Feb 19, 22; m 42; c 2. DAIRY CHEMISTRY. Educ: Tex Technol Col, BS, 47, MS, 49; NC State Col, PhD, 56. Prof Exp: Vet voc teacher, Knox County Voc Sch, Tex, 47; instr dairy mfg, Tex Technol Col, 47-50; res asst dairy chem, NC State Col, 50-52; from asst prof to assoc prof dairy mfg, 52-64, PROF DAIRY MFG, MISS STATE UNIV & DAIRY CHEMIST, AGR EXP STA, 64- Mem: Am Dairy Sci Asn. Res: Milk plant sanitation; oxidized flavor development in milk; body and texture defects in cheddar cheese; frozen cultures; chocolate milk; flavor components of milk and cheese; new products developed from sweet whey. Mailing Add: Miss State Univ Col of Agr Drawer J C Mississippi State MS 39762

CARDWELL, PAUL H, b Metamora, Mich, Aug 24, 12; m 39; c 4. COLLOID CHEMISTRY. Educ: Cent Mich Teacher's Col, AB, 35; Univ Mich, PhD(chem), 41. Prof Exp: Chief chemist, Dowell Inc, 41-52, dir lab, 52-54; tech specialist, Dow Chem Co, Mich, 54-65, asst mgr apparatus & instruments, 65-67, mgr apparatus & instruments bus, 67-69; DIR RES, DEEPSEA VENTURES INC, 69- Mem: AAAS; Am Chem Soc; Am Inst Aeronaut & Astronaut; Am Inst Mining, Metall & Petrol Eng; Electrochem Soc. Res: Contact angles; colloid chemistry; resins; corrosion inhibitors; chemical removal of aluminum from oil wells; reproducible contact angles on reproducible metal surfaces; extractive metallurgy. Mailing Add: Zanoni VA 23191

CARDWELL, VERNON BRUCE, b Ft Morgan, Colo, Oct 8, 36; m 54; c 4. AGRONOMY, CROP PHYSIOLOGY. Educ: Colo State Univ, BS, 58, MS, 61; Iowa State Univ, PhD(crop prod), 67. Prof Exp: Res asst agron, Colo State Univ, 58-60, asst agronomist, 60-64; instr agron, Iowa State Univ, 64-67; asst prof, 67-73, ASSOC PROF AGRON, UNIV MINN, ST PAUL, 73- Mem: Am Soc Agron; Crop Sci Soc Am; Coun Agr Sci & Technol. Res: Seed physiology; seed quality; low temperature germination corn & soybeans. Mailing Add: Dept of Agron Univ of Minn St Paul MN 55108

CAREFOOT, THOMAS HENRY, invertebrate physiology, see 12th edition

CAREN, ROBERT POSTON, b Columbus, Ohio, Dec 25, 32; m 63. PHYSICS. Educ: Ohio State Univ, BS, 53, MS, 54, PhD(physics), 61. Prof Exp: Sr physicist, NAm Aviation Inc, Ohio, 59-60; instr physics, Ohio State Univ, 60-61; res scientist & sr mem res lab, 62-68, mgr infrared progs lab, 68-70, dir eng sci lab, 70-75, DIR PALO ALTO RES LABS, LOCKHEED MISSILES & SPACE CO, 75- Concurrent Pos: Lectr, Univ Santa Clara. Mem: Am Phys Soc; Am Asn Physics Teachers; assoc fel Am Inst Aeronaut & Astronaut; Am Defense Preparedness Asn. Res: Development of advanced infrared sensor systems and subsystems; radiation heat transfer theory; nuclear effects on materials; experimental low temperature research. Mailing Add: 658 Toyon Pl Palo Alto CA 94306

CARES, WILLIAM RONALD, b Dearborn, Mich, 41; m; c 1. SURFACE CHEMISTRY, CHEMICAL KINETICS. Educ: Case Inst Technol, BSChE, 63; Univ Ill, MS, 65, PhD(chem), 69. Prof Exp: Res fel, Rice Univ, 69-71; res chemist, US Naval Res Lab, 71-73; res chemist catalysis, Petro-Tex Chem Corp, 73-76; SR RES CHEMIST, PULLMAN-KELLOGG, 76- Mem: Am Chem Soc; Am Inst Chem Engrs; The Chem Soc; Catalysis Soc. Res: Kinetics and mechanics of heterogeneous catalytic reactions. Mailing Add: Pullman-Kellogg R&D Ctr 16200 Park Row-Indust Park Ten Houston TX 77084

CARESS, EDWARD ALAN, b Columbus, Nebr, Feb 6, 36; m 61; c 3. ORGANIC CHEMISTRY. Educ: Dartmouth Col, AB, 58; Univ Rochester, PhD(org chem), 63. Prof Exp: Fel, Univ Rochester, 63; res assoc org chem, Mass Inst Technol, 63-65; asst prof, 65-70, ASSOC PROF CHEM, GEORGE WASHINGTON UNIV, 70-, ASST DEAN GRAD SCH ARTS & SCI, 71- Mem: Am Chem Soc; The Chem Soc. Res: Organic reaction mechanisms; photochemistry of organic compounds; resolution of organic compounds through use of platinum complexes; structure determination of natural products. Mailing Add: Dept of Chem George Washington Univ Washington DC 20006

CARET, ROBERT LAURENT, b Biddeford, Maine, Oct 7, 47; m 69. ORGANIC CHEMISTRY, ENVIRONMENTAL CHEMISTRY. Educ: Suffolk Univ, BA, 69; Univ NH, PhD(org chem), 74. Prof Exp: Res assoc, Bio-Res Inst, 67-69; teacher chem, Rumford High Sch, 69-70; teaching asst, Univ NH & Suffolk Univ, 66-72; vis asst prof, 74-75, INSTR CHEM, TOWSON STATE COL, 75- Concurrent Pos: Lectr, Suffolk Univ, 72-73. Mem: Sigma Xi; Am Chem Soc; Am Asn Univ Prof. Res: Stereochemistry, conformational analysis; nuclear magnetic resonance including Carbon-13 nmr and their use in the study of organosulfur compounds. Mailing Add: Dept of Chem Towson State Col Towson MD 21204

CARETTO, ALBERT A, JR, b Baldwin, NY, May 16, 28; m 60; c 2. NUCLEAR CHEMISTRY, PHYSICAL CHEMISTRY. Educ: Rensselaer Polytech Inst, BS, 50; Univ Rochester, PhD(chem), 54. Prof Exp: Nuclear chemist, Brookhaven Nat Lab, 54-56; nuclear chemist, Univ Calif, Berkeley, 56-57, res chemist, Livermore, 58-59; asst prof chem, 57-58, from asst prof to assoc prof, 59-67, chmn dept, 70-74, PROF CHEM, CARNEGIE-MELLON UNIV, 67- Concurrent Pos: Sabbatical Award, Europ Orgn Nuclear Res, 64-65 & Europ Ctr Nuclear Res, Genoa, Switz, 74-75. Mem: AAAS; Am Chem Soc; Am Phys Soc. Res: Nuclear reactions induced with high energy particles; nuclear spectroscopy; radiochemical effects of recoil atoms. Mailing Add: Dept of Chem Carnegie-Mellon Univ Pittsburgh PA 15213

CAREW, DAVID P, b Monson, Mass, Oct 21, 28; m 51; c 3. PHARMACY. Educ: Mass Col Pharm, BS, 52, MS, 54; Univ Conn, PhD(pharm), 58. Prof Exp: Asst instr pharmacog, Univ Conn, 54-57; from asst prof to assoc prof, 57-65, PROF PHARMACOG, UNIV IOWA, 65-, ASST DEAN, COL PHARM, 75- Concurrent Pos: Collab scientist, UN Cannabis Res Prog; mem adv panel, US Pharmacopaeia, 71- Mem: AAAS; Am Soc Pharmacog (vpres, 64-65, pres, 65-66); Am Pharmaceut Asn; fel Acad Pharmaceut Sci; Int Plant Tissue Cult Asn. Res: Natural product research; products with therapeutic activity; plant tissue culture and biosynthesis; antibiotics. Mailing Add: Col of Pharm Univ of Iowa Iowa City IA 52242

CAREW, JOHN, b New York, NY, Feb 12, 20; m 44; c 4. HORTICULTURE. Educ: Pa State Col, BS, 40; Cornell Univ, PhD(veg crops), 47. Prof Exp: Asst prof veg crops, Cornell Univ, 47-52, prof, 52-55; PROF HORT, MICH STATE UNIV, 55-, CHMN DEPT, 62- Prof Exp: Vis scientist, Orgn Econ Coop & Develop, Eng, 61-62; mem joint adv coun, Univ Nigeria, 62-65. Mem: Fel AAAS; fel Am Soc Hort Sci (pres-elect, 64-65, pres, 65-66); Int Soc Hort Sci. Res: Horticultural science. Mailing Add: Dept of Hort Mich State Univ East Lansing MI 48823

CAREW, LYNDON BELMONT, JR, b Lynn, Mass, Nov 27, 32; m 60; c 2. ANIMAL NUTRITION, BIOCHEMISTRY. Educ: Univ Mass, BS, 55; Cornell Univ, PhD(animal nutrit), 61. Prof Exp: Res asst poultry nutrit, Cornell Univ, 55-58, res assoc, 58-59, res asst, 59-61; tech dir, Colombian Nat Poultry Prog & Animal Nutrit Lab, Colombian Agr Prog, Rockefeller Found, Bogota, 61-65; sr res assoc poultry sci, Cornell Univ, 65-66; head poultry res sect, Hess & Clark Div, Richardson Merrill Corp, Ohio, 66-69; assoc prof animal sci & nutrit, 69-75, PROF ANIMAL SCI & NUTRIT, UNIV VT, 75- Concurrent Pos: Mem tech comt, Animal Nutrit Res Coun, 72- Mem: AAAS; Am Inst Nutrit; Poultry Sci Asn; Soc Exp Biol & Med; NY Acad Sci. Res: Essential fatty acid deficiency and endocrine effect; fatty acid and phospholipid metabolism; metabolizable energy; energy metabolism; general poultry nutrition; nutritive properties of fats; Lat Am poultry science; light and pineal biochemistry. Mailing Add: Univ of Vt Dept of Animal Sci Biores Lab 655 Spear St Burlington VT 05401

CAREY, ANDREW GALBRAITH, JR, b Baltimore, Md, Apr 11, 32; m 57; c 2. BIOLOGICAL OCEANOGRAPHY. Educ: Princeton Univ, AB, 55; Yale Univ, PhD(zool), 62. Prof Exp: Asst prof oceanog, 61-71, ASSOC PROF OCEANOG, ORE STATE UNIV, 71- Concurrent Pos: Marshal fel, Denmark, 70. Mem: AAAS; Am Soc Limnol & Oceanog; Am Soc Zool; Ecol Soc Am; Marine Biol Asn UK. Res: Marine benthic ecology; community, energetics, deep sea; invertibrate zoology; polar ecology. Mailing Add: Sch of Oceanog Ore State Univ Corvallis OR 97331

CAREY, BENJAMIN WATSON, b Pleasant Hill, Ill, Feb 26, 07; m 35; c 1. PEDIATRICS. Educ: Univ Ill, BS, 26; Harvard Univ, MD, 32. Prof Exp: Intern, Johns Hopkins Hosp, 32-33; intern & resident, Children's Hosp, Mass, 33-35; instr bact & pediat, Harvard Med Sch, 35-38; asst prof pediat, Wayne Univ, 38-41; asst dir, Lederle Labs, Am Cyanamid Co, 41-46, dir, 46-54, med dir, 54-68; admin asst med affairs, Pascack Valley Hosp, Westwood, NJ, 68-72; RETIRED. Concurrent Pos: Dir bact lab, Children's Hosp, Mass, 35-38; dir bact & chem lab, Children's Hosp, Mich, 38-41. Mem: AMA; Am Acad Pediat; Soc Pediat Res; NY Acad Sci. Res: Medical bacteriology; human clinical research in chemotherapy; infectious diseases; serum and vaccine production. Mailing Add: 142 Second Ave Westwood NJ 07675

CAREY, BERNARD JOSEPH, b Pittsburgh, Pa, Feb 28, 41; m 66; c 2. COMPUTER SCIENCE. Educ: Univ Pittsburgh, BS, 62; Univ Calif, Santa Barbara, MS, 69, PhD(elec eng comput sci), 71. Prof Exp: Design engr, Guid & Control, Litton Industs, 63-68; res asst speech processing, Univ Calif, Santa Barbara, 69-71; ASST PROF COMPUT SCI, UNIV CONN, 71- Concurrent Pos: Consult microprocessing syst, Rogers Corp, 75- Mem: Asn Comput Mach; Inst Elec & Electronics Engrs. Res: Computer speech and signal processing; digital systems design; applications of microprocessors. Mailing Add: U-157 Univ of Conn Storrs CT 06268

CAREY, DAVID CROCKETT, b Montclair, NJ, Oct 2, 39; m 69. HIGH ENERGY PHYSICS, PARTICLE PHYSICS. Educ: Mass Inst Technol, BS, 62; Univ Mich, MS, 64, PhD(physics), 67. Prof Exp: Technician, Forrestal Res Ctr, Princeton-Penn Accelerator, 59-60; instr physics & math, Upsala Col, 62; res assoc physics, City Col New York, 67-69; PHYSICIST, NAT ACCELERATOR LAB, 69- Mem: AAAS; Am Phys Soc. Res: Theoretical and experimental investigations into strong interaction dynamics; design of particle accelerators and external beams. Mailing Add: Nat Accelerator Lab Batavia IL 60510

CAREY, FRANCIS ARTHUR, b Philadelphia, Pa, May 28, 37; m 63; c 3. ORGANIC CHEMISTRY. Educ: Drexel Univ, BS, 59; Pa State Univ, PhD(org chem), 63. Prof Exp: NIH fel, Harvard Univ, 63-64; asst prof, 66-71, ASSOC PROF CHEM, UNIV VA, 71- Mem: Am Chem Soc. Res: Stereochemistry; generation of reactive intermediates; organo-silicon chemistry. Mailing Add: Dept of Chem Univ of Va Charlottesville VA 22903

CAREY, GEORGE WARREN, b New York, NY, Jan 1, 27; m 50; c 4. URBAN GEOGRAPHY, ANALYTICAL STATISTICS. Educ: Columbia Univ, AB, 48, AM, 49, EdD(geog), 64. Prof Exp: Pop Instr Mat Proj grant, Teachers Col, Columbia Univ, 64-65; US Dept Water Resources Res grants, 67-69 & 70-72; instr geog, Univ PR, Mayaguez, 61-62; instr, Teachers Col, Columbia Univ, 63-64, from asst prof to assoc prof urban geog, 64-68; prof urban geog in planning, Livingston Col, Rutgers Univ, 68-74, actg dean, Livingston Col, 73-74; PROF URBAN STUDIES, RUTGERS UNIV, NEWARK, 74- Mem: Asn Am Geog; Regional Sci Asn. Res: Interface between urban, social, economic and managerial systems with environmental ecological systems. Mailing Add: Dept of Urban Studies Rutgers Univ Newark NJ 07102

CAREY, JOHN HUGH, b Windsor, Ont, Jan 28, 47; m 70; c 2. PHOTOCHEMISTRY. Educ: Univ Windsor, BSc, 70, MSc, 72; Carleton Univ, PhD(chem), 74. Prof Exp: NAT RES COUN CAN FEL, CAN CENTRE INLAND WATERS, 74- Mem: Chem Inst Can. Res: Photochemistry of natural waters and of solution-sediment interface; photochemical aspects of pollution and water treatment processes. Mailing Add: Can Centre Inland Waters 867 Lakeshore Rd Burlington ON Can

CAREY, LARRY CAMPBELL, b Coal Grove, Ohio, Nov 5, 33; m 56; c 4. SURGERY. Educ: Ohio State Univ, BSc, 55, MD, 59. Prof Exp: Intern surg, New York Hosp, Cornell Univ, 59-60; resident, Marquette Integrated Residency Prog, 60-64, chief admin resident, Marquette Univ, 64-65, from asst prof to prof surg, Sch Med, Univ Pittsburgh, 68-74; PROF SURG & CHMN DEPT, COL MED, OHIO STATE UNIV, 75- Concurrent Pos: Markle Scholar acad med, 65-70; William S Middleton lectr, Wis State Med Soc, 65; consult, Milwaukee County Gen Hosp, Wis, 65-, Milwaukee Lutheran Hosp, 65-, Columbia Hosp, 65- & St Luke's Hosp, 68-; asst clin prof, Boston Univ, 66-67. Mem: AAAS; fel Am Col Surg; AMA; Soc Univ Surg; Soc Surg Alimentary Tract. Res: Pancreatic physiology; bioelectric phenomena; shock. Mailing Add: Dept of Surg of Col of Med Ohio State Univ Columbus OH 43210

CAREY, PAUL L, b Arrowsmith, Ill, Nov 4, 23; m 48; c 5. BIOCHEMISTRY. Educ: Ill Wesleyan Univ, BS, 48; Kansas State Col, MS, 50; Purdue Univ, PhD(biochem), 58. Prof Exp: Jr biochemist, Smith, Kline & French Labs, 50-53; asst, Purdue Univ, 53-54; mgr spec chows lab, Ralston Purina, 58-62; chemist, Penick & Ford, Ltd, 62-65; biochemist, R J Reynolds Co, 65-68, group leader, 68-70; ASSOC SCIENTIST, RALSTON PURINA CO, 70- Concurrent Pos: Instr, Winston Salem State Teachers Col, 67. Mem: AAAS; Am Chem Soc. Res: Nutrition; chemistry of biological materials; process research and development. Mailing Add: Cent Res Dept Ralston Purina Co 900 Checkerboard Sq Plaza St Louis MO 63188

CAREY, WARREN F, microbiology, see 12th edition

CARFAGNO, DANIEL GAETANO, b Syracuse, NY, Aug 17, 35; m 58; c 2. PHYSICAL CHEMISTRY, INORGANIC CHEMISTRY. Educ: Le Moyne Col, NY, BS, 57; Syracuse Univ, PhD(phys chem), 65. Prof Exp: Sr res chemist, 65-67, res group leader, 67-69, RES SPECIALIST, MONSANTO RES CORP, MIAMISBURG, OHIO, 69- Mem: Am Chem Soc. Res: Phase equilibria; x-ray diffraction; differential thermal analysis; purification of alkaline earth metals; physical properties of plutonium and its compounds; isotopic fuels; chemical vapor deposition; calorimetric measurements. Mailing Add: Monsanto Res Corp Mound Lab PO Box 32 Miamisburg OH 45342

CARGES, GERARD LEO, organic chemistry, physical chemistry, see 12th edition

CARGILL, DAVID INNES, b Perth, Scotland, July 5, 38; m 61; c 5. BIOCHEMISTRY. Educ: St Andrews Univ, BSc, 60, PhD(steroid metab), 64. Prof Exp: Asst lectr biochem, Queens Col, St Andrews Univ, 61-64; res assoc steroid metab, Col Physicians & Surgeons, Columbia Univ, 64-66; scientist, Warner-Lambert Res Inst, 66-67; sr biochemist, Ciba Pharmaceut Co, 67-71, MGR ATHEROSCLEROSIS RES, CIBA-GEIGY CORP, 71- Res: Platelet physiology; atherosclerosis; steroid hormone metabolism. Mailing Add: Ciba-Geigy Corp MR2 108 Ardsley NY 10502

CARGILL, ROBERT LEE, JR, b Marshall, Tex, Sept 11, 34; m 65; c 3. ORGANIC CHEMISTRY. Educ: Rice Univ, BA, 55; Mass Inst Technol, PhD(org chem), 60. Prof Exp: NIH fel org chem, Univ Calif, Berkeley, 60-62; from asst prof to assoc prof chem, 67-73, PROF CHEM, UNIV SC, 73- Mem: Am Chem Soc; The Chem Soc. Res: Photochemistry; isomerization reactions, especially in small ring compounds; structure and synthesis of natural products; synthesis and properties of highly strained molecules. Mailing Add: Dept of Chem Univ of SC Columbia SC 29208

CARGO, DAVID GARRETT, b Pittsburgh, Pa, Dec 28, 24; m 49; c 6. ZOOLOGY. Educ: Univ Pittsburgh, BS, 49, MS, 50. Prof Exp: Biologist, State of Md, 50-60; RES ASSOC MARINE BIOL, CHESAPEAKE BIOL LAB, UNIV MD, SOLOMONS, 60- Mem: Am Soc Zool. Res: Biology and ecology of marine invertebrates, especially Crustacea and coelenterates. Mailing Add: Chesapeake Biol Lab Univ of Md Box 38 Solomons MD 20688

CARGO, DOUGLAS BRUCE, b Sewickley, Pa, Aug 6, 43; m 70; c 3. ENVIRONMENTAL SCIENCES, GEOGRAPHY. Educ: Clarion State Col, BS, 61; Kent State Univ, MA, 69; Univ Chicago, PhD(geog), 75. Prof Exp: Res analyst planning, Seven County Land Use & Transp Study, Cleveland, Ohio, 65-67; instr, Eastview Jr High Sch, Bath, Ohio, 67-69; instr geog, Univ Akron, 69-71; res asst, Univ Chicago, 71-75; ASST PROF ENVIRON SCI, UNIV TEX, DALLAS, 75- Concurrent Pos: Jr planner, Tri-County Regional Planning Comn, Akron, Ohio, 65-69; consult, Areawide Waste Treat Mgt Planning Tech Sem, Region VI, Environ Protection Agency, 75- Res: Solid waste generation rates, land use and the analysis of environmental pollutants through space and time. Mailing Add: Dept of Environ Sci Univ of Tex at Dallas Box 688 Richardson TX 75080

CARGO, GERALD THOMAS, b Dowagiac, Mich, Mar 2, 30; m 56; c 2. MATHEMATICAL ANALYSIS. Educ: Univ Mich, BBA, 52, MS, 53, PhD(math), 59. Prof Exp: Asst math, Willow Run Res Ctr, Univ Mich, 55-56; asst prof, 59-63, ASSOC PROF MATH, SYRACUSE UNIV, 63- Concurrent Pos: Nat Res Coun-Nat Bur Stand resident res assoc, 61-62; NSF res grant, 63- Mem: Am Math Soc; Math Asn Am. Res: Boundary behavior of analytic functions; inequalities and convex functions; Hardy classes and Nevanlinna theory; Tauberian theorems; mathematical microbiology; zeros of polynomials. Mailing Add: 200 Carnegie Hall Syracuse Univ Syracuse NY 13210

CARHART, HOMER WALTER, organic chemistry, see 12th edition

CARHART, RAYMOND, audiology, acoustics, deceased

CARHART, RICHARD ALAN, b Evanston, Ill, Aug 30, 39; m 60; c 3. THEORETICAL HIGH ENERGY PHYSICS, MATHEMATICAL PHYSICS. Educ: Northwestern Univ, BA, 60; Univ Wis, MA, 62, PhD(physics), 64. Prof Exp: Res assoc physics, Univ Wis, 64 & Brookhaven Nat Lab, 64-66; asst prof, 66-70, ASSOC PROF PHYSICS, UNIV ILL, CHICAGO CIRCLE, 70- Mem: Am Phys Soc. Res: Parity violations in quantum electro-dynamics; nucleon electromagnetic form factors and other topics in theoretical elementary particle physics; nonlinear ordinary and partial differential equations of physics. Mailing Add: Dept of Physics Box 4348 Univ of Ill at Chicago Circle Chicago IL 60680

CARIANI, ANTHONY R, geology, deceased

CARIS, JOHN CLAYTON, physics, see 12th edition

CARITHERS, JEANINE RUTHERFORD, b Boone, Iowa, Sept 26, 33; m 53; c 3. ANATOMY, ENDOCRINOLOGY. Educ: Iowa State Univ, BS, 56, MS, 65; Univ Mo, PhD(anat), 68. Prof Exp: Asst prof, 68-72, ASSOC PROF ANAT, IOWA STATE UNIV, 72- Mem: Am Asn Anatomists; Am Asn Vet Anatomists; World Asn Vet Anatomists; Sigma Xi; AAAS. Res: Ultrastructure of liver and hypophysis in stress-prone pigs; endocrine and neuroendocrine responses of adrenals and hypophysis in states of stress and altered water balance. Mailing Add: Dept of Vet Anat Iowa State Univ Ames IA 50010

CARITHERS, WILLIAM CORNELIUS, JR, b Toccoa, Ga, Nov 20, 42; m 65; c 2. EXPERIMENTAL HIGH ENERGY PHYSICS. Educ: Mass Inst Technol, BS, 64; Yale Univ, MS, 66, PhD(physics), 68. Prof Exp: Res assoc physics, Columbia Univ, 68-70, asst prof, 70-73; asst prof physics, Univ Rochester, 73-75; FEL DIV PHYSICS, LAWRENCE BERKELEY LAB, 75- Concurrent Pos: Alfred P Sloane Found fel, 72-74. Res: High energy electron-positron annihilation physics. Mailing Add: Lawrence Berkeley Lab Univ of Calif Berkeley CA 94720

CARL, JAMES DUDLEY, b Centralia, Ill, June 4, 35; m 61. GEOCHEMISTRY. Educ: Mo Sch Mines, BS, 57; Univ Ill, MS, 60, PhD(geol), 61. Prof Exp: Asst prof geol, Cent Mo State Col, 61-63 & Ill State Univ, 63-68; ASSOC PROF GEOL, STATE UNIV NY COL POTSDAM, 68- Mem: Geol Soc Am; Geochem Soc; Mineral Soc Am. Res: Petrology; mineralogy; geochemical and structural studies of igneous and metamorphic rocks from coastal Maine and the Northwest Adirondacks, New York; major and trace element variation. Mailing Add: Dept of Geol State Univ of NY Potsdam NY 13676

CARL, PHILIP LOUIS, b Cleveland, Ohio, Dec 6, 39; m 64. MOLECULAR GENETICS. Educ: Harvard Univ, BA, 61; Univ Calif, Berkeley, MS, 63, PhD(biophys), 68. Prof Exp: Jane Coffin Childs Mem Fund fel med res, 68-70; ASST PROF MICROBIOL, UNIV ILL, URBANA, 70- Mem: AAAS; Am Soc Microbiol. Res: Molecular biology of DNA replication. Mailing Add: Dept of Microbiol Univ of Ill Urbana IL 61801

CARLANDER, KENNETH DIXON, b Gary, Ind, May 25, 15; m 39. FISHERIES. Educ: Univ Minn, BS, 36, MS, 38, PhD(zool), 43. Prof Exp: Lab technician, Work Proj Admin, Univ Minn, 36-38; fishery biologist, Minn State Dept Conserv, 38-46; from asst prof to assoc prof zool, 46-57, PROF ZOOL, IOWA STATE UNIV, 57- Concurrent Pos: Leader, Iowa Coop Fishery Res Unit, 46-65; consult, Ford Found, Egypt, 65-66. Mem: AAAS; Am Fisheries Soc (pres, 60); Am Soc Limnol & Oceanog; Am Soc Ichthyol & Herpet; Biomet Soc. Res: Fishery biology; fish population estimation; limnology; age and growth of fishes. Mailing Add: Dept of Animal Ecol Iowa State Univ Ames IA 50011

CARLBERG, DAVID MARVIN, b Los Angeles, Calif, Feb 9, 34; m 62; c 2. MICROBIOLOGY, VIROLOGY. Educ: Univ Calif, Los Angeles, BA, 56, PhD(microbiol), 63. Prof Exp: Chemist, Riker Labs, Inc, 56-58; head formula off, Rexall Drug & Chem Co, 58; res engr, McDonnell Douglas Corp, 63-65; mem staff, Hughes Aircraft Co, 65-66; from asst prof to assoc prof, 66-75, PROF MICROBIOL,

CALIF STATE UNIV LONG BEACH, 75- Mem: Am Soc Microbiol; Am Inst Aeronaut & Astronaut; AAAS. Res: Microbial genetics; collection, detection and analysis of microorganisms in air and on surfaces; bio-instrumentation; exobiology. Mailing Add: Dept of Microbiol Calif State Univ Long Beach CA 90840

CARLBORG, FRANK WILLIAM b Chicago, Ill, Nov 23, 28; m 56; c 3. STATISTICS. Educ: Ripon Col, BA, 50; Univ Ill, MS, 54; Univ Chicago, PhD(statist), 64. Prof Exp: Mathematician, Aerial Measurements Lab, Northwestern Univ, 54-55; consult opers res, Booz-Allen & Hamilton, Inc, 55-57; asst prof math, Rockford Col, 57-61; instr statist, Grad Sch Bus, Univ Chicago, 63-64; assoc prof math, Northern Ill Univ, 64-67; INDEPENDENT STATIST CONSULT, 67- Mem: Am Statist Asn; Inst Math Statisticians. Res: Experimental design and analysis of experimental results. Mailing Add: 400 S Ninth St St Charles IL 60174

CARLE, KENNETH ROBERTS, b Keene, NH, Sept 16, 29; m 57; c 4. PHYSICAL ORGANIC CHEMISTRY. Educ: Middlebury Col, AB, 51; Univ NH, MS, 53; Del Univ, PhD(chem), 55. Prof Exp: Asst, Gen Chem Lab, Univ NH, 51-52 & Qual Chem Lab, Del Univ, 52-53; res chemist org chem, Am Cyanamid Co, 55-59; assoc prof chem, 59-62, head dept, 62-69, PROF CHEM, HOBART & WILLIAM SMITH COL, 62- Concurrent Pos: Instr, Bridgeport Eng Inst, 57-; Fulbright-Hays lectr, Univ Manila, 66-67; vis res prof, Silliman Univ, Philippines, 74-75. Mem: AAAS; Am Chem Soc. Res: Organic synthesis; polymers; kinetics; rubber chemistry; structure determination of myeloma proteins. Mailing Add: Dept of Chem Hobart & William Smith Col Geneva NY 14456

CARLETON, ALBERT EUGENE, agronomy, plant breeding, see 12th edition

CARLETON, BLONDEL HENRY, b Portland, Ore, Dec 8, 04; m 35; c 5. PHYSIOLOGY. Educ: Univ Ore, BA, 26; Univ Rochester, PhD(physiol), 36. Prof Exp: Asst zool, Univ Ore, 26-27; asst physiol, Sch Med, Univ Rochester, 33-36; instr biol & physics, Ga Teachers Col, 36-38; from asst prof to prof zool, 38-75, chmn fac biol, 67-70, EMER PROF ZOOL, UNIV PORTLAND, 75- Mem: Sigma Xi; Am Asn Biol Teachers. Res: Excitability of amphibian muscle; narcosis and excitability in nerve; oxygen metabolism of muscle and nerve. Mailing Add: 6705 N Wilbur Ave Portland OR 97217

CARLETON, HERBERT RUCK, b Rockville Centre, NY, Dec 5, 28; m 51; c 3. OPICAL PHYSICS, SOLID STATE SCIENCE. Educ: Univ Southern Calif, BA, 58; Cornell Univ, PhD(theoret physics), 64. Prof Exp: Test methods engr, Sperry Gyroscope Corp, 49-54; design engr, Gilfillan Bros, Inc, 54-55; sr engr, Canoga Corp, 55-56; prin engr, Bendix Pac Corp, Calif, 56-58; staff mem, Res Ctr, Sperry Rand Corp, 62-67; assoc prof mat sci, 67-73, joint assoc prof mat sci & elec sci, 73-75, JOINT PROF MAT SCI & ELEC SCI, STATE UNIV NY, STONY BROOK, 75- Concurrent Pos: Consult, Bendix Pac Corp, 59-60 & Defense Systs Dept, Gen Elec Corp, 60-62; vis scientist, IBM San Jose Res Lab, 74, consult, 74-75. Mem: Optical Soc Am; Am Phys Soc; sr mem, Inst Elec & Electronics Eng. Res: Ultrasonic amplification in piezoelectric semiconductors; optical properties of materials; strain-optic properties; non-crystalline solids; elastic properties of crystals; brillouin scattering; hypersonics. Mailing Add: Dept of Mat Sci State Univ of NY Stony Brook NY 11790

CARLETON, NATHANIEL PHILLIPS, b Burlington, Vt, Mar 16, 29; m 51; c 4. ASTROPHYSICS. Educ: Harvard Univ, AB, 51, AM, 52, PhD, 56. Prof Exp: From instr to asst prof physics, 56-62, LECTR, HARVARD UNIV, 62-, PHYSICIST, SMITHSONIAN ASTROPHYS OBSERV, 62- Mem: Fel Am Phys Soc; Am Astron Soc. Res: Atomic physics; physics of planetary atmospheres and the interstellar medium. Mailing Add: Perkin 307 60 Garden St Harvard Univ Cambridge MA 02138

CARLETON, RALPH KIMBALL, inorganic chemistry, deceased

CARLETON, RICHARD ALLYN, b Providence, RI, Mar 15, 31; m 54; c 4. CARDIOLOGY. Educ: Dartmouth Col, AB, 52; Dartmouth Med Sch, cert, 53; Harvard Med Sch, MD, 55. Prof Exp: Intern med, Boston City Hosp, 55-56, asst resident, 56-57, sr resident, 57-58, sr med resident, Metab Sect, 59-60; from asst prof to assoc prof, Med Sch, Univ Ill, 62-68; assoc dir cardiol, Sect Cardiorespiratory Dis, Rush-Presby-St Luke's Med Ctr, 62-68, dir cardiol, 68-72; prof med, Rush Med Col, 68-72; prof, Univ Calif, San Diego, 72-74; PROF MED & CHMN DEPT, DARTMOUTH MED SCH, 74- Concurrent Pos: Res fel, Harvard Med Sch, 58-59; Burton E Hamilton res fel, Cardiol Div, Thorndike Mem Labs, Boston City Hosp, Mass, 58-59; teaching fel, Sch Med, Tufts Univ, 59-60; asst med, Harvard Med Sch, 56-58; from asst attend physician to assoc attend physiciand, Rush-Presby-St Luke's Med Ctr, 62-67, attend physician, 67-; fel coun cardiol, Am Heart Asn. Mem: Fel Am Col Cardiol; Am Soc Clin Invest; fel Am Col Physicians; Asn Univ Cardiologists. Res: Indicator dilution methods; ventricular volume and function under autonomic influence; clinical and physiologic aspects of cardiac pacing. Mailing Add: Dept of Med Dartmouth Med Sch Hanover NH 03755

CARLEY, DAVID DON, b Battle Creek, Mich, July 2, 35. THEORETICAL PHYSICS. Educ: Western Mich Univ, BS, 57; Univ Mich, MS, 58; Univ Fla, PhD(physics), 63. Prof Exp: Proj engr, Noise & Vibration Lab, Gen Motors Corp, 58-60; res assoc, Univ Fla, 64; asst prof, 64-67, ASSOC PROF PHYSICS, WESTERN MICH UNIV, 67- Mem: Am Phys Soc. Res: Statistical mechanics; theory of fluids. Mailing Add: Dept of Physics Western Mich Univ Kalamazoo MI 49001

CARLEY, DAVID WILCOX, b Galesburg, Ill, June 21, 22; m 45; c 2. ANALYTICAL CHEMISTRY. Educ: Knox Col (Ill), MA, 44; Univ Ill, MS, 47, PhD(anal chem), 52. Prof Exp: Res assoc, George Univ, 44; chem process engr, C F Braun Co, 47-49; sales mgr, Wilson Meyer Co, 52-59; assoc prof chem, 60-66, PROF CHEM, RIPON COL, 66-, CHMN DEPT, 68- Concurrent Pos: NSF fel, 62, consult, 63 & assoc prog dir, Inst Sect, 63-64; vchmn, State of Wis Air Pollution Control Adv Coun, 67-; vis scientist, Cambridge Univ, 68; fel Univ Colo, 69; dir, NSF Summer Insts & Col Sci Improv Plan, Ripon Col; consult, Photoengravers; cong sci counr, Am Chem Soc, 74- Mem: AAAS; Am Chem Soc; Am Inst Chemists. Res: Hydrogen bonding in DNA molecule; development and use of instrumental methods. Mailing Add: Dept of Chem Ripon Col Ripon WI 54971

CARLEY, HAROLD EDWIN, b Syracuse, NY, July 3, 42; m 65; c 3. PLANT PATHOLOGY, SOIL SCIENCE. Educ: Cornell Univ, BS, 64; Univ Idaho, MS, 66; Univ Minn, St Paul, PhD(plant path), 69. Prof Exp: Proj leader agr bactericide-viricide develop, 69-71, GROUP LEADER FUNGICIDES, ROHM & HAAS CO, 71- Mem: AAAS; Am Phytopath Soc. Res: Allelopathy; biological and cultural control of soil-borne plant diseases; chemotherapy of phytopathogenic organisms. Mailing Add: Rohm & Haas Co Spring House PA 19477

CARLILE, CLAYTON GEORGE, b Meadville, Pa, July 13, 32; m 54; c 3. ORGANIC CHEMISTRY. Educ: Vanderbilt Univ, BA, 54, PhD(org chem), 61. Prof Exp: Res chemist, Org Chem Div, Am Cyanamid Co, 60-67, sr chemist, 67-68; assoc prof, 68-70, PROF CHEM, RADFORD COL, 70- Mem: Am Chem Soc. Res: Mechanisms of organic reactions; vapor phase and high pressure reactions; light absorbing compounds; process research. Mailing Add: Dept of Chem Radford Col Radford VA 24141

CARLIN, CHARLES HERRICK, b Rockford, Ill, Jan 25, 39; m 65; c 2. ORGANIC CHEMISTRY. Educ: Carthage Col, AB, 61; Johns Hopkins Univ, MA, 63, PhD(org chem), 66. Prof Exp: Asst prof chem, 66-67, ASSOC PROF CHEM, CARLETON COL, 67- Concurrent Pos: Sci adv, Food & Drug Admin, 69- Mem: AAAS; Am Chem Soc; Sigma Xi. Res: Mechanism of action of pesticides and trace metals in ecological systems. Mailing Add: Dept of Chem Carleton Col Northfield MN 55057

CARLIN, FRANCES, b Springfield, Mass, Aug 21, 06. FOOD CHEMISTRY. Educ: Columbia Univ, BS, 31, MA, 33; Cornell Univ, MS, 43; Iowa State Univ, PhD(foods), 47. Prof Exp: Head phys educ, St Catherine Col, 31-32; asst buyer, R H Macy, NY, 33-34; head health & phys educ, St Mary-of-the-Woods, 34-38; instr food & nutrit & head phys educ, Georgetown High Sch & Jr Col, 38-41; asst chem & foods, NY State Col Home Econ, 41-43; asst prof foods & nutrit, Hood Col, 43-45; from asst prof to assoc prof, 45-53, PROF FOOD & NUTRIT & FOOD TECHNOL, IOWA STATE UNIV, 54- Mem: AAAS; Poultry Sci Asn; Inst Food Technologists; Am Home Econ Asn. Res: Effect of freezing on tenderness and on ice crystal formation in poultry after various periods of aging; vitamin research; effect of processing food on safety functional properties and flavor; development of new products. Mailing Add: Dept Food Nutrit Food Technol 108 MacKay Hall Iowa State Univ Ames IA 50010

CARLIN, RICHARD LEWIS, b Boston, Mass, July 28, 35; m 59; c 2. INORGANIC CHEMISTRY. Educ: Brown Univ, BS, 57; Univ Ill, MS, 59, PhD, 60. Prof Exp: From instr to asst prof chem, Brown Univ, 60-67; assoc prof, 67-70, PROF CHEM, UNIV ILL, CHICAGO CIRCLE, 70- Mem: Am Chem Soc; Am Phys Soc; The Chem Soc. Res: Electronic structure of transition metal compounds; synthetic inorganic chemistry; magnetism and spectroscopy; magnetism at low temperatures. Mailing Add: Dept of Chem Univ of Ill Chicago Circle Chicago IL 60680

CARLIN, ROBERT BURNELL, b St Paul, Minn, Nov 13, 16; m 52; c 5. ORGANIC CHEMISTRY. Educ: Univ Minn, BCh, 37, PhD(org chem), 41. Prof Exp: Asst, Univ Minn, 37-40; Lalor Found fel, Univ Ill, 41-42, instr org chem, 42-43; instr, Univ Rochester, 43-46; assoc prof, 46-52, head dept chem, 60-67 & assoc dean, Col Eng & Sci, 67-70, BECKER PROF ORG CHEM, CARNEGIE-MELLON UNIV, 52- Concurrent Pos: Consult, Koppers Co, Inc, 52-; Reilly lectr, Univ Notre Dame, 56; chmn, Gordon Res Conf Org Reactions & Processes, 56; mem comt, Off Ord Res, 56-59; eval postdoctoral fel applns, Nat Acad Sci-Nat Res Coun, 55-58, NSF, 55-59 & Air Force Off Sci Res, 66-68. Mem: Fel AAAS; Am Chem Soc. Res: Molecular rearrangements, such as, benzidine and alkyl group migrations; Fischer indole synthesis; polymer synthesis and structure; synthesis of bicyclic N heterocycles. Mailing Add: Dept of Chem Mellon Inst Carnegie-Mellon Univ Pittsburgh PA 15213

CARLISLE, ALAN, b Blackpool, Eng, Feb 11, 24; m 57; c 3. ECOLOGY. Educ: Univ Col NWales, BSc, 48; Aberdeen Univ, PhD(forestry), 54. Prof Exp: Res fel forestry, 50-56; res officer, Oxford Univ, 56-58; head woodlands sect, Merlewood Res Sta, Nature Conserv, UK, 58-68; HEAD TREE BIOL, PETAWAWA FOREST EXP STA, CAN FORESTRY SERV, 68- Res: Ecology and silviculture of Scots pine; rainfall interception in woodlands; oak taxonomy; nutrient cycles in woodlands; chemistry of precipitation; tree nutrition; effects of forest practices; management of woodland reserves; tree improvement; conservation. Mailing Add: Petawawa Forest Exp Sta Chalk River ON Can

CARLISLE, DAVID B, b Salford, Eng, Mar 12, 26. MARINE BIOLOGY, FRESH WATER BIOLOGY. Educ: Oxford Univ, BA, 47, MS, 51, DPhil(zool), 54, DSc(zool), 63. Prof Exp: Zoologist, Marine Biol Asn UK, 51-62; head lab res div, Anti Locust Res Ctr, 62-69; prof biol & chmn dept, Trent Univ, 69-72; ADV WATER QUAL RES, DEPT ENVIRON, INLAND WATERS DIRECTORATE, GOVT CAN, 72- Mem: Fel Brit Inst Biol; fel Zool Soc London; fel Linaean Soc London; fel Am Anthrop Asn; fel AAAS. Res: Comparative endocrinology of arthropods; tunicate biology; physiological ecology; environment induced changes in the endocrine systems of arthropods; environmental biology. Mailing Add: Inland Waters Directorate Place Vincent Massey Hull PQ Can

CARLISLE, DONALD, b Vancouver, BC, June 21, 19; m 44; c 4. ECONOMIC GEOLOGY. Educ: Univ BC, BASc, 42, MASc, 44; Univ Wis, PhD(geol), 50. Prof Exp: Geologist, Consol Mining & Smelting Co, Can, 44-46; lectr geol, 49-50, from instr to assoc prof, 51-64, assoc dean grad div, 67-74, PROF GEOL, UNIV CALIF, LOS ANGELES, 64- Mem: Geol Soc Am; Soc Econ Geol; Am Inst Mining, Metall & Petrol Eng. Res: Mineral economics and economic theory; mineral deposits and geochemistry; areal studies. Mailing Add: Dept of Geol Univ of Calif Los Angeles CA 90024

CARLISLE, FRANK JEFFERSON, JR, b Fargo, NDak, Feb 8, 20; m 44; c 5. SOIL SCIENCE. Educ: NDak State Col, BS, 42; Univ Wis, MS, 47; Cornell Univ, PhD, 54. Prof Exp: Soil scientist, Soil Conserv Serv, USDA, 47-54, soil correlator, 54-60; head soil surv lab, Nebr, 60-61, asst dir soil classification & correlation, 61-73; ASST DIR SOIL SURV INVEST, SOIL CONSERV SERV, USDA, 73- Mem: Soil Sci Soc Am; Am Soc Agron. Res: Soil classification, morphology and genesis. Mailing Add: 6501 Otis St Hyattsville MD 20784

CARLISLE, GENE OZELLE, b Bivins, Tex, Feb 11, 39; m 61; c 1. INORGANIC CHEMISTRY. Educ: E Tex State Univ, BSc, 61, MSc, 65; N Tex State Univ, PhD(chem), 69. Prof Exp: Teacher high sch, Tex, 61-62; instr chem, Texarkana Col, 62-65, asst prof physics, 65-66; instr chem, N Tex State Univ, 66-67; res assoc, Univ NC, 69-70; asst prof chem, 70-74, ASSOC PROF CHEM, W TEX STATE UNIV, 74- Mem: Am Chem Soc; Am Phys Soc; Am Inst Physics. Res: Synthesis of transition metal complexes and applications of information gained from spectral and magnetic studies of these complexes to structural and bonding problems in coordination chemistry. Mailing Add: Dept of Chem W Tex State Univ Canyon TX 79015

CARLISLE, JOHN GRIFFIN, JR, b Port Washington, NY, May 25, 11; m 43; c 3. MARINE BIOLOGY. Educ: Loyola Univ (Calif), BS, 36. Prof Exp: Jr biologist, Marine Fisheries, 45-51, MARINE BIOLOGIST, MARINE FISHERIES, CALIF DEPT FISH & GAME, 51- Mem: Am Fisheries Soc; fel Am Inst Fishery Res Biologists. Res: Marine ecology; life histories of fishes and molluscs; marine habitat development. Mailing Add: 350 Golden Shore Long Beach CA 90802

CARLISLE, VICTOR WALTER, b Bunnell, Fla, Oct 3, 22; m 50; c 3. SOIL MORPHOLOGY. Educ: Univ Fla, BSA, 47, MS, 53, PhD(soils), 62. Prof Exp: Asst soil surveyor, Fla Agr Exp Sta, 47-54; asst soils, Inst Food & Agr Sci, 54-60, from instr to asst prof, 60-67, asst chemist, Agr Exp Sta, 62-67, assoc prof & assoc chemist, 67-74, PROF SOIL SCI, INST FOOD & AGR SCI, UNIV FLA, 74- Concurrent Pos: Consult, US AID, Univ Costa Rica, 66, Battelle Mem Inst subcontract, 67 & Jamaica Sch Agr, 70; course coordr, Orgn Trop Studies, Inc, Costa Rica, 68. Mem: Soil Sci Soc Am; Int Soc Soil Sci; Clay Minerals Soc; Am Asn Quaternary Environ. Res: Soil genesis, classification and mapping; source of parent materials of soils; weathering; soil mineralogy. Mailing Add: Inst of Food & Agr Sci Univ of Fla Gainesville FL 32611

CARLITZ, LEONARD, b Philadelphia, Pa, Dec 26, 07; m 31; c 2. MATHEMATICS. Educ: Univ Pa, AB, 27, AM, 28, PhD(math), 30. Prof Exp: Nat Res fel math, Calif Inst Technol, Univ Pa & Cambridge Univ, 30-32; from asst prof to prof, 32-64, JAMES B DUKE PROF MATH, DUKE UNIV, 64- Concurrent Pos: Mem, Inst Adv Study, 35-36. Mem: Am Math Soc; Math Asn Am; Soc Indust & Appl Math. Res: Theory of numbers; arithmetic of polynomials and power series; combinatorial analysis; special functions. Mailing Add: Dept of Math Duke Univ Durham NC 27706

CARLOCK, HENRY ARTHUR, b Alexandria, Ohio, Sept 30, 05; m 36; c 1. NUCLEAR PHYSICS. Educ: Denison Univ, BS, 28; Ohio State Univ, MS, 31. Prof Exp: Asst physics, Ohio State Univ, 30-31 & Union Univ (NY), 31-32; TV engr, Eng Lab, Gen Elec Co, 36-38; prof physics, Miss Col, 38-41 & La Col, 45-46; prof physics, Miss Col, 46-73, prof advan electronics eng, 70-73, EMER PROF PHYSICS, MISS COL, 73- Concurrent Pos: White fel, Cornell Univ, 34-36, fel, 35-36. Mem: Am Phys Soc. Res: Linear accelerator for positive ions and ion sources; control of currents in discharge tubes by means of magnetic fields; electrodless discharges; television research; velocity modulated oscillators; effects of x-radiation and ultrasonics. Mailing Add: Dept of Physics Miss Col 508 W College Clinton MS 39056

CARLON, HUGH ROBERT, b Camden, NJ, Oct 4, 34; m 59; c 4. ATMOSPHERIC PHYSICS. Educ: Drexel Inst, BS, 57; George Washington Univ, MS, 71. Prof Exp: US Army, 57-, staff officer, Off Chief Engrs, 57-58, chem engr, Army Chem Ctr, Md, 56-60, PHYSICIST, EDGEWOOD ARSENAL, US ARMY, 60- Concurrent Pos: Pres, Trans-Harford Corp, 70-73, pres, Harford Serv Corp, 73-; mem, Inter-Union Comn Radio Meteorol, 75- Honors & Awards: First Prize, US Armed Forces Mgt Essay Contest, 72. Mem: Sigma Xi; Am Inst Physics; Optical Soc Am; Am Inst Chem Eng. Res: Atmospheric physics, especially infrared transmission; instruments for detection and measurement of infrared spectral bands over long optical paths; detection of trace atmospheric gases and pollutants; infrared sources, optical filters; infrared emissions by water aerosols. Mailing Add: 11 McGregor Way Glennwood Bel Air MD 21014

CARLOTTI, RONALD JOHN, b Martins Ferry, Ohio, Sept 20, 42; m 69; c 2. NUTRITIONAL BIOCHEMISTRY. Educ: Ohio State Univ, BS, 64; WVa Univ, MS, 66, PhD(nutrit biochem), 70. Prof Exp: Res asst biochem, Col Agr, WVa Univ, 66-70; res assoc enzymol, Univ Iowa, 71-72, asst res scientist infant nutrit, 72-73, res assoc enzymol, 73-74; NUTRITIONIST HUMAN NUTRIT, RES DEPT, KELLOGG CO, 74- Mem: Am Chem Soc; AAAS; Inst Food Technologists; Am Meat Sci Asn. Res: Carbohydrates, lipids, amino acids and proteins, trace elements, vitamins, composition of foods; infant nutrition; growth and development of poultry; isolation, purification and immunochemical characterization of lactate dehydrogenase from Morris hepatomas. Mailing Add: Res Dept Kellogg Co Battle Creek MI 49016

CARLOZZI, MICHAEL, b New York, NY, Jan 28, 16; m 42; c 3. CLINICAL PHARMACOLOGY. Educ: Columbia Univ, BS, 37; Long Island Col Med, MD, 40. Prof Exp: Pvt pract, 45-47; assoc med dir, Merck & Co, 47-51; med dir, Chas Pfizer & Co, Inc, 51-72; ASST PROF MED, ALBERT EINSTEIN COL MED, 72- Mem: AAAS; AMA; Am Acad Dermatol; Am Geriat Soc; NY Acad Sci. Res: Conception and evaluation of new therapeutic agents; endocrinology; chemotherapy of bacterial and malignant diseases. Mailing Add: 48 Benedict Pl Pelham NY 10803

CARLQUIST, PHILIP RICH, b Salt Lake City, Utah, Mar 25, 10; m 33; c 3. MICROBIOLOGY. Educ: Univ Utah, BA, 35; Univ Calif, MPH, 38; Yale Univ, PhD(bact), 51. Prof Exp: Bacteriologist, Utah State Health Dept, 35-37, 38-39; dir state health lab, Dept Health, Wyo, 39-41; chief div bact, Army Med Sch, US Army, 46-48, bacteriologist, Europe, 50-54, Fourth Army Med Lab, 54-59 & Madigan Army Hosp, Wash, 59-65; MICROBIOLOGIST, DeKALB GEN HOSP & PHYSICIAN'S LAB, 65- Concurrent Pos: Mem comt Enterobacteriacaea, Int Asn Microbiol Socs, 47- Mem: Am Pub Health Asn; assoc mem Am Soc Clin Path; Brit Soc Gen Microbiol; Am Soc Microbiol. Res: Diagnostic medical bacteriology, particularly entericcs and anaerobes. Mailing Add: DeKalb Gen Hosp Physician's Lab PO Box 1307 Decatur GA 30031

CARLQUIST, SHERWIN, b Los Angeles, Calif, July 7, 30. BOTANY. Educ: Univ Calif, BA, 52, PhD(bot), 56. Prof Exp: Asst prof bot, Grad Sch, Claremont Cols, 56-61, assoc prof, 61-66; PROF BOT, CLAREMONT GRAD SCH & UNIV CTR, 66- Honors & Awards: Gleason Prize, NY Bot Garden, 67. Mem: Bot Soc Am; Am Soc Plant Taxon. Res: Comparative anatomy of flowering plants; Compositae, Rapateaceae; problems of insular floras and faunas. Mailing Add: Rancho Santa Ana Bot Garden Claremont CA 91711

CARLS, RALPH A, b Ringtown, Pa, Aug 9, 38; m 59; c 2. BACTERIOLOGY. Educ: Mansfield State Col, BS, 60; Okla State Univ, MS, 64; Univ Wis, PhD(bact), 71. Prof Exp: Teacher pub schs, 60-66; PROF MICROBIOL, EDINBORO STATE COL, 70- Concurrent Pos: Consult, Am Sterilizer Co, 71- Mem: Am Soc Microbiol. Res: Microbial ecology, primarily concerned with Bdelloribio and spherotilus life cycles and growth characterization; contamination control and biological indicators. Mailing Add: Dept of Biol Edinboro State Col Edinboro PA 16444

CARLSEN, RICHARD CHESTER, b San Francisco, Calif, Feb 5, 40; m 69; c 1. NEUROPHYSIOLOGY. Educ: Univ Calif, Berkeley, AB, 63; Univ Ore, PhD(physiol), 73. Prof Exp: RES ASSOC NEUROPHYSIOL, DUKE UNIV MED CTR, 73- Concurrent Pos: Muscular Dystrophy Asn fel physiol & pharmacol, Duke Univ, 74-76; NIH fel, 75. Mem: Am Physiol Soc; Soc Neurosci; Sigma Xi. Res: Development and plasticity of synaptic connections in the spinal cord; trophic relationship between nerve and muscle. Mailing Add: Physiol & Pharmacol Box 3709 Duke Univ Med Ctr Durham NC 27710

CARLSON, ALBERT DEWAYNE, JR, b Colfax, Iowa, Dec 8, 30; m 56; c 2. INVERTEBRATE PHYSIOLOGY. Educ: Univ Iowa, BA, 52, MS, 59, PhD(zool), 60. Prof Exp: ASSOC PROF BIOL, STATE UNIV NY, STONY BROOK, 60- Res: Neural and cellular control of luminescent responses of larval and adult fireflies; neurobiology and behavior of fireflies and aquatic invertebrates. Mailing Add: Dept of Cellular & Comp Biol State Univ of NY Stony Brook NY 11794

CARLSON, ANTON BURCHARD, analytical chemistry, see 12th edition

CARLSON, ARTHUR, JR, b Buffalo, Kans, June 5, 22; m 53; c 3. VETERINARY PHARMACOLOGY. Educ: Kans State Univ, DVM, 50; Univ Mo, MS, 67. Prof Exp: Pvt pract, Humboldt, Kans, 50-62; res assoc pharmacol res, Haver-Lockhart Labs, Kansas City, 62-64, assoc dir pharmacol res, 64-65, dir pharmacol res, 65-68, vpres pharmacol res, 68-74; DIR QUAL ASSURANCE, BAYNET CORP, SHAWNEE, KANS, 74- Mem: Am Vet Med Asn; Am Pharmaceut Asn; Animal Health Inst; Am Soc Vet Physiol & Pharmacol. Res: Calcium and amino acid metabolism; Escherichia coli endotoxin, vaccine adjuvants, prolonged-release oral medications, and antiviral agents. Mailing Add: Baynet Corp Shawnee KS 66203

CARLSON, ARTHUR STEPHEN, b Brooklyn, NY, Oct 24, 19; m 41; c 2. PATHOLOGY. Educ: Brooklyn Col, AB, 41; Cornell Univ, MD, 52; Am Bd Path, Nat Bd Med Exam, dipl. Prof Exp: Asst bacteriologist, Brooklyn Col, 41-42; instr path, Med Col, Cornell Univ, 53-56, lectr, 56-69; assoc pathologist, 56-74, CHIEF PATHOLOGIST, COMMUNITY HOSP AT GLEN COVE, 74-; CLIN ASST PROF PATH, MED COL, CORNELL UNIV, 69- Concurrent Pos: From intern to resident, NY Hosp, 52-55, provisional asst, 55-57; res fel, NY Heart Asn, 55-56; pres, NY State Bd Med Examr; inspector, Inspection & Acrreditation Prog, Am Asn Blood Banks. Mem: AAAS; Col Am Path; Am Asn Path & Bact; Int Acad Path. Res: Pathological and bacteriological biochemistry; morphological pathology. Mailing Add: Community Hosp Glen Cove NY 11542

CARLSON, BILLE CHANDLER, b Jamaica Plain, Mass, June 27, 24; m 47; c 2. APPLIED MATHEMATICS. Educ: Harvard Univ, BA & MA, 47; Oxford Univ, PhD(physics), 50. Prof Exp: Staff mem radiation lab, Mass Inst Technol, 43-44; instr physics, Princeton Univ, 50-52, res assoc, 52-54; from asst prof to prof physics, 54-65, PROF PHYSICS & MATH, AMES LAB, IOWA STATE UNIV, 65- Concurrent Pos: Sr res fel math, Calif Inst Technol, 62-63; visitor, Poincare Inst, Univ Paris, 71-72. Mem: Fel Am Phys Soc; Am Math Soc; Soc Indust & Appl Math; Math Asn Am. Res: Special functions, particularly elliptic integrals and R-functions. Mailing Add: Depts of Physics & Math Iowa State Univ Ames IA 50011

CARLSON, BRUCE ARNE, b St Paul, Minn, Apr 8, 46; m 68. ORGANIC CHEMISTRY, ORGANOMETALLIC CHEMISTRY. Educ: Cornell Univ, AB, 68; Purdue Univ, PhD(org chem), 73. Prof Exp: RES CHEMIST ORG CHEM, CENT RES & DEVELOP DEPT, E I DU PONT DE NEMOURS & CO, INC, 73- Mem: AAAS; Am Chem Soc; Fedn Am Scientists; Sigma Xi. Res: Exploratory chemistry of organoboranes and boron hydrides; heterocyclic chemistry based on HCN and unusual diazo and diazonium ion chemistry. Mailing Add: Cent Res & Develop Dept Exp Sta E I du Pont de Nemours & Co Inc Wilmington DE 19898

CARLSON, BRUCE MARTIN, b Gary, Ind, July 11, 38; m 68; c 1. ANATOMY. Educ: Gustavus Adolphus Col, BA, 59; Cornell Univ, MS, 61; Univ Minn, MD & PhD(anat), 65. Prof Exp: From asst prof to assoc prof, 66-75, PROF ANAT, UNIV MICH, ANN ARBOR, 75- Concurrent Pos: US Acad Sci exchange fel, Inst Develop Biol, Moscow, USSR, 65-66 & Inst Physiol, Prague, Czech, 71, 72, 74 & 75; Fulbright fel, Hubrecht Lab, Utrecht, Neth, 73-74. Honors & Awards: Newcomb-Cleveland Award, AAAS, 74. Mem: AAAS; Am Asn Anat; Am Soc Zool; Am Soc Ichthyol & Herpet; Soc Develop Biol. Res: Regeneration; muscle transplantation; bone induction. Mailing Add: Dept of Anat Univ of Mich Ann Arbor MI 48104

CARLSON, CARL WILBURN, b Clifton, Kans, Mar 25, 19; m 45; c 2. SOIL CONSERVATION. Educ: Kans State Univ, BS, 49, MS, 50. Prof Exp: Agronomist, Kans State Univ, 49-53; soil scientist, USDA & Univ Wyo, Riverton, 53-54, USDA, Mandan, NDak, 54-61, asst dir soil & water conserv res div, Agr Res Serv, 61-71, asst to dep adminr soil, water & eng, Off Adminr, 71-72, ASST ADMINR SOIL, WATER & AIR SCI, AGR RES SERV, USDA, 72- Honors & Awards: Cert Merit Award, USDA, 58. Mem: Fel AAAS; fel Am Soc Agron; Soil Sci Soc Am; Soil Conserv Soc Am. Res: Soil and water conservation; moisture conservation as influenced by cultural practices; agricultural research administration. Mailing Add: Rm 330 Admin Bldg Agr Res Serv USDA Washington DC 20250

CARLSON, CHARLES MERTON, b Duluth, Minn, Apr 3, 34; wid. THEORETICAL CHEMISTRY. Educ: Univ Calif, Riverside, AB, 56; Univ Utah, PhD(chem), 60. Prof Exp: Staff scientist, Metall Dept, Brookhaven Nat Lab, NY, 60-62; res specialist, Solid State Physics Group, Boeing Sci Res Labs, 62-70; VIS SCHOLAR, COMPUT SCI GROUP, UNIV WASH, 71- Mem: Am Phys Soc; Am Chem Soc; Soc Indust & Appl Math. Res: Basic theoretical research in quantum chemistry and statistical mechanics as applied to molecules and solids. Mailing Add: 6507 1/2 Parker Ct NW Seattle WA 98117

CARLSON, CHARLES WENDELL, b Eaton, Colo, May 15, 21; m 43; c 5. ANIMAL NUTRITION. Educ: Colo State Univ, BS, 42; Cornell Univ, MSA, 48, PhD(animal nutrit), 49. Prof Exp: Asst poultry husb, Cornell Univ, 46-49; from asst prof to assoc prof, 49-56, PROF POULTRY HUSB, S DAK STATE UNIV, 56-, LEADER POULTRY RES, 67- Concurrent Pos: Res assoc, Wash State Univ, 62-63; prin non-ruminant nutritionist, Coop States Res Serv, USDA, Washington, DC, 75-76. Honors & Awards: Nat Turkey Fedn res award, 61. Mem: Fel AAAS; Am Chem Soc; fel Poultry Sci Asn (pres, 70-71); Am Inst Nutrit. Res: Poultry; biochemistry; amino acid and energy metabolism; growth factors; calcification. Mailing Add: Dept of Animal Sci SDak State Univ Brookings SD 57006

CARLSON, CLARENCE ALBERT, JR, b Moline, Ill, Apr 3, 37; m 58; c 2. FISH BIOLOGY. Educ: Augustana Col, Ill, AB, 58; Iowa State Univ, MS, 60, PhD(zool), 63. Prof Exp: Asst prof biol, Augustana Col, Ill, 62-66; asst prof fishery biol & asst leader NY Coop Fishery Unit, Cornell Univ, 66-72; ASSOC PROF FISHERY BIOL, COLO STATE UNIV, 72-, CHMN FISHERY MAJOR, 72- Mem: Am Fisheries Soc; Ecol Soc Am; Am Inst Biol Sci; AAAS. Res: Ecology of fisheres, aquatic radioecology; river ecology; environmental effects of power generation. Mailing Add: Dept of Fishery & Wildlife Biol Colo State Univ Ft Collins CO 80523

CARLSON, DANA PETER, b Red Wing, Minn, Oct 31, 31; m 56; c 4. ORGANIC CHEMISTRY. Educ: Univ Minn, BS, 53; Carnegie Inst Technol, MS, 56, PhD(chem), 57. Prof Exp: From res chemist, Plastics Dept, to sr res chemist, 57-71, RES ASSOC, E I DU PONT DE NEMOURS & CO, 71- Mem: Am Chem Soc. Res: Fluorocarbon chemistry; synthesis of monomers and polymers. Mailing Add: E I du Pont de Nemours & Co PO Box 1217 Parkersburg WV 26101

CARLSON, DAVID EMIL, b Weymouth, Mass, Mar 5, 42; m 66; c 2. SOLID STATE PHYSICS. Educ: Rensselaer Polytech Inst, 63; Rutgers Univ, PhD(physics), 68. Prof Exp: Res scientist physics, US Army Nuclear Effects Lab, Edgewood Arsenal, 68-69; MEM TECH STAFF PHYSICS, RCA LABS, NJ, 70- Mem: Am Phys Soc; Am Ceramic Soc; Sigma Xi. Res: Ion motion in glasses and insulators; thin film photovoltaic devices. Mailing Add: RCA Labs Princeton NJ 08540

CARLSON, DAVID HILDING, b Chicago, Ill, Nov 10, 36; m 59; c 3. MATHEMATICS. Educ: San Diego State Col, AB, 57; Univ Wis, MS, 59, PhD(math), 63. Prof Exp: Instr math, Univ Wis, Milwaukee, 62-63; from asst prof to assoc prof, 63-73, PROF MATH, ORE STATE UNIV, 73- Concurrent Pos: Fulbright-Hays lectr, Univ of Repub Uruguay, 65-66; vis prof, Kent State Univ, 70-71. Mem: Am Math Soc; Math Asn Am. Res: Matrix theory, especially inertia theory and location of eigenvalues. Mailing Add: Dept of Math Ore State Univ Corvallis OR 97331

CARLSON, DAVID STEN, b New Bedford, Mass, Aug 30, 48; m 70; c 1.

ANTHROPOLOGY, DENTAL RESEARCH. Educ: Univ Mass, Amherst, BA, 70, MA, 72, PhD(anthrop), 74. Prof Exp: Teaching fel anthrop, Univ Mass, Amherst, 70-74, lectr, 73-74; ASST PROF ANTHROP, WAYNE STATE UNIV, 74- Concurrent Pos: Marshall fel, Am Scandanavian Found, 73; assoc, Sch Med, Wayne State Univ, 74-; scholar, Univ Mich, 75- Mem: AAAS; Am Asn Phys Anthropologists; Human Biol Coun. Res: Analysis of the phylogenetic and ontogenetic factors influencing the form and function of the musculoskeletal system in general and the craniofacial complex in particular. Mailing Add: Ctr for Human Growth & Dev Univ of Mich Ann Arbor MI 48105

CARLSON, DONALD EARLE, continuum mechanics, see 12th edition

CARLSON, DONALD EUGENE, b Superior, Wis, Feb 27, 38; m 61; c 2. RADIOBIOLOGY, IMMUNOBIOLOGY. Educ: Univ Wis, BS, 60, MS, 62; Univ Tenn, PhD(radiation biol), 68. Prof Exp: Nat Cancer Inst fel, Oak Ridge Assoc Univs, 68-69; asst prof radiol, 69-75, ASSOC PROF RADIOL, UNIV TEX HEALTH SCI CTR, 75- Mem: AAAS; Radiation Res Soc; Am Inst Biol Sci; Radiol Soc N Am; Los Alamos Meson Physics Facility Users Group. Res: Effects of ionizing radiation on immune processes; kinetics of antibody producing cells; biological effects of negative pi mesons, neutrons and alpha particles. Mailing Add: Univ of Tex Health Sci Ctr 5323 Harry Hines Dallas TX 75235

CARLSON, DOUGLAS W, b Jamestown, NDak, Jan 7, 30; m 54; c 2. PHYSICAL CHEMISTRY. Educ: Univ NDak, BS, 55; Univ Del, MS, 57, PhD(phys chem), 59. Prof Exp: Fel, Princeton Univ, 59-61; res chemist, Minn Mining & Mfg Co, Minn, 61-64; res chemist, E I du Pont de Nemours & Co, Del, 64-67, tech rep, 67-70; DIR RES, SPRINGFIELD RES CTR, DAYCO CORP, 70- Mem: Am Chem Soc. Res: Polymer chemistry; photochemistry. Mailing Add: 1839 E Swallow St Springfield MO 65804

CARLSON, EARL JOHN, b Two Harbors, Minn, July 7, 14; m 41; c 4. POLYMER CHEMISTRY. Educ: Univ Minn, BCh, 37. Prof Exp: Jr res chemist, 38-40, develop chemist, 40-42, from res chemist to sr res chemist, 42-56, sect leader, 56-60, res assoc, B F Goodrich Co, 60-75; RETIRED. Mem: Am Chem Soc. Res: Plasticizer synthesis; plastics sales development; condensation polymerization; vinyl polymerization; stereo specific polymerizations; textile chemistry; propellent chemistry; ablators; rubber chemistry and physics. Mailing Add: 2516 Ridgewood Rd Akron OH 44313

CARLSON, EDWARD H, b Lansing, Mich, Apr 29, 32; m 60; c 3. SOLID STATE PHYSICS. Educ: Mich State Univ, BS, 54, MS, 56; Johns Hopkins Univ, PhD(physics), 59. Prof Exp: Instr physics, Johns Hopkins Univ, 59-60; NSF fel, State Univ Leiden, 60-61; asst prof, Univ Ala, 61-65; from asst prof to assoc prof, 65-74, PROF PHYSICS, MICH STATE UNIV, 74- Mem: Am Phys Soc. Res: Rare earth spectra; electron and nuclear magnetic spin resonance; ordered magnetic states; atomic and molecular structure. Mailing Add: Dept of Physics Mich State Univ East Lansing MI 48823

CARLSON, ELMER CARL, b Seattle, Wash, Jan 21, 18; m 41; c 2. ECONOMIC ENTOMOLOGY. Educ: State Col Wash, BS, 40, MS, 43. Prof Exp: Asst entom & asst entomologist, State Col Wash, 43-46; sr lab technician, 46-71, SPECIALIST, AGR EXP STA, UNIV CALIF, DAVIS, 71- Mem: Entom Soc Am. Res: Biology and control of insects and mites affecting vegetables seed and oil crops; interrelationships of pest populations, plant damage and biology; plant resistance to insects, especially its nature. Mailing Add: Dept of Entom Agr Exp Sta Univ of Calif Davis CA 95616

CARLSON, ELOF AXEL, b Brooklyn, NY, July 15, 31; m 59; c 5. GENETICS. Educ: NY Univ, BA, 53; Ind Univ, PhD, 58. Prof Exp: Lectr genetics, Ind Univ, 57-58 & Queen's Univ, 58-60; asst prof zool, Univ Calif, Los Angeles, 60-65, assoc prof, 65-68; prof biol, 68-75, DISTINGUISHED TEACHING PROF BIOL, STATE UNIV NY, STONY BROOK, 75- Mem: Fel AAAS; Genetics Soc Am. Res: Gene structure and function; chemical and radiation induced mutagenesis; mosaicism in Drosophila; history of genetics. Mailing Add: Dept of Biol State Univ of NY Stony Brook NY 11794

CARLSON, EMIL HERBERT, b Portland, Ore, Oct 25, 29; m 51; c 4. ORGANIC CHEMISTRY. Educ: Willamette Univ, BS, 51; Carnegie Inst Technol, MS, 54, PhD(chem), 56. Prof Exp: Asst chem lab, Carnegie Inst Technol, 51-52, asst org qual lab, 52-53, sr asst, 53-54; res chemist, 5S-64, prod specialist org div, Mo, 64-65, mkt analyst, 65-70, RES SPECIALIST, AGR DIV, MONSANTO CO, 70- Mem: Am Chem Soc; NY Acad Sci; Sigma Xi. Res: Agricultural chemicals; functional fluids; intermediate chemicals. Mailing Add: R R 1 Box A-20 Muscatine IA 52761

CARLSON, ERIC DUNGAN, b Kansas City, Mo, Jan 19, 29; m 70. ASTRONOMY. Educ: Wash Univ (St Louis), AB, 50; Northwestern Univ, MS, 65, PhD(astron), 68. Prof Exp: Mgt trainee printing, R R Donnelley & Sons Co, 56-58; asst to ed-in-chief encycl writing, Consol Bk Publs, 58-61; SR ASTRONR, ADLER PLANETARIUM, 68-; ASST PROF ASTRON RES, DEARBORN OBSERV, NORTHWESTERN UNIV, 68- Mem: Am Astron Soc. Res: Radial velocities; peculiar emissionline stars. Mailing Add: Adler Planetarium 1300 S Lake Shore Dr Chicago IL 60605

CARLSON, ERIC THEODORE, b Westbrook, Conn, Aug 22, 22; m 50; c 1. PSYCHIATRY. Educ: Wesleyan Univ, BA, 44; Cornell Univ, MD, 50. Prof Exp: Intern int med, New York Hosp, 50-51, asst res psychiat, 51-55, psychiatrist out-patients, 55-56, asst attend psychiatrist, 52-60, assoc attend psychiatri st, 60-70, ATTEND PSYCHIATRIST, NEW YORK HOSP, 70-; CLIN PROF PSYCHIAT, MED COL, CORNELL UNIV, 70- Concurrent Pos: Asst psychiat, Med Col, Cornell Univ, 52-53, instr, 53-58, from clin asst prof to clin assoc prof, 58-70; consult psychiatrist, Hosp Spec Surg, New York. Mem: AAAS; Am Psychiat Asn; Int Soc Hist Med; Am Asn Hist Med. Res: Medical education; development of psychiatric thought. Mailing Add: 60 Sutton Pl S New York NY 10022

CARLSON, ERNEST HOWARD, b Seattle, Wash, Dec 23, 33. GEOLOGY, MINERALOGY. Educ: Univ Wash, BSc, 56; Univ Colo, MSc, 60; McGill Univ, PhD(crystal growth), 66. Prof Exp: Geologist, US Geol Surv, 58-60; instr geol, Villanova Univ, 65-66; asst prof, 66-75, ASSOC PROF GEOL, KENT STATE UNIV, 75- Concurrent Pos: Vis prof, Pahlavi Univ, Shiraz, Iran, 70-71. Mem: Am Crystallog Asn; Geol Soc Am; Mineral Soc Am; Nat Asn Geol Teachers. Res: Experimental mineralogy; exploration geochemistry. Mailing Add: Dept of Geol Kent State Univ Kent OH 44242

CARLSON, FRANCIS DEWEY, b Syracuse, NY, June 29, 21; m 50; c 3. BIOPHYSICS. Educ: Johns Hopkins Univ, AB, 42; Univ Pa, PhD(biophys), 49. Prof Exp: Asst, Electroacoust Lab, Harvard Univ, 42-43, res assoc, 43-46; asst, Univ Pa, 46-49; from instr to assoc prof biophys, 49-60, chmn dept, 56-74, PROF BIOPHYS, JOHNS HOPKINS UNIV, 60- Concurrent Pos: NSF sr fel, 61-62. Mem: AAAS; Am Phys Soc; Biophys Soc. Res: Mechano-chemistry of muscular contraction; molecular biology; neurophysiology. Mailing Add: Dept of Biophys Johns Hopkins Univ Baltimore MD 21218

CARLSON, GARY, b Los Angeles, Calif, Mar 6, 28; m 54; c 5. COMPUTER SCIENCES. Educ: Univ Calif, Los Angeles, BS, 56, MA, 58, PhD(indust psychol), 62. Prof Exp: Res asst, Western Data Processing Ctr, Univ Calif, Los Angeles, 58-59; sr opers res specialist, Indust Dynamics Dept, Hughes Aircraft Co, Calif, 59-61; with, Info Systs, Inc, 61-63; dir, Comput Res Ctr, Brigham Young Univ, 63-70, DIR COMPUT SERV, BRIGHAM YOUNG UNIV, 70- Concurrent Pos: Mem, Utah State Citizens Comt Rev State Comput Orgn, 68-69. Mem: Asn Comput Mach; AAAS. Res: Computer administration; computer performance measurement and monitoring. Mailing Add: Comput Serv Brigham Young Univ Provo UT 84602

CARLSON, GARY ALDEN, b Hastings, Nebr, Apr 18, 41; m 62; c 2. PHYSICAL CHEMISTRY. Educ: Univ Idaho, BS, 63; Univ Calif, Berkeley, PhD(phys chem), 66. Prof Exp: MEM TECH STAFF PHYS CHEM, SANDIA LABS, 66- Mem: Am Chem Soc; Am Phys Soc. Res: Free radical studies; rapid scan infrared spectroscopy; high temperature chemistry; detonation chemistry; exploding wires; pulsed electron beam energy deposition; plasma spectroscopy; nonequilibrium phase changes; liquid metal fast breeder reactor; safety research. Mailing Add: Div 5423 Sandia Labs PO Box 5800 Albuquerque NM 87115

CARLSON, GARY P, b Buffalo, NY, Feb 21, 43; m 68; c 1. TOXICOLOGY, PHARMACOLOGY. Educ: St Bonaventure Univ, BS, 65; Univ Chicago, PhD(pharmacol), 69. Prof Exp: From asst prof to assoc prof pharmacol, Univ RI, 74-75; ASSOC PROF TOXICOL, PURDUE UNIV, 75- Mem: Soc Toxicol; Am Soc Pharmacol & Exp Therapeut; Am Indust Hyg Asn; NY Acad Sci. Res: Toxicity of pesticides, drugs, chemicals; drug metabolism. Mailing Add: Dept of Pharmacol & Toxicol Purdue Univ West Lafayette IN 47907

CARLSON, GERALD EUGENE, b Wausa, Nebr, Oct 28, 32; m 54; c 3. ENVIRONMENTAL PHYSIOLOGY, AGRONOMY. Educ: Iowa State Univ, BS, 58, MS, 59; Pa State Univ, PhD(forage physiol), 63. Prof Exp: Res agronomist, Crop Res Div, Agr Res Serv, USDA, 63-66, res leader, Humid Pasture & Range Invests, 66-72, LAB CHIEF, LIGHT & PLANT GROWTH LAB, BELTSVILLE AGR RES CTR, USDA, 72- Concurrent Pos: Vis foreign scientist, Japanese Inst Sci & Technol, 70. Mem: Am Soc Agron; Crop Sci Soc Am; Am Soc Photobiol. Res: Identification of factors limiting light utilization by plants including the effect of temperature and carbon dioxide on photosynthesis, respiration and translocation. Mailing Add: Bldg 046-A Beltsville Agr Res Ctr Beltsville MD 20705

CARLSON, GERALD LEROY, b Kane, Pa, July 13, 32; m 59; c 5. PHYSICAL CHEMISTRY. Educ: Grove City Col, BS, 54; Univ Pittsburgh, MS, 57, PhD(chem), 60. Prof Exp: Res assoc, Mellon Inst, 54-60, res fel, 60-61; res chemist, Res Labs, Alcoa Aluminum Co Am, 61-62; res fel, Mellon Inst, 62-67 & Mellon Inst Sci, Carnegie-Mellon Univ, 67-74; SR SCIENTIST, WESTINGHOUSE RES LABS, 74- Mem: Soc Appl Spectros; Am Chem Soc; Coblentz Soc. Res: Molecular structure; molecular spectroscopy; mass spectrometry. Mailing Add: Westinghouse Res Labs Beulah Rd Pittsburgh PA 15235

CARLSON, GERALD LOWELL, b Jackson, Minn, Dec 2, 29; m 56; c 2. BIOLOGICAL CHEMISTRY. Educ: Luther Col, Iowa, AB, 51; Univ Ga, MS, 56; Mass Inst Technol, PhD(biochem), 61. Prof Exp: Donner Found res fel med, 61-62; Given Found res, fel, 62-63; asst prof biochem, 63-69, ASSOC PROF BIOCHEM, MED CTR, UNIV ALA, 69- Concurrent Pos: Lalor Found res grant, 63-64. Mem: AAAS; Am Soc Zool; Soc Gen Physiol. Res: Mechanisms of enzyme action; germ cell differentiation; autoimmunity; biochemistry of regeneration and metamorphosis. Mailing Add: Dept of Biochem Univ of Ala Med Ctr Birmingham AL 35203

CARLSON, GLENN RICHARD, b New Haven, Conn, Oct 22, 45; m 67; c 2. SYNTHETIC ORGANIC CHEMISTRY. Educ: Bates Col, BS, 67; Mich State Univ, PhD(org chem), 71. Prof Exp: Res assoc org chem, Univ Chicago, 71-73; RES CHEMIST, ROHM AND HAAS CO, 73- Mem: Am Chem Soc. Res: Synthesis of biologically active molecules; development of new synthetic methods in organic chemistry. Mailing Add: Rohm and Haas Co Spring House PA 19477

CARLSON, GORDON ANDREW, b Jamestown, NY, Jan 11, 17; m 41; c 3. INORGANIC CHEMISTRY. Educ: Wittenberg Col, AB, 40; Ohio State Univ, PhD(chem), 44. Prof Exp: Res chemist, Columbia Chem Div, Pittsburgh Plate Glass Co, Ohio, 44-50, asst chief chemist, WVa, 50-51 & Columbia-Southern Chem Co, 51-55, asst dir res, 55-62; dept head inorg res, 62-65, SR RES ASSOC, CHEM DIV, BARBERTON CHEM TECH CTR, PITTSBURGH PLATE GLASS CO, 65- Mem: Am Chem Soc; Electrochem Soc. Res: Chlor-alkali cells; coordination compounds; rare earths; alkalies and chlorine; industrial electrochemistry; metal halides and inorganic chlorinations; chemical metallurgy. Mailing Add: 254 Tanglewood Trail Wadsworth OH 44281

CARLSON, GUSTAF HARRY, b Bjornfall, Sweden, Dec 18, 04; nat 41. ORGANIC CHEMISTRY. Educ: Clark Univ, AB, 26, MA, 27; Harvard Univ, MA, 30, PhD(chem), 32. Prof Exp: Asst to Prof Arthur Michael, 32-39; res chemist, Merck & Co, NJ, 39-40, Lederle Labs, NY, 40-43, Wyeth Inst, 43-46 & R P Scherer Corp, 46-52, vpres, 52-63, EXEC VPRES, R P SCHERER CORP, 63- Mem: AAAS; Am Chem Soc; NY Acad Sci. Res: Mechanism of organic reactions; production of riboflavin, intermediates, vitamins, pharmaceuticals and penicillin. Mailing Add: R P Scherer Corp 9425 Grinnel Detroit MI 48213

CARLSON, GUSTAV GUNNAR, b Gwinn, Mich, Nov 21, 09; m 33; c 2. ANTHROPOLOGY. Educ: Northern Mich Univ, AB, 32; Univ Mich, MA, 34, PhD(anthrop & sociol), 36. Prof Exp: Instr sociol, Eastern Mich Univ, 34; instr, Univ Mich, 35-46; from instr to asst prof anthrop, Univ Cincinnati, 36-43; chief intel div, Off War Info, Kunming, 43-46; assoc prof, 46-52, head dept sociol & anthrop, 61-69, PROF ANTHROP, UNIV CINCINNATI, 52-, HEAD DEPT, 69- Concurrent Pos: Consult supvr mus, Ohio Works Progress Admin, 38-40; mem bd trustees, Ach Fund, 47-60; chmn all-univ fac, Univ Cincinnati, 62-64; Taft grant Europ ethnol mus res, Univ Cincinnati, 63; Miami Purchaser Asn grant Ohio Valley archaeol res, 65-71; Cincinnati Health & Welfare Coun grant urban studies res, 67-69; ed, Bull Cent States Anthrop Soc, 71-76 & Proceedings Cent States Anthrop Soc, 75-77. Res: European ethnology; Far Eastern cultures; anthropology of complex societies; primitive art. Mailing Add: 6404 Edwood Ave Cincinnati OH 45224

CARLSON, HARVE J, b Jerome, Idaho, June 10, 11; m 37; c 3. BACTERIOLOGY, VIROLOGY. Educ: Univ Wash, BS, 34; Univ Mich, MSPH, 40, DPH, 43. Prof Exp: Lab technician, Idaho State Dept Health, 36-39, asst bacteriologist, 39-40; asst bacteriologist, Univ Mich, 40-41, res assoc, 41-42, instr bact, 42-43; asst prof pediat res, Western Reserve Univ, 46-51; biologist, Off Naval Res, Calif, 51-56, sci liaison off, Eng, 56-58, head microbiol br, 58-59; prog dir facil & spec prog, Div Biol & Med Sci, NSF, 59-60, from dep asst dir to dir div, 60-72; SCI & EDUC CONSULT, 72- Mem: Am Soc Microbiol; Am Soc Exp Biol & Med. Res: Antibiotics; ultraviolet irradiation; airborne organisms; oligiodynamics of metals; effects on bacteria and viruses; effect of chemical agents of poliomyelitis virus; research administration. Mailing Add: 5510 Hoover St Bethesda MD 20034

CARLSON, HUGH DOUGLAS, b Haileybury, Ont, Aug 29, 22; m 50; c 2. ECONOMIC GEOLOGY. Educ: Queen's Univ (Ont), BSc, 49, PhD(geol), 53; Univ Toronto, MASc, 50. Prof Exp: Resident geologist, Ont Dept Mines, 53-56; consult geologist, 56-57; asst prof geol, Univ SDak, 57-60, assoc prof, 60-63; res geologist, Ont Dept Mines, 63-67; CONSULT GEOLOGIST, 67- Mem: AAAS; Geol Asn Can; Can Inst Mining & Metall; Mineral Asn Can. Mailing Add: 110 Martin St Porcupine ON Can

CARLSON, IRVING THEODORE, b Colbert, Wash, July 14, 26; m 52; c 2. PLANT BREEDING. Educ: Wash State Univ, BS, 50, MS, 52; Univ Wis, PhD(agron), 55. Prof Exp: Asst wheat breeding, Wash State Univ, 50-52; asst forage grass breeding, Univ Wis, 52-55; asst wheat breeding, Wash State Univ, 55-56; asst prof field crops, NC State Col, 56-60; assoc prof agron, 60-73, PROF AGRON, IOWA STATE UNIV, 73- Mem: Am Soc Agron. Res: Breeding and evaluation of forage crops, including orchard grass, reed canary grass, smooth bromegrass, tall fescue, alfalfa and birdsfoot trefoil. Mailing Add: Dept of Agron Iowa State Univ Ames IA 50010

CARLSON, JAMES ANDREW, b Lewiston, Idaho, Nov 14, 46. PURE MATHEMATICS. Educ: Univ Idaho, BS, 67; Princeton Univ, PhD(math), 71. Prof Exp: Asst prof math, Stanford Univ, 71-73 & Brandeis Univ, 73-75; ASST PROF MATH, UNIV UTAH, 75- Res: Several complex variables, particularly value distribution theory, and complex algebraic geometry. Mailing Add: Dept of Math Univ of Utah Salt Lake City UT 84112

CARLSON, JAMES C, b Mankato, Minn, Jan 4, 28; m 52; c 3. RADIOLOGICAL PHYSICS. Educ: Univ Minn, AB, 51, MS, 56. Prof Exp: Cancer res scientist, Roswell Park Mem Inst, 54-57; RADIOL PHYSICIST, HACKLEY HOSP, MUSKEGON, MICH, 57- Concurrent Pos: Consult, Vet Admin Hosp, Wood, Wis, 60-; pres, Mich Med Software, Inc, 70- Mem: Radiol Soc NAm; Soc Nuclear Med; Health Physics Soc; Am Asn Physicists Med. Res: Radiation measuring instrumentation; computer programming for nuclear medicine; hospital computer systems. Mailing Add: Nuclear Med Dept Hackley Hosp Muskegon MI 49443

CARLSON, JAMES GORDON, b Port Allegany, Pa, Jan 24, 08; m 36; c 3. CYTOLOGY, RADIOBIOLOGY. Educ: Univ Pa, AB, 30, PhD(cytol), 35. Prof Exp: Demonstr biol, Bryn Mawr Col, 30-31, instr, 31-35; instr zool, Univ Ala, 35-39, from asst prof to assoc prof, 39-46; sr biologist, NIH, USPHS, Md, 46-47; prof zool, Univ Tenn, Knoxville, 47-62, head dept zool & entom, 47-67, dir, Inst Radiation Biol, 55-75, ALUMNI DISTINGUISHED SERV PROF, UNIV TENN, KNOXVILLE, 62- Concurrent Pos: Rockefeller fel, Carnegie Inst, Cold Spring Harbor Biol Lab & Univ Mo, 40-41; USPHS spec fel, Univ Heidelberg, 64-65; spec consult, USPHS, 43-46, 47-48; consult, Oak Ridge Nat Lab, Tenn, 47-; mem comt biol & agr fels, Nat Res Coun, 49-52. Mem: AAAS (vpres, 55); Radiation Res Soc; Am Soc Cell Biol; Am Inst Biol Sci. Res: Effects of fixatives on staining reactions; orthopteran cytology; effects of ultraviolet and ionizing radiations on chromosomes and cell division; effect of chemical agents on chromosomes and cell division. Mailing Add: Dept of Zool & Entom Univ of Tenn Knoxville TN 37916

CARLSON, JAMES H, b Cleveland, Ohio, June 10, 35; m 62; c 2. GENETICS. Educ: Fenn Col, BS, 58; Ohio State Univ, MS, 60, PhD(genetics), 63. Prof Exp: Asst instr zool, Ohio State Univ, 61-63; asst prof biol, Fairleigh Dickinson Univ, 63-66; assoc prof, 66-70, chmn dept, 70-74, PROF BIOL, RIDER COL, 70- Mem: AAAS; Genetics Soc Am; Am Genetic Asn. Res: Penetrance, expressivity and chromosomal control of a wing venation mutant system in Drosophila melanogaster. Mailing Add: Dept of Biol Rider Col Trenton NJ 08602

CARLSON, JAMES ROY, b Windsor, Colo, Mar 9, 39; m 61; c 2. BIOCHEMISTRY. Educ: Colo State Univ, BS, 61; Univ Wis-Madison, MS, 64, PhD(biochem), 66. Prof Exp: Asst prof, 66-71, ASSOC PROF ANIMAL SCI, WASH STATE UNIV, 71-, CHMN GRAD PROG NUTRIT, 73- Mem: Am Inst Nutrit. Res: Nutritional biochemistry, involving amino acid and protein metabolism and enzyme adaptation in response to nutritional factors. Mailing Add: Dept of Animal Sci Wash State Univ Pullman WA 99163

CARLSON, JOHN BERNARD, b Virginia, Minn, Jan 23, 26; m 49. PLANT ANATOMY. Educ: St Olaf Col, BA, 50; Iowa State Col, PhD(plant anat), 53. Prof Exp: Instr bot, Iowa State Col, 54; from asst prof to assoc prof, 54-67, PROF BOT, UNIV MINN, DULUTH, 67- Mem: Bot Soc Am. Res: Developmental morphology of soybeans and other plants; floral anatomy of wild rice. Mailing Add: Dept of Biol Univ of Minn Duluth MN 55812

CARLSON, JOHN W, b Topeka, Kans, Nov 10, 40; m 61; c 2. MATHEMATICS. Educ: Kans State Univ, BS, 63, MS, 64; Univ Mo-Rolla, PhD(math), 70. Prof Exp: Instr math, Washburn Univ, 64-67; instr, Univ Mo-Rolla, 67-70; asst prof, 70-74, ASSOC PROF, KANS STATE TEACHERS COL, 74- Mem: Am Math Soc; Math Asn Am. Res: Topology; quasi-uniform spaces. Mailing Add: Dept of Math Kans State Teachers Col Emporia KS 66801

CARLSON, JON FREDERICK, b Newport News, Va, July 3, 40; m 66; c 2. ALGEBRA. Educ: Old Dom Col, BA, 62; Univ Va, MA, 65, PhD(math), 67. Prof Exp: Instr math, Univ Va, 67-68; ASST PROF MATH, UNIV GA, 68- Mem: AAAS; Am Math Soc. Res: Quadratic form theory and finite group theory; finite group representations. Mailing Add: Dept of Math Univ of Ga Athens GA 30601

CARLSON, KEITH DOUGLAS, b Los Angeles, Calif, Feb 28, 33. PHYSICAL CHEMISTRY. Educ: Univ Redlands, BS, 54; Univ Kans, PhD(chem), 60. Prof Exp: Fulbright scholar theoret chem, Oxford Univ, 60-62, hon Am Ramsey fel, 61-62; from asst prof to assoc prof, 62-71, PROF CHEM, CASE WESTERN RESERVE UNIV, 71- Concurrent Pos: Consult, Gen Elec Co, 74- Mem: Am Phys Soc; Am Chem Soc. Res: High temperature chemistry; quantum theory of molecular structure; molecular beam kinetics. Mailing Add: Dept of Chem Case Western Reserve Univ Cleveland OH 44106

CARLSON, KEITH J, b White Bear Lake, Minn, June 3, 38; m 65. GEOLOGY. Educ: Gustavus Adolphus Col, BS, 60; Iowa State Univ, MS, 62; Univ Chicago, PhD(paleozool), 66. Prof Exp: Asst prof, 67-74, ASSOC PROF GEOL, GUSTAVUS ADOLPHUS COL, 74- Mem: Soc Vert Paleont; Soc Study Evolution. Res: Vertebrate paleontology. Mailing Add: Dept of Geol Gustavus Adolphus Col St Peter MN 56082

CARLSON, KENNETH THEODORE, b Douglas, NDak, June 18, 21; m 49; c 3. CHEMISTRY. Educ: Minot State Col, BS, 47; Colo State Col, MA, 53. Prof Exp: Instr high schs, NDak, 47-48, 49-51, prin, 51-54; asst prof physics & chem, 54-55, from asst prof to assoc prof chem, 55-69, PROF SCI, MAYVILLE STATE COL, 69-, CHMN DEPT SCI, 55- Mem: Am Chem Soc; Nat Sci Teachers Asn. Res: History of chemistry, especially as it pertains to chemical education in the United States. Mailing Add: Dept of Sci Mayville State Col Mayville ND 58257

CARLSON, KERMIT HOWARD, b Jamestown, NY, Dec 31, 13; m 46; c 2. MATHEMATICS. Educ: Upsala Col, BA, 39; Univ Iowa, MS, 41; Univ Wis, PhD, 54. Prof Exp: Instr math, Mich State Univ, 51-54; PROF MATH, VALPARAISO UNIV, 54- Mem: Am Math Soc; Math Asn Am. Res: Area theory; harmonic functions; calculus of variations; topology. Mailing Add: Dept of Math Valparaiso Univ Valparaiso IN 46383

CARLSON, KRISTIN ROWE, b Minneapolis, Minn, July 31, 40. PSYCHOPHARMACOLOGY. Educ: Univ Mich, BA, 62; McGill Univ, MA, 63, PhD(psychol), 66. Prof Exp: NIMH fel, Univ Waterloo, 66-68; asst prof pharmacol & psychol, 68-75; RES ASST PROF PHARMACOL, SCH MED, UNIV PITTSBURGH, 75- Mem: Sigma Xi; Soc Neurosci; NY Acad Sci; Am Psychol Asn. Res: Psychopharmacology of morphine and methadone dependence; primate learning. Mailing Add: 620 Scaife Hall Dept of Pharmacol Univ of Pittsburgh Sch of Med Pittsburgh PA 15261

CARLSON, LESTER WILLIAM, b Warren, Wis, Sept 12, 33; m 54; c 5. PLANT PATHOLOGY, SILVICULTURE. Educ: Carroll Col, Wis, BS, 55; Okla State Univ, MS, 59; Univ Wis, PhD(plant path), 63. Prof Exp: Asst prof plant path, SDak State Univ, 63-66; forest pathologist, 66-74, PROJ LEADER, CAN FORESTRY SERV, 74- Mem: Can Phytopath Soc; Can Inst Forestry. Res: Plant disease control; nursery cultural practices; tree seedling physiology; forest regeneration and reclamation. Mailing Add: Can Forestry Serv 5320 122nd St Edmonton AB Can

CARLSON, LEWIS JOHN, b Valley City, NDak, Oct 4, 24; m 50; c 2. PAPER CHEMISTRY. Educ: Jamestown Col, BS, 47; Univ Iowa, MS, 50. Prof Exp: Teacher high sch, NDak, 47-48; res chemist, Rayonier, Inc, 50-60; sr chemist, Minn Mining & Mfg Co, 60-62; res chemist, 62-70, SUPVR COATING APPLN RES, CROWN ZELLERBACH CORP, 70- Mem: Am Chem Soc; Tech Asn Pulp & Paper Indust. Res: Wood chemistry; carbohydrate; organic synthesis; polymers; paper coatings and treatments. Mailing Add: 232 NW 19th Camas WA 98607

CARLSON, LOREN DANIEL, physiology, deceased

CARLSON, MARGARET JOYCE, b Kankakee, Ill, Nov 30, 11. MICROBIOLOGY. Educ: Univ Ill, BS, 41; Mich State Univ, MS, 47; Univ Md, PhD(microbiol), 59. Prof Exp: Bacteriologist, Mich State Dept Health Labs, 41-49 & Vet Admin Hosp, Houston, Tex, 49-50; supvr bact med, Walter Reed Med Ctr, Washington, DC, 50-55; prof adminr microbiol, 59-73, MEM STAFF CLIN & PHYSIOL SCI PROGS, NAT INST GEN MED SCI, NIH, 73- Mem: Am Soc Microbiol; Am Pub Health Asn. Res: Medical and public health microbiology with emphasis on immunological aspects of host-parasite relationships. Mailing Add: Nat Inst of Gen Med Sci NIH Bethesda MD 20014

CARLSON, MARVIN PAUL, b Creston, Iowa, Sept 27, 35; div; c 3. GEOLOGY. Educ: Univ Nebr, BS, 57, MS, 63, PhD(geol), 69. Prof Exp: Stratigrapher, 58-63, PRIN GEOLOGIST, CONSERV & SURV DIV, UNIV NEBR, 63-, ASST DIR, 70- Concurrent Pos: Res chmn, Interstate Oil Compact Comn, 71; adv comt mem, Nat Gas Surv, Fed Power Comn, 75. Mem: Fel Geol Soc Am; AAAS; Sigma Xi. Res: Remote sensing of natural resources; land use planning; long range effect of man on natural resource systems. Mailing Add: 113 NH Univ of Nebr Lincoln NE 68588

CARLSON, MERLE WINSLOW, organic chemistry, see 12th edition

CARLSON, MILDRED V, b Arthur, Iowa. BIOCHEMISTRY, ORGANIC CHEMISTRY. Educ: State Col Iowa, BA, 65; Univ Minn, Minneapolis, PhD(biochem), 70. Prof Exp: ASST PROF CHEM, RUSSELL SAGE COL, 69- Mem: AAAS; Am Chem Soc; Grad Women in Sci. Res: Hormonal control of enzyme activities. Mailing Add: Dept of Chem Russell Sage Col Troy NY 12180

CARLSON, NORMAN ARTHUR, b Geneseo, Ill, Aug 5, 39; m 67; c 1. ORGANIC CHEMISTRY. Educ: Augustana Col, Ill, BA, 61; Univ Wis-Madison, MS, 63; Univ Mich, PhD(org chem), 67. Prof Exp: CHEMIST, ORG CHEM DEPT, E I DU PONT DE NEMOURS & CO, INC, WILMINGTON, 67- Mem: Am Chem Soc. Mailing Add: Sharpless Rd Hockessin DE 19707

CARLSON, OSCAR VERDELL, b Sioux Rapids, Iowa, Dec 12, 31; m 54; c 3. ENTOMOLOGY, INVERTEBRATE ZOOLOGY. Educ: Buena Vista Col, BS, 58; Iowa State Univ, MS, 60, PhD(entom), 67. Prof Exp: Instr biol, Wartburg Col, 60-62 & Estherville Jr Col, 62-64; ASSOC PROF BIOL, STOUT STATE UNIV, 68- Mem: Entom Soc Am. Res: Mating and oviposition of Empoasca fabae. Mailing Add: Dept of Biol Stout State Univ Menomonie WI 54751

CARLSON, PAUL ROLAND, b St Paul, Minn, Nov 23, 33; m 57; c 2. MARINE GEOLOGY. Educ: Gustavus Adolphus Col, BA, 55; Iowa State Univ, MS, 57; Ore State Univ, PhD(oceanog), 67. Prof Exp: Res asst eng geol, Soil Eng Lab, Iowa State Univ, 55-57; geologist, US Army Corps Engrs, 57-58; instr geol, Gustavus Adolphus Col, 58-59; teacher gen sci, Cleveland Consol Schs, 59-61; instr geol, Pac Lutheran Univ, 61-63; res assoc geol oceanog, Ore State Univ, 63-67; GEOLOGIST, US GEOL SURV, 67- Mem: Geol Soc Am; Am Geophys Union; Soc Econ Paleont & Mineral. Res: Sedimentology; submarine canyons; processes of sedimentation; estuaries, especially San Francisco Bay; acoustical subbottom profiling; oceanographic factors affecting estuaries. Mailing Add: US Geol Surv 345 Middlefield Rd Menlo Park CA 94025

CARLSON, PETER S, genetics, see 12th edition

CARLSON, PHILIP R, b Evanston, Ill, June 2, 31; m 54; c 3. MATHEMATICS. Educ: Bethel Col (Minn), BS, 53; Univ Minn, BS, 57, MS, 65, PhD, 71. Prof Exp: Teacher high sch, Minn, 57-59; instr math, 60-67, from asst prof to assoc prof, 67-71, PROF MATH, BETHEL COL (MINN), 71- Concurrent Pos: Programmed Course in Algebra for High Sch Teachers, 60-66; dir, Job Corps Math Proj Minn Nat Lab, 67-68; vis scholar math, Cambridge Univ (Eng), 73-74. Mem: Math Asn Am; Nat Coun Teachers of Math. Mailing Add: Dept of Math Bethel Col St Paul MN 55101

CARLSON, PHILLIP RICHARD, b Gig Harbor, Wash, Oct 28, 11; m 38; c 3. PHYSICS. Educ: Col Puget Sound, BS, 35; Univ Wash, PhD(physics), 40. Prof Exp: Prof physics, Pasadena Col, 39-41; res engr, Lockheed Aircraft Corp, 41-45; prof physics & head dept sci, Pasadena Col, 45-52; mgr dept, Air Force Studies, Lockheed Aircraft Corp, 52-62, div engr adv studies & applns, Lockheed-Calif Co, 62, res dir aircraft, 63-65, RES ADV, LOCKHEED-CALIF CO, LOCKHEED AIRCRAFT CORP, 65- Mem: Am Phys Soc; Opers Res Soc Am. Res: Photonuclear effect; heat transmission in aircraft; sound control in aircraft; photo-disintegration of atomic nuclei; operations research; research management. Mailing Add: 1531 N Michigan Ave Pasadena CA 91104

CARLSON, RAYMOND ELIAS, organic chemistry, see 12th edition

CARLSON, RICHARD EUGENE, b Wausa, Nebr, Oct 19, 40; m 60; c 2. AGRONOMY. Educ: Univ Nebr, BS, 67; Iowa State Univ, MS, 69, PhD(agr climat), 71. Prof Exp: Researcher & teaching asst, Iowa State Univ, 67-68, NDEA fel, 68-71; photointerpreter, Purdue Univ, 71; asst prof, 71-75, ASSOC PROF AGR CLIMAT, IOWA STATE UNIV, 75- Mem: Am Soc Agron; Crop Sci Soc Am; Soil Sci Soc Am; Am Meteorol Soc. Res: Remote sensing and agriculture; microclimate studies on corn and soybeans with emphasis on moisture stress and photosynthesis. Mailing Add: Dept of Agron Iowa State Univ Ames IA 50010

CARLSON, RICHARD FREDERICK, b St Paul, Minn, June 19, 36; m 57; c 3. PHYSICS. Educ: Univ Redlands, BS, 57; Univ Minn, MS, 62, PhD, 64. Prof Exp: Asst res physicist, Univ Calif, Los Angeles, 64, asst prof physics, 64-67; asst prof, 67-71, coordr dept, 70-72, ASSOC PROF PHYSICS, UNIV REDLANDS, 71- Mem: Am Phys Soc; Am Sci Affil. Res: Experimental nuclear physics; nuclear structure; few nucleon problem. Mailing Add: Dept of Physics Univ of Redlands Redlands CA 92373

CARLSON, RICHARD OSCAR, b Flushing, NY, June 2, 26; m 52; c 3. PHYSICS. Educ: Columbia Univ, AB, 47, AM, 49, PhD(physics), 52. Prof Exp: Asst, Columbia Univ, 47-50, physicist, Hudson Labs, 52-54; PHYSICIST, GEN ELEC CORP RES & DEVELOP CTR, 54- Mem: Fel Am Phys Soc; Electrochem Soc; sr mem Inst Elec & Electronics Eng. Res: Processing of silicon devices and integrated circuits; experimental transport measurements on pure and doped semiconductor crystals; semiconductor injection lasers; thermoelectric energy conversion; tuning capacitors; silicon solar cells. Mailing Add: Gen Elec Corp R&D Ctr PO Box 8 Schenectady NY 12301

CARLSON, RICHARD RAYMOND, b Chicago, Ill, Sept 15, 23; m 51; c 3. NUCLEAR PHYSICS. Educ: Univ Chicago, PhD(physics), 51. Prof Exp: Res assoc physics, Univ Chicago, 51; from asst prof to assoc prof, 51-63, PROF PHYSICS, UNIV IOWA, 63- Concurrent Pos: Guggenheim fel, Oxford Univ, 59; mem staff, Los Alamos Sci Lab, Univ Calif, 56; physicist, Oak Ridge Nat Lab, 52. Mem: Am Phys Soc. Mailing Add: Dept of Physics Univ of Iowa Iowa City IA 52241

CARLSON, RICHARD VERNER, nuclear chemistry, physical chemistry, see 12th edition

CARLSON, ROBERT BRUCE, b Virginia, Minn, Sept 15, 38; m 60. ENTOMOLOGY. Educ: Univ Minn, Duluth, BS, 60; Mich State Univ, MS, 62, PhD(entom), 65. Prof Exp: Assoc insect ecologist, NCent Forest Exp Sta, US Forest Serv, 65-67; asst prof entom, 68-71, ASSOC PROF ENTOM, NDAK STATE UNIV, 71- Mem: AAAS; Ecol Soc Am; Entom Soc Am; Entom Soc Can. Res: Ecology. Mailing Add: Dept of Entom NDak State Univ Fargo ND 58102

CARLSON, ROBERT FRITZ, b Sweden, June 27, 09; nat US; m 49; c 3. HORTICULTURE. Educ: Univ Minn, BS, 44; Mich State Univ, MS, 49, PhD, 52. Prof Exp: Res assoc pomol, NY Exp Sta, Cornell Univ, Geneva, NY, 44-46; from asst prof to assoc prof, 46-66, PROF HORT, MICH STATE UNIV, 66- Concurrent Pos: Vis prof, Univ Uppsala, Lund & Agr Exp Stas, Sweden, 53; agr res adv to Univ Ryukyus, Okinawa, Japan, 56-58; speaker, Apricot Symp, Yugoslavia, 68 & Int Hort Congs, Israel, 70 & Poland, 74; consult, US AID to Uruguay, 75. Honors & Awards: Stark Award, Am Soc Hort Sci, 66; Paul Shepard Award, Am Pomol Soc, 73; Gold Medal, Mass Hort Soc & Wilder Award, Am Pomol Soc, 74. Mem: Fel Am Soc Hort Sci; Am Pomol Soc; Int Hort Soc. Res: Rootstock physiology; dormancy problems in tree fruits; rootstocks for various tree fruits. Mailing Add: Dept of Hort Mich State Univ East Lansing MI 48824

CARLSON, ROBERT GEORGE, b Grand Rapids, Mich, Apr 1, 22; m 42; c 2. VETERINARY PATHOLOGY. Educ: Mich State Univ, DVM, 52; Purdue Univ, MS, 54, PhD(vet path), 56; Am Col Lab Animal Med, dipl; Am Col Vet Path, dipl. Prof Exp: Instr vet sci, Purdue Univ, 52-56, asst prof, 56; res assoc, 56-61, SECT HEAD PATH, UPJOHN CO, 61-, INT ADV, JAPAN UPJOHN, LTD, 74- Mem: Am Vet Med Asn; Am Col Vet Path; Am Col Lab Animal Med; Soc Toxicol. Res: Veterinary pathology; toxicology and pathology research. Mailing Add: Pharmaceut Res & Develop Upjohn Co Kalamazoo MI 49001

CARLSON, ROBERT GIDEON, b Chicago, Ill, Feb 4, 38; m 62; c 2. ORGANIC CHEMISTRY. Educ: Univ Ill, BS, 59; Mass Inst Technol, PhD(org chem), 63. Prof Exp: From asst prof to assoc prof chem, 63-72, PROF CHEM, UNIV KANS, 72- Concurrent Pos: Alfred P Sloan Found res fel, 70-72. Mem: Am Chem Soc; The Chem Soc. Res: Synthetic organic chemistry; structure and synthesis of natural products; highly strained ring systems; photochemical reactions of organic compounds. Mailing Add: Dept of Chem Univ of Kans Lawrence KS 66044

CARLSON, ROBERT KENNETH, b Chicago, Ill, July 7, 28; m 54; c 2. INORGANIC CHEMISTRY. Educ: Northwestern Univ, PhB, 54. Prof Exp: Asst chemist, Great Lakes Carbon Corp, 49-50, chemist, 53-57; res chemist, Borg-Warner Corp, 57-59; mat engr, Chance Vought Corp, 59-60, res scientist, 60-62; sr scientist, Res Ctr, Ling-Temco-Vought Corp, 62-63, head mat sci, 63-64, prog mgr, Carbon & Graphite Div, 64-67; V PRES & GEN MGR, POCO GRAPHITE INC, UNION OIL CO, CALIF, 67- Concurrent Pos: Prin investr, US Air Force Indust res grant, 61-62. Mem: Sigma Xi; Am Chem Soc; fel Am Inst Chemists. Res: Synthesis of new carbon and graphite products; high temperature refractory oxides for rocket and missile application; synthesis of salts of organo phosphoric acids. Mailing Add: Poco Graphite Union Oil Co Calif PO Box 2121 Decatur TX 76234

CARLSON, ROBERT LEONARD, b Duluth, Minn, Feb 26, 32; m 58; c 2. PHOTOGRAPHIC CHEMISTRY. Educ: Univ Minn, BA, 58; Univ Ill, PhD(inorg chem), 62. Prof Exp: PROD DEVELOP MGR, 3M CO, 62- Res: Metal ion complexes and molecular complexes in non-aqueous solvents; electrochemistry; photoconductors; photographic chemistry; color photographics imaging systems. Mailing Add: 2243 Berland Pl St Paul MN 55119

CARLSON, ROBERT M, b Cokato, Minn, Oct 30, 40; m 62; c 2. ORGANIC CHEMISTRY. Educ: Univ Minn, BChem, 62; Princeton Univ, PhD(chem), 66. Prof Exp: Fel, Harvard Univ, 65-66; from asst prof to assoc prof org chem, 66-74, PROF CHEM, UNIV MINN, DULUTH, 74- Mem: Am Chem Soc. Res: Total synthesis of natural products; new synthetic methods. Mailing Add: Dept of Chem Univ of Minn Duluth MN 55812

CARLSON, ROBERT MARVIN, b Denver, Colo, Mar 18, 32; m 62; c 1. SOIL CHEMISTRY, PLANT NUTRITION. Educ: Colo State Univ, BS, 54; Univ Calif, Berkeley, PhD(soil sci), 62. Prof Exp: Soil scientist, USDA, 52-54; sr lab technician, Univ Calif, Berkeley, 55-62; prog specialist, Ford Found, Arg, 62-65; asst pomologist, 65-71, ASSOC POMOLOGIST, UNIV CALIF, DAVIS, 71- Concurrent Pos: Vis res prof, Nat Univ South, Arg, 62-65. Mem: AAAS; Am Soc Agron; Soil Sci Soc Am; Am Chem Soc. Res: Analytical chemistry. Mailing Add: Dept of Pomol Univ of Calif Davis CA 95616

CARLSON, ROBERT WARNER, b Waseca, Minn, Oct 26, 41; m 63; c 2. AERONOMY, ATOMIC PHYSICS. Educ: Calif State Polytech Col, San Luis Obispo, BS, 63; Univ Southern Calif, PhD(physics), 70. Prof Exp: Res scientist, Jet Propulsion Lab, Calif Inst Technol, 63-66; res asst physics, 65-70, RES PHYSICIST, UNIV SOUTHERN CALIF, 70- Concurrent Pos: Consult, Jet Propulsion Lab, Calif Inst Technol, Gen Dynamics Corp & var indust orgn, 67- Mem: AAAS; Optical Soc Am; Am Geophys Union. Res: Atomic and molecular physics and their application to atmospheric and astrophysical problems. Mailing Add: Dept of Physics Univ of Southern Calif Los Angeles CA 90007

CARLSON, ROGER, b St Joseph, Mo, Aug 15, 37; m 60; c 1. MATHEMATICAL STATISTICS. Educ: Univ Kansas City, BS, 59, MA, 60; Harvard Univ, PhD(statist), 64. Prof Exp: Asst prof, 63-69, ASSOC PROF MATH, UNIV MO-KANSAS CITY, 69- Mem: Am Statist Asn. Res: Foundations of statistical inference; applications of mathematics to social sciences. Mailing Add: Dept of Math Univ of Mo Kansas City MO 64110

CARLSON, RONALD H, b Worcester, Mass, Nov 11, 38; m 60; c 3. INDUSTRIAL CHEMISTRY. Educ: Univ Mass, BS, 60; Univ Ill, MS, 62, PhD(inorg chem), 65. Prof Exp: From res chemist to sr res chemist, Hooker Chem Corp, 64-71; SR RES CHEMIST, FMC CORP, 71- Mem: Am Chem Soc; The Chem Soc. Res: Metal coordination chemistry; electrochemistry; inorganic phosphate and synthetic organophosphorus chemistry; heterocyclic chemistry; environmental chemistry; catalysis; inorganic coatings; spectroscopy. Mailing Add: FMC Corp Indust Chem Div Princeton NJ 08540

CARLSON, ROY DAVID, b Duluth, Minn, Dec 22, 14; m 47; c 2. POULTRY HUSBANDRY. Educ: Univ Minn, BS, 47. Prof Exp: Mem staff, Swift & Co, 36-42; exec secy, NDak Poultry Improv Bd, 46-48; poultry husbandman, Nat Poultry Improv Plan, Agr Res Ctr, Agr Res Serv, USDA, 48-58; dir div poultry industs, 58-67, dep comnr, 67-70, DIR DIV POULTRY INDUSTS, MINN DEPT AGR, 71- Mem: Poultry Sci Asn. Res: Egg marketing and poultry improvement. Mailing Add: Minn Dept of Agr 530 State Off Bldg St Paul MN 55155

CARLSON, ROY DOUGLAS, b San Bernardino, Calif, Dec 29, 44; m 66; c 2. BIOPHYSICS. Educ: Pomona Col, BA, 66; Univ Wis-Madison, MA, 68, PhD(biophys), 73. Prof Exp: NIH FEL, UNIV TENN-OAK RIDGE GRAD SCH BIOMED SCI & BIOL DIV, OAK RIDGE NAT LAB, 73- Mem: Biophys Soc; Sigma Xi. Res: X-ray scattering studies and physical biochemistry of chromatin and natural and synthetic nucleic acids; macromolecular structure and assembly. Mailing Add: Univ Tenn-Oak Ridge Grad Sch Biomed Sci Biol Div Oak Ridge Nat Lab Oak Ridge TN 37830

CARLSON, SHELDON DUANE, organic chemistry, see 12th edition

CARLSON, STANLEY DAVID, b St Paul, Minn, Sept 4, 34; m 58, 69; c 4. ENTOMOLOGY, PHYSIOLOGY. Educ: Univ Minn, BS, 56; Univ Nebr, MS, 61; Kans State Univ, PhD(entom, physiol), 65. Prof Exp: Res entomologist, Stored Prod Insects Res Br, Agr Res Serv, USDA, Kans, 59-64; asst prof entom, Va Polytech Inst, 65-67; NIH spec fel physiol, Karolinska Inst, Sweden, 67-69 & biol, Yale Univ, 69-70; asst prof entom, Univ Ill, Urbana, 70-71; asst prof, 71-76, ASSOC PROF ENTOM, UNIV WIS-MADISON, 76- Concurrent Pos: Entomologist, Colo Dept Agr, 59; USDA res grant, 65-; consult, Panogen Div, Morton Chem Co, 61-62. Mem: AAAS; Entom Soc Am; Am Inst Biol Sci; Scand Physiol Soc. Res: Sensory physiology of insects; physiology of insect vision and microspectrophotometry of insect visual pigments; ultrastructure (SEM, TEM, HVEM), of compound eye and optic tract. Mailing Add: Dept of Entom Univ of Wis Madison WI 53706

CARLSON, STEVEN ALLEN, b Baltimore, Md, Nov 3, 43; m 68; c 1. ORGANIC CHEMISTRY. Educ: Principia Col, BS, 65; Mass Inst Technol, PhD(org chem), 69. Prof Exp: Sr chemist, Lexington Res Labs, Itek Corp, Mass, 69-71; MGR RES & DEVELOP, ARKWRIGHT-INTERLAKEN, INC, 71- Mem: Am Chem Soc; Am Inst Chemists; Am Soc Photog Scientist & Engrs. Res: Photochemistry of anthraquinone; delayed thermal fluorescence; organic dry-imaging reactions; photochemistry of diazo compounds; coating of polyester film. Mailing Add: Arkwright-Interlaken Inc Fiskeville RI 02823

CARLSON, THOMAS ARTHUR, b Waterbury, Conn, Apr 1, 28; m 50. MOLECULAR PHYSICS Educ: Trinity Col, Conn, BS, 50; Johns Hopkins Univ, MA, 51, PhD(chem), 54. Prof Exp: SR RES STAFF MEM, OAK RIDGE NAT LAB, 54- Concurrent Pos: Guggenheim fel, 67; ed, J Electron Spectroscopy, 72- Mem: Am Chem Soc; fel Am Phys Soc; AAAS. Res: Electron spectroscopy; Auger and electron shake-off phenomena; hot atom chemistry; atomic physics. Mailing Add: Physics Div Oak Ridge Nat Lab PO Box X Oak Ridge TN 37830

CARLSON, TOBY NAHUM, b Brooklyn, NY, Nov 4, 36; m 69; c 2. METEOROLOGY. Educ: Mass Inst Technol, BS, 58, MS, 60; Univ London, PhD(meteorol), 65. Prof Exp: Res meteorologist, Weather Serv, Inc, on contract to Air Force Cambridge Res Lab, 61-62, Nat Hurricane Res Lab, Nat Oceanic & Atmospheric Admin, Environ Res Lab, 65-74; ASSOC PROF METEOROL, PA STATE UNIV, 74- Honors & Awards: Distinguished Authorship, Nat Oceanic & Atmospheric Admin, 74. Mem: Am Meteorol Soc. Res: Modelling urban heat fluxes, heatbudget surface temperature; investigation of aerosol transport and effects of aerosol layer on solar radiation balance over equatorial Atlantic. Mailing Add: Dept of Meteorol Pa State Univ University Park PA 16802

CARLSON, WALTER ERIK, animal nutrition, see 12th edition

CARLSON, WAYNE R, b Moline, Ill, Dec 31, 40; m. CYTOGENETICS. Educ: Rockford Col, BA, 62; Ind Univ, Bloomington, MA, 67, PhD(zool), 68. Prof Exp: ASSOC PROF GENETICS, UNIV IOWA, 68- Concurrent Pos: NSF res grant, 75-77. Res: Cytogenetics of maize, especially the B chromosome. Mailing Add: Dept of Bot Univ of Iowa Iowa City IA 52240

CARLSON, WILLARD EMMETT, b Ft Dodge, Iowa, May 11, 23; m 50; c 4. PAPER CHEMISTRY. Educ: Iowa State Univ, BS, 47; Lawrence Col, MS, 49. Prof Exp: Dir res, Whiting-Plover Paper Co, 52-57; from supvr res & develop, Cent Tech Dept to mgr cent res, St Regis Paper Co, 57-64, mgr develop, 64-66, TECH DIR, KRAFT DIV, ST REGIS PAPER CO, 66- Mem: Tech Asn Pulp & Paper Indust. Res: Pulp and paper research; coatings; natural and synthetic fibrous web formation; fiber bonding; water removal; elastic and resilient properties of fibrous webs; Kraft odor abatement. Mailing Add: St Regis Paper Co 2400 Gulf Life Tower Jacksonville FL 32207

CARLSON, WILLIAM DWIGHT, b Denver, Colo, Nov 5, 28; m 50; c 2.

VETERINARY RADIOLOGY, RADIATION BIOLOGY. Educ: Colo State Univ, DVM, 52, MS, 56; Univ Colo, PhD(radiol), 58; Am Bd Vet Radiol, dipl, 62. Prof Exp: Vet, 52-53; asst prof dept med, Colo State Univ, 53-55, Am Vet Med Asn fel, 55-57, assoc prof & radiologist, Col Vet Med, 57-62, prof radiol, 62-68, chmn dept radiol & radiation biol, 64-68, dir radiol health animal res lab, Univ-USPHS, 62-68, pres bd trustees, Colo State Univ Res Found, 66-68; PRES, UNIV WYO, 68- Concurrent Pos: Civilian adv, US Army Gen Command & Staff Sch, Ft Leavenworth, Kans, 69-72; consult, USPHS, 62-, Vet Admin Med Facil, 68-, Am Inst Biol Sci, 69-70, NASA, 69- & Surg Gen, US Air Force, 70-; mem bd comnr, Nat Comn on Accrediting, 69-72; mem nat adv coun health res facil, NIH, 69-73; secy adv comt coal mine safety res, Dept Interior, 71; mem bd visitors, Air Univ, 73- Mem: AAAS; Am Soc Nuclear Med; Am Vet Radiol Soc; Am Vet Med Asn; Radiol Soc NAm. Res: Diagnostic, therapeutic, radioactive isotopes; lower animals. Mailing Add: Box 3434 Univ Sta Laramie WY 82070

CARLSON, WILLIAM H, b Kingston, Pa, Aug 20, 41; m 63; c 2. HORTICULTURE. Educ: Pa State Univ, BS, 63, MS, 64, PhD(hort), 66. Prof Exp: Asst prof floricult, 66-71, ASSOC PROF HORT, MICH STATE UNIV, 71-, EXTEN SPECIALIST HORT, 66- Concurrent Pos: Exec secy, Bedding Plants, Inc, 70- Honors & Awards: Kenneth Post Award, Am Soc Hort Sci, 68. Mem: Am Soc Hort Sci; Int Soc Hort Sci. Res: Plant nutrition of Rosa hybrida, Chrysanthemum morifolium; physiology of Lilium longifolium, Petunia hybrida and Euphorbia pulcherrima. Mailing Add: Dept of Hort Mich State Univ East Lansing MI 48823

CARLSON, WILLIAM SAMUEL, b Ironwood, Mich, Nov 18, 05; m 32; c 1. GEOLOGY, METEOROLOGY. Educ: Univ Mich, AB, 29, MS, 32, PhD(geol), 38. Hon Degrees: LLD, Dickinson Col, 48, Univ Mich, 50, Univ Del & Middlebury Col, 51 & Bowling Green State Univ, 64; DSc, Alfred Univ, 53; LHD, Univ Cincinnati, 70; Dr Arts and Letters, Univ Toledo, 72. Prof Exp: Asst geol, Univ Mich, 26-27, 29-30, asst instr, 32-34; prin high sch, Mich, 34-37; asst prof educ, Univ Minn, 37-39, assoc prof, 39-41, dir admissions & records, 41; pres, Univ Del, 46-50, Univ Vt, 50-52 & State Univ NY, 52-58; pres, 58-72, EMER PRES, UNIV TOLEDO, 72- Concurrent Pos: Aerologist, Univ Mich Greenland Exped, 28-29, leader, 30-31. Mem: AAAS. Res: Glaciology, geology and meteorology of Greenland; aerology near bodies of ice. Mailing Add: 4047 Newcastle Dr Sylvania OH 43560

CARLSON, WILLIAM THEODORE, b Lyons, Nebr, Mar 27, 33; m 60; c 2. INFORMATION SCIENCE. Educ: Univ Nebr, BS, 57, MS, 59. Prof Exp: Asst & instr agron, Univ WVa, 59-62; asst analyst, 62-63, chief plant sci, 63-66, CHIEF AGR & ALLIED APPL SCI, SCI INFO EXCHANGE, SMITHSONIAN INST, 66- Concurrent Pos: Adv, Agr Res Inst-Nat Acad Sci, 66- Mem: Am Soc Agron; Am Soc Info Sci. Res: Forage management; weed control; information storage and retrieval techniques; agricultural and plant sciences; environmental biology; water resources; pesticides. Mailing Add: Smithsonian Inst Sci Info Exch 1730 M St NW Washington DC 20036

CARLSSON, ERIK, b Vingaker, Sweden, Mar 31, 24; m 51; c 2. RADIOLOGY. Educ: Karolinska Inst, Sweden, MD, 52, PhD, 70. Prof Exp: Mem staff, Radiol Dept, Karolinska Sjukhuset, Stockholm, Sweden, 54-57, pediat roentgenol, 55-57 & thorax roentgenol, 60-62; from asst prof to assoc prof radiol, Sch Med, Wash Univ, 58-64; assoc prof radiol, head cardiovasc radiol & staff mem, Cardiovasc Res Inst, Med Ctr, 64-68, PROF RADIOL, UNIV CALIF, SAN FRANCISCO, 68- Concurrent Pos: NIH training grant & consult. Mem: Asn Univ Radiol; NY Acad Sci. Res: Diagnostic and cardiovascular radiology. Mailing Add: Univ of Calif Med Ctr San Francisco CA 94143

CARLSTEAD, EDWARD MEREDITH, b Chillicothe, Mo, Aug 1, 25; m 50; c 3. METEOROLOGY, COMPUTER SCIENCE. Educ: Univ Calif, Los Angeles, BS, 46; US Naval Postgrad Sch, MS, 53. Prof Exp: Analyst, Joint Weather Bur-Air Force-Navy Anal Ctr, Washington, DC, 49-51; chief analyst, Joint Numerical Weather Prediction Unit, Md, 56-59; res meteorologist, Fleet Numerical Weather Facility, 59-62; officer in chg opers anal, Pac Command Detachment, Naval Command Systs Support Activity, 62-65; chief sci serv div, Pac Regional Hq, Nat Weather Serv, Environ Sci Serv Admin, 65-71, METEOROLOGIST IN CHG, NAT WEATHER SERV FORECAST OFF, NAT OCEANIC & ATMOSPHERIC ADMIN, HONOLULU, 71- Mem: Am Meteorol Soc; Sigma Xi. Res: Programming of digital computers to analyze and predict wind and temperature fields in the tropics; improvement of forecasting techniques; physical oceanography; operations analysis. Mailing Add: Nat Weather Serv Forecast Off Honolulu Int Airport Honolulu HI 96819

CARLSTON, RICHARD CHARLES, inorganic chemistry, solid state physics, see 12th edition

CARLSTONE, DARRY SCOTT, b Vinita, Okla, May 15, 39; m 62; c 2. PARTICLE PHYSICS. Educ: Univ Okla, BS, 61; Purdue Univ, MS, 64, PhD(physics), 68. Prof Exp: Asst prof, 67-72, ASSOC PROF PHYSICS, CENT STATE UNIV, OKLA, 72- Mem: Am Phys Soc; Am Asn Physics Teachers. Res: Theoretical particle physics with special interest in symmetry principles. Mailing Add: Dept of Physics Cent State Univ Edmond OK 73034

CARLTON, BRUCE CHARLES, b Burrillville, RI, Aug 3, 35; m 56; c 2. GENETICS. Educ: Univ NH, BS, 57; Mich State Univ, MS, 58, PhD(genetics), 60. Prof Exp: Asst hort, Mich State Univ, 57-59; USPHS trainee, Stanford Univ, 60-62; from asst prof to assoc prof biol, Yale Univ, 62-71; PROF BIOCHEM & MICROBIOL, UNIV GA, 71- Concurrent Pos: USPHS res grant, 63-l NSF res grant, 65-; fel, Silliman Col, 69-71. Mem: Am Soc Microbiol; Genetics Soc Am. Res: Mechanisms of genetic control through studies of mutationally-altered proteins and structure and function of circular DNA elements. Mailing Add: 523 Biol Sci Bldg Univ of Ga Athens GA 30602

CARLTON, RICHARD WALTER, b Nov 23, 42; US citizen; m 66; c 2. SEDIMENTARY PETROLOGY. Educ: Wash State Univ, BS, 65; Ore State Univ, MS, 68, PhD(geol), 72. Prof Exp: GEOLOGIST, OHIO GEOL SURV, 70- Mem: Soc Econ Paleontologists & Mineralogists; Int Asn Sedimentologists. Res: Chlorite-illite rations used in conjunction with stratigraphic and petrographic studies to reconstruct the paleogeography of Lower Silurian clastic rocks in Ohio. Mailing Add: Dept of Natural Resources Div of Geol Surv Columbus OH 43221

CARLTON, ROBERT AUSTIN, b Brownsville, Tenn, Apr 30, 27; m 50; c 2. ZOOLOGY. Educ: Lambuth Col, BS, 50; George Peabody Col, MA, 51; Ala Polytech Inst, PhD(zool), 58. Prof Exp: Instr biol, Northeast Miss Jr Col, 51-54; zool, Ala Polytech Inst, 54-56; assoc prof biol, Delta State Col, 56-64; PROF BIOL, LAMBUTH COL, 64- Mem: Nat Asn Biol Teachers. Res: Vertebrate ecology; natural history of vertebrates. Mailing Add: Lambuth Col Jackson TN 38302

CARLTON, TERRY SCOTT, b Peoria, Ill, Jan 29, 39; m 60; c 2. THEORETICAL CHEMISTRY. Educ: Duke Univ, BS, 60; Univ Calif, Berkeley, PhD(phys chem), 63. Prof Exp: Asst prof chem, 63-69, ASSOC PROF CHEM, OBERLIN COL, 69- Mem: Am Phys Soc. Res: Quantum theory. Mailing Add: Dept of Chem Oberlin Col Oberlin OH 44074

CARLTON, VIRGINIA, b Rosebud, Tex, Mar 20, 18. MATHEMATICS. Educ: Centenary Col, BS, 39; Tulane Univ, MA, 40; Northwestern Univ, PhD, 59. Prof Exp: Teacher high sch, La, 40-41; instr math, Wesleyan Col, 41-46, prof & head dept, 50-55; asst prof, Centenary Col, 46-48; assoc prof, Northwestern State Col, 48-50; lectr, Northwestern Univ, 55-57; PROF MATH, CENTENARY COL, 57-, HEAD DEPT, 57- Concurrent Pos: Ford fel, Fund Advan Educ, 53-54; Fulbright lectr, Ghana, 63-64 & Liberia, 70-72. Mem: Am Math Soc; Soc Indust & Appl Math; Math Asn Am. Res: Teaching of mathematics. Mailing Add: Dept of Math Centenary Col Shreveport LA 71104

CARLTON, WILLIAM HERBERT, b Statesboro, Ga, Oct 6, 40; m 62; c 3. MEDICAL PHYSICS. Educ: Emory Univ, BS, 62, MS, 64; Rutgers Univ, PhD(radiation biophys), 69; Am Bd Health Physics, dipl, 70; Am Bd Radiol, dipl & cert radiol physics, 73. Prof Exp: Lectr radiation sci, Rutgers Univ, 64-69; asst prof, 69-74, ASSOC PROF RADIOL, MED COL GA, 74- Concurrent Pos: Consult Colgate-Palmolive Res Lab, 68-69 & Vet Admin Hosp, Augusta, Ga, 70- Mem: Am Asn Physicists in Med; Soc Nuclear Med; Health Physics Soc. Res: Low energy x-ray absorption; bioelectric potentials of root tips; lacrimal scanning. Mailing Add: Dept of Radiol Med Col of Ga Augusta GA 30902

CARLTON, WILLIAM MARION, botany, see 12th edition

CARLTON, WILLIAM WALTER, b Owensboro, Ky, June 17, 29; m 55; c 2. VETERINARY PATHOLOGY, VETERINARY TOXICOLOGY. Educ: Univ Ky, BS, 53, MS, 56; Fla Southern Col, BS, 54; Auburn Univ, DVM, 60; Purdue Univ, PhD(vet path), 63; Am Col Vet Path, dipl. Prof Exp: Instr path, Purdue Univ, 60-62; asst prof, Mass Inst Technol, 62-65; assoc prof, 65-68, PROF PATH, SCH VET MED, PURDUE UNIV, WEST LAFAYETTE, 68- Mem: Am Inst Nutrit; Am Asn Avian Path; Am Soc Exp Path; Int Acad Path; Soc Toxicol. Res: Nutritional and toxicological diseases, especially of cardiovascular and nervous systems. Mailing Add: Dept of Vet Microbiol Purdue Univ Sch of Vet Med West Lafayette IN 47906

CARLUCCI, ANGELO FRANCIS, b Plainfield, NJ, Feb 17, 31. MICROBIAL ECOLOGY, MARINE MICROBIOLOGY. Educ: Rutgers Univ, BS, 53, MS, 56, PhD(microbiol), 59. Prof Exp: Asst microbiol, Tela RR Co, United Fruit Co, Honduras, 59-61; jr res biologist, Scripps Inst Oceanog, Univ Calif, San Diego, 61-62, postgrad res biologist, 62-64, asst res biologist, Inst Marine Resources, 64-70, ASSOC RES MICROBIOLOGIST, INST MARINE RESOURCES, UNIV CALIF, SAN DIEGO, 71-, LECTR, 70- Mem: AAAS; Am Soc Microbiol; Am Soc Limnol & Oceanog; Sigma Xi. Res: General marine microbiology; nitrogen cycle in the sea; vitamins and metabolites in marine ecology; survival of bacteria in seawater; inhibitory properties of seawater. Mailing Add: Inst of Marine Resources Univ of Calif at San Diego La Jolla CA 92093

CARLUCCIO, LEEDS MARIO, b Leominster, Mass, Sept 12, 36; m 61; c 3. BOTANY, MORPHOLOGY. Educ: Mass Col Pharm, BS, 58, MS, 60; Cornell Univ, PhD(paleobot), 66. Prof Exp: Instr gen bot, Cornell Univ, 64-66; from asst prof to assoc prof, 66-72, PROF GEN BOT, CENT CONN STATE COL, 72- Concurrent Pos: Mem Int Orgn Paleobot. Mem: Bot Soc Am; Sigma Xi; Am Asn Univ Prof. Res: Anatomy and morphology of the progymnosperms of Devonian floras. Mailing Add: Dept of Biol Cent Conn State Col New Britain CT 06050

CARLYLE, DAVID WESLEY, b Johnson Co, Mo, June 9, 38; m 59; c 3. INORGANIC CHEMISTRY. Educ: Univ Mo, BS, 60; Iowa State Univ, PhD(chem), 68. Prof Exp: Chem engr, Procter & Gamble Mfg Co, Kans, 60-62; teacher high sch, Raytown, Mo, 62-64; res assoc chem, Colo Univ, 68-69; asst prof, Tex Tech Univ, 69-74; TEACHER, ODESSA HIGH SCH, MO, 74- Mem: Am Chem Soc; AAAS. Res: Mechanism of inorganic reactions in aqueous solution. Mailing Add: R R 1 Box 115D Odessa MO 64076

CARMACK, MARVIN, b Dana, Ind, Sept 1, 13; m 60. ORGANIC CHEMISTRY. Educ: Univ Ill, AB, 37; Univ Mich, MS, 39, PhD(org chem), 40. Prof Exp: Asst org chem, Univ Ill, 40-41; Towne instr, Univ Pa, 41-44, from asst prof to prof, 44-53; PROF ORG CHEM, IND UNIV, BLOOMINGTON, 53- Concurrent Pos: Guggenheim fel, Swiss Fed Inst Technol, 49-50; Fulbright res scholar, Commonwealth Sci & Indust Res Orgn, Melbourne, 60-61. Mem: AAAS; Am Chem Soc; NY Acad Sci; The Chem Soc; Swiss Chem Soc. Res: Natural products; organic sulfur chemistry; heterocyclic compounds. Mailing Add: Dept of Chem Ind Univ Bloomington IN 47401

CARMACK, ROBERT, b Winslow, Ariz, Feb 24, 34; m 59; c 4. ANTHROPOLOGY. Educ: Univ Calif, Los Angeles, BA, 60, MS, 62, PhD(anthrop), 65. Prof Exp: Lectr social & introductory anthrop & Mesoamerica & asst coordr Latin Am studies, Peace Corps, Univ Calif, Los Angeles, 63-64; asst prof anthrop, Ariz State Univ, 64-66, fac res grant, 65-66; Ford Found Foreign Area fel, 66-67; asst prof, Univ Calif, San Diego, 67-70; ASSOC PROF ANTHROP & ASSOC CTR INTER-AM STUDIES, STATE UNIV NY ALBANY, 70- Mem: Fel Am Anthrop Asn; Royal Anthrop Inst Gt Brit & Ireland. Res: Social anthropology, especially traditional political systems; middle-America, especially the Maya. Mailing Add: Dept of Anthrop State Univ of NY Albany NY 12203

CARMAN, GLENN ELWIN, b Waterloo, Iowa, June 8, 14; m 41; c 2. ENTOMOLOGY. Educ: State Univ Iowa, BS, 36; Cornell Univ, PhD(entom), 42. Prof Exp: Res entomologist, Rohm & Haas Co, 42-43; from jr entomologist to assoc entomologist, Agr Exp Sta, 43-53, chmn dept, Univ Calif, 63-69, ENTOMOLOGIST, CITRUS RES CTR & AGR EXP STA, UNIV CALIF, RIVERSIDE, 53-, PROF ENTOM, 63- Concurrent Pos: VChmn & mem, Bd Dirs, Indust Comt Citrus Additives & Pesticides, Inc, 68- Mem: AAAS; Entom Soc Am. Res: Insect toxicology; economic entomology; biology and control of insect pests; evaluation of application equipment; reentry studies; snail control on citrus crops. Mailing Add: 5368 Pinehurst Dr Riverside CA 92504

CARMAN, JOHN HOMER, b Pittsburgh, Pa, Sept 27, 35; m 56; c 4. PETROLOGY, GEOCHEMISTRY. Educ: Allegheny Col, BS, 57; NMex Inst Mining & Technol, MS, 60; Pa State Univ, PhD(petrol & mineral), 69. Prof Exp: Res asst geol, Stanford Univ, 65-68; asst prof, 68-74, ASSOC PROF GEOL, UNIV IOWA, 74- Mem: Geol Soc Am; Mineral Soc Am; Am Geophys Union. Res: Experimental phase equilibrium studies bearing on the natural history of rocks and minerals; graphical studies concerning the chemical evolution of igneous rocks and minerals. Mailing Add: Dept of Geol Univ of Iowa Iowa City IA 52242

CARMAN, MAX FLEMING, JR, b Lansing, Mich, Oct 2, 24; m 47; c 2. GEOLOGY. Educ: Univ Calif, Los Angeles, AB, 48, PhD(geol), 54. Prof Exp: Asst, Univ Calif, Los Angeles, 51-54; asst prof, Univ Houston, 54-57; prof geol, Petrobras Petrol Co, Brazil, 57-59; assoc prof geol, 59-64, assoc dean col arts & sci, 61-65, PROF GEOL,

UNIV HOUSTON, 64- Concurrent Pos: NSF & Fulbright sr res fel, Victoria, NZ, 63. Mem: Geol Soc Am; Mineral Soc Am; Geochem Soc; Brazilian Geol Soc. Res: Petrology; petrography; areal and structural geology. Mailing Add: Dept of Geol Univ of Houston Houston TX 77004

CARMAN, PHILIP DOUGLAS, b Ottawa, Can, Oct 28, 16; m 51; c 3. OPTICS. Educ: Univ Toronto, BA, 40; Univ Rochester, MSc, 51. Prof Exp: Optical instruments, Res Enterprises Ltd, 40; from jr physicist to SR RES OFFICER, NAT RES COUN CAN, 41- Mem: Fel Optical Soc Am; Can Asn Physicists; Can Inst Surv. Res: Performance of photographic and photogrammetric systems; optical instrument design and testing. Mailing Add: Nat Res Coun Montreal Rd Ottawa ON Can

CARMEAN, WILLARD HANDY, b Philadelphia, Pa, Jan 4, 22; m 49; c 3. FOREST SOILS. Educ: Pa State Univ, BS, 43; Duke Univ, MF, 47, PhD(forest soils), 53. Prof Exp: Res forester soil res & forest surv, Pacific Northwest Forest Exp Sta, 46-51; soil scientist forest soils, Cent States Forest Exp Sta, 53-67, PROJ LEADER, N CENT FOREST EXP STA, 67- Mem: AAAS; Soc Am Foresters; Soil Sci Soc Am; Ecol Soc Am; Am Soil Conserv Soc. Res: Relations between tree growth and factors of soil and topography. Mailing Add: N Cent Forest Exp Sta Univ of Minn Folwell Ave St Paul MN 55108

CARMEL, RALPH, b Riga, Latvia, Aug 8, 40; US citizen; m 67; c 2. HEMATOLOGY. Educ: Yeshiva Univ, BA, 59, BHL, 59; NY Univ, MD, 63. Prof Exp: USPHS res fel hemat, Mt Sinai Sch Med, 66-68; res assoc, Aerospace Med Lab, Lackland AFB, 68-70; Wellcome fel, St Mary's Hosp Med Sch, London, 71; from asst prof to assoc prof med, Wayne State Univ, 72-75; chief hemat, Grace Hosp, Detroit, 75; ASSOC PROF MED, SCH MED, UNIV SOUTHERN CALIF, 75- Mem: Am Soc Hemat; Am Soc Clin Nutrit; Am Inst Nutrit; Am Fedn Clin Res; AAAS. Res: Megaloblastic anemia and folic acid and vitamin B-12 metabolism, with special interest in the transport of vitamin B-12 and in the proteins binding vitamin B-12. Mailing Add: Univ Southern Calif Sch of Med 2025 Zonal Ave Los Angeles CA 90033

CARMELI, MOSHE, b June 15, 33; US citizen; m 61; c 3. THEORETICAL PHYSICS. Educ: Hebrew Univ, Jerusalem, MSc, 60; Israel Inst Technol, PhD(physics), 64. Prof Exp: Lectr physics, Israel Inst Technol, 64; res assoc physics, Lehigh Univ, 64-65; res assoc physics, Univ Md, 65-67, asst prof, 67-68; res physicist, US Air Force, Wright-Patterson AFB, 67-69, sr scientist, 69-72; assoc prof, 72-74, PROF PHYSICS, BEN GURION UNIV, 74-, HEAD DEPT, 73- Mem: Fel Am Phys Soc; AAAS; Am Asn Univ Profs; Sigma Xi; Int Soc Gen Relativity & Gravitation. Res: General relativity and gauge theory. Mailing Add: Dept of Physics Ben Gurion Univ Beer Sheva Israel

CARMER, SAMUEL GRANT, b Buffalo, NY, Dec 19, 32; m 60; c 3. BIOMETRICS. Educ: Cornell Univ, BS, 54; Univ Ill, Urbana, MS, 58, PhD(agron), 61. Prof Exp: Res fel biomath, NC State Col, 61-62; from asst prof to assoc prof, 62-71, PROF BIOMET, UNIV ILL, URBANA-CHAMPAIGN, 71- Mem: Am Soc Agron; Biomet Soc; Crop Sci Soc Am. Res: Biomathematics; teaching graduate level courses and providing individual advice on problems of experimental design; statistical analysis and data processing by computer. Mailing Add: Dept of Agron Univ of Ill Urbana IL 61801

CARMICHAEL, BYRON M, physics, mathematics, see 12th edition

CARMICHAEL, DAVID JAMES, b Casterton, Australia, June 2, 36; m 68; c 1. PROTEIN CHEMISTRY. Educ: Univ Melbourne, BAgrSci, 60; Univ Nottingham, PhD(food sci), 66. Prof Exp: Technologist, Kraft Foods Ltd, 60-63; demonstr food sci, Univ Nottingham, 63-66; Am Dent Asn fel, Northwestern Univ, 66-67; Helen Hay Whitney Found fel, 67-68; assoc prof, 68-73, PROF DENT, UNIV ALTA, 73- Concurrent Pos: Med Res Coun Can grant, 73-76; vis scientist, Med Res Coun Can, 75; vis scientist, Monash Univ, Clayton, Australia, 75-76. Mem: Brit Biochem Soc; Int Asn Dent Res. Res: Collagen research; biochemical characterization of normal and lathyritic dentin matrix collagen; immunohistochemical study of collagen fibrillogenesis. Mailing Add: Fac of Dent Univ of Alta Edmonton AB Can

CARMICHAEL, HALBERT HART, b St Louis, Mo, Aug 29, 37; m 58; c 2. CHEMICAL KINETICS. Educ: Univ Tenn, BS, 59; Univ Calif, PhD(chem), 63. Prof Exp: Nat Bur Stand fel, 63-64; asst prof, 64-69, ASSOC PROF CHEM, NC STATE UNIV, 69- Mem: Am Chem Soc. Res: Photochemistry and radiation chemistry of gases. Mailing Add: Dept of Chem NC State Univ Raleigh NC 27607

CARMICHAEL, HUGH, b Scotland, Nov 10, 06; m 37; c 4. COSMIC RAY PHYSICS. Educ: Univ Edinburgh, BSc, 29; Cambridge Univ, PhD(physics), 36, MA, 39. Prof Exp: Demonstr physics, Cambridge Univ, 37-44; sr prin sci officer exp physics, Brit Ministry Supply, Atomic Energy Mission, Can, 44-50; prin res officer & head gen physics br, Atomic Energy Can, Ltd, Chalk River, 50-71; RETIRED. Concurrent Pos: Fel, St John's Col, Cambridge Univ, 36-40; chmn working group interplanetary disturbances & mem, Inter-Union Comn Solar Terrestrial Physics, Int Coun Sci Unions, 67-73. Mem: Am Geophys Union; fel Royal Soc Can; Can Asn Physicists. Res: Cosmic radiation; balloon measurements; bursts; time variations; radiation monitors; latitude survey; reactor control; health monitoring; aerial prospecting; quartz fibre instruments; microbalances; electrometers. Mailing Add: 9 Beach Ave Deep River ON Can

CARMICHAEL, IAN STUART, b London, Eng, Mar 29, 30; m 70; c 4. PETROLOGY, GEOCHEMISTRY. Educ: Cambridge Univ, BA, 54; Univ London, PhD(geol), 59. Prof Exp: Lectr geol, Imp Col, Univ London, 58-63; assoc prof, 65-67, PROF GEOL, UNIV CALIF, BERKELEY, 67-, CHMN DEPT GEOL & GEOPHYSICS, 72- Concurrent Pos: Sr foreign scientist, NSF, Univ Chicago, 63; Miller res prof, Miller Inst Sci Res, 67-68; ed-in-chief, Contrib to Mineral & Petrol, 74- Mem: Am Geophys Union; Mineral Soc Am; Mineral Soc Gt Brit & Ireland. Res: Origin and cooling history of igneous rocks. Mailing Add: Dept of Geol Univ of Calif Berkeley CA 94720

CARMICHAEL, J W, JR, b Lamesa, Tex, Feb 9, 40. PHYSICAL CHEMISTRY, INORGANIC CHEMISTRY. Educ: Eastern NMex Univ, BS, 61; Univ Ill, MS, 63, PhD(phys chem), 65. Prof Exp: Asst chem, Univ Ill, 62-65; asst prof, Univ Ark, 65-70; asst prof, 70-73, ASSOC PROF CHEM, XAVIER UNIV LA, 73- Mem: Am Chem Soc; Am Crystallog Asn; The Chem Soc; Nat Sci Teachers Asn. Res: Use of x-ray diffraction techniques to determine the crystal structure of transition-metal complexes; investigation of the manner in which changes in structure affect the electronic spectra. Mailing Add: Dept of Chem Xavier Univ of La New Orleans LA 70125

CARMICHAEL, JACK B, physical chemistry, see 12th edition

CARMICHAEL, JOHN WILLIAM, medical mycology, see 12th edition

CARMICHAEL, LELAND E, b Huntington Park, Calif, June 15, 30; m 57; c 1. VETERINARY MICROBIOLOGY. Educ: Univ Calif, AB, 52, DVM, 56; Cornell Univ, PhD(virol), 59. Prof Exp: Asst bact, State Univ NY Vet Col, Cornell Univ, 56-59, res assoc virol, 59-63, asst prof infectious dis & John M Olin Chair, 63-69, PROF VET VIROL & JOHN M OLIN PROF VIROL & VET MICROBIOL, STATE UNIV NY VET COL, CORNELL UNIV, 69- Concurrent Pos: NIH grant, 61-63; Scientific dir, James A Baker Inst for Animal Health, 75- Honors & Awards: Gaines Award, Am Vet Med Asn, 75. Mem: Am Vet Med Asn; NY Acad Sci; US Livestock Sanit Asn; Am Soc Microbiol. Res: Infectious diseases of domestic animals, principally viral and mycoplasmal diseases; infectious hepatitis; immunology; pathogenesis. Mailing Add: Vet Virus Res Inst Snyder Hall SUNY Vet Col Cornell Univ Ithaca NY 14850

CARMICHAEL, LEONARD, psychology, deceased

CARMICHAEL, LYNN PAUL, b Louisville, Ky, Sept 15, 28; m 54; c 3. FAMILY MEDICINE. Educ: Univ Louisville, MD, 52; Am Bd Family Pract, dipl, 70. Prof Exp: Intern, Brooke Army Hosp, 52-53; resident gen pract, Dade County Hosp, 54-55; clin instr, 56-64, asst prof med & pediat, 65-68, asst dean, Sch Med, 66-67, assoc prof family med, 68-72, PROF FAMILY MED & CHMN DEPT, UNIV MIAMI, 72- Concurrent Pos: USPHS res fel pediat, Harvard Med Sch, 63-64; Secy, Family Pract Comt, AMA, 68-71. Mem: AAAS; Soc Teachers Family Med (pres, 67-); fel Am Acad Family Physicians; Asn Am Med Cols; Can Col Family Physicians. Res: Family health care; primary care; medical education. Mailing Add: Dept of Family Med Univ of Miami Box 520875 Miami FL 33152

CARMICHAEL, RALPH HARRY, b Freetown, Ind, Jan 20, 23; m 42; c 1. PHARMACOLOGY. Educ: Ind Univ, BS, 49; Butler Univ, MS, 56. Prof Exp: Assoc chemist, 50-56, supvr clin chem, 56-62, dept head clin chem & hemat, 62-70, DEPT HEAD DRUG METAB, LILLY LAB CLIN RES, ELI LILLY & CO, 70- Mem: Am Chem Soc; Am Asn Clin Chemists; NY Acad Sci. Res: Drug metabolism. Mailing Add: 2732 Parkwood Dr Indianapolis IN 46224

CARMICHAEL, RICHARD DUDLEY, b High Point, NC, Mar 13, 42; m 67. MATHEMATICAL ANALYSIS. Educ: Wake Forest Col, BS, 64; Duke Univ, AM, 66, PhD(math), 68. Prof Exp: Asst prof math, Va Polytech Inst & State Univ, 68-71; ASSOC PROF MATH, WAKE FOREST UNIV, 71- Mem: Am Math Soc; Math Asn Am; Soc Indust & Appl Math; Calcutta Math Soc. Res: Theory of distributions; complex variables. Mailing Add: Dept of Math Wake Forest Univ Winston-Salem NC 27109

CARMICHAEL, ROBERT STEWART, b Toronto, Ont, Jan 11, 42; m 67. GEOPHYSICS, GEOLOGY. Educ: Univ Toronto, BASc, 63; Univ Pittsburgh, MS, 64, PhD(earth & planetary sci), 67. Prof Exp: Teaching asst geol, Univ Pittsburgh, 63-64 & 66-67; fel geophys, Osaka Univ, 67-68; geophysicist, Explor & Prod Res Ctr, Shell Develop Co, 68-72; ASST PROF GEOL, MICH STATE UNIV, 72- Concurrent Pos: Visitor, NATO Advan Study Inst, Univ Newcastle, 67. Mem: Am Geophys Union; Sigma Xi; Soc Explor Geophys; Soc Terrestrial Magnetism & Elec. Res: Rock and paleo-magnetism; computer analysis; properties of earth materials; high-pressure geophysics; exploration geophysics; tectonics. Mailing Add: Dept of Geol Mich State Univ East Lansing MI 48824

CARMICHAEL, STEPHEN WEBB, b Detroit, Mich, July 17, 45; m 70. HUMAN ANATOMY. Educ: Kenyon Col, AB, 67; Tulane Univ, PhD(anat), 71. Prof Exp: Asst prof biol, Delgado Col, 69-71; from instr to asst prof anat, 71-75, ASSOC PROF ANAT, SCH MED, WVA UNIV, 75- Concurrent Pos: Fel, Giorgio Cini Found, Milan, 73. Mem: Am Asn Anat; Electron Micros Soc Am; Am Soc Cell Biol. Res: Morphological aspects of secretion; adrenal medullary cytology. Mailing Add: Dept of Anat WVa Univ Med Ctr Morgantown WV 26506

CARMIN, ROBERT LEIGHTON, b Muncie, Ind, Nov 28, 18; m 40; c 2. GEOGRAPHY. Educ: Ohio Univ, BS, 40; Univ Nebr, MA, 42; Univ Chicago, PhD, 53. Prof Exp: Fulbright res scholar, Nat Univ Cuyo, 58; instr geog, Mich State Univ, 42-44; cartogr, Off Strategic Serv, 44-45; asst prof geog, Mich State Univ, 47-50; from asst prof to prof, Univ Ill, Urbana, 51-62; head Latin Am Studies Unit, Lang Develop Br, US Off Educ, 62; DEAN COL SCI & HUMANITIES, BALL STATE UNIV, 62- Concurrent Pos: Consult cartog, Cowles Comn Res Econ, Univ Chicago, 46 & 47; consult geog & cartog, Spencer Press, 54-62; mem comt geog, Nat Acad Sci-Nat Res Coun, 61-66; consult, Latin Am, US Off Educ, 61, Fels, 64, Geog Sec Teachers Summer Inst, 65; consult, Peace Corps, 64, lectr, 64 & 66; consult, NSF, 64-75; consult AID, Brazil, 65 & Costa Rica, 65; pres, Ind Partners of the Americas, 65 & 66, chmn, Educ Comt, 67-; consult int affairs comt, Asn State Cols & Univs, 66- Mem: Asn Am Geog; Nat Coun Geog Educ; Latin Am Studies Asn (vpres, 60-61, pres, 61); Asn Brazilian Geog. Res: Frontier settlements and towns in Latin America. Mailing Add: Col of Sci & Humanities Ball State Univ Muncie IN 47306

CARMODY, DONALD RICHARD, physical organic chemistry, see 12th edition

CARMODY, GEORGE R, b Brooklyn, NY, Mar 29, 38; m 62; c 3. POPULATION GENETICS. Educ: Columbia Univ, AB, 60, PhD(zool), 67. Prof Exp: Teaching asst zool, Columbia Univ, 61-62; USPHS fel, Univ Chicago, 67-68; asst prof, 69-74, ASSOC PROF BIOL, CARLETON UNIV, 74-, ASSOC CHMN, DEPT OF BIOL, 73- Concurrent Pos: Ford Found genetics training grant fel, Univ Chicago, 68-69. Mem: AAAS; Genetics Soc Am; Soc Study Evolution; Genetics Soc Can. Res: Speciation in Drosophila; protein polymorphisms; maintenance of genetic variability in natural populations; genetic variability in cave-dwelling organisms. Mailing Add: Dept of Biol Carleton Univ Ottawa ON Can

CARMON, JAMES LAVERN, b Ga, May 7, 26; m 46; c 1. STATISTICS, GENETICS. Educ: Univ Ga, BSA, 48; Univ Md, MS, 50; NC State Col, PhD(statist), 55. Prof Exp: Instr, animal husb, Univ Ga, 50-55, from asst prof to assoc prof, 55-58, statistician, Col Exp Sta, 56-59, assoc dir, Inst Statist, 59, DIR COMPUT CTR, UNIV GA, 59-, ASST VCHANCELLOR COMPUT SYSTS, 68- Concurrent Pos: NIH res fel statist, Va Polytech Inst, 63-64. Mem: Am Soc Animal Sci; Am Statist Asn. Res: Population genetics and statistics. Mailing Add: Off of Comput Activ Univ of Ga Athens GA 30602

CARMONY, DONALD DUANE, b Indianapolis, Ind, Sept 16, 35; m 61; c 2. HIGH ENERGY PHYSICS. Educ: Ind Univ, BS, 56; Univ Calif, Berkeley, PhD(physics), 62. Prof Exp: Res physicist, Lawrence Radiation Lab, Univ Calif, Berkeley, 61-62, res physicist & lectr, Univ Calif, Los Angeles, 62-63, res physicist, Univ Calif, San Diego, 63-66; assoc prof, 66-75, PROF PHYSICS, PURDUE UNIV, 75- Concurrent Pos: Alexander von Humboldt sr scientist award, 72-73. Mem: Am Phys Soc. Res: Experimental high energy physics; neutrino interactions in big bubble chamber; high energy hadrom collisions; study of pion-pion interactions and strange meson resonances. Mailing Add: Dept of Physics Purdue Univ Lafayette IN 47907

CARNAHAN, CHALON LUCIUS, b Beverly, Mass, Sept 17, 33; m 60; c 2.

GROUNDWATER HYDROLOGY. Educ: Calif Inst Technol, BS, 55; Univ Calif, Berkeley, MS, 58; Univ Nev, Reno, PhD(hydrol), 75. Prof Exp: Radiol chemist, US Naval Radiol Defense Lab, 57-62; sr radiochemist, Hazleton-Nuclear Sci Corp, 62-65, mgr phys chem dept, 65-67; sr assoc scientist, Isotopes, a Teledyne Co, 67-69, scientist & group leader, 69-70, Teledyne Isotopes, 70-71; RES ASSOC, DESERT RES INST, UNIV NEV, 71- Mem: AAAS; Am Geophys Union. Res: Physical and chemical effects of underground nuclear explosions; nuclear reactions; thermodynamics of irreversible processes in flow through porous media; contamination transport in ground water. Mailing Add: 1593 Benton St Sunnyvale CA 94087

CARNAHAN, HOWARD LEON, b Erie, Kans, Mar 25, 20; m 45, 73; c 2. GENETICS, PLANT BREEDING. Educ: Kans State Col, BS, 42; Univ Minn, MS, 47, PhD(plant genetics), 49. Prof Exp: Asst, Univ Minn, 42, res assoc, 46-48; asst prof agron, Pa State Univ, 49-52; res agronomist, US Regional Pasture Res Lab, 53-57, agronomist in chg, 57-60; res agronomist, Agr Res Serv, USDA, 60-65; dir res, Arnold-Thomas Seed Serv, 65-69; DIR PLANT BREEDING, CALIF COOP RICE RES FOUND, 69- Mem: Am Soc Agron; Crop Sci Soc Am; Am Genetic Asn. Res: Forage plant breeding; cytogenetics; agronomy; rice breeding. Mailing Add: Calif Coop Rice Res Found PO Box 306 Biggs CA 95917

CARNAHAN, JAMES ELLIOT, b Kaukauna, Wis, Jan 26, 20; m 44; c 4. ENTOMOLOGY, ENVIRONMENTAL BIOLOGY. Educ: Univ Wis, BS, 42, MS, 44, PhD(org chem), 46. Prof Exp: Asst chem warfare agents, Univ Wis, 42-46; res chemist, 46-56, RES SUPVR, E I DU PONT DE NEMOURS & CO, 56- Honors & Awards: Hoblitzelle Nat Award Res Agr Sci, Tex Res Found, 65. Mem: Am Chem Soc; Am Soc Plant Physiol; Am Soc Biol Chem; Entom Soc Am. Res: Organic synthesis; catalytic chemistry; polymer chemistry; biological nitrogen fixation; plant biochemistry; agricultural chemicals, drugs and air pollution effects on plants. Mailing Add: E I du Pont de Nemours & Co Wilmington DE 19898

CARNAHAN, ROBERT EDWARD, b Kaukauna, Wis, Jan 5, 25; m 50; c 3. ORGANIC CHEMISTRY. Educ: Univ Wis, BS, 47; Univ Ill, MS, 48, PhD(org chem), 50. Prof Exp: Res chemist, Pfizer, Inc, 50-55, patent agent, 55-60; patent agent, 60-69, DIR PATENTS, MEAD JOHNSON & CO, 69- Mem: Am Chem Soc. Res: Medicinal chemistry. Mailing Add: Mead Johnson & Co Evansville IN 47721

CARNALL, WILLIAM THOMAS, b Denver, Colo, May 23, 27; m 50; c 3. PHYSICAL CHEMISTRY. Educ: Colo State Univ, BS, 50; Univ Wis, PhD(chem), 54. Prof Exp: Asst, Univ Wis, 52-54; SR CHEMIST, ARGONNE NAT LAB, 54- Concurrent Pos: Sigma Xi fel, Munich, 61-62. Mem: Am Chem Soc. Res: Chemistry of the actinide elements; theory of lanthanide and actinide element spectra; chemistry and spectrophotometry in molten salt systems. Mailing Add: 5333 Seventh Ave La Grange IL 60525

CARNEIRO, ROBERT LEONARD, b New York, NY, June 4, 27. ANTHROPOLOGY. Educ: Univ Mich, BA, 49, MA, 52, PhD(anthrop), 57. Prof Exp: Instr anthrop, Univ Wis, 56-57; from asst cur to assoc cur SAm ethnol, 57-69, CUR S AM ETHNOL, AM MUS NATURAL HIST, 69- Concurrent Pos: Lectr anthrop, Hunter Col, 63-64 & Columbia Univ, 64-65; vis assoc prof, Univ Calif, Los Angeles, 68; vis prof, Pa State Univ, 73. Mem: AAAS; Am Anthrop Asn; Am Ethnol Soc; Soc Am Archaeol. Res: Cultural evolution, including the reconstruction of sequences and the history of evolutionism; the origin and development of the state; cultural ecology of Amazonia, especially the effects of subsistence. Mailing Add: Am Mus of Natural Hist Central Park W at 79th St New York NY 10024

CARNELL, PAUL HERBERT, b Oakfield, Wis, May 27, 17; m 42; c 5. PETROLEUM CHEMISTRY. Educ: Albion Col, AB, 39; Western Reserve Univ, PhD(chem), 43. Hon Degrees: DrLaws, Alderson Broaddus Col, 73. Prof Exp: Res chemist, Phillips Petrol Co, 43-47; dir res, Leonard Refining, Inc, 47-48; asst prof, Marietta Col, 48-49; asst prof, Albion Col, 48-52, head chem dept, 52-66; mem staff, 66-68, ASST DIR, DIV INSTL DEVELOP, BUR POST-SEC EDUC, US OFF EDUC, DEPT HEALTH, EDUC & WELFARE, 68- Concurrent Pos: Res Corp grant, 52; res assoc, Yale Univ, 59-60; NIH res grants, 59-; Am Chem Soc vis scientist, 59-; assoc prof dir acad years insts, NSF, 64-65. Mem: Am Chem Soc; NY Acad Sci. Res: Petroleum chemistry; Treating with hydrofluoric acid, lubricating oils, fuel oils, fuel oil stability, and viscosity; synthetic membranes. Mailing Add: Bur Post-Sec Educ US Off of Educ HEW Washington DC 20201

CARNES, DAVID LEE, JR, b Youngstown, Ohio, Mar 16, 46; m 71. COMPARATIVE PHYSIOLOGY. Educ: Allegheny Col, BS, 68; Rice Univ, MA, 74, PhD(biol), 75. Prof Exp: Res assoc biol, Rice Univ, 74-75; RES ASSOC PHARMACOL, UNIV MO-COLUMBIA & HARRY S TRUMAN VET ADMIN HOSP, 75- Mem: AAAS. Res: Interrelationship of Parathyroid hormone, calcitronin and vitamin D, as they relate to the control of calcium metabolism. Mailing Add: Dept of Pharmacol Med Sci Bldg Univ of Mo Columbia MO 65201

CARNES, JAMES EDGAR, b Harlington, Tex, Sept 20, 33; m 55; c 2. HUMAN ANATOMY. Educ: Arlington State Col, BA, 64; NTex State Univ, MA, 66, PhD(biol), 72. Prof Exp: Instr biol, Univ Tex, Arlington, 66-67; instr microbiol, Baylor Dent Col, 67-69; ASST PROF ANAT, TEX COL OSTEOP MED, N TEX STATE UNIV, 72- Mem: Am Soc Microbiol; AAAS; Am Asn Anatomists. Res: Ultrastructural cytochemistry and histochemistry; enzyme localization studies. Mailing Add: PO Box 13046 NTex State Univ Denton TX 76203

CARNES, JAMES EDWARD, b Cumberland, Md, Sept 27, 39. SOLID STATE ELECTRONICS. Educ: Pa State Univ, BS, 61; Princeton Univ, MA, 67, PhD(electron device physics), 70. Prof Exp: MEM TECH STAFF SOLID STATE ELECTRONICS, RCA LABS, 69- Concurrent Pos: Lectr, Short Course Prog, Univ Calif, Los Angeles, 73- Mem: Inst Elec & Electronics Engrs. Res: Experimental and analytical studies of metal-oxide-silicon integrated circuits, specifically charge-coupled devices including understanding of physics of operation and optimum design for imaging, memory and signal processing applications. Mailing Add: RCA Labs Princeton NJ 08540

CARNES, JOSEPH JOHN, b Rock Island, Ill, Aug 3, 17; m 42; c 5. ORGANIC CHEMISTRY. Educ: St Ambrose Col, BS, 39; Univ Iowa, MS, 41, PhD(org chem), 43. Prof Exp: Chemist, Nat Defense Res Comt contract, Univ Iowa, 42-43; sr chemist, Am Cyanamid Co, 43-51, group leader, 51-54, sect mgr, 54-59, dir appl res, 59-63, dir contract res, 63-68; dir advan planning, 68-69, VPRES, NEW ENG INST, 69- Mem: Fel AAAS; Am Chem Soc; fel Am Inst Chemists. Res: Synthetic organic chemistry; process development; ring closure of N-chloroamines; nitrogen mustards; chemical warfare agents; surface active agents; organic phosphorus insecticides; new product development; paper and industrial chemicals; rocket propellant chemistry; energy conversion. Mailing Add: New Eng Inst PO Box 308 Ridgefield CT 06877

CARNES, WILLIAM HENRY, b Ft Worth, Tex, Nov 2, 09; m 50. PATHOLOGY. Educ: Columbia Univ, AB, 32; Johns Hopkins Univ, MD, 36. Prof Exp: Asst res & path, City Hosps, Baltimore, Md, 36-38; assoc path, Sch Med, Johns Hopkins Univ, 38-39; instr, Columbia Univ, 39-41; from asst prof to assoc prof, Stanford Univ, 41-47; from asst prof to assoc prof, Johns Hopkins Univ, 47-51; from assoc prof to prof, Stanford Univ, 51-56; prof path & head dept, Univ Utah, 56-68; PROF PATH, UNIV CALIF, LOS ANGELES, 68- Mem: AAAS; Am Asn Path & Bact; Am Soc Cell Biol; Am Soc Exp Path; Soc Exp Biol & Med. Res: Experimental pathology. Mailing Add: Dept of Path Univ of Calif Sch of Med Los Angeles CA 90024

CARNEVALE, EDMUND HENRY, b Beverly, Mass, Oct 13, 24; m 47; c 10. PHYSICS. Educ: Boston Col, BS, 51, MS, 52; Cath Univ, PhD(physics), 58. Prof Exp: Res asst, Boston Univ, 52-53; group leader, US Naval Ord Lab, 53-58; sr scientist, Avco Corp, 58-60; staff scientist & vpres, Parametrics, Inc, 60-68, PRES, PANAMETRICS, INC, 68- Mem: Acoust Soc Am. Res: Ultrasonic propagation studies in gases, liquids and solids as a function of temperature and pressure. Mailing Add: Panametrics Inc 221 Crescent St Waltham MA 02154

CARNEY, ALBERT STRICKER, b Grantsville, Md, June 8, 16; m 43; c 3. ORGANIC CHEMISTRY. Educ: Juniata Col, BS, 38; Univ Pa, MS, 39; Pa State Univ, PhD(chem), 55. Prof Exp: Instr, 44-55, asst prof, 55-57, ASSOC PROF CHEM, PA STATE UNIV, 57- Concurrent Pos: Consult, Suter Chem Co, 44-51; instr chem, Altoona Hosp, Sch Nursing & Lewistown Hosp, Sch Nursing, 52- Mem: AAAS; Am Chem Soc; Nat Sci Teachers Asn; fel Am Inst Chem. Res: Teaching general, analytical and organic chemistry. Mailing Add: Altoona Campus RD 4 Altoona PA 16601

CARNEY, EDWARD J, b Rochester, NY, May 15, 29; m 51; c 5. STATISTICS, COMPUTER SCIENCES. Educ: Univ Rochester, AB, 51, MS, 58; Iowa State Univ, PhD(statist), 67. Prof Exp: Asst factory eng dept, Bausch & Lomb Optical Co, 54-58; staff asst area commun proj, Gen Dynamics/Electronics, 58-59; instr indust eng, Iowa State Univ, 59-63, asst prof indust eng & statist, 63-67, assoc prof statist, 67-74, PROF COMPUT SCI & STATIST, UNIV RI, 74- Mem: Inst Math Statist; Am Statist Asn; Asn Comput Mach. Res: Design of experiments; variances of variance component estimates; statistical computations. Mailing Add: Comput Lab Univ of RI Kingston RI 02881

CARNEY, GEORGE OLNEY, b Clinton, Mo, Mar 23, 42; m 71. HISTORICAL GEOGRAPHY, CULTURAL GEOGRAPHY. Educ: Cent Mo State Univ, BA & BSE, 64, MA, 65; Okla State Univ, PhD(hist), 72. Prof Exp: Asst prof soc sci, Mo Western Col, 65-68; ASST PROF GEOG, OKLA STATE UNIV, 70- Concurrent Pos: Mem, John Edwards Mem Found, 70- & Okla Humanities Task Force State Based Prog, Nat Endowment Humanities, 72- Mem: Fel Royal Geog Soc; Asn Am Geog; Orgn Am Historians. Res: Spatial and temporal dimensions of American popular music. Mailing Add: Dept of Geog Okla State Univ 301 Home Econ E Stillwater OK 74074

CARNEY, GORDON C, b Glasgow, Scotland, Sept 15, 34; m 60; c 2. INVERTEBRATE PHYSIOLOGY. Educ: Univ Durham, BSc, 57, MSc, 60; Univ Minn, PhD(entom), 64. Prof Exp: Jr res officer, Atomic Energy Can, Ltd, 63-64, asst res officer, 64-66; asst prof biol, Bowling Green State Univ, 66-69; lectr biol, 69-74, SR LECTR CELL PHYSIOL, BRISTOL POLYTECH, 74- Mem: Estuarine & Brackish Waters Sci Asn. Res: Metabolism of invertebrate mitochondria; physiological effects of pollutants on aquatic invertebrates; binding of heavy metals by marine algae. Mailing Add: Dept of Sci Bristol Polytechnic Bristol England

CARNEY, JAMES JOSEPH, b Duluth, Minn, Feb 23, 12; m 38; c 1. ORGANIC CHEMISTRY. Educ: Col St Thomas, BS, 33; Catholic Univ, MS, 34; Univ Minn, PhD(org chem), 42. Prof Exp: Asst prof chem, Col St Thomas, 34-42; chemist, Merck & Co, Inc, 42-45, factory dept head, 45-48, dept head sterile tech, 48-49, mgr antibiotics prod, 49-50, asst dir com develop, 50-53, mgr prod coord, 54-55 & mkt develop, 55-57; chmn dept chem, 70-75, PROF CHEM, COL ST THOMAS, 57- Mem: Am Chem Soc; The Chem Soc. Res: Structure and chemical behavior of dyestuff from furfural, aniline and aniline hydrochloride; atabrine; riboflavin; vitamin B1; antibiotics. Mailing Add: 1440 Randolph Ave Apt 204 St Paul MN 55105

CARNEY, RICHARD WILLIAM JAMES, b Novelty, Mo, June 19, 34; m 57; c 3. ORGANIC CHEMISTRY. Educ: McPherson Col, BS, 57; Iowa State Univ, MS, 61, PhD(org chem), 62. Prof Exp: Sr chemist, 62-73, SR RES CHEMIST, MEDICINAL CHEM, CHEM RES DIV, CIBA PHARMACEUT CO, 73- Concurrent Pos: Vis scientist, Basel, 68. Mem: Am Chem Soc; Am Inst Chem; NY Acad Sci. Res: Diterpenes; chemistry of heterocyclics. Mailing Add: Chem Res Div CIBA Pharmaceut Co 556 Morris Ave Summit NJ 07901

CARNEY, ROBERT GIBSON, b Ann Arbor, Mich, Apr 25, 14; m 39; c 4. DERMATOLOGY. Educ: Univ Mich, AB, 35, MD, 39. Prof Exp: Intern med, 39-40, resident dermat, 40-43, assoc, 46-47, from asst prof to assoc prof, 47-54, PROF DERMAT, UNIV IOWA HOSPS, 61- Concurrent Pos: Consult, Comt Rev, US Pharmacopeia, 50-; consult, Am Acad Dermat Adv Comt, Food & Drug Admin, 61- Mem: Am Dermat Asn; Am Acad Dermat; Soc Invest Dermat. Res: Topical therapy and pharmaceuticals; incontinentia pigmenti. Mailing Add: Dept of Dermat Univ of Iowa Hosp Iowa City IA 52240

CARNEY, ROSE AGNES, b Chicago, Ill. PHYSICS. Educ: DePaul Univ, BS, 42, MS, 46; Ill Inst Technol, PhD, 61. Prof Exp: Asst metal labs, Univ Chicago, 42-43; instr, Army Spec Training Prog, De Paul, 43-44, instr physics, 44-46; chmn physics & math, St Xavier Col, Ill, 46-48; assoc prof physics, 48-59, PROF MATH & CHMN DEPT, ILL BENEDICTINE COL, 59-, CHMN DIV NATURAL SCI, 69- Concurrent Pos: Consult, Argonne Nat Lab, 62, 63, 64. Mem: Am Asn Physics Teachers; Math Asn Am. Res: Molecular spectroscopy; mathematical physics. Mailing Add: Dept of Math Ill Benedictine Col Lisle IL 60532

CARNEY, THOMAS PATRICK, organic chemistry, see 12th edition

CARNOW, BERTRAM WARREN, b Philadelphia, Pa, June 19, 22; m 60; c 3. ENVIRONMENTAL HEALTH, THORACIC DISEASES. Educ: NY Univ, BA, 47; Chicago Med Sch, MB & MD, 51. Prof Exp: Intern, Cook County Hosp, Chicago, 51-52; resident cardiol, Michael Reese Hosp, Chicago, 52-53, resident clin internal med, 53-55; physician pvt pract, 55-69; clin assoc, 64-65, asst clin prof, 65-67, from asst prof to assoc prof, 67-70, PROF PREV MED & COMMUNITY HEALTH, SCH MED, UNIV ILL, CHICAGO, 70-, DIR & PROF OCCUP & ENVIRON MED, SCH PUB HEALTH, 72- Concurrent Pos: Consult & attend physician, Michael Reese Hosp, 55-72 & Univ Ill Hosp, 70-; chest consult, Union Health Serv, 57-; med dir, Chicago Lung Asn, 69-; dir, Environ Health Resource Ctr, Ill Inst Environ Qual, 70- Mem: AAAS; Am Thoracic Soc; fel Am Pub Health Asn; fel Am Col Chest Physicians; fel Royal Soc Health. Res: Effects of environmental hazards on health, including air pollution morbidity and mortality; health and energy; occupational diseases including pneumoconiosis and noise; natural history and levels of prevention chronic pulmonary diseases. Mailing Add: Dept of Occup & Environ Med Univ of Ill Sch Pub Health Chicago IL 60680

CARNS, HARRY ROBERT, b Cedar Rapids, Iowa, Nov 17, 17; m 37; c 3. PLANT PHYSIOLOGY. Educ: Univ Iowa, BS, 39, MS, 41; Univ Calif, Los Angeles, PhD(bot), 51. Prof Exp: Plant physiologist, Cotton Field Sta, USDA & Delta Br Exp Sta, Miss, 50-57; plant physiologist, Dept Bot, Univ Calif, Los Angeles, 57-61; leader cotton physiol invests, Plant Indust Sta, USDA, 61-72, CHMN, PLANT PHYSIOL INST, BELTSVILLE AGR RES CTR, USDA, MD, 72- Mem: AAAS; Am Soc Plant Physiol; Am Inst Biol Sci; NY Acad Sci; Am Soc Photobiol. Res: Physiology of the cotton plant; plant growth regulators; foliar and fruit abscission; phytotoxicity of agricultural chemicals; defoliation; bioconversion to fuels. Mailing Add: Plant Physiol Inst USDA Beltsville Agr Res Ctr West Beltsville MD 20705

CARO, JOSEPH HENRY, b Boston, Mass, Sept 20, 20; m 55; c 3. ENVIRONMENTAL SCIENCES. Educ: Northeastern Univ, BS, 43. Prof Exp: Physicist optics, Bur Ord, US Dept Navy, 43-49; chemist, Fertilizer Lab, 49-65 & Soils Lab, 65-72, RES CHEMIST, AGR CHEM MGT LAB, AGR ENVIRON QUAL INST, AGR RES SERV, USDA, 72- Mem: Am Chem Soc; Asn Off Anal Chem. Res: Studies of environmental fate of pest control chemicals and analytical methodology of chemical residues. Mailing Add: Rm 227 Bldg 007 Beltsville Agr Res Ctr W Beltsville MD 20705

CARO, LUCIEN G, b Toulon, France, July 5, 28; US citizen; m 54; c 2. MOLECULAR BIOLOGY. Educ: Univ Tulane, BS, 57; Yale Univ, PhD(biophys), 59. Prof Exp: NSF fel & guest investr cell biol, Rockefeller Inst, 59-61, Helen Hay Whitney fel & res assoc, 61-64; biophysicist, Oak Ridge Nat Lab, 64-70; PROF MOLECULAR BIOL, UNIV GENEVA, 70- Concurrent Pos: NSF res grant, 64; vis investr, Inst Molecular Biol, Univ Geneva, Switz, 62-64. Mem: AAAS; Biophys Soc; Am Soc Microbiol; Genetics Soc Am. Res: Control of DNA replication; episomes; bacterial conjugation; electron microscopy; autoradiography. Mailing Add: Inst of Molec Biol Univ Geneva 30 quai E Ansermet Geneva Switzerland

CAROFF, LAWRENCE JOHN, b Beaverdale, Pa, Aug 26, 41; c 3. ASTROPHYSICS. Educ: Swarthmore Col, BS, 62; Cornell Univ, PhD(appl physics), 67. Prof Exp: RES SCIENTIST ASTROPHYS, AMES RES CTR, NASA, 67- Mem: Am Astron Soc. Res: Theoretical research into nature of quasars and active galaxies; cosmotology; galactic infrared sources; infrared observations from aircraft and ground observations of galactic H II regions, stars and planets. Mailing Add: Ames Res Ctr MS 245-3 NASA Moffett Field CA 94035

CAROL, BERNARD, b New York, NY; c 4. MATHEMATICAL STATISTICS, COMPUTERS. Educ: Columbia Univ, MA, 49. Prof Exp: Statistician, Metrop Life Ins Co, 49-61; biostatistician & dir comput ctr, NY Med Col, 61-68; biomathematician, Dept Biostatist, Montefiore Hosp & Med Ctr, 68-71; DIR BIOSTATIST, AYERST LABS, AM HOME PROD, 71- Mem: Inst Math Statist; Biomet Soc; Am Statist Asn; Asn Comput Mach; Opers Res Soc Am. Res: Mathematical statistics applied to medicine marketing, insurance, design of computer files, validation of aptitude tests, sampling company files, teaching with heavy use of computers. Mailing Add: 15 Linden St Great Neck NY 11021

CAROLIN, VALENTINE MOTT, JR, b Sayville, NY, Aug 23, 18; m 52; c 2. INSECT ECOLOGY. Educ: Syracuse Univ, BS, 39, MS, 42. Prof Exp: Scout, Bur Entom & Plant Quarantine, USDA, NJ, 39; survr, Radio Corp Am Commun, Inc, NY, 39-40; asst, State Univ NY Col Forestry, Syracuse Univ, 40-42; entomologist, Bur Entom & Plant Quarantine, USDA, 46-54, Pac Northwest Forest & Range Exp Sta, US Forest Serv, 54-74, SUPVRY RES ENTOMOLOGIST, PAC NORTHWEST FOREST & RANGE EXP STA, US FOREST SERV, USDA, 74- Mem: Entom Soc Am; Soc Am Foresters. Res: Biological control factors, especially parasites of forest insects; ecology of spruce budworm; western hemlock looper; techniques for field fumigation of European pine shoot moth. Mailing Add: 9030 SE Mill St Portland OR 97216

CAROME, EDWARD F, b Cleveland, Ohio, May 22, 27; m 51; c 6. PHYSICS. Educ: John Carroll Univ, BS & MS, 51; Case Inst Technol, PhD(physics), 54. Prof Exp: From asst prof to prof physics, John Carroll Univ, 54-68; liaison scientist, London Br, Off Naval Res, 68-69; PROF PHYSICS, JOHN CARROLL UNIV, 69- Mem: Am Phys Soc; Acoustical Soc Am; Seismol Soc Am. Res: Theoretical nuclear structure; theoretical and experimental studies of ultrasonic waveguides; absorption and dispersion of ultrasound in liquids; propagation of acoustic transients; laser induced effects in liquids and solids. Mailing Add: Dept of Physics John Carroll Univ Cleveland OH 44118

CARON, AIMERY PIERRE, b Paris, France, Apr 20, 30; US citizen; m 56; c 1. PHYSICAL CHEMISTRY. Educ: Univ Calif, Los Angeles, BS, 55; Univ Southern Calif, MS, 58, PhD(crystallog chem), 62. Prof Exp: Lab asst, Univ Southern Calif, 55, US Air Force & Army fel, 62-63; mem res staff phys chem, Space Mat Lab, Northrop Corp, 63-66; asst prof chem, Univ Mass, 66-68; mgr, C & M Caron, Inc, VI, 68-69; admin officer, 69-73, ASST TO PRES & ASSOC PROF CHEM, COL VI, 73- Mem: Am Crystallog Asn. Res: Determination of molecular structures by x-ray diffraction; inorganic syntheses; infrared spectroscopy; general physical chemistry; scientific computer programming. Mailing Add: Col of VI Box 1826 St Thomas VI 00801

CARON, DEWEY MAURICE, b North Adams, Mass, Dec 25, 42; m 65. ENTOMOLOGY, APICULTURE. Educ: Univ Vt, BS, 64; Univ Tenn, Knoxville, MS, 66; Cornell Univ, PhD(entom), 70. Prof Exp: Instr entom, Cornell Univ, 68, admin asst, Dept Entom, 69-70; ASSOC PROF APICULT, UNIV MD, COLLEGE PARK, 70- Mem: Entom Soc Am; Am Inst Biol Sci; Bee Res Asn. Res: Pollination ecology; biology and behavior of bees and wasps. Mailing Add: Dept of Entom Univ of Md College Park MD 20742

CARON, WILFRID M, b L'Islet, Que, Sept 27, 12; m 47. SURGERY. Educ: Laval Univ, BA, 34, MD, 39. Prof Exp: Intern, St Sacrement Hosp, 39-40; resident, Peter Bent Brigham Hosp, Boston, 40-41; fel, St Louis Univ Hosp, 41-44; asst, 44-49, assoc prof, 49-63, PROF SURG & HEAD DEPT, LAVAL UNIV, 63- Concurrent Pos: Marlow lectr, Acad Med Toronto, 64; vis prof, Univ Western Ont; chief dept surg, Centre Hosp, Laval Univ; consult, St Michel Archange, Jeffery Hales & St Sacrement Hosps. Mem: Fel Am Col Surgeons; fel Am Surg Asn; Can Med Asn; Can Asn Clin Surgeons; Royal Soc Med. Mailing Add: Dept of Surg Laval Univ Fac of Med Ste-Foy PQ Can

CARONE, FRANK, b New Kensington, Pa, Nov 28, 27; m 52; c 5. PATHOLOGY. Educ: WVa Univ, AB, 48; Yale Univ, MD, 52. Prof Exp: Instr, Sch Med, Yale Univ, 59-60; from asst prof to prof, 60-69, MORRISON PROF PATH & DEP CHMN DEPT, SCH MED, NORTHWESTERN UNIV, CHICAGO, 69- Concurrent Pos: Life Inst Med Res Fund res fel, 57-59; Markle scholar, 64-69; attend, Wesley Mem Hosp, Chicago, 61- Mem: Am Fedn Clin Res; Int Acad Path; Am Soc Exp Path. Res: Renal pathophysiology employing micropuncture techniques; light and electron microscopic study of human renal disease. Mailing Add: Dept of Path Northwestern Univ Med Sch Chicago IL 60611

CAROSELLI, NESTOR EDGAR, b Fall River, Mass, Sept 23, 13; m 42; c 2. PLANT PATHOLOGY. Educ: Univ RI, BS, 37, MS, 40; Brown Univ, PhD(bot), 54. Prof Exp: Asst plant pathologist, Bartlett Tree Res Lab, 40-42, from assoc plant to plant pathologist, 46-54; asst prof plant path, 55-57, assoc prof, 57-59, PROF BOT, UNIV RI, 60- Res: Chemotherapy of vascular diseases of plants; physiology of fungi. Mailing Add: Dept of Bot Univ of RI Kingston RI 02881

CAROSELLI, REMUS FRANCIS, b Providence, RI, Oct 4, 16; m 48; c 3. TEXTILE CHEMISTRY. Educ: Univ RI, BS, 37. Prof Exp: Process control technologist, Ashton Plant, Owens-Corning Fiberglas Corp, 41-46, chief chemist, 46-48, proj mgr textile res, 48-50, asst res mgr, 50-57, lab mgr textile process & prod develop, 57-60, mgr textile prod develop lab, 60-73; PRES, R F CAROSELLI PROD DEVELOP SERV, 74- Mem: Am Chem Soc; Asn Textile Chemists & Colorists. Res: Textiles, plastics, coatings and research and development organization. Mailing Add: 230 Colonel John Gardner Rd Narragansett RI 02882

CAROTHERS, JAMES (EDWARD), b Iowa Falls, Iowa, Mar 27, 23; m 45; c 4. PHYSICS. Educ: Univ Calif, PhD(physics), 52. Prof Exp: ASSOC DIR NUCLEAR TESTING, LAWRENCE RADIATION LAB, UNIV CALIF, 70- Mem: Am Nuclear Soc. Res: Neutron physics; physics of critical systems. Mailing Add: Lawrence Radiation Lab Univ of Calif Bldg 121 Rm 1041 Livermore CA 94550

CAROTHERS, STEVEN WARREN, b Prescott, Ariz, Dec 19, 43; m; c 3. ECOLOGY. Educ: Northern Ariz Univ, BS, 66, MS, 69; Univ Ill, Urbana, PhD(ecol), 74. Prof Exp: Asst ornithologist, Mus Northern Ariz, 66-67; teaching asst biol, Northern Ariz Univ, 67-68; asst cur zool, Mus Northern Ariz, 69-70; instr ornith, Northern Ariz Univ, 70; teaching asst biol, Univ Ill, 70-71; cur zool, 71-74, CUR BIOL, MUS NORTHERN ARIZ, 74- Mem: Am Ornithologists Union; Am Soc Mammalogists; Cooper Ornith Soc; Wilson Ornith Soc. Res: Work with federal agencies and private institutions at designing proper land-use management plans particularly those affecting non-game wildlife. Mailing Add: Mus of Northern Ariz Colton Res Ctr Flagstaff AZ 86001

CAROTHERS, ZANE BLAND, b Philadelphia, Pa, Nov 7, 24; m 52; c 2. BOTANY. Educ: Temple Univ, BS, 50, MEd, 52; Univ Michigan, PhD(bot), 58. Prof Exp: Instr bot, Univ Ky, 57-59; asst prof, 59-64, ASSOC PROF BOT, UNIV ILL, URBANA-CHAMPAIGN, 64-, ASSOC HEAD DEPT, 70- Mem: AAAS; Bot Soc Am; Torrey Bot Club. Res: Cell ultrastructure; developmental and systematic anatomy. Mailing Add: Dept of Bot Univ of Ill Urbana-Champaign Urbana IL 61801

CAROVILLANO, ROBERT L, b Newark, NJ, Aug 2, 32; m 52; c 3. SPACE PHYSICS, ASTROPHYSICS. Educ: Rutgers Univ, AB, 54; Ind Univ, PhD(physics), 59. Prof Exp: From asst prof to assoc prof, 59-67, PROF PHYSICS, BOSTON COL, 67-, CHMN DEPT, 69- Concurrent Pos: Vis fac, Mass Inst Technol, 67-68; assoc ed, Cosmic Electrodynamics, 69-72 & Rev Geophys-Space Physics, 72-76. Mem: AAAS; Am Phys Soc; Am Geophys Union (secy magnetospheric sect, 70-); Am Asn Physics Teachers; NY Acad Sci. Res: Theoretical studies on the solar wind, the magnetosphere, and the ionosphere and auroras. Mailing Add: Dept of Physics Boston Col Chestnut Hill MA 02167

CAROW, JOHN, b Ladysmith, Wis, Aug 26, 13; m 42; c 2. FORESTRY. Educ: Univ Mich, BSF, 37, MF, 38. Prof Exp: Field asst forest surv, Appalachian Forestry Exp Sta, 37-38, jr forester, 39-42; forest statistician, Am Paper & Pulp Asn, 38-39; shelterbelt asst, US Forest Serv, 39; res forester, Southeastern Forestry Exp Sta, 42-46; from instr to assoc prof, 47-62, PROF FOREST MGT, UNIV MICH, 62- Mem: Am Foresters. Res: Forest mensuration, inventory techniques, logging cost analysis; forest management. Mailing Add: Sch Natural Resources Univ of Mich Ann Arbor MI 48104

CAROZZI, ALBERT VICTOR, b Geneva, Switz, Apr 26, 25; nat US; m 49; c 2. GEOLOGY. Educ: Univ Geneva, MS, 47, DSc(geol mineral), 48. Prof Exp: Lectr spec geol, Univ Geneva, 48-53, asst prof, 53-57; assoc prof geol, 57-59, assoc mem, Ctr Advan Study, 69-70, PROF GEOL, UNIV ILL, URBANA-CHAMPAIGN, 59- Concurrent Pos: Asst vis prof, Univ Ill, 55-56; Am Asn Petrol Geol distinguished lectr, 59; adv, Govt Ivory Coast, Africa, 60-; corresp mem, Int Comt Hist of Geol Sci, 68-; consult adv, Petroleo Brasileiro SAm, Brazil, 69- & Philippine Oil Develop Co, Manila, 70- Honors & Awards: Davy Award, Univ Geneva, 49, 54; Plantamour-Prevost Award, 55. Mem: Fel Geol Soc Am; Am Asn Petrol Geol; Soc Econ Paleont & Mineral; Hist Sci Soc. Res: Sedimentary petrology, such as models of deposition of carbonate rocks and sandstones, experimental studies on porosity in carbonate rocks; history of geology; oil exploration. Mailing Add: Dept of Geol 250 Nat Hist Bldg Univ of Ill Urbana-Champaign Urbana IL 61801

CARP, OSCAR, surgery, deceased

CARP, RICHARD IRVIN, b Philadelphia, Pa, May 10, 34; m 60; c 1. MICROBIOLOGY, VIROLOGY. Educ: Univ Pa, BA, 55, VMD, 58, PhD(microbiol), 62. Prof Exp: Asst virol, Wistar Inst, 58-62, fel, 62-63, assoc, 64-68; virol res dir, Alembic Chem Co, India, 63-64; MEM STAFF, INST BASIC RES MENT RETARDATION, 68- Concurrent Pos: Vis prof, Div Biochem Virol, Col Med, Baylor Univ, 66- Mem: Am Soc Microbiol. Res: Slow infections of the central nervous system with particular interest in scrapie and the search for the causes of multiple sclerosis and amyotrophic lateral sclerosis; viruses that cause birth defects and mental retardation, such as the cytomegaloviruses. Mailing Add: Inst of Basic Res Ment Retardation 1050 Forest Hill Rd Staten Island NY 10314

CARPELAN, LARS HJALMAR, b Calif, June 9, 13; m 54; c 3. ECOLOGY. Educ: San Jose State Col, AB, 34; Stanford Univ, PhD, 53. Prof Exp: Res assoc algal physiol, Hopkins Marine Sta, Stanford Univ, 53-54; res assoc zoo-fisheries, Univ Calif, Los Angeles, 54-56; from asst prof to assoc prof, 56-71, PROF BIOL, UNIV CALIF, RIVERSIDE, 71- Mem: Ecol Soc Am; Am Soc Limnol & Oceanog. Res: Aquatic biology; brackish waters. Mailing Add: Dept of Biol Univ of Calif Riverside CA 92507

CARPENDER, JAMES WOOD JOHNSON, b New Brunswick, NJ, Nov 18, 11; m 46; c 3. RADIOLOGY. Educ: Yale Univ, BA, 34; Columbia Univ, MD, 38. Prof Exp: From assoc prof to prof radiol, Univ Chicago, 51-66; CO-CHMN DEPT RADIOL, GUTHRIE CLIN LTD, 66- Mem: AAAS; Radiol Soc NAm; AMA; Am Roentgen Ray Soc; Am Radium Soc. Res: Radiation therapy. Mailing Add: Dept of Radiol Guthrie Clin Ltd Sayre PA 18840

CARPENTER, ADELAIDE TROWBRIDGE CLARK, b Athens, Ga, June 24, 44. GENETICS. Educ: NC State Univ, BS, 66; Univ Wash, MS, 69, PhD(genetics), 72. Prof Exp: NIH fel cytogenetics, Univ Wis-Madison, 72-74; res assoc, Dept Anat, Duke Univ, 74-75, asst adj prof, 75-76, asst med res prof, 76; ASST PROF CYTOGENETICS, DEPT BIOL, UNIV CALIF, SAN DIEGO, 76- Mem: Genetics Soc Am; Am Soc Naturalists; Am Soc Cell Biol; AAAS. Res: Analysis of meiotic mutants in Drosophila melanogaster females, particularly recombination-defectives, by electron microscopy and effects on somatic crossing-over and chromosome maintenance. Mailing Add: Dept of Biol Univ of Calif at San Diego La Jolla CA 92037

CARPENTER, ALDEN B, b Newton, Mass, Feb 24, 36; m 61; c 2. GEOCHEMISTRY. Educ: Harvard Univ, AB, 57, PhD(geol), 63. Prof Exp: From asst prof to assoc prof geol, 63-73, PROF GEOL, UNIV MO-COLUMBIA, 73- Mem: Mineral Soc Am; Mineral Asn Can; Geochem Soc; Soc Econ Paleontologists & Mineralogists; Am Asn Petrol Geologists. Res: Mineralogy; mineral equilibria; geochemistry of subsurface waters. Mailing Add: Dept of Geol Univ of Mo Columbia MO 65202

CARPENTER, ANNA-MARY, b Ambridge, Pa, Jan 14, 16. ANATOMY. Educ: Geneva Col, AB, 36; Univ Pittsburgh, MS, 37, PhD(microtech), 40; Univ Minn, MD, 58. Hon Degrees: DSc, Geneva Col, 68. Prof Exp: Asst, Univ Pittsburgh, 38-40; instr lab tech, Moravian Col Women, 41-42; chmn biol curricula, Scranton-Keystone Jr Col, 42-44; res assoc path dept, Children's Hosp, Pittsburgh, 44-53; lectr mycol, Sch Med, Univ Pittsburgh, 46-53 & Western Reserve Univ, 53-54; from instr to assoc prof, 54-65, PROF ANAT, SCH MED, UNIV MINN, MINNEAPOLIS, 65- Mem: AAAS; Histochem Soc (secy, 74-75, treas, 75-76); Am Asn Anat; Int Soc Human & Animal Mycol; Int Soc Stereology (secy-treas, 72-79). Res: Mycology; histochemistry; quantitation. Mailing Add: 153 Orlin Ave SE Minneapolis MN 55414

CARPENTER, BARRY KEITH, b Hastings, Eng, Feb 13, 49; m 74, ORGANIC CHEMISTRY, ORGANOMETALLIC CHEMISTRY. Educ: Warwick Univ, BSc, 70; Univ Col, Univ London, PhD(chem), 73. Prof Exp: NATO fel, Yale Univ, 73-75; ASST PROF CHEM, CORNELL UNIV, 75- Mem: The Chem Soc; Am Chem Soc. Res: Mechanistic organic and mechanistic organometallic chemistry. Mailing Add: Dept of Chem Cornell Univ Ithaca NY 14853

CARPENTER, BRUCE H, b Rapid City, SDak, Feb 5, 32; m 52; c 2. PLANT PHYSIOLOGY. Educ: Calif State Col, Long Beach, AB, 57, MA, 58; Univ Calif, Los Angeles, PhD(bot), 62. Prof Exp: Instr biol, 57-59, from asst prof to assoc prof, 62-71, chmn dept, 67-72, PROF BIOL, CALIF STATE UNIV, LONG BEACH, 71-, ASSOC VPRES ACAD AFFAIRS-ACAD PERSONNEL, 72- Concurrent Pos: Nat Sci Found res grant, 63- Mem: AAAS; NY Acad Sci; Am Soc Plant Physiol. Res: Plant photoperiodism and Circadian rhythms. Mailing Add: Dept of Biol Calif State Univ 6101 S Seventh St Long Beach CA 90840

CARPENTER, CAROLYN VIRUS, b Chicago, Ill, Jan 1, 40; m 64; c 3. BIOCHEMISTRY. Educ: Univ Ill, Urbana, BS, 63; Univ Ill, Chicago Circle, PhD(biochem), 68. Prof Exp: Fel biochem, 68-75, RES ASSOC BIOCHEM & MOLECULAR BIOL, NORTHWESTERN UNIV, 75- Mem: Am Chem Soc. Res: Mechanism of action of anti-metabolites; specificity profiles of the enzymes involved in cell wall biosynthesis in several Gram-positive bacteria. Mailing Add: Dept of Biochem & Molec Biol Northwestern Univ Evanston IL 60201

CARPENTER, CHARLES, b Newark, NJ, July 17, 08; m 36; c 3. CHEMISTRY. Educ: Syracuse Univ, BS, 29, MS, 31; Darmstadt Tech Univ, Dr Ing, 33. Prof Exp: Asst chem, Carnegie Inst Technol, 34-36; chief chemist, asst dir & tech dir, Herty Found Lab, 36-39; tech dir & gen supt, Southland Paper Mills, 39-47; asst to vpres, NY & Pa County, 47-50; vpres, White Star Paper Co, 51; CONSULT, 52- Mem: Soc Am Foresters; Tech Asn Pulp & Paper Indust. Res: Cellulose, wood and pulp chemistry. Mailing Add: 2345 Wildwood Dr Montgomery AL 36111

CARPENTER, CHARLES C J, b Savannah, Ga, Jan 5, 31; m 58; c 3. INTERNAL MEDICINE, INFECTIOUS DISEASE. Educ: Princeton Univ, AB, 52; Johns Hopkins Univ, MD, 56. Prof Exp: From asst prof to prof med, Johns Hopkins Univ, 62-73; PROF MED & CHMN DEPT, CASE WESTERN RESERVE UNIV, 73- Concurrent Pos: Res career develop award, Johns Hopkins Univ, 64-69; mem US deleg, US-Japan Coop Med Sci Prog, 65-, chmn cholera panel, 66-73; mem cholera adv comt, NIH, 66-73; mem expert adv panel bact dis, WHO, 67- Mem: Am Soc Clin Invest; Infectious Dis Soc Am; Asn Profs Med; Am Fedn Clin Res; Asn Am Physicians. Res: Pathogenesis and pathophysiology of bacterial infections. Mailing Add: Univ Hosps Dept of Med Adelbert Rd Cleveland OH 44106

CARPENTER, CHARLES CONGDEN, b Denison, Iowa, June 2, 21; m 47; c 3. ZOOLOGY. Educ: Univ Northern Mich, BA, 43; Univ Mich, MS, 47, PhD(zool), 51. Prof Exp: Instr zool, Univ Mich, 51-52 & Wayne State Univ, 52; from instr to assoc prof, 53-66, PROF ZOOL, UNIV OKLA, 66-, CUR REPTILES, MUS ZOOL, 54- Concurrent Pos: NY Zool Soc grant-in-aid, Jackson Hole Res Sta, 51; NSF grant, 56-75; treas, Grassland Res Found, 58-62; mem, Galapagos Int Sci Proj, 64; mem, Sci Adv Comt, Charles Darwin Found for Galapagos Islands. Mem: Am Soc Zoologists; Ecol Soc Am; Am Soc Ichthyol & Herpet; Am Soc Mammal; fel Animal Behav Soc (secy, 65-68). Res: Ecology and behavior of vertebrates; herpetology; dynamics of populations and space relationships of reptiles and amphibians; comparative ecology and behavior. Mailing Add: Dept of Zool Univ of Okla Norman OK 73069

CARPENTER, CHARLES PATTEN, b Sellersville, Pa, July 5, 10; m 34; c 2. TOXICOLOGY, BACTERIOLOGY. Educ: Franklin & Marshall Col, BS, 31; Univ Pa, AM, 34, PhD(med sci), 37. Prof Exp: Instr bact, Hyg Dept, Med Sch, Univ Pa, 36-39, asst prof pub health & prev med, Lab, 39-40; Union Carbide indust fel, Mellon Inst, 40-46, sr fel, 46-56, from asst admin fel to admin fel, 56-75, ADV FEL, CARNEGIE-MELLON UNIV, MELLON INST RES, 75- Mem: AAAS; Am Chem Soc; Am Soc Toxicol; Am Indust Hyg Asn. Res: Toxicity of synthetic organic chemicals; industrial hygiene. Mailing Add: Mellon Inst Carnegie-Mellon Univ Pittsburgh PA 15213

CARPENTER, DAVID FRANCIS, b Springfield, Mass, Dec 24, 45; m 68; c 2. MICROBIAL PHYSIOLOGY. Educ: Univ Vt, BA, 67; Univ NH, PhD(microbiol), 71. Prof Exp: Nat Acad Sci-Nat Res Coun assoc, Washington, DC, 71-73; RES MICROBIOLOGIST, FOOD SCI LAB, US ARMY NATICK DEVELOP CTR, 73- Mem: Am Soc Microbiol; Soc Indust Microbiol; Sigma Xi. Res: Abatement of military pollutants; prevention of microbial deterioration of military material; microbial fermentations for optimizing biodegradation of recalcitrant molecules. Mailing Add: Pollution Abatement Div Food Sci Lab Army Dev Ctr Natick MA 01760

CARPENTER, DAVID O, b Fairmont, Minn, Jan 27, 37; m 61; c 2. NEUROPHYSIOLOGY, BIOPHYSICS. Educ: Harvard Univ, BA, 59, MD, 64; Prof Exp: Med officer neurophysiol, Lab Neurophysiol, NIMH, 65-72; CHMN NEUROBIOL DEPT, ARMED FORCES RADIOBIOL RES INST, 73- Concurrent Pos: Fel neurophysiol, Sch Med, Harvard Univ, 64-65. Mem: Am Physiol Soc; Soc Gen Physiol; Soc Neurosci; Int Brain Res Orgn; NY Acad Sci. Res: Electrogenic sodium pumps in Aplysia neurons; neurotransmitter substances in Aplysia; ionic basis of action potentials in invertebrate neurons; significance of cell size in spinal motor neurons; supraspinal control mechanisms. Mailing Add: Neurobiol Dept Armed Forces Radiobiol Res Inst Bethesda MD 20014

CARPENTER, DELMA RAE, JR, b Salem, Va, Apr 15, 28; m 52; c 3. PHYSICS ENGINEERING. Educ: Roanoke Col, BS, 49; Cornell Univ, MS, 51; Univ Va, PhD(physics), 57. Prof Exp: Instr physics, 51-53, from asst prof to assoc prof, 56-62, proj dir res labs, 60-74, dep dir, 63-65, head dept physics, 69-74, PROF PHYSICS, VA MIL INST, 63-, DIR RES LABS, 65- Concurrent Pos: Res assoc, US Army Ord Contract, 53, 54, 56 & 60; consult, US Army Res Off, 71-75; chmn bd trustees, Sci Mus Va, 73-76. Mem: Fel AAAS; Am Asn Physics Teachers; Soc Res Adminr. Res: Ordnance development and design; heat transfer in satellite instruments; physics teaching demonstrations. Mailing Add: Dept of Physics Va Mil Inst Lexington VA 24450

CARPENTER, DEWEY KENNETH, b Omaha, Nebr, June 30, 28; m 55; c 3. PHYSICAL CHEMISTRY, POLYMER CHEMISTRY. Educ: Syracuse Univ, BS, 50; Duke Univ, AM, 52, PhD(chem), 55. Prof Exp: Res assoc chem, Cornell Univ, 55-56; res assoc & instr, Duke Univ, 56-58; from asst prof to assoc prof chem, Ga Inst Technol, 58-69; assoc prof, 69-74, PROF CHEM, LA STATE UNIV, BATON ROUGE, 74- Concurrent Pos: Vis fel, Dartmouth Col, 67-68. Mem: Am Chem Soc; Am Sci Affil; Sigma Xi; Biophys Soc; Am Crystallog Asn. Res: Physical chemistry of high polymers; physical chemistry of micromolecules in solution. Mailing Add: Dept of Chem La State Univ Baton Rouge LA 70803

CARPENTER, DOROTHY IRENE, b South Bend, Ind, Aug 12, 15. MATHEMATICS. Educ: Ashland Col, AB, 37; Univ Mich, MA, 44. Prof Exp: Instr math, Denison Univ, 46-53; ASSOC PROF MATH, ASHLAND COL, 53-, CHMN DEPT, 70- Mem: Math Asn Am. Res: History of mathematics. Mailing Add: 407 Claremont Ave Ashland OH 44805

CARPENTER, DWIGHT WILLIAM, b Paducah, Ky, July 25, 36; m 58; c 3. PHYSICS. Educ: Univ Ky, BS, 58; Univ Ill, Urbana-Champaign, MS, 59, PhD(physics), 65. Prof Exp: Res assoc physics, Univ Ill, Urbana-Champaign, 64-66; asst prof, Duke Univ, 66-72; ASSOC PROF PHYSICS, LIMESTONE COL, 72- Mem: Am Phys Soc; Am Asn Physics Teachers; Math Asn Am. Res: Elementary particle physics. Mailing Add: Div of Sci & Math Limestone College Gaffney SC 29340

CARPENTER, EDWARD J, b Buffalo, NY, Mar 28, 42; m 67; c 2. BIOLOGICAL OCEANOGRAPHY. Educ: State Univ NY Col Fredonia, BS, 64; NC State Univ, MS, 67, PhD(zool), 70. Prof Exp: NSF fel biol oceanog, Woods Hole Oceanog Inst, 70-71, from asst scientist to assoc scientist, 71-75; ASSOC PROF BIOL, STATE UNIV NY STONY BROOK, 75- Concurrent Pos: Mem, US-USSR Comt Ocean Pollution, 73- Mem: Am Soc Limnol & Oceanog; Estuarine Res Fedn; Sigma Xi; Phycol Soc Am. Res: Nitrogen cycling in marine environment; physiology of nitrogen incorporation by algae; denitrification and nitrogen fixation; entrainment of plankton through coastal power plants; effects of pollutants on microorganisms. Mailing Add: Marine Sci Res Ctr State Univ of NY Stony Brook NY 11794

CARPENTER, EDWIN DAVID, b Great Falls, Mont, Aug 3, 32. ORNAMENTAL HORTICULTURE. Educ: Wash State Univ, BS, 57; Mich State Univ, MS, 62, PhD(hort), 64. Prof Exp: Sr exp aide, Coastal Wash Res & Exten Unit, Wash State Univ, 57-60; asst prof, 64-70, ASSOC PROF ORNAMENTAL HORT, UNIV CONN, 70-, CHMN DEPT PHYS SCI, 71- Concurrent Pos: Mem, Northeast Regional Tech Comn, Phys Hort Introd, USDA, 65-, chmn, 72-74. Mem: Am Soc Hort Sci; Am Hort Soc; Int Plant Propagators Soc; Am Soc Bot Gardens & Arboretums; Int Soc Hort Sci. Res: Plant anatomy, taxonomy and ecology. Mailing Add: Dept of Plant Sci Univ of Conn Storrs CT 06268

CARPENTER, ERWIN LOCKWOOD, physical chemistry, deceased

CARPENTER, ESTHER, b Meriden, Conn, June 4, 03. ZOOLOGY. Educ: Ohio Wesleyan Univ, AB, 25, Univ Wis, MS, 27; Yale Univ, PhD(zool), 32. Hon Degrees: DSc, Ohio Wesleyan Univ, 56. Prof Exp: Asst zool, Univ Wis, 25-27; lab technician, Yale Univ, 28-29; lab technician biol, Albertus Magnus Col, 30-32; lab technician, Dept Embryol, Carnegie Inst, 32-33; lab technician, 33-34, from instr to prof, 34-63, Myra MSampson prof, 63-68, EMER PROF ZOOL, SMITH COL, 68- Concurrent Pos: Instr, Albertus Magnus Col, 33-34; Howald Scholar, Ohio State Univ, 42-43; res, Strangeways Res Lab, Eng, 53-54, 61; Sophia Smith fel, 73- Mem: AAAS; Am Soc Cell Biol; Am Soc Zool; Soc Develop Biol; Am Asn Anat. Res: Experimental embryology; vital staining and transplantation in amphibia; regeneration in Eisenia foetida; tissue culture-differentiation of avian thyroid and femora; differentiation and physiological activities of embryonic thyroid glands in vitro. Mailing Add: 55 Prospect St Northampton MA 01060

CARPENTER, FRANCES LYNN, b Oklahoma City, Okla, Feb 14, 44. ECOLOGY, EVOLUTION. Educ: Univ Calif, Riverside, BA, 66; Univ Calif, Berkeley, PhD(zool), 72. Prof Exp: ASST PROF ECOL, UNIV CALIF, IRVINE, 72- Concurrent Pos: Am Philos Soc res grant, 74. Mem: Ecol Soc Am; Am Soc Naturalists; Soc Study Evolution; Cooper Ornith Soc; Am Ornithologists Union. Res: Energetics of plant-pollinator coevolved relationships; pollination strategies; behavior and resource partitioning in avian nectar-eaters; territoriality; comparison of generalist and specialist adaptive strategies in plants, birds, insects. Mailing Add: Dept of Ecol & Evolutionary Biol Univ of Calif Irvine CA 92717

CARPENTER, FRANK GRANT, b Toledo, Ohio, Oct 8, 23; m 51; c 2. PHYSIOLOGY. Educ: Ohio State Univ, BSc, 48; Columbia Univ, PhD, 51. Prof Exp: Asst physiol, Ohio State Univ, 48 & Columbia Univ, 49-51; instr, Univ Rochester, 52-54; asst prof, Med Col, Cornell Univ, 54-57; from asst prof to assoc prof, Dartmouth Med Sch, 57-67; ASSOC PROF PHARMACOL, UNIV ALA, BIRMINGHAM, 67- Concurrent Pos: Fel, Univ Rochester, 51-54. Mem: Am Physiol Soc; Harvey Soc; Am Soc Pharmacol & Exp Therapeut. Res: Anesthesia; nerve metabolism; autonomic neuroeffectors. Mailing Add: Dept of Pharmacol Univ of Ala Birmingham AL 35294

CARPENTER, FRANK MORTON, b Boston, Mass, Sept 6, 02. ENTOMOLOGY. Educ: Harvard Univ, AB, 26, MS, 27, ScD, 29. Prof Exp: Nat Res Coun fel zool, 28-31, assoc entom, 31-32, from asst prof to prof, 36-39, chmn dept, 53-59, prof entom & Agassiz prof zool, 36-69, Fisher Prof natural hist, 69-72, asst cur invert paleont, Mus Comp Zool, 32-36, cur fossil insects, 36-72, EMER FISHER PROF NATURAL HIST, HARVARD UNIV & EMER CUR FOSSIL INSECTS, MUS COMP ZOOL, 72- Concurrent Pos: Assoc, Carnegie Inst, 31-32; ed, Psyche, 46- Honors & Awards: Paleont Soc Medal, 75. Mem: Fel Am Acad Arts & Sci. Res: Paleoentomology and insect evolution; North American Neuroptera; Permian insects of Kansas and Oklahoma; Carboniferous insects of North America and Europe. Mailing Add: Dept of Biol Harvard Univ Cambridge MA 02138

CARPENTER, FREDERICK HILTMAN, b Cortez, Colo, June 8, 18; m 43; c 4. BIOCHEMISTRY. Educ: Stanford Univ, AB, 40, AM, 41, PhD(biochem), 44. Prof Exp: Lab asst biochem, Stanford Univ, 40-43, actg instr, 42; asst, Med Col, Cornell Univ, 43-45, res assoc, 45-48; from asst prof to assoc prof, 49-62, PROF BIOCHEM, UNIV CALIF, BERKELEY, 62-, DEAN DEPT BIOL SCI, 72- Concurrent Pos: Spec consult, NIH, 46-47, mem biochem study sect, 65-69; Royal Victor fel & Rockefeller fel, 48; Guggenheim fel, 64; vis sci, Technische Hochschule, Aachen, WGer, 71-72.

Mem: AAAS; Am Chem Soc; Am Soc Biol Chem. Res: Chemistry and synthesis of penicillin; chemistry of nucleic acid; action of mustard gas; synthesis, degradation and isolation of peptides; theory and practice of chromatography; biological activity in insulin and leucine amino peptidase. Mailing Add: Dept of Biochem Univ of Calif Berkeley CA 94720

CARPENTER, GAIL ALEXANDRA, b New York, NY, Dec 23, 48. BIOMATHEMATICS. Educ: Univ Colo, Boulder, BA, 70; Univ Wis-Madison, MA, 72, PhD(math), 74. Prof Exp: INSTR APPL MATH, MASS INST TECHNOL, 74- Mem: Am Math Soc. Res: Mathematical biology, using methods of topological dynamics; study of excitable membrane and network phenomena. Mailing Add: Dept of Math Rm 2-336 Mass Inst of Technol Cambridge MA 02139

CARPENTER, GARY GRANT, b San Francisco, Calif, Aug 25, 29. PEDIATRICS. Educ: Rutgers Univ, AB, 56; Jefferson Med Col, MD, 60. Prof Exp: From instr to asst prof, Sch Med, Temple Univ, 65-68; ASSOC PROF PEDIAT, JEFFERSON MED COL, 68- Concurrent Pos: Training fel metab & amino acids, St Christopher's Hosp Children, Pa, 62-63; asst prog dir, clin res ctr, St Christopher's Hosp Children, 65-68. Mem: AAAS; Ny Acad Sci. Res: Amino acid metabolism; cytogenetics. Mailing Add: Dept of Pediat Jefferson Med Col Philadelphia PA 19107

CARPENTER, GENE BLAKELY, b Evansville, Ind, Dec 15, 22; m 49; c 2. PHYSICAL CHEMISTRY, CRYSTALLOGRAPHY. Educ: Univ Louisville, BA, 44; Harvard Univ, MA, 45, PhD(phys chem), 47. Prof Exp: Nat Res fel, Calif Inst Technol, 48-49; from instr to assoc prof chem, 49-63, PROF CHEM, BROWN UNIV, 63- Concurrent Pos: Guggenheim fel, Univ Leeds, 56-57; vis prof, State Univ Groningen, 63-64; Fulbright-Hays lectr, Univ Zagreb, 71-72. Mem: AAAS; Am Chem Soc; Am Crystallog Asn. Res: Crystal structure by x-ray diffraction. Mailing Add: Dept of Chem Brown Univ Providence RI 02912

CARPENTER, GENE PAUL, b Fergus Falls, Minn, July 27, 32; m 59; c 2. ECONOMIC ENTOMOLOGY. Educ: Okla State Univ, BSc, 55; Ore State Univ, MSc, 61, PhD(entom), 63. Prof Exp: Field res rep, Agr Chem Div, Geigy Chem Corp, NY, 63-66; ASST RES PROF ENTOM, RES & EXTEN CTR, UNIV IDAHO, 66- Mem: AAAS; Entom Soc Am. Res: Insect vectors of plant viruses; biology and control of the alfalfa weevil and potato wireworms. Mailing Add: Res & Exten Ctr Univ of Idaho Rte 1 Box 186 Kimberly ID 83341

CARPENTER, IRVIN WATSON, JR, b Washington, DC, Nov 29, 23; m 48; c 3. PLANT TAXONOMY. Educ: Purdue Univ, BSF, 48, MS, 50, PhD(bot), 52. Prof Exp: Instr forestry, Purdue Univ, 52-53; from asst prof to prof biol, 53-72, CHMN BIOL, APPALACHIAN STATE UNIV, 72- Mem: Bot Soc Am; AAAS. Res: Flora of southern Appalachians. Mailing Add: Dept of Biol Appalachian State Univ Boone NC 28606

CARPENTER, JACK WILLIAM, b Worthington, Ohio, July 17, 25; m 50; c 2. THEORETICAL PHYSICS. Educ: Mass Inst Technol, BS, 51, MS, 52, PhD(physics), 57. Prof Exp: Proj scientist & sr physicist, Allied Res Assoc, Inc, 57-58, chief proj scientist, 58; vpres geophys div & sr physicist, Am Sci & Eng, Inc, Cambridge, 58-69; PRES & DIR, VISIDYNE, INC, MASS, 69- Res: Plasma physics; nuclear weapons effects; magnetohydrodynamics; nuclear physics; infrared spectroscopy.

CARPENTER, JAMES ANDREW, JR, b New Orleans, La, Sept 10, 37; m 58; c 2. ANIMAL BREEDING. Educ: Abilene Christian Col, BS, 59; Tex A&M Univ, MS, 61, PhD(animal breeding), 71. Prof Exp: Animal husbandman beef & sheep, Res Farm, Tex Technol Col, 63-67; teaching asst animal breeding, Tex A&M Univ, 67-71; instr animal sci, Tarleton State Col, 71-72; ASST PROF ANIMAL SCI, COLO STATE UNIV, 73- Concurrent Pos: Livestock consult, Booz-Allen & Hamilton Int Consults, 75-76. Mem: Am Soc Animal Sci; Coun Agr Sci & Technol. Res: Applied research in range nutrition and management; breeding research includes crossbreeding, energetic efficiency and systems analysis. Mailing Add: Dept of Animal Sci Colo State Univ Ft Collins CO 80523

CARPENTER, JAMES HEISKELL, oceanography, see 12th edition

CARPENTER, JAMES WILLIAM, b Shamokin, Pa, Sept 7, 35; m 58; c 3. ORGANIC CHEMISTRY. Educ: Lebanon Valley Col, BS, 60; Univ Nebr, MS, 63, PhD(org chem), 65. Prof Exp: Instr chem, Univ Nebr, 64-65; CHEMIST, EASTMAN KODAK CO, 65- Mem: Soc Photog Sci & Eng. Res: Photographic chemistry. Mailing Add: 120 Campfire Henrietta NY 14467

CARPENTER, JAMES WOODFORD, b Union, Ky, Jan 6, 22; m 49; c 2. ANIMAL SCIENCE. Educ: Univ Ky, BS, 52, MS, 53; Univ Fla, PhD(meat technol), 59. Prof Exp: Plant mgr, S4-59, asst meat scientist, 59-66, assoc meat scientist, 66-71, PROF & MEAT SCIENTIST, MEAT LAB, UNIV FLA, 71- Honors & Awards: Educ & Cult Medal, Govt S Vietnam, 70. Mem: Am Soc Animal Sci; Am Meat Sci Asn. Res: Meat animal carcass evaluation, especially quality and palatability factors. Mailing Add: Dept of Animal Sci Univ of Fla Gainesville FL 32611

CARPENTER, JOHN BARTLEY, plant pathology, see 12th edition

CARPENTER, JOHN HAROLD, b Owatonna, Minn, May 1, 29; m 53; c 2. HIGH TEMPERATURE CHEMISTRY. Educ: Macalester Col, BA, 51; Purdue Univ, MS, 53, PhD(chem), 55. Prof Exp: Chemist, Univ Calif, 54-55, res chemist, Lawrence Radiation Lab, 55-68; assoc prof chem, 68-71, PROF CHEM, ST CLOUD STATE UNIV, 71-, CHMN CHEM DEPT, 73- Mem: Am Chem Soc; fel Am Inst Chemists. Res: Structures, vapor pressures and thermodynamics of high melting inorganic compounds. Mailing Add: Dept of Chem St Cloud State Univ St Cloud MN 56301

CARPENTER, JOHN MELVIN, b Terre Haute, Ind, May 21, 10; m 46; c 3. ZOOLOGY. Educ: Univ Tex, BA, 36, MA, 40, PhD(zool), 46. Prof Exp: Res assoc entom, Clayton Found, Univ Tex, 38-45, tutor zool, Univ Tex, 42-45, instr, 45-46; asst prof, Univ Tenn, 46-53; head dept zool, 53-63, chmn biol sci div, 54-55, chmn dept zool, 63-65, PROF BIOL, UNIV KY, 53- Concurrent Pos: Res awards, Clayton Found Res, 44, Univ Ky Faculty Res Fund, 53-58, 65, Nat Sci Found, 56, 64-66, NIH & AEC, 62-65. Mem: Fel AAAS; Am Inst Biol Sci; Am Genetic Asn; Am Soc Zool; Ecol Soc Am. Res: Natural populations and speciation in Drosophila; gen-ecology; biotic potential; interspecific and intraspecific competition; radiation effects. Mailing Add: Dept of Zool Univ of Ky Lexington KY 40506

CARPENTER, JOHN RICHARD, b Galveston, Tex, May 20, 38; m 58; c 3. GEOCHEMISTRY. Educ: Rice Univ, BA, 59; Fla State Univ, MS, 62, PhD(geochem geol), 64. Prof Exp: Asst geol, Fla State Univ, 63-64; geologist, Fla Geol Surv, 64; US Naval Oceanog Off, 64-66; asst prof geol, 66-70, actg head dept, 69-70, ASSOC PROF GEOL, UNIV SC, 70-, DIR GRAD STUDIES & ASST CHMN DEPT GEOL, 74- Concurrent Pos: Sr col consult, 72-74, sr staff mem, Earth Sci Teacher Prep Proj, 73- Mem: Mineral Soc Am; Geochem Soc; Int Asn Geochem & Cosmochem; Nat Asn Geol Teachers. Res: Element distribution in coexisting phases of metamorphic and igneous rocks; geochemistry and petrogenesis of ultramafic rocks; structural control of metamorphic mineral assemblages; geochemistry of opaque minerals; alternative structure modes for earth science education. Mailing Add: Dept of Geol Univ of SC Columbia SC 29208

CARPENTER, LAWRENCE EDWARD, b Spring Green, Wis, Nov 20, 15; m 40; c 4. BIOCHEMISTRY. Educ: Univ Wis, BS, 39, PhD(biochem), 43; Northwestern Univ, MS, 41. Prof Exp: Biochemist, Pillsbury Mills, Inc, 43-45; asst prof biochem, Univ Minn, 45-49, assoc prof animal husb, 49-53; EXEC DIR, DISTILLERS FEED RES COUN, 53- Mem: Am Chem Soc; Am Dairy Sci Asn; Am Soc Animal Sci; Poultry Sci Asn. Res: Swine nutrition; chemistry; nutrition and foods. Mailing Add: Distillers Feed Res Coun 1435 Enquirer Bldg Cincinnati OH 45202

CARPENTER, LEE GRAYDON, b Milestone, Sask, Aug 15, 25; US citizen; m; c 2. INDUSTRIAL CHEMISTRY. Educ: Univ Sask, BSc, 49, MSc, 51; Columbia Univ, PhD(phys chem), 56. Prof Exp: Reduction engr, Aluminum Co Can, 56-57; sect head alkali metal salts, Am Potash & Chem Corp, 57-60; dept mgr radiation & radiochem, Aerojet-Gen Corp, 60-70; SR ENGR, ALUMINUM CO AM, 73- Mem: Am Chem Soc; Chem Inst Can; AAAS. Res: Recovery of aluminum from domestic ores and treatment of waste gases and residues resulting from recovery methods. Mailing Add: Alcoa Tech Ctr Alcoa Center PA 15069

CARPENTER, MALCOLM BRECKENRIDGE, b Montrose, Colo, July 7, 21; m 49; c 3. ANATOMY, PHYSIOLOGY. Educ: Columbia Univ, BA, 43; Long Island Col Med, MD, 47; Am Bd Psychiat & Neurol, dipl, 55. Prof Exp: Asst neurol, 47 & 48-50, from instr to assoc prof anat, 53-62, PROF ANAT, COLUMBIA UNIV, 62- Concurrent Pos: Fel neurol, Columbia Univ, 48-50; Markle scholar med sci, 53-58; surg intern, Bellevue Hosp, NY, 47-48; asst res neurologist, Neurol Inst, NY, 50 & 52-53; consult, Nat Inst Neurol Dis & Blindness, 62-66 & 68-72; mem, Inst Brain Res Orgn; ed, Neurology, 63-72, Am J Anat, 68-74, J Comp Neurol, 71 & Neurobiology, 71. Mem: Am Asn Anat; Asn Res Nerv & Ment Dis; Am Neurol Asn; Am Acad Neurol; NY Acad Med. Res: Neuroanatomic, neurophysiologic and neuropathologic study of motor disturbances, particularly those due to disease of basal ganglia. Mailing Add: 630 W 168th St New York NY 10032

CARPENTER, MARTHA STAHR, b Bethlehem, Pa, May 29, 20; m 51; c 1. ASTRONOMY. Educ: Wellesley Col, BA, 41; Univ Calif, MA, 43, PhD, 45. Prof Exp: Asst astron, Univ Calif, 41-44; instr, Wellesley Col, 45-47; from asst prof to assoc prof, Cornell Univ, 47-54; res grant radio astron, Australian Commonwealth Sci & Indust Res Org, 54-55; res assoc, Ctr Radiophysics & Space Res, Cornell Univ, 55-69; LECTR ASTRON, UNIV VA, 69-, ASSOC PROF ASTRON, 73- Mem: AAAS; Am Astron Soc; Am Asn Variable Star Observers (2nd vpres, 48-49, lst vpres, 49-51, pres, 51-54). Res: Galactic structure; radio astronomy. Mailing Add: Dept of Astron Univ of Va 416 Cabell Hall Charlottesville VA 22901

CARPENTER, MARY PITYNSKI, b Detroit, Mich, Feb 20, 26; m 47; c 3. BIOCHEMISTRY. Educ: Wayne Univ, BS, 46; Univ Mich, MA, 48, PhD(zool), 52. Prof Exp: Res assoc zool, Univ Mich, 51-53; res assoc biochem, Okla Med Res Found, 54-58, biochemist, 58-66; from asst prof to assoc prof biochem, 65-69, PROF BIOCHEM & MOLECULAR BIOL, SCH MED, UNIV OKLA, 69- Concurrent Pos: Assoc mem, Okla Med Res Found, 69- Mem: AAAS; Am Soc Biol Chem; Am Chem Soc; Am Inst Nutrit; Brit Biochem Soc. Res: Biochemistry of mammalian testis; metabolism of unsaturated fatty acids and prostaglandins; mixed function oxidases; function of vitamin E. Mailing Add: 1218 Cruce Norman OK 73069

CARPENTER, PAUL GERSHOM, b Salem, Ore, Jan 16, 14; m 37; c 2. ORGANIC CHEMISTRY. Educ: Willamette Univ, AB, 35; Ore State Col, MS, 37; Univ Wis, PhD(org chem), 41. Prof Exp: Asst, Ore State Col, 35-37 & Univ Wis, 37-38; asst prof chem, Willamette Univ, 38-39; asst, Wis Alumni Res Found, 39-40 & Univ Wis, 40-41; res chemist, Hercules Powder Co, 41-42, asst group leader, 42-45; sales res investr, Phillips Petrol Co, 45-47, chief prod res sect, 47-51; mgr synthetic rubber res br, Copolymer Rubber & Chem Corp, 51-56, vpres res & develop, 56-60, pres & chief exec off, 60-69; pres & chief exec off, Polyform, Inc, 70-73; DEVELOP MGR, DRACO INC, 73- Concurrent Pos: Chem & petrol consult, 69- Mem: Am Chem Soc. Res: Hydrogenation; organic synthesis; polymerization; petroleum production; synthetic rubber. Mailing Add: 1325 Marilyn Dr Baton Rouge LA 70815

CARPENTER, PAUL NATHANIEL, b Patten, Maine, June 9, 10; m 37. AGRONOMY. Educ: Bates Col, BS, 33; Univ Maine, MS, 49. Prof Exp: Instr sci, Aroostook Cent Inst, 34-37 & Skowhegan High Sch, 37-41; prin, Bridgewater Class Acad, 41-43; instr physics, Univ Maine, 43-44, farm labor supv, Exten Serv, 44-45, asst agronomist & agr chemist, Agr Exp Sta, 46, assoc agronomist, 56-60, SPECTROSCOPIST, AGR EXP STA, UNIV MAINE, 60- Mem: Fel Am Inst Chem; Am Chem Soc; Am Soc Agron; Soc Appl Spectroscopy. Res: Soil analysis; plant analysis and nutrition. Mailing Add: 406 Deering Hall Univ of Maine Orono ME 04473

CARPENTER, PHILIP LEWIS, b Newtown, Conn, Mar 17, 12; m 35; c 1. MICROBIOLOGY, IMMUNOLOGY. Educ: Middlebury Col, BS, 33; Brown Univ, MS, 34; Univ Wis, PhD(agr bact), 37. Prof Exp: Asst comp anat, Brown Univ, 34-35; asst bact, Univ Wis, 36-37; from instr to asst prof, Iowa State Col, 37-42; from instr to prof, 42-75, head dept, 45-70, EMER PROF BACT, UNIV RI, 75- Concurrent Pos: Fac fel, Ford Found, 51-52; vis prof, Univ Melbourne, 70. Mem: AAAS; Am Soc Microbiol; Am Acad Microbiol; NY Acad Sci; Brit Soc Immunol. Res: Immunology and serology; microbiology of lake sediments; serology of dysentery bacteria; physiology and classification of the coliform bacteria. Mailing Add: 42 Upper College Rd Kingston RI 02881

CARPENTER, RAYMOND, mathematics, see 12th edition

CARPENTER, RAYMOND ALLISON, b Hamilton, Ont, Can, July 12, 21; US citizen; m 49; c 4. MOLECULAR SPECTROSCOPY. Educ: Univ Tenn, BS, 50, MS, 51, PhD(physics), 65. Prof Exp: Res analyst, Douglas Aircraft Co, Inc, 51-52; physicist, Naval Ord Lab, 52-54, 55, electronic scientist, 54-55; classification analyst, 55-68, SR CLASSIFICATION ANALYST, DIV CLASSIFICATION, US AEC, 68- Mem: Am Phys Soc. Res: Infrared spectroscopy; underwater acoustics; electrical measurements; atomic and molecular physics. Mailing Add: 902 S Warfield Dr Mt Airy MD 21771

CARPENTER, RAYMOND T, b Topeka, Kans, Jan 14, 29; m 53; c 1. NUCLEAR PHYSICS. Educ: Univ Kans, BS, 54, MS, 56; Northwestern Univ, PhD(nuclear physics), 62. Prof Exp: Asst instr physics, Univ Kans, 54-56; asst instr physics, Northwestern Univ, 56-58; physicist, Argonne Nat Lab, 58-62; asst prof physics, 62-65, ASSOC PROF PHYSICS, UNIV IOWA, 65- Mem: AAAS. Res: Experimental nuclear structure physics of light nuclei, using accelerated lithium and helium ions as projectiles; gamma ray spectroscopy. Mailing Add: Dept of Physics Univ of Iowa Iowa City IA 52240

CARPENTER, RICHARD A, b Kansas City, Mo, Aug 22, 26; m 48; c 3. SCIENCE

ADMINISTRATION. Educ: Univ Mo, BS, 48, MA, 49. Prof Exp: Chemist, Shell Oil Co, 49-51; asst mgr, Midwest Res Inst, 51-58; mgr, Callery Chem Co, 58-64; sr specialist sci & tech, Libr Cong, 64-69, chief, Environ Policy Div, Cong Res Serv, 69-72; EXEC DIR COMN ON NAT RESOURCES, NAT RES COUN, NAT ACAD SCI, 72- Mem: Fel AAAS; Am Chem Soc; fel Am Inst Chem; Sci Res Soc Am; Ecol Soc Am. Res: Air pollution; environmental chemistry; technology assessment; boron chemistry; rocket propellants. Mailing Add: Comn Nat Resources Nat Res Coun 2101 Constitution Ave NW Washington DC 20418

CARPENTER, ROBERT DEAN, b Paris, Mo, Dec 12, 24; m 51; c 4. ANALYTICAL CHEMISTRY. Educ: Iowa State Univ, BS, 52. Prof Exp: Chemist, Com Solvents Corp, 52-54; from chemist to res chemist, 54-59, sr scientist, 65-72, ASSOC PRIN SCIENTIST CHEM, PHILIP MORRIS USA, 72- Res: Chemistry of tobacco and cigarette smoke; analysis of components and elucidation of mechanisms. Mailing Add: Philip Morris Res Ctr Box 26583 Richmond VA 23261

CARPENTER, ROBERT FRANCIS, b Schenectady, NY, Oct 28, 25; m 61; c 2. MOLECULAR PHYSICS. Educ: Union Univ, NY, BS, 49; Ohio State Univ, MSc, 52, PhD(physics), 63. Prof Exp: Instr physics, Kenyon Col, 60-63; asst prof, Denison Univ, 63-64; asst prof, Dayton Campus, Miami-Ohio State Univs, 63-69; MEM STAFF, AIR FORCE FLIGHT DYNAMICS LAB, WRIGHT PATTERSON AFB, 69- Mem: Am Phys Soc; Am Asn Physics Teachers. Res: Vibrational spectra of polyatomic molecules; line widths in nuclear magnetic resonance; spectroscopic gas temperature measurements. Mailing Add: 3431 Hickory Ct Xenia OH 45385

CARPENTER, ROBERT HALSTEAD, b Pasadena, Calif, June 24, 14; m 41; c 2. GEOLOGY. Educ: Stanford Univ, AB, 40, MA, 43, PhD(geol), 48. Prof Exp: Geologist, Anaconda Copper Mining Co, Mont, 42-44 & Int Smelting & Ref Co, Utah, 44-46; from asst prof to prof geol, Colo Sch Mines, 47-74; PROF GEOL, UNIV GA, 74- Concurrent Pos: Consult geologist, NY & Honduras Rosario Mining Co, Honduras, 49-52, Molybdenum Corp Am, 53-, Thomp Creek Coal & Coke, 53-, Utah Construct Co, 54 & UN Spec Fund Proj, Baldwin Mines, Burma, 63; Fulbright res scholar, italy, 57-58; pres, Int Mineral Eng, 63- Mem: Geol Soc Am; Soc Econ Geologists; Am Inst Mining, Metall & Petrol Eng. Res: Economic and structural geology; geochemistry of ore deposits. Mailing Add: Dept of Geol Univ of Ga Athens CA 30602

CARPENTER, ROBERT LELAND, b St Louis, Mo, June 27, 42; m 65; c 4. BIOPHYSICS. Educ: Univ Mo-Rolla, BS, 65; Univ Tenn, PhD(molecular biol), 75. Prof Exp: Res asst plasma physics, Los Alamos Sci Lab, 65-66; engr, McDonnell Douglas Corp, 66-70; SR STAFF INHALATION TOXICOL, LOVELACE BIOMED & ENVIRON RES INST, 75- Mem: AAAS; Am Chem Soc. Res: Inhalation toxicology of airborne effluents and aerosolized consumer products; aerosol physics. Mailing Add: Inhalation Toxicol Res Inst Lovelace Found Med Educ & Res Albuquerque NM 87115

CARPENTER, ROBERT RAYMOND, b Connellsville, Pa, Mar 6, 33; m 54; c 2. INTERNAL MEDICINE. Educ: Univ Pittsburgh, BS, 54; Univ Rochester, MD, 57. Prof Exp: Intern, King County Hosp, 57-58; resident, Sch Med, Univ Wash, 58-60; clin investr & attend physician, Lab Clin Invest, Nat Inst Allergy & Infectious Dis, 62-63, actg chief clin immunol, 63-64; asst prof med, Baylor Col Med, 64-68; from asst prof to assoc prof med & community med, Univ Pittsburgh, 68-72; PROF INTERNAL MED & DIR PRIMARY CARE-COMMUNITY MED, UNIV MICH ANN ARBOR, 72- Concurrent Pos: Markle sholar, 64; consult infectious dis, Georgetown Univ Serv, DC Gen Hosp, 62-63 & Montefiore Hosp, Pittsburgh, 68-71; consult immunol, Vet Admin Hosp, Houston, 64-68, dir immunol & infectious dis, 68; attend physician, Ben Taub Gen Hosp, Houston, 64-68; dir health care, Western Pa Regional Med Prog, Pittsburgh, 68-72. Mem: Am Asn Immunol; Am Col Physicians; Am Fedn Clin Res; Am Hosp Asn; Am Col Prev Med. Res: Immunology; infectious disease; health care. Mailing Add: Primary Care-Community Med Univ of Mich Med Ctr Ann Arbor MI 48104

CARPENTER, ROGER EDWIN, b Tucson, Ariz, Oct 13, 35; m 73; c 2. ZOOLOGY. Educ: Univ Ariz, BA, 57; Univ Calif, Los Angeles, PhD(zool), 63. Prof Exp: Assoc biol, Univ Calif, Riverside, 61-63; from asst prof to assoc prof zool, 63-70, chmn dept, 71-75, PROF ZOOL, SAN DIEGO STATE UNIV, 70- Mem: AAAS; Am Soc Mammal; Am Soc Zoologists. Res: Environmental physiology of vertebrates. Mailing Add: Dept of Zool San Diego State Univ San Diego CA 92182

CARPENTER, ROLAND LEROY, b Los Angeles, Calif, Apr 26, 26; m 51; c 1. ASTRONOMY, ASTROPHYSICS. Educ: Los Angeles State Col, BA, 51; Univ Calif, Los Angeles, MA, 64, PhD, 66. Prof Exp: Electronics technician, Collins Radio Co, 52-55, engr digital commun, 55-57, group supvr, 57-59; res engr, 59-62, scientist, 62-68, MEM STAFF LUNAR & PLANETARY SCI SECT, JET PROPULSION LAB, CALIF INST TECHNOL, 68-; MEM FACULTY, DEPT PHYSICS, CALIF STATE COL, LOS ANGELES, 68-, ASSOC PROF PHYSICS, 74- Mem: AAAS; Am Astron Soc; Royal Astron Soc. Res: Radar astronomy, especially studies of nearer planets by earth-based radar; galaxies; planetary astronomy. Mailing Add: Dept of Physics Calif State Col 5151 State College Dr Los Angeles CA 90032

CARPENTER, ROSE MARIE, b Chippewa Falls, Wis, Oct 3, 27; m 45; c 1. ZOOLOGY. Educ: Midwestern State Univ, BS, 63, MS, 69. Prof Exp: TEACHER BIOL, WICHITA FALLS INDEPENDENT SCH DIST, 63- Mem: Am Arachnological Soc; Soc Vert Paleontologists; Nat Asn Biol Teachers. Res: Distribution of American jumping spiders; extinct Pleistocene vertebrate faunas in Texas. Mailing Add: S H Rider High Sch 4611 Cypress Wichita Falls TX 76310

CARPENTER, ROY, US citizen. MARINE CHEMISTRY, GEOCHEMISTRY. Educ: Wash & Lee Univ, BS, 61; Univ Calif, San Diego, PhD(chem), 68. Prof Exp: Asst prof marine chem & geochem, 68-73, ASSOC PROF OCEANOG, UNIV WASH, 73- Mem: Am Chem Soc. Res: Chemical reactions in the oceans and marine sediments. Mailing Add: Dept of Oceanog Univ of Wash Seattle WA 98105

CARPENTER, RUSSELL LE GRAND, b Meriden, Conn, Nov 7, 01; m 29; c 2. RADIOBIOLOGY, HISTOLOGY. Educ: Tufts Col, BS, 24; Harvard Univ, PhD(zool), 28. Prof Exp: Instr anat, Col Physicians & Surgeons, Columbia Univ, 28-31, assoc, 31-35, asst prof, 35-38; prof, 38-68, EMER PROF ZOOL, TUFTS UNIV, 68-; LECTR OPHTHAL, HARVARD MED SCH, 53-; RES BIOLOGIST, BUR RADIOL HEALTH, USPHS, 70- Concurrent Pos: Mem corp, Marine Biol Lab, Woods Hole, 33-; instr, Harvard Univ, 46-53; mem teaching staff, Lancaster Course Ophthal, 52; consult microwave radio-biol, USPHS, 69-70; consult, Retina Found. Mem: Am Asn Anat; Am Soc Zoologists; Asn Res Vision & Ophthal; fel Am Acad Arts & Sci; Int Microwave Power Inst. Res: Anatomy and histology of the vertebrate eye; biological effects of microwave radiation with particular reference to the eye. Mailing Add: Bur of Radiol Health 109 Holton St Winchester MA 01890

CARPENTER, SAMMY, b Bolckow, Mo, July 20, 28; m 50; c 2. ORGANIC CHEMISTRY. Educ: Northwest Mo State Col, AB & BS, 50; Univ Mo, PhD(chem), 58. Prof Exp: Asst chem, Univ Mo, 54-57, asst instr, 58; res chemist, Celanese Corp Am, 58-64; asst prof, 64-69, ASSOC PROF CHEM, NORTHWEST MO STATE UNIV, 69-, CHMN DEPT, 68- Concurrent Pos: Guest chemist, Nat Bur Stand, 52-54. Mem: Am Chem Soc; Sigma Xi; AAAS. Res: Substituted styrenes; organometallics and epoxides. Mailing Add: Dept of Chem Northwest Mo State Univ Maryville MO 64468

CARPENTER, STANLEY JOHN, b Mansfield, Ohio, Feb 12, 36; m 63; c 3. CYTOLOGY. Educ: Oberlin Col, AB, 58; Univ Iowa, PhD(zool), 64. Prof Exp: Trainee path, 64-66, instr anat & cytol, 66-67, asst prof, 67-73, ASSOC PROF ANAT, DARTMOUTH MED SCH, 73- Res: Cell fine structure; choroid plexus, placenta, embryo. Mailing Add: Dept of Anat Dartmouth Med Sch Hanover NH 03755

CARPENTER, STEVE HAYCOCK, b Cedar City, Utah, May 15, 38; m 60; c 2. SOLID STATE PHYSICS. Educ: Univ Utah, BS, 59, PhD(physics), 64. Prof Exp: Asst elec & magnetic lab, Univ Utah, 60-62; res physicist, Aerojet-Gen Corp Div, Gen Tire & Rubber Co, 64-65; asst prof physics & metall, 65-68, assoc prof physics, 68-72, PROF PHYSICS & METALL, UNIV DENVER, 72- Mem: Am Phys Soc. Res: Defects, including impurities, and their interactions in crystals by means of internal friction measurements. Mailing Add: Dept of Physics Univ of Denver Denver CO 80210

CARPENTER, T J, b Middlebourne, WVa, Jan 9, 27; m 49; c 3. ORGANIC CHEMISTRY, ANALYTICAL CHEMISTRY. Educ: WVa Univ, BS, 49, MS, 51. Prof Exp: Plant chemist, Corning Glass Works, Pa, 54-56, supv chem eng, NY, 56-57, mfg eng consult, 57-61, mgr chem & metall eng dept, Tech Staff Div, 61-64; sr mem tech staff, Signetics Corp, 64-65, sect head, Process Improv, 65-66; mgr process technol dept, 66-71, DIR PROCESS TECHNOL, CORNING GLASS WORKS, 71- Mem: Am Chem Soc; Am Ceramic Soc; Am Inst Chem Eng. Res: Chemical process engineering. Mailing Add: Corning Glass Works Bldg 8 Fifth Floor Corning NY 14830

CARPENTER, WILL DOCKERY, b Moorhead, Miss, July 13, 30; m; c 1. PLANT PHYSIOLOGY. Educ: Miss State Univ, BS, 52; Purdue Univ, MS, 56, PhD(plant physiol), 58. Prof Exp: Plant biochemist, Inorg Chem Div, Monsanto Chem Co, 58-60, market develop dept, Agr Div, Monsanto Co, 60-75, mgr dept, 68-70, dir dept, 71-75, DIR PROD DEVELOP, MONSANTO AGR PROD CO, 75- Mem: Weed Sci Soc Am (treas, 75). Res: Respiration and carbohydrate metabolism; soil fertility; herbicides; agriculture. Mailing Add: Monsanto Prod Develop 800 N Lindbergh Blvd St Louis MO 63166

CARPENTER, WILLIAM GRAHAM, b West Liberty, WVa, May 7, 31; m 65; c 3. POLYMER CHEMISTRY. Educ: WVa Wesleyan Col, BS, 53; Univ Md, MS, 56, PhD(org chem), 60. Prof Exp: Res chemist, Stamford Res Labs, Am Cyanamid Co, 59-63; sr chemist, Plastics Div, Moorehead-Patterson Res Ctr, Am Mach & Foundry Co, 63-64 & Org Div, Cent Res Labs, Interchem Corp, Bloomfield, 64-67; SR CHEMIST, NL INDUSTRIES, INC, 67- Mem: Am Chem Soc; Soc Plastics Engrs. Res: Fiber process development; polyamides containing phosphorus; addition polymerization; synthesis and evaluation of high temperature polymers; compounding epoxy resins; block copolymers; epoxide polymerization; water soluble polymers, thickeners and retention aids. Mailing Add: 39 Pinehurst Dr Cranbury NJ 08512

CARPENTER, WILLIAM JOHN, b Pittsburgh, Pa, Sept 15, 27; m 52; c 3. FLORICULTURE. Educ: Univ Md, BS, 49; Mich State Univ, PhD(hort), 53. Prof Exp: From asst prof to assoc prof hort, Kans State Univ, 53-66; PROF HORT, MICH STATE UNIV, 68- Mem: Am Soc Hort Sci. Res: Measuring and programming of the greenhouse environment; supplemental lighting of greenhouse crops; floriculture crop physiology. Mailing Add: Dept of Hort Mich State Univ East Lansing MI 48823

CARPENTER, ZERLE LEON, b Thomas, Okla, July 21, 35; m 58; c 2. ANIMAL SCIENCE, FOOD SCIENCE. Educ: Okla State Univ, BS, 57; Univ Wis, MS, 60, PhD(animal sci), 62. Prof Exp: Res assoc, Univ Wis, 58-62; from asst prof to assoc prof, 62-70, PROF ANIMAL SCI, TEX A&M UNIV, 70- Mem: Am Meat Sci Asn; Am Soc Animal Sci; Inst Food Tech. Res: Determination of histological, biochemical and physical characteristics of beef, pork and lamb muscle as related to quantitative and qualitative components of meat animal species. Mailing Add: Dept of Animal Sci Tex A&M Univ College Station TX 77843

CARPER, WILLIAM ROBERT, b Syracuse, NY, Feb 8, 35; m 66; c 3. ENZYMOLOGY. Educ: State Univ NY Albany, BS, 60; Univ Miss, PhD(chem), 63. Prof Exp: Welch Fund fel chem, Texas A&M Univ, 63-65; asst prof, Calif State Col Los Angeles, 65-67; assoc prof, 67-70, PROF BIOCHEM, WICHITA STATE UNIV, 70- Concurrent Pos: NIMH spec fel, Univ S Fla, 72-73. Mem: Am Chem Soc; The Chem Soc; Asn Biol Chemists; Sigma Xi. Res: Structure and function of proteins. Mailing Add: Dept of Chem Wichita State Univ Wichita KS 67208

CARPINO, LOUIS ALBERT, b Des Moines, Iowa, Dec 13, 27; m 58; c 6. CHEMISTRY. Educ: Iowa State Col, BS, 50; Univ Ill, MS, 51, PhD(org chem), 53. Prof Exp: Assoc prof org chem, 54-67, PROF CHEM, UNIV MASS, AMHERST, 67- Mem: Am Chem Soc; The Chem Soc; Soc Ger Chem. Res: Small-ring heterocycles; non-benzenoid aromatic systems; new amino-protecting groups; organo-nitrogen and organo-sulfur chemistry. Mailing Add: Dept of Chem Univ of Mass Amherst MA 01002

CARR, ALBERT A, b Covington, Ky, Dec 20, 30; m 56; c 4. ORGANIC CHEMISTRY, MEDICINAL CHEMISTRY. Educ: Xavier Univ, BS, 53, MS, 55; Univ Fla, PhD(org chem), 58. Prof Exp: SECT HEAD ORG CHEM, MERRELL-NAT LABS, 58- Mem: Am Chem Soc. Res: Pharmaceuticals; psychotherapeutics; design preparation and characterization of psychotherapeutic, cardiovascular and antiallergy agents. Mailing Add: 8505 Brent Dr Cincinnati OH 45231

CARR, ARCHIE FAIRLY, JR, b Mobile, Ala, June 16, 09; m 37; c 5. ZOOLOGY. Educ: Univ Fla, BS, 33, MS, 34, PhD(herpet), 37. Prof Exp: Asst biol, 33-37, instr biol sci, 38-40, asst prof, 40-44, from assoc prof to prof, 45-59, grad res prof, 59-73, GRAD PROF ZOOL, UNIV FLA, 73- Concurrent Pos: Mem, Univ Fla exped, Mex, 39-40, 41, Shire Valley Surv, Nyasaland, Africa, 52, Am Philos Soc exped, Trinidad & Costa Rica, 53, Univ Fla-Fla State Mus exped, Panama & Costa Rica, 54, Nat Sci Found, Cent Am, 55-65, Brazil, West Africa, Portugal, Azores, 56, Mex, Spain, SAfrica, Argentina, Chile & Costa Rica, 57, 58, Off Naval Res, Nat Sci Found, Leeward Islands of Hawaiian Archipelago, 62, EAfrica & Madagascar, 63; prof, Escuela Agricola Panamericana, 45-49; biologist, United Fruit Co, 49; res assoc, Am Mus Natural Hist, 49-; assoc, Fla State Mus, 53-; tech adv to faculty sci & letters, Univ Costa Rica, 56-57; tech dir, Caribbean Conserv Corp, 59-; head marine turtle group, Survival Serv Comn, Int Union Conserv Nature, 63- Honors & Awards: Elliot Medal, Nat Acad Sci, 55; Burroughs Award, 56. Mem: Am Soc Ichthyol & Herpet (vpres, 41); Am Soc Nat. Res: Zoogeography of turtles; ecology of reptiles and amphibians of Florida; tropical natural history; ecology and migration of sea turtles. Mailing Add: Dept of Zool Univ of Fla Gainesville FL 32601

CARR, CHARLES JELLEFF, b Baltimore, Md, Mar 27, 10; m 32; c 3. PHARMACOLOGY, CHEMISTRY. Educ: Univ Md, BS, 33, MS, 34, PhD(pharmacol), 37. Hon Degrees: DSc, Purdue Univ, 64. Prof Exp: From asst prof to prof pharmacol, Univ Md, 37-55; prof, Purdue Univ, 55-57; head pharmacol unit, Psychopharmacol Serv Ctr, NIMH, 57-63; chief sci anal br, Life Sci Div, Army Res Off, chief res & develop hqs, US Dept Army, Va, 63-67; DIR LIFE SCI RES OFF, FEDN AM SOCS EXP BIOL, 67- Concurrent Pos: Adj prof, Univ Md, 57- Mem: Soc Pharmacol & Exp Therapeut; Am Chem Soc; Am Pharmaceut Asn; NY Acad Sci; Am Col Neuropsychopharmacol. Res: Carbohydrate metabolism; general anesthetic agents; hypotensive alkylnitrites; psychopharmacology; genetic basis for drug metabolic effects. Mailing Add: Fedn of Am Socs for Exp Biol 9650 Rockville Pike Bethesda MD 20014

CARR, CHARLES WILLIAM, b Minneapolis, Minn, July 20, 17; m 45; c 2. BIOCHEMISTRY. Educ: Univ Minn, BChem, 38, MS, 39, PhD(phys chem), 43. Prof Exp: Jr chemist, 39-43, res fel, 43-46, from instr to assoc prof, 46-64, PROF BIOCHEM, UNIV MINN, MINNEAPOLIS, 64- Mem: Am Chem Soc; Am Soc Biol Chem; Soc Exp Biol & Med; NY Acad Sci; Am Soc Cell Biol. Res: Membrane structure and permeability; ion binding with proteins and other biopolymers; ionic effects on enzymes. Mailing Add: Dept of Biochem Univ of Minn Minneapolis MN 55455

CARR, CLIDE ISOM, b Creston, Mont, June 9, 20; m 45; c 3. PHYSICAL CHEMISTRY, RESEARCH ADMINISTRATION. Educ: Univ Mont, BA, 42; Univ Calif, PhD(chem), 49. Prof Exp: Asst, Univ Calif, 46-49; res chemist, Gen Labs, US Rubber Co, 49-55, Naugatuck Chem, 55-56 & Calif Res Corp, 56-57; res chemist & group leader, Res Ctr, US Rubber Co, 57-60, dept mgr, Fiber Res, 60-61 & Elastomer Res, 61-66, Mgr, Tire Eng, 66-69, MGR ELASTOMER RES, UNIROYAL RES CTR, UNIROYAL INC, 69- Mem: AAAS; Am Chem Soc; Sigma Xi. Res: Management of tire technology and elastomer applications; plastics; fibers; rubber; tires; polymers. Mailing Add: Oxford Mgt & Res Ctr Uniroyal Inc Middlebury CT 06749

CARR, DANIEL OSCAR, b Kansas City, Kans, May 1, 34. BIOCHEMISTRY. Educ: Univ Mo-Kansas City, BS, 56; Iowa State Univ, PhD(biochem), 60. Prof Exp: From instr to asst prof, 63-69, ASSOC PROF BIOCHEM, UNIV KANS MED CTR, KANSAS CITY, 69- Concurrent Pos: USPHS fel, Univ Kans Med Ctr, Kansas City, 60-63. Mem: AAAS; Am Chem Soc; Am Soc Biol Chemists. Res: Biosynthetic mechanisms; catalyses by flavoproteins. Mailing Add: Dept of Biochem Univ Kans Med Ctr Kansas City KS 66103

CARR, DAVID HARVEY, b Southport, Eng, Jan 10, 28; Can citizen; m 51; c 1. ANATOMY, CYTOGENETICS. Educ: Univ Liverpool, MB, ChB, 50, DSc, 70. Prof Exp: Lectr, Univ Western Ont, 58-61, from asst prof to assoc prof, 61-67; assoc prof, 67-70, PROF ANAT, McMASTER UNIV, 70- Concurrent Pos: Can Soc Study Fertil res award, 61. Mem: Am Asn Anat; Can Fedn Biol Soc; Genetics Soc Am. Res: Chromosome studies in clinical syndromes and spontaneous abortions. Mailing Add: Dept of Anat McMaster Univ Col of Health Sci Hamilton ON Can

CARR, DAVID TURNER, b Richmond, Va, Mar 12, 14; div; c 1. INTERNAL MEDICINE, ONCOLOGY. Educ: Med Col Va, MD, 37; Univ Minn, MS, 47. Prof Exp: Intern, Grady Hosp, Ga, 37-38, asst resident, 38-39; chest serv, Bellevue Hosp, New York, 40-41, chief resident, 41-42; physician, Mt Morris Tuberc Hosp, 42-43; from asst prof to assoc prof med, Mayo Med Sch, 53-64, chmn dept oncol & dir, Mayo Comphrensive Cancer Ctr, 74-75, PROF MED, MAYO MED SCH, 64-, CONSULT, MAYO CLIN, 47- Mem: Am Lung Asn (vpres, 71-72); Am Cancer Soc; fel Am Col Physicians; Am Thoracic Soc (vpres, 63-64); Int Asn Study Lung Cancer (vpres, 74-75). Mailing Add: Mayo Clin Rochester MN 55901

CARR, DONALD DEAN, b Fredonia, Kans, Mar 28, 31; m 55; c 4. GEOLOGY. Educ: Kans State Univ, BS, 53, MS, 58; Ind Univ, Bloomington, PhD(geol), 69. Prof Exp: Teacher pub sch, 59-60; geologist, Humble Oil & Ref Co, 60-62 & Geosci Div, Tex Instruments Co, 62-63; GEOLOGIST, INDUST MINERALS SECT, IND GEOL SURV, 63- Mem: Geol Soc Am; Am Asn Petrol Geologists; Am Inst Mining, Metall & Petrol Engrs. Res: Sedimentology; stratigraphy; geomorphology; economic geology of coal and industrial minerals. Mailing Add: Ind Geol Surv 611 N Walnut Grove Bloomington IN 47401

CARR, DONALD EATON, b Los Angeles, Calif, Oct 17, 03; m 34; c 1. CHEMISTRY. Educ: Univ Calif, BS, 30. Prof Exp: Res chemist, Union Oil Co, Calif, 30-38, from res supvr to res mgr, 38-47; consult, R R Collier Corp, 47; consult, Phillips Petrol Co, 47, assoc dir fuels & lubricants div, Res Dept, 48-50, asst dir res, 50-65; INDEPENDENT CONSULT & WRITING, 65- Concurrent Pos: Mem, Am Petrol Inst, 45- Honors & Awards: Ord Award, Soc Automotive Engrs. Mem: Am Chem Soc; Soc Automotive Engrs. Res: Lubricants; engines; utilization of fuels and lubricants in aviation and automotive engines, including jet and rocket propelled. Mailing Add: 1433 Shawnee Bartlesville OK 74003

CARR, DUANE TUCKER, b Gunnison, Colo, July 6, 32; m 54; c 2. CHEMISTRY. Educ: Western State Col Colo, BA, 54; Purdue Univ, PhD(phys chem), 62. Prof Exp: Instr chem, Wabash Col, 60-61; asst prof 61-67, ASSOC PROF CHEM, COE COL, 67- Concurrent Pos: Asst prof chem, Agr Col, Haile Selassie Univ, 68-70. Mem: AAAS; Am Chem Soc. Res: Molecular structure, carbon-13 splittings in fluorine magnetic resonance spectra. Mailing Add: Dept of Chem Coe Col Cedar Rapids IA 52402

CARR, EDWARD ALBERT, b Cranston, RI, Mar 3, 22; m 52; c 2. PHARMACOLOGY, INTERNAL MEDICINE. Educ: Brown Univ, AB, 42; Harvard Univ, MD, 45. Prof Exp: Intern, RI Hosp, 45-46; asst resident internal med, Cushing Hosp, 48; instr, Harvard Med Sch, 49-51; resident internal med, Pa Hosp, 51-52; from asst prof to prof internal med & pharmacol, Univ Mich, Ann Arbor, 53-74, dir, Upjohn Ctr Clin Pharmacol, 66-74; prof med & Pharmacol & chmn dept Pharmacol, Univ Louisville, 74-76; PROF MED & PHARMACOL & CHMN DEPT PHARMACOL & THERAPEUT, STATE UNIV NY BUFFALO, 76- Concurrent Pos: Res fel pharmacol, Harvard Med Sch, 48-49; exchange fel, St Bartholomew's Hosp, London, 52-53; mem pharmacol-toxicol comt, Nat Inst Gen Med Sci, 70-74; mem revision comt, US Pharmacopeia, 65-75. Mem: AAAS; Am Col Physicians; Am Soc Pharmacol & Exp Therapeut; Am Soc Clin Pharmacol & Therapeut (pres, 74-75); Am Thyroid Asn. Res: Clinical pharmacology; drug allergy; thyroid; radioisotopes. Mailing Add: Dept of Pharmacol & Therapeut State Univ NY Buffalo NY 14214

CARR, EDWARD FRANK, b St Johnsbury, Vt, Aug 18, 20; m 54; c 3. PHYSICS Educ: Mich State Univ, BS, 43, PhD(physics), 54. Prof Exp: Asst physics, Mich State Univ, 48-53, instr, 53-54; asst prof, St Lawrence Univ, 54-57; from asst prof to assoc prof, 57-69, PROF PHYSICS, UNIV MAINE, ORONO, 69- Mem: Am Phys Soc; Am Asn Physics Teachers. Res: Liquid crystals; microwave dielectric measurements. Mailing Add: Dept of Physics Univ of Maine Orono ME 04473

CARR, EDWARD MARK, b Warsaw, Poland, Sept 16, 18; nat US; m 44; c 2. ANALYTICAL CHEMISTRY. Educ: Univ Minn, BCh, 49, MS, 52. Prof Exp: Res chemist process develop, E I du Pont de Nemours & Co, 52-53; res chemist, Toni Div, Gillette Co, 53-56, res supvr anal chem sect, 56-61, res assoc, 61-64; mgr anal chem group, Noxzema Chem Co, 64-69, MGR ANAL CHEM, NOXELL CORP, 69- Mem: Am Chem Soc; Soc Cosmetic Chemists. Res: Analytical chemistry of cosmetic materials and products; physical chemistry of surface active agents; emulsion polymerization. Mailing Add: 9605 Labrador Lane Cockeysville MD 21030

CARR, GEORGE LEROY, b Upperco, Md, Dec 11, 27; m 52; c 3. PHYSICS, SCIENCE EDUCATION. Educ: Western Md Col, BS, 48, MEd, 59; Cornell Univ, PhD(sci educ, physics), 69. Prof Exp: Teacher high sch, Md, 48-51; chmn dept sci, Milford Mill High Sch, 53-61; instr physics, Cornell Univ, 63-64; chmn dept sci, Pikesville High Sch, Md, 64-65; asst prof educ & physics, Western Md Col, 65-66; from asst prof to assoc prof, 66-70, PROF PHYSICS, UNIV LOWELL, 70- Concurrent Pos: Consult, Phys Sci Study Comt, 57-61; adv, Univ Lowell Prof Improv Prog, 69-71. Mem: AAAS; Am Asn Physics Teachers; Am Geophys Union. Res: Problem-solving in physics; environmental applications of physics; teaching of physics. Mailing Add: 2 Gifford Lane Chelmsford MA 01824

CARR, GERALD DWAYNE, b Pasco, Wash, Apr 1, 45; m 68. SYNTHETIC BOTANY, EVOLUTION. Educ: Eastern Wash State Col, BA, 68; Univ Wis-Milwaukee, MS, 70; Univ Calif, Davis, PhD(bot), 75. Prof Exp: Assoc bot, Univ Calif, Davis, 74-75; ASST PROF BOT, UNIV HAWAII, 75- Concurrent Pos: Consult bot, Kuakini Hosp & Home, 76. Mem: Soc Study Evolution; Am Soc Plant Taxonomists; Bot Soc Am; Int Asn Plant Taxon. Res: Biosynthetic and evolutionary studies of the Pacific states and Hawaiian tarweeds; biosystematics and cytogenetics of the Hawaiian flora. Mailing Add: Dept of Bot Univ of Hawaii Honolulu HI 96822

CARR, HERMAN YAGGI, b Alliance, Ohio, Nov 28, 24; m 59; c 2. CONDENSED STATE PHYSICS. Educ: Harvard Univ, BS, 48, MA, 49, PhD(physics), 53. Prof Exp: From asst prof to assoc prof, 52-64, PROF PHYSICS, RUTGERS COL, 64- Concurrent Pos: Guggenheim fel, 67. Mem: Fel Am Phys Soc; Am Asn Physics Teachers. Res: Nuclear magnetic resonance; physics of fluids; phase transitions. Mailing Add: Dept of Physics Rutgers The State Univ New Brunswick NJ 08903

CARR, HOWARD EARL, b Headland, Ala, Sept 16, 15; m 39; c 2. PHYSICS. Educ: Auburn Univ, BS, 36; Univ Va, AM, 39, PhD(physics), 41. Prof Exp: Asst, Univ Va, 39-40; from asst to assoc prof physics, Univ SC, 41-44; physicist, Ford, Bacon, Davis Corp, Manhattan Dist, Oak Ridge, Tenn, 44; asst prof physics, US Naval Acad, 46-48; assoc prof, 48-53, PROF PHYSICS & HEAD DEPT, AUBURN UNIV, 53- Mem: AAAS; fel Am Phys Soc; Am Asn Physics Teachers. Res: Isotope separation; thermal diffusion in liquids; mass spectrography; negative ions. Mailing Add: Dept of Physics Auburn Univ Auburn AL 36830

CARR, JAMES DAVID, b Ames, Iowa, Apr 3, 38; m 68; c 1. ANALYTICAL CHEMISTRY. Educ: Iowa State Univ, BS, 60; Purdue Univ, PhD(anal chem), 66. Prof Exp: Technician gas chromatog, Ivorydale Labs, Procter & Gamble Co, 60; res fel chem, Univ NC, 65-66; asst prof, 66-70, ASSOC PROF CHEM, UNIV NEBR-LINCOLN, 71- Concurrent Pos: Vis assoc prof chem, Purdue Univ, 74-75. Mem: Am Chem Soc. Res: Coordination chain reactions; kinetics and mechanism of ligand exchange reactions; optically active multidentate ligands; trace analysis by kinetic means; water quality analysis and treatment. Mailing Add: Dept of Chem Univ of Nebr Lincoln NE 68508

CARR, JEROME BRIAN, b Syracuse, NY, Dec 17, 38; m 61; c 3. OCEANOGRAPHY, GEOPHYSICS. Educ: St Louis Univ, BS, 61; Boston Col, MS, 65; Rensselaer Polytech Inst, PhD(oceanog), 71. Prof Exp: Phys oceanogr, US Naval Oceanog Off, Md, 61-62; oceanogr-geophysicist, Sperry Rand Res Ctr, Mass, 63-66; oceanogr, Gen Dynamics/Electronics, NY, 66; vis asst prof oceanog & geol, Purdue Univ, 66-67; oceanogr, Raytheon Corp, RI, 67-68 & Hazeltine Corp, Braintree, 69-70; environ consult, 70-71; environ scientist, Lowell Technol Inst Res Found, 71-72 & Anal Systs Eng Corp, 72-74; PRES & TECH DIR, CARR RES LAB, INC, 74- Mem: AAAS; Am Geophys Union; Am Inst Mining, Metall & Petrol Engrs. Res: Study of space-time variations of the ocean's thermal structure and relations to underwater acoustics; global tectonics; developing a trophic state index for lake management and reclamation. Mailing Add: 17 Waban St Wellesley MA 02181

CARR, JOHN B, b Olympia, Wash, Aug 29, 37; m 60; c 4. MEDICINAL CHEMISTRY. Educ: St Martin's Col, BSc, 59; Univ Wash, PhD(med chem), 63. Prof Exp: NIH fel med chem, Univ Kans, 63-64; ORG RES CHEMIST, SHELL DEVELOP CO, 64- Res: Lipio metabolism; animal physiology and growth; parasitology; insecticides; herbicides. Mailing Add: Shell Develop Co PO Box 4248 Modesto CA 95352

CARR, JOHN FRANK, b Johnson City, Tenn, June 28, 29; m 50; c 4. FISHERIES. Educ: Mich State Univ, BS, 56, MS, 58. Prof Exp: Asst fishery biol, Mich State Univ, 56-57, instr, 57-60; instr, Great Lakes Fishery Lab, US Fish & Wildlife Serv, 60-66, chief limnol prog, 66-69, asst lab dir, 69-73, GREAT LAKES LIAISON OFFICER, NAT MARINE FISHERIES SERV, NAT OCEANIC & ATMOSPHERIC ADMIN, 73- Mem: Am Soc Limnol & Oceanog; Am Fisheries Soc. Res: Institutional arrangements for optical utilization and management of Great Lakes fisheries. Mailing Add: Nat Marine Fisheries Serv PO Box 648 Ann Arbor MI 48107

CARR, JOHN HALDEN, b El Centro, Calif, Feb 3, 21; m 44; c 1. BACTERIOLOGY. Educ: Kans State Teachers Col, Emporia, BS, 48; Kans State Univ, MS, 52, PhD(bact), 57. Prof Exp: Assoc prof bact, 63-67, PROF MICROBIOL, FRESNO STATE COL, 67- Concurrent Pos: Grant, NIH, 57-60. Mem: Am Soc Microbiol. Res: Animal virology, interference phenomenon and immunity to viruses and mechanism involved. Mailing Add: Dept of Biol Fresno State Col Fresno CA 93726

CARR, JOHN WEBER, III, b Durham, NC, May 16, 23; m 49; c 3. MATHEMATICS. Educ: Duke Univ, BS, 43; Mass Inst Technol, MS, 49, PhD(math), 51. Prof Exp: Res assoc, Mass Inst Technol, 51-52; res mathematician, Univ Mich, 52-55, asst & assoc prof, 55-59; assoc prof & dir res comput ctr, Univ NC, 59-62, prof, 62-63; assoc prof, 63-66, chmn grad group comput & info sci, 66-73, PROF COMPUT & INFO SCI, MOORE SCH ENG, UNIV PA, 66- Concurrent Pos: Lectr math, Univ Mich, 53-55; ed, Comput Reviews, 59-63. Mem: Am Math Soc; Asn Comput Mach (pres, 57-58). Res: Solution of Schrodinger equation; computing machinery logic and programming theory; numerical analysis. Mailing Add: Moore Sch of Elec Eng Univ of Pa Philadelphia PA 19104

CARR, JULIUS JAY, b Brooklyn, NY, Feb 24, 13; m 39; c 1. BIOCHEMISTRY. Educ: Brooklyn Col, BS, 34; NY Univ, MS, 40, PhD(biochem), 52. Prof Exp: Technician clin chem, Jewish Hosp, Brooklyn, 34-36, asst lab dir, 36-39; jr chemist,

Bellevue Hosp, 39-42; sr chemist, Metropolitan Hosp, 46-52; assoc chemist, Mt Sinai Hosp, 52-56; asst prof pharmacol & biochem, NY Med Col, 56-59, asst prof biochem, 59-61; CHIEF BIOCHEMIST, METHODIST HOSP, BROOKLYN, 60- Mem: AAAS; Am Chem Soc; Am Asn Clin Chem; NY Acad Sci. Res: Reactions of aromatic nitrocompounds with creatinine and related substances; absorption and metabolism of salicylates and other drugs; methodology in clinical chemistry; acid phosphatase in lipoidosis. Mailing Add: Methodist Hosp 506 Sixth St Brooklyn NY 11215

CARR, LAURENCE A, b Ann Arbor, Mich, Mar 21, 42; m 64; c 1. PHARMACOLOGY. Educ: Univ Mich, BS, 65; Mich State Univ, MS, 67, PhD(pharmacol), 69. Prof Exp: Asst, Mich State Univ, 66; asst prof, 69-75, ASSOC PROF PHARMACOL, UNIV LOUISVILLE, 75- Honors & Awards: Bristol Lab Award, 65. Mem: AAAS; Soc Neurosci; Int Soc Neuroendocrinol. Res: Brain catecholamines and reproductive hormones; effects of amphetamines on the release and metabolism of brain catecholamine. Mailing Add: Dept of Pharmacol Univ Louisville Health Sci Ctr Louisville KY 40201

CARR, LAWRENCE J, organic chemistry, see 12th edition

CARR, MALCOLM WALLACE, b New York, NY, Oct 8, 99; m 42. ANATOMY, SURGICAL PATHOLOGY. Educ: Univ Pa, DDS, 22; Am Bd Oral Surg, dipl. Prof Exp: Instr oral surg, Col Physicians & Surgeons, Columbia Univ & clin asst vis oral surgeon, Vanderbilt Clin, 23-25; from asst attend to assoc attend, Fifth Ave Hosp, 23-36; assoc prof oral surg, NY Med Col & assoc oral surgeon, Flower & Fifth Ave Hosps, 36-65; CONSULT, 65- Concurrent Pos: Assoc, St Mary's Hosp, Children, 33-35; dir & vis oral surgeon, Metrop Hosp, 34-55, consult, 55-; dir & vis oral surgeon, Knickerbocker Hosp, 35-41, consult, 41-; lectr, Grad Sch Med, Univ Pa, 36-41, 46-55, lectr, Sch Dent, 38-41; attend oral surgeon & dir outpatinet clin, St Luke's Hosp, 40-62, consult, 62-; consult-instr, Bur Med & Surg, US Dept Navy, St Albans, NY, 48-55; attend oral surgeon, NY Polyclin Med Sch & Hosp, 49-55, prof oral surg, 49-55; dir dent & vis oral surgeon, Bird S Coler Mem Hosp, 51-55, consult, 55-; consult, Metrop Med Ctr, NY Col Med, 56- Honors & Awards: Hon fel, Royal Col Surg, Eng; comdr, Order Knights Hosp St John Jerusalem, 75. Mem: AAAS; fel Am Col Dent (pres, 45); AMA; Am Soc Anesthesiol; NY Acad Med. Res: Acute infections of face and neck; fractures of the Maxillae; oral surgery; neoplastic diseases. Mailing Add: 52 E 61st St New York NY 10021

CARR, MICHAEL H, b Leeds, Eng, May 26, 35; US citizen; m 61; c 5. ASTROGEOLOGY, GEOCHEMISTRY. Educ: Univ London, BSc, 56; Yale Univ, MS, 57, PhD(geol), 60. Prof Exp: Res assoc geophys, Univ Western Ont, 60-62; geologist, 62-74, CHIEF, BR ASTROGEOL STUDIES, US GEOL SURV, 74- Mem: AAAS; Geol Soc Am; Am Geophys Union. Res: Geology of the moon and the planets; lunar stratigraphy; chemistry of the lunar regolith; geologic history of Mars; Martian volcanism; planetary exploration; trace element geochemistry; electron microprobe analysis. Mailing Add: US Geol Surv 345 Middlefield Rd Menlo Park CA 94025

CARR, MICHAEL JOHN, b Portland, Maine, Nov 12, 46; m 72. GEOLOGY. Educ: Dartmouth Col, AB, 69, MS, 71, PhD(geol), 74. Prof Exp: ASST PROG GEOL, RUTGERS UNIV, 74- Mem: Geol Soc Am; Am Geophys Union. Res: The relation of volcanos and active faults to underthrusting at convergent plate margins, with special emphasis on the Central American convergent plate margin. Mailing Add: Dept of Geol Rutgers Univ New Brunswick NJ 08903

CARR, PAUL HENRY, b Boston, Mass, May 12, 35; m 60; c 3. SOLID STATE PHYSICS. Educ: Mass Inst Technol, BS, 57, MS, 61; Brandeis Univ, PhD(solid state physics), 66. Prof Exp: Res physicist, Calibration Ctr, Redstone Arsenal, Ala, 61-62; res physicist, 62-68, SUPVRY RES PHYSICIST, AIR FORCE CAMBRIDGE RES LAB, 68- Mem: Am Phys Soc; Sigma Xi; sr mem Inst Elec & Electronics Eng. Res: Electron paramagnetic resonance; vacuum standards and measurements; nonlinear phenomena in microwave ultrasonics; microwave phonons; kilomegacycle ultrasonics; microwave frequency elastic surface waves. Mailing Add: Air Force Cambridge Res Lab Bedford MA 01730

CARR, RAYMOND NIEL, b Goodland, Kans, Nov 1, 41; m 64; c 2. MATHEMATICAL STATISTICS, BIOSTATISTICS. Educ: Southwestern Col, Kans, BA, 63; Kans State Univ, MS, 65, PhD(math statist & probability), 69. Prof Exp: Asst prof statist, Univ Del, 68-74; MGR, SCHERING-PLOUGH CORP, 74- Concurrent Pos: Spec consult, Nat Air Pollution Control Admin, Cincinnati, Ohio, 68-70; Univ Del Res Found grant, 69-70; signature & propagation lab, Aberdeen Res & Develop Ctr, 71. Mem: Math Asn Am; Biomet Soc; Inst Math Statist; Am Statist Asn; fel Royal Statist Soc. Res: Stochastic models of biological systems; probability theory; decision theory; pattern recognition. Mailing Add: Schering-Plough Corp 60 Orange St Bloomfield NJ 07003

CARR, RICHARD DEAN, b Columbus, Ohio, June 29, 29; m 53; c 4. DERMATOLOGY. Educ: Ohio State Univ, BA, 51, MD, 54. Prof Exp: Asst prof dermat, 63-65, dir div, 67-69, ASSOC PROF DERMAT, COL MED, OHIO STATE UNIV, 65- Concurrent Pos: Consult, Dayton Vet Admin Hosp, 63- Mem: AMA; Am Acad Dermat; Soc Invest Dermat; Am Col Physicians; Am Asn Prof Dermat. Res: Clinical dermatology; causes of vasculitis of the skin; percutaneous absorption of topical corticosteroids. Mailing Add: 1840 Zollinger Rd Columbus OH 43221

CARR, ROBERT H, b Ames, Iowa, June 3, 35; m 62. PHYSICS. Educ: Cornell Univ, BA, 57; Iowa State Univ, PhD(solid state physics), 63. Prof Exp: Res officer physics, Commonwealth Sci & Indust Res Orgn, Australia, 63-64; from asst prof to assoc prof, 64-72, PROF PHYSICS, CALIF STATE UNIV, LOS ANGELES, 72- Mem: Am Asn Physics Teachers; Cryogenic Soc Am (past pres). Res: Measurement of low temperature characteristics of materials. Mailing Add: Dept of Physics Calif State Univ Los Angeles CA 90032

CARR, ROBERT JOSEPH, b Milwaukee, Wis, Mar 27, 31. NUCLEAR CHEMISTRY. Educ: Univ Calif, Los Angeles, BS, 51; Univ Calif, Berkeley, PhD(chem), 56. Prof Exp: From instr to asst prof chem, Wash State Univ, 55-61; chemist, Shell Develop Co, 61-67; prof chem, Merritt Col, Oakland, 67-70; PROF CHEM, COL ALAMEDA, 70- Mem: Am Chem Soc. Res: Nuclear reactions and spectroscopy; activation analysis; Mössbauer effect spectroscopy; application of radio-isotopes to problems in physical and analytical chemistry. Mailing Add: Dept of Chem Col of Alameda Alameda CA 94501

CARR, ROBERT WILSON, JR, b Montpelier, Vt, Sept 7, 34; m 58; c 3. CHEMICAL KINETICS, CHEMICAL ENGINEERING. Educ: Norwich Univ, BS, 56; Univ Vt, MS, 58; Univ Rochester, PhD(phys chem), 62. Prof Exp: Asst prof, 65-69, ASSOC PROF CHEM ENG, UNIV MINN, MINNEAPOLIS, 69- Concurrent Pos: Res fel, Harvard Univ, 63-65, lectr chem, 64-; asst ed, J Phys Chem, 70-; NSF fel, Cambridge Univ, 71-72; Hon Ramsay Mem fel, 71-72. Mem: AAAS; Am Chem Soc; Sigma Xi. Res: Gas kinetics; photochemistry; energy transfer; unimolecular reactions; atomic and free radical reactions. Mailing Add: Dept of Chem Eng & Mat Sci Univ of Minn Minneapolis MN 55455

CARR, ROGER BYINGTON, b Haverhill, Mass, Mar 25, 36; m 60; c 2. ASTRONOMY, PHYSICS. Educ: Bucknell Univ, BS, 63; Univ Fla, MS, 65, PhD(physics, astron), 67. Prof Exp: From instr to asst prof astron, Carleton Col, 67-72; ASST PROF PHYSICS, ALFRED UNIV, 72- Mem: Am Astron Soc; Am Meteorol Soc; Sigma Xi. Res: Atmospheric extinction; eclipsing binary stars. Mailing Add: Dept of Physics Alfred Univ Alfred NY 14802

CARR, RONALD E, b Newark, NJ, Sept 17, 32; m 57; c 2. OPHTHALMOLOGY, VISUAL PHYSIOLOGY. Educ: Princeton Univ, AB, 54; Johns Hopkins Univ, MD, 58; NY Univ, MS, 63. Prof Exp: Intern, Cornell-Bellevue Med Serv, 58-59; resident ophthal, Med Ctr, NY Univ, 59-62; clin assoc ophthal, NIH, Md, 63-64, actg assoc ophthalmologist, 64-65; asst prof, 65-69, ASSOC PROF OPHTHAL, MED CTR, NY UNIV, 69- Concurrent Pos: Fel ophthal, Med Ctr, NY Univ, 62-63; attend physician, Univ Hosp, NY, 65-; chief ophthal serv, Goldwater Mem Hosp, 67-; dir retinal clin, Bellevue Hosp, 70- Mem: Am Acad Ophthal & Otolaryngol; Am Col Surg; Asn Res Ophthal; Am Ophthalmol Soc. Res: Clinical applications of visual electrophysiology; hereditary diseases of the retina; drug induced retinal degenerations. Mailing Add: Dept of Ophthal NY Univ Med Ctr New York NY 10016

CARR, RONALD IRVING, b Toronto, Ont, May 17, 35; m 68; c 2. IMMUNOLOGY, BIOCHEMISTRY. Educ: Univ Toronto, BA, 58, MD, 62; Rockefeller Univ, PhD(life sci), 69. Prof Exp: Intern, St Michael's Hosp, Toronto, 62-63; asst resident physician, Rockefeller Univ Hosp, 64-69; MEM STAFF, DEPT ALLERGY & CLIN IMMUNOL, NAT JEWISH HOSP, 70-; ASST PROF MED, UNIV COLO, DENVER, 70- Concurrent Pos: Sr res fel, Dept Allergy & Clin Immunol, Nat Jewish Hosp, Denver, 69-70; mem, Pulmonary Allergy & Clin Immunol Adv Comt, Food & Drug Admin, 73- Mem: Am Rheumatism Asn; Can Soc Immunol; Am Asn Immunol; Am Acad Allergy; Am Fedn Clin Res. Res: Nucleic acid antibodies in disease; origin of circulating nucleic acids; immune complex disease due to orally ingested substances; therapy of immunologic diseases; allergic disease, asthma. Mailing Add: Dept of Allergy & Clin Immunol Nat Jewish Hosp Denver CO 80206

CARR, RUSSELL L K, b Wakefield, Mich, Apr 26, 26; wid; c 2. ORGANIC CHEMISTRY. Educ: Ohio State Univ, BSc, 51, PhD(chem), 55. Prof Exp: From chemist to sr chemist, 55-64, res supvr, 64-66, sect mgr res, 66-70, MGR RES, HOOKER CHEM CORP, 70- Mem: Am Chem Soc. Res: Organic fluorine chemistry; fluorinations; hydrogenations, catalysis; organic phosphorus chemistry; reactions of elemental phosphorus and of esters of phosphorus acids; synthesis of phosphines; polymer chemistry. Mailing Add: Hooker Chem & Plastics Corp Res Ctr, Long Rd Grand Island NY 14072

CARR, SCOTT BLIGH, b Linton, Ky, Oct 11, 34; m 56; c 2. ANIMAL NUTRITION. Educ: Western Ky State Col, BS, 56; Univ Ky, MS, 63, PhD(dairy nutrit), 67. Prof Exp: Exten specialist forages, 67-71, ASST PROF RUMINANT NUTRIT & SCIENTIST-IN-CHARGE FORAGE TESTING LAB, VA POLYTECH INST & STATE UNIV, 71- Mem: Am Dairy Sci Asn; Am Soc Animal Sci. Res: Forages; dairy and beef nutrition. Mailing Add: Dept of Animal Nutrit Va Polytech Inst & State Univ Blacksburg VA 24060

CARR, THOMAS DEADERICK, b Ft Worth, Tex, Jan 2, 17; m 61; c 1. PHYSICS, ASTRONOMY. Educ: Univ Fla, BS, 37, MS, 39, PhD(physics), 58. Prof Exp: Physicist & head blast measurement sect, Ballistic Res Lab, Aberdeen Proving Ground, Md, 40-45; civilian scientist with US Navy Bur Ord, 46; physicist & head antenna & propagation sect & staff mem directorate range develop, Air Force Missile Test Ctr, Patrick Air Force Base, Fla, 50-56; from asst prof to assoc prof, 58-69, PROF PHYSICS & ASTRON, UNIV FLA, 69- Concurrent Pos: Mem, Int Astron Union. Mem: Am Phys Soc; Am Astron Soc; Am Geophys Union. Res: Radio astronomy; geophysics; radio propagation; cosmic radiation; x-ray diffraction; blast measurements; ballistics; guided missile instrumentation. Mailing Add: Dept of Physics & Astron Univ of Fla Gainesville FL 32603

CARR, WALTER JAMES, JR, b Knob Noster, Mo, May 6, 18; m 53; c 2. PHYSICS. Educ: Mo Sch Mines, BS, 40; Stanford Univ, EE, 42; Carnegie Inst Technol, DrSc(physics), 50. Prof Exp: From physicist to adv physicist, 42-65, mgr theoret physics dept, 65-70, CONSULT MAGNETISM, WESTINGHOUSE RES LABS, 70- Mem: Am Phys Soc; Inst Elec & Electronics Engrs. Res: Solid state physics; ferromagnetism; superconductivity. Mailing Add: Westinghouse Res Lab Beulah Rd Pittsburgh PA 15235

CARR, WILLIAM EDWARD STATTER, b New Smyrna Beach, Fla, Oct 6, 35; m 55; c 4. MARINE ECOLOGY. Educ: Stetson Univ, BS, 59; Duke Univ, PhD(zool), 65. Prof Exp: NIH physiol trainee, 61-63; US Fish & Wildlife Serv training grant, 63-65; asst prof, 65-74, ASSOC PROF ZOOL, UNIV FLA, 74- Concurrent Pos: Marine consult estuarine ecol, Fla Power Corp res grant, 70- Mem: AAAS. Res: Chemoreception in marine animals; estuarine ecology and the impact of man. Mailing Add: Dept of Zool Univ of Fla Gainesville FL 32611

CARRABINE, JOHN ANTHONY, b Cleveland, Ohio, Nov 8, 28; m 51; c 6. X-RAY CRYSTALLOGRAPHY, INORGANIC CHEMISTRY. Educ: John Carroll Univ, BS, 51, MS, 53; Case Western Reserve Univ, PhD, 70. Prof Exp: Chemist, Thompson Prods, Inc, 51-56; supvr phys res, Brush Beryllium Co, 56-66; asst prof, 66-74, ASSOC PROF CHEM, JOHN CARROLL UNIV, 74- Concurrent Pos: Lectr, Eve Col, John Carroll Univ, 56-66. Mem: Am Chem Soc. Res: Crystal and molecular structures of metal complexes of biochemically important substances; physical metallurgy of beryllium. Mailing Add: Dept of Chem John Carroll Univ North Park & Miramar Blvd Cleveland OH 44118

CARRAHER, CHARLES EUGENE, JR, b Des Moines, Iowa, May 8, 41; m 63; c 4. POLYMER CHEMISTRY, PHYSICAL CHEMISTRY. Educ: Sterling Col, BA, 63; Univ Mo-Kansas City, PhD(chem), 68. Prof Exp: Teaching asst chem, Univ Mo, 63-67; asst prof, 67-71, ASSOC PROF CHEM, UNIV S DAK, 71-, DIR GEN CHEM PROG, 67- Concurrent Pos: Petrol Res Fund grant, 68-; Nat Sci Found grant, 70-73. Mem: AAAS; Am Inst Chem. Res: Preparation and characterization of phosphorus containing extended polar polymers and organometallic polymers; practical and theoretical molecular weight distribution determinations. Mailing Add: Dept of Chem Univ of SDak Vermillion SD 57069

CARRANO, ANTHONY VITO, b New York, NY, Mar 22, 42; m 64; c 2. CYTOGENETICS, BIOPHYSICS. Educ: Rensselaer Polytech Inst, BS, 64; Univ Calif, Berkeley, MB, 70, PhD(biophys), 72. Prof Exp: Fel, Div Biol & Med Res, Argonne Nat Lab, Ill, 72-73; BIOPHYSICIST, BIOMED DIV, LAWRENCE LIVERMORE LAB, UNIV CALIF, 73- Concurrent Pos: Adj asst prof, Sch Med, Univ Calif, Davis, 74. Mem: Am Soc Cell Biol; Am Soc Human Genetics; Radiation Res Soc; Genetics Soc Am; AAAS. Res: Flow microfluorometric and cytophotometric analysis of chromosomes; mechanisms of chromosomal aberration production;

persistence of chromosomal damage in proliferating cells; chromosome structure; medical genetics. Mailing Add: Biomed Div L-523 Lawrence Livermore Lab Livermore CA 94550

CARRANO, RICHARD ALFRED, b Bridgeport, Conn, Oct 1, 40; div; c 2. PHARMACOLOGY, TOXICOLOGY. Educ: Univ Conn, BS, 63, MS, 65, PhD(pharmacol & biochem), 67. Prof Exp: Res pharmacologist, ICI US, Inc, 66-67, supvr gen pharmacol, 68-74; MGR PHARMACOL & TOXICOL, ADRIA LABS, INC, 74- Mem: Am Chem Soc; Am Pharmaceut Asn; NY Acad Sci. Res: Development of new drugs by guiding the studies to demonstrate efficacy and to prove safety in animals for extrapolation to man, administration of the relationships between regulatory and clinical research groups which are necessary to obtain this goal. Mailing Add: Adria Labs Inc 1105 Market St Wilmington DE 19899

CARRASCO, PEDRO, b Madrid, Spain, Sept 20, 21; Mex citizen; m 49; c 2. ANTHROPOLOGY. Educ: Columbia Univ, PhD(anthrop), 53. Prof Exp: Guggenheim Found fel, Peabody Mus, Harvard Univ, 53; from asst to assoc prof anthrop, Univ Calif, Los Angeles, 57-67; PROF ANTHROP, STATE UNIV NY STONY BROOK, 67- Concurrent Pos: Am Coun Learned Socs fel, Archivo Indias, Seville, 63; NSF fel, Mex arch, 66-69. Mem: Am Anthrop Asn; Am Ethnol Soc. Res: Mesoamerican ethnology and ethnohistory. Mailing Add: Dept of Anthrop State Univ of NY Stony Brook NY 11790

CARRASQUER, GASPER, b Valenica, Spain, Dec 21, 25; nat US; c 3. MEDICINE. Educ: Univ Valenica, MD, 51. Prof Exp: Intern, North Hudson Hosp, NJ, 53; resident internal med, City Hosp New York, 54-55 & Louisville Gen Hosp, Ky, 55-56; from instr to assoc prof med, 59-70, PROF EXP MED, SCH MED, UNIV LOUISVILLE, 71-, ASSOC PHYSIOL, 61- Concurrent Pos: Fel, Sch Med, Univ Louisville, 56-59; res fel, Am Heart Asn, 58-60, adv res fel, 60-62; USPHS career develop award, 67-72; estab investr, Am Heart Asn, 62-67. Mem: Am Physiol Soc; Am Biophys Soc; Soc Exp Biol & Med. Res: Renal physiology; ion transport; concentration mechanism of the urine. Mailing Add: Dept of Med Univ of Louisville Louisville KY 40201

CARRAWAY, KERMIT LEE, b Utica, Miss, Mar 1, 40; m 62; c 1. BIOCHEMISTRY. Educ: Miss State Univ, BS, 62; Univ Ill, PhD(org chem), 66. Prof Exp: NIH res fel biochem, Univ Calif, 66-68; from asst prof to assoc prof, 68-75, PROF BIOCHEM, OKLA STATE UNIV, 75- Concurrent Pos: Mem, Molecular Cytol Study Sect, NIH, 75- Mem: AAAS; Am Chem Soc; Am Soc Biol Chemists. Res: Membrane biochemistry; protein chemistry. Mailing Add: Dept of Biochem Okla State Univ Stillwater OK 74074

CARREA, RAUL, b Buenos Aires, Arg, Jan 26, 17; 42, 56; c 6. NEUROSURGERY. Educ: Univ Buenos Aires, MD, 41. Prof Exp: Asst neurol, Columbia Univ, 45-47, 52-53, res neurosurgeon, Neurol Inst, 46-47; chief neurosurgeon, Inst Exp Med, 4858, PROF NEUROSURG, UNIV BUENOS AIRES, 58- Concurrent Pos: Chief neurosurgeon, Neurosurg Serv, Children's Hosp, 54; dir neurol res ctr, Di Tella Inst, 63; ed-in-chief, J Med & J Acta Neurol Latinoam; ed, Bull, Arg Neurosurg Asn, 67-; ed, World Fedn Neurol Surgeons, 69; ed in chief, Child's Brain, 74- Mem: Arg Neurosurg Asn (pres, 65); Int Soc Pediat Neurosurg (pres, 73); Am Neurol Asn; Am Asn Anatomists; Int Col Surgeons. Res: Cerebellar physiology; child neurosurgery. Mailing Add: Ctr for Neurol Invest Gallo 1330 Buenos Aires Argentina

CARREGAL, ENRIQUE JOSE ALVAREZ, b Padron, Spain, July 2, 32; US citizen; m 60; c 2. MEDICINE, PHYSIOLOGY. Educ: Univ Santiago, MD, 54, PhD(med), 60. Prof Exp: House doctor, Hosp Clin, Santiago, 54-55; instr physiol, Univ Santiago Med Dch, 55; fel physiol, Span Pharmacol Inst Madrid, 56-59; fel radiol, Univ Madrid, 57-59; fel neurophysiol, Coun Sci Invest, 60; res fel, Inst Med Res, Huntington Mem Hosp, Pasadena, Calif, 60-63; neurophysiologist, Stanford Res Inst, 63-65; assoc prof physiol, 6669, ASSOC PROF NEUROSURG, SCH MED, UNIV SOUTHERN CALIF, 70- Concurrent Pos: Consult, Huntington Mem Hosp, Pasadena, Calif & City of Hope Nat Med Ctr, Duarte. Mem: AAAS; Am Physiol Soc; Soc Neurosci; Span Soc Physiol Sci. Res: Neurophysiology of pain and sensory mechanism, neural control of respiration; synaptology. Mailing Add: Dept of Neurosurg Univ of Southern Calif Sch of Med Los Angeles CA 90033

CARREIRA, LIONEL ANDRADE, b Worcester, Mass, Nov 1, 44; m 70; c 1. PHYSICAL CHEMISTRY, MOLECULAR SPECTROSCOPY. Educ: Worcester Polytech Inst, BS, 66; Mass Inst Technol, PhD(phys chem), 69. Prof Exp: Fel chem, Univ Fla, 69-71; res assoc, Univ SC, 71-73; instr, 73-75, ASST PROF CHEM, UNIV GA, 75- Mem: Am Chem Soc; Sigma Xi. Res: Studies of far infrared and Raman spectra of simple polyatomic molecules having large amplitude oscillations, quasi-linear molecules with anomalously low frequency bending modes and other molecules with unusual vibrational potential functions. Mailing Add: Dept of Chem Univ of Ga Athens GA 30602

CARRELL, HORACE LYNN, physical chemistry, x-ray crystallography, see 12th edition

CARRELL, JOHN CRAIG, chemical physics, see 12th edition

CARRERA, GUILLERMO MANUEL, b Vieques, PR, Jan 3, 13; m 45; c 2. PATHOLOGY. Educ: Univ PR, BS, 34; Tulane Univ, MD, 37. Prof Exp: Instr, 45-53, ASSOC PROF PATH, TULANE UNIV, 53-, HEAD DEPT PATH, OCHSNER MED CTR, 54- Mem: AAAS; Am Asn Exp Path; Soc Exp Biol & Med; Am Asn Path & Bact; AMA. Res: Pathogenesis and pathology of experimental visceral larva migrans, amebiasis and parasitic infections; cancer. Mailing Add: Dept of Path Ochsner Med Ctr New Orleans LA 70121

CARRICK, WAYNE LEE, b Benton, Ark, Feb 23, 27; m 49; c 5. ORGANIC CHEMISTRY, POLYMER CHEMISTRY. Educ: Univ Ark, BS, 52, MS, 53, PhD(phys org chem), 55. Prof Exp: Res chemist, 54-56, group leader, 56-65, ASSOC DIR RES & DEVELOP, UNION CARBIDE CORP, 65- Mem: Am Chem Soc. Res: Reactions mechanisms; kinetics; catalysis; organic compounds of transition metals; high polymers. Mailing Add: 23 Hamlin Rd East Brunswick NJ 08816

CARRICO, JAMES LEON, b Sanger, Tex, Nov 22, 06; m 33; c 3. CHEMISTRY. Educ: NTex State Teachers Col, AB, 27, BS, 29; Univ Tex, AM, 31; Calif Inst Technol, PhD(chem), 35. Prof Exp: Prof chem, NTex State Teachers Col, 31-32; prof physics & chem, Lamar Col, 35-36; res consult, Tex Co, 36-47; from assoc prof to prof chem, 37-74, dir dept, 42-69, EMER PROF CHEM, N TEX STATE UNIV, 74- Mem: AAAS; Am Chem Soc. Res: Reactions of hydrocarbons in radio frequency plasma. Mailing Add: Dept of Chem NTex State Univ Denton TX 76203

CARRICO, JOHN P, physics, see 12th edition

CARRICO, ROBERT JOSEPH, b Mishawaka, Ind, Dec 27, 38; m 76. BIOCHEMISTRY. Educ: Purdue Univ, BS, 64; Univ Wis, PhD(biochem), 68. Prof Exp: Res asst biochem, Univ Wis, 68-69; fel, Univ Gothenburg, 69-71; RES SCIENTIST BIOCHEM, MILES LABS, INC, 71- Concurrent Pos: Fel biochem, Albert Einstein Col Med, 71-72. Res: Applications of immobilized enzymes in clinical chemistry; activities of modified substrates and cofactors with enzymes; monitoring of therapeutic drugs in blood. Mailing Add: Miles Labs Inc 1127 Myrtle St Elkhart IN 46514

CARRIEL, JONATHAN TURNER, chemistry, see 12th edition

CARRIER, ERNEST BERNARD, bacteriology, biochemistry, see 12th edition

CARRIER, OLIVER, JR, b Detroit, Mich, Apr 10, 23; m 44; c 5. PHYSIOLOGY, PHARMACOLOGY. Educ: Col Charleston, BS, 60; Univ Miss, PhD(physiol, biophys), 64. Prof Exp: From instr to asst prof, 64-67, ASSOC PROF PHARMACOL, UNIV TEX MED SCH, SAN ANTONIO, 68- Mem: AAAS; Am Chem Soc; Am Physiol Soc; Am Soc Pharmacol & Exp Therapeut. Res: Properties of vascular smooth muscle; biophysics; effects of oxygen, pH, electrolytes and vasoactive drugs. Mailing Add: 7703 Floyd Curl Dr San Antonio TX 78229

CARRIER, STEVEN THEODORE, b Havre, Mont, Nov 8, 38; m 63; c 2. MEDICAL STATISTICS, BIOSTATISTICS. Educ: Calif Polytech State Univ, BS, 67; Tex A&M Univ, PhD(statist), 75. Prof Exp: Analyst math, Naval Weapons Ctr, Calif, 67-68; statistician, 71-75, GROUP LEADER STATIST ANAL, LEDERLE LABS, AM CYANAMID CO, 75- Mem: Am Statist Asn; Biometrics Soc. Res: New statistical methods for the design and analysis of clinical trial data. Mailing Add: 132 N Lincoln Ave Pearl River NY 10965

CARRIERE, RITA MARGARET, b Toronto, Ont, Can, Apr 25, 30. HISTOLOGY. Educ: McGill Univ, BSc, 50, MSc, 54, PhD, 60. Prof Exp: Lectr histol, Univ Montreal, 55-57, asst prof histol, embryol & endocrinol, 57-60; ASST PROF ANAT, STATE UNIV NY DOWNSTATE MED CTR, 60- Res: Liver growth and differentiation; intestinal epithelium; dynamics of cell populations. Mailing Add: Dept of Anat State Univ NY Downstate Med Ctr Brooklyn NY 11203

CARRIGAN, RICHARD ALFRED, b Somerville, Mass, May 11, 06; m 31; c 1. ENVIRONMENTAL CHEMISTRY. Educ: Univ Fla, BS, 32; Cornell Univ, PhD(soil chem), 48. Prof Exp: High sch teacher, Fla, 32-33; lab supvr, Magnolia Petrol Co, Tex, 33-38; from asst chemist to assoc chemist exp sta, Univ Fla, 38-45, from assoc biochemist to biochemist, 45-51, prof soils, Col Agr, 48-51; supvr anal chem, Armour Res Found, Ill, 51-60; prog dir sci facil eval group, Div Inst Prog, 60-64, staff assoc, 64-70, prog mgr, Div Inst Develop, 70-71, PROG MGR, DIV ADVAN ENVIRON RES & TECHNOL, NAT SCI FOUND, 71- Concurrent Pos: Specialist anal chem, Union of Burma Appl Res Inst, 54-55. Mem: Fel AAAS; Am Chem Soc; Am Inst Chem; Soil Sci Soc Am; Soc Environ Geochem & Health. Res: Chemistry of minor elements in soils; spectrographic analysis; pasture fertility; chemistry of cobalt in soils; spectroscopy of plasmas; federal grant administration. Mailing Add: 2475 Virginia Ave NW Apt 304 Washington DC 20037

CARRIGAN, RICHARD ALFRED, JR, b Miami, Fla, Feb 17, 32; m 54; c 2. ELEMENTARY PARTICLE PHYSICS. Educ: Univ Ill, BS, 53, MS, 56, PhD(physics), 62. Prof Exp: Jr physicist, Firestone Tire & Rubber Co, 53; res physicist, Carnegie Inst Technol, 61-64, asst prof physics, 64-68; PHYSICIST, FERMI NAT ACCELERATOR LAB, 68-, DIR PERSONNEL SERV, 72- Concurrent Pos: Guest res assoc, Brookhaven Nat Lab, 62-63, guest asst physicist, 63-67; consult, Am Inst Res, 63-64; sr Fulbright fel, Deutsches Elektron-Synchrotron, 67-68. Mem: AAAS; Am Phys Soc; Sigma Xi. Res: Experimental elementary particle physics; particle scattering; magnetic monopole hypothesis. Mailing Add: Fermi Nat Accelerator Lab Box 500 Batavia IL 60510

CARRIKER, MELBOURNE ROMAINE, b Santa Marta, Colombia, Feb 25, 15; US citizen; m 43; c 4. MALACOLOGY, MARINE BIOLOGY. Educ: Rutgers Univ, BS, 39; Univ Wis, PhM, 40, PhD(invert zool), 43. Hon Degrees: DSc, Beloit Col, 68. Prof Exp: Asst zool, Univ Wis, 39-43; instr, Rutgers Univ, 46-47, asst prof, 47-54; assoc prof, Univ NC, 54-61; fisheries res biologist & chief shellfish mortality prog, Biol Lab, US Bur Com Fisheries, Md, 61-62; dir systs-ecol prog, Marine Biol Lab, 62-72, investr, 72-73; PROF, COL MARINE STUDIES, UNIV DEL, 73- Concurrent Pos: Mem ornith exped, 34; trustee, Int Oceanog Found, 64-; mem corp, Marine Biol Lab; adv comt biol study of plant site, Conn Yankee Atomic Power Plant, 65-75; adv comt oceanic biol to Off Naval Res, Am Inst Biol Sci, 66-69; adj prof, Univ RI, 65-73 & Boston Univ, 68-73; bioinstrumentation coun, 69-72; mem, President's Fed Water Pollution Control Adv Bd, 69-72; assoc mus comp zool, Harvard Univ, 67-73; res fel, Acad Natural Sci, Philadelphia, 68-; mem adv bd, Quarterly Rev Biol, 68-; mem ecol adv comt, Environ Protection Agency Sci Adv Bd, 75-; assoc, Del Mus Natural Hist, 75- Mem: AAAS; Am Asn Limnol & Oceanog; Ecol Soc Am; Atlantic Estuarine Res Soc (pres, 61); Nat Shellfisheries Asn (pres, 57-59). Res: Anatomy, histology, behavior, ecology and physiology of gastropods; ecology of bivalve larvae; estuarine and marine ecology; mechanisms of penetration of calcareous substrata by invertebrates, especially of boring gastropods. Mailing Add: Col of Marine Studies Univ of Del Lewes DE 19958

CARRIKER, ROY C, b Spokane, Wash, July 21, 37; m 65. PHYSICS. Educ: Wash State Univ, BS, 60; Trinity Col, Conn, MS, 63; Univ Conn, PhD(physics), 68; Harvard Univ, MBA, 76. Prof Exp: Sr engr, Pratt & Whitney Aircraft Div, United Aircraft Corp, 60-65, res engr, United Aircraft Res Labs, 65-66, sr res engr, 66-68, supvr optical technol, 68-69, chief electronic instrumentation, 69-70; tech dir, Precision Metals Div, Hamilton Watch Co, 70-72; dir eng & develop, Hamilton Technol, Inc, 72-74; pres, R C Carriker & Assoc, 74-76; GEN MGR, SERMETEL INC, 76- Res: Solid state physics, especially rare earth-transition metal alloys, transport properties and superconductivity; optics, especially spectroscopy of flames and plasmas, coherent optics and holography; instrumentation techniques. Mailing Add: Sermetel Int Hq Limerick Rd Limerick PA 19468

CARRILLO, ANGEL LUIS, biochemistry, pathology, see 12th edition

CARRINGTON, ELSIE REID, b Philadelphia, Pa, Sept 19, 11; m 43. OBSTETRICS & GYNECOLOGY. Educ: Wheaton Col, Ill, AB, 33; Temple Univ, MD, 41, MS, 49; Am Bd Obstet & Gynec, dipl. Prof Exp: Assoc prof, Sch Med, Temple Univ, 58-61; res prof, 61-67, PROF OBSTET & GYNEC & CHMN DEPT, MED COL PA, 67- Mem: AMA; Am Asn Obstet & Gynec; Asn Prof Gynec & Obstet; Am Col Obstet & Gynec. Res: Carbohydrate metabolism in pregnancy; fetal and neonatal welfare; placental function; monitor study of fetal and neonatal instantaneous heart rate patterns. Mailing Add: Dept of Gynec & Obstet Med Col of Pa Philadelphia PA 19129

CARRINGTON, THOMAS JACK, b Amarillo, Tex, June 1, 29; m 56; c 2. GEOLOGY. Educ: Univ Ky, BS, 58, MS, 60; Va Polytech Inst, PhD(geol), 65. Prof Exp: Petrol consult, Ky, 59; asst prof geol, Birmingham-Southern Col, 61-65, actg chmn dept, 61-62 & 63-65, chmn, 65-67, assoc prof, 65-67; PROF GEOL & HEAD

DEPT, AUBURN UNIV, 67- Concurrent Pos: Dir, Nat Sci Found undergrad res participation prog & equip grant, Birmingham-Southern Col, 63-64. Mem: AAAS; Geol Soc Am; Nat Asn Geol Teachers. Res: Stratigraphy, origin and areal distribution of geologic rock formations in the Talladega Group, Alabama; geologic structural development and economic mineral deposits in Alabama, particularly the Piedmont area. Mailing Add: Dept of Geol Auburn Univ 8080 Haley Center Auburn AL 36830

CARRINGTON, TUCKER, b Cincinnati, Ohio, Oct 19, 27; m 57; c 3. PHYSICAL CHEMISTRY. Educ: Univ Va, BS, 48; Calif Inst Technol, PhD(chem), 52. Prof Exp: Instr chem, Yale Univ, 52-54; phys chemist, Nat Bur Standards, 56-68, Nat Acad Sci-Nat Res Coun resident res assoc, 56-57; PROF CHEM, YORK UNIV, 68- Honors & Awards: Silver Medal, Combustion Inst, 62. Mem: Am Phys Soc; Chem Inst Can. Res: Energy transfer between resolved quantum states of small molecules in gases. Mailing Add: Dept of Chem York Univ Toronto ON Canada

CARRITT, DAYTON ERNEST, b Boston, Mass, Mar 12, 15; m 39; c 1. CHEMISTRY, OCEANOGRAPHY. Educ: RI State Col, BS, 37; Harvard Univ, PhD(anal chem), 48. Prof Exp: Chem technician, Woods Hole Oceanog Inst, 37-38; res chemist, Bur Ships, US Navy Dept, Woods Hole, 41-42; instr chem, RI State Col, 42-43; scientist, Manhattan Dist, Los Alamos, 43-46; instr chem, Scripps Inst, Univ Calif, 47-51; assoc prof oceanog, Johns Hopkins Univ, 51-60; from assoc prof to prof, Mass Inst Technol, 60-69; Gold Key prof chem oceanog, Nova Univ, 69-71; PROF MARINE SCI, UNIV MASS & DIR, INST MAN & ENVIRON, 71- Concurrent Pos: With Am Dynamics Int, Inc. Mem: AAAS; Am Chem Soc; Am Geophys Union; Am Soc Limnol & Oceanog. Res: Chemical oceanography; theory of the action of anti-fouling paints; atomic weight studies; polarography; plutonium chemistry; chemical properties of sea water. Mailing Add: Dept of Marine Sci Univ of Mass Amherst MA 01002

CARROCK, FREDERICK E, b Utica, NY, Oct 9, 31; m 59; c 1. PHYSICAL CHEMISTRY, ORGANIC CHEMISTRY. Educ: Syracuse Univ, BA, 54, MS, 56; State Univ NY, Syracuse, PhD(chem), 59. Prof Exp: Asst, State Univ NY, Syracuse, 54-56; res chemist, US Rubber Co, 59-61; supvr lab develop res, Rexall Chem Co, 61-65, asst mgr process develop res, 65-70, DIR PROD DEVELOP, DART INDUSTS INC, 70- Mem: Am Chem Soc; fel Am Inst Chemists. Res: Free radical and polymer chemistry. Mailing Add: 151 Albright Lane Paramus NJ 07652

CARROLL, ARTHUR GEORGE, b Sheridan, Wyo, Sept 18, 17; m 44; c 3. PLANT PHYSIOLOGY. Educ: Univ Wyo, BS, 48, MS, 49; Univ Iowa, PhD(bot), 53. Prof Exp: Prof biol, Nebr State Teachers Col, 52-61; from asst prof to assoc prof, 61-69, PROF BOT, OKLA STATE UNIV, 69- Mem: Am Soc Plant Physiologists; Sigma Xi; Am Inst Biol Sci. Res: Plant growth. Mailing Add: Sch of Biol Sci Okla State Univ Stillwater OK 74074

CARROLL, BENJAMIN, chemistry, see 12th edition

CARROLL, BENJAMIN L, b Jasper Co, Tex, Nov 15, 37; m 58; c 3. PHYSICAL CHEMISTRY, APPLIED MATHEMATICS. Educ: McNeese State Col, BS, 59; Iowa State Univ, PhD(phys chem), 63. Prof Exp: Asst prof chem, McNeese State Col, 65-66; SCI SYSTS SUPVR, LOCKHEED ELECTRONICS CO, 66- Concurrent Pos: Consult, Aptech, Inc, Tex, 68-69. Mem: Fel Am Inst Chem; Am Phys Soc; Am Chem Soc. Res: Molecular structure studies by electron diffraction and calculation of atomic energy levels of transition metals; theoretical scattering calculations. Mailing Add: Lockheed Electronics Co C-31 16811 El Camino Real Houston TX 77058

CARROLL, BERNARD JAMES, b Sydney, Australia, Nov 21, 40; m 66; c 2. EXPERIMENTAL PSYCHIATRY, PSYCHOPHARMACOLOGY. Educ: Univ Melbourne, BSc, 61, MB, BS, 64, DPM, 69, PhD(endocrinol), 72. Prof Exp: Resident med officer, Royal Melbourne Hosp, 65-66, sr house physician, 66-67; clin supvr psychiat, Univ Melbourne, 67-68, med res fel, 68-69; sr res officer, Nat Health & Med Res Coun, 69-70, res fel, 70-71; asst prof psychiat, Univ Pa, 71-73; assoc prof, 73-76, PROF PSYCHIAT, UNIV MICH, ANN ARBOR, 76-, RES SCIENTIST, MENT HEALTH RES INST, 73- Concurrent Pos: Royal Australian Col Physicians traveling fel endocrinol, 70; Nat Health & Med Res Coun Australia Charles J Martin overseas fel, 71; mem extramural grants comt, Ill Dept Ment Health, 73-, chmn, 76-; consult, Intramural Prog Rev, 75; assoc ed, Psychoneuroendocrinology, 75- Mem: Int Soc Psychoneuroendocrinol; Int Col Neuropsychopharmacol; AAAS; Soc Biol Psychiat; Am Psychosom Soc. Res: Clinical psychobiology of depression and mania; clinical and experimental psychopharmacology; behavioral pharmacology; psychoendocrinology; neuroendocrinology. Mailing Add: Ment Health Res Inst Univ of Mich Ann Arbor MI 48109

CARROLL, BURT HARING, b Tenafly, NJ, Mar 20, 96; m 24; c 3. CHEMISTRY. Educ: Cornell Univ, ChB, 17; Univ Wis, PhD(chem), 22. Prof Exp: Mem staff, Arthur D Little, Inc, 17; assoc chemist, Nat Bur Standards, 22-29, chemist, 29-33; chemist, Eastman Kodak Co, 33-45, supvry scientist, 45-51, sr res assoc, 52-61; PROF PHOTOG, ROCHESTER INST TECHNOL, 62- Honors & Awards: Jansen Medal, French Photog Soc; Henderson Medal, Royal Photog Soc, 51; Centennial Medal, Sch Eng, Columbia Univ, 64. Mem: Am Chem Soc; fel Optical Soc Am; hon mem Soc Photog Sci & Eng; Royal Photog Soc; Ger Photog Soc. Res: Optical sensitization; photographic emulsion making; photographic theory. Mailing Add: Sch of Photog Arts & Sci Rochester Inst of Technol One Lomb Memorial Dr Rochester NY 14623

CARROLL, CATHERINE, b Salmon Arm, BC, Oct 27, 18. NUTRITION. Educ: McGill Univ, BHS, 39; Univ Chicago, SM, 54; Mich State Univ, PhD(nutrit), 60. Prof Exp: Dietitian, Royal Inland Hosp, Kamloops, Can, 41-50; dietitian, Vancouver Gen Hosp, 50-53, asst dir dietetics, 54-57; assoc prof, 60-66, PROF NUTRIT, UNIV ARK, FAYETTEVILLE, 66- Mem: AAAS; Am Dietetic Asn; Am Inst Nutrit; Can Dietetic Asn. Res: Interrelationships in carbohydrate and lipid metabolism; amino acid imbalance; ethanol metabolism and liver lipids. Mailing Add: Dept of Home Econ Univ of Ark Fayetteville AR 72701

CARROLL, CHARLES OWEN, theoretical physics, see 12th edition

CARROLL, DANA, b Palm Springs, Calif, Sept 2, 43; m 66; c 2. MOLECULAR BIOLOGY. Educ: Swarthmore Col, BA, 65; Univ Calif, Berkeley, PhD(chem), 70. Prof Exp: Fel cell biol, Beatson Inst Cancer Res, 70-72; fel develop biol, Carnegie Inst Washington Dept Embryol, 72-75; ASST PROF MICROBIOL, MED CTR, UNIV UTAH, 75- Mem: Sigma Xi. Res: Sequence organization in specific, isolated genes of Xenopus, and its relation to their evolution and developmental regulation. Mailing Add: Dept of Microbiol Univ of Utah Med Ctr Salt Lake City UT 84132

CARROLL, DOUGLAS GORDON, b Baltimore, Md, Jan 31, 15; m 48; c 4. PHYSIOLOGY. Educ: Yale Univ, BA, 37; Johns Hopkins Univ, MD, 42; Am Bd Internal Med, dipl, 51. Prof Exp: Asst med, 42-48, instr environ med & med, 50-53, asst prof med, 53-59, ASSOC PROF MED, JOHNS HOPKINS UNIV, 60-; CHIEF PHYSIOL MED & REHAB, BALTIMORE CITY HOSP, 60- Mem: AMA; Am Soc Clin Invest; Am Clin & Climat Asn. Res: Human pulmonary physiology. Mailing Add: Baltimore City Hosp Baltimore MD 21224

CARROLL, EDWARD JAMES, JR, b San Diego, Calif, Dec 25, 45; m 68; c 2. DEVELOPMENTAL BIOLOGY, BIOCHEMISTRY. Educ: Sacramento State Col, BA, 68; Univ Calif, Davis, PhD(biochem), 72. Prof Exp: Fel develop biol, Scripps Inst Oceanog, 72-75; ASST PROF ZOOL, UNIV MD, COLLEGE PARK, 75- Mem: AAAS; Am Chem Soc; Am Soc Cell Biol; Am Soc Zoologists; Soc Develop Biol. Res: Biochemistry of fertilization and activation of embryonic metabolism. Mailing Add: Dept of Zool Univ of Md College Park MD 20742

CARROLL, EDWARD MAJOR, b Corsicana, Tex, Dec 30, 16; m 41; c 2. MATHEMATICS EDUCATION. Educ: Bishop Col, BS, 39; Columbia Univ, MA, 52, EdD(math educ), 64. Prof Exp: Asst to pres, Bishop Col, 39-52, asst prof math & physics, 52-57; teacher math, Dwight Morrow High Sch, Englewood, NJ, 58-65; PROF MATH & MATH EDUC, NY UNIV, 65- Concurrent Pos: Vis prof, Univ Wis-Madison, 64; consult-writer, Educ Serv Inc, Mass, 65; mem comt examr, Nat Teacher Exam, Educ Testing Serv, NJ, 69-73. Mem: Am Math Soc; Am Educ Res Asn; Math Asn Am; Nat Coun Teachers Math; AAAS. Res: Learning and teaching of mathematics in early childhood. Mailing Add: 23 Press Bldg NY Univ New York NY 10003

CARROLL, F IVY, b Norcross, Ga, Mar 28, 35; m 57; c 2. ORGANIC CHEMISTRY. Educ: Auburn Univ, BS, 57; Univ NC, PhD(chem), 61. Prof Exp: Sr chemist, 60-67, group leader, 67-71, ASST DIR, RES TRIANGLE INST, 71- Mem: Am Chem Soc; The Chem Soc. Res: Synthesis of organic compounds including the study of interesting reactions and physical properties, particularly carbon-13 NMR, of compounds thus formed; synthesis of biotransformation products of steroids, quinidine and 5,5-dialkylbarbituric acids. Mailing Add: Res Triangle Inst Research Triangle Park NC 27709

CARROLL, FLOYD DALE, b Mt Clare, Nebr, Jan 8, 14; m 44; c 1. ANIMAL HUSBANDRY. Educ: Univ Nebr, BS, 37; Univ Md, MS, 39; Univ Calif, PhD(animal nutrit), 48. Prof Exp: Prin technician, 48-49, from instr to assoc prof, 49-64, PROF ANIMAL HUSB, UNIV CALIF, DAVIS, 64- Concurrent Pos: Fulbright fel, Univ Ceylon, 55-56. Res: Vitamin B synthesis in the horse; nutrition in range beef production; beef production under hot climatic conditions; dwarfism in beef cattle; factors affecting beef carcass grades; composition and palatability. Mailing Add: Dept of Animal Sci Univ of Calif Davis CA 95616

CARROLL, FRANCIS W, b Philadelphia, Pa, Aug 22, 32; m 59; c 4. MATHEMATICAL ANALYSIS. Educ: St Joseph's Col, Pa, BS, 54; Purdue Univ, MS, 56, PhD(math), 59. Prof Exp: Asst prof math, Purdue Univ, 59-60, Mich State Univ, 60 & Univ Wisconsin-Milwaukee, 60-61; from asst prof to assoc prof, 61-75, PROF MATH, OHIO STAE UNIV, 75- Mem: Am Math Soc; Math Asn Am; AAAS. Res: Complex analysis; topological groups; functional equations. Mailing Add: Dept of Math Ohio State Univ Columbus OH 43210

CARROLL, GEORGE C, b Alton, Ill, Feb 11, 40; m 68; c 2. MYCOLOGY. Educ: Swarthmore Col, BA, 62; Univ Tex, PhD(bot), 66. Prof Exp: Asst prof, 67-72, ASSOC PROF BIOL, UNIV ORE, 72- Concurrent Pos: Vis prof, Swiss Fed Inst Technol, Inst Gen Bot, 73-74. Mem: Mycol Soc Am; Bot Soc Am; Am Soc Microbiol; Soc Gen Microbiol; Brit Mycol Soc. Res: Ultrastructure of spore formation in fungi; evolution of the fungi; ecology of fungi in terrestrial ecosystems; microbiology of the coniferous forest canopy. Mailing Add: Dept of Biol Univ of Ore Eugene OR 97403

CARROLL, GERALD V, b Meriden, Conn, Apr 9, 21. GEOLOGY. Educ: Lehigh Univ, BA, 43; Yale Univ, PhD(geol), 52. Prof Exp: Instr, Lehigh Univ, 50-51; instr, Trinity Col, Conn, 51-53; asst prof geol, Dartmouth Col, 53-54; assoc prof, Agr & Mech Col, Univ Tex, 54-59; vis assoc prof, Pa State Univ, 59-61; assoc prof, 61-65, PROF GEOL, GEORGE WASHINGTON UNIV, 65- Mem: AAAS; Geol Soc Am. Res: Petrology. Mailing Add: Dept of Geol George Washington Univ Washington DC 20006

CARROLL, HAROLD WILSON, b Phoenix, Ariz, Nov 9, 21; m 46; c 1. NUTRITION, BIOCHEMISTRY. Educ: Univ Calif, AB, 49, PhD(nutrit), 54. Prof Exp: Asst nutrit, Univ Calif, 52-54, res specialist, Agr Exp Sta, 54-55; radiol biologist, US Navy Radiol Defense Lab, 55-70, res physiologist, 70-72, CHIEF, PHYSIOL DIV, NAVAL MED FIELD RES LAB, 72- Mem: Am Inst Nutrit. Res: Radiobiology; enzymes; gerontology; environmental acclimatization; enzyme changes with physical exercise. Mailing Add: Physiol Div Naval Med Field Res Lab Camp Lejeune NC 28542

CARROLL, HARVEY FRANKLIN, b New Haven, Conn, Aug 25, 39. PHYSICAL CHEMISTRY. Educ: Hunter Col, AB, 61; Cornell Univ, PhD(phys chem), 69. Prof Exp: Sr chemist, Chem Div, Uniroyal Inc, Conn, 68-69; asst prof, 69-75, ASSOC PROF CHEM, KINGSBOROUGH COMMUNITY COL, CITY UNIV NY, 75- Mem: Am Chem Soc. Res: High temperature gas phase chemical kinetics; shock tubes; biomolecular reactions; unimolecular reactions. Mailing Add: Dept of Phys Sci Kingsborough Community Col Brooklyn NY 11235

CARROLL, JAMES BARR, b Chicago, Ill, Mar 25, 29; m 52; c 6. SCIENCE ADMINISTRATION, SCIENCE POLICY. Educ: Brown Univ, ScB, 52, MS, 57; Univ Conn, PhD(physics), 67. Prof Exp: Res engr, United Aircraft Res Labs, 55-68; staff scientist, Royal Prod Co, Litton Industs, 68-72; consult, Off Planning & Mgt, US Environ Protection Agency, 72-74; dir, Off Energy Supply Progs, Fed Energy Admin, 74; CONSULT, 74- Concurrent Pos: Lectr, Trinity Col, Conn, 65-66 & Univ Conn, 68. Mem: Am Phys Soc. Res: Technology assessment; computer modeling techniques; energy research. Mailing Add: 63 Pippin Dr Glastonbury CT 06033

CARROLL, JAMES JOSEPH, b Scranton, Pa, Dec 12, 35; m 60; c 1. BIOCHEMISTRY, CLINICAL CHEMISTRY. Educ: Univ Scranton, BS, 57; Pa State Univ, MS, 60, PhD(biochem), 62; NAT Registry Clin Chem, cert. Prof Exp: Scientist, Sandoz Pharmaceut, 62-65; scientist, Warner-Lambert Res Inst, 65-71, sr scientist, 71-73, SR RES ASSOC, WARNER-LAMBERT CO, INC, 73- Mem: Am Chem Soc; Am Asn Clin Chemists; Asn Clin Scientists. Res: Cholesterol metabolism; drug metabolism; creation and development of enzyme assays for use in diagnostics reagents; immuno-enzyme assays for isoenzymes. Mailing Add: Warner-Lambert Co Inc Morris Plains NJ 07950

CARROLL, JUNE STARR, b Lincoln, Nebr, June 23, 21. PHYSICAL GEOGRAPHY, GEOGRAPHY OF WESTERN UNITED STATES. Educ: Univ Calif, Los Angeles, BA, 46, MA, 48. Prof Exp: PROF GEOG, LOS ANGELES CITY COL, 47- Mem: AAAS; Asn Am Geog; Am Geog Soc; Artic Inst NAm; Nat Coun Geog Educ. Mailing Add: Dept of Earth Sci Los Angeles City Col 855 N Vermont Ave Los Angeles CA 90029

CARROLL, KEITH JAMES, physics, see 12th edition

CARROLL, KENNETH GIRARD, b Pittsburgh, Pa, Feb 18, 14; m 45; c 1. PHYSICS.

Educ: Carnegie Inst Technol, BS, 34, MS, 35; Yale Univ, PhD(physics), 39. Prof Exp: Instr physics, NC State Col, 39-40; physicist, Nat Adv Comt Aeronaut, Va & Ohio, 40-44, Elastic Stop Nut Corp Am, NJ, 44-46 & res lab, US Steel Corp, 46-54; head physics sect, Int Nickel Co Res Lab, 54-61; sr scientist, Sperry Rand Res Ctr, 61-67 & NASA Electronics Res Ctr, 67-70; vis scientist, Forsyth Dent Ctr, Boston, 70-71; PHYSICIST, ARGONNE NAT LAB, 74- Concurrent Pos: Nat Acad Sci-Nat Res Coun vis scientist, US Army Natick Labs, 65-66. Mem: Am Phys Soc; Electron Micros Soc; Sigma Xi. Res: Biophysics. Mailing Add: 57 S Seventh Ave La Grange IL 60525

CARROLL, KENNETH KITCHENER, b Carrolls, NB, Mar 9, 23; m 50; c 3. BIOCHEMISTRY. Educ: Univ NB, BSc, 43, MSc, 46; Univ Toronto, MA, 46; Univ Western Ont, PhD(med), 49. Prof Exp: From asst prof to prof med res & actg head dept, 54-68, PROF BIOCHEM, DEPT MED RES, UNIV WESTERN ONT, 68- Concurrent Pos: Can Life Ins Off Asn fel, Dept Med Res, Univ Western Ont, 49-52; Merck & Co fel, Dept Chem, Cambridge Univ, 52-53, Agr Res Coun Can fel, 53-54; hon secy, Can Fedn Biol Socs, 67-71; mem coun on atherosclerosis, Am Heart Asn. Mem: AAAS; Am Oil Chem Soc; Am Soc Biol Chem; Chem Inst Can; Can Physiol Soc. Res: Lipid metabolism, cholesterol; atherosclerosis; bacterial lipids; mammary cancer; polyprenols. Mailing Add: Dept of Biochem Univ of Western Ont London ON Can

CARROLL, MARCUS NEWMAN, JR, b Atlantic City, NJ, Nov 29, 24; m 48; c 4. PHYSIOLOGY, PHARMACOLOGY. Educ: Johns Hopkins Univ, AB, 49; Univ Fla, PhD(pharmacol), 52. Prof Exp: Res assoc pharmacol, Univ Fla, 51-52; instr, Sch Med, Marquette Univ, 52-53; physiologist & pharmacologist, Miles-Ames Res Labs, 53-58; neuropharmacologist, A H Robins Co, 58-60; asst prof pharmacol & neurol sci, Med Col Va, 60-62, dir res coord, Dept Neurol Sci, 61-62; assoc prof physiol & pharmacol, Ohio State Univ, 63-64; chief div pharmacol, Brookdale Hosp Ctr, Brooklyn, NY, 64-72; ASST DIR TOXICOL, AFFILIATED MED RES INC, 72- Concurrent Pos: Lederle fac award, 60; Nat Inst Neurol Dis & Blindness career develop award, 61; consult, Warner-Lambert Res Inst, NJ, 61-62; head sect pharmacodynamics & neuropharmacol, Strasenburg Labs, NY, 62-63; consult, Carroll Assocs, 62-; sr tech specialists, NAm Aviation, Inc, Ohio, 63-64. Mem: Am Soc Pharmacol & Exp Therapeut; Am Acad Neurol; Aerospace Med Asn. Res: Neurophysiology of centrally acting skeletal muscle relaxants; effects of drugs on central nervous system evoked arrhythmias and cerebrogenic pathological processes; biochemical and biophysical alterations evoked by aerospace stress; effects of drugs in one hundred per cent oxygen under various pressures. Mailing Add: Affiliated Med Res Inc PO Box 57 Princeton NJ 08540

CARROLL, MURRAY NORMAN, b Calgary, Alta, June 14, 23; m 52. PHYSICAL CHEMISTRY. Educ: Univ BC, BA, 47; McGill Univ, PhD(chem), 52. Prof Exp: Res chemist, Celanese Corp Am, 51-55 & Borden Chem Co, 55-61; assoc chemist, Div Indust Res, Wash State Univ, 62-67; sect head, Forest Prod Lab, Can Dept Forestry & Rural Develop, 67-70; RES MGR, WOOD PROD DIV, CAN DEPT ENVIRON, 70- Mem: Forest Prod Res Soc. Res: Synthetic textiles; pulp and paper; adhesive chemistry; wood products. Mailing Add: Western Forest Prod Lab 6620 NW Marine Dr Vancouver BC Can

CARROLL, NICHOLAS VINCENT, biochemistry, see 12th edition

CARROLL, PAUL JOSEPH, b Lawrence, Mass, May 14, 44; m 68; c 3. LOW TEMPERATURE PHYSICS. Educ: Lowell Technol Inst, BS, 65; Georgetown Univ, MS, 71, PhD(physics), 72. Prof Exp: RES PHYSICIST, NAVAL COASTAL SYST LAB, 71- Res: Applications of superconducting magnetic measurement systems. Mailing Add: Naval Coastal Syst Lab Code 792 Panama City FL 32401

CARROLL, ROBERT BAKER, b Tuscaloosa, Ala, Apr 8, 21; m 41, 51; c 4. ORGANIC CHEMISTRY, PLANT PATHOLOGY. Educ: Univ Ala, BS, 42; La State Univ, BS, 47, MS, 49, PhD(plant path), 51. Prof Exp: Asst jr pyrometrist, Ensley-Fairfield Works, US Steel Corp, Ala, 41; asst res chemist, Swann Chem Co, 41-42; asst res chemist, Exp Sta, Hercules Powder Co, Del, 42-43, explosives chemist, Radford Ord Works, Va, 42-43; asst plant physiologist, Boyce Thompson Inst Plant Res, NY, 50-52; microbiologist, Biochem Res Lab, Borden Co, 52-53; independent consult microanal, 53-68; PRES, O D V, INC, 68- Mem: AAAS; Am Ord Asn; Sigma Xi; Weed Sci Soc Am; Am Inst Chemists. Res: Food and flavors; pharmaceuticals; pesticides; instrumentation; plant physiology; agronomy; alcoholic beverages; plastics; oils and fats; microanalytical methods using all forms of chromatography; forensic chemistry, especially analysis of drugs and narcotics; field identification of narcotics and dangerous drugs. Mailing Add: ODV Inc PO Box 305 South Paris ME 04281

CARROLL, ROBERT LEON, b Three Lakes, Wash, Jan 15, 10; m 44; c 2. PHYSICS, MATHEMATICS. Educ: Fairmont State Col, AB, 33; WVa Univ, MS, 40, PhD(physics), 44. Prof Exp: Mem staff eletronics, Field Exp Sta, Mass Inst Technol, 44-45; assoc proj leader, Bur Standards, Wash, DC, 45-46; head dept physics, Fairmont State Col, 46-56; chief engr, US Naval Test Pilot Sch, Md, 56-58; aerodyn specialist, Bell Aircraft Corp, NY, 58-59; mem advan physics sect, Am Mach & Foundry Co, Va, 59-62; opers analyst, Opers Eval Group, Off Chief Naval Opers, Pentagon, 62-63; mem staff advan res, Melpar Inc, Va, 63-64; opers analyst, Combat Opers Res Group, Ft Belvoir, Va, 64-65; HEAD DEPT MATH & PHYSICS, BAPTIST COL CHARLESTON, 65- Res: Release and control of nuclear energy by disruption of the nucleus at low temperature. Mailing Add: Dept of Physics Baptist Col PO Box 10087 Charleston SC 29411

CARROLL, ROBERT LYNN, b Kalamazoo, Mich, May 5, 38; m 74; c 2. VERTEBRATE PALEONTOLOGY. Educ: Mich State Univ, BS, 59; Harvard Univ, MA, 61, PhD(biol), 63. Prof Exp: Nat Res Coun Can fel, 62-63; NSF fel, 63-64; assoc cur geol, Redpath Mus, 64-69, assoc prof vert paleont, 69-74, PROF VERT PALEONT, McGILL UNIV, 74-, CUR, REDPATH MUS, 69- Mem: Soc Vert Paleont; Soc Study Evolution; Am Soc Zool; Paleont Soc; Linnean Soc London. Res: Anatomy and phylogeny of Carboniferous, Permian and Triassic amphibians and reptiles—labyrinthodonts, microsaurs, captorhinomorphs and eosuchians; origin of lizards and Lissamphibia. Mailing Add: Redpath Mus McGill Univ PO Box 6070 Sta A Montreal PQ Can

CARROLL, ROBERT WAYNE, b Chicago, Ill, May 10, 30; m 57, 74; c 2. MATHEMATICS. Educ: Univ Wis, BS, 52; Univ Md, PhD(math), 59. Prof Exp: Res aeronaut scientist, Nat Adv Comt Aeronaut, Ohio, 52-54; Nat Sci Found fel, 59-60; asst prof math, Rutgers Univ, 60-63, assoc prof, 63-64; assoc prof, 64-67, PROF MATH, UNIV ILL, URBANA-CHAMPAIGN, 67- Concurrent Pos: Nat Sci Found res grant, 63-70. Mem: Math Soc Am. Res: Partial differential equations; functional analysis; differential geometry; mathematical physics; Lie groups. Mailing Add: Dept of Math Univ of Ill Urbana IL 61801

CARROLL, ROBERT WILLIAM, b Geneva, NY, Jan 16, 38. PHYSICAL CHEMISTRY. Educ: Hobart Col, BS, 59; Fordham Univ, PhD(phys chem), 65. Prof Exp: Res chemist, Inmont Corp, 65-68; sr scientist photoconductor res & develop, Optonetics, Ind, 69-70; ASST PROF CHEM, HERBERT H LEHMAN COL, 71- Mem: Am Chem Soc; Sigma Xi; AAAS. Res: Kinetics of hydrocarbon pyrolysis; field emission and field ionization microscopy. Mailing Add: Dept of Chem Herbert H Lehman Col Bronx NY 10468

CARROLL, SAMUEL EDWIN, b Thamesville, Ont, Dec 20, 28; m; c 2. CARDIOVASCULAR SURGERY, THORACIC SURGERY. Educ: Univ Western Ont, BA, 56, MD, 53. Prof Exp: Trainee surg, Victoria Hosp, 53-58; registr, Edgware Gen Hosp, London, Eng, 59-60; CARDIOVASC SURGEON, DEPT VET AFFAIRS, WESTMINSTER HOSP, 63-; CHIEF SURGEON, ST JOSEPH'S HOSP, 66-; PROF SURG, UNIV WESTERN ONT, 66- Concurrent Pos: Res fel hypothermia, Surg Lab, Children's Hosp Med Ctr, Boston, Mass, 61; Ont Heart Found res fel, 61-, grant in aid, 64-; res asst, St Mark's Hosp, London, Eng, 59-60; examr, Royal Col Physicians & Surgeons Can, 68- Mem: Fel Am Col Cardiol; fel Am Col Chest Physicians; fel Am Col Surg; NY Acad Sci; Can Cardiovasc Soc. Res: Hypothermia; small vessel anastomosis; surgical shock. Mailing Add: St Joseph's Hosp London ON Can

CARROLL, THOMAS JOSEPH, b Pittsburgh, Pa, April 26, 12. PHYSICS. Educ: Univ Pittsburgh, AB, 32; Yale Univ, PhD(physics), 36. Prof Exp: Lab asst physics, Yale Univ, 32-36; prof math & physics, Col New Rochelle, 36-41; radio engr, Signal Corps Labs, Ft Monmouth, NJ, 41-43, physicist, Off Chief Signal Officer, Washington, DC, 43-46, Bur Standards, 46-51, Lincoln Lab, Mass Inst Technol, 51-58 & Bendix Radio Co, 58-68; res prof, Dept Elec Eng, George Washington Univ, 68-70; CONSULT, 70- Concurrent Pos: Mem, Int Sci Radio Union. Mem: Am Phys Soc; Inst Elec & Electronics Engrs; Optical Soc Am; Acoust Soc Am; Am Asn Physics Teachers. Res: Microwave radio propagation; twilight scatter propagation; molecular spectra; Faraday effect in molecular spectra. Mailing Add: 162 Lake Shore Rd Brighton MA 02135

CARROLL, THOMAS WILLIAM, b Los Angeles, Calif, Aug 22, 32; m 52; c 4. PLANT PATHOLOGY, ELECTRON MICROSCOPY. Educ: Calif State Polytech Col, BS, 54; Univ Calif, Davis, MS, 62, PhD(plant path), 65. Prof Exp: NIH res plant pathologist, 65-66; from asst prof to assoc prof bot, 66-74, PROF PLANT PATH, MONT STATE UNIV, 75- Mem: Am Phytopath Soc. Res: Plant virology. Mailing Add: Dept of Plant Path Mont State Univ Bozeman MT 59715

CARROLL, VERN, b Brooklyn, NY, Sept 2, 33; m 61; c 4. ANTHROPOLOGY. Educ: Yale Univ, BA, 59, MA, 62; Cambridge Univ, BA, 61, MA, 66; Univ Chicago, PhD(anthrop), 66. Prof Exp: From asst prof to assoc prof anthrop, Univ Wash, 66-72; assoc prof, 72-75, PROF ANTHROP, UNIV MICH, ANN ARBOR, 75- Concurrent Pos: Vis asst researcher, Pac & Asian Ling Inst, Univ Hawaii, 67-68; Am Coun Learned Socs fel & NSF grant Nukuoro ling, 67-68; Nat Inst Ment Health spec fel, 70-71; sr fel, East-West Ctr, Honolulu, 72-73. Mem: Fel Am Anthrop Asn; fel Rpyal Anthrop Inst Gt Brit & Ireland; Polynesian Soc; Asn Social Anthrop Oceania. Res: Theoretical cultural anthropology, especially the structure of cultures as information systems. Mailing Add: 560 S First Ann Arbor MI 48103

CARROLL, WALTER WILLIAM, b Chicago, Ill, June 25, 15; m; c 2. CANCER, SURGERY. Educ: Northwestern Univ, MD, 41, MS, 44. Prof Exp: From instr to prof surg, Med Sch, Northwestern Univ, 42-74; MED DIR, ST VINCENT MED CTR, LOS ANGELES, 74- Concurrent Pos: Surg attend staff, Passavant Mem Hosp Exp Study, Nat Res Coun, 42-45; assoc div surg, Cook County Hosp & staff physician, Commonwealth Edison Co, 44-46; courtesy surg staff, St Joseph Hosp; assoc dir, Joint Comn Accreditation of Hosps, 70- Mem: Am Col Surg; Am Geriat Soc; Am Radium Soc; Am Pub Health Asn; Int Soc Surg. Res: Nerve repair; wound healing; soft tissue tumor surgery; venous surgery. Mailing Add: St Vincent Med Ctr 2131 W Third St Los Angeles CA 90057

CARROLL, WILLIAM ROBERT, b Logan, Utah, June 9, 16; m 41; c 4. BIOCHEMISTRY. Educ: Swarthmore Col, AB, 38; Harvard Univ, AM, 40, PhD(physiol), 42. Prof Exp: Scientist dir, Nat Inst Arthritis, Metab & Digestive Dis, 48-71; TEACHER SCI, BALLOU HIGH SCH, WASHINGTON, DC, 71- Concurrent Pos: Res fel, Med Col, Cornell Univ, 46-48. Mem: AAAS; Am Soc Biol Chem; Nat Sci Teachers Asn. Res: Influence of estrogen on the metabolism of the uterus; enzyme reactions; metabolism of amino acids; effect of x-rays on proteins; physical chemistry of proteins and nucleic acids. Mailing Add: 4802 Broad Brook Dr Bethesda MD 20014

CARRUTH, BETTY RUTH, b Comanche, Tex. NUTRITION. Educ: Tex Tech Univ, BS, 65, MS, 68; Univ Mo, PhD(human nutrit, sociol), 74. Prof Exp: Instr nutrit, Tex Tech Univ, 68-71; res dietician, Med Ctr & instr med dietetics, Dept Human Nutrit, Foods & Food Systs Mgt, Univ Mo, 71; ASST PROF NUTRIT, UNIV MINN, ST PAUL, 74- Mem: Am Dietetic Asn; Soc Nutrit Educ; Am Pub Health Asn; Sigma Xi. Res: Planning services for high-risk groups in the community; surveying medical, social and educational professionals serving the adolescent population; determination of significant nutritional and socio-health needs of youth; attitude measurement. Mailing Add: Dept of Food Sci & Nutrit Univ of Minn St Paul MN 55101

CARRUTH, JAMES HARVEY, b Baton Rouge, La, Aug 17, 38; m 65; c 2. PURE MATHEMATICS. Educ: La State Univ, BS, 61, MS, 63, PhD(math), 66. Prof Exp: Asst prof, 66-69, ASSOC PROF MATH, UNIV TENN, KNOXVILLE, 70- Mem: Am Math Soc; Math Asn Am. Res: Topological semigroups. Mailing Add: Dept of Math Univ of Tenn Knoxville TN 37916

CARRUTH, KAYLA BERNARD, b Natchitoches, La, Oct 4, 42; m 65; c 2. NUTRITION. Educ: Northwestern State Univ, BS, 63; Univ Tenn, PhD(nutrit), 75. Prof Exp: Asst mgr food serv, La State Univ, 64-65, asst nutrit specialist, La Coop Exten Serv, 65-66; instr nutrit, 66-67, ASST PROF & LEADER HEALTH & NUTRIT, TENN AGR EXTEN SERV, UNIV TENN, KNOXVILLE, 67- Mem: Am Dietetic Asn; Am Home Econ Asn. Res: Solutions used in total parenteral nutrition with specific interest in the sulfur-containing amino acids contained in the nutrient solutions. Mailing Add: Univ of Tenn PO Box 1071 Knoxville TN 37901

CARRUTH, LAURENCE ADAMS, b Mansfield, Mass, June 11, 07; m 37; c 2. ENTOMOLOGY. Educ: Univ Mass, BS, 29; SDak State Col, MS, 31; Cornell Univ, PhD(entomology), 35. Prof Exp: Asst entom & zool, SDak State Col, 29-31; asst biol, Cornell Univ, 31-35, from instr to prof entom, 35-73, head dept, 49-67, EMER PROF ENTOM, UNIV ARIZ, 73- Concurrent Pos: From asst to assoc, NY Exp Sta, Geneva, 35-41. Mem: Fel Am Soc Agr Sci; Entom Soc Am; Entom Soc Can. Res: Insects of the Southwest; insects of economic importance. Mailing Add: 2904 E Kleindale Rd Tucson AZ 85716

CARRUTH, PHILIP WILKINSON, b Cleveland, Ohio, July 19, 14; m 42; c 2. MATHEMATICS. Educ: Hamilton Col, AB, 36; Syracuse Univ, MA, 37; Univ Ill, PhD(math), 41. Prof Exp: Asst math, Univ Ill, 37-41, instr, 45-47; from asst to assoc prof, Swarthmore Col, 47-66; PROF MATH, MIDDLEBURY COL, 66- Mem: Am Math Soc. Res: Valuation theory; ordinal number theory. Mailing Add: Dept of Math Middlebury Col Middlebury VT 05753

CARRUTH, WILLIS LEE, b Summit, Mass, Feb 21, 09; m 37; c 2. PHYSICAL CHEMISTRY. Educ: Asbury Col, BA, 35; Univ Ky, MS, 38. Prof Exp: Instr chem, Asbury Col, 35-36 & Univ SDak, 36-38; instr chem & physics, Lewis & Clark Col, 38-39, from asst prof to prof chem, 39-44; prof, Nebr Wesleyan Univ, 44-46; assoc prof math, Col Puget Sound, 46-47, from asst prof to prof chem, 47-58; mem staff, Res & Develop, Appl Physics Corp, 58-74; TECH DIR, TINSLEY REPLICATION GROUP, TINSLEY LABS, INC, 74- Concurrent Pos: Registr & admin secy fac, Lewis & Clark Col, 42-44; consult scientist, Varian Assocs, 74- Mem: Fel AAAS; Am Chem Soc; Optical Soc Am. Res: Spectroscopy; instrumental methods of analysis. Mailing Add: 2120 Maginn Dr Glendale CA 91202

CARRUTHERS, CHRISTOPHER, b Motherwell, Scotland, Mar 17, 09; nat US; m 40, 59; c 4. CANCER. Educ: Syracuse Univ, BS, 33, MS, 35; Univ Iowa, PhD(biochem), 38. Prof Exp: Asst chem, Syracuse Univ, 33-35; asst chem, Univ Iowa, 35-38; res assoc, Barnard Free Skin & Cancer Hosp, 42-48; mem div cancer res, Med Sch, Wash Univ, 48-53; assoc cancer res scientist, 53-66, PRIN CANCER RES SCIENTIST, ROSWELL PARK MEM INST, 66- Concurrent Pos: Res fel, Barnard Free Skin & Cancer Hosp, 38-42; instr, Sch Med, Wash Univ, 41-44; assoc res prof, Med Sch, State Univ, NY Buffalo, 56- Mem: Fel AAAS; fel NY Acad Sci; Am Soc Biol Chem; Am Asn Cancer Res; Soc Exp Biol & Med. Res: Polarography in biochemistry and cancer research; biochemistry of epidermis, cell particulates and carcinogenesis; immunology and biochemistry of epidermal proteins and of membranes of various types of rat mammary carcinomas and in rat liver carcinogenesis. Mailing Add: Roswell Park Mem Inst Orchard Park Labs Orchard Park NY 14127

CARRUTHERS, GEORGE ROBERT, astrophysics, see 12th edition

CARRUTHERS, JOHN ROBERT, b Toronto, Ont, Sept 12, 35; m 57; c 2. MATERIALS SCIENCE. Educ: Univ Toronto, BASc, 59, PhD(metall), 66; Lehigh Univ, MS, 61. Prof Exp: Mem tech staff semiconductor mat, Bell Tel Labs, 59-63; lectr mat sci, Univ Toronto, 64-65, asst prof, 65-67; mem tech staff crystal chem, Bell Tel Labs, 67-75, HEAD CRYSTAL GROWTH & GLASS RES & DEVELOP DEPT, BELL LABS, MURRAY HILL, NJ, 75- Concurrent Pos: Consult, Off Applns, NASA, 74- Mem: AAAS; Am Phys Soc. Res: Crystal growth and evaluation, especially influences of fluid convection and phase equilibria on crystal growth, space processing, optical fibers for communications. Mailing Add: Bell Labs Inc Murray Hill NJ 07974

CARRUTHERS, PETER A, b Lafayette, Ind, Oct 7, 35; m 55, 69; c 3. THEORETICAL PHYSICS. Educ: Carnegie Inst Technol, BS & MS, 57; Cornell Univ, PhD(theoret physics), 61. Prof Exp: From asst prof to prof physics, Cornell Univ, 61-73; LEADER THEORET DIV, LOS ALAMOS SCI LAB, UNIV CALIF, 73- Concurrent Pos: NSF fel, 60-61; Sloan res fel, 63-65; vis assoc prof, Calif Inst Technol, 65, vis prof, 69-70; NSF sr fel, Univ Rome, 67-68; mem, Bd Dirs & trustee, Aspen Ctr Physics, 75-; mem, Physics Adv Panel, NSF, 75- Mem: Fel Am Phys Soc; AAAS. Res: Theory of strong interactions of elementary particles; symmetries of elementary particles. Mailing Add: Los Alamos Sci Lab UC PO Box 1663 Los Alamos NM 87544

CARRYER, HADDON MCCUTCHEN, b Unionville, Mo, Aug 25, 14; m 41; c 3. INTERNAL MEDICINE, ALLERGY. Educ: Drake Univ, BA, 35; Northwestern Univ, BM, 38, MS & MD, 39; Univ Minn, PhD(med), 48. Prof Exp: Asst med, Mayo Grad Sch, Univ Minn, 42-43, consult, 43-, from instr to assoc prof, Mayo Med Sch, 46-73, PROF MED, MAYO MED SCH, 73- Concurrent Pos: Consult, Div Med, Mayo Clin, 43- Mem: Fel Am Acad Allergy; fel Am Col Physicians; AMA. Res: Executive health periodic examinations. Mailing Add: Mayo Med Sch Rochester MN 55901

CARSKI, THEODORE ROBERT, b Baltimore, Md, June 22, 30; m 54; c 4. IMMUNOLOGY, MICROBIOLOGY. Educ: Johns Hopkins Univ, AB, 52; Univ Md, MD, 56; Am Bd Microbiol, dipl, 65. Prof Exp: Med intern, Univ Md, 56-57; sr asst surgeon, Commun Dis Ctr, Ala, 57-60; dir med res, Baltimore Biol Lab Div, B-D Labs, Inc, 60-68; dir microbiol, 68-70, from asst dir to dir, Huntingdon Res Ctr, 70-74, assoc med dir, 74-75, CORP MED DIR, BECTON, DICKINSON & CO, 75- Mem: Am Soc Microbiol; NY Acad Sci. Res: Fluorescent antibody techniques; mycoplasma; medical immunology; bacteriology and virology. Mailing Add: Becton Dickinson & Co Box 243 Cockeysville MD 21030

CARSOLA, ALFRED JAMES, b Los Angeles, Calif, June 6, 19; m 47; c 7. OCEANOGRAPHY, MARINE GEOLOGY. Educ: Univ Calif, Los Angeles, AB, 42; Univ Southern Calif, MS, 47; Scripps Inst, Univ Calif, PhD(oceanog), 53. Prof Exp: Asst, Univ Southern Calif, 46-47; geophysicist & oceanographer, Electronics Lab, US Navy, 47-60; staff scientist, 60-67, HEAD OCEANICS DIV, LOCKHEED-CALIF CO, 67- Concurrent Pos: Instr, San Diego State Col, 56-59, Loyola Univ, 63, Univ San Diego, 59 & Univ Calif, Los Angeles, 60-61, 62. Mem: AAAS; Soc Econ Paleont & Mineral; Am Geophys Union; fel Geol Soc Am; Am Soc Limnol & Oceanog. Res: Marine sediments; arctic marine seafloor, bathymetry and geomorphology; Seamount sediments; nearshore physical marine processes; micropaleontology; physical oceanography and acoustical oceanography. Mailing Add: 3569 Addison St San Diego CA 92106

CARSON, ALBERT B, b Alpena Pass, Ark, Dec 7, 08; m 36; c 1. MATHEMATICS. Educ: Univ Ark, BS, 32; Vanderbilt Univ, MA, 33; Univ Chicago, PhD(math), 41. Prof Exp: Prof physics & math, Boise Jr Col, Idaho, 35-37; instr math, Ill Inst Technol, 38-40; instr, La State Univ, 40-42; assoc prof, 46-51, prof, 51-75, EMER PROF MATH, US AIR FORCE INST TECHNOL, 75-, HEAD DEPT, 51- Mem: AAAS; Am Soc Eng Educ; Am Math Soc; Math Asn Am. Res: Analogue of Green's Theorem for multiple integral problems of the calculus of variations. Mailing Add: Dept of Math US Air Force Inst Technol Wright-Patterson AFB Dayton OH 45433

CARSON, CHARLES E, geology, agronomy, see 12th edition

CARSON, CHESTER CARROL, b Passaic, NJ, Nov 21, 18; m 47; c 2. ANALYTICAL CHEMISTRY. Educ: Newark Col Eng, BS, 41; Rensselaer Polytech Inst, MS, 59. Prof Exp: Control chemist, Rayon Prod, Celanese Corp Am, Md, 41-42; CHEMIST, LARGE STEAM TURBINE-GENERATOR DEPT, GEN ELEC CO, 46- Mem: Am Soc Testing & Mat; Am Chem Soc; fel Am Inst Chemists. Res: Sampling and analysis of atmospheric particulates at gas turbine sites; sampling and analysis of pyrolysates detected in large gas-cooled generators; effect and measurements of hydrogen in steel, of hydrogen, oxygen and nitrogen in steel and metals; problems in gases; immediate detection of overheating; gas-cooled electrical machines. Mailing Add: Mat & Processes Lab Gen Elec Co Schenectady NY 12345

CARSON, EUGENE WATSON, JR, b Cumberland, Va, Mar 27, 39; m 60; c 3. AGRONOMY, PLANT BIOCHEMISTRY. Educ: Va Polytech Inst & State Univ, BS, 61, MS, 63; NC State Univ, PhD(soil sci), 66. Prof Exp: Va Coun Hwy Invest & Res asst turf & hwy, Va Polytech Inst & State Univ, 60-63; res asst plant physiol, NC State Univ, 63-66; assoc prof, 66-72, PROF AGRON, VA POLYTECH INST & STATE UNIV, 72- Mem: Am Soc Agron; Crop Sci Soc Am. Res: Plant ecology. Mailing Add: Dept of Agron Va Polytech Inst & State Univ Blacksburg VA 24060

CARSON, FREDERICK WALLACE, b Quincy, Mass, Mar 18, 40; m 69. ORGANIC CHEMISTRY, BIOCHEMISTRY. Educ: Mass Inst Technol, BS, 61; Washington Univ, MA, 63; Univ Chicago, PhD(chem), 65. Prof Exp: NIH fel org chem, Princeton Univ, 65-66; asst prof org chem & biochem, Ind Univ, Bloomington, 66-70; asst prof, 70-71, ASSOC PROF ORG CHEM & BIOCHEM, AM UNIV, 72- Concurrent Pos: Grants, Washington Heart Asn, 72, NIMH, 73, US Dept Interior, 75, Petrol Res Fund, Biomed Sci & Res Corp. Mem: AAAS; Am Chem Soc; The Chem Soc; NY Acad Sci. Res: Biochemical mechanisms; model enzyme systems; enzyme kinetics; stereochemistry; organosulfur chemistry. Mailing Add: Dept of Chem American Univ Washington DC 20016

CARSON, GEORGE STEPHEN, b Lakewood, Ohio, Dec 7, 48; m 69; c 1. MATHEMATICS. Educ: Univ Tenn, Knoxville, BS, 70; Univ Calif, Riverside, PhD(math), 75. Prof Exp: LECTR MATH, CALIF STATE COL, SAN BERNARDINO, 75- Mem: Am Math Soc; Math Asn Am; AAAS; Sigma Xi. Res: Approximation theory; numerical analysis. Mailing Add: Dept of Math Calif State Col San Bernardino CA 92407

CARSON, GEORGE WALTER, b Salem, Ind, May 26, 04; m 27; c 3. MATHEMATICS. Educ: Hanover Col, AB, 27; Univ Ill, AM, 35. Prof Exp: Head dept math, Pikeville Col, 35-41; prof, Grove City Col, 42-58; assoc prof, Univ Redlands, 58-61; prof, 61-71, EMER PROF MATH, CALIF STATE POLYTECH UNIV, 71- Res: Mathematics education. Mailing Add: 234D Paseo Quinta Green Valley AZ 85614

CARSON, HAMPTON LAWRENCE, b Philadelphia, Pa, Nov 5, 14; m 37; c 2. EVOLUTIONARY BIOLOGY. Educ: Univ Pa, AB, 36, PhD(zool), 43. Prof Exp: Instr zool, Univ Pa, 38-42; from instr to prof, Wash Univ, 43-70; PROF GENETICS, UNIV HAWAII, 71- Concurrent Pos: Mem Wheelock Exped, Labrador, 34; prof biol, Univ Sao Paulo, 51; Fulbright res scholar, Univ Melbourne, 61. Mem: Genetics Soc Am; Soc Study Evolution (pres, 71); Am Soc Naturalists (pres, 73); AAAS; Am Soc Zoologists. Res: Population genetics; genetic systems and their relation to evolution; cytogenetics and evolution of drosophila and other insects. Mailing Add: Dept of Genetics Univ of Hawaii Honolulu HI 96822

CARSON, J DAVID, b Lehi, Utah, Dec 19, 17; m 42; c 6. GENETICS. Educ: Colo State Univ, BS, 47; Univ Calif, PhD(genetics), 53. Prof Exp: Asst & assoc poultry sci & animal genetics, Utah State Univ, 52-64; PROF BIOL SCI & CHMN, UNIV OF THE PACIFIC, 64- Concurrent Pos: Consult, West-Line Breeders, Wash, 56-58; NIH res grant, 61. Mem: Genetics Soc Am; Poultry Sci Asn. Res: Genetics of reproduction in domestic turkeys as influenced through hybridization and natural selection; genetic forces regulating serum cholesterol levels in white mice; dietary interactions. Mailing Add: Dept of Biol Sci Univ of the Pacific Stockton CA 95204

CARSON, JAMES ESTLE, b Canton, Ohio, Dec 21, 21; m 48; c 2. METEOROLOGY. Educ: Kent State Univ, BS, 43; Univ Chicago, SM, 48, PhD(meteorol), 60. Prof Exp: Res asst meteorol, Univ Chicago, 47-51; asst prof, Rutgers Univ, 51-53; meteorologist, US Army Natick Res & Develop Command, 53-55; asst prof physics, Iowa State Univ, 55-61; METEOROLOGIST, ARGONNE NAT LAB, 61- Mem: Am Meteorol Soc; Sigma Xi; Air Pollution Control Asn. Res: Meteorology; micrometeorology; turbulent transfer and diffusion; thermal pollution; cooling tower effluents. Mailing Add: Environ Statements Proj Bldg 11 Argonne Nat Lab Argonne IL 60439

CARSON, JAMES ROLLAND, b Omaha, Nebr, Feb 21, 14; m 47; c 4. POULTRY SCIENCE. Educ: Ore State Col, BS, 37; Cornell Univ, MS, 42, PhD(animal breeding), 49. Prof Exp: Asst, Cornell Univ, 40-42 & 46-48; asst prof poultry sci, Univ Conn, 48-55, assoc prof, 55-59; regional coordr, NCent Regional Poultry Breeding Proj, USDA, 59-62; PROF POULTRY HUSB, PURDUE UNIV, 62- Mem: AAAS; Poultry Sci Asn; Am Genetic Asn. Res: Poultry management and physiology. Mailing Add: Poultry Bldg Purdue Univ West Lafayette IN 47906

CARSON, JOHNNY LEE, b Asheville, NC, Feb 6, 49; m 73. ENVIRONMENTAL BIOLOGY, CYTOPATHOLOGY. Educ: Western Carolina Univ, BS, 71; Univ NC, PhD(bot), 75. Prof Exp: RES ASSOC PATH, SCH MED, UNIV NC, CHAPEL HILL, 75- Mem: Sigma Xi. Res: Cytopathological studies of effects of environmental pollutants in mammals; cytology-ecology of algae and fungi. Mailing Add: Dept of Path Sch of Med Univ NC Chapel Hill NC 27514

CARSON, MERL J, PEDIATRICS. Educ: Univ NC, AB, 34; Vanderbilt Univ, MD, 38; Am Bd Pediat, dipl, 43. Prof Exp: Intern, Rochester Gen Hosp, NY, 38-39; jr resident, St Louis Children's Hosp, Mo, 39-40, asst resident, 40-41; pvt pract, NC, 41-42; lectr & instr pediat, Sch Med, Univ NC, 42-45; instr, Sch Med, Wash Univ, 45-47, asst prof, 47, asst dean chg postgrad affairs, 47-50; prof pediat, Univ Southern Calif, 50-60; DIR PEDIAT & ELECTROENCEPHALOG LAB, ORANGE COUNTY GEN HOSP, 60-63, 64-; MED DIR, CHILDRENS HOSP ORANGE COUNTY, 64-; PROF PEDIAT, UNIV CALIF, IRVINE, 67- Concurrent Pos: Chief pediat serv, St Agnes Hosp, Raleigh, NC, 42-45 & St Louis County Hosp, 46; asst physician, St Louis Children's Hosp, 45-50; sect head, St Louis City Hosp, Denver, Colo & Orange, Calif; asst neurol, Inst Neurol, Eng, 57-58; chief res neurol & electroencephalog & clin prof pediat, Med Ctr, Univ Calif, Los Angeles, 60-63 & 64-; consult pediatrician, Ft MacArthur & Orthop Hosps, Los Angeles & Nat Jewish Hosp, Denver; mem rev bd, Neurol & Sensory Dis Inst, NIH, DC. Mem: Soc Pediat Res; Am Acad Pediat; Am Pediat Soc; AMA. Res: Pediatric neurology. Mailing Add: 1771 Sirrine Dr Santa Ana CA 97200

CARSON, PAUL LANGFORD, US citizen. MEDICAL PHYSICS, ULTRASONIC RESEARCH. Educ: Colo Col, BS, 65; Univ Ariz, PhD(physics), 72. Prof Exp: Instr, 71-73, ASST PROF RADIOL, UNIV COLO MED CTR, DENVER, 71- Mailing Add: Dept of Radiol Univ of Colo Med Ctr Denver CO 80220

CARSON, PAUL LLEWELLYN, b Ames, Iowa, Mar 27, 19; m 53; c 3. SOIL FERTILITY. Educ: Northwest Mo State Teachers Col, BS, 41; Iowa State Col, MS, 47. Prof Exp: Teacher pub schs, Mo, 41; agronomist, 48-69, PROF AGRON, S DAK STATE UNIV, 69- Concurrent Pos: Adv, Rockefeller Found, Colombia, SA, 67-68. Mem: Am Soc Agron; Soil Sci Soc Am. Res: Soil management; soil testing to determine fertilizers needed for farmers soils. Mailing Add: Dept of Plant Sci S Dak State Univ Brookings SD 57006

CARSON, ROBERT CLELAND, b Akron, Ohio, Mar 11, 24; m 51; c 4. MATHEMATICS. Educ: Purdue Univ, BS, 48, MS, 50; Univ Wis, PhD(math), 53. Prof Exp: Asst prof math & statist, Lehigh Univ, 53-57; asst prof math, Western Reserve Univ, 57-58; mathematician, Goodyear Aircraft Corp, Ohio, 58-63; coordr res, 63-71, asst dean grad studies, 68-71, ASSOC PROF MATH, UNIV AKRON, 63- Mem: Am Math Soc; Am Statist Asn; Soc Indust & Appl Math. Res: Variational methods and stochastic processes. Mailing Add: 1537 Maple St W Barberton OH 44203

CARSON, ROBERT JAMES, III, b Lexington, Va. GEOLOGY, GEOMORPHOLOGY. Educ: Cornell Univ, AB, 63; Tulane Univ, MS, 67; Univ Wash, PhD(geol), 70. Prof Exp: Geologist, Texaco Inc, La, 63-67 & Dept Ecol, Univ Wash, 69-70; MEM STAFF GEOL, NC STATE UNIV, 70- Mem: AAAS; Geol Soc Am; Glaciol Soc; Am Quaternary Asn. Res: Quaternary geology of the Olympic Peninsula, Washington. Mailing Add: Dept of Geosci NC State Univ Raleigh NC 27607

CARSON, STANLEY FREDERICK, b San Francisco, Calif, Oct 4, 12; m 44; c 3. MICROBIOLOGY. Educ: Stanford Univ, AB, 34, PhD(microbiol), 41. Prof Exp: Sr microbiologist, Merck & Co, 42-45 & Wyeth Inst Appl Biochem, Pa, 45-46; sr microbiologist, 47-48, prin biologist, 48-67, from asst dir to assoc dir, 48-67, DEP DIR BIOL DIV, OAK RIDGE NAT LAB, 67- Concurrent Pos: E R Squibb lectr, Rutgers Univ, 58; Haskin Labs fel, Stanford Univ, 41-42; mem, NSF Adv Panel Molecular Biol, 55-57, Adv Panel Spec Facilities & Progs, 61-; vpres, Microbial Metab Div, Int Cong Microbiol, Italy, 53; assoc ed, J Bact, 51-56 & Bact Rev, 58-64; hon res prof, Univ Ga, 60-; prof biomed sci, Univ Tenn, 67- Mem: Fel AAAS; Am Chem Soc; Am Soc Biol Chemists; Am Soc Microbiol; Environ Mutagen Soc. Res: Biochemistry and physiology of microorganisms; intermediary metabolism; enzyme mechanisms; isotopic tracers. Mailing Add: Biol Div Oak Ridge Nat Lab PO Box Y Oak Ridge TN 37830

CARSON, STEVEN, b Brooklyn, NY, Oct 17, 25; m 48; c 2. PHARMACOLOGY. Educ: Wash Univ, BS, 48; NY Univ, MS, 50, PhD(biol pharmacol), 58. Prof Exp: Asst, Pub Health Res Inst, NY, 48-49; chief chemist, NY Med Col, 49-51; pharmacologist, Endo Labs, Inc, 51-58, vpres & sci dir, Food & Drug Res Labs, Inc, 59-72; vpres & dir sci affairs, Biomet Testing, Inc, 72-75; INDEPENDENT CONSULT, 75- Concurrent Pos: Adj assoc prof, St John's Univ; bd forum advan toxicol, Univ Tenn. Mem: NY Acad Sci; fel Soc Cosmetic Chemists; Am Soc Pharmacol & Exp Therapeut; fel Royal Soc Health; fel Soc Cosmetic Chemists. Res: Pharmacology of central nervous depressants and stimulants; pharmacological properties of marine products and marine inhabitants; toxicological evaluations of food additives; drug supplements; respiratory physiology and pharmacology in laboratory animals exposed to environmental irritant; clinical pharmacology. Mailing Add: Biomet Testing Inc PO Box 373 Ryder Station Brooklyn NY 11234

CARSON, THEOPHILUS ROOSEVELT, b Eastman, Ga, Jan 23, 24; m 49; c 3. BIOLOGY. Educ: Morgan State Col, BS, 50. Prof Exp: Med technician, Vet Admin Hosp, Perry Point, Md, 54-56; biologist, Dept Army, Edgewood Arsenal, Md, 56-58, pharmacologist, 58-63; pharmacologist, 63-72, PHARMACOLOGIST FOOD ADDITIVES SAFETY & COSMETIC FORMULATION, FOOD & DRUG ADMIN, DEPT HEALTH, EDUC & WELFARE, 72- Mem: AAAS; Am Chem Soc; NY Acad Sci. Res: Inhalation toxicology of agents; dermatologic and inhalation research to aid in regulating drugs and cosmetics. Mailing Add: FDA BF/ADS/DT HFF-152 200 C St SW Washington DC 20204

CARSTEA, DUMITRU DUMITRU, b Comuna Paduroiu, Romania, Mar 22, 30; US citizen; m 56; c 4. ENVIRONMENTAL SCIENCES, SOIL CHEMISTRY. Educ: MS & BS, Agr Inst, Bucharest, 54; Ore State Univ, MS, 65, PhD(soil chem, clay mineral), 67. Prof Exp: Res scientist, Romanian Acad Sci, 54-60; res asst clay mineral & soil chem, Ore State Univ, 61-66; res scientist, proj leader, Can Dept Agr, 66-67; soil scientist, chemist, & hydrologist, US Geol Surv, 67-68, res hydrologist & proj leader, 68-70; ENVIRON TECH STAFF, MITRE CORP, WASHINGTON OPERS, 74- Mem: Int Asn Study Clays; Int Soc Soil Sci; Am Geophys Union; Clay Minerals Soc; Soil Sci Soc Am. Res: Clay mineralogy; hydrology; water quality; sedimentation; soil genesis-classification; estuarine sedimentation; environmental analysis and planning; energy analysis. Mailing Add: 12801 Point Pleasant Dr Fairfax VA 22030

CARSTEN, ARLAND L, b Hastings, Minn, Apr 17, 30; m 69. RADIOBIOLOGY, HEALTH PHYSICS. Educ: Mankato State Col, BSc, 53, MS, 56; Univ Rochester, PhD(biol), 57. Prof Exp: From asst to sr assoc radiation biol, Univ Rochester, 55-64; assoc health physicist, 57-62, vis assoc biologist, 62-64, assoc scientist, 66-70, SCIENTIST, MED RES CTR, BROOKHAVEN NAT LAB, 70-; ASSOC PROF PATH, STATE UNIV NY STONY BROOK, 73- Concurrent Pos: Res fel, Royal Dent Col, Copenhagen, Denmark, 65-66; health physics fel adv, AEC, 60-62; lectr, Am Inst Biol Sci Prog, Med Educ Nat Defense, 61-63; res assoc neurol, Columbia Univ, 64-72; res assoc, Lerner Marine Lab, Bimini, Bahamas. Mem: AAAS; Radiation Res Soc; Health Physics Soc; Am Soc Hemat; Int Soc Exp Hemat. Res: Acute and late effects of ionizing radiation on mammals, particularly on hematopoietic and nervous tissues; protection against and recovery from radiation injury to man. Mailing Add: Med Res Ctr Brookhaven Nat Lab Upton NY 11973

CARSTEN, MARY E, b Berlin, Ger, Mar 2, 22; nat US; m 64. BIOCHEMISTRY. Educ: NY Univ, AB, 46, MS, 68, PhD(biochem), 51. Prof Exp: Instr, NY Univ, 52-53; res assoc dept microbiol, Col Physicians & Surg, Columbia Univ, 53-55; asst res physiol chemist, Dept Physiol Chem, 56-61, assoc res biochemist, Depts Biol Chem & Med, 61-63, assoc prof, Depts Physiol, Obstet & Gynec, 63-70, PROF OBSTET & GYNEC, SCH MED, UNIV CALIF, LOS ANGELES, 70- Concurrent Pos: Nat Found Infantile Paralysis fel, 54-55; Am Cancer Soc fel, Univ Calif, Los Angeles, 55-57; USPHS res career develop award, 64-69 & 69-74; estab investr, Los Angeles County Heart Asn, 61-64. Honors & Awards: Res Award, Los Angeles County Heart Asn, 62, 63, 64. Mem: Hon mem Soc Gynecol Invest; Am Soc Biol Chem; Am Chem Soc; NY Acad Sci; Am Physiol Soc. Res: Ion exchange chromatography; amino acids; protein chemistry; immunochemistry; skeletal, heart and smooth muscle proteins and calcium transport; myometrial and uterine physiology. Mailing Add: Dept of Obstet & Gynec Univ of Calif Sch of Med Los Angeles CA 90024

CARSTENS, ALLAN MATLOCK, b Aurora, Ill, Jan 14, 39; m 60; c 3. OPERATIONS RESEARCH. Educ: Univ NMex, BS, 61, MS, 63, PhD, 67. Prof Exp: Instr math, Wash State Univ, 66-67, asst prof, 67-70; from asst prof to assoc prof, Mankato State Col, 70-73; SR PROGRAMMER, SPERRY RAND CORP, UNIVAC, 73- Mem: Asn Comput Mach; AAAS; Am Math Soc; Math Asn Am. Res: Generalized topological spaces; convergence algebras; products of pretopologies; structures of the lattices of pretopologies, pseudotopologies, limitierungen. Mailing Add: UNIVAC Sperry Rand Corp 2276 Highcrest Dr Roseville MN 55113

CARSTENS, HERMAN PAUL, b Chicago, Ill, Dec 24, 10; m 43; c 4. MEDICINE. Educ: Univ Chicago, SB, 32; Univ Ill, MS, 39, MD, 41; Univ Pa, cert, 54. Prof Exp: Res internal med, Grad Sch, Univ Ill, 40; intern, Garfield Park & Cook County Hosps, 40-41; resident med, Cook County Hosp, 41-42; asst internal med, Col Med, Univ Ill, 43-44, clin instr, 45-53, asst prof physiol, 57; assoc internal med, 57-67, ASST PROF MED, MED SCH, NORTHWESTERN UNIV CHICAGO, 67- Concurrent Pos: Fel path, Cook County Hosp, Chicago, Ill, 45-46; asst internal med, Col Med, Univ Ill, 40-41; assoc prof, Cook County Postgrad Sch, 44-47; assoc attend, Cook County Hosp, 44-47, mem attend staff, 61-; active staff, Lutheran Gen Hosp, Park Ridge, 61-; attend staff, Northwest Community Hosp, Arlington Heights, 61-; assoc attend, Ill Masonic Hosp, Chicago, 47-61, mem consult staff, 61-; attend, Vet Admin Res Hosp, 74- Mem: Am Soc Internal Med; Am Col Chest Physicians; Am Thoracic Soc; Am Rheumatism Asn; AMA. Res: Rheumatism. Mailing Add: 1430 Arlington Heights Rd Arlington Heights IL 60004

CARSTENS, JOHN C, b Chicago, Ill, Oct 8, 37; m 67. PHYSICS. Educ: Monmouth Col, Ill, AB, 59; Univ Mo-Rolla, PhD(physics), 66. Prof Exp: Fel, Univ Mo-Rolla, 66-67; asst prof physics, Western Ill Univ, 67-68; ASST PROF CLOUD PHYSICS & SR INVESTR, CLOUD PHYSICS CTR, UNIV MO-ROLLA, 68- Concurrent Pos: Resident res assoc, Argonne Lab, 66-67. Res: Mass and heat transport problems in cloud physics; droplet growth. Mailing Add: Dept of Physics Norwood Hall Univ of Mo Rolla MO 65401

CARSWELL, ALLAN IAN, b Toronto, Ont, Oct 4, 33; m 56; c 3. OPTICAL PHYSICS. Educ: Univ Toronto, BApplSci, 56, MA, 57, PhD(physics), 60. Prof Exp: Nat Res Coun Can fel, Inst Theoret Physics, Amsterdam, 60-61; sr mem sci staff plasma physics, RCA Victor Co, 61-65, dir optical & microwave physics lab, 65-68; PROF PHYSICS, YORK UNIV, 68-, DIR GRAD PROG PHYSICS, 71- Mem: Am Phys Soc; Can Asn Physicists; Can Aeronaut & Space Inst; Optical Soc Am; Asn Prof Engrs. Res: Laser systems and applications; atmospheric optics; lidar; light scattering. Mailing Add: Dept of Physics York Univ Toronto ON Can

CARTA, GUY RODNEY, microbiology, see 12th edition

CARTAN, FRED O, chemistry, see 12th edition

CARTE, IRA F, b Winona, WVa, Jan 21, 38; m 62; c 1. GENETICS. Educ: Va Polytech Inst, BS, 63, MS, 66, PhD(genetics), 68. Prof Exp: GENETICIST, PERDUE FARMS, INC, 68- Mem: Poultry Sci Asn; Genetics Soc Am; Am Genetic Asn. Res: Population genetics of chickens. Mailing Add: Primary Breeder Dept Perdue Farms Inc Salisbury MD 21801

CARTEN, FREDERICK HOWARD, organic chemistry, see 12th edition

CARTER, ALBERT SMITH, b Conemaugh, Pa, July 2, 03; m 43; c 2. CHEMISTRY. Educ: Carnegie Inst Technol, BS, 24; Univ Wis, MS, 25, PhD(org chem), 27. Prof Exp: Asst chem, Univ Wis, 24-27; res chemist, Jackson Lab, E I du Pont de Nemours & Co, Inc, 27-41, asst mgr, Louisville works, 41-45, div head, Jackson Lab, 46-48, asst dir, 48-57, res dir elastomer chem dept, 57-63; CONSULT, 63- Concurrent Pos: Sci consult, US Dept Commerce, Germany, 46. Honors & Awards: Mod Pioneer Award, Nat Asn Mfrs, 40. Mem: AAAS; Am Chem Soc; Am Inst Chem Engrs; Am Soc Chem Indust; fel Am Inst Chemists. Res: Chemistry of furfural and acetylene; synthetic rubber; ethyl pimelic acid; polymer chemistry. Mailing Add: 5550 Bayview Dr Ft Lauderdale FL 33308

CARTER, ANNE COHEN, b New York, NY, Nov 27, 19; m 47; c 2. ENDOCRINOLOGY. Educ: Wellesley Col, BA, 41; Cornell Univ, MD, 44; Am Bd Internal Med, dipl. Prof Exp: From instr to asst prof med, Med Col, Cornell Univ, 46-55; from asst prof to assoc prof, 55-68, PROF MED, STATE UNIV NY DOWNSTATE MED CTR, 67- Concurrent Pos: Fel, Russell Sage Inst Path; mem, Cancer Clin Training Comt, Nat Cancer Inst, 71-74, mem, Cancer Control Treatment & Rehab Rev Comt, 74- Mem: AAAS; Soc Exp Biol & Med; Harvey Soc; Endocrine Soc; Am Fedn Clin Res. Res: Metabolism. Mailing Add: Dept of Med State Univ NY Downstate Med Ctr Brooklyn NY 11230

CARTER, ASHLEY, b Glen Ridge, NJ, June 27, 24; m 72; c 3. ELECTROPHYSICS, ACOUSTICS. Educ: Harvard Univ, AB, 45; Brown Univ, ScM, 50, PhD(physics), 63. Prof Exp: Res assoc underwater explosives, Woods Hole Oceanog Inst, 46-47; underwater acoustics, res anal group, Brown Univ, 51-53; mem tech staff, 53-65, head eng mech & physics dept, 65-71, HEAD ELEC PROTECTION DEPT, BELL TEL LABS, 71- Concurrent Pos: Lectr, Fairleigh Dickinson Univ, 72- & Drew Univ, 75- Mem: Am Phys Soc; Acoust Soc Am; Inst Elec & Electronics Engrs. Res: Electromagnetic interference; electrical protection techniques; physics of gas discharge; optical waveguides; propagation of waves in inhomogeneous media; underwater acoustics. Mailing Add: 420 River Rd Chatham NJ 07928

CARTER, BETTINA BUSH (MRS DANIEL F JACKSON), b Woburn, Mass, Sept 4, 10; m 29, 51; c 3. IMMUNOBIOLOGY, ENVIRONMENTAL BIOLOGY. Educ: Univ Mich, AB, 29, MS, 45; Univ Pittsburgh, PhD(bact, biochem), 51. Prof Exp: Jr lab technician div labs, NY State Dept Health, 37-40; chief serologist, La State Health Labs, 40-41; chief serologist, Ky State Health Dept, 41-43; chief serologist, Ill Br Lab, Chicago, 43-44; res immunologist, Inst Path, WPa Hosp, Pittsburgh, 44-53; asst prof microbiol, Col Med, State Univ NY Syracuse, 53-55; from asst prof biol to assoc prof biol, Univ Western Mich, 55-59; res assoc, Sch Med, Univ Louisville, 59-63, assoc prof natural sci, Univ, 60-63; from adj assoc prof bact to prof bact, Syracuse Univ, 65-70; prof biol & chmn dept sci & math, Cazenovia Col, 70-73; adj prof, Div Environ Technol, Fla Int Univ, 74-75; EDUC CONSULT, WARD'S NATURAL SCI ESTAB, 69- Concurrent Pos: Lectr, Duquesne Univ, 46-49; serologist to dist atty, Pa, 52. Honors & Awards: Gerber Award, 56; Zonta Int Woman's Year, 75. Mem: AAAS; Am Chem Soc; NY Acad Sci. Res: RH factor; effect of nutrition on antibody formation; serology of syphilis; complement fixation; algal immune patterns; environment; effect of plant auxins on animal cells. Mailing Add: 5220 SW 60 Pl Miami FL 33155

CARTER, BRIAN GEOFFREY, b Manchester, Eng, May 3, 35; Can citizen; m 58; c 2. IMMUNOBIOLOGY. Educ: Univ London, BSc, 56, MSc, 57, PhD(immunol), 61. Prof Exp: Instr pediat, Univ Pa, 64-65; asst prof microbiol & immunol, McGill Univ, 68-69; ASST PROF IMMUNOL, UNIV MAN, 69- Concurrent Pos: USPHS trainee, 62-64; Nat Multiple Sclerosis Soc fel, 66-68. Mem: Can Soc Immunol; Brit Soc Immunol; Brit Soc Immunol; Am Asn Immunol. Res: Cellular aspects of antibody heterogeneity; differentiation of immunocompetent cells; cellular mechanisms underlying the synthesis of different immunoglobulin classes. Mailing Add: Dept of Immunol Univ of Man Winnipeg MB Can

CARTER, CAROL SUE, b San Francisco, Calif, Dec 25, 44; m 70; ETHOLOGY. Educ: Drury Col, BA, 66; Univ Ark, Fayetteville, PhD(zool), 69. Prof Exp: Instr biol, Drury Col, BA, 66; NIH trainee, Mich State Univ, 69-70; NIH fel, WVa Univ, 70-71; adj asst prof biol, 70-72; res fel psychopharmacol, Ill Dept Ment Health, 72-74; ASST PROF, DEPT PSYCHOL, ECOL, ETHNOL & EVOLUTION, UNIV ILL, CHAMPAIGN, 74- Mem: AAAS; Soc Neurosci; Animal Behav Soc. Res: Mechanisms regulating mammalian reproductive behavior; utilization of endocrine, neuroendocrine and pharmacological techniques to study the physiological basis of behavior. Mailing Add: 829 Psychol Bldg Univ of Ill Champaign IL 61820

CARTER, CHARLES CONRAD, b Seattle, Wash, July 20, 24; m 48; c 4. NEUROLOGY. Educ: Reed Col, BA, 46; Univ Ore, MD, 48. Prof Exp: Intern, Good Samaritan Hosp, 48-49; resident neurol, Med Sch, Wash Univ, 54-56; from asst prof to assoc prof neurol, 62-70, actg head div, 74-75, PROF NEUROL, HEALTH SCI CTR, UNIV ORE, 70- Concurrent Pos: Nat Inst Neurol Dis & Stroke grant, Sch Med, Wash Univ, 69-70; vis assoc prof med, Sch Med, Wash Univ, 69-70. Mem: AMA; Am Acad Neurol. Res: Cerebral blood flow. Mailing Add: 16686 SW Maple Circle Lake Oswego OR 97304

CARTER, CHARLES EDWARD, b Boise, Idaho, Aug 25, 19; m 46; c 2. BIOCHEMISTRY. Educ: Reed Col, BA, 41; Cornell Univ, MD, 44. Prof Exp: Prin biochemist, Oak Ridge Nat Lab; asst prof med, Sch Med, Western Reserve Univ, 50-53; assoc prof pharmacol, Sch Med, Yale Univ, 53-57, prof, 57-64; dir dept, 64-72, PROF PHARMACOL, SCH MED, CASE WESTERN RESERVE UNIV, 64- Mem: Am Soc Biol Chem. Res: Biochemistry of nucleic acids. Mailing Add: Dept of Pharmacol Case Western Reserve Univ Sch of Med 2109 Adelbert Rd Cleveland OH 44106

CARTER, CLAUDE FRANCIS, b Hillsdale, Mich, Dec 24, 05; m 29; c 1. PHYSICS. Educ: Hillsdale Col, BS, 29; Univ Mich, MS, 36. Prof Exp: Assoc prof physics, Hillsdale Col, 42-46, assoc prof, 46-71, EMER ASSOC PROF NATURAL SCI & PHYSICS, UNIV MIAMI, 71- Mem: Electron Micros Soc Am; Am Asn Physics Teachers. Res: Electron microscopy; nuclear emulsions. Mailing Add: 5970 SW 46th St Miami FL 33155

CARTER, CLINT EARL, b Durant, Okla, Apr 28, 41; m 62; c 2. ANIMAL PHYSIOLOGY, PARASITOLOGY. Educ: La Sierra Col, BA, 65; Loma Linda Univ, MA, 67; Univ Calif, Los Angeles, PhD(zool), 71. Prof Exp: Teaching assoc microbiol, San Bernardino Valley Jr Col, 66-67; teaching asst gen biol & cell & comp physiol, Univ Calif, Los Angeles, 67-68; NIH teaching assoc fel gen zool & cell physiol, Univ Mass, 71-72; ASST PROF BIOL, VANDERBILT UNIV, 72- Mem: Am Soc Parasitologists; AAAS. Res: Energy metabolism of parasitic helminths. Mailing Add: Dept of Biol Vanderbilt Univ Nashville TN 37235

CARTER, CURTIS HAROLD, b Scott, Ga, Oct 1, 15; m 38; c 5. INTERNAL MEDICINE. Educ: Univ Ga, MD, 38, BS, 39; Am Bd Internal Med, dipl, 53. Prof Exp: Intern Univ Hosp, Augusta, Ga, 38-39, resident, 47-50; assoc med off, Vet Admin, 39-40; assoc med & instr, Postgrad Sch, Sch Med, Univ Tex M D Anderson Hosp & Tumor Inst, Houston, 50-51; from asst prof to assoc prof, 51-57, clin prof med, 68, assoc dean clin sci, 68-72, dean sch med, 72-75, PROF MED, SCH MED, MED COL GA, 57- Concurrent Pos: Pvt pract, 51-55; consult, Milledgeville State Hosp, Ga, 52-59, Vet Admin Hosp, 56-67 & Ga State Training Sch, Gracewood, 57-60; vis assoc prof, Sch Med, Univ Colo, 56; chg, Pulmonary Dis Div, Eugene Talmadge Mem Hosp, Augusta, 56-68, attend physician, 56-; dir educ internal med, Mem Hosp, Chatham County, Savannah, 68. Mem: AAAS; AMA; fel Am Col Physicians; fel Am Col Chest Physicians; NY Acad Sci. Res: Pulmonary disease, especially emphysema and tuberculosis. Mailing Add: Sch of Med Med Col of Ga Augusta GA 30902

CARTER, DAVID, b Brooklyn, NY, April 1, 20; m 43; c 2. PHYSICS. Educ: City Col, BEE, 45; Stanford Univ, PhD(physics), 51. Prof Exp: Tutor, City Col, 43-44; radio engr, Hallicrafters, Inc, Ill, 45-46; elec engr, Aladdin Radio Industs, 46; instr elec eng, Ill Inst Technol, 46-47; asst physics, Stanford Univ, 47-49; asst prof elec eng, NY Univ, 51-52; sr res engr, Gen Dynamics Div, Convair, 52-55; PROF PHYSICS, SAN JOSE STATE COL, 55- Concurrent Pos: Instr, Radio-TV Inst, NY, 45 & Am TV Inst Technol, Chicago, 46; electronics engr, Argonne Nat Lab, 46-47; lectr, San Diego State Col, 53 & Univ Calif, Los Angeles, 53-55; consult, Varian Assocs, 56- Mem: Inst Elec & Electronics Eng. Res: Microwave tubes and antennas. Mailing Add: Dept of Physics San Jose State Col San Jose CA 95114

CARTER, DAVID L, b Cleveland, Ohio, June 27, 33; m 57; c 2. SOLID STATE SCIENCE. Educ: Ohio State Univ, BSc & MSc, 56; Columbia Univ, PhD(physics), 62. Prof Exp: Res asst microwave physics, Radiation Lab, Columbia Univ, 58-62; res assoc solid state physics, Univ Pa, 62-64; vis prof, Physics Lab, Ecole Normale Superieure, Paris, 64-66; mem tech staff solid state microwave physics, Physics Res Br, Tex Instruments, Inc, 66-68, mgr electron transport physics br, 68-71; staff scientist, Chief Tech Off, Singer Co, 71-74; MGR, ALLOY RECEPTOR RES, XEROX CORP, 74- Concurrent Pos: Consult, Nuclear Res Assocs, 61-62, TRG Inc, 61-64 & Philco Res Labs, 62-64; chmn, Conf Physics Semimetals & Narrow Gap Semiconductors, Dallas, 70. Mem: Am Phys Soc. Res: Microwave interactions with solids; microwave solid state masers; narrow gap semiconductors; electro-optics technology; technical management. Mailing Add: Xerox Corp 800 Phillips Rd Webster NY 14580

CARTER, DAVID LAVERE, b Tremonton, Utah, June 10, 33; m 53; c 3. SOIL CHEMISTRY. Educ: Utah State Univ, BS, 55, MS, 57; Ore State Univ, PhD(soil sci), 61. Prof Exp: Lab technician soil sci, Utah State Univ, 52-54; phys sci aid, Soil Conserv Serv, USDA, 54-55; instr, Ore State Univ, 56; soil scientist, Soil & Water Conserv Res Div, 56-60, res soil scientist & line proj leader, 60-65, res soil scientist, 65-75, SUPVRY SOIL SCIENTIST, AGR RES SERV, USDA, 75- Honors & Awards: Emmett J Culligan Award, World Water Soc, 75. Mem: AAAS; Am Soc Agron; Soil Sci Soc Am; Int Soc Soil Sci. Res: Chemistry; applied statistics; plant physiology; salt and ion movement through soils; soil and water pollution control. Mailing Add: Snake River Conserv Res Ctr Agr Res Serv USDA Rte 1 Box 186 Kimberly ID 83341

CARTER, DAVID MARTIN, b Doniphan, Mo, June 10, 36; m 61; c 3. DERMATOLOGY. Educ: Dartmouth Col, AB, 58; Harvard Med Sch, MD, 61; Yale Univ, PhD(biol), 71. Prof Exp: Intern med & surg, Med Ctr, Univ Rochester, 61-62, asst resident med, 62-63; surgeon, Venereal Dis Br, Commun Dis Ctr, USPHS, 63-65; resident dermat, Hosp Univ Pa, 65-67; asst prof, 70-73, ASSOC PROF DERMAT, SCH MED, YALE UNIV, 73- Concurrent Pos: Med investr, Howard Hughes Med Inst, 70- Mem: Am Acad Dermat; Soc Invest Dermat. Res: Defenses of cutaneous cells against ultraviolet irradiation. Mailing Add: Dept of Dermat Yale Univ Sch of Med New Haven CT 06510

CARTER, DAVID SOUTHARD, b Victoria, BC, Mar 25, 26; US citizen; m 49; c 4. MATHEMATICS. Educ: Univ BC, BA, 46, MA, 48; Princeton Univ, PhD(math physics), 52. Prof Exp: Asst math physics, Princeton Univ, 49-52; mem staff, Los Alamos Sci Lab, Calif, 52-58; NSF res grant math & instr math sci, NY Univ, 57-58; vis assoc prof, Univ Wash, 58 & Univ Calif, Berkeley, 59-61; PROF MATH, ORE STATE UNIV, 61-, ACTG CHMN DEPT, 69- Concurrent Pos: Consult, Lockheed Missiles & Space Co, 59- Mem: Am Math Soc. Res: Applied analysis; hydrodynamics; statistical mechanics; control and stability theory; numerical analysis and computation. Mailing Add: Dept of Math Ore State Univ Corvallis OR 97331

CARTER, EARL THOMAS, b Baltimore, Md, July 7, 22; m 47; c 3. PHYSIOLOGY. Educ: Northwestern Univ, BS, 34, MD, 48, MS, 50; Univ Tex, PhD(physiol), 55. Prof Exp: Staff physician, Chicago Munic Tuberc Sanitarium, 50-51; third yr resident med, Chicago Wesley Mem Hosp, 55-56; asst prof physiol & prev med, Ohio State Univ, 56-60; asst prof med, 60-73, PROF PREV MED, MAYO CLIN & MAYO GRAD SCH MED & CHMN DIV, CLIN, 73- Concurrent Pos: Staff physician, Univ Hosp & Ohio Tuberc Hosp, 56-60; consult, Mayo Clin, 60- Res: Environmental medicine and environmental physiology. Mailing Add: 200 First St SW Rochester MN 55901

CARTER, EDWARD PENDLETON, b Boston, Mass, Dec 19, 11; m 36; c 1. BOTANY. Educ: Univ Md, BS, 36, MS, 39. Prof Exp: Agent, Bur Plant Indust, Soils & Agr Eng, USDA, Exp Sta, Md, 36-38, jr pathologist, 38-42, asst pathologist, 42-47, asst physiologist, 47-49, plant pathologist, Insecticide Sect, Livestock Br, Prod & Mkt Admin, 49-53, pathologist in charge fungicide & herbicide testing lab, 53-60, fungicide & hematocide sect, 60-61, asst chief staff officer for plant biol, 61-66, chief staff officer, Agr Res Serv, 66-70, CHIEF STAFF OFFICER, PESTICIDES REGULATIONS DIV, FUNGICIDES & NEMATICIDES EVAL, ENVIRON PROTECTION AGENCY, USDA, 70- Mem: AAAS; Am Phytopath Soc; Bot Soc Am. Res: Mycology and physiology of heating of moist grain in storage; organic chemistry of pigments of sorghum plant; physiology of weed control and development of new herbicides and applications. Mailing Add: 5505 42nd Ave Hyattsville MD 20781

CARTER, ELIZABETH FRANCIS, b Chicago, Ill, Mar 27, 43. ANTHROPOLOGY. Educ: Univ Chicago, BA, 65, PhD(Near Eastern lang & civilizations), 71. Prof Exp: ASST PROF ANTHROP, UNIV ORE, 73- Mem: Am Anthrop Asn; Soc Am Archaeol; Archaeol Inst Am; Am Orient Soc. Res: Proto-historic and early historic Iran. Mailing Add: Dept of Anthrop Univ of Ore Eugene OR 97403

CARTER, ELMER BUZBY, b Brooklyn, NY, Mar 28, 30; m 57; c 2. COMPUTER SCIENCE. Educ: Haverford Col, SB, 53; Fla State Univ, MS, 60, PhD(physics), 62. Prof Exp: Res assoc physics, Rice Univ, 62-63, asst prof, 63-65; asst prof, Strasbourg Univ, 65-66; ASSOC PROF PHYSICS, TRINITY UNIV, 66- Concurrent Pos: NSF comput sci resident, Syst Develop Corp, Santa Monica, Calif & Sch Archit & Urban Planning, Univ Calif, Los Angeles, 70-71. Mem: AAAS; Am Phys Soc; Asn Comput Mach. Res: Urban studies. Mailing Add: Dept of Comput & Info Sci Trinity Univ San Antonio TX 78284

CARTER, FAIRIE LYN, b Biloxi, Miss, Oct 1, 26. ANALYTICAL CHEMISTRY. Educ: Miss State Col for Women, BS, 48; Univ NC, MA, 50. Prof Exp: Asst anal chem, Univ NC, 48-50; asst cur limnol, Acad Nat Sci Philadelphia, 50-54; assoc chemist, E Reg Res Ctr, 55-57, res chemist, S Reg Res Ctr, 57-65, RES CHEMIST, WOOD PROD INSECT LAB, US FOREST SERV, USDA, 65- Mem: Am Chem Soc; Am Oil Chem Soc; Sci Res Soc Am; Entom Soc Am. Res: Insect biochemistry; insecticides; wood extractives; composition of natural products; amino acids; lipids; chromatography. Mailing Add: Wood Prod Insect Lab US Forest Serv PO Box 2008 Gulfport MS 39501

CARTER, FORREST LEE, organic chemistry, see 12th edition

CARTER, FREDERICK J, b Vernon, NY, Dec 16, 29; m 61; c 5. MATHEMATICS. Educ: Le Moyne Col, NY, BS, 56; Univ Detroit, MA, 58. Prof Exp: Instr math, Le Moyne Col, NY, 58-63; from instr to asst prof, 63-70, chmn dept, 70-74, ASSOC PROF MATH, ST MARY'S UNIV, TEX, 70-, UNDERGRAD ADV, 74- Mem: Am Math Soc; Math Asn Am. Res: Integral transforms and distribution theory. Mailing Add: Dept of Math St Mary's Univ San Antonio TX 78284

CARTER, GEORGE EMMITT, JR, b Fayetteville, NC, Jan 18, 46; m 70; c 1. PLANT PHYSIOLOGY. Educ: Wake Forest Univ, BS, 68, MA, 70; Clemson Univ, PhD(plant physiol), 73. Prof Exp: ASST PROF PLANT PHYSIOL, CLEMSON UNIV, 73- Mem: Am Soc Plant Physiologists; Am Phytopath Soc; AAAS. Res: Mechanism of dormancy in plants; quantitative analysis of plant growth regulators; peach tree physiology. Mailing Add: Dept of Plant Path & Physiol Clemson Univ Clemson SC 29631

CARTER, GEORGE FRANCIS, b San Diego, Calif, Apr 6, 12; m 39; c 3. CULTURAL GEOGRAPHY. Educ: Univ Calif, Berkeley, AB, 34, PhD(geog), 42. Prof Exp: Cur anthrop, San Diego Mus, 34-38; asst geog, Univ Calif, 38-40 & 41-42; assoc soc sci analyst, Latin Am Div, Off Strategic Servs, DC, 42-43; analyst, 45; from instr to prof geog, Johns Hopkins Univ, 43-67; DISTINGUISHED PROF GEOG, TEX A&M UNIV, 67- Concurrent Pos: Asst, San Diego State Col, 37-38, instr, 40-41; John Simon Guggenheim fel field work, San Diego, 53-54; Am Philos Soc grant field work, Nev; Wenner-Gren Found grant field work, San Diego; fel, Inter-Am Inst. Mem: Asn Am Geogr; Am Geog Soc; Am Antiq Soc. Res: Pre Columbian discoveries of America and their cultural influences; the antiquity of man in America. Mailing Add: Dept of Geog Tex A&M Univ College Station TX 77843

CARTER, GEORGE H, b Dobbs Ferry, NY, June 16, 16; m 46; c 3. PSYCHIATRY. Educ: Williams Col, BA, 38; Harvard Univ, MD, 43. Prof Exp: From instr to asst prof prev med, 50-53, from instr to asst prof psychiat, 53-59, ASSOC PROF PSYCHIAT, SCH MED, BOSTON UNIV, 59- Concurrent Pos: Resident fel psychiat, Harvard Med Sch, 48, teaching fel, 48-49; Commonwealth Fund fel, 66-67; asst, Univ Hosp, 50-55, asst vis physician, 55-59, vis physician, 59-; chmn residency training psychiat, Med Ctr, Boston Univ, 71-74 & curriculum coordr, 74-75, clin dir family ther, 74-75; asst chief inpatient serv psychiat, Bradford Vet Admin Hosp, 75- Mem: Am Med Soc; Am Psychiat Asn; Am Psychoanal Asn. Res: Psychotherapy; psychoanalytic theory. Mailing Add: 10 Garden Terr Cambridge MA 02138

CARTER, GERALD BATE, b Belle Fourche, SDak, Aug 26, 21; m 44; c 3. ANALYTICAL CHEMISTRY. Educ: Univ Kansas, BS, 43, MS, 48. Prof Exp: Army ord civilian chemist, Sunflower Ord Works, 43-46; chemist, US Potash Co, 46-47; from proj leader to sect leader anal res, Houston Res & Develop Lab, Indust Chem Div, Shell Chem Co, 48-68, DEPT HEAD ANAL, BIOL SCI RES CTR, SHELL DEVELOP CO, 68- Mem: Am Chem Soc; Sigma Xi. Res: Separation techniques and instruments; spectroscopy; chromatography; physical measurement; radiochemistry. Mailing Add: 209 Robin Hood Dr Modesto CA 95350

CARTER, GESINA C, b Nootdorp, Netherlands, Dec 15, 39; US citizen; m 62; c 3. SOLID STATE PHYSICS. Educ: Univ Mich, BS, 60; Carnegie Inst Technol, MS, 62, PhD(physics), 65. Prof Exp: Res assoc acoust, Cath Univ, 65-66; PHYSICIST, NAT BUR STANDARDS, 66- Concurrent Pos: Student, Phys Spectrometry Lab, Fac Sci, Univ Grenoble, 75-76. Mem: Am Phys Soc; Am Inst Physics; Am Soc Info Sci; NY Acad Sci; Am Soc Testing & Mat. Res: Magnetism in metals; fermi surfaces and electronic behavior in metals; transport properties in metals; nuclear magnetic resonance in metals and alloys; evaluation of phase diagrams. Mailing Add: Nat Bur Standards Washington DC 20234

CARTER, GILES FREDERICK, b Lubbock, Tex, Mar 22, 30; m 54; c 3. METALLURGICAL CHEMISTRY, ARCHAEOLOGICAL CHEMISTRY. Educ: Tex Tech Univ, BS, 49; Univ Calif, PhD(chem), 53. Prof Exp: Asst chem, Univ Calif, 49-50, chemist radiation lab, 50-52; res chemist, E I du Pont de Nemours & Co, 52-63, staff scientist, 63-67; dir state tech serv, 67-70, ASSOC PROF CHEM, EASTERN MICH UNIV, 70- Mem: Am Chem Soc; Am Numis Soc; Am Soc Metals;

Sigma Xi. Res: X-ray fluorescence analyses of ancient coins; diffusivity measurements in metals; equilibria and general physical chemistry of molten salts. Mailing Add: 1303 Grant St Ypsilanti MI 48197

CARTER, H KENNON, b Athens, Ga, Apr 23, 41; m 62; c 3. EXPERIMENTAL NUCLEAR PHYSICS. Educ: Univ Ga, BS, 63; La State Univ, MS, 65; Vanderbilt Univ, PhD(physics), 69. Prof Exp: Asst prof physics, Furman Univ, 68-74; SCIENTIST, OAK RIDGE ASSOC UNIVS-UNIV ISOTOPE SEPARATOR, OAK RIDGE, TENN, 74- Concurrent Pos: Fac res assoc, Univ Isotope Separator, 72-74. Mem: Am Phys Soc. Res: Nuclear spectroscopy; study of nuclei far from stability with an isotope separator on-line to a heavy-ion cyclotron. Mailing Add: Bldg 6000 Oak Ridge Nat Lab Oak Ridge TN 37830

CARTER, HARRY HART, b Ponca, Nebr, Mar 14, 21; m 46; c 7. PHYSICAL OCEANOGRAPHY. Educ: US Coast Guard Acad, BS, 43; Scripps Inst Oceanog, MS, 48. Prof Exp: Res assoc, 63-68, RES SCIENTIST, CHESAPEAKE BAY INST, JOHNS HOPKINS UNIV, 68- Concurrent Pos: Pres, Hydrocon, Inc, 70-; prof part-time, Marine Sci, State Univ NY, Stony Brook, 75- Mem: AAAS; Sigma Xi. Res: Estuarine circulation and mixing; coastal oceanography. Mailing Add: Chesapeake Bay Inst Johns Hopkins Univ Baltimore MD 21218

CARTER, HARRY NELSON, b Halleyville, Okla, Mar 7, 12; m 39. MATHEMATICS. Educ: Northeastern Okla State Col, BS, 40; Univ Colo, MS, 50. Prof Exp: Teacher pub schs, Ark & Okla, 36-41; instr math, Spartan Sch Aeronaut, 41-45; from instr to assoc prof, 45-61, asst dean eng, 58-59, actg dean, 59-60, men's counr, 60-62, dean students, 62-70, PROF MATH, UNIV TULSA, 61-, COORDR STUDENT SERV, 70- Mem: AAAS; Am Soc Eng Educ; Am Math Soc; Math Asn Am; NY Acad Sci. Res: Summation of divergent series; numerical solution of differential equations. Mailing Add: 3739 S Fulton Ave Tulsa OK 74135

CARTER, HARVEY PATE, b Friendship, Tenn, Dec 15, 27; m 51; c 1. COMPUTER SCIENCE. Educ: David Lipscomb Col, BA, 49; Vanderbilt Univ, MA, 50, PhD(math), 59. Prof Exp: Instr math, David Lipscomb Col, 50-51, assoc prof, 53-57; sr mathematician, Oak Ridge Nat Lab, 57-68, asst dir math div, 64-68; assoc prof math, Univ Tenn, Knoxville, 68-70; dir math div, Oak Ridge Nat Lab, 69-73; DIR COMPUT SCI DIV, UNION CARBIDE NUCLEAR DIV, UNION CARBIDE CORP, 73- Mem: Asn Comput Mach. Res: Applications of computer techniques to scientific and engineering problems; management of computer facilities. Mailing Add: Union Carbide Nuclear Div PO Box X Oak Ridge TN 37830

CARTER, HERBERT EDMUND, b Morresville, Ind, Sept 25, 10; m 33; c 2. CHEMISTRY, BIOCHEMISTRY. Educ: DePauw Univ, AB, 30; Univ Ill, AM, 31, PhD(chem), 34. Hon Degrees: ScD, DePauw Univ, 51, Univ Ill, 74, Univ Ind, 74; DHL, Thomas Jefferson Univ, 75. Prof Exp: From instr to prof chem, Univ Ill, Urbana-Champaign, 32-67, head dept, 54-67, acting dean grad col, 63-65, vchancellor acad affairs, 67-71; COORDR INTERDISCIPLINARY PROGS, UNIV ARIZ, 71- Concurrent Pos: Mem div chem & chem technol, Nat Res Coun, 50-55, mem-at-large & mem exec comt, 57-59; mem mgt comt, Gordon Res Conf, 53-56, mem coun, 59-; chmn biochem study sect, NIH, 54-56, chmn biochem training grant comt, 58-61; mem nat comt, Int Union Pure & Appl Chem, 55-62, chmn, 60-62; mem bd sci counr, Nat Heart Inst, 57-59; mem nat comt, Int Union Biochem, 62-65; mem, President's Comt Nat Medal Sci, 63-66; chmn sect biochem, Nat Acad Sci, 63-66, mem coun, 66-69; mem Nat Sci Bd, 64-, chmn, 70-74; mem bd trustees, Nutrit Found, 70- Honors & Awards: Lilly Award, 43; Nichols Medal, 65; Award Lipid Chemistry, Am Oil Chem Soc, 66; Kenneth A Spencer Award, 68; Alton E Bailey Award, 70. Mem: Nat Acad Sci; AAAS; Am Chem Soc; Am Soc Biol Chemists (pres, 56-57); Am Oil Chem Soc. Res: Fatty acid metabolism; biochemistry of amino acids; chemistry of streptomycin and other antibiotic substances; biochemistry of complex lipids of plants and animals; structure of bacterial lipopolysaccharides. Mailing Add: Admin Bldg Univ of Ariz Tucson AZ 85721

CARTER, HOWARD PAYNE, b Houston, Tex, Sept 9, 21; m 52; c 2. ZOOLOGY. Educ: Tuskegee Inst, BS, 41; Columbia Univ, MA, 48; Univ Calif, MA, 57; Univ Wis, PhD(protozool), 64. Prof Exp: Instr, 50-54, from asst prof to assoc prof, 54-70, PROF BIOL, TUSKEGEE INST, 70-, DEAN COL ARTS & SCI, 68- Concurrent Pos: Res assoc, Carver Res Found, 58- Mem: AAAS; Nat Inst Sci; Soc Protozool; Am Soc Parasitol; Am Micros Soc. Res: Infraciliature of ciliated protozoa; morphology of parasitic and freeliving protozoa. Mailing Add: PO Box 773 Tuskegee Institute AL 36088

CARTER, HUBERT KENNON, b Athens, Ga, Apr 23, 41; m 62; c 3. EXPERIMENTAL NUCLEAR PHYSICS. Educ: Univ Ga, BS, 63; La State Univ, MS, 65; Vanderbilt Univ, PhD(physics), 69. Prof Exp: Asst prof physics, Furman Univ, 68-74; SCIENTIST PHYSICS, UNIV ISOTOPE SEPARATOR, OAK RIDGE ASSOC UNIVS, 74- Concurrent Pos: Mem fac res, Univ Isotope Separator, Oak Ridge Assoc Univs, 72-74. Mem: Am Phys Soc; AAAS. Res: The study of nuclei far from stability with an isotope separator on-line to a heavy-ion cyclotron. Mailing Add: Bldg 6000 Oak Ridge Nat Lab Oak Ridge TN 37830

CARTER, IRVING DOYLE, b Brewer, Maine, July 4, 18; m 39; c 2. CHEMISTRY. Educ: Northeastern Univ, BS, 42. Prof Exp: Anal chemist, 42-45, res chemist, 45-53, APPL STATISTICIAN, AM CYANAMID CO, 53- Mem: Am Soc Qual Control; Am Statist Asn. Res: Statistical techniques for the design and analysis of experiments. Mailing Add: Cent Res Div Am Cyanamid Co 1937 W Main St Stamford CT 06904

CARTER, JACK FRANKLIN, b Lodgepole, Nebr, Oct 1, 19; m 41; c 5. AGRONOMY. Educ: Univ Nebr, BS, 41; State Col Wash, MS, 47; Univ Wis, PhD(agron, plant path), 50. Prof Exp: Res & teaching fel agron, State Col Wash, 41 & 46-47; res asst agron & plant path, Univ Wis, 47-50; assoc prof & assoc agronomist, 50-59, PROF AGRON & AGRONOMIST, N DAK STATE UNIV, 59-, CHMN DEPT, 60- Mem: Am Soc Agron; Crop Sci Soc Am (pres, 72-73). Res: Forage crop production and management; pasture research; forage crop diseases. Mailing Add: 1345 11th St N Fargo ND 58102

CARTER, JACK LEE, b Kansas City, Kans, Jan 23, 29; m 51; c 3. BOTANY. Educ: Kans State Teachers Col, BS, 50, MS, 54; Univ Iowa, PhD(bot), 60. Prof Exp: Assoc prof biol & head dept, Northwestern Col, Iowa, 55-58 & Simpson Col, 60-62; assoc prof & coordr res & inst grants, Kans State Teachers Col, 62-66; assoc dir biol sci curric study, Univ Colo, 66-68; PROF BIOL, COLO COL, 68- Concurrent Pos: Ed, Am Biol Teacher, 70-74. Mem: AAAS; Nat Asn Biol Teachers; Bot Soc Am; Am Soc Plant Taxonomists; Am Inst Biol Sci. Res: Developmental and systematic botany. Mailing Add: Dept of Biol Colo Col Colorado Springs CO 80903

CARTER, JAMES CEDRIC, b Wingate, Ind, Oct 25, 05; m 34; c 1. PLANT PATHOLOGY. Educ: Purdue Univ, BS, 28, MS, 32, PhD(plant path), 34. Prof Exp: Teacher high sch, Ind, 28-30; asst plant path, Purdue Univ, 31-34; asst botanist, Ill Natural Hist Surv, 34-46, assoc botanist, 46-47, plant pathologist, 47-74; head sect bot & plant path, 55-74, prof plant path, 55-74, EMER PROF PLANT PATH, UNIV ILL, URBANA-CHAMPAIGN, 74- Honors & Awards: Citations, Nat Arborist Asn, 45 & Midwest chap, Int Shade Tree Conf, 67. Mem: AAAS; Am Phytopath Soc; Mycol Soc Am; Am Inst Biol Sci; Int Shade Tree Conf. Res: Diffusable nature of the inhibitory agent produced by fungi. Mailing Add: 709 E Main St Crawfordsville IN 47933

CARTER, JAMES CLARENCE, b New York, NY, Aug 1, 27. THEORETICAL PHYSICS. Educ: Spring Hill Col, BS, 52; Fordham Univ, MS, 53; Catholic Univ, PhD(physics), 56; Woodstock Col, STL, 59. Prof Exp: From instr to asst prof, 60-66, acad vpres, 70-74, ASSOC PROF PHYSICS, LOYOLA UNIV, LA, 66-, PRESIDENT, 74- Concurrent Pos: NSF exten grant, 63-64; NSF res grant, Loyola Univ, La, 66-70; Nat Acad Sci-Nat Res Coun resident res assoc, Nat Bur Standards, 64-65. Mem: Am Phys Soc; Am Asn Physics Teachers. Res: Nuclear structure; symmetries of elementary particles. Mailing Add: President Loyola Univ New Orleans LA 70118

CARTER, JAMES CLYDE, b Hardin, Mo, Aug 19, 31; m 56; c 2. INORGANIC CHEMISTRY. Educ: Univ Okla, BS, 53; Univ Mich, MS, 55, PhD(chem), 61. Prof Exp: Instr chem, Univ Mich, 60-61; asst prof, 61-64, ASSOC PROF CHEM, UNIV PITTSBURGH, 64- Concurrent Pos: Consult, Kelsey-Hayes Co, 61-62; mem comt inorg nomenclature, Nat Acad Sci-Nat Res Coun. Mem: AAAS; Am Chem Soc; Am Inst Aeronaut & Astronaut. Res: Boron hydrides and their derivates; selective reducing agents; ion propulsion systems; halogens; nuclear magnetic resonance spectrometry; vibrational spectroscopy; alkaloid chemistry; immunology and cancer chemotherapy. Mailing Add: Dept of Chem Univ of Pittsburgh Pittsburgh PA 15213

CARTER, JAMES EDWARD, b Great Neck, NY, Nov 3, 29; m 51; c 3. OBSTETRICS & GYNECOLOGY, PATHOLOGY. Educ: Univ Vt, BA, 51; NY Med Col, MD, 55. Prof Exp: Assoc prof, 70-73, PROF OBSTET & GYNEC & PATH, SCH MED, IND UNIV, INDIANAPOLIS, 73-, ASSOC DEAN STUDENT AFFAIRS, 73- Mem: AAAS; Am Col Obstet & Gynec; Electron Micros Soc Am. Mailing Add: Ind Univ Sch of Med 1100 W Michigan St Indianapolis IN 46202

CARTER, JAMES HORACE, organic chemistry, polymer chemistry, see 12th edition

CARTER, JAMES M, b Cass Co, Mich, Dec 7, 21; m 51; c 4. VETERINARY PHYSIOLOGY. Educ: Mich State Univ, DVM, 51; Purdue Univ, MS, 63, PhD(physiol), 65. Prof Exp: Pvt vet practr, Ind, 51-61; Am Vet Med Asn fel, 61-63, asst, 63-65, asst prof, 65-68, ASSOC PROF VET PHYSIOL, PURDUE UNIV, WEST LAFAYETTE, 68- Mem: Am Vet Med Asn; Am Soc Vet Physiol & Pharmacol. Res: Renal and respiratory physiology, especially fluid and electrolyte and acid-base balance studies. Mailing Add: Sch of Vet Med Purdue Univ West Lafayette IN 47906

CARTER, JAMES ROLAND, b Fulton, Ky, Aug 26, 42; m 69; c 2. PHARMACOLOGY. Educ: Murray State Univ, BA, 64; Univ Tenn, MS, 65, PhD(pharmacol), 67, MD, 70. Prof Exp: Instr pharmacol, Univ Tenn, 67-70; intern internal med, Vanderbilt Univ Hosp, 70-71; resident anesthesiol, Sch Med, Univ Ala, Birmingham, 71-73; asst prof anesthesiol & pharmacol, 73-75, ASSOC PROF ANESTHESIOL, COL MED, UNIV TENN, MEMPHIS, 75- Res: Drug metabolism; endocrinology; clinical pharmacology; anesthesiology. Mailing Add: 774 West Dr Memphis TN 38112

CARTER, JOHN C H, b Toronto, Ont, Sept 16, 25; m 49. MARINE BIOLOGY, FRESH WATER BIOLOGY. Educ: Univ Toronto, BA, 48; McGill Univ, MSc, 61, PhD(zool), 63. Prof Exp: Res fel, Systs & Ecol Prog, Marine Biol Lab, 63-65; ASSOC PROF BIOL, UNIV WATERLOO, 65- Mem: Am Soc Limnol & Oceanog. Res: Hydrography and plankton of landlocked fiords of Canadian Arctic; Arctic plankton; ecology of estuarine and freshwater copepods. Mailing Add: Dept of Biol Univ of Waterloo Waterloo ON Can

CARTER, JOHN HAAS, b Trevorton, Pa, Oct 25, 30; m 54; c 3. OPTOMETRY. Educ: Pa State Col Optom, OD, 53; Univ Ind, MS, 59, PhD, 62. Prof Exp: Asst div optom, Univ Ind, 57-59, res asst, 59-62; res assoc prof physiol optics, Pa Col Optom, 62-66, res prof, 66-70; dir visual sci div, 71-73, PROF PHYSIOL OPTICS, MASS COL OPTOM, 70-, ACAD DEAN, 73- Mem: Fel AAAS; fel Am Acad Optom; Am Optom Asn. Res: Physiological optics; infrared self-recording refractionometer; nature of stimulus to accomodative mechanism of the eye. Mailing Add: Mass Col Optom 424 Beacon St Boston MA 02115

CARTER, JOHN LEMUEL, JR, b Clarksville, Tex, Mar 17, 20; m 47; c 3. PHYSICS. Educ: Baylor Univ, BA, 41; Brown Univ, MSc, 43; Cornell Univ, PhD(physics), 53. Prof Exp: Engr sound div, US Naval Res Lab, 43-45; engr, US Naval Underwater Sound Reference Lab, 45-47; asst physics, Cornell Univ, 47-52; engr measurements lab, Gen Elec Co, Mass, 52-55, physicist, Hanford Atomic Prod Oper, 55-58, supvr theoret physics res unit, Hanford Labs, 58-63, tech specialist, 63-64; RES ASSOC, REACTOR PHYSICS DEPT, PAC NORTHWEST LAB, BATTELLE MEM INST, 65- Mem: Am Phys Soc; Am Nuclear Soc. Res: Nuclear reactor physics; solid state theory; digital computer codes. Mailing Add: 78 McMurray Rd Richland WA 99352

CARTER, JOHN LYMAN, b Sisseton, SDak, Mar 14, 34; m 63. INVERTEBRATE PALEONTOLOGY. Educ: Univ NDak, BS, 59; Univ Cincinnati, PhD(geol), 66. Prof Exp: Res assoc paleont & cur, Univ Ill, Urbana-Champaign, 66-72; ASSOC CUR, CARNEGIE MUS NATURAL HIST, 72- Mem: Paleont Soc; Soc Syst Zool; Brit Paleont Asn; Int Paleont Union; Soc Econ Paleont & Mineral. Res: Late Paleozoic Brachiopoda and biostratigraphy. Mailing Add: Carnegie Mus of Nat Hist 4400 Forbes Ave Pittsburgh PA 15213

CARTER, JOHN NEWTON, b Columbia, Mo, Jan 24, 21; m 48; c 3. AGRONOMY. Educ: Univ Mo, BS, 43; Univ Ill, MS, 48, PhD(soil fertil), 50. Prof Exp: Spec res asst agron, Univ Ill, 47-50; prin soil scientist, Battelle Mem Inst, 51-55; SOIL SCIENTIST SOIL & WATER CONSERV, AGR RES SERV, USDA, 55- Mem: Am Soc Agron; Soil Sci Soc Am; Am Soc Sugar Beet Technologists. Res: Sugarbeet nitrogen nutrition; soil fertility; soil nitrogen transformations; soil organic matter; plant nutrition. Mailing Add: Snake River Conserv Res Sta USDA-ARS Rt 1 Box 186 Kimberly ID 83341

CARTER, JOHN ROBERT, b Buffalo, NY, Apr 21, 17; m 43; c 2. PATHOLOGY. Educ: Hamilton Col, BS, 39; Univ Rochester, MD, 43. Prof Exp: Asst path, Univ Iowa, 44, from instr to prof, 44-59; prof path & oncol & chmn dept, Med Ctr, Univ Kans, 59-66; PROF PATH, CHMN DEPT & DIR INST PATH, CASE WESTERN RESERVE UNIV, 66- Concurrent Pos: Dir path, Univ Hosps, Cleveland, 66- Mem: Am Soc Clin Path; Am Soc Exp Path; Col Am Path; Am Asn Path & Bact (treas, 61-62, secy-treas, 62-65, vpres, 66-67, pres, 67-68); Int Acad Path. Res: Blood coagulation. Mailing Add: Inst of Path Case Western Reserve Univ Cleveland OH 44106

CARTER, JOHN VERNON, b Boise, Idaho, Dec 21, 40; m 64; c 3. BIOPHYSICAL CHEMISTRY. Educ: Whittier Col, BA, 62; Purdue Univ, PhD(org chem), 67. Prof Exp: Res scientist org chem, Koppers Co, Inc, Pa, 67-68; res fel, Univ Pittsburgh, 68-

70; from asst prof to assoc prof chem, Adams State Col, 70-74; RES ASSOC CHEM, UNIV MINN, MINNEAPOLIS, 74- Mem: Am Chem Soc. Res: Protein formation in solution. Mailing Add: Dept of Chem Univ of Minn Minneapolis MN 55455

CARTER, JOSEPH OMER, mathematics, see 12th edition

CARTER, KENNETH, b Morecambe, Eng, Feb 6, 14; m 40, 70; c 3. MEDICINE, PHARMACEUTICAL CHEMISTRY. Educ: Univ London, MRCS & LRCP, 47. Prof Exp: Res chemist, Glaxo Labs, Eng, 38-42, med dir, Buenos Aires, 48-50, dir, Qual Control & Prod, 49-57; res liaison officer & secy, Therapeut Res Corp Gt Brit, 46-48; dir develop, Smith Kline & French, Pa, 52-54, med dir & dir develop, Eng, 54-57; sci dir, Ames Co, Ind, 57-58, vpres res & med affairs, 58-62, Miles Labs, Inc, 62-65; vpres & dir med affairs, Syntex Int, 65-71, CORP VPRES REGULATORY AFFAIRS, SYNTEX CORP, 71- Concurrent Pos: House physician, Postgrad Med Sch, Univ London, 47-48; mem, Royal Col Surgeons; mem, Bd Dirs, Royal Soc Med Found, Inc, 68- Mem: Pharmaceut Soc Gt Brit; fel Royal Soc Med; Brit Med Asn; Brit Harvein Soc. Res: Administration of research, development and medical operations; disciplines necessary for developing new diagnostic and therapeutic agents, such as chemistry, biochemistry, immunology, toxicology, pharmacology, clinical pharmacology and clinical research; government regulatory affairs. Mailing Add: 24612 Olive Tree Lane Los Altos Hills CA 94022

CARTER, KENNETH NOLON, b Columbia, SC, Oct 2, 25; m 54; c 1. ORGANIC CHEMISTRY. Educ: Erskine Col, AB, 47; Vanderbilt Univ, MS, 49, PhD(chem), 51. Prof Exp: HEAD DEPT CHEM, PRESBY COL, 51-, CHARLES A DANA PROF, 70- Mem: Am Chem Soc. Res: Organic synthesis; molecular rearrangements. Mailing Add: Dept of Chem Presby Col Clinton SC 29325

CARTER, LARK POLAND, b Lytton, Iowa, June 26, 30; m 54; c 2. AGRONOMY. Educ: Iowa State Univ, BS, 53, MS, 56, PhD(agron), 60. Prof Exp: Asst prof agron, 60-62, asst dean, Col Agr, 62-65, ASSOC DEAN COL AGR & ASST DIR AGR EXP STA, MONT STATE UNIV, 65- Mem: Am Soc Agron. Res: Field crop production; forage management; seed certification; grass and legume seed production. Mailing Add: Col of Agr Mont State Univ Bozeman MT 59715

CARTER, LELAND LAVELLE, b Oberlin, Kans, Nov 27, 37; m 58; c 4. REACTOR PHYSICS, NUCLEAR ENGINEERING. Educ: Northwest Nazarene Col, BA, 61; Univ Wash, MS, 64, PhD(nuclear eng), 69. Prof Exp: Physicist, Gen Elec Co, 62-63; physicist, Pac Northwest Labs, Battelle Mem Inst, 64-65; nuclear engr, 69-73, ALT GROUP LEADER NEUTRON TRANSPORT, LOS ALAMOS SCI LAB, UNIV CALIF, 74- Mem: Am Nuclear Soc. Res: Application of the Monte Carlo method to solve three-dimensional particle transport problems on the digital computer, including adjoint simulation, neutron cross-sections, nonlinear radiative transport, criticality and unbiased sampling schemes. Mailing Add: Group TD-6 Los Alamos Sci Lab Los Alamos NM 87545

CARTER, LOREN SHELDON, b Nampa, Idaho, Jan 10, 39; m 68. PHYSICAL CHEMISTRY, INORGANIC CHEMISTRY. Educ: Ore State Univ, BS, 61, MS, 66; Wash State Univ, PhD(phys chem), 70. Prof Exp: Reactor engr, Phillips Petrol Co, 61-63; asst prof, 70-74, ASSOC PROF CHEM, BOISE STATE COL, 74- Mem: AAAS; Am Chem Soc. Res: Surface adsorption on clays; transference number measurements using a centrifuge in organic solvents. Mailing Add: Dept of Chem Boise State Col Boise ID 83707

CARTER, MARY EDDIE, b Americus, Ga, March 14, 25. ORGANIC CHEMISTRY. Educ: La Grange Col, BA, 46; Univ Fla, MS, 49; Univ Edinburgh, PhD, 56. Prof Exp: Instr chem, La Grange Col, 46-47; microscopist, Calloway Mills, 47-48; textile chemist, Southern Res Inst, 49-51; chemist, West Point Mfg Co, 51-53; res chemist, Am Viscose Corp, 56-62, res assoc, FMC Corp, Am Viscose Div, Res & Develop, 62-71; CHIEF TEXTILES & CLOTHING LAB, SMNRD, AGR RES SERV, US DEPT AGR, 71- Mem: Am Chem Soc; Am Asn Textile Chemists & Colorists; Inter-Soc Color Coun; Fiber Soc; Sigma Xi. Res: Naturally occurring polymers for fiber, food and feed, including safety, nutrition and processing aspects. Mailing Add: PO Box 19687 New Orleans LA 70179

CARTER, MARY KATHLEEN, b Franklinton, La, July 11, 22. PHARMACOLOGY. Educ: Tulane Univ, BA, 49, MS, 53; Vanderbilt Univ, PhD(pharmacol), 55. Prof Exp: Asst pharmacol, Tulane Univ, 49-52; asst, Vanderbilt Univ, 52-53; res assoc, Med Ctr, Univ Kans, 55-57; from instr to assoc prof, 57-73, PROF PHARMACOL, MED SCH, TULANE UNIV, 73- Concurrent Pos: USPHS fel, Med Ctr, Univ Kans, 55-57, sr res fel, Med Sch, Tulane Univ, 57-61. Mem: AAAS; Soc Exp Biol & Med; Am Soc Pharmacol & Exp Therapeut; Am Soc Nephrology. Res: Renal pharmacology; transport of electrolytes and sugars; cholinesterases. Mailing Add: Dept of Pharmacol Tulane Univ Sch of Med New Orleans LA 70112

CARTER, MASON CARLTON, b Wash, DC, Jan 14, 33; m 53. PLANT PHYSIOLOGY, FORESTRY. Educ: Va Polytech Inst, BS, 55, MS, 57; Duke Univ, PhD(forestry), 59. Prof Exp: Asst plant physiol, Va Polytech Inst, 55-56; res forester, Southeastern Forest Exp Sta, USDA, 59-60; asst prof forestry, Auburn Univ, 60-66, assoc prof forestry & bot, 66-70, alumni assoc prof, 70-72, alumni prof, 72-73; PROF FORESTRY & CONSERV & CHMN DEPT, PURDUE UNIV, WEST LAFAYETTE, 73- Mem: AAAS; Am Soc Plant Physiol; Soc Am Foresters. Res: Mechanisms of herbicidal action; physiology of woody plants. Mailing Add: Dept of Forestry & Conserv Purdue Univ West Lafayette IN 47906

CARTER, MELVIN K, b Alameda, Calif, Dec 10, 37; m 62; c 3. PHYSICAL CHEMISTRY, ENVIRONMENTAL CHEMISTRY. Educ: San Jose State Col, BA, 59, BS, 60; Univ Wash, PhD(phys chem), 66. Prof Exp: Chemist, Shell Develop Co, 66-72; dir, CSW Enterprises, 69-72; assoc prof chem, Porterville Col, 72; sr res fel, Nat Res Coun, Ames Res Ctr, 72-73; MGR APPLN LAB, DOHRMANN DIV, ENVIROTECH CORP, SANTA CLARA, 73- Concurrent Pos: Dir res & develop, Mat Metrics Inc, 73; chmn bd, Actinic Spectra Inc, 75-; prof chem, De Anza Col, 75- Mem: Am Chem Soc; Am Soc Test & Mat; Instrument Soc Am. Res: New analytical pollution measuring instrumentation; production of synthetic oil and coal from vegetation cellulose; high energy density solar power storage. Mailing Add: 19807 Colby Ct Saratoga CA 95070

CARTER, MELVIN WINSOR, b Phoenix, Ariz, Jan 22, 28; m 50; c 5. APPLIED STATISTICS, BIOMETRICS. Educ: Ariz State Univ, BS, 53; NC State Col, MS, 54, PhD, 56. Prof Exp: Asst nutrit, NC State Col, 52-56; asst prof statist, Purdue Univ, 56-58; exp statist, NC State Col, 58-61; assoc prof, 61-66, PROF STATIST, BRIGHAM YOUNG UNIV, 66- Concurrent Pos: Vis prof, NC State Univ, 67-68 & Texas A&M Univ, 71; biostatistician, Upjohn Co, 75. Mem: Am Statist Asn; Biomet Soc. Res: Biometrics; nutrition; endocrinology; biochemistry; methodology for the analysis of unbalanced designs. Mailing Add: Dept of Statist Brigham Young Univ Provo UT 84601

CARTER, NEVILLE LOUIS, b Los Angeles, Calif, Aug 21, 34; m 56; c 3. GEOLOGY, GEOPHYSICS. Educ: Pomona Col, AB, 56; Univ Calif, Los Angeles, MA, 58, PhD(geol), 63. Prof Exp: Res geophysicist, Univ Calif, Los Angeles, 61-63, asst, 63; res geologist, Shell Develop Co, 63-66; assoc prof geol & geophys, Yale Univ, 66-71; PROF GEOPHYS, EARTH & SPACE SCI, STATE UNIV NY STONY BROOK, 71- Mem: Am Geophys Union; Geol Soc Am. Res: Tectonophysics; experimental and natural deformation of rocks and minerals. Mailing Add: Dept of Earth & Space Sci State Univ of NY Stony Brook NY 11794

CARTER, ORWIN LEE, b Geneseo, Ill, Aug 22, 42; m 66; c 2. PHYSICAL CHEMISTRY. Educ: Univ Iowa, BS, 64; Univ Ill, MS, 65, PhD(x-ray crystallog), 67; Rider Col, MBA, 75. Prof Exp: Asst prof phys chem, US Mil Acad, 67-70; from scientist to sr scientist, Rohm & Haas Co, 70-74; DIR PROD DEVELOP, MICROMEDIC SYSTS, INC, 74- Mem: Am Chem Soc; Am Asn Clin Chemists. Res: Immunoassay by radioisotope techniques; enzymatic assays in clinical chemistry. Mailing Add: 102 Witmer Rd Horsham PA 19044

CARTER, OWEN, JR, b Sligo, La, Aug 18, 15; m 44; c 3. PHYSICAL CHEMISTRY. Educ: Centenary Col, BS, 35; Univ Wis, PhD(phys chem), 39. Prof Exp: Sr chemist, Procter & Gamble Co, 39-41; res assoc, Nat Defense Res Comt Proj, Univ Wis, 41-44; SR CHEMIST, PROCTER & GAMBLE CO, 44- Mem: Am Chem Soc; Am Acad Dermat; Soc Invest Dermat; Sigma Xi. Res: Molecular-kinetic study of proteins; gelatinization of high polymers; fats and oils; soaps and detergents; toxicology and safety evaluation. Mailing Add: 6309 Lisbon Cincinnati OH 45213

CARTER, PAUL BEARNSON, b Spanish Fork, Utah, Feb 17, 18; m 42; c 1. BACTERIOLOGY. Educ: Univ Utah, BS, 48, MS, 50, PhD(bact), 55. Prof Exp: Instr bact, Univ Utah, 50-55, res instr anat, 55-56; asst prof biol, 56-62, ASSOC PROF BIOL, UTAH STATE UNIV, 62- Mem: Am Soc Microbiol; NY Acad Sci; Sigma Xi. Res: Pathogenic bacteriology; medical mycology; immunology; inflammation. Mailing Add: Dept of Biol Utah State Univ Logan UT 84322

CARTER, PAUL RICHARD, b St Louis, Mo, Apr 14, 22; m 44; c 2. MEDICINE, SURGERY. Educ: Union Col, Nebr, BA, 44; Loma Linda Univ, MD, 48; Am Bd Surg, dipl; Am Bd Thoracic Surg, dipl. Prof Exp: Surg resident, Los Angeles County Gen Hosp, 50-52, 54-56; thoracic surg resident, Olive View Sanitarium, Calif, 56-57; head physician surg, Los Angeles County Gen Hosp, 57-66; assoc prof surg, Univ Calif-Calif Col Med, 66-68; assoc prof in residence, 68-70, PROF SURG, COL MED, UNIV CALIF, IRVINE, 70-; CHIEF SURG, ORANGE COUNTY MED CTR, 70- Concurrent Pos: Fulbright scholar, Oxford Univ, 59; assoc prof surg, Sch Med, Loma Linda Univ, 62-66; chief surg, Rancho Los Amigos Hosp, 66-68; sr attend surgeon, Los Angeles County Gen Hosp & White Mem Hosp, Los Angeles; mem sr staff, Intercommunity Hosp, Covina; mem active staff, Queen of the Valley Hosp, West Covina. Mem: Fel Am Col Surg; Am Col Chest Physicians; Am Asn Thoracic Surg; Soc Thoracic Surg. Res: Mesenteric vascular occlusion; subphrenic abscess; volvulus of the colon and gall-bladder; pseudotumors of the lung; traumatic thoracic injuries; gallstone obstruction; surgical significance of sternal fracture; rupture of the bronchus; segmental reversal of small intestine; use of the Celestin tube for inoperable carcinoma of the esophagus and cardia; bronchotomy; diaphragmatic hernia. Mailing Add: 227 W Badillo Covina CA 91723

CARTER, PAUL RICHARD, b Brooklyn, NY, Apr 26, 27; m 50; c 3. Educ: Brooklyn Col, BS, 50; Calif Inst Technol, PhD(chem), 53. Prof Exp: Res chemist titanium div, Nat Lead Co, 52-55; HEAD SURFACE CHEM SECT, RES LAB, US STEEL CORP, 56- Mem: Electrochem Soc. Res: Corrosion; surface treatment of metals; electrochemistry; kinetics. Mailing Add: 469A Newell Lane Trafford PA 15085

CARTER, RICHARD LESTON, b Boston, Mass, Apr 30, 22; m 54. MATHEMATICAL STATISTICS. Educ: Mass Inst Technol, SB, 47; Southern Methodist Univ, MS, 54; Univ NC, PhD(math statist), 57. Prof Exp: Chem engr, Carter's Ink Co, Mass, 47-49; res assoc geochem, Boston Univ, 49-51; statist engr, Res Inc, Tex, 51-54; asst, Univ NC, 55-57; assoc prof indust eng, Ill Inst Technol, 57-60; PROF MGT, RENSSELAER POLYTECH INST, 60- Concurrent Pos: Asst dir, N C Opers Anal Unit, USAF, 56-57; statist consult, Parmly Found Auditory Res, Ill, 58-60. Mem: Inst Math Statist; Am Statist Asn; Inst Indust Eng; Biomet Soc; Inst Mgt Sci. Res: Mathematical statistics and programming; design and analysis of experiments; applications to geochemistry, mineral exploration and psychoacoustics; industrial engineering; operations research; computers and data processing. Mailing Add: Sch of Mgt Rensselaer Polytechnic Inst Troy NY 12181

CARTER, RICHARD POWELL, JR, inorganic chemistry, see 12th edition

CARTER, RICHARD THOMAS, b Portland, Ore, Apr 4, 36; m 57; c 3. PARASITOLOGY, INVERTEBRATE ZOOLOGY. Educ: Portland State Univ, BS, 63; Ore State Univ, MA, 67, PhD(parasitol), 73. Prof Exp: Instr biol, Portland State Col, 62-63; asst prof, 68-73, ASSOC PROF BIOL, PAC UNIV, 73-, CHMN DEPT, 72- Mem: Sigma Xi. Res: Investigations into life stage development in Nanophyetus salmincola as related to life history. Mailing Add: Dept of Biol Pac Univ Forest Grove OR 97116

CARTER, ROBERT CLIFTON, b Gate City, Va, June 29, 09; m 40. ANIMAL BREEDING, GENETICS. Educ: Va Polytech Inst, BS, 31; Iowa State Col, MS, 39, PhD(animal breeding), 56. Prof Exp: County agr agent, State Agr Exten Serv, Va, 31-40 & 42-47; secy, Va Cattleman's Asn, 47-48; PROF ANIMAL SCI, VA POLYTECH INST & STATE UNIV, 48- Honors & Awards: Distinguished Serv Award, Am Soc Animal Sci, 70, Animal Breeding & Genetics Award, 74. Mem: Genetics Soc Am; Am Genetics Asn (secy, 67-69, vpres, 70, pres, 71); Am Inst Biol Sci; fel AAAS; Am Soc Animal Sci. Res: Estimates of genetic parameters; comparison of breeding systems; systems of animal production; beef cattle and sheep. Mailing Add: 25 Agnew Hall Va Polytech Inst & State Univ Blacksburg VA 24061

CARTER, ROBERT DUNCAN, b Twin Falls, Idaho, Mar 2, 29; m 50; c 3. ENTOMOLOGY. Educ: Univ Calif, BS, 50, PhD(entom), 54. Prof Exp: Entomologist, Bur Entom, Calif State Dept Agr, 50; entomologist, 55-70, MGR ENTOM RES, AGR RES CTR, DEL MONTE CORP, 70- Mem: Entom Soc Am (sec-treas, 72-77); Am Phytopath Soc. Res: Arthropods in relation to plant disease; agricultural entomology. Mailing Add: Box 36 San Leandro CA 94577

CARTER, ROBERT ELDRED, b Minneapolis, Minn, July 14, 23; m 46; c 3. PEDIATRICS, HEMATOLOGY. Educ: Univ Minn, BS, 45, MB, 46, MD, 48. Prof Exp: From instr to asst prof pediat, Univ Chicago, 56-59; from asst prof to prof, Univ Iowa, 59-67, from asst dean to assoc dean, 61-67; prof pediat, dean & dir, Sch Med, Univ Miss, 67-70; DEAN MED EDUC PROG, UNIV MINN, DULUTH, 70- Concurrent Pos: John & Mary R Markle scholar med sci, 57-62. Mem: Soc Pediat Res. Res: Biological effects of ionizing radiations; general hematology and bone marrow function. Mailing Add: Med Educ Prog Univ of Minn Duluth MN 55812

CARTER, ROBERT EMERSON, b Philadelphia, Pa, Feb 3, 20; m 47; c 11. PHYSICS. Educ: Washington Col, BS, 42; Univ Ill, MS, 47. Prof Exp: Jr physicist, Off Sci Res &

Develop, Purdue Univ, 42-43, instr eng sci & war mgt training, 42-43; jr physicist, Los Alamos Sci Lab, Univ Calif, 43-45; asst nuclear res, Univ Ill, 46-48; mem staff, Los Alamos Sci Lab, 48-63; CHMN PHYS SCI DEPT, ARMED FORCES RADIOBIOL RES INST, 63- Concurrent Pos: Consult, AEC, 47-48, Dept Army, 59-63; instr, Los Alamos Grad Ctr, Univ NMex, 56-63; mem, Subcomt Res Reactors Comt Phys Sci, Nat Acad Sci-Nat Res Coun, 64-69. Mem: Am Nuclear Soc; Am Phys Soc; AAAS. Res: Nuclear physics; molecular and atomic physics; mathematical physics; electronics; radiation physics; medical physics; radiation biology; on-line digital computers. Mailing Add: 9512 Edgeley Rd Bethesda MD 20014

CARTER, ROBERT EVERETT, b Jamaica, NY, Dec 3, 37; m 62; c 3. PHYSICAL ORGANIC CHEMISTRY. Educ: Columbia Col, BA, 58; Calif Inst Technol, PhD(org chem), 62; Gothenburg Univ, Fil Dr, 70. Prof Exp: NIH fel gen med sci, Inst Org Chem, Gothenburg Univ, 62-65, lectr org chem, 64-65; consult, A B Hassle Co, 65-67; res assoc, 68-70, DOCENT, DIV ORG CHEM, LUND INST TECHNOL, 70- Res: Kinetic deuterium effects; applications of nuclear magnetic resonance spectroscopy to physical organic chemistry. Mailing Add: Div of Org Chem Chem Ctr Lund Inst of Technol Lund Sweden

CARTER, ROBERT LEONIDAS, b Atlanta, Ga, Aug 31, 09; m 34; c 2. SOIL FERTILITY. Educ: Univ Ga, BS, 31, MS, 33. Prof Exp: Ed adv, USDA, 35-46; SOIL SCIENTIST, GA COASTAL PLAIN EXP STA & PROF AGRON, UNIV GA, 46- Mem: Am Soc Agron; Am Chem Soc; Soil Conserv Soc Am. Res: Soil classification and survey; soil conservation and soil fertility research; soil and plant micronutrient research. Mailing Add: 1207 N Central Ave Tifton GA 31794

CARTER, ROBERT SAGUE, b Poughkeepsie, NY, Nov 15, 25; m 49; c 4. NUCLEAR PHYSICS. Educ: Princeton Univ, AB, 48; Harvard Univ, MA, 49, PhD(physics), 52. Prof Exp: Assoc physicist, Brookhaven Nat Lab, 52-56; physicist, Westinghouse Res Labs, 56-58; physicist, 58-69, CHIEF REACTOR RADIATION DIV, NAT BUR STANDARDS, 69- Mem: Am Phys Soc. Res: Elementary particle physics; neutron physics. Mailing Add: 14710 Pettit Way Potomac MD 20854

CARTER, ROY MERWIN, b Mauston, Wis, Jan 5, 13; m 40; c 2. WOOD TECHNOLOGY. Educ: Univ Minn, BSF, 35; Mich State Univ, MS, 39. Prof Exp: Asst, Mich State Univ, 36-38; dist forester, Wis State Dept Conserv, 38-39; instr & exten forester, Univ Wis, 39-42; chief procurement inspector woodcraft, Army Air Force, 42-43; wood process engr, Fairchild Aircraft Corp, 43-44; forest utilization specialist, US Forest Serv, 44-48; PROF WOOD TECHNOL, SCH FOREST RESOURCES, NC STATE UNIV, 48- Mem: Soc Am Foresters; Forest Prod Res Soc (2nd vpres, 49, 1st vpres, 50, pres, 51). Res: Wood moisture relations; wood preservation and finishing; gluing; plant operations and processes for wood industries. Mailing Add: Dept of Wood & Paper Sci NC State Univ Sch Forest Res Raleigh NC 27607

CARTER, SIDNEY, b Boston, Mass, Dec 8, 12; m 45; c 3. NEUROLOGY. Educ: Dartmouth Col, AB, 34; Boston Univ, MD, 38. Prof Exp: Intern, St Mary's Hosp, Waterbury, Conn, 38-39; resident psychiat, Westboro State Hosp, Mass, 39-40; resident neurol, Boston City Hosp, 40-42; asst, Harvard Med Sch, 41-42; from instr to assoc prof, 47-60, PROF NEUROL, COLUMBIA UNIV, 60-, CHIEF DIV CHILD NEUROL, COLUMBIA-PRESBY MED CTR, 54- Concurrent Pos: Asst, Div Neuropsychiat, Montefiore Hosp, 46-47, adj attend neurologist, 48-52; from asst attend neurologist to attend neurologist, Columbia-Presby Hosp, 48-; attend neurologist, Lawrence Hosp, Bronxville, 52-; consult, Morristown Mem Hosp, NJ, 53- Mem: Am Psychiat Asn; Am Acad Neurol; AMA; Am Acad Cerebral Palsy; Int League Against Epilepsy. Res: Epilepsy; pediatric neurology. Mailing Add: 710 W 168th St New York NY 10032

CARTER, STEFAN A, b Warsaw, Poland, Mar 25, 28; Can citizen; m 58; c 2. PHYSIOLOGY. Educ: Univ Man, MD & BSc, 54, MSc, 56. Prof Exp: Lectr, 58-59, asst prof, 59-67, ASSOC PROF PHYSIOL, FAC MED, UNIV MAN, 67- Concurrent Pos: Clin res fel, NY Hosp-Cornell Med Ctr, 56-67; Nat Res Coun Can fel physiol, Mayo Found, Univ Minn, 57-58; Man Heart Found grant-in-aid, 59-; dir cardiovasc sect, Clin Invest Unit, St Boniface Gen Hosp, 58-; dir long term anticoagulant & peripheral vascular dis clin, St Boniface Hosp, 67-; mem coun circulation, Am Heart Asn, 69. Mem: Can Physiol Soc; Can Soc Clin Invest; Can Cardiovasc Soc. Res: Vasomotor regulation in diabetes; function and elasticity of medium-sized arteries and arterial pressure pulses in health and arterial diseases; arterial occlusive disease in the limbs. Mailing Add: Dept of Physiol Fac of Med Univ of Man Winnipeg MB Can

CARTER, STEPHEN KEITH, b New York, NY, Oct 30, 37; m 66; c 2. INTERNAL MEDICINE, ONCOLOGY. Educ: Columbia Col, AB, 59; New York Med Col, MD, 63. Prof Exp: Intern med, Lenox Hill Hosp, 63-64, resident, 64-66, chief resident, 66-67; spec asst to sci dir chemother, 67-68, chief cancer ther eval br, 68-73, assoc dir cancer ther eval, 73-74, DEP DIR DIV CANCER TREATMENT, NAT CANCER INST, 74- Concurrent Pos: Exec secy, Clin Trials Task Force, Nat Cancer Inst, 67, Nat Brain Tumor Study Group, 69- Mem: NY Acad Sci. Res: Monitoring clinical trials of new chemical agents developed to fight cancer. Mailing Add: Div of Cancer Treatment Nat Cancer Inst Rm 3A51 Bldg 31 Bethesda MD 20014

CARTER, WALTER HANSBROUGH, JR, b Winchester, Va, Mar 20, 41; m 63; c 2. MATHEMATICAL STATISTICS. Educ: Univ Richmond, BS, 63; Va Polytech Inst, MS, 66, PhD(statist), 68. Prof Exp: Asst prof, 68-74, ASSOC PROF BIOMET, MED COL VA, VA COMMONWEALTH UNIV, 74- Res: Estimation from combinations of distributions; test theory; legal application of statistical methods; grouped regression; multiple response ridge analysis. Mailing Add: Dept of Biomet Med Col of Va Richmond VA 23219

CARTER, WILLIAM ALFRED, b Ada, Okla, Sept 16, 35. ZOOLOGY, ORNITHOLOGY. Educ: ECent State Col, BS, 57; Okla State Univ, MS, 60, PhD(zool), 65. Prof Exp: Instr biol, Northwestern State Col, Okla, 63-64; from asst prof to assoc prof, 64-71, PROF BIOL, E CENT OKLA STATE UNIV, 71-, CHMN DEPT, 72- Concurrent Pos: Res assoc ornith, Stovall Mus, Univ Okla, 71-; Okla coordr, US Fish & Wildlife Breeding Bird Surv, 74-; consult, Environ Impact Assessment, Williams Bros Eng Co, 75 & 76. Mem: Am Ornithologists Union; Am Soc Ichthyologists & Herpetologists; Cooper Ornith Soc; Wilson Ornith Soc; Sigma Xi. Res: Ecology and distribution of birds and herps. Mailing Add: Dept of Biol ECent Okla State Univ Ada OK 74820

CARTER, WILLIAM CASWELL, b Waterville, Maine, Jan 16, 17; m 42; 57; c 3. MATHEMATICS. Educ: Colby Col, AB, 38; Harvard Univ, PhD(math), 47. Prof Exp: Mathematician, Ballistic Res Lab, Aberdeen, 47-52; instr math, Univ Md, 47-51, Johns Hopkins Univ, 51-52; mathematician, Computer Dept, Raytheon Mfg Co, 52-55; dept mgr systs anal, Datamatic, Minn-Honeywell, 55-59; sr engr, Res Div, Int Bus Mach Corp, 59-61, data systs div, 61-66, staff mem, Int Bus Mach Res Ctr, 66-74, MGR, IBM CORP, 74- Concurrent Pos: Instr, Boston Univ, 52-58. Mem: Am Math Soc; Asn Comput Mach; Soc Indust & Appl Math; Inst Elec & Electronics Eng. Res: Fault tolerant computer system design and design methods; combinatorial math; computer systems design and analysis; logic design and methods of logic design; computer systems programming. Mailing Add: 3 Shagbark Lane Woodbury CT 06798

CARTER, WILLIAM DOUGLAS, b Keene, NH, Apr 24, 26; m 50; c 4. GEOLOGY. Educ: Dartmouth Col, AB, 49. Prof Exp: Geol field asst, Permafrost, US Geol Surv, 48-49, petrol res, 50, geologist uranium explor prog, 51-57, mining geologist & tech adv to Govt Chile, AID, 57-62, commodity geologist, Light Metals & Indust Minerals Br, Resources Res Group, 62-65, geol coordr remote sensing eval & coord staff, 65-67, chmn mineral & land resources working group, Earth Resources Observ Systs Prog, 67-70; asst mgr, Applications Res, EROS Prog, 70-74, RES SCIENTIST, EROS PROG, US GEOL SURV, 75- Concurrent Pos: Lectr remote sensing, Am Asn Petrol Geologists Continuing Educ Prog, 75-76. Mem: Fel Geol Soc Am; Soc Econ Geologists; Am Asn Petrol Geologists; Am Soc Photogrammetry; Yugoslav Acad Sci. Res: Use of satellite data in studies of the tectonics and location of ore deposits in the United States and Andes Mountain region, South America; mineral deposits, chiefly uranium, copper, silica, phosphate and potash; photogeology; remote sensing; space applications. Mailing Add: 1925 Newton Square Bldg E2 Reston VA 22091

CARTER, WILLIAM EARL, b Dayton, Ohio, Apr 29, 27; m 54; c 3. ETHNOLOGY, APPLIED ANTHROPOLOGY. Educ: Muskingum Col, AB, 49; Boston Univ, STB, 55; Columbia Univ, MA, 58, PhD(anthrop), 63. Prof Exp: Instr social probs, Nat Sch Social Work, Bolivia, 60-61; asst prof anthrop, Univ Fla, 62-65; assoc prof & Peace Corps area studies coordr, Univ Wash, 65-66; assoc prof, 67-69, PROF ANTHROP, UNIV FLA, 69-, DIR LATIN AM STUDIES, 68- Concurrent Pos: Instr anthrop, Brooklyn Col, 62; res assoc, Bur Appl Social Res, Columbia Univ, 62-63; US Off Educ Fulbright Hays fel, Guatemala, 65-66; fac res award, Univ Fla, 67-; NSF fel, Bolivia, 68-69; consult, US Off Educ & Foreign Area Fel Prog, 69-71; mem exec bd, Consortium Latin Am Studies & Progs, 71-74. Honors & Awards: Citation, Partners of Alliance, 70. Mem: Fel Am Anthrop Asn; Latin Am Studies Asn; fel Soc Appl Anthrop. Res: Agrarian reform; religion; Aymara ethnography; shifting cultivation in Guatemala; Bolivian national character; chronic cannabis use in Costa Rica. Mailing Add: 1011 N W 21st St Gainesville FL 32601

CARTER, WILLIAM EUGENE, b Steubenville, Ohio, Oct 16, 39; m 61; c 3. GEODESY. Educ: Univ Pittsburgh, BS, 61; Ohio State Univ, MS, 65; Univ Ariz, PhD(civil eng), 73. Prof Exp: Chief astron surv team, 1381st Geodetic Surv Squad, US Air Force, 61-63, chief, Astron Surv Br, 1st Geodetic Surv Squad, 66-69, res geodesist lunar laser ranging, Air Force Cambridge Res Lab, 69-72; LURE PROJ MGR, INST ASTRON, UNIV HAWAII, 72- Concurrent Pos: Mem, Lunar Ranging Exp Team, 75- Mem: Am Soc Civil Engrs; Am Cong Surv & Mapping; Int Asn Geodesy; Am Soc Photogram. Res: Application of lunar laser ranging techniques to geodetic and geophysical investigations; research and development of advanced geodetic field instrumentation, stressing automation of measuring fuctions and data handling. Mailing Add: Inst for Astron Univ of Hawaii Kula HI 96790

CARTER, WILLIAM HAROLD, b Houston, Tex, Nov 17, 38; m 61; c 2. OPTICAL PHYSICS. Educ: Univ Tex, BS, 62, MS, 63, PhD(elec eng), 66. Prof Exp: Res assoc elec eng, Univ Tex, 63-66, instr, 66; contract monitor optics, US Army Mil Intel Corps, 67-69; res assoc physics, Univ Rochester, 69-70; RES PHYSICIST OPTICS, OPTICAL SCI DIV, NAVAL RES LAB, 71- Concurrent Pos: Asst prof lectr, George Washington Univ, 68-69, assoc prof lectr, 71- Mem: Inst Elec & Electronics Engrs; Am Phys Soc; Am Optical Soc. Res: Electromagnetic wave propagation; holography; digital image processing; coherence theory; angular spectrum representation; radio astronomy; laser resonater theory; scattering theory; microscopy; electromagnetic beam theory. Mailing Add: Code 5531 Naval Res Lab Washington DC 20375

CARTER, WILLIAM WALTON, physics, see 12th edition

CARTER, WILLIAM WHITNEY, b El Reno, Okla, May 8, 41; m 70. PHYTOPATHOLOGY, NEMATOLOGY. Educ: Southwestern State Univ, Okla, BS, 63; Univ Ariz, MS, 65, PhD(phytopath), 73. Prof Exp: Plant pathologist, USDA, Phoenix, Ariz, 65-72, res plant pathologist, College Station, Tex, 72-75, RES PLANT PATHOLOGIST POST HARVEST PATH, USDA, WESLACO, TEX, 75- Concurrent Pos: Mem grad fac, Tex A&M Univ, 74-75. Mem: Am Phytopath Soc; Soc Nematologists; Cotton Dis Coun. Res: Determine efficiency of treatments for decay control; affects of harvesting practices, storage, transportation and market on quality of vegetables and citrus. Mailing Add: Market Qual Res Lab Box 267 Weslaco TX 78596

CARTERETTE, EDWARD CALVIN HAYES, b Mt Tabor, NC, July 10, 21; m 55; c 1. PSYCHOACOUSTICS, NEUROPSYCHOLOGY. Educ: Univ Chicago, AB, 49; Harvard Univ, AB, 52; Ind Univ, MA, 54, PhD(psychol), 57. Prof Exp: Res staff mem acoustics lab, Mass Inst Technol, 51-52; from instr to asst prof psychol, Univ Calif, Los Angeles, 56-63, assoc prof exp psychol, 63-68; vis assoc prof psychol, Univ Calif, Berkeley, 65-66; PROF EXP PSYCHOL, UNIV CALIF, LOS ANGELES, 68- Concurrent Pos: NSF fel, Royal Inst Technol, Stockholm, Sweden & Cambridge Univ, Eng, 60-61; NSF sr fel, Inst Math Studies in Soc Sci, Stanford Univ, 64-65; rev ed, J Auditory Res, 60-69; assoc ed, Perception & Psychophysics, 71-; mem, Brain Res Inst, Univ Calif, Los Angeles, 74- Mem: Fel Acoust Soc Am; fel AAAS; fel Am Psychol Asn; Soc Exp Psychologists; Psychonomic Soc. Res: Psychoacoustics, hearing and speech perception; neuropsychology; mathematical models of cognitive processes. Mailing Add: Dept of Psychol Univ of Calif Los Angeles CA 90024

CARTIER, GEORGE ETIENNE, polymer chemistry, see 12th edition

CARTIER, GEORGE THOMAS, b Scranton, Pa, Jan 26, 24; m 46; c 2. APPLIED CHEMISTRY. Educ: Haverford Col, AB, 49. Prof Exp: Asst org chem, Smith Kline & French, 49-51; tech liaison & mkt develop, Quaker Chem Prods, 51-55; lab dir, Res & Develop Qual Control, A M Collins Div, Int Paper Co, 55-61; consult chemist, Web Processing, 61-65; pres, Keystone Filter Media Co, 65-74; CONSULT, 74- Concurrent Pos: Mem, Franklin Inst. Mem: Am Chem Soc; Tech Asn Pulp & Paper Indust; Asn Consult Chemists & Chem Engrs; Filtration Soc. Res: High polymer applications research; decorative and functional paper and paper board research and development; market development; fibrous filter media; high efficiency filter construction. Mailing Add: 311 Middle Rd Falmouth ME 04105

CARTIER, JEAN JACQUES, b Beauharnois, Que, Apr 1, 27; m 54. ENTOMOLOGY. Educ: Univ Montreal, BA, 48, BS, 52, MS, 53; Kans State Univ, PhD(entom), 56. Prof Exp: Res officer, Res Sta, 53-69, RES COORD ENTOM, RES BR, CENT EXP FARM, CAN DEPT AGR, 69- Mem: Entom Soc Am; Entom Soc Can. Res: Crop plants resistance to insects; biology of aphids, particularly aphid biotypes. Mailing Add: K W Weatby Bldg Cent Exp Farm Can Dept of Agr Ottawa ON Can

CARTIER, PETER G, b Green Bay, Wis. PHYSICAL CHEMISTRY. Educ: Lawrence Univ, BA, 68; Cornell Univ, PhD(phys chem), 73. Prof Exp: SR CHEMIST, ROHM AND HAAS CO, 73- Mem: Am Chem Soc. Res: Surface chemistry; the development and characterization of adsorbents and ion exchange resins. Mailing Add: 13 Chelfield Rd Glenside PA 19038

CARTLEDGE, FRANK, b Emory, Ga, Aug 26, 38; m 61. ORGANIC CHEMISTRY. Educ: King Col, BA, 60; Iowa State Univ, PhD(org chem), 64. Prof Exp: Nat Sci Found res fel organosilicon chem, Univ Sussex, 64-65; fel, Univ Göttingen, 65-66; asst prof, 66-74, ASSOC PROF CHEM, LA STATE UNIV, BATON ROUGE, 74- Mem: Am Chem Soc. Res: Synthesis and kinetics of reactions of group four organic compounds. Mailing Add: Dept of Chem La State Univ Baton Rouge LA 70803

CARTLEDGE, GROVES HOWARD, b Gainesville, Ga, Feb 10, 91; m 18; c 2. CHEMISTRY Educ: Davidson Col, AB & AM, 11; Univ Chicago, PhD(chem), 16. Hon Degrees: ScD, Davidson Col, 37. Prof Exp: Actg prof chem, Davidson Col, 12-13; prof chem & physics, Presby Col, SC, 13-17; assoc prof chem, Davidson Col, 17-18; chief chemist, Island Refining Corp, 19-20; from assoc prof to prof chem, Johns Hopkins Univ, 20-31; prof & head dept, Univ Buffalo, 31-45; prof chem & dean fac, King Col, 46-51; group leader, Chem Div, Oak Ridge Nat Lab, 51-61, consult, 61-71; RETIRED. Concurrent Pos: Smyth lectr, Columbia Theol Sem, 47. Honors & Awards: Whitney Award, Nat Asn Corrosion Engrs, 66; Herty Medal, Am Chem Soc, 67, Lind Lectr Award, 71. Mem: Am Chem Soc. Res: Physical chemistry; electrochemistry. Mailing Add: 124 Franklin St Brevard NC 28712

CARTMILL, MATT, b Los Angeles, Calif; m 71. PHYSICAL ANTHROPOLOGY. Educ: Pomona Col, BA, 64; Univ Chicago, MA, 66, PhD(primate evolution), 70. Prof Exp: Assoc anat, Med Ctr, 69-70, asst prof, 70-74, sociol & anthrop, Duke Univ, 70-74, ASSOC PROF ANTHROP, DUKE UNIV & ANAT, MED CTR, 74- Concurrent Pos: NIH career develop award, 75. Mem: Am Anthrop Asn; Am Asn Phys Anthrop; Am Soc Mammal; Soc Study Evolution; Australian Soc Mammal. Res: Primate evolution. Mailing Add: Dept of Anat Duke Univ Med Ctr Durham NC 27710

CARTON, CHARLES ALLAN, b New York, NY, Feb 28, 20; m 57; c 2. NEUROSURGERY. Educ: Yale Univ, BA, 41; Columbia Univ, MD, 44; Am Bd Neurol Surg, dipl, 55. Prof Exp: Intern med & surg, Bellevue Hosp, New York, 44-45; asst resident neurol & neuropath, Neurol Inst, NY, 47-48, fel neurosurg, 49, asst resident, 50-51, chief resident, 52; asst prof, Baylor Col Med, 53-56; asst clin prof, Albert Einstein Col Med, 56-58, assoc clin prof, 58-61; assoc clin prof & mem attend staff, 61-74, CLIN PROF NEUROSURG, MED CTR, UNIV CALIF, LOS ANGELES, 74- Concurrent Pos: Asst resident, Presby Hosp, New York, 49; chief neurosurg, Vet Admin Hosp, Houston, 53; asst attend neurosurgeon, Jefferson Davis Hosp, 54; asst neurosurgeon, Methodist Hosp & Clin; asst neurosurgeon, M D Anderson Hosp, 55; exec officer, Div Neurosurg, Montefiore Hosp, New York, 56, chief, 58; vis neurosurgeon, Bronx Munic Hosp Ctr, 58; vis neurosurgeon, Morrisania Hosp, 59; adj, Cedars Lebanon Hosp, Los Angeles, 61, asst, 63; asst, Mt Sinai Hosp, 64; attend & co-chief neurosurg, Cedars-Sinai Med Ctr, 71-75. Mem: AAAS; Am Asn Neuropath; Am Acad Neurol; fel Am Col Surg; AMA. Res: Cerebrovascular disease aneurysm and small vessel anastomosis; hydrocephalus. Mailing Add: 465 N Roxbury Dr Beverly Hills CA 90210

CARTON, EDWIN BECK, organic chemistry, polymer chemistry, see 12th edition

CARTON, ROBERT WELLS, b Chicago, Ill, Nov 22, 20; m 47; c 4. INTERNAL MEDICINE. Educ: Princeton Univ, AB, 42; Northwestern Univ, MD, 46. Prof Exp: From clin asst to prof med, Col Med, Univ Ill, 51-70; PROF MED & ASSOC DEAN, RUSH MED COL, 70- Mem: Fel Am Col Physicians. Res: Correlation of pulmonary structure and function; chronic bronchopulmonary infections; pulmonary function testing. Mailing Add: Presby-St Luke's Hosp 1753 W Congress Pkwy Chicago IL 60612

CARTWRIGHT, BRIAN GRANT, b Seattle, Wash, May 29, 47. ASTROPHYSICS. Educ: Yale Univ, BS, 67; Univ Chicago, SM, 68, PhD(physics), 71. Prof Exp: Res assoc physics, Enrico Fermi Inst, Univ Chicago, 71-72; ASST RES PHYSICIST, SPACE SCI LAB, UNIV CALIF, BERKELEY, 73- Mem: Am Phys Soc. Res: High energy particle astrophysics and nucleosynthesis. Mailing Add: Dept of Physics Univ of Calif Berkeley CA 94720

CARTWRIGHT, DAVID CHAPMAN, b Minneapolis, Minn, Dec 2, 37; m 65; c 2. CHEMICAL PHYSICS, MOLECULAR PHYSICS. Educ: Hamline Univ, BS, 62; Calif Inst Technol, MS, 63, PhD(chem physics, physics), 68. Prof Exp: NATOfel atmospheric physics, Max Planck Inst Extraterrestrial Physics, 67-68; vis asst prof physics, Univ Colo, Boulder, 68-69; mem tech staff, Space Physics Lab, Aerospace Corp, 69-74; staff member, Theoret Div, 74-75, ALTERNATE GROUP LEADER, LASER THEORY GROUP, THEORET DIV, LOS ALAMOS SCI LAB, 75- Concurrent Pos: Session chmn lab exp, Int Asn Geomagnetism & Aeronomy, Am Geophys Union, 74-79; consult atmospheric physics, Aerospace Corp, El Segundo, Calif, 74- Mem: Am Phys Soc; Am Geophys Union. Res: Electron impact processes in atoms and molecules; structure of simple molecules; auroral and ionospheric processes; gas laser processes; theory of low energy electron scattering by atoms and molecules; physics of auroras and airglow. Mailing Add: Theoret Div MS 228 Los Alamos Sci Lab PO Box 1663 Los Alamos NM 87545

CARTWRIGHT, GEORGE EASTMAN, b Lancaster, Wis, Dec 1, 17; m 48; c 5. INTERNAL MEDICINE. Educ: Univ Wis, BA, 39; Johns Hopkins Univ, MD, 43. Prof Exp: Intern, Hopkins Hosp, 43, asst resident, 44; resident med, Salt Lake Gen Hosp, 44-45; from instr to assoc prof, 45-58, PROF MED, COL MED, UNIV UTAH, 58-, CHMN DEPT INTERNAL MED, 67- Mem: Am Asn Physicians; Soc Exp Biol & Med; Am Soc Clin Invest. Res: Hematology. Mailing Add: Dept of Internal Med Univ of Utah Col of Med Salt Lake City UT 84132

CARTWRIGHT, KEROS, b Los Angeles, Calif, July 25, 34; m 62; c 4. HYDROGEOLOGY. Educ: Univ Calif, Berkeley, AB, 59; Univ Nev, MS, 61; Univ Ill, PhD(geol), 73. Prof Exp: Asst hydrol, US Geol Surv, 60-61; res asst, 61-63, from asst geologist to assoc geologist, 63-74, GEOLOGIST-IN-CHARGE, HYDROGEOLOGY & GEOPHYSICS SECT, ILL GEOL SURV, 74- Mem: Am Geophys Union; Geol Soc Am; Am Inst Mining, Metall & Petrol Engrs. Mailing Add: Ill State Geol Surv Urbana IL 61801

CARTWRIGHT, OSCAR LING, b Sharpsville, Pa, Apr 12, 00; m 28. ENTOMOLOGY. Educ: Allegheny Col, BS, 23; Ohio State Univ, MS, 25. Prof Exp: From asst entomologist to acting assoc entomologist, Agr Exp Sta, Clemson Col, 25-34, assoc entomologist, 34-35 & 47-48; sr asst sanitarian, US Pub Health Serv, SC & Tenn, 45-47; from assoc cur to cur insects, 48-70, EMER ENTOMOLOGIST, DIV COLEOPTERA, DEPT ENTOM, US NAT MUS, 70- Mem: Fel Entom Soc Am; Soc Syst Zool. Res: Fauna of South Carolina; biology and taxonomy of Scarabaeidae, especially subfamily Aphodiinae. Mailing Add: Rm NHB-169 Dept of Entom Mus Nat Hist Smithsonian Inst Washington DC 20560

CARTWRIGHT, RICHARD VANCE, physical organic chemistry, polymer chemistry, see 12th edition

CARTWRIGHT, THOMAS CAMPBELL, b York, SC, Mar 8, 24; m 46; c 4. ANIMAL BREEDING. Educ: Clemson Col, BS, 48; Tex A&M Univ, MS, 49, PhD, 54. Prof Exp: Animal husbandman & geneticist, Agr Exp Sta, 52-58, PROF ANIMAL BREEDING, TEX A&M UNIV, 58- Mem: AAAS; Am Genetic Asn; Am Soc Animal Sci. Res: Population genetics; genetics of cattle. Mailing Add: Dept of Animal Sci Tex A&M Univ College Station TX 77843

CARTWRIGHT, THOMAS EDWARD, b Monessen, Pa, April 27, 16; m 41; c 4. BIOPHYSICS. Educ: Univ Pittsburgh, BS, 39, MS, 49, PhD(biophys), 54. Prof Exp: Research asst virol, 47-49, res assoc biophys, 51-59, asst prof, 59-61, ASSOC PROF BIOPHYS, UNIV PITTSBURGH, 61- Mem: AAAS; Biophys Soc; Nat Sci Teachers Asn. Res: Biophysical properties of viruses and toxins. Mailing Add: Dept of Biophys Univ of Pittsburgh Pittsburgh PA 15213

CARTY, ARTHUR JOHN, b Hookergate, Eng, Sept 12, 40; m 67; c 1. INORGANIC CHEMISTRY, ORGANOMETALLIC CHEMISTRY. Educ: Univ Nottingham, BSc, 62, PhD(chem), 65. Prof Exp: Asst prof chem, Mem Univ Nfld, 65-67; asst prof, 67-69, ASSOC PROF CHEM, UNIV WATERLOO, 69- Concurrent Pos: Royal Soc Nuffield Found fel, 74; actg dir, Guelph-Waterloo Ctr Grad Studies in Chem, 75-76. Mem: Am Chem Soc; The Chem Soc. Res: Acetylenic phosphines as ligands in organometallic chemistry; structural elucidation via infrared, Mössbauer, mass spectra and x-ray crystallography; metal-phosphorus bond reactions of coordinated phosphines; acetylene oligomerisation; binding of heavy metal pollutants at biologically important sites. Mailing Add: Dept of Chem Univ of Waterloo Waterloo ON Canada

CARTY, DANIEL T, b Greenville, Tex, Aug 19, 35; m 59; c 2. ORGANIC CHEMISTRY, POLYMER CHEMISTRY. Educ: Univ Calif, Riverside, BA, 61; Univ Hawaii, MS, 63; Stanford Univ, PhD(chem), 68. Prof Exp: Chemist, Stanford Res Inst, 63-65; res assoc, Inst Org Chem, Karlsruhe Tech Univ, Ger, 67-68; chemist, Rohm and Haas Co, Bristol, Pa, 68-72; sr res assoc, Fibers Pioneering Res, 72-76; SR RES ASSOC, PLASTICS PIONEERING RES, 76- Mem: Am Chem Soc. Res: Homopolymers and copolymers of acrylates and their alloys with other polymeric systems; chemistry of novel small ring hydrocarbons and valence bond isomers of benzene; chemical modification of nylon and polyester fibers. Mailing Add: 850 Stump Rd Chalfont PA 18914

CARUBELLI, RAOUL, b Cordoba, Arg, June 17, 29; US citizen; m 59; c 2. BIOCHEMISTRY. Educ: Cordoba Nat Univ, PhD(biochem), 60. Prof Exp: Lab instr anal chem drugs & quant anal & biol chem, Cordoba Nat Univ, 53-56; res assoc biochem, Okla Med Res Found, 57-59, biochemist, 60-64, sr investr, 64-65, assoc, 65-67, assoc mem, 67-73; from asst prof to assoc prof, 63-70, PROF BIOCHEM, SCH MED, UNIV OKLA, 70-; MEM STAFF, OKLA MED RES FOUND, 73- Concurrent Pos: USPHS res grants, 62-; NIH res career develop award, 68-72; vis investr, Max Planck Inst Virus Res, Ger, 63-64. Mem: AAAS; Am Chem Soc; Am Soc Biol Chem. Res: Biological and carbohydrate chemistry; chemistry and metabolism of carbohydrates and glycoproteins; electrolytes in biological fluids. Mailing Add: Biomembrane Res Lab Okla Med Res Found 825 NE 13 St Oklahoma City OK 73104

CARUCCIO, FRANK THOMAS, b New York, NY, Sept 7, 35; m 63; c 1. GROUNDWATER GEOLOGY, ENVIRONMENTAL GEOLOGY. Educ: City Col NY, BS, 58; Pa State Univ, MS, 63, PhD(geol), 67. Prof Exp: Res assoc geol, Pa State Univ, 67-68; asst prof, State Univ NY Col New Paltz, 68-70, assoc prof, 70-71; ASSOC PROF GEOL, UNIV SC, 71- Concurrent Pos: State Univ NY Res Found grant, 69; consult geologist, Res Ctr, Uniroyal, Inc, NY, 69-71; Environ Protection Agency res grant, 73-75 & 75-76, demonstration grant, 75-77. Mem: Am Geophys Union; Am Water Works Asn; Water Resources Asn; Water Pollution Control Fedn; Soc Environ Geochem & Health. Res: Groundwater, its occurrence, movement and quality in relation to the hydrogeologic environment; interactions of pollutants within the hydrogeologic regime and their affects on the ground water potability. Mailing Add: Dept of Geol Univ of SC Columbia SC 29208

CARUOLO, EDWARD VITANGELO, b Providence, RI, Nov 1, 31; m 53; c 4. ANIMAL PHYSIOLOGY, MEDICAL PHYSIOLOGY. Educ: Univ RI, BS, 53; Univ Conn, MS, 55; Univ Minn, PhD(physiol), 63. Prof Exp: Instr physiol, Univ Minn, 60-63; asst prof, 63-68, ASSOC PROF EXP PHYSIOL, NC STATE UNIV, 68-, CHMN FAC PHYSIOL, 74- Mem: Am Dairy Sci Asn; Nat Mastitis Coun. Res: Physiology of milk secretion and ejection in health and disease; general experimental animal physiology; animal models in byssinosis. Mailing Add: Grinnells Animal Health Lab NC State Univ Raleigh NC 27607

CARUSO, FRANK SAN CARLO, b Hartford, Conn, Aug 29, 36; m 55, 73; c 3. PHARMACOLOGY. Educ: Trinity Col, Conn, BS, 58; Univ Rochester, MS, 61, PhD(pharmacol), 63. Prof Exp: Sr pharmacologist, 63-65, asst dir pharmacol-cardiovasc res, 65-66, coordr biol screening, 67-72, asst dir clin pharmacol, 72-74, DIR CLIN ANALGESIC RES, BRISTOL LABS, 74- Mem: AAAS; NY Acad Sci; Am Soc Clin Pharmacol & Therapeut; Int Asn Study Pain. Res: Sodium fluoride effects on renal function in dogs; pharmacological and toxicological effects on blood pressure, heart rate and respiration; narcotics and analgesics, narcotics and antagonists. Mailing Add: Bristol Labs Thompson Rd Syracuse NY 13201

CARUSO, SEBASTIAN CHARLES, b Jamestown, NY, March 7, 26. ENVIRONMENTAL CHEMISTRY. Educ: Alfred Univ, BA, 49; Univ Pittsburgh, PhD, 54. Prof Exp: Asst, Univ Pittsburgh, 49-50, res asst, 50-54; fel, 54-66, SR FEL, CARNEGIE-MELLON UNIV, 66- Mem: AAAS; Am Chem Soc; Am Soc Testing & Mat. Res: Lipid chemistry; micro-isolation and identification of organic substances obtained from natural sources; biochemistry of water pollution; gas chromatography; chemical analysis of surface waters and air. Mailing Add: Carnegie-Mellon Univ 4400 Fifth Ave Pittsburgh PA 15213

CARUTHERS, JERALD WAYNE, underwater acoustics, physical oceanography, see 12th edition

CARUTHERS, JOHN QUINCY, b Columbia, Tenn, Feb 28, 13; m 49; c 3. BACTERIOLOGY. Educ: Hampton Inst, BS, 33; Iowa State Univ, MS, 41. Prof Exp: Instr bact, Atlanta Col Mortuary Sci, 38-62, pres, 53-62; ASST PROF BIOL, SPELLMAN COL, 62- Concurrent Pos: Instr pub schs, Ga, 33-59. Mem: Am Soc Microbiol; Nat Sci Teachers Asn. Res: Physiological bacteriology; antibiotic resistance in staphylococci. Mailing Add: 668 Fielding Ln SW Atlanta GA 30311

CARUTHERS, LEO THOMAS, JR, bSheffield, Ala, July 12, 25; m 52; c 4. HEALTH PHYSICS. Educ: Univ Richmond, BS, 53. Prof Exp: Field rep, USAEC, Tenn, 54-57, health physicist, Savannah River Opers Off, SC, 57-58; RADIATION PROTECTION OFFICER, NC STATE UNIV, 58- Concurrent Pos: US AEC fel, Vanderbilt Univ, 53-54. Mem: Health Physics Soc. Mailing Add: 214 David Clark Labs NC State Univ Raleigh NC 27607

CARVAJAL, FERNANDO, b San Jose, Costa Rica, June 4, 13; nat US; m 41; c 4. MICROBIOLOGY. Educ: Univ Costa Rica, BS, 38; Cornell Univ, MS, 42; La State

Univ, PhD(mycol), 43. Prof Exp: Res mycologist & dir field studies, Inst del Cafe, San Jose, Costa Rica, 39; res mycologist, Exp Sta, La State Univ, 43-44; sr res mycologist, Schenley Res Labs, 44-49, head div microbiol, 49-54; co-head div microbiol, Schering Corp, 54-57, head div microbiol & vpres, Formet Labs, 57-61; dir res & labs, Arroyo Pharmaceut Corp, 61-72; MGR FERMENTATION PROD, UPJOHN MFG CORP, PR, 72- Concurrent Pos: Fel, US Off Educ. Honors & Awards: Cert Merit, La State Univ, 42. Mem: Am Soc Microbiol; Mycol Soc Am; Am Chem Soc. Res: Bacteriology; studies on genetics, mutations, physiology and fermentation of fungi and bacteria; plant pathology; antibiotics; discovery of the sexual state of Colletotrichum falcatum; synthesis of steroids; bacteriophages; production of vitamins by microbial fermentation. Mailing Add: Upjohn Mfg Co PO Box 11307 Barceloneta PR 00617

CARVELL, KENNETH LLEWELLYN, b N Andover, Mass, May 1, 25. FOREST ECOLOGY. Educ: Harvard Univ, BA, 49; Yale Univ, MF, 50; Duke Univ, DF, 53. Prof Exp: Assoc forester, Agr Exp Sta & assoc prof silvicult, 53-64, PROF FOREST ECOL & FOREST ECOLOGIST, WVA UNIV, 64- Res: Forest regeneration methods and improvement cuttings for hardwood forests; use of chemical tree poisons in forest improvement work; ecological effects of herbicides on electric transmission line rights-of-way. Mailing Add: Dept of Forest Ecol WVa Univ Morgantown WV 26505

CARVER, ALFRED CURTIS, mathematics, see 12th edition

CARVER, DAVID HAROLD, b Boston, Mass, Apr 18, 30; m 63; c 3. PEDIATRICS, INFECTIOUS DISEASES. Educ: Harvard Col, AB, 51; Duke Univ, MD, 55. Prof Exp: Asst prof pediat, microbiol & immunol, Albert Einstein Col Med, 63-66; assoc prof pediat, 66-73, PROF PEDIAT, SCH MED, JOHNS HOPKINS UNIV, 73-, ASSOC PROF MICROBIOL, 67- Concurrent Pos: Res fel pediat, Med Sch, Western Reserve Univ, 56-58; USPHS res fel, Children's Hosp Med Ctr & Harvard Med Sch, 61-63; mem comt infectious dis, Am Acad Pediat, 73- Honors & Awards: Schaffer Award, Johns Hopkins Univ Hosp, 73. Mem: Soc Pediat Res; Infectious Dis Soc Am; Am Pediat Soc; Am Acad Pediat. Res: Virology. Mailing Add: CMSC 1109 Johns Hopkins Univ Hosp Baltimore MD 21205

CARVER, DONALD S, nutrition, see 12th edition

CARVER, EUGENE ARTHUR, b Randolph, Vt, Dec 28, 44; m 70; c 2. METEORITICS. Educ: Univ Md, BS, 67; Univ Chicago, MS, 70, PhD(chem), 74. Prof Exp: Res assoc meteoritics, Goddard Space Flight Ctr, NASA, 72-74; ED CHEM, CHEM ABSTR SERV, INC, 74- Mem: Meteoritic Soc. Res: Solar system history as inferred from meteorites and lunar samples with emphasis on fission-track dating and geochronology. Mailing Add: 5060 Wyandot Pl Hilliard OH 43026

CARVER, GARY PAUL, b Brooklyn, NY, 42; m 75. EXPERIMENTAL SOLID STATE PHYSICS. Educ: Clarkson Col Technol, BS, 63; Cornell Univ, PhD(physics), 70. Prof Exp: Res asst, Lab Atomic & Solid State Physics, Cornell Univ, 66-69; RES PHYSICIST SOLID STATE, NAVAL SURFACE WEAPONS CTR, WHITE OAK LAB, 69- Mem: Am Phys Soc; AAAS. Res: Study of phenomena which determine electrical response of semiconductors to high electric fields via high field transport measurements of hot electron effects, energy loss mechanisms and magnetophonon resonances. Mailing Add: Rm 2-004 White Oak Lab Naval Surface Weapons Ctr Silver Spring MD 20910

CARVER, GEORGE EVANS, JR, b Scottdale, Pa, Nov 10, 19; m 55. GEOLOGY. Educ: Univ WVa, AB, 42; Univ Okla, MS, 47. Prof Exp: Geologist, Kerr-McGee Oil Industs, Inc, 47-49 & United Carbon Co, 49-52; chief geologist, Decem Drilling Co, 52-56; eastern div geologist, United Carbon Co, 56-57, staff geologist, 57-59, sr staff geologist, Ashland Oil Co, 59-64, mgr explor, 64-67, mgr for explor, 67-72, SR EXPLOR ADV, ASHLAND OIL CO, INC, 72- Mem: Am Asn Petrol Geologists; fel Geol Soc Am. Mailing Add: 6130 Del Monte Dr Houston TX 77027

CARVER, JAMES EDWARD, JR, population genetics, see 12th edition

CARVER, JOHN GUILL, b Mt Juliet, Tenn, Feb 10, 24; m 56; c 4. PHYSICS ENGINEERING. Educ: Ga Inst Technol, BS, 50; Yale Univ, MS, 51, PhD(physics), 55. Prof Exp: Res asst physics, Yale Univ, 51-55; sr engr, Aircraft Nuclear Propulsion Dept, Gen Elec Co, 55-56, task leader, 56-60, specialist advan reactor physics develop, Atomic Power Equip Dept, 60-62, mgr fuels & irradiations physics, 62-67; mgr advan res & technol, Space Div, NAm Rockwell Corp, 67-70, SUPVR ADVAN RES, SPACE DIV, ROCKWELL INTERNAT CORP, 70- Concurrent Pos: Chmn, Tech Comt Sensor Systs, Am Inst Aeronaut & Astronaut, 74-75. Mem: Fel AAAS; assoc fel Am Inst Aeronaut & Astronaut; Am Phys Soc; Sigma Xi. Res: Particle accelerators; neutron cross sections; nuclear shielding; neutron spectrometry; physics of plutonium fuel cycle; research reactor operation. Mailing Add: 3079 Pinewood St Orange CA 92665

CARVER, MICHAEL JOSEPH, b Omaha, Nebr, Apr 4, 23; m 48; c 3. BIOCHEMISTRY. Educ: Creighton Univ, BS, 47, MS, 48; Univ Mo, PhD(biochem), 52. Prof Exp: Instr chem, Creighton Univ, 48-49; supvr anal dept, Cudahy Labs, 52-56; assoc res prof, 56-66, PROF BIOCHEM, UNIV NEBR MED CTR, OMAHA, 66-, ASST DEAN COL MED, 70- Mem: AAAS; Am Chem Soc; Am Inst Chem; Soc Exp Biol & Med; Am Soc Neurochem. Res: Cytochemistry; electrophoresis; protein chemistry; neurochemistry; psycotropic drugs; mental retardation. Mailing Add: 553 S 90th Omaha NE 68114

CARVER, ROBERT E, b Kansas City, Mo, Jan 6, 31; m 57; c 2. SEDIMENTARY PETROLOGY, NATURAL RESOURCES. Educ: Mo Sch Mines, BSMinE, 53; Univ Mo, AM, 59, PhD(geol), 61. Prof Exp: Geologist, Bellaire Res Labs, Texaco, Inc, 61-64; asst prof, 64-71, ASSOC PROF GEOL & ASST HEAD DEPT, UNIV GA, 71- Concurrent Pos: Vis prof, Univ Rio Grande do Sul, Brazil, 72. Mem: Fel Geol Soc Am; Am Asn Petrol Geologists; Soc Econ Paleontologists & Mineralogists; Int Asn Sedimentol. Res: Sedimentary petrology; industrial mineralogy; hydrogeology; geology and natural resources of the southeast Atlantic coastal plain. Mailing Add: Dept of Geol Univ Ga Athens GA 30602

CARVER, ROBERT (GLYNN), botany, see 12th edition

CARVER, THOMAS RIPLEY, b Rochester, NY, Mar 6, 29; m 51; c 2. PHYSICS. Educ: Harvard Univ, AB, 50; Univ Ill, PhD(physics), 54. Prof Exp: Instr, Univ Ill, 54; from instr to assoc prof, 54-67, PROF PHYSICS, PRINCETON UNIV, 67- Concurrent Pos: John Simon Guggenheim fel, 64-65; consult, Texaco Develop Corp, 66-; sr vis res scientist, Royal Radar Estab, Malvern, Eng, 74-75; dir, Triphibian Res Corp, Calif, 75- Mem: Fel Am Phys Soc; Am Asn Physics Teachers; AAAS. Res: Nuclear magnetic resonance and paramagnetic spectroscopy; solid state and low temperature physics; optical orientation and spectroscopy. Mailing Add: J Henry Labs Phys Jadwin Hall Box 708 Princeton Univ Princeton NJ 08540

CARVER, WILLIAM ANGUS, b Fairforest, SC, Aug 6, 95; m 36; c 1. PLANT BREEDING. Educ: Clemson Col, BS, 21; Univ Wis, MS, 22; Iowa State Col, PhD(genetics), 25. Prof Exp: Asst genetics, Univ Wis, 21-22 & Iowa State Col, 22-25; asst cotton specialist, State Agr Exp Sta, Univ Fla, 25-32, from assoc agronomist to agronomist, 32-64; dir peanut seed res, Gold Kist Peanut Growers, 64-67, adv, Gold Kist Peanuts, 67-75; RETIRED. Honors & Awards: Golden Peanut Award, Nat Peanut Coun, 63. Res: Genetic studies in maize and cotton; crop improvement by breeding in peanuts and grasses. Mailing Add: 605 NE Seventh Terr Gainesville FL 32601

CARY, ARTHUR SIMMONS, b Sacramento, Calif, Nov 30, 25; m 57; c 3. HIGH ENERGY PHYSICS. Educ: Fisk Univ, BA, 49, MA, 51; Univ Calif, Riverside, MA & PhD(physics), 69. Prof Exp: Instr phys sci, Dillard Univ, 54-56; assoc prof physics, Tenn State Univ, 56-63; from asst prof to assoc prof, Harvey Mudd Col, 69-74; ASST PROF PHYSICS, CALIF POLYTECH STATE UNIV, 74- Mem: Am Phys Soc. Res: High energy experimental physics; nuclear emulsion; bubble chamber. Mailing Add: Dept of Physics Calif Polytech State Univ San Luis Obispo CA 93407

CARY, BOYD BALFORD, JR, b Enid, Okla, Oct 29, 23; m 57; c 2. ACOUSTICS, FLUID PHYSICS. Educ: Univ Md, BS, 47, MS, 48, PhD(physics), 54. Prof Exp: Res asst, Inst Fluid Dynamics & Appl Math, Univ Md, 54-56; physicist, Space Sci Lab, Space Tech Div, Gen Elec Co, Pa, 56-64; sr res staff mem, Gen Dynamics Corp, 64-71; lead scientist, Tracor Inc, 71-74; CONSULT, 75- Mem: Am Phys Soc; Sigma Xi. Res: Physics of fluids; analytical studies of nonlinear acoustical propagation; shock wave structure in nitrogen and air; experimental study of two-dimensional compression of jet mixing; acoustical applications of ferrofluids. Mailing Add: 20 Audubon Ct Short Hills NJ 07078

CARY, JOHN W, soil physics, see 12th edition

CASABELLA, PHILIP A, b Albany, NY, Feb 18, 33; m 60; c 2. PHYSICS. Educ: Rensselaer Polytech Inst, BS, 54, MS, 57; Brown Univ, PhD(physics), 59. Prof Exp: Res assoc physics, Brown Univ, 59-60; from asst prof to assoc prof, 61-69, PROF PHYSICS, RENSSELAER POLYTECH INST, 69-, CHMN DEPT PHYSICS & ASTRON, 70- Mem: Fel Am Phys Soc. Res: Nuclear magnetic resonance and pure quadruple resonance in solids. Mailing Add: Dept of Physics Rensselaer Polytech Inst Troy NY 12181

CASADY, ALFRED JACKSON, b Milton, Iowa, Feb 16, 16; m 41; c 2. AGRONOMY, PLANT BREEDING. Educ: Kans State Univ, BS, 48, MS, 50, PhD(agron), 62. Prof Exp: Agronomist plant breeding, Agr Res Serv, USDA, Kans State Univ, 49-70; PROF AGRON, KANS STATE UNIV, 70- Mem: Am Soc Agron. Res: Breeding and selection of cereal crops for improved yield; insect and disease resistance. Mailing Add: Dept of Agron Kans State Univ Manhattan KS 66506

CASADY, ROBERT BARNES, b Los Angeles, Calif, Nov 29, 17; m 43; c 3. PHYSIOLOGY, RESEARCH ADMINISTRATION. Educ: Univ Calif, BS, 41, PhD(comp physiol), 48. Prof Exp: Instr zool & jr zoologist, Exp Sta, Univ Calif, 48-50; asst prof animal indust, NC State Univ, 50-57; supt, US Rabbit Exp Sta, Animal Res Div, Calif, 57-65, res animal husbandman, Animal Husb Div, Agr Res Ctr, 65-67, ASST DIR INT PROGS DIV, AGR RES CTR, USDA, 67- Concurrent Pos: Res assoc, Inst Co-op Res, Johns Hopkins Univ, 53-54. Mem: AAAS. Res: Rabbit physiology; nutrition; genetics and management. Mailing Add: Agr Res Serv USDA Hyattsville MD 20782

CASAGRANDE, DANIEL JOSEPH, b Bridgeport, Conn, Jan 25, 45; m 64; c 2. ORGANIC GEOCHEMISTRY. Educ: Univ Scranton, BSc, 66; Pa State Univ, PhD(org chem), 70. Prof Exp: Fel org geochem, Univ Calgary, 71; UNIV PROF ORG GEOCHEM, GOVERNORS STATE UNIV, 71- Concurrent Pos: Consult, Village Park Forest South, Ill, 72-, Armour Dial, Inc, 72 & Izaak Walton League, 73. Mem: Am Chem Soc; Geol Soc Am; The Chem Soc; Geochem Soc; Am Water Works Asn. Res: Organic geochemistry of sulfur, metals, amino acids, porphyrins in sediments, petroleum, peat, lignite and coal; advanced water-waste treatment including the recycling of water for municipal use. Mailing Add: Col of Environ & Appl Sci Governors State Univ Park Forest South IL 60466

CASAGRANDE, JOSEPH BARTHOLOMEW, b Cincinnati, Ohio, Feb 14, 15; m 45, 69; c 4. ANTHROPOLOGY. Educ: Univ Wis, BA, 38; Columbia Univ, PhD(anthrop), 51. Prof Exp: Instr anthrop, Univ Rochester, 49-50; mem staff, Soc Sci Res Coun, 50-60; prof anthrop & head dept, Univ Ill, Champaign-Urbana, 60-67, PROF ANTHROP & DIR CTR INT COMP STUDIES, UNIV ILL, URBANA, 67- Concurrent Pos: Mem behav sci fel comt, NIMH, 60-62, mem behav sci study sect, 64-69; mem adv panel anthrop, NSF, 62-64; dir, Soc Sci Res Coun, 63-65; NSF grant field res, Ecuador & Guggenheim fel, Ecuador, 66-67; mem joint comt foreign area fel prog, Am Coun Learned Soc-Soc Sci Res Coun, 69-73; NSF grant archival res, Ecuador, Spain & US, 70-73. Mem: Am Anthrop Asn (pres-elect & pres 71-73); Am Ethnol Soc (vpres & pres, 62-64); Soc Appl Anthrop; Latin Am Studies Asn; Royal Anthrop Soc Gt Brit & Ireland. Res: Field and archival research on the position of the Indian in Ecuadorian society from colonial times to the present; ethnolinguistics and psycholinguistics; the history of anthropology. Mailing Add: Dept of Anthrop 109 Davenport Hall Univ Ill Urbana IL 61801

CASALI, LIBERTY, b Vivian, WVa, Oct 15, 11. CHEMISTRY. Educ: Duke Univ, BS, 33; Univ Colo, PhD(phys chem), 52. Prof Exp: Teacher pub schs, WVa, 33-42; asst chemist, Gen Elec Co Labs, Mass, 42-45; asst gen qual phys chem lab, Wheaton Col, Mass, 45-47; instr math, Univ Colo, 48-50, asst chem, 47-51; from instr to asst prof chem, Russell Sage Col, 52-61; assoc prof chem & physics, Winthrop Col, 61-65 & Bridgewater Col, 65-66; assoc prof, 67-72, PROF CHEM, MADISON COL, VA, 72- Concurrent Pos: Abstractor, Chem Abstr. Mem: Am Chem Soc. Res: Heats of reaction of some fluroolefins; molecular spectra; history of science. Mailing Add: Dept of Chem Madison Col Harrisonburg VA 22801

CASALS, JORDI, b Viladrau, Spain, May 15, 11; nat US; m 41; c 1. VIROLOGY. Educ: Instituto Nacional, BS, 28; Univ Barcelona, MD, 34. Prof Exp: Asst resident physician, Med Sch Hosp, Univ Barcelona, 34-36; asst path, Sch Med, Cornell Univ, 37-38; asst, Dept Path & Bact, Rockefeller Inst, 38-52, virus res prog, Rockefeller Found, 52-69; PROF EPIDEMIOL, YALE UNIV, 69- Concurrent Pos: Consult, Walter Reed Army Med Ctr; mem, Surgeon Gen's Virus Comn, Japan, 47. Mem: AAAS; fel Soc Exp Biol & Med; fel NY Acad Sci. Res: Arthropod-borne virus infections. Mailing Add: Dept of Epidemiol Yale Univ New Haven CT 06520

CASANOVA, JOSEPH, b Stafford Springs, Conn, May 31, 31; m 56; c 2. ORGANIC CHEMISTRY. Educ: Mass Inst Technol, SB, 53; Carnegie Inst Technol, MS, 56, PhD(org chem), 57. Prof Exp: Res chemist, E I du Pont de Nemours & Co, Del, 57; res chemist & dir res, Chem Warfare Lab, Army Chem Ctr, 58-59; NIH res fel, Harvard Univ, 59-61; asst prof, 61-63, assoc prof, 64-74, PROF CHEM, CALIF STATE COL, LOS ANGELES, 74- Concurrent Pos: Vis prof, Ind Univ, 65-66; Fulbright scholar, Dept Org Chem, Univ Lund, Sweden, 70-71. Mem: Am Chem Soc; The Chem Soc. Res: Sulfonium salts; organic boron compounds; isocyanides;

electroorganic chemistry. Mailing Add: Dept of Chem Calif State Col Los Angeles CA 90032

CASARETT, ALISON PROVOOST, b New York, NY, Apr 17, 30; c 2. RADIOBIOLOGY. Educ: St Lawrence Univ, BS, 51; Univ Rochester, MS, 53, PhD(radiation biol), 57. Prof Exp: Res assoc radiation biol, Univ Rochester, 53-58, instr, 58-63; asst prof, 63-69, ASSOC PROF PHYSICAL BIOL, CORNELL UNIV, 69-, ASSOC DEAN GRAD SCH, 73- Mem: Health Physics Soc; Radiation Res Soc. Res: Physiological, endocrinological and pathological effects of radiation on mammals. Mailing Add: Sage Grad Ctr Cornell Univ Ithaca NY 14853

CASARETT, GEORGE WILLIAM, b Rochester, NY, Aug 17, 20; m 44; c 2. PATHOLOGY. Educ: Univ Rochester, PhD(anat), 52. Prof Exp: Jr res assoc, 41-43, asst path, Manhattan Proj, 43-47, chief path unit, Atomic Energy Proj, 47-49, asst chief radiation tolerance sect, 49-52, scientist, 52-62, res assoc radiation ther, 59-68, from instr to asst prof, 53-59, prof radiation biol, 63-66, PROF RADIATION BIOL & BIOPHYS, SCH MED, UNIV ROCHESTER, 66-, RADIOL, 68-, CHIEF RADIATION PATH SECT, 62- Concurrent Pos: Consult, Surgeon Gen, US Army, 61-70, sci comt effects atomic radiation, UN, 63-, US Armed Forces Radiobiol Res Inst, 64-70 & coun, Am Asn Advan Aging Res; mem, US deleg, UN Conf Peaceful Uses Atomic Energy, Switz, 55, subcomt effects radiation, Nat Acad Sci, 56-64, subcomt nat comt radiation protection, 56-, adv comt to Fed Radiation Coun, 66-70, nat coun radiation protection, 62-, comt on radiation effects, 73- & Int Comn Radiol Protection, 67-70, chmn comt on biol effects of radiation, Nat Acad Sci; mem cancer res training comt, Nat Cancer Inst, 69- Honors & Awards: Award & Silver Medal, Am Roentgen Ray Soc, 59; Award, Radiol Soc NAm, 59. Mem: AAAS; Am Asn Anat; Am Soc Exp Path; fel Am Geront Soc; fel NY Acad Sci. Res: Oncology; pathology and hematology of radiations; radioactive substances; radiation biology; gerontology; sterility. Mailing Add: Dept Radiation Biol & Biophys Univ of Rochester Sch of Med Rochester NY 14642

CASARETT, LOUIS J, pharmacology, toxicology, deceased

CASASSA, EDWARD FRANCIS, b Portland, Maine, Nov 10, 24; m 54; c 3. PHYSICAL CHEMISTRY. Educ: Univ Maine, BS, 45; Mass Inst Technol, PhD(phys chem), 53. Prof Exp: Chemist, E I du Pont de Nemours & Co, 45-48; asst, Mass Inst Technol, 49-52; proj assoc chem, Univ Wis, 52-56; fel, 56-59, SR FEL, MELLON INST, 59-, PROF CHEM, CARNEGIE-MELLON UNIV, 67- Concurrent Pos: Lectr, Univ Pittsburgh, 56-59; asst ed, J Polymer Sci, 65-69, assoc ed, 69- Mem: AAAS; Am Phys Soc; Am Chem Soc. Res: Physical chemistry of polymers and proteins; statistical mechanics; light scattering. Mailing Add: Dept of Chem Carnegie-Mellon Univ 4400 Fifth Ave Pittsburgh PA 15213

CASAZ, GERONIMO, b Stamford, Tex, Sept 30, 24; c 3. MICROBIOLOGY, CLINICAL PATHOLOGY. Educ: Roosevelt Univ, BSc, 46; Univ Chicago, MSc, 52; Loyola Univ Chicago, PhD(microbiol), 63; Am Bd Bioanal, cert, 72. Prof Exp: Tech dir microbiol & immunohemat, Ill Masonic Med Ctr, Chicago, 63-67; TECH DIR LABS & ASST PROF CLIN PATH, CHICAGO COL OSTEOPATH MED & CHICAGO OSTEOPATH HOSP, 67- Concurrent Pos: Mem curriculum bd, Moraine Valley Community Col, 72-74; mem health care licensure comt, State of Ill, 73-74; mem, Ill Health Resources Coun, 74- Mem: Am Osteopath Col Path; Am Asn Microbiol; Am Asn Clin Chem; Am Asn Univ Prof; Am Asn Bioanalysts. Res: Hospital epidemiology. Mailing Add: 14655 S Ridge Ave Orland Park IL 60462

CASCARANO, JOSEPH, b Brooklyn, NY, Oct 26, 28; m 54; c 2. CELL PHYSIOLOGY. Educ: NY Univ, BA, 50, MS, 53, PhD(biol), 56. Prof Exp: US Pub Health Serv fels, Univ Minn, 56-57, NY Univ, 57-58, instr path, Sch Med, 58-59, asst prof, 59-65; assoc prof zool, 65-70, PROF CELL BIOL, UNIV CALIF, LOS ANGELES, 70- Honors & Awards: Distinguished Teaching Award, Univ Calif, Los Angeles, 70. Mem: AAAS; Am Physiol Soc; Histochem Soc; Soc Exp Biol & Med; NY Acad Sci. Res: Cation transport; cell permeability; anaerobic energy metabolism; relation of cell function to energy metabolism; mitochondrial biogenesis; metabolic regulation, heart metabolism; acclimation and adaptation of animals to altitude. Mailing Add: Dept of Zool Univ of Calif Los Angeles CA 90024

CASCIANO, DANIEL ANTHONY, b Buffalo, NY, Mar 1, 41; m 64; c 2. CELL BIOLOGY. Educ: Canisius Col, BS, 62; Purdue Univ, PhD(cell biol), 71. Prof Exp: Res asst tissue cult, Roswell Park Mem Inst, 63-64; res asst cell biol, Purdue Univ, 65-66, asst microbiol, 69; investr biochem, Univ Tenn, 71-73; RES BIOLOGIST MUTAGENESIS, NAT CTR TOXICOL RES, 73- Res: Development of mammalian somatic cell systems capable of in vitro metabolism of promutagens and procarcinogens. Mailing Add: Div of Mutagenic Res Nat Ctr of Toxicol Res Jefferson AR 72079

CASCIATO, RONALD J, cell biology, biochemistry, see 12th edition

CASE, ARTHUR ADAM, b Manhattan, Kans, Dec 3, 10; m 40; c 4. VETERINARY MEDICINE. Educ: Kans State Col, BS, 37, MS, 39, DVM, 42; Am Bd Vet Toxicol, dipl, 72. Prof Exp: Asst parasitol, Kans State Col, 35-41; from instr to asst prof path, Ohio State Univ, 42-47; assoc prof, 47-51, PROF VET MED & SURG & EXT VET, SCH VET MED, UNIV MO, 51-, CHIEF TOXICOL VET CLIN STAFF, 67- Concurrent Pos: Exten vet, Vet Continuing Educ, 73. Mem: Fel AAAS; Am Vet Med Asn; Am Asn Vet Clinicians; Am Asn Path & Bact; Am Soc Parasitol. Res: Parasitology; pathology; infectious diseases of swine; toxicology, especially poisonous plants; white snake-root poisoning, green acorn-oak leaf poisoning, bullnettle nighshade poisoning and mold toxicities. Mailing Add: Vet Teaching Hosp Univ of Mo Columbia MO 65202

CASE, CARL TYLER, b Louisville, Ky, Oct 17, 39; m 62; c 2. PLASMA PHYSICS. Educ: Univ Louisville, BEE, 62, MS, 63; Air Force Inst Technol, PhD(plasma physics), 70. Prof Exp: Physicist, Air Force Cambridge Res Labs, 63-66; chief, plasma devices group, Air Force Avionics Lab, 68-71; prof physics, 72-74, DEP HEAD DEPT PHYSICS, AIR FORCE INST TECHNOL, 74- Concurrent Pos: Consult, Air Force Aerospace Res Labs, 68-74, Asst Chief Staff Air Force Studies & Anal, 72-, Air Force Avionics Lab, 72- & Air Force Weapons Lab, 73- Mem: Am Phys Soc. Res: Analysis of wave-plasma interactions, including ionospheric propagation, pulse dispersion, and plasma diagnostics; analysis of plasma kinetic theory, correlation theory and the many-body problem. Mailing Add: Dept of Physics Air Force Inst Technol AFIT/ENP Wright-Patterson AFB OH 45433

CASE, CHARLES CALVIN, b Tulare, Calif, Oct 20, 22; c 2. ANTHROPOLOGY. Educ: Univ Calif, Los Angeles, BA, 47; Univ Southern Calif, MA, 61; Univ Ore, PhD(anthrop), 69. Prof Exp: Teaching fel anthrop, Univ Ore, 62-63; asst prof anthrop, Ariz State Col, 63-65 & Northern Ariz Univ, 66-70; ASSOC PROF ANTHROP, US INT UNIV, SAN DIEGO, 70- Mem: Fel Am Anthrop Asn; Can Anthrop Asn. Res: Underground water tunnels of pre-Roman Italy; marriage and migration patterns in an early American family. Mailing Add: Dept of Anthrop US Int Univ 10455 Pomerado Rd San Diego CA 92131

CASE, CLINTON MEREDITH, b Oregon City, Ore, Nov 20, 40; m 65; c 2. SURFACE PHYSICS, HYDROLOGY. Educ: Linfield Col, BA, 63; Univ Nev, Reno, MS, 67, PhD(physics), 70. Prof Exp: Res assoc hydrol, 70-72, asst res prof, 73-75, ASSOC RES PROF HYDROL, CTR WATER RESOURCES RES, DESERT RES INST, UNIV NEV, RENO, 75- Concurrent Pos: Fulbright scholar, 63. Mem: Am Geophys Union; Am Phys Soc; Am Vacuum Soc; Am Asn Physics Teachers. Res: Statistical mechanics of cooperative phenomena, including surface effects; theoretical groundwater hydrology; experimental solid state physics. Mailing Add: Ctr for Water Resources Res Desert Res Inst Univ of Nev Syst Reno NV 89507

CASE, JAMES EDWARD, b Mountain View, Ark, Feb 15, 33; m 72; c 4. GEOLOGY, GEOPHYSICS. Educ: Univ Ark, BS, 53, MS, 54; Univ Calif, Berkeley, PhD(geol), 63. Prof Exp: Instr geol & math, Lamar State Col, 54-55; geologist, US Geol Surv, 55, geologist & geophysicist, 56-58, geophysicist, 58-65, chief Denver area pub unit & geologist, 66; assoc prof geol & geophys, Tex A&M Univ, 66-69 & Univ Mo, 69-71; GEOPHYSICIST, US GEOL SURV, CALIF, 71- Mem: Fel Geol Soc Am; Am Geophys Union; Seismol Soc Am; Soc Explor Geophys; Am Asn Petrol Geol. Res: Gravity; magnetic data; correlation of geophysical data with geologic structure; tectonics of northern South America and Caribbean region; geophysical expression of mafic and ultramafic belts. Mailing Add: US Geol Surv 345 Middlefield Rd Menlo Park CA 94025

CASE, JAMES FREDERICK, b Bristow, Okla, Oct 27, 26; m 50; c 3. COMPARATATIVE PHYSIOLOGY. Educ: Johns Hopkins Univ, PhD(biol), 51. Prof Exp: Physiologist avian physiol, USDA, 51-52; insect physiologist, Med Labs, Army Chem Ctr, Md, 55-57, from asst prof to assoc prof zool, Univ Iowa, 57-63; assoc prof, 63-69, PROF NEUROBIOL, UNIV CALIF, SANTA BARBARA, 69- Res: Invertebrate physiology. Mailing Add: 1119 Biol Sci Unit II Univ of Calif Santa Barbara CA 93018

CASE, JAMES HUGHSON, b Franklinville, NY, May 25, 28; m 51; c 3. TOPOLOGY. Educ: Ala Polytech Inst, BS, 50; Tulane Univ, PhD(math), 54. Prof Exp: Asst prof math, Univ Utah, 54-59; asst prof, Univ Rochester, 59-61; assoc prof, 61-68, PROF MATH, UNIV UTAH, 68- Concurrent Pos: Consult, Gen Dynamics/Electronics, 60-61. Mem: Am Math Soc. Res: Automata theory; general topology. Mailing Add: Dept of Math Univ of Utah Salt Lake City UT 84112

CASE, KENNETH MYRON, b New York, NY, Sept 23, 23. PHYSICS. Educ: Harvard Univ, SB, 45, MA, 46, PhD(physics), 48. Prof Exp: Mem staff physics, Los Alamos Sci Lab, 44-45; Nat Res fel, 45-48; fel, Inst Adv Study, 48-50, asst prof physics, Univ Mich, 50-52, from assoc prof to prof chem, 53-67; PROF PHYSICS, ROCKEFELLER UNIV, 67- Concurrent Pos: Consult, Los Alamos Sci Lab, 48-; res assoc, Radiation Lab, Univ Calif, 49 & Univ Rochester, 50. Res: Neutron diffusion; field theory; relativistic wave equations. Mailing Add: Dept of Physics Rockefeller Univ New York NY 10021

CASE, LLOYD ALLEN, theoretical physics, see 12th edition

CASE, MARVIN THEODORE, b Anna, Ill, Dec 20, 34; m 58; c 2. VETERINARY PATHOLOGY, TOXICOLOGY. Educ: Univ Ill, BS, 57, DVM, 59, MS, 64, PhD(vet path), 68. Prof Exp: Vet, Ill State Dept Agr, 61-62; from instr to asst prof vet path, Col Vet Med, Univ Ill, 62-69; head path-toxicol sect, 69-71, MGR PATH, RIKER LABS, 71- Concurrent Pos: Dep vet Los Angeles County, Calif & assoc prof comp med, Univ Southern Calif, 69-71. Mem: AAAS; Am Vet Med Asn; NY Acad Sci; Int Acad Path. Res: Neoplasia; toxicology; animal diseases; teratology. Mailing Add: Riker Labs Bldg 218-2 3-M Ctr St Paul MN 55101

CASE, MARY ELIZABETH, b Crawfordsville, Ind, Dec 10, 25. MICROBIAL GENETICS. Educ: Maryville Col, Tenn, BA, 50; Univ Tenn, MS, 50; Yale Univ, PhD(bot), 57. Prof Exp: Res assoc genetics, Yale Univ, 57-72; ASSOC PROF ZOOL, UNIV GA, 72- Mem: Bot Soc Am; Am Genetics Soc. Res: Genetics of microorganisms; neurospora crassa. Mailing Add: Dept of Zool Univ of Ga Athens GA 30602

CASE, NORMAN MONDELL, b Milton, Ore, Oct 12, 17. HUMAN ANATOMY. Educ: Col Med Evangelists, BS, 49; Univ Southern Calif, MS, 54; Loma Linda Univ, PhD(anat), 58. Prof Exp: Tech asst anat, 50-54, from asst instr to asst prof, 54-75, ASSOC PROF ANAT, LOMA LINDA UNIV, 75- Mem: Am Asn Anat; Electron Micros Soc Am; assoc AMA; Royal Micros Soc. Res: Electron microscopy of the optic lobe in Octopus. Mailing Add: Dept of Anat Loma Linda Univ Loma Linda CA 92354

CASE, ROBERT B, b Columbus, Ohio, July 19, 20; m 58; c 1. CARDIOLOGY, PHYSIOLOGY. Educ: Ohio Wesleyan Univ, BA, 43; Mass Inst Technol, BS, 43; Columbia Univ, MD, 48. Prof Exp: Res assoc cardiac physiol, Sch Pub Health, Harvard Univ, 52-54; DIR LAB EXP CARDIOL, ST LUKE'S HOSP, 56-, ASSOC PROF MED, COL PHYSICIANS & SURGEONS, COLUMBIA UNIV, 72- Concurrent Pos: Nat Heart Inst res fel, 52-54; NY Heart Asn sr res fel, 56-61; USPHS res career develop award, 62-67; chief cardiac consult clin, City Health Dept, New York, 62-70; asst clin prof med, Col Physicians & Surgeons, Columbia Univ, 65-72. Mem: Am Physiol Soc; Am Fedn Clin Res; Am Col Chest Physicians. Res: Cardiac physiology; coronary artery disease. Mailing Add: St Luke's Hosp 421 W 113th St New York NY 10025

CASE, RONALD MARK, b Wausau, Wis, Oct 7, 40; m 63; c 2. WILDLIFE ECOLOGY. Educ: Ripon Col, AB, 62; Univ Ill, Urbana, MS, 64; Kans State Univ, PhD(biol), 71. Prof Exp: Instr ecol, Univ Mo, 71-72; asst prof, 72-75, ASSOC PROF WILDLIFE, UNIV NEBR, 75- Concurrent Pos: Consult ecol, Midwest Res Inst, 71-72, Mits Kawamoto & Assocs, 72-73. Mem: Wildlife Soc; Ecol Soc Am; Am Inst Biol Sci; Am Soc Naturalists; Am Ornithologists Union. Res: Pocket gophers and grasslands; population dynamics and damages to forage; bioenergetics of Bobwhites. Mailing Add: Dept of Poultry & Wildlife Sci Univ of Nebr Lincoln NE 68583

CASE, VERN WESLEY, b Kitchener, Ont, May 7, 35; m 58; c 3. SOIL FERTILITY. Educ: Univ BC, BSA, 58; Cornell Univ, MSc, 61. Prof Exp: Res officer res br, Can Dept Agr, 61-68; agr specialist, Res & Develop, Int Minerals & Chem Corp, 68-73; instr soils & plant nutrit, Triton Col, 73-74; MGR AGR ANAL & AGRON, INT MINERALS & CHEM CORP, 74- Mem: Am Soc Agron; Agr Inst Can; Can Soc Soil Sci; Int Soc Soil Sci. Res: Fertilizer evaluation and plant nutrition research; cropping systems, land reclamation and environmental concerns in agriculture. Mailing Add: IMC Plaza Int Minerals & Chem Corp Libertyville IL 60048

CASEBIER, RONALD LEROY, organic chemistry, see 12th edition

CASELLA, ALEXANDER JOSEPH, b Taylor, Pa, Aug 10, 39; m 66; c 2. BIOPHYSICS. Educ: Villanova Univ, BS, 61; Drexel Inst Technol, MS, 64; Pa State Univ, PhD(physics), 69. Prof Exp: Physicist, Frankford Arsenal, Philadelphia, 61-65; asst prof physics, Jacksonville Univ, 69-74; ASST PROF PHYS SCI, SANGAMON

STATE UNIV, 74- Mem: AAAS; Am Asn Physics Teachers. Res: Lasers; biophysics of visual systems. Mailing Add: Dept of Phys Sci Sangamon State Univ Springfield IL 62703

CASELLA, CLARENCE J, b New York, NY, Nov 9, 29; m 59; c 3. GEOLOGY. Educ: Hunter Col, BA, 56; Columbia Univ, PhD(geol), 62. Prof Exp: Lectr geol, Hunter Col, 58-60 & Brooklyn Col, 60-61; asst prof, Villanova Univ, 61-65; asst prof, Dept Earth Sci, 65-68, ASSOC PROF, DIV GEOL, NORTHERN ILL UNIV, 68- Mem: AAAS; Am Geophys Union; Geol Soc Am. Res: Structural petrology of metamorphic rocks; lunar geology. Mailing Add: Dept of Geol Northern Ill Univ De Kalb IL 60115

CASELLA, JOHN FRANCIS, b Oneida, NY, June 9, 44; m 68; c 2. POLYMER CHEMISTRY. Educ: Clarkson Col Technol, BS, 66, MS, 71, PhD(phys chem), 73. Prof Exp: Res assoc phys chem of proteins, Grad Dept Biochem, Brandeis Univ, 72-74; RES SCIENTIST POLYMER CHARACTERIZATION, ETHICON, INC, 74- Concurrent Pos: Am Cancer Soc fel, 74. Mem: Am Chem Soc; AAAS; NY Acad Sci. Res: Physical chemistry of macromolecules; solution properties of synthetic polymers and how they correlate with processability and final properties of the polymer. Mailing Add: Ethicon Inc Somerville NJ 08876

CASELLA, RUSSELL CARL, b Framingham, Mass, Nov, 6, 29; m 52; c 2. PHYSICS. Educ: Mass Inst Technol, BS, 51; Univ Ill, MS, 53, PhD(physics), 56. Prof Exp: Physicist, Air Force Cambridge Res Ctr, 51-52; asst, Univ Ill, 53-54, from res asst to assoc, 54-58; physicist, Int Bus Machines Res Lab, 58-65; physicist, 65-69, ELEM PARTICLE THEORIST, REACTOR RADIATION THEORY DIV, NAT BUR STANDARDS, 69- Honors & Awards: Silver Medal Award, US Dept Com, 73. Mem: Am Phys Soc. Res: Theoretical high energy physics; theoretical solid state physics. Mailing Add: 1485 Dunster Lane Potomac MD 20854

CASERIO, FREDERICK F, JR, b Los Angeles, Calif, Sept 19, 28; m 57; c 2. PHYSICAL ORGANIC CHEMISTRY. Educ: Univ Calif, Los Angeles, BS, 51, PhD, 54. Prof Exp: Res asst, Calif Inst Technol, 56-57; res chemist, Am Potash & Chem Corp, 56-60; res chemist, Nat Eng Sci Co, 60-61; group leader, Magna Corp, 61-62; dir petrochem res, Richfield Oil Corp, 62-69 & Union Oil Co, 69-72; TECH DIR, MAGNA CORP, 72- Mem: Am Chem Soc. Res: Oil and water treating research. Mailing Add: 11808 S Bloomfield Ave Santa Fe Springs CA 90670

CASERIO, MARJORIE C, b London, Eng, Feb 26, 29; nat US; m 57; c 2. ORGANIC CHEMISTRY. Educ: Univ London, BSc, 50; Bryn Mawr Col, MA, 51, PhD(chem), 56. Prof Exp: Assoc chemist, Fulma Res Inst, Eng, 52-53; from asst to instr chem, Bryn Mawr Col, 53-56; fel, Calif Inst Tech, 56-64; from asst prof to assoc prof, 65-71, PROF CHEM, UNIV CALIF, IRVINE, 71- Concurrent Pos: John S Guggenheim Found fel, 73-76. Honors & Awards: Garvan Medal, Am Chem Soc, 75. Mem: Am Chem Soc; The Chem Soc; Sigma Xi. Res: Reaction mechanisms in organic chemistry. Mailing Add: Dept of Chem Univ of Calif Irvine CA 92664

CASETTI, EMILIO, b Rome, Italy, Jan 31, 28; m 53. GEOGRAPHY. Educ: Univ Rome, DrLegge, 51; Northwestern Univ, PhD(geog), 64. Prof Exp: Asst prof math methods in geog, Univ Toronto, 64-66; asst prof math methods in geog, Ohio State Univ, 66-67, assoc prof, 67-71, prof, 71-73, EMER PROF GEOG, OHIO STATE UNIV, 73- Mem: Am Statist Asn; Regional Sci Asn; Economet Soc; Asn Am Geog. Res: Classification problems; mathematical model of economic development mathematical methods for optimal investment allocation; interrelationships among economic and demographic variables. Mailing Add: Dept of Geog Ohio State Univ 1775 S College Rd Columbus OH 43210

CASEY, ADRIA CATALA, b Havana, Cuba, Apr 24, 34; m 66; c 2. ORGANIC CHEMISTRY, MEDICINAL CHEMISTRY. Educ: Univ Havana, BS, 56; Univ Miami, MS, 62; Clarkson Col Technol, PhD(org chem), 64. Prof Exp: Asst prof chem, Univ Villanueva, Cuba, 56-60; instr, Clarkson Col Technol, 64-65; res chemist, Textile Fibers, E I du Pont de Nemours & Co, Inc Exp Sta, Del, 65-66; mem staff chem, NEng Inst, 66-73; asst prof chem, Univ Bridgeport, Conn, 73-75; SR RES CHEMIST, STAUFFER CHEM CO, 75- Mem: AAAS; Am Chem Soc; The Chem Soc; Int Soc Heterocyclic Chem. Res: Synthesis; organic synthesis of specialty chemicals. Mailing Add: Stauffer Chem Co Dobbs Ferry NY 10522

CASEY, CHARLES P, b St Louis, Mo, Jan 11, 42; m 68. ORGANIC CHEMISTRY. Educ: St Louis Univ, BS, 63; Mass Inst Technol, PhD(org chem), 68. Prof Exp: Fel org chem, Harvard Univ, 67-68; asst prof, 68-74, ASSOC PROF ORG CHEM, UNIV WIS, MADISON, 74- Honors & Awards: Eastman Kodak Award, Mass Inst Technol, 67. Mem: Am Chem Soc; The Chem Soc. Res: Mechanism of organometallic reactions; metal carbene complexes; homogeneous catalysis. Mailing Add: Dept of Chem Univ of Wis Madison WI 53706

CASEY, DONALD JAMES, organic chemistry, see 12th edition

CASEY, HAROLD W, b Reuter, Mo, Sept 24, 32; m 55; c 2. VETERINARY PATHOLOGY. Educ: Univ Mo-Columbia, BS, 54, DVM, 57; Tulane Univ, MPH, 58; Univ Calif, Davis, PhD(comp path), 65. Prof Exp: US Air Force, 58-, vet, Nouasseur AFB, 58-59, asst chief vet serv, Morocco, 59-61, res scientist biol div, Hanford Atomic Works, 61-63, chief anat path sect, Air Force Sch Aerospace Med, 65-67, chief cytolpath br, 67-70, chief vet path br, 70-71, chief gen vet path br, 71-74, CHMN DEPT VET PATH, ARMED FORCES INST PATH, 74- Mem: Int Acad Path; AAAS; Am Vet Med Asn; Am Col Vet Path. Res: Comparative pathology, including ultrastructure and the biological effects of ionizing radiation with special interest in non-human primate pathology. Mailing Add: Dept of Vet Path Armed Forces Inst of Path Washington DC 20306

CASEY, HELEN LILES, b Greer, SC, Oct 5, 22; m 43; c 2. BACTERIOLOGY, IMMUNOLOGY. Educ: Univ SC, BS, 43; 51; Fla State Univ, MS, 53; Purdue Univ, PhD(immunol), 58. Prof Exp: Lab technician bact, Shannon WTex Mem Hosp, San Angelo, Tex, 43-45; lab technician, Doctor's Lab, San Angelo, 45; med technician, Alvin Mem Hosp, Alvin, Tex, 46-47; head technician, Clin Lab, McLeod Infirmary, Florence, SC, 47-48; asst chief & med technician serol, Venereal Dis Res Lab, USPHS, Durham, NC, 48-50, med bacteriologist in chg venereal dis serol lab, Chamblee, Ga, 50-54, chief virus serol lab & supvry med bacteriologist, Viral & Rickettsial Dis Sect, Lab Br, Commun Dis Ctr, Atlanta, 57-61, chief immuno-serol unit, Virol Sect, 61-72, CHIEF AUTOIMMUNE DIS LAB, VIRAL IMMUNOL BR, LAB BR, CTR DIS CONTROL, 72-; ASSOC CLIN PROF, DEPT MED, RHEUMATOL & IMMUNOL DIV, MED SCH, EMORY UNIV, 73- Concurrent Pos: Fel rheumatic dis, Southwestern Med Sch, Dallas, 71-72. Mem: AAAS; Sigma Xi; Am Soc Microbiol; Am Pub Health Asn. Res: Viral antigen-antibody reactions in relation to diagnostic work; possible viral etiology of rheumatoid arthritis; detection of various types immune-complexes in serum or body fluids of patients with rheumatic diseases; production of antigens for newer viruses; crude versus purified antigens for diagnosis of viral diseases. Mailing Add: Virol Sect Ctr for Dis Control Atlanta GA 30333

CASEY, JAMES JOSEPH, inorganic chemistry, see 12th edition

CASEY, JAMES PATRICK, b Syracuse, NY, Aug 5, 15; m 41; c 2. CHEMISTRY. Educ: Syracuse Univ, BS, 37; State Univ NY, MS, 47. Prof Exp: Head paper serv lab, A E Staley Mfg Co, 37-46; assoc prof pulp & paper mfg, State Univ NY, 46-51; dir tech serv, A E Staley Mfg Co, 51-56, dir res, 56-59; vpres res & develop, Union Starch & Ref Co, 59-67, vpres mkt & develop, Union Div, 67-70, vpres res, Marschall Div, 70-76, CONSULT, MILES LABS, 76- Mem: Am Chem Soc; fel Tech Asn Pulp & Paper Indust; Am Asn Cereal Chem; Indust Res Inst. Res: Paper; starch; textiles; adhesives; consumer products; foods; synthetic resin emulsions; vegetable oils; chemical technology; enzymes; food cultures; microbiology; fermentation. Mailing Add: 1521 Dogwood Dr Elkhart IN 46514

CASEY, JEREMIAH P, organometallic chemistry, biophysical chemistry, see 12th edition

CASEY, JOHN ADDIS, b Frostburg, Md, Jan 3, 41; m 64; c 3. PHYSICS. Educ: Loyola Col, Md, BS, 62; Mich State Univ, MS, 64, PhD(physics), 67. Prof Exp: Chmn dept, 68-71, ASSOC PROF PHYSICS, CARTHAGE COL, 67-, DIR ACAD SERV & REGISTR, 71- Mem: AAAS; Am Phys Soc; Am Asn Physics Teachers. Res: Nuclear magnetic resonance in magnetically ordered crystals; pulsed nuclear magnetic resonance in liquids. Mailing Add: Carthage Col Kenosha WI 53140

CASEY, JOHN EDWARD, JR, b Cranston, RI, Dec 2, 30; m 56; c 2. ORGANIC CHEMISTRY. Educ: Providence Col, BS, 52, MS, 57. Prof Exp: Control chemist, Allied Chem Corp, 52; asst chem, Providence Col, 55-57; org chemist, Smith Kline & French Labs, 57-64; process chemist, Geigy Chem Corp, RI, 64-66; develop chemist, Rohm and Haas Co, 66-71; mgr process develop, 71-75, MGR CHEM RES & DEVELOP, YARDNEY ELEC CORP, 75- Concurrent Pos: Mem sci adv bd, Environ Defence Fund. Mem: Am Chem Soc. Res: Correlation of chemical structure and biological activity; natural products; waste treatment; environmental chemistry; improvement of the silver-zinc, nickel-cadmium, nickel-zinc and nickel-hydrogen alkaline batteries. Mailing Add: 10 Laurel Wood Dr North Stonington CT 06359

CASEY, KENNETH L, b Ogden, Utah, Apr 16, 35; m 58; c 3. NEUROPHYSIOLOGY, NEUROLOGY. Educ: Whitman Col, BA, 57; Univ Wash, MD, 61. Prof Exp: Intern, NY Hosp-Cornell Med Ctr, 61-62; res assoc psychol, McGill Univ, 64-66; from asst prof to assoc prof physiol, 66-74, resident neurol, 71-74, assoc prof physiol, neurol & neurophysiol, 74-75, PROF PHYSIOL & ASSOC PROF NEUROL, UNIV MICH, ANN ARBOR, 75- Mem: AAAS; Am Physiol Soc; Soc Neurosci; Am Acad Neurol. Res: Neurophysiological correlates of behavior; neurophysiology of limbic system; somatosensory neurophysiology and neural mechanism of pain sensation. Mailing Add: Dept of Physiol Med Sci Bldg II Univ of Mich Ann Arbor MI 48109

CASEY, MARTHA L, b Gaffney, SC, June 15, 42; m 68; c 1. BIOCHEMISTRY. Educ: Bryn Mawr Col, AB, 64; Mass Inst Technol, PhD(org chem), 68. Prof Exp: Res assoc org chem, 68-74, SPECIALIST ACAD PLANNING, CHANCELLOR'S OFF, UNIV WIS-MADISON, 74- Concurrent Pos: Lectr chem dept, Univ Wis-Madison, 74- Res: Biosynthetic studies using carbon-13 nuclear magnetic resonance. Mailing Add: 14 Farley Ave Madison WI 53705

CASH, DEWEY BYRON, b Wadley, Ala, Dec 22, 30; m 54; c 2. MATHEMATICAL ANALYSIS. Educ: Auburn Univ, BS, 55, MEd, 57, MS, 64. Prof Exp: Teacher high schs, Fla & Ga, 55-58; ASSOC PROF MATH, COLUMBUS COL, 58- Mem: Math Asn Am. Res Differential equations. Mailing Add: Dept of Math Columbus Col Columbus GA 31907

CASH, RICHARD ALAN, b Milwaukee, Wis, June 9, 41. TROPICAL PUBLIC HEALTH. Educ: Univ Wis-Madison, BS, 63; NY Univ, MD, 66; Johns Hopkins Univ, MPH, 73. Prof Exp: Intern surg, Bellevue Hosp, New York, 66-67; res physician cholera res, Pakistan SEATO Cholera Res Lab, Dacca, Bangladesh, 67-70; resident internal med, 70-71, fel infectious dis, 71-73, ASST PROF INTERNAL MED, UNIV MD HOSP, 73-, ASST PROF SOCIAL & PREV MED, 75- Concurrent Pos: Consult var found & orgn, 74- Mem: AAAS; Soc Epidemiol Res; Royal Soc Trop Med; NY Acad Sci. Res: Expanding oral therapy use in treating diarrhea; evaluation of nutritional deficiencies, especially in less developed societies and evaluating methods of supplement delivery; development and evaluation of rural health systems. Mailing Add: Dept of Social & Prev Med Univ Md Hosp 31 S Greene St Baltimore MD 21201

CASH, ROWLEY VINCENT, b Waterville, NY, June 7, 17; m 51; c 3. HISTORY OF CHEMISTRY. Educ: Colgate Univ, AB, 39; Ind Univ, PhD(chem), 52. Prof Exp: Asst chem, Ind Univ, 39-42; head dept, Olivet Col, 42-51; from asst prof to assoc prof, 51-57, chmn dept, 69-75, PROF CHEM, CENT CONN STATE COL, 57- Mem: Fel AAAS; Hist Sci Soc; Am Chem Soc; fel Am Inst Chemists; Soc Hist Technol. Res: History of chemistry; contributions by Scandinavian chemists. Mailing Add: 70 Pendleton Rd New Britain CT 06053

CASH, WILLIAM DAVIS, b Chesnee, SC, Feb 23, 30; m 61; c 3. BIOLOGICAL CHEMISTRY. Educ: Univ NC, BS, 51, PhD(pharmaceut chem), 54. Prof Exp: Res assoc biochem, Med Col, Cornell Univ, 54, from instr to assoc prof, 56-68; DIR BIOCHEM, CIBA-GEIGY CORP, 68- Mem: AAAS; Am Chem Soc; Am Soc Biol Chemists; NY Acad Sci. Res: Diabetes; prostaglandin biochemistry; platelet biochemistry; neurochemistry. Mailing Add: Ciba-Geigy Corp Ardsley NY 10502

CASHEL, MICHAEL, b Worthington, Minn, Feb 18, 37; m 62; c 3. BIOCHEMISTRY, GENETICS. Educ: Amherst Col, AB, 59; Western Reserve Univ, MD, 63; Univ Wash, PhD(genetics), 68. Prof Exp: HEAD, SECT REGULATORY CELL PHYSIOL, LAB MOLECULAR BIOL, NAT INST NEUROL DIS & STROKE, 63-65 & 67- Mem: Am Soc Biol Chem; Am Soc Microbiol. Res: Regulation of RNA synthesis; transcription specificity of RNA polymerase and nucleotide synthesis; protein synthesis; metabolic dormancy and cell physiology. Mailing Add: Molecular Genetics Bldg 6 Rm 335 Nat Inst Neurol Dis & Stroke Bethesda MD 20014

CASHMAN, ROBERT JOSEPH, b Wilmington, Ohio, Sept 27, 06; m 40; c 2. PHYSICS. Educ: Bethany Col, WVa, AB, 28, ScD, 53; Northwestern Univ, AM, 30, PhD(physics), 35. Prof Exp: Asst physics, 29-30, instr, 30-36, asst prof, 36-40, assoc prof 41-47, PROF PHYSICS, NORTHWESTERN UNIV, 47- Concurrent Pos: Physicist, Nat Defense Res Comt Contract, Mich, 41; dir res, Nat Defense Res Comt Contract, Northwestern Univ, 41-45; contracts, US Navy, 45-, US Army, 62-64 & US Air Force, 62- Honors & Awards: Certs Commendation, US Army & US Navy. Mem: Fel Am Phys Soc; fel Optical Soc Am. Res: Photo-electricity; photoconductive cells; thalofide; lead sulfide; temperature measurement; gas analysis; solid state physics. Mailing Add: Dept of Phys Northwestern Univ Evanston IL 60201

CASHWELL, EDMOND DARRELL, b Groveland, Fla, Feb 14, 20; m 45; c 1. APPLIED MATHEMATICS. Educ: Univ Wis, PhD(math), 49. Prof Exp: Jr physicist,

AEC, Metal Lab, Univ Chicago, 44-45 & Clinton Labs, 45-46; instr math, Univ Wis, 49 & Ohio State Univ, 49-51; MEM STAFF MATH, LOS ALAMOS SCI LAB, 51- Mem: Am Math Soc; Am Nuclear Soc. Res: The Monte Carlo method, especially as applied to transport problems in physics; differential equations; number theory; probability theory and random processes. Mailing Add: Los Alamos Sci Lab PO Box 1663 Los Alamos NM 87545

CASIDA, JOHN EDWARD, b Phoenix, Ariz, Dec 22, 29; m 56; c 2. TOXICOLOGY. Educ: Univ Wis, BS, 51, MS, 52, PhD(entom, biochem), 54. Prof Exp: Res asst, Univ Wis, 46-53; med entomologist, Camp Detrick, Md, 53; from asst prof to prof entom, Univ Wis, 54-63; PROF ENTOM & INSECT TOXICOLOGIST, UNIV CALIF, BERKELEY, 64- Concurrent Pos: Haight travel fel, 58-59; Guggenheim fel, 70-71; int res award pesticide chem, Am Chem Soc, 70. Honors & Awards: Medal, Seventh Int Cong Plant Protection, Paris, 70. Mem: AAAS; Am Chem Soc; Entom Soc Am. Res: Pesticide chemistry; comparative biochemistry. Mailing Add: Div of Entom Univ of Calif Berkeley CA 94720

CASIDA, LESTER EARL, b Chula, Mo, Apr 9, 04; m 27; c 3. REPRODUCTIVE PHYSIOLOGY. Educ: Northeast Mo State Teachers Col, BS, 26; Univ Mo, AM, 27, PhD(animal breeding), 32. Prof Exp: Asst animal husb, Univ Mo, 27-29; actg prof biol sci, Ariz State Teachers Col, 29-30; actg prof agr, Ark State Teachers Col, 30-31; Nat Res Coun fel, Univ Wis, 32-34, from asst prof to prof genetics, 34-74; RETIRED. Honors & Awards: Borden Award, Am Dairy Sci Asn, 54; Morrison Award, Am Soc Animal Sci, 59, Animal Physiol & Endocrinol Award, 65; Master & Pioneer Medal, 5th Int Cong Animal Reprod, 64; Order Cavalier Ufficiale, Repub of Italy, 66; Carl G Hartman Award, Soc Study Reprod, 75. Mem: Am Soc Animal Sci (vpres, 55, pres, 56); Am Dairy Sci Asn; Soc Exp Biol & Med; Endocrine Soc; Am Asn Anat. Res: Endocrinology. Mailing Add: 4229 Mandan Circle Madison WI 53711

CASIDA, LESTER EARL, JR, b Columbia, Mo, Aug 25, 28; m 53; c 2. MICROBIOLOGY, ECOLOGY. Educ: Univ Wis, BS, 50, MS, 51, PhD(bact), 53. Prof Exp: Bacteriologist, Abbott Labs, 51 & Pabst Labs, 53-54; res biochemist, Charles Pfizer & Co, Inc, 54-57; asst prof bact, 57-62, assoc prof microbiol, 62-66, PROF MICROBIOL, PA STATE UNIV, 66- Mem: Am Soc Microbiol; Am Chem Soc; Brit Soc Gen Microbiol. Res: Microbial physiology and ecology; industrial microbiology. Mailing Add: Dept of Microbiol S-101 Frear Bldg Pa State Univ University Park PA 16802

CASILLAS, EDMUND RENE, b Westwood, Calif, Nov 24, 38; m 62; c 2. BIOCHEMISTRY. Educ: Calif State Univ, BA, 59; Ore State Univ, PhD(biochem), 68. Prof Exp: Res assoc biochem, Ore State Univ, 64-68; fel, Ore Regional Primate Res Ctr, 69-71, res scientist reproductive physiol, 71-76; ASSOC PROF CHEM, NMEX STATE UNIV, 76- Mem: Am Soc Biol Chemists; Soc Study Reproduction; Am Chem Soc; Sigma Xi. Res: Reproductive processes in male animal with emphasis on spermatozoan maturation in the epididymis and regulatory mechanisms involved in spermatozoa physiology and metabolism. Mailing Add: Dept of Chem NMex State Univ Box 3C Las Cruces NM 88003

CASJENS, SHERWOOD REID, b Kesley, Iowa, July 10, 45; m 72. BIOCHEMISTRY. Educ: Mich State Univ, BS & MS, 67; Stanford Univ, PhD(biochem), 72. Prof Exp: ASST PROF MOLECULAR BIOL, SCH MED, UNIV UTAH, 74- Concurrent Pos: NIH res grant, 75-78. Res: Structure and assembly of viruses; genetics and biochemistry of bacteriophage P22 morphogenesis. Mailing Add: Dept of Microbiol Univ of Utah Med Ctr Salt Lake City UT 84132

CASKEY, ALBERT LEROY, b Wichita, Kans, Nov 26, 31; m 57; c 2. ANALYTICAL CHEMISTRY. Educ: Southeast Mo State Col, BS, 53; Iowa State Col, MS, 55, PhD(anal chem), 61. Prof Exp: Asst prof chem, Southeast Mo State Col, 58-61, assoc prof, 61-64; ASSOC PROF CHEM, SOUTHERN ILL UNIV, CARBONDALE, 64- Concurrent Pos: Nat Sci Found grants, 63-71; Off Water Resources Res grants, 65-70. Mem: AAAS; Am Chem Soc; Soc Appl Spectros. Res: Chelation systems; substituent effects in chelation; water pollution; spectrophotometric methods; ion exchange; inorganic reactions. Mailing Add: Dept of Chem Southern Ill Univ Carbondale IL 62901

CASKEY, CHARLES (DIRXON), JR, b Fannin Co, Tex, Oct 26, 07; m 31; c 3. ANIMAL NUTRITION. Educ: Okla Agr & Mech Col, BS, 29; Cornell Univ, PhD, 40. Prof Exp: Lab asst agr chem, Okla Agr & Mech Col, 29; chief chemist, State Bd Agr, Okla, 29-37; dir southern states labs, 40-43, V PRES IN CHARGE RES, CO-OP MILLS, INC, 43- Mem: Am Chem Soc; Poultry Sci Asn; Am Dairy Sci Asn. Res: Manganese requirements in poultry. Mailing Add: 701 Morningside Dr Towson MD 21204

CASKEY, CHARLES THOMAS, b Lancaster, SC, Sept 22, 38; m 60; c 2. HUMAN GENETICS. Educ: Duke Univ, MD, 63. Prof Exp: Intern & resident, Sch Med, Duke Univ, 63-65; res assoc, Nat Heart & Lung Inst, 65-67, sr investr, 65-70, head med genetics, 70-71; HEAD MED GENETICS, BAYLOR COL MED, 71- Concurrent Pos: Howard Hughes investr, Baylor Col Med, 71; genetics adv, Nat Heart & Lung Inst, 75-79. Mem: Am Soc Genetics; Am Soc Biol Chem; Fedn Am Socs Exp Biol; Am Soc Clin Invest. Res: The mechanism of polypeptide chain determination; somatic cell genetics, inborn error of metabolism and medical genetics. Mailing Add: Baylor Col of Med 1200 Moursund Ave Houston TX 77025

CASKEY, JAMES EDWARD, JR, b Heath Springs, SC, Mar 13, 18; m 42; c 2. METEOROLOGY. Educ: Furman Univ, BS, 39; Duke Univ, MA, 40. Prof Exp: Observer, US Weather Bur, Charleston, SC, 41, meteorologist, Washington, DC, 46; asst prof physics, Furman Univ, 46-48; meteorologist, US Weather Bur, Washington, DC, 48-65; chief sci info & doc, Environ Sci Serv Admin, Nat Oceanog & Atmospheric Admin, Rockville, Md, 66-70, dir environ sci info ctr, 71-74; DIR PUBL & TECH ED, AM METEOROL SOC, 74- Concurrent Pos: Ed, Monthly Weather Rev, 48-68; US mem comn bibliog & publ, World Meteorol Orgn, 54-59. Mem: AAAS; Am Meteorol Soc; Am Geophys Union. Res: Synoptic and dynamic meteorology; climatology and forecasting. Mailing Add: 21 Flume Rd Magnolia MA 01930

CASLER, DAVID ROBERT, b Norwich, NY, May 25, 20; m 42; c 2. PHARMACEUTICS. Educ: Union Col, NY, BS, 41. Prof Exp: Apprentice, Lathrop's Pharm, 41-42, pharmacist, 42, 46-47; from res asst to res assoc, 47-61, assoc mem, 61-63, res pharmacist, 63-74, SR RES PHARMACIST, STERLING-WINTHROP RES INST, 74-, GROUP LEADER CLIN PACKAGING & PROD DEVELOP PILOT LAB, 64- Concurrent Pos: Chmn flavor test panel, Sterling-Winthrop Res Inst, 59-67. Mem: Am Pharmaceut Asn; Acad Pharm Sci. Res: Flavors; flavor test panel methodology; perfumery; liquid processing and processing equipment; surfactants and emulsions; packaging equipment. Mailing Add: Prod Develop Div Sterling-Winthrop Res Inst Rensselaer NY 12144

CASO, LOUIS VICTOR, b Union City, NJ, July 6, 24. IMMUNOLOGY, HISTOLOGY. Educ: Manhattan Col, BS, 47; Columbia Univ, AM, 49; Rutgers Univ, BS, 54, PhD(zool), 58. Prof Exp: Instr biol sci, Col Pharm, Rutgers Univ, 49-54, asst zool, Univ, 55-58; asst res prof pharmacol, Sch Med, George Washington Univ, 59-60; asst prof anat, Col Med, Ohio State Univ, 61-64; asst prof histol, 64-68, ASSOC PROF ANAT SCI, SCH DENT, TEMPLE UNIV, PA, 68- Concurrent Pos: NIH grant, 62-63; Am Cancer Soc grants, 63-64 & 73-74. Mem: Am Asn Anat; NY Acad Sci; Tissue Cult Asn. Res: Effects of viral hemagglutination on human erythrocyte antigens; effects of homologous antiserum on cell growth and morphology; immunology of cancer; action of phytohemagglutinin and neuraminidase on cell growth; antigenicity; cytochemistry. Mailing Add: Sch of Dent Temple Univ Philadelphia PA 19140

CASO, MARGUERITE MIRIAM, b Union City, NJ, Mar 2, 19. CHEMISTRY. Educ: Col Mt St Vincent, AB, 40; Fordham Univ, MS, 55, PhD(chem), 58. Prof Exp: Chemist, Colgate-Palmolive-Peet Co, 41-42; chemist, Nat Starch Prod Co, 42-44; teacher sci, Holy Cross Acad, New York, 46-53; lectr chem, Fordham Univ, 54-55; instr, 55-60, asst prof, 60-69, ASSOC PROF CHEM, COL MT ST VINCENT, 69- Concurrent Pos: Dir undergrad res prog, Nat Sci Found, 62- Mem: AAAS; Am Chem Soc. Res: Analytical and inorganic chemistry; nonaqueous solvents; infrared spectroscopy; complexes. Mailing Add: Dept of Chem Col of Mt St Vincent Riverdale NY 10471

CASOLA, ARMAND RALPH, b Newark, NJ, Feb 21, 18; m 54; c 5. PHARMACEUTICAL CHEMISTRY, MEDICINAL CHEMISTRY. Educ: City Col NY, BS, 40; Fordham Univ, MS, 50, PhD(org chem), 56. Prof Exp: Res chemist, Am Cyanamid Co, 54-59; sr scientist, Strasenburgh Labs, 59-65; chemist, Div New Drugs, Bur Med, 65-66, supvry chemist, Div Dent & Surg Adjs, Off New Drugs, 66-70, chief chemist, Bur Drugs, 70-72, SUPVRY CHEMIST, DIV ANTI INFECTIVE DRUG PRODS, BUR DRUGS, FOOD & DRUG ADMIN, 72- Mem: AAAS; Am Chem Soc; NY Acad Sci; Acad Pharmaceut Sci. Res: Synthetic organic chemistry; intermediates; chemotherapeutics; analgesics; hypotensives.

CASON, JAMES, JR, b Murfreesboro, Tenn, Aug 30, 12; m 35; c 2. ORGANIC CHEMISTRY. Educ: Vanderbilt Univ, AB, 34; Univ Calif, MS, 35; Yale Univ, PhD(org chem), 38. Prof Exp: Asst & Nat Cancer Inst fel, Harvard Univ, 38-40; instr org chem, DePaul Univ, 40-41; from instr to asst prof, Vanderbilt Univ, 41-45; from asst prof to assoc prof, 45-52, PROF ORG CHEM, UNIV CALIF, BERKELEY, 52- Mem: Am Chem Soc. Res: Structure and synthesis of natural products; branched-chain acids; cyclic reaction intermediates. Mailing Add: Dept of Chem Univ of Calif Berkeley CA 94720

CASON, JAMES LEE, b Shongaloo, La, Feb 22, 22; m 46; c 2. DAIRY HUSBANDRY, NUTRITION. Educ: La Polytech Inst, BS, 48; Mich State Univ, MS, 50; NC State Col, PhD(dairy husb, nutrit), 56. Prof Exp: Actg asst prof dairy husb, La Polytech Inst, 49; instr, Univ Ark, 50-53, asst prof, 53-54; asst prof, Rutgers Univ, 56-59; assoc prof, Univ Md, 59-68, prof, 68-70; PROF AGR & HEAD DEPT, COL PURE & APPL SCI, NORTHEAST LA UNIV, 69- Mem: Am Soc Animal Sci; Biomet Soc; Am Dairy Sci Asn. Res: Dairy cattle nutrition, particularly force utilization. Mailing Add: Dept of Agr Sch of Pure & Appl Sci Northeast La Univ Monroe LA 71201

CASON, LOUIS FORESTER, organic chemistry, see 12th edition

CASON, NEAL M, b Chicago, Ill, July 26, 38; c 4. HIGH ENERGY PHYSICS. Educ: Ripon Col, AB, 59; Univ Wis, MS, 61, PhD(physics), 64. Prof Exp: Res assoc high energy physics, Univ Wis, 64-65; from instr to asst prof, 65-69, ASSOC PROF PHYSICS, UNIV NOTRE DAME, 70- Mem: Am Phys Soc. Res: Multi-pion production at high energy; associated production of strange particles; meson spectroscopy. Mailing Add: Dept of Physics Univ of Notre Dame Notre Dame IN 46556

CASORSO, DONALD ROY, b Vancouver, BC, Dec 23, 27; m 55; c 6. PATHOLOGY. Educ: Univ BC, BSA, 49, MSA, 51; Univ Toronto, DVM & VS, 55; Univ Conn, PhD, 59; Univ London, MRCVS, 67. Prof Exp: Animal pathologist, Rockefeller Found, 58-64; PATHOLOGIST, ANIMAL HEALTH RES & DEVELOP, DOW CHEM CO, 64- Concurrent Pos: Prof, Nat Vet Col, Colombia; mem ed comn, Pan-Am Vet & Zootech Cong; mem int comt, Animal Health Inst. Mem: Am Vet Med Asn; NY Acad Sci; World Path Asn; World Parasitol Asn; World Vet Med Asn. Res: Avian diseases; international research in animal disease, nutrition and management; supervising and visiting the research of investigators on each continent. Mailing Add: Agr Dept Dow Chem Co PO Box 1706 Midland MI 48640

CASPARI, ERNST WOLFGANG, b Berlin, Ger, Oct 24, 09; nat US; m 38. DEVELOPMENTAL GENETICS, BEHAVIORAL GENETICS. Educ: Univ Göettingen, PhD(zool), 33; Wesleyan Univ, MA, 50. Prof Exp: Asst zool, Univ Göettingen, 33-35; asst microbiol, Univ Istanbul, 35-38; fel biol, Lafayette Col, 38-41, asst prof, 41-44; asst prof zool, Univ Rochester, 44-45, res assoc, 45-46; assoc prof biol, Wesleyan Univ, 46-47; res assoc genetics, Carnegie Inst Technol, 47-49; prof biol, Wesleyan Univ, 49-60; prof, 60-75, chmn dept, 60-65, EMER PROF BIO, UNIV ROCHESTER, 75- Concurrent Pos: Ctr Advan Study Behav Sci fel, Stanford Univ, 56-57; ed, Advan Genetics & Genetics, 68-72; guest prof, Inst Genetics, Univ Giessen, Ger, 75-76. Mem: Am Soc Naturalists (vpres, 61); Genetics Soc Am (treas, 51-53, vpres, 65, pres, 66); Soc Study Evolution; fel Am Acad Arts & Sci; Behav Genetics Asn. Res: Physiological genetics; genetics of behavior; genetic transformation in Ephestia; effect of base analogues on development. Mailing Add: Dept of Biol Univ of Rochester Rochester NY 14627

CASPARI, MAX EDWARD, b Frankfurt-on-Main, Ger, Mar 17, 23; nat US; m 51; c 3. SOLID STATE PHYSICS. Educ: Wesleyan Univ, AB, 48; Mass Inst Technol, PhD(physics), 54. Prof Exp: Asst insulation res lab, Mass Inst Technol, 48-54; from instr to assoc prof, 54-64, PROF PHYSICS, UNIV PA, 64- Concurrent Pos: Chmn dept physics, Univ Pa, 68-73. Mem: Am Phys Soc. Res: Hyperfine interactions; perturbed angular correlations; magnetism; surfaces. Mailing Add: 1520 Spruce St Philadelphia PA 19102

CASPARIAN, SARKIS MANOUG, b Gurun, Armenia, Nov 29, 06; nat US; m 39; c 1. ORGANIC CHEMISTRY, ANALYTICAL CHEMISTRY. Educ: Northeastern Univ, BChE, 31; Columbia Univ, MA, 42. Prof Exp: Asst, Columbia Univ, 40-42; res chemist, Allied Chem & Dye Corp, 42-44; organizer & head anal dept, Dighton Dye Plant, 44-52; head anal dept, Dighton Works, Arnold, Hoffman & Co, Inc, 52-63; SUPVR, ANAL & CONTROL LABS, ICI ORGANICS, INC, 63- Mem: AAAS; Am Chem Soc. Res: Ultraviolet and visible range spectrophotometry applied to qualitative and quantitative analytical problems; Anthraquinone and other dyestuffs and intermediates. Mailing Add: 51 Dix Ave Johnston RI 02919

CASPE, SAUL, b New York, NY, Mar 1, 05; m; c 2. BIOCHEMISTRY. Educ: Polytech Inst Brooklyn, BS, 30;xColumbia 30. Prof Exp: Asst nutrit chemist, Col Physicians & Surgeons, Columbia Univ, 21-23; develop chemist, H A Metz Labs, NY,

24-30; res engr & fel, Casa Biochemica, France, 30-31; consult chemist, 32-35; assoc dir res, Philip Morris & Co, 35-55; res biochemist, Jewish Mem Hosp, NY, 58-60; CONSULT RES BIOCHEMIST, 60- Concurrent Pos: Field interviewer, Lea-Mendota Res Group, 66-75, field rep, Int Med Serv Am, 75; consult chemist, Lab Indust Hyg, 67-; assoc Hodgkins dis study, St Vincent's Hosp, NY. Mem: Am Chem Soc; fel Am Inst Chem. Res: Enzyme chemistry and physiology; creatine metabolism; organic synthesis of tertiary amines; sulphur drugs; biologicals of value in wound healing. Mailing Add: PO Box 608 Grand Cent Sta New York NY 10017

CASPER, BARRY MICHAEL, b Knoxville, Tenn, Jan 21, 39; m 61. THEORETICAL PHYSICS. Educ: Swarthmore Col, BA, 60; Cornell Univ, PhD(theoret physics), 66. Prof Exp: Consult, Inst Defense Anal, 62-63; instr physics, Cornell Univ, 65-66; asst prof 66-71, ASSOC PROF PHYSICS, CARLETON COL, 71- Concurrent Pos: Nat coun mem, Fedn Am Scientist, 71-75; chmn, Forum Physics & Soc, Am Phys Soc, 74; res fel, Prog Sci & Int Affairs, Harvard Univ, 75-76; humanities fel, Rockefellar Found, 75-76; NSF fel, Sci Appl to Societal Probs, 76-77. Mem: Am Phys Soc; Am Asn Physics Teachers. Res: Elementary particle research; arms control research; scientist and public policy; history of science. Mailing Add: Dept of Physics Carleton Col Northfield MN 55057

CASPER, BERENICE MARGARET, b Bellwood, Nebr. URBAN GEOGRAPHY. Educ: Univ Nebr, BSc 40, MA, 45, EdD(geog), 59. Prof Exp: Teacher geog, Nebr Pub Schs, 24-46; instr econ geog, Univ Pa, 46-47; prof geog, 47-74, EMER PROG GEOG, TRENTON STATE COL, 74- Concurrent Pos: Instr geog, Nebr Wesleyan Univ, 42-43 & Univ Mo, 47-48; chmn dept geog, Trenton State Col, 66-72; writer, Stark Enterprise, 74- Mem: Am Geog Soc; Asn Am Geog. Res: Geography of Anglo-America; urban geography of New Jersey; geographic education, especially curriculum development; analysis of urban census data; comparative analysis of Guyots elementary text books with those of today; locational factors in ecomonic development of New Jersey 1900-1975. Mailing Add: 446 Ewingville Rd Trenton NJ 08638

CASPER, JOHN MATTHEW, b Middletown, Pa, Jan 28, 46; m 71; c 1. PHYSICAL CHEMISTRY. Educ: Univ Scranton, BS, 67; Univ SC, PhD(chem), 71. Prof Exp: Res assoc chem, Univ Md, 71-73; ASST PROF CHEM, POLYTECH INST NY, 73- Concurrent Pos: Consult, Church & Dwight Co, Inc, 75- Mem: Am Chem Soc; Coblentz Soc; Soc Appl Spectros. Res: Application of vibrational spectroscopy to studies of chemical structure and bonding. Mailing Add: Dept of Chem Polytech Inst of NY Brooklyn NY 11201

CASPER, KARL JOSEPH, b Kansas City, Mo, Dec 4, 32; m 54, 68; c 1. PHYSICS. Educ: Ohio State Univ, BS, 53, PhD(physics), 60. Prof Exp: From instr to asst prof physics, Western Reserve Univ, 60-67; assoc prof, 67-74, PROF PHYSICS, CLEVELAND STATE UNIV, 74- Mem: Am Phys Soc. Res: Nuclear spectroscopy; angular correlations; solid state radiation detectors. Mailing Add: Dept of Physics Cleveland State Univ Cleveland OH 44115

CASPERS, HORST J, b Reitsch, Ger, Mar 3, 25; m 61. PHYSICAL CHEMISTRY, ANALYTICAL CHEMISTRY. Educ: Univ Munich, dipl, 55, Dr rer Nat (phys chem), 58. Prof Exp: Res scientist spectros, Siemens & Halske Res Labs, 59-61; supvr anal chem, Res Div, Am Stand Corp, NB, 61-68; mgr anal res sect, 68-71, mgr inorg res sect, 71-74, MGR ANAL TECH SUPPORT SECT, STAUFFER CHEMS, 75- Mem: Am Chem Soc; Soc Ger Chem. Res: Kinetic studies of donor-acceptor complexes of polymers with Friedel-Crafts catalysts; investigation of polarized molecules by ultraviolet spectroscopy; low temperature infrared spectroscopy of semiconductor materials; research on clay-organic compounds. Mailing Add: 12 Wildwood Rd Saddle River NJ 07458

CASPERS, HUBERT HENRI, b Oostkamp, Belg, June 5, 29; US citizen; m 59; c 2. SOLID STATE PHYSICS, SPECTROSCOPY. Educ: Univ Calif, Los Angeles, BA, 53, MA, 58, PhD(spectros), 62. Prof Exp: RES PHYSICIST & HEAD LUMINESCENT MAT SECT, US NAVAL ELECTRONICS LAB CTR, SAN DIEGO, 62- Mem: Am Phys Soc. Res: Rare earth solid state spectroscopy; infrared and Raman spectroscopy on solids. Mailing Add: Naval Electronics Lab Ctr Luminescent Mat Sect San Diego CA 92152

CASPI, ELIAHU, b Warsaw, Poland, June 10, 13; US citizen; m 48; c 1. ORGANIC CHEMISTRY, BIOCHEMISTRY. Educ: Clark Univ, PhD(org chem), 55. Prof Exp: Scientist, Anal Chem, Israel Stand Inst, 43-50; from staff res scientist to sr res scientist, 51-70, PRIN RES SCIENTIST, WORCESTER FOUND EXP BIOL, 71- Concurrent Pos: US Pub Health Serv res career prog award, 63-72. Mem: Fel AAAS; Am Chem Soc; Am Soc Biol Chemists; The Chem Soc. Res: Bioorganic chemistry of natural products; biosynthesis and metabolism of hormones; biosynthesis of sterols in relation to cancer. Mailing Add: Worcester Found for Exp Biol Shrewsbury MA 01545

CASS, CAROL E, b Lexington, Ky, Oct 18, 42; m 66. CELL BIOLOGY, BIOCHEMISTRY. Educ: Univ Okla, BS, 63, MS, 65; Univ Calif, Berkeley, PhD(cell biol), 71. Prof Exp: Fel cancer chemother, 70-73, ASST PROF BIOCHEM, CANCER RES UNIT, UNIV ALTA, 74- Mem: Am Soc Cell Biol; Can Soc Cell Biol; Am Asn Cancer Res; AAAS. Res: Nucleoside transport in mammalian cells, biochemical and biological mechanisms of anti-neoplastic agents; mechanisms of resistance to anti-neoplastic agents. Mailing Add: Cancer Res Unit McEachern Lab Univ of Alta Edmonton AB Can

CASS, DAVID D, b Indianapolis, Ind, Sept 1, 38; m 66. PLANT EMBRYOLOGY. Educ: Butler Univ, BS, 61; Univ Okla, PhD(bot), 67. Prof Exp: US Pub Health Serv fel, Univ Calif, Berkeley, 66-68, lectr biol, 68-69; asst prof, 69-74, ASSOC PROF BOT, UNIV ALTA, 74- Mem: Can Bot Asn; Bot Soc Am. Res: Vascular plant embryology, especially fertilization mechanisms and early embryo development; determining roles of embryo sac transfer cells; histochemistry; electron microscopy. Mailing Add: Dept of Bot Univ of Alta Edmonton AB Can

CASS, JULES SILAND, b New York, NY, Oct 7, 14; m 40; c 1. LABORATORY ANIMAL MEDICINE. Educ: Ohio State Univ, DVM, 36, MSc, 38; Am Col Lab Animal Med, dipl, 61. Prof Exp: Res fel, Bur Biol, Wildlife Res Sta, 36-38; field vet, Bur Animal Indust, USDA, Pa, 38-39, vet meat inspector, Minn, 39-45; res fel entom, Univ Minn, 45-47, USPHS grant, 47-48; instr vet anat & histol, Sch Vet Med, 48-49; biologist, Tech Develop Div, USPHS, 49-51; asst prof indust health, Kettering Lab, Col Med, Univ Cincinnati, 51-62; chief, Res Lab Animal Med & Care, 62-65, CHIEF RES LAB ANIMAL MED, SCI & TECHNOL, RES SERV, DEPT MED & SURG, VET ADMIN, WASHINGTON, DC, 65- Concurrent Pos: Fel, Univ Minn, 41; prin investr, Water Toxicity Proj, Ohio River Valley Water Sanit Comn, 51-56; biologist, Minn State Conserv Dept, 54-57; mem, Bd Govrs, Inst Lab Animal Resources & chmn, Comt Animal Prod, Nat Res Coun, 56-59; mem, Coun, Am Col Lab Animal Med, 57-58, Bd Dirs, 64, chmn, Comt Curric & Training, 64; consult, Animal Care Lab, Pan-Am Sanit Bur, WHO, Mex, 59; mem, Inter-govt Ad Hoc Comt Minimum Standards, 67 & Comt Lab Animal Lit, Inst Lab Animal Resources, 67-71; rep, Study Sect Primate Res, NIH, 64, mem, Toxicol Study Group, 67-72; vet liaison mem, Div Biol & Agr, Nat Res Coun, 68-; vis scientist, Armed Forces Inst Path, 68- mem, Va Comn Safety & Occup Health, chmn, Subcomt Control Infections, 73- Mem: Am Asn Lab Animal Sci (pres, 57-58); Conf Res Workers Animal Dis; Am Col Vet Toxicologists; Am Vet Med Asn; Sigma Xi. Res: Study of laboratory animals; biological model in biomedical research, its characteristics, biological quality and state and its interaction with factors and procedures in the laboratory environment; encephalitedes; bacterial diseases in wild animals; toxicity of chemicals in water to man and other warm-blooded animals. Mailing Add: Vet Admin Dept Med & Surg Res Serv Washington DC 20420

CASS, WILLIAM EMERSON, b Richford, Vt, Mar 5, 14; m 38; c 4. ORGANIC POLYMER CHEMISTRY. Educ: Univ Vt, PhB, 35; NY Univ, PhD(org chem), 38. Prof Exp: Instr org chem, NY Univ, 38-42; res assoc, Res Lab, Gen Elec Co, 42-50, mgr, Org Chem Sect, 50-53, mgr, New Prod Develop Lab, 53-57; staff mem, Arthur D Little, Inc, 57-59; mgr chem res, Owens-Corning Fiberglas Corp, 59-64; assoc prof, 64-75, PROF CHEM, NORTHEASTERN UNIV, 75-, EXEC OFFICER DEPT, 69- Mem: AAAS; Am Chem Soc (sec, 66-69); Sigma Xi. Res: Addition and condensation polymers; mechanism of polymerization; reactions of organic peroxides; reinforced plastics. Mailing Add: Dept of Chem Northeastern Univ Boston MA 02115

CASSADY, GEORGE, b Los Angeles, Calif, Aug 9, 34; m 57; c 3. MEDICINE, PEDIATRICS. Educ: Duke Univ, MD, 58. Prof Exp: Intern pediat, Duke Univ Med Ctr, 58-59, resident pediat, 60; clin assoc, Med Invest & Genetic Unit, Nat Inst Dent Res, 60-62; sr resident pediat, Children's Hosp Med Ctr, Boston, Mass, 62-63; from asst prof to assoc prof pediat, 64-70, PROF PEDIAT, MED CTR, UNIV ALA, BIRMINGHAM, 70-, DIR NEWBORN DIV, 65- Concurrent Pos: Fel pediat cardiol, Duke Hosp, Durham, NC, 59; teaching fel, Harvard Med Sch, 62-63; res fel newborn physiol, Boston Lying-In Hosp & Harvard Med Sch, 63-64; mem bd cert, Am Acad Pediat; chmn maternal & child health comt, State Ala. Mem: Am Soc Human Genetics; fel Am Acad Pediat; Soc Pediat Res; Am Fedn Clin Res; NY Acad Sci. Mailing Add: Dept of Pediat Univ of Ala Birmingham AL 45233

CASSADY, JOHN MAC, b Vincennes, Ind, Aug 16, 38; m 59; c 4. ORGANIC CHEMISTRY, MEDICINAL CHEMISTRY. Educ: DePauw Univ, BA, 60; Case Western Reserve Univ, MS, 62, PhD(org chem), 64. Prof Exp: Res assoc, Case Western Reserve Univ, 64-65; NIH fel, Univ Wis, 65-66; from asst prof to assoc prof, 66-74, PROF MED CHEM, SCH PHARM & PHARMACOL SCI, PURDUE UNIV, 74- Mem: AAAS; Am Chem Soc; The Chem Soc; Acad Pharmaceut Sci; Am Soc Pharmacog. Res: Isolation and structure elucidation of tumor inhibitors from plants; synthesis of potential tumor inhibitors including ergolines, lactones and anthracyclonones. Mailing Add: Dept of Med Chem Sch of Pharm Purdue Univ West Lafayette IN 47907

CASSADY, WILLIAM EMMETT, molecular biology, see 12th edition

CASSAN, STANLEY MORRIS, b Montreal, Que, Nov 3, 36; US citizen; m 61. PULMONARY DISEASES, INTERNAL MEDICINE. Educ: McGill Univ, BSc, 58, MD, CM, 62; Univ Minn, PhD(pulmonary med, path), 70. Prof Exp: ASST PROF MED & CHIEF RESPIRATORY INTENSIVE CARE UNIT, MED CTR, UNIV CALIF, LOS ANGELES, 72- Concurrent Pos: NIH fel pulmonary med, Med Ctr, Stanford Univ, 71-72. Honors & Awards: Mayer Res Award, Am Col Chest Physicians, 70. Mem: Am Soc Clin Res; fel Am Col Chest Physicians; fel Am Col Physicians; Am Lung Asn. Res: Morphometry in pulmonary disease; glycolysis and redox ratio in human erythrocytes; pollutants and pulmonary carcinoma. Mailing Add: Div of Pulmonary Dis Univ of Calif Los Angeles CA 90024

CASSARD, DANIEL WATERS, b Grand Rapids, Mich, July 30, 23; m 48; c 4. ANIMAL SCIENCE. Educ: Univ Calif, BS, 47, PhD(genetics), 52. Prof Exp: From instr to asst prof animal husb, Univ Calif, 52-56; from assoc prof to prof, Univ Nev, Reno, 56-72; asst sci dir agr, Pfizer Int, Inc, 72-73, sci dir agr, Pfizer Latin Am, 73-74, VPRES AGR DEVELOP, PFIZER AFRICA-MID EAST, PFIZER CORP, 74- Concurrent Pos: Fulbright lectr, Agrarian Univ, Peru, 63. Mem: Am Soc Animal Sci. Res: Animal growth; genetics of sheep; livestock production. Mailing Add: Pfizer Corp PO Box 30340 Nairobi Kenya

CASSARETTO, FRANK PHILIP, b Chicago, Ill, Dec 5, 06; m 31; c 3. CHEMISTRY. Educ: Loyola Univ, Ill, BS, 30, PhD(chem), 48; Univ Chicago, MS, 39. Prof Exp: Instr math, 30-32, instr chem, 32-42, from assoc prof to prof, 46-75, EMER PROF CHEM, LOYOLA UNIV CHICAGO, 75- Mem: Am chem Soc. Res: Inorganic electrodeposition; physical Grignard potentials. Mailing Add: 6300 Sheridan Rd Chicago IL 60660

CASSATT, JAMES C, inorganic chemistry, biological chemistry, see 12th edition

CASSATT, WAYNE A, JR, nuclear chemistry, see 12th edition

CASSEDAY, JOHN HERBERT, b Pasadena, Calif, Aug 11, 34; m 65; c 2. NEUROSCIENCES, PSYCHOLOGY. Educ: Univ Calif, Riverside, BA, 60; Ind Univ, MA, 63, PhD(psychol), 70. Prof Exp: LECTR, DEPT PSYCHOL, DUKE UNIV, 72-, ASST PROF OTOLARYNGOL, DEPT SURG, MED CTR, 72- Concurrent Pos: USPHS trainee, Duke Univ, 70-72, NSF grant, 72-75, NIH grant, 75- Mem: AAAS; Acoust Soc Am; Soc Neurosci. Res: Hearing. Mailing Add: Lab of Otolaryngol Duke Univ Med Ctr Durham NC 27710

CASSEL, D KEITH, b Bader, Ill, July 23, 40; m 65; c 2. SOIL PHYSICS. Educ: Univ Ill, BS, 63; Univ Calif, Davis, MS, 64, PhD(soil physics), 68. Prof Exp: Lab technician, Univ Calif, Davis, 65-68; assoc prof soil physics, NDak State Univ, 68-74; ASSOC PROF SOIL PHYSICS, NC STATE UNIV, 74- Mem: Soil Sci Soc Am; Soil Conserv Soc Am; Int Soc Soil Sci; Can Soc Soil Sci; Am Soc Agron. Res: Water and solute movement in unsaturated soils; tillage and modification of hardpan soils. Mailing Add: Dept of Soil Sci NC State Univ Raleigh NC 27607

CASSEL, DAVID GISKE, b Ainsworth, Nebr, Dec 12, 39; m 66; c 2. PHYSICS. Educ: Calif Inst Technol, BS, 60; Princeton Univ, MA, 62, PhD(physics), 65. Prof Exp: NSF fel, Europ Orgn Nuclear Res, Geneva, Switz, 65-66; asst prof, 66-71, ASSOC PROF PHYSICS, CORNELL UNIV, 71- Mem: Am Phys Soc. Res: Pi meson form factor; neutral K meson decays; symmetries of electromagnetic interactions; neutral K meson photoproduction; inelastic electron scattering. Mailing Add: Lab of Nuclear Studies Cornell Univ Ithaca NY 14850

CASSEL, DAVID WAYNE, b Toronto, Can, Sept 21, 36; US citizen; m 60; c 3. ALGEBRA. Educ: Greenville Col, BS, 59; Syracuse Univ, MA, 62, PhD(math), 67. Prof Exp: Instr math, Messiah Col, 62-64; Danforth teacher, Syracuse Univ, 65-67; from asst prof to assoc prof, 67-73, PROF MATH, CHMN DEPT, REGISTR & ASST TO DEAN, MESSIAH COL, 73- Mem: Am Math Soc; Math Asn Am. Res: Homological algebra; structure of projective modules as it relates to the base ring. Mailing Add: Dept of Math Messiah Col Grantham PA 17027

CASSEL, HANS MAURICE, b Berlin, Ger, Jan 19, 91; nat US; m 40; c 1. SURFACE CHEMISTRY. Educ: Univ Berlin, PhD, 14. Prof Exp: Asst prof phys chem, Berlin Inst Technol, 27-33; res assoc chem, Stanford Univ, 34-35; res, Great Western Electrochem Co, 36, Colgate-Palmolive-Peet Co, 37-40 & Dearborn Chem Co, Chicago, 42-44; consult phys chemist, Gen Ceramics Co, NJ, 44-46; phys chemist, US Bur Mines, 47-63; CONSULT PHYS CHEMIST, 63- Concurrent Pos: Emer prof, Tech Univ Berlin, 54- Mem: Am Phys Soc; fel Am Inst Chem. Res: Adsorption of gases and organic vapors on mercury; kinetics of steam generation; thermodynamics of adsorbed phases; flame propagation through dust clouds; sonic method of particle size determination; induction periods in thermal explosions. Mailing Add: 7135 Collins Ave Miami Beach FL 33141

CASSEL, JAMES MARTIN, b New Kensington, Pa, Nov 21, 18; m 44; c 4. DENTAL MATERIALS. Educ: Washington & Jefferson Col, BS, 42; Georgetown Univ, MS, 55, PhD(chem), 68. Prof Exp: Leather technologist, 46-47, res chemist leather, 47-62, res chemist, polymer characterization, 62-68, CHIEF DENT & MED MAT SECT, NAT BUR STANDARDS, 68-, PROG MGR, SYNTHETIC IMPLANTS, 75- Honors & Awards: Alsop Award, Am Leather Chem Asn, 59. Mem: Am Chem Soc; Am Leather Chem Asn; Int Asn Dent Res. Res: Chemical and physical studies of natural polymers; dental materials and synthetic surgical implants. Mailing Add: Dent & Med Mat Sect Nat Bur Standards Washington DC 20234

CASSEL, JOHN CHARLES, b Johannesburg, SAfrica, Oct 19, 21; US citizen; c 4. EPIDEMIOLOGY. Educ: Univ Witwatersrand, BSc, 41, MB, BCh, 45; Univ NC, MPH, 53. Prof Exp: Resident house physician & surgeon, Gen Hosp, Johannesburg, SAfrica, 46; med officer, Training Scheme for Health Personnel, Johannesburg, 47-48; med officer chg, Polela Health Centre, 48-52; from assoc prof to prof, 54-75, chmn dept, 59-75, ALUMNI DISTINGUISHED PROF EPIDEMIOL, SCH PUB HEALTH, UNIV NC, CHAPEL HILL, 75- Concurrent Pos: Consult, NIMH, 65-; assoc ed, Am J Epidemiol, 67-70; consult, Coop Drug Study in Hypertensive Dis, USPHS, 67-; chmn epidemiol & dis control study sect, NIH, 68-69; mem exec comt, Coun Epidemiol, Am Heart Asn, 68-71; consult, Div Environ Health Sci, Nat Environ Health Sci Ctr, 68-; mem adv bd, Inst Environ Health Studies, Univ NC, 68-; chmn policy adv bd, Multiple Risk Factor Intervention Trial, Nat Heart & Lung Inst, 72-; mem comt epidemiol & vet follow-up study, Nat Acad Sci; mem bd overseers, Am J Epidemiol; mem heart spec proj comt, NIH; mem bd collabr, Health Probs of Urban Environ; consult, Am Health Found. Mem: Int Soc Cardiol; Int Epidemiol Asn; Soc Epidemiol Res; fel Am Pub Health Asn; Am Col Prev Med. Res: Health consequences of rapid culture change, especially cardiovascular disease; epidemiology health services research. Mailing Add: Dept of Epidemiol Univ NC Sch of Pub Health Chapel Hill NC 27514

CASSEL, JOSEPH FRANKLIN, b Reading, Pa, July 9, 16; m 43; c 4. ORNITHOLOGY, ECOLOGY. Educ: Wheaton Col, BS, 38; Cornell Univ, MS, 41; Univ Colo, PhD(zool), 52. Prof Exp: From instr to asst prof zool, Colo State Univ, 46-50; from asst prof to assoc prof, 50-64, chmn dept, 50-64, PROF ZOOL, NDAK STATE UNIV, 64-, CHMN DEPT, 69- Concurrent Pos: NSF sci fac fel, Mus Comp Zool, Harvard Univ, 63-64. Mem: Sigma Xi; Am Ornithologists Union; Wilson Ornith Soc; Cooper Ornith Soc; Am Soc Mammal. Res: Population dynamics; speciation; creationism and evolution; birds and mammals of North Dakota. Mailing Add: Dept of Zool NDak State Univ Fargo ND 58102

CASSEL, WILLIAM ALWEIN, b Philadelphia, Pa, Mar 25, 24; m 49; c 2. MICROBIOLOGY. Educ: Philadelphia Col Pharm, BS, 46, MS, 47; Univ Pa, PhD(microbiol), 52. Prof Exp: Jr bacteriologist, Philadelphia Gen Hosp, 47-48, lab tech asst, 48-50; asst instr microbiol, Univ Pa, 50-51; res assoc, Hahnemann Med Col, 53 52-53; from asst prof to assoc prof, 53-69, PROF MICROBIOL, EMORY UNIV, 69- Concurrent Pos: Med fac award, Lederle Labs, 55-58; USPHS res career develop award, 60-65. Mem: AAAS; Am Soc Microbiol; Soc Exp Biol & Med; Am Asn Immunol; Tissue Culture Asn. Res: Variation in viruses; relation of viruses to cancer; microbial cytology; animal virology; oncolytic viruses; genetics. Mailing Add: Dept of Microbiol Emory Univ Atlanta GA 30322

CASSELBERRY, SAMUEL EMERSON, b Wayne, Pa, June 19, 43; m 64; c 2. ANTHROPOLOGY, ARCHAEOLOGY. Educ: Univ Pittsburgh, BA, 65; Pa State Univ, MA, 68, PhD(anthrop), 71. Prof Exp: Dir archaeol field sch, Pa State Univ, 68-69, instr anthrop, 69-70; asst prof, 70-72, ASSOC PROF ANTHROP, MILLERSVILLE STATE COL, 72-, CHMN DEPT, 76- Mem: Fel AAAS; fel Am Anthrop Asn. Res: Proxemics; settlement patterns; paleodemography; peasantry; kinship terminology. Mailing Add: 1501 Carlton Dr Lancaster PA 17601

CASSELL, GAIL HOUSTON, b Alexander City, Ala, Jan 25, 46; m 67; c 1. COMPARATIVE MEDICINE, INFECTIOUS DISEASES. Educ: Univ Ala, BS, 69; Univ Ala, Birmingham, MS, 71, PhD(microbiol), 73. Prof Exp: Biol technician virol & drug metab, Southern Res Inst, 64-67; res assoc molecular biol, Univ Ala, 67-68; res asst, 68-70, instr, 70-73, ASST PROF COMP MED, UNIV ALA, BIRMINGHAM, 73-, ASST PROF MICROBIOL, 74- Concurrent Pos: NIH grants, 73-77; Vet Admin grants, 73-78; resident microbiol, Vet Admin Hosp, Birmingham, 70- Mem: Am Soc Microbiol; Recticuloendothel Soc; Am Thoracic Soc. Res: Host-parasite relationships in mycoplasmal diseases and the phagocytic cell in host resistance. Mailing Add: Dept of Comp Med Univ of Ala Birmingham AL 35294

CASSELMAN, WARREN GOTTLIEB BRUCE, b Vancouver, BC, July 26, 21; m 50; c 2. PHARMACOLOGY, PUBLIC HEALTH ADMINISTRATION. Educ: Univ BC, BA, 43, MA, 44; Univ Toronto, MD, 49, PhD(physiol), 52. Prof Exp: Asst, Banting & Best Dept Med Res, Univ Toronto, 49-52, res assoc, 52-55, assoc prof, 55-58, assoc prof histol, Dept Anat, 57-58; sr res histochemist, Dept Neuropath, NY State Psychiat Inst, 58-59; assoc mem in chg div cell biol, Inst Muscle Dis, Inc, NY, 59-61; head training sect, Geigy Chem Corp, 61-65; prof pharmacol, Univ Toronto, 65-66; med dir, Pharmaceut Div, Geigy (Can) Ltd, 65-70, head clin pharmacol unit, 70; clin pharmacol adv, Drug Adv Bur, 70-72, dir bur, 72-73, dir gen, Drugs Directorate, 73-74, SR ADV MED, HEALTH PROTECTION BR, DEPT NAT HEALTH & WELFARE, CAN, 74-, SR MED ADV, INT HEALTH SERVS, 75- Concurrent Pos: Merck fel natural sci, Cytol Lab, Oxford Univ, 53-54; res assoc psychiat, Col Physicians & Surgeons, Columbia Univ, 59-65; Merck lectr, Univs BC, Alta & Sask, 61. Honors & Awards: Starr Medal, Univ Toronto, 57. Mem: Am Physiol Soc; Can Physiol Soc; Pharmacol Soc Can; Can Med Asn. Res: Clinical pharmacology; pharmacokinetics; histochemistry; cell biology. Mailing Add: 112 Helena St Ottawa ON Can

CASSELS, DONALD ERNEST, b Ellendale, NDak, Sept 8, 06; m 38. MEDICINE, PEDIATRICS. Educ: Univ NDak, BA, 32, BS, 34; Harvard Univ, MD, 36; Am Bd Pediat, dipl & cert cardiol. Prof Exp: From instr to assoc prof pediat, 40-54, PROF PEDIAT, MED SCH, UNIV CHICAGO, 54- Mem: Am Pediat Soc; Soc Pediat Res; Am Acad Pediat; AMA; Am Heart Asn. Res: Cardio-vascular and pulmonary physiology and clinical investigation. Mailing Add: Wyler Children's Hosp Univ of Chicago Chicago IL 60637

CASSEN, BENEDICT, physics, deceased

CASSEN, PATRICK MICHAEL, b Chicago, Ill, May 13, 40; m 65; c 1. GAS DYNAMICS, MAGNETOHYDRODYNAMICS. Educ: Univ Mich, BAeroE, 62, MS, 63, PhD(aeronaut & astronaut eng), 67. Prof Exp: RES SCIENTIST, SPACE SCI DIV, AMES RES CTR, NASA, 67- Mem: Am Geophys Union. Res: Magnetohydrodynamic boundary layers; magnetospheric physics; origin and evolution of the solar system. Mailing Add: Space Sci Div Ames Res Ctr NASA Moffett Field CA 94035

CASSEN, THOMAS JOSEPH, b Chicago, Ill. PHYSICAL CHEMISTRY. Educ: Polytech Inst Brooklyn, BS, 61, PhD(phys chem), 66. Prof Exp: Res assoc chem, Univ Calif, Riverside, 66-68 & Univ Ariz, 68-70; NIH spec fel, 70; coordr, Gen Chem Labs, Univ Ariz, 71; ASST PROF CHEM, UNIV GA, 72- Mem: Am Chem Soc; AAAS. Res: Molecular spectroscopy. Mailing Add: Dept of Chem Univ of Ga Athens GA 30602

CASSENS, PATRICK, b Litchfield, Ill, Oct 21, 38; m 62; c 2. MATHEMATICAL ANALYSIS. Educ: St Louis Univ, BS, 60, MS, 62, PhD(math), 66. Prof Exp: Teaching asst math, St Louis Univ, 62-64; from instr to asst prof, Univ Mo, St Louis, 64-68; asst prof, 68-70, ASSOC PROF MATH, STATE UNIV NY COL, OSWEGO, 70-, ASSOC DEAN ARTS & SCI, 73- Mem: AAAS; Am Math Soc; Math Asn. Res: Summability theory and natural boundaries of functions; 2-metric geometry. Mailing Add: Dept of Math State Univ of NY Col Oswego NY 13126

CASSENS, ROBERT G, b Morrison, Ill, June 10, 37; m 60; c 1. BIOCHEMISTRY. Educ: Univ Ill, BS, 59; Univ Wis, MS, 61, PhD(biochem), 63. Prof Exp: From asst prof to assoc prof, 64-71, PROF MEAT & ANIMAL SCI, UNIV WIS, MADISON, 71- Concurrent Pos: Fulbright grant, Australia, 63. Mem: Inst Food Technol; Am Soc Animal Sci. Res: Meat science; examination of muscle ultrastructure as effected by rate of postmortem glycolysis; function of zinc in muscle; effect of temperature on rigor mortis and associated changes; histochemistry of fiber types; myoglobin localization; phosphorylase fluorescent antibody. Mailing Add: Muscle Biol Lab Univ of Wis Madison WI 53706

CASSERBERG, BO R, b Halsingborg, Sweden, Oct 3, 41; m 63; c 2. PHYSICS. Educ: Univ Minn, Minneapolis, BPhys, 64; Princeton Univ, PhD(physics), 68. Prof Exp: Asst prof, 68-74, ASSOC PROF PHYSICS, UNIV MINN, DULUTH, 74- Res: Hyperfine structure; magnetic resonance. Mailing Add: Dept of Phys Univ of Minn Duluth MN 55812

CASSIDY, CARL EUGENE, b Salineville, Ohio, Dec 4, 24; m 61; c 2. ENDOCRINOLOGY, INTERNAL MEDICINE. Educ: Kenyon Col, AB, 46; Western Reserve Univ, MD, 48. Prof Exp: Asst, 54-56, clin instr, 56-58, instr, 58-59, sr instr, 59-62, from asst prof to assoc prof, 62-73, CLIN PROF MED, SCH MED, TUFTS UNIV, 73-; PHYSICIAN-IN-CHIEF, MED CTR WESTERN MASS, 72- Concurrent Pos: Res fel, New Eng Ctr Hosp, 54-56; trainee, Nat Inst Arthritis & Metab Dis, 54-56; asst physician, New Eng Ctr Hosp, 56-68, physician, 68-71. Mem: Endocrine Soc; AMA; Am Thyroid Asn; Am Col Physicians. Res: Diseases of endocrine glands; antithyroid drugs; radioiodine; hormones of reproductive system; pituitary hormones. Mailing Add: Med Ctr of Western Mass 759 Chestnut St Springfield MA 01107

CASSIDY, HAROLD GOMES, b Havana, Cuba, Oct 17, 06; nat US; m 34. ORGANIC CHEMISTRY, SCIENCE WRITING. Educ: Oberlin Col, AB, 390, AM, 32; Yale Univ, PhD, 39. Hon Degrees: DSc, St Thomas Inst, 72. Prof Exp: Res instr chem, Oberlin Col, 32-33; res chemist, Wm S Merrell Co, Ohio, 33-36; instr chem, Oberlin Col, 36-37; from instr to prof, 38-72, EMER PROF CHEM, YALE UNIV, 72- Concurrent Pos: Chmn, Gordon Conf Separation & Purification, 56 & Comt Grants-in-Aid, Nat Exec Bd, Soc Sigma Xi, 69; Nat Sigma Xi lectr, 60 & 65; Ayd lectr, 62; Korzybski Mem lectr, 62; Res Soc spec lectr, 63; Danforth vis lectr, Asn Am Cols Arts Prog, 68 & 71; Sigma Xi centennial lectr, Ohio State Univ, 70; seminar leader libr arts educ, Danforth Workshop, 62-65; sr fel sci, Ctr Advan Studies, Wesleyan Univ, 65-66; consult, Improv Sci Educ in India, 69 & Nat Humanities Fac, 70; prof-at-large, Hanover Col, 72-; Green Honors Chair prof, Tex Christian Univ, 74; assoc ed, Am J Sci. Mem: Fel AAAS; Am Chem Soc. Res: Oxidation-reduction; chromatography; cybernetics; science education. Mailing Add: Rte 2 Box 251 Hanover IN 47243

CASSIDY, JAMES EDWARD, b Springfield, Mass, Aug 17, 28; m 59; c 2. METABOLISM, ANALYTICAL CHEMISTRY. Educ: Univ Mass, BS, 49; Univ Vt, MS, 54; Rensselaer Polytech Inst, PhD(chem), 58. Prof Exp: Asst chemist eng exp sta, Univ NH, 49-50; asst dept chem, Univ Vt, 53-54; asst, Rensselaer Polytech Inst, 54-57; res chemist, Am Cyanamid Co, 58-64; res assoc, 64-69, GROUP LEADER, CIBA-GEIGY CORP, 69- Mem: Am Chem Soc; Sigma Xi; NY Acad Sci. Res: Metabolism of pesticides and herbicides in soil, plants and animals. Mailing Add: 503 Tangle Dr Jamestown NC 27282

CASSIDY, JAMES T, b Oil City, Pa, Sept 10, 30; m 55; c 2. RHEUMATOLOGY. Educ: Univ Mich, BS, 53, MD, 55. Prof Exp: From instr to assoc prof internal med, 62-73, PROF MED & PEDIAT, MED SCH, UNIV MICH, ANN ARBOR, 73- Concurrent Pos: Fel, Rackham Arthritis Res Unit, Med Sch, Univ Mich, Ann Arbor, 61-62; Arthritis Found fel, 63-66. Mem: Soc Pediat Res; Am Asn Immunol; Am Rheumatism Asn; Am Fedn Clin Res. Res: Immunology. Mailing Add: R4633 Kresge Med Res Bldg Univ of Mich Med Sch Ann Arbor MI 48109

CASSIDY, PATRICK EDWARD, b East Moline, Ill, Nov 8, 37; m 61; c 1. CHEMISTRY. Educ: Univ Ill, BS, 59; Univ Iowa, MS, 62, PhD(chem), 63. Prof Exp: Fel, Univ Ariz, 63-64; mem tech staff polymer chem, Sandia Corp, 64-66; sr scientist-group leader, Tracor, Inc, 66-69, asst to dir res lab, 69-71; ASST PROF CHEM, SOUTHWEST TEX STATE UNIV, 71- Concurrent Pos: Vpres, Tex Res Inst, 75- Mem: Am Chem Soc; Soc Plastics Engrs. Res: Polyphenylene; polyphenylethers; ferrocenes and other organo-metallics; thermal analyses of polymers; epoxy resin modifications; urethanes; high-temperature polymers; adhesives; coupling agents; phenolics; permeation through polymers; heterocyclic polymers. Mailing Add: Dept of Chem Southwest Tex State Univ San Marcos TX 78666

CASSIDY, SAMUEL H, b Escanaba, Mich, Oct 3, 22; m 56. MATHEMATICS. Educ: Univ Minn, Minneapolis, BA, 48; Univ Mich, MS, 50. Prof Exp: Instr math, Napa Col, 54-57; mathematician, US Naval Radiol Defense Lab, 57-69; MATHEMATICIAN, NAVAL UNDERSEA CTR, 69- Mem: AAAS. Res: Computer systems analysis. Mailing Add: Naval Undersea Ctr San Diego CA 92107

CASSIDY, WILLIAM ARTHUR, b New York, NY, Jan 3, 28; m 61; c 3. GEOCHEMISTRY, GEOLOGY. Educ: Univ NMex, BS, 52; Pa State Univ, PhD(geochem), 61. Prof Exp: Mem staff seismic comput, Superior Oil Co Calif, 52-53; res scientist meteoritics, Lamont Geol Observ, Columbia Univ, 61-67; ASSOC PROF GEOL, DEPT EARTH & PLANETARY SCI, UNIV PITTSBURGH, 68-

Concurrent Pos: Prin investr, NSF grants res on meteorites & meteorite craters, Arg & Chile, 61-68. Mem: Am Geophys Union; Meteoritical Soc. Res: Origin and evolution of planetary and subplanetary bodies; element abundances and fractionations; meteorites and meteorite craters; primitive planetary atmospheres; lunar & planetary surface phenomena. Mailing Add: Dept of Earth & Planetary Sci 506 Langley Hall Univ of Pitts Pittsburgh PA 15260

CASSIE, ROBERT MACGREGOR, b Lowville, NY, May 22, 35; m 71. GEOLOGY. Educ: St Lawrence Univ, BS, 56; Univ Wis, PhD(geol), 65. Prof Exp: Sr res mineralogist, Chem Div, Pittsburgh Plate Glass Co, 62-65, res supvr, Minerals Res, 65-66; asst prof geol, Col Wooster, 66-67; asst prof, 67-73, ASSOC PROF GEOL, STATE UNIV NY COL BROCKPORT, 73- Concurrent Pos: Sr fel, Geophys Lab, Carnegie Inst, 69-70. Mem: Geol Soc Am; Am Geophys Union; Nat Asn Geol Teachers; AAAS. Res: Petrology and structure of metamorphic and igneous terrains, especially western Connecticut; silicate phase equilibria. Mailing Add: Dept of Earth Sci State Univ NY Coll Brockport NY 14420

CASSIN, JOSEPH M, b Lowell, Mass, July 21, 28; m 69. MICROBIAL ECOLOGY. Educ: Cath Univ Am, AB, 52; St John's Univ, AM, 58; Howard Univ, MS, 64; Fordham Univ, PhD(microbial ecol), 68. Prof Exp: Teacher, sec schs, 52-67; ASST PROF BIOL, ADELPHI UNIV, 67-, ASST PROF MICROBIAL ECOL, INST MARINE SCI, 68- Concurrent Pos: Instr, Sci Honors Prog, Sch Eng, Columbia Univ, 59-60; inst dir biol sci curric study, NY Archdiocese Sci Coun, 64-65; res asst microbial ecol, Fordham Univ, 66-68. Mem: Phycol Soc Am; Soc Protozoologists; Ecol Soc Am; Am Inst Biol Sci; Am Soc Limnol & Oceanog. Res: Phytoplankton physiology and taxonomy. Mailing Add: Dept of Biol Adelphi Univ Garden City NY 11530

CASSIN, PATRICIA ERLBAUM, phycology, cell physiology, see 12th edition

CASSIN, SIDNEY, b Mass, June 8, 28; m 50; c 4. PHYSIOLOGY. Educ: NY Univ, BA, 50; Univ Tex, MA, 54, PhD(physiol), 57. Prof Exp: From instr to assoc prof physiol, 57-68, PROF PHYSIOL, COL MED, UNIV FLA, 68- Concurrent Pos: NIH spec fel, Nuffield Inst Med Res, Oxford Univ, 62-63. Mem: AAAS; Am Physiol Soc; Soc Exp Biol & Med. Res: Respiratory physiology; neonatal anoxia; fetal and newborn pulmonary circulation; renal physiology. Mailing Add: Dept of Physiol Col of Med Univ of Fla Gainesville FL 32601

CASSINELLI, JOSEPH PATRICK, b Cincinnati, Ohio, Aug 23, 40; c 3. ASTROPHYSICS. Educ: Xavier Univ, Ohio, BS, 62; Univ Ariz, MS, 65; Univ Wash, PhD(astron), 70. Prof Exp: Res asst, Kitt Peak Nat Observ, 63-65; res engr aerospace sci, Boeing Co, Wash, 65-66; res assoc, Joint Inst Lab Astrophys, Colo, 70-72; ASST PROF ASTRON, UNIV WIS-MADISON, 72- Concurrent Pos: Vis scientist, Space Res Lab, Astron Inst Utrecht, 75. Mem: Am Astron Soc; Int Astron Union. Res: Theoretical studies of the structure of the stellar atmosphere of hot stars; radiative transfer in stellar atmospheres and studies of stellar coronae and stellar winds. Mailing Add: Washburn Obersv Univ of Wis Madison WI 53706

CASSMAN, MARVIN, biochemistry, see 12th edition

CASTAGNA, FRANK, b Spokane, Wash, June 2, 30; m 63. MATHEMATICS. Educ: Eastern Wash State Col, BA, 55; Univ Wash, BS, 58, MA, 63, PhD(math), 67. Prof Exp: Teacher, Kent Schs, Wash, 55-57 & Seattle Schs, 58-59; res engr math anal prog, Boeing Co, 63-65; res instr, NMex State Univ, 67-68; asst prof math, Wayne State Univ, 68-74; ASSOC PROF MATH, FRAMINGTON STATE COL, 74- Mem: Am Math Soc; Math Asn Am. Res: Algebra; Abelian groups. Mailing Add: Dept of Math Framington State Col Framington MA 01701

CASTAGNOLI, NEAL, JR, b Los Angeles, Calif, Sept 6, 36; m 57; c 5. ORGANIC CHEMISTRY, MEDICINAL CHEMISTRY. Educ: Univ Calif, Berkeley, BS, 59, MA, 61, PhD(chem), 64. Prof Exp: NIH fel, Higher Inst Health, Italy, 64-65; fel, Imp Col, Univ London, 65-67; asst prof, 67-73, ASSOC PROF CHEM & PHARMACEUT CHEM, MED CTR, UNIV CALIF, SAN FRANCISCO, 73- Mem: Am Chem Soc; The Chem Soc; NY Acad Sci. Res: Synthesis, metabolism and pharmacological activity of centrally active compounds. Mailing Add: Dept of Chem Univ of Calif Med Ctr 926 Med Sci Bldg San Francisco CA 94122

CASTAGNOZZI, DANIEL M, b Newark, NJ, Aug 30, 15; m 41; c 4. FORESTRY. Educ: Univ Mich, BSF, 52; State Univ NY Col Forestry, Syracuse Univ, MF, 57. Prof Exp: From instr to asst prof mensuration & silvicult, 56-61, assoc prof, 61-71, PROF FORESTRY, NY STATE RANGER SCH, STATE UNIV NY COL ENVIRON SCI & FORESTRY, SYRACUSE UNIV, 71-, DIR, FOREST TECHNOL PROG, 72- Res: Surveying; accounting. Mailing Add: Forest Technol Prog SUNY Col Environ Sci & Forestry Wanakena NY 13695

CASTALDI, COSMO RAYMOND, b Sudbury, Ont, Nov 12, 20; m 51; c 4. DENTISTRY. Educ: Univ Toronto, DDS, 44; Northwestern Univ, MSD, 51. Prof Exp: Demonstr children's dent, Northwestern Univ, 51-52; asst prof, Ind Univ, 52-56; prof, Univ Alta, 56-65; prof, Dept Restorative Dent, Univ Man, 65-69; PROF PEDIAT DENT & HEAD DEPT, SCH DENT MED, UNIV CONN, 69- Mem: Int Asn Dent Res. Res: Growth of well children; handicapped children; calcification in biological sciences; dental epidemiology. Mailing Add: Dept of Pediat Dent Univ of Conn Health Ctr Hartford CT 06112

CASTANEDA, ALDO RICARDO, b Genoa, Italy, July 17, 30; US citizen; m 56; c 3. THORACIC SURGERY, CARDIOVASCULAR SURGERY. Educ: San Carlos Univ, Guatemala, MD, 56; Univ Minn, PhD(surg), 63, MS, 64. Prof Exp: From instr to prof surg, Univ Minn, Minneapolis, 63-72; PROF CARDIOVASC SURG & CARDIOVASC SURGEON-IN-CHIEF, HARVARD MED SCH, 72- Concurrent Pos: Mem adv coun, Cardiovasc Surg Coun, Am Heart Asn, 68. Mem: AAAS; Am Asn Thoracic Surg; Am Col Cardiol; Am Col Surg; Am Surg Asn. Res: Cardiac physiology; extracorporeal circulation and its biologic effects; combined cardiopulmonary transplantation. Mailing Add: Dept of Cardiovasc Surg Harvard Med Sch Boston MA 02115

CASTANER, DAVID, b New York, NY, Aug 4, 34; m 62; c 3. SYSTEMATIC BOTANY. Educ: City Col New York, BS, 61; Iowa State Univ, MS, 63, PhD(plant path), 65. Prof Exp: Asst plant path, Iowa State Univ, 61-65; from asst prof to assoc prof, 66-73, PROF BOT & CUR HERBARIUM, CENT MO STATE UNIV, 73- Mem: Mycol Soc Am; Bot Soc Am; Am Soc Plant Taxon. Res: Flora of Johnson County, Missouri; genus Carex. Mailing Add: Dept of Biol Cent Mo State Univ Warrensburg MO 64093

CASTANERA, ESTHER GOOSSEN, b Winnipeg, Man, July 19, 20; nat US. BIOCHEMISTRY, NUTRITION. Educ: Univ Man, BSc, 42; Univ Calif, PhD(animal nutrit), 54. Prof Exp: Chemist nitrocellulose & small arms ammunition, Defence Industs, Ltd, Can, 42-44; chemist, Can Breweries Ltd & Graham's Dried Foods, Ltd, 44-45; technician hematol & serol, Can Red Cross Blood Transfusion Serv, 46-48; asst pentosuria, Dept Biol Chem, Sch Med, Creighton Univ, 48-49, Univ Calif, 49-54; asst microenzyme methods, Dept Pharmacol, Sch Med, Univ Wash, 54-55; biochemist cellulose metabolism in rats, Chem Div, US Army Med Nutrit Lab, Colo, 56-57; res biochemist pyridoxine & pregnancy, Univ Calif, 57-58; biochemist res assoc pharmacol, Bio-Med Div, US Radio Defense Lab, 58-60; res biochemist, Univ Calif, Berkeley, 61-65; res chemist, Calif State Dept Pub Health, 65-67; RES BIOCHEMIST, MED CTR, UNIV CALIF, SAN FRANCISCO, 67- Mem: AAAS; fel Am Inst Chem; Am Chem Soc. Res: Radioactive tracers in metabolism; microbiochemical techniques; enzymology. Mailing Add: 1417 Grizzly Peak Blvd Berkeley CA 94708

CASTANO, JOHN ROMAN, b New York, NY, June 10, 26; m 51; c 1. GEOLOGY, GEOCHEMISTRY. Educ: City Col NY, BS, 48; Northwestern Univ, MS, 50. Prof Exp: Asst, Northwestern Univ, 48-50; from geol trainee to staff geologist, 50-75, PROJECT LEADER GEOCHEM SERV, SHELL DEVELOP CO, 75- Mem: Geol Soc Am; Am Asn Petro Geologists; Soc Econ Paleontologists & Mineralogists. Res: Petrology and genesis of ancient and recent sand bodies; environment of deposition of sedimentary iron ores; coal petrology; organic geochemistry. Mailing Add: Shell Develop Co Box 481 Houston TX 77001

CASTATER, ROBERT DEWITT, b Janesville, Wis, Jan 27, 22; m 44; c 2. OPERATIONS ANALYSIS. Educ: Milton Col, BS, 47; State Univ Iowa, MS, 52. Prof Exp: OPERS ANALYST & DIV CHIEF, STRATEGIC AIR COMMAND, OFFUTT AFB, 57- Res: Bomber penetration analysis; war planning; analysis; war games; operational test and evaluation; electronic counter measures. Mailing Add: 11421 Hickory Rd Omaha NE 68144

CASTELFRANCO, PAUL ALEXANDER, b Florence, Italy, Oct 16, 21; nat US; m 54; c 2. PLANT PHYSIOLOGY. Educ: Univ Calif, AB, 43, MS, 50, PhD(agr chem), 54; Harvard Univ, STB, 57. Prof Exp: Asst, Dept Agr Biochem, Univ Calif, 51-54, jr res biochemist, 55-56; USPHS fel biochem, Med Sch, Tufts Col, 57-58; from asst botanist to assoc botanist, 58-65, assoc prof, 65-70, PROF BOT, UNIV CALIF, DAVIS, 70- Concurrent Pos: Guggenheim Mem Found fel, 73-74. Mem: Am Soc Plant Physiologists. Res: Fat and ethanol metabolism; chlorophyll biosynthesis; fate and mode of action of herbicides. Mailing Add: Dept of Bot Univ of Calif Davis CA 95616

CASTELL, JOHN DANIEL, b Guelph, Ont, Apr 19, 43; m 66, 75; c 4. MARINE SCIENCES, NUTRITIONAL BIOCHEMISTRY. Educ: Dalhousie Univ, BSc Hons, 65, MSc, 66; Ore State Univ, PhD(food sci), 70. Prof Exp: Fel, Hormel Inst, Univ Minn, 70; RES SCIENTIST MARINE NUTRIT, DEPT ENVIRON, FISHERIES & MARINE SERV, 71- Mem: Nat Shellfish Asn; World Maricult Soc. Res: Nutritional requirements of marine species of commercial interest for aquaculture with a main emphasis on lobsters and minor emphasis on oysters and salmonids. Mailing Add: Fisheries & Marine Serv Dept Environ PO Box 429 Halifax NS Can

CASTELLAN, GILBERT WILLIAM, b Denver, Colo, Nov 21, 24; m 56; c 4. PHYSICAL CHEMISTRY. Educ: Regis Col, Colo, BS, 45; Cath Univ, PhD(chem), 49. Hon Degrees: ScD, Regis Col, Colo, 67. Prof Exp: AEC fel theoret physics, Univ Ill, 49-50; from instr to prof chem, 50-69, asst head dept, 63-65, assoc dean phys sci & eng, grad sch, 69-74, PROF CHEM, UNIV MD, COLLEGE PARK, 69-, ASSOC CHMN DEPT, 74- Concurrent Pos: Consult, Naval Res Lab, 56-63 & Melpar, Inc, 63-67; NSF fel, Max Planck Inst Phys Chem, Göttingen, 62-63. Mem: Am Phys Soc; Electrochem Soc; Am Chem Soc. Res: Chemical relaxation; electrochemical thermodynamics and kinetics. Mailing Add: Dept of Chem Univ of Md College Park MD 20742

CASTELLANI, ANTHONY GEORGE, microbiology, see 12th edition

CASTELLANI, MARIA, b Milano, Italy, July 29, 96; nat US. MATHEMATICS. Educ: Univ Rome, DrMath, 19, Actuary, 20, privat dozent, 40; Univ Geneva, privat dozent, 32. Prof Exp: Consult statistician, Censur Bur, Italy, 20; actuary, Nat Security Fund, 21-24; ed in chief, Social Security Rev, 25-30; actuary, Int Labor Off, League of Nations, Switz, 30-32; chief actuary, Nat Security Fund, Italy, 32-44; assoc prof math, Univ Rome, 42-43; instr, US Armed Forces Inst, 44; prof, Univ Kansas City, 46-59; prof, 59-65, EMER PROF MATH, FAIRLEIGH DICKINSON UNIV, 65- Concurrent Pos: Lectr, Univ Geneva, 32-36. Mem: Am Statist Asn; Am Math Soc; Math Asn Am; Inst Math Statist; Int Cong Actuaries. Res: Theory of random variables according to statistics and insurance; frequency curves and convergency in probabilities; geometry; functional analysis. Mailing Add: Via Paisiello 47 Rome Italy

CASTELLANO, GABRIEL ANGELO, virology, see 12th edition

CASTELLI, JOHN P, b Boston, Mass, Nov 7, 16; m 51; c 4. RADIO ASTRONOMY, SOLAR PHYSICS. Educ: Boston Coll, AB, 38, MA, 39. Prof Exp: Radio engr, 46-49, electronics scientist, 49-55, SUPVRY RES ELECTRONICS ENGR, RADIO ASTRON BR, AIR FORCE CAMBRIDGE RES LABS, 55-, CHIEF SOLAR SECT, TRANS-IONOSPHERIC PROPAGATION BR, 65- Concurrent Pos: Mem Comn V, Int Union Radio Sci. Honors & Awards: Tech Achievement Award, Off Aerospace Res, 68. Mem: Inst Elec & Electronic Engrs; Am Astron Soc; Int Astron Union; Int Union Radio Sci. Res: Application of solar radio burst spectral parameters in prediction and warning of proton events and other geophysical phenomena. Mailing Add: 125 Hillside Ave Arlington MA 02174

CASTELLI, MARY ROSE, plant evolution, systematic botany, see 12th edition

CASTELLI, WALTER ANDREW, b Iquique, Chile, May 5, 29; m 53; c 2. ANATOMY. Educ: Univ Chile, DDS, 57; Univ Mich, MS, 64. Prof Exp: Instr anat, Concepcion Univ, 57-62, instr anat & res assoc physiopath, 64-66; asst prof, 66-70, ASSOC PROF ANAT, UNIV MICH, ANN ARBOR, 70- Concurrent Pos: W K Kellogg Found fel anat, Univ Mich, 62-64. Mem: Int Asn Dent Res. Res: Gross anatomy; vascular studies; head and neck area; experimental tooth transplants in monkeys. Mailing Add: Dept of Anat 3725 Med Sci II Univ of Mich Ann Arbor MI 48104

CASTELLINO, FRANCIS JOSEPH, b Pittston, Pa, Mar 7, 43; m 65; c 2. BIOCHEMISTRY. Educ: Univ Scranton, BS, 64; Univ Iowa, MS, 66, PhD(biochem), 68. Prof Exp: NIH fel biochem, Duke Univ, 68-70; asst prof, 70-74, ASSOC PROF BIOCHEM, UNIV NOTRE DAME, 74- Res: Structure function relationships in proteins. Mailing Add: Dept of Chem Univ of Notre Dame Notre Dame IN 46556

CASTELLION, ALAN WILLIAM, b North Tonawanda, NY, May 1, 34; m 61; c 2. PHARMACOLOGY. Educ: Univ Buffalo, BS, 56; Univ Utah, PhD(pharmacol), 64. Prof Exp: Teaching asst pharmacog, Univ Utah, 56-58, pharm, 58-59 & pharmacol, 59-63, res asst, 63-64; sr pharmacologist, Neuropharmacol Sect, Dept Neurol & Cardiol, Res & Develop Div, Smith Kline & French Labs, 66-67, sr scientist, Dept Pharmacol, 67-69; CHIEF, SECT PHARMACOL, NORWICH PHARMACAL CO, 70- Concurrent Pos: Smith Kline & French res fel, Col Med, Univ Utah, 64-66.

Honors & Awards: Lunsford Richardson Pharm Award, 64. Mem: AAAS; Am Pharmaceut Asn; sr mem Acad Pharmaceut Sci; NY Acad Sci. Res: Basic pharmacology of the central nervous, autonomic and cardiovascular systems; psychopharmacology; drug screening and development. Mailing Add: Bradley Hill Rd Oxford NY 13830

CASTELLION, GEORGE AUGUSTUS, physical chemistry, see 12th edition

CASTELLO, ROBERT ANTHONY, b Bridgeport, Conn, Oct 22, 30; m 58; c 3. PHARMACEUTICAL CHEMISTRY. Educ: Univ Conn, BS, 52, MS, 59; Univ Mich, PhD(pharmaceut chem), 62. Prof Exp: Res assoc pharmaceut res, Res Labs, 62-70, mgr process eng & develop, 70-73, SR MGR PHARMACEUT QUAL CONTROL, MERCK SHARP & DOHME, 73- Mem: Am Pharmaceut Asn; Acad Pharmaceut Sci; NY Acad Sci; Am Inst Chem. Res: Rheology as applied to pharmaceutical product development. Mailing Add: Merck Sharp & Dohme West Point PA 19486

CASTEN, RICHARD FRANCIS, b New York, NY, Nov 1, 41; m 64. NUCLEAR PHYSICS. Educ: Col of the Holy Cross, BS, 63; Yale Univ, MS, 64, PhD(physics), 67. Prof Exp: Fel nuclear physics, Niels Bohr Inst, Univ Copenhagen, 67-69 & Los Alamos Sci Lab, 69-71; from asst physicist to assoc physicist, 71-76, PHYSICIST, BROOKHAVEN NAT LAB, 76- Mem: Sigma Xi; NY Acad Sci; AAAS; Am Phys Soc. Res: Nuclear structure, especially collective modes such as pairing, vibrational, rotational excitations, Coulomb excitation, one and two nucleon transfer reactions, gamma ray deexcitations, Nilsson, Coriolis models. Mailing Add: Physics Dept Brookhaven Nat Lab Upton NY 11973

CASTEN, RICHARD G, b Philadelphia, Pa, May 14, 43; m 70. APPLIED MATHEMATICS. Educ: Temple Univ, AB, 65; Calif Inst Technol, PhD(appl math), 70. Prof Exp: Math technician magnetohydrodyn, Gen Elec Co, 64-65; math technician satellite orbits, Aerospace Corp, 66; res fel appl math, Calif Inst Technol, 70; ASST PROF MATH, PURDUE UNIV, WEST LAFAYETTE, 70- Mem: Soc Indust & Appl Math. Res: Biomathematics. Mailing Add: Dept of Math Purdue Univ West Lafayette IN 47907

CASTENHOLZ, RICHARD WILLIAM, b Chicago, Ill, May 9, 31; m 54; c 2. BOTANY, MICROBIOLOGY. Educ: Univ Mich, BS, 52; Wash State Univ, PhD(bot), 57. Prof Exp: Asst bot, Wash State Univ, 53-57; from instr to assoc prof, 57-69, PROF BIOL, UNIV ORE, 69- Concurrent Pos: John Simon Guggenheim fel, 70-71. Mem: Ecol Soc Am; Phycol Soc Am; Am Soc Limnol & Oceanog; Am Soc Microbiol; Brit Soc Gen Microbiol. Res: Physiology and ecology of marine and freshwater algae; biology of thermophilic microorganisms. Mailing Add: Dept of Biol Univ of Ore Eugene OR 97403

CASTER, KENNETH EDWARD, b New Albany, Pa, Jan 26, 08; m 33. GEOLOGY, PALEONTOLOGY. Educ: Cornell Univ, AB, 29, MS, 31, PhD(stratig), 33. Prof Exp: Asst entom, Cornell Univ, 28-30, asst geol, 29-30, instr paleont, 30-32, instr geol, 32-35; researcher, Paleont Res Inst, NY, 35-36; cur paleont, Mus, 36-40, from asst prof to assoc prof, 40-52, PROF GEOL, UNIV CINCINNATI, 52-, FEL, GRAD SCH, 37- Concurrent Pos: Asst head dept sci, NY State Norm Sch, Geneseo, 35-36; Nat Res Coun grant-in-aid, 35-37; trustee, Paleont Res Inst, NY, 39-, vpres, 41-43 & 65, pres, 44, 45, 51-54 & 66; trustee, Cushman Found, 51-56; vis prof awards, US Dept State, 45-47; prof, Sao Paulo, 45-48; mem, Pan Am Cong Mining, Eng & Geol, 47 & US comn geol, 60-64; Guggenheim fel geol studies, SAm, 47-48, SAfrica, 54-55, Australia & NZ, 56; Fulbright vis fel, Univ Tasmania, 55-56 & Univ Cologne, 64; German Res Asn grant, 64; US rep, Int Paleont Union, 60-; US off del, Int Geol Cong, 60 & 64. Honors & Awards: Gondwana Medal, India, 55. Mem: AAAS; Am Asn Petrol Geol; fel Paleont Soc (secy, 47-55, vpres, 57, pres, 59); fel Geol Soc Am; Soc Vert Paleont. Res: Invertebrate paleontology; paleozoic stratigraphy; paleogeography; Southern Hemisphere historical geology; early echinoderm history. Mailing Add: 425 Riddle Rd Cincinnati OH 45220

CASTER, WILLIAM OVIATT, b Topeka, Kans, Dec 7, 19; m 43; c 4. NUTRITION. Educ: Univ Wis, BA, 42, MS, 44; Univ Minn, PhD(physiol chem), 48. Prof Exp: Asst inorg chem, Univ Wis, 42-43; chemist, Lab Physiol Hyg, Univ Minn, 44-46, chemist, Physiol Hyg, 46-47, asst prof physiol chem, 51-63; biochemist, Nutrit Unit, USPHS, 48-51; assoc prof, 63-70, PROF NUTRIT, UNIV GA, 70- Concurrent Pos: USPHS fel, Nat Heart Inst, 56-61. Mem: AAAS; Am Chem Soc; Am Soc Biol Chemists; Am Inst Nutrit; Brit Nutrit Soc. Res: Human nutrition; water and electrolyte metabolism; radiobiology; chemistry and function of cardiovascular system. Mailing Add: 368 Dawson Hall Sch of Home Econ Univ of Ga Athens GA 30602

CASTILLO, JESSICA MAGUILA, b Zamboanga City, Philippines, Jan 6, 38. INSECT PATHOLOGY, NEMATOLOGY. Educ: Univ Philippines, BS, 58; Univ Mass, MS, 65; Univ Md, PhD(plant path), 68. Prof Exp: Asst instr zool, Univ Philippines, 58-62; res assoc nematol, Univ Mass, 68-70; assoc prof biol, Mindanao State Univ, 70-72; res assoc nematol, Univ Mass, 72-74; asst dir microbiol, Quipse Labs, Ill, 74; FEL, BOYCE THOMPSON RES INST, 75- Mem: AAAS; Am Inst Biol Sci; Soc Nematologists; Soc Invert Path. Res: Culture in vitro of the fungus Coelomomyces punctatus, an obligate parasite of the larvae of Anopheles quadrimaculatus; mosquito tissue culture work. Mailing Add: Boyce Thompson Inst Plant Res 1086 N Broadway Yonkers NY 10701

CASTLE, EDWARD SEARS, physiology, deceased

CASTLE, GORDON BENJAMIN, b Portland, Ind, Aug 10, 06; m 31; c 2. ZOOLOGY. Educ: Wabash Col, AB 28; Univ Calif, AM, 30, PhD(zool), 34. Prof Exp: Asst biol, Termite Invest Comt, Univ Calif, 28-31; from instr to prof, 34-62, acting head dept, 37-38, chmn dept, 38-49, dir biol sta, 38-62, sr acad dean & dean col arts & sci, 49-52, dean grad sch, 52-57, acting pres, dept 8-59, chmn zool, 62-64, vpres univ, 64-67, PROF ZOOL, ARIZ STATE UNIV, 62- Mem: AAAS; Ecol Soc Am; Am Soc Zoologists. Res: Social and aquatic insects. Mailing Add: Dept of Zool Ariz State Univ Tempe AZ 85281

CASTLE, JOHN EDWARDS, b Minneapolis, Minn, June 10, 19; m 42; c 3. APPLIED CHEMISTRY. Educ: Carleton Col, BA, 40; Univ Wis, PhD(org chem), 44. Prof Exp: Asst chem, Univ Wis, 40-42, tech asst, Off Sci Res & Develop Contract, 42-43; chemist cent res dept, 43-50, res supvr, 50-61, asst lab dir, 61-63, assoc dir mat res, 63-65, lab dir, Electrochem Dept, 65-68, asst res dir, 68-72, mgr environ prods sect, Indust Chem Dept, 72-74, RES MGR, ENERGY & MAT DEPT, E I DU PONT DE NEMOURS & CO, INC, 74- Mem: Am Chem Soc; Am Inst Chemists; Chem Soc London; Am Inst Chem Engrs. Res: Contact catalysis; organic synthesis; polymers; biochemistry; coal gasification and liquefaction. Mailing Add: Energy & Mat Dept Du Pont Co Wilmington DE 19898

CASTLE, JOHN GRANVILLE, JR, b Buffalo, NY, Sept 9, 24; m 46; c 5. MAGNETIC RESONANCE, PHYSICAL OPTICS. Educ: State Univ NY Buffalo, BA, 47; Yale Univ, PhD(physics), 50. Prof Exp: Instr physics, State Univ NY Buffalo, 50-51; assoc res physicist, Cornell Aeronaut Lab, 51-52; res assoc, State Univ NY Buffalo, 53-55; res physicist, Westinghouse Res Lab, 55-58, adv physicist, Westinghouse Res & Develop Ctr, 58-67; prof elec eng & res assoc, Learning Res & Develop Ctr, Univ Pittsburgh, 67-69; dir div natural sci & math, 70-71, PROF PHYSICS, UNIV ALA, HUNTSVILLE, 69- Concurrent Pos: Vis scientist, High Energy Physics Lab, Stanford Univ, 70-71; clin prof therapeut radiol, Med Ctr, Univ Ala, Birmingham, 70-; consult, US Army Missile Command, 72- Mem: AAAS; Am Phys Soc. Res: Cryogenic engineering; solid state physics; medical instrumentation; educational media; microwave and optical properties of solids; infrared properties of gases; biomedical engineering. Mailing Add: 4018 Heatherhill Rd SE Huntsville AL 35802

CASTLE, MANFORD C, b Coeburn, Va, Apr 27, 42; m 69; c 2. PHARMACOLOGY. Educ: Berea Col, BA, 65; Univ Kans, PhD(pharmacol), 72. Prof Exp: Vol chem teacher, Peace Corps, Univ Antioquia, Colombia, 66-68; clin instr pharmacol, Univ Kans, 72-73; staff assoc & NIH fel, Nat Inst Gen Med Sci, 73-75; STAFF ASSOC PHARMACOL, NAT CANCER INST, 75- Mem: AAAS. Res: Metabolism and disposition of drugs, particularly cardiac glycosides and cancer chemotherapeutic agents; alteration of these processes by other drugs; methods involve use of radioactive tracers and high-pressure liquid chromatography. Mailing Add: Lab of Chem Pharmacol Div Cancer Treat Nat Cancer Inst Bethesda MD 20014

CASTLE, PETER MYER, b Detroit, Mich, Apr 19, 40; m 61; c 3. CHEMICAL PHYSICS, HIGH TEMPERATURE CHEMISTRY. Educ: Univ Mich, BS, 62; Purdue Univ, PhD(phys chem), 70. Prof Exp: Vis asst prof phys chem, Purdue Univ, 70; SR ENGR PHYS CHEM, WESTINGHOUSE RES LABS, 70- Concurrent Pos: Instr night sch, Community Col Allegheny County, Boyce Campus, 74- Mem: Am Soc Mass Spectrometry. Res: Studies of infrared, visible and near ultraviolet laser photolysis reactions for synthetic applications and understanding of fundamental photophysical processes; high temperature mass spectrometric investigations of liquid vapor and solid-vapor equilibria in fused salt systems. Mailing Add: Westinghouse Res Labs Beulah Rd Pittsburgh PA 15235

CASTLE, RAYMOND NIELSON, b Boise, Idaho, June 24, 16; m 37; c 8. CHEMISTRY. Educ: Univ Idaho, BS, 39; Univ Colo, MA, 41, PhD(chem), 44. Prof Exp: Asst chem, Univ Colo, 39-42; instr, Univ Idaho, 42-43; instr, Univ Colo, 43-44; res engr, Battelle Mem Inst, 44-46; asst prof chem, Univ NMex, 46-47, pharmaceut chem, 47-50, from assoc prof to prof chem, 50-70, chmn dept, 63-70; PROF CHEM, BRIGHAM YOUNG UNIV, 70- Concurrent Pos: Res fel, Univ Va, 52-53; ed, J Heterocyclic Chem. Mem: Am Chem Soc; fel The Chem Soc. Res: Organic chemistry; optical crystallography; optical crystallographic properties of organic compounds; synthesis of cinnolines; pyridazines; pyridines; other related heterocycles of medicinal interest. Mailing Add: Dept of Chem Brigham Young Univ Provo UT 84601

CASTLE, RICHARD THOMAS, b Urbana, Ohio, Apr 11, 34; m 55; c 1. PHYSICS. Educ: Otterbein Col, BS, 56; Ohio Univ, MS, 61, PhD(physics), 63. Prof Exp: Res assoc, Ohio Univ, 60-63; sr scientist, 63-70, DIV CHIEF, BATTELLE MEM INST, 70- Res: Nuclear spectroscopy and structure; atomic processes and ionization phenomena. Mailing Add: Battelle Mem Inst 505 King Ave Columbus OH 43201

CASTLE, ROBERT O, b Berkeley, Calif, Dec 31, 26; m 54; c 2. GEOLOGY, TECTONICS. Educ: Stanford Univ, BS, 48; McGill Univ, MSc, 49; Univ Calif, Los Angeles, PhD(geol), 64. Prof Exp: Instr geol, Univ Mass, 49-50; geol field asst, 50-51, geologist, 51-52, GEOLOGIST, US GEOL SURV, 54- Mem: Geol Soc Am; Asn Eng Geologists. Res: Tectonics engineering geology; igneous petrology; contemporary crustal deformation. Mailing Add: 730 Torreya Ct Palo Alto CA 94303

CASTLE, WILLIAM BOSWORTH, b Cambridge, Mass, Oct 21, 97; m 33; c 2. CLINICAL MEDICINE, HEMATOLOGY. Educ: Harvard Univ, MD, 21; FRACP; FRCP, 64. Hon Degrees: SM, Yale Univ, 33; MD, Univ Utrecht, 36; SD, Univ Chicago, 52; LLD, Jefferson Med Col, 64; DSc, Harvard Univ, 64, Univ Pa, 66, Marquette Univ, 69, Mt Sinai Sch Med, 72; LHD, Boston Col, 66; hon FRCPS(C), 65; hon FRCP(E), 67. Prof Exp: Intern, Mass Gen Hosp, 21-23; asst physiol, Sch Pub Health, 23-25, instr, Med Sch, 24-25, asst med, 25-29, from instr to prof, 29-57, George Richards Minot prof, 57-63, Francis Weld Peabody fac prof, 63-68, EMER FRANCIS WELD PEABODY FAC PROF MED, HARVARD UNIV, 68- Concurrent Pos: Dir anemia comn, Rockefeller Found, PR, 31; assoc dir, Thorndike Mem Lab, Boston City Hosp, 32-48, dir lab, 48-63, dir second & fourth med serv, 40-63, hon dir lab, 63-; distinguished physician, Vet Admin, 68-72; sr physician, Vet Admin Hosp, West Roxbury, Mass, 72-73, consult, 73-75. Honors & Awards: Phillips Prize, Am Col Physicians, 32; Perpetual Student, Med Col, St Bartholomew's Hosp, London, 70. Mem: Nat Acad Sci; AAAS; master Am Col Physicians; fel Am Inst Nutrit, 73; hon mem Am Soc Hemat. Res: Clinical medicine; etiology and therapy of pernicious anemia, sprue; anemias in pregnancy; hookworm anemia; relation asmotic fragility to cell shape; hemolytic effects of oxidants; mechanism of hemolytic anemias; splenic filtration; bolld viscosity of sicklemia; vitamin B12 intrinsic factor relations. Mailing Add: 22 Irving St Brookline MA 02146

CASTLEBERRY, GEORGE E, b Herring, Okla, Mar 20, 18; m 44. INORGANIC CHEMISTRY. Educ: Southwestern State Col, Okla, BS, 39; Univ Okla, MS, 51, PhD(sci educ), 68. Prof Exp: Prin & teacher, E Walnut Schs, Okla, 40-41; teacher, Clinton High Sch, 41-42, teacher & asst prin, 45-58; PROF CHEM, SOUTHWESTERN STATE COL, OKLA, 58- Mem: Am Chem Soc; Nat Sci Teachers Asn. Res: Science education; graduate education of secondary school science teachers of Oklahoma public schools; orthoaminobenzenethiol as a reagent for the gravimetric determination of selenium in compounds. Mailing Add: Dept of Chem Southwestern State Col Weatherford OK 73096

CASTLEMAN, ALBERT WELFORD, JR, b Richmond, Va, Jan 7, 36; m 76; c 2. CHEMICAL PHYSICS. Educ: Rensselaer Polytech Inst, BChE, 57; Polytech Inst Brooklyn, MSc, 63, PhD, 69. Prof Exp: Engr, Olin Mathieson Chem Corp, 57-58; engr, Brookhaven Nat Lab, 58-66, scientist, 66-75; PROF CHEM, UNIV COLO, 75- Concurrent Pos: Adj prof atmospheric chem, Dept Earth & Space Sci & Mech, State Univ NY Stony Brook, 73-75; fel, Coop Inst Res Environ Sci, Univ Colo, 75-; consult, Mfg Chemists Asn; consult, Adv Comt Reactor Safeguard, AEC, 58-75, Reactor Safety Div, US Nuclear Regulatory Comn, 75-; adv, Nat Ctr Atmospheric Res, 75-; mem, Subcomn Ions, Aerosols & Radioactivity, Int Comn Atmospheric Electricity, Int Asn Meteorol & Atmospheric Physics. Mem: Am Chem Soc; Am Phys Soc; Am Geophys Union; Int Asn Meteorol & Atmospheric Physics (secy, 74-75). Res: Nucleation phenomena, molecular properties of small clusters; kinetics of association reactions, statistical mechanics; aerosol and surface chemistry; atmospheric chemistry. Mailing Add: Dept of Chem Univ of Colo Boulder CO 80309

CASTLEMAN, BENJAMIN, b Everett, Mass, May 17, 06; m 35; c 3. PATHOLOGY. Educ: Harvard Univ, BA, 27; Yale Univ, MD, 31. Hon Degrees: MD, Gothenburg Univ, 66. Prof Exp: Instr path, 35-42, assoc, 43-48, asst prof, 48-53, clin prof, 53-61, prof, 61-70, Shattuck prof path anat, 70-72, EMER SHATTUCK PROF PATH ANAT, HARVARD MED SCH, 72-; SR CONSULT PATH, MASS GEN HOSP, 74- Concurrent Pos: Asst pathologist, Mass Gen Hosp, 35-42, pathologist, 42-43, chief dept path, 53-74; consult pathologist, hosps. Mem: Am Asn Path & Bact; Am Soc Clin Path; Am Soc Exp Path; Am Acad Arts & Sci; Int Acad Path. Res: Hypertension; parathyroid and thymus glands; myasthenia gravis; pulmonary disease. Mailing Add: Dept of Path Mass Gen Hosp Boston MA 02114

CASTLES, THOMAS R, b St Louis, Mo, Oct 27, 37; m 59; c 3. PHARMACOLOGY. Educ: Grinnell Col, BA, 59; Univ Iowa, MS, 62, PhD(pharmacol), 65. Prof Exp: From assoc pharmacologist to sr pharmacologist, 67-70, PRIN PHARMACOLOGIST, MIDWEST RES INST, 70- Mem: Am Soc Pharmacol & Exp Therapeut; Soc Am Toxicol. Res: Thiazide diuretics; aldosterone; carbonic anhydrose inhibitors; pharmacology and toxicology of antimalarial compounds; analgesic tolerance. Mailing Add: Midwest Res Inst 425 Volker Blvd Kansas City MO 64110

CASTLETON, KENNETH BITNER, b Salt Lake City, Utah, July 29, 03; m 31; c 4. SURGERY. Educ: Univ Utah, AB, 23, Univ Pa, MD, 27; Univ Minn, PhD(surg), 33; Am Bd Surg, dipl. Prof Exp: Instr anat, 33-34, phys diagnosis, 35-36, surg anat, 38-43, assoc clin prof surg, 43-62, prof surg & dean, Col Med, 62-69, vpres med affairs, 69-71, EMER PROF SURG, UNIV UTAH, 71- Concurrent Pos: Pvt pract, 33-62. Mem: Fel Am Col Surg. Res: Clinical surgery; gastroenterology; experimental physiology. Mailing Add: 1235 E 2nd South 303 Salt Lake City UT 84102

CASTNER, HENRY WALKER, b Louisville, Ky, May 3, 32; m 64; c 1. GEOGRAPHY. Educ: Centre Col, BA, 55; Vanderbilt Univ, BMEng, 55; Univ Pittsburgh, MA, 60; Univ Wis, PhD(geog), 64. Prof Exp: Div salesman, Westinghouse Elec Corp, 55; from asst prof to assoc prof, 64-71, PROF GEOG, QUEEN'S UNIV, ONT, 71- Honors & Awards: Cert of Merit, Am Cong Surveying & Mapping, 74. Mem: Asn Am Geog; Am Geog Soc; Can Asn Geog; Soc Hist Discovery. Res: Cartography. Mailing Add: Dept of Geog Queen's Univ Kingston ON Can

CASTNER, THEODORE GRANT, JR, b Orange, NJ, June 17, 30; m 67; c 2. SOLID STATE PHYSICS. Educ: Cornell Univ, BEng, 53; Univ Ill, MS, 55, PhD(physics), 58. Prof Exp: Physicist, Gen Elec Res Labs, 58-63; assoc prof, 63-70, PROF PHYSICS, UNIV ROCHESTER, 70- Concurrent Pos: Guggenheim fel, Swiss Fed Inst Technol, 69-70. Mem: Fel Am Phys Soc; Am Asn Phys Teachers. Res: Electron spin resonance phenomena in solids; impurities in insulators, semiconductors; cooperative phenomena in magnetic materials; spin-lattice relaxation; dielectric phenomena and Metal-Insulator transition in doped semiconductors. Mailing Add: Dept of Physics & Astron Univ of Rochester Rochester NY 14627

CASTO, BRUCE CORDELL, animal virology, see 12th edition

CASTO, CLYDE CHRISTY, b Edmond, Okla, Oct 8, 12; m 42; c 2. ANALYTICAL CHEMISTRY. Educ: Cent State Col, BS, 36; Univ Mich, MS, 48. Prof Exp: Anal chemist, George W. Gooch Lab, Los Angeles, 36-38, anal chemist, Commercial Lab, 38-40; res chemist, Consol Electrodyn Corp, Calif, 40-41; anal chemist, George W. Gooch Lab, Los Angeles, 41-42; anal res chemist, Basic Magnesium, Nev, 42-44; head anal res & develop, Tenn Eastman Corp, Oak Ridge, 44-46; res chemist, Rayon Dept, E I du Pont de Nemours & Co, 48-50, res supvr textile fibers dept, Nylon Res Div, 50-56; sect head anal instrumental, Am Potash & Chem Corp, Nev, 56-57; lab dir, vpres & partner, George W. Gooch Labs, Calif, 57-59; staff chemist, Aerojet-Gen Corp, 59-60; mgr phys & anal chem dept, Rocket Power, Inc, Ariz, 60-64; CHEMIST, OFF STATE CHEMIST, 64- Mem: Am Chem Soc. Res: Light metals; uranium chemistry; nylon and polyesters; chemicals and high energy fuels; solid propellants; agricultural chemicals. Mailing Add: 2149 Nicklaus Dr Mesa AZ 85205

CASTON, J DOUGLAS, b Ellenboro, NC, June 16, 32; m 58; c 2. BIOCHEMISTRY, EMBRYOLOGY. Educ: Lenoir-Rhyne Col, BA, 54; Univ NC, MA, 58; Brown Univ, PhD(biol), 61. Prof Exp: From sr instr to asst prof anat, 63-71, ASSOC PROF ANAT, SCH MED, CASE WESTERN RESERVE UNIV, 71- Concurrent Pos: Fel develop biol, Carnegie Inst, 61-63. Mem: Biophys Soc; Am Soc Gen Physiol; Soc Develop Biol. Res: Control of metabolic pathways during development; catecholamine synthesis; protein and ribonucleic acid synthesis. Mailing Add: Dept of Anat Sch of Med Case Western Reserve Univ Cleveland OH 44106

CASTON, RALPH HENRY, b Akron, Ohio, Nov 22, 15; m 46; c 3. PHYSICS. Educ: Univ Akron, MS, 39; Univ Notre Dame, PhD(physics), 42. Prof Exp: Mem staff, Nat Defense Res Comt Proj, Radiation Lab, Mass Inst Technol, 42-45; MEM TECH STAFF, KIMBERLY-CLARK RES & DEVELOP LAB, 45- Mem: Am Phys Soc. Res: Physics of rubber; radar; general physics in the paper industry; operations research; computer process control. Mailing Add: Kimberly-Clark Res & Develop Lab Neenah WI 54956

CASTOR, CECIL WILLIAM, b Detroit, Mich, Oct 9, 25; m 48; c 3. INTERNAL MEDICINE, RHEUMATOLOGY. Educ: Univ Mich, MD, 51; Am Bd Internal Med, dipl, 58. Prof Exp: Intern & resident internal med, 51-55; from instr to assoc prof internal med, 55-67, PROF INTERNAL MED, MED SCH, UNIV MICH, ANN ARBOR, 67- Concurrent Pos: Arthritis & Rheumatism Found fel, 55-57, sr investr, 57-62; USPHS career res develop award, 63-67; mem study sect gen med, NIH, 70-74; fel comt & prof educ comt, Arthritis Found. Mem: Fel Am Col Physicians; Am Soc Clin Invest; Am Fedn Clin Res; Am Rheumatism Asn; Tissue Cult Asn. Res: Regulation of connective tissue cells, in vitro and in vivo, especially on the hormonal control of mucopolysaccharide and collagen synthesis. Mailing Add: Dept of Internal Med Univ of Mich Med Ctr Ann Arbor MI 48104

CASTOR, JOHN I, b Fresno, Calif, Jan 5, 43. ASTROPHYSICS. Educ: Fresno State Col, BS, 61; Calif Inst Technol, PhD(astron), 67. Prof Exp: Res fel physics, Calif Inst Technol, 66-67; res assoc astrophys, 67-69, asst prof, 69-72, ASSOC PROF PHYSICS & ASTROPHYS, UNIV COLO, BOULDER, 72-, FEL JOINT INST LAB ASTROPHYS, 70- Mem: Fel Royal Astron Soc; Am Astron Soc; Int Astron Union. Res: Stellar interiors; pulsating stars; radiative transfer; stellar radiation hydrodynamics. Mailing Add: Joint Inst Lab Astrophys Univ of Colo Boulder CO 80309

CASTOR, LAROY NORTHROP, b Philadelphia, Pa, Sept 27, 24; m 53; c 4. CELL BIOLOGY. Educ: Mass Inst Technol, BS, 48; Univ Pa, PhD(biophys), 54. Prof Exp: Am Cancer Soc fel, Univ Toronto, 55; fel, Univ Pa, 56-57; assoc biophysics, Johnson Res Found, 57-63; sr investr cell biol, Biochem Res Found, Del, 63-66; ASST MEM, INST CANCER RES, 66- Mem: Am Soc Cell Biol; Tissue Cult Asn; Am Asn Cancer Res. Res: Photochemical action spectra of cytochrome oxidases; growth control in cell culture. Mailing Add: Inst for Cancer Res 7701 Burholme Ave Fox Chase Philadelphia PA 19111

CASTOR, WILLIAM STUART, JR, inorganic chemistry, physical chemistry, see 12th edition

CASTRILLON, JOSE P A, b Buenos Aires, Arg, Jan 4, 26; m 64; c 2. ORGANIC CHEMISTRY, RADIOCHEMISTRY. Educ: Univ Buenos Aiers, Dr(chem), 51. Prof Exp: Scientist chem, Atanor, SAM, Arg, 50-52, Squibb, SA, 53-55, Arg AEC, 56-64 & PR Nuclear Ctr, 64-73; DIV HEAD PHYS SCI, PR NUCLEAR CTR, 73- Concurrent Pos: Fel Arg Coun, Univ Colo, 60-61; prof org chem, Univ Asuncion, 71-72. Mem: Am Chem Soc; Arg Chem Soc. Res: New solvents and solutes in liquid scintillation counting; synthesis of sulfur heterocycles as potential schistosomicides, trypanosomicides and antitumor agents; conformational studies of flexible ring systems. Mailing Add: PR Nuclear Ctr San Juan PR 00935

CASTRO, ALBERT JOSEPH, b Santa Clara, Calif, May 13, 16; m 37; c 3. ORGANIC CHEMISTRY. Educ: San Jose State Col, AB, 39; Stanford Univ, AM, 42, PhD(org chem), 45. Prof Exp: Res chemist, Calif Res Corp, 44-47; asst prof chem, Univ Ariz, 47-49; from asst prof to assoc prof, 49-58, PROF CHEM, SAN JOSE STATE UNIV, 58- Concurrent Pos: Asst, Univ Santa Clara, 49; consult, Nassau Chems, Inc, 49-53; vis lectr, Johns Hopkins Univ, 54-55. Mem: Fel Am Inst Chemists; Am Chem Soc; Brit Chem Soc. Res: Composition of ether solutions of the Grignard reagent; synthesis and structure of compounds related to chelidonine; reaction of epoxides and ammonia; synthesis of arylethylenes; Friedel-Crafts reaction; structure of compounds from Serratia marcescens; chemistry of pyrroles; compounds of possible pharmacological value. Mailing Add: Dept of Chem San Jose State Univ San Jose CA 95192

CASTRO, ALBERTO, b San Salvador, El Salvador, Nov 15, 33; m 56; c 5. BIOCHEMISTRY, ENDOCRINOLOGY. Educ: Univ Houston, BS, 58; Univ El Salvador, PhD(biol chem), 62. Prof Exp: From asst prof to prof microbiol & biochem, Univ El Salvador, 58-63, dir grad res & coordr-dir gen studies & preprof curric, 65-66; asst prof pediat, Sch Med, Univ Ore, 70-73; ASSOC PROF MED & PATH & DIR HORMONE RES LAB, SCH MED, UNIV MIAMI, 73- Concurrent Pos: Prof basic sci & head dept, Univ El Salvador, 65-68; NIH sr res fel diabetes & metab, Sch Med, Univ Ore, 66-70; dir endocrinol lab, United Med Lab, 70-73; consult, Union Carbide Corp, NY, 73-; sr scientist, Papanicolaou Cancer Res Inst, 73-75. Mem: AAAS; Am Chem Soc; Am Inst Chem; NY Acad Sci; fel Royal Soc Trop Med & Hyg. Res: Carbohydrate biochemistry, metabolism and modes of action; hormone mechanisms and the interrelationship to hypertension and primary aldosteronism. Mailing Add: Dept of Med Univ of Miami Sch of Med Miami FL 33136

CASTRO, ANTHONY EDWARD, b New York, NY, Mar 4, 35; m 67; c 2. VIROLOGY, IMMUNOLOGY. Educ: NY Univ, BA, 55, MS, 62; Purdue Univ, PhD(virol), 71. Prof Exp: Asst prof virol, Purdue Univ, 70-71; ASST PROF VIROL & IMMUNOL, UNIV MINN, MINNEAPOLIS, 71- Concurrent Pos: Minn Med Found grant, Med Sch, Univ Minn, Minneapolis, 73-74. Mem: AAAS; Am Soc Microbiol. Res: Virological and immunological aspects of carcinogenesis; slow virus diseases of man and animals. Mailing Add: 1181 Raleigh St St Paul MN 55108

CASTRO, ANTHONY J, b Chicago, Ill, Nov 30, 30; m 60; c 3. ORGANIC POLYMER CHEMISTRY. Educ: Univ Chicago, MS, 58, PhD(chem), 62; John Marshall Law Sch, JD, 75. Prof Exp: Res supvr polymer chem, Continental Can Co, 62-65; MGR POLYMER CHEM, ARMAK CO, McCOOK, 65- Mem: AAAS; NY Acad Sci; Am Chem Soc; Soc Plastics Engrs. Res: Investigation into areas of unconventional polymer formation; preparation of novel polymeric materials from conventional polymerision techniques. Mailing Add: 127 S Cuyler Ave Oak Park IL 60302

CASTRO, CHARLES E, b Santa Clara, Calif, Nov 17, 31; m 53; c 7. ORGANIC CHEMISTRY. Educ: San Jose State Col, AB, 53; Univ Calif, Davis, PhD(phys org chem), 57. Prof Exp: Res chemist, Shell Develop Co, Calif, 57-60; from asst chemist to assoc chemist, 60-70, CHEMIST & PROF BIOL CHEM, UNIV CALIF, RIVERSIDE, 70- Concurrent Pos: NSF res grant, 61-; NIH res grant, 63- Mem: Am Chem Soc. Res: Reactions of organics with transition metal species; elmintics; mode of action; mechanisms of biochemical transformations; hemes and hemeproteins. Mailing Add: Dept of Nematol Univ of Calif Riverside CA 92502

CASTRO, GEORGE, b Los Angeles, Calif, Feb 23, 39; m 63; c 4. RESEARCH ADMINISTRATION, PHYSICAL CHEMISTRY. Educ: Univ Calif, Los Angeles, BS, 60; Univ Calif, Riverside, PhD(chem), 65. Prof Exp: Fel chem, Univ Pa, 65-67 & Calif Inst Technol, 67-68; staff mem res, 68-69, proj leader org photoconductors, 69-73, mgr org solids, 73-75, MGR PHYS SCI, IBM SAN JOSE RES LAB, 75- Res: Electronic properties of organic solids. Mailing Add: IBM Res Lab Monterey & Cottle Rds San Jose CA 95193

CASTRO, GILBERT ANTHONY, b Port Arthur, Tex, Apr 24, 39; m 61; c 2. PHYSIOLOGY, MICROBIOLOGY. Educ: Lamar State Col, BS, 61; Univ Ark, MS, 63; Univ Tex, PhD(microbiol), 66. Prof Exp: NIH fel zool, Univ Mass, 66-68; from asst prof to assoc prof parasitol & lab pract, Med Ctr, Univ Okla, 71-72; ASSOC PROF PHYSIOL, UNIV TEX MED SCH HOUSTON, 72- Mem: Am Physiol Soc; Am Soc Parasitol; Soc Exp Biol & Med. Res: Intestinal physiology; host-parasite relationships; physiology and immunology of host-parasite systems; pathogenesis of gastrointestinal parasites. Mailing Add: Dept of Physiol Univ of Tex Med Sch Houston TX 77025

CASTRO, PETER, b Mayagüez, PR, July 20, 43; US citizen. MARINE ZOOLOGY, PARASITOLOGY. Educ: Univ PR, Mayagüez, BS, 64; Univ Hawaii, MS, 66, PhD(zool), 69. Prof Exp: Res assoc, Dept Marine Sci, Univ PR, Mayagüez, 70; instr, Dept Biol Sci, Univ PR, Rio Piedras, 70-71; asst prof, 72-75, ASSOC PROF BIOL SCI, CALIF STATE POLYTECH UNIV, POMONA, 75- Concurrent Pos: Vis investr, Hopkins Marine Sta, Stanford Univ, 71-82 & Smithsonian Trop Res Inst, 74; lectr, Moss Landing Marine Labs, 73. Mem: AAAS; Am Soc Zoologists. Res: Ecological, physiological and behavioral aspects of marine symbioses. Mailing Add: Dept of Biol Sci Calif State Polytech Univ Pomona CA 91768

CASWELL, HAL, b Los Angeles, Calif, Apr 27, 49. ECOLOGY, MATHEMATICAL BIOLOGY. Educ: Mich State Univ, BS, 71, PhD(zool), 74. Prof Exp: NSF fel zool, Mich State Univ, 71-74, res assoc, 74-75; ASST PROF BIOL SCI, UNIV CONN, 75- Mem: Ecol Soc Am; Brit Ecol Soc. Res: Mathematical analysis of ecological processes; population and community theory; plant-herbivore interactions. Mailing Add: Biol Sci Group Univ of Conn Storrs CT 06268

CASWELL, HERBERT HALL, JR, b Marblehead, Mass, May 21, 23; m 48; c 6. ECOLOGY, ORNITHOLOGY. Educ: Harvard Univ, SB, 48; Univ Calif, Los Angeles, MA, 50; Cornell Univ, PhD(zool), 56. Prof Exp: PROF BIOL, EASTERN MICH UNIV, 55-, HEAD DEPT, 74- Mem: Ecol Soc Am; Wilson Ornith Soc; Am Ornith Union; Am Inst Biol Sci; Sigma Xi. Res: Terrestrial ecology. Mailing Add: Dept of Biol Eastern Mich Univ Ypsilanti MI 48197

CASWELL, LYMAN RAY, b Omaha, Nebr, Sept 29, 28; m 64; c 3. ORGANIC CHEMISTRY. Educ: Ind Univ, BS, 49, MA, 50; Mich State Univ, PhD(org chem), 56. Prof Exp: Asst prof chem, Ohio Northern Univ, 55-56; head chem dept, upper Iowa Col, 56-61; from asst prof to assoc prof, 61-68, actg chmn dept, 67-70, PROF CHEM, TEX WOMEN'S UNIV, 68- Mem: AAAS; Am Chem Soc; Am Inst Chemists; Soc Appl Spectros. Res: Heterocyclic compounds; cyclic imbides and anhydrides; ultraviolet visible and fluorescence spectra. Mailing Add: Dept of Chem Tex Woman's Univ Denton TX 76204

CASWELL, RANDALL SMITH, b Eugene, Ore, Feb 7, 24; m 45; c 6. PHYSICS. Educ: Mass Inst Technol, SB, 47, PhD(physics), 51. Prof Exp: Assoc prof physics, Univ Ky, 50-52; res partic solid state physics, Oak Ridge Nat Lab, 52; physicist neutron physics, 52-69, DEP DIR, CTR RADIATION RES, NAT BUR STAND, 69- Concurrent Pos: Adj prof physics, Am Univ, 57-71. Mem: Fel Am Phys Soc; Radiation Res Soc. Res: Neutron cross sections; physics and dosimetry. Mailing Add: Ctr for Radiation Res Nat Bur Stand Washington DC 20234

CASWELL, ROBERT LITTLE, b San Francisco, Calif, Jan 27, 18; m 57; c 1. PESTICIDE CHEMISTRY. Educ: Univ Calif, Berkeley, BS, 39. Prof Exp: Chemist pesticides, Agr Res Serv, USDA, 41-42 & 46-64, asst chief staff, Off Enforcement Chem, Pesticides Regulation Div, 64-70, Off Pesticides, 70-72, CHEMIST, CRITERIA & EVAL DIV, OFF PESTICIDES, ENVIRON PROTECTION AGENCY, 72- Mem: AAAS; Am Chem Soc; Asn Off Anal Chem. Res: Development of chemical methods of analysis for pesticides. Mailing Add: 11626 35th Pl Beltsville MD 20705

CASWELL, WILLIAM BRADFORD, JR, hydrogeology, see 12th edition

CATACOSINOS, PAUL ANTHONY, b New York, NY, Sept 29, 33; m 58; c 3. STRATIGRAPHY, GEOLOGY. Educ: Univ NMex, BA, 57, MA, 62; Mich State Univ, PhD(geol), 72. Prof Exp: Explor geologist, Mountain Fuel Supply Co, 62-66; explor geologist, Consumers Power Co, 67-69; from instr to asst prof, 69-75, ASSOC PROF GEOL, DELTA COL, 75- Mem: Am Asn Petrol Geologists; fel Geol Soc Am; Soc Explor Paleontologists & Mineralogists; Sigma Xi; fel AAAS. Res: Origin and evolution of the Michigan Basin as a structure; detailed analysis of the Cambrian and Ordovician stratigraphic sections. Mailing Add: Dept of Geol Delta Col University Center MI 48710

CATALANOTTO, FRANK ALFRED, b Brooklyn, NY, Aug 4, 44; m 69; c 1. DENTAL RESEARCH, PEDODONTICS. Educ: NJ Col Dent, DMD, 68; Harvard Univ, cert pedodontics, 71. Prof Exp: Instr pedodontics, Sch Dent Med, Harvard Univ, 69-71; assoc epidemiologist, Naval Dent Res Inst, Great Lakes, Ill, 72-74; ASST PROF PEDODONTICS, HEALTH CTR, UNIV CONN, FARMINGTON, 74- Concurrent Pos: Asst pedodontics, Childrens Hosp Med Ctr, Boston, 69-71; Nat Inst Dent Res career develop award, 75. Mem: Am Dent Asn; Int Asn Dent Res; Am Acad Pedodontics. Res: Oral sensation and perception, including taste, smell, touch; physiology and biochemistry of taste and taste bud function. Mailing Add: Dept of Pediat Dent Univ of Conn Health Ctr Farmington CT 06032

CATALFOMO, PHILIP, b Providence, RI, Dec 27, 31; m 62; c 2. PHARMACOGNOSY. Educ: Providence Col, BS, 53; Univ Conn, BS, 58; Univ Wash, MS, 60, PhD(pharmacog), 63. Prof Exp: From asst prof to prof pharmacog, Sch Pharm, Ore State Univ, 63-75, head dept, 66-75; PROF PHARMACOG & DEAN, SCH PHARM, UNIV MONT, 75- Concurrent Pos: Am Found Pharmaceut Educ Gustavus A Pfeiffer Mem res fel, 69-70. Mem: AAAS; Am Soc Pharmacog; Am Pharmaceut Asn; Acad Pharmaceut Sci. Res: Investigation of higher plants and fungi for pharmacologically active components; secondary metabolism of marine fungi and mycorrhizal fungi. Mailing Add: Sch of Pharm Univ of Mont Missoula MT 59801

CATALINE, ELMON LAMONT, b Flint, Mich, June 14, 12; m 34; c 2. PHARMACY. Educ: Univ Mich, BS, 34, MS, 35, PhD(pharm, chem), 38. Prof Exp: Asst prof pharm, Univ Toledo, 37-40; from instr to prof & dean col, 40-69, EMER PROF PHARM & EMER DEAN COL PHARM, UNIV NMEX, 72- Mem: Am Pharmaceut Asn. Res: Organic arsenicals; phytochemistry; emulsifying agents; pharmaceutical history. Mailing Add: PO Box 492 Chama NM 87520

CATANA, ANTHONY J, JR, b Trenton, NJ, Aug 17, 31; m 53; c 2. PLANT ECOLOGY. Educ: Rutgers Univ, BS, 53; Univ Wis, MS, 54, PhD(bot), 60. Prof Exp: Instr bot & zool, Univ Wis, 58-60, instr microbiol, 60, asst prof bot & zool, 60-63; from asst prof to assoc prof ecol, Albion Col, 63-69, dir, Ott Biol Preserve, 64-69, assoc dean fac, 66-69; prof biol & dean, Doane Col, 69-70, vpres acad affairs, 70-74; VPRES ACAD AFFAIRS & PROF BIOL, YANKTON COL, 74- Concurrent Pos: Johnson's Wax award, 62-; mem, Nebr Environ Control Coun, 72-74. Mem: Fel AAAS; Am Inst Biol Sci; Ecol Soc Am. Res: Spatial distribution of plants and ecological sampling methods; distribution of soil bacteria and actinomycetes with grazing; plant community studies in Michigan. Mailing Add: Yankton Col Yankton SD 57078

CATANZARO, EDWARD JOHN, b Jamaica, NY, Nov 4, 33; m 62; c 2. GEOCHEMISTRY. Educ: Brooklyn Col, BS, 55; Univ Wyo, MA, 57; Columbia Univ, PhD(geochem), 62. Prof Exp: Geologist, US Geol Surv, Washington, DC, 62-63; res chemist, Nat Bur Standards, 63-69; asst prof geol, Southampton Col, 69-71; ASSOC PROF CHEM, FAIRLEIGH DICKINSON UNIV, 71- Prof Exp: Vis sr res assoc, Lamont-Doherty Geol Observ, NY, 70- Mem: Geol Soc Am; Geochem Soc; Am Geophys Union. Res: Atomic weights of chemical elements; natural isotopic variations in strontium and lead; geochemistry and petrology of Precambrian rocks; mass spectrometric techniques and development; trace metal concentrations and movements in natural waters. Mailing Add: Dept of Chem Fairleigh Dickinson Univ Teaneck NJ 07666

CATCHINGS, ROBERT MERRITT, III, b Washington, DC, Apr 2, 42; m 70. SOLID STATE PHYSICS. Educ: Univ Mich, BS, 64; Wayne State Univ, MS, 66, PhD(physics), 70. Prof Exp: ASST PROF PHYSICS, HOWARD UNIV, 70- Res: Magnetization, electrical conductivity and nuclear magnetic resonance studies of amorphous solids. Mailing Add: Dept of Physics Howard Univ Washington DC 20001

CATCHPOLE, HUBERT RALPH, b London, Eng, May 13, 06; nat US. PHYSIOLOGY. Educ: Cambridge Univ, BA, 28; Univ Calif, PhD(physiol), 33. Prof Exp: Asst, Univ Calif, 34-35 & Cutter Labs, 35-36; from instr to asst prof physiol, Yale Univ, 36-43, Commonwealth Fund fel, 41-43; assoc path, 46-47, asst prof, 47-51, res assoc prof, 51-61, RES PROF PATH, UNIV ILL COL MED, 61- Mem: Assoc Am Physiol Soc; assoc Endocrine Soc; assoc Brit Soc Endocrinol. Res: Physiology of reproduction, aeroembolism and capillary vessels; lactogenic and gonadotrophic hormones; biophysics of connective tissue and ionic distribution; histochemistry of connective tissue. Mailing Add: Dept of Path Univ of Ill Col of Med Chicago IL 60612

CATE, JAMES RICHARD, JR, b Winters, Tex; m 64; c 2. ENTOMOLOGY. Educ: Tex A&M Univ, BS, 67, MS, 68; Univ Calif, Berkeley, PhD(entom), 75. Prof Exp: Res asst entom, Tex A&M Univ, 66-68, res assoc, 68-71; res asst, Univ Calif, Berkeley, 71-72, res assoc, 72-74; ASST PROF ENTOM, TEX A&M UNIV, 74- Mem: Entom Soc Am; Entom Soc Can; Int Orgn Biol Control; AAAS. Res: Population ecology and biological control of insect pests of grain sorghum and cotton. Mailing Add: Dept of Entom Tex A&M Univ College Station TX 79927

CATE, ROBERT BANCROFT, JR, soil science, see 12th edition

CATE, THOMAS RANDOLPH, b Nashville, Tenn, Feb 19, 35; m 56; c 5. VIROLOGY, CLINICAL MEDICINE. Educ: Vanderbilt Univ, BA, 56, MD, 59. Prof Exp: Asst prof med, Sch Med, Washington Univ, 66-68; assoc prof med, Duke Univ, 68-75; ASSOC PROF MICROBIOL & MED, BAYLOR COL MED, 75- Concurrent Pos: Consult, Microbiol Labs, Barnes Hosp, 66-68. Mem: Infect Dis Soc Am; Am Soc Microbiol; Am Thoracic Soc; Am Fedn Clin Res. Res: Pathogenesis of respiratory virus infections with a major focus on mechanisms of resistance. Mailing Add: Dept of Microbiol Baylor Col of Med 1200 Moursund Houston TX 77025

CATENHUSEN, JOHN ALFONS, b Tecumseh, Nebr, June 17, 09; m 41; c 2. PLANT ECOLOGY. Educ: Univ Wis, PhB, 37, PhM, 39, PhD(bot), 48. Prof Exp: Biologist, Arboretum & Wildlife Refuge, Univ Wis, 41-43; asst div mgr, Trop Plantation, Haiti, 43-44; res botanist, Firestone Plantations Co, Liberia, WAfrica, 44-51; assoc prof biol & head dept, Col of Steubenville, 52-55; PROF BIOL & HEAD DIV SCI & MATH, HILLSDALE COL, 55- Mem: AAAS. Res: Plant taxonomy; animal ecology; plant breeding; Hevea brasiliensis. Mailing Add: Baw Beese Heights Rte 3 Hillsdale MI 49242

CATER, CARL MALCOM, b Brookshire, Tex, Aug 8, 25; m 49; c 2. BIOCHEMISTRY. Educ: Tex A&M Univ, BS, 49, MS, 63, PhD(nutrit, biochem), 68. Prof Exp: Asst mgr, Cypress Farms, Inc, Tex, 49-50; operator rice farm, 52-54; asst mgr, Cater Prod, Inc, 54-61; res asst biochem & nutrit, 61-64, from instr to asst prof biochem & biophys, 65-71, asst prof soil & crop sci, 69-71, from asst head to actg head oilseed prod div, 66-70, assoc res chemist, 71-75, HEAD OILSEED PROD DIV, FOOD PROTEIN RES & DEVELOP CTR, TEX A&M UNIV, 71-, ASSOC PROF BIOCHEM & BIOPHYS & ASSOC PROF SOIL & CROP SCI, 71-, RES CHEMIST, 75- Mem: Am Chem Soc; Am Oil Chemists Soc; Inst Food Technologists; Am Asn Cereal Chemists. Res: Chemistry of proteins; utilization of oilseed proteins as human food; human nutrition; chemistry of cottonseed pigments. Mailing Add: Food Protein Res & Develop Ctr Tex A&M Univ College Station TX 77843

CATER, EARLE DAVID, b San Antonio, Tex, Apr 4, 34; m 63. PHYSICAL CHEMISTRY. Educ: Trinity Univ, BS, 54; Univ Kans, PhD(chem), 60. Prof Exp: Resident assoc chem, Argonne Nat Lab, 58-60, res assoc, 60-61; asst prof, 61-67, ASSOC PROF CHEM, UNIV IOWA, 67- Concurrent Pos: Consult, Argonne Nat Lab, 61-; vis chemist, Oxford Univ, 71. Mem: AAAS; Am Chem Soc. Res: High temperature physical chemistry; thermodynamics of vaporization processes; solid phase reactions at high temperatures; matrix isolation spectroscopy of high temperature molecules; high temperature mass spectrometry. Mailing Add: Dept of Chem Univ of Iowa Iowa City IA 52240

CATER, FRANK SYDNEY, b Chicago, Ill, June 27, 34. MATHEMATICS. Educ: Univ Southern Calif, BA, 56, MA, 57, PhD(math), 60. Prof Exp: Asst prof math, Univ Ore, 60-65; from asst prof to assoc prof, 65-70, PROF MATH, PORTLAND STATE UNIV, 70- Concurrent Pos: Referee, Math Mag, 72- Mem: Am Math Soc; Math Asn Am. Res: Analysis and real variables. Mailing Add: Dept of Math Portland State Univ Portland OR 97207

CATES, DAVID MARSHALL, b Salisbury, NC, Jan 7, 22; m 48; c 2. POLYMER CHEMISTRY, TEXTILE CHEMISTRY. Educ: NC State Col, BS, 49, MS, 51; Princeton Univ, PhD(chem), 55. Prof Exp: PROF TEXTILE CHEM, NC STATE UNIV, 55- Mem: Am Chem Soc; Fiber Soc; Am Asn Textile Chemists & Colorists. Res: High polymers and textile fibers; absorption chemistry. Mailing Add: 3 David Clark Lab NC State Univ Raleigh NC 27607

CATES, GEOFFREY WILLIAM, b Toronto, Ont, Oct 4, 23; m 48; c 5. CLINICAL PATHOLOGY. Educ: Univ Toronto, MD, 47. Prof Exp: Labs dir, Saskatoon City Hosp, 58-63, lectr, 56-63, assoc prof & asst dir labs, Univ Hosp, 63-67, PROF PATH, UNIV SASK, 67-, ASSOC DIR LABS UNIV HOSP, 67- Mem: Am Soc Clin Path; Can Asn Path; Can Med Asn; Int Acad Path. Res: Practice of clinical pathology in relationship to patient care and undergraduate teaching of morbid anatomy. Mailing Add: Univ Hosp Saskatoon SK Can

CATES, HARRY LOUIS, JR, b Trumbull, Tex, Sept, 16, 21; m 45; c 3. PLASTICS CHEMISTRY. Educ: Eastern Tex State Teachers Col, BA, 41; Univ Tex, MS, 48; Ohio State Univ, PhD(org Chem), 51. Prof Exp: Res chemist, 51-56, sales technologist, 57-60, mkt develop supvr, 60-63, dist sales mgr, 63-66, lab dir, Plastics Dept, 66-68, prod mgr flurocarbons div, 68-70, res mgr plastic prod div, 70-74, RES CONSULT, PLASTICS DEPT, E I DU PONT DE NEMOURS & CO, 75- Mem: Am Chem Soc; Soc Plastics Eng. Res: Engineering plastics; research consultant, based on marketing experience, in search of new business for plastics. Mailing Add: Plastics Dept E I du Pont de Nemours & Co Wilmington DE 19898

CATES, LINDLEY A, b Chicago, Ill, Nov 20, 32; m 57; c 2. PHARMACEUTICAL CHEMISTRY, ORGANIC CHEMISTRY. Educ: Univ Minn, BS, 54; Univ Colo, MS, 58, PhD(pharmaceut chem), 61. Prof Exp: Instr pharm, Univ Colo, 58-61; from asst prof to assoc prof, 61-68, PROF PHARMACEUT CHEM, SCH PHARM, UNIV HOUSTON, 68-, CHMN MED CHEM & PHARMACOG, 73- Concurrent Pos: Prin investr, NIH grant, 62-; Robert A Welch Found grant, 69-74. Mem: Am Pharmaceut Asn; Am Chem Soc. Res: Synthesis of organophosphorus compounds as potential chemotherapeutics. Mailing Add: Col of Pharm Univ of Houston Houston TX 77004

CATES, VERNON E, b Parsons, Kans, Feb 17, 31; m 66; c 2. ANALYTICAL CHEMISTRY, INORGANIC CHEMISTRY. Educ: Kans State Univ, BS, 53, MS, 56, PhD(chem), 62. Prof Exp: Teacher high sch, Kans, 56; instr chem, Centenary Col, 56-59; from asst prof to assoc prof, ETex State Univ, 62-67; PROF CHEM & CHMN DEPT, DALLAS BAPTIST COL, 67- Mem: Am Chem Soc. Res: Gas chromatography; analytical oxidizing agents; analytical methods; teaching innovations. Mailing Add: Dallas Baptist Col Dept of Chem Box 21206 Dallas TX 75211

CATHCART, JAMES BACHELDER, b Berkeley, Calif, Nov 22, 17; m 44; c 2. GEOLOGY. Educ: Univ Calif, AB, 39. Prof Exp: Asst geol, Univ Calif, 39-41; geologist, Quicksilver Invests, 42-43, Bauxite Invests, Ark, 43-45 & Potash Invests, NMex, 46-47, in charge phosphate invests, Southeast US, 47-53, sr geologist, Trace Elements Res & Resource Group, 53-57, COMMODITY GEOLOGIST, US GEOL SURV, 57, PHOSPHATE RES, S AM, 66- Mem: Geol Soc Am; Soc Econ Geologists; Am Asn Petrol Geologists; Soc Econ Paleontologists & Mineralogists. Res: Economic geology of non-metallic rocks, potash, phosphate, bauxite. Mailing Add: 17225 W 16th Pl Golden CO 80401

CATHCART, JOHN ALMON, b Sparta, Ill, Jan 17, 16; m 41; c 2. ORGANIC CHEMISTRY. Educ: Monmouth Col, Ill, BS, 37; Ohio State Univ, PhD(org chem), 41. Prof Exp: Asst chem, Ohio State Univ, 37-41; Anna Fuller Fund fel, Ohio State Univ, 41-42; from instr to asst prof chem & math, Monmouth Col, 42-45; res chemist, 45-51, develop engr, 51-60, patent search specialist, 60-64, supvr info sect, Patent Dept, 64-74, PATENT SEARCH SPECIALIST, EASTMAN KODAK CO, 74- Mem:

Am Chem Soc. Res: Polynuclear hydrocarbons; polymers; organic synthesis, patents. Mailing Add: 65 Kemphurst Rd Rochester NY 14612

CATHCART, JOHN VARN, b St George, SC, Nov 28, 23; m 66. PHYSICAL CHEMISTRY. Educ: Clemson Col, BS, 47; Univ Va, PhD(chem), 51. Prof Exp: CHEMIST, PHYS CHEM METAL SURFACES, OAK RIDGE NAT LAB, 51- Mem: AAAS; Am Soc Metals; Electrochem Soc Inc; Am Inst Mining, Metall & Petrol Engrs. Res: Oxidation of metal surfaces. Mailing Add: Clinch View Lane RFD 17 Knoxville TN 37921

CATHER, JAMES NEWTON, b Carthage, Mo, Mar 17, 31; m 51; c 2. ZOOLOGY. Educ: Southern Methodist Univ, BS, 54, MS, 55; Emory Univ, PhD, 58. Prof Exp: Instr biol, Emory Univ, 58; from instr to assoc prof zool, 58-73, PROF ZOOL, UNIV MICH, 73-, LEADER DEPT EXP BIOL, 75- Concurrent Pos: Upjohn fac fel, 65; instr embryol, Marine Biol Lab, 66-67; vis assoc prof, Ore Inst Marine Biol, 69. Mem: AAAS; Malacol Soc London; Am Soc Zoologists; Soc Develop Biol. Res: Cellular differentiation; genetics of development; molluscan development; invertebrate embryology. Mailing Add: Div of Biol Sci Dept of Exp Biol Univ of Mich Natural Sci Bldg Ann Arbor MI 48104

CATHERINO, HENRY ALVES, analytical chemistry, see 12th edition

CATHEY, EVERETT HENRY, b Little Rock, Ark, Jan 9, 31; m 61; c 2. SCIENCE EDUCATION, GEOPHYSICS. Educ: Univ Ark, BA, 56, MS, 60; Univ Ariz, PhD(geophys, educ), 76. Prof Exp: Instr & res assoc physics & math, Univ Ark, 58-62; asst prof physics, Ft Hays Kans State Col, 62-64; assoc prof geol & physics, Amarillo Col, 67-69; PROF SCI, ARIZ COL TECHNOL, 69-, CHMN GEN STUDIES, 75- Mem: Soc Explor Geophysicists; Tree-Ring Soc. Res: Science administration; paleoclimatology; aeronomy. Mailing Add: Ariz Col of Technol Star Rte Box 97 Winkelman AZ 85292

CATHEY, HENRY M, horticulture, plant physiology, see 12th edition

CATHEY, LECONTE, b Statesville, NC, Oct 18, 23; m 48; c 5. RADIATION PHYSICS. Educ: Davidson Col, BS, 47; Emory Univ, MS, 48; Univ NC, PhD(physics), 52. Prof Exp: Sr res physicist, Savannah River Lab, 52-65; PROF PHYSICS, UNIV SC, 66- Concurrent Pos: Mem grants rev comt, Dept Health, Educ & Welfare & Environ Protection Agency, 71. Mem: Am Phys Soc; Inst Elec & Electronic Engrs. Res: Instrumentation; radiation measurement; Mössbauer spectroscopy. Mailing Add: 1225 Belt Line Blvd Columbia SC 29205

CATHEY, WILLIAM NEWTON, b Pulaski, Tenn, Feb 20, 39; m 63. SOLID STATE PHYSICS. Educ: Univ Tenn, BS, 61, MS, 62, PhD(solid state physics), 66. Prof Exp: Fel, Nat Res Coun Can, 66-67; asst prof, 67-74, ASSOC PROF PHYSICS, UNIV NEV, RENO, 74- Concurrent Pos: Consult, Reno Metall Ctr, US Bur Mines, Nev. Res: Electronic structure of metals using Mössbauer effect; electrotransport of impurities in metals. Mailing Add: Dept of Physics Univ of Nev Reno NV 89507

CATHLES, LAWRENCE MACLAGAN, III, b Brooklyn, NY, Feb 9, 43; m 74. GEOPHYSICS. Educ: Princeton Univ, AB, 65, PhD(geophys), 71. Prof Exp: SR GEOPHYSICIST, LEDGEMONT LAB, KENNECOTT COPPER CORP, 71- Concurrent Pos: Mem adv coun, Dept Geol & Geophys Sci, Princeton Univ, 73- Mem: Am Geophys Union; AAAS. Res: Earth's viscosity structure inferred from isostatic rebound phenomena; physics and chemistry of copper sulfide leaching from waste dumps; physics and chemistry of igneous intrusive environments. Mailing Add: Ledgemont Lab Kennecott Copper Corp 128 Spring St Lexington MA 02173

CATHOU, RENATA EGONE, b Milan, Italy, June 21, 35; US citizen; m 59. IMMUNOLOGY. Educ: Mass Inst Technol, BS, 57, PhD(biochem), 63. Prof Exp: Res assoc phys chem, Mass Inst Technol, 62-65, fel, 64-65; res assoc biochem, Med Sch, Harvard Univ, 65-69, instr med, 69-70; asst prof, 70-73, ASSOC PROF BIOCHEM & PHARMACOL, SCH MED, TUFTS UNIV, 73- Concurrent Pos: Am Heart Asn grant, 69-; Nat Inst Allergy & Infect Dis grant, 71-; sr investr, Arthritis Found, 70-75. Mem: Biophys Soc; AAAS; Am Asn Immunol; Am Chem Soc; Am Soc Biol Chem. Res: Structure, conformation and function of biological macromolecules in the immune response. Mailing Add: Dept of Biochem & Pharmacol Tufts Univ Med Sch Boston MA 02111

CATIGNANI, GEORGE LOUIS, b Nashville, Tenn, Apr 9, 43; m 72. NUTRITIONAL BIOCHEMISTRY. Educ: Vanderbilt Univ, BA, 67, PhD(biochem), 74. Prof Exp: Res assoc toxicol, Ctr Environ Toxicol, Dept Biochem, Vanderbilt Univ, 74-75; STAFF FEL NUTRIT BIOCHEM, LAB NUTRIT & ENDOCRINOL, NAT INST ARTHRITIS, METAB & DIGESTIVE DIS, 75- Mem: Assoc Am Inst Nutrit; AAAS. Res: Mechanism of action of vitamin E; identification, characterization and function of specific cytoplasmic tocopherol binding protein and its role in vitamin E action. Mailing Add: Bldg 10 Rm 5N106 NIAMDD NIH Bethesda MD 20014

CATLETT, DUANE STEWART, b Fremont, Nebr, July 13, 40; m 61; c 1. PHYSICAL CHEMISTRY, CHEMICAL METALLURGY. Educ: Nebr Wesleyan Univ, BA, 63; Iowa State Univ, PhD(phys chem), 67. Prof Exp: Asst prof chem, Minot State Col, 67-68; asst prof, Pac Lutheran Univ, 68-70; staff mem, 70-74, ALT GROUP LEADER MAT TECHNOL GROUP, LOS ALAMOS SCI LAB, 74- Mem: Am Vacuum Soc; Am Soc Metals; Am Chem Soc. Res: Gas-solid kinetics and its theory of the rate-determining processes, such as nucleation and diffusion; thermodynamics of solution; thin film deposition processes; hot-atom chemistry and radiolysis of condensed media. Mailing Add: Group CMB-6/ms-770 Los Alamos Sci Lab Los Alamos NM 87545

CATLIN, B WESLEY, b Mt Vernon, NY, June 26, 17. MICROBIOLOGY. Educ: Univ Calif, Los Angeles, AB, 42, MA, 44, PhD(microbiol), 47. Prof Exp: Med lab technician, Challis Clin Lab, Calif, 39-41; med lab technician, Seaside Mem Hosp, 42; chief lab technologist, Santa Barbara Gen Hosp, 43; teaching asst bact, Univ Calif, Los Angeles, 43-47, res assoc, 47-49; res assoc genetics, Carnegie Inst Wash, 49-50; from asst prof to assoc prof, 50-65, PROF MICROBIOL, MED COL WIS, 65- Mem: AAAS; Am Soc Microbiol; Soc Study Evolution; Genetics Soc Am; Brit Soc Gen Microbiol. Res: Neisseriaceae; genetic transformation. Mailing Add: Med Col of Wis Milwaukee WI 53233

CATLIN, DONALD E, b Erie, Pa, Apr 29, 36; m 61; c 2. MATHEMATICS. Educ: Pa State Univ, BS, 58, MA, 61; Univ Fla, PhD(math), 65. Prof Exp: Instr math, Univ Fla, 64-65; ASST PROF MATH & STATIST, UNIV MASS, 65-, VCHMN DEPT MATH, 70- Concurrent Pos: Mem, Inst Fundamental Studies. Mem: Am Math Soc; Math Asn Am. Res: Orthomodular lattice theory and quantum logics; functional analysis; mechanics. Mailing Add: Dept of Math Univ of Mass Amherst MA 01002

CATLIN, FRANCIS I, b Hartford, Conn, Dec 6, 25; m 48; c 3. OTOLARYNGOLOGY. Educ: Johns Hopkins Univ, MD, 48, ScD, 59. Prof Exp: Intern, Union Mem Hosp, 48-49; intern otolaryngol, Johns Hopkins Hosp, 50 & 52-53, asst resident, 53-54 & 55, otolaryngologist, 56-72; from instr to asst prof otolaryngol, Johns Hopkins Univ, 56-63, asst prof audiol & speech, Sch Hyg & Pub Health, 60-70, assoc prof otolaryngol, Sch Med, 63-72, sci dir, Info Ctr Hearing, Speech & Dis Human Commun, 68-72, assoc prof audiol & speech, Pub Health Admin, 70-72; PROF OTORHINOLARYNGOL & COMMUN SCI, BAYLOR COL MED, 72-; CHIEF-OF-SERV, DEPT OTOLARYNGOL, TEX CHILDREN'S & ST LUKE'S EPISCOPAL HOSPS, HOUSTON, 72- Concurrent Pos: Consult, Vet Admin Hosp, Perry Point, 61-72; spec consult, Neurol & Sensory Dis Serv Prog, US Dept Health, Educ & Welfare, 63-64; mem communicative dis res training comt, Nat Inst Neurol Dis & Blindness, 63-65; mem prof adv coun, Nat Easter Seal Soc Crippled Children & Adults, 73- Honors & Awards: Cert audiol, Am Speech & Hearing Asn, 65; Award of Merit, Am Acad Ophthal & Otolaryngol, 74. Mem: AMA; fel Am Acad Opthal & Otolaryngol; fel Am Laryngol, Rhinol & Otol Soc; Am Broncho-Esophagol Asn; Am Speech & Hearing Asn. Res: Hearing problems in audiology including instrumentation; infectious diseases involving the paranasal sinuses. Mailing Add: St Luke's-Tex Children's Hosps PO Box 20269 Houston TX 77025

CATLIN, PETER BOSTWICK, b Ross, Calif, Sept 22, 30; m 52; c 3. PLANT PHYSIOLOGY, POMOLOGY. Educ: Univ Calif, BS, 52, MS, 55, PhD(plant physiol), 58. Prof Exp: Asst pomologist, 58-64, ASSOC POMOLOGIST, UNIV CALIF, DAVIS, 64-, LECTR POMOL, 69- Mem: Am Soc Hort Sci. Res: Physiology and biochemistry of plant growth and development; respiratory metabolism; chemical taxonomy. Mailing Add: Dept of Pomol Univ of Calif Davis CA 95616

CATLIN, SETH, b Portland, Ore, Aug 30, 39; m 63. MATHEMATICAL LOGIC. Educ: Univ Ore, BA, 61; Portland State Univ, MST, 65; Ariz State Univ, PhD(math), 68. Prof Exp: Asst prof math, Southern Ore Col, 68-69; ASSOC PROF MATH, EASTERN ORE COL, 69- Mem: Math Asn Am. Res: Regressive functions of order greater than or equal to one. Mailing Add: Dept of Math Eastern Ore Col La Grande OR 97850

CATO, BENJAMIN RALPH, JR, b Belmont, NC, Aug 24, 25; m 48; c 3. MATHEMATICS. Educ: Duke Univ, AB, 48, AM, 50. Prof Exp: Instr math & physics, Univ Ariz, 50-52; instr math, Univ Md, 52-55; from asst prof to assoc prof, 55-74, PROF MATH, COL WILLIAM & MARY, 74- Concurrent Pos: Assoc prof, Summer Inst High Sch Teachers, Nat Sci Found, 59-, assoc dir, 60-68, dir, 68- Mem: Am Math Soc; Math Asn Am. Res: Partial differential equation; integral equations; matrices; analysis; linear operators. Mailing Add: Dept of Math Col of William & Mary Williamsburg VA 23185

CATON, ROY DUDLEY, JR, b Fresno, Calif, June 7, 30. ELECTROANALYTICAL CHEMISTRY. Educ: Fresno State Col, BS, 52, MA, 53; Ore State Univ, PhD(chem), 63. Prof Exp: Chemist, Chem & Radiol Labs, US Army Chem Ctr, Md, 57; asst prof, 62-69, ASSOC PROF CHEM, UNIV NMEX, 69- Concurrent Pos: Petrol Res Fund grant, 63-64. Mem: Am Chem Soc; Electrochem Soc. Res: Polarography and coulometry of metals in fused salt media; electrode potentials of metals in fused alkali metaphosphates; ion exchange and liquid chromatography detectors. Mailing Add: Dept of Chem Univ of NMex Albuquerque NM 87131

CATRAMBONE, JOSEPH ANTHONY, SR, b Chicago, Ill, Sept 21, 24; m 51; c 7. MATHEMATICS, DATA PROCESSING. Educ: St Benedict's Col, BS, 47; Univ Maine, MA, 53. Prof Exp: Instr math & chem, Damar Acad, 47-50; mathematician adv bd on simulation secretariat, Univ Chicago, 53-54, syst res, 54-56, group leader, Inst Syst Res, 56-58, sr mathematician & asst dir labs appl sci, 58-61; sr scientist sci comput ctr, Serv Bur Corp, 62-63; opers analyst res inst, Ill Inst Technol, 63-68; dir, admin data processing, 68-72, ASST VPRES DATA PROCESSING, UNIV ILL, CHICAGO CIRCLE, 72- Concurrent Pos: Instr math, Walton Col Com, 52-, Loyola Univ, 56- & Chicago Jr Col, 57-58. Mem: Am Math Soc. Res: Military weapons systems analyses and evaluations; operations research; applied mathematics. Mailing Add: 3023 N 77th Ct Elmwood Park IL 60635

CATRAVAS, GEORGE NICHOLAS, b Argostoli, Greece, June 22, 16; US citizen. ORGANIC CHEMISTRY, BIOCHEMISTRY. Educ: Univ Athens, DCh, 37; Univ Leeds, PhD(org chem), 47; Sorbonne, DSc, 53. Prof Exp: Instr org chem, Nat Univ Athens, 37-40; res chemist, Lever Bros & Unilever, Eng, 47-49; in charge res, Nat Ctr Sci Res, France, 50-54; Foreign Oper Admin-Nat Acad Sci fel, Univ Chicago, 54-56, asst prof, 56-63; head biochem res, Technicon Corp, 63-66; proj dir, Molecular Biol Exp, 66-72, CHIEF DIV NEUROCHEM, ARMED FORCES RADIOBIOL RES INST, 72- Concurrent Pos: Prof lect, Am Univ, 67- & adj prof, 72- Honors & Awards: Except Serv Civilian Award, Gold Medal, Defense Nuclear Agency, 73. Mem: AAAS; Radiation Res Soc; NY Acad Sci; The Chem Soc. Res: Intermediary metabolism of lipids; control mechanisms; membranes; action of ionizing radiations on cell constituents; mammalian central nervous system; opiates; microwaves. Mailing Add: Armed Forces Radiobiol Res Inst Defense Nuclear Agency Bethesda MD 20014

CATSIFF, EPHRAIM HERMAN, PHYSICAL CHEMISTRY, POLYMER CHEMISTRY. Educ: Pa State Col, BS, 45; Univ Southern Calif, MS, 48; Polytech Inst Brooklyn, PhD(polymer chem), 52. Prof Exp: Res asst chem, Princeton Univ, 51-53, res assoc polymers, 53-57; chemist, Shell Develop Co, Calif, 57-62; supvr phys properties sect, Res Dept, Thiokol Chem Corp, 62-67, head polymer physics & instrumental res, 67-69, head explor polymer res, 69-74; GROUP LEADER POLYMER PHYSICS, CIBA-GEIGY CORP, 74- Mem: Am Chem Soc; Soc Rheol; Soc Plastics Engrs; NY Acad Sci; Sigma Xi. Res: Creep and stress relaxation of crystalline and amorphous polymers; swelling behavior of filled rubbers; Mullins effect; melt viscometry; epoxy resins characterization. Mailing Add: Plastics & Additives Div Ciba-Geigy Corp Ardsley NY 10502

CATSIMPOOLAS, NICHOLAS, b Athens, Greece, Feb 9, 31; US citizen; m 59; c 3. BIOPHYSICS. Educ: Athens Univ, BS, 55 Univ Tenn, MS, 62, PhD(biochem), 64. Prof Exp: Biochemist, King Gustaf V Res Inst, Sweden, 59-60; sr res scientist, Res Ctr, Cent Soya Co , Chicago, 65-73; adj assoc prof, Stritch Sch Med, Loyola Univ, 72-73; ASSOC PROF MASS INST TECHNOL, 73- Mem: AAAS; Am Soc Biol Chem; NY Acad Sci; Am Chem Soc; Biophys Soc. Res: Macromolecular and cell biophysics; electrophoresis; immunophysics. Mailing Add: Biophys Lab Dept of Nutrit & Food Sci Mass Inst of Technol Rm 56-307 Cambridge MA 02139

CATTANI, RAY AUGUST, b San Bernadino, Calif, Feb 18, 30; m 54; c 4. AGRICULTURAL CHEMISTRY, SOIL CHEMISTRY. Educ: Brigham Young Univ, BA, 57; Ore State Univ, 57-60, MS, 60; Univ Ariz, PhD(agr chem), 63. Prof Exp: Asst soil chem, Ore Satae Univ, 57-60; NSF res asst agr chem, Univ Ariz, 60-63; mem chem fac, Phoenix Col, 63-66, Mesa Community Col, 66-67, acad dean, 67-73; EXEC DEAN, SCOTTSDALE COMMUNITY COL, 73- Concurrent Pos: Consult & examr, NCent Asn Comn Higher Educ, 69- Mem: AAAS; Am Soc Plant Physiol; Am Chem Soc; Am Inst Chem. Res: Inorganic and organic chemistry; quantitative analysis. Mailing Add: Scottsdale Community Col Chaparral & Pima PO Box Y Scottsdale AZ 85252

CATTELL, MCKEEN, b Garrison, NY, Nov 17, 91; m 22; c 4. PHARMACOLOGY.

Educ: Columbia Univ, BS, 14; Harvard Univ, AM, 17, PhD(physiol), 20, MD, 24. Hon Degrees: DSc, Univ Antioquia, Columbia. Prof Exp: Teaching fel physiol, Harvard Med Sch, 14-17, teaching fel pharmacol, 20-24; from instr to asst prof physiol, 24-36, assoc prof pharmacol in charge dept, 36-43, prof & head dept, 43-56, prof clin pharmacol, 56-59, EMER PROF CLIN PHARMACOL, MED COL, CORNELL UNIV, 59-; ED, J CLIN PHARMACOL, 61- Concurrent Pos: Mem div biol & agr, Nat Res Coun, 37-46, vchmn, 41-43; mem pharmacol study sect, USPHS, 46-47; mem teaching mission, Austria, 47, Columbia, 48 & Japan, 50; chmn adv comt chem-biol coord ctr, Nat Res Coun & physiol-pharmacol sub-comt, 49-50, mem div med sci, 50-53; vis prof, Univ Tokyo, 59. Mem: AAAS; Am Soc Pharmacol & Exp Therapeut (treas, 44-47, pres, 51); fel NY Acad Med; fel NY Acad Sci; Am Col Clin Pharmacol & Chemother (pres, 63-64). Res: Traumatic shock; physiological effects of hydrostatic pressure; nerve-muscle physiology; mechanism of digitalis action. Mailing Add: Dept of Pharmacol Cornell Univ Med Col New York NY 10021

CATTERSON, ALLEN DUANE, b Denver, Colo, June 26, 29; m 50; c 3. PREVENTIVE MEDICINE, AEROSPACE MEDICINE. Educ: Univ Colo, BA, 51, MD, 55; Ohio State Univ, MS, 61. Prof Exp: Gen pract, Colo, 58-59; resident aviation med, Ohio State Univ, 59-61; chief resident, Lovelace Found Med Educ & Res, 61-62; actg asst chief, Ctr Med Opers Off, NASA Manned Spacecraft Ctr, 62-63, assoc chief, 63-64, asst to chief, Ctr Med Prog, 64-65, chief flight med br, Ctr Med Off, 65-66, dep dir med res & opers, 66-71; PRES, AEROSPACE MED CONSULT, 71- Concurrent Pos: Mem air traffic controller career comt, US Dept Transp, 69-70. Honors & Awards: Except Serv Award, NASA, 69; Melbourne W Boynton Award, Am Astronaut Soc, 71. Mem: Fel Aerospace Med Asn; fel Am Col Prev Med. Res: Environmental physiology. Mailing Add: Aerospace Med Consult PO Box 60385 Houston TX 77205

CATTO, PETER JAMES, b Boston, Mass, Mar 23, 43; m 69; c 1. PLASMA PHYSICS. Educ: Mass Inst Technol, BS & MS, 67; Yale Univ, PhD(eng & appl sci), 72. Prof Exp: Mem sch natural sci, Inst Advan Study, Princeton, NJ, 71-73; ASST PROF PLASMA PHYSICS, UNIV ROCHESTER, 73- Concurrent Pos: Mem, Theory Group, Lab Laser Energetics, Univ Rochester, 73-; consult, Nuclear Div, Oak Ridge Nat Lab, 74- Mem: Am Phys Soc. Res: Propagation and stability of waves in magnetized and unmagnetized plasmas with a particular interest in drift instabilities in sheared magnetic fields. Mailing Add: Dept of Mech & Aerospace Sci Univ of Rochester Rochester NY 14627

CATTOLICO, ROSE ANN, b Philadelphia, Pa, July 2, 43. DEVELOPMENTAL BIOLOGY. Educ: Temple Univ, BA, 65, MA, 67; State Univ NY Stony Brook, PhD(develop biol), 73. Prof Exp: Res assoc biol, Brookhaven Nat Lab, 67-68; fel biol, McGill Univ, 73-75; ASST PROF BOT, UNIV WASH, 75- Concurrent Pos: Res grant, Univ Wash, 75. Mem: Soc Develop Biol; Soc Plant Physiologists; AAAS; Asn Women Sci. Res: Control of organelle biogenesis, nucleic acid metabolism during growth and morphogenesis of synchronized unicellular algal cells. Mailing Add: Dept of Bot Univ of Wash Seattle WA 98195

CATURA, RICHARD CLARENCE, b Arkansas, Wis, July 31, 35; m 59; c 3. PHYSICS, X-RAY ASTRONOMY. Educ: Univ Minn, BS, 57; Univ Calif, Los Angeles, MS, 59, PhD(physics), 62. Prof Exp: Res physicist, Univ Calif, Los Angeles, 62-63; res assoc elem particle physics, Princeton Univ, 63-66; RES SCIENTIST, LOCKHEED PALO ALTO RES LAB, 66- Concurrent Pos: Math consult, Radioisotope Serv, Vet Admin, Los Angeles, Calif, 59-62. Mem: Am Phys Soc; Am Astron Soc. Res: Nuclear physics; transport of high intensity charged particle beams; properties of wide-gap spark chambers; elementary particle physics; investigation of solar and stellar x-ray emissions. Mailing Add: Lockheed Palo Alto Res Lab Dept 52-14 Bldg 202 3251 Hanover St Palo Alto CA 94304

CATY, JEAN LOUIS, b Matheson, Ont, Mar 12, 43; m 69. GEOLOGY, SEDIMENTOLOGY. Educ: Univ Montreal, BS, 67, MS, 70, PhD(geol), 76. Prof Exp: ASST PROF EARTH SCI, UNIV QUE, CHICOUTIMI, 72- Mem: Geol Asn Can; Can Inst Mining & Metall. Res: Stratigraphy, structural geology, geochemistry and sedimentology of the northern edge of the Archean Greenstone Belt, Chibougamau, Province of Quebec. Mailing Add: Dept of Earth Sci Univ of Que Chicoutimi PQ Can

CATZ, BORIS, b Russia, Feb 25, 23; US citizen; m; c 4. MEDICINE. Educ: Nat Univ Mex, BS, 41, MD, 47; Univ Southern Calif, MS, 51. Prof Exp: Intern, Gen Hosp, Mexico City, 45-46; adj prof, Sch Med, Nat Univ Mex, 47-48; instr, 52-54, asst clin prof, 54-59, ASSOC CLIN PROF MED, UNIV SOUTHERN CALIF, 59- Concurrent Pos: Practicing physician, Los Angeles, 51-; chief thyroid clin, Los Angeles County Hosp, 59-69, sr consult, 70- Mem: AAAS; fel Am Col Physicians; Soc Exp Biol & Med; Endocrine Soc; Soc Nuclear Med. Res: Thyroid disease. Mailing Add: Suite 404 435 N Roxbury Dr Beverly Hills CA 90210

CATZ, CHARLOTTE SCHIFRA, b Paris, France; US citizen. PEDIATRICS, TERATOLOGY. Educ: Univ Buenos Aires, MD, 52. Prof Exp: Staff physician, Pediat Serv, Hosp Fernandez, Buenos Aires, Arg, 51-56, supvr, Pediat Tuberc Clin, 55-56; teaching asst pediat, Stanford Univ, 59-61; asst dir, Pediat Out-Patient Clin, Palo Alto Med Ctr, Calif, 63-66; from asst prof to assoc prof pediat, State Univ NY Buffalo, 66-75; PEDIAT MED OFFICER, PREGNANCY & INFANCY BR, NAT INST CHILD HEALTH & HUMAN DEVELOP, 75- Concurrent Pos: Pvt pract pediat, Inst Padua, Buenos Aires, 53-56; actg instr, Sch Med, Stanford Univ, 63-64; assoc attend physician, Children's Hosp, Buffalo, NY, 69-75, clin dir, Birth Defects Ctr, 70-75; Fulbright sr res scholar & guest prof, Univ Rene Descartes, Paris, 73-74; Nat Inst Child Health & Human Develop liaison officer to comt on drugs, Am Acad Pediat, 75- Mem: Am Acad Pediat; AAAS; Am Soc Pharmacol & Exp Therapeut; Asn Ambulatory Pediat Serv; Pan Am Med Asn. Mailing Add: Pregnancy & Infancy Br Nat Inst of Child Health & Human Develop Landow C709 Bethesda MD 20014

CAUDLE, DANNY DEARL, b Maud, Tex, Oct 8, 37; m 64; c 4. PHYSICAL CHEMISTRY, CORROSION. Educ: Centenary Col, BS, 61; Univ Okla, PhD(phys chem), 66. Prof Exp: Res asst chem res inst, Univ Okla, 62-66; res scientist, 66-72, CORROSION & CHEM SPECIALIST, PROD DEPT, CONTINENTAL OIL CO, 72- Mem: Am Chem Soc; Nat Asn Corrosion Eng; Soc Petrol Eng. Res: Oil field chemical problems, including corrosion, emulsion and paraffin problems. Mailing Add: 9231 Rentur Dr Houston TX 77031

CAUGHEY, JOHN LYON, JR, b Rochester, NY, May 30, 04; m 37; c 1. MEDICAL EDUCATION. Educ: Harvard Univ, AB, 25, MD, 30; Columbia Univ, MScD, 35. Prof Exp: Intern med, Presby Hosp, New York, 30-32, from asst resident to resident, 32-37; from asst to assoc, Columbia Univ, 35-45; asst dean, 45-48, asst prof med, 45-48, assoc prof clin med, 48-69, PROF MED & MED EDUC, SCH MED, CASE WESTERN RESERVE UNIV, 69-, ASSOC DEAN, 48- Concurrent Pos: Asst physician, Presby Hosp, New York, 37-45; tech aide, Comt Med Res, Off Sci Res & Develop, 43-45. Res: Constitutional medicine; comprehensive health services. Mailing Add: Sch of Med Case Western Reserve Univ Cleveland OH 44106

CAUGHEY, THOMAS KIRK, b Scotland, Oct 22, 27; m 52; c 4. APPLIED MECHANICS. Educ: Glasgow Univ, BSc, 48; Cornell Univ, MME, 52; Calif Inst Technol, PhD(eng sci), 54. Prof Exp: Instr appl mech, 52-54, from asst prof to assoc prof, 55-62, PROF APPL MECH, CALIF INST TECHNOL, 62- Concurrent Pos: Consult engr, Jas Howden & Co, Scotland, 49-51, 54-55, NAm Electronics, 55-58, Aeronaut Eng Res Inc, 58-62, Inca Inc, 62-68, Jet Propulsion Lab, Calif Inst Technol, 69- & Tetra-Tech Inc, 70- Mem: AAAS; Soc Indust & Appl Math; Seismol Soc Am. Res: Non-linear mechanics; vibrations; acoustics; electronics; applied mathematics; classical physics. Mailing Add: Div of Eng & Appl Sci Thomas Lab Calif Inst of Technol Pasadena CA 91109

CAUGHEY, WINSLOW SPAULDING, b Antrim, NH, Nov 9, 26; m 52; c 4. BIOCHEMISTRY. Educ: Univ NH, BS, 48, MS, 49; Johns Hopkins Univ, PhD(chem), 53. Prof Exp: Sr res asst physiol chem, Med Sch, Johns Hopkins Univ, 53-54, res assoc chem, 54-56; pres, Monadnock Res Inst, 56-59; from asst prof to assoc prof physiol chem, Med Sch, Johns Hopkins Univ, 59-67; prof chem, Univ SFla, 67-69 & Ariz State Univ, 69-73; PROF BIOCHEM & CHMN DEPT, COLO STATE UNIV, 73- Concurrent Pos: Chmn subcomt porphyrins, Div Chem & Chem Technol, Nat Acad Sci-Nat Res Coun; Lederle med fac award, 63-66. Mem: AAAS; Am Chem Soc; NY Acad Sci; Am Soc Biol Chem; The Chem Soc. Res: Inorganic biochemistry; mechanisms of enzymic reactions; hemeproteins and metalloporphyrins; reactions of oxygen in biological systems. Mailing Add: Dept of Biochem Colo State Univ Ft Collins CO 80521

CAUGHLAN, CHARLES NORRIS, b Pullman, Wash, Jan 20, 15; m 36; c 4. PHYSICAL CHEMISTRY. Educ: Univ Wash, BS, 36, PhD(chem), 41. Prof Exp: Instr chem, Mont State Col, 41-44; chemist, Eastman Kodak Co, 44-46; from asst prof to assoc prof, 46-51, head dept chem, 67-73, PROF CHEM, MONT STATE UNIV, 51- Mem: Am Chem Soc; Am Crystallog Asn. Res: Hydrogen bonds in acetoxime; Raman spectroscopy; dielectric properties of polymers and titanium compounds; structures; metal alkoxides; x-ray diffraction; structures of organic phosphates, organic titanates, organic vanadates and natural products. Mailing Add: Dept of Chem Mont State Univ Bozeman MT 59715

CAUGHLAN, GEORGEANNE ROBERTSON, b Montesano, Wash, Oct 25, 16; m 36; c 4. PHYSICS, ASTROPHYSICS. Educ: Univ Wash, BS, 37, PhD(physics, 64. Prof Exp: From instr to assoc prof, 57-74, PROF PHYSICS, MONT STATE UNIV, 74- Mem: Am Phys Soc; Am Astron Soc; Am Asn Physics Teachers; Int Astron Union; Sigma Xi. Res: Nuclear astrophysics and the analysis of synthesis of elements in the stars. Mailing Add: Dept of Physics Mont State Univ Bozeman MT 59715

CAUGHLAN, JOHN ARTHUR, b Pittsfield, Ill, Apr 29, 21; m 42; c 4. ORGANIC CHEMISTRY. Educ: Univ Ill, BS, 42; Univ Nebr, MS, 44. Prof Exp: Lab foreman, Tenn Eastman Corp, 44-46; res chemist, 46-50, develop chemist, 50-54, head evaluation sect, 54-56, staff asst to dir prod develop, 56-57, asst dir prod develop, 57-60, mgr process chem, 60-62, asst to dir res, 62-67, asst dir res, 67-69, ASST DIR, CHEM GROUP RES & DEVELOP, MALLINCKRODT CHEM WORKS, 69- Mem: Am Chem Soc; Commercial Develop Asn; Am Inst Chem; Soc Photog Sci & Eng. Res: Amino ketones; opium alkaloids; organic chemistry; columbium; tantalum; industrial chemicals. Mailing Add: Mallinckrodt Chemical Works 3600 N Second St St Louis MO 63160

CAUL, JEAN FRANCES, b Cleveland, Ohio, Aug 19, 15. FOOD SCIENCE. Educ: Lake Erie Col, AB, 37; Ohio State Univ, MA, 38, PhD(physiol chem), 42. Prof Exp: Res chemist, Borden Co, NY, 42-44; sr proj leader, Arthur D Little, Ind, 44-67; distinguished prof, 67-70, PROF FOODS & NUTRIT, COL HOME ECON, KANS STATE UNIV, 70- Concurrent Pos: Vis instr, Inst Food Sci, Giessen, Ger, 60. Mem: Am Chem Soc; Inst Food Technol; NY Acad Sci. Res: Food technology; flavor measurement; consumer product testing; catfish flavor and preflavoring. Mailing Add: Dept of Foods & Nutrit Col of Home Econ Kans State Univ Manhattan KS 66506

CAULDER, JERRY DALE, b Gideon, Mo, Nov 7, 42; m 63; c 1. WEED SCIENCE. Educ: Southeast Mo State Univ, BS & BA, 64; Univ Mo, MS, 66, PhD(agron), 69. Prof Exp: Res asst weed sci, Univ Mo, 66-70; mkt develop specialist, 69-71, mgr, Colombia, SA, 71-73, develop assoc, 73, tech mgr herbicides, 73-74, NEW PROD MGR, MONSANTO CO, 74- Res: Coordination of the discovery, development and manufacture of herbicides and plant growth regulators. Mailing Add: Monsanto Co 800 N Lindbergh St Louis MO 63166

CAULEY, DARRELL LEE, b Washington, DC, Sept 18, 41; m 68; c 3. WILDLIFE ECOLOGY. Educ: Univ Cincinnati, BA, 65, MS, 70; Mich State Univ, PhD(wildlife ecol), 74. Prof Exp: Pres, Tech Environ Serv, Inc, 71-74; ENVIRON SPECIALIST, BROWN & ROOT, INC, 74- Mem: Wildlife Soc. Res: Ecological relationships of urban flora and fauna. Mailing Add: 6114 Pine Cove Houston TX 77018

CAULFIELD, DANIEL FRANCIS, b Brooklyn, NY, Aug 4, 35; m 60; c 2. POLYMER CHEMISTRY. Educ: Brooklyn Col, BS, 57; Polytech Inst Brooklyn, PhD(chem), 62. Prof Exp: Res Assoc chem, Polytech Inst Brooklyn, 62; fel Cornell Univ, 62-65; RES CHEMIST, FOREST PROD LAB, US FOREST SERV, 65- Mem: Am Chem Soc; Tech Asn Pulp & Paper Indust. Res: Light scattering; scattering and diffraction of x-rays. Mailing Add: Forest Prod Lab US Forest Serv Madison WI 53705

CAULFIELD, HENRY JOHN, optics, see 12th edition

CAULFIELD, JAMES BENJAMIN, b Minneapolis, Minn, Jan 1, 27; m 50; c 3. MEDICINE, PATHOLOGY. Educ: Miami Univ, BA, 47; Univ Ill, BS, 48, MD, 50. Prof Exp: Vis investr, Rockefeller Inst Med Res, 55-56; from instr to asst prof path, Med Ctr, Univ Kans, 56-59; asst prof, 59-70, ASSOC PROF PATH, HARVARD MED SCH, 70- Concurrent Pos: USPHS fel, 56-58; asst path, Mass Gen Hosp, 59-69, assoc pathologist, 69-; pathologist, Shrine Burns Inst, Boston, Mass, 70- Mem: Int Acad Path. Res: Electron microscopy; spontaneous and induced alterations in fine structure of cells. Mailing Add: Dept of Path Mass Gen Hosp Boston MA 02114

CAUNA, NIKOLAJS, b Riga, Latvia, Apr 4, 14; US citizen; m 42. ANATOMY, CELL BIOLOGY. Educ: Riga Univ, MD, 42; Univ Durham, MSc, 54, DSc, 61. Prof Exp: Student demonstr anat, Univ Riga, 35-42, lectr, 42-44; med practitioner, WGer, 44-46; lectr anat, Baltic Univ, Ger, 46-48; from lectr to reader, Durham, Eng, 48-61; PROF ANAT, SCH MED, UNIV PITTSBURGH, 61-, CHMN DEPT, 75- Concurrent Pos: Res grants, Royal Soc Eng, 59-61, Am Cancer Soc, 62-63 & USPHS, 62- Mem: Am Asn Anat; Histochem Soc; Am Soc Cell Biol; Anat Soc Gt Brit & Ireland; fel Royal Micros Soc. Res: Development and evolution of tetrapod limbs; development, structure and function of the peripheral receptor organs; control mechanism of the autonomic nervous system; fine structure and functions of the human nasal respiratory mucosa; urticaria. Mailing Add: Sch of Med Univ of Pittsburgh Pittsburgh PA 15261

CAUSA, ALFREDO G, b Montevideo, Uruguay, June 25, 28; US citizen. POLYMER SCIENCE, TEXTILES. Educ: Sch Chem & Chem Eng, Montevideo, BSc, 58; Case

Inst Technol, MS, 62; Univ Akron, PhD(polymer sci), 68. Prof Exp: Chemist textile chem, SAm Subsidiaries, Courtaulds, Ltd, 52-58; res chemist indust fibers, Textile Fibers Div, Can Indust Ltd, 61-64; res chemist polymers, Tarrytown Tech Ctr, Union Carbide Corp, 68-70; PRIN CHEMIST TIRE REINFORCING TECHNOL, GOODYEAR TIRE & RUBBER CO, 70- Mem: Am Chem Soc; AAAS. Res: Failure modes in tires and other fiber-reinforced composites; fiber fracture; polymer and fiber microstructure; chemistry of fiber finishes and adhesives; chemistry and physics of interfaces. Mailing Add: Goodyear Tire & Rubber Co Plant 1 Dept 469B 1144 E Market St Akron OH 44316

CAUSEY, GEORGE DONALD, b Baltimore, Md, July 9, 26; m 61; c 4. AUDIOLOGY, SPEECH PATHOLOGY. Educ: Univ Md, BA, 50, MA, 51; Purdue Univ, PhD(audiol, speech path), 54. Prof Exp: Asst chief audiol clin, DC Health Dept, 54-55; chief acoust res audiol, Vet Benefits Off, 55-64; CHIEF CENT AUDIOL & SPEECH PATH PROG, VET ADMIN HOSP, 64- Concurrent Pos: Res prof, Univ Md, 56-, dir biocommun lab, 67-; consult, Fairfax County Health Dept, Va, 60-67, Subcomt Antitrust & Monopoly, Senate Comt on Judiciary, 61-62 & Fed Trade Comn, 64-; mem standing comt med care & hosps, Pub Health Adv Coun, Govt DC, 66-; mem comt hearing, bioacoust & biomech, Nat Acad Sci-Nat Res Coun, 67-; mem comt hearing aids, Am Nat Standards Inst, 67-; consult, Md Dept Pub Health, 67- Mem: Fel Am Speech & Hearing Asn; Acoust Soc Am. Res: Hearing impairment and measurement techniques; hearing aids. Mailing Add: 3504 Dunlop Chevy Chase MD 20015

CAUSEY, MILES KEITH, b Monroe, La, Dec 23, 40; m 63; c 1. WILDLIFE BIOLOGY. Educ: La State Univ, BS, 62, MS, 64, PhD(entom), 68. Prof Exp: Asst prof, 68-75, ASSOC PROF ZOOL & ENTOM, AUBURN UNIV, 75- Mem: AAAS; Wildlife Soc; Am Soc Mammal. Res: Wildlife biology and conservation, especially pesticide-wildlife relationships and environmental degradation. Mailing Add: Dept of Zool & Entom Auburn Univ Auburn AL 36830

CAUSEY, NELL BEVEL, b Trenton, Tenn, Dec 8, 10; m 38. INVERTEBRATE ZOOLOGY. Educ: Col of Ozarks, BS, 31; Univ Ark, MA, 37; Duke Univ, PhD(animal ecol), 40. Prof Exp: Instr zool, Univ Ark, 43-44, marine ecologist, Marine Lab, Duke Univ, 44; instr zool, Univ Ark, 45-48; pvt investr, 48-64; from asst prof to assoc prof, 64-71, PROF ZOOL, LA STATE UNIV, 71- Res: Ecology and taxonomy of Diplopoda. Mailing Add: Dept of Zool La State Univ Baton Rouge LA 70803

CAUSEY, WILLIAM MCLAIN, b Cleveland, Miss, Feb 6, 38. MATHEMATICAL ANALYSIS. Educ: Univ Miss, BS, 60, MA, 62; Univ Kans, PhD(math), 66. Prof Exp: Asst prof math, Miss State Univ, 66-67; asst prof, Univ Cincinnati, 67-68; assoc prof, 68-73, PROF MATH, UNIV MISS, 73- Mem: Am Math Soc; London Math Soc. Res: Complex variables; univalent functions. Mailing Add: Dept of Math Univ of Miss University MS 38677

CAUTHER, SALLY EUGENIA, b Montgomery, Ala, Oct 1, 32. BIOCHEMISTRY. Educ: Abilene Christian Col, BS, 53; La State Univ, Baton Rouge, MS, 57; Oxford Univ, PhD(biochem), 65. Prof Exp: Assoc prof chem, Abilene Christian Col, 56-67; asst prof, Tex Tech Univ, 67-68; PROF CHEM, NORTHEAST LA UNIV, 68- Mem: AAAS; Am Chem Soc; Am Soc Microbiol. Res: Vitamins and coenzymes, especially biotin, folic acid and vitamin B12; methionine biosynthesis in bacteria; enzymology, especially atropinesterase and methyltransferase; azolesterases related to schizophrenia. Mailing Add: Dept of Chem Northeast La Univ Monroe LA 71201

CAUWENBERG, WINFRED JOSEPH, b Green Bay, Wis, Apr 18, 96; m 39; c 1. CHEMISTRY. Educ: Univ Wis, BS, 19; Univ Buffalo, AM, 28; Columbia Univ, PhD(chem), 30. Prof Exp: Res chemist, Newport Co, 19-24; res chemist, Nat Aniline & Chem Co, 24-28; res chemist, Titanium Pigment Co, NY, 30-32; res chemist, United Color & Pigment Co, NJ, 32-36; res chemist, Va Chem Corp, 36-38, dir res, 38-44; dir res develop, Piney River Works, Calco Chem Div, Am Cyanamid Co, 44-48, mgr titanium dept, 48-54, tech dir, Pigments Div, 54-61, consult, 62-73; RETIRED. Mem: AAAS; Am Chem Soc; fel Am Inst Chem. Res: Dyestuffs; pigments. Mailing Add: PO Box 413 Amherst VA 24521

CAVA, MICHAEL PATRICK, b Brooklyn, NY, Feb 13, 26; m 51; c 1. ORGANIC CHEMISTRY. Educ: Harvard Univ, BS, 46; Univ Mich, MS, 48, PhD(chem), 51. Prof Exp: Fel, Harvard Univ, 51-53; from asst prof to prof chem, Ohio State Univ, 53-65; prof, Wayne State Univ, 65-69; PROF CHEM, UNIV PA, 69- Mem: Am Chem Soc. Res: Natural products chemistry; strained ring systems. Mailing Add: Dept of Chem Univ of Pa Philadelphia PA 19104

CAVAGNA, GIANCARLO ANTONIO, b Milano, Italy, June 3, 38; m 64; c 2. ORGANIC CHEMISTRY, PAPER CHEMISTRY. Educ: Univ Pavia, PhD(org chem), 61. Prof Exp: Org res chemist, Inst Carlo Erba Therapeut Res, Italy, 61-63; res chemist, 63-74, SR RES CHEMIST, RES CTR, WESTVACO CORP, 74- Mem: Am Chem Soc; Tech Asn Pulp & Paper Indust. Res: Organic and colloid chemistry of the papermaking process; chemistry of wood by-products. Mailing Add: Westvaco Corp Res Ctr Johns Hopkins Rd Laurel MD 20810

CAVAGNOL, JERRY CHARLES, chemistry, see 12th edition

CAVALIERE, ALPHONSE RALPH, b New Haven, Conn, Jan 23, 37; m 61; c 3. BOTANY. Educ: Ariz State Univ, BS, 60, MS, 62; Duke Univ, PhD(mycol), 65. Prof Exp: Vis asst prof bot, Duke Univ, 65-66; asst prof biol, 66-69, ASSOC PROF BIOL, GETTYSBURG COL, 69-, CHMN DEPT, 75- Concurrent Pos: Assoc mem, Surtsey Res Soc, Iceland, 65- Mem: Mycol Soc Am; Am Inst Biol Sci. Res: Fungi of Iceland; mycological research on the new volcanic upthrust, Surtsey; marine fungi of Eastern US. Mailing Add: Dept of Biol Gettysburg Col Gettysburg PA 17325

CAVALIERI, DONALD JOSEPH, b New York, NY, May 5, 42; m 70; c 1. DYNAMIC METEOROLOGY. Educ: City Col New York, BS, 64; Queens Col, New York, MA, 67; NY Univ, PhD(meteorol), 74. Prof Exp: Physicist, US Naval Appl Sci Lab, 66-67; from instr to asst prof physics, State Univ NY, 67-70; NAT RES COUN RES ASSOC METEOROL, NAT GEOPHYS & SOLAR-TERRESTRIAL DATA CTR, NAT OCEANOG & ATMOSPHERIC ADMIN, 74- Mem: Am Geophys Union; Am Meteorol Soc. Res: Statistical association of stratospheric and ionospheric planetary-scale waves. Mailing Add: Environ Data Serv Nat Ocean & Atmos Admin Boulder CO 80302

CAVALIERI, LIEBE FRANK, b Philadelphia, Pa, Aug 26, 19; c 3. PHYSICAL BIOCHEMISTRY. Educ: Univ Pa, BS, 43, MS, 44, PhD(chem), 45. Prof Exp: Asst instr, Univ Pa, 43; fel amino compounds & sugars, Ohio State Univ, 45; fel, 46-48, from asst to assoc, 48-60, assoc dir, 61-68, asst prof biochem, 52-54, assoc prof, 54-60, PROF BIOCHEM, SLOAN-KETTERING DIV, CORNELL UNIV, 60-, MEM, 60- Concurrent Pos: Fel, Columbia Univ, 50. Mem: AAAS; Am Chem Soc; Am Soc Biol Chem; Harvey Soc. Res: Macromolecular structure of nucleic acids; DNA replication. Mailing Add: Div Genetics Walker Lab Sloan-Kettering Inst Cancer Res 145 Boston Post Rd Rye NY 10580

CAVALIERI, RALPH R, b New York, NY, Jan 15, 32; m 57; c 1. ENDOCRINOLOGY, NUCLEAR MEDICINE. Educ: NY Univ, BA, 52, MD, 56; Am Bd Internal Med, 65. Prof Exp: Intern med, Third Div, Bellevue Hosp, 56-57, resident, 57-59; mem staff nuclear med, US Naval Hosp, Bethesda, Md, 59-61; NATO fel biochem, Nat Inst Med Res, Eng, 61-62; USPHS spec fel nuclear med, Johns Hopkins Univ Hosp, 62-63; CHIEF RADIOISOTOPE SERV, VET ADMIN HOSP, 63-; ASSOC PROF MED & RADIOL, MED CTR, UNIV CALIF, SAN FRANCISCO, 71- Concurrent Pos: Asst clin prof med & radiol, Med Ctr, Univ Calif, San Francisco, 63-71. Mem: Soc Nuclear Med; Endocrine Soc; Am Fedn Clin Res; Am Soc Clin Invest. Res: Thyroid physiology and biochemistry; endocrine control of metabolism; application of radioisotope techniques to medicine. Mailing Add: Vet Admin Hosp 42nd Ave & Clement San Francisco CA 94121

CAVALLITO, CHESTER JOHN, b Perth Amboy, NJ, May 7, 15; m 40; c 3. ORGANIC CHEMISTRY, PHYSIOLOGICAL CHEMISTRY. Educ: Rutgers Univ, BS, 36; Ohio State Univ, AM, 38, PhD(chem), 40. Prof Exp: Asst entom, NJ Exp Sta, 36; res chemist, Goodyear Tire & Rubber Co, 40-41, Winthrop Chem Co, 42-46 & Sterling-Winthrop Res Inst, 46-50; res dir, Irwin, Neisler & Co, 51-63, vpres & dir res, Neisler Labs, Inc, 63-66; prof med chem, Sch Pharm, Univ NC, Chapel Hill, 66-70; EXEC VPRES, AYERST LABS, 70- Concurrent Pos: Lectr pharmacol, Univ Ill, 62- Mem: Fel AAAS; Am Chem Soc; Am Soc Microbiol; fel NY Acad Sci; Am Soc Pharmacol & Exp Therapeut. Res: Medicinals, synthetic and natural; mechanisms of drug action; research administration. Mailing Add: Ayerst Labs 685 Third Ave New York NY 10017

CAVANAGH, DENIS, b Paisley, Scotland, Dec 27, 23; US citizen; m 51; c 3. OBSTETRICS & GYNECOLOGY. Educ: Univ Glasgow, MB, ChB, 52; FRCOG. Prof Exp: Mike Hogg Award, Postgrad Sch Med, Univ Tex, 59; from asst prof to prof obstet & gynec, Univ Miami, 59-66; prof & chmn dept, Sch Med, St Louis Univ, 66-71; prof, Univ Tasmania, 71-72; PROF GYNEC & OBSTET & CHMN DEPT, SCH MED, ST LOUIS UNIV, 73- Mem: AAAS; AMA; fel Am Col Surg; fel Am Col Obstet & Gynec; Am Gynec Soc. Res: Diagnosis and treatment of gynecological cancer; clinical and laboratory aspects of septic shock; eclamptogenic toxemia. Mailing Add: Dept of Gynec & Obstet St Louis Univ Sch of Med St Louis MO 63104

CAVANAGH, PAUL R, organic chemistry, medicinal chemistry, see 12th edition

CAVANAGH, TIMOTHY D, b Berkeley, Calif, Aug 16, 38; m 63; c 2. MATHEMATICS. Educ: Sacramento State Col, AB, 59, MA, 62; Ohio State Univ, PhD(math educ), 65. Prof Exp: Assoc prof, 65-74, PROF MATH, UNIV NORTHERN COLO, 74- Concurrent Pos: Vis prof, Univ Col, Galway, Ireland, 70-71. Mem: Math Asn Am; Am Educ Res Asn. Res: Mathematics education. Mailing Add: Dept of Math Univ of Northern Colo Greeley CO 80639

CAVANAH, LLOYD (EARL), b Keytesville, Mo, Sept 18, 19; m 48; c 2. AGRONOMY. Educ: Univ Mo, BS, 48, MS, 50. Prof Exp: From instr to assoc prof, 48-74, PROF AGRON, UNIV MO-COLUMBIA, 74-, SUPT DEPT RES FARM, 73- Concurrent Pos: Exec secy-treas, Mo Seed Improv Asn, Univ Mo, 55-62. Mem: Am Soc Agron. Res: Quality seeds; factors that affect the cleanliness and germination of seeds. Mailing Add: Dept of Agron 135 Mumford Hall Univ of Mo Columbus MO 65201

CAVANAUGH, CHARLES JOHNSON, SR, b Leesville, La, Jan 23, 11; m; c 5. PHYSIOLOGY. Educ: La Col, BA, 32; Univ Tenn MS, 34. Prof Exp: Teaching fel zool, Univ Tenn, 32-34; head sci dept, Ark high sch, 34-35; teaching fel zool, NY Univ, 35-37; instr biol, Hofstra Col, 37-40; asst prof, 40-41; prof, Union Univ, 42-44; PROF BIOL LA COL, 45- Mem: AAAS. Res: Endocrinology; congenital anomalies. Mailing Add: Dept of Biol La Col Pineville LA 71360

CAVANAUGH, DANIEL JAMES, pharmacology, see 12th edition

CAVANAUGH, ROBERT J, b Scranton, Pa, Nov 11, 42; m 64; c 3. ORGANIC CHEMISTRY. Educ: Carnegie-Mellon Univ, BS, 64; Univ Pittsburgh, PhD(org chem), 67. Prof Exp: RES CHEMIST, E I DU PONT DE NEMOURS & CO, INC, 69- Mem: Am Chem Soc. Res: Reactions of enamines; catalytic oxidation; polymer synthesis. Mailing Add: 4 Foxboro Dr Vienna WV 26105

CAVANAUGH, ROBERT MORRIS, chemistry, see 12th edition

CAVAZOS, LAURO FRED, b Kingsville, Tex, Jan 4, 27; m 54; c 10. ANATOMY. Educ: Tex Tech Col, BA, 49, MA, 51; Iowa State Univ, PhD(physiol), 54. Prof Exp: Lab asst, Tex Tech Col, 49-51; res asst, Iowa State Col, 51-54; from instr to assoc prof anat, Med Col Va, 54-64; prof anat & chmn dept, 64-72, assoc dean sch med, 72-73, actg dean sch med, 73-75, DEAN SCH MED, TUFTS UNIV, 75- Mem: Am Asn Anat; Asn Am Med Cols. Res: Physiology; histochemistry, electron microscopy and biochemistry of male reproductive system; hormonal factors; fine structure of cells of steroid secretion. Mailing Add: Off of the Dean Tufts Univ Sch of Med Boston MA 02111

CAVE, MAC DONALD, b Philadelphia, Pa, May 14, 39; m 63; c 2. ANATOMY, CELL BIOLOGY. Educ: Susquehanna Univ, BA, 61; Univ Ill, MS, 63, PhD(anat), 65. Prof Exp: Am Cancer Soc Swed-Am Exchange fel, Inst Genetics, Univ Lund, 65-66; USPHS fel, Max Planck Inst Biol, 66-67; asst prof anat & cell biol, Sch Med, Univ Pittsburgh, 67-72; ASSOC PROF ANAT, UNIV ARK MED SCI, 72- Mem: AAAS; Am Asn Anat; Am Soc Cell Biol. Res: The replication and structure of the genetic machinery, synthesis of chromosomal proteins and nucleic acids, organization and localization of genes coding for ribosomal RNA, their amplification during oogenesis and their expression during early embryogenesis. Mailing Add: Dept of Anat Univ of Ark Med Ctr Little Rock AR 72201

CAVE, WILLIAM THOMPSON, b Winnipeg, Man, June 8, 17; US citizen; m 41; c 2. PHYSICAL CHEMISTRY. Educ: Univ Man, BSc, 39; Oxford Univ, DPhil(phys chem), 48. Prof Exp: Res chemist, Shawinigan Chems Co, 44-46 & 48-51; res mgr, Cent Res Dept, Monsanto Chem Co, 51-68, tech dir, Monsanto Co, 68-70, DIR NUCLEAR OPERS, MOUND LAB, MONSANTO RES CORP, 70- Concurrent Pos: Mem inspection bd, UK & Can, 40-44. Mem: Fel AAAS; Am Chem Soc; Soc Appl Spectros. Res: Spectroscopy; organic chemistry; instrumentation. Mailing Add: Monsanto Res Corp Mound Lab PO Box 32 Miamisburg OH 45342

CAVELL, RONALD GEORGE, b Sault St Marie, Ont, Oct 15, 38; m 60; c 2. INORGANIC CHEMISTRY. Educ: McGill Univ, BSc, 58; Univ BC, MSc, 60, PhD(inorg chem), 62; Cambridge Univ, PhD, 64. Prof Exp: From asst prof to assoc prof, 64-74, PROF CHEM, UNIV ALTA, 74- Mem: Chem Inst Can; The Chem Soc; Am Chem Soc. Res: X-ray and ultraviolet photoelectron spectroscopy; chemistry of simple and complex transition metal halides; particularly fluorides; halogen and perfluoroalkyl derivatives of phosphorus. Mailing Add: Dept of Chem Univ of Alta Edmonton AB Can

CAVENDER, JAMES C, b Tuxedo, NY, July 27, 36; m 64; c 2. MYCOLOGY. Educ: Union Col, BS, 58; Univ Wis, MS, 61, PhD(bot), 63. Prof Exp: Res assoc bact, Univ Wis, 63-64; asst prof biol, Wabash Col, 64-69; asst prof, 69-70, ASSOC PROF BOT, OHIO UNIV, 71- Mem: AAAS; Bot Soc Am; Mycol Soc Am. Res: Ecology, taxonomy and morphology of cellular slime molds. Mailing Add: Dept of Bot Ohio Univ Athens OH 45701

CAVENDER, JAMES VERE, JR, b San Antonio, TX, Oct 10, 22; m 45; c 2. ORGANIC POLYMER CHEMISTRY. Educ: Agr & Mech Col, Tex, BSc, 48. Prof Exp: Res chemist, Monsanto Chem Co, Mo, 48-51, res chemist, Tex, 51-54, res group leader, 54-74, SR PROCESS SPECIALIST, MONSANTO CO, 74- Mem: Am Chem Soc. Res: Linear polyolefin process development; Ziegler chemistry; organic synthesis; organic medicinals. Mailing Add: Monsanto Co PO Box 1311 Texas City TX 77590

CAVENESS, FIELDS EARL, b Long Beach, Calif, Feb 3, 28; m 49; c 3. NEMATOLOGY. Educ: Chico State Col, BA, 52; Ore State Univ, PhD(plant path), 56. Prof Exp: Nematologist, Beet Sugar Develop Found, Ft Collins, Colo, 56-58; assoc nematologist, Dept Plant Path, SDak State Univ, 58-59; nematologist, Int Develop Serv, Washington, DC, 59-62, US AID, Lagos, Nigeria, 62-65 & USDA, Shafter, Calif, 65-69; NEMATOLOGIST, INT INST TROP AGR, IBADAN, NIGERIA, 69- Mem: Soc Nematol; Nigerian Soc Plant Protection (vpres, 70-71, pres, 71-72); Orgn Trop Am Nematol; Asn Advan Agr Sci Africa. Res: Breeding plants for resistance to nematodes; chemical control of nematodes; biology of nematodes. Mailing Add: Int Inst of Trop Agr PMB 5320 Ibadan Nigeria

CAVENESS, WILLIAM FIELDS, b Zebulon, NC, Sept 13, 08; m 61. NEUROPHYSIOLOGY, CLINICAL NEUROLOGY. Educ: Univ NC, AB, 29; McGill Univ, MD, 43. Prof Exp: Asst prof neurol, Col Physicians & Surgeons, Columbia Univ, 53-63, assoc prof, 63-70; assoc dir collab & field res, 65-69, CHIEF LAB EXP NEUROL, NAT INST NEUROL DIS & STROKE, 69- Concurrent Pos: Rep, Educ & Cult Exchange Prog, US State Dept, Latin Am, 62. Mem: Am Neurol Asn; Asn Res Nerv & Ment Dis; Am Epilepsy Soc (pres, 61). Res: Central nervous system development; convulsive disorders; cranio-cerebral trauma. Mailing Add: Bldg 36 Rm 4A-27 Nat Inst Neurol Dis & Stroke Bethesda MD 20014

CAVENY, ELMER LEONARD, b NC, May 26, 07; m 30; c 1. PSYCHIATRY. Educ: Emory Univ, MD, 30; Am Bd Psychiat, dipl; Am Bd Prev Med, dipl. Prof Exp: Asst clin prof psychiat, Womens Med Col, Philadelphia, 48-51; clin prof, Georgetown Univ, 52-55; prof psychiat & neurol & chmn dept, 55-58, PROF PSYCHIAT, MED COL ALA, 59- Concurrent Pos: Chief neuropsychiat treatment & training ctr, US Naval Hosp, Philadelphia, 48-51, chief neuropsychiat serv, Nat Naval Med Ctr, Bethesda, 51-53, head neuropsychiat br, Bur Med & Surg, US Navy Dept, Washington, DC, 53-55; pvt pract, 59-63. Mem: Fel Am Col Physicians; Am Psychoanal Asn; fel Am Psychiat Asn; AMA. Res: Psychological structure of man through psychiatry and psychoanalysis. Mailing Add: 3516 Robin Dr Birmingham AL 35223

CAVERS, PAUL BRETHEN, b Toronto, Ont, Jan 18, 38; m 61; c 3. PLANT ECOLOGY, WEED SCIENCE. Educ: Ont Agr Col, BSA, 60; Univ Wales, PhD(weed ecol), 63; DIC, Univ London, 72. Prof Exp: Lectr, 63-64, asst prof, 64-69, ASSOC PROF ECOL, UNIV WESTERN ONT, 69- Concurrent Pos: Chmn subcomt life hist studies, Can Weed Comt; grant selection comt pop biol, Nat Res Coun Can, 73-76. Mem: Weed Sci Soc Am; Ecol Soc Am; Brit Ecol Soc; Agr Inst Can; Can Bot Asn(pres, 73-74). Res: Comparative ecology of closely related species living in the same area including the following genera: Rumex, Polygonum, Plantago and Melilotus; seed dispersal and dormancy; seedling establishment; dynamics of plant populations. Mailing Add: Dept of Plant Sci Univ of Western Ont London ON Can

CAVERT, HENRY MEAD, b Minneapolis, Minn, Mar 30, 22; m; c 3. PHYSIOLOGY, MEDICAL SCHOOL ADMINISTRATION. Educ: Univ Minn, MD, 51, PhD(physiol), 52. Prof Exp: From asst to assoc prof physiol, 51-68, asst dean med sch, 57-64, PROF PHYSIOL, MED SCH, UNIV MINN, MINNEAPOLIS, 68-, ASSOC DEAN, 64- Concurrent Pos: Am Heart Asn res fel, 51-54, estab investr, 54-57; Nat Heart Inst spec res fel biochem & vis prof biochem, Sch Med, Univ Edinburgh, 61-62; mem heart prog proj comt, Nat Heart & Lung Inst, 66-69, consult, 69- Mem: AAAS; Am Physiol Soc; Asn Am Med Cols. Res: Transport of sugars and amino acids across muscle cell membranes; effects of muscle activity on transmembrane transport; physician manpower needs and supply. Mailing Add: 151 Owre-Jackson Univ of Minn Med Sch Minneapolis MN 55455

CAVES, THOMAS COURTNEY, b Pryor, Okla, Apr 8, 40; m 64; c 1. THEORETICAL CHEMISTRY. Educ: Univ Okla, BS, 62; Columbia Univ, PhD(chem physics), 68. Prof Exp: NASA res fel atomic physics, Harvard Col Observ, 68-69, res assoc, 69; asst prof, 69-74, ASSOC PROF CHEM, NC STATE UNIV, 74- Mem: Am Phys Soc. Res: Ab initio calculation of atomic and molecular properties; upper and lower bounds to calculated properties; semiempirical Green's function method for calculating atomic properties. Mailing Add: Dept of Chem NC State Univ Raleigh NC 27607

CAVEY, MICHAEL JOHN, b Elkhorn, Wis, Oct 8, 46. BIOLOGICAL STRUCTURE, EMBRYOLOGY. Educ: Univ Va, BA, 68; Univ Wash, MS, 71, PhD(zool), 73. Prof Exp: RES SCIENTIST ANAT, SCH MED, UNIV SOUTHERN CALIF, 74- Concurrent Pos: NIH fel, 75. Mem: Sigma Xi; Am Soc Zoologists; Western Soc Naturalists. Res: Fine structure and differentiation of contractile tissues and intercellular junctions; morphogenetic movements in marine invertebrates and their subcellular mechanisms. Mailing Add: Dept of Anat Univ Southern Calif Sch Med Los Angeles CA 90033

CAVIN, WILLIAM PINCKNEY, b Spartanburg, SC, June 2, 25; m 50; c 2. ORGANIC CHEMISTRY. Educ: Wofford Col, AB, 45; Duke Univ, AM, 46; Univ NC, PhD(chem), 53. Prof Exp: Asst, Duke Univ, 45-46; from instr to prof, 46-62, JOHN M REEVES PROF CHEM, WOFFORD COL, 62-, CHMN DEPT, 71- Concurrent Pos: NSF fac fel-vis prof chem, Brown Univ, 65-66. Mem: AAAS; Am Chem Soc. Res: Isotope effect of carbon-14 in organic reactions; kinetics and mechanisms of organic reactions. Mailing Add: Dept of Chem Wofford Col Spartanburg SC 29301

CAVINESS, BOBBY FORRESTER, b Asheboro, NC, Mar 24, 40; m 61; c 1. COMPUTER SCIENCE, APPLIED MATHEMATICS. Educ: Univ NC, Chapel Hill, BS, 62; Carnegie-Mellon Univ, MS, 64, PhD(math), 68. Prof Exp: Asst prof math, Duke Univ, 67-70; asst prof comput sci, Univ Wis-Madison, 70-75; ASSOC PROF COMPUT SCI, ILL INST TECHNOL, 75- Concurrent Pos: NSF grant, 69-70 & 71; assoc ed, Asn Comput Mach J Transactions Math Software, 75- Mem: Asn Comput Mach; Math Asn Am; Soc Indust & Appl Math; AAAS; Fedn Am Scientists. Res: Symbolic and algebraic manipulation; recursive systems for symbolic mathematics; list processing. Mailing Add: Dept of Comput Sci Ill Inst of Technol Chicago IL 60616

CAVINESS, CHARLES E, plant breeding, plant genetics, see 12th edition

CAVINESS, VERNE STRUDWICK, JR, b Raleigh, NC, July 25, 34; m 62; c 2. NEUROLOGY, NEUROPATHOLOGY. Educ: Duke Univ, BA, 56; Oxford Univ, DPhil(exp path), 60; Harvard Univ, MD, 62. Prof Exp: ASST PROF NEUROL, HARVARD MED SCH, 71- Concurrent Pos: NIH spec res fel, Harvard Med Sch, 69-71; asst neurologist, Mass Gen Hosp, 71-; sr investr, E K Shriver Inst, 71- Mem: AAAS; Am Neurol Asn; Am Acad Neurol. Res: Developmental neuroanatomy and neuropathology. Mailing Add: E K Shriver Inst 200 Trapelo Rd Waltham MA 02154

CAVITT, STANLEY BRUCE, b Red Oak, Tex, Apr 5, 34; m 65; c 1. PETROLEUM CHEMISTRY. Educ: NTex State Col, BA, 56, MS, 57; Univ Tex, PhD(chem), 61. Prof Exp: Chemist, Am Oil Co, 61-63; res chemist, 63-69, PROJ CHEMIST, JEFFERSON CHEM CO, INC, 69- Mem: Am Chem Soc; The Catalysis Soc. Res: Catalyst research; petrochemicals; organic synthesis. Mailing Add: Jefferson Chem Co Inc PO Box 4128 Austin TX 78765

CAVONIUS, CARL RICHARD, b Santa Barbara, Calif, Dec, 23, 32. NEUROSCIENCE. Educ: Wesleyan Univ, BA, 53; Brown Univ, MSc, 61, PhD(psychol), 62. Prof Exp: USPHS fel, Brown Univ, 62-63; res scientist, Human Sci Res, Inc, 63-65; dir, Eye Res Found, 65-71; Von Humboldt fel, Univ Munich, 71-73; J McKeen Cattell Fund fel, Cambridge Univ, 73-74; chief sci officer lab med phys, Univ Amsterdam, 74-75; PROF NEUROPHYS, UNIV DORTMUND, W GERMANY, 76- Concurrent Pos: From asst prof to assoc prof, Sch Med, Univ Md, 64-70; res dir, Inst Pedestrian Res, 73-74. Mem: Am Psychol Asn; Optical Soc Am; Psychonomic Soc; Exp Psychol Soc. Res: Sensory coding and processing; human psychophysics, especially visual; human factors and applied physiology.

CAWEIN, MADISON JULIUS, b Bloomfield, NJ, Jan 31, 26; m 69; c 4. HEMATOLOGY, CLINICAL PHARMACOLOGY. Educ: Harvard Univ, BA, 49; Tulane Univ, MD, 54; Univ Minn, MS, 59. Prof Exp: Assoc prof med, Med Sch, Univ Ky, 60-66; dir clin pharmacol, Eaton Labs, 66-68; assoc prof clin pharmacol, Univ Tenn, 68-70; DIR CLIN PHARMACOL, MERRELL-NAT LABS, 70- Concurrent Pos: Consult, Vet Admin Hosp, Lexington, Ky, 60-66, USPHS Hosp, Lexington, 60-66, St Mary's Hosp, Lexington, 59-66, Univ Ky Hosp, 60-66, Vet Admin Hosp, Johnson City, Tenn, 68-70 & Univ Tenn Mem Hosp, Knoxville, 68-70; lectr, Roswell Park Mem Inst, 64. Mem: Am Soc Hemat; Am Soc Clin PHarmacol & Therapeut; Am Fedn Clin Res; NY Acad Sci. Res: Genetics of erythrocyte enzymes; hemoglobinopathy; leukemoid reactions; pharmacogenetics; drug interactions; carcinogenesis; pharmacology of 1-dihydroxyphenylalanine. Mailing Add: Merrell-Nat Labs 110 E Amity Rd Cincinnati OH 45215

CAWLEY, ALLAN JOSEPH, veterinary medicine, see 12th edition

CAWLEY, EDWARD PHILIP, b Jackson, Mich, Sept 1, 12; m 39; c 2. DERMATOLOGY. Educ: Univ Mich, AB, 36, MD, 40; Am Bd Dermat, dipl, 47. Prof Exp: Asst prof dermat, Med Sch, Univ Mich, 48-51; PROF DERMAT & CHMN DEPT, SCH MED, UNIV VA, 51- Concurrent Pos: Dir, Am Bd Dermat, 58-67, pres. Mem: Am Dermat Asn; Soc Invest Dermat; AMA; NY Acad Sci. Res: Medical mycology and dermatopathology. Mailing Add: Dept of Dermat Univ of Va Sch of Med Charlottesville VA 22901

CAWLEY, EDWARD T, b Chicago, Ill, Mar 13, 31; m 55; c 5. PLANT ECOLOGY. Educ: Northern Ill State Teachers Col, BS, 53; Univ Wis, MS, 58, PhD(bot), 60. Prof Exp: From asst prof to assoc prof, 60-71, PROF BIOL, LORAS COL, 71- Concurrent Pos: Mem Iowa State Preserves Adv Bd, past chmn, 63-68. Mem: Ecol Soc Am. Res: Fresh water ecology-diversity indices. Mailing Add: Dept of Biol Loras Col Dubuque IA 52003

CAWLEY, JOHN JOSEPH, b Somerville, Mass, Sept 18, 32; m 56; c 5. PHYSICAL ORGANIC CHEMISTRY. Educ: Boston Col, BS, 55; Harvard Univ, MA, 57, PhD(chem), 61. Prof Exp: Fel, Univ Wash, 60-61; asst prof, 61-69, ASSOC PROF CHEM, VILLANOVA UNIV, 69- Mem: Am Chem Soc; The Chem Soc. Res: Chromic acid oxidations; hydrolytic processes in concentrated acid media; chemistry of acetals, especially of dioxolans and dioxoles. Mailing Add: Dept of Chem Villanova Univ Villanova PA 19085

CAWLEY, LEO PATRICK, b Oklahoma City, Okla, Aug 11, 22; m 48; c 3. PATHOLOGY, IMMUNOCHEMISTRY. Educ: Okla State Univ, BS, 48; Univ Okla, MD, 52; Am Bd Path, dipl, 57, cert clin chem, 65. Prof Exp: Intern path, Wesley Hosp, Wichita, Kans, 52-53, resident, 53-54; resident, Wayne County Gen Hosp, Eloise, Mich, 54-57; clin pathologist & assoc dir labs, 57-69, DIR LABS, WESLEY MED CTR, 69- Concurrent Pos: Sci dir, Wesley Med Res Found. Mem: AAAS; Am Soc Clin Pathologists; Am Soc Human Genetics; Am Soc Exp Path; Am Chem Soc. Res: Biochemistry; characterization of proteins other than hemoglobin of human erythrocytes, using methods based on column chromatography, electrophoresis, analytic electrophoresis, immunoelectrophoresis and gas chromatography. Mailing Add: Wesley Med Res Found 550 N Hillside Wichita KS 67214

CAWLEY, ROBERT, b Scranton, Pa, Jan 29, 36; m 58; c 4. THEORETICAL PHYSICS, THEORETICAL MECHANICS. Educ: Mass Inst Technol, BS, 58, MS, 60, PhD(physics), 65. Prof Exp: Asst prof physics, Clarkson Col Technol, 65-67; res physicist, Nuclear Physics Div, 67-73, RES PHYSICIST PRIMAL THER, NUCLEAR PHYSICS BR, NAVAL SURFACE WEAPONS CTR, 73- Mem: Am Phys Soc. Mailing Add: Nuclear Physics Br Naval Surface Weapons Ctr Silver Spring MD 20910

CAYEN, MITCHELL NESS, b Montreal, Que, Nov 6, 38; m 67; c 2. BIOCHEMISTRY. Educ: McGill Univ, BSc, 59, MSc, 51, PhD(agr chem), 65. Prof Exp: RES BIOCHEMIST & HEAD METAB SECT, AYERST RES LABS, 65- Concurrent Pos: Fel coun arteriosclerosis, Am Heart Asn. Mem: Am Soc Pharmacol & Exp Therapeut; AAAS; Can Biochem Soc; NY Acad Sci; Sigma Xi. Res: Lipid metabolism; pathogenesis of atherosclerosis; drug metabolism. Mailing Add: 5625 Hudson Ave Cote St Luc Montreal PQ Can

CAYER, DAVID, b Hartford, Conn, Nov 5, 13; m 42; c 2. GASTROENTEROLOGY. Educ: Duke Univ, AB, 35, MD, 38; Am Bd Internal Med, dipl, 46; Am Bd Gastroenterol & Am Bd Nutrit, dipl, 51. Prof Exp: Res physician, NC Sanitarium Tuberc, 39; from intern to resident, Med Sch, Duke Univ, 39-42, asst physiol & pharm, 41, instr med, 42-44; asst prof med, 45-49, assoc prof, 49-53, prof gastroenterol, 53-56, PROF MED, BOWMAN GRAY SCH MED, WAKE FOREST UNIV, 56- Concurrent Pos: Consult, US Vet Admin, NC, 48 & Nat Cancer Inst. Honors & Awards: Silver Medal, AMA, 51, Billings Medal, 51 & 71, Gold Medal, 71. Mem: AAAS; fel AMA; fel Am Col Physicians; Am Gastroscopic Soc; fel Am Col Clin Pharmacol & Chemother. Res: Digestive diseases. Mailing Add: 2240 Cloverdale Ave Winston-Salem NC 27103

CAYFORD, AFTON HERBERT, b Hollywood, Calif, Dec 15, 29; m 51; c 4. MATHEMATICAL ANALYSIS, NUMBER THEORY. Educ: La Verne Col, BA, 51;

Univ Calif, Los Angeles, MA, 58, PhD(math), 61. Prof Exp: Mem tech staff, Hughes Aircraft Co, 56-58; instr math, Univ BC, 59-62; res scientist, Jet Propulsion Lab, Calif Inst Technol, 62-63; asst prof, 62-70, ASSOC PROF MATH, UNIV BC, 70- Mem: AAAS; Am Math Soc; Math Asn Am; Can Math Cong; Sigma Xi. Res: Analytic number theory and properties of certain classes of entire functions of complex variable with special characteristics of finite point sets; growth rate. Mailing Add: Dept of Math Univ of BC Vancouver BC Can

CAYLE, THEODORE, b New York, NY, Mar 1, 28; m 49; c 4. ENZYMOLOGY, INDUSTRIAL MICROBIOLOGY. Educ: Brooklyn Col, BA, 49, MA, 51; Univ Ill, PhD(plant physiol, biochem), 56. Prof Exp: Asst, Brooklyn Col, 49-51; asst, Univ Ill, 51-53, 55-56; instr bot, Univ Wash, 56-58; biochemist, Wallerstein Co, 58-62, mgr biochem res, 62-65, asst dir res, 65-70, dir res, 70-73; pres, CCF Consult Corp, 74-75; VPRES TECH DIR, DAIRYLAND FOOD LABS, INC, 75- Mem: Am Dairy Sci Asn; Am Chem Soc; Am Asn Cereal Chem; fel Am Inst Chem; Am Soc Microbiol. Res: Dairy chemistry. Mailing Add: Dairyland Food Labs 620 Progress Ave Waukesha WI 53186

CAYWOOD, STANLEY WILLIAM, JR, b Akron, Ohio, Aug 10, 24; m 59. ORGANIC CHEMISTRY. Educ: Harvard Univ, AB, 48, AM, 49; Univ NH, MS, 50; Cornell Univ, PhD(org chem), 53. Prof Exp: RES CHEMIST, ELASTOMERS DEPT, E I DU PONT DE NEMOURS & CO, INC, 53- Concurrent Pos: Vis prof, Clarkson Col Technol, 64-65. Res: Polymer synthesis, elastomer testing, evaluation and processing. Mailing Add: 115 Watford Rd Westgate Farms Wilmington DE 19808

CAYWOOD, THOMAS E, b Lake Park, Iowa, May 9, 19; m 41; c 3. OPERATIONS RESEARCH. Educ: Cornell Col, AB, 39; Northwestern Univ, MA, 40; Harvard Univ, PhD(math), 47. Prof Exp: Tutor math, Northwestern Univ, 39-40; tutor, Harvard Univ, 41-42, spec res assoc physics, 42-45, asst, 46; sr mathematician & coordr res, Inst Air Weapons Res, Ill, 47-52; supvr opers res, Armour Res Found, 52-53; partner, Caywood-Schiller Assoc, 53-61, Peat, Marwick, Caywood, Schiller & Co, 62-66 & Caywood-Schiller Assoc, 66-70; VPRES, CAYWOOD-SCHILLER DIV, A T KEARNEY CO, 71- Concurrent Pos: Lectr sch bus, Univ Chicago, 53-57; consult, Off Asst Secy Defense, 53-60 & opers eval group, US Navy, 60-61; pres, Invests Opers Mex, SA, 59-60; mem alumni bd dir, Cornell Col, 62-65, trustee, 64-, pres bd trustees, 70-72; mem statist comt, Nat Acad Sci-Nat Res Coun, 60-63; mem defense sci bd & chmn ord panel, Off Dir Defense Res & Eng, 60-64; chmn task force, Gun Systs Acquisition, Defense Sci Bd, 75; ed, Opers Res, Opers Res Soc Am, 61-68. Honors & Awards: George E Kimball Medal, Opers Res Soc Am, 74. Mem: Am Math Soc; Am Inst Indust Eng; Opers Res Soc Am (vpres, 68-69, pres, 69-70); Math Asn Am; Inst Mgt Sci. Res: Applied mathematics; operational research; theory of games; probability; evaluation engineering. Mailing Add: Caywood Schiller Div A T Kearney Co 100 S Wacker Dr Chicago IL 60606

CAZEAU, CHARLES J b Rochester, NY, June 25, 31; m 60; c 3. GEOLOGY. Educ: Univ Notre Dame, BS, 54; Fla State Univ, MS, 55; Univ NC, PhD(geol), 62. Prof Exp: Explor Geologist, Humble Oil & Refining Co, Tex, 55-56, 57-58; asst prof geol, Clemson Col, 60-63; asst prof, 63-67, ASSOC PROF GEOL, STATE UNIV NY BUFFALO, 67- Concurrent Pos: Sigma Xi study grant, 64-65; consult, Union Camp Corp, 74-75. Mem: Am Asn Petrol Geologists; Soc Econ Paleontologists & Mineralologists; AAAS; Nat Asn Geol Teachers. Res: Detrital mineralogy; recent sediments; Triassic and Pleistocene geology; geoarchaeology of West Mexico. Mailing Add: Dept of Geol State Univ of NY Buffalo NY 14214

CAZES, JACK chromatography, see 12th edition

CAZIER, MONT ADELBERT, b Cardston, Alta, May 27, 11; US citizen; m 55; c 3. SYSTEMATIC ENTOMOLOGY. Educ: Univ Calif, BS, 35, PhD(entom), 42. Prof Exp: Asst cur entom, Dept Insects & Spiders, Am Mus Natural Hist, 41-43, assoc cur, 46-51, cur, 52-59, chmn dept, 46-59, founder & dir, Southwestern Res Sta, 55-59, resident dir, 50-62; prof entom, 62-68, PROF ZOOL, ARIZ STATE UNIV, 68- Concurrent Pos: Res entomologist, Univ Calif, 62-64. Res: Bionomic entomology; biology; ecology and behavior studies in aculeate Hymenoptera, Coleoptera and Diptera. Mailing Add: Dept of Zool Ariz State Univ Tempe AZ 85281

CAZIN, JOHN, JR, b Wheeling, WVa, July 2, 29; m 53; c 3. MICROBIOLOGY, MEDICAL MYCOLOGY. Educ: Univ NC, BS, 52, MS, 54, PhD(bact), 57. Prof Exp: Instr, 57-58, assoc, 58-59, from asst prof to assoc prof, 59-72, PROF MICROBIOL, UNIV IOWA, 72- Mem: Am Soc Microbiol; Mycol Soc Am; Int Soc Human & Animal Mycol; fel Am Acad Microbiol; Med Mycol Soc of the Americas. Res: Medical mycology; immunology of deep seated mycoses; medical bacteriology. Mailing Add: Dept of Microbiol Univ of Iowa Iowa City IA 52240

CCARELLI, ANTHONY JOSEPH, b New York, NY, Aug 11, 44; m 68; c 1. MOLECULAR BIOLOGY, MOLECULAR GENETICS. Educ: Cornell Univ, BS, 66; Loma Linda Univ, MS, 68; Calif Inst Technol, PhD(biophys), 74. Prof Exp: Fel molecular biol, Am Cancer Soc, 74-76; ASST PROF BIOL, LOMA LINDA UNIV, 76- Mem: Am Soc Microbiol. Res: Investigation of the early stages of infection with the single-stranded DNA bacteriophage ØX174, particularly the mechanisms of initiation and synthesis of the complementary strand in parental replicative forms. Mailing Add: Dept of Biol Loma Linda Univ Loma Linda CA 92354

CEASAR, GERALD P, b New York, NY, Jan 8, 40; m 67. CHEMICAL PHYSICS. Educ: Manhattan Col, BS, 62; Columbia Univ, PhD(chem), 67. Prof Exp: Air Force Off Sci Res fel chem, Calif Inst Technol, 68-69; NATO & Ramsay Mem fel, Univ Bristol & Oxford Univ, 69; asst prof, Univ Rochester, 69-74; SCIENTIST, JOSEPH C WILSON CTR TECHNOL, XEROX CORP, 74- Mem: Am Phys Soc; Am Chem Soc. Res: Polarized optical spectroscopy; magnetic resonance spectroscopy; liquid crystals; surface physics and chemistry; photoelectron and auger spectroscopies; electronic structure of molecules and solids; solid state chemistry; xerography. Mailing Add: Joseph C Wilson Ctr for Technol Xerox Corp Webster NY 14580

CEBALLOS, RICARDO, b Cadiz, Spain, Jan 12, 30; m 57; c 3. PATHOLOGY, NEUROLOGY. Educ: Univ Madrid, MD, 53. Prof Exp: Intern, Dept Physiol, Med Sch, Univ Madrid, 47-49, intern, Dept Internal Med, 49-52, rotating intern, 52-53; assoc prof path, Clin Concepcion, Madrid, 55-57; instr, Med Ctr, Univ Ala, 58, asst prof, 58-60; asst pathologist & dir labs, Hotel Dieu Hosp, Kingston, Ont, 60-63; asst dir, 63-64, DIR ANAT PATH, UNIV HOSP & HILLMAN CLIN, BIRMINGHAM, 64-; PROF PATH & ASSOC PROF NEUROL, MED CTR, UNIV ALA, 69- Concurrent Pos: Fel, Menendez & Pelayo Univ, Spain, 53; fel path, Inst Clin & Med Invest, 53-54; Doherty Found fel histochem & path, Med Sch, Univ Ala, 55; Marquesa de Pelayo Found fel, Spain, 56; Med Res Coun Can fel, 62; intern, Dept Path, Gen Hosp, Prov Diputacion of Madrid, Spain, 50-53; secy, Int Cong Internal Med, 56; consult, Vet Admin Hosp, Tuskegee, Ala, 58-60 & 64-67; lectr, Med Sch, Queen's Univ, Ont, 60-63; assoc prof path & instr neurol & med, Med Ctr, Univ Ala, 64-69; consult, WHO. Mem: Int Acad Path; AMA; Latin Am Soc Path. Res: Pituitary changes in head trauma; neuropathology; hyperparathyroidism. Mailing Add: Dept of Path Univ of Ala Med Ctr Birmingham AL 35233

CEBRA, JOHN JOSEPH, b Philadelphia, Pa, May 7, 34; m 56; c 4. IMMUNOBIOLOGY, IMMUNOCHEMISTRY. Educ: Univ Pa, AB, 55; Rockefeller Inst, PhD(immunochem), 60. Prof Exp: Nat Found fel immunol, Weizmann Inst Sci, 60-61; from instr to assoc prof microbiol, Col Med, Univ Fla, 61-67; assoc prof biol, 67-69, PROF BIOL, JOHNS HOPKINS UNIV, 69- Concurrent Pos: NIH career develop award, 64-67; vis prof immunochem, St Mary's Hosp Med Sch, London, 66-67; mem study sect, Nat Inst Allergy & Infectious Dis, 67-71; instr-in-charge physiol course, Marine Biol Lab, Woods Hole, Mass, 72-76. Honors & Awards: Eli Lilly Award Microbiol & Immunol, Am Soc Microbiol, 68. Mem: Am Asn Immunologists; Am Soc Microbiol. Res: Protein chemistry; structure of immunoglobulins; interaction of antigen, antibody and complement components; cellular synthesis of immunoglobulin polypeptide chains. Mailing Add: Mergenthaler Lab for Biol Johns Hopkins Univ Baltimore MD 21218

CEBULL, STANLEY EDWARD, b Albany, Calif, June 1, 34; m 56; c 3. STRUCTURAL GEOLOGY. Educ: Univ Calif, Berkeley, AB, 57, MA, 58; Univ Wash, PhD(geol), 67. Prof Exp: Geologist, Tex Petrol Co, Venezuela, 58-62; ASST PROF GEOSCI, TEX TECH UNIV, 67- Mem: AAAS; Am Asn Petrol Geologists; Geol Soc Am. Res: Tectonic geology. Mailing Add: Dept of Geosci Texas Tech Univ Lubbock TX 79409

CECCHI, JOSEPH LEONARD, b Chicago, Ill, Mar 29, 47; m 70; c 1. PLASMA PHYSICS, ATOMIC PHYSICS. Educ: Knox Col, AB, 68; Harvard Univ, AM, 69, PhD(physics), 72. Prof Exp: Res assoc atomic physics, Argonne Nat Lab, 67; RES STAFF MEM PLASMA PHYSICS, PLASMA PHYSICS LAB, PRINCETON UNIV, 72- Mem: Am Phys Soc. Res: Controlled thermonuclear fusion by the Tokamak method of magnetic confinement, with specific interests in plasma stability, impurity transport, and fundamental atomic processes. Mailing Add: Plasma Physics Lab Princeton Univ PO Box 451 Princeton NJ 08540

CECH, FRANKLIN CHARLES, b Cleveland, Ohio, Nov 26, 19; m 42; c 4. FOREST GENETICS. Educ: Univ Ohio, AB, 42; Mont State Univ, BSF, 49, MF, 53; Tex A&M Univ, PhD(physiol, genetics), 57. Prof Exp: Nurseryman, Mont State Univ, 49-53; silviculturist, Tex Forest Serv, 55-57; res silviculturist, Int Paper Co, 57-64; assoc prof, 64-69, PROF FOREST GENETICS, WVA UNIV, 69- Mem: Soc Am Foresters. Res: Forest genetics and tree improvement with southern pines and northern hardwoods; silvicultural research in regeneration of species. Mailing Add: Div of Forestry WVa Univ Morgantown WV 26506

CECH, IRINA, b Moscow, USSR, Feb 6, 39. ENVIRONMENTAL SCIENCES, MEDICAL ECOLOGY. Educ: Univ Tex, PhD(community health, environ sci), 73. Prof Exp: From engr to sr engr hydrologist, Hydrometeorol Inst, Bratislava, Czech, 65-69; res asst, 69-73, ASST PROF ENVIRON SCI, SCH PUB HEALTH, UNIV TEX HEALTH SCI CTR, 73- Concurrent Pos: Asst prof environ planning, Dept Archit, Rice Univ, 74- Res: Health consequences of man-made climatic and hydrologic modifications; water related diseases; water resources development and management. Mailing Add: Univ Tex Health Sci Ctr Sch of Pub Health PO Box 20186 Houston TX 77025

CECH, JOSEPH JEROME, JR, b Berwyn, Ill, Dec 5, 43; m 67; c 2. PHYSIOLOGICAL ECOLOGY. Educ: Univ Wis-Madison, BS, 66; Univ Tex, Austin, MA, 70, PhD(zool), 73. Prof Exp: ASST PROF FISHERIES BIOL, UNIV CALIF, DAVIS, 73- Mem: Am Inst Biol Sci; Ecol Soc Am; Am Fisheries Soc; Sigma Xi. Res: Investigatiions in the physiological adjustments and adaptations of marine and freshwater fishes to their environments with emphasis on respiratory, circulatory and hematological responses to extreme environments or environmental changes. Mailing Add: Dept of Animal Physiol Univ of Calif Davis CA 95616

CECICH, ROBERT ALLEN, b Chicago, Ill, July 15, 41. PLANT ANATOMY. Educ: Northern Ill Univ, BS, 63; Iowa State Univ, PhD(bot), 74. Prof Exp: Technician genetics, 65-69, BOTANIST PLANT ANAT, INST FOREST GENETICS, 69- Mem: Bot Soc Am; Int Soc Plant Morphologists. Res: Anatomical and ultrastructural aspects of apical meristem differentiation of vegetable or reproductive structures. Mailing Add: Inst of Forest Genetics Box 898 Rhinelander WI 54501

CECIL, DAVID ROLF, b Tulsa, Okla, July 12, 35; m 58; c 1. MATHEMATICS. Educ: Univ Tulsa, BA, 58; Okla State Univ, MS, 60, PhD(math), 62. Prof Exp: Sales engr fluid dynamics, Black, Sivalls & Bryson, 57-58; sr res mathematician, Atlantic Ref Co, 62; asst prof math, North Tex State Univ, 62-69; prof, Butler Univ, 69-70; assoc prof, 70-73, PROF MATH, TEX A&I UNIV, 73- Mem: Am Math Soc. Res: Vector lattices; topological algebra; group generalizations. Mailing Add: Dept of Math Tex A&I Univ Kingsville TX 78363

CECIL, HELENE CARTER, b Tunkhannock, Pa, Jan 25, 33; m 54; c 1. REPRODUCTIVE PHYSIOLOGY. Educ: Univ Md, BS, 63, PhD(poultry physiol), 68. Prof Exp: RES BIOLOGIST, USDA, 57- Mem: Am Physiol Soc; Am Chem Soc; Poultry Sci Asn; AAAS. Res: Avian reproductive enviro- nmental pollutants on reproduction. Mailing Add: Avian Physiol Lab Animal Physiol & Genetics Inst USDA Beltsville MD 20705

CECIL, JACK T, virology, tissue culture, see 12th edition

CECIL, OLIN B, physical chemistry, see 12th edition

CECIL, SAM REBER, b San Francisco, Calif, Feb 22, 16; m 47; c 2. FOOD SCIENCE. Educ: Milligan Col, BS, 37; Univ Ga, MSA, 54. Prof Exp: Asst biochem, Sch Med, Vanderbilt Univ, 37-39, asst nutrit, 40-41; asst food technologist, 41-43, assoc food technologist, 43-58, food scientist, 58-67, PROF FOOD SCI RES, AGR EXP STA, UNIV GA, 67- Concurrent Pos: Jr food technologist, Ore State Col, 56-58; mem sci adv coun, Refrigeration Res Found, 74-77. Mem: AAAS; Inst Food Technologists; Am Chem Soc; Am Oil Chem Soc. Res: Canning and freezing of fruits and vegetables; storage of canned foods, military and civil defense rations; effects of newer cultural practices on processing and product quality of peanuts. Mailing Add: Dept of Food Sci Ga Sta Univ of Ga Col Agr Experiment GA 30212

CECIL, THOMAS E, b Louisville, Ky, Dec 13, 45; m 71; c 1. GEOMETRY. Educ: Col Holy Cross, AB, 68; Brown Univ, PhD(math), 73. Prof Exp: ASST PROF MATH, VASSAR COL, 73- Concurrent Pos: NSF basic res grant, 75. Mem: Am Math Soc; Math Asn Am; Sigma Xi. Res: Taut immersions of manifolds. Mailing Add: Box 308 Vassar Col Poughkeepsie NY 12601

CECILIA, MARY, b Muscatine, Iowa, Oct 15, 05. TAXONOMY. Educ: St Ambrose Col, BA, 35; Univ Iowa, MS, 37, PhD, 50. Prof Exp: Teacher high sch, 25-40; PROF BIOL, MUNDELEIN COL, 40- Concurrent Pos: Chmn dept biol, Mundelein Col, 40-65; vis prof dept oral path & res, Sch Dent, Loyola Univ, Ill, 65-73. Mem: AAAS; Bot Soc Am; Mycol Soc Am. Res: Taxonomy of fungi; taxonomy and cytology of the basidiomycetes. Mailing Add: 6363 Sheridan Rd Mundelein Col Chicago IL 60660

CEDAR, WARREN RICHARD, b Chicago, Ill, June 28, 20; c 4. DENTISTRY. Educ: Northwestern Univ, DDS, 43. Prof Exp: Pvt pract, 45-70; PROF OPER DENT, DENT SCH, NORTHWESTERN UNIV, CHICAGO, 71-, CHMN DEPT, 74- Mem: Am Dent Asn; fel Am Col Dent. Mailing Add: Dept of Oper Dent Northwestern Univ Dent Sch Chicago IL 60611

CEDER, JACK G, b Spokane, Wash, Aug 25, 33; m 55; c 2. PURE MATHEMATICS. Educ: Univ Wash, BS, 55, MS, 57, PhD(math), 59. Prof Exp: From asst prof to assoc prof, 59-74, PROF MATH, UNIV CALIF, SANTA BARBARA, 74- Mem: Am Math Soc. Res: Abstract topological spaces; real functions. Mailing Add: Dept of Math Univ of Calif Santa Barbara CA 93106

CEDERBERG, JAMES W, b Oberlin, Kans, Mar 16, 39. MOLECULAR PHYSICS, PHYSICS. Educ: Univ Kans, AB, 59; Harvard Univ, AM, 60, PhD(physics), 63. Prof Exp: Lectr & res fel physics, Harvard Univ, 63-64; asst prof, 64-68, ASSOC PROF PHYSICS, ST OLAF COL, 68- Concurrent Pos: NSF sci fac fel, Duke Univ, 69-70. Mem: Am Phys Soc; Am Asn Physics Teachers. Res: Molecular beams; rotational magnetic moments and magnetic interactions in molecules; molecular hyperfine interactions. Mailing Add: Dept of Physics St Olaf Col Northfield MN 55057

CEDERQUIST, DENA CAROLINE, b Madrid, Iowa, Aug 29, 10. NUTRITION, BIOCHEMISTRY. Educ: Iowa State Col, BS, 31, MS, 37; Univ Wis, PhD(nutrit & biochem), 45. Prof Exp: Asst dietitian, Monmouth Mem Hosp, NJ, 33-35; instr dietetics, Kans State Col, 37-41; instr dietetics, Univ Wis, 41-42; from asst prof to assoc prof foods & nutrit, 44-56, prof & head dept, 56-70, PROF FOOD SCI & HUMAN NUTRIT, MICH STATE UNIV, 70- Mem: AAAS; Am Dietetic Asn; Am Home Econ Asn; Am Chem Soc; Am Inst Nutrit. Res: Food requirements of women; protein requirements of college age women; nutritive value of proteins; nutritive value of soybean protein; weight control studies. Mailing Add: Dept of Food Sci & Human Nutrit Mich State Univ Col Human Ecol East Lansing MI 48823

CEDERSTROM, DAGFIN JOHN, hydrology, see 12th edition

CEFOLA, MICHAEL, b Barile, Italy, Oct 22, 08; US citizen; m 48. NUCLEAR CHEMISTRY, INORGANIC CHEMISTRY. Educ: City Col BS, 33;New York, BS, 33; NY Univ, PhD(chem), 41. Prof Exp: Instr microchem, NY Univ, 41-42; instr chem, City Col New York, 41-42; res assoc, Univ Chicago, 42-44; mem staff, Radiation Lab, Mass Inst Technol, 44-45; microchemist, Socony-Vacuum Oil Co, NY, 45-47; chemist, Gen Elec Co, 47-50; from assoc prof to prof, 50-72, EMER PROF CHEM, FORDHAM UNIV, 72- Mem: Am Chem Soc; hon mem Am Microchem Soc; fel NY Acad Sci. Res: Ultramicrochemistry in radiochemistry; high vacuum technique; chemistry of chelate compounds. Mailing Add: Dept of Chem Fordham Univ New York NY 10458

CEGLOWSKI, WALTER STANLEY, b Newark, NJ, Nov 24, 32; m 64; c 1. MICROBIOLOGY. Educ: Univ Vt, BS, 54; Rutgers Univ, MS, 56, PhD(dairy microbiol), 62. Prof Exp: Lab asst rickettsial dis, Walter Reed Army Inst Res, DC, 56-58; res asst dairy bact, Rutgers Univ, 58-62; res assoc immunol, Plum Island Animal Dis Lab, USDA, 62-65; fel dept microbiol, Sch Med, Temple Univ, 65-67, asst prof & dir immunoserol lab, 67-70; ASSOC PROF MICROBIOL, PA STATE UNIV, 70- Mem: Am Soc Microbiol; Am Asn Immunol. Res: Applied microbiology and immunology. Mailing Add: Dept of Microbiol Pa State Univ University Park PA 16802

CEITHAML, JOSEPH JAMES, b Chicago, Ill, May 23, 16; m 42; c 2. BIOCHEMISTRY. Educ: Univ Chicago, BS, 37, PhD(biochem), 41. Prof Exp: Res assoc biochem, 41-45, from asst prof to assoc prof, 46-58, PROF BIOCHEM, UNIV CHICAGO, 58-, DEAN STUDENTS, DIV BIOL SCI, 51- Mem: AAAS; Asn Am Med Cols; Am Soc Biol Chem. Res: Isolation of anterior pituitary hormones; metabolism of the malaria parasite; isolation and study of plant enzymes; biochemical genetics. Mailing Add: Dept of Biochem Univ of Chicago 950 E 59th St Chicago IL 60637

CELANDER, DAVID ROBERT, b Des Moines, Iowa, Oct 12, 23; m 46; c 3. BIOCHEMISTRY, PHYSIOLOGY. Educ: Drake Univ, BA, 46; Univ Iowa, MS, 49, PhD(biochem), 52. Prof Exp: Teacher, Iowa Pub Schs, 47-48; res asst biochem, Univ Iowa, 49-52; res assoc physiol, Med Br, Univ Tex, 52-55, asst prof biochem, 55-61; presiding prof biochem & physiol, 61-63, PROF BIOCHEM & CHMN DEPT, COL OSTEOP MED & SURG, 63- Concurrent Pos: Consult, Nat Osteopath Bd Exam; ed hematol sect, Nuclear Eng Int J Cyclic Res. Honors & Awards: Rutherford Medal, 67. Mem: Fel Int Cardiovasc Soc; Am Chem Soc; NY Acad Sci; Brit Biochem Soc; fel Royal Soc Med. Res: Amino acid metabolism in mice and rats; chemical and physiological significance of plasma fibrinolytic systems in man and other species; Se-75 and I-125 labeled proteins. Mailing Add: Dept of Biochem Col of Osteop Med & Surg Des Moines IA 50312

CELANDER, EVELYN FAUN, b Ottumwa, Iowa, Nov 4, 26; m 46; c 3. BIOCHEMISTRY. Educ: Drake Univ, BA, 48; Col Osteop Med & Surg, MS, 67. Prof Exp: Reader & res asst, State Univ Iowa & Univ Tex Med Br, 48-55; res technician, Univ Tex Med Br, 55-59, res assoc physiol, 59-61; from instr to asst prof, 61-71, ASSOC PROF BIOCHEM, COL OSTEOP MED & SURG, 71- Res: Control and function of fibrinolytic enzyme system in health and disease; biosynthesis of radioactive protein substances; computer technology as applied to information retrieval, data processing and instruction. Mailing Add: Dept of Biochem Col of Osteop Med & Surg Des Moines IA 50312

CELAURO, FRANCIS L, b Jersey City, NJ, Sept 12, ll; m 40; c 3. MATHEMATICS. Educ: NY Univ, AB, 37, AM, 39, PhD(math), 52. Prof Exp: Instr math, NY Univ, 37-40; asst prof, Loyola Col, 40-43; mathematician, Nat Bur Standards, 43-45; asst prof math, Lehigh Univ, 45-49; prof, East Tenn State Col, 54-57; prof, Cent Mich Univ, 57-62; PROF MATH, GEORGE PEABODY COL, 62- Concurrent Pos: NSF award, Princeton Univ, 60. Mem: Math Asn Am. Res: Applied mathematics; psychology of mathematics learning at college level; factors associated with retention of college mathematics. Mailing Add: Dept of Math George Peabody Col for Teachers Nashville TN 37203

CELENTANO, VINCENT DOMINIC, organic chemistry, biochemistry, see 12th edition

CELESIA, GASTONE G, b Genoa, Italy, Nov 22, 33. NEUROLOGY, NEUROPHYSIOLOGY. Educ: Univ Genoa, MD, 59; McGill Univ, MS, 65. Prof Exp: Resident neurol, Montreal Neurol Inst, 62-65; from asst prof to prof neurol, Med Ctr, Univ Wis-Madison, 66-76, dir lab EEG & clin neurophysiol, 70-76; PROF NEUROL & VCHMN DEPT, ST LOUIS UNIV, 76- Concurrent Pos: Fel neurophysiol, Med Ctr, Univ Wis-Madison, 60-62 & Montreal Neurol Inst, 64-65; demonstr, McGill Univ, 63-65; clin investr, Vet Admin Hosp, 66-70, consult, 70-76; consult, Cent Wis Colony, 70-76. Mem: Am Acad Neurol; Soc Neurosci; Am Epilepsy Soc; Am EEG Soc; AMA. Res: Electroencephalography; auditory cortex; sensory system. Mailing Add: Dept of Neurol St Louis Univ 1221 S Grand Blvd St Louis MO 63104

CELESTE, JACK RICHARD, organic chemistry, see 12th edition

CELESTE, VINCENT, b Brooklyn, NY, Apr 13, 40; m 60. MATHEMATICS. Educ: Polytech Inst Brooklyn, BS, 60, MS, 63, PhD(math), 66. Prof Exp: Asst proj engr, Eclipse-Pioneer Div, Bendix Corp, 60-62; mathematician, Systs Develop Corp, 62-63; instr, Polytech Inst Brooklyn, 64-68, ASST PROF MATH, POLYTECH INST NEW YORK, 68- Mem: Am Math Soc. Res: Abstract harmonic analysis; general measure theory; point set topology; group theory. Mailing Add: Dept of Math Polytech Inst of New York Brooklyn NY 11201

CELIANO, ALFRED, b Orange, NJ, Aug 8, 28. CHEMICAL KINETICS, PHYSICAL INORGANIC CHEMISTRY. Educ: Seton Hall Univ, AB, 49; Catholic Univ, STL, 53; Fordham Univ, MS, 56, PhD(chem), 59. Prof Exp: PROF CHEM & CHMN DEPT, SETON HALL UNIV, 59- Mem: AAAS; Am Chem Soc. Res: The stability and kinetics of formation and substitution of inorganic complexes. Mailing Add: Dept of Chem Seton Hall Univ South Orange NJ 07079

CELIK, HASAN ALI, b Bozkir, Turkey. ALGEBRA. Educ: Middle East Tech Univ, BS, 64, MS, 65; Univ Calif, Santa Barbara, PhD(math), 71. Prof Exp: Asst math, Middle East Tech Univ, 64-66; asst prof, 71-75, ASSOC PROF MATH, CALIF STATE POLYTECH UNIV, POMONA, 75- Mem: Am Math Soc; Math Asn Am. Res: Non-associative algebra including flexible-antiflexible algebras; Jordan-Lie-Alternative Rings. Mailing Add: Dept of Math Calif State Polytech Univ Pomona CA 91768

CELIS, TEODORO F R, b La Paz, Entre Rios, Arg, Apr 22, 30; m 63; c 2. MOLECULAR BIOLOGY, MICROBIOLOGY. Educ: Univ Buenos Aires, MD, 57, PhD(biochem), 64. Prof Exp: Res assoc microbiol, Univ Buenos Aires, 60-62, lectr, 62-64, asst prof, 64-66; asst prof microbiol, 72-75, ASSOC PROF MICROBIOL, MED SCH, NY UNIV, 75- Concurrent Pos: Trainee genetics, Med Sch, NY Univ, 68-70, NIH res grant, 73- Mem: Am Soc Microbiol. Res: Transport of basic amino acids on escherichia coli. Mailing Add: Dept of Microbiol NY Univ Med Sch New York NY 10016

CELITANS, GERARD JOHN, b Riga, Latvia, Feb 1, 37; US citizen; m 66. BIOPHYSICS, NUCLEAR PHYSICS. Educ: Univ New South Wales, BSc, 59, PhD(nuclear & radiation chem), 63. Prof Exp: Res officer, Australian Atomic Energy Res Estab, Southerland, NSW, 63-64; res scientist, New Eng Inst Med Res, Conn, 64-66; assoc prof chem & physics, Hiram Scott Col, 66-67; assoc prof chem, Fla Atlantic Univ, 67-68; ASSOC PROF BIOENG, UNIV TEX MED SCH SAN ANTONIO, 68- Concurrent Pos: Vis prof & res assoc, New Eng Inst Med Res, 66- Mem: AAAS; Am Chem Soc; Am Phys Soc; Am Asn Physics Teachers. Res: Low energy positronium; gamma-ray interaction and spectroscopy; sub-nanosecond electronics. Mailing Add: Dept of Bioeng Univ of Tex Med Sch San Antonio TX 78229

CELLA, JOHN ANTHONY, organic chemistry, see 12th edition

CELLA, RICHARD JOSEPH, JR, b Philadelphia, Pa, Nov 2, 42; m 64; c 2. POLYMER CHEMISTRY, PHYSICAL CHEMISTRY. Educ: Univ Del, BS, 64; Cornell Univ, PhD(phys chem), 69. Prof Exp: RES CHEMIST, ELASTOMER CHEM DEPT, E I DU PONT DE NEMOURS & CO, INC, 69- Mem: Am Chem Soc; Sigma Xi. Res: X-ray diffraction studies of polymer structure; morphology and solid state characterization of polymers; physical and mechanical properties of elastomers; adhesives. Mailing Add: Elastomer Chem Dept Exp Sta E I Du Pont de Nemours & Co Wilmington DE 19898

CELLARIUS, RICHARD ANDREW, b Oakland, Calif, July 28, 37; m 59; c 2. BOTANY, BIOPHYSICS. Educ: Reed Col, BA, 58; Rockefeller Univ, PhD(biol), 65. Prof Exp: USPHS fel, 65-66; asst prof bot, Univ Mich, 66-72; MEM FAC, EVERGREEN STATE COL, 72- Mem: AAAS; Am Soc Photobiol; Am Soc Plant Physiol; Am Inst Biol Sci. Res: Light reactions of photosynthesis, photochemistry, plant physiology, photobiology. Mailing Add: Evergreen State Col Olympia WA 98505

CELLI, VITTORIO, b Parma, Italy, Aug 13, 36; m 62; c 1. SOLID STATE PHYSICS. Educ: Univ Pavia, DSc(physics), 58. Prof Exp: Res assoc solid state physics, Univ Ill, 59-61, asst prof, 61-62; asst res physicist, Univ Calif, San Diego, 62-64; lectr & asst theoret physics, Univ Bologna, 64-66; assoc prof physics, 66-69, PROF PHYSICS, UNIV VA, 69- Concurrent Pos: Fulbright scholar, 59-62; docent, Univ Rome, 65; prof extraordinary, Univ Trieste & guest scientist, Int Ctr Theoret Physics, Trieste, 73-74. Mem: Am Phys Soc; Ital Phys Soc. Res: Theoretical physics; theoretical solid state physics; surface science. Mailing Add: 210 Magnolia Dr Charlottesville VA 22901

CELMER, WALTER DANIEL, b Plymouth, Pa, Sept 13, 25; m 46; c 4. BIO-ORGANIC CHEMISTRY. Educ: Bucknell Univ, BS, 47; Univ Ill, PhD(biochem), 50. Prof Exp: Res chemist, 50-54, res supvr, 54-61, res mgr, 61-72, RES ADV, PFIZER INC, 72- Concurrent Pos: Ed, Antimicrobial Agents & Chemother, 72- Mem: Am Chem Soc; Am Soc Biol Chem; Am Soc Microbiol; Marine Technol Soc. Res: Discovery from microbial sources of novel chemical entities possessing biological activities, especially antibiotics and to elucidate their structures, stereochemistry, biogenesis, as well as to prepare and study their synthetic modifications. Mailing Add: Cent Res Pfizer Inc Groton CT 06340

CELMINS, AIVARS KARLIS RICHARDS, b Riga, Latvia, Apr 10, 27; m 55; c 3. APPLIED MATHEMATICS. Educ: Hannover Tech Univ, BS, 50, MS, 53; Clausthal Tech Univ, PhD(geophys), 57. Prof Exp: Mathematician, Seismos GmbH, Ger, 53-57; asst sect head explor geophys, Petrobras-DEPEX, Brazil, 57-58, sect head gravity, 59-61; sr mathematician, Inst Instrumental Math, Bonn, Ger, 61-64; RES MATHEMATICIAN, US ARMY BALLISTIC RES LABS, 64- Concurrent Pos: Instr, Univ Del, 65-70. Mem: Asn Comput Mach; Europ Asn Explor Geophysicists; Ger Soc Appl Math & Mech. Res: Numerical mathematics and its applications to technical and physics problems, particularly in the fields associated with fluid dynamics, engineering and exploration geophysics. Mailing Add: Appl Math & Sci Lab Ballistic Res Labs Aberdeen Proving Ground MD 21005

CELS, ROBERT, b Levuka, Fiji Islands, Jan 3, 23; US citizen; m 64; c 2. PHYSICAL CHEMISTRY, ELECTROCHEMISTRY. Educ: Univ NZ, BS, 45; Case Western Reserve Univ, PhD(electrochem), 55. Prof Exp: Chemist, Vacuum Oil Co, NZ, 47-50; chemist, Werner G Smith, Ohio, 50-51; chemist, Harshaw Chem Co, 51-52; res chemist, Kemet Dept, Linde Co, Union Carbide Corp, 54-65; RES SPECIALIST, RES CTR, BABCOCK & WILCOX CO, 65- Mem: Electrochem Soc; Sigma Xi. Res: Solid electrolyte tantalum capacitors; tantalum analyses for trace impurities; chemisorption of gases by alkaline earth metal films; silicon and germanium films prepared by high vacuum evaporation; silicon pressure transducers and silicon devices; corrosion research at high temperatures and pressures using electrochemical techniques. Mailing Add: 3390 Mogadore Rd Apt 5 Mogadore OH 44260

CEMBER, HERMAN, b Brooklyn, NY, Jan 14, 24; m 43; c 2. RADIOBIOLOGY, HEALTH PHYSICS. Educ: City Col New York, BS, 49; Univ Pittsburgh, MS, 52, PhD(biophys), 60; Environ Engrs Intersoc Bd, dipl. Prof Exp: Res assoc health physics, Grad Sch Pub Health, Univ Pittsburgh, 50-54, from asst prof to assoc prof indust hyg, 54-60; asst prof indust health, Col Med, Univ Cincinnati, 60-64; PROF ENVIRON HEALTH, TECH INST & LECTR RADIOL, SCH MED, NORTHWESTERN UNIV, EVANSTON, 64- Concurrent Pos: Health physics consult, Carnegie Inst Technol, 52-60; radiol safety officer, Univ Pittsburgh, 52-60; lectr, US Naval Training Prog, Westinghouse Atomic Power Div, 52-55; tech expert occup health, Int Labour Off, Switz, 61-62; vis prof indust health, Col Med, Univ Cincinnati, 64-; Fulbright vis prof environ health, Hadassah Med Sch, Hebrew Univ Jerusalem, 72-73. Mem: AAAS; Radiation Res Soc; Health Physics Soc; Am Acad Environ Engrs; fel Am Pub Health Asn. Res: Biological effects of radiation; experimental lung cancer. Mailing Add: Technol Inst Northwestern Univ Evanston IL 60201

CENCE, ROBERT J, b Cleveland, Ohio, July 16, 30; m 54; c 2. PHYSICS. Educ: Univ Calif, Berkeley, AB, 52, PhD(physics), 59. Prof Exp: Res assoc physics, Lawrence Radiation Lab, Univ Calif, 59-63, instr, Univ Calif, Berkeley, 62-63; assoc prof, 63-74, PROF PHYSICS & ASTRON, UNIV HAWAII, 74- Mem: Am Phys Soc. Res: Elementary particle physics. Mailing Add: Dept of Physics Univ of Hawaii Honolulu HI 96822

CENCI, HARRY JOSEPH, b Brooklyn, NY, Oct 2, 30; m 54; c 6. POLYMER CHEMISTRY. Educ: Brooklyn Col, BS, 53; Univ Pa, MS, 55, PhD(org chem), 57. Prof Exp: Head synthesis lab, 57-74, PROJ LEADER INDUSTRIAL COATINGS, ROHM AND HAAS CO, 74- Mem: Am Chem Soc. Res: Shythesis and application of industrial coatings; polymer synthesis; modifiers for polyvinyl chloride; emulsion and solution polymerization; physical organic chemistry; reactions of cis and trans cyclohexane -1, 3- diols; synthesis and reactions of 1, 2- difluoro-1, 2-dicyanoethylene; photochemistry. Mailing Add: Indust Coatings Dept Rohm and Haas Co Springhouse PA 19477

CENEDELLA, RICHARD J, b Pittsburgh, Pa, Jan 12, 39; m 64; c 1. BIOCHEMISTRY, PHARMACOLOGY. Educ: Pa State Univ, BS, 61; Jefferson Med Col, PhD(biochem), 66. Prof Exp: Res assoc, 65-68, ASST PROF PHARMACOL, SCH MED, W VA UNIV, 68- Res: Fatty acid metabolism; biochemistry of prostaglandins; chemotherapy and metabolism of the malarial parasite; lipid pharmacology; mechanism of action of hypolipemic drugs; pharmacology of the prostaglandins. Mailing Add: Dept of Pharmacol WVa Univ Med Ctr Morgantown WV 26506

CENGEL, JOHN ANTHONY, b East Chicago, Ind, July 21, 36; m 62; c 3. PHYSICAL ORGANIC CHEMISTRY. Educ: Purdue Iniv, BS, 58, PhD(phys chem), 65; Ore State Univ, MS, 59. Prof Exp: From proj chemist to sr proj chemist, 65-70, SR RES SCIENTIST, AMOCO CHEM CORP, STAND OIL CO IND, 70- Concurrent Pos: AEC grant, 63-65. Mem: Am Chem Soc; Inst Chem Eng; Soc Plastics Indust; Am Soc Test & Mat. Res: Infrared spectroscopy; metal carbonyl chemistry; polymer development; cellular plastics; viscous polymers; petroleum additives. Mailing Add: Dept of Res & Develop PO Box 400 Amoco Chem Corp Naperville IL 60540

CENTER, ELIZABETH M, b Sterling, Ill, Aug 11, 28; m 51; c 1. GENETICS. Educ: Augustana Col, AB, 50; Stanford Univ, PhD(genetics), 57. Prof Exp: Res asst anat, 51-61, res assoc, 61-69, instr biol sci, 68-71, LECTR BIOL SCI, STANFORD UNIV, 71- Mem: AAAS; Am Soc Zoologists; Genetics Soc Am; Am Soc Anat; Sigma Xi. Res: Developmental genetics. Mailing Add: Dept of Biol Sci Stanford Univ Stanford CA 94305

CENTER, ROBERT E, b Breslau, Ger, Nov 12, 35; m 59; c 2. LASERS, CHEMICAL PHYSICS. Educ: Univ Sydney, BSc, 56, BE, 58, MEngSc, 59, PhD(gas dynamics), 63. Prof Exp: Scientist gas dynamics, Jet Propulsion Lab, Calif Inst Technol, 63-67; prin res scientist, Avco Everett Res Lab, 67-75; PRIN RES SCIENTIST, MATH SCI NORTHWEST, 75- Mem: Am Phys Soc. Res: Scattering of fast electrons in gases; vibrational relaxation in anharmonic diatomic molecules under conditions of thermal nonequilibrium; high temperature gas phase kinetics; infrared laser development. Mailing Add: Math Sci Northwest PO Box 1887 Bellevue WA 98009

CENTIFANTO, YSOLINA M, b Panama, Aug 12, 28; US citizen; m 53; c 3. BACTERIOLOGY. Educ: Univ Panama, BS, 51; Western Reserve Univ, MS, 54; Univ Fla, PhD(bact), 64. Prof Exp: Asst prof physiol & genetics, Univ Panama, 55-56; res biologist, Kodak Trop Res Lab, Panama, 57; res biologist, Eastman Kodak Res Lab, NY, 58-61; res assoc virol, Ophthal Div, 64-65, from instr to asst prof, 65-72, ASSOC PROF OPHTHAL & MICROBIOL, COL MED, UNIV FLA, 72- Mem: AAAS; Am Chem Soc; Am Soc Microbiol; Asn Res Vision & Ophthal. Res: Viral and bacterial infections of the eye; pathogenesis of disease and its relation to host defense mechanisms, such as interferon and antibody synthesis; carcinoma of the prostate; studies on recurrent viral infections. Mailing Add: Dept of Ophthal Univ of Fla Gainesville FL 32601

CENTNER, ROSEMARY LOUISE, b Newport, Ky, Sept 23, 26. ORGANIC CHEMISTRY. Educ: Our Lady of Cincinnati Col, BA, 47; Univ Cincinnati, MS, 49. Prof Exp: Libr asst tech libr, 49-52, br librn, 52-56, tech librn, 56-66, mgr tech info serv, 66-72, mgr div info consults, 72-73, MGR, NDA COORD, PROCTER & GAMBLE CO, 73- Mem: AAAS; Am Chem Soc; Am Soc Infor Sci; Spec Libr Asn. Res: Scientific information; information retrieval; technical translation; analgetic effects of salicylic acid derivatives. Mailing Add: Spec Prods Technol Div Procter & Gamble Co Miami Valley Labs PO Box 39175 Cincinnati OH 45247

CENTOFANTI, LOUIS F, b Youngstown, Ohio, July 25, 43; m 63; c 3. INORGANIC CHEMISTRY. Educ: Youngstown State Univ, BS, 65; Univ Mich, MS, 67, PhD(chem), 68. Prof Exp: Fel chem, Univ Utah, 68-69; asst prof, Emory Univ, 69-73; SR RES CHEMIST, MONSANTO CO, 73- Mem: Am Chem Soc. Res: Synthetic and physical inorganic chemistry. Mailing Add: Monsanto Co 800 N Lindbergh St Louis MO 63166

CENTORINO, JAMES JOSEPH, b Salem, Mass, July 18, 23; m 46; c 5. PHYSICAL GEOGRAPHY. Educ: Boston Univ, AB, 50, AM, 51. Prof Exp: Teacher jr high sch, Mass, 53-S6; from instr to asst prof, 56-62, ASSOC PROF GEOG & EARTH SCI, SALEM STATE COL, 62- Concurrent Pos: Dir earth sci div, NSF In-Serv Inst, 59-67 & Coop Col-Sch Sci Prog, 71; Consult, Soc Studies Curric Projs, 62, Boston Univ Mapping Serv, 63 & Mass Coastal Environ, Mass Coastal Zone Mgt, 75. Mem: Royal Can Geog Soc; Int Oceanog Found; The Oceanic Soc; Nat Geog Soc. Res: Air pollution in the Boston area; locational patterns of the New England shoe industry; local conservation problems. Mailing Add: Dept of Geog Salem State Col Salem MA 01970

CENTURY, BERNARD, b Chicago, Ill, June 15, 28; m 58; c 4. PHARMACOLOGY. Educ: Univ Chicago, PhB, 46, BS & MS, 51, PhD(pharmacol), 53. Prof Exp: Biochemist, Biochem Res Lab, Elgin State Hosp, 54-59, res assoc, 59-74, actg dir, 69-74; asst prof biol chem, Univ Ill Col Med, 62-74; MEM STAFF, DEPT BIOCHEM, MICHAEL REESE MED CTR, 74- Concurrent Pos: NSF fel pharmacol, Univ Chicago, 53-54. Mem: AAAS; Am Chem Soc; Am Inst Nutrit; Am Soc Pharmacol & Exp Therapeut; Am Asn Clin Chem. Res: Biochemistry of mental diseases; relationships of dietary lipids to vitamin E deficiency; effect of diet lipids on metabolism of essential fatty acids and tissue compositions; effect of diet lipids on pharmacological responses. Mailing Add: Dept of Biochem Michael Reese Med Ctr Chicago IL 60616

CEPONIS, MICHAEL JOHN, b Brooklyn, NY, June 25, 16; m 51; c 6. PLANT PATHOLOGY. Educ: Cornell Univ, BS, 50, MS, 53. Prof Exp: Plant pathologist, Tischler Res Serv, NJ, 53-55; from asst plant pathologist to plant pathologist, 55-63, RES PLANT PATHOLOGIST, AGR RES SERV, USDA, NEW BRUNSWICK, NJ, 63- Mem: Am Phytopath Soc; Am Soc Hort Sci. Res: Market diseases of fruits and vegetables. Mailing Add: 42 Fieldstone Dr Somerville NJ 08876

CEPRINI, MARIO Q, b Jamaica, NY, Sept 30, 25; m 49; c 3. ORGANIC CHEMISTRY. Educ: Queens Col, NY, BS, 47; St John's Univ, NY, MS, 61, PhD(org chem), 67. Prof Exp: Jr chemist, NY State Racing Comn Lab, 47-52, chemist, 52-60; jr chemist, Hoffmann-LaRoche Inc, NJ, 60-64, assoc chemist, 64-65; sr chemist, 67-69, group leader metallo-org, 70-72, group leader process develop, 72-75, SR RES SCIENTIST, HEYDEN DIV, TENNECO CHEM, INC, GARFIELD, NJ, 70-, GROUP LEADER POLYMER ADDITIVES, 75-, SR CHEMIST, INTERMEDIATES DIV, PISCATAWAY, 70- Mem: Am Chem Soc. Res: Peptide and amino acid chemistry; antibiotics; organic synthesis; polymer additives; process development. Mailing Add: 513 Ocean Point Ave Cedarhurst NY 11516

CERAME-VIVAS, MAXIMO JOSE, marine ecology, biological oceanography, see 12th edition

CERANKOWSKI, LEON DENNIS, b Philadelphia, Pa, July 31, 40; m 63. PHYSICAL CHEMISTRY. Educ: Drexel Univ, BS, 63; Princeton Univ, MA, 65, PhD(chem), 69. Prof Exp: Chemist, USDA, 61-64; instr phys chem, Princeton Univ, 67-69; SCIENTIST, POLYMER LAB, POLAROID CORP, 69- Mem: AAAS; Am Chem Soc. Res: Physical chemistry of electrolyte and macromolecular solutions. Mailing Add: Polymer Lab Polaroid Corp 730 Main St Cambridge MA 02139

CERBULIS, JANIS, b Smiltene, Latvia, Dec 5, 13; nat US; m 52; c 3. BIOCHEMISTRY, AGRICULTURE. Educ: Acad Agr, Jelgava, PhD(agr), 44; Univ Pa, MS, 57; Rutgers Univ, PhD(biochem), 67. Prof Exp: Adminr & instr hort, Baltic Univ, Ger, 47-49; res chemist, Stephen F Whitman & Sons, Pa, 51-55 & Borden Chem Co, 55-56; RES CHEMIST, MILK PROPERTIES LAB, EASTERN REGIONAL RES CTR, AGR RES SERV, USDA, 56- Concurrent Pos: USPHS maintenance grant, 61-63; Am Cancer Soc res grant-in-aid, 63; abstractor, Chem Abstracts, 58- Mem: Am Chem Soc; Ger Soc Fat Res; Sigma Xi. Res: Growing of grass seeds, vegetable seeds and medicinal plants; food value of grass; phenolic resins; earthworm chemical composition; whey composition; lactose chemistry; milk proteins and lipids; milk clotting; milk composition and relationship among milk constituents. Mailing Add: RD 2 Boyertown PA 19512

CERCONE, NICHOLAS JOSEPH, b Pittsburgh, Pa, Dec 18, 46. COMPUTER SCIENCE. Educ: Col Steubenville, BS, 68; Ohio State Univ, MS, 70; Univ Alta, PhD(comput sci), 75. Prof Exp: Programmer design automation, Int Bus Mach Corp, 68-69; instr comput sci, Ohio State Univ, 70-71; instr comput, Int Bus Mach Corp, 71-72; asst prof comput sci, Old Dom Univ, 75-76; ASST PROF COMPUT SCI, COMPUT SCI PROG, SIMON FRASER UNIV, 76- Mem: Asn Comput Mach; Inst Elec & Electronics Engrs; Royal Photographic Soc. Res: Artificial intelligence; natural language processing; representation of knowledge. Mailing Add: Comput Sci Prog Simon Fraser Univ Burnaby BC Can

CEREFICE, STEVEN A, b Newark, NJ, Aug 10, 43; m 67. ORGANIC CHEMISTRY, ORGANOMETALLIC CHEMISTRY. Educ: Rutgers Univ, BA, 65; Columbia Univ, MA, 66, PhD(org chem), 69. Prof Exp: RES CHEMIST ORG CHEM, AMOCO CHEM CORP, 70- Mem: AAAS; Am Chem Soc. Res: Organic synthesis; photochemistry of hydrocarbons; small ring compounds; bicyclic-polycyclic hydrocarbons; transition metal-catalyzed cycloaddition and electrocyclic reaction of polycyclic hydrocarbons; catalysis and mechanisms studies. Mailing Add: 27 W 161 95th St Naperville IL 60540

CERETTI, ELENA, b Mendoza, Arg, Nov 19, 33; wid. CARDIOLOGY, ELECTROPHYSIOLOGY. Educ: Univ Cuyo, MD, 59. Prof Exp: Sr instr physiol, Fac Med, Univ Cuyo, 65-68; asst prof, 68-70, ASSOC PROF BIOPHYS, FAC MED, UNIV SHERBROOKE, 70- Concurrent Pos: Res fel heart physiol, Fac Med, Univ Cuyo, 59-63; Arg Nat Coun Sci & Technol Invest fel, 60-62, grant, 66-67, fel heart electrophysiol, Univ Southern Calif, 63-65; Med Res Coun Ottawa grants, 67-70; Que Health Found grant, 69-70; Can Heart Found fel, 69-70. Mem: Biophys Soc; Arg Cardiol Soc. Res: Electrical activity of the heart and ionic changes; anoxia and ischemia; atrioventricular conduction; membrane impedance and cytoplasmic resistivity in skeletal and cardiac muscle. Mailing Add: Dept of Biophys Univ of Sherbrooke Fac of Med Sherbrooke PQ Can

CERIMELE, BENITO JOSEPH, b Cincinnati, Ohio, May 11, 36; m 63; c 4. BIOMATHEMATICS, COMPUTER SCIENCES. Educ: Xavier Univ, Ohio, BS, 57, MS, 59; Univ Cincinnati, PhD(math), 63. Prof Exp: Reactor physicist, Gen Elec Co, Ohio, 57-59; from instr to asst prof math, Xavier Univ, Ohio, 62-66; NIH fel biomath, NC State Univ, 66-68, asst prof, 68-70; SR SYSTS ANALYST, LILLY RES LABS, ELI LILLY & CO, 70- Concurrent Pos: Adj asst prof biomath, Sch Med, Univ NC, Chapel Hill, 69-70. Mem: Math Asn Am; Asn Comput Mach. Res: Deterministic and stochastic modeling of biological systems; neurodynamics; signal processing; systems theory; computer process control. Mailing Add: Lilly Res Labs Eli Lilly & Co Indianapolis IN 46206

CERINI, COSTANTINO PETER, b Philadelphia, Pa, Nov 19, 31; m 60; c 2. VIROLOGY, IMMUNOLOGY. Educ: La Salle Col, BS, 53; Lehigh Univ, MS, 60, PhD(virol), 64. Prof Exp: Res virologist, 64-70, GROUP LEADER, LEDERLE LABS DIV, AM CYANAMID CO, 70- Mem: AAAS; Am Soc Microbiol; Tissue Cult Asn; NY Acad Sci. Res: Virus immunology; rubella, mumps, rabies, measles, influenza. Mailing Add: Lederle Labs Pearl River NY 10965

CERLON, PETER JOHN, metallurgy, inorganic chemistry, see 12th edition

CERNOSEK, STANLEY FRANK, JR, b Shiner, Tex, Dec 19, 40; m 68; c 1. IMMUNOCHEMISTRY. Educ: Pan Am Col, BA, 63; Univ Tex, Austin, PhD(chem), 69. Prof Exp: Res assoc peptide chem, Sch Med, Univ Pittsburgh, 68-70; fel biochem, Grad Dept Biochem, Brandeis Univ, 70-73; ASST PROF DEPT BIOCHEM, COL

MED, UNIV ARK, 73- Concurrent Pos: Consult, Nat Ctr Toxicol Res, 73- & Bur Vet Med, Food & Drug Admin, 74- Mem: Am Chem Soc; The Chem Soc; Biophys Soc, Sigma Xi. Res: Synthetic protein and peptide syntheses; sequential polypeptide syntheses; physical biochemistry; development and applications of radioimmunoassays for chemical carcinogens, hormones and environmental toxicans. Mailing Add: Dept of Biochem Univ of Ark Col of Med Little Rock AR 72201

CERNUSCHI, FELIX, b Montevideo, Uruguay, May 17, 08; m 47; c 2. PHYSICS, ASTROPHYSICS. Educ: Univ Buenos Aires, CE, 32; Cambridge Univ, PhD(physics), 38. Prof Exp: Res worker, Cordoba Observ, Argentina, 38-39; prof phys sci, Nat Univ Tucuman, Argentina, 39-43; res assoc astrophys, Harvard Col Observ, 44-46; sci adv phys sci, UNESCO, France, 47-48; invited prof physics, Univ PR, 49-50; PROF ASTRON, UNIV OF THE REPUB, URUGUAY, 50-, PHYSICS, 55-, DIR DEPT ASTRON & PHYSICS, 55- & FAC HUMANITIES & SCI, 57- Concurrent Pos: Argentine Asn Adv Sci fel, 38-39; Guggenheim fel, 44-46; prof physics & dir fac eng, Univ Buenos Aires, 57-68; mem, Radioastronomy Comn, Univ Buenos Aires, 59-63 & Nat Coun Sci & Tech Invests, Argentina, 59-; invited lectr, Inter-Am Conf Physics, Brazil, 63 & Inter-Am Conf Sci & Technol, DC, 64. Honors & Awards: Phys Sci Prize, Buenos Aires City, 65. Mem: Fel AAAS; fel Am Phys Soc; Am Astron Soc; Argentine Nat Acad Sci; Argentine Physics Asn. Res: Statistical mechanics and its application to liquid state, solid state and astrophysics; interstellar matter; polarization of stellar light; cosmogony; methods for teaching science and organization of universities; transition solid-liquid; solar energy applications. Mailing Add: Reconquista 398 Apt 406 Montevideo Uruguay

CERNY, JOSEPH, III, b Montgomery, Ala, Apr 24, 36; m 59; c 2. NUCLEAR CHEMISTRY. Educ: Univ Miss, BS, 57; Univ Calif, Berkeley, PhD(nuclear chem), 61. Prof Exp: From asst prof to assoc prof, 61-71, PROF CHEM, UNIV CALIF, BERKELEY & RES CHEMIST, LAWRENCE BERKELEY LAB, 71-, CHMN DEPT CHEM, 75- Concurrent Pos: Consult, US Army Res Off, 63-; Guggenheim fel, Oxford Univ, 69-70; vis fel, Australian Nat Univ, Canberra, 75. Honors & Awards: E O Lawrence Award, US AEC, 74. Mem: Am Chem Soc; Am Phys Soc; Fedn Am Sci; AAAS. Res: Low energy nuclear physics utilizing stripping or pickup reactions to investigate reaction mechanisms and nuclear spectroscipy; two-particle transfer reactions; studies of isobaric analogue states and exotic nuclei. Mailing Add: Lawrence Berkeley Lab Bldg 88 Berkeley CA 94720

CERNY, LAURENCE CHARLES, b Cleveland, Ohio, Mar 5, 29; m 55; c 3. BIOPHYSICAL CHEMISTRY. Educ: Case Inst Technol, BS, 51, MS, 53; State Univ Ghent, PhD(phys chem), 56. Prof Exp: Asst prof chem, John Carroll Univ, 56-60; PROF CHEM, UTICA COL, 60- Concurrent Pos: Fel, Med Sch, Univ Minn, 58-59; res assoc hemodynamics, St Vincent Hosp, 59-60; estab investr, Masonic Med Res Lab, 62-67, career develop award, 67-; US-Czech exchange fel, Nat Acad Sci. 67. Mem: Fel Am Inst Chem; Biophys Soc; Am Chem Soc; Soc Rheol; Am Heart Asn. Res: Hemorheology; hemodynamics; polymers; chemical education; plasma expanders. Mailing Add: Masonic Med Res Lab Utica NY 13501

CERON, GABRIEL, b Bogota, Colombia. ANATOMY, TERATOLOGY. Educ: Nat Univ Colombia, MD, 64; Univ Fla, PhD, 69. Prof Exp: Asst prof anat, Nat Univ Colombia, 70-73; ASST PROF ANAT, JEFFERSON MED COL, THOMAS JEFFERSON UNIV, 73- Concurrent Pos: Consult, Dept Surg, US Naval Hosp, Philadelphia, 73- Mem: Teratol Soc; Soc Develop Biol. Res: Experimental morphology; control mechanisms of morphogenesis; molecular and cellular basis of teratogenesis. Mailing Add: Jefferson Med Col 1020 Locust St Philadelphia PA 19107

CERRONI, ROSE E, b Weirton, WVa, Mar 29, 30. BIOLOGY, PHYSIOLOGY. Educ: Col of Steubenville, BS, 52; Vanderbilt Univ, MA, 55, PhD(biol), 59. Prof Exp: Instr nursing sci, Wheeling Hosp Sch Nursing, 52-53; instr microbiol, Vanderbilt Univ, 55-56; NIH fel, Carlsberg Biol Inst, 60-61; res assoc, Temple Univ, 61-62; asst prof biol, West Liberty State Col, 62-67; PROF BIOL, COL STEUBENVILLE, 67- Mem: Am Inst Biol Sci. Res: Physiology and ultra structure of the nucleus of acanthamoeba radioautographic studies on tetrahymena pyriformis to determine aspects of the mechanism of heat-induced division sychrony. Mailing Add: Dept of Biol Col of Steubenville Steubenville OH 43952

CERTAINE, JEREMIAH, mathematics, see 12th edition

CERUTTI, PETER A, b Zurich, Switz, Oct 8, 31; US & Swiss citizen; m 72; c 1. BIOCHEMISTRY. Educ: Univ Zurich, MD, 56, PhD(org chem), 63. Prof Exp: Asst prof biochem sci, Princeton Univ, 66-70; PROF BIOCHEM SCI & CHMN DEPT, UNIV FLA, 71- Concurrent Pos: Fel org chem, Swiss Sci Found, 63-64; fel, NIH, 64-66, res grant, 66-; AEC res grant, 69-, travel grant to Int Cong Biochem, Montreux, Switz, 70; Nat Acad Sci travel grant to Int Cong Biophys, Moscow, USSR, 72. Mem: Am Chem Soc; Am Soc Biol Chemists; Am Soc Photobiol; NY Acad Sci; Biophys Soc. Res: Molecular biology of DNA repair in bacterial and eukaryotic cells. Mailing Add: Dept of Biochem Box 724 JHMHC Univ of Fla Gainesville FL 32610

CERVENKA, JAROSLAV, b Prague, Czech, Mar 15, 33; US citizen; m 59; c 2. MEDICAL GENETICS, CANCER. Educ: Charles Univ, Czech, MD, 58; Czech Acad Sci, CSc(med genetics), 68. Prof Exp: Act chief genetics, Lab Plastic Surg, Prague, 67-68; asst prof, 68-71, ASSOC PROF GENETICS, SCH DENT, UNIV MINN, 71- Concurrent Pos: Staff consult, Sch Dent, Univ Minn, 66-, mem staff, Grad Sch, 69-, Genetic Clin, Health Sci Ctr, 72- Mem: Am Soc Human Genetics; Int Dermatoglyphics Asn. Res: Cytogenetics of cancer; genetics of congenital abnormalities; prenatal diagnosis of genetic disease; clinical cytogenetics. Mailing Add: Div of Oral Path & Hum Gen Univ of Minn Sch of Dent Minneapolis MN 55455

CERVONI, PETER, b Jamaica, NY, Mar 4, 31; m 64; c 3. PHARMACOLOGY. Educ: St John's Univ, NY, BS, 52; Univ Wash, MS, 55, PhD(pharmacol), 57. Prof Exp: Asst pharmacol, Univ Wash, 52-57; pharmacologist, US Army Chem Warfare Labs, 57-59; asst prof pharmacol, Univ Miss, Jackson, 60-61; instr, State Univ NY Downstate Med Ctr, 66-70, sect head cardiovasc/autonomic pharmacol, 70-72, DIR & RES SCIENTIST, USV PHARMACEUT CORP, 72- Concurrent Pos: Life Ins Med Res Fund fel, 59-60. Mem: AAAS; Am Soc Pharmacol & Exp Therapeut; NY Acad Sci. Mailing Add: Div of Biol Res & Develop USV Pharmaceut Corp Tuckahoe NY 10707

CERWONKA, ROBERT HENRY, b Endicott, NY, Mar 16, 31; m 57; c 2. MARINE ECOLOGY. Educ: State Univ NY, Albany AB, 53, MA, 57; Univ Conn, PhD, 68. Prof Exp: From instr to assoc prof, 59-69, PROF BIOL, STATE UNIV NY COL POTSDAM, 69- Mem: Ecol Soc Am; Am Soc Limnol & Oceanog; Am Ornith Union; Sigma Xi. Res: Filtering rates of bivalve mollusks. Mailing Add: Dept of Biol State Univ of NY Col Potsdam NY 13676

CERYCH, JOHN Z, inorganic chemistry, see 12th edition

CESARE, FRANK CHARLES, organic chemistry, polymer chemistry, see 12th edition

CESARI, LAMBERTO, b Bologna, Italy, Sept 23, 10; m 39. MATHEMATICAL ANALYSES. Educ: Univ Pisa, PhD(math), 33. Prof Exp: Asst prof math, Univ Rome, 37-39; assoc prof, Univ Pisa, 39-42; from assoc prof to prof, Univ Bologna, 42-48; vis prof, Inst Adv Study, 48, Univ Calif, 49 & Univ Wis, 50; vis prof, Purdue Univ, 50, prof, 52-60; PROF MATH, UNIV MICH, 60-, R L WILDER PROF, 75- Mem: Am Math Soc; Math Asn Am; Soc Indust & Appl Math; Math Union Italy. Res: Real functions; calculus of variations; surface area theory; asymptotic behavior of differential equations; numerical analysis; ordinary and partial differential equations; optimal control theory; nonlinear analysis. Mailing Add: Dept of Math Univ of Mich Ann Arbor MI 48104

CESCAS, MICHEL PIERRE, b Bordeaux, France, Sept 23, 36; m 68; c 2. SOIL CHEMISTRY, SOIL FERTILITY. Educ: Laval Univ, BSc, 60; Univ Ill, Urbana, MSc, 65, PhD(agron), 68. Prof Exp: Res asst agron, Univ Ill, Urbana, 60-61, 63-68; from asst prof to assoc prof soil chem, 68-74, PROF SOIL CHEM, LAVAL UNIV, 74- Mem: AAAS; Am Soc Agron; Soil Sci Soc Am; Int Soc Soil Sci; Am Chem Soc. Res: Electron probe analysis of soils and related materials; phosphorus evolution in soils; rock phosphate transformation with time; analytical chemistry methodology in soil testing and analysis; adsorption phenomena in soils; optimization of fertilization of important corps; soil pollution by agricultural practices and inorganic compounds. Mailing Add: Dept of Soils Laval Univ Fac of Agr Quebec PQ Can

CESCON, LAWRENCE A, organic chemistry, see 12th edition

CESSNA, JOHN CURTIS, b Johnstown, Pa, Apr 15, 26; m 47; c 2. ORGANIC CHEMISTRY. Educ: Augustana Col, BA, 50; Iowa State Col, MS, 52. Prof Exp: Res chemist, Nat Carbon Res Lab, 52-64; sr res chemist, Consumer Prod Div, 64-72, SR RES CHEMIST, BATTERY PRODS DIV, UNION CARBIDE CORP, 72- Mem: Am Chem Soc; Electrochem Soc; Nat Asn Corrosion Eng. Res: Corrosion studies relating to batteries and coolants; basic and applied research on batteries. Mailing Add: Battery Prods Div Union Carbide Corp 12900 Snow Rd Parma OH 44130

CESSNA, LAWRENCE C, JR, b Cumberland, Md, Feb 1, 39; m 60; c 2. POLYMER PHYSICS, CHEMICAL ENGINEERING. Educ: Johns Hopkins Univ, BES, 61; Rensselaer Polytech Inst, PhD(polymer sci), 65. Prof Exp: Res assoc polymer sci, Rennselaer Polytech Inst, 65-66; res engr, 66-70, sr res engr, 70-73, res scientist, 73, MGR MAT SCI DIV, HERCULES RES CTR, 73- Mem: Soc Chem Indust; Soc Plastics Engr; Am Chem Soc. Res: Ultimate properties of high polymers; physical and mechanical behavior of polymers and polymer based composite materials; chemical engineering fundamentals. Mailing Add: 111 Neptune Dr Newark DE 19711

CETAS, ROBERT CHARLES, b Harbor Springs, Mich, Feb 23, 22; m 47; c 2. PLANT PATHOLOGY. Educ: Mich State Col, BS, 47; Cornell Univ, PhD(plant path), 52. Prof Exp: From asst prof to assoc prof, 52-71, PROF PLANT PATH, STATE UNIV NY COL AGR & EXP STA, CORNELL UNIV, 71- Mem: AAAS; Am Inst Biol Sci; Am Phytopath Soc; Potato Asn Am; NY Acad Sci. Res: Evaluation of fungicides as foliar sprays on potatoes and other vegetables and as seedpiece treatments on potatoes; evaluation of potatoes for scab and leaf roll resistance. Mailing Add: Long Island Veg Res Farm Cornell Univ 39 Sound Av RR 1 Riverhead NY 11901

CEVALLOS, WILLIAM HERNAN, b Quito, Ecuador, Mar 11, 32; US citizen; m 56; c 4. BIOCHEMISTRY, METABOLISM. Educ: Mt St Mary's Col, Md, BS, 54; St John's Univ, NY, MS, 56; Georgetown Univ, PhD(biochem), 60. Prof Exp: Asst microbiol, St John's Univ, NY, 54-56, lab instr histol, 55-56; asst biochem, Georgetown Univ, 56-59, lab instr chem, Sch Nursing, 57-59; res assoc biochem, Sinai Hosp, Baltimore, Md, 59-64; res assoc, Smith Kline & French Labs, 64-68; RES ASSOC BIOCHEM, DIV RES, LANKENAU HOSP, 68- Concurrent Pos: NIH fel, 60-62, grant, 62-64. Mem: NY Acad Sci. Res: Lipid metabolism related to atherosclerosis and heart disease; hormonal and chemotherapeutic control of cholesterol metabolism; relationship of circulating and dietary lipids to platelets and thrombosis. Mailing Add: Lankenau Hosp Lancaster & City Line Aves Philadelphia PA 19151

CEVASCO, ALBERT ANTHONY, b New York, NY, Sept 4, 40; m 63; c 1. ORGANIC CHEMISTRY. Educ: Manhattan Col, BS, 62; Fordham Univ, PhD(org chem), 68. Prof Exp: Asst chem, Fordham Univ, 62-67; res chemist, Explor Res & Develop Dept, Bound Brook, 67-69, res chemist, Decision Making Systs, 69-70, process res chemist, Dyes Tech Dept, 70-75, SR RES CHEMIST, AGR RES CTR, AM CYANAMID CO, PRINCETON, NJ, 75- Mem: Am Chem Soc. Res: Heterocyclic syntheses; organic luminescers; dye and brightener intermediates process research and development; process research and development of organic agricultural and animal health products. Mailing Add: 19 Shearn Dr Middlesex NJ 08846

CEZAIRLIYAN, ARED, b Istanbul, Turkey, May 9, 34; US citizen; m 70. THERMAL PHYSICS, MATERIALS SCIENCE. Educ: Robert Col, Istanbul, BSME, 57; Purdue Univ, Lafayette, MSME, 60, PhD, 63. Prof Exp: PHYSICIST, NAT BUR STANDARDS, 63- Concurrent Pos: Consult, Thermophys Properties Res Ctr, Purdue Univ, 68-; mem, Int Orgn Comt, Europ Thermophys Properties Conf, 74- Honors & Awards: Spec Act Award, US Dept Com, 70, Silver Medal Award, 75. Mem: AAAS; Am Phys Soc; Am Inst Aeronaut & Astronaut; Sigma Xi. Res: High temperature thermophysics; material properties; calorimetry; high temperature thermometry; high speed pyrometry; optics; transient measurement techniques. Mailing Add: Phys Chem Div Nat Bur of Standards Washington DC 20234

CHA, CHUL YUNG, physical chemistry, see 12th edition

CHA, MOON HWA, b Ann Arbor, Mich, Apr 25, 31; m 53; c 1. PHYSICS. Educ: Univ Md, BS, 56, PhD(physics), 64. PHYSICIST, US NAVAL ORD LAB, 56- Mem: Am Phys Soc. Res: Elementary particle physics. Mailing Add: Naval Surface Weapons 4-282 Naval Ordnance Lab Silver Spring MD 20910

CHA, SUNGMAN, b Chungpyong, Korea, Mar 1, 28; m 60; c 3. BIOCHEMISTRY, PHARMACOLOGY. Educ: Yonsei Univ, Korea, MD, 54; Univ Wis, PhD(pharmacol), 63. Prof Exp: Asst pharmacol, Univ Wis, 59-61, trainee, 61-63; asst prof, 63-68, ASSOC PROF MED SCI, BROWN UNIV, 68- Mem: Am Soc Biol Chem. Res: Metabolism of nucleotides including analogues; mechanism of phosphorylation at the substrate level. Mailing Add: Div of Biomed Sci Brown Univ Providence RI 02912

CHABAI, ALBERT JOHN, b Mont, Feb 1, 29; m 58; c 6. PHYSICS. Educ: Mont State Col, BS, 51; Lehigh Univ, MS & PhD(physics), 58. Prof Exp: Mem physics res staff, 58-65, div supvr, Adv Systs Develop, 65-71, DIV SUPVR, SOLID DYNAMICS RES, SANDIA CORP, 71- Mem: AAAS. Res: Fluid dynamics. Mailing Add: Sandia Corp Div 5166 Albuquerque NM 87115

CHABRECK, ROBERT HENRY, b Lacombe, La, Mar 18, 33; m 54; c 4. WILDLIFE MANAGEMENT, ECOLOGY. Educ: La State Univ, Baton Rouge, MS, 57, PhD(bot), 70. Prof Exp: Refuge biologist, La Wildlife & Fisheries Comn, 57-59, res biologist, 59-66, res supvr, 66-67; asst leader, La Coop Wildlife Res Unit, 67-72;

ASSOC PROF FORESTRY & WILDLIFE MGR, LA STATE UNIV, BATON ROUGE, 72- Mem: Wildlife Soc; Ecol Soc Am; Int Union Conserv Nature & Natural Resources. Res: Wetland ecology, especially wetland management, life history studies of American alligator and waterfowl. Mailing Add: Sch of Forestry & Wildlife Mgt La State Univ Baton Rouge LA 70803

CHACE, FENNER ALBERT, JR, b Fall River, Mass, Oct 5, 08; m 34; c 1. ZOOLOGY. Educ: Harvard Univ, AB, 30, AM, 31, PhD(biol), 34. Prof Exp: Asst cur mar invert, Mus Comp Zool, Harvard Univ, 34-42, cur Crustacea, 42-46, cur marine invert, US Nat Mus, 46-63; SR ZOOLOGIST, SMITHSONIAN INST, 63- Concurrent Pos: Asst biol, Harvard Univ, 35, Agassiz fel, 35-39, tutor biol, 40-41. Mem: AAAS; Am Soc Limnol & Oceanog; Soc Syst Zool. Res: Taxonomy, morphology and distribution of decapod Crustacea. Mailing Add: Dept Invert Zool Smithsonian Inst Washington DC 20560

CHACE, FREDERIC MASON, b Swansea, Mass, Aug 24, 06; m 43. MINING GEOLOGY. Educ: Brown Univ, PhB, 29, MA, 32; Harvard Univ, PhD(geol), 47. Prof Exp: Geologist, Bendigo Mines Ltd, Australia, 34-36; consult geologist, Mining Co of Oruro, Boliva, 37; geologist, Cerro de Pasco Copper Corp, Peru, 38-40; indust specialist, War Prod Bd, Washington, DC, 42-45; commodity geologist, US Geol Surv, 46-47, staff geologist, 47-51; regional geologist, M A Hanna Co, Minn & Mich, 51-52; mining geologist, Gold Fields Am Develop Co, Ltd, 52-55; asst dir explor, M A Hanna Co, 55-61, chief geologist, Hanna Mining Co, 61-68, vpres geol & explor, 68-74; CONSULT, 74- Mem: Am Inst Min, Metall & Petrol Eng; Soc Econ Geol; Geol Soc Am; Can Mining Inst. Res: Structural control of ore deposits; paragenesis of ore minerals; mineral exploration and mineral economics; gold, copper and iron ore deposits. Mailing Add: 23838 Duffield Rd Shaker Heights OH 44122

CHACHARONIS, PETER, b Zanesville, Ohio, Dec 9, 25. PROTOZOOLOGY. Educ: Marshall Col, BA, 48; Ohio State Univ, MA, 50, PhD(zool), 54. Prof Exp: Asst zool, Ohio State Univ, 49-54; instr biol, Monticello Col, 54-58, chmn dept sci, 55-58; res assoc pharmacol, Sci Assocs, 58-64; instr sci & math, Monticello Col, 64-66, head div, 64-71, assoc prof, 66-71; PROF BIOL SCI & CHMN DIV HEALTH & LIFE SCI, LEWIS & CLARK COMMUNITY COL, 71- Mem: AAAS; Soc Protozoologists. Res: Ecology of the protozoa; pharmacology. Mailing Add: Div Health & Life Sci Lewis & Clark Community Col Godfrey IL 62035

CHACKERIAN, CHARLES, JR, b San Francisco, Calif, Feb 6, 35; c 2. PHYSICAL CHEMISTRY, MOLECULAR SPECTROSCOPY. Educ: Univ Calif, Berkeley, BS, 58; Univ Wash, PhD(chem), 64. Prof Exp: RES SCIENTIST, AMES RES CTR, NASA, 64- Mem: Am Phys Soc. Res: Shock waves; high temperature chemical kinetics and molecular relaxation; infrared spectroscopy related to planetary and stellar spectroscopy. Mailing Add: Ames Res Ctr NASA Moffett Field CA 94035

CHACKO, GEORGE KUTTY, b Kottarakkara, India, Feb 15, 33; m 62; c 3. BIOCHEMISTRY. Educ: Univ Col, Trivandrum, India, BSc, 56; Maharaja's Col, Ernakulam, MSc, 58; Univ Ill, Urbana, PhD(food chem, biochem), 66. Prof Exp: Fel biochem, Univ Wash, 66-67; fel, Univ Ariz, 67-68; res asst prof, 68-74, RES ASSOC PROF BIOCHEM & PHYSIOL, MED COL PA, 74- Mem: Am Chem Soc; Am Oil Chemists Soc; AAAS. Res: Biochemical characterization of axon plasma membranes; structure and function of platelet complex lipids. Mailing Add: Dept of Biochem & Physiol Med Col of Pa Philadelphia PA 19129

CHACON, RAFAEL VAN SEVEREN, b El Salvador, Cent Am, June 10, 31; nat US; m 53; c 4. MATHEMATICS. Educ: Univ Rochester, BS, 51; Syracuse Univ, PhD(math), 56. Prof Exp: Instr, Ohio State Univ, 56-58; asst prof math, Univ Wis, 58-61; assoc prof, Brown Univ, 61-64 & Ohio State Univ, 64-69; prof, Univ Minn, Minneapolis, 69-74; PROF MATH, UNIV BC, 74- Mem: Am Math Soc. Res: Probability theory; ergodic theory; functional analysis. Mailing Add: Dept of Math Univ of BC Vancouver BC Can

CHADAM, JOHN MARTIN, mathematical physics, see 12th edition

CHADDE, FRANK ERNEST, b Chicago, Ill, June 23, 29; m 53; c 4. ANALYTICAL CHEMISTRY. Educ: Univ Ill, BS, 51. Prof Exp: Control chemist, 51-52, anal chemist, 54-63, mgr anal res dept, 63-67, mgr chem control dept, 67-69, MGR ANAL LAB, ABBOTT LABS, 69- Mem: AAAS; Am Chem Soc; Acad Pharmaceut Sci. Res: Ultra violet spectroscopy; colorimetry; titrimetric analysis; polarography. Mailing Add: D-866 Anal Labs Abbott Lab North Chicago IL 60064

CHADER, GERALD JOSEPH, b Buffalo, NY, Apr 15, 37; c 3. BIOCHEMISTRY. Educ: Univ Buffalo, BA, 59; Univ Louisville, PhD(biochem), 66. Prof Exp: High sch teacher, NY, 59-60; instr, Med Sch & tutor, Dept Biochem & Molecular Biol, Harvard Univ, 69-71; BIOCHEMIST, NAT EYE INST, 71- Concurrent Pos: Fel biochem, Sch Med, Univ Louisville, 66-67; fel biol chem, Harvard Med Sch, 67-69. Mem: AAAS. Res: Mechanism of action of hormones, involving hormone-receptor interactions and study of these interactions in relation to retinal function. Mailing Add: Lab of Vision Res Nat Eye Inst Bethesda MD 20014

CHADWICK, CLAUDE SIMPSON, zoology, see 12th edition

CHADWICK, DAVID HENRY, b Sutton, NH, Aug 8, 18; m 44; c 3. INDUSTRIAL ORGANIC CHEMISTRY. Educ: Univ NH, BS, 40, MS, 42; Univ Ill, PhD(org chem), 46. Prof Exp: From asst to instr chem, Univ NH, 40-42; asst, Univ Ill, 42-46; res chemist, Monsanto Chem Co, 46-49, res group leader, 49-59; asst dir res, 59-67, DIR PROCESS RES DEPT, MOBAY CHEM CO, 67- Concurrent Pos: Anal lab control supvr, Clinton Eng Works, Tenn Eastman Corp, 42-46; res chemist, Nat Defense Res Comt Contract, Monsanto Chem Co, 43; instr chem, Univ Ill, 46. Mem: Am Chem Soc. Res: Molecular rearrangements; stable vinyl alcohols; addition of Grignard reagents to aronitriles; synthetic detergents; organophosphorus compounds; isocyanates; tall oil; polyurethanes; polyesters; polycarbonates. Mailing Add: Process Res Dept Mobay Chem Corp New Martinsville WV 26155

CHADWICK, GEORGE F, b Buffalo, NY, July 11, 30; m 52; c 4. PLASTICS CHEMISTRY, SOLID STATE PHYSICS. Educ: Univ Buffalo, BA, 51; Pa State Univ, MA, 56. Prof Exp: Asst petrol, Pa State Univ, 51-54; chemist, Durez Plastics Div, Hooker Chem Corp, 54-57; res scientist, 57-64, res supvr, 64-65, develop supvr, 65-67, MGR DEVELOP-ELECTRONICS, AIRCO ELECTRONICS, 67- Mem: Am Chem Soc; Soc Plastics Engrs; Inst Elec & Electronics Engrs; Fedn Socs Paint Technol; fel Brit Plastics Inst. Res: Plastics; rheology; conductivity of fine dispersed conductive materials in plastics composites. Mailing Add: Airco Electronics Packard Rd & 47th St Niagara Falls NY 14302

CHADWICK, HAROLD KING, b Bay Shore, NY, May 28, 30; m 55; c 5. FISHERIES. Educ: Cornell Univ, BS, 52; Univ Mich, MS, 56. Prof Exp: FISHERY BIOLOGIST, CALIF DEPT FISH & GAME, 56- Mem: Am Fisheries Soc. Res: Investigation of Sacramento River striped bass population including harvest and natural mortality rates and strength of year classes; reservoir ecology; fisheries management. Mailing Add: Calif Dept of Fish & Game 3900 N Wilson Way Stockton CA 95205

CHADWICK, JUNE MARIE, b Fredericton, NB, Aug 3, 28; m 62; c 1. MICROBIOLOGY. Educ: Univ NB, BSc, 47; Queen's Univ, Ont, MA, 50; Univ London, PhD(bact), 61. Prof Exp: Res officer insect path, Can Dept Agr, 50-62; asst prof, 64-72, ASSOC PROF MICROBIOL & IMMUNOL, QUEEN'S UNIV, ONT, 72- Mem: Soc Invert Path; Am Soc Microbiol; Can socSoc Microbiol. Res: Immunity in insects as an aspect of comparative immunology; cellular responses and humoral responses in both natural and acquired immunity. Mailing Add: Dept of Microbiol & Immunol Queen's Univ Kingston ON Can

CHADWICK, ROBERT AULL, b Milwaukee, Wis, May 4, 29; m 51; c 2. GEOLOGY. Educ: Princeton Univ, AB, 51; Univ Wis, PhD(geol), 56. Prof Exp: Geologist, Eagle-Picher Co, 56-59; asst prof geol, Long Beach State Col, 59-61; from asst prof to assoc prof, 61-73, PROF GEOL, MONT STATE UNIV, 73- Concurrent Pos: NSF grants attend, Conf Struct & Origin Volcanic Rocks, Wayne State Univ, 62 & Geol of Scand, Int Field Inst, 63; consult, Mont Power Co, 73; res grants, Mont Power Co, 74, US Environ Protection Agency & US Geol Surv, 75. Mem: Fel Geol Soc Am; Am Geophys Union. Res: Petrology; economic geology; Tertiary volcanic rocks of southwestern Montana; geology of pegmatites; paleomagnetism of volcanic rocks; geology of Montana coal deposits; geothermal energy potential of Montana. Mailing Add: Dept of Earth Sci Mont State Univ Bozeman MT 59715

CHADWICK, ROBERT WILLIAM, toxicology, pharmacology, see 12th edition

CHAE, KUN, b Seoul, Korea, Apr 13, 44; m 73; c 1. MEDICINAL CHEMISTRY, BIOCHEMISTRY. Educ: Seoul Nat Univ, BS, 66; Univ NC, Chapel Hill, MS, 71, PhD(med chem), 73. Prof Exp: Vis fel, 73-74, STAFF FEL, NAT INST ENVIRON HEALTH SCI, RESEARCH TRIANGLE PARK, 74- Mem: Am Chem Soc; Sigma Xi. Res: Organic synthesis and metabolic studies of glyceryl ether lipids; synthesis of chlorinated aromatics; toxicological and metabolic studies of polychlorinated biphenyls. Mailing Add: 321 Bywood Dr Durham NC 27705

CHAET, ALFRED BERNARD, b Boston, Mass, June 7, 27; m 50; c 3. PHYSIOLOGY. Educ: Univ Mass, BS, 49, MS, 50; Univ Pa, PhD(zool), 53. Prof Exp: Asst, Univ Pa, 51-53; instr zool, Univ Maine, 53-56; asst prof physiol, Sch Med, Boston Univ, 56-58; from assoc prof to prof biol, Am Univ, 58-66; PROF BIOL, UNIV WFLA, 66-, PROVOST, 67- Concurrent Pos: Res, Marine Biol Lab, Woods Hole, 49, 51-53 & 55-58 & corp mem; res assoc, Boston City Hosp, 56-; NIH spec fel & vis scholar, Scripps Inst, Calif, 64-65; assoc dean sci, Univ WFla, 66-67. Mem: AAAS; Am Soc Zool; Soc Gen Physiol; Am Soc Physiol; Soc Biophys. Res: Shedding substance in starfish-invertebrate neurohormone; absorption properties of echinoderm tube feet; adhering mechanisms in invertebrates; toxic factor in heat death; thiaminase. Mailing Add: Gamma Col Univ of WFla Pensacola FL 32504

CHAFETZ, HARRY, b New York, NY, Oct 27, 29; m 55; c 2. ORGANIC CHEMISTRY. Educ: City Col New York, BS, 50; Pa State Univ, PhD(chem), 54. Prof Exp: From chemist to res chemist, 56-61, group leader, 61-70, RES ASSOC, BEACON RES LABS, TEXACO INC, 70- Mem: Am Chem Soc. Res: Organic synthesis; petrochemicals; free radical chemistry; halogenetion and oxidation reactions; hydrocarbon chemistry. Mailing Add: 4 Robin Lane Poughkeepsie NY 12603

CHAFETZ, LESTER, pharmaceutical chemistry, see 12th edition

CHAFETZ, MORRIS EDWARD, b Worcester, Mass, Apr 20, 24; m 46; c 3. PSYCHIATRY. Educ: Tufts Univ, MD, 48; Am Bd Psychiat & Neurol, dipl, 56. Prof Exp: Dir alcohol clin, Mass Gen Hosp, 57-68, dir acute psychiat serv, 61-68, psychiatrist, 64-70, dir clin psychiat serv, 68-70; actg dir, Div Alcohol Abuse & Alcoholism, 70-71, DIR, NAT INST ALCOHOL ABUSE & ALCOHOLISM, NIMH, 71- Concurrent Pos: Mem ad hoc rev bd res in alcoholism, NIMH, 58-61; mem subcomt alcoholism, Mass Ment Health Planning Proj, 63-; assoc clin prof psychiat, Harvard Med Sch, 68-70. Mem: Group Advan Psychiat. Res: Treatment, prevention and dynamics of alcoholism and alcohol-related disorders; psychiatric care of urban poor. Mailing Add: Nat Inst on Alcohol Abuse NIMH 5600 Fishers Lane Rockville MD 20852

CHAFFEE, ELEANOR, b Cambridge, Mass, Oct 9, 34. PHYSICAL INORGANIC CHEMISTRY. Educ: Mt Holyoke Col, BA, 56; Harvard Univ, MAT, 62; Wellesley Col, MA, 67; Brown Univ, PhD(chem), 71. Prof Exp: Res physicist photoconductivity, Stanford Res Inst, 56-57; res physicist phys chem, Arthur D Little, Inc, 59-61; teacher chem, Am High Sch, Lugano, Switz, 62-63 & Lexington High Sch, Mass, 63-65; fel chem, State Univ NY Buffalo, 71-72; RES CHEMIST INORG CHEM, EASTMAN KODAK CO, NY, 72- Res: Kinetics and mechanisms of oxidation of substrates, their catalysis by metals, and their complexes; peroxide chemistries. Mailing Add: Res Lab Eastman Kodak Co Rochester NY 14650

CHAFFEE, ELMER FENN, b Omak, Wash, May 6, 11; m 37; c 4. IMMUNOLOGY. Educ: Univ Idaho, BS, 38; Duke Univ, PhD(microbiol), 52. Prof Exp: Bacteriologist, State Dept Health, Idaho, 38-42; chief dept serol, Army Med Serv Grad Sch, Walter Reed Army Med Ctr, US Army, 48-50, chief serol sect, Trop Res Med Lab, PR, 52-55, exec officer, 5th US Army Med Lab, Mo, 55-61, mem staff immunol & bact br, Armed Forces Inst Path, 61-62, chief, 63-66; ASSOC PROF, DEPT PARASITOL & LAB PRACT, UNIV NC, CHAPEL HILL, 66- Mem: Am Soc Trop Med & Hyg; NY Acad Sci. Res: Serodiagnosis of schistosomiasis, trypanosomiasis and syphilis; immunology of schistosomiasis, trypanosomiasis and leishmaniasis; factors affecting immune hemolysis. Mailing Add: Dept of Parasitol & Lab Pract Univ of NC Chapel Hill NC 27514

CHAFFEE, MAURICE AHLBORN, b Wilkes-Barre, Pa, Jan 10, 37; m 59; c 2. ECONOMIC GEOLOGY, GEOCHEMISTRY. Educ: Colo Sch Mines, Geol E, 59; Univ Ariz, MS, 64, PhD(econ geol, mineral), 67. Prof Exp: Mine geologist, NJ Zinc Co, Va, 60-62; GEOLOGIST, US GEOL SURV, 67- Mem: Geol Soc Am; Soc Econ Geol; Asn Explor Geochem; Sigma Xi. Res: Geology and hydrothermal alteration of mineral deposits; trace element chemistry related to mineral deposits; development of new methods and concepts for application to trace element chemistry in mineral exploration. Mailing Add: Explor Res Br US Geol Surv Fed Ctr Denver CO 80225

CHAFFEE, ROBERT GIBSON, b Rutland, Vt, Mar 6, 14; m 38; c 4. PALEONTOLOGY. Educ: Dartmouth Col, AB, 36; Univ Pa, MS, 41; Columbia Univ, PhD, 52. Prof Exp: Field asst, Am Mus Natural Hist, NY, 26-28; asst cur geol & paleont, Acad Natural Sci, Pa, 38-42; photogrammetrist, Alaskan Br, US Geol Surv, 42-44; cur geol, 48-68, DIR, DARTMOUTH COL MUS, 68- Mem: Soc Vert Paleont; Geol Soc Am. Res: Vertebrate paleontology; stratigraphy. Mailing Add: Dartmouth Col Mus Hanover NH 03755

CHAFFEE, ROWAND R J, b El Paso, Tex, Nov 4, 25; m 53; c 6. PHYSIOLOGY, BIOCHEMISTRY. Educ: Univ NMex, BS, 46, BA, 51, MS, 52; Harvard Univ,

PhD(biol sci), 57. Prof Exp: Asst prof zool, Univ Redlands, 58-59; NIH fel cellular physiol, Univ Calif, Berkeley, 59-60; asst prof zool, Univ Calif, Riverside, 60-64; mem staff cellular physiol, Los Alamos Sci Lab, 64-69; PROF ERGONOMICS, UNIV CALIF, SANTA BARBARA, 69- Mem: AAAS; Am Soc Mech Engrs; Am Physiol Soc; Soc Exp Biol & Med; Brit Ergonomics Res Soc. Res: Biochemistry of hibernating; cold and heat acclimation; cellular physiology; mitosis; enzyme kinetics; primate temperature; altitude acclimation biochemistry. Mailing Add: Dept of Ergonomics Univ of Calif Santa Barbara CA 93106

CHAFFIN, TOMMY L, b Dallas, Tex, Apr 11, 43; m 65. ORGANIC CHEMISTRY. Educ: Okla State Univ, BS, 65; Univ Ill, PhD(org chem), 69. Prof Exp: Sr res chemist, 69-75, TECH MGR, 3M CO, ST PAUL, 75- Mem: Am Chem Soc. Res: Stereospecific alkylations; anionic block copolymers; Friedel-Crafts chemistry; photographic chemistry. Mailing Add: 2002 Fairmeadows Rd Stillwater MN 55082

CHAGANTI, RAJU SREERAMA KAMALASANA, b Samalkot, India, Mar 12, 33; m 66; c 2. GENETICS. Educ: Andhra Univ, India, BSc, 54, MSc, 55; Harvard Univ, PhD(biol), 64. Prof Exp: Demonstr & res asst bot, Andhra Univ, India, 55-61, lectr, 61-67; mem sci staff, Med Res Coun Radiobiol Unit, Harwell, Eng, 67-71; res assoc & assoc investr, Lab Human Genetics, New York Blood Ctr, 71-76; HEAD LAB GENETICS & ASSOC ATTEND PATHOLOGIST, MEM SLOAN-KETTERING CANCER CTR, 76- Concurrent Pos: Vis asst prof genetics, Grad Sch Med Sci, Cornell Univ, 74-; consult, Lab Human Genetics, New York Blood Ctr, 76- Mem: Am Soc Human Genetics; Genetics Soc Am; AAAS; Sigma Xi; Indian Soc Genetics & Plant Breeding. Res: Human and mammalian genetics; genetic control of chromosome form and behavior at mitosis and meiosis; chromosome change in neoplastic cells. Mailing Add: Mem Sloan-Kettering Cancer Ctr 1275 York Ave New York NY 10021

CHAGNON, ANDRE, b Montreal, Que, Aug 16, 32; m 58; c 2. VIROLOGY, TISSUE CULTURE. Educ: Univ Montreal, BA, 53, BSc, 57, PhD(microbiol), 60. Prof Exp: RES ASST VIROL, INST MICROBIOL & HYG, UNIV MONTREAL, 60- Concurrent Pos: Partic, Int Conf Rubella Immunization, 69. Mem: Soc Cryobiol; Tissue Cult Asn; Fr-Can Asn Advan Sci; Can Soc Cell Biol; Can Soc Microbiol. Mailing Add: Inst of Microbiol & Hyg Univ of Montreal CP100 Ville de Laval PQ Can

CHAGNON, NAPOLEON ALPHONSEAU, b Port Austin, Mich, Aug 27, 38; m 60; c 2. ANTHROPOLOGY. Educ: Univ Mich, BA, 60, MA, 63, PhD(anthrop), 66. Prof Exp: Res assoc human genetics, Med Sch, Univ Mich, 66-72, asst prof anthrop, 67-72; ASSOC PROF ANTHROP, PA STATE UNIV, 72- Concurrent Pos: AEC & NSF grants, Univ Mich, 66-72; Wenner-Gren Found Anthrop Res grant, 69-70; NSF filming grants, Brandeis Univ & Univ Mich, 71-73. Honors & Awards: Golden Eagle Award, CINE, 70 & 71; Grand Prize, Brussels Film Festival, 70; Grand Prize, Int Festival Sci & Educ Films, Padua, Italy, 70; Blue Ribbon, Am Film Festival, 72 & 74; Prize, Int Festival Ethnic Sociol Films, Venice, 72. Mem: Fel Am Anthrop Asn; fel Am Asn Phys Anthrop. Res: The demographic, political, economic and social aspects of tribal warfare; the demographic basis of social organization; the use of ethnographic film in education; cultural evolution; South American native peoples. Mailing Add: Dept of Anthrop 409 Soc Sci Bldg Pa State Univ University Park PA 16802

CHAGNON, PAUL ROBERT, b Woonsocket, RI, Nov 11, 29. NUCLEAR PHYSICS. Educ: Col of the Holy Cross, BS, 50; Johns Hopkins Univ, PhD(physics), 55. Prof Exp: Assoc, Univ Mich, 55, instr, 55-57, asst prof physics, 57-63; asst prof, 63-69, PROF PHYSICS, UNIV NOTRE DAME, 69- Mem: Am Phys Soc. Res: Nuclear spectroscopy. Mailing Add: Dept of Physics Univ of Notre Dame Notre Dame IN 46556

CHAH, CHEONG CHOO, b Seoul, Korea, May 21, 42; m 70; c 1. ANIMAL NUTRITION. Educ: Kon-Kuk Univ, Korea, BSA, 66; Malling Agr Col, Denmark, dipl, 67; Univ Guelph, MSc, 71; SDak State Univ, Brookings, PhD(animal nutrit), 74. Prof Exp: Asst animal sci, Kon-kuk Univ, 64-66; livestock technician animal sci, Univ Guelph, 68-69, res asst poultry sci, 69-71; res asst animal sci, SDak State Univ, 72-75; RES ASSOC FOODS & NUTRIT, UNIV GA, 75- Mem: Poultry Sci Asn; Am Inst Nutrit; AAAS; Nutrit Today Soc; Sigma Xi. Res: Identification of the macro- and micro-mineral elements involved in the development of cardiovascular-related diseases. Mailing Add: Dept of Foods & Nutrit Sch of Home Econ Univ of Ga Athens GA 30602

CHAHINE, MOUSTAFA TOUFIC, b Beirut, Lebanon, Jan 1, 35; US citizen; m 60; c 1. FLUID PHYSICS, ATMOSPHERIC PHYSICS. Educ: Univ Wash, BS, 56, MS, 57; Univ Calif, Berkeley, PhD(fluid physics), 60. Prof Exp: Res assoc fluid dynamics, Univ Calif, Berkeley, 58-60, teaching assoc aeronaut, 59-60; STAFF SCIENTIST, JET PROPULSION LAB, CALIF INST TECHNOL, 60- Concurrent Pos: Vis scientist, Mass Inst Technol, 69-70; mem earth surv panel, NASA, 69-; assoc prof, Am Univ Beirut, 71-72; sci consult, Naval Postgrad Sch, Monterey, Calif, 74- Honors & Awards: Exceptional Sci Achievement Medal, NASA, 69. Mem: Am Inst Physics; Am Meteorol Soc. Res: Thermodynamics and statistical fluid physics; strong shock waves and remote sensing of planetary atmospheres; long range numerical weather prediction. Mailing Add: Jet Propulsion Lab Calif Inst of Technol Pasadena CA 91103

CHAI, AN-TI, b Honan, China, 39. MOLECULAR SPECTROSCOPY. Educ: Nat Taiwan Univ, BS, 61; Kans State Univ, MS, 66, PhD(physics), 68. Prof Exp: Asst prof physics, Mich Technol Univ, 68-73; INTERIM ASST PROF, INTERDISCIPLINARY CTR AERONOMY & ATMOSPHERIC SCI, UNIV FLA, 74- Mem: Am Phys Soc; Am Asn Physics Teachers; AAAS. Res: Atomic spectroscopy; direct and diffused solar irradiance measurements; radiative transfer; related atmospheric studies. Mailing Add: Dept of Physics & Astron Univ of Fla Gainesville FL 32611

CHAI, CHEN KANG, b Hopeh, China, Feb 14, 16; m 54; c 2. GENETICS. Educ: Army Vet Col, China, DVM, 37; Mich State Col, MS, 49, PhD(animal breeding), 51. Prof Exp: Asst, Mich State Col, 49-51; US Dept State fel, 51-52, res fel, 52-55, res assoc, 56, staff scientist, 57-67, SR STAFF SCIENTIST, JACKSON LAB, 67- Concurrent Pos: Vis fel, Mass Inst Technol, 52-53; Guggenheim fel, 62-63. Mem: Fel AAAS; Am Genetic Asn; Genetics Soc Am; Biomet Soc; Am Asn Phys Anthrop. Res: Quantitative genetics; genetic study of endocrine variation; mouse and rabbit genetics; genetic variations in Taiwan aborigines. Mailing Add: Jackson Lab Bar Harbor ME 04609

CHAI, WINCHUNG A, b Hunan, China, Aug 21, 39; US citizen; m 69; c 1. MATHEMATICAL ANALYSIS, APPLIED MATHEMATICS. Educ: Wittenberg Univ, BA, 60; NY Univ, MS, 64; Polytech Inst Brooklyn, PhD(math), 68. Prof Exp: Mathematician, Am Tel & Tel Co, 60-63 & Aerospace Res Ctr, Gen Precision Inc, NJ, 63-68; ASSOC PROF MATH, MONTCLAIR STATE COL, 68- Mem: Am Math Soc; Math Asn Am; Soc Indust & Appl Math; Asn Comput Mach. Res: Applied mathematics and computer science. Mailing Add: Dept of Math Montclair State Col Upper Montclair NJ 07043

CHAIKEN, JAN MICHAEL, b Philadelphia, Pa, Oct 19, 39; m 39; c 2. MATHEMATICS. Educ: Carnegie Inst Technol, BS, 60; Mass Inst Technol, PhD(math), 66. Prof Exp: From instr to asst prof math, Cornell Univ, 64-68; RESEARCHER, RAND CORP, 68- Concurrent Pos: Res assoc, Mass Inst Technol, 67-68; adj assoc prof, Univ Calif, Los Angeles, 72-; assoc ed, Opers Res, 75- Mem: AAAS; Am Math Soc; Opers Res Soc Am. Res: Allocation of urban services. Mailing Add: Rand Corp 1700 Main St Santa Monica CA 90406

CHAIKEN, ROBERT FRANCIS, b Brooklyn, NY, Dec 19, 28; m 53; c 3. CHEMICAL PHYSICS. Educ: Univ Ill, BS, 49; Brooklyn Polytech Inst, MS, 58; Univ Calif, Riverside, PhD(chem), 66. Prof Exp: Res chemist, Tracerlab Inc, 50-51 & US Testing Co, 51-53; res assoc, George Wash Univ, 53-57; res assoc, Aerojet-Gen Corp, 57-59, tech specialist, 59-61, tech consult, 61-68; sr chem physicist, Stanford Res Inst, 68-70; RES CHEMIST, PITTSBURGH MINING & SAFETY RES CTR, US BUR MINES, 70- Concurrent Pos: Lectr, Univ Pittsburgh, 72- Mem: AAAS; Am Chem Soc; Combustion Inst (asst treas, 74-). Res: Solid state chemistry; organic semiconductors; high temperature and pressure reaction kinetics; theory of combustion and detonation processes; high speed photography. Mailing Add: Pittsburgh Mining & Safety Res Ctr US Bur of Mines 4800 Forbes Ave Pittsburgh PA 15213

CHAIKIN, LAWRENCE, b New York, NY, Mar 14, 14; m 43; c 2. ORAL SURGERY. Educ: NY Univ, BA, 34, DDS, 37. Prof Exp: From instr to asst prof, 38-68, ASSOC PROF ORAL SURG, COL DENT, NY UNIV, 69- Concurrent Pos: Assoc vis oral surgeon, Bellevue Hosp; lectr & clinician; attend oral surgeon, NShore Hosp, Manhasset, NY. Mem: Fel Am Col Dent. Res: Local anesthesia in dentistry. Mailing Add: 15 Bond St Great Neck NY 11021

CHAIKIN, SAUL WILLIAM, b New York, NY, Dec 25, 21; m 50; c 2. CHEMISTRY. Educ: Brooklyn Col, BA, 43; Univ Chicago, MS, 48, PhD(chem), 48. Prof Exp: Asst chem, Toxicity Lab, Univ Chicago, 43-45; res assoc, Univ Calif, Los Angeles, 48-49; asst prof anal chem, Univ WVa, 49-51; sr chemist, Stanford Res Inst, 51-64, mgr surface chem sect, Electronic Mat Dept, 64-68; mgr chem res dept, Memorex Corp, 68-69; DIR RES, XIDEX CORP, 69- Concurrent Pos: Res Corp grantee, 50. Mem: Am Chem Soc; Sigma Xi. Res: Development of analytical methods; organic reductions with complex metal hydrides; chelate formation; organoboron chemistry; vapor pressure determination; chromatography; microscopy. Mailing Add: 305 Soquel Way Sunnyvale CA 94086

CHAIT, ARNOLD, b New York, NY, Jan 20, 30; m 65; c 3. MEDICINE. Educ: NY Univ, BA, 51; Univ Utrecht, MD, 57; Am Bd Radiol, dipl, 63. Prof Exp: Intern, Kings County Hosp, NY, 58-59, resident radiol, 59-62; from instr to assoc prof, State Univ NY Downstate Med Ctr, 62-67; from asst prof to assoc prof, 67-74, PROF RADIOL, UNIV PA, 74- Concurrent Pos: Attend physician, Philadelphia Vet Admin Hosp, 69-; consult, Children's Hosp Philadelphia, 74- Mem: Inter-Am Col Radiol; Soc Cardiovasc Radiol; Radiol Soc NAm; Am Roentgen Ray Soc; fel Am Col Radiol. Res: Cardiovascular radiology. Mailing Add: Dept of Radiol Hosp of the Univ of Pa Philadelphia PA 19104

CHAIT, EDWARD MARTIN, b Brooklyn, NY, May 8, 42; m 66; c 2. ANALYTICAL CHEMISTRY, MASS SPECTROMETRY. Educ: Cornell Univ, AB, 64; Purdue Univ, PhD(anal chem), 68. Prof Exp: Applns supvr, 67-74; PROD MGR INSTRUMENTS, E I DU PONT DE NEMOURS & CO, INC, 74- Mem: Am Chem Soc; Am Soc Testing & Mat. Res: Organic mass spectrometry; molecular spectroscopy; field ionization phenomena. Mailing Add: 1500 S Shamrock Ave Monrovia CA 91016

CHAKERIAN, GULBANK DONALD, b Parlier, Calif, Dec 21, 33; m 58; c 2. MATHEMATICS. Educ: Univ Calif, Berkeley, AB, 55, PhD(math), 60. Prof Exp: Instr math, Calif Inst Technol, 60-63; lectr, 63-64, asst prof, 64-69, PROF MATH, UNIV CALIF, DAVIS, 69- Mem: Am Math Soc; Math Asn Am. Res: Integral geometry and convex bodies. Mailing Add: Dept of Math Univ of Calif Davis CA 95616

CHAKKALAKAL, DENNIS ABRAHAM, b Irinjalakuda, India, Mar 16, 39; m 70. THEORETICAL PHYSICS. Educ: Madras Univ, BSc, 58; Marquette Univ, MS, 62; Washington Univ, PhD(physics), 68. Prof Exp: ASSOC PROF PHYSICS, SOUTHERN UNIV, 69- Concurrent Pos: NIH fel biophys, Rensselaer Polytech Inst, 76. Mem: Am Phys Soc; Sigma Xi. Res: Study of symmetrical nuclear matter and neutron-star structure using many-body theory; the biophysics research: modelling of bone as a viscoelastic composite- is in the beginning stages. Mailing Add: Dept of Physics Southern Univ Box 9365 Baton Rouge LA 70813

CHAKO, NICHOLAS, b Hotove, Albania, Nov 11, 10; nat US; m 52; c 1. MATHEMATICS. Educ: France, BS, 28; Johns Hopkins Univ, PhD(physics), 34; Sorbonne, DSc, 66. Prof Exp: Prof math & physics & head dept, State Gym, Albania, 36-37; staff mem, Crufts Lab & tutor physics, Harvard Univ, 38-40; staff mem, Spectros Lab, Mass Inst Technol, 40-41; assoc prof physics, Kans State Univ, 46-47 & Ala Polytech Inst, 47-49; Fulbright exchange prof, State Univ Utrecht, 50-51; guest prof math & physics, Chalmers Univ Technol, Sweden, 51-52; res assoc, Inst Math Sci, NY Univ, 53-56; ASSOC PROF MATH, QUEENS COL, NY, 56- Concurrent Pos: Lectr, Ill Inst Technol, 40-41; consult, Balkan Affairs, Off Strategic Serv, DC, 42-45; consult & math physicist, Russell Elec Co, Ill, 42-46; Fulbright grant to Holland, 50-51; lectr, Univ Lund, 52; prof, French Atomic Energy Ctr, Saclay & Univ Paris, 66-67; vis lectr, Laval Univ, 68. Honors & Awards: Annual Prize, Royal Soc Eng & Chalmers Alumni Asn Sweden, 52. Mem: Am Phys Soc; Inst Elec & Electronics Engrs; Am Math Soc; NY Acad Sci; Acoust Soc Am. Res: Absorption of light by organic compounds; geometrical and electron optics; crystal vibrations and acoustics fields; diffraction; special functions; asymptotic integration. Mailing Add: Dept of Math Queens Col Flushing NY 11367

CHAKRABARTI, CHUNI LAL, b Patuakhali, India, Mar 1, 20; m 62; c 1. ANALYTICAL CHEMISTRY, INORGANIC CHEMISTRY. Educ: Univ Calcutta, BSc, 41; Univ Birmingham, MSc, 60; Queen's Univ, Belfast, PhD(chem), 62; FRIC, 63. Prof Exp: Supvr chemist, Metal & Steel Factory, Govt India, 41-45; chemist-in-charge, Mines & Indust Dept, Govt Burma, 45-52, chief chemist & mgr, Mineral Resources Develop Corp, 52-59; vis asst prof & fel, La State Univ, 63-65; group leader res ctr, Noranada Mines Ltd, 65; asst prof, 65-67, ASSOC PROF ANAL & INORG CHEM, CARLETON UNIV, CAN, 67- Mem; Fel Chem Inst Can; Am Chem Soc; Brit Soc Anal Chem. Res: Atomic-absorption, atomic-fluorescence and emission spectroscopy; determination of ultratrace elements in air and water; speciation and complexation of trace metals in the natural environment; electroanalytical techniques for characterization and quantitation of trace metal species. Mailing Add: Dept of Chem Carleton Univ Colonel By Dr Ottawa ON Can

CHAKRABARTI, SIBA GOPAL, b Rangpur, EPakistan, July 1, 23; m 59; c 2. BIOCHEMISTRY. Educ: Univ Calcutta, BSc, 45; Rutgers Univ, MS, 60, PhD(biochem), 63. Prof Exp: Chemist, Bhartia Elec Steel Co, India, 45-47 & Burn & Co, Ltd, 47-53; assoc chemist, Sam Tour & Co, Inc, NY, 53-54; res assoc biochem, Sloan-Kettering Inst Cancer Res, 54-58; res asst, Rutgers Univ, 58-60; clin chemist,

Middlesex Gen Hosp, NJ, 60-62 & St Joseph's Hosp, Hamilton, Ont, 63-65; asst prof dermat, Med Ctr, Univ Mich, Ann Arbor, 68-73; ASSOC PROF DERMAT, COL MED, HOWARD UNIV, 73- Concurrent Pos: NIH res trainee dermat, Univ Mich, 65-68. Mem: Am Chem Soc; Soc Invest Dermat; NY Acad Sci. Res: Epidermal protein synthesis and characterization of epidermal pre-keratins; control mechanisms in epidermal differentiation. Mailing Add: Dept of Dermat Howard Univ Col of Med Washington DC 20001

CHAKRABARTY, ANANDA MOHAN, b Sainthia, India, Apr 4, 38; m 65; c 2. MICROBIAL GENETICS. Educ: St Xavier's Col, India, BSc, 58; Calcutta Univ, MSc, 60, PhD(biochem), 65. Prof Exp: Sr sci officer biochem, Calcutta Univ, 64-65; res assoc, Univ-Ill, Urbana, 65-71; STAFF MICROBIOLOGIST, GEN ELEC CO, 71- Honors & Awards: Scientist of the Year Award, Industrial Res Mag, 75. Mem: Am Soc Microbiol; AAAS; Genetics Soc Am. Res: Evolution and application of hydrocarbon degradative plasmids in Pseudomonas; molecular cloning and genetic engineering with plasmids. Mailing Add: Res & Develop Ctr Gen Elec Co PO Box 8 Schenectady NY 12301

CHAKRABARTY, MANOJ R, b Bajitpur, Bangladesh, Jan 1, 33; m 59; c 1. PHYSICAL INORGANIC CHEMISTRY. Educ: Univ Calcutta, BS, 51, MS, 54; Univ Toronto, PhD(inorg chem), 62. Prof Exp: Instr chem, Bengal Eng Col, India, 55-56; lectr, Indian Sch Mines, 56-57; sessional instr, Univ Alta, 57-59; assoc res scientist mat chem, Ont Res Found, 62-63; from asst prof to assoc prof chem, 63-69, PROF CHEM, MARSHALL UNIV, 69- Mem: Am Chem Soc. Res: Coordination compounds; inorganic reaction kinetics; nuclear and solid state chemistry. Mailing Add: Dept of Chem Marshall Univ Huntington WV 25701

CHAKRABARTY, RAMESWAR PRASAD, b Sylhet, Bangladesh, June 2, 35; Indian citizen; m 71; c 1. STATISTICS. Educ: Univ Gauhati, India, BS, 55, MS, 57; Tex A&M Univ, PhD(statist), 68. Prof Exp: Lectr statist, Univ Gauhati, 58-59; statistician, Tea Res Asn of India, 59-64; vis asst prof, Univ Ga, 67-69; statistician, Fertilizer Asn of India, 69-71; ASST PROF STATIST, UNIV GA, 71- Mem: Int Asn Surv Statist; Am Statist Asn; Biomet Soc; Sigma Xi. Res: Ratio methods of estimation, variance estimation in surveys; jack-knife statistics; statistical consulting on designs of surveys, experiments and data analysis. Mailing Add: Dept of Statist Univ of Ga Athens GA 30602

CHAKRABORTY, PRABIR KUMAR, b Calcutta, India, Jan 1, 36; m 63; c 3. PHYSIOLOGY, NEUROENDOCRINOLOGY. Educ: Calcutta Univ, BS, 55; Nat Dairy Res Inst, Bangalore, Indian dairy dipl, 57; Ore State Univ, MS, 70, PhD(reproductive physiol, animal sci), 71. Prof Exp: Supvr, Cent Cattle Res & Breeding Sta, Haringhata, WBengal, India, 57-63; asst res officer, Nat Diary Res Inst, Karnal, Haryana, India, 63-67; res asst physiol, Dept Animal Sci, Ore State Univ, 67-71; NIH res assoc neuroendocrinol, Dept Animal Sci, Wash State Univ, 71-74; ASST PROF REPRODUCTIVE ENDOCRINOL, UNIV ORE HEALTH SCI CTR, 74- Mem: Am Soc Animal Sci; Soc Study Reproduction; Endocrine Soc. Res: Endocrine and neuroendocrine control of reproduction in laboratory and domestic animals. Mailing Add: Dept of Surg Univ of Ore Health Sci Ctr Portland OR 97201

CHAKRABORTY, RANAJIT, b Calcutta, India, Apr 17, 46; m 74. POPULATION GENETICS, HUMAN GENETICS. Educ: Indian Statist Inst, Calcutta, BStatist, 67, MStatist, 68, PhD(biostatist), 71. Prof Exp: From lectr to sr lectr statist, Indian Statist Inst, Calcutta, 71-72; vis consult genetics, Pop Genetics Lab, Univ Hawaii, 72-73; res assoc pop genetics, Health Sci Ctr, 73, ASST PROF POP GENETICS, GRAD SCH & ASST PROF HUMAN ECOL, SCH PUB HEALTH, UNIV TEX, HOUSTON, 73- Mem: Am Soc Human Genetics; Am Soc Phys Anthropologists; AAAS. Res: Statistical methods for genetic determination of quantitative traits; analysis of pedigree data for detection and estimation of familial aggregation of various disorders; mathematical theories of molecular evolution and population dynamics. Mailing Add: Ctr for Demog & Pop Genetics 1100 Holcombe Blvd Rm 1109 Houston TX 77030

CHAKRAVARTI, ANINDA KUMAR, b Varanasi, India, Dec 3, 26; m 57; c 1. AGRICULTURAL GEOGRAPHY. Educ: Univ Allahabad, BA, 49, MA, 53; Univ Wis, PhD(geog), 67. Prof Exp: Asst prof geog, Univ Allahabad, 53-60; asst prof, 65-70, ASSOC PROF GEOG, UNIV SASK, 70- Concurrent Pos: Can Coun, Ottawa fel, 72-73. Mem: Asn Am Geogr; Can Asn Geogr; AAAS. Res: Geography of South Asia; agricultural geography; climatology; food grain deficiency problems in India. Mailing Add: Dept of Geog Univ of Sask Saskatoon SK Can

CHAKRAVARTI, DIPTIMAN, b Sylhet, Assam, Sept 19, 28; m 55; c 2. RADIOCHEMISTRY, FOOD CHEMISTRY. Educ: Univ Calcutta, BSc, 48; Univ Mass, MS, 51; Wash State Univ, PhD(animal s ci), 55. Prof Exp: Food technologist fisheries res, Col Fisheries, Univ Wash, 57-58, from res instr to res assoc prof radiochem, Lab Radiation Biol, 58-66; mgr life sci advan mkt develop, Corning Glass Works, NY, 66-67, head life sci dept & mgr com develop, 67-69; PRES & CHIEF EXEC OFF, INNOVA CORP, 69- Concurrent Pos: Consult indust orgns, 56-; head anal radio chem div, Univ Wash, 61-66; dir, Oper Interface, Am Chem Soc Environ Conf, 70, chmn centennial comt, Puget Sound Sect, Am Chem Soc, 75-76; mem bd dirs, Inst Technol Corp; mem bd trustees, Pac Sci Ctr Found, Seattle. Mem: Am Chem Soc; Inst Food Technol; NY Acad Sci. Res: Food technology; radiochemical analysis of environmental biological samples, methodology and microanalytical techniques; electrochemistry and metal process quality control; industrial ion-exchange processes; oil pollution control in industrial and marine environment. Mailing Add: Innova Corp 444 Ravenna Blvd Seattle WA 98115

CHAKRAVARTI, KALIDAS, b Gobindapur, India. PHYSICAL CHEMISTRY, SURFACE CHEMISTRY. Educ: Univ Calcutta, BS, 57, MS, 59, PhD(chmem), 64. Prof Exp: Res assoc Univ Calcutta, 64-65; res fel, Univ Minn, Minneapolis, 65-66; res assoc, Lehigh Univ, 67-69; SR SCIENTIST, ALLIED CHEM CORP, 69- Mem: Am Chem Soc; Am Phys Soc; fel Am Inst Chemists. Res: Physical chemistry of macromolecules and biopolymers; colloid and surface chemistry; synthetic fibers and synthetic polymers; fiber surface science; adhesion and bonding of elastomers; fiber finishes. Mailing Add: Allied Chem Corp Fibers Tech Ctr PO Box 31 Petersburg VA 23803

CHAKRIN, ALAN LEONARD, b New York, NY, Mar 7, 40; m 64; c 1. CLINICAL CHEMISTRY, TOXICOLOGY. Educ: Brooklyn Col, BA, 61; Univ Chicago, PhD(biochem), 69. Prof Exp: Res assoc biochem, Univ Chicago, 69-70; instr, 70-74, ASSOC PATH, MED SCH, NORTHWESTERN UNIV, CHICAGO, 74-; BIOCHEMIST, NORTHWESTERN MEM HOSP, 70- Mem: Am Chem Soc; Am Asn Clin Chemists. Res: Applications of digital computers in clinical laboratories; new methods in clinical toxicology and drug screening. Mailing Add: Wesley Pavillion Northwestern Mem Hosp Chicago IL 60611

CHAKRIN, LAWRENCE WILLIAM, b Brooklyn, NY, Oct 21, 38; m 64. PHARMACOLOGY. Educ: Long Island Univ, BSc, 62; Univ Minn, PhD(neuropharmacol), 67. Prof Exp: USPHS fel, Cambridge, Eng, 67-68; RES PHARMACOLOGIST & ASST DIR BIOL RES, SMITH KLINE & FRENCH LABS, 68- Mem: Am Soc Pharmacol & Exp Therapeut; Am Soc Neurochem. Res: Respiratory pharmacology; pharmacology of immediate hypersensitivity reactions; central nervous system neuropharmacology; cholinergic mechanisms. Mailing Add: Smith Kline & French Labs 1500 Spring Garden St PO Bx 7929 Philadelphia PA 19101

CHALAMALASETTY, VENKATESWARA RAO, b Bantumelli, India, Dec 26, 41; m 71; c 2. REPRODUCTIVE ENDOCRINOLOGY. Educ: Sri Venkateswara Univ, BVSc, 64; Wash State Univ, MS, 66, PhD(animal sci), 69. Prof Exp: Res asst animal sci, Wash State Univ, 66-69; res fel biochem & instr urol, Albert Einstein Col Med, 69-70; res fel reproductive endocrinol, Med Col, Cornell Univ, 70-72; ASST PROF OBSTET & GYNEC & FAC ASSOC BIOCHEM, SCH MED, UNIV LOUISVILLE, 72- Mem: Am Soc Biol Chemists; Endocrine Soc; Soc Study Reproduction; Am Fertil Soc; AAAS. Res: Mechanism of action of prostaglandins and gonadotropins in Corpus Luteum; identification of receptors for these agents in luteal cell membranes and molecular mechanisms involved beyond receptor binding. Mailing Add: Dept of Obstet & Gynec Univ of Louisville Sch of Med Louisville KY 40202

CHALFANT, RICHARD BRUCE, b Akron, Ohio, Aug 15, 29; m 53; c 3. ENTOMOLOGY. Educ: Univ Akron, BS, 54; Univ Wis, MS, 56, PhD(entom), 59. Prof Exp: Asst prof entom, Sch Agr, NC State Univ, 59-66; ASSOC PROF ENTOM, GA COASTAL PLAIN EXP STA, UNIV GA, 66- Mem: Entom Soc Am. Res: Biology, control and management of insects affecting vegetable crops; host-plant resistance. Mailing Add: Ga Coastal Plain Exp Sta Univ of Ga Tifton GA 31794

CHALGREN, STEVE DWAYNE, b Ft Dodge, Iowa, Jan 3, 40; m 62; c 2. MICROBIOLOGY, VIROLOGY. Educ: Univ Iowa, BA, 61; Univ Mo, MS, 65, PhD(microbiol, virol), 68. Prof Exp: ASSOC PROF BIOL, RADFORD COL, 68- Concurrent Pos: Microbiol consult, Labs, Radford Community Hosp, Va, 68- Mem: Am Soc Microbiol. Res: Diagnostic microbiology. Mailing Add: Dept of Biol Radford Col Box 597 Radford VA 24141

CHALK, RONALD C, organic chemistry, see 12th edition

CHALKLEY, DONALD THOMAS, b Lake Charles, La, Feb 8, 20; m 42, 66; c 6. EMBRYOLOGY, SCIENCE POLICY. Educ: Oberlin Col, BA, 42; Amherst Col, MA, 47; Princeton Univ, PhD(zool, embryol), 50. Prof Exp: Mem staff, Dept Biol, Princeton Univ, 50; asst prof, Univ Notre Dame, 50-56; asst prof, Nat Cancer Inst, USPHS, 56-59, exec secy, Path Study Sect, Div Res Grants, 59-66, Reproductive Biol Study Sect, 66-67, spec asst to dir, 67-72, chief, Inst Rels Br, 72-74, DIR OFF PROTECTION RES RISKS, NIH, 74- Concurrent Pos: From asst referral officer to assoc referral officer, Div Res Grants, USPHS, 62-67; asst to vchancellor, Univ Miss Med Ctr, 65-66. Res: Research and training grant administration; policy development and implementation; legal and ethical aspects of research. Mailing Add: NIH-USPHS Bethesda MD 20014

CHALKLEY, G ROGER, b Sleaford, Eng, June 28, 39; m 62; c 3. BIOCHEMISTRY. Educ: Oxford Univ, BA, 61, MA & DPhil(chem), 64. Prof Exp: Res fel biol, Calif Inst Technol, 64-67; from asst prof to assoc prof biochem, 67-73, PROF BIOCHEM, ENDOCRINOL & GENETICS, UNIV IOWA, 73- Res: Structure and function of chromosomal nucleoproteins; mode of action of steroid hormones; interaction of carcinogens with nuclear material. Mailing Add: Dept of Biochem Univ of Iowa Iowa City IA 52240

CHALKLEY, ROGER, b Cincinnati, Ohio, June 21, 31. MATHEMATICS, ALGEBRA. Educ: Univ Cincinnati, ChE, 54; AM, 56, PhD(math), 58. Prof Exp: Instr math, Univ Cincinnati, 57-58; mathematician, Oak Ridge Nat Lab, 58-59; asst prof math, Knox Col, Ill, 60-62; asst prof, 62-64, ASSOC PROF MATH, UNIV CINCINNATI, 64- Mem: Am Math Soc; Math Asn Am. Res: Real analysis; algebraic differential equations; number theory. Mailing Add: Dept of Math Univ of Cincinnati Cincinnati OH 45221

CHALLICE, CYRIL EUGENE, b London, Eng, Jan 17, 26; m 51. PHYSICS, BIOPHYSICS. Educ: Univ London, BSc, 46, PhD(physics), 49, DSc(biophys), 75; Imp Col, Univ London, ARCS, 46, DIC, 49; Univ Alta, PEng, 74. Prof Exp: Biophysicist, Nat Inst Med Res, Eng, 49-52 & Wright-Fleming Inst Microbiol, 52-54; biophysicist & lectr physics, St Mary's Hosp Med Sch, London, 54-57; from asst prof to assoc prof physics, 57-63, head dept, 63-71, PROF PHYSICS, UNIV CALGARY, 63-, VDEAN, FAC ART & SCI, 73- Concurrent Pos: NIH fel, 56; NY State Dept Health fel, 58-59; chmn, Nat Comt Biophys, Can, 75- Mem: AAAS; Biophys Soc; Electron Micros Soc Am; NY Acad Sci; fel Brit Inst Physics. Res: Structure of biological systems using electron microscopy, electron diffraction and x-ray diffraction; structure and function of the heart. Mailing Add: Dept of Physics Univ of Calgary Calgary AB Can

CHALLIFOUR, JOHN LEE, b Bristol, Eng, June 13, 39; m 67. MATHEMATICAL PHYSICS. Educ: Univ Calif, Berkeley, BA, 60; Cambridge Univ, PhD(theoret physics), 63. Prof Exp: Instr math, Princeton Univ, 63-65, lectr, 65-66; asst prof physics, Brandeis Univ, 66-68; ASSOC PROF MATH & PHYSICS, IND UNIV, BLOOMINGTON, 68- Concurrent Pos: Vis prof, Univ Göttingen, 70-71 & Univ Bielefeld, 75-76. Mem: Am Math Soc; Am Phys Soc. Res: Axiomatic quantum field theory; theory of distributions; partial differential operators. Mailing Add: Dept of Phys Ind Univ Bloomington IN 47401

CHALLINOR, DAVID, b New York, NY, July 11, 20; m 52; c 4. FOREST ECOLOGY. Educ: Harvard Univ, BA, 43; Yale Univ, MF, 59, PhD(forest ecol), 66. Prof Exp: Forestry asst, Conn Agr Exp Sta, 59-60; dep dir, Peabody Mus Natural Hist, Yale Univ, 60-66; spec asst trop biol, Mus Nat Hist, Smithsonian Inst, 66-67, dep dir, 67-69, dir off int activ, 69-71, ASST SECY SCI, SMITHSONIAN INST, 71- Mem: Soc Am Foresters; Ecol Soc Am; Wildlife Soc. Res: Tree-soil interactions in temperate and tropical forest environments. Mailing Add: Off of Asst Secy Sci Smithsonian Inst Washington DC 20650

CHALLONER, DAVID REYNOLDS, b Appleton, Wis, Jan 31, 35; m 58; c 3. INTERNAL MEDICINE, ENDOCRINOLOGY. Educ: Lawrence Col, BS, 56; Harvard Univ, MD, 61. Prof Exp: Intern, Columbia-Presby Hosp, 61-62, asst resident, 62-63; res assoc, Lab Metab, Nat Heart Inst, 63-65; chief resident, King County Hosp, Univ Wash, 65-66, USPHS spec fel endocrinol, 66-67; from asst prof to prof med & biochem & asst chmn dept med, Sch Med, Ind Univ, Indianapolis, 67-75; PROF INTERNAL MED & DEAN, ST LOUIS UNIV SCH MED, 75- Mem: Am Fedn Clin Res (pres); Endocrine Soc; Am Physiol Soc; Am Diabetes Asn; Am Soc Clin Invest. Res: Control mechanisms in intermediary and oxidative metabolism. Mailing Add: St Louis Univ Sch of Med 221 N Grand Blvd St Louis MO 63103

CHALMERS, BRUCE, b London, Eng, Oct 15, 07; nat US; m 38; c 5. APPLIED PHYSICS Educ: Univ London, BSc, 29, PhD, 32, DSc, 41. Hon Degrees: AM, Harvard Univ, 53. Prof Exp: Lectr physics & math, Sir John Cass Inst, London, 32-38; physicist, Tin Res Inst, 38-41; sr sci officer, Brit Ministry of Supply, 41-44; head metall div, Royal Aircraft Estab, Farnborough, 44-46; Atomic Energy Res Estab, Harwell, 46-48; prof phys metall, Univ Toronto, 48-53; McKAY PROF METALL, HARVARD UNIV, 53- Concurrent Pos: Marberg lectr, 60; Australian Inst Metals lectr, 63; master, John Winthrop House, Harvard Univ, 64-74. Honors & Awards: Albert Sauveur Award, Am Soc Metals, 61; Clamer Medal, Franklin Inst, 64. Mem: Nat Acad Sci; fel Am Acad Arts & Sci; hon mem French Metall Soc, Indian Inst Metals & Japan Inst Metals. Res: Processes of solidification; plastic deformation; structure of grain boundaries; fracture of metals and other substances. Mailing Add: Div of Eng & Appl Physics Harvard Univ Cambridge MA 02138

CHALMERS, JOHN HARVEY, JR, b St Paul, Minn, Mar 5, 40. BIOCHEMICAL GENETICS, MUSICAL ACOUSTICS. Educ: Stanford Univ, AB, 62; Univ Calif, San Diego, PhD(biol), 68. Prof Exp: NIH-USPHS fel genetics, Univ Wash, 68-71; NIH-USPHS trainee, Univ Calif, Berkeley, 71-73; res fel appl microbiol, Merck Sharp & Dohme Res Labs, 73-75; ASST PROF BIOCHEM, BAYLOR COL MED, TEX MED CTR, 76- Concurrent Pos: Ed-publ, Xenharmonikon, 73- Mem: Sigma Xi; AAAS; Genetics Soc Am; Am Soc Microbiol. Res: Biochemical genetics; multienzyme complexes; organelle and mitochondrial genetics; industrial microbiology; fungal genetics; musical acoustics and experimental music. Mailing Add: Dept of Biochem Baylor Col of Med Tex Med Ctr Houston TX 77025

CHALMERS, JOSEPH STEPHEN, b Detroit, Mich, Feb 26, 38; m 62; c 3. THEORETICAL PHYSICS. Educ: Wayne State Univ, BS, 60, MA, 62, PhD(physics), 67. Prof Exp: ASSOC PROF PHYSICS, UNIV LOUISVILLE, 67- Mem: Am Phys Soc. Res: Theory of scattering of elementa ry particles by optical potentials; paramagnetic resonance of free radicals. Mailing Add: Dept of Physics Univ of Louisville Louisville KY 40208

CHALMERS, ROBERT ANTON, b Wildwood, NJ, Nov 4, 30; m 56; c 2. NUCLEAR PHYSICS. Educ: Princeton Univ, AB, 52; Northwestern Univ, PhD(physics), 63. Prof Exp: RES SCIENTIST, LOCKHEED MISSILES & SPACE CO, INC, 63- Mem: Am Phys Soc. Res: Low energy nuclear physics research with electrostatic accelerator and development of computer-oriented laboratory instrumentation. Mailing Add: Lockheed Missiles & Space Co Inc Palo Alto CA 94304

CHALMERS, ROBERT KENNY, b Kent, Ohio, Jan 26, 37; m 58; c 4. PHARMACY, PHARMACOLOGY. Educ: Ohio Northern Univ, BS, 58; Purdue Univ, MS, 60, PhD(pharmacol), 61. Prof Exp: From asst prof to prof pharmacol, 61-69, PROF CLIN PHARM, SCH PHARM & PHARMACAL SCI, PURDUE UNIV, WEST LAFAYETTE, 69- Concurrent Pos: USPHS grant, 63-64. Mem: Am Soc Pharmacol & Exp Therapeut; Am Pharmaceut Asn. Res: Pharmacy services to optimize drug therapy. Mailing Add: Dept of Clin Pharm Purdue Univ West Lafayette IN 47906

CHALMERS, THOMAS CLARK, b Forest Hills, NY, Dec 8, 17; m 42; c 4. INTERNAL MEDICINE, GASTROENTEROLOGY. Educ: Columbia Col, MD, 43; Am Bd Internal Med, dipl, 50. Prof Exp: Intern med, Presby Hosp, New York, 43-44; res fel, Malaria Res Unit, Goldwater Mem Hosp, NY Univ, 44-45; resident, 2nd & 4th Med Serv, Boston City Hosp, 45-47, out-patient physician, 47-48; dir hepatitis study, Comn Liver Dis, Armed Forces Epidemiol Bd, 51-53; chief med serv, Lemuel Shattuck Hosp, Boston, 55-68; asst chief med dir res & educ, Vet Admin, DC, 68-70; prof med, Sch Med, George Washington Univ, 70-73; PRES, MT SINAI MED CTR, PROF MED & ACTG CHMN DEPT MED EDUC, PRES & DEAN, MT SINAI SCH MED, 73- Concurrent Pos: Asst, Harvard Med Sch, 47-49, from instr to asst clin prof, 49-61, lectr, 61-; pvt pract internal med, Cambridge, Mass, 47-53; asst physician, Thorndike Mem Lab, Boston City Hosp, 47-53, assoc vis physician, 2nd & 4th Med Serv, 55-68; jr physician, Mt Auburn Hosp, Cambridge, 47-53, assoc vis physician, 55-68; lectr, Sch Med, Tufts Univ, 55-61, prof, 61-68; consult, Faulkner Hosp, Jamaica Plain, Mass, 55-68; assoc staff, New Eng Hosp Ctr, Boston, 55-68; mem training comt, Nat Heart Inst, 61-65, mem spec rev panel, Coronary Drug Proj, 66-69, mem diet-heart rev panel, 68, mem policy bd, Urokinase-Streptokinase Pulmonary Embolism Trial, 68-; mem cancer chemother collab prog rev comt, Nat Cancer Inst, 65-66; mem comt epidemiol & vet follow-up studies, Nat Acad Sci-Nat Res Coun, 65-69, 72-, mem ad hoc comt hepatitis-associated antigen tests, 70-71; mem subcomt on liver, Adv Comt Gen Med, Army Surg Gen Adv Comt, 65-72; mem sci adv comt, Pharmaceut Mfrs Asn Found, 66-68, mem adv comt to fac develop awards in clin pharmacol, 66-70; mem coop studies eval comt, Vet Admin, 70-74; assoc dir clin care & dir clin ctr, NIH, 70-73; chmn nat coop Crohn's dis study adv bd, Nat Inst Arthritis, Metab & Digestive Dis, 71-; mem hyper-immune gamma globulin trials policy bd, Nat Heart & Lung Inst, 72-75; chmn rev panel new drug regulation, Food & Drug Admin, 75- Mem: Inst of Med of Nat Acad Sci; Am Asn Study Liver Dis (pres, 59); Am Clin & Climat Asn; Am Col Physicians; Am Gastroenterol Asn (pres, 69). Res: Epidemiology of acute hepatitis; clinical epidemiology. Mailing Add: Mt Sinai Sch of Med New York NY 10029

CHALQUEST, RICHARD ROSS, b Denver, Colo, Nov 4, 29; m 54; c 7. VETERINARY MEDICINE, MICROBIOLOGY. Educ: Wash State Univ, BS, 51, DVM, 57; Cornell Univ, MS, 59, PhD(path), 60. Prof Exp: Asst prof vet microbiol, Wash State Univ, 60-62; res vet, Pfizer Inc, 62-63, mgr agr res & develop, 63-65, dir, 65-72; PROF AGR & DIR DIV AGR, ARIZ STATE UNIV, 72- Concurrent Pos: Mem gov bd, Agr Res Inst, Nat Res Coun, Nat Acad Sci, 67-70. Mem: Am Asn Avian Path; Am Vet Med Asn; Poultry Sci Asn; Am Asn Vet Parasitol. Res: Agricultural research. Mailing Add: Div of Agr Ariz State Univ Tempe AZ 85281

CHALUPA, LEO M, b Ger, Mar 28, 45; US citizen; m 66. NEUROSCIENCES, VISION. Educ: Queens Col, BA, 66; City Univ New York, PhD(neuropsychol), 70. Prof Exp: Res physiologist psychol, Brain Res Inst, Univ Calif, Los Angeles, 70-75; ASST PROF PSYCHOL, UNIV CALIF, DAVIS, 75- Concurrent Pos: Fel, Brain Res Inst, Univ Calif, Los Angeles, 70-72. Honors & Awards: US-USSR Scientist Exchange Award, Nat Acad Sci, 74. Mem: AAAS; Soc Neurosci; Am Psychol Asn. Res: Visual neurophysiology and neuropsychology, brain-behavior correlates of learning and attention. Mailing Add: Dept of Psychol Univ of Calif Davis CA 95616

CHALUPA, WILLIAM VICTOR, b New York, NY, Dec 11, 37; m 60; c 2. ANIMAL NUTRITION. Educ: Rutgers Univ, BS, 58, MS, 59, PhD(nutrit), 62. Prof Exp: Asst dairy sci, Rutgers Univ, 58-59, asst instr, 59-62, res fel nutrit, 62-63; from asst prof to assoc prof, Clemson Univ, 63-71; MGR RUMEN METABOLIC RES, SMITH KLINE ANIMAL HEALTH PROD DIV, 71- Concurrent Pos: Vis scientist, USDA, Md, 69-70; adj assoc prof, Sch Vet Med, Univ Pa, 75- Mem: AAAS; Am Soc Animal Sci; Am Dairy Sci Asn; Am Inst Nutrit. Res: Energy and protein utilization of foodstuffs by ruminant and monogastric animals; biochemistry of rumen metabolism. Mailing Add: Smith Kline Animal Health Prods Div 1600 Paoli Pike West Chester PA 19380

CHAMBERLAIN, CHARLES CALVIN, b Evart, Mich, Jan 23, 20; m 46; c 2. ANIMAL NUTRITION. Educ: Mich State Univ, BS, 41, MS, 48; Iowa State Univ, PhD, 59. Prof Exp: From instr to assoc prof animal nutrit, 49-71 & from asst animal husbandman to assoc animal husbandman, 52-71, PROF ANIMAL NUTRIT, UNIV TENN, 71- Mem: Am Soc Animal Sci. Res: Cattle and swine. Mailing Add: Dept of Animal Sci Univ of Tenn Knoxville TN 37916

CHAMBERLAIN, CHARLES CRAIG, b Milford, Utah, June 1, 33; m 59; c 4. MEDICAL PHYSICS, RADIOBIOLOGY. Educ: Univ Calif, Los Angeles, BA, 59, MA, 61, PhD(med physics), 67. Prof Exp: Instr, 67-70, ASST PROF RADIOL, STATE UNIV NY UPSTATE MED CTR, 70- Mem: Health Physics Soc; Am Asn Physicists in Med; Tissue Cult Asn. Res: Effects of heat and radiation on mammalian cells in culture; etiology of human leukemia and lymphoma. Mailing Add: Dept of Radiol State Univ NY Upstate Med Ctr Syracuse NY 13210

CHAMBERLAIN, CHARLES KENT, b Cedar City, Utah, Nov 19, 37; m 58; c 2. GEOLOGY, INVERTEBRATE PALEONTOLOGY. Educ: Brigham Young Univ, BS, 64, MS, 66; Univ Wis-Madison, PhD(geol), 70. Prof Exp: Explor geologist, Texaco Inc, 69-71, sr geologist, 71-72; asst prof geol, Ohio Univ, 72-76; ASST PROF GEOL, UNIV NEV, LAS VEGAS, 76- Mem: Geol Soc Am; Soc Econ Paleontologists & Mineralogists; Paleont Soc; Am Asn Petrol Geologists. Res: Trace fossils in environments of deposition and in determining soft-bodied parts of fossil communities. Mailing Add: Dept of Geol Univ of Nev Las Vegas NV 89109

CHAMBERLAIN, DAVID LEROY, JR, b Kansas City, Kans, Sept 2, 17; m 45; c 3. ORGANIC CHEMISTRY, PHYSICAL CHEMISTRY. Educ: Univ Kans, BS, 44, MS, 50; Univ Southern Calif, PhD(org chem), 53. Prof Exp: Lab asst, Univ Kans, 43-44; chem engr & tech asst to opers, Rayon Mfg, E I du Pont de Nemours & Co, 44-46; lab asst, Univ Kans, 46-47; res org chemist, Callery Chem Co, 53-56; sr org chemist, Stanford Res Inst, 56-72; consult, 72-73; RES ASSOC, NAT FOREST PROD ASN-NAT BUR OF STAND, 73- Mem: Am Chem Soc; fel Am Inst Chem; Combustion Inst; Sigma Xi. Res: Thermal degradation of organic materials; chemistry of organic-inorganic interfaces; chemistry of fire retardant materials; fire test methods for materials. Mailing Add: 12209 Pawnee Dr Gaithersburg MD 20760

CHAMBERLAIN, DONALD WILLIAM, b Green Bay, Wis, Nov 28, 05; m 45; c 3. PLANT PATHOLOGY. Educ: St Norbert Col, BA, 29; Univ Wis, MA, 32, PhD(plant path), 43. Prof Exp: Instr hort, Univ Wis, 43-45; asst agron, Agr Exp Sta, Univ Ky, 45-46; assoc pathologist, 46-56, PATHOLOGIST, CROPS RES DIV, AGR RES SERV, USDA, 56-; PROF PLANT PATH, UNIV ILL, URBANA-CHAMPAIGN, 75- Concurrent Pos: From asst prof to assoc prof plant path, Univ Ill, Urbana-Champaign, 56-75. Mem: AAAS; Am Phytopath Soc. Res: Bacterial and fungus diseases of soybean; disease resistance; occurrance of races of Pseudomonas glycinea in Illinois; resistance to Phytophthora rot in soybean as expressed in roots or stems. Mailing Add: 2022 Boudreau Dr Urbana IL 61801

CHAMBERLAIN, ERLING WILLIAM, b Oslo, Norway, Jan 5, 34; US citizen; m 57. MATHEMATICS. Educ: Columbia Univ, AB, 55, MA, 56, PhD(math), 61. Prof Exp: From asst prof to assoc prof, 62-70, PROF MATH, UNIV VT, 70- Concurrent Pos: NSF res grants, 63-70. Mem: Am Math Soc. Res: Asymptotic theory of ordinary differential equations in the complex domain, especially with regard to factorization of differential operators. Mailing Add: Dept of Math Univ of Vt Burlington VT 05401

CHAMBERLAIN, JACK G, b Detroit, Mich, May 15, 33; m 55; c 4. DEVELOPMENTAL ANATOMY. Educ: Occidental Col, BA, 55; San Diego State Col, MA, 57; Univ Calif, Berkeley, PhD(anat), 62. Prof Exp: Instr anat, Univ Calif, Berkeley, 62-63; instr, Univ Mich Sch Med, 63-64; asst prof, Univ Calif, San Francisco, 64-72; ASSOC PROF ANAT & CHMN DEPT, UNIV OF THE PAC SCH DENT, 72- Concurrent Pos: Fac grant, Univ Calif, 62-64; Rackham fac & local cancer res grants, Univ Mich, 63-64, fac grant, 64-65, USPHS grant, 65-70. Mem: Soc Develop Biol; Am Asn Anat. Res: Normal and abnormal developmental biology, especially pathogenesis of central nervous system abnormalities induced by antivitamins; scanning electron microscopy. Mailing Add: Dept of Anat Univ of the Pac San Francisco CA 94115

CHAMBERLAIN, JAMES LUTHER, b West Chester, Pa, May 16, 25; m 51; c 3. ZOOLOGY. Educ: Cornell Univ, BS, 48; Univ Mass, MS, 51; Univ Tenn, PhD(zool), 57. Prof Exp: Asst proj leader, WVa Conserv Comn, 48-49; field agt, US Fish & Wildlife Serv, Mass, 51; instr zool, State Teachers Col, NY, 52-53; res assoc, La State Univ, 55-57; assoc prof biol, Randolph-Macon Woman's Col, 57-69; chmn div sci & math, 69-74, PROF BIOL, UTICA COL, 74- Concurrent Pos: Res grant, US Forest Serv. Mem: Ecol Soc Am; Am Soc Mammal; Am Soc Ichthyologists & Herpetologists; Am Ornithologists Union; Wildlife Soc. Res: Marsh ecology; vertebrate ecology. Mailing Add: Div of Sci & Math Utica Col Utica NY 13502

CHAMBERLAIN, JOHN PAUL, b Chicago, Ill, Nov 4, 43; m 67; c 1. DEVELOPMENTAL BIOLOGY. Educ: Princeton Univ, AB, 65; Univ Miami, PhD(biol), 70. Prof Exp: Fel biochem & microbiol, Univ Wash, 70-73; ASST PROF BIOL, UNIV MICH, ANN ARBOR, 73- Mem: Soc Develop Biol. Res: Gene regulation and genetic control of embryonic development. Mailing Add: Div of Biol Sci Univ of Mich Ann Arbor MI 48109

CHAMBERLAIN, JOSEPH MILES, b Peoria, Ill, July 26, 23; m 45; c 3. ASTRONOMY. Educ: US Merchant Marine Acad, BS, 44; Bradley Univ, BA, 47; Columbia Univ, AM, 50, EdD, 62. Prof Exp: Instr nautical sci, US Merchant Marine Acad, 47-50, asst prof astron & meteorol, 50-52; asst astronomer, Am Mus-Hayden Planetarium, NY, 52-53, chmn & astronomer, 53-64; asst dir, Am Mus Natural Hist, 64-68; DIR, ADLER PLANETARIUM, 68-; PROF ASTRON, NORTHWESTERN UNIV, 68- Concurrent Pos: Instr, Naval Reserve Officer Sch, NY, 54-55; consult, Norman Porter & Assocs, NY, 54-; lectr, Nat Artists Corp, NY, 55-57; instr, Hunter Col, 64-68; prof lectr, Univ Chicago, 68- Mem: AAAS; Am Astron Soc; Am Asn Mus (vpres, 71-74, pres, 74-75); Am Polar Soc. Res: Determination of geodetic coordinates by astronomic methods; planetarium education and administration. Mailing Add: Adler Planetarium 1300 S Lake Shore Dr Chicago IL 60605

CHAMBERLAIN, JOSEPH WYAN, b Boonville, Mo, Aug 24, 28; m 49; c 3. AERONOMY, ASTRONOMY. Educ: Univ Mo, AB, 48, AM, 49; Univ Mich, MS, 51, PhD(astron), 52. Prof Exp: Proj scientist aurora & airglow, US Air Force Cambridge Res Ctr, 51-53; res assoc Yerkes Observ, Univ Chicago, 53-55, from asst prof to prof, 55-62; assoc dir planetary sci div, Kitt Peak Nat Observ, 62-70, astronr, 70-71; dir, Lunar Sci Inst, 71-73; PROF SPACE PHYSICS & ASTRON, RICE UNIV, 73- Concurrent Pos: Mem exec comt, Assembly Math Phys Sci, Nat Acad Sci-Nat Res Coun, 72-; ed, Rev Geophys & Space Physics, 74-, sect chmn, Nat Acad Sci. Mem: Nat Acad Sci; AAAS; Am Astron Soc; Am Phys Soc; Am Geophys Union. Res: Planetary atmospheres; aurora and airglow; aeronomy of the stratosphere. Mailing Add: Dept Space Physics & Astron Rice Univ Houston TX 77001

CHAMBERLAIN, MALCOLM, b Binghamton, NY, June 7, 25; m 50; c 2. ORGANIC CHEMISTRY. Educ: Bowdoln Col, BS, 47; Mass Inst Technol, PhD(org chem), 51. Prof Exp: Res chemist, 51-54, proj leader, 54, group leader, 55, asst lab dir cellulose & plastics lab, 56-60, staff asst exec res, 61-67, admin asst human health res, 67-69, ASST MGR CORP RES & DEVELOP INDUST RELS, DOW CHEM CO, 69-

Mem: AAAS; Am Chem Soc. Res: Synthetic organic chemistry; autoxidation of polymers; plastics; carbohydrate derivatives. Mailing Add: 1414 Crescent Dr Midland MI 48640

CHAMBERLAIN, NUGENT FRANCIS, b Henderson, Tex, Mar 10, 16; m 43; c 3. SPECTROSCOPY, ELECTRON MICROSCOPY. Educ: Agr & Mech Col Tex, ChE, 38. Prof Exp: Tech trainee, Humble Oil & Ref Co, 38, from jr chemist to chemist, 39-41, from res chemist to sr res chemist, 41-59, sr res chem engr, 59-61, res specialist, 61-63, res assoc, 63-66, res assoc Esso Res & Eng Co, 66-69, sr res assoc, 69-74, SR RES ASSOC, BAYTOWN RES & DEVELOP DIV, EXXON RES & ENG CO, 74- Honors & Awards: Southeastern Tex Sect Award & Southwest Regional Award, Am Chem Soc, 69. Mem: Am Chem Soc; fel Am Inst Chemists; Electron Micros Soc Am. Res: Nuclear magnetic resonance spectroscopy; electron microscopy; neutron activation analysis. Mailing Add: Exxon Res & Eng Co Baytown Res & Dev Div Box 4255 Baytown TX 77520

CHAMBERLAIN, OWEN, b San Francisco, Calif, July 10, 20; m 43; c 4. EXPERIMENTAL HIGH ENERGY PHYSICS. Educ: Dartmouth Col, AB, 41, Unviv Chicago, PhD(physics), 49. Prof Exp: From instr to assoc prof, 48-58, PROF PHYSICS, UNIV CALIF, BERKELEY, 58- Concurrent Pos: Civilian physicist, Manhattan Dist, Berkeley & Los Alamos, 42-46; Guggenheim fel, 57-58; Loeb lectr, Harvard Univ, 59. Honors & Awards: Nobel Prize, 59. Mem: Nat Acad Sci; fel Am Phys Soc; fel AAAS. Res: Fission; Alphaparticle decay; neutron diffraction in liquids; high energy nucleon scattering; antinucleons. Mailing Add: Dept of Physics Univ Oof Calif Berkeley CA 94720

CHAMBERLAIN, PHYLLIS IONE, b Belfast, NY, Oct 22, 38. INORGANIC CHEMISTRY. Educ: Houghton Col, BS, 60; State Univ NY Buffalo, PhD(chem kinetics), 68. Prof Exp: Teacher pub sch, NY, 60-62; ASST PROF CHEM, ROBERTS WESLEYAN COL, NY, 67- Mem: Am Chem Soc; Am Sci Affil. Res: Kinetics and mechanisms of the reactions of transition metal complexes of organic ligands. Mailing Add: Roberts Wesleyan Col 2301 Westside Dr Rochester NY 14624

CHAMBERLAIN, RICHARD HALL, radiology, deceased

CHAMBERLAIN, ROBERT ENGLISH, b Utica, NY, Jan 15, 21; m 48; c 2. MICROBIOLOGY. Educ: Rensselaer Polytech Inst, BS, 44, MS, 47; Univ Mich, PhD(bact), 52; Am Bd Med Microbiol, dipl. Prof Exp: Asst, Behr Manning Corp, NY, 44; lab asst zool, Rensselaer Polytech Inst, 46-47; instr, Univ Mich, 51-52; head microbiol sect, Smith Kline & French Labs, 52-61; sr res microbiologist, Nat Drug Co, 61-65; CHIEF MED MICROBIOL SECT, NORWICH PHARMACAL CO, 65- Mem: Am Soc Microbiol. Res: Hypersensitivity; antimicrobial agents. Mailing Add: Norwich Pharmacal Co Norwich NY 13815

CHAMBERLAIN, ROY WILLIAM, b Stanton, Calif, July 24, 16; m 44; c 1. MEDICAL ENTOMOLOGY, VIROLOGY. Educ: Mont State Col, BS, 42; Johns Hopkins Univ, ScD(parasitol), 49. Prof Exp: Chief arbovirus vector lab, 49-67, chief arbovirus infections unit, 67-68, DEP CHIEF VIROL DIV, CTR DIS CONTROL, USPHS, 68- Mem: Am Soc Trop Med & Hyg; Sigma Xi; Am Mosquito Control Asn. Res: Arthropod transmission of virus diseases of man and animals; behavior of arboviruses in insects and vertebrates. Mailing Add: Virol Div Ctr for Dis Control Atlanta GA 30333

CHAMBERLAIN, THEODORE KLOCK, b Detroit, Mich, July 18, 30. GEOLOGY, OCEANOGRAPHY. Educ: Univ NMex, BS, 52; Scripps Inst Oceanog, Univ Calif, MS, 53, PhD(oceanog), 60. Prof Exp: Res marine geologist, Scripps Inst Calif, 56-60; oceanogr, Tokyo Univ Fisheries, Tokyo, 60-62; assoc prof & asst chmn dept oceanog, Univ Hawaii, 62-67; sr oceanogr & mgr sci div, Ocean Sci & Eng Inc, 67-71; dir, Chesapeake Res Consortium, Johns Hopkins Univ, 72-75; PROF & HEAD EARTH RESOURCES DEPT, COLO STATE UNIV, 75- Concurrent Pos: Nat Acad Sci-Nat Res Coun fel, 60-62. Mem: Am Soc Limnol & Oceanog; AAAS; Royal Siam Soc. Res: Littoral processes, physical limnology, marine placers, exploration geology; earth resources development. Mailing Add: Dept of Earth Resources Colo State Univ Ft Collins CO 80523

CHAMBERLAIN, VIRGIL RALPH, b Calgary, Alta, Feb 5, 17; US citizen; m 49; c 4. GEOLOGY, ENGINEERING. Educ: Mont Sch Mines, BS, 38; Stanford Univ, MS, 49. Prof Exp: Jr geologist, US Geol Surv, 38-39; geologist, Anaconda Copper Co, 39-42; party chief, Union Mines Develop, US Army Engrs, 42-45; geologist, Irvine E Stewart & Assocs, 46-47; geologist & stratigrapher, Gen Petrol Corp, 49-50; CONSULT GEOLOGIST & ENGR, 50- Mem: AAAS; Am Asn Petrol Geol; Geol Soc Am; NY Acad Sci; Int Acad Law & Sci. Res: Subsurface stratigraphy; mineral exploration; oil production; evaluations. Mailing Add: PO Box 2341 Great Falls MT 59403

CHAMBERLAIN, WILLIAM MAYNARD, b Montreal, Que, Apr 7, 38; m 61; c 2. AQUATIC BIOLOGY. Educ: Univ Toronto, BS, 61, PhD(zool), 68. Prof Exp: Res assoc, Limnol Res Ctr, Univ Minn, Minneapolis, 67-69; ASST PROF LIFE SCI, IND STATE UNIV, 69- Concurrent Pos: Partic, Gordon Res Conf, 66 & 70 & Int Symp Eutrophication, 67. Mem: Am Soc Limnol & Oceanog; Ecol Soc Am. Res: Nutrient circulation studies on phosphorus in aquatic ecosystems; assay procedures for available phosphorus; temperature effects on cladoceran populations; nutrient cycles. Mailing Add: Dept of Life Sci Ind State Univ Terre Haute IN 47809

CHAMBERLAND, BERTRAND LEO, b Manchester, NH, Mar 17, 34; m 65; c 3. INORGANIC CHEMISTRY. Educ: St Anselms Col, AB, 55; Col of the Holy Cross, MS, 56; Univ Pa, PhD(inorg chem), 60. Prof Exp: Res chemist, Nat Carbon Co, Ohio, 56-57 & E I du Pont de Nemours & Co, Inc, 60-69; ASSOC PROF CHEM, UNIV CONN, 69- Mem: Am Chem Soc; The Chem Soc. Res: Inorganic synthesis of molecular and solid state compounds; preparation, characterization and crystal growth of solid state materials; high pressure synthesis and reactions. Mailing Add: Dept of Chem Univ of Conn Storrs CT 06268

CHAMBERLAND, MAURICE R, organic chemistry, see 12th edition

CHAMBERLIN, EARL MARTIN, b Cochranville, Pa, Dec 4, 14; m 47; c 7. ORGANIC CHEMISTRY. Educ: Philadelphia Col Pharm & Sci, ScB, 36; Boston Univ, AM, 37; Harvard Univ, AM, 44, PhD(org chem), 46. Prof Exp: Asst org chem, Merck & Co, 38-40, admin asst to dir res, 40-41, res chemist, 46-47, staff asst to vpres & sci dir, 48-49, res chemist, 49-59, mgr develop res, 59-68, DIR SYNTHETIC PREP LAB, PROCESS RES, MERCK & CO, 69- Mem: Am Chem Soc; fel Am Inst Chemists; The Chem Soc. Res: Synthetic organic chemistry; alkaloids; fatty acids; quinones; organic therapeutic agents; steroidal hormones. Mailing Add: 2028 Hilltop Rd Westfield NJ 07090

CHAMBERLIN, HARRIE ROGERS, b Cambridge, Mass, June 13, 20; m; c 4. PEDIATRICS. Educ: Harvard Univ, AB, 42, MD, 45. Prof Exp: Instr pediat, Sch Med, Yale Univ, 51-52; from instr to assoc prof, 53-70, PROF PEDIAT & DIR DIV FOR DIS OF DEVELOP & LEARNING, CHILD DEVELOP INST, SCH MED, UNIV NC, CHAPEL HILL, 70- Concurrent Pos: Mem, White House Ad Hoc Adv Comt Ment Retardation, 63-65; consult, Div Hosp & Med Facil, USPHS, 65-68. Mem: AMA; Am Acad Pediat; assoc Am Acad Neurol; Am Pediat Soc; Am Acad Cerebral Palsy. Res: Mental retardation; developmental neurology in the infant and child. Mailing Add: Dept of Pediat Univ of NC Chapel Hill NC 27514

CHAMBERLIN, HENRY HOWARD, b Palmerton, Pa, Mar 8, 13; m 48; c 1. FORESTRY. Educ: Pa State Col, BSF, 39; Yale Univ, MF, 40. Prof Exp: Forester, Pa Turnpike Comn, 41; instr forestry, Univ La, 41-45; head dept, 45-74, PROF FORESTRY, UNIV ARK, MONTICELLO, 45- Mem: Soc Am Foresters. Res: Private forest land ownership. Mailing Add: Dept of Forestry Univ of Ark Monticello AR 71655

CHAMBERLIN, HOWARD ALLEN, b Dayton, Ohio, Nov 17, 21; m 46; c 3. POLYMER CHEMISTRY, TEXTILE CHEMISTRY. Educ: Ohio State Univ, BSc, 44; NC State Univ, MSc, 68. Prof Exp: Chemist, Monsanto Cent Res Labs, Ohio, 44-51; chemist, Res & Develop Labs, Chemstrand Corp, Ala, 52-55, develop group leader, Acrylic Fiber Plant, 55-59, res chemist, Chemstrand Res Ctr, Monsanto Co, 60-67; RES CHEMIST, HOOKER RES CTR, HOOKER CHEM CORP, 67- Mem: Am Chem Soc. Res: Polymer process development and improvement; synthetic fiber technical assistance and development in acrylic fiber plant; application of polymerization technology to textile fabric finishes; research approaches. Mailing Add: Hooker Chem Corp Res Ctr Box 8 MPO Niagara Falls NY 14302

CHAMBERLIN, JAMES WESLEY, b Kansas City, Mo, July 4, 33; m 56; c 4. ORGANIC CHEMISTRY. Educ: Univ Kansas City, BS, 55; Iowa State Univ, MS, 58; Stanford Univ, PhD, 63. Prof Exp: Res assoc org chem, Iowa State Univ, 58-59; org chemist, Stanford Res Inst, 59-61; sr org chemist, Chem Res Div, 63-70, sr org chemist, Biol & Microbiol Prod Res Div, 70-71, RES SCIENTIST, BIOL & MICROBIOL PROD RES DIV, ELI LILLY & CO, 71- Mem: Am Chem Soc. Res: Chemistry of organic natural products; application of organic chemistry to the study of biological processes; medicinal chemistry. Mailing Add: Chem Res Div Eli Lilly & Co Res Labs Indianapolis IN 46206

CHAMBERLIN, JOHN MACMULLEN, b Old Hickory, Tenn, Oct 12, 35; m 65; c 1. INORGANIC CHEMISTRY, PHYSICAL CHEMISTRY. Educ: Western Ky Univ, BS, 57; Duke Univ, MA, 61, PhD(inorg chem), 64. Prof Exp: ASSOC PROF INORG & PHYS CHEM, WESTERN KY UNIV, 64-, COORDR MED TECHNOL, 65-, ASST TO DEAN, COL APPL ARTS & HEALTH, 72- Res: Potentiometric and polarographic determination of stability constants of complex ions in molten salts; corrosion of metals in molten salts. Mailing Add: Dept of Chem Western Ky Univ Bowling Green KY 42101

CHAMBERLIN, MICHAEL JOHN, biochemistry, see 12th edition

CHAMBERLIN, RICHARD ELIOT, b Cambridge, Mass, Mar 20, 23; m 53; c 4. MATHEMATICS. Educ: Univ Utah, AB, 43; Harvard Univ, AM, 47, PhD(math), 50. Prof Exp: Asst prof math, 49-51, from asst prof to assoc prof, 52-70, PROF MATH, UNIV UTAH, 70- Mem: Am Math Soc. Res: Algebraic topology; critical points of polynomials. Mailing Add: Dept of Math Univ of Utah Salt Lake City UT 84112

CHAMBERLIN, THOMAS LELAND, b Hamilton, Ohio, Oct 12, 46. PETROLEUM GEOLOGY. Educ: Mich State Univ, East Lansing, BS, 68; Univ Ill, Urbana, MS, 71, PhD(geol), 75. Prof Exp: GEOLOGIST PETROL GEOL, TEXACO, INC, 75- Concurrent Pos: Res asst, Ill State Geol Surv, 69-74. Mem: Sigma Xi; Am Asn Petrol Geologists; Soc Econ Paleontologists & Mineralogists. Res: The regional geologic analysis of selected areas to be evaluated in terms of hydrocarbon accumulation and production. Mailing Add: Texaco Inc 3350 Wilshire Blvd PO Box 3756 Los Angeles CA 90051

CHAMBERLIN, THOMAS WILSON, SR, b Gays, Ill, May 23, 14; m 36; c 2. GEOGRAPHY OF THE SOVIET UNION. Educ: Eastern Ill State Teachers Col, BEd, 36; Clark Univ, MA, 37, PhD(geog), 46. Hon Degrees: PedD, Eastern Ill Univ, 56. Prof Exp: Asst prof geog, ETenn State Teachers Col, 38-42; asst prof, Northern Ill State Teachers Col, 46-47; from asst prof to assoc prof, 47-53, head dept, 47-54, prof, 53-54, acad dean, 54-70, spec asst to provost, 70-71, PROF GEOG, UNIV MINN, DULUTH, 70- Mem: Asn Am Geog; Am Geog Soc; Nat Coun Geog Educ. Res: Regional geography of western Europe and the Soviet Union. Mailing Add: Dept of Geog Univ of Minn Duluth MN 55812

CHAMBERS, ALFRED HAYES, b Reading, Pa, Nov 15, 14; m 45. PHYSIOLOGY. Educ: Swarthmore Col, AB, 36; Univ Pa, PhD(physiol), 42. Prof Exp: Asst instr physiol, Univ Pa, 37-42, instr, 42-45, assoc, 45-47, asst prof, 47-48; from asst prof to assoc prof, 48-70, PROF PHYSIOL & BIOPHYSICS, UNIV VT, 70- Concurrent Pos: NIH spec fel, 61, 62. Mem: Am Physiol Soc. Res: Metabolism; respiration; hearing. Mailing Add: Dept of Physiol Univ of Vt Burlington VT 05401

CHAMBERS, BARBARA MAE FROMM, b Syracuse, NY, Nov 23, 40; m 62; c 4. MATHEMATICS. Educ: Univ Ala, BS, 62, MA, 64, PhD(math), 69. Prof Exp: Asst math, Univ Ala, 62-69; consult, Washington, DC, 69-70; lectr, Univ Va, Fairfax, 70-71; ASST PROF MATH, GEORGE MASON UNIV, 71- Concurrent Pos: NASA fel, 63; adv, Northern Va Sci Fair, 72- Mem: Math Asn Am. Res: Analytic function theory. Mailing Add: Dept of Math George Mason Univ Fairfax VA 22030

CHAMBERS, CECIL WILLIAM, b Boardman, Ohio, June 9, 09; m 37; c 3. BACTERIOLOGY. Educ: Ohio State Univ, AB, 32. Prof Exp: Chemist, Youngstown Sheet & Tube Co, 36-37; gen bacteriologist, USPHS, 39-40; serologist, Ohio State Dept Health, 40-42; res bacteriologist, USPHS, 42-50, in-charge bact studies germicides, 50-65, in-charge waste treatability & effluent disinfection activity, Fed Water Pollution Control Admin, 66-72 & US Environ Protection Agency, 73-75; CONSULT, 75- Honors & Awards: Bronze Medal, US Environ Protection Agency, 75. Mem: Am Soc Microbiol; Water Pollution Control Fedn. Res: Effect of environmental factors on the bactericidal efficiency of free chlorine, ozone, iodine, quaternary ammonium compounds, and silver; development of germicide test methods; adaption of bacteria to degrade aromatic compounds; applied research in disinfection of wastewater. Mailing Add: 1576 Collinsdale St Cincinnati OH 45230

CHAMBERS, CHARLES MCKAY, JR, b Hampton, Va, June 22, 41; m 62; c 4. ACADEMIC ADMINISTRATION, MATHEMATICS. Educ: Univ Ala, BS, 62, MS, 63, PhD(physics), 64; George Washington Univ, JD, 75. Prof Exp: Aerospace engr, Marshall Space Flight Ctr, NASA, 62-63; res assoc physics, Univ Ala, 63-64; NSF res fel, Harvard Univ, 64-65; from asst prof to assoc prof math, Univ Ala, 65-69; charter officer & dir, Univ Assocs, Inc, 69-72; ASSOC DEAN, GEORGE WASHINGTON UNIV, 72- Concurrent Pos: Dir, NSF grant Ala math talent search; consult, NASA, US Air Force, Salk Inst, US Cong, US Off Educ, NSF, Am Asn Higher Educ & Coun Grad Schs US. Mem: AAAS; Am Math Soc; Am Asn Physics Teachers; NY Acad Sci; Am Asn Univ Adminrs. Res: Development and administration of interdisciplinary programs; abstract harmonic analysis; group representation theory. Mailing Add: PO Box 138 Benjamin Franklin Sta Washington DC 20044

CHAMBERS, DAVID SMITH, b Clarksville, Tex. APPLIED STATISTICS. Educ: Univ Tex, AB, 39, MBA, 47. Prof Exp: Spec instr math, Univ Tex, 38-41; asst state supvr div wage-hour, US Dept Labor, Miss, 41-42; regional examr, Off Price Admin, Tex, 42-43; instr aeronaut eng, Univ Tex, 43-44, instr appl math & astron, 44-46, instr bus statist, 46-47; from asst prof to assoc prof, 47-58, PROF STATIST, UNIV TENN, KNOXVILLE, 58- Concurrent Pos: Mem bd dirs, Engr Joint Coun, 72-74. Honors & Awards: Eugene Grant Award, Am Soc Qual Control. 70. Mem: AAAS; Am Soc Qual Control (pres, 71-72); Am Statist Asn; Am Soc Test & Mat. Res: Quality control. Mailing Add: Dept of Statist Univ of Tenn Knoxville TN 37916

CHAMBERS, DERRELL LYNN, b Los Angeles, Calif, Feb 3, 34; m 54; c 3. ENTOMOLOGY. Educ: Whittier Col, BA, 55; Ohio State Univ, MS, 57; Ore State Univ, PhD, 65. Prof Exp: Biol aide entom res div, Whittier Lab, Agr Res Serv, USDA, 55; forestry aide, US Forest Serv, Ohio, 56-57; asst zool & entom, Ohio State Univ, 56-57; entomologist, Entom Res Div, Mexican Fruit Flies Invest Lab, Agr Res Ser, USDA, Mexico City, 57-59, entomologist, Pioneering Res Lab Insect Physiol, Md, 59-61; asst, Sci Res Inst, Ore State Univ, 61-64; sr entomologist, Entom Res Div, Arid Areas Citrus Insects Invest Lab, Agr Res Serv, Calif, 65-67, invests leader, Hawaiian Fruit Flies Invest Lab, Honolulu, 68-72, LAB DIR INSECT ATTRACTANTS & BIOL RES LAB, AGR RES SERV, USDA, 72- Concurrent Pos: Consult, various govt agencies and progs. Mem: Am Inst Biol Sci; Entom Soc Am. Res: Insect physiology and behavior. Mailing Add: Insect Attract & Biol Res Lab PO Box 14565 Gainesville FL 32604

CHAMBERS, DOYLE, b Line, Ark, Mar 24, 18; m 43; c 3. ANIMAL GENETICS, ANIMAL BREEDING. Educ: La State Univ, BS, 40, MS, 47; Oklahoma State Univ, PhD(animal breeding), 50. Prof Exp: Asst animal sci, La State Univ, 40-42, instr, 45-47; from assoc prof to prof, Okla State Univ, 50-62, assoc dir, Agr Exp Sta, 62-64; DIR AGR EXP STA, LA STATE UNIV, BATON ROUGE, 64- Concurrent Pos: Don M Tyler distinguished prof, Okla State Univ, 61. Mem: Fel AAAS; Am Soc Animal Sci. Res: Inheritance of quantitative traits of economic importance in beef cattle and swine, including growth and carcass traits, maternal traits and efficiency of feed use; dwarfism and cancer eye studies. Mailing Add: Agr Exp Sta La State Univ Baton Rouge LA 70803

CHAMBERS, EDWARD LUCAS, b Manhattan, NY, Jan 27, 17; m 54; c 1. CELL PHYSIOLOGY. Educ: Princeton Univ, BA, 38; NY Univ, MD, 43. Prof Exp: Intern, 2nd Div, Bellevue Hosp, 43-44, asst resident, 4th Div, 44; asst prof anat, Sch Med, Johns Hopkins Univ, 50-52; assoc prof, Sch Med, Univ Ore, 53-54; assoc prof physiol, 54-63, chmn grad prog cellular & molecular biol, 67-71, prof physiol & biochem, 63-73, PROF PHYSIOL & BIOPHYS, SCH MED, UNIV MIAMI, 73- Concurrent Pos: Mem corp, Marine Biol Lab, Woods Hole, Mass. Mem: Soc Gen Physiol; Am Asn Anat; Am Physiol Soc; Am Soc Cell Biol. Res: Ion exchanges between cell and environment; cell activation; cell metabolism; micromanipulation. Mailing Add: Dept of Physiol & Biophys Univ of Miami Sch of Med PO Box 520875 Miami FL 33152

CHAMBERS, HOWARD WAYNE, b Buda, Tex, Dec 27, 39. TOXICOLOGY. Educ: Tex A&M Univ, BS, 61, MS, 63; Univ Calif, Berkeley, PhD(entom), 66. Prof Exp: Res entomologist, Univ Calif, 66-68; asst prof entom, 68-74, ASSOC PROF ENTOM, MISS STATE UNIV, 74- Mem: Am Chem Soc; Entom Soc Am. Res: Insecticide chemistry; mechanism of action and metabolism of insecticides; insecticide synergists; degradation of pesticides by soil fungi; resistance to insecticides. Mailing Add: PO Drawer EM Miss State Univ State College MS 39762

CHAMBERS, JACK VIRGIL, b Waco, Tex, Dec 18, 17; m 42; c 5. PHYSICAL GEOGRAPHY. Educ: Univ Chicago, BS, 42, MS, 47; Bristol Univ, PhD, 70. Prof Exp: Geogr, US Bur Census, 48-50; geogr, US Army Natick Labs, 50-71; ASSOC PROF GEOG, INDIANA UNIV PA, 71- Concurrent Pos: Pa Coun Geog Educ, 72. Mem: Asn Am Geogr; fel Am Geog Soc. Res: Tropical microclimatology; tropical topoclimatology; geography of color; geographical characteristics and the frequencies and patterns of mesoprecipitation in the Gatun Drainage Basin of Panama. Mailing Add: Dept of Geog Indiana Univ of Pa Indiana PA 15701

CHAMBERS, JAMES Q, b Kansas City, Mo, Jan 14, 38; m 64. ANALYTICAL CHEMISTRY. Educ: Princeton Univ, AB, 59; Univ Kans, PhD(chem), 64. Prof Exp: Asst prof chem, Univ Colo, 64-69; asst prof, 69-71, ASSOC PROF CHEM, UNIV TENN, 71- Res: Mechanisms and kinetics of electrode reactions; electrosynthetic methods; electron paramagnetic resonance; ultraviolet and visible spectroscopy of radical ions. Mailing Add: Dept of Chem Univ of Tenn Knoxville TN 37916

CHAMBERS, JAMES RICHARD, b Birmingham, Ala, Aug 20, 14; m 39; c 1. ORGANIC CHEMISTRY. Educ: Columbia Union Col, BA, 39; Western Reserve Univ, MS, 49; Tex A&M Univ, PhD(org chem), 58. Prof Exp: Asst chem, Columbia Union Col, 39-41; prin, Nashville Jr Acad, 41-42; chief chemist, Pennzoil Co, 42-46; head chem dept, Atlantic Union Col, 46-54; instr chem, Southwestern Jr Col, 54-56; asst, Agr & Mech Col Tex, 56-58; head sci dept, Southwestern Jr Col, 58-60; assoc prof, 60-61, PROF CHEM, WALLA WALLA COL, 61- Mem: Am Chem Soc. Res: Organic synthesis; organophosphorus chemistry; biochemistry. Mailing Add: Dept of Chem Walla Walla Col College Place WA 99324

CHAMBERS, JAMES VERNON, b Pekin, Ill, Mar 12, 35; m 57; c 1. FOOD SCIENCE. Educ: Ohio State Univ, BSc, 61, MSc, 66, PhD(food sci), 72. Prof Exp: Head microbiologist div qual control, Ross Labs Div, Abbott Labs, 61-68; lab dir, Div Food, Dairies & Drugs, Ohio Dept Agr, 69-71; asst prof food sci, Univ Wis, River Falls, 72-74; ASST PROF & EXTEN SPECIALIST DAIRY MFG, PURDUE UNIV, 74- Concurrent Pos: Consult microbiol food waste treatment & mgt pract, Borden Dairy, 72-75, Assoc Milk Producers, 73-74, Mid Am Dairyman Inc, 73-, Wise Potato Chip Co, 75-, Level Valley Dairy, 75- & Nafziger Ice Cream Co, 75- Mem: Inst Food Technologists; Am Soc Microbiol: Water Pollution Control Fedn; Am Dairy Sci Asn; Int Asn Milk Food & Environ Sanitarians Inc. Res: Translocation of Xanthine oxidase from milk into the blood and possible effect on cholesterol accumulation; up grading whey for human consumption and improving the economic value of whey. Mailing Add: Smith Hall Purdue Univ West Lafayette IN 47907

CHAMBERS, JOHN EDWARD, b Sand Springs, Okla, Nov 4, 32; m; c 2. PHYCOLOGY, ELECTRON MICROSCOPY. Educ: Univ Tulsa, BS, 58; Univ Tex, MA, 62; Univ Kans, PhD(bot), 66. Prof Exp: Palynologist, Cities Serv Res & Develop Co, 56-58; instr biol, Southwest Tex State Col, 61-63; asst prof, Univ Southwestern La, 66-69; assoc prof, ETex State Univ, 69-72; PROF BIOL, COL ENVIRON & APPL SCI, GOVERNORS STATE UNIV, 72- Mem: Am Phycol Soc; Int Phycol Soc. Res: Ultrastructure of red algae; algal ecology; ultrastructure of latex cells; aquatic biology. Mailing Add: Col of Environ & Appl Sci Governors State Univ Park Forest South IL 60466

CHAMBERS, JOHN MCKINLEY, b Toronto, Ont, Apr 28, 41; m 71. STATISTICS, COMPUTER SCIENCE. Educ: Univ Toronto, BSc, 63; Harvard Univ, AM, 65, PhD(statist), 66. Prof Exp: MEM TECH STAFF STATIST, BELL TEL LABS, 66- Concurrent Pos: Vis lectr math, Imp Col, Univ London, 66-67; vis lectr statist, Harvard Univ, 69 & Princeton Univ, 71; assoc ed, J Am Statist Asn, 71- Mem: Am Statist Asn; fel Royal Statist Soc; Asn Comput Mach; Brit Comput Soc. Res: Numerical and non-numerical methods of scientific computation, particularly as applied to statistical data analysis. Mailing Add: Bell Tel Labs Murray Hill NJ 07974

CHAMBERS, JOHN WILLIAM, b Richmond, Ky, Nov 7, 29; m 65; c 2. ENDOCRINOLOGY, PHARMACOLOGY. Educ: Eastern Ky State Col, BS, 58; Vanderbilt Univ, PhD(pharmacol), 65. Prof Exp: From res asst to instr, 64-68, ASST PROF PHARMACOL, MED COL VA, 70- Concurrent Pos: NIH-USPHS fel, 65-66. Mem: AAAS; Am Soc Pharmacol & Exp Therapeut; assoc Am Chem Soc; NY Acad. Sci. Res: Effects of hormones and drugs on amino acid transport and metabolism, on membrane function and on enzyme activity. Mailing Add: Dept of Pharmacol Med Col of Va Richmond VA 23298

CHAMBERS, KENTON LEE, b Los Angeles, Calif, Sept 27, 29; m 58; c 2. BOTANY. Educ: Whittier Col, AB, 50; Stanford Univ, PhD(biol), 56. Prof Exp: Actg instr biol sci, Stanford Univ, 54-55; from instr to asst prof bot, Yale Univ, 56-60; assoc prof, 60-65, PROF BOT, ORE STATE UNIV, 65-, CUR HERBARIUM, 60- Concurrent Pos: Prog dir syst biol, NSF, 67-68. Mem: AAAS; Am Soc Plant Taxonomists; Soc Study Evolution; Bot Soc Am; Asn Trop Biol. Res: Taxonomy and biosystematics of angiosperms, especially Compositae; flora of Oregon. Mailing Add: Dept of Bot Ore State Univ Corvallis OR 97331

CHAMBERS, LEE MASON, b St Marys, WVa, Mar 17, 36; m 60; c 1. ANALYTICAL CHEMISTRY, ELECTROCHEMISTRY. Educ: Marshall Univ, BS, 58; Univ Ill, MS, 60, PhD(anal chem), 63. Prof Exp: Staff chemist, Ivorydale Tech Ctr, 62-66, group leader res, 66-70, MEM STAFF ANAL SERV ADMIN, IVORYDALE TECH CTR, PROCTER & GAMBLE CO, 70- Mem: Am Chem Soc; Soc Appl Spectros; Am Oil Chemist's Soc. Res: Development of chemical and instrumental methods for analysis. Mailing Add: Ivorydale Tech Ctr Procter & Gamble Co Cincinnati OH 45217

CHAMBERS, LESLIE ADDISON, b Mystic, Iowa, Oct 11, 05; m 30; c 4. BIOPHYSICS, ZOOLOGY. Educ: Tex Christian Univ, BS, 27, MS, 28; Princeton Univ, PhD(biol), 30. Prof Exp: Asst prof biol, Tex Christian Univ, 30-32; Johnson Found fel med physics, Univ Pa, & instr pediat sch med, 32-36, assoc med physics & lectr biophys, 36-42, assoc pediat, 36-46, from asst prof to assoc prof biophys, 42-46; chief phys defense div, Biol Labs, Chem Corps, US Army, 46-51; dir res, R A Taft Sanit Eng Ctr, USPHS, 51-56, dir res, Los Angeles Air Pollution Control Dist, 56-60; prof biol & dir Hancock Found, Univ Southern Calif, 60-68; PROF ENVIRON HEALTH, UNIV TEX SCH PUB HEALTH HOUSTON, 68-, DIR INST ENVIRON HEALTH, 69-71. Concurrent Pos: Consult, USPHS, 56-, NSF, 63- & Pan Am Health Orgn, 75-; adj prof, Environ Sci & Eng Dept, Rice Univ, 74- Mem: AAAS; Am Soc Microbiol; Am Physiol Soc; Am Chem Soc; fel NY Acad Sci. Res: Biological and chemical properties of viruses; physical fractionation of rickettsiae; physical properties of bacterial cells; environmental health; air pollution; aerobiology; marine biology. Mailing Add: Univ of Tex Sch of Pub Health PO Box 20168 Houston TX 77025

CHAMBERS, RALPH ARNOLD, b Harlan, Ky, Sept 5, 33; m 53; c 2. POLYMER CHEMISTRY. Educ: Presby Col, SC, BS, 59; Vanderbilt Univ, PhD(chem), 63. Prof Exp: From chemist to sr chemist, 62-67, sr res chemist, 67-71, res assoc, 71-75, SUPT DEVELOP & CONTROL, TENN EASTMAN CO, 75- Mem: Am Chem Soc; Sigma Xi; Tech Asn Pulp & Paper Indust. Res: Chemistry of cellulose and its derivatives. Mailing Add: Tenite Plastics Div Tenn Eastman Co Kingsport TN 37660

CHAMBERS, RICHARD, b London, Eng, Apr 22, 23; m 56. NEUROLOGY. Educ: Oxford Univ, BA, 44, BM & BCh, 47, MA, 48; FRCP(C), 59. Prof Exp: Instr, Univ Toronto, 57-60; assoc prof, NJ Col Med & Dent, 60-61, prof, 61-66; PROF NEUROL, JEFFERSON MED COL, 66- Concurrent Pos: Fel neurol, Harvard Med Sch, 51-53 & 56; scholar, Royal Col Physicians, Eng, 55-56. Mem: Am Acad Neurol; Am Asn Neuropath. Res: Cortical physiology; fructose metabolism; virus encephalitis; peripheral neuropathy. Mailing Add: Dept of Neurol Jefferson Med Col Philadelphia PA 19107

CHAMBERS, RICHARD LEE, b Algona, Iowa, Feb 26, 47; m 70. MARINE GEOCHEMISTRY, SEDIMENTOLOGY. Educ: Univ Mont, BA, 70, MS, 71; Mich State Univ, PhD(geol), 75. Prof Exp: RES SCIENTIST CHEM SEDIMENTOLOGY, GREAT LAKES ENVIRON RES LAB, DEPT OF COM, 73- Mem: Am Asn Petrol Geologists; Am Quaternary Asn. Res: Evaluation of the chemical and physical mobility of phosphorus using in-situ spiked samples with control samples; evaluate the rate of phosphorus cycling at the sediment-water interface. Mailing Add: NOAA Great Lakes Environ Res Lab 2300 Washtenaw Ave Ann Arbor MI 48104

CHAMBERS, ROBERT J, b Atlanta, Ga, Sept 23, 30; m 58; c 2. ASTRONOMY. Educ: Univ Wash, BSc, 52; Univ Calif, Berkeley, PhD(astron), 64. Prof Exp: From instr to assoc prof, 67-73, PROF ASTRON, POMONA COL, 73-, DIR, BRACKETT OBSERV, 64- Concurrent Pos: NSF sci fac fel, Univ Calif, Berkeley, 69-70. Mem: AAAS; Am Astron Soc; assoc Am Soc Mech Engrs. Res: Galactic star clusters; astronomical instrumentation. Mailing Add: F P Brackett Observ Pomona Col Claremont CA 91711

CHAMBERS, ROBERT ROOD, b Lincoln, Nebr, May 23, 23; m 65; c 3. ORGANIC CHEMISTRY. Educ: Univ Nebr, AB, 44; Univ Ill, PhD(org chem), 47; DePaul Univ, JD, 51. Prof Exp: Res chemist, Sinclair Res Inc, 47-50, group leader, 50-52, div dir, 52-59, tech mgr, 59-66, vpres, 67-68, pres, 68, vpres res, Sinclair Oil Corp, 68-69, VPRES, ATLANTIC RICHFIELD CO, 69- Mem: Am Chem Soc; Brit Chem Soc. Res: Petroleum; petrochemicals. Mailing Add: Atlantic Richfield Co 1500 Market St Philadelphia PA 19101

CHAMBERS, ROBERT WARNER, b Oakland, Calif, Oct 27, 24; m 49; c 3. BIOCHEMISTRY. Educ: Univ Calif, AB, 49, PhD(biochem), 54. Prof Exp: Asst biochem, Univ Calif, 51-54; Life Ins Med Res Fund, Res Coun fel, 54-56; from instr to assoc prof, 56-69, PROF BIOCHEM, SCH MED, NY UNIV, 69- Concurrent Pos: NSF grant, 58-62; Nat Inst Gen Med Sci grant, 60-78; New York Health Res Coun grant, 64-70; Am Cancer Soc grant, 70-72; Nat Cancer Inst grant, 74-78; career scientist, Health Res Coun City New York, 62-72; mem subcomt purines & pyrimidines, Comt Biol Chem, Nat Acad Sci-Nat Res Coun, 62, chmn, 64; mem comt res etiology of cancer, Am Cancer Soc, 66-69; mem postdoctoral fel comt, NSF, 67; Nat Acad Sci-Nat Res Coun & Polish Acad Sci exchange scientist, 68. Mem: AAAS; Am Chem Soc; The Chem Soc; Am Soc Biol Chem; Harvey Soc. Res: Photochemistry of nucleic acids; structure-action relationships in transfer RNA; synthesis of oligonucleotides; mutagenesis and carcinogenesis. Mailing Add: 35 Crest Dr Tarrytown NY 10591

CHAMBERS, VAUGHAN CRANDALL, (JR), b Philadelphia, Pa, June 14, 25; m 48; c

4. ORGANIC CHEMISTRY. Educ: Swarthmore Col, AB, 47; Mass Inst Technol, PhD(org chem), 50. Prof Exp: Chemist, Org Photog Mat, 50-59, sr chemist, 59-60, res assoc, 60-61, res supvr, 61-64, RES MGR PHOTO PROD DEPT, E I DU PONT DE NEMOURS & CO, INC, 64- Mem: Am Chem Soc; Soc Photog Sci & Eng. Res: Physical organic study of small carbocyclic compounds; photography, sensitizing dyes; polymers for use in photographic products. Mailing Add: Photo Prod Dept Nemours Bldg E I du Pont de Nemours & Co Inc Wilmington DE 19898

CHAMBERS, VELMA CATHERINE, b Atkinson, Nebr, Oct 27, 09. MICROBIOLOGY. Educ: Univ Wash, BS, 42, MS, 48, PhD(microbiol), 54. Prof Exp: Res instr microbiol, 54-64, res asst prof, 64-71, RES ASSOC PROF MICROBIOL, UNIV WASH, 71- Mem: Am Soc Cell Biol; Am Soc Microbiol; NY Acad Sci. Res: Extraneural growth of poliomyelitis virus; propagation of viruses in tissue culture; virus induced tumors; transplantation immunity; electron microscopy. Mailing Add: Dept of Microbiol Univ of Wash Seattle WA 98105

CHAMBERS, VIVIAN MURRAY, biology, see 12th edition

CHAMBERS, WILBERT FRANKLIN, b Cameron, WVa, Feb 26, 23; m 57; c 2. NEUROANATOMY. Educ: WVa Univ, BS, 46, MS, 47; Univ Wis, PhD(anat), 52. Prof Exp: Instr anat, Univ Pittsburgh, 49-52; fel, Vanderbilt Univ, 52-54, instr, 54-55; from instr to assoc prof, Univ Vt, 55-67; assoc prof, 67-71, PROF ANAT, DARTMOUTH MED SCH, 71- Mem: Am Asn Anat; Am Soc Zool. Res: Morphology of primate brain; nervous system function in altered endocrine states; the subcommissural organ and water metabolism; effect of amphethamines on the reticular activation response; gonadatrophic centers of the hypothalamus; the thyroid release factors region of the hypothalamus. Mailing Add: Dept of Anat-Cytol Dartmouth Med Sch Hanover NH 03755

CHAMBERS, WILLIAM EDWARD, b Ravenswood, WVa, Aug 14, 33; m 55; c 3. ANALYTICAL CHEMISTRY. Educ: Marshall Univ, BS, 55; Univ Ill, MS, 57, PhD(anal chem), 60. Prof Exp: Anal chemist, Parma Res Center, Union Carbide Corp, 59-63, supvr anal div, 63-68, asst dir advan technol projs, Carbon Prod Div, 68-73, DIR CARBON FIBER DEVELOP, CARBON PROD DIV, UNION CARBIDE CORP, 73- Mem: Soc Appl Spectros; Am Chem Soc. Res: Flame spectroscopy; nuclear magnetic resonance. Mailing Add: Union Carbide Corp Carbon Prod Div PO Box 6116 Cleveland OH 44101

CHAMBERS, WILLIAM HYLAND, b St Louis, Mo, July 30, 22; m 45; c 4. NUCLEAR SCIENCE. Educ: Cornell Univ, BA, 43, MS, 48; Ohio State Univ, PhD(physics), 50. Prof Exp: Asst, Cornell Univ, 46-48; asst, Ohio State Univ, 48-49; MEM STAFF & GROUP LEADER, LOS ALAMOS SCI LAB, 50- Concurrent Pos: With AEC Combined Opers Planning Group, Oak Ridge, 67. Mem: Fel AAAS; Am Phys Soc; Am Geophys Union; Am Nuclear Soc. Res: Nuclear weapon development; detection of nuclear detonations in space; solar flare x-rays; soft x-ray spectrometry; nuclear safeguards and arms control; nuclear material detection and identification. Mailing Add: 336 Andanada Los Alamos NM 87544

CHAMBERS, WILLIAM L, organic chemistry, polymer chemistry, see 12th edition

CHAMBLEE, DOUGLAS SCALES, b Zebulon, NC, Jan 4, 21; m 49; c 3. AGRONOMY. Educ: NC State Univ, BS, 44, MS, 47; Iowa State Univ, PhD(agron), 49. Prof Exp: Res instr agron, 43-47, from asst prof, to assoc prof, 48-60, PROF AGRON, NC STATE UNIV, 60- Concurrent Pos: Mem, NC State Mission, Peru, 61-64; agent div forage crops, USDA. Mem: Fel Am Soc Agron. Res: Moisture requirements of alfalfa; grass mixtures; fertility and management of permanent pastures. Mailing Add: Dept of Crop Sci NC State Univ Raleigh NC 27607

CHAMBLISS, GLENN HILTON, b Jasper, Tex, Feb 14, 42; m 65; c 1. MICROBIOLOGY. Educ: Univ Tex, Austin, BA, 65; Miami Univ, MA, 67; Univ Chicago, PhD(microbiol), 72. Prof Exp: Fel, Jane Coffin Childs Mem Fund Med Res, 71-73; fel, Phillipe Found, 73-74; ASST PROF BACT, UNIV WIS-MADISON, 74- Concurrent Pos: Fel microbiol, Inst Microbiol, Univ Paris-Sud, 72-74. Mem: Am Soc Microbiol; Fedn Europ Biol Socs; NY Acad Sci. Res: Examination of the involvement of translational controls in regulating temporal gene expression during bacterial sporulation. Mailing Add: Dept of Bact Univ of Wis Madison WI 53706

CHAMBLISS, KEITH WAYNE, b Flora, Ill, Dec 16, 26; m 56; c 3. IMMUNOCHEMISTRY. Educ: Univ Ill, AB, 48; Univ Ind, AM, 50, PhD(biochem), 52; Am Bd Clin Chem, dipl, 61. Prof Exp: Res immunochemist serol, US Army Med Serv Grad Sch, 52-53; res biochemist agr chem, Univ Wyo, 55-56; clin chemist, Toledo Hosp, Ohio, 56-61; sect head immunol, Ames Res Lab Div, Miles Labs Inc, 62-73; DIR IMMUNOCHEM RES & DEVELOP, DADE DIV, AM HOSP SUPPLY CORP, 73- Mem: Am Chem Soc; Am Asn Clin Chem. Res: Antibody-antigen reactions and/or radionuclides applied to diagnostic tests. Mailing Add: DADE Div Am Hosp Supply Corp PO Box 520672 Miami FL 33152

CHAMBLISS, OYETTE LAVAUGHN, b Chapman, Ala, Nov 4, 36; m 61; c 2. VEGETABLE CROPS. Educ: Auburn Univ, BS, 58, MS, 62; Purdue Univ, PhD(plant breeding), 66. Prof Exp: Res horticulturist, Veg Breeding Lab, Crops Res Div, Agr Res Serv, USDA, 66-70; ASSOC PROF VEG CROPS, AUBURN UNIV, 70- Honors & Awards: Marion Meadows Award, Am Soc Hort Sci, 66, Asgrow Award, 72. Mem: Am Soc Hort Sci; Am Hort Soc; AAAS; Am Inst Biol Sci. Res: Developing insect resistant varieties and investigating nature of resistance in cucumber and cowpea. Mailing Add: Dept of Hort Auburn Univ Auburn AL 36830

CHAMEIDES, WILLIAM LLOYD, b New York, NY, Nov 21, 49; m 69. AERONOMY. Educ: State Univ NY Binghamton, BA, 70; Yale Univ, MPh, 73, PhD(atmospheric sci), 74. Prof Exp: Res investr, 74-75, ASST RES SCIENTIST, SPACE PHYSICS RES LAB, UNIV MICH, 75- Mem: Am Geophys Union. Res: Theoretical studies of the physical and chemical phenomena of planetary atmospheres with specific attention to the processes that control the abundances of trace gases in the earth's lower atmosphere. Mailing Add: Space Physics Res Lab Univ of Mich 2455 Hayward Ann Arbor MI 48105

CHAMELIN, ISIDOR MARIE, b Jersey City, NJ, Oct 10, 04; m 31; c 2. CHEMISTRY. Educ: City Col New York, BS, 26, ChE, 35, NY Univ, MS, 28; Columbia Univ, ScD(biochem), 40. Prof Exp: Technician, Mt Sinai Hosp, New York, 22-23, sr technician, 24-28; technician, Rockefeller Inst, 23-24; asst, Washington Sq Col, NY Univ, 28-29; asst, City Col New York, 29-40, sr chemist, Goldwater Mem Hosp, 39-45; dir res, Rystan Co, Inc, 45-50; consult chemist, 50-54; assoc dir biores lab & asst prof chem, Fla Southern Col, 54-56; dir PML Labs, res & develop div, & toxicologist med exam off, Sarasota County, 56-64; BIOCHEMIST, ORANGE MEM HOSP, 64- Concurrent Pos: Chief chemist, Morrisania Hosp, New York, 28-39; chemist, Jewish Mem Hosp, 38-39; Lakeland Found fel, Univ Chicago, 41-45; toxicologist, Med Exam Off, Orange County & Osceola County, Fla, 64-; adj prof toxicol, Fla Technol Univ, 70- Mem: Fel AAAS; Am Chem Soc; NY Acad Sci; fel Am Inst Chemists. Res: Chemurgy; plant and animal pigments; hormones; pharmacology; physiology; enzymes; toxicology. Mailing Add: Orange Mem Hosp Lab Orlando FL 32806

CHAMOT, WALTER M, b New Bedford, Mass, Oct 7, 28; m 49; c 3. POLYMER CHEMISTRY, COLLOID CHEMISTRY. Educ: Univ Chicago, MS, 49; Ill Inst Technol, PhD(eng), 68. Prof Exp: Group leader org synthesis, Nalco Chem Co, Ill, 57-59, coagulation, 59-63, polymer synthesis, 59-64; WATER & WASTE TREATMENT CONSULT, 64-; ASSOC PROF CHEM, CITY COLS CHICAGO, 64- Honors & Awards: Edward Bartow Award, 63. Mem: AAAS; Sigma Xi; Am Inst Chem. Res: Biocides; antioxidants; polymers. Mailing Add: 4243 Raymond Ave Brookfield IL 60513

CHAMPAGNE, JOSEPH R, food science, food technology, see 12th edition

CHAMPE, SEWELL PRESTON, b Montgomery, WVa, Nov 24, 32; m 59, 69; c 2. MOLECULAR BIOLOGY, GENETICS. Educ: Mass Inst Technol, SB, 54; Purdue Univ, PhD(biophys), 59. Prof Exp: Am Cancer Soc fel, Purdue Univ, 59-61, from asst prof to assoc prof biol, 61-69; PROF MICROBIOL, RUTGERS UNIV, 69- Concurrent Pos: NIH res career award, 63- Res: Bacteriophage structure and genetics; chemical mutagenesis; protein synthesis. Mailing Add: Inst of Microbil Rutgers Univ New Brunswick NJ 08903

CHAMPION, KENNETH STANLEY WARNER, b Parramatta, Australia, Dec 7, 23; m 48; c 4. PHYSICS. Educ: Univ Sydney, BSc, 44; Univ Birmingham, PhD, 51. Prof Exp: Asst lectr physics, Univ Queensland, 46-49; hon res fel, Univ Birmingham, 51-52; res assoc, Mass Inst Technol, 52-54; asst prof, Tufts Univ, 54-59; sect chief, Space Physics Labs, 59-63, CHIEF, ATMOSPHERIC STRUCT BR, AIR FORCE CAMBRIDGE RES LABS, 63- Concurrent Pos: Res assoc, Comput Ctr, Mass Inst Technol, 56-59; consult, Photochem Lab, Geophys Res Directorate, Air Force Cambridge Ctr, 56-59, Guenter Loeser lectr, 62; vis prof, Univ Adelaide, 64; chmn working group 4, Comt Space Res, Int Coun Sci Unions, 74- Mem: Am Geophys Union; Am Phys Soc; Am Meteorol Soc; fel Brit Phys Soc. Res: Plasma physics; effects of electromagnetic and magnetic fields; measurement of cross sections for atomic processes; upper atmosphere physics; properties and processes of the atmosphere; model atmospheres. Mailing Add: 6 Rolfe Rd Lexington MA 02173

CHAMPION, LLOYD REGINALD, genetics, see 12th edition

CHAMPION, WILLIAM (CLARE), b Rockford, Ill, Mar 12, 30; m 63. ORGANIC CHEMISTRY. Educ: Univ Ill, BS, 52; Cornell Univ, PhD(org chem), 58. Prof Exp: Asst org chem, Cornell Univ, 54-57; fel, Iowa State Univ, 58-59; from asst prof to assoc prof, 59-73, PROF ORG CHEM, COLO COL, 73- Mem: Am Chem Soc; Am Crystallog Asn. Res: Synthetic organic chemistry; x-ray crystallography. Mailing Add: Dept of Chem Colo Col Colorado Springs CO 80903

CHAMPLIN, ARTHUR KINGSLEY, b Portland, Maine, Nov 30, 38; m 66; c 2. REPRODUCTIVE BIOLOGY, GENETICS. Educ: Williams Col, BA, 61, MA, 63; Univ Rochester, PhD(biol), 69. Prof Exp: NIH trainee biol, Univ Rochester, 65-69; fel biol & reproduction physiol, Jackson Lab, 69-71; ASST PROF BIOL, COLBY COL, 71- Mem: Soc Study Reproduction; Am Soc Zoologists. Res: Genetic and environmental factors affecting mammalian reproduction and early development. Mailing Add: Dept of Biol Colby Col Waterville ME 04901

CHAMPLIN, WILLIAM G, b Rogers, Ark, Sept 10, 23; m 51; c 1. MEDICAL MICROBIOLOGY. Educ: Northeastern State Univ, Okla, BS, 48; Univ Ark, MS, 65, PhD(microbiol), 71. Prof Exp: SUPV MICROBIOLOGIST, VET ADMIN HOSP, 53- Concurrent Pos: Consult, Antaeus Lineal Res Assoc & Washington Regional Hosp, 71-75; guest lectr immunol, Univ Ark, 71-75. Mem: Am Soc Microbiol; Am Soc Clin Pathologists; Sigma Xi. Res: Rapid diagnosis of viral diseases by cell culture and indirect immunofluorescence of clinical material. Mailing Add: Vet Admin Hosp Fayetteville AR 72701

CHAMPNEY, WILLIAM SCOTT, b Cleveland, Ohio, Jan 15, 43; m 66; c 2. BIOCHEMICAL GENETICS. Educ: Univ Rochester, AB, 65; State Univ NY Buffalo, PhD(biol), 70. Prof Exp: Instr microbiol, Col Med, Univ Calif, Irvine, 70-72; ASST PROF BIOCHEM, UNIV GA, 72- Mem: Am Soc Microbiol. Res: Genetics of bacterial ribosomes; protein-nucleic acid interactions. Mailing Add: Grad Studies Res Bldg Dept of Biochem Univ of Ga Athens GA 30602

CHAN, ALLAN P, b Montreal, Que, Oct 2, 21; m 48. RESEARCH MANAGEMENT, PLANT SCIENCE. Educ: McGill Univ, BSc, 44, MSc, 46; Ohio State Univ, PhD(floricult), 49. Prof Exp: Res asst, Macdonald Col, McGill Univ, 44-46; officer-in-chg floricult res, Cent Exp Farm, 47-59; head ornamental plant sect, 59-62, from asst dir to assoc dir, Inst, 62-65, DIR, PLANT RES INST, 65- Mem: Am Inst Biol Sci; Am Soc Hort Sci; Int Soc Hort Sci; Agr Inst Can; Am Asn Bot Gardens & Arboretums. Res: Floriculture research; plant physiology and anatomy; controlled environment facilities. Mailing Add: Plant Res Inst Central Exp Farm Ottawa ON Can

CHAN, AN SOO, b Port Dickson, Malaysia, July 14, 35. HISTOLOGY, EMBRYOLOGY. Educ: Queen's Univ, Kingston, Can, BA, 63; Univ Ottawa, PhD(histol & embryol), 69. Prof Exp: Res asst path, Hosp for Sick Children, Toronto, 63-65; demonstr histol & embryol, Univ Ottawa, 66-69; res fel exp path, Hosp for Sick Children & Univ Toronto, 69-71; ASST PROF ANAT, COL MED, HOWARD UNIV, 71- Mem: Am Asn Anatomists; Am Soc Cell Biol; Am Soc Zoologists; Electron Micros Soc Am; Teratology Soc. Res: Structure, function and embryology of thyroid, parathyroid and ultimobranchial glands in the vertebrate species and the effects of hormonal and nutritional variations on these glands. Mailing Add: Dept of Anat Col of Med Howard Univ 520 W St NW Washington DC 20059

CHAN, BOCK G, b Kwantung, China, June 15, 35; US citizen; m 68. PHYTOCHEMISTRY. Educ: Cornell Univ, PhD(plant physiol), 70. Prof Exp: Experimentalist mineral nutrit biochem, Dept Pomol, Cornell Univ, 60-61 & 63-67, res assoc photosynthesis, Dept Veg Crops, 70-71; fel biochem, 71-73, RES PLANT PHYSIOLOGIST, WESTERN REGIONAL RES CTR, AGR RES SERV, USDA, 73- Mem: Sigma Xi; Phytochem Soc NAm; Am Soc Plant Physiol. Res: Phytochemical basis and the physiology of plant resistance to insects, other pests and pathogens, with special emphasis on the economical crop plants; investigation into emzymology and biosynthetic pathways of the biologically active compounds. Mailing Add: Western Regional Res Ctr Agr Res Serv USDA 800 Buchanan Albany CA 94706

CHAN, CHEUNG-KING, b Swatow, China, Mar 8, 22; m 53; c 2. BOTANY, HORTICULTURE. Educ: Lingnan Univ, BSc, 45; Pa State Univ, MSc, 62, PhD(bot), 65. Prof Exp: From instr to asst prof hort, Lingnan Univ, 48-52; asst prof, SChina Agr Col, 53-56; chmn dept biol, Chung Chi Col, Hong Kong, 59-67; ASSOC PROF BOT, PA STATE UNIV, ALTOONA, 67- Mem: Bot Soc Am; Int Soc Plant Morphologists. Res: Developmental morphology in Ipomoea reptans Poir. and other species in the

Convolvulaceae; orchid culture. Mailing Add: Dept of Bot Pa State Univ Altoona PA 16603

CHAN, CHIA HWA, b Shanghai, China, Apr 28, 36; m 61; c 3. HIGH ENERGY PHYSICS, THEORETICAL PHYSICS. Educ: Univ London, BSc, 58, PhD(physics), 62. Prof Exp: Vis asst prof, Mid East Tech Univ, Ankara, 62-64; asst res physicist, Univ Calif, San Diego, 64-66; asst prof physics, Purdue Univ, 66-70; chmn dept, 72-74, ASSOC PROF PHYSICS, UNIV ALA, HUNTSVILLE, 70- Mem: Am Phys Soc; Am Asn Physics Teachers. Mailing Add: Dept of Physics Univ of Ala Huntsville AL 35807

CHAN, CHIU YEUNG, b Hong Kong, Feb 28, 41; Brit citizen; m 70; c 1. MATHEMATICAL ANALYSIS, APPLIED MATHEMATICS. Educ: Univ Hong Kong, BSc, 64 & 64; Univ Ottawa, MSc, 67; Univ Toronto, PhD(math), 69. Prof Exp: Asst prof, 69-74, ASSOC PROF MATH, FLA STATE UNIV, 74- Mem: Am Math Soc. Res: Partial differential equations, including Stefan problems, theory of the Sturm type, nuclear reactor kinetics, heat conduction, heat radiation, and demography. Mailing Add: Dept of Math Fla State Univ Tallahassee FL 32306

CHAN, DAVID SIUPOON, b Hong Kong, July 23, 40; m 75. BIOCHEMISTRY. Educ: San Jose State Univ, BA, 64, MS, 70; Univ Southern Miss, PhD(biochem), 73. Prof Exp: Biochemist, Stanford Res Inst, 67-70; res fel, 73-75, RES ASSOC, HARVARD MED SCH, 75-; ASST BIOCHEMIST, McLEAN HOSP, 73- Mem: Fedn Am Scientists; AAAS. Res: Structure and function relationship of central nervous system myelin proteolipid, amino acid sequence. Mailing Add: Biol Res Lab McLean Hosp 115 Mill St Belmont MA 02178

CHAN, EDDIE CHIN SUN, b Singapore, May 13, 31; Can citizen; m 59. MICROBIAL PHYSIOLOGY EDUniv Tex, El Paso, BA, 54; Univ Tex, Austin, MA, 57; Univ Md, PhD(microbiol), 60. Prof Exp: Nat Res Coun Can fel, 60-62; asst prof microbiol & biochem, Univ NB, 62-65; asst prof, 65-68, ASSOC PROF MICROBIOL & IMMUNOL, McGILL UNIV, 68- Concurrent Pos: Nat Res Coun Can & Med Res Coun Can grants. Mem: AAAS; Am Soc Microbiol; Can Soc Microbiol. Mailing Add: Dept of Microbiol & Immunol McGill Univ Montreal PQ Can

CHAN, FRANK LAI-NGI, analytical chemistry, deceased

CHAN, HAK-FOON, b Hong Kong, Oct 10, 42; m 68; c 1. AGRICULTURAL CHEMISTRY. Educ: Chung Chi Col, Chinese Univ, Hong Kong, dipl sci, 64; Bowling Green State Univ, MA, 68; Univ Mich, PhD(org chem), 71. Prof Exp: Fel chem, Univ Rochester, 71-73; SR CHEMIST FUNGICIDE & BIOCIDE, ROHM & HAAS CO, 73- Mem: Am Chem Soc; The Chem Soc. Res: To prepare for evaluation organic compounds which possess fungicidal, bactericidal and biocidal activities. Mailing Add: Res Lab Rohm & Haas Co Norristown & McKean Rds Spring House PA 19477

CHAN, HARVEY THOMAS, JR, b Astoria, Ore, Mar 5, 40; m 66; c 3. FOOD SCIENCE. Educ: Ore State Univ, BSc, 63, PhD(food sci), 69; Univ Hawaii, MSc, 66. Prof Exp: Res asst food sci, Univ Hawaii, 63-65 & Ore State Univ, 65-68; FOOD TECHNOLOGIST, AGR RES SERV, USDA, 68- Concurrent Pos: Affil grad fac, Dept Food Sci, Univ Hawaii, 69- Mem: Inst Food Technologists (secy-treas, 73-74 & 76); Am Chem Soc; Sigma Xi. Res: Chemical and biochemical composition of tropical fruits and vegetables; process and product development of tropical fruits and vegetables; changes in nutrients and biochemical constituents during processing. Mailing Add: Hawaii Fruit Lab Agr Res Serv USDA Univ of Hawaii 1920 Edmondson Rd Honolulu HI 96822

CHAN, JAMES C, b Hong Kong, Nov 20, 37; m 64. MICROBIOLOGY. Educ: Int Christian Univ Tokyo, BA, 61; Univ Rochester, PhD(microbiol), 66. Prof Exp: Sr res scientist, Squibb Inst Med Res, 66-68; asst prof med, Sch Med, Ind Univ, Indianapolis, 68-71; asst prof biochem virol, Baylor Col Med, 71-72; ASST PROF VIROL & ASST VIROLOGIST, CANCER CTR, UNIV TEX M D ANDERSON HOSP & TUMOR INST, 72- Concurrent Pos: Ind Univ fac res grant & Little Red Door, Inc cancer res grants, 69 & 70; Am Cancer Soc inst grant, 68-70; St Joseph Co, Inc cancer res grant, 71-72. Mem: AAAS; Am Soc Microbiol. Res: Biochemical changes in cells following infection by tumor viruses, such as polyoma and Shope fibroma viruses; virus induced tumor; host resistance; studies of viral involvement in neoplastic diseases in animals and man. Mailing Add: Dept of Virol Cancer Ctr Univ Tex M D Anderson Hosp & Tumor Inst Houston TX 77025

CHAN, KAI CHIU, b Canton, China, May 16, 34; US citizen; m 66; c 2. DENTISTRY. Educ: Chiba Univ, Japan, BS, 58; Tokyo Med & Dent Univ, DDS, 62; Univ Iowa, MS, 64, DDS, 67. Prof Exp: From instr to asst prof, 64-71, ASSOC PROF OPER DENT, COL DENT, UNIV IOWA, 71- Mem: Am Asn Dent Schs; Am Dent Asn; Int Asn Dent Res; Acad Oper Dent. Res: Burnished amalgam surfaces; retention pins; dental cements. Mailing Add: Col of Dent Univ of Iowa Iowa City IA 52242

CHAN, KWING LAM, b Hong Kong, Oct 13, 49; m 73. ASTROPHYSICS. Educ: Univ Calif, Berkeley, BA, 70; Princeton Univ, MA, 72, PhD(physics), 74. Prof Exp: RES ASSOC ASTROPHYS, THOMAS J WATSON RES CTR, IBM CORP, 74- Mem: Am Astron Soc; Am Phys Soc. Res: Fluid dynamics and radiative transfer. Mailing Add: Thomas J Watson Res Ctr IBM Corp PO Box 218 Yorktown Heights NY 10598

CHAN, LAI KOW, b Hong Kong, Nov 5, 40; m 67. STATISTICS. Educ: Hong Kong Baptist Col, BSc, 62; Univ Western Ont, MA, 64, PhD(statist), 66. Prof Exp: Instr statist, Univ Toronto, 65-66; asst prof, 66-69, ASSOC PROF STATIST, UNIV WESTERN ONT, 69- Concurrent Pos: Hon res fel, Univ Col, London, 73-74. Res: Linear estimation based on order statistics. Mailing Add: Dept of Math Univ of Western Ont London ON Can

CHAN, LEE-NIEN LILLIAN, b Hong Kong, Sept 28, 41; m 69; c 2. DEVELOPMENTAL BIOLOGY, PHYSIOLOGY. Educ: Acadia Univ, BSc, 63; Univ Wis, MS, 66; Yale Univ, PhD(biol), 71. Prof Exp: Assoc in res, Yale Univ, 67, res assoc molecular biophys & biochem, 71; res assoc biol, Mass Inst Technol, 71-73; ASST PROF PHYSIOL, UNIV CONN HEALTH CTR, 73- Mem: Soc Develop Biol. Res: Development of eukaryotes; regulation of specific gene expression; erythropoiesis; differentiation of red cell membrane. Mailing Add: Dept of Physiol Univ of Conn Health Ctr Farmington CT 06032

CHAN, LELAND, b Los Angeles, Calif, May 2, 41. ENVIRONMENTAL HEALTH. Educ: Calif State Univ, Los Angeles, BS, 64; Univ Calif, Los Angeles, MS, 66, DrPH(environ health), 76. Prof Exp: Lab dir environ health, Sch Pub Health, Univ Calif, Los Angeles, 65-66; environ health specialist res hosp, Commissioned Corps, NIH, 66-68; ASST PROF ENVIRON HEALTH SCI & MICROBIOL, CALIF STATE UNIV, LOS ANGELES, 72- Concurrent Pos: Mem, Adv Comt Sanitarian Standards & Environ Health Prog, Calif State Bd Health, 75-77. Mem: Am Conf Govt Indust Hygienists; Nat Environ Health Asn. Res: Staphylococcal food intoxications; environmental control in medical facilities; bacterial analysis of water; environmental bacteriology. Mailing Add: Dept of Microbiol & Pub Health Calif State Univ Los Angeles CA 90032

CHAN, LOCK LIM, Can citizen. POLYMER CHEMISTRY, PAPER CHEMISTRY. Educ: Chung Chi Col, Hong Kong, BSc, 62; Dartmouth Col, MA, 64; State Univ NY Col Forestry, PhD(polymer chem), 69. Prof Exp: Nat Res Coun Can fel, Univ Montreal, 69-70; GROUP LEADER, RES & DEVELOP LAB, BORDEN CHEM CO (CAN) LTD, 70- Mem: Am Chem Soc; sr mem Chem Inst Can; assoc Tech Asn Pulp & Paper Indust. Res: Synthesis and characterization of polymers; solution properties of polymers; anionic block copolymerization; condensation and emulsion polymerization; chemical additive in paper industry; polyelectrolytes. Mailing Add: Borden Chem Co (Can) Ltd 595 Coronation Dr West Hill ON Can

CHAN, MAUREEN GILLEN, b Brooklyn, NY, June 2, 39; m 63. POLYMER CHEMISTRY. Educ: Chestnut Hill Col, BS, 61; Stevens Inst Technol, MS, 65. Prof Exp: Sr tech aide chem, 62-64, assoc mem tech staff, 64-74, MEM TECH STAFF, BELL LABS, 74- Mem: Am Chem Soc; Soc Plastics Engrs; Am Soc Testing & Mat. Res: Stabilization of high polymers. Mailing Add: Bell Labs Murray Hill NJ 07974

CHAN, MING SUI MICHAEL, b Hong Kong, Feb 17, 24; Can citizen. FOOD TECHNOLOGY, BIOCHEMISTRY. Educ: Univ Wash, BS, 49; Univ Mass, MS, 52, PhD(food technol), 56. Prof Exp: Res instr, Univ Mass, 56; chemist, Tech Sta, Fisheries Res Bd, Can, 56-57, assoc scientist, 58-61; chemist, Fish Processing Exp Plant, Indust Develop Serv, Can Dept Fisheries, 57-58; res officer, Plant Res Inst, Can Dept Agr, 61-62, res officer, Food Res Inst, 62-64; sr food scientist, New Brunswick Res & Productivity Coun, 64-67, actg head food sci dept, 66-67; sr res chemist, 67-75, SR RES SCIENTIST, RES & DEVELOP DIV, KRAFTCO CORP, 75- Concurrent Pos: Can Patents & Develop Corp award, 63-64; res grants, Atlantic Develop Bd, Ont, 65-66 & Sugar Res Found, 65-67. Mem: Inst Food Technologists; Can Inst Food Technologists; Am Oil Chemists Soc; fel Royal Soc Health; Brit Soc Chem Indust. Res: Oilseed; dairy protein; new product developments in cheese and dairy products; nutritional high protein foods; product and process developments in meat and dairy technology; food nutrition, toxicology, microbiology and dehydration. Mailing Add: Kraftco Corp Res & Develop Div 801 Waukegan Rd Glenview IL 60025

CHAN, MOSES HUNG-WAI, b Hsi-an, China, Nov 23, 46; m 72; c 1. LOW TEMPERATURE PHYSICS. Educ: Bridgewater Col, BA, 67; Cornell Univ, MS, 70, PhD(exp physics), 74. Prof Exp: Teaching asst physics, Cornell Univ, 67-69; asst lectr, Univ Hong Kong, 69-70; res asst, Cornell Univ, 70-73; RES ASSOC PHYSICS, DUKE UNIV, 73-, INSTR PHYSICS, 75- Mem: Am Phys Soc; Sigma Xi. Res: Low temperature properties of quantum fluids and solids, in particular, superfluidity of very thin films on substrate and dielectric behavior of liquid helium 4. Mailing Add: Dept of Physics Duke Univ Durham NC 27706

CHAN, PAULA PUI-YING, b Hong Kong. NUMERICAL ANALYSIS. Educ: Chinese Univ Hong Kong, BS, 65; Case Western Reserve Univ, MA, 67, PhD(math), 70. Prof Exp: Asst prof, 70-75, ASSOC PROF MATH, CLEVELAND STATE UNIV, 75- Mem: Am Math Soc. Mailing Add: Dept of Math Cleveland State Univ Cleveland OH 44115

CHAN, PETER SINCHUN, b Kwang-tung, China, Dec 1, 38; m 64; c 3. PHARMACOLOGY, BIOCHEMISTRY. Educ: Nat Taiwan Univ, BSc, 61; Univ Cincinnati, MSc, 64; Ind Univ, Bloomington, PhD(pharmacol, org chem), 67. Prof Exp: Res fel med chem, Univ Mich, 67-68; instr pharmacol, Med Br, Univ Tex, Galveston, 68-69, asst prof, 69-70; res pharmacologist, 70-73, group leader & prin res pharmacologist, 73-75, HEAD DEPT CARDIOVASC-RENAL PHARMACOL, LEDERLE LABS DIV, AM CYCNAMID CO, 75- Mem: Am Soc Pharmacol & Exp Therapeut; Soc Exp Biol & Med; Am Heart Asn; Am Chem Soc. Res: Cardiovascular and renal pharmacology; pharmacological and biochemical approaches to mechanisms of drug actions, drug design and regulation of enzyme systems of pharmacologic importance. Mailing Add: Lederle Labs Div American Cyanamid Co Pearl River NY 10965

CHAN, PHILLIP C, b Amoy, China, June 14, 28; US citizen; m 65; c 1. CHEMISTRY. Educ: Monmouth Col, Ill, BS, 52; Columbia Univ, MA, 53, PhD(chem), 57. Prof Exp: Fel, Sch Med, Johns Hopkins Univ, 57-59; Jan Coffin Childs fel, Max Planck Inst Cell Chem, Ger, 59-60; asst prof, 60-67, ASSOC PROF BIOCHEM, STATE UNIV NY DOWNSTATE MED CTR, 67- Mem: Am Chem Soc; Harvey Soc; Am Soc Biol Chem. Res: Free radical reactions in biological systems. Mailing Add: Dept of Biochem State Univ of NY Downstate Med Ctr Brooklyn NY 11203

CHAN, PO CHUEN, b Ichong, Hupeh, China, May 13, 35; m 61; c 2. CELL BIOLOGY. Educ: Int Christian Univ, Tokyo, BA, 60; Columbia Univ, MA, 63; NY Univ, PhD(biol), 67. Prof Exp: Res assoc, Sloan-Kettering Inst Cancer Res, 67-70; ASSOC MEM, AM HEALTH FOUND, 70- Mem: AAAS; Am Soc Hemat; Am Soc Zool; NY Acad Sci; Tissue Cult Asn. Res: Cell production and kinetics; mechanism of chemical carcinogenesis. Mailing Add: Am Health Found Hammond House Rd Valhalla NY 10595

CHAN, RAYMOND KAI-CHOW, b Hong Kong, Oct 10, 33; m 60; c 2. PHYSICAL CHEMISTRY. Educ: Univ Toronto, BA, 58, PhD(phys chem), 61. Prof Exp: Nat Res Coun Can fel, 61-62; asst prof, 62-71, ASSOC PROF CHEM, UNIV WESTERN ONT, 71- Mem: Chem Inst Can; NAm Thermal Anal Soc. Res: Molecular solid phase transitions under pressure, supercooled liquids and glassy plastic crystals by thermally stimulated depolarization, dielectric properties and differential thermal analysis; air pollution, especially sulfur dioxide limestone/dolomite reaction using thermogravimetry. Mailing Add: Dept of Chem Univ of Western Ont London ON Can

CHAN, SAMUEL H P, b Nanking, China, Aug 1, 41; m 71. BIOCHEMISTRY, MOLECULAR BIOLOGY. Educ: Int Christian Univ, Tokyo, BA, 64; Univ Rochester, PhD(biochem), 69. Prof Exp: Asst, Univ Rochester, 64-69; res assoc biochem & fel, Cornell Univ, 69-71; ASST PROF BIOCHEM, SYRACUSE UNIV, 71- Mem: AAAS; Am Chem Soc; NY Acad Sci; Sigma Xi. Res: Physical, chemical and enzymatic properties of cytochromes and cytochrome oxidase; resolution and reconstitution of inner mitochondrial membrane; mechanism of oxidative phosphorylation. Mailing Add: Dept of Biol Syracuse Univ Syracuse NY 13210

CHAN, SHUNG KAI, b Hong Kong, Mar 31, 35; m 63. BIOCHEMISTRY. Educ: WVa Univ, AB, 56; Univ Wis, PhD(biochem), 62. Prof Exp: Asst res chemist, Samuel Robert Noble Found, 58-59; sr res biochemist, Abbott Labs, 62-66; asst prof, 66-70, ASSOC PROF BIOCHEM, MED CTR, UNIV KY, 70- Mem: AAAS; Am Chem Soc; Brit Biochem Soc; Am Soc Biol Chem. Res: Glycoproteins, structure and function; glycoproteins, biosynthesis and regulation. Mailing Add: Dept of Biochem Univ of Ky Med Ctr Lexington KY 40506

CHAN, SUNNEY IGNATIUS, b San Francisco, Calif, Oct 5, 36; m 64; c 1. BIOPHYSICAL CHEMISTRY, BIOPHYSICS. Educ: Univ Calif, Berkeley, BS, 57, PhD(chem), 60. Prof Exp: NSF fel physics, Harvard Univ, 60-61; asst prof chem, Univ Calif, Riverside, 61-63; from asst prof to assoc prof chem physics, 63-68, PROF CHEM PHYSICS, CALIF INST TECHNOL, 68- Concurrent Pos: Consult, Unified Sci Assocs, 63-64, Div Gen Med Sci, USPHS, 70-74, Procter & Gamble Co & Merck Sharp & Dohme Res Labs, 74-; Sloan fel, 65-67; Guggenheim Mem fel, 68-69; mem ed rev bd, Appl Spectros, 68-70; assoc ed, Ann Rev Magnetic Resonance, 70; Reilly lectr, Univ Notre Dame, 73. Mem: AAAS; Am Phys Soc; Am Chem Soc; Am Soc Biol Chemists; fel NY Acad Sci. Res: Physical methods for the determination of molecular structure; applications of magnetic resonance spectroscopy to biological problems, particularly membrane structure and function, protein-lipid interactions and mechanisms of ion and electron transport. Mailing Add: Dept of Chem Calif Inst of Technol Pasadena CA 91125

CHAN, TAK-HANG, b Hong Kong, June 28, 41. ORGANIC CHEMISTRY. Educ: Univ Toronto, BSc, 62; Princeton Univ, MA, 63, PhD(chem), 65. Prof Exp: Asst prof, 66-71, ASSOC PROF CHEM, McGILL UNIV, 71- Concurrent Pos: Res fel, Harvard Univ, 65-66. Mem: Am Chem Soc; The Chem Soc; AAAS. Res: Synthesis of complicated natural products; structural determination of natural products; mechanisms of organic reactions; new synthetic methods with silicon, sulfur and phosphorus compounds. Mailing Add: Dept of Chem McGill Univ Montreal PQ Can

CHAN, TEH-SHENG, b Taiwan, China. HUMAN GENETICS, PHYSIOLOGY. Educ: Nat Taiwan Univ, MD, 63; Yale Univ, PhD(molecular biol), 69. Prof Exp: Intern, Nat Taiwan Univ Hosp, 62-63; ASST PROF PHYSIOL, SCH MED, UNIV CONN, FARMINGTON, 73- Concurrent Pos: Helen Hay Whitney fel, Rockefeller Univ, 69-71 & Mass Inst Technol, 71-73. Res: Somatic cell genetics; inborn errors of purine metabolism; viral gene transfer in mammalian cells. Mailing Add: Dept of Physiol Univ of Conn Health Ctr Farmington CT 06032

CHAN, WAH YIP, b Shanghai, China, Dec 1, 32; nat US; m 61; c 2. PHARMACOLOGY. Educ: Univ Wis, BA, 56; Columbia Univ, PhD(pharmacol), 61. Prof Exp: From res assoc biochem to assoc prof pharmacol, 60-76, PROF PHARMACOL, MED COL, CORNELL UNIV, 76- Concurrent Pos: USPHS res career develop award, 68-73; Hirschl career scientist award, 73-; mem basic pharmacol adv comt, Pharmaceut Mfrs Asn Found, 73- Mem: Soc Exp Biol & Med; Am Soc Pharmacol & Exp Therapeut; Soc Study Reproduction; Harvey Soc; NY Acad Sci. Res: Pharmacology of neurohypophysial hormones and polypeptides; renal pharmacology; uterine pharmacology. Mailing Add: Dept of Pharmacol Cornell Univ Med Col New York NY 10021

CHAN, YAU WA, b Canton, China, Nov 30, 24; US citizen; m 51; c 3. PHYSICS. Educ: Lingnam Univ, BS, 50, MS, 52; Univ Calif, Berkeley, PhD(physics), 62. Prof Exp: Demonstr physics, Univ Hong Kong, 53-58; res physicist, Univ Calif, Berkeley, 62-63; res assoc physics, Brookhaven Nat Lab, 63-65, from asst physicist to assoc physicist, 65-74; MEM FAC, SCI CTR PHYSICS, UNIV HONG KONG, 74- Concurrent Pos: Demonstr, Univ Hong Kong, 53-58; lectr, Chu Hai Col, Hong Kong, 57-58; sr lectr, Chinese Univ Hong Kong, 71- Mem: Am Asn Physics Teachers; Am Phys Soc. Res: Atomic beams; magnetic resonance measurements of nuclear spin, moment and hyperfine structure; interaction of charged particles with coherent radiation; secular equations. Mailing Add: Sci Ctr Physics Univ Hong Kong Shatin New Terr Hong Kong

CHAN, YICK-KWONG, b China, Oct 14, 35; m 69. BIOMETRY. Educ: Taiwan Prov Col Agr, BS, 55; Univ Minn, MS, 60, PhD(biostatist), 66. Prof Exp: Asst prof pub health, Yale Univ, 66-68; ASSOC PROF BIOMET & HEAD REGIONAL STATIST OFF SOUTHEASTERN CANCER STUDY GROUP, EMORY UNIV, 68- Mem: Am Pub Health Asn; Am Statist Asn; Inst Math Statist; Biomet Soc. Res: Application of statistics to clinical trials, epidemiology and bioassays; epidemic simulations. Mailing Add: Dept of Biometry Emory Univ Sch of Med Atlanta GA 30322

CHANANA, ARJUN DEV, b Lyallpur, Punjab, India, Nov 6, 30; m 63; c 2. EXPERIMENTAL PATHOLOGY, SURGERY. Educ: Univ Rajasthan, MB & BS, 55; FRCS(E) & FRCS, 60. Prof Exp: Internship, residencies & postdoctoral work surg, 55-60; chief resident surg, Bolton Dist Gen Hosp, Eng, 60-63; asst scientist, 63-66, assoc scientist, 66-69, actg head div hemat, 68-69, SCIENTIST, MED DEPT & HEAD DIV EXP PATH, BROOKHAVEN NAT LAB, 69- Concurrent Pos: Asst attend physician, Hosp of Med Res Ctr, Brookhaven Nat Lab, 63-65, assoc attend physician, 65-70, attend physician, 70-, chief of staff, 74-; assoc prof, Health Sci Ctr, State Univ NY Stony Brook, 70-; res consult surg, Nassau County Med Ctr, 70- Mem: Am Soc Hemat; Soc Exp Biol & Med; Transplantation Soc; Radiation Res Soc; Am Soc Exp Path. Res: Lymphocytopoiesis; transplantation immunology; leukemia; extracorporeal irradiation of blood and lymph. Mailing Add: Med Dept Brookhaven Nat Lab Upton NY 11973

CHANCE, BRITTON, b Wilkes-Barre, Pa, July 24, 13; m 38, 56; c 12. BIOPHYSICS, BIOCHEMISTRY. Educ: Univ Pa, BS, 35, MS, 36, PhD(phys chem), 40; Univ Cambridge, PhD(physiol), 42, DSc, 52. Hon Degrees: MD, Karolinska Inst, Sweden, 62; DSc, Med Col Ohio, 74. Prof Exp: Actg dir, Johnson Res Found, 40-41, asst prof biophys, 41-49, prof, 49-64, DIR JOHNSON RES FOUND, UNIV PA, 49-, JOHNSON PROF BIOPHYS & PHYS BIOCHEM, SCH MED, 64- Concurrent Pos: Investr, Off Sci Res & Develop, 41; res assoc, Radiation Lab, Mass Inst Technol, 41-42, group leader, 41-45, assoc div head, 42-45, mem vis comt, 54-56; sci consult to attache for res, USN, London, 48; consult, NSF, 51-56, mem physics panel comt on growth, 52-57; mem vis comt, Bartol Res Found, 55-59; Harvey lectr, 54; Phillips lectr, 56 & 65; Pepper lectr, 57; mem, President's Sci Adv Comt, 59; mem comn blood & blood derivatives, Am Red Cross & Nat Res Coun; exchange scholar, USSR, 63; foreign fel, Churchill Col, Univ Cambridge, 66; Keilin lectr, 66; Hackett lectr, 66; Redfearn lectr, 70; mem adv coun, Nat Inst Alcoholism, 71-75; vpres, Int Union Pure & Appl Biophys, 72-75, pres, 75- Honors & Awards: Presidential Cert of Merit, 50; Paul Lewis Award, 50; Morlock Award, 61; Genootschaps Medal, Dutch Biochem Soc, 65; Franklin Medal, 66; Harrison Howe Award, Rochester Chap, Am Chem Soc, 66, Philadelphia Chap Award, 69, Nichols Award, NY Chap, 70; Pa Award for Excellence, 68; Heineken Medal, Royal Neth Acad Sci & Lett, 70; Gairdner Award, 72; Post-Cong Festschrift, 73; Semmelweis Medal, 74; Nat Medal Sci, 74. Mem: Nat Acad Sci; fel Am Phys Soc; Am Soc Biol Chem; Am Chem Soc; Am Acad Arts & Sci. Res: Automatic ship steering; photoelectric control units; radar timing and computing devices; sensitive spectrophotometers; enzyme-substrate compounds and reaction mechanisms of catalases; peroxidases, dehydrogenases; cytochromes; kinetics of multienzyme systems; oscillating enzyme systems; cation accumulation; cell oxygen requirements; in vivo readout of enzyme function. Mailing Add: Johnson Res Found Univ of Pa Sch of Med Philadelphia PA 19174

CHANCE, CHARLES JACKSON, b Belen, NMex, Apr 18, 14; m 36; c 2. ZOOLOGY, FISHERIES. Educ: Berea Col, AB, 35; Univ Tenn, MS, 42. Prof Exp: Aquatic biologist, Fisheries Mgt, Tenn Conserv Dept, 40-48; aquatic biologist, 48-58, chief aquatic biologist, 58-60, CHIEF FISHERIES & WATERFOWL RESOURCES BR, TENN VALLEY AUTHORITY, 60- Mem: AAAS; Am Inst Biol Sci; Am Fisheries Soc; Wildlife Soc. Res: Fisheries populations and management. Mailing Add: 4304 Gaines Rd Knoxville TN 37918

CHANCE, CHARLES MARION, b Easton, Md, Dec 31, 19; m 43; c 2. DAIRY NUTRITION. Educ: Univ Md, BS, 41; Va Polytech Inst, MS, 49; Mich State Col, PhD(dairy cattle nutrit), 52. Prof Exp: Asst, Va Polytech Inst, 47-48; asst, Mich State Col, 49-52; asst prof res & exten, Cornell Univ, 52-59, assoc prof, 59; dir dairy indust rels, Agway Inc, 59-65; ASSOC PROF DAIRY SCI, UNIV MD, 65- Mem: Am Dairy Asn. Res: Relation of cow families to milk production and reproductive efficiency; influence of antibiotics on synthesis and digestion in the rumen. Mailing Add: 3711 Marlbrough Way College Park MD 20740

CHANCE, NORMAN ALLEE, b Lynn, Mass, Sept 13, 27; m 49; c 3. CULTURAL ANTHROPOLOGY. Educ: Earlham Col, AB, 51; Cornell Univ, PhD, 57. Prof Exp: Res assoc anthrop, Cornell Univ, 56-57; asst prof, Univ Okla, 57-62, asst prof psychiat & consult, Sch Med, 58-62; assoc prof anthrop, McGill Univ, 62-68, dir prog anthrop & develop, 65-68; PROF ANTHROP, UNIV CONN, 68- Concurrent Pos: Russell Sage Found fel, 59-61; res fel, Sch Pub Health, Harvard Univ, 60-61. Mem: AAAS; fel Am Anthrop Asn; Am Ethnol Soc; Arctic Inst NAm. Res: Economic, social and political development in North America and People's Republic of China. Mailing Add: Dept of Anthrop Univ of Conn Box U-158 Storrs CT 06268

CHANCE, ROBERT L, b Detroit, Mich, Feb 1, 24; m 52; c 4. ANALYTICAL CHEMISTRY, INORGANIC CHEMISTRY. Educ: Wayne State Univ, BS, 48. Prof Exp: Sr res chemist, 49-54, group leader inorg anal chem, 54-61, SR RES CHEMIST, GEN MOTORS RES LABS, 62- Mem: Am Chem Soc; Nat Asn Corrosion Engrs; Am Soc Testing & Mat. Res: General and automotive corrosion research; electrochemical polarization techniques; surface analysis; engine coolants; inhibitor systems; coatings; alloy compositions; metallurgical structure; metal deformation; heat treatment; cavitation phenomena; corrosion surveys. Mailing Add: Gen Motors Res Labs Phys Chem Dept Gen Motors Tech Ctr Warren MI 48090

CHANCE, RONALD E, b Lapeer, Mich, Jan 17, 34; m 61; c 1. BIOCHEMISTRY, METABOLISM. Educ: Purdue Univ, BS, 56, MS, 59, PhD(biochem), 62. Prof Exp: Asst prof biochem, Purdue, 62-63; sr biochemist, 63-69, RES SCIENTIST, ELI LILLY & CO, 69- Concurrent Pos: USDA grant, 62-63. Mem: AAAS; Am Chem Soc; Am Soc Animal Sci. Res: Animal nutrition, amino acid requirements and interrelationships in Chinook salmon and weanling pigs; chemistry of protein hormones; isolation, purification, characterization and chemistry of protein hormones. Mailing Add: Dept of Biochem Lilly Res Labs Indianapolis IN 46206

CHANCE, RONALD RICHARD, b Memphis, Tenn, July 24, 47; m 67; c 2. SOLID STATE CHEMISTRY. Educ: Delta State Univ, BS, 70; Dartmouth Col, PhD(chem), 74. Prof Exp: STAFF CHEMIST, ALLIED CHEM CORP, 74- Mem: Am Phys Soc; Am Chem Soc; AAAS. Res: Optical and electrical properties of polydiacetylenes; theory of fluorescence and energy transfer in layered systems; photoconduction and exciton dynamics in molecular crystals. Mailing Add: Mat Res Ctr Allied Chem Corp Morristown NJ 07960

CHANDAN, RAMESH CHANDRA, b Lahore, Pakistan, July 5, 34; m 60; c 3. FOOD TECHNOLOGY, BIOCHEMISTRY. Educ: Panjab Univ, India, BSc, 53, Hons, 55, MSc, 56; Univ Nebr, Lincoln, PhD(dairy mfg), 63. Prof Exp: Lectr chem, Panjab Univ, India, 56-57; lectr dairy chem, Nat Dairy Res Inst, India, 57-59; from asst to assoc dairy sci, Univ Nebr, Lincoln, 59-66; scientist dairy technol, Unilever, Ltd, Eng, 67-69; mgr res & develop, Dairylea Coop, Inc, 70-74; VPRES, WHEY PROD & TECH SERV, PURITY CHEESE CO, ANDERSON CLAYTON FOODS, 74- Mem: Am Dairy Sci Asn; Inst Food Technologists; Am Chem Soc; fel Am Inst Chemists; Am Mgt Asn. Res: Physico-chemical properties of milk; lipases; lysozymes; lipids; milk protein; microbial chemistry and enzymes; diary product technology; cultures; cultured products; food product development; whey protein manufacturing technology, quality control, cheese product development; whey fractionation. Mailing Add: 865 Mayer Lane Mayville WI 53050

CHANDER, JAGDISH, b Toba Tek Singh, India, Mar 7, 33; m 65; c 2. EXPERIMENTAL NUCLEAR PHYSICS. Educ: DAV Col, Jullundur, India, BSc, 52; Panjab Univ, India, BA, 53 & 54; Univ Rajasthan, MSc, 56; Univ Erlangen, Dr rer nat(physics), 61. Prof Exp: Demonstr chem, DAV Col, Jullundur, 52-53; demonstr, DSD Col, Gurgaon, 53-54; lectr physics, DAV Col, Jullundur, 56-68; lectr, Birla Sci Col, Pilani, 62 & Panjab Univ, 62-66; from asst prof to assoc prof, 66-70, PROF PHYSICS, UNIV WIS-STEVENS POINT, 70- Concurrent Pos: Res assoc, Siemens Res Lab, Erlangen, Ger, 61; dir, NSF Undergrad Res Partic, 68-71; acad year exten grant col teachers, NSF, 69-70. Mem: Am Phys Soc. Res: Low energy nuclear physics; semiconductor radiation detectors. Mailing Add: Dept of Physics & Astron Univ of Wis Stevens Point WI 54481

CHANDLER, A BLEAKLEY, b Augusta, Ga, Sept 11, 26; m; c 2. PATHOLOGY. Educ: Med Col Ga, MD, 48. Prof Exp: Intern, Baylor Univ Hosp, 48-49; resident path, Univ Hosp, Augusta, Ga, 49-50; from asst prof to assoc prof, 53-62, PROF PATH, MED COL GA, 62-, CHMN DEPT, 75- Concurrent Pos: Nat Cancer Inst trainee path, Med Col Ga, 50-51; Commonwealth Fund res fel, Inst Thrombosis Res, Norway, 63-64. Mem: Am Asn Hist Med; Int Acad Path; Am Soc Exp Path; Am Path & Bact. Res: In vitro thrombosis; experimental and human cardiovascular pathology. Mailing Add: Dept of Path Med Col of Ga Augusta GA 30902

CHANDLER, ALBERT MORRELL, b Pontiac, Mich, July 10, 32; m 59; c 3. BIOCHEMISTRY. Educ: Wayne State Univ, BS, 55, PhD(biochem), 62. Prof Exp: Pharmacist, Mich, 51-56; res assoc anat, Wayne State Univ, 63-66; from asst prof to assoc prof, 66-73, PROF BIOCHEM, SCH MED, UNIV OKLA, 73- Concurrent Pos: NIH fels, Ohio State Univ, 61-63. Mem: AAAS; Am Chem Soc. Res: Plasma protein metabolism; hepatic nucleic acid metabolism; regulation of hexosamine synthesis. Mailing Add: 2600 Abbey Rd Oklahoma City OK 73120

CHANDLER, ALFRED BERTRAM, b Geelong, Australia, Nov 30, 16; nat US; m 41; c 1. INORGANIC CHEMISTRY, ANALYTICAL CHEMISTRY. Educ: Univ Tenn, BS, 47. Prof Exp: Chemist, Aluminum Co Am, Tenn, 40-43; chemist-spectroscopist, Tenn Eastman Co, 43-47; chief chemist, Foote Mineral Co, 47-65; managing dir, Int Div, Andrew S McCreath & Son, Inc, 65-68; SCIENTIST, ROY F WESTON, INC, 68- Concurrent Pos: Dir, Andrew S McCreath & Son, Inc, 61- Mem: Am Chem Soc; Am Soc Testing & Mat; Am Inst Mining, Metall & Petrol Eng. Res: Instrumentation; emission and x-ray spectroscopy; flame photometry; atomic absorption; environmental science; waste water treatability. Mailing Add: Process Develop Lab Roy F Weston Inc Weston Way West Chester PA 19380

CHANDLER, ARTHUR CECIL, JR, b Hinton, WVa, Feb 14, 33; m 57; c 4. MEDICINE, OPHTHALMOLOGY. Educ: Fla Southern Col, AB, 53; Univ Tenn, MS, 55; Duke Univ, MD, 59; Am Bd Ophthal, dipl, 65. Prof Exp: Instr surg, Sch Med, Stanford Univ, 63-65; assoc, 65-66, asst prof, 66-70, ASSOC PROF OPHTHAL,

SCH MED, DUKE UNIV, 70-, ASSOC ANAT, 67- Concurrent Pos: Consult, Durham Vet Admin Hosp, 65-, chief ophthal, 66-; consult, Watts Hosp & Sea Level Hosp. Mem: AAAS; fel Am Col Surg; AMA; Am Asn Ophthal. Res: Prevention and treatment of amblyopia ex anopsia. Mailing Add: Box 3802 Duke Univ Med Ctr Durham NC 27710

CHANDLER, CARL DAVIS, JR, b Pulaski, Va, Jan 25, 44; m 66; c 2. ANALYTICAL CHEMISTRY. Educ: Emory & Henry Col, BS, 65; ETenn State Univ, MA, 67; Va Polytech Inst & State Univ, PhD(chem), 73. Prof Exp: ANAL GROUP AREA SUPVR CHEM, RADFORD ARMY AMMUNITION PLANT, HERCULES INC, 67- Concurrent Pos: Lectr short course liquid chromatography, Am Chem Soc, 73-; consult, US Army Res Off, Picatinny Arsenal. Mem: Am Chem Soc. Res: Investigation of column efficiency improvements in gas and liquid chromatography and analysis of propellants and explosives. Mailing Add: Radford Army Ammunition Plant Hercules Inc Radford VA 24141

CHANDLER, CLAY MORRIS, b McKenzie, Tenn, Nov 2, 27; m 56; c 2. ZOOLOGY. Educ: Bethel Col, Tenn, BS, 50; George Peabody Col, MA, 54; Ind Univ, PhD(zool), 65. Prof Exp: Asst prof biol, WGa Col, 59-61; prof zool, Bethel Col, Tenn, 65-70; PROF BIOL, MID TENN STATE UNIV, 70- Mem: AAAS; Am Soc Limnol & Oceanog; NAm Benthological Soc. Res: Ecology and systematics of freshwater triclad Turbellaria. Mailing Add: Dept of Biol Box 121 Mid Tenn State Univ Murfreesboro TN 37130

CHANDLER, COLSTON, b Boston, Mass, June 7, 39; m 65; c 3. THEORETICAL PHYSICS. Educ: Brown Univ, ScB, 61; Univ Calif, Berkeley, PhD(physics), 67. Prof Exp: Asst prof, 66-68, 70-73, ASSOC PROF PHYSICS, UNIV N MEX, 73- Mem: AAAS; Am Phys Soc; Sigma Xi. Res: Nonrelativistic quantum mechanical scattering theory. Mailing Add: Dept of Physics & Astron Univ of NMex Albuquerque NM 87131

CHANDLER, DAVID, b New York, NY, Oct 15, 44; m 66; c 2. CHEMICAL PHYSICS, STATISTICAL MECHANICS. Educ: Mass Inst Technol, SB, 66; Harvard Univ, PhD(chem physics), 69. Prof Exp: Res chemist, Univ Calif, San Diego, 69-70; asst prof, 70-75, ASSOC PROF CHEM, UNIV ILL, URBANA, 75- Concurrent Pos: Sloan fel, 72-76. Mem: Am Inst Physics; Am Phys Soc; Am Chem Soc. Res: Statistical mechanics of strongly interacting systems near and at equilibrium; phase transitions; structure of simple and complex molecular liquids; ionic solutions. Mailing Add: Dept of Chem Univ Ill Sch of Chem Sci Urbana IL 61801

CHANDLER, DAVID CULBERTSON, b Walnut Grove, Minn, July 11, 06; m 35; c 2. ZOOLOGY. Educ: Greenville Col, AB, 29; Univ Mich, AM, 30, PhD(zool), 34. Prof Exp: Instr zool, Univ Ark, 34-35; prof biol, McMurray Col, Tex, 35-36; instr zool, Univ Ark, 36-38; asst prof, Franz Theodore Stone Lab, Ohio State Univ, 38-44, assoc prof, 44-47, prof hydrobiol, 47-49; prof limnol, Cornell Univ, 49-53; prof zool, Biol Sta, 51-59, prof zool, Univ, 53-74, dir, Great Lakes Res Div, 60-74, EMER PROF ZOOL, UNIV MICH, ANN ARBOR, 74- Concurrent Pos: Mem comt water pollution, Comt Instnl Coop, chmn, 63-69, mem Great Lakes comt, chmn, 65-69; mem planning comt eutrophication, Nat Acad Sci, 65-67, mem nat sea-grant comt, 65-68; mem steering comt, Int Great Lakes Study Group, 65-; mem, Am Inst Biol Sci adv comt oceanic biol to Off Naval Res, 66-69; mem, Great Lakes Found, pres, 66-69; mem steering comt, Int Field Year of Great Lakes, Int Hydrol Decade, 66-; consult comt pub responsibilities, Am Inst Biol Sci, 67-; mem adv comt, Orgn Trop Studies, 68- Mem: AAAS; Ecol Soc Am; Am Micros Soc (vpres, 50, pres, 5; Am Soc Limnol & Oceanog (vpres, 49, pres, 57); Int Asn Great Lakes Res (pres, 67). Res: Limnology; plankton of streams and lakes; biological productivity of the Great Lakes. Mailing Add: 2980 Crayton Rd Naples FL 33940

CHANDLER, DEAN WESLEY, b Chicago, Ill, June 5, 44. PHYSICAL CHEMISTRY, ANALYTICAL CHEMISTRY. Educ: Harvard Univ, BA, 67; Northwestern Univ, PhD(chem), 73. Prof Exp: Lectr chem, State Univ NY Albany, 73-74, vis asst prof, 74-75; ASST PROF CHEM, WILLIAMS COL, 75- Mem: Am Chem Soc; Am Inst Physics. Res: Development of new physical techniques which might prove useful as tools for chemical analysis. Mailing Add: Dept of Chem Williams Col Williamstown MA 01267

CHANDLER, DONALD ERNEST, b San Bernardino, Calif, Nov 22, 25; m 46; c 3. PHYSICS. Educ: US Naval Acad, BS, 46; Univ Calif, Los Angeles, MA, 54, PhD(physics), 58. Prof Exp: Tech officer, Electronic Systs, Off Naval Res, US Navy, 55-57, physicist, Armed Forces Spec Weapons Proj, 57-59, proj officer, Adv Res Proj Agency, 59-63, sr prog officer, Electronics Lab, 63-66, proj mgr, Advan Res Proj Agency, 66-68; mem tech staff, 68-71, MGR PHYS SCI, TEMPO, GEN ELEC CO, 71- Mem: Acoustical Soc Am; Am Geophys Union. Res: Propagation of acoustic waves in media exhibiting relaxation; propagation of electromagnetic waves in ionized gases. Mailing Add: Tempo Gen Elec Co 816 State St Santa Barbara CA 93103

CHANDLER, FRANCIS WOODROW, JR, b Milledgeville, Ga, July 25, 43; m 68; c 2. VETERINARY PATHOLOGY. Educ: Univ Ga, BS, 66, DVM, 67, PhD(vet path), 73; Am Col Vet Pathologists, dipl. Prof Exp: Vet med officer, Venereal Dis Res Lab, Ctr Dis Control, 67-70; resident, Dept Vet Path, Col Vet Med, Univ Ga, 70-73; DEP DIR PATH DIV & CHIEF HISTOPATH LAB, BUR LABS, CTR DIS CONTROL, USPHS, 73- Mem: Int Acad Path; NY Acad Sci; Am Asn Lab Animal Sci; Sigma Xi. Res: Animal models of venereal diseases; pneumocystis carinii pneumonia; deep mycoses; anemia and neoplasia; surface ultrastructure of cells; rabies. Mailing Add: Path Div Bldg 1/2301 Ctr for Dis Control Atlanta GA 30333

CHANDLER, HORACE W, b Brooklyn, NY, May 23, 27; m 54; c 2. PHYSICAL CHEMISTRY, CHEMICAL ENGINEERING. Educ: Cornell Univ, BChE, 50; NY Univ, MChE, 55; Columbia Univ, DrEngSc, 60. Prof Exp: Tech trainee, Gen Chem Div, Allied Chem & Dye Corp, 50-51; chem engr, Gen Aniline & Film Corp, 51-53; res engr, Columbia Mineral Beneficiation Labs, 53-58; tech coordr, Radiation Applns, Inc, 58-60; tech adv to pres, Isomet Corp, 60-68; CHMN DEPT PHYS SCI & MATH, BERGEN COMMUNITY COL, 68- Concurrent Pos: Lectr, Stevens Inst Technol, 60. Mem: Am Inst Chem Eng; Am Chem Soc. Res: Radiation, polymer and high temperature chemistry; life support systems. Mailing Add: Dept of Physical Sci & Math Bergen Community Col Paramus NJ 07652

CHANDLER, J RYAN, b Charleston, SC, July 30, 23; m 41; c 3. SURGERY, OTOLARYNGOLOGY. Educ: Duke Univ, MD, 47; Am Bd Otolaryngol, dipl. Prof Exp: PROF OTOLARYNGOL, SCH MED, UNIV MIAMI, 52-, CHMN DEPT, 72- Concurrent Pos: Consult, Vet Admin Hosp, Coral Gables, Fla, 57 & NIH Commun Dis Training Grant Comt, 60-64. Mem: Am Soc Head & Neck Surg; Soc Head & Neck Surg; Am Laryngol Asn; Am Acad Ophthal & Otolaryngol; Am Col Surg. Res: Otology and head and neck cancer surgery; otolaryngology teaching. Mailing Add: Dept of Otolaryngol Univ of Miami Sch of Med Miami FL 33152

CHANDLER, KIRBY, b New York, NY, Apr 1, 27; m 48. PHYSICAL ANTHROPOLOGY. Educ: Univ Wash, BA, 57, PhD(primatology), 66. Prof Exp: Asst prof anthrop, Univ Wash, 60-64; asst prof, 64-66, chmn dept social sci, 66-69, PROF PHYS ANTHROP, SHORELINE COMMUNITY COL, 70- Mem: Am Anthrop Asn. Res: Primate taxonomy and ecology. Mailing Add: Shoreline Community Col 16101 Greenwood N Seattle WA 98133

CHANDLER, LELAND, b Rising Sun, Ind, June 12, 24; m 45; c 3. ENTOMOLOGY. Educ: Hanover Col, AB, 49; Purdue Univ, MS, 51, PhD, 55. Prof Exp: From instr to assoc prof entom, 53-62, asst head dept, 65-69, cur entom mus, 57-67, PROF ENTOM, PURDUE UNIV, 62- Concurrent Pos: Consult, US AID-Purdue/Vicosa Proj, Brazil, 73; entomologist, Nat Ctr Res on Rice & Beans, Goiania, Brazil, 75- Mem: Entom Soc Am. Res: Biosystematics of bees-wasps. Mailing Add: Dept of Entom Purdue Univ West Lafayette IN 47906

CHANDLER, LOUIS, b Rumania, Jan 15, 22; nat US; m 43. BIOPHYSICS, RADIATION PHYSICS. Educ: Univ Chicago, BS, 43; Univ Ill, MS, 47, PhD(biophysics), 54. Prof Exp: Instr physics, Univ Chicago, 43-44; physicist, Manhattan Dist, Chicago, Oak Ridge, Calif, 44-46; asst physics, Univ Ill, 46-48; physicist, Anderson Phys Lab, 48-49; chief physicist, Swift & Co, 55-57; from asst prof to assoc prof, Univ Ill, 57-67; PROF ENVIRON SCI, RUTGERS UNIV, 67-, DIR GRAD PROG RADIOL HEALTH, 68- Concurrent Pos: Pres, Radiation Control, Inc, 59-67; res prof, Univ Tokyo, 63-64; fel adv, AEC, 68- Mem: Am Phys Soc; Radiation Res Soc; Am Asn Physics Teachers; Health Phys Soc; Am Asn Physicists in Med. Res: Effect of small dose radiation on cell growth; radiation protection legislation; applications of Mössbauer effect. Mailing Add: Radiation Sci Ctr Rutgers Univ New Brunswick NJ 08903

CHANDLER, MICHAEL LYNN, b Bloomington, Ind, Aug 6, 45; m 65; c 2. PHARMACOLOGY, BIOCHEMISTRY. Educ: Ind Univ, AB, 67, MS, 70, PhD(pharmacol), 73. Prof Exp: Res assoc biochem, Fels Res Inst, 71-74; GROUP LEADER PHARMACOL, DOW CORNING CORP, 74- Res: Insulin catabolism and enzyme kinetics; pharmacokinetics and drug metabolism; pharmacology of organosilicon compounds; central nervous system and cardiovascular pharmacology. Mailing Add: Biosci Res Dow Corning Corp Midland MI 48640

CHANDLER, RAY JAMES, b Tioga, Tex, June 3, 24; m 41; c 6. CORROSION. Educ: Southern Methodist Univ, BS, 58. Prof Exp: Res chemist, 51-75, SR RES CHEMIST, MOBIL RES & DEVELOP CORP, 75- Mem: Nat Asn Corrosion Engrs; Sigma Xi. Res: Inhibition of corrosion or prevention, including studies of cathodic protection, corrosion inhibitors, coatings and special metals to control corrosion. Mailing Add: Mobil Res & Develop Corp Box 900 FRL Dallas TX 75221

CHANDLER, REGINALD FRANK, b Edmonton, Alta, Sept 18, 41; m 62; c 2. MEDICINAL CHEMISTRY. Educ: Univ Alta, BSc, 62, MSc, 65; Univ Sydney, PhD(pharmaceut chem), 69. Prof Exp: Lectr phytochem, Univ Sydney, 65-68; asst prof, 68-74, ASSOC PROF PHARM, DALHOUSIE UNIV, 74- Concurrent Pos: Secy-treas, Asn Fac of Pharm Can, 72-74. Mem: Acad Pharmaceut Sci; Am Soc Pharmacog; Chem Inst Can; Am Chem Soc; Int Pharmaceut Fedn. Res: Chemistry of the natural products of Sanguinaria canadensis and Calotropis procera; a study of sterols possessing antihyper cholesteremic activity. Mailing Add: Col of Pharm Dalhousie Univ Halifax NS Can

CHANDLER, RICHARD EDWARD, b Ft Pierce, Fla, Sept 9, 37; m 61; c 1. MATHEMATICS. Educ: Fla State Univ, BS, 59, MS, 60, PhD(math), 63. Prof Exp: Res assoc, Duke Univ, 63-65; asst prof math, 65-67, ASSOC PROF MATH, NC STATE UNIV, 67- Mem: Math Soc; Math Asn Am. Res: Point-set and algebraic topology, especially fixed point theory and rings of continuous functions. Mailing Add: Dept of Math NC State Univ Raleigh NC 27607

CHANDLER, ROBERT FLINT, JR, b Columbus, Ohio, June 22, 07; m 31; c 3. AGRONOMY. Educ: Univ Maine, BS, 29; Univ Md, PhD(pomol), 34. Hon Degrees: LLD, Univ Maine, 51; DH, Cent Luzon State Univ, 71; ScD, Univ Notre Dame, 71, Univ Philippines, 72, Univ NH, 72 & Univ Md, 75; LittD, Univ Singapore, 71. Prof Exp: Horticulturist, State Dept Agr, Maine, 29-31; asst, Exp Sta, Univ Md, 31-34; Nat Res Found fel forestry, Univ Calif, 34-35; from asst prof to prof forest soils, Cornell Univ, 35-47; dean col agr & dir exp sta, Univ NH, 47-50, pres, 50-54; asst dir agr, Rockefeller Found, 54-57, assoc dir agr sci, 57-65, spec field staff mem, Asian Veg Res & Develop Ctr, Taiwan, 65-75; CONSULT INT AGR, 75- Concurrent Pos: Vis prof, Agr Mech Col Tex, 40; soil scientist, Rockefeller Found, 46-47; dir, Int Rice Res Inst, Laguna, Philippines; trustee, China Found Educ & Cult, mem adv bd, Mekong Comt. Honors & Awards: Gold Medal Award, Govt India, 66; Sitari-I-Imtiaz Award, Govt Pakistan, 68. Mem: Fel Am Acad Arts & Sci; Am Soc Agron; Crop Sci Soc Am; Soc Int Develop. Res: Chemical composition of forest tree leaves and litter; vegetation as soil-forming factor; potassium nutrition of alfalfa; agricultural education and research in the Far East; administration of scientific research. Mailing Add: Petersham Rd Templeton MA 01468

CHANDLER, ROGER EUGENE, b Elmira, NY, Sept 16, 34; m 56; c 2. ORGANIC CHEMISTRY. Educ: Colgate Univ, AB, 56; Mass Inst Technol, PhD(org chem), 61. Prof Exp: Res chemist, Esso Res & Eng Co, 61-64, proj leader detergent additives, 64-66, res coordr, Paramins Div, Enjay Chem Co, 66-67, sect head detergent additives, 67-68, sect head res dept, Esso Petrol Co, Eng, 68-70, head engine testing activity, Enjay Additives Lab, 70-73, ADV ENVIRON CONSERV, PUB AFFAIRS DEPT, EXXON CORP, 73- Mem: Am Chem Soc; Soc Automotive Engrs. Res: Lubricating oil additives; environmental conservation. Mailing Add: 809 Village Green Westfield NJ 07090

CHANDLER, WEBSTER A, plant pathology, see 12th edition

CHANDLER, WILLIAM DAVID, b Brantford, Ont, Jan 29, 40; m 61; c 4. PHYSICAL ORGANIC CHEMISTRY. Educ: Queen's Univ, Ont, BSc, 62, PhD(chem), 65. Prof Exp: NATO fel, Pa State Univ, 65-66 & Rutgers Univ, 66-67; from asst prof to assoc prof, 67-74, PROF CHEM & HEAD DEPT, UNIV REGINA, 74- Mem: AAAS; fel Chem Inst Can; Am Chem Soc. Res: Organic synthesis; diphenyl ether chemistry; natural product chemistry of algae. Mailing Add: Dept of Chem Univ of Regina Regina SK Can

CHANDRA, G RAM, b India, Feb 10, 33; m. BIOCHEMISTRY. Educ: Agra Univ, BS, 51, MS, 53; Univ Alta, PhD(biochem), 62. Prof Exp: Res asst biochem, Indian Agr Res Inst, 55-59; scientist, Res Inst Advan Studies, 62-65; scientist, Mich State Univ, 65-66; SCIENTIST, AGR RES SERV, USDA, 66- Res: Biological volatiles; senescence; plant nutrition; hormones. Mailing Add: USDA Agr Res Serv Beltsville MD 20705

CHANDRA, JAGDISH, b Hyderabad, India, Oct 11, 35; US citizen; m 61; c 3. MATHEMATICS. Educ: Osmania Univ, India, BA, 55, MA, 57; Rensselaer Polytech Inst, PhD(math), 65. Prof Exp: Instr math, Rensselaer Polytech Inst, 65-66; res mathematician, US Army Arsenal, Watervliet, 66-70; chief appl math br, 70-73, assoc dir math div, 73-74, DIR MATH DIV, US ARMY RES OFF, DURHAM, 74-

Concurrent Pos: Adj asst prof, Rensselaer Polytech Inst, 66-70, Union Col, 68-69; adj assoc prof, Duke Univ, 74- Mem: Am Math Soc; Soc Indust & Appl Math. Res: Mathematical analysis, nonlinear differential and integral equation; operator inequalities. Mailing Add: US Army Res Off PO Box 12211 Research Triangle Park NC 27709

CHANDRA, KAILASH, b Kanpur, UP, India, Aug 20, 38; m 61; c 3. PHYSICS. Educ: Agr Univ, BSc, 56, MSc, 58; Gorakhpur Univ, PhD(physics), 67. Prof Exp: Lectr physics, KK Degree Col, India, 58-60; res fel, Coun Sci & Indust Res, 60-62; lectr, Gorakhpur Univ, 62-68; res assoc, Univ Ga, 68-69; assoc prof, 69-73, PROF PHYSICS, SAVANNAH STATE COL, 73- Mem: Am Asn Physics Teachers; Am Phys Soc. Res: Molecular spectroscopy; competency based education in physics. Mailing Add: 410 Paradise Dr Savannah GA 31406

CHANDRA, PURNA, b Khatauli, India, June 23, 29; m 64. AGRICULTURAL MICROBIOLOGY. Educ: Agra Univ, BSc, 49, MSc, 51; Ore State Univ, PhD(bact), 58. Prof Exp: Instr bact, Ore State Univ, 58-59; lectr, Univ Bagdad, 59-60; microbiologist, Can Dept Agr Res Sta, Sask, 60-63; asst prof biol, Mt Allison Univ, 63-67; assoc prof, 67-70, PROF BIOL, LAKE SUPERIOR STATE COL, 70- Mem: AAAS; Am Soc Microbiol; Soil Sci Soc Am; Am Inst Biol Sci. Res: Decomposition of organic matter; nitrogen transformation in soils; biocidal effect on soil microorganisms; soil respiration studies; effects of fertilizers on soil bacteria. Mailing Add: Dept of Biol Sci Lake Superior State Col Sault Ste Marie MI 49783

CHANDRAN, KRISHNAN BALA, b Madurai City, India, May 16, 44; m 72; c 1. BIOMECHANICS, MECHANICAL ENGINEERING. Educ: Madras Univ, BS, 63; Wash Univ, MS, 69, DSc(biomed eng), 72. Prof Exp: Tool engr, Hindustan Motors Ltd, Calcutta, India, 66-67; ASST PROF ORTHOP, BIOMECH LAB, MED SCH, TULANE UNIV, 74- Mem: Am Acad Mechanics; Soc Eng Sci. Res: Investigations on traumatology of the human head; hemodynamics at arterial curvature sites in relation to the origin of atherosclerosis; left ventricular dynamics with emphasis on the non invasive diagnosis of myocardial infarction. Mailing Add: Div of Orthop Biomech Lab Tulane Univ Med Sch New Orleans LA 70112

CHANDRAN, SATISH RAMAN, b Mar 2, 35; US citizen; m 66; c 2. ENTOMOLOGY, PARASITOLOGY. Educ: Univ Kerala, BS, 55, MS, 58; Univ Ill, Urbana, PhD(entom), 66. Prof Exp: Res assoc entom, Univ Ill, Urbana, 65-66; from instr to asst prof biol, Univ Ill, Chicago Circle, 66-72; ASSOC PROF BIOL, KENNEDY-KING COL, 72- Mem: Entom Soc Am; Am Inst Biol Sci; Nat Asn Biol Teachers. Res: Developmental morphology due to thermal stress in mosquitoes; microsomal oxidase activity in insects; toxicity of the photoisomere of cyclodiene insecticides of freshwater animals; lepidopteran morphology. Mailing Add: 817 S Loomis Chicago IL 60607

CHANDRASEKHAR, BELLUR S, b Bangalore, India, May 24, 28; m 55; c 2. PHYSICS. Educ: Mysore Univ, BSc, 47; Univ Delhi, MSc, 49; Oxford Univ, DPhil(physics), 52. Prof Exp: Res assoc physics, Univ Ill, 52-54; res physicist, Westinghouse Res Labs, 54-59, fel physicist, 59-61, sect mgr cryophysics, 61-63; prof physics, 63-67, chmn physics dept, 65-67, chmn biol dept & dean sci, 69-70, PERKINS PROF PHYSICS, CASE WESTERN RESERVE UNIV, 67- Concurrent Pos: Vis scientist, Oxford, 54-55; sr vis res fel, Imp Col, London, 61; consult, Bell Tel Labs, 65, 66 & 68, Argonne Nat Lab, 65-68, Lewis Res Ctr, 66 & NSF Inst Progs, 71. Mem: AAAS; fel Am Phys Soc; fel Brit Inst Physics & Phys Soc. Res: Liquid helium; superconductivity; electronic properties of solids. Mailing Add: Dept of Sci Case Western Reserve Univ Cleveland OH 44106

CHANDRASEKHAR, SUBRAHMANYAN, b Lahore, India, Oct 19, 10; nat US; m 36. ASTRONOMY, ASTROPHYSICS. Educ: Univ Madras, BA, 30; Cambridge Univ, PhD(theoret physics), 33, ScD(astrophys), 42. Prof Exp: Fel, Trinity Col, Cambridge Univ, 33-37; res assoc, 36-37, from asst prof to prof, 37-45, DISTINGUISHED SERV PROF, YERKES OBSERV, UNIV CHICAGO, 46- Concurrent Pos: Ed, Astrophys J, 52-71. Honors & Awards: Bruce Gold Medal, Astron Soc Pac, 52; Gold Medal, Royal Astron Soc, 53; Rumford Medal, Am Acad Arts & Sci, 57; Royal Medal, Royal Soc, 62; Nat Medal of Sci, 66; Henry Draper Medal, 71. Mem: Nat Acad Sci; Am Acad Arts & Sci; Am Astron Soc; Royal Astron Soc; Royal Soc. Res: Internal constitution of stars; white dwarfs; dynamics of stellar systems; theory of stellar atmospheres; radiative transfer; hydrodynamics and hydromagnetics; general relativity. Mailing Add: Lab for Astrophys & Space Res 933 E 56th St Chicago IL 60637

CHANDROSS, EDWIN A, b Brooklyn, NY, Oct 13, 34; m 61; c 2. ORGANIC CHEMISTRY, PHYSICAL CHEMISTRY. Educ: Mass Inst Technol, BS, 55; Harvard Univ, MA, 57, PhD(org chem), 60. Prof Exp: MEM TECH STAFF CHEM, BELL TEL LABS, 59- Concurrent Pos: Instr, Rutgers Univ, 62-63. Mem: AAAS; Am Chem Soc; The Chem Soc. Res: Photochemistry; fluorescence spectroscopy; chemiluminescence. Mailing Add: Bell Tel Labs Murray Hill NJ 07974

CHANDROSS, RONALD JAY, b New York, NY, Mar 21, 35; m 59; c 2. BIOPHYSICS. Educ: Polytech Inst Brooklyn, BS, 56; Mass Inst Technol, PhD(phys chem), 61. Prof Exp: Sr physicist, Gen Dynamics/Astronaut, 61-63; res chemist, Gen Chem Div, Allied Chem Corp, 63-66; fel biol, Mass Inst Technol, 66-68; res fel, Dept Surg, Mass Gen Hosp, 68-69; RES ASSOC, LAB REPRODUCTIVE BIOL, UNIV NC, CHAPEL HILL, 69- Mem: Am Crystallog Asn; Am Phys Soc; AAAS. Res: X-ray, electron diffraction; electron microscopy; structures of biological importance; steroid structure-activity relationships. Mailing Add: 111 Swing Bldg Univ of NC Div of Health Affairs Chapel Hill NC 27514

CHANEY, ALLAN HAROLD, b Kerrville, Tex, Dec 11, 23; m 48; c 2. ZOOLOGY. Educ: Tulane Univ, BS, 46, MS, 49, PhD(zool), 58. Prof Exp: Instr comp anat, Tulane Univ, 50; from asst prof to assoc prof zool, Ark Polytech Col, 53-59; prof & head dept, Del Mar Col, 59-63; assoc prof, 63-66, PROF BIOL, TEX A&I UNIV, 66- Mem: Am Soc Syst Zool; Am Soc Ichthyologists & Herpetologists. Res: Herpetology; systematics; marine biology. Mailing Add: 1208 W Richard Kingsville TX 78363

CHANEY, CHARLES LESTER, b Denver, Colo, Dec 21, 30; m 61; c 1. ANALYTICAL CHEMISTRY. Educ: Univ Wash, BS, 58. Prof Exp: Chemist, US Bur Mines, Nev, 58-60 & Md, 60-62; res asst anal chem, 62-67, STAFF ASSOC, GULF GEN ATOMIC CO, 67-; OWNER & MGR, SPECTRA/IRCO, 71- Mem: Am Chem Soc; Soc Appl Spectros; Am Soc Testing & Mat. Res: Design, construct spectroscopy equipment and instrumentation; spectrographic analysis. Mailing Add: 2987 Governor Dr San Diego CA 92122

CHANEY, DAVID WEBB, b Cleveland, Ohio, Dec 19, 15; m 38; c 1. ORGANIC CHEMISTRY. Educ: Swarthmore Col, AB, 38; Univ Pa, MS, 40, PhD(org chem), 42. Prof Exp: Asst sect leader, Am Viscose Corp, 42-52; sr res group leader, Chemstrand Corp, 52-53, asst dir res, 53-58, exec dir res, 58-60, vpres & exec dir, Chemstrand Res Ctr, Inc, 60-65, tech dir new prod & basic res, 65-67; DEAN SCH TEXTILES, NC STATE UNIV, 67- Mem: AAAS; Am Soc Eng Educ; Am Chem Soc; Am Asn Textile Chem & Colorists. Res: Organic synthetic fluorine compounds, especially fluorovinyls; vinyl copolymers; condensation polymers; synthetic fibers. Mailing Add: Sch of Textiles NC State Univ Raleigh NC 27607

CHANEY, EDWARD L, b Vicksburg, Miss, Jan 15, 43; m 64; c 3. RADIOLOGICAL PHYSICS, HEALTH PHYSICS. Educ: Millsaps Col, BS, 65; Univ Tenn, Knoxville, PhD(radiation, health physics), 69. Prof Exp: Fel, Univ Western Ont, 70-71; asst prof radiol, Univ Colo Med Ctr, Denver, 71-75; MEM FAC RADIOL, MEM MED CTR, SAVANNAH, GA, 75- Mem: Am Asn Physicists in Med; Am Col Radiol. Res: Medical physics. Mailing Add: Dept of Radiol Mem Med Ctr PO Box 6688 Sta C Savannah GA 31405

CHANEY, GEORGE L, b Coffeyville, Kans, Mar 16, 30; m 52; c 4. MATHEMATICS. Educ: Univ Kans, BS, 53, PhD(math educ), 67; Kans State Col Pittsburg, MS, 59. Prof Exp: High sch teacher, Kans, 53-56; teacher math, Coffeyville Community Col, 56-61; prof, Kans State Col Pittsburg, 62-64 & 66-68; PROF MATH, OTTAWA UNIV, KANS, 68- Mem: Math Asn Am. Mailing Add: Dept of Math Ottawa Univ Ottawa KS 66067

CHANEY, ROBERT BRUCE, JR, b Helena, Mont, Aug 22, 32; m 63; c 2. AUDIOLOGY. Educ: Univ Mont, BA, 58, MA, 60; Stanford Univ, PhD(audiol), 65. Prof Exp: Audiologist, Vet Admin, 60-62; res psychologist, US Navy Electronics Lab, Calif, 63-64; asst prof, 64-70, ASSOC PROF AUDIOL, UNIV MONT, 70- Mem: Acoust Soc Am; Am Speech & Hearing Asn. Res: Speech and hearing science; acoustical theory of speech production; neurological basis of speech perception. Mailing Add: Speech & Hearing Clin Univ of Mont Missoula MT 59801

CHANEY, ROBIN W, b Cleveland, Ohio, Dec 13, 38. NUMERICAL ANALYSIS, OPERATIONS RESEARCH. Educ: Ohio State Univ, BS, 60, PhD(math), 64. Prof Exp: Asst prof math, Western Wash State Col, 64-67 & Univ Calif, Santa Barbara, 67-69; PROF MATH, WESTERN WASH STATE COL, 69- Concurrent Pos: NSF fel, 66-67; vis prof, Chalmers Inst Technol, Sweden, 72-73. Mem: AAAS; Am Math Soc; Soc Indust & Appl Math; Inst Mgt Sci. Res: Convergence analysis of optimization algorithms. Mailing Add: Dept of Math Western Wash State Col Bellingham WA 98225

CHANEY, STEPHEN GIFFORD, b Ware Co, Pa, Feb 8, 44; m 68. BIOCHEMISTRY. Educ: Duke Univ, BS, 66; Univ Calif, Los Angeles, PhD(biochem), 70. Prof Exp: Am Cancer Soc fel microbiol, Sch Med, Washington Univ, 70-72; ASST PROF BIOCHEM, SCH MED, UNIV NC, CHAPEL HILL, 72- Res: Nucleic acid research; control mechanisms. Mailing Add: Dept of Biochem Univ of NC Chapel Hill NC 27514

CHANEY, WILLIAM R, b McAllen, Tex, Dec 2, 41; m 68; c 1. FORESTRY, PLANT PHYSIOLOGY. Educ: Tex A&M Univ, BS, 64; Univ Wis, PhD(tree physiol), 69. Prof Exp: Asst prof, 70-73, ASSOC PROF TREE PHYSIOL, PURDUE UNIV, WEST LAFAYETTE, 73- Concurrent Pos: Fel, Dept Forestry, Univ Wis, 69-70. Mem: Am Soc Plant Physiol; Ecol Soc Am; Soc Am Foresters. Res: Water relations of woody species, particularly studies of transpiration; air pollution effects on woody plants. Mailing Add: Dept of Forestry & Nat Resources Purdue Univ West Lafayette IN 47906

CHANG, ALBERT YEN, b China, Apr 15, 36; m 64. BIOCHEMISTRY. Educ: Nat Taiwan Univ, BS, 58; Univ Calif, Berkeley, MA, 62; Univ Ill, Urbana, PhD(biochem), 67. Prof Exp: SR RES SCIENTIST DIABETES RES, UPJOHN CO, 67- Mem: AAAS; Am Chem Soc; Am Diabetes Asn. Res: Etiology of diabetes; mechanism of protein synthesis; nucleic acid sequence; regulation of enzymes in mammalian systems. Mailing Add: Diabetes Res Upjohn Co Kalamazoo MI 49001

CHANG, BERKEN, b Berkeley, Calif, Apr 17, 35; m 59; c 3. ATOMIC PHYSICS. Educ: Calif Inst Technol, BS, 58, MS, 60; Univ Calif, Berkeley, PhD(physics), 67. Prof Exp: Mem tech staff, Hughes Aircraft Co, 60-61; consult, Bell Tel Labs, 67-68; res assoc, NASA Ames Res Ctr, 68-70; asst prof physics, 70-74, ASSOC PROF PHYSICS, CALIF STATE UNIV, LOS ANGELES, 74- Mem: Am Phys Soc; Optical Soc Am. Res: Optical pumping and spectroscopy; nuclear moments; collisional processes. Mailing Add: Dept of Physics Calif State Univ Los Angeles CA 90032

CHANG, BOMSHIK, b Inchon, Korea, Feb 6, 31; m; c 2. MATHEMATICS. Educ: Seoul Nat Univ, BA, 54, MA, 56; Univ BC, PhD, 59. Prof Exp: From lectr to asst prof, 58-69, ASSOC PROF MATH, UNIV BC, 69- Mem: Am Math Soc; Math Asn Am; Can Math Cong. Res: Group theory; Lie algebra. Mailing Add: Dept of Math Univ of BC Vancouver BC Can

CHANG, CATHERINE TEH-LIN, b China. PHOTOCHEMISTRY, PHOTOGRAPHIC CHEMISTRY. Educ: Nat Taiwan Univ, BS, 58; Washington Univ, PhD(chem), 64. Prof Exp: Res chemist, 64-69, sr res chemist, 69-75, RES ASSOC, PHOTO PROD DEPT, E I DU PONT DE NEMOURS & CO, INC, 75- Mem: Am Chem Soc; Soc Photog Scientists & Engrs. Res: Stereochemical course of the diazomethane-carbonyl reaction; acid-catalyzed cyclization of farnesol; photopolymerization; photoimaging systems. Mailing Add: Photo Prod Dept Exp Sta E I du Pont de Nemours & Co Inc Wilmington DE 19898

CHANG, CHAE HAN JOSEPH, b Seoul, Korea, July 7, 29; US citizen; m 56; c 4. RADIOLOGY. Educ: Severance Union Med Col, MD, 53; Am Bd Radiol, dipl, 59; Nagoya Univ, PhD(med sci), 66. Prof Exp: Resident radiol, Emory Univ Hosp, 55-58, instr, Sch Med, 58-59; chief, Man Mem Hosp, WVa, 59-63; from assoc prof to prof, Sch Med, WVa Univ, 64-70; PROF RADIOL, SCH MED, UNIV KANS, 70- Mem: AMA; fel Am Col Radiol; Radiol Soc NAm; Am Roentgen Ray Soc; Asn Univ Radiol. Res: Roentgenological measurement of the right descending pulmonary artery in normal state and pulmonary hypertension. Mailing Add: Dept of Radiol Univ of Kans Med Ctr Kansas City KS 66103

CHANG, CHANG-SUN, b Hangchow, China, Dec 28, 34; m 59; c 2. APPLIED MECHANICS. Educ: Univ Ill, BS, 55; Univ Minn, PhD(mech), 61. Prof Exp: Civil engr, Ill State Hwy Dept, 55; engr, Res Labs Div, Bendix Corp, 59-61, sr engr, 61-64; res specialist, Lockheed Missiles & Space Co, 64-67, staff scientist, 67-70; prin engr, Bieber-Chang Assocs, 70-71; PRES, APPL DYNAMICS RES CORP, 71- Mem: Soc Eng Sci; Am Inst Aeronaut & Astronaut. Res: Experimental mechanics; aeroelastic testing; damping studies; real-time dynamic data acquisition and analysis; advanced electromechanical devices; dynamic control; structural and material damping; piezoelectric transducers; thermoelasticity. Mailing Add: Appl Dynamics Res Corp 117 Sahara Off Park Huntsville AL 35801

CHANG, CHARLES C, b Liaoning, China, May 28, 39; m 73. PHYSICAL CHEMISTRY. Educ: Taiwan Nat Normal Univ, BS, 66; Univ Wis, Milwaukee, MS, 69; Johns Hopkins Univ, PhD(phys chem), 72. Prof Exp: Fel catalysis, Johns Hopkins Univ, 72-73; ASSOC SR RES CHEMIST, GEN MOTORS RES LAB, 73- Mem: Am Chem Soc. Res: Heterogeneous catalysis on interactions of gases with the surface of metal catalysts. Mailing Add: 13136 Wales Huntington Woods MI 48070

CHANG, CHARLES HUNG, b Szechwan, China, Apr 4, 25; US citizen; m 56; c 3. SYNTHETIC ORGANIC CHEMISTRY. Educ: Nat Cent Polytech Col, China, dipl, 45; Wayne State Univ, PhD(org chem), 59. Prof Exp: Chem engr, Taiwan Fertilizer Co, 47-54; res assoc chem, Wayne State Univ, 58-61; res chemist, GAF Corp, 61-65, res specialist dyestuffs, 65-73; RES ASSOC DYESTUFFS, SCOTT ALTHOUSE RES CTR, CROMPTON & KNOWLES CORP, 73- Mem: Am Chem Soc. Res: Design and synthesis of organic dyestuffs and intermediates. Mailing Add: 36 Dorchester Dr Wyomissing Hills PA 19610

CHANG, CHARLES YU-CHUN, b Harbin, China, Sept 18, 41; US citizen; m 69; c 1. PHARMACEUTICAL CHEMISTRY. Educ: Tamkang Col, Taiwan, BS, 64; Am Univ, MS, 74. Prof Exp: Res chemist, Kingdom Pharmaceut Co, Taiwan, 65-67; instr tech & mil Mandarin Chinese, Berlitz Lang Inst, Washington, DC, 67-69; DIR QUAL CONTROL, MIFFLIN McCAMBRIDGE CO, RIVERDALE, 69- Mem: Nat Asn Pharmaceut Mfrs. Res: Large scale purification and disinfection of water; origination of rapid pharmaceutical tests and analyses; activation parameters for hydrogen-deuterium exchanges. Mailing Add: 4106 Quintana St Hyattsville MD 20782

CHANG, CHEN CHUNG, b Tientsin, China, Oct 13, 27; nat US; m 51; c 4. MATHEMATICAL LOGIC. Educ: Harvard Univ, AB, 49; Univ Calif, PhD(math), 55. Prof Exp: Lectr math, Univ Calif, 54-55; instr, Cornell Univ, 55-56; asst prof, Univ Southern Calif, 56-58; ASST PROF MATH, UNIV CALIF, LOS ANGELES, 58- Concurrent Pos: NSF sr fel, Inst Advan Study, 62-63; Fulbright sr res fel, UK, 66-67; vis fel, All Souls Col, Oxford Univ, 66-67; consult ed, J Symbolic Logic, Asn Symbolic Logic, 68-; mem US nat comt, Int Union Hist & Philos of Sci, 70-72; ed, Ann Math Logic, 70- Mem: Am Math Soc; Asn Symbolic Logic. Res: Logic; set theory; abstract algebra Mailing Add: Dept of Math Univ of Calif Los Angeles CA 90024

CHANG, CHIN-CHUAN, b Laiyang, China, Oct 5, 25; m 45; c 3. ENDOCRINOLOGY, REPRODUCTIVE PHYSIOLOGY. Educ: Cath Univ Peiping, BS, 48; NY Univ, MA, 61; Univ Wis-Madison, PhD(endocrinol, reprod physiol), 67. Prof Exp: Instr biol, Nat Defense Med Col, Taiwan, 55-58; instr zool, Nat Taiwan Univ, 58-63, assoc prof, 63-65; res assoc endocrinol, 69-70, STAFF SCIENTIST, BIOMED DIV, POP COUN, ROCKEFELLER UNIV, 71- Concurrent Pos: Res fel med, Pop Coun, Rockefeller Univ, 67-69. Mem: AAAS; Am Fertil Soc; Endocrine Soc; NY Acad Sci. Res: Relationship between blastocyst and endometrium in process of implantation and requirements of ovarian hormones for formation of deciduomata; endocrine activity of steroids released through dimethylpolysiloxane membrane; more effective methods in contraception. Mailing Add: Pop Coun Rockefeller Univ York Ave & 66th St New York NY 10021

CHANG, CHING MING, b Nanking, China, Oct 13, 35; m 64; c 2. FLUID PHYSICS, ENGINEERING SCIENCE. Educ: Aachen Tech Univ, Dipl Ing, 62, DrIng(fluid physics), 67. Prof Exp: Res assoc, Shock Tube Lab, Inst Mech, Aachen Tech Univ, 62-64, instr, 64-67, res assoc, 67; vis asst prof eng mech, NC State Univ, 68-70, asst prof, 70-73; sr engr, 73-75, consult, 75, SUPVR, LINDE DIV, UNION CARBIDE CORP, 75- Concurrent Pos: Deleg, Int Coun Sci Unions, Madrid, 65, London, 67; asst ed, Plasma Physics, 71-72; adj assoc prof eng sci, aerospace & nuclear eng, State Univ NY Buffalo, 75- Mem: AAAS; Am Phys Soc; Am Soc Mech Engrs; Nat Soc Prof Engrs. Res: Heat transfer; energy conversion; electrohydrodynamics; applied mechanics; particle physics; thermal sciences and air pollution. Mailing Add: Linde Div Union Carbide Corp PO Box 44 Tonawanda NY 14150

CHANG, CHING-JEN, b Keelung, Taiwan, Feb 2, 41; m 67; c 1. PHYSICAL ORGANIC CHEMISTRY. Educ: Tunghai Univ, Taiwan, BS, 63; Marquette Univ, MS, 67; Univ Calif, Berkeley, PhD(org chem), 71. Prof Exp: Res fel chem, Univ Fla, 71-73; SR CHEMIST COATINGS, ROHM & HAAS CO, 73- Mem: Am Chem Soc. Res: Carbanions; ion-pairs structures; water soluble polymers; hydrophobic association; ionic association; polyurethanes; thickeners. Mailing Add: Res Labs Rohm & Haas Co Norristown & McKean Rds Spring House PA 19477

CHANG, CHING-JER, b Hsin-chu, Taiwan, Oct 17, 42. NATURAL PRODUCTS CHEMISTRY, MEDICINAL CHEMISTRY. Educ: Nat Taiwan Cheng Kung Univ, BS, 65; Ind Univ, PhD(org chem), 72. Prof Exp: Res asst chem, Nat Taiwan Univ, 66-67; teaching asst, NMex Highlands Univ, 68; res & teaching asst, Ind Univ, 68-72; res assoc, 72-73, ASST PROF MED CHEM & PHARMACOG, PURDUE UNIV, WEST LAFAYETTE, 73- Mem: Am Chem Soc; The Chem Soc; Am Soc Pharmacog; Phytochem Soc NAm. Res: Structure elucidation, stereochemistry, biosynthesis and partial synthesis of natural products; interaction of small molecules and drugs with macromolecules; biomedical application of spectroscopy. Mailing Add: Dept of Med Chem & Pharmacog Purdue Univ Sch of Pharm West Lafayette IN 47906

CHANG, CHIN-HAI, b Tainan, Taiwan. CANCER, IMMUNOCHEMISTRY. Educ: Nat Taiwan Univ, BS, 65; Wash Univ, PhD(develop biochem), 71. Prof Exp: Fel physiol chem, Roche Inst Molecular Biol, 71-73; ASST PROF CANCER RES, SCH MED, TUFTS UNIV, 73- Concurrent Pos: Consult immunochem, Leary Lab, 73- Mem: AAAS. Res: Biochemical and immunological studies of placental proteins of relevance to onco-developmental gene expression and reproductive physiology; radioimmunoassays; organ-specificity and physiological roles of alkaline phosphatases. Mailing Add: Dept of Path & Cancer Res Ctr Tufts Univ Sch of Med Boston MA 02111

CHANG, CHU HUAI, b Fukien, China, Oct 1, 17; m 59; c 2. RADIOLOGY. Educ: St John's Univ, China, BS, 41, MD, 44. Prof Exp: From instr to asst prof radiol, Sch Med, Yale Univ, 54-62; assoc prof, 62-67, PROF RADIOL, COL PHYSICIANS & SURGEONS, COLUMBIA UNIV, 67- Concurrent Pos: Res fel radiol, Sch Med, Univ Calif, 47-49; res fel, Sch Med, Yale Univ, 50-54. Mem: AAAS; Am Col Radiol; Asn Univ Radiol. Res: Radiation therapy and radiobiology. Mailing Add: Col Physicians & Surgeons Columbia Univ New York NY 10032

CHANG, CHUAN CHUNG, b Tainan, Formosa, Nov 28, 38; m 63; c 2. PHYSICS. Educ: Rensselaer Polytech Inst, BS, 62; Cornell Univ, PhD(physics), 67. Prof Exp: Res asst physics, Cornell Univ, 63-67; MEM TECH STAFF, BELL LABS, 67- Res: Surface physics; crystallography; electron diffraction; electron spectroscopy; electronics materials. Mailing Add: 2C-136 Bell Labs Murray Hill NJ 07974

CHANG, CLIFFORD WAH JUN, b Honolulu, Hawaii, July 25, 38. ORGANIC CHEMISTRY Educ: Univ Southern Calif, BS, 60; Univ Hawaii, PhD(chem), 64. Prof Exp: Jr chemist, Cyclo Chem Corp, Calif, 59; asst marine chemist, Hawaii Marine Lab, 61; fel chem, Univ Ga, 64-68; asst prof, 68-73, ASSOC PROF CHEM, UNIV W FLA, 73- Mem: Am Chem Soc; The Chem Soc. Res: Structural determinations and synthesis of natural products. Mailing Add: Dept of Chem Univ of WFla Pensacola FL 32504

CHANG, DONALD CHOY, b Kwangtung, China, Aug 28, 42. BIOPHYSICS, CELL PHYSIOLOGY. Educ: Nat Taiwan Univ, BS, 65; Rice Univ, MA, 67, PhD(physics), 70. Prof Exp: Res assoc physics, Rice Univ, 70-74; instr, 73-74, ASST PROF BIOPHYS, BAYLOR COL MED, 74- Concurrent Pos: Welch Found fel biophys, Baylor Col Med, 70-73; adj asst prof physics, Rice Univ, 74- Mem: AAAS; Am Phys Soc; Biophys Soc. Res: Nuclear magnetic resonance studies of cellular water and ions; cellular transport. Mailing Add: Dept of Physics Rice Univ Houston TX 77001

CHANG, EUGENE Y C, polymer chemistry, see 12th edition

CHANG, FA YAN, b Shantung, China, May 5, 32; m 60; c 2. WEED SCIENCE, PLANT PHYSIOLOGY. Educ: Nat Taiwan Univ, BSc, 53; Univ Alta, MSc, 66, PhD(plant sci), 69. Prof Exp: Asst agronomist, Taiwan Tobacco Res Inst, Repub China, 54-64; instr weed sci, Univ Alta, 69-70; res assoc, Univ Guelph, 70-74; HERBICIDE EVAL OFFICER, CAN DEPT AGR, 74- Concurrent Pos: Nat executive, Can Weed Comt, 75-; mem, Plant Growth Regulator Working Group. Mem: Agr Pesticide Soc Can; Weed Sci Soc Am; Can Soc Plant Physiol. Res: Weed control; herbicide physiology and antidotes. Mailing Add: Plant Prod Div Agr Can 930 Carling Ave Ottawa ON Can

CHANG, FRANKLIN, b Princeton, NJ, Feb 12, 42; m 67; c 1. PHYSIOLOGY, ENTOMOLOGY. Educ: Univ Md, BS, 63; Univ Ill, PhD(entom), 69. Prof Exp: Asst prof biol, Alma Col, Mich, 69-70; asst prof entom, 70-74, ASSOC PROF ENTOM, UNIV HAWAII, 74- Mem: AAAS; Entom Soc Am; Am Inst Biol Sci. Res: Carbohydrate and lipid metabolism in insects. Mailing Add: Dept of Entom Univ of Hawaii 2500 Dole St Honolulu HI 96822

CHANG, FRANKLIN SHIH CHUAN, b Nanking, China, Dec 30, 15; US citizen; m 38, 69; c 4. POLYMER CHEMISTRY. Educ: Purdue Univ, MS, 49; Univ Md, PhD(chem), 52. Prof Exp: Group leader anal & phys chem, Mystik Adhesive Prod, Inc, 52-60; MGR POLYMER PHYSICS, INGERSOLL RES CTR, BORG-WARNER CORP, 60- Mem: Am Chem Soc. Res: Physical testing of viscoelastic materials; adhesion and adhesives; infrared spectrometry; molecular structure; submicron particle size distribution measurements; polymer physics. Mailing Add: Ingersoll Res Ctr Borg-Warner Wolf & Algonquin Rds Des Plaines IL 60018

CHANG, FREDERIC CHEWMING, b San Francisco, Calif, Aug 5, 05; m 32; c 3 ORGANIC CHEMISTRY. Educ: Harvard Univ, MA, 40, PhD(org chem), 41. Prof Exp: From instr to asst prof chem, Lingnan Univ, 30-38, cur dept, 32-38; res assoc, Stanford Univ, 41-42; spec res assoc, Off Sci Res & Develop, Harvard Univ, 42-46; prof chem & chemn dept, Lingnan Univ, 46-51; lectr chem & res assoc path, 51-59, prof pharmacog, Col Pharm, 59-72, PROF BIOCHEM, COL BASIC MED SCI, UNIV TENN, MEMPHIS, 72- Mem: AAAS; Am Chem Soc; The Chem Soc; Am Soc Pharmacol; Phytochem Soc NAm Res: Naphthoquinones; antimalarials; steroids; medicinal plants. Mailing Add: Dept of Biochem Univ of Tenn Ctr for Health Sci Memphis TN 38163

CHANG, GEORGE K, low temperature physics, space science, see 12th edition

CHANG, GEORGE WASHINGTON, b Madison, Wis, Feb 22, 42. MICROBIAL PHYSIOLOGY, MICROBIAL GENETICS. Educ: Princeton Univ, AB, 63; Univ Calif, Berkeley, PhD(biochem), 67. Prof Exp: Vis scientist, Lab Molecular Biol, NIH, 67-68; NIH fel biochem, 68-70, ASST PROF FOOD MICROBIOL, NUTRIT SCI DEPT, UNIV CALIF, BERKELEY, 70- Mem: Am Soc Microbiol; Inst Food Technologists; Environ Mutagen Soc. Res: Microbial genetics and physiology; food microbiology and fish spoilage; human intestinal microflora and microecology; genetic engineering and enzyme technology; mycotPxins and environmental mutagens. Mailing Add: Dept of Nutrit Sci Univ of Calif Berkeley CA 94720

CHANG, HAN-CHUAN LIU, b Kwangsi, China; US citizen; m 63; c 2. MOLECULAR SPECTROSCOPY. Educ: Tunghai Univ, Taiwan, BS, 60; Univ Southern Calif, MS, 64, PhD(physics), 73. Prof Exp: Teaching asst physics, Tunghai Univ, Taiwan, 60-61; res asst neurophysiol, Med Sci Res, Rancho Los Amigos Hosp, Downey, Calif, 64-66; MEM TECH STAFF, LOGICON, SAN PEDRO, 73- Res: Optics; energy sources. Mailing Add: 3711 Toland Ave Los Alamitos CA 90720

CHANG, HERBERT HANG SHING, organic chemistry, physical organic chemistry, see 12th edition

CHANG, HOU-MIN, b Chiayi, Taiwan, Aug 29, 38; m 66; c 2. WOOD CHEMISTRY. Educ: Nat Taiwan Univ, BA, 62; Univ Wash, MS, 66, PhD(wood chem), 68. Prof Exp: Fel, 68-69, vis asst prof, 69, asst prof, 70-73, ASSOC PROF WOOD & PAPER SCI, NC STATE UNIV, 73- Mem: Am Chem Soc; Am Tech Asn Pulp & Paper Indust. Res: Species variation in wood lignins; isolation and characterization of cellulase lignin; characterization of residual lignin in kraft pulps of various yield; delignification by oxygen and alkali; lignin biodegradation. Mailing Add: Dept of Wood & Paper Sci NC State Univ Raleigh NC 27607

CHANG, HOWARD HOW CHUNG, b Honolulu, Hawaii, Nov 16, 22; m 52; c 1. THEORETICAL PHYSICS. Educ: Calif Inst Technol, BS, 44; Univ Calif, MA, 49; Harvard Univ, PhD(physics), 55. Prof Exp: Instr physics, Clarkson Tech Inst, 49-50; instr math, Univ Hawaii, 50-52; mem tech staff, Microwave Lab, Gen Elec Co, 55-57; physicist, Rand Corp, Calif, 57-58 & Hughes Aircraft Co, 58-61; sr math physicist, Stanford Res Inst, 61-75; STAFF SCIENTIST, LOCKHEED MISSILES & SPACE CO, 75- Concurrent Pos: Liaison physicist, Off Naval Res, London, 69-70. Mem: Am Phys Soc; AAAS. Res: Electromagnetic theory; plasma physics; controlled fusion; solar energy; macroscopic applications of superconductivity; energy storage; energy economics. Mailing Add: 337 Los Altos Ave Los Altos CA 94022

CHANG, HSIEN-HSIN, b Yuan-lin, China, June 16, 42; m 69. ORGANIC CHEMISTRY. Educ: Nat Taiwan Univ, BS, 65; Univ Miami, MS, 69; Univ Wash, PhD(org chem), 74. Prof Exp: Teaching assoc chem, Univ Wash, 73-74; RES ASSOC BIOCHEM, MOLECULAR & CELL BIOL, CORNELL UNIV, 74- Mem: Am Chem Soc. Res: Bacterial metabolism of coenzymes; synthesis of biologically significant compounds. Mailing Add: Biochem Molecular & Cell Biol Savage Hall Cornell Univ Ithaca NY 14853

CHANG, HSING-TZE RUAN, b Tainan, Taiwan, Apr 15, 39; US citizen; m 67; c 3. PROTEIN CHEMISTRY, NUTRITIONAL BIOCHEMISTRY. Educ: Nat Taiwan Norm Univ, BS, 63; Miss State Univ, MS, 66; Univ Calif, Berkeley, PhD(nutrit), 75. Prof Exp: Res asst org chem, Inst Chem, Acad Sinica, Taiwan, 62-64; res asst protein chem, Dept Microbiol, Columbia Univ, 66-67; biochemist, Arequipa Found, Berkeley, 67-68; res asst nutrit & food, Western Regional Res Lab, USDA, Berkeley, 70-75; CHEMIST, INST MED SCI, PAC MED CTR, 75- Res: Application of biochemical expertise in protein chemistry to the solution of scientific problems of medical and nutritional interest. Mailing Add: Inst of Med Sci Pac Med Ctr 2200 Webster St San Francisco CA 94115

CHANG, IK-CHIN, b Suncheon, Korea, Jan 5, 16; m 42; c 3. IMMUNOLOGY. Educ: Yonsei Univ, Korea, MD, 40, DmSc(microbiol), 60; Johns Hopkins Univ, MPH, 54. Prof Exp: Dep dir bioprod, Nat Health Inst, Korea, 45-56; assoc prof microbiol,

Yonsei Univ Col, 57-60; prof, Cath Med Col, Korea, 60-66; res assoc, Sch Pub Health, Univ Pittsburgh, 66-69; BACTERIOLOGIST BIO-PROD, BUR DIS CONTROL & LAB SERV, MICH DEPT PUB HEALTH, 69- Mem: Am Soc Microbiol; Am Asn Immunol. Res: Establishment of standard procedure in toxic factors in Pertussis vaccine. Mailing Add: Mich State Dept Pub Health Bur Dis Control 3500 N Logan St Lansing MI 48914

CHANG, I-LOK, b Amoy, China, July 9, 43; US citizen; m 70; c 1. MATHEMATICAL ANALYSIS. Educ: Calif Inst Technol, BS, 65; Cornell Univ, PhD(math), 71. Prof Exp: From instr to asst prof, 70-76, ASSOC PROF MATH, AM UNIV, 76- Mem: Am Math Soc; AAAS. Res: Functions of a complex variable; numerical analysis. Mailing Add: Dept of Math Statist Comput Sci Am Univ Washington DC 20016

CHANG, IN-KOOK, b Choonchun, Korea, Aug 24, 43; m 71; c 2. PLANT PHYSIOLOGY. Educ: Seoul Nat Univ, BS, 66; Va Polytech Inst & State Univ, MS, 70; Univ Chicago, PhD(biol), 74. Prof Exp: ASST PROF PLANT PHYSIOL, VA POLYTECH INST & STATE UNIV, 75- Mem: AAAS; Am Soc Plant Physiologists; Weed Sci Soc Am. Res: Plant growth and development; effects of growth regulators on crop productivity; plant hormone metabolism; mode of action of herbicides, degradation of herbicides and growth regulators in plants and soils. Mailing Add: Dept of Plant Path & Physiol Va Polytech Inst & State Univ Blacksburg VA 24061

CHANG, IRENE CHING LAI, b Peiping, China, July 14, 16; m 41; c 2. NUTRITION. Educ: Nanking Univ, BS, 39, MS, 42; State Univ Wash, PhD(foods nutrit), 49. Prof Exp: Lab instr gen org & anal chem, Nanking Univ, 39-42, instr chem, 42-45; res asst foods & nutrit, State Univ Wash, 46-48; res assoc, Syracuse Univ, 48-51, from asst prof to assoc prof, 51-56; res scientist, Joseph E Seagram & Sons, 56-64; RES ASSOC BIOCHEM, NC STATE UNIV, 65- Mem: Inst Food Technologists; Am Chem Soc. Res: Mineral metabolism and nutrition. Mailing Add: 516 Emerson Dr Raleigh NC 27609

CHANG, JACK CHE-MAN, b Shanghai, China, Nov 19, 41; m 65. ANALYTICAL CHEMISTRY. Educ: Asbury Col, BA, 61; Univ Ill, Urbana, MS, 63, PhD(chem), 65. Prof Exp: Res assoc electrochem luminescence, Mass Inst Technol, 66-67; SR RES CHEMIST, EASTMAN KODAK CO, 67- Res: Electrochemistry of organic compounds. Mailing Add: Eastman Kodak Co Res Labs 343 State St Rochester NY 14650

CHANG, JAMES C, b Shanghai, China, Aug 8, 30; m 70. PHYSICAL CHEMISTRY, INORGANIC CHEMISTRY. Educ: Mt Union Col, BS, 57; Univ Calif, Los Angeles, PhD(chem), 64. Prof Exp: Res chemist, E R Squibb & Sons Div, Olin Mathieson Chem Co, 59-62; asst prof chem, State Col Iowa, 64-67; from asst prof to assoc prof, 67-74, actg head dept, 75-76, PROF CHEM, UNIV NORTHERN IOWA, 74- Concurrent Pos: Vis prof, Univ Copenhagen, 69-70. Mem: AAAS; Am Chem Soc. Res: Inorganic synthesis and substitution reactions of inorganic complex compounds. Mailing Add: Dept of Chem Univ of Northern Iowa Cedar Falls IA 50613

CHANG, JAW-KANG, b Cholon, S Vietnam, Aug 11, 42; Taiwan citizen; m 62. BIOCHEMISTRY. Educ: Nat Taiwan Univ, BS, 65; Univ NB, Fredericton, PhD(org chem), 69. Prof Exp: Res fel peptide chem, Inst Biomed Res, Univ Tex, Austin, 69-73; SR DEVELOP CHEMIST, DEPT BIO-PROD, BECKMAN INSTRUMENTS INC, 73- Res: Synthetic chemistry of natural products; alkaloids, peptides and proteins. Mailing Add: Beckman Instruments Inc Dept of Bio-prod 1117 California Ave Palo Alto CA 94304

CHANG, JEFFREY PEH-I, b Changteh, China, Oct 10, 17; US citizen; c 2. CELL BIOLOGY. Educ: Nat Cent Univ, Chungking, BS, 41; Univ Ill, MS, 46, PhD(zool), 49. Prof Exp: From asst prof to assoc prof, Univ Tex Postgrad Sch Med, 55-62; from asst biologist to assoc biologist, Univ Tex M D Anderson Hosp & Tumor Inst Houston, 55-64, actg chief sect exp path, 59-64, biologist & prof biol, 64-72; PROF CELL BIOL, UNIV TEX MED BR GALVESTON, 72- Concurrent Pos: Spec consult, Nat Cancer Inst, 58-61 & Sch Aerospace Med, Brooks AFB, 62-64; consult, Univ Tex M D Anderson Hosp & Tumor Inst Houston, 72-; res consult, Nat Sci Coun, Repub China, 74-; academician, Academia Sinica, Repub China. Mem: AAAS; Acad Sinica; Am Soc Exp Path; Am Asn Cancer Res; Histochem Soc. Res: Ultrastructural and histochemical studies of cells and tissues; initiation and mechanism of carcinogenesis; tumor production and biology. Mailing Add: Dept of Human Biol Chem & Genet Univ of Tex Med Br Galveston TX 77550

CHANG, JEN HU, b Ningpo, China, Mar 21, 28; m 57. GEOGRAPHY. Educ: Chekiang Univ, China, BA, 49; Clark Univ, PhD, 54. Prof Exp: Res assoc climat, Harvard Univ, 54-57; assoc prof geog, 66-71, PROF GEOG, UNIV HAWAII, 71-; CLIMATOLOGIST, EXP STA, HAWAIIAN SUGAR PLANTERS ASN, 57- Mem: Am Meteorol Soc; Asn Am Geogr. Res: Geography of Asia; regional and agricultural climatology. Mailing Add: Dept of Geog Univ of Hawaii Honolulu HI 96822

CHANG, JOHN WAN-YUIN, b Peiping, China, May 19, 11. BIOCHEMISTRY, NUTRITION. Educ: Army Vet Col, China, DVM, 32; Va Polytech Inst, MS, 49; Ala Polytech Inst, MS, 51; Agr & Mech Col Tex, PhD(biochem), 55. Prof Exp: Res asst, Ala Polytech Inst, 50; res asst, Agr & Mech Col Tex, 51-55; res assoc, Univ Md, 56; assoc prof chem, Wilmington Col, 56-58, prof, 59-61; CHIEF ENVIRON CHEM DIV, BUR LABS, MD STATE DEPT HEALTH & MENT HYG, 61- Mem: Am Pub Health Asn; AAAS; Am Chem Soc; Asn Off Anal Chemists; Air Pollution Control Asn. Res: Microbiological assays of amino acids; studies on gossypol; chemical index of protein; organic and analytical chemistry on food, drug, water, air and textiles; microbiology; radiology. Mailing Add: Md State Dept Health & Ment Hyg 16 E 23rd St Lab Admin Baltimore MD 21218

CHANG, JOSEPH YUNG, b Nanking, China, Jan 30, 32; nat US; m 63. PHYSICAL CHEMISTRY. Educ: Taiwan Col Eng, BS, 53; Univ Notre Dame, MS, 57, PhD(phys chem), 58. Prof Exp: Res assoc & fel, Univ Notre Dame, 58-61; proj scientist, Res Div, Philco Corp, 61; assoc res scientist, NY Univ, 61-63; RES SCIENTIST, GRUMMAN AEROSPACE CORP, 63- Mem: AAAS; Am Phys Soc; Am Chem Soc; Am Nuclear Soc. Res: Radiation chemistry and effects on solid state materials; radiation conversion of wastes; solar energy conversion. Mailing Add: Res Ctr Plant 26 Grumman Aerospace Corp Bethpage NY 11714

CHANG, KENNETH SHUEH-SHEN, b Taipei, Taiwan, Jan 3, 29; m 52; c 5. MICROBIOLOGY, ONCOLOGY. Educ: Nat Taiwan Univ, MD, 51; Univ Tokyo, DMSc, 60. Prof Exp: Asst microbiol, Col Med, Nat Taiwan Univ, 51-55, lectr, 55-59, from assoc prof to prof, 59-67; sr virologist, Flow Labs, 67-69, med officer, Lab Biol, 69-70, HEAD, SECT VIRAL ONCOGENESIS, LAB CELL BIOL, NAT CANCER INST, 70- Concurrent Pos: WHO fel, Commonwealth Serum Labs & Dept Microbiol, Univ Melbourne, 56; Nat Acad Sci fel, Dept Med Microbiol & Immunol, Univ Calif, Los Angeles, 62-64; head Bacillus Calmette-Guerin Vaccine Lab, Taiwan Serum & Vaccine Inst, 57-58; head microbiol & serol sect, Dept Clin Path, Nat Taiwan Univ Hosp, 58-67; round table discussant, Comn Bacillus Calmette-Guerin Vaccine, Int Union Against Tuberc, Can, 61. Mem: Am Asn Cancer Res; Am Soc Microbiol; Tissue Cult Asn; Am Asn Immunologists. Res: Cancer virology and immunology; interferon; influenza; clinical microbiology. Mailing Add: Lab of Cell Biol Nat Cancer Inst Bethesda MD 20014

CHANG, KERN KO NAN, b Shanghai, China, Sept 9, 18; m 48; c 3. ELECTRONIC PHYSICS. Educ: Nat Cent Univ, China, BS, 40; Univ Mich, MS, 48; Polytech Inst Brooklyn, DEE, 54. Prof Exp: Engr, Cent Radio Mfg Co, 40-45, mem staff, Radio Corp Am Labs, 48-62, head microwave solid-state device group, 62-67, HEAD, RCA LABS, 67- Honors & Awards: RCA Labs Achievement Awards, 56, 60, 64 & 67. Res: Magnetrons; traveling wave tubes; beam-focusing devices; parametric amplifiers; tunnel diode amplifiers and converters; solid-state microwave devices; light emitting and avalanche devices; display systems. Mailing Add: RCA Labs Princeton NJ 08540

CHANG, KUANG-CHOU, b Taipei, Taiwan, Jan 2, 49; m. PHYSICAL ORGANIC CHEMISTRY. Educ: Tunghai Univ, Taiwan, BS, 70; Univ Minn, PhD(org chem), 75. Prof Exp: Res asst hydrogen bonding, Univ Minn, 71-74; res fel solution kinetics, Brandeis Univ, 74-76; MEM STAFF, BELL LABS, 76- Res: Spectroscopic studies of reaction kinetics and structure determination. Mailing Add: Bell Labs Murray Hill NJ 07974

CHANG, KUEI-SHENG, b An-yang, Ho-nan, China, Aug 13, 21; US citizen; m 56; c 2. GEOGRAPHY. Educ: Nat Cent Univ, China, AB, 45; Univ Mich, AM, 50, PhD(geog), 55. Prof Exp: Lectr geog, Univ Mich, Ann Arbor, 55-56; asst prof, Wis State Univ-Oshkosh, 56-59; from asst prof to assoc prof, Wayne State Univ, 59-66; ASSOC PROF GEOG, UNIV WASH, 66- Concurrent Pos: Soc Sci Res Coun res award, 62; Fulbright Exchange Prog res award, Taiwan, 62; vis assoc prof, Nat Taiwan Univ, 62-63. Mem: Asn Am Geogr; Asn Asian Studies; Soc Hist Discoveries. Res: Industrial geography of China and history of geographical thought and exploration. Mailing Add: Dept of Geog Univ of Wash Seattle WA 98105

CHANG, KUO WEI, b Shanghai, China, Nov 21, 38; m 63; c 1. BIOMEDICAL ENGINEERING. Educ: Nat Taiwan Univ, BS, 60; Univ Cincinnati, MS, 63; Princeton Univ, MA, 67, PhD(aerospace sci), 69. Prof Exp: Engr, First Naval Shipyard, Chinese Navy, 60-62; res asst aerospace eng, Univ Cincinnati, 62-64; asst res aerospace sci, Princeton Univ, 64-68; sr scientist physics, 68-71, mgr biosensors dept, 71-75, TECH ADV, SPACE SCI DIV, WHITTAKER CORP, 75- Concurrent Pos: Pres, Indust & Biomed Sensors Corp, Lexington, 75- Mem: Am Phys Soc; Asn Advan Med Instrumentation; Inst Elec & Electronic Engrs. Res: Aerospace sciences; plasma physics; magnetohydrodynamics; biophysics; fluid mechanics; kinetic theory of gases; electronics. Mailing Add: Whittaker Corp Space Sci Div 335 Bear Hill Rd Waltham MA 02154

CHANG, KWANG-CHIH, b Peking, China, Apr 15, 31; US citizen; m 57; c 2. ANTHROPOLOGY, ARCHAEOLOGY Educ: Nat Taiwan Univ, BA, 54; Harvard Univ, PhD(anthrop), 60. Hon Degrees: MA, Yale Univ, 69. Prof Exp: Lectr anthrop, Harvard Univ, 60-61; instr, Yale Univ, 61-63, from asst prof to assoc prof, 63-69, chmn dept, 70-73, PROF ANTHROP, YALE UNIV, 69- Concurrent Pos: Am Coun Learned Socs fel, Yale Univ, 62-63 & 67-68, NSF res grants, 64-74; Nat Endowment for Humanities grant, 75-77. Mem: Academia Sinica; Am Anthrop Asn; Asn Asian Studies. Res: Prehistory and early history of China and Southeast Asia. Mailing Add: Dept of Anthrop Yale Univ New Haven CT 06520

CHANG, KWANG-POO, b Taipei, Taiwan, Nov 12, 42; m 72; c 1. PARASITOLOGY. Educ: Nat Taiwan Univ, BSc, 65; Univ Guelph, MS, 68, PhD(biol), 72. Prof Exp: Fel parasitol, 72-74, res assoc, 74-76, ASST PROF PARASITOL, ROCKEFELLER UNIV, 76- Mem: Am Soc Microbiol; Am Soc Parasitologists; Sigma Xi. Res: Biology of invertebrates; intercellular symbiosis, including structural and physiological aspects of cellular interactions between prokaryotic symbiotes and insect or protozoan hosts; intracellular parasitism and mammalian Leishmaniasis. Mailing Add: Rockefeller Univ New York NY 10021

CHANG, LOUIS WAI-WAH, b Hong Kong, July 1, 44; Brit citizen; m 68; c 1. EXPERIMENTAL PATHOLOGY. Educ: Univ Mass, Amherst, BA, 66; Tufts Univ, MS, 69; Univ Wis-Madison, PhD(path), 72. Prof Exp: Instr path, 72-73, dir path lab, 72-75, ASST PROF PATH, UNIV WIS-MADISON, 73- Concurrent Pos: NIH & NSF res grants, 73- Mem: Am Asn Pathologists & Bacteriologists; Am Asn Neuropath; Am Asn Exp Path; AAAS; Nat Soc Histotechnol. Res: Environmental toxicology; heavy metal toxicology; anesthetic toxicity; histochemistry; electron microscopy; developmental biology. Mailing Add: Dept of Path Univ of Wis Madison WI 53706

CHANG, LUCY MING-SHIH, b China, Aug 20, 42; US citizen. BIOCHEMISTRY. Educ: Western Reserve Univ, AB, 64, Ind Univ, PhD(biochem), 68. Prof Exp: Res assoc biochem, Univ Ky, 68-70, asst prof, 70-72; ASST PROF BIOCHEM, UNIV CONN, 72- Mem: AAAS; Am Soc Biol Chem. Res: Enzymatic synthesis of DNA in eukaryotic cells. Mailing Add: Dept of Biochem Univ of Conn Health Ctr Farmington CT 06032

CHANG, LUKE LI-YU, b Honan, China, Sept 18, 35; US citizen; m 59; c 2. MINERALOGY. Educ: Nat Taiwan Univ, BS, 57; Univ Chicago, PhD(geophys sci), 63. Prof Exp: Sr scientist mineral & ceramic sci, Tem-Pres Res, Inc, 63-67; asst prof geol, Cornell Univ, 67-70; assoc prof, 70-75, PROF GEOL, MIAMI UNIV, 75- Concurrent Pos: Contrib ed, Phase Diagrams Ceramist, Am Ceramic Soc & Nat Bur Standards, 74- Mem: Fel Mineral Soc Am; Am Ceramic Soc; Geochem Soc; Mineral Soc Gt Brit; Can Mineral Soc. Res: Mineral synthesis and equilibrium relations in the systems of carbonates, sulfides, and oxides; crystal chemistry of tungstates. Mailing Add: Dept of Geol Miami Univ Oxford OH 45056

CHANG, MARTIN MING YANG, physical chemistry, polymer chemistry, see 12th edition

CHANG, MEI LING (WU), b Kiangsi, China, Mar 1, 15; US citizen; m 53. NUTRITIONAL BIOCHEMISTRY. Educ: Ginling Col, China, BS, 38; Ore State Univ, MS, 49, PhD(nutrit), 51. Prof Exp: Res asst histochem of brain, Sch Med, Wash Univ, 51-53; res assoc nutrit, Ore State Univ, 53-54; res assoc nutrit biochem, Univ Ill, 54-62; RES CHEMIST, HUMAN NUTRIT DIV, USDA, 62- Mem: Am Inst Nutrit Res: Vitamin metabolism; histochemistry of brain; dietary effect on enzyme system; carbohydrate metabolism. Mailing Add: Nutrit Inst Agr Res Serv USDA Beltsville MD 20705

CHANG, MIN CHUEH, b Taiyuan, China, Oct 10, 08; US citizen; m 48; c 3. PHYSIOLOGY. Educ: Tsing Hua, China, BSc, 33; Univ Edinburgh, dipl Agr, 39; Univ Cambridge, PhD, 41, ScD, 69. Prof Exp: Res fel, Sch Agr, Univ Cambridge, 41-45; res assoc, 45-56, sr scientist, 56-70, PRIN SCIENTIST, WORCESTER FOUND EXP BIOL, 71-; ADJ PROF BIOL, BOSTON UNIV, 74- Concurrent Pos: Assoc prof, Boston Univ, 51-61, res prof, 61-74. Honors & Awards: Ortho Award, 50; Lasker Award, 54; Ortho Medal, 61; Hartman Award, 70; Marshall Medal, 71. Mem: AAAS; Am Physiol Soc; Am Asn Anatomists; Am Acad Arts & Sci; Soc Study

Reproduction. Res: Physiology of reproduction; mammalian germ cells; animal husbandry. Mailing Add: Worcester Found for Exp Biol 222 Maple Ave Shrewsbury MA 01545

CHANG, MINGTEH, b Fukien, China, Jan 18, 39; m 70; c 1. FOREST HYDROLOGY. Educ: Nat Chung-Hsing Univ, Taiwan, BS, 60; Pa State Univ, MS, 68; WVa Univ, PhD(forest hydrol), 73. Prof Exp: Watershed technologist, Mountainous Agr Resources Develop Bur, Govt of Taiwan, 61-64; teaching asst surv, Chung-Hsing Univ, 64-67; res assoc hydrol, Water Res Inst, WVa Univ, 73-75; ASST PROF FOREST HYDROL & SOIL PHYSICS, SCH FORESTRY, STEPHEN F AUSTIN STATE UNIV, 75- Concurrent Pos: Co-prin investr, Water Res Inst, WVa Univ, 73- Mem: Am Geophys Union; Soc Am Foresters; Soil Conserv Soc Am. Res: Quantitative analysis and interpretation of hydrologic and climatologic data; physical and physiologic processes of the soil-plant-atmosphere system in forest environment. Mailing Add: Sch of Forestry Box 6109 Stephen F Austin State Univ Nacogdoches TX 75961

CHANG, MONICA LIU, biomathematics, see 12th edition

CHANG, NADA, b Beograd, Yugoslavia, Oct 18, 40; US citizen; m 69; c 2. ENDOCRINOLOGY. Educ: Western Col Women, BA, 65; Univ Ky, PhD(anat), 70. Prof Exp: Fel endocrinol, Sch Vet Med, Univ Mo, 70-73; fel, Med Sch, Univ Ky, 73-75; ASST PROF ANAT, MED SCH, UNIV ARIZ, 75- Mem: Am Asn Anatomists. Res: Central nervous system control of prolactin secretion and lactation; endogenous factors involved in the modification of the same. Mailing Add: Dept of Anat Ariz Med Ctr Univ of Ariz Tucson AZ 85724

CHANG, NGEE PONG, b Singapore, Dec 24, 40; m 65; c 2. THEORETICAL HIGH ENERGY PHYSICS Educ: Ohio Wesleyan Univ, BA, 59; Columbia Univ, PhD(physics), 63. Prof Exp: Res assoc physics, Columbia Univ, 62-63; res fel, Inst Advan Study, 63-64; res assoc, Rockefeller Univ, 64-65; vis prof, 65-66, PROF PHYSICS, CITY COL NEW YORK, 66- Mem: Am Phys Soc. Res: Field theory of weak interactions; symmetries; kinematics at infinite momentum; impact parameter representation; infinite energy scattering; non-abelian gauge theories; quark dynamics. Mailing Add: Dept of Physics City Col of New York New York NY 10031

CHANG, PAULINE (WUAI) KIMM, b Shanghai, China, Jan 19, 26; nat US; m 52; c 2. ORGANIC CHEMISTRY. Educ: Wellesley Col, BA, 49; Univ Mich, MS, 50, PhD(chem), 55. Prof Exp: From res asst to res assoc, 55-66, SR RES ASSOC PHARMACOL, SCH MED, YALE UNIV, 66- Mem: Am Chem Soc; The Chem Soc. Res: Synthetic organic chemistry; medicinal chemistry; synthesis of labeled compounds. Mailing Add: Dept of Pharmacol Yale Univ Sch of Med New Haven CT 06510

CHANG, PEI WEN, b China, Apr 26, 23; nat US; m 51; c 3. VETERINARY PATHOLOGY. Educ: Mich State Univ, DVM, 51; Univ RI, MS, 60; Yale Univ, PhD, 65. Prof Exp: Gen practitioner, Ind, 51-53; area veterinarian, Ind Livestock Sanit Bd, 53-55; asst prof & asst res prof animal path, 55-60, assoc prof, 60-66, PROF ANIMAL PATH, UNIV RI, 66- Concurrent Pos: Danforth fel, 61-63. Mem: Am Vet Med Asn. Res: Animal virology; characterization of animal viruses. Mailing Add: Dept of Animal Path Univ of RI Kingston RI 02881

CHANG, PERKINS PERN KENG, biochemistry, carbohydrate chemistry, see 12th edition

CHANG, PETER HON, b Shanghai, China, Feb 19, 41; US citizen; m 68; c 2. BIOENGINEERING. Educ: Univ Calif, Berkeley, BS, 64, MS, 66, PhD(bioeng), 70. Prof Exp: Res engr, Sondell Sci Instruments, Calif, 65-70; staff scientist, Hewlett Packard Co, 70-71; PROD MGR INSTRUMENTATION, MILES LABS, INC, 71- Concurrent Pos: Mem bd dirs, VIR Electronics, Ltd, Hong Kong, 73- & Nike Enterprise Ltd, 75- Mem: Inst Elec & Electronics Engrs. Res: Glucose controlled insulin infusion system—an instrumentation system to simulate pancreatic endocrine function of a normal subject for diabetic research. Mailing Add: Miles Labs Inc 1127 Myrtle St Elkhart IN 46514

CHANG, POTTER CHIEN-TIEN, b Canton, China, Mar 21, 34; US citizen; m 61; c 2. BIOSTATISTICS. Educ: Nat Taiwan Univ, BS, 58; Univ Minn, MS, 66, PhD(biometry), 68. Prof Exp: ASST PROF BIOSTATIST, SCH PUB HEALTH, UNIV CALIF, LOS ANGELES, 68- Concurrent Pos: Consult, Shock Res Unit, Ctr Critical Care, Sch Med, Univ Southern Calif, 72- Mem: Am Statist Asn; Biometric Soc. Res: Statistical methodology. Mailing Add: Div of Biostatist Univ of Calif Sch of Pub Health Los Angeles CA 90024

CHANG, RAYMOND, b Hong Kong, Mar 6, 39; m 68. PHYSICAL CHEMISTRY Educ: Uni0iv London, BSc, 62; Yale Univ, MS, 63, PhD(phys chem), 66. Prof Exp: Res fel, Wash Univ, 66-67; asst prof chem, Hunter Col, 67-68; ASST PROF CHEM, WILLIAMS COL, 68- Mem: Am Chem Soc Res: Electron spin resonance and nuclear magnetic resonance; chemical kinetics of fast reactions; photochemistry. Mailing Add: Dept of Chem Williams Col Williamstown MA 01267

CHANG, REN-FANG, b Nanking, China, Jan 14, 38; m 68. THERMODYNAMICS, LASERS. Educ: Taiwan Nat Univ, BS, 60; Univ Md, PhD(physics), 68. Prof Exp: Res assoc physics, 68-71, ASST PROF PHYSICS & ASTRON, UNIV MD, 71- Honors & Awards: NASA Apollo Achievement Award. Mem: AAAS; Am Phys Soc; Sigma Xi. Res: Measurements on scattered laser light from samples undergoing phase transitions; studies of optical properties of retro reflector. Mailing Add: Dept of Physics & Astron Univ of Md College Park MD 20742

CHANG, RICHARD KOUNAI, b Hong Kong, June 22, 40; m 61; c 3. SOLID STATE PHYSICS, QUANTUM ELECTRONICS. Educ: Mass Inst Technol, BS, 61; Harvard Univ, MS, 62, PhD(solid state physics), 65. Prof Exp: Res fel solid state physics, Harvard Univ, 65-66; asst prof, 66-69, ASSOC PROF ENG & APPL SCI, YALE UNIV, 70- Concurrent Pos: Consult, Mithras, Inc, Sanders Inc, 65-70. Mem: Fel Am Phys Soc. Res: Nonlinear optics; Raman spectroscopy; solid state laser emission; air pollution monitoring. Mailing Add: Becton Ctr 15 Prospect St Yale Univ New Haven CT 06520

CHANG, ROBERT CHI-HENG, b Taichung, Taiwan, May 16, 37; US citizen; m 65; c 2. POLYMER CHEMISTRY. Educ: Tunghai Univ, BS, 60; Okla State Univ, MS, 65; Univ Akron, PhD(polymer chem), 72. Prof Exp: Scientist polymer synthesis, B F Goodrich Chem Co, 65-69; SR SCIENTIST POLYMER SYNTHESIS, OWENS-CORNING FIBERGLAS CORP, 72- Mem: Am Chem Soc. Res: Structure-property correlation of polymeric materials used to treat glass and the glass fiber reinforced plastics obtained from them, including reinforcement for thermoplastics, thermoset and elastomers. Mailing Add: Owens-Corning Fiberglas Corp Tech Ctr Rte 16 Granville OH 43023

CHANG, ROBERT SHIHMAN, b China, July 26, 22; nat US; m 51; c 4. VIROLOGY. Educ: St John's Univ, China, MD, 46; Harvard Univ, DSc, 52. Prof Exp: Assoc prof microbiol, Harvard Univ, 54-68; PROF MED MICROBIOL, SCH MED, UNIV CALIF, DAVIS, 68- Mem: AAAS; Soc Exp Biol & Med. Res: Tissue culture; cell physiology; genetics. Mailing Add: Sch of Med Univ of Calif Davis CA 95616

CHANG, SEA BONG, biochemistry, deceased

CHANG, SEN-DOU, b Shaohsing, Chekiang, China, Aug 16, 28; US citizen; m 69. GEOGRAPHY OF CHINA, URBAN GEOGRAPHY. Educ: Chi-nan Univ, Shanghai, BA, 49; Univ Wis, MA, 55; Univ Wash, PhD(geog), 61. Prof Exp: Asst prof geog, San Fernando Valley State Col, 62-66; assoc prof, 66-71, PROF GEOG, UNIV HAWAII, 71- Concurrent Pos: Soc Sci Res Coun grant, Hong Kong & Tokyo, 61-62. Mem: Asn Am Geog; Am Geog Soc; Asn Asian Studies; Am Soc Photogrammetry. Res: Transformation of economic landscape and spatial organizations of traditional and contemporary China. Mailing Add: Dept of Geog Univ of Hawaii Honolulu HI 96822

CHANG, SHAO-CHIEN, b Mar 1, 30; Can citizen; m; c 2. MATHEMATICS. Educ: Taiwan Norm Univ, BSc, 56; Carleton Univ, Ont, MSc, 63, PhD(math), 68. Prof Exp: Teacher, Taipei Chien Kuo High Sch, 55-56; supvr, Chnong Hwa NTS Sch, KL Malaysia, 57-62; sessional lectr math, Carleton Univ, Ont, 63-65; lectr, 65-67, asst prof, 67-73, ASSOC PROF MATH, BROCK UNIV, 73- Mem: Am Math Soc; Math Asn Am; Can Math Cong. Res: Mathematical logic, especially syntactical transforms; summability, especially classical and functional analytical method; sequences. Mailing Add: Dept of Math Brock Univ St Catharines ON Can

CHANG, SHAU-JIN, b Kiangsu, China, Jan 7, 37; m 64; c 2. ELEMENTARY PARTICLE PHYSICS, MATHEMATICAL PHYSICS. Educ: Taiwan Univ, BS, 59; Tsing Hua Univ, Taiwan, MS, 61; Harvard Univ, PhD(physics), 67 Prof Exp: Mem physics, Inst Advan Study, 67-69; from asst prof to assoc prof, 69-74, PROF PHYSICS, UNIV ILL, URBANA, 74- Concurrent Pos: Mem, Inst Advan Study, Princeton, 72-73; Alfred P Sloan fel, 72-74; vis physicist, Fermi Nat Accelerator Lab, 75. Mem: Am Phys Soc; Chinese Astron Soc. Res: Various theoretical topics in quantum field theory and elementary particle physics. Mailing Add: Dept of Physics Univ of Ill Urbana IL 61801

CHANG, SHAW FAI, b Anhwei, China, July 26, 33; US citizen; m 60; c 4. BIOCHEMICAL PHARMACOLOGY. Educ: Taiwan Col Eng, BS, 55; Univ Minn, MS, 62. Prof Exp: Jr biochemist chem carcinogenesis, Vet Admin Hosp, 62-64; associated res biochemist drug metab, Parke, Davis & Co, 64-69; sr biochem pharmacologist, 69-71, RES SPECIALIST DRUG METAB, RIKER LABS, INC, 3M CO, 71- Mem: Am Chem Soc; AAAS; Sigma Xi. Res: Study of the physical translocation and biochemical modification of central nervous system and anti-inflammatory agents in laboratory animals and human volunteers; development of extremely sensitive analytical method in quantitating drug residues in biological fluids. Mailing Add: Riker Labs Inc 3M Ctr Bldg 218-2 St Paul MN 55101

CHANG, SHEN CHIN, b China, Oct 1, 14; US citizen; m 53; c 2. ENTOMOLOGY, BIOCHEMISTRY. Educ: Chekiang Univ, BS, 38; Ore State Univ, PhD(entom), 52. Prof Exp: Res asst prof insect toxicol, Univ Ill, 54-62; RES CHEMIST, INSECT CHEMOSTERILANTS LAB, AGR RES SERV, USDA, 62- Mem: AAAS; Entom Soc Am; Am Chem Soc. Res: Chemical and biochemical studies of pyrethrins; mode of action of chemosterilants; discover/co-discoverer of phosphoramides, methylmelamines, dithiobiurets and dithiazolium salts as insect chemosterilants. Mailing Add: Insect Chemosterilants Lab USDA Agr Res Center E Beltsville MD 20705

CHANG, SHIH LU, b China, Dec 28, 13; nat US; m 48; c 2 MICROBIOLOOLOGY, PUBLIC HEALTH. Educ: Yale-in-China Med Col, MD, 35; Harvard Univ, DPH, 41; Am Bd Microbiol, dipl. Prof Exp: Res fel sanit eng, Harvard Univ, 41-44, res fel comp path & trop med, 42-46, fac instr sanit biol & instr comp path & trop med, 44-46, from asst prof to assoc prof sanit biol, 46-54; sr surgeon, Microbiol Sect, Robert A Taft Sanit Eng Ctr, USPHS, 54-59, med dir, 59-70; chief etiology, Div Water Hyg, Water Supply Res Lab, Nat Environ Res Ctr, 70-75, CHIEF ETIOLOGY, HEALTH EFFECTS RES LAB, ENVIRON PROTECTION AGENCY, 75- Concurrent Pos: Responsible envestr projs, Off Sci Res & Develop, Nat Res Coun, Off Qm Gen & Surgeon Gen Off, USPHS, 41-54. Mem: AMA; Am Pub Health Asn; Am Acad Microbiol; AAAS; Royal Soc Health. Res: Bacteriology and parasitology in relation to public health and sanitary engineering, especially in field of disinfection and on epidemiology of amebic meningoencephalitis and recreational waters. Mailing Add: 4806 Beverly Hills Dr Cincinnati OH 45226

CHANG, SHIH-YUNG, b Hopei, China, Feb 10, 38; m 64; c 1. QUANTUM CHEMISTRY. Educ: Nat Taiwan Univ, BS, 60; Kans State Univ, MS, 65; Univ Wash, PhD(phys chem), 69. Prof Exp: Res assoc chem, Univ Calif, Santa Barbara, 69-70; NIH fel, Johns Hopkins Univ, 70-71; mem sci staff, Wolf Res & Develop Corp, 71-73; SR PHYS CHEMIST, ENVIROSPHERE CO, EBASCO SERV INC, 73- Concurrent Pos: Adj asst prof, Dept Pharmacol, Mt Sinai Sch Med, City Univ New York, 76- Mem: Am Phys Soc; Sigma Xi. Res: Ab-initio study of molecular wave functions and physical properties; perturbation theory of molecular polarizabilities, force constants and dipole moments; perturbational approach to the determination of molecular electrostatic interaction potential for drug design. Mailing Add: Envirosphere Co EBASCO 19 Rector St New York NY 10006

CHANG, SHU-PEI, b Sinwui, China, Oct 11, 22; US citizen; m 69; c 1. POLYMER CHEMISTRY, INDUSTRIAL ORGANIC CHEMISTRY. Educ: Nat Checkiang Univ, BS, 45; Univ Louisville, PhD(chem), 63. Prof Exp: Chem engr & soil scientist, Taiwan Sugar Co, 46-59; res fel, Univ Louisville, 63-64; RES CHEMIST, NEW CROPS, NORTHERN REGIONAL RES CTR, USDA, 64- Mem: Am Chem Soc; Am Oil Chemist's Soc. Res: Plasticizers, lubricants and extenders from new seed oils or their derived fatty acids; addition and condensation polymerizations; application of calorimetric, chromatographic and spectroscopic methods to analysis. Mailing Add: 1815 N University St Peoria IL 61604

CHANG, SHU-SING, b Shanghai, China, Feb 18, 35; m 60; c 2. PHYSICAL CHEMISTRY. Educ: Univ Taiwan, BS, 56; Univ Mich, MS, 59, PhD(chem), 62. Prof Exp: Chemist, Cent Res Lab, Allied Chem Corp, 61-63; chemist, Inorg Solids Div, 63-64, CHEMIST, POLYMER DIV, NAT BUR STANDARDS, 64- Mem: Am Phys Soc; Am Chem Soc. Res: Thermodynamic properties of globular molecules, plastic crystals, vitreous state and polymers. Mailing Add: Polymer Div Nat Bur of Standards Washington DC 20234

CHANG, STEPHEN SZU SHIANG, b Peiking, China, Aug 15, 18; nat US; m 52. FOOD CHEMISTRY. Educ: Nat Chi-nan Univ, BS, 41; Kans State Univ, MS, 49; Univ Ill, PhD(food chem), 52. Prof Exp: Res chemist, Universal Pharmaceut Corp, China, 41-44; assoc engr, Nat Resources Comn China, 44-46; supt prod, Chinchow Pulp & Paper Mill, 46-47; res assoc food chem, Univ Ill, 52-55; res chemist, Swift & Co, 55-57; sr res chemist, A E Staley Mfg Co, 57-60; assoc prof food sci, 60-62,

PROF FOOD SCI, RUTGERS UNIV, 62- Honors & Awards: Spec Award, Potato Chip Inst Int; Putnam Food Award, Putnam Publ Co; Am Oil Chem Soc Achievement Award & Bailey Award. Mem: Inst Food Technologists; Am Chem Soc; Am Oil Chem Soc (pres, 70); NY Acad Sci. Res: Flavor stability of fats and oils; chemical reactions involved in the processing of edible fats and oils; mechanisms of the autoxidation of unsaturated fatty acids; chemistry of food emulsifiers; chemistry of food flavors; isolation and identification of flavor compounds in foods. Mailing Add: Dept of Food Sci Rutgers Univ New Brunswick NJ 08903

CHANG, SUK CHUL, b Hamhung, Korea, June 19, 23; m 67; c 2. PATHOLOGY, CYTOLOGY. Educ: Seoul Nat Univ, MD, 48; Washington Univ, PhD(path anat), 56. Prof Exp: Intern, 34th Gen Hosp, US Army, Korea, 48-49; assistantship, Seoul Univ Hosp, 49-52; instr path, Col Med, Univ Utah, 57-59; lectr, Univ Toronto, 59-61; assoc prof, Post-Grad Sch Med, Univ Tex & assoc pathologist, M D Anderson Hosp & Tumor Inst, 62-68, dir cytol, Sch Cytotechnol, 64-68; assoc prof, 69-72, PROF PATH, HAHNEMANN MED COL, 72-, DIR DIV CYTOPATH & SCH CYTOTECHNOL, 70- Concurrent Pos: Vis pathologist, Toronto Gen Hosp, 59-61. Mem: Col Am Path; Am Soc Cytol; Am Soc Clin Path. Res: Anatomic pathology; exfolitive cytology; cancer research in lung and female reproductive organs; epidemiology; early detection of cancer. Mailing Add: Div of Cytopath Hahnemann Med Col Philadelphia PA 19102

CHANG, SUN-YUNG ALICE, b Ci-an, China, Mar 24, 48; m 73. MATHEMATICAL ANALYSIS. Educ: Nat Taiwan Univ, BA, 70; Univ Calif, Berkeley, PhD(math), 74. Prof Exp: Asst prof math, State Univ NY Buffalo, 74-75; HEDRICK ASST PROF MATH, UNIV CALIF, LOS ANGELES, 75- Mem: Am Math Soc. Res: Investigation of behavior of analytic functions in the complex plane and in several complex variables, approximation of bounded functions by analytic functions. Mailing Add: Dept of Math Univ of Calif Los Angeles CA 90024

CHANG, TAI YUP, b Korea, Oct 25, 33; m 62; c 2. AIR POLLUTION, QUANTUM CHEMISTRY. Educ: Seoul Nat Univ, BS, 58, MS, 60; Univ Wis-Madiosn, PhD(theoret chem), 66. Prof Exp: Res fel theoret chem, Univ Wis-Madison, 67 & Harvard Univ, 67-69; RES SCIENTIST, CHEM DEPT, RES STAFF, FORD MOTOR CO, 69- Mem: Am Phys Soc; Air Pollution Control Asn; Am Meteorol Soc Res: Atomic and molecular physics; development and application of the perturbation theory; air pollution modeling, urban air quality modeling including atmospheric dispersion and photochemistry; global balance of trace gases. Mailing Add: Chem Dept Res Staff Ford Motor Co PO Box 2053 Dearborn MI 48121

CHANG, TE WEN, b Nanchang, China, Oct 12, 20; US citizen; m 52; c 5. INFECTIOUS DISEASES, VIROLOGY. Educ: Nat Cent Univ, AB, 41, MD, 45. Prof Exp: Resident & asst med, Nat Cent Univ Hosp, 46-49; intern, St Joseph Hosp, Kansas City, 50; res fel virol, Univ Kans, 50-52; fel med, Mass Mem Hosp & Boston Univ, 52-57; instr med, 58-59, asst prof med & microbiol, 60-67, ASSOC PROF MED, NEW ENG MED CTR, TUFTS UNIV, 68- Concurrent Pos: Asst physician, New Eng Med Ctr Hosp, 60; assoc prof sch med, Tufts Univ, 68. Mem: AAAS; Fedn Clin Invest; Am Soc Microbiol; Infectious Dis Soc Am; Am Pub Health Asn. Res: Treatment and prevention of viral diseases in man. Mailing Add: Dept of Med New Eng Med Ctr Hosp Boston MA 02111

CHANG, TED TEH-LIANG, b Tainan, Taiwan, China, Oct 6, 35; m 60; c 3. PHYSICAL CHEMISTRY, ANALYTICAL CHEMISTRY. Educ: Nat Univ Taiwan, BS, 57; Univ Va, MS, 63, PhD(chem), 65. Prof Exp: Asst lectr anal chem, Chen-Kung Univ, Taiwan, 59-61; fel, Calif Inst Technol, 65-66; res chemist, Am Cyanamid Co, 66-71; SECT LEADER, WYETH LABS INC, 71- Mem: AAAS; Am Chem Soc; Am Soc Mass Spectrometry; fel Am Inst Chemists. Res: Mass spectrometry; gel permeation chromatography; atomic absorption spectroscopy; polarography; spectrophotometry gas chromatography. Mailing Add: 428 Falcon Rd Audubon PA 19407

CHANG, THOMAS MING SWI, b Swatow, China, Apr 8, 33; Can citizen; m 58; c 4. PHYSIOLOGY, MEDICAL RESEARCH. Educ: McGill Univ, BSc, 57, MD, CM, 61, PhD(physiol), 65. Prof Exp: Intern, Montreal Gen Hosp, 61-62; sessional lectr, 64-65, lectr, 65-66, from asst prof to assoc prof, 66-72, prof physiol, 72-75, PROF MED & DIR ARTIFICIAL ORGAN RES UNIT, McGILL UNIV, 75-; MED RES COUN CAN ASSOC, 68- Concurrent Pos: Med Res Coun Can fel, 62-65, scholar, 65-68. Mem: Am Physiol Soc; Can Physiol Soc; Can Nephrology Soc; Europ Dialysis & Transplant Soc; fel Royal Col Physicians Can. Res: Artificial and biological membranes; artificial cells and kidney; microencapsulation of enzymes, detoxicants and other biologically active materials for biomedical research and clinical applications. Mailing Add: Dept of Physiol McGill Univ Montreal PQ Can

CHANG, TIEN-DING, b Chwansha, China, Oct 13, 21; m 51; c 2. GENETICS. Educ: Chekiang Univ, BS, 44; Univ Minn, MS, 60; Iowa State Univ, PhD(genetics), 63. Prof Exp: Teacher high schs, China, 44-46; agronomist, Taiwan Agr Res Inst, 46-57; cytogeneticist, Dept Med Genetics, Children's Hosp Winnipeg, 64-69; res assoc genetics, Univ Mo-Columbia, 69-74; RES ASSOC CELLULAR & BIOCHEM GENETICS, SLOAN-KETTERING INST CANCER RES, 74- Concurrent Pos: Damon Runyon Mem Fund vis investr, Biol Div, Oak Ridge Nat Lab, 63-64; cytogeneticist, Jenkins Found for Res, Salinas, Calif, 73-74. Mem: Genetics Soc Am; Genetics Soc Can; NY Acad Sci. Res: Plant and mammalian cytogenetics. Mailing Add: Sloan-Kettering Inst 145 Boston Post Rd Rye NY 10580

CHANG, TIMOTHY SCOTT, b Shaowu, China, May 30, 25; m 55; c 4. BACTERIOLOGY, POULTRY PATHOLOGY. Educ: Fukien Christian Univ, BA, 46; Duke Univ, MDiv, 51; NC State Univ, BS, 52; Ohio State Univ, MS, 53, PhD(poultry path, bact), 57. Prof Exp: Teacher pub sch, China, 46-48; asst poultry sci, Ohio State Univ, 51-57; dir bact res lab, Whitmoyer Labs, Inc, Rohm and Haas Co, Pa, 57-65; group leader vet microbiol, Vet Res Div, Norwich Pharmacal Co, NY, 65-66, sect chief, 66-69; mgr diagnostic reagents, Burroughs Wellcome Co, 69-70; dir animal technol dept, S B Penick & Co, CPC Int, Inc, 70-72; ASSOC PROF POULTRY SCI, MICH STATE UNIV, 71- Mem: Poultry Sci Asn; Am Soc Microbiol; World Poultry Sci Asn; Am Asn Avian Pathologists; NY Acad Sci. Res: Bursa of facricius in antibody production; poultry disease and nutriton; general disinfectant for use in poultry and animal fields; waste and waste utilization; avian microbiology. Mailing Add: Dept of Poultry Sci Mich State Univ East Lansing MI 48824

CHANG, TSUEN KUNG, b Canton, China, May 6, 10; US citizen; m 43; c 3. GEOGRAPHY. Educ: Lingnan Univ, BA, 32; Yenching Univ, MA, 36; Univ Nebr, PhD(geog), 54. Prof Exp: Lectr Oriental studies, Pomona Col, 37-38; instr, Univ Hawaii, 40-42; instr geog, Univ Okla, 50-51 & Carrol Col, 55-56; from asst prof to assoc prof, 56-59, PROF GEOG, UNIV WIS-STEVENS POINT, 59- Mem: Asn Am Geogr. Res: Geography and ethnography of China; geography of domesticated plants and animals; dispersal of food tubers in Eastern Asia. Mailing Add: Dept of Geog Univ of Wis Stevens Point WI 54481

CHANG, TSUENG-HSING, genetics, see 12th edition

CHANG, WEN-HSUAN, b Tsingtao, China, Mar 28, 26; m 59; c 1. ORGANIC CHEMISTRY. Educ: Fu Jen Univ, China, BSc, 48; Wesleyan Univ, MA, 56; Northwestern Univ, PhD, 59. Prof Exp: Assoc engr, Taiwan Agr & Chem Works, Formosa, 49-54; sr res chemist, 58-65, from res assoc to sr res assoc, 65-69, scientist, 69-71, SR SCIENTIST, PPG INDUSTS, 71- Mem: Am Chem Soc; NY Acad Sci; The Chem Soc. Res: Organic chemical synthesis; reaction mechanisms; kinetics; polymer synthesis. Mailing Add: PPG Industs PO Box 9 Rosanna Dr Allison Park PA 15101

CHANG, WILLIAM WEI-LIEN, b Taipei, Taiwan, Feb 7, 33; m 65; c 3. HISTOLOGY, PATHOLOGY. Educ: Nat Taiwan Univ, MD, 58; Ohio State Univ, MSc(path), 66; McGill Univ, PhD(anat), 70. Prof Exp: Asst pharmacol, Nat Taiwan Univ, 60-61; rotating intern med, Buffalo Gen Hosp, 61-62; from asst resident to resident path, Ohio State Univ, 62-66; asst prof, 70-73, ASSOC PROF ANAT, MT SINAI SCH MED, 73- Concurrent Pos: Ont Heart Asn fel, Dept Path, Queen's Univ, Ont, 66-67; Nat Cancer Inst res grant, Dept Anat, Mt Sinai Sch Med, 74-77. Mem: Am Asn Anat; Am Soc Cell Biol. Res: Cell population kinetics of colon and salivary glands; chemical carcinogenesis of colon. Mailing Add: Dept of Anat Mt Sinai Sch Med Fifth Ave & 100th St New York NY 10029

CHANG, YEONG-JEN PETER, b Szechuan, China, June 19, 46; m 70; c 1. PHYSICAL CHEMISTRY. Educ: Nat Taiwan Univ, BS, 69; Princeton Univ, MA, 72, PhD(polymer phys chem), 74. Prof Exp: Chemist, 74-75, SR CHEMIST, TECH CTR, UNION CARBIDE CORP, 75- Mem: Am Chem Soc. Res: Structure-property relations of coatings materials, including rheological, rheo-optical, morphological and thermomechanical aspects; the permeation of water and salts through polymer membranes. Mailing Add: Union Carbide Tech Ctr South Charleston WV 25303

CHANG, YI-CHI, b Peiping, China; US citizen. PHARMACOLOGY. Educ: Southeast Mo State Col, BS, 61; Univ Conn, MS, 68, PhD(pharmacol), 69. Prof Exp: From instr to asst prof pharmacol, Bowman Gray Sch Med, Wake Forest Univ, 69-75; UNIT LEADER CARDIOVASC PHARMACOL, NORWICH PHARMACAL CO DIV, MORTON-NORWICH PROD, INC, 75- Mem: AAAS; Sigma Xi. Res: Autonomic, cardiovascular and cerebrovascular pharmacology; pharmacology of native peptide hormones and peptide antagonists. Mailing Add: Res & Develop Norwich Pharmacal Co PO Box 191 Norwich NY 13815

CHANG, YI-HAN, b Peking, China, May 26, 33; US citizen; m 63. BIOCHEMICAL PHARMACOLOGY, IMMUNOLOGY. Educ: Stetson Univ, BS, 56; Pa State Univ, PhD(org chem), 61; Univ Conn, PhD(pharmacol), 65. Prof Exp: Res chemist, 60-63, RES BIOCHEM PHARMACOLOGIST, PFIZER, INC, 63- Mem: Am Soc Pharmacol & Exp Therapeut; Am Chem Soc. Res: Anti-inflammatory and immunosuppressive agents; etiology and pathology of rheumatoid arthritis and hypersensitivity diseases; in vitro, in vivo models of cell-mediated hypersensitivity; drug metabolism. Mailing Add: Dept of Pharmacol Pfizer Med Res Labs Groton CT 06340

CHANG, YUNG-FENG, b Taiwan, China, Nov 15, 35; m 68. BIOCHEMSITRY, MICROBIOLOGY. Educ: Nat Taiwan Univ, BS, 58, MS, 60; Univ Pittsburgh, PhD(biochem), 66. Prof Exp: Res asst biochem, Nat Taiwan Univ, 61-62; res asst, Univ Pittsburgh, 62-66; res assoc, Sch Med, 66-70, asst prof histol & embryol, Sch Dent, 70-71, asst prof microbiol, 71-74, ASSOC PROF MICROBIOL, SCH DENT, UNIV MD, BALTIMORE, 74- Mem: Am Chem Soc; Am Soc Microbiol. Res: Lysine metabolism and regulation of enzyme in Pseudomonas putida; isoenzymes of glutaric semialdehyde dehydrogenase; cariostatic effects of amino acids on dental plaque formation and cell wall biosynthesis of streptococcus mutans; lysine metabolism in the mammalian brain; developmental aspects of the alternative catabolic pathways of lysine in the rat. Mailing Add: Dept of Microbiol Univ of Md Sch of Dent Baltimore MD 21201

CHANG, YU-WEI, b Chen-Kiang, Kiangsu, China, July 14, 19; m 53; c 4. ORGANIC CHEMISTRY. Educ: Nat Cent Univ, China, BS, 40; Univ Wash, PhD, 53. Prof Exp: Res fel, Columbia Univ, 52-56; res assoc, Worcester Found Exp Biol, 56-58; res fel org chem, Harvard, 58-60; RES CHEMIST, JACKSON LAB, E I DU PONT DE NEMOURS & CO, INC, 60- Res: Synthetic organic chemistry; reaction mechanisms. Mailing Add: Jackson Lab PO Box 525 E I du Pont de Nemours & Co Inc Wilmington DE 19899

CHANG-FANG, CHUEN-CHUEN, b Taipei, Taiwan, June 21, 30; US citizen; m 61; c 3. PARTICLE PHYSICS. Educ: Nat Taiwan Univ, BS, 53; Univ SC, MS, 57; Duke Univ, PhD(physics), 61. Prof Exp: Teaching asst physics, Prov Tainan Eng Col, 53-55; instr, Univ Miami, 60-61; sr physicist, Controls for Radiation, 61-62; asst prof physics, Southern Ill Univ, Carbondale, 66-72; vis assoc prof, Nat Taiwan Univ, 72-73; ASST PROF PHYSICS, SOUTHERN ILL UNIV, CARBONDALE, 73- Mem: Am Phys Soc; Am Asn Physics Teachers. Res: Phenomenological analysis of anti-proton-proton interaction at 2.32 GeV/c. Mailing Add: Dept of Physics & Astron Southern Ill Univ Carbondale IL 62903

CHANGNON, STANLEY ALCIDE, JR, b Donovan, Ill, Apr 14, 28; m 50; c 3. CLIMATOLOGY. Educ: Univ Ill, BS, 51, MS, 56. Prof Exp: Proj supvr cloud physics, Univ Ill, 52-54; assoc scientist, 55-65, prof climat scientist, 65-70, actg head, 70-71, HEAD ATMOSPHERIC SCI SECT, ILL STATE WATER SURV, URBANA, 71- Concurrent Pos: NSF res grants, 65-; mem nat comt, Comt Weather Info for Agr, 65; mem, Nat Comt Severe Local Storms & Nat Comt Weather Modification. Honors & Awards: Robert Horton Award, Am Geophys Union, 64; Award, Bldg Res Inst, 65. Mem: Am Meteorol Soc; Am Geophys Union. Res: Climatology of Lake Michigan, Illinois and Middle West; severe storms; physical geography; weather modification; agrometeorology; irrigation; urban industrial effects on precipitation. Mailing Add: Atmospheric Sci Sect Univ Ill 271A Water Resrcs Urbana IL 61801

CHANGUS, GEORGE WILLIAM, b Ely, Nev, Oct 17, 12; m 42; c 3. PATHOLOGY Educ: Univ Calif, AB, 35, MA, 38; St Louis Univ, MD, 43; Am Bd Nuclear Med, dipl, 73 Prof Exp: Asst physiol, Univ Calif, 35-40; res path, US Marine Hosp, New Orleans, 46-47; instr, Sch Med, Tulane Univ, 47-50; from asst to assoc prof, Sch Med, Univ Tenn, 50-52; spec res fel, Mem Ctr for Cancer & Allied Dis, NY, 52-59; assoc dir path, Norwalk Hosp, 52-59; DIR LABS & CHMN DEPT PATH, MERCY HOSP & MED CTR, 59- Concurrent Pos: Clin prof path, Univ Ill Med Ctr, 70- Concurrent Pos: Am Cancer Soc fel, Tulane Univ, 47-49; Ewing fel, Mem Hosp, New York, 49-50. Mem: Am Soc Clin Pathologists; Col Am Pathologists; Int Acad Path; Am Asn Bacteriologists & Pathologists. Res: Radioactive isotope tracer studies; lipid metabolism; histochemical and chemical pathology in relation to cancer; new concept on the genesis of Tietze's Syndrome; virus-like particles in lethal midline granuloma not known previously; macrophage origin in granulation tissue; experimental induction of malakoplakia in guinea pigs. Mailing Add: Mercy Hosp & Med Ctr Dept Path Stevenson Expressway at King Dr Chicago IL 60616

CHANIN, LORNE MAXWELL, b Roland, Man, Aug 14, 27; m 50; c 3. PLASMA PHYSICS. Educ: Univ Man, BS, 49; Univ NMex, MS, 51; Univ Pittsburgh, PhD(physics), 59. Prof Exp: Res engr, Westinghouse Res Labs, 51-59; res scientist & sect head, Honeywell Res Ctr, 59-65; assoc prof, 65-68, PROF ELEC ENG, UNIV MINN, MINNEAPOLIS, 68- Concurrent Pos: Consult, US Bur Mines. Mem: Am Phys Soc; Inst Elec & Electronics Engrs; Europ Phys Soc. Res: Atomic physics; interactions and reactions involving electrons, ions and excited atoms; plasma physics; electrical discharges; gaseous electronics. Mailing Add: Dept of Elec Eng Univ of Minn Minneapolis MN 55455

CHANIN, MARTIN, b Newark, NJ, Jan 24, 22; m 46; c 2. CHEMISTRY. Educ: Univ Pa, BA, 42; Univ Mich, MS, 44, PhD(chem), 46. Prof Exp: Jr chemist, Hoffmann-La Roche, NJ, 42; asst, Univ Mich, 43-47; asst prof chem, Evansville Col, 47-50; assoc prof, Detroit Inst Technol, 50-53; res biochemist, City of Hope Med Ctr, 53-56; assoc prof, Memphis State Univ, 56-57; res chemist, Humko Co, 57-59; prof textile chem, Clemson Univ, 59-62; prof chem & chmn dept sci, Florence State Col, 62-63; fel pharmacol, Med Sch, Vanderbilt Univ, 63-64; assoc prof chem, Tenn State Univ, 64-68; res biochemist, Meharry Med Col, 68-69; DYE CHEMIST, KITAN-DIMONA, ISRAEL, 70- Mem: AAAS; Am Chem Soc. Res: Synthetic drugs; serology; arthritis; synthetic analgesics; fatty acids; dye chemistry. Mailing Add: 3810 Richland Ave Nashville TN 37205

CHANIOTIS, BYRON NICOLAS, medical entomology, see 12th edition

CHANLEY, JACOB DAVID, b New York, NY, Jan 8, 18; m 42; c 2. BIOCHEMISTRY. Educ: City Col New York, BS, 38; Univ Chicago, MS, 40; Harvard Univ, MA, 42, PhD(chem), 44. Prof Exp: Asst, Mt Sinai Hosp, 40-41; res worker, War Prod Bd, Columbia Univ, 44-45; res chemist, Mt Sinai Hosp, 45-67; ASSOC PROF BIOCHEM, SCH MED & GRAD SCH, CITY UNIV NEW YORK, 67- Concurrent Pos: Fel chem, Harvard Univ, 41-44; Rockefeller traveling fel, Oxford Univ, 49-50; Dazian Found fel, 49-50; sr res assoc, Mt Sinai Hosp, 67-70. Mem: AAAS; Am Chem Soc; The Chem Soc; Am Soc Biol Chemists. Res: Chemistry; steroid saponins of marine origin; metabolism of adrenergic transmitters; enzyme kinetics. Mailing Add: Mt Sinai Sch of Med Fifth Ave & 100th St New York NY 10029

CHANMUGAM, GANESAR, b Colombo, Ceylon, Oct 24, 39; m 66; c 2. THEORETICAL ASTROPHYSICS. Educ: Univ Ceylon, BSc, 61; Cambridge Univ, BA, 63; Brandeis Univ, PhD(physics), 70. Prof Exp: Instr physics, Univ Mass, Amherst, 63-64; res fel astrophys, Inst Astrophys, Univ Liege, Belg, 69-71; res assoc, 71-72, ASST PROF PHYSICS & ASTRON, LA STATE UNIV, BATON ROUGE, 72- Mem: Am Phys Soc; Am Astron Soc; Royal Astron Soc. Res: Physics of dense matter with astrophysical applications to white dwarfs; neutron stars. Mailing Add: Dept of Physics & Astron La State Univ Baton Rouge LA 70803

CHANNAPRAGADA, RAO, applied mathematics, see 12th edition

CHANNEL, LAWRENCE EDWIN, b Boise, Idaho, Mar 17, 27; m 47; c 5. PHYSICS. Educ: Pasadena Col, AB, 50; Univ Calif, Los Angeles, MA, 55. Prof Exp: Physicist, US Naval Ord Test Sta, Pasadena, Calif, 53-57; opers analyst, Mass Inst Technol physics eval group, US Dept Navy, Washington, DC, 57-63; res & develop scientist, 63-67, dept mgr, 67-69, DIV MGR, LOCKHEED-CALIFORNIA CO, 69- Concurrent Pos: Asst physics, Univ Calif, Los Angeles, 54; part-time instr, Pasadena Col, 64-65 & 69. Mem: Acoust Soc Am; Am Inst Aeronaut & Astronaut; Am Helicopter Soc. Res: Underwater acoustics; military operations research. Mailing Add: 2829 N Lake Ave Altadena CA 91001

CHANNELL, ROBERT BENNIE, b Gallman, Miss, July 4, 24. SYSTEMATIC BOTANY. Educ: Miss State Col, BS, 47, MS, 49; Duke Univ, PhD, 55. Prof Exp: Pub sch teacher, 47-49; instr bot, Miss State Col, 49-51; asst, Duke Univ, 51-54; asst cur, Herbarium, 54-55; botanist, Gray Herbarium-Arnold Arboretum, Harvard, 55-57; from asst prof to assoc prof biol, 57-69, chmn dept, 63-73, PROF BIOL, VANDERBILT UNIV, 69- Mem: Am Soc Plant Taxon; Int Asn Plant Taxon. Res: Conventional and experimental taxonomy of vascular plants, particularly Compositae and Cyperaceae; flora of the Southeastern United States. Mailing Add: Dept of Biol Substa B Vanderbilt Univ Nashville TN 37203

CHANNING, CORNELIA POST, b Boston, Mass, Apr 23, 38. ENDOCRINOLOGY. Educ: Hood Col, BA, 61; Harvard Univ, MA, 63, PhD(biochem), 66. Prof Exp: Fel reproductive physiol, Sch Vet Med, Univ Cambridge, 65-67; from instr to asst prof physiol, Sch Med, Univ Pittsburgh, 67-73; ASSOC PROF PHYSIOL, SCH MED, UNIV MD, BALTIMORE, 73- Concurrent Pos: Nat Inst Child Health & Human Develop fel, 65-67, res grant, 68-76, contract grant, 70-71; Am Cancer Soc grant, 68; Pop Coun New York res grant, 72-76; WHO grant, 75; mem bladder-prostate adv comt, Nat Cancer Inst, 72-73; mem study sect on molecular cytol, NIH, 74-77. Honors & Awards: Newcomb Cleveland Prize, AAAS, 69. Mem: Am Physiol Soc; Soc Study Reproduction; Endocrine Soc; Soc Exp Biol & Med; Tissue Cult Asn. Res: Reproductive physiology and biochemistry of the mammalian ovary; use of tissue culture as a method for studying the mechanisms of luteinization in the rhesus monkey, pig and human; control of ovarian steroidogenesis; control of oocyte meiosis. Mailing Add: Dept of Physiol Univ of Md Sch of Med Baltimore MD 21201

CHANOCK, ROBERT MERRITT, b Chicago, Ill, July 8, 24; m 48; c 2. PEDIATRICS, BACTERIOLOGY. Educ: Univ Chicago, BS, MD, 47. Prof Exp: Asst prof res pediat, Sch Med, Univ Cincinnati, 54-56; asst prof epidemiol, Sch Pub Health, Johns Hopkins Univ, 56-57; MED DIR, LAB INFECT DIS, NAT INST ALLERGY & INFECT DIS, 57-, VCHMN, BD VACCINE DEVELOP, 64-; PROF CHILD HEALTH & DEVELOP, SCH MED, GEORGE WASHINGTON UNIV, 71- Concurrent Pos: Nat Res Coun fel, 50-51; Nat Found Infantile Paralysis fel, 51-52; sr res fel, USPHS, 56-57; virologist, Children's Hosp DC, 57-; mem int nomenclature comt myxoviruses, 7th & 8th Int Microbiol Cong; mem, Armed Forces Epidemiol Bd, Comn Acute Respiratory Dis, assoc mem, Comn Influenza; dir int ref ctr respiratory viruses, WHO, 62-; mem int comt nomenclature bacteria, Int Asn Microbiol Soc, 69-72; clin prof, Georgetown Univ, 70-71. Honors & Awards: E Mead Johnson Award Pediat Res, 64; USPHS Meritorious Serv Medal, 65; Squibb Award, Infect Dis Soc Am, 69; USPHS Distinguished Serv Medal, 71. Mem: Nat Acad Sci; Soc Pediat Res; Am Soc Microbiol; Am Epidemiol Soc; Am Soc Clin Invest. Res: Respiratory virus diseases; virus microbiology. Mailing Add: Lab of Infect Dis Nat Inst Allergy & Infect Dis Bethesda MD 20014

CHAN-PALAY, VICTORIA, b Singapore, Oct 9, 45; m; c 2. NEUROSCIENCES. Educ: Smith Col, AB, 65; Tufts Univ, PhD, 69; Harvard Med Sch, MD, 75. Prof Exp: Res asst biophys, New Eng Med Ctr, Boston, 65-66; teaching asst histol & gross anat, Sch Med, Tufts Univ, 68-69; teaching asst gross anat & neuroanat, 70-71, instr neurobiol, 71-72, ASST PROF NEUROBIOL, HARVARD MED SCH, 72- Honors & Awards: William F Milton Fund Award, Harvard Univ, 73; Leon Reznick Mem Prize, Harvard Med Sch, 75. Mem: Am Asn Anat; Am Asn Cell Biol; Soc Neurosci; Am Asn Univ Women; Am Med Women's Asn. Mailing Add: Dept of Neurobiol Harvard Med Sch Boston MA 02115

CHANT, DONALD A, b Toronto, Ont, Sept 30, 28; m 59; c 1. ENTOMOLOGY. Educ: Univ BC, BA, 50, MA, 52; Univ London, PhD(zool), 56. Prof Exp: Res officer, Res Inst, Can Dept Agr, 56-60, dir entom & plant path, Res Lab, 60-64; chmn dept biol control, Univ Calif, Riverside, 64-67; chmn dept zool, 67-75, VPRES & PROVOST, UNIV TORONTO, 75- Concurrent Pos: Mem, Nat Comt Pesticide Use in Agr, Can, 61-64 & subcomt insect control, Nat Acad Sci-Nat Res Coun, 64- Mem: Entom Soc Can; Can Soc Zool (pres, 74-75); fel Royal Entom Soc. Res: Acarology; taxonomy and ecology of predacious phytoseiid mites; principles of predation; biological control. Mailing Add: Simcoe Hall Univ of Toronto Toronto ON Can

CHANTELL, CHARLES J, b Chicago, Ill, May 19, 31; m 57; c 2. VERTEBRATE ANATOMY. Educ: Univ Ill, BS, 61; Univ Notre Dame, MS, 63, PhD(biol), 65. Prof Exp: ASSOC PROF BIOL, UNIV DAYTON, 65- Mem: Soc Vert Paleont; Am Soc Ichthyologists & Herpetologists; Soc Study Amphibians & Reptiles. Res: Evolution, phylogeny and osteology of the lower vertebrates. Mailing Add: Dept of Biol Univ of Dayton Dayton OH 45409

CHAO, CHONG-YUN, b Kunming, China, July 5, 31; m 56; c 2. MATHEMATICS. Educ: Univ Iowa, BA, 53, MS, 54; Univ Mich, PhD(math), 61. Prof Exp: Instr math, Coe Col, 54-56; mathematician, Res Ctr, Int Bus Mach Corp, 61-63; assoc prof math, 63-66, PROF MATH, UNIV PITTSBURGH, 66- Mem: Am Math Soc; Math Asn Am. Res: Lie algebras, groups and graphs. Mailing Add: Dept of Math Univ of Pittsburgh Pittsburgh PA 15213

CHAO, EDWARD CHING-TE, b Soochow, China, Nov 30, 19; nat US; m 42; c 3. GEOLOGY. Educ: Nat Southwest Assoc Univ, BS, 41; Univ Chicago, PhD(geol), 48. Prof Exp: Jr geologist, Geol Surv Szechuan, China, 41-45; fel petrol & geochem, Univ Chicago, 48-49; GEOLOGIST, US GEOL SURV, WASHINGTON, DC, 49- Mem: Geol Soc Am; Mineral Soc Am; Geochem Soc. Res: Petrology; geochemistry; mineralogy; impact metamorphism; lunar petrology and geology. Mailing Add: US Geol Surv Nat Ctr 12201 Sunrise Valley Dr Reston VA 22092

CHAO, FU-CHUAN, b Hong Kong, Feb 8, 19; US citizen; m 47; c 3. BIOCHEMISTRY Educ: Lingnan Univ, BA, 41; Univ Calif, Berkeley, PhD(biochem), 51. Prof Exp: Res fel biochem, Univ Calif, Berkeley, 51-52, jr res biochemist, 52-53; res instr biochem, Univ Utah, 53-55; res assoc, Stanford Univ, 55-61, biophys chemist, Stanford Res Inst, 61-74; RES CHEMIST, UNIV SAN FRANCISCO, 74- Concurrent Pos: Am Heart Asn fel, 55-57. Mem: AAAS; Am Chem Soc; fel Am Inst Chemists; NY Acad Sci; Am Soc Microbiol. Res: Stability of ribonucleoproteins in solution; purification of viruses by chemical methods; prognosis of cancer; metabolism of marijuana. Mailing Add: 1524 Channing Ave Palo Alto CA 94303

CHAO, JIA-ARNG, b Hunan, China, July 8, 41; m 67; c 1. MATHEMATICS. Educ: Nat Taiwan Normal Univ, BS, 64; Nat Tsing Hua Univ, MS, 66; Washington Univ, PhD(math), 72. Prof Exp: Teaching & res asst math, Washington Univ, 66-71, instr, 72; ASST PROF MATH, UNIV TEX, AUSTIN, 72- Mem: Am Math Soc. Res: Harmonic analysis on local fields. Mailing Add: Dept of Math Univ of Tex Austin TX 78712

CHAO, JING, b Chekiang, China, Nov 7, 24; m 52; c 1. PHYSICAL CHEMISTRY. Educ: Nat Cent Univ, China, BS, 47; Carnegie Inst Technol, BS, 61, PhD(phys chem), 62. Prof Exp: Asst chem engr, Hsinchu Res Inst, Chinese Petrol Corp, 47-53; assoc chem engr, Union Indust Res Inst, Ministry Econ Affairs, China, 53-57; phys chemist, Thermal Res Lab, Dow Chem Co, 61-69; sr thermodynamicist, 69-73, ASST DIR MOLECULAR THERMODYN, THERMODYN RES CTR, TEX A&M UNIV, 73- Mem: Am Chem Soc; Sigma Xi. Res: Thermochemistry; applied thermodynamics; collection, analysis of chemical data and critical evaluation of physical and thermodynamic properties for chemical substances. Mailing Add: Thermodyn Res Ctr Tex Eng Exp Sta Tex A&M Univ College Station TX 77843

CHAO, JOWETT, b Chekiang, China, Nov 16, 15; nat US; m 39; c 1. PARASITOLOGY. Educ: WChina Union Univ, BS, 39; Nat Cent Univ, China, MS, 42; Cornell Univ, PhD(entom), 49. Prof Exp: Instr gen zool, WChina Union Univ, 39-40; entomologist, Ministry Agr & Forestry, China, 42-44; teacher chem, Putney Sch, Vt, 47-48; chemist, Vitaminerals, Inc, Calif, 48-54; from asst res zoologist to assoc res zoologist, 54-65, RES ZOOLOGIST, UNIV CALIF, LOS ANGELES, 65- Concurrent Pos: Nat Found Infantile Paralysis fel, 57; lectr, Immaculate Heart Col, Calif, 58-59; Inter-Am Prog Trop Med fel, La State Univ, 61; mem int panel workshop malaria immunol, Walter Reed Army Inst Res, 63; consult, Ore State Univ, 75. Mem: AAAS; Am Soc Trop Med & Hyg; Soc Protozool; Am Soc Parasitol; Am Mosquito Control Asn. Res: In vitro culture of plasmodium; internal microorganisms; in vitro culture of insect cells and the insect phase of blood parasites; reptilian hemogregarine life cycle and transmission; invertebrate tissue culture. Mailing Add: Dept of Zool Univ of Calif Los Angeles CA 90024

CHAO, LI-PEN, b Hunan, Repub of China, Apr 14, 33; m 67; c 2. NEUROCHEMISTRY, BIOCHEMISTRY. Educ: Chung Hsing Univ, Taiwan, BS, 57; Univ Minn, MS, 64, PhD(biochem), 67. Prof Exp: Teaching asst plant physiol, Chung Hsing Univ, Taiwan, 59-60; res asst biochem, Univ Minn, 61-67; asst res biochemist, Sch Med, Univ Calif, San Francisco, 67-70; asst res neurologist, 70-74, ASSOC RES NEUROLOGIST, SCH MED, UNIV CALIF, LOS ANGELES, 74- Mem: AAAS; Am Chem Soc; Sigma Xi; Am Soc Neurochem; Int Soc Neurochem. Res: Chemistry of enzymes and proteins—isolation, purification and characterization by various chemical and physical methods; chemistry of cholinergic nerve transmission and localization of cholinergic neurons. Mailing Add: Dept of Neurol Univ of Calif Sch of Med Los Angeles CA 90024

CHAO, MIN-TE, b Hupei, China, Jan 23, 38; US citizen; m 65; c 2. STATISTICS. Educ: Nat Taiwan Univ, BS, 61; Univ Calif, Berkeley, MA, 65, PhD(statist), 67. Prof Exp: MEM TECH STAFF, BELL LABS, 68- Concurrent Pos: Assoc ed, Bull Acad Sinica, 75- Mem: Am Statist Asn. Res: Large sample theory; stochastic approximating; models for error process in telecommunication applications; traffic theory. Mailing Add: Rm 2E-420 Bell Labs Holmdel NJ 07733

CHAO, MOU SHU, b Changsha, China, Nov 20, 24; nat US; m 68. ELECTROCHEMISTRY, ANALYTICAL CHEMISTRY. Educ: Nat Cent Univ, China, BS, 47; Univ Ill, MS, 57, PhD(anal chem), 61. Prof Exp: Asst chem, Nat Cent Univ, China, 47-49; technician, Taiwan Fertilizer Co, 50-53, asst engr, 53-55; chemist, Spec Serv Lab, 60-63, res chemist, Electro & Inorg Res Lab, 63-72, RES SPECIALIST, INORG LAB, DOW CHEM CO, 72- Mem: Am Chem Soc; Electrochem Soc; Am Inst Chemists. Res: Chemical instrumentation; electrode kinetics. Mailing Add: Cent Res-Inorg Lab Dow Chem Co Midland MI 48640

CHAO, TSAI HSIANG, b Canton, China, Oct 24, 12; US citizen; m 39; c 2. ORGANIC CHEMISTRY. Educ: Ind Univ, BS, 35, AM, 36; Purdue Univ, PhD(org chem), 40. Prof Exp: Asst chief chemist, Org Sect, Res Dept, Calco Chem Div, Am Cyanamid Co, 44-52, sr res chemist, Org Chem Div, 52-75; RETIRED. Mem: Fel AAAS; NY Acad Sci; Am Chem Soc. Res: Chlorination and utilization of chlorohydrocarbons; rubber chemicals and dye intermediates; hydrogenation of aromatics; textile chemicals. Mailing Add: 7 Monroe St Somerville NJ 08876

CHAO, TSUN TIEN, b Anhwei, China, July 23, 18; US citizen; m 47; c 3. GEOCHEMISTRY. Educ: Nat Cent Univ, China, BS, 42; Ore State Univ, MS, 59, PhD(soil chem), 60. Prof Exp: Jr soil chemist, Nat Agr Res Bur, China, 44-45, 46-48; assoc agronomist, Taiwan Sugar Exp Sta, 48-52, soil technologist, 52-56; res fel soil chem, Ore State Univ, 56-59, res assoc, 59-61, asst prof, 61-64; from assoc soil chemist to soil chemist, Pineapple Res Inst, 64-67; RES CHEMIST, US GEOL SURV, 67- Mem: Soil Sci Soc Am; Am Soc Agron; Geochem Soc; AAAS. Res: Geochemistry of heavy metals in the weathering zone; research in methods of chemical analysis for geochemical exploration. Mailing Add: US Geol Surv Federal Ctr Denver CO 80225

CHAO, TYNG TSAIR, b Kan-su Prov, China, Oct 14, 27; US citizen; m 62; c 3. CHEMISTRY. Educ: Nat Taiwan Univ, BS & MS, 57; Va Polytech Inst, PhD(soil chem), 61. Prof Exp: Asst chem res, NC State Univ, 61; prof phys sci, 61-62, PROF CHEM, FAYETTEVILLE STATE UNIV, 62- Mem: Am Chem Soc; Am Soc Agron. Res: Nitrogen transformation through biological and chemical reactions. Mailing Add: Dept of Chem Fayetteville State Univ Fayetteville NC 28301

CHAP, JAMES JOHN, b Springfield, Mass, May 30, 09; m 43; c 2. CHEMISTRY Educ: Bates Col, AB, 31; Mass State Col, MS, 32, PhD(org chem), 34. Prof Exp: Jr chemist, NY Customs Lab, US Food & Drug Admin, 36-44, 46-47, asst to chief div labs, Bur Customs, Washington, DC, 47-49, from asst chief to chief org div, NY Customs Lab, 49-74; RETIRED. Mem: AAAS; Am Chem Soc. Res: Phenobarbital synthesis; analysis of iron in pharmaceuticals; gas chromatograph and infrared analysis of narcotics; analysis of surfactants and petroleum products. Mailing Add: 14 Victoria Dr Nanuet NY 10954

CHAPARAS, SOTIROS D, b Lowell, Mass, May 4, 29; m 56; c 2. MICROBIOLOGY, IMMUNOLOGY. Educ: Northeastern Univ, BS, 51; Univ Mass, MS, 53; St Louis Univ, PhD(microbiol), 58; Am Bd Med Microbiol, dipl. Prof Exp: Instr microbiol, Sch Med, St Louis Univ, 58-59; instr, Sch Med, Univ Southern Calif, 59-60; sr scientist, 60-68, CHIEF SECT MYOBACT & FUNGAL ANTIGENS, BUR BIOLOGICS, FOOD & DRUG ADMIN, 68- Concurrent Pos: Lectr, Children's Hosp, DC, 61-63, Howard Univ, 62- & NIH Fac, 62- Mem: Am Asn Immunol; Am Thoracic Soc; Am Soc Microbiol. Res: Pathogenicity, immunity and hypersensitivity in tuberculosis; standardization of skin test reagents; immunity in cancer; immunology of fungi. Mailing Add: Bur of Biologics Food & Drug Admin Bethesda MD 20014

CHAPAS, RICHARD BERNARD, b Cleveland, Ohio, Dec 5, 45; m 68; c 2. PHOTOGRAPHIC CHEMISTRY. Educ: St Vincent Col, BS, 68; Univ Ill, PhD(chem), 72. Prof Exp: SR RES CHEMIST, EASTMAN KODAK CO, 72- Concurrent Pos: Mem adj fac, Rochester Inst Technol, 73- Mem: Am Chem Soc; Soc Photog Scientists & Engrs. Res: Photographic systems development and analysis; light sensitive materials production. Mailing Add: 15 Saddlehorn Dr Rochester NY 14626

CHAPCO, WILLIAM, genetics, biometry, see 12th edition

CHAPEL, JAMES L, b Ont, Jan 1, 20; US citizen; m 52; c 3. CHILD PSYCHIATRY. Educ: Univ Toronto, MD, 54. Prof Exp: Assoc child psychiat, Univ Iowa Psychopath Hosp, 66; PROF PSYCHIAT & CHIEF SECT CHILD PSYCHIAT, COL MED, UNIV MO-COLUMBIA, 66- Concurrent Pos: Consult child psychiat, Bd Control, Iowa, 66; consult, Bd Training Schs, Mo, 67- & Boone & Calloway County Juv Courts, 67- Mem: Am Psychiat Asn. Res: Diagnosis, evaluation and treatment of learning disabilities such as perceptual deficits and cognitive processes; the biological and biochemical basis of sexual deviancies. Mailing Add: Univ of Mo Med Ctr 807 Stadium Rd Columbia MO 65201

CHAPERON, EDWARD ALFRED, b Burlington, Vt, Sept 24, 30; m 62; c 2. IMMUNOLOGY, MICROBIOLOGY. Educ: LeMoyne Col, BS, 57; Marquette Univ, MS, 59; Univ Wis-Madison, PhD(zool), 65. Prof Exp: NIH fel immunol, Med Ctr, Univ Colo, Denver, 65-68; asst prof, 68-71, ASSOC PROF MICROBIOL, SCH MED, CREIGHTON UNIV, 71- Mem: Am Asn Immunol; Am Soc Microbiol; Reticuloendothelial Soc. Res: Cellular immunology. Mailing Add: Dept of Med Microbiol Creighton Univ Sch of Med Omaha NE 68178

CHAPIN, CHARLES EDWARD, b Porterville, Calif, Oct 25, 32; m 58; c 3. GEOLOGY. Educ: Colo Sch Mines, Geol Engr, 54, DSc(geochem), 65. Prof Exp: Asst prof geol, Univ Tulsa, 64-65; asst prof, NMex Inst Mining & Technol, 65-68, assoc prof & head geosci dept, 68-70; GEOLOGIST, NMEX BUR MINES, 70- Mem: Geol Soc Am; Soc Econ Paleontologists & Mineralogists; Am Geochem Soc. Res: Volcanology; mineral deposits; sedimentary petrology. Mailing Add: NMex Bur of Mines Socorro NM 87801

CHAPIN, DOUGLAS SCOTT, b Muskegon, Mich, July 14, 22; m 44; c 4. PHYSICAL CHEMISTRY Educ: Kans State Col, BS, 44; Ill Inst Technol, MS, 48; Ohio State Univ, PhD(chem), 54. Prof Exp: Researcher, Nat Bur Standards, 47; asst, Cryogenic Lab, Ohio State Univ, 48-51; sr cryogenic oper, Herrick L Johnston, Inc, 52; asst, Cryogenic Lab, Ohio State Univ, 53-54; cryogenic engr, Herrick L Johnston, Inc, 54; from asst prof to assoc prof chem, Univ Ariz, 54-66; assoc prog dir grad fels & traineeships, 66-68, prog dir, 68-73, HEAD FELS & TRAINEESHIPS SECT, NAT SCI FOUND, 73- Concurrent Pos: Staff mem, Lincoln Lab, Mass Inst Technol, 63-64. Mem: AAAS; Faraday Soc; Am Chem Soc. Res: Low temperature kinetics; ortho-parahydrogen catalysis; separation of ortho-parahydrogen; low temperature thermodynamics of gases absorbed on solids; solid state chemistry of transition metal oxides. Mailing Add: Div of Grad Educ in Sci Nat Sci Found Washington DC 20550

CHAPIN, EARL CONE, b Farmington, Ill, Feb 5, 19; m 45; c 3. ORGANIC CHEMISTRY. Educ: Univ Ill, BS, 41; Pa State Univ, MS, 42, PhD(org chem), 44. Prof Exp: Res chemist, Monsanto Chem Co, 44-50, res group leader, 51-62; chmn dept phys sci, 62-68, dean sch arts & sci, 68-74, PROF CHEM, WESTERN NEW ENGLAND COL, 62- Concurrent Pos: Consult, Spencer Chem Div, Gulf Oil Co, Kans & Arthur D Little, Mass. Mem: Am Chem Soc; Am Asn Physics Teachers; Brit Chem Soc; Swiss Chem Soc. Res: Synthetic polymers; organic synthesis. Mailing Add: Sch of Arts & Sci Western New England Col Springfield MA 01119

CHAPIN, EDWARD WILLIAM, JR, b Baltimore, Md, May 28, 43. MATHEMATICAL LOGIC. Educ: Trinity Col, Conn, BS, 65; Princeton Univ, MA, 67, PhD(math), 69. Prof Exp: ASST PROF MATH, UNIV NOTRE DAME, 69- Mem: AAAS; Am Math Soc; Math Asn Am; Asn Symbolic Logic; Soc Indust & Appl Math. Res: Algebraic structure of deductive systems; non-standard measure theory; stochastic models of set theory; modal logic. Mailing Add: 740 E Woodside South Bend IN 46614

CHAPIN, JOHN LADNER, b New York, NY, Sept 1, 16; m 50; c 3. RESPIRATORY PHYSIOLOGY. Educ: Denison Univ, AB, 39; Univ Rochester, PhD(physiol), 50. Prof Exp: Instr physiol, Sch Med, La State Univ, 50; from instr to asst prof, Sch Med, Univ Colo, 51-65; PROF PHYSIOL, INST CHILD STUDY, UNIV MD, COLLEGE PARK, 65- Mem: AAAS; Am Ornithologists Union; Am Physiol Soc; NY Acad Sci; Am Inst Biol Sci. Res: Respiratory and environmental physiology; man and his environment; physical aspects of human development. Mailing Add: 2400 Parker Ave Wheaton MD 20902

CHAPIN, JOHN STILLMAN, soil chemistry, see 12th edition

CHAPLER, CHRISTOPHER KEITH, b Des Moines, Iowa, July 19, 40; m 61; c 3 MEDICAL PHYSIOLOGY. Educ: Drake Univ, BA, 62, MA, 64; Univ Fla, PhD(physiol), 67. Prof Exp: Can Heart Found fel, 67; asst prof, 68-73, ASSOC PROF PHYSIOL, QUEEN'S UNIV, ONT, 73- Mem: Am Physiol Soc; Can Physiol Soc; fel Am Col Sports Med; Fedn Am Socs Exp Biol; AAAS. Res: Muscle metabolism; cardiovascular physiology. Mailing Add: Dept of Physiol Queen's Univ Kingston ON Can

CHAPLIN, HUGH, JR, b New York, NY, Feb 4, 23; m 45; c 4. MEDICINE. Educ: Princeton Univ, AB, 43; Columbia Univ, MD, 47. Prof Exp: From intern to resident, Mass Gen Hosp, 47-50; physician in chg, Clin Ctr Blood Bank, NIH, 53-55; from instr to asst prof, 55-62, dir student health serv, 56-57, assoc dean, Med Sch, 57-63, assoc prof & dir, Johnson Insts Rehab, 64-65, PROF MED & PREV MED & KOUNTZ PROF PREV MED, MED SCH, WASH UNIV, 65- Concurrent Pos: Fel, Brit Postgrad Med Sch, London, 51-53; Nat Inst Arthritis & Metab Dis trainee, Med Sch, Wash Univ, 55-56; Commonwealth Fund res fel, Wright-Fleming Inst, London, 62-63; mem, Am Bd Internal Med, 56. Mem: Am Soc Hematol; Am Soc Clin Invest; Am Fedn Clin Res; Asn Am Physicians; Royal Soc Med. Res: Hematology; secondary hemolytic anemia. Mailing Add: 159 Linden Ave St Louis MO 63105

CHAPLIN, JAMES FERRIS, b Lamar, SC, Nov 10, 20; m 45; c 2. GENETICS. Educ: Clemson Col, BS, 48; NC State Col, MS, 52, PhD(plant breeding, genetics), 59. Prof Exp: Asst agronomist, Clemson Col, 48-57; res agronomist, Pee Dee Exp Sta, 57-65, sr scientist, Oxford Tobacco Res Lab, 65-67, leader tobacco breeding & dis invests, Plant Sci Res Div, 67-72, DIR, OXFORD TOBACCO RES LAB, USDA, 72-; PROF CROP SCI, NC STATE UNIV, 65- Mem: Am Soc Agron; Am Genetics Asn. Res: Tobacco breeding and genetics for disease resistance, chemical constituents and agronomic characteristics. Mailing Add: USDA Tobacco Res Lab Agr Res Serv Southern Region Oxford NC 27565

CHAPLIN, MICHAEL H, b Olney, Ill, Aug 22, 43; m 65; c 1. PLANT PHYSIOLOGY, NUTRITION. Educ: Univ Ky, BS, 65; Rutgers Univ, MS, 66; Mich State Univ, PhD(hort), 68. Prof Exp: Dir plant anal lab, 68-74, ASSOC PROF HORT, ORE STATE UNIV, 68- Mem: Am Soc Hort Sci. Res: Biologically active compounds to control tree growth and fruit maturation processes; nutritional status of economic crops as it affects quality, yield and maturation processes. Mailing Add: Dept of Hort Ore State Univ Corvallis OR 97331

CHAPLIN, ROBERT LEE, JR, b Savannah, Ga, Mar 22, 23; m 56; c 1. PHYSICS. Educ: Clemson Univ, BS, 48; NC State Col, MS, 53, PhD(physics), 62. Prof Exp: Instr physics, NC State Col, 56-60; res asst, Univ NC, 60-61, fel, 61-62; from asst prof to assoc prof, 62-74, PROF PHYSICS, CLEMSON UNIV, 74- Concurrent Pos: Res partic, Oak Ridge Nat Lab, 63- Mem: AAAS; Am Phys Soc. Res: Defect state of metal crystals as produced by electron irradiation damage and removed by thermal annealing. Mailing Add: Dept of Physics Clemson Univ Clemson SC 29631

CHAPLINE, WILLIAM RIDGELY, (JR), b Lincoln, Nebr, Jan 10, 91; m 21; c 2. RANGE MANAGEMENT, FORESTRY. Educ: Univ Nebr, BS. Prof Exp: Grazing asst, US Forest Serv, 13-14, examr grazing, 15-20, from inspector to sr inspector in charge grazing res, 20-35, chief div range res, 35-52; chief forest conserv sect, Food & Agr Orgn, UN, Rome, 52-54; prof int course on pastures, Inst Interam Agr Sci, Montevideo, 54-55; consult, Ministry Agr, Arg & Chile, 55, Inst Interam Agr Sci, Peru, 55 & Govt of Spain, 56-57; consult watershed mgt, Charles Lathrop Pack Forestry Found, 57-59; CONSULT, 59- Concurrent Pos: Spec investr, USDA, 17-18; consult, Civilian Conserv Corp, 34 & 36, US Navy, 44, Forest Serv, USDA, 67-69, Am Univ African Drought Proj, 74- & Vols Int Tech Assistance; coordr spec course, Colo State Univ, 67. Honors & Awards: Cert Merit, Soc Range Mgt, 67. Mem: AAAS; Soc Foresters; Soc Range Mgt; hon mem Grassland Soc Southern Africa. Res: Forest, range and watershed management; ecology; range plants; livestock production; erosion control; range re-seeding; worldwide rangeland management and improvement. Mailing Add: 4225 43rd St NW Washington DC 20016

CHAPMAN, ALAN THEODORE, chemistry, see 12th edition

CHAPMAN, ALBERT LEE, b Anderson, Mo, Nov 5, 33; m 56; c 4. ANATOMY. Educ: Univ Mo, AB, 56, MS, 59; Univ Nebr, PhD(anat), 62. Prof Exp: From instr to assoc prof, 62-74, PROF ANAT, UNIV KANS MED CTR, KANSAS CITY, 74- Concurrent Pos: Spec fel, Viral Lymphoma & Leukemia Br, Nat Cancer Inst, 69- Mem: Am Asn Anat; AAAS; NY Acad Sci; Soc Exp Biol & Med; Electron Micros Soc Am. Res: Electron microscopic studies of lymphatic tissues, especially leukemic tissue and virus. Mailing Add: Dept of Anat Univ Kans Med Ctr Kansas City KS 66103

CHAPMAN, ALBERT SIMONDS, b Pasadena, Calif, Mar 4, 24; m 51; c 3. GEOGRAPHY. Educ: Stanford Univ, BA, 48; Northwestern Univ, MS, 50, PhD, 54. Prof Exp: Asst prof geog, Mich State Univ, 53; lectr, Univ Minn, Duluth, 54; asst prof geog, Univ Ohio, 54-57; geog attache, Am Embassy, New Delhi, India, 57-61, Paris, 61-64; assoc prof geog, Univ Ga, 64-67; GEOG ATTACHE, AM EMBASSY, BONN, 67- Mem: Asn Am Geogr. Res: Transportation; South Asia and Middle East. Mailing Add: Geog Attache American Embassy Box 190 Bonn West Germany

CHAPMAN, ARTHUR BARCLAY, b Windermere, Eng, Oct 25, 08; nat US; m 34; c 3. ANIMAL BREEDING. Educ: State Col Wash, BS, 30; Iowa State Col, MS, 31; Univ Wis, PhD(genetics), 35. Prof Exp: Asst genetics, Univ Wis, 31-36; Nat Res fel agr, Iowa State Col & Chicago, 36-37; from instr to prof genetics, 37-75, EMER PROF GENETICS, UNIV WIS-MADISON, 75- Concurrent Pos: Rockefeller Found res & teaching award, Poland, 60; ed, J Am Soc Animal Sci, 61-63; Fulbright & Guggenheim Mem Found fels, NZ, 66-67 & Midwest Univ Consortium Int Activities-AID, Indonesia, 73. Honors & Awards: Animal Breeding & Genetics Award, Am Soc Animal Sci, 68, Morrison Award, 74. Mem: Fel AAAS; Biomet Soc; Genetics Soc Am; fel Am Soc Animal Sci (vpres, 63-64, pres, 64-65); Am Dairy Sci Asn. Res: Animal breeding; genetic effects of irradiation. Mailing Add: 1117 Risser Rd Madison WI 53705

CHAPMAN, ARTHUR OWEN, b Heber, Utah, Mar 16, 13; m 41; c 2. HISTOLOGY, PATHOLOGY. Educ: Brigham Young Univ, AB, 41; Univ Kans, MA, 49; Univ Nebr, PhD(med sci), 53. Prof Exp: Asst instr zool, Univ Kans, 46-47, asst instr anat, 47-49;

instr, Col Med, Univ Nebr, 49-53, asst prof, 53-59; assoc prof zool, 59-67, PROF ZOOL, BRIGHAM YOUNG UNIV, 67- Mem: Am Asn Anatomists. Res: Nervous system; irradiation effects on the brain and on embryos; thiamin deficiency and mercury effects on the brain. Mailing Add: Dept of Zool 259 WIDB Brigham Young Univ Provo UT 84601

CHAPMAN, ASTRID GRØNNEBERG, b Skotselv, Norway, Feb 3, 38. BIOCHEMISTRY. Educ: Univ Calif, Los Angeles, BS, 62, PhD(biochem), 67. Prof Exp: Lab asst biochem, Inst Enzyme Res, Madison, Wis, 58-59 & 62-63; res assoc, Inst Cancer Res, Norway Radiation Hosp, Oslo, 67-68; fel biochem, Vet Admin Hosp, Los Angeles, 68-69; asst res chemist, Univ Calif, Los Angeles, 69-74; asst res chemist, Brain Res Unit, Univ Hosp, Lund, Sweden, 75; ASST RES CHEMIST, UNIV CALIF, LOS ANGELES, 75- Mem: Sigma Xi. Res: Regulation of energy metabolism; correlation between in vitro kinetic regulatory properties of key enzymes and observed physiological variations in metabolite concentrations. Mailing Add: Dept of Chem Univ of Calif Los Angeles CA 90024

CHAPMAN, CARL HALEY, b Steelville, Mo, May 29, 15; m 42; c 2. ANTHROPOLOGY, ARCHAEOLOGY. Educ: Univ Mo, AB, 39; Univ NMex, MA, 46; Univ Mich, PhD(anthrop), 59. Prof Exp: Res asst, Archaeol Res Exp Sta, Univ Mo, 38-40, dir archaeol, Univ Mo-Columbia, 41, dir mus anthrop, 46-50 & 51-55, dir, Am Archaeol, 46-50, 51-56 & 60-65, instr anthrop, 46-50, from asst prof to assoc prof, 51-60, PROF ANTHROP, UNIV MO-COLUMBIA, 60-, DIR ARCHAEOL RES ACTIV, 65- Concurrent Pos: Nat Park Serv grants archaeol invests, Univ Mo, 52-60, NSF grant, Osage Prehist, 61-63 & Nat Endowment Humanities grant archaeol, 70-72; expert witness, Indian Claims Sect, US Justice Dept, 60-65. Honors & Awards: Distinguished Serv Award, Soc Am Archaeol, 75 Memm: Fel AAAS; fel Am Anthrop Asn; Soc Am Archaeol; Soc Hist Archaeol; Am Ethnol Soc. Res: Archaeology of Missouri, in the eastern United States from the earliest periods to historical times; ethnohistory and archaeology of the Osage Indian tribe. Mailing Add: 205 Swallow Hall Univ of Mo Columbia MO 65201

CHAPMAN, CARL JOSEPH, b New York, NY, July 4, 39; m 65. SYSTEMATIC BOTANY. Educ: Univ NH, BS, 61, PhD(bot), 65. Prof Exp: Fel, Univ NH, 65; instr bot, Univ RI, 65-66; res assoc, Univ Fla, 66-67; asst prof biol, 67-73, ASSOC PROF BIOL, CONCORD COL, 73- Mem: Am Orchid Soc; Bot Soc Am; Sigma Xi. Res: Floristic studies; scanning electron microscopy studies of fern spores. Mailing Add: Dept of Biol Concord Col Athens WV 24712

CHAPMAN, CARLETON ABRAMSON, b Groveton, NH, Oct 14, 11; m 40; c 1. GEOLOGY. Educ: Univ NH, BS, 33; Harvard Univ, AM, 35, PhD(petrog), 37. Prof Exp: Asst petrog, Harvard Univ, 34-35, instr, 35-37; instr geol, 37-39, assoc, 39-42, from asst prof to assoc prof, 45-48, PROF GEOL, UNIV ILL, URBANA-CHAMPAIGN, 48- Mem: AAAS; fel Geol Soc Am; fel Mineral Soc Am; Am Geochem Soc; fel Geol Soc London. Res: Petrology; igneous and metamorphic geology; mineralogy; structural geology. Mailing Add: Dept of Geol Univ of Ill Urbana IL 61822

CHAPMAN, CARLETON BURKE, b Sycamore, Ala, June 11, 15; m 40; c 3. MEDICINE. Educ: Davidson Col, AB, 36; Oxford Univ, BS, 38; Harvard Univ, MD, 41, MPH, 44; Am Bd Internal Med, dipl, 48; Am Bd Cardiovasc Dis, 53. Hon Degrees: MA, Dartmouth Col, 68; LLD, Davidson Col, 68. Prof Exp: From intern to resident med, Boston City Hosp, Mass, 41-44; from instr to asst prof med, Med Sch, Univ Minn, 47-53; prof, Southwestern Med Sch, Univ Tex, 53-66; dean, Dartmouth Med Sch, 66-73, vpres, 72-73; PRES, COMMONWEALTH FUND, NEW YORK, 73- Concurrent Pos: Rockefeller Found fel, Sch Pub Health, Harvard Univ, 44; asst, Harvard Med Sch, 43; consult, Surgeon-Gen, US Army, 44 & US Vet Admin, 47-; fel, Coun Clin Cardiol, Am Heart Asn. Mem: AAAS; Am Acad Arts & Sci; Am Soc Clin Invest; fel Am Col Cardiol; Am Heart Asn (pres, 64-65). Res: Cardiovascular disease. Mailing Add: Thetford VT 05074

CHAPMAN, CHARLES R, b Bismarck, NDak, June 29, 25; m 52; c 2. ECOLOGY. Educ: Utah State Univ, BS, 50; Univ Wyo, MS, 52. Prof Exp: Fishery res biologist, US Fish & Wildlife Serv, 50-51; fishery mgt supvr, Ohio Div Wildlife, 53-55; fishery res biologist, Bur Com Fisheries, US Fish & Wildlife Serv, 55-57, supvry fishery res biologist, Bur Sport Fisheries & Wildlife, 57-59, fishery biologist, 59-62, supvry fishery res biologist, Bur Com Fisheries, 62-69, fishery mgt biologist, Bur Com Fisheries, Div Resource Mgt, 69-70, fish & wildlife adminr, Bur Sport Fisheries & Wildlife, US Dept Interior, 70-73; resource analyst, Nat Oceanic & Atmospheric Admin, 73-74; MGR AQUATIC ECOSYSTS PROG, OFF BIOL SERV, US FISH & WILDLIFE SERV, 74- Concurrent Pos: Consult, Bur Sport Fisheries & Wildlife, 62-, Tex Water Pollution Control Bd, 63-64 & La Wildlife & Fisheries Comn, 64-65; mem estuarine tech coord comt, Gulf States Marine Fisheries Comt, 64-70. Mem: Am Fisheries Soc. Res: Estuarine ecology; coastal zone management; marine fishery biology; coastal water development planning; oil pollution; coastal ecosystems; power plants; stream altevation. Mailing Add: Fish & Wildlife Serv US Dept of the Interior Washington DC 20240

CHAPMAN, CLARK RUSSELL, b Palo Alto, Calif, May 13, 45; m 66. PLANETARY SCIENCES. Educ: Harvard Col, AB, 67; Mass Inst Technol, MS, 68, PhD(planetary sci), 72. Prof Exp: Res scientist astrosci, Res Inst, Ill Inst Technol, 71-72; RES SCIENTIST, PLANETARY SCI INST, SCI APPLN, INC, 72- Concurrent Pos: Consult, Task Group Mercury Nomenclature, Int Astron Univ, 74-75 & Lunar Sci Inst Coun, 75-; mem sci adv group inner solar syst, NASA, 75- Mem: AAAS; Am Astron Soc; Am Geophys Union; Meteoritical Soc. Res: Spectrophotometry of planets, satellites, asteroids and compositional interpretation; analysis of telescopic and spacecraft imagery of planetary atmospheres and surfaces; cratering and impact-erosional processes; planetary accretion; Jupiter's atmospheric circulation; comparative planetology. Mailing Add: 6160 N Montebella Tucson AZ 85704

CHAPMAN, DAVID J, b Kingston, WI, Dec 12, 39; m 63; c 1. PLANT BIOCHEMISTRY, PHYCOLOGY. Educ: Univ Auckland, BSc, 60; Univ Calif, PhD(marine biol), 65. Prof Exp: Res assoc marine biol, Scripps Inst, Calif, 65-66; res assoc biol, Brookhaven Nat Lab, 66-67; asst prof, Univ Chicago, 68-73; ASSOC PROF BIOL, UNIV CALIF, LOS ANGELES, 73- Concurrent Pos: Ger Acad Exchange award, Univ Saarlandes, 70. Mem: Am Chem Soc; Linnean Soc London; Am Soc Plant Physiol; Bot Soc Am; Phycol Soc Am. Res: Algal, chloroplast and natural product biochemistry; chemical taxonomy. Mailing Add: Dept of Biol Univ of Calif Los Angeles CA 90024

CHAPMAN, DAVID MACLEAN, b Thunder Bay, Ont, Jan 3, 35. ANATOMY, ZOOLOGY. Educ: Univ Man, BSc, 59, MSc, 61; Univ Cambridge, PhD(zool), 64. Prof Exp: PROF ANAT, MED COL, DALHOUSIE UNIV, 64- Mem: Can Soc Zool; Marine Biol Asn UK. Res: Electron microscopy of cnidarians. Mailing Add: Dept of Anat Dalhousie Univ Halifax NS Can

CHAPMAN, DEREK D, b Lincoln, Eng, Feb 13, 32; m 58; c 2. ORGANIC CHEMISTRY. Educ: Univ Nottingham, BSc, 53, PhD(org chem), 56. Prof Exp: Lectr chem, SE Essex Tech Col, Eng, 60-61; sr res chemist, 62-66, RES ASSOC, EASTMAN KODAK CO, 66- Concurrent Pos: Fel, Univ Rochester, 56-59, Hull Univ, 59-60 & Univ Rochester, 61-62. Mem: Am Chem Soc; Soc Chem Indust; Brit Chem Soc. Res: Synthesis of alkaloids; structure determination of antibiotics; synthesis of photographically useful compounds. Mailing Add: Eastman Kodak Co Res Labs 1669 Lake Ave Rochester NY 14650

CHAPMAN, DONALD HARDING, b Saginaw, Mich, Nov 14, 04; m 32. GEOMORPHOLOGY, GLACIAL GEOLOGY. Educ: Univ Mich, AB, 27, AM, 28, PhD(geol), 31. Prof Exp: Instr geol, Dartmouth Col, 28-30 & Univ Mich, 31; from instr to prof, 31-74, in chg geog & meteorol, 36-58, EMER PROF GEOL, UNIV NH, 74- Concurrent Pos: Asst prof, La State Univ, 37-38; Fulbright scholar, Norway, 50. Honors & Awards: Medal of Order of St Olav, Norway, 58. Mem: Am Meteorol Soc; Geol Soc Am. Res: Glacial physiography; meteorology; climatology; water resources. Mailing Add: Dept of Earth Sci James Hall Univ of NH Durham NH 03824

CHAPMAN, DONALD WALLACE, fisheries, see 12th edition

CHAPMAN, DOUGLAS GEORGE, b Provost, Alta, Mar 20, 20; nat US; m 43; c 3. BIOMETRICS. Educ: Univ Sask, BA, 39; Univ Calif, MA, 40, PhD(math), 49; Univ Toronto, MA, 44. Prof Exp: Meteorologist, Meteorol Serv Can, 41-46; asst prof math, Univ BC, 46-48; asst math statist, Univ Calif, 48-49; from asst prof to prof math, 49-68, dir ctr quantitative sci, 68-71, DEAN COL FISHERIES, UNIV WASH, 71- Concurrent Pos: Guggenheim fel, Oxford Univ, 54-55; vis prof, NC State Univ, 58-59 & Univ Calif, San Diego, 63-64; adv NPac Fur Seal Comn, 59-; chmn spec study group, Int Whaling Comn, 61-64, sci comn, 65-74; mem, Wash State Census Bur, 65-68; mem comn nat statist, Nat Acad Sci, 71-74 & ocean affairs bd, 72-76; chmn comn sci adv, Marine Mammal Comn, 74- Mem: Fel Inst Math Statist; Biomet Soc; fel Am Statist Asn; Am Fisheries Soc; AAAS. Res: Mathematical statistics theory; population estimation; population dynamics. Mailing Add: Col of Fisheries Univ of Wash Seattle WA 98195

CHAPMAN, DOUGLAS WILFRED, b Can, Sept 21, 21; nat US; m 53; c 2. ORGANIC CHEMISTRY. Educ: Univ Calif, BS, 48; Mass Inst Technol, PhD(chem), 51. Prof Exp: Res chemist, 51-65, RES ASSOC, MALLINCKRODT CHEM WORKS, 65- Mem: Am Chem Soc. Res: Organic synthesis; x-ray contrast media. Mailing Add: 48 Greendale Dr St Louis MO 63121

CHAPMAN, FLOYD BARTON, b Zaleski, Ohio, Apr 1, 11. HORTICULTURE Educ: Ohio State Univ, AB, 32, AM, 33, PhD(zool), 38 Prof Exp: Asst bot, Ohio State Univ, 32-35; field ecologist, State Div Conserv, Ohio, 35-38, game mgt technician, 38-40; asst biologist, Fish & Wildlife Serv, US Dept Interior, 40-42; forest game technician, State Div Wildlife, Ohio, 45-47, from asst chief to assoc chief game sect, 47-57; res ecologist, Malabar Farm, Friends of the Land, Ohio, 58-63; HORTICULTURIST, ADJUNCTIVE THER DEPT, HARDING HOSP, 63- Concurrent Pos: Mem, Nat Coun Rehab & Ther Through Hort. Mem: Am Soc Mammal; Asn Interpretive Naturalists; Am Hort Soc; fel Royal Hort Soc. Res: Ecology of white-tailed deer, tree squirrels and ruffed grouse; ecology of forest game animals; nature interpretation; environmental ecology; seed germination in alpine plants. Mailing Add: Adjunctive Ther Dept Harding Hosp Worthington OH 43085

CHAPMAN, FRANCIS WORTHINGTON, JR, analytical chemistry, petroleum chemistry, see 12th edition

CHAPMAN, GARY ADAIR, b Corvallis, Ore, Aug 30, 37; m 60; c 2. AQUATIC BIOLOGY, WATER POLLUTION. Educ: Ore State Univ, BSc, 59, MSc, 65, PhD(fisheries), 69. Prof Exp: Agr res technician, USDA, 60-62; FISHERIES BIOLOGIST, US ENVIRON PROTECTION AGENCY, 68- Honors & Awards: Superior Serv Medal, US Environ Protection Agency, 72. Mem: Am Fisheries Soc; Sigma Xi. Res: Pollution effects on aquatic organisms; salmonid biology and toxicity bioassays; chemistry and toxicity of heavy metals in water. Mailing Add: Western Fish Toxicol Sta 1350 SE Goodnight Ave Corvallis OR 97330

CHAPMAN, GARY ALLEN, b Bryan, Tex, Mar 25, 38; m 62; c 1. SOLAR PHYSICS. Educ: Univ Ariz, BS, 60, PhD(astron), 68. Prof Exp: Asst astronr, Univ Hawaii, 68-69; MEM TECH STAFF SOLAR PHYSICS, AEROSPACE CORP, 69- Mem: AAAS; Am Astron Soc; Sigma Xi. Res: Fine structure of solar magnetic fields and solar faculae, the effect of faculae on solar oblateness measurements, and the study of solar x-rays. Mailing Add: Bldg A-1 Aerospace Corp PO Box 92957 Los Angeles CA 90009

CHAPMAN, GEORGE BUNKER, b Bayonne, NJ, June 10, 25. CYTOLOGY. Educ: Princeton Univ, AB, 50, AM, 52, PhD, 53. Prof Exp: Asst instr biol, Princeton Univ, 50-52, asst res, 52-53, res assoc, 54-56; asst prof zool, Harvard, 56-60; assoc prof anat, Med Col, Cornell Univ, 60-63; PROF BIOL & CHMN DEPT, GEORGETOWN UNIV, 63- Concurrent Pos: Res biologist, Labs Div, Radio Corp of Am, 53-56. Mem: Am Soc Microbiol; Electron Micros Soc Am; Am Asn Anatomists; Bot Soc Am; NY Acad Sci. Res: Bacteriophagy; bacterial and general cytology; fine structure of human skin, eye, uterus, and gallbladder; electron microscopy of coelenterates and protozoa. Mailing Add: Dept of Biol Georgetown Univ Washington DC 20001

CHAPMAN, GEORGE HERBERT, b London, Eng, Apr 14, 96; nat US; m 21, 52; c 1. BIOCHEMICAL PHARMACOLOGY. Prof Exp: Dir, Clin Res Lab, Bronx, NY, 24-74; RES ASSOC, BIOL RES INST, WHITE PLAINS, 74- Mem: Fel AAAS; Am Soc Microbiol; fel NY Acad Sci. Res: Bacteriology and infection of staphylococci and coliform bacteria; basic biochemical changes in disease; pharmacology of nucleic acids: pH-Eh relationships. Mailing Add: Rte 2 Colchester CT 06415

CHAPMAN, HAROLD CLYDE, b Kalamazoo, Mich, Dec 1, 21; m 44; c 4. ENTOMOLOGY. Educ: Mich State Univ, BS, 48, MS, 50; Rutgers Univ, PhD, 59. Prof Exp: Asst, Mich State Univ, 49-50; forest entomologist, Div Forest Insects, 50-51, med entomologist, Entom Res Div, Insects Affecting Man & Animals Res Br, Fla, 51-52, Nev, 58-61 & Calif, 61-64, LOCATION LEADER, GULF COAST MOSQUITO RES LAB, AGR RES SERV, USDA, 64- Concurrent Pos: Assoc, Rutgers Univ, 55-58. Mem: Soc Invert Path; Am Mosquito Control Asn (pres, 75-76). Res: Biological and ecological control research on mosquitoes. Mailing Add: USDA Gulf Coast Mosquito Res Lab McNeese State Univ Chennault Campus Lake Charles LA 70601

CHAPMAN, HERBERT L, JR, b Kansas City, Mo, July 15, 23; m 46; c 4. ANIMAL NUTRITION. Educ: Univ Fla, BSA, 48, MSA, 51; Iowa State Univ, PhD(animal nutrit), 55. Prof Exp: Asst animal husbandman, 51-53, from asst animal nutritionist to animal nutritionist, 55-65, PROF ANIMAL SCI & HEAD RANGE CATTLE STA, UNIV FLA AGR EXP STA, 65- Mem: Fel AAAS; Am Soc Animal Sci. Res: Mineral requirement and interrelations in beef cattle; vitamin interrelations; nutritional requirements of all classes of beef cattle; pasture crop evaluation; chemical residues in beef cattle. Mailing Add: Range Cattle Sta Univ of Fla Ona FL 33865

CHAPMAN, HOMER DWIGHT, b Darlington, Wis, Oct 4, 98; m 28. SOIL FERTILITY Educ: Univ Wis, BS, 23, MS, 25, PhD(soil chem), 27. Hon Degrees: LLD, Univ Calif, 69. Prof Exp: Asst soil surv, Geol Natural Hist Surv, Calif, 23-25; asst soils, Univ Wis, 25-27; from asst chemist to chemist, Citrus Exp Sta, 27-67, chmn dept, 38-61, from assoc prof to prof agr chem, 38-44, prof soils & nutrit, 44-67, EMER CHEMIST, CITRUS EXP STA & EMER PROF SOILS & PLANT NUTRIT, UNIV CALIF, RIVERSIDE, 67- Concurrent Pos: Spec studies, Mediter Countries, 52, SAfrica & Japan, 57; consult, Brazil, 54, Japan, 60 & Aerojet, 75-; specialist, Int Coop Admin, Chile, 57; vis prof, India, 61-62. Mem: Fel Am Soc Agron; Soil Sci Soc Am; Am Soc Hort Sci; Int Soc Soil Sci; Int Soc Citricult (secy-treas). Res: Soil chemistry; citrus nutrition; development of diagnostic methods; literature reviewing and consulting advisory on research. Mailing Add: 830 S University Ave Riverside CA 92507

CHAPMAN, JEFFERSON, b Kinston, NC, Mar 13, 43; m 65; c 2. ANTHROPOLOGY, ARCHAEOLOGY. Educ: Yale Univ, BA, 65; Brown Univ, MAT, 68; Univ NC, Chapel Hill, MA, 73, PhD(anthrop), 75. Prof Exp: Teacher hist & anthrop, Webb Sch, Knoxville, 65-67, chmn social studies dept, 68-71; dir excavations, 70-75, RES ASST PROF ANTHROP, UNIV TENN, KNOXVILLE, 75- Mem: AAAS; Am Anthrop Asn; Asn Field Archaeol; Soc Am Archaeol. Res: Early archaic period in Eastern North America; paleoethnobotany of eastern woodlands. Mailing Add: Dept of Anthrop Univ of Tenn Knoxville TN 37916

CHAPMAN, JOE ALEXANDER, b Westpoint, Tenn, Oct 23, 19; m 41; c 3. PLANT ECOLOGY. Educ: Carson Newman Col, BS, 40; Peabody Col, MA, 47; Univ Tenn, PhD, 57. Prof Exp: High sch teacher, 41-42 & 42-44; from asst prof to assoc prof biol, 47-51, PROF BIOL & HEAD DEPT, CARSON NEWMAN COL, 53- Mem: AAAS; Ecol Soc Am; Am Bot Soc. Res: Taxonomy. Mailing Add: Carson Newman Col Jefferson City TN 37760

CHAPMAN, JOHN DONALD, b Estevan, Sask, Feb 18, 41; m 63; c 4. BIOPHYSICS, RADIATION BIOPHYSICS. Educ: Univ Sask, BSc, 63, MSc, 65; Pa State Univ, PhD(biophys), 67. Prof Exp: Nat Res Coun Can fel, 67-68; RES OFFICER MED BIOPHYS, ATOMIC ENERGY CAN, LTD, 68- Concurrent Pos: Assoc ed, Radiation Res, 75-78; vis lectr, Univ Calif, Berkeley, 75-76. Mem: AAAS; Biophys Soc; Radiation Res Soc; Brit Inst Radiol. Res: Radiation biology; in vitro mammalian cell biology; cellular and molecular biophysics; somatic cell mutation; biochemistry of mammalian cell cycle. Mailing Add: c/o Med Biophys Br Atomic Energy of Can Ltd Pinawa MB Can

CHAPMAN, JOHN DONERIC, b Eng, July 24, 23; m 45; c 1. GEOGRAPHY. Educ: Oxford Univ, BA, 47, MA, 49; Univ Wash, PhD, 58. Prof Exp: Asst prof geol & geog, Univ BC, 47-55, assoc prof geog, 55-62, head dept geog, 68-74, PROF GEOG, UNIV BC, 62- Mem: Asn Am Geogr; Can Asn Geogr. Res: Geography of energy and manufacturing. Mailing Add: Dept of Geog Univ of BC Vancouver BC Can

CHAPMAN, JOHN E, b Springfield, Mo, July 5, 31. PHARMACOLOGY. Educ: Southwest Mo State Col, BS(educ) & BS(biol, chem), 54; Univ Kans, MD, 58. Prof Exp: Instr pharmacol, Med Ctr, Univ Kans, 61-62, asst prof, 62-67, asst acad dean, Med Sch, 63-65, assoc dean, 65-67; assoc dean educ, 67-72, actg dean, 72-73, actg vchancellor med affairs, 73-75, DEAN, VANDERBILT UNIV SCH MED, 75- Concurrent Pos: Chmn & bd dirs, Health Educ Media Asn, Coun Deans. Mem: Am Soc Pharmacol & Exp Therapeut. Res: Role of humoral agents in central nervous system activity; medical school and medical center administration. Mailing Add: Sch of Med Vanderbilt Univ Nashville TN 37203

CHAPMAN, JOHN HERBERT, b London, Ont, Aug 28, 21; m 49; c 5. PHYSICS. Educ: Univ Western Ont, BSc, 48; McGill Univ, MSc, 49, PhD(physics), 51. Hon Degrees: LLD, Laurentian Univ, 67; DEng, Waterloo Univ, 69. Prof Exp: Sect leader ionospheric physics, Radio Physics Lab, Defence Res Telecommun Estab, 50-57, supt commun lab, 57-59, dep chief supt, 59-69; ASST DEP MINISTER SPACE PROG, DEPT COMMUN, 70- Concurrent Pos: Mem, Can Nat Comt Radio Sci, Int Sci Radio Union, 58-, secy, 64. Honors & Awards: Dellinger Gold Medal, Int Union Radio Sci, 66; Charles Chree Medal & Prize, Phys Soc & Inst Physics, London, 67; McCurdy Award, Can Aeronaut & Space Inst, 67. Mem: Fel Inst Elec & Electronics Engrs; Am Geophys Union; Royal Soc Can; Can Asn Physicists; fel Can Aeronaut & Space Inst. Res: Ionospheric propagation; aurora; ionospheric winds and tides; radio propagation; space research. Mailing Add: Dept of Commun Rm 2016 300 Slater St Ottawa ON Can

CHAPMAN, JOHN JUDSON, b Valdosta, Ga, May 10, 18; m 47; c 4. GEOLOGY. Educ: Univ Ill, PhD(geol), 53. Prof Exp: Topog engr, US Geol Surv, 41-44 & 46-47; sr geologist, Creole Petrol Corp, 48-51; asst prof geol & head dept, Southern State Col, 53-57, prof geol & chmn div natural sci, 57-68; PROF EARTH SCI & HEAD DEPT, W CAROLINA UNIV, 68- Mem: AAAS; Geol Soc Am; Am Asn Petrol Geologists; Am Asn Geol Teachers; Am Inst Prof Geologists. Res: Physical stratigraphy; petroleum geology. Mailing Add: Dept of Earth Sci Western Carolina Univ Cullowhee NC 28723

CHAPMAN, JOHN S, b Sweetwater, Tex, Jan 30, 08; m 32; c 1. MEDICINE. Educ: Southern Methodist Univ, BA & BS, 27, MA, 28; Univ Tex, MD, 32. Prof Exp: Clin asst prof med, Univ Tenn, 43; clin asst, 43-45, from clin instr to clin assoc prof, 45-52, PROF MED, UNIV TEX HEALTH SCI CTR DALLAS, 52- Prof Exp: Tuberc consult, Vet Admin Hosp, Dallas, 46; civilian consult, Brooke Army Hosp, 64; consult, Bur Radiol Health & ETex Tuberc Hosp, Tyler; ed-in-chief, Arch Environ Health, AMA, 71. Mem: AAAS; Am Col Physicians; Am Thoracic Soc. Res: Eosinophilic leucocyte; mycobacteria; juvenile tuberculosis. Mailing Add: 5323 Harry Hines Blvd Dallas TX 75235

CHAPMAN, JUDITH-ANNE WILLIAMS, b Timmins, Ont, Aug 17, 49; m; c 1. MEDICAL STATISTICS. Educ: Univ Waterloo, BS, 71, PhD(statist), 74. Prof Exp: NAT CANCER INST CAN FEL STATIST, UNIV WATERLOO, 74- Mem: Can Oncol Soc; Am Statist Asn. Res: Use of statistics in analysing cancer mortality and incidence data; regional differences in Ontario; different statistics for significance tests. Mailing Add: 11 Dayman Ct Kitchener ON Can

CHAPMAN, KENNETH REGINALD, b Croydon, Eng, Apr 10, 24; m 46; c 5. EXPERIMENTAL NUCLEAR PHYSICS. Educ: Univ London, BSc, 50 & 51, MSc, 60, PhD(nuclear physics), 65. Prof Exp: Develop engr, Mullard Radio Valve Co, UK, 40-47; res demonstr physics, Univ Col, Univ Leicester, 51-56; res assoc, Univ Birmingham, 56-66; ACCELERATOR PHYSICIST, FLA STATE UNIV, 66- Mem: Brit Inst Physics; Brit Inst Elec Eng. Res: Accelerator and nuclear physics; ion source research; accelerator development. Mailing Add: Dept of Physics Fla State Univ Tallahassee FL 32306

CHAPMAN, KENT M, b Minneapolis, Minn, Oct 28, 28; m 51; c 4. BIOPHYSICS. Educ: Univ Minn, BA, 49, MS, 53, PhD(biophys), 62. Prof Exp: From asst prof to assoc prof physiol, Univ Alta, 58-65; ASSOC PROF MED SCI, BROWN UNIV, 65-, CHMN NEUROSCI SECT, DIV BIOL & MED SCI, 73- Mem: AAAS; Biophys Soc; Am Physiol Soc; Soc Neurosci. Res: Neurophysiology; sensory transduction and encoding; electrophysiology; insect mechanoreceptors; invertebrate neurophysiology. Mailing Add: Div of Biol & Med Sci Box G Brown Univ Providence RI 02912

CHAPMAN, LORING FREDERICK, b Los Angeles, Calif, Oct 4, 29; m 54; c 3. BEHAVIORAL BIOLOGY, MEDICAL PHYSIOLOGY. Educ: Univ Nev, BS, 50; Univ Chicago, PhD(biopsychol), 55. Prof Exp: Res asst, Dept Med, Univ Chicago, 52-55; from asst prof to assoc prof psychol, Dept Med, Col Med, Cornell Univ, 54-61; assoc prof med psychol, Dept Psychiat, Univ Calif, Los Angeles, 61-65; prof, Univ Ore, 65-66 & Georgetown Univ, 66-67; PROF BEHAV BIOL & CHMN DEPT, SCH MED, UNIV CALIF, DAVIS, 67- Nat Acad Sci award, 57, 59 & 61; Wilson prize, 58; partic, Skin Conf, 59; vis scientist, Univ Sao Paulo, 59; partic, Ciba Found Conf, London, Eng, 60; consult, Nat Inst Neurol Dis, 62-; USPHS career develop award, 64; dir res, Fairview Hosp, 65-66; chief lab & behav biol br, NIH, 66-67; mem, Comt, Nat Inst Neurol Dis & Blindness, 67-68 & Res & Training Comt, Nat Inst Child Health & Human Develop, 68-73; Commonwealth Fund award & vis scientist, Univ Col, Univ London, 70; consult, NASA, 73- & Calif Med Facility, Vacaville, 74- Mem: Am Physiol Soc; Am Neurol Asn; Royal Soc Med; Aerospace Med Asn. Res: Brain function; biology of behavior; mental retardation; mental illness and deviant behavior; pain; epilepsy; computer analysis of electroencephalography; coding principles in neural systems; memory; psychopharmacology; biorhythms; aerospace physiology; environmental physiology. Mailing Add: Dept of Behav Biol Sch of Med Univ of Calif Davis CA 95616

CHAPMAN, LYMAN JOHN, b Agincourt, Ont, Aug 5, 08; m 40; c 1. GEOMORPHOLOGY, CLIMATOLOGY. Educ: Ont Agr Col, BSA, 30; Mich State Univ, MS, 38. Prof Exp: Soil surveyor, Ont Agr Col, 30-32; physiographer, Ont Res Found, 32-75; RETIRED. Concurrent Pos: Can Dept Agr grant, 63-65. Mem: Agr Inst Can. Res: Glacial geology; soils and climate; agriculture and forestry. Mailing Add: RR 1 Thornbury ON Can

CHAPMAN, ORVILLE LAMAR, b New London, Conn, June 26, 32; m 55; c 2. ORGANIC CHEMISTRY. Educ: Va Polytech Inst, BS, 54; Cornell Univ, PhD(chem), 57. Prof Exp: From instr to assoc prof, 57-64, PROF CHEM, IOWA STATE UNIV, 64- Concurrent Pos: Alfred P Sloan Found res fel, 61- Honors & Awards: Award in Pure Chem, Am Chem Soc, 68; Founders Prize, Tex Instrument Found, 74. Mem: Nat Acad Sci; Am Chem Soc; The Chem Soc. Res: Organic photochemistry; metal-organic complexes; natural products; insect, fish and mammalian pheromones. Mailing Add: Dept of Chem Iowa State Univ Ames IA 50010

CHAPMAN, PAUL JONES, b Cazadero, Calif, Sept 9, 00; m 25. ENTOMOLOGY. Educ: Ore State Univ, BS, 22; Cornell Univ, PhD(entom), 28. Prof Exp: Field asst entom & plant path, Cornell Univ, 23, exten instr, 24-28; entomologist, Va Truck Exp Sta, 28-30; prof entom, Univ & mem staff, Geneva Exp Sta, 30-68, head dept entom, 48-65, EMER PROF ENTOM, CORNELL UNIV, 68- Honors & Awards: Am Entom Soc Gold Medal & Award, 40, 42. Mem: AAAS; Am Entom Soc (vpres, 53). Res: Biology, ecology and control of insect and mite fruit pests; Tortricid fauna of apple; insecticide research, especially on petroleum oils; taxonomy of Psocoptera. Mailing Add: State Agr Exp Sta Geneva NY 14456

CHAPMAN, R KEITH, b Oak Lake, Man, Oct 31, 16; m 42; c 3. ENTOMOLOGY. Educ: Ont Agr Col, BSA, 40; Univ Wis, PhD(entom), 49. Prof Exp: Lectr entom & zool, Ont Agr Col, 40-45; asst entom, 45-47, from instr to assoc prof, 47-59, PROF ENTOM, UNIV WIS-MADISON, 59- Mem: Entom Soc Am; Potato Asn Am; Am Phytopath Soc. Res: Insect transmission of plant disease; vegetable insects and their control. Mailing Add: Dept of Entom Univ of Wis Col Agr & Life Sci Madison WI 53706

CHAPMAN, RANDOLPH WALLACE, b Lebanon, NH, Jan 2, 07; m 50; c 1. GEOLOGY. Educ: Univ NH, BS, 29; Harvard Univ, AM, 32, PhD(geol), 34. Prof Exp: Asst physiog, Harvard Univ, 30-31, asst geol, 31-34; instr, Vassar Col, 34-37; asst prof, Marshall Col, 37-39; from instr to assoc prof petrog, Johns Hopkins Univ, 39-52; res geologist, US Geol Surv, 52-54; prof geol, Trinity Col, Conn, 54-67; prof geol, Col Petroleum & Minerals, Saudi Arabia, 67-76. Concurrent Pos: Fulbright award, Gt Brit, 49-50; Smith-Mundt award, Univ Libya, 60-61; US State Dept grant, Univ Baghdad, 64-65. Mem: Fel Geol Soc Am; fel Mineral Soc Am; fel Am Geophys Union; fel geochem Soc; fel Geol Soc London. Res: Petrography and structure of rocks in New Hampshire and Maryland; petrology of batholith, Boulder, Montana; petrology of volcanic rocks of central Connecticut; geomorphology of eastern Saudi Arabia. Mailing Add: PO Box 785 Roswell NM 88201

CHAPMAN, RICHARD ALEXANDER, b Teague, Tex, Sept 24, 32; m 54; c 4. SEMICONDUCTORS. Educ: Rice Univ, BA, 54, MA, 55, PhD, 57. Prof Exp: Asst physics, Rice Univ, 54-56; physicist, Vallecitos Atomic Lab, Gen Elec Corp, 57-59; physicist, 59-64, mgr mat physics br, 64-71, MGR INFRARED DEVICES BR, CENT RES LAB, TEX INSTRUMENTS INC, 71- Concurrent Pos: Mem bd gov, Rice Univ, 75-79 Mem: Fel Am Phys Soc. Res: Optoelectronic devices and physics; photovoltaic and photoconductive infrared detectors; light emitting diodes and electrochromic digital displays; photoluminescence and infrared properties of impurities in semiconductors; semiconductor materials. Mailing Add: 7240 Briarcove Dr Dallas TX 75240

CHAPMAN, RICHARD ALEXANDER, b Akron, Ohio, July 5, 18; m 42; c 1. PLANT PATHOLOGY. Educ: Kent State Univ, BS, 40; Univ Ill, PhD(plant path), 48. Prof Exp: Asst plant pathologist, Crop Protection Inst, Conn Exp Sta, 48-50; assoc plant pathologist, Agr Exp Sta, 50-54, asst prof bot, Univ, 53-54, assoc prof plant path, 59-60, chmn dept plant path, 63-68, chmn dept bot, 65-68, PROF PLANT PATH, UNIV KY, 60-, PLANT PATHOLOGIST, AGR EXP STA, 54- Mem: Fel AAAS; Bot Soc Am; Am Phytopath Soc; Soc Nematol. Res: Plant nematology and pathology. Mailing Add: Dept of Plant Path Univ of Ky Rm S-305 Agr Sci Ctr Lexington KY 40506

CHAPMAN, RICHARD DAVID, b Atlanta, Ga, June 25, 28; m 55; c 4. ORGANIC CHEMISTRY. Educ: Univ Fla, BS, 49; Northwestern Univ, MS, 51, PhD(org chem), 54. Prof Exp: Sr chemist, Chemstrand Corp, Monsanto Co, 53-58, group leader, 58-60, res chemist, Chemstrand Res Ctr, 60-69, RES SPECIALIST, MONSANTO TEXTILES CO, 69- Res: Polymer chemistry; condensation polymers. Mailing Add: Tech Ctr Monsanto Textiles Co PO Box 12830 Pensacola FL 32575

CHAPMAN, ROBERT DEWITT, b Erie, Pa, July 13, 37; m 64; c 2. ASTROPHYSICS. Educ: Pa State Univ, BS, 59; Harvard Univ, PhD(astron), 65. Prof Exp: Asst prof astron, Univ Calif, Los Angeles, 64-67; HEAD COMPUT SECT & ASTRONR, GODDARD SPACE FLIGHT CTR, NASA, 67- Honors & Awards: Sci Writing Award, Am Inst Physics-US Steel Found, 74. Mem: Am Astron Soc; Int Astron Union. Res: Structure of the outermost layers of the sun and stars and related physical problems. Mailing Add: Lab Solar Physics & Astrophys Code 681 NASA Goddard Space Flight Ctr Greenbelt MD 20771

CHAPMAN, ROBERT EARL, JR, b Borger, Tex, Jan 11, 41; m 61; c 2. BIOPHYSICS. Educ: Yale Univ, MS, 66, PhD, 68. Prof Exp: Fel biophys, Univ Calif, San Diego, 68; CHIEF PHOTOG SCIENTIST, UNICOLOR DIV, PHOTOSCI INC, 68- Res: Photographic science; photo chemistry. Mailing Add: Unicolor Div Photosci Inc 7200 Huron River Dr Dexter MI 48130

CHAPMAN, ROBERT MILLS, b Chicago, Ill, Aug 29, 18; m 45, 68; c 1. GEOLOGY. Educ: Northwestern Univ, BS, 41. Prof Exp: Geologist, Alaskan Br, 42-47, in charge Fairbanks off, 47-55 & geochem explor br, Colo, 55-61, in charge off, College, Alaska, 61-69, GEOLOGIST, ALASKAN GEOL BR, US GEOL SURV, 70- Mem: Fel Geol Soc Am; Soc Econ Geol; Am Asn Petrol Geol; fel Arctic Inst NAm. Res: Alaskan geology; structure and stratigraphy of interior region; geochemical exploration; mineral deposits. Mailing Add: US Geol Surv 345 Middlefield Rd Menlo Park CA 94025

CHAPMAN, ROBERT PRINGLE, b Hartland, NB, Sept 12, 26; m 50; c 3. PHYSICS. Educ: Mt Allison Univ, BS, 47; McGill Univ, MS, 49; Univ Sask, PhD(physics), 53. Prof Exp: Sci officer, Defence Res Estab Atlantic, 53-73, HEAD OCEAN ACOUSTICS SECT, DEFENCE RES ESTAB PAC, 73- Mem: Acoustical Soc Am. Res: Underwater acoustics; physics of aurora borealis; mass spectroscopy. Mailing Add: Defence Res Estab Pac Victoria BC Can

CHAPMAN, ROSS ALEXANDER, b Oak Lake, Man, Dec 10, 13; m 42; c 2. FOOD SCIENCE. Educ: Univ Toronto, BSA, 40; McGill Univ, MSc, 41, PhD(chem), 44. Hon Degrees: DSc, Univ Guelph, 72. Prof Exp: Asst prof, MacDonald Col, McGill Univ, 44-48; head food sect, Food & Drug Div, Dept Nat Health & Welfare, Ottawa, 48-55, head food sect, WHO, Geneva, 55-57, asst dir sci serv, 58-63, asst dir gen food, 63-65, asst dep minister food & drugs, 65-71, spec adv, Off Dep Minister Health, 71-72, dir-gen int health serv, 72-73; CONSULT FOOD LEGIS & CONTROL, S PAC COMN, NEW CALEDONIA, 75- Concurrent Pos: Can deleg, Food & Agr Orgn/WHO Prog Food Standards & Additives; head Can deleg, UN Comt Narcotic Drugs, 70-; mem deleg, UN Conf Protocol Psychotropic Substances, 71. Honors & Awards: Underwood-Prescott Award, Mass Inst Technol, 73. Res: Methods of determination of thiamine and riboflavin; ferric thiocyanate methods for fat peroxides and its application to milk powders and fats and oils; methods for determination of antioxidants in fat and oils; behavior of antioxidants; methods for arsenic in fruits and vegetables and tocopherol in butter fat; food and drug legislation. Mailing Add: Suite 2205 505 St Laurent Blvd Ottawa ON Can

CHAPMAN, RUSSELL LEONARD, b Brooklyn, NY, May 30, 46; m 69; c 2. PHYCOLOGY. Educ: Dartmouth Col, AB, 68; Univ Calif, Davis, MS, 70, PhD(bot), 73. Prof Exp: ASST PROF PHYCOL, LA STATE UNIV, BATON ROUGE, 73- Mem: Phycol Soc Am; Bot Soc Am; Int Phycol Soc; Brit Phycol Soc; Electron Micros Soc Am. Res: Cytology and ultrastructure of Cephaleuros and related Chroolepidaceaen green algae including Phycopeltis and Trentopohlia. Mailing Add: Dept of Bot La State Univ Baton Rouge LA 70803

CHAPMAN, SEVILLE, b Seville, Spain, Nov 4, 12; US citizen; m 42; c 3. PHYSICS. Educ: Univ Calif, AB, 34, PhD(physics), 38. Prof Exp: Asst, Univ Calif, 35-37; instr physics, Univ Kans, 38-41; from instr to asst prof, Stanford Univ, 41-48; from prin physicist to chief scientist, Cornell Aeronaut Lab, 48-70; DIR, NY STATE ASSEMBLY SCI STAFF, 71- Mem: AAAS; Am Geophys Union; Am Inst Aeronaut & Astronaut; Am Phys Soc; Inst Elec & Electronics Engrs. Res: Relation of technology to problems of society; administration of research; design of electronic, missile and physical systems; arms control and disarmament; gaseous electronics; atmospheric physics; electrification of liquids by spraying. Mailing Add: 94 Harper Rd Buffalo NY 14226

CHAPMAN, STANLEY LANE, soil mineralogy, soil genesis, see 12th edition

CHAPMAN, STEPHEN R, b San Francisco, Calif, Oct 21, 36; m 58; c 2. POPULATION GENETICS, AGRONOMY. Educ: Univ Calif, Davis, BS, 59, MS, 63, PhD(genetics), 66. Prof Exp: Lab tech agron, Univ Calif, Davis, 60-66; asst prof agron & genetics, Univ Mont, 66-70; ASSOC PROF AGRON & GENETICS, MONT STATE UNIV, 70- Mem: Am Soc Agron. Res: Population genetics and ecology of forage grasses. Mailing Add: Dept of Plant & Soil Sci Mont State Univ Bozeman MT 59715

CHAPMAN, THOMAS EVERETT, b Globe, Ariz, Feb 16, 39; m 65; c 2. VETERINARY PHYSIOLOGY. Educ: Univ Calif, Davis, BS, 62, DVM, 64, PhD(physiol), 69. Prof Exp: Asst prof, 69-73, ASSOC PROF PHYSIOL, KANS STATE UNIV, 73- Res: Water metabolism in ruminants and poultry; gluconeogenesis and glucose utilization in ruminants; propionate production and propionate-lactate interrelations in ruminants. Mailing Add: Dept of Physiol Sci Col Vet Med Kans State Univ Manhattan KS 66502

CHAPMAN, THOMAS SHELBY, b McAlester, Okla, Nov 1, 05; m 31; c 1. PHYSICAL CHEMISTRY. Educ: Rice Inst, BA, 28, MS, 31; Cornell Univ, PhD(phys chem), 38. Prof Exp: Asst, Rice Inst, 27-31; high sch teacher, Okla, 31-36; res engr, Humble Oil & Refining Co, Tex, 38-42; army inspector, Maumelle Ord Works, 42-43; procurement planning officer, Off Chief of Ord, 43; tech officer, Chicago Area Off, Manhattan Dist, 43-46, chief opers br, Res Div, 46; chief opers br, AEC, 47; asst lab dir, Argonne Nat Lab, 47-51; dir health, physics & med sect, Dow Chem Co, Colo, 51-61, tech info officer, 57-67; CHIEF REGULATION & APPRAISAL DIV TECH INFO, ENERGY RES & DEVELOP ADMIN, 67- Mem: Am Chem Soc. Res: Colloid suspensions; plant pigments; oil well drilling fluids; analysis of oil well cores and cuttings by chemical and physical methods; hydrocarbon phase studies; behavior of fluids in oil field reservoirs; colloidal carbohydrates; radiation and nuclear sciences. Mailing Add: 438 E Tennessee Ave Oak Ridge TN 37830

CHAPMAN, TOBY MARSHALL, b Chicago, Ill, Nov 16, 38; m 61; c 2. BIO-ORGANIC CHEMISTRY, POLYMER CHEMISTRY. Educ: Univ Ill, BS, 60; Polytech Inst Brooklyn, PhD(chem), 65. Prof Exp: Res fel biol chem, Harvard Med Sch, 65-67; asst prof, 67-74, ASSOC PROF CHEM, UNIV PITTSBURGH, 74- Mem: Am Chem Soc. Res: New methods of peptide and oligonucleotide synthesis; determination of biopolymer structure; vinyl copolymerization; synthesis of phosphorus containing polymers; synthesis of sugar phosphates. Mailing Add: Dept of Chem Univ of Pittsburgh Pittsburgh PA 15213

CHAPMAN, VERNE M, b Sacramento, Calif, Oct 4, 38; m 68. GENETICS. Educ: Calif State Polytech Col, BS, 60; Ore State Univ, MS, 63, PhD(genetics), 65. Prof Exp: Asst prof biol, Millersville State Col, 65-66; fel, Jackson Lab, 66-68; res fel, Yale Univ, 68-72; SR CANCER RES SCIENTIST, ROSWELL PARK MEM INST, 72-; ASST PROF GENETICS, STATE UNIV NY BUFFALO, 73- Mem: AAAS; Genetics Soc Am. Res: Biochemical genetics and developmental genetics of the mouse. Mailing Add: Dept of Molecular Biol Roswell Park Mem Inst Buffalo NY 14263

CHAPMAN, WARREN HOWE, b Chicago, Ill, Oct 30, 25; c 5. UROLOGY. Educ: Mass Inst Technol, BS, 46; Univ Chicago, MD, 52. Prof Exp: Intern, St Luke's Hosp, Chicago, 52-53; resident urol, Univ Chicago Clins & instr, Sch Med, 53-57; res assoc, Western Wash State Col, 63-66; clin assoc, Dept Surg, Affil Hosps, 57-62, clin instr, 62-66, from asst prof to assoc prof urol, 66-73, PROF UROL, UNIV WASH, 73-, ADMIN OFF DEPT, 66- Mem: AMA; Am Urol Asn; Am Asn Cancer Res; Asn Am Med Cols; Soc Univ Urol. Res: Renal and adrenal hypertension; carcinoma of the bladder. Mailing Add: Dept of Urol RL10 Univ of Wash Seattle WA 98105

CHAPMAN, WILLIAM FRANK, b Hanover, NH, July 26, 44; m 67. GLACIAL GEOLOGY. Educ: Univ NH, BS, 66; Univ Mich, MS, 68, PhD(geol), 72. Prof Exp: ASST PROF GEOL, SLIPPERY ROCK STATE COL, 71- Mem: Geol Soc Am; Am Quaternary Asn; Sigma Xi. Res: Stufy of the glacial geology of the Oil City Quadrangle, Pennsylvania. Mailing Add: Dept of Geol Slippery Rock State Col Slippery Rock PA 16057

CHAPMAN, WILLIE LASCO, JR, b Chattanooga, Tenn, Dec 17, 28; m 58; c 2. COMPARATIVE PATHOLOGY. Educ: Univ Tenn, BS, 50; Auburn Univ, DVM, 57; Colo State Univ, MS, 63; Univ Wis, PhD(vet sci), 68. Prof Exp: Pvt pract, 57-62; resident radiol, Colo State Univ, 62-63; asst prof vet med & surg, Sch Vet Med, Univ Ga, 63-64; res fel comp path, Univ Wis, 64-67; asst prof path, 67-71, assoc prof med & surg & head dept, 71-75, ASSOC PROF PATH, COL VET MED, UNIV GA, 75- Concurrent Pos: NIH spec fel, Dept Path & Regional Primate Res Ctr, Univ Wis, 65-67. Mem: Am Vet Med Asn; Am Animal Hosp Asn; Am Asn Lab Animal Sci; Int Acad Path. Res: Thymus; neoplasms; mechanisms of immunity to blood parasites; chemotherapy of Leishmaniasis Mailing Add: Dept of Path Univ of Ga Col of Vet Med Athens GA 30602

CHAPMAN, WILLIS HARLESTON, b Liberty, SC, Jan 23, 16; m 42; c 1. AGRONOMY. Educ: Clemson Univ, BS, 36; NC State Univ, MS, 38. Prof Exp: Asst agronomist, NC State Univ, 38-42; from asst agronomist to assoc agronomist, 42-53, AGRONOMIST, INST FOOD & AGR SCI, AGR RES & EDUC CTR, UNIV FLA, 53-, PROF AGRON & HEAD DEPT, UNIV, 59- Mem: Fel Am Soc Agron. Res: Small grain breeding; research administration. Mailing Add: PO Box 470 Quincy FL 32351

CHAPPEL, CLIFFORD, b Guelph, Ont, Aug 23, 25; m 54, 70; c 4. BIOLOGY. Educ: Ont Vet Col, DVM, 50; McGill Univ, MSc, 53, PhD(invest med), 59. Prof Exp: Assoc dir biol res, Ayerst Res Labs, 53-65; PRES & DIR, BIO-RES LABS, LTD, 65- Concurrent Pos: Lectr, Dept Invest Med, McGill Univ. Mem: Soc Exp Biol & Med; Am Soc Pharmacol; Pharmacol Soc Can. Res: Pharmacology; toxicology. Mailing Add: Bio-Res Labs Ltd 265 Hymus Blvd Pointe Claire PQ Can

CHAPPEL, SAMUEL ESTELLE, b Abingdon, Va, Apr 18, 31. RADIATION PHYSICS. Educ: Va State Col, BS, 52; Pa State Univ, MS, 59, PhD(physics), 62. Prof Exp: Physicist radiation, 62-72, PROG COORDR INFO PROG, NAT BUR STAND, 72- Mem: Am Phys Soc; Am Soc Test & Mat; AAAS. Res: Interaction of ionizing of matter including radiation effects and dosimetry; scientific and technical information management. Mailing Add: Nat Bur of Stand Rm A935 Admin Bldg Washington DC 20234

CHAPPELEAR, JOHN EMERSON, b Enid, Okla, Feb 13, 29; m 55; c 4. APPLIED MATHEMATICS, PHYSICS. Educ: Univ Okla, BS, 50; Ind Univ, MS, 52, PhD(physics), 54. Prof Exp: From physicist to sr physicist, 54-64, res assoc, 64-66, sr res assoc, 66-74, SR STAFF MATHEMATICIAN, SHELL DEVELOP CO, 74- Concurrent Pos: Lectr, Univ Houston, 68-70 & Rice Univ, 70-71. Mem: Soc Petrol Engrs. Res: Computer science; petroleum reservoir engineering; numerical solution of systems of non-linear partial differential equations Mailing Add: 9714 Rice Ave Houston TX 77035

CHAPPELL, CHARLES FRANKLIN, b St Louis, Mo, Dec 7, 27; m 51; c 3. METEOROLOGY, ATMOSPHERIC PHYSICS. Educ: Wash Univ, BS, 49; Colo State Univ, MS, 66, PhD(atmospheric sci), 71. Prof Exp: Flight test engr, MacDonnell Aircraft Corp, 50-55; anal & forecasting, Nat Weather Serv, 56-67; res meteorologist, Dept Atmospheric Sci, Colo State Univ, 67-70; assoc prof meteorol, Utah State Univ, 70-72; res meteorologist, Off Weather Modification, 72-73, SR SCIENTIST & DEP DIR, ATMOSPHERIC PHYSICS & CHEM LAB, ENVIRON RES LABS, NAT OCEANIC & ATMOSPHERIC AGENCY, 73- Concurrent Pos: Adj prof, Utah State Univ, 72-74; assoc mem grad fac, Colo State Univ, 73-, vis assoc prof, 74. Mem: Am Geophys Union; Am Meteorol Soc; Sigma Xi; Weather Modification Asn. Res: Parameterization of severe convection in mesoscale numerical models, genesis and organization of mesoscale convective systems, and mesoscale modeling of snowfall over mountainous terrain. Mailing Add: Atmospheric Physics & Chem Lab 30th & Marine Boulder CO 80302

CHAPPELL, CHARLES RICHARD, b Greenville, SC, June 2, 43; m 68; c 1. MAGNETOSPHERIC PHYSICS. Educ: Vanderbilt Univ, BA, 65; Rice Univ, PhD(space sci), 69. Prof Exp: Consult, Lockheed Palo Alto Res Lab, 68, assoc res scientist, 68-70, res scientist, 70-73, staff scientist, 73-74; CHIEF MAGNETOSPHERIC & PLASMA PHYSICS BR, SPACE SCI LAB, MARSHALL SPACE FLIGHT CTR, NASA, 74- Concurrent Pos: Assoc res physicist, Univ Calif, San Diego, 73-74. Mem: Am Geophys Union; AAAS; Int Asn Geomagnetism & Aeronomy; Int Union Radio Sci. Res: Space science; study of low energy particle population of the magnetosphere; the plasmasphere and ionosphere; magnetospheric convection. Mailing Add: 2803 Downing Court Huntsville AL 35801

CHAPPELL, ELIZABETH, b Chicago, Ill, Jan 26, 27; m 73; c 2. PHARMACOLOGY. Educ: St Xavier Col, BS, 45; Univ Ill, MS, 47, PhD(pharmacol), 61. Prof Exp: Instr sci, St Xavier Col, 47-48; res asst pharmacol, 48-53, chem, 56-57, sr pharmacologist, 60-74, REGULATORY AFFAIRS MGR, ABBOTT LABS, 74- Res: General pharmacology, including blood coagulation, enzyme induction, antidiabetic drugs and general screening procedures; enzymology; biogenic amines and cyclic amp. Mailing Add: Abbott Labs North Chicago IL 60064

CHAPPELL, GARY S, medicinal chemistry, see 12th edition

CHAPPELL, GEORGE A, physical organic chemistry, computer science, see 12th edition

CHAPPELL, GUY LEE MONTY, b Marysville, Ohio, Aug 23, 40; m 62; c 2. RUMINANT NUTRITION, PHYSIOLOGY. Educ: Va Polytech Inst, MA, 65, PhD(ruminant nutrit), 66. Prof Exp: Asst prof, 66-70, ASSOC PROF SHEEP EXTEN & RES, UNIV KY, 70- Mem: Am Soc Animal Sci. Res: Factors affecting cellulose digestibility in the ruminant; factors affecting roughage utilization ruminants; high energy rations for early weaned lambs; intensive sheep production. Mailing Add: Dept of Animal Sci Univ of Ky Lexington KY 40506

CHAPPELL, RICHARD LEE, b Buffalo, NY, Mar 9, 38; m 68; c 1. NEUROBIOLOGY. Educ: Princeton Univ, BSE, 62; Johns Hopkins Univ,

PhD(biophys), 70. Prof Exp: Test engr nuclear eng, Naval Reactors, US AEC, 62-66; ASSOC PROF BIOL SCI, HUNTER COL, 70- Mem: AAAS; Asn Res Vision & Ophthal; Inst Elec & Electronics Engrs; Soc Neurosci; Soc Gen Physiologists. Res: Electrophysiology; pharmacology; neuroanatomy and information processing with emphasis on the retina and visual systems. Mailing Add: Dept of Biol Sci Hunter Col 695 Park Ave New York NY 10021

CHAPPELL, STERLING FRANK, III, organic chemistry, see 12th edition

CHAPPELL, WILBERT, chemistry, see 12th edition

CHAPPELL, WILLIAM ADRIAN, b Belvidere, NC, Oct 22, 25; m 52; c 3. VIROLOGY. Educ: Univ NC, AB, 47, MSPH, 51, PhD, 64. Prof Exp: Microbiologist, Virus & Rickettsia Div, Ft Detrick, 51-66; chief arbovirus ref lab, Nat Commun Dis Ctr, 66-70, CHIEF, VIRAL & RICKETTSIAL PROD BR, BIOL PROD DIV, CTR DIS CONTROL, 70- Mem: Sigma Xi; Am Soc Microbiol; Am Soc Trop Med & Hyg; Tissue Cult Asn. Res: Medical virology; aerobiology, viruses and bacteria; combined infections; arboviruses; Bedsonia agents; viral, rickettsial and mycoplasmal reagent development and production. Mailing Add: Biol Prod Div Ctr for Dis Control Atlanta GA 30333

CHAPPELL, WILLIAM EVERETT, horticulture, see 12th edition

CHAPPELLE, DANIEL EUGENE, b Washington, DC, Mar 15, 33; m 56; c 3. FOREST ECONOMICS. Educ: Colo State Univ, BS, 56; Duke Univ, MF, 59; State Univ NY Col Forestry, Syracuse Univ, PhD(forestry econ), 65. Prof Exp: Res forester econ, Southeastern Forest Exp Sta, US Forest Serv, 56-61; res asst forestry econ, State Univ NY Col Forestry, Syracuse Univ, 61-64; economist, Pac Northwest Forest & Range Exp Sta, US Forest Serv, 64-66, prin economist, 66-68; assoc prof resource econ, Mich State Univ, 68-71, PROF RESOURCE ECON, MICH STATE UNIV, 71- Mem: Am Econ Asn; Am Statist Asn; Regional Sci Asn. Res: Natural resource economics; regional economics and regional science; land use modeling. Mailing Add: Dept of Resource Develop 323 Natural Resources Bldg Mich State Univ East Lansing MI 48824

CHAPPELLE, EMMETT W, b Phoenix, Ariz, Oct 24, 25; m 47; c 4. BIOCHEMISTRY, PHOTOBIOLOGY Educ: 0Uiv Calif, BA, 50; Univ Wash, MS, 54. Prof Exp: Instr biochem, Meharry Med Col, 50-52; res assoc, Stanford Univ, 55-58; scientist biochem, Res Inst Advan Studies, 58-63; biochemist, Hazleton Labs, 63-66; exobiologist, 66-70, astrochemist, 70-73, PHOTOBIOLOGIST, GODDARD SPACE FLIGHT CTR, NASA, 73- Concurrent Pos: Consult, Appl Magnetics Corp, 73-74; NASA fel, Johns Hopkins Univ, 75- Mem: Am Chem Soc; NY Acad Sci; Am Soc Photobiol. Res: Metabolism of iron in mammalian systems; methods for quantitative assay of proteins and amino acids; carbon monoxide utilization by green plants; bioluminescence; methods for microbial determination; interstellar molecules. Mailing Add: 2502 Allendale Rd Baltimore MD 21216

CHAPPELOW, CECIL CLENDIS, JR, b Kansas City, Mo, Apr 12, 28; m 47; c 5. ORGANIC CHEMISTRY. Educ: Univ Southern Calif, BA, 51; Univ Mo-Kansas City, MA, 57, PhD, 68. Prof Exp: Sr chemist, Chem Div, 50-66, prin chemist, 66-68, head org & polymeric mat sect, 68-70, HEAD ORG & POLYMER CHEM, MIDWEST RES INST, 70- Mem: Am Chem Soc; Sigma Xi. Res: Synthesis, physicochemical characterization; structure-property correlation; substituted ureas and sulfamides; organometallic monomers and polymers; epoxidation and hydroxylation reactions; hydrogen-fluoride catalyzed condensations; metal chelating agents; biomedical polymer systems; cellulose conversion and utilization; corrosion protective coatings. Mailing Add: Midwest Res Inst Phys Sci Div 425 Volker Blvd Kansas City MO 64110

CHAPPLE, ELIOT D, anthropology, psychology, see 12th edition

CHAPPLE, WILLIAM DISMORE, b Boston, Mass, July 24, 36; m 63; c 1. NEUROPHYSIOLOGY, COMPARATIVE PHYSIOLOGY. Educ: Harvard Univ, BA, 58; Syracuse Univ, MA, 60; Stanford Univ, PhD(biol), 65. Prof Exp: Res biologist, Control Systs Lab, Stanford Res Inst, 63-65; NATO fel, Cambridge Univ & Bristol Univ, 65-66; asst prof biol, 66-70, ASSOC PROF BIOL, UNIV CONN, 70- Mem: Fel AAAS; Brit Soc Exp Biol. Res: Comparative neurophysiology; neural basis of patterned movement. Mailing Add: Biol Sci Group U-42 Univ of Conn Storrs CT 06268

CHAPPLE, WILLIAM MASSEE, b Billings, Mont, Apr 15, 34; m 56; c 3. STRUCTURAL GEOLOGY, TECTONICS. Educ: Calif Inst Technol, BS, 56, MS, 57, PhD(geol), 64. Prof Exp: From instr to assoc prof, 61-72, PROF GEOL, BROWN UNIV, 72- Mem: AAAS; Geol Soc Am; Am Geophys Union. Res: Plate tectonics; formation of orogenic belts; application of continuum mechanics to the study of the mechanical origin of geologic structures. Mailing Add: Dept of Geol Sci Brown Univ Providence RI 02912

CHAR, DONALD F B, b Honolulu, Hawaii, Mar 25, 25; m 51; c 5. PUBLIC HEALTH, PEDIATRICS. Educ: Temple Univ, MD, 50. Prof Exp: Intern, Atlantic City Hosp, 50-51; resident pediat, St Christopher's Hosp Children, 53-56; dir med educ, Kauikeolani Children's Hosp, 56-59 & 62-65; from instr to asst prof pediat, Med Sch, Univ Wash, 59-62; DIR STUDENT HEALTH & PROF PEDIAT, UNIV HAWAII, 65- Concurrent Pos: Fel pediat cardiol, St Christopher's Hosp Children, 55-56; res instr, Med Sch, Univ Wash, 62-63; lectr, East-West Training Prog Med Practitioners, Apia, WSamoa, 65. Mem: Am Acad Pediat; Am Col Health Asn. Res: Immune status in adrenalectomized animals; drug use and abuse on college campuses; sex behavior of college students; health care of foreign students. Mailing Add: Student Health Serv Univ Hawaii 1710 East West Rd Hon-lulu HI 96822

CHAR, WALTER F, b Honolulu, Hawaii, May 27, 20; m 48; c 3. PSYCHIATRY. Educ: Temple Univ, MD, 45; Am Bd Psychiat & Neurol, cert psychiat, 52, cert child psychiat, 60. Prof Exp: Intern, Med Ctr, Temple Univ, 45-46; resident psychiat, Univ Pittsburgh, 48-49; resident, Med Ctr, Temple Univ, 49-52, assoc prof psychiat, Sch Med & dir child psychiat, Med Ctr, 52-62; chmn dept psychiat, 67-69, PROF PSYCHIAT, SCH MED, UNIV HAWAII, 67- Concurrent Pos: Pvt pract psychiat & psychoanal, 52-; consult psychiat, Child & Family Serv, Honolulu, 62-, Cath Social Serv & Army Tripler Gen Hosp, 65- Mem: Am Psychiat Asn; Am Psychoanal Asn; Am Acad Child Psychiat; Am Orthopsychiat Asn; Am Col Psychoanal. Res: Evaluation of admission procedure of medical students; transcultural psychiatry. Mailing Add: Univ of Hawaii Sch of Med 3675 Kilauea Ave Honolulu HI 96816

CHARACHE, PATRICIA, b Newark, NJ, Dec 26, 29; m 51; c 1. MEDICINE. Educ: Hunter Col, BA, 52; NY Univ, MD, 57. Prof Exp: Intern med, Baltimore City Hosps, 57-58; res assoc immunol, Childrens Hosp & Harvard Univ, 62-64; from instr to asst prof med, 64-73, ASSOC PROF MED & LAB MED, JOHNS HOPKINS UNIV, 73-, ASST PROF MICROBIOL, 70-, PHYSICIAN, HOSP, 67-, PATHOLOGIST, 73- Concurrent Pos: USPHS fel res med, Univ Pa, 58-59, Porter fel, 59-60; USPHS fel infect dis, Johns Hopkins Univ, 60-62; res career develop award, Nat Inst Allergy & Infect Dis, 69- Mem: Am Soc Clin Pharmacol; Reticuloendothelial Soc; Infect Dis Soc Am; Am Soc Microbiol; Fedn Clin Res. Res: Immunology and infectious disease, emphasizing bacterial-host interaction; microbiology. Mailing Add: Dept of Lab Med Johns Hopkins Hosp Baltimore MD 21205

CHARACHE, SAMUEL, b New York, NY, Jan 12, 30; m 51; c 1. MEDICINE. Educ: Oberlin Col, BA, 51; NY Univ, MD, 55; Am Bd Internal Med, dipl & cert hemat. Prof Exp: Clin assoc, NIH, 56-58; resident med, Hosp Univ Pa, 58-60; asst prof, 66-70, ASSOC PROF MED, SCH MED, JOHNS HOPKINS UNIV, 70- Concurrent Pos: USPHS, 56-58; fel hemat, Johns Hopkins Hosp, 60-62 & 64-66; fel biol, Mass Inst Technol, 62-64. Mem: Am Soc Clin Invest; Am Soc Hemat. Res: Hematology. Mailing Add: Dept of Med Johns Hopkins Hosp Baltimore MD 21205

CHARALAMPOUS, FRIXOS C, b Cyprus, Jan 15, 20; nat US. BIOCHEMISTRY. Educ: Univ Athens, MD, 46; Harvard Univ, DSc(nutrit & biochem), 49. Prof Exp: Fels, Nat Heart Inst, Md, 49-50, Nat Cancer Inst, 50-52; assoc biochem, 52-54, from asst prof to assoc prof, 54-64, PROF BIOCHEM, UNIV PA, 64- Mem: AAAS; Soc Biol Chem. Res: Biochemistry of growth and cellular differentiation; mechanism of induction of respiratory enzymes in yeast. Mailing Add: Dept of Biochem Univ of Pa Philadelphia PA 19104

CHARAP, STANLEY H, b Brooklyn, NY, Apr 21, 32; m 55; c 2. SOLID STATE SCIENCE. Educ: Brooklyn Col, BS, 53; Rutgers Univ, PhD(physics), 59. Prof Exp: Asst physics, Rutgers Univ, 53-55, instr, 57-58; physicist, Res Ctr, Int Bus Mach Corp, 58-64; res scientist, Res Div, Am Stand Corp, 64-65, supvr solid state physics, 65-66, mgr physics & electronics, 66-68; assoc prof, 68-71, PROF ELEC ENG, CARNEGIE-MELLON UNIV, 71- Concurrent Pos: Consult, Westinghouse Res Labs, 69-74. Mem: Am Phys Soc; Inst Elec & Electronics Engrs; Magnetics Soc; Sigma Xi. Res: Magnetic domains, magnetic hysteresis models; theory of solid state; magnetism of solids. Mailing Add: Dept of Elec Eng Carnegie Mellon Univ Pittsburgh PA 15213

CHARBENEAU, GERALD T, b Mt Clemens, Mich, July 22, 25; m 47; c 3. DENTISTRY. Educ: Univ Mich, DDS, 48, MS, 49. Prof Exp: Teaching fel dent, Sch Dent, Univ Mich, Ann Arbor, 48-49; from instr to assoc prof, 49-65, PROF DENT, SCH DENT, UNIV MICH, ANN ARBOR, 65-, CHMN DEPT OPER DENT, 69- Mem: Am Dent Asn; Int Asn Dent Res; Am Acad Restorative Dent; Acad Oper Dent. Res: Operative and general restorative dentistry with specific relationship of dental materials to these clinical areas. Mailing Add: 2062 Dent Sch of Dent Univ of Mich Ann Arbor MI 48109

CHARBONNEAU, LARRY FRANCIS, b Faribault, Minn, Aug 14, 39; m 66; c 2. ORGANIC POLYMER CHEMISTRY. Educ: Mankato State Col, BS, 64; Univ Ill, Urbana-Champaign, MS, 69, PhD(org chem), 72. Prof Exp: Chem technician polymer chem, 3M Co Cent Res, 60-62, chemist photochem, 64-67; ASSOC SR RES CHEMIST ORG POLYMER CHEM, GEN MOTORS RES LABS, WARREN, 72- Mem: AAAS; Am Chem Soc; Soc Photog Scientists & Engrs. Res: Synthesis of high performance polymers to meet the perceived future needs of the automotive industry. Mailing Add: 65 Stratford Lane Rochester MI 48063

CHARBONNIER, FRANCIS MARCEL, b Monaco, Apr 28, 27; nat US; m 52; c 6. PHYSICS. Educ: Polytech Sch Paris, Dipl d'Ing, 49; Univ Wash, PhD(physics), 52. Prof Exp: Engr, Militaire de l'Armement, France, 52-55; sr physicist, Linfield Res Inst, 56-57, tech asst to dir, 57-59, asst dir, 59-62; dir res & develop div, Field Emission Corp, 62-64, vpres & res dir, 64-69, asst to pres, 69-74; MEM STAFF, McMINNVILLE DIV, HEWLETT-PACKARD CO, 74- Mem: Am Phys Soc; Inst Elec & Electronics Engrs. Res: Field emission; electron physics, optics and devices; radiation sources. Mailing Add: Hewlett-Packard Co McMinnville Div 500 Links St McMinnville OR 97128

CHARD, CHESTER STEVENS, b New York, NY, Sept 15, 15; m 53; c 6. ANTHROPOLOGY. Educ: Harvard Univ, AB, 37; Univ Calif, PhD(anthrop), 53. Prof Exp: Instr anthrop, Univ Wash, 54-55; vis asst prof, 55-56; lectr, 58-59, from asst prof to prof, 59-72, EMER PROF ANTHROP, UNIV WIS-MADISON, 72- Mem: Fel AAAS; fel Am Anthrop Asn; Soc Am Archaeol; Arctic Inst NAm; fel Anthrop Inst Gt Brit & Ireland. Res: Prehistory and culture history of Soviet Union and northeastern Asia; Old Worlds origins and relationships of New World cultures. Mailing Add: Dept of Anthrop Univ of Wis Madison WI 53706

CHARDON, ROLAND E, b Boston, Mass, Apr 15, 29. CULTURAL GEOGRAPHY Educ: Univ Minn, BA, 51; Fla State Univ, MS, 54; Univ Minn, PhD(geog), 61 Prof Exp: Instr geog, State Univ Iowa, 56-58; res contractor, Nat Acad Sci-Nat Res Coun, 58-59; asst prof, Ohio State Univ, 59-63; asst prof, Vanderbilt Univ, 63-66, assoc prof geog, 66-68, dir, Vanderbilt-in-France, 65-66; ASSOC PROF GEOG, LA STATE UNIV, BATON ROUGE, 68- Concurrent Pos: Actg dir, Grad Ctr Latin Am Studies, Vanderbilt Univ, 64-65; contract researcher, NASA, 71; consult var state & pvt planning agencies, 71-75; adj assoc prof planning, Dept Archit & Archit Eng, Sch Eng & Environ Design, Univ Miami, 74-75. Mem: Asn Am Geogr; Am Geog Soc; Latin Am Studies Asn; Sigma Xi. Res: Cultural-historical geography, with emphasis on urban, economic and coastal geography and problems, especially Anglo-America, Mexico, the Caribbean and Brazil. Mailing Add: Dept of Geog & Anthrop La State Univ Baton Rouge LA 70803

CHAREN, GEORGE, b Newark, NJ, June 18, 13; m 43; c 3. PHYSICS, CHEMISTRY. Educ: Columbia Univ, AB, 34, MA, 37; Univ Colo, EdD(sci), 62. Prof Exp: Pub sch instr, NJ, 38-41; head dept develop & prod fermentations, Apex Chem Co, 41-46 & S B Penick & Co, 46-48; high sch instr, NY, 56-58; instr phys sci, Univ Colo, 59-60; assoc prof sci & coordr physics, Jersey City State Col, 60-67; PROF PHYS SCI & MATH & DEAN INSTR, BERGEN COMMUNITY COL, 67- Concurrent Pos: Mem inserv inst, Esso Found & NY Univ, 57-58 & Newark Col Eng, 61-62; mem univ phys sci study comt physics, 62-63; mem acad year inst, NSF, Univ Colo, 58-59. Mem: AAAS; Nat Sci Teachers Asn; Am Asn Physics Teachers. Res: Nuclear physics. Mailing Add: Bergen Community Col 400 Paramus Rd Paramus NJ 07652

CHARETTE, LAURENT A, b Gogama, Ont, Jan 2, 24; m 48; c 4. ANIMAL PRODUCTION, ANIMAL BREEDING. Educ: Ont Agr Col, BSA, 48; Univ Minn, St Paul, MS, 51, PhD(animal breeding), 57. Prof Exp: Res officer animal sci, Exp Farm, Can Dept Agr, Ont, 48-62; dir dept animal sci, 62-74, PROF ANIMAL SCI, FAC AGR & FOOD SCI, LAVAL UNIV, 62- Mem: Can Soc Animal Prod; Am Soc Animal Sci. Res: Effects of sex and age of castration in swine; crossbreeding beef breeds; crossbreeding dairy and beef cattle. Mailing Add: Dept of Animal Sci Laval Univ Fac of Agr & Food Sci Quebec PQ Can

CHARGAFF, ERWIN, b Austria, Aug 11, 05; US citizen; m 29; c 1. BIOCHEMISTRY. Educ: Univ Vienna, Dr phil(chem), 28. Prof Exp: Milton Campbell res fel, Yale Univ, 28-30; asst dept bact & pub health, Univ Berlin, 30-33; res assoc, Pasteur Inst, 33-34; res assoc, 35-38, from asst prof to prof, 38-74, chmn dept, 70-74, EMER PROF BIOCHEM, COL PHYSICIANS & SURGEONS, COLUMBIA UNIV, 74- Concurrent Pos: Guggenheim fels, 49 & 57-58; vis prof, Wenner Gren Inst, Univ

Stockholm, 49, Univs Rio de Janeiro, Sao Paulo & Recife, 59, Cornell Univ & Univs Naples & Palermo, 66 & Biol Sta, Naples, 69; Harvey lectr, Rockefeller Univ, 56; Plenary Cong lectr, Int Biochem Cong Vienna, 58; vis lectr, Univs Tokyo, Kyoto, Sendai & others, 58; Jesup lectr, Columbia Univ, 59; first K A Forster lectr, Mainz, 68 & Miescher Mem lectr, Basel, 69; mem comt growth, Nat Res Coun, 52-54; mem adv coun biol, Oak Ridge Nat Lab, 58-67; Albert Einstein chair, Col France, 65- Honors & Awards: Pasteur Medal, Paris, 49; Carl Neuberg Medal, 58; Soc Chem Biol Medal, Paris, 61; Charles Leopold Mayer Prize, Acad Sci, Paris, 63; Dr H P Heineken Prize, Royal Neth Acad Sci, 64; Bertner Found Award, 65; Gregor Mendel Medal, Halle, Ger, 73; Nat Medal Sci, 74. Mem: Nat Acad Sci; fel Am Acad Arts & Sci; for mem Royal Swed Physiol Soc; Ger Acad Sci. Res: Lipids; lipoproteins; blood coagulation; metabolism of amino acids and inositol; chemistry and biosynthesis of nucleic acids and nucleoproteins; phosphotransferases and other enzymes. Mailing Add: 350 Cent Park W New York NY 10025

CHARGOIS, DEBORAH MAJEAU, b New Orleans, La, Nov 8, 40; m 69. MEDICAL PHYSIOLOGY. Educ: St Mary's Col, BA, 63; La State Univ, MS, 67, PhD(physiol), 69. Prof Exp: Instr otorhinolaryngol, Med Sch, 68-69, asst prof physiol, 69-71, clin asst prof physiol & biochem, 71-74, asst prof physiol, Sch Dent, 74-75, DIR DENT PHYSIOL, SCH DENT, LA STATE UNIV, NEW ORLEANS, 75- Concurrent Pos: Space biol fel, Wallop's Island Base, NASA, 69, spec lectr, Missile Test Facil, 69-71; trop med fel, Med Sch, La State Univ, 73. Honors & Awards: First Award for Sci Merit, Am Speech & Hearing Asn, 70. Mem: Am Physiol Soc; Soc Neurosci; Sigma Xi. Res: Recording electrical potentials of the mammalian retina in response to discrete monochromatic stimuli. Mailing Add: Dept of Physiol La State Univ Sch of Dent New Orleans LA 70119

CHARKES, N DAVID, b New York, NY, Aug 13, 31; m 53; c 3. NUCLEAR MEDICINE. Educ: Columbia Univ, AB, 52; Washington Univ, MD, 55. Prof Exp: USPHS fel arthritis & metab dis, 60-61; assoc radiol, 62-66, clin asst prof, 66, assoc prof radiol & med, 66-71, PROF RADIOL & ASSOC PROF MED, SCH MED, TEMPLE UNIV, 71-, DIR DEPT NUCLEAR MED, HOSP, 66- Concurrent Pos: Dir radioisotope unit, North Div, Albert Einstein Med Ctr, 62-66; consult nuclear med, Walson Army Hosp, Ft Dix, NJ, 67- & Vet Admin Hosp, Wilmington, Del; Fogarty Sr Internat fel, 76. Mem: Soc Nuclear Med; Am Fedn Clin Res. Mailing Add: Temple Univ Sch of Med Broad & Ontario Sts Philadelphia PA 19140

CHARKEY, LOWELL WILLIAM, b Denver, Colo, Mar 4, 11; m 35; c 2. NUTRITIONAL BIOCHEMISTRY. Educ: Colo State Univ, BS, 33; Univ Colo, MS, 37; Cornell Univ, PhD(animal nutrit), 45. Prof Exp: Bench chemist, Great Western Sugar Co, 33-36; from instr to prof chem, 36-74, chemist, Exp Sta, 36-43 & 45-74, EMER PROF BIOCHEM, COLO STATE UNIV, 74- Concurrent Pos: Res instr, Cornell Univ, 43-45; Fulbright lectr, Rangoon, 56-57; biochemist & nutritionist, UN Food & Agr Orgn, Malaya, 59-60. Mem: Fel AAAS; Am Inst Nutrit; Am Chem Soc; Soc Exp Biol & Med; Poultry Sci Asn. Res: Carotene and tocopherol analysis; egg quality and composition; poultry nutrition; unidentified growth factors; vitamins and stress factors in amino acid metabolism; magnesium; thyroid. Mailing Add: Dept of Biochem Colo State Univ Ft Collins CO 80521

CHARKOUDIAN, JOHN CHARLES, b Springfield, Mass, July 29, 41. INORGANIC CHEMISTRY, PHOTOGRAPHIC CHEMISTRY. Educ: Bates Col, BS, 63; Babson Col, MBA, 66; Boston Univ, MS, 67; Va Polytech Inst & State Univ, PhD(phys inorg chem), 70. Prof Exp: SCIENTIST PHOTOG CHEM, POLAROID CORP, 70- Mem: AAAS; Am Chem Soc; Soc Photog Scientists & Engrs. Res: Mechanism of photographic development electron spin resonance applied to photographic science; fast reaction kinetics; stopped flow and temperature jump kinetics. Mailing Add: Polaroid Corp 750 Main St Cambridge MA 02139

CHARLANG, GISELA WOHLRAB, b Berlin, Ger, May 31, 38; US citizen; m 69 MICROBIALAL PHYSIOLOGY, MICROBIAL ECOLOGY. Educ: Univ Chicago, BA, 61, MS, 62, PhD(bot), 64. Prof Exp: NSF fel, Univ Liverpool, 64-65; instr bot, Univ Chicago, 65-66; asst prof biol, Grand Valley State Col, 66-68; res assoc, Univ Ill, Urbana, 68-69; res fel, 69-73, ASSOC SCIENTIST, DIV BIOL, CALIF INST TECHNOL, 73- Mem: AAAS; Asn Women in Sci; Am Asn Univ Women. Res: Water metabolism and requirements of Neurospora crassa; effects of low water activity environments on membranes; evolution of xerotolerance. Mailing Add: Div of Biol Calif Inst of Technol Pasadena CA 91125

CHARLAP, LEONARD STANTON, b Wilmington, Del, Aug 1, 38. MATHEMATICS. Educ: Mass Inst Technol, BS, 59; Columbia Univ, PhD(math), 62. Prof Exp: Mem, Inst Advan Study, 62-64; from asst prof to assoc prof math, Univ Pa, 64-69; assoc prof, 69-70, PROF MATH, STATE UNIV NY STONY BROOK, 70- Mem: Am Math Soc. Res: Differential geometry; differential topology; homological algebra; flat Riemannian manifolds. Mailing Add: Dept of Math State Univ of NY Stony Brook NY 11790

CHARLEBOIS, CLARENCE THOMAS, b Rouyon, Que, Apr 20, 49. NEUROSCIENCE, SCIENCE POLICY. Educ: McGill Univ, BSc & dipl ethology, 70. Prof Exp: Researcher neurosci, Inst Exp Psychol, Oxford Univ, 70-73; SCI ADV, SCI COUN CAN, 75- Mem: Can Pub Health Asn. Res: Neurophysiological control of hunger and thirst; regulation of environmental food contaminants. Mailing Add: Sci Coun of Can 150 Kent St Ottawa ON Can

CHARLES, DONALD FOSTER, b Iloilo, Philippines, Oct 16, 22; m 44; c 3. PHYSICAL CHEMISTRY. Educ: Univ Calif, AB, 47. Prof Exp: RES CHEMIST, CALIF & HAWAIIAN SUGAR REFINING CO, 47- Mem: Am Chem Soc. Res: Physical properties of sucrose solutions; technology of sugar refining; chemistry of cane sugar colorants Mailing Add: C & H Sugar Refining Co Crockett CA 94525

CHARLES, GEORGE WILLIAM, b Columbus, Ohio, Dec 24, 15; m 39; c 2. PHYSICS. Educ: Ohio State Univ, BA, 37, PhD(physics), 47. Prof Exp: Asst physics, Ohio State Univ, 38-42; instr, NC State Col, 42-44; physicist, Naval Ord Lab, 44-46; asst prof physics, Univ Okla, 47-52; sr res physicist, Mound Lab, 52-54; PHYSICIST, OAK RIDGE NAT LAB, 54- Mem: Optical Soc Am. Res: Atomic spectroscopy; physical optics; heat; spectra of columbium and molybdenum in the extreme ultraviolet; spectra of polonium and of rare earths. Mailing Add: Oak Ridge Nat Lab Physics Div 840 W Outer Dr Oak Ridge TN 37830

CHARLES, JERRY THOMAS, organic chemistry, see 12th edition

CHARLES, RICHARD LLOYD, analytical chemistry, medicinal chemistry, see 12th edition

CHARLES, ROBERT GEORGE, physical chemistry, see 12th edition

CHARLES, ROBERT WILSON, b Altoona, Pa, Sept 1, 45. GEOCHEMISTRY. Educ: Bucknell Univ, BS, 67; Mass Inst Technol, PhD(geol), 72. Prof Exp: Fel geochem, Univ BC, 73-74; STAFF MEM GEOCHEM, LOS ALAMOS SCI LAB, 74- Mem: AAAS; Mineral Soc Am; Am Geophys Union. Res: Experimental determination of phase equilibria suitable to describe mineral assemblages found in nature; experimentation involves the routine use of high pressure-temperature hydrothermal equipment to duplicate natural conditions. Mailing Add: Los Alamos Sci Lab CNC-11 M/S 514 Univ of Calif Los Alamos NM 87545

CHARLESWORTH, LLOYD JAMES, JR, b Allentown, Pa, Feb 17, 34; m 66; c 1. SEDIMENTOLOGY. Educ: Lehigh Univ, BA, 56; Univ Mich, MS, 58, PhD(geol), 68. Prof Exp: Sr geologist, NJ Bur Geol & Topog, 61-67; res assoc geol, Univ Mich, 67-68; sr geologist, NJ Bur Navig, 68; vis scholar, Univ Mich, 68-69; ASSOC PROF GEOL & DIR SUBSURFACE DATA CTR, UNIV TOLEDO, 69- Mem: AAAS; Geol Soc Am; Soc Econ Paleontologists & Mineralogists; Sigma Xi; Int Asn Great Lakes Res. Res: Sedimentology, geomorphology, and process-response conditions of sedimentation in Holocene paludal, coastal and nearshore marine sedimentary environments, particularly New Jersey, Lake Erie and south Florida regions. Mailing Add: Dept of Geol Univ of Toledo Toledo OH 43606

CHARLIER, ROGER HENRI, b Antwerp, Belg, Nov 10, 21; nat US; m 58; c 2. GEOLOGY, GEOGRAPHY. Educ: Colonial Univ, Belg, BPol & Admin Sci, 40; Free Univ Brussels, MPolSc, 41, MS, 45; Univ Liege, BS(geol) & BS(geog), 43; Univ Erlangen, PhD(phys geog), 47; Indust Col Armed Forces, dipl, 53; McGill Univ, cert, 53; Univ Paris, LittD(cult geog), 57, ScD(geol, oceanog), 58. Prof Exp: Prof geol, Col Baudouin, Belg, 41-42; personal student asst geol, Univ Liege, 43-44; press corresp, 45-51; assoc prof geog & chmn dept, Poly Univ, 51-52; prof phys sci & chmn dept, Finch Col, 52-55; chmn dept geol & geog, Hofstra Univ, 55-58; adj prof geol, Univ Paris, 58-59; vis prof educ, Univ Minn, 59-60; prof geol & geog, Parsons Col, 60-61; dir bur educ travel & study abroad, 61-63, chmn area earth sci, 62-63, coordr earth sci progs, 63-66, vchmn dept geog & environ studies, 67-71, PROF GEOG, GEOL & OCEANOG, NORTHEASTERN ILL UNIV, 61- Concurrent Pos: Dept dir, UNRRA, 46-47; res analyst, US Govt, 47, 49 & 50; bursar, Carnegie Corp, 53; vis lectr, NY Univ, 53-58, Hunter Col, 57-58 & Univ Aix-Marseille, 58-60; Fr Govt spec fel, 58-59; resident scholar, Northeastern Ill Univ, 62-65; vis prof, Western NMex Univ, 62-; NSF grants, 63, 64; vis prof, De Paul Univ, 64-66; vis scientist, Govt SAfrica, 68, Rhodesia, 68, Romania, 68, Israel, 69; exchange sr scientist, NSF, 68, Romanian Acad Sci, 68, Nat Res Coun Romania, 70 & Int Res & Exchange Comn, 69; sr NATO fel & grant, 70; prof oceanog, Fac Sci, Univ Bordeaux, 70-74; extraordinary prof, Flemish Free Univ Brussels, 71-; exec dir, Inst Develop Riverine & Estuarine Systs, 74-76; sr Fulbright fel, 76. Honors & Awards: Francois Franck Prize, 39; Belg Govt Prize, 39; City of Antwerp Prize, 39; Cross of the Rhine, Belg; Gold Medal for Excellence in Educ, Belg; Medal for Touristic Merit, Belg; Knight, Order of Acad Palms, France; Comdr, Arts-Sci-Lett, France; Knight, Order of Leopold, Belg; Gold Medal Advan of Progress, France. Mem: Asn Am Geogr; Nat Asn Geol Teachers; fel Geol Soc Am; Marine Technol Soc; Am Soc Oceanog. Res: Oceanography; coastal erosion; sedimentary processes; applications of statistics to the domain of the earth sciences; regional geography of Europe. Mailing Add: 4055 N Keystone Ave Chicago IL 60641

CHARLTON, DAVID BERRY, b Vancouver, BC, Jan 26, 04; nat US; m 30; c 4. CHEMISTRY, BACTERIOLOGY. Educ: Univ BC, BA, 25; Cornell Univ, MS, 29; Iowa State Col, PhD(bact), 33. Prof Exp: Asst bacteriologist, Portland, Ore, 26-28; instr bact, Ore State Col, 29-31 & Univ Nebr, 31-32; owner & dir, Charlton Labs, 34-71, CONSULT, MEI-CHARLTON INC, 71- Concurrent Pos: Instr bact, Ore State Col, 34-36. Mem: AAAS; Am Chem Soc; Am Pub Health Asn. Res: Food bacteriology; sanitary bacteriology; chlorine compounds as germicides. Mailing Add: MEI-Charlton Inc 2340 SW Canyon Rd Portland OR 97201

CHARLTON, GORDON RANDOLPH, b Newport News, Va, Aug 30, 37; m 65; c 1. HIGH ENERGY PHYSICS, RESEARCH ADMINISTRATION. Educ: Ohio State Univ, BSc, 57; WVa Univ, MSc, 60; Univ Md, PhD(physics), 66. Prof Exp: Res asst high energy physics, Univ Md, 62-66; physicist, Ecole Polytechnique, Paris, 66-69; asst physicist, High Energy Physics Div, Argonne Nat Lab, 69-72; res assoc, Stanford Linear Accelerator Ctr, 72-73; syst mgr Physics Dept, Univ Toronto, 73-75; PHYSICIST HIGH ENERGY PHYSICS, DIV PHYS RES, ENERGY RES & DEVELOP ADMIN, 75- Mem: Am Phys Soc. Res: Experimental high energy physics with emphasis on the bubble chamber technique; computer controlled film-measuring machines and data processing. Mailing Add: Div of Phys Res Energy Res & Develop Admin Washington DC 20545

CHARLTON, HARVEY JOHNSON, b Dillwyn, Va, Aug 18, 34; m; c 3. MATHEMATICS. Educ: Va Polytech Inst, BS, 60, MS, 62, PhD(math), 66. Prof Exp: Proj physicist, Atomic Energy Div, Babcock Wilcox Co, Va, 57-59; instr math, Va Polytech Inst, 60-66; ASST PROF MATH, NC STATE UNIV, 66- Mem: Am Math Soc; Asn Symbolic Logic. Res: Modern topology. Mailing Add: Dept of Math NC State Univ Box 5126 Raleigh NC 27607

CHARLTON, JAMES LESLIE, b London, Ont, Dec 12, 42; m 64; c 2. PHOTOCHEMISTRY, ORGANIC CHEMISTRY. Educ: Univ Western Ont, BSc, 65, PhD(chem), 68. Prof Exp: Nat Res Coun Can fel, Calif Inst Technol, 68-70; asst prof, 70-74, ASSOC PROF CHEM, UNIV MAN, 74- Mem: Am Chem Soc; Can Inst Chem. Res: Photochemistry of quinones; spectroscopy of deuterated aromatics; photochemical synthesis. Mailing Add: Dept of Chem Univ of Man Winnipeg MB Can

CHARLTON, LOWELL ARTHUR, theoretical nuclear physics, see 12th edition

CHARLTON, RALPH WOODARD, organic chemistry, see 12th edition

CHARMAN, HOWARD PRENTIS, b San Diego, Calif, June 15, 41; m 61; c 2. MEDICAL RESEARCH. Educ: Univ Calif, Berkeley, BS, 63; Univ Southern Calif, MD, 67. Prof Exp: Intern, Los Angeles County-Univ Southern Calif Med Ctr, 67-68, resident path, 68-71; SCIENTIST CANCER RES, SCH MED, UNIV SOUTHERN CALIF, 71- Mem: AAAS; Am Soc Microbiol. Res: Immunochemistry and natural history of onco RNA viruses. Mailing Add: 870 New Mark Esplanade Rockville MD 20850

CHARMATZ, RICHARD, b New York, NY, Aug 9, 36. MICROPALEONTOLOGY. Educ: NY Univ, BA, 57, MS, 61, PhD(geol), 67. Prof Exp: Res asst, 57-61, sci asst, 61-66, ASST CUR, DEPT MICROPALEONT, AM MUS NATURAL HIST, 67- Concurrent Pos: Adj asst prof, Newark Col Arts & Sci, Rutgers Univ, 68- Mem: Am Asn Petrol Geologists; Geol Soc Am; Am Paleont Soc; Soc Econ Paleontologists & Mineralogists. Res: Taxonomic and nomenclatural research on Foraminifera and Ostracoda, emphasizing stratigraphic ranges of index forms. Mailing Add: Dept of Micropaleontology Am Mus of Natural Hist New York NY 10024

CHARMS, BERNARD, b Carrollton, Ohio, Nov 20, 27; m 50, 66. CARDIOPULMONARY PHYSIOLOGY. Educ: Ohio State Univ, BA, 47; Case Western Reserve Univ, MD, 51; Am Bd Internal Med, dipl, 58. Prof Exp: Dir cardiopulmonary lab, 58-67, dir pulmonary res, Mt Sinai Hosp, 67-72; sr clin instr med,

Case Western Reserve Univ, 58-74. Concurrent Pos: NIH grant; sr vis physician & physician, Out-Patient Cardiac Clin, Mt Sinai Hosp, 55- Mem: AAAS; Am Heart Asn; fel Am Col Cardiol; fel Am Col Physicians; fel Am Col Chest Physicians. Res: Pulmonary physiology; biochemistry; cardiology. Mailing Add: 11900 Shaker Blvd Cleveland OH 44120

CHARNES, ABRAHAM, b Hopewell, Va, Sept 4, 17; m 50; c 3. MATHEMATICS, ECONOMICS. Educ: Univ Ill, AB, 38, MS, 39, PhD(math), 47. Prof Exp: Off Naval Res fel, Univ Ill, 47-48; from asst prof to assoc prof math, Carnegie Inst Technol, 48-52, assoc prof indust admin, 52-55; prof math & dir res dept transportation & indust mgt, Purdue Univ, 55-57; res prof appl math & econ, Northwestern Univ, 57-68; Walter P Murphy prof, 68-73, JESSE H JONES PROF BIOMATH & MGT SCI, PROF MATH, GEN BUS & COMPUT SCI & DIR CTR CYBERNETIC STUDIES, UNIV TEX, AUSTIN, 73- Concurrent Pos: Ed, J Inst Mgt Sci, 54- Mem: Fel AAAS; fel Economet Soc; Opers Res Soc Am; Asn Comput Mach; Inst Mgt Sci (vpres, 58, pres, 60). Res: Topological algebra; functional analysis; differential equations; aerodynamics; hydrodynamic theory of lubrication; statistics; extremal problems in theory of inequalities; game theory; mathematical theory of management science; biomathematics. Mailing Add: Dept of Math Univ of Tex Austin TX 78712

CHARNEY, ELLIOT, b New York, NY, June 1, 22; m 47; c 3. CHEMICAL PHYSICS. Educ: City Col New York, BS, 42; Columbia Univ, PhD, 56. Prof Exp: Res chemist, Manhattan Proj, 42-45, tech adv, 45-48; consult, US AEC, 48-50; consult writer, Kellex Corp, 50-54; res scientist, Lab Phys Biol, 56-72, CHIEF SECT SPECTROS & STRUCT, NIH, 72- Concurrent Pos: Asst, Columbia Univ, 51-55; vis scientist, Univ Oxford, 62-63; vis fac assoc, Dartmouth Col, 74. Mem: Am Phys Soc; NY Acad Sci. Res: Infrared and ultraviolet spectroscopy; optical rotatory dispersion; structure and interactions of molecules in condensed phases; electro-optic properties; biopolymers. Mailing Add: NIH Bldg 2 Room B1-03 Bethesda MD 20014

CHARNEY, JESSE, b New York, NY, Dec 26, 17; m 55. BIOCHEMISTRY. Educ: City Col New York, BS, 38; NY Univ, MS, 39. Prof Exp: Res chemist, Schwarz Labs, 39-42; jr chemist, Edgewood Arsenal, 42-44; res chemist, John Wyeth & Bro, 44-47; res assoc, Sharp & Dohme, Inc, 47-52, Merck Sharp & Dohme Res Labs, MercK & Co, 52-61; HEAD DEPT BIOCHEM, INST MED RES, 61- Concurrent Pos: Vis asst prof, Univ Pa, 62- Mem: Am Soc Biol Chem; Am Chem Soc. Res: Isolation of antibiotics; fermentation factors; purification and biochemical characterization of viruses; tumor immunochemistry. Mailing Add: Inst for Med Res Sheridan & Copewood Sts Camden NJ 08103

CHARNEY, JULE GREGORY, b San Francisco, Calif, Jan 1, 17; m 46; c 3. METEOROLOGY. Educ: Univ Calif, Los Angeles, BA, 38, MA, 40, PhD(meteorol), 46. Hon Degrees: DSc, Univ Chicago, 70. Prof Exp: Asst math, Univ Calif, 38-40, instr physics-meteorol, 42-43, lectr, 43-46; res assoc, Univ Chicago, 46-47; Nat Res fel, Oslo, 47-48; staff mem, Inst Advan Study, 48-52, long term mem, 52-56; prof meteorol, 56-66, SLOAN PROF METEOROL, MASS INST TECHNOL, 66-, CHMN DEPT, 74- Concurrent Pos: Vis lectr, Univ Chicago, 50-55; Woods Hole Assocs lectr, Woods Hole Oceanog Inst, 54; chmn US nat comt global atmospheric res prog, Nat Acad Sci, 68-71; overseas fel, Churchill Col, Eng, 72- Honors & Awards: Meisinger Award, Am Meteorol Soc, 49, Rossby Res Medal, 64; Losey Award, Am Inst Aeronaut & Astronaut, 57; Symons Mem Gold Medal, Royal Meteorol Soc, 61; Hodgkins Medal, Smithsonian Inst, 69; World Meteorol Orgn Prize, 71. Mem: Nat Acad Sci; Am Math Soc; Am Meteorol Soc; fel Am Geophys Union; fel Am Acad Arts & Sci. Res: Study of the general circulation of the coupled atmosphere-ocean system; use of topological-dynamical methods for the direct calculation of climate, with special application to desert-monsoon circulations. Mailing Add: Rm 54-1424 Mass Inst of Technol Cambridge MA 02139

CHARNEY, MICHAEL, b New York, NY, Aug 6, 11; m 41, 66; c 5. PHYSICAL ANTHROPOLOGY. Educ: Univ Tex, Austin, BA, 34; Univ Colo, Boulder, PhD(anthrop), 69. Prof Exp: Chief lab serv clin path, Station Hosp, Camp Gordon, Ga, 45-46; chief bacteriologist cancer res, Longevity Res Found, NY, 47-48; dir clin path, Hackensack Bio-Chem Lab, NJ, 46-65; asst prof anthrop, Idaho State Univ, 68-72; ASSOC PROF ANTHROP, COLO STATE UNIV, 71- Concurrent Pos: Co-dir, Forensic Sci Lab, 73 & assoc prof zool, 74-, Colo State Univ; dep coroner, Larimer County, Colo, 75- Mem: Fel Royal Anthrop Inst Gt Brit; Am Asn Phys Anthropologists; Am Acad Forensic Sci; Soc Study Human Biol; Soc Study Social Biol. Res: Population genetics study of a triracial hybrid population on the Costa Chica of Mexico, containing the only living, visibly-Negroid descendants of African negro slaves brought to Mexico in the 16th and 17th centuries; forensic anthropology. Mailing Add: Dept of Anthrop Colo State Univ Ft Collins CO 80523

CHARNEY, WILLIAM, b Russia, Jan 10, 18; nat US; m 47; c 2. MICROBIOLOGY. Educ: Johns Hopkins Univ, BA, 40; Rutgers Univ, PhD(microbiol), 53. Prof Exp: Bacteriologist, Rare Chems, Inc, 46-50; microbiologist, 53-70, assoc dir microbiol develop, 70-73, DIR MICROBIOL DEVELOP, SCHERING CORP, 73- Mem: Am Soc Microbiol; Am Chem Soc; NY Acad Sci. Res: Microbial transformation of steroids; vitamin B-12; antibiotics. Mailing Add: 110 Christopher St Montclair NJ 07042

CHARNICKI, WALTER FRANCIS, b Haverhill, Mass, Mar 6, 21; m 46; c 2. PHARMACEUTICAL CHEMISTRY. Educ: Mass Col Pharm, BS, 43, MS, 48; Purdue Univ, PhD(pharmaceut chem), 51. Prof Exp: Control chemist, E L Patch Co, 43, 46; instr, Franklin Tech Inst, 47-48; res assoc, Merck Sharp & Dohme Res Lab, 51-59; dir prod develop, 59-71, VPRES RES, DEVELOP & PROD, DORSEY LABS, SANDOZ INC, 71- Concurrent Pos: Retail pharmacist, 43, 46-48. Mem: Am Chem Soc; Am Pharmaceut Asn; NY Acad Sci. Res: Synthesis of organic medicinal agents; sterile and tablet pharmaceuticals. Mailing Add: Dorsey Labs Sandoz Inc PO Box 83288 Lincoln NE 68501

CHARNOCK, JOHN S, b Adelaide, Australia, Dec 29, 30; m 57; c 2. PHARMACOLOGY, BIOCHEMISTRY. Educ: Univ Adelaide, BSc, 55, PhD(med sci), 60. Prof Exp: Trainee biol, Commonwealth Sci & Indust Res Orgn, 48-55; res asst med sci, Univ Adelaide, 56-60; instr physiol, Vanderbilt Univ, 63; sr lectr cell physiol, Univ Adelaide, 64-68; assoc prof, 68-71, PROF PHARMACOL, UNIV ALTA, 71-, CHMN DEPT, 72- Concurrent Pos: Res fel biochem, Res Inst, McGill-Montreal Gen Hosp, 61-62; C J Martin fel, Nat Health & Med Res Coun Australia, 61-63. Mem: Overseas mem, Brit Biochem Soc. Res: Cell physiology, especially the action of drugs and hormones on cellular function; membrane metabolism and ion transport mechanisms; oxidative phosphorylation. Mailing Add: Dept of Pharmacol Univ of Alta Edmonton AB Can

CHARNY, EUGENE JOSEPH, b Philadelphia, Pa, Dec 31, 27; m 53; c 3. PSYCHOANALYSIS. Educ: Swarthmore Col, BA, 50; Univ Pa, MD, 54; Pittsburgh Psychoanal Inst, dipl, 67. Prof Exp: Intern, Grad Hosp, Univ Pa, 54-55; clin dir, State Hosp, Mayview, Pa, 60-61; instr, 61-64, ASST PROF PSYCHIAT, WESTERN PSYCHIAT INST, MED SCH, UNIV PITTSBURGH, 64- Concurrent Pos: Teaching fel psychiat, Western Psychiat Inst, Univ Pittsburgh, 55-58; teaching analyst, Pittsburgh Psychoanal Inst. Mem: AAAS; Am Psychoanal Asn; Am Psychiat Asn; Am Psychosom Soc. Res: Human communication; linguistic-kinesic analysis of psychotherapy films; general systems theory; psychoanalytic theory. Mailing Add: 3700 Fifth Ave Pittsburgh PA 15213

CHARON, NYLES WILLIAM, b Minneapolis, Minn, Sept 13, 43; m 69. MICROBIOLOGY. Educ: Univ Minn, BA, 65, MS, 69, PhD(microbiol), 72. Prof Exp: Teaching res asst microbiol, Univ Minn, 65-72; fel biol sci, Stanford Univ, 72-74; ASST PROF MICROBIOL, WVA UNIV, 74- Mem: AAAS; Am Soc Microbiol; Sigma Xi; Am Leptospirosis Res Conf. Res: Late gene regulation in bacteriophage lambda; biochemical and genetic studies of the spirochete Leptospira, their relative sensitivity to ultraviolet-light irradiation, DNA repair mechanisms, means for motility and isoleucine biosynthesis. Mailing Add: Dept of Microbiol WVa Univ Med Ctr Morgantown WV 26506

CHARPIE, ROBERT ALAN, b Cleveland, Ohio, Sept 9, 25; m 47; c 4. THEORETICAL PHYSICS. Educ: Carnegie Inst Technol, BS, 48, MS, 49, DSc(theoret physics), 50. Hon Degrees: PhD, Denison Univ, 65. Prof Exp: Physicist, Westinghouse Elec Corp, 47-50; physicist, Oak Ridge Nat Lab, 50-55, asst dir, 55-61, dir, Reactor Div, 58-61; mgr advan develop, Union Carbide Corp, 61-63, gen mgr develop dept, 63-64, dir technol, 64-66, pres electronics div, 66-68; pres, Bell & Howell Co, 68-69; PRES CABOT CORP, 69- Concurrent Pos: Asst, US Mem Seven-Nation Adv Comt, Int Conf Peaceful Uses Atomic Energy, 55, coordr, US Fusion Res Exhib, 58, secy gen adv comt, AEC, 59-63; ed-in-chief, Proc Int Conf, 55; gen ed, Int Monogr Ser on Nuclear Energy, 55-60; ed, J Nuclear Energy, 55-60; mem, Oak Ridge Bd Ed, 57-61; mem adv comn UN sci activities, State Dept, 61-; mem panel, Civilian Technol Pakistan, President's Sci Adv Comn, 61-, mem panel oceanog, President's Sci Adv Comt, 65; trustee, Carnegie Inst Technol, 62- Honors & Awards: Award, US Chamber Com, 55. Mem: Nat Acad Eng; fel Am Nuclear Soc; fel Am Phys Soc; fel NY Acad Sci; Sigma Xi. Res: Theoretical, nuclear and reactor physics. Mailing Add: Cabot Corp 125 High St Boston MA 02110

CHART, JEROME JAMES, b Wis, Oct 22, 22; m 53; c 6. ENDOCRINOLOGY. Educ: Univ Wis, PhD(zool), 55. Prof Exp: Sr endocrinologist, Ciba Pharmaceut Co, 54-58, assoc dir endocrinol, 58-62, asst dir biol res, 62-69, dir endocrinol, 69-71, DIR METAB DIS, CIBA-GEIGY CORP, 71- Res: Adrenal physiology; control of carbohydrate and salt metabolism. Mailing Add: Ciba-Geigy Corp Ardsley NY 10502

CHARTERS, ALEXANDER CRANE, aeronautical engineering, see 12th edition

CHARTERS, ELAINE MARY, b Springfield, Ohio, Aug 15, 32. ZOOLOGY, ENDOCRINOLOGY. Educ: Our Lady Cincinnati Col, BA, 54; Cath Univ Am, MA, 60, PhD(zool), 63. Prof Exp: Med technologist, Mercy Hosp, Springfield, Ohio, 55-56; from instr to assoc prof biol, 62-68, PROF BIOL, EDGECLIFF COL, 68- Mem: Nat Asn Biol Teachers; Am Inst Biol Sci. Res: Effects of anoxia on liver function of rats; effects of upper lethal temperature on thyroid of tadpoles. Mailing Add: Edgecliff Col Dept of Biol 2220 Victory Pkwy Cincinnati OH 45206

CHARTOCK, MICHAEL ANDREW, b Palo Alto, Calif, May 25, 43; m 71; c 1. ECOLOGY, SCIENCE POLICY. Educ: Univ Calif, Berkeley, AB, 65; San Jose State Univ, MA, 71; Univ Southern Calif, PhD(biol), 72. Prof Exp: ASST PROF ZOOL, UNIV OKLA, 71- Concurrent Pos: Res fel sci & pub policy, Univ Okla, 71-; consult, BDM Corp, 74-75. Mem: AAAS; Ecol Soc Am; Am Inst Biol Sci. Res: Technology assessment and policy oriented research in energy development including coal, oil, gas, uranium, geothermal and other sources; energy flow in aquatic ecosystems; role of detritus in coral reefs. Mailing Add: Dept of Zool Univ of Okla Norman OK 73069

CHARTON, MARVIN, b Brooklyn, NY, May 1, 31; m 55; c 3. PHYSICAL ORGANIC CHEMISTRY. Educ: City Col New York, BS, 53; Brooklyn Col, MA, 56; Stevens Inst Technol, PhD(chem), 62. Prof Exp: Res chemist, Evans Res & Develop Corp, 55-56; instr, 56-61, from asst prof to assoc prof, 61-67, PROF CHEM, PRATT INST, 67-, CHMN DEPT, 69- Concurrent Pos: Fel, Intrasci Res Found, 69- Mem: AAAS; Am Chem Soc; Brit Chem Soc; NY Acad Sci. Res: Linear free energy relationships in organic chemistry; quantitative treatment of proximity effects. Mailing Add: Dept of Chem Pratt Inst Ryerson St Brooklyn NY 11205

CHARTRAND, GARY, b Sault Ste Marie, Mich, Aug 24, 36; m 68. MATHEMATICS. Educ: Mich State Univ, BS, 58, MS, 60, PhD(math), 64. Prof Exp: From asst prof to assoc prof, 64-70, PROF MATH, UNIV WESTERN MICH, 70- Concurrent Pos: Res grants, US Air Force Off Sci Res, Univ Mich, 65-66 & NIMH, Res Ctr Group Dynamics, 66; NSF grant, 68-69; Off Naval Res fel, 70-71; vis math scholar, Univ Calif, Santa Barbara, 70-71. Mem: Am Math Soc; Math Asn Am. Res: Theory of graphs; connectivity and line-connectivity; graphical partitions; traversability; line, total and permutation graphs; planarity; colorability; graphs and matrices; reconstruction of graphs. Mailing Add: Dept of Math Western Mich Univ Kalamazoo MI 49001

CHARTRAND, MARK RAY, III, b Miami, Fla, Aug 2, 43; m 75. ASTRONOMY. Educ: Case Inst Technol, BS, 65; Case Western Reserve Univ, PhD(astron), 70. Prof Exp: Asst to dir astron, Ralph Mueller Planetarium, Cleveland Mus Natural Hist, 65-66; asst astronr & dir educ, 70-74, CHMN & ASSOC ASTRON, HAYDEN PLANETARIUM, AM MUS, 74- Concurrent Pos: Adj asst prof, Fordham Univ, Lincoln Ctr Campus, 72- Mem: AAAS; Am Astron Soc; Int Soc Planetarium Educ. Res: Galactic structure; photoelectric and photographic photometry. Mailing Add: Am Mus-Hayden Planetarium 81st & Central Park W New York NY 10024

CHARVONIA, DAVID ALAN, b Denver, Colo, July 19, 29; m 53; c 2. DYNAMICS. Educ: Univ Colo, BS, 51; Purdue Univ, MS, 53, PhD(propulsion), 59. Prof Exp: Div mgr & other positions in electronics & space syst, Aerojet-Gen Corp, 61-68; sr staff mem radar syst, ITT Gilfillan, 68; vpres & tech dir res & develop admin, Telluron, 68-72; staff specialist res & develop prog admin, 72-75, actg asst dir electronic technol, 75, SPEC ASST TO DEP DIR RES & EXPLOR DEVELOP, OFF DIR DEFENSE RES & ENG, DEPT DEFENSE, WASHINGTON, DC, 75- Res: Exploratory development pertinent to military applications. Mailing Add: 9119 Hamilton Dr Fairfax VA 22030

CHARYULU, KOMANDURI K N, b Hanamkonda, India, May 24, 24; US citizen; m 44; c 6. ONCOLOGY, RADIOLOGY. Educ: Andhra Univ, BSc, 45, MD, 51. Prof Exp: Radium registr & asst surgeon radiother, Radium Inst & Cancer Hosp, India, 55-57; tutor radiol, Osmania Med Col, India, 57-58; hon clin asst, London Hosp, Eng, 59-60, registr, 60-61; locum consult, St Mary's Hosp, Portsmouth, Eng, 62; asst prof radiol, Univ Minn Hosps, Minneapolis, 64-67, assoc prof, 67-70; PROF RADIOL & DIR RADIATION THER, SCH MED, UNIV MIAMI, 70- Concurrent Pos: Spec vis res fel, Mem Hosp & Sloan-Kettering Cancer Inst, 62-63. Mem: Radiation Res Soc; Am Soc Therapeut Radiol; Am Soc Clin Oncol; Am Col Radiol; Am Radium Soc. Res: Oxygenation of tissues and study of radiation sensitivity; modification of radio sensitivity by heat and microwaves; endocrine relationships in carcinoma; dose distribution in electron and x-ray therapeutic regimens. Mailing Add: Radiation Ther Div Univ Miami Sch of Med Miami FL 33152

CHASALOW, IVAN G, b New York, NY, Mar 6, 30; m 54; c 6. OPERATIONS RESEARCH. Educ: Mass Inst Technol, BS, 51; Columbia Univ, MA, 52, PhD(chem), 57. Prof Exp: Opers analyst, opers eval group, Mass Inst Technol, 56-59; OPERS ANALYST, BELL TEL LABS, 59- Mem: Am Chem Soc; Opers Res Soc Am. Res: Military operations research and applied research in underwater sound. Mailing Add: 3C328B Bell Tel Labs Whippany NJ 07981

CHASANOV, MARTIN GERSON, b Philadelphia, Pa, Aug 23, 27; m 54; c 3. PHYSICAL CHEMISTRY. Educ: Univ Del, BS, 49, MS, 50, PhD(chem), 52. Prof Exp: Res chemist, Rohm and Haas Co, 52-54; chemist, Chem Corps, US Army, 54-56; staff chemist, Mil Prod, Int Bus Mach Corp, 56-59; CHEMIST, ARGONNE NAT LAB, 59- Mem: Am Chem Soc; Sigma Xi. Res: Chemical kinetics; heterogeneous catalysis; nuclear radiation effects on electronic materials; nuclear dosimetry; thermodynamics; surface chemistry; semiconductors; ion exchange; high temperature chemistry; nuclear reactor materials. Mailing Add: Agronne Nat Lab Argonne IL 60439

CHASAR, DWIGHT WILLIAM, organic chemistry, spectroscopy, see 12th edition

CHASE, CARL TRUEBLOOD, physics, see 12th edition

CHASE, CHARLES ELROY, JR, b Lyndonville, Vt, May 16, 29; m 54; c 2. LASERS. Educ: Mass Inst Technol, BS, 50; Camridge Univ, PhD(physics), 54. Prof Exp: Staff mem physics, Lincoln Lab, Mass Inst Technol, 54-63 & Francis Bitter Nat Magnet Lab, 63-75; PRES, TACHISTO INC, 75- Concurrent Pos: Fulbright award, Univ Leiden, 62-63; adj assoc prof physics, Boston Univ, 70-71. Mem: Fel Am Phys Soc. Res: Low temperature physics; ultrasonics in solids and liquids; nonlinear optics; plasmas; quantum electronics; laser development. Mailing Add: 39 Edgewater Dr Waltham MA 02154

CHASE, DAN L, b Mt Vernon, Ohio, Jan 12, 16; m 40; c 3. ANALYTICAL CHEMISTRY. Educ: Ohio State Univ, BA, 39. Prof Exp: Gen analyst, Repub Steel Co, Ohio, 39-41; res engr, 41-50, asst chief anal div, 50-58, supvr anal chem div, 58-75, SR CHEMIST, COLUMBUS LABS, BATTELLE MEM INST, 75- Mem: Am Chem Soc; Sigma Xi. Res: Analytical methods. Mailing Add: Anal & Phys Chem Sect Battelle Mem Inst 505 King Ave Columbus OH 43201

CHASE, DAVID MARION, b Denver, Colo, Jan 20, 30; m 63; c 1. THEORETICAL PHYSICS, APPLIED PHYSICS. Educ: Univ Colo, BS, 51; Princeton Univ, AM, 53, PhD(physics), 55. Prof Exp: Staff mem, Los Alamos Sci Lab, 54-57; sr physicist, TRG, Inc, Melville, 57-68; SR SCIENTIST, BOLT BERANEK & NEWMAN, INC, 68- Concurrent Pos: Vis asst prof, Iowa State Univ, 56. Mem: Am Phys Soc; Acoust Soc Am. Res: General relativity; nuclear models and scattering; acoustics; turbulence; stochastic investment analysis. Mailing Add: 14 Pinckney St Boston MA 02114

CHASE, FRANCIS EDWARD, b Frankford, Ont, Sept 16, 14; m 40; c 2. SOIL MICROBIOLOGY. Educ: Ont Agr Col, BSA, 38; Univ Toronto, MSA, 40; McGill Univ, PhD(soil microbiol), 51. Prof Exp: Agr asst, Mold Count Tomato Prod, Dominion Dept Agr, Toronto, 38; asst med bact, Banting Inst, Toronto, 39-41; agr asst soil microbiol & antibiotics, Dominion Dept Agr, Ottawa, 41-44; from asst prof to assoc prof soil microbiol, 44-56, chmn dept microbiol, 67-71, PROF SOIL MICROBIOL, ONT AGR COL, UNIV GUELPH, 56- Concurrent Pos: Ed, Can J Microbiol, 60-70. Mem: Soil Sci Soc Am; Am Soc Agron; Can Soil Sci Soc; Can Soc Microbiologists; Agr Inst Can. Res: Effect of silicic acid on the growth of the tubercle bacillus; grouping of soil microorganisms according to nutritional requirements; use of sulphite waste liquor as a medium for production of antibiotics; occurrence of salmonella in eggs and fowl; studies on nitrification, nitrogen fixation and soil respiration in agricultural and forest soils. Mailing Add: Dept of Environ Biol Ont Agr Col Univ of Guelph Guelph ON Can

CHASE, FRED LEROY, b Dedham, Mass, Nov 30, 14; m 45; c 2. CHEMISTRY, RESEARCH ADMINISTRATION. Educ: Harvard Univ, AB, 37; Mass Inst Technol, ScD(chem eng), 42. Prof Exp: Instr chem eng, Mass Inst Technol, 39-40; res chemist, Dewey & Almy Chem Div, 41-44, lab mgr, 44-58, asst dir res, Container & Chem Spec Div, 58-62, mgr compounding tech ctr, Overseas Chem Div, Eng, 62-64, asst dir res, Container & Chem Spec Div, 64, res dir can & drum sealing compounds, Dewey & Almy Chem Div, 64-70, assoc dir res, 70-73, ASSOC DIR RES, INDUSTRIAL CHEM GROUP, W R GRACE & CO, 73- Mem: Am Chem Soc. Res: Rubber; colloid chemistry; canning technology; sealing compounds for containers for food preservation and industrial packaging. Mailing Add: 30 Lake Shore Dr Arlington MA 02174

CHASE, GARY ANDREW, b New York, NY, Jan 5, 45; m 68; c 1. HUMAN GENETICS, STATISTICS. Educ: Harvard Univ, AB, 66; Johns Hopkins Univ, PhD(statist), 70. Prof Exp: NIH fel, Sch Med, Johns Hopkins Univ, 70-71, ASST PROF MED & BIOSTATIST, JOHNS HOPKINS UNIV, 71-, STATISTICIAN, LIPID RES CLIN, 72- Concurrent Pos: Consult, Nat Heart & Lung Inst Collab Lipid Res Prog, 72-; investr human genetics, Howard Hughes Med Inst, 73- Mem: Am Statist Asn; Am Soc Human Genetics. Res: Statistical methods in human genetics. Mailing Add: Div of Genetics Johns Hopkins Hosp Baltimore MD 21205

CHASE, GERALD ROY, b Janesville, Wis, Oct 16, 38; m 61, 73; c 4. BIOSTATISTICS. Educ: Beloit Col, BS, 61, Stanford Univ, MS, 63, PhD(statist), 66. Prof Exp: From asst prof to assoc prof statis & community health, Univ Mo-Columbia, 66-73; vis scientist, environ biomet, Nat Inst Environ Health Sci, 73-74; ASSOC PROF STATIST & COMMUNITY HEALTH, UNIV MO-COLUMBIA, 74- Mem: Inst Math Statist; Am Statist Asn; Biomet Soc. Res: Applications of statistics in health related fields. Mailing Add: Dept of Statist Univ of Mo Columbia MO 65201

CHASE, GRAFTON D, b NJ, May 2, 21; m 53; c 3. PHYSICAL CHEMISTRY. Educ: Philadelphia Col Pharm, BSc, 43; Temple Univ, MA, 51, PhD(chem), 55. Prof Exp: Instr chem, Philadelphia Col Pharm, 46-48; res scientist, Johnson & Johnson, 48-49; from instr to assoc prof, 49-65, PROF CHEM, PHILADELPHIA COL PHARM & SCI, 65- Concurrent Pos: Assoc canning technologist, Crown Can Co, 45-47; consult, Clinica Quintero Venezuela, 55-56; ed, Remington's Pharmaceut Sci. Mem: Am Chem Soc. Res: Fundamentals of radiochemistry; investigation of antigen-antibody interactions and radioimmunoassay. Mailing Add: Dept of Chem Philadelphia Col of Pharm & Sci Philadelphia PA 19104

CHASE, HAROLD FREDERICK, b Nanuet, NY, June 20, 12; m 40; c 3. ANESTHESIOLOGY. Educ: Colby Col, BS, 33; Boston Univ, MD, 38. Prof Exp: From instr to asst prof pharmacol, Med Sch, Wayne State Univ, 39-44; from asst prof to assoc prof, Western Reserve Univ, 44-48; Commonwealth Fund fel, Hartford Hosp, 48, resident anesthesia, 48-49; anesthesiologist in-chg, Univ Va Hosp, 49-55; prof clin & res anesthesiol, Thomas Jefferson Univ, 55-65; anesthesiologist, Lankenau Hosp, Philadelphia, Pa, 65-68; PROF CLIN & RES ANESTHESIOL, THOMAS JEFFERSON UNIV, 68- Concurrent Pos: Consult, Univ Hosp, Cleveland, 45-48. Res: Pharmacology of curare and curare-like drugs and anesthetic drugs; clinical usefulness of newer analgesics; carbon dioxide absorption; relationship of intrathoracic and intercranial pressures; humidity in anesthesia systems. Mailing Add: 1732 Old Gulph Rd Villanova PA 19085

CHASE, HARRISON VERNON, b Big Rapids, Mich, Aug 17, 13; m 32; c 1. GEOGRAPHY. Educ: Univ Mich, AB, 35, MA, 39. Prof Exp: High sch & jr col teacher, Mich, 37-42; instr geog, Syracuse Univ, 42-43; res analyst, Off Strategic Serv, DC, 43-45; asst prof geog, 47-65, chmn interdiv prog social sci, 69-71, instrnl chmn social sci, 69-74, ASSOC PROF GEOG, FLA STATE UNIV, 65-, DIR PROG SOCIAL SCI, 74- Res: Geography of Japan; political geography; critical resources. Mailing Add: Dept of Geog Fla State Univ Tallahassee FL 32306

CHASE, HELEN CHRISTINA (MATULIC), b New York, NY, Mar 21, 17; m 42. BIOSTATISTICS. Educ: Hunter Col, AB, 38; Columbia Univ, MSc, 51; Univ Calif, DrPH, 61. Prof Exp: Jr statistician, NY State Dept Health, 48-50, from biostatistician to prin biostatistician, 50-63; chief mortality statist br, Nat Ctr Health Statist, US Dept Health, Educ & Welfare, 63-65, statistician, Off Health Statist Anal, 65-69; dir res, Asn Schs Allied Health Prof, 69-71; staff assoc biostatist, Inst Med, Nat Acad Sci, 71-72; STATISTICIAN, OFF RES & STATIST, SOCIAL SECURITY ADMIN, US DEPT HEALTH, EDUC & WELFARE, 73- Concurrent Pos: Lectr biostatist, Grad Sch Nursing, Cath Univ Am, 66-; consult, US Dept Health, Educ & Welfare, 61-62, White House Conf Food, Nutrit & Health, 69, Nat Acad Sci, 70 & Maternal & Child Health Proj, George Washington Univ, 70-; mem radiation bioeffects & epidemiol adv comt, Food & Drug Admin, 71-75. Mem: Fel AAAS; fel Am Pub Health Asn; fel Am Statist Asn; Soc Epidemiol Res; Pop Asn Am. Res: Public health; health planning; epidemiology; infant mortality. Mailing Add: 6417 15th St Alexandria VA 22307

CHASE, HERMAN BURLEIGH, b New Hampton, NH, May 7, 13; m 37, 67; c 5. ANIMAL GENETICS. Educ: Dartmouth Col, AB, 34; Univ Chicago, PhD(zool), 38. Prof Exp: Asst zool, Univ Chicago, 35-38; instr, Univ Ill, 38-41, assoc, 41-45, asst prof, 45-48; assoc prof biol, 48-52, chmn dept, 63-67, dir, Inst Life Sci, 67-75, PROF BIOL, BROWN UNIV, 52- Concurrent Pos: Nat Cancer Inst spec fel, Radiobiol Unit, Mt Vernon Hosp, Eng, 56-57; USPHS spec fel, Commonwealth Sci & Indust Res Orgn, Australia, 64-65 & 72-73. Res: Genetics Soc Am; Am Soc Zoologists; Soc Study Evolution; Soc Develop Biol; Radiation Res Soc. Res: Genetics of mice; physiological genetics; radiation biology. Mailing Add: Biomed Div Brown Univ Providence RI 20912

CHASE, IVAN DMITRI, b Syracuse, NY, Feb 1, 43. ETHOLOGY. Educ: Univ SC, BS, 65; Harvard Univ, MA, 70, PhD(sociol), 72. Prof Exp: Vis fel math, Dartmouth Col, 71-73, vis scholar math & soc sci, 73-74, vis scholar sociol, 74-75; HON FEL ZOOL, UNIV WIS, MADISON, 75- Concurrent Pos: Fel Soc Sci Res Coun, 71-73; fel, Harry Frank Guggenheim Found, 75-76. Mem: Animal Behav Soc; Am Sociol Asn. Res: The social processes used by animals in the formation and maintenance of dominance hierarchies. Mailing Add: Dept of Zool Birge Hall Univ of Wis Madison WI 53706

CHASE, JOHN WALTER, physical chemistry, nuclear chemistry, see 12th edition

CHASE, JOHN WILLIAM, b Baltimore, Md, May 30, 44; m 67. BIOCHEMICAL GENETICS, MOLECULAR BIOLOGY. Educ: Drew Univ, BA, 66; Johns Hopkins Univ, PhD(biochem), 71. Prof Exp: Res fel biol chem, Harvard Med Sch, 71-74, res assoc, 74-75; ASST PROF MOLECULAR BIOL, ALBERT EINSTEIN COL MED, 75- Concurrent Pos: NIH fels, 71-72 & 74-75; Am Cancer Soc fel, 72-73. Mem: Am Soc Microbiol. Res: Enzymology of DNA replication, recombination and repair including biochemical and genetic studies of the functions of nucleases in these processes. Mailing Add: Albert Einstein Col of Med 1300 Morris Park Ave Bronx NY 10461

CHASE, LARRY EUGENE, b Wadsworth, Ohio, Sept 23, 43; m 64; c 2. ANIMAL NUTRITION, ANIMAL PHYSIOLOGY. Educ: Ohio State Univ, BS, 66; NC State Univ, Raleigh, MS, 69; Pa State Univ, University Park, PhD(animal nutrit), 75. Prof Exp: Dairy supvr, NC Dept Agr, Willard, 68-69; res aide animal nutrit & physiol, Pa State Univ, 69-74; ASST PROF ANIMAL SCI, CORNELL UNIV, 75- Mem: Am Dairy Sci Asn; Am Soc Animal Sci. Res: Improvement of intake and utilization of foodstuffs by ruminants with emphasis on forage utilization and nitrogen metabolism. Mailing Add: 102 Sharlene Rd Ithaca NY 14850

CHASE, LEE MACARTHUR, physics, see 12th edition

CHASE, LLOYD FREMONT, JR, b San Francisco, Calif, Feb 1, 31; m 52; c 3. PHYSICS. Educ: Stanford Univ, BS, 53, PhD(physics), 57. Prof Exp: Res assoc physics, Stanford Univ, 57-58; res scientist, Res Labs, Lockheed Missiles & Space Co, 58-64, SR STAFF SCIENTIST & SR MEM LABS, LOCKHEED PALO ALTO RES LAB, 64- Concurrent Pos: Res collabr, Brookhaven Nat Lab, 61-62; vis sr res officer, Univ Oxford, 62-63. Mem: Am Phys Soc. Res: Low energy nuclear physics; nuclear structure and nuclear reaction mechanisms; space physics. Mailing Add: O R G N Lockheed 52-10 Bldg 203 3251 Hanover St Palo Alto CA 94304

CHASE, LLOYD LEE, b Milwaukee, Wis, Oct 24, 39; m 68. PHYSICS. Educ: Univ Ill, BS, 61; Cornell Univ, PhD(physics), 66. Prof Exp: Mem tech staff, Bell Tel Labs, NJ, 66-69; asst prof physics, 69-74, ASSOC PROF PHYSICS, IND UNIV, BLOOMINGTON, 74- Mem: Am Phys Soc. Res: Electron spin resonance; optical pumping in solids; Raman scattering. Mailing Add: Dept of Physics Ind Univ Bloomington IN 47401

CHASE, NORMAN E, b Cincinnati, Ohio, June 29, 26; m 54; c 2. MEDICINE. Educ: Univ Cincinnati, BS, 49, MD, 53. Prof Exp: Instr radiol, Col Physicians & Surgeons, Columbia Univ, 59-61, assoc, 61; assoc attend, 61, from asst prof to assoc prof, 61-66, PROF RADIOL, MED CTR, NY UNIV, 66-, CHMN DEPT, 69- Concurrent Pos: Dir radiol, Bellevue Hosp, NY, 64-74, assoc dir, 74-; sr consult, Manhattan Vet Admin Hosp, 69- Mem: AMA; Am Soc Neuroradiol (secy-treas, 62, pres elect, 71); Asn Univ Radiol; NY Acad Sci; Am Col Radiol. Res: Cerebrovascular disease; radiology; neuroradiology. Mailing Add: NY Univ Med Ctr 550 First Ave New York NY 10012

CHASE, RANDOLPH MONTIETH, JR, b Brooklyn, NY, Aug 10, 28; m 55; c 2. MEDICINE, IMMUNOLOGY. Educ: NY Univ, AB, 50, MD, 58. Prof Exp: Asst med, Sch Med, NY Univ, 59-61, instr, 61-62; asst physician, Rockefeller Inst Hosp, 62-64; asst prof, 64-70, ASSOC PROF MED, SCH MED, NY UNIV, 70-, DIR MICROBE LAB, UNIV HOSP, 64- Concurrent Pos: Nat Inst Allergy & Infect Dis fel, Rockefeller Inst, 62-64; clin asst, Bellevue Hosp, 62-64, asst attend physician, 62-, asst vis physician, 65- Mem: AAAS; Transplantation Soc. Res: Infectious disease and allergy, including the effect of antibiotics on streptococcal cell wall, particularly the effect of prophylactic antibiotics in RHD and the appearance of given streptoccal strains in SBE; cross reacting antigens existing between bacteria and mammalicin tissues, especially tissue transplants. Mailing Add: NY Univ Med Ctr 560 First Ave New York NY 10016

CHASE, RICHARD CONANT, elementary particle physics, see 12th edition

CHASE, RICHARD GOLDEN, organic chemistry, see 12th edition

CHASE, RICHARD L, b Perth, Australia, Dec 25, 33; m 65; c 2. GEOLOGY. Educ: Univ Western Australia, BSc, 56; Princeton Univ, PhD(geol), 63. Prof Exp: Geologist, WAustralian Petrol Ltd, 54-55; asst geologist, Geosurv Australia Ltd, 56-57; sr asst geologist, Ministry Mines, Que, 59; geologist, Ministry Mines & Hydrocarbons, Venezuela, 60-61; Ford Found fel, Woods Hole Oceanog Inst, 63-64, asst scientist, 64-68; asst prof geol, 68-74, ASSOC PROF GEOL SCI, UNIV BC, 74- Mem: AAAS; Geol Soc Am; Am Geophys Union; Australian Geol Soc; Geol Asn Can. Res: Petrology and structural geology; origin of oceanic igneous rocks; history of ocean floors; marine geology, geotectonics and petrology in the northeastern Pacific, Caribbean, Mid-Atlantic Ridge, Mediterranean Sea and Red Sea. Mailing Add: Dept of Geol Univ of BC Vancouver BC Can

CHASE, ROBERT A, b Keene, NH, Jan 6, 23; m 46; c 3. RECONSTRUCTIVE SURGERY. Educ: Univ NH, BS, 45; Yale Univ, MD, 47; Am Bd Surg, dipl, 55; Am Bd Plastic Surg, dipl, 60. Prof Exp: Intern, New Haven Hosp, 47-48; asst resident surg path, bact & cancer clin & asst surg path & bact, Sch Med, Yale Univ, 48-49; asst resident, New Haven Hosp, 49-50, sr asst resident surg, 52-53, chief res surgeon, 53-54; res plastic surgeon, Univ Pittsburgh Hosp, 57-59; from asst prof to assoc prof surg, Sch Med, Yale Univ, 59-63; prof surg & chmn dept, Sch Med, Stanford Univ, 63-73; PRES & DIR, NAT BD MED EXAMR, 73- Concurrent Pos: Teaching fel plastic surg, Univ Pittsburgh Hosp, 57-59; asst, Sch med, Yale Univ, 53-54; attend surgeon, US Vet Admin Hosp, West Haven, Conn, 59-62, consult, 62-63; attend surg, Grace New Haven Community Hosp, 59-63; consult, Christian Med Col & Hosp, India, 62 & Vet Admin Hosp, Palo Alto, Calif; mem med staff, Santa Clara County Hosp; mem, Plastic Surg Res Coun. Mem: Inst of Med of Nat Acad Sci; Asn Am Med Cols; fel Am Col Surg; Am Soc Plastic & Reconstruct Surg; Am Soc Surg of the Hand. Res: Nerve pedicle regeneration in anterior ocular chamber; objective evaluation of palatopharyngeal function. Mailing Add: Nat Bd of Med Examr 3930 Chestnut St Philadelphia PA 19104

CHASE, ROBERT SILMON, JR, b Abington, Pa, June 9, 30; m 55; c 4. VERTEBRATE ZOOLOGY. Educ: Haverford Col, AB, 52; Univ Ark, MS, 55; Bryn Mawr Col, PhD(biol), 67. Prof Exp: From instr to assoc prof biol, 58-74, asst dean, 61-64, dean studies, 68-70, dean col, 69-72, provost, 70-72, PROF BIOL, LAFAYETTE COL, 74- Mem: Fel AAAS. Res: Amphibian development, vertebrate behavior and ecology. Mailing Add: Dept of Biol Lafayette Col Easton PA 18042

CHASE, SHERRET SPAULDING, b Toledo, Ohio, June 30, 18; m 43; c 5. GENETICS, BOTANY. Educ: Yale Univ, BS, 39; Cornell Univ, PhD(bot cytol, genetics), 47. Prof Exp: Assoc prof bot, Iowa State Col, 47-54; res geneticist & mgr foreign seed opers, DeKalb Agr Asn, Inc, 54-66, dir, Dekalb-Italiana, 65-66; Bullard fel, Bot Mus, Harvard Univ, 66-67, Cabot fel, Forest Res, 67-69, res assoc econ bot, Bot Mus, 69-70; PROF BIOL, STATE UNIV NY OSWEGO, 70- Concurrent Pos: Pres, Catskill Ctr Conserv & Develop, Inc; dir, Old Mill East Meredith Corp, 73- Mem: AAAS; Bot Soc Am; Am Soc Agron; Genetics Soc Am. Res: Plant breeding; cytotaxonomy of Najas; parthenogenesis in maize; corn breeding; forest genetics. Mailing Add: Shokan NY 12481

CHASE, THEODORE, JR, b Boston, Mass, Aug 20, 38; m 65; c 1. ENZYMOLOGY. Educ: Harvard Univ, AB, 60; Univ Calif, Berkeley, PhD(biochem), 66. Prof Exp: Res assoc biol, Brookhaven Nat Lab, 67-69; asst prof, 69-74, ASSOC PROF BIOCHEM & MICROBIOL, RUTGERS UNIV, 74- Mem: Am Soc Microbiol; Am Chem Soc; Am Ornith Union; Brit Ornith Union. Res: Mechanism of enzyme action; practical utilization of enzymes and enzyme inhibitors. Mailing Add: Dept of Biochem & Microbiol Cook Col Rutgers Univ New Brunswick NJ 08903

CHASE, THOMAS NEWELL, b Westfield, NJ, May 23, 32; m 59; c 2. NEUROLOGY, NEUROPHARMACOLOGY. Educ: Mass Inst Technol, BS, 54; Yale Univ, MD, 62. Prof Exp: Engr, Singer Mfg Co, Conn, 54-55; res technician, Col Physicians & Surgeons, Columbia Univ, 57-58; intern internal med, Yale-New Haven Med Ctr, 62-63; from asst resident to resident neurol, Mass Gen Hosp, 63-66; guest worker, Lab Clin Sci, 66-68, chief neurol unit, NIMH, 68-74, chief exp therapeut, 70-74; CHIEF LAB NEUROPHARMACOL & DIR INTRAMURAL RES, NAT INST NEUROL DIS & STROKE, 74- Concurrent Pos: Fel neuropath, Mass Gen Hosp & Harvard Sch Med, 64-65; Nat Inst Neurol Dis & Stroke spec fel, 66-68; clin assoc prof neurol, Sch Med, Georgetown Univ, 71-; mem neurol adv comt, Food & Drug Admin; res group Huntington's chorea, World Fedn Neurol. Honors & Awards: Winternitz Prize Path, Yale Univ, 60, Ramsay Prize Clin Med, 61; Dipl of Merit, Govt Bolivia, 74. Mem: AAAS; Am Soc Neurochem; Soc Neurosci; Asn Res Nerv & Ment Dis; Am Acad Neurol. Res: Neuropharmacology; clinical and experimental neurology; neurochemistry; neurohumoral mechanisms; research administration. Mailing Add: Nat Inst Neurol Dis & Stoke 9000 Rockville Pike Bethesda MD 20014

CHASE, VERNON LINDSAY, b Baltimore, Md, Mar 20, 20; m 43; c 4. TEXTILE CHEMISTRY. Educ: Western Md Col, BA, 41. Prof Exp: Res chemist electrochem, Am Smelting & Refining Co, 41-46; res chemist org chem, Ridbo Labs, 47; dir res & develop textile colors, Color & Chem Div, Interchem Corp, 48-57, prog mgr org coatings, Cent Res Labs, 58-66; RES ASSOC POLYMERS, TECH CTR, J P STEVENS & CO, INC, 67- Mem: Am Chem Soc; Am Asn Textile Chemists & Colorists. Res: Polymeric systems for elastic fabrics; acrylic binders for pigment printing and non-woven fabrics; foam-backcoating; systems for carpet backing and flame retardancy; dyeing systems for new fiber blends. Mailing Add: J P Stevens & Co Inc Tech Ctr 141 Lanza Ave Garfield NJ 07026

CHASE, WILLIAM HENRY, b Montreal, Que, June 15, 27. IMMUNOPATHOLOGY, NEUROPATHOLOGY. Educ: McGill Univ, BSc, 48, MD & CM, 52. Prof Exp: Resident path, Vancouver Gen Hosp, 52-56; from asst prof to assoc prof, 58-74, PROF PATH, UNIV BC, 74- Concurrent Pos: Res fel anat, Univ Chicago, 56-58. Res: Electron microscopy. Mailing Add: Dept of Path Univ of BC Vancouver BC Can

CHASENS, ABRAM I, b Woodbine, NJ, Sept 7, 12; m 42. PERIODONTICS, DENTISTRY. Educ: Temple Univ, DDS, 36; NY Univ, cert, 52; Am Bd Periodont, dipl, 54. Prof Exp: Asst prof periodont & oral med, Col Dent, NY Univ, 53-57; PROF PERIODONT & ORAL MED & CHMN DEPT, SCH DENT, FAIRLEIGH DICKINSON UNIV, 57- Concurrent Pos: Consult, Muhlenberg Hosp, Plainfield, NJ, 54- Honors & Awards: Samuel Charles Miller Mem Award, 71; Hirschfeld Medal, Northeastern Soc Periodont, 72. Mem: Fel Am Col Dent; fel Am Acad Oral Med (pres, 67-68); Am Dent Asn; Am Acad Periodont; fel Royal Soc Health. Res: Periodontology; oral medicine; occlusion diseases and disturbances of the temporomandibular joint; periodontal surgery. Mailing Add: Sch of Dent Fairleigh Dickinson Univ Hackensack NJ 07601

CHASIN, LAWRENCE ALLEN, b Willimantic, Conn, July 2, 41; m 61, 75; c 2. BIOCHEMICAL GENETICS. Educ: Brown Univ, BS, 62; Mass Inst Technol, PhD(biol), 67. Prof Exp: Res assoc microbiol, Lab Enzymol, Ctr Nat Sci Res, 66-68; sr instr cell genetics, Univ Colo Med Ctr, 68-70; asst prof, 70-75, ASSOC PROF BIOL SCI, COLUMBIA UNIV, 75- Concurrent Pos: Mem genetics study sect, Div Res Grants, NIH, 75-79. Mem: AAAS; Genetics Soc Am; Am Soc Cell Biol. Res: Application of somatic cell genetic techniques to the study of the regulation of gene expression; biochemical characterization of regulatory variants of cultured mammalian cells. Mailing Add: Dept of Biol Sci Columbia Univ New York NY 10027

CHASIN, MARK, b New York, NY, Feb 20, 42; m 63; c 3. BIOCHEMISTRY, ENZYMOLOGY. Educ: Cornell Univ, AB, 63; Mich State Univ, PhD(biochem), 67. Prof Exp: Sr res investr biochem pharmacol, Squibb Inst Med Res, 67-74; group leader molecular biol, Ortho Res Found, 74-75, SECT HEAD BIOCHEM RES, ORTHO PHARMACEUT CORP, 75- Mem: AAAS; Am Chem Soc; NY Acad Sci; Am Soc Biol Chemists. Res: Enzymology and enzyme inhibitors; enzymology concerned with 3', 5'-cyclic adenosine monophosphate; reproductive research. Mailing Add: Ortho Pharmaceut Corp Raritan NJ 08869

CHASIN, WERNER DAVID, b Danzig, Feb 29, 32; US citizen; m 63; c 3. OTOLARYNGOLOGY. Educ: Harvard Univ, AB, 54; Tufts Univ, MD, 58. Prof Exp: Rotating intern, Mt Sinai Hosp, New York, 58-59; resident otolaryngol, Mass Eye & Ear Infirmary, 59-62, asst otolaryngologist, 62-64; chief otolaryngol, Beth Israel Hosp, 64-68; CHMN DEPT OTOLARYNGOL, SCH MED, TUFTS UNIV, 68-; OTOLARYNGOLOGIST-IN-CHIEF, TUFTS-NEW ENG MED CTR, 68-; CHIEF OTOLARYNGOL, USPHS HOSP, BRIGHTON, 74- Concurrent Pos: Vis surgeon, Boston City Hosp, 68-; secy-treas, New Eng Otolaryngol Soc, 69-72. Mem: Fel Am Acad Ophthal & Otolaryngol. Mailing Add: New Eng Med Ctr Hosp 171 Harrison Ave Boston MA 02111

CHASIS, HERBERT, b New York, NY, Nov 9, 05; m 43; c 2. MEDICINE. Educ: Syracuse Univ, AB, 26; NY Univ, MD, 30, ScD(med), 37. Prof Exp: From instr to assoc prof, 35-64, PROF MED, COL MED, NY UNIV, 64-, ATTEND PHYSICIAN, UNIV HOSP & CLIN, 55- Concurrent Pos: Asst vis physician, Bellevue Hosp, 38-43, assoc attend physician, 44-54, attend physician, 57-; chief, Cardiac Clin, French Hosp, 46-64, consult physician, 64-; consult, St Lukes Hosp, Newburg, NY & Vet Admin, 51-; consult physician, Phelps Mem Hosp, Tarrytown, NY. Mem: Am Soc Clin Invest; Am Physiol Soc; Harvey Soc; Soc Exp Biol & Med; fel Am Col Physicians. Res: Cardiovascular and renal physiology; physiological and clinical investigation of renal and hypertensive disease in man. Mailing Add: Dept of Med NY Univ Med Sch New York NY 10016

CHASMAN, CHELLIS, b New York, NY, Feb 11, 32. NUCLEAR PHYSICS. Educ: Harvard Univ, BA, 53; Columbia Univ, PhD(physics), 61. Prof Exp: SR SCIENTIST PHYSICS, BROOKHAVEN NAT LAB, 61- Mem: Am Phys Soc. Mailing Add: Brookhaven Nat Lab Upton NY 11973

CHASON, JACOB LEON, b Monroe, Mich, May 12, 15; m 42; c 3. PATHOLOGY. Educ: Univ Mich, AB, 37, MD, 40. Prof Exp: PROF PATH, SCH MED, WAYNE STATE UNIV, 51-, CHMN DEPT, 67- Concurrent Pos: Chief path, Detroit Gen Hosp, 66-, pres, 70-72; assoc dean, Wayne State Univ Sch Med, 70-72. Mem: Am Soc Clin Path; Col Am Path; Am Acad Neurol; Int Acad Path. Res: Pathology of the nervous system. Mailing Add: Dept of Path Wayne State Univ Sch of Med Detroit MI 48201

CHASSAN, JACOB BERNARD, b New York, NY, Oct 16, 16; m 52; c 3. PSYCHIATRY, STATISTICS. Educ: City Col New York, BS, 39; George Washington Univ, MA, 49, PhD, 58. Prof Exp: Statistician, Air Tech Serv Command, Wright Field, Ohio, 42-48; statistician, Med Statist Div, Off Surgeon Gen, US Dept Army, 46-47, chief health reports br, 47-49; chief div tuberc, USPHS, 49-50; chief clin eval & follow-up studies unit, Dept Med & Surg, Vet Admin, 50-53; sci analyst, Mass Inst Tech Opers Eval Group, 53-55; chief statistician, St Elizabeths Hops, DC, 55-60; math statistician, Off Educ, US Dept Health, Educ & Welfare, 60-61; head statistician, Hoffmann-La Roche, Inc, 61-66; dir statist serv, Sandoz, Inc, NJ, 66-67; assoc dir med res planning, 67-69, psychiat res planner, 69-73, CLIN RES SCIENTIST, HOFFMANN-LA ROCHE INC, 73- Concurrent Pos: Mem fac, USDA Grad Sch; asst clin prof psychiat, George Washington Univ, 61-73, spec lectr psychiat & behav sci, Sch Med, 73-; lectr biostatist, Seton Hall, 64-65; clin assoc prof statist in psychiat, Med Col, Cornell Univ, 71-; fac mem, NY Ctr Psychoanal Training, 72-; pvt pract psychother. Mem: Fel AAAS; fel Am Statist Asn; Am Acad Psychother; Am Asn Marriage & Family Counrs; Math Asn Am. Res: Design of clinical research; applied statistics in epidemiology, psychiatry and psychoanalysis; mathematical statistics. Mailing Add: Hoffmann-La Roche Inc Nutley NJ 07110

CHASSON, ROBERT LEE, b Cincinnati, Ohio, May 30, 19; m 42; c 2. EXPERIMENTAL PHYSICS. Educ: Univ Calif, Berkeley, AB, 40, AM, 50, PhD(physics), 51. Prof Exp: Sr res asst, Elec Resistance Welding Processes, Lockheed Aircraft Corp, 40-42; sr eng aide radar, SigC, Spec Tech Sch, 42-43; asst physics, Univ Calif, 46-50, jr physicist, Off Naval Res & AEC Cosmic-Ray Proj, 50-51; from asst prof to prof, Univ Nebr, 51-62, chmn dept, 56-62; PROF PHYSICS & CHMN DEPT, UNIV DENVER, 62-, DIR PHYSICS RES, DENVER RES INST, 62- Concurrent Pos: Fac res fel, Nebr Res Coun, 54; mem comt cosmic rays, Int Geophys Year & Int Geophys Coop, 55-; fel mem vis scientist prog, Am Inst Physics, 58-; Guggenheim fel, sr vis fel, UK Dept Sci & Indust Res & vis prof, Imp Col, London, 62-63; fac res lectr, Univ Denver, 66-67; mem rep, Univ Corp Atmospheric Res, 66-, trustee, 70-, vchmn trustees, 71-72; mem atmospheric sci adv panel, NSF, 68-70; vis prof, Space Res Lab, Astron Inst, Univ Utrecht, 70; resident vis scientist, Haleakala Observ, Univ Hawaii, 70; mem geophys res bd, Nat Acad Sci-Nat Res Coun, 73- Mem: Fel AAAS; fel Phys Soc; Am Geophys Union. Res: Cosmic rays; fields and particles in space. Mailing Add: Dept of Physics & Astronomy Univ of Denver Denver CO 80210

CHASSON, ROBERT MORTON, b St Louis, Mo, Mar 3, 30; m 54; c 4. PLANT PHYSIOLOGY. Educ: Univ Mo, AB, 52, PhD(plant physiol), 59. Prof Exp: Asst prof biol, Ill State Univ, 59-60; from asst prof to assoc prof bot, Iowa State Univ, 60-65; ASSOC PROF BOT, ILL STATE UNIV, 65- Concurrent Pos: NSF res grant, 62-64; Univ Found res grant, Ill State Univ, 66-71. Mem: Bot Soc Am; Am Soc Plant Physiol. Res: Cytological and biochemical changes associated with aging of plant storage tissues. Mailing Add: Dept of Biol Sci Ill State Univ Normal IL 61761

CHASSY, BRUCE MATTHEW, b Ft Jackson, SC, Oct 22, 42; m 64; c 2. BIOCHEMISTRY, ORGANIC CHEMISTRY. Educ: San Diego State Col, AB, 62; Cornell Univ, NIH fel & PhD(biochem), 66. Prof Exp: Fel biochem, Albert Einstein Med Ctr, 65-67; fel biochem, 68-69, RES CHEMIST, NAT INST DENT RES, 69-

Concurrent Pos: Prof lectr, Am Univ, 69-72. Mem: AAAS; NY Acad Sci; Am Soc Biol Chem; Am Soc Microbiol. Res: Enzyme mechanisms and specificity; plasmids; nucleic acids; nucleotides. Mailing Add: Nat Inst Dent Res NIH Bldg 30 Bethesda MD 20014

CHASTAIN, BENJAMIN BURTON, b Tuscaloosa, Ala, Dec 21, 36; m 62; c 2. INORGANIC CHEMISTRY. Educ: Birmingham-Southern Col, BS, 56; Columbia Univ, MA, 57, PhD(inorg chem), 67. Prof Exp: Assoc chemist, Southern Res Inst, 57-58; pub sch teacher, Ala, 58-59; from instr to assoc prof chem, 59-70, PROF CHEM, SAMFORD UNIV, 70- Concurrent Pos: Mem secretary's adv comt coal mine safety res, US Dept Interior, 71-74. Mem: Am Chem Soc. Res: Synthesis and electronic structures of coordination complexes on transition metals; crystal and molecular structures of complexes of biological interest. Mailing Add: Dept of Chem Samford Univ Birmingham AL 35209

CHASTAIN, MARIAN FAULKNER, b Sept 9, 22; US citizen; m 56; c 2. FOOD CHEMISTRY, NUTRITION. Educ: Cedar Crest Col, BS, 44; Fla State Univ, MS, 53, PhD(food, nutrit), 55. Prof Exp: Jr chemist, Hoffmann-La Roche, Inc, NJ, 44-52; asst prof foods & nutrit, Purdue Univ, 55-56; ASSOC PROF FOODS & NUTRIT, AUBURN UNIV, 56-59, 62- Mem: AAAS; Am Home Econ Asn; Am Dietetic Asn; Inst Food Technol. Res: Improvement of nutritional value of proteins; effects of microwave heating on palatability and nutritive value; oxidative changes in stored foods. Mailing Add: 1104 S Gay St Auburn AL 36830

CHASTEEN, NORMAN DENNIS, b Flint, Mich, Oct 6, 41; m 67. BIOINORGANIC CHEMISTRY. Educ: Univ Mich, AB, 65; Univ Ill, Urbana-Champaign, MS, 66, PhD(chem), 69. Prof Exp: NIH fel, 69-70; asst prof chem, Lawrence Univ, 70-74; ASSOC PROF CHEM, UNIV NH, 74- Mem: AAAS; Am Chem Soc; NY Acad Sci. Res: Magnetic resonance and biological molecules; trace metals in marine environments. Mailing Add: Dept of Chem Univ of NH Durham NH 03824

CHATELAIN, JACK ELLIS, b Ogden, Utah, July 17, 22; m 46; c 2. THEORETICAL PHYSICS. Educ: Utah State Univ, BS, 47, MS, 48; Lehigh Univ, PhD(physics), 57. Prof Exp: Instr physics, Univ Wyo, 50-52; physicist, Dugway Proving Ground, 53; instr physics, Lehigh Univ, 53-57; physicist, Phillips Atomic Energy Div, Phillips Petrol Co, 58; from asst prof to assoc prof physics, 57-71, PROF PHYSICS, UTAH STATE UNIV, 71- Concurrent Pos: Physicist, White Sands Proving Ground, 54-55; sci specialist & consult, Edgerton, Germeshausen & Grier, Inc, Nev, 61-63. Mem: Am Phys Soc. Res: Theoretical aspects of radiation and its interaction with matter. Mailing Add: Dept of Physics Utah State Univ Logan UT 84321

CHATLAND, HAROLD, b Hamilton, Can, Nov 13, 11; nat US; m 37; c 3. MATHEMATICS. Educ: McMaster Univ, BA, 34; Univ Chicago, MS, 35, PhD(math), 37. Prof Exp: Asst prof math, Univ Mont, 39-46 & Ohio State Univ, 46-49; prof, Univ Mont, 49-59, dean col arts & sci, 54, dean fac, 56-57, acad vpres, 57-59; eng specialist, Sylvania Elec Prod, Inc, 59-62; acad dean, Western Wash State Col, 62-64; sr eng specialist, Electronic Defense Labs, 64-68; mem staff, Sylvania Electronics Systs, 68-71; CONSULT, 71- Mem: Soc Indust & Appl Math; Math Asn Am. Res: Number theory; decision-making techniques. Mailing Add: 10566 Blandor Way Los Altos Hills CA 94022

CHATTEN, LESLIE GEORGE, b Calgary, Alta, May 10, 20; m 43; c 3. PHARMACEUTICAL CHEMISTRY. Educ: Univ Alta, BSc, 47, MSc, 49; Ohio State Univ, PhD(pharmaceut chem), 61. Prof Exp: Head pharmaceut chem sect, Food & Drug Directorate, Dept Nat Health & Welfare, Govt Can, 49-61; assoc prof, 61-65, PROF PHARMACEUT CHEM, UNIV ALTA, 65- Concurrent Pos: Mem comt assay tablets & capsules, Brit Pharmacopoeia, 53-63 & comt org synthetic substances, 58-63; vis scientist, Med Res Coun Can, 70-71. Mem: Am Chem Soc; Can Pharmaceut Asn; Acad Pharmaceut Sci; Chem Inst Can; Sigma Xi. Res: Qualitative and quantitative pharmaceutical chemistry; application of nonaqueous titrimetry to analysis of drugs and pharmaceuticals; identification of organic medicinal agents; polarography; absorption spectrophotometry; fluorimetry; high performance liquid chromatography. Mailing Add: Fac of Pharm & Pharmaceut Sci Univ of Alta Edmonton AB Can

CHATTERJEE, PRONOY KUMAR, b Varanasi, India, Oct 26, 36; m 62; c 1. POLYMER CHEMISTRY, PHYSICAL CHEMISTRY. Educ: Banaras Hindu Univ, BS, 56, MS, 58; Calcutta Univ, PhD(chem), 63, DSc(chem), 74. Prof Exp: Sr res asst polymers, Indian Asn Cultivation Sci, Calcutta, 59-63; Nat Acad Sci-Nat Res Coun res assoc chem, Southern Regional Res Labs, USDA, La, 63-65; res assoc polymers, Princeton Univ, 65-66; SR RES CHEMIST, PERSONAL PROD CO DIV, JOHNSON & JOHNSON, 66- Honors & Awards: P B Hofman Res Scientist Award, Johnson & Johnson, 73; Educ Serv Award, Plastic Inst Am, 74. Mem: AAAS; Am Chem Soc; fel Am Inst Chemists; Tech Asn Pulp & Paper Indust; Int Confederation Thermal Anal. Res: Rubber vulcanization mechanism; analysis of rubber chemicals; heterogeneous reaction kinetics; thermal analysis; reaction mechanism of polysulfides; polymerization; inter-fiber bonding mechanism of cellulose; characterization and preparation of chemically modified wood pulp. Mailing Add: Res & Eng Personal Prod Co Div Johnson & Johnson Milltown NJ 08850

CHATTERJEE, RAMANANDA, b India, Mar 1, 36; m 65; c 2. SOLID STATE PHYSICS. Educ: Calcutta Univ, BS, 54, MS, 56, PhD(physics), 63. Prof Exp: Fel, Inst Theoret Physics, Univ Alta, 63-65; from asst prof to assoc prof physics, 65-74, PROF PHYSICS, UNIV CALGARY, 74- Res: Theoretical aspects of electron paramagnetic resonance and electron-nuclear double resonance spectrum. Mailing Add: Dept of Physics Univ of Calgary Calgary AB Can

CHATTERJEE, SAMPRIT, b Calcutta, India, June 3, 38; m. STATISTICS, OPERATIONS RESEARCH. Educ: Univ Calcutta, BS & MS, 60; Univ Cambridge, DStat, 62; Harvard Univ, PhD(statist), 66. Prof Exp: Instr statist, Boston Univ, 62-63; res asst, Harvard, 63-65; asst prof statist & opers res, 66-69, assoc prof statist, 69-74, PROF STATIST, NY UNIV, 74- Concurrent Pos: Consult, Mass Ment Health Ctr, 63-65; res fel, Univ Col, London, 73-74; res scholar, Int Inst Appl Systs Anal, Vienna, 74. Mem: Am Statist Asn; Biomet Soc; Royal Statist Soc; Opers Res Soc Am; AAAS. Res: Environmental problems; ecology; linear models; sample survey; systems analysis; public policy. Mailing Add: Dept of Quantitative Anal NY Univ New York NY 10006

CHATTERS, ROY (MILTON), biology, see 12th edition

CHATTERTON, BRIAN DOUGLAS EYRE, b Khartoum, Sudan, July 31, 43; Irish citizen; m 70; c 2. PALEONTOLOGY. Educ: Trinity Col, Dublin, BA, 65; Australian Nat Univ, PhD(paleont), 70. Prof Exp: Sr demonstr paleont, Australian Nat Univ, 69-70; ASST PROF PALEONT, UNIV ALTA, 70- Mem: Geol Asn Can; Paleont Soc; Palaeont Asn; Int Palaeont Asn. Res: Systematics, paleoecology and biostratigraphy of conodont faunas from Western and Arctic Canada; ontogenetic, paleoecologic and systematic studies of trilobites from Western and Arctic Canada; Devonian paleontology. Mailing Add: Dept of Geol Univ of Alta Edmonton AB Can

CHATTERTON, NEIL ELLIS, operations research, applied physics, see 12th edition

CHATTERTON, ROBERT TREAT, JR, b Catskill, NY, Aug 9, 35; m 56; c 4. ENDOCRINOLOGY, BIOCHEMISTRY. Educ: Cornell Univ, BS, 58, PhD(physiol biochem), 63; Univ Conn, MS, 60. Prof Exp: Res fel biol chem, Harvard Univ, 63-65; res assoc, Div Neoplastic Med, Montefiore Hosp & Med Ctr, 65-70; asst prof obstet & gynec & physiol, 70-72, ASSOC PROF OBSTET & GYNEC, PHYSIOL & BIOL CHEM, UNIV ILL MED CTR, 72- Mem: AAAS; Endocrine Soc; Soc Study Reproduction; Soc Gynec Invest; Am Physiol Soc. Res: Physiology, biochemistry and histology of mammary gland development and ovarian function; steroid biosynthesis; autoradiography. Mailing Add: Dept of Obstet & Gynec Univ of Ill at the Med Ctr Chicago IL 60680

CHATTHA, MOHINDER SINGH, b Gurdaspur, India, Oct 15, 40; m 67; c 2. POLYMER CHEMISTRY. Educ: Sikh Nat Col, India, BSc, 60; Banaras Hindu Univ, MSc, 63; Tulane Univ, PhD(chem), 71. Prof Exp: Teacher sci, Khalsa High Sch, Wadala, India, 60-61; lectr chem, Sikh Nat Col, 63-65; res fel, Coun Sci & Indust Res, New Delhi, 65-67; fel, Tulane Univ, 71-72; Nat Res Coun fel polymer chem, Wright-Patterson AFB, Ohio, 72-73; SR RES SCIENTIST POLYMER CHEM, RES & DEVELOP, FORD MOTOR CO, 73- Mem: Am Chem Soc. Res: Organophosphorus chemistry; synthesis and chemistry of new polymers of potential use in coatings, adhesives and elastomers. Mailing Add: Polymer Sci Res PO Box 2053 Ford Motor Co Dearborn MI 48121

CHATTORAJ, SATI CHARAN, b WBengal, India, Aug 1, 34; m 61; c 1. REPRODUCTIVE ENDOCRINOLOGY. Educ: Univ Calcutta, BS, 54, MS, 56; Boston Univ, PhD(biochem), 65. Prof Exp: Asst prof obstet & gynec, Chicago Med Sch, 65-69; asst prof, 69-71, ASSOC PROF BIOCHEM OBSTET & GYNEC & DIR RES LABS, SCH MED, BOSTON UNIV, 71- Concurrent Pos: Nat Inst Child Health & Human Develop grant, Sch Med, Boston Univ, 74-77; asst dir gynecic endocrinol, Michael Reese Hosp & Med Ctr, 65-69. Honors & Awards: Morris L Parker Award, Chicago Med Sch, 67. Mem: AAAS; Am Inst Biol Sci; Endocrine Soc; Soc Study Reproduction; Am Chem Soc. Res: Reproductive endocrinology; control of steroidogenesis during pregnancy; contraception; breast and endometrial cancer. Mailing Add: Dept of Obstet & Gynec Boston Univ Sch of Med Boston MA 02118

CHATTORAJ, SHIB CHARAN, b Kendua, West-Bengal, India, Sept 17, 24; nat US; m 64; c 2. INORGANIC CHEMISTRY, ANALYTICAL CHEMISTRY. Educ: Univ Calcutta, BSc, 43; Brigham Young Univ, MS, 57; Univ Cincinnati, PhD(inorg & anal chem), 64. Prof Exp: Inspector chem, Chief Inspectorate Mil Explosives, Govt India, 45-46; demonstr, Asutosh Col, Calcutta, 46-55; Ohio State Res Found vis res assoc, Aerospace Res Lab, Chem Res Lab, Wright-Patterson AFB, 64-66, Univ Cincinnati res assoc, Ceramics & Graphite Br, Mat Lab, 66-67, mat engr, 67-70, mat res engr, Advan Metall Studies Br, Metals & Ceramics Div, 70-2, res scientist, Aerospace Res Labs, 72, res chemist, 72-73. Concurrent Pos: Am Chem Soc-Asia Found grants, 62. Honors & Awards: Invention Award, US Air Force, 69, patent award, 72. Mem: AAAS; Am Chem Soc; Sigma Xi; Am Ceramic Soc; fel Am Inst Chemists. Res: Vapor pressure composition study by vacuum line techniques; Grignard reagents; gas chromatography; coordination compounds; metal beta-ketoenolates; metal alkoxides; boron suboxides; stress corrosion; trace metals analysis. Mailing Add: 1350 Redbud Dr Fairborn OH 45324

CHATURVEDI, RAM PRAKASH, b India, Dec 15, 31; m 64; c 1. ATOMIC PHYSICS. Educ: Agra Univ, BSc, 53, MSc, 55; Univ BC, PhD(nuclear spectros), 63. Prof Exp: Lectr physics, Agra Univ, 55-59; fel & lectr, Panjab Univ, India, 63-64; fel, State Univ NY Buffalo, 64-65; assoc prof, 65-70, PROF PHYSICS, STATE UNIV NY COL CORTLAND, 70- Concurrent Pos: Mem users group, Oak Ridge Nat Lab, 70. Mem: Am Phys Soc; Am Asn Physics Teachers. Res: Nuclear level schemes. Mailing Add: Dept of Physics State Univ of NY Col Cortland NY 13045

CHATURVEDI, RAMA KANT, b Kanker, India, July 7, 33; m 58; c 3. BIO-ORGANIC CHEMISTRY. Educ: Agra Univ, India, BSc, 54, MSc, 56, PhD(chem), 60. Prof Exp: Lectr chem, H B Technol Inst, India, 60-64 & St Stephen's Col, Univ Delhi, 64-65; res assoc, Ind Univ, 65-66; res assoc biochem, Yale Univ, 66-70; ASSOC PROF CHEM, GTR HARTFORD COMMUNITY COL, 70- Concurrent Pos: Fulbright Award, US Educ Found in India, 65; asst instr, Yale Univ, 70-74; lectr, Cent Conn State Col, 73 & 75- Mem: AAAS; Am Chem Soc; Nutrit Today Soc. Res: Mechanism of acyl transfer reactions with special attention to the formation and partitioning of tetrahedral intermediates involved in these reactions. Mailing Add: 39 Beacon St Newington CT 06111

CHAU, ALFRED SHUN-YUEN, b Hong Kong, Nov 20, 41; Can citizen; m 66; c 1. ANALYTICAL CHEMISTRY. Educ: Univ BC, BSc, 61; Carleton Univ, MSc, 66. Prof Exp: Pesticide analyst, Dept Agr, Ottawa, 65-70; chemist, 70-73, HEAD SPEC SERV SECT, DEPT ENVIRON, CAN, 73- Concurrent Pos: Sci reviewer, J Asn Off Anal Chemists, 70-; sci authority, Off Res Subventions, Inland Waters Dir, Environ Can, 74- Mem: Am Chem Soc; Asn Off Anal Chemists. Res: Development of methodology for the analysis and positive confirmation of organic pollutants, particularly biocides, by chemical derivatization-gas chromatographic techniques. Mailing Add: Water Qual Br Can Ctr Inland Water PO Box 5050 Burlington ON Can

CHAU, CHEUK-KIN, b Hong Kong, Sept 25, 41; m 68. SOLID STATE PHYSICS, CRYOGENICS. Educ: MacMurray Col, BA, 62; Univ Ill, Urbana, MS, 63, PhD(physics), 68. Prof Exp: Vis asst prof, Ill Inst Technol, 68-69, asst prof physics, 69-75; VIS ASSOC PROF PHYSICS, CALIF STATE UNIV, CHICO, 75- Mem: Am Phys Soc; Am Asn Physics Teachers; Sigma Xi. Res: Apply cryogenic technology to investigate heat transport, electrical transport, optics and radiation damages in condensed materials such as insulators, semiconductors, semi-metal and superconductors. Mailing Add: Dept of Physics Calif State Univ Chico CA 95926

CHAU, MICHAEL MING-KEE, b Hong Kong, Nov 14, 47; m 75. PHYSICAL ORGANIC CHEMISTRY. Educ: Calif State Univ Fresno, BS, 71; Univ Ill, Urbana, PhD(chem), 75. Prof Exp: RES ASSOC ORG CHEM, TEX TECH UNIV, 75- Mem: Am Chem Soc. Res: Mechanistic studies of organosulfur compounds; anchimeric acceleration of bond homolyses. Mailing Add: Dept of Chem Tex Tech Univ Lubbock TX 79409

CHAU, RAYMOND YING PUI, b Hong Kong, Aug 15, 38; c 1. TOXICOLOGY, PHARMACOLOGY. Educ: Univ Ore, BS, 63; Univ Kans, MS, 65; New York Med Col, PhD(pharmacol), 69. Prof Exp: ASST PROF PHARMACOL, NEW YORK MED COL, 71-; SR RES TOXICOLOGIST, LEDERLE LABS, AM CYANAMID CO, 74- Concurrent Pos: Pres, Int Found for Educ, Health & Res, Inc, 72-; vpres, New York Inst Biblical Studies, 73- Mem: AAAS; Soc Comp Ophthal; Int Study Group Res Cardiac Metab. Res: Exploration of the pathogenesis of myocardial infarction, necrosis and healing; beneficial influence of various drugs upon the development and course of cardiac injury. Mailing Add: 58-44 75th St Elmhurst NY 11373

CHAU, YIU-KEE, b Canton, China, Dec 6, 27; m 56; c 2. CHEMICAL OCEANOGRAPHY, LIMNOLOGY. Educ: Lingnan Univ, BS, 49; Univ Hong Kong, MS, 61; Univ Liverpool, PhD(chem oceanog), 65. Prof Exp: Sci master, Mid Sch, Macau, 50-54; res officer chem oceanog, Fisheries Res Unit, Univ Hong Kong, 54-59; assoc prof chem, Chung Chi Col, Hong Kong, 59-68; RES SCIENTIST, CAN CTR INLAND WATERS, 68- Concurrent Pos: UNESCO fel, Commonwealth Sci & Indust Res Orgn, Australia, 57; Brit Coun fel, Univ Liverpool, 63-65. Mem: AAAS; assoc Royal Australian Chem Inst; Am Chem Soc; Am Soc Limnol & Oceanog. Res: Chemical and biological processes of trace metals in the aquatic environment; interaction of metals and organics in natural water; atomic absorption spectroscopy. Mailing Add: Can Ctr for Inland Waters Burlington ON Can

CHAUBAL, MADHUKAR GAJANAN, b Nasik City, India, May 15, 30; m 66. PHARMACOGNOSY, CHEMISTRY. Educ: Univ Poona, BSc, 51; Univ Bombay, BSc, 54; Univ Toronto, MSc, 60; Univ RI, PhD(pharmaceut sci), 64. Prof Exp: Demonstr pharm, Dept Chem Technol, Univ Bombay, 56-58; demonstr pharm, Univ Toronto, 58-60; res asst, Univ RI, 60-64; from asst prof to assoc prof, 64-74, PROF PHARMACOG, UNIV OF THE PAC, 74- Mem: AAAS; Am Pharmaceut Asn; Am Soc Pharmacog; NY Acad Sci; fel Am Inst Chem. Res: Phytochemistry of medicinal plants; biosynthesis of plant constituents; essential oils from plants; constituents of the Saururaceae and pharmacology of the constituents. Mailing Add: Sch of Pharm Univ of the Pac 751 Brookside Rd Stockton CA 95207

CHAUBE, SHAKUNTALA, experimental embryology, see 12th edition

CHAUDHARI, BIPIN BHUDHARLAL, b Taloda, India, Aug 31, 35; m 65. MEDICINAL CHEMISTRY. Educ: Univ Bombay, BS, 57 & 59; Univ Iowa, MS, 62, PhD(med chem), 65. Prof Exp: Anal chemist, Oriental Chem Industs, Bombay, India, 59-60; res assoc chem, Ill Inst Technol, 65; res chemist, Bauer & Black-Polyken Res Ctr, Kendall Co, 65-69; SR RES CHEMIST, MED CHEM SECT, BIOMED RES LABS, ICI US INC, 69- Mem: Am Chem Soc. Mailing Add: ICI US Inc Biomed Res Labs Concord Pike & New Murphy Wilmington DE 19897

CHAUDHARY, RABINDRA KUMAR, b India; Can citizen. MEDICAL MICROBIOLOGY. Educ: Bihar Vet Col, BVSc, 58; Guelph Univ, MSc, 65; Univ Ottawa, PhD(microbiol), 71. Prof Exp: Res scientist microbiol, Bell & Craig Pharmaceut Co, 66-67; res assoc, Univ Ottawa, 71-75; BIOLOGIST, LAB CTR DIS CONTROL, BUR VIRAL DIS, 75- Mem: Am Soc Microbiologists. Res: Biophysical studies of hepatitis B antigen. Mailing Add: Lab Ctr Dis Control Bur Viral Dis Tunney's Pasture Ottawa ON Can

CHAUDHARY, SOHAN SINGH, b Pakistan, Oct 15, 30; m 61; c 4. PHYSICAL ORGANIC CHEMISTRY. Educ: Univ Punjab, India, BSc, 52, Hons, 54, MSc, 56; Univ Calif, Davis, PhD(phys org chem), 65. Prof Exp: Chemist, Indian Coun Med Res, 56; res asst, Coun Sci Indust Res, India, 56-60, sr res asst, 60-61; res asst aroma of Calif wines, Univ Calif, Davis, 61-63, teaching asst chem, 63-65; NIH fel syntheses & reaction mech, Case Inst Technol, 65-66; res chemist, Res Dept, 66-70, PROJ SCIENTIST, ARCO/POLYMER, INC, 70- Mem: Am Chem Soc; Am Soc Enol; NY Acad Sci; Sigma Xi. Res: Isolation and structural elucidation of active principals of herbs; organic syntheses and reaction mechanism; thermoplastics. Mailing Add: ARCO/Polymer Inc Res Dept 440 College Park Dr Monroeville PA 15146

CHAUDHRY, ANAND P, b WPunjab, India, Oct 19, 22; US citizen; m 57; c 2. ORAL PATHOLOGY. Educ: Panjab Univ, India, BS, 42, BDS, 47; Univ Mich, MS, 53; Univ Minn, PhD(path), 56; Am Bd Oral Path, dipl. Prof Exp: From instr to prof, Univ Minn, 56-61; prof path & chmn dept, Sch Dent, Univ Pittsburgh, 61-66; PROF PATH, STATE UNIV NY BUFFALO, 67- Concurrent Pos: Consult, Children's Hosp, Millard Fillmore Hosp, Buffalo Gen Hosp & Vet Admin Hosp. Mem: Am Acad Oral Path; Am Asn Dent Res. Res: Experimental carcinogenesis of the oral cavity, skin and the salivary glands; experimental teratology with special interest in the study of embryogenesis and pathogenesis of cleft lip and palate. Mailing Add: Dept of Path State Univ of NY Sch of Med Buffalo NY 14207

CHAUDHURI, TUHIN, b Bengal, India, Jan 3, 42; m 69. NUCLEAR MEDICINE, HEMATOLOGY. Educ: Univ Calcutta, MB & BS, 64. Prof Exp: Intern med, Med Col & Hosp, Univ Calcutta, 64-65; resident surg, RKM Seva Pratishthan, Calcutta, 65-66; res assoc radiol & nuclear med, Yale Univ, 66-67, ASST PROF NUCLEAR MED & RADIOL & DIR, NUCLEAR MED, UNIV IOWA, 69- Concurrent Pos: Fel nuclear med, Yale Univ, 67-68; James Picker fel, Nat Acad Sci-Nat Res Coun, 67-70; fel, Donner Lab, Univ Calif, Berkeley, 68-69; guest scientist, Lawrence Radiation Lab, Berkeley, 68-69. Mem: AAAS; Am Heart Asn; Am Fedn Clin Res; Biophys Soc; Soc Nuclear Med. Res: Early detection of cancer and its therapy; C-14 breath analysis study; evaluation of splenic and hepatic function by multiple radio labeled erythrocytes; hemoglobin and nonhemoglobin heme kinetic study; blood flow studies in different organs. Mailing Add: 3111 Juniper Dr Iowa City IA 52240

CHAUFFE, LEROY, b Freetown, La, Dec 26, 36; m 69. ORGANIC CHEMISTRY. Educ: Xavier Univ, La, BS, 59; Howard Univ, MS, 64; Univ Calif, Davis, PhD(chem), 66. Prof Exp: Chemist, Shell Chem Co, 65-66; fel, Univ Calif, Davis, 66-67; asst prof chem, State Col Long Beach, 67-68; asst prof, 68-74, assoc dean instr, 70-71, actg dean grad studies, 71-72, ASSOC PROF CHEM, CALIF STATE UNIV, HAYWARD, 74- Mem: Am Chem Soc; The Chem Soc. Res: Physical organic chemistry; biochemical kinetics and mechanism. Mailing Add: Dept of Chem Calif State Univ Hayward CA 94542

CHAUNCEY, HOWARD H, biological chemistry, see 12th edition

CHAUVIN, ROBERT S, b West Beekmantown, NY, Nov 20, 20; m 46; c 2. GEOLOGY, GEOGRAPHY. Educ: NY Univ, BS, 43; Columbia Univ, MA, 50, PhD(geol, geog), 55. Prof Exp: Head dept geol & geog, 50-68, dean sci, 68-70, assoc prof geol, 50-55, PROF GEOL, STETSON UNIV, 55-, DEAN COL LIB ARTS, 70- Mem: Am Geog Soc. Res: Arctic flora, fauna and glaciation. Mailing Add: Col of Lib Arts Stetson Univ De Land FL 32721

CHAVE, KEITH ERNEST, b Chicago, Ill, Jan 18, 28; m 51; c 2. GEOLOGY. Educ: Univ Chicago, PhB, 48, MS, 51, PhD, 52. Prof Exp: Asst, State Geol Surv, Ill, 48; res geologist, Calif Res Corp, 52-59; from asst prof to prof geol, Lehigh Univ, 59-67, assoc dir marine sci ctr, 62-67; PROF OCEANOG, UNIV HAWAII, 67-, CHMN DEPT, 70- Concurrent Pos: Trustee, Bermuda Biol Sta, 62-68; Alexander von Humboldt sr scientist, Univ Kiel, 73-74. Mem: AAAS; Soc Econ Paleont & Mineral; Geochem Soc; Am Geophys Union; Am Soc Limnol & Oceanog. Res: Geochemistry; marine geology. Mailing Add: Dept of Oceanog Univ of Hawaii Honolulu HI 96822

CHAVEL, ISAAC, b Louisville, Ky, Apr 2, 39. GEOMETRY. Educ: Brooklyn Col, BA, 61; NY Univ, MS, 64; Yeshiva Univ, PhD(math), 66. Prof Exp: Teaching asst math, Brooklyn Col, 61-64; asst prof, Univ Minn, 66-70; asst prof, 70-73, ASSOC PROF MATH, CITY COL NEW YORK, 73- Mem: Am Math Soc. Res: Interplay of Riemannian geometry with mathematical analysis, especially as relates to the Laplace operator. Mailing Add: Dept of Math City Col of New York New York NY 10463

CHAVEZ, HENRY BERNARD, plant pathology, see 12th edition

CHAVIN, WALTER, b US, Dec 6, 25 ENDOCRINOLOGY, RADIOBIOLOGY. Educ: City Col New York, BS, 46; NY Univ, MS, 49, PhD(zool), 54. Prof Exp: Instr biol, City Col New York, 46-47; asst, NY Aquarium, 48-49; instr zool, Univ Ariz, 49-51; res specialist, Am Mus Natural Hist, 51-53; assoc prof endocrinol & radiobiol, 53-68, prof biol, 68-74, PROF BIOL & RADIOL, WAYNE STATE UNIV, 74- Concurrent Pos: Res assoc, Div Biol & Med Res, Argonne Nat Lab, 55, 56, consult, 57-58; Sigma Xi fac res award, 68; coordr, US-Japan Seminar Responses of Fish to Environ Changes, Tokyo, 70; consult, Great Lakes Lab, Bur Com Fisheries, US Dept Interior. Mem: Fel AAAS; Am Soc Zool; Endocrine Soc; Radiation Res Soc; Soc Exp Biol & Med. Res: Comparative endocrinology; pigment cell physiology; environmental biology; radiation biology. Mailing Add: Depts of Biol & Radiol Wayne State Univ Detroit MI 48202

CHAVKIN, LEONARD THEODORE, b New York, NY, Dec 2, 25; m 44; c 4. PHARMACY. Educ: Columbia Univ, BS, 44; Philadelphia Col Pharm, MS, 47; NY Univ, PhD, 60. Prof Exp: Assoc pharm, Col Pharm, Columbia Univ, 47-49, assoc prof, 49-59; dir develop, 59-63, dir res develop, 63, asst vpres res develop, 63-67, VPRES RES & DEVELOP, BRISTOL-MYERS CO, 67- Mem: AAAS; Am Pharmaceut Asn; Soc Cosmetic Chem. Res: Industrial pharmaceutical manufacturing. Mailing Add: Res & Develop Dept Bristol-Myers Co 1350 Liberty Ave Hillside NJ 07207

CHAWNER, WILLIAM DONALD, b Calif, May 13, 03; m; c 2. PETROLEUM GEOLOGY. Educ: Occidental Col, BS, 25; Calif Inst Technol, MS, 34; La State Univ, PhD(geol), 37. Prof Exp: Geologist, Atlantic Refining Co, 27-33, State Geol Surv, La, 34-36, Australasian Petrol Co, 36-41, Carter Oil Co, 41-58 & Humble Oil & Refining Co, 58-68; geol consult, 68-75; RETIRED. Mem: Am Asn Petrol Geol; fel Geol Soc Am; Am Inst Mining, Metall & Petrol Eng; Am Inst Prof Geol. Res: Petroleum geology; stratigraphy; mineral exploration; uranium deposits. Mailing Add: PO Box 608 Del Mar CA 92014

CHAYES, FELIX, b New York, NY, May 10, 16; m 41. PETROLOGY. Educ: NY Univ, BA, 36; Columbia Univ, MA, 39, PhD(geol), 42 Prof Exp: Chemist, Gillis & Pawel Metal Co, NC, 41; bus analyst, War Prod Bd, Washington, DC, 41-42; chemist-petrographer, US Bur Mines, 42-46; mineralogist, Mass Inst Technol, 46-47; PETROLOGIST, GEOPHYS LAB, CARNEGIE INST, 47- Mem: Fel Mineral Soc Am; fel Geol Soc Am; Am Geophys Union. Res: Petrology of alkali intrusion of Bancroft, Ontario; quantitative analysis by fragment counting; precision of linear analysis; application of statistics to chemical petrography; petrography of granite; geochemical data. Mailing Add: Carnegie Inst Geophys Lab 2801 Upton St NW Washington DC 20008

CHAYKIN, STERLING, b New York, NY, Sept 18, 29; m 54; c 4. BIOCHEMISTRY. Educ: NY Univ, AB, 50; Univ Wash, PhD(biochem), 54. Prof Exp: Runyon fel, Harvard, 56-59; from asst prof to assoc prof biochem, 59-69, chmn dept biochem & biophys, 68-70, assoc dean resident instr, Col Agr & Environ Sci, 70-74, PROF BIOCHEM, UNIV CALIF, DAVIS, 74- Concurrent Pos: Fulbright fel, Ger, 66-67; Guggenheim fel, 66-67. Mem: AAAS; Am Chem Soc; Am Soc Biol Chemists. Res: Enzymology; biochemistry of development, vitamins and mammalian genetics. Mailing Add: Dept of Biochem & Biophys Univ of Calif Davis CA 95616

CHAYKOVSKY, MICHAEL, b Mayfield, Pa, Sept 19, 34; m 66; c 2. ORGANIC CHEMISTRY, BIOCHEMISTRY. Educ: Pa State Univ, BS, 56; Univ Mich, MS, 59, PhD(org chem), 61. Prof Exp: Res fel chem, Harvard Univ, 61-64; asst prof, State Univ NY Buffalo, 64-65; res fel, Labs of Chem, Nat Inst Arthritis & Metab Dis, 65-66; sr res chemist, Hoffmann-La Roche Inc, 66-69; RES ASSOC CHEM, HARVARD MED SCH, 69- Mem: Am Chem Soc. Res: Synthetic organic chemistry; steroids, terpenes, sulfur ylides and carbanions; photochemistry; organophosphorous chemistry; heterocycles; carbenes; nitrenes; diborane reductions; cancer chemotherapy. Mailing Add: Labs of Bio-org Chem Farber Cancer Ctr Harvard Med Sch 35 Binney St Boston MA 02132

CHAZIN, ROBERT L, topology, computer science, see 12th edition

CHEADLE, VERNON IRVIN, b Salem, SDak, Feb 6, 10; m 39; c 1. BOTANY. Educ: Miami Univ, AB, 32; Harvard Univ, AM, 34, PhD(biol bot), 36. Hon Degrees: LLD, Miami Univ & Univ RI, 64. Prof Exp: Austin teaching fel bot, Harvard Univ, 33-36; from instr to asst prof, RI State Col, 36-42, prof & head dept, 42-52, dir grad div, 43-52; botanist, Exp Sta & prof bot, Univ Calif, Davis, 52-62, chmn dept, 52-60; actg vchancellor, 61-62, CHANCELLOR, UNIV CALIF, SANTA BARBARA, 62- Concurrent Pos: Fulbright fel, 59. Mem: Fel AAAS; fel Am Acad Arts & Sci; Am Soc Plant Taxon; Bot Soc Am (pres, 61); Torrey Bot Club. Res: Anatomy; morphology of vascular plants. Mailing Add: Off of the Chancellor Univ of Calif Santa Barbara CA 93106

CHEAL, MARYLOU, b St Clair Co, Mich. PSYCHOBIOLOGY. Educ: Oakland Univ, BA, 69; Univ Mich, Ann Arbor, PhD(psychobiol), 73. Prof Exp: Res investr taste regeneration, Dept Zool, 73-75, RES INVESTR TASTE DEVELOP, DEPT ORAL BIOL, SCH DENT, UNIV MICH, ANN ARBOR, 75-, LECTR PSYCHOL, 73- Mem: AAAS; Sigma Xi; Sci Neurosci; Europ Chemoreception Orgn. Res: Motivational aspects of social and feeding behavior as influenced by chemical stimuli; regeneration of taste function; development of gustatory and olfactory systems; preference behavior. Mailing Add: Dept of Oral Biol Sch of Dent Univ of Mich Ann Arbor MI 48109

CHEATHAM, WILLIAM J, b Jackson, Tenn, June 18, 25; m 48; c 2. PATHOLOGY. Educ: Vanderbilt Univ, BA, 47, MD, 50. Prof Exp: From intern to resident path, Vanderbilt Univ Hosp, 50-53; Nat Found Infantile Paralysis fel, Children's Med Ctr, Mass, 53-54; asst pathologist, Children's Med Ctr, Mass, 54-55; from asst prof to assoc prof, 57-70, PROF PATH, SCH MED, VANDERBILT UNIV, 70- Mem: AMA; Int Acad Path; Am Soc Exp Path; Am Asn Path & Bact; NY Acad Sci. Res: Virology. Mailing Add: Dept of Path Vanderbilt Univ Sch Med Nashville TN 37232

CHEATUM, ELMER (PHILIP), ecology, see 12th edition

CHEATUM, EVELYN LEONARD, b Lerado, Kans, Oct 31, 09; m 39; c 2. WILDLIFE PATHOLOGY. Educ: Univ Mich, AB, 34, MS, 35, PhD(zool), 48. Prof Exp: Assoc prof biol, John Tarleton Col, 37-38; wildlife pathologist, Bur Game, State Conserv Dept, NY, 39-43, sr game pathologist, Bur Fish & Wildlife Invests, 46-50, chief bur game, 52-59, asst dir, Div Fish & Game, 59-64, asst comnr fish & game, 64-68; DIR INST NATURAL RESOURCES, UNIV GA, 68-, PROF FOREST RESOURCES, 69- Concurrent Pos: Leader, Mont Coop Wildlife Res Unit; assoc prof, Univ Mont, 50-52; consult mem, Atomic Safety Licensing Bd Panel, Nuclear Regulatory Comn, 72-; mem marine resources adv comt to Coastal Plains Regional Comn, US Dept

Com, 72- Honors & Awards: Nash Conserv Award, 54. Mem: AAAS; Wildlife Soc (treas, 52-53, vpres, 53-54, pres, 60-62). Res: Zoopathology; parasitology; physiology of reproduction; medical parasitology; wildlife management; administration of natural resources. Mailing Add: 100 Torrey Pine Pl Athens GA 30601

CHEAVENS, THOMAS HENRY, b Dallas, Tex, May 19, 30; m 55; c 3. INDUSTRIAL CHEMISTRY, PETROLEUM CHEMISTRY. Educ: Univ Tex, BS, 50, PhD(chem), 55. Prof Exp: Res chemist, Org Pigments, Res Div, Am Cyanamid Co, 55-56, develop chemist, Org Chem Div, 56, develop chemist, Intermediates & Rubber Chem, 57, group leader, process develop dept, NJ, 58-61, group leader, Refinery Catalysts Group, Conn, 61-67; supvr indust catalysts res, W R Grace & Co, 67-70, dir indust catalysts res dept, 71-75; SR PROJ CHEMIST, DIV FOSSIL ENERGY RES, US ENERGY RES & DEVELOP ADMIN, WASHINGTON, DC, 75- Mem: Am Chem Soc. Res: Catalysis in fossil fuel production and refining, industrial chemical processes. Mailing Add: 14114 Burntwoods Rd Glenwood MD 21738

CHEEKE, PETER ROBERT, b Duncan, BC, Oct 19, 41; m 70. ANIMAL SCIENCE, NUTRITION. Educ: Univ BC, BSA, 63, MSA, 65; Ohio State Univ, PhD(animal nutrit), 69. Prof Exp: Asst prof, 69-74, ASSOC PROF ANIMAL SCI, ORE STATE UNIV, 74- Res: Nutritional roles of vitamin E and selenium; amino acid requirements of swine; nutrition of rabbits. Mailing Add: Dept of Animal Sci Ore State Univ Corvallis OR 97330

CHEEMA, MOHINDAR SINGH, b Sialkot, Panjab, India, Jan 15, 29; m 51; c 4. MATHEMATICS. Educ: Univ Panjab, BA, 48, MA, 50; Univ Calif, Los Angeles, MA, 60, PhD(math), 61. Prof Exp: Res scholar math, Univ Panjab, 51-54, instr, 54-58; jr res mathematician, Univ Calif, Los Angeles, 58-61; from asst prof to assoc prof, 61-68, PROF MATH, UNIV ARIZ, 68- Mem: Am Math Soc; Math Asn Am; Asn Comput Mach. Res: Number theory; pure mathematics; numerical analysis. Mailing Add: Dept of Math Univ of Ariz Tucson AZ 85721

CHEEMA, ZAFRULLAH K, b Gakkhar, WPakistan, Apr 21, 34; US citizen; m 62; c 2. ORGANIC CHEMISTRY, PHARMACEUTICAL CHEMISTRY. Educ: Panjab Univ, WPakistan, BPharm, 54; Univ Tübingen, Dr rer nat(chem), 57. Prof Exp: Res assoc, Univ Kans, 57-58 & Univ Tübingen, 58-59; chmn dept chem, Knoxville Col, 59-63, assoc prof chem, 61-63; adv chemist, Health Dept, Govt Pakistan, 60-61; lectr chem, Univ Del, 63-64; res chemist, Gen Chem Div, Allied Chem Corp, 64-65, sr res chemist agr div, 66, res assoc, Plastics Div, 66-69, tech supvr res, 69-71; group leader, Am Hoechst Corp, 71-75, MGR RES, KEUFFEL & ESSER CO, 75- Concurrent Pos: Consult, Oak Ridge Nat Lab, 62-63; res fel, Univ Tenn, 63; adj prof & chmn chem dept, Fairleigh Dickinson Univ, 66-67. Mem: Am Chem Soc; NY Acad Sci. Res: Terpenes; conformational analysis; reaction mechanisms such as epimerization and pinnacol rearrangement; synthesis of pesticides; oxidation; photopolymers, diazo, graphic arts and engineering product development. Mailing Add: 20 Whippany Rd Morristown NJ 07960

CHEER, CLAIR JAMES, b Lakewood, Ohio, May 16, 37; m 62; c 2. ORGANIC CHEMISTRY. Educ: Kenyon Col, BA, 59; Wayne State Univ, PhD(org chem), 64. Prof Exp: Res asst chem, Parma Res Lab, Union Carbide Corp, 59; res assoc org chem, Frank J Seiler Res Lab, off aerospace res, US Air Force Acad, 64-67; res assoc chem, Univ Ariz, 67-68; asst prof, 68-74, ASSOC PROF CHEM, UNIV RI, 74- Mem: Am Chem Soc; The Chem Soc; Am Inst Physics; Am Crystallog Asn; NY Acad Sci. Res: Organic reactions and mechanisms; organic synthetic methods; natural products; use of x-ray crystallography in solution of organic chemical problems, especially geometrical relationships in trigonal-pyramidal centers. Mailing Add: Dept of Chem Univ of RI Kingston RI 02881

CHEETHAM, ALAN HERBERT, invertebrate paleontology, see 12th edition

CHEEVER, CHARLES LYLE, b Muskegon, Mich, May 1, 28; m; c 3. INDUSTRIAL HYGIENE, ENVIRONMENTAL HEALTH. Educ: Mich State Univ, BSME, 50; Harvard Univ, MS, 57. Prof Exp: Engr, Gen Gas & Light Co, Mich, 50-51; indust hyg engr, Mich Dept Health, 53-56; INDUST HYGIENIST, ARGONNE NAT LAB, 57-, ASSOC DIR, OCCUP HEALTH & SAFETY DIV, 74- Mem: Am Indust Hyg Asn; Am Acad Indust Hyg; Am Conf Govt Indust Hygienists; Health Physics Soc. Res: Health hazards, especially identification, behavior of and quantification of various toxic air pollutants; aerosol characteristics and collection efficiencies for air pollution control. Mailing Add: Bldg 14 Argonne Nat Lab Argonne IL 60439

CHEEVER, FRANCIS SARGENT, b Wellesley, Mass, Aug 20, 09; m 42; c 4. MICROBIOLOGY. Educ: Harvard Univ, AB, 32, MD, 36. Hon Degrees: DSc, Waynesburg Col, 60. Prof Exp: Intern, Presby Hosp, New York, 36-38; res fel bact & immunol, Harvard Med Sch, 38-39, from asst instr to instr, 39-41, assoc instr, 46, asst prof, 46-50; prof microbiol, Grad Sch Pub Health, Univ Pittsburgh, 50-58, prof, Sch Med, 58-74, dean sch, 58-69, lectr bact, Med Sch, 51-58, vchancellor health professions, 67-74, pres univ, 70-74; DIR ADMIS, HARVARD MED SCH, 75- Concurrent Pos: Mem enteric dis comn, Armed Forces Epidemiol Bd, 52-, mem bd, 64; vis prof microbiol & molecular genetics, Harvard Med Sch, 74- Mem: AAAS; Am Soc Microbiol; Am Soc Trop Med & Hyg; Am Epidemiol Soc; Am Asn Immunol (secy-treas, 54, pres, 63-64). Res: Epidemiology and immunity in bacterial and virus diseases of man and animals. Mailing Add: Harvard Med Sch Boston MA 02115

CHEEVER, GORDON DALE, physical chemistry, inorganic chemistry, see 12th edition

CHEFURKA, WILLIAM, b Brandon, Man, July 24, 24; m 46; c 3. BIOCHEMISTRY, PHYSIOLOGY. Educ: Univ Man, BSc, 46; Mont State Col, MSc, 49; Harvard Univ, PhD(biol), 53. Prof Exp: Asst physics, Brandon Col, 43-44, asst chem, 44-45; asst biophys, Mass Inst Technol, 49-50; teaching fel physiol, Harvard Univ, 51-52; RES OFFICER, RES INST, CAN DEPT AGR, 53- Concurrent Pos: Hon lectr, Univ Western Ont, 53- Mem: Am Soc Gen Physiol; NY Acad Sci; Can Soc Zoologists; Can Biochem Soc. Res: Comparative biochemistry, especially biochemistry of intermediary metabolism and biochemical mechanism of action of toxic agents; physiology of invertebrates including insects. Mailing Add: Res Inst Can Dept of Agr Univ Sub PO London ON Can

CHEIN, ORIN NATHANIEL, b New York, NY, Aug 29, 43; m 65; c 1. MATHEMATICS. Educ: NY Univ, BA, 64, MS, 66, PhD(math), 68. Prof Exp: Asst prof, 68-71, ASSOC PROF MATH, TEMPLE UNIV, 71- Mem: Am Math Soc. Res: Automorphisms of free and free metabelian groups; Moufang loops. Mailing Add: Dept of Math Temple Univ Philadelphia PA 19122

CHEITLIN, MELVIN DONALD, b Wilmington, Del, Mar 25, 29; m 52; c 3. INTERNAL MEDICINE, CARDIOLOGY. Educ: Temple Univ, BA, 50, MD, 54. Prof Exp: Chief cardiovasc serv, Madigan Gen Hosp, US Army, 60-64, Tripler Gen Hosp, 64-68 & Letterman Gen Hosp, 68-71, asst clin prof med, Univ Calif, San Francisco, 69-71, chief cardiovasc serv, Walter Reed Army Hosp, 71-74; PROF MED, UNIV CALIF, SAN FRANCISCO, 74- Concurrent Pos: Consult, Queen's Hosp, Honolulu, 65-68 & Letterman Gen Hosp, 75-; fel, Coun Clin Cardiol, Am Heart Asn, 66-68; cardiovasc consult to surgeon gen, Walter Reed Gen Hosp, 71-74; assoc dir cardiopulmonary unit, San Francisco Gen Hosp, 74- Mem: AMA; fel Am Col Physicians; fel Am Col Cardiol; Am Heart Asn; Am Fedn Clin Res. Mailing Add: 224 Castenada Ave San Francisco CA 94116

CHELIKOWSKY, JOSEPH R, b N Tonawanda, NY, Apr 13, 07; m 36; c 2. GEOLOGY. Educ: Cornell Univ, AB, 31, AM, 32, PhD(geol), 35. Prof Exp: Asst instr geol, Cornell Univ, 31-35; instr, Case Western Reserve Univ, 35-36; asst prof, 37-46, PROF GEOL, KANS STATE UNIV, 46- Concurrent Pos: Assoc ed, Geology. Mem: Am Asn Petrol Geol; fel Geol Soc Am. Res: Economic geology; structural geology; stratigraphy and sedimentation. Mailing Add: Dept of Geol Kans State Univ Manhattan KS 66504

CHELLEW, NORMAN RAYMOND, b Aurora, Minn, June 17, 17; m 43. INORGANIC CHEMISTRY. Educ: Univ Minn, BS, 39. Prof Exp: Teacher, Breitung Twp Schs Mich, 40-41; chemist & supvr, E I du Pont de Nemours & Co, 41-46; jr chemist, 47-49, ASSOC CHEMIST, ARGONNE NAT LAB, 49- Mem: Sci Res Soc Am; Am Chem Soc; Am Nuclear Soc. Res: Research and development of inorganic chemistry; radiochemistry; development of fission product monitors for nuclear reactor systems. Mailing Add: Chem Eng Bldg 205 Argonne Nat Lab 9700 S Cass Ave Argonne IL 60439

CHELLO, PAUL LARSON, b New Haven, Conn, Nov 11, 42. BIOCHEMICAL PHARMACOLOGY. Educ: Johns Hopkins Univ, BA, 64; Univ Vt, PhD(pharmacol), 71. Prof Exp: Res assoc molecular therapeut, 74-75, ASSOC, SLOAN KETTERING INST CANCER RES, 75- Mem: Am Asn Cancer Res. Res: Cancer chemotherapy; mechanisms of drug resistance; mechanism of action of antifolates; folic acid, vitamin B12 and methionine metabolism. Mailing Add: Sloan Kettering Cancer Ctr Walker Lab 145 Boston Post Rd Rye NY 10580

CHEMBURKAR, PRAMOD BHAURAO, b Bombay, India, Sept 3, 32; m 61; c 2. PHARMACEUTICAL SCIENCES, CHEMICAL ENGINEERING. Educ: Univ Bombay, BS, 53; Gujarat Univ, India, BPharm, 55, MPharm, 58; Univ Fla, PhD(pharmaceut sci, chem eng), 67. Prof Exp: Work mgr & chief chemist, Pharmaceut Lab, Indoco Remedies Ltd, India, 58-62; sr res pharmacist, Wyeth Labs, 67-68; sect head, Pharm Chem Dept, William H Rorer, Inc, 68-75; UNIT HEAD, WYETH LABS, 75- Mem: AAAS; Am Pharmaceut Asn; Acad Pharmaceut Sci; Indian Pharmaceut Asn. Res: Chemical kinetics connected with stability of drugs; diffusion of drugs through polymeric membranes; study of absorption, metabolism and excretion of drugs and its correlation with in vitro tests; development of pharmaceutical dosage formulations. Mailing Add: Wyeth Labs Inc PO Box 8299 Philadelphia PA 19101

CHEMERDA, JOHN MARTIN, b New Castle, Pa, Dec 20, 14; m 39; c 1. ORGANIC CHEMISTRY. Educ: Pa State Univ, BS, 35; Univ Mich, MS, 37, PhD(chem), 39. Prof Exp: Jr res chemist, Shell Petrol Refinery, 35-36; asst, Univ Mich, 39-40, res fel, 41-42, res assoc, 42-44; asst chem, Calif Inst Technol, 40-41 & Inst, Johns Hopkins Univ, 44-46; res chemist, 46-58, from dir to sr dir process res, 58-69, EXEC DIR PROCESS RES, MERCK SHARP & DOHME RES LABS, MERCK & CO, INC, 69- Mem: Am Chem Soc. Res: Synthetic organic chemistry. Mailing Add: Merck Sharp & Dohme Res Labs Rahway NJ 07065

CHEMSAK, JOHN A, b Ambridge, Pa, Feb 19, 32; m 59; c 3. ENTOMOLOGY. Educ: Pa State Univ, BS, 54, MS, 56; Univ Calif, Berkeley, PhD(entom), 61. Prof Exp: Biol aideplant path, USDA, 56; from jr res entomologist to asst res entomologist, 61-62, SPEC- IALIST ENTOM, UNIV CALIF, BERKELEY, 62- Res: Biosystematics and zoogeography of North and central American cerambycid beetles; biologies of insects. Mailing Add: Div of Entom 201 Wellman Hall Univ of Calif Berkeley CA 94720

CHEN, ABEL JER-JIUNN, b Taipei, Taiwan, Dec 4, 41; m 67; c 2. SPACE PHYSICS, AERONOMY. Educ: Nat Taiwan Univ, BS, 64; Rice Univ, MS, 68, PhD(space sci), 70. Prof Exp: Fel & res assoc space physics, Univ Alta, 70-72; Nat Res Coun-Nat Acad Sci resident res assoc, Goddard Space Flight Ctr, NASA, 72-74; SR PHYSICIST AERONOMY & SPACE PHYSICS, AIKEN INDUST, INC, 74- Mem: AAAS; Am Geophys Union. Res: Magnetospheric convection; plasmaphere dynamics; ion-compositions in the ionosphere; atmosphere, ionosphere, magnetosphere coupling processes and correlation between severe weather and ionospheric disturbance. Mailing Add: Aiken Indust Inc 7411 50th Ave College Park MD 20740

CHEN, AN-BAN, b Chiayi, Taiwan, Oct 10, 42; m 69; c 2. SOLID STATE PHYSICS. Educ: Taiwan Norm Univ, BS, 66; Col William & Mary, MS, 69, PhD(physics), 71. Prof Exp: Res assoc, Col William & Mary, 71-72; res assoc, Case Western Reserve Univ, 72-74; ASST PROF PHYSICS, AUBURN UNIV, 74- Mem: Am Phys Soc. Res: Study of electronic structures and related properties in crystalline solids and in disordered condensed matters. Mailing Add: Dept of Physics Auburn Univ Auburn AL 36830

CHEN, ANDREW TAT-LENG, b Kedah, Malaysia, June 17, 38; m 70; c 2. CYTOGENETICS, GENETICS. Educ: Chinese Univ Hong Kong, BSc, 61; McGill Univ, MSc, 63; Univ Western Ont, PhD(human cytogenetics), 69. Prof Exp: Assoc psychiat, Emory Univ, 69-70, asst prof, 70-72; asst prof genetics & pediat & dir cytogenetics lab, Med Col Va, 72-74; CHIEF CELLULAR GENETICS LAB, CTR DIS CONTROL, 74- Concurrent Pos: Res geneticist, Ga Ment Health Inst, 69-72. Mem: AAAS; Am Soc Human Genetics. Res: Human cytogenetics; tissue culture of mammalian cells. Mailing Add: Cellular Genetics Lab Path Div Ctr for Dis Control Atlanta GA 30333

CHEN, BANG-YEN, b Ilan, Taiwan, Oct 3, 43; m 68; c 2. GEOMETRY. Educ: Tamkang Col, Taiwan, BS, 65; Nat Tsinghua Univ, MS, 67; Univ Notre Dame, PhD(math), 70. Prof Exp: Res assoc math, 70-72, ASSOC PROF MATH, MICH STATE UNIV, 72- Concurrent Pos: Ed, Tamkang J Math, 70- Mem: Am Math Soc. Res: Differential geometry, global analysis and algebraic geometry. Mailing Add: Dept of Math Mich State Univ East Lansing MI 48824

CHEN, CATHERINE S H, b Chungking, China; US citizen. POLYMER CHEMISTRY. Educ: Barnard Col, BA, 50; Columbia Univ, MA, 52; Polytech Inst New York, PhD(polymer chem), 55. Prof Exp: Res assoc, Dept Chem, Columbia Univ, 54-56, Celanese Res Co, Celanese Corp, 64-70; sr res chemist, Cent Res Div, Am Cyanamid Co, 56-64; res scientist, Corp Res & Develop Lab, Singer Co, 72-75; RES ASSOC, CENT RES DIV, MOBIL RES & DEVELOP CORP, 75- Mem: Am Chem Soc; Soc Petrol Engrs. Res: All aspects of polymer science; tertiary oil recovery chemicals; surfactants and mobility control agents, including polymers. Mailing Add: Mobil Res & Develop Corp PO Box 1025 Princeton NJ 08540

CHEN, CHANG-HWEI, b Taipei, Taiwan, Oct 12, 41; m 70; c 2. PHYSICAL CHEMISTRY. Educ: Nat Taiwan Univ, BS, 64; Univ Conn, PhD(phys chem), 70.

Prof Exp: Res assoc chem, Univ Pittsburgh, 70-73; fel, 73-74, res scientist I, 74-75, RES SCIENTIST II CHEM, DIV LABS & RES, NY STATE DEPT HEALTH, 75- Mem: Am Chem Soc. Res: Physical chemical properties of macromolecular solutions; thermodynamics and transport processes; model membranes and structure of water. Mailing Add: Div of Labs & Res NY State Dept of Health Albany NY 12201

CHEN, CHAO LING, b Tsaotun, Taiwan, Sept 28, 37; m 66. VETERINARY PHYSIOLOGY, ENDOCRINOLOGY. Educ: Nat Taiwan Univ, DVM, 60; Iowa State Univ, MS, 66; Mich State Univ, PhD(neuroendocrinol), 69. Prof Exp: Instr histol, Taipei Med Col, 61-63; asst vet, US Med Res Unit, 63-64; res asst reproductive physiol, Iowa State Univ, 64-66; asst neuroendocrinol, Mich State Univ, 66-69; asst prof, 69-74, ASSOC PROF, DEPT PHYSIOL, COL VET MED, KANS STATE UNIV, 74- Mem: Soc Study Reproduction; Am Soc Animal Sci; Endocrine Soc; Int Soc Neuroendocrinol. Res: Hypothalamic regulation of the pituitary gonadotropin secretion by hypothalamic stimulation, lesion and sectioning and by radioimmunoassay; relationship between the uterus and luteal function. Mailing Add: Dept of Physiol Sci Col Vet Med Kans State Univ Manhattan KS 66502

CHEN, CHARLES CHIN-TSE, b Taipei, Taiwan, May 22, 29. SOLID STATE PHYSICS. Educ: Nat Taiwan Univ, BS, 51; Univ Md, PhD(physics), 62. Prof Exp: Instr physics, Nat Taiwan Univ, 51-57; asst, Univ Md, 57-62, res assoc, 63; asst prof, 63-67, ASSOC PROF PHYSICS, OHIO UNIV, 67- Mem: Am Phys Soc; Phys Soc Japan. Res: Statistical mechanics. Mailing Add: Dept of Physics Ohio Univ Athens OH 45701

CHEN, CHEN HO, b Kiangsu, China, Sept 9, 29; US citizen; m 60; c 3. PLANT CYTOLOGY, PLANT PHYSIOLOGY. Educ: Nat Taiwan Univ, BS, 54; La State Univ, MS, 60; SDak State Univ, PhD(plant sci), 64. Prof Exp: Fel, SDak State Univ, 63-64 & Argonne Nat Lab, US Atomic Energy Comn, 64-65; lectr & head biol, Hong Kong Baptist Col, 65-68; from asst prof to assoc prof, 68-75, PROF BOT & BIOL, SDAK STATE UNIV, 75- Mem: Bot Soc Am; Tissue Cult Asn. Res: Development of tissue culture techniques for use in breeding monocotyledonous species. Mailing Add: Dept of Bot & Biol SDak State Univ Brookings SD 57006

CHEN, CHIADAO, b Peiping, China, Dec 7, 13; nat US. BIOCHEMISTRY. Educ: Shanghai Univ, BS, 36; Dresden Tech Univ, Dipl Ing, 39; Berlin Tech, DSc(biochem), 41. Prof Exp: Res chemist, Ciba Pharmaceut, Inc, 42-44; sr res fel, Univ Pittsburgh, 44-48; res assoc & asst prof biochem, Univ Notre Dame, 48-49; assoc prof, 49-67, PROF BIOCHEM, MED SCH, NORTHWESTERN UNIV, CHICAGO, 67- Concurrent Pos: Biochemist & sect chief, Vet Admin Res Hosp, 54-56. Mem: AAAS; Am Chem Soc; NY Acad Sci; Am Soc Biol Chem; Brit Biochem Soc. Res: Steroid chemistry; oxygenation mechanism; chemical carcinogenesis; liver function. Mailing Add: Dept of Biochem Northwestern Univ Med Sch Chicago IL 60611

CHEN, CHIH SHAN, b Chekiang, China, Oct 1, 29; m 62; c 1. STRATIGRAPHY, SEDIMENTARY PETROLOGY. Educ: Nat Taiwan Univ, BS, 54; Fla State Univ, MS, 60; Northwestern Univ, PhD(geol), 64. Prof Exp: Asst geol, Nat Taiwan Univ, 54-57 & Fla State Univ, 60; res geologist, Pure Oil Co, 64-65; RES SCIENTIST, UNION OIL CO CALIF, 65- Mem: Am Asn Petrol Geol; Soc Econ Paleontologists & Mineralogists; Sigma Xi; Geol Soc Am. Res: Sedimentary petrology; regional lithostratigraphic analysis; clay mineralogy; petrographic study of reservoir rocks. Mailing Add: Union Oil Co of Calif Res Ctr PO Box 76 Brea CA 92621

CHEN, CHIN HSIN, b Che-Kiang, China, Feb 7, 43; m 69. SYNTHETIC ORGANIC CHEMISTRY. Educ: Tunghai Univ, Taiwan, BS, 64; Okla State Univ, PhD(org chem), 71. Prof Exp: Fel org chem, Ohio State Univ, 71-72 & Harvard Univ, 72-73; SR RES CHEMIST, SYNTHETIC ORG CHEM, EASTMAN KODAK CO RES LAB, 73- Mem: Am Chem Soc; Sigma Xi. Res: Exploratory organic synthesis in the field of organosulfur and organophosphorus heterocyclic chemistry. Mailing Add: Eastman Kodak Co Res Lab 1669 Lake Ave Rochester NY 14650

CHEN, CHING-CHIH, b Foochow, China, Sept 3, 37; US citizen; m 61; c 3. INFORMATION SCIENCE, SCIENTIFIC BIBLIOGRAPHY. Educ: Nat Taiwan Univ, BA, 59; Univ Mich, AMLS, 61; Case Western Reserve Univ, PhD(info sci), 74. Prof Exp: Serv librn, Univ Mich, 61-62; sci ref librn, Windsor Pub Libr, Ont, 62; ref librn, McMaster Univ, 62-63, head sci librn, 63-64; sr sci librn, Univ Waterloo, 64-65, head librn, Eng, Math & Sci Libr, 65-68; assoc head sci libr, Mass Inst Technol, 68-71; asst prof, 71-75, ASSOC PROF LIBR SCI, SIMMONS COL, 75- Concurrent Pos: Consult, Sci & Tech Doc Ctr, Nat Sci Coun, Repub of China, 73- Mem: Am Soc Info Sci; Med Libr Asn; Spec Libr Asn; Am Asn Univ Prof. Res: Scientific management, especially the application of modern analytic techniques in library problems; systems analysis; biomedical, scientific and technical library and information services and systems. Mailing Add: Sch of Libr Sci Simmons Col 300 The Fenway Boston MA 02115

CHEN, CHONG MAW, b Taoyuan, Taiwan; m 66; c 2. PLANT BIOCHEMISTRY. Educ: Taiwan Norm Univ, BS, 58; Univ Kans, MA, 64, PhD(plant biochem), 67. Prof Exp: Teaching asst biol, Taiwan Norm Univ, 61-62; fel biochem, McMaster Univ, 67-69; res fel, Roche Inst Molecular Biol, 69-71; asst prof life sci, 71-73, ASSOC PROF LIFE SCI, UNIV WIS-PARKSIDE, 73- Mem: AAAS; Am Chem Soc; Am Soc Plant Physiol. Res: Mechanism of action of plant hormones; nucleic acid and protein biosynthesis. Mailing Add: Sci Div Univ of Wis-Parkside Kenosha WI 53140

CHEN, CHUNG WEI, b Hunan, China, Dec 25, 21; US citizen; m 65; c 2. APPLIED STATISTICS. Educ: Nat Chengchi Univ, China, BA, 46, LLB, 48; La State Univ, MBA, 54, PhD(statist), 57. Prof Exp: Secy, Human Prov Govt, 46-47; judge, Hunan Changsha Dist Ct, 47-48; statistician, Haloid-Xerox, Inc, 56-61, chief statistician, Xerox Corp, 61-72, MGR STATIST SERV, XEROX CORP, 72- Concurrent Pos: Instr, Rochester Inst Technol, 59-67. Mem: Am Statist Asn; Am Econ Asn; Opers Res Soc Am. Mailing Add: 310 Orchard Park Blvd Rochester NY 14609

CHEN, CHUNG-HO, b Kaohsiung, Taiwan, Dec 1, 37; US citizen; m 65; c 4. BIOCHEMISTRY. Educ: Chung-Hsing Univ, BS, 62; Okla State Univ, PhD(biochem), 69. Prof Exp: Res asst pharmacol, Kaohsiung Med Col, 63-64; asst biochem, Okla State Univ, 64-69; ASST PROF OPHTHALMIC BIOCHEM, SCH MED, JOHNS HOPKINS UNIV, 73- Concurrent Pos: Fel biochem, Okla State Univ, 69; fel, Sch Med, Johns Hopkins Univ, 69-73, consult ophthal, 72-73; Nat Eye Inst res grant, 75-; Am Diabetes Asn grant, 75- Honors & Awards: Mayers Award, 75. Mem: AAAS; Am Chem Soc; Asn Res Vision & Ophthal. Res: Ocular biochemistry; enzymatic activities in mitochondria; metabolic regulations and disorders; angiogenesis; membrane chemistry; active transport. Mailing Add: Dept of Ophthal Johns Hopkins Univ Med Sch Baltimore MD 21205

CHEN, DAVID HOU-CHUNG, b Quayang, June 8, 41; US citizen; m 71; c 1. MEDICAL PARASITOLOGY, IMMUNOLOGY. Educ: Univ Md, BS, 67; Am Univ, MS, 72; NY Univ, PhD(med sci), 74. Prof Exp: Res entomologist, Walter Reed Army Inst Res, 67-69; trainee, Dept Prev Med, Sch Med, NY Univ, 71-74; STAFF FEL, PARASITE IMMUNOL, LAB PARASITIC DIS, NAT INST ALLERGY & INFECTIOUS DIS, 74- Mem: Am Soc Trop Med & Hyg. Res: Mechanisms of parasite induced immunity; malaria in rodent and primate models. Mailing Add: NIH Lab of Parasitic Dis Nat Inst Allergy & Infectious Dis Bethesda MD 20014

CHEN, DAVIDSON TAH-CHUEN, b Wenling, China, Apr 1, 42; US citizen; m 66; c 2. PHYSICAL OCEANOGRAPHY, REMOTE SENSING. Educ: Nat Taiwan Univ, BS, 63; Univ Calif, Berkeley, MS, 66; NC State Univ, PhD(phys oceanog), 72. Prof Exp: Consult engr struct dynamics, John A Blume & Assoc Engrs, 66-67; Nat Res Coun resident res assoc phys oceanog & remote sensing, Wallops Flight Ctr, NASA, 72-74; PHYS OCEANOGR, NAVAL RES LAB, 74- Mem: Am Geophys Union. Res: Developing mathematical models, based upon fluid dynamics, for physical parameters which describe important and meaningful phenomena geophysically; developing microwave remote sensors for inferring the measurements of these physical parameters. Mailing Add: Code 7112 C Naval Res Lab Washington DC 20375

CHEN, EDWARD CHUCK MING, physical chemistry, see 12th edition

CHEN, EDWIN HUNG-TEH, b Tainan, Taiwan, Aug 23, 34; US citizen; m 66; c 2. BIOSTATISTICS. Educ: Nat Taiwan Univ, BS, 57; Mich State Univ, MS, 64; Univ Calif, Los Angeles, PhD(biostatist), 69. Prof Exp: Systs analyst statist, Union Am Comput Corp, 64-65; asst res statistician & lectr biostatist, Univ Calif, Los Angeles, 69-72; ASSOC PROF BIOMET, UNIV ILL MED CTR, 72- Mem: Am Statist Asn; Inst Math Statist; Biomet Soc; AAAS. Res: Data analysis; robust statistical procedures; statistical computing; computer simulation. Mailing Add: Sch of Pub Health Univ of Ill PO Box 6998 Chicago IL 60680

CHEN, FRANCIS F, b Canton, China, Nov 18, 29; US citizen; m 56; c 3. PHYSICS. Educ: Harvard Univ, AB, 50, MA, 51, PhD(physics), 54. Prof Exp: Res assoc physics, Brookhaven Nat Lab, 53-54; res staff mem, Plasma Physics Lab, Princeton Univ, 54-64, res physicist, 64-69, sr res physicist, 69; PROF ENG & APPL SCI, UNIV CALIF, LOS ANGELES, 69- Concurrent Pos: Attend physicist, Nuclear Res Ctr, Fontenay, France, 62-63; chmn fusion adv comt, Elec Power Res Inst, 73- Mem: Fel Am Phys Soc; sr mem Inst Elec & Electronics Engrs. Res: Basic plasma physics; fusion reactors; laser-plasma interactions. Mailing Add: Boelter Hall 7731 Univ of Calif Los Angeles CA 90024

CHEN, FREEMAN PHILIP, b San Fernando, Trinidad, Feb 27, 47; US citizen; m 70; c 1. CHEMICAL PHYSICS. Educ: Brooklyn Polytech Inst, BS, 69; State Univ NY Stony Brook, MS, 73, PhD(chem physics), 75. Prof Exp: VIS PROF & TEACHING ASSOC CHEM, STATE UNIV NY BUFFALO, 75- Res: Studies of molecules and molecular crystals using Stark spectroscopy; laser Raman and electronic spectroscopy of solid state organic charge transfer complexes. Mailing Add: Dept of Chem Acheson Hall State Univ of NY Buffalo NY 14214

CHEN, HAROLD H, polymer chemistry, biochemistry, see 12th edition

CHEN, HARRY WU-SHIONG, b Kaohsiung, Taiwan, June 17, 37; m 69. CELL BIOLOGY, BIOCHEMISTRY. Educ: Nat Chengchi Univ, BA, 60; Univ Kans, PhD(biochem), 69. Prof Exp: Fel, 69-70, assoc staff scientist, 70-74, STAFF SCIENTIST, JACKSON LAB, 74- Res: Regulation of cell growth; function of cholesterol in the surface membranes of mammalian cells. Mailing Add: Jackson Lab Bar Harbor ME 04609

CHEN, HO SOU, b Taiwan, Nov 24, 32; US citizen; c 2. SOLID STATE PHYSICS. Educ: Nat Taiwan Univ, BS, 56; Brown Univ, MS, 63; Harvard Univ, PhD(appl physics), 67. Prof Exp: Mem tech staff mat, Bell Tel Labs, 68-70; asst prof chem, Yeshiva Univ, 70-71; physicist, Allied Chem Corp, 71-72; MEM TECH STAFF MAT, BELL LABS, 72- Res: Structure, thermal and physical properties of metallic glasses. Mailing Add: Bell Labs Murray Hill NJ 07974

CHEN, HUBERT JAN-PEING, b Kiangsu Prov, China, Oct 29, 42; m; c 1. STATISTICS. Educ: Nat Taiwan Univ, BA, 67; Univ Rochester, MA, 71, PhD(statist), 74. Prof Exp: Asst teacher statist, Nat Taiwan Univ, 67-69; teaching asst, Univ Rochester, 69-72; lectr, Ohio State Univ, 72-73; ASST PROF STATIST, MEMPHIS STATE UNIV, 73- Concurrent Pos: Mem staff, Math Rev, 72- Mem: Am Statist Asn; Inst Math Statist; Am Soc Qual Control. Res: Estimation of ranked parameters; ranking and selections; multiple comparisons; simulation and Monte Carlo methods. Mailing Add: Dept of Math Sci Memphis State Univ Memphis TN 38152

CHEN, INAN, b Tainan, Taiwan, Oct 9, 33; m 59; c 2. SOLID STATE PHYSICS, SPECTROSCOPY. Educ: Nat Taiwan Univ, BS, 56; Tsinghua Univ, China, MS, 58; Univ Mich, PhD(nuclear sci), 64. Prof Exp: Res assoc nuclear sci, Univ Mich, 64-65; from scientist to sr scientist, 65-71, PRIN SCIENTIST, RES LABS, XEROX CORP, 71- Mem: Am Phys Soc; Soc Photog Scientists & Engrs. Res: Theoretical studies of electronic states in solids; theoretical studies of electronic properties of amorphous and molecular solids, and their applications to electrophotographic processes. Mailing Add: Res Labs Xerox Corp Xerox Sq Rochester NY 14644

CHEN, I-NGO, b Canton, China, July 13, 34; c 1. COMPUTER SCIENCE. Educ: Nat Taiwan Univ, BSc, 55, Chiao Tung Univ, MSc, 63; Univ Ottawa, PhD(elec eng), 69. Prof Exp: ASSOC PROF COMPUT SCI, UNIV ALTA, 68- Mem: Inst Elec & Electronics Engrs; Asn Comput Mach; Can Info Processing Soc. Res: Automata theory; learning machines; modeling of human behavior. Mailing Add: Dept of Comput Sci Univ of Alta Edmonton AB Can

CHEN, I-WEN, b Tokyo, Japan, Aug 3, 34; US citizen; m 64; c 3. RADIOBIOLOGY, BIOCHEMISTRY. Educ: Nat Taiwan Univ, BS, 56; Univ Pa, PhD(biochem), 64. Prof Exp: Asst prof, 66-72, ASSOC PROF RADIOL, UNIV CINCINNATI, 72- Concurrent Pos: NIH fel, 64-66; adj affil mem, Jewish Hosp, Cincinnati, 73- Mem: AAAS; Radiation Res Soc. Res: Biochemical effect of radiation; radioimmunoassays for various hormones. Mailing Add: E561 Sch of Med Univ of Cincinnati Cincinnati OH 45267

CHEN, JAMES CHE WEN, b Taipei, Taiwan, Nov 21, 30; m 59; c 3. DEVELOPMENTAL BIOLOGY, BOTANY. Educ: Taiwan Prov Norm Univ, BS, 55; Univ Pa, PhD(bot), 63. Prof Exp: Asst instr bot, Taiwan, 56-58; asst instr biol, Univ Pa, 61-62; res assoc, Princeton Univ, 63-65; from asst prof to assoc prof, Washington & Jefferson Col, 65-68; ASSOC PROF BOT, DOUGLASS COL, RUTGERS UNIV, 68- Mem: AAAS; Soc Develop Biol; Bot Soc Am. Res: Investigations of growth and differentiation of plants and quantitative description of growth processes of plants. Mailing Add: Dept of Biol Sci Douglass Col Rutgers Univ New Brunswick NJ 08903

CHEN, JAMES L, b Nanking, China, Nov 3, 14; nat US; m 46; c 2. PHARMACEUTICAL CHEMISTRY. Educ: Sino-French Univ, China, BS, 39; Purdue Univ, MS, 40, PhD, 47. Prof Exp: Chemist, Merck & Co, 42-45; res chemist, Arlington Chem Co, 47-49; res assoc, 49-69, RES FEL, E R SQUIBB & SONS, 69- Mem: Am Pharmaceut Asn; NY Acad Sci; fel Am Inst Chem. Res: Phytochemistry;

biological adhesives; pharmaceutical research; analytical chemistry. Mailing Add: 30 Fairview Ave East Brunswick NJ 08816

CHEN, JAMES PAI-FUN, b Fengyuan, Taiwan, May 1, 29; nat US; m 64; c 3. BIOCHEMISTRY, IMMUNOLOGY. Educ: Houghton Col, BS, 55; St Lawrence Univ, MS, 57; Pa State Univ, PhD(biochem), 62. Prof Exp: Instr chem, Houghton Col, 60-62, assoc prof, 62-64; res assoc, Div Exp Med, Col Med, Univ Vt, 64-65; res assoc med, Sch Med, State Univ NY Buffalo, 65-68; res asst prof internal med, Univ Tex Med Br Galveston, 68-70, asst prof human genetics, 70-75; sr res assoc, Biomed Res Div, NASA Johnson Space Ctr, 75-76; RES ASSOC PROF, MEM RES CTR, UNIV TENN, KNOXVILLE, 76- Mem: AAAS; Am Asn Immunologists. Res: Structures of human IgM and IgA; genetic control of antibody synthesis and hemostatic mechanism of coagulation. Mailing Add: Mem Res Ctr Univ of Tenn Knoxville TN 37920

CHEN, JAMES RALPH, b Kingston, Jamaica, Sept 22, 39; US citizen; m; c 3. PARTICLE PHYSICS, ATOMIC PHYSICS. Educ: Brandeis Univ, BA, 62; Harvard Univ, MA, 64, PhD(physics), 69. Prof Exp: Res asst physics, Harvard Univ, 67-68; res assoc & fel, Univ Pa, 68-69, asst prof, 69-73; ASST PROF PHYSICS, STATE UNIV NY COL GENESEO, 72- Concurrent Pos: Res consult, Los Alamos Sci Lab, 72, spokesman-scientist, Los Alamos Meson Physics Facil, 74-75; spokesman-scientist, Brookhaven Nat Lab, 72-73; State Univ NY Res Found res fels, 74-76. Mem: Am Phys Soc; AAAS; Am Asn Physics Teachers. Res: Particle physics experiments; atomic physics x-ray spectroscopic calculations and measurements. Mailing Add: 6 Elm St Geneseo NY 14454

CHEN, JANE LEE, b Kweiyang, China, Aug 6, 40; m 64; c 1. PLANT PHYSIOLOGY. Educ: Radcliffe Col, BA, 62; Univ Calif, Berkeley, MA, 64, Univ Calif, Los Angeles, PhD(plant sci), 66. Prof Exp: Res fel bot sci, Univ Calif, Los Angeles, 67-69; asst prof biol sci, 69-72, ASSOC PROF BIOL SCI, SAN JOSE STATE UNIV, 72- Mem: AAAS; Am Soc Plant Physiol. Res: Nature and function of chloroplast ribosomes; nature of products formed by isolated chloroplasts incubated for protein synthesis; effect of light on protein and RNA synthesis by isolated chloroplasts. Mailing Add: Dept of Biol San Jose State Univ San Jose CA 95192

CHEN, JOHN HENG, b Peking, China, Aug 15, 34; US citizen; m 63; c 2. BIOCHEMISTRY. Educ: Nat Chung-Hsing Univ, BA, 59; Boston Univ, MS, 63; State Univ NY Stony Brook, PhD(biochem), 68. Prof Exp: Instr chem, Nat Chung-Hsing Univ, 59-60; teaching fel biochem, Sch Med, Boston Univ, 62-63; assoc in res, Rutgers Univ, 64-65; instr, State Univ NY Stony Brook, 67-68; fel biochem & med, Yale Univ, 68-70, res assoc biophys, 70-71; ASSOC PROF BIOCHEM, COL PHYSICIANS & SURGEONS, COLUMBIA UNIV, 71- Mem: Am Chem Soc; Am Inst Biol Sci; Asn Res Vision & Ophthal; AAAS. Res: Molecular biology, genetic engineering, structure and function of mammalian messenger RNA, nucleic acids chemistry and metabolism. Mailing Add: Col of Physicians & Surgeons Columbia Univ New York NY 10032

CHEN, JOSEPH CHENG YIH, b Nanking, China, Nov 12, 33; nat US; m 59; c 1. THEORETICAL PHYSICS, THEORETICAL CHEMISTRY. Educ: St Anselm's Col, BA, 57; Univ Notre Dame, PhD(theoret chem), 61. Prof Exp: Res assoc theoret chem, Brookhaven Nat Lab, 61-63, assoc chemist, 63-65; vis fel, Joint Inst Lab Astrophys, Univ Colo, 65-66; from asst prof to assoc prof physics, 66-74, PROF PHYSICS, UNIV CALIF, SAN DIEGO, 74- Concurrent Pos: Vis lectr, Univ Manchester, 64; Nordic Inst Theoret Atomic Physics vis prof, Univ Oslo, 70. Mem: Fel Am Phys Soc. Res: Scattering problem; reaction theories; atomic and molecular systems; many body theory. Mailing Add: Dept of Physics Univ of Calif La Jolla CA 92037

CHEN, JOSEPH H, b Tientsin, China, Mar 22, 31; m 56; c 2. PHYSICS. Educ: St Procopius Col, BS, 54; Univ Notre Dame, PhD(physics), 58. Prof Exp: Instr physics, Univ Notre Dame, 57-58; asst prof, 58-64, ASSOC PROF PHYSICS, BOSTON COL, 64- Mem: Am Phys Soc. Res: Dielectric relaxation phenomena; electron spin resonance and optical properties of solids; Mössbauer effect and high pressure studies in solids. Mailing Add: Dept of Physics Boston Col Chestnut Hill MA 02167

CHEN, JOSEPH KE-CHOU, b Hupei, China, May 27, 36; m 67; c 2. MEDICAL MICROBIOLOGY. Educ: Nat Taiwan Univ, BS, 59; Univ Pittsburgh, PhD(microbiol), 72. Prof Exp: Res asst microbiol, Sch Med, Univ Pittsburgh, 66-72, res assoc biochem, 72-74; CANCER RES SCIENTIST VIRAL ONCOL, ROSWELL PARK MEM INST, 75- Mem: Am Soc Microbiol; Am Chem Soc. Res: Study of biochemical and biological aspects of interferon. Mailing Add: Roswell Park Mem Inst 666 Elm St Buffalo NY 14263

CHEN, KO KUEI, b Shanghai, China, Feb 26, 98; nat US; m 29; c 2. PHARMACOLOGY. Educ: Univ Wis, BS, 20, PhD(physiol), 23; Johns Hopkins Univ, MD, 27. Hon Degrees: ScD, Philadelphia Col Pharm, 46, Univ Wis, 52 & Ind Univ, 71. Prof Exp: Sr asst, Peking Union Med Col, China, 23-25; res pharmacologist, Univ Wis, 25-26; assoc pharmacol, Johns Hopkins Univ, 27-29; dir pharmacol res, Eli Lilly & Co, 29-63; prof pharmacol, 37-68, EMER PROF PHARMACOL, SCH MED, IND UNIV, INDIANAPOLIS, 68- Concurrent Pos: Lectr, Univ Wis, 39, 52, 53, Ind Univ, 49, Vanderbilt Univ, 65, Dartmouth Col, 70, Med Br Univ Tex, 70, Univ Kans, 71, Univ Calif, San Francisco, 72, 73, Univ Ill, 75 & Univ Helsinki, 75; consult med serv, Marion County Gen Hosp, 48-; spec consult & mem pharmacol & exp therapeut study sect, NIH, 50-55, 57-60; chmn bd, Fedn Am Socs Exp Biol, 54; deleg & secy to gen assembly, US Nat Comt, Int Union Physiol Sci, 56-59; mem drug res bd, Nat Acad Sci-Nat Res Coun, 64-67; ed, Soc Hist, Am Soc Pharmacol & Exp Therapeut, 69. Honors & Awards: Remington Honor Medal, 65. Mem: AAAS; Am Soc Pharmacol & Exp Therapeut (treas, 47-50, pres, 53); Am Pharmaceut Asn; Am Physiol Soc; AMA. Res: Ephedrine and vasoconstrictors; Chinese herbs; Senecio alkaloids; cyanide antidotes; synthetic analgesics; genetic pharmacology. Mailing Add: 7975 Hillcrest Rd Indianapolis IN 46240

CHEN, KUO-MEI, b Nanking, China, Oct 25, 47; m 73; c 1. CHEMICAL PHYSICS. Educ: Tunghai Univ, Taiwan, BS, 69; Princeton Univ, PhD(chem), 75. Prof Exp: RES ASSOC CHEM, AMES LAB, ENERGY RES & DEVELOP ADMIN, 75- Mem: Am Chem Soc. Res: Laser isotope separation and nonlinear optics. Mailing Add: Ames Lab Iowa State Univ Ames IA 50011

CHEN, KUO-TSAI, b Chekiang, China, July 15, 23; m 53; c 3. MATHEMATICS. Educ: Southwest Assoc Univ, China, BS, 46; Columbia Univ, Phd(math), 50. Prof Exp: Instr math, Princeton Univ, 50-51; res assoc, Univ Ill, 51-52; lectr, Univ Hong Kong, 52-58; assoc prof, Tech Inst Aeronaut, Brazil, 58-60, prof, 60-61; from assoc prof to prof, Rutgers Univ, 62-65; prof, State Univ NY Buffalo, 65-67; PROF MATH, UNIV ILL, URBANA-CHAMPAIGN, 67- Concurrent Pos: Mem, Inst Advan Study, 60-62. Mem: Am Math Soc. Res: Differential topology and global analysis on path spaces. Mailing Add: Dept of Math Univ of Ill Urbana IL 61801

CHEN, KWAN-YU, b Shanghai, China, Aug 29, 30; m 61; c 3. ASTRONOMY. Educ: Ill Inst Technol, BS, 53, MS, 56; Univ Pa, PhD(astron), 63. Prof Exp: Asst prof astron, 63-68, asst prof astron & phys sci, 68-69, ASSOC PROF ASTRON & PHYS SCI, UNIV FLA, 69- Mem: AAAS; Am Astron Soc; Int Astron Union. Res: Photometric study of variable stars. Mailing Add: Dept of Physics & Astron Univ of Fla Gainesville FL 32601

CHEN, LINDA LI-YUEH HUANG, b Tokyo, Japan, Mar 22, 37; m 61; c 2. BIOCHEMISTRY, NUTRITION. Educ: Nat Taiwan Univ, BS, 59; Univ Louisville, PhD(biochem), 64. Prof Exp: Res assoc biochem, Univ Louisville, 64-66; asst prof nutrit, 67-72, ASSOC PROF NUTRIT, UNIV KY, 72- Mem: AAAS; Am Inst Chemists; Am Chem Soc; Am Inst Nutrit; Nutrit Today Soc. Res: Vitamin K and oxidative phosphorylation; physiological function of vitamin E and tissue antioxidant status; fish tempeh combination for human food. Mailing Add: Dept of Nutrit & Food Sci Univ of Ky Lexington KY 40506

CHEN, LO-CHAI, b Kwang-Tung, China, Dec 9, 39; m 65. ICHTHYOLOGY. Educ: Nat Taiwan Univ, BS, 61; Univ Alaska, MS, 65; Univ Calif, San Diego, PhD(marine biol), 69. Prof Exp: Asst prof, 67-72, ASSOC PROF ZOOL, SAN DIEGO STATE UNIV, 72- Mem: Am Soc Ichthyologists & Herpetologists. Res: Systematics and zoogeography of fishes, especially Sebastes Scorpaenidae; growth and meristic determination in fishes. Mailing Add: Dept of Zool San Diego State Univ San Diego CA 92182

CHEN, MING CHIH, b China, Aug 15, 20; m 43; c 3. ORGANIC CHEMISTRY. Educ: Fukien Christian Univ, BS, 42; Univ Buffalo, PhD, 50. Prof Exp: Asst, Univ Buffalo, 48-50; res fel, Purdue Univ, 50-52; res assoc, O-Cel-O Div, Gen Mills Inc, 52-57; group leader, Simoniz Co, 57-62; mgr urethane develop, Sheller Labs, Sheller Mfg Corp, 62-67, dir prod develop, 67-73, DIR MFG DEVELOP, SHELLER GLOBE CORP, 73- Mem: Chem Soc; Soc Plastics Engrs; Soc Plastics Indust; Soc Automotive Engrs; NY Acad Sci. Res: Cyanogen in the formation of oxamidine; halogenated hydrocarbon for non-inflammable hydraulic fluid and additives; cellulose and its derivatives; organic coating and polymeric foams. Mailing Add: Sheller Globe Corp 1641 Porter St Detroit MI 48216

CHEN, MIN-SHIH, b Kiangsu, China, Nov 13, 42; m 69; c 1. THEORETICAL HIGH ENERGY PHYSICS. Educ: Nat Taiwan Univ, BSc, 64; Yale Univ, MPhil, 67, PhD(physics), 70. Prof Exp: Res assoc physics, Brookhaven Nat Lab, 70-72; res assoc, Stanford Linear Accelerator Ctr, 72-74; lectr, 74-75, SCHOLAR PHYSICS, UNIV MICH, ANN ARBOR, 75- Mem: Am Phys Soc. Res: Phenomenology in multiparticle production and weak interactions; atomic physics and computational science. Mailing Add: Dept of Physics Univ of Mich Ann Arbor MI 48109

CHEN, MOU-SHAN, physical chemistry, chemical engineering, see 12th edition

CHEN, PAUL EAR, b Hangchow, China, June 29, 25; US citizen; m 49; c 2. APPLIED MECHANICS, APPLIED MATHEMATICS. Educ: Chiao Tung Univ, BS, 47; Purdue Univ, MS, 53; Washington Univ, DSc(appl mech), 62. Prof Exp: Instr civil eng, Nat Taiwan Univ, 47-51; struct engr, Mississippi Valley Struct Steel Co, 53-55; sr struct engr, Sverdrup & Parcel Eng Co, 55-59; proj engr, Cent Res Dept, Monsanto Co, 59-66, res scientist, Adv Res Proj Agency Proj, 66-70; SUPVR ADVAN PHYS DESIGN, BELL LABS, NAPERVILLE, 70- Concurrent Pos: Affil prof mat sci, Washington Univ, 66-70; dir bd, Unisysts, Inc, 69-; adj prof, Ill Inst Technol, 75- Mem: AAAS; Am Soc Civil Eng; Am Soc Mech Eng; Am Soc Eng Educ; Soc Rheol. Res: Structural engineering. Mailing Add: 22W131 Glen Park Glen Ellyn IL 60137

CHEN, PHILIP STANLEY, JR, b St Johns, Mich, July 3, 32; m 55; c 2. PHARMACOLOGY, PHYSIOLOGY. Educ: Clark Univ, BA, 50; Univ Rochester, PhD(pharmacol), 54. Prof Exp: Res assoc, Atomic Energy Proj, Univ Rochester, 50-54, jr scientist, 54-56; sr asst scientist, Nat Heart Inst, 56-59; asst prof radiation biol & biophys & pharmacol, Univ Rochester, 59-66; Guggenheim fel, Copenhagen, Denmark, 66-67; grants assoc, 67-68, spec asst to asst dir prog planning & eval, 68-70, chief spec projs br, Off Prog Anal, 70-71, chief anal & eval br, Prog Planning & Eval, 71-72, assoc dir prog planning & eval br, Nat Inst Gen Med Sci, 72-74, ASST DIR INTRAMURAL AFFAIRS, NIH, 74- Concurrent Pos: NSF fel, Copenhagen, Denmark, 54-55. Mem: Am Chem Soc; Am Physiol Soc; Radiation Res Soc. Res: Radioactive tracer techniques in biology; microanalytical chemistry; bone and mineral metabolism; vitamin D; renal excretion. Mailing Add: Bldg 1 Rm 103 Off of Dir NIH Bethesda MD 20014

CHEN, PING-FAN, b Kiangyin, China, May 13, 17; m 47; c 3. GEOLOGY. Educ: Nat Cent Univ, China, BS, 38; Univ Cincinnati, MS, 57; Va Polytech Inst, PhD(geol), 59. Prof Exp: From jr to sr geologist, Nat Geol Surv China, 38-46; from sr geologist to chief petrol geol, Chinese Petrol Corp, 46-55; petrol geologist & head petrol div, 60-66, STRATIGRAPHER, WVA GEOL & ECON SURV, 67- Concurrent Pos: Adj prof geol, WVa Univ, 75- Res: Geological exploration work for petroleum, ground water, coal and other mineral deposits in China; detail stratigraphic and structural geology studies in Central Appalachian for oil and gas possibilities in Lower Paleozoic rocks. Mailing Add: 1277 Dogwood Morgantown WV 26505

CHEN, RAYMOND F, biological chemistry, see 12th edition

CHEN, ROBERT CHIA-HUA, b Shanghai, China, Oct 26, 46; m 71; c 1. COMPUTER SCIENCE, ELECTRICAL ENGINEERING. Educ: Rensselaer Polytech Inst, BEE, 66; Mass Inst Technol, SM, 68; Carnegie-Mellon Univ, PhD(comput sci), 74. Prof Exp: Eng programmer comput sci, Burroughs Corp, 68-69; ASST PROF COMPUT & INFO SCI, UNIV PA, 74- Concurrent Pos: Staff engr, Burroughs Corp, 74- Mem: Asn Comput Mach; Inst Elec & Electronics Engrs. Res: Concurrency and modularity in computer systems. Mailing Add: Dept of Comput & Info Sci Univ of Pa Philadelphia PA 19174

CHEN, ROBERT LONG WEN, b Shanghai, China, Aug 23, 25; US citizen; m 58; c 2. PLASMA PHYSICS. Educ: Nanking Univ, BS, 47; Univ Syracuse, MA, 57, PhD(physics), 60. Prof Exp: Res assoc physics, Syracuse Univ, 60-61, asst prof, 61-62; res assoc, Goddard Inst Space Studies, 62-64; assoc prof, 64-70, PROF PHYSICS, EMORY UNIV, 70- Mem: Am Phys Soc. Res: Nonequilibrium statistical mechanics; plasma theory; molecular biophysics. Mailing Add: Dept of Physics Emory Univ Atlanta GA 30322

CH'EN, SHANG-YI, b China, Mar 4, 10; nat US; m 31; c 5. SPECTROSCOPY. Educ: Yenching Univ, BS, 32, MS, 34; Calif Inst Technol, PhD(physics), 40. Prof Exp: Asst, Inst Physics, Nat Acad, Peiping, 34-37; Norman Bridge Physics Lab, Calif Inst Technol, 38-39; lectr physics, Yenching Univ, 39-41, asst prof optics & spectros, 41-43, prof optics & chmn dept physics, 43-46; res prof, Inst Physics, Peiping Nat Acad, 46-49; from assoc prof to prof physics, 49-75, EMER PROF PHYSICS, UNIV ORE, 75- Concurrent Pos: Res grants, NSF, 52-75, Off Ord Res, 57-60 & Air Force Off Sci Res, 63-70; vis prof, High Pressure Lab, Nat Ctr Sci Res, Bellevue, France, 61 & Clarendon Lab, Oxford, 68; assoc ed, J Quant Spectros & Radiative Transfer, 72-75. Mem: Fel Am Phys Soc; Am Asn Physics Teachers; French Phys Soc. Res: Atomic spectroscopy; spectral line shape; pressure and temperature effects. Mailing Add: Dept of Physics Univ of Ore Eugene OR 97403

CHEN, SHAO LIN, b China, Aug 15, 18; nat; m 50; c 2. BIOCHEMISTRY. Educ: Nanking Univ, BS, 40; Cornell Univ, PhD(plant physiol, biochem), 49. Prof Exp: Fel plant biochem, Carnegie Inst, 49-50; res assoc, Am Smelting & Refining Co, 50-52; biochemist, Red Star Yeast Co, 52-56, sr scientist, 56-60; dir microbiol chem lab, 60-65, DIR BIOCHEM LAB, UNIVERSAL FOODS CORP, 66- Concurrent Pos: Res fel, China Found, 43; vis assoc prof, Univ Wis-Milwaukee, 65-67. Mem: AAAS; Am Chem Soc; Inst Food Technol; fel Royal Soc Health; Sigma Xi. Res: Fermentation; enzymology; metabolism; cereal chemistry; food technology. Mailing Add: Universal Foods Corp Biochem Lab 325 N 27th St Milwaukee WI 53208

CHEN, SHEPLEY S, b Taipei, Taiwan, Mar 28, 38; m 67; c 1. BIOLOGY, PLANT PHYSIOLOGY. Educ: Nat Taiwan Univ, BSc, 61; Harvard Univ, PhD(biol), 66. Prof Exp: Res assoc biochem, Mich State Univ-AEC Plant Res Lab, 65-69; asst prof, 69-73, ASSOC PROF BIOL SCI, UNIV ILL, CHICAGO CIRCLE, 73- Concurrent Pos: Jane Coffin Childs Mem Fund med res fel, 65-66; Nat Coun Sci Develop lectr, Repub of China, 67 & 69. Mem: Am Soc Plant Physiologists; Bot Soc Am. Res: Mechanism of seed dormancy; biochemistry of seed germination; plant growth hormones; protein synthesis in cultured plant cells. Mailing Add: Dept of Biol Sci Univ of Ill at Chicago Circle Chicago IL 60680

CHEN, SHI-HAN, b Che-Kiang, China, June 29, 36; US citizen; m 68; c 2. BIOCHEMICAL GENETICS, MEDICAL GENETICS. Educ: Taiwan Norm Univ, BS, 59; Nat Taiwan Univ, MS, 63; Univ Tex, Austin, PhD(zool), 68. Prof Exp: Instr biol, Taiwan Norm Univ, 63-64; teaching & res asst, Univ Tex, Austin, 64-68; res assoc & fel, King County Blood Bank, 69-71; res asst prof, 72-74, RES ASSOC PROF, DEPT PEDIAT, SCH MED, UNIV WASH, 74- Mem: Am Soc Human Genetics; AAAS. Res: Genetic variation of enzyme systems in man; biochemical causes of immunodeficiency diseases. Mailing Add: Dept of Pediat Univ of Wash Sch of Med Seattle WA 98195

CHEN, SHUI-CHIN, b Kaohsiung, Taiwan, Oct 15, 39; m 64; c 1. CLINICAL CHEMISTRY, TOXICOLOGY. Educ: Nat Taiwan Univ, BSc, 62; Mass Inst Technol, MSc, 65, PhD(biochem), 66. Prof Exp: Clin chemist, US Naval Med Res Unit, Taipei, 62-63; clin chemist & supvr, Robert B Brigham Hosp & Mass Gen Hosp, 64-67; chief clin chemist, Bionetics Res Labs, Litton Indust, 67-68; dir core lab, clin res & assoc dir clin lab, Childrens Hosp, 68-69; asst prof pediat, George Washington Univ, 68-70; asst prof, 69-71, ASSOC PROF PATH, MED COL OHIO, 71-, DIR CLIN CHEM & TOXICOL, MED COL HOSP, 69- Mem: AAAS; Am Asn Clin Chem; Am Chem Soc; Asn Clin Sci; Soc Acad Clin Lab Physicians & Scientists. Res: Methodology development; diagnostic enzymology; drug residue in the human organs and tissues; methodology of rapid emergency toxic and drug determination. Mailing Add: Med Col of Ohio PO Box 6190 Toledo OH 43614

CHEN, SOW-HSIN, b Taiwan, Mar 5, 35; m 61; c 3. EXPERIMENTAL FLUID PHYSICS, RADIATION PHYSICS. Educ: Nat Taiwan Univ, BS, 56; Nat Tsing-Hua Univ Taiwan, MS, 58; Univ Mich, MNS, 62; McMaster Univ, PhD(physics), 64. Prof Exp: Asst reactor physics, Argonne Nat Lab, 59-60; asst prof physics, Univ Waterloo, 64-67, assoc prof, 67-68; res assoc, Harvard Univ, 67-68; from asst prof to assoc prof nuclear eng, 68-74, PROF NUCLEAR ENG, MASS INST TECHNOL, 74- Concurrent Pos: Res assoc & fel, Atomic Energy Res Estab, Harwell, Eng, 64-65; vis scientist, Solid State Sci Div, Argonne Nat Lab, 75. Mem: Fel Am Phys Soc; Am Nuclear Soc; AAAS. Res: Study of molecular dynamics in solids and fluids by thermal neutron and laser light scattering. Mailing Add: 24-209 Dept of Nuclear Eng Mass Inst of Technol Cambridge MA 02139

CHEN, SOW-YEH, b Chang-Hwa, Taiwan, Aug 28, 39; m 72; c 1. ORAL PATHOLOGY. Educ: Nat Taiwan Univ, BMD, 65; Univ Ill Med Ctr, MS, 70, PhD(path), 72. Prof Exp: Nat Inst Dent Res spec fel, 71-73; ASST PROF PATH, SCH DENT, TEMPLE UNIV, 73- Mem: Am Acad Oral Path; Int Asn Dent Res; Sigma Xi; AAAS. Res: Ultrastructural study of tumors in the oral region and cell biology of oral squamous cell carcinoma. Mailing Add: 3223 N Broad St Philadelphia PA 19140

CHEN, STEPHEN P K, b Shanghai, China, June 19, 42. CHEMISTRY. Educ: Chung Chi Col, Hong Kong, Dipl, 63; Univ Wis, PhD(chem), 68. Prof Exp: SR RES CHEMIST, EASTMAN KODAK CO, 68- Mem: Am Chem Soc; Soc Rheol; Am Phys Soc. Res: Rheological behavior of polymers; molecular mobilities in both rubbery and glassy polymers; polymer physical chemistry. Mailing Add: Eastman Kodak Co Kodak Park B-81 Rochester NY 14650

CHEN, STEPHEN SHIOWSHIUNG, b Lotung, Taiwan, Sept 9, 38; US citizen; m 68; c 1. PHYSICAL CHEMISTRY. Educ: Taiwan Norm Univ, BS, 63; Tex A&M Univ, PhD(phys chem), 71. Prof Exp: Fel chem, Mobil Oil Found, 71-73; CHEM PHYSICIST, THERMODYN RES CTR, TEX A&M UNIV, 73- Mem: Am Chem Soc; Sigma Xi. Res: Applications of equation of state to pure compounds and mixtures; chemical thermodynamics; phase equilibria; potential barrier to internal rotation; statistical thermodynamic methods; computer applications. Mailing Add: Thermodyn Res Ctr Tex A&M Univ College Station TX 77843

CHEN, STEVE SHIH-CHIEH, b Taipei, Repub of China, Aug 8, 34; US citizen; m 61; c 3. AGRICULTURAL BIOCHEMISTRY. Educ: Nat Taiwan Univ, BS, 56; WVa Univ, MS, 61, PhD(agr biochem), 65. Prof Exp: Lab technician res, US Navy Med Res Unit No 2, 57-59; asst res, Grad Sch, WVa Univ, 59-64; sr scientist, Clinton Corn Prod Co, Div, Standard Brands, 64-67; res chemist, Ralston Purina Co, 67-68; dir mktg, Ralston Purine Eastern, 68-69; COUNTRY DIR/TAIWAN, AM SOYBEAN ASN, 69- Mem: Am Chem Soc; Am Oil Chemist's Soc; China Nutrit Soc; Chinese Inst Food Sci & Technol; Sigma Xi. Res: Soy protein foods, soybean oil and animal feed nutrition. Mailing Add: 14 79 St 4th Rd Tien-Mou Taipei Republic of China

CHEN, TCHAW-REN, b Kaohsiung, Taiwan, June 2, 35; US citizen; m 68; c 2. CELL BIOLOGY, CYTOGENETICS. Educ: Nat Taiwan Univ, BA, 58; Yale Univ, PhD(biol), 67. Prof Exp: Res staff ichthyol, Taiwan Fisheries Res Inst, Keelung, 58; res asst ichthyol, Inst Fisheries Biol, Nat Taiwan Univ, 60, asst zool, 60-61; cur asst ichthyol, Dept Zool, Yale Univ, 61-63; technician I & II ichthyol, Univ Calif, Santa Barbara, 63-67; res assoc, Biol Dept, Yale Univ, 67-71; ASST PROF BIOL, GRAD SCH BIOMED SCI, UNIV TEX HEALTH SCI CTR HOUSTON, 72- Mem: Am Soc Ichthyologists & Herpetologists; AAAS; Am Soc Cell Biol; Am Genetic Asn; Genetics Soc Can. Res: Chemically, physically, and virally induced human cell alteration in vitro; cytogenetics of human and other vertebrate cells in vitro and in natural populations; factors influencing cell growth of in vitro cells. Mailing Add: Grad Sch of Biomed Sci Univ of Tex Health Sci Ctr Houston TX 77025

CHEN, TIEN CHI, b Hong Kong, Nov 12, 28; m 67. COMPUTER SCIENCE. Educ: Brown Univ, SB, 50; Duke Univ, MA, 52, PhD(physics), 57. Prof Exp: Assoc physicist res ctr, 56-58, staff mathematician, 58-59, mathematician data systs div, 59-61, develop engr, 61-62, sr programmer & mgr problem-oriented programming, 62-67, tech staff mem, Advan Comput Systs, 67-68, STAFF MEM RES CTR, INT BUS MACH CORP, 68- Concurrent Pos: Vis scientist, Univ Uppsala, 65-66; mgr technol & systs, Int Bus Mach San Jose Res Lab, 73. Mem: Am Phys Soc; Asn Comput Mach; Inst Elec & Electronics Eng. Res: Eigenvalue problems in chemical physics; digital computer design and applications; numerical analysis; computer algorithms; magnetic bubbles. Mailing Add: IBM San Jose Res Lab 5600 Cottle Rd San Jose CA 95193

CHEN, TSANG JAN, b Taiwan, China, Nov 13, 34; m 61; c 3. POLYMER CHEMISTRY. Educ: Nat Taiwan Univ, BS, 57; NDak State Univ, PhD(polymer chem), 67. Prof Exp: SR CHEMIST, RES LABS, EASTMAN KODAK CO, 67- Res: Emulsion polymerization; anionic polymerization. Mailing Add: Eastman Kodak Co Res Labs 343 State St Rochester NY 14650

CHEN, TSEH-AN, b Shanghai, China, Oct 26, 28; US citizen; m 54; c 4. BOTANY. Educ: Nat Taiwan Univ, BS, 51; Univ Wis, MS, 53; Univ NH, PhD(bot), 62. Prof Exp: Res analyst bot, Univ NH, 60-62; from asst prof to assoc prof biol, Fairleigh Dickinson Univ, 62-69; assoc prof entom & econ zool, 69-74, PROF PLANT PATHOL, RUTGERS UNIV, 74-, DIR GRAD PROG NEMATOL, 70- Concurrent Pos: Res grants, NIH, 64-66, NJ Dept Health, 65 & NSF, 70-; res assoc, Cornell Univ, 63-66. Mem: Am Phytopath Soc; Soc Nematologists; Am Soc Microbiol. Res: Interaction of soil fungi and plant parasitic nematodes on development of root rots; mechanism of plant virus transmission by nematodes; myco-plasma-like organisms that cause plant diseases; corn stunt spiro-plasma. Mailing Add: Dept of Plant Biol Rutgers Univ New Brunswick NJ 08903

CHEN, TSONG MENG, b Yunlin, Taiwan, July 1, 35; US citizen; m 63; c 3. PLANT PHYSIOLOGY. Educ: Nat Taiwan Univ, BS, 58, MS, 62; Univ Calif, Davis, PhD(plant physiol), 66. Prof Exp: Res assoc plant physiol, Mich State Univ, 66-68 & Univ Ga, 68-70; assoc prof crop physiol, Nat Taiwan Univ, 70-72; res assoc plant physiol, Univ Ga, 72-73; BIOLOGIST PLANT PHYSIOL, UNION CARBIDE CORP, 73- Mem: Am Soc Plant Physiologists; Weed Sci Soc Am; Am Agron Soc. Res: Natural and synthetic plant growth regulators which will increase the yield of the major agronomic crops. Mailing Add: Res & Develop Union Carbide Corp PO Box 8361 South Charleston WV 25303

CHEN, TU, b I-Lan, Taiwan, Mar 19, 35; US citizen; m 61; c 2. MAGNETISM, MATERIALS SCIENCE. Educ: Cheng Kung Univ, Taiwan, BS, 58; Univ Minn, Minneapolis, MS, 64, PhD(metall eng, mat sci), 67. Prof Exp: Teacher math & physics, Nan-Yiang Girls High Sch, Taiwan, 61-62; staff engr mat sci res, IBM Corp, 67-68; prin scientist mat sci, Corp Res Ctr, Northrop Corp, 68-71; res scientist, 71-75, PRIN SCIENTIST MAT SCI, XEROX PALO ALTO RES CTR, 75- Mem: Am Asn Crystal Growth; Inst Elec & Electronics Engrs; Am Chem Soc; Mat Res Soc. Res: Solid state physics and magnetism; crystal growth, structure and phase equilibria; inorganic chemistry. Mailing Add: Gen Sci Lab Xerox Palo Alto Res Ctr 3333 Coyote Hill Rd Palo Alto CA 94304

CHEN, TUAN WU, b Taiwan, China, Mar 26, 36; m 64; c 2. THEORETICAL PHYSICS. Educ: Nat Taiwan Univ, BS, 58; Tsing Hua Univ, Taiwan, MS, 60; Syracuse Univ, PhD(physics), 66. Prof Exp: Instr physics, Univ Guelph, 67-68; asst prof, 68-73, ASSOC PROF PHYSICS, NMEX STATE UNIV, 73- Concurrent Pos: Fel physics, Univ Toronto, 66-68; vis scientist, Los Alamos Sci Lab, 74- Mem: Am Phys Soc. Res: Formalism of asymptotic quantum field theory in functional derivatives; effect of masses of gauge fields on chiral symmetry; high energy behaviors of scattering processes; nucleus scattering at medium energy. Mailing Add: Dept of Physics NMex State Univ Las Cruces NM 88001

CHEN, TZE TUAN, cytology, protozoology, see 12th edition

CHEN, WILLIAM KWO-WEI, b Shanghai, China, July 18, 28; US citizen; m 51; c 2. POLYMER CHEMISTRY, CHEMICAL ENGINEERING. Educ: Mass Inst Technol, BSc, 51; Polytech Inst Brooklyn, PhD(chem), 58. Prof Exp: Chem engr, Quaker Chem Prod Co, 51-52; chem engr, Am Mach & Foundry Co, 52-53, sr chem engr, 53-54, group leader plastic prod, 54-56, sect mgr desalination & membranes, 56-61, asst mgr chem lab, 61-63, mgr liquid processing lab, 63-64; mgr res & develop, Celanese Plastic Co, 64-68, VPRES AMCEL CO & GEN EXPORT MGR, CELANESE PLASTIC CO, 68- Mem: Am Chem Soc; Soc Plastics Eng. Res: Membranes, films and battery separators; desalination and liquid processing; plastic and resins; marketing; international trade. Mailing Add: 102 Pinegrove Rd Berkeley Heights NJ 07922

CHEN, YI-DER, b Taipei, Taiwan, May 29, 40; m 69. THEORETICAL CHEMISTRY, BIOPHYSICS. Educ: Nat Taiwan Univ, BS, 63; Nat Tsing Hua Univ, MS, 65; Pa State Univ, PhD(chem), 70. Prof Exp: Res chemist biophys Univ Calif, Santa Cruz, 69-72; RES ASSOC THEORET BIOL, NIH, 72- Mem: Biophys Soc; Am Chem Soc. Res: Statistical mechanics of nonspherical molecules; fluctuations and noise in chemical and biological systems; theoretical studies of membrane transports, muscle contractions and conformational transitions in biopolymers. Mailing Add: Bldg 2 Rm 319 NIH Bethesda MD 20014

CHEN, YOK, b Soochow, China, July 27, 31; US citizen; m 61; c 3. PHYSICS. Educ: Univ Wis, BSc, 52; Purdue Univ, PhD(physics), 65. Prof Exp: Electron microscopist, Univ Chicago, 53-55; physicist, Hoffman Semiconductor Prod, Evanston, Ill, 55-58; RES PHYSICIST, OAK RIDGE NAT LAB, 65- Mem: Am Phys Soc; Am Ceramic Soc. Res: Radiation damage in semiconductors; impurities and defects in insulators; radiation effects and ion implantation in oxides; optical and magnetic resonance spectroscopy. Mailing Add: Solid State Div Oak Ridge Nat Lab Oak Ridge TN 37830

CHEN, YU WHY, b Nantungchow, China, Apr 1, 10; m 38; c 1. MATHEMATICS. Educ: Univ Göttingen, PhD, 34. Prof Exp: Prof math, Peking Univ, 36-45; res assoc, NY Univ, 46-49; res fel, Inst Adv Study, 49-50; assoc prof, Univ Okla, 50-52; from assoc prof to prof, Wayne State Univ, 52-65; PROF MATH, UNIV MASS, AMHERST, 65- Mem: Am Math Soc. Res: Partial differential equations and applications. Mailing Add: Dept of Math Univ of Mass Amherst MA 01002

CHEN, YUH-CHING, b Fukien, China, May 20, 30; m 55; c 3. PURE MATHEMATICS. Educ: City Univ New York, PhD(math), 66. Prof Exp: Asst math, Taiwan Norm Univ, 54-59; asst, Nanyang Univ, Singapore, 59-60; teacher high sch, Malaya, 60-62; asst prof, Univ Minn, Morris, 64-65 & Wesleyan Univ, 66-71; ASST PROF MATH, FORDHAM UNIV, 71-, CHMN DEPT, 74- Mem: Sigma Xi. Res: Homology and homotopy theories and their applications. Mailing Add: Dept of Math Fordham Univ Bronx NY 10458

CHEN, YUNG MING, b China, Dec 30, 35; US citizen; m 62. APPLIED MATHEMATICS. Educ: Univ Md, BS, 56; Drexel Inst, MS, 58; Univ Calif, Berkeley, MA, 60; NY Univ, PhD(appl math), 63. Prof Exp: Res engr, Radio Corp Am, NJ, 56-58; asst, Electronics Res Labs, Univ Calif, Berkeley, 58-59; asst, Dept Math, 59-60; asst appl math, Courant Inst Math Sci, NY Univ, 60-63; asst prof, Purdue Univ, 63-65; assoc prof, Univ Fla, 65-67; assoc prof, 67-72, PROF APPL MATH, STATE

UNIV NY STONY BROOK, 72- Mem: Am Math Soc; Am Phys Soc; Inst Elec & Electronics Engrs; Soc Eng Sci; Soc Indust & Appl Math. Res: Wave propagation; approximation methods in initial and boundary value problems; numerical analysis and stochastic process. Mailing Add: Dept of Appl Math State Univ of NY Stony Brook NY 11790

CHENERY, PETER JASPERSEN, b Chicago, Ill, May 26, 19; m 41; c 3. SCIENCE ADMINISTRATION, INFORMATION SCIENCE. Educ: Harvard Univ, BS, 40. Prof Exp: Asst aeronaut eng, Mass Inst Technol, 40-42; from asst prod engr to eng sect head, Sperry Gyroscope Co, 42-44, 46-54, dir res & develop, Wright Mach Co div, Sperry Rand Corp, 54-63; DIR, NC BD SCI & TECHNOL, 63- Concurrent Pos: Pres, Asn Sci Info Dissemination Ctrs, 72-74; chmn, Southern Interstate Nuclear Bd, 73-75. Honors & Awards: US Naval Ord Develop Award, 46. Mem: AAAS; Inst Elec & Electronics Eng; Am Soc Info Sci. Res: Research administration; industrial utilization of new technology; instrumentation and control; mechanized information retrieval. Mailing Add: NC Bd of Sci & Technol Box 12235 Research Triangle Park NC 27709

CHENEY, CHARLES BROOKER, b New Haven, Conn, Mar 2, 12; m 34; c 5. OBSTETRICS & GYNECOLOGY. Educ: Yale Univ, BA, 34, MD, 41; Am Bd Obstet & Gynec, dipl, 52. Prof Exp: Asst surg obstet & gynec, 41-43, asst obstet & gynec, 46-47, instr, 47-49, clin instr, 49-52, asst clin prof, 52-69, sr clin assoc, 69-74, ASSOC CLIN PROF OBSTET & GYNEC, SCH MED, YALE UNIV, 74-; ASST ASSOC CHIEF OBSTET & GYNEC & ATTEND OBSTETRICIAN & GYNECOLOGIST, YALE-NEW HAVEN HOSP, 57- Concurrent Pos: Pres, Conn Med Exam Bd. Mem: AMA; Am Col Obstet & Gynec. Mailing Add: 111 Park St New Haven CT 06511

CHENEY, ELLIOTT WARD, (JR), b Gettysburg, Pa, June 28, 29; m 52; c 3. MATHEMATICS. Educ: Lehigh Univ, BA, 51; Univ Kans, PhD(math), 57. Prof Exp: Instr math, Univ Kans, 52-56; design specialist & mathematician, Convair-Astronaut Div, Gen Dynamics Corp, 56-59; mem tech staff, Space Tech Labs, Inc, 59-61; asst prof math, Iowa State Univ, 61-63; from asst prof to assoc prof, Univ Calif, Los Angeles, 62-65; assoc prof, 65-66, PROF MATH, UNIV TEX, AUSTIN, 66- Concurrent Pos: Vis assoc prof, 64; guest prof, Univ Lund, 66-67; vis prof, Mich State Univ, 69-70; assoc ed, J Approximation Theory. Mem: Am Math Soc; Math Asn Am; Soc Indust & Appl Math (ed, J Numerical Anal). Res: Approximation theory; linear inequalities; numerical analysis. Mailing Add: Dept of Math Univ of Tex Austin TX 78712

CHENEY, ERIC SWENSON, b New Haven, Conn, Nov 17, 34; m 58; c 4. ECONOMIC GEOLOGY. Educ: Yale Univ, BS, 56, PhD(geol), 64. Prof Exp: Instr sci, Southern Conn State Col, 63-64; asst prof geol, 64-69, ASSOC PROF GEOL, UNIV WASH, 69- Concurrent Pos: Consult, Maine Geol Surv, 63-69, Amoco Mining Co, 70-71, Tex Gulf Inc, 72, Crangesellschart-USA, 74 & Continental Oil Co Inc, 75. Mem: AAAS; Geol Soc Am; Am Inst Mining, Metall & Petrol Eng; Sigma Xi; Soc Econ Geologists. Res: Geology and geochemistry of ore deposits; mineral and energy resources. Mailing Add: Dept of Geol Sci Univ of Wash Seattle WA 98195

CHENEY, FREDERICK WYMAN, b Ayer, Mass, Jan 17, 35; m 59; c 2. ANESTHESIOLOGY. Educ: Tufts Univ, BS, 56, MD, 60. Prof Exp: Instr anesthesiol, 64-66, from asst prof to assoc prof, 66-74, PROF ANESTHESIOL, SCH MED, UNIV WASH, 74- Concurrent Pos: NIH res fel, Sch Med, Univ Wash, 66-67. Mem: Am Physiol Soc; Am Soc Anesthesiologists; Am Thoracic Soc; Soc Critical Care Med. Res: Pulmonary edema; pulmonary vasculature; clinical respiratory care. Mailing Add: Dept Anesthesiol Mail Stop RN10 Univ of Wash Sch of Med Seattle WA 98195

CHENEY, HORACE BELLATTI, b Emerson, Iowa, Dec 15, 13; m 40; c 3. SOIL FERTILITY. Educ: Iowa State Univ, BS, 35; Ohio State Univ, PhD(soil fertility), Prof Exp: Jr soil surveyor, Soil Conserv Serv, USDA, Iowa, 35-37; exten assoc agron, Iowa State Univ, 37-39; res assoc, Ohio State Univ, 39-41; exten & res asst prof agron, econ & sociol, Iowa State Univ, 41-43, exten asst prof agron, 43-44, from assoc prof to prof, 45-52; PROF SOILS & HEAD DEPT, ORE STATE UNIV, 52- Mem: Fel AAAS; Soil Sci Soc Am (pres, 64); fel Soil Conserv Soc Am; fel Am Soc Agron (pres, 73). Res: Soil management. Mailing Add: Dept of Soil Sci Ore State Univ Corvallis OR 97331

CHENEY, LEE CANNON, b Centerville, Utah, Aug 24, 08; m 43; c 4. MEDICINAL CHEMISTRY. Educ: Univ Utah, AB, 31, MS, 34; Iowa State Col, PhD(org chem), 38. Prof Exp: Res chemist, Parke, Davis & Co, 38-45; res chemist, Bristol Labs, Bristol-Myers Co, 45-50, dir org chem res, 50-73; RETIRED. Honors & Awards: Syracuse Sect Award, Am Chem Soc, 64. Mem: Am Chem Soc. Res: Dibenzofuran derivatives; vitamin K; antispasmodics; analgesics; biotin-type compounds; antihistamine agents; sympatholytic agents; ganglionic blocking agents; antibiotics; diuretics; hypoglycemics; aminopenicillanic acid derivatives; cephalosporin derivatives. Mailing Add: Woodchuck Hill Rd Fayetteville NY 13066

CHENEY, MONROE G, b Ft Worth, Tex, Mar 10, 19; m 52; c 3. PHYSICS, GEOPHYSICS. Educ: Rice Univ, BA, 41; Univ Tex, MA, 50; Columbia Univ, MA, 52. Prof Exp: Radio physicist, US Naval Res Lab, DC, 42-47; asst prof physics, Hardin-Simmons Univ, 47-49; physicist, US Bur Mines, Pa, 52; reservoir engr, Anzac Oil Corp, Tex, 52-55 & Socony Mobil, Venezuela, 55-59; asst prof physics, Arlington State Col, 59-67, ASST PROF PHYSICS, UNIV TEX, ARLINGTON, 67- Concurrent Pos: Mem, Byrd Antarctic Exped, 46-47; consult, Anzac Oil Corp, 55-74. Mem: Am Inst Mining, Metall & Petrol Eng; Am Geophys Union. Res: Ground constants at radio frequency; resaturation of oil by gas at elevated pressures; seismic surface-wave models; fracturing massive limestone with light crude oil; seismic waves in ice and snow. Mailing Add: 1217 W Cedar Arlington TX 76012

CHENG, CHENG-YIN, b Shinpu, Formosa, Jan 29, 30; nat US; m 61. SOIL CHEMISTRY, ANALYTICAL CHEMISTRY. Educ: Berea Col, BS, 55; Univ Ill, MS, 56, PhD(soil chem), 60. Prof Exp: Instr chem, Wilson Col, 60-61; from asst prof to assoc prof, Shippensburg State Col, 61-65; from asst prof to assoc prof, Ithaca Col, 65-68; PROF CHEM, EAST STROUDSBURG STATE COL, 68- Mem: Am Chem Soc; fel Am Inst Chemists; Am Soc Agron. Res: Cation diffusion in soils; radioisotope technology; instrumental methods of analysis; precipitation from homogeneous solution; ion selective electrodes; air and water quality. Mailing Add: Dept of Chem East Stroudsburg State Col East Stroudsburg PA 18301

CHENG, CHIA-CHUNG, b China, May 5, 25; m 53; c 4. ORGANIC CHEMISTRY. Educ: Chekiang Univ, BS, 48; Univ Tex, MA, 51, PhD, 54. Prof Exp: Res assoc org res, NMex Highlands Univ, 54-57 & Princeton Univ, 57-59; head, Cancer Chemother Sect, 59-66, HEAD MED CHEM SECT, MIDWEST RES INST, 66- Concurrent Pos: Mem med chem A study sect, NIH, 73-77. Honors & Awards: Sci Award, Midwest Res Inst, 74. Mem: AAAS; Am Chem Soc; NY Acad Sci; The Chem Soc. Res: Synthesis, identification and reaction mechanism study of organic compounds; synthesis and evaluation of antimetabolites, antibiotics, alkylating agents, vitamin analogs and natural products. Mailing Add: Midwest Res Inst Kansas City MO 64110

CHENG, CHIANG-SHUEI, b Sinchu, Taiwan, Sept 22, 35; US citizen; m 60; c 1. PLASMA PHYSICS. Educ: Nat Taiwan Univ, BS, 58; Nat Tsing Hua Univ, MS, 60; Lehigh Univ, PhD(physics), 68. Prof Exp: Instr physics, Lehigh Univ, 65-69; assoc prof, 69-73, PROF PHYSICS, EAST STROUDSBURG STATE COL, 73- Mem: Am Phys Soc; Sigma Xi. Res: Scattering function and transport theory of a plasma. Mailing Add: Dept of Physics East Stroudsburg State Col East Stroudsburg PA 18301

CHENG, DAVID, b Chungking, China, July 21, 41; US citizen; m 63; c 2. EXPERIMENTAL PHYSICS. Educ: Univ Calif, Berkeley, BS, 62, MS, 63, PhD(exp high energy physics), 65. Prof Exp: Physicist exp physics, Lawrence Radiation Lab, Univ Calif, 65-67; assoc physicist, Brookhaven Nat Lab, 67-70; mem tech staff, Bell Labs, 70-74; MEM RES STAFF, XEROX RES CTR, 74- Mem: Optical Soc Am; Am Phys Soc. Res: High speed and high density optical recording technique; laser physics and applications; integrated circuit pattern generation techniques; laser and CCD image recording; acousto-optical devices and electro optics. Mailing Add: Xerox Res Ctr 3333 Coyote Hill Rd Palo Alto CA 94304

CHENG, FRANK HSIEH FU, b Shanghai, China, Nov 16, 23; US citizen; m 58; c 3. BIOCHEMISTRY, IMMUNOCHEMISTRY. Educ: St John's Univ, China, BS, 46; Univ Tenn, MS, 50; Ind Univ, PhD(biochem), 57. Prof Exp: Chemist & assoc supt pharmaceut lab, T W Wu & Co, China, 46-49; asst, Ind Univ, 53-56; biochemist, Toledo Hosp Inst Med Res, Ohio, 58-63; asst prof, 64-69, ASSOC PROF RADIOL, UNIV IOWA, 69- Concurrent Pos: Fel, Univ Wis, 56-58. Mem: AAAS; Am Chem Soc; Am Acad Allergy; Soc Nuclear Med. Res: Radiobiology. Mailing Add: Dept of Nuclear Med Univ of Iowa Hosp Iowa City IA 52242

CHENG, FRED FA WU, b China, Dec 28, 18. ORGANIC CHEMISTRY, ANALYTICAL CHEMISTRY. Educ: Ga Inst Technol, MS, 50; Emory Univ, PhD(chem), 58. Prof Exp: Res chemist, Tenn Corp Res Labs, 51-57 & Atlas Chem Industs, Inc, 58-72, RES CHEMIST, ICI UNITED STATES INC, 72- Mem: Am Chem Soc; Am Microchem Soc. Res: Infrared analysis; oil seed proteins; microanalysis; polarography. Mailing Add: ICI United States Inc Concord Pike & New Murphy Rd Wilmington DE 19897

CHENG, GEORGE CHIWO, b China, Sept 7, 29; US citizen; m 70. BIOMEDICAL ENGINEERING, ELECTRICAL ENGINEERING. Educ: South China Univ, BA, 53; Mont State Univ, MS, 60. Prof Exp: Res assoc, Electronics Res Lab, Mont State Univ, 59-61; asst prof elec eng, Southeastern Mass Tech Inst, 61-62; sr res scientist, 63-68, head biomath div, Nat Biomed Res Found, 68-71; consult, Toshiba Res & Develop Ctr, Japan, 71-72; assoc prof, Univ Fla, 73-74; CONSULT AUTOMATIC MED DATA PROCESSING & INT TECHNOL EXCHANGE, 74- Concurrent Pos: Vis lectr, Southeastern Mass Tech Inst, 62-; managing ed, Pattern Recognition Soc Jour. Mem: AAAS; Am Soc Eng Educ; Pattern Recognition Soc. Res: Information processing; pattern recognition; microelectronics and billiongate computer design; nervous system simulation; automatic medical data processing; pictorial data processing by computers; international technology exchange. Mailing Add: Rogers Rd M102 Athens GA 30601

CHENG, HAZEL PEI-LING, b Hong Kong; Can citizen. GASTROENTEROLOGY, CELL BIOLOGY. Educ: McGill Univ, BSc, 67, MSc, 69, PhD(anat), 72. Prof Exp: Lectr anat, McGill Univ, 72-73; res staff cell biol, Yale Univ, 73-75; ASST PROF ANAT, UNIV ILL, URBANA, 75- Concurrent Pos: Adj res assoc cell biol, Rockefeller Univ, 73. Mem: Am Asn Anatomists; Am Asn Cell Biologists. Res: Biological mechanism involved in the origin, differentiation and renewal of the four main epithelial cell types in the mouse small intestine. Mailing Add: Sch of Basic Med Sci Med Sci Bldg Univ Ill Urbana IL 61801

CHENG, HWEI-HSIEN, b Shanghai, China, Aug 13, 32; US citizen; m 62; c 2. SOIL CHEMISTRY, BIOCHEMISTRY. Educ: Berea Col, BA, 56; Univ Ill, MS, 58, PhD(agron), 61. Prof Exp: Res assoc soil chem, Univ Ill, 61-62; soil biochem, Iowa State Univ, 62-63, asst prof, 64-65; Fulbright res scholar & collab soil chem, soils res ctr, State Agr Univ, Belgium, 63-64; asst prof soils, 65-71, ASSOC PROF SOILS, WASH STATE UNIV, 71- Concurrent Pos: Guest scientist, Jülich Nuclear Res Ctr, Ger, 72-73. Mem: AAAS; fel Am Inst Chem; Soil Sci Soc Am; Am Chem Soc; Am Soc Agron. Res: Fractionation and distribution of nitrogen in soils; methods for use of nitrogen-15 in soils research; movement and transformation of pesticides in soils; soil organic matter turnover. Mailing Add: Dept of Agron & Soils Wash State Univ Pullman WA 99163

CHENG, KUANG LU, b Yangchow, China, Sept 14, 19; nat US. ANALYTICAL CHEMISTRY. Educ: Northwestern Col, BS, 41; Univ Ill, MS, 49, PhD(soil chem), 51. Prof Exp: Fel, Univ Ill, 51-52; microchemist, Com Solvents Corp, 52-53; asst prof chem, Univ Conn, 53-55; engr, Westinghouse Elec Corp, 55-57; assoc dir res, Metals Div, Kelsey-Hayes Co, 57-59; mem tech staff, Labs, Radio Corp Am, 59-66; PROF CHEM, UNIV MO-KANSAS CITY, 66- Mem: Fel AAAS; Am Chem Soc; Electrochem Soc; Soc Appl Spectros. Res: Photoelectron spectroscopy; ion selective electrodes; ligand chromatography. Mailing Add: Dept of Chem Univ of Mo Kansas City MO 64110

CHENG, KUO-JOAN, b Taiwan, China, Dec 9, 40; m 68; c 1. AGRICULTURAL MICROBIOLOGY. Educ: Nat Taiwan Univ, BS, 63; Univ Sask, MS, 66, PhD(microbiol), 69. Prof Exp: Res asst dairy & food sci, Univ Sask, 64-66; Nat Res Coun Can fel microbiol, MacDonald Col, McGill Univ, 69-71; RUMINANT MICROBIOLOGIST, ANIMAL SCI SECT, RES STA, CAN DEPT AGR, 71- Mem: Am Soc Microbiol; Can Soc Microbiol. Res: Coliform bacteria in Canadian diary products; degradation of rutin and related flavonoids by anaerobic rumen bacteria; studies on the localization and the role of periplasmic enzymes in bacterial cell; microbiology and biochemistry of digestion in the rumen. Mailing Add: Animal Sci Sect Res Sta of Can Dept of Agr Lethbridge AB Can

CHENG, LANNA, b Singapore, Apr 27, 41; m 69. MARINE ZOOLOGY, ENTOMOLOGY. Educ: Univ Singapore, BSc, 63, hons, 64, MSc, 66; Oxford Univ, DPhil(entom), 69. Prof Exp: Nat Res Coun Can fel biol, Univ Waterloo, 68-69; Nat Res Coun Can fel biol, 69-72, ASST RES BIOLOGIST, SCRIPPS INST OCEANOG, UNIV CALIF, SAN DIEGO, 72- Mem: Int Ecol Soc; fel Royal Entom Soc London; Brit Ecol Soc; Am Soc Limnol & Oceanog; Entom Soc Am. Res: Taxonomic and biological studies of aquatic insects especially Gerridae, Hemiptera; population ecology; host-parasite relationships; physiological adaptations of insects to marine environments; ecology of marine insects; pleuston; animals of the sea-air interface. Mailing Add: Scripps Inst of Oceanog Univ of Calif San Diego La Jolla CA 92093

CHENG, LAWRENCE KAR-HIU, b Hong Kong, July 25, 47. PHARMACEUTICS. Educ: Univ Calif, Santa Barbara, BS, 71; State Univ NY Buffalo, PhD(pharmaceut), 75. Prof Exp: ANAL CHEMIST PHARMACEUT, STUART PHARMACEUT, ICI US INC, 75- Mem: Am Pharmaceut Asn; Am Chem Soc. Res: Analytical methods development for drugs and metabolites in the biological tissues and dosage forms; pharmacokinetics of drugs and other pharmacologically active chemicals;

bioequivalency studies of new formulations. Mailing Add: Stuart Pharmaceut ICI US Inc CR & DL Wilmington DE 19897

CHENG, LI-JEN, b Chekiang, China, Mar 18, 32; m 66; c 2. EXPERIMENTAL SOLID STATE PHYSICS. Educ: Ordnance Eng Col, BS, 56; Nat Tsing Hua Univ, MS, 61; Rensselaer Polytech Inst, PhD(solid state physics), 66. Prof Exp: Fel & asst res officer solid state sci, Chalk River Nuclear Labs, 66-69; prof physics, Chung Chen Inst Technol, 70-72; vis sr scientist solid state physics, NY Univ, 72-73; SR FEL SOLID STATE PHYSICS, INST STUDY DEFECTS IN SOLIDS, STATE UNIV NY ALBANY, 73- Concurrent Pos: Assoc physicist, Inst Nuclear Energy Res, 70-72. Honors & Awards: Chuang Shon-Kent Sci Award of 72, Chuang Mem Found, 73. Mem: Am Phys Soc; Sigma Xi. Res: Semi-conductors and metals; electronic materials; lattice defects; radiation effects; ion implantation; atomic diffusion; optical and electrical properties; positron annihilation; interfaces; device physics; nuclear science; instrumentation. Mailing Add: Inst for Study of Defects in Solids Dept of Physics State Univ of NY Albany NY 12222

CHENG, PING YAO, chemistry, see 12th edition

CHENG, SHU-SING, b Kwangtung, China, Sept 7, 23. ORGANIC CHEMISTRY, MICROBIOLOGY. Educ: Nat Col Pharm, Nanking, BS, 47; Univ NC, MS, 59, PhD(pharmaceut chem), 61. Prof Exp: Res assoc virol, Sch Med, Univ NC, 60-62; res fel steroid biochem, Worcester Found Exp Biol, 62-63; res asst prof med chem & biochem, Univ NC, 63-65; HEAD MICROBIOL RES, KENDALL RES CTR, KENDALL CO, 66- Mem: AAAS; Am Chem Soc; Am Pharmaceut Asn; Am Soc Microbiol. Res: Design and synthesis of organic compounds of medicinal interest; chemistry and biochemistry of steroids; tissue culture and microbiology; screening of antimicrobial agents; process research in cold sterilization and monitoring devices. Mailing Add: Kendall Res Ctr 411 Lake Zurich Rd Barrington IL 60010

CHENG, SZE-CHUH, b Soochow, China, Nov 11, 21; US citizen; m; c 3. NEUROCHEMISTRY, BIOLOGY. Educ: Southwest Assoc Univ, China, BSc, 43; Brown Univ, MSc, 49; Univ Pa, PhD(gen physiol), 54. Prof Exp: Res asst gen physiol, Tsinghua Univ, Peking, 43-47; res assoc neurophysiol, Rockefeller Univ, 55-60; sr res scientist neurochem, NY State Psychiat Inst, 60-64; sr res scientist neurochem, NY State Res Inst Neurochem & Drug Addiction, 65-71; ASSOC PROF ANESTHESIOL, SCH MED, NORTHWESTERN UNIV, CHICAGO, 71- Concurrent Pos: Fel, McCollum-Pratt Inst, Johns Hopkins Univ, 53-55; Pub Health Serv spec fel, Agr Res Coun Inst Animal Physiol, Babraham, Eng, 64-65. Mem: Am Soc Neurochem; Soc Neurosci. Res: Metabolism and function of nervous tissue including drug effects. Mailing Add: Dept of Anesthesia Northwestern Univ Sch of Med Chicago IL 60611

CHENG, TAI CHUN, b Shanghai, China; US citizen; m 64; c 2. ORGANIC POLYMER CHEMISTRY, ORGANOMETALLIC CHEMISTRY. Educ: Tunghai Univ, BS, 60; Wash State Univ, PhD(chem), 68. Prof Exp: Res assoc chem, Univ Notre Dame, 67-68; res scientist, 68-75, SR RES SCIENTIST POLYMERS, FIRESTONE TIRE & RUBBER CO, 75- Mem: Am Chem Soc; Sigma Xi. Res: Anionic polymerization and Ziegler-Natta polymerization of olefin and diolefin; synthesis of specialty rubber such as phosphazene polymer; new polymers and monomers synthesis; mechanism study of radical anion and aryl halides; synthesis and kinetic study of organoboron compounds. Mailing Add: Cent Res Labs Firestone Tire & Rubber Co Akron OH 44303

CHENG, THOMAS CLEMENT, b Nanking, China, Nov 5, 30; nat US; m 56; c 3. BIOLOGY. Educ: Wayne State Univ, AB, 52; Univ Va, MS, 56, PhD(biol), 58. Prof Exp: Lab asst, Wayne State Univ, 52; lab instr, Univ Va, 55-57, instr biol, 57-58; asst prof, Univ Md, 58-59; from asst prof to assoc prof, Lafayette Col, 59-64; chief parasitol & immunol, Northeast Marine Health Sci Lab, USPHS, 64-65; from assoc prof to prof, Univ Hawaii, 65-69; PROF PHYSIOL, DIR INST PATHOBIOL & CTR FOR HEALTH SCI, LEHIGH UNIV, 69- Prof Exp: Sr investr, Mt Lake Biol Sta, 63; vis scientist, Pac Biomed Res Ctr, Hawaii, 63-64; mem, Surgeon Gen Comn Food Protection, 64-65; adj prof, Univ RI, 64-65; instr-chg, NSF US-Japan Coop Res Training Prog Biol, 68; mem study sect, Environ. Biol & Chem Prog, USPHS, 69-72; ed, J Invert Path, 69-, Current Topics in Comp Pathobiol, 70-73 & Comp Pathobiol, 73-; asst ed, J Parasitol, 70-74 . Honors & Awards: Fleming Award, 58; Phi Sigma Award, 59; Darbaker Award, 61; Jones Teaching Award, 62. Mem: Fel AAAS; Soc Exp Biol & Med; Am Soc Parasitol; Soc Invert Path; Am Micros Soc (vpres, 75-75). Res: Experimental invertebrate pathology; general parasitology; physiology of host-parasite relationships; biochemistry of protozoa; helminths of mollusks; comparative serology and immunology; transplantation immunity; histochemistry; radioisotope tracers; epidemiology of parasitic diseases; marine ecology. Mailing Add: Inst for Pathobiol Lehigh Univ Bethlehem PA 18015

CHENG, TIEN-HSI, b Foochow, China, Apr 5, 12; nat US; m 41; c 2. ZOOLOGY, ENTOMOLOGY. Educ: Fukien Christian Univ, BS, 33; State Univ NY, BA, 33; Dickinson Col, PhB, 37; Ohio State Univ, MSc, 38, PhD(entom, zool), 39. Prof Exp: Instr biol, Trinity Col, Foochow & prov high sch, Changchow, 33-36; asst prof zool & entom, Lingnan Univ, 40-42, assoc prof & dir agr biol res lab, 42-45; res fel, Univ Minn, 46-47; res assoc, Crop Protection Inst, Univ NH, 48-49; from asst prof to prof zool, 49-72, head dept, 67-70, EMER PROF ZOOL, PA STATE UNIV, UNIVERSITY PARK, 72- Concurrent Pos: Sr tech adv, Chinese Relief & Rehab Admin, UN, 46, mem econ comn, Asia & Far East, 47-48; Asia Found vis prof, Chung Chi & Baptist Cols, Hong Kong, 60; Griswold lectr, Cornell Univ, 62; mem comt sci develop Communist China, NSF, 65-69. Mem: AAAS; Entom Soc Am; Am Inst Biol Sci. Res: Insecticides; livestock insect control; automatic spraying devices; biological control; physiological and histological effects of antibiotics; television teaching. Mailing Add: Dept of Biol Pa State Univ University Park PA 16802

CHENG, TSUNG O, b Shanghai, China, Mar 30, 25; nat US; m; c 2. CARDIOLOGY. Educ: St Johns Univ, China, BS, 47; Pa Med Sch, China, MD, 50; Univ Pa, MMedSc, 56; Am Bd Internal Med, dipl, 61; Am Bd Cardiovasc Dis, dipl, 63. Prof Exp: Intern, Hosp St Barnabas, NJ, 50-51; resident internal med, Cook County Hosp, Ill, 52-55; asst cardiol, Mass Gen Hosp, 56-57; asst physician, Cardiac Clin, Johns Hopkins Hosp, 57-59; dir cardiopulmonary lab, Brooklyn Hosp & asst prof med, State Univ NY Downstate Med Ctr, 59-66; dir cardiovasc lab & chief cardiol, Vet Admin Hosp, Brooklyn, 66-70; assoc prof, Sch Med, 70-72, PROF MED, SCH MED & DIR CARDIAC CATHETERIZATION LAB, MED CTR, GEORGE WASHINGTON UNIV, 72- Concurrent Pos: Fel Northwestern Univ, 54-55; fel cardiol, Sch Med, George Washington Univ, 55-56; res fel cardiopulmonary physiol, Johns Hopkins Univ & Hosp, 57-59, Am Heart Asn advan res fel, 58-59; consult & chief pediat cardiac clin, Cumberland Hosp, 63-66; asst vis physician, Kings County Hosp, 64-70; fel coun clin cardiol, Am Heart Asn, 64-; clin asst prof, State Univ NY Downstate Med Ctr, 66-70; consult, Beth Israel Hosp, New York, 70-; chief cardiol, DC Gen Hosp, 71-72. Mem: Fel Am Col Physicians; fel Am Col Cardiol; AMA; fel Am Col Chest Physicians; fel Int Col Angiol. Res: Clinical investigations in cardiopulmonary pathophysiology in health and diseases. Mailing Add: Med Ctr George Washington Univ Washington DC 20037

CHENG, WU-CHIEH, physical chemistry, see 12th edition

CHENG, YUNG-CHI, b London, Eng, Dec 29, 44; Chinese citizen; m 69; c 1. BIOCHEMICAL PHARMACOLOGY. Educ: Tunghai Univ, Taiwan, BSc, 66; Brown Univ, PhD(biochem pharmacol), 72. Prof Exp: Fel pharmacol, Yale Univ, 72-73, res assoc, 73-74; ASST PROF PHARMACOL, STATE UNIV NY BUFFALO, 74-, SR CANCER RES SCIENTIST, ROSWELL PARK MEM RES INST, 74- Mem: Am Soc Microbiol; Am Soc Pharmacol & Exp Therapeut. Res: Development and use of nucleoside analogs for cancer and viral chemotherapy. Mailing Add: Dept of Exp Therapeut Roswell Park Mem Res Inst Buffalo NY 14263

CHENHALL, ROBERT GENE, b Maurice, Iowa, Jan 24, 23; m 44, 72; c 3. ANTHROPOLOGY, ARCHAEOLOGY. Educ: Calif State Univ, San Diego, BA, 46; Ariz State Univ, MA, 65, PhD(anthrop), 72. Prof Exp: Sr systs analyst, Price Waterhouse & Co, 51-55; treas & dir, Fisher Contracting Co, 55-63; corp controller, Del E Webb Corp, 63-66; fac assoc anthrop, Ariz State Univ, 65-66, lectr, 66-67, mgr data processing, Inspiration Consol Copper Co, 67-68; asst prof anthrop, Ark Archeol Surv & Univ Ark, Fayetteville, 69-72, ASST PROF ANTHROP, EXEC DIR MUS DATA BANK COORD COMT & RES ASSOC, MUS, UNIV ARK, FAYETTEVILLE, 72- Concurrent Pos: Ed, Newslett Comput Archaeol, 65-71; Wenner-Gren Found Anthrop Res grants, Ark Archeol Surv, Univ Ark, Fayetteville, 70-73, NSF grant, 72-74, Smithsonian Inst Mus Act grant, Mus, 72-74; organizer, Archaeol Data Bank Conf, 71; consult, NMex Data Bank Consortium, 71-72; chmn, Mus Data Bank Study Group, 72- Mem: AAAS; fel Am Anthrop Asn; Soc Am Archaeol. Res: Applications of computers in museum cataloging and in archaeological research; prehistory of the southwestern United States; archaeological theory. Mailing Add: Univ of Ark Mus Fayetteville AR 72701

CHENIAE, GEORGE MAURICE, b Mounds, Ill, Aug 27, 28; m 52; c 3. PLANT BIOCHEMISTRY. Educ: Univ Ill, BS, 50; NC State Col, MS, 57, PhD(plant physiol), 59. Prof Exp: Asst, Oak Ridge Nat Lab, 50-52; Nat Sci fel, 59-60; res scientist, Res Inst Advan Study, 60-75; PROF AGRON, UNIV KY, 75- Concurrent Pos: Adv ed, Am Soc Plant Physiologists, 70- Mem: Am Soc Plant Physiologists; Am Soc Biol Chemists. Res: Lipid metabolism; respiration; photosynthesis. Mailing Add: Dept of Agron Univ of Ky Lexington KY 40506

CHENICEK, ALBERT GEORGE, b Chicago, Ill, Dec 15, 13; m 43; c 3. POLYMER CHEMISTRY. Educ: Univ Chicago, BS, 34, PhD(org chem), 37. Prof Exp: Res chemist, Columbia Chem Div, Pittsburgh Plate Glass Co, Ohio, 37-41 & Interchem Corp, 41-53; actg dir develop, Standard Coated Prods, Inc, 53-56; mgr res, Stoner-Mudge Co, 56-61; PRES, UNIFILM CORP, 61- Mem: AAAS; Am Chem Soc. Res: Chlorination of organic compounds; synthetic resins; coated fabrics; protective coatings; inks. Mailing Add: Unifilm Corp 60 Cornell Blvd Somerville NJ 08876

CHENOT, CHARLES FREDERIC, b Canton, Ohio, Sept 16, 38; m 62; c 4. SOLID STATE CHEMISTRY. Educ: Col Wooster, BA, 60; Univ Cincinnati, PhD(phys chem), 64. Prof Exp: ENG SPECIALIST, CHEM & METALL DIV, GTE SYLVANIA INC, 64- Mem: Am Ceramic Soc; Am Chem Soc; Electrochem Soc. Res: Physical chemical studies of diffusion in the solid state; research and development of solid state luminescent materials including phase equilibrium relationships, solid state reactions and associated luminescence spectroscopy. Mailing Add: Chem & Metall Div GTE Sylvania Inc Towanda PA 18848

CHENOWETH, MAYNARD BURTON, b Chicago, Ill, Nov 25, 17; m 40; c 5. Educ: Columbia Univ, BA, 38; Cornell Univ, MD, 42. Prof Exp: Asst prof pharmacol, Cornell Univ, 46-48; assoc prof, Univ Mich, 48-53; RES SCIENTIST PHARMACOL, DOW CHEM CO, 53- Mem: Am Soc Pharmacol & Exp Therapeut; Soc Toxicol; Soc Exp Biol & Med. Res: Study of new drugs or chemicals in man. Mailing Add: Dow Chem Co Midland MI 48640

CHENOWETH, PHILIP ANDREW, b Chicago, Ill, Aug 21, 19; m 52; c 2. GEOLOGY, STRATIGRAPHY. Educ: Columbia Univ, BA, 46, MA, 47, PhD, 49. Prof Exp: Instr geol, Amherst Col, 49-51; sr geologist, Sinclair Oil & Gas Co, 51-54; assoc prof geol, Univ Okla, 54-60; staff geologist, Sinclair Oil & Gas Co, 60-65, res assoc, Sinclair Res Ctr, 65-68; CONSULT GEOLOGIST, 54-60, 68- Honors & Awards: Prof Award, Univ Okla, 55. Mem: AAAS; fel Geol Soc Am; Am Asn Petrol Geologists; Am Inst Prof Geologists; Am Inst Mining, Metall & Petrol Eng. Res: Petroleum and exploration geology. Mailing Add: 702 Petrol Club Bldg Sixth at Boulder Tulsa OK 74119

CHENOWETH, WILLIAM LYMAN, b Wichita, Kans, Sept 16, 28; m 55; c 4. GEOLOGY. Educ: Univ Wichita, AB, 51; Univ NMex, MS, 53. Area geologist, Div Raw Mat, AEC, 55-58, chief geol engr, Sect Off, Flagstaff, Ariz, 58-62, proj geologist, Resource Appraisal Br, 62-70, Chief, 70-74, STAFF GEOLOGIST, US ENERGY RES & DEVELOP ADMIN, 75- Mem: Geol Soc Am; Am Inst Mining, Metall & Petrol Eng; Am Asn Petrol Geologists. Res: Uranium geology; Mesozoic stratigraphy; depositional environment of uranium ore deposits. Mailing Add: 707 Brassie Dr Grand Junction CO 81501

CHEO, PEN CHING, b Ho-Fei, China, Mar 28, 19; m 49. PLANT VIROLOGY. Educ: Nanking Univ, China, BS, 41; WVa Univ, MS, 49; Univ Wis, PhD(plant path), 51. Prof Exp: Fel, Plant Indust Sta, USDA, 51-53; res assoc, Univ RI, 53-55; fel, Tree Fruit Exp Sta, Wash State Col, 55-57; res fel, Div Biol, Calif Inst Technol, 57-61; asst plant pathologist, Wash State Univ, Wenatchee, 61-66; plant pathologist, Dept Arboreta & Bot Gardens, 66, CHIEF RES DIV, DEPT ARBORETA & BOT GARDENS, LOS ANGELES STATE & COUNTY ARBORETUM, 66- Mem: Am Phytopath Soc. Mailing Add: Los Angeles State & Co Arboretum 301 N Baldwin Ave Arcadia CA 91006

CHEO, PETER K, b Nanking, China, Feb 2, 30; US citizen; m 56; c 4. PHYSICS. Educ: Aurora Col, BS, 51; Va Polytech Inst, MS, 53; Ohio State Univ, PhD(physics), 64. Prof Exp: Instr physics, Bethany Col, 54-57; asst prof, Aurora Col, 57-61; mem prof staff, Bell Tel Labs, 63-70; mgr laser appln res, Aerojet-Gen Corp, 70-71; SR RES SCIENTIST, UNITED TECHNOLOGIES RES CTR, 71- Mem: Am Phys Soc; Optical Soc Am; sr mem Inst Elec & Electronics Eng. Res: Laser and device research; optical communication; non-linear and coherent phenomena; atomic and molecular spectroscopy; integrated optics. Mailing Add: 86 Waldridge Rd West Hartford CT 06119

CHEPENIK, KENNETH PAUL, b Jacksonville, Fla, Mar 14, 38; m 63; c 2. DEVELOPMENTAL BIOLOGY. Educ: Univ Fla, BSAdv, 61, MS, 65, PhD(human anat, biochem, physiol), 68. Prof Exp: Instr anat, Bowman Gray Sch Med, Wake Forest Univ, 68-70, asst prof, 70-73; ASSOC PROF ANAT, JEFFERSON MED COL, THOMAS JEFFERSON UNIV, 73- Mem: Sigma Xi; NY Acad Sci; Teratology Soc; Am Asn Anatomists; Soc Develop Biol. Res: Biochemical mechanisms underlying normal and abnormal mammalian embryogenesis. Mailing Add: Dept of Anat Jefferson Med Col Philadelphia PA 19107

CHER, MARK, b Buenos Aires, Arg, June 14, 32; nat US; m 56; c 4. PHYSICAL CHEMISTRY. Educ: Calif Inst Technol, BS, 54; Harvard Univ, AM, 55, PhD(chem), 58. Prof Exp: Instr chem, Univ Calif, Los Angeles, 57-59, asst prof, 59-60; res specialist, Atomics Int, Calif, 60-63; mem tech staff, NAm Aviation Sci Ctr, 63-69; assoc prof chem & chmn dept, Wis State Univ, River Falls, 69-71; sr chemist, Addressograph-Multigraph Corp, 71-74; MGR QUAL ASSURANCE, ROCKWELL INT AIR MONITORING CTR, 74- Mem: Am Chem Soc; Am Phys Soc. Res: Fast reaction kinetics; atoms and free radicals in gas phase; application of ultrasonics to chemical kinetics; shock waves and detonations; photochemistry; radiation chemistry; air pollution monitoring. Mailing Add: Rockwell Int Air Monitor Ctr 2421 W Hillcrest Newbury Park CA 91320

CHERASKIN, EMANUEL, b Philadelphia, Pa, June 9, 16; m 44; c 1. ORAL MEDICINE. Educ: Univ Ala, AB, 39, MA, 41, DMD, 52; Univ Cincinnati, MD, 43; Am Bd Oral Med, dipl. Prof Exp: Intern med, Hartford Munic Hosp, 43-44; resident, St Mary's Hosp, Ind, 46-47; asst prof physiol, 50-52, assoc prof oral med, Sch Dent, 52-56, prof oral surg & oral med & chmn dept, 56-62, PROF ORAL MED & CHMN DEPT, MED CTR, UNIV ALA, BIRMINGHAM, 62- Concurrent Pos: Consult, Vet Admin Hosps, 52- & Southeastern Area, Vet Admin, 55-64; mem staff, Univ Hosp, Univ Ala & Hillman Clin, 57- Mem: Am Acad Oral Med; Am Dent Asn; AMA; hon mem Circle Odontol Paraguay; hon mem Dom Odontol Soc. Res: Predictive medicine; nutrition; metabolism. Mailing Add: Univ of Ala University Sta Birmingham AL 35294

CHERAYIL, GEORGE DEVASSIA, b Kothamangalam, India, Dec 17, 29; m 57; c 3. BIOCHEMISTRY, CHEMISTRY. Educ: Univ Madras, BSc, 49, Hons, 52, MA, 55; St Louis Univ, PhD(biochem), 62. Prof Exp: Demonstr chem, St Xavier's Col, India, 49-50; lectr, Fatima Mata Nat Col, India, 52-54; lectr, Med Col, Univ Mysore, 54-55; chmn dept, Andhra Loyola Col, 55-58; ASST PROF PATH, MED COL WIS, 62- Mem: AAAS; Am Chem Soc; Am Soc Neurochem. Res: Metabolism of steroids, especially bile acids; lipid metabolism in central nervous system disorders; biochemistry of phospholipids, sphingolipids and glycolipids. Mailing Add: Dept of Path Med Col of Wis 8700 W Wisconsin Ave Milwaukee WI 53226

CHERBAS, LUCY FUCHSMAN, b New York, NY, May 24, 43; m 68; c 1. DEVELOPMENTAL BIOLOGY. Educ: Swarthmore Col, BA, 64; Harvard Univ, PhD(biol), 71. Prof Exp: Res fel biol, Harvard Univ, 70-73; res fel biochem, Cambridge Univ, 73-74; res fel biol, Mass Inst Technol, 74-75; RES FEL BIOL, HARVARD UNIV, 76- Mem: Soc Develop Biol. Res: Mechanism of heme requirement for globin synthesis; action of ecdysone on cultured Drosophila cells. Mailing Add: Biol Labs Harvard Univ 16 Divinity Ave Cambridge MA 02138

CHERBAS, PETER THOMAS, b Bryn Mawr, Pa, Mar 26, 46; m 68; c 1. DEVELOPMENTAL BIOLOGY, INSECT PHYSIOLOGY. Educ: Harvard Col, BA, 67; Harvard Univ, PhD(biol), 73. Prof Exp: Res fel genetics, Cambridge Univ, 73-74; RES FEL BIOL, HARVARD UNIV, 74- Mem: Genetics Soc Am. Res: Biochemical studies of the mechanism of action of the insect steroid hormone ecdysone; genetic studies of the activity of this hormone in cell cultures and in polytene chromosomes. Mailing Add: Biol Labs Harvard Univ 16 Divinity Ave Cambridge MA 02138

CHERENACK, PAUL FRANCIS, b Hazleton, Pa, June 19, 42. MATHEMATICS. Educ: Villanova Univ, BS, 63; Univ Pa, PhD(math), 68. Prof Exp: Teaching asst, Univ Pa, 63-68; instr, Villanova Univ, 68; asst prof, Ind Univ, Bloomington, 68-74; MEM FAC MATH, UNIV CAPE TOWN, SAFRICA, 74- Mem: Am Math Soc; Math Asn Am. Res: Nature of singularities on algebraic varieties via their analytic homotopy groups; algebraic geometry. Mailing Add: Dept of Math Univ Cape Town Private Bag Rondebasch South Africa

CHERIAN, SEBASTIAN K, b Palai, India, Sept 23, 38; m 67; c 2. ZOOLOGY. Educ: Univ Kerala, BSc, 59; Duquesne Univ, MS, 63; St Bonaventure Univ, PhD(physiol), 67. Prof Exp: USPHS res grant, 61-63; teaching asst physiol, Duquesne Univ, 60-63 & zool, St Bonaventure Univ, 63-66; asst prof biol, St Francis Col, Pa, 66-69; asst prof, 69-74, ASSOC PROF BIOL, JAMESTOWN COL, 74-, ACTG CHMN DEPT, 70- Mem: Am Soc Zoologists. Res: Ovarian and uterine responses to exogenous hormone administration in the immature rat; concentration and distribution of uterine glycogen in unilaterally pregnant rats. Mailing Add: Dept of Biol Jamestown Col Jamestown ND 58401

CHERIN, PAUL, b Brooklyn, NY, Oct 14, 34; m 57; c 3. SOLID STATE SCIENCE. Educ: Brooklyn Col, BS, 55; Polytech Inst Brooklyn, PhD(phys chem), 63. Prof Exp: Staff scientist, Int Bus Mach Corp, 60-61; scientist, 62-66, SR SCIENTIST, XEROX CORP, 66- Mem: Am Phys Soc; Am Chem Soc; Am Crystallog Asn; Sigma Xi; NY Acad Sci. Res: X-ray diffraction; relation of crystallographic properties of solids with their physical properties. Mailing Add: Xerox Corp 800 Philips Rd Bldg 218 Webster NY 14580

CHERKIN, ARTHUR, b Latrobe, Pa, July 24, 13; m 43; c 2. BIOCHEMISTRY. Educ: Univ Calif, Los Angeles, BA, 33, PhD(biochem), 53. Prof Exp: Res chemist, Max Factor & Co, Calif, 33; co-owner, Synthetics, 34; chemist, Don Baxter, Inc, 34-36, chief chemist, 36-43, dir res, 43-48, vpres, 48-63; chief psychobiol res lab, 65-75, DIR, CLIN & RES CTR AGING, VET ADMIN HOSP, 75- Concurrent Pos: Res chemist, Univ Calif, Los Angeles, 33, lectr anesthesiol, Sch Med, 66-, res biochemist, Dept Psychiat, 73-; res fel, Calif Inst Technol, 62-66; consult, Don Baxter, Inc, 63-65. Mem: Geront Soc; Int Brain Res Orgn; Soc Neurosci; NY Acad Sci; Am Soc Pharmacol & Exp Therapeut. Res: Amino acids and peptides; pyrogens; parenteral alimentation and therapy; mechanism of general anesthesia; learning and memory; experimental amnesias; memory enhancement. Mailing Add: Clin & Res Ctr on Aging Vet Admin Hosp Sepulveda CA 91343

CHERKOFSKY, SAUL CARL, b Lynn, Mass, June 2, 42; m 62; c 3. ORGANIC CHEMISTRY. Educ: Mass Inst Technol, BS, 63; Harvard Univ, MA, 64, PhD(org chem), 67. Prof Exp: RES CHEMIST, CENT RES DEPT, E I DU PONT DE NEMOURS & CO, INC, 66- Mem: Am Chem Soc. Res: Anthracene photodimers; carbonium ion rearrangements; bicyclobutane synthesis and polymers; aromatic substitutions; heterocyclic chemistry; medicinal chemistry. Mailing Add: 1013 Woodstream Dr Ramblewood Wilmington DE 19810

CHERLIN, GEORGE (YALE), b New Haven, Conn, Feb 21, 24; m 45; c 2. MATHEMATICS. Educ: Rutgers Univ, MSc, 49, PhD(math), 51. Prof Exp: Asst instr, Rutgers Univ, 47-51; asst mathematician, Mutual Benefit Life Ins Co, 51-62; vpres-actuary, Nat Health & Welfare Retirement Asn, Inc, 62-72; ASSOC MATHEMATICIAN, MUTUAL BENEFIT LIFE INS CO, 72- Mem: Casualty Actuarial Soc; Soc Actuaries; Am Acad Actuaries. Res: Actuarial science; mathematical logic; complex variable. Mailing Add: 9 Porter Pl Newark NJ 07112

CHERMACK, EUGENE E A, b New York, NY, Aug 31, 34; m 56; c 4. PHYSICAL METEOROLOGY. Educ: Queens Col, NY, BS, 56; Univ Wash, BS, 58; NY Univ, MS, 62, PhD(meteorol), 70. Prof Exp: Weather officer, US Air Force, 56-59; asst res scientist, NY Univ, 59-62, instr meteorol, 62-67; ASSOC PROF METEOROL, STATE UNIV NY COL OSWEGO, 67- Mem: Am Meteorol Soc; Am Geophys Union; Optical Soc Am; Am Inst Physics; Int Asn Gt Lakes Res. Res: Development of indirect sounding techniques applied to the atmosphere; infrared radiation and atmospheric optics; surface temperature of lakes and rivers. Mailing Add: Dept of Earth Science State Univ of NY Col Oswego NY 13126

CHERMOCK, RALPH LUCIEN, b Pittsburgh, Pa, Aug 25, 18; m 43; c 1. ECOLOGY. Educ: Univ Pittsburgh, BS, 39; Duquesne Univ, MS, 41; Cornell Univ, PhD(entom), 47. Prof Exp: Instr biol, Beaver Col, 41-42; prof, Univ Ala, 47-66, dir mus natural hist, 61-66; dept chmn, Parsons Col, 66-73; CHIEF ENVIRON DIV, GEOL SURV ALA, 73- Honors & Awards: US Dept Army Spec Citation, 72. Mem: Sigma Xi; AAAS; Am Inst Biol Sci; Lepidopterists Soc; Lepidopterists Found. Res: Environmental studies as related to industrial or municipal development; taxonomy of lepidoptera; zoogeography. Mailing Add: PO Drawer 0 University AL 35486

CHERMS, FRANK LLEWELLYN, JR, b Warwick, RI, June 4, 30; m 52; c 2. POULTRY PHYSIOLOGY. Educ: Univ RI, BS, 52; Univ NH, MS, 54; Univ Md, PhD, 58. Prof Exp: Instr & asst geneticist, Univ NH, 54-55; from asst prof to prof, Univ Wis, 57-69; REPROD PHYSIOLOGIST, NICHOLAS TURKEY BREEDING FARMS, INC, 69- Mem: AAAS; Poultry Sci Asn; Soc Study Reprod; World Poultry Sci Asn. Res: Improving the reproductive performance of the turkey through breeding and physiology. Mailing Add: 1212 Apple Tree Ct Sonoma CA 95476

CHERN, BERNARD, theoretical physics, see 12th edition

CHERN, MING-FEN MYRA, b Kiangsu Prov, China, Apr 19, 46; m 69; c 1. BIOSTATISTICS, POPULATION GENETICS. Educ: Nat Taiwan Univ, BS, 67; Univ Minn, MS, 70, PhD(biomet), 73. Prof Exp: Res fel biol res, 72-73, res assoc, 73-74, ASST PROF HEALTH COMPUT SCI & BIOMET, UNIV MINN, 74- Mem: Am Soc Human Genetics; Am Statist Asn; Biomet Soc; AAAS; Sigma Xi. Res: The application of principles in biostatistics, population genetics and computer science on genetic modeling and diagnostic aids in medicine. Mailing Add: Box 511 Mayo Mem Bldg Univ of Minn Minneapolis MN 55455

CHERN, SHIING-SHEN, b Kashing, China, Oct 26, 11; US citizen; m 39; c 2. GEOMETRY. Educ: Nankai Univ, China, BS, 30; Tsing Hua Univ, China, MS, 34; Univ Hamburg, DSc, 36. Hon Degrees: LLD, Chinese Univ Hong Kong & DSc, Univ Chicago, 69. Prof Exp: Prof math, Tsing Hua Univ, China, 37-43, Acad Sinaca, 46-48 & Univ Chicago, 49-59; PROF MATH, UNIV CALIF, BERKELEY, 60- Concurrent Pos: Guggenheim fel, 54-55 & 67; colloquium lectr, Am Math Soc, 60. Honors & Awards: Chauvenet Prize, Math Asn Am, 70. Mem: Nat Acad Sci; Am Acad Arts & Sci; Am Math Soc (vpres, 63-64); Math Asn Am; Acad Sinaca. Res: Differential geometry; integral geometry; topology. Mailing Add: Dept of Math Univ of Calif Berkeley CA 94720

CHERNESKY, MAX ALEXANDER, b Inglis, Man, Aug 1, 38; m 65. MEDICAL VIROLOGY. Educ: Univ Guelph, BS, 65; Univ Toronto, MD, 67; Univ BC, PhD(virol), 69. Prof Exp: Instr, 70-72, ASST PROF PEDIAT, McMASTER UNIV, 72- Mem: Am Soc Microbiol; Am Soc Trop Med & Hyg; Can Soc Microbiol. Res: Pathogenesis of viral infections. Mailing Add: Dept of Pediat Fac of Med McMaster Univ Hamilton ON Can

CHERNETSKI, KENT EUGENE, zoology, neurophysiology, see 12th edition

CHERNIACK, LOUIS, b Winnipeg, Man, Nov 23, 08. PULMONARY DISEASES. Educ: Univ Man, MD, 32, BSc, 34; FRCP(C); FRCP. Prof Exp: Asst prof, 50-60, ASSOC PROF INTERNAL MED, FAC MED, UNIV MAN, 60-, MEM, JOINT RESPIRATORY PROG, 68- Concurrent Pos: Physician, Health Sci Ctr, 47- Mem: Fel Am Col Physicians; fel Am Col Chest Physicians; Am Thoracic Soc; Can Med Asn; Can Thoracic Soc. Res: Cinical investigation. Mailing Add: Respiratory Ctr Health Sci Ctr 668 Bannatyne Ave Winnipeg MB Can

CHERNIACK, NEIL S, b Brooklyn, NY, May 28, 31; m 55; c 3. INTERNAL MEDICINE. Educ: Columbia Univ, AB, 52; State Univ NY, MD, 56. Prof Exp: Intern med, Univ Ill Med Ctr, 56-57, res fel, 57-58, resident, 60-62; res fel pulmonary dis, Columbia Univ, 62-64; from asst prof to assoc prof med, Univ Ill Med Ctr, 64-69; assoc prof, 69-73, PROF MED & ASSOC DIR PULMONARY SERV, UNIV PA, 73- Concurrent Pos: Consult, Chicago State Tuberc Sanitarium, Ill, 64-69; assoc attend physician, Cook County Hosp, 65-69; assoc attend physician & sr res assoc, Michael Reese Hosp, 67-69; sr attend physician, Philadelphia Gen Hosp, Pa, 69- Mem: Am Thoracic Soc; Am Physiol Soc; Am Soc Clin Invest. Res: Effect of acceleration on the lung; control of ventilation and circulation; pulmonary disease; oxygen and carbon dioxide stores of the body; bioengineering. Mailing Add: Dept of Med Univ of Pa Philadelphia PA 19041

CHERNIACK, REUBEN MITCHELL, b Can, June 15, 24; m 52; c 3. MEDICINE. Educ: Univ Man, MD, 48, MSc, 51; FRCP(C). Prof Exp: Lectr, Dept Physiol & Med Res, 54-56, from asst prof to assoc prof med, 56-66, PROF MED, FAC MED, UNIV MAN, 66- Concurrent Pos: Fel med, Columbia Univ, 52-54; Life Ins fel & Markle scholar, 54; dir, Cardiorespiratory Unit, Winnipeg Gen Hosp, 54, Inhalation Ther Unit, 61, consult physician in respiratory dis, 58, dir respiratory div, Clin Invest Unit; dir, Joint Respiratory Prog, Sanatorium Bd Man & Univ Man; med dir, D A Stewart Ctr Study & Treat Respiratory Dis; consult, Man Rehab Hosp & Munic Hosps, 58. Honors & Awards: Prowse Prize, 52; Drewery Prize, 53. Mem: Fel Am Col Physicians; Am Soc Clin Invest; Am Physiol Soc; Can Col Physicians & Surg; Can Soc Clin Invest (secy, 61-63, pres, 63, past pres, 64). Res: Internal medicine; respiratory function and diseases. Mailing Add: Respiratory Lab Ward F-2 Winnipeg Gen Hosp Winnipeg MB Can

CHERNIAK, EUGENE ANTHONY, b Windsor, Ont, Dec 17, 30; m 55; c 2. PHYSICAL CHEMISTRY. Educ: Queen's Univ, Ont, BA, 53, MA, 56; Univ Leeds, PhD(radiation chem), 59. Prof Exp: Sci master, Pickering Col, Can, 53-55; Nat Res Coun Can fel, 59-60; lectr chem, Carleton Univ, Can, 60-61, asst prof, 61-65; chmn dept, 65-69, PROF CHEM, BROCK UNIV, 65- Mem: AAAS; Chem Inst Can; Faraday Soc; The Chem Soc. Res: Chemical kinetics; photochemistry (including flash photochemistry); radiation chemistry. Mailing Add: 7 Wychwood Rd St Catharines ON Can

CHERNIAK, ROBERT, b New York, NY, June 26, 36; m 61; c 2. BIOCHEMISTRY. Educ: City Col New York, BS, 59; Duke Univ, PhD(biochem), 64. Prof Exp: Arthritis Found fel biochem, Univ Newcastle, 64 & Albert Einstein Col Med, 65; assoc res biologist, Sterling-Winthrop Res Inst Div, Sterling Drug, Inc, 66-68; asst prof, 68-73, ASSOC PROF CHEM, GA STATE UNIV, 73- Res: Structure and biosynthesis of polysaccharides of biological origin. Mailing Add: Dept of Chem Ga State Univ University Plaza Atlanta GA 30303

CHERNICK, CEDRIC LOUIS, inorganic chemistry, physical chemistry, see 12th edition

CHERNICK, SIDNEY SAMUEL, b Winnipeg, Man, Mar 6, 21; US citizen; m 47; c 3. BIOCHEMISTRY. Educ: Univ Calif, Los Angeles, AB, 43; Univ Calif, MA, 45, PhD(physiol), 48. Prof Exp: Physiologist, Med Sch, Univ Calif, 48-51; prof pharmacol & physiol, NDak State Col, 51-52; scientist, 52-63, SCIENTIST DIR, NAT INST ARTHRITIS, METAB & DIGESTIVE DIS, 63- Concurrent Pos: Vis scientist, Med Clin, Munich, 63-64. Honors & Awards: Purkinje Medal, Czech Med Soc, 69. Mem: Soc Exp Biol & Med; Am Soc Biol Chem. Res: Metabolic defects in endocrine and nutritional diseases; in vitro metabolism; diabetes. Mailing Add: Lab Nutrit & Endocrinol Nat Inst Arthritis Metab & Digestive Dis Bethesda MD 22014

CHERNICK, VICTOR, b Winnipeg, Man, Dec 31, 35; m 57; c 4. PEDIATRICS, PHYSIOLOGY. Educ: Univ Man, MD, 59; Am Bd Pediat, dipl, 65. Prof Exp: Rotating intern, Winnipeg Gen Hosp, Man, 59-60; Nat Inst Neurol Dis & Blindness perinatal fel, Johns Hopkins Univ, 60, from jr asst resident to asst resident pediat, 60-62, univ fel environ med & pediat, 62-64, from instr to asst prof pediat, 64-66; from asst prof to assoc prof pediat & physiol, 66-71, PROF PEDIAT & HEAD DEPT, UNIV MAN, 71- Concurrent Pos: Resident, Vet Admin Hosp, Baltimore, 64, attend physician, 64-65, consult, 66; chief respiratory dis & perinatal physiol, Children's Hosp, 67-71; pediatrician in chief, Health Sci Ctr, 71- Honors & Awards: Queen Elizabeth II Scientist Award, 67; Medal, Can Pediat Soc, 70. Mem: Fel Am Thoracic Soc; fel Am Acad Pediat; Soc Pediat Res; Can Soc Clin Invest; Am Physiol Soc. Res: Pulmonary physiology; neonatology. Mailing Add: Children's Ctr 685 Bannatyne Ave Winnipeg MB Can

CHERNICK, WARREN SANFORD, b Providence, RI, Oct 6, 29. PHARMACOLOGY. Educ: RI Col Pharm, BS, 52; Philadelphia Col Pharm, MS, 54, DSc, 56. Prof Exp: From instr to asst prof pharmacol, Philadelphia Col Pharm, 54-64; from asst prof to assoc prof, 64-68, PROF PHARMACOL & CHMN DEPT, HAHNEMANN MED COL & HOSP, 68- Concurrent Pos: Res assoc, Children's Hosp & Med Sch, Univ Pa, 57-64. Mem: Am Pharmaceut Asn; Am Soc Pharmacol & Exp Therapeut. Res: Salivary secretion; psychopharmacology; cystic fibrosis. Mailing Add: Dept of Pharmacol Hahnemann Med Col & Hosp Philadelphia PA 19102

CHERNIN, ELI, b New York, NY, Sept 12, 24; m 56; c 3. MEDICAL PARASITOLOGY, TROPICAL PUBLIC HEALTH. Educ: City Col New York, BS, 44; Univ Mich, MA, 48; Johns Hopkins Univ, ScD(parasitol), 51. Hon Degrees: AM, Harvard Univ, 70. Prof Exp: Asst zool, Univ Mich, 47-48; asst parasitol, Sch Med, Johns Hopkins Univ, 48-49, 50-51 & Sch Hyg & Pub Health, 49-51; res assoc, 51-52, from instr to assoc prof, 52-69, PROF TROP PUB HEALTH, SCH PUB HEALTH, HARVARD UNIV, 70- Concurrent Pos: Consult, Ludlow Mfg & Sales Co, Calcutta & Boston, 51-53, trop med & parasitol study sect, USPHS, 66-70 & parasitic dis panel, US-Japan Coop Med Sci Prog, 70-74; sr res fel, USPHS, 56-61, career develop award, 61-64, res career award, 64-; China Med Bd-La State Univ travel fel, Cent Am, 57; asst ed, Jour Parasitol, 68-72; adv, WHO, 70; mem ed bd, Am J Trop Med & Hyg, 71-72; consult, Ctr Dis Control, USPHS, PR, 71-72, mem training grant comt, Nat Inst Allergy & Infectious Dis, 72-73; vchmn & mem bd dir, Coun Biol Ed, 75- Honors & Awards: Bailey K Ashford Res Award, Am Soc Trop Med & Hyg, 61. Mem: AAAS; Am Soc Parasitol; Am Soc Trop Med & Hyg; Royal Soc Trop Med & Hyg. Res: Incidence and epidemiology of malaria and other parasitic infections in industrial workers in India; biology and biological control of disease-carrying snails, especially of vectors of human schistosomiasis; transmission and biology of filariasis. Mailing Add: Dept of Trop Pub Health Harvard Sch of Pub Health Boston MA 02115

CHERNOFF, AMOZ IMMANUEL, b Malden, Mass, Mar 17, 23; m 53; c 3. HEMATOLOGY. Educ: Yale Univ, BS, 43, MD, 47. Prof Exp: Intern med, Mass Gen Hosp, 47-48; asst resident, Barnes Hosp, St Louis, Mo, 48-49; res fel hemat, Michael Reese Hosp, Chicago, 49-51; from instr to asst prof, Wash Univ, 51-56; assoc prof, Duke Univ, 56-58; res prof, 58-64, DIR, MEM RES CTR, UNIV TENN, KNOXVILLE, 64-, PROF MED, SCH MED, 75- Concurrent Pos: Asst dir hemat res lab, Michael Reese Hosp, 50-51; Am Col Physicians fel med, Sch Med, Wash Univ, 51-52, USPHS fel, 52-53; consult, City Hosp, St Louis, 52-56; asst physician, Barnes Hosp, 52-56; chief hemat sect, Vet Admin Hosp, Durham, NC, 56-58; mem cancer chemother study sect, USPHS, 58-63; USPHS res career award, 62- Mem: AAAS; fel Am Col Physicians; Am Soc Human Genetics; Int Soc Hemat; Am Soc Clin Invest. Res: Hemolytic anemias; abnormal hemoglobins; biochemical genetics. Mailing Add: Univ of Tenn Mem Res Ctr Knoxville TN 37920

CHERNOFF, HERMAN, b New York, NY, July 1, 23; m 47; c 2. MATHEMATICAL STATISTICS. Educ: City Col New York, BS, 43; Brown Univ, ScM, 45, PhD(appl math), 48. Prof Exp: Res assoc, Cowles Comn Res Econ, Chicago, 47-49; asst prof statist & math, Univ Ill, 49-52; from assoc prof to prof statist, Stanford Univ, 52-74; PROF APPL MATH, MASS INST TECHNOL, 74- Concurrent Pos: Mem exec comt, Assembly Math & Phys Sci, Nat Res Coun, 74-76. Mem: Am Acad Arts & Sci; Am Math Soc; Inst Math Statist; Am Statist Asn. Res: Statistical problems in econometrics; sequential design of experiments; rational selection of decision functions; large sample theory; pattern recognition. Mailing Add: Dept of Math 2-381 Mass Inst Technol Cambridge MA 02139

CHERNOFF, HYMAN MORDECAI, internal medicine, deceased

CHERNOSKY, EDWIN JASPER, b Rosenberg, Tex, May 21, 14; m 43; c 3. ENVIRONMENTAL PHYSICS, SPACE PHYSICS. Prof Exp: Chemist, Champion Paper Co, Tex, 37-42; observer, Huancayo Magnetic Observ, Peru, Carnegie Inst, 42-45, res geophysicist, 45-46; mem tech staff physics res dept, Naval Ord Lab, Washington, DC, 46-52; physicist & actg chief geomagnetics unit, 52-55; chief geomagnetic activity sect, 55-67, RES PHYSICIST, AIR FORCE CAMBRIDGE RES LABS, 67- Concurrent Pos: Deleg, Int Asn Geomagnetism & Aeronomy, Toronto, 57, Helsinki, 60, Berkeley, 63, Zurich, 67, Madrid, 69, Moscow, 71, Kyoto, 73 & Grenoble, 75, mem comn IV, IX & lunar variations, chmn interdiv comn of hist; mem organizing comt & deleg, Int Symp Equatorial Aeronomy, Huaychulo, Peru, 62, San Jose do Campos, Brazil, 65 & Ahmedabad, India, 69. Mem: AAAS; Am Geophys Union; Am Phys Soc; Inst Elec & Electronics Engrs; Sigma Xi. Res: Solar geomagnetic relationships, morphology of solar activity and geomagnetic variations, characterization of geomagnetic time variations, recurrence phenomena of geomagnetic variations; 22 year geomagnetic cycle, dichotomy in geomagnetic activity; solar-terrestrial physics. Mailing Add: 48 Berkley St Waltham MA 02154

CHERNOV, HARVEY IRWIN, b Providence, RI, Mar 31, 36; m 57; c 2. PHARMACOLOGY. Educ: RI Col Pharm, BSc, 57; Univ Wis, MS, 59; Univ Iowa, PhD(pharmacol), 64. Prof Exp: Pharmacologist, Sterling-Winthrop Res Inst, 59-61; sect head neuro-pharmacol, Ciba Pharmaceut Co, 64-72; WITH NEUROPHARMACOL DIV, US FOOD & DRUG ADMIN, 72- Mem: Am Soc Pharmacol & Exp Therapeut. Res: Analgesics, biotransformation and mechanism of action; central nervous system stimulants and depressants; neurophysiology. Mailing Add: Neuropharmacol Div US Food & Drug Admin Rockville MD 20852

CHERNY, WALTER B, b Montreal, Que, Apr 13, 26; nat US; m 55; c 2. OBSTETRICS & GYNECOLOGY. Educ: McGill Univ, BSc, 48, MD, CM, 50. Prof Exp: Instr obstet & gynec, Sch Med, Duke Univ, 55, assoc, 55-57, from asst prof to prof, 57-70; DIR RESIDENCY, POST-GRAD TRAINING IN OBSTET & GYNEC, GOOD SAMARITAN HOSP, 70- Concurrent Pos: Chief serv obstet, Lincoln Hosp, 58; consult, Watts Hosp, 58-70. Mem: AAAS; AMA; Am Col Obstet & Gynec. Res: Obstetrics and gynecology affecting physical and emotional health; reproductive physiology. Mailing Add: Dept of Obstet & Gynec Good Samaritan Hosp Phoenix AZ 85006

CHERRINGTON, ERNEST HURST, JR, b Westerville, Ohio, Sept 10, 09; m; c 2. ASTRONOMY. Educ: Ohio Wesleyan Univ, AB, 31, MS, 32; Univ Calif, PhD(astrophysics), 35. Prof Exp: Instr math & astron, Syracuse Univ, 35-36; asst astronomer, Perkins Observ, Ohio Wesleyan Univ, 36-46, from instr to asst prof astron, 36-46; assoc prof physics & asst dean, Centenary Col, 46-47, head dept physics & astron & dean, 47-48; prof astron, Univ Akron, 48-67, dean, Col Lib Arts, 48-60, dir, Grad Studies, 55-60, dean, Grad Sch, 60-67; prof astron, Hood Col, 67-75; RETIRED. Concurrent Pos: From instr to asst prof physics, Ohio State Univ, 36-46. Mem: Am Astron Soc; Royal Astron Soc Can. Res: Dynamics of particles in comet tails; spectrophotometry of solar radiation; Be Stars; the moon. Mailing Add: Rte 5 Box 242 Frederick MD 21701

CHERRINGTON, VIRGIL ARTHUR, bacteriology, deceased

CHERRY, DONALD STEPHEN, b Paterson, NJ, Sept 23, 43; m 66; c 1. AQUATIC ECOLOGY. Educ: Furman Univ, BSc, 65; Clemson Univ, MSc, 70, PhD(zool), 73. Prof Exp: Teacher biol & football coach, J L Mann High Sch, SC, 65-68; instr human ecol, Clemson Univ, 72-73; res assoc & fel, 73-74, ASST PROF BIOL, CTR ENVIRON STUDIES, VA POLYTECH INST & STATE UNIV, 74- Concurrent Pos: Consult, Dept Microbiol, Clemson Univ, 73; investr, Facil Use Agreement, Savannah River Proj, 72-75, co-investr, AEC contract, 73-75; consult, Am Elec Power Serv Corp, Canton, Ohio, 74-75, co-investr, 74-76. Mem: Ecol Soc Am; Am Water Works Asn; Int Water Resources Asn. Res: Impact of power production discharges upon aquatic food chains in the drainage systems by site-specific field laboratory and field biomonitoring activities. Mailing Add: Ctr Environ Studies Dept of Biol Va Polytech Inst & State Univ Blacksburg VA 24061

CHERRY, EDWARD TAYLOR, b Gainesboro, Tenn, Nov 12, 41; m 67; c 3. ENTOMOLOGY. Educ: Tenn Polytech Inst, BS, 63; Univ Tenn, MS, 66, PhD(entom), 70. Prof Exp: Res assoc appl entom, Miss State Univ, 70-71; asst prof agr biol, Univ Tenn, 71-74; RES SPECIALIST PLANT PROTECTANTS, CIBA-GEIGY CORP, 74- Mem: Entom Soc Am; Sigma Xi. Res: Chemical insecticides both from a potential nature and extension of existing compounds; subject areas of pest management, biological control. Mailing Add: Ciba-Geigy Corp Box 11422 Greensboro NC 27409

CHERRY, JAMES DONALD, b Summit, NJ, June 10, 30; m 54; c 3. PEDIATRICS, INFECTIOUS DISEASES. Educ: Springfield Col, BS, 53; Univ Vt, MD, 57; Am Bd Pediat, dipl, 62. Prof Exp: Intern pediat, Boston City Hosp, 57-58, asst resident, 58-59; resident, Kings County Hosp, Brooklyn, NY, 59-60; instr, Col Med, Univ Vt, 60-61; NIH fel med, Harvard Med Sch & Thorndike Mem Lab, 61-62; from asst prof to assoc prof, Med Sch, Univ Wis, 63-66; from assoc prof to prof pediat, Sch Med, St Louis Univ, 68-73, assoc prof microbiol, 68-73, vchmn dept pediat, 70-73; PROF PEDIAT, SCH MED, UNIV CALIF, LOS ANGELES, 73- Concurrent Pos: Asst attend physician, Mary Fletcher Hosp & DeGoesbriand Hosp, Burlington, Vt, 61-62; asst pediat, Boston City Hosp, 61-62; assoc attend physician, Madison Gen Hosp, 62-67; dir, John A Hartford Res Found, 62-67; Markle scholar, 64; mem med staff, Cardinal Glennon Mem Hosp Children, 66-67; vis worker, Common Cold Res Unit & Clin Res Ctr, Salisbury, Eng, 69. Mem: Am Soc Microbiol; Am Fedn Clin Res; Am Acad Pediat; AAAS; Soc Pediat Res. Res: Clinical manifestations of viral diseases; viral vaccines; interaction of infectious agents in the pathogenesis of disease. Mailing Add: Div Infectious Dis Dept Pediat Univ Calif Sch Med Los Angeles CA 90024

CHERRY, JOE H, b Newbern, Tenn, June 3, 34; m 55; c 3. PLANT PHYSIOLOGY, BIOCHEMISTRY. Educ: Univ Tenn, BS, 57; Univ Ill, MS, 59, PhD(agron, biochem), 61. Prof Exp: Res assoc biochem, Seed Protein Pioneering Res Lab, USDA, La, 61-62; from asst prof to assoc prof hort, 62-67, PROF HORT, PURDUE UNIV, 67- Mem: AAAS; Am Soc Plant Physiol; Am Soc Biol Chemists. Res: Nucleic acid metabolism during seed germination and plant growth; induction of enzymes during cell differentation; effects of ionizing radiation on plants; mechanism of action of plant hormones. Mailing Add: Dept of Hort Purdue Univ West Lafayette IN 47907

CHERRY, JOHN PAUL, b Rhinebeck, NY, Jan 31, 41; m 64; c 2. FOOD BIOCHEMISTRY, AGRICULTURAL BIOCHEMISTRY. Educ: Furman Univ, BS, 63; WVa Univ, MS, 66; Univ Ariz, PhD(genetics & biochem), 71. Prof Exp: Res assoc biochem, Tex A&M Univ, 72-73; asst prof food sci, Univ Ga, 73-75; SUPVRY RES CHEMIST/RES LEADER FOOD BIOCHEM, SOUTHERN REGIONAL RES CTR, AGR RES SERV, USDA, 76- Concurrent Pos: Res chemist, Southern Res Ctr, Agr Res Serv, USDA, 70-72; res grant, Univ Ga, 74-75. Mem: Inst Food Technologists; Am Chem Soc; Am Asn Cereal Chemists; Am Peanut Res & Educ Asn; Sigma Xi. Res: Discovery, isolation, fractionation and purification of proteins from conventional and nonconventional food sources; characterization of biochemical, functional and nutritional properties of proteins for use as food and feed ingredients. Mailing Add: USDA Southern Regional Res Ctr PO Box 19687 New Orleans LA 70179

CHERRY, LEONARD VICTOR, b Los Angeles, Calif, May 3, 23; m 53; c 1. SOLID STATE PHYSICS. Educ: City Col New York, BS, 47; Duke Univ, PhD(chem), 53. Prof Exp: Prin chemist, Battelle Mem Inst, 51-53; from asst prof to assoc prof chem, Hampton Inst, 53-57; instr & res assoc, Univ Pittsburgh, 57-61; asst prof, 61-65, ASSOC PROF PHYSICS, FRANKLIN & MARSHALL COL, 65- Mem: AAAS; Am Phys Soc; Am Asn Physics Teachers; Fedn Am Sci. Res: Kerr effect in aromatic fluorine compounds; physicochemical problems connected with photoengraving and electrophotography; properties of intermetallic compounds; Mössbauer effect. Mailing Add: Dept of Physics Franklin & Marshall Col Lancaster PA 17604

CHERRY, MARIANNA, b Hartford, Conn, Dec 28, 24. IMMUNOGENETICS. Educ: Wheaton Col, BA, 46; Bryn Mawr Col, MA, 51; Yale Univ, PhD(biophys chem), 64. Prof Exp: Res biologist, Univ Calif, San Diego, 62; vis asst prof chem, Mt Holyoke Col, 62-63; asst prof chem, State Univ NY Albany, 63-65; from assoc staff scientist to staff scientist, 65-75, SR STAFF SCIENTIST, JACKSON LAB, 75- Mem: Transplantation Soc; AAAS. Res: Genetic control in mice of histocompatibility antigens and other cell membrane alloantigens; genetic control of immune response. Mailing Add: Jackson Lab Bar Harbor ME 04609

CHERRY, WILLIAM BAILEY, b Bowling Green, Ky, Apr 27, 16; m 44; c 2. BACTERIOLOGY. Educ: Western Ky State Teachers Col, BS, 37; Univ Ky, MS, 42; Univ Wis, PhD(bact), 49; Am Bd Med Microbiol, dipl, 62. Prof Exp: Bacteriologist, Univ Ky, 41-43; asst prof bact, Univ Tenn, 49-51; bacteriologist, 51-70, SCIENTIST DIR, CTR DIS CONTROL, USPHS, 70- Concurrent Pos: Assoc prof, Sch Pub Health, Univ NC. Honors & Awards: Meritorious Serv Award, USPHS, 63; Kimble Award, 67; P R Edwards Award, 68; Difco Award, 74. Mem: AAAS; Am Soc Microbiol; fel Am Acad Microbiol; Fedn Am Scientists. Res: Enteric bacteriology; bacteriology of anthrax; listeriosis; fluorescent antibody techniques. Mailing Add: 2857 Talisman Ct NE Atlanta GA 30345

CHERRY, WILLIAM HENRY, b New York, NY, Oct 9, 19; m 47; c 3. PHYSICS. Educ: Mass Inst Technol, BS, 41; Princeton Univ, MA, 48, PhD, 58. Prof Exp: RES PHYSICIST, LABS, RCA CORP, 41- Honors & Awards: Levy Medal, Franklin Inst. Mem: Am Phys Soc; Inst Elec & Electronics Engrs. Res: Electrodynamics in magnetron, betatron and velocity modulated tubes; gas and ultra high frequency discharge; colorimetry in color television; time division multiplex, communications systems; information theory; multiplex, color television systems; secondary electron emission from surfaces bombarded by positrons. Mailing Add: RCA Corp Labs Princeton NJ 08540

CHERTKOFF, MARVIN JOSEPH, b Baltimore, Md, Nov 20, 30; m 63. PHARMACEUTICAL CHEMISTRY, PHYSICAL PHARMACY. Educ: Univ Md, BS, 51, MS, 54; Purdue Univ, PhD(pharm), 58. Prof Exp: Chemist chem warfare labs, Army Chem Ctr, Md, 56; mgr tech unit, Qual Stand Sect, Merck Sharp & Dohme, 58-59, corp trainee, 59-61, mgr qual control, Pharm Prod, 61-63, supt, Sterile Opers, 63-65, qual motivation coordr, 65-67; planning mgr, Hoffmann-La Roche Inc, 67-68; dir bus develop, Givaudan Corp, 68-69, dir, Aroma Chem Div, 69-70, vpres, Div, 70-73; PRES, BIOZEST LABS, 73- Mem: Am Chem Soc; Am Pharmaceut Asn; Am Soc Qual Control. Res: Quality control and pharmaceutical production; physical pharmacy. Mailing Add: 202 Beechwood Dr Ridgewood NJ 07450

CHERTOCK, GEORGE, b New York, NY, Aug 1, 14; m 37; c 2. PHYSICS. Educ: City Col New York, BS, 39; George Washington Univ, MA, 43; Cath Univ Am, PhD(physics), 52. Prof Exp: Phys sci aide to physicist, Bur Standards, 40-46; PHYSICIST, NAVAL SHIP RES & DEVELOP CTR, 46- Mem: Acoust Soc Am. Res: Structure of stars; response of elastic and plastic structures to underwater explosions and to pressure waves; shock waves in air and water; underwater acoustics. Mailing Add: Naval Ship Res & Develop Ctr Washington DC 20007

CHERTOK, BENSON T, b Laconia, NH, May 15, 35; m 61; c 2. NUCLEAR PHYSICS, PARTICLE PHYSICS. Educ: Mass Inst Technol, SB, 57, SM, 60; Boston Univ, PhD(physics), 64. Prof Exp: From asst prof to assoc prof, 66-75, PROF PHYSICS, AM UNIV, 75- Concurrent Pos: Guest worker, Nat Bur Stand, 64-70; NSF res grant, 67-; AEC grant, 68-74; vis scientist, Stanford Linear Accelerator Ctr, 70-71, 73-74. Mem: Am Phys Soc. Res: Electron and photon interaction with nuclei; the nucleon and few nucleon problems. Mailing Add: Dept of Physics Am Univ Washington DC 20016

CHERTOK, ROBERT JOSEPH, b Spartanburg, SC, Sept 5, 35; m 59; c 2. PHYSIOLOGY. Educ: Univ SC, BS, 57; Univ Miami, PhD(renal physiol), 65. Prof Exp: Sr scientist, Lawrence Livermore Lab, Univ Calif, 64-73; assoc prof, Jackson State Univ, 73-75; ASSOC PROF, COMP ANIMAL RES LAB, UNIV TENN, OAK RIDGE, 76- Mem: Am Physiol Soc; Am Soc Nephrol; Soc Exp Biol & Med. Res: Renal transport; metabolism of environmental pollutants. Mailing Add: Univ Tenn Comp Animal Res Lab 1299 Bethel Valley Rd Oak Ridge TN 37830

CHERVENICK, PAUL A, b Pittsburgh, Pa, Apr 20, 32; m 54; c 3. HEMATOLOGY, INTERNAL MEDICINE. Educ: Univ Pittsburgh, BS, 57, MD, 61. Prof Exp: From intern to resident med, Univ Pittsburgh, 61-64; fel hemat, Univ Utah, 64-67; asst prof med, Rutgers Med Sch, 67-69; assoc prof, 69-73, PROF MED, SCH MED, UNIV PITTSBURGH, 73- Concurrent Pos: Mem site vis team, NIH; Leukemia Soc Am scholar, 70. Mem: Am Soc Clin Invest; Am Soc Clin Res; Am Soc Hemat; fel Am Col Physicians. Res: Study of factors controlling the proliferation and maturation of blood leukocytes in health and diseases, such as leukemia; proliferation of cells in an in vitro culture system. Mailing Add: 931 Scaife Hall Univ of Pittsburgh Sch of Med Pittsburgh PA 15261

CHERVENKA, CHARLES HENRY, b Howe, Okla, July 3, 21; m 47; c 2. BIOCHEMISTRY. Educ: Univ Okla, BS, 49, MS, 50; Univ Wash, PhD(biochem), 55. Prof Exp: Anal org chemist, Dow Chem Co, 50-52; res chemist, Reichhold Chems, Inc, 55-56; res assoc enzyme chem, Palo Alto Med Res Found, 56-60; APPLN CHEMIST, BECKMAN INSTRUMENTS, INC, 60- Mem: AAAS; Am Chem Soc. Res: Structure of proteins and enzymes; biomedical instrumentation. Mailing Add: Beckman Instruments Inc Appln Res Dept 1117 Calif Ave Palo Alto CA 94304

CHESBRO, WILLIAM RONALD, b Cohoes, NY, Oct 6, 28; m 50; c 2. MICROBIOLOGY. Educ: Ill Inst Technol, BS, 51, MS, 55, PhD(bact), 59. Prof Exp: Plant bacteriologist, Wanzer Dairy, Ill, 51-55; asst bacteriologist, Am Meat Inst Found, 55-59; from asst prof to assoc prof, 59-68, PROF MICROBIOL, UNIV NH, 68- Concurrent Pos: NIH res grants, 60- Mem: Am Soc Microbiol. Res: Microbial physiology and pathogenic microbiology. Mailing Add: Dept of Microbiol Univ of NH Durham NH 03824

CHESEMORE, DAVID LEE, b Janesville, Wis, Nov 3, 39; m 61; c 1. WILDLIFE ECOLOGY. Educ: Univ Wis-Stevens Point, BS, 61; Univ Alaska, College, MS, 67; Okla State Univ, PhD(wildlife ecol), 75. Prof Exp: Res asst, Wildlife Res Unit, Univ Alaska, College, 61-63, res asst, Dept Wildlife Mgt, 63-64; forester, US Peace Corps, Nepalese Forestry Dept, Birganj, 64-65; res biologist, US Fish & Wildlife Serv, Univ Alaska, College, 67-68; res asst, Wildlife Res Unit, Okla State Univ, 68-72; ASST PROF BIOL, CALIF STATE UNIV, FRESNO, 72- Mem: Wildlife Soc; Am Soc Mammalogists; Sigma Xi. Res: Population dynamics and ecology of big game; canid ecology; biometrics and computer applications to field ecology; ecology of rare and endangered species; aspects of scientific photography. Mailing Add: Dept of Biol Calif State Univ Fresno CA 93740

CHESHIRE, DAVID LYNN, elementary particle physics, nuclear physics, see 12th edition

CHESICK, JOHN POLK, b New Castle, Ind, Aug 8, 33; m 56; c 2. PHYSICAL CHEMISTRY. Educ: Purdue Univ, BS, 54; Harvard Univ, PhD(chem), 57. Prof Exp: From instr to asst prof chem, Yale Univ, 57-62; assoc prof, 62-71, PROF CHEM, HAVERFORD COL, 71- Mem: Am Chem Soc; Am Phys Soc. Res: Chemical kinetics; gas reactions; photochemistry. Mailing Add: Dept of Chem Haverford Col Haverford PA 19041

CHESKY, JEFFREY ALAN, b Lynn, Mass, May 11, 46; m 70; c 1. MUSCULAR PHYSIOLOGY, GERONTOLOGY. Educ: Cornell Univ, AB, 67; Univ Miami, PhD(physiol, biophys), 74. Prof Exp: NIH TRAINEE & RES INSTR PHYSIOL & BIOPHYS, SCH MED, UNIV MIAMI, 74- Mem: Geront Soc; AAAS; Am Physiol Soc; Sigma Xi. Res: Physiology of aging; age related changes in cardiac and skeletal muscle; alterations in contractile proteins. Mailing Add: 9352 SW 77th Ave Miami FL 33156

CHESLER, DAVID ALAN, b New York, NY. MEDICAL PHYSICS. Educ: Mass Inst Technol, SB, 55, SM, 55, ScD(elec eng), 60. Prof Exp: Engr, Gen Tel & Electronics, 60-70; ASSOC PHYSICIST RADIOL, MASS GEN HOSP, 70- Concurrent Pos: NIH res fel, Mass Gen Hosp, 70-72. Mem: Sigma Xi; Inst Elec & Electronics Engrs. Res: Tomography and computer processing of medical x-ray and radionuclide images. Mailing Add: Physics Res Lab Mass Gen Hosp Fruit St Boston MA 02114

CHESLEY, LEON CAREY, b Montrose, Pa, May 22, 09; m 34; c 5. BIOCHEMISTRY. Educ: Duke Univ, PhD(physiol), 32. Prof Exp: Asst biophysicist, Mem Hosp, New York, 32-35; biochemist, Margaret Hague Maternity Hosp, 35-53; assoc prof, 53-59, PROF OBSTET & GYNEC, STATE UNIV NY DOWNSTATE MED CTR, 59- Concurrent Pos: Instr eve session, City Col New York, 33-35. Mem: AAAS; Am Physiol Soc; hon mem Am Gynec Soc. Res: Toxemias of pregnancy. Mailing Add: Dept of Obstet & Gynec State Univ NY Downstate Med Ctr Brooklyn NY 11203

CHESNIN, LEON, b New York, NY, Mar 28, 19; m 40; c 4. SOIL CHEMISTRY, PLANT NUTRITION. Educ: Univ Ky, BS, 40; Rutgers Univ, PhD(soils), 48. Prof Exp: Jr soil surveyor, Soil Conserv Serv, USDA, Ind, 41-42; chemurgic res supvr, Joseph E Seagram & Sons, Inc, Ky, 42-44; asst, NJ Exp Sta, 44-47; asst prof agron, 47-54, ASSOC PROF AGRON & AGRONOMIST, UNIV NEBR, LINCOLN, 54- Concurrent Pos: Consult fertilizer co; mem, Nat Micronutrient Comt, Am Coun Fertilizer Appln. Mem: Am Soc Agron; Soil Sci Soc Am. Res: Micronutrients for crop production; micronutrient and major nutrient interrelations for crop growth; nutrition of crop varieties; influence of soil management on nutrient availability; disposal of animal, municipal and industrial wastes in soil for crop production as influenced by soil management and waste management practices. Mailing Add: Dept of Agron Univ of Nebr Lincoln NE 68583

CHESNUT, DONALD BLAIR, b Richmond, Ind, Dec 27, 32; m 54. PHYSICAL CHEMISTRY. Educ: Duke Univ, BS, 54; Calif Inst Technol, PhD(chem), 58. Prof Exp: Res assoc & instr physics, Duke Univ, 57-58; res chemist, Cent Res Dept, Exp Sta, E I du Pont de Nemours & Co, Inc, 58-65; assoc prof, 65-71, PROF CHEM, DUKE UNIV, 71- Mem: Am Phys Soc. Res: Quantum mechanics; magnetic resonance. Mailing Add: Dept of Chem Duke Univ Durham NC 27706

CHESNUT, THOMAS LLOYD, b Pulaski, Miss, June 14, 42; m 61; c 2. ENTOMOLOGY, FRESH WATER ECOLOGY. Educ: Miss State Univ, BS, 65, MS, 66, PhD(entom), 69. Prof Exp: Res technologist insect ecol, Boll Weevil Res Lab, Agr Res Serv, 64-68; asst prof biol sci, Fla Technol Univ, 69-70, dir fresh water ecol, 70-74; dir, Off Res Serv, 74-75, DIR, OFF GRAD STUDIES, GA COL, 75- Mem: Entom Soc Am. Res: Effects of competitive displacement between natural populations of insects; biological control of insect populations using insect parasites; effects of eutrophication on the aquatic habitat. Mailing Add: Off of Grad Studies Ga Col Milledgeville GA 31061

CHESNUT, WALTER G, b Montclair, NJ, July 20, 28; m 51; c 5. ENGINEERING PHYSICS. Educ: Lehigh Univ, BS, 50, MS, 52; Univ Rochester, PhD(physics), 56. Prof Exp: Res assoc high energy physics, Cosmotron Dept, Brookhaven Nat Lab, 56-58; sr assoc appl physics, G C Dewey Corp, NY, 58-62; sr physicist, Radio Physics Lab, 62-63, STAFF SCIENTIST, STANFORD RES INST, 63- Concurrent Pos: Consult, Stanford Res Inst, 58-61 & Los Alamos Sci Lab, 68-; vis lectr elec eng, Stanford Univ, 73. Mem: AAAS; Am Phys Soc. Res: Utilization of radio and radar waves as a probe of the upper atmosphere or space environments. Mailing Add: Stanford Res Inst 333 Ravenswood Ave Menlo Park CA 94025

CHESNUTT, CLARENCE, JR, b Winchester, Tenn, Nov 11, 25; m 57; c 4. BIOLOGY, CHEMISTRY. Educ: Univ Tenn, BS, 50; Mich State Univ, MS, 53; Ore State Univ, PhD(nutrit physiol, biochem), 56. Prof Exp: Res assoc radiation physiol, Inst Atomic Energy Comn, Univ Tenn, 56; assoc prof dairying, Mid Tenn State Col, 56-58 & Univ Calif, Berkeley, 58-62; prof biol & chmn dept natural sci, Campbellsville Col, 62-65; PROF BIOL & CHMN DEPT BIOL SCI, BAPTIST COL CHARLESTON, 65- Concurrent Pos: Dir, Cent Ky Regional Sci Fair, 62-; chmn, State Sci Youth Activities Comt, 63- Mem: Am Dairy Sci Asn; Nat Sci Teachers Asn. Res: Ruminant nutrition, plant and animal genetics; plant physiology and radiation biology including mutagenic effects of ionizing radiation on laboratory plants and animals. Mailing Add: Dept of Biol Sci Baptist Col PO Box 10087 Charleston SC 29411

CHESNUTWOOD, CHARLES MARK, b Johnstown, Pa, May 6, 16; m 42; c 2. GEOGRAPHY. Educ: Pa State Teachers Col, E Stroudsburg, BS, 38; Univ Utah, MS, 50; Pa State Univ, PhD, 54. Prof Exp: Instr geog, Univ Utah, 49-50 & Pa State Univ, 50-54; asst prof phys sci, Univ Fla, 54-55; intel officer, Cent Intel Agency, 55-61; sr engr, Palo Alto Div, Itek Corp, 61-64; instr geog, Univ Utah, 64-67; AEROSPACE TECHNOLOGIST, NASA MANNED SPACECRAFT CTR, HOUSTON, 67- Concurrent Pos: Mem, Int Geog Cong. Mem: Asn Am Geog; Am Soc Photogram. Res: Photographic interpretation. Mailing Add: 1908 S Palm Ct Pasadena TX 77502

CHESS, KARIN V T, b Hobbs, NMex, Dec 17, 39; m 60. MATHEMATICS. Educ: Univ Kans, BS, 62, MA, 64, PhD(math), 68. Prof Exp: Instr, Univ Kans, 64-65 & 68-69; ASST PROF MATH, WIS STATE UNIV, EAU CLAIRE, 69- Mem: Am Math Soc; Math Asn Am. Res: Generalized nilpotent groups. Mailing Add: Dept of Math Wis State Univ Eau Claire WI 54701

CHESSER, NANCY JEAN, b Albany, NY, Aug 31, 46; m 75. EXPERIMENTAL SOLID STATE PHYSICS. Educ: Cornell Univ, BA, 67; State Univ NY Stony Brook, PhD(physics), 72. Prof Exp: Asst physicist & instr physics, Ames Lab & Iowa State Univ, 72-73, assoc physicist & asst prof, 73-75; RESIDENT RES ASSOC PHYSICS, FELTMAN RES LAB, 75- Mem: Am Phys Soc. Res: Investigation of the dynamical properties of various solids through inelastic and quasielastic neutron scattering. Mailing Add: Reactor Bldg Nat Bur of Standards Washington DC 20234

CHESSICK, RICHARD D, b Chicago, Ill, June 2, 31; m 53; c 3. PSYCHIATRY. Educ: Univ Chicago, PhB, 49, SB & MD, 54. Prof Exp: Asst chief psychiat, USPHS Hosp, Lexington, Ky, 58-60; staff psychiatrist, Michael Reese Hosp, Chicago, 60-61; instr psychiat, 60-61; assoc, 61-62, from asst prof to assoc prof, 62-72, PROF PSYCHIAT, NORTHWESTERN UNIV, EVANSTON, 73- Concurrent Pos: Pvt pract, 60-; chief, Vet Admin Res Hosp, Chicago, 61-65, assoc dir res training prog, 64; sr attend psychiatrist, Evanston Hosp & Northwestern Mem Hosp, Chicago. Honors & Awards: Merck Award, 54. Mem: Fel Am Psychiat Asn; Am Psychosom Soc; Am Acad Psychother; fel Am Orthopsychiat Asn; Asn Advan Psychother. Res: Psychotherapy. Mailing Add: Suite 628 636 Church St Evanston IL 60201

CHESSIN, HENRY, b Cleveland, Ohio, Dec 8, 19; m 50; c 3. PHYSICS. Educ: Western Reserve Univ, BS, 47; Purdue Univ, MS, 50; Polytech Inst Brooklyn, PhD(physics), 59. Prof Exp: Instr physics, Polytech Inst Brooklyn, 51-57; res physicist, US Steel Corp, 57-64; PROF PHYSICS, STATE UNIV NY ALBANY, 64- Mem: Am Crystallog Asn; Am Inst Mining, Metall & Petrol Engrs; Am Phys Soc. Res: X-ray crystallography; crystal physics. Mailing Add: Dept of Physics State Univ of NY Albany NY 12203

CHESSIN, HYMAN, b Cleveland, Ohio, Sept 27, 20; m 42; c 4. PHYSICAL CHEMISTRY. Educ: Western Reserve Univ, BS, 47, MS, 49, PhD(phys chem), 51. Prof Exp: Anal chemist, Harshaw Chem Co, 41-43, 46-47; asst phys chem, Western Reserve Univ, 47-50; asst prof phys & anal chem & res assoc, Kenyon Col, 50-52; asst prof phys chem, Univ Ark, 52-54; dir res, Vander Horst Corp, NY, 54-62; sr res chemist, 62-68, res assoc, 68-73, SR RES ASSOC, M&T CHEM, INC, FERNDALE, 73- Mem: Am Chem Soc; Electrochem Soc; Am Electroplaters Soc. Res: Electrochemistry; physical chemistry. Mailing Add: 7146 Heather Heath Lane West Bloomfield MI 48033

CHESSIN, MEYER, b New York, NY, Feb 5, 21; m 45; c 5. PLANT PHYSIOLOGY, ENVIRONMENTAL MANAGEMENT. Educ: Univ Calif, BS, 41, PhD(plant physiol), 50. Prof Exp: Asst bot, Univ Calif, 47-48; from instr to assoc prof, 49-61, PROF BOT, UNIV MONT, 51- Concurrent Pos: Res fel, USPHS, Rothamsted Exp Sta, Eng, 56-57; travel awards, Int Photobiol Cong, Copenhagen, 60, Oxford, 64; AEC fel, Univ Minn, 64-65; consult, Nat Libr Med, 70; Nat Acad Sci mem sci exchange, Romania, 71 & 75. Mem: AAAS; Am Phytopath Soc. Res: Virology; biological effects of radiation. Mailing Add: Dept of Bot Univ of Mont Missoula MT 59801

CHESSMORE, ROY A, b Oklahoma City, Okla, Sept 23, 13; m 40; c 2. PLANT BREEDING. Educ: Okla Agr & Mech Col, BS, 41, MS, 48; Univ Nebr, PhD(plant breeding), 52. Prof Exp: Agronomist, Soil Conserv Serv, USDA, 41-42; supt exp sta, Okla Agr & Mech Col, 42, agronomist, 46-52; plant breeder, Noble Found, 52-58, agronomist, 58-62, asst dir, Agr Div, 63-65; DIR KERR FOUND, 65- Mem: Am Soc Agron; Am Soc Range Mgt. Res: Supervise cattle crossbreeding; pasture management; channel catfish breeding and selection. Mailing Add: Kerr Found Box 588 Poteau OK 74953

CHESTER, ARTHUR NOBLE, b Seattle, Wash, Aug 5, 40; m 69. THEORETICAL PHYSICS. Educ: Univ Tex, Austin, BS, 61; Calif Inst Technol, PhD(theoret physics), 65. Prof Exp: Physicist, Bell Tel Labs, NJ, 65-69; mem tech staff physics, 69-71, head chem laser sect, 71-73, mgr, Laser Dept, 73-75, ASST DIR, HUGHES RES LABS, 75- Concurrent Pos: Consult, Dept Defense Adv Group, 75- Honors & Awards: A A Bennett Math Award, 58. Mem: AAAS; Am Phys Soc; Inst Elec & Electronics Eng; Optical Soc Am. Res: Elementary particle theory; x-ray camera tubes; field quenching of photoluminescence; electrophoresis in gas discharges; gas laser physics; chemical lasers. Mailing Add: Hughes Res Labs Malibu CA 90265

CHESTER, ARTHUR WARREN, b Brooklyn, NY, Jan 9, 40; m 61; c 3. INORGANIC CHEMISTRY. Educ: Brooklyn Col, BS, 61; Mich State Univ, PhD(inorg chem), 66. Prof Exp: Res chemist, 66-70, SR RES CHEMIST, 70- Mem: Am Chem Soc; The Chem Soc. Res: Reactions of metal complexes in non-aqueous solvents; metals in hydrocarbon autoxidations; catalysis by inorganic solids; zeolite chemistry and catalysts; cracking catalysts. Mailing Add: Mobil Res & Develop Corp Paulsboro NJ 08046

CHESTER, BRENT, b New York, NY, Apr 19, 42; m 67; c 2. CLINICAL MICROBIOLOGY. Educ: City Col New York, BS, 62; Long Island Univ, MS, 70; NY Univ, PhD(microbiol), 75. Prof Exp: Jr bacteriologist, Bur Labs, New York City Dept Health, 65-68; supvr clin microbiol res, Kings County Res Lab, New York, 68-70; sr supvr microbiol, Mt Sinai Serv, Elmhurst Hosp, New York, 70-74; hosp microbiologist, Vet Admin Hosp, Gainesville, Fla, 74-75, HOSP MICROBIOLOGIST, VET ADMIN HOSP, MIAMI, 75- Mem: Am Soc Microbiol. Res: Isolation and identification procedures for bacteria in clinical specimens with emphasis on Yersinia, Klebsiella and nonfermentative Bacilli. Mailing Add: Vet Admin Hosp Lab Serv 1201 NY 16th St Miami FL 33155

CHESTER, CLARENCE LUCIAN, b Cabot, Vt, Jan 5, 15; m 47; c 2. PATHOLOGY, ANATOMY. Educ: Univ Vt, BS, 37, MD, 40. Prof Exp: Intern, Burbank Hosp, Mass, 40-42; res path, St Elizabeth's Hosp, 47-48; asst path med col, Tufts Col, 48-49, instr, 53, lectr path & bact, 53-55; pathologist, Vet Admin Hosp, Little Rock, Ark, 55-59 & Portland, Ore, 59-74; asst clin prof path, Med Sch, Univ Ore, 62-74; CHIEF ANAT PATH SERV, VET ADMIN HOSP, PORTLAND, 74-; ASSOC PROF PATH, MED SCH, UNIV ORE, 74- Concurrent Pos: Asst pathologist, Mt Auburn Hosp, 49-53, assoc pathologist, 53-55. Mem: AAAS; NY Acad Sci; Col Am Path; Asn Advan Med Instrumentation. Res: Cancer. Mailing Add: Anat Path Serv Vet Admin Hosp Sam Jackson Park Portland OR 97207

CHESTER, CLIVE RONALD, b Brooklyn, NY, Apr 6, 30. MATHEMATICS. Educ: NY Univ, AB, 50, MS, 51, PhD(math), 55. Prof Exp: Asst math, Inst Math Sci, NY Univ, 51-56; instr, Queens Col, NY, 56-61; from asst prof to assoc prof, Polytech Inst Brooklyn, 61-69; ASSOC PROF MATH, NY INST TECHNOL, 69- Mem: Am Math Soc. Res: Applied mathematics; wave propagation; partial differential equations. Mailing Add: Dept of Math NY Inst of Technol 268 Wheatley Rd Old Westbury NY 11568

CHESTER, DANIEL LEON, b Albany, Calif, Feb 26, 43. COMPUTER SCIENCES. Educ: Univ Calif, Berkeley, BA, 66, MA, 68, PhD(math), 73. Prof Exp: ASST PROF MATH & COMPUT SCI, UNIV TEX, AUSTIN, 73- Mem: Asn Comput Mach; Asn Comput Ling; Asn Symbolic Logic; Am Math Soc. Res: Computer generation and understanding of mathematical discourse; natural language question answering systems; artificial intelligence. Mailing Add: Dept of Comput Sci Univ of Tex Austin TX 78712

CHESTER, EDWARD M, b Queens Co, NY, Jan 12, 12; m 38; c 2. INTERNAL MEDICINE. Educ: NY Univ, BS, 32; Iowa State Univ, MD, 36; Am Bd Internal Med, dipl, 44. Prof Exp: From clin instr to sr clin instr, 43-49, asst clin prof, 49-54, assoc prof, 54-74, PROF MED, CASE WESTERN RESERVE UNIV, 74- Concurrent Pos: Vis in med & med dir out-patient clins, Cleveland Metrop Gen Hosp, 59-63; dir ambulatory teaching clin, Case Western Reserve Univ. Mem: Am Heart Asn; Am Fedn Clin Res; fel Am Col Physicians; Am Diabetes Asn. Res: Vertebral column acromegaly; infarction of the lung; lung abscess secondary to aseptic pulmonary infarction; diabetes and medical education. Mailing Add: Cleveland Metrop Gen Hosp 3395 Scranton Rd Cleveland OH 44109

CHESTER, MARVIN, b New York, NY, Dec 29, 30; div; c 2. SOLID STATE PHYSICS, CRYOGENICS. Educ: City Col New York, BS, 52; Calif Inst Technol, PhD(physics), 61. Prof Exp: From asst prof to assoc prof, 61-74, PROF PHYSICS, UNIV CALIF, LOS ANGELES, 74- Mem: Am Phys Soc. Res: Configurational emf, a transport property in semiconducting materials; second sound, a thermal wave which may exist in certain solid at low temperatures; electric field effects on the infrared absorption of silicon. Mailing Add: Dept of Physics Univ of Calif Los Angeles CA 90024

CHESTNUT, ALPHONSE F, b Stoughton, Mass, Nov 20, 17; m 43; c 2. MARINE ECOLOGY, ZOOLOGY. Educ: Col William & Mary, BSc, 41; Rutgers Univ, MSc, 43, PhD, 49. Prof Exp: Asst zool, Rutgers Univ, 41-43; res assoc oyster culture, NJ Agr Exp Sta, 43-48; specialist, Inst Marine Sci, 48-49, asst to dir, 49-55, assoc prof, 49-59, PROF OYSTER CULTURE, INST MARINE SCI, UNIV NC, 59-, DIR INST, 55- Mem: Am Soc Limnol & Oceanog; Nat Shellfisheries Asn (vpres, 51-53, pres, 53); Sigma Xi; Atlantic Estuarine Res Soc (secy-treas, 52-53). Res: Food and feeding mechanism of lamellibranchia; estuarine ecology of pelecypods. Mailing Add: Inst of Marine Sci Univ of NC Morehead City NC 28557

CHESTON, CHARLES EDWARD, b Princeton, NJ, Nov 23, 11; m 38; c 2. FORESTRY. Educ: Syracuse Univ, BS, 33; Yale Univ, MF, 40. Prof Exp: Asst forester, NJ, 33-42; PROF FORESTRY & CHMN DEPT, UNIV OF THE SOUTH, 42- Concurrent Pos: Asst, Univ Mich, 57-58; mem, Tenn Conserv Comn, 63- Mem: Soc Am Foresters. Res: Forest economics; management of hardwood forest lands; rehabilitation of devastated forest lands on the Cumberland plateau. Mailing Add: Dept of Forestry Univ of the South Sewanee TN 37375

CHESTON, WARREN BRUCE, b Rochester, NY, Mar 15, 26; m 50; c 4. PHYSICS. Educ: Harvard Univ, BS, 47; Univ Rochester, PhD(physics), 51. Prof Exp: Asst prof physics, Washington Univ, 51-52; from assoc prof to prof, Univ Minn, Minneapolis, 53-71, dir space sci ctr, 65-68, dean inst technol, 68-71; CHANCELLOR, UNIV ILL, CHICAGO CIRCLE, 71- Concurrent Pos: Fulbright lectr, Univ Utrecht, 58-59; dept sci attache, Am Embassy, London, 63-65. Mem: Am Phys Soc. Res: Theoretical nuclear and meson physics. Mailing Add: Off of the Chancellor Univ of Ill at Chicago Circle Chicago IL 60680

CHETSANGA, CHRISTOPHER J, b Chetsanga Village, Rhodesia, Aug 22, 35; m 70. MOLECULAR BIOLOGY, BIOCHEMISTRY. Educ: Pepperdine Univ, BS, 64; Univ Toronto, MS, 67, PhD(molecular biol), 69. Prof Exp: Tutor biochem, Harvard Univ, 70-72; asst prof, 72-75, ASSOC PROF MOLECULAR BIOL, UNIV MICH, DEARBORN, 75- Concurrent Pos: Fel biochem, Harvard Univ, 69-72. Mem: AAAS; Soc Develop Biol; Fedn Am Scientists. Res: Alterations in the structural integrity of DNA from postmitotic heart and brain cells of aging mice and their effect on gene functions. Mailing Add: Dept of Natural Sci Univ of Mich Dearborn MI 48128

CHEUNG, AUGUSTINE Y, b Hong Kong, Feb 11, 47. RADIOLOGICAL HEALTH, MICROWAVE ENGINEERING. Educ: Univ Md, BS, 69, MS, 71, PhD(electrophys), 73. Prof Exp: From res asst to res assoc, 69-75, ASST PROF, INST FLUID DYNAMICS & APPL MATH, UNIV MD, COLLEGE PARK, 75- Mem: Am Phys Soc; Inst Elec & Electronics Engrs; Microwave Power Inst. Res: Therapeutic applications of microwave power and the prevention of microwave related health hazards. Mailing Add: Inst of Fluid Dynamics & Appl Math Univ of Md College Park MD 20742

CHEUNG, HERBERT CHIU-CHING, b Canton, China, Dec 19, 33; nat US; m 66; c 2. BIOPHYSICS, PHYSICAL CHEMISTRY. Educ: Rutgers Univ, AB, 54, PhD(phys chem, physics), 60; Cornell Univ, MS, 56. Prof Exp: Asst chem, Cornell Univ, 54-56; asst scientist, Fundamental Res Labs, US Steel Corp, 56-57; asst instr chem, Rutgers Univ, 58-60; res chemist res & develop dept, Am Viscose Div, FMC Corp, 60-63 & Gen Chem Div Res Labs, Allied Chem Corp, 63-66; sr fel biophys, Cardiovasc Res Inst, Med Ctr, Univ Calif, San Francisco, 66-69; assoc prof biophys, 69-73 & biomath, 73-74, ASSOC PROF BIOCHEM, 69-, PROF BIOMATH & HEAD BIOPHYS SECT, UNIV ALA MED CTR, 74- Concurrent Pos: Lectr eve div, Pa Mil Col, 61-63; USPHS res career develop award, 71-76. Mem: Am Chem Soc; Biophys Soc; Am Soc Biol Chemists. Res: Molecular basis of contractility; relationship between macromolecular conformation and biological function; fluorescence spectroscopy. Mailing Add: Dept of Biomath Univ of Ala Univ Sta Birmingham AL 35233

CHEUNG, PAUL JAMES, b Hong Kong, May 6, 42; Brit citizen; m 69; c 1. MARINE BIOLOGY. Educ: Iona Col, BS, 66; NY Univ, MS, 69, PhD(marine biol), 73. Prof Exp: RES STAFF MARINE FOULING, OSBORN LABS MARINE SCI, NEW YORK ZOOL SOC, 67- Mem: NY Acad Sci; Am Soc Zoologists. Res: Mass rearing of marine invertebrates; evaluation of anti-fouling paints; biogenesis and biochemistry of barnacle adhesive and neuroendocrine physiology of barnacle. Mailing Add: Osborn Labs of Marine Sci W Eighth St Coney Island Brooklyn NY 11224

CHEUNG, PETER PAK LUN, b China, Feb 2, 39; m 65; c 1. DENTAL MATERIALS. Educ: Colo State Univ, BS, 64; Okla State Univ, PhD(chem), 67. Prof Exp: Instr chem, Okla State Univ, 67-68; investr, N J Zinc Co, 68-69; res chemist, 69-70, SR CHEMIST, PENWALT CHEM CORP, 70- Mem: Am Chem Soc; Am Dent Asn. Res: Chemistry of titantium dioxide; surface chemistry; dental materials; organic resin composite materials; alginate and silicone impression materials; physical properties of materials; x-ray opacity of materials; physical properties of polymeric composite materials. Mailing Add: Pennwalt Corp 900 First Ave King of Prussia PA 19406

CHEUNG, TSUN-SUNG HARRY, b Hong Kong, Sept 9, 32; m 61; c 2. MARINE BIOLOGY. Educ: Univ Hong Kong, BSc, 55, MSc, 60; Univ London, BA, 60; Univ Glasgow, PhD(zool), 64. Prof Exp: Demonstr zool, Univ Hong Kong, 55-58, asst res officer marine biol, 58-60, demonstr physiol, 60-61 & tutor extra-mural studies, 64-65; asst prof marine sci, Univ Miami, 65-69; sr biologist, Inmont Marine Lab, 69-70; res assoc biol, 70-74, ASSOC PROF BIOL, TRENT UNIV, 70- Concurrent Pos: Hong Kong Auxiliary Med Serv. Mem: Am Inst Biol Sci. Res: Growth, reproduction, development and culture of decapod crustaceans and functions of morphogenetic hormones controlling these physiological processes; prawns of Solenocera subnuda; Metapenaeus ensis; crabs of Carcinus maenas; Menippe mercenaria; physiological ecology of Orconectes. Mailing Add: Dept of Biol Trent Univ Peterboro ON Can

CHEUNG, WAI YIU, b Canton, China, July 15, 33; US citizen; m 62; c 3. BIOCHEMISTRY. Educ: Chung Hsing Univ, Taiwan, BS, 56; Univ Vt, MS, 60; Cornell Univ, PhD(biochem), 64. Prof Exp: USPHS trainee, 64-67; asst mem, 67-70, assoc mem, 70-71, MEM, LAB BIOCHEM, ST JUDE CHIREN'S RES HOSP, 71- Concurrent Pos: Assoc prof biochem, Univ Tenn Ctr Health Sci; USPHS res career develop award, 71-76. Mem: AAAS; Am Chem Soc; Am Soc Biol Chem. Res: Biological regulation; mechanism of hormonal action; cyclic nucleotides. Mailing Add: Lab of Biochem St Jude Children's Res Hosp Memphis TN 38101

CHEVALIER, PETER ANDREW, b Chicago, Ill, Mar 4, 40; m 64; c 2. CARDIOPULMONARY PHYSIOLOGY, PHYSIOLOGY. Educ: Univ Minn, BA, 62, PhD(physiol), 67. Prof Exp: Asst prof physiol, Univ Del, 67-73; ASSOC CONSULT & ASST PROF PHYSIOL, MAYO MED SCH, 73- Mem: AAAS; Am Physiol Soc; Am Thoracic Soc. Res: Pulmonary ventilation and perfusion; dynamic regional lung mechanics; Roentgen videodensitometry and dynamic three-dimensional reconstruction techniques. Mailing Add: Dept of Physiol & Biophys Mayo Med Sch Rochester MN 55901

CHEVALIER-SKOLNIKOFF, SUZANNE, b Berkeley, Calif, Jan 12, 38; m 65; c 2. PHYSICAL ANTHROPOLOGY, PRIMATOLOGY. Educ: Univ Calif, Berkeley, PhD(anthrop), 71. Prof Exp: LECTR ANTHROP, STANFORD UNIV, 70- Concurrent Pos: NIMH res fel, Langley Porter Neuropsychiat Inst, 71-; corresp ed med primatology, Am Anthrop Asn Group Med Anthrop, 72- Mem: Am Anthrop Asn. Res: Medical primatology; sex role differentiation in nonhuman primates with implications for man; visual and tactile communication in Macaca arctoides, and its ontogenetic development; sexual behavior of male and female stumptail monkeys and implications for understanding the sexual behavior of other primates. Mailing Add: Dept of Anthrop Stanford Univ Stanford CA 94305

CHEVILLE, NORMAN F, b Rhodes, Iowa, Sept 30, 34; m 58; c 4. VETERINARY PATHOLOGY. Educ: Iowa State Univ, DVM, 59; Univ Wis, MS, 63, PhD(path), 64. Prof Exp: Res vet virol, US Army Biol Lab, 59-61; proj assoc path, Univ Wis, 61-63; PATHOLOGIST, NAT ANIMAL DIS LAB, USDA, 63-, PROF VET PATH, IOWA STATE UNIV, 69- Concurrent Pos: Res pathologist, Nat Inst Med Res, London, 68; Mem: Am Vet Med Asn; Am Col Vet Path (secy-treas, 74-); Conf Res Workers Animal Dis. Res: Pathology of disease of infectious origin and domesticated animals. Mailing Add: Nat Animal Dis Lab Iowa State Univ Ames IA 50010

CHEVONE, BORIS IVAN, b Lynn, Mass, Aug 25, 43; m 69; c 1. INSECT PHYSIOLOGY. Educ: Univ Mass, Amherst, BA, 65, MS, 68; Univ Minn, St Paul, PhD(entom), 74. Prof Exp: RES SCIENTIST VIROL, UNIV MINN, ST PAUL, 74- Mem: Entom Soc Am; Sigma Xi. Res: Physiology of insect vectors of plant viruses. Mailing Add: Dept of Plant Path Univ of Minn St Paul MN 55108

CHEVRETTE, JOSEPH EDGAR, b St Simon, Can, Aug 28, 12; m 40; c 3. AGRONOMY. Educ: Univ Montreal, BA, 32; Laval Univ, BSA, 36; Cornell Univ, PhD(plant breeding), 41. Prof Exp: Prof farm crops, Ecole Superieure Agr, 41-56; res officer forage crops breeding & mgt, Exp Farm, 56-62; head crops sci dept, 62-71, PROF SPEC CROPS, LAVAL UNIV, 62- Concurrent Pos: Mem, Corp Prof Agriculturists, Que. Mem: Can Soc Agron; Agr Inst Can; Can Asn Adv Sci. Res: Crop management. Mailing Add: Dept of Crop Sci Fac of Agr Univ of Laval Quebec PQ Can

CHEVRIER, JEAN-CLAUDE JACQUES, b Paris France, Feb 27, 39; m 60; c 1. SOLID STATE PHYSICS. Educ: George Washington Univ, BA, 60; Univ Va, MS, 62, PhD(physics), 64. Prof Exp: From res assoc to instr physics, Univ Va, 64-66; res physicist, Plastics Dept, 66-73, PROD MGR, FABRICS & FINISHES DEPT, E I DU PONT DE NEMOURS & CO, 73- Mem: Am Phys Soc; Sci Res Soc Am. Res: Electrical and mechanical properties of metals and alloys; single crystals; superconductivity; electrical properties of high polymers. Mailing Add: Brandywine Bldg 1213 Wilmington DE 19898

CHEW, FRANCES SZE-LING, b Los Angeles, Calif, May 11, 48. BIOLOGY. Educ: Stanford Univ, AB, 70; Yale Univ, PhD(biol), 74. Prof Exp: Fel, Dept Biol Sci, Stanford Univ, 74-75; ASST PROF BIOL, TUFTS UNIV, 75- Concurrent Pos: Grant, Am Philos Soc, 75. Mem: Soc Study Evolution; Lepidopterists Soc; AAAS. Res: Plant-herbivore coevolution, especially evolution of pierid butterflies and their cruciferous foodplants. Mailing Add: Dept of Biol Tufts Univ Medford MA 02155

CHEW, FRANK, b San Francisco, Calif, Aug 17, 16; m 46; c 3. OCEANOGRAPHY. Educ: Univ Calif, Los Angeles, BA, 43, MA, 50; Univ Miami, PhD(oceanog), 73. Prof Exp: Asst prof oceanog, Univ Miami, 52-59; res assoc, Gulf Coast Lab, Miss, 59-62; res scientist, Lockheed-Calif Co, 62-65 & Bissett-Berman Corp, 65-66; RES OCEANOGR, ATLANTIC OCEANOG & METEOROL LABS, NAT OCEANIC & ATMOSPHERIC ADMIN, 66- Mem: Am Meteorol Soc; Am Geophys Union. Res: Accelerative process in ocean currents and its relation to sea level change. Mailing Add: 901 Sistina Ave Coral Gables FL 33146

CHEW, GEOFFREY FOUCAR, b Washington, DC, June 5, 24; m 45, 71; c 4. THEORETICAL HIGH ENERGY PHYSICS. Educ: George Washington Univ, BS, 44; Univ Chicago, PhD(physics), 48. Prof Exp: Jr theoret physicist, Los Alamos Sci Lab, NMex, 44-46; Nat Res fel, Univ Chicago, 46-48; theoret physics res radiation lab, Univ Calif, 48-49, asst prof physics, 49-50; from asst prof to prof, Univ Ill, 50-57; theoret group leader, 70-74, PROF PHYSICS, UNIV CALIF, BERKELEY, 57-, CHMN DEPT, 74- Concurrent Pos: Fulbright lectr, Les Houches, 53, 60 & 65; fel, Churchill Col, Cambridge, 62-63; lectr, Tata Inst, Nainital, 69; vis prof, Princeton Univ, 70-71; consult, Los Alamos & Brookhaven Nat Labs. Honors & Awards: Hughes Prize, Am Phys Soc, 62; E O Lawrence Award, 69. Mem: Nat Acad Sci; fel Am Phys Soc; Am Acad Arts & Sci. Res: Theoretical particle physics; scattering matrix theory; strong interactions. Mailing Add: 10 Maybeck Twin Dr Berkeley CA 94708

CHEW, KENNETH KENDALL, b Red Bluff, Calif, Oct 29, 33; m 58; c 2. MARINE BIOLOGY. Educ: Chico State Col, BA, 55; Univ Wash, MS, 58, PhD(fisheries), 61. Prof Exp: Sr fisheries biologist, Fisheries Res Inst, 61-62, from res asst prof to res assoc prof, 62-67, assoc prof, 67-71, PROF FISHERIES, UNIV WASH, 71- Mem: Am Fisheries Soc; Nat Shellfisheries Asn; Am Soc Zoologists; Am Soc Ichthyologists & Herpetologists; Marine Biol Asn U K. Res: Shellfish biology; marine ecology; growth, condition and survival of Pacific oysters; shellfish toxicity studies in Washington and Southeast Alaska; clam and mussel culture studies; ecological baseline studies. Mailing Add: Col of Fisheries Univ of Wash Seattle WA 98105

CHEW, ROBERT MARSHALL, b Wheeling, WVa, Oct 7, 23; m 46; c 3. ECOLOGY. Educ: Washington & Jefferson Col, BS, 44; Univ Ill, MS, 46, PhD(zool, animal ecol), 48. Prof Exp: Teaching asst zool, Univ Ill, 44-46; asst prof biol, Lawrence Col, 48-52; instr zool, 52-55, from asst prof to assoc prof biol, 55-66, PROF BIOL, UNIV SOUTHERN CALIF, 66- Concurrent Pos: Consult, Northrop Space Labs, 62-66 & Lab Nuclear Med & Radiation Biol, Univ Calif, Los Angeles, 65- Mem: AAAS; Ecol Soc Am; Am Soc Mammalogists; Am Ornithologists Union; Am Physiol Soc. Res: Wild mammals; water metabolism; desert ecology. Mailing Add: Dept of Biol Sci Univ of Southern Calif Los Angeles CA 90007

CHEW, VICTOR, b Djakarta, Indonesia, July 9, 23; m 49; c 3. MATHEMATICAL STATISTICS. Educ: Univ Western Australia, BSc, 44; Univ Melbourne, BA, 53. Prof Exp: Tech officer, Commonwealth Sci & Indust Res Orgn, Melbourne, Australia, 47-49, res officer, Sydney, 53-55; instr math, Univ Melbourne, 49-53; asst math statist, Univ Fla, 55-56, asst prof, 56-57; asst statistician, NC State Col, 57-60; math statistician, US Naval Weapons Lab, 60-62; sr engr, RCA Serv Co, 62-70; MATH STATISTICIAN, BIOMET SERV STAFF, AGR RES SERV, USDA, 71- Concurrent Pos: Lectr, Am Univ, 60-61, 62-63, Johns Hopkins Univ, 61-62, Brevard Eng Col, 63-67 & Fla State Univ, 67-70. Mem: Inst Math Statist; Biomet Soc; Am Statist Asn; Int Asn Statist in Phys Sci; fel Royal Statist Soc. Res: Experimental designs; mathematical modeling; statistical inference; regression analysis; biometry. Mailing Add: Biomet Serv Staff Southern Region Agr Res Serv USDA 215 Rolfs Hall Univ of Fla Gainesville FL 32611

CHEW, WILLIAM HUBERT, JR, b Macon, Ga, Sept 21, 33; m 57; c 3. INTERNAL MEDICINE, INFECTIOUS DISEASES. Educ: Med Col Ga, MD, 58. Prof Exp: Asst med, Sch Med, Tufts Univ, 62-63, instr, 63-64; from instr to assoc prof, 64-72, chief infectious dis sect, 67-70, dir & coord, Physician Augmentation Prog, 70-75, PROF MED, MED COL GA, 72- Concurrent Pos: Res fel infectious dis, Pratt Clin, New Eng Ctr Hosp, 63-64; Markle scholar acad med, 64; dir, NIH Training Grant, 67. Mem: Am Fedn Clin Res. Res: Candida-host defense interactions, specifically in what alterations in host defense occur to permit either local or disseminated candidiasis. Mailing Add: Sch of Med Med Col of Ga Augusta GA 30902

CHEY, TONG CHULL, b Seoul, Korea, Sept 29, 31; m 62; c 1. PHYSICS. Educ: Seoul Nat Univ, BS, 56; Pa State Univ, MS, 61, PhD(physics), 63. Prof Exp: Instr physics, Pa State Univ, 63-64; asst prof, Wash State Col, 64-66; asst prof, 66-67, ASSOC PROF PHYSICS, MARIETTA COL, 67- Mem: Am Phys Soc; Am Asn Physics Teachers. Res: High pressure physics; first-order phase transitions under high pressures. Mailing Add: Dept of Physics Marietta Col Marietta OH 45750

CHHABRA, RAJENDRA S, b India, Mar 4, 39; m 66; c 2. PHARMACOLOGY, TOXICOLOGY. Educ: Vet Col, Mhow, India, BVSc&AH, 62; Univ London, PhD(pharmacol), 70. Prof Exp: Res asst pharmacol, Vet Col, Mhow, India, 62-64, asst res officer, 64-66; res pharmacologist, Biorex Labs, London, Eng, 66-67; vis assoc pharmacol, 70-73, SR STAFF FEL, PHARMACOL BR, NAT INST ENVIRON HEALTH SCI, 73- Mem: Soc Toxicol; Am Soc Pharmacol & Exp Therapeut. Res: Transport and biotransformation of foreign chemicals; species and strain variations in enzymatic biotransformation of xenobiotics. Mailing Add: Pharmacol Br Nat Inst of Environ Health Sci Research Triangle Park NC 27709

CHHEDA, GIRISH B, b Kutch, India, Mar 4, 34; m 62. MEDICINAL CHEMISTRY, BIOCHEMISTRY. Educ: Univ Bombay, BSc, 55, BS, 57; Univ Mich, MS, 59; State Univ NY Buffalo, PhD(med chem), 63. Prof Exp: Apprentice drug anal, Glaxo Labs, Bombay, India, 56; pharmaceut chemist, Castophene Mfg Co, 57; fel, State Univ NY Buffalo, 63-64; cancer res scientist, 64-65, sr cancer res scientist, 65-68, assoc cancer res scientist, 68-73, PRIN CANCER RES SCIENTIST, ROSWELL PARK MEM INST, 73- Concurrent Pos: Res prof, Niagara Univ, 68- & State Univ NY Buffalo, 68. Mem: Am Chem Soc. Res: Anti-metabolites and enzyme inhibitors; synthesis of oligonucleotides; investigation of human urinary nucleic acid constituents. Mailing Add: 200 Donna Lea Williamsville NY 14221

CHI, BENJAMIN, b Tientsin, China, June 18, 33; US citizen. PHYSICS. Educ: Antioch Col, BS, 55; Rensselaer Polytech Inst, PhD(physics), 62. Prof Exp: Fel & lectr physics, Western Reserve Univ, 62-63, instr, 63-65; asst prof, 65-69, chmn dept, 70-73, ASSOC PROF PHYSICS, STATE UNIV NY ALBANY, 69- Mem: AAAS; Am Phys Soc; Asn Comput Mach; Am Asn Physics Teachers. Res: Structure of atomic nuclei; methods in undergraduate physics teaching; computer applications in theoretical physics. Mailing Add: Dept of Physics State Univ of NY Albany NY 12203

CHI, CHE, b Peking, China, Feb 6, 49. NEUROSCIENCE. Educ: Nat Taiwan Univ, BS, 70; Okla State Univ, MS, 72; Univ Wis, PhD(neurosci), 76. Prof Exp: Res asst, Okla State Univ, 71-72, Univ Wis, 72-76; RES FEL, UNIV MINN, MINNEAPOLIS, 76- Mem: Sigma Xi; Entom Soc Am. Res: Scanning and transmission electron microscopy of the house fly ommatidium and first optic neuropile. Mailing Add: Dept of Neurol Univ of Minn Minneapolis MN 55455

CHI, CHIEN CHEN, b Shang-tung, China, Feb 16, 15; m 50; c 4. PLANT PATHOLOGY. Educ: Western China Union Univ, BS, 42; Nanking Univ, MS, 47; Univ Wis, PhD(plant path), 59. Prof Exp: Plant pathologist sugar cane, Taiwan Sugar Exp Sta, 47-56; proj assoc plant path, Univ Wis, 59-60; fel, 60-62; PLANT PATHOLOGIST FORAGE CROPS, CAN DEPT AGR, 62- Mem: Am Phytopath Soc. Res: Diseases of forage crops host-pathogen relationships of soil-inhabiting fungi in forage crops. Mailing Add: Ottawa Res Sta Can Dept of Agr Ottawa ON Can

CHI, CHRISTINA HADINATA, b Semarang, Indonesia, Apr 12, 43; US citizen; m 68; c 1. PHARMACOLOGY. Educ: Gadjah Mada State Univ, Indonesia, BS, 65; Univ Pittsburgh, PhD(pharmacol), 73. Prof Exp: NIH FEL PHARMACOL, UNIV PITTSBURGH, 74- Res: Characterization of d,l-methadone uptake by rat lung; fluorescence assay for d,l-methadone; narcotic-analgesics. Mailing Add: 1504 Murray Ave Pittsburgh PA 15217

CHI, DONALD NAN-HUA, b Medan, Indonesia, June 28, 39; US citizen; m 68; c 1. APPLIED MATHEMATICS. Educ: Willamette Univ, BA, 62; Carnegie Inst Technol, MS, 65; Univ Pittsburgh, PhD(math), 70. Prof Exp: Instr math, Univ Pittsburgh, 68-70, res physicist, 70-75; SUPVRY RES PHYSICIST NUMERICAL ANAL & AERODYNAMICS, PITTSBURGH MINING & SAFETY RES CTR, US BUR MINES, 75- Concurrent Pos: Mem comt Masters & PhD degrees, Dept Mech Eng, Univ Pittsburgh, 73-; reviewer, Appl Mech Rev, Am Soc Mech Engrs, 74- Mem: Soc Indust & Appl Math. Res: Transient phenomena of flame propagation in coal mine networks via computer simulation; numerical techniques in solving gas-dynamic and heat transfer problems and in solving nonlinear algebraic equations and nonlinear least square problems. Mailing Add: 1504 Murray Ave Pittsburgh PA 15217

CHI, KUO-RUEY, b Foochow, China, Jan 20, 24; m 54; c 1. CROP BREEDING, GENETICS. Educ: Nat Taiwan Univ, BS, 52; Va Polytech Inst, MS, 62; Iowa State Univ, PhD(crop breeding), 65. Prof Exp: F rom asst prof to assoc prof biol, Hardin-Simmons Univ, 64-70, prof biol, 70-73; FEL BIOCHEM, SCH HYG & PUB HEALTH, JOHNS HOPKINS UNIV, 73- Res: Quantitative genetics on corn. Mailing Add: Dept of Biochem Sch Pub Health Johns Hopkins Univ Baltimore MD 21205

CHI, LOIS WONG, b Foochow, Fukien, China; m 45; c 3. PARASITOLOGY. Educ: Wheaton Col, Ill, BS, 45; Univ Southern Calif, MS, 48, PhD, 53. Prof Exp: Res assoc & instr, Loma Linda Univ, 52-56; from instr to assoc prof biol, Immaculate Heart Col, 57-66, chmn dept, 63-66; assoc prof biol sci, 66-70, chmn dept, 71-72, PROF BIOL SCI, CALIF STATE COL, DOMINGUEZ HILLS, 70- Concurrent Pos: Mem adv coun, Allergy & Infectious Dis, NIH, 73-74. Mem: Am Soc Parasitol; Am Soc Trop Med & Hyg; AAAS; NY Acad Sci. Res: Host and parasite relationship; control of Schistosoma parasites through snail vector Oncomelania; hybrid snail reproduction and susceptibility to schistosome infection. Mailing Add: Dept of Biol Sci Calif State Col Dominguez Hills CA 90747

CHIA, FU-SHIANG, b Shantung, China, Jan 15, 31; m 63; c 2. ZOOLOGY. Educ: Taiwan Norm Univ, BS, 55; Univ Wash, MS, 62, PhD(zool), 64. Prof Exp: Lab instr biol, Tunghai Univ, 57-58; asst zool, Univ Wash, 58-64; asst prof life sci, Sacramento State Col, 64-66; sr res officer zool, Univ Newcastle, 66-69; assoc prof, 69-75, PROF ZOOL, UNIV ALTA, 75- Mem: Am Soc Zoologists. Res: Developmental biology;

marine invertebrate zoology. Mailing Add: Dept of Zool Univ of Alta Edmonton AB Can

CHIAD, TANG, b Nanking, China, Apr 15, 40; m 67; c 1. ATOMIC PHYSICS. Educ: Tunghai Univ, Taiwan, BS, 64; Kans State Univ, MS, 69, PhD(physics), 74. Prof Exp: Res asst, Physics Inst, Chinese Acad Sinica, 65-67; asst physics, Kans State Univ, 68-73; RES ASSOC NUCLEAR CHEM, CYCLOTRON INST, TEX A&M UNIV, 73- Mem: Am Phys Soc; Am Asn Physics Teachers; Sigma Xi. Res: Study of electron transfer and inner shell ionization in heavy ion-atom collisions. Mailing Add: Cyclotron Inst Tex A&M Univ College Station TX 77843

CHIAKULAS, JOHN JAMES, b Chicago, Ill, Aug 3, 15; m 49; c 1. ANATOMY. Educ: Northwestern Univ, BS, 40, MA, 47; Univ Chicago, PhD(zool), 51. Prof Exp: Spec lectr zool, Grinnell Col, 46-47; instr biol, Roosevelt Univ, 51-52; asst prof anat, Chicago Col Optom, 52-53; instr, 53-55, assoc, 56-57, from asst prof to assoc prof, 58-66, PROF ANAT, CHICAGO MED SCH, 66- Mem: Int Soc Chronobiol; Am Soc Zool; Am Asn Anat; Soc Develop Biol. Res: Wound healing; tissue specificity; organ regeneration; biorythmicity. Mailing Add: Dept of Anat Univ Health Sci Chicago Med Sch Chicago IL 60612

CHIAMORI, NEIL Y, biochemistry, see 12th edition

CHIANELLI, RUSSELL ROBERT, b Newark, NJ, May 22, 44; m 71; c 3. PHYSICAL CHEMISTRY. Educ: Polytech Inst Brooklyn, BS, 70, PhD(chem), 74. Prof Exp: RES CHEMIST, EXXON RES & ENG CO, 73- Mem: AAAS; NY Acad Sci; Am Chem Soc; Am Phys Soc; Electrochem Soc. Res: Physics and chemistry of solids particularly transition metal chalcogenides, metallic non-metals and related compounds which are studied with x-ray crystallography, optical microscopy and spectroscopy. Mailing Add: Exxon Res & Eng Co Corp Res Labs Linden NJ 07036

CHIANG, ANNE, b Canton, China, Oct 3, 42; US citizen; m 67; c 1. PHYSICAL CHEMISTRY, DISPLAY TECHNOLOGY. Educ: Nat Taiwan Univ, BS, 64; Univ Southern Calif, PhD(phys chem), 68. Prof Exp: Sr chemist photochem & colloid chem, Memorex Corp, 69-71; MEM RES STAFF, XEROX PALO ALTO RES CTR, 72- Mem: Am Chem Soc; Soc Photographic Scientists & Engrs. Res: Viscoelasticity of polymers; non-aqueous colloid chemistry; photochemistry of coordination compounds; material research of display devices. Mailing Add: Xerox Palo Alto Res Ctr 3333 Coyote Hill Rd Palo Alto CA 94304

CHIANG, CHIN LONG, b Ningpo, China, Nov 12, 16; US citizen; m 45; c 3. BIOSTATISTICS. Educ: Tsing Hua Univ, China, BA, 40; Univ Calif, Berkeley, MA, 48, PhD(statist), 53. Prof Exp: Instr pub health, 53-55, from asst prof to assoc prof biostatist, 55-66, PROF BIOSTATIST, UNIV CALIF, BERKELEY, 66-, CO-CHMN GROUP, 72- Concurrent Pos: Consult, State Dept Health, Calif, 58 & 64, State Dept Hyg, 61-62, NY State Dept Health, 62, Nat Inst Neurol Dis & Stroke & Nat Ctr Health Serv Res & Develop, 73-; spec res fel, Nat Heart Inst, 59-60; vis asst prof, Univ Mich, 59; vis lectr, Univ Minn, Minneapolis, 60, 61; spec consult, Nat Vital Statist, 62-64, WHO, 70, 71, 73 & 74; vis assoc prof, Univ NC, 63, vis prof, 70; Fulbright fel, Gt Brit, 64; vis prof, Yale Univ, 65 & 66, Emory Univ, 67, Univ Pittsburgh, 68, Univ Wash, 69, Univ Tex, 73 & Vanderbilt Univ, 75; assoc ed, Biometrics, 70-75. Mem: AAAS; Am Pub Health Asn; Am Statist Asn; Inst Math Statist; Biomet Soc. Res: Stochastic studies of the life table; competing risks; illness and death processes; stochastic processes; birth-illness-death process. Mailing Add: 844 Spruce St Berkeley CA 94707

CHIANG, HUAI C, b Sunkiang, China, Feb 15, 15; m 46; c 3. ENTOMOLOGY. Educ: Tsing Hua Univ, China, BS, 38; Univ Minn, MS, 46, PhD(entom), 48. Prof Exp: Asst entom, Tsing Hua Univ, China, 38-40, instr, 40-44; asst, 45-48, res fel, 48-53, from asst prof to assoc prof biol, 54-60, PROF BIOL, UNIV MINN, ST PAUL, 60- Concurrent Pos: Guggenheim fel, 56-57. Mem: Entom Soc Am; Ecol Soc Am; Can Entom Soc; fel Royal Entom Soc London. Res: Insect biology and ecology. Mailing Add: Dept of Entom Univ of Minn St Paul MN 55108

CHIANG, JOSEPH FEI, b Hunan, China, Feb 22, 38; m 63; c 1. PHYSICAL CHEMISTRY, APPLIED PHYSICS. Educ: Tunghai Univ, Taiwan, BS, 60; Cornell Univ, MS, 64, PhD(phys chem), 67. Prof Exp: Asst prof chem, Hudson Valley Community Col, 64-65; fel, Cornell Univ, 67-68; asst prof, 68-74, ASSOC PROF CHEM, STATE UNIV NY COL ONEONTA, 74- Concurrent Pos: NIH fel, 75. Mem: Am Chem Soc; Am Phys Soc. Res: Use of electron diffraction technique and spectroscopic techniques to study molecular structures in gas phase. Mailing Add: Dept of Chem State Univ of NY Col Oneonta NY 13820

CHIANG, KWEN-SHENG, b Shanghai, China, Feb 12, 39. MOLECULAR BIOLOGY. Educ: Nat Taiwan Univ, BS, 59; PhD(biochem Princeton Univ, PhD(biochem sci), 65. Prof Exp: Res assoc biochem, Princeton Univ, 65-66; asst prof, 66-72, ASSOC PROF BIOPHYS, UNIV CHICAGO, 72- Concurrent Pos: USPHS res career develop award, 70-75; ed bd, Plant Sci Letters J, 73-; vis prof, Inst Bot, Acad Sinica, 74-75. Mem: Genetics Soc Am; Am Soc Microbiol; Biophys Soc; Am Soc Cell Biol. Res: Molecular mechanisms of meiosis and sexual reproduction; molecular biology of cellular organelles; chloroplast and mitochondria; biochemical mechanisms of non-Mendelian genetics. Mailing Add: Dept of Biophys & Theoret Biol Univ of Chicago Chicago IL 60637

CHIANG, MORGAN S, b Kiangsu, China, Dec 30, 26; m 65; c 1. GENETICS. Educ: Nat Taiwan Univ, BSc, 50; McGill Univ, MSc, 59; Tex A&M Univ, PhD(genetics), 65. Prof Exp: Asst genetics, Nat Taiwan Univ, 50-55 & lectr, 55-57; res asst, Jackson Mem Lab, 59-61; RES SCIENTIST, RES STA, CAN DEPT AGR, 65- Mem: Agr Inst Can; Genetics Soc Can; Can Soc Hort Sci. Res: Discovery of mutant Careener in mice; breeding cabbage variety resistant to clubroot disease; development of hybrid cabbage variety Chateauguay; cabbage pollen physiology; breeding grain corn resistant to corn borer. Mailing Add: Can Dept of Agr PO Box 457 Res Sta St Jean PQ Can

CHIANG, SCHUMANN, b Shanghai, China, May 27, 44; Can citizen; m 70; c 1. CHEMISTRY. Educ: Queen's Univ, Ont, BSc, 68; Univ Waterloo, PhD(polymer chem), 74. Prof Exp: RES SCIENTIST, FIBERGLAS CAN LTD, 74- Mem: Chem Inst Can; Fedn Soc Coating Technol. Mailing Add: Fiberglas Can Ltd PO Box 3005 Sarnia ON Can

CHIANG, TZU SUNG, b Taiwan, China, Feb 18, 35; m 62; c 3. PHARMACOLOGY. Educ: Nat Taiwan Univ, BS, 58, MS, 61; Univ Kans, PhD(pharmacol), 67. Prof Exp: Teaching asst pharmacol, Nat Taiwan Univ, 58-64; teaching asst, Sch Med, Univ Kans, 64-67; instr, Sch Med, Univ Okla, 67-68; sr scientist, Alcon Labs, Inc, 68-70; ASST PROF PHARMACOL & ASST RES PROF OPHTHAL, MED COL GA, 70- Concurrent Pos: Fel, Sch Med, Univ Okla, 67-68. Mem: NY Acad Sci; Am Soc Pharmacol & Exp Therapeut. Res: Autonomic, cardiovascular and ocular pharmacology; neuropharmacology. Mailing Add: Box 121 Med Col of Ga Augusta GA 30902

CHIANG, YUEN-SHENG, b Tsingtao, China, Feb 2, 36. PHYSICAL CHEMISTRY. Educ: Nat Taiwan Univ, BS, 56; Univ Louisville, MChE, 60; Princeton Univ, PhD(phys chem), 64. Prof Exp: Res assoc phys chem, Princeton Univ, 63; from scientist to sr scientist, Xerox Corp, NY, 64-69; MEM TECH STAFF, RCA LABS, 69- Mem: AAAS; Am Chem Soc; Electron Micros Soc Am; Electrochem Soc; Inst Elect & Electronics Engrs. Res: Physics and chemistry of surfaces; crystal growth and dislocation studies; electron paramagnetic resonance; electron microscopy; organic semiconductors; metal physics; ultra high vacuum technology. Mailing Add: RCA Labs Princeton NJ 08540

CHIAO, RAYMOND YU, b Hong Kong, Oct 9, 40; US citizen; m 68; c 2. PHYSICS. Educ: Princeton Univ, AB, 61; Mass Inst Technol, PhD(physics), 65. Prof Exp: Asst prof physics, Mass Inst Technol, 65-67; asst prof, 67-70, ASSOC PROF PHYSICS, UNIV CALIF, BERKELEY, 70- Concurrent Pos: Alfred P Sloan fel, 67-72. Mem: Am Phys Soc. Res: Lasers; non-linear optics; spontaneous and stimulated Brillouin scattering; stimulated Raman scattering; self-trapping of optical beams; superconductivity; astrophysics. Mailing Add: Dept of Physics Univ of Calif Berkeley CA 94720

CHIAO-YAP, LUNG WEN, b Nanking, China. THEORETICAL NUCLEAR PHYSICS. Educ: Nat Taiwan Univ, BS, 56; Univ Calif, Berkeley, PhD(chem), 61. Prof Exp: Res assoc chem, Brookhaven Nat Lab, 61-62; res assoc nuclear data, Nat Acad Sci, 62-63; asst prof, 64-69, ASSOC PROF PHYSICS, GEORGETOWN UNIV, 69- Res: Nuclear collective model. Mailing Add: Dept of Physics Georgetown Univ Washington DC 20057

CHIARAPPA, LUIGI, b Rome, Italy, Dec 12, 25; US citizen; m 51; c 3. PHYTOPATHOLOGY. Educ: Univ Florence, Gen Agr Laurea, 50; Univ Calif, PhD(plant path), 58. Prof Exp: Agronomist tech comn, Ital Colonization in Chile, 50-51; entomologist, Di Giorgio Corp, 52-55, plant pathologist, Res Dept, 59-62; from tropical plant pathologist to SR PLANT PATHOLOGIST, RES DEPT, FOOD & AGR ORGN, UN, 62- Concurrent Pos: Vis prof, Univ Calif, Davis, 75-76. Mem: Am Phytopath Soc; Int Orgn Citrus Virol. Res: Diseases of fruit, nut and vine crops; epidemiology; disease loss appraisal. Mailing Add: Via San Lucio 38 Rome Italy

CHIARODO, ANDREW, b New York, NY, June 26, 34; m 69. DEVELOPMENTAL BIOLOGY, CELL BIOLOGY. Educ: Fordham Univ, AB, 56, MS, 59; Washington Univ, PhD(zool), 63. Prof Exp: NIH fel, Med Col, Cornell Univ, 63; from instr to assoc prof biol, Georgetown Univ, 63-73; grants assoc, Div Res Grants, NIH, 73-74; PROG DIR, NAT ORGAN SITE PROGS, NAT CANCER INST, 74- Mem: AAAS; Soc Develop Biol; Am Soc Zoologists. Mailing Add: Nat Cancer Inst Bethesda MD 20014

CHIASSON, BERTRAND ARNOLD, b New Waterford, NS, Dec 16, 46; US citizen; m 73. ORGANIC CHEMISTRY. Educ: St Louis Univ, BS, 69; Mass Inst Technol, PhD(org chem), 73. Prof Exp: ORG CHEMIST, UNION CARBIDE CORP, 73- Mem: Am Chem Soc. Res: Plant growth regulators. Mailing Add: Union Carbide Corp PO Box 8361 South Charleston WV 25303

CHIASSON, LEO PATRICK, b Cheticamp, NS, May 14, 18; m 48; c 5. GENETICS. Educ: St Francis Xavier Univ, BA, 38, BSc, 40; Univ Toronto, PhD(genetics), 44. Prof Exp: Assoc prof biol, 44-49, PROF BIOL, ST FRANCIS XAVIER UNIV, 49- Concurrent Pos: Assoc scientist, Fisheries Res Bd Can, 44-55; sr researcher zool, Columbia Univ, 63-64. Mem: AAAS; Genetics Soc Am; Genetics Soc Can. Res: Tomato species hybrids; relative growth in mice; blood groups and dermatoglyphics in Micmac Indians; distribution of scallops; species hybrids of Abies; mutagenic effects of heat in bacteria. Mailing Add: Dept of Biol St Francis Xavier Univ Antigonish NS Can

CHIASSON, ROBERT BRETON, b Griggsville, Ill, Oct 9, 25; m 44; c 8. COMPARATIVE ANATOMY, COMPARATIVE ENDOCRINOLOGY. Educ: Ill Col, AB, 49; Univ Ill, MS, 50; Stanford Univ, PhD(biol sci), 56. Prof Exp: Spec supvr, Ill State Mus, 50-51; from instr zool to prof biol sci, 51-75, PROF VET SCI, UNIV ARIZ, 75- Concurrent Pos: Fulbright lectr, Univ Sci & Technol, Ghana, 69-70. Mem: Am Soc Zoologists; Am Physiol Soc; Int Soc Stereology; Soc Vert Paleont; World Asn Vet Anatomists. Res: Regulation of pituitary function in the chicken and anatomy of vertebrates; anatomy of the eye of doves and skin of snakes. Mailing Add: Dept of Vet Sci Univ of Ariz Tucson AZ 85721

CHIASSON, WILLIAM JOSEPH, organic chemistry, industrial chemistry, see 12th edition

CHIAZZE, LEONARD, JR, b Falconer, NY, June 19, 34; m 54; c 4. BIOSTATISTICS, EPIDEMIOLOGY. Educ: Univ Buffalo, BS, 55, MBA, 57; Univ Pittsburgh, ScD(biostatist), 64. Prof Exp: Asst health serv officer, Nat Cancer Inst, 57-60, sr asst health serv officer, 60-63, health serv officer, 63-64, scientist, 64-66; res assoc ctr pop res & dir div biostatist & epidemiol, Sch Med, Georgetown Univ, 66-68, dir cerebrovasc dis follow-up & surveillance syst, 68-71, ASSOC PROF COMMUNITY MED & INT HEALTH, SCH MED, GEORGETOWN UNIV, 68-, DIR, GRAD PROG BIOSTATIST, 71- Concurrent Pos: Chief biomet br, Nat Cancer Inst, 75- Mem: Fel Am Pub Health Asn; Am Statist Asn; Pop Asn Am; Soc Epidemiol Res; Int Epidemiol Soc. Res: Chronic disease epidemiology, especially cancer; morbidity survey and case register methodologies; population research; delivery of medical services; clinical trials. Mailing Add: Dept Community Med & Int Health Georgetown Univ Sch of Med 3900 Reservoir Rd Washington DC 20007

CHIBA, MIKIO, b Miyagi-Ken, Japan, Aug 4, 29; m 56; c 2. ANALYTICAL CHEMISTRY. Educ: Hokkaido Univ, MSc, 53, DSc(anal chem), 62. Prof Exp: Chemist, Hokkaido Police Hq, 54-56; res chemist, Sci Police Res Inst, 56-64; res officer, 64-66, RES SCIENTIST, CAN DEPT AGR, 66- Concurrent Pos: Nat Res Coun Can fel, 62-64; hon res prof, Brock Univ, 73- Mem: Am Chem Soc; NY Acad Sci; Chem Inst Can; Chem Soc Japan; Food Hyg Soc Japan. Res: Method development for pesticide analysis and to find better ways of applying pesticides. Mailing Add: Can Dept of Agr PO Box 185 Vineland Station ON Can

CHIBNIK, SHELDON, b New York, NY, Dec 20, 25; m 45; c 2. ORGANIC CHEMISTRY. Educ: Cornell Univ, AB, 44; Polytech Inst Brooklyn, MS, 51; Temple Univ, PhD(chem), 55. Prof Exp: Lab instr chem, Hunter Col, 46-48; from chemist to sr chemist, Nat Lead Co, 48-55, group leader indust finishes, 55-61; sr res chemist, Mobil Chem Co, Edison, 61-70 & Mobil Res & Develop Corp, Paulsboro, 71-75, ASSOC CHEM, MOBIL RES & DEVELOP CORP, 75- Mem: Am Chem Soc. Res: Polymer chemistry; protective coatings; monomer synthesis; liquid phase oxidations; heterogeneous catalysis; lubricating oil and fuel additives. Mailing Add: 7 Glen View Pl Cherry Hill NJ 08034

CHIBURIS, EDWARD FRANK, b Omaha, Nebr, July 31, 33; m 54; c 5. GEOPHYSICS. Educ: Tex A&M Univ, BS, 60, MS, 62; Ore State Univ, PhD(geophys), 65. Prof Exp: Res geophysicist, Seismic Data Lab, Teledyne, Inc, 65-

68 & dir res, 68-69; ASSOC PROF GEOPHYS & ASST DIR, MARINE SCI INST, UNIV CONN, 69- Concurrent Pos: Geophys consult, Teledyne, Inc, 69- Mem: Am Geophys Union; Seismol Soc Am; Soc Explor Geophys. Res: Hypocenter location techniques; seismic network and array analyses; focal mechanism studies; gravity and magnetic methods; crustal studies; computer modeling; geophysical data processing. Mailing Add: Marine Sci Inst Univ of Conn Groton CT 06340

CHIBUZO, GREGORY ANENONU, b Abor, Nigeria, July 25, 43; m 72; c 1. VETERINARY ANATOMY. Educ: Tuskegee Inst, BS, 68, DVM, 70, MS, 75. Prof Exp: Instr, 70-73, ASST PROF VET ANAT, SCH VET MED, TUSKEGEE INST, 73- Concurrent Pos: Consult bio-med studies, Southern Voc Col, Tuskegee, 74-75. Mem: World Asn Vet Anatomists; Am Asn Vet Anatomists; Soc Study Reproduction; Am Vet Med Asn. Res: Influence of neurohumoral substances on uterine motility; effect of progesterone and/or estrogen on the integrity of intra-ovarian vascular growth and distribution at birth, maturity and menopause. Mailing Add: Dept of Anat Tuskegee Inst Sch of Vet Med Tuskegee Institute AL 36088

CHICHESTER, CLINTON OSCAR, b New York, NY, Feb 11, 25; m 47; c 3. FOOD TECHNOLOGY. Educ: Mass Inst Technol, SB, 49; Univ Calif, MS, 51, PhD, 54. Prof Exp: From asst prof to prof food technol, Univ Calif, Davis, 53-70, chmn dept, 67-70; PROF FOOD & RESOURCE CHEM, INT CTR MARINE RESOURCE DEVELOP, UNIV RI, 70- Concurrent Pos: Mem, NIH, 57-, Coun Foods & Nutrit, AMA, 62- & Space Sci Bd, Nat Acad Sci, 63-; on leave as vpres res, Nutrit Found, New York, 72-74. Honors & Awards: Bernardo O'Higgins Award, Govt Chile, 69; Medal, Czech Acad Sci. Mem: Inst Food Technologists; Am Chem Soc; Optical Soc Am; Am Inst Chem Eng; Am Soc Biol Chemists. Res: Pigment biochemistry; food processing. Mailing Add: Dept of Food & Resource Chem Univ of RI Kingston RI 02881

CHICHESTER, FREDERICK WESLEY, b New Haven, Conn, Apr 28, 29; m 58; c 4. SOIL CHEMISTRY, PLANT PHYSIOLOGY. Educ: Colo State Univ, BS, 52; Univ Conn, MS, 60; Ore State Univ, PhD(soils), 66. Prof Exp: Asst soils, Ore State Univ, 63-66; RES SCIENTIST, AGR RES SERV, USDA, 66- Mem: Am Soc Agron; Soil Sci Soc Am. Res: Role of soil organo-mineral colloids in plant nutrient cycling; soil nitrogen transformations; effects of agricultural chemicals use and animal wastes on water quality; transport mechanisms of potential agricultural water pollutants. Mailing Add: North Appalachian Exp Watershed NCent Region Agr Res Serv USDA PO Box 478 Coshocton OH 43812

CHICHESTER, LYLE FRANKLIN, b Albany, NY, Nov 5, 31; m 54; c 3. ZOOLOGY. Educ: Univ Conn, BS, 54, PhD(zool), 68; Cent Conn State Col, MS, 61. Prof Exp: From instr to assoc prof biol, 60-74, PROF BIOL, CENT CONN STATE COL & CHMN DEPT BIOL SCI, 74- Mem: AAAS. Res: Distribution, ecology and systematics of terrestrial slugs, especially introduced European species; application of biochemical methods to systematic problems involving mollusks. Mailing Add: Dept of Biol Sci Cent Conn State Col New Britain CT 06050

CHICK, ERNEST WATSON, b Lynchburg, Va, Feb 20, 28; m 52; c 5. MYCOLOGY, PREVENTIVE MEDICINE. Educ: Duke Univ, BA, 49, MD, 53. Prof Exp: Epidemic intel serv officer, Commun Dis Ctr, Kansas City Field Sta, 54-56; assoc path, Duke Univ, 59-61; asst chief lab serv, Durham Vet Admin Hosp, NC, 62-63; from assoc prof to prof med, WVa Univ, 63-69, chmn div prev med, 63-69; assoc dir, Tobacco & Health Res Inst, 71-74, PROF COMMUNITY MED & MED & CO-DIR MYCOL PROG, COL MED, UNIV KY, 69-, ACTG CHMN DEPT CELL BIOL, 75- Concurrent Pos: Clin investr, Durham Vet Admin Hosp, 59-61; mem, Vet Admin Fungus Dis Study Group, 59-63, consult, 63-69, chmn, 70-; mem, Govt Task Force Health, WVa, 65-66; mem standards & exam comt pub health & med mycol, Am Bd Med Microbiol, 71- Mem: Fel Am Col Chest Physicians; Am Thoracic Soc; Am Soc Exp Path; Int Soc Human & Animal Mycol; Med Mycol Soc of the Americas (pres, 72). Res: Epidemiology; pathogenesis and treatment of mycotic infections. Mailing Add: Dept of Community Med Univ of Ky Col of Med Lexington KY 40506

CHICK, THOMAS WESLEY, b Martin, Tenn, May 1, 40; m 72; c 3. PULMONARY PHYSIOLOGY. Educ: Univ Cent Ark, BS, 61; Univ Ark, Little Rock, MD, 65. Prof Exp: From intern to resident internal med, Univ Tex Southwestern Med Sch Dallas, 65-68, fel pulmonary med, 68-70, from instr to asst prof internal med, 70-72; ASST PROF INTERNAL MED, SCH MED, UNIV N MEX, 72- Res: Clinical pulmonary physiology of obstructive airway disease; effects of bronchodilators, oxygen and exercise. Mailing Add: Vet Admin Hosp 2100 Ridgecrest Dr SE Albuquerque NM 87108

CHICKERING, ARTHUR MERTON, zoology, deceased

CHICKOS, JAMES S, b Buffalo, NY, Oct 27, 41; m 66; c 2. ORGANIC CHEMISTRY, PHYSICAL ORGANIC CHEMISTRY. Educ: Univ Buffalo, BA, 63; Cornell Univ, PhD(org chem), 66. Prof Exp: NIH vis fel, Princeton Univ, 66-67; NIH fel & res assoc, Univ Wis, 67-69; ASST PROF ORG CHEM, UNIV MO-ST LOUIS, 69- Mem: Am Chem Soc; The Chem Soc. Res: Role of tautomeric catalysis in organic and enzymatic systems; small ring, non-benzenoid aromatics; chemistry for the non-major. Mailing Add: Dept of Chem Univ of Mo St Louis MO 63121

CHICKS, CHARLES HAMPTON, b Sandpoint, Idaho, Nov 10, 30; m 56; c 4. MATHEMATICS. Educ: Linfield Col, BA, 53; Univ Ore, MA, 56, PhD(math), 60. Prof Exp: Adv res engr, Sylvania Electronic Defense Labs, Gen Tel & Electronics Corp, 60-62, engr specialist, 62-69; SR MEM TECH STAFF, ESL INC, 69- Concurrent Pos: Lectr, Univ Santa Clara, 64- Mem: Am Math Soc. Res: Periodic automorphisms on banach algebras; military operations research; arms control and disarmament. Mailing Add: Tech Staff ESL Inc 495 Java Dr Sunnyvale CA 94086

CHICO, RAYMUNDO JOSE, b Hernando, Arg, Sept 17, 30; m 59; c 4. MINING GEOLOGY, ECONOMIC GEOLOGY. Educ: Univ Cordoba, dipl geol, 53; Mo Sch Mines, MS, 58; Harvard Univ, MA, 63. Prof Exp: Asst, Univ Cordoba, 53; geologist, Direccion Gen de Ingenieros, Arg, 54; geologist, Peruvian Mines, Cerro de Pasco Corp, 54-55, off geologist, NY, 58-59; geol engr, Four Corners Uranium Corp, Colo, 56-57; consult, Nat Lead Co, Arg, 59, Int Basic Econ Corp, NY, 60 & Air Force Cambridge Res Labs, 60; geologist, Ltd War Lab, Aberdeen, Md, 63-65; oceanogr, Nat Oceanog Data Ctr, DC, 65-66; consult econ mining & eng geol indust, US & Latin Am, 66-68; PRES, RAYMUNDO J CHICO, INC, 68- Concurrent Pos: Guest crystallog lab, Johns Hopkins Univ, 65; US deleg, NATO Advan Study Inst Uranium, London, 72. Honors & Awards: State of Md Gov Citation, 69. Mem: Am Inst Mining, Metall & Petrol Eng; Geol Soc Am; Sigma Xi; Am Mining Cong. Res: Engineering earth sciences; inter-American mineral industry; applied geology and mining for business development and economic growth; ore genesis; minerals; field economic geology. Mailing Add: 9600 E Grand Ct Englewood CO 80110

CHICOINE, LUC, b Montreal, Que, Apr 19, 29; m 53; c 1. PEDIATRICS. Educ: Univ Montreal, BA, 48, MD, 53; FRCP(C), 60. Prof Exp: From asst prof to assoc prof, 61-71, PROF PEDIAT, UNIV MONTREAL, 71-, CHMN DEPT, 75- Concurrent Pos: Dir, Poison Control Ctr. Mem: Am Acad Pediat; Can Med Asn; Can Pediat Soc. Res: Pediatric water and electrolyte problems. Mailing Add: Ste Justine Hosp 3175 Cote Ste Catherine Montreal PQ Can

CHICOYE, ETZER, b Jacmel, Haiti, Nov 4, 26; US citizen; m 54; c 2. FOOD CHEMISTRY. Educ: Univ Haiti, BS, 48; Univ Wis-Madison, MS, 54, PhD(food sci), 68. Prof Exp: Grader & cup tester, Nat Coffee Bur, Haiti, 48-52; res asst, Univ Wis-Madison, 52-54, 64-67; anal chemist, Chicago Pharmacol Co, 54-56; res chemist, Julian Labs, Ill, 56-64; chem res supvr, 67-72, MGR RES, MILLER BREWING CO, 72- Mem: Am Chem Soc; Inst Food Technol; Master Brewers Asn Am; Am Soc Brewing Chemists. Res: Steroid chemistry; steroid hormones and vitamin D; cholesterol and degradation products of cholesterol in food; brewing chemistry; flavor chemistry. Mailing Add: Miller Brewing Co Res Lab 4000 W State St Milwaukee WI 53208

CHIDAMBARASWAMY, JAYANTHI, b Hamsavaram, India, Nov 14, 27; m 46; c 5. MATHEMATICS. Educ: PR Col, Kakinada, India, BA, 50; Andhra Univ, India, MA, 53; Univ Calif, Berkeley, PhD(math), 64. Prof Exp: Lectr math, Andhra Univ, India, 53-62; teaching asst, Univ Calif, Berkeley, 62-64, instr, 64-65; asst prof, Univ Kans, 65-66; assoc prof, 66-69, PROF MATH, UNIV TOLEDO, 69- Mem: Am Math Soc; London Math Soc; Indian Math Soc. Res: Theory of numbers. Mailing Add: Dept of Math Univ of Toledo Toledo OH 43606

CHIDDIX, MAX EUGENE, b Palestine, Ill, Apr 13, 18; m 44; c 2. ORGANIC CHEMISTRY. Educ: Ill State Norm Univ, BEd, 40; Univ Ill, PhD(org chem), 43. Prof Exp: Res chemist, Gen Aniline & Film Corp, 43-49, group leader, 50-53, res fel, 53-55, prog mgr acetylene derivatives res, 55-60 & chem & polymer res, 60-68; chief chemist amiben, 68-69, CHIEF PROJ CHEMIST, GAF CORP, 69- Mem: AAAS; NY Acad Sci; Am Chem Soc. Res: Polymers; plastics; resins; polyamide fibers and film; acetylene and textile chemicals; surfactants; alkylphenols; chelating agents; corrosion inhibitors; reactive dyes; lube oil additives; bactericides; fungicides; herbicides; analytical methods; pollution control. Mailing Add: Apt 103 310 Waco St League City TX 77573

CHIDESTER, ALFRED HERMAN, b East Moline, Ill, Sept 23, 14; m 37; c 3. GEOLOGY. Educ: Augustana Col, AB, 42; Univ Chicago, PhD, 59. Prof Exp: Geologist, US Geol Surv, 43-46; instr mapping, Univ Chicago, 47-48; geologist, Br Rocky Mountain Natural Resources, 48-72, CHIEF BR LATIN AM & AFRICAN GEOL, US GEOL SURV, 72- Mem: Mineral Soc Am; Geol Soc Am; Soc Econ Geologists; Am Geophys Union; Geochem Soc. Res: Geology and petrology of talc and asbestos bearing ultramafic rocks of Vermont; petrology and mineralogy; structure; pyrite, sulfure and barite resources of Japan; astronaut training in geology; regional geology of Liberia. Mailing Add: US Geol Surv Nat Ctr Reston VA 22092

CHIDLEY, BRUCE GORDON, b Toronto, Ont, Can, Dec 29, 32; m 60; c 3. PHYSICS. Educ: McMaster Univ, BSc, 53, MSc, 54; Univ Sask, PhD, 56. Prof Exp: Res physicist, Carnegie Inst Technol, 56-59; RES PHYSICIST, CHALK RIVER NAT LAB, ATOMIC ENERGY CAN, LTD, 59- Res: Accelerator and reactor physics. Mailing Add: Chalk River Nat Lab Atomic Energy of Can Ltd Chalk River ON Can

CHIDSEY, CHARLES AUGUSTUS, b Detroit, Mich, Sept 11, 29; m 54; c 3. INTERNAL MEDICINE, CLINICAL PHARMACOLOGY. Educ: Trinity Col, BS, 50; Columbia Univ, MD, 54. Prof Exp: Intern, First Med Div, Bellevue Hosp, 54-55, asst resident, 55-56; asst resident, Grace-New Haven Hosp, 58-59; clin assoc, Nat Heart Inst, 59-61, sr investr, 61-65; PROF MED & PHARMACOL, UNIV COLO MED CTR, DENVER, 65- Concurrent Pos: Fel, Bellevue Hosp, 56-58; mem coun high blood pressure res, Am Heart Asn. Mem: Cardiac Muscle Soc; Am Soc Clin Pharmacol & Therapeut; Am Soc Pharmacol & Exp Therapeut; Am Physiol Soc; Am Soc Clin Invest. Res: Clinical and cardiovascular pharmacology; myocardial biochemistry in heart failure; adrenergic mechanisms. Mailing Add: Dept Med & Pharmacol Univ Colo Med Ctr 4200 E 9th Ave Denver CO 80220

CHIDSEY, JANE LOUISE, b Wilkes-Barre, Pa, Apr 1, 08. PHYSIOLOGY. Educ: Wellesley Col, AB, 29; Brown Univ, AM, 31; Cornell Univ, PhD(physiol), 34. Prof Exp: Demonstr biol, Brown Univ, 29-31; Coxe fel, Yale Univ, 34-35; instr zool, Smith Col, 35-39; from asst prof to prof biol, 39-74, Fund Adv Educ fac fel, 52-53, actg dean, 61-62, EMER PROF BIOL, WHEATON COL, MASS, 74- Concurrent Pos: Mem corp, Mt Desert Island Biol Lab. Mem: AAAS; Am Inst Biol Sci. Res: Carbohydrate and fat metabolism; general animal and cellular physiology. Mailing Add: Dept of Biol Wheaton Col Norton MA 02766

CHIEN, CHIH-YUNG, b Chungking, China, Aug 5, 39; m 63; c 3. HIGH ENERGY PHYSICS. Educ: Nat Taiwan Univ, BS, 60; Yale Univ, MS, 63, PhD(physics), 66. Prof Exp: Asst res physicist, Univ Calif, Los Angeles, 66-67, asst prof in residence, 67-68, asst prof, 68-69; asst prof, 69-73, ASSOC PROF PHYSICS, JOHNS HOPKINS UNIV, 73- Mem: Am Phys Soc. Res: Experimental research on the structure and interactions of elementary particles. Mailing Add: Dept of Physics Johns Hopkins Univ Baltimore MD 21218

CHIEN, HWEI CHIU, biochemistry, food science, see 12th edition

CHIEN, JAMES C W, b Shanghai, China, Nov 4, 29; US citizen; m 53; c 3. PHYSICAL CHEMISTRY. Educ: St John's Univ, China, BS, 49; Univ Ky, MS, 51; Univ Wis, PhD(phys chem), 54. Prof Exp: Sr res chemist, Hercules Powder Co, Del, 54-69; PROF CHEM, UNIV MASS, AMHERST, 69- Mem: Am Chem Soc. Res: Photochemistry; energy transfer; nuclear magnetic resonance; radiation and quantum chemistry; polymerization; oxidation; ultraviolet spectroscopy; electron spin resonance. Mailing Add: Dept of Chem Univ of Mass Amherst MA 01002

CHIEN, PING-LU, b China, Nov 5, 28; m 57; c 3. ORGANIC CHEMISTRY. Educ: Nat Taiwan Univ, BS, 52; Univ Kans, PhD(org chem), 64. Prof Exp: Chemist, Union Indust Res Inst, 55-59; res assoc med chem, Univ Kans, 64-65; assoc chemist, 65-68, SR CHEMIST, MIDWEST RES INST, 68- Mem: Am Chem Soc. Res: Synthetic organic and medicinal chemistry; absorption spectroscopy. Mailing Add: Midwest Res Inst 425 Volker Blvd Kansas City MO 64110

CHIEN, SEN HSIUNG, b Taiwan, China, Aug 31, 41; m 70; c 1. SOIL CHEMISTRY. Educ: Nat Taiwan Univ, BS, 63; Univ NH, MS, 68; Iowa State Univ, PhD(soil chem), 72. Prof Exp: Assoc, Iowa State Univ, 72-73 & Washington Univ, 73-75; RES CHEMIST SOIL CHEM, INT FERTILIZER DEVELOP CTR, 75- Mem: Soil Sci Soc Am; Am Soc Agron; Int Soil Sci Soc. Res: Dissolution of phosphate rock in relation to the utilization of phosphate rock for direct application to soils. Mailing Add: Int Fertilizer Develop Ctr Florence AL 35630

CHIEN, SHU, b Peiping, China, June 23, 31; m 57; c 2. PHYSIOLOGY. Educ: Nat Taiwan Univ, MB, 53; Columbia Univ, PhD(physiol), 57. Prof Exp: Intern, Taiwan Univ Hosp, 52-53; asst, 54-56, from instr to assoc prof, 56-69, PROF PHYSIOL, COL PHYSICIANS & SURGEONS, COLUMBIA UNIV, 69- Mem: AAAS; Harvey

Soc; Instrument Soc Am; Soc Exp Biol & Med; NY Acad Sci. Res: Blood viscosity; red cell membrane; microcirculation; blood flow and volume; hemorrhage; endotoxin shock; body fluids; autonomic nervous system. Mailing Add: Dept of Physiol Columbia Univ Col Physicians & Surgeons New York NY 10032

CHIERICI, GEORGE J, b Napa, Calif, Nov 8, 26; m 56; c 4. PROSTHODONTICS. Educ: Univ Pac, DDS, 50. Prof Exp: Asst clin prof, 63-68, asst prof, 68-70, ASSOC PROF OROFACIAL ANOMALIES, UNIV CALIF, SAN FRANCISCO, 70- Mem: Int Asn Dent Res; Am Cleft Palate Asn. Res: Normal and abnormal growth and development; morphologic and physiologic interrelationships in the orofacial complex. Mailing Add: Dept of Orofacial Anomalies Univ of Calif Sch of Dent San Francisco CA 94143

CHIGA, MASAHIRO, b Tokyo, Japan, Mar 6, 25; nat US; m 62; c 2. PATHOLOGY. Educ: Univ Tokyo, MD, 50. Prof Exp: Asst, Inst Infectious Dis, Univ Tokyo, 54; resident, Med Ctr, Univ Kans, 54-58; from asst prof to assoc prof path, Sch Med, Univ Utah, 60-69; assoc prof, 69-72, PROF PATH, SCH MED, UNIV KANS, 72- Concurrent Pos: Childs Mem Fund fel, 56-57; Nat Found fel, Metab Lab, Univ Utah, 58-60. Mem: Int Acad Path; Am Soc Exp Path. Res: Viral and rickettsial infection; experimental oncology; enzymes. Mailing Add: Dept of Path Univ of Kans Med Ctr Kansas City KS 66103

CHIGNELL, COLIN FRANCIS, b London, Eng, Apr 7, 38; US citizen; m 66; c 2. ORGANIC CHEMISTRY, PHARMACOLOGY. Educ: Univ London, BPharm, 59, PhD(med chem), 62. Prof Exp: Vis fel, Nat Inst Arthritis, Metab & Digestive Dis, 62-65; vis assoc, 65-70, RES PHARMACOLOGIST, NAT HEALTH & LUNG INST, 70- Concurrent Pos: Res assoc, Nat Inst Gen Med Sci, 66-69. Honors & Awards: J J Abel Prize, Am Soc Pharmacol & Exp Therapeut, 73. Mem: Am Chem Soc; Am Soc Pharmacol & Exp Therapeut; Am Soc Biol Chemists; NY Acad Sci; Soc Exp Biol & Med. Res: Spectroscopic studies of drug interactions with biological systems at a molecular level; structure and function of biological membranes and their interaction with drug molecules; mechanisms of drug toxicity. Mailing Add: Molec Pharm Sect Pulmonary Br Nat Heart & Lung Inst Bethesda MD 20014

CHIH, CHUNG-YING, b China, Dec 11, 16; nat US; m 55. PHYSICS. Educ: Nat Tsing Hua Univ, China, BSc, 37; Univ Calif, Berkeley, PhD(physics), 54. Prof Exp: Instr physics, Fukien Med Col, 37-40; assoc prof, Fukien Teachers Col, 40-44; prof, Nat Chi-nan Univ, 44-45 & Kiangsu Col, 45-48; physicist radiation lab, Univ Calif, 53-54; from asst prof to prof physics, Middlebury Col, 54-68; SCI CONSULT, 68- Concurrent Pos: NSF res grant, 57-60. Mem: Am Phys Soc. Res: Neutron proton scattering; elementary particles. Mailing Add: PO Box 2556 Noble Sta Bridgeport CT 06608

CH'IH, JOHN JUWEI, b Tsingtao, China, Oct 29, 33; US citizen; m 62; c 1. BIOLOGICAL CHEMISTRY. Educ: Southern Ill Univ, BA, 60; Univ Del, MS, 63; Thomas Jefferson Univ, PhD(biochem), 68. Prof Exp: Res technician, Biochem Res Found, Newark, Del, 60-63; clin chemist, St Mary's Hosp, Philadelphia, 63-65; teaching asst biochem, Thomas Jefferson Univ, 65-68, instr, 68-69; sr instr biol chem, 69-71, asst prof, 71-76, ASSOC PROF BIOL CHEM, HAHNEMANN MED COL, 76- Mem: Am Soc Biol Chemists; Am Asn Clin Chemists; Sigma Xi; AAAS; Am Chem Soc. Res: Regulatory mechanisms in nucleic acids and protein biosynthesis of the eukaryotic cells; biogenesis of mammalian cell organelles. Mailing Add: Dept of Biol Chem Hahnemann Med Col Philadelphia PA 19102

CHIHARA, CAROL JOYCE, b New York, NY, Oct 31, 41; m 64; c 1. DEVELOPMENTAL GENETICS. Educ: Univ Calif, Berkeley, BA, 62, PhD(develop genetics), 72; San Francisco State Univ, MA, 67. Prof Exp: NIH fel genetics, Cambridge Univ, 72-73; RESEARCHER DEVELOP BIOL, UNIV CALIF, BERKELEY, 74- Concurrent Pos: Lectr cell & molecular biol, San Francisco State Univ, 74. Mem: AAAS. Res: Genetic control mechanisms in Drosophila as expressed by the phenomena of transdetermination in imaginal discs; effect of environmental factors and hormones as well as developmental capacities of the discs. Mailing Add: Dept of Molecular Biol Univ of Calif Berkeley CA 94720

CHIHARA, THEODORE SEIO, b Seattle, Wash, Mar 14, 29; m 56; c 5. MATHEMATICAL ANALYSIS. Educ: Seattle Univ, BS, 51; Purdue Univ, MS, 53, PhD(math), 55. Prof Exp: Asst math, Purdue Univ, 52-55; from asst prof to prof, Seattle Univ, 55-69, actg head dept, 58-59, head, 59-66; prof assoc, Univ Alta, 69-70; vis prof, Univ Victoria, BC, 70-71; PROF MATH & CHMN DEPT, PURDUE UNIV, CALUMET CAMPUS, 71- Mem: Am Math Soc; Math Asn Am; Soc Indust Appl Math. Res: Theory of orthogonal polynomials; moment problems; special functions. Mailing Add: Dept of Math Purdue Univ Hammond IN 46322

CHILCOTE, DAVID OWEN, b Sask, Can, July 8, 31; m 53; c 2. c 2. AGRONOMY, PLANT PHYSIOLOGY. Educ: Ore State Univ, BS, 53, MS, 57; Purdue Univ, PhD(agron), 61. Prof Exp: From instr to assoc prof farm crops, 53-70, PROF AGRON CROP SCI, ORE STATE UNIV, 70- Mem: Am Soc Plant Physiol; Am Soc Agron; Crop Sci Soc Am. Res: Herbicides and growth regulators; crop physiology and ecology. Mailing Add: Dept of Agron Crop Sci Ore State Univ Corvallis OR 97331

CHILCOTE, MAX ELI, b Bemidji, Minn, Sept 1, 17; m 43; c 3. CLINICAL CHEMISTRY, LABORATORY MEDICINE. Educ: Univ Minn, BS, 38, MS, 41; Univ Mich, PhD(biol chem), 44. Prof Exp: Asst biochem, Univ Mich, 40-44; instr, Med Sch, Loyola Univ, Ill, 44-46; Nutrit Found res fel, Pa State Col, 46-48; from asst prof to assoc prof biochem, Sch Med State Univ NY Buffalo, 48-60, asst dir biochem lab, 59-66, dir clin biochem, 66-69, assoc dir, 69-70, DIR ERIE COUNTY LABS, 70- Concurrent Pos: Clin assoc prof, 60-70, clin prof biochem, 70- & Dept of Pathol, Sch Med, State Univ NY Buffalo, 74- Honors & Awards: Educ Award, Am Asn Clin Chem, 75. Mem: Am Asn Clin Chem; Can Soc Clin Chem; Am Chem Soc; Acad Clin Lab Physicians & Scientists (pres, 73-75). Res: Clinical chemistry. Mailing Add: Erie County Labs 462 Grider St Buffalo NY 14215

CHILCOTE, WILLIAM W, b Washington, Iowa, Mar 6, 18; m 46; c 2. PLANT ECOLOGY. Educ: Iowa State Col, BS, 43, PhD(bot), 50. Prof Exp: Instr forestry, Iowa State Col, 46-50; asst prof bot & asst ecologist, 50-56, assoc prof bot & assoc ecologist, 56-62, PROF BOT, AGR EXP STA, ORE STATE UNIV, 62- Concurrent Pos: Fulbright grant, Finland, 58-59. Mem: Ecol Soc Am; Soc Am Foresters. Res: Forest and range ecology; autecology; community dynamics. Mailing Add: Dept of Bot Ore State Univ Corvallis OR 97331

CHILCOTT, JOHN HENRY, b Evanston, Ill, Feb 21, 24; m 50; c 3. ANTHROPOLOGY. Educ: Harvard Univ, AB, 48; Univ Colo, MEd, 52, PhD(educ), 58. Prof Exp: Asst pres, Menlo Sch & Col, 56-58; asst prof educ, Univ Calif, Santa Barbara, 58-63; assoc prof, 63-70, PROF ANTHROP & EDUCATION, UNIV ARIZ, 70- Concurrent Pos: Fulbright lectr, Trujillo, 65-66; ed, Coun Anthrop & Educ, 73- Mem: Am Anthrop Asn; Soc Appl Anthrop; Am Ethnol Soc; Am Educ Res Asn. Res: Education of minority group members in the United States; directed culture change. Mailing Add: Dept of Anthrop Univ of Ariz Tucson AZ 85721

CHILD, CHARLES GARDNER, III, b New York, NY, Feb 1, 08; m 41; c 6. SURGERY. Educ: Yale Univ, AB, 30; Cornell Univ, MD, 34; Am Bd Surg, dipl, 42. Prof Exp: Attend surgeon, NY Hosp, 47-53; assoc prof clin surg, Med Col, Cornell Univ, 47-53; prof surg & chmn dept, Med Sch, Tufts Univ, 53-58; chmn dept, 59-74, PROF SURG, MED SCH, UNIV MICH, ANN ARBOR, 59- Concurrent Pos: Surgeon in chief, Boston Dispensary & surgeon, Boston Floating Hosp, Infants & Children, 53-58; dir first surg serv, Boston City Hosp, 54-58; consult to Surg Gen, Liver Study Sect, US Army, 56-62; consult, Vet Admin, 57-62, mem adv comt res, Dept Med & Surg, 60-68; mem, Am Bd Surg, 59-65, chmn, 64-65; mem surg study sect, NIH, 59-63, mem gen clin res ctr comt, 63-66; ed, J Surg Res, 60-66; mem surg test comt, Nat Bd Med Exam, 61-68, mem ad hoc comt study & improv part II exam, 62-64, mem at large, 65-; mem halothane anesthesia comt, NSF, 63-67; mem sci adv comt, United Health Founds, 63-, mem grants rev subcomt, 63-; mem study comt, USPHS Hosps, 65; mem bd med, Nat Acad Sci, 67-; chmn differences in postoperative mortality policy comt, Nat Res Coun, 70-; actg dir, Study Instnl Differences in Postoperative Mortality, Nat Acad Sci-Inst Med-Nat Res Coun, 71- Mem: Am Surg Asn; Soc Clin Surg; fel Am Col Surg; Soc Surg Alimentary Tract; NY Acad Sci. Res: Portal hypertension, liver diseases and pancreatic duodenal physiology and neoplasia. Mailing Add: Dept of Surg Univ of Mich Hosp Ann Arbor MI 48104

CHILD, FRANK MALCOLM, b Jersey City, NJ, Nov 30, 31; m 60; c 3. CELL BIOLOGY. Educ: Amherst Col, AB, 53; Univ Calif, PhD(zool), 57. Prof Exp: Instr zool, Univ Chicago, 57-60, asst prof, 60-65; assoc prof biol, 65-73, PROF BIOL, TRINITY COL, CONN, 73-, CHMN DEPT, 74- Mem: Soc Protozool; Am Soc Cell Biol; Am Soc Zoologists. Res: Protozoan physiology; developmental and cellular biology; cilia and flagella. Mailing Add: Dept of Biol Trinity Col Hartford CT 06106

CHILD, HARRY RAY, b Bedford, Ind, Oct 30, 28; m 54; c 1. SOLID STATE PHYSICS, MAGNETISM. Educ: Univ Tex, BS, 56; Univ Tenn, PhD(physics), 65. Prof Exp: PHYSICIST, OAK RIDGE NAT LAB, 56- Mem: Am Phys Soc. Res: Neutron scattering studies of solids, mostly magnetic properties. Mailing Add: Solid State Div Oak Ridge Nat Lab PO Box X Oak Ridge TN 37830

CHILD, JEFFREY JAMES, b Gateshead, Eng, June 26, 36; m 58; c 2. MICROBIOLOGY. Educ: Univ Durham, BSc, 58, PhD, 62. Prof Exp: Fel, Prairie Regional Lab, Nat Res Coun Can, 62-63; asst lectr biol, Univ Salford, 63-64, lectr microbiol, 64-67; ASSOC RES OFFICER, PRAIRIE REGIONAL LAB, NAT RES COUN CAN, 67- Concurrent Pos: With Commonwealth Sci & Indust Res Orgn, Australia, 74-75. Mem: Can Soc Microbiol; Brit Soc Gen Microbiol. Res: Physiology of fungi, especially nutrition and reproduction; microbiological degradation of natural products; symbiotic nitrogen fixation. Mailing Add: Nat Res Coun Prairie Regional Lab Saskatoon SK Can

CHILD, PROCTOR LOUIS, b Brooklyn, NY, Nov 29, 25; m 52; c 3. PATHOLOGY. Educ: Long Island Col Med, MS, 49. Prof Exp: Intern, St Agnes Hosp, White Plains, NY, US Army, 49-50, gen med officer, 155th Sta Hosp, Japan, 50, battalion & regimental surgeon, 1st Cavalry Div, Korea, 50-51, gen med officer, Army Hosp, Camp Pickett, Va, 51-52, path resident, Fitzsimons Gen Hosp, Denver, 52-56, chief path serv, 5th Gen Hosp, Stuttgart, 56-58, 130th Sta Hosp, Heidelberg, 58-60, chief path, William Beaumont Gen Hosp, El Paso, 60-64, mem geog path div, Armed Forces Inst Path, 64-66, chief viro-path br & asst chief geog path div, 67; ASSOC PROF PATH, SCH MED, TEMPLE UNIV, 68- Concurrent Pos: Mem staff, Roxborough Mem Hosp, Philadelphia, 70- Mem: AMA; Am Soc Clin Path; Col Am Path; Int Acad Path. Res: Geographic pathology and infectious disease, especially viro-pathology and study of hemorrhagic fevers. Mailing Add: 307 Colket Ln Wayne PA 19087

CHILD, RALPH GRASSING, b New York, NY, Oct 7, 19; m 44; c 3. MEDICINAL CHEMISTRY. Educ: Hofstra Col, BA, 41; George Washington Univ, MA, 48; Univ Iowa, PhD(org chem), 50. Prof Exp: Res chemist, 50-74, SR RES CHEMIST, LEDERLE LABS, AM CYANAMID CO, 74- Mem: Am Chem Soc. Res: Medicinal organic chemistry; chemotherapy of virus and neoplastic diseases; chemistry of antibacterial compounds; immune response inhibitors. Mailing Add: Lederle Labs Pearl River NY 10965

CHILD, WILLIAM CLARK, JR, b Elizabeth, NJ, Sept 2, 27; m 60; c 3. PHYSICAL CHEMISTRY. Educ: Oberlin Col, AB, 50; Univ Wis, PhD(chem), 55. Prof Exp: Proj assoc chem, Univ Wis, 55-56; from instr to assoc prof, 56-69, PROF CHEM, CARLETON COL, 69- Concurrent Pos: NSF fac fel, Univ Calif, Santa Barbara, 71-72. Mem: Am Chem Soc. Res: Thermodynamics, titration calorimetry. Mailing Add: Dept of Chem Carleton Col Northfield MN 55057

CHILDERS, CLIFFORD W, physical chemistry, see 12th edition

CHILDERS, H MALCOLM, physics, see 12th edition

CHILDERS, NORMAN FRANKLIN, b Moscow, Idaho, Oct 29, 10; m 71; c 3. HORTICULTURE. Educ: Univ Mo, BS, 33, MS, 34; Cornell Univ, PhD(pomol), 37. Prof Exp: Asst pomol, Cornell Univ, 34-37; asst prof hort, Ohio State Univ, 37-44, assoc, Ohio Exp Sta, 39-44; asst dir & sr plant physiologist, PR Exp Sta, USDA, 44-47; prof, res specialist & chmn dept hort & forestry, 48-66, BLAKE PROF HORT, AGR EXP STA, RUTGERS UNIV, NEW BRUNSWICK, 66- Mem: AAAS; Am Soc Plant Physiologists; Am Soc Agron; fel Am Soc Hort Sci. Res: Photosynthesis; transpiration; respiration; nutrition of fruits and other horticultural plants; tropical and temperate pomology; tropical vegetables. Mailing Add: Dept of Hort & Forestry Rutgers Univ New Brunswick NJ 08903

CHILDERS, RAY FLEETWOOD, b Los Angeles, Calif, Apr 16, 45; m 71. PHARMACEUTICAL CHEMISTRY, ANALYTICAL CHEMISTRY. Educ: Univ Calif, Los Angeles, BS, 67; Ind Univ, PhD(inorg chem), 72. Prof Exp: Res assoc biophys chem, Ind Univ, 72-74; SR ANAL CHEMIST, ELI LILLY & CO, 74- Mem: Am Chem Soc. Res: Automated flow analysis of pharmaceutics; activity and structure of biological molecules using nuclear magnetic resonance; pharmaco-kinetics. Mailing Add: Eli Lilly & Co Indianapolis IN 46206

CHILDERS, RICHARD LEE, b Birmingham, Ala, Dec 10, 30; m 60; c 2. PHYSICS. Educ: Presby Col, SC, BS, 53; Univ Tenn, MS, 56, PhD(particle physics), 62. Prof Exp: Res assoc physics, Univ Tenn, 61-63; asst prof, 63-66, ASSOC PROF PHYSICS, UNIV SC, 66- Concurrent Pos: Consult, Neutron Physics Div, Oak Ridge Nat Lab, 62-64; dir, Honors Prog, 67-69. Mem: Am Phys Soc; Am Asn Physics Teachers; Am Inst Physics. Res: Nuclear physics using nuclear track emulsion; acoustics of musical instruments, especially guitars and other string instruments. Mailing Add: Dept of Physics Univ of SC Columbia SC 29208

CHILDERS, ROBERT LEE, b Parkersburg, WVa, May 3, 36; m 62; c 2. PHOTOGRAPHIC CHEMISTRY. Educ: WVa State Col, BS, 60, MS, 61; Ohio State Univ, PhD(org chem), 65. Prof Exp: Sr res chemist, 65-71, RES ASSOC, EASTMAN KODAK CO, 71- Mem: Am Chem Soc; Soc Photog Sci & Eng. Res: Chemistry of photographic emulsions and photographic processing. Mailing Add: Res Labs B-59 Eastman Kodak Co Rochester NY 14650

CHILDERS, ROBERT WAYNE, b Ft Worth, Tex, May 25, 37; m 62; c 3. THEORETICAL PHYSICS. Educ: Howard Payne Col, BA, 60; Vanderbilt Univ, PhD(physics), 63. Prof Exp: Res fel, Argonne Nat Lab, 63-65; asst prof, 65-72, ASSOC PROF PHYSICS, UNIV TENN, KNOXVILLE, 65- Concurrent Pos: Consult, Oak Ridge Nat Lab, 66- Mem: Am Phys Soc. Res: Elementary particle physics; theory of infinitely rising Regge trajectories and local duality; Veneziano model; quantum field theory; symmetries of elementary particles. Mailing Add: Dept of Physics & Astron Univ of Tenn Knoxville TN 37916

CHILDERS, RODERICK W, b Paris, France, 31; m; c 1. MEDICINE, CARDIOLOGY. Educ: Univ Dublin, BA, 53, MD, 54, MA, 58; Am Bd Cardiovasc Dis, cert, 69. Prof Exp: Intern med, St Andrew's Hosp, London, Eng, 54-55; intern surg, May Day Hosp, Croydon, 55-56; chief cardiologist, Royal City of Dublin Hosp, Ireland, 59-63; asst prof cardiol, 63-69, ASSOC PROF MED, UNIV CHICAGO, 69-, HEAD HEART STA, 66- Concurrent Pos: Fel, Harvard Univ & WRoxbury Vet Admin Hosp, Mass, 58-59; res assoc nutrit, Sch Pub Health, Harvard Univ, 59-62; med tutor cardiol, Med Sch, Univ Dublin, 59-63; cardiac asst, Nat Children's Hosp, Dublin & vis pediat cardiologist, Rotunda Maternity Hosp, 59-63. Mailing Add: Dept of Med Univ of Chicago Chicago IL 60637

CHILDERS, WALTER ROBERT, b Kelowna, BC, Mar 29, 16; m 53; c 2. PLANT BREEDING, PLANT GENETICS. Educ: McGill Univ, BSc, 38; Univ Wis, MS, 47, PhD(plant breeding & genetics), 51. Prof Exp: Asst corn & soybeans, Forage Crop Div, 38-40, CHIEF FORAGE SECT, OTTAWA RES STA, CAN DEPT AGR, 46- Mem: Am Soc Agron; Agr Inst Can; Can Soc Genetics; Can Soc Agron; Can Phytopath Soc. Res: Orchard and brome grass; timothy; cytology and genetics; alfalfa. Mailing Add: Ottawa Res Sta Can Dept of Agr Ottawa ON Can

CHILDRESS, CHARLES CURTIS, b Pittsburg, Kans, Nov 10, 39; m 60; c 3. BIOCHEMISTRY, BIOMEDICAL ENGINEERING. Educ: Col Great Falls, BS, 61; Kans State Univ, MS, 64; Johns Hopkins Univ, PhD(biochem), 69. Prof Exp: Res asst biol, Kans State Univ, 62-64; res biochemist, US Army Chem Res & Develop Labs, Md, 64-67; biochemist, NIH, 67-69; dir of labs, Upsher Labs, Kansas City, Mo, 69-71; pres, Midwest Sci Labs, Inc, 71-72; PRES, MIDWEST SCI INSTRUMENTS, INC, 72-; PRES, LIEBRA SYSTS, INC, 74- Honors & Awards: Qual Performance Award, US Army, 65. Mem: AAAS; Am Chem Soc; Am Asn Clin Chem; Am Asn Off Anal Chemists; Am Acad Clin Toxicol. Res: Plant physiology; insect biochemistry; clinical biochemistry; computerized diagnosis; laboratory computer systems. Mailing Add: 1203 Willow Dr Olathe KS 66061

CHILDRESS, DENVER RAY, b Alcoa, Tenn, Feb 5, 37; m 57; c 2. PURE MATHEMATICS. Educ: Maryville Col, Tenn, BS, 59; Univ Tenn, Knoxville, MMath, 64, EdD(math educ), 75. Prof Exp: Teacher math, Powell High Sch, Knox County, Tenn, 59-60, Maryville Jr High Sch, Maryville, Tenn, 60-62 & Maryville High Sch, 62-65; asst prof, 67-76, ASSOC PROF MATH, CARSON-NEWMAN COL, 76- Mem: Nat Coun Teachers Math; Math Asn Am. Res: Factors which influence student ratings of mathematics teaching and teachers, especially attitudes. Mailing Add: Rte 1 Box 46 New Market TN 37820

CHILDRESS, DUDLEY STEPHEN, b Cass Co, Mo, Sept 25, 34; m 59; c 2. BIOMEDICAL ENGINEERING. Educ: Univ Mo-Columbia, BS, 57, MS, 58; Northwestern Univ, PhD(elec eng), 67. Prof Exp: From instr to asst prof elec eng, Univ Mo-Columbia, 59-63; res asst, Physiol Control Syst Lab, Northwestern Univ, Evanston, 64-66, asst prof, 66-72, ASSOC PROF ELEC ENG, TECHNOL INST & ASSOC PROF ORTHOPED SURG, MED SCH, NORTHWESTERN UNIV, CHICAGO, 73-, DIR, PROSTHETICS RES LAB, 71-, CO-DIR, REHAB ENG PROG, 72- Concurrent Pos: Consult, Social & Rehab Serv, Div Int Activities, Dept Health, Educ & Welfare, 68; consult, Crippled Children's Bur, Mich Dept Health, 69-72; mem comt prosthetics res & develop, Nat Acad Sci-Nat Res Coun, 69-72, mem sub-comt design, 70-73, chmn upper-extremity prosthetics panel, 71-; Nat Inst Gen Med Sci res career develop award, 70-75; mem appl physiol & bioeng study sect, NIH, 74-76. Mem: AAAS; Inst Elec & Electronics Eng; Biomed Eng Soc. Res: Rehabilitation engineering, design and development of modern technological systems for disabled people and scientific approach to analysis and description of problems of these people. Mailing Add: Prosthetics Res Lab Northwestern Univ Rm 1441 345 E Superior St Chicago IL 60611

CHILDRESS, EVELYN TUTT, b Joplin, Mo, Feb 8, 26; m 67. MICROBIOLOGY, IMMUNOLOGY. Educ: Lincoln Univ, Mo, BS, 47; Univ Mich, MS, 48, MS, 56; Stanford Univ, PhD(med microbiol), 67. Prof Exp: Instr biol, Fla Agr & Mech Univ, 48-49; instr, Lincoln Univ, Mo, 49-52, asst prof, 52-63; mem fac, Fullerton Jr Col, 67-69; asst prof, 69-72, ASSOC PROF BIOL, CALIF STATE COL, DOMINGUEZ HILLS, 72- Mem: Am Soc Microbiol; Sigma Xi; AAAS; Am Asn Univ Prof. Res: Aging in the immune system; origin of naturally occuring antibodies, particularly with their specificity and the question of necessity of antigenic stimulation for their appearance.

CHILDRESS, JAMES J, b Kokomo, Ind, Nov 17, 42; m 67. COMPARATIVE PHYSIOLOGY, BIOLOGICAL OCEANOGRAPHY. Educ: Wabash Col, BA, 64; Stanford Univ, PhD(biol), 69. Prof Exp: ASST PROF ZOOL, UNIV CALIF, SANTA BARBARA, 69- Concurrent Pos: Prin investr, NSF grant, 70-72. Mem: AAAS; Am Soc Zoologists; Am Soc Limnol & Oceanog. Res: Ecological physiology of marine invertebrates and fishes; respiratory physiology; DDT and deep sea trophic structure; effects of hydrostatic pressure on organisms. Mailing Add: Dept of Zool Univ of Calif Santa Barbara CA 93106

CHILDRESS, NOEL A, b Lafayette Co, Miss, Jan 12, 20; m 46; c 2. GEOMETRY. Educ: Univ Miss, BAE, 41, MA, 47; Univ Fla, PhD(math), 54. Prof Exp: Pub sch teacher, Miss, 41-42; from instr to assoc prof, 48-57, PROF MATH, UNIV MISS, 57- Mem: Math Asn Am; Am Math Soc. Res: Projective transformations in algebraic geometry; involutions. Mailing Add: Dept of Math Univ of Miss University MS 38677

CHILDRESS, SCOTT JULIUS, b Greenville, SC, Apr 6, 26. ORGANIC CHEMISTRY. Educ: Furman Univ, BS, 47; Univ NC, PhD(chem), 51. Prof Exp: Res chemist catalysis, Tenn Eastman Co, 51-52 & pharmaceut, Wallace & Tiernan, Inc, 52-58; res chemist pharmaceut, 59-60, group leader, 60-61, mgr med chem sect, 61-68, asst to vpres res & develop, 68-74, ASST VPRES RES & DEVELOP, WYETH LABS, 74- Mem: Am Chem Soc. Res: Design, synthesis and testing of organic compounds of possible therapeutic value. Mailing Add: Wyeth Labs PO Box 8299 Philadelphia PA 19101

CHILDRESS, WILLIAM STEPHEN, b Houston, Tex, Oct 5, 34; m 60; c 2. FLUID MECHANICS, APPLIED MATHEMATICS. Educ: Princeton Univ, BSE, 56, MSE, 58; Calif Inst Technol, PhD(aeronaut, math), 61. Prof Exp: Assoc res Scientist, Jet Propulsion Lab, 61-64; res assoc magneto-fluid dynamics, Courant Inst Math Sci, 64-66, asst prof, 66-70, ASSOC PROF MATH, NY UNIV, 70- Concurrent Pos: Assoc, Inst Henri Poincare, Univ Paris, 67-68. Mem: Am Math Soc; Soc Indust & Appl Math. Res: Singular perturbation problems in fluid dynamics and applied mathematics; magnetohydrodynamics; dynamo theory of geomagnetism; viscous flow theory. Mailing Add: Dept of Math Col of Arts & Sci NY Univ New York NY 10003

CHILDS, DANA PITT, b Herington, Kans, Mar 24, 26; m 50; c 2. ENTOMOLOGY. Educ: Kans State Univ, BS, 49. Prof Exp: RES ENTOMOLOGIST, AGR RES SERV, USDA, 49- Mem: Entom Soc Am. Res: Biological and chemical control of insect pests attacking stored products. Mailing Add: Agr Res Serv USDA PO Box 10125 Richmond VA 23240

CHILDS, DONALD RAY, b Lynn, Mass, May 31, 30; m 64; c 2. MATHEMATICAL PHYSICS. Educ: Univ NH, BS, 52, MS, 54; Vanderbilt Univ, PhD(physics), 58. Prof Exp: Scientist, Westinghouse Elec Corp, 57-59; sr scientist, Allied Res Assocs, 59-60; prin res scientist, Avco-Everett Res Lab, 60-64; sr scientist, Lab of Electronics, 64-66 & Quincy Div, Elec Boat Co, 66-69; PHYSICIST, NAVAL UNDERWATER SYST CTR, 69- Concurrent Pos: Spec lectr, Northeastern Univ, 67-69. Res: Multivariate analysis; signal processing; non-linear mechanics; non-linear control systems; non-linear differential equations. Mailing Add: Naval Underwater Systs Ctr Middletown RI 02840

CHILDS, DONALD SMYTHE, JR, b Syracuse, NY, July 5, 16; m 42; c 6. NUCLEAR MEDICINE. Educ: Haverford Col, BA, 38; Yale Univ, MD, 42; Univ Minn, MSc, 49. Prof Exp: Fel radiol, Mayo Found, Univ Minn, 46-49; from instr to prof clin radiol, Mayo Grad Sch Med, Univ Minn, 69-74, PROF RADIOL, MAYO MED SCH, 74-, CONSULT, MAYO CLIN, 49-, HEAD DEPT THERAPEUT RADIOL, 53- Mem: AAAS; Radiol Soc NAm; AMA; Radiation Res Soc. Res: Medical applications of ionizing radiations. Mailing Add: Dept of Therapeut Radiol Mayo Clin Rochester MN 55901

CHILDS, GEORGE RICHARD, b Terra Alta, WVa, Oct 9, 24; m 49; c 4. NUTRITION. Educ: WVa Univ, BS, 50; Purdue Univ, MS, 51; Univ Md, PhD(nutrit), 64. Prof Exp: Poultry res specialist, Cent Soya Co, Inc, 51-61; animal husbandman, Bur Commercial Fisheries, 61-64; mgr poultry res, 64-68, DIR FEED RES, CENT SOYA CO, INC, 68- Mem: Poultry Sci Asn. Res: Amino acid requirements of growing chicks and laying hens. Mailing Add: Res Dept Cent Soya Co Inc 1230 N Second St Decatur IN 46733

CHILDS, JAMES FIELDING LEWIS, b Tucson, Ariz, Jan 3, 10; m 36; c 4. PLANT PATHOLOGY. Educ: Univ Calif, BS, 37, PhD(plant path), 41. Prof Exp: Agent, USDA, 41-43, from asst pathologist to prin pathologist, 43-66, RES PATHOLOGIST, AGR RES SERV, USDA, 66- Concurrent Pos: Consult, Egypt, 55, Morroco, 59, Surinam, 63 & Sudan, 64; ed proc, Int Orgn Citrus Virol, 66. Mem: Am Phytopath Soc; Int Orgn Citrus Virol. Res: Etiology, virus indexing procedures and programs; control of virus and other diseases of citrus. Mailing Add: Agr Res Serv USDA 2120 Camden Rd Orlando FL 32803

CHILDS, LINDSAY NATHAN, b Boston, Mass, Apr 17, 40. MATHEMATICS. Educ: Wesleyan Univ, BA, 62; Cornell Univ, PhD(math), 66. Prof Exp: Asst prof math, Northwestern Univ, 66-68; asst prof, 68-71, ASSOC PROF MATH, STATE UNIV NY ALBANY, 71- Mem: Am Math Soc; Math Asn Am. Res: Algebra. Mailing Add: Dept of Math State Univ of NY Albany NY 12203

CHILDS, RICHARD FRANCIS, b Battlecreek, Mich, Sept 20, 18; m 58; c 3. PHARMACEUTICAL CHEMISTRY. Educ: Olivet Col, BA, 41; Univ Wis, BS, 54, MS, 55; Univ Ariz, PhD, 62. Prof Exp: Chemist, Cleaver-Brooks Co, Wis, 42-48 & Armour Co, Ill, 48-50; from instr to asst prof, 55-70, ASSOC PROF PHARM, UNIV ARIZ, 55- Concurrent Pos: Dir, Regional Sci Fair, 58 & 64; NSF sci fac fel, 59-60. Mem: Am Pharmaceut Asn. Res: Development of undergraduate laboratory equipment; chemical analysis of pharmaceutical drugs. Mailing Add: Col of Pharm Univ of Ariz Tucson AZ 85721

CHILDS, RONALD FRANK, b Liss, Eng, Nov 30, 39; m 65; c 2. PHYSICAL ORGANIC CHEMISTRY. Educ: Bath Univ Technol, BSc, 63; Univ Nottingham, PhD(org chem), 66. Prof Exp: Fel, Univ Calif, Los Angeles, 66-68; asst prof, 68-72, ASSOC PROF CHEM, McMASTER UNIV, 72- Mem: Am Chem Soc; The Chem Soc; sr mem Chem Inst Can. Res: Physical organic chemistry, particularly thermal and photochemical rearrangements of carbonium ions. Mailing Add: Dept of Chem McMaster Univ Hamilton ON Can

CHILDS, WILLIAM HENRY, b Princeton, Ill, Jan 1, 07; m 34; c 4. POMOLOGY. Educ: Univ Ill, BS, 30, MS, 31; Cornell Univ, PhD(pomol), 40. Prof Exp: Instr hort univ & asst agr exp sta, 31-38, from asst prof hort & asst horticulturist to prof hort & horticulturist, 40-74, actg chmn dept hort univ, 60-62, EMER PROF HORT, WVA AGR EXP STA, WVA UNIV, 72- Mem: Am Soc Hort Sci. Res: Small fruits; cultural studies with strawberries, raspberries, blackberries and grapes; blueberry selection, hybridization and propagation. Mailing Add: 936 Virginia St Apt 309 Dunedin FL 33528

CHILDS, WILLIAM JEFFRIES, b Boston, Mass, Nov 9, 26; m 51; c 2. ATOMIC PHYSICS. Educ: Harvard Univ, AB, 48; Univ Mich, MS, 49, PhD(physics), 56. Prof Exp: PHYSICIST, ARGONNE NAT LAB, 56- Concurrent Pos: Vis prof, Univ Bonn, Ger, 72-73. Mem: Am Phys Soc; Optical Soc Am. Res: Atomic-beam magnetic resonance; hyperfine structure; laser spectroscopy. Mailing Add: Argonne Nat Lab 9700 S Cass Ave Argonne IL 60439

CHILDS, WILLIAM VES, b Cale Ark, Sept 14, 35; m 62; c 3. FLUORINE CHEMISTRY, ELECTROCHEMISTRY. Educ: Southern State Col, BS, 56; Univ Ark, MS, 60, PhD(phys chem), 63. Prof Exp: SR CHEMIST, PHILLIPS PETROL CO, 62- Concurrent Pos: Vis scientist, Univ Tex, 69-70; consult, NIH, Lung & Heart Inst, 75. Mem: AAAS; Am Chem Soc (treas, 75); Am Inst Chem Engrs. Res: Fluorine chemistry; kinetics; computer simulations; application of small, dedicated computers to data acquisition and processing; synthetic electrochemistry. Mailing Add: 1504 Harris Dr Bartlesville OK 74003

CHILGREEN, DONALD RAY, b Jenkins, Ky, Nov 8, 39; m 64; c 3 3. SOIL MICROBIOLOGY. Educ: Marion Col, AB, 64; Kans State Univ, MS, 67, PhD(microbiol), 74. Prof Exp: Asst prof biol, Marion Col, 67-68; instr microbiol, Kans State Univ, 68-69; asst prof, 70-75, ASSOC PROF BIOL, MARION COL, 75- Mem: Am Soc Microbiol. Res: Diversity of the indigenous thermophilic microorganism in prairie soils; growth curves and respiratory activity of thermophilic bacteria from soil. Mailing Add: Dept of Biol Marion Col 4201 S Washington St Marion IN 46952

CHILGREN, JOHN DOUGLAS, b New Ulm, Minn, Sept 14, 43. PHYSIOLOGY. Educ: Gonzaga Univ, BS, 65; Wash State Univ, MS, 68, PhD(zoophysiol), 75. Prof Exp: Res assoc psychobiol, US Army Human Eng Lab, 68-69; opers & training adminr, Eighth US Army, UN Command, 69-70; ASST PROF ZOOL, ORE STATE UNIV, 75- Mem: Am Inst Biol Sci; Am Soc Zoologists; Am Ornithologists Union; Cooper Ornith Soc. Res: Ecological energetics of vertebrates; vertebrate annual cycles and periodicities; reproductive endocrinology and behavioral correlates in birds, especially passerines. Mailing Add: Dept of Zool Ore State Univ Corvallis OR 97331

CHILTON, BRUCE L, b Buffalo, NY, June 14, 35. MATHEMATICS. Educ: Univ Buffalo, BA, 58, MA, 60; Univ Toronto, PhD(math), 62. Prof Exp: Asst prof math, State Univ NY Buffalo, 62-68; dean dept, 68-74, ASSOC PROF MATH, STATE UNIV NY COL FREDONIA, 68- Mem: Math Asn Am; Am Math Soc. Res: Geometry, especially properties of regular and semiregular figures in Euclidean n-spaces. Mailing Add: Dept of Math State Univ of NY Fredonia NY 14063

CHILTON, FRANK, theoretical physics, see 12th edition

CHILTON, JOHN MORGAN, b Tuscaloosa, Ala, Apr 20, 21; m 45; c 3. INORGANIC CHEMISTRY. Educ: Univ Ala, AB, 42; Univ Va, MS, 47, PhD(chem), 50. Prof Exp: Jr chemist, Gen Anal Lab, Tenn Valley Authority, 42-45; asst prof anal chem, Ala Polytech Inst, 49-51; chemist, Ionic Develop Group, Anal Chem Div, 51-54, CHEMIST, CHEM DEVELOP SECT, CHEM TECH DIV, OAK RIDGE NAT LAB, 54- Mem: Am Chem Soc. Res: Structure of metal ions in solution; spectrophotometric methods of analysis; chemistry of actinides. Mailing Add: 104 Wedgewood Dr Oak Ridge TN 37830

CHILTON, NEAL WARWICK, b New York, NY, June 24, 21; m 47; c 5. ORAL MEDICINE. Educ: City Col New York, BSc, 39; NY Univ, DDS, 43; Columbia Univ, MSc, 46; Am Bd Endodont & Am Bd Periodont, dipl. Prof Exp: Intern, Lincoln Hosp, New York, 43; from instr to asst clin prof pharmacol & therapeut, NY Univ, 44-50; res assoc dent & asst prof dent pub health pract, Columbia Univ, 49-54; from asst prof to assoc prof periodont, Sch Dent, 52-63, assoc prof prev med, Sch Med, 59-66, CLIN PROF PERIODONT, SCH DENT, TEMPLE UNIV, 63-, PROF ORAL MED, SCH MED, 66- Concurrent Pos: Lectr, Seton Hall Univ, 47-52 & Temple Univ, 52-53; guest lectr, Evans Dent Inst, Univ Pa, 50-; clin prof, Univ Kansas City, 54-55; res assoc, Fac Med, Columbia Univ, 57-70, sr res assoc biostatist, Sch Pub Health & sr res assoc prev med, Sch Dent & Oral Surg, 70-; asst prof, Grad Sch Med, Univ Pa, 57-70, lectr, Sch Dent Med, 70-; asst chief, Bur Dent Health, State Dept Health, NJ; consult, Coun Dent Therapeut, Am Dent Asn & Surgeon Gen, USPHS; mem comt res manpower, Nat Inst Dept Res & mem dent study sect, Div Res Grants, NIH. Mem: Am Asn Endodont; fel Am Col Dent; Am Acad Periodont; Am Acad Oral Path; Int Asn Dent Res. Res: Diseases of the mouth and gums, etiology, pathology, treatment and prevention; dental public health; design and analysis of agents in clinical trial; clinical therapeutics.

CHILTON, ST JOHN POINDEXTER, b Philadelphia, Pa, Feb 3, 09; m 35. PLANT PATHOLOGY. Educ: La State Univ, BS, 35, MS, 36; Univ Minn, PhD(plant path), 38. Prof Exp: Instr, Univ Minn, 37-38; agent, USDA, 38-40; from asst prof to assoc prof, 40-48, from asst pathologist to pathologist, 42-50, PROF BOT & PLANT PATH & CHMN DEPT, LA STATE UNIV, BATON ROUGE, 50-, PLANT PATHOLOGIST & HEAD AGR EXP STA, 50- Mem: AAAS; Am Soc Sugar Cane Tech (past pres); Int Soc Sugar Cane Tech. Res: Genetics of fungi; sugar cane breeding and pathology. Mailing Add: 3617 Hyacinth Ave Baton Rouge LA 70808

CHILTON, WILLIAM SCOTT, b Philadelphia, Pa, Aug 29, 33; m 65; c 2. ORGANIC CHEMISTRY. Educ: Duke Univ, BS, 55; Univ Ill, Urbana-Champaign, PhD(org chem), 63. Prof Exp: Asst prof, 63-68, ASSOC PROF CHEM, UNIV WASH, 68- Concurrent Pos: Sci adv, Food & Drug Admin, 67- Mem: AAAS; Am Chem Soc; Ger Chem Soc; The Chem Soc. Res: Structure of natural products; new naturally occurring amino acids; synthesis of higher carbon sugars; applications of circular dichroism. Mailing Add: Dept of Chem Univ of Wash Seattle WA 98105

CHIMENTI, FRANK A, b Erie, Pa, May 3, 39; m 64; c 2. MATHEMATICAL ANALYSIS. Educ: Gannon Col, BA, 61; John Carroll Univ, MS, 63; Pa State Univ, PhD(math), 70. Prof Exp: Res asst appl math, Lord Mfg Co, Pa, 63-65 & Ord Res Lab, Pa State Univ, 65-67; asst prof, 69-74, ASSOC PROF MATH, STATE UNIV NY COL FREDONIA, 74- Mem: Math Asn Am; Am Math Soc. Res: General topology; convergence of sequences of sets; multivalued functions; convergence formulas. Mailing Add: Dept of Math State Univ of NY Fredonia NY 14063

CHIMOSKEY, JOHN EDWARD, b Traverse City, Mich, Apr 15, 37; div; c 2. PHYSIOLOGY. Educ: Univ Mich, MD, 63. Prof Exp: Intern internal med, Univ Calif, 64; USPHS fel, Harvard Med Sch, 64-66 & Retina Found, Boston, 66-67; assoc prof physiol, Hahnemann Med Col, 69-70; actg instr dermat, Med Ctr, Stanford Univ, 70-71; asst prof bioeng, Univ Wash, 71-75; ASSOC PROF PHYSIOL & SURG & DIR, TAUB LABS MECH CIRCULATORY SUPPORT, BAYLOR COL MED, 75- Concurrent Pos: Guest scientist, US Naval Air Develop Ctr, 69-70; NIH spec fel, Hahnemann Med Col, 70; guest lectr, Hahnemann Med Col & Calif Col Podiatric Med, 70-71; NIH spec fel & Dermat Found fel, Stanford Univ, 70-71; NIH grant, 73-75; adj assoc prof bioeng, Rice Univ, 75- Mem: AAAS; Am Physiol Soc. Res: Cardiovascular physiology. Mailing Add: Taub Labs Baylor Col Med Houston TX 77025

CHIN, BYONG HAN, b Shanghai, China, Nov 27, 34; US citizen; m 61; c 3. BIOCHEMISTRY. Educ: Yonsei Univ, Korea, BS, 57; Univ Hawaii, MS, 64, PhD(hort), 67. Prof Exp: Marine biochemist, Hawaii Marine Lab, Univ Hawaii, 63-64; FEL, MELLON INST, CARNEGIE-MELLON UNIV, 67- Mem: Am Chem Soc; Soc Toxicol. Res: The metabolism of pesticides by plants and animals; bioassay of poisonous fishes; methodology development in clinical biochemistry. Mailing Add: Carnegie-Mellon Univ Mellon Inst 4400 Fifth Ave Pittsburgh PA 15213

CHIN, DER-TAU, b Chekiang, China, Sept 14, 39. ELECTROCHEMISTRY, CHEMICAL ENGINEERING. Educ: Chung Yuan Col Sci & Eng, BS, 62; Tufts Univ, MS, 65; Univ Pa, PhD(chem eng), 69. Prof Exp: Process engr, Taiwan Sugar Corp, 62-63; sci programmer, US Air Force Cambridge Res Labs, 65; sr res engr, Res Labs, Gen Motors Corp, 69-75; ASSOC PROF CHEM ENG, CLARKSON COL TECHNOL, 75- Honors & Awards: Young Author's Award, Electrochem Soc, 72. Mem: Electrochem Soc; Am Inst Chem Eng. Res: Electrolytic mass transfer; electrochemical study of flow turbulence; electrochemical machining; high current density electrode process; electrochemical waste treatment; potential and current distribution in electrochemical systems; rotating disk and spherical electrode. Mailing Add: Dept of Chem Eng Clarkson Col Technol Potsdam NY 13676

CHIN, EDWARD, b Boston, Mass, Sept 4, 26; m 52; c 4. BIOLOGY. Educ: Harvard Univ, BS, 48; Univ NH, MS, 53; Univ Wash, PhD, 61. Prof Exp: Biol aide clam invest, US Fish & Wildlife Serv, 49-51; fishery res biologist king crab studies, 54-55 & gulf shrimp studies, 55-61; asst sci dir, US Prog Biol Int Indian Ocean Exped, 62-65; assoc prof biol, Tex A&M Univ, 65-68; dir biol oceanog prog, NSF, 68-70; ASSOC DIR, INST NATURAL RESOURCES, UNIV GA, 70- Mem: Am Fisheries Soc; Am Inst Fishery Res Biol; Brit Marine Biol Asn. Res: Marine invertebrates; marine ecology. Mailing Add: Inst of Natural Resources Univ of Ga Athens GA 30601

CHIN, JANE ELIZABETH HENG, b Augusta, Ga, Nov 20, 33; m 60; c 2. PHARMACOLOGY. Educ: Univ Ga, BS, 54; Univ Mich, MS, 56, PhD(pharmacol), 60. Prof Exp: USPHS fel neuropharmacol, Univ Ill, 59-60; USPHS fel neuropharmacol, Sch Med, 60-68, USPHS fel biosci, Univ, 68-71, INSTR PHARMACOL, SCH MED, STANFORD UNIV, 71- Mem: NY Acad Sci; Am Soc Pharmacol & Exp Therapeut. Res: Neuropharmacology; pain and analgesics; physiological mechanisms in brain; central modulation of afferent inputs; psychopharmacology; neuropsychology; biophysics; high pressure physiology; drug effects on membranes. Mailing Add: Dept of Pharmacol Stanford Univ Sch of Med Palo Alto CA 94305

CHIN, SEE LEANG, b Padang Rengas, Malaya, May 24, 42; Can citizen; m 71. LASERS. Educ: Nat Taiwan Univ, BSc, 64; Univ Waterloo, MSc, 66, PhD(physics), 69. Prof Exp: Teacher math, Hua Lian High Sch, Taiping, Malaya, 60; fel physics, 69-70, res assoc, 71-72, ASST PROF PHYSICS, LAVAL UNIV, 72- Concurrent Pos: Sci consult, K A Mace Ltd, Kitchener, Ont, 73-; vis scientist, Nuclear Study Ctr, Saclay, France, 75. Mem: Can Asn Physicists; Optical Soc Am. Res: Pulse compression in a dye medium; interaction of high power lasers with atoms and molecules. Mailing Add: Dept of Physics Laval Univ Quebec PQ Can

CHIN, TOM DOON YUEN, b Kwangtung, China, May 29, 22; US citizen; m 50; c 2. EPIDEMIOLOGY. Educ: Univ Mich, MD, 46; Tulane Univ, MPH, 50; Am Bd Prev Med & Am Bd Med Microbiol, dipl. Prof Exp: Intern, Binghamton City Hosp, 47; intern, Western Pa Hosp, 47-48; resident, Sea View Hosp, 48-49; dir health unit, State Dept Health, La, 50-51; asst chief epidemiol, Kansas City Field Sta, USPHS, 54-64, chief, 64-66, dir ecol invests prog, Ctr Dis Control, Kansas City, 67-73; PROF HUMAN ECOL & COMMUNITY HEALTH, UNIV KANS MED CTR, KANSAS CITY, 73-, CHMN DEPT, 74- Mem: Fel Am Col Prev Med; fel Am Pub Health Asn; Am Epidemiol Soc; Soc Epidemiol Res; Infectious Dis Soc Am. Res: Infectious diseases. Mailing Add: Dept Human Ecol & Community Health Univ of Kans Med Ctr Kansas City KS 66103

CHIN, WEI TSUNG, b Loo-Yee, China, July 5, 28; m 63; c 1. AGRICULTURAL CHEMISTRY. Educ: Nat Taiwan Univ, BS, 52, MS, 57; Va Polytech Inst, PhD(agron), 62, MS, 63. Prof Exp: Res chemist pesticides, Niagara Chem Div, FMC Corp, 62-66 & Uniroyal Chem Corp, 66-72; SR RES CHEMIST, DIAMOND SHAMROCK CORP, 73- Mem: Am Chem Soc. Res: Chemistry of soils, fertilizers and pesticides. Mailing Add: Diamond Shamrock Corp T R Evans Res Ctr Painesville OH 44077

CHIN, YEH-HAO, b Oct 17, 43; Chinese citizen. COMPUTER SCIENCES. Educ: Nat Taiwan Univ, BS, 66; Univ Tex, MS, 70, PhD(elec eng), 72. Prof Exp: Res asst comput sci, Univ Tex, 69-72; programmer analyst, CDC Sunnyvale Div, 72-73; researcher, Telecommun Labs Taiwan, 73-74; ASST PROF COMPUT SCI, NORTHWESTERN UNIV, EVANSTON, 74- Mem: Inst Elec & Electronics Engrs; Asn Comput Mach. Res: Data base systems; file structure; memory hierarchy. Mailing Add: Dept of Comput Sci Northwestern Univ Evanston IL 60201

CHINARD, FRANCIS PIERRE, b Berkeley, Calif, June 30, 18; m 43; c 3. PHYSIOLOGICAL CHEMISTRY, INTERNAL MEDICINE. Educ: Univ Calif, AB, 37; Johns Hopkins Univ, MD, 41. Prof Exp: Intern, Presby Hosp, New York, 41-42; Nat Res Coun fel, Rockefeller Inst, 45-46, asst Rockefeller Inst Hosp, 46-49; instr med & physiol chem, Sch Med, Johns Hopkins Univ, 49-51, asst prof physiol chem, 51-56, asst prof med, 52-56, assoc prof physiol chem & med, 59-63; prof exp med, Fac Med, McGill Univ, 63-64; prof med, Sch Med, NY Univ, 64-68; prof med & chmn dept, 68-74, PROF EXP MED, NJ MED SCH, COL MED & DENT NJ, 75- Concurrent Pos: Markle scholar, 49-54; asst chief med, Baltimore City Hosps, Md, 53-60, physiciant in chief, 60-63; chief med, Goldwater Mem Hosp, 66-68; adj prof, NY Univ, 68-70; career scientist, Health Res Coun, New York. Mem: Am Chem Soc; Am Soc Biol Chem; Am Soc Clin Invest; Soc Exp Biol & Med; fel Am Col Physicians. Res: Membrane permeability; renal and pulmonary physiology and metabolism. Mailing Add: Col of Med & Dent of NJ NJ Med Sch 100 Bergen St Newark NJ 07103

CHINAS, BEVERLY NEWBOLD, b Minden, Nebr; m 69; c 2. CULTURAL ANTHROPOLOGY, ETHNOLOGY. Educ: Fresno State Col, BA, 63; Univ Calif, Los Angeles, MA, 65, PhD(anthrop), 68. Prof Exp: Asst prof anthrop, 68-70, ASSOC PROF ANTHROP, CALIF STATE UNIV, CHICO, 70- Mem: Fel Am Anthrop Asn. Res: Peasant communities; Latin America; women's roles cross-culturally; economic anthropology; field research on Isthmus Zapotecs, Oaxaca, Mexico. Mailing Add: Dept of Anthrop Calif State Univ Chico CA 95926

CHINEA, JOSE JUAN, b Bayamon, PR, Dec 20, 44; m 74; c 1. HISTOLOGY, DENTISTRY. Educ: Univ PR, San Juan, BS, 65, DMD, 69; Ind Univ, Indianapolis, MS, 74. Prof Exp: Instr histol, Sch Dent, Univ PR, 69-72; teaching asst, Sch Med, Ind Univ, 73-74; ASST PROF HISTOL, SCH DENT, UNIV PR, SAN JUAN, 74- Res: Histological aspects of tonsils under electron microscopy and the immunological aspects of the tonsils; congenital malformations of the face and oral structures. Mailing Add: Basic Sci Dept Sch of Dent Univ of PR GPO Box 5067 San Juan PR 00936

CHING, HILDA, b Honolulu, Hawaii, June 30, 34; m 60; c 3. PARASITOLOGY. Educ: Ore State Univ, BA, 56, MS, 57; Univ Nebr, PhD(zool), 59. Prof Exp: Asst parasitol, Agr Exp Sta, Univ Hawaii, 59-60; RES ASSOC, DEPT ZOOL, UNIV BC, 60- Mem: Am Soc Parasitol. Res: Trematodes of fishes and birds. Mailing Add: Dept of Zool Univ of BC Vancouver BC Can

CHING, JASON KWOCK SUNG, b Honolulu, Hawaii, Dec 18, 40; m 64; c 3. ENVIRONMENTAL SCIENCE. Educ: Univ Hawaii, BS, 62; Pa State Univ, MS, 64; Univ Wash, PhD(meteorol), 74. Prof Exp: Res asst meteorol, Pa State Univ, 62-64, Woods Hole Oceanog Inst, 64-66 & Univ Wash, 66-70; res meteorologist, Barbados Oceanog & Meteorol Anal Proj, 70-75, METEOROLOGIST, METEOROL DIV, AIR RES LAB, ENVIRON RES LABS, NAT OCEANIC & ATMOSPHERIC ADMIN, 75- Mem: Am Meteorol Soc; Sigma Xi. Res: Numerical-theoretical models of the dynamics, thermodynamics and transport characteristics in the planetary boundary layer of the atmosphere. Mailing Add: Meteorol Div Environ Sci Res Lab Environ Protection Agency Research Triangle Park NC 27711

CHING, MELVIN CHUNG HING, b Honolulu, Hawaii, Feb 11, 35; m 65; c 2. ANATOMY. Educ: Univ Nebr, AB, 57, MSc, 60; Univ Calif, Berkeley, PhD(anat), 71. Prof Exp: Instr, 71-73, ASST PROF ANAT, SCH MED & DENT, UNIV ROCHESTER, 73- Mem: Am Asn Anat; Sigma Xi. Res: Neuroendocrinology; endocrinology; neuroendocrine control mechanisms; hypothalamic-pituitary-thyroid gonadal axis. Mailing Add: Dept of Anat Univ Rochester Sch Med & Dent Rochester NY 14642

CHING, TE MAY, b Soochow, China, Jan 9, 23; US citizen; m 46; c 2. PLANT PHYSIOLOGY. Educ: Nat Cent Univ, China, BS, 44; Mich State Col, MS, 50; Mich State Univ, PhD(cytol), 54. Prof Exp: Asst wood chem, Nat Cent Univ, China, 44-48; asst plant anat, hist & cytol, Mich State Univ, 50-52, asst plant physiol & cytol, 52-54, instr, 54-56; from asst prof to assoc prof seed physiol, 56-71, PROF SEED PHYSIOL, ORE STATE UNIV, 71- Mem: AAAS; Am Soc Plant Physiol; Genetics Soc Am; Am Soc Agron; Am Oil Chem Soc. Res: Seed physiology; cytology; lipid metabolism; structure and function of cellular organelles; developmental biology. Mailing Add: Dept of Crop Sci Ore State Univ Corvallis OR 97331

CHINN, AUSTIN BROCKENBROUGH, b Warsaw, Va, May 8, 08; m 38; c 3. MEDICINE. Educ: Univ Va, MD, 32. Prof Exp: Instr med, George Washington Univ, 36-38, assoc, 38-41; asst clin prof, Sch Med, Western Reserve Univ, 46-53, assoc prof, 53-62, assoc dean, 60-62; CHIEF GERONT BR, DIV CHRONIC DIS, USPHS, 62- Concurrent Pos: Prof & dir rehab res & training ctr, Univ Southern Calif, 67-69. Mem: AMA; Am Col Physicians; Am Fedn Clin Res. Res: Gastrointestinal disease. Mailing Add: 422 E Beverley St Staunton VA 24401

CHINN, CLARENCE EDWARD, b Cheney, Wash, Dec 1, 25; m 53; c 3. CHEMISTRY. Educ: Walla Walla Col, BA, 51; Ore State Col, MS, 53, PhD(soils), 56; Univ Tenn, PhD(inorg chem), 69. Prof Exp: From asst prof to assoc prof chem & math, Southern Missionary Col, 56-67; assoc prof, 67-74, PROF CHEM, WALLA WALLA COL, 74- Mem: Am Chem Soc. Res: Soil moisture measurement; effect of herbicides on plant enzymes; solvent extraction of metal chelates. Mailing Add: Dept of Chem Walla Walla Col College Place WA 99324

CHINN, HERMAN ISAAC, b Connellsville, Pa, Apr 8, 13; m 45; c 4. BIOCHEMISTRY. Educ: Pa State Col, BS, 34; Northwestern Univ, MS, 35, PhD(biochem), 38. Prof Exp: Instr biochem, Med Sch, Northwestern Univ, 38-42; prin chemist, Fla State Bd Health, 46-47; chief dept biochem, Sch Aviation Med, 47-55; sci liaison officer, Off Naval Res, London, 55-57; biochemist, Air Force Off Sci Res, 57-60; dep sci attache, Am Embassy, Ger, 60-63; sci officer, Off Int Sci Affairs, US Dept State, DC, 63-65; sci attache, Am Embassy, Tehran, Iran, 65-67; sci officer, US Dept State, 67-70; sci attache, Am Embassy, Stockholm, Sweden, 70-73; sci attache, Am Embassy, Tel Aviv, Israel, 73-75; CONSULT, US DEPT STATE, WASHINGTON, DC, 75- Mem: AAAS; Soc Exp Biol & Med; Am Physiol Soc; Am Soc Pharmacol & Exp Therapeut; Am Chem Soc. Res: Biochemistry of the eye; aviation physiology; motion sickness. Mailing Add: 9907 Wildwood Rd Kensington MD 20795

CHINN, LELAND JEW, b Sacramento, Calif, Oct 19, 24; m 59; c 1. MEDICINAL CHEMISTRY. Educ: Univ Calif, BS, 48; Univ Wis, PhD(chem), 51. Prof Exp: Asst to prof, Univ Wis, 48-51; res chemist, 52-70, group leader, 70-72, RES FEL, G D SEARLE & CO, 72- Concurrent Pos: Vis scientist, Univ Southern Calif, 68. Mem: AAAS; Am Chem Soc. Res: Natural products; stereochemistry of polycyclic compounds; medicinal chemistry. Mailing Add: 6141 Elm St Morton Grove IL 60053

CHINN, PHYLLIS ZWEIG, b Rochester, NY, Sept 26, 41; m 68. MATHEMATICS. Educ: Brandeis Univ, BA, 62; Harvard Univ, MAT, 63; Univ Calif, San Diego, MA, 66; Univ Calif, Santa Barbara, PhD(math), 69. Prof Exp: Teacher jr high sch, Mass, 63-64; instr math, Mass State Col Salem, 64; asst prof math, Towson State Col, 69-75; ASST PROF MATH, HUMBOLDT STATE UNIV, 75- Mem: Math Asn Am; Nat Coun Teachers Math. Res: Graph reconstruction problems; frequency partition of graphs; means of improving the teaching of mathematics to prospective teachers; graph coloring problems; coding; discovery learning of mathematics; graphical operations and properties. Mailing Add: Dept of Math Humboldt State Univ Arcata CA 95521

CHINN, STANLEY H F, b Vancouver, BC, Apr 9, 14; m 44; c 2. SOIL MICROBIOLOGY. Educ: Iowa State Col, BS, 40, MSc, 42, PhD, 46. Prof Exp: Lectr, Univ Sask, 49-51; BACTERIOLOGIST, RES BR, CAN DEPT AGR, 51- Concurrent Pos: Adj prof, Univ Sask. Mem: Am Soc Microbiol; Can Soc Phytopath; Can Soc Microbiol. Res: Soil microbiology as related to common rootrot of wheat. Mailing Add: Res Br Can Dept of Agr Univ Campus Saskatoon SK Can

CHINNERY, MICHAEL ALISTAIR, b London, Eng, Sept 27, 33; m 64. GEOPHYSICS, SEISMOLOGY. Educ: Cambridge Univ, BA, 57, MA, 61; Univ Toronto, MA, 59, PhD(geophys), 62. Prof Exp: Geophysicist, Seismog Serv, Ltd, Eng, 57-58; geophysicist, Hunting Surv Corp, 59; lectr geophys, Univ Toronto, 61-62; instr, Univ BC, 62-63, asst prof, 63-65; res assoc geol & geophys, Mass Inst Technol, 65-66; from assoc prof to prof, Dept Geol Sci, Brown Univ, 66-73; GROUP LEADER, LINCOLN LAB, MASS INST TECHNOL, 73- Concurrent Pos: Consult geophysicist, 60-; mem subcomt gravity, Nat Res Coun Can, 63-; assoc ed, J Geophys Res, Am Geophys Union, 70- Mem: Am Geophys Union; Seismol Soc Am; fel Royal Astron Soc Can. Res: Displacements and stresses in faulting; strength of earth's crust; earthquake mechanism; geotectonics; elasticity theory; earthquake risk; seismic discrimination. Mailing Add: Lincoln Lab 42 Carleton St Mass Inst Technol Cambridge MA 02142

CHINNICI, JOSEPH (FRANK) PETER, b Philadelphia, Pa, Oct 12, 43; m 65; c 3. GENETICS, EVOLUTIONARY BIOLOGY. Educ: La Salle Col, AB, 65; Univ Va, PhD(biol), 70. Prof Exp: ASST PROF BIOL, VA COMMONWEALTH UNIV, 70- Honors & Awards: Andrew Fleming Award, Univ Va, 70. Mem: Genetics Soc Am; AAAS; Soc Study Evolution; Sigma Xi. Res: Genetic control of crossing-over in Drosophila melanogaster; development of sexual isolation in Drosophila and its genetic control; frequency-dependent selection; environmental effects on crossing-over; genetic aspects of human birth weight. Mailing Add: Dept of Biol Va Commonwealth Univ Richmond VA 23284

CHINNOCK, ROBERT FISHER, medicine, pediatrics, deceased

CHINOWSKY, WILLIAM, b New York, NY, Feb 24, 29; m 50; c 2. PHYSICS. Educ: Columbia Univ, AB, 49, AM, 51, PhD(physics), 55. Prof Exp: Res assoc physics, Brookhaven Nat Lab, 54-56, assoc physicist, 56-61; assoc prof, 61-67, PROF PHYSICS, UNIV CALIF, BERKELEY, 67- Mem: Am Phys Soc. Res: High energy physics. Mailing Add: Dept of Physics Univ of Calif Berkeley CA 94720

CHIOLA, VINCENT, b Bayonne, NJ, May 7, 22; m 52; c 2. INORGANIC CHEMISTRY. Educ: Wagner Col, BS, 47; Univ Tex, MA, 50. Prof Exp: Chemist, Gen Aniline & Film Corp, 50-51; asst, Plastics Lab, Princeton Univ, 51; eng chemist, Chem & Metall Div, 51-60, from develop engr to adv develop engr, 60-68, sect head, Chem Develop Lab, 68-69, SECT HEAD, CHEM & METALL DIV, PHOSPHOR DEVELOP LAB & PHOSPHOR PILOT PLANT, GTE SYLVANIA INC, 69- Mem: AAAS; Am Chem Soc; Am Inst Chem; Electrochem Soc. Res: Allyl compounds reaction rates; polyurethane polymerization; protective coatings; chemistry of tungsten and molybdenum; electronic grade chemicals; inorganic luminescent chemicals; phosphors. Mailing Add: Chem & Metall Div GTE Sylvania Inc Towanda PA 18848

CHIONG, MIGUEL ANGEL, b Havana, Cuba, Dec 20, 25; Can citizen. INTERNAL MEDICINE, CARDIOLOGY. Educ: Univ Havana, MD, 50; Queen's Univ, Ont, MSc, 64, PhD(cardiovasc physiol), 65; FRCPS(C), 63. Prof Exp: ASSOC PROF MED & PHYSIOL, FAC MED, QUEEN'S UNIV, ONT, 66- Concurrent Pos: Sr fel, Ont Heart Found, 66-; attend staff, Kingston Gen Hosp, 66-; mem Am group, Int Group Study Myocardial Metab, 70- Mem: AAAS; Can Med Asn; Can Physiol Soc; Can Cardiovasc Soc; Am Col Cardiol. Res: Study of myocardial metabolism and hemodynamics in response to cardiovascular drugs and hypoxia or ischemia in isolated hearts and in patients with coronary artery disease. Mailing Add: Dept of Med Etherington Hall Queen's Univ Kingston ON Can

CHIOTTI, PREMO, b Cuba, Ill, Feb 18, 11; m 42. PHYSICAL CHEMISTRY. Educ: Univ Ill, BS, 38; Iowa State Univ, PhD(phys chem), 50. Prof Exp: Supt finishing dept, Otsego Falls Paper Mills, Inc, Mich, 39-42 & Manhattan Project, US Army, 42-46; jr chemist, 44-50, from asst prof to assoc prof, 50-61, PROF CHEM & SR CHEMIST, IOWA STATE UNIV, 61-, SR METALLURGIST, 62- Mem: AAAS; Metall Soc; Am Chem Soc; Am Soc Metals. Res: Physical and chemical metallurgy; thermodynamic properties and high temperature properties of metals and alloys. Mailing Add: RFD 3 Ames IA 50010

CHIOU, GEORGE CHUNG-YIH, b Taoyuan, Taiwan, July 11, 34; US citizen; m 61; c 2. PHARMACOLOGY, BIOCHEMISTRY. Educ: Nat Taiwan Univ, BS, 57, MS, 60; Vanderbilt Univ, PhD(pharmacol), 67. Prof Exp: Pharmacist, William Pharmaceut Works, Taiwan, 60-61; pharmacist, Chinese Air Force Hosp, 61-62; instr pharmacol, China Med Col, Taiwan, 62-64; from res asst to res assoc, Col Med, Vanderbilt Univ, 64-68; fel pharmacol, Univ Iowa, 68-69; asst prof, 69-73, ASSOC PROF PHARMACOL, COL MED, UNIV FLA, 73- Concurrent Pos: NIH health sci advan award pharmacol, Vanderbilt Univ, 67-68; NIH res grant, Univ Fla, 69-71, Nat Inst Neurol Dis & Stroke res grant, 71-74. Mem: NY Acad Sci; Am Soc Pharmacol & Exp Therapeut; Sigma Xi. Res: Autonomic pharmacology; calcium antagonists; neurochemistry; enzymology; structure-activity relationships of cholinergic and cholinolytic agents; enzyme kinetics of acetylcholinesterase and butyrylcholinesterase; action mechanisms of nicotinic responses; nature of cholinergic receptor; cytolysis of neuroblastomas. Mailing Add: Dept of Pharmacol & Therapeut Univ of Fla Col of Med Gainesville FL 32610

CHIOU, WIN LOUNG, b Hsinchu, Taiwan, Aug 29, 38; m 63; c 2. PHARMACOLOGY. Educ: Nat Taiwan Univ, BS, 61; Univ Calif, San Francisco, PhD(pharmaceut chem), 69. Prof Exp: From res assoc to asst prof pharm, Wash State Univ, 69-71; asst prof pharm, 71-73, assoc prof pharm & occup & environ med, 73-75, DIR CLIN PHARMACOKINETICS LAB, COL PHARM, UNIV ILL MED CTR, 75-, PROF PHARM, 76- Concurrent Pos: Consult, Med Lett on Drugs & Therapeut, 74-75. Mem: Acad Pharmaceut Sci; Am Pharmaceut Asn. Res: Biopharmaceutics and pharmacokinetics of drugs such as aspirin; propranolol and digoxin; renal function; formulation of dosage forms; antacid-drug and soft-drink-drug interactions; toxicity of aerosol propellants. Mailing Add: Univ of Ill Med Ctr 833 S Wood St Chicago IL 60612

CHIPAULT, JACQUES ROBERT, b La Vernelle, France, May 13, 14; nat US; m 41; c 2. BIOCHEMISTRY. Educ: Carleton Col, BA, 35; Univ Minn, MS, 41, PhD(biochem), 46. Prof Exp: Res fel, Hormel Inst, Univ Minn, 46-48, res assoc, 48-51, from asst prof to assoc prof biochem, 51-60, PROF BIOCHEM, UNIV MINN, 60- Mem: AAAS; Am Chem Soc; Am Oil Chem Soc; Coblentz Soc. Res: Fat antioxidants; fat oxidation; fat metabolism: infrared spectroscopy; effect of high-energy radiation on fats; lipid deterioration; lipids of the eye; intestinal sterols and bile acids. Mailing Add: Hormel Inst 801 16th Ave NE Austin MN 55912

CHIPLEY, JOHN RAYMOND, b Athens, Ga, Nov 10, 44; m 70. MICROBIOLOGY. Educ: Univ Ga, BS, 66, MS, 67, PhD(microbiol), 69. Prof Exp: Asst prof microbiol, Exp Sta, Univ Ga, 69-71; asst prof, 71-73, ASSOC PROF MICROBIOL, OHIO STATE UNIV, 71- Concurrent Pos: Reviewer, J Food Sci, 73- & J Asn Off Anal Chemists, 75- Mem: Am Soc Microbiol; Sigma Xi; Asn Off Anal Chemists. Res: Studies of transport mechanisms and physiology of bacteria; identification of toxic metabolites; antigenic properties of cell components; production of single-cell protein; mycology and public health; steroid metabolism by bacteria. Mailing Add: Ohio State Univ Dept of Microbiol Columbus OH 43210

CHIPLEY, ROBERT MACNEILL, b Cincinnati, Ohio, Nov 20, 39; m 67; c 1. ORNITHOLOGY. Educ: Yale Univ, BA, 61; Cornell Univ, PhD(ecol), 74. Prof Exp: Sci ed, Dover Publ, 65-68; STAFF ECOLOGIST, THE NATURE CONSERVANCY, 74- Mem: Ecol Soc Am; Soc Syst Zool; Am Ornithologists Union; Cooper Ornith Soc. Res: Inventory and preservation of natural areas; conservation of endangered species. Mailing Add: 3220 Lothian Rd 203 Fairfax VA 22030

CHIPMAN, DANIEL MYRON, b Ames, Iowa, July 26, 45; m 69; c 1. QUANTUM CHEMISTRY. Educ: Iowa State Univ, BS, 67; Univ Wis, PhD(phys chem), 72. Prof Exp: Asst chem, Univ Wis, 67-72; lectr, Univ Calif, Santa Barbara, 72-73; vis asst prof, Univ Colo, 73-75; ASST PROF CHEM, UNIV IOWA, 75- Concurrent Pos: Res chemist, Univ Calif, Santa Barbara, 72; test consult, Am Col Testing Prog, 75-; textbook rev, Harper & Row Publ Inc, 75- Res: Calculation and interpretation of the electronic structure and properties of atoms and molecules, particularly as it relates to chemical bonding and intermolecular forces. Mailing Add: Dept of Chem Univ of Iowa Iowa City IA 52242

CHIPMAN, DAVID MAYER, b New York, NY, Oct 7, 40; m 62; c 3. ENZYMOLOGY, BIO-ORGANIC CHEMISTRY. Educ: Columbia Univ, BA, 62, PhD(org chem), 65. Prof Exp: Nat Acad Sci-Nat Res Coun res fel biophys, Weizmann Inst, 65-66, NIH res fel, 66-67; asst prof chem, Mass Inst Technol, 67-71; SR LECTR BIOCHEM, BEN GURION UNIV OF THE NEGEV, 71-, CHMN DEPT BIOL, 74- Mem: AAAS; Am Chem Soc; Israel Biochem Soc. Res: Mechanisms of chemical reactions of biological interest; enzyme mechanisms; physical organic chemistry. Mailing Add: Ben Gurion Univ of the Negev PO Box 653 Beersheva Israel

CHIPMAN, DAVID RANDOLPH, b Atlanta, Ga, Jan 23, 28; m 52; c 2. X-RAY CRYSTALLOGRAPHY. Educ: Mass Inst Technol, BS, 49, ScD(metall), 55; Univ Ill, MS, 50. Prof Exp: Fulbright fel, Stuttgart Tech Univ, Ger, 55-56; SOLID STATE PHYSICIST, US ARMY MAT & MECH RES CTR, 56- Mem: Am Phys Soc; Am Crystallog Asn. Res: X-ray and neutron diffraction studies of the structure of metals and alloys; magnetic studies of hard magnetic materials; x-ray studies of amorphous metals. Mailing Add: Army Mat & Mech Res Ctr Watertown MA 02172

CHIPMAN, E W, b Bridgetown, NS, Nov 15, 13; m 47; c 5. BOTANY, HORTICULTURE. Educ: McGill Univ, BSc, 39. Prof Exp: Overseer agr dept, United Fruit Co, Mass, 39-42; RES SCIENTIST, CAN DEPT AGR, 46- Mem: Am Soc Hort Sci; Agr Inst Can; Can Soc Hort Sci. Res: Vegetable research. Mailing Add: Res Sta Can Dept of Agr Kentville NS Can

CHIPMAN, GARY RUSSELL, b Berlin, Wis, Feb 27, 43; m 67; c 2. POLYMER

CHEMISTRY. Educ: Univ Wis, BS, 65; Univ Mich, MS, 67, PhD(org chem), 70. Prof Exp: RES CHEMIST, AMOCO CHEM CORP, 70- Mem: Am Chem Soc. Res: Product and process research on random olefin copolymers, polyesters, and block copolymers. Mailing Add: Amoco Chem Corp Amoco Res Ctr PO Box 400 Naperville IL 60540

CHIPMAN, ROBERT K, b New York, NY, Nov 16, 31; m 54; c 2. ZOOLOGY. Educ: Amherst Col, AB, 53; Tulane Univ, MS, 58, PhD(zool), 63. Prof Exp: From instr to asst prof biol, State Univ NY Col Plattsburgh, 61-62; from asst prof to assoc prof zool, Univ Vt, 62-68, asst dean grad sch, 67-68; chmn dept, 68-74, PROF ZOOL, UNIV RI, 68- Mem: Am Soc Mammal; Am Soc Zool; Soc Study Reprod. Res: Morphological variation of fish; vertebrate ecology; rodent population dynamics; physiology of reproduction of rodents. Mailing Add: Dept of Zool Univ of RI Kingston RI 02881

CHIPMAN, WILMON B, b Reading, Mass, July 6, 32; m 60; c 3. ORGANIC CHEMISTRY, BIOCHEMISTRY. Educ: Harvard Univ, AB, 54; Dartmouth Col, AM, 56; Univ Ill, PhD(org chem), 60. Prof Exp: Instr chem, Colby Col, 60-62, asst prof, 62-65; assoc prof, 65-67, PROF CHEM, BRIDGEWATER STATE COL, 67-, CHMN DEPT, 65- Concurrent Pos: Vis lectr, NSF Inst, Mt Hermon Sch, 61-64; NSF vis lectr, 63-65; partic, Heterocyclic Chem Conf, Univ Mont, 65-; lectr, NSF in-serv inst, Bridgewater State Col, 66-71; consult, Encyclop Britannica Films, 66- Mem: Am Chem Soc. Res: Heterocyclic chemistry; infrared and ultraviolet spectroscopy; transannular interactions; natural products; nuclear magnetic resonance spectrometry. Mailing Add: 64 Pleasant Dr Bridgewater MA 02324

CHIQUOINE, A DUNCAN, b Upland, Pa, May 3, 26; m 50; c 4. CELL BIOLOGY. Educ: Swarthmore Col, AB, 47; Cornell Univ, PhD(zool), 52. Prof Exp: Asst, Cornell Univ, 47-52; instr anat, Univ Wash, 52-53; from instr to asst prof biol, Princeton Univ, 53-57; from asst prof to assoc prof anat, Wash Univ, 57-67; PROF BIOL & CHMN DEPT, HAMILTON COL, 72- Mem: AAAS; Histochem Soc. Res: Histochemistry and cytochemistry; mammalian embryology; electron microscopy of mammalian embryos; germ cells. Mailing Add: Dept of Biol Hamilton Col Clinton NY 13323

CHIRIGOS, MICHAEL ANTHONY, b Wierton, WVa; Sept 14, 24; c 3. BIOCHEMISTRY. Educ: Western Md Col, BS, 52; Univ Del, MS, 54; Rutgers Univ, PhD(biochem), 57. Prof Exp: Asst bact, Univ Del, 52-54; fel, Nat Heart & Lung Inst, 57-59, pharmaceut chemist, 59-67; head viral chemother sect, Drug Eval Br, 66-67, HEAD VIRUS & DIS MODIFICATION SECT, NAT CANCER INST, 67-, ASSOC BR CHIEF VIRAL BIOL BR, 70- Mem: AAAS; Am Asn Cancer Res; Soc Exp Biol & Med. Res: Chemotherapy of cancer and oncogenic viruses; oncogenic virology; transport mechanisms in vitro and in vivo. Mailing Add: Viral Biol Br Nat Cancer Inst Bethesda MD 20014

CHIRIKJIAN, JACK G, b Dec 10, 40; US citizen; m 64; c 2. BIOLOGICAL CHEMISTRY. Educ: Trenton State Col, BA, 63; Rutgers Univ, MS, 66, PhD(biochem), 69. Prof Exp: Nat Cancer Inst fel biochem, Princeton Univ, 69-71, res assoc, 71-72; ASST PROF BIOCHEM, SCHS MED & DENT, GEORGETOWN UNIV, 72-, LEUKEMIA SOC AM SCHOLAR, 75- Concurrent Pos: Consult, Schwarz-Mann Div, Becton Dickenson Co, 73-75 & Pharmacia Fine Chem, 74- Mem: Am Soc Biol Chemists; Am Soc Microbiol; Am Chem Soc; Sigma Xi. Res: Studies dealing with protein-nucleic acid interactions using viral DNA polymerases, restriction endonucleases and tRNA synthetases as model systems; emphasis is placed on structure and function of the various enzymes. Mailing Add: Dept of Biochem Georgetown Univ Schs of Med & Dent Washington DC 20007

CHIRINO, FERNANDO PORFIRIO, b Havana, Cuba, Sept 15, 18; US citizen; m 40; c 2. MEDICINE. Educ: Tulane Univ, BS, 40, MD, 43; Univ Havana, MD, 46. Prof Exp: Indust physician, Hershey Corp, Cuba, 46-59; PROF MED, SCH MED, TULANE UNIV, 63- Res: Internal medicine; geriatrics. Mailing Add: Dept of Med Tulane Univ Sch of Med New Orleans LA 70112

CHISCON, J ALFRED, b Kingston, Pa, Feb 18, 33; m 69. GENETICS, EVOLUTION. Educ: Bloomsburg State Col, BS, 54; Purdue Univ, MS, 56, PhD(biol), 61. Prof Exp: From instr to assoc prof, 61-70, PROF BIOL, PURDUE UNIV, 70- Concurrent Pos: Carnegie fel, 68-69. Mem: AAAS. Res: Social impact of biology; molecular evolution; nucleic acid reassociation studies; DNA relationships between primates; rate of nucleotide sequence change during evolution. Mailing Add: Dept of Biol Sci Purdue Univ Lafayette IN 47907

CHISCON, MARTHA OAKLEY, b Chicago, Ill, Aug 27, 35; m 69; c 2. IMMUNOBIOLOGY. Educ: Western Ill Univ, BSEd, 56; Purdue Univ, PhD(immunobiol), 71. Prof Exp: High sch teacher biol, chem & physics, Ill & Alaska, 56-64; instr biol, 64-70, NIH fel immunol, 71-72, ASST PROF BIOL, PURDUE UNIV, 72- Mem: Soc Develop Biol; AAAS. Res: Functional development of the immune response and associated areas of immunobiology. Mailing Add: Dept of Biol Sci Purdue Univ West Lafayette IN 47907

CHISHOLM, ALEXANDER, organic chemistry, see 12th edition

CHISHOLM, ALEXANDER JAMES, b Minnedosa, Man, July 28, 41; m 66; c 3. CLOUD PHYSICS. Educ: Univ Alta, BSc, 62; McGill Univ, MSc, 66, PhD(radar meteorol), 70. Prof Exp: Meteorol officer forecast div, Can Meteorol Serv, 62-66, meteorologist, 66-70, res scientist, Res & Training Div, 70-74, ACTG CHIEF CLOUD PHYSICS RES DIV, ATMOSPHERIC ENVIRON SERV, 74- Honors & Awards: President's Prize, Can Meteorol Soc, 74. Mem: Am Meteorol Soc; Royal Meteorol Soc; Can Meteorol Soc. Res: Hailstorm airflow; hail suppression concepts; rainfall enhancement of cumuliform clouds by weather modification. Mailing Add: Atmospheric Environ Serv 4905 Dufferin St Downsview Toronto ON Can

CHISHOLM, JAMES JOSEPH, b Natick, Mass, May 29, 36; c 4. OPTICS, SPECTROSCOPY. Educ: Boston Col, BS, 58; Univ Rochester, MS, 65. Prof Exp: Res physicist, Air Force Cambridge Res Ctr, 57-58; from physicist to prog mgr spectrophotom, 58-75, DIR SPECTROPHOTOM RES & DEVELOP, BAUSCH & LOMB INC, 75- Mem: Optical Soc Am. Res: Monochromators, spectrophotometers, refractometers, radiometry and photometry; color measurements; atomic absorption spectroscopy; enzyme and kinetic measurements; ultraviolet and infrared optics; light sources and detectors; interferometry; diffraction gratings; photometric standards. Mailing Add: 34 Sawmill Dr Penfield NY 14526

CHISHOLM, RODERICK G, b Superior, Wis, Sept 21, 13. GEOPHYSICS. Educ: DePaul Univ, BA, 33; Univ St Tomas, Manila, MS, 41, PhD(math physics), 49; St Paul Col, Manila, dipl, 50; La Salle Col, MA, 52. Hon Degrees: DMathSci, Col Los Angeles, Philippines, 58. Prof Exp: Mem fac, Christian Bros Col High Sch, 33-35; mem fac, De La Salle Inst, 35-38; teacher physics high sch, 38-40, prin, 40-41 & 46-48, dean sch com & eng, vpres & pro-dir, De La Salle Col, 49-51, dean sch eng, 52-54, sch lib arts, 54-56, dean col, 56-58; chmn dept physics, St Mary's Col, Minn, 58-60, chmn dept math, 60-63, assoc prof, 62-63, dean grad studies, 64-65, dir instnl res, 65-68, vpres acad affairs & dean col, 66-68; HEAD DIV PHYS OCEANOG, MARGARITA MARINE RES STA & HEAD DEPT OCEANOG, 68- Concurrent Pos: Consult, Robot Statist, Inc, PI, 54-58, Philippine Govt, 56-58, Pac Glass Co, 56-58 & Benber Mines, Inc, 57-58; exec vpres, Philippine Accrediting Asn Schs, Cols & Univs, 57-58; scholar, Oak Ridge Inst Nuclear Studies; NSF grants, Am Univ & Univ Calif, Los Angeles; head radiol defense serv, Mobile Support Area 1, State of Minn, 58-60; lectr eng anal lab, Int Bus Mach Corp, 60-62, consult, 64-; radiol defense officer, City of Winona, Minn, 62-63. Mem: AAAS; Am Soc Eng Educ; Am Phys Soc; Am Math Soc; Math Asn Am. Res: Military applications of probability theory; radiological defense studies; fallout studies over Venezuela. Mailing Add: Colegio La Salle Apartado 61223-706 Caracas Venezuela

CHISHOLM, RONALD C, physical chemistry, see 12th edition

CHISHOLM, SALLIE WATSON, b Marquette, Mich, Nov 5, 47. AQUATIC ECOLOGY. Educ: Skidmore Col, BA, 69; State Univ NY Albany, PhD(biol), 74. Prof Exp: RES ASSOC PHYTOPLANKTON ECOL, INST MARINE RESOURCES, UNIV CALIF, SAN DIEGO, 74- Mem: Am Soc Limnol & Oceanog; Ecol Soc Am; Phycological Soc Am; Int Asn Limnol. Res: Diel periodicity in physiological processes in phytoplankton. Mailing Add: Inst of Marine Resources Univ of Calif at San Diego La Jolla CA 92093

CHISLER, JOHN ADAM, b Daybrook, WVa, Feb 25, 37; m 59; c 2. BACTERIOLOGY, GENETICS. Educ: Ohio State Univ, BSc, 59, MSc, 61, PhD(plant path), 62. Prof Exp: Asst prof bot, Marshall Univ, 62-66; PROF BIOL & CHMN DIV SCI & MATH, GLENVILLE STATE COL, 66- Mem: AAAS; Am Phytopath Soc. Res: Physiology of plant pathogens-respiration of the tomato Fusarium wilt organism; chemical interactions of fungi in plants. Mailing Add: Div of Sci & Math Glenville State Col Glenville WV 26351

CHISM, GRADY WILLIAM, III, b Tampa, Fla, June 18, 46; m 72; c 1. FOOD SCIENCE. Educ: Univ Fla, BS, 68; Univ Mass, Amherst, PhD(food sci), 73. Prof Exp: Fel food sci, Cook Col, Rutgers Univ, 73-74; ASST PROF FOOD SCI, OHIO STATE UNIV & OHIO RES & DEVELOP CTR, 74- Mem: Inst Food Technologists; Sigma Xi. Res: Biochemical control mechanisms of enzymatic processes important in the development and maintenance of quality in plant tissues used as food; enzymatic regulation of cytokinin levels in tomato fruits. Mailing Add: Dept of Food Sci & Nutrit Ohio State Univ Columbus OH 43210

CHISOLM, JAMES JULIAN, JR, b Baltimore, Md, July 24, 21; m 48; c 2. MEDICINE. Educ: Princeton Univ, AB, 44; Johns Hopkins Univ, MD, 46; Am Bd Pediat, dipl, 52. Prof Exp: Intern pediat, Johns Hopkins Hosp, 46-47, asst resident, Johns Hopkins Hosp & asst in pediat, Sch Med, Johns Hopkins Univ, 48; sr asst resident, Babies Hosp, New York, 50-51; resident, Johns Hopkins Hosp, 51-52, asst pediat, Sch Med, Johns Hopkins Univ, 53-55, from instr to asst prof, 55-63, ASSOC PROF PEDIAT, SCH MED, JOHNS HOPKINS UNIV, 63-, PEDIATRICIAN, JOHNS HOPKINS HOSP, 52- Concurrent Pos: Fel pediat, Sch Med, Johns Hopkins Univ, 51-53; hosp physician, Baltimore City Hosp, 53-56, asst chief hosp physician, 56-61, assoc chief hosp physician, 61-; mem panel on lead, Nat Res Coun, 70-71; pediat consult, USPHS, 71. Mem: AAAS; Soc Pediat Res; Am Acad Pediat; Am Pediat Soc. Res: Renal tubular function; biochemical effects of lead poisoning; porphyrin metabolism. Mailing Add: Dept of Pediat Baltimore City Hosp Baltimore MD 21224

CHITHARANJAN, DAKSHINAMURTHY, b Madras, India, May 25, 40; m 64; c 2. ORGANIC CHEMISTRY. Educ: Annamalai Univ, Madras, BSc, 60, MSc, 61; Wayne State Univ, PhD(org chem), 69. Prof Exp: Asst prof, 68-70, ASSOC PROF CHEM, UNIV WIS-STEVENS POINT, 70-, DIR MED TECHNOL, 74- Mem: Am Chem Soc; Sigma Xi. Res: Synthesis of 4-amino-4, 6-dideoxy hexoses and unsaturated derivatives of carbohydrates. Mailing Add: Dept of Chem Univ of Wis Stevens Point WI 54481

CHITTENDEN, FAYETTE DUDLEY, b New Haven, Conn, Aug 11, 02; m 32; c 2. CHEMISTRY. Educ: Yale Univ, BS, 23, PhD(chem eng), 26. Prof Exp: Asst, Yale Univ, 23-25; chemist, Gen Labs, Uniroyal, Inc, 26-31, asst develop mgr, Gen Prods Div, 31-42, tech supvr, GRS Plant, WVa, 42-44, tech coordr, Synthetic Rubber Div, 44-45, factory mgr, Naugatuck Plant, 45-46, develop mgr, Chem Div, 46-51, factory mgr, Painesville Plant, 51-55, prod mgr, Naugatuck Chem Div, 55-58, mgr plastics opers, 58-60, gen mgr, Naugatuck Chem Div & Vpres, 60-65, vpres tech & mfg staff, 65-68; RETIRED. Mem: Am Chem Soc. Res: Chemistry of rubber and textiles; plastics. Mailing Add: 3A Harbour Village Branford CT 06405

CHITTICK, DONALD ERNEST, b Salem, Ore, May 3, 32; m 57; c 2. PHYSICAL CHEMISTRY. Educ: Willamette Univ, BS, 54; Ore State Univ, PhD(chem), 60. Prof Exp: Instr chem, Univ Puget Sound, 58-59, from asst prof to assoc prof, 59-68; PROF CHEM, GEORGE FOX COL, 68-, CHMN SCI & MATH, 74- Mem: Am Chem Soc; fel Am Inst Chem; Creation Res Soc. Res: Photochemistry; electrochemistry; programmed instruction. Mailing Add: Rte 2 Box 194 Newberg OR 97132

CHITTIM, RICHARD LEIGH, b Easthampton, Mass, Dec 2, 15; m 49; c 3. MATHEMATICS. Educ: Bowdoin Col, AB, 41; Oxford Univ, BA, 50, MA, 55. Prof Exp: From instr to assoc prof, 42-63, PROF MATH, BOWDOIN COL, 63- Concurrent Pos: Dir, NSF In-Serv Inst, 59-61; NSF fac fel, Univ London, 61-62. Mem: Am Math Soc. Res: Algebra and analysis; complex function theory. Mailing Add: Dept of Math Bowdoin Col Brunswick ME 04011

CHITTY, DENNIS HUBERT, b Bristol, Eng, Sept 18, 12; m 36; c 3. POPULATION ECOLOGY. Educ: Univ Toronto, BA, 35; Oxford Univ, MA, 47, DPhil(zool), 49. Prof Exp: From res officer to sr res officer ecol, Bur Animal Pop, 35-61; PROF ZOOL, UNIV BC, 61- Concurrent Pos: NSF sr foreign scientist fel, Smith Col, 68-69. Mem: Ecol Soc Am; Brit Ecol Soc (vpres, 60); Can Soc Zool; fel Royal Soc Can. Res: Regulation of numbers in natural populations; control of rats; history and principles of scientific methodology. Mailing Add: Dept of Zool Univ of BC Vancouver BC Can

CHITWOOD, HENRY CADY, b Washington Co, Md, Sept 18, 11; m 38; c 2. ORGANIC CHEMISTRY. Educ: WVa Univ, AB, 32; Johns Hopkins Univ, PhD(chem), 34. Prof Exp: Res chemist, Carbide & Carbon Chem Corp, WVa, 34-38 & Nat Carbon Co, Ohio, 35-36; group leader res, 48-53, asst dir res org chem, 53-55, asst dir res biol chem, 55-60, sr res consult, 60-67, ASST DIR UNIV RELS, UNION CARBIDE CHEM CO, NEW YORK, 67- Mem: AAAS; Am Chem Soc. Res: Glyoxalidines; electro-organic chemistry; catalysis; aliphatic syntheses; oxidation of organic compounds; nitrogen heterocyclics; ethylene oxide. Mailing Add: 19 Ledge Rd Old Greenwich CT 06870

CHITWOOD, HOWARD, b Creekmore, Ky, Feb 4, 32; m 53; c 2. MATHEMATICS. Educ: Carson-Newman Col, BS, 53; Univ Fla, MS, 55; Univ Tenn, PhD, 71. Prof

Exp: PROF MATH, CARSON-NEWMAN COL, 57- Mem: Math Asn Am. Res: Differential equations. Mailing Add: Rt 1 Laurel Hills Jefferson City TN 37760

CHITWOOD, JAMES LEROY, b St Petersburg, Fla, Mar 17, 43; m 64; c 2. ORGANIC CHEMISTRY. Educ: Emory Univ, BS, 65; Univ Calif, Berkeley, PhD(org chem), 68. Prof Exp: Res chemist, 68-69, sr res chemist, 70-72, res assoc, 72-74, div head phys & anal chem res, 74, staff asst to exec vpres develop, Res & Develop Admin, 74-75, DIR CHEM RES DIV, TENN EASTMAN CO, 76- Mem: AAAS; Am Chem Soc; Sigma Xi; Am Asn Textile Chemists & Colorists; Am Inst Chemists. Res: Organic synthesis; reaction mechanisms; radiation induced reactions; structure-property relationships. Mailing Add: Res Labs Tenn Eastman Co Kingsport TN 37662

CHITWOOD, MAY BELLE HUTSON, b Lubbock, Tex, Sept 17, 08; div; c 2. PARASITOLOGY, NEMATOLOGY. Educ: Univ Md, BA, 58. Prof Exp: Collabr nematol, Bulb Pest Lab, USDA, NY, 37-47, vet parasitologist, 59-61, res parasitologist, Beltsville Parasitol Lab, 61-64, res fel parasitologist, 64-65, sr res parasitologist, 65-73; in chg primate parasite registry, Primate Res Ctr, Univ Calif, Davis, 71-75; RETIRED. Mem: Am Soc Parasitol; Soc Syst Zool; Wildlife Dis Asn; Am Micros Soc; Am Soc Trop Med & Hyg. Res: Systematic relationships of parasitic nematodes. Mailing Add: 7560 Pindell School Rd Fulton MD 20759

CHIU, ANDREW TAK-CHAU, b Hong Kong, Sept 1, 45; Brit citizen; m 72; c 1. PHARMACOLOGY. Educ: Olivet Col, Mich, BA, 70; Univ Va, PhD(pharmacol), 74. Prof Exp: Fel, 74-75, RES SCIENTIST BIOCHEM PHARMACOL, PAPANICOLAOU CANCER RES INST, 75-, ADJ ASST PROF, 75- Res: Renal hypertension, specifically the Renin-angiotensin system; metabolisms of angiotensins and prostaglandins. Mailing Add: 7701 Camino Real A218 Miami FL 33143

CHIU, CHARLES BIN, b Foochow, China, May 19, 40; m 64; c 1. ELEMENTARY PARTICLE PHYSICS, HIGH ENERGY PHYSICS. Educ: Seattle Pac Col, BSc, 61; Univ Calif, Berkeley, PhD(physics), 66. Prof Exp: Fel theoret particle physics, Theoret Group, Lawrence Radiation Lab, Calif, 65-67; vis scientist theory div, Europ Orgn Nuclear Res, Switz, 67-68; sr res fel, Cavendish Lab, Cambridge, 68-69; res fel, Calif Inst Technol, 69-70, Tolman sr res fel, 70-71; asst prof physics, 71-74, ASSOC PROF PHYSICS, UNIV TEX, AUSTIN, 74- Concurrent Pos: Co-organizer, Conf Phenomenology Particle Physics, 71. Mem: Am Phys Soc; Sigma Xi; Am Asn Univ Prof; AAAS. Res: Participation of measurements on particle scattering cross sections; high energy scattering phenomenon; Regge theory; theory of strong interactions and general theory of elementary particle physics. Mailing Add: Dept of Physics Univ of Tex Austin TX 78712

CHIU, CHIN-SHAN, b Cheh-Chiang, China, Mar 6, 46. ATMOSPHERIC PHYSICS, CLOUD PHYSICS. Educ: Nat Taiwan Norm Univ, BS, 68; NMex Inst Mining & Technol, MS, 71, PhD(physics), 75. Prof Exp: RES SCIENTIST ATMOSPHERIC ELEC, INST ATMOSPHERIC SCI, S DAK SCH MINES & TECHNOL, 74- Mem: Am Geophys Union; Sigma Xi. Res: Use of computers to do numerical simulation study in order to predict the effects of atmospheric electricity on convective clouds, and vice versa. Mailing Add: Inst of Atmospheric Sci SDak Sch of Mines & Technol Rapid City SD 57701

CHIU, G C, b Kweiyang, China, Apr 16, 12; m 57; c 1. MEDICINE. Educ: Tung-chi Univ, China, MD, 37; Univ Hamburg, MD, 39. Prof Exp: Lectr, NIH, China, 46-49; pvt pract, Shanghai, 49-50; SR PHYSICIAN, MED RES DIV, LILLY RES LABS, 52-; ED-IN-CHIEF, CARDIOVASC ABSTR INDEX CARDS, MED ABSTR CO, IND, 63- Concurrent Pos: Res fel cardiol, State Univ NY Upstate Med Ctr, 51-52; ed, Stand Ger-Chinese Dict, 40-50; mem spec staff, Ministry Health, Nanking, China, 46-49; mem, State Bd Health & Educ, Kweichow, 46-49; chief med officer & cardiologist, Shanghai-Nanking RR, 46-49; mem coun arteriosclerosis, Am Heart Asn, 55- Mem: Am Heart Asn; Am Diabetes Asn; NY Acad Sci. Res: Cardiovascular diseases, especially atherosclerosis. Mailing Add: Med Res Div Lilly Res Labs Indianapolis IN 46206

CHIU, HONG-YEE, b Shanghai, China, Oct 4, 32; m 66; c 3. ASTROPHYSICS. Educ: Okla State Univ, BSc, 56; Cornell Univ, PhD(physics), 59. Prof Exp: Nat Acad Sci res assoc physics, Theoret Div, NASA, 60-62, PHYSICIST, GODDARD INST SPACE STUDIES, 62-; ADJ PROF PHYSICS, CITY COL OF NEW YORK, 66- Concurrent Pos: Mem, Inst Advan Study, 59-61; asst prof physics, Yale Univ, 61-62; adj asst prof, Columbia Univ, 62-65, adj assoc prof, 65- Mem: Am Phys Soc; Am Astron Soc; fel Royal Astron Soc. Res: Nuclear physics; strange particles; stellar evolution; neutrino astrophysics; x-ray astronomy. Mailing Add: Goddard Inst for Space Studies 2880 Broadway New York NY 10025

CHIU, JEN, b China, June 22, 24; US citizen; m; c 4. ANALYTICAL CHEMISTRY, POLYMER CHEMISTRY. Educ: Hunan Univ, BS, 46; Univ Ill, MS, 59, PhD(anal chem), 61. Prof Exp: Res chemist, 60-65, sr res chemist, 65-72, RES ASSOC, E I DU PONT DE NEMOURS & CO, INC, 72- Mem: Am Chem Soc; Int Confedn Thermal Anal; NAm Thermal Anal Soc (vpres, 75); Sigma Xi. Res: Analytical research in polymer characterization using chromatographic, spectroscopic, and thermal methods. Mailing Add: E I du Pont de Nemours & Co Inc Exp Sta Wilmington DE 19898

CHIU, JEN-FU, b Taiwan, China, Sept 30, 40; m 70; c 1. BIOCHEMISTRY. Educ: Taipei Med Col, Taiwan, BPharm, 64; Nat Taiwan Univ, MSc, 67; Univ BC, PhD(biochem), 72. Prof Exp: Teaching asst chem, Taipei Med Col, 65-67; lectr biochem, Chung Shan Med Col, 67-68; proj investr, Univ Tex M D Anderson Hosp & Tumor Inst Houston, 72-74, asst biochemist, 74-75; ASST PROF BIOCHEM, SCH MED, VANDERBILT UNIV, 75- Concurrent Pos: Rosalie B Hite fel, Univ Tex, 72-73; mem med educ comt, Vanderbilt Univ, 75- Mem: Am Soc Cell Biol; Am Asn Cancer Res; Can Biochem Soc; Biophys Soc. Res: Role of nuclear nonhistone proteins in the regulation of genetic activity; biochemistry of chromatin; macromolecular mechanism of carcinogenesis; antibodies to human tissue or tumor associated antigens. Mailing Add: Dept of Biochem Vanderbilt Univ Sch of Med Nashville TN 37232

CHIU, LUE-YUNG CHOW, b Kiang-su, China, Sept 14, 31; m 60; c 2. QUANTUM CHEMISTRY, ATOMIC PHYSICS. Educ: Nat Taiwan Univ, BS, 52; Bryn Mawr Col, MA, 54; Yale Univ, PhD(phys chem), 57. Prof Exp: Fel atomic physics, Yale Univ, 57-60; res physicist, Radiation Lab, Columbia, 60-62; res assoc, Lab Molecular Struct & Spectra, Univ Chicago, 62-63 & Inst for Studies of Metals, 63-64; res asst prof chem, Cath Univ Am, 64-65; Nat Res Coun-Nat Acad Sci sr resident res assoc, Lab Theoret Studies, Goddard Space Flight Ctr, NASA, 65-67, sr resident res assoc, Astrochem Sect, 67-68; ASSOC PROF QUANTUM CHEM, HOWARD UNIV, 68- Mem: Am Phys Soc. Res: Photochemistry; atomic beam magnetic resonances; theoretical studies on magnetic interactions in molecules; atomic and molecular scattering; interaction of radiation with atoms and molecules. Mailing Add: Dept of Chem Howard Univ Washington DC 20001

CHIU, PETER JIUNN-SHYONG, b Miao-Li, Taiwan, June 9, 42; m 67; c 2. PHARMACOLOGY. Educ: Taipei Med Col, BS, 64; Nat Taiwan Univ, MS, 66; Columbia Univ, PhD(pharmacol), 72. Prof Exp: Res fel nephrology, Sch Med, Univ Pa, 72-74; SR SCIENTIST RENAL PHARMACOL, RES DIV, SCHERING CORP, 74- Mem: Int Soc Nephrology; Am Soc Nephrology; Am Fedn Clin Res; AAAS. Res: Effect of antihypertensive agents on fluid and electrolytes balance; renal handling of antibiotics and their effects on kidney function; evaluation of new natriuretic agents. Mailing Add: Schering Corp Res Div 60 Orange St Bloomfield NJ 07003

CHIU, TAI-WOO, b China, May 18, 44. PHYSICAL CHEMISTRY. Educ: Chinese Univ Hong Kong, BSc, 66; Univ Miami, PhD(phys chem), 71. Prof Exp: Res assoc polymers, Mat Res Ctr, Lehigh Univ, 71-73 & Univ Cincinnati, 73-74; PHYS CHEMIST RUBBER RES, POLYFIBRON DIV, W R GRACE & CO, 75- Mem: Am Chem Soc; Sigma Xi. Res: Surface properties of nitrite rubber. Mailing Add: 55 Hayden Ave Lexington MA 02173

CHIU, TIN-HO, b Kiang Si Prov, China, June 22, 42; m 69; c 2. SURFACE CHEMISTRY. Educ: Chung Chi Col, Chinese Univ Hong Kong, BSc, 64; Lehigh Univ, PhD(chem), 72. Prof Exp: High sch teacher chem, Fai Yuen Col, Hong Kong & sci teacher biol, South Western Col, Hong Kong, 64-66; teaching asst chem, Lehigh Univ, 67-69, res asst, Ctr Surface & Coatings Res, 69-72; RES ASSOC MICROCALORIMETRY, AVCO EVERETT RES LAB, SUBSID AVCO CORP, 72- Mem: Am Chem Soc; Am Soc Artifical Internal Organs. Res: The interaction of solid/gas, solid/liquid, and gas/liquid including surface, biophysical, colloid, corrosion, polymer and pollution chemistry. Mailing Add: 22 Bolton St Reading MA 01867

CHIU, WAN-CHENG, b Meihsien, China, Nov 1, 19; US citizen; m 54; c 3. METEOROLOGY. Educ: Nat Cent Univ, China, BS, 41; NY Univ, MS, 47, PhD(meteorol), 51. Prof Exp: Technician, Fukein Weather Bur, China, 41-42; teacher math, Pungshan Model Sch, China, 42-43; asst teacher meteorol, Nat Cent Univ, China, 44-45; res assoc, NY Univ, 51-54, from assoc meteorologist to meteorologist, 54-60, res scientist, 60-61; PROF METEOROL, UNIV HAWAII, 61- Concurrent Pos: Vis scientist, Nat Ctr Atmospheric Res, 67-68, sr fel, 75. Mem: Am Meteorol Soc; Am Geophys Union; Royal Meteorol Soc. Res: Atmospheric energy and circulation; study effect of Southeastern Pacific sea surface warming on the momentum and heat transport in the atmosphere. Mailing Add: Dept of Meteorol Univ of Hawaii Honolulu HI 96822

CHIU, YAM-TSI, b Canton, China, Sept 5, 40; US citizen; m 68. ATMOSPHERIC PHYSICS. Educ: Yale Univ, BS, 61, MS, 63, PhD(physics), 65. Prof Exp: Res assoc elem particle physics, Yale Univ, 65 & Enrico Fermi Inst Nuclear Studies, Univ Chicago, 65-67; mem tech staff, 67-74, STAFF SCIENTIST, AEROSPACE CORP, 74- Concurrent Pos: Mem Upper Atmosphere Comn, Int Asn Geomagnetism & Aeronomy, 72-73. Mem: Am Phys Soc; Am Geophys Union; Am Meteorological Soc. Res: Atmospheric dynamics; ionospheric and solar coronal dynamics-nonlinear waves in gas dynamic and magneto gas dynamic media. Mailing Add: Space Physics Lab Aerospace Corp Box 92957 Los Angeles CA 90009

CHIU, YING-NAN, b Canton, China, Nov 25, 33; m 60; c 2. PHYSICAL CHEMISTRY. Educ: Berea Col, BA, 55; Yale Univ, MS, 56, PhD(phys chem), 60 Prof Exp: Fel chem, Columbia Univ, 60-62; res assoc physics, Univ Chicago, 62-64; from asst prof to assoc prof, 64-70, PROF QUANTUM CHEM, CATH UNIV AM, 70-, CHMN CHEM DEPT, 72- Concurrent Pos: A P Sloan fel, Harvard & Princeton Univs, 69-71. Mem: Am Chem Soc. Res: Spin-spin and spin-orbit interaction in molecules; quantum theory of molecular structure and valence; theory of optical and magnetic resonance spectra, molecular interactions, molecular quantum mechanics; energy transfer. Mailing Add: Dept of Chem Cath Univ Am Washington DC 20017

CHIVERS, HUGH JOHN, b Frome, Eng, June 26, 32; m 56; c 2. PHYSICS. Educ: Univ Manchester, BSc, 56, PhD(physics), 59. Prof Exp: Res asst radio astron, Univ Manchester, 60-61; proj leader ionospheric physics, Environ Sci Serv Admin, 61-62, sect chief, 62-64, dir, Space Disturbance Monitoring Sta, 65-67, mem staff, Tropospheric Wave Propagation, Colo, 67-68; MEM STAFF, DEPT APPL ELECTROPHYSICS, UNIV CALIF, SAN DIEGO, 68- Concurrent Pos: Mem Comn III, Int Sci Radio Union, 61- Mem: Am Geophys Union. Res: Ionospheric research using scintillations and absorption caused by charged particles from space. Mailing Add: Dept of Appl Electrophysics Univ of Calif at San Diego La Jolla CA 92037

CHIVIAN, JAY SIMON, b Newark, NJ, Mar 17, 31; m 56; c 2. PHYSICS. Educ: Franklin & Marshall Col, BS, 52; Lehigh Univ, MS, 54, PhD(physics), 60. Prof Exp: Res asst, Brookhaven Nat Lab, 56; instr physics, Lafayette Col, 59-60; res physicist, Texas Instruments, Inc, 60-67; res scientist, Ling-Temco-Vought Res Ctr, 67-71; SR SCIENTIST, ADVAN TECHNOL CTR, INC, 71- Mem: Am Phys Soc; Inst Elec & Electronics Engrs; Laser Inst Am. Res: Laser applications, holography and nonlinear optics; physical electronics; plasma and neutron physics; thermionic energy conversion; signal detection in antisubmarine warfare; thin film physics; infrared detection; quantum electronics. Mailing Add: Advan Technol Ctr Inc PO Box 6144 Dallas TX 75222

CHIVUKULA, RAMAMOHANA RAO, b Vijayavada, India, June 20, 33; m 57; c 3. MATHEMATICS. Educ: Andra Univ, BA, 53, MA, 55, PhD(math), 60; Univ Ill, PhD, 62. Prof Exp: Lectr math, Andhra Univ, 53-59 & Univ Mich, 62-63; asst prof, 63-71, ASSOC PROF MATH, UNIV NEBR, LINCOLN, 71- Mem: Am Math Soc; Math Asn Am; Indian Math Soc. Res: Functional analysis. Mailing Add: Dept of Math Univ of Nebr Lincoln NE 68508

CHIYKOWSKI, LLOYD NICHOLAS, b Garson, Man, July 26, 29; m 53; c 4. ENTOMOLOGY. Educ: Univ Man, BSA, 53, MSc, 54; Univ Wis, PhD(entomol), 58. Prof Exp: Asst entomol, Univ Wis, 54-58; res officer, Plant Res Inst, 58-67, res scientist, Cent Exp Farm, Cell Biol Res Inst, 67-71, RES SCIENTIST, CENT EXP FARM, CHEM & BIOL RES INST, CAN DEPT AGR, 71- Mem: Entom Soc Am; Entom Soc Can; Agr Inst Can; Can Phytopath Soc. Res: Leafhopper transmission of plant viruses. Mailing Add: Chem & Biol Res Inst Res Br Canada Dept of Agr Ottawa ON Can

CHIZINSKY, WALTER, b Springfield, Mass, Nov 27, 26; m 53; c 3. REPRODUCTIVE PHYSIOLOGY, VERTEBRATE BIOLOGY. Educ: Univ Mass, BS, 49; Univ Chicago, SM, 50; NY Univ, PhD(biol), 60. Prof Exp: Instr & assoc prof biol, Bennett Col, NY, 54-67, chmn dept sci & math, 60-67; from assoc prof to prof biol, 67-74, acad dean, 70-74, DISTINGUISHED PROF SCI, BRIARCLIFF COL, 74- Concurrent Pos: Shell Merit fel, Stanford Univ, 69- Mem: AAAS; Am Asn Univ Prof; Am Asn Sex Educr & Counr; Soc Sci Study Sex; Sex Info & Educ Coun US. Res: Human sexuality, especially behavior; anatomy, physiology and behavior of sex and reproduction, especially in vertebrates. Mailing Add: Dept of Biol Briarcliff Col Briarcliff Manor NY 10510

CHLOUPEK, FRANK J, physical organic chemistry, see 12th edition

CHMURA, CAROL A, b Chicopee, Mass, Sept 22, 39. GEOLOGY. Educ: Smith Col, BA, 61; Univ NMex, MS, 63; Stanford Univ, PhD(geol), 70. Prof Exp: Lectr geol, Muhlenberg Col, 67-72; sr paleontologist, Mobile Oil Corp, 72-74; RES GEOLOGIST

PALYNOLOGY GEOL, CHEVRON OIL FIELD RES CO, 74- Concurrent Pos: Res assoc geol, Lehigh Univ, 67-72. Mem: Sigma Xi; Am Asn Stratig Palynologists. Res: Palynology and sedimentology. Mailing Add: Chevron Oil Field Res Co PO Box 446 La Habra CA 90631

CHMURA, NORMAN WALTER, b Cleveland, Ohio, May 28, 28; m 53; c 2. MICROBIOLOGY. Educ: Western Reserve Univ, BS, 49; Univ NH, MS, 55; Univ Md, PhD(microbiol), 58. Prof Exp: Res asst bovine mastitis & disinfectants, Univ NH, 53-55; asst bacter & virol, Univ Md, 55-57; res bacteriologist meat microbiol & food poisoning, Swift & Co, 57-59; from asst prof to assoc prof biol sch, Carnegie Inst Technol, 59-65; assoc prof biol & chmn dept, 65-70, prof biol & provost, 70-74, MARY HELEN MARKS PROF BIOL, CHATHAM COL, 74- Mem: AAAS; Am Soc Microbiol. Res: Microbial associations and genetics; ultraviolet resistance and sensitivity. Mailing Add: Dept of Biol Chatham Col Pittsburgh PA 15232

CHMURNY, ALAN BRUCE, b Oak Park, Ill, May 31, 44; m 66; c 1. ORGANIC CHEMISTRY. Educ: Univ Ill, Urbana, BS, 66; Univ Calif, Los Angeles, PhD(org chem), 71. Prof Exp: NIH fel, Mass Inst Technol, 72-73; RES SCIENTIST ORG CHEM, PFIZER INC, GROTON, 73- Mem: Am Chem Soc. Res: Application of enzyme catalyzed reactions to the production of fine organic chemicals on an industrial scale. Mailing Add: 6 Ferry View Dr Gales Ferry CT 06335

CHO, ALFRED CHIH-FANG, b Shanghai, China, Dec 31, 21; US citizen; m 57; c 2. ACOUSTICS, AEROSPACE SCIENCES. Educ: Univ Shanghai, BSc, 43; Univ Tex, MA, 50, PhD(physics), 58. Prof Exp: Asst engr, Shanghai Tel Co, 44-48; teaching fel physics, Univ Tex, 49-53, spec instr math, 53-57; sr struct engr, Gen Dynamics/Ft Worth, 57-60, sr physicist, 60-62; sr tech specialist acoust & vibration, res & eng, Space Div, NAm Rockwell, 62-63, supvr, 63-67, mem tech staff, 67-69; sr res specialist antisubmarine warfare & shock & vibration dir, Lockheed-Calif Co, 69-71; eng supvr specialist, Litton Ship Syst, 71-72; mem tech staff, Hughes Aircraft, Space & Commun Group, 72-73; MEM TECH STAFF, SPACE DI, ROCKWELL INT, 73- Concurrent Pos: Adj prof physics, Tex Christian Univ, 58-62; translr, Am Inst Physics, 64-69. Mem: Acoust Soc Am; Brit Acoust Soc. Res: Acoustics and vibration of aircraft and launch vehicle; dynamic responses of spacecraft; aeromechanics in aeronautics; pulse statistical analysis in architectural acoustics. Mailing Add: 6263 S Roundhill Dr Whittier CA 90601

CHO, ALFRED Y, b Peking, China, July 10, 37; US citizen; m 68; c 3. SURFACE PHYSICS, SEMICONDUCTORS. Educ: Univ Ill, BSEE, 60, MSEE, 61, PhD(elec eng), 68. Prof Exp: Res physicist, Ion Physics Corp, 61-62; mem tech staff, TRW-Space Tech Labs, 62-65; res asst, Univ Ill, Urbana, 65-68; MEM TECH STAFF MOLECULAR BEAM EPITAXY, BELL LABS, 68- Mem: NY Acad Sci; Am Phys Soc; Am Vacuum Soc; Electrochem Soc; Sigma Xi. Res: Surface structure reconstructions of III-V compounds; crystal growth by molecular beam epitaxy; preparation of various microwave and optoelectronic devices. Mailing Add: Bell Labs IC-217 Murray Hill NJ 07974

CHO, ARTHUR KENJI, b Oakland, Calif, Nov 7, 28; m 53; c 2. PHARMACOLOGY, ORGANIC CHEMISTRY. Educ: Univ Calif, BS, 52; Ore State Univ, MS, 53; Univ Calif, Los Angeles, PhD(chem), 58. Prof Exp: Asst res pharmacologist, Univ Calif, Los Angeles, 58-61; res chemist, Don Baxter Inc, 61-65; res pharmacologist, Nat Heart & Lung Inst, 65-70; assoc prof, 70-74, PROF PHARMACOL, UNIV CALIF, LOS ANGELES, 74- Mem: Am Chem Soc; Am Soc Pharmacol & Exp Therapeut. Res: Drug metabolism; adrenergic mechanisms. Mailing Add: Dept of Pharmacol Med Sch Univ of Calif Los Angeles CA 90024

CHO, BYUNG-RYUL, b Seoul, Korea, Feb 3, 26; m 48; c 3. VETERINARY MEDICINE, MICROBIOLOGY. Educ: Seoul Nat Univ, DVM, 50; Univ Minn, MS, 59, PhD(vet microbiol), 61. Prof Exp: Instr animal infectious dis, Col Vet Med, Seoul Nat Univ, 57-58, from asst prof to assoc prof poultry dis, 61-64; from res assoc to asst prof, 64-72, ASSOC PROF AVIAN TUMORS, WASH STATE UNIV, 72- Mem: AAAS; Am Asn Avian Pathologists; Am Soc Microbiologists; Conf Res Workers Animal Dis. Res: Viral diseases of poultry, particularly in avian tumor research. Mailing Add: Dept of Vet Microbiol Wash State Univ Pullman WA 99163

CHO, CHUNG WON, b Seoul, Korea, Feb 7, 31; m 58; c 2. MOLECULAR PHYSICS. Educ: Seoul Nat Univ, BSc, 53; Univ Toronto, MA, 55, PhD(physics), 58. Prof Exp: Asst, Univ Toronto, 58; from asst prof to assoc prof, 58-67, PROF PHYSICS, MEM UNIV NFLD, 67- Concurrent Pos: Vis assoc prof, Pa State Univ, 66-68. Mem: Europ Phys Soc; Can Asn Physicists; Am Phys Soc; Optical Soc Am. Res: Lasers; laser spectroscopy; stimulated laser light scattering; pressure-induced infrared absorption; lidars. Mailing Add: Dept of Physics Mem Univ of Nfld St John's NF Can

CHO, HAN-RU, b Peking, China, Feb 4, 45; m 71; c 1. DYNAMIC METEOROLOGY. Educ: Nat Taiwan Univ, BS, 67; Univ Ill, MS, 70, PhD(atmospheric sci), 72. Prof Exp: Res assoc meteorol, Lab Atmospheric Res, Univ Ill, 72-74, asst prof, 74; ASST PROF METEOROL, DEPT PHYSICS, UNIV TORONTO, 74- Mem: AAAS; Am Meteorol Soc; Can Meteorol Soc; Can Asn Physicists. Res: Interactions of cumulus cloud ensembles with large scale convective weather systems and the theory of cumulus parameterization. Mailing Add: Dept of Physics Univ of Toronto Toronto ON Can

CHO, KON HO, b Kangwha, Korea, May 3, 37; m 68. PHYSICAL CHEMISTRY, BIOCHEMISTRY. Educ: Seoul Nat Univ, BS, 62; Auburn Univ, MS, 65; Princeton Univ, MA & PhD(chem), 70. Prof Exp: Proj engr, Hyosung Moolsan Co, Ltd, 62-63; MEM RES STAFF, WESTERN ELEC CO, INC, 69- Mem: Am Chem Soc. Res: Mechanisms of enzyme catalysis; conformation and conformational changes of proteins; radiation chemistry of polymers. Mailing Add: Western Elec Co Inc PO Box 900 Princeton NJ 08540

CHO, YOUNG WON, b Seoul, Korea, Mar 3, 31; US citizen; m 59; c 1. CLINICAL PHARMACOLOGY, CARDIOLOGY. Educ: Seoul Nat Univ, MD, 56; Emory Univ, MS, 62. Prof Exp: Intern internal med, Vanderbilt Univ Hosp, 56-57; resident, Baptist Hosp, Nashville, Tenn, 57-60; instr physiol, Sch Med, Emory Univ, 60-64; chief div cardiol & asst cardiologist, Philadelphia Gen Hosp, 64-66; assoc dir clin pharmacol, Mead Johnson Res Ctr, 66-68; asst dir clin pharmacol, William S Merrell Co, Ohio, 68-69, group dir cardiovasc clin res, 69-71; ASSOC PROF PHARMACOL & MED, MED CTR, UNIV NEW ORLEANS, 71- Concurrent Pos: NIH cardiovasc res trainee, Emory Univ, 60-64; sr res fel pharmacol, Mead Johnson Res Ctr, 66-68; prin investr, NIH grants, Emory Univ, 63-64 & Philadelphia Gen Hosp, 65-66; instr, Sch Med, Univ Pa, 64-70; consult, US Vet Admin Hosp, Castle Point, NY, 69- Mem: Fel Am Col Angiol; Am Heart Asn; Am Fedn Clin Res; Am Therapeut Soc; fel Am Col Chest Physicians. Res: During various forms of cardiovascular shocks, cardia mitochondrial respiratory enzyme systems lose their activities, which is followed by inactivation of cardiac myosin ATPase activity and repolarization activity by actin. Mailing Add: Dept of Pharmacol Univ of New Orleans Med Ctr New Orleans LA 70122

CHOATE, JERRY RONALD, b Bartlesville, Okla, Mar 21, 43; m 63; c 1. MAMMALOGY, EVOLUTIONARY BIOLOGY. Educ: Kans State Col, Pittsburg, BA, 65; Univ Kans, PhD(zool), 69. Prof Exp: Asst prof biol, Univ Conn, 69-71; DIR, MUS HIGH PLAINS, FT HAYS, KANS STATE COL, 71- Concurrent Pos: Guest lectr, Yale Univ, 70; spec consult, Synecol Corp, 70- Mem: AAAS; Am Soc Mammalogists (recording sec, 74-); Soc Study Evolution; Soc Syst Zool. Res: Systematics, biogeography and natural history of mammals; speciation and evolutionary biology of insectivores, bats and rodents. Mailing Add: Mus of the High Plains Ft Hays Kans State Col Hays KS 67601

CHOBANIAN, ARAM V, b Pawtucket, RI, Aug 10, 29; m 55; c 3. CARDIOVASCULAR DISEASES, INTERNAL MEDICINE. Educ: Brown Univ, AB, 51; Harvard Med Sch, MD, 55. Prof Exp: Intern & chief resident med, Univ Hosp, Boston, 55-59; assoc prof, 60-71, PROF MED, SCH MED, BOSTON UNIV, 71-, DIR CARDIOVASCULAR INST, MED CTR, 73-, DIR HYPERTENSION CTR, 75- Concurrent Pos: NIH cardiovascular fel, Univ Hosp, Boston, 59-62; Nat Heart & Lung Inst grants, Sch Med, Boston Univ, 68-; lectr, Harvard Med Sch, 71; mem coun arteriosclerosis and high blood pressure res, Am Heart Asn; mem cardiovasc & renal adv comt, Food & Drug Admin, 75-; mem hypertension & arteriosclerosis adv comt, Nat Heart & Lung Inst, 75- Mem: Am Soc Clin Invest; Am Heart Asn; Am Physiol Soc; Am Fedn Clin Res. Res: Hypertension; arterial metabolism; body cholesterol metabolism. Mailing Add: Dept of Med Boston Univ Sch of Med Boston MA 02118

CHOBOTAR, BILL, b Vita, Man, Sept 2, 34; m 56; c 3. ZOOLZOOLOGY, ANIMAL PARASITOLOGY. Educ: Walla Walla Col, BA, 63, MA, 65; Utah State Univ, PhD(zool), 69. Prof Exp: Asst prof biol, 68-72, ASSOC PROF BIOL, ANDREWS UNIV, 72- Concurrent Pos: Alexander von Humboldt fel, Univ Bonn, 74. Mem: AAAS; Am Soc Parasitol; Soc Protozool; Wildlife Dis Asn. Res: Biology, development and pathology of parasitic infections, especially of parasitic protozoa; life cycles and fine structure of coccidia. Mailing Add: Dept of Biol Andrews Univ Berrien Springs MI 49104

CHOBY, EDWARD GEORGE, JR, organic chemistry, see 12th edition

CHOCK, ERNEST PHAYNAN, b Medan, Indonesia, Oct 27, 37; US citizen; m 65; c 2. PHYSICAL INORGANIC CHEMISTRY, POLYMER CHEMISTRY. Educ: Univ Calif, Santa Barbara, MA, 63, PhD(surface chem), 66. Prof Exp: Asst chem, Univ Calif, Santa Barbara, 61-66; consult, Bell & Howell Res Ctr, Calif, 66-67; researcher dept physics, Univ Calif, Santa Barbara, 67-68 & Los Angeles, 67-70, RES PHYSICIST, UNIV CALIF, LOS ANGELES, 70- Mem: Am Chem Soc; Am Phys Soc. Res: General material science, inorganic syntheses, formation of single crystals, thin films and ultrafine particles; physical properties of solids by magnetic susceptibility, magnetic resonances, conductivity, surface adsorption and catalytic activity; polymerization of organic and inorganic molecules. Mailing Add: Dept of Physics Univ of Calif Los Angeles CA 90024

CHOCK, JAN SUN-LUM, b Honolulu, Hawaii, Jan 22, 44; m 68. PHYCOLOGY. Educ: Univ Hawaii, BA, 67; Univ NH, PhD(bot), 75. Prof Exp: Res asst marine phycol, Univ Hawaii, 67-68; inspector, Animal & Plant Health Inspection Serv, Agr Res Serv, USDA, 68-70; ASST PROF BIOL, KING'S COL, PA, 75- Mem: AAAS; Phycol Soc Am. Res: Marine algal ecology, specifically in the estuarine intertidal environment; productivity of brown algae and salt marsh grass. Mailing Add: Dept of Biol King's Col Wilkes Barre PA 18711

CHODOROW, MARVIN, b Buffalo, NY, July 16, 13; m 37; c c 2. ACOUSTICS, ELECTRONICS. Educ: Univ Buffalo, AB, 34; Mass Inst Technol, PhD(physics), 39. Hon Degrees: DL, Univ Glasgow, 72. Prof Exp: Asst, Carnegie Inst, 39-40; res assoc physics, Pa State Univ, 40-41; instr, City Col, 41-43; res physicist, Gen Elec Industs, Conn, 43; sr proj engr, Sperry Gyroscope Corp, NY, 43-47; assoc prof physics, 47-54, chmn dept appl physics, 62-69, PROF APPL PHYSICS & ELEC ENG, STANFORD UNIV, 54-, DIR MICROWAVE LAB, 59- Concurrent Pos: Vis lectr, Ecole Normale Superiure, Univ Paris, 55-56; Fulbright fel, Univ Cambridge, 62-63; vis res assoc, Univ Col, UnUniv London, 69-70; consult, Rand Corp & Lincoln Lab; mem, US Nat Comt, Int Sci Radio Union; mem adv comt on USSR & Eastern Europe, Nat Acad Sci, 69-71, chmn, 71-73. Honors & Awards: Baker Award, Inst Elec & Electronics Engrs, 62. Mem: Nat Acad Sci; Nat Acad Eng; fel Am Phys Soc; fel Inst Elec & Electronics Engrs; fel Am Acad Arts & Sci. Res: Electronic devices and microwave acoustics. Mailing Add: Microwave Lab Stanford Univ Stanford CA 94305

CHODOS, ARTHUR A, b New York, NY, Oct 27, 22; m 43; c 2. GEOCHEMISTRY, ANALYTICAL CHEMISTRY. Educ: City Col NY, BS, 43; Polytech Inst Brooklyn, MS, 69. Prof Exp: Org chemist, Fed Telecommun Lab, NJ, 43-48; anal chemist, Gen Serv Admin, Fed Supply Serv, NY, 48-50; chemist spectros, US Geol Surv, 50-51, spectroscopist, Anal Labs Br, Colo, 51-52; sr spectroscopist, 52-75, SR RES SPECTROSCOPIST, DIV GEOL SCI, CALIF INST TECHNOL, 75- Concurrent Pos: Consult, US Geol Surv, 54-64; co-investr lunar samples. Mem: Microbeam Anal Soc (treas, Electron Probe Anal Soc Am, 68-69, pres, 71); Soc Appl Spectros; Geochem Soc. Res: Application of instrumental analysis to problems in geochemistry and petrology; electron microprobe analysis of lunar materials Mailing Add: Dept of Geol 170-25 Calif Inst of Technol Pasadena CA 91125

CHODOS, ROBERT BRUNO, b Gap, Pa, July 12, 18; m 50; c 4. MEDICINE. Educ: Franklin & Marshall Col, BS, 39; Univ Pa, MD, 43; Am Bd Internal Med, dipl, 53; Am Bd Nuclear Med, dipl, 72. Prof Exp: Intern, US Naval Hosp, Philadelphia, 43-44; asst med, Sch Med, Boston Univ, 47-49, instr, 52-53; from instr to assoc prof, State Univ NY Upstate Med Ctr, 53-69; PROF MED & RADIOL, ALBANY MED COL, 69-; HEAD NUCLEAR MED DIV, RADIOL DEPT, ALBANY MED CTR, 69- Concurrent Pos: Fel med, Evans Mem Hosp, Boston, 47-48, asst resident, 48-49; resident, Cushing Vet Admin Hosp, Framingham, 49-50, physician med serv & radioisotope unit, 50-52; resident, Vet Admin Hosp, Boston, 52-53, asst chief med & dir radioisotope unit, 53-57, assoc chief staff, 57-65, chief radioisotope serv, 57-69; from asst attend physician to assoc attend physician, State Univ NY Upstate Med Ctr, 53-59; attend physician med & radiol, Albany Med Ctr, 69-; chief nuclear med serv, Vet Admin Hosp, 69-71, physician, 71-75; assoc ed, J Nuclear Med, 70-75. Mem: AMA; fel Am Col Physicians; Am Fedn Clin Res; Soc Nuclear Med. Res: Radioisotopes; human iron absorption; pathologic physiology of anemias; human thyroid metabolism; renal function in hypertension; blood volume in man in health and disease. Mailing Add: Nuclear Med Div Albany Med Ctr Hosp Albany NY 12208

CHODROFF, SAUL, b Brooklyn, NY, Apr 29, 14; m 39; c 2. CHEMISTRY. Educ: Brooklyn Col, BS, 38, MA, 42; Polytech Inst Brooklyn, PhD(org chem), 48. Prof Exp: Asst res chemist, Weiss & Downs, Inc, NY, 39-40; res chemist, Nat Oil Prods Co, NJ, 41-48; assoc res dir, Nopco Chem Co, 48-50; res dir, Norda Essential Oil & Chem Co, 50-65, VPRES RES & DEVELOP, NORDA INC, 65- Honors & Awards: Am Inst Chem Award. Mem: Am Chem Soc. Res: Chemistry of organic sulfur; catalytic oxidation; organic synthesis in vitamins, hormones, pharmaceuticals; essential

CHODROFF

oils, aromatics, perfumes and flavors. Mailing Add: Norda Inc 475 Tenth Ave New York NY 10018

CHOE, BYUNG-KIL, b Taegu, Korea, Feb 15, 33; m 60; c 2. MICROBIOLOGY, CELL BIOLOGY. Educ: Kyungpook Nat Univ, Korea, MD, 58, MS, 60; Ind Univ, Bloomington, PhD(microbiol), 70. Prof Exp: Res assoc med microbiol, Med Sch, Kyungpook Nat Univ, Korea, 59-62; res assoc microbial genetics, Karolinska Inst, 63-68; res assoc microbiol, Ind Univ, Bloomington, 70-71; res asst prof immunol, Med Sch, State Univ NY, Buffalo, 71-73; ASST PROF IMMUNOL, MED SCH, WAYNE STATE UNIV, 73- Concurrent Pos: Boswell fel, NY Res Found, 71-72. Mem: Am Soc Microbiol; Tissue Cult Asn; Am Soc Cell Biol. Res: Growth regulation of mammalian cells; tumor immunology. Mailing Add: Dept of Immunol & Microbiol Wayne State Univ Med Sch Detroit MI 48202

CHOE, HYUNG TAE, b Seoul, Korea, Apr 27, 27; m 64; c 2. PLANT PHYSIOLOGY, HORTICULTURE. Educ: Seoul Nat Univ, BS, 54, MSc, 57; Ohio State Univ, PhD(plant physiol, hort), 63. Prof Exp: Instr agr, UNESCO Fundamental Educ Ctr, Ministry of Educ, Korea, 56-59; from asst prof to assoc prof, 63-72, PROF BIOL, MANKATO STATE UNIV, 72- Concurrent Pos: Res biologist & NSF fel, Thimann Labs, Univ Calif, Santa Cruz, 73-74. Mem: AAAS; Am Soc Plant Physiol; Am Soc Hort Sci; Int Soc Hort Sci; Bot Soc Am. Res: Senescence of isolated chloroplasts and metabolic functions during the senescence in oat leaves; the effect of growth regulators, light qualities on the senescence of chloroplasts and leaves of oat seedlings. Mailing Add: Dept of Biol Mankato State Univ Mankato MN 56001

CHOE, JAE YNE, b Korea, Mar 27, 39. PHARMACOLOGY. Educ: Seoul Nat Univ, BS, 61, MS, 66; Univ NC, Chapel Hill, PhD(pharmacol), 73. Prof Exp: Instr pharmacol, Sch Pharm, Won Kwang Univ, 66-70; fel, Sch Med, Northwestern Univ, 73-75; RES ASSOC, DENT RES CTR, UNIV NC, CHAPEL HILL, 75- Res: Androgen metabolism in rodent submaxillary gland and kinetics on the metabolites binding with receptors in the gland. Mailing Add: Dent Res Ctr Univ of NC Chapel Hill NC 27514

CHOGUILL, HAROLD SAMUEL, b Humboldt, Kans, Jan 12, 07; m ; c 1. CHEMISTRY Educ: Col of Emporia, AB, 27; Univ Kans, AM, 31, PhD(chem), 38. Prof Exp: Teacher high sch, Kans, 27-30, prin, 28-30; instr math, Garden City Jr Col, 32-37; instr chem, Independence Jr Col, 37-43, ground instr, Civilian Pilot Training, 37-43; PROF CHEM, FT HAYS KANS STATE COL, 46- Concurrent Pos: NSF res fel, Univ Col, London, 59-60; vis lectr, Div Chem Educ, Am Chem Soc. Mem: AAAS; Am Chem Soc; The Chem Soc. Res: Iodination of phenols and ethers; kinetics of deiodination; halogenated nitroparaffins. Mailing Add: Dept of Chem Fort Hays Kans State Col Hays KS 67601

CHOI, BYUNG HO, b Hwang Hae Do, Korea, Oct 16, 28; m; c 2. PATHOLOGY, NEUROPATHOLOGY. Educ: Yonsei Univ, Korea, MD, 53, MSD, 63. Prof Exp: Asst in path, Yonsei Univ, Korea, 53-54; resident, Inst Path, Western Reserve Univ, 54-57, demonstr, 56-57; from instr to assoc prof, Yonsei Univ, Korea, 59-65; res assoc path, Albany Med Col, 65-67, asst prof, 67-69; assoc prof path, Sch Med, St Louis Univ, 69-72; ASSOC PROF PATH, SCH MED & DENT, UNIV ROCHESTER, 72- Concurrent Pos: Univ fel, Western Reserve Univ, 57-59; consult pathologist, Cuyahoga County Hosp, Cleveland, 57-59; attend pathologist, Vet Admin Hosp, Albany, NY, 68-69; assoc pathologist, Cardinal Glennon Mem Hosp, St Louis, 69-72; pathologist, Firmin-Desloge Hosp, 69-72; sr assoc pathologist, Strong Mem Hosp, Rochester, 72- Mem: Am Soc Clin Path; Col Am Path; Am Asn Neuropath. Res: Kinetics of cell proliferation in arterial intimal cells under different experimental conditions; developmental neurobiology. Mailing Add: 15 Ross Brook Dr Rochester NY 14625

CHOI, DUK-IN, b Inchon, Korea, Apr 30, 36; m 65; c 1. PLASMA PHYSICS. Educ: Seoul Nat Univ, Korea, BS, 59; Univ Colo, PhD(physics), 68. Prof Exp: Fel statist mech, Univ Brussels, 68-70; res assoc, Ctr Statist Mech, Univ Tex, Austin, 70-73; RES SCI ASSOC PLASMA PHYSICS, FUSION RES CTR, UNIV TEX, AUSTIN, 73- Mem: Am Phys Soc. Res: Theory and application of plasma turbulence in the thermonuclear fusion research. Mailing Add: Fusion Res Ctr Univ of Tex Austin TX 78712

CHOI, KEEWHAN, b Seoul, Korea, Jan 26, 31; US citizen; m 56; c 3. STATISTICS. Educ: Case Inst, BS, 58; Harvard Univ, PhD(statist), 63. Prof Exp: Asst prof statist, Cornell Univ, 62-66; res assoc, Atomic Bomb Casualty Comn, 68-72; PROF MATH, GA STATE UNIV, 72- Concurrent Pos: NSF fel, 75-; consult clin cancer invest rev comt, Nat Cancer Inst, 75- Mem: Math Asn Am; Inst Math Statist; Am Statist Asn. Res: Classification; pattern recognition; application of statistics to biological societal problems. Mailing Add: Dept of Math Ga State Univ Atlanta GA 30303

CHOI, NUNG WON, b Pyong Yong, Korea, Nov 1, 31; m 62; c 3. EPIDEMIOLOGY, ONCOLOGY. Educ: Seoul Nat Univ, MD, 58; Univ Minn, MPH, 61, PhD(epidemiol), 66. Prof Exp: Assoc prof social & prev med, 66-72, PROF EPIDEMIOL & CHIEF SECT EPIDEMIOL & BIOSTATIST, FAC MED, UNIV MAN, 72-; DIR EPIDEMIOL & BIOSTATIST, MAN CANCER TREAT & RES FOUND, 68- Concurrent Pos: Med fel epidemiol, Univ Minn, 60-65, fel epidemiol & med statist, Mayo Grad Sch Med, 65-66; USPHS traineeship epidemiol, 62-65; Am Pub Health Asn associateship for prep vital & health statist monogr, 65-66; Can Nat Health Dept positionship epidemiol, 66-72; mem adv comt statist studies, Nat Cancer Inst Can, 69-72; mem nat adv comt epidemiol, Dept Nat Health & Welfare, 74- Mem: Fel Am Pub Health Asn; Am Acad Neurol; Can Pub Health Asn; NY Acad Sci; Am Soc Trop Med & Hyg. Res: Epidemiologies of toxoplasmosis; brain tumor, cardiovascular, cerebrovascular and other neurological diseases; cancer of gastrointestinal tract; childhood and female breast and genital tract congenital malformation; cancer epidemiology. Mailing Add: Dept of Social & Prev Med Univ of Man Winnipeg MB Can

CHOI, RICHARD PARK, organic chemistry, food chemistry, deceased

CHOI, SANG-IL, b Korea, Sept 1, 31; m 61; c 1. SOLID STATE PHYSICS, THEORETICAL CHEMISTRY. Educ: Seoul Nat Univ, BSc, 53; Brown Univ, PhD(chem), 61. Prof Exp: Res assoc theoret chem, Univ Chicago, 61-63; from asst prof to assoc prof, 63-72, PROF PHYSICS, UNIV NC, CHAPEL HILL, 72- Mem: Am Chem Soc; fel Am Phys Soc. Res: Transport phenomena of gases; semiclassical scattering theory; physics of organic solids; theoretical study of solid electrolytes and organic crystals. Mailing Add: Dept of Physics & Astron Univ of NC Chapel Hill NC 27514

CHOI, SOON DAL, electronics, electromagnetics, see 12th edition

CHOI, SUNG CHIL, b Seoul, Korea, Dec 30, 30; m 67. BIOSTATISTICS. Educ: Univ Wash, Seattle, BS, 57, MA, 60; Univ Calif, PhD(biostatist), 66. Prof Exp: Mathematician, Boeing Co, 59-62; mem tech staff, Aerospace Corp, 63-64; asst prof math, Calif State Col, 64-66; mem sci staff, Measurement Anal Corp, 66-67; asst prof, 67-73, ASSOC PROF BIOSTATIST, MED SCH, WASH UNIV, 73- Mem: Am Pub Health Asn; Am Statist Asn; Biomet Soc. Res: Biostatistical methodology. Mailing Add: Dept of Prev Med Wash Univ Med Sch St Louis MO 63110

CHOI, YONG CHUN, b Sunchun, Korea, Dec 25, 35; m 67; c 1. BIOCHEMISTRY, ORGANIC CHEMISTRY. Educ: Seoul Nat Univ, MD, 59; Univ Rochester, PhD(biochem), 67. Prof Exp: Res assoc, 66-68, instr, 68-69, ASST PROF PHARMACOL, BAYLOR COL MED, 69-; MEM, SLOAN-KETTERING INST, 74- Res: Molecular biology of mammalian cell nuclei; mechanism of ribosome synthesis; primary structure of high molecular weight RNA of mammalian cell nucleoli. Mailing Add: Sloan-Kettering Inst 145 Boston Post Rd Rye NY 10880

CHOLAK, JACOB, b Elsass, Russia, Aug 23, 00; nat US; m 32; c 2. ANALYTICAL CHEMISTRY. Educ: Univ Cincinnati, ChE, 24; Am Acad Sanit Engrs, dipl; Am Bd Indust Hyg, dipl. Prof Exp: Asst appl physiol, Col Med, Univ Cincinnati, 26-40, res assoc, 40-42, from asst prof to prof, 42-65, prof environ health eng, 65-70, EMER PROF, ENVIRON HEALTH ENG, UNIV CINCINNATI, 70- Honors & Awards: Eminent Chemist Award, Am Chem Soc, 55. Mem: Am Chem Soc; Am Indust Hyg Asn; Air Pollution Control Asn. Res: Industrial hygiene; atmospheric pollution; field surveys; analytical chemistry; instrumental and physical analytical methods. Mailing Add: 3115 S Whitetree Circle Cincinnati OH 45236

CHOLLET, RAYMOND, b Flushing, NY, Oct 4, 46; m 69; c 2. PLANT PHYSIOLOGY. Educ: Colgate Univ, AB, 68; Univ Ill, MS, 69, PhD(bot), 72. Prof Exp: Res assoc plant physiol, Univ Ill, 71-72; RES SCIENTIST PLANT PHYSIOL, E I DU PONT DE NEMOURS & CO, INC, 72- Mem: Am Soc Plant Physiologists. Res: Photosynthetic and photorespiratory carbon metabolism in higher plants. Mailing Add: Cent Res & Develop Dept Exp Sta E I du Pont de Nemours & Co Inc Wilmington DE 19898

CHOMAN, BOHDAN RUSSELL, b Cass Twp, Pa, May 17, 26; m 56; c 2. MICROBIOLOGY. Educ: Pa State Univ, BS, 49, MS, 50, PhD(bact), 54. Prof Exp: Res biochemist, Union Carbide Co, 50-57; res assoc, Lever Bros Co, NJ, 57-63; microbiol expert, Lederle Labs, 63-67, SUPT, LEDERLE LABS, AM CYANAMID CO, 67- Mem: Am Chem Soc; Am Soc Microbiol; fel Am Acad Microbiol. Res: Immunology; virology; biologicals. Mailing Add: 574 Blauvelt Dr Oradell NJ 07649

CHOMCHALOW, NARONG, b Bangkok, Thailand, Aug 26, 35; m 60; c 3. AGRONOMY, PLANT GENETICS. Educ: Kasetsart Univ, Bangkok, BS, 57; Univ Hawaii, MS, 61; Univ Chicago, PhD(bot), 64. Prof Exp: Instr agr, Kasetsart Univ, 57-59; res asst bot, Univ Chicago, 61-64; asst prof biol, Northern Ill Univ, 64-66; RES OFFICER AGR, APPL SCI RES CORP THAILAND, BANGKOK, 66-, RES DIR, AGR PROD RES INST, 71- Mem: Agr Sci Soc Thailand; Sigma Xi; Orchid Soc Thailand; Sci Soc Thailand. Res: Agronomic investigation and plant improvement of kenaf, jute, peanut, sunflower, banana, basil, Japanese mint and other essential oil crops; genetics and cytogenetics research on banana, mint and Ocimum species. Mailing Add: Appl Sci Res Corp 196 Phahonyothin Rd Bang Khen Bangkok 9 Thailand

CHONACKY, NORMAN J, b Cleveland, Ohio, Oct 14, 39; m 64; c 2. PHYSICS. Educ: John Carroll Univ, BS, 61; Univ Wis, PhD(physics), 67. Prof Exp: Asst prof, 67-74, ASSOC PROF PHYSICS, SOUTHERN CONN STATE COL, 74- Mem: Am Asn Physics Teachers. Res: Physics of organic and biologically important materials, including water. Mailing Add: Dept of Physics Southern Conn State Col New Haven CT 06515

CHONG, BERNI PATRICIA, b West Palm Beach, Fla, Aug 29, 45; m 68; c 2. SYNTHETIC ORGANIC CHEMISTRY. Educ: Swarthmore Col, AB, 67; Univ Mich, PhD(org chem), 71. Prof Exp: Fel org synthesis, Cornell Univ, 71-72; fel org chem, Univ Calif, San Diego, 72-73; RES CHEMIST POLLUTION CONTROL, ROHM AND HAAS CO, 74- Mem: Am Chem Soc; The Chem Soc. Res: Synthesis of adsorbants and extractants for pollution control. Mailing Add: Rohm and Haas Co Spring House PA 19477

CHONG, CLYDE HOK HEEN, b Honolulu, Hawaii, Mar 6, 33; m 64; c 3. ANALYTICAL CHEMISTRY, INORGANIC CHEMISTRY. Educ: Wabash Col, BA, 54; Mich State Univ, PhD(chem), 58. Prof Exp: Asst chem, Mich State Univ, 54-57; group leader develop, 61-68, RES SPECIALIST, MOUND LAB, 68- Mem: Am Chem Soc. Res: Some reactions and properties of tetralithium peroxydiphosphate tetrahydrate. Mailing Add: Mound Lab Miamisburg OH 45342

CHONG, DELANO PUN, b Colon, Panama, Nov 21, 36; US citizen PHYSICAL CHEMISTRY. Educ: Univ Calif, Berkeley, BS, 58; Harvard Univ, AM, 59, PhD(chem), 63. Prof Exp: NATO res fel chem, Oxford Univ, 63-64; proj assoc, Univ Wis, 64-65; asst prof, 65-70, ASSOC PROF CHEM, UNIV BC, 70- Res: Theoretical chemistry. Mailing Add: Dept of Chem Univ of BC Vancouver BC Can

CHONG, JOSHUA ANTHONY, b Kingston, Jamaica, May 15, 43; US citizen; m 68; c 2. SYNTHETIC ORGANIC CHEMISTRY. Educ: Univ Calif, Berkeley, BS, 67; Univ Mich, PhD(org chem), 71. Prof Exp: Fel org synthesis, Cornell Univ, 71-74; RES CHEMIST HEALTH PROD, ROHM AND HAAS CO, 74- Mem: Am Chem Soc. Res: Synthesis of physiologically active compounds. Mailing Add: Rohm and Haas Co Spring House PA 19477

CHOONG, ELVIN T, b Jakarta, Indonesia, Oct 20, 32; US citizen; m 60; c 2. WOOD TECHNOLOGY, FORESTRY. Educ: Mont State Univ, BS, 56; Yale Univ, FM, 58; State Univ NY Col Forestry, Syracuse, PhD(wood technol), 62. Prof Exp: Instr wood sci & technol, State Univ NY Col Forestry, Syracuse, 61-63; asst prof, Humboldt State Col, 64-65; from asst prof to assoc prof, 65-73, PROF FORESTRY, LA STATE UNIV, 73- Concurrent Pos: Vis res fel, Commonwealth Sci & Indust Res Orgn, Australia, 72. Mem: Forest Prod Res Soc; Soc Wood Sci & Technol. Res: Physical and mechanical properties of wood and wood products; nondestructive testings of wood; wood drying and preservation; wood quality. Mailing Add: Sch of Forestry La State Univ Baton Rouge LA 70803

CHOONG, HSIA SHAW-LWAN, b China, July 12, 45; m 71; c 1. ORGANIC CHEMISTRY, PROTEIN CHEMISTRY. Educ: Nat Taiwan Univ, BS, 66; Mass Inst Technol, PhD(chem), 71. Prof Exp: Teaching asst chem, Mass Inst Technol, 67-68, res asst, 68-71; sr chemist, Gen Foods Corp, 72-73; MEM TECH STAFF, HEWLETT PACKARD CO, 73- Mem: Am Chem Soc; Sigma Xi. Res: Liquid crystals for displays and electronic devices. Mailing Add: Hewlett Packard Co 1501 Page Mill Rd Palo Alto CA 94304

CHOPOORIAN, JOHN ANDREW, inorganic chemistry, see 12th edition

CHOPPIN, GREGORY ROBERT, b Eagle Lake, Tex, Nov 9, 27; m 51; c 4. INORGANIC CHEMISTRY, NUCLEAR CHEMISTRY. Educ: Loyola Univ La, BS, 49; Univ Tex, PhD(chem), 53; Loyola Univ, DSc, 69. Prof Exp: Mem staff, Radiation

Lab, Univ Calif, 53-56; from asst prof to assoc prof, 56-63, PROF CHEM, FLA STATE UNIV, 63-, CHMN DEPT, 68- Concurrent Pos: Vis scientist, Ctr Study Nuclear Energy, Belgium, 62-63; Fulbright lectr, Uruguay, 65 & Portugal, 69. Mem: AAAS; Am Chem Soc. Res: Nuclear chemistry; physical chemistry of the actinides and lanthanides; structure of water and aqueous solutions Mailing Add: Dept of Chem Fla State Univ Tallahassee FL 32306

CHOPPIN, PURNELL WHITTINGTON, b Baton Rouge, La, July 4, 29; m 59; c 1. VIROLOGY, INTERNAL MEDICINE. Educ: La State Univ, MD, 53; Am Bd Internal Med, dipl. Prof Exp: Intern ward med, Barnes Hosp & Sch Med, Wash Univ, 53-54, asst resident, 56-57; vis investr, 57-59, res assoc, 59-60, from asst prof to assoc prof, 60-70, PROF VIROL & MED, ROCKEFELLER UNIV, 70- Concurrent Pos: Nat Found fel, 57-59; asst physician, Rockefeller Univ Hosp, 57-60, res assoc physician, 60-62, from assoc physician to physician, 62-70, sr physician, 70-; mem virol study sect, NIH, 68-72, chmn, 75- Mem: Am Soc Microbiol; Am Soc Cell Biol; Am Asn Immunol; Am Soc Clin Invest; Am Cancer Soc. Res: Animal virology; myxoviruses; virus multiplication; virus structure. Mailing Add: Rockefeller Univ New York NY 10021

CHOPRA, BALDEO K, b Multan, W Pakistan, Aug 10, 42; m 64; c 1. PLANT PATHOLOGY, MICROBIOLOGY. Educ: Benares Hindu Univ, BSc, 60, MSc, 62; Auburn Univ, PhD(plant path, microbiol), 68. Prof Exp: Res asst plant path, Univ Allahabad, 62-65; res asst, Auburn Univ, 65-68, tech res asst, 68-69; assoc prof biol, Prairie View Agr & Mech Col, 69-73; ASSOC PROF BIOL, J C SMITH UNIV, 73- Concurrent Pos: Proj leader, USDA soybean study grant, Prairie View Agr & Mech Col, 69-72. Mem: Am Phytopath Soc; Am Soc Microbiol. Res: Mycology; taxonomy of Aspergilli and Xylaria; interaction of pesticides and soil microflora; nutritional physiology of Fusarium and Aspergillus flavus. Mailing Add: Dept of Biol J C Smith Univ Charlotte NC 28216

CHOPRA, DEV RAJ, b Jullundur City, Punjab, Apr 14, 30; m 56; c 2. EXPERIMENTAL ATOMIC PHYSICS, SURFACE PHYSICS. Educ: Punjab Univ, MS, 52; Univ Nebr, MA, 60; NMex State Univ, PhD(physics), 64. Prof Exp: Demonstr physics, Punjab Univ, 51-52, lectr, 52-58; asst, Univ Nebr, 58-60; res assoc, NMex State Univ, 60-64; assoc prof, 64-71, res grants, 66, 67-, 70-71, PROF PHYSICS, EAST TEX STATE UNIV, 71- Mem: Am Asn Physics Teachers; Am Phys Soc; Sigma Xi. Res: Soft x-ray spectroscopy; valence band studies of transition and rare-earth elements using photo-absorption and appearance potential spectroscopy techniques. Mailing Add: Dept of Physics East Tex State Univ Commerce TX 75428

CHOPRA, DHARAM PAL, b India, Feb 2, 44; m 68; c 1. CELL BIOLOGY. Educ: Univ Delhi, BS, 63; Univ London, MS, 67; Univ Newcastle, Eng, PhD(cell biol, path), 71. Prof Exp: Res assoc cellular develop biol, Univ Newcastle, Eng, 67-71; asst prof dermat, Skin & Cancer Hosp, Temple Univ, 71-74; SR SCIENTIST CELL BIOL, SOUTHERN RES INST, 74- Concurrent Pos: Nat Cancer Inst & Nat Inst Arthritis & Metab Dis grant, Temple Univ, 71-72. Mem: AAAS; Soc Invest Dermat; Tissue Cult Asn; Am Asn Cell Biol; Brit Soc Develop Biol. Res: Regulation of growth and differentiation in normal and neoplastic tissues. Mailing Add: Kettering-Meyer Lab Southern Res Inst 2000 9th Ave S Birmingham AL 35209

CHOPRA, DHARAM-VIR, b Jullundur, India, Oct 15, 30; m 69. STATISTICS, MATHEMATICS. Educ: Panjab Univ, MA, 53; Univ Mich, MS, 61, MA, 63; Univ Nebr, PhD(statist), 68. Prof Exp: Lectr math, DAV Col, India, 53-59; instr, Univ Nebr, 63-66; asst prof, SC State Col, 66-67; asst prof, 67-71, ASSOC PROF MATH, WICHITA STATE UNIV, 71- Concurrent Pos: Statist consult dent sch, Univ Nebr, 64-65. Mem: Am Statist Asn; Inst Math Statist; Math Asn Am. Res: Design of experiments and their analysis; combinatorial mathematics; application of statistics to social sciences; psychology. Mailing Add: Dept of Math Wichita State Univ Wichita KS 67208

CHOPRA, INDER JIT, b Gujranwala, India, Dec 15, 39; m 66; c 3. ENDOCRINOLOGY, INTERNAL MEDICINE. Educ: All India Inst Med Sci, New Delhi, MB, BS, 61, MD, 65; Am Bd Internal Med, dipl, 72, cert endocrinol, 73. Prof Exp: Intern, All India Inst Med Sci, New Delhi, 62, resident med, 63-65; res officer, Indian Coun Med Res, New Delhi, 66; registr med, All India Inst Med Sci, 66-67; resident med, Queen's Med Ctr, Honolulu, Hawaii, 67-68; fel endocrinol, Harbor Gen Hosp, Sch Med, Univ Calif, Los Angeles, Torrance, 68-71; asst prof, 71-74, ASSOC PROF MED, SCH MED, UNIV CALIF, LOS ANGELES, 74- Concurrent Pos: Staff physician, Harbor Gen Hosp, Torrance, 71-72 & Ctr Health Sci, Univ Calif, Los Angeles, 72-; NIH res career develop award, 72-77 & grant, 72-78. Mem: Sigma Xi; Am Soc Clin Invest; Am Thyroid Asn; Endocrine Soc; fel Am Col Physicians. Res: Thyroid physiology and disease; nature of biologically active thyroid hormones, thyroid hormone metabolism, pathogenesis of Graves' disease, nature of thyroid stimulators, radioimmunoassay, pituitary-thyroid axis. Mailing Add: Dept of Med Ctr Health Sci Univ of Calif Sch of Med Los Angeles CA 90024

CHOPRA, KULDIP P, b Srinagar, Kashmir, Mar 25, 32; m 68. ENVIRONMETNAL PHYSICS, SPACE PHYSICS. Educ: Univ Delhi, BSc, 51, MSc, 53, PhD(physics), 60. Prof Exp: Res assoc & res asst, Univ Md, 57-58; vis asst prof physics & res scientist, Univ Southern Calif, 58-60; res asst prof astronaut, Polytech Inst Brooklyn, 60-63; sr scientist, Melpar Inc, 63-65, head space physics, 64-65; assoc prof atmospheric sci, Univ Miami, 65-67; prof appl physics, Nova Univ, 67-69; PROF PHYSICS, OLD DOM UNIV, 69- Concurrent Pos: Vis prof, Va Inst Marine Sci, 71-; sci consult, NASA Wallops Flight Ctr, 75- Honors & Awards: Melpar Award, 64. Mem: AAAS; fel Am Inst Physics; Am Inst Aeronaut & Astronaut; Am Geophys Union; Am Meteorol Soc. Res: Cosmical magnetism; space vehicles in ionized media; plasma physics and magneto-fluidynamics; low speed aerodynamics; mesometeorology; mathematical physics; urban coastal environment, atmospheric and oceanic flow problems caused by islands. Mailing Add: Dept of Physics Old Dom Univ PO Box 6173 Norfolk VA 23508

CHOPRA, NAITER MOHAN, b Amritsar, India, Nov 23, 23; m 53; c 4. ORGANIC CHEMISTRY. Educ: Punjab Univ, Lahore, BSc, 44, MSc, 45; Trinity Col, Dublin, PhD(chem), 55. Prof Exp: Demonstr chem, Forman Christian Col, Lahore, 46-47; asst lectr, Allahabad Agr Inst, India, 47-49; res fel, Univ Toronto, 55-57; res officer, Can Dept Agr Res Sta, Man, 57-65; assoc prof chem, 65-67, PROF CHEM, NC AGR & TECH STATE UNIV, 67-, DIR TOBACCO RES PROJ, 67- Mem: Am Chem Soc. Res: Chemistry of wheat rust and pesticides; chemistry of viricides and antibiotics which could be employed in cure of cereal disease; breakdown of pesticides in tobacco and cigaret smokes. Mailing Add: Dept of Chem NC Agr & Tech State Univ Greensboro NC 27411

CHOQUETTE, PHILIP WHEELER, b Utica, NY, Aug 16, 30; m 59; c 2. GEOLOGY. Educ: Allegheny Col, BS, 52; Johns Hopkins Univ, MA, 54, PhD(geol), 57. Prof Exp: Geologist, US Geol Surv, 56-58; from res geologist to adv res geologist, 58-72, SR RES GEOLOGIST, MARATHON OIL CO, 72- Mem: AAAS; Geol Soc Am; Am Asn Petrol Geologists; Soc Econ Paleontologists & Mineralogists; Am Inst Prof Geologists. Res: Physical stratigraphy, petrology and geochemistry of sedimentary carbonate rocks; diagenesis and porosity in limestones and dolomites. Mailing Add: Denver Res Ctr Marathon Oil Co PO Box 269 Littleton CO 80120

CHORBAJIAN, TORCOM, plant physiology, see 12th edition

CHORNOCK, FRANCIS WILLIAM, b Munkacs, Austria-Hungary, Nov 3, 14; m 38; c 2. BIOCHEMISTRY. Educ: Pa State Univ, BS, 35; MS, 37, PhD(biochem), 40. Prof Exp: Asst chemist, Med Clin, Univ Pa, 39-42; chief chemist, Bryn Mawr Hosp, 41-45, res chemist, Off Sci Res & Develop Proj, 42-45; res chemist, Swift & Co, 46-52; res biochemist, Res Lab, Commercial Solvents Corp, Ind, 52-59; clin biochemist, Hertzler Clin, 59-61; PROF BIOCHEM & CHMN DEPT, KIRKSVILLE COL OSTEOP MED, 61-, DIR CLIN LAB, KIRKSVILLE OSTEOP HOSP, 68- Mem: AAAS; Am Chem Soc; Am Asn Clin Chem; NY Acad Sci. Res: Clinical biochemistry; nutritional biochemistry; limitation of certain methods used in the estimation of vitamin A and its precursors; a plasma substitute from red blood cells. Mailing Add: Dept of Biochem Kirksville Col of Osteop Med Kirksville MO 63501

CHORPENNING, FRANK WINSLOW, b Marietta, Ohio, Aug 17, 13; m 42; c 4. MICROBIOLOGY, IMMUNOBIOLOGY. Educ: Marietta Col, AB, 39; Ohio State Univ, MSc, 50, PhD(microbiol), 63. Prof Exp: Admin asst typhus comn, US Army, Philippines & Japan, 45-46, bacteriologist, 4th Army Area Lab, 48-49; serologist & dir blood bank, Med Lab, Ger, 52-55; chief clin path & dir blood bank, Brooke Army Hosp, 55-61; from asst prof to assoc prof, 63-73, PROF MICROBIOL & IMMUNOL, OHIO STATE UNIV, 73- Concurrent Pos: Mem coop study group, WHO, 53-55. Mem: AAAS; fel Am Acad Microbiol; Am Soc Microbiol; Am Asn Immunol. Res: Naturally occuring antibodies and development of natural immunity; immunochemical specificity of bacterial antigens. Mailing Add: Dept of Microbiol Ohio State Univ Columbus OH 43210

CHORTYK, ORESTES T, organic chemistry, see 12th edition

CHORVAT, ROBERT JOHN, b Chicago, Ill, Aug 16, 42; m 64; c 3. MEDICINAL CHEMISTRY. Educ: Ill Benedictine Col, BS, 64; Ill Inst Technol, PhD(org Chem), 68. Prof Exp: SR RES CHEMIS, G D SEARLE & CO, 68- Mem: Am Chem Soc. Res: Four-membered, single phosphorus atom heterocycles; synthesis; chemistry; physical properties; synthesis of steroids and nucleo-hetero steroids. Mailing Add: G D Searle & Co PO Box 5110 Chicago IL 60680

CHOSY, JULIUS J, b Columbus, Ohio, May 20, 31; m 56. INTERNAL MEDICINE, PSYCHOSOMATIC MEDICINE. Educ: Ohio State Univ, BA, 52, MD, 55. Prof Exp: Asst prof, 64-70, ASSOC PROF MED, UNIV WIS-MADISON, 70- Concurrent Pos: Fel hemat, 61-62; NIH spec res fel, 62-64, career develop award, 64-69. Mem: Am Psychosom Soc; Soc Psychophysiol Res; Am Fedn Clin Res. Res: Psychophysiology, particularly catecholamines in relation to anxiety, depression, and performance effectiveness. Mailing Add: 733 Seneca Pl Madison WI 53711

CHOU, CHEN-LIN, b Kiangsu, China, Oct 8, 43; m 70; c 1. GEOCHEMISTRY, METEORITICS. Educ: Nat Taiwan Univ, BS, 65; Univ Pittsburgh, PhD(geochem), 71. Prof Exp: Scholar geochem, Univ Calif, Los Angeles, 71-72, asst res geochemist, 72-75; sr res assoc geol, 75-76, ASST PROF GEOL, UNIV TORONTO, 76- Concurrent Pos: Lectr earth sci, Calif State Univ, Fullerton, 73-74. Mem: Geochem Soc; Meteoritical Soc; Am Geophys Union; Geol Soc Am; Int Asn Geochem & Cosmochem. Res: Elemental abundances and origin of meteorites; chemistry of lunar samples; geochemistry of Archean rocks; neutron activation analysis. Mailing Add: Dept of Geol Univ of Toronto Toronto ON Can

CHOU, CHING-CHUNG, b Taipei, Taiwan, June 25, 32; m 62; c 3. PHYSIOLOGY, INTERNAL MEDICINE. Educ: Nat Taiwan Univ, BM, 58; Northwestern Univ, MS, 64; Univ Okla, PhD(physiol), 66. Prof Exp: Intern, Nat Taiwan Univ Hosp, Taipei, 57-58; resident radiol, Chinese Air Force Gen Hosp, 59-60; intern, Washington Hosp Ctr, DC, 60-61; resident internal med, Northwestern Univ, 61-64; instr physiol, Univ Okla, 65-66; from asst prof to assoc prof, 66-73, PROF PHYSIOL & MED, MICH STATE UNIV, 73- Mem: Am Fedn Clin Res; NY Acad Sci; Am Physiol Asn; Am Gastroenterol Asn. Res: Gastroenterology; gastrointestinal physiology; cardiovascular physiology. Mailing Add: Dept of Physiol Mich State Univ East Lansing MI 48823

CHOU, DAVID YUAN PIN, b Shantung, China, Mar 5, 22; US citizen; m 53; c 3. PHYSICAL CHEMISTRY. Educ: Tokyo Inst Technol, BE, 48; Ohio State Univ, PhD(soil chem), 54. Prof Exp: Assoc prof chem, St Augustine's Col, 54-56; GLENN FRYE PROF CHEM & HEAD DEPT, LENOIR RHYNE COL, 56- Concurrent Pos: Am Chem Soc Petrol Res Fund fac award advan sci study, 63; res fel, Univ Kans, 64-65. Honors & Awards: R M Bost Distinguished Prof Award, 71. Mem: AAAS; Am Chem Soc; fel Am Inst Chem. Res: Nonaqueous solvents; phase equilibria; colloidal properties of silicates. Mailing Add: 768 Eighth St N E Hickory NC 28601

CHOU, SHELLEY NIEN-CHUN, b Chekiang, China, Feb 6, 24; nat US; m 56; c 3. NEUROSURGERY. Educ: St John's Univ, China, BS, 46; Univ Utah, MD, 49; Univ Minn, MS, 54, PhD, 64. Prof Exp: Intern, Providence Hosp, 49-50; univ fel, Univ Minn, 50-53, res fel, 53-55, AEC grant, 54-55; clin asst neurosurg, Univ Utah, 56-58; vis scientist, NIH, 59-60; from instr to assoc prof, 60-68, PROF NEUROSURG, MED SCH, UNIV MINNEAPOLIS, 68-, HEAD DEPT, 74- Concurrent Pos: Mem, Am Bd Neurol Surg, 74- Mem: Soc Neurol Surg; Am Col Surg; Soc Nuclear Med; Neurosurg Soc Am; Am Acad Neurol Surg. Res: Experimental neurosurgery and neurophysiology; isotopic tracer investigation in neurophysiology. Mailing Add: Dept of Neurosurg Univ of Minn Med Sch Minneapolis MN 55455

CHOU, SHIH-TOON, veterinary medicine, biochemistry, see 12th edition

CHOU, TING-CHAO, b Taiwan, Sept 9, 38; m 65; c 2. PHARMACOLOGY, CHEMOTHERAPHY. Educ: Kaohsiung Med Col, Taiwan, BS, 61; Nat Taiwan Univ, MS, 65; Yale Univ, PhD(pharmacol), 70. Prof Exp: Teaching asst pharmacol, Col Med, Nat Taiwan Univ, 64-65; res asst, Yale Univ, 65; fel, Sch Med, Johns Hopkins Univ, 69-72; ASSOC PHARMACOL, SLOAN-KETTERING INST CANCER RES, 72-; ASST PROF PHARMACOL, SLOAN-KETTERING DIV, GRAD SCH MED SCI, CORNELL UNIV, 73- Mem: AAAS; Am Asn Cancer Res; Am Soc Pharmacol Exp Therapeut. Res: Pharmacology and biochemistry of cancer chemotherapeutic agents; enzyme kinetics; theoretical biology of dose-effect relationships. Mailing Add: Sloan-Kettering Inst Cancer Res Lab of Pharmacol 410 E 68th St New York NY 10021

CHOU, TSU-TEH, b Shanghai, China, Mar 10, 34. ELEMENTARY PARTICLE PHYSICS, HIGH ENERGY PHYSICS. Educ: Nat Taiwan Univ, BS, 56; Tsing Hua Univ, MS, 58; Univ Iowa, PhD, 65. Prof Exp: Res assoc, State Univ NY Stony Brook, 65-70; asst prof physics, Univ Denver, 70-74; ASST PROF PHYSICS, UNIV GA, 74- Mem: Am Phys Soc. Mailing Add: Dept of Physics Univ of Ga Athens GA 30602

CHOUDARY, JASTI BHASKARARAO, b Jalipudi, India, Jan 15, 33; m 73; c 2. REPRODUCTIVE PHYSIOLOGY, ENDOCRINOLOGY. Educ: Madras Vet Col,

BVSc, 54; Kans State Univ, MS, 64, PhD, 66. Prof Exp: Vet surgeon, India, 54-61; res asst reproductive physiol, Kans State Univ, 63-66; Ford Found fel, Univ Kans Med Ctr, Kansas City, 66-68; sect head reproductive physiol, William S Merrell Co Div, Richardson-Merrell, Inc, 68-71; RES SCIENTIST ENDOCRINOL & METAB REGULATION, G D SEARLE & CO, 71- Mem: Soc Study Reproduction; Endocrine Soc; Am Fertil Soc; Am Asn Anat; NY Acad Sci. Res: In vitro storage of spermatozoa; pituitary-ovarian relationships; ovarian follicular development and atresia; luteotropic mechanisms; luteolytic mechanisms; control of fertility; drugs; endocrinology of pregnancy; biology of gonadal steroids; biosynthesis of progesterone; radioimmunoassays of steroid hormones. Mailing Add: Searle Labs PO Box 5110 Chicago IL 60680

CHOUDHURY, ABDUL LATIF, b Dacca, Bangladesh, Jan 1, 33; m 60; c 2. THEORETICAL HIGH ENERGY PHYSICS. Educ: Univ Dacca, BS, 53, MS, 54; Free Univ Berlin, PhD(theoret physics), 60. Prof Exp: Asst to prof physics, Univ Dacca, 55, sr lectr, 61-66 & 68, reader, 69; asst to prof theoret physics, Free Univ Berlin, 58-60; res fel, Fritz-Haber Inst Ger, 60; Brit Coun-Colombo Plan res asst, Imp Col, Univ London, 60-61; consult, Nat Bur Standards-Univ Dacca proj, 63-66; assoc prof, 66-73, PROF PHYSICS & MATH, ELIZABETH CITY STATE UNIV, 73- Concurrent Pos: Vis physicist, Int Ctr Theoret Physics, Italy, 68. Mem: Am Phys Soc; Am Asn Univ Prof. Res: Quantum field theory; atomic physics; symmetry principles and group theoretical approach to particle physics; magnetic properties of the proposed charmed particles. Mailing Add: Dept of Phys Sci & Math Box 287 Elizabeth City State Univ Elizabeth City NC 27909

CHOUDHURY, DEO C, b Darbhanga, India, Feb 1, 27; m 63; c 1. NUCLEAR PHYSICS. Educ: Univ Calcutta, BS, 45, MS, 47; Univ Calif, Los Angeles, PhD(physics), 59. Prof Exp: Asst physics, Univ Rochester, 55-56 & Univ Calif, Los Angeles, 56-59; asst prof, Univ Conn, 59-62; assoc prof, 62-67, PROF PHYSICS, POLYTECH INST NY, 67- Mem: Am Phys Soc; NY Acad Sci; Indian Phys Soc. Res: Theoretical investigations of nuclear structure and reactions; theory of B-decay; theory of strong interactions in nuclear physics. Mailing Add: Dept of Physics Polytech Inst of NY 333 Jay St Brooklyn NY 11201

CHOUDRY, AMAR, b New Delhi, India, Sept 8, 38; m 67. NUCLEAR PHYSICS. Educ: Univ Delhi, BSc, 56, MSc, 58; Columbia Univ, PhD(physics), 67. Prof Exp: Asst physics, Univ Colo, 60-61 & Columbia Univ, 61-67; ASST PROF PHYSICS, UNIV RI, 67- Concurrent Pos: Fac res fels, Argonne Nat Lab, 68 & Univ RI Res Comt, 58-59; NASA res grant on sec electron conduction. Comput systs consult, Celanese Corp, NY, 67-68; electronic systs consult, Enthone, Inc, 69- Mem: AAAS; Am Phys Soc. Res: High energy and neutron physics; radiation damage; application of computers to scientific problems. Mailing Add: Dept of Physics Univ of RI Kingston RI 02881

CHOUKAS, NICHOLAS C, b Chicago, Ill, Sept 5, 23; m 51; c 5. ORAL SURGERY. Educ: Loyola Univ, Ill, DDS, 50, MS, 58; Am Bd Oral Surg, dipl, 61. Prof Exp: Fel, 53, from instr to assoc prof oral surg, 56-70, assoc prof oral biol & chmn dept oral surg, 63-69, PROF ORAL BIOL, GRAD SCH & PROF ORAL & MAXILLOFACIAL SURG, SCH DENT, LOYOLA UNIV CHICAGO, 69- Concurrent Pos: Attend oral surgeon, Hines Vet Admin Hosp, 58-60, consult, 60-; NIH teachers training grant, 61, res grant, 61-64; chief oral surg, Loyola Univ Hosp, 69- Mem: Am Soc Oral Surg; fel Int Asn Oral Surg; fel Am Col Dent; fel Int Col Dent; fel Am Col Stomatologic Surgeons. Res: Growth changes of the temporomandibular joint and mandible of the Macca rhesus monkey following various experimentally induced environments. Mailing Add: Sch of Dent Loyola Univ Med Ctr Maywood IL 60153

CHOULES, GEORGE LEW, b Salt Lake City, Utah, Mar 5, 33; m 60; c 3. BIOCHEMISTRY. Educ: Utah State Agr Col, BS, 55, MS, 57; Johns Hopkins Univ, PhD(biochem), 64. Prof Exp: Fel protein chem, Univ Calif, San Diego, 64-67; ASST RES PROF BIOL, UNIV UTAH, 67- Mem: AAAS; Biophys Soc; Am Soc Microbiol. Res: Protein chemistry; nitrogen fixation; subcellular particles; structure of antibodies; structure of membranes. Mailing Add: Dept of Biol Univ of Utah Salt Lake City UT 84112

CHOULIS, NICOLAS HELIAS, b Athens, Greece, Nov 17, 30; m 67; c 2. PHARMACEUTICAL CHEMISTRY, PHARMACEUTICS Educ: Athens Tech, BPharm, 59; Univ London, PhD(pharm chem), 64. Prof Exp: Vis teacher pharm chem, Chelsea Sch Pharm, London, 64; res assoc, Univ Kans, 64-65; asst prof pharm, Tex Southern Univ, 65-66, assoc prof, 66-67 & 68-70; dir res & develop & qual control, Minerva Pharmaceut Indust, 67-68; assoc prof, 70-74, PROF PHARMACEUT CHEM & PHARMACEUT, MED CTR, WVA UNIV, 74- Concurrent Pos: Mead Johnson Labs res award, 66-67; mem, US Pharmacopeia; consult, World Health Orgn, 72. Mem: AAAS; Am Chem Soc; Am Pharmaceut Asn; Am Inst Chem; The Chem Soc. Res: Analytical and medicinal chemistry; chromatographic procedures; drug abuse control. Mailing Add: Dept of Pharm WVa Univ Med Ctr Morgantown WV 26506

CHOVER, JOSHUA, b Detroit, Mich, Mar 26, 28; m 52. MATHEMATICS. Educ: Univ Mich, PhD(math), 52. Prof Exp: Res mathematician, Bell Tel Labs, 52-56; from instr to assoc prof, 56-65, PROF MATH, UNIV WIS-MADISON, 65- Concurrent Pos: Mem, Inst Advan Study, 55-56. Mem: Am Math Soc. Res: Probability and analysis. Mailing Add: Dept of Math Univ of Wis Madison WI 53706

CHOVITZ, BERNARD H, b Norfolk, Va, Nov 10, 24; m 49; c 3. GEODESY, CARTOGRAPHY. Educ: Col William & Mary, BS, 44; Harvard Univ, MA, 47. Prof Exp: Mathematician, Army Map Serv, 48-60, Geod Intel, Mapping Res & Develop Agency, 60-61; prin scientist, Autometric Oper, Raytheon Co, 61-64; geodesist, Off Res & Develop, US Coast & Geod Surv, 64-65; geodesist, Environ Sci Serv Admin, 65-70; geodesist, 70-74, DIR GEOD RES & DEVELOP LAB, NAT OCEANIC & ATMOSPHERIC ADMIN, 74- Concurrent Pos: Pres sect II, Int Asn Geod, 75-79. Honors & Awards: Meritorious Civilian Serv Award, US Dept Army, 57. Mem: Am Math Asn; Am Geophys Union; Int Asn Geod. Res: Mathematical analysis of map projections; determination of size of earth; application of artificial satellites to the determination of the earth's gravitational field and its size and shape. Mailing Add: 8813 Clifford Chevy Chase MD 20015

CHOVNICK, ARTHUR, b New York, NY, Aug 2, 27; m 49; c 2. GENETICS. Educ: Ind Univ, AB, 49, MA, 50; Ohio State Univ, PhD(genetics), 53. Prof Exp: Asst physiol genetics, Ohio State Univ, 51-53; instr zool, Univ Conn, 53-57, asst prof genetics, 57-59; asst dir, Long Island Biol Asn, 59-60, dir, 60-62; PROF GENETICS, UNIV CONN, 62- Concurrent Pos: Assoc ed, Genetics, 72-76; mem genetics study sect, Div Res Grants, NIH, 72-76. Mem: AAAS; Genetics Soc Am; Am Genetic Asn; Am Soc Naturalists; Am Soc Cell Biol. Res: Gene structure, function, mutation, mechanism of recombination. Mailing Add: Dept of Genetics & Cell Biol Univ of Conn Storrs CT 06268

CHOW, ALFRED WEN-JEN, b Peiping, China, Jan 11, 24; m 57. MEDICINAL CHEMISTRY, ORGANIC CHEMISTRY. Educ: Univ Mich, BS, 46; Univ Minn, PhD(med chem), 50. Prof Exp: Res chemist, Bjorksten Res Lab, 51 & Kremers Urban Co, 52-56; sr res chemist, Pabst Res Lab, 56-59; sr med chemist, Smith Kline & French Lab, 59-71, SR INVESTR, SMITH KLINE CORP, 71- Mem: AAAS; Am Chem Soc; The Chem Soc. Res: Preparation of organic compounds of antimicrobial and antiviral interest; synthetic organic chemistry; animal health products; ruminant nutrition. Mailing Add: Smith Kline Corp 1600 Paoli Pike West Chester PA 19380

CHOW, ARTHUR, b Meaford, Ont, Oct 23, 36; m 63. ANALYTICAL CHEMISTRY. Educ: Univ Toronto, BSc, 61, MA, 62, PhD(anal chem), 66. Prof Exp: Lectr chem, Univ Toronto, 62-64, instr, 64-66; Nat Res Coun Can fel, 66-68; asst prof chem, 68-72, ASSOC PROF CHEM, UNIV MAN, 72- Mem: Am Chem Soc; Chem Inst Can. Res: Analytical chemistry of the noble metals; activation analysis, separation and isolation of the noble metals; analysis of inorganic pollutants. Mailing Add: Dept of Chem Univ of Manitoba Winnipeg MB Can

CHOW, BACON FIELD, nutrition, deceased

CHOW, BRYANT, b Peking, China, Dec 24, 36; nat US; m 59; c 2. MATHEMATICS, STATISTICS. Educ: Franklin & Marshall Col, BS, 59; Rutgers Univ, MS, 61; Va Polytech Inst, PhD(statist), 66. Prof Exp: Instr statist, Va Polytech Inst, 64-65; asst prof appl math statist, Rutgers Univ, 65-69; ASSOC PROF MATH, UNIV SOUTHWESTERN LA, 69- Mem: Am Soc Qual Control; Am Statist Asn; Biomet Soc. Res: Non-parametric statistics; biometrics. Mailing Add: Dept of Math Univ of Southwestern La Lafayette LA 70501

CHOW, CHE CHUNG, b Shanghai, China, Feb 6, 35; m 60. PHYSICAL CHEMISTRY. Educ: Univ Hong Kong, BSc, 58; Brown Univ, PhD(chem), 64. Prof Exp: Asst chem, Brown Univ, 58-63; res chemist, Cent Res Dept, E I du Pont de Nemours & Co, Inc, Del, 63-69, chemist, East Lab, NJ, 69-70; SCIENTIST, XEROGRAPHIC TECHNOL DEPT, XEROX CORP, 70- Mem: Am Chem Soc. Res: Chemical reactions in shock waves; molecular energy transfer; growth kinetics of inorganic fibers; electrical and magnetic properties of solids; magneto-optics; laser-magnetic memories and imaging systems; coagulation mechanism of polymer solutions; mechanism of detonation in solid explosives; imaging and reproduction systems. Mailing Add: Xerographic Technol Dept Xerox Corp 800 Phillips Rd W147 Webster NY 14580

CHOW, CHING KUANG, b Chiayi, Taiwan, May 7, 40; m 67; c 2. NUTRITIONAL BIOCHEMISTRY. Educ: Nat Taiwan Univ, BS, 63; Univ Ill, Urbana, MS, 66, PhD(nutrit sci), 69. Prof Exp: From res asst to res assoc nutrit biochem, Univ Ill, Urbana, 64-71; assoc res scientist biochem, NY Univ Med Ctr, 69-70; Nat Vitamin Found fel biochem, 71-72, ASST RES BIOCHEMIST, UNIV CALIF, DAVIS, 72- Mem: Am Inst Nutrit; NY Acad Sci. Res: Metabolism and function of vitamin E; intrinsic antioxidant defense mechanism; nutritional and environmental stresses on normal and pathological processes. Mailing Add: Calif Primate Res Ctr Univ of Calif Davis CA 95616

CHOW, CHRISTOPHER N, b Nanking, China, Dec 23, 46. ELECTROOPTICS. Educ: Calif Lutheran Col, BA, 68; Univ Minn, PhD(physics), 74. Prof Exp: SR PHYSICIST ELECTROOPTICS, 3M CO, 74- Res: Electrophotography; electro-optical information processing and display; laser technology. Mailing Add: 3M Ctr 235-2E St Paul MN 55101

CHOW, FU HO CHEN, b Hangchow, China, Apr 26, 16; US citizen; m 50. BIOCHEMISTRY, PHYSICAL CHEMISTRY. Educ: Ginling Col, China, BS, 38; Mich State Univ, MS, 50; Colo State Univ, PhD(chem, biochem), 57. Prof Exp: Teacher, World High Sch, China, 38-46; instr chem, Ginling Col, China, 46-48; asst enzyme biochem, Univ Wis, 51-53; from asst chemist to assoc chemist, 57-73, CHEMIST, DEPT PATH, COLO STATE UNIV, 73- Mem: Am Chem Soc. Res: Effect of chemical and physical conditions on association or dissociation of polyelectrolytes in urine and its correlation with formation of urinary calculi; dietary manipulation in preventing urinary calculi formation. Mailing Add: Dept of Path Colo State Univ Ft Collins CO 80521

CHOW, KAO LAING, b Tientsin, China, Apr 21, 18; US citizen; m 64. NEUROSCIENCES, NEUROANATOMY. Educ: Yenching Univ China, BA, 43; Harvard Univ, PhD(psychol), 50. Prof Exp: Asst, Yerkes Labs Primate Biol, 47-54; asst prof physiol, Univ Chicago, 54-60; assoc prof med, 60-65, PROF NEUROL, MED SCH, STANFORD UNIV, 65- Concurrent Pos: Mem, Int Brain Res Orgn. Mem: AAAS; Am Physiol Soc; Soc Neurosci. Res: Neurophysiology of learning and vision; neuroanatomy of vision and central nervous system. Mailing Add: Dept of Neurol Stanford Univ Sch Med Palo Alto CA 94305

CHOW, LAURENCE CHUNG-LUNG, b Taipei, Taiwan, Feb 8, 43; US citizen; m 67; c 3. PHYSICAL CHEMISTRY, DENTAL RESEARCH. Educ: Chen-Kung Univ, Taiwan, BS, 64; Georgetown Univ, PhD(chem), 70. Prof Exp: Res assoc, 69-75, CHIEF RES SCIENTIST, DENT CHEM DIV, AM DENT ASN HEALTH FOUND RES UNIT, NAT BUR STANDARDS, 76- Mem: Am Chem Soc; Int Asn Dent Res. Res: Dental research; dental caries prevention; topical fluoridation of teeth; thermodynamics; solution theories; transport of ions through membranes. Mailing Add: Am Dent Asn Health Found Res Unit Nat Bur of Standards Washington DC 20234

CHOW, LOUISE TSI, b Hunan, China, Sept 30, 43; m 74. MOLECULAR BIOLOGY. Educ: Nat Taiwan Univ, BS, 65; Calif Inst Technol, PhD(chem), 73. Prof Exp: Res fel biochem, Univ Calif Med Ctr, San Francisco, 73-74; res fel chem, Calif Inst Technol, 74-75; RES FEL ELECTRON MICROS, COLD SPRING HARBOR LAB, 75- Res: Electron microscope studies of gene arrangements in bacterial and viral chromosomes. Mailing Add: Cold Spring Harbor Lab PO Box 100 Cold Spring Harbor NY 11724

CHOW, PAO LIU, b Fukien, China, Nov 28, 36; m 65; c 2. APPLIED MECHANICS, APPLIED MATHEMATICS. Educ: Cheng Kung Univ, Taiwan, BS, 59; Rensselaer Polytech Inst, MS, 64, PhD(mech), 67. Prof Exp: Asst prof math, Rensselaer Polytech Inst, 66-67; asst prof, NY Univ, 67-74; ASSOC PROF MATH, WAYNE STATE UNIV, 74- Mem: Soc Indust & Appl Math. Res: Elasticity; fluid dynamics; wave propagation in random medium. Mailing Add: Dept of Math Wayne State Univ Detroit MI 48202

CHOW, PAUL C, b Peking, China, Aug 1, 26; m 65; c 3. SOLID STATE PHYSICS, THEORETICAL PHYSICS. Educ: Univ Calif, Berkeley, BA, 60; Northwestern Univ, PhD(physics), 65. Prof Exp: Res assoc physics, Northwestern Univ, 65; res assoc, Univ Southern Calif, 65-66, asst prof, 66-67; res scientist, Univ Tex, 67-68; ASSOC PROF PHYSICS, CALIF STATE UNIV, NORTHRIDGE, 68- Concurrent Pos: Vis asst prof, Univ Tex, 70. Mem: Am Phys Soc. Res: Theoretical solid state physics. Mailing Add: Dept of Physics & Astron Calif State Univ Northridge CA 91324

CHOW, RICHARD H, b Vancouver, BC, Sept 6, 24; nat US; m 48; c 5. PHYSICS.

Educ: Univ BC, BA, 47, MA, 49; Univ Calif, Los Angeles, PhD(nuclear physics), 55. Prof Exp: Asst physics, Univ BC, 46-49; from asst to assoc, Univ Calif, Los Angeles, 50-54; asst res officer reactor physics, Atomic Energy Can, Ltd, 54-58; from asst prof to assoc prof, 58-65, PROF PHYSICS, CALIF STATE UNIV, LONG BEACH, 65- Concurrent Pos: Richland fac appointment, NWCol & Univ Asn Sci, 68-69. Mem: Am Asn Physics Teachers; Am Phys Soc; Am Nuclear Soc. Res: Reactor physics; nuclear physics. Mailing Add: Dept of Physics & Astron Calif State Univ Long Beach CA 90840

CHOW, SUI-WU, b Nanking, China, Nov 12, 26; US citizen; m 53; c 2. ORGANIC CHEMISTRY. Educ: St John's Univ, Minn, BA, 49; Duquesne Univ, MS, 51; Mass Inst Technol, PhD, 54. Prof Exp: Res chemist, Atlantic Ref Co, 54-56; res assoc, Univ Ill, 56-57; RES SCIENTIST CHEM & PLASTICS, UNION CARBIDE CORP, 57- Mem: Am Chem Soc. Res: Organic and polymer chemistry. Mailing Add: Bound Brook Tech Ctr R&D Union Carbide Corp River Rd Bound Brook NJ 08805

CHOW, TAI-LOW, b Yencheng, China, May 18, 37; m 63; c 2. ASTROPHYSICS. Educ: Nat Taiwan Univ, BS, 58; Case Western Reserve Univ, MS, 63; Univ Rochester, PhD(physics, astron), 70. Prof Exp: Instr physics, Cheng Kung Univ, 60-61; asst prof, 69-73, ASSOC PROF PHYSICS & CHMN DEPT, CALIF STATE COL, STANISLAUS, 73- Mem: AAAS; Am Phys Soc; Am Astron Soc; Am Asn Physics Teachers. Res: Scattering of high-energy electrons by nucleons; x-ray and gamma-ray astronomy; electron gas in ultraintense magnetic fields and the astrophysical applications; high-velocity neutral hydrogen gases at high galactic latitudes; interstellar medium; pulsar and gravitation radiations Mailing Add: Dept of Physics Calif State Col Turlock CA 95380

CHOW, TSAIHWA JAMES, b Shanghai, China, Oct 13, 24; US citizen. ANALYTICAL CHEMISTRY, GEOCHEMISTRY. Educ: Nat Chiaotung Univ, BS, 46; Wash State Univ, MS, 49; Univ Wash, PhD(anal chem), 53. Prof Exp: Instr, Nat Chiaotung Univ, 46-47; res asst, Univ Wash, 50-52, asst, 52-53, res assoc oceanog, 53-55; res fel geochem, Calif Inst Technol, 55-60; MEM STAFF, SCRIPPS INST OCEANOG, 60- Concurrent Pos: Vis researcher, Royal Inst Technol, Sweden, 62-63. Mem: AAAS; NY Acad Sci; Am Chem Soc; Am Geophys Union. Res: Microanalytical chemistry; mass spectrometry; chemical oceanography; trace elements in the sea; strontium-calcium ratio in marine organisms; geochemistry of lead isotopes; geochronology; solution chemistry; lead pollution. Mailing Add: Scripps Inst of Oceanog Univ of Calif La Jolla CA 92037

CHOW, TSENG YEH, b Shanghai, China, July 22, 21. MATHEMATICS Educ: Nat Chiao-Tung Univ, BS, 42; Mass Inst Technol, SM, 48; Cornell Univ, PhD(math), 53. Prof Exp: From asst prof to assoc prof math, Rensselaer Polytech Inst, 52-63; assoc prof, 64-67, PROF MATH, CALIF STATE UNIV, SACRAMENTO, 67- Mem: Am Math Soc; Math Asn Am. Res: Mathematical analysis; applied mathematics. Mailing Add: Dept of Math Calif State Univ Sacramento CA 95819

CHOW, TSU LING, b China, Apr 4, 15; nat US; m 50. ANIMAL VIROLOGY. Educ: Nat Cent Univ, China, BVS, 40; Mich State Univ, PhD(animal path), 50. Prof Exp: Asst, Nat Cent Univ, China, 40-42; vet, Ministry Agr & Forestry, 42-45; res assoc, Univ Wis, 50-53; assoc prof path & bact, Sch Vet Med, 54-69, PROF PATH & BACT, SCH VET MED, COLO STATE UNIV, 69- Mem: Am Vet Med Asn; Am Soc — Microbiol; Conf Res Workers Animal Dis; NY Acad Sci; Int Acad Path. Res: Virus diseases of domestic animals such as rinderpest, contagious ecthyma, vesicular stomatitis; infectious bovine rhinotracheitis; blue tongue. Mailing Add: 923 Valleyview Rd Ft Collins CO 80521

CHOW, TSU-SEN, b China, Nov 8, 39; US citizen; m 67; c 1. APPLIED MATHEMATICS, MECHANICS. Educ: Cheng Kung Univ, Taiwan, BS, 62; Rensselaer Polytech Inst, MS, 66; Carnegie-Mellon Univ, PhD(math, mech), 68. Prof Exp: Teaching asst thermodyn, Nat Cheng Kung Univ, 63-64; res assoc polymer/composite, Univ NC, Chapel Hill, 68-72; SCIENTIST MAT, WILSON CTR TECHNOL, XEROX CORP, 72- Mem: Am Phys Soc; Soc Rheology; Am Acad Mech. Res: Mechanical and thermal properties of polymers and composites; rheology and stability of disperse systems; adhesion and surface sciences; statistical and continuous mechanics. Mailing Add: Wilson Ctr Technol Xerox Corp 800 Phillips Rd Webster NY 14580

CHOW, WEN MOU, b Peiping, China, Apr 2, 19; m 51; c 2. COMPUTER SCIENCE. Educ: Chiao-Tung Univ, Shanghai, BS, 41; Mass Inst Technol, MS, 42, DSc, 45. Prof Exp: Asst prof, Lafayette Col, 47-51 & Cath Univ, 51-52; chem engr, Calco Chem Div, Am Cyanamid Co, 45-47 & 52-56; statistician, Union Carbide Plastics Co, 56-68; Mead Johnson prof mgt & chmn dept, Univ Evansville, 68-69; PROF QUANT METHODS, CALIF STATE UNIV, FULLERTON, 69- Mem: Asn Comput Mach; Inst Mgt Sci; Opers Res Soc Am; Am Statist Asn. Res: Applied mathematics; digital computers; applied statistics; operations research. Mailing Add: Dept of Quant Methods Calif State Univ Fullerton CA 92634

CHOW, YUAN LANG, b Formosa, May 28, 29; m 58; c 2. ORGANIC CHEMISTRY Educ: Nat Taiwan Univ, BSc, 51; Duquesne Univ, PhD(chem), 57. Prof Exp: Engr, Chinese Petrol Co, 51-54; res fel, Ill Inst Technol, 57-58; res fel, Royal Inst Technol, Sweden, 58-59, asst lectr, 59-61; res assoc, Imp Col, Univ London, 61-62; lectr, Univ Singapore, 62-63; asst prof org chem, Univ Alta, 63-66; assoc prof, 66-69, PROF CHEM, SIMON FRASER UNIV, 69- Mem: Fel Chem Inst Can; The Chem Soc; Am Chem Soc. Res: Photochemistry in solution, synthetic and mechanistic studies of free radical reaction; carcinogen chemistry. Mailing Add: Dept of Chem Simon Fraser Univ Burnaby BC Can

CHOW, YUAN SHIH, b Hupeh, China, Sept 1, 24; m 63; c 3. MATHEMATICS. Educ: Chekiang, China, BS, 49; Univ Ill, MA, 55, PhD(math), 58. Prof Exp: Res assoc, Univ Ill, 58-59; staff mathematician, Inst Bus Mach Corp, 59-62; vis assoc prof statist, Columbia Univ, 62-63; from assoc prof to prof, Purdue Univ, 63-68; PROF STATIST, COLUMBIA UNIV, 68- Mem: Am Math Soc; Soc Indust & Appl Math; Inst Math Statist. Res: Probability; Fourier series. Mailing Add: Dept of Math Statist Columbia Univ New York NY 10027

CHOW, YUTZE, b Shanghai, China, Sept 29, 27; m 54; c 3. THEORETICAL PHYSICS, MATHEMATICS. Educ: Chin-Kung Univ, Taiwan, 52; Univ RI, MS, 57; Brandeis Univ, MA, 64, PhD(physics), 65. Prof Exp: Res asst elec eng, Univ RI, 55-57; asst prof, Inst Tech Aeronaut, Brazil, 57-61; asst prof elec eng, Univ Waterloo, 61-62; teaching asst physics, Brandeis Univ, 62-63, res asst, 63-64; from asst prof to assoc prof, 64-67, PROF PHYSICS, UNIV WIS-MILWAUKEE, 67- Concurrent Pos: Vis assoc prof, Syracuse Univ, 57. Mem: Am Phys Soc. Res: Quantum field theory and elementary particle symmetries; Lie groups and Lie algebras. Mailing Add: Dept of Physics Univ of Wis Milwaukee WI 53201

CHOWDHURY, AJIT KUMAR, b Calcutta, India, Apr 6, 28; m 61; c 2. REPRODUCTIVE PHYSIOLOGY, REPRODUCTIVE ENDOCRINOLOGY. Educ: Univ Calcutta, BSc, 51, MSc, 53, PhD(reprod physiol), 58. Prof Exp: Asst prof pharmacol, Bengal Vet Col, Calcutta, 54-64, prof physiol, 64-68; asst mem reprod endocrinol, Albert Einstein Med Ctr, Philadelphia, 68-71; ASST PROF REPROD BIOL & ENDOCRINOL, UNIV TEX MED SCH HOUSTON, 71-, ASSOC MEM, GRAD SCH BIOMED SCI, 74- Concurrent Pos: Pop Coun grant, Albert Einstein Med Ctr, 61-63. Mem: Soc Study Reprod; Am Soc Andrology; Indian Physiol Soc. Res: Morphology and kinetics of spermatogenesis in man, monkey and rodents; endocrine control and effect of noxious agents. Mailing Add: Dept Reprod Med & Biol Univ of Tex Med Sch Houston TX 77025

CHOWDHURY, DIPAK KUMAR, b Jhargram, India, Nov 15, 36; m 65; c 1. GEOPHYSICS. Educ: Indian Inst Technol, Kharagpur, BS, 56, MTechnol, 58; Tex A&M Col, PhD(geophys), 61. Prof Exp: Lectr geophys, Indian Sch Mines & Appl Geol, Dhanabad, 61-63, asst prof, 63-67; Nat Res Coun fel, Univ BC, 67-69; physicist, Shell Develop Co, Tex, 69-70; ASST PROF GEOL, IND UNIV, FT WAYNE, 70-, CHMN DEPT EARTH & SPACE SCI, 74- Concurrent Pos: Geophys consult, Law Eng & Testing Co, Ga, 70- Mem: Am Geophys Union. Res: Interpretation of self potential data for tabular shaped ore bodies; elastic wave propagations along layers in two-dimensional models; elastic wave velocities and attenuations in rocks and plastics; variations in azimuths and incident angles due to dipping interfaces; studies on earthquake seismology. Mailing Add: Dept of Earth & Space Sci Ind Univ Ft Wayne IN 46805

CHOWDHURY, IKBALUR RASHID, b Dacca, Pakistan, Feb 12, 39; m 67; c 1. SOIL FERTILITY, AGRONOMY. Educ: Univ Dacca, BSc, 60, MS, 63; NDak State Univ, PhD(soil fertil), 70. Prof Exp: Sr res asst bact, Pak-SEATO Cholera Res Inst, 64; sr lectr soil microbiol, Univ Dacca, 64-65; res asst soil fertil, NDak State Univ, 65-70, fel soil chem, 70-72; asst prof agr, 72-75, ASSOC PROF AGR, LINCOLN UNIV, 75- Mem: Am Soc Agron; Soil Sci Soc Am; Int Soc Soil Sci. Res: Soil testing; soil chemistry; plant physiology; environmental quality; soybean fertilization with respect to oil and protein. Mailing Add: 1403 Chestnut St Jefferson City MO 65101

CHOWDHURY, MRIDULA, b Calcutta, India, Feb 22, 38; m 61; c 2. REPRODUCTIVE BIOLOGY. Educ: Presidency Col, Calcutta, BS, 57; Calcutta Univ, MS, 59, PhD(reproductive physiol), 68. Prof Exp: NIH fel reproductive biol, Albert Einstein Med Ctr, 69-71; res assoc, 71-74, SR RES SCIENTIST REPRODUCTIVE BIOL, SCH MED, UNIV TEX, 74- Mem: AAAS. Res: Feedback control of gonadotropins in male with special emphasis on the control mechanism of synthesis of gonadotropins in the pituitary. Mailing Add: Dept of Reproductive Med & Biol Sch of Med Univ of Tex Houston TX 77025

CHOWDHURY, TUSHAR KUMAR, b Burdwan, India, Dec 22, 36; m 59; c 3. BIOPHYSICS, PHYSIOLOGY. Educ: Univ Calcutta, BSc, 56; Mont State Univ, MS, 61; State Univ NY Buffalo, PhD(biophys), 65. Prof Exp: Analyzer, Indian Sci & Eng Works, 56-58; res asst phys chem, Bose Inst, 58-59; teaching asst physics, Mont State Univ, 59-61; res asst biophys, State Univ NY Buffalo, 61-64, res instr, 65-66, res asst prof, 66; asst prof physiol, George Washington Univ, 66-69; assoc prof, 69-73, PROF PHYSIOL & BIOPHYS, MED SCH, UNIV OKLA, 73-, DIR INTERDISCIPLINARY BIOPHYS PROG, 69- Concurrent Pos: United Health Found fel, 65-66; grants, USPHS, 66-68, NSF, 67-71, Wash Heart Asn, 68-69, Okla Heart Asn, 71-72, Am Cancer Soc, 69-71, Am Heart Asn, 73-75. Mem: Biophys Soc; Soc Exp Biol & Med; Am Physiol Soc; Sigma Xi (secy-treas, 68-69); NY Acad Sci. Res: Mechanisms of the regulation of growth in living cells; regulation of body fluid; electrochemical properties of cancer cells. Mailing Add: Dept of Physiol & Biophys Univ of Okla Health Sci Ctr Oklahoma City OK 73190

CHOWN, EDWARD HOLTON, b Kingston, Ont, Feb 24, 32; m 65; c 2. PETROLOGY, STRUCTURAL GEOLOGY. Educ: Queen's Univ, Ont, BSc, 55; Univ BC, MASc, 57; Johns Hopkins Univ, PhD(geol), 63. Prof Exp: Geologist, Dept Natural Resources, Que, 63-66; asst prof geol, Loyola Col Montreal, 66-70, assoc prof, 70-75; ASSOC PROF SCI, UNIV QUE, CHICOUTIMI, 75- Mem: Geol Soc Am; Geol Asn Can; Can Inst Mining & Metall. Res: Proterozoic clastics and carbonates in central Quebec; differentiation and alteration of gabbroic sill complex and associated dyke swarm weathered profile beneath carbonate sequence; shock features in carbonates. Mailing Add: Dept of Earth Sci Univ of Que Chicoutimi PQ Can

CHOWN, H BRUCE, b Winnipeg, Man, Nov 10, 93; m 22, 49; c 5. MEDICINE, GENETICS. Educ: McGill Univ, BA, 14; Univ Man, MD, 22. Hon Degrees: DSc, Univ Man, 63. Prof Exp: PROF PEDIAT, UNIV MAN, 48- Mem: Royal Soc Can; Am Pediat Soc; Am Soc Human Genetics; Can Pediat Soc; Genetics Soc Can (pres, 64). Res: Cause and prevention of erythroblastosis; genetics of human blood groups. Mailing Add: Dept of Pediat Univ of Man Winnipeg MB Can

CHOYKE, WOLFGANG JUSTUS, b Berlin, Ger, July 24, 26; nat US; m 49; c 2 SOLID STATE PHYSICS. Educ: Ohio State Univ, BSc, 48, PhD(physics), 52. Prof Exp: Res physicist, 52-60, fel physicist, 60-62, ADV PHYSICIST, WESTINGHOUSE RES LABS, 63- Concurrent Pos: Adj prof physics, Univ Pittsburgh, 74- Mem: Am Phys Soc. Res: Experimental nuclear physics; radiative recombination processes; optical absorption; silicon carbide; reflectivity and ellipsometry of metals and semiconductors; ion implantation; ion beam simulation of neutron damage in solids. Mailing Add: Westinghouse Res Labs Beulah Rd Churchill Boro Pittsburgh PA 15235

CHRAPLIWY, PETER STANLEY, b Pulaski, Wis, Oct 20, 23; m 47; c 4. HERPETOLOGY. Educ: Univ Kans, BA, 53, MA, 56; Univ Ill, PhD(zool), 64. Prof Exp: From instr to asst prof, 60-64, ASSOC PROF BIOL, UNIV TEX, EL PASO, 64- Mem: Am Soc Ichthyologists & Herpetologists. Res: Taxonomy of modern amphibians and reptiles; biology of desert flora and fauna Mailing Add: Dept of Biol Sci Univ of Tex El Paso TX 79968

CHRAPLYVY, ZENOBIUS VOLODYMYR, b Ukraine, Mar 15, 04; nat nat US; m 47; c 2. PHYSICS. Educ: Univ Lemberg, PhD(physics), 32. Prof Exp: Prof physics, Univ Lemberg, 39-41; assoc prof, UNRRA Int Univ, Munich, 45-47; prof, Ukranian Inst Technol, 47-48; prof, 48-72, EMER PROF PHYSICS, ST LOUIS UNIV, 72- Mem: Am Phys Soc. Res: Quantum mechanics; electrodynamics; particle theory. Mailing Add: Dept of Physics St Louis Univ 221 N Grand Blvd St Louis MO 63103

CHRENKO, RICHARD MICHAEL, b Gillette, NJ, July 16, 30; m 63; c 2 SPECTROSCOPY, SOLID STATE PHYSICS. Educ: NY Univ, BA, 52; Harvard Univ, MA, 54. Prof Exp: Physicist, Knolls Atomic Power Lab, 56, physicist, Res Lab, 56-65, PHYSICIST, RES & DEVELOP CTR, GEN ELEC CO, 65- Res: Molecular and solid state spectroscopy; nuclear physics and reactors; physical properties of diamonds; residual stress. Mailing Add: Gen Elec Res & Develop Ctr PO Box 8 Schenectady NY 12301

CHREPTA, STEPHEN JOHN, b Watervliet, NY, May 19, 21. ANALYTICAL CHEMISTRY. Educ: Fordham Univ, MS, 51. Prof Exp: Instr, 46-53, ASST PROF CHEM & DEAN, ST BASIL'S COL, 53- Mem: AAAS; Am Chem Soc. Res:

Spectroscopy; x-ray diffraction. Mailing Add: Dept of Chem St Basil's Col Stamford CT 06902

CHRESTENSON, HUBERT EDWIN, b Grandview, Wash, Oct 21, 27; m 47; c 3. MATHEMATICS. Educ: State Col Wash, BA, 49, MA, 51; Univ Ore, PhD(math), 53. Prof Exp: Instr math, Purdue Univ, 53-54; asst prof, Whitman Col, 54-57; ASST PROF MATH, REED COL, 57- Mem: Am Math Soc; Math Asn Am. Res: Fourier analysis. Mailing Add: Dept of Math Reed Col Portland OR 97202

CHRETIEN, MAX, b Basel, Switz, Feb 29, 24; m 58; c 1. PHYSICS. Educ: Univ Basel, Switz, PhD(physics), 49. Prof Exp: Fels, Univ Birmingham, Eng, 51-53 & Columbia Univ, 53-54, from instr to asst prof, 53-58, ASSOC PROF HIGH ENERGY PHYSICS, BRANDEIS UNIV, 58- Concurrent Pos: Swiss Nat Fund fel, 63 & 69. Mem: Am Physical Soc. Res: Theory of elementary particles. Mailing Add: 39 Florence Rd Waltham MA 02154

CHRETIEN, MICHEL, b Shawinigan, Que, Mar 26, 36; m 60; c 2. ENDOCRINOLOGY. Educ: Univ Montreal, BA, 55, MD, 60; McGill Univ, MSc, 62. Prof Exp: Med Res Coun Can fel, McGill Univ, 60-62; resident med, Peter Bent Brigham Hosp, 62-63, clin fel endocrinol, 63-64; Childs Mem Fund fel biochem, Hormone Res Lab, Univ Calif, Berkeley, 64-66, Med Res Coun Can fel, 66-67; SR INVESTR ENDOCRINOL, CLIN RES INST MONTREAL, 67-; PROF MED, UNIV MONTREAL & HOTEL-DIEU HOSP, 67- Concurrent Pos: Res fel endocrinol, Hotel Dieu Hosp, 60-62; asst prof exp med, McGill Univ, 67- Honors & Awards: Basic Res Award, Asn Fr Speaking Physicians Can, 71. Mem: Can Soc Clin Invest; Endocrine Soc; NY Acad Sci; Can Biomed Soc; AAAS. Res: Purification, isolation, chemical characterization and physiology of pituitary hormones related to the lipolytic and the melanophore stimulating activities. Mailing Add: Clin Res Inst Montreal 110 Pine Ave W Montreal PQ Can

CHRIEN, ROBERT EDWARD, b Cleveland, Ohio, Apr 15, 30; m 53; c 4. NUCLEAR PHYSICS. Educ: Rensselaer Polytech Inst, BS, 52; Case Inst Technol, MS, 55, PhD(physics), 58. Prof Exp: Res assoc physics, 57-59, from asst physicist to physicist, 59-72, SR PHYSICIST, BROOKHAVEN NAT LAB, 72- Concurrent Pos: Past secy & past chmn nuclear cross sect adv comt, US AEC; mem, Nuclear Data Comt, Nuclear Energy Agency, 72-, Tech Adv Panel, Los Alamos Meson Physics Facil, 72-, Nuclear Physics Div, Comt Nuclear Data, Am Phys Soc, 73 & Nuclear Data Comt, Energy Res & Develop Admin, 75- Mem: Fel Am Phys Soc; AAAS; Sigma Xi; NY Acad Sci. Res: Neutron physics; neutron resonance parameters; capture gamma rays from neutron resonances; neutron total cross sections; applications of on-line computers in nuclear physics; nuclear detectors; photonuclear reactions; intermediate energy physics. Mailing Add: Dept of Physics Brookhaven Nat Lab Upton NY 11973

CHRISMAN, CHARLES LARRY, b St Joseph, Mo, Mar 1, 41; m 62; c 2. CYTOGENETICS. Educ: Univ Mo-Columbia, BS, 67, MS, 69, PhD(cytogenetics), 71. Prof Exp: ASST PROF GENETICS & CYTOGENETICS, PURDUE UNIV, 71- Mem: AAAS; Am Soc Animal Sci; Am Genetic Asn; Genetics Soc Can; Genetics Soc Am. Res: Cytogenetic investigations into the effects of chemicals and drugs on the chromosomes of domestic animals, utilizing tissue cultures; investigations of animals with birth defects as possibly caused by chromosomal aberrations. Mailing Add: Dept of Animal Sci Lilly Hall Purdue Univ West Lafayette IN 47907

CHRISMAN, NOEL JUDSON, b Visalia, Calif, July 30, 40; m 62; c 1. MEDICAL ANTHROPOLOGY. Educ: Univ Calif, Riverside, BA, 62, Univ Calif, Berkeley, PhD(anthrop), 66, MPH, 67. Prof Exp: NIMH fel pub health, Sch Pub Health, Univ Calif, Berkeley, 66-67; asst prof anthrop, Pomona Col, 67-73; ASST PROF COMP NURSING CARE SYSTS, SCH NURSING, UNIV WASH, 73- Concurrent Pos: Nat Endowment Humanities Younger Humanist fel, Univ Calif, Berkeley, 71. Mem: Fel Am Anthrop Asn; fel Soc Appl Anthrop; Coun Anthrop & Educ. Res: Urban studies, particularly the role of social networks in urban adaptation; clan and ethnic variation in health seeking behaviors; working class Americans; network analysis. Mailing Add: Sch of Nursing Univ of Wash Seattle WA 98195

CHRISPEELS, MAARTEN JAN, b Kortenberg, Belg, Feb 10, 38; m 66. PLANT PHYSIOLOGY. Educ: Univ Ghent, Engr, 60; Univ Ill, Urbana, PhD(agron), 64. Prof Exp: Res asst agron, Univ Ill, 63-64; res assoc plant biochem, Res Inst Advan Studies, 64-65; plant res lab, AEC, 65-67; res assoc microbiol, Purdue Univ, 67; asst prof biol, 67-74, ASSOC PROF BIOL, UNIV CALIF, SAN DIEGO, 74- Concurrent Pos: John S Guggenheim Found fel, 74-75. Mem: AAAS; Am Soc Plant Physiol. Res: Biochemistry of plant development, especially germination; structure-function relationships in cells (secretion, lysosomes, protein bodies). Mailing Add: Dept of Biol Univ Calif San Diego PO Box 109 La Jolla CA 92037

CHRIST, ADOLPH ERVIN, b Reedley, Calif, May 13, 29; m 51; c 4. CHILD PSYCHIATRY. Educ: Univ Calif, Berkeley, AB, 51; Univ Calif, San Francisco, MD, 54; Am Bd Psychiat & Neurol, dipl, 62, cert child psychiat, 67. Prof Exp: Intern univ hosp, Univ Calif, San Francisco, 54-55; resident psychiat, Langley Porter Neuropsychol Inst, 58-60, child psychiat, 60-62; instr med sch, Univ Wash, 62-65, asst prof, 65-67, dir psychol inst Scivia children, 62-67; asst prof child psychiat & dir inpatient child psychiat, Albert Einstein Col Med, 68-69; dir training clin psychol, 69-72, EXEC DIR DIV CHILD PSYCHIAT, STATE UNIV NY DOWNSTATE MED CTR, 72- Mem: Am Psychiat Asn; Am Orthopsychiat Asn; Am Col Psychiat; World Fedn Ment Health. Res: Treatment of childhood psychosis, especially milieu treatment. Mailing Add: 853 Seventh Ave Apt 6C New York NY 10019

CHRIST, CHARLES LOUIS, b Baltimore, Md, Mar 12, 16; m 38. PHYSICAL CHEMISTRY Educ: Jo3hns Hopkins Univ, PhD(chem), 40. Prof Exp: Jr instr chem, Johns Hopkins Univ, 36-40, instr, 42-45; chemist, Gen Elec Co, 40-41; instr chem, Wesleyan Univ, 41-42; consult, Rheem Res Corp, 45-46; group leader x-ray crystallog, Am Cyanamid Corp, 46-49; staff assoc, 49-56, res physicist, 59-60, chief br exp geochem & mineral, 60-62, RES PHYSICIST, US GEOL SURV, 62- Concurrent Pos: Mem nat comt crystallog, Nat Res Coun-Nat Acad Sci, 55-59; prof lectr, George Washington Univ, 56-, consult, 59-; res fel geol sci, Harvard Univ, 59-60; vis prof geol, Univ Mo-Columbia, 68-69. Honors & Awards: Rockefeller Pub Serv Award, 59. Mem: AAAS; Am Chem Soc; Mineral Soc Am; Am Crystallog Asn; Mineral Soc Gt Brit & Ireland. Res: Crystal structure determination, especially of minerals; crystal chemistry; geochemistry. Mailing Add: US Geol Surv 345 Middlefield Rd Menlo Park CA 94025

CHRIST, DARYL DEAN, b Buffalo Center, Iowa, Nov 3, 42; m 66; c 1. PHARMACOLOGY, PHYSIOLOGY. Educ: Univ Iowa, BS, 64; Loyola Univ Chicago, PhD(pharmacol), 69. Prof Exp: NIH fel, Loyola Univ Chicago, 69-70; asst prof, 71-75, ASSOC PROF PHARMACOL, SCH MED, UNIV ARK, LITTLE ROCK, 75- Res: Effects of general anesthetics on ganglionic transmission; pharmacology and physiology of synaptic transmission. Mailing Add: Dept of Pharmacol Univ of Ark Med Ctr Little Rock AR 72201

CHRIST, JOHN CONRAD, b Waumandee, Wis, May 17, 08; m 32; c 2. BIOLOGY. Educ: NCent Col, BA, 31; Northwestern Univ, MA, 38; Univ Bari, PhD, 60. Prof Exp: Teacher high sch, Ill, 31-36; from asst prof to assoc prof biol, 46-73, head div sci & math, 53-71, dean sch natural sci, 71-73, EMER PROF BIOL, PERU STATE COL, 73- Res: Hemorocallis hybridization; aphid control; yellow birch; nesting habits of eastern phoebe; rusts. Mailing Add: Box 162 Peru State Col Peru NE 68421

CHRISTE, KARL OTTO, b Ulm, Ger, July 24, 36; nat US; m 62; c 3. INORGANIC CHEMISTRY, PHYSICAL CHEMISTRY. Educ: Stuttgart Tech Univ, BS, 57, MS, 60, PhD(inorg chem), 61. Prof Exp: Teaching asst anal chem, Stuttgart Tech Univ, 58-60, res assoc inorg polymer chem, 60-61; mem res staff, Stauffer Chem Co, 62-67; MEM TECH STAFF, ROCKETDYNE DIV, ROCKWELL INT, 67- Mem: Am Chem Soc. Res: Inorganic and organic fluorine chemistry; high energy oxidizers and explosives; structural studies; vibrational spectroscopy; low-temperature matrix isolation; chlorinated hydrocarbons; inorganic polymers; boron hydrides; water treatment chemicals. Mailing Add: Rocketdyne Div Rockwell Int 6633 Canoga Ave Canoga Park CA 91304

CHRISTENA, RAY CLIFFORD, b Indianapolis, Ind, Aug 17, 15; m 47; c 5. INDUSTRIAL CHEMISTRY. Educ: Univ Tex, BS, 39; Ind Univ, MA, 48, PhD(phys chem), 51. Prof Exp: Lab asst, Allison Div, Gen Motors Corp, 40-42 & Calco Div, Am Cyanamid, Ind, 46-51; res chemist, Benger Lab, E I du Pont de Nemours & Co, 51-58, Thiokol Chem Corp, NJ, 58-62, Univ Utah, 62-63 & Celanese Corp Am, 63-67; GROUP LEADER RES & DEVELOP, CHEMICALS DIV, VULCAN MAT CO, 67- Mem: Am Chem Soc; Soc Rheol. Res: Research and development in area of sodium hydroxide, chlorine and clorinated hydrocarbons. Mailing Add: Res Dept Vulcan Mat Co PO Box 545 Wichita KS 67201

CHRISTENBERRY, GEORGE ANDREW, b Macon, Ga, Sept 3, 15; m 37; c 3. MYCOLOGY, ACADEMIC ADMINISTRATION. Educ: Furman Univ, BS, 36; Univ NC, AM, 38, PhD(bot), 40. Prof Exp: Asst, Univ NC, 36-40; asst prof biol, Meredith Col, 40-41, assoc prof & head dept, 41-43, prof, 43; instr pre-flight prog, Furman Univ, 43-44, instr voc appraiser, Vet Guid, 46, assoc prof biol, 46-48, prof & dean men's col, 48-53; pres, Shorter Col, Ga, 53-58; admin dir, Furman Univ, 58-64; prof biol & chmn dept, Ga Col Milledgeville, 64-65, dean, 65-70; PRES, AUGUSTA COL, 70- Concurrent Pos: Carnegie Found grant, 48; bd dir, Asn State Cols & Univs, 73- Mem: Am Inst Biol Sci. Res: Taxonomic study of Mucorales in southeastern United States. Mailing Add: Augusta Col 2500 Walton Way Augusta GA 30904

CHRISTENS, JOHN MANUEL, physical chemistry, chemical engineering, see 12th edition

CHRISTENSEN, ALBERT KENT, b Washington, DC, Dec 3, 27; m 52; c 5. ANATOMY, CELL BIOLOGY. Educ: Brigham Young Univ, AB, 53; Harvard Univ, PhD(biol), 58. Prof Exp: NIH fel, Col Med, Cornell Univ, 58-59; NIH fel, Harvard Med Sch, 59-60, instr anat, 60-61; from asst prof to assoc prof, Sch Med, Stanford Univ, 61-71; PROF ANAT & CHMN DEPT, SCH MED, TEMPLE UNIV, 71- Mem: Am Soc Cell Biol; Am Asn Anat; Soc Study Reprod; Am Soc Zool; Electron Micros Soc Am. Res: Ultrastructure and cytochemistry of steroid-secreting cells; reproductive biology; macrophages. Mailing Add: Dept of Anat Sch of Med Temple Univ Philadelphia PA 19140

CHRISTENSEN, BERT EINAR, b Duluth, Minn, Oct 20, 04; m 32; c 3. SYNTHETIC ORGANIC CHEMISTRY. Educ: Wash State Univ, BS, 27; Univ Wash, PhD(chem), 32. Prof Exp: Chemist, Atmospheric Nitrogen Corp, Syracuse, 27-28; from instr to prof chem, 31-70, chmn dept, 56-70, EMER PROF CHEM, ORE STATE UNIV, 70- Mem: AAAS; Am Chem Soc. Res: Organic synthesis of quinazoline, purine, pyrimidine and other related heterocyclic compounds. Mailing Add: 337 NW 23rd Corvallis OR 97330

CHRISTENSEN, BURGESS NYLES, b San Francisco, Calif, Oct 4, 40; m 61; c 2. NEUROPHYSIOLOGY. Educ: Univ Utah, BA, 63, PhD(biophys, bioeng), 67. Prof Exp: Fel physiol, Univ Utah, 67-68; fel, Yale Univ, 68-70; ASST PROF NEUROSCI, BROWN UNIV, 70- Mem: AAAS; Soc Neurosci. Res: Synaptic transmission in the central nervous system. Mailing Add: Dept of Med Sci Brown Univ Providence RI 02912

CHRISTENSEN, BURTON GRANT, b Waterloo, Iowa, Apr 8, 30; m 61; c 2. BIO-ORGANIC CHEMISTRY, MEDICINAL CHEMISTRY. Educ: Iowa State Univ, BS, 52; Harvard Univ, AM, 54, PhD(chem), 56. Prof Exp: Chemist, 56-71, asst dir, 71-73, DIR SYNTHETIC CHEM RES, MERCK, SHARP & DOHME RES LABS, 73- Mem: Am Chem Soc; The Chem Soc. Res: Organic chemistry, with emphasis on antibacterial synthesis. Mailing Add: 2665 Skytop Dr Scotch Plains NJ 07076

CHRISTENSEN, CARL JOSEPH, b Provo, Utah, Apr 3, 01; m 24; c 4. PHYSICAL CHEMISTRY. Educ: Brigham Young Univ, BS, 23; Univ Wis, MS, 25; Univ Calif, PhD(phys chem), 29. Prof Exp: Instr chem, Brigham Young Univ, 23-24 & 25-27; asst, Univ Wis, 24- 25; res investigator, Crocker Res Lab, San Francisco, 27-28; res phys chemist, Bell Tel Labs, 29-46; prof, 46-49,xEMER 46-69, EMER PROF CHEM & CERAMICS, UNIV UTAH, 69- Concurrent Pos: Dean sch mines & mineral indust, 46-53, coordr coop res & dir eng exp sta, 53-66; dir & vpres, Lemco Corp, Salt Lake City; assoc ed, Ann Review Phys Chem, 55-75. Mem: Am Chem Soc; Am Phys Soc; Am Ceramic Soc. Res: Fluorescence; microphone carbon resistance fluctuations in granular electrical resistors; television; deposited carbon electrical resistors; ceramics; semiconductors; piezo-electric crystals; magnetic metals; flow and creep of solid materials; preservation of building stone; food preservation by gamma irradiation. Mailing Add: 2597 Sherwood Dr Salt Lake City UT 84108

CHRISTENSEN, CHARLES RICHARD, b Florence, Ala, Oct 18, 38; m 59; c 3. OPTICAL PHYSICS, SOLID STATE PHYSICS. Educ: Vanderbilt Univ, BE, 60; Calif Inst Technol, PhD(chem), 66. Prof Exp: Res chemist, 67-68; RES PHYSICIST, PHYS SCI LAB, US ARMY MISSILE COMMAND, 69- Mem: Am Phys Soc; Am Chem Soc; Optical Soc Am. Res: Coherent optics with applications in signal processing and correlation, coherent optical imaging, optical synthetic aperture methods; magnetic resonance and magnetic materials. Mailing Add: 1400 Levert Ave Athens AL 35611

CHRISTENSEN, CHRISTIAN MARTIN, b Ft Collins, Colo, Sept 24, 46; m 68; c 2. ECONOMIC ENTOMOLOGY. Educ: Rutgers Univ, BS, 68; Purdue Univ, MS, 70, PhD(entom), 74. Prof Exp: LIVESTOCK ENTOM EXTEN SPECIALIST, UNIV KY, 74- Mem: Entom Soc Am. Res: Practical application techniques and control methods for controlling livestock insect pests. Mailing Add: Dept of Entom Univ of Ky Lexington KY 40506

CHRISTENSEN, CLYDE MARTIN, b Sturgeon Bay, Wis, Aug 8, 05; m 35; c 3. PLANT PATHOLOGY. Educ: Univ Minn, BS, 29, MS, 30, PhD(plant path), 37. Prof Exp: Instr plant path, Univ Minn, 30-32 & 34-40, asst prof, 40-47; asst pathologist, Forest Prod Lab, US Forest Serv, Wis, 43; assoc prof plant path, 47-48, PROF PLANT PATH, UNIV MINN, ST PAUL, 48- Concurrent Pos: Spec temp sci aide,

Rockefeller Found Agr Prog, Mex, 59-63 & 64; consult, Cargill, Inc, 62- Mem: Am Asn Cereal Chemists; Mycol Soc Am; Am Phytopath Soc; Am Soc Microbiol. Res: Deterioration of stored grains caused by molds; mycotoxins. Mailing Add: Dept of Plant Path Univ of Minn St Paul MN 55101

CHRISTENSEN, DAVID EMUN, b Ashland, Wis, Feb 17, 21; m 46; c 4. ACADEMIC ADMINISTRATION, GEOGRAPHY. Educ: Univ Chicago, MA, 48, PhD(geog), 56. Prof Exp: From asst prof to assoc prof geog, Fla State Univ, 48-62; assoc prof, 62-67, asst dean col lib arts & sci, 66-70, PROF GEOG, SOUTHERN ILL UNIV, CARBONDALE, 67-, ASSOC DEAN COL LIB ARTS & SCI, 70- Mem: Asn Am Geogr; Nat Coun Geog Educ; Inst Brit Geog. Res: Urban planning and geography. Mailing Add: Col of Lib Arts & Sci Southern Ill Univ Carbondale IL 62901

CHRISTENSEN, EARL MARTIN, botany, see 12th edition

CHRISTENSEN, EDWARD RICHARDS, b Salt Lake City, Utah, Dec 21, 24; m 49; c 1. PETROLEUM CHEMISTRY. Educ: Univ Utah, BS, 48, MS, 49; Union Col, MS, 74. Prof Exp: Chemist, Beacon Res Labs, 49-56, group leader fuels process develop, 56-58, admin asst to dir res, 58-59, asst supvr petrochem res, 60-65, supvr, 65-67, res dir chem & process res, 67-75, ASST MGR RES & TECH DEPT, BEACON RES LABS, TEXACO, INC, 75- Mem: Am Chem Soc; Am Inst Chem Eng; Sigma Xi. Res: Synthesis and applications testing of chemicals derived from petroleum and development of pilot scale processes for their manufacture; development of new or improved catalysts and processes for the conversion of crude oil to marketable products; development of improved methods and equipment for carrying out process development. Mailing Add: Beacon Res Labs Texaco Inc PO Box 509 Beacon NY 12508

CHRISTENSEN, ELEANOR, b Berkeley, Calif, May 6, 29; m 54. PHYSIOLOGY. Educ: Stanford Univ, AB, 49, PhD(biol sci), 54. Prof Exp: Asst gen physiol, Stanford Univ, 52, photophysiol, 53; res fel neurophysiol, Calif Inst Technol, 53-54, res fel immunol, 54-55; lectr cell physiol, Univ San Diego, 59 & res assoc carbohydrate metab, 60-62, lectr gen physiol, 64-65, LECTR BACT PHYSIOL, SAN DIEGO STATE UNIV, 65- Mem: AAAS; Soc Gen Physiol. Res: General and cellular physiology; photophysiology; neurc carbohydrate metabolism; immunology. Mailing Add: 6361 Rockhurst Dr San Diego CA 92120

CHRISTENSEN, ERIC, b Brooklyn, NY, Sept 18, 14; m 39; c 2. BIOLOGY. Educ: NY State Col Forestry, BS, 47; Hofstra Col, MA, 58. Prof Exp: Asst & tech specialist, Brookhaven Nat Lab, 47-53; assoc prof, 55-67, PROF BIOL, STATE UNIV NY AGR & TECH COL FARMINGDALE, 67- Mem: Bot Soc Am. Res: Radiobiology; teaching sciences; Ginkgo biloba. Mailing Add: Dept of Biol State Univ of NY Agr & Techn Col Farmingdale NY 11735

CHRISTENSEN, FRITJOF ERNEST, b La Crosse, Wis, Apr 14, 07; m 33; c 1. PHYSICS. Educ: Augsburg Col, BA, 28; Univ Minn, MA, 45. Prof Exp: Prin high sch, Minn, 28-30, instr, 30-42; supvr pre-radar, Ill Inst Technol, 42-43; instr physics, Univ Minn, 43-45, prin lab attend & instr, 45-53; assoc prof, 53-72, EMER ASSOC PROF PHYSICS, ST OLAF COL, 72- Concurrent Pos: Physicist, Radioisotope Res Lab, Vet Admin, Minn; lectr, Augsburg Col, 47-49; proj dir, Ctr Ed Apparatus in Physics, NY, 63-65. Honors & Awards: Distinguished Serv Citation, Am Asn Physics Teachers, 73. Mem: AAAS; Am Phys Soc; Am Asn Physics Teachers. Res: Transverse wave apparatus; calcite crystal apparatus; radioactivity decay; Wilson cloud chamber; model mass spectrometer; biophysics; usage of the oscilloscope. Mailing Add: 33 Lincoln Lane Northfield MN 55057

CHRISTENSEN, GEORGE CURTIS, b New York, NY, Feb 21, m 47; c 4. VETERINARY ANATOMY. Educ: Cornell Univ, DVM, 49, MS, 50, PhD(mammal anat, higher educ), 53. Prof Exp: Instr vet anat, Cornell Univ, 49-53; assoc prof, Iowa State Univ, 53-58; prof vet anat & head dept, Purdue Univ, 58-63; dean col vet med & dir vet med res inst, 63-65, ACAD AFFAIRS, IOWA STATE UNIV, 65- Concurrent Pos: Consult, NIH & NCent Asn Cols & Sec Schs; mem comt educ & res, Nat Acad bd dirs, Quad Cities Grad Ctr, Iowa State Hyg Lab & Ctr Res Libr, Chicago; mem, Iowa Dept Health; mem exec comt, Nat Asn State Univs & Land-Grant Cols, 73-, chmn coun acad affairs, 73-74; vpres, Mid-Am State Univs Asn, 75-76. Mem: Am Asn Anatomists; Am Vet Med Asn; Conf Res Workers Animal Dis; Am Asn Vet Anatomists (pres, 63); World Asn Vet Anatomists (vpres, 63-65). Res: College administration; comparative cardiovascular anatomy and physiology; comparative vasculature of urogenital and central nervous systems; history of veterinary medical education; higher education. Mailing Add: 111 Beardshear Hall Iowa State Univ Ames IA 50010

CHRISTENSEN, GERALD M, b Pocatello, Idaho, Dec 23, 28; m 49; c 2. BIOCHEMISTRY, RADIOBIOLOGY. Educ: Univ Utah, BS, 51; Emory Univ, PhD, 58. Prof Exp: Fel, Virus Lab, Univ Calif, Berkeley, 58-59; sr biochemist, Boeing Co, Wash, 59-64; asst prof radiol, 64-69, ASSOC PROF RADIOL, UNIV WASH, 69- Res: Biological effects of ionizing radiation; enzyme and protein chemistry; amino acid and nucleotide metabolism. Mailing Add: Dept of Radiol Univ of Wash Seattle WA 98105

CHRISTENSEN, GLEN C, b Nevada City, Calif, Sept 13, 24; m 51; c 2. WILDLIFE MANAGEMENT. Educ: Univ Nev, BS, 50, MS, 52. Prof Exp: Wildlife biologist, Nev Fish & Game Comn, 52-56, wildlife mgr, 56-58, res, India, 59-61, wildlife specialist res & mgt, 61-71, CHIEF OF GAME MGT, NEV DEPT FISH & GAME COMN, 71- Concurrent Pos: Wildlife Mgt Inst grant-in-aid, 51-52; lectr, Univ Nev, 64-66, mem adj staff, 64-75. Res: Development of the field of exotic game bird introductions into game deficient habitat of western United States; research on this problem in the arid regions of India, Pakistan and Afghanistan; development of a more precise deer harvest management program for Nevada. Mailing Add: Nev Dept of Fish & Game PO Box 10678 Reno NV 89502

CHRISTENSEN, GLENN MARVIN, b Duluth, Minn, Jan 17, 30; m 53; c 3 BIOCHEMISTRY. Educ: Univ Minn, BA, 52, PhD(biochem), 57. Prof Exp: Instr elem biochem, Univ Minn, Duluth, 55-57, from asst prof to assoc prof chem, 57-66; RES BIOCHEMIST, NAT WATER QUALITY LAB, US ENVIRON PROTECTION AGENCY, 66- Concurrent Pos: NIH fel, 64. Mem: AAAS; Am Chem Soc; Am Fisheries Soc. Res: Carbohydrate aspects of biological chemistry; normal and abnormal carbohydrate metabolism; endocrinology; biomedical research; aquatic life; biochemical toxicology. Mailing Add: 6201 Longdon Blvd Duluth MN 55804

CHRISTENSEN, HALVOR NIELS, b Cozad, Nebr, Oct 24, 15; m 39; c 3. BIOCHEMISTRY, BIOPHYSICS. Educ: Kearney State Col, BS, 35; Purdue Univ, MS, 37; Harvard Univ, PhD(biol chem), 40. Prof Exp: Fel chem, Harvard Univ, 40-41; res biochemist, Lederle Labs, 41-42; instr biol chem, Harvard Med Sch, 42-44; dir chem labs, Mary-Imogene Bassett Hosp, Cooperstown, NY, 44-47; asst prof biochem, Harvard Med Sch, 47-49; prof biochem & nutrit & head dept, Sch Med, Tufts Univ, 49-55; chmn dept biol chem, 55-70, PROF BIOL CHEM, UNIV MICH, ANN ARBOR, 55- Concurrent Pos: Dir lab biochem res, Childrens Hosp, Boston, 47-49; Guggenheim fel, Carlsberg Lab, Copenhagen, 52; consult, NIH, 61-68; Nobel guest prof, Univ Uppsala, 68. Mem: Fel Am Acad Arts & Sci; Am Soc Biol Chem; Am Chem Soc; Am Inst Nutrit; Biophys Soc. Res: Amino acid transport; intravenous amino acid nutrition; peptide metabolism and antibiotics. Mailing Add: Dept of Biol Chem Univ of Mich Ann Arbor MI 48109

CHRISTENSEN, HERBERT EDWARD, b Berkeley, Calif, May 31, 22; m 70; c 5. TOXICOLOGY, OCCUPATIONAL HEALTH. Educ: Univ Calif, Berkeley, AB, 43, MPH, 49; Univ Calif, San Francisco, MS, 49; Univ Cincinnati, DSc(indust health), 64; Am Bd Indust Hyg, dipl. Prof Exp: Pharmacologist, Western Regional Res Labs, USDA, 49-50; chief toxicol br, Army Environ Health Lab, Army Chem Ctr, Md, 51-55; fel indust toxicol, Kettering Lab, Univ Cincinnati, 55-58; toxicologist, Armed Forces Inst Path, Walter Reed Army Med Ctr, 58-62, chief toxicol div, US Army Environ Hy Agency, Edgewood Arsenal, 62-66, exec officer & chief chem div, 3rd US Army Med Lab, Ft McPherson, Ga, 67; asst dir res admin, Ctr Res Pharmacol & Toxicol, Sch Med & assoc prof indust toxicol, Sch Pub Health, Univ NC, Chapel Hill, 68-69; staff indust toxicologist, Standard Oil Co, Calif, 69-70; consult, indust hyg, toxicol & air pollution, 70-71; ed, Toxic Substances List, 71-75, chief toxicity & res anal br, Off Res & Standards Develop, 71-74, dep asst dir, Off, 72-74, actg dir, 74-75, DEP DIR, DIV CRITERIA DOCUMENTATION & STANDARDS DEVELOP & ED, REGISTRY OF TOXIC EFFECTS OF CHEM SUBSTANCES, NAT INST OCCUP SAFETY & HEALTH, 75- Concurrent Pos: Consult, Environ Health Ctr, Naval Ord Systs Command, Crane Naval Ammunition Depot, 66-71, Schuyler Develop Corp, 67-75 & WHO, Nepal, 73. Mem: Fel AAAS; Soc Toxicol; Am Indust Hyg Asn; Am Acad Indust Hyg; Conf Govt Indust Hygienists; Sigma Xi. Res: Toxicology of insecticides; insect repellents; economic poisons; industrial chemicals; carcinogens; effects pf putrefaction on animal tissues; forensic toxicology; preparation of occupational health standards. Mailing Add: 1299 Glen Ave Berkeley CA 94708

CHRISTENSEN, HOWARD ANTHONY, b Oakland, Calif, Apr 26, 28; m 52. MEDICAL ENTOMOLOGY, MEDICAL PARASITOLOGY. Educ: San Francisco State Col, BA, 63, MA, 66; Univ Calif, Davis, PhD(entom), 68. Prof Exp: MED ENTOMOLOGIST, GORGAS MEM LAB, 68- Mem: Am Soc Parasitol; Entom Soc Am; Am Soc Trop Med & Hyg; Am Mosquito Control Asn. Res: Bionomics of arthropod vectors of medical importance and the epidemiology of arthropod borne diseases; leishmaniasis. Mailing Add: Gorgas Mem Lab PO Box 2016 Balboa Heights CZ

CHRISTENSEN, HOWARD DIX, b Logan, Utah, Mar 30, 40; m 66; c 2. NEUROPHARMACOLOGY. Educ: Univ Nev, Reno, BS, 61; Univ Calif, Los Angeles, PhD(biophys, nuclear med), 66. Prof Exp: Fel pharmacol, Univ Calif, San Francisco, 66-68 & Columbia Univ, 68-69; pharmacologist, Res Triangle Inst, 69-74; ASSOC PROF PHARMACOL, SCHS MED & DENT, UNIV OKLA, 74- Mem: AAAS; NY Acad Sci; Sigma Xi; Soc Neurosci; Int Asn Study Pain. Res: Central nervous system mechanisms of cannabinoids, anticonvulsants, barbiturates and contraceptive steroids; analysis of brain stem regulatory mechanism. Mailing Add: 416 NW 17th Oklahoma City OK 73103

CHRISTENSEN, JAMES, b Ames, Iowa, Jan 4, 32; m 58; c 3. PHYSIOLOGY, GASTROENTEROLOGY. Educ: Univ Nebr, BS, 53, MS & MD, 57. Prof Exp: From instr to assoc prof, 65-72, PROF INTERNAL MED, UNIV IOWA, 72- Concurrent Pos: Lectr, Univ Alta, 65-66; Markle scholar, 65-70; USPHS career develop award, 69-74; mem eval comt gastroenterol, Vet Admin Res, 69- Mem: Am Fedn Clin Res; Am Gastroenterol Asn; Am Soc Clin Invest. Res: Gastrointestinal motility; autonomic and smooth muscle physiology. Mailing Add: Dept of Internal Med Univ of Iowa Hosps Iowa City IA 52242

CHRISTENSEN, JAMES BOYD, b Utah, Dec 24, 21; m 47; c 2. ANTHROPOLOGY. Educ: Univ Utah, BS, 47, MS, 48; Northwestern Univ, PhD(anthrop), 52. Prof Exp: From instr to assoc prof anthrop, Wayne State Univ, 52-66, PROF ANTHROP, WAYNE STATE UNIV, 66-, CHMN DEPT & DIR SOC SCI, 71- Concurrent Pos: Consult, US Govt, Africa; NSF sr fel, 59-60, scholar, 62-63. Mem: Fel AAAS; fel Am Anthrop Asn; fel African Studies Asn; Royal African Soc; Int African Inst. Res: Cultural anthropology; subsaharan Africa; Negro in new world; social organization; culture change; primitive religion. Mailing Add: Dept of Anthrop Wayne State Univ Detroit MI 48202

CHRISTENSEN, JAMES ROGER, b Des Moines, Iowa, Oct 28, 25; m 51; c 2. VIROLOGY. Educ: Iowa State Col, BS, 49; Cornell Univ, PhD(biochem), 53. Prof Exp: Fel biophysics, Sch Med, Univ Colo, 53-55; from instr to assoc prof, 55-70, actg chmn, 67-70, PROF MICROBIOL, SCH MED & DENT, UNIV ROCHESTER, 70- Mem: Am Soc Microbiologists. Res: Bacteriophage; host-controlled modification; gene function; mutual exclusion. Mailing Add: Dept of Microbiol Univ of Rochester Sch of Med Rochester NY 14642

CHRISTENSEN, JOHN, b Mahtowa, Minn, June 19, 08; m 32, 64; c 3. BIOCHEMISTRY. Educ: Union Col, Nebr, BA, 39; Univ Nebr, MA, 46; Mich State Univ, PhD, 56. Prof Exp: Teacher, Shelton Acad, 39-41; asst chem, Union Col, Nebr, 41-45; from instr to asst prof chem, Emmanuel Missionary Col, 45-55; prof, 55-75, EMER PROF CHEM, SOUTHERN MISSIONARY COL, 75- Mem: Am Chem Soc RES: Periodate oxidations of sugars and related organic compounds; chemistry of 1,3-diketones. Mailing Add: Box 507 Collegedale TN 37315

CHRISTENSEN, JOHN BERT, b Richfield, Utah, Sept 11, 27; m 52; c 3. ANATOMY. Educ: Brigham Young Univ, BS, 54; Univ Okla, MS, 55, PhD(med sci), 58. Prof Exp: Asst dept anat, Univ Okla, 54-58, instr, 58-59; from asst prof to assoc prof, George Washington Univ, 59-73; PROF & DIR EDUC RESOURCES, UNIV MO-KANSAS CITY SCH MED & SCH DENT, 73- Concurrent Pos: Picker x-ray fel, 58-59; vis prof, Univ Va, 70-75. Mem: Am Asn Anatomists. Res: Normal ossification; calcification of bone and cartilage; development of multimedia teaching material. Mailing Add: Univ Mo Sch of Med 2411 Holmes Kansas City MO 64108

CHRISTENSEN, LARRY WAYNE, b Elkhart, Ind, Sept 1, 43; m 65; c 2. ORGANIC CHEMISTRY. Educ: Goshen Col, BA, 65; Purdue Univ, PhD(chem), 69. Prof Exp: ASSOC PROF CHEM, HOUGHTON COL, 69- Concurrent Pos: Vis prof, Purdue Univ, 70- Mem: Am Chem Soc; Am Sci Affil. Res: Mechanistic and synthetic organo-sulfur chemistry; applications of mass spectrometry to organic chemistry. Mailing Add: Dept of Chem Houghton Col Houghton NY 14744

CHRISTENSEN, LAURITZ ROYAL, b Everson, Wash, Aug 2, 14; m 60; c 3. PATHOLOGY. Educ: Univ Wash, BS, 36; St Louis Univ, PhD(bact), 41. Prof Exp: Fel med, Col Med, NY Univ, 41-43, from instr to assoc prof bact, 42-59, assoc prof path, 59-67; PROF MED BIOPHYS & DIR DIV LAB ANIMAL SCI, FAC MED, UNIV TORONTO, 67- Concurrent Pos: Nat Res Coun fel, 41-42; dir, Berg Inst Sci Serv, 53-67; mem bd gov, Inst Lab Animal Resources, Nat Res Coun, 54-67 & resources panel, Can Coun Animal Care. Honors & Awards: Lasker Award, Lasker

Found, 49; Griffin Award , Am Asn Lab Animal Sci, 73. Mem: Am Asn Lab Animal Sci (pres, 63); Am Soc Microbiologists; fel NY Acad Sci; Can Asn Lab Animal Sci (pres, 68-70); Can Microbiol. Res: Pathology of infectious disease; mouse and rabbit pox; latent infections of animal populations; laboratory animal disease. Mailing Add: Div of Lab Animal Sci Fac of Med Univ of Toronto Toronto ON Can

CHRISTENSEN, LEONARD, b Cloquet, Minn, 13; m 62. MEDICINE, SURGERY. Educ: Ore State Univ, BS, 38; Univ Ore, MD, 41, MS, 49; Am Bd Ophthal, dipl, 50. Prof Exp: From asst prof to assoc prof, 50-72, PROF OPHTHAL, MED SCH, UNIV ORE, 72- Concurrent Pos: Heed fel, 49-50. Mem: AMA; Am Acad Ophthal & Otolaryngol; Am Ophthal Soc. Res: Ophthalmology, particularly ophthalmological pathology. Mailing Add: Dept of Ophthal Univ of Ore Med Sch Portland OR 97201

CHRISTENSEN, MARK NEWELL, b Green Bay, Wis, July 16, 30; m 55. GEOLOGY. Educ: Univ Alaska, BS, 52; Univ Calif, PhD, 59. Prof Exp: From actg instr to instr geol, Univ Calif, Berkeley, 59-60, from asst prof to assoc prof, 60-74; PROF GEOL & GEOPHYS, UNIV CALIF, SANTA CRUZ, 74-, CHANCELLOR, 74- Concurrent Pos: Asst dean, Col Letters & Sci, Univ Calif, Berkeley, 65-74. Mem: AAAS; Geol Soc Am; Am Geophys Union. Res: Structural geology; deformation of rocks; stratigraphy; tectonics. Mailing Add: 292 Cent Serv Bldg Univ House Univ Calif Meyer Dr Santa Cruz CA 95064

CHRISTENSEN, MARTHA, b Ames, Iowa, Jan 4, 32. MYCOLOGY, ECOLOGY. Educ: Univ Nebr, BSc, 53; Univ Wis, MS, 56, PhD(bot), 60. Prof Exp: Proj assoc mycol, Univ Wis, 60-63; asst prof, 63-68, ASSOC PROF BOT, UNIV WYO, 68- Honors & Awards: Weber-Ernst Award, 53. Mem: Mycol Soc Am; Ecol Soc Am; Brit Mycol Soc. Res: Ecology and taxonomy of soil microfungi. Mailing Add: Dept of Bot Univ of Wyo Laramie WY 82070

CHRISTENSEN, NED JAY, b Clarkston, Utah, June 23, 29; m 51; c 3. AUDIOLOGY. Educ: Brigham Young Univ, BA, 54, MA, 55; Pa State Univ, PhD, 59. Prof Exp: Instr, Pa State Univ, 55-59; asst prof speech & clin supvr, WVa Univ, 59-62; from asst prof to assoc prof, 62-74, asst dir, 62-70, PROF SPEECH, UNIV ORE, 74-, DIR SPEECH PATH-AUDIOL PROG, 72- Mem: Am Speech & Hearing Asn. Res: Audiological rehabilitation. Mailing Add: Dept of Speech Path-Audiol Univ of Ore Col of Educ Eugene OR 97403

CHRISTENSEN, NIELS GUNNAR, b Los Angeles, Calif, June 10, 24; m 61; c 1. ELECTROOPTICS. Educ: Okla State Univ, BS, 48, MS, 50. Prof Exp: Res physicist, Frost Geophys Corp, Okla, 50; mathematician, Douglas Aircraft Co, 54; group head guided missile div, Firestone Tire & Rubber Co, 54-56; proj engr space technol labs, TRW, Inc, 56-58, dept mgr commun satellite, 59-60; reliability systs mgr, AC Electronics Div, Gen Motors Corp, Wis, 60-61, qual assurance dir, 62-63, engr dir, Apollo Guid Systs, 64-65, inertial instruments, 66-68; mgr, Electro-Optical Labs, 68-70, opers mgr, 70-75, DIV MGR, ELECTRO-OPTICAL DIV, HUGHES AIRCRAFT CO, 75- Res: Electro-optical sensors for night vision; missile guidance; space applications; electronic countermeasures; infrared surveillance. Mailing Add: Hughes Aircraft Bldg 119-2 PO Box 90515 Los Angeles CA 90009

CHRISTENSEN, NIKOLAS IVAN, b Madison, Wis, Apr 11, 37; m m 60; c 1. GEOPHYSICS, MINERALOGY. Educ: Univ Wis, BS, 59, MS, 61, PhD(geol), 63. Prof Exp: Res fel geophys, Harvard Univ, 63-64; assoc prof geol, Univ Southern Calif, 64-67; PROF GEOL, UNIV WASH, 67- Mem: Fel Geol Soc Am; Am Geophys Union; Seismol Soc Am. Res: Elasticity of rocks and minerals; crystal physics, nature of the earth's interior. Mailing Add: Dept of Geol Univ of Wash Seattle WA 98105

CHRISTENSEN, NORMAN LEROY, JR, b Fresno, Calif, Dec 28, 46; m 68; c 2. PLANT ECOLOGY. Educ: Calif State Univ, Fresno, BA, 68, MA, 70; Univ Calif, Santa Barbara, PhD(biol), 73. Prof Exp: ASST PROF BOT, DUKE UNIV, 73- Mem: Ecol Soc Am. Res: Effects of disturbance on plant community structure and function; effects of fire on community nutrient relations; forest demography. Mailing Add: Dept of Bot Duke Univ Durham NC 27706

CHRISTENSEN, ODIN DALE, b Duluth, Minn, Dec 12, 47; m 71. MINERALOGY. Educ: Univ Minn, BA, 70; Stanford Univ, PhD(geol), 75. Prof Exp: ASST PROF GEOL, UNIV NDAK, 75- Mem: Geol Soc Am; Mineral Soc Am; Soc Mining Engrs; Am Geophys Union. Res: Defining the mineralogic and petrographic changes which occur during diagenesis of pelitic rocks and using these criteria as indicators of low-grade metamorphism. Mailing Add: Dept of Geol Univ of NDak Grand Forks ND 58201

CHRISTENSEN, ROBERT LEE, b Orange, NJ, July 23, 29; m 52; c 3. ATOMIC PHYSICS, COMPUTER SCIENCE. Educ: Princeton Univ, AB, 50, MA, 54, PhD(physics), 57. Prof Exp: Asst reactor dept, Brookhaven Nat Lab, 53; asst physics dept, Princeton Univ, 57-58; physicist res lab, Int Bus Mach Corp, 58-61, eng mgt, 61-65; vpres, Quantum Sci Corp, 66-67; VPRES, BECKER SECURITIES CORP, 68- Mem: Am Phys Soc; Inst Elec & Electronics Engrs. Res: Nuclear and atomic resonance; ultra-high vacuum; photoelectric effect; application of digital computers to image and language processing; management information systems, applied statistics and econometrics. Mailing Add: Becker Securities Corp 55 Water St New York NY 10041

CHRISTENSEN, SABINUS HOEGSBRO, b New York, NY, May 20, 15; m 46. PHYSICS. Educ: Pratt Inst, BME, 41; Harvard Univ, MS, 48, ScD(math physics), 51. Prof Exp: Lab asst, Pratt & Whitney Aircraft Div, United Aircraft Corp, 33-37; jr res engr, Sperry Gyroscope Co, 41-42; res engr, Fairchild Aircraft Corp, 42-44; physicist in chg aerodynamic res, Carrier Corp, 44-47; res assoc prof, Ga Inst Technol, 51-53; prof & chmn dept physics, Clark & Morehouse Cols, 53-60; prof sci & math, Deep Springs Col, 60-61; prof physics, Hobart & William Smith Cols, 61-64; prof & chmn dept, Bard Col, 64-66; chmn dept, 69-75, PROF PHYSICS, LINCOLN UNIV, PA, 66- Concurrent Pos: Consult, Oak Ridge Assoc Univs, 63- & Goddard Space Flight Ctr, NASA. Mem: Am Asn Physics Teachers; Am Phys Soc; Sigma Xi. Res: Radiation physics. Mailing Add: Dept of Physics Lincoln University PA 19352

CHRISTENSEN, STANLEY HOWARD, b Boone, Iowa, Mar 6, 35; m 56; c 4. SOLID STATE PHYSICS. Educ: Iowa State Univ, BS, 57; Cornell Univ, PhD(physics), 63. Prof Exp: Jr physicist, Iowa State Univ, 57; asst prof, 63-67, ASSOC PROF PHYSICS, KENT STATE UNIV, 67- mem: Am Phys Soc; Am Asn Physics Teachers. Res: Electron paramagnetic resonance; laser light scattering. Mailing Add: Dept of Physics Kent State Univ Kent OH 44242

CHRISTENSEN, THOMAS GASH, b Richmond, Va, Sept 16, 44. EXPERIMENTAL BIOLOGY. Educ: Rutgers Univ, BS, 66; Univ Vt, PhD(bot), 71. Prof Exp: Res assoc med biochem, Sch Med, Univ Vt, 71-74; RES ASSOC PULMONARY PATH, SCH MED, BOSTON UNIV, 74- Concurrent Pos: Vis lectr biol, Univ Vt, 71-74. Mem: Sigma Xi; AAAS. Res: Pathogenesis of chronic bronchitis and emphysema; histology and secretory products of respiratory epithelium; mechanism of transbronchial passage of aerosolized medications. Mailing Add: Boston Univ Sch of Med Boston MA 02118

CHRISTENSEN, VALDEMAR J, inorganic chemistry, see 12th edition

CHRISTENSON, CHARLES O, b Oakland, Calif, Sept 17, 36; m 60; c 4. TOPOLOGY. Educ: Univ Kans, BA, 58, MA, 60; NMex State Univ, PhD(math), 64. Prof Exp: Asst prof, 64-74, ASSOC PROF MATH, UNIV IDAHO, 74- Mem: AAAS; Am Math Soc; Math Asn Am. Res: Knot theory; study of the knotting number; topology of manifolds. Mailing Add: Dept of Math Univ of Idaho Moscow ID 83843

CHRISTENSON, DONALD ROBERT, b Terry, Mont, Mar 31, 37; m 59; c 3. SOIL SCIENCE, PLANT PHYSIOLOGY. Educ: Mont State Col, BS, 60, MS, 64; Mich State Univ, PhD(soil sci), 68. Prof Exp: Asst prof, 68-73, ASSOC PROF SOIL SCI, MICH STATE UNIV, 73- Honors & Awards: Scarseth Award, Am Soc Agron, 67. Mem: Am Soc Agron; Potato Asn Am. Res: Soil-plant nutrient relationships; response of plants to applied nutrients; soil test correlations; mechanisms of nutrient release from soil minerals. Mailing Add: Dept of Crop & Soil Sci Mich State Univ East Lansing MI 48824

CHRISTENSON, GEORGE LAURENT, bacteriology, deceased

CHRISTENSON, PAUL JOHN, b Watervliet, NY, Aug 13, 21; m 48; c 9. MEDICINE, PREVENTIVE MEDICINE. Educ: Siena Col, NY, BS, 43; Marquette Univ, MD, 46; Columbia Univ, MPH, 55. Prof Exp: Pvt pract med, NY, 49-52; dir pub health, Tri County Unit, Va, 52-53 & Mich, 53-54; dir pub health, Utica City Dept Health, NY, 55-57; mem med staff, Eaton Labs, Norwich Pharmacol Col, 57-59, dir clin res, 59-60, med dir, Norwich Prod Div, 60-65; med dir, Quinton Div, Merck & Co, Inc, NJ, 65-69; med dir, S E Massengill Co, 69-71; med dir, Semed Pharmaceut, 69-71; MED DIR, OTTAWA COUNTY HEALTH DEPT, 71- Concurrent Pos: Chief med examr, Ottawa Co, Mich; mem Mich Gov comt med manpower. Mem: AAAS; Am Pub Health Asn; Am Col Prev Med; World Med Asn; AMA. Res: Clinical pharmaceutical research, especially chemotherapeutic agents; public health. Mailing Add: 414 Washington Grand Haven MI 49417

CHRISTENSON, ROGER MORRIS, b Sturgeon Bay, Wis, Sept 28, 19; m 47; c 3. CHEMISTRY. Educ: Univ Wis, BS, 41, MS, 42, PhD(food chem), 44. Prof Exp: Asst, Univ Wis, 41-44; res chemist, 44-52, res supvr, Synthetic Vehicles, 52-56, mgr new prods res, 58-65, DIV DIR RESIN RES, PPG INDUSTS, 65- Mem: Am Chem Soc; Fedn Socs Paint Technol. Res: Polymer chemistry related to coatings and adhesives; thermosetting and thermoplastic acrylic polymers, alkyd and epoxy resins, melamine and urea resins, polymer dispersions, oils and fatty acids, separation processes, electrodeposition of paints, pressure sensitive acrylic adhesives, radiation curing of polymers, applied coating technology. Mailing Add: PPG Industs Res Ctr Springdale PA 15144

CHRISTIAN, CHARLES DONALD, b Parker, Kans, Nov 28, 30; m 56; c 3. OBSTETRICS & GYNECOLOGY. Educ: Univ Kans, AB, 52; Duke Univ, PhD, 56, MD, 58. Prof Exp: Instr anat, Sch Med, Duke Univ, 55-59; resident obstet & gynec, Columbia-Presby Med Ctr, 59-62; asst prof, Col Med, Univ Fla, 62-64; assoc prof obstet & gynec & dir endocrine div, Med Ctr, Duke Univ, 64-69; PROF OBSTET & GYNEC & HEAD DEPT, SCH MED, UNIV ARIZ, 69- Concurrent Pos: Macy Found fel, Columbia-Presby Med Ctr, 59-62. Res: Neuro-endocrine relationships in reproductive processes. Mailing Add: Dept of Obstet & Gynec Univ of Ariz Sch of Med Tucson AZ 85724

CHRISTIAN, CHARLES L, b Wichita, Kans, July 10, 26; m 54; c 2. IMMUNOLOGY. Educ: Univ Wichita, BS, 49; Western Reserve Univ, MD, 53. Prof Exp: From instr to prof med, Col Physicians & Surgeons, Columbia Univ, 58-70; PROF MED, COL MED, CORNELL UNIV, 70-; PHYSICIAN IN CHIEF, HOSP SPEC SURG, 70- Concurrent Pos: Consult, USPHS, 63- Mem: AAAS; Soc Exp Biol & Med; Am Asn Immunol. Res: Immunochemistry; experimental pathology. Mailing Add: The Hosp for Spec Surg 535 E 70th St New York NY 10021

CHRISTIAN, CURTIS GILBERT, b Norwich, NY, Nov 16, 17; m 46; c 2. ORGANIC CHEMISTRY. Educ: Univ Calif, BS, 49, MS, 50. Prof Exp: Res chemist, Ansco Div, Gen Aniline & Film Corp, 50-52 & Union Oil Co, Calif, 52-55; res assoc, Gasparcolor, Inc, 55-58, prod mgr, 58-60; mgr photo prod pilot plant, 60-63, mgr photo prod tech serv, 63-66, mgr tech serv, Photo Film Div, 66-70, mgr prof photog prod lab, Photog Prod Div, 70-72, SR SPECIALIST, PHOTOG PROD DIV, MINN MINING & MFG CO, 72- Mem: Am Chem Soc; Soc Photog Sci & Eng; Royal Photog Soc; Soc Motion Picture & TV Engrs. Res: Photographic chemistry; phosphorus chemistry; organic synthesis. Mailing Add: Box 3064 St Paul MN 55165

CHRISTIAN, ERMINE AMERICA, b Macon, Ga, Nov 16, 22. MATHEMATICS. Educ: Mercer Univ, AB, 42; Univ Md, MBA, 60; Cath Univ Am, MSE, 70. Prof Exp: Chemist, E I du Pont de Nemours & Co, Ind, 42; aerodynamicist, Curtiss-Wright Airplane Co, NY, 44-45; res assoc, Underwater Explosives Res Lab, Mass, 45-47; mathematician, Naval Ord Lab, 47-58, physicist, 58-74, PHYSICIST, US NAVAL SURFACE WEAPONS CTR, WHITE OAK, 74- Mem: Acoust Soc Am; Am Geophys Union; Marine Technol Soc. Res: Underwater explosions phenomena. Mailing Add: Naval Surface Weapons Ctr Silver Spring MD 20910

CHRISTIAN, GARY DALE, b Eugene, Ore, Nov 25, 37; m 61; c 2. ANALYTICAL CHEMISTRY. Educ: Univ Ore, BS, 59; Univ Md, MS, 62, PhD(anal chem), 64. Prof Exp: Res anal chemist, Walter Reed Army Inst Res, 61-67; from asst prof to assoc prof, Univ Ky, 67-72; PROF CHEM, UNIV WASH, 72- Concurrent Pos: Asst prof, Univ Md, 65-66; guest lectr, Walter Reed Army Inst Res, 66; consult electroanal chem, 68-; consult, Miles Labs, Inc, 68-72 & Beckman Instruments, Inc, 70- Mem: AAAS; Am Chem Soc; Soc Appl Spectros; fel Am Inst Chemists. Res: Atomic absorption spectroscopy; electroanalytical chemistry; fluorometric analysis; gas chromatography of trace elements; clinical chemistry; enzyme assay; catalytic analysis; drug analysis; environmental analysis. Mailing Add: Dept of Chem Univ of Wash Seattle WA 98195

CHRISTIAN, HOWARD HARRIS, b Philadelphia, Pa, Sept 30, 26; m 55. BIOLOGY, DATA PROCESSING. Educ: DePauw Univ, BA, 50; Univ Mass, MA, 52. Prof Exp: Instr biol, Susquehanna Univ, 53-54; res biochemist, Smith Kline & French Labs, 54-57; lit scientist, 57-65, SUPVR TECH RECORDS SECT, WYETH LABS, 65- Res: Biological sciences; pharmaceutics; experimental pharmacology and therapeutics. Mailing Add: Tech Rec Sect Sci Info Sect Wyeth Labs Box 8299 Philadelphia PA 19101

CHRISTIAN, HOWARD J, b Cambridge, Mass, Sept 17, 23; m 45; c 8. PATHOLOGY. Educ: Tufts Univ, BS, 49, MD, 52. Prof Exp: Intern path, Boston City Hosp, 52-53, resident, 53-56; from instr to asst prof, 56-70, ASSOC PROF PATH, SCH MED, TUFTS UNIV, 70-; PATHOLOGIST, CARNEY HOSP, BOSTON, 57- Concurrent Pos: Instr, Sch Med, Boston Univ, 54-56; asst pathologist, St Elizabeth's Hosp, Brighton, Mass, 56-57; jr vis pathologist, Lemuel Shattuck Hosp, Boston, 57-; attend pathologist, Vet Admin Hosp, 61-; hon sr lectr, Aberdeen Univ, 63-64; Commonwealth Fund fel, 63- Mem: AMA; Int Acad Path. Res: Electron microscopy; cancer and diabetes. Mailing Add: Carney Hosp 2100 Dorchester Ave Boston MA 02124

CHRISTIAN, JAMES A, b Kansas City, Mo, June 20, 35; m 55; c 3. BOTANY, GENETICS. Educ: Univ Mo-Columbia, BS, 58, MA, 62, PhD(biosysts), 71. Prof Exp: Teacher, St Clair Sch Dist, Mo, 58-60 & Mehlville Sch Dist, 60-61; asst bot, Univ Mo, 61-64; asst prof biol, Tarkio Col, 64-66; asst prof, 66-74, ASSOC PROF BOT, LA TECH UNIV, 74- Mem: Am Soc Plant Taxonomists; Int Asn Plant Taxonomists; Bot Soc Am; Torrey Bot Club. Res: Biosystematics and genetics of the genus Lupinus; taxonomy and phylogeny of vascular plants. Mailing Add: Dept of Bot & Bact La Tech Univ Ruston LA 71270

CHRISTIAN, JAMES ALVIN, food science, see 12th edition

CHRISTIAN, JERRY DALE, physical chemistry, thermodynamics, see 12th edition

CHRISTIAN, JOE CLARK, b Marshall, Okla, Sept 12, 34; m 60; c 2. MEDICAL GENETICS. Educ: Okla State Univ, BS, 56; Univ Ky, MS, 59, PhD(genetics), 60, MS, 64. Prof Exp: From intern to resident internal med, Vanderbilt Univ Hosp, 64-66; from asst prof to assoc prof, 66-74, PROF MED GENETICS, MED SCH, IND UNIV, INDIANAPOLIS, 74- Mem: Am Soc Human Genetics; Am Fedn Clin Res; Am Oil Chem Soc. Res: Quantitative genetics of vascular diseases; clinical genetics. Mailing Add: Dept of Med Genetics Ind Univ Med Sch Indianapolis IN 46202

CHRISTIAN, JOHN DONALD, organic chemistry, see 12th edition

CHRISTIAN, JOHN EDWARD, bionucleonics, pharmaceutical chemistry, see 12th edition

CHRISTIAN, JOHN JERMYN, b Scranton, Pa, Apr 12, 17; m 42, 58; c 1. ENDOCRINOLOGY, PATHOBIOLOGY. Educ: Princeton Univ, AB, 39; Johns Hopkins Univ, ScD, 54. Prof Exp: Asst res pharmacologist, Wyeth Inst Appl Biochem, 48-51; head animal labs, US Naval Med Res Inst, 51-56, physiologist exp med, 56-59; assoc prof comp path, Univ Pa, 59-62; mem div endocrinol, Albert Einstein Med Ctr, 62-69, PROF BIOL SCI, STATE UNIV NY BINGHAMTON, 69- Concurrent Pos: Res assoc pathobiol, Johns Hopkins Univ, 54-59; assoc dir, Penrose Res Lab, Philadelphia Zool Soc, 59-62; mem ad hoc comt comp path, Nat Res Coun, Nat Acad Sci, 63-70. Honors & Awards: Mercer Award, Ecol Soc Am, 57. Mem: AAAS; Am Inst Biol Sci; Am Ornithologists Union; Soc Exp Biol & Med; Wildlife Dis Asn (vpres, 63-64, pres, 65-67). Res: Relationship of population density and social factors to endocrine adaptive mechanisms; reproduction; adrenal cortex; pathogenesis of renal disease. Mailing Add: Dept of Biol Sci State Univ of NY Binghamton NY 13901

CHRISTIAN, JOSEPH RALPH, b Chicago, Ill, June 15, 20; m 44; c 2. PEDIATRICS. Educ: Loyola Univ, Ill, MD, 44; Am Bd Pediat, dipl, 50. Prof Exp: Clin asst pediat, Stritch Sch Med, Loyola Univ, Ill, 48-50, clin instr, 50-51, asst clin prof, 51-52, from asst prof to prof, 53-61, asst chmn dept & dir res, 53-61; prof, Col Med, Univ Ill, 61-71; PROF PEDIAT & CHMN DEPT, RUSH MED COL, 71-; CHMN DEPT PEDIAT, PRESBY-ST LUKE'S HOSP, 61- Concurrent Pos: Dir med educ & sr pediatrician, Mercy Hosp, 48-61, dir pediat cardiac clin & cardiac in patient serv, 48-61, dir pediat out patient clin, 49-61, dir pediat residency training prog, 54-61; attend pediatrician, Loyola Serv, La Rabida Sanitarium, 48-61; chief pediat, Lewis Mem Hosp, 51-61; sr attend pediatrician, Cook County Hosp, 59-65. Mem: Fel Am Acad Pediat; fel Am Col Chest Physicians; fel Am Col Physicians; Am Fedn Clin Res; Am Therapeut Soc. Res: Infant nutrition; fluid and electrolyte balance; accidental poisoning in children; pediatric cardiology. Mailing Add: Rush-Presby-St Luke's Med Ctr 1753 Congress Pkwy Chicago IL 60612

CHRISTIAN, PAUL JACKSON, b Barre, Vt, Sept 9, 20; m 46; c 4. SYSTEMATIC ENTOMOLOGY. Educ: Wheaton Col, AB, 47; Univ Kans, PhD(entom), 52. Prof Exp: Asst instr biol, Univ Kans, 48-51; from asst prof to assoc prof, Univ Louisville, 52-61; assoc prof, 61-63, PROF BIOL, BETHEL COL, MINN, 63-, CHMN DEPT, 73- Concurrent Pos: Consult, Louisville & Jefferson County Dept Pub Health, 60-61. Mem: Soc Study Evolution; Soc Syst Zool. Res: Classification of leaf hoppers. Mailing Add: Dept of Biol Bethel Col St Paul MN 55101

CHRISTIAN, ROBERT ROLAND, b Meriden, Conn, Nov 10, 24. MATHEMATICS. Educ: Yale Univ, BS, 47, MA, 49, PhD(math), 54. Prof Exp: Instr math, New Haven Col, 46-49; asst instr, Yale Univ, 48-49; instr, Clark Univ, 49-51; from asst to asst prof, 52-62, ASSOC PROF MATH, UNIV BC, 62- Concurrent Pos: Pvt sch teacher, 47; vis lectr, Univ Ill, 59-60. Mem: Am Math Soc; Math Asn Am. Res: Mathematics education; integration; partially ordered vector spaces. Mailing Add: Dept of Math Univ of BC Vancouver BC Can

CHRISTIAN, ROBERT THOMAS, b Lansing, Mich, Jan 31, 24; m 64. VIROLOGY. Educ: Univ Notre Dame, BS, 50; Mich State Univ, MS, 52, PhD(virol), 56. Prof Exp: Bacteriologist, Mich State Dept Health, 50; asst poultry diagnostician, Mich State Univ, 52-55, asst pathogenic bact & immunol, 56; res assoc obstet & gynec, Med Sch, Univ Mich, 57-67, acting dir, Reuben Peterson Mem Res Lab, 64-67; ASSOC PROF ENVIRON HEALTH, KETTERING LAB, SCH MED, UNIV CINCINNATI, 67- Mem: Am Soc Microbiol; Tissue Culture Asn; NY Acad Sci. Res: Early phases of DNA virus replication; viral flora of female genital tract; poultry respiratory viruses; lyophilization of microorganism and tissue cells; in vitro toxicity testing, in vitro carcinogenesis of environmental agents and cell biological effects of these agents. Mailing Add: Dept Environ Hlth Kettering Lab Sch of Med Univ of Cincinnati Cincinnati OH 45267

CHRISTIAN, ROBERT VERNON, JR, b Wichita, Kans, Mar 1, 19; m 44; c 3. CHEMISTRY. Educ: Munic Univ Wichita, BS, 40; Iowa State Col, PhD(chem), 46. Prof Exp: Asst, Nat Defense Res Comt Proj, Iowa State Col, 42-43; from asst prof to assoc prof, 46-60, PROF CHEM, WICHITA STATE UNIV, 60- Mem: AAAS; Am Chem Soc. Res: Mass spectroscopy; volatile metal chelates; metal dithiocarbamates; trace metal analysis; chemical instrumentation. Mailing Add: Dept of Chem Wichita State Univ Wichita KS 67208

CHRISTIAN, ROSS EDGAR, b DuBois, Pa, Nov 1, 25; m 46; c 2. PHYSIOLOGY, GENETICS. Educ: Pa State Univ, BS, 47; Univ Wis, MS, 49, PhD(genetics), 51. Prof Exp: Asst genetics, Univ Wis, 47-50, instr & agt, Bur Dairy Indust, USDA, 50-51; asst prof animal husb, Wash State Univ, 51-56; from asst prof to assoc prof, 56-67, PROF ANIMAL SCI, UNIV IDAHO, 67- Mem: AAAS; Am Soc Animal Sci; Am Dairy Sci Asn; Am Genetic Asn. Res: Sterility in farm animals; genetics of fertility. Mailing Add: Dept of Animal Sci Univ of Idaho Moscow ID 83843

CHRISTIAN, SAMUEL TERRY, b Huntington, WVa, Dec 4, 37; m 58; c 3. BIOCHEMISTRY, ORGANIC CHEMISTRY. Educ: Marshall Univ, BA, 60; Univ Tenn, PhD(biochem), 66. Prof Exp: Res biochemist, Addiction Res Ctr, NIMH, Ky, 68-69, chief biochem pharmacol sect, 69-72; CHIEF NEUROSCI PROG, NEUROCHEM SECT & ASSOC PROF PSYCHIAT & BIOCHEM, MED CTR, UNIV ALA, BIRMINGHAM, 72- Concurrent Pos: Fel med chem, Univ Tenn, Memphis, 66-67; fel pharmacol, Univ Ky, 67-68; consult, NMiss Res Found, 66-; adj asst prof, Depts Community Med Pharmaceut Chem, Univ Ky, 69- Mem: AAAS; NY Acad Sci; Am Soc Biol Chem; Am Soc Neurochem; Soc Neurosci. Res: Neurochemistry and molecular pharmacology; investigation of events produced by psychoactive agents on central nervous system macromolecular or subcellular organelles and their relevance to behavioral or physiological parameters; basic neurochemistry of the brain and its relevance to brain function. Mailing Add: Neurosci Prog Univ Ala Med Ctr Univ Sta-CDLD Birmingham AL 35294

CHRISTIAN, SHERRIL DUANE, b Estherville, Iowa, Sept 28, 31; m 56; c 3. PHYSICAL CHEMISTRY. Educ: Iowa State Univ, BS, 52, PhD, 56. Prof Exp: From asst prof to prof, 56-69, asst dean, Col Arts & Sci, 63-66, chmn dept chem, 68-69, GEORGE LYNN CROSS RES PROF CHEM, UNIV OKLA, 69- Concurrent Pos: Okla Found res award, 56-57; guest prof chem, Univ Oslo, 66-67 & 74-75; res grants, NSF, Off Saline Water, Dept Interior & PRF Res Corp. Mem: Am Chem Soc; The Chem Soc. Res: Physical chemistry of molecular complexes; spectral and thermodynamic properties of hydrogen-bonded and charge-transfer complexes; effect of solvents on complex equilibria; effects of pressure on confirmational equilibria. Mailing Add: Univ of Okla Dept of Chem 620 Parrington Oval Rm 211 Norman OK 73069

CHRISTIAN, WALTER, b Chicago, Ill, Mar 5, 40; m 64; c 3. MICROBIOLOGY, IMMUNOLOGY. Educ: Univ Ill, BS, 62; Univ Mich, MPH, 67, PhD(epidemiol sci), 74. Prof Exp: Res asst microbiol, Univ Mich, 62-64; lab technician med technol, US Army Hosp, Sandia Base, Albuquerque, NMex, 64-66; res asst epidemiol, Sch Pub Health, Univ NMex, 67-70; ASST LAB MGR LAB PROCEDURES, UPJOHN CO, 74- Mem: Am Soc Microbiol. Res: Auto-immune antibodies as applied to clinical applications; rapid diagnostic procedures as currently employed in microbiology. Mailing Add: Lab Procedures Upjohn Co 6330 Variel Ave Woodland Hills CA 91364

CHRISTIAN, WAYNE GILLESPIE, b King City, Mo, Oct 28, 18; m 43; c 2. GEOPHYSICS. Educ: WTex State Univ, BS, 39; Univ Denver, MS, 48, EdD(sci ed), 51. Prof Exp: Mus technician, WTex State Teachers Col, 36-37, hist geol lab supvr, 36-39, field & lab supvr paleont & archeol, Mus, 39-40; chief computer, Western Geophys Los Angeles, 40-44; supt schs, Mo, 43-44 & 46-48; teacher pub schs, 44-46; dept dir, Colo State Home Children, 51-52; GEOPHYSICIST, SUN OIL CO, 52- Concurrent Pos: Supvr, Ground Water Surv & Mineral Surv, US Dept Interior & State Tex, 37-39; instr & teaching fel, Univ Denver, 48-56. Mem: Soc Vert Paleont; Soc Explor Geophys Mailing Add: Sun Oil Co Sci & Technol 503 N Central Expressway Richardson TX 75080

CHRISTIANO, JOHN G, b Falerna, Italy, Aug 29, 17; nat US; m 43; c 3. APPLIED MATHEMATICS. Educ: Univ Pittsburgh, BS, 39, MS, 42, PhD(math), 50. Prof Exp: From instr to assoc prof math, Univ Pittsburgh, 42-59; asst prof & actg head dept, Duquesne Univ, 46-47; PROF MATH, NORTHERN ILL UNIV, 59- Mem: Fedn Am Scientists; Math Asn Am; Am Math Soc. Res: Mathematics; mechanics. Mailing Add: Dept of Math Northern Ill Univ DeKalb IL 60115

CHRISTIANS, CHARLES J, b Parkersburg, Iowa, Apr 15, 34; m 57; c 2. ANIMAL BREEDING. Educ: Iowa State Univ, BS, 55; NDak State Univ, MS, 58; Okla State Univ, PhD(animal breeding), 62. Prof Exp: Asst animal husb, NDak State Univ, 56-58 & Okla State Univ, 58-61; from asst prof to assoc prof, Miss State Univ, 61-64; from asst prof to assoc prof, 64-71, PROF ANIMAL HUSB, UNIV MINN, ST PAUL, 71- Mem: Am Soc Animal Sci. Res: Beef breeding; factors affecting various beef carcass traits. Mailing Add: Animal Husb Sci & Agr Exten Dept of Agr Univ of Minn St Paul MN 55101

CHRISTIANSEN, C ARTHUR, plant anatomy, see 12th edition

CHRISTIANSEN, E A, b Shellbrook, Sask, Sept 20, 28; m 54. GEOLOGY. Educ: Univ Sask, BSA, 52, MSc, 56; Univ Ill, PhD(geol), 59. Prof Exp: Asst res officer, 59-63, ASSOC RES OFFICER GEOL, SASK RES COUN, UNIV SASK, 63-, ADJ PROF GEOL SCI, 74- Mem: Fel Geol Soc Am. Res: Glacial and groundwater geology; occurrence of groundwater in drift, Tertiary and upper Cretaceous sediments. Mailing Add: Sask Res Coun Univ of Sask Saskatoon SK Can

CHRISTIANSEN, FRANCIS WYMAN, b Richfield, Utah, Feb 19, 12; m 33; c 6. GEOLOGY. Educ: Univ Utah, BS, 35, MS, 37; Princeton Univ, PhD(struct ecol geol), 48. Prof Exp: Mining geologist, Sierra Consol Mines, Inc, Nev, 37-38; instr geol, Univ Utah, 39-40; indust specialist, War Prod Bd, Washington, DC, 42, asst dep dir mining div, 42-43, chief metals sect, 43-46; from asst prof to assoc prof, 46-58, PROF ECON & STRUCT GEOL, UNIV UTAH, 58- Mem: AAAS; Geol Soc Am; Am Geophys Union; Am Inst Mining, Metall & Petrol Engrs; Am Asn Petrol Geologists. Res: Structural geology and ore deposit; polygonal yielding of tabular bodies; magasutures of the earth and continental genesis. Mailing Add: Dept of Geol & Geophys Sci Univ of Utah Salt Lake City UT 84112

CHRISTIANSEN, GORDON SECRIST, biochemistry, see 12th edition

CHRISTIANSEN, JAMES BRACKNEY, b Alden, Minn, Mar 14, 11; m 37; c 2. BIOLOGICAL CHEMISTRY. Educ: Carroll Col, Wis, BA, 32; Univ Wis, MA, 34, PhD(biol chem), 39. Prof Exp: Res assoc, Larrowe Div, Gen Mills Inc, Mich, 39-50; PROF & HEAD DEPT CHEM, BUENA VISTA COL, 54- Concurrent Pos: Pvt res, 50- Mem: Fel AAAS; Am Chem Soc; World Poultry Sci Asn; Poultry Sci Asn; Am Dairy Sci Asn. Res: Chemical measurements of vitamins and other nutrient factors; nutritional requirements of poultry and dogs; livestock feed. Mailing Add: Dept of Chem Buena Vista Col Storm Lake IA 50588

CHRISTIANSEN, JAMES LEARNED, b Detroit, Mich, Oct 20, 40; m 68. ECOLOGY, HERPETOLOGY. Educ: Buena Vista Col, BA, 62; Univ Utah, MS, 65; Univ NMex, PhD(biol), 69. Prof Exp: Asst, Univ Utah, 62-64; asst, Univ NMex, 64-65, asst cur herpet div, Mus Southwestern Biol, 65-67, asst chg zool II lab prog, 67-68; instr biol, Univ Albuquerque, 68-69; asst prof, 69-75, ASSOC PROF BIOL, DRAKE UNIV, 75- Concurrent Pos: Asst, Univ Utah, Exped to Cent Am with Dr John M Legler, 62 & leader, Exped to Mex, 65. Mem: Am Soc Ichthyologists & Herpetologists; Soc Study Amphibians & Reptiles; Am Inst Biol Sci; Brit Herpet Soc. Res: Reproduction and distribution of turtles and lizards; turtles and water pollution in Iowa. Mailing Add: Dept of Biol Drake Univ Des Moines IA 50311

CHRISTIANSEN, KENNETH ALLEN, b Chicago, Ill, June 24, 24; m 47; c 4. EVOLUTIONARY BIOLOGY, SPELEOLOGY. Educ: Boston Univ, BA, 48; Harvard Univ, PhD(biol), 51. Prof Exp: Asst prof biol, Am Univ Beirut, 51-54; instr, Smith Col, 54-55; from asst prof to assoc prof, 55-62, PROF BIOL, GRINNELL COL, 62- Concurrent Pos: Correspondent, Mus Paris. Mem: Fel AAAS; Soc Study Evolution; Soc Syst Zool; fel Nat Speleol Soc. Res: Taxonomy and evolution; Collembola. Mailing Add: Dept of Biol Grinnell Col Grinnell IA 50112

CHRISTIANSEN, MARJORIE MINER, b Canton, Ill, Feb 28, 22; m 51; c 1. NUTRITION, BIOCHEMISTRY. Educ: Univ NMex, BS, 49, MA, 55; Utah State Univ, PhD(nutrit, biochem), 67. Prof Exp: Chemist, Carnegie-Ill Steel Co, Ind, 42-44; control chemist, Blockson Chem Co, Ill, 44-47; asst dietitian, St Joseph Hosp, Albuquerque, NMex, 48-50; instr sci, Regina Sch Nursing, 50-64, instr nutrit, 52-64, proj dir utilization of basic sci prin in solving nursing care probs, 66-69; PROF HOME ECON, MADISON COL, VA, 69- Concurrent Pos: USPHS div nursing training grant, 66-68; proj dir dietary sem, Va Regional Med Prog proj grant, 73- Mem: Am Dietetic Asn; Am Home Econ Asn; Soc Nutrit Educ; Nutrit Today Soc. Res: Serum alpha-tocopherol, cholesterol and lipids in adults on self-selected diets at different levels of polyunsaturated fat. Mailing Add: Dept of Home Econ Madison Col Harrisonburg VA 22801

CHRISTIANSEN, MERYL NAEVE, b Gooselake, Iowa, Sept 5, 25; m 50. PLANT PHYSIOLOGY. Educ: Univ Ark, BS, 50, MS, 55; NC State Univ, PhD(crop sci), 60. Prof Exp: Asst, Univ Ark, 51-54; agronomist crops res div, USDA, 55-58; asst, NC State Univ, 58-60; plant physiologist, Crops Res Div, USDA, 60-73, PLANT PHYSIOLOGIST & CHIEF PLANT STRESS LAB, PLANT PHYSIOL INST, AGR RES SERV, USDA, 73- Mem: Am Soc Plant Physiol; Crop Sci Soc Am; Phytochem Soc NAm; NY Acad Sci. Res: Seed germination physiology; environmental influences on seedling development and metabolism. Mailing Add: Plant Stress Lab Physiol Inst USDA Beltsville Agr Res Ctr Beltsville MD 20705

CHRISTIANSEN, PAUL ARTHUR, b Mitchell Co, Iowa, June 7, 32; m 55; c 2. BOTANY, PLANT ECOLOGY. Educ: Univ Iowa, BA, 59; Univ Ore, MS, 64; Iowa State Univ, PhD(plant ecol), 67. Prof Exp: Teacher, Humboldt Community Schs, Iowa, 59-64; asst prof biol, 67-74, ASSOC PROF BIOL, CORNELL COL, 74- Concurrent Pos: Vis prof, Ore State Univ, 69-70. Mem: AAAS; Am Inst Biol Scientists; Ecol Soc Am. Res: Establishment of prairie species; management of natural areas. Mailing Add: Dept of Biol Cornell Col Mt Vernon IA 52314

CHRISTIANSEN, RICHARD LOUIS, b Denison, Iowa, Apr 1, 35; m 56; c 3. ORTHODONTICS, PHYSIOLOGY. Educ: Univ Iowa, DDS, 59; Ind Univ, Bloomington, MSD, 64; Univ Minn, Minneapolis, PhD(physiol), 70. Prof Exp: Intern dent, USPHS Hosp, San Francisco, 59-60; chief dent officer, USPHS Outpatient Clin, St Louis, 60-62; NIH trainee, Sch Dept, Ind Univ, Bloomington, 62-64; staff orthodontist, Oral Med & Surg Br, Nat Inst Dent Res, 64-66; staff lectr orthod, Sch Dent, Univ Minn, Minneapolis, 66-70; PRIN INVESTR, ORAL MED & SURG BR, NAT INST DENT RES, 70-, CHIEF CRANIOFACIAL ANOMALIES PROG BR, 73- Concurrent Pos: NIH res fel physiol, Univ Minn, Minneapolis, 66-70; mem numerous state of the art planning comt, Nat Inst Dent Res, 69-; vis prof orthod, Sch Dent, Georgetown Univ, 70-; vis lectr, Univ Md, 70- Mem: Am Dent Asn; Am Asn Orthod; Int Asn Dent Res; Int Union Physiol Sci. Res: Craniofacial malformations; oral physiology, especially intra-oral pressures and motor function, hemodynamics of oral-facial tissues, equilibrium of the dentition and biophysics of orthodontic tooth movement. Mailing Add: Craniofacial Anomalies Prog Br Nat Inst Dent Res Bethesda MD 20014

CHRISTIANSEN, ROBERT GEORGE, b Sangudo, Alta, Apr 23, 24; m 48. ORGANIC CHEMISTRY. Educ: Univ Alta, BSc, 46, MSc, 48; Univ Wis, PhD(chem), 52 Prof Exp: 6ctr org chem, Univ Alta, 46-48; asst, Univ Wis, 48-50; res assoc, 51-61, SR RES ASSOC & GROUP LEADER, STERLING-WINTHROP RES INST, 61- Mem: Am Chem Soc. Res: Steroids; medicinal chemistry; modified steroidal hormones. Mailing Add: Sterling-Winthrop Res Inst Rensselaer NY 12144

CHRISTIANSEN, ROBERT LORENZ, b Kingsburg, Calif, June 13, 35; m 62; c 3. GEOLOGY. Educ: Stanford Univ, BS, 56, MS, 57, PhD(geol), 61. Prof Exp: Geologist explor geol, Utah Construct & Mining Co, 57-58; geologist mineral, Stanford Res Inst, 60-61; GEOLOGIST VOLCANIC GEOL & PETROL, US GEOL SURV, 61- Mem: AAAS; Geol Soc Am; Mineral Soc Am; Am Geophys Union. Res: Igneous petrology; volcanology; geothermal energy; geology of cordilleran region of the United States. Mailing Add: US Geol Surv 345 Middlefield Rd Menlo Park CA 94025

CHRISTIANSEN, WAYNE ARTHUR, b Ft Collins, Colo; c 2. RADIO ASTRONOMY. Educ: Univ Colo, BS, 62; Univ Calif, Santa Barbara, MA, 66, PhD(physics), 68. Prof Exp: Scientist physics, E G & G Inc, 62-64; fel astrophys, Joint Inst Lab Astrophys, Univ Colo, 68-70; ASST PROF ASTRON, UNIV NC, CHAPEL HILL, 70- Mem: Am Astron Soc; Royal Astron Soc. Res: Origin and evolution of radio galaxies and quasars and the interaction between these objects and their surroundings. Mailing Add: Dept of Physics & Astron Univ of NC Chapel Hill NC 27514

CHRISTIANSON, DONALD DUANE, b Fertile, Minn, May 26, 31; c 4. PLANT BIOCHEMISTRY. Educ: Concordia Col, BA, 55; NDak State Univ, MS, 57. Prof Exp: PRIN RES CHEMIST BIOCHEM, NORTHERN REGIONAL LAB, USDA, 57- Honors & Awards: Cert of Merit, Northern Regional Lab, USDA, 70. Mem: Int Asn Cereal Chemists; Am Chem Soc. Res: Determine enzymes and nonenzymic lipid oxidation processes in wheat and corn that contribute to off-flavors and loss in nutritional quality. Mailing Add: 1010 N Summit Blvd Peoria IL 61606

CHRISTIANSON, GEORGE, b Volga, SDak, May 7, 17; m 47; c 2. FOOD SCIENCE. Educ: SDak State Col, BS, 39; Univ Tenn, MS, 40; Univ Minn, MS, 51, PhD(biochem), 53. Prof Exp: Asst blood lipids, Univ Minn, 41-42, asst dairy chem, 50-53; res chemist cereal chem, Gen Mills, Inc, 46-48; res scientist meats & meat prod, Rath Packing Co, 53-63; RES ASSOC, JAMES FORD BELL TECH CTR, GEN MILLS, INC, 63- Mem: Inst Food Technologists. Res: Milk stability; meat preservation studies; cereal chemistry; freezing and storage of meats; meat processing equipment; ready to eat cereals; physical chemistry of sugars; confectionary development; cereal snack development. Mailing Add: 18210 30 Place N Wayzata MN 55319

CHRISTIANSON, JOHN DEAN, b Chester, Iowa, Sept 30, 41; m 65. PLANT ECOLOGY. Educ: Upper Iowa Univ, BS, 63; Mankato State Col, MS, 65; Rutgers Univ, PhD(plant ecol), 69. Prof Exp: Instr high sch, Iowa, 63-66; ASST PROF BIOL, WAGNER COL, 69- Mem: AAAS; Am Inst Biol Sci; Torrey Bot Club. Mailing Add: Dept of Biol Wagner Col Staten Island NY 10301

CHRISTIANSON, LEE (EDWARD), b Dayton, Ohio, May 5, 40; m 63; c 2. MAMMALOGY, ECOLOGY. Educ: Univ NDak, BS, 63; Southern Ill Univ, MA, 65; Univ Ariz, PhD(zool), 67. Prof Exp: Asst prof, 67-72, ASSOC PROF BIOL SCI, UNIV OF THE PAC, 72- Mem: Am Soc Mammalogists; Soc Syst Zool. Res: Mammalian systematics and ecology. Mailing Add: Dept of Biol Sci Univ of the Pac Stockton CA 95204

CHRISTIE, ALISTAIR D, b Calcutta, India, Mar 3, 33; Can citizen; m 55; c 2. ATMOSPHERIC PHYSICS, METEOROLOGY. Educ: St Andrews Univ, BSc, 55; Univ Toronto, MA, 56; Univ London, PhD(atmospheric physics), 65. Prof Exp: Meteorologist, 56-61, RES METEOROLOGIST, ATMOSPHERIC ENVIRON SERV, DEPT OF TRANSP, CAN, 65- Concurrent Pos: Mem working group, Int Comn Meteorol of Upper Atmosphere, 67- Mem: Am Meteorol Soc; Royal Meteorol Soc; Can Meteorol Soc. Res: Meteorological exchange processes; atmospheric tracers; noctilucent clouds; photochemistry and dynamics of the stratosphere and mesosphere; D-region electron density variations. Mailing Add: Atmospheric Environ Serv DOE 4905 Dufferin St Toronto ON Can

CHRISTIE, BERTRAM RODNEY, b Moorefield, Ont, Mar 22, 33; m 60; c 3. CROP BREEDING. Educ: Ont Agr Col, BSA, 55, MSA, 56; Iowa State Univ, PhD(crop breeding), 59. Prof Exp: From asst prof to assoc prof crop sci, Ont Agr Col, 59-70, PROF CROP SCI, ONT AGR COL, UNIV GUELPH, 70- Mem: Am Soc Agron; Agr Inst Can. Res: Forage crop breeding. Mailing Add: Dept of Crop Sci Ont Agr Col Univ of Guelph Guelph ON Can

CHRISTIE, BRUCE ROBERT, b Colac, Australia, Sept 22, 32; div; c 4. VETERINARY PATHOLOGY, ANTHROPOLOGY. Educ: Univ Sydney, BVSc, 57; Mich State Univ, MS, 67, PhD(path), 69. Prof Exp: Lectr vet sci, Longerenong Col, Dooen, Australia, 56-60; res officer, Cameron Lab, Werribee, 61-64; sr pathologist, Attwood Res Lab, Westmeadows, 69-72; sr scientist path, Ill Inst Technol Res, 72-73; DIR PATH, SPECIALIZED CTR RES, RANCHO LOS AMIGOS HOSP, 73- Concurrent Pos: Consult, Asian Develop Bank, Rizal, Philippines, 74-; asst clin prof, Sch Med, Univ Southern Calif, 75- Mem: Royal Col Vet Surgeons; Australian Col Vet Scientists; Australian Vet Asn; Am Vet Med Asn. Res: Quantitative pathology of the respiratory tract; comparative pathology; animal models of human disease; rural folk music dance and legend in Karnataka, India; folk arts and artisans of Mexico. Mailing Add: Rancho Los Amigos Hosp Med Sci Bldg 7601 E Imperial Hwy Downey CA 90242

CHRISTIE, DAN EDWIN, mathematics, deceased

CHRISTIE, JOHN MCDOUGALL, b Calcutta, India, Dec 4, 31; m 57; c 3. STRUCTURAL GEOLOGY, ELECTRON MICROSCOPY. Educ: Univ Edinburgh, BSc, 53, PhD, 56. Prof Exp: Instr geol, Pomona Col, 56-58; from asst prof to assoc prof, 58-68, PROF GEOL, UNIV CALIF, LOS ANGELES, 68- Concurrent Pos: Guggenheim fel, 64-65; hon fel, Australian Nat Univ, 64-65. Mem: Geol Soc Am; Am Geophys Union; Electron Micros Soc Am. Res: Structural geology and petrology; electron microscopy of minerals; experimental deformation of minerals and rocks. Mailing Add: Dept of Geol Univ of Calif Los Angeles CA 90024

CHRISTIE, JOSEPH HERMAN, b Magnolia, Ark, Aug 30, 37. ANALYTICAL CHEMISTRY. Educ: Rensselaer Polytech Inst, BS, 59; La State Univ, MS, 62; Colo State Univ, PhD(chem), 74. Prof Exp: Mem tech staff chem, Rockwell Int, 62-69; asst prof, Colo State Univ, 74-75; SUPVRY CHEMIST, US GEOL SURV, 75- Concurrent Pos: Fac affil chem, Colo State Univ, 75- Mem: Am Chem Soc; Electrochem Soc; Soc Appl Spectros. Res: Electrochemical and spectroscopic trace analysis; computer applications in analytical chemistry and instrumentation. Mailing Add: Br Anal Lab US Geol Surv Menlo Park CA 94025

CHRISTIE, PETER ALLAN, b Englewood, NJ, Feb 2, 40; m 62. ORGANIC POLYMER CHEMISTRY. Educ: Juniata Col, BS, 62; Univ Del, PhD(org chem), 67. Prof Exp: RES CHEMIST POLYMER SYNTHESIS, RES & DEVELOP CTR, ARMSTRONG CORK CO, 67- Mem: Am Chem Soc. Res: Heterocyclic synthesis; organophosphorus chemistry; application of nuclear magnetic resonance spectroscopy to stereochemistry; condensation, addition, and ring-opening polymerization; reactions on polymers; organic-inorganic polymer systems. Mailing Add: Res & Develop Ctr Armstrong Cork Co Lancaster PA 17604

CHRISTIE, STEPHEN ROLLAND, b Dunedin, Fla, Nov 5, 29; m 68; c 1. PLANT VIROLOGY. Prof Exp: Lab technician food sci, Dept Food Technol, 58-60, lab technician plant path, 60-64, sr lab technician plant virol, 64-68, electron micros technician, 68-73, PLANT PATHOLOGIST PLANT VIROL, UNIV FLA, 73- Mem: Am Phytopath Soc. Res: Hybridizes nicotiana species for use in plant virus research; characterizes new plant viruses by host ranges, serology, and electron microscopy; tests crops for plant viruses; identifies them and considers control. Mailing Add: Dept of Plant Path Univ of Fla Plant Virus Lab Gainesville FL 32611

CHRISTIE, WARNER HOWARD, b Brooklyn, NY, Oct 29, 29; m 66; c 3. MASS SPECTROMETRY. Educ: Univ Miami, Fla, BS, 51, MS, 53; Univ Fla, PhD(chem), 58. Prof Exp: CHEMIST, MASS SPECTROMETRY, ANAL CHEM DIV, OAK RIDGE NAT LAB, 59- Mem: Am Chem Soc. Res: Synthesis and reactions of organic fluorine containing materials; isotope exchange reactions of fluorocarbons; mass spectrometry; application of small computers to mass spectrometry; ion microprobe mass analysis. Mailing Add: 952 W Outer Dr Oak Ridge TN 37830

CHRISTINSEN, JAMES EDWARD, organic chemistry, see 12th edition

CHRISTMAN, ADAM A, b Shannon, Ill, Dec 11, 95; m 23; c 2. PHYSIOLOGICAL CHEMISTRY. Educ: Grinnell Col, BS, 17; Univ Ill, MS, 20, PhD(biol chem), 22. Prof Exp: Chemist, Hercules Powder Co, 17-19; teaching asst, Univ Ill, 19-21; instr, 22-25, from asst prof to prof, 25-64; EMER PROF BIOL CHEM, MED SCH, UNIV MICH & CONSULT, 64- Mem: Am Soc Biol Chemists; Soc Exp Biol & Med. Res: Purine metabolism; methods for uric acid and allantoin; studies on methylated purines; determination of carbon monoxide in blood and air; histidine compounds in muscle extracts; distribution of anserine and carnosine in muscle of white rat, comparative levels in skeletal muscle of other animals. Mailing Add: Dept of Biol Chem Med Sci Bldg Univ of Mich Med Sch Ann Arbor MI 48104

CHRISTMAN, ARTHUR CASTNER, JR, b North Wales, Pa, May 11, 22; m 45; c 6. PHYSICS. Educ: Pa State Univ, BS, 44, MS, 50. Prof Exp: Instr physics, George Washington Univ, 48-51; physicist opers res off, Johns Hopkins Univ, 51-58; sr physicist, Stanford Res Inst, 58-62, head opers res group, 62-64, mgr opers eval dept, 65-66, mgr opers res dept, 66-69, dir opers res dept, 69-71, dir tactical systs, 71-75; SCI ADV, HQ US ARMY TRAINING & DOCTRINE COMMAND, 75- Concurrent Pos: Consult, US Navy, 50-51. Mem: Am Phys Soc; Sigma Xi; Opers Res Soc Am; fel AAAS. Res: Operations research; systems analysis; weapons, information, traffic, postal and health systems; analytic modeling; simulation; field experimentation; reconnaissance; surveillance; target acquisition; interdiction; close support; air defense; countermeasures; x-rays.

CHRISTMAN, DAVID R, b Columbus, Ohio, Oct 14, 23; m 52; c 3. ORGANIC CHEMISTRY, PHARMACEUTICAL CHEMISTRY. Educ: Ohio State Univ, BSc, 47; Carnegie Inst Technol, MSc, 50, DSc(chem), 51. Prof Exp: Asst, Carnegie Inst Technol, 47-51; assoc chemist, 51-64, CHEMIST, BROOKHAVEN NAT LAB, 64- Concurrent Pos: Lectr, Columbia Univ, 58-64. Mem: Am Chem Soc; The Chem Soc; Soc Nuclear Med. Res: Organic radioactivity analysis and syntheses; organic radiation chemistry; organic radiopharmaceuticals with isotopes of short half-life; data processing. Mailing Add: Dept of Chem Brookhaven Nat Lab Upton NY 11973

CHRISTMAN, EDWARD ARTHUR, b Lakewood, Ohio, Aug 3, 43. RADIATION CHEMISTRY. Educ: Ohio Univ, BS, 65; Rutgers Univ, MS, 74, PhD(radiation sci), 76. Prof Exp: Mech engr aerospace, Missile Syst Div, Avco Corp, Mass, 65-71; INSTR RADIATION SCI, RUTGERS UNIV, BUSCH CAMPUS, 73- Mem: Sigma Xi. Res: Radiation chemistry of tritium, heavy ions and radiological health and protection. Mailing Add: Radiation Sci Doolittle Hall Rutgers Univ Busch Campus New Brunswick NJ 08903

CHRISTMAN, JUDITH KERSHAW, b Teaneck, NJ, Apr 8, 41; m 59. BIOCHEMISTRY, MOLECULAR BIOLOGY. Educ: NY Univ, AB, 62; Columbia Univ, PhD(biochem), 67. Prof Exp: Res fel, Nucleic Acid Dept, NY Blood Ctr, 67-71; asst mem dept enzym, Inst Muscle Dis, 71-74; asst prof, 74-75, ASSOC PROF MOLECULAR BIOL, DEPT PEDIAT, MT SINAI SCH MED, 75- Mem: Am Soc Biol Chemists; Am Soc Cell Biol; Sigma Xi; Harvey Soc; AAAS. Res: Biochemical basis of regulatory mechanisms involved in cell differentiation, oncogenesis; regulation of biosynthesis of mRNA, proteins; chemical and immunological characterization of plasminogen activators from transformed cells; determination of role of proteases in tumorigenicity; replication of Reovirus in mammalian cells. Mailing Add: Dept of Pediat Mt Sinai Sch of Med New York NY 10029

CHRISTMAN, LUTHER PARMALEE, b Summit Hill, Pa, Feb 26, 15; m 39; c 3. ANTHROPOLOGY. Educ: Temple Univ, BS, 48, MEd, 52; Mich State Univ, PhD(sociol, anthrop), 65. Prof Exp: Instr nursing, Cooper Hosp Sch Nursing, 48-53; dir, Yankton State Hosp, SDak, 53-56; consult, Mich Dept Health, 56-63; assoc prof, Sch Nursing, Univ Mich, 63-67, res assoc, Inst Social Res, 64-67 & Bur Hosp Admin, 66-67; prof sociol, Col Arts & Sci & prof & dean, Col Nursing, Vanderbilt Univ, 67-72; PROF & DEAN COL NURSING & ALLIED HEALTH SCI, RUSH UNIV, 72-, VPRES NURSING AFFAIRS, RUSH-PRESBY-ST LUKE'S MED CTR, 72- Concurrent Pos: Consult community serv res, NIMH, 63-66; mem nursing panel, Nat Comn Study Nursing & Nursing Educ, 68-70, mem panel consults, Comt Nursing, 71-73, mem, White House Conf Children, 70; mem exchange med info rev group, Vet Admin, 71-; mem comt memberships, Nat Inst Med, 72-75, comt educ health prof, 73-74 & ad hoc mem recruitment comt, 75- Mem: Nat Inst Med; AAAS; Soc Gen Syst Res; Am Sociol Asn; Am Acad Nursing. Res: Effective models for delivering care to patients; facilitation of interdisciplinary collaboration; clinical nursing practice. Mailing Add: Col Nursing & Allied Health Sci Rush Univ 1725 W Harrison Chicago IL 60612

CHRISTMAN, ROBERT ADAM, b Ann Arbor, Mich, May 16, 24; m 53; c 4. GEOLOGY. Educ: Univ Mich, BS, 46, MS, 47; Princeton Univ, PhD(geol), 50. Prof Exp: Geologist, US Geol Surv, 50-54; asst prof geol, Cornell Univ, 54-60; ASSOC PROF GEOL, WESTERN WASH STATE COL, 60- Mem: AAAS; Nat Asn Geol Teachers; Geol Soc Am. Res: Petrology; mineralogy; earth science for teachers. Mailing Add: Dept of Geol Western Wash State Col Bellingham WA 98225

CHRISTMAN, RUSSELL FABRIQUE, b June 20, 36; m 58; c 3. CHEMISTRY. Educ: Univ Fla, BS, 58, MS, 60, PhD(chem), 62. Prof Exp: Res asst prof sanit chem, Univ Wash, 62-66, asst prof civil eng, 66-68, assoc prof appl sci, 68-74, asst to provost & dir div environ affairs, 70-74; PROF ENVIRON SCI, UNIV NC, CHAPEL HILL, 74- Mem: AAAS; Water Pollution Control Fedn; Am Water Works Asn; Am Soc Limnol & Oceanog. Res: Chemical structures of natural product organic materials in water; methods of organic analysis in water samples; mechanisms of colloidal destabilization with hydrolysis products of aluminum III. Mailing Add: Dept of Environ Sci & Eng Univ of NC Chapel Hill NC 27514

CHRISTMANN, MARVIN HENRY, b Petrel, NDak, July 3, 34; m 60; c 2. SOLID STATE PHYSICS, MATHEMATICS. Educ: SDak Sch Mines & Technol, BS, 56; Iowa State Univ, MS, 60. Prof Exp: Consult, Minneapolis-Honeywell Res Ctr, 56-57; res physicist, 60-61 & 63-66, sr res physicist, 66-71, RES SPECIALIST, CENT RES LABS, MINN MINING & MFG CO, 71- Mem: Inst Elec & Electronic Engrs; Metall Soc; Am Inst Mining, Metall & Petrol Engrs. Res: Determine by experimental optical and electrical methods the complicated dependence of perfection and properties on the degree of supersaturation of films grown epitaxially by vacuum vapor deposition. Mailing Add: Cent Res Labs Minn Min & Mfg Co PO Box 33221 St Paul MN 55133

CHRISTMAS, ELLSWORTH P, b Warrick Co, Ind, Nov 5, 35; m 58; c 2. AGRONOMY. Educ: Purdue Univ, BS, 58, MS, 61, PhD(agr ed, agron), 64. Prof Exp: Teacher sec sch, Ind, 58-60; from asst prof to assoc prof, 64-74, PROF AGRON, PURDUE UNIV, 74-, ASST DIR, IND COOP EXT SERV, 74- Concurrent Pos: Agronomist, Int Progs Agr, Purdue-Brazil Proj, Brazil, 69-73. Mem: Am Soc Agron; Soil Sci Soc Am; Crop Sci Soc Am. Res: Teaching methods in agriculture; soil characterization and conservation. Mailing Add: Coop Ext Serv Agr Admin Bldg Purdue Univ West Lafayette IN 47906

CHRISTOFFERSEN, DONALD JOHN, b Ogema, Wis, July 27, 34; m 51; c 4. ANALYTICAL CHEMISTRY. Educ: Wis State Univ-Stevens Point, BS, 56; Univ Wis, PhD(anal chem), 66. Prof Exp: Chemist, Pure Oil Co, 61-63, group leader gas chromatography, 63-65, sr res chemist, 64-69, SUPVR, SPECTRAL ANAL CHEM, UNION OIL CO, 69- Mem: Am Chem Soc. Res: Gas chromatography with petroleum oriented applications. Mailing Add: Union Oil Co Res Ctr Box 76 Brea CA 92621

CHRISTOFFERSEN, RALPH EARL, b Elgin, Ill, Dec 4, 37; m 61; c 1. PHYSICAL CHEMISTRY. Educ: Cornell Col, BS, 59; Ind Univ, PhD(phys chem), 64. Prof Exp: NIH fel quantum chem, Univ Nottingham, 64-65 & Iowa State Univ, 65-66; from asst prof to assoc prof phys chem, 66-72, PROF PHYS CHEM, UNIV KANS, 72- Concurrent Pos: Alfred P Sloan res fel, 71-73; consult, Upjohn Co, Mich & Argonne Nat Lab, Ill; consult, Xerox Corp, 75- Honors & Awards: Am Inst Chemists Award, 59. Mem: AAAS; Am Inst Chemists; Am Chem Soc; Am Phys Soc. Res: Quantum chemistry; theory of chemical bonds; relativistic effects; evaluation of molecular integrals; ab initio calculations on large molecules. Mailing Add: Dept of Chem Univ of Kans Lawrence KS 66044

CHRISTOFFERSON, ERIC, b Newburyport, Mass, May 29, 39; m 61; c 2. GEOLOGICAL OCEANOGRAPHY. Educ: Princeton Univ, AB, 61; Univ RI, PhD(oceanog), 73. Prof Exp: Instr oceanog, Univ RI, 73-74, res assoc, 73-75; ASST PROF GEOL, RUTGERS UNIV, 75- Mem: Sigma Xi; Am Geophys Union. Res: Geologic history of the Caribbean Sea. Mailing Add: Dept of Geol Rutgers Univ New Brunswick NJ 08903

CHRISTOFFERSON, GLEN DAVIS, b Tacoma, Wash, Feb 7, 31; m 51; c 2. PHYSICAL CHEMISTRY. Educ: Univ Wash, BS, 53; Univ Calif, Los Angeles, PhD(chem), 58. Prof Exp: Asst, Univ Calif, Los Angeles, 53-57; res chemist, Calif Res Corp, 57-63, sr res chemist, 64-68, SR RES ASSOC, CHEVRON RES CO, 68- Mem: Soc Appl Spectros; Am Chem Soc; Am Crystallog Asn. Res: Organic and inorganic crystal structure determination; x-ray emission and absorption spectroscopy; x-ray low angle scattering; electron diffraction and microscopy. Mailing Add: Chevron Res Co 576 Standard Ave Richmond CA 94802

CHRISTOPH, FRANCIS THEODORE, JR, b Alexandria, La, Jan 1, 43; m 68. TOPOLOGY. Educ: St Peter's Col, NJ, BS, 64; Rutgers Univ, MS, 66, PhD(math), 69. Prof Exp: ASST PROF MATH, TEMPLE UNIV, 69- Mem: Math Asn Am; Am Math Soc. Res: Decompositions and extensions of topological semigroups; embedding topological semigroups in topological groups. Mailing Add: Dept of Math Temple Univ Philadelphia PA 19122

CHRISTOPH, ROY JAY, biology, see 12th edition

CHRISTOPHER, EVERETT PERCY, b Hamilton, NY, May, 10, 04; m 30, 70; c 4. HORTICULTURE. Educ: RI State Col, BS, 26, MS, 30; Cornell Univ, PhD, 34. Prof Exp: From instr to prof hort, 27-69, prof plant & soil sci, 69-71, actg asst dean, Col Agr, 40-42, vdean, 42-59, assoc dean, 62-69, exten horticulturist, 27-62, pomologist, Agr Exp Sta, 40-62, assoc dir, 62-69, actg dean & dir, 70, EMER PROF PLANT & SOIL SCI, UNIV RI, 71- Concurrent Pos: Mem sci adv comt, Refrig Res Found, 57-72; res adv, Off Rural Develop, AID, SKorea, 71-73. Mem: AAAS; Am Soc Hort Sci. Res: Pomology; soil management; pruning; storage; care and feeding of garden plants. Mailing Add: Dept Plant & Soil Sci Col Resource Develop Univ of RI Kingston RI 02881

CHRISTOPHER, JOHN, b Chicago, Ill, Oct 15, 23; m 47; c 3. MATHEMATICS. Educ: Knox Col, AB, 46; Univ Ore, MA, 50, PhD(math), 52. Prof Exp: Instr math, Univ Ore, 50-52 & Knox Col, 52-54; asst prof, Univ of the Pac, 54-55; instr, Fresno State Col, 55-56; sr mathematician, Electrodata Corp, Calif, 56-58; asst prof math, Sacramento State Col, 58-60; dir comput ctr, Univ Nebr, 60-63; assoc prof, 63-67, PROF MATH, CALIF STATE UNIV, SACRAMENTO, 67- Res: Number theory and numerical analysis. Mailing Add: Dept of Math Sacramento State Col 600 Jay St Sacramento CA 95819

CHRISTOPHER, ROBERT PAUL, b Cleveland, Ohio, Apr 27, 32; m 62; c 3. PHYSICAL MEDICINE & REHABILITATION. Educ: Northwestern Univ, BS, 54; St Louis Univ, MD, 59. Prof Exp: US Off Voc Rehab fel phys med & rehab, Univ Mich, Ann Arbor, 60-63, from instr to asst prof, 63-67; assoc prof, 67-71, PROF PHYS MED & REHAB, UNIV TENN, MEMPHIS, 71-; ASSOC MED DIR, LES PASSEES REHAB CTR, 70- Concurrent Pos: Chief phys med & rehab, Vet Admin Hosp, Ann Arbor, 63-67; consult, St Jude Children's Res Hosp, 67-, Le Bonheur Children's Hosp, 68-, Vet Admin Hosps, Memphis, 67- & Nashville, 70-, Coun Med Educ, AMA, 69- & Comn Accreditation Rehab Facil, 73- Mem: Fel Am Acad Phys Med & Rehab; Am Cong Rehab Med; Am Asn Electromyog & Electrodiag. Res: Electrodiagnostic studies in collagen vascular disease; habilitation programs for brain damaged children; primary muscle disease. Mailing Add: Div of Phys Med & Rehab Univ Tenn 800 Madison Ave Memphis TN 38163

CHRISTOPHERSON, WILLIAM MARTIN, b Salt Lake City, Utah, July 2, 16; m 43; c 1. MEDICINE. Educ: Univ Louisville, MD, 42. Prof Exp: Ewing fel, Mem Cancer Ctr, NY, 49-50; from asst prof to assoc prof, 50-56, chmn dept, 56-74, PROF PATH, SCH MED, UNIV LOUISVILLE, 56- Concurrent Pos: Pathologist, Louisville Gen Hosp, 56-; consult, Med Div, Nat Cancer Inst, Vet Admin Hosp & Ireland Army Hosp, Ky; mem adv comt & spec consult, Cancer Control Prog, USPHS. Mem: Int Acad Path (past pres); Am Soc Cytol (past pres); Soc Exp Path; Am Asn Cancer Educ (past pres); Am Cancer Soc. Res: Cancer. Mailing Add: Dept of Path Univ Louisville Health Sci Ctr Louisville KY 40201

CHRISTOPHOROU, LOUCAS GEORGIOU, b Limassol, Cyprus, Jan 21, 37; m 63; c 2. ATOMIC PHYSICS, MOLECULAR PHYSICS. Educ: Nat Univ Athens, BSc, 60; Manchester, dipl adv physics, 61, PhD(physics), 63, DSc(physics), 69. Prof Exp: Sr res physicist, Health Physics Div, Oak Ridge Nat Lab, 63-64; from asst prof to assoc prof, 64-68, PROF PHYS, UNIV TENN, KNOXVILLE, 69-; HEAD ATOMIC & MOLECULAR RADIATION PHYSICS GROUP, HEALTH PHYSICS DIV, OAK RIDGE NAT LAB, 66- Concurrent Pos: Consult, Oak Ridge Nat Lab, 64-66. Mem: Am Phys Soc; Radiation Res Soc; Health Physics Soc; AAAS. Res: Radiation physics; chemical physics; low-energy electron-molecule interactions; photophysical processes. Mailing Add: Health Physics Div Oak Ridge Nat Lab PO Box X Oak Ridge TN 37830

CHRISTOPOULOS, GEORGE NICK, b Kokkino, Greece, Sept 13, 38; US citizen; m 64; c 2. TOXICOLOGY. Educ: Roosevelt Univ, BS, 66; Univ Ill Med Ctr, MS, 71, PhD(med chem), 72. Prof Exp: Lab technician, Dearborn Chem Co, 60-63, Div Corn Prod, Best Foods, 63-65 & Durkee's Famous Food, 65-66; CHIEF TOXICOLOGIST, COOK COUNTY CORONER, 66- Concurrent Pos: Consult, Dept Pub Health, State of Ill, 74-75. Mem: Am Chem Soc; NY Acad Sci; Am Acad Forensic Sci. Res: Isolation and identification of toxic gases other than carbon monoxide in fire victims. Mailing Add: 1828 W Polk St Chicago IL 60612

CHRISTY, JOHN HARLAN, b Ft Smith, Ark, Aug 13, 37; m 60; c 2. MATHEMATICS. Educ: Mass Inst Technol, BS, 59; Vanderbilt Univ, PhD(math), 64. Prof Exp: Res asst physics, Los Alamos Sci Lab, 59-60; asst prof math, Southwestern at Memphis, 63-66; assoc prof, Hendrix Col, 66-67; PROF MATH & CHMN DEPT, TEX WOMAN'S UNIV, 67- Concurrent Pos: Vis prof, Univ Ark, 64. Mem: Math Asn Am. Res: Topological dynamics; expansive transformation groups. Mailing Add: Dept of Math Tex Woman's Univ Denton TX 76204

CHRISTY, NICHOLAS PIERSON, b Morristown, NJ, June 18, 23; m 47; c 2. MEDICINE. Educ: Yale Univ, AB, 45; Columbia Univ, MD, 51; Am Bd Internal Med, 58. Prof Exp: From instr to assoc prof med, 56-65, assoc clin prof, 65-72, PROF MED, COL PHYSICIANS & SURGEONS, COLUMBIA UNIV, 72-, CHMN DEPT MED, ROOSEVELT HOSP, 70- Concurrent Pos: Markle scholar, 56; asst physician, Presby Hosp, 54-60, asst attend physician, 60-62, assoc attend physician, 62-; asst vis physician, Francis Delafield Hosp, 54- & first med div, Bellevue Hosp, 58-; ed, J Clin Endocrin & Metab, 63-67. Honors & Awards: Borden Award, 51. Mem: AAAS; Am Physiol Soc; Soc Exp Biol & Med; Fedn Clin Res; Endocrine Soc. Res: Clinical disorders of the adrenal cortex; adrenal cortical physiology of animals and man; mechanisms of action of adrenal cortical and gonadal steroids at a cellular level; metabolism of estrogens in hepatic disease. Mailing Add: Dept of Med Roosevelt Hosp 428 W 59th St New York NY 10019

CHRISTY, ROBERT FREDERICK, b Vancouver, BC, May 14, 16; nat US; m 73; c 3. THEORETICAL PHYSICS, THEORETICAL ASTROPHYSICS. Educ: Univ BC, BA, 35, MA, 37; Univ Calif, PhD(theoret physics), 41. Prof Exp: Instr physics, Ill Inst Technol, 41-42; res assoc, Univ Chicago, 42-43 & AEC, Los Alamos, NMex, 43-46, chmn fac, 69-70, from asst prof to assoc prof, 46-50, PROF PHYSICS, CALIF INST TECHNOL, 50-, VPRES & PROVOST, 70- Concurrent Pos: Chmn comt nat acad to surv risks of nuclear power. Mem: Int Astron Union; Am Phys Soc; Am Astron Soc; Nat Acad Sci. Res: Cosmic rays; nuclear physics; astrophysics; variable stars. Mailing Add: Calif Inst of Technol Pasadena CA 91125

CHRISTY, ROBERT WENTWORTH, b Chicago, Ill, Nov 2, 22. PHYSICS. Educ: Univ Chicago, MS, 49, PhD(physics), 53. Prof Exp: Consult, Motorola, Inc, 52-53;

from instr to assoc prof, 53-62, chmn dept, 62-67, PROF PHYSICS, DARTMOUTH COL, 62- Concurrent Pos: Consult, Space Technol Labs, Inc, 58-68. Mem: AAAS; Am Phys Soc; Am Asn Physics Teachers. Res: Ionic crystals; plastic flow; thermoelectric power; color centers; luminescence; thin films; metal optics. Mailing Add: Dept of Physics & Astron Dartmouth Col Hanover NH 03755

CHROMEY, FRED CARL, b Philadelphia, Pa, June 30, 18; m 43; c 4. APPLIED PHYSICS. Educ: St Josephs Col, Pa, BS, 40; Cornell Univ, PhD(physics), 44. Prof Exp: Asst physics, Ind Univ, 40-42 & Cornell Univ, 42-44; res assoc, Mass Inst Technol contract, res physicist, Los Alamos Sci Lab, NMex, 44-46; RES PHYSICIST, E I DU PONT DE NEMOURS & CO, INC, 46- Concurrent Pos: Instr univ exten, Purdue Univ, 41-42, supvr studies, 41. Mem: Am Phys Soc. Res: Measurement of radio-activity; cyclotron construction and operation; cosmic rays; theory of scattering by colored bodies; viscoelasticity; digital computer programs; statistical experimental designs; explosion hazards. Mailing Add: 6 N Cliffe Dr Wycliffe Wilmington DE 19809

CHRONIC, HALKA (PATTISON), atmospheric sciences, geology, see 12th edition

CHRONIC, JOHN, b Tulsa, Okla, June 3, 21; m 48; c 4. INVERTEBRATE PALEONTOLOGY, STRATIGRAPHY. Educ: Univ Tulsa, BS, 42; Univ Kans, MS, 47; Columbia Univ, PhD(geol), 49. Prof Exp: Instr geol, Univ Mich, 49-50; from asst prof to assoc prof, 50-65, PROF GEOL, UNIV COLO, BOULDER, 65- Concurrent Pos: Exchange lectr, Univ Edinburgh, 58-59; NSF lectr, State Univ NY Col Oneonta, 62; prof & chmn dept, Haile Sellassie Univ, 65-66 & Australian Nat Univ, 69-70. Mem: AAAS; Geol Soc Am; Soc Econ Paleontologists & Mineralogists; Am Asn Petrol Geologists; Paleont Soc. Res: Colorado geology; economic geology. Mailing Add: 971 Sixth St Boulder CO 80302

CHRYSANT, STEVEN GEORGE, b Gargaliani, Greece, Feb 22, 34; US citizen; m 69; c 2. CARDIOVASCULAR DISEASES. Educ: Univ Athens, MD, 59, PhD(biochem), 68. Prof Exp: From instr to asst prof med, Stritch Sch Med, Loyola Univ, 70-72; asst prof, 72-75, ASSOC PROF MED, HEALTH SCI CTR, UNIV OKLA, 76- Concurrent Pos: Res assoc nephrology, Hines Vet Admin Hosp, Ill, 70-71, staff physician, 71-72, asst sect chief renal hypertension & in-chg-hemodialysis unit, 72; dir hypertension screening & treatment prog, Oklahoma City Vet Admin Hosp, 72-; mem coun kidney dis, Am Heart Asn. Honors & Awards: Young Investr Travel Award, Am Soc Nephrology, 72; Gold Medal, Univ Patrae, Greece, 74; Physician's Recognition Award, AMA, 76. Mem: Am Heart Asn; Am Soc Nephrology; Int Soc Nephrology; Soc Exp Biol & Med; Sigma Xi. Res: Systemic and renal hemodynamics in both animals and humans as this pertains to the cause and treatment of hypertension. Mailing Add: Vet Admin Hosp 921 NE 13th St Oklahoma City OK 73104

CHRYSOCHOOS, JOHN, b Icaria, Greece, Feb 27, 34; nat US; m 64; c 2. PHYSICAL CHEMISTRY. Educ: Athens Tech Univ, dipl, 57; Univ BC, MSc, 62, PhD(phys chem), 64. Prof Exp: Instr chem, Univ BC, 60-64; res fel chem physics, Harvard Univ, 64-65; res assoc biophys, Michael Reese Hosp, 65-66; res assoc phys chem, Ill Inst Technol, 66-67; asst prof, 67-71, ASSOC PROF CHEM, UNIV TOLEDO, 71- Mem: AAAS; Am Chem Soc; Radiation Res Soc. Res: Molecular luminescence; spectroscopy; lasers; flash spectroscopy; radiation chemistry. Mailing Add: Dept of Chem Univ of Toledo Toledo OH 43606

CHRYSSANTHOU, CHRYSSANTHOS, b Thessalonika, Greece, Oct 15, 25; m 58; c 2. EXPERIMENTAL PATHOLOGY. Educ: Univ Thessalonika, MD, 53. Prof Exp: Pathologist, Gynec-Obstet Clin, Med Sch, Univ Thessalonika, 53-54; instr path, Sch Med, NY Univ, 57; fel, 58-59, res assoc path, 59-60, assoc exp path, 60-63, assoc pathologist, 64-68, PATHOLOGIST, DEPT LABS & RES, BETH ISRAEL MED CTR, 68-, ASSOC DIR DEPT, 64-; ASSOC PROF PATH, MT SINAI SCH MED, CITY UNIV NEW YORK, 67- Concurrent Pos: Vis prof, Claude Bernard Inst, Univ Montreal, 66; NIH res grants hypertension, 67-72; US Naval res grants decompression sickness & shock, Off Naval Res, 68-76; lectr, Sch Nursing, Beth Israel Med Ctr. Honors & Awards: Physician's Recognition Award, AMA, 69 & 72; Am Soc Clin Pathologists-Col Am Pathologists Award, 75. Mem: Soc Exp Biol & Med; Am Soc Exp Path; Am Asn Path & Bact; Am Asn Cancer Res; NY Acad Sci. Res: Hypertension; mechanism and prevention of decompression sickness; pathophysiology of vasoactive polypeptides; tolerance and addiction to narcotics; endotoxin induced reactions; experimental tumor chemotherapy; coagulation-fibrinolysis; laser photocoagulation. Mailing Add: Dept Labs & Res Beth Israel Med Ctr 10 Nathan D Perlman Pl New York NY 10003

CHRZANOWSKI, FRANCIS ALAN, b Camden, NJ, Feb 1, 45; m 68; c 3. PHYSICAL PHARMACY. Educ: Philadelphia Col Pharm & Sci, BSc, 68, MSc, 72, PhD(phys pharm), 75. Prof Exp: Asst prof pharm, Mass Col Pharm, 74-75; ASST PROF PHARM, COL PHARM & ALLIED HEALTH PROF, NORTHEASTERN UNIV, 75- Mem: Am Pharmaceut Asn. Res: Analysis of drugs in biological fluids; pharmacokinetics of theophylline; metal amino acid complexation. Mailing Add: Col Pharm & Allied Health Prof Northeastern Univ Boston MA 02115

CHU, BENJAMIN PENG-NIEN, b Shanghai, China, Mar 3, 32; m 59; c 3. PHYSICAL CHEMISTRY. Educ: St Norbert Col, BS, 55; Cornell Univ, PhD(phys chem), 59. Prof Exp: Res asst nuclear eng, Brookhaven Nat Lab, NY, 57; res assoc chem, Cornell Univ, 58-62; from asst prof to assoc prof, Univ Kans, 62-68; PROF CHEM, STATE UNIV NY STONY BROOK, 68- Concurrent Pos: Sloan res fel, 66-68; Guggenheim fel, 68-69. Mem: Am Chem Soc; Am Phys Soc. Res: Critical phenomena; molecular configuration and dynamics of macromolecules in solution; structure of non-crystalline media; light scattering and small angle x-ray scattering; ion exchange. Mailing Add: Dept of Chem State Univ of NY Stony Brook NY 11794

CHU, CHI HSUIN ULLI, b Hankow, China, Jan 20, 08; nat US; m 58; c 2. ANATOMY. Educ: Nanking Univ, BS, 31; Yenching Univ, MS, 33; Washington Univ, PhD(anat), 47. Prof Exp: Instr, Nanking Univ, 33-35; asst prof anat, Med Col, Lingnan Univ, 37-40; from assoc prof to prof, Med Col, Nat Cent Univ, 41-45; asst res prof, Sch Med, Univ Utah, 52-55; sr cancer res scientist, Roswell Park Inst, 55-57; asst prof anat, 57-73, ASSOC PROF ANAT, CASE WESTERN RESERVE UNIV, 73- Concurrent Pos: Nat Cancer Inst fel, 48-50. Mem: AAAS; Am Soc Cell Biol; Am Asn Anatomists; Am Asn Cancer Res; Histochem Soc. Res: Histo-chemistry of nervous tissues; cytochemistry of Golgi apparatus; cytogenesis of connective tissue. Mailing Add: Dept of Anat Case Western Reserve Univ Cleveland OH 44106

CHU, CHIA-KUN, b Shanghai, China, Aug 14, 27; m 52; c 3. APPLIED MATHEMATICS. Educ: Chiao-Tung Univ, China, BS, 48; Cornell Univ, MME, 50; NY Univ, PhD(math), 59. Prof Exp: Develop engr, Gen Elec Co, 50-53; asst prof mech eng, Stevens Inst Technol, 53-57; assoc prof eng sci, Pratt Inst, 57-59; assoc prof aero eng, NY Univ, 59-63; vis res assoc, Plasma Res Lab, 63-65, assoc prof, 65-68, PROF ENG SCI, COLUMBIA UNIV, 68- Concurrent Pos: Guggenheim fel, 71-72. Mem: Am Math Soc; Am Inst Aeronaut & Astronaut; Am Phys Soc. Res: Fluid dynamics; plasma physics; numerical methods and computing. Mailing Add: Sch of Eng & Appl Sci Columbia Univ New York NY 10027

CHU, CHING-WU, b Hoo-nan, China, Dec 2, 41; US citizen; m 68; c 1. SOLID STATE PHYSICS. Educ: Cheng-Kung Univ, BS, 62; Fordham Univ, MS, 65; Univ Calif, San Diego, PhD(physics), 68. Prof Exp: Mem tech staff physics, Bell Labs, NJ, 68-70; from asst prof to assoc prof, 70-75, PROF PHYSICS, CLEVELAND STATE UNIV, 75- Concurrent Pos: Vis staff mem, Los Alamos Sci Lab, 75-76. Mem: Am Phys Soc. Res: High pressure and low temperature study of solids with emphasis in superconductivity and magnetism. Mailing Add: Dept of Physics Cleveland State Univ Cleveland OH 44115

CHU, EDITH JU-HWA, b Tai Tsang, China, May 18, 08; m 48; c 1. ORGANIC CHEMISTRY. Educ: Cent Univ, China, BS, 30; Univ Mich, MS, 34, ScD(org chem), 36. Prof Exp: Prof org chem, Nat Univ Peking, 36-47; vis scientist univs res insts, Brit Coun, 47; PROF CHEM, IMMACULATE HEART COL, 48- Mem: Fel AAAS; Am Chem Soc. Res: Pinacol-pinacolone rearrangement; vitamin K related compounds; microbiological work; porphyrins. Mailing Add: Dept of Chem Immaculate Heart Col Los Angeles CA 90027

CHU, ELIZABETH WANN, b Shanghai, China, Oct 29, 21; US citizen; m 46; c 1. CYTOLOGY, PATHOLOGY. Educ: Univ Hong Kong, BS & BM, 46; Shanghai Med Col, MD, 46; Am Bd Path, dipl, 65. Prof Exp: Resident internal med, Beekman Hosp, New York, 48-49; resident internal med, Cambridge City Hosp, Mass, 50-51 & path, 51-52; resident, Boston City Hosp, 52-53; med officer cytol, Washington Cytol Unit, 56-58, MED OFFICER CYTOL, PATH LAB, NAT CANCER INST, 58- Mem: Am Soc Cytol. Res: Exfoliative cytology, its value in experimental carcinogenesis, metastases and endocrine factors. Mailing Add: Path Lab Nat Cancer Inst Bethesda MD 20014

CHU, EN LUNG, b Yangchow, China, Oct 1, 08; m 30; c 5. ELECTRODYNAMICS. Educ: Chiao Tung Univ, BS, 30; Stanford Univ, PhD(physics), 51. Prof Exp: Student engr, Hangchow Elec Works, China, 30-32; oper engr, Zakow Power Plant, 32-33; instr physics, Chiao Tung Univ, 33-34; from asst to res assoc, Inst Physics, Chinese Acad Sci, 34-46; res assoc, Microwave Lab, Stanford Univ, 49-60, staff mem, Stanford Linear Accelerator Ctr, 61-70, engr-physicist, 71-74; RETIRED. Concurrent Pos: Fel, Kewilin Lab, 39-44; assoc prof, Kwangsi Univ, 41-43. Mem: Am Phys Soc; Chinese Inst Elec Eng. Res: Analytical electrodynamics. Mailing Add: Stanford Linear Accelerator Ctr Stanford Univ Stanford CA 94305

CHU, ERNEST HSIAO-YING, b Haining, China, June 3, 27; US citizen; m 54; c 3. GENETICS. Educ: St John's Univ, China, BS, 47; Univ Calif, Berkeley, MS, 51, PhD(genetics), 54. Prof Exp: From res asst to res assoc bot, Yale Univ, 54-59, lectr anat, Sch Med, 58-59; biologist genetics, Oak Ridge Nat Lab, 59-72; PROF HUMAN GENETICS, MED SCH, UNIV MICH, ANN ARBOR, 72- Concurrent Pos: Prof zool & biomed sci, Univ Tenn, Knoxville, 67-72. Mem: Genetics Soc Am; Am Soc Human Genetics; Am Soc Cell Biol; Tissue Cult Asn; Environ Mutagen Soc. Res: Somatic cell genetics; mammalian cytogenetics; radiation biology. Mailing Add: Dept of Human Genetics Univ of Mich Med Sch Ann Arbor MI 48109

CHU, FLORENCE CHIEN-HWA, b China, May 20, 18; nat US; m 43; c 3. RADIOLOGY. Educ: Nat Med Col, Shanghai, MD, 42; Am Bd Radiol, dipl, 50. Prof Exp: Clin asst radiation ther, Mem Hosp Cancer & Allied Dis, 50-53; res assoc, Radiobiol Sec, Sloan Kettering Inst, 55; from asst attend radiation therapist to assoc attend radiation therapist, 55-69, ATTEND RADIATION THERAPIST, MEM SLOAN-KETTERING CANCER CTR, 69- Concurrent Pos: Fel radiol, City Hosp, New York, 47-48; fel, Mem Hosp Cancer & Allied Dis, 49-50; res fel, Radiobiol Sect, Sloan Kettering Inst, 53-54; instr radiol, Med Col Cornell Univ, 56-61, asst prof clin radiol, 61-69, assoc prof radiol, 69-73, clin prof, 73-; assoc attend radiologist, New York Hosp, 70- Mem: AMA; Am Col Radiol; Radiol Soc NAm; Am Radium Soc. Res: Ionizing radiation. Mailing Add: Mem Sloan-Kettering Cancer Ctr 1275 York Ave New York NY 10021

CHU, FUN SUN, b China, May 7, 33; m 58; c 3. BIOCHEMISTRY. Educ: Nat Chung-Hsin Univ, BS, 54; WVa Univ, MS, 59; Univ Mo, PhD(biochem), 64. Prof Exp: Res assoc, Food Res Inst, Univ Chicago, 63-67; asst prof, Food Res Inst, 67-72, ASSOC PROF, DEPT FOOD SCIENCE & FOOD RES INST, UNIV WIS-MADISON, 72- Mem: Am Soc Microbiol. Res: Protein chemistry; biochemistry of microorganisms; biochemistry of microbial toxins. Mailing Add: Food Res Inst Univ of Wis 1925 Willow Dr Madison WI 53706

CHU, KAI-CHING, b Szechwan, China, Nov 19, 44; m 72; c 1. APPLIED MATHEMATICS, SYSTEM THEORY. Educ: Nat Taiwan Univ, BS, 66; Harvard Univ, MS, 68, PhD(appl math), 71. Prof Exp: Res asst appl math, Harvard Univ, 68-71; mathematician, Systs Control, Inc, 71-72; res staff mem appl math, 73-75, ACTG MGR SOCIAL SCI GROUP, DEPT GEN SCI, T J WATSON RES CTR, IBM CORP, 75- Concurrent Pos: Assoc ed, Trans on Automatic Control, 71- Mem: Inst Elec & Electronics Engrs. Res: Decision and control theories, optimization and estimation techniques; game theory; computer applications to urban and industrial problems. Mailing Add: T J Watson Res Ctr PO Box 218 Yorktown Heights NY 10598

CHU, KEH-CHANG, b Feng-Yang, China, May 19, 33; m 63; c 2. NUCLEAR MAGNETIC RESONANCE, RADIATION PHYSICS. Educ: Nat Taiwan Univ, BS, 55; Univ Mich, MS, 62, PhD(nuclear sci), 67. Prof Exp: Instr nuclear sci, Nat Tsing Hua Univ, 60-62; asst prof, 68-74, ASSOC PROF PHYSICS, WESTERN ILL UNIV, 74- Mem: Am Phys Soc. Res: Electron spin resonance; electron nuclear double resonance; radiation effects in solids. Mailing Add: Dept of Physics Western Ill Univ Macomb IL 61455

CHU, LUKE LO-HWA, b Chungking, China, July 1, 39; US citizen; m 66; c 2. BIOCHEMISTRY, ENDOCRINOLOGY. Educ: Nat Taiwan Univ, BS, 62; Univ Calif, San Francisco, PhD(biochem), 67. Prof Exp: Res biochemist, Univ Calif, San Francisco, 67-68, res fel endocrinol, 68-70; res assoc, 70-72, RES BIOCHEMIST, VET ADMIN HOSP, 72-; ASST PROF BIOCHEM, UNIV MO-KANSAS CITY, 73- Concurrent Pos: Instr, Univ Mo-Kansas City, 70-73. Mem: Am Chem Soc; AAAS; Endocrine Soc. Res: Mechanism of hormone actions; hormone production; mineral metabolism. Mailing Add: Vet Admin Hosp 4801 Linwood Blvd Kansas City MO 64128

CHU, SHERWOOD CHENG-WU, b Shanghai, China, Aug 30, 37; US citizen; m 59; c 1. MATHEMATICS. Educ: Harvard Univ, BA, 59; Univ Md, MA, 61, PhD(math), 63. Prof Exp: Res asst, Inst Fluid Dynamics & Appl Math, Univ Md, 60-63; Nat Acad Sci-Nat Res Coun resident res assoc, US Naval Ord Lab, 63-64; mem tech staff, Bellcomm, Inc, 64-66; asst prof math, Univ Del, 66-68; mem tech staff, Bellcomm, Inc, Washington, DC, 68-72; res mathematician, NIH, 72-76, GEN ENGR, US DEPT TRANSP, 76- Concurrent Pos: Asst prof lectr, George Washington Univ, 65-66. Mem: Am Math Soc; Soc Indust & Appl Math. Res: Applied mathematics. Mailing Add: 7012 Marbury Rd Bethesda MD 20034

CHU, SHIH-HSI, organic chemistry, see 12th edition

CHU, SHIRLEY SHAN-CHI, b Peiping, China, Feb 16, 29; US citizen; m 54; c 3. SOLID STATE SCIENCE. Educ: Taiwan Nat Univ, BS, 51; Duquesne Univ, MS, 54; Univ Pittsburgh, PhD(phys chem), 61. Prof Exp: Res assoc x-ray crystallog, Crystallog Lab, Univ Pittsburgh, 61-67; asst prof, 68-73, ASSOC PROF ELECTRON SCI, SOUTHERN METHODIST UNIV, 68- Mem: Am Crystallog Asn; Am Chem Soc; Inst Elec & Electronics Engrs. Res: X-ray crystallography; crystal structures of organic compounds and electronic materials by x-ray diffraction; crystallographic computer programming; characterization of electronic materials. Mailing Add: Inst of Technol Southern Methodist Univ Dallas TX 75275

CHU, SHU-YUAN, b Shanghai, China, July 5, 39; m 69. THEORETICAL PHYSICS. Educ: Nat Taiwan Univ, BS, 60; Univ Calif, Berkeley, PhD(physics), 66. Prof Exp: Physicist, Lawrence Radiation Lab, Univ Calif, Berkeley, 66-67; asst res physicist, Univ Calif, Riverside, 67-69; ASST PROF PHYSICS, IND UNIV, BLOOMINGTON, 69- Concurrent Pos: Vis mem fac, Univ Calif, Riverside, 75-76. Mem: Am Phys Soc. Res: Theoretical particle physics. Mailing Add: Dept of Physics Ind Univ Bloomington IN 47401

CHU, SOU YIE, b Taipei, Taiwan, Feb 17, 42; m 67; c 1. DRUG METABOLISM. Educ: Nat Taiwan Univ, BS, 64; Univ Ill, Chicago, PhD(pharmaceut chem), 70. Prof Exp: Res asst chem pharmacol, Univ Ill Med Ctr, 70, trainee, 70-74; vis fel, Nat Cancer Inst, 72-73, Nat Inst Arthritis, Metab & Digestive Dis spec fel, Roche Inst, 73-74; PHARMACOLOGIST, DRUG METAB DEPT, ABBOTT LABS, 74- Mem: AAAS; Am Chem Soc. Res: Synthesis of biologically active compounds; analysis of drugs in biological fluids; pharmacokinetic studies. Mailing Add: Drug Metab Dept Abbott Labs North Chicago IL 60064

CHU, TSANN MING, b Kaoh-siung, Formosa, Apr 18, 38; m 67. BIOCHEMISTRY. Educ: Nat Taiwan Univ, BS, 61; NC State Univ, MS, 65; Pa State Univ, PhD(biochem), 67. Prof Exp: Clin chemist, Buffalo Gen Hosp, 69-70; sr cancer res scientist & asst dir clin chem, 70-71, prin cancer res scientist, 71-72, ASSOC CHIEF CANCER RES SCIENTIST & DIR CLIN CHEM, ROSWELL PARK MEM INST, 72- Concurrent Pos: Res fel biochem, Med Found Buffalo, 67-69; United Health Found Western NY fel, 68-69; asst prof exp path, State Univ NY Buffalo, 71-74, assoc prof, 74- Mem: AAAS; Am Chem Soc. Res: Biochemistry of steroids and its conjugates; clinical endocrinology; clinical enzymology; tumor antigen and antibody. Mailing Add: Roswell Park Mem Inst 666 Elm St Buffalo NY 14203

CHU, VICTOR FU HUA, b Hankow, China, Jan 22, 18; nat US; m 47; c 3. PHYSICAL CHEMISTRY. Educ: Cent China Univ, BS, 38; Yale Univ, PhD(phys chem), 50. Prof Exp: Chemist, Chungking Saltpeter Ref, 38-39; chemist & plant supt, Kweichow Saltpeter Ref, 39-42; chemist, Hunan Oil Ref, 42-43; plant supt, Pai-Yeh Oil Ref, 43-44; instr chem, Cent China Univ, 44-47; from res chemist to sr res chemist, 50-65, res assoc, 65-75, RES FEL, PHOTO PRODS DEPT, E I DU PONT DE NEMOURS & CO, INC, 75- Honors & Awards: Journal Award, Photog Soc Am, 53. Mem: Am Chem Soc. Res: Color photography; conductance of electrolytes; photographic chemistry and systems. Mailing Add: 2502 Garth Rd Chalfonte Wilmington DE 19810

CHU, VINCENT HAO KWONG, b Shanghai, China, Oct 20, 18; m 50; c 1. INORGANIC CHEMISTRY, PHYSICAL CHEMISTRY. Educ: Sun Yat-Sen Univ, BSc, 44; Lehigh Univ, MSc, 57, PhD(inorg chem), 62. Prof Exp: Asst engr, Cent Indust Res Inst, China, mfg head, Taiwan Camphor Bur & supt res & mfg, Taipei Chem Works, 44-52; ENGR RAW MAT, HOMER RES LABS, BETHLEHEM STEEL CORP, 57- Mem: AAAS; Am Chem Soc; Am Inst Mining, Metall & Petrol Engrs; Sigma Xi. Res: Organo-metal compound; coordination in aprotic media; reduction kinetics of iron ore; iron ore agglomeration; crystal field theory; powder metallurgy; pyrometallurgy; coal gasification. Mailing Add: 1310 Woodland Circle Bethlehem PA 18017

CHU, WEI-KAN, b Yunnan, China, Apr 1, 40. EXPERIMENTAL ATOMIC PHYSICS, ENGINEERING PHYSICS. Educ: Cheng-Kung Univ, BS, 62; Baylor Univ, 63, MS, 65, PhD(physics), 69. Prof Exp: Fel, Baylor Univ, 69-72; res fel, Calif Inst Technol, 72-73, sr res fel, 73-75; ENG ION IMPLANTATION & BACKSCATTERING, IBM CORP, 75- Mem: Am Phys Soc; Electrochem Soc; Sigma Xi; Bohmische Phys Soc. Res: Energy loss of ions in matter; ion beam surface layer analysis; ion implantation in semiconductors; thin film interactions. Mailing Add: IBM Corp D171 300-095 East Fishkill Hopewell Junction NY 12533

CHU, WILLIAM HOW-JEN, b Shanghai, China, July 9, 37; US citizen; c 3. POLYMER PHYSICS. Educ: Nat Taiwan Univ, BS, 59; Univ Mass, MS, 65 & 67, PhD(chem), 69. Prof Exp: Sr phys chemist, Plastic Coating Corp, 62-67; RES STAFF MEM POLYMER PHYSICS, INT BUS MACH CORP, 69- Mem: Am Chem Soc. Res: Structural-property relationship of polymers. Mailing Add: K42/282 Int Bus Mach Corp 5600 Cottle Rd San Jose CA 95193

CHU, WILLIAM PETER, b Kumming, China, June 6, 43; US citizen; m 70; c 1. OPTICAL PHYSICS. Educ: Polytech Inst Brooklyn, BS, 65; Univ Rochester, PhD(physics), 70. Prof Exp: Res assoc chem, 70-73, RES ASST PROF PHYSICS, OLD DOM UNIV, 74- Mem: Optical Soc Am. Res: Optical properties of stratospheric aerosols including remote sensing with space craft. Mailing Add: 609 Old Dominion Rd Yorktown VA 23692

CHU, WILLIAM TONGIL, b Seoul, Korea, Apr 16, 34; m 62; c 2. RADIATION PHYSICS. Educ: Carnegie Inst Technol, BS, 57, MS, 59, PhD(physics), 63. Prof Exp: Res assoc high energy physics, Brookhaven Nat Lab, 63-64; asst prof physics, Ohio State Univ, 64-70; asst prof radiol, 71-75, ASSOC PROF RADIOL, LOMA LINDA UNIV, 75- Concurrent Pos: Res collabr, Brookhaven Nat Lab, 72-73. Mem: Am Phys Soc; Radiation Res Soc; Am Asn Physicists Med; Sigma Xi; AAAS. Res: Experimental elementary particle physics; radiation physics and radiation biology. Mailing Add: 621 E Mariposa Dr Redlands CA 92373

CHU, WILLIAM WEI-LING, b Shanghai, China, June 27, 38; US citizen; m 70. APPLIED MECHANICS, APPLIED MATHEMATICS. Educ: Northeastern Univ, MS, 65; Cornell Univ, PhD(theoret & appl mech), 69. Prof Exp: Engr, Beacon Brass Co, Mass, 62-63; ASST PROF MECH ENG, NORTHEASTERN UNIV, 69- Concurrent Pos: Assoc, Abacus Intersysts, Inc, 69- Mem: Am Acad Mech; Am Soc Mech Engrs. Res: Numerical method for the solution of problems in three dimensional elasticity; bond stresses in fiber reinforced composites; stress concentrations in composite materials. Mailing Add: 34 Sweetwater Ave Bedford MA 01730

CHU, YANG-MING, b Hankow, China, May 5, 22; m 48; c 4. MICROBIOLOGY, IMMUNOLOGY. Educ: Univ Wash, BS, 58; George Washington Univ, MS, 64, PhD(microbiol), 67. Prof Exp: Res asst, 64-67, asst res prof, 67-68, ASST PROF MICROBIOL, SCH MED, GEORGE WASHINGTON UNIV, 68- Concurrent Pos: USPHS grant, 68-71. Res: Immune response studied at cellular and molecular level. Mailing Add: Dept of Microbiol George Washington Univ Sch Med Washington DC 20005

CHU, YAW-EN, b Taipei, Taiwan, Sept 28, 37; m 66; c 2. PLANT GENETICS. Educ: Chung-Shin Univ, BS, 61; Tokyo Univ, MS, 65, PhD(plant genetics), 68. Prof Exp: Fel appl genetics, Nat Inst Genetics, Japan, 68-69; assoc res fel plant genetic, Acad Sinica, Taiwan, 69-72, head lab, Inst Bot, 71-72; asst res prof, Univ Utah, 72-74, res assoc, 74-76; SR PLANT GENETICIST, GREENFIELD RES LAB, ELI LILLY & CO, 76- Concurrent Pos: Vis assoc prof, Inst Food Crops, Chung-Shing Univ, 71; asst ed newsletter, Soc Advan Breeding Res in Asia & Oceania, 70-72. Mem: Sigma Xi; Genetic Soc Am; Crop Soc Am; Japanese Soc Genetics; Chinese Soc Agron. Res: Genetical studies on plant tissue, cell and protoplast culture; application of tissue culture techniques for plant breeding. Mailing Add: Dept of Biol Univ of Utah Salt Lake City UT 84112

CHU, YUNG YEE, b Hangchow, China, Aug 18, 33; m 67. NUCLEAR CHEMISTRY. Educ: Nat Taiwan Univ, BSc, 54; Univ Calif, Berkeley, PhD(chem), 60. Prof Exp: Res assoc nuclear chem, 59-61, assoc chemist, 61-65, CHEMIST, BROOKHAVEN NAT LAB, 65- Mem: Am Chem Soc; Am Phys Soc. Res: Nuclear fission; high energy nuclear reactions; nuclear spectroscopy. Mailing Add: Chem Dept Brookhaven Nat Lab Upton NY 11973

CHUANG, HANSON YII-KUAN, b Nanking, China, Sept 24, 35; US citizen; m 66; c 2. BIOCHEMISTRY, PATHOLOGY. Educ: Nat Taiwan Univ, BS, 58; Univ NC, PhD(biochem), 68. Prof Exp: Res asst chem, Acad Sinica, China, 60-63; res assoc path, 72-73, instr path & biochem, 73-74, ASST PROF PATH & BIOCHEM, UNIV NC, CHAPEL HILL, 74- Concurrent Pos: Res fel physiol chem, Johns Hopkins Univ, 68-71. Mem: Am Chem Soc. Res: Enzyme and protein isolation, purification and characterization; protein biosynthesis; drug metabolism; metabolism and function of biogenic and cholinergic amines; blood enzymology; platelet function in blood. Mailing Add: Dept of Path 811 Preclin Bldg Univ of NC Chapel Hill NC 27514

CHUANG, RONALD YAN-LI, b Szuchuan, China, Feb 12, 40; m 67; c 2. BIOCHEMISTRY, ONCOLOGY. Educ: Nat Taiwan Univ, BS, 61; Univ Calif, Davis, MS, 66, PhD(biochem), 71. Prof Exp: Res assoc cancer res, Columbia Univ, 71-72; ASST PROF MED, PHYSIOL & PHARMACOL, MED CTR, DUKE UNIV, 72- Concurrent Pos: NIH fel, Col Physicians & Surgeons, Columbia Univ, 71-72; chemist, Vet Admin Hosp, Durham, NC, 73-; NIH res grant, Med Ctr, Duke Univ, 74-77. Mem: AAAS. Res: Study of the control mechanism of gene expression in leukemic cells. Mailing Add: Hemat Div Dept of Med Duke Univ Med Ctr Durham NC 27710

CHUANG, TSAN IANG, b Hsin-Chu, Taiwan, Apr 21, 33; US citizen; m 58; c 3. SYSTEMATIC BOTANY. Educ: Taiwan Normal Univ, BS, 56; Nat Taiwan Univ, MS, 59; Univ Calif, Berkeley, PhD(bot), 66. Prof Exp: Asst res fel & cur herbarium, Inst Bot, Academia Sinica, Taiwan, 59-62; asst prof, Univ RI, 66-67; asst prof, 67-71, ASSOC PROF BOT & CUR HERBARIUM, DEPT BIOL SCI, ILL STATE UNIV, 71- Concurrent Pos: NSF res grants, 72 & 75. Mem: Bot Soc Am; Am Soc Plant Taxonomists; Int Asn Plant Taxon; Int Orgn Plant Biosystematists. Res: Systematics and evolution of genera cordylanthus, castilleia, orthocarpus; cytotaxonomy of umbelliferae; pollen morphology and its taxonomic significance of hydrophyllaceae, Campanulaceae and scrophulariaceae. Mailing Add: Dept of Biol Sci Ill State Univ Normal IL 61761

CHUBB, FRANCIS LEARMONTH, b Que, June 26, 13; m 44. ORGANIC CHEMISTRY. Educ: McGill Univ, BSc, 35; Univ Southern Calif, MSc, 49, PhD, 52. Prof Exp: Chemist, Dom Oilcloth & Linoleum Co, 35-46; lab asst, Univ Southern Calif, 46-50; Can Cancer Soc fel, Univ Alta, 51-54; sr chemist, Merck & Co, Ltd, 54-55; dir org chem res, Frank W Horner Ltd, 55-75; RES ASSOC CHEM, McGILL UNIV, 75- Mem: Am Chem Soc; fel Chem Inst Can. Res: Pharmaceutical chemistry; five and six-membered heterocyclic compounds; six-membered alicyclic compounds. Mailing Add: Dept of chem McGill Univ Montreal PQ Can

CHUBB, MICHAEL, b Eastbourne, Eng, Dec 10, 31; US citizen; m 56; c 1. RESOURCE GEOGRAPHY. Educ: Univ Toronto, BScF, 55; Mich State Univ, MS, 64, PhD(geog), 67. Prof Exp: Field officer, Ont Ministry Natural Resources, 55-63; researcher recreation resources, Mich Dept Natural Resources, 65-66; asst prof resource develop, 66-69, assoc prof park & recreation resources & dir recreation research & planning unit, 69-71, ASSOC PROF GEOG, MICH STATE UNIV, 71- Concurrent Pos: Asst ed, Can J Forest Res, 70-; mem panel recreation & aesthetics nat water qual criteria proj, Nat Acad Sci, 71-73; consult, Bur Planning, Minn Dept Natural Resources, Waterways Div, Mich Dept Natural Resources & Planning Comn, City of Lansing; fac developer, NSF Proj Developing MS Specialization in Land Use Analysis, 75- Mem: Nat Recreation & Parks Asn; Soc Park & Recreation Educr; Soc Am Foresters; Asn Am Geogr. Res: Interrelation between biological and social aspects of natural resource management, particularly the physical and psychological aspects of the carrying capacity of land for recreation; user studies; recreation resource planning. Mailing Add: Dept of Geog Mich State Univ East Lansing MI 48824

CHUBB, TALBOT ALBERT, b Pittsburgh, Pa, Nov 5, 23; m; c 4. GEOPHYSICS, ASTROPHYSICS. Educ: Princeton Univ, AB, 44; Univ NC, PhD(physics), 50. Prof Exp: HEAD, UPPER AIR PHYSICS BR, NAVAL RES LAB, 59- Honors & Awards: E O Hulburt Award, Naval Res Lab, 63. Mem: Am Geophys Union; Am Astron Soc; Am Phys Soc; Int Solar Energy Soc. Res: Aeronomy; optical geophysics; x-ray astronomy; solar thermal processes for energy recovery. Mailing Add: Code 7120 Naval Res Lab Washington DC 20375

CHUBER, STEWART, b Queens Village, NY, Dec 22, 30; m 53; c 2. PETROLEUM GEOLOGY. Educ: Colo Sch Mines, GeolE, 52; Stanford Univ, MS, 53, PhD(geol), 61. Prof Exp: Subsurface geologist, Magnolia Petrol Corp, 53-54; field geologist, Mobil Oil Co Can, Ltd, 54-56, party chief, 56; subsurface geologist, Western Div, Mobil Oil Co, Calif, 57-60 & Franco Western Oil Co, Calif & Tex, 61-65; consult geologist, 65-68; div geologist, Buttes Gas & Oil Co, 68-70; consult geologist, 71; VPRES, CANTRELL, WHEELER, LEWIS & CHUBER GEOLOGISTS & ENGRS, 71-; CONSULT, 74- Mem: Am Asn Petrol Geol; Geol Soc Am; Soc Econ Paleontologists & Mineralogists. Res: Stratigraphy; stratigraphic nomenclature; geologic history, especially Permian and Pennsylvanian cyclic sedimentation. Mailing Add: Ste 1002 711 Polk St Houston TX 77002

CHUCKROW, VICKI G, b Brooklyn, NY, July 26, 41. MATHEMATICS. Educ: City Col NY, BS, 62; NY Univ, MS, 64, PhD(math), 66. Prof Exp: Asst prof, 66-72, ASSOC PROF MATH, CITY COL NY, 72- Concurrent Pos: NSF grant, 66-67. Mem: Am Math Soc; Math Asn Am. Res: Riemann surfaces; Schottky groups, a special subclass of Kleinian groups. Mailing Add: Dept of Math City Col of NY New York NY 10031

CHUDD, CLETUS CHARLES, b Cleveland, Ohio, May 5, 11. ORGANIC CHEMISTRY. Educ: Univ Dayton, BS, 35; Western Reserve Univ, PhD(chem), 52. Prof Exp: Teacher private sch, Hawaii, 35-45 & Ohio, 45-46; instr, 47-50, chmn dept,

55-64, PROF CHEM, UNIV DAYTON, 52- Honors & Awards: Distinguished Serv Prof, Univ Dayton, 75. Mem: Am Chem Soc. Res: Conjugated and non-conjugated diene systems; organic synthesis; polymers. Mailing Add: Dept of Chem Univ of Dayton Dayton OH 45469

CHUI, CHARLES KAM-TAI, b Macao, China, May 7, 40; m 64; c 2. MATHEMATICS. Educ: Univ Wis-Madison, BS, 62, MS, 63, PhD(math), 67. Prof Exp: Asst prof math, State Univ NY Buffalo, 67-70; assoc prof, 70-74, PROF MATH, TEX A&M UNIV, 74- Mem: Math Asn Am; Am Math Soc. Res: Analysis; approximation theory. Mailing Add: Dept of Math Tex A&M Univ College Station TX 77843

CHUI, SIU-TAT, b Hong Kong, Apr 20, 49; Chinese citizen. THEORETICAL SOLID STATE PHYSICS. Educ: McGill Univ, BSc, 69; Princeton Univ, PhD(physics), 72. Prof Exp: Instr physics, Princeton Univ, 72-73; mem tech staff, Bell Tel Lab, 73-75; ASST PROF PHYSICS, STATE UNIV NY ALBANY, 75- Mem: Am Phys Soc; Sigma Xi. Res: Superconductivity; one-dimensional physics; lattice dynamics; metal-insulator transition. Mailing Add: Dept of Physics State Univ NY Albany NY 12222

CHULICK, EUGENE THOMAS, b Jackson, Calif, Jan 8, 44; m 66. NUCLEAR CHEMISTRY Educ: Univ of the Pac, BS, 65; Wash Univ, PhD(nuclear chem), 69. Prof Exp: Res assoc nuclear chem, Cyclotron Inst, Tex A&M Univ, 69-74, instr, Physics Dept, 74; SR RES CHEMIST, BABCOCK & WILCOX, LYNCHBURG RES CTR, 74- Concurrent Pos: Tex Res Coun fel, 70-; Tech expert, Int Atomic Energy Agency, 73-74. Mem: AAAS; Am Chem Soc; Am Phys Soc; Am Nuclear Soc. Res: Delayed neutrons from fission products; nuclear reactions; internal ionization during beta decay; radiochemistry of pressurized water reactors. Mailing Add: Babcock & Wilcox Lynchburg Res Ctr PO Box 1260 Lynchburg VA 24505

CHULSKI, THOMAS, b Grand Rapids, Mich, Aug 6, 21; m 47; c 6. ANALYTICAL CHEMISTRY. Educ: Mich State Univ, PhD(chem), 53 Prof Exp: Chemist, Lindsay Chem Co, 47-50; chemist, Upjohn Co, 53-63; ASSOC PROF CHEM, FERRIS STATE COL, 63- Mem: AAAS; Am Chem Soc; NY Acad Sci. Res: Analytical chemistry of drugs in biological systems; drug dosage in relation to metabolism and mode of degradation. Mailing Add: Dept of Phys Sci Ferris State Col Big Rapids MI 49307

CHUN, ALEXANDER HING CHINN, b Wahiawa, Hawaii, Jan 15, 28; m 57; c 3. PHYSICAL PHARMACY. Educ: Purdue Univ, BS, 54, MS, 56, PhD, 59. Prof Exp: Pharmacist, 58-62, assoc res fel, 71, PHARMACOLOGIST, ABBOTT LABS, 62- Mem: Am Pharmaceut Asn; Am Chem Soc; fel Acad Pharmaceut Sci. Res: Drug absorption and distribution; chemical pharmacology; application of physical and colloidal chemistry to pharmaceuticals; biopharmaceutics. Mailing Add: 1908 Linden Ave Waukegan IL 60085

CHUN, BYUNGKYU, b Korea, Apr 10, 28; US citizen; m 57; c 2. SURGICAL PATHOLOGY. Educ: Seoul Nat Univ, MD, 52. Prof Exp: From instr to asst prof, 61-69, ASSOC PROF PATH, MED CTR, GEORGETOWN UNIV, 69- Concurrent Pos: Consult, Glen Dale Hosp, Glen Dale, Md, 64 & Children's Hosp, DC, 65. Honors & Awards: Achievement Award, Angiol Res Found, 66. Mem: Col Am Path. Res: Oncology. Mailing Add: Dept of Path Georgetown Univ Med Ctr Washington DC 20007

CHUN, EDWARD HING LOY, b Wahiawa, Hawaii, Nov 2, 30. CHEMISTRY. Educ: Univ Hawaii, BA, 52, MA, 54; Harvard Univ, PhD(chem), 58. Prof Exp: Res assoc biol, Mass Inst Technol, 58-63; res investr chem, Univ Pa, 63-68, res investr, Sch Vet Med, 68-70, asst prof biochem, Sch Vet Med, 70-73; CHEMIST, ABBOTT LABS, 74- Mem: Am Chem Soc. Res: Biophysical chemistry; molecular structure of biological macromolecules. Mailing Add: Abbott Labs 14th & Sheridan North Chicago IL 60064

CHUN, KEE WON, b Korea; US citizen. THEORETICAL PHYSICS. Educ: Univ Pa, AB, 51, PhD(physics), 60; Princeton Univ, AM, 55. Prof Exp: Res assoc theoret physics, Columbia Univ, 59-62 & Yale Univ, 62-65; assoc prof physics, 65-69, PROF PHYSICS, UNIV NEW HAVEN, 69-, CHMN DEPT, 71- Mem: Am Phys Soc; Am Asn Physics Teachers. Res: Field theory and nuclear physics; quantizied field in general theory of relativity and astrophysics. Mailing Add: Dept of Physics Univ of New Haven West Haven CT 06516

CHUN, PAUL W, b Repub of Korea, Dec 14, 28; US citizen; m 64; c 1. PHYSICAL BIOCHEMISTRY. Educ: Northwestern State Col, Okla, 55; Okla State Univ, MS, 57; Univ Mo-Columbia, PhD(biochem), 65. Prof Exp: Chemist, Worcester Found Exp Biol, 61-63; fel biochem, Univ Mo-Columbia, 65-66; fel, 66-67, asst prof, 67-70, ASSOC PROF BIOCHEM, UNIV FLA, 70- Mem: AAAS; Am Chem Soc; Am Soc Biol Chemists; Biophys Soc. Res: Protein-protein interaction; molecular exclusion. Mailing Add: Dept of Biochem Univ of Fla Gainesville FL 32601

CHUN, RAYMOND WAI MUN, b Honolulu, Hawaii, Jan 21, 26; m 60; c 3. PEDIATRICS, NEUROLOGY. Educ: St Joseph's Col, BS, 51; Georgetown Univ, MD, 55. Prof Exp: Intern med, Philadelphia Gen Hosp, Pa, 55-56; resident pediat, Univ Hosp, Georgetown Univ, 56-58; resident neurol, 58-59 & 60-61, from asst prof to assoc prof, 61-72, PROF PEDIAT NEUROL, UNIV WIS-MADISON, 72- Concurrent Pos: Fel pediat neurol, Columbia-Presby Hosp, New York, 59-60; consult, Wis Diag Ctr, 61- Mem: Fel Am Acad Pediat; Am Acad Neurol. Res: Epilepsy; neurophysiology; clinical research; transillumination of skull of infants. Mailing Add: Dept of Pediat Univ Wis Univ Hosp Madison WI 53706

CHUNG, ALBERT EDWARD, b Jamaica, West Indies, Dec 18, 36; m 58; c 3. BIOCHEMISTRY. Educ: Univ West Indies, BS, 57, MS, 59; Johns Hopkins Univ, PhD(biochem), 62. Prof Exp: Res fel, Harvard Univ, 62-64; asst prof biochem, Med Ctr, Univ Colo, 64-67; assoc prof, 67-74, PROF BIOCHEM, UNIV PITTSBURGH, 74- Concurrent Pos: Fulbright sr res scholar, France, 74-75. Mem: AAAS; Am Chem Soc; Am Soc Biol Chemists. Res: Regulation of enzymes; control of coenzyme levels in biological systems; metabolism and function of cyclopropane compounds; cellular differentiation. Mailing Add: Dept of Biochem Fac Arts & Sci Univ of Pittsburgh Pittsburgh PA 15213

CHUNG, CHIN SIK, b Taejon, Korea, May 6, 24; nat US; m 57; c 3. HUMAN GENETICS, BIOSTATISTICS. Educ: Ore State Univ, BS, 51; Univ Wis, MS, 53, PhD, 57. Prof Exp: Asst genetics, Univ Wis, 52-57, res assoc med genetics, 57-61; vis scientist, NIH, 61- 6 64, res biologist, 64-65; PROF PUB HEALTH, UNIV HAWAII, 65-, PROF GENETICS, 69- Concurrent Pos: Res grants, NSF & NIH; chmn, Dept Pub Health Sci, Sch Pub Health, Univ Hawaii, 73-; consult, NIH, 74. Mem: Am Soc Human Genetics; Am Statist Asn; Biomet Soc; Am Pub Health Asn; Soc Study Social Biol. Res: Population genetics of humans; biometrical genetics; epidemiological genetics. Mailing Add: Dept Pub Health Sci Sch Pub Hlth Univ of Hawaii Honolulu HI 96822

CHUNG, CHOONG WHA, b Korea, Aug 14, 18; nat US; m 49; c 2. MICROBIOLOGY, MEDICINE. Educ: Keio Univ, Japan, MD, 46; Rutgers Univ, PhD(microbiol), 53. Prof Exp: Asst microbiol, Med Sch, Seoul Nat Univ, 46-49; res fel microbial biochem, Inst Microbiol, Rutgers Univ, 49-53; instr biochem, Dept Pediat, Sch Med, Johns Hopkins Univ, 53-55; instr, Med Sch, Univ Colo, 55-56; instr chem, Ind Univ, 57-60; chief med chem sect, Med Lab Br, Nat Commun Dis Ctr, Atlanta, Ga, 60-67, ASST CHIEF DERMAL TOXICITY BR, DIV TOXICOL, FOOD & DRUG ADMIN, USPHS, 67- Mem: AAAS; Am Chem Soc; Sigma Xi; Am Soc Microbiol; Brit Biochem Soc. Res: Delayed and immediate hypersensitivity; chemical allergens; photoallergy; skin biochemistry; immunological aspects of neoplasia. Mailing Add: Dermal Toxicity Br Div Toxicol Food & Drug Admin 200 C St SW Washington DC 20204

CHUNG, DAVID YIH, b Shanghai, China, Nov 14, 36; m 67; c 1 SOLID STATE PHYSICS, LOW TEMPERATURE PHYSICS. Educ: Nat Taiwan Univ, BSc, 58; Univ BC, MSc, 62, PhD(physics), 66. Prof Exp: Res physicist, Heat Div, Nat Bur Standards, 66-67; asst prof, 67-71, ASSOC PROF PHYSICS, HOWARD UNIV, 71- Concurrent Pos: Vis prof, Dept Appl Physics & Electronics, Univ Durham, Eng & Brit Sci Res Coun vis fel, 73-74. Mem: Am Phys Soc. Res: Second sound propagation in solids; superfluid flow in liquid helium; ultrasonic waves in solids; optical property of solids; non-linear acoustics in liquids and solids; magnetic properties of solids at low temperature. Mailing Add: Dept of Physics Howard Univ Washington DC 20001

CHUNG, ED BAIK, b Seoul, Korea, Mar 16, 28; US citizen; m 58; c 5. MEDICINE, PATHOLOGY. Educ: Severance Union Med Col, MD, 51; Georgetown Univ, MS, 56, PhD(path), 58; Am Bd Path, cert anat path & clin path. Prof Exp: Resident path, Med Ctr, Georgetown Univ, 54-58, from instr to asst prof, 58-63; assoc prof, 64-70, PROF PATH, COL MED, HOWARD UNIV, 70- Concurrent Pos: Attend pathologist, Howard Univ Hosp, 64-; consult path, Glenn Dale Hosp, Md, 68-74 & Coroner's Off, Washington, DC, 69-71; spec res fel, Nat Cancer Inst, 71-72. Mem: AMA; Int Acad Path; Am Soc Clin Path; Col Am Path; NY Acad Sci. Res: Orthopedic and renal pathology; oncologic pathology; tumor immunology. Mailing Add: Dept of Path Howard Univ Col of Med Washington DC 20001

CHUNG, FAN RONG KING, b China, Oct 9, 49; m 70; c 1. MATHEMATICS. Educ: Nat Taiwan Univ, BS, 70; Univ Pa, MA, 72, PhD(math), 74. Prof Exp: MEM TECH STAFF MATH, BELL TEL LABS, 74- Mem: Am Math Soc. Res: Combinatories; graph theory; switching networks; mathematical algorithms. Mailing Add: Bell Tel Labs 600 Mountain Ave Murray Hill NJ 07974

CHUNG, IN-CHO, systematic botany, see 12th edition

CHUNG, JIWHEY, b Korea, Feb 26, 36; m 64; c 2. BIOCHEMISTRY, FOOD SCIENCE. Educ: Seoul Nat Univ, BS, 61; Univ Ill, MS, 66; Univ Tenn, PhD(biochem), 69. Prof Exp: Res assoc, Albert Einstein Med Ctr, 69-71; RES ASSOC BIOCHEM, UNIV CHICAGO, 71- Concurrent Pos: NIH fel, 71-72 & spec res fel, 72-73; Ill & Chicago Heart Asn grant-in-aid, 75. Mem: Sigma Xi. Res: Lipid metabolism; mechanism of action of lipoprotein lipase and lecithin cholesterol acyl transferase; synthesis and metabolism of plasma; very low density lipoproteins. Mailing Add: 2849 Farmington Rd Northbrook IL 60069

CHUNG, KAI LAI, b Shanghai, China, Sept 19, 17; nat US; m; c 3. MATHEMATICS. Educ: Princeton Univ, MA & PhD(math), 47. Prof Exp: Instr math, Princeton Univ, 47-48; asst prof, Cornell Univ, 48-50, vis assoc prof, 51-52; from assoc prof to prof, Syracuse Univ, 53-61; PROF MATH, STANFORD UNIV, 61- Concurrent Pos: Vis prof, Columbia Univ, 50-51 & 59, Univ Chicago, 56-57, Univ Strasbourg, 68-69 & Swiss Fed Inst Technol, 70; G A Miller vis prof, Univ Ill, 70-71; Guggenheim fel, 75-76; fel, Churchill Col, Cambridge Univ, 76; ed, Zeitschrift für Wahrscheinlichkeitstheorie und Verwandte Gebiete. Mem: Am Math Soc; Inst Math Statist. Res: Probability. Mailing Add: Dept of Math Stanford Univ Stanford CA 94305

CHUNG, KUK PYO, b Kyungpuk, Korea, Nov 15, 35; US citizen. ASTROPHYSICS, PARTICLE PHYSICS. Educ: Yale Univ, BA, 59; Princeton Univ, MA, 61, PhD(physics), 69. Prof Exp: Asst prof physics, State Univ NY, 62-68; res assoc, Princeton Univ, 69-70; res scientist, Watson Res Ctr, Int Bus Mach Corp, 70-72; resident res assoc astrophys, Columbia Univ, 72-73; ASST PROF PHYSICS, LEHMAN COL, 73- Concurrent Pos: Consult, McNulty Assoc, 67-69. Mem: Am Phys Soc. Res: Theoretical work on gravitational radiations from black holes; dispersion-theoretic approach to pion-pion interactions. Mailing Add: Herbert Lehman Col Bedford Park Blvd Bronx NY 10468

CHUNG, KWOK-LEUNG, b Hong Kong, July 5, 32; Can citizen; m 66. MICROBIOLOGY, BIOCHEMISTRY. Educ: Univ Man, BSc, 58, MSc, 61, PhD(microbiol), 65. Prof Exp: Lectr microbiol, Univ Man, 53-54; ASST PROF MICROBIOL, QUEEN'S UNIV, ONT, 54- Concurrent Pos: Nat Res Coun Can res grants, 64- Mem: Can Soc Microbiol. Res: Cell wall replication and synthesis of cell wall material in bacteria. Mailing Add: Dept of Microbiol Queen's Univ Kingston ON Can

CHUNG, KYUNG WON, b Seoul, Korea, Aug 15, 38; m 66; c 2. REPRODUCTIVE ENDOCRINOLOGY. Educ: Yonsei Univ, Korea, BS, 64, MS, 66; St Louis Univ, MS, 69; Univ Okla, PhD(anat), 71. Prof Exp: Instr biol, Yonsei Univ, 66; fel endocrinol, Hershey Med Ctr, Pa State Univ, 71-72; instr, 72-75, ASST PROF ANAT, STATE UNIV NY DOWNSTATE MED CTR, 75- Mem: Sigma Xi; Am Asn Anatomists. Res: Ultrastructural and biochemical studies on the testes in mice and rats with testicular feminization. Mailing Add: 59 Maxwell Rd Garden City NY 11530

CHUNG, OKKYUNG KIN, b Seoul, Korea, Apr 11, 36; m 61; c 2. CEREAL CHEMISTRY. Educ: Ewha Womens Univ, Korea, BS, 59; Kans State Univ, MS, 65, PhD(grain sci), 73. Prof Exp: Res asst lipids, Kans State Univ, 64-66, res assoc lipids & surfactants, 73-74; RES CHEMIST LIPIDS & SURFACTANTS, US GRAIN MKT RES CTR, AGR RES SERV, USDA, 74- Mem: Am Asn Cereal Chemists; Sigma Xi. Res: Functionality of wheat flour lipids and lipid related surfactants in breadmaking; interaction of lipids and surfactants with other flour components during processing into bread; fractionation of wheat flour components. Mailing Add: US Grain Mkt Res Ctr 1515 College Ave Manhattan KS 66502

CHUNG, RONALD ALOYSIUS, b Christiana, Jamaica, Sept 30, 36; m 63; c 3. FOOD SCIENCE. Educ: Col Holy Cross, BSc, 59; Purdue Univ, MS, 61, PhD(food technol), 63. Prof Exp: Asst food technol, Purdue Univ, 61-63; PROF FOOD SCI & NUTRIT & PROG COORDR, TUSKEGEE INST, 63- Concurrent Pos: Res assoc, Carver Res Found, 63- Honors & Awards: Res Award, Poultry Sci Asn, 66. Mem: Fel AAAS; Am Chem Soc; Inst Food Technol; Poultry Sci Asn. Res: Lipid metabolism in poultry and livestock; biochemical activity of food additives in mammalian cell systems; new food product development; dietary related toxemia in pregnancy. Mailing Add: Sch Appl Sci Food Sci & Nutrit Tuskegee Institute AL 36088

CHUNG, SUH URK, b Pyongyang, Korea, Nov 11, 36; US citizen; m 63; c 2. HIGH ENERGY PHYSICS. Educ: Univ Fla, BS, 61; Univ Calif, Berkeley, PhD(physics), 66. Prof Exp: Assoc physicist, 66-74, PHYSICIST, BROOKHAVEN NAT LAB, 74- Mem: Am Phys Soc. Res: Experimental high energy physics. Mailing Add: Dept of Physics Brookhaven Nat Lab Upton NY 11973

CHUNG, VICTOR, b Vancouver, BC, Sept 7, 40; m 64; c 1. PHYSICS. Educ: Mass Inst Technol, SB, 61, SM, 62; Univ Calif, Berkeley, PhD(physics), 66. Prof Exp: Asst res physicist, Univ Calif, San Diego, 66-68; ASST PROF PHYSICS, CITY COL NEW YORK, 68- Mem: Am Phys Soc. Res: Theoretical high energy physics; quantum electrodynamics. Mailing Add: Dept of Physics City Col of New York New York NY 10031

CHUNG-PHILLIPS, ALICE, b Kowloon, Hong Kong, Feb 8, 38. THEORETICAL CHEMISTRY, QUANTUM CHEMISTRY. Educ: Univ Chicago, BS, 59, MS, 60, PhD(phys chem), 63. Prof Exp: Res assoc quantum chem, Ill Inst Technol, 63-64; res chemist & lectr, Univ Calif, Davis, 64-66; resident res assoc, Argonne Nat Lab, 66-67, vis scientist, 67-69; asst prof chem, 69-73, ASSOC PROF CHEM, MIAMI UNIV, 73- Mem: Am Chem Soc; Am Phys Soc. Res: Theoretical studies of molecular structure and spectra. Mailing Add: Dept of Chem Miami Univ Oxford OH 45056

CHUPKA, WILLIAM ANDREW, b Pittston, Pa, Feb 12, 23; m 55; c 2. PHYSICAL CHEMISTRY. Educ: Univ Scranton, BS, 43; Univ Chicago, MS, 49, PhD(chem), 51. Prof Exp: Instr chem, Harvard Univ, 51-54; consult, Argonne Nat Lab, 52-54, assoc physicist, 54-68, sr physicist, 68-75; PROF CHEM, YALE UNIV, 75- Concurrent Pos: Guggenheim fel, 61-62. Mem: Am Chem Soc; fel Am Phys Soc. Res: Mass spectrometry; molecular and atomic structure; high temperature thermodynamics; chemical kinetics; photoionization; vacuum ultraviolet spectroscopy; photoelectron spectroscopy; ion-molecular reactions. Mailing Add: Sterling Chem Lab Yale Univ New Haven CT 06520

CHUPP, EDWARD LOWELL, b Lincoln, Nebr, May 14, 27; m 50; c 3. PHYSICS. Educ: Univ Calif, Berkeley, AB, 50, PhD(physics), 54. Prof Exp: Staff mem physics, Lawrence Radiation Lab, Univ Calif, 54-59; unit chief geospace physics, Aerospace Div, Boeing Co, 59-62; assoc prof physics, 62-67, PROF PHYSICS, UNIV NH, 67- Concurrent Pos: Consult, Lawrence Radiation Lab, Univ Calif, 59-63 & Geophys Corp Am, Mass, 63-; mem, State of NH Radiation Adv Comt, 66; consult, Solar Physics Subcomt, NASA, 67; mem sci baloon panel, Nat Ctr Atmospheric Res, 70; NATO sr res fel, Max Planck Inst, Munich, 70, Alexander von Humboldt sr award, 72-73, Fulbright-Hayes hon sr fel, Max Planck Inst Extraterrestrial Physics, 72-73. Honors & Awards: Exceptional Sci Achievement Medal, NASA, 72. Mem: Fel Am Phys Soc; Am Geophys Union; Am Asn Physics Teachers; Am Astron Soc. Res: Cosmic radiation time variations; gamma ray spectroscopy and astronomy; neutron and gamma ray detectors; solar flare physics; solar terrestrial relations; measurements of atomic transition probabilities. Mailing Add: Dept of Physics Univ of NH Durham NH 03824

CHUPP, JOHN P, organic chemistry, see 12th edition

CHURCH, ALONZO, b Washington, DC, June 14, 03; m 25; c 3. MATHEMATICS. Educ: Princeton Univ, AB, 24, PhD(math), 27. Hon Degrees: DSc, Case Western Reserve Univ, 69. Prof Exp: Nat Res fel math, Harvard Univ, 27-28, Univ Göttingen, 28-29 & Univ Amsterdam, 29; from asst prof to prof, Princeton Univ, 29-67; PROF PHILOS & MATH, UNIV CALIF, LOS ANGELES, 67- Concurrent Pos: Ed, Asn Symbolic Logic Jour. Mem: AAAS; Am Math Soc; Asn Symbolic Logic; Am Acad Arts & Sci; Brit Acad. Res: Mathematical logic. Mailing Add: Dept of Philos Univ of Calif Los Angeles CA 90024

CHURCH, BROOKS DAVIS, b Youngstown, Ohio, May 6, 18; c 2. MICROBIOLOGY. Educ: Univ Mich, BS, 47, MS, 52, PhD(bact), 55. Prof Exp: Res bacteriologist, Ft Detrick, Md, 49-50; asst med, Univ Chicago, 50-52; sr res assoc, Warner Res Inst, 55-60; asst prof, Univ Wash, 60-62; from asst prof to assoc prof microbiol, Univ Minn, 62-66; sr microbiologist, North Star Res Inst, 66-72; sr microbiologist, Denver Res Inst, 72-76, PROF BIOL SCI, UNIV DENVER, 72- Concurrent Pos: Mem comt safe drinking water, Nat Acad Sci, 76-77. Mem: AAAS; Am Soc Microbiol. Res: Biochemistry of bacterial spores and cell walls; fungal digestion of food processing wastes; biosynthesis of heteropolysaccharides. Mailing Add: Dept of Biol Sci Univ Park Univ of Denver Denver CO 80210

CHURCH, CHARLES ALEXANDER, JR, b Rock Hill, SC, Oct 29, 32; m. MATHEMATICS. Educ: Va Polytech Inst, BS, 57; Duke Univ, PhD(math), 65. Prof Exp: Instr math, Roanoke Col, 59-62; asst prof, WVa Univ, 65-66; asst prof biomet, Med Col Va, 66-67; ASSOC PROF MATH, UNIV NC, GREENSBORO, 67- Mem: AAAS; Am Math Soc; Math Asn Am. Res: Combinatorial mathematics. Mailing Add: Dept of Math Univ of NC Greensboro NC 27412

CHURCH, CHARLES HENRY, b Phoenix, Ariz, May 15, 29; m 56; c 3. OPTICAL PHYSICS. Educ: Mo Sch Mines, BS, 50; Pa State Univ, MS, 51; Univ Mich, PhD(physics), 59. Prof Exp: Assayer anal chem, Bradley Mining Co, Idaho, 50; asst physics, Pa State Col, 50-51; jr physicist, Res Labs Div, Gen Motors Corp, 51-52; res assoc, Eng Res Inst, Univ Mich, 56-58; assoc physicist, Cent Res Lab, Crucible Steel Co Am, 58-59; sr physicist, Westinghouse Res Labs, 59-63, fel physicist, 63-65, advan physicist, 65-68; staff specialist laser technol & systs, Advan Res Projs Agency, US Dept Defense, Va, 68-71, dep dir, Res & Develop Ctr, Thailand, 71, staff specialist laser technol & syst, Va, 71-72, asst dir target acquisition, identification & advan delivery concepts, 72-74, sr prog mgr, 74-75; ASST DIR ARMY RES TECHNOL, OFF DEP CHIEF STAFF RES, DEVELOP & ACQUISITION, US ARMY, 75- Concurrent Pos: Teacher, Univ Calif Far East Exten, 53-54; lectr, Univ Pittsburgh, 58-59 & Carnegie Inst Technol, 64. Mem: AAAS; Am Phys Soc; fel Optical Soc Am; Inst Elec & Electronics Eng; Am Inst Aeronaut & Astronaut. Res: Lasers; atomic and molecular plasmas; optical intrumentation and design; spectroscopy; weapons systems; solid state physics. Mailing Add: US Army-Army Res Technol Hq DA DAMA-ARZ-E Pentagon Washington DC 20310

CHURCH, CLIFFORD CARL, b Carmen, Okla, Oct 4, 99; m 27; c 2. GEOLOGY, PALEONTOLOGY. Educ: Univ Okla, AB, 23; Stanford Univ, MA, 25. Prof Exp: Heavy mineral res, Marland Oil Co, San Francisco, 26; micropaleontologist, Assoc Oil Co, 26-48; sr micropaleontologist, Getty Oil Co, 48-60; CONSULT MICROPALEONTOL, 60-; CUR PALEONTOL, KERN COUNTY OIL MUS, 74- Concurrent Pos: Optical worker, Calif Acad Sci, 43-45; lectr, Stanford Univ, 46-48; consult, Western Geothermal, Inc, 62. Mem: Am Asn Petrol Geol; Paleont Soc; Soc Econ Paleontologists & Mineralogists (vpres, 30, pres, 52); fel Geol Soc Am; fel Am Inst Prof Geologists. Res: Micropaleontology; malacology; mineralogy; Mesozoic and Cenozoic Foraminifera in California. Mailing Add: 15 Montrose St Bakersfield CA 93305

CHURCH, DAVID CALVIN, b Iola, Kans, Nov 1, 25; m 52. ANIMAL NUTRITION, ANIMAL PHYSIOLOGY. Educ: Kans State Univ, BS, 50; Univ Idaho, MS, 52; Okla State Univ, PhD(animal nutrit), 56. Prof Exp: From asst prof to assoc prof, 56-70, PROF ANIMAL NUTRIT, ORE STATE UNIV, 70- Mem: Am Soc Animal Sci; Am Dairy Sci Asn. Res: Nutrition of the ruminant animal; rumen physiology; feedstuff and forage evaluation. Mailing Add: Dept of Animal Sci Ore State Univ Corvallis OR 97331

CHURCH, EUGENE LENT, b Yonkers, NY, July 30, 25; m 48; c 2. NUCLEAR PHYSICS, OPTICAL PHYSICS. Educ: Princeton Univ, AB, 48; Harvard Univ, PhD(physics), 53. Prof Exp: Res assoc, Princeton Univ, 48; res assoc, Brookhaven Nat Lab, 50-52; guest scientist, Argonne Nat Lab, 52-55; guest scientist, Brookhaven Nat Lab, 55-59; Secy of Army fel, Univ Inst Theoret Physics, Copenhagen, 59-61; guest scientist, Brookhaven Nat Lab, 61-71; PHYSICIST, FRANKFORD ARSENAL, US ARMY, 71- Concurrent Pos: Mem solid state adv panel, Nat Res Coun, 73- Mem: Fel Am Phys Soc; sr mem Inst Elec & Electronics Engr; Optical Soc Am; Am Vacuum Soc; Sigma Xi. Res: Experimental and theoretical nuclear structure and laser physics. Mailing Add: Frankford Arsenal Philadelphia PA 19137

CHURCH, GEORGE LYLE, b Boston, Mass, Dec 19, 03; m 34; c 1. BOTANY. Educ: Mass Col, BS, 25; Harvard Univ, AM, 27, PhD(bot), 28. Prof Exp: Teaching fel bot, Harvard Univ, 26-28; instr, Brown Univ & RI Col Pharm, 28-34; from asst prof to prof bot, 34-59, cur herbarium, 40-72, chmn dept bot, 58-66, Stephen Olney prof, 59-72, EMER PROF BOT, BROWN UNIV, 72- Mem: AAAS; Bot Soc Am; Soc Develop Biol; Soc Study Evolution; Am Soc Plant Taxonomists. Res: Cytology and taxonomy; cytotaxonomy and cytogenetics of Gramineae. Mailing Add: Div of Biomed Sci Brown Univ Providence RI 02912

CHURCH, GILBERT, b New London, Conn, Aug 4, 14. ZOOLOGY, EMBRYOLOGY Educ: Uni0v Chicago, PhB, 48; Stanford Univ, PhD(embryol), 54. Prof Exp: Res assoc, Stanford Univ, 50-55; asst prof & vis chmn, Univ Calif Field Staff Med Ed, Univ Indonesia, Djakarta, 55-59, assoc prof & adv, Univ Ky Contract Team, Inst Technol, Bandung, 59-61; assoc prof zool, San Fernando Valley State Col, 61-64, assoc dean, Sch Letters & Sci, 64-65; MGR DIR, ANTHROPOSOPHIC PRESS, INC, 65- Concurrent Pos: NSF grant, 60-62. Mem: AAAS; Am Soc Ichthyologists & Herpetologists. Res: Auxetic growth and related physiological problems in amphibia; reproductive cycles of tropical amphibia including physiological and endocrine relationships. Mailing Add: 258 Hungry Hollow Rd Spring Valley NY 10977

CHURCH, JOHN ARMISTEAD, b Richmond, Va, Apr 3, 37; m 64; c 3. CELLULOSE CHEMISTRY, PHYSICAL CHEMISTRY. Educ: Univ Va, BA, 59; Inst Paper Chem, MS, 61, PhD(org chem), 64. Prof Exp: Sr scientist, 64-70, RES ASSOC, PRINCETON LAB, AM CAN CO, 70- Mem: Am Chem Soc. Res: Cellulose and carbohydrate chemistry; preparation, characterization and utilization of polysaccharide graft copolymers; reaction kinetics; autoxidation and degradation of carbohydrates, cellulose and paper; pulp and paper science and technology. Mailing Add: Princeton Lab Am Can Co PO Box 50 Princeton NJ 08540

CHURCH, JOHN PHILLIPS, b Columbus, Ohio, July 14, 34; m 59; c 2. REACTOR PHYSICS. Educ: Univ Cincinnati, ChE, 57; Univ Fla, MSc, 60, PhD(nuclear eng), 63. Prof Exp: Chem engr, Nat Cash Register Co, 58-59; RES ENGR, E I DU PONT DE NEMOURS & CO, INC, 63- Mem: Am Nuclear Soc. Res: Reactor kinetics; core design; reactor safety analysis. Mailing Add: Savannah River Lab Aiken SC 29801

CHURCH, LARRY B, b St Louis, Mo, Apr 19, 39; m 62. NUCLEAR CHEMISTRY, RADIOCHEMISTRY. Educ: Univ Rochester, BS, 61; Carnegie Inst Technol, MS, 64, PhD(chem), 66. Prof Exp: Fel chem, Univ Calif, Irvine, 66-68; asst prof, State Univ NY Buffalo, 68-73; ASSOC PROF CHEM, REED COL, 73- Mem: Am Chem Soc. Res: Nuclear reactions; hot-atom chemistry. Mailing Add: Dept of Chem Reed Col Portland OR 97202

CHURCH, LLOYD EUGENE, b Littleton, WVa, Sept 25, 19; m 64. ANATOMY. Educ: WVa Univ, AB, 42; Univ Md, DDS, 44; George Washington Univ, MS, 51, PhD(anat), 59. Prof Exp: Intern oral surg, Bellevue Hosp, New York, 44-45; resident & instr, Med Col Va, 45-46; clin instr, George Washington Univ, 52-62; sr res scientist, Nat Biomed Res Found, 63-67; ASSOC RES PROF ANAT, SCH MED, GEORGE WASHINGTON UNIV, 67- Concurrent Pos: Mem staff, Dept Oral Path, Armed Forces Inst Path, DC, 59-62, vis scientist, 62; mem attend staff, Suburban Montgomery County Gen & Prince Georges Gen Hosps. Mem: Fel AAAS; Am Asn Anat; fel Am Col Dent; Am Acad Oral Path; Am Acad Oral Roentgenol. Res: Anatomy of temporo-mandibular joint; salivary glands; growth and function of mandible. Mailing Add: Dept of Anat George Washington Univ Sch Med Washington DC 20006

CHURCH, NORMAN STANLEY, b Alta, Apr 3, 29. ENTOMOLOGY. Educ: Univ Alta, BSc, 50; Mont State Col, MS, 54; Cambridge Univ, PhD, 58. Prof Exp: Res officer insect physiol, Lethbridge, Alta, 50-63 & Saskatoon, Sask, 63-67, RES SCIENTIST, CAN AGR RES STA, 67- Mem: AAAS; Can Soc Zoologists; Entom Soc Can. Res: Insect metamorphosis, hyper-metamorphosis and diapause; temperature relations of insects; ecology of blister beetles and wireworms; wireworm physiology and behavior. Mailing Add: Agr Can Res Sta Univ Campus Saskatoon SK Can

CHURCH, PHIL EDWARDS, b Berwyn, Ill, Apr 8, 02; m 33; c 4. CLIMATOLOGY. Educ: Univ Chicago, BA, 23; Clark Univ, AM, 32, PhD(climatol), 37. Prof Exp: Teacher pub schs, Ill, 25-35; from instr to assoc prof meteorol & geog, 35-49, chmn dept atmospheric sci, 49-68, prof, 49-73, EMER PROF ATMOSPHERIC SCI, UNIV WASH, 73- Concurrent Pos: Res assoc meteorol & oceanog, Univ Chicago, 41-44; meteorologist, Manhattan Dist, 43-45, Dupont-Hanford Works, 45-47 & Gen Elec-Hanford Works, 47. Mem: AAAS (pres, Pac Div, 62-65); Am Meteorol Soc (vpres, 64-66); Am Geophys Union. Res: Microclimatology; meteorology. Mailing Add: Dept of Atmospheric Sci Univ of Wash Seattle WA 98195

CHURCH, PHILIP THROOP, b Winchester, Conn, Mar 18, 31; m 54; c 3. TOPOLOGY. Educ: Wesleyan Univ, BA, 53; Harvard Univ, MA, 54; Univ Mich, PhD, 59. Prof Exp: From asst prof to assoc prof, 58-65, PROF MATH, SYRACUSE UNIV, 65- Concurrent Pos: NSF grants, 59-62 & 63-69; mathematician, Inst Defense Anal, 62-63; mem, Inst Advan Study, 62 & 65-66; NSF sr fel, 65-66; ed topol, Transactions of & Memoirs of Am Math Soc, 74-77. Mem: Am Math Soc; Math Asn Am. Res: Singularities of differentiable maps. Mailing Add: Dept of Math Syracuse Univ Syracuse NY 13210

CHURCH, ROBERT BERTRAM, b Calgary, Alta, May 7, 37; m 57; c 2. DEVELOPMENTAL BIOLOGY. Educ: Univ Alta, BSc, 62, MSc, 63; Univ Edinburgh, PhD(animal genetics), 65. Prof Exp: NIH fel cancer biol, Med Sch, Univ Wash, 65-66, res assoc, 66-67; mem fac develop biol, 67-68, assoc prof, 68-69, PROF MED BIOCHEM & HEAD DIV, UNIV CALGARY, 69- Concurrent Pos: Nat Res Coun grant develop biol, Univ Calgary, 67-; pres, Church Livestock Consult, Ltd, 70-; Med Res Coun grant, 72- Mem: Soc Study Reprod; AAAS; Genetics Soc Am; Can Biochem Soc. Res: Analysis of genetic transcription in developing mammalian systems utilizing biochemical parameters; synthesis of ribonucleic acid and control at the transcription and translational levels; evolution of desoxyribonucleic acid base sequences in mammals; bovine embryo transplants. Mailing Add: Div of Med Biochem Univ of Calgary Fac of Med Calgary AB Can

CHURCH, ROBERT FITZ (RANDOLPH), b Philadelphia, Pa, Mar 27, 30; m 53; c 3. ORGANIC CHEMISTRY. Educ: Amherst Col, BA, 51; Univ Mich, MS, 61, PhD(chem), 62. Prof Exp: Sales serv rep, Am Cyanamid Co, 51-53, chemist, 55-57; instr, Univ Mich, 60-61; res chemist, Am Cyanamid Co, 61-62; RES CHEMIST, LEDERLE LABS, PEARL RIVER, NY, 63- Concurrent Pos: Instr, Bridgeport Eng Inst, 56-57 & 62-68. Mem: Am Chem Soc; Sigma Xi. Res: Chemistry of steroids and other natural products; pharmaceutical and synthetic organic chemistry; chemistry of small ring compounds; process research. Mailing Add: 201 Sheephill Rd Riverside CT 06878

CHURCH, RONALD L, b Monterey, Calif, Nov 19, 30; m 56; c 3. ZOOLOGY, POLLUTION BIOLOGY. Educ: San Jose State Col, BA, 53; Univ Calif, Berkely, MA, 61, PhD(zool), 64. Prof Exp: Asst prof biol, San Jose State Col, 63-65; asst prof, Univ Nev, Reno, 65-70, res assoc, Desert Res Inst, 68-70; environ specialist, Div Water Qual, State Water Resources Control Bd, 70-72, ENVIRON SPECIALIST, NORTH COAST REGION, CALIF REGIONAL WATER QUAL CONTROL BD, 72- Concurrent Pos: NSF grant, 65-66; adj prof, Pac Marine Sta, Univ of the Pac, 74- Mem: AAAS; Ecol Soc Am; Wildlife Soc; Am Ornith Union; Am Soc Mammal. Res: Water quality biology; vertebrate biology; animal ecology. Mailing Add: Calif Region Water Qual Control Bd 1000 Coddingtown Ctr Santa Rosa CA 95401

CHURCH, SHEPARD EARLL, (JR), b Syracuse, NY, Dec 4, 21; m 43; c 3. PHYSICAL CHEMISTRY. Educ: NY State Col Forestry, Syracuse Univ, BS, 43, MS, 50; Syracuse Univ, PhD(chem), 58. Prof Exp: Res chemist, Rohm and Haas Co, Pa, 43-47; from instr to assoc prof pulp & paper technol, State Univ NY Col Forestry, 47-57; tech serv rep, Rayonier, Inc, 57-59, mgr tech serv, 59-69, GEN MGR TECH SERV & PROD DEVELOP, ITT RAYONIER, INC, 69- Mem: Am Chem Soc; Tech Asn Pulp & Paper Indust; Can Pulp & Paper Asn. Res: Pulp and paper technology; cellulose chemistry; kinetics of organic redox reactions; photochemistry; urea formaldehyde and malamine formaldehyde resins; plywood adhesives; paper wet strength. Mailing Add: ITT Rayonier Inc 605 Third Ave New York NY 10016

CHURCH, STANLEY EUGENE, b Oakland, Calif, Sept 13, 43; m 66; c 2. GEOCHEMISTRY. Educ: Univ Kans, BS, 65, MS, 67; Univ Calif, Santa Barbara, PhD(geochem), 70. Prof Exp: Res asst geochem, Univ Calif, Santa Barbara, 69-70; res assoc, Carnegie-Mellon Univ, 70-71; Nat Res Coun Assoc, Johnson Space Ctr, NASA, 71-73; res assoc geochem, Univ Calif, Santa Barbara, 73-75; RES ASSOC GEOCHEM, HASLER RES CTR, APPL RES LAB, GOLETA, CA, 75- Mem: Geol Soc Am; Am Geophys Union; Geochem Soc; Sigma Xi. Res: The application of isotopic and trace element geochemical methods to the problem of the genesis of calc-alkaline magma and the evolution of the earth's mantle; trace element abundance distributions and the effect of micrometeorite bombardment of the lunar soils to form agglutinites. Mailing Add: Hasler Res Ctr Appl Res Lab Goleta CA 93017

CHURCH, WILLIAM RICHARD, b Tonyrefail, Glamorgan, UK, July 10, 36; m 63; c 4. GEOLOGY. Educ: Univ Wales, BSc, 57, PhD(geol), 61. Prof Exp: Boese fel, Columbia Univ, 61-62; lectr, 62-63, asst prof, 63-68, ASSOC PROF GEOL, UNIV WESTERN ONT, 68- Mem: Fel Geol Soc London; fel Geol Asn Can; Geol Soc Am; Can Inst Mining & Metall. Res: Structural geology of northeast Newfoundland, northwest Ireland, Huronian rocks north of Lake Huron; eclogites, Ireland; eclogites, ariegites and lherzolites, Pyrenees; peridotites, northeast Newfoundland. Mailing Add: Dept of Geol Univ of Western Ont London ON Can

CHURCHER, CHARLES STEPHEN, b Aldershot, Eng, Mar 21, 28; Can citizen; m 59; c 3. VERTEBRATE PALEONTOLOGY, MAMMALOGY. Educ: Univ Natal, BSc, 50, Hons, 52, MSc, 54; Univ Toronto, PhD, 57. Prof Exp: Lectr zool, 57-59, from asst prof to assoc prof, 60-70, PROF ZOOL, UNIV TORONTO, 70-, ASSOC PROF DENT, 69- Concurrent Pos: Res assoc, Royal Ont Mus, Toronto, 59-; consult, Geol Surv Can, 66- Mem: Am Soc Mammalogists; Soc Vert Paleont; Can Soc Zoologists; Australian Mammal Soc. Res: Pleistocene mammals, especially Canadian and African. Mailing Add: Dept of Zool Univ of Toronto Toronto ON Can

CHURCHILL, ALGERNON COOLIDGE, b Aug 15, 37; US citizen; m 59; c 3. PHYCOLOGY, PLANT ECOLOGY. Educ: Harvard Univ, BA, 59; Univ Ore, MS, 63, PhD(biol), 68. Prof Exp: From instr to asst prof, 66-74, ASSOC PROF BIOL, ADELPHI UNIV, 74- Concurrent Pos: Sr investr, NY Ocean Sci Lab grant, 70-71. Mem: AAAS; Phycol Soc Am; Am Inst Biol Sci; Am Soc Limnol & Oceanog; Int Phycol Soc. Res: Physiological ecology of algae; growth and differentiation of marine algae; culturing of algae. Mailing Add: Adelphi Inst of Marine Sci Adelphi Univ Garden City NY 11530

CHURCHILL, ARTHUR VICTOR, organic chemistry, see 12th edition

CHURCHILL, BRUCE WENZEL, b Belvidere, Ill, Sept 21, 21; m 46; c 3. MICROBIOLOGY. Educ: Northern Ill State Teachers Col, BE, 43; Univ Wis, MA, 47, PhD(bot & plant physiol), 49. Prof Exp: MICROBIOLOGIST, UPJOHN CO, 49- Mem: Soc Indust Microbiol. Res: Strain selection and biochemistry of fermentation of antibiotic producing microorganisms. Mailing Add: Res Dept Upjohn Co 7000 Portage St Kalamazoo MI 49001

CHURCHILL, CONSTANCE LOUISE, b Los Angeles, Calif, May 10, 41. ORGANIC CHEMISTRY. Educ: Baylor Univ, BS, 63, PhD(org chem), 69. Prof Exp: From asst prof to assoc prof, 68-73, PROF CHEM, DAKOTA STATE COL, 73-, CHMN, DIV SCI, MATH & HEALTH SERV, 74- Mem: AAAS; Am Chem Soc; The Chem Soc; Sigma Xi. Res: Decomposition of ozonides; water pollution. Mailing Add: Dept of Chem Dakota State Col Madison SD 57042

CHURCHILL, DEWEY ROSS, JR, b Blackwell, Okla, May 19, 26; m 53; c 4. PHYSICS. Educ: Univ Kans, BS, 48, MS, 53. Prof Exp: Comput-observer geophys, Seismograph Serv Corp, 50-53; physicist, Res & Develop Lab, Phillips Petrol, 54-55; staff mem physics, Los Alamos Sci Lab, 55-60; sr res engr, 60-61, RES SCIENTIST THEORET PHYSICS, LOCKHEED MISSLES & SPACE CO, 61- Mem: AAAS; Am Phys Soc. Res: Radiative processes in molecules and atoms; radiative transfer; nuclear weapons effects; shock hydrodynamics; ion ballistics; aeronomy. Mailing Add: 1711 Karameos Dr Sunnyvale CA 94087

CHURCHILL, DON W, b Seattle, Wash, Feb 5, 30; m 55; c 3. PSYCHIATRY. Educ: Lawrence Col, BS, 51; Univ Wis, MS, 56, MD, 57; Am Bd Psychiat & Neurol, dipl psychiat, 63, dipl child psychiat, 66. Prof Exp: Instr path, Univ Wis, 53-56; intern, King County Hosp, Seattle, 57-58; resident psychiat, Cincinnati Gen Hosp, 58-60, fel child psychiat, 60-62; from asst prof to assoc prof, 64-73, PROF PSYCHIAT, MED CTR, IND UNIV-PURDUE UNIV, INDIANAPOLIS, 73-, DIR, RILEY CHILD GUID CLIN, 73- Concurrent Pos: Consult, Ind Boys' Sch, 64-69 & Ind Sch for Deaf,

69- Mem: AAAS; Am Psychiat Asn. Res: Pathology of connective tissue and histochemistry of mucomucopolysaccharides; childhood psychosis; language development and its relation to performance and cognitive levels; relationship of success-failure level to affect mood and specific behavioral measures; language and learning abilities. Mailing Add: Dept of Psychiat Ind Univ Sch of Med Indianapolis IN 46202

CHURCHILL, DONALD, physical chemistry, see 12th edition

CHURCHILL, EDMUND, b Chelsea, Mass, Nov 10, 12; m 38; c 1. MATHEMATICS. Educ: Univ Md, BS, 43; Columbia Univ, AM, 44. Prof Exp: Instr math, Rutgers Univ, 45-47; from asst prof to prof, Antioch Col, 47-70, dir anthrop res proj, 50-70; MEM STAFF, WEBB ASSOCS, 70- Concurrent Pos: Lectr, Grad Ctr, Ohio State Univ, Wright Field, 48. Mem: Am Math Soc; Am Statist Asn; Am Asn Phys Anthrop; Am Anthrop Asn. Res: Statistics; anthropometry. Mailing Add: Webb Assocs Yellow Springs OH 45387

CHURCHILL, EDWARD DELOS, surgery, deceased

CHURCHILL, HELEN MAR, b Lawrence, Kans, Nov 14, 07. BIOLOGY. Educ: Univ Kans, AB, 28; Univ Mich, MA, 36, PhD(zool), 51. Prof Exp: Instr biol, William Jewell Col, Mo, 31-36 & Hibbing Jr Col, Minn, 37-38; teaching asst, Univ Mich, 39-40, 41, teaching fel, 40, 41-42; bacteriologist, Patuxent Res Refuge, US Fish & Wildlife Serv, Md, 42-45; protozoologist, NIH, 45; instr biol, Cornell Col, 45-47; from asst prof to assoc prof, 47-73, EMER ASSOC PROF BIOL, HOLLINS COL, 73- Concurrent Pos: Res parasitologist, Univ Philippines, 64-65. Mem: AAAS. Res: Germ cell cycle; digenetic trematode; cercaricides; preventive ointments and fabrics for control of schistosomiasis; methods for culturing parasites. Mailing Add: 515 Strand Rd NE Roanoke VA 24012

CHURCHILL, JOHN ALVORD, b Boston, Mass, Mar 25, 20; c 1. NEUROLOGY, PEDIATRIC NEUROLOGY. Educ: Trinity Col, BS, 42; Univ Pa, MD, 45. Prof Exp: Kirby-McCarthy fel, Johnston Found, Univ Pa, 46-50, asst prof neurol, Sch Med, 49-50; mem staff, Hartford Hosp, 50-53; assoc, Henry Ford Hosp, 53-60; chief dept child neurol, Lafayette Clin & assoc prof neurol, Wayne State Univ, 60-67; head sect neuropediat, Perinatal Br, Nat Inst Neurol Dis & Blindness, 67-72; PROF NEUROL & CHIEF CHILD NEUROL, SCH MED, WAYNE STATE UNIV, 72- Concurrent Pos: Res consult, pediat, Guys Hosp & Spastic Soc, UK, 60; mem comt ment retardation, Dept Ment Health State of Mich, 64-67; res fel bot, Mus Natural Hist, Smithsonian Inst, 70-; res fel, Cranbrook Inst Sci, 72- Mem: AAAS; Am Acad Neurol; Asn Res Nerv & Ment Dis; Am Acad Cerebral Palsy. Res: Relationships of perinatal events to neurological disabilities in the child. Mailing Add: Dept of Neurol Wayne State Univ Sch Med Detroit MI 48201

CHURCHILL, JOHN WILBUR, chemistry, see 12th edition

CHURCHILL, LYNN, b Sacramento, Calif, Mar 27, 47. NEUROCHEMISTRY. Educ: Univ Houston, BS, 69; Univ Calif, Irvine, PhD(biol sci), 73. Prof Exp: Teaching asst psychobiol, Univ Calif, Irvine, 69-73; res assoc, 73-74, NIH FEL PHARMACOL, SCH MED, UNIV WIS-MADISON, 74- Mem: Soc Neurosci. Res: In vivo and in vitro turnover rates and biogenesis of sodium-plus-potassium ion-activated adenosine triphosphatase subunits, plasma membrane proteins in the electric organ of electrophorus electricus. Mailing Add: Dept of Pharmacol Med Sch Univ of Wis 387 Med Sci Bldg Madison WI 53706

CHURCHILL, MELVYN ROWEN, b London, Eng, June 2, 40; m 66; c 2. INORGANIC CHEMISTRY, CRYSTALLOGRAPHY. Educ: Univ London, BSc, 61, PhD(inorg chem), 64. Prof Exp: From instr to assoc prof chem, Harvard Univ, 64-71; prof, Univ Ill, Chicago Circle, 71-75; PROF CHEM, STATE UNIV NY BUFFALO, 75- Concurrent Pos: Alfred P Sloan fel, 68-70; assoc ed, Inorg Chem, 70-. Mem: Am Chem Soc; Am Crystallog Asn; Brit Chem Soc. Res: Crystallographic studies on inorganic compounds particularly organometallic and transition metal complexes; synthetic organometallic chemistry. Mailing Add: Dept of Chem State Univ NY Buffalo NY 14214

CHURCHILL, PAUL CLAYTON, b Carson City, Mich, Feb 23, 41; m 70. PHYSIOLOGY. Educ: Univ Mich, Ann Arbor, BS, 63, PhD(physiol), 69. Prof Exp: NSF fel pharmacol, Univ Lausanne, 69-70; asst prof physiol, Sch Med, Univ Mich, Ann Arbor, 70-72; ASST PROF PHYSIOL, SCH MED, WAYNE STATE UNIV, 72- Mem: Am Physiol Soc; Int Soc Nephrology; Am Soc Nephrology. Res: Physiology of kidney, excretory function, transport; control or renin secretion from kidney. Mailing Add: Dept of Physiol Wayne State Univ Sch Med Detroit MI 48201

CHURCHILL, RALPH JOHN, b Pittsburgh, Pa, July 16, 44; m 66; c 1. WATER CHEMISTRY, ENVIRONMENTAL CHEMISTRY. Educ: Univ Ky, BS, 66; Univ Houston, MS, 70; Univ Calif, Berkeley, PhD(civil eng), 73. Prof Exp: Engr pollution control, Shell Oil Co, 66-71; consult, Eng-Sci Inc, 73-75; GROUP LEADER WATER RES, TRETOLITE DIV, PETROLITE CORP, 75- Mem: Am Inst Chem Engr; Am Water Works Asn; Water Pollution Control Fedn. Res: Water and wastewater investigation; oil-water separation; water and wastewater treatment technology; mineral scale deposition-inhibition; municipal and industrial wastewater management. Mailing Add: Tretolite Div Petrolite Corp 369 Marshall Ave St Louis MO 63119

CHURG, JACOB, b Dolhinow, Poland, July 16, 10; US citizen; m 42; c 2. PATHOLOGY. Educ: Univ Wilno, MD, 33, DMedSc, 36. Prof Exp: Asst path, Sch Med, Univ Wilno, 34-36; asst bact, 38, fel path, 41-43, res assoc, 46-62, assoc attend pathologist, 62-75, ATTEND PATHOLOGIST, MT SINAI HOSP, 76-; PROF PATH & RES PROF COMMUNITY MED, MT SINAI SCH MED, 66- Concurrent Pos: Resident, Beth Israel Hosp, Newark, NJ, 39-40; pathologist, Barnert Mem Hosp, Paterson, 46-; lectr, NJ Col Med, 58- Mem: Am Asn Path & Bact; Am Soc Exp Path; NY Acad Sci. Res: Vascular diseases; renal structure and diseases; pneumoconioses. Mailing Add: 711 Ogden Ave Teaneck NJ 07666

CHURGIN, JAMES, b New York, NY, Sept 3, 28; m 55; c 3. OCEANOGRAPHY. Educ: City Col New York, BS, 50; WVa Univ, MS, 54. Prof Exp: Staff & div geologist, Vitro Minerals Corp, 54-59; oceanogr, Naval Oceanog Off, 59-60; chief serv br, 61-69, chief applns, 69-74, DIR DATA SERV DIV, NAT OCEANOG DATA CTR, 74- Mailing Add: Nat Oceanog Data Ctr Washington DC 20235

CHURKIN, MICHAEL, JR, b San Francisco, Calif, Jan 6, 32; m 60; c 2. GEOLOGY. Educ: Univ Calif, Berkeley, BA, 57, MA, 58; Northwestern Univ, PhD, 61. Prof Exp: Fel geol, Columbia Univ, 61-62; GEOLOGIST, ALASKAN GEOL BR, US GEOL SURV, 62- Mem: AAAS; Geol Soc Am. Res: Paleozoic stratigraphy and structure of Alaska; tectonic history of north Pacific and Arctic; geology of North-East USSR; graptolites. Mailing Add: US Geol Surv Menlo Park CA 94025

CHURNEY, LEON, b Philadelphia, Pa, Dec 23, 10; m 46; c 1. PHYSIOLOGY. Educ: Univ Pa, AB, 32, PhD(zool), 39. Prof Exp: Instr zool, Univ Pa, 34-39 & 40-41; from asst prof to assoc prof, 46-64, PROF PHYSIOL, SCH MED, UNIV NEW ORLEANS, 64- Concurrent Pos: Harrison fel, Univ Pa, 39-40. Mem: Am Physiol Soc. Res: Cell division; physico-chemical properties of nucleus; physiology of cell cortex; electrophysiology; muscle. Mailing Add: Univ of New Orleans Sch of Med New Orleans LA 70112

CHUSED, THOMAS MORTON, b St Louis, Mo, Mar 29, 40; m 65; c 2. IMMUNOLOGY. Educ: Harvard Col, BA, 62; Harvard Med Sch, MS, 67. Prof Exp: Clin assoc immunol, Arthritis Br, Nat Inst Arthritis & Metab Dis, 69-71; sr staff fel, 71-74, SR INVESTR IMMUNOL, LAB MICROBIOL & IMMUNOL, NAT INST DENT RES, 74- Mem: Am Asn Immunologists; Am Rheumatism Asn. Res: Understanding the mechanism of autoimmune disease in New Zealand mice and applying this to human autoimmune disease. Mailing Add: Bldg 10 Rm 2B10 Nat Inst Health Bethesda MD 20014

CHUSID, JOSEPH GEORGE, b Newark, NJ, Aug 23, 14; m 42; c 2. NEUROLOGY. Educ: Univ Pa, AB, 34, MD, 38; Am Bd Psychiat & Neurol, dipl. Prof Exp: Asst instr neurol, Univ Pa, 41-45; resident & fel, Neuropsychiat Inst, Univ Ill, 46-47; from asst attend neurologist to attend neurologist, 48-59, assoc dir dept neurol & neurosurg, 59-64, chief neurol serv, 64-70, DIR DEPT NEUROL, ST VINCENT'S HOSP & MED CTR, 70- Concurrent Pos: Clin assoc prof med, NJ Col Med & Dent, 61-72; assoc clin prof neurol, Col Physicians & Surgeons, Columbia Univ, 65-, assoc attend neurologist, Columbia Presby Med Ctr, 65- Mem: Soc Biol Psychiat; fel Am Acad Neurol; Asn Res Nerv & Ment Dis; Soc Exp Biol & Med; Am Epilepsy Soc. Res: Neurological sciences; cortical connections and functions in the monkey; chronic experimental epilepsy; electroencephalographic studies on humans; cerebral angiography; clinical neurology; effects of major cerebral arterial ligations in monkeys; neuropharmacology; effects of metals on the brain. Mailing Add: St Vincent's Hosp & Med Ctr 145 W 11th St New York NY 10011

CHUTE, HAROLD LEROY, b Winnipeg, Ont, Sept 4, 21; nat US; m 54; c 3. ANIMAL PATHOLOGY. Educ: Univ Toronto, DVM, 49, DVSc, 55; Ont Vet Col, VS, 49; Ohio State Univ, MSc, 53. Prof Exp: Animal pathologist, Nova Scotia Agr Col, 49; asst prof, 49-52, assoc prof & assoc animal pathologist, 52-55, PROF ANIMAL PATH & ANIMAL PATHOLOGIST, UNIV MAINE, ORONO, 55-, DIR DEVELOP, 69- Concurrent Pos: Dir develop, Pullorum-Typhoid Agency, 58-69. Mem: AAAS; fel Am Vet Med Asn; Am Soc Microbiol; Am Asn Avian Path (pres, 63). Res: Virology in poultry diseases; veterinary bacteriology. Mailing Add: Off Dir Develop Alumni Ctr Univ of Maine Orono ME 04473

CHUTE, JOHN LAWRENCE, JR, b Biddeford, Maine, June 12, 38; m 60; c 2. GEOLOGY, OCEANOGRAPHY. Educ: Univ Notre Dame, BS, 60; Ind Univ, MA, 62; Columbia Univ, PhD(geol), 69. Prof Exp: ASST PROF GEOL, LEHMAN COL, 63-; RES ASSOC, LAMONT-DOHERTY GEOL OBSERV, 69- Mem: AAAS; Am Geophys Union. Res: Physical oceanography of coastal waters, especially the Hudson Estuary and New York Bight; lunar studies, including the Apollo lunar heat flow experiments. Mailing Add: 680 Piermont Ave Piermont NY 10968

CHUTE, ROBERT MAURICE, b Naples, Maine, Feb 13, 26; m 46; c 2. ENVIRONMENTAL PHYSIOLOGY, LIMNOLOGY. Educ: Univ Maine, AB, 50; Johns Hopkins Univ, ScD, 53. Prof Exp: From instr to asst prof biol, Middlebury Col, 53-58; asst prof, San Fernando Valley State Col, 58-60; assoc prof & chmn, Lincoln Univ, Pa, 60-61; prof & chmn, 62-69, PROF BIOL & ACTG CHMN DEPT, BATES COL, 75- Mem: AAAS. Res: Human ecology; estimation of cultural impact on lake systems; individual characteristics of Maine lakes. Mailing Add: Dept of Biol Bates Col Lewistown ME 04240

CHUTE, WALTER JOHN, b Brooklyn Corner, Kings Co, NS, March 22, 14; m 46; c 1. ORGANIC CHEMISTRY. Educ: Acadia Univ, BSc, 39; Univ Toronto, MA, 40, PhD(org chem), 43. Prof Exp: Demonstr chem, Univ Toronto, 39-40; assoc prof, 43-46, PROF CHEM, DALHOUSIE UNIV, 46- Mem: Fel Chem Inst Can. Res: Organic chemistry of nitrogen compounds; heterocyclic compounds. Mailing Add: Dept of Chem Dalhousie Univ Halifax NS Can

CHVAPIL, MILOS, b Kladno, Czech, Sept 29, 28; m 53; c 2. PHYSIOLOGICAL CHEMISTRY, EXPERIMENTAL PATHOLOGY. Educ: Charles Univ, Prague, MD, 52, DSc(exp path), 66; Czech Acad Sci, PhD(biochem), 55. Prof Exp: Scientist, Inst Indust. Hyg & Occup Dis, Prague, 52-68; fel sci, Max Planck Inst Protein & Leather Res, Ger, 68-69; vis prof med, Sch Med, Univ Miami, 69-70; PROF SURG BIOL, COL MED, UNIV ARIZ, 70- Concurrent Pos: Consutl scientist, Ministry of Light Indust & Ministry of Food Indust, 58-68; fel surg, Med Sch, Univ Ore, 63; assoc prof exp path, Med Sch, Charles Univ, Prague, 65. Honors & Awards: Laureate of State Prize, Czech Repub, 66; Ministry of Health Award & Sci Bd, Czech Acad Sci Award, 68. Mem: Am Soc Exp Path. Res: Connective tissue physiology; biochemistry control of collagen biosynthesis by chelating agents; oxygen effect on collagen synthesis. Mailing Add: Dept of Surg Univ of Ariz Col of Med Tucson AZ 85721

CHYATTE, SAMUEL BARUCH, b New York, NY, Jan 9, 36; m 57; c 4. PHYSICAL MEDICINE. Educ: Lehigh Univ, BA, 57; Jefferson Med Col, MD, 61. Prof Exp: Intern, York Gen Hosp, Pa, 61-62; from resident to chief resident phys med & rehab, Highland View Hosp, Cleveland, Ohio, 62-65; from asst prof to assoc prof, 65-72, PROF PHYS MED & REHAB, SCH MED, EMORY UNIV, 72-, DIR TRAINING, REGIONAL REHAB RES & TRAINING CTR, 65-, CHIEF DEPT PHYS MED, UNIV HOSP, 67-, DIR MUSCLE DIS CLIN, 68- Mem: AMA; Am Acad Phys Med & Rehab; Am Cong Rehab Med; Asn Acad Physiatrists; Int Rehab Med Asn. Res: Disorders of the neuromuscular system. Mailing Add: Rm 256 Woodruff Mem Bldg Emory Univ Sch of Med Atlanta GA 30322

CHYLEK, PETR, b Ostrava, Czech, Nov 6, 37; US citizen. ATMOSPHERIC PHYSICS, OPTICS. Educ: Charles Univ, Prague, dipl, 66; Univ Calif, Riverside, PhD(physics), 70. Prof Exp: Res assoc physics, Ind Univ, Bloomington, 70-72; fel, Nat Ctr Atmospheric Res, Boulder, 72-73; asst prof, State Univ NY Albany, 73-75; ASSOC PROF GEOSCI, PURDUE UNIV, 75- Mem: Am Geophys Union; Am Meteorol Soc; Am Optical Soc. Res: Light scattering in the atmosphere, effects of air pollution on climate, radiative transfer and atmospheric optics. Mailing Add: Dept of Geosci Purdue Univ West Lafayette IN 47907

CHYNOWETH, ALAN GERALD, b Harrow, Eng, Nov 18, 27; m 50; c 2. SOLID STATE PHYSICS, MATERIALS SCIENCE. Educ: King's Col, Univ London, BSc, 48, PhD(physics), 50. Prof Exp: Demonstr physics, King's Col, Univ London, 48-50; res fel, Nat Res Coun Can, 50-52; mem tech staff, 53-60, head crystal electronics res dept, 60-65, asst dir mat res, 65-73, DIR MAT RES, BELL LABS, 73- Concurrent Pos: Surv dir, Nat Acad Sci Comt on Surv Mat Sci & Eng, 70-75; mem & panel chmn, Nat Acad Sci Comt on Mineral Resources & Environ, 73-75. Honors & Awards: Baker Prize, Inst Elec & Electronics Engrs, 67. Mem: Fel Am Phys Soc; fel Brit Inst Physics; Inst Elec & Electronics Engrs; Am Inst Mining, Metall & Petrol Engrs; Mat Res Soc. Res: Transport properties of semiconductors and insulators; electrical breakdown; ferroelectrics; tunnelling; solid state plasmas; materials research; national materials policies. Mailing Add: Bell Labs Murray Hill NJ 07974

CHYNOWETH, DAVID PAUL, microbiology, environmental biology, see 12th edition

CHYTIL, FRANK, b Prague, Czech, Aug 28, 24; m 49; c 3. BIOCHEMISTRY. Educ: Col Chem Tech, Prague, Ing, 49, PhD(biochem), 52. Prof Exp: Res biochemist, Charles Univ, Prague, 49-51; Czech Acad Sci res fel biochem, Inst Human Nutrit, Prague, 52-55; sr scientist physiol, Czech Acad Sci, 56-62, sr scientist microbiol, 63-64; sr res fel biochem, Brandeis Univ, 64, sr res assoc, 65-66; from asst prof to assoc prof, 66-75, PROF BIOCHEM, SCH MED, VANDERBILT UNIV, 75- Concurrent Pos: Sect head, Southwest Found Res & Educ, 66-72. Mem: Am Inst Nutrit; Am Soc Biol Chem; Endocrine Soc; Am Chem Soc. Res: Metabolic regulations. Mailing Add: Dept of Biochem Vanderbilt Univ Sch Med Nashville TN 37232

CIABATTONI, JOSEPH, organic chemistry, see 12th edition

CIACCIO, EDWARD I, b Brooklyn, NY, Oct 23, 25; m 55; c 3. BIOCHEMISTRY. Educ: Cornell Univ, MNS, 55, PhD(biochem, physiol), 59. Prof Exp: Chemist, Lever Bros, 49-50, supvr prod control, 52-53; res asst biochem, Cornell Univ, 54-58; res assoc, Merck Sharp & Dohme Res Labs Div, Merck & Co Inc, 59-65; asst prof, 65-69, ASSOC PROF PHARMACOL, HAHNEMANN MED COL, 69- Mem: AAAS; Am Chem Soc; NY Acad Sci; Brit Biochem Soc. Res: Biochemical differentiation of normal tissues during different functional states and compared to tumor tissues; intermediary metabolism and enzymology; enzyme inhibitors; mechanism of action and effect on metabolism in vivo. Mailing Add: Dept of Pharmacol Hahnemann Med Col Philadelphia PA 19102

CIACCIO, LEONARD LOUIS, physical chemistry, analytical chemistry, see 12th edition

CIALDELLA, CATALDO, b Rochester, NY, Aug 26, 26; m 59; c 4. ORGANIC CHEMISTRY, POLYMER CHEMISTRY. Educ: Clarkson Col Technol, BS, 50; Case Inst Technol, MS, 54, PhD(org chem), 56. Prof Exp: Chemist, Bausch & Lomb Optical Co, 50-52; res chemist, Esso Res & Eng Co, 56-58, res dir, 58-60, dir res & develop, 60-61; VPRES, HYSOL DIV, DEXTER CORP, 61- Mem: Am Chem Soc. Res: Polymers, especially epoxy, urethane silicones. Mailing Add: Hysol Div Dexter Corp 211 Franklin St Olean NY 14760

CIALONE, JOSEPH C, horticulture, see 12th edition

CIANCIO, SEBASTIAN GENE, b Jamestown, NY, June 21, 37; m 63; c 1. PHARMACOLOGY, PERIODONTOLOGY. Educ: Univ Buffalo, DDS, 61; Am Bd Periodont, dipl. Prof Exp: Fel pharmacol & periodont, 63-65, from asst prof to assoc prof periodont & chmn dept, 65-73, assoc prof pharmacol, Sch Med, 65-73, PROF PERIODONT-ENDODONT & CHMN DEPT, SCH DENT, STATE UNIV NY BUFFALO, 73-, CLIN PROF PHARMACOL, SCH MED, 73- Concurrent Pos: Res grants, United Health Found of Western NY, 65-66 & 70-71, Nat Inst Dent Res, 67-69 & Merrill Nat Labs, 73-; consult, Vet Admin Hosp, Buffalo, 70-, Pharmaceut Mfrs Asn, 73- & US Pharmacopae & Nat Formulary, 75- Mem: Am Dent Asn; Am Acad Periodont; Int Asn Dent Res. Res: Papain induced changes in rabbit tissues; local hemostasis; principal fibers of the periodontal ligament; plaque control agents; periodontal observations in twins; acid mucopolysaccharides in gingivitis and periodontitis. Mailing Add: Dept Periodont-Endodont State Univ NY Sch Dent Buffalo NY 14214

CIAPETTA, FRANK GEORGE, chemistry, see 12th edition

CIAUDELLI, JOSEPH PETER, organic chemistry, see 12th edition

CIBILS, LUIS ANGEL, b Yuty, Paraguay, Mar 22, 27; m 61; c 3. OBSTETRICS & GYNECOLOGY. Educ: Col San Jose, BA, 42; Univ Paraguay, MD, 50. Prof Exp: Scholar, Span Inst Cult, 51-52; vis asst gynec, Univ Paris, 52-53; res fel obstet & gynec, Univ of Repub Uruguay, 57-60; res consult, Western Reserve Univ, 60, from instr to asst prof, 60-66; assoc prof, 66-70, PROF OBSTET & GYNEC, UNIV CHICAGO, 70- Honors & Awards: Found Prize, Am Asn Obstet & Gynec, 65. Res: Physiology of reproduction. Mailing Add: Dept of Obstet & Gynec Univ of Chicago Chicago IL 60637

CIBULA, ADAM BURT, b Salem, Ohio, June 4, 34. BIOLOGY, ENTOMOLOGY. Educ: Kent State Univ, BS, 56, MA, 58; Ohio State Univ, PhD(entom), 65. Prof Exp: Asst biol, Kent State Univ, 56-58, instr, 58-60; asst zool, Ohio State Univ, 60-64; asst prof, 64-69, ASSOC PROF BIOL, KENT STATE UNIV, 69- Mem: AAAS; Entom Soc Am; Nat Asn Biol Teachers; Am Inst Biol Sci. Res: Insect nutrition; relationship of free amino acid composition in plants and insect growth and development. Mailing Add: Dept of Biol Sci Kent State Univ Kent OH 44242

CICCARELLI, ROGER N, b Rochester, NY, Dec 23, 34; m 60; c 2. POLYMER CHEMISTRY. Educ: St Bonaventure Univ, BS, 56; Syracuse Univ, MS, 58, PhD(org chem), 61; State Univ NY, PhD(org chem), 61. Prof Exp: Chemist, Exxon Res & Eng Co, 61-62; scientist, 62-70, SR SCIENTIST, XEROX CORP, 70- Mem: Am Chem Soc; Sigma Xi; The Chem Soc. Res: Synthetic and polymer chemistry; structure versus electrical properties of polymers; toners and carriers; electrostatics; photoconductors; reaction mechanisms; rocket fuels. Mailing Add: 145 Hibiscus Dr Rochester NY 14618

CICCHINELLI, ALEXANDER L, b Waterford, NY, Oct 13, 34; m 59; c 7. BIOSTATISTICS, ACADEMIC ADMINISTRATION. Educ: Cornell Univ, AB, 56; Univ Mich, MA, 59, PhD(biostatist), 62. Prof Exp: Asst prof biostatist, Western Reserve Univ, 62-64; dir comput ctr, Clark Col Technol, 64-70; DIR ANAL STUDIES & MGT ANAL, CENT ADMIN, STATE UNIV NY, 70- Concurrent Pos: Univ wide coordr State Univ NY & mem consult comt, Nat Ctr Higher Educ Mgt. Res: Management of higher education; faculty productivity analysis; program cost estimation and analysis. Mailing Add: State Univ NY Cent Admin 99 Washington Ave Albany NY 12210

CICCONE, PATRICK EDWIN, b Newark, NJ, Nov 20, 44; m 67; c 1. PSYCHOPHARMACOLOGY, PSYCHIATRY. Educ: Harvard Col, BA, 66; Univ Pa, MD, 70; Hosp Univ Pa, residency dipl psychiat, 74. Prof Exp: Staff psychiatrist, Erich Lindemann Ment Health Ctr, 74-75; ASST DIR & CLIN INVESTR PSYCHOPHARMACOL, McNEIL LABS, INC, 75- Concurrent Pos: Clin assoc psychiat, Mass Gen Hosp, 74-75; dir & staff psychiatrist, Revere Community Ment Health Ctr, 74-75; instr psychiat, Med Sch, Univ Pa, 76- Mem: Am Psychiat Asn; Intra-Sci Res Found. Res: Writing and development of clinical research drug protocols and monitoring the conduct of clinical investigations to prove the safety and efficacy of psychotropic compounds. Mailing Add: McNeil Labs Inc Camp Hill Rd Ft Washington PA 19034

CICERO, THEODORE JAMES, b Niagara Falls, NY, Aug 14, 42; m 66; c 3. BIOCHEMICAL PHARMACOLOGY. Educ: Villanova Univ, BS, 64; Purdue Univ, MS, 66, PhD(physiol psychol), 69. Prof Exp: Fel neurochem, 68-70, asst prof, 70-74, ASSOC PROF NEUROPHARMACOL, SCH MED, WASH UNIV, 74-, ASSOC PROF NEUROBIOL, 76- Concurrent Pos: Consult, Nat Inst Alcoholism & Alcohol Abuse, 70-74 & NIMH, 72-73. Mem: Soc Neurosci; Soc Biol Psychiat; Am Psychopath Asn; Soc Neurochem. Res: Neurochemical, neurobiological and neuroendocrinological correlates of tolerance to and dependence on narcotics and alcohol; developmental neurochemistry. Mailing Add: Dept of Psychiat Sch Med Wash Univ 4940 Audubon Ave St Louis MO 63110

CICERONE, RALPH JOHN, b New Castle, Pa, May 2, 43; m 67; c 1. AERONOMY, ATMOSPHERIC CHEMISTRY. Educ: Mass Inst Technol, SB, 65; Univ Ill, MS, 67, PhD(elec eng & physics), 70. Prof Exp: Physicist, US Dept Com, 67; res asst aeronomy, Univ Ill, 67-70; ASSOC RES SCIENTIST AERONOMY, SPACE PHYSICS RES LAB, UNIV MICH, ANN ARBOR, 70- Concurrent Pos: Consult, Aeronomy Corp, 70-71 & NSF, 75; lectr & asst prof elec eng, Univ Mich, Ann Arbor, 73-75. Mem: AAAS; Am Geophys Union; Am Meteorol Soc. Res: Theoretical and experimental studies of the earth's upper atmosphere, including the stratosphere, mesosphere and ionosphere. Mailing Add: Space Physics Res Lab Univ of Mich Ann Arbor MI 48109

CICHOCKI, FREDERICK PAUL, b Detroit, Mich, Dec 28, 43; m 66; c 3. EVOLUTIONARY BIOLOGY, ICHTHYOLOGY. Educ: Univ Miami, BS, 66, MS, 68; Univ Mich, PhD(zool), 76. Prof Exp: Instr biol, Kalamazoo Col, 73-74; ASST PROF ZOOL, UNIV GUELPH, 76- Mem: Am Soc Ichthyologists & Herpetologists; Soc Syst Zool. Res: Evolution of reproductive strategies of fishes; comparative morphology and cladistic history of fishes. Mailing Add: Dept of Zool Univ of Guelph Guelph ON Can

CICHOWSKI, ROBERT STANLEY, b Lakewood, Ohio, Feb 28, 42; m 66; c 3. PHYSICAL CHEMISTRY, CERAMICS. Educ: Purdue Univ, BSChE, 64; State Univ NY Col Ceramics, Alfred Univ, PhD(ceramic sci), 68. Prof Exp: Res chemist, Phillips Petrol Co, Okla, 68-71; lectr, 71, asst prof, 71-75, ASSOC PROF CHEM, CALIF POLYTECH STATE UNIV, SAN LUIS OBISPO, 75- Mem: Am Chem Soc; Catalysis Soc; Am Ceramic Soc; Nat Sci Teachers Asn. Res: Chemistry of glazes; surface chemistry of solids and solid state chemistry related to heterogeneous catalysis. Mailing Add: Dept of Chem Calif Polytech State Univ San Luis Obispo CA 93407

CIECIUCH, RONALD FRANK WALTER, physical chemistry, organic chemistry, see 12th edition

CIEGLER, ALEX, b Yonkers, NY, Apr 25, 24; m 55; c 4. MICROBIOLOGY. Educ: NY Univ, BA, 49; Univ Ky, MS, 51; Univ Wis, PhD(bact), 56. Prof Exp: Bacteriologist, US Army Chem Corps Biol Labs, Ft Detrick, 51-53; res microbiology, 56-70, HEAD BIOL PROD INVESTS, FERMENTATION LAB, NORTHERN REGIONAL RES LAB, USDA, 70- Mem: Am Soc Microbiol. Res: Fermentation; microbiological production of carotenoids, pigments organic acids; freeze drying of microorganisms; mycotoxins; medical enzymes; microbial transformations. Mailing Add: 2610 W Greenbriar Lane Peoria IL 61614

CIERESZKO, LEON STANLEY, b Holyoke, Mass, July 30, 17; m 43; c 1. BIOCHEMISTRY. Educ: Mass State Col, BS, 39; Yale Univ, PhD(physiol chem), 42. Prof Exp: Lab asst physiol chem, Yale Univ, 39- 42; res biochemist, Med Res Div, Sharp & Dohme, Inc, Pa, 42-45; instr biol chem, Sch Med, Univ Utah, 45-46; instr chem, Univ Ill, 46-48; from asst prof to assoc prof, 48-56, chmn dept, 69- 70, PROF CHEM, UNIV OKLA, 56- Concurrent Pos: Fulbright fel, Zool Sta, Naples, Italy, 55-56; res fel, Yale Univ, 59; off partic, US Prog Biol, Int Indian Ocean Exped, 63; fel & vis investr, Friday Harbor Labs, Seattle, 64; consult prof biochem, Sch Med, Univ Okla, 67-; vis investr, Dept Marine Sci, Univ PR, 70 & 74-75, Caribbean Res Inst, VI, 71 & Marine Sci Inst, Univ Tex, 74; consult, Caribbean Res Inst, 70-; vis lectr, Col VI, 71. Mem: AAAS; Am Chem Soc; Geochem Soc; The Chem Soc. Res: Comparative biochemistry; chemistry of natural products from marine animals; chemistry of coelenterates and their zooxanthellae; toxic substances of marine origin; biogeochemistry of coral reefs. Mailing Add: Dept of Chem Univ of Okla 620 Parrington Oval 211 Norman OK 73069

CIFELLI, RICHARD, b Newark, NJ, Apr 26, 23; m 51; c 4. INVERTEBRATE PALEONTOLOGY. Educ: Univ Mont, BA, 47; Univ Calif, MA, 51; Harvard Univ, PhD, 59. Prof Exp: Geologist, Phillips Petrol Co, 51-55; instr geol, Brown Univ, 57-59; CUR, NAT MUS NATURAL HIST, 59- Concurrent Pos: Dir, Cushman Found Foraminiferal Res. Res: Paleont Soc; Soc Syst Zool; Soc Study Evolution. Res: Paleontology, speciality foraminifera; stratigraphy; oceanography. Mailing Add: Nat Mus of Natural Hist Washington DC 20560

CIFONELLI, JOSEPH ANTHONY, b Utica, NY, Mar 19, 16; m 49; c 2. CARBOHYDRATE CHEMISTRY. Educ: Univ Minn, PhD(biochem), 52. Prof Exp: Res assoc carbohydrate chem, Univ Minn, 53; res assoc pediat, Bobs Roberts Hosp, 53-64, RES ASSOC BIOCHEM, LA RABIDA INST, 64-; PROF PEDIAT, UNIV CHICAGO, 71- Concurrent Pos: Assoc prof pediat, Univ Chicago, 67-71. Mem: Am Chem Soc; Am Soc Biol Chemists. Res: Biosynthesis of mucopolysaccharides; enzyme chemistry. Mailing Add: Dept of Pediat Pritzker Sch of Med Univ of Chicago Chicago IL 60637

CIFTAN, MIKAEL, b Istanbul, Turkey, Aug 12, 35; US citizen; m 57; c 1. THEORETICAL PHYSICS. Educ: Robert Col, Istanbul, BSc, 57; Mass Inst Technol, MSc, 59; Duke Univ, PhD(physics), 67. Prof Exp: Res physicist, Raytheon Co, 60-65; res assoc physics, Ind Univ, Bloomington, 68-70; physicist, US Army Res Off-Durham, 70-74, MEM RES STAFF, US ARMY RES OFF-RESEARCH TRIANGLE PARK, 74- Concurrent Pos: Mem res staff, Duke Univ, 70-74. Mem: Am Phys Soc; NY Acad Sci. Res: Symmetry principles in physics and representation theory; elementary particle phenomenology; special functions via group theory; theoretical and experimental laser physics, spectroscopy; biomethematical approach to experimental histocompatibility and genetic mapping; theoretical biology; solid state physics. Mailing Add: US Army Res Off PO Box 12211 Research Triangle Park NC 27709

CIGNETTI, JESS A, b Oklahoma, Pa, Sept 18, 41; m 63; c 1. SCIENCE EDUCATION. Educ: Slippery Rock State Col, BS, 62; Duquesne Univ, MS, 65; Ohio State Univ, PhD(sci educ), 70. Prof Exp: Sci teacher, Penn Hills High Sch, 62-64 & Gateway Schs, 64-67; admin asst, Ohio State Univ, 67-68; TEACHER PHYS SCI, CALIFORNIA STATE COL, PA, 68- Concurrent Pos: Consult local schs, 74- Mem: Nat Sci Teachers Asn. Res: Interdisciplinary sciences; curriculum for nonscience major in colleges. Mailing Add: New Sci Bldg 214 California State Col California PA 15419

CIHONSKI, JOHN LEO, b Butler, Pa, Oct 24, 48; m 72. ANALYTICAL CHEMISTRY. Educ: Slippery Rock State Col, BA, 71; Tex A&M Univ, PhD(chem), 75. Prof Exp: Res assoc chem eng, Tex A&M Univ, 75; ANAL CHEMIST WATER RECOVERY & INORG ANAL EL PASO PROD CO, 75- Mem: Am Chem Soc; AAAS. Res: Water recovery and disposal; industrial heterogeneous catalysis with continued interests in ozone research; organometallic photochemistry and molecular spectroscopy. Mailing Add: El Paso Prod Co PO Box 3986 Odessa TX 79760

CILLEY, JONATHAN HUBBARD, b Arlington, Md, Aug 14, 16; m 40; c 3. BIOCHEMISTRY. Educ: Wheaton Col, BS, 38; Northwestern Univ, PhD(physiol chem), 47. Prof Exp: Lab asst, Wheaton Col, 35-38, asst, 38; lab asst, Med Sch, Northwestern Univ, 38-47, asst, Off Sci Res & Develop Contract, 42-43, instr chem, Sch Nursing, 44-45; from instr to asst prof, 47-64, ASSOC PROF BIOCHEM, SCH MED, TEMPLE UNIV, 64-, DIR STUDENT LABS, 68- Mem: Am Chem Soc; Am Asn Clin Chem; Asn Multidiscipline Educ in Health Sci. Res: High altitude physiology; factors influencing an increase of urinary amino nitrogen; clinical chemistry. Mailing Add: Dept of Student Labs Temple Univ Sch of Med Philadelphia PA 19140

CIMINERA, JOSEPH LOUIS, b Philadelphia, Pa, Dec 4, 17; m 45; c 2. PHARMACY, BIOSTATISTICS. Educ: Philadelphia Col Pharm & Sci, BSc, 38; Villanova Univ, MSc, 60. Hon Degrees: DSc, Philadelphia Col Pharm & Sci, 69. Prof Exp: Pharmaceut worker, 31-38; res assoc, 38-49, mgr statist serv, 49-70, assoc dir, 70-71, res fel, 71-72, SR RES FEL, MERCK SHARP & DOHME RES LABS, 72- Concurrent Pos: Lectr, Philadelphia Col Pharm & Sci, 53-; lectr & head dept statist, Grad Sch, Villanova Univ, 53; mem clin guidelines comt, FDA-Pharmaceut Mfrs Asn, 70-; mem panel on drug bioavailability studies, US Pharmacopoeia, 70- Mem: Am Pharmaceut Asn; Am Statist Asn; Biomet Soc. Res: Biometry; pharmaceutical chemistry. Mailing Add: Merck Sharp & Dohme Res Lab West Point PA 19486

CINADER, BERNHARD, b Vienna, Austria, Mar 30, 19; c 1. IMMUNOCHEMISTRY. Educ: Univ London, BSc, 45, PhD, 48, DSc, 58. Prof Exp: Asst, Lister Inst Prev Med, Eng, 45-46, Brit Mem fel, 49-53, Agr Res Coun grantee, 53-56; fel immunochem, Inst Path, Western Reserve Univ, 48-49; prin sci officer, Dept Exp Path, Agr Res Coun Inst Animal Physiol, Eng, 56-58; head subdiv immunochem div biol, Ont Cancer Inst, 58-69; assoc prof, 58-69, PROF MED BIOPHYS, UNIV TORONTO, 69-, PROF MED GENETICS, 69-, PROF DEPT CLIN BIOCHEM, 70-, DIR INST IMMUNOL, 71- Concurrent Pos: Lectr & medallist, Fr Soc Biol Chemists, 54; joint chmn sect immunochem, Int Cong Biochem, Belgium, 55, dir, Austria, 58; spec lectr, Pasteur Inst, 60; pub lectr, Univ Col, Univ London, 63; Enrique E Ecker lectr, Western Reserve Univ, 64; vis prof, Univ Alta, 68 & Univ Manitoba; Pfizer fel, Inst Clin Res, Montreal, 72; A Harrington lectr, State Univ NY, Buffalo, 74; guest lectr, Acads Sci, Czech & Hungary & Acad Med, Rumania, 74. Mem grant panel epidemiol, immunol, microbiol, path & virol, Nat Cancer Inst; mem comt on antilymphocytic serum, Med Res Coun, chmn grants panel immunol & transplantation, 70; mem expert adv panel on immunol, WHO, 70-, mem task force standardization immune reagents, 73-; mem spec comt clin immunol, Royal Col Physicians & Surgeons; pres, Int Union Immunol Socs, 69-74; chmn immunol comt, Biol Coun Can; mem, WHO Human Reprod Task Force Adv Bd, Immunol-Med Univ SC, Centre Immunol, Buffalo, IUIS-WHO Inst, Amsterdam, 71-; mem, Can Sci Deleg to USSR, 75. Mem: Am Asn Immunologists; Brit Soc Immunologists; Can Soc Immunologists (pres, 67); fel Royal Soc Can; Can Fedn Biol Socs (vchmn, 75-76, chmn, 76-77). Res: Immunochemistry and genetics of mammalian polymorphic proteins; antibodies to enzymes; regulation of the immune response; acquired immunological tolerance; allotypes; antibody synthesis; tumor immunology. Mailing Add: Inst Immunol Med Sci Bldg Univ of Toronto Toronto ON Can

CINCOTTA, JOSEPH JOHN, b Queens, NY, Sept 15, 31; m 55; c 3. ANALYTICAL CHEMISTRY. Educ: Columbia Univ, BS, 53; City Univ New York, MS, 66. Prof Exp: Anal chemist, Am Molasses Co, 53-59; anal res chemist, Am Cyanamid Co, 59-68 & M W Kellogg Co, 68-69; SR ANAL CHEMIST, CONTINENTAL OIL CO, 69- Mem: Am Chem Soc; fel Am Inst Chemists. Res: Gas chromatography; coulometry; spectrophotometry; atomic absorption spectrometry and liquid chromatography. Mailing Add: 5025 Cloudburst Hill Columbia MD 21044

CINES, MARTIN R, chemistry, see 12th edition

CINLAR, ERHAN, b Divrigi, Turkey, May 28, 41; US citizen. MATHEMATICAL STATISTICS, OPERATIONS RESEARCH. Educ: Univ Mich, BSE, 63, MA, 64, PhD(indust eng), 65. Prof Exp: From asst prof to assoc prof opers res, Northwestern Univ, 65-71; vis prof, Stanford Univ, 71-72; PROF OPERS RES, NORTHWESTERN UNIV, EVANSTON, 72- Concurrent Pos: Assoc ed, J Stochastic Processes & Their Applns, Math of Opers Res & Mgt Sci. Mem: Am Math Soc; fel Inst Math Statist; Inst Mgt Sci. Res: Stochastic processes; theory of regeneration; Markov processes and boundary theory; Markov renewal theory; point processes; random measures; random sets. Mailing Add: Dept of Indust Eng Northwestern Univ Evanston IL 60201

CINO, PAUL MICHAEL, b New York, NY, Dec 8, 46; m 73. INDUSTRIAL MICROBIOLOGY. Educ: Hunter Col, BA, 68; Rutgers Univ, MS, 70, PhD(microbiol), 73. Prof Exp: Res fel, Waksman Inst Microbiol, Rutgers Univ, 73-75; RES INVESTR MICROBIOLOGIST, E R SQUIBB & SONS INC, 75- Mem: Am Soc Microbiol; Soc Indust Microbiol. Res: Microbial bioconversions and enzymatic synthesis of microbial products. Mailing Add: Squibb Inst for Med Res Georges Rd New Brunswick NJ 08903

CINOTTI, ALFONSE A, b Jersey City, NJ, Jan 1, 23; m 46; c 5. MEDICINE, OPHTHALMOLOGY. Educ: Fordham Univ, BS, 43; Long Island Univ, MD, 46; Am Bd Ophthal, dipl. Prof Exp: From asst prof surg & ophthal to assoc prof ophthal, 57-73, PROF OPHTHAL & CHMN DEPT, COL MED NJ, 73- Concurrent Pos: Assoc examr, Am Bd Ophthal, 53-63; dir resident training ophthal, New York Eye & Ear Infirmary, 55-63, asst dir inst ophthal, 57-, attend ophthalmologist & dir glaucoma, 63-, consult glaucoma, 70-; chmn eye health screening prog, State of NJ, 58-; dir ophthal, Jersey City Med Ctr, 63; mem bd joint comn, Allied Health Personnel Ophthal. Mem: Fel Am Col Surg; Asn Res Vision & Ophthal; Am Asn Ophthal; Am Acad Ophthal & Otolaryngol. Res: Influence of hormones and methods of case finding of glaucoma; vitreous and retinal detachment; diabetic cataracts; scanning electron microscopy of human cataract; trace metals in retinitis pigmentosa; color field studies in retinitis pigmentosa; comparative studies of biomicroscopy of cataracts in vivo with scanning electron microscopy. Mailing Add: Dept of Ophthal Col of Med of NJ 100 Bergen St Newark NJ 07103

CINOTTI, WILLIAM RALPH, b Jersey City, NJ, Sept 14, 26; m 54; c 2. PROSTHODONTICS, PERIODONTICS. Educ: Georgetown Univ, BS, 46, DDS, 51; NY Univ, cert periodont, 64. Prof Exp: Intern oral surg, Martland Med Ctr, 51-52; pvt pract, 52-60; from clin instr to clin asst prof dent, Sch Dent, Seton Hall Univ, 60-67; clin assoc prof, 67-68, assoc prof, 68-71, PROF DENT, NJ DENT SCH, COL MED & DENT NJ, 71- Concurrent Pos: Dir dept dent serv, Hudson County, NJ, 70- Mem: Am Prosthodont Soc; fel Am Col Dent; Am Geriat Soc; Royal Soc Health; fel Int Col Dent. Res: Termporomandibular joint dysfunction; desensitization of teeth; psychologic evaluation of dental patients; evaluation of efficacy of denture adhesive and denture cleanser. Mailing Add: NJ Dent Sch Col of Med & Dent of NJ Jersey City NJ 07307

CINQUINA, CARMELA LOUISE, b Philadelphia, Pa, Mar 11, 36. BACTERIOLOGY. Educ: West Chester State Teachers Col, BS, 57; Villanova Univ, MS, 63; Rutgers Univ, PhD(bact), 68. Prof Exp: Instr biol & phys educ, York Jr Col, 57-61; from instr to asst prof biol, 61-68, assoc prof, 68-69, PROF BACT, WEST CHESTER STATE COL, 69- Mem: AAAS; Am Soc Microbiol. Res: The effect of caffeine on growth in bacterial cells wiparticularly on morphological changes, variation in lipid composition and changes in cyclic adenosine monophosphate phosphodiesterase activity. Mailing Add: Dept of Biol West Chester State Col West Chester PA 19380

CINTI, DOMINICK LOUIS, b Wilkes-Barre, Pa, Dec 16, 39; m 67. PHYSIOLOGY, PHARMACOLOGY. Educ: Univ Scranton, BS, 61; Jefferson Med Col, MS, 66, PhD(physiol), 68; Drexel Inst Technol, MSBmE, 69. Prof Exp: Sr med technician, Jefferson Med Col, 62-63, asst, 64-66; instr physiol, Sch Med, Temple Univ, 66-67; Nat Heart Inst fel, 67-69; res assoc pharmacol, Sch Med, Yale Univ, 69-72; vis scientist, Karolinska Inst, Stockholm, 72-73; ASST PROF PHARMACOL, HEALTH CTR, UNIV CONN, FARMINGTON, 73- Res: Hepatic membrane bound enzymes, such as mixed function oxidase system; microsomal electron transport system. Mailing Add: Dept of Pharmacol Univ of Conn Health Ctr Farmington CT 06032

CIOCCO, ANTONIO, biochemistry, deceased

CIOFFI, PAUL PETER, b Cervinara, Italy, June 29, 96; nat US; m 26; c 2. MAGNETISM. Educ: Cooper Union, BS, 19, EE, 22; Columbia Univ, AM, 24. Prof Exp: Mem tech staff, Bell Tel Labs, 17-61; consult, Arnold Eng Co, Ill, 61-65; MAGNETICS TECHNOL CONSULT, 65- Mem: AAAS; fel Am Phys Soc; Inst Elec & Electronics Engrs; NY Acad Sci; Sigma Xi. Res: Magnetic measurements, materials and circuit; electromagnets for intense magnetic fields. Mailing Add: 132 Kent Pl Blvd Summit NJ 07901

CIORDIA, HONORICO, b Vega Baja, PR, Aug 18, 20; m 43; c 2. PARASITOLOGY. Educ: Univ Tenn, BA, 47, MS, 48, PhD, 52. Prof Exp: Instr zool & entom, Univ Tenn, 48-49, asst, 48-52; assoc parasitologist, Agr Exp Sta, Univ PR, 52-53; instr, Univ Tenn, 53-54, res assoc zool & entom, 54-55; PARASITOLOGIST, VET SCI RES DIV, AGR RES SERV, USDA, 55- Mem: Am Soc Parasitologists; Am Micros Soc. Res: Cytotaxonomy of helminth parasites; relationship between irradiation and parasitism; ecology of preparasitic nematodes; control of internal parasites of cattle and sheep. Mailing Add: Vet Sci Res Div Agr Res Serv US Dept of Agr Experiment GA 30212

CIPERA, JOHN DOMINIK, b Czech, Aug 7, 23. ORGANIC BIOCHEMISTRY. Educ: Tech Univ, Czech, Ing, 48; Univ Toronto, MSA, 51; McGill Univ, PhD(chem), 54. Prof Exp: Asst chem, Col Forestry, State Univ NY, 55-56; res assoc, Univ Pittsburgh, 56-58; RES SCIENTIST, ANIMAL RES INST, CAN DEPT AGR, 58- Honors & Awards: Eddy Found Award, 52. Mem: Am Chem Soc; Biochem Soc; Chem Inst Can. Res: Organic chemistry of naturally occurring polymers; peptides; glycosaminoglycans; chemistry and physiology of connective tissues; role of organic matrix in calcification processes. Mailing Add: 830 Maplecrest Ave Ottawa ON Can

CIPOLLA, SAM J, b Chicago, Ill, July 24, 40; m 66; c 2. ATOMIC PHYSICS, NUCLEAR PHYSICS. Educ: Loyola Univ Chicago, BS, 62; Purdue Univ, MS, 65, PhD(nuclear physics), 69. Prof Exp: Asst, Purdue Univ, 62-69, res assoc, 69; asst prof, 69-72, ASSOC PROF PHYSICS, CREIGHTON UNIV, 72- Concurrent Pos: Res partic, Oak Ridge Nat Lab, 71-; Cottrell Col Sci Grant, Res Corp, USA, 72-75; consult, Omaha Pub Power Dist, 74- Mem: Am Phys Soc; Am Asn Physics Teachers. Res: Radioactivity measurement; operation of nuclear instrumentation; nuclear spectroscopy measurements; radioactive source preparation; vacuum technology; ion-induced inner-shell ionization measurements in atoms. Mailing Add: Dept of Physics Creighton Univ Omaha NE 68178

CIPOLLARO, ANTHONY C, medicine, deceased

CIPORIN, LEON, polymer chemistry, see 12th edition

CIPPARONE, JOSEPH ROBERT, b Windsor, Ont, Aug 29, 28; US citizen; m 53; c 5. PATHOLOGY. Educ: Univ Detroit, BS, 48; Marquette Univ, MD, 52. Prof Exp: ASSOC PATHOLOGIST, ST LAWRENCE HOSP, 59-; ASSOC PROF PATH, COL HUMAN MED, MICH STATE UNIV, 68- Mem: Col Am Path; NY Acad Sci; Am Soc Clin Path; Am Acad Forensic Sci; Pan-Am Med Asn. Res: Pathology of the placenta. Mailing Add: Dept of Path St Lawrence Hosp Lansing MI 48914

CIPRIANI, CIPRIANO, b Venezia, Italy, Aug 25, 23; m 54; c 5. INDUSTRIAL CHEMISTRY. Educ: Univ Bologna, Italy, PhD(indust chem), 49. Prof Exp: Res chemist, Snia Viscosa, Italy, 48-52 & Courtaulds Can Ltd, Ont, 52-57; sr res chemist, Celanese Corp Am, 57-62; supvr res, Fibers Div, Allied Chem Corp, 62-68, dep sci dir, Allied Chem SA, Brussels, 68-70, technol mkt specialist, Corp Res & Develop, 70-72; PRIN ENGR, FIBER DIV, FMC CORP, 72- Mem: Am Chem Soc; Fiber Soc. Res: Research and development, manufacture and applications of man-made fibers; technology utilization, particularly licensing. Mailing Add: 9 Sunderland Dr Morristown NJ 07960

CIPRIANO, LEONARD FRANCIS, b New York, NY, Feb 26, 38; m 62; c 1. PHYSIOLOGY. Educ: City Col New York, BS, 59; Univ Calif, Berkeley, PhD(physiol), 70. Prof Exp: Res physiologist, US Army Res Inst Environ Med, Natick, Mass, 70-72; LAB DIR CARDIOVASC & PULMONARY PHYSIOL, LOVELACE FOUND MED EDUC & RES, ALBUQUERQUE, N MEX, 72- Concurrent Pos: Instr, Calif State Col, Bakersfield, 73-75. Mem: Am Physiol Soc; Can Physiol Soc; AAAS; Sigma Xi. Res: Cellular and systemic physiology; acclimatization and adaptation to altitude; thermoregulation; exercise physiology; pulmonary and cardiovascular physiology; man's interaction with the environment. Mailing Add: 43850 Fenner Ave Lancaster CA 93534

CIRCLE, SIDNEY JOSEPH, protein chemistry, food chemistry, see 12th edition

CIRIACKS, KENNETH W, b West Bend, Wis, May 7, 38; m 65; c 2. GEOLOGY. Educ: Univ Wis, BS, 58; Columbia Univ, PhD(geol), 62. Prof Exp: Res scientist, Pan Am Petrol Corp Res Ctr, Stand Oil Co Ind, 62-65, sr res scientist, 65-71, staff res scientist, Res Ctr, 71-72, proj geologist, 72-73, dist geologist, 73-75, DIV GEOLOGIST, AMOCO PROD CO, 76- Concurrent Pos: NSF fel, Columbia Univ, 62. Mem: Geol Soc Am; Am Paleont Soc; Asn Econ Paleont & Mineral; Brit Paleont Asn. Res: Late Paleozoic biostratigraphy; taxonomy, evolution and ecology of fossil and living pelecypods; geological aspects of physical and biological processes in modern marine carbonate environments. Mailing Add: Amoco Prod Co Security-Life Bldg Denver CO 80202

CIRIACY, EDWARD W, b Philadelphia, Pa, Feb 12, 24; c 4. FAMILY MEDICINE. Educ: Pa State Col, BS, 48; Temple Univ, MD, 52. Prof Exp: Intern, Frankford Hosp, Philadelphia, 52-53, resident surg, Frankford & Temple Univ Hosps, 53-54; pvt pract, 54-71; PROF FAMILY PRACT & HEAD DEPT, UNIV MINN MINNEAPOLIS, 71- Concurrent Pos: Mem bd & mem res & develop comt, Am Bd Family Pract, 72-75, mem recert exam panel, 74-76; mem adv bd, Mod Med Publ, 74- Mem: Am Acad Family Physicians; AMA; Asn Am Med Cols; Pan-Am Med Asn; Soc Teachers Family Med. Mailing Add: Dept Fam Pract & Community Hlth Univ Minn Mayo Mem Bldg Box 381 Minneapolis MN 55455

CIRIC, JULIUS, b Kragujevac, Yugoslavia, Nov 10, 22; nat US; m 56; c 4. PHYSICAL CHEMISTRY, CHEMICAL ENGINEERING. Educ: Darmstadt Tech Univ, Dipl, 49; Univ Toronto, MASc, 52, PhD, 56. Prof Exp: Instr, Univ Toronto, 50-52; jr chem engr, Ont Paper Co, 52-53; chem engr, Shawinigan Chems, Ltd, 56-57; asst res specialist, Sch Chem, Rutgers Univ, 57-59 & Ont Res Found, 59-62; SR RES CHEMIST, MOBIL RES & DEVELOP CORP, 62- Mem: Am Chem Soc. Res: Ion-exchange; chemical engineering unit operations; inorganic preparative and physical chemistry. Mailing Add: Mobil Res & Develop Corp Paulsboro NJ 08066

CIRILLO, VINCENT PAUL, b New York, NY, Oct 16, 25; m 49; c 4. BIOCHEMISTRY. Educ: Univ Buffalo, BA, 47; NY Univ, MS, 52; Univ Calif, Los Angeles, PhD(biol), 53. Prof Exp: Asst biol, Univ Buffalo, 46-47 & NY Univ, 48-50; asst zool, Univ Calif, Los Angeles, 50-53; asst prof prev med & pub health, Sch Med, Univ Okla, 53-56; sr res microbiologist, Anheuser-Busch, Inc, Mo, 56-59; asst prof microbiol, Col Med & Dent, Seton Hall Univ, 59-62, assoc prof biochem, 62-64; assoc prof, 64-69, PROF BIOCHEM, STATE UNIV NY STONY BROOK, 69- Mem: Am Soc Biol Chemists; Am Soc Microbiol; Am Soc Cell Biol; Soc Gen Physiol. Res: Membrane structure; chemistry and mechanisms of sugar transport. Mailing Add: Dept of Biochem State Univ of NY Stony Brook NY 11790

CIRINO, ELIZABETH FAHEY, b Taunton, Mass, Oct 28, 17; m 53; c 1. MARINE ECOLOGY. Educ: Mass State Teachers Col, BS, 40; Boston Univ, AM, 51, PhD(biol), 58. Prof Exp: From instr to assoc prof, 52-59, PROF BIOL, BRIDGEWATER STATE COL, 59- Mem: AAAS; NY Acad Sci; Am Soc Limnol & Oceanog; Phycol Soc Am; Ecol Soc Am. Res: Intertidal and subtidal communities. Mailing Add: Dept of Biol Sci Bridgewater State Col Bridgewater MA 02324

CISIN, IRA HUBERT, b New York, NY, Sept 1, 19; m 46; c 3. STATISTICS. Educ: NY Univ, BS, 39; Am Univ, MA, 51, PhD(statist), 57. Prof Exp: Asst res dir, Samuel E Gill, 41-42; res technician, Off War Info, 42 & Res Br, War Dept, 42-45; res assoc, Columbia Broadcasting Syst, 45-46; co-chief prof staff attitude, Res Br, Defense Dept, 46-52; sr res scientist, Human Resources Res Off, George Washington Univ, 52-53, dir res, motivation, morale & leadership, 53-54, adv res design, 54-59; res specialist, Calif Dept Pub Health, 59-62; PROF SOCIOL & DIR SOCIAL RES GROUP, GEORGE WASHINGTON UNIV, 62- Concurrent Pos: Consult, Bur Social Sci Res, 54-, Calif Dept Pub Health, 62-, Columbia Broadcasting Syst, 64-, Social Res Group, Univ Calif, Berkeley, 70- & Nat Inst Drug Abuse, 74-; mem, Surg Gen Adv Comt TV & Social Behav, 69-70. Mem: AAAS; Am Asn Pub Opinion Res; Am Statist Asn; Inst Math Statist. Res: Development of mathematical models and improvement of measurement techniques for social science; devising statistical procedures for application in problem areas not previously amenable to quantitative approaches. Mailing Add: Dept of Sociol George Washington Univ Washington DC 20037

CISKOWSKI, JOSEPH M, b Bridgeport, Conn, May 1, 17; m 38; c 3. ORGANIC CHEMISTRY. Educ: Univ Ill, BS, 38, MS, 39. Prof Exp: Chem engr, Merck & Co, Inc, 39-41, res chemist, 41-42 & 45-49, proj engr, 49-52, mgr chem eng design group, 52-56, dir eng develop technol, 56-57, dir prod develop, 57-63; dir res ctr, Stauffer Chem Co, Calif, 63-65; vpres res, Celotex Div, 65-67, PRES, JIM WALTER RES CORP, 67- Mem: Am Chem Soc; Am Inst Chem Engrs; Inst Food Technologists. Res: Process and product research and development in organic, industrial and agricultural chemicals. Mailing Add: Jim Walter Res Corp 10301 Ninth St N St Petersburg FL 33702

CISLAK, FRANCIS EDWARD, b Chicago, Ill, Oct 3, 05; m. ORGANIC CHEMISTRY. Educ: Univ Chicago, BS, 25, MS, 26; Northwestern Univ, PhD(chem), 29. Prof Exp: Res chemist, Reilly Tar & Chem Corp, 29-37, dir res, 37-73; RETIRED. Concurrent Pos: Adv, NSF, 58-59. Mem: AAAS (vpres, 57); Am Chem Soc; Am Inst Chem Engrs; The Chem Soc; Brit Soc Chem Indust. Res: Coal tar chemistry; insecticides; fungicides; pyridine chemistry. Mailing Add: 5331 N Kenwood Ave Indianapolis IN 46208

CITRON, IRVIN MEYER, b Atlanta, Ga, May 5, 24; m 65; c 1. ANALYTICAL CHEMISTRY, SCIENCE EDUCATION. Educ: Hebrew Univ, Jerusalem, 58; Emory Univ, MS, 61; NY Univ, PhD(sci educ), 69. Prof Exp: Chem lab asst, Israel Defense Dept Labs, Weizmann Inst Sci, 54-56; res asst inorg chem, Hebrew Univ, Jerusalem, 56-58; res asst anal chem, Emory Univ, 58-61; asst prof anal & inorg chem, Troy State Col, 61-62; asst prof, 62-69, asst chmn dept chem, 67-69, ASSOC PROF ANAL & INORG CHEM, FAIRLEIGH DICKINSON UNIV, 69- Concurrent Pos: Deleg, Colloquium Spectroscopicum Int, Ottawa, 67 & Spectros Symp Can, 69. Mem: Am Chem Soc; Soc Appl Spectros; Nat Sci Teachers Asn; Nat Asn Res Sci Teaching. Res: Use of organometallic complexes for analytical purposes; methods of teaching science at high school and college levels. Mailing Add: Dept of Chem Fairleigh Dickinson Univ Rutherford NJ 07070

CITRON, JOEL DAVID, b Brooklyn, NY, Apr 19, 41. ORGANIC CHEMISTRY, POLYMER CHEMISTRY. Educ: Polytech Inst Brooklyn, BS, 62; Univ Calif, Davis, PhD(org chem), 67. Prof Exp: Teaching assoc chem, Univ Calif, Davis, 67-68; RES CHEMIST, ELASTOMER CHEM DEPT, E I DU PONT DE NEMOURS & CO, INC, 69- Mem: Am Chem Soc. Res: Elastomers. Mailing Add: Elastomer Chem Dept E I du Pont de Nemours & Co Inc Wilmington DE 19898

CIULA, RICHARD PAUL, b Lorain, Ohio, Dec 8, 33; m 59; c 3. ORGANIC CHEMISTRY. Educ: Bowling Green State Univ, BA, 55; Univ Calif, MS, 57; Univ Wash, PhD, 60. Prof Exp: From asst prof to assoc prof, 60-68, chmn dept, 66-71, PROF CHEM, CALIF STATE UNIV, FRESNO, 68- Mem: Am Chem Soc; Am Inst Chemists; The Chem Soc. Res: Synthesis and properties of small ring compounds, particularly in the cyclobutane series; kinetics and mechanism of the nitrile exchange reaction; synthesis of bicyclic amines. Mailing Add: Dept of Chem Calif State Univ Fresno CA 93740

CIULLO, ROBERT HENRY, b Arlington, Mass, May 14, 28; m 58; c 1. ZOOLOGY. Educ: Boston Col, BS, 53, MS, 56; Univ NH, PhD(zool), 68. Prof Exp: Teacher high sch, Mass, 57-59; from asst prof to assoc prof, 59-72, coordr, Div Sci & Math, 71-73, PROF BIOL, NASSON COL, 72- Concurrent Pos: Biol consult & res dir, Var Water Qual Monitoring Progs, 70- Mem: Am Soc Zoologists; Am Soc Cell Biol; Am Inst Biol Sci; AAAS. Res: Histology and histochemistry of vertebrate organisms, utilizing saltwater minnow Fundulus heteroclitus; ecology; water pollution. Mailing Add: 306 Main St Springvale ME 04083

CIVAN, MORTIMER M, b New York, NY, Nov 13, 34; m 61; c 2. PHYSIOLOGY. Educ: Columbia Univ, AB, 55, MD, 59. Prof Exp: Intern, Presby Hosp, 59-60, asst resident, 60-62; staff assoc biophys, NIH, 62-64; instr med, Harvard Med Sch, 65-68, assoc, 68-69, asst prof, 69-72; ASSOC PROF PHYSIOL & ASSOC PROF MED, SCH MED, UNIV PA, 72- Concurrent Pos: USPHS clin & res fel, Mass Gen Hosp & Harvard Univ, 64-65 & USPHS spec fel, Weizmann Inst Sci, 70-71; Am Heart Asn grant-in-aid, 71-73; NSF grant, 73-; NIH grant, 74-; asst, Mass Gen Hosp, 65-72; estab investr, Am Heart Asn, 71- Mem: Am Soc Clin Invest; Am Physiol Soc; Biophys Soc; Am Soc Nephrology; Soc Gen Physiol. Res: Kinetics of muscle contraction; transport of solutes and water across membranes. Mailing Add: Dept of Physiol Richards Bldg G4 Univ of Pa Sch of Med Philadelphia PA 19174

CIVEN, MORTON, b Boston, Mass, June 20, 29; m 54; c 2. BIOCHEMISTRY. Educ: Harvard Univ, MSc, 53, PhD(biochem), 57. Prof Exp: Fel, Harvard Med Sch, 58-59; USPHS fel, Nat Inst Med Res, London, 59-61; RES BIOCHEMIST, US VET ADMIN HOSP, LONG BEACH, 62- Concurrent Pos: Asst prof biochem, Univ Southern Calif, 62-68, adj asst prof, 68-; adj assoc prof physiol, Univ Calif, Irvine, 72- Mem: AAAS; Brit Biochem Soc; Am Soc Biol Chemists. Res: Mechanism of enzyme induction, especially effects of peptide hormones on target cell membranes; mechanisms of action of gonadotropins; enzymatic regulation of amino acid metabolism; biochemistry of adrenal cells; effect of toxic chemicals on adrenocortical secretion. Mailing Add: Med Res 151 Vet Admin Hosp Long Beach CA 90801

CIVEROLO, EDWIN LOUIS, b Los Angeles, Calif, July 24, 40; m 67; c 1. PLANT PATHOLOGY, PLANT VIROLOGY. Educ: Univ Calif, Riverside, BA, 62, PhD(plant path), 67. Prof Exp: Res asst plant path, Univ Calif, Riverside, 62-67; PLANT PATHOLOGIST, CROPS RES DIV, PLANT INDUST STA, AGR RES SERV, USDA, 67- Mem: Am Phytopath Soc; Am Inst Biol Sci. Res: Bacteriophage; phytopathogenic bacteria. Mailing Add: Plant Indust Sta Agr Res Serv USDA Beltsville MD 20705

CIVIN, PAUL, b Rochester, NY, April 29, 19; m 39; c 2. MATHEMATICS. Educ: Univ Buffalo, BA, 39; Duke Univ, MA, 41, PhD(math), 42. Prof Exp: Instr math, Univ Mich, 42-43 & Univ Buffalo, 43-46; from asst prof to assoc prof, 46-57, PROF MATH, UNIV ORE, 57- Concurrent Pos: Mem, Inst Advan Study, 53-54; vis res prof, Univ Fla, 60-61; vis prof, Copenhagen Univ, 61-62 & 68-69; consult, Pres, Univ Ore, 73- Mem: Am Math Soc. Res: Fourier series; topology; two-to-one mappings of manifolds; Banach algebra. Mailing Add: Dept of Math Univ of Ore Eugene OR 97403

CIZEK, LOUIS JOSEPH, b New York, NY, Apr 11, 16; m 41; c 2. PHYSIOLOGY. Educ: Fordham Univ, BS, 37; Columbia Univ, MD, 41. Prof Exp: Intern med serv, Beekman Hosp, NY, 41-42; from instr to asst prof, 46-56, ASSOC PROF PHYSIOL, COL PHYSICIANS & SURGEONS, COLUMBIA UNIV, 56- Concurrent Pos: Managing ed, Proc, Soc Exp Biol & Med. Mem: Fel AAAS; Am Physiol Soc; Harvey Soc; Soc Exp Biol & Med (secy-treas); Am Soc Zool. Res: Water and electrolyte balance. Mailing Add: Dept of Physiol Columbia Univ Col Physicians & Surgeons New York NY 10032

CLAASSEN, CARL ERNEST, agronomy, see 12th edition

CLAASSEN, HOWARD HUBERT, b Hillsboro, Kans, Apr 10, 18; m 41; c 3. CHEMICAL PHYSICS. Educ: Univ Okla, PhD(physics), 49. Prof Exp: Asst prof physics, Univ Okla, 49-52; asst prof, Wheaton Col, 52-64, prof & chmn sci div, 64-66; sr physicist, Chem Div, Argonne Nat Lab, 66-71; PROF PHYSICS & CHMN SCI DIV, WHEATON COL, 71- Concurrent Pos: Res assoc, Ohio State Univ, 50-51; consult chem div, Argonne Nat Lab. Honors & Awards: Rosenberger Medal, Univ Chicago, 64. Mem: AAAS; fel Am Phys Soc. Res: Raman and infrared absorption spectra of molecules. Mailing Add: Dept Physics Wheaton Col Wheaton IL 60187

CLAASSEN, RICHARD STRONG, b Ithaca, NY, May 10, 22; m 45; c 3. SOLID STATE PHYSICS. Educ: Cornell Univ, AB, 43; Columbia Univ, MA, 47; Univ Minn, PhD(physics), 50. Prof Exp: Asst, Substitute Alloy Material Labs, 44-46 & Univ Minn, 47-50; physicist, 51-53, supvr, 53-57, mgr phys sci res dept, 57-60, dir phys res, 60-68, dir electronic component develop, 68-75, DIR MATERIALS & PROCESSES, SANDIA LABS, 75- Concurrent Pos: Chmn Nat Sci Seminar, 63; mem, Rocky Mountain Sci Coun, 61-, chmn, 65-66; mem solid state sci panel, Nat Acad Sci-Nat Res Coun, 65-, chmn, 74, mem nat material adv bd, 73-; panel chmn surv mat sci, Nat Acad Sci, 71. Mem: Fel Am Phys Soc. Res: Physics of solids; research and development administration. Mailing Add: Sandia Labs Albuquerque NM 87111

CLABAUGH, STEPHEN EDMUND, b Carthage, Tex, Apr 2, 18; m 45; c 3. GEOLOGY. Educ: Univ Tex, BS, 40, MA, 41; Harvard Univ, PhD(geol), 50. Prof Exp: Asst geol, Univ Tex, 40-41; geologist, US Geol Surv, 42-54; from asst prof to assoc prof, 47-55, chmn dept, 62-66, PROF GEOL, UNIV TEX, AUSTIN, 55- Concurrent Pos: Nat Res Coun fel, Harvard Univ, 46-47; Piper prof geol, 58. Mem: AAAS; fel Geol Soc Am; Am Geophys Union; fel Mineral Soc Am; Geochem Soc. Res: Geology of Montana corundum deposits; tungsten deposits of Osgood Range, Nevada; igneous and metamorphic rocks of Cornudas Peaks; Texas and New Mexico, and Christmas Mountains, Texas; vermiculite deposits and metamorphic rocks, central Texas; volcanic rocks of western Texas and Mexico. Mailing Add: Dept of Geol Sci Univ of Tex Austin TX 78712

CLABEAUX, MARIE STRIEGEL, b Buffalo, NY, July 5, 41. PHYSICAL ANTHROPOLOGY. Educ: State Univ NY Buffalo, BA, 63, MA, 66, PhD(phys anthrop), 67. Prof Exp: Lectr anthrop, Lehman Col, 66-67, from instr to asst prof, 67-73; ASSOC PROF ANTHROP & CHMN DEPT, STATE UNIV NY COL BUFFALO, 73- Concurrent Pos: Consult, WNED, Buffalo, NY, 65-66; dir, NSF Instructional Equip Grant, 69-71, recipient, NSF Sci Instructional Equip Grant, 74-76. Mem: AAAS; fel Am Asn Phys Anthrop; Am Anthrop Asn. Res: Palaeopathology of the pre-Columbian Indians of North America; physical and cultural environmental determinants of disease patterns in skeletal populations; methodology of analysis of osteoarthritis. Mailing Add: Dept of Anthrop State Univ NY Col Buffalo NY 14222

CLABOUGH, JEANNE WHITAKER, b Farmville, Va, Sept 2, 41; m 62. ANATOMY. Educ: Longwood Col, BA, 64; Med Col Va, PhD(anat), 69. Prof Exp: Instr, 68-69, ASST PROF ANAT, MED COL VA, 69- Mem: Am Asn Anat. Res: Neuroendocrinology, the mammalian pineal organ; electron microscopy; interaction of light, retina, sympathetic nerves, pineal and gonads. Mailing Add: Dept of Anat Med Col of Va Richmond VA 23219

CLADIS, JOHN BAROS, b Dawson, NMex, June 21, 22; m 47; c 4. NUCLEAR PHYSICS, PLASMA PHYSICS. Educ: Univ Colo, BS, 44; Univ Calif, Berkeley, PhD(nuclear physics), 52. Prof Exp: Asst, Univ Colo, 46-47; physicist, Lawrence Radiation Lab, Univ Calif, 48-52, mem res staff, Los Alamos Sci Lab, 52-55; SR STAFF SCIENTIST, LOCKHEED MISSILES & SPACE CO, 55-, DIR THEORET SPACE PHYSICS, LOCKHEED PALO ALTO RES LAB, 65- Mem: Am Phys Soc; Am Geophys Union. Res: High energy nuclear scattering experiments; nuclear weapons diagnostic measurements; Van Allen radiation belt measurements; plasma-magnetic field interactions; magnetospheric physics. Mailing Add: Lockheed Palo Alto Res Lab Dept 52-12 Bldg 202 3251 Hanover St Palo Alto CA 94304

CLADIS, PATRICIA ELIZABETH RUTH, b Shanghai, China, July 13, 37; US citizen; m 62; c 2. CHEMICAL PHYSICS, FLUID DYNAMICS. Educ: Univ BC, BA, 59; Univ Toronto, MA, 60; Univ Rochester, PhD(physics), 68. Prof Exp: Meteorologist, Govt Can, 59-62; programmer-analyst, Katz, Casciato & Shapiro, Ltd, Can, 62; instr physics, Western Conn State Col, 63-64; res asst, Univ Rochester, 64-68, consult, 68; consult, Univ Toronto, 68-69; research, Fac Sci, Lab Physics of Solids, Univ Paris South, Orsay, 69-70, sr researcher, 70-73; MEM TECH STAFF, BELL LABS, 73- Mem: Am Phys Soc. Res: Liquid crystals, macromolecules and defects; static and dynamic properties of liquid crystals. Mailing Add: Bell Labs 600 Mountain Ave Murray Hill NJ 07970

CLAESSENS, PIERRE, b Brussels, Belg, Sept 5, 39; m 68; c 1. ELECTROCHEMISTRY. Educ: Univ Louvain, Lic, 63, Dr(electrochem), 67. Prof Exp: Asst prof chem, Univ Montreal, 67-68; res chemist, 68-70, group leader, 70-73, HEAD DEPT, NORANDA RES CTR, 73- Mem: Electrochem Soc; Nat Asn Corrosion Engrs. Res: Electrodeposition of metals; cathodic process; study of the physical properties of solutions. Mailing Add: Noranda Res Ctr 240 Hymus Blvd Pointe Claire PQ Can

CLAFF, CHESTER ELIOT, JR, b Brockton, Mass, Apr 17, 28; m 52; c 2. ORGANIC CHEMISTRY. Educ: Mass Inst Technol, BS, 50, PhD(org chem), 53. Prof Exp: Chemist polymerization, B B Chem Co, Inc, 55-60; gen mgr, Mark Co, 60-64; vpres, M B Claff & Sons, Inc, 64-71; TRANSL MGR, LINGUISTIC SYSTS, INC, 74- Concurrent Pos: Res assoc, Mass Inst Technol; partic, Rubber Reserve Prog, Reconstruction Corp, 53-55. Res: Preparation and reaction of organosodium compounds; leather technology; acrylic polymerization technology. Mailing Add: 300 Oak St Brockton MA 02401

CLAFLIN, ALICE J, b River Falls, Wis, Feb 12, 32. IMMUNOBIOLOGY. Educ: Northern State Col, BS, 53; Univ Wis, PhD(med genetics), 70. Prof Exp: Instr med, 70-73, RES ASST PROF SURG, SCH MED, UNIV MIAMI, 73- Mem: Tissue Cult Asn. Res: Immune mechanisms of tumor-bearing animals; cellular and humoral immune response with immunosuppressive therapy and transplantation immunology. Mailing Add: Dept of Surg Univ of Miami Sch of Med Miami FL 33152

CLAFLIN, ROBERT MALDEN, b Flint, Mich, Nov 11, 21; m 57; c 3. VETERINARY PATHOLOGY. Educ: Mich State Univ, DVM, 52; Purdue Univ, MS, 56, PhD(vet path), 58. Prof Exp: Instr res animal dis, 52-58, assoc prof vet path, 58-59, PROF & HEAD DEPT VET MICROBIOL, PATH & PUB HEALTH, SCH VET MED, PURDUE UNIV, 59- Mem: Am Vet Med Asn; Conf Res Workers Animal Dis; Int Acad Path. Res: Etiology, pathology and epizoology of respiratory diseases of swine, particularly atrophic rhinitis and mucosal diseases of cattle. Mailing Add: Vet Microbiol Path & Pub Health Purdue Univ West Lafayette IN 47906

CLAFLIN, TOM O, b Ripon, Wis, Apr 1, 39; m 61; c 1. BIOLOGY. Educ: Northern State Col, BS, 61; Univ SDak, MA, 63, PhD(zool), 66. Prof Exp: From asst prof to assoc prof, 66-74, PROF BIOL, UNIV WIS-LA CROSSE, 69- Mem: AAAS; Am Fisheries Soc. Res: Ecology of the benthos of river and lake systems. Mailing Add: Dept of Biol Univ of Wis La Crosse WI 54601

CLAGETT, CARL OWEN, b Lebanon, Pa, Jan 25, 13; m; c 3. BIOCHEMISTRY. Educ: Pa State Univ, BS, 39; Univ Wis, MS, 41, PhD(biochem), 47. Prof Exp: Assoc prof biochem & agr chemist, NDak Agr Col, 47-49, prof, 50-56; PROF BIOCHEM, PA STATE UNIV, 56- Mem: AAAS; Am Chem Soc; Am Soc Plant Physiologists; Am Soc Biol Chemists. Res: Plant enzymes and metabolism; carrier proteins; riboflavin binding protein. Mailing Add: Dept of Biochem Pa State Univ University Park PA 16802

CLAGETT, DONALD CARL, b Madison, Wis, Dec 31, 39; m 68; c 4. ORGANIC CHEMISTRY. Educ: Pa State Univ, BS, 61; Yale Univ, MS, 63, PhD(chem), 66. Prof Exp: Asst res scientist, NY Univ, 67-68; asst prof chem, Northeastern Univ, 68-73; group leader, 73-75, SR GROUP LEADER, DEWEY & ALMY DIV, W R GRACE & CO, 75- Mem: AAAS; Am Chem Soc; The Chem Soc; NY Acad Sci. Res: Chemistry of small ring organic compounds; chemistry of nucleic acids; chemical mutagens; chemistry of arthropod venoms; rubber latex formulations; powdered coatings; urethane foam systems. Mailing Add: Dewey & Almy Div W R Grace & Co 55 Hayden Ave Lexington MA 02173

CLAGETT, OSCAR THERON, b Jamesport, Mo, Oct 19, 08; m 34; c 6. THORACIC SURGERY. Educ: Univ Colo, MD, 33; Univ Minn, MS, 38. Hon Degrees: DSc, Univ Colo, 62. Prof Exp: Intern, Univ Gen Hosp, Univ Colo, 33-34; pvt pract surg & med, Colo, 34-35; asst surgeon, 38-40, CONSULT GEN SURG, SECT THORACIC, DEPT SURG, MAYO CLIN, 40-; PROF SURG, MAYO GRAD SCH MED, UNIV MINN, 50- Concurrent Pos: Assoc prof, Mayo Found, Univ Minn, 46-50. Honors & Awards: Norlin Medal, Univ Colo, 47. Mem: Hon mem Royal Col Surg Ireland; Am Col Surg; Am Asn Thoracic Surg; Am Surg Asn; Soc Clin Surg. Mailing Add: Mayo Clinic 200 First St SW Rochester MN 55901

CLAGUE, WILLIAM DONALD, b Mobile, Ala, Nov 29, 20; m 44; c 2. SCIENCE EDUCATION. Educ: Bridgewater Col, AB, 41; Univ Va, MEd, 52, EdD, 60. Prof Exp: Teacher high sch, Ala, 41-43; from asst prof to assoc prof chem, Bridgewater Col, 43-60, dean students, 52-66, prof natural sci, 60-66; PROF EDUC & DEAN GRAD & PROF STUDIES, LA VERNE COL, 66-, VPRES ACAD AFFAIRS, 75- Res: Choline; methods of laboratory instruction in college chemistry; sources of teaching personnel for church related colleges; development leading to accreditation of a new institution of higher education in a frontier. Mailing Add: La Verne Col 1950 Third St La Verne CA 91750

CLAIBORNE, IMOGENE B, b Lynchburg, Va, Aug 28, 06. INORGANIC CHEMISTRY, ANALYTICAL CHEMISTRY. Educ: Randolph Macon Women's Col, AB, 29; Duke Univ, AM, 32. Prof Exp: Anal chemist, Mead Corp, 29-31; head sci & math dept, Sullins Col, 34-42; asst prof math & physics, Catawba Col, 42-44; asst prof chem, Richmond Prof Inst, 44-46; from instr to asst prof, 46-71, EMER ASST PROF, RANDOLPH-MACON WOMEN'S COL, 71- Concurrent Pos: Instr, Lynchburg Gen Hosp, 51-54; grant, Duke Univ, 54; Danforth Found fel, 55; NSF grant, 56; grant, Oak Ridge Inst Nuclear Studies, 59. Mem: Am Chem Soc. Res: Rates of reaction; standardization of enzymes and analyzation of organic compounds. Mailing Add: 2413 Terrell Pl Lynchburg VA 24503

CLAIBORNE, LEWIS T, JR, b Holly Grove, Ark, Sept 17, 35; m 62. ACOUSTICS. Educ: Baylor Univ, BS, 57; Brown Univ, PhD(physics), 61. Prof Exp: Res assoc physics, Brown Univ, 61-62; res physicist, 62-69, br mgr, Cent Res Labs, 69-74, LAB DIR, ADVAN TECHNOL LAB, CENT RES LABS, TEX INSTRUMENTS, INC, 75- Mem: Am Inst Physics; Inst Elec & Electronics Engrs. Res: Ultrasonic attenuation; lattice-electron interactions in both normal and superconducting metals; surface devices; charge-coupled devices. Mailing Add: Tex Instruments Inc PO Box 5936 Dallas TX 75222

CLAMAN, HENRY NEUMANN, b New York, NY, Dec 13, 30; m 56; c 3. INTERNAL MEDICINE, IMMUNOLOGY. Educ: Harvard Univ, AB, 52; NY Univ, MD, 55. Prof Exp: Intern, Barnes Hosp, St Louis, Mo, 55-56, asst resident, 56-57; from asst resident to resident, Mass Gen Hosp, Boston, 57-61; from instr to assoc prof, 62-73, assoc dean fac affairs, 69-71, PROF MED & MICROBIOL, UNIV COLO MED CTR, DENVER, 73- Concurrent Pos: Fel allergy, Sch Med, Univ Colo, 61-62; consult, Fitzsimons Gen Hosp, 68-; mem immunobiol study sect, NIH, 68-72, mem allergy immunol res comn, 73- Mem: Fel Am Acad Allergy; Am Asn Immunol; Soc Exp Biol & Med. Res: Immunological tolerance to protein antigens; roles of the thymus and bone marrow cells in immunocompetence; effect of corticosteroids on immunocompetence; graft-versus-host reactions; cell interaction in immune responses; immunology of contact allergy. Mailing Add: Univ of Colo Med Ctr Denver CO 80220

CLAMANN, H PETER, b Berlin, Ger, Nov 18, 39; US citizen; m 67; c 2. BIOMEDICAL ENGINEERING, PHYSIOLOGY. Educ: St Mary's Univ, Tex, BS, 61; Johns Hopkins Univ, PhD(biomed eng), 68. Prof Exp: Res physiologist, Walter Reed Army Inst Res, 68-70; instr physiol, Harvard Med Sch, 72-73; ASST PROF PHYSIOL, MED COL VA, 73- Concurrent Pos: Res fel neurophysiol, Harvard Med Sch, 70-72. Mem: AAAS; Biomed Eng Soc; Soc Neurosci; NY Acad Sci; Inst Elec & Electronics Engrs. Res: Neurophysiology and electromyography; control of voluntary muscle. Mailing Add: 4001 Laurelwood Rd Richmond VA 23234

CLAMBEY, GARY KENNETH, b Fergus Falls, Minn, Feb 27, 45; m 69; c 1. PLANT ECOLOGY. Educ: NDak State Univ, BS, 67, MS, 69; Iowa State Univ, PhD(bot), 75. Prof Exp: Instr nat sci, Fergus Falls Jr Col, Minn, 68-69; specialist prev med, US Army Med Dept, 69-71; ASST PROF BOT, N DAK STATE UNIV, 74- Mem: AAAS; Am Inst Biol Sci; Ecol Soc Am; Am Soc Limnol & Oceanog; Int Asn Ecol. Res: Analysis of plant community structure and dynamics in forest and wetland vegetation. Mailing Add: Dept of Bot NDak State Univ Fargo ND 58102

CLAMPITT, BERT HOWARD, physical chemistry, see 12th edition

CLAMPITT, PHILIP THEODORE, b Marshalltown, Iowa, Feb 17, 30; m 60; c 2. ZOOLOGY, ANIMAL ECOLOGY. Educ: Cornell Col, BA, 52; Univ Iowa, MS, 60, PhD(zool), 63. Prof Exp: Teacher, High Sch, Iowa, 55-57; teaching asst zool, Univ Iowa, 57-63; asst prof, Grand Valley State Col, 63-68; assoc zoologist, 68-73; ZOOLOGIST, CRANBROOK INST SCI, 73- Concurrent Pos: Lectr, Oakland Univ, 69-74, adj assoc prof biol sci, 74- Mem: AAAS; Am Soc Zool; Ecol Soc Am; Am Inst Biol Sci. Res: Ecology of aquatic invertebrates; malacology. Mailing Add: Cranbrook Inst Sci 500 Lone Pine Rd Bloomfield Hills MI 48013

CLANCY, CARL FRANCIS, b Boston, Mass, Jan 14, 10; m 37; c 3. BACTERIOLOGY. Educ: Mass State Col, BS, 33, MS, 36; Yale Univ, PhD(bact), 42. Prof Exp: Asst, Storrs Exp Sta, Conn, 36-39; bacteriologist, Col Physicians & Surgeons, Columbia Univ, 42-43 & Lederle Labs, NY, 43-47; assoc, Thomas Jefferson Univ, 47-63, assoc prof microbiol, 63-75; RETIRED. Concurrent Pos: Assoc, Off Sci Res & Develop, 44; bacteriologist, Pa Hosp, 47-63. Mem: Am Soc Microbiol. Res: Serology; salmonellas; streptococci grouping; Rorientalis vaccine; Newcastle virus vaccine; wound infections; acid fast bacilli; antimicrobials and chemotherapy. Mailing Add: Dept of Microbiol Thomas Jefferson Univ Philadelphia PA 19107

CLANCY, CLARENCE WILLIAM, b Cincinnati, Ohio, Sept 18, 05; m 39; c 2. GENETICS. Educ: Univ Ill, BS, 30, MS, 32; Stanford Univ, PhD(biol), 40. Prof Exp: Asst biol, Stanford Univ, 36-40; from instr to assoc prof zool, 40-58, prof biol, 58-74, EMER PROF BIOL, UNIV ORE, 74- Concurrent Pos: USPHS spec res fel, Zool Inst, Univ Zurich, 60-61. Mem: Fel AAAS; Genetics Soc Am; Am Soc Zoologists; Am Genetic Asn. Res: Developmental and physiological genetics of Drosophila. Mailing Add: Dept of Biol Univ of Ore Eugene OR 97403

CLANCY, EDWARD PHILBROOK, b Beloit, Wis, July 3, 13; m 43; c 5. PHYSICS. Educ: Beloit Col, BS, 35; Harvard Univ, AM, 37, PhD(physics), 40. Prof Exp: Instr physics, Harvard Univ, 37-43; asst prof, Hamilton Col, 43-44; res assoc, Underwater Sound Lab, Harvard Univ, 44-45, lectr, 46; from asst prof to assoc prof, 46-57, PROF PHYSICS, MT HOLYOKE COL, 57- Mem: Am Phys Soc; Am Asn Physics Teachers. Res: Radiation physics; optics. Mailing Add: Dept of Physics Mt Holyoke Col South Hadley MA 01075

CLANCY, JOHN, b Dungarvan, Ireland, Oct 27, 22; US citizen; m 52; c 6. PSYCHIATRY. Educ: Nat Univ Ireland, MB & ChB, 46; FRCPS(C). Prof Exp: Intern med, St Vincents Hosp, Dublin, Ireland, 46; pvt pract, 47-51; resident psychiat, Iowa, 51-54; dir psychiat, Union Hosp, Moosejaw, Sask, 55-59; from asst prof to assoc prof, 59-66, PROF PSYCHIAT, UNIV IOWA, 66- Concurrent Pos: Mem Gov Comn Alcoholism, Iowa, 60-61 & 66-; consult, Vet Admin Hosp, Iowa City, 66- Mem: AMA; fel Am Psychiat Asn; Am Psychopath Asn. Res: Psychopathology and treatment of alcoholism; psychophysiological relationships; psychotherapy. Mailing Add: Dept of Psychiat Univ Iowa Psychopathic Hosp Iowa City IA 52242

CLANCY, RICHARD L, b Hardy, Iowa, Dec 26, 33; m 56; c 2. PHYSIOLOGY. Educ: Univ Minn, BA, 56, MSc, 61; Univ Kans, PhD(physiol), 65. Prof Exp: Asst prof physiol, Ohio State Univ, 67-69; ASSOC PROF PHYSIOL, SCH MED, UNIV KANS, 69- Concurrent Pos: Nat Heart Inst fel, 65-67. Mem: AAAS; Am Physiol Soc. Res: Acid-base and cardiovascular physiology. Mailing Add: Dept of Physiol Univ Kans Med Ctr Kansas City KS 66103

CLANDININ, DONALD ROBERT, b Vandura, Sask, Jan 19, 14; m 38; c 3. POULTRY NUTRITION. Educ: Univ BC, BSA, 35, MSA, 36; Univ Wis, PhD(biochem, poultry), 48. Prof Exp: Poultry geneticist, Govt Alta, 36-38; from lectr to assoc prof poultry nutrit, 38-53, PROF POULTRY NUTRIT, UNIV ALTA, 53- Mem: AAAS; fel Poultry Sci Asn; World Poultry Sci Asn; Animal Nutrition Res Coun; fel Agr Inst Can. Res: Nutrient requirements of chickens and turkeys; factors affecting protein quality. Mailing Add: Dept of Animal Sci Univ Alta Poultry Div Edmonton AB Can

CLANTON, DONALD CATHER, b Belle Fourche, SDak, Dec 22, 26; m 50; c 2. ANIMAL NUTRITION. Educ: Colo State Univ, BS, 49; Mont State Univ, MS, 54, PhD(animal nutrit), 57. Prof Exp: From asst prof to assoc prof, 58-66, PROF ANIMAL SCI, UNIV NEBR, 66- Mem: AAAS; Am Soc Animal Sci; Am Soc Range Mgt. Res: Ruminant nutrition, particularly nutrition of reproduction. Mailing Add: North Platte Exp Sta Univ of Nebr Box 429 North Platte NE 69101

CLANTON, DONALD HENRY, b Hickory, NC, Sept 3, 26; m 49; c 3. MATHEMATICS. Educ: Baylor Univ, BS, 50, MA, 52; Auburn Univ, PhD(math), 64. Prof Exp: Teacher high sch, Tex, 50-51; instr math, Allen Mil Acad, 51-53, Univ SC, 53-56 & Auburn Univ, 56-60; Oak Ridge Inst Nuclear Studies fel, Oak Ridge Nat Lab, Tenn, 60-62; from asst prof to assoc prof, 62-70, PROF MATH, FURMAN UNIV, 70- Concurrent Pos: Eve instr, Baylor Univ, 50-51. Mem: Math Asn Am; Am Math Soc. Res: Characteristic roots and values, and inclusion regions of matrices. Mailing Add: Dept of Math Furman Univ Greenville SC 29613

CLANTON, UEL S, JR, b Brownwood, Tex, June 23, 31; m 56; c 2. GEOCHEMISTRY, ASTROGEOLOGY. Educ: Univ Tex, Austin, BS, 55, MA, 60, PhD(geol), 68. Prof Exp: Chief computer, United Geophys Corp, 55-56; teaching asst, Univ Tex, Austin, 58-61, res scientist, Environ Health Eng Res Lab, 62-63; PHYS SCIENTIST & GEOLOGIST, GEOL BR, NASA JOHNSON SPACE CTR, 63- Mem: AAAS; Geol Soc Am; Microbeam Anal Soc. Res: Vapor-phase crystallization in lunar breccias; morphology and chemistry of impact and volcanic glassy droplets; faulting and subsidence along the Texas Gulf Coast. Mailing Add: Geol Br NASA Johnson Space Ctr TN 6 Houston TX 77058

CLAPHAM, WENTWORTH B, JR, b New York, NY, Mar 20, 42. ECOLOGY, ENVIRONMENTAL MANAGEMENT. Educ: Amherst Col, BA, 63; Univ Chicago, PhD(evolutionary biol), 68. Prof Exp: Asst prof geol, 68-74, sr res assoc, Syst Res Ctr, 74-75, ASSOC PROF SYSTS ENGR, CASE WESTERN RESERVE UNIV, 75- Concurrent Pos: Mem tech adv comt eval math models in hwy planning, Ohio Environ Protection Agency, 74-; partic modeling food production, Mesarovic-Pestel Global Modeling Proj, Club of Rome. Mem: Ecol Soc Am; Soc Study Evolution; Am Soc Limnol & Oceanog; Soc Gen Systs Res. Res: Developing approaches to understanding structure and dynamics of human ecosystems; application to food-producing ecosystems in the context of global population and resource demands. Mailing Add: Systs Res Ctr Case Western Reserve Univ Cleveland OH 44106

CLAPP, CHARLES EDWARD, b Holden, Mass, Aug 29, 30; m 53; c 4. SOIL BIOCHEMISTRY. Educ: Univ Mass, BS, 52; Cornell Univ, MS, 54, PdD(soil chem), 57. Prof Exp: Asst soil chemist, Cornell Univ, 52-56; org chemist, Agr Res Serv, USDA, 56-61; asst prof, 61-69, ASSOC PROF SOIL SCI, UNIV MINN, ST PAUL, 69-; RES CHEMIST, AGR RES SERV, USDA, 61- Mem: Am Chem Soc; Am Soc Agron; Soil Sci Soc Am; Int Soil Sci Soc. Res: Chemistry of soil organic matter; clay-organic complexes; electrophoresis; polysaccharide chemistry; ethylenimine chemistry; viscosity; soil structure; sludge and waste water chemistry. Mailing Add: Agr Res Serv US Dept of Agr 329 Soil Sci Bldg Univ of Minn St Paul MN 55108

CLAPP, JAMES R, b Siler City, NC, Sept 3, 31; m 53; c 2. INTERNAL MEDICINE. Educ: Univ NC, MD, 57. Prof Exp: Intern & resident med, Parkland Mem Hosp, Dallas, Tex, 57-59; investr kidney & electrolytes, NIH, 61-63; assoc, 63-66, asst prof, 66-70, ASSOC PROF INTERNAL MED, SCH MED, DUKE UNIV, 70- Concurrent Pos: USPHS trainee, 59-61 & grant, 63-; fel renol, Southwestern Med Sch, Univ Tex, 59-61; estab investr, Am Heart Asn. Mem: Am Physiol Soc; Am Fedn Clin Res. Res: Renal physiology and pathophysiology. Mailing Add: Box 3014 Duke Univ Med Ctr Durham NC 27706

CLAPP, JOHN GARLAND, JR, b Greensboro, NC, Oct 27, 36; m 59; c 3. AGRONOMY. Educ: NC State Univ, BS, 59, MS, 61, PhD(crop sci), 69. Prof Exp: Asst agr exten agent, NC State Univ, 61-62; exten agronomist, Clemson Univ, 62-63 & NC State Univ, 63-75; SCI AGRONOMIST, ALLIED CHEM CORP, 75- Concurrent Pos: Exten agronomist, Nat Soybean Resource Comt, 70- Honors & Awards: Geigy Award in Agron, Am Soc Agron, 72; Meritorious Serv Award, Am Soybean Asn, 74. Mem: Am Soc Agron. Res: Applied on-farm evaulation of fertilizers, herbicides, growth regulators, nematocides, plant population and tillage methods for soybean production. Mailing Add: PO Box 2120 Houston TX 77001

CLAPP, LEALLYN BURR, b Paris, Ill, Oct 13, 13; m 40. ORGANIC CHEMISTRY. Educ: Eastern Ill Univ, BEd, 35; Univ Ill, AM, 39, PhD(chem), 41. Hon Degrees: PdD, Eastern Ill Univ, 56; LLD, RI Col, 64. Prof Exp: Instr high sch, Ill, 35-38; asst chemist, Univ Ill, 39-41; from instr to assoc prof org chem, 41-56, exec officer dept chem, 55-59, PROF ORG CHEM, BROWN UNIV, 56- Honors & Awards: Sci Apparatus Makers Award in Chem Educ, Am Chem Soc, 76. Mem: Am Chem Soc. Res: Chemistry of ethylenimines and other heterocyclic nitrogen compounds. Mailing Add: Dept of Chem Brown Univ Providence RI 02912

CLAPP, NEAL K, b Shelby Co, Ind, Oct 14, 28; m 53; c 3. RADIOBIOLOGY, PATHOLOGY. Educ: Purdue Univ, BS, 50; Ohio State Univ, DVM, 60; Colo State Univ, MS, 62, PhD, 64. Prof Exp: Instr surg, Vet Clins, Colo State Univ, 60-61; RADIATION PATHOLOGIST, OAK RIDGE NAT LAB, 64- Mem: AAAS; Am Vet Med Asn; Radiation Res Soc; Am Asn Cancer Res. Res: Radiation pathology; chemical carcinogenesis. Mailing Add: Biol Div Oak Ridge Nat Lab Y-12 Oak Ridge TN 37830

CLAPP, PHILIP CHARLES, b Belleville, Ont, Oct 14, 35; US citizen; m 61; c 3. SOLID STATE PHYSICS. Educ: Queen's Univ, BS, 57; Mass Inst Technol, PhD(physics), 63. Prof Exp: Lectr magnetism, Mass Inst Technol, 63; physicist 63-75, HEAD PHYSICS & METALLURGY GROUP, LEDGEMONT LAB, KENNECOTT COPPER CORP, 75- Concurrent Pos: Sr vis scientist, Oxford Univ, 69-70; vis prof, Nat Comn Atomic Energy, Buenos Aires, Arg, 72; adj prof physics, Boston Col, 73- Mem: AAAS; Metallurgical Soc; Inst Solar Energy Soc; Am Phys Soc; Can Asn Physicists. Res: Alloy research; Martensitic phase transformations; solar energy; theories of order-disorder phenomena; low temperature physics and magnetism; biophysics; psychology. Mailing Add: 121 Sudbury Rd Concord MA 01742

CLAPP, RICHARD CROWELL, organic chemistry, see 12th edition

CLAPP, ROGER EDGE, b Cleveland, Ohio, Oct 9, 10; m 57; c 2. THEORETICAL PHYSICS. Educ: Harvard Univ, AB, 41, AM, 42, PhD(physics), 49. Prof Exp: Mem staff microwaves, Radiation Lab, Mass Inst Technol, 42-46, AEC fel, 49-50; sr physicist, Snow & Schule, Inc, 50-52; consult, Ultrasonic Corp, 52-57; staff consult, Adv Industs, Inc, 57-61; STAFF CONSULT, AIR TECH CORP, 61-; PRES, BASIC RES ASSOC INC, 70- Concurrent Pos: Res contractor, US Off Naval Res, 52-54; consult, Airborne Instruments Lab Div, Cutler-Hammer, Inc, NY, 59- & Carter's Ink Co, 66- Mem: AAAS; Am Phys Soc; Am Geophys Union; Inst Elec & Electronics Eng. Res: Nuclear three-body problem; radar ground reflections; electromagnetic radiation from nuclear detonations in the lower atmosphere; electron and muon structure; gravitational theory. Mailing Add: 19 Copley St Cambridge MA 02183

CLAPP, ROGER WILLIAMS, JR, b Tampa, Fla, Aug 31, 29; m 59; c 4. PHYSICS. Educ: Davidson Col, BS, 50; Univ Va, MS, 52, PhD(physics), 54. Prof Exp: Res physicist, Army Missile Command, Redstone Arsenal, 56-63; asst prof, 63-66, ASSOC PROF PHYSICS, UNIV S FLA, 66- Mem: Am Phys Soc; Am Asn Physics Teachers. Res: Surface physics; thin films; history of physics. Mailing Add: Dept of Physics Univ of SFla Tampa FL 33620

CLAPP, WILLIAM LEE, b Memphis, Tenn, Feb 16, 43; m 65; c 2. ANALYTICAL

CHEMISTRY. Educ: Wake Forest Col, BS, 64; Duke Univ, MA, 66, PhD(chem), 69. Prof Exp: Res chemist anal chem, R J Reynolds Tobacco Co, 68-69; chem officer, Weapons Develop & Eng Lab, Edgewood Arsenal, US Army, 69-71; res chemist anal chem, 71-72, SECT HEAD PROJ MGT, R J REYNOLDS TOBACCO CO, 72- Mem: Am Chem Soc; Sigma Xi. Res: Research project management; analytical methods for analysis of pesticides in tobacco. Mailing Add: Res Dept R J Reynolds Tobacco Co Winston-Salem NC 27102

CLAPPER, MUIR, b Detroit, Mich, May 26, 13; m 62. INTERNAL MEDICINE. Educ: Wayne State Univ, AB, 33, MD, 36, MS, 40. Prof Exp: From instr to assoc prof, 40-53, PROF MED, SCH MED, WAYNE STATE UNIV, 53- Concurrent Pos: Consult, Dearborn Vet Hosp, 51-, Detroit Mem Hosp, 57-, Jennings Hosp, 60- & Harper Hosp, 62-; attend, Detroit Gen Hosp. Mem: AMA; Am Heart Asn; Am Col Physicians; Asn Univ Cardiol; Am Col Cardiol. Res: Cardiology. Mailing Add: Wayne State Univ Sch of Med Detroit MI 48201

CLAPPER, THOMAS WAYNE, b McKean, Pa, Oct 15, 15; m 41; c 3. ORGANIC CHEMISTRY. Educ: St Vincent Col, BS, 37; Pa State Univ, MS, 38, PhD(org chem), 42. Prof Exp: Res chemist, Pharmaceut Div, Calco Chem Div, Am Cyanamid Co, 40-44, from asst chief chemist to chief chemist, Pharmaceut Dept, 44-48, prod mgr, 48-51, tech dir, Atomic Energy Div, 51-52, gen supt, Chem Processing Plant, 52, asst gen mgr, 52-53; asst to gen mgr, Atomic Energy Div, Phillips Petrol Co, 54; plant mgr, Calera Ref, Chem Construct Corp, 54-56; res mgr, Am Potash & Chem Corp, 56-63, tech dir res, 63-68; DIR RES, KERR-MCGEE CORP, 68- Mem: Am Chem Soc; Electrochem Soc; Am Tech Asn Pulp & Paper Indust; Am Soc Metals; Am Inst Mining, Metall & Petrol Engrs. Res: Sulfa drugs; vitamins; chemical processing of uranium; cobalt; high energy fuels; electrochemistry; rare earths; boron compounds; maganese metal and compounds. Mailing Add: 12104 Camelot Pl Oklahoma City OK 73120

CLAPPER, WILLIAM EVERETT, microbiology, see 12th edition

CLARDY, JON CHRISTEL, b Washington, DC, May 16, 43; m 66; c 2. STRUCTURAL CHEMISTRY. Educ: Yale Univ, BS, 64; Harvard Univ, PhD(chem), 69. Prof Exp: Instr chem, 69-70, from asst prof to assoc prof, 70-75, PROF CHEM, IOWA STATE UNIV, 75- Concurrent Pos: Camille & Henry Dreyfus Found fel, 72; Alfred P Sloan Found fel, 73. Mem: Am Chem Soc; Am Crystallog Asn. Res: Application of x-ray and neutron diffraction to problems of biological and chemical interest. Mailing Add: Dept of Chem Iowa State Univ Ames IA 50011

CLARDY, LEROY, b Ft Worth, Tex, July 16, 10; m 38; c 1. PHYSICAL CHEMISTRY. Educ: Tex Christian Univ, BS, 31, MS, 34. Prof Exp: Anal chemist, Armour & Co, Tex, 34-36; chief chemist, Terrell's Labs, 36-37; chemist, Swift & Co, 37-43, physicist, 43-70, mgr control eng div, Eng Res Dept, 70-75; CONSULT, 75- Mem: Inst Elec & Electronics Engr. Res: Application of instrumentation and automatic control systems to meat packing and allied processes. Mailing Add: 835 Edgewater Dr Naperville IL 60540

CLARE, STEWART, b Montgomery Co, Mo, Jan 31, 13; m 36. ZOOLOGY, BIOCHEMISTRY. Educ: Univ Kans, BA, 35; Iowa State Univ, MS, 37; Univ Chicago, PhD(zool), 49. Prof Exp: Tech consult, White-Fringed Beetle Proj, Bur Entom & Plant Quarantine, US Civil Serv Comn, 41-42, instr meteorol, Army Air Force Weather Sch, 42-43; res biologist, Midwest Res Inst, Mo, 45-46; mem spec res proj, Univ Mo-Kansas City, Midwest Res Inst & Kansas City Art Inst, 46-49; instr zool, Univ Alta, 49-50, asst prof zool & lectr sci of color, 50-53; asst prof physiol & pharmacol, Kansas City Col Osteop & Surg, 53; lectr, Univ Adelaide, 54-55; sr res officer entom, Ministry Agr & Gezira Res Sta, Sudan Govt, NAfrica, 55-56; sr entomologist, Klipfontein Org Prod Corp, SAfrica, 57; prof biol & head dept, Union Col, Ky, 58-59, chmn div sci, 59-61; prof biol & head dept, Mo Valley Col, 61-62; Buckbee Found prof biol & lectr, Eve Col, Rockford Col, 62-63; prof biochem & chmn dept & mem res div, Kansas City Col Osteop & Surg, 63-67; prof biol, 67-72, dir biol res, 72-74, EMER PROF BIOL, COL EMPORIA, 74-; RES BIOLOGIST & CONSULT, 74- Concurrent Pos: Res & study grants, Alta Res Coun, 51-53, Union Col, Ky, 59-61, Mo Valley Col, 61-62, Rockford Col, 62-63, Adirondack Res Sta, 63-66, NIH, 63-65 & Col Emproia, 67-74; consult, Vols for Tech Assistance, 62- & Info Resource, Nat Referral Ctr for Sci & Technol, Libr of Cong, 70- Mem: NY Acad Sci; Am Entom Soc; Brit Asn Advan Sci; Arctic Inst NAm; Nat Asn Biol Teachers. Res: Comparative physiology-biochemistry; circulation of the Arthropoda; trace elements in invertebrates; capillary movement in porous materials; gums, extractives and extraneous materials of plants; biometeorology; chromatology; history of science. Mailing Add: 4000 Charlotte St Kansas City MO 64110

CLARENBURG, RUDOLF, b Utrecht, Holland, May 3, 31; US citizen; m 59; c 1. PHYSIOLOGICAL CHEMISTRY. Educ: Univ Utrecht, Drs, 59, DSc(chem), 65. Prof Exp: Res physiologist, Univ Calif, Berkeley, 59-66; assoc prof, 66-74, PROF PHYSIOL, KANS STATE UNIV, 74- Mem: AAAS; Am Physiol Soc; NY Acad Sci. Res: Lipid metabolism; transport across biological membranes. Mailing Add: Dept of Physiol Sci VMS Bldg Kans State Univ Manhattan KS 66502

CLARIDGE, CHARLES ALFRED, b Victoria, BC, Sept 5, 21; m 47; c 4. MICROBIOLOGY. Educ: Univ BC, BA, 43; Iowa State Col, PhD(physiol bact), 53. Prof Exp: Microbiologist, Merck & Co, Inc, 54-56; microbiologist, Fisheries Res Bd Can, 56-59; MICROBIOLOGIST, BRISTOL LAB, INC DIV, BRISTOL-MYERS CO, 59- Mem: AAAS; Am Soc Microbiol; Am Chem Soc. Res: Microbial metabolism; microbial transformation of organic compounds; antibiotic fermentations; mutational biosynthesis of new anti- biotics by idiotrophic cultures. Mailing Add: Bristol Labs Inc Box 657 Syracuse NY 13201

CLARK, A GAVIN, b Warrington, Eng, Nov 18, 38. MICROBIOLOGY. Educ: Univ Edinburgh, BSc, 63, PhD(microbiol), 66. Prof Exp: Res asst bact, Med Sch, Univ Edinburgh, 61-62, res asst microbiol, Sch Agr, 63-66; fel soil sci, Univ Alta, 66-68; ASST PROF MICROBIOL, UNIV TORONTO, 68- Mem: Brit Soc Appl Bact; Brit Soc Gen Microbiol; Can Soc Pub Health. Res: Agrobacterium species; pathogenicity, taxonomy, host response to infection; Myxobacteria; taxonomy, mechanism of encystment and energetics of growth; pathogenicity of Vibro parahaemolyticus. Mailing Add: Dept of Microbiol & Parasitol Univ of Toronto St George Campus Toronto ON Can

CLARK, ALBERT F, biochemistry, see 12th edition

CLARK, ALFRED, physical chemistry, see 12th edition

CLARK, ALFRED, JR, b Elizabethton, Tenn, May 5, 36; m 60; c 1. APPLIED MATHEMATICS. Educ: Purdue Univ, BS, 58; Mass Inst Technol, PhD(appl math), 63. Prof Exp: NSF fel, 63-64; from asst prof to assoc prof, 64-74, PROF MECH & AEROSPACE SCI, UNIV ROCHESTER, 74-, CHMN DEPT, 72- Concurrent Pos: Vis fel, Joint Inst Lab Astrophys, Univ Colo, 70-71. Mem: Am Astron Soc; Int Astron Union; Am Phys Soc; Int Asn Gt Lakes Res; Am Geophys Union. Res: Astrophysical and geophysical fluid dynamics; solar physics; dynamics of large lakes. Mailing Add: Dept of Mech & Aerospace Sci Univ of Rochester Rochester NY 14627

CLARK, ALLAN H, b Cincinnati, Ohio, July 16, 35; m 61; c 3. MATHEMATICS. Educ: Mass Inst Technol, BS, 57; Princeton Univ, MA, 59, PhD(math), 61. Prof Exp: From instr to prof math, Brown Univ, 61-75; DEAN, SCH SCI, PURDUE UNIV, WEST LAFAYETTE, 75- Concurrent Pos: NSF grant, 62-; vis mem, Inst Advan Study, Princeton Univ, 65-66; vis prof, Math Inst, Aarhus Univ, 70-71. Mem: Am Math Soc. Res: Algebraic topology. Mailing Add: Sch of Sci Purdue Univ West Lafayette IN 47906

CLARK, ALLEN KEITH, b Bridgeton, NJ, June 25, 33; m 57; c 2. ORGANIC CHEMISTRY. Educ: Catawba Col, AB, 55; Univ NC, PhD(org chem), 60. Prof Exp: From asst prof to assoc prof, 62-66, actg chmn dept, 68-69, PROF CHEM, OLD DOM UNIV, 66-, CHMN DEPT, 69-, ASST PROVOST, 72- Concurrent Pos: Sigma Xi res grant-in-aid, 63-64. Mem: Am Chem Soc; Am Inst Chemists. Res: Chemistry of ferrocene; aromatic nitroso compounds. Mailing Add: 1050 Manchester Ave Norfolk VA 23508

CLARK, ALLEN LEROY, b Delaware, Iowa, Sept 29, 38; m 55; c 3. ECONOMIC GEOLOGY, GEOCHEMISTRY. Educ: Iowa State Univ, BS, 61; Univ Idaho, MS, 63, PhD(geol), 68. Prof Exp: Instr geol, Univ Idaho, 65-66; geologist, Bear Creek Mining Co, Kennecott Copper Corp, 66-67; res geologist, 67-72, CHIEF, OFF RESOURCE ANAL, US GEOL SURV, 72- Concurrent Pos: Co-investr, Apollo 12, NASA, 70- Mem: Am Asn Petrol Geol; Geol Soc Am; Soc Econ Geologists; Int Asn Genesis of Ore Deposits. Res: Economics of international development; economic analysis of exploration and resource availability; wallrock alteration and trace element distributions associated with base metal deposits; platinum group metals distribution in ultramafic rocks; structural analysis of lunar samples. Mailing Add: US Geol Surv 12201 Sunrise Valley Dr Reston VA 22091

CLARK, ALLEN VARDEN, b Attleboro, Mass, Nov 1, 41; m 63; c 3. FOOD SCIENCE. Educ: Mass Inst Technol, BS, 63, MS, 65, PhD(food sci & technol), 69. Prof Exp: RES SCIENTIST, CORP RES & DEVELOP LAB, THE COCA-COLA CO, 68- Mem: Am Chem Soc; Inst Food Technologists. Res: Isolation and characterization of pigments from protein carbonyl browning systems. Mailing Add: 2139 East Lake Rd NE Atlanta GA 30307

CLARK, ALTON HAROLD, b Bangor, Maine, Oct 10, 39; m 61; c 2. SOLID STATE PHYSICS. Educ: Univ Maine, BA, 61; Univ Wis, Madison, MS, 63; Cornell Univ, PhD(physics), 67. Prof Exp: Physicist, Sprague Elec Co, 66-68; asst prof, 68-73, ASSOC PROF PHYSICS, UNIV MAINE, 73- Concurrent Pos: Vis scientist, Xerox Palo Alto Res Ctr, 74-75. Mem: Am Phys Soc. Res: Electrical and optical properties of crystaline and amorphous semiconductors. Mailing Add: Dept of Physics Univ of Maine Orono ME 04473

CLARK, ALVIN JOHN, b Oak Park, Ill, Apr 13, 33; div. GENETICS, BACTERIOLOGY. Educ: Univ Rochester, BS, 55; Harvard Univ, PhD(chem), 59. Prof Exp: Asst prof bact, 62-64, asst prof bact & molecular biol, 64-67, assoc prof, 67-72, PROF MOLECULAR BIOL, 72- Concurrent Pos: Am Cancer Soc fel, 59-61; fel, Yale Univ, 62; John Simon Guggenheim Mem Found fel, 69. Mem: Am Soc Microbiol; Genetics Soc Am. Res: Enzymological and genetic analysis of genetic recombination; bacterial conjugation. Mailing Add: Dept of Molecular Biol Univ of Calif Berkeley CA 94720

CLARK, ANDREW HILL, historical geography, cultural geography, deceased

CLARK, ARMIN LEE, b Huntington, WVa, June 16, 28; m 51; c 2. GEOLOGY. Educ: Marshall Univ, BA, 51; Ohio State Univ, MSc, 58; Univ Tenn, PhD(geol), 73. Prof Exp: Teacher sci, Cabell & Mingo Counties, WVa Schs, 58-60 & Jefferson County, Ky Schs, 60-61; ASSOC PROF GEOL, MURRAY STATE UNIV, 61- Mem: Geol Soc Am; Nat Asn Geol Teachers. Res: Petrology of the Eocene sediments in western Kentucky and Tennessee. Mailing Add: 1504 Oak Dr Murray KY 42071

CLARK, ARNOLD FRANKLIN, b Madison, Wis, Apr 27, 16; div; c 3. ENERGY CONVERSION, SOLAR PHYSICS. Educ: Swarthmore Col, AB, 37; Ind Univ, AM, 39, PhD(physics), 41. Prof Exp: Asst math & physics, Univ Wis, 37-38; asst physics, Ind Univ, 38-41; res fel & physicist, Univ Calif, 41-46; res assoc, Univ Rochester, 46-47, asst prof physics, 47-49; asst prof physics, Carnegie Inst Technol, 49-54; PHYSICIST, LAWRENCE LIVERMORE LAB, UNIV CALIF, 54- Mem: Fel Am Phys Soc; Sigma Xi; Int Solar Energy Soc. Res: Nuclear physics; engineering physics; design of electromagnetic accelerators; nuclear emulsions; radiation effects; cloud chambers; solar energy research and development; shallow solar ponds. Mailing Add: Lawrence Livermore Lab Livermore CA 94550

CLARK, ARNOLD M, b Philadelphia, Pa, Jan 28, 16; m 53; c 2. GENETICS. Educ: Pa State Col, AB, 37; Univ Pa, MA, 39, PhD(zool), 43. Prof Exp: Res biologist, Smyth Labs, Philadelphia, 40-46; asst instr zool, Univ Pa, 44-45; instr, Philadelphia Col Pharm, 45-46; from instr to assoc prof, 46-56, PROF BIOL, UNIV DEL, 56- Concurrent Pos: Radiation biologist, Brookhaven Nat Lab, 53-54. Mem: AAAS; Am Soc Naturalists; Radiation Res Soc; Am Genetics Soc; Genetic Asn Am. Res: Genetics of Habrobracon; toxicological studies on insecticides; resins; pharmaceuticals; gene dosage; radiation damage; action of cell poisons; oxygen poisoning; genetics of aging; studies of genetic mosaics in insects and their use in the analysis of development and behavior; analysis of chromosomal aberrations in man. Mailing Add: Sch of Arts & Sci Univ of Del Newark DE 19711

CLARK, ARTHUR EDWARD, b Scranton, Pa, July 9, 32; m 58; c 4. SOLID STATE PHYSICS. Educ: Univ Scranton, BS, 54; Univ Del, MS, 56; Cath Univ, PhD(physics), 60. Prof Exp: RES PHYSICIST, US NAVAL ORD LAB, 59- Mem: AAAS; Am Phys Soc. Res: Magnetic, elastic and magnetoelastic properties of solids; ultrasonics and hypersonics. Mailing Add: US Naval Surface Weapons Ctr White Oak Lab Silver Spring MD 20910

CLARK, ARTHUR RANDOLPH, b Bethlehem, Pa, Nov 3, 11; m 39; c 1. CHEMISTRY. Educ: Franklin & Marshall Col, BS, 34; Pa State Col, MS, 35, PhD(metallo-org), 38. Prof Exp: Asst, Pa State Col, 34-38; res chemist, Rohm and Haas, Mallinckrodt Chem Works, Hanovia Chem & Mfg Co, 38-50; PRES, FOXLYN LABS, FOXLYN FARMS CO, 50- Mem: AAAS; Am Chem Soc; fel Am Inst Chemists. Res: Agricultural, medicinal, and industrial chemical products. Mailing Add: Foxlyn Labs PO Box 471 Farmingdale NJ 07727

CLARK, BARRY GILLESPIE, b Happy, Tex, Mar 5, 38; m 63; c 4. ASTRONOMY. Educ: Calif Inst Technol, BS, 59, PhD(astron), 64. Prof Exp: Asst scientist, 64-69, SCIENTIST, NAT RADIO ASTRON OBSERV, 69- Mem: Am Astron Soc. Res: Radio astronomy interferometry and array design and use. Mailing Add: VLA Proj Nat Radio Astron Observ PO Box O Socorro NM 87801

CLARK, BENJAMIN EDWARD, b Southampton, NY, Oct 3, 14; m 47; c 2. AGRONOMY. Educ: Cornell Univ, BS, 40, MS, 46; Mich State Col, PhD(hort), 49. Prof Exp: From asst prof to assoc prof, 48-56, head dept, 52-68, PROF SEED INVESTS, AGR EXP STA, NY STATE COL AGR & LIFE SCI, CORNELL UNIV, 56-; ASST DIR, NY STATE AGR EXP STA, GENEVA, 68- Mem: Am Soc Agron; Am Soc Hort Sci. Res: Seed germination. Mailing Add: 75 Highland Ave Geneva NY 14456

CLARK, BENJAMIN FRANKLIN, organic chemistry, see 12th edition

CLARK, BENTON C, b Oklahoma City Okla, Aug 4, 37; m 65; c 2. GEOCHEMISTRY, BIOPHYSICS. Educ: Univ Okla, BS, 59; Univ Calif, MA, 61; Columbia Univ, PhD(biophys), 69. Prof Exp: Res asst radiation instrumentation res & develop, Los Alamos Sci Lab, 59-60; assoc electronic speech recognition, Advan Systs Develop Div, Int Bus Mach Corp, 61; sr staff scientist res & develop, Avco Corp, 68-71; SR RES SCIENTIST, MARTIN-MARIETTA CORP, 71- Concurrent Pos: Dep team leader, NASA Viking Inorg Chem Team, 72- Mem: AAAS; Am Geophys Union. Res: Radiobiological effects of very soft x-rays; space radiation research; detection of life on Mars; geochemical analysis of planetary surfaces by x-ray fluorescence spectrometry. Mailing Add: 6752 S Lamar Littleton CO 80123

CLARK, BILL PAT, b Bartlesville, Okla, May 15, 39. SEMICONDUCTORS, SOLID STATE PHYSICS. Educ: Okla State Univ, BS, 61, MS, 64, PhD(physics), 68. Prof Exp: Asst physics, Okla State Univ, 61-68; res fel, Dept Theoret Physics, Univ Warwick, Eng, 68-69; sr mem tech staff, Booz Allen Appl Res, 69-70; SR MEM TECH STAFF & ANALYST, COMPUT SCI CORP, 70- Mem: AAAS; Am Phys Soc. Res: Ultraviolet reflectivity and band structure; elementary particles; quantum field theory; instabilities and transport properties in solids; mathematical models; systems analysis; computer simulations of solar cells, injection lasers, and other devices. Mailing Add: Box 336 Dewey OK 74029

CLARK, BRIAN ROGER, b Minneapolis, Minn, Oct 26, 37; m 72. BIOCHEMISTRY. Educ: Pomona Col, BA, 61; Univ Calif, Los Angeles, PhD(biochem), 74. Prof Exp: FEL & RES BIOCHEMIST, DEPT PSYCHIAT, SCH MED, UNIV CALIF, LOS ANGELES, HARBOR GEN HOSP CAMPUS, 74- Mem: Sigma Xi; AAAS; NY Acad Sci. Res: Effects of psychiatric disorders and alcoholism on regulation of the metabolism of one-carbon fragments, including transmethylation, de novo synthesis of purines and thymidylic acid synthesis. Mailing Add: Harbor Gen Hosp 1000 W Carson St Torrance CA 90509

CLARK, BRUCE R, b Pittsburgh, Pa, June 17, 41; m 67. GEOLOGY. Educ: Yale Univ, BS, 63; Stanford Univ, PhD(geol), 68. Prof Exp: Res assoc geol, Stanford Univ, 67-68; asst prof, 68-73, ASSOC PROF GEOL, UNIV MICH, ANN ARBOR, 73- Concurrent Pos: Vis lectr, Monash Univ, Australia, 74-75; mem, US Nat Comt on Rock Mech. Mem: Geol Soc Am; Am Geophys Union; Int Soc Rock Mech. Res: Structural geology; rock mechanics; experimental deformation of rocks at high pressure and temperature. Mailing Add: Dept of Geol & Mineral 1008 CC Little Bldg Univ of Mich Ann Arbor MI 48104

CLARK, BURR, JR, b Howell, Mich, Jan 20, 24; m 47; c 3. AGRICULTURAL BIOCHEMISTRY. Educ: Mich State Univ, BS, 52; Univ NH, MS, 60; WVa Univ, PhD(agr biochem), 66. Prof Exp: Self-employed in agr, 47-49 & 52-56; from asst ed to assoc ed, 63-69, SR ED BIOCHEM, CHEM ABSTRACTS SERV, 69- Mem: AAAS; Am Chem Soc. Res: Volatile fatty acids metabolism in ruminants and forage quality in relation to volatile fatty acids metabolism in ruminants. Mailing Add: 200 Larrimer Ave Worthington OH 43085

CLARK, BYRON BRYANT, b Temple, Tex, Apr 5, 08; m 31; c 3. PHARMACOLOGY. Educ: Baylor Univ, AB, 30; Univ Iowa, MS, 32, PhD, 34. Prof Exp: Asst biochem, Univ Iowa, 30-31; asst path chemist, Gen Hosp, Univ Iowa, 31-36; from instr to assoc prof physiol & pharmacol, Albany Med Col, 36-47; prof pharmacol & chmn dept, Med Sch, Tufts Univ, 47-57; dir pharmacol & chemother, Mead Johnson & Co, Ind, 57-62, vpres res ctr, 62-68; DIR PHARMACOL & TOXICOL PROGS, NAT INST GEN MED SCI, 68- Concurrent Pos: Consult pharmacologist, Albany Hosp, NY, 37-47 & New Eng Ctr Hosp, 47-57; mem comt drug safety, Drug Res Bd, Nat Acad Sci-Nat Res Coun. Mem: Fel AAAS; Am Soc Pharmacol & Exp Therapeut; Soc Exp Biol & Med; Soc Toxicol; NY Acad Sci. Res: Insulin and carbohydrate metabolism; ethyl and methyl alcohol pharmacology and metabolism; drugs on blood and hemoglobin; gastric secretion; antacids; antispasmodics; autonomic drugs; cardiovascular drugs. Mailing Add: 5101 River Rd Bethesda MD 20016

CLARK, C ELMER, b Tooele, Utah, Mar 5, 21; m 51; c 4. PHYSIOLOGY, BIOCHEMISTRY. Educ: Utah State Univ, BS, 50; Univ Md, MS, 60, PhD(poultry physiol), 62. Prof Exp: Asst prof poultry sci, Utah State Univ, 52-57 & Univ Md, 57-61; assoc prof, 62-70, asst dir agr exp sta, 70-75, ASSOC DIR AGR EXP STA, UTAH STATE UNIV, 75-, PROF POULTRY SCI, 70- Mem: AAAS; Poultry Sci Asn; World Poultry Sci Asn; Soc Exp Biol & Med. Res: Neurohumoral factors in ovulation in chickens; environmental-physiology relationships in the avian species. Mailing Add: Agr Exp Sta Utah State Univ Logan UT 84322

CLARK, CARL CYRUS, b Manila, Philippines, Apr 23, 24; US citizen; m 47; c 4. BIOPHYSICS. Educ: Worcester Polytech Inst, BS, 44; Columbia Univ, PhD(zool), 50. Prof Exp: Res assoc physiol & infrared spectrophotom, Med Col, Cornell Univ, 47-51; asst prof zool, Univ Ill, 51-55; head biophys div, Naval Aviation Med Acceleration Lab, 55-61; mgr, Lift Sci Dept, Martin Co, 61-66; assoc chief, Sci & Technol Div, Libr of Cong, 66-68; chief, Task Group on Indust Self-Regulation, Nat Comn Prod Safety, 68-70; staff consult prod safety, Prod Eval Technol Div, Nat Bur Standards, 70-72; head dept life sci, Worcester Polytech Inst, 72-74; EXEC DIR, COMN ADVAN PUB INTEREST ORGN, MONSOUR MED FOUND, 74- Concurrent Pos: From assoc physiol to asst prof, Sch Med, Univ Pa, 55-61; pres, Safety Systs Co, 68- Mem: AAAS; Systs Safety Soc; Am Inst Aeronaut & Astronaut; Aerospace Med Asn; Human Factors Soc. Res: Infrared spectrophotometry and x-ray diffraction studies of biochemicals; microbiospectrophotometry; human centrifuge dynamic flight simulation; airbag restraint development; flight physiology; auto and home safety; information systems; public interest organization. Mailing Add: 23 Seminole Ave Baltimore MD 21228

CLARK, CARL HERITAGE, b Los Angeles, Calif, Nov 18, 25; m 48; c 1. PHARMACOLOGY. Educ: State Col Wash, BS & DVM, 47; Ohio State Univ, MS, 49, PhD(physiol), 53. Prof Exp: Asst, State Col Wash, 45-47; instr physiol, Ohio State Univ, 47-53; assoc head, Dept Animal Dis Res, 60-67, PROF PHYSIOL & PHARMACOL & HEAD DEPT, SCH VET MED, AUBURN UNIV, 53- Mem: Am Vet Med Asn; Am Soc Vet Physiologists & Pharmacologists (pres, 61). Res: Nervous control of the motility of the stomach in ruminant animals; pharmacology of veterinary analeptics and anesthetics; pharmacology of body fluids and fluid therapy. Mailing Add: Dept of Physiol & Pharmacol Auburn Univ Sch of Vet Med Auburn AL 36830

CLARK, CARROLL THOMAS, b Louisa, Va, Feb 26, 23; m 47; c 4. ORGANIC CHEMISTRY. Educ: Univ Va, BS, 49, PhD(org chem), 57; Georgetown Univ, MS, 53. Prof Exp: Res biochemist, NIH, 49-53; asst prof chem, Univ Ga, 57-63; assoc prof, The Citadel, 63-68; PROF CHEM & CHMN DEPT, BELHAVEN COL, 68- Mem: Am Chem Soc. Res: Organic synthesis; mechanisms of reactions. Mailing Add: Dept of Chem Belhaven Col Jackson MS 39202

CLARK, CHARLES AUSTIN, b Owego, NY, Dec 18, 15; m 37; c 4. ORGANIC CHEMISTRY. Educ: Cornell Univ, BS, 37. Prof Exp: Bacteriologist, NY State Dept Health Labs, 37-42; org res chemist, 42-50, MGR ORG PREP LAB UNIT, ANSCO DIV, GAF CORP, 50- Mem: Am Chem Soc; Soc Photog Scientists & Engrs; Am Inst Chemists. Res: Optical photographic sensitizing dyes and related intermediates. Mailing Add: 14 Westwood Ct Binghamton NY 13905

CLARK, CHARLES CHRISTOPHER, b Erie, Pa, Feb 17, 43; m. BIOCHEMISTRY. Educ: Gannon Col, BA, 65; Northwestern Univ, PhD(biochem), 70. Prof Exp: ASST PROF BIOCHEM, MED SCH, UNIV PA, 72- Concurrent Pos: NIH grant, Univ Wash, 70-72 & res career develop award, 75-80. Mem: Am Chem Soc. Res: Structure and biosynthesis of connective tissue macromolecules with particular emphasis on interstitial collagens, and basement membrane collagen and noncollagen glycoprotein(s). Mailing Add: Dept Biochem & Biophys Anat-Chem Bldg Univ Pa Med Sch Philadelphia PA 19174

CLARK, CHARLES KITTREDGE, b Berkeley, Calif, Oct 15, 06; m 35; c 2. NATURAL PRODUCTS CHEMISTRY. Educ: Stanford Univ, AB, 28, MA, 29; Univ Fla, PhD(chem), 40. Prof Exp: Res chemist, Hercules Powder Co, 30-31; res chemist petrochems, Shell Develop Co, 32; res chemist naval stores, USDA, 33-37; res chemist tall oil, Quaker Chem Prods Corp, 40-42; res chemist terpenes, Naval Stores Div, Glidden Co, 43-45; res chemist wood prods, Weyerhaeuser Timber Co, 46-47; res chemist naval stores, Crosby Chem Inc, 48-52; res chemist wood prods, Crossett Co, 53-62; res chemist chem prod div, Union Camp Corp, 62-73; RETIRED. Mem: Am Chem Soc. Res: Rosin; terpenes and tall oil products. Mailing Add: 231 Andover Dr Savannah GA 31405

CLARK, CHARLES LESTER, b San Jose, Calif, Nov 17, 17; m 40; c 3. MATHEMATICS. Educ: Stanford Univ, AB, 39, MA, 40; Univ Va, PhD(math), 44. Prof Exp: Asst math, Stanford Univ, 39-40; instr, Univ Va, 42-44; from asst prof to prof, Ore State Col, 44-57; dir comput ctr, 61-71, dir inst res, 64-71, PROF MATH, CALIF STATE UNIV, LOS ANGELES, 57-, HEAD DEPT, 57-64 & 71- Concurrent Pos: Vis prof, Univ Va, 55-56; consult to educ, indust & legislative groups, 58- Mem: AAAS; Am Math Soc; Math Asn Am. Res: Topology; analysis; arc reversing transformations. Mailing Add: Dept of Math Calif State Univ Los Angeles CA 90032

CLARK, CHARLES MALCOLM, JR, b Greensburg, Ind, Mar 12, 38; m 63; c 2. MEDICINE. Educ: Ind Univ, AB, 60, MD, 63. Prof Exp: Intern, St Vincent's Hosp, Indianapolis, Ind, 63-64; resident internal med, Ind Univ, 64-65; staff assoc metab res, Nat Inst Arthritis & Metab Dis, 67-69; res & educ assoc, 69-71, clin investr endocrinol & metab, 71-74, ASSOC CHIEF STAFF EDUC, VET ADMIN HOSP, 74-; ASST PROF MED & PHARMACOL, MED SCH, IND UNIV, INDIANAPOLIS, 69- Concurrent Pos: Clin fel diabetes, Joslin Clin, Boston, Mass, 65-66, NIH fel, Joslin Res Lab, 66-67; fel med, Peter Bent Brigham Hosp, Boston, 66-67. Mem: AMA; Am Diabetes Asn; Endocrine Soc; fel Am Col Physicians; Am Fedn Clin Res. Res: Metabolic control mechanisms; regulation of intermediary metabolism in fetal development. Mailing Add: Vet Admin Hosp 1481 W Tenth St Indianapolis IN 46202

CLARK, CHESTER WILLIAM, b San Francisco, Calif, July 18, 06; m 30; c 2. PHYSICS. Educ: Univ Calif, BS, 27, MS, 29; State Univ Leiden, PhD(physics), 35. Prof Exp: Res chemist, Standard Oil Co, Calif, 29-33; instr chem, Univ Calif, 35-37 & San Francisco Jr Col, 37-41; res assoc, Johns Hopkins Univ, 46-47; low temp consult & physicist, Naval Res Lab, US Army, Washington, DC 47, sci asst to dir, Ballistic Res Lab, Aberdeen Proving Ground, 47-51, chief res & develop div & dep comdr, Picatinny Arsenal, 51-54, chief ord res & develop, 55-61, dir army res, 61-63, Commanding Gen, US Army-Japan, 63-65; vpres res, Res Triangle Inst, NC, 65-71; SPECIAL SCI & TECHNOL ASST TO AMBASSADOR, AM EMBASSY, TAIWAN, 73- Concurrent Pos: Mem div chem & chem technol, Nat Res Coun; Am Comnr, Joint US-Rep China Comn Rural Reconstruction, Taipei, 73- Mem: Soc Rheol; Am Phys Soc. Res: Low temperature specific heats of inorganic substances; superconductivity of metals and alloys; preparation of interstitial alloys; attainment of temperatures below 1K. Mailing Add: American Embassy Taipei Taiwan

CLARK, CLARENCE FLOYD, b Briceton, Ohio, May 26, 12; m 34; c 1. ZOOLOGY. Educ: Miami Univ, BS, 34; Ohio State Univ, MS, 42. Prof Exp: Teacher, High Sch, Ohio, 35-37; fish mgt agent, Ohio Div Wildlife, 37-57, asst supvr fish mgt, 57-60, supvr fish invest, 60-63, asst supvr fish mgt, 63-69; asst fisheries res, Sch Natural Resources, Ohio State Univ, 69-71; CONSULT ENVIRON & FISHERIES PROPAGATION & MGT PROBS, 71- Concurrent Pos: Consult, Environ Consult, 72-75 & Dept Nat Resources, Ohio Div Wildlife, 74-75. Mem: Fel Am Inst Fishery Res Biol; Am Fisheries Soc; Int Acad Fishery Sci; Am Soc Ichthyologists & Herpetologists. Res: Management and propagation of northern pike, minnows, muskelluge, creek chubs; status of freshwater naiads in relation to environmental impacts from man made changes in the environment. Mailing Add: 3835 Fairlington Dr Columbus OH 43220

CLARK, CLIFTON BOB, b Ft Smith, Ark, July 8, 27; m 50; c 3. SOLID STATE PHYSICS. Educ: Univ Ark, BA, 49, MA, 51; Univ Md, PhD(physics), 57. Prof Exp: Asst prof sci & math, Florence State Teachers Col, Ala, 50-51; asst prof physics, US Naval Acad, 51-55; physicist, US Naval Res Lab, 55-56; assoc prof physics, US Naval Acad, 56-57; from assoc prof to prof, Southern Methodist Univ, 57-65, chmn dept, 62-65; head dept, 65-75, PROF PHYSICS, UNIV NC, GREENSBORO, 65- Mem: AAAS; Am Asn Physics Teachers; Am Phys Soc; Sigma Xi. Res: Electron-lattice interactions in metals; lattice dynamics. Mailing Add: 800 Montrose Dr Greensboro NC 27410

CLARK, CORODON SCOTT, b Rochester, NY, Feb 1, 38. PUBLIC HEALTH, ENVIRONMENTAL HEALTH ENGINEERING. Educ: Antioch Col, BS, 61; Johns Hopkins Univ, MS, 63, PhD(eng sci), 65. Prof Exp: Sanit engr, Ohio River Valley Water Sanit Comn, 65-67; sr res assoc, 67-70, ASST PROF ENVIRON HEALTH, UNIV CINCINNATI, 71- Concurrent Pos: Consult, Am Pub Works Asn, 69, Ohio River Valley Water Sanit Comn, 71-73 & Environ Protection Agency, Cincinnati, 72-74. Mem: Am Pub Health Asn; Am Water Works Asn; Water Pollution Control Fedn. Res: Epidemiologic-Serologic study of health risks of wastewater exposure; sources of lead in pediatric lead absorption; development of potability indicator for direct reuse water. Mailing Add: Dept of Environ Health Kettering Lab 3223 Eden Ave Cincinnati OH 45267

CLARK, CROSMAN JAY, b Jackson, Mich, Mar 6, 25; m 45; c 3. MATHEMATICS. Educ: Okla State Univ, BA, 46, MS, 48, PhD(math), 53. Prof Exp: Res engr, Curtiss-

Wright Corp, 48; instr elec eng, Ohio State Univ, 49; instr & res asst, Okla State Univ, 49-52; mathematician, Stanolind Gas & Oil Co, 51; res mathematician, Continental Oil Co, 53-56; mathematician, Lockheed Missile & Space Corp, 56-61; eng specialist, Sylvania Electronic Defense Lab, Gen Tel & Electronics Corp, 61-64; staff scientist, Apparatus Res Dept, Tex Instruments Inc, Dallas, 64-66; dir advan studies, Northrop Corp, Calif, 66-70; vpres res & develop, Underwater Sci, Inc, Calif, 70-72; PRES, INTERSCI SYSTS, INC, 72- Concurrent Pos: Instr, Okla State Univ, 46-48 & 50-53, Foothill Col, 61-63 & Univ Santa Clara, 64; lectr, Calif State Col, Hayward, 71-72; consult, Nat Endowment Arts, 72-; adj prof cybernetic syst, San Jose State Univ, 73- Mem: Environ Design Res Asn; Soc Gen Systs Res; Math Asn Am; Soc Indust & Appl Math. Res: Topology; set theory; logic; wave theory; automata; operations research; stochastic processes; decision theory; general systems theory; mathematical-statistical model building; behavioral cybernetics; humanistic measures of design; business-management systems. Mailing Add: 19200 Shubert Dr Saratoga CA 95070

CLARK, DALE ALLEN, b Munden, Kans, Sept 14, 22; m 48; c 2. BIOCHEMISTRY. Educ: Hastings Col, BA, 44; Univ Colo, MA, 47; Univ Utah, PhD(biochem), 50. Prof Exp: Asst prof biochem, Sch Med, Univ Okla, 50-52, assoc prof, 53-54; res biochemist, Vet Admin Hosp, Dallas, 54-59; BIOCHEMIST, CLIN SCI DIV, US AIR FORCE SCH AEROSPACE MED, BROOKS AFB, 59- Mem: Fel AAAS; Am Chem Soc; Am Oil Chemists Soc; Soc Exp Biol & Med. Res: Atherosclerosis; steroid and sterol metabolism. Mailing Add: Clin Path Br US Air Force Sch Aerospace Med Brooks AFB TX 78235

CLARK, DAVID BARRETT, b Glen Ellyn, Ill, Nov 1, 13; m 48; c 2. NEUROLOGY, NEUROPATHOLOGY. Educ: Univ Chicago, PhD(neuroanat), 40, MD, 46. Prof Exp: Asst anat, Univ Chicago, 38-47; intern med, Johns Hopkins Hosp, 47-48, asst resident neurol, 48-49; Fulbright lectr, Nat Hosp, Queen Sq, Eng, 50-51; from asst prof to assoc prof neurol & pediat, Johns Hopkins Hosp, 51-65; PROF NEUROL & CHMN DEPT, SCH MED, UNIV KY, 65- Concurrent Pos: Fel neurol, Johns Hopkins Hosp, 49-50; attend neurologist, Baltimore City Hosp, 51-59, Johns Hopkins Hosp, 51-65 & Rosewood State Training Sch, 52-65; adv, Epilepsy Found, 66-; Teale lectr, Royal Col Physicians, 67-; mem field study sect, Nat Inst Neurol Dis & Blindness, 58-60, prog proj study sect, 60-64, neurol res & training sect, 65-69 & residency rev comt, 67- Mem: Am Acad Neurol; Am Neurol Asn; Am Asn Neuropath; AMA; Royal Soc Med. Res: Pediatric neurology; pathology of cerebral birth injuries; developmental defects of the central nervous system. Mailing Add: Univ of Ky Col of Med Lexington KY 40506

CLARK, DAVID C, b Woodland, Calif, May 19, 41. MATHEMATICAL ANALYSIS. Educ: Calif Inst Technol, BS, 63; Stanford Univ, MS, 65, PhD(math), 67. Prof Exp: Vis mem, Courant Inst Math Sci, 67-68; asst prof math, Rutgers Univ, 68-72; ASST PROF MATH, UNIV PR, MAYAGUEZ, 72- Mem: Am Math Soc. Res: Ordinary and partial differential equations. Mailing Add: Dept of Math Univ of PR Mayaguez PR 00708

CLARK, DAVID DELANO, b Austin, Tex, Feb 10, 24; m 49; c 3. NUCLEAR PHYSICS. Educ: Univ Calif, Berkeley, AB, 48, PhD(physics), 53. Prof Exp: Asst physics, Univ Calif, Berkeley, 48-51, physicist, 53; res assoc, Brookhaven Nat Lab, 53-55; from asst prof to assoc prof eng physics, 55-64, PROF APPL PHYSICS, CORNELL UNIV, 64-, DIR WARD LAB NUCLEAR ENG, 60- Concurrent Pos: Euratom fel, Italy, 62; Guggenheim fel, Niels Bohr Inst, Copenhagen, Denmark, 68-69. Mem: Am Phys Soc; Am Nuclear Soc; AAAS. Res: Nuclear structure physics, especially isomers; nuclear instrumentation; reactor physics. Mailing Add: Ward Lab of Nuclear Eng Cornell Univ Ithaca NY 14853

CLARK, DAVID ELLSWORTH, b Paso Robles, Calif, Nov 22, 22; m 47; c 2. ORGANIC CHEMISTRY. Educ: Univ Redlands, BA, 47; Stanford Univ, MS, 48, PhD(org chem), 53. Prof Exp: Instr chem, Fresno State Col, 50-51 & 53-54, from asst prof to prof, 54-65; assoc acad planning, Chancellor's Off, Calif State Cols, 65-67; acad admin internship, Brown Univ, 67-68; ASSOC VPRES ACAD AFFAIRS, CALIF STATE UNIV, FRESNO, 70- Concurrent Pos: NSF sci fac fel, Harvard Univ, 62-63. Mem: Am Chem Soc. Res: Organic synthesis; chemical therapeutics. Mailing Add: 1456 E Browning Fresno CA 93710

CLARK, DAVID GORDON, b Helena, Ark, July 27, 16; m 44. PHYSICS. Educ: Park Col, AB, 38; Agr & Mech Col Tex, MS, 40; Pa State Col, PhD(physics), 47. Prof Exp: Instr physics, Pa State Col, 42-47; asst prof, 47-51, ASSOC PROF PHYSICS, UNIV NH, 51- Res: Underwater sound transmission. Mailing Add: Dept of Physics Univ of NH Durham NH 03824

CLARK, DAVID LEE, b Detroit, Mich, Apr 7, 39; m 67; c 1. ANIMAL BEHAVIOR, ANATOMY. Educ: Kalamazoo Col, BA, 62; Univ Okla, MS, 63; Mich State Univ, PhD(zool), 67. Prof Exp: ASST PROF ANAT, COL MED, OHIO STATE UNIV, 68- Concurrent Pos: Great Lakes Cols Asn teaching fel, Kenyon Col, 67-68. Mem: AAAS; Am Inst Biol Sci; Aerospace Med Asn; Am Soc Zool; Animal Behav Soc. Res: Ontogeny of the vestibular system of equilibrium in the rat to determine the anatomical correlates of vestibular behavior during postnatal development; effects of 2g environment on vestibular system. Mailing Add: Dept of Anat Ohio State Univ Columbus OH 43210

CLARK, DAVID LEIGH, b Albuquerque, NMex, June 15, 31; m 51; c 4. PALEONTOLOGY. Educ: Brigham Young Univ, BS, 53, MS, 54; Univ Iowa, PhD(geol), 57. Prof Exp: Geologist, Standard Oil Co, 54; asst geol, Columbia Univ, 54-55; asst, Univ Iowa, 55-57; asst prof, Southern Methodist Univ, 57-59; from asst prof to assoc prof, Brigham Young Univ, 59-63; assoc prof geol, 63-68, prof geol & geophys, 68-74, chmn dept, 71-74, W H TWENHOFEL PROF, UNIV WIS-MADISON, 74- Concurrent Pos: Sr Fulbright fel & vis prof, Univ Bonn, 65-66. Mem: Geol Soc Am; Am Asn Petrol Geologists; Paleont Soc; Soc Econ Paleontologists & Mineralogists. Res: Cretaceous cephalopods; Mesozoic stratigraphy; Paleozoic and Mesozoic conodonts. Mailing Add: Dept of Geol & Geophys Weeks Hall Univ of Wis Madison WI 53706

CLARK, DAVID SEDGEFIELD, b St Stephen, NB, Nov 13, 29; m 52; c 3. FOOD MICROBIOLOGY. Educ: McGill Univ, MS, 53, PhD(physiol), 57. Prof Exp: Lectr agr bact, MacDonald Col, McGill Univ, 53-57; tech sales, Buckman Labs Can, Ltd, 57-58; from asst res officer to assoc res officer, 58-70, SR RES OFFICER, NAT RES COUN CAN, 70- Concurrent Pos: Secy-treas, Int Standing Comn Microbiol Specifications Foods, Int Asn Microbiol Socs; hon mem expert panel on food microbiol & hyg, WHO; assoc ed, Can J Microbiol, 73- Mem: Inst Food Technologists; Can Soc Microbiol; Can Inst Food Technologists. Res: Industrial fermentations; bacterial physiology; meat microbiology; microbiol methodology; effect of the gaseous environment on microorganisms. Mailing Add: 1 Kaymar Dr Ottawa ON Can

CLARK, DAVID THURMOND, b Topeka, Kans, Aug 3, 25; m 47; c 2. PARASITOLOGY. Educ: Univ Nebr, BA, 49, MA, 51; Univ Ill, PhD, 55. Prof Exp: Asst zool, Univ Nebr, 49-51 & Univ Ill, 51-55; from instr to prof microbiol & pub health, Mich State Univ, 56-65, asst vpres res & develop, 65-69; staff assocuniv sci & develop, Instnl Rels, NSF, Washington, DC, 69-70; DEAN GRAD STUDIES & RES, PORTLAND STATE UNIV, 70- Concurrent Pos: Mem subcomt prenatal & postnatal mortality in swine, Comt Animal Health, Nat Res Coun, 59- Mem: Am Soc Parasitologists; Am Micros Soc; NY Acad Sci; Soc Protozool. Res: Parasites of wildlife; physiological studies on nematodes; immunology of parasitic infections; science policy. Mailing Add: Dept of Grad Studies Portland State Univ Portland OR 97207

CLARK, DONALD ELDON, physical chemistry, nuclear chemistry, see 12th edition

CLARK, DONALD GREGORY, biochemistry, molecular biology, see 12th edition

CLARK, DONALD RAY, JR, b Garrett, Ind, Jan 20, 40; m 58; c 3. POLLUTION BIOLOGY. Educ: Univ Ill, Urbana, BS, 61; Tex A&M Univ, MS, 64; Univ Kans, PhD(zool), 68. Prof Exp: Asst prof wildlife sci, Tex A&M Univ, 68-72; RES BIOLOGIST, PATUXENT WILDLIFE RES CTR, US FISH & WILDLIFE SERV, DEPT INTERIOR, 72- Concurrent Pos: Prin investr, Tex Agr Exp Sta, 68-72. Mem: Ecol Soc Am; Am Soc Ichthyologists & Herpetologists; Soc Study Amphibians & Reptiles; Am Soc Mammalogists; AAAS. Res: Understanding the relationships between environmental contaminants and declining populations of bats. Mailing Add: Patuxent Res Ctr US Fish & Wildlife Res Ctr Laurel MD 20811

CLARK, DOUGLAS NAPIER, b New York, NY, Jan 24, 44; m 68. MATHEMATICAL ANALYSIS. Educ: Johns Hopkins Univ, AB, 64, PhD(math), 67. Prof Exp: Fel math, Univ Wis-Madison, 67-68; asst prof, Univ Calif, Los Angeles, 68-73; ASSOC PROF MATH, UNIV GA, 73- Concurrent Pos: Res assoc, Off Naval Res, Nat Res Coun, 67-68. Mem: Am Math Soc. Res: Operator theory and its applications to complex variables, chiefly Toeplitz and Hankel matrices and the study of invariant subspaces; analytic functions in polydiscs; interpolation problems. Mailing Add: Dept of Math Univ of Ga Athens GA 30602

CLARK, DUNCAN WILLIAM, b New York, NY, Aug 31, 10; m 43, 71; c 3. MEDICINE. Educ: Fordham Col, AB, 32; Long Island Col Med, MD, 36. Prof Exp: Intern, Brooklyn Hosp, 36-38; resident med, Kings County Hosp, 38-40; dir student health, Long Island Col Med, 41-49, instr med, 42-47, from asst dean to assoc dean, 43-48, asst prof med, actg chmn prev med & dean, 48-50; PROF ENVIRON MED & COMMUNITY HEALTH & CHMN DEPT, STATE UNIV NY DOWNSTATE MED CTR, 51- Concurrent Pos: Fel, Yale Univ, 40-41; traveling fel, WHO, 52; Commonwealth Found fel, 61; vis prof, Univ Birmingham, 61; consult, Health Serv Res Study Sect, US Pub Health Serv, 61-65 & 73-76, consult & chmn community health res training, 65-69, consult, Nat Ctr Health Serv Res & Develop, 70-73, mem tech adv group on med care effectiveness, 70-72, res scientist, Fel Rev Comt, 70-72; consult, Nat Acad Sci-Nat Res Coun, 65-68. Mem: Am Col Prev Med; Harvey Soc; Am Pub Health Asn; Asn Teachers Prev Med (pres, 53-56); NY Acad Med (vpres, 76-). Res: Medical education; public health and medical care; preventive medicine. Mailing Add: Dept of Environ Sci State Univ NY Downstate Med Ctr Brooklyn NY 11203

CLARK, EDGAR WILLIAM, b Royal Oak, Mich, June 28, 22; m 45; c 4. INSECT PHYSIOLOGY, FOREST ENTOMOLOGY. Educ: Univ Calif, AB, 47, PhD(entom), 50. Prof Exp: Asst res zoologist, Univ Calif, Los Angeles, 50-53; entomologist, Entom Res Div, USDA, 53-61; PRIN INSECT PHYSIOLOGIST, INSECT PHYSIOL & BIOCHEM, DIV FOREST PROTECTION RES, US FOREST SERV, 61- Concurrent Pos: Adj assoc prof, Duke Univ; mem tech comt, Southern Forest Dis & Insect Res Coun; consult, Food & Agr Orgn UN. Mem: AAAS; Entom Soc Am. Res: Insect physiology, biochemistry and ecophysiology; insecticidal and biological control of forest insects; insect nutrition, chemostimulants, host resistance, host plant chemistry, insect reproduction and tropical forest entomology. Mailing Add: Forest Sci Lab US Forest Serv Box 12254 Research Triangle Park NC 27709

CLARK, EDWARD ALOYSIUS, b Jersey City, NJ, Jan 28, 34; m 55; c 4. THEORETICAL PHYSICS. Educ: Col Holy Cross, BS, 55; Fordham Univ, MS, 60, PhD(physics), 66. Prof Exp: From instr to assoc prof physics, 60-70, chmn dept, 66-70, asst to pres, Univ, 70-71, acad vpres, Brooklyn Ctr, 74-75, PROF PHYSICS, LONG ISLAND UNIV, BROOKLYN CTR, 70-, DEAN COL LIB ARTS & SCI, 71-, PRES, BROOKLYN CTR, 75- Concurrent Pos: Consult, State NY Dept Educ, 66- Mem: Am Phys Soc; Am Asn Physics Teachers. Res: Calculation of molecular vibration-rotation spectra. Mailing Add: Long Island Univ Brooklyn Ctr Brooklyn NY 11201

CLARK, EDWARD MAURICE, b Edinburgh, Scotland, Feb 16, 20; m 44; c 1. PLANT BREEDING. Educ: Univ Minn, BS, 49, MS, 55, PhD(plant breeding), 56. Prof Exp: From asst botanist to assoc botanist, 56-62, ASSOC PROF BOT, AUBURN UNIV, 62- Mem: AAAS; Genetics Soc Am; Am Genetics Asn. Res: Interspecific relationships within Vicia; corncytogenetics; genetics and cytogenetics of Vicia and Zea; Arachis diseases, including Cercospora. Mailing Add: Dept of Bot Auburn Univ Auburn AL 36830

CLARK, EDWARD SHANNON, b Schenevus, NY, Apr 26, 30. POLYMER CHEMISTRY. Educ: Union Univ, NY, BS, 51; Univ Calif, PhD(chem), 56. Prof Exp: Res phys chemist, E I du Pont de Nemours & Co, Inc, 55-62, sr res chemist, 62-72; PROF, DEPT CHEM & METALL ENG, UNIV TENN, KNOXVILLE, 72- Concurrent Pos: Fulbright scholar, Aarhus Univ, 62-63. Mem: Am Chem Soc; Am Crystallog Asn; Am Phys Soc; Soc Plastics Eng; Am Inst Chem Engr. Res: Structure-property relationships in polymers. Mailing Add: Dept of Chem & Metall Eng Univ of Tenn Knoxville TN 37916

CLARK, ELOISE ELIZABETH, b Grundy, Va, Jan 20, 31. BIOCHEMISTRY, BIOPHYSICS. Educ: Mary Washington Col, BA, 51; Univ NC, PhD(zool), 58. Prof Exp: Instr biol, Women's Col, Univ NC, Greensboro, 52-53; res asst physiol, Univ NC, Chapel Hill, 53-55; from instr to assoc prof biol & biochem, Columbia Univ, 59-69; prog dir, Develop Biol Prog, 69-70, prog dir, Biophysic Prog, 70-73, sect head, Molecular Biol Sect, 71-73; div dir, Biol & Med Sci Div, 73-75, DEP ASST DIR, BIOL, BEHAV & SOC SCI DIRECTORATE, NSF, 75- Mem: AAAS; Am Soc Cell Biol; Soc Gen Physiologists (secy, 65-67); Biophysical Soc. Res: Physical biochemistry of muscle protein and enzymes; science adminstration. Mailing Add: Dep Asst Dir Biol Behav & Soc Sci Nat Sci Found Washington DC 20550

CLARK, ERVIL DELWYN, b Angwin, Calif, Jan 23, 27; m 48; c 2. BIOLOGY. Educ: Pac Union Col, BA, 50, MA, 55; Ore State Univ, PhD(radiation biol), 71. Prof Exp: Instr sci & math, San Diego Union Acad, 51-56; from instr to assoc prof, 56-73, PROF BIOL & CHMN DEPT, PAC UNION COL, 73- Concurrent Pos: Instr anat, physiol & microbiol, Paradise Valley Sch Nursing, 53-54. Mem: Ecol Soc Am; Am Ornith Union. Res: Plant ecology; ecology of California North Coast Ranges. Mailing Add: Dept of Biol Pac Union Col Angwin CA 94508

CLARK, EUGENIE, b New York, NY, May 4, 22; div; c 4. ZOOLOGY. Educ: Hunter Col, BA, 42; NY Univ, MS, 46, PhD(zool), 50. Prof Exp: Asst ichthyol, Scripps Inst Oeanog, 46-47 & NY Zool Soc, 47-48; asst animal behavior, Am Mus Natural Hist, 48-49, res assoc, 50-66; dir marine biol, Cape Haze Marine Lab, 55-66; assoc prof biol, City Col New York, 66-67; assoc prof, 69-73, PROF ZOOL, UNIV MD, COLLEGE PARK, 73- Concurrent Pos: AEC fel, 50; Fulbright scholar, Egypt, 51; Saxton fel & Breadloaf Writer's fel, 52; instr, Hunter Col, 54. Honors & Awards: Gold Medal, Soc Women Geogr, 75. Mem: Am Soc Ichthyologists & Herpetologists; Soc Women Geogr. Res: Ichthyology; reproductive behavior of fishes; morphology and taxonomy of plectognath fishes; isolating mechanisms of poeciliid fishes; behavior of sharks; Red Sea fishes. Mailing Add: Dept of Zool Univ of Md College Park MD 20742

CLARK, EVELYN GENEVIEVE, b Brooksville, Ky, Jan 8, 22. MICROBIOLOGY. Educ: Georgetown Col, AB, 47; Univ Ky, MS, 53. Prof Exp: From instr to asst prof, 47-64, ASSOC PROF BIOL, GEORGETOWN COL, 64- Concurrent Pos: Consult microbiologist, Cent Baptist Hosp, Ky, 53-64; bacteriologist, Ford Mem Hosp, Ky, 67-71. Mem: AAAS; Nat Asn Biol Teachers; Am Soc Microbiol; Am Inst Biol Sci. Res: Clinical mycology. Mailing Add: Dept of Biol Georgetown Col Georgetown KY 40324

CLARK, FLORA MAE, b Houston Co, Ala, Nov 19, 33. GENETICS. Educ: Ala Col, BS, 60, MAT, 65; Univ Tenn, PhD(zool), 70. Prof Exp: Teacher biol, Rehobeth High Sch, Ala, 60-62 & Dependent Educ Group, US Army, France, 62-64; instr, Jacksonville State Univ, 65-67; res assoc genetics, Univ Ga, 70-71; ASSOC PROF BIOL, COLUMBUS COL, 71- Concurrent Pos: Consult, St Francis Hosp Lab, 75. Mem: AAAS; Am Inst Biol Sci. Res: Evolution of water snakes; karotypes, similarity of plasma proteins, hemoglobins, behavior and thermoregulation and distribution. Mailing Add: Fac Off Bldg Columbus Col Columbus GA 31907

CLARK, FLOYD BRYAN, b Akron, Ohio, Nov 15, 25; m 49; c 2. FORESTRY, RESEARCH ADMINISTRATION. Educ: Purdue Univ, BSF, 49; Univ Mo, MSF, 54; Univ Southern Ill, PhD(bot), 68. Prof Exp: Res forester silvicult, 49-59, proj leader, 59-68, asst dir, NCent Forest Exp Sta, 68-74, DIR, NORTHEASTERN FOREST EXP STA, US FOREST SERV, 74- Mem: Am Soc Foresters. Res: Forest management; silviculture; silvics; natural and artificial regeneration. Mailing Add: Northeastern Forest Exp Sta 6816 Market St Upper Darby PA 19082

CLARK, FRANCIS EUGENE, soil microbiology, see 12th edition

CLARK, FRANCIS JOHN, b Chicago, Ill, May 30, 33. NEUROPHYSIOLOGY. Educ: Northwestern Univ, BSEE, 56; Purdue Univ, MSEE, 57, PhD(elec eng), 65. Prof Exp: Engr, Cook Res Labs, Ill, 57-59; proj mgr, Advan Res Dept, Sunbeam Corp, 59-61; instr elec eng, Purdue Univ, West Lafayette, 64, asst prof, Sch Elec Eng & Dept Vet Anat, 64-68, assoc prof, 68-72; ASSOC PROF PHYSIOL & BIOPHYS, UNIV NEBR MED CTR, OMAHA, 72- Concurrent Pos: Nat Inst Neurol Dis & Stroke spec fel, Nobel Inst Neurophysiol, Karolinska Inst, Sweden, 69-71. Mem: AAAS; Am Phys Soc; Soc Neurosci. Res: Mammalian nervous system. Mailing Add: Dept Physiol & Biophys Univ Nebr Med Ctr Omaha NE 68105

CLARK, FRANCIS MATTHEW, b Augusta, Ill, Sept 26, 00; m 25. MICROBIOLOGY. Educ: Univ Ill, BS, 23, MS, 26, PhD(bact), 33. Prof Exp: Asst soil fertility, 23-25, asst soil biol, 25-29, asst bact, 29-33, instr, 33-37, assoc, 37-39, from asst prof to assoc prof, 39-57, prof, 57-70, EMER PROF BACT, UNIV ILL, URBANA-CHAMPAIGN, 70- Mem: Am Soc Microbiologists; Am Chem Soc; Inst Food Technologists. Res: Thermophilic organisms and food spoilage; bacterial nutrition; yeast nutrition; vitamin and nitrogen requirements. Mailing Add: 726 S Foley St Champaign IL 61820

CLARK, FRANK EUGENE, b St Louis, Mo, Sept 16, 19; m 48. MATHEMATICS. Educ: Dartmouth Univ, BA, 41; Duke Univ, MA, 46, PhD(math), 48. Prof Exp: Vis instr math, Duke Univ, 47-48; instr, Tulane Univ, 48-50; from asst prof to assoc prof, 50-61, PROF MATH, RUTGERS UNIV, 61-, CHMN DEPT, UNIV COL, 59- Mem: Am Math Soc; Math Asn Am; Soc Indust & Appl Math. Res: Algebraic inequalities; linear programming. Mailing Add: Dept of Math Univ Col Rutgers Univ New Brunswick NJ 08903

CLARK, FRANK RINKER, geology, deceased

CLARK, FRANK S, b San Mateo, Calif, Sept 27, 33. ORGANIC CHEMISTRY. Educ: Stanford Univ, BS, 55; Purdue Univ, PhD(org chem), 60. Prof Exp: Res chemist, 60-68, RES SPECIALIST, MONSANTO CO, 68- Mem: AAAS; Am Chem Soc. Res: Development of high temperature gas turbine engine oils; organic synthesis; synthesis of synthetic lubricants; boundary lubrication, corrosion and wetting of synthetic lubricants. Mailing Add: Corp Res Dept Monsanto Co 800 N Lindbergh Blvd St Louis MO 63166

CLARK, GEOFFREY ANDERSON, b Philadelphia, Pa, Aug 17, 44; m 68. ANTHROPOLOGY, ARCHAEOLOGY. Educ: Univ Ariz, BA, 66, MA, 67; Univ Chicago, PhD(anthrop), 71. Prof Exp: ASST PROF ANTHROP, ARIZ STATE UNIV, 71- Concurrent Pos: Nat Park Serv grant for report test excavations, Williams AFB, Ariz, 73; ed, Anthrop Res Papers, 74- Mem: Fel AAAS; fel Am Anthrop Asn; fel Royal Anthrop Inst; Soc Am Archaeologists; SAfrican Archaeol Soc. Res: Old World prehistory with specialization in Pleistocene and post-Pleistocene ages of Western Europe, especially France and Spain; early man; Puebloan societies of American southwest; statistical applications in archaeology. Mailing Add: Dept of Anthrop Ariz State Univ Tempe AZ 85281

CLARK, GEORGE, b Sunnyside, Wash, May 18, 05; m 34; c 4. NEUROPHYSIOLOGY. Educ: Univ Wash, BS, 36; Northwestern Univ, MS, 37, PhD(neurophysiol), 39. Prof Exp: Instr neurol, Northwestern Univ, 39-40; assoc anat, Med Col SC, 40-42, asst prof, 42; asst prof psychobiol, Yerkes Labs Primate Biol, 43-47; assoc prof neuroanat, Chicago Med Sch, 47-53; assoc prof physiol, Univ Buffalo, 53-59; res physiologist, US Army Med Res Lab, Ky, 59-61 & Inst Environ Med, Mass, 61-64; res physiologist, Civil Aeromed Res Inst, Fed Aviation Agency, 64-70; VIS PROF ANAT, MED UNIV SC, 70-; MEM STAFF VET ADMIN HOSP, 70- Mem: Am Acad Neurol; Am Asn Anat; Soc Exp Biol & Med; Am Physiol Soc; Biol Stain Comn. Res: Functions of hypothalamus; temperature regulation; staining mechanisms. Mailing Add: Vet Admin Hosp Charleston SC 29403

CLARK, GEORGE ALFRED, JR, b Camden, NJ, May 6, 36; m 61. ORNITHOLOGY. Educ: Amherst Col, BA, 57; Yale Univ, PhD(biol), 64. Prof Exp: Res assoc zool, Univ Wash, 63-64, actg instr, 64-65; asst prof, 65-70, ASSOC PROF ZOOL, UNIV CONN, 70- Mem: Soc Study Evolution; Am Ornith Union; Am Soc Naturalists; Cooper Ornith Soc; Wilson Ornith Soc. Res: Integumental structure, behavior and evolution of birds. Mailing Add: Biol Sci Group Univ of Conn Storrs CT 06268

CLARK, GEORGE HOWARD, genetics, see 12th edition

CLARK, GEORGE RICHMOND, II, b Princeton, Maine, Mar 23, 38; m 61; c 2. PALEOECOLOGY, MARINE BIOLOGY. Educ: Cornell Univ, AB, 61; Calif Inst Technol, MS, 66, PhD(geobiol), 69. Prof Exp: Asst prof geol, Univ NMex, 69-74; ASST PROF GEOL, STATE UNIV NY COL GENESEO, 74- Mem: AAAS; Am Soc Limnol & Oceanog; Soc Econ Paleont & Mineral; Geol Soc Am; Paleont Soc. Res: Growth lines; environmental variations in shell morphology; marginal calcification in invertebrates; heavy metal uptake by marine invertebrates. Mailing Add: Dept of Geol Sci State Univ NY Col Geneseo NY 14454

CLARK, GEORGE STUART, physical organic chemistry, see 12th edition

CLARK, GEORGE WHIPPLE, b Evanston, Ill, Aug 31, 28; m 54; c 2. PHYSICS. Educ: Harvard Univ, AB, 49; Mass Inst Technol, PhD(physics), 52. Prof Exp: From instr to assoc prof, 52-65, PROF PHYSICS, MASS INST TECHNOL, 65- Concurrent Pos: Guggenheim fel & Fulbright res scholar, 63. Mem: Am Phys Soc; Am Acad Arts & Sci. Res: Cosmic rays; x-ray astronomy. Mailing Add: Dept of Physics Mass Inst Technol Cambridge MA 02139

CLARK, GERALD ROBERT, b Parr, Alta, Mar 17, 18; US citizen; m 51; c 6. PSYCHIATRY. Educ: Univ Ore, BA, 42, MD, 45; Harvard Univ, MPH, 51; Am Bd Prev Med, dipl pub health, 53; Am Bd Psychiat & Neurol, dipl, 62. Prof Exp: Dir, State Dept Health, NMex, 53-55; psychiatrist, Norristown State Hosp, Pa, 55-58; supt, Somerset State Hosp, Pa, 58-59; dir psychiat clins, Jefferson Med Col, 59-60; supt, Elwyn Sch, 60-66, PRES, ELWYN INST, 66- Concurrent Pos: Sr physician, Childrens Hosp of Philadelphia, 67-; prof psychiat & pediat, Univ Pa, 70- Honors & Awards: US Physicians Award, President Nixon, AMA & Comn Handicapped, 70; Gold Achievement Awards, Am Acad Achievement, 71 & Am Psychiat Inst, 72. Mem: Fel Am Col Physicians; fel Am Col Psychiatrists. Res: Mental retardation; vocational training and rehabilitation. Mailing Add: 77 Washington Rd Elwyn PA 19063

CLARK, GLEN W, b Newdale, Idaho, Apr 17, 31; m 51; c 5. PARASITOLOGY, PROTOZOOLOGY. Educ: Ricks Col, BS, 56; Utah State Univ, MS, 58; Univ Calif, Davis, PhD(zool), 62. Prof Exp: Instr life sci, Am River Jr Col, 62-64; from asst prof to assoc prof, 64-71, PROF ZOOL, CENT WASH STATE COL, 71- Mem: Soc Protozool; Wildlife Dis Asn. Res: Blood parasites of the class Aves; coccidial parasites of reptiles. Mailing Add: Dept of Biol Cent Wash State Col Ellensburg WA 98926

CLARK, GLENN R, b Reidsville, NC, Dec 29, 36; m 66. HUMAN ANATOMY, VERTEBRATE MORPHOLOGY. Educ: Wake Forest Univ, BS, 58, Bowman Gray Sch Med, MD, 63. Prof Exp: Instr biol, Wake Forest Univ, 64-65; instr gross anat, Basic Health Sci Div, Emory Univ, 65-66, instr microanat, 66-70; asst prof anat & physiol & dir biosci, 70-71, ASST DIR, DIV ALLIED HEALTH PROGS, BOWMAN GRAY SCH MED, 71-; CHMN DEPT ALLIED HEALTH TECHNOL, FORSYTH TECHNOL INST, 71- Mem: AMA; Am Soc Zoologists; Asn Schs Allied Health Prof. Res: Human microanatomy. Mailing Add: Div of Allied Health Progs Bowman Gray Sch of Med Winston-Salem NC 27103

CLARK, GORDON MURRAY, b Montreal, Que, July 9, 25; m 51; c 6. RADIATION BIOLOGY. Educ: Sir George Williams Univ, BSc, 48; McGill Univ, MSc, 51; Emory Univ, PhD(radiation biol), 54. Prof Exp: Asst, McGill Univ, 50-51 & Univ Miami, 51-52; res, Emory Univ, 54-55 & Univ Mich, 55-57; asst prof zool & radiation biol, 57-64, assoc prof radiobiol, 64-68, PROF RADIOBIOL, UNIV TORONTO, 68- Mem: Soc Protozool; Royal Astron Soc Can; Can Genetic Soc. Res: Radiation effects on the cellular level; chemistry of the cell; dose and dose-rate studies in plant and animal cells; fallout studies; hyperbaric oxygen studies; x-ray and gamma ray irradiation; molecular biology; radiation injury and recovery; molecular level. Mailing Add: Dept of Zool Univ of Toronto Toronto ON Can

CLARK, GRADY WAYNE, b Candler, NC, Nov 29, 22; m 52. PHYSICS. Educ: Clemson Col, BS, 44; Univ Va, PhD(physics), 51. Prof Exp: Res engr, Univ Va, 51-52, Linde Air Prod, 52-55 & Va Inst Sci Res, 55-58; res engr, 58-59, LAB HEAD, OAK RIDGE NAT LAB, 59- Mem: Am Phys Soc; Am Soc Metals; Am Ceramics Soc. Res: Crystal physics; crystal growth; eutetic solidification; biomagnetism. Mailing Add: Oak Ridge Nat Lab Oak Ridge TN 37830

CLARK, HADDEN, polymer chemistry, physical chemistry, see 12th edition

CLARK, HARLAN EUGENE, b Bloomington, Ill, July 29, 41; m 67; c 2. PHYSICAL CHEMISTRY, ANALYTICAL CHEMISTRY. Educ: Univ Ill, BS, 63; Univ Mich, MS, 64, PhD(chem), 68. Prof Exp: Sr res chemist, Res Ctr, Sherwin-Williams Co, 68-73, SR SCIENTIST, SHERWIN-WILLIAMS CHEM, CHICAGO, 73- Mem: Am Chem Soc; Fine Particle Soc. Res: Molecular vibrations; infrared and Raman spectroscopy; titanium dioxide pigments; zinc and barium chemicals. Mailing Add: 413 Todd St Park Forest IL 60466

CLARK, HAROLD ARTHUR, b East Jordan, Mich, Apr 10, 10; m 38; c 2. POLYMER CHEMISTRY. Educ: Mich State Univ, BS, 31. Prof Exp: Analyst, Monolith Portland Midwest Co, 37-40 & Dow Chem Co, 42-43; anal supt, 43-48, res chemist, 48-53, res group leader, 53-60, res supvr, 60-64, tech dir, Eng Prod Div, 64-69, asst dir corp develop, 69-70, res scientist, 70-75, CONSULT, DOW CORNING CORP, 75- Mem: Am Chem Soc; Sigma Xi. Res: Silicone resins and polymers; abrasion resistant coatings for plastics and solar collectors. Mailing Add: 1718 SE 39th St Cape Coral FL 33904

CLARK, HAROLD EUGENE, b Sunderland, Mass, Feb 21, 06; m 38; c 2. PLANT PHYSIOLOGY. Educ: Mass Col, BS, 28; Rutgers Univ, MS, 31, PhD(plant physiol), 33. Prof Exp: Nat Res Coun fel, Yale Univ & Univ Conn Exp Sta, 33-35; assoc biochemist, Exp Sta, Pineapple Res Inst, Hawaii, 35-38, head dept physiol & soil, 38-46, head dept chem, 46-47; from assoc prof to prof plant physiol, 47-74, EMER PROF PLANT PHYSIOL, RUTGERS UNIV, 74- Mem: AAAS; Am Soc Plant Physiol; Am Chem Soc; Bot Soc Am. Res: Mineral nutrition. Mailing Add: 24 E Lawrence St Milltown NJ 08850

CLARK, HELEN EDITH, b Edam, Sask, Feb 4, 12; nat US. NUTRITION. Educ: BHSc, Univ Sask, 39; Iowa State Col, MS, 45, PhD(nutrit), 50. Prof Exp: Res assoc, Iowa State Col, 45-50; from asst prof to assoc prof nutrit, Kans State Col, 50-54; assoc prof, 54-59, PROF NUTRIT, PURDUE UNIV, 59- Honors & Awards: Borden Award, Am Home Econ Asn, 68. Mem: Am Bd Nutrit; AAAS; Am Home Econ Asn; Am Dietetic Asn; Am Inst Nutrit. Res: Factors influencing utilization of proteins and amino acids; amino acid requirements of man. Mailing Add: Apt a 414 Vine St West Lafayette IN 47906

CLARK, HERBERT MOTTRAM, b Derby, Conn, Sept 3, 18. RADIOCHEMISTRY. Educ: Yale Univ, BS, 40, PhD(phys chem), 44. Prof Exp: Lab asst, Yale Univ, 40-42, instr chem, 42-46; asst prof phys chem, 46-49, assoc prof phys nuclear chem, 49-51, PROF PHYS NUCLEAR CHEM, RENSSELAER POLYTECH INST, 51- Concurrent Pos: Res assoc, Monsanto Chem Co & Clinton Labs, Oak Ridge, 46-47;

consult, Union Carbide & Carbon Corp, 55-58 & US AEC, 65-70; mem subcomt radiochem, Comt Nuclear Sci, Nat Acad Sci-Nat Res Coun, 61-72; consult radiol health, USPHS, 67-69. Mem: Fel AAAS; fel NY Acad Sci; Am Chem Soc; Am Phys Soc; Am Soc Eng Educ. Res: Radiochemistry applied to the environment; radiological health, nuclear power, nuclear medicine; radiochemical separations; Mössbauer spectrometry. Mailing Add: Dept of Chem Rensselaer Polytech Inst Troy NY 12181

CLARK, HOWARD CHARLES, b Auckland, NZ, Sept 4, 29; m 54; c 2. INORGANIC CHEMISTRY. Educ: Univ Auckland, BSc, 51, MSc, 52, PhD, 54; Cambridge Univ, PhD, 57, ScD(chem), 72. Prof Exp: Jr lectr chem, Univ Auckland, 54-55; res fel, Cambridge Univ, 55-57; from asst prof to prof, Univ BC, 57-65; SR PROF INORG CHEM, UNIV WESTERN ONT, 65-, HEAD DEPT, 67- Concurrent Pos: Mem chem grant selection comt, Nat Res Coun Can, 68-70, chmn, 70; chmn comt chem dept chmn, Ont Univs, 69-; consult, E I du Pont de Nemours & Co, Inc, Del, 70-71. Honors & Awards: Noranda Lectr Award, Chem Inst Can. Mem: The Chem Soc; Am Chem Soc; fel Chem Inst Can; fel Royal Soc Can. Res: Chemistry of inorganic fluorides and organometallic compounds; coordination chemistry; organometallic and coordination compounds. Mailing Add: Dept of Chem Univ of Western Ont London ON Can

CLARK, HOWARD CHARLES, JR, b Wichita, Kans, June 4, 37; m 57; c 5. GEOPHYSICS, GEOENVIRONMENTAL SCIENCE. Educ: Univ Okla, BS, 59; Stanford Univ, MS, 65, PhD(geophys), 67. Prof Exp: Teaching asst, Stanford Univ, 59-60; instr physics, Kansas City Jr Col, 61-62; res asst geophys, Stanford Univ, 65; instr geol, Menlo Col, 62-66; asst prof, 66-73, ASSOC PROF GEOL, RICE UNIV, 73- Mem: Geol Soc Am; Am Geophys Union; Soc Explor Geophys. Res: Marine geophysics; paleomagnetism. Mailing Add: Dept of Geol Rice Univ Houston TX 77001

CLARK, HOWARD GARMANY, b Birmingham, Ala, Feb 25, 28; m 47; c 3. BIOMEDICAL ENGINEERING. Educ: Howard Col, AB, 47; Univ Notre Dame, MS, 49; Univ Md, PhD(org chem), 54. Prof Exp: Chemist, Chemstrand Corp, 54-59; group leader, Peninsular Chem Res, Inc, 60; sr chemist, Camille Dreyfus Lab, Res Triangle Inst, 60-67; assoc prof textiles, Clemson Univ, 67-68; assoc prof biomed eng, 68-75, PROF BIOMED ENG & MECH ENG, DUKE UNIV, 75- Mem: AAAS; Am Chem Soc. Res: Medical applications of polymer materials; coagulation of blood. Mailing Add: Dept of Biomed Eng Duke Univ Durham NC 27706

CLARK, HOWARD SELBY, b Portsmouth, Ohio, Aug 29, 06; m 40; c 3. CHEMISTRY. Educ: Ohio State Univ, BA, 37. Prof Exp: Asst microanalyst, Res Lab, Merck & Co, Inc, 39-43; from assoc chemist to chemist, Ill Geol Surv, 43-53; DIR, CLARK MICROANAL LAB, 47- Mem: AAAS; Am Chem Soc; fel Am Inst Chemists. Res: Organic microanalytical chemistry for the determination of purity or identity. Mailing Add: Clark Microanal Lab PO Box 69 Urbana IL 61801

CLARK, HOWELL R, b Dexter, Ky, Aug 9, 26; m 58; c 2. INORGANIC CHEMISTRY, ANALYTICAL CHEMISTRY. Educ: Murray State Univ, BS, 56; Vanderbilt Univ, MS, 58, PhD(inorg chem), 70. Prof Exp: Chemist, Shell Oil Co, 58-63; from asst prof to assoc prof 63-73, PROF CHEM, MURRAY STATE UNIV, 73- Mem: Am Chem Soc. Res: Kinetics and mechanisms of fluoride hydrolysis; catalysis and synthetic processes based on hard and soft acid-base interactions. Mailing Add: Dept of Chem Murray State Univ Murray KY 42072

CLARK, HUGH, b Pawling, NY, Apr 15, 14; m 39; c 4. EMBRYOLOGY. Educ: Clark Univ, AB, 34; Univ Mich, PhD(zool), 41. Prof Exp: Asst zool, Univ Mich, 35-39; prof physiol, Des Moines Still Col Osteop, 39-45; assoc zool, Univ Iowa, 45-46, asst prof, 46-47; from asst prof to assoc prof, 47-59, PROF ZOOL, UNIV CONN, 59-, ASSOC DEAN, GRAD SCH, 63- Concurrent Pos: Dir mus reconstruct, Southern Ill State Norm Univ, 39. Mem: AAAS; Am Soc Zoologists; Soc Develop Biol. Res: Embryology of hemipenis in North American snakes; homogamy in the earthworm; respiration in reptile embryos; nitrogen metabolism in embryonic development; factors in embryonic differentiation. Mailing Add: Dept of Biol Sci Univ of Conn Storrs CT 06268

CLARK, HUGH KIDDER, b St Louis, Mo, Jan 22, 18; m 42; c 2. REACTOR PHYSICS. Educ: Oberlin Col, AB, 39; Cornell Univ, PhD(phys chem), 43. Prof Exp: Res assoc, Radio Res Lab, Harvard Univ, 43-45; res chemist, 45-62, RES ASSOC, E I DU PONT DE NEMOURS & CO, INC, 62- Mem: AAAS; Am Chem Soc; fel Am Nuclear Soc. Res: X-ray crystallography; radar direction finding; spinning of synthetic fibers; nuclear reactor physics; criticality safety. Mailing Add: Savannah River Lab E I du Pont de Nemours & Co Inc Aiken SC 29801

CLARK, IRWIN, b Boston, Mass, Apr 28, 18; m 49; c 4. BIOCHEMISTRY. Educ: Harvard Univ, AB, 39; Columbia Univ PhD(biochem), 50. Prof Exp: Head isotope dept, Merck Inst, 51-59; from asst prof to prof biochem, Col Physicians & Surgeons, Columbia Univ, 59-71; prof biochem & surg, Sch Med, Univ NC, 71-74; PROF SURG (BIOCHEM), COL MED & DENT NJ-RUTGERS MED SCH, 74- Concurrent Pos: Sr Fulbright scholar, Cambridge Univ, 50-51; res career develop award, 62-69; consult, Merck, Sharpe & Dohme, 59-60 & Squibb Inst Med Res, 60-62; vis prof, Rice Univ, 63; ed, Proc, Soc Exp Biol & Med, 67- Mem: Soc Exp Biol & Med; Am Soc Biol Chem; Am Inst Nutrit; Endocrine Soc; Brit Biochem Soc. Res: Metabolism of bone; hormonal effects on bone; inorganic biochemistry. Mailing Add: Dept of Surg Col Med Dent NJ-Rutgers Med Sch Piscataway NJ 08854

CLARK, JACK L, b Canton, Ill, Jan 15, 41; m 64; c 2. ANIMAL HUSBANDRY. Educ: Univ Ill, Urbana-Champaign, BS, 62; Mont State Univ, MS, 65; Univ Mo-Columbia, PhD(animal husb), 68. Prof Exp: ASST PROF ANIMAL HUSB, UNIV MO-COLUMBIA, 68- Mem: Am Soc Animal Sci. Res: Beef cattle nutrition; body composition studies with cattle and hogs. Mailing Add: 125 Mumford Univ of Mo Columbia MO 65201

CLARK, JAMES BENNETT, b Shamrock, Tex, Aug 24, 23; m 44; c 2. MICROBIAL PHYSIOLOGY. Educ: Univ Tex, BA, 47, MA, 48, PhD(bact), 50. Prof Exp: Res asst bact, Univ Tex, 46-48, res scientist bact genetics, 48-50; asst prof biol, Univ Houston, 50-51; ASST PROF MICROBIOL, UNIV OKLA, 51- Concurrent Pos: Consult, McDonnell Douglas Corp, Mo, 69- Mem: Am Soc Microbiol; Am Acad Microbiol. Res: Control mechanisms during morphogenesis in procaryotic cells. Mailing Add: Dept of Bot & Microbiol Univ of Okla Norman OK 73069

CLARK, JAMES D'ARGAVILLE, b Beith, Scotland, Jan 21, 01; nat US, m 30. PULP & PAPER TECHNOLOGY. Educ: Univ Cape Town, BSc, 18; Univ London, BSc, 22; Lawrence Col, PhD(chem), 41. Prof Exp: Apprentice, Masson Scott & Co, London, 22 & Hendon Paper Works Co, Ltd, 23; asst elec engr, Edward Lloyd, Ltd, 24-25; chief chemist, Bowater's Paper Mills, Ltd, 26-30; develop engr, Mead Corp, Ohio, 30-32; tech dir, Scott Paper Co, Pa, 32-35, eng & tech mgr, 35-39; res assoc, Inst Paper Chem, Lawrence Col, 41-42; CONSULT, 46-; HON RES ASSOC, WESTERN WASH STATE COL, 64- Concurrent Pos: Prof pulp & paper sci, Ore State Univ, 59-63. Honors & Awards: Gold Medal, Tech Asn Pulp & Paper Indust, 63; hon mem Can Pulp & Paper Tech Asn; Papermakers Asn Gt Brit & Ireland; Australian Pulp & Paper Indust Tech Asn. Res: Fundamental properties of pulps; cellulose-water relationships; pulp, paper and board manufacture; paper and board quality evaluation; manufacture of structural boards by dry processes and non-woven fabrics. Mailing Add: Chuckanut Point Bellingham WA 98225

CLARK, JAMES DERRELL, b Atlanta, Ga, Mar 8, 37; m 60; c 3. LABORATORY ANIMAL MEDICINE. Educ: Univ Ga, DVM, 61, MS, 64; Tulane Univ, DSc(microbiol), 69. Prof Exp: Asst prof lab animal med, Sch Med, Tulane Univ, 67-72; res asst vet med, 61-62, DIR LAB ANIMAL MED, COL VET MED, UNIV GA, 72- Concurrent Pos: Vet consult, Audubon Park Zoo, New Orleans, 66-72; consult, Am Asn Accreditation Lab Animal Care, 70- Mem: Am Vet Med Asn; Am Asn Lab Animal Sci; Am Asn Zoo Vets; Am Asn Vet Med Cols; Am Col Lab Animal Med. Res: Study the effects of mycotoxins upon the health of laboratory and domestic animals; development of new and innovative teaching methods. Mailing Add: Lab Animal Med Univ of Ga Col of Vet Med Athens GA 30602

CLARK, JAMES EDWARD, b Elkins, WVa, Nov 19, 26; m 49; c 3. INTERNAL MEDICINE. Educ: WVa Univ, AB, 48; Jefferson Med Col, MD, 52; Am Bd Internal Med, dipl, 59. Prof Exp: Intern, Jefferson Med Col Hosp, 52-53, resident med, 53-55, chief resident, 55-56, asst, 56-58, instr, 58-62, assoc, 62-64; asst prof clin med, Jefferson Med Col & dir artificial kidney unit & dialysis unit, Hosp, 64-68; CHIEF MED, CROZER-CHESTER MED CTR, 68-; PROF MED, HAHNEMANN MED COL, 69- Concurrent Pos: Vis lectr, Univ Pa; lectr, US Naval Hosp, Philadelphia; courtesy med staff, Pa Hosp, 62-; consult, Jefferson Med Col Med Serv, Philadelphia Gen Hosp & Riddle Mem Hosp; chmn nat adv coun, Nat Kidney Dis Found, 62-64; chmn pharm comt, Riddle Mem Hosp, 62-64; dir health serv, Swarthmore Col; med dir, The Franklin Mint, Franklin Ctr, Pa; mem bd dirs, Kidney Found Southeastern Pa, Inc. Mem: AAAS; fel Am Col Physicians; AMA; NY Acad Sci; Am Heart Asn. Res: Kidney disease and electrolyte metabolism. Mailing Add: Crozer-Chester Med Ctr 15th & Upland Sts Chester PA 19113

CLARK, JAMES HENRY, b Earlington, Ky, June 17, 32; m 57; c 2. REPRODUCTIVE PHYSIOLOGY, ENDOCRINOLOGY. Educ: Western Ky State Univ, BS, 59; Purdue Univ, MS, 66, PhD(endorrinol), 68. Prof Exp: Instr develop biol, Purdue Univ, 64-68, asst prof endocrinol, 70-73; ASSOC PROF CELL BIOL, BAYLOR COL MED, 73- Concurrent Pos: NIH fel biochem endocrinol, Univ Ill, Urbana, 68-70; NIH res grant, 70; Am Cancer Soc res grant, 72; mem rev panel for contraceptive devices, NIH, 73, mem endocrinol study sect, 74-; mem ed bd, Endocrinol, 74- & Biol of Reprod, 74- Mem: AAAS; Endocrine Soc; Soc Study Reprod. Res: Mechanism of steroid hormone action; the control of reproductive function and the control of hormone indued growth. Mailing Add: Dept of Cell Biol Baylor Col of Med Houston TX 77025

CLARK, JAMES PERRY, organic chemistry, see 12th edition

CLARK, JAMES WILLIAM, b Beaumont, Tex, Nov 28, 24; m 49; c 3. DENTISTRY. Educ: Univ Tex, DDS, 47; Univ Toronto, dipl, 52; Am Bd Periodont, dipl, 55. Prof Exp: Pvt pract, 47-64; assoc prof periodont, Med Ctr, Univ Ala, 64-69; PROF PERIODONT & CHMN DEPT, COL DENT, UNIV TENN, MEMPHIS, 69- Concurrent Pos: Co-founder, Rowe Smith Mem Found; ed-in-chief, Clin Dent; consult, Memphis Vet Admin Hosp. Mem: Am Acad Periodont. Res: Clinical investigation of etiology and treatment of periodontal disease. Mailing Add: Dept of Periodont Univ of Tenn Col of Dent Memphis TN 38163

CLARK, JASPER ARNOLD, b Kansas City, Mo, Sept 4, 09; m 31; c 2. BOTANY. Educ: William Jewell Col, AB, 31; Univ Okla, MS, 33; Univ Mo, PhD(bot), 56. Prof Exp: Asst bot, Univ Okla, 31-33 & Univ Mo, 34-35; prof biol, Hannibal-LaGrange Col, 35-38; chmn natural sci div, 65-75, PROF BIOL SCI, SOUTHWEST BAPTIST COL, 38- Concurrent Pos: Instr bot, Univ Mo, 56; NSF partic, Ore Inst Marine Biol, 58; res partic, Okla State Univ, 61; environ consult, Clark, Dietz & Assocs-Engrs, Inc, 72-75. Mem: Am Inst Biol Sci; Am Soc Plant Physiologists. Res: Drought resistance by plants; plant physiology. Mailing Add: 233 W College Bolivar MO 65613

CLARK, JEFFREY LEE, b Ft Wayne, Ind, Feb 6, 41. BIOCHEMISTRY. Educ: Univ Mich, BS, 63; Univ Chicago, PhD(biochem), 69. Prof Exp: ASST PROF BIOL CHEM, UNIV CALIF, IRVINE, 71- Concurrent Pos: NIH fel, Univ Calif, San Diego, 69-71. Res: Control of cell growth in culture; origin and nature of serum growth factors; roles of polyamines in regulation of cell growth; regulation of polyamine metabolism. Mailing Add: Dept of Biol Chem Univ of Calif Col of Med Irvine CA 92717

CLARK, JIMMY DORRAL, b Hobart, Okla, Feb 21, 39. MYCOLOGY. Educ: Wayne State Univ, BS, 65, MS, 66; Univ Calif, Berkeley, PhD(bot), 72. Prof Exp: Res assoc, Univ Calif, Berkeley, 72-74; ASST PROF BIOL, UNIV KY, 75- Mem: Mycol Soc Am; Bot Soc Am; Am Inst Biol Sci. Res: Genetic control of gametic and agametic cell fusion in the true slime molds and the associated physiological aspects of the compatible and incompatible reactions. Mailing Add: Sch of Biol Sci Univ of Ky Lexington KY 40506

CLARK, JOAN ROBINSON, b Madison, Wis, Jan 22, 20; wid. CRYSTALLOGRAPHY. Educ: Barnard Col, BA, 45; Johns Hopkins Univ, PhD(crystallog), 58. Prof Exp: Jr sci aide phys chem, Eastern Regional Res Lab, USDA, 43; math asst, Manhattan Proj, Carbide & Carbon Chem Corp, 45; jr proj engr develop eng, Brown Instruments Div, Minneapolis-Honeywell Regulator Corp, 46-49; asst physics, Inst for Cancer Res, 49-53; mathematician, 53-56, PHYSICIST CRYSTALLOG, US GEOL SURV, DEPT INTERIOR, 56- Concurrent Pos: Fulbright res scholar, Univ Sydney, 62; co-investr, Apollo Lunar Samples, NASA, 69- Mem: AAAS; fel Geol Soc Am; fel Am Mineral Soc; Am Crystallog Asn; Am Phys Soc. Res: X-ray diffraction studies of crystal structure; borates and other inorganics. Mailing Add: US Geol Surv Washington DC 20242

CLARK, JOE HALLER, b Nashville, Ark, Nov 26, 13; m 47; c 3. RESEARCH ADMINISTRATION. Educ: Univ Tex, AB, 35, AM 37; Univ Ill, PhD(org chem), 40. Prof Exp: Res chemist, Eastman Kodak Co, 40-42 & Am Cyanamid Co, 42-58; dir tech info serv, 58-71, dir med res coord, 71-75, DIR ADMIN SERV, RES & DEVELOP, LEDERLE LABS, 75- Mem: AAAS; Am Chem Soc; Am Soc Info Sci; Drug Info Asn. Res: Chemistry of marihuana; synthetic organic chemistry; technical information; pharmaceutical research. Mailing Add: 25 Clinton Pl Woodcliff Lake NJ 07675

CLARK, JOHN, b Chicago, Ill, July 26, 09; m 57. PETROLOGY. Educ: Univ Ill, BS, 31; Univ Pittsburgh, MS, 32; Princeton Univ, AM, 34, PhD(geol), 35. Prof Exp: Docent, Princeton Univ, 35-37; assoc prof geol & cur vert paleont, Tex Tech Col, 37; instr & cur, Univ Colo, 37-39, temp dir mus, 38; asst cur vert paleont, Carnegie Mus, 39-46, assoc cur, 46-47, cur phys geol, 47-48; vpres, Cent Asiatic Res Found, 48-53; res assoc, Yellowstone-Bighorn Res Asn, 53-56; cur geol, Cleveland Mus Natural Hist, 56-57; asst prof geol eng, SDak Sch Mines & Technol, 57-61; dir res, Black Hills Clay

Prods Co, 61-63; assoc cur sedimentary petrol, Field Mus Natural Hist, 63-74, cur, 74; RETIRED. Concurrent Pos: Lectr, Univ Mich, 49-50 & Univ Chicago, 53. Mem: Soc Vert Paleont; fel Geol Soc Am; fel Paleont Soc. Res: Sedimentary petrology; fluvial sedimentation; desert geologic processes; vertebrate paleontology. Mailing Add: 3B Colony House Apts Franklin TN 37064

CLARK, JOHN BELL, physical chemistry, biochemistry, see 12th edition

CLARK, JOHN DESMOND, b London, Eng, Apr 10, 16; m 38; c 2. ARCHAEOLOGY, ETHNOGRAPHY. Educ: Cambridge Univ, BA, 37, MA, 42, PhD(archaeol), 50, ScD(prehist archaeol), 75. Prof Exp: Dir, Rhodes-Livingstone Mus, Northern Rhodesia, 37-61; PROF ANTHROP, UNIV CALIF, BERKELEY, 61-, CUR PREHIST ARCHAEOL, LOWIE MUS ANTHROP, 75- Concurrent Pos: Dir antiquities, Northern Rhodesia Nat Monuments Comn, 48-61; adv, Dept Archaeol, Dundo Mus, Angola, 51-; Wenner-Gren Found Anthrop Res grants, Cent Africa, 54 & 62, Kalambo Falls Prehist Site, 56, Exten Media Film Unit Univ, Univ Calif, 65-68, Berkeley Off, Comn Nomenclature, Pan-African Cong Prehist, 68- & Conf Probs Nubian Prehist, Dallas, 70; mem permanent comt, Pan-African Cong Prehist & Study Quaternary, 55-, ed, Proc 3rd Cong, 59, vpres, Cong, 59-, chmn, Sect Prehist Archaeol, 63 & 67, chmn, Comt Atlas African Prehist & mem, Comn Nomenclature, 63- & joint ed, Bull, 67-; Oxford Univ Boise Fund grant, Kalambo Falls Prehist Site, 59-60; NSF grants, Kalambo Falls Prehist Site, 62-64, Lake Malawi Rift, 65-70, Cahora Bassa, Mozambique, 67 & Ethiopia, 74-75; Univ Calif, Los Angeles, African Studies Ctr grants, Kalambo Falls Prehist Site, 62, Latamne Site, Syria, 63 & Repub Niger, 70 & Near Eastern Ctr grant, Latmne Site, 65; Univ Calif, Berkeley, Inst Social Sci grant edge damage study, 67; consult antiquities in Malawi, UNESCO, 68-72; mem, Viking Fund Medal Comt, Wenner-Gren Found, 70-; Brit Acad grant, Air & Tenere, 70, Sudan, 73. Mem: Fel AAAS; fel Am Acad Arts & Sci; fel Am Anthrop Asn; fel African Studies Asn; Am Quaternary Asn. Res: Prehistoric archaeology and cultural evolution of early man in Africa, particularly south of the Sahara; origins of domestication and agriculture in Sahara. Mailing Add: Dept of Anthrop Univ of Calif Berkeley CA 94720

CLARK, JOHN FRANCIS BULLOCK, b Perth, NB, Apr 20, 11; m 41; c 2. CLINICAL CHEMISTRY, ANALYTICAL CHEMISTRY. Educ: Acadia Univ, BS, 34; Univ Mich, MA, 34. Hon Degrees: ScD, Gt Lakes Col, 58. Prof Exp: Technician med, Gen Hosp, NB, 34-35; teacher, Pub Schs, Mich, 36-41; anal chemist, Chevrolet Labs, Mich, 41-46; prof chem & head dept, Col Pharm, Detroit Inst Technol, 46-51; anal chemist, Chevrolet Lab, Mich, 51-54; asst prof anal chem, Broome Tech Inst, NY, 54-56; prof clin & biol sci, Broome Tech Community Col, 56-73, head dept, 57-67, dir div health sci & fed grants adminr, 67-73; RETIRED. Concurrent Pos: Instr, Gt Lakes Col, 41-45; consult, Detroit Clin Lab, 41 & Detroit Mfg Lab, 41-; dir eve div chem, Detroit Inst Technol, 46-51; secy-treas, NY State Registr Med Technol, Inc, 61-68, exec secy, 68-75, emer secy, 75-; dir area, Fed Water Qual Proj, 64; chmn educ comt, Tri County Respiratory Asn, 69-73; chmn, Broome County Health Planning Comt, 70-72. Mem: Am Asn Bioanalysts; Am Pub Health Asn; Asn Schs Allied Health Prof; NY Acad Sci. Res: Promotion of better instruction in basic sciences at junior college level. Mailing Add: c/o Spanish Lakes 12 Margarita Ln Port St Lucie FL 33452

CLARK, JOHN FULMER, b Reading, Pa, Dec 12, 20; m 43, 74; c 2. PLANETARY SCIENCES. Educ: Lehigh Univ, BS, 42, EE, 47; George Washington Univ, MS, 46; Univ Md, PhD(physcis), 56. Prof Exp: Engr, US Naval Res Lab, 42-47; asst prof elec eng, Lehigh Univ, 47-48; unit head ionospheric physics, US Naval Res Lab, 48-54, br head atmospheric elec, 54-58; dir physics & astron progs, NASA, 58-63, dep assoc adminr space sci & appln & chmn space sci steering comt, 63-65, DIR, GODDARD SPACE FLIGHT CTR, NASA, 65- Concurrent Pos: Lectr, George Washington Univ, 56-58, res assoc atmospheric physics, Grad Coun, 60-66; mem, Joint Comt Atmospheric Elec, Int Union Geod & Geophys, 60-63; mem, US Nat Comt, Int Sci Radio Union, 62-64; indust & prof adv coun, Pa State Univ, 63-65; mem vis comt physics, Lehigh Univ, 66-74; mem, Comt Fed Labs, 71-; mem, Md Gov Sci Adv Coun, 72- Honors & Awards: NASA Medals for Distinguished Serv, Outstanding Leadership & Except Serv; Collier Trophy, Nat Aeronaut Asn, 75. Mem: Sr mem Inst Elec & Electronics Engr; fel Am Astron Soc; Am Geophys Union; Am Inst Aeronaut & Astronaut; Sigma Xi. Res: Space applications—earth resources survey, meteorology, communications; upper atmospheric physics; ionospheric physics; atmospheric potential gradient and polar conductivities; algebraic ring theory; radar beacon development. Mailing Add: Goddard Space Flight Ctr NASA Greenbelt MD 20771

CLARK, JOHN HARLAN, b Helena, Mont, Aug 30, 48; m 67; c 1. FISHERIES. Educ: Carrol Col, BA, 71; Colo State Univ, MS, 74, PhD(fisheries), 75. Prof Exp: FISHERIES BIOLOGIST III, ALASKA DEPT FISH & GAME, 75- Mem: Am Fisheries Soc; Nat Audubon Soc; Sigma Xi. Res: Production capabilities of lakes used for nursery areas by sockeye salmon. Mailing Add: 333 Raspberry Rd Anchorage AK 99502

CLARK, JOHN JEFFERSON, b Shrewsburg, Mass, Dec 30, 22; m 43; c 3. VETERINARY PATHOLOGY. Educ: Univ Ga, DVM, 53; Univ Minn, PhD(vet path), 59. Prof Exp: Res fel vet path, Univ Minn, 53-57; res assoc, 57-70, RES SECT HEAD VET PATH, UPJOHN CO, 70- Mem: Am Vet Med Asn; Am Soc Clin Path. Res: Cancer embryology, histology, chemotherapy and immunity; immunology, bacteriology and pathology of leptospirosis and other infectious diseases; parasitology of domesticated animals; hematology; clinical chemistry. Mailing Add: Upjohn Co Kalamazoo MI 49001

CLARK, JOHN KAPP, b Williamsport, Pa, July 17, 14; m 42; c 5. PHYSIOLOGY, PHARMACOLOGY. Educ: Trinity Col, BS, 36; Univ Pa, MD, 40. Hon Degrees: DSc, Trinity Col, 64. Prof Exp: Asst instr med, 43-47, from instr to asst prof, 47-54, ASSOC PROF MED, SCH MED, UNIV PA, 54-, CHIEF REGIONAL MED PROG, 67- Concurrent Pos: Univ res fel pharmacol, Sch Med, Univ Pa, 46-47, NIH res fel, 47-49; dir res, Smith Kline & French Labs, 51-61, vpres res & develop, 61-66; attend physician & chief renal sect, Hosp Univ Pa. Mem: AAAS; fel Am Col Physicians; Am Fedn Clin Res; AMA; Am Physiol Soc. Res: Clinical and renal physiology; fluid and electrolyte metabolism; health care delivery. Mailing Add: Univ Pa Unit Regional Med Prog 36th & Hamilton Walk/GA Philadelphia PA 19174

CLARK, JOHN MAGRUDER, JR, b Ithaca, NY, June 10, 32; m 57; c 2. BIOCHEMISTRY. Educ: Cornell Univ, BS, 54; Calif Inst Technol, PhD(biochem), 58. Prof Exp: From instr to asst prof, 58-66, ASSOC PROF BIOCHEM, UNIV ILL, URBANA, 66- Mem: Am Soc Biol Chemists; Am Chem Soc. Res: Enzymology related to protein biosynthesis. Mailing Add: Dept of Biochem Univ of Ill Urbana IL 61801

CLARK, JOHN S, b Tientsin, N China, Oct 1, 25; m 51; c 2. SOIL CHEMISTRY. Educ: Univ BC, BSA, 48, MSA, 50; Cornell Univ, PhD(soil chem), 53. Prof Exp: Res officer soil chem, Can Dept Agr, 53-55; asst prof, Cornell Univ, 55-56 & Univ BC, 56-60; res officer, 60-69, DIR SOIL RES INST, CAN DEPT AGR, 69- Res: Ion equilibria and phosphate fixation in soils; mineralogical and chemical properties of soil clays. Mailing Add: Soil Res Inst Can Dept of Agr Ottawa ON Can

CLARK, JOHN WALTER, b Lockhart, Tex, Apr 7, 35. THEORETICAL NUCLEAR PHYSICS, THEORETICAL ASTROPHYSICS. Educ: Univ Tex, BS, 55, MA, 57; Wash Univ, PhD(physics), 59. Prof Exp: Res assoc physics, Wash Univ, 59; NSF fel, Princeton Univ, 59-61; assoc res scientist, Denver Div, Martin Co, 61; NATO fel, 62-63; from asst prof to assoc prof, 63-72, PROF PHYSICS, WASH UNIV, 72- Concurrent Pos: Alfred P Sloan Found fel, 65-67; guest prof, Swed Univ Abo, 71-72. Mem: Fel Am Phys Soc. Res: Quantum mechanics of many-body systems; nuclear interactions and nuclear structure; hypernuclear physics; neutron stars; quantum fluids and solids; theoretical neurophysics. Mailing Add: Dept of Physics Wash Univ St Louis MO 63130

CLARK, JOHN WHITCOMB, b Walkerton, Ind, Aug 14, 18; m 61. RADIOLOGY. Educ: Harvard Univ, MD, 43. Prof Exp: From instr to prof radiol, Univ Ill, 48-70, PROF RADIOL, RUSH MED COL, 70- Concurrent Pos: Resident, Presby-St Luke's Hosp, 46-49, from asst attend radiologist to attend radiologist, 49-; assoc scientist, Argonne Nat Lab, 52-56, consult, 56-62. Mem: AMA; fel Am Col Radiol; Radiation Res Soc; Radiol Soc NAm; Am Roentgen Ray Soc. Res: Clinical radiology; radiobiology. Mailing Add: Dept of Radiol Presby-St Luke's Hosp 1753 W Congress Pkwy Chicago IL 60612

CLARK, JOSEPH CLYDE, geology, see 12th edition

CLARK, JULIA BERG, b Moline, Ill, June 7, 40; m 63; c 2. BIOCHEMISTRY, PHARMACOLOGY. Educ: Radcliffe Col, BA, 62; Ind Univ, PhD(biochem), 66. Prof Exp: Res assoc, 70-71, ASST PROF PHARMACOL, SCH MED, IND UNIV, INDIANAPOLIS, 71- Concurrent Pos: USPHS fels, Harvard Univ, 66-67, NIH, 67-68, Ind Univ, 70-71; Nat Inst Arthritis & Metab Dis fel, 67-69. Res: Enzymology; endocrinology. Mailing Add: Dept of Pharmacol Ind Univ Med Sch Indianapolis IN 46202

CLARK, JUNIUS MANSON, b Vicksburg, Miss, Sept 5, 40; m 64; c 2. IMMUNOLOGY, MICROBIOLOGY. Educ: Tex A&M Univ, BS, 63; Univ Tex Med Br Galveston, PhD(microbiol), 67. Prof Exp: ASST PROF MICROBIOL, JEFFERSON MED COL, 69- Concurrent Pos: USPHS fel, Sch Med, Johns Hopkins Univ, 67-69. Mem: Reticuloendothelial Soc; Am Soc Microbiol. Res: Inflammation and the role of the mast cell and Basophil in resistance to infectious and noxious agents. Mailing Add: Jefferson Med Col 1020 Locust St Philadelphia PA 19107

CLARK, KENNETH COURTRIGHT, b Austin, Tex, Sept 30, 19; m 47; c 2. Educ: Univ Tex, BA, 40; Harvard Univ, AM, 41, PhD(physics), 47. Prof Exp: Tutor physics, Harvard Univ, 41-42, spec res assoc, Nat Defense Res Comt Proj, Electro-Acoustic Lab, 41-45, instr physics, 47-48; from asst prof to assoc prof, 48-60, PROF PHYSICS, UNIV WASH, 60- Concurrent Pos: Actg head physics div & res assoc prof, Geophys Inst, Univ Alaska, 57-58; consult, US Agency Int Develop, India, 64 & 66 & Battelle Northwest Labs, Battelle Mem Inst, 71-; prog dir aeron, NSF, Washington, DC, 69-70; mem adv comt, Geophys Inst, Univ Alaska, 73- Mem: Fel Am Phys Soc; fel Optical Soc Am; Am Asn Physics Teachers; Am Geophys Union. Res: Auroral physics; afterglow spectra; laboratory aeronomy; extreme ultraviolet spectroscopy; optical excitation processes in upper atmosphere. Mailing Add: Dept of Physics Univ of Wash Seattle WA 98195

CLARK, KENNETH FREDERICK, b Liverpool, Eng, Apr 4, 33; m 62. ECONOMIC GEOLOGY. Educ: Univ Durham, BSc, 56, Univ NMex, MS, 62, PhD(geol), 66. Prof Exp: Geologist, Anglo Am Corp of SAfrica, Ltd, 56-60; asst prof geol sci, Cornell Univ, 66-71; ASSOC PROF GEOL, UNIV IOWA, 71- Concurrent Pos: Consult, Mexican Govt, 69- Mem: Geol Soc Am; Am Inst Mining, Metall & Petrol Eng. Res: Exploration and development of mineral resources; geophysical and geochemical exploration. Mailing Add: Dept of Geol Univ of Iowa Iowa City IA 52242

CLARK, KERRY BRUCE, b Woodbury, NJ, Aug 22, 45; m 66; c 1. INVERTEBRATE ZOOLOGY, MARINE ECOLOGY. Educ: Rutgers Univ, BA, 66; Univ Conn, MS, 68, PhD(invert zool), 71. Prof Exp: Asst prof, 71-75, ASSOC PROF BIOL SCI, FLA INST TECHNOL, 75- Mem: Ecol Soc Am; Am Inst Biol Sci; Am Soc Zoologists; Sigma Xi. Res: Taxonomy and ecology of opisthobranch molluscs; ecology of fouling communities; structure of benthic marine communities. Mailing Add: Dept of Biol Sci Fla Inst of Technol Melbourne FL 32901

CLARK, LARRY P, b Chicago, Ill, June 18, 36; m 59; c 2. ORGANIC CHEMISTRY, MEDICINAL CHEMISTRY. Educ: Univ Mich, BS, 58; Univ Notre Dame, PhD(org chem), 66. Prof Exp: Control chemist, Abbott Labs, 58-62 & Miles Labs, 62-63; res assoc org chem, Col Pharm, Univ Mich, 66-67; asst prof chem, Univ Pittsburgh, Bradford, 67-69; group leader drug anal, Pharmaco Inc, Schering Corp, 69-71, GROUP LEADER DRUG ANAL, PLOUGH, INC, SCHERING-PLOUGH CORP, 71- Mem: AAAS; Am Chem Soc. Res: Analysis of drugs in dosage forms and in vivo; determination of drug stability. Mailing Add: Res Dept Plough Inc 3022 Jackson Ave Memphis TN 38101

CLARK, LEIGH BRUCE, b Seattle, Wash, Sept 9, 34; c 1. SPECTROCHEMISTRY. Educ: Univ Calif, Berkeley, BS, 57; Univ Wash, PhD(chem), 63. Prof Exp: Asst prof, 64-72, ASSOC PROF CHEM, UNIV CALIF, SAN DIEGO, 72- Res: Molecular spectroscopy; reflection spectroscopy of molecular crystals. Mailing Add: Dept of Chem Univ of Calif at San Diego La Jolla CA 92037

CLARK, LELAND CHARLES, JR, b Rochester, NY, Dec 4, 18; m 39; c 4. BIOCHEMISTRY. Educ: Antioch Col, BS, 41; Univ Rochester, PhD(biochem), 44. Prof Exp: Chmn biochem dept, Fels Res Inst, 44-58; asst prof biochem, Antioch Col, 44-56, prof, 56-58; from assoc prof to prof surg, Med Ctr, Univ Ala, 58-68; PROF RES PEDIAT, CHILDREN'S HOSP RES FOUND, MED COL, UNIV CINCINNATI, 68- Concurrent Pos: Sr res assoc surg & pediat, Univ Cincinnati, 55-58; consult, Wright-Patterson AFB, 56-58 & NIH, 61-; NIH res career award, 62-68; vis prof, Cardiovasc Res Inst San Francisco, 67; Edmund Hall lectr, Univ Louisville Chap Sigma Xi, 67; fel, coun cerebrovascular dis, Am Heart 67-; ed, Symp Oxygen Transport. Mem: AAAS; Am Chem Soc; fel Am Inst Am Physiol Soc; Am Heart Asn. Res: Vitamin, steroid and oxygen metabolism; polarography; cardiovascular disease; hydrogen and oxygen electrode in diagnosis; ion exchange resins in biology; glucose electrode; surgical monitoring; intermediary metabolism and synthesis of psychotomimetric drugs; fluorocarbon liquid breathing; artificial blood. Mailing Add: Div of Neurophysiol Children's Hosp Res Found Elland Ave & Cincinnati OH 45229

CLARK, LEWIS EDWIN, agronomy, see 12th edition

CLARK, LEWIS JESSE, inorganic chemistry, deceased

CLARK, LINCOLN DUFTON, b Andover, Mass, Jan 18, 23; m 49; c 2. MEDICINE. Educ: Harvard Univ, MD, 47. Prof Exp: Asst instr psychiat, Med Sch, Harvard Univ,

49-50, asst physician, 51-53; from asst prof to assoc prof, 55-64, PROF PSYCHIAT & DIR BEHAV SCI LAB, COL MED, UNIV UTAH, 64- Concurrent Pos: NIMH res career award, 63-67 & res scientist award, 67-72; asst physician, Mass Gen Hosp, 51-53; sci assoc, Roscoe B Jackson Mem Lab; chmn adv comt preclin psychopharmacol, NIMH, 61-66, res scientist, 68- Mem: Am Psychiat Asn; Am Col Neuropsychopharmacol. Res: Experimental psychiatry; psychopharmacology; animal behavior. Mailing Add: 50 N Medical Dr Salt Lake City UT 84110

CLARK, LLOYD ALLEN, b North Battleford, Sask, Mar 17, 32; m 55; c 4. ECONOMIC GEOLOGY, GEOCHEMISTRY. Educ: Univ Sask, BE, 54, MSc, 55; McGill Univ, PhD(geol), 59. Prof Exp: Fel geochem, Carnegie Inst Geophys Lab, 58-60; from asst prof to assoc prof econ geol, McGill Univ, 60-70; CHIEF GEOCHEM RES & LAB DIV, KENNECOTT EXPLOR, INC, 70- Concurrent Pos: Nat Res Coun Can res grants, 60-70, fel, 65-66; guest investr, Univ Tokyo & Univ Florence, 65-66. Mem: Soc Econ Geol; fel Geol Asn Can; fel Mineral Asn Can; Can Inst Mining & Metall; Soc Geol Appl Mineral Depostis. Res: Phase equilibrium and related studies in synthetic systems; studies of naturally occurring minerals and ores to yield quantitative information about environment of ore formation. Mailing Add: Kennecott Explor Inc 2300 W 1700 South Salt Lake City UT 84104

CLARK, LORIN DELBERT, geology, see 12th edition

CLARK, LOUIS WATTS, b Comanche, Okla, July 2, 16; m 42; c 11. PHYSICAL ORGANIC CHEMISTRY. Educ: Univ Okla, BS, 42, MS, 44; Kans State Univ, PhD(chem), 50. Prof Exp: Asst prof chem, Utah State Univ, 50-51; res chemist, Spreckels Sugar Co, Calif, 51; asst prof chem, St Mary's Col, Calif, 51-52; assoc prof, Col St Elizabeth, 52-53; prof, Panhandle Agr & Mech Col, 53-54; assoc prof, St Joseph Col, Md, 54-58, St Bonaventure, 58-59 & St Mary of the Plains Col, 59-60; PROF CHEM, WESTERN CAROLINA UNIV, 60- Concurrent Pos: Consult chemist, Spreckels Sugar Co, 51-52 & Vick Chem Co, 53. Mem: Am Chem Soc. Res: Chemical kinetics; mechanisms. Mailing Add: Dept of Chem Western Carolina Univ Cullowhee NC 28723

CLARK, MALCOLM A, b Ottawa, Ont, May 30, 25; m 47; c 2. SPACE PHYSICS, AEROSPACE TECHNOLOGY. Educ: Queen's Univ, Ont, BSc, 47, MSc, 48; Mass Inst Technol, PhD(physics), 52. Prof Exp: Mem tech staff acoustics res, Bell Tel Labs, Inc, 52-54; assoc res officer, Physics Div, Atomic Energy Can Ltd, 54-62; staff scientist, Space Physics Lab, 62-69, systs eng dir, Space Exp Support Prog, 69-72, GROUP DIR, SPACE TEST TECHNOL DIV, AEROSPACE CORP, 72- Res: Cosmic ray balloon experiments; beta and gamma ray spectroscopy; short nuclear lifetimes; night airglow and aurora; ultraviolet dayglow; far ultraviolet sky radiance; experimental spacecraft. Mailing Add: Off for Technol Aerospace Corp Box 95085 Los Angeles CA 90045

CLARK, MALCOLM JOHN ROY, b Bournemouth, Eng, May 22, 44; m 68. ENVIRONMENTAL CHEMISTRY. Educ: Univ Victoria, BSc, 66; Univ NB, PhD(chem), 71. Prof Exp: BR ENVIRON CHEMIST, BC WATER RESOURCES SERV, 71- Concurrent Pos: Vis scientist, Univ Victoria, 75- Mem: Am Chem Soc; Am Fisheries Soc; Am Soc Limnol & Oceanog; Am Soc Testing & Mat; Can Inst Chem. Res: Investigation of various problems in pollution and environmental chemistry, particularly regarding metals, color and dissolved gases; environmental data storage and retrieval. Mailing Add: 336 Foul Bay Rd Victoria BC Can

CLARK, MALCOLM MALLORY, b Palo Alto, Calif, Sept 21, 31; m 57; c 5. GEOLOGY. Educ: Univ Calif, Berkeley, BS, 57; Stanford Univ, PhD(geol), 67. Prof Exp: Engr, Temescal Metall Corp, 57-60; lab mgr, Dumont Mfg Co, 60-61; mfg mgr, Monitor Plastics Co, 61-63; GEOLOGIST, US GEOL SURV, 67- Mem: Geol Soc Am; Am Soc Photogram; Glaciol Soc; Am Soc Testing & Mat. Res: Glacial geology; glaciation of the Sierra Nevada, California; weathering of granite; geology of active faults; remote sensing. Mailing Add: US Geol Surv 345 Middlefield Rd Menlo Park CA 94025

CLARK, MARION THOMAS, b Hapeville, Ga, Aug 20, 17; m 42; c 3. CHEMISTRY. Educ: Emory Univ, AB, 38, AM, 39; Univ Va, PhD(org chem), 46. Prof Exp: Instr chem, Emory Jr Col, 39-43; asst, Univ Va, 43-46; assoc prof, Birmingham-Southern Col, 46-48; asst prof, Emory Univ, 48-51; asst & actg chmn, Univ Rel Div, Oak Ridge Inst Nuclear Studies, 51-53; assoc prof chem, Emory Univ, 53-63; PROF CHEM, AGNES SCOTT COL, 63- Mem: Am Chem Soc; Sigma Xi. Res: Synthesis of quinoline methanols and aminomethyl benzyl alcohols as possible antimalarials. Mailing Add: Dept of Chem Agnes Scott Col Decatur GA 30030

CLARK, MARY ELEANOR, b San Francisco, Calif, Apr 28, 27. ZOOLOGY, COMPARATIVE PHYSIOLOGY. Educ: Univ Calif, Berkeley, AB, 49, MA, 51, PhD(zool), 60. Prof Exp: USPHS fel zool, Bristol Univ, 61-63; Sci Res Coun Gt Brit fel, Bristol & Newcastle Univs, 63-66; vis prof zool, Univ Lund, 67; res asst organismic biol, Univ Calif, Irvine, 67-68; NSF fel environ health eng, Calif Inst Technol, 68-69; asst prof, 69-70, ASSOC PROF BIOL, SAN DIEGO STATE UNIV, 70- Mem: AAAS. Res: Tissue culture of normal and malignant mouse fibroblasts; polychaete neurosecretion; monoamine histochemistry; regeneration; biochemistry of amino acids in marine invertebrates; polychaete and oligochaete osmoregulation; marine pollution and kelp-bed ecology. Mailing Add: Dept of Biol San Diego State Univ San Diego CA 92115

CLARK, MARY, JANE, b McKeesport, Pa, Sept 18, 25. BIOCHEMISTRY. Educ: Univ Pittsburgh, BS, 47, PhD(biochem), 57. Prof Exp: Jr chemist, Koppers Co, Inc, Pa, 52-54; res assoc biochem, Sch Pub Health, Univ Pittsburgh, 54-58; res assoc biochem, Col Physicians & Surgeons, Columbia Univ, 58-64; asst prof, NY Med Col, Flower & Fifth Ave Hosps, 64-68; ASSOC PROF BIOCHEM, JERSEY CITY STATE COL, 68-, ASSOC PROF CHEM & CHMN DEPT, 73- Mem: Am Chem Soc. Res: Synthesis of peptides and their enzymic hydrolysis; aromatic biosynthesis in bacteria; general intermediary metabolism of amino acids and derivatives; coenzymes in intermediary metabolism; folic acid and leukemia. Mailing Add: Dept Biochem Jersey City State Col 2039 Kennedy Blvd Jersey City NJ 07305

CLARK, MARY MARGARET, b Amarillo, Tex, Jan 9, 25. CULTURAL ANTHROPOLOGY. Educ: Southern Methodist Univ, BS, 45; Univ Calif, PhD(anthrop), 57. Prof Exp: Res anthropologist, USPHS, DC, 57-58; field anthropologist & lectr pub health, Sch Pub Health, Univ Calif, 58-59; res anthropologist, Langley Porter Neuropsychiat Inst, 60-73, lectr psychiat, Sch Med, 66-70, PROF ANTHROP, SCH MED, UNIV CALIF, SAN FRANCISCO, 70-, CHMN MED ANTHROP PROG, 73- Concurrent Pos: Lectr, San Francisco State Col, 67; vis prof, Univ Calif, Berkeley, 67-68; prin investr, Nat Inst Child Health & Human Develop Res Grant, 67- Mem: Fel AAAS; fel Am Anthrop Asn; fel Soc Appl Anthrop; fel Geront Soc (vpres, 74-75). Res: Applied anthropology in medicine and public health; social gerontology and psychiatry; culture and personality of adult life. Mailing Add: Univ Calif Sch of Med 1320 Third Ave Med Anthrop Prog San Francisco CA 94143

CLARK, MELVILLE, b Syracuse, NY, Dec 19, 21. PHYSICS, ELECTRICAL ENGINEERING. Educ: Mass Inst Technol, SB, 43; Harvard Univ, AM, 47, PhD(physics), 49. Prof Exp: Mem staff microwaves, Radiation Lab, Mass Inst Technol, 42-45; mem staff electronics, Los Alamos Sci Lab, 45-46; mem staff reactors, Brookhaven Nat Lab, 49-53; mem staff neutronics, Radiation Lab, Univ Calif, 53-55; assoc prof nuclear eng, Mass Inst Technol, 55-62; sr eng specialist, Appl Res Lab, Sylvania Elec Prod, Inc, 62-64; sr consult scientist, Res & Advan Develop Div, Avco Corp, 64-65; pres, Meldor Corp, 65-67; sr scientist, NASA Electronics Ctr, 67-70; sr develop engr, Thermo Electron Eng Corp, 70-73; MEM STAFF, COMBUSTION ENG, 73- Concurrent Pos: Lectr, United Shoe Mach Corp, 56; consult, Raytheon Mfg Co, 55-58 & Arthur D Little, 57-58; pres, Melville Clark Assocs, 55-; vpres, Clark Music Co, 57-60; dir, 416 S Salina St Corp, 57-60; vpres & dir, Meldor Corp, 60-65. Mem: AAAS; Am Phys Soc; Acoust Soc Am; Inst Elec & Electronics Engrs; Am Inst Physics. Res: Microwave radiation; quantum mechanics; nuclear and plasma physics; reactor engineering; neutral particle transport; musical acoustics; ionospheric propagation; speech research; electric space propulsion; auditory perception. Mailing Add: 8 Richard Rd Wayland MA 01778

CLARK, MERVIN LESLIE, b Baltimore, Md, May 18, 21; m 49; c 4. MEDICINE. Educ: Va Polytech Inst, BS, 42; Northwestern Univ, MD, 48. Prof Exp: From asst prof to assoc prof med, 62-69, actg dir div clin pharmacol, 70-75, PROF MED, SCH MED, UNIV OKLA, 69- Concurrent Pos: Res fel, Exp Therapeut Unit, Sch Med, Univ Okla, 55-56; chief med serv & dir res unit, Cent State Griffin Mem Hosp, Norman, Okla, 56-; mem sci rev panel, Drug Interactions Eval Prog, Am Pharmaceut Asn, 74. Mem: Am Soc Pharmacol & Exp Therapeut; Am Col Neuropsychopharmacol. Res: Clinical pharmacology and therapeutics; psychopharmacology. Mailing Add: Sch of Med Univ of Okla Oklahoma City OK 73104

CLARK, MILTON B, b Cleveland, Ohio, Dec 22, 17; m 57; c 1. PHYSICAL CHEMISTRY. Educ: Western Reserve Univ, AB, 41, MS, 52, PhD(phys chem), 56. Prof Exp: Inorg chem analyst, Harshaw Chem Co, 46-53, develop engr electrochem, Nat Carbon Co Div, 56-60, RES CHEMIST, PARENA TECH CTR, UNION CARBIDE CORP, 60- Mem: Electrochem Soc. Res: Electrochemistry; primary and secondary cells. Mailing Add: 6603 Greenbrier Dr Brecksville OH 44141

CLARK, MINOR E, b Waddy, Ky, Aug 5, 13; m 38. FISHERIES. Educ: Eastern State Col, BS, 34. Prof Exp: Chief biologist, Div Game & Fish, Ky State Dept Fish & Wildlife, 37-40, supt fisheries, 40-52, dir div fisheries, 52-53, from asst comnr to comnr, 53-71; RETIRED. Mem: Am Fisheries Soc; Wildlife Soc. Mailing Add: 411 Wapping St Frankfort KY 40601

CLARK, NANCY BARNES, b Hamden, Conn, July 1, 39; m 61. COMPARATIVE ENDOCRINOLOGY, COMPARATIVE PHYSIOLOGY. Educ: Mt Holyoke Col, BA, 61; Columbia Univ, MA, 62, PhD(endocrinol), 65. Prof Exp: Asst prof zool, 65-70, ASSOC PROF BIOL, UNIV CONN, 70- Mem: AAAS; Am Soc Zool; Endocrine Soc. Res: Parathyroid function, and calcium and phosphate regulation in non-mammalian vertebrates; comparative studies of thyroid function. Mailing Add: Biol Sci Group Univ of Conn Storrs CT 06268

CLARK, NATHAN EDWARD, b Milford, Conn, Feb 26, 40. MARINE METEOROLOGY, FISHERIES. Educ: Brown Univ, ScB, 62; Mass Inst Technol, PhD(oceanog), 67. Prof Exp: METEOROLOGIST, SOUTHWEST FISHERIES CTR, NAT MARINE FISHERIES SERV, NAT OCEANIC & ATMOSPHERIC ADMIN, 67- Concurrent Pos: Res assoc, Scripps Inst Oceanog, 74- Mem: Am Meteorol Soc; Am Geophys Union. Res: Large scale air-sea heat transfer processes and fluctuations in the North Pacific Ocean; effects of large scale changes of ocean and atmosphere on northeastern Pacific fisheries. Mailing Add: Southwest Fisheries Ctr Box 271 La Jolla CA 92038

CLARK, NERI ANTHONY, b New Haven, Conn, Apr 19, 18; m 43; c 1. AGRONOMY. Educ: Univ Md, BS, 54, PhD(agron), 59. Prof Exp: From asst prof to assoc prof, 58-70, PROF AGRON, UNIV MD, COLLEGE PARK, 70- Mem: Am Soc Agron. Res: Agronomy; forage management. Mailing Add: 1305 Millgrove Pl Silver Spring MD 20904

CLARK, ORRIN H, b North Adams, Mass, June 2, 09; m 42; c 1. PHYSICS. Educ: Columbia Univ, AB, 30, MA, 31; NY Univ, PhD(physics), 37. Prof Exp: Instr physics, Newark Col Eng, 39-40; lectr, Columbia Univ, 40-41; from asst dir to dir, Works Control Lab, Corning Glass Works, 41-47; res assoc physics, Socony Mobil Oil Co, Inc, NJ, 47-50, supvr, Catalysis Res Sect, Res & Develop Dept, 50-64; from assoc prof to prof, 64-74, EMER PROF PHYSICS, SAN DIEGO STATE COL, 74- Mem: AAAS; Am Phys Soc; Soc Rheol. Res: Basic and theoretical lubrication; strength of glass; nuclear physics; geometrical optics; multiple scattering of neutrons; catalysis. Mailing Add: Dept of Physics San Diego State Col San Diego CA 92115

CLARK, PATRICIA ANN, b Dubuque, Iowa, Mar 28, 40. PHYSICAL CHEMISTRY. Educ: Univ NC, Greensboro, AB, 62; Univ Mass, PhD(chem), 67. Prof Exp: Fel, Cornell Univ, 67-68; asst prof, 68-73, ASSOC PROF CHEM, VASSAR COL, 73- Mem: AAAS; Am Phys Soc; Am Chem Soc. Res: Molecular electronic absorption spectroscopy; semi-empirical molecular orbital theory calculations; molecular charge-transfer complexes. Mailing Add: Dept of Chem Vassar Col Poughkeepsie NY 12601

CLARK, PATRICIA ANN ANDRE, b Leadville, Colo, Mar 23, 38; m 60; c 1. ASTROPHYSICS. Educ: Mass Inst Technol, SB, 61, SM, 64; Univ Rochester, PhD(astrophys), 69. Prof Exp: Instr eng, Univ Rochester & instr math, Rochester Inst Technol, 73-74; ASST CHIEF, INT FIELD YEAR GREAT LAKES, US ENVIRON PROTECTION AGENCY, 74- Mem: Am Astron Soc. Res: Modeling of chemical and physical processes in the Great Lakes. Mailing Add: 210 Chelmsford Rd Rochester NY 14618

CLARK, PATRICK JOSEPH, b Sheboygan, Wis, Oct 9, 42. ANALYTICAL CHEMISTRY. Educ: Univ Wis-Madison, BS, 65; Univ Va, PhD(chem), 76. Prof Exp: Res engr, Allen-Bradley Co, 65-67; res specialist natural prod, Dept Pharm, Univ Wis, 67-69; res specialist, Dept Chem, Univ Va, 69-71; RES ASSOC CHEM, TEX A&M UNIV, 76- Res: Application of charged particle activation analysis to geological samples; application of optical emission spectrometry using the inductively-coupled high-frequency plasma arc to trace element analysis in lignite. Mailing Add: Dept of Chem Tex A&M Univ College Station TX 77843

CLARK, PAUL ENOCH, b Cambridge, Ohio, Nov 24, 05; m 41; c 2. PHYSICAL CHEMISTRY. Educ: Muskingum Col, AB, 27; Ohio State Univ, MS, 31, PhD(phys chem), 38. Prof Exp: From instr to assoc prof chem, Muskingum Col, 27-42, prof chem & chmn div nat sci, 42-43; prof chem & head dept, Washington & Jefferson Col, 43-49; CHEMIST, APPL PHYSICS LAB, JOHNS HOPKINS UNIV, 49- Concurrent Pos: Chemist, Armstrong Cork Co, 44-46. Honors & Awards: Cert of Achievement Award, Int Publ Competition, Soc Tech Commun, Boston, 72, Award of Merit, Houston, 73. Mem: Am Chem Soc; fel Am Inst Chemists. Res: Heat capacities of oils; photovoltaic effect; threshold values for silver, gold and copper electrodes in electrolytic solutions; monomolecular surface films. Mailing Add: 2505 Eccleston St Silver Spring MD 20902

CLARK, RALPH B, b Farmington, Utah, Sept 12, 33; m 57; c 5. PLANT PHYSIOLOGY, PLANT BIOCHEMISTRY. Educ: Brigham Young Univ, BS, 57; Utah State Univ, MS, 59; Univ Calif, Los Angeles, PhD(plant sci), 62. Prof Exp: Res assoc, Ore State Univ, 63; res chemist, NCent Region, Agr Res Serv, USDA, 63-75; MEM FAC, DEPT AGRON, UNIV NEBR-LINCOLN, 75- Mem: AAAS; Am Soc Plant Physiologists; Am Soc Agron; Crop Sci Soc Am; Soil Sci Soc Am. Res: Physiology and biochemistry of mineral nutrition and metabolism in sorghum. Mailing Add: Dept of Agron Univ of Nebr Lincoln NE 68583

CLARK, RALPH M, b Stowe, Vt, Dec 12, 26; m 59; c 1. ZOOLOGY. Educ: Univ Vt, AB, 50, MEd, 51; Wash State Univ, MS, 61; Univ Mass, PhD(zool), 63. Prof Exp: Teacher, Vt High Sch, 51-56, USAF High Sch, Ger, 56-58 & NY High Sch, 58-59; from asst prof to assoc prof, 63-72; PROF BIOL, STATE UNIV NY COL PLATTSBURGH, 72- Concurrent Pos: NY Res Found grants-in-aid, 63-66. Mem: AAAS; Am Soc Cell Biol. Res: Developmental biology; effects of carcinogenic agents on chick embryos; wound healing processes in pupae of the silkmoth, Hyalophora cecropia. Mailing Add: Div of Sci & Math State Univ of NY Col Plattsburgh NY 12901

CLARK, RALPH O, b Broken Bow, Nebr, Aug 17, 12; m 34. CHEMISTRY. Educ: Univ Nebr, BS, 34, MS, 35. Prof Exp: Control chemist, Kendall Ref Co, 36-37; head anal sect, 37-55, ASST DIR, GULF RES & DEVELOP CO, 55-, TECH ASSOC, 61- Mem: Am Chem Soc; Am Soc Testing & Mat; Am Petrol Inst. Res: Microanalysis; polarography as applied to analysis; chromatography of petroleum products; absorption analysis; analysis by x-ray fluorescence; emission spectroscopy; process analyzers. Mailing Add: 7126 Shannon Rd Verona PA 15147

CLARK, RANDOLPH LEE, b Hereford, Tex, July 2, 06; m 32; c 2. SURGERY. Educ: Univ SC, BS, 27; Med Col Va, MD, 32; Univ Minn, MSc, 38. Hon Degrees: DSc, Med Col Va, 54. Prof Exp: Intern, Garfield Mem Hosp, Washington, DC, 33; chief resident, Am Hosp, Paris, France, 33-35; first asst, Mayo Clin, Minn, 35-39, asst surgeon, 39; chief surgeon, Shands Clin, Jackson, Miss, 39-42; dir & surgeon-in-chief, 46-68, PROF SURG, UNIV TEX M D ANDERSON HOSP & TUMOR INST, 65-, PRES, UNIV TEX SYST CANCER CTR, 68- Concurrent Pos: Actg dean, Univ Tex Grad Sch Biomed Sci, 48-50, prof surg, 48-65; consult, Surgeon Gen, US Dept Air Force, 48-53 & consult med div, Oak Ridge Inst Nuclear Studies, 50-56; med ed, Cancer Bull; co-ed, Bk of Health, 53- & Year Bk of Cancer, 56-; mem, Nat Adv Cancer Coun, 61-65 & President's Comn Heart Dis, Cancer & Stroke, 64-65; co-chmn senate panel, Consult for Conquest Cancer, 70-71; mem, President's Cancer Panel, 72-; chmn comt int collab activities, Int Union Against Cancer, 75- Honors & Awards: Nat Award, Am Cancer Soc, 64; Distinguished Serv Award, Am Col Surgeons, 69; Rodman E & Thomas G Sheen Award, AMA, 74. Mem: Fel AMA; Am Cancer Soc (vpres & pres-elect, 75-); fel Am Col Surg; Asn Am Cancer Insts (pres, 61-62 & 75-). Res: Cancer. Mailing Add: Univ of Tex Syst Cancer Ctr M D Anderson Hosp & Tumor Inst Houston TX 77030

CLARK, RAYMOND DONALD, b Whitehouse, Tenn, July 28, 32; m 55; c 3. ORGANIC CHEMISTRY. Educ: David Lipscomb Col, BA, 54; Vanderbilt Univ, MA, 56, PhD(org chem), 58. Prof Exp: Res chemist, 58-62, chemist, Org Chem Div, 62-63, SR RES CHEMIST, RES DIV, TENN EASTMAN CO, 63- Mem: Am Chem Soc. Res: Ketenes and their derivatives; rearrangements and pyrolyses; organic phosphites and phosphates; polyurethanes and extenders. Mailing Add: 532 Dogwood Dr Kingsport TN 37663

CLARK, RAYMOND GEORGE, b Elmhurst, NY, July 9, 19; m 45; c 3. CHEMISTRY. Educ: Polytech Inst Brooklyn, BChE, 47. Prof Exp: Group leader, Panelyte Div, St Regis Paper Co, 47-53; supvr, Phenolic Plastics Lab, 53-56, dir, 56-64, tech dir, Phenolic Plastics Div, 64-65, mgr resin res & develop, Molding Compound Div, 65-66, TECH DIR, CARTERET PLANT, REICHHOLD CHEM, INC, 66- Mem: Am Chem Soc; fel Am Inst Chem; Soc Plastics Eng. Res: Phenolic resins and homologues. Mailing Add: Reichhold Chem PO Box 38 Middlesex Ave Carteret NJ 07008

CLARK, RAYMOND LOYD, b Tacoma, Wash, Jan 23, 35; m 55; c 4. PLANT PATHOLOGY. Educ: Wash State Univ, BS, 57, PhD(plant path), 61. Prof Exp: RES PLANT PATHOLOGIST, CROPS RES DIV, AGR RES SERV, USDA, 61- Mem: Am Phytopath Soc. Res: Diplodia stalk rot of corn; tomato fruit rots; leptosphaerulina leafspot on alfalfa; root knot nematode; disease resistance; evaluation of foreign and wild germplasm. Mailing Add: NCgnt Regional Plant Intro Sta USDA Agr Res Serv Ames IA 50010

CLARK, RICHARD BENNETT, b Charleston, WVa, Nov 1, 20; m 49; c 2. CHEMISTRY. Educ: WVa Univ, BS, 42; Yale Univ, PhD(chem), 51. Prof Exp: Chemist, 50-54, sr chemist, 54-65, DEPT SUPT ORG CHEM DEVELOP, TENN EASTMAN CO, 65- Mem: Am Chem Soc. Res: Processes for manufacture of organic chemicals with particular reference to dyestuffs and chemicals used in photographic processing. Mailing Add: 1821 E Sevier Ave Kingsport TN 37664

CLARK, ROBERT, b New York, NY, Apr 19, 33; m 55; c 3. PSYCHOPHARMACOLOGY. Educ: Columbia Col, AB, 54; Columbia Univ, AM, 55, PhD(psychol), 58. Prof Exp: Asst psychol, Columbia Univ, 54-58, lectr, 56-58; res psychologist, Walter Reed Army Inst Res, 58-62; dir behav res, Labs, Va, 62-63; res psychologist, E I du Pont de Nemours & Inc, 63-66, res supvr, Stine Labs, 66-71, RES ASSOC PSYCHOL, STINE LAB, E I DU PONT DE NEMOURS & CO, INC, 71- Mem: AAAS; Am Psychol Asn; Am Soc Pharmacol & Exp Therapeut. Mailing Add: E I du Pont de Nemours & Co Inc Stine Lab PO Box 30 Newark DE 19711

CLARK, ROBERT A, b Boston, Mass, Aug 4, 39; m 65; c 1. ORGANIC CHEMISTRY. Educ: Mass Inst Technol, BS, 61; Univ Md, PhD(org chem), 66. Prof Exp: Res assoc org chem, Univ Wis, 65-66; asst prof chem, Clarkson Col Technol, 66-74; MEM FAC, DEPT CHEM, ROCHESTER INST TECHNOL, 74- Mem: Am Chem Soc. Res: Physical organic chemistry. Mailing Add: Dept of Chem Rochester Inst of Technol Rochester NY 14623

CLARK, ROBERT ALFRED, b Boston, Mass, Oct 28, 08; m 37; c 3. PSYCHIATRY. Educ: Harvard Univ, AB, 30, MD, 34. Prof Exp: Resident physician neurol, Boston City Hosp, 34-35; intern med, Univ Hosps, Cleveland, 35-37; resident physician psychiat, Boston Psychopathic Hosp, 37-39; sr physician, RI State Hosp, 39-42; sr physician, Western Psychiat Inst, Pittsburgh, 42-44, clin dir, 44-55; clin dir, Friends Hosp, 55-58; med dir, 58-70, chief outpatient serv, 70-71, asst to med dir med educ & res, 71-74, PSYCHIATRIST, NORTHEAST COMMUNITY MENT HEALTH CTR, 74-; CLIN PROF, HAHNEMANN MED COL, 74-; DIR STUDENT TRAINING, FRIENDS HOSP, 68- Concurrent Pos: Rockefeller fel, C G Jung Inst, Zurich, 48-49, Bollingen fel, 54; from instr to assoc prof, Sch Med, Univ Pittsburgh, 42-55; asst, Harvard Med Sch, 38-39; asst prof, Jefferson Med Col, 61-; clin assoc prof, Hahnemann Med Col, 70-74. Mem: Life fel Am Psychiat Asn. Res: Psychiatry and religion. Mailing Add: Friends Hosp Adams Ave & Roosevelt Blvd Philadelphia PA 19124

CLARK, ROBERT ALFRED, b Smith Center, Kans, Aug 18, 24; m 52; c 2. METEOROLOGY, CIVIL ENGINEERING. Educ: Kans State Univ, BS, 48; Tex A&M Univ, MS, 59, PhD(meteorol), 64. Prof Exp: Hydraul engr, US Bur Reclamation, 48-50, 52-60; from assoc prof to prof meteorol, Tex A&M Univ, 60-73; ASSOC DIR, OFF HYDROL, NAT WEATHER SERV, 73- Mem: Am Meteorol Soc; Am Soc Civil Eng; Am Geophys Union; Royal Meteorol Soc. Res: Hydrology; hydrometeorology; physical meteorology. Mailing Add: Off of Hydrol Nat Weather Serv Silver Spring MD 20910

CLARK, ROBERT AMOS, b Oswego, NY, Jan 14, 42; m 65; c 3. INFECTIOUS DISEASES. Educ: Syracuse Univ, AB, 63; Columbia Univ, MD, 67. Prof Exp: Intern med, Univ Wash, 67-68; asst resident, Columbia Presby Med Ctr, 68-69; clin assoc, Nat Inst Allergy & Infectious Dis, 69-71, sr staff fel, 71-72; instr, 72-73, ASST PROF MED, UNIV WASH, 73- Concurrent Pos: Consult infectious dis, Nat Naval Med Ctr, 70-72; chief resident, Harborview Med Ctr, Univ Wash, 72-73, attend physician, 73-; Nat Cancer Inst res career develop award, 75. Mem: Am Col Physicians; Am Fedn Clin Res; AAAS. Res: Mechanisms of host defense against microorganisms and neoplasia, with emphasis on the function of polymorphonuclear leukocytes. Mailing Add: Dept of Med Rm 16 Univ of Wash Seattle WA 98195

CLARK, ROBERT ARTHUR, b Melrose, Mass, May 3, 23; m 66. APPLIED MATHEMATICS. Educ: Duke Univ, AB, 44; Mass Inst Technol, MS, 46, PhD(math), 49. Prof Exp: Instr math, Mass Inst Technol, 46-49, res assoc, 49-50; from instr to prof, Case Inst Technol, 50-67, PROF MATH, CASE WESTERN RESERVE UNIV, 67- Mem: Am Math Soc; Math Asn Am; Soc Indust & Appl Math. Res: Elasticity, shell theory; asymptotic theory of differential equations. Mailing Add: Dept of Math Case Western Reserve Univ Cleveland OH 44106

CLARK, ROBERT BECK, b Rock Springs, Wyo, July 18, 41; m 59; c 4. ELEMENTARY PARTICLE PHYSICS. Educ: Yale Univ, BA, 63, MPhil, 67, PhD(physics), 68. Prof Exp: Fac assoc physics, Ctr Particle Theory, Univ Tex, Austin, 68-70, asst prof, Dept Physics, 70-73; asst prof, 73-76, ASSOC PROF PHYSICS, TEX A&M UNIV, 76- Mem: Am Phys Soc; Am Asn Physics Teachers. Res: Investigations of the electromagnetic and weak interactions of elementary particles. Mailing Add: Dept of Physics Tex A&M Univ College Station TX 77843

CLARK, ROBERT DAVID, genetics, animal science, see 12th edition

CLARK, ROBERT H, b Winnipeg, Man, Dec 25, 21; m 43; c 2. HYDROLOGY. Educ: McGill Univ, BEng, 43, MEng, 45. Prof Exp: Demonstr civil eng, McGill Univ, 43-44; hydraul design engr, Hydraul Div, Dominion Eng Works, Ltd, Montreal, 46-48; lectr civil eng, Univ Man, 48-52, asst prof, 52; chief hydraul engr, Red River Invest, Winnipeg, Can Dept Resources & Develop, 50-53, asst chief water resources br, 53-57; chief hydraul engr, Can Dept Northern Affairs & Nat Resources, 57-66, chief planning div, Can Dept Energy, Mines & Resources, 66-68, spec adv, 68-74, DIR, INLAND WATERS BR, DEPT ENVIRON, 74- Concurrent Pos: Mem, Prairie Prov Water Bd, 53-69; mem, Souris-Red River Eng Bd, 53-; mem subcomt hydrol, Nat Res Coun Can, 57-66; mem, Greater Winnipeg Floodway Adv Bd, 62-69; secy, Can Nat Comt, Int Hydrol Decade, 64-68; chmn working group on guide & tech regulations, Comn Hydrometeorol, World Meteorol Orgn, 64-72 & mem adv working group, 68-72; chmn, Can Sect, Int Great Lakes Working Comt, 65-; chmn, Atlantic Tidal Power Eng & Mgt Comt, 66-70; chmn, Can Sect, Int Niagara Bd Control, 69-; mem for Can, Int Lake Superior Bd of Control & Int Niagara Comt, 69- Mem: Am Geophys Union; Int Asn Sci Hydrol. Res: Snowmelt floods and river flow under ice conditions; water resources planning and development. Mailing Add: 1461 McRobie Ave Ottawa ON Can

CLARK, ROBERT KENLEY, b Williamsburg, Va, June 24, 18; m 43; c 5. PHYSICS. Educ: Univ Mont, BA, 39; Univ Ill, MS, 41, PhD(physics), 49. Prof Exp: Asst physics, Univ Ill, 39-43, res assoc res & develop betatron, 43-45; physicist radiol physics, Med Sch, Columbia Univ, 45-50; assoc physicist, Argonne Nat Lab, 50-54; physicist, Vet Admin Res Hosp, 54-63; sr vpres, Human Systs Inst, 63-65, dir, 63-68; sr res physicist, 68-70, MGR APPL RES, MOSLER SAFE CO, 70- Concurrent Pos: Assoc radiol, Northwestern Med Sch; consult, Med Sch, Univ Ill, 63-68, mem fac, 65-68; dir & pres, Radiation Control, 59-68 & Appl Sci Enterprise, Inc, 64-68; dir, AND, Inc, 64-68. Mem: AAAS; Am Phys Soc; Am Asn Physics Teachers; Am Mgt Asn; Inst Elec & Electronics Engrs. Res: Systems physics; security products and systems; medical physics; application of feedback theory to human behavior. Mailing Add: 834 Holyoke Cincinnati OH 45240

CLARK, ROBERT KINGSBURY, JR, biochemistry, see 12th edition

CLARK, ROBERT LONG, b Tekamah, Nebr, June 15, 15; m 41; c 3. MEDICINAL CHEMISTRY. Educ: Park Col, AB, 37; Univ Nebr, MA, 39, PhD(chem), 42. Prof Exp: Sr chemist, 42-64, RES FEL, MERCK & CO, INC, 64- Mem: Am Chem Soc. Res: Synthetic organic compounds; analgesics; anticonvulsants; arsenicals; chemotherapeutic compounds; veterinary medicinals; anti-inflammatory compounds. Mailing Add: 119 Grove Ave Woodbridge NJ 07095

CLARK, ROBERT M, b Canton, Ohio. PATHOLOGY. Educ: Muskingum Col, BS, 44; Case Western Reserve Univ, MD, 48. Prof Exp: From instr to asst prof, Case Western Reserve Univ, 50-62; ASSOC PROF PATH, UNIV MIAMI, 62-; CHIEF LAB PATH, VET ADMIN HOSP, MIAMI, 62- Concurrent Pos: Staff physician, Vet Admin Hosp, Cleveland, Ohio, 50-62. Mem: Am Soc Clin Path; Int Acad Path. Res: Blood coagulation. Mailing Add: Sch of Med Univ of Miami Coral Gables FL 33124

CLARK, ROBERT PAUL, b Jackson, Mich, July 19, 35; m 59; c 4. ANALYTICAL CHEMISTRY, PHYSICAL CHEMISTRY. Educ: Univ Mich, BS, 57; Univ Ill, MS, 59, PhD(anal chem), 62. Prof Exp: Staff mem, Power Sources Div, 62-69, MEM TECH STAFF, EXPLOR BATTERY DIV, SANDIA LABS, 69- Mem: AAAS; Am Chem Soc; Am Inst Chemists; Int Confederation Thermal Anal; NAm Thermal Anal Soc. Res: Batteries; energy conversion; thermal batteries; power sources; electrochemistry; fused salts; phase equilibria; thermal analysis; calorimetry. Mailing Add: Div 2523 Sandia Labs Albuquerque NM 87115

CLARK, ROBERT VERNON, b PEI, Aug 9, 26; m 54; c 4. PLANT PATHOLOGY. Educ: McGill Univ, BSc, 49, MSc, 52; Univ Wis, PhD(plant path), 56. Prof Exp: Res officer, Plant Path, Bot & Plant Path Lab, 49-59, Cereal Dis, Res Br, Genetics & Plant Breeding Inst, 59-64, RES OFFICER, CEREAL DIS, RES BR, OTTAWA RES STA, CAN DEPT AGR, 64- Mem: Can Phytopath Soc (secy-treas, 62-64). Res: Plant pathology and mycology of cereal diseases. Mailing Add: Res Br Ottawa Res Sta Can Dept of Agr Ottawa ON Can

CLARK, ROGER WILLIAM, b Oxford, Nebr, Nov 23, 42; m 67; c 2. MOLECULAR

BIOLOGY, GENETICS. Educ: Colo State Univ, BS, 65, MS, 67; Univ Ill, PhD(genetics), 71. Prof Exp: INSTR MOLECULAR GENETICS, SCH MED & DENT, UNIV ROCHESTER, 74- Concurrent Pos: NIH fel gen med sci, Fla State Univ, 71-72; Nat Cancer Inst fel, Univ Tex, Houston, 72-74. Res: Chromosome structure; biochemistry and biophysics of nucleic acids. Mailing Add: Dept of Exp Radiol Univ Rochester Sch Med & Dent Rochester NY 14642

CLARK, RONALD DAVID, b Leeds, Eng, July 18, 38; m 64; c 1. ORGANIC CHEMISTRY, ACADEMIC ADMINISTRATION. Educ: Univ Leeds, BSc, 59, PhD(chem), 62. Prof Exp: Fel, Univ Nebr, 62-65; sci officer chem, Radiochem Ctr, UK Atomic Energy Authority, 65-67; assoc prof chem, Jamestown Col, 67-75; DEAN, SCH NATURAL & SOCIAL SCI, KEARNEY STATE COL, 75- Mem: Am Chem Soc; The Chem Soc. Res: Steroid synthesis; synthesis of small ring compounds and isotopically labelled polypeptides. Mailing Add: Sch Natural & Social Sci Kearney State Col Kearney NE 68847

CLARK, RONALD DUANE, b Hollywood, Calif, Nov 21, 38; m 67; c 3. ORGANIC CHEMISTRY, ENVIRONMENTAL CHEMISTRY. Educ: Univ Calif, Los Angeles, BS, 60; Univ Calif, Riverside, PhD(org chem), 64. Prof Exp: Fel, Mich State Univ, 64-65; sr res chemist, Standard Oil Co, 65-69; asst prof, 69-73, ASSOC PROF CHEM, NMEX HIGHLANDS UNIV, 73- Mem: Am Chem Soc; The Chem Soc. Res: Stereochemistry; catalysis; atmospheric reactions of particulates. Mailing Add: Dept of Chem NMex Highlands Univ Las Vegas NM 87701

CLARK, RONALD GREY, b Norfolk, Va, Jan 16, 38; c 3. NEUROANATOMY. Educ: Johns Hopkins Univ, BA, 60; George Washington Univ, MS, 63, PhD(anat), 66. Prof Exp: Chief sect histopath, Lab Path, NIH, 62-70; ASSOC PROF NEUROANAT, DEPT BIOL STRUCT, UNIV MIAMI, 70-, ACTG CHMN DEPT, 74- Mem: Am Asn Anatomists; Soc Neurosci; Am Soc Cell Biol. Res: The rhesus monkey on the pathogenesis of brain tumors, cellular response to positive contrast agents used in myelography and ventriculography and mechanisms of hydrocephalus. Mailing Add: Dept of Biol Struct Sch of Med Univ of Miami PO Box 520875 Biscayne Annex Miami FL 33152

CLARK, RONALD HERSHEL, b Memphis, Tenn, Mar 15, 45; m 66; c 2. ELECTROMAGNETICS, LOW TEMPERATURE PHYSICS. Educ: Ark State Univ, BS, 66; Fla State Univ, PhD(physics), 71. Prof Exp: RES PHYSICIST, NAVAL COASTAL SYSTS LAB, 71- Res: Development of superconducting magnetic measuring devices and the implementation of these devices into magnetic measuring systems. Mailing Add: Code 792 Naval Coastal Systs Lab Panama City FL 32401

CLARK, RONALD JENE, b Hutchinson, Kans, July 11, 32; m 56; c 2. INORGANIC CHEMISTRY. Educ: Univ Kans, BS, 54, PhD(chem), 58. Prof Exp: Asst, Univ Kans, 54-57; res chemist, Linde Co, NY, 58-61; instr & assoc, Iowa State Univ, 61-62; from instr to assoc prof, 62-72, PROF CHEM, FLA STATE UNIV, 72- Mem: Am Chem Soc; The Chem Soc; AAAS; Sigma Xi. Res: Metal coordination compounds-phosphorus triflouride substitution products of metal carbonyls; lower oxidation state compounds of transition and acetinide elements. Mailing Add: Dept of Chem Fla State Univ Tallahassee FL 32306

CLARK, RONALD KEITH, b Los Angeles, Calif, Aug 28, 41; m 60; c 2. PHYSICAL CHEMISTRY. Educ: Univ Calif, Riverside, BA, 63, PhD(phys chem), 66. Prof Exp: NIH fel, Cornell Univ, 66-67; chemist, 67-74, SR RES CHEMIST, SHELL DEVELOP CO, 74- Mem: AAAS; Am Chem Soc; Soc Petrol Engrs. Res: Statistical thermodynamics and mechanics of nonelectrolyte solutions; theory of phase transitions, particularly in the critical region; phase behavior of polymer solutions. Mailing Add: Shell Develop Co PO Box 481 Houston TX 77001

CLARK, SALEM THOMAS, b Mokane, Mo, June 26, 27; m 47; c 4. PHYSICAL CHEMISTRY. Educ: Univ Mo, BS, 54, MS, 56, PhD(phys chem), 58. Prof Exp: Res chemist, Linde Co, 57-59; infrared spectroscopist, Celanese Corp, 59-60; prod mgr, Union Carbide Corp, 60-63; anal group leader, Celanese Fibers Co, 63-66, tech planning coordr, 66-69; DIR UNIV RELS, CELANESE CORP, 69- Mem: Am Chem Soc; Instrument Soc Am; Am Soc Qual Control. Res: Gas chromatography; instrumental analysis. Mailing Add: Celanese Corp 1211 Ave of Americas New York NY 10036

CLARK, SAM LILLARD, JR, b St Louis, Mo, June 9, 26; m 74; c 3. ANATOMY. Educ: Harvard Univ, MD, 49. Prof Exp: Intern med, Mass Gen Hosp, 49-50; from instr to assoc prof anat, Wash Univ, 54-68; PROF ANAT & CHMN DEPT, MED SCH, UNIV MASS, 68- Concurrent Pos: Nat Res Coun fel med sci biochem & nutrit, Med Sch, Vanderbilt Univ, 50-52; Palmer sr res fel; USPHS sr res fel; USPHS career develop award anat, Wash Univ. Mem: AAAS; Asn Am Med Cols; Am Asn Anat; Am Soc Cell Biol; Am Soc Exp Path. Res: Cellular differentiation; immunology; electron microscopy of tissues, relating structure to function. Mailing Add: Dept of Anat Univ of Mass Med Sch Worcester MA 01605

CLARK, SAMUEL FRIEND, b Danville, Ky, Jan 16, 14; m 46; c 2. ORGANIC CHEMISTRY. Educ: Univ WVa, AB, 34, MS, 37; Univ NC, PhD(org chem), 39. Prof Exp: Asst org chem, Univ WVa, 34-36, Johns Hopkins Univ, 36-37 & Univ NC, 37-38; res chemist, Union Carbide Chem Co, WVa, 39-42, group leader, 42-46; from assoc prof to prof chem, Univ Miss, 46-63, head ord chem div, 46-55, chmn, 55-63; chmn dept chem & phys sci, 63-67, PROF CHEM, FLA ATLANTIC UNIV, 63-, CHMN DEPT, 68- Concurrent Pos: Summer sr scientist, Union Carbide Nuclear Co, Tenn, 51-56; temporary chem liaison & field rep, NSF-AID Prog to Asn Cent Am Univs, Costa Rica, 67-68; US AID consult, Pedag Insts & Simon Bolivar Univ, Venezuela, 69. Mem: Am Chem Soc. Res: Constitution of natural tannins; synthesis of vinyl monomers and polymers, azo dyes and insecticides; reaction kinetics in radiology; molecular rearrangements. Mailing Add: Dept of Chem Fla Atlantic Univ Boca Raton FL 33432

CLARK, SANDRA HELEN BECKER, b Kansas City, Mo, July 27, 38; m 55; c 2. GEOLOGY. Educ: Univ Idaho, BS, 63, MS, 64, PhD(geol), 68. Prof Exp: Geologist, Cominco Am, Inc, 66-67; geologist, Alaska Mineral Resources Br, Calif, 67-72, STAFF GEOLOGIST, OFF MINERAL RESOURCES, US GEOL SURV, 72- Concurrent Pos: Coord staff mem, Alaska Natural Gas EIS Task Force, Dept of Interior, Washington, DC, 74-75, partic, Mgr Develop Prog, 75-76. Mem: AAAS; Geol Soc Am. Res: Geologic mapping; field and laboratory studies of structure and petrology of metamorphic and igneous terranes in northern Idaho, east central and south central Alaska. Mailing Add: 11910 Barrel Cooper Ct Reston VA 22092

CLARK, SHELDON LEWIS, organic chemistry, see 12th edition

CLARK, SIDNEY GILBERT, b Wolfsburg, Pa, Sept 2, 30; m 57; c 2. ORGANIC CHEMISTRY. Educ: Juniata Col, BS, 53; Pa State Univ, MS, 56, PhD(org chem), 58. Prof Exp: Res chemist, 57-62, res specialist, 62-64, group leader sulfonation-sulfation, 64-66, sr res group leader, 66-69, SECT MGR, INORG RES DEPT, MONSANTO CO, 69- Mem: Am Chem Soc; Sigma Xi; Am Oil Chemists' Soc. Res: Alkylation; sulfonation; ethoxylation; surfactant intermediates; anionic and nonionic surfactants; enzymes. Mailing Add: Res & Develop Dept Detergents & Phosphates Div Monsanto Co 800 N Lindbergh Blvd St Louis MO 63166

CLARK, STEPHEN DARROUGH, b Seattle, Wash, Apr 10, 45; m 68; c 2. ORGANIC CHEMISTRY. Educ: Seattle Univ, BS, 68; Mass Inst Technol, PhD(org chem), 72. Prof Exp: Fel chem, Syntex Res Inc, Calif, 72-73; res chemist, Arapahoe Chem Inc, 73-75, GROUP LEADER CHEM, ARAPAHOE-NEWPORT DIV, SYNTEX INC, 75- Mem: Am Chem Soc; Sigma Xi. Res: Process development research in pharmaceutical intermediates and finished drugs. Mailing Add: Arapahoe Chem Inc Rockhill Lab Newport TN 37821

CLARK, SYDNEY P, JR, b Philadelphia, Pan, July 26, 29; m 63; c 4. GEOPHYSICS. Educ: Harvard Univ, AB, 51, MA, 53, PhD(geol), 55. Prof Exp: Res fel geophys, Harvard Univ, 55-57; geophysicist, Geophys Lab, Carnegie Inst, 57-62; WEINBERG PROF GEOPHYS, YALE UNIV, 62-, DIR GRAD STUDIES, 74- Concurrent Pos: Fulbright scholar, Australian Nat Univ, 63. Mem: Fel Am Geophys Union. Res: Terrestrial and lunar heat flow; high pressure phase equilibria; constitution of earth's interior. Mailing Add: Box 2161 Yale Sta New Haven CT 06520

CLARK, THOMAS ALAN, b Leicestershire, Eng, Mar 14, 38; m 60; c 2. ASTRONOMY. Educ: Univ Leeds, BSc, 59, PhD(cosmic ray physics), 63. Prof Exp: Fel physics, Univ Calgary, 62-64, sessional lectr, 64-65; vis scientist, Defence Res Telecommun Estab, Defence Res Bd Can, 65; lectr physics, Univ Col, Univ London, 66-69, tutor, 68-69; asst prof, 70, ASSOC PROF PHYSICS, UNIV CALGARY, 70- Mem: Royal Astron Soc Can; assoc Brit Inst Physics & Phys Soc; Am Astron Soc; Can Astron Soc. Res: Infrared astronomy; far infrared solar spectral studies by balloon-borne and mountain altitude instrumentation; stratospheric emission; spectral measurements in the Far Infra Red as an aid in pollution studies; Infra Red Solar measurements. Mailing Add: Dept of Physics Univ of Calgary Calgary AB Can

CLARK, THOMAS ARVID, b Durango, Colo, Aug 23, 39; m 67. RADIO ASTRONOMY, GEODESY. Educ: Univ Colo, BS, 61, PhD(astrophys), 67. Prof Exp: Staff scientist astron, Boulder Labs, Environ Sci Serv Admin, Colo, 61-66 & NASA Marshall Space Flight Ctr, 66-68; RADIO ASTRONR, NASA GODDARD SPACEFLIGHT CTR, 68-; ASSOC PROF ASTRON, UNIV MD, 69- Concurrent Pos: Exec vpres, Radio Amateur Satellite Corp, 74-; mem study group 2nd radio astron, Int Consultative Radio Comn, 74-; mem radio astron subcomt, Comt Radio Frequencies, Nat Acad Sci-Nat Res Coun, 74-; co-chmn serv working group radio astron, Fed Commun Comn, 75- Mem: Int Astron Union; Am Astron Soc; Int Sci Radio Union; Am Geophys Union; AAAS. Res: Development of very long baseline interferometry for high accuracy astronomical and geophysical measurements; millimeter and infrared astronomy; cometary physics; development of astronomical instrumentation; radio frequency spectrum management; education. Mailing Add: Radio & Infrared Astron Br Code 693 NASA Goddard Space Flight Ctr Greenbelt MD 20771

CLARK, THOMAS HENRY, b London, Eng, Dec 3, 93; m 27; c 1. STRATIGRAPHY, PALEONTOLOGY. Educ: Harvard Univ, AB, 17, AM, 21, PhD(geol, paleont), 23. Prof Exp: Asst geol, Harvard Univ, 15-20, instr, 20-24; from asst prof to assoc prof paleont, 24-29, Logan prof, 29-62, prof, 62-64, dir univ mus, 25-52, chmn dept geol sci, 52-59, EMER PROF PALEONT, McGILL UNIV, 64- Concurrent Pos: Geologist, Geol Surv Can, 28-31 & 35 & Dept Mines, Que, 38-64; consult, Dept Natural Resources, Que, 64-69; consult geologist, 62-; adv geol, Redpath Mus, 64- Honors & Awards: Logan Medal, 71. Mem: Paleont Soc; Geol Soc Am; Royal Soc Can (pres sect IV, 53-54); Geol Asn Can (pres, 58-59); Can Inst Mining & Metal. Res: Invertebrate paleontology; Paleozoic stratigraphy. Mailing Add: Dept of Geol Sci McGill Univ Montreal PQ Can

CLARK, TREVOR H, b Haviland, Kans, July 16, 09; m 33; c 1. PHYSICS. Educ: Friends Univ, AB, 30, Univ Mich, MS, 33. Prof Exp: Serv mgr, Geo E Marshall Co, Kans, 29-32; serviceman, Int Radio, Mich, 33-34; lab asst, Univ Mich, 34; engr, Radio Corp Am Mfg Co, NJ, 34-38; engr, Les Laboratoires LMT, Int Tel & Tel Co, France, 38-40, dept head, Fed Telecommunication Labs, NJ, 40-45, dir head, 45-47, mgr tech servs, 47-48, mgr eng servs & special projs, 49-51; asst to pres, Fed Tel & Radio Crop, 48-49; assoc dir, Southwest Res Inst, 51-55; asst to eng mgr, Air Arm Div, Westinghouse Elec Corp, 55-61, mgr underwater launch prog, Aerospace Div, 61-64, mgr deep submergence prog, 64-65, mgr prog opers, 65-67, asst div mgr, 67-69, mgr info serv, 70-71; CONSULT, 71- Mem: Fel Inst Elec & Electronics Engr; Acoustical Soc Am; assoc fel Am Inst Aeronaut & Astronaut; Am Inst Physics. Res: High vacuum; sound; thermionics; microwaves; wave propagation; telephony; electron emission; navigation; vacuum tubes; photocells; multipliers; beam tubes; switching systems; direction finders; antennas; communications; countermeasures; radar; research administration. Mailing Add: 2 Cove Pl Whitehall Beach RFD 5 Box 234C Annapolis MD 21401

CLARK, TRUMAN BENTON, b Pine City, Minn, Jan 13, 28; m 51; c 3. INSECT PATHOLOGY. Educ: Univ Minn, BA, 51, MS, 53, PhD(zool), 58. Prof Exp: Res specialist parasitol, Minn State Bd Health, 56-58; res asst, Univ Minn, 58-59; asst prof zool, Iowa State Univ, 59-61; specialist I insect path, Calif Pub Health Dept, 61-66 & Univ Calif, Berkeley, 66-67; res entomologist, Entom Res Div, 67-70, RES ENTOMOLOGIST, AGR RES SERV, USDA, 72- Concurrent Pos: Assoc prof biol, Calif State Univ, Fresno, 70-72. Mem: Soc Protozoologists; Soc Invert Path; Int Orgn Biol Control; Am Mosquito Control Asn. Res: Microbial control of mosquitoes and honey bee pathology and disease control. Mailing Add: Agr Res Serv USDA Bioenviron Bee Lab Beltsville MD 20705

CLARK, VIRGINIA, b Grand Rapids, Mich, Nov 18, 28; m 53. BIOSTATISTICS. Educ: Univ Mich, BA, 50, MA, 51; Univ Calif, Los Angeles, PhD(biostatist), 63. Prof Exp: Statistician, Gen Elec Co, 51-54; appl mathematician, Econ Res Proj, Harvard Univ, 54-56, res assoc biostatist, Med Sch, 60-61; statistician, Systs Lab Corp, 56-57; from asst prof to assoc prof biostatist, 63-74, assoc prof biomath, 71-74, PROF BIOSTATIST & PROF BIOMATH, SCH PUB HEALTH, UNIV CALIF, LOS ANGELES, 74- Mem: Inst Math Statist; Biomet Soc; fel Am Statist Asn. Mailing Add: Dept of Biostatist Univ of Calif Los Angeles CA 90024

CLARK, WALLACE HENDERSON, JR, b LaGrange, Ga, May 16, 24; m; c 5. MEDICINE, PATHOLOGY. Educ: Tulane Univ, BS, 44, MD, 47. Prof Exp: From instr to prof path, Sch Med, Tulane Univ, 49-62; asst prof, Harvard Univ, 62-68, assoc clin prof, 68-69; PROF PATH, SCH MED, TEMPLE UNIV, 69-, CHMN DEPT, 74- Concurrent Pos: Markle scholar, 54-60; consult pathologist, Orleans Parish Coroner's Off, 50-52, 54-56 & Armed Forces Inst Path, Washington, DC; assoc path, Mass Gen Hosp, 62-68. Mem: Am Asn Path & Bact; Am Asn Cancer Res; Am Soc Exp Path; Am Acad Dermat. Res: Dermal pathology; electron microscopy; correlation of ultrastructural changes with known changes in cellular function; tumor progression in human neoplastic systems; immunology and fine structure of primary human cutaneous malignant melanomas; induction of animal model of human malignant melanoma in the guina pig. Mailing Add: Health Sci Ctr Dept of Path Temple Univ Sch of Med Philadelphia PA 19140

CLARK, WALTER, b Hendon, Eng, Nov 9, 99; m; c 3. CHEMISTRY. Educ: Univ London, BSc, 21, MSc, 22, PhD(chem), 23. Prof Exp: Res phys chemist, Brit Photog Res Asn, 22-27; asst, Sci Mus, London, 27-28; dir res, Kodak Ltd, 28-31, asst to vpres in charge res, Eastman Kodak Co, 31-47, head appl photog div, Res Lab, 47-69; CONSULT, 69- Concurrent Pos: Secy, Int Cong Photog, 28; ed, Var Photog & Graphic Arts Publ. Mem: Fel Soc Motion Picture & TV Engrs; fel Photog Soc Am; fel Soc Photog Sci & Eng; Am Soc Photogram (vpres, 46); hon fel Royal Photog Soc. Res: Photographic theory; photography and its applications; graphic arts; photomicrography; photographic chemistry; infrared; document reproduction; tropical deterioration; conservation of photographs. Mailing Add: 94 Southern Pkwy Rochester NY 14618

CLARK, WALTER ERNEST, b Stuart, Va, Sept 25, 16; m 39; c 3. APPLIED CHEMISTRY. Educ: Va Mil Inst, BS, 37; George Washington Univ, MA, 39; Univ Wis, PhD(chem & chem eng), 49. Prof Exp: Instr chem, Va Mil Inst, 39-41; asst prof chem eng, Mo Sch Mines, 49-51; sr res chemist, 51-56, GROUP LEADER, OAK RIDGE NAT LAB, 56- Concurrent Pos: Lectr, Univ Tenn, 63-64; Fulbright lectr, Tribhuvan Univ, Nepal, 67-68. Mem: Am Chem Soc; fel Am Inst Chem. Res: Electrochemistry; polarography; corrosion; nuclear fuel processing; nuclear waste disposal. Mailing Add: 386 East Dr Oak Ridge TN 37830

CLARK, WALTER LEIGHTON, III, b Springfield, Pa, Feb 3, 21; m 45; c 1. FOOD SCIENCE. Educ: Pomona Col, BA, 42; Georgetown Univ, MS, 46; Cornell Univ, PhD(biochem), 53. Prof Exp: Res assoc, Food Sci & Tech Div, NY State Agr Exp Sta, Cornell Univ, 48-50, asst, 50-53, asst prof biochem, Grad Sch Nutrit, 53-56; sr res biologist, Dept Nutrit & Food Technol, Lederle Labs, Agr Ctr, Am Cyanamid Co, 56-58, group leader food res, 58-61, group leader food res & develop, 61-64; tech mgr new prod develop refrig foods, Res Ctr, Pillsbury Co, 65-67; assoc dir res-explor, Quaker Oats Co, Ill, 67-73; CORP DIR SCI & NUTRIT, HUNT-WESSON FOODS, INC, 73- Mem: Am Chem Soc; Inst Food Technologists; Am Asn Cereal Chemists. Res: Teaching food science; food additives; diet and nutrition; food microbiology; protein sources and technology; food fabrication; heavy metals; food regulations; tomato processing; edible fats and oils; enzyme technology; research management. Mailing Add: Hunt-Wesson Foods Inc 1645 W Valencia Dr Fullerton CA 92634

CLARK, WARREN PARKER, physics, deceased

CLARK, WARREN S, JR, dairy microbiology, nutrition, see 12th edition

CLARK, WESLEY GLEASON, b Wadsworth, Ohio, July 1, 33; m 65; c 3. PHARMACOLOGY. Educ: Univ Colo, BA, 55, MS, 58; Univ Utah, PhD(pharmacol), 62. Prof Exp: Instr, 62-63, asst prof, 63-72, ASSOC PROF PHARMACOL, UNIV TEX HEALTH SCI CTR DALLAS, 72- Concurrent Pos: USPHS grant, Nat Inst Allergy & Infectious Dis, 64-66 & Nat Inst Neurol Dis & Stroke, 70- Mem: Am Physiol Soc; Am Soc Pharmacol & Exp Therapeut; Soc Exp Biol & Med; Soc Neurosci; NY Acad Sci. Res: Neuropharmacology; effects of drugs on thermoregulation; bacterial pyrogens; food poisoning; vomiting. Mailing Add: Dept of Pharmacol Univ Tex Health Sci Ctr Dallas TX 75325

CLARK, WILLIAM ARTHUR, b Chicago Heights, Ill, Nov 6, 23; m 46; c 3. BACTERIOLOGY. Educ: Univ Colo, BA, 50; Cornell Univ, MS, 51, PhD(bact), 53. Prof Exp: Asst prof bact, Cornell Univ, 53-54; asst cur, Am Type Cult Collection, 54-60, dir, 60-73; MEM FAC, DEPT MICROBIOL, QUEENSLAND UNIV, 73- Concurrent Pos: Exec secy, Int Comn Syst Bact. Mem: Am Soc Microbiol; Brit Soc Gen Microbiol. Res: Taxonomy of myxobacteria: preservation of bacteriophage and other microorganisms; taxonomy of bacteriophage. Mailing Add: Dept of Microbiol Queensland Univ Brisbane Queensland 4067 Australia

CLARK, WILLIAM ARTHUR VALENTINE, b Christchurch, NZ, Mar 21, 38; m 70. URBAN GEOGRAPHY. Educ: Univ Canterbury, BA, 60, MA, 61; Univ Ill, Urbana, PhD(geog), 64. Prof Exp: Lectr geog, Univ Canterbury, 64-66; asst prof, Univ Wis-Madison, 66-68, assoc prof geog & urban planning, 68-70; assoc prof, 70-72, PROF GEOG, UNIV CALIF, LOS ANGELES, 72- Concurrent Pos: Vis prof geog, Univ Auckland, 74. Mem: NZ Geog Soc; Asn Am Geogr; Regional Sci Asn. Res: Models of urban spatial structure and spatial behavior; particularly models which focus on intra-urban population migration. Mailing Add: Dept of Geog Univ of Calif Los Angeles CA 90024

CLARK, WILLIAM DEAN, b Guthrie, Okla, Feb 1, 36; m 54; c 2. MATHEMATICAL ANALYSIS. Educ: Cent State Col, Okla, BSEd, 62; Univ Tex, Austin, MA, 64, PhD(math), 68. Prof Exp: NSF Acad Year Inst partic, Univ Tex, Austin, 62-63; assoc prof math, 66-74, res grant, 68-69, PROF MATH, STEPHEN F AUSTIN STATE UNIV, 74- Concurrent Pos: NSF res grant, 70-73. Mem: Soc Indust & Appl Math; Math Asn Am. Res: Summability of series. Mailing Add: Dept of Math Stephen F Austin State Univ Nacogdoches TX 75961

CLARK, WILLIAM DEMPSEY, b Buffalo, NY, July 15, 21; m 47; c 3. PHYSICAL CHEMISTRY. Educ: Univ Miami, BS, 50, MS, 51; Univ Ore, MS, 54, PhD(phys chem), 58. Prof Exp: Staff mem, Los Alamos Sci Lab, NMex, 56-61; proj mgr, Phys Sci Corp, 61-62; consult, Aerosols, Mass Spectrometry, Radiation, 62-63; staff mem, Hughes Aircraft Co, 63-64; eng specialist, Garret Corp, 64-65; tech dir, Dyna-Therm Corp, 65-66; founder, 63, PRES, ROPAT CO, 68-; PRES, ROPAT-CASLON INC, 68- Concurrent Pos: Consult, Pub Utilities & Foreign Co. Mem: AAAS; Am Chem Soc; Fedn Am Scientists; Am Inst Aeronaut & Astronaut; Am Inst Physics; Brit Inst Physics. Res: Chemical kinetics; mass and optical spectroscopy; high temperature measurement; gravitation; nuclear propulsion; cryogenics; specialized analytical and space instrumentation; energy systems development; aerosols; horology; nuclear weapons and reactors. Mailing Add: 12616 Chadron St Hawthorne CA 90250

CLARK, WILLIAM DONALD KENNEDY, physical chemistry, see 12th edition

CLARK, WILLIAM EDWIN, b Brunswick, Ga, Dec 7, 34; m 60; c 1. MATHEMATICS. Educ: Sam Houston State Col, BA, 60; Tulane Univ, PhD(math), 64. Prof Exp: Ford Found res fel math, Calif Inst Technol, 64-65; from asst prof to assoc prof, Univ Fla, 65-70; PROF MATH, UNIV S FLA, 70- Concurrent Pos: NSF res grant, 67- Mem: Am Math Soc; Math Asn Am. Res: Arithmetic coding theory. Mailing Add: Dept of Math Univ of SFla Tampa FL 33620

CLARK, WILLIAM GILBERT, b Kansas City, Mo, Apr 17, 09; m 37, 54; c 2. PHYSIOLOGY, BIOCHEMISTRY. Educ: Univ Tex, AB, 31; Calif Inst Technol, PhD(physiol), 37. Prof Exp: Asst, Univ Tex, 27-31; res assoc biochem, Scripps Inst Oceanog, 37-38; res assoc physiol, Scripps Clin & Res Found, 38-39; instr physiol zool, Univ Minn, 39-40; asst prof space med & physiol, Univ Southern Calif, 43-46, Fowler res fel pharmacol, 46-47; res assoc physiol & pharmacol, Scripps Clin & Res Found, 47-49; ASSOC CLIN PROF BIOL CHEM, CTR HEALTH SCI, UNIV CALIF, LOS ANGELES, 49-; CHIEF, PSYCHOPHARMACOL RES LABS, VET ADMIN HOSP, SEPULVEDA, 60- Concurrent Pos: Chief physiol & pharmacol, Res Labs, Vet Admin Ctr, Los Angeles, 49-60. Honors & Awards: Walter Reed Mem Award, 53. Mem: Fel AAAS; Soc Exp Biol & Med; Am Physiol Soc; Am Soc Pharmacol & Exp Therapeut; Am Soc Biol Chemists. Res: Neuropharmacology; psychopharmacology; physiology; biochemistry; neurotransmitters; histamine; behavior sciences; mental retardation; psychiatry. Mailing Add: Psychopharmacol Res Lab Vet Admin Hosp Sepulveda CA 91343

CLARK, WILLIAM GILBERT, b Los Angeles, Calif, May 26, 30; m 59; c 2. MAGNETIC RESONANCE, EXPERIMENTAL SOLID STATE PHYSICS. Educ: Stanford Univ, BS, 52; Cornell Univ, PhD(physics), 61. Prof Exp: Asst, Stanford Univ, 52-53; asst, Cornell Univ, 54-60; jr res physicist, Univ Calif, San Diego, 60-62, asst res physicist, 62-64; from asst prof to assoc prof, 64-73, PROF PHYSICS, UNIV CALIF, LOS ANGELES, 73- Concurrent Pos: Nat Ctr Sci Res fel, Fac Sci, Orsay, France, 69-70, exchange prof, 70; assoc prof, Sci & Med Univ, Univ Grenoble, France, 75-76. Mem: Am Phys Soc. Res: Low temperature physics; physical properties of pseudo one-dimensional solids at very low temperatures. Mailing Add: Dept of Physics Univ of Calif Los Angeles CA 90024

CLARK, WILLIAM GLENN, b Sesser, Ill, Aug 15, 15. ALGEBRA. Educ: Union Col, Ky, AB, 36; Univ Ky, MA, 38, PhD(math), 42. Prof Exp: Res analyst, Army Security Agency, Washington, DC, 46-47; from asst prof to assoc prof 47-57, PROF PHYSICS & MATH, MT UNION COL, 57-, HEAD DEPT MATH, 63- Mem: AAAS; Math Asn Am. Res: Theory of ideals in quaternion algebras; generalized quaternion algebras and universal hermitian forms. Mailing Add: Dept of Math Mt Union Col Alliance OH 44601

CLARK, WILLIAM JESSE, b Salt Lake City, Utah, Sept 29, 23; m 51; c 2. LIMNOLOGY, AQUATIC ECOLOGY. Educ: Utah State Univ, BS, 50, MS, 56, PhD(aquatic biol), 58. Prof Exp: From asst prof to assoc prof biol, 57-68, assoc prof biol & wildlife sci, 68-75, ASSOC PROF WILDLIFE & FISHERIES SCI, TEX A&M UNIV, 75- Mem: Am Soc Limnol & Oceanog; Ecol Soc Am; Int Asn Theoret & Appl Limnol. Res: Ecology of ponds; regional limnology; limnology and ecology of rivers. Mailing Add: Dept of Wildlife & Fisheries Sci Tex A&M Univ College Station TX 77843

CLARK, WILLIAM KEMP, b Dallas, Tex, Sept 2, 25; m; c 6. NEUROSURGERY. Educ: Univ Tex, BA, 45, MD, 48. Prof Exp: From asst prof to assoc prof, 56-69, PROF SURG, UNIV TEX HEALTH SCI CTR DALLAS, 69-, CHMN DIV NEUROSURG, 56- Concurrent Pos: Dir serv neurosurg, Parkland Mem Hosp, 56-; attend neurol surg, Vet Admin Hosp, 56-, consult, 64-; consult, Children's Med Ctr, 56- Res: Injuries to the nervous system; ultrastructure of nervous system. Mailing Add: Div of Neurosurg Univ of Tex Health Sci Ctr Dallas TX 75235

CLARK, WILLIAM MELVIN, JR, b Baldwin, Kans, Apr 17, 22; m 45; c 2. MEDICAL EDUCATION ADMINISTRATION. Educ: Baker Univ, AB, 46; Univ Chicago, MD, 49; Am Bd Pediat, dipl. Prof Exp: From instr to assoc prof, 54-67, PROF PEDIAT, SCH MED, UNIV ORE HEALTH SCI CTR, 67-, ASSOC MED DIR, UNIV HOSPS & CLINS, 72- Mem: Am Acad Pediat; Am Acad Neurol. Res: Pediatric neurology. Mailing Add: Univ Ore Health Sci Ctr Portland OR 97201

CLARK, WILLIAM RICHMOND, b Poultney, Vt, Feb 16, 22; m 50. BIOCHEMISTRY. Educ: Haverford Col, BS, 48; Ind Univ, MA, 50; Georgetown Univ, PhD(biochem), 56. Prof Exp: Asst microbiol, Med Ctr, Ind Univ, 50-52; res chemist biochem, Walter Reed Army Inst Res, 52-56; asst prof chem, Bethany Col, 56-58; prof & head dept, Parsons Col, 58-60, chmn div sci, 59-60; res chemist, US Agr Res Serv, Washington, DC, 60-63; prof, Mary Baldwin Col, 63-67; assoc prof, Bethany Col, 67-74; MEM FAC, DEPT CELL BIOL, SCH MED, UNIV CALIF, LOS ANGELES, 74- Concurrent Pos: Iowa Heart Asn res grant, 59. Mem: AAAS; Am Chem Soc. Res: Immunochemistry; bacterial enzymes; purification and characterization of high molecular weight substances; tissue thromboplastins. Mailing Add: Dept of Cell Biol Univ Calif Sch Med Los Angeles CA 90024

CLARK, WILLIAM THOMAS, JR, b Covington, Ky, June 20, 31; m 63; c 2. GEOGRAPHY OF LATIN AMERICA. Educ: Univ Ky, BS, 53, MA, 55, PhD(geog educ), 67. Prof Exp: Instr geog, Ala Polytech Inst, 57-59; asst prof, Indiana State Col, Pa, 60-62; assoc prof, 64-75, PROF GEOG, MOREHEAD STATE UNIV, 75- Mem: Nat Coun Geog Educ. Res: Physical geography, including meteorology and climatology; Middle America; Europe. Mailing Add: Dept of Geog Morehead State Univ Box 688 Morehead KY 40351

CLARK, WILSON FARNSWORTH, b Schenectady, NY, Feb 25, 21; m 45; c 4. ENVIRONMENTAL SCIENCES, CONSERVATION. Educ: Middlebury Col, BA, 42; Cornell Univ, PhD(conserv educ), 49. Prof Exp: Asst prof & exten conservationist, Cornell Univ, 49-54; PROF & HEAD DIV SCI & MATH, EASTERN MONT COL, 54- Concurrent Pos: Res chemist, Manhattan Proj, 45-47; mem, Mont State Bd Natural Resources & Conserv. Mem: Soc Am Foresters; Wildlife Soc; Soil Conserv Soc Am; Conserv Educ Asn (pres, 65-69). Res: Broad interpretation of methods, results, development, concepts and implications of environmental science and education. Mailing Add: Div of Sci & Math Eastern Mont Col Billings MT 59101

CLARKE, ALEXANDER MALLORY, b Richmond, Va, Mar 29, 36; m 59; c 3. BIOPHYSICS, BIOMEDICAL ENGINEERING. Educ: Va Mil Inst, BS, 58; Univ Va, MS, 60, PhD(physics), 63. Prof Exp: Asst prof, 64-69, ASSOC PROF BIOPHYS, MED COL VA, 69- Concurrent Pos: Mem subcomt ocular effects, Am Nat Stand Inst, 69-; mem tech comt, 76, Int Electrotech Comn, 73- Mem: Asn Res Vision & Ophthal; Am Phys Soc. Res: Biomedical instrumentation; physical chemistry of macromolecules; effects of intense optical sources on the pupil and retina. Mailing Add: Dept of Biophys Box 877 Med Col of Va Richmond VA 23298

CLARKE, ALLEN BRUCE, b Saskatoon, Sask, Sept 8, 27; nat US; m 49; c 3. MATHEMATICS. Educ: Univ Sask, BA, 47; Brown Univ, MS, 49, PhD(math), 51. Prof Exp: From instr to prof math, Univ Mich, 51-67; PROF MATH & CHMN DEPT, WESTERN MICH UNIV, 67- Concurrent Pos: Fulbright lectr, Univ Turku & Abo Acad, Finland, 59-60. Mem. Am Math Soc; Inst Math Statist; Math Asn Am. Res: Probability theory; theory of waiting lines. Mailing Add: Dept of Math Western Mich Univ Kalamazoo MI 49008

CLARKE, ARTHUR A, mathematics, see 12th edition

CLARKE, ARTHUR HADDLETON, JR, b Danvers, Mass, July 12, 26; m 49; c 1. MALACOLOGY. Educ: Univ Boston, AB, 52; Cornell Univ, MS, 58; Harvard Univ, PhD(biol), 60. Prof Exp: Chem engr, Raytheon Mfg Co, Mass, 50-55; cur mollusks, Cornell Univ, 55-56; marine biologist, Columbia Univ, 57-59; cur mollusks, 59-67, HEAD INVERT ZOOL, NAT MUS CAN, 67- Concurrent Pos: Gen consult malacol, Govt Can, 59-; assoc ed, Malacologia & The Nautilus; publ ed, Am Malacol Union, 72-75. Mem: Am Malacol Union (pres, 67-68). Res: North American freshwater mollusks, especially Unionidae; arctic and eastern North American marine mollusks; world-wide abyssal marine mollusks; Monoplacophora. Mailing Add: Nat Mus of Natural Sci Nat Mus of Can Ottawa ON Can

CLARKE, BRUCE LESLIE, b Toronto, Ont, Aug 4, 42; m 68; c 2. THEORETICAL CHEMISTRY. Educ: Univ Toronto, BSc, 65; Univ Chicago, PhD(chem), 69. Prof Exp: Res asst chem, Univ Calif, Santa Cruz, 69-70; ASST PROF CHEM, UNIV ALTA, 70- Mem: Am Phys Soc. Res: Diagrammatic stability analysis of oscillatory chemical reaction systems; topological properties of self-organizing chemical networks; non-equilibrium statistical mechanics, fluctuations, phase transitions and critical phenomena. Mailing Add: Dept of Chem Univ of Alta Edmonton AB Can

CLARKE, CHARLES HENRY DOUGLAS, b Kerwood, Ont, June 14, 09; m 38; c 3. BIOLOGY. Educ: Univ Toronto, BScF, 31, PhD(biol), 35. Prof Exp: Asst, Univ Toronto, 31-35; field asst, Nat Mus Can, 35-37; mammalogist, Nat Parks Bur, Can Dept Mines & Resources, 38-44; mammalogist, Res Div, Dept Lands & Forests, Ont, 44-46, supvr wildlife mgt, 46-59, chief fish & wildlife br, 60-71; RETIRED. Concurrent Pos: Asst, Univ Toronto, 36-37; wildlife consult, Can Aid Kenya Game Dept, 65, Temporary Comn of Future of Adirondacks, NY, 69 & Tanzania Nat Parks, 72-74. Mem: Am Soc Mammal; Wildlife Soc (pres, 53); Am Ornith Union; fel Arctic Inst NAm. Mailing Add: 26 Lockie Ave Agincourt ON Can

CLARKE, DAVID BRUCE, b Newburgh, NY, Feb 19, 42; m 67; c 2. INDUSTRIAL ORGANIC CHEMISTRY. Educ: Dartmouth Col, AB, 64; Univ Rochester, PhD(chem), 69. Prof Exp: Res chemist additive synthesis, Shell Oil Co-Shell Develop, 69-75; SR RES CHEMIST SYNTHETIC LUBRICANTS, TENNECO CHEM CO, 75- Mem: Am Chem Soc. Res: Research and development of lubricants based on synthetic fluids. Mailing Add: Tenneco Chem Co Turner Pl Piscataway NJ 08854

CLARKE, DAVID HARRISON, b Jamestown, NY, Aug 14, 30; m 52; c 3. HUMAN PHYSIOLOGY. Educ: Springfield Col, BS, 52, MS, 53; Univ Ore, PhD(phys educ), 59. Prof Exp: Asst prof phys educ, Univ Calif, Berkeley, 58-64; from assoc prof to PROF PHYS EDUC, UNIV MD, COLLEGE PARK, 64-, DIR GRAD STUDIES, DEPT PHYS EDUC, 73- Concurrent Pos: Chmn res sect, Nat Col Phys Educ Asn, 65-66. Mem: Asn Health, Phys Educ & Recreation; Am Acad Phys Educ; Am Col Sports Med. Res: Physiology of exercise, especially in the area of muscular fatigue and the strength debt of exercise. Mailing Add: 11402 Hennessey Dr Beltsville MD 20705

CLARKE, DONALD DUDLEY, b Kingston, BWI, Mar 20, 30; US citizen; m 53; c 7. BIOCHEMISTRY. Educ: Fordham Univ, BS, 50, MS, 51, PhD(org Chem, enzym), 55. Prof Exp: Nat Res Coun Can fel, Banting Inst, 55-57; res scientist neurochem, NY State Psychiat Inst, 57-62; assoc prof, 62-70, PROF BIOCHEM, FORDHAM UNIV, 70- Concurrent Pos: Res assoc biochem, Col Physicians & Surgeons, Columbia Univ, 59-61; adj assoc prof, Fordham Univ, 61-62; spec fel, NIH, 72-73; mem adv comt, NIMH, 73-77. Mem: AAAS; Am Chem Soc; Am Soc Biol Chemists; Int Soc neurochem; Am Soc Neurochem. Res: Neurochemistry, glutamic acid metabolism and related compounds with special reference to brain; mold metabolites, structure and biosynthesis. Mailing Add: Dept of Chem Forham Univ New York NY 10458

CLARKE, DONALD WALTER, b Vermilion, Alta, Apr 12, 20; m 51. BIOCHEMISTRY. Educ: Univ Alta, BSc, 41, MSc, 43; Calif Inst Technol, PhD(chem), 51. Prof Exp: Asst fuels, Res Coun Alta, 41-43; from asst prof to assoc prof, 51-64, prof, Banting & Best Dept Med Res, 64-68, PROF PHYSIOL, UNIV TORONTO, 68- Mem: Can Physiol Soc. Res: Intermediate metabolism of carbohydrates; protein complexes with other larger molecules; electrophoresis. Mailing Add: Dept of Physiol Univ of Toronto Med Sci Bldg Toronto ON Can

CLARKE, DUANE GROOKETT, b Philadelphia, Pa, Jan 7, 18; m 49. CHEMISTRY. Educ: Fla Southern Col, BS, 40; Pa State Univ, MS, 42, PhD(org chem), 44. Prof Exp: From res chemist to develop chemist, 43-67, head pollution abatement lab, 67-69, ASST TO MGR WATER & AIR CONSERV, ROHM AND HAAS CO, 70- Mem: Am Chem Soc. Res: Heavy hydrocarbons synthesis and properties; synthesis of insecticides, fungicides and monomers; process development; pollution abatement. Mailing Add: Eng Div Rohm and Haas Co PO Box 584 Bristol PA 19007

CLARKE, EDWARD NIELSEN, b Providence, RI, Apr 25, 25; m 49; c 4. RESEARCH ADMINISTRATION. Educ: Brown Univ, BS, 45, Phd(physics), 51, Harvard Univ, MS, 47, MES, 48. Prof Exp: Physicist solid state physics, res lab, Sylvania Elec Prod, 50-56 & Sperry Semiconductor Div, Sperry Rand Corp, 56-59; founder & vpres opers, Nat Semiconductor Corp, 59-64, vpres corp develop & diversification, 64-65; ASSOC DEAN FAC & DIR RES, WORCESTER POLYTECH INST, 65- Concurrent Pos: Consult, Semiconductor Indust; mem, Nat Coun Univ Res Adminr; tri-col res coordr, Clark Univ, Holy Cross Col & Worcester Polytech Inst, 74-; mgr, Urban Technol Syst Backup Site, 74- Mem: AAAS; Sigma Xi; Am Phys Soc; Inst Elec & Electronics Engrs; Am Soc Eng Educ. Res Semicondictor device research; technology transfer. Mailing Add: 85 Richards Ave Paxton MA 01612

CLARKE, ERIC THACHER, physics, see 12th edition

CLARKE, ERNEST MAURICE, b St John, NB, May 12, 11. PHYSICS. Educ: St Francis Xavier Univ, Can, BSc, 32; Laval Univ, DSc(physics), 56. Prof Exp: Prof eng, St Patrick's Col, Can, 32-34; head dept, 54-70, PROF PHYSICS, ST FRANCIS XAVIER UNIV, 42- Mem: Am Phys Soc; Am Asn Physics Teachers; Can Asn Physicists. Res: Mass spectrometry; ionization and dissociation energies; electronics. Mailing Add: Dept of Physics St Francis Xavier Univ Antigonish NS Can

CLARKE, FRANK ELDRIDGE, b Brunswick, Md, Dec 26, 13; m 34; c 2. HYDROLOGY. Educ: Western Md Col, AB, 35. Prof Exp: Head chem process br, Chem Eng Lab, US Naval Eng Exp Sta, 51-57, head chem engr div, 57-61; res engr, US Geol Surv, 61-62, chief water qual res, 62, chief gen hydrol br, 62-65, asst chief hydrologist, 65-67, assoc chief hydrologist, 67-68, asst dir, 68-71; dep undersecy, US Dept Interior, 71-72, SR SCIENTIST, US GEOL SURV, US DEPT INTERIOR, 72- Concurrent Pos: Consult, US State Dept, 62- & Univ Queensland, 67- Honors & Awards: Am Chem Soc Cert Merit, 53; Superior Achievement Award, US Navy, 56; Am Soc Testing & Mat Merit Award, 61; Max Hecht Award, 64; Cert Award, Gordon Res Conf, 66; US Dept Interior Distinguished Serv Award. Mem: AAAS; Am Inst Chem Eng; Am Chem Soc; Am Soc Testing & Mat (pres, 74-75). Res: Corrosion and encrustation mechanisms; geochemical controls of water quality; environmental sciences; corrosion and encrustation processes in water wells; systems for environmental impact assessment. Mailing Add: 165 Williams Dr Annapolis MD 21401

CLARKE, FRANK HENDERSON, b Newcastle, NB, Can, Dec 6, 27; m 54; c 2. ORGANIC CHEMISTRY, MEDICINAL CHEMISTRY. Educ: Univ NB, BSc, 49, MS, 50; Harvard Univ, PhD(org chem), 54. Prof Exp: Fel, Columbia Univ, 53-55; sr res chemist, med chem, Schering Corp, 55-62; res supvr, Geigy Chem Corp, 62-65 assoc dir med chem, 65-67, dir med chem, 67-71, DIR MED CHEM, CIBA-GEIGY CORP, 71- Concurrent Pos: Ed-in-chief, Ann Reports Med Chem, Am Chem Soc, 75- Mem: AAAS; Harvey Soc; NY Acad Sci; Pharmaceut Mfgrs Asn. Res: Design and synthesis of medicinal agents. Mailing Add: 14 Long Pond Rd Windmill Farms Armonk NY 10504

CLARKE, FRANK RUSSELL, experimental psychology, see 12th edition

CLARKE, GARRY K C, b Hamilton, Ont, Oct 6, 41. SEISMOLOGY, GLACIOLOGY. Educ: Univ Alta, BSc, 63; Univ Toronto, MA, 64, PhD(physics), 67. Prof Exp: ASSOC PROF GEOPHYS, UNIV BC, 67- Concurrent Pos: Partic glaciol expeds, Can & Greenland. Mem: Soc Explor Geophys; Am Geophys Union; Seismol Soc Am; Inst Elec & Electronics Engrs. Res: Geophysical applications of statistical communication theory; glacier flow theory. Mailing Add: Dept of Geophys Univ of BC Vancouver BC Can

CLARKE, GARY ANTHONY, b Washington, DC, May 31, 46; m 75. MICROBIAL PHYSIOLOGY. Educ: Indiana Univ, Pa, BS, 68; St Bonaventure Univ, PhD(biol), 73. Prof Exp: ASST PROF BIOL, ROANOKE COL, 73- Mem: Am Soc Microbiol. Res: Chemostatic growth of nitrogen fixing bacteria and the effect of environmental variables on this growth; carbon dioxide fixation in achlorophyllous angiosperms. Mailing Add: Dept of Biol Roanoke Col Salem VA 24153

CLARKE, GEORGE, b Readfield, Maine, Mar 10, 15; m 38; c 4. ANALYTICAL CHEMISTRY. Educ: Univ Maine, AB, 36. Prof Exp: Chemist, 43-51, group leader anal chem, reactor testing sta, Chem Processing Plant, 51-54, res chemist, 54-62, group leader anal chem, 62-70, GROUP HEAD ANAL CHEM, CENT RES DIV, AM CYANAMID CO, 70- Mem: Am Chem Soc; Am Microchem Soc. Mailing Add: Stamford Res Lab Am Cyanamid Co 1937 W Main St Stamford CT 06904

CLARKE, GEORGE A, b New York, NY, Apr 4, 33; m 66; c 2. PHYSICAL CHEMISTRY. Educ: City Col New York, BS, 55; Pa State Univ, PhD(phys chem), 60. Prof Exp: Res assoc theoret chem, Columbia Univ, 60-62; asst prof chem, State Univ NY Buffalo, 62-68; assoc prof, Drexel Inst Technol, 68-70 & Drexel Univ, 70-71; ASSOC PROF CHEM, UNIV MIAMI, 71- Concurrent Pos: USPHS grants, 64-66; co-sr investr, AEC Proj Grant, 64-68. Mem: Am Chem Soc; Am Phys Soc. Res: Intermolecular interactions in gaseous and condensed media; studies on molecular complexes; inter-and intramolecular energy transfer processes; electrolyte effects on solvent structure and reactive species; approximation methods in quantum chemistry. Mailing Add: Dept of Chem Univ of Miami Coral Gables FL 33124

CLARKE, GEORGE LEONARD, biology, see 12th edition

CLARKE, HANS THACHER, organic chemistry, deceased

CLARKE, JAMES SPENCER, b Chicago, Ill, June 28, 18; m 49; c 2. SURGERY. Educ: Harvard Univ, BS, 40, MD, 43. Prof Exp: From instr to asst prof surg, Univ Chicago, 53-56; from res asst prof to res assoc prof, Med Ctr, Univ Calif, Los Angeles, 56-63; prof & chmn dept, Sch Med, Univ NMex, 63-66; PROF IN RESIDENCE SURG, SCH MED, UNIV CALIF, LOS ANGELES, 69- Mem: AMA; Am Col Surg; Soc Univ Surg; Soc Exp Biol & Med; Am Surg Asn. Res: Gastrointestinal physiology. Mailing Add: Dept of Surg Ctr for Health Sci Univ of Calif Sch of Med Los Angeles CA 90024

CLARKE, JOHN, b Cambridge, Eng, Feb 10, 42. PHYSICS. Educ: Cambridge Univ, BS, 64, MS, 68, PhD(physics), 68. Prof Exp: Scholar, 68-69, from asst prof to assoc prof, 69-73, PROF PHYSICS, UNIV CALIF, BERKELEY, 73- Concurrent Pos: Alfred P Sloan fel, 70-72. Mem: Am Phys Soc. Res: Superconductivity; Josephson tunneling application to measurement of low voltages and magnetic fields and detection of electromagnetic radiation; experimental and theoretical study of low frequency electrical noise in solids. Mailing Add: Dept of Physics Univ of Calif Berkeley CA 94720

CLARKE, JOHN, b Belfast, Northern Ireland, Jan 5, 43; Can & Brit citizen; m 68; c 1. HISTORICAL GEOGRAPHY. Educ: Queen's Univ, Belfast, BA, 65; Univ Man, MA, 67; Univ Western Ont, PhD(geog), 70. Prof Exp: Asst prof, 71-75, ASSOC PROF GEOG, CARLETON UNIV, 75- Concurrent Pos: Can Coun fel & Nat Adv Comt Geog Res award grant, Univ Western Ont, 70-71; Can Coun res grant, Carleton Univ, 71-72. Mem: Can Asn Geogr; Am Asn Geogr. Res: Historical geography of Canada; quantitative methods and historical geography. Mailing Add: Dept of Geog Carleton Univ Ottawa ON Can

CLARKE, JOHN F, b Hempstead, NY, Sept 5, 39. PHYSICS. Educ: Fordham Univ, BS, 61; Mass Inst Technol, MS, 64, PhD(nuclear eng), 66. Prof Exp: Res staff mem plasma physics, 66-73, group leader confinement physics, 73-74, DIR THERMONUCLEAR DIV, OAK RIDGE NAT LAB, 74- Concurrent Pos: Mem fusion power coord comt, Div Controlled Thermonuclear Res, Energy Res & Develop Admin, 74; mem, Joint US/USSR Fusion Power Coord Comt, 74. Mem: Am Phys Soc; fel AAAS. Res: Thermonuclear fusion; plasma heating and confinement; MHD equilibrium; neutral particle and heavy ion transport in plasmas. Mailing Add: Oak Ridge Nat Lab PO Box Y Bldg 9201-2 Oak Ridge TN 37830

CLARKE, JOHN FREDERICK GATES, b Victoria, BC, Feb 22, 05; nat US; m 29; c 2. ENTOMOLOGY. Educ: State Col Wash, PhC, 26, BS, 30, MS, 31; Univ London, PhD(entom), 53. Prof Exp: Pharmacist, Red Cross Pharm, Wash, 26-29; fel, State Col Wash, 30-32, instr, 33-35; assoc entomologist, Bur Entom & Plant Quarantine, USDA Nat Mus, 36-40, entomologist, 40, entom res br, Agr Res Serv, 53-54; cur insects, 54-63, chmn dept entom, 63-66, SR ENTOMOLOGIST, SMITHSONIAN INST, 66- Mem: Fel Royal Entom Soc London. Res: Insect gall-formation; life histories and habits of Microlepidoptera; morphology and classification of Microlepidoptera. Mailing Add: Dept of Entom Smithsonian Inst Washington DC 20560

CLARKE, JOHN FREDERICK GATES, JR, b Pullman, Wash, Nov 11, 33; m 56; c 2. ANALYTICAL CHEMISTRY. Educ: Wash State Univ, BS, 55; Purdue Univ, MS, 58, PhD(anal chem), 60. Prof Exp: Res chemist, 60-70, SR RES CHEMIST & GROUP LEADER GAS CHROMATOGRAPHY SECT, HERCULES RES CTR, HERCULES, INC, 70- Mem: Am Chem Soc. Res: Automation of analytical instrumentation, particularly gas chromatography. Mailing Add: 2610 Belaire Dr Montclare Newport DE 19804

CLARKE, JOHN ROSS, b Martinsville, Va, Mar 23, 41; m 65; c 2. SOLID STATE ELECTRONICS. Educ: Univ Va, BEE, 64, MS, 67, PhD(elec eng), 70. Prof Exp: Elec engr, Warrenton Training Ctr, 64-65; PHYSICIST, EASTMAN KODAK CO RES LABS, 70- Res: Vacuum deposition preparation and characterization of photoconductive films. Mailing Add: Eastman Kodak Co Res Labs 1669 Lake Ave Rochester NY 14650

CLARKE, JOY HAROLD, b Lafayette, Ind, June 12, 99; m 22; c 2. HORTICULTURE. Educ: Purdue Univ, BSA, 21; Univ Del, MS, 23; Columbia Univ, PhD(bot), 42. Prof

Exp: Asst horticulturist, Exp Sta, Univ Del, 21-23; from instr to prof pomol, Rutgers Univ, 23-46, from asst pomologist to assoc pomologist, NJ Exp Sta, 23-46; gen mgr, Cranguyma Farms, 46-54; NURSERYMAN, HORT CONSULT & WRITER, 54- Concurrent Pos: Owner, Clarke Nursery. Mem: Am Soc Hort Sci; Royal Hort Soc. Res: Culture and breeding of small fruits; rhododendrons and azaleas. Mailing Add: Rte 1 Box 168 Long Beach WA 98631

CLARKE, LEMUEL FLOYD, b Newton, Utah, June 21, 06; m 27; c 3. ZOOLOGY. Educ: Utah State Col, BS, 27, MS, 31; Univ Chicago, PhD(exp embryol), 35. Prof Exp: Field asst, bur entom, USDA, Utah, 27; teacher, high sch, 27-29; asst entom, exp sta, Utah State Col, 30-31; asst zool, Univ Chicago, 31-33; sr teaching asst, Univ Rochester, 33-34; instr, agr col br, Utah State Col, Cedar City, 34-35; from instr to assoc prof, 35-45, head dept zool & physiol, 52-68, PROF ZOOL, UNIV WYO, 45-, DIR PREMED CURRIC, 37-, ASSOC DEAN COL HEALTH SCI, 68- Concurrent Pos: Dir, Jackson Hole Biol Res Sta, 53-72. Mem: AAAS; Am Soc Zoologists; AMA; Am Asn Med Cols; Am Soc Allied Health Professions. Res: Potentialities of embryonic germ areas; insects related to alfalfa seed production; sage grouse reproductive cycles. Mailing Add: Col of Health Sci Univ of Wyo Laramie WY 82071

CLARKE, LILIAN A, b Humboldt, Iowa, Aug 9, 15; m 42. INORGANIC CHEMISTRY. Educ: Grinnell Col, BA, 36; Pa State Univ, MS, 39, PhD(inorg chem), 42. Prof Exp: Instr chem, Pa State Univ, 37-42; ASST PROF CHEM, VILLANOVA UNIV, 63- Concurrent Pos: Consult, Berks Assoc, Inc, 68- & Basic Inc, 69- Mem: AAAS; Am Chem Soc. Res: Studies of properties of surface active agents; identification of chemical and biological warfare agents; rerefining of crankcase oils; water pollution abatement; solvent extraction of lubricating oils; chemistry of carbon black. Mailing Add: Dept of Chem Villanova Univ Villanova PA 19085

CLARKE, NORMAN ARTHUR, b Waterbury, Conn, May 20, 22; m 47; c 3. MICROBIOLOGY. Educ: Univ Conn, BS, 47; Yale Univ, PhD(microbiol), 51; Am Bd Microbiol, dipl. Prof Exp: CHIEF, MICROBIOL CONTROL BR, HEALTH EFFECTS RES LAB, ENVIRON PROTECTION AGENCY, 50- Mem: AAAS; Am Soc Microbiol; fel Am Pub Health Asn; Am Water Works Asn; Am Acad Microbiol. Res: Enteric viruses; chemical and physical disinfection; water and sewage treatment. Mailing Add: Environ Protection Agency 26 St Clair St Cincinnati OH 45268

CLARKE, RAY ALLEN, chemistry, see 12th edition

CLARKE, RICHARD HENRY, b Worcester, Mass, Dec 6, 42; m 66; c 4. PHYSICAL CHEMISTRY. Educ: Boston Col, BS, 65; Univ Pa, PhD(phys chem), 69. Prof Exp: Fel chem, Univ Chicago, 69-71; asst prof, 71-72, ASSOC PROF CHEM, BOSTON UNIV, 72- Concurrent Pos: NIH fel, 70; Alfred P Sloan Found res fel, 72. Mem: Am Phys Soc. Res: Molecular electronic spectroscopy and magnetic resonance of biological molecules and molecular systems. Mailing Add: Dept of Chem Boston Univ Boston MA 02215

CLARKE, RICHARD PENFIELD, b Baltimore, Md, Jan 30, 19; m 46; c 4. PHYSICAL CHEMISTRY Educ: Princeton Univ, AB, 41, MA, 48, PhD(phys chem), 50. Prof Exp: Asst, Manhattan Proj, Princeton Univ, 43-45; chemist, Air Reduction Co, 46; asst, Princeton Univ, 47-50; chemist, res dept, Standard Oil Co Ind, 50-51; proj leader, Exp, Inc, 51-56; mgr res, Okonite Co, 56-59; vpres, Hasche Eng Co, Tenn, 59-64; from vpres to pres, Kalamazoo Spice Extraction Co, 64-70; CONSULT CHEM ENGR, FUEL GAS PROD & ECON, 70- Concurrent Pos: Vpres, Thermo Tile Corp, 53-59, pres, 59-69. Mem: Am Chem Soc; Nat Soc Prof Engrs; NY Acad Sci. Res: Dielectric increments of amino acid and polypetide solutions; kinetics; decomposition diborane; heterogeneous catalysis; oxidation hydrocarbons; combustion; chemical engineering. Mailing Add: 2126 Benjamin Ave Kalamazoo MI 49001

CLARKE, ROBERT ALMA, b American Fork, Utah, June 26, 11; m 36; c 5. PHYSICS, MATHEMATICS. Educ: Brigham Young Univ, BS, 33, MS, 34; Calif Inst Technol, PhD(physics), 37. Prof Exp: Teacher physics & math, 37-40, chmn tech div, 41-45, dean fac, 47-67, part-time teacher, 45-60, PROF PHYSICS & MATH, WEBER STATE COL, 60-, ADMIN VPRES, 67- Mem: AAAS; Am Asn Physics Teachers. Res: Principles of physics; analytical mechanics; calculus. Mailing Add: Weber State Col 3750 Harrison Blvd Ogden UT 84403

CLARKE, ROBERT FRANCIS, b Portsmouth, Va, Oct 8, 19; m 47; c 2. ZOOLOGY, ANIMAL BEHAVIOR. Educ: Kans State Teachers Col, BSEd, 55, MS, 57; Okla Univ, PhD(zool), 63. Prof Exp: Instr educ, 56-58, from instr to assoc prof biol, 58-68, PROF BIOL, EMPORIA KANS STATE COL, 68-, CHMN DEPT, 72- Mem: Am Soc Ichthyol & Herpet. Res: Display behavior of lizards, particularly the family Iguanidae; ecology of reptiles; color change in gravid lizards. Mailing Add: Div of Biol Sci Emporia Kans State Col Emporia KS 66801

CLARKE, ROBERT LA GRONE, b Tullahoma, Tenn, Mar 10, 17; m 43; c 2. PHARMACEUTICAL CHEMISTRY. Educ: Ga Inst Technol, BS, 38; Emory Univ, MS, 39; Univ Wis, PhD(org chem), 47. Prof Exp: Instr chem, Young Harris Jr Col, Ga, 39-40; ORG CHEMIST, STERLING-WINTHROP RES INST, 47- Mem: Am Chem Soc. Res: Ketene acetals; nitrogen heterocycles; steroids; mercurial diuretics; alkaloids. Mailing Add: Sterling-Winthrop Res Inst Rensselaer NY 12144

CLARKE, ROBERT LEE, b Vermilion, Alta, Apr 17, 22; m 45; c 4. NUCLEAR PHYSICS. Educ: Univ Alta, BSc, 43; McGill Univ, PhD(physics), 48. Prof Exp: Asst physics, Nat Res Coun, Can, 43-45; from asst res officer to assoc res officer, Atomic Energy Can, Ltd, 48-68; PROF PHYSICS, CARLETON UNIV, 68- Mem: Am Phys Soc; Can Asn Physicist. Res: Nuclear reactions; neutrons, low to medium energy; radiography; medical physics. Mailing Add: Dept of Physics Carleton Univ Ottawa ON Can

CLARKE, ROBERT TRAVIS, b Brooklyn, NY, Nov 17, 37; m 58; c 4. PALYNOLOGY, PALEONTOLOGY. Educ: Univ Okla, MS, 61, PhD(geol), 63. Prof Exp: Res tech palynology, Socony Mobil Oil Co, Inc, 63-67, SR RES GEOLOGIST, FIELD RES LAB, MOBIL OIL CORP, 67- Mem: Geol Soc Am; Brit Paleont Asn; Am Asn Stratig Palynologists. Res: Research in the field of palynology applicable to resolving problems in stratigraphic zonation and correlation; paleoecologic and environmental interpretations and age determinations. Mailing Add: Field Res Lab Mobil Res & Develop Corp PO Box 900 Dallas TX 75221

CLARKE, ROY SLAYTON, JR, b Philadelphia, Pa, Jan 23, 25; m 51; c 3. GEOCHEMISTRY. Educ: Cornell Univ, AB, 49; George Washington Univ, MS, 57. Prof Exp: Chemist, USDA, 49-51; res assoc chem, George Washington Univ, 51-52; chemist, USDA, 52-53; anal chemist, US Geol Surv, 53-57; chemist, Div Meteorites, 57-66, assoc cur, 66-70, CUR, NAT MUS NATURAL HIST, 70- Mem: AAAS; Am Chem Soc; Mineral Soc Am; Geochem Soc. Res: Chemical analysis of meteorites and minerals. Mailing Add: US Nat Mus Washington DC 20560

CLARKE, THEODORE HUBER, chemistry, see 12th edition

CLARKE, THOMAS ARTHUR, b Peoria, Ill, Aug 13, 40. OCEANOGRAPHY, ECOLOGY. Educ: Univ Chicago, BS, 62; Univ Calif, San Diego, PhD(oceanog), 68. Prof Exp: Asst prof oceanog, 68-74, ASST MARINE BIOLOGIST, HAWAII INST MARINE BIOL, 68-, ASSOC PROF OCEANOG, UNIV HAWAII, 74- Mem: AAAS; Ecol Soc Am; Am Soc Limnol & Oceanog; Brit Ecol Soc. Res: Behavior and population dynamics of pomacentrid fish; shark ecology; fisheries ecology. Mailing Add: Hawaii Inst Marine Biol PO Box 1067 Univ of Hawaii Kaneohe HI 96744

CLARKE, W T W, b Toronto, Ont, Nov 6, 20; m 47; c 4. INTERNAL MEDICINE. Educ: Univ Toronto, MD, FRCP(C), 50. Prof Exp: PROF MED, UNIV TORONTO, 66- Concurrent Pos: Dep physician-in-chief, Toronto Gen Hosp, 69-74; mem drug qual & therapeut comt, Govt of Ont. Mem: Fel Am Col Physicians; Am Soc Nephrology; Am Diabetes Asn; Can Diabetic Asn; Can Med Asn. Res: Diabetes; nephrology. Mailing Add: Fac of Med Univ of Toronto Toronto ON Can

CLARKE, WILBUR BANCROFT, b Colon, Panama, July 22, 29; US citizen; m 59; c 1. ORGANIC CHEMISTRY. Educ: Xavier Univ, BS, 50, MS, 53; Univ Ind, PhD(chem), 62. Prof Exp: Instr chem, Xavier Univ, 50-53; asst neurol, US Army Chem Ctr, Md, 53-55; asst chem, Ind Univ, 56-58; chemist, Northern Regional Labs, 58-59; PROF CHEM, SOUTHERN UNIV, 60-, CHMN DEPT, 70- Concurrent Pos: Res grants, NSF & Sigma Xi, 63-64, NIH, 64-66; res specialist, Miss Test Facil, NASA-Gen Elec Co, 66 & 67; fel, La State Univ, 67-68; consult, NSF, 64. Mem: AAAS; Am Chem Soc; Brit Chem Soc. Res: Heterocyclics; preparation and elucidation of antiviral and anti-carcinogenic agents; infrared spectroscopy. Mailing Add: Dept of Chem Southern Univ Baton Rouge LA 70813

CLARKE, WILLIAM CAREY, cultural geography, see 12th edition

CLARKE, WILLIAM DIXON, oceanography, marine biology, see 12th edition

CLARKE, WILLIAM HENRY, astrophysics, see 12th edition

CLARKE, WILLIAM JAMES, b Seattle, Wash, July 16, 19; m 42; c 1. VETERINARY MEDICINE. Educ: Wash State Univ, BS, 43, DVM, 44, PhD(vet sci), 62. Prof Exp: Asst chief vet, res lab, Swift & Co, Ill, 44; dep state vet, Wash State Dept Agr, 44-45; owner, vet hosp, Wash, 45-56; res pathologist & sr scientist, biol lab, Gen Elec Co, 56-65; res assoc & mgr path sect, Pac Northwest Lab, 65-72, MGR BIOL, ECOL & MED SCI DEPT, COLUMBUS LAB, BATTELLE MEM INST, 72- Concurrent Pos: NIH fel, Wash State Univ, 60-62, adj assoc prof, 63-72 & Univ Wash, 67-72. Mem: AAAS; Am Vet Med Asn; Am Col Vet Toxicol; Health Physics Soc; Radiation Res Soc. Res: Research pathology; radiobiology; toxicology; bioengineering; research management. Mailing Add: Battelle Mem Inst Columbus Lab 505 King Ave Columbus OH 43201

CLARKE, WINSTON BROMLEY, physics, see 12th edition

CLARKSON, ALLEN BOYKIN, JR, b Augusta, Ga, July 1, 43; m 67; c 1. PARASITOLOGY. Educ: Univ of the South, BS, 65; Univ Ga, PhD(zool), 75. Prof Exp: Instr biol, Univ Ga, 72-74; FEL PARASITOL, ROCKEFELLER UNIV, 74- Mem: AAAS; Soc Protozoologists. Res: Parasitic protozoa, particularly trypanosomatids; carbohydrate pathways for energy production, especially as relates to microbodies, antigenic variations, immunological response and host parasite interactions at the parasite cell surface. Mailing Add: Dept of Parasitol Rockefeller Univ New York NY 10021

CLARKSON, BAYARD D, b New York, NY, July 15, 26; c 4. HEMATOLOGY, ONCOLOGY. Educ: Yale Univ, BA, 48; Columbia Univ, MD, 52. Prof Exp: From intern to resident, NY Hosp, 52-58; instr clin med, Med Col, Cornell Univ, 58-62, from asst prof to assoc prof med, 62-74; res fel, 58-59, res assoc, 59-61, assoc, 61-65, assoc mem, 65-71, MEM, SLOAN-KETTERING INST CANCER RES, 71-; PROF MED, MED COL, CORNELL UNIV, 74- Concurrent Pos: Mem adv comt ther, Am Cancer Soc, 65-68; mem pharmacol B study sect, NIH, 67-71; mem bd trustees, Cold Spring Harbor Lab, 68; mem ed adv bd, Cancer Res, 70; mem chemother adv comt, Nat Cancer Inst, 71-74; attend physician & chief hemat & lymphoma serv, Mem Hosp, NY, 71-; assoc ed, Cancer Res, 73-76. Mem: NY Acad Sci; Am Asn Cancer Res; fel Am Col Physicians; Am Soc Clin Oncol (pres, 73-74); Am Soc Clin Invest. Res: Cancer chemotherapy; leukemia; cell kinetics and regulation of cell growth as related to control of cancer. Mailing Add: Mem Sloan-Kettering Cancer Ctr 1275 York Ave New York NY 10021

CLARKSON, DONALD R, b Milford, Conn, Oct 27, 28; m 51; c 1. MATHEMATICS. Educ: Southern Conn State Col, BS, 58; Univ Conn, MS, 60, PhD(math, educ), 68. Prof Exp: PROF MATH & MATH EDUC, UNIV BRIDGEPORT, 63- Mem: AAAS; Math Asn Am. Res: Mathematics education. Mailing Add: 920 Robert Treat Dr Milford CT 06460

CLARKSON, JACK E, b Provo, Utah, June 17, 36; m 63; c 3. ANALYTICAL CHEMISTRY, NUCLEAR CHEMISTRY. Educ: Bringham Young Univ, BS, 60; Univ Calif, Berkeley, PhD(chem), 65. Prof Exp: SR CHEMIST, LAWRENCE LIVERMORE LAB, UNIV CALIF, 65- Mem: Am Chem Soc. Res: Trace impurity analysis by gas chromatography; gas chromatographic analysis of explosives; gel permeation chromatography of polymers and explosives; helium ionization detector utilization part-per-million analysis; high pressure liquid chromatography. Mailing Add: Lawrence Livermore Lab Univ Calif PO Box 808 L-404 Livermore CA 94550

CLARKSON, JAMES DAVID, b Cincinnati, Ohio, Dec 18, 32; m 57; c 1. GEOGRAPHY. Educ: Univ Chicago, PhD(geog), 67. Prof Exp: Actg asst prof geog, Univ & asst res geogr, Soc Sci Res Inst, Univ Hawaii, 66-67; from asst prof to assoc prof, Univ, 67-74, staff assoc, Ctr S & Southeast Asian Studies, 67-70, PROF GEOG, UNIV MICH, ANN ARBOR, 74-, RES ASSOC, CTR POP PLANNING, SCH PUB HEALTH, 70- Concurrent Pos: Assoc univ seminars, Columbia Univ, 67-68; mem, Southeast Asia Develop Adv Group, Asia Soc & AID, 66-; adv bd mil personnel supplies, Nat Res Coun-Nat Acad Sci, 67-70. Res: Tropical agriculture; Southeast Asia, especially Malaysia; problems of world food supply; cultural ecology. Mailing Add: Dept of Geog Univ of Mich Ann Arbor MI 48104

CLARKSON, MERTON ROBERT, b Ferndale, Wash, July 25, 08; m 30; c 3. VETERINARY MEDICINE. Educ: Wash State Univ, BS & DVM, 30; Georgetown Univ, LLB, 42. Prof Exp: Jr vet, meat inspection div, Bur Animal Indust, USDA, SDak, 30-32, NY, 32-35, inspector, Ind, 35-39, asst chief, trade label sect, 39-42 & war food admin, 42-44, chief, 44, asst chief, meat inspection div, 44-47, chief inspection & quarantine div, Agr Res Admin, 47-51, asst to adminr, defense progs, Agr Res Serv, 51-52, dep adminr, 52-59, assoc adminr, 59-64; CONSULT, 64- Concurrent Pos: Consult, Nat Security Resources Bd, 50-51; mem comt animal health, Nat Acad Sci-Nat Res Coun, chmn, 63-69, mem agr bd, vchmn, 65-70; dir, Am Vet Med Asn, 66-71, trustee, Prof liability ins trust, 66-71 & group ins trust, 67-; dir, Bur Vet Med, Food & Drug Admin, 66-, dir, prof exam serv, 70-71; mem adv coun, NY State Vet Col; dir, Monadnock Community Col. Honors & Awards: Distinguished

Serv Award, USDA, 56; Prizes, Am Vet Med Asn, 62 & 72; Award, Am Meat Inst, 64. Mem: Fel AAAS; Am Vet Med Asn (pres elect, 63, pres, 64); Am Soc Animal Sci; US Animal Health Asn; Conf Res Works Animal Dis. Res: Prevention, control and eradication of diseases and pests of livestock, crops and agricultural products with inspection and research programs to improve and safeguard the quality of agricultural products. Mailing Add: PO Box 388 Peterborough NH 03458

CLARKSON, QUENTIN DEANE, b Eugene, Ore, Sept 26, 25; m 47; c 4. BIOSTATISTICS. Educ: Univ Ore, BS, 49, MS, 50; Ore State Col, PhD(bot), 55. Prof Exp: Instr bot, Ore State Col, 52-54; from asst prof to assoc prof, Portland State Col, 55-65; assoc prof, NC State Univ, 65-69; assoc prof neurol, Med Sch, Univ Ore, 69-74; MEM FAC, DEPT SOCIAL WORK, PORTLAND STATE UNIV, 74- Res: Biostatistics; systematics. Mailing Add: Dept of Social Work Portland State Univ Portland OR 97207

CLARKSON, ROBERT BRECK, b Buffalo, NY, Apr 19, 43; m 65; c 2. PHYSICAL CHEMISTRY, SURFACE CHEMISTRY. Educ: Hamilton Col, BA, 65; Princeton Univ, MA, 68, PhD(chem), 69. Prof Exp: ASST PROF PHYS CHEM, UNIV WIS-MILWAUKEE, 69- Mem: Am Chem Soc; Am Phys Soc; Catalysis Soc. Res: Applications of electron paramagnetic resonance and nuclear magnetic resonance characterization of gas-solid interactions. Mailing Add: Dept of Chem Univ of Wis Milwaukee WI 53201

CLARKSON, ROY BURDETTE, b Cass, WVa, Oct 25, 26; m 52; c 3. BOTANY. Educ: Davis & Elkins Col, BS, 51; WVa Univ, MA, 54, PhD, 60. Prof Exp: Teacher, pub sch, WVa, 51-56; from instr to assoc prof, 56-69, assoc chmn dept, 69-74, actg chmn, 74-75, PROF BIOL, W VA UNIV, 69-, CUR HERBARIUM, 75- Res: Plant taxonomy and geography; chemosystematics, especially comparison of macromolecules utilizing electrophoretic and serologic techniques. Mailing Add: Dept of Biol WVa Univ Morgantown WV 26506

CLARKSON, THOMAS WILLIAM, b UK, Aug 1, 32; m 57; c 2. TOXICOLOGY. Educ: Univ Manchester, BSc, 53, PhD(biochem), 56. Prof Exp: Med Res Coun fel, Univ Manchester, 56-57; instr radiation biol, Univ Rochester, 57-61, asst prof, 61-62; sci officer, Med Res Coun, UK, 62-64; sr fel, Weizmann Inst, 64-65; assoc prof, biophys, pharmacol & radiation biol, 65-71, PROF RADIATION BIOL, BIOPHYS, PHARMACOL & TOXICOL, UNIV ROCHESTER, 71-, DIR, ENVIRON HEALTH SCI CTR, 75- Concurrent Pos: Mem comt food protection, Nat Acad Sci-Nat Acad Eng, 73-76, subcomt toxicol, 72-76; mem toxicol adv bd, Food & Drug Admin, 75-77; mem toxicol study sect, NIH, 76-77. Mem: AAAS; Health Physics Soc; Brit Pharmacol Soc; Soc Toxicol; The Chem Soc. Res: Cellular physiology; reabsorption mechanisms in intestine and kidney; heavy metal toxicology; action of metals on cellular level in intestine, kidney and red blood cells. Mailing Add: Sch of Med Univ of Rochester Rochester NY 14642

CLARKSON, VERNON A, b Tacoma, Wash, Apr 19, 21; m 47; c 3. HORTICULTURE. Educ: Wash State Univ, BS, 49; Ore State Univ, MS, 51. Prof Exp: Asst, Inst Hort, Ore State Univ, 50-57; horticulturist, Plastics Div, Union Carbide Corp, 57-60 & Olefins Div, 60-70; HORTICULTURIST, AGR CHEM DIV, CIBA-GEIGY CHEM CORP, 70- Mem: Am Soc Hort Sci. Res: Agricultural uses of plastic materials; agricultural pesticides. Mailing Add: Ciba-Geigy Res Farm Star Rte Livingston NY 12541

CLARY, WARREN POWELL, b Lewellen, Nebr, Sept 8, 36; m 57; c 3. PLANT ECOLOGY. Educ: Univ Nebr, BS, 58; Colo State Univ, MS, 61, PhD, 72. Prof Exp: RANGE SPECIALIST, ROCKY MT FOREST & RANGE EXP STA, US FOREST SERV, 60- Mem: Soc Range Mgt. Res: Effects of site and woody overstory on herbage production. Mailing Add: Rocky Mountain Forest & Range Exp Sta Flagstaff AZ 86001

CLASE, HOWARD JOHN, b Salisbury, Eng, June 14, 38; m 63; c 2. INORGANIC CHEMISTRY. Educ: Cambridge Univ, BA, 60, PhD(inorg chem), 63, MA, 65. Prof Exp: Fel, McMaster Univ, 63-65; asst chem, Univ Oulu, 65-66; tutorial fel, Univ Sussex, 66-68; asst prof, 68-74, ASSOC PROF CHEM, MEM UNIV NF, 74- Mem: The Chem Soc; Chem Inst Can. Res: Application of Raman spectroscopy to problems in inorganic chemistry. Mailing Add: Dept of Chem Mem Univ Nfld St John's NF Can

CLASEN, RAYMOND ADOLPH, b Chicago, Ill, June 28, 26; m 50; c 1. PATHOLOGY. Educ: Univ Ill, BS, 50, MD, 52. Prof Exp: Intern, Cook County Hosp, 52-53; resident, Presby Hosp, Chicago, 53-57; Nat Inst Neurol Dis & Blindness fel, 57-59, asst prof 59-73, ASSOC PROF PATH, RUSH-PRESBY-ST LUKE'S MED CTR, CHICAGO, 73- Mem: AAAS; Am Asn Neuropath; Am Asn Pathologists & Bacteriologists; Am Soc Exp Path; Int Acad Path. Res: Experimental neuropathology. Mailing Add: 3440 Parthenon Way Olympia Fields IL 60461

CLASS, CALVIN MILLER, b Baltimore Co, Md, Jan 27, 24; m 48; c 1. NUCLEAR PHYSICS. Educ: Johns Hopkins Univ, AB, 43, PhD(physics), 51. Prof Exp: Physicist hydrodyn, Nat Adv Comt Aeronaut, 44-46; asst nuclear physics, Johns Hopkins Univ, 49-52; from instr to assoc prof, 52-63, PROF PHYSICS, RICE UNIV, 63- Concurrent Pos: Guggenheim fel, 55-56. Mem: Fel Am Phys Soc; Ital Phys Soc. Res: Spectroscopy of light and medium nuclei. Mailing Add: Dept of Physics Rice Univ Houston TX 77001

CLASS, JAY BERNARD, b Baltimore, Md, Apr 14, 28; m 58; c 2. ORGANIC CHEMISTRY. Educ: Univ Md, BS, 49; Pa State Univ, PhD(org chem), 52. Prof Exp: Res chemist, 52-66, RES SUPVR, RES CTR, HERCULES INC, 66- Mem: Am Chem Soc; Sigma Xi. Res: Rubber chemicals; synthetic rubber; rosin and fatty acid derivatives; resins and plastics; organic synthesis. Mailing Add: Hercules Res Ctr Hercules Inc Wilmington DE 19899

CLASSEN, HAROLD ARTHUR, b Gilman, Ill, Aug 15, 21; m 52; c 2. ECONOMIC GEOGRAPHY, GEOGRAPHY OF ASIA. Educ: Ill State Univ, BEd, 42; Univ Nebr, PhD(geog), 55. Prof Exp: Teaching asst geog, Univ Nebr, 47-49; instr, Univ Nev, 49-51; teaching asst, Univ Nebr, 51-53; PROF GEOG, UNIV WIS-LA CROSSE, 54- Mem: Asn Am Geogr; Nat Coun Geog Educ. Res: Minerals industries. Mailing Add: Dept of Geog Univ of Wis La Crosse WI 54601

CLATOR, IRVIN GARRETT, b Huntington, WVa, Nov 2, 41. NUCLEAR PHYSICS, EXPLOSIVES. Educ: WVa Univ, BS, 63, MS, 65, PhD(physics), 69. Prof Exp: From physicist to res physicist, US Naval Weapons Lab, 65-70; asst prof, 70-74, ASSOC PROF PHYSICS, UNIV NC, WILMINGTON, 74-, CHMN DEPT, 71- Mem: AAAS. Res: Neutron induced reaction in the 10 to 20 mev energy range with medium A nuclei; explosive material properties. Mailing Add: Dept of Physics Univ of NC PO Box 3725 Wilmington NC 28401

CLATWORTHY, WILLARD HUBERT, b Auxier, Ky, Oct 16, 15; m 44; c 2. STATISTICS. Educ: Berea Col, BA, 38; Univ Ky, MA, 40; Univ NC, PhD(math statist), 52. Prof Exp: Asst, Univ Ky, 38-40; prof, Louisburg Col, 40-42; tool designer, Wright Automatic Packing Mach Co, 42 & Bell Aircraft Corp, 42-43; instr math, Wayne State Univ, 46-49; instr math & probability, Nat Bur Standards, 52-55; statistician, Bettis Atomic Power Div, Westinghouse Elec Corp, 55-62; PROF STATIST, STATE UNIV NY BUFFALO, 62- Mem: Fel AAAS; Inst Math Statist; fel Am Statist Asn; Int Asn Statist in Phys Sci; fel Royal Statist Soc. Res: Mathematics of statistical design of experiments; combinatorial aspects of design of experiments; regression and least squares; analysis of variance. Mailing Add: 378 Cottonwood Dr Williamsville NY 14221

CLAUDE, PHILIPPA, b New York, NY, Jan 21, 36. CELL BIOLOGY, NEUROBIOLOGY. Educ: Cornell Univ, BA, 57; Univ Pa, PhD(zool), 68. Prof Exp: Instr, Harvard Med Sch, 69-72, prin res assoc neurobiol, 72-75; ASST SCIENTIST, WIS REGIONAL PRIMATE RES CTR, 75- Concurrent Pos: NIH fel, Harvard Med Sch, 69-72, NIH spec fel, 72-73. Mem: Am Soc Cell Biol; Soc Neurosci; Electron Micros Soc Am. Res: Intercellular junctions; neuronal development and neurospecificity; neuronal tissue culture. Mailing Add: Wis Regional Primate Res Ctr 1223 Capitol Ct Madison WI 53706

CLAUNCH, ROBERT THOMAS, organic chemistry, inorganic chemistry, see 12th edition

CLAUS, CARL JACOB, electrophotography, see 12th edition

CLAUS, GEORGE, b Budapest, Hungary, Jan 16, 32; US citizen; m. MICROBIOLOGY, MARINE BIOLOGY. Educ: Eötvös Lorand Univ, Dipl bot, 55; Univ Budapest, Dr rer nat, 56. Prof Exp: Instr plant taxon, Eötvös Lorand Univ, 53-56; Rockefeller Found res fel ecol, Univ Vienna, 56-58; asst cur limnol, Acad Natural Sci, Philadelphia, 58-59; instr microbiol & Am Rheumatism Asn res fel, Med Ctr, NY Univ, 59-64; assoc prof biol, Fla State Univ, 64-65; chief microbiol, Repub Aviation Div, Fairchild Hiller Corp, 65-68; VPRES, OFFSHORE SEA DEVELOP CORP, 68- Concurrent Pos: Res grants, USPHS, 59-64, Lederle Labs, 61-64 & NASA, 62-63; mem, Coun Sci & Indust Res, Pretoria, SAfrica, 60-61; vis asst res prof, Fordham Univ, 62-64; assoc prof, Long Island Univ, 65-69. Mem: NY Acad Sci; Int Phycol Soc; Phycol Soc Am; Int Asn Theoret & Appl Limnol; Am Inst Biol Sci. Res: Medical microbiology; exo-biology; ultra-structure of the blue-green algae, water resources and microbial pollution of waters. Mailing Add: Offshore Sea Develop Corp 99 Nassau St New York NY 10038

CLAUS, GEORGE WILLIAM, b Council Bluffs, Iowa, Aug 15, 36; m 58; c 3. MICROBIOLOGY. Educ: Iowa State Univ, BS, 59, PhD(physiol bact), 64. Prof Exp: Bacteriologist & biochemist, US Army Med Unit, Ft Detrick, Md, 64-66; asst prof microbiol, Pa State Univ, University Park, 66-73; ASST PROF MICROBIOL, VA POLYTECH INST & STATE UNIV, 73- Concurrent Pos: Consult, Hoffmann-La Roche, Inc, 74- Mem: AAAS; Am Soc Microbiol; Brit Soc Gen Microbiol; Sigma Xi; Electron Micros Soc Am. Res: Physiology and fine-structure of acetic acid bacteria; intracytoplasmic membrane development in nonphotosynthetic gram-negative bacteria; limited oxidation of Gluconobacter; biochemical cytology. Mailing Add: Dept of Biol Va Polytech Inst & State Univ Blacksburg VA 24061

CLAUS, THOMAS HARRISON, b Kansas City, Mo, Jan 17, 43; m 66. MEDICAL PHYSIOLOGY. Educ: Wheaton Col, Ill, BS, 64; Univ Ill, Chicago, PhD(biochem), 70. Prof Exp: Res assoc physiol, 69-73, ASST PROF PHYSIOL, SCH MED, VANDERBILT UNIV, 73- Mem: Am Chem Soc; Brit Biochem Soc. Res: Mechanism of intracellular protein degradation and of hormonal regulation of hepatic carbohydrate metabolism. Mailing Add: Dept of Physiol Vanderbilt Univ Sch of Med Nashville TN 37232

CLAUS, WILBUR SCHEIRICH, b Spencer Co, Ind, Mar 26, 11; m 38; c 4. CHEMISTRY. Educ: Wheaton Col, BS, 31; Mich State Col, MS, 34; Iowa State Col, PhD(plant chem), 37. Prof Exp: Sr indust fel, Mellon Inst, 37-41; res chemist, Campbell Taggart Res Corp, Mo, 41-43; group leader, Gen Mills, Inc, Minn, 43-46; res chemist, Brown & Bigelow, 47; dir res & develop, Albers Milling Co, 47-53; head cereal res div, 53-55, head cereal & frozen food res, 55-59, asst dir res, 59-66, dir basic res, 66-72, DIR COORD & PLANNING, RES LAB, CARNATION CO, 72- Concurrent Pos: Dir, Campbell Taggart Res Corp; mem bd dirs, League Int Food Educ, 70- Mem: Am Asn Cereal Chemists (pres, 66-67); Inst Food Technologists; Am Chem Soc; Am Soc Bakery Engrs. Res: Biochemistry; nutrition; flavor; microbiology; analytical research. Mailing Add: 2706 W Appalachian Ct Westlake Village CA 91361

CLAUSEN, CHRIS ANTHONY, b New Orleans, La, Dec 7, 40; m 62; c 2. INORGANIC CHEMISTRY. Educ: La State Univ, Baton Rouge, BS, 63; La State Univ, New Orleans, PhD(inorg chem), 69. Prof Exp: Chemist, Standard Oil Co Calif, 63-66; asst prof, 69-72, ASSOC PROF CHEM, FLA TECHNOL UNIV, 72- Concurrent Pos: AEC res grant, 69-70; tour speaker, Am Chem Soc, 74. Honors & Awards: Excellence in Sci Res Award, Sigma Xi, 69. Mem: Am Chem Soc; Soc Appl Spectros. Res: Mössbauer spectroscopy as applied in bonding and structural studies; molecular vibrations in the far infrared; coordination chemistry in marine environments. Mailing Add: Dept of Chem Fla Technol Univ PO Box 25000 Orlando FL 32816

CLAUSEN, CONRAD DUANE, b Takoma Park, Md, May 18, 43; m 75. INVERTEBRATE ZOOLOGY. Educ: Columbia Union Col, BA, 66; Loma Linda Univ, PhD(biol), 72. Prof Exp: ASST PROF BIOL, LOMA LINDA UNIV, 72- Mem: Am Inst Biol Sci; Sigma Xi. Res: The use of periodical growth lines found in invertebrates in geochronometry, ecology and paleoecology. Mailing Add: Dept of Biol Loma Linda Univ Loma Linda CA 92354

CLAUSEN, ERIC NEIL, b Ithaca, NY, July 2, 43. GEOMORPHOLOGY. Educ: Columbia Univ, BS, 65; Univ Wyo, PhD(geol), 69. Prof Exp: Asst prof geol, 68-75, ASSOC PROF EARTH SCI & DIR ACAD COMPUT SERV, MINOT STATE COL, 75- Mem: Geol Soc Am; Nat Asn Geol Teachers; Am Statist Asn; Opers Res Soc Am; Am Asn Quaternary Environ. Res: Badland geomorphology; quantitative techniques in geology; earth science education. Mailing Add: Dept of Earth Sci Minot State Col Minot ND 58701

CLAUSEN, HARRY JOHN, anatomy, see 12th edition

CLAUSEN, JEWELL JOHANNA, b Chicago, Ill, Nov 8, 31; m 56. ECOLOGY. Educ: Butler Univ, BA, 52; Univ Wis-Madison, PhD(bot), 55. Prof Exp: Berliner res fel, Am Asn Univ Women, Copenhagen, 55-56; res assoc, dept microbiol, Univ Minn, 56-61; part-time instr zool, Univ Wis Ctr Syst, 63-64, summer proj assoc, dept forestry, 63-66, asst prof bot, 66-70, assoc prof, Marathon County Campus, 70-71; ECOLOGIST, NICOLET COL & TECH INST, 71- Mem: AAAS; Ecol Soc Am. Mailing Add: Nicolet Col & Tech Inst Box 518 Rhinelander WI 54501

CLAUSEN, KNUD ERIK, b Saksk bing, Denmark, Nov 20, 27; US citizen; m 56.

FOREST GENETICS. Educ: Univ Minn, MS, 59, PhD(forest genetics), 61. Prof Exp: Asst forester, H rsholm Forestry Dist, Denmark, 50-51; field asst, Gävleborgs Läns Skogsvardsstyrelse, Sweden, 51; trainee, Skoghallsverken, Uddeholm Co, 51; field asst, genetics, Forest Res Inst, Sweden, 52-53; res asst, Univ Wis, 54; field asst genetics, Forest Res Inst Sweden, 55-56; res forester, 61-62, GENETICIST, N CENT FOREST EXP STA, US FOREST SERV, 62- Concurrent Pos: Nat Acad Sci travel grant, Poland & Ger Acad Exchange Serv study grant, 70; consult, Inst Forest Improvement, Sweden, 70. Mem: AAAS; Am Genetic Asn; Soc Study Evolution. Res: Genetics of Betula. Mailing Add: NCent Forest Exp Sta Inst of Forest Genetics Rhinelander WI 54501

CLAUSEN, ROBERT THEODORE, b New York, NY, Dec 26, 11; m 42; c 4. PLANT TAXONOMY. Educ: Cornell Univ, AB, 33, AM, 34, PhD(plant taxon), 37. Prof Exp: Asst, Univ, 33-35, from asst to asst prof, Baily Hortorium, 35-41, from asst prof to assoc prof, Univ, 41-49, PROF BOT, CORNELL UNIV, 49- Concurrent Pos: Collabr, USDA, 43. Mem: AAAS; Bot Soc Am; Am Soc Plant Taxonomists; Torrey Bot Club; Am Fern Soc (pres, 39-42). Res: Taxonomy of vascular plants; phytogeography; monograph of Ophioglossaceae; Crassulaceae; Sedum; ecology. Mailing Add: Wiegand Herbarium 462 Mann Libr Cornell Univ Ithaca NY 14850

CLAUSER, JOHN FRANCIS, b Pasadena, Calif, Dec 1, 42; m 64. QUANTUM MECHANICS. Educ: Calif Inst Technol, BS, 64; Columbia Univ, MA, 66, PhD(physics), 69. Prof Exp: Res physicist quantum physics, Univ Calif & Lawrence Berkeley Lab, 69-75, RES PHYSICIST, UNIV CALIF & LAWRENCE LIVERMORE LAB, 75- Mem: Am Phys Soc. Res: Reconciliation of everyday notions of objectivity, space and time with the observed and predicted behavior of quantum mechanical systems. Mailing Add: 35 Kazar Ct Moraga CA 94556

CLAUSER, MILTON JOHN, b Santa Monica, Calif, June 17, 40; m 61; c 2. PLASMA PHYSICS. Educ: Mass Inst Technol, SB, 61; Calif Inst Technol, PhD(physics), 66. Prof Exp: NSF res fel physics, Munich Tech Univ, 66-67; MEM TECH STAFF, SANDIA LABS, 67- Mem: Am Phys Soc. Res: Electron and ion beam fusion; target behavior; laser created plasmas. Mailing Add: Sandia Labs Div 5241 Albuquerque NM 87115

CLAUSS, JAMES K, b Cambridge, Mass, Aug 3, 20; m 44; c 3. ANALYTICAL CHEMISTRY, PHYSICAL CHEMISTRY. Educ: Reed Col, BA, 41; Ore State Univ, MS, 43, PhD(anal chem), 55. Prof Exp: Sr analyst, Gen Chem Div, Allied Chem Corp, 43-46, chief chemist, 46-52; from assoc chemist to chemist, Stanford Res Inst, 54-58; chief chemist display devices, Electron Tube Div, Litton Indust Inc, 58-64; MEM SR STAFF CHEM, DEPT RES & DEVELOP, SIGNETICS CORP, 64- Mem: Electrochem Soc; Sigma Xi. Res: Analytical methods; plant stream and effluent analyzers; emission spectrography trace elements; vacuum technology; cathode-ray tube materials and fabrication techniques; xerographic materials; electrophoretic coatings; integrated circuit process development. Mailing Add: 744 Coastland Dr Palo Alto CA 94303

CLAUSS, ROY H, b Ill, Feb 8, 23; m 45; c 3. CARDIOVASCULAR SURGERY. Educ: Northwestern Univ, BS, 43, MD, 46; Am Bd Surg, dipl, 56; Am Bd Thoracic Surg, dipl, 57. Prof Exp: Instr surg, Col Physicians & Surgeons, Columbia Univ, 55-57; asst, Harvard Med Sch, 57-58; asst prof, Col Med, Univ Cincinnati, 58-60; assoc prof, Sch Med, NY Univ, 60-69; PROF SURG, NEW YORK MED COL, 69- Concurrent Pos: Am Trudeau Soc teaching fel, 54-55; USPHS spec fel, Harvard Med Sch, 57-58; attend surgeon, Flower & Fifth Ave Hosps; vis surgeon, Metrop & Coler Hosps; consult, NY Vet Med Admin Hosp; mem coun circulation, Am Heart Asn. Mem: Am Col Chest Physicians; Am Heart Asn; Am Col Surg; Am Asn Thoracic Surg; Int Cardiovasc Soc. Res: Cardiovascular and pulmonary surgery and physiology. Mailing Add: New York Med Col Flower & Fifth Ave Hosps New York NY 10029

CLAUSSEN, DENNIS LEE, b Pender, Nebr, Sept 23, 41; m 64; c 2. PHYSIOLOGICAL ECOLOGY. Educ: Pomona Col, BA, 63; Univ Calif, Riverside, MA, 66; Univ Mont, PhD(zool), 71. Prof Exp: Entomologist, Nutrilite Prod Inc, Calif, 66-68; res asst biol control, Univ Calif, Riverside, 68; asst prof, 71-75, ASSOC PROF ZOOL, MIAMI UNIV, 75- Mem: AAAS; Sigma Xi; Am Soc Zoologists. Res: The metabolism, thermal relations and water relations of amphibians and reptiles. Mailing Add: Dept of Zool Miami Univ Oxford OH 45056

CLAUSSEN, WALTER FREDERICK, b Rock Rapids, Iowa, June 25, 16; m 45; c 1. PHYSICAL CHEMISTRY. Educ: Univ Ill, AB, 36, PhD(phys chem), 39. Prof Exp: Res asst, Univ Ill, 36-37, asst instr, 37-39; res chemist, Texas Co, 39-45, SAM Labs, Carbide & Carbon Chem Corp, NY, 45-46 & WVa, 46-47; asst res prof, Univ Ill, 47-51; res chemist, Corning Glass Works, 51-56, Res Lab, Gen Elec Co, 56-65 & Ill State Water Surv, 65-69; RES CONSULT, 69- Mem: AAAS; Am Chem Soc. Res: Infrared absorption spectra; structures of watery substances; petroleum chemistry; glassy oxides; super pressure technique; phase changes, kinetics and thermodynamics. Mailing Add: 2017 Cureton Dr Urbana IL 61801

CLAUS-WALKER, JACQUELINE LUCY, b Paris, France, Dec 13, 15; US citizen; m 65. ENDOCRINOLOGY. Educ: Univ Paris, BA, 35; Sorbonne, MS, 46; Union Col, BS, 51; Univ Houston, MS, 55; Baylor Univ, PhD(physiol), 66. Prof Exp: Lab technician, Robert Packer Hosp, 47-48; chief clin path & chief pharmacist, James Walker Mem Hosp, 51-53; res asst, Univ Tex M D Anderson Hosp & Tumor Inst, 55-58; res asst, 61-66, from instr to asst prof physiol chem, Dept Biochem & Dept Rehab, 66-71, ASSOC PROF PHYSIOL, DEPTS REHAB & PHYSIOL & ASST PROF BIOCHEM, BAYLOR COL MED, 71-, DIR NEUROENDOCRINE LAB, TEX INST REHAB & RES, 66- Mem: AAAS; Am Pharmaceut Asn; Am Soc Hosp Pharmacists; Endocrine Soc; Fr Soc Therapeut & Pharmacodyn. Res: Androgens and estrogen excretion in man; interrelation of adrenal and thyroid function in rats; endocrine function in man with section of the cervical spinal cord. Mailing Add: Tex Inst for Rehab & Res 1333 Moursund Ave Houston TX 77025

CLAUSZ, JOHN CLAY, b Hackensack, NJ, Oct 5, 40; m 63; c 2. MYCOLOGY. Educ: Ohio Wesleyan Univ, BA, 62; Univ NC, Chapel Hill, MA, 66, PhD(bot), 70. Prof Exp: ASST PROF BIOL, ST ANDREWS PRESBY COL, 69- Mem: Mycol Soc Am; Am Inst Biol Sci; Sigma Xi; Nat Sci Teachers Asn. Res: Physiology and ecology of fungi, especially the aquatic fungi; lipids in water molds. Mailing Add: Dept of Biol St Andrews Presby Col Laurinburg NC 28352

CLAVAN, WALTER, b Philadelphia, Pa, Apr 6, 21; m 45; c 2. ANALYTICAL CHEMISTRY. Educ: Univ Pa, BS, 42, MS, 47, PhD(anal chem), 49. Prof Exp: Anal chemist, E J Lavino & Co, 42-44; group leader, anal dept, 49-70, dir anal serv, 70-74, MGR, ANAL CHEM DEPT, KING OF PRUSSIA TECHNOL CTR, PENNWALT CORP, 74- Concurrent Pos: Lectr, eve div, La Salle Col, 60-70. Mem: AAAS; Am Chem Soc; Geochem Soc; Sigma Xi. Mailing Add: Benson E 620 Jenkintown PA 19046

CLAVEAU, ROSARIO, b Chicoutimi, Que, Dec 13, 24; m 51; c 2. HEMATOLOGY, INTERNAL MEDICINE. Educ: Chicoutimi Sem, BA, 45; Laval Univ, MD, 50; FRCP(C). Prof Exp: CHIEF DEPT HEMAT, HOPITAL DE CHICOUTIMI, LAVAL UNIV, 60-, HON PROF, FAC MED, 64- Mem: Am Soc Hemat; fel Am Col Physicians; NY Acad Sci; Can Med Asn. Mailing Add: Hemat Lab Hopital de Chicoutimi Chicoutimi PQ Can

CLAVERAN, RAMON A, b Aguascalientes, Mex, July 16, 34; m 61; c 2. RANGE SCIENCE. Educ: Nat Sch Agr, Mex, BS, 58; Univ Ariz, MS, 64, PhD(range mgt), 67. Prof Exp: Mem staff exten & admin animal prod, Fondo de Garantia, Bank Mex, 58-62, mem staff res & develop projs, 67-74; RESEARCHER & CONSULT PASTURE & ANIMAL PROD, NAT INST AGR INVEST, 74- Concurrent Pos: Teaching range mgt, Nat Sch Agr, Mex, 68-75; head, Forage Dept, Nat Inst Agr, 74; ed jour, Mex Soc Animal Prod, 69- & Latin Am Soc Animal Prod, 75; adv, Secy Agr Mex. Mem: Mex Soc Animal Prod (pres, 70-72); Latin Am Soc Animal Prod (vpres, 75); Soc Range Mgt; Brit Grassland Soc; Australian Trop Grassland Soc. Res: Beef and milk production systems in arid range, temperate pastures and tropical grasslands. Mailing Add: Nat Inst Agr Invest Apartado Postal 41-781 Mexico 10 DF Mexico

CLAWSON, ALBERT J, b Curtis, Nebr, Feb 15, 24; m 48; c 6. ANIMAL NUTRITION. Educ: Univ Nebr, BS, 49; Kans State Univ, MS, 51; Cornell Univ, PhD(nutrit), 55. Prof Exp: Asst, Kans State Univ, 49-51; animal husbandman, Exp Sta, North Platte, Nebr, 51-52; asst, Cornell Univ, 52-55; assoc prof, 55-69, PROF ANIMAL SCI, NC STATE UNIV, 69- Concurrent Pos: Res fel, Centro Internacional Agricultura Tropical, Cali, Colombia, 72-73. Mem: Am Inst Nutrit; Am Soc Animal Sci. Res: Nutrient requirements for reproduction in swine; amino acid requirements of pig as determined by manipulation of dietary ingredients; indirect methods of determining live animal composition and factors influencing carcass composition. Mailing Add: Dept of Animal Sci NC State Univ Raleigh NC 27607

CLAWSON, DAVID KAY, b Salt Lake City, Utah, Aug 8, 27; m 52; c 2. ORTHOPEDIC SURGERY. Educ: Harvard Univ, MD, 52; Am Bd Orthop Surg, dipl, 61. Prof Exp: Intern surg, Stanford Univ Hosp, 52-53, resident, 53-54, resident orthop, 54-55; resident, San Francisco City & County Hosp, 55-56; resident orthop, Stanford Univ Hosp, 56-57; asst prof surg, Univ Calif, Los Angeles, 58; head div surg & from asst prof to assoc prof surg, 58-65, PROF ORTHOP & CHMN DEPT, UNIV WASH, 65- Concurrent Pos: Nat Found Infantile Paralysis fels orthop, 55-57 & advan orthop, 57-58; hon sr registr, Royal Nat Orthop Hosp & clin res asst, Univ London, 57-58. Mem: Am Acad Orthop Surg; Royal Soc Med; Asn Bone & Joint Surgeons; fel Am Geriat Soc; AMA. Res: Infections of bone and joints; bone implants; health care delivery; orthopedic manpower. Mailing Add: Dept of Orthop Univ of Wash Seattle WA 98105

CLAWSON, ROBERT CHARLES, b South Haven, Mich, May 10, 29; m 58; c 5. HISTOLOGY, EMBRYOLOGY. Educ: Spring Hill Col, BS, 50; St Louis Univ, MS, 53; Loyola Univ Chicago, PhD(anat), 62. Prof Exp: From instr to asst prof anat, Stritch Sch Med, 64-68; asst prof, 68-70, ASSOC PROF ANAT, SCH MED, LA STATE UNIV, SHREVEPORT, 70- Concurrent Pos: NIH fel, Loyola Univ Chicago, 62-64. Mem: Am Soc Zool; Am Asn Anat; Pan Am Asn Anat. Res: Endocrinology; histochemistry; embryology and electron microscopy; primordial germ cells and effects of corticosteroids on growth and liver glycogen in the chick embryo. Mailing Add: Dept of Anat La State Univ Sch of Med Shreveport LA 71103

CLAXTON, WILLIAM EUGENE, b Hartville, Mo, Oct 27, 23; m 52; c 4. MATHEMATICS, STATISTICS. Educ: Harvard Univ, BS, 48; Univ Cincinnati, MS, 50. Prof Exp: Res physicist, 51-73, res assoc, 73-75, SR RES ASSOC, CENT RES LAB, FIRESTONE TIRE & RUBBER CO, 75- Mem: Soc Comput Simulations; Am Chem Soc. Res: Applications of experimental design and statistical analysis; APL scientific computer applications; development of special purpose computers; instrument development; microprocessor applications. Mailing Add: Cent Res Lab Firestone Tire & Rubber Co Akron OH 44317

CLAY, CLARENCE SAMUEL, b Kansas City, Mo, Nov 2, 23; m 45; c 4. GEOPHYSICS. Educ: Kans State Univ, BA, 47, MS, 48; Univ Wis, PhD(physics), 51. Prof Exp: Asst prof physics, Univ Wyo, 50-51; res physicist, Carter Oil Co, 51-55; sr res scientist, Hudson Lab, Columbia Univ, 55-67; PROF GEOL & GEOPHYS, UNIV WIS-MADISON, 68- Mem: Fel Acoust Soc Am; Am Geophys Union; Soc Explor Geophys. Res: Wave propagation in inhomogeneous media; scattering at rough interfaces; statistical geophysical measurements; marine geophysics; ocean acoustics. Mailing Add: Weeks Hall Dept Geol & Geophys Univ of Wis Madison WI 53706

CLAY, FORREST PIERCE, JR, b Sutherland, Va, Nov 15, 27. PHYSICS. Educ: Randolph-Macon Col, BS, 48; Univ Va, MS, 50, PhD(physics), 52. Prof Exp: With Atlantic Res Corp, 52; asst prof physics, Georgetown Univ, 52-54 & Rutgers Univ, 54-61; from asst prof to assoc prof, 61-70, PROF PHYSICS, OLD DOM UNIV, 70- Concurrent Pos: Vis prof, Randolph-Macon Col, 57-58. Mem: Am Phys Soc; Am Asn Physics Teachers; Sigma Xi. Res: Positronium decay; microwave transmission; quadrupole mass spectrometry electronics; development of flight mass spectrometer. Mailing Add: Dept of Physics Old Dom Univ Norfolk VA 23508

CLAY, GEORGE A, b Cambridge, Mass, June 24, 38; m 65; c 3. NEUROPHARMACOLOGY. Educ: Dartmouth Col, AB, 61; Boston Univ, MA, 64, PhD, 68. Prof Exp: Nat Heart & Lung Inst fel, 67-68, Nat Inst Gen Med Sci Res Assocs Pharmacol Training Prog fel, 68-70; asst prof pharmacol, Bowman Gray Sch Med, 70-72; from res investr to sr res investr, 72-74, GROUP LEADER CENT NERV SYST PHARMACOL, G D SEARLE & CO, 74- Mem: AAAS. Res: Biochemical mechanisms of action of drugs affecting the central nervous system. Mailing Add: Searle Labs PO Box 5110 Chicago IL 60680

CLAY, JAMES RAY, b Burley, Idaho, Nov 5, 38; m 59; c 3. MATHEMATICS. Educ: Univ Utah, BS, 60; Univ Wash, MS, 62, PhD(math), 66. Prof Exp: Assoc engr, Boeing Co, 60-63; phys scientist, US Govt, 64-66; from asst prof to assoc prof, 66-74, assoc head dept, 69-74, PROF MATH, UNIV ARIZ, 74- Concurrent Pos: Humboldt Found spec award, 72; guest prof, Univ Tübingen, 72-73 & Univ London, 73. Mem: Am Math Soc; Math Asn Am. Res: Abstract algebra; computer science; algebraic structures arising from endomorphism and mappings of groups. Mailing Add: Dept of Math Univ of Ariz Tucson AZ 85721

CLAY, JAMES WILLIAM, b Crum, WVa, Apr 26, 33; m 60; c 4. GEOGRAPHY, CARTOGRAPHY. Educ: Marshall Univ, BS, 56, MA, 63; Univ NC, Chapel Hill, PhD(geog), 68. Prof Exp: Oil scout & geologist, Mobil Oil Venezuela, 57-60; geophysicist geol, Western Exploration, 60-62; oceanogr geol, US Hydrographic Off, US Navy, 62-63; asst prof geog, NC Col, 63-65; instr, 65-69, PROF GEOG & EARTH SCI, UNIV NC, 69- Res: Natural environment of North America; regional analysis of physiography; geology; soils, climate and vegetation; interrelationship of these components. Mailing Add: Dept of Geog & Earth Sci Univ of NC Charlotte NC 28223

CLAY, JOHN PAUL, b Dawson, Ga, Oct 2, 10; m 33; c 3. PHYSICAL CHEMISTRY. Educ: Univ Ga, BS, 32, MS, 34; Columbia Univ, PhD(chem), 38. Prof Exp: Asst chem, Univ Ga, 32-34 & Columbia Univ, 34-38; res chemist, Interchem Corp, 38; from instr to assoc prof, 39-55, PROF CHEM, LEHMAN COL, 55-, CHMN DEPT, 68- Concurrent Pos: Chem adv, Europ Command, 49-50; sci dir, Dugway Proving Ground, 52-55. Res: Complex ion exchange; catalysis; aerosols; fuels; wood chemistry; lignin; reaction rates; organo-phosphorous compounds. Mailing Add: Dept Chem Herbert H Lehman Col Bedford Park Blvd W Bronx NY 10468

CLAY, MARY ELLEN, b Freeport, Ohio, July 28, 40. ENTOMOLOGY, BIOLOGY. Educ: Muskingum Col, BS, 63; Ohio State Univ, MS, 66, PhD(entom), 69. Prof Exp: Res assoc mosquito biol, 69-73, lectr introd biol prog, 73-74, ASST PROF ENTOM, OHIO STATE UNIV, 74- Mem: Am Inst Biol Sci; Am Mosquito Control Asn; Entom Soc Am. Res: Insect biology; structural and functional aspects of the mosquito crop; insect diapause; mosquito neuroendocrine system; internal and external factors controlling diapause processes. Mailing Add: Dept of Entom Ohio State Univ Columbus OH 43210

CLAY, MICHAEL M, b Cleveland, Ohio, Aug 10, 20; m 55; c 2. PHARMACOLOGY. Educ: Ohio State Univ, BA, 41, PhD(pharmacol), 53; Univ Toledo, BS, 50, MS, 51. Prof Exp: Res assoc endocrinol, Ohio State Univ, 52-53; from asst to assoc prof pharmacol, Col Pharm, Columbia Univ, 53-64; guest prof, Med Clin, Univ Münster, 64-68; PROF PHARMACOL, UNIV HOUSTON, 68- Mem: Am Pharmaceut Asn; Am Geront Soc; NY Acad Sci. Res: Connective tissue physiology; cardiovascular disease. Mailing Add: Dept of Pharmacol Univ of Houston Houston TX 77004

CLAY, ROBERT EDWARD, b New Ulm, Minn, July 16, 33; m 58; c 4. MATHEMATICAL LOGIC. Educ: St Edward's Univ, BS, 55; Univ Notre Dame, MS, 58, PhD(math), 61. Prof Exp: Asst prof math, Univ Notre Dame, 61-62 & San Jose State Col, 62-63; asst prof math, Univ Notre Dame, 63-68; assoc prof math, Prescott Col, Ariz, 68-69; ASSOC PROF MATH, UNIV NOTRE DAME, 69- Concurrent Pos: Vis assoc prof, Fulbright-Hays lectureship, Univ Col Sci Educ, Ghana, 71-72; adv bd mem, Portage Twp, Ind, 71-75. Mem: Am Math Soc. Res: Logical system of Lesniewski and its application to the foundations of mathematics. Mailing Add: Dept of Math Univ of Notre Dame Notre Dame IN 46556

CLAY, WALLACE GORDON, physics, see 12th edition

CLAY, WILLIAM MARION, b Myers, Ky, Oct 3, 06; m 29, 42; c 2. ICHTHYOLOGY, HERPETOLOGY. Educ: Transylvania Col, AB, 27; Univ Mich, AM, 33, PhD(zool), 37. Prof Exp: From instr to asst prof biol, Transylvania Col, 27-32; asst zool, Univ Mich, 32-36; from instr to prof biol, Univ, 36-74, head dept, 56-63, Tom Wallace Prof Conserv, 63-74, founder & dir, Inst, 60-63, SR RES ASSOC, POTAMOLOGICAL INST, 63-, EMER PROF BIOL, GRAD SCH, UNIV LOUISVILLE, 74- Concurrent Pos: Asst, US Geol Surv, 45. Mem: Fel AAAS; Am Fisheries Soc; Soc Study Evolution; Soc Syst Zool; Am Soc Ichthyol & Herpet. Res: Distribution of Kentucky fishes; amphibians and reptiles; taxonomy of water snakes; ecology of freshwater fishes; conservation of biological resources. Mailing Add: Dept of Biol Grad Sch Univ of Louisville Louisville KY 40208

CLAYBAUGH, GLENN ALAN, b Lincoln, Nebr, Dec 10, 27; m 50; c 2. BACTERIOLOGY. Educ: Univ Nebr, BSc, 49; Mich State Univ, MSc, 50; Iowa State Univ, PhD(dairy bact), 53. Prof Exp: Sr bacteriologist, 53-60, sect leader, 60-63, prod mgr, 63-65, assoc, Mkt Div, 65-69, mkt dir, 69-72, DIR PROF SERV, MEAD JOHNSON LABS, MEAD JOHNSON & CO, 72- Mem: AAAS; Am Soc Microbiol; Am Dairy Sci Asn; Am Chem Soc; fel Am Pub Health Asn. Res: Infant feeding; dairy and food bacteriology; antibiotics and non-sporulating anaerobic bacteria in dairy products; quality control of milk products, infant formulas and other specialized food products; hospital consulting. Mailing Add: Mead Johnson Labs Mead Johnson & Co Evansville IN 47721

CLAYBERG, CARL DUDLEY, b Tacoma, Wash, Mar 1, 31; m 56; c 2. GENETICS. Educ: Univ Wash, BS, 54; Univ Calif, PhD(genetics), 58. Prof Exp: Asst genetics, Univ Calif, 54-56; asst geneticist, Conn Agr Exp Sta, 57-61, assoc geneticist, 61-74; ASSOC PROF HORT & FORESTRY, KANS STATE UNIV, 74- Mem: AAAS; Am Genetic Asn; Genetics Soc Am; Bot Soc Am; Am Soc Hort Sci. Res: Plant genetics, especially in Gesneriaceae, Lycopersicon, and Phaseolus. Mailing Add: Dept of Hort & Forestry Kans State Univ Manhattan KS 66506

CLAYBROOK, JAMES RUSSELL, b Cleburne, Tex, Aug 24, 36; m 63. BIOLOGICAL CHEMISTRY. Educ: Univ Tex, BS, 57, PhD(chem), 63. Prof Exp: Asst prof biochem, Med Sch, Univ Ore, 63-66; asst scientist, Ore Regional Primate Res Ctr, 63-66; NIH res fel microbiol, Univ Ill, Urbana-Champaign, 66-68; researcher, Int Lab Genetics & Biophys, Naples, Italy, 68-69; ASSOC PROF PHYSIOL, MED COL OHIO, TOLEDO, 69- Mem: AAAS; Am Chem Soc; Soc Gen Physiol. Res: Role of nucleic acids in development and cell differentiation. Mailing Add: Dept of Physiol Med Col of Ohio Box 6190 Toledo OH 43614

CLAYCOMB, CECIL KEITH, b Twin Falls, Idaho, Oct 19, 20; m 43; c 2. BIOCHEMISTRY. Educ: Univ Ore, PhD(biochem), 51. Prof Exp: AEC res asst, Med Sch, 49-51, asst prof, 51-61, PROF BIOCHEM, SCH DENT, UNIV ORE HEALTH SCI CTR, 61-, HEAD DEPT, 51-, ASST TO PRES HEALTH SCI CTR FOR MINORITY AFFAIRS, 75- Concurrent Pos: Vis res scientist, Inst Dent Res, United Dent Hosp, New South Wales, Australia, biol sci coordr, 72- Mem: AAAS; Am Chem Soc; NY Acad Sci; Int Asn Dent Res. Res: Oral collagen metabolism using proline labeled with tritium and/or radiocarbon. Mailing Add: Dept Biochem Sch Dent Univ Ore Health Sci Ctr Portland OR 97201

CLAYCOMB, WILLIAM CREIGHTON, b Cincinnati, Ohio, Dec 20, 42; m 65; c 1. BIOLOGICAL CHEMISTRY. Educ: Ind Univ, AB, 66, PhD(pharmacol), 69. Prof Exp: ASST PROF CELL BIOPHYS, BAYLOR COL MED, 72- Concurrent Pos: Res fel biol chem, Harvard Med Sch, 69-72 & NIH fel differentiation, 70-72. Mem: AAAS; Soc Develop Biol; Am Soc Cell Biol; Am Soc Zool; Am Soc Biol Chemists. Res: Regulation of cell differentiation and cell proliferation; developmental biology; genetic regulation; endocrinology. Mailing Add: Dept of Cell Biophys Baylor Col of Med Houston TX 77025

CLAYDON, THOMAS JOSEPH, bacteriology, see 12th edition

CLAYMAN, BRUCE PHILIP, b New York, NY, Sept 2, 42; m 62; c 1. SOLID STATE PHYSICS. Educ: Rensselaer Polytech Inst, BS, 64; Cornell Univ, PhD(physics), 69. Prof Exp: Asst prof, 68-73, ASSOC PROF PHYSICS, SIMON FRASER UNIV, 73- Mem: Am Phys Soc; Can Asn Physicists. Res: Far-infrared spectroscopic study of pure and doped crystalline materials at low temperature. Mailing Add: Dept of Physics Simon Fraser Univ Burnaby BC Can

CLAYPOOL, DON PEARSON, b Salt Lick, Ky, Sept 20, 19; m 43; c 4. ORGANIC CHEMISTRY. Educ: Tulane Univ, BS, 46; Univ Ky, MS, 50, PhD(chem), 52. Prof Exp: Instr chem, Morehead State Col, 46-47 & Univ Ky, 47-52; res chemist, Monsanto Chem Co, WVa, 52-56; assoc prof, 56-63, PROF CHEM, MEMPHIS STATE UNIV, 63- Mem: AAAS; Am Chem Soc (treas, 61-62). Res: Fats and oils; organosulfur compounds; sulfonium salts; reactions of dimethyl sulfoxide; organonitrogen compounds; other syntheses and mechanisms of reactions. Mailing Add: Dept of Chem Memphis State Univ Memphis TN 38111

CLAYPOOL, LAWRENCE LEONARD, b Pueblo, Colo, Dec 30, 07; m 29, 41; c 4. POMOLOGY. Educ: Univ Calif, BS, 28; State Col Wash, PhD(hort), 35. Prof Exp: Asst horticulturist, State Col Wash, 30-34; supvr, Prod Credit Corp, Wash, 34-37; from instr to prof pomol, 37-74, from jr pomologist to pomologist, Exp Sta, 37-74, EMER PROF POMOL, UNIV CALIF, DAVIS, 74- Concurrent Pos: Prof, Univ Ankara, Turkey, 57-58. Mem: Fel AAAS; fel Am Soc Hort Sci; Int Soc Hort Sci. Res: Maturity, handling and physiology of fruits in relation to fresh shipment, canning and drying; mechanical harvesting. Mailing Add: Col of Agr & Environ Sci Univ of Calif Davis CA 95616

CLAYTON, ANTHONY BROXHOLME, b Solihull, Eng, Jan 14, 40; m 63; c 4. ORGANIC CHEMISTRY. Educ: Univ Aston, ARIC, 62; Univ Birmingham, PhD(org chem), 65. Prof Exp: Res assoc, Cornell Univ, 65-67; RES CHEMIST, HERCULES INC, 67- Mem: Am Chem Soc; The Chem Soc. Res: Organic Fluorine chemistry; agricultural chemistry; general organic chemistry. Mailing Add: Hercules Inc Res Ctr Wilmington DE 19899

CLAYTON, CARLYLE NEWTON, b Liberty, SC, Dec 20, 12; m 35; c 2. PLANT PATHOLOGY. Educ: Clemson Agr Col, BS, 34; Univ Wis, PhD(plant path, physiol), 40. Prof Exp: Asst plant path, Univ Wis, 35-40; asst & assoc plant pathologist, SC Truck Exp Sta, 40-45; assoc prof, 45-50, PROF PLANT PATH, NC STATE UNIV, 50- Mem: AAAS; Am Phytopath Soc. Res: Fruit diseases. Mailing Add: Dept of Plant Path NC State Univ Raleigh NC 27607

CLAYTON, CHARLES CURTIS, b Minneapolis, Minn, June 20, 20; m 43; c 3. BIOCHEMISTRY. Educ: Univ Wis, BS, 42, MS, 47, PhD(biochem), 49. Prof Exp: Chemist, Ord Dept, Badger Ord Works, 44-45; from asst prof to assoc prof, 49-64, PROF BIOCHEM, MED COL VA, 64-, ASST DEAN BASIC SCI, 73- Mem: Am Inst Nutrit; NY Acad Sci. Res: Experimentally induced cancer; action of chemotherapeutic agents. Mailing Add: Dept of Biochem Med Col of Va Richmond VA 23298

CLAYTON, DALE LEONARD, b Harrisville, Mich, Apr 16, 39; m 61; c 2. ANIMAL BEHAVIOR. Educ: Andrews Univ, BA, 62; Loma Linda Univ, MA, 64; Mich State Univ, PhD(zool), 68. Prof Exp: Teaching asst biol sci, Mich State Univ, 64-67, dir human biol labs, 67-69, from instr to asst prof physiol, 67-69, NIH gen med res grant, 68-69; ASST PROF BIOL, WALLA WALLA COL, 70- Mem: AAAS; Am Soc Zoologists; Animal Behav Soc. Res: Physiological basis of animal behavior, especially the interaction of internal circadian rhythms, homeostatic mechanisms, development rates and external variables of photoperiod, twilight, temperature and other environmental perturbations. Mailing Add: Dept of Biol Walla Walla Col College Place WA 99324

CLAYTON, DAVID WALTON, b Leicester, Eng; m 55; c 3. ORGANIC CHEMISTRY. Educ: Univ London, BSc, 45, MSc, 51; Cambridge Univ, BA, 50, PhD(org chem), 53. Prof Exp: Res asst chem, Brit Leather Mfrs Res Asn, London, 45-48; Nat Res Coun Can fel, 53-55; sr scientific officer, Radiochem Ctr, United Kingdom Atomic Energy Authority, Amersham, Eng, 55-58; res chemist, 58-68, DIR PROCESS RES DIV, PULP & PAPER RES INST CAN, 68- Mem: The Chem Soc; Chem Inst Can; Am Chem Soc; Can Pulp & Paper Asn. Res: Chemistry of alkaline pulping; chemistry of pulp bleaching; carbohydrate and wood chemistry. Mailing Add: PO Box 358 24 Westwood Dr Hudson PQ Can

CLAYTON, DONALD DELBERT, b Shenandoah, Iowa, Mar 18, 35. ASTROPHYSICS, NUCLEAR PHYSICS. Educ: Southern Methodist Univ, BS, 56; Calif Inst Technol, MS, 59, PhD(physics), 62. Prof Exp: Res fel physics, Calif Inst Technol, 61-63; asst prof space sci, 63-65, from assoc prof to prof physics & space sci, 65-75, A H BUCHANAN PROF ASTROPHYS, RICE UNIV, 75- Concurrent Pos: Alfred P Sloan res fel, 66-68; mem panel astrophys & relativity, astron & physics surv comt, Nat Res Coun, 70- Mem: Fel Am Phys Soc; Am Geophys Union; Am Astron Soc. Res: Nucleosynthesis; space science; stellar evolution; geochemistry. Mailing Add: Dept of Space Physics & Astron Rice Univ Houston TX 77001

CLAYTON, FRANCES ELIZABETH, b Texarkana, Tex, Nov 6, 22. GENETICS. Educ: Tex State Col Women, BA, 44; Univ Tex, MA, 47, PhD(zool, genetics), 51. Prof Exp: Teacher, high sch, Ark, 44-45; tutor & fel, Univ Tex, 45-50; instr zool, Univ Ark, 50-51 & Univ Tex, 51-52, Hite fel, 52-53, res scientist, Genetics Found, 53-54; from asst prof to assoc prof, 54-62, PROF ZOOL, UNIV ARK, FAYETTEVILLE, 62- Concurrent Pos: Vis colleague, Univ Hawaii, 63-64. Mem: Soc Exp Biol & Med; Am Soc Zoologists; Genetics Soc Am; Am Genetic Asn; Am Soc Cell Biol. Res: Developmental and irradiation genetics; cytogenetics in Drosophila. Mailing Add: Dept of Zool Univ of Ark Fayetteville AR 72701

CLAYTON, GLEN TALMADGE, b Elmo, Ark, Jan 30, 29; m 50; c 2. PHYSICS. Educ: Univ Ark, BS, 53, MS, 54; Univ Mo, PhD, 60. Prof Exp: Instr physics, Univ Ark, 53-54; asst prof, William Jewell Col, 54-56; instr, Univ Mo, 56-58; res assoc, Argonne Nat Lab, 58-60; from asst prof to prof, Univ Ark, Fayetteville, 60-74; PROF PHYSICS & DEAN SCH SCI & MATH, STEPHEN F AUSTIN STATE UNIV, 74- Mem: Am Asn Physics Teachers; Am Phys Soc. Res: X-ray; neutron diffraction; electronics. Mailing Add: Sch of Sci & Math Stephen F Austin State Univ Nacogdoches TX 75961

CLAYTON, JAMES OLIVER, b San Francisco, Calif, Aug 20, 05; m 35; c 2. CHEMISTRY. Educ: Univ Calif, BS, 27, PhD(chem), 31. Prof Exp: Head dept chem, Col of St Teresa, 32-35; res chemist, Naval Res Lab, Washington, DC, 35-36, Standard Oil Co Calif, 36-45, Calif Res Corp, 45-65 & Chevron Res Co, 65-70; RETIRED. Mem: AAAS; Am Chem Soc. Res: Physical and chemical research on petroleum products; lubricating oil additives; heat capacity and entropy of carbon monoxide and nitrogen at low temperatures. Mailing Add: 1082 Miller Ave Berkeley CA 94708

CLAYTON, JAMES WALLACE, b New Westminster, BC, Nov 4, 33; m 57; c 4. BIOCHEMISTRY. Educ: Univ BC, BA, 55; Univ Sask, PhD(phys & org chem), 62. Prof Exp: Chemist wood pulp bleaching, Res Div, MacMillan & Bloedel, Ltd, BC, 56-58 & Grain Res Lab, Bd Grain Comnrs, Can, 62-66; CHEMIST, FRESHWATER INST, ENVIRON CAN, FISHERIES & MARINE SERV, 67- Mem: Chem Inst Can. Res: Protein chemistry of freshwater fish; genetics of fish proteins and enzymes. Mailing Add: Freshwater Inst Environ Can 501 University Crescent Winnipeg MB Can

CLAYTON, JOHN CHARLES (HASTINGS), b Pittston, Pa, June 15, 24; m 68. PHYSICAL INORGANIC CHEMISTRY. Educ: St Joseph's Col, Philadelphia, BS, 49; Univ Pa, MS, 50, PhD(chem), 53. Prof Exp: Asst chem, Univ Pa, 52, fel, 53-54;

SR RES CHEMIST, BETTIS ATOMIC POWER LAB, WESTINGHOUSE ELEC CORP, 54- Mem: AAAS; Am Chem Soc; Am Ceramic Soc; Am Nuclear Soc; Sigma Xi. Res: Inorganic chemistry of solids; physical and inorganic chemistry of nuclear materials. Mailing Add: Bettis Atomic Power Lab Westinghouse Elec Corp PO Box 79 West Mifflin PA 15122

CLAYTON, JOHN MARK, b Kevil, Ky, Aug 6, 45; m 70; c 2. MEDICINAL CHEMISTRY, MEDICAL RESEARCH. Educ: Tenn Technol Univ, BS, 68; Univ Tenn, PhD(pharmaceut sci), 71. Prof Exp: Res assoc drug design, Dept Chem, Pomona Col, 71-72; res biologist chem carcinogenesis, Nat Ctr Toxicol Res, Food & Drug Admin, 72-73; clin res assoc, 74-75, DIR CLIN & REGULATORY SERV, PLOUGH, INC, 75- Concurrent Pos: Asst prof pharmacol, Col Med, Univ Ark, 73-; asst prof molecular biol, Col Pharm, Univ Tenn, 74. Mem: Acad Pharmaceut Sci; Am Chem Soc; Am Inst Chemists. Res: Quantitative structure-activity relationship approach to drug design; methodologies of human prophetic patch testing; evaluations of sunscreens. Mailing Add: Plough Inc 3030 Jackson Ave Memphis TN 38151

CLAYTON, JOHN WESLEY, JR, b Philadelphia, Pa, Sept 1, 24. TOXICOLOGY, RESPIRATORY PHYSIOLOGY. Educ: Wheaton Col, AB, 48; Univ Pa, AM, 50, PhD(parasitol), 54. Prof Exp: Toxicologist, Haskell Lab, E I du Pont de Nemours & Co, Inc, 54-60, asst dir labs, 60-69; dir environ sci lab, Hazleton Labs, TRW Inc, 69-71; dir toxicol ctr, Univ Wis, 71-73; chief toxicol br & dir health effects div, US Environ Protection Agency, 73-74; PROF PHARMACOL & TOXICOL & DIR TOXICOL PROG, UNIV ARIZ, 74- Concurrent Pos: Mem tech adv bd, State Air Pollution Control Bd, Va, 70-71 & policy bd, Nat Ctr Toxicol Res, 73-74; consult, Wis Dangerous Substances Bd, 72-73, Reese & Schluechter, Ill, 72-73, Kennecott Copper Corp, 74- & Food & Drug Admin, 74-; mem sci adv bd, Food & Drug Admin; mem adv bd, Inhalation Toxicol Res Inst, Electronic Resources Develop Agency; mem ed bd, Am Indust Hyg Asn, 67-; mem ed bd, Soc Toxicol, 69-, chmn tech comt, 70-71. Mem: Am Indust Hyg Asn; Soc Toxicol. Res: Toxicology of fluorocarbons, including cardiac effects; action of fluoro-olefins on renal function; pyrolysis products of fluoropolymers. Mailing Add: Toxicol Prog Biol Sci W Univ of Ariz Tucson AZ 85721

CLAYTON, NEAL, b Ripley, Miss, Aug 23, 13; m 40; c 1. SEISMOLOGY, GEOPHYSICS. Educ: Miss Col, BA, 34; La State Univ, MS, 37. Prof Exp: Eng trainee, Schlumberger Well Surv, Tex, 37; seismic helper, Humble Oil Co, 37, seismic computer, 37-39 & seismic Explor, Inc, 40-41; assoc physicist, US Navy, 42-43; seismic observer, Magnolia Petrol Co, 43-44; seismic party chief, NAm Geophys Co, 44-46 & Repub Explor Co, 46-51; asst mgr domestic opers, Century Geophys Corp, 51-54; pres & supvr, Liberty Explor Co, 54-56; mem staff & dist geophysicist, Sohio Petrol Co, La, 56-60; supvr, Index Explor Co, Tex, 60-61; Gulf Coast geophysicist, Cosden Petrol Co, 61-64; CONSULT GEOPHYSICIST, 64- Mem: Soc Explor Geophys; Am Asn Petrol Geologists; Soc Petrol Engrs. Res: Application of seismology to exploration for oil and gas; geophysical prospecting. Mailing Add: PO Box 2811 Corpus Christi TX 78403

CLAYTON, PAULA JEAN, b St Louis, Mo, Dec 1, 34; m 58; c 3. PSYCHIATRY. Educ: Univ Mich, Ann Arbor, BS, 56; Washington Univ, MD, 60. Prof Exp: Intern, St Luke's Hosp, St Louis, 60-61; asst resident & chief resident psychiat, Barnes & Renard Hosps, St Louis, 61-65; from instr to assoc prof psychiat, 65-74, PROF PSYCHIAT, SCH MED, WASHINGTON UNIV, 74- Concurrent Pos: Consult psychiatrist, Malcolm Bliss Ment Health Ctr, St Louis, 72- & dir training & res, 75; dir, Barnes & Renard Hosp Psychiat Inpatient Serv, 75- Mem: Fel Am Psychiat Asn; Psychiat Res Soc; Asn Res Nerv Ment Dis; Am Psychopath Asn; Soc Biol Psychiat. Res: Studies dealing with nosology, course and treatment of patients with psychiatric diagnosis; also the symptomatology and course of normal bereavement. Mailing Add: Dept of Psychiat Washington Univ Sch of Med 4940 Audubon Ave St Louis MO 63110

CLAYTON, RAYMOND BRAZENOR, b Manchester, Eng, Sept 16, 25; m 62; c 2. BIOCHEMISTRY, ENDOCRINOLOGY. Educ: Univ Manchester, BSc, 49, MSc, 50, PhD(chem), 52. Prof Exp: Res fel org chem, Univ Manchester, 52-53; res fel biochem, Univ Chicago, 53-54, res fel chem, Harvard Univ, 54-55; Imp Chem Indust fel, Oxford Univ, 55-56; res dir, Manchester Cancer Res Trust Fund, 56-58; res fel chem, Harvard Univ, 59-63; assoc prof, 63-68, PROF BIOCHEM, DEPT PSYCHIAT & BEHAV SCI, STANFORD UNIV, Concurrent Pos: Hon lectr, Univ Manchester, 56-58; estab investr, Am Heart Asn, 60-65. Mem: AAAS; Am Soc Biol Chemists; Endocrine Soc; Am Soc Zoologists; The Chem Soc. Res: Steroid biosynthesis; comparative aspects of steroid and terpenoid metabolism; genetics of steroid hormone metabolism and hormonal effects; action of steroid hormones in the central nervous system. Mailing Add: Dept of Psychiat & Behav Sci Stanford Univ Stanford CA 94305

CLAYTON, ROBERT ALLEN, b Milwaukee, Wis, Nov 21, 22; m 47; c 3. BIOCHEMISTRY. Educ: Univ Wis, BS, 49, MS, 51, PhD(biochem), 53. Prof Exp: Instr, Univ Wis, 52-53; asst prof biochem, Med Sch, George Washington Univ, 53-56; res assoc, Am Tobacco Co, 56-57, sr res assoc, 57-58, head biochem sect, 59; head, Fundamental Food Res Dept, Gen Mills, Inc, 59-61 & Explor Food Res Dept, 61-63, dir food sci activity, 64-67; DIR RES, ANHEUSER-BUSCH, INC, 67- Mem: Am Chem Soc. Res: Plant biochemistry; research administration. Mailing Add: Anheuser-Busch Inc 721 Pestalozzi St St Louis MO 63118

CLAYTON, ROBERT NORMAN, b Hamilton, Ont, Mar 20, 30; m 71; c 1. GEOCHEMISTRY. Educ: Queen's Univ, Ont, BSc, 51, MS, 52; Calif Inst Technol, PhD(chem), 55. Prof Exp: Res fel geochem, Calif Inst Technol, 55-56; asst prof geochem, Pa State Univ, 56-58; from asst prof to assoc prof chem, 58-66, master phys sci col div, assoc dean col & assoc dean phys sci, 69-72, PROF CHEM & GEOPHYS, UNIV CHICAGO, 66- Concurrent Pos: Guggenheim fel, 64-65; Sloan fel, 64-66. Mem: AAAS; Am Geophys Union; Meteoritical Soc. Res: Natural variations of stable isotope abundances. Mailing Add: Enrico Fermi Inst Univ of Chicago Chicago IL 60637

CLAYTON, RODERICK KEENER, b Tallin, Estonia, Mar 29, 22; US citizen; m 44; c 2. BIOPHYSICS. Educ: Calif Inst Technol, BS, 47, PhD(physics, biol), 51. Prof Exp: Merck fel, Stanford Univ, 51-52; assoc prof physics, US Naval Post Grad Sch, 52-57; NSF sr fel, 57-58; sr biophysicist, biol div, Oak Ridge Nat Lab, 58-62; vis prof microbiol, Dartmouth Col Med Sch, 62-63; sr investr, C F Kettering Res Lab, 63-66; PROF BIOL & BIOPHYS, CORNELL UNIV, 66- Concurrent Pos: Lalor fel, Woods Hole Marine Biol Lab, 55; consult, Firestone Res & Develop Lab, 56-57. Mem: AAAS; Am Soc Biol Chemists; Biophys Soc; Soc Gen Physiol; Am Soc Plant Physiol. Res: Physical aspects of photosynthesis; biochemistry of photosynthetic bacteria. Mailing Add: Cornell Univ Dept of GDP Plant Sci Bldg Ithaca NY 14853

CLAYTON, WILLIAM HOWARD, b Dallas, Tex, Aug 16, 21; m 64; c 2. PHYSICAL OCEANOGRAPHY. Educ: Bucknell Univ, BSc, 49; Tex A&M Univ, PhD(phys oceanog), 56. Prof Exp: Instr physics, Bucknell Univ, 47-49; asst, Ohio State Univ, 49 & Univ NMex, 50; asst oceanog & meteorol, 50-51 & oceanog, 51-54, assoc & instr math, 54-56, micrometeorologist & prin investr oceanog & meteorol res for US Air Force & US Army, 56-58, from asst prof to assoc prof, 58-65, dir micrometeorol res, 58-61, PROF OCEANOG & METEOROL, TEX A&M UNIV, 65-, ASSOC DEAN COL GEOSCI, 70-, PROVOST, MOODY COL MARINE SCI & MARITIME RESOURCES, 74- Concurrent Pos: Vis prof, Univ Hawaii, 63-64. Mem: Am Meteorol Soc; Am Geophys Union. Res: Micrometeorology; numerical analysis; water level variations; air-sea interchange; oceanographic and meteorological instrumentation; machine computational methods. Mailing Add: Moody Col Marine Sci & Maritime Resources PO Box 1675 Galveston TX 77550

CLAYTON, WILLIAM JOSEPH, b Greenville, SC, Dec 17, 13; m 40; c 5. CHEMISTRY. Educ: Southeastern State Col, BA, 32; Duke Univ, MA, 35, PhD(phys chem), 37. Prof Exp: Instr chem, Southeastern State Col, 32-34; res chemist, gen labs, US Rubber Co, NJ, 37-45; tech dir, Minn Mining & Mfg Co, 45-55; vpres, consumer prods, res & develop, Int Latex Corp, 55-63 & Mobil Plastics Develop Corp, 63-66, VPRES RES & DEVELOP, PLASTICS DIV, MOBIL CHEM CO, 66- Mem: Am Chem Soc. Res: Development of rubber products; latex products; equilibrium in solutions of cadmium and zinc oxalates. Mailing Add: Plastics Div Mobil Chem Co Macedon NY 14522

CLAYTON-HOPKINS, JUDITH ANN, b Santa Monica, Calif, Sept 17, 39; c 1. ENDOCRINOLOGY. Educ: Univ Calif, Los Angeles, BA, 60, MA, 63, PhD(zool), 66. Prof Exp: Staff physiologist, Worcester Found Exp Biol, 68-70; res assoc, Dept Biol Sci & Dept Nutrit Sci, Univ Conn, Storrs, 70-71, asst prof physiol, 71-72; DIR, RADIOISOTOPE/ENDOCRINE LAB, DEPT PATH, ST FRANCIS HOSP, HARTFORD, 72-; ASST PROF LAB MED, HEALTH CTR, UNIV CONN, FARMINGTON, 73- Concurrent Pos: Res assoc NIH grant, Harvard Med Sch & Beth Israel Hosp, Boston, 67-68; consult, Automation Med Lab Sci Rev Comt, NIH, 73-77; consult radioimmunoassay, Bur Med Devices & Diag Prods, Food & Drug Admin, 76-; mem comt radionuclides, Am Asn Clin Chem & ed, Selected Methods Clin Chem, 74- Mem: Am Soc Zool; Endocrine Soc; NY Acad Sci; Am Physiol Soc; Am Asn Clin Chem. Res: Mechanisms of gonadotropin action, ovulation; radioimmunochemistry; biochemical and molecular mechanisms of hormone action. Mailing Add: Dept of Path St Francis Hosp 114 Woodland St Hartford CT 06105

CLEARE, HENRY MURRAY, b Dalton, Ga, Aug 5, 28; m 50; c 3. PHYSICS. Educ: Ga Inst Technol, BS, 51. Prof Exp: Physicist, 51-58, res physicist, 58-62, RES ASSOC, RES LABS, EASTMAN KODAK CO, 62-, HEAD RADIOGRAPHY DEPT, 64- Mem: Health Physics Soc. Res: Image-forming properties of medical and industrial radiographic systems; properties of quantum limited radiographic systems; photographic radiation dosimetry; photographic effects of radiations in space. Mailing Add: Res Labs Eastman Kodak Co Rochester NY 14650

CLEARFIELD, ABRAHAM, b Philadelphia, Pa, Nov 9, 27; m 49; c 2. INORGANIC CHEMISTRY. Educ: Temple Univ, BA, 48, MA, 50; Rutgers Univ, PhD(phys chem, crystallog), 54. Prof Exp: Assoc chemist, Titanium Alloy Mfg Div, Nat Lead Co, 54-56, sr chemist, 56-58, asst chief chem res, 58-63; from asst prof to prof chem, Ohio Univ, 63-74; assoc prog dir thermodyn, NSF, 74-75; PROF CHEM, TEX A&M UNIV, 76- Concurrent Pos: Lectr, Niagara Univ, 57-60; consult, Bio-Rad Labs, Tizon Chem Co & Magnesium Elektron, Manchester, Eng, 75- Mem: Am Chem Soc; Am Crystallog Asn. Res: Chemistry of transition metals, especially titanium and zirconium; inorganic synthesis and structure; x-ray diffraction and crystal structure, inorganic ion exchangers. Mailing Add: Dept of Chem Tex A&M Univ College Station TX 77843

CLEARY, JAMES WILLIAM, b Evanston, Ill, Apr 13, 26; m 60; c 4. ORGANIC CHEMISTRY, POLYMER CHEMISTRY. Educ: Loyola Univ, Ill, BS, 50; State Col Wash, MS, 53, PhD, 56. Prof Exp: Sr exp aide agr chem, State Col Wash, 51-53, asst chem, 53-55; res chemist, 56-68, SR RES CHEMIST, PHILLIPS PETROL CO, 68- Mem: AAAS; Am Chem Soc. Res: Peppermint oil; a-ketobutyrolactones; synthesis and modification of synthetic rubber; sulfur compounds; polyolefins; polymerization catalysts; pyridine polymers; condensation polymers; polyamides; polyesters; polyvinyl pyridines. Mailing Add: 1215 S Dewey Ave Bartlesville OK 74003

CLEARY, LAURENCE TWOMEY, b Andover, Mass, 23; m 47; c 4. TEXTILES, CHEMISTRY. Educ: Tufts Col, BS, 47; Inst Textile Technol, MS, 49; Columbia Univ, MA, 52, PhD(chem eng), 54. Prof Exp: Res engr, synthetic fiber process develop, 53-59, sr res engr & group supvr, 59, res supvr, 59-63, tech serv rep, 63-68, TECH SERV SPECIALIST, E I DU PONT DE NEMOURS & CO, 69- Mem: AAAS; Am Chem Soc. Res: Solution properties of surface active agents; fire retardant investigations of cellulose structures; synthetic fiber process development; new process scouting; product improvement and development. Mailing Add: 323 Spalding Rd Wilmington DE 19803

CLEARY, PAUL PATRICK, b Watertown, NY, July 9, 41; m 63; c 2. MOLECULAR BIOLOGY, MEDICAL MICROBIOLOGY. Educ: Univ Cincinnati, BS, 65; Univ Rochester, MS, 69, PhD(microbiol genetics), 71. Prof Exp: ASST PROF MICROBIOL & PEDIAT, UNIV MINN, MINNEAPOLIS, 72- Concurrent Pos: Trainee biol sci, Univ Calif, Santa Barbara, 71-72. Mem: Am Soc Microbiol; Sigma Xi. Res: Genetic structure and regulation of the biotin gene cluster in Escherichia coli; regulation of the arabinose operon in Escherichia coli; genetic determinants for resistance to phagocytosis in group A streptococci. Mailing Add: Dept of Microbiol Box 196 Univ of Minn Minneapolis MN 55400

CLEARY, ROBERT WILLIAM, pharmaceutical chemistry, see 12th edition

CLEARY, STEPHEN FRANCIS, b New York, NY, Sept 28, 36; m 59; c 3. BIOPHYSICS, RADIOBIOLOGY. Educ: NY Univ, BSChE, 58, PhD(biophys), 64; Univ Rochester, MS, 60. Prof Exp: Res engr sterio-specific polymers, Texus-US Chem Co, 58-59; teaching asst radiation biol, Inst Environ Med, NY Univ, 60-62, res assoc biophys, 62-64; asst prof, 64-67, ASSOC PROF BIOPHYS, MED COL VA, VA COMMONWEALTH UNIV, 67- Concurrent Pos: Consult, Environ Biophys Br, Nat Inst Environ Health Sci, NIH, 74- Mem: Biophys Soc; NY Acad Sci; Am Soc Photobiol. Res: Biological effects of non-ionizing radiation; lasers, light, microwave and radiofrequency radiation; effects of ionizing radiation on the mammalian eye; structural bonding forces in viruses. Mailing Add: Dept of Biophys Box 877 Med Col of Va Richmond VA 23298

CLEARY, TIMOTHY JOSEPH, b Philadelphia, Pa, Aug 8, 42; m 65; c 2. MEDICAL MICROBIOLOGY. Educ: Mt St Mary's Col, BS, 64; Univ Cincinnati, MS, 68, PhD(microbiol), 69; Am Bd Med Microbiol, dipl, 74. Prof Exp: Asst prof biol, Duquesne Univ, 69-71; fel, Ctr Dis Control, 71-73; ASST PROF PATH, UNIV MIAMI, 73- Mem: Am Soc Microbiol; Sigma Xi. Res: Antimicrobial drug interactions and assays; serological procedures for the detection of Neisseria gonorrhoeae antibodies. Mailing Add: Dept of Path Univ of Miami Jackson Mem Hosp Miami FL 33161

CLEARY, WILLIAM JAMES, b St Louis, Mo, Dec 10, 43; m 67; c 1. MARINE

GEOLOGY, SEDIMENTARY PETROLOGY. Educ: Southern Ill Univ, BA, 65; Duke Univ, MA, 67; Univ SC, PhD(geol), 72. Prof Exp: Geologist, Pan Am Petrol Corp, 67-68; ASST PROF GEOL, UNIV NC, WILMINGTON, 72-, RES ASSOC MARINE SCI, 74- Mem: Geol Soc Am; Sigma Xi; Soc Econ Paleontologists & Mineralogists; Am Asn Geol Teachers. Res: Continental margin sedimentation off Southeastern United States; barrier island sedimentation; turbidite sedimentation on the Hatteras Abyssal Plain. Mailing Add: Dept of Marine Sci Box 3725 Univ of NC Wilmington NC 28401

CLEAVER, CHARLES E, b Paris, Ky, Mar 14, 38; m 59; c 3. MATHEMATICS. Educ: Eastern Ky Univ, BS, 60; Univ Ky, MS, 63, PhD(math), 68. Prof Exp: Instr math, Murray State Univ, 62-64, asst prof, 64-65; asst prof, 68-73, ASSOC PROF MATH, KENT STATE UNIV, 73, ASST CHMN MATH, 75- Mem: Am Math Soc; Math Asn Am. Res: Operator theory in Banach spaces; functional analysis. Mailing Add: Dept of Math Kent State Univ Kent OH 44242

CLEAVER, FRANK L, b Palm, Pa, Feb 3, 25; m 53; c 2. MATHEMATICS. Educ: Pa State Univ, BS, 48; Univ Miami, MS, 55; Tulane Univ, PhD(math), 60. Prof Exp: Instr math, Tulane Univ, 55-60; asst prof, Univ Ky, 61-62; from asst prof to assoc prof, 60-66, PROF MATH, UNIV S FLA, 66- Mem: Am Math Soc; Math Asn Am. Res: Topology and geometry of numbers. Mailing Add: Dept of Math Univ of S Fla Tampa FL 33620

CLEAVER, FREDERICK CHARLES, b Everett, Wash, June 27, 16; m 41; c 3. FISH BIOLOGY. Educ: Univ Wash, BS, 41, PhD(fisheries), 67. Prof Exp: Fishery biologist, Wash State Dept Fish, 42-48; fishery res biologist, US Fish & Wildlife Serv, 48-51 & 56-58; fishery res biologist, Ore Fish Comn, 51-52, dir res, 52-54, asst dir, 54-56; fishery res biologist & chief Seattle marine invest, US Fish & Wildlife Serv, 58-68; DIR COLUMBIA FISHERIES OFF, BUR COM FISHERIES, NAT MARINE FISHERIES SERV, NAT OCEANIC & ATMOSPHERIC ADMIN, US DEPT COM, 68- Concurrent Pos: Lectr, Univ Wash, 56- Mem: Am Inst Fishery Biol. Res: Biological fishery research and administration; improving yields of fish and shellfish stocks. Mailing Add: Columbia Fisheries Off Bur Com Fisheries 811 NE Oregon St PO Box 4332 Portland OR 97208

CLEAVER, JAMES EDWARD, b Portsmouth, Hants, Eng, May 17, 38; m 64. CANCER. Educ: St Catharine's Col, Cambridge Univ, BA, 61, PhD(radiobiol), 64. Res fel neurosurg, Mass Gen Hosp, Boston & surg, Harvard Med Sch, 64-66; asst res biophysicist, lab radiobiol, 66-68, from asst prof to assoc prof radiobiol, 68-74, PROF RADIOL, SCH MED, UNIV CALIF, SAN FRANCISCO, 74- Honors & Awards: Research Award, Radiation Res Soc, 73. Mem: Radiation Res Soc; Biophys Soc. Res: Effects of ultraviolet light on mammalian cells and mechanisms of recovery from radiation damage; dermatology; mutagenesis; xeroderma pigmentosum; radiobiology of tritium decays. Mailing Add: Lab of Radiobiol Univ of Calif Sch of Med San Francisco CA 94122

CLEAVES, ARTHUR BAILEY, b North Scituate, Mass, Dec 6, 05; m 35; c 2. GEOLOGY. Educ: Brown Univ, PhB, 27, AM, 29; Univ Toronto, MA, 30; Harvard Univ, AM, 32, PhD(geol), 33. Prof Exp: Instr geol & paleont, Lafayette Col, 33-36; jr geologist, Pa Topog & Geol Surv, 36-37, asst geologist, 37-38; chief geologist, Pa Turnpike Comn, 38-41; architect-engr, Caribbean, 41-42; consult engr, Venezuela, 42; spec eng geologist, Cananea Consol Copper Co, Mex, 43; field serv consult, Off Sci Res & Develop, Wash, 43-45; consult, City of Philadelphia Water Comn & pvt firms, 46; from assoc prof to prof geol, 46-74, EMER PROF EARTH & PLANETARY SCI, WASH UNIV, 74- Concurrent Pos: Chief geologist, Manu-mine Res & Develop Co, Pa, 54-55; hwy res engr, Bur Pub Roads, 59; land planner, Vet Admin, Chicago & St Louis Dists, 59-; consult geol engr, Parson Brinkerhoff Quade & Douglas, 40- & Wabash Drilling Co, 46-; foreign off consult, Ammann & Whitney, 54-; consult, Leo A Daly Co, 56-; consult engr, Calif Div Hwys, 57, 59 & 61 & World Bank, Lebanon, 59; consult engr, Rome Off, Lublin McGaughey Assocs; chmn spec adv bd slope stability, Bldg Res Adv Bd, Nat Acad Sci-Nat Res Coun; mem bd consults, Tenn Valley Authority, 52-, Electroconsult, Italy, 62- & SADE, 64-; with traffic comn, University City, Mo, 62-; chief geologist, Exped, Ed Greenland, 33; co-organizer exped, Matto Grosso, Brazil, 34, Antigua, St Lucia, Trinidad, Brit Guiana, Surinam, Venezuela, New Caledonia & Solomon Islands. Honors & Awards: Presidental Cert Merit, 49. Mem: Fel Geol Soc Am; fel Am Soc Civil Engrs; Nat Soc Prof Eng; Am Geochem Soc; Asn Eng Geol. Res: Pennsylvania; geology in engineering. Mailing Add: Dept of Earth Sci Wash Univ St Louis MO 63130

CLEAVES, DUNCAN WORSTER, b Bangor, Maine, Aug 23, 19; m; c 3. PHYSICAL CHEMISTRY. Educ: Brown Univ, BS, 40; Univ Calif, PhD(chem), 51. Prof Exp: Jr chemist, E I du Pont de Nemours & Co, 40-41; jr engr, Chile Explor Co, 41-44; jr chemist, Shell Develop Co, 44-46 & Int Minerals & Chem Corp, 46-48; asst, Univ Calif, 48-50; instr chem, Univ Ore, 50-51; sr res engr, Titanium Metals Corp, 51-56; assoc prof chem & physics, southern regional div, Univ Nev, 56-63; sr res scientist, astropower lab, Douglas Aircraft Co, Inc, 63-64; instr sci, high sch, 64-68; instr chem, Santa Rosa Jr Col, 68; PROF CHEM & MATH, CERRO COSO COMMUNITY COL, 68- Mem: Am Chem Soc. Res: Polymerization of alkenes in sulfuric acid; vacuum fusion analysis; lactam formation of glutamic acid. Mailing Add: Cerro Coso Community Col Ridgecrest CA 93555

CLEAVES, EMERY TAYLOR, b Easton, Pa, May 11, 36; m 60; c 4. GEOMORPHOLOGY, ENVIRONMENTAL GEOLOGY. Educ: Harvard Col, BA, 60; John Hopkins Univ, MA, 64, PhD(geog), 73. Prof Exp: Assoc geologist, 63-65, geologist IV, 65-73, ASST DIR, MD GEOL SURV, 73- Mem: Fel Geol Soc Am. Res: Chemical weathering of crystalline rocks; role of chemical weathering in landform development; landform mapping and its application to environmental geology. Mailing Add: Md Geol Surv 214 Latrobe Hall Johns Hopkins Univ Baltimore MD 21218

CLEBSCH, EDWARD ERNST COOPER, b Clarksville, Tenn, June 6, 29; m 56; c 3. PLANT ECOLOGY. Educ: Univ Tenn, AB, 55, MS, 57; Duke Univ, PhD(bot), 60. Prof Exp: Res assoc, 60-63, asst prof, 63-66, ASSOC PROF BOT, UNIV TENN, KNOXVILLE, 66- Mem: Fel AAAS; Ecol Soc Am; Bot Soc Am; Asn Trop Biol. Res: Mineral cycling in Southern Appalachian ecosystems; radiation ecology; physiological ecology of Arctic-Alpine plants; flora and vegetation of Tennessee; mountain environments; ecology of the Aleutian Islands. Mailing Add: Dept of Bot Univ of Tenn Knoxville TN 37916

CLECKLEY, JAMES JENNINGS, b Bamberg, SC, Feb 17, 14; m 42; c 3. MEDICINE. Educ: The Citadel, BS, 35; Univ SC, MD, 39. Prof Exp: Assoc, 47-49, from asst prof to assoc prof, 49-56, chmn dept, 56-68, PROF PSYCHIAT, MED UNIV SC, 56- Concurrent Pos: Consult, US Naval Hosp, Charleston, 49- & Vet Admin Hosp. Mem: AMA; Am Psychiat Asn. Mailing Add: Dept of Psychiat Med Univ of SC Charleston SC 29401

CLEE, THOMAS EDWARD, b Pembroke, Ont, May 29, 45; m 71; c 1. EXPLORATION GEOPHYSICS. Educ: Univ Toronto, BSc, 67, MSc, 68, PhD(geophys), 73. Prof Exp: RES GEOPHYSICIST EXPLOR GEOPHYS, WESTERN GEOPHYS CO, 73- Mem: Soc Explor Geophysicists. Res: Digital signal processing; spectral analysis. Mailing Add: PO Box 2469 Houston TX 77001

CLEEK, GEORGE KIME, b Warm Springs, Va, Aug 27, 26; m 48; c 3. CHEMISTRY. Educ: Va Polytech Inst, BS, 48. Prof Exp: Res chemist, Nitrogen Div, Allied Chem Corp, 48-55, res chemist, Tech Serv, 55-63, res chemist, Appln Res, 63-66, tech serv prod supvr, Plastics Div, 66-69; mgr res lab, Caradco Div, Scovill Mfg Co, 69-70; develop chemist, Chem Div, 70-73, MGR RES & DEVELOP LAB CHEM-RESINS, GA-PAC CORP, 73- Mem: Am Chem Soc; Forest Prod Res Soc. Res: Ureaformaldehyde resins. Mailing Add: Develop Lab Chem Div Ga-Pac Corp 2883 Miller Rd Decatur GA 30032

CLEEK, GIVEN WOOD, b Warm Springs, Va, Nov 6, 16; m 41; c 3. CHEMISTRY, GLASS TECHNOLOGY. Educ: George Washington Univ, BS, 54. Prof Exp: Lab apprentice, Nat Adv Comt Aeronaut, 35-36; lab apprentice, 36-41, glassworker, 46-49, technologist, 49-57, phys chemist, 58-67, RES CHEMIST, NAT BUR STANDARDS, 67- Honors & Awards: Silver Medal, US Dept Commerce, 73. Mem: Am Chem Soc; Am Ceramic Soc; Optical Soc Am; Am Soc Test & Mat. Res: Development of special optical glasses having higher refractive indices and special dispersions; development of infrared transmitting glasses; determination of physical properties of glass as a function of chemical composition. Mailing Add: 5512 N 24th St Arlington VA 22205

CLEELAND, CHARLES SAMUEL, b Jacksonville, Ill, Sept 23, 38; m 65; c 1. NEUROPSYCHOLOGY, PSYCHOLOGY. Educ: Wesleyan Univ, BS, 60; Wash Univ, PhD(psychol), 66. Prof Exp: Instr psychiat, Med Sch, Univ Mo, 64-65; from instr to asst prof, 66-72, ASSOC PROF NEUROL, MED SCH, UNIV WIS, MADISON, 72- Concurrent Pos: Consult, Dept Ment Health, State of Wis, 68- & Dept Corrections, 72- Mem: AAAS; Soc Psychophysiol Res; Soc Neurosci; Pavlovian Soc NAm. Mailing Add: Dept of Neurol Univ of Wis Ctr for Health Sci Madison WI 53706

CLEERE, ROY LEON, b Madisonville, Tex, Dec 20, 05; m 31. PUBLIC HEALTH. Educ: Agr & Mech Col Tex, BS, 27; Univ Tex, MD, 29; Johns Hopkins Univ, MPH, 36; Am Bd Prev Med & Pub Health, cert, 50. Prof Exp: Intern, Kansas City Gen Hosp, 29-30; resident, Presby Hosp, 30-31; pvt pract, Denver, Colo, 31-35; exec dir, State Dept Pub Health, Colo, 35-74; asst prof pub health & lab diag, Sch Med, Univ Colo, Denver, 47-74; DIR, COLO-WYO REGIONAL ARTHRITIS DEPT, 74- Concurrent Pos: Deleg, World Health Assembly, Geneva, 51. Mem: Fel Am Pub Health Asn; AMA. Mailing Add: 4210 E 11th Ave Denver CO 80220

CLEGG, DAVID JOHN, b Bramhall, UK, Oct 4, 31; m 57; c 3. TOXICOLOGY, TERATOLOGY. Educ: Univ Col North Wales, BSc, 53, MSc, 56. Prof Exp: Jr sci officer serol, Blood Transfusion Serv, Manchester, Eng, 54-57; exp biologist, Benger Labs, Fisons Ltd, Cheshire, Eng, 57-62; chief exp biologist, Brit Indust Biol Res Asn, 62-65; sci adv, Pesticide Toxicol & Gen Teratology, Can Dept Nat Health & Welfare, Health Protection Br, 65-73, proj dir, Int Proj Food Irradiation, 74-75, HEAD PESTICIDE UNIT, DIV TOXICOL EVAL, HEALTH PROTECTION BR, CAN DEPT NAT HEALTH & WELFARE, 75- Concurrent Pos: Consult joint meeting, Food & Agr Orgn Working Party on Pesticide Residues & WHO Expert Comt on Pesticide Residues, 68, 71 & 72; mem ad hoc drafting groups, Codex Comt on Pesticide Residues, 69; consult, Joint Food & Agr Orgn/WHO Expert Comt on Food Additives, 70. Mem: Soc Develop Biol; Teratology Soc; Can Asn Res Toxicol; Brit Inst Biol. Res: Naturally occurring and induced congenital malformations; pesticide toxicology. Mailing Add: Health Protection Br Nat Health & Welfare Ottawa ON Can

CLEGG, JAMES S, b Aspinwall, Pa, July 27, 33; m 58; c 3. PHYSIOLOGY, BIOCHEMISTRY. Educ: Pa State Univ, BS, 57; Johns Hopkins Univ, PhD(biol), 61. Prof Exp: Res assoc biol, Johns Hopkins Univ, 61-62; asst prof zool, 62-64, assoc prof biol, 64-70, PROF BIOL, UNIV MIAMI, 70- Concurrent Pos: Wilson fel, 58-59. Mem: AAAS; Am Soc Zoologists; Am Soc Cell Biol. Res: Comparative biochemistry; mechanisms of cryptobiosis; structure and role of water in cellular metabolism; biophysics and biochemistry of dried but viable organisms. Mailing Add: Dept of Biol Univ of Miami Coral Gables FL 33124

CLEGG, LAWRENCE FRANK LEVEY, b Brighton, Eng, Apr 10, 13; m 41; c 2. BACTERIOLOGY, CHEMISTRY. Educ: Univ Toronto, BSA, 35; Univ London, PhD(chem) & Imp Col, dipl, 38, DSc(microbiol), 57. Prof Exp: Res bacteriologist, Ministry Agr & Fisheries, 37-42; asst adv dairy bacteriologist, Harper Adams Agr Col, 42-45, adv dairy bacteriologist, 45-47; prov adv bacteriologist, Ministry of Agr & Fisheries, 47-51; dep head dept bact, Nat Inst Res Dairying, Reading, 51-58; chmn dept food sci, 58-74, PROF DAIRY FOOD SCI, UNIV ALTA, 58- Concurrent Pos: Civilian with Ministry Agr & Fisheries. Mem: Can Soc Microbiol; Am Dairy Sci Asn; Brit Soc Dairy Technol; Brit Soc Appl Bact; Brit Soc Gen Microbiol. Res: Bacteriology of water and shellfish; dairy bacteriology; spores; chemical sterilization; keeping quality of milk. Mailing Add: Dept of Food Sci Univ of Alta Edmonton AB Can

CLEGG, MOSES TRAN, animal physiology, see 12th edition

CLEGG, ROBERT EDWARD, b Providence, RI, July 29, 14; m 41; c 3. BIOCHEMISTRY. Educ: Univ RI, BS, 36; NC State Col, MS, 39; Iowa State Univ, PhD(phys chem, nutrit), 48. Prof Exp: Assoc prof, 48-54, PROF CHEM, KANS STATE UNIV, 54- Mem: AAAS; Am Chem Soc; Am Soc Biol Chem. Res: Hormone influence on lipoprotein level; application of trace techniques to biochemical problems; enzyme kinetics; in vivo protein formation. Mailing Add: Dept of Biochem Kans State Univ Manhattan KS 66502

CLEGG, THOMAS BOYKIN, b Emory University, Ga, Jan 6, 40; m 68. EXPERIMENTAL NUCLEAR PHYSICS. Educ: Emory Univ, BA, 61; Rice Univ, MA, 63, PhD(physics), 65. Prof Exp: Res assoc physics, Rice Univ, 65 & Univ Wis-Madison, 65-68; asst prof, 68-72, ASSOC PROF PHYSICS, UNIV NC, CHAPEL HILL, 68- Concurrent Pos: Fulbright grant, 75; vis physicist, Ctr Nuclear Res, Saclay, France, 75-76; res assoc, Inst Fundamental Electronics, Univ Paris, 76. Mem: Am Phys Soc. Res: Elastic scattering of polarized protons and deuterons from nuclei; development of polarized ion beams for nuclear physics experiments. Mailing Add: Dept of Physics Phillips Hall Univ of NC Chapel Hill NC 27514

CLEGG, WILLIAM JOSIAH, b Social Circle, Ga, Apr 15, 17; m 46; c 2. INDUSTRIAL ORGANIC CHEMISTRY. Educ: Emory Univ, BA, 39, MA, 40; Univ Tex, PhD(org chem), 47. Prof Exp: Lab instr chem, Emory Univ, 38-40 & Univ Tex, 40-44; res chemist, Nat Aniline & Chem Co, Inc, NY, 44-45; lab instr chem, Univ Tex, 45-47; assoc chemist, 47-50, chemist, 50-61, SR CHEMIST, TENN EASTMAN CO, 61- Mem: Am Chem Soc. Res: Pharmaceutical chemicals and intermediates; aliphatic organic chemicals; physical measurements on organic chemicals; aromatic organic chemicals; dyes; chromatography. Mailing Add: Bldg 267 Tenn Eastman Co Kingsport TN 37662

CLEGHORN, ROBERT ALLEN, b Cambridge, Mass, Oct 6, 04; m 32; c 3. PSYCHIATRY. Educ: Univ Toronto, MD, 28; Aberdeen Univ, DSc(psychiat), 32; FRCPS, 65. Prof Exp: Jr rotating intern, Toronto Gen Hosp, 28-29; demonstr physiol, Aberdeen Univ, 29-32; demonstr med & asst attend physician, Toronto Gen Hosp, 33-46; from asst to prof psychiat, 46-64, prof & chmn dept, 64-70, EMER PROF PSYCHIAT, McGILL UNIV, 71- Concurrent Pos: Dir therapeut res lab, Allan Mem Inst, 46-64, dir, 64-70, hon consult, 70-; res assoc, Harvard Univ Med Sch, 53-54; psychiatrist in chief, Royal Victoria Hosp, 64-70. Mem: Am Psychiat Asn; Can Physiol Soc; Can Med Asn; Can Psychiat asn; fel Royal Col Psychiatrists. Res: Physiology and clinical aspects of the adrenal cortex; clinical endocrinology; autonomic nervous system in adrenal insufficiency; shock and blood substitutes; physiological correlates and psychoanalytic studies in psychosomatic states and psychopharmacology; study of schizo-affective psychoses and lithium therapy. Mailing Add: Allan Mem Inst 1025 Pine Ave W Montreal PQ Can

CLELAND, CHARLES EDWARD, b Kane, Pa, Feb 2, 36. ARCHAEOLOGY, CULTURAL ANTHROPOLOGY. Educ: Denison Univ, BA, 58; Univ Ark, MS, 60; Univ Mich, MA, 64, PhD(anthrop), 66. Prof Exp: From asst prof to assoc prof, 65-73, PROF ANTHROP, MICH STATE UNIV, 73-, CUR MUS, 65- Mem: Am Anthrop Asn; Soc Am Archaeol; Soc Hist Archaeol (pres, 73-74). Res: Subsistence and settlement studies of prehistoric Great Lakes Indians; archaeology of historic sites; ethnohistory of Indians of the eastern United States. Mailing Add: Mich State Univ Mus East Lansing MI 48824

CLELAND, CHARLES FREDERICK, b Indianapolis, Ind, July 1, 39. PLANT PHYSIOLOGY. Educ: Wabash Col, BA, 61; Stanford Univ, PhD(plant physiol), 67. Prof Exp: NSF fels, 66-68; Milton Fund res grant, Harvard Univ, 68; asst prof biol, Harvard Univ, 68-73, lectr, 73-74; vis asst prof bot, Univ NC, Chapel Hill, 74-75; PLANT PHYSIOLOGIST, RADIATION BIOL LAB, SMITHSONIAN INST, 75- Concurrent Pos: NSF res grants, 69 & 72. Mem: Bot Soc Am; Am Soc Plant Physiologists; AAAS; Phycol Soc Am; Japanese Soc Plant Physiologists. Res: Plant growth and development, especially hormonal basis for the photoperiodic control of flowering; biology and physiology of Lemnaceae. Mailing Add: Radiat Biol Lab Smithsonian Inst 12441 Parklawn Dr Rockville MD 20852

CLELAND, GEORGE HORACE, b Pasadena, Calif, July 19, 21; m 53; c 1. ORGANIC CHEMISTRY. Educ: Occidental Col, BA, 42; Calif Inst Technol, PhD(chem), 51. Prof Exp: Res assoc, Nat Defense Res Comn, 43-45; res chemist org chem, Naval Ord Test Sta, 52-54; assoc prof, 54-70, PROF CHEM, OCCIDENTAL COL, 70- Mem: AAAS; Am Chem Soc. Res: Reactions of amino acids; reaction mechanisms. Mailing Add: Dept of Chem Occidental Col Los Angeles CA 90041

CLELAND, JOHN W, b New Concord, Ohio, Oct 29, 21; m 47; c 2. EXPERIMENTAL SOLID STATE PHYSICS. Educ: Monmouth Col, BS, 43; Purdue Univ, MS, 49. Prof Exp: PHYSICIST, SOLID STATE DIV, OAK RIDGE NAT LAB, 49- Mem: Am Meteorol Soc; Sigma Xi; Am Asn Crystal Growers; AAAS. Res: Electrical properties of semiconductors; radiation effects research; low temperature work, thermal properties; reactor research; crystal growth materials. Mailing Add: 7101 Stockton Dr Knoxville Tn 37919

CLELAND, MARSHALL ROBERT, b Vermillion, SDak, Feb 9, 26; m 48; c 4. NUCLEAR PHYSICS. Educ: Univ SDak, BA, 47; Wash Univ, St Louis, PhD(physics), 51. Prof Exp: Staff scientist, Nat Bur Stand, 51-52; staff scientist, Nuclear Res & Develop, Inc, Mo, 52-53; pres high voltage accelerator develop, Teleray Corp, Mo, 53-58; VPRES ACCELERATOR DEVELOP & MANUFACTURE, RADIATION DYNAMICS, INC, 58-, TECH DIR, 74- Concurrent Pos: Mem, Nat Acad Sci-Nat Res Coun, Panel Adv to Ctr Radiation Res, Nat Bur Stand, 65-, chmn, 68 & 69. Mem: Am Phys Soc. Res: Industrial and medical uses of radiation; development of high voltage particle accelerators; electron and ion sources. Mailing Add: Radiation Dynamics Inc 1800 Shames Dr Westbury NY 11590

CLELAND, ROBERT E, b Baltimore, Md, Apr 30, 32; m 57; c 2. PLANT PHYSIOLOGY. Educ: Oberlin Col, AB, 53; Calif Inst Technol, PhD(biochem), 57. Prof Exp: USPHS fels plant physiol, Lund, 57-58 & King's Col, London, 58-59; asst prof bot, Univ Calif, Berkeley, 59-64; assoc prof, 64-68, PROF BOT, UNIV WASH, 68- Concurrent Pos: Guggenheim fel, Univ Leeds, 67-68. Mem: Am Soc Plant Physiol; Bot Soc Am; Am Soc Cell Biol. Res: Mechanism of auxin action; cell wall metabolism. Mailing Add: Dept of Bot Univ of Wash Seattle WA 98105

CLELAND, ROBERT LINDBERGH, b St Francis, Kans, June 10, 27; m 56; c 4. PHYSICAL CHEMISTRY. Educ: Agr & Mech Col Tex, BS, 49; Mass Inst Technol, SM, 51, PhD(phys chem), 56. Prof Exp: Res asst chem, Mass Inst Technol, 48-50, res employee, div indust coop, 50-52, res asst, 52-56; res assoc chem, Cornell Univ, 56-58; from asst prof to assoc prof, 60-71, PROF CHEM, DARTMOUTH COL, 71- Concurrent Pos: Fulbright res scholar, State Univ Leiden, 58-59; USPHS spec res fel, Retina Found, Mass, 59-60; res fel, Univ Uppsala, 68; vis prof, Univ Strasbourg, 68-69; assoc ed, Macromolecules, 74- Mem: AAAS; Am Chem Soc; The Chem Soc. Res: Physical chemistry of solutions of ionic polysaccharides; liquid transport in membranes. Mailing Add: Dept of Chem Dartmouth Col Hanover NH 03755

CLELAND, WILFRED EARL, b St Francis, Kans, Aug 10, 37; m 66; c 2. ELEMENTARY PARTICLE PHYSICS. Educ: Agr & Mech Col Tex, BS, 59; Yale Univ, MS, 60, PhD(physics), 64. Prof Exp: Instr physics, Yale Univ, 63-64; vis scientist, Europ Orgn Nuclear Res, Geneva, 64-67; from asst prof to assoc prof physics, Univ Mass, Amherst, 67-70; ASSOC PROF PHYSICS, UNIV PITTSBURGH, 70- Concurrent Pos: NATO fel, 64-65; NSF fel, 65-66. Mem: Am Phys Soc. Res: Studies of interactions of elementary particles using electronic techniques at high energy accelerators, including muonium, lambda decay, muon tridents and K Mesons systems. Mailing Add: Dept of Physics Univ Pittsburgh 4200 Fifth Ave Pittsburgh PA 15260

CLELAND, WILLIAM WALLACE, b Baltimore, Md, Jan 6, 30; m 67. BIOCHEMISTRY. Educ: Oberlin Col, AB, 50; Univ Wis, MS, 53, PhD(biochem), 55. Prof Exp: NSF fel, Univ Chicago, 57-59; from asst prof to assoc prof, 59-66, PROF BIOCHEM, UNIV WIS-MADISON, 66- Mem: Am Chem Soc; Am Soc Biol Chemists. Res: Use of enzyme kinetics to deduce enzymatic mechanisms. Mailing Add: Dept of Biochem Univ of Wis Madison WI 53706

CLELLAND, RICHARD COOK, b Camden, NY, Aug 23, 21; m 63; c 2. APPLIED STATISTICS. Educ: Hamilton Col, BA, 44; Columbia Univ, AM, 49; Univ Pa, PhD(statist), 56. Prof Exp: Instr math, Syracuse Univ, 46-47; instr, Hamilton Col, 50-53; from asst prof to assoc prof statist, 56-66, chmn dept statist & opers res, 66-71, actg dean, Wharton Sch, 71-72, PROF STATIST, UNIV PA, 66-, ASSOC DEAN, WHARTON SCH, 75- Mem: Fel Am Statist Asn; Inst Math Statist; Opers Res Soc Am; Am Math Soc. Res: Experimental design; statistical methodology; operations research. Mailing Add: Dept of Statist & Opers Res E-111 Dietrich Hall Univ of Pa Philadelphia PA 19174

CLEM, JOHN RICHARD, b Waukegan, Ill, Apr 24, 38; m 60; c 1. PHYSICS. Educ: Univ Ill, Urbana, BS, 60, MS, 62, PhD(physics), 65. Prof Exp: Res assoc physics, Univ Md, 65-66; vis res fel, Tech Univ Munich, 66-67; from asst prof to assoc prof, 67-75, PROF PHYSICS, IOWA STATE UNIV, 75- Concurrent Pos: Consult, Argonne Nat Lab, 71-; vis staff mem, Los Alamos Sci Lab, 71-; Fulbright-Hayes sr res scholar, Inst Solid Bodies Res, Julich, Ger, 74-75. Mem: AAAS; Am Phys Soc. Res: Theoretical research in solid state physics, low-temperature solid state physics; superconductivity; superfluidity. Mailing Add: Dept of Physics Iowa State Univ Ames IA 50010

CLEM, JUDY ROBERTA, b Atlanta, Ga, Nov 28, 38; m 75. PARASITOLOGY, LABORATORY MEDICINE. Educ: Birmingham-Southern Col, BA, 60; Univ NC, MPH, 67, PhD(parasitol & lab pract), 74. Prof Exp: Med technologist, Peace Corps, Thailand, 61-63 & Huntsville Hosp, Ala, 64-66; instr, 67-74, ASST PROF PARASITOL, SCH PUB HEALTH, UNIV NC, CHAPEL HILL, 74- Mem: Am Soc Microbiol; Am Pub Health Asn; Am Soc Clin Pathologists; Sigma Xi; Conf State & Prov Pub Health Lab Dir. Res: Immunoserology of parasitic infections, particularly Angiostrongylus cantonensis, with emphasis on exploring host-parasite relationships and developing diagnostic tests. Mailing Add: Dept of Parasitol Sch of Pub Health Univ of NC Chapel Hill NC 27514

CLEM, LEROY HOFFMAN, meteorology, see 12th edition

CLEM, LESTER WILLIAM, b Frederick, Md, June 23, 34; m 57; c 5. IMMUNOLOGY, IMMUNOCHEMISTRY. Educ: Western Md Col, BS, 56; Univ Del, MS, 60; Univ Miami, PhD(microbiol), 63. Prof Exp: From instr to assoc prof microbiol, 64-71, PROF IMMUNOL & MED MICROBIOL, COL MED, UNIV FLA, 71- Concurrent Pos: Res assoc immunol, Variety Children's Res Found, Miami, 63-66; WHO consult immunol, India, 70; instr physiol course, Marine Biol Lab, Woods Hole, 72 & 74. Mem: AAAS; Am Asn Immunol; Soc Exp Biol & Med; Am Asn Zool; Am Soc Microbiol. Res: Phylogenetic development of immunological competency and immunoglobulin structure and function; cholera immunity. Mailing Add: Dept of Immunol & Med Microbiol J Hillis Miller Med Ctr Univ of Fla Gainesville FL 32601

CLEM, WILLIAM HENRY, b Champaign, Ill, Dec 8, 32; m 59; c 1. ENDODONTICS, ORAL MICROBIOLOGY. Educ: Northwestern Univ, DDS, 56; Univ Wash, MSD, 65. Prof Exp: Instr endodont, Dent Sch, Northwestern Univ, 60-63; asst prof, Sch Dent, Univ Wash, 65-67; ASSOC PROF ENDODONT, COL DENT, UNIV ILL, 67- Concurrent Pos: Nat Inst Dent Res fel, Univ Wash, 63-65; res assoc & consult, USPHS Hosp, Seattle, 65-; consult, Coun Dent Educ, Am Dent Asn, 74- Mem: Am Dent Soc; Am Soc Microbiol; Am Asn Endodont. Res: Antibacterial factors in saliva; role of peroxidase; glutamic acid isotope transport in lactobacillus; dental caries in rat; pathogenic potential of root canal bacteria, streptococci; post-treatment endodontic pain. Mailing Add: Dept of Endodontia Univ of Ill Sch of Dent Chicago IL 60612

CLEMANS, GEORGE BURTIS, b Huntington, NY, May 11, 38; m 61. ORGANIC CHEMISTRY, BIOCHEMISTRY. Educ: Va Polytech Inst, BS, 60; Duke Univ, MA, 63, PhD(chem), 64. Prof Exp: Res assoc chem, Ind Univ, 64-65; res assoc chem, Univ Ark, 65-66, asst prof, 66-67; asst prof, 67-74, ASSOC PROF CHEM, BOWLING GREEN STATE UNIV, 74- Res: Stereospecific reactions of dicyclopentadiene derivatives; synthesis of diterpenes. Mailing Add: Dept of Chem Bowling Green State Univ Bowling Green OH 43402

CLEMANS, KERMIT GROVER, b Adrian, NDak, Apr 14, 21; m 44; c 3. MATHEMATICAL STATISTICS. Educ: Jamestown Col, BS, 43; Univ Minn, MA, 48; Univ Ore, PhD(math statist), 53. Prof Exp: Instr math, Willamette Univ, 48-50 & Univ Ore, 50-53; math & statist consult, US Naval Test Sta, 53-59; dean div sci & technol, 60-67, PROF MATH STUDIES, UNIV SOUTHERN ILL, EDWARDSVILLE, 59- Mem: AAAS; Am Math Soc; Math Asn Am; Inst Math Statist. Res: Nonparametric statistics; extreme value statistics. Mailing Add: Fac of Math Studies Southern Ill Univ Edwardsville IL 62025

CLEMANS, STEPHEN D, b Gloversville, NY, Apr 1, 39; m 61; c 2. STRUCTURAL CHEMISTRY. Educ: Rensselaer Polytech Inst, BS, 61, MS, 64, PhD(org chem), 67. Prof Exp: Asst res chemist, 63-69, res chemist, 69-76, SR RES CHEMIST & GROUP LEADER, STERLING WINTHROP RES INST, RENSSELAER, 76- Concurrent Pos: Fel, Harvard Univ, 67. Mem: Am Chem Soc. Res: Pharmaceutical research; structure determination by spectroscopic methods of compounds of pharmaceutical interest. Mailing Add: Box 149 RD 4 Troy NY 12180

CLEMENCE, GERALD MAURICE, astronomy, deceased

CLEMENCY, CHARLES V, b New York, NY, Feb 12, 29; m 54; c 3. GEOCHEMISTRY, CLAY MINERALOGY. Educ: Polytech Inst Brooklyn, BS, 50; NY Univ, MS, 58; Univ Ill, PhD(geol), 61. Prof Exp: Chemist, Sylvania Elec Prod Inc, 53-58; asst prof, 61-67, ASSOC PROF GEOL, STATE UNIV NY BUFFALO, 67- Concurrent Pos: Ford Found for study grant, Brazil, 69. Mem: Mineral Soc Am; Geol Soc Am; Geochem Soc; Clay Minerals Soc. Res: Analytical geochemistry; rock weathering; low temperature water-rock interactions. Mailing Add: Dept of Geol Sci State Univ of NY Buffalo NY 14226

CLEMENS, ANTON HUBERT, b Gerolstein, Ger, Nov 19, 28; m 52; c 3. BIOINSTRUMENTATION. Educ: Polytech Inst Bingen, MScEE, 51. Prof Exp: Europ res rep, Picker Int Corp, Switz, 59-62, consult res & develop nuclear instr patient monitoring, 62-65; dir instrument res & develop, Ames Co Div, 65-72, DIR LIFE SCI INSTRUMENTS, MILES LABS, INC, 72- Mem: Asn Advan Med Instrumentation; Instrument Soc Am; Ger Soc Nuclear Med; Ger Soc Data Processing & Automation Med. Res: Patient monitoring; nuclear medicine systems; spectrophotometry; laboratory automation; research instrumentation for endocrinology and metabolism; artificial endocrine pancreas. Mailing Add: 3435 Calumet Ave Elkhart IN 46514

CLEMENS, CARL FREDERICK, b Elkland, Pa, Nov 24, 24; m 53; c 8. CHEMISTRY. Educ: Ohio Univ, BS, 55; Univ Rochester, MBA, 66. Prof Exp: Asst biochem, Univ Rochester, 52-53; jr chemist, Haloid Co, 55-56, chemist, Haloid-Xerox, Inc, 56-58, proj chemist, 58-59; sr proj chemist, 59-64, scientist, 64-66, mgr instrumental anal, 66-67, mgr mat appln & develop, 67-69, tech prog mgr, Advan Develop Dept, 69-72, TECHNOL PROG MGR, XEROGRAPHIC TECHNOL DEPT, XEROX CORP, 72- Mem: Am Chem Soc; Am Mgt Asn; fel Am Inst Chemists. Res: Xerography; application of polymer science to xerographic materials; polymer characterization and analytical chemistry; technical management. Mailing Add: 2448 Lake Rd Ontario NY 14519

CLEMENS, CHARLES HERBERT, b Dayton, Ohio, Aug 15, 39; m 66; c 2. MATHEMATICS. Educ: Holy Cross Col, AB, 61; Univ Calif, Berkeley, PhD(math), 66. Prof Exp: From asst prof to assoc prof math, Columbia Univ, 70-75; ASSOC

PROF MATH, UNIV UTAH, 75- Mem: Am Math Soc. Res: Topology of algebraic varieties. Mailing Add: Dept of Math Univ of Utah Salt Lake City UT 84112

CLEMENS, DAVID HENRY, b Newton, Mass, Nov 8, 31; m 53; c 3. ORGANIC CHEMISTRY. Educ: Middlebury Col, AB, 53; Univ Wis, PhD(chem), 57. Prof Exp: Head ion exchange & pollution control synthesis, 57-75, PROJ LEADER COATINGS RES, ROHM AND HAAS CO, 75- Mem: Am Chem Soc; The Chem Soc. Res: Organic synthesis; vinyl polymers; plasticizers; ion exchange; adsorbents; membrane processes; organic coatings. Mailing Add: Spring House Res Labs Rohm and Haas Co Spring House PA 19477

CLEMENS, DONALD F, inorganic chemistry, see 12th edition

CLEMENS, HOWARD PAUL, b Arthur, Ont, May 31, 23; nat US; m; c 4. ZOOLOGY. Educ: Univ Western Ont, BS, 46, MS, 47; Ohio State Univ, PhD(zool), 49. Prof Exp: Asst zool, Univ Western Ont, 47; from instr to assoc prof, 49-72, PROF ZOOL, UNIV OKLA, 72- Mem: AAAS; Am Soc Limnol & Oceanog; Am Fisheries Soc; Am Soc Zoologists; Am Soc Ichthyologists & Herpetologists. Res: Limnology; fishery biology; aquatic invertebrates; fish endocrines. Mailing Add: Dept of Zool Univ of Okla Norman OK 73069

CLEMENS, JAMES ALLEN, b Windsor, Pa, Feb 4, 41; m 64; c 2. NEUROENDOCRINOLOGY, NEUROPHYSIOLOGY. Educ: Pa State Univ, BS, 63, MS, 65; Mich State Univ, PhD(physiol), 68. Prof Exp: RES ASSOC, ELI LILLY & CO, 69- Concurrent Pos: Nat Inst Neurol Dis & Blindness fel, Univ Calif, Los Angeles, 68-69. Mem: AAAS; Endocrine Soc; Soc Neurosci; Int Soc Psychoneuroendocrinol. Res: Neural mechanisms that control anterior pituitary hormone secretion. Mailing Add: Dept of Physiol Res Lilly Res Labs Indianapolis IN 46206

CLEMENS, LAWRENCE MARTIN, b Chicago, Ill, Nov 14, 37; m 63. ORGANIC CHEMISTRY. Educ: Ill Inst Technol, BS, 59; Carnegie Inst Technol, PhD(org chem), 64. Prof Exp: NIH fel chem, Univ Notre Dame, 63-64; res chemist, Archer Daniels Midland Co, 64-69; SR CHEMIST, MINN MINING & MFG CO, 69- Mem: Am Chem Soc. Res: Biphenyl isomerism; charge-transfer complexing; reaction kinetics. Mailing Add: Minn Mining & Mfg Co 3M Ctr St Paul MN 55101

CLEMENS, STANLEY RAY, b Souderton, Pa, May 11, 41; m 62; c 2. MATHEMATICS. Educ: Bluffton Col, AB, 63; Ind Univ, MA, 65; Univ NC, PhD(math), 68. Prof Exp: Instr, Univ NC, 67-68; ASST PROF MATH, ILL STATE UNIV, 68- Mem: Am Math Soc; Math Asn Am. Res: Topology. Mailing Add: Dept of Math Ill State Univ Normal IL 61761

CLEMENS, WILLIAM ALVIN, b Berkeley, Calif, May 15, 32; m 55; c 4. PALEONTOLOGY. Educ: Univ Calif, Berkeley, BA, 54, PhD(pa- leont), 60. Prof Exp: NSF fel, 60-61; from asst prof & asst cur to assoc prof zool & assoc cur fossil higher vert, Univ Kans, 61-67; assoc prof, 67-71, PROF PALEONT, UNIV CALIF, BERKELEY, 71- Concurrent Pos: NSF fel, 68-69; John Simon Guggenheim Found fel, 74. Mem: Soc Syst Zool; Soc Vert Paleont; Geol Soc Am; Palaeont Asn; Zool Soc London. Res: Evolution of Mesozoic and Cenozoic mammals. Mailing Add: Dept of Paleont Univ of Calif Berkeley CA 94720

CLEMENS, WILLIAM BRYSON, b Milton, Pa, Sept 28, 16; m 42. BACTERIOLOGY. Educ: Bucknell Univ, BS, 37, MS, 38. Prof Exp: High sch head dept sci, Pa, 38-42; sr supvr, Acid Dept, Pa Ord Works, US Rubber Co, 42-43, supvr acid area 3, Kankakee Ord Works, 43-45; high sch head dept sci, Ga, 45-46; from instr to prof bact, 46-73, chmn dept sci, 58-61, sci bldg coordr, 62-64, exec chmn div & chmn dept biol, 64-68, EMER PROF BACT, STATE UNIV NY COL CORTLAND, 73- Mem: AAAS; Am Nature Study Soc; Nat Asn Biol Teachers; Nat Sci Teachers Asn; Am Forestry Asn. Res: Subject matter and methods used in introductory bacteriology courses in colleges and universities. Mailing Add: E River Rd RD 1 Cortland NY 13045

CLEMENT, ANTHONY CALHOUN, b Spartanburg, SC, Nov 24, 09; m 59. EMBRYOLOGY. Educ: Univ SC, BS, 30; Princeton Univ, AM, 33, PhD(biol), 35. Prof Exp: From asst prof to prof biol, Col Charleston, 30-49; assoc prof, 49-56, PROF BIOL, EMORY UNIV, 56- Concurrent Pos: Guggenheim fel, 54-55; prog dir develop biol, NSF, 58-59, mem adv panel develop biol, 59-62; trustee, Marine Biol Lab, Woods Hole, 64-72; vis assoc, Calif Inst Technol, 67. Mem: AAAS; Am Soc Zoologists; Soc Develop Biol. Res: Molluscan embryology; cytoplasmic localization and embryonic determination; deletion experiments. Mailing Add: Dept of Biol Emory Univ Atlanta GA 30322

CLEMENT, DUNCAN, b Pittsfield, Mass, Oct 22, 17. SCIENCE ADMINISTRATION. Educ: Mt St Mary's Col, Md, BS, 40; Harvard Univ, MA, 42, PhD(bot), 48. Prof Exp: Econ botanist, Atkins Garden, Harvard Univ, Cuba, 48-53, in chg, 49-53, dir, 53-60, dir, US, 61-63; consult, Off Int Sci Activities, NSF, 62-63, prog dir, 63-64, head sci liaison staff, Costa Rica, 64-68, prof assoc, Off Int Progs, 68-74; SCI ATTACHE, US EMBASSY, MADRID, SPAIN, 74- Res: International science cooperation. Mailing Add:

CLEMENT, GERALD EDWIN, b Austin, Minn, Nov 5, 35; m 57; c 2. CLINICAL CHEMISTRY. Educ: Univ Minn, BA, 57; Purdue Univ, PhD(org chem), 61; Am Bd Clin Chemists, cert, 75. Prof Exp: NIH fel, Northwestern Univ, 61-63; asst prof org chem, Harpur Col, 63-68; assoc prof chem, Kenyon Col, 68-74; CLIN CHEMIST, ALLENTOWN & SACRED HEART HOSP, 74- Concurrent Pos: Fel, Hahnemann Hosp, 72-74. Mem: Am Asn Clin Chemists; Am Chem Soc; Am Soc Biol Chemists. Res: Organic reaction mechanisms, especially acid, base and enzyme catalysis. Mailing Add: Allentown & Sacred Heart Hosp 1200 S Cedar Crest Blvd Allentown PA 18105

CLEMENT, JACOB JAMES, b Chicago, Ill, Feb 28, 44. RADIOBIOLOGY. Educ: St Mary's Col, BA, 66; Roosevelt Univ, MS, 69; Univ NC, MSPH, 70, PhD(radiol health), 73. Prof Exp: Instr, 73-75, ASST PROF RADIOBIOL, DEPT THERAPEUT RADIOL, UNIV MINN HOSPS, 75- Mem: Radiation Res Soc. Res: Radiation effects on tumors, growth response, cellularity, vascularity, oxygenation and cell kinetics; radiation-induced cell synchrony in vivo; combined chemotherapy and radiotherapy. Mailing Add: Dept of Therapeut Radiol Univ of Minn Hosps Minneapolis MN 55455

CLEMENT, JOHN REID, JR, b East Spencer, NC, Apr 14, 21; m 45; c 2. PHYSICS. Educ: Catawba Col, AB, 43. Prof Exp: Proj physicist low temp thermomet & calorimet, 46-53, head cryogenic properties, matter & cryogenic devices sect, 53-59, asst head, Cryogenics Br, 59-62, head, 62-69, ASSOC SUPT SOLID STATE DIV, US NAVAL RES LAB, 69- Mem: Fel Am Phys Soc; Sigma Xi. Res: High voltage electricity; low temperature thermometry and calorimetry; cryogenic devices; high magnetic fields. Mailing Add: US Naval Res Lab Washington DC 20390

CLEMENT, MAURICE JAMES, b Vancouver, BC, Sept 11, 38. THEORETICAL ASTROPHYSICS. Educ: Univ BC, BSc, 60, MSc, 61; Univ Chicago, PhD(astrophys), 65. Prof Exp: Asst prof, 67-74, ASSOC PROF ASTRON, UNIV TORONTO, 74- Concurrent Pos: Fel, Princeton Univ, 65-66; Nat Res Coun Can fel, 66-67. Mem: Am Astron Soc. Res: Equilibrium and stability of rotating stars; differential rotation and meridian circulation in stars. Mailing Add: Dept of Astron Univ of Toronto Toronto ON Can

CLEMENT, PAUL ARNOLD, b Denmark, Oct 29, 16; nat US; m 38; c 3. MATHEMATICS. Educ: State Col Wash, BA, 38, MA, 40; Univ Calif, Los Angeles, PhD(math), 49. Prof Exp: Asst, Univ Calif, Los Angeles, 40-42; asst instr math & physics, Civil Serv, US Air Force, 42; asst, Univ Calif, Los Angeles, 46-49; from instr to asst prof, 49-58, ASSOC PROF MATH, WASH STATE UNIV, 58- Concurrent Pos: NSF guest worker, Nat Bur Stand, DC, 57. Mem: Am Math Soc; Math Asn Am; Soc Indust & Appl Math. Res: Theory of functions; number theory; modern geometry. Mailing Add: Dept of Math Wash State Univ Pullman WA 99163

CLEMENT, ROBERT ALTON, b Brockton, Mass, Aug 12, 29; m 55; c 4. INDUSTRIAL ORGANIC CHEMISTRY. Educ: Mass Inst Technol, BS, 50; Univ Calif, Los Angeles, PhD(chem), 54. Prof Exp: Proj assoc chem, Univ Wis, 54-55; from instr to asst prof, Univ Chicago, 55-62; RES CHEMIST, E I DU PONT DE NEMOURS & CO, INC, 62- Mem: Am Chem Soc; The Chem Soc. Res: Synthetic methods in organic chemistry; polymer chemistry. Mailing Add: Cent Res & Develop Dept Exp Sta E I du Pont de Nemours & Co Inc Wilmington DE 19898

CLEMENT, ROLAND CHARLES, b Fall River, Mass, Nov 22, 12; m 47; c 3. ORNITHOLOGY. Educ: Brown Univ, AB, 49; Cornell Univ, MS, 50. Prof Exp: Staff biologist, 58-66, V PRES, NAT AUDUBON SOC, 66- Concurrent Pos: Secy for the Americas, Int Coun Bird Preserv, 74- Mem: Am Ornith Union; Wilson Ornith Soc; Wildlife Soc. Res: Resources conservation; faunal investigations in centra; Ungava, Labrador. Mailing Add: Nat Audubon Soc 950 Third Ave New York NY 10022

CLEMENT, WILLIAM GLENN, b Denver, Colo, Apr 11, 31; m 52; c 2. GEOPHYSICS, ELECTRICAL ENGINEERING. Educ: Stanford Univ, BS, 56, PhD(geophys), 63. Prof Exp: Geophysicst, Pan Am Petrol Corp, Tex, 63-65, sr geophysicist, Res Lab, Okla, 65-68, staff res engr, 68-74, SR RES ENGR, AMOCO PROD RES CTR, 74- Concurrent Pos: Part-time instr, Sch Earth Sci, Tulsa Univ, 69. Mem: Soc Explor Geophys; Europ Asn Explor Geophys. Res: Application of communication theory to the analysis of seismic and potential field data. Mailing Add: Amoco Prod Res Ctr 4502 E 41st St Tulsa OK 74115

CLEMENT, WILLIAM H, b Johnstown, Pa, Dec 17, 31; m 65; c 2. ORGANIC CHEMISTRY. Educ: Gettysburg Col, AB, 54; Univ Del, MS, 57, PhD(chem), 60. Prof Exp: Chemist, E I du Pont de Nemours & Co, 55 & Pittsburgh Plate Galss Co, 56-57; res chemist, Gulf Oil Corp, 59-62; asst prof chem, Waynesburg Col, 62-63; NSF fel, Univ Buffalo, 63; USDA fel, Univ Cincinnati, 63-65; asst prof, Ithaca Col, 65-70, prof, 70-71; RES ASSOC ENVIRON SCI, RUTGERS UNIV, 71- Mem: Am Chem Soc; Sigma Xi. Res: Oxidation reactions and mechanisms; organometallics; catalysis; organic synthesis; environmental chemistry; water quality. Mailing Add: Dept of Environ Sci Rutgers Univ New Brunswick NJ 08903

CLEMENT, WILLIAM MADISON, JR, b Rome, Ga, Dec 15, 28; m 50. GENETICS. Educ: Univ Ga, BSA, 50; Univ Calif, PhD, 58. Prof Exp: Lab technician agron, Univ Calif, 52-57; res geneticist, Forage & Range Br, Agr Res Serv, USDA, 58-65; ASSOC PROF CYTOL & CYTOGENETICS, VANDERBILT UNIV, 65- Concurrent Pos: Asst prof, Univ Minn, 61-65. Mem: Crop Sci Soc Am; Am Genetics Soc; Genetics Soc Can; Am Soc Naturalists. Res: Cytogenetics of medicago sativa and related species; chromosome structure. Mailing Add: Dept of Gen Biol Vanderbilt Univ Nashville TN 37203

CLEMENTE, CARMINE DOMENIC, b Penns Grove, NJ, Apr 29, 28; m 68. NEUROANATOMY. Educ: Univ Pa, AB, 48, MS, 50, PhD, 52. Prof Exp: From instr to assoc prof, 52-63, chmn dept, 63-73, PROF ANAT, SCH MED, UNIV CALIF, LOS ANGELES, 63- Concurrent Pos: Gianinni Found fel, 53-54; hon res assoc, Univ Col, Univ London, 53-54; consult, Vet Admin Hosp, Sepulveda, Calif; ed, Gray's Anat & Exp Neurol; vis scientist & spec consult, NIH, 58; mem, Biol Stain Comn; mem subcomt neuropath, Nat Acad Sci; mem med adv bd, Bank of Am-Gianinni Found; mem admin bd, Asn Am Med Cols, 73- Honors & Awards: Res Award, Pavlovian Soc NAm, 68; Award of Merit in Sci, Nat Paraplegic Found, 73. Mem: Am Asn Anat (vpres, 70-72, pres-elect, 75); Am Acad Neurol; Am Physiol Soc; Am Acad Cerebral Palsy; Pavlovian Soc NAm (vpres, 70-72, pres, 72-73). Res: Regeneration of nerve fibers; effects of x-irradiation on brain; neurocytology; basic neurology; sleep and wakefulness. Mailing Add: Dept of Anat Univ of Calif Sch of Med Los Angeles CA 90024

CLEMENTS, BURIE WEBSTER, b Pierce, Fla, Dec 16, 27; m 52; c 3. MEDICAL ENTOMOLOGY. Educ: Univ Fla, BS, 54, MSA, 56. Prof Exp: Entomologist, stored prod insects, Ga Lab, USDA, 56-58, Savannah, Ga Lab, 58-60; Vero Beach Lab, Entom Res Ctr, 60-64, ENTOMOLOGIST & HEAD FLY CONTROL RES SECT, W FLA ARTHROPOD RES LAB, DIV HEALTH, 64- Mem: Entom Soc Am; Am Mosquito Control Asn; Sigma Xi. Res: Evaluation of insecticides in the control of stable flies and sand flies. Mailing Add: WFla Arthropod Res Lab Dept of Health & Rehab Serv Div of Health PO Box 2326 Panama City FL 32401

CLEMENTS, GEORGE FRANCIS, b Colfax, Wash, Apr 17, 31; m 52; c 4. MATHEMATICS. Educ: Univ Wis, BSME, 53; Syracuse Univ, MA, 57, PhD(math), 62. Prof Exp: Asst math, Syracuse Univ, 55-62; asst prof appl math, 62-68, ASSOC PROF MATH, UNIV COLO, BOULDER, 68- Mem: Am Math Soc. Res: Combinatorial theory. Mailing Add: Dept of Math Univ Colo Boulder CO 80302

CLEMENTS, GERALD RICHARD, b Hillsboro, Ind, June 1, 20. EXPERIMENTAL PATHOLOGY. Educ: Ind Univ, AB, 49; Northwestern Univ, MS, 53, PhD(path), 56. Prof Exp: HEAD DEPT PATH & TOXICOL, ARMOUR PHARMACEUT CO, 51- Mem: Soc Toxicol; Am Cancer Soc. Res: Toxicology and pathology of pharmaceutical preparations; spontaneous diseases of laboratory animals; care of laboratory animals. Mailing Add: Dept of Path & Toxicol Armour Pharmaceut Co Box 511 Kankakee IL 60901

CLEMENTS, JOHN ALLEN, b Auburn, NY, May 16, 23; m 49; c 2. PHYSIOLOGY. Educ: Cornell Univ, MD, 47. Prof Exp: Res asst physiol, Med Sch, Cornell Univ, 47-49; physiologist, Med Labs, Army Chem Ctr, Md, 51-62, asst chief clin invest br, 52-61; res assoc physiol & assoc prof pediat, 61-64, PROF PEDIAT & AM HEART ASN CAREER INVESTR, SCH MED, UNIV CALIF, SAN FRANCISCO, 64- Concurrent Pos: Res assoc physiol, Sch Med, Univ Pa, 52-58; lectr, Sch Med, Johns Hopkins Univ, 55-61; consult, Baltimore City Hosp, 57-61, Roswell Park Mem Inst, 58-61 & Surgeon Gen, USPHS, 64-68; sci counr, Nat Heart & Lung Inst, 72; assoc ed, Am Rev Respiratory Dis, 73- Honors & Awards: Modern Med Distinguished Achievement Award, 73. Mem: Nat Acad Sci; Am Physiol Soc; NY Acad Sci. Res: Biophysics; respiration; membrane and cardiovascular physiology. Mailing Add: Cardiovasc Res Inst Univ of Calif Med Ctr San Francisco CA 94143

CLEMENTS, JOHN HERBERT, b Alma, Ont, Apr 2, 07; nat US; m 36; c 5. PHYSICS. Educ: Univ Toronto, BA, 27, MA, 28; Univ Chicago, PhD(physics), 35. Prof Exp: Asst demonstr, Univ Toronto, 27-28; instr physics, Mt Allison Univ, 28-30; asst, Univ Chicago, 31-34; engr & br off mgr, Schlumberger Well Surv Corp, Houston, 36-44; instr physics, Ohio Wesleyan Univ, 44-45; asst prof, SDak State Col, 45-48; prof, 48-74, head dept, 48-68, EMER PROF PHYSICS, E TEX STATE UNIV, 74- Mem: AAAS; Am Phys Soc. Res: Band spectroscopy; temperature variation method to assist in vibrational analyses of complex molecular spectra; absorption spectrum of sulphur dioxide. Mailing Add: Dept of Physics ETex State Univ Commerce TX 75428

CLEMENTS, JOHN RICHARD, b Welland, Ont, Feb 21, 31; m 60; c 5. ENVIRONMENTAL SCIENCES. Educ: Univ Mich, BS, 57, MF, 58, PhD(forest ecol), 63. Prof Exp: Compiler forest inventory, Ont Paper Co, 55-56; res officer forest ecol, Can Dept Forestry, 62-66, res scientist, Petawawa Forest Exp Sta, 66-73; ENVIRON ADV, NAT ENERGY BD, 73- Mem: AAAS; Ecol Soc Am; Soc Am Foresters; Can Inst Forestry; Can Bot Asn. Res: Rainfall patterns within forests; influence of forest structure and silviculture on penetration of rainfall through forest vegetation. Mailing Add: Environ Group Nat Energy Bd Trebla Bldg 473 Albert St Ottawa ON Can

CLEMENTS, REGINALD MONTGOMERY, b Vancouver, BC, Apr 13, 40; m 64. PLASMA PHYSICS. Educ: Univ BC, BASc, 63, MASc, 64; Univ Sask, PhD(plasma physics), 67. Prof Exp: Fel plasma physics, Univ Alta, 67-68; asst prof, 68-73, ASSOC PROF PLASMA PHYSICS, UNIV VICTORIA, 73- Mem: Can Asn Physicists. Res: Gas discharge physics; plasma diagnostic methods, specifically electrostatic and rf probes; microwave-plasma interactions; electrical phenomena in combustion. Mailing Add: Dept of Physics Univ of Victoria Victoria BC Can

CLEMENTS, RICHARD GERALD, b Indianapolis, Ind, Apr 28, 31; m 56; c 4. SOIL SCIENCE, ECOLOGY. Educ: Purdue Univ, BS, 56; Univ Ga, PhD(soil sci), 66. Prof Exp: Soil scientist, Standard Fruit & Steamship Co, 56-58, dir qual control, 58-61, agronomist, 61-63; res asst soils, Univ Ga, 63-66, res assoc ecol, 66-69; DIR TERRESTRIAL ECOL, PR NUCLEAR CTR, AEC, 69- Concurrent Pos: Mem comt, US Int Biol Prog, Trop Biome, 70-71; mem nat comt, Man & Biosphere, US Directorate Trop Forests, 75- Mem: Am Soc Agron; Ecol Soc Am; Am Inst Biol Sci. Res: Tropical terrestrial ecology involving hydrology, climatology, soils, plant and animal ecology and limnology; ecosystem analysis and modelling. Mailing Add: PR Nuclear Ctr Caparra Heights Sta San Juan PR 00935

CLEMENTS, ROBERT LAWRENCE, b Howell, Mich, Jan 3, 25; m 54; c 1. FOOD BIOCHEMISTRY. Educ: Mich State Univ, BS, 51; Ohio State Univ, PhD(agr biochem), 55. Prof Exp: Asst instr agr biochem, Ohio State Univ, 53-54, fel, 55-56; from jr biochemist to asst biochemist, plant biochem, Univ Calif, 56-62; asst dir res labs, inst nutrit & food technol, Ohio State Univ, 62-66, assoc prof hort, Ohio State Univ & Ohio agr Res & Develop Ctr, 66-68; RES CHEMIST, USDA SOFT WHEAT QUAL LAB, AGR RES SERV, 68- Mem: Fel AAAS; Am Soc Plant Physiol; Am Asn Cereal Chemists; Inst Food Technologists; NY Acad Sci. Res: Cereal chemistry; proteins and carbohydrates of wheat; bioanalytical methodoly. Mailing Add: USDA Soft Wheat Qual Lab Crops Res Div Agr Res Serv Wooster OH 44691

CLEMENTSON, GERHARDT C, b Blackearth, Wis, May 3, 17; m 43; c 3. COMPUTER SCIENCES, OPERATIONS RESEARCH. Educ: US Mil Acad, BS, 42; Calif Inst Technol, MS, 45; Mass Inst Technol, MSAE, 48, ScD, 50. Prof Exp: Asst prof elec eng, US Air Force Inst Technol, 54-55, prof aeronaut & head dept, US Air Force Acad, 55-61; asst to exec vpres-tech, Space & Info Syst Div, NAm Aviation, Inc, 64-68; dir opers res, Auto-Tronix Universal Corp, 68-70; PROF COMPUT & MGT SCI & CHMN DEPT MGT SCI, METROP STATE COL, 70- Concurrent Pos: Mem, Order of Daedalians, 60. Mem: Opers Res Soc Am. Res: Guidance and control; education; management sciences. Mailing Add: 6423 Sycamore St Littleton CO 80120

CLEMENTZ, DAVID MICHAEL, b Cleveland, Ohio, Sept 4, 45; m 67; c 2. CLAY MINERALOGY, SOIL SCIENCE. Educ: Univ Ariz, BS, 67; Purdue Univ, MS, 69; Mich State Univ, PhD(clay mineral), 73. Prof Exp: RES CHEMIST SURFACE CHEM, CHEVRON OIL FIELD RES CO, 73- Mem: Clay Minerals Soc; Soil Sci Soc Am; Am Chem Soc. Res: Surface chemistry of minerals especially clays; clay organic interactions. Mailing Add: Chevron Oil Field Res Co PO Box 446 La Habra CA 90631

CLEMETSON, CHARLES ALAN BLAKE, b Canterbury, Eng, Oct 31, 23; m 47; c 4. OBSTETRICS & GYNECOLOGY. Educ: Oxford Univ, BM & BCh, 48, MA, 50; FACOG, FRCOG, FRCSC. Prof Exp: Res asst obstet, Univ Col Hosp, London, 50-52, lectr obstet & gynec, 56-58; house surgeon, London Hosps, 52-54; registr, Ashton-under-Lyne Hosp, 54-56; asst prof, Univ Sask Hosp, 58-61; asst prof, Med Ctr, Univ Calif, San Francisco, 61-67; lectr maternal health, Univ Calif, Berkeley, 63-67; assoc prof, 67-72, PROF OBSTET & GYNEC, STATE UNIV NY DOWNSTATE MED CTR, 72-; DIR DEPT OBSTET & GYNEC, METHODIST HOSP BROOKLYN, 67- Mem: Fel Am Col Obstet & Gynec. Res: Obstetric and menstrual physiology. Mailing Add: Dept of Obstet & Gynec Methodist Hosp of Brooklyn Brooklyn NY 11215

CLEMMENS, RAYMOND LEOPOLD, b Baltimore, Md, Apr 2, 22; m 52; c 3. MEDICINE, PEDIATRICS. Educ: Loyola Col, Md, BS, 47; Univ Md, MD, 51; Am Bd Pediat, dipl, 56. Prof Exp: From instr to assoc prof, 54-70, PROF PEDIAT, UNIV MD, BALTIMORE, 70- Concurrent Pos: Pediat consult, USPHS & Md State Health Dept, 62-; mem adv comt, Crippled Children's Prog, Md State Health Dept, 62-; mem adv coun ment hyg, State Bd Health & Ment Hyg, 62-67; mem task force, Nat Inst Neurol Dis & Stroke, 65- Mem: Fel Am Acad Pediat; Am Pediat Soc; Am Acad Ment Deficiency. Res: Handicapped children's diagnostic and evaluation clinic. Mailing Add: Dept of Physiol Univ of Md Hosp Baltimore MD 21201

CLEMMONS, JACKSON JOSHUA WALTER, b Beloit, Wis, Mar 24, 23; m 52; c 4. BIOCHEMISTRY, PATHOLOGY. Educ: Univ Wis, BS, 48, MS, 49, PhD, 56; Western Reserve Univ, MD, 59; Am Bd Path, dipl, 64. Prof Exp: Res assoc path, Univ Wis, 51-56; univ fel, Western Reserve Univ, 57-60, Helen Hay Whitney fel, 60-62; asst prof, 62-64, ASSOC PROF PATH, SCH MED, UNIV VT, 64- Mem: Am Soc Exp Path; NY Acad Sci; Int Acad Path; Am Asn Clin Chem. Mailing Add: Dept of Path Univ of Vt Med Sch Burlington VT 05401

CLEMMONS, JOHN B, b Rome, Ga, Apr 11, 18; m 47; c 2. MATHEMATICS, PHYSICS. Educ: Morehouse Col, AB, 37; Univ Atlanta, MS, 39. Prof Exp: Admin prin high sch, Ky, 40-43; asst prin high sch, Md, 43-47; assoc prof, 47-74, PROF MATH, SAVANNAH STATE COL, 74-, HEAD DEPT MATH & PHYSICS, 47- Concurrent Pos: Lectr, Univ Southern Calif, 50; chmn acad adv comt to Bd Regents, State of Ga. Mem: Am Math Soc; Nat Inst Sci. Res: Special properties of convex sets. Mailing Add: Dept of Math & Physics Savannah State Col Savannah GA 65717

CLEMONS, RUSSELL EDWARD, b Warner, NH, Oct 1, 30; m 68. ECONOMIC GEOLOGY. Educ: Univ NMex, BS, 60, MS, 62; Univ Tex, Austin, PhD(geol), 66. Prof Exp: From instr to asst prof geol, Univ Tex, Arlington, 65-69; assoc prof, 69-74, PROF GEOL, N MEX STATE UNIV, 74- Mem: Fel Geol Soc Am; Am Asn Petrol Geologists; Am Inst Prof Geologists. Res: Areal field geology and geologic mapping in Mexico and New Mexico; igneous petrography and associated mineral deposits of plutons in northeastern Mexico and southern New Mexico. Mailing Add: Dept of Earth Sci N Mex State Univ Las Cruces NM 88003

CLEMSON, HARRY C, b Pawtucket, RI, Aug 12, 34; m 57; c 3. BIOCHEMISTRY, CLINICAL CHEMISTRY. Educ: Univ RI, BS, 58, PhD(pharmaceut chem), 66; Univ NH, MS, 61. Prof Exp: Instr chem, Rochester Inst Technol, 60-63; USPHS fel pharmacol, Yale Univ, 66-68; asst prof med chem, Northeastern Univ, 68-70; HEAD BIOCHEM, LYNN HOSP, 70- Mem: AAAS; Am Chem Soc; Am Asn Clin Chemists; Am Inst Chemists. Res: Diagnostic enzymology; drug analysis. Mailing Add: Path Lab Lynn Hosp 212 Boston St Lynn MA 01904

CLENCH, HARRY KENDON, b Ann Arbor, Mich, Aug 12, 25; m 47, 67; c 1. ENTOMOLOGY. Educ: Univ Mich, BS, 50, MS, 51. Prof Exp: ASSOC CUR INSECTS, CARNEGIE MUS NATURAL HIST, 51- Mem: Soc Syst Zool; Asn Trop Biol; Lepidop Soc (pres, 74-75). Res: Systematics and zoogeography of butterflies. Mailing Add: Sect of Insects Carnegie Mus of Natural Hist Pittsburgh PA 15213

CLENCH, MARY HEIMERDINGER, b Louisville, Ky, Jan 18, 32; m 67. ORNITHOLOGY. Educ: Wheaton Col, Mass, BA, 53; Yale Univ, MS, 55 & 59, PhD(biol), 64. Prof Exp: Instr biol & conserv, Am Mus Natural Hist, 55-56; instr biol, Wittenberg Col, 56-57; asst zool, Yale Univ, 57-58, instr, 58-59, asst, 60; asst ornith, Peabody Mus, Yale Univ, 59-63; from asst to asst cur, 63-67, ASSOC CUR ORNITH, CARNEGIE MUS NATURAL HIST, 68- Concurrent Pos: Frank M Chapman Mem Fund grant, Am Mus Natural Hist, 64-65; asst prof, Univ Fla, 67-68. Mem: Fel Am Ornith Union; Wilson Ornith Soc; Cooper Ornith Soc; Brit Ornith Union. Res: Avian anatomy; pterylography of passerines; passerine systematics and taxonomy; migration patterns; other banding-related studies. Mailing Add: Carnegie Mus of Natural Hist Pittsburgh PA 15213

CLENCH, WILLIAM JAMES, b Brooklyn, NY, Oct 24, 97; m 24; c 2. ZOOLOGY. Educ: Mich State Col, BS, 21; Harvard Univ, MS, 23; Univ Mich, PhD, 53. Hon Degrees: Dsc, Mich State Univ, 53. Prof Exp: Custodian, Grand Rapids Pub Mus, 25-26; lectr zool, 30-36, cur mollusks, 26-66, HON CUR MALACOL, MUS COMP ZOOL, HARVARD UNIV, 66- Concurrent Pos: Res assoc, Bishop Mus, 41 & Am Mus Natural Hist, New York, 47-; mem, Pac Sci Cong, 46; field expeds, Ky, Tenn, Ala, Ga, Fla, West Indies & Hawaiian Islands; collab, Univ Fla, 53-; sr res assoc, mus zool, Ohio State Univ, 68-71, cur mollusks, 71-, adj prof zool, 71-74. Spec ed, Webster's New Int Dict, 32-34; ed, Johnsonia & Occasional Paper on Mollusks, 41-; collab, World Bk Encycl, 44-46; spec ed, Encycl Americana, 53- Mem: Am Malacol Union (pres, 34); Conchol Soc Gt Brit & Ireland. Res: Taxonomy of western Atlantic mollusks; medically important mollusks; marine boring mollusks; land mollusks of western Pacific Islands and of the West Indies. Mailing Add: 26 Rowena St Dorchester MA 02124

CLENDENIN, MARTHA ANNE, b Salem, Ohio, Jan 26, 44. NEUROSCIENCES. Educ: Med Col Va, Richmond, BS, 65, MS, 70, PhD(anat), 72. Prof Exp: NIH fel, Sweden, 72-73; ASST PROF ANAT, EASTERN VA MED SCH, 73- Honors & Awards: Dorothy Briggs Mem Award for Res, Am Phys Ther Asn, 72. Mem: Soc Neurosci; Sigma Xi; Am Phys Ther Asn; AAAS. Res: Investigations of normal mechanisms of movement and posture utilizing intra and extracellular recording techniques in animals and electromyography in human subjects. Mailing Add: Dept of Anat Eastern Va Med Sch PO Box 1980 Norfolk VA 23501

CLENDENING, JOHN ALBERT, b Martinsburg, WVa, Mar 6, 32; m 54; c 3. PALYNOLOGY. Educ: WVa Univ, BS, 58, MS, 60, PhD(palynology), 70. Prof Exp: Asst coal geologist, WVa Geol & Econ Surv, 60-64, palynologist, 64-66, coal geologist & palynologist, 66-68; geologist, Pan Am Petrol Corp, 68-71; sr geologist, 71-75, SR STAFF PALEONTOLOGIST, AMOCO PROD CO, 75- Mem: Fel Geol Soc Am; Am Asn Stratig Palynologists. Res: Applied Paleozoic stratigraphic palynology with emphasis on the Pennsylvanian and Permian systems of North America. Mailing Add: Amoco Prod Co Box 3092 Houston TX 77001

CLENDENNING, WILLIAM EDMUND, b Waynesburg, Pa, June 23, 31; m 58; c 4. DERMATOLOGY. Educ: Allegheny Col, BS, 52; Jefferson Med Col, MD, 56. Prof Exp: Instr dermat, Case Western Reserve Univ, 60-61; sr investr, Dermat Br, Nat Cancer Inst, 61-63; sr instr dermat, Case Western Reserve Univ, 63-66, asst prof, 66-67; clin assoc prof, 67-72, PROF CLIN DERMAT, DARTMOUTH MED SCH, 72- Concurrent Pos: Nat Cancer Inst fel, Sch Med, Case Western Reserve Univ, 63-66 & 67-69; mem gen med A study sect, NIH, 65-69; assoc ed, J Invest Dermat, 67-72; mem, Path & Chemother Panels-Mycosis Fungoides Coop Study Group, 72-; mem ed adv bd, Contact Dermatitis, 74. Mem: Am Acad Dermat; Soc Invest Dermat; Am Dermat Asn; Am Soc Dermatopath; Am Fedn Clin Res. Res: Immunoglobulin E in atopic dermatitis; mycosis fungoides; contact dermatitis. Mailing Add: Dartmouth Med Sch 2 Maynard Rd Hanover NH 03755

CLENDENON, NANCY RUTH, b Ahoskie, NC, July 27, 33. NEUROCHEMISTRY. Educ: Old Dom Univ, BS, 62; Univ NC, Chapel Hill, PhD(biochem), 68. Prof Exp: Instr, 69-71, ASST PROF NEUROCHEM, COL MED, OHIO STATE UNIV, 71- Concurrent Pos: Res fel neurochem, Ohio State Univ Hosp, 68-69; Ohio State Univ Res Found-Charles R Kistler Mem Found, 69- Mem: AAAS; Int Soc Neurochem; Am Chem Soc; Soc Neurosci; Am Soc Neurochem. Res: Cerebral hydrolytic enzyme localization in subcellular fractions of normal and pathologic tissues; combined modalities in the treatment of experimentally induced brain tumors in rats and their effects on lysosomal enzymes; response of hydrolytic enzymes in dog spinal cord and cerebrospinal fluid to experimental injury. Mailing Add: 10573 Riverside Dr Powell OH 43065

CLENDINNING, ROBERT ANDREW, bSchenectady, NY, Dec 12, 31; m 57; c 2. ORGANIC CHEMISTRY, POLYMER CHEMISTRY. Educ: Union Col, BS, 53; Rensselaer Polytech Inst, PhD(org chem), 59. Prof Exp: Asst, Rensselaer Polytech Inst, 53-58; chemist, 58-68, PROJ SCIENTIST, UNION CARBIDE CHEM & PLASTICS, 68- Mem: Am Chem Soc. Res: Synthetic organic chemistry; monomer synthesis and polymerization; mechanism of polymerization reactions; condensation polymerization reactions; polyolefin and olefin copolymers research and development; polymer evaluation. Mailing Add: 48 Crest Rd New Providence NJ 07974

CLERMONT, YVES WILFRED, b Montreal, Can, Aug 14, 26; m 50; c 3. MICROSCOPIC ANATOMY. Educ: Univ Montreal, BSc, 49; McGill Univ, PhD(anat), 53. Prof Exp: From lectr to assoc prof, 53-63, chmn dept, 75, PROF ANAT, McGILL UNIV, 63- Concurrent Pos: Anna Fuller fel, 54-55; Lalor Found award, 62; consult, WHO, Geneva, 72- Mem: Am Asn Anat (2nd vpres, 70-72); Soc Study Reproduction; Can Asn Anat. Res: Histology and histophysiology of mammalian testes; cytological studies of spermatogenesis with light and electron microscopes. Mailing Add: 567 Townshend St Lambert PQ Can

CLEROUX, ROBERT A, operations research, mathematical statistics, see 12th edition

CLESCERI, LENORE STANKE, b Chicago, Ill, Aug 9, 35; m 57; c 5. BIOCHEMISTRY, MICROBIOLOGY. Educ: Loyola Univ, BS, 57; Marquette Univ, MS, 61; Univ Wis, PhD(biochem), 63. Prof Exp: NIH res fel, 63-65; asst prof, 66-74, ASSOC PROF MICROBIAL BIOCHEM, RENSSELAER POLYTECH INST, 74- Mem: AAAS; Am Chem Soc; Am Soc Microbiol. Res: Kinetics of microbial growth; microbial ecology; inorganic nitrogen metabolism. Mailing Add: Dept of Biol Rensselaer Polytech Inst Troy NY 12181

CLEVELAND, ANNE STACK, b Chicago, Ill, Oct 7, 08; m 40; c 2. BIOCHEMISTRY. Educ: Univ Chicago, SB, 30, SM, 48, PhD(org chem), 54. Prof Exp: From asst to res assoc instr cancer res, Univ Chicago, 47-55; res biochemist, Univ Calif, 55-56; instr biochem, Med Sch, Northwestern Univ, 56-57; res biochemist cancer res, Donner Lab, Univ Calif, Berkeley, 57-74; consult, Endocrine Div, Alta Bates Hosp, Berkeley, 74-75; BIOCHEMIST, FAC MED, ATMA JAYA CATH UNIV, JAKARTA, INDONESIA, 76- Concurrent Pos: Commonwealth fel, Gt Brit, 61-62; biochemist in chg metab res lab, Chicago Wesley Mem Hosp, 56-57; consult, Cancer Res Inst, Sch Med, Univ Calif, San Francisco, 60-66; consult, St Joseph Hosp, Kaohsiung, Taiwan, 67-69; vis prof, Dept Chem, Fu Jen Univ, Taiwan, 71-74; Bur Med consult, US Naval Med Res Unit-2, Taiwan, 71-74. Mem: AAAS; Sigma Xi; Am Chem Soc; Endocrine Soc; Brit Soc Endocrinol. Res: Steroid studies in cancer and pituitary-adrenal pathologies; chromatography; carbohydrates; enzymes; clinical chemistry. Mailing Add: Fac of Med Atma Jaya Cath Univ Jakarta Indonesia

CLEVELAND, BRUCE TAYLOR, b Boston, Mass, Aug 6, 37; m 64. NUCLEAR PHYSICS. Educ: Johns Hopkins Univ, PhD(physics), 70. Prof Exp: Lectr physics, State Univ NY Buffalo, 70-72; res assoc, Columbia Univ, 72-75; SR RES ASSOC CHEM, BROOKHAVEN NAT LAB, 75- Mem: Am Phys Soc. Res: Experimental nuclear physics. Mailing Add: Dept of Chem Brookhaven Nat Lab Upton NY 11973

CLEVELAND, ELONZA ALEXANDER, JR, b Water Valley, Miss, May 2, 15; m 44; c 1. ORGANIC CHEMISTRY. Educ: Memphis State Col, BS, 36; Univ Ill, MS, 37. Prof Exp: RES CHEMIST, CPC INT, INC, 38-40 & 42- Concurrent Pos: Dir, Ital Res Lab, CPC Int, Inc, Milan, 49-51 & 57-61. Mem: Am Chem Soc; Am Asn Cereal Chemists. Res: Synthesis; carbohydrate chemistry; proteins and amino acids; catalysis; analytical methods; wet-milling grains. Mailing Add: Moffett Tech Ctr CPC Int Inc Box 345 Argo IL 60501

CLEVELAND, ERNEST LYNN, b Lebanon, Ky, June 15, 16; m 46; c 3. PHYSICS. Educ: Univ Ky, AB, 38, MS, 40; Pa State Univ, PhD(physics), 50. Prof Exp: Instr physics, Pa State Univ, 41-42; asst prof, Mo Sch Mines, 49-51; from asst prof to prof, 51-74, assoc physicist, 51-57, EMER PROF PHYSICS, NMEX STATE UNIV, 74- Mem: Fel AAAS; Optical Soc Am; Am Asn Physics Teachers. Res: Precipitation static; infrared radiation; solar energy; physics education at high school and college level. Mailing Add: Dept of Physics NMex State Univ Las Cruces NM 88001

CLEVELAND, GREGOR GEORGE, b San Jose, Calif, Sept 21, 48; m 69; c 1. BIOPHYSICS, MAGNETIC RESONANCE. Educ: La Tech Univ, BS, 70; Rice Univ, MA, 73, PhD(physics), 75. Prof Exp: Instr physics, Univ Houston, Downtown, 75; ASST PROF PHYSICS, UNIV NC, GREENSBORO, 75- Mem: AAAS; Am Phys Soc; Biophys Soc. Res: Investigation of the physicochemical state of water and ions in various biological or model systems utilizing pulsed nuclear magnetic resonance. Mailing Add: Dept of Physics Univ of NC Greensboro NC 27412

CLEVELAND, JAMES PERRY, b Charlotte, NC, Feb 20, 42; m 65. ORGANIC CHEMISTRY. Educ: Ga Inst Technol, BS, 63, PhD(org chem), 67. Prof Exp: NIH fel org chem, Ore State Univ, 69-70; MEM STAFF, RES LABS, TENN EASTMAN CO, 70- Mem: Am Chem Soc; The Chem Soc. Res: Mechanisms of reactions of organo-phosphorus and organo-sulfur compounds. Mailing Add: Res Labs Tenn Eastman Co Kingsport TN 37660

CLEVELAND, JESSE MARVIN JR, b Newnan, Ga, July 3, 29; m 51; c 1. INORGANIC CHEMISTRY. Educ: Ga Inst Technol, BS, 51; Univ Colo, MS, 55, PhD(inorg chem), 59. Prof Exp: Chemist, Phillips Chem Co, 51-52, Dow Chem Co, 52-57 & Gen Elec Co, 59-63; sr develop specialist, 63-65, sr res chemist, 65-68, ASSOC SCIENTIST, ROCKY FLATS DIV, DOW CHEM CO, 68- Mem: Am Chem Soc. Res: Chemistry of actinide elements; solvent extraction; nonaqueous solvents; technical writing; ceramic fabrication process; coordination chemistry; actinide compound syntheses; chemical kinetics. Mailing Add: Rocky Flats Div Dow Chem Co PO Box 888 Golden CO 80401

CLEVELAND, JOHN H, b Bloomington, Ind, Nov 12, 32; m 53; c 3. ECONOMIC GEOLOGY. Educ: Univ Ind, BA, 54, PhD(geol), 63; Univ Wis, MS, 61. Prof Exp: From asst prof to assoc prof, 62-70, PROF GEOL, IND STATE UNIV, 70- Mem: AAAS; Nat Asn Geol Teachers; Mineral Soc Am; Geol Soc Am. Res: Geology and geochemistry of magnesite deposits; lead-zinc deposits of the upper Mississippi Valley; economic geology of Indiana; computer utilization. Mailing Add: Dept of Geol Ind State Univ Terre Haute IN 47809

CLEVELAND, MERRILL L, b Orleans, Ind, Feb 29, 28; m 50; c 1. ENTOMOLOGY. Educ: Ind State Teachers Col, BS, 50; Purdue Univ, MS, 54, PhD, 64. Prof Exp: Pub sch teacher, 50-52; entomologist, Fruit Insects Sect, Entom Res Div, 55-68, asst chief fruit insects res br, 68-72, STAFF SCIENTIST, NAT PROG STAFF, FRUIT & VEG INSECTS, AGR RES SERV, USDA, 72- Mailing Add: 1220 Burton St Silver Spring MD 20910

CLEVELAND, RICHARD WARREN, b Santa Ana, Calif, Dec 4, 24; m 52; c 2. PLANT BREEDING. Educ: Univ Calif, BS, 49, PhD(genetics), 53. Prof Exp: Asst agron, Univ Calif, 49-53; from asst prof to assoc prof, 53-74, PROF AGRON, PA STATE UNIV, STATE COLLEGE, 74- Mem: Am Soc Agron. Res: Breeding and cytogenetics of forage plants. Mailing Add: 361 Laurel Lane State College PA 16801

CLEVELAND, THOMAS HILBURN, b Anniston, Ala, Dec 1, 19; m 50; c 4. ORGANIC CHEMISTRY. Educ: Birmingham Southern Col, BS, 42. Prof Exp: From chemist to group leader, Monsanto Chem Co, 47-54; group leader res, 54-69, ASST DIR RES, MOBAY CHEM CO, 69- Mem: Am Chem Soc; Soc Plastics Engrs. Res: Isocyanate and isocyanate polymer chemistry; polyether and polyester chemistry; polycaronate polymer chemistry. Mailing Add: 41 Oriole Dr New Martinsville WV 26155

CLEVELAND, WILLIAM SWAIN, b Sussex, NJ, Jan 24, 43; c 2. STATISTICS, AIR POLLUTION. Educ: Princeton Univ, AB, 65; Yale Univ, MS, 67, PhD(statist), 69. Prof Exp: Asst prof statist, Univ NC, Chapel Hill, 69-72; MEM TECH STAFF STATIST, BELL LABS, 72- Concurrent Pos: Mem comt seasonal adjust, Census Bur, 75- Mem: Am Statist Asn; Air Pollution Control Asn. Res: Analysis of air pollution data; graphical methods in statistics; time series analysis; seasonal adjustment of economic data. Mailing Add: Bell Labs 600 Mountain Ave Murray Hill NJ 07974

CLEVELAND, WILLIAM WEST, b Pleasant Shade, Tenn, June 1, 21; m 53. PEDIATRICS. Educ: Harvard Univ, BS, 43; Vanderbilt Univ, MD, 50. Prof Exp: Res chemist, Synthetic Rubber Indust, 43; intern, Vanderbilt Hosp, 51, resident, 52-56; resident, St Louis Children's Hosp, 56; from instr to assoc prof, 56-66, PROF PEDIAT, SCH MED, UNIV MIAMI, 67-, CHMN DEPT, 69- pediat endocrinol, Johns Hopkins Hosp, 58-60. Res: Pediatric Mailing Add: Dept of Pediat Univ of Miami Sch of Med Miami FL 33136

CLEVEN, GALE W, astronomy, economic geology, see 12th edition

CLEVENGER, IMA FUCHS, b Mayfield, Okla, Jan 24, 03; m 35; c 4. SPEECH PATHOLOGY, AUDIOLOGY. Educ: Abilene Christian Col, BA, 24; Univ Iowa, MA, 28; Univ Okla, PhD(speech), 55. Prof Exp: Prof speech & Eng & chmn dept, Oklahoma City Univ, 46-56; prof & dir dept, 56-71, EMER PROF SPEECH PATH & AUDIOL, ABILENE CHRISTIAN COL, 71-; ASST DIR, SCH SPEECH PATH, LANG & AUDIOL, UNIV PR, SAN JUAN, 71- Concurrent Pos: Consult, WTex Rehab Ctr, 57-71 & Abilene State Sch, 62-71. Mem: Am Speech & Hearing Asn; Speech Asn Am; Coun Except Children. Mailing Add: Sch of Speech Path Lang & Audiol Univ of PR GPO 5067 San Juan PR 00936

CLEVENGER, RICHARD LEE, b Columbus, Ind, May 16, 31. ORGANIC CHEMISTRY, BIOCHEMISTRY. Educ: Univ Ind, BS, 59; Univ Louisville, PhD(org chem), 63. Prof Exp: Instr chem, Univ Louisville, 62-63; asst prof, 63-65, ASSOC PROF CHEM, E TEX STATE UNIV, 65-, DIR FORENSIC CHEM, 73- Mem: AAAS; Am Chem Soc; Am Acad Forensic Sci. Res: Thiadiazoles; hydrazones; geometrical isomerism; cyclopentane derivatives; azasteroids; neuro-biochemistry; B vitamins; drug metabolites; forensic chemistry. Mailing Add: 112 Briarwood Dr Commerce TX 75428

CLEVENGER, SARAH, b Indianapolis, Ind, Dec 19, 26. PLANT BIOCHEMISTRY. Educ: Miami Univ, AB, 47; Ind Univ, PhD(plant physiol), 57. Prof Exp: Asst prof biol, Berea Col, 57-59, Wittenburg Col, 59-60, Eastern Ill Univ, 60-61 & Berea Col, 61-63; asst prof, 63-66, ASSOC PROF LIFE SCI, IND STATE UNIV, 66- Concurrent Pos: Contrib ed, Book Forum, 74- Mem: AAAS; Am Asn Plant Physiol; Int Asn Plant Taxon; Phytochem Soc NAm (secy, 67-68); Am Inst Biol Sci. Res: Flower pigments; gene control of flower pigments and simulation of population biology. Mailing Add: 717 S Henderson St Bloomington IN 47401

CLEVER, HENRY LAWRENCE, b Mansfield, Ohio, June 14, 23; m 56; c 1. PHYSICAL CHEMISTRY. Educ: Ohio State Univ, BSc, 45, MS, 49, PhD(chem), 51. Prof Exp: Jr chemist, Shell Develop Co, 45-47; asst, Res Found, Ohio State Univ, 47-50; instr & res assoc, Duke Univ, 51-54; from instr to assoc prof, 54-65, PROF CHEM, EMORY UNIV, 65- Concurrent Pos: Partic, Oak Ridge Nat Lab, 57; res assoc, Univ Mich, 63-64; res assoc, Polymer Res Inst, Univ Mass, Amherst, 72-73. Mem: AAAS; Am Chem Soc. Res: Thermodynamics; solubility; surface tension; heat capacity; thermal properties; light scattering. Mailing Add: Dept of Chem Emory Univ Atlanta GA 30322

CLEVER, ULRICH, genetics, deceased

CLEWE, THOMAS HAILEY, b San Francisco, Calif, Sept 9, 25; m 52; c 3. INTERNAL MEDICINE, MEDICAL RESEARCH. Educ: Stanford Univ, BS, 49, MD, 55. Prof Exp: Res assoc reprod, Stanford Univ, 55-56; instr anat, Sch Med, Yale Univ, 57-58; asst res prof obstet & gynec & anat, Sch Med, Univ Kans, 58-61; asst res prof obstet & gynec, Sch Med, Vanderbilt Univ, 61-66; res assoc, Div Reprod Physiol, Delta Regional Primate Res Ctr, 66-73; asst clin res dir, 73-75, ASSOC CLIN RES DIR, SQUIBB INST MED RES, 75- Concurrent Pos: Pop Coun med res fel, Sch Med, Yale Univ, 56-57; assoc prof anat, Tulane Univ, 66-73, clin assoc prof obstet & gynec, Med Sch, 68-73. Honors & Awards: Rubin Award, Am Fertil Soc, 59. Mem: AMA; Am Asn Anat; Am Fertil Soc; Brit Soc Study Fertil; Soc Study Reproduction. Res: Structure and function of mammalian oviduct; physiology of early stages of mammalian reproduction; fertility and sterility; comparative reproduction, especially of primates; fertility control devices; physiology of skin; cardiovascular physiology. Mailing Add: Squibb Inst Med Res PO Box 4000 Princeton NJ 08540

CLEWELL, ANDRE F, b Canton, Ohio, Mar 27, 34; m 60; c 2. SYSTEMATIC BOTANY, PLANT ECOLOGY. Educ: Oberlin Col, BA, 56; Kent State Univ, MA, 57; Univ Ind, PhD(bot), 63. Prof Exp: From instr to asst prof, 62-67, ASSOC PROF BIOL SCI, FLA STATE UNIV, 67-; BOTANIST, TALL TIMBERS RES STA, 66- Concurrent Pos: NSF res grant, 63-65. Mem: AAAS; Am Soc Plant Taxon; Soc Study Evolution; Int Soc Plant Taxon. Res: Systematics of Lespedeza and Eupatorium; description and successional relationships of north Florida vegetation types. Mailing Add: Dept of Biol Sci Fla State Univ Tallahassee FL 32306

CLEWELL, DAYTON HARRIS, b Berwick, Pa, Dec 15, 12; m 38; c 2. PHYSICS. Educ: Mass Inst Technol, BS, 33, PhD(physics), 36. Prof Exp: Physicist, C K Williams Co, 35-38 & Magnolia Petrol Co, 38-42, supvr physics res, 42-46, from asst dir to dir field res labs, 46-56; gen mgr res dept, 56-62, gen mgr res & eng, 62-64, SR VPRES, MOBIL OIL CORP, 64- Mem: Am Phys Soc; Soc Explor Geophys; Am Asn Petrol Geologists; Inst Elec & Electronics Engrs. Res: Design of spectrophotometers; gravity meters; seismographs; terrestrial magnetism; propagation of electric waves through earth; research administration in areas of petroleum exploration, production, processing and product development; engineering administration. Mailing Add: Mobil Oil Corp 150 E 42nd St New York NY 10017

CLEWELL, DON BERT, b Dallas, Tex, Sept 5, 41; m 68. BIOCHEMISTRY, MICROBIOLOGY. Educ: Johns Hopkins Univ, AB, 63; Ind Univ, Indianapolis, PhD(biochem), 67. Prof Exp: Biologist, Univ Calif, San Diego, 69-70; asst prof, 70-73, ASSOC PROF ORAL BIOL, SCH DENT & MICROBIOL, SCH MED, UNIV MICH, ANN ARBOR, 73- Concurrent Pos: Nat Cancer Inst fel molecular genetics, Univ Calif, San Diego, 67-69. Mem: AAAS; Am Chem Soc; Biophys Soc; Am Soc Biol Chem; Int Asn Dent Res. Res: Molecular biology; molecular genetics; nucleic acid chemistry. Mailing Add: Dept of Oral Biol Univ of Mich Sch of Dent Ann Arbor MI 48104

CLIBURN, JOSEPH WILLIAM, b Hazlehurst, Miss, Jan 20, 26; m 50; c 2. ZOOLOGY, BOTANY. Educ: Millsaps Col, BS, 47; Univ Southern Miss, MA, 53; Univ Ala, PhD(zool), 60. Prof Exp: Instr, pub schs, Miss, 47-53; instr biol, Copiah-Lincoln Jr Col, 53-55; instr, pub schs, Miss, 55-58; assoc prof, 60-64, PROF BIOL, UNIV SOUTHERN MISS, 64- Concurrent Pos: Prof zool, Gulf Coast Res Lab, 70- Mem: Am Soc Ichthyologists & Herpetologists; Soc Study Amphibians & Reptiles. Res: Taxonomy and zoogeography of southeastern amphibians, reptiles and fishes. Mailing Add: Dept of Biol Univ of Southern Miss Hattiesburg MS 39401

CLICK, ROBERT EDWARD, b Wenatchee, Wash, Mar 22, 37; m 56; c 2. IMMUNOBIOLOGY. Educ: Wash State Univ, BS, 60; Univ Calif, Berkeley, PhD(biochem), 64. Prof Exp: NIH fels, Columbia Univ, 64-65 & Sloan-Kettering Cancer Inst, 65-66; NIH fel, Univ Wis, 66-68, asst prof immunol, 68-72; scientist, Wis

Alumni Res Inst, Madison, 72-73; assoc, Sloan-Kettering Cancer Inst, 73-74; ASST PROF IMMUNOL, DEPT LAB MED & PATH, UNIV MINN, MINNEAPOLIS, 74- Mem: AAAS; Soc Develop Biol; Soc Plant Physiol; Am Asn Immunol. Res: Genetic control of immune responses, primarily in vitro. Mailing Add: Dept of Lab Med & Path Univ of Minn Minneapolis MN 55455

CLIFF, FRANK SAMUEL, b Carson City, Nev, Apr 3, 28; m 55; c 3. VERTEBRATE ZOOLOGY. Educ: Stanford Univ, AB, 51, PhD, 54. Prof Exp: Herpetologist, Sefton-Stanford exped, Gulf of Calif, 52-53; asst comp anat & gen biol, Natural Hist Mus, Stanford Univ, 52, 53-56; from instr to asst prof, Colgate Univ, 56-59; from asst prof to assoc prof, 59-69, PROF COMP ANAT & GEN BIOL, CALIF STATE UNIV, CHICO, 69- Mem: Am Soc Ichthyologists & Herpetologists. Res: Reptiles of islands adjacent to Baja California and Mexico; reptiles of western North America and Mexico; insular evolution; osteology of reptiles. Mailing Add: Dept of Biol Calif State Univ Chico CA 95929

CLIFFORD, ALAN FRANK, b Natick, Mass, June 8, 19; m 49; c 2. INORGANIC CHEMISTRY, FLUORINE CHEMISTRY. Educ: Harvard Univ, AB, 41; Univ Del, MS, 47, PhD(inorg chem), 49. Prof Exp: Anal chemist & lab supvr, Kankakee Ord Works, Ill, 41-43; asst, Manhattan Dist Proj, Radiation Lab, Univ Chicago, 43; anal res chemist, Clinton Labs, Tenn, 43-44; res chemist, Hanford Eng Works, Wash, 44-45; develop chemist, Exp Sta, E I du Pont de Nemours & Co, Del, 45-47; instr chem, Univ Del, 47-49; asst prof inorg chem, Ill Inst Technol, 49-51; from asst prof to assoc prof, Purdue Univ, 53-66; PROF INORG CHEM & HEAD DEPT CHEM, VA POLYTECH INST & STATE UNIV, 66- Concurrent Pos: Guggenheim fel, Cambridge Univ, 51-53; mem subcomt solubility data, Comn Equilibrium Data, Int Union Pure & Appl Chem, 73- Mem: AAAS; Am Chem Soc; fel NY Acad Sci; The Chem Soc. Res: Rare earth, inorganic fluoride chemistry; hydrogen fluoride system; acid theory; inorganic polymers; oxidations in liquid ammonia; multiple bonding in organic compounds; hypofluorites; Mössbauer spectrometry of rare earths and biological materials. Mailing Add: Dept of Chem Va Polytech Inst & State Univ Blacksburg VA 24061

CLIFFORD, ALFRED HOBLITZELLE, b St Louis, Mo, July 11, 08; m 42; c 2. MATHEMATICS. Educ: Yale Univ, AB, 29; Calif Inst Technol, PhD(math), 33. Prof Exp: Mem, Inst Advan Study, 33-36, asst, 36-38; instr math, Mass Inst Technol, 38-41, asst prof, 41-42; assoc prof math, Johns Hopkins Univ, 46-55; prof, 55-74, EMER PROF MATH, NEWCOMB COL, TULANE UNIV, 74- Mem: Am Math Soc; Math Asn Am. Res: Algebraic theory of semigroups and ordered groups. Mailing Add: Dept of Math Tulane Univ New Orleans LA 70118

CLIFFORD, CHARLES E, nuclear physics, see 12th edition

CLIFFORD, DONALD H, b Burlington, Vt, June 7, 25; m 54; c 4. VETERINARY SURGERY. Educ: Univ Montreal, DVM, 50; Univ Minn, MPH, 55, PhD(vet med), 59; Am Col Lab Animal Med, dipl, 60; Am Col Vet Surgeons, dipl. Prof Exp: Intern vet med, Angell Mem Hosp, Boston, Mass, 50-51, intern res, 51-52; from instr to assoc prof vet surg, Col Vet Med, Univ Minn, 52-65; med assoc lab animal care, Brookhaven Nat Lab, 62-63; assoc prof med, sci & technol, Col Med, Baylor Univ, 65-71, lectr exp surg, 66-71; chief animal res facil, Vet Admin Hosp, Houston, Tex, 65-71; DIR DIV LAB ANIMAL MED, MED COL OHIO AT TOLEDO, 71- Concurrent Pos: Consult, dept pharmac- ol, Boston Univ, 51 & St Paul Como Zoo; ed, Minn Vet, 64-65; mem subcomt dog & cat standards, Nat Res Coun, Nat Acad Sci; exec comt & exam comt, Am Col Lab Med; consult, Am Asn Accreditation of Lab Animal Care. Mem: Fel Am Col Vet Surgeons; Am Vet Med Asn; Am Asn Lab Animal Sci; Am Asn Lab Animal Practitioners. Res: Veterinary surgery, especially comparative restraint and anesthesiology in laboratory, zoological and domestic animals; pathology and surgery of the canine mouth and esophagus; pathogenesis and treatment of achalasia of the esophagus in dogs and cats; effect of acupuncture on the cardiovascular system of dogs. Mailing Add: Div Lab Animal Med Med Col Ohio Box 6190 Toledo OH 43614

CLIFFORD, GEORGE O, b Akron, Ohio, Apr 30, 24; m 48; c 3. INTERNAL MEDICINE, HEMATOLOGY. Educ: Tufts Univ, MD, 49; Am Bd Internal Med, dipl, 57. Prof Exp: Intern Henry Ford Hosp, Detroit, 49-50; resident med, Detroit Receiving Hosp, 54-55; from instr to assoc prof, Col Med, Wayne State Univ, 55-63; assoc prof, Med Col, Cornell Univ, 63-72; PROF MED & CHMN DEPT, SCH MED, CREIGHTON UNIV, 72- Concurrent Pos: Res fel, Detroit Receiving Hosp, 50-52; res fel, Med Sch, Univ NC, 54; Markel scholar, 59; dir blood bank med lab serv, Mem Hosp, Sloan-Kettering Inst, 63-72, dir hemat, 63-70, assoc chmn dept med, 70-72. Mem: AMA; Am Fedn Clin Res; Am Soc Hemat; NY Acad Sci; fel Am Col Physicians. Res: Clinical and research hematology; hemaglobiopathies; megaloblastic anemias; leukemias. Mailing Add: Dept of Med Creighton Univ Sch of Med Omaha NE 68131

CLIFFORD, HOWARD JAMES, b Binghamton, NY, May 29, 39; m 62; c 3. PHYSICAL CHEMISTRY, MOLECULAR SPECTROSCOPY. Educ: Univ NMex, BS, 63; Wash State Univ, PhD(chem), 71. Prof Exp: Lab technician, Vet Admin Hosp, Albuquerque, 63-64; instr, Regina Sch Nursing, Univ Albuquerque, 64-66; teaching asst chem, Univ NMex, 66-67; res asst, Wash State Univ, 67-70; ASST PROF CHEM, UNIV PUGET SOUND, 70- Concurrent Pos: Lab technician, Med Sch, Univ NMex, 65. Mem: AAAS; Am Chem Soc. Res: Synthesis, purification and spectroscopic investigation of six-coordinated metal complexes containing ions from the second and third transition series group VIII elements; examination of effects due to differences in molecular symmetries by using absorption, emission, optical rotatory dispersion and circular dichroism spectroscopy. Mailing Add: Dept of Chem Univ of Puget Sound Tacoma WA 98416

CLIFFORD, HUGH FLEMING, b Warren, Pa, Dec 9, 31; m 61; c 2. LIMNOLOGY, INVERTEBRATE ZOOLOGY. Educ: Mich State Univ, BS, 58, MS, 59; Ind Univ, PhD(limnol), 65. Prof Exp: Fishery biologist, Mo Conserv Comn, 60-62; asst prof, 65-71, ASSOC PROF ZOOL, UNIV ALTA, 71- Mem: Am Soc Limnol & Oceanog; Ecol Soc Am; Am Fisheries Soc; Can Soc Zoologists. Res: Stream limnology and ecology of mayflies. Mailing Add: Dept of Zool Univ of Alta Edmonton AB Can

CLIFFORD, JOSEPH MICHAEL, b Oak Park, Ill, Nov 11, 25; m 59; c 5. PHYSICS. Educ: Harvard Univ, AB, 48; Lehigh Univ, MS, 49, PhD(physics), 55. Prof Exp: Analyst, Inst Res, Lehigh Univ, 53-54; sci warfare adv, Off Asst Secy Defense Res & Eng, US Dept Defense, 55; mem staff, weapons systs eval group, Inst Defense Anal, 56-61; chief missions anal, Martin Co, Colo, 61-63; HEAD OPERS ANAL DEPT, AEROSPACE CORP, SAN BERNARDINO, CALIF, 63- Mem: Opers Res Soc Am; Am Inst Aeronaut & Astronaut. Res: Operations research; mathematical physics. Mailing Add: 30907 Rue de la Pierre Palos Verdes Peninsula CA 90274

CLIFFORD, PAUL CLEMENT, b Bismarck, NDak, Nov 23, 10; m 36; c 7. STATISTICS, QUALITY CONTROL. Educ: Columbia Univ, AB, 31, AM, 34. Prof Exp: Instr math, Columbia Univ, 32-35; from instr to assoc prof, 35-57, chmn dept, 63-72, PROF MATH, MONTCLAIR STATE COL, 57- Concurrent Pos: Lectr, NY Univ, 44-45, Newark Col Eng, 46-47, Rutgers Univ, 48-64, Univ Mich, 57-69 & Univ Wis, 62-70; indust consult, Nat Broadcasting Co Continental Classroom, 44-49, instr, 61-62; indust consult, UN, 53, Int Coop Admin, 54-59 & AID, 59-64. Honors & Awards: Shewhart Medal, Am Soc Qual Control, 65 & Ott Award, 73. Mem: Fel AAAS; fel Am Soc Qual Control; fel Am Statist Asn; Inst Math Statist; hon mem Europ Orgn Qual Control. Res: Application of statistics and industrial quality control. Mailing Add: Dept of Math Montclair State Col Upper Montclair NJ 07043

CLIFFTON, EUGENE EVERETT, b Landsdale, Pa, June 26, 11; m 41; c 5. SURGERY, ENZYMOLOGY. Educ: Lafayette Col, BS, 33; Yale Univ, MD, 37. Prof Exp: Instr surg, New York Hosp, Cornell Univ, 43-44; Am Cancer Soc sr fel, Yale Univ, 47-49, asst prof surg, asst prof, sect oncol, 50-52; asst prof clin surg, Med Col, Cornell Univ, 52-61; from asst to assoc, 52-60, ASSOC MEM, SLOAN-KETTERING INST CANCER RES, 60-; ASSOC PROF CLIN SURG, MED COL, CORNELL UNIV, 61- Concurrent Pos: Consult, Newington Vet Admin Hosp, 47-52, Meriden Hosp, Conn, 50-53, Hosp Spec Surg, New York, 58- & French & Polyclin Hosps; assoc surgeon, New Haven Hosp, 47-52; vis surgeon, Bellevue Hosp, New York, 52-; asst attend surgeon, James Ewing Hosp, New York, 52-; asst attend surgeon, New York Hosp, 52-58, assoc attend surgeon, 58-; from asst attend surgeon to assoc attend surgeon, Thoracic Surg Serv, Mem Hosp Cancer & Allied Dis, New York, 53-65, attend surgeon & assoc chief thoracic surg, 65- Mem: AAAS; Soc Univ Surgeons; AMA. Res: Enzymes; therapy; diagnostic tests for cancer; thrombosis and hemorrhage. Mailing Add: 449 E 68th St New York NY 10021

CLIFT, CECIL WILLIAM, b Patoka, Ind; m 37; c 4. AGRONOMY. Educ: Purdue Univ, BSA, 35; Univ Ill, MS, 43. Prof Exp: Jr asst agronomist, Soil Conserv Serv, USDA, Ind, 38-40; instr agron, Tuskegee Inst, 40-44, asst prof soil fertil & plant physiol, 46-52; dir agr, 52-57, chmn div sci, 58-60, prof sci, chmn terminal ed & dir work study prog, 60-64, PROF CHEM & GEN SCI, 64-, CHMN DIV PHYS & LIFE SCI, 67- Concurrent Pos: Assoc, George Washington Carver Found, 46-52. Res: Plant nutrition; agricultural chemistry and engineering; plant physiology; geology. Mailing Add: Div of Phys & Life Sci Jarvis Christian Col PO Box 368 Hawkins TN 75765

CLIFT, WILLIAM ORRIN, b Flint, Mich, Mar 27, 14; m 51; c 2. GEOLOGY. Educ: Univ Mich, BS, 38; Columbia Univ, PhD, 56. Prof Exp: Paleontologist, Sinclair Refining Co, Venezuela, 46-49, stratigr-paleontologist, Sinclair Petrol Co, 49, chief geologist, Ethiopia, 49-52, gen supt, 53-56, mgr, Sinclair Somal Corp, Somalia, 56-59, pres, Sinclair & BP Explor Co, NY, 59-62, vpres, Sinclair Int Oil Co, 62-69; vpres, Podesta, Meyers, Rominger & Clift, Inc, 69-74; DIR EXPLOR SCI, FOREST OIL CORP, DENVER, 74- Concurrent Pos: Chmn bd, Sinclair Mediter Petrol Co; pres & dir, Sinclair Somal Corp & Sinclair Libyan Oil Co; dir, Sinclair Venezuelan Oil Co. Mem: Fel Geol Soc Am; Am Petrol Geologists; Paleont Soc. Res: Eocene stratigraphy and paleontology. Mailing Add: 25 Wedge Way Littleton CO 80123

CLIFTON, CARL MOORE, b Greenup Co, Ky, Sept 18, 14; m 47; c 3. DAIRY HUSBANDRY. Educ: Eastern Ky State Col, BS, 36; Univ Ky, MS, 39; Ohio State Univ, PhD(dairy sci), 54. Prof Exp: Supt dairy, Univ Ky, 45-46, field agt, 46-50; co-op agt, Dairy Husb Res Br, Agr Res Serv, USDA, 50-55; asst prof dairy, Univ Minn, 55-59; asst prof, 60-65, ASSOC PROF DAIRY, UNIV GA, 65- Mem: Am Dairy Sci Asn. Res: Dairy cattle breeding and management. Mailing Add: Dept of Animal & Dairy Sci Univ of Ga Athens GA 30602

CLIFTON, DAVID GEYER, b Pomeroy, Ohio, Mar 20, 24; m 56. PHYSICAL CHEMISTRY. Educ: Miami Univ, Ohio, BA, 48, MA, 50; Ohio State Univ, PhD, 55. Prof Exp: Res chemist, Film Dept, E I du Pont de Nemours & Co, Inc, 55-56; staff mem, Los Alamos Sci Lab, Univ Calif, 57-64; sr res chemist, Gen Motors Defense Res Labs, 64-68; STAFF MEM, LOS ALAMOS SCI LABS, UNIV CALIF, 68- Mem: Am Chem Soc; Am Inst Physics. Res: Thermodynamics; rocket propellant systems; aerophysics. Mailing Add: Los Alamos Sci Labs Univ of Calif Los Alamos NM 87544

CLIFTON, HUGH EDWARD, b Prospect, Ohio, July 29, 34; m 57; c 3. GEOLOGY. Educ: Ohio State Univ, BSc, 56; Johns Hopkins Univ, PhD(geol), 63. Prof Exp: GEOLOGIST, US GEOL SURV, 63- Mem: AAAS; Geol Soc Am; Soc Econ Paleont & Mineral; Int Soc Sedimentol. Res: Sedimentary petrography; sedimentology. Mailing Add: US Geol Surv 345 Middlefield Rd Menlo Park CA 94025

CLIFTON, JAMES ALBERT, b Fayetteville, NC, Sept 18, 23; m 49; c 3. INTERNAL MEDICINE. Educ: Vanderbilt Univ, BA, 44, MD, 47; Am Bd Internal Med, dipl, 55; Am Bd Gastroenterol, dipl, 62. Prof Exp: Intern, Univ Hosps, Univ Iowa, 47-48, resident dept med, 48-51; mem staff, Vet Admin Hosp, Tenn, 52-53; assoc internal med, 53-54, from asst prof to assoc prof, 54-63, vis prof dept physiol, 64, vchmn dept med, 67-70 & head dept, 70-75, PROF MED, UNIV IOWA, 63-, ACTG HEAD DEPT, 75- Concurrent Pos: Fel med, Mass Mem Hosp, 55-56; NIH spec res fel, 55-56; attend physician, Vet Admin Hosp, Iowa, 53-; consult, Surgeon Gen, USPHS, 64-; mem, Nat Adv Arthritis & Metab Dis, 70-73. Mem: AAAS; AMA; Am Col Physicians; Am Gastroenterol Asn (pres, 70-71); Asn Profs Med (secy-treas, 73-75). Res: Patho-physiology of the gastrointestinal system; liver disease and mechanisms of intestinal absorption. Mailing Add: Dept of Med Univ of Iowa Iowa City IA 52241

CLIFTON, JAMES ALFRED, b St Louis, Mo, Jan 6, 27; m 47; c 4. APPLIED ANTHROPOLOGY, CULTURAL ANTHROPOLOGY. Educ: Univ Chicago, PhB, 50; San Francisco State Col, MA, 57; Univ Ore, PhD(anthrop), 60. Prof Exp: Asst prof anthrop, Univ Colo, 60-62; from asst prof to assoc prof, anthrop, Univ Kans, 62-69; asst dean, col community sci, 70-72, PROF MODERNIZATION PROCESSES & ANTHROP & DIR SUMMER SESSIONS & JAN INTERIM, UNIV WIS-GREEN BAY, 70-, DIR APPL ANTHROP, UNIV YEAR FOR ACTION PROG, 71- Concurrent Pos: Consult, Bur Indian Affairs, 64, High Coun Cent Am Univs, 67 & UN Ctr for Housing, Bldg & Planning, 68; NSF vis sci lectr, Midwest Lib Arts Cols, 64-; NSF sci fac fel, 65-66; lectr & partic, Brazilian Planning Cong, 70; NSF vis lectr, Wis Cols, 70- Mem: Fel Am Anthrop Asn; fel Soc Appl Anthrop; Am Ethnol Asn. Res: Ute Indians, Spanish-American and Anglos of southwest Colorado; Klamath Indians; Santiago slum community; Potawatomi Indians. Mailing Add: Dir Action Progs Univ of Wis-Green Bay Green Bay WI 54302

CLIFTON, KELLY HARDENBROOK, b Spokane, Wash, July 22, 27; m 49; c 3. EXPERIMENTAL BIOLOGY. Educ: Univ Mont, BA, 50; Univ Wis, MS, 51, PhD(zool), 55. Prof Exp: Am Cancer Soc res fel, Children's Cancer Res Found, 55-56, res assoc exp path, 56-59; res fel, dept path, Harvard Univ, 57-59; from asst prof to assoc prof, dept radiol, 59-69, prof radiol & path, 69-75, PROF HUMAN ONCOL & RADIOL, MED SCH, UNIV WIS-MADISON, 75-, ASST DEAN PRE-MED AFFAIRS, 72- Mem: Am Asn Cancer Res; Soc Exp Biol & Med; Am Soc Exp Path; Radiation Res Soc. Res: Endocrine oncogenesis; radiobiology; physiologic feedback mechanisms in general. Mailing Add: Dept of Human Oncol Med Sch Univ Wis-Madison Madison WI 53706

CLIFTON, YEATON HOPLEY, b Camden, NJ, Oct 9, 22; m 59; c 2. MATHEMATICS. Educ: Columbia Univ, BS, 54, PhD(math), 61. Prof Exp: Instr math, Columbia Univ, 58-59 & Mass Inst Technol, 59-61; asst prof, Univ Calif, Los Angeles, 61-67; assoc ed, 67-74, ADMIN ED MATH REV, UNIV MICH, 74- Mem: Am Math Soc; Ling Soc Am. Res: Differential geometry; geometric integration theory; comparative Indo-European linguistics. Mailing Add: Math Rev 611 Church St Ann Arbor MI 48104

CLIMENHAGA, JOHN LEROY, b Delisle, Sask, Nov 7, 16; m 43; c 2. ASTROPHYSICS. Educ: Univ Sask, BA, 45, MA, 49; Univ Mich, MA, 56, PhD, 60. Prof Exp: Instr physics, Regina Col, 46-48; from asst prof to assoc prof, 49-63, head dept physics, 56-69, dean Fac Arts & Sci, 69-72, PROF PHYSICS, UNIV VICTORIA, BC, 63- Concurrent Pos: Mem, Nat Comt Can, Int Astron Union, 67-71; mem scholar comt, Nat Res Coun, 69-72, mem radio & elec eng div adv bd, 71-74. Mem: Am Astron Soc; Am Asn Physics Teachers; Can Asn Physicists; Royal Astron Soc Can; Int Astron Union. Res: Abundance ratio C-12/C-13 in carbon stars; line blanketing and micro-turbulence in late type stars; cometary spectra. Mailing Add: Dept of Physics Univ of Victoria Victoria BC Can

CLINCH, NORMAN FREDERICK, b Tunbridge Wells, Eng, Apr 3, 41; m 63; c 1. PHYSIOLOGY, BIOPHYSICS. Educ: Univ London, BS, 62, PhD(physiol), 65. Prof Exp: From lectr to asst prof physiol, Univ Alta, 65-69; asst prof, 69-74, ASSOC PROF PHYSIOL, UNIV MAN, 74- Res: Physiology and biophysics of skeletal muscle, especially active state and excitation-contraction coupling processes; electrical properties of biological membranes. Mailing Add: Dept of Physiol Univ of Man 770 Bannatyne Ave Winnipeg MB Can

CLINE, ALAN KAYLOR, mathematics, see 12th edition

CLINE, ATHOL L, biochemistry, see 12th edition

CLINE, DOUGLAS, b York, Eng, Aug 28, 34. NUCLEAR PHYSICS. Educ: Univ Manchester, BSc, 57, PhD(physics), 63. Prof Exp: Res fel physics, Univ Manchester, 60-63; res assoc, 63-65, asst prof, 65-70, ASSOC PROF PHYSICS, UNIV ROCHESTER, 70- Concurrent Pos: Mem prog adv comt, Brookhaven Nat Lab, 74- Mem: Am Phys Soc. Res: Heavy ion physics, including coulomb excitation, electromagnetic moments and transition strengths, over vertical fields; collective model interpretation of nuclear properties. Mailing Add: Nuclear Structure Res Lab Univ of Rochester Rochester NY 14627

CLINE, EDWARD TERRY, b Ischua, NY, Aug 20, 14; m 39; c 3. ORGANIC CHEMISTRY. Educ: Antioch Col, BS, 36; Ohio State Univ, PhD(chem), 39. Prof Exp: Asst, Ohio State Univ, 36-39; res org chemist, 39-66, RES ASSOC, E I DU PONT DE NEMOURS & CO, INC, 66- Mem: Am Chem Soc; Sigma Xi. Res: Textile treatments; new textile fibers; polymers; specialty film; chemical development. Mailing Add: 18 Crestfield Rd Wilmington DE 19810

CLINE, GEORGE BRUCE, b McConnellsburg, Pa, Sept 30, 36; m 61. PHYSIOLOGY, BIOPHYSICS. Educ: Juniata Col, BS, 58; State Univ NY, PhD(physiol), 67. Prof Exp: Consult molecular anat sect, Oak Ridge Nat Lab, 64-66, res assoc, 66-67; from asst prof to assoc prof, 67-75, PROF BIOL, UNIV ALA, BIRMINGHAM, 75-, CHMN DEPT, 70- & ASST PROF PHYSIOL & BIOPHYS, SCH MED, 67- Concurrent Pos: Consult, Electro-Nucleonics, Inc, NJ, 67-; vis prof, Univ Brussels, 73-74. Mem: AAAS; Am Soc Zool; Soc Study Reproduction. Res: Development of zonal centrifuge separation methods for animal, microbial and insect viruses from culture fluids, tissue homogenates and natural waters; characterization of phase-specific antigens from mammalian embryonic and fetal cells. Mailing Add: Dept of Biol Univ Col Univ of Ala Univ Sta Birmingham AL 35294

CLINE, JACK HENRY, b Columbus, Ohio, Feb 27, 27; m 48; c 7. ANIMAL NUTRITION. Educ: Ohio State Univ, BS, 50, MS, 52, PhD(animal sci), 56. Prof Exp: Asst animal nutrit, Exp Sta, 51-56, from instr to assoc prof, 56-69, PROF ANIMAL SCI, OHIO STATE UNIV, 69- Concurrent Pos: With Ohio Agr Res & Develop Ctr, Wooster. Mem: Am Soc Animal Sci. Res: Ruminant and non-ruminant nutrition; feeding and metabolism studies with sheep and swine; mineral metabolism. Mailing Add: Dept of Animal Sci Ohio State Univ Columbus OH 43210

CLINE, JAMES E, b Detroit, Mich, Mar 10, 31; m 53; c 4. EXPERIMENTAL NUCLEAR PHYSICS. Educ: Univ Mich, BSE, 53, MS, 54, PhD(physics), 58. Prof Exp: Res assoc synchrotron proj, Univ Mich, 54-57; physicist, Atomic Energy Div, Phillips Petrol Co, 57-64, group leader, Decay Schemes Group, 64-66; group leader exp physics, Nat Reactor Test Sta, Aerojet Nuclear Co, 66-70, sect chief, 70-73; LAB DIR, NUCLEAR ENVIRON SERVS, DIV SCI APPLICATIONS, INC, 73- Concurrent Pos: Lectr nuclear physics, Idaho State Univ, 58-59 & Univ Idaho, 58-73; asst prof & collab physics, Utah State Univ, 65-73. Mem: Am Phys Soc; Am Nuclear Soc; Inst Elec & Electronics Eng. Res: Gamma ray spectroscopy and automatic data analysis, measurement and study of various molecular forms of radioiodine in ventilation air in nuclear power plants; study of transuranics in Radwaste systems. Mailing Add: Nuclear Environ Servs No 3 Choke Cherry Rd Suite 100 Rockville MD 20850

CLINE, JAMES EDWARD, b Glens Falls, NY, Nov 13, 13; m 37; c 3. OPERATIONS RESEARCH. Educ: Cornell Univ, AB, 34; Harvard Univ, AM, 35, PhD(phys chem), 37. Prof Exp: Asst photochem, Harvard Univ, 37-41; asst chemist, Tenn Valley Authority, Wilson Dam, Ala, 41-42, assoc chemist, 42-45; chief res chemist, Beacon Co, Boston, 45-48; phys chemist & proj leader, Mass Inst Technol, 48-52; engr specialist, Sylvania Elec, 52-58; eng group leader, Raytheon Mfg Co, 58-61; staff scientist, Kearfott Semiconductor Corp, 61-62; eng specialist, Sylvania Electronic Syst Div, Gen Tel & Electronics Corp, 62-64; eng specialist aerospace technol, Electronic Res Ctr, NASA, 64-70; OPERS RES, US DEPT TRANSP, CAMBRIDGE, 70- Mem: Inst Elec & Electronics Engrs; Transp Res Forum. Res: Instrumentation for vapor trace detection; computer interfacing; microelectronics. Mailing Add: 23 Stetson St Brookline MA 02146

CLINE, MARLIN GEORGE, b Bertha, Minn, Dec 31, 09; m 36; c 3. SOIL MORPHOLOGY. Educ: NDak State Col, BS, 35; Cornell Univ, PhD(soils), 42. Hon Degrees: DSc, NDak State Univ & Trinity Col, Dublin, 65. Prof Exp: Jr soil surveyor, 35-38, assoc soil scientist, 41-42, soil scientist, 44-45, AGT CORRELATION SOILS, USDA, 46-; EMER PROF SOIL SCI, STATE UNIV NY COL AGR & LIFE SCI, CORNELL UNIV, 74- Concurrent Pos: From instr to assoc prof soils, State Univ NY Col Agr, Cornell Univ, 42-46, head dept agron, 63-70; soil scientist, Econ Coop Admin, Brit Africa, 49; Cornell contract, Philippines, 54-56; mem US mission soil & water, USSR, 58. Mem: Soil Sci Soc Am; Am Soc Agron; Brit Soc Soil Sci; Int Soc Soil Sci. Res: Morphology, genesis and cartography of soils; soil management. Mailing Add: Dept of Agron State Univ NY Col Agr & Life Sci Ithaca NY 14850

CLINE, MICHAEL CASTLE, b Richmond, Va, Apr 3, 45; m 69; c 1. FLUID DYNAMICS. Educ: Va Polytech Inst & State Univ, BS, 67; Purdue Univ, MS, 68, PhD(mech eng), 71. Prof Exp: Nat Acad Sci res assoc, Langley Res Ctr, NASA, 71-73; STAFF MEM FLUID DYNAMICS, LOS ALAMOS SCI LAB, 73- Mem: Am Inst Aeronaut & Astronaut. Res: Computation of fluid flows in ducts. Mailing Add: Theoret Div Los Alamos Sci Lab Univ of Calif Los Alamos NM 87545

CLINE, MORRIS GEORGE, b Los Angeles, Calif, Aug 10, 31; m 59; c 4. PLANT PHYSIOLOGY, ECOLOGY. Educ: Univ Calif, Berkeley, BS, 53; Brigham Young Univ, MS, 61; Univ Mich, PhD(bot), 64. Prof Exp: Asst plant physiologist, Colo State Univ, 64-66; res fel environ plant physiol, Calif Inst Technol, 66-68; asst prof, 68-74, ASSOC PROF BOT, OHIO STATE UNIV, 74- Mem: Am Soc Plant Physiologists. Res: Effects of temperature, low and high intensity visible and ultraviolet radiation on plants; photoreversal of ultraviolet damage; effects of soil environment on root growth of mountain shrubs; hormone action and nucleic acid metabolism. Mailing Add: Fac of Bot Ohio State Univ Columbus OH 43210

CLINE, RANDALL EUGENE, b Marietta, Ohio, Oct 4, 31; m 56; c 3. APPLIED MATHEMATICS. Educ: Marietta Col, BA, 53; Purdue Univ, MS, 55, PhD(math), 63. Prof Exp: Res asst & instr math, Purdue Univ, 55-59; res assoc, Inst Sci & Technol, Mich, 59-63, assoc res mathematician, 63-65, res mathematician, 65-68; assoc prof, 68-72, PROF MATH & COMPUT SCI, UNIV TENN, KNOXVILLE, 72- Concurrent Pos: Mem staff, Math Res Ctr, US Army-Univ Wis, 64-65. Mem: AAAS; Asn Comput Mach; Soc Indust & Appl Math. Res: Matrix theory; algorithms. Mailing Add: Depts of Math & Comput Sci Univ of Tenn Knoxville TN 37916

CLINE, RICHARD EMORY, organic chemistry, see 12th edition

CLINE, SYLVIA GOOD, b Atlantic City, NJ, Dec 27, 28; m 50; c 3. CELL PHYSIOLOGY, CYTOLOGY. Educ: Bryn Mawr Col, PhD(biol), 65. Prof Exp: Res asst physiol chem, Univ Pa, 50-51; res asst biol, Bryn Mawr Col, 60-64, fel, 65-66; res assoc biochem, Queens Col, NY, 66-67, lectr chem & biochem, 67-68; asst prof, 68-71, ASSOC PROF BIOL, QUEENSBOROUGH COMMUNITY COL, 71-, ASST DEAN INSTR & ACAD AFFAIRS, 72- Concurrent Pos: Guest investr, Biochem Cytol Lab, Rockefeller Univ, 66-67. Mem: AAAS; Am Chem Soc; Am Soc Cell Biol; Soc Protozool. Res: Glucose metabolism; RNA catabolism; growth of protozoa. Mailing Add: Dept of Biol Queensborough Community Col Bayside NY 11364

CLINE, THOMAS L, b Peiping, China, May 14, 32; US citizen; m 54; c 3. PHYSICS. Educ: Hiram Col, BA, 54; Mass Inst Technol, PhD(physics), 61. Prof Exp: Res assoc, lab nuclear sci, Mass Inst Technol, 60-61; PHYSICIST, SPACE SCI LAB, GODDARD SPACE FLIGHT CTR, NASA, 61- Concurrent Pos: Mem solar physics subcomt, Space Sci Steering Comt, DC, 62-64; actg chief, high energy astrophys prog, NASA, DC, 70-71, asst head, cosmic radiation br, 70- Mem: AAAS; fel Am Phys Soc; Am Astron Soc. Res: Cosmic rays; solar particle production; astrophysics. Mailing Add: Lab High-Energy Astrophys Goddard Space Flight Ctr NASA-Code 661 Greenbelt MD 20771

CLINE, WARREN KENT, b Bluefield, Va, July 28, 21; m 46; c 2. ORGANIC CHEMISTRY. Educ: Va Polytech Inst, BS, 42; Ohio State Univ, PhD(chem), 50. Prof Exp: Chemist, Gen Chem Defense Corp, 42-43; res chemist, Magnolia Petrol Co, 43-46; asst, Ohio State Univ, 46-50; res org chemist, Olin Mathieson Chem Corp, 50-59, res assoc, 59-60, supvr res group, Film Opers, 60-69, MGR POLYMER RES GROUP, FILM DIV, OLIN CORP, 69- Mem: AAAS; Am Chem Soc; fel Am Inst Chem. Res: Organic synthesis; polymer chemistry. Mailing Add: Olin Corp Film Div Box 200 Pisgah Forest NC 28768

CLINESCHMIDT, BRADLEY VAN, b Redding, Calif, Dec 11, 41. NEUROPHARMACOLOGY. Educ: Ore State Univ, BS, 64; Univ Wash, PhD(pharmacol), 68. Prof Exp: Res assoc pharmacol, Exp Therapeut Br, Nat Heart & Lung Inst, 69-71, sr staff fel, 71-72; res fel, 72-75, DIR NEUROPSYCHOPHARMACOL, MERCK INST THERAPEUT RES, 75- Mem: Am Soc Pharmacol & Exp Therapeut. Res: Neuropharmacological, behavioral and neurochemical actions of drugs affecting the central nervous system. Mailing Add: Merk Inst Therapeut Res West Point PA 19486

CLINGMAN, WILLIAM HERBERT, JR, b Grand Rapids, Mich, May 5, 29; m 51; c 2. PHYSICAL CHEMISTRY. Educ: Univ Mich, BS, 51; Princeton Univ, MA & PhD(phys chem), 54. Prof Exp: Res chemist, Am Oil Co, 54-57, group leader, 57-59; sect head, Tex Instruments Inc, 59-61, dir energy res lab, 61-62, mgr res exploitation dept, 62-64, corporate res & develop mkt, 64-67; PRES, W H CLINGMAN & CO, 67- Mem: Am Chem Soc; Inst Elec & Electronics Eng. Res: Radiation chemistry; catalysis; organic reaction mechanism; energy conversion; thermodynamics. Mailing Add: W H Clingman & Co 2001 Bryan St Suite 2265 Dallas TX 75201

CLINNICK, MANSFIELD, b Somerville, NJ, Jan 21, 22; m 42; c 6. MATHEMATICS. Educ: Calif State Polytech Col, BS, 47; Univ Calif, Berkeley, AB, 53, MA, 55. Prof Exp: Prin programmer, Lawrence Radiation Lab, Univ Calif, 55-58; proj mgr, Broadview Res Corp, 58-60; from asst prof to assoc prof math, Calif State Polytech Col, 60-67; MATHEMATICIAN-PROGRAMMER, LAWRENCE RADIATION LAB, UNIV CALIF, BERKELEY, 67- Mem: Am Math Soc; Math Asn Am; Asn Comput Mach. Res: Digital computer programming; algebra. Mailing Add: Math & Comput Group Lawrence Radiation Lab Univ Calif Berkeley CA 94720

CLINTON, CHARLES ANTHONY, b Wittier, Calif, June 22, 39; m 67; c 1. ETHNOGRAPHY. Educ: Western Wash State Col, BA, 62; Univ Kans, MA, 66; Wash State Univ, PhD(anthrop), 73. Prof Exp: Instr anthrop, Western Wash State Col, 67-69, NSF prin investr urban anthrop, 71-73; ON-SITE RESEARCHER ANTHROP & EDUC, NAT INST EDUC CONTRACT, ABT ASSOC, INC, 73- Mem: Am Anthrop Asn; Soc Appl Anthrop; Coun Anthrop & Educ. Res: Ethnographic research designed to show the inter-relationships between a rural county, its school system and a period of planned educational change. Mailing Add: Box 409 Lewisport KY 42351

CLINTON, RAYMOND OTTO, b Burbank, Calif, Apr 12, 18; m 38; c 2. MEDICINAL CHEMISTRY. Educ: Calif Inst Technol, BS, 40; Univ Calif, Los Angeles, MA, 41, PhD(org chem), 43. Prof Exp: Jr chemist, C F TenEyck & Co, Calif, 35-38; asst, Calif Inst Technol, 36-40; asst inorg chem, Univ Calif, Los Angeles, 41; chemist, Union Oil Co, Calif, 41-42; instr org chem, Marymount Col, 42; sr res chemist, Gasparcolor, Inc, Calif, 46-47 & Sterling-Winthrop Res Inst, 43-46, group leader, 47-56, head org sect, 56-60, dir res admin, 60-67; asst to chmn, 67-74, DIR CORP SERV, STERLING DRUG , INC, 74- Concurrent Pos: Chemist, Black Diamond Oil Co, Calif, 39-40; res chemist, Nat Defense Res Comt proj, Univ Calif, Los Angeles, 41-43; adj prof Rensselaer Polytech Inst, 52-62. Mem: Am Chem Soc; NY Acad Sci; The Chem Soc. Res: Chemistry of plant pigments; antitubercular agents; local anesthetics; antomalarials; sulfur-containing amines; anti-virus agents; color photography dyes; flavonones and related plant pigments; synthesis of acids related to vitamin A; steroids; structural activity relationships. Mailing Add: Sterling Drug, Inc 90 Park Ave New York NY 10016

CLINTON, WILLIAM L, b St Louis, Mo, Sept 17, 30; m 52; c 9. THEORETICAL PHYSICS. Educ: St Louis Univ, PhD(chem), 59. Prof Exp: Teacher chem, St Louis

Univ, 55-58; res assoc, Brookhaven Nat Lab, 58-60; asst prof chem, 60-63, from asst prof to assoc prof physics, 64-70, PROF PHYSICS, GEORGETOWN UNIV, 70- Concurrent Pos: Consult, Nat Bur Standards. Mem: Am Phys Soc. Res: Application of quantum mechanics to molecular physics and theoretical chemistry; electronic structure. Mailing Add: Dept of Physics Georgetown Univ Washington DC 20007

CLIPPINGER, FRANK WARREN, JR, b Appleton, Wis, Oct 27, 25; m 50; c 2. ORTHOPEDIC SURGERY. Educ: Drury Col, Mo, AB, 48; Wash Univ, St Louis, MD, 52. Prof Exp: From instr to assoc prof, 57-70, PROF ORTHOP SURG, SCH MED, DUKE UNIV, 70- Concurrent Pos: Chmn, Prosthetics Res & Develop Comt, Nat Acad Sci, 75-76; med dir, Duke Hosp W-Duke Univ Med Sch, 75-, dir rehab, Duke Univ Med Ctr, 75-; chmn, Orthop Surg Adv Coun, Am Col Surgeons, 73-76; chmn, Orthop Sect, Southern Med Asn, 73-74. Mem: Am Orthop Asn; Am Acad Orthop Surgeons; Am Col Surgeons; Am Soc Surg Hand. Res: Research and development of artificial limbs; socket design and sensory feedback mechanism. Mailing Add: Box 3435 Duke Univ Med Ctr Durham NC 27710

CLISE, RONALD LEO, b Westernport, Md, Aug 19, 23; m; c 3. GENETICS, ZOOLOGY. Educ: Marietta Col, BA, 49; Mich State Univ, MS, 52; Western Reserve Univ, PhD(genetics), 60. Prof Exp: From instr to assoc prof, 55-72, chmn dept, 60-62, PROF BIOL, CLEVELAND STATE UNIV, 72- Mem: AAAS. Res: Population studies with Drosophila. Mailing Add: Dept of Biol Cleveland State Univ Euclid Ave at 24th St Cleveland OH 44115

CLITHEROE, H JOHN, b Hornchurch, Eng, Jan 2, 35; m 68; c 1. ENDOCRINOLOGY, GYNECOLOGY. Educ: Univ Sheffield, BSc, 59, MIBiol & PhD(med), 62; Royal Soc Health, FRSH, 74. Prof Exp: Johnson & Johnson fel, Rutgers Univ, 62-63; NIH fel pharmacol, 63-65, asst prof, 66-70, CLIN ASST PROF OBSTET & GYNEC, NJ COL MED, 70-; PROF BIOL SCI, STATEN ISLAND COMMUNITY COL, CITY UNIV NEW YORK, 75- Concurrent Pos: Consult cytol, St Elizabeth's Hosp, Elizabeth, NJ & St Vincent's Med Ctr, New York, 67-75; assoc prof biol sci, Staten Island Community Col, New York Univ, 70-75. Mem: Brit Soc Endocrinol; Brit Inst Biol; Pan-Am Cancer Cytol Soc; NY Acad Sci; World Population Soc. Res: Uterine physiology; cytology of reproductive organs; sex selection of offspring; sexual therapy. Mailing Add: Dept of Biol Sci Staten Island Community Col Staten Island NY 10301

CLIVER, DEAN OTIS, b Berwyn, Ill, Mar 2, 35; m 60; c 4. VIROLOGY. Educ: Purdue Univ, BS, 56, MS, 57; Ohio State Univ, PhD(agr), 60. Prof Exp: Fel, Ohio State Univ, 60; resident res assoc virus serol, US Army Chem Corps Biol Labs, Ft Detrick, Md, 61-62; res assoc virol, food res inst & dept microbiol, Univ Chicago, 62-66; asst prof, 66-67, ASSOC PROF VIROL, FOOD RES INST & DEPT BACT, UNIV WIS-MADISON, 67- Concurrent Pos: Resident res assoc, Nat Acad Sci-Nat Res Coun, 61-62; consult, WHO, 69- Mem: Am Soc Microbiol. Res: Food research; stuies on virus contamination of foods and water; animal virology. Mailing Add: Food Res Inst Univ of Wis 1925 Willow Dr Madison WI 53706

CLOAK, FRANK THEODORE, JR, anthropology, see 12th edition

CLODMAN, JOSEPH, b Toronto, Ont, June 9, 17; m 41; c 2. METEOROLOGY. Educ: Univ Toronto, BA, 41, MA, 48; NY Univ, PhD(meteorol), 61. Prof Exp: Forecaster, 43-52, res aviation meteorologist, 52-61, supvr synoptic res, 61-69, SUPT FORECAST RES SECT, METEOROL SERV CAN, 69- Concurrent Pos: Mem comn synoptic meteorol working group, World Meteorol Orgn, 58-60. Honors & Awards: Darton Prize, Royal Meteorol Soc, 58. Mem: Fel Royal Meteorol Soc. Res: Aviation meteorology, particularly aircraft turbulence; clear air turbulence; mesometeorology; short range forecasting. Mailing Add: Atmospheric Environ Serv 4905 Dufferin St Downsview ON Can

CLOGSTON, ALBERT MCCAVOUR, b Boston, Mass, July 13, 17; m; c 2. SOLID STATE PHYSICS. Educ: Mass Inst Technol, SB, 38, PhD(physics), 41. Prof Exp: Teaching fel physics, Mass Inst Technol, 38-41, mem staff, radiation lab, 41-46; res physicist, Bel Tel Labs, 46-63, asst dir mat res lab, 63-65, dir, phys res lab, 65-71; vpres res, Sandia Labs, 71-73; EXEC DIR RES, PHYSICS & ACAD AFFAIRS DIV, BELL TEL LABS, 73- Mem: Nat Acad Sci; fel Am Phys Soc. Res: Magnetism; theory of metals; superconductivity; nuclear magnetic resonance; alloys and intermetallic compounds. Mailing Add: Bell Tel Labs Murray Hill NJ 07974

CLOKE, PAUL LEROY, b Orono, Maine, Feb 6, 29; m 55; c 2. GEOCHEMISTRY. Educ: Harvard Univ, AB, 51; Mass Inst Technol, PhD(geol), 54. Prof Exp: Res geologist, mining & explor geol, Anaconda Co, 54-57; res fel geochem, Harvard Univ, 57-59; from asst prof to assoc prof, 59-69, PROF GEOCHEM, UNIV MICH, ANN ARBOR, 69- Concurrent Pos: Chemist, Dept Sci & Indust Res, Gracefield, NZ, 66-67; ed, The Geochem News, 65-73. Mem: AAAS; Soc Econ Geologists; Geochem Soc; Am Inst Mining, Metall & Petrol Engrs; Mineral Soc Can. Res: Application of chemistry to geologic problems, such as hydrothermal solutions, ore deposits, mineral-solution equilibria at low to high temperature and pressure, and recent sediments. Mailing Add: Dept of Geol & Mineral Univ of Mich Ann Arbor MI 48104

CLONEY, RICHARD ALAN, b Port Angeles, Wash, Feb 12, 30; m 52; c 3. DEVELOPMENTAL BIOLOGY. Educ: Humboldt State Col, AB, 52, MA, 54; Univ Wash, PhD(zool), 59. Prof Exp: NIH fel anat, Sch Med, 59-61, from asst prof to assoc prof zool, 61-72, PROF ZOOL, UNIV WASH, 72- Concurrent Pos: Consult, Develop Biol Sect, Educ Develop Ctr, Mass; NSF grant. Mem: Am Soc Zoologists; Soc Develop Biol; Am Soc Cell Biol. Res: Electron microscopic, cinematographic and experimental analyses of metamorphosis in ascidians; microfilaments and morphogenesis; intracytoplasmic movements; muscle differentiation. Mailing Add: Dept of Zool Univ of Wash Seattle WA 98105

CLONEY, ROBERT DENNIS, b Boston, Mass, May 6, 27. PHYSICAL CHEMISTRY. Educ: Spring Hill Col, BS, 52; Cath Univ Am, PhD(chem), 57; Woodstock Col, STB, 61. Prof Exp: Instr, pvt sch, NY, 52-53; res assoc chem, Woodstock Col, 57-62; from instr to asst prof, 62-71, ASSOC PROF CHEM, FORDHAM UNIV, 71- Mem: AAAS; Am Phys Soc; Am Chem Soc; Am Asn Jesuit Sci. Res: Quantum chemistry; molecular structure. Mailing Add: Dept of Chem Fordham Univ Bronx NY 10458

CLOOS, ERNST, geology, geophysics, deceased

CLOPTON, JOHN RAYMOND, biochemistry, see 12th edition

CLORE, WALTER JOSEPH, b Tecumseh, Okla, July 1, 11; m 34; c 3. HORTICULTURE. Educ: Okla Agr & Mech Col, BS, 33; State Col Wash, PhD, 47. Prof Exp: Asst hort, 37-46, assoc horticulturist, 46-51, HORTICULTURIST, IRRIG AGR RES & EXTEN CTR, WASH STATE UNIV, 51- Mem: Am Soc Hort Sci. Res: Production of grapes for juice and wine and asparagus production under irrigation. Mailing Add: Irrig Agr Res & Exten Ctr Wash State Univ Prosser WA 99350

CLOSE, DONALD ALAN, b Tucson, Ariz, Nov 19, 46. NUCLEAR PHYSICS. Educ: Hastings Col, BA, 68; Univ Kans, MA, 70, PhD(physics), 72. Prof Exp: NSF presidential internship, 72-73, staff mem, 73, STAFF PHYSICIST NUCLEAR PHYSICS, LOS ALAMOS SCI LAB, UNIV CALIF, 73- Mem: Am Phys Soc; Sigma Xi. Res: Muonic atoms, gamma-ray spectroscopy and nuclear structure; proton induced x-ray fluorescence. Mailing Add: Los Alamos Sci Lab Los Alamos NM 87545

CLOSE, DONALD HENRY, b Milwaukee, Wis, Nov 15, 37; m 69; c 3. OPTICS. Educ: Univ Kans, BS, 60; Calif Inst Technol, MS, 62; PhD(elec eng), 65. Prof Exp: Engr, US Naval Ord Test Sta, 60-61; mem tech staff, 65-72, sr mem tech staff, 72-73, HEAD HOLOGRAM OPTICS SECT, HUGHES RES LAB, HUGHES AIRCRAFT CO, 73- Mem: Am Phys Soc; Optical Soc Am; Soc Photo-Optical Instrumentation Engrs. Res: Design and fabrication of optical elements with unique capabilities using holographic techniques; optical and digital information processing. Mailing Add: Hughes Res Lab 3011 Malibu Canyon Rd Malibu CA 90265

CLOSE, PERRY, b Chicago, Ill, May 20, 21; m 61; c 2. GENETICS, PHYSIOLOGY. Educ: Univ Calif, Berkeley, AB, 47, MA, 48; Univ Tex, PhD(zool), 55. Prof Exp: Asst prof biol, Univ Southwestern La, 48-49; asst human genetics, Univ Tex, 54-55; aviation physiologist, med dept, US Naval Air Sta, Va, 55-57 & US Naval Sch Aviation Med, 57-62; mem res staff, Northrop Space Labs, Calif, 62-64; res biologist, Vet Admin Hosp, Long Beach, Calif, 64-65; head life sci, Chrysler Space Div, 65-68; PROF BIOL, CITY COL SAN FRANCISCO, 68- Mem: Am Soc Human Genetics; assoc fel Aerospace Med Asn; Soc Study Social Biol. Res: Genetics og longevity; hereditary deafness; low pressure and impact patho-physiology and the effects of radiation in combination with aerospace stresses; evaluation of radiotherapy procedures; genetics of mental retardation. Mailing Add: 272 Dennis Dr Daly City CA 94015

CLOSE, RICHARD THOMAS, b New York, NY, Dec 24, 34; m 58; c 7. COMPUTER SCIENCE, ELECTROMAGNETIC THEORY. Educ: Iona Col, BS, 56; Cath Univ Am, PhD(physics), 67. Prof Exp: Res physicist, Naval Res Lab, Washington, DC, 59-68; assoc prof physics, St Bonaventure Univ, 68-71; DIR COMPUT CTR, STATE UNIV NY AGR & TECH COL, ALFRED UNIV, 71- Concurrent Pos: Lectr, Univ Md, 60-68. Mem: Am Phys Soc; Am Asn Physics Teachers; Asn Comput Mach. Res: Antenna research with particular emphasis on frequency independent scanning arrays; intense electron beam studies including both theoretical and experimental studies; solution of elliptic partial differential equations using various numerical methods; environmental remote sensing and its military uses. Mailing Add: Comput Ctr State Univ of NY Agr & Tech Col Alfred NY 14802

CLOSE, WARREN JAMES, b Big Rock, Ill, July 24, 20; m 43; c 3. PHARMACEUTICAL CHEMISTRY. Educ: DePauw Univ, AB, 42; Univ Wis, MS & PhD(org Chem), 46. Prof Exp: Res chemist, 46-61, head org chem res, 61-65, dir chem res, 65-70, dir exp chem, 70-75, DIR SCI LIAISON, ABBOTT LABS, 75- Mem: Am Chem Soc; AAAS; Am Soc Microbiol; Am Soc Pharmacog. Res: Drugs affecting the central nervous system. Mailing Add: Sci Div Abbott Labs Chicago IL 60064

CLOSMANN, PHILIP JOSEPH, b New Orleans, La, July 28, 25; m 56; c 6. PHYSICS, CHEMICAL PHYSICS. Educ: Tulane Univ, BE, 44; Mass Inst Technol, SM, 48; Calif Inst Technol, MS, 50; Rice Inst, PhD(physics), 53. Prof Exp: PHYSICIST, SHELL DEVELOP CO, 53- Mem: Am Phys Soc; Am Inst Chem Engrs; Soc Petrol Engrs. Res: Fluid flow; heat flow; low temperature physics; fluidized solids. Mailing Add: 27 Williamsburg Houston TX 77024

CLOSS, GERHARD LUDWIG, b Wuppertal, Ger, May 1, 28. ORGANIC CHEMISTRY. Educ: Univ Tübingen, Dipl Chem, 53, PhD(chem), 55. Prof Exp: Fel, Harvard Univ, 55-57; from asst prof or assoc prof, 57-63, PROF CHEM, UNIV CHICAGO, 63- Concurrent Pos: A P Sloan Found fel, 62-66. Honors & Awards: James Flack Norris Award, Am Chem Soc, 74. Mem: Nat Acad Sci; AAAS; Am Chem Soc; The Chem Soc; Am Acad Arts & Sci. Res: Chemistry of reactive intermediates; carbenes; carbanions; porphyrins; magnetic resonance. Mailing Add: Dept of Chem Univ of Chicago Chicago IL 60637

CLOSSON, WILLIAM DEANE, b Barryton, Mich, Feb 3, 34; m 69. ORGANIC CHEMISTRY. Educ: Wayne State Univ, BS, 56; Univ Wis, PhD(org chem), 60. Prof Exp: Asst org chem, Univ Wis, 56-60; NSF res fel, Har- vard Univ, 60-61; from instr to asst prof, 61-66; assoc prof, 66-71, PROF CHEM, STATE UNIV NY ALBANY, 71- Concurrent Pos: Res grants, Petrol Res Fund, 62-65, USPHS, 63-66 & 67- & NSF, 65-67; Alfred P Sloan res fel, 68-70; Nat Acad Sci-Nat Res Coun travel grant, IVPAC Cong, Jerusalem, 75. Mem: AAAS; Am Chem Soc; The Chem Soc. Res: Solvolytic reactions of organic compounds; electronic absorption spectra of ketones, esters and alkyl azides; reactions of organic anion radiacls; mecuration reactions and silver pi complexes of substituted alkenes. Mailing Add: Dept of Chem State Univ NY at Albany Albany NY 12222

CLOTFELTER, BERYL EDWARD, b Prague, Okla, Mar 23, 26; m 51; c 3. PHYSICS. Educ: Okla Baptist Univ, BS, 48; Univ Okla, MS, 49, PhD(physics), 53. Prof Exp: Instr math & physics, Okla Baptist Univ, 49-50; res physicist, Phillips Petrol Co, 53-55; assoc prof physics, Univ Idaho, 55-56; from asst prof to assoc prof physics, Okla Baptist Univ, 56-63; assoc prof, 63-68, PROF PHYSICS, GRINNELL COL, 68- Concurrent Pos: NSF sci fac fel, 68-69. Mem: Am Phys Soc; Am Asn Physics Teachers. Res: Astrophysics; cosmology. Mailing Add: Dept of Physics Grinnell Col Grinnell IA 50112

CLOTHIER, GALEN EDWARD, b Stafford, Kans, Nov 7, 33; m 55; c 3. CELL BIOLOGY, DEVELOPMENTAL BIOLOGY. Educ: Fresno State Col, AB, 55; Ore State Univ, MS, 57, PhD(biol), 60. Prof Exp: Asst prof zool, Los Angeles State Col, 60-62; from asst prof to assoc prof, 62-68, PROF BIOL, CALIF STATE UNIV, SONOMA, 68- Mem: Am Soc Zoologists; AAAS; Sigma Xi. Res: Physiology of mitosis; developmental biology of sea urchins. Mailing Add: Dept of Biol Calif State Univ, Sonoma Rohnert Park CA 94928

CLOTHIER, RONALD RAYMOND, b Cimarron, Kans, June 8, 24; m 46; c 1. MAMMALOGY. Educ: Fresno State Col, AB, 48; Univ Mont, MA, 50; Univ NMex, PhD, 57. Prof Exp: Asst zool, Univ Mont, 48-50; asst biol, Univ NMex, 50-52; asst prof biol, Kans Wesleyan Univ, 52-55; ASST PROF ZOOL, ARIZ STATE UNIV, 55- Mem: Am Soc Mammalogists; Wildlife Soc; Soc Syst Zool. Res: Mammal life history; mammalian systematics. Mailing Add: Dept of Zool Ariz State Univ Tempe AZ 85281

CLOTHIER, WILLIAM DELBERT, b Potlatch, Idaho, Feb 11, 25; m 48; c 3. FISH & WILDLIFE MANAGEMENT, AQUATIC BIOLOGY. Educ: Mont State Col, BS, 51, MS, 52. Prof Exp: Fisheries res fieldman, Mont Fish & Game Dept, 49-50, lab supvr, 50, jr fishery biologist, 51-52; leader fishery invests, SDak Dept Game, Fish & Parks, 53-58; proj leader, Coastal Rivers Invests, Ore State Fish Comn, 58-60, water resources analyst, 60-62, asst state fisheries dir, 62-65; aquatic biologist, Nat Coastal

Pollution Res Prog, Marine Sci Ctr, Pac Northwest Water Lab, Fed Water Qual Admin, 65-69, regional res & develop prog specialist, 69-70, regional res & monitoring rep, Environ Protection Agency, 71-73, CHIEF REGION X SILVICULT PROJ, ENVIRON PROTECTION AGENCY, 73- Mem: Am Fisheries Soc; Inst Fishery Res Biol. Res: Loss and movement of trout in irrigation diversions; ecological field investigations involving warm water fishes and salmonids; administration in sport and commerical fisheries, marine and freshwater; commercial fisheries, marine and fresh water; water pollution control. Mailing Add: Environ Protection Agency 1200 Sixth Ave Seattle WA 98101

CLOUD, PRESTON E, JR, b West Upton, Mass, Sept 26, 12; m 72; c 3. GEOLOGY, INVERTEBRATE PALEONTOLOGY. Educ: George Washington Univ, BS, 38; Yale Univ, PhD, 40. Prof Exp: Instr geol, Mo Sch Mines, 40-41; Sterling res fel, Yale Univ, 41-42; geologist, US Geol Surv, 42-61, chief, paleont & stratig br, 49-59; prof geol & geophys, Univ Minn, 61-65, chmn dept, 61-63, head, sch earth sci, 62-63; prof geol, Univ Calif, Los Angeles, 65-68; PROF BIOGEOL, UNIV CALIF, SANTA BARBARA, 68-; GEOLOGIST, US GEOL SURV, 74- Concurrent Pos: Asst prof, Harvard Univ, 46-48; mem exec comt, earth sci div, Nat Res Coun, 52-56; lectr, Univ Tex, 62. Del, Pac Sci Cong, NZ, 49 & Philippines, 53, Int Geol Cong, Algiers, 52 & Norden, 60. Honors & Awards: Morrison Prize, NY Acad, 40; Rockefeller Pub Serv Award, 56; Distinguished Serv Award, US Dept Interior, 59; Medal, Paleont Soc Am, 71; Am Philos Soc Award, 73; Lucius Wilbur Cross Medal, Yale Univ Grad Sch, 73. Mem: Nat Acad Sci, (mem coun, 72-75, exec comt, 73-75); AAAS; Geol Soc Am; Am Paleont Soc; Soc Study Evolution. Res: Paleoecology; carbonate rocks; sedimentary and organic processes in geology; mineral resources. Mailing Add: Dept of Geol Univ of Calif Santa Barbara CA 93106

CLOUD, WILLIAM K, b Tucson, Ariz, May 7, 10; m 40; c 4. SEISMOLOGY, MECHANICAL ENGINEERING. Educ: Univ Ariz, BS, 34. Prof Exp: Irrig engr, Exten Serv, Univ Ariz, 35-37; state engr, Agr Adjust Admin, Ariz, 37-42; geophysicist, US Coast & Geod Surv, 46-52, chief seismol field surv, 52-71; ASSOC RES SEISMOLOGIST, UNIV CALIF, BERKELEY, 71- Concurrent Pos: Treas, Earthquake Eng Res Inst, 65. Honors & Awards: Colbert Medal, Soc Am Mil Engrs, 58. Mem: Seismol Soc Am; Soc Am Mil Engrs. Res: Engineering seismology; natural and artificial earthquake vibrations of engineering significance; vibration characteristics of structures. Mailing Add: 1920 Eighth Ave San Francisco CA 94116

CLOUD, WILLIAM MAX, b Wilmot, Kans, Mar 27, 23; m 47; c 3. ATOMIC SPECTROSCOPY. Educ: Southwestern Col, BA, 47; Univ Wis, MS, 49, PhD(physics), 55. Prof Exp: Instr physics, Southwestern Col, 49-50, asst prof physics & counr men, 50-53; from asst prof to assoc prof, Kans State Teachers Col, 55-62; assoc prof, 62-66, PROF PHYSICS & CHMN DIV PRE-ENG STUDIES, EASTERN ILL UNIV, 66- Mem: Am Asn Physics Teachers; Am Phys Soc. Res: High resolution atomic spectroscopy using an atomic beam. Mailing Add: Dept of Physics Eastern Ill Univ Charleston IL 61920

CLOUGH, DONALD J, b Toronto, Ont, Jan 9, 31; m 51; c 8. OPERATIONS RESEARCH. Educ: Univ Toronto, BASc, 54, MBA, 58. Prof Exp: Demonstr fluid mech, Univ Toronto, 55-56, from instr to lectr, 57-59, from asst prof to assoc prof opers res, 61-67; chmn dept, 69-74, PROF OPERS RES, UNIV WATERLOO, 68- Concurrent Pos: Secy-treas & dir, Systs Eng Assocs Ltd, 65-66, pres, 67-; consult, Ont Govt, 58- & Can Govt, 65-; mem comt inspection, Can Govt Specifications Bd, 65-68; mem grant selection comt comput sci, Nat Res Coun Can, 66-68; dir, NATO Conf Manpower Planning Models, Cambridge, UK, 71. Honors & Awards: Gold Medal, Can Oper Res Soc, 63-67 & 68. Mem: AAAS; Am Inst Indust Eng; Can Oper Res Soc; Opers Res Soc Am; Inst Mgt Sci. Res: Computer models of industrial and educational systems; sampling statistics. Mailing Add: Dept of Mgt Sci Univ of Waterloo Waterloo ON Can

CLOUGH, FRANCIS BOWMAN, b Boise, Idaho, Feb 4, 24; m 62. PHYSICAL INORGANIC CHEMISTRY. Educ: Univ Wyo, BS, 44; Princeton Univ, MA, 48, PhD(chem), 51. Prof Exp: Chemist org res, Distillation Prod, Inc, 44-46; asst, Princeton Univ, 46-50; from instr to asst prof chem, Va Polytech Inst & State Univ, 50-55; asst prof, 55-59, ASSOC PROF CHEM, STEVENS INST TECHNOL, 59- Mem: AAAS; Am Chem Soc; fel Am Inst Chemists. Res: Optical rotatory power; solvent interactions; energy transfer in inorganic reactions. Mailing Add: Dept of Chem Stevens Inst of Technol Hoboken NJ 07030

CLOUGH, GARRETT CONDE, b Mystic, Conn, Dec 25, 31; m 61; c 2. ECOLOGY, CONSERVATION. Educ: Union Col, BS, 53; Univ Mich, MS, 54; Univ Wis, PhD(zool), 62. Prof Exp: Asst zool, Univ Wis, 58-62; asst prof, Dalhousie Univ, 61-63; NIMH res fel, Norweg State Game Res Inst, 63-64; vis res fel ecol, Cornell Univ, 64-65; assoc prof zool, Univ RI, 65-74; SCI DIR, CTR NATURAL AREAS, 74- Concurrent Pos: Nat Res Coun Can grant, 61-63; vis res prof, Univ Oslo, 69-70; NIH spec res fel, 69-70. Mem: Am Inst Biol Sci; Sigma Xi; Am Soc Mammalogists; Ecol Soc Am; Animal Behav Soc. Res: Ecology and animal behavior; population dynamics; arctic ecology. Mailing Add: Ctr for Natural Areas 1525 New Hampshire Ave NW Washington DC 20036

CLOUGH, JOHN WENDELL, b Oak Bluffs, Mass, Jan 3, 42; m 68; c 1. GEOPHYSICS. Educ: Northeastern Univ, BS, 65; Univ Wis, Madison, MS, 70, PhD(geophys), 74. Prof Exp: Proj assoc geophys, Geophys & Polar Res Ctr, Univ Wis, 74-75; SCI DIR & ASST PROF GEOPHYS, ROSS ICE SHELF PROJ, UNIV NEBR, LINCOLN, 75- Mem: Sigma Xi; Am Geophys Union; Soc Explor Geophys; Int Glaciol Soc. Res: Radar echo sounding of polar ice thickness; geophysical survey of Ross Ice Shelf, Antartica. Mailing Add: Ross Ice Shelf Proj Mgt Off Univ of Nebr 135 Bancroft Hall Lincoln NE 68588

CLOUGH, OLIVER WENDELL, b NS, Nov 2, m 34; c 2. DENTISTRY. Educ: Dalhousie Univ, BS, 29, DDS, 32; Univ Rochester, MS, 34. Prof Exp: Pvt pract, NS, 34-36; from instr to prof oper dent, 36-72, assoc dean admin, 70-72, EMER PROF OPER DENT, MED COL VA, 72- Mem: Am Dent Asn; Am Col Dentists; Int Asn Dent Res. Res: Antibacterial action of saliva; tooth development. Mailing Add: 7104 Pinetree Rd Richmond VA 23229

CLOUGH, PHILIP JAMES, b Lewistown, Maine, Sept 3, 20; m 44. PHYSICAL INORGANIC CHEMISTRY. Educ: Bowdoin Col, BS, 43; Middlebury Col, MS, 45. Prof Exp: Instr chem, Bowdoin Col, 43-44; res chemist, Reynolds Metals Co, Mass, 45-46; res chemist & proj leader, Nat Res Corp, 46-51, asst dir res, 62-65, gen mgr, metallized prod div, 65-68, mgr, Norton Co, 68-69, tech dir, 69; dir tech develop, Gorham Res Corp, 69-71, VPRES, GORHAM INT INC, 71- Mem: Vacuum Metallizers Asn (pres, 69). Res: Vacuum reduction and purification of metals; continuous metallic coatings for decorative purposes; surface coatings for high temperature corrosion and erosion prevention; surface coating flexible webs; fire retardancy of cellulosics. Mailing Add: Gorham Int Inc Gorham ME 04038

CLOUGH, ROBERT RAGAN, b Sibley, Iowa, Feb 25, 42. MATHEMATICS. Educ: Univ Md, AB, 64; Northwestern Univ, MS, 66, PhD(math), 67. Prof Exp: Asst prof math, Univ Notre Dame, 67-73; SYSTS REP, BURROUGHS CORP, 73- Mem: Am Math Soc; Math Asn Am. Res: Algebraic topology; fiber spaces; homotopy theory. Mailing Add: Burroughs Corp 324 S Michigan Ave Chicago IL 60604

CLOUGH, STUART BENJAMIN, b Tisbury, Mass, Mar 16, 37; m 61; c 2. PHYSICAL CHEMISTRY, POLYMER CHEMISTRY. Educ: Univ Mass, Amherst, BS, 59, PhD(chem), 66; Univ Del, MChE, 61. Prof Exp: Chemist, Dewey & Almy Chem Div, W R GRace Co, 61-62; res fel chem, Univ Mass, Amherst, 62-65; chemist, US Army Natick Labs, 65-68 & US Army Mat & Mech Res Ctr, 68-70; asst prof, 70-74, ASSOC PROF CHEM, UNIV LOWELL, 74- Mem: Am Chem Soc; Am Phys Soc. Res: Structure and physical properties of bulk polymers. Mailing Add: Dept of Chem Univ of Lowell Lowell MA 01854

CLOUGH, STUART CHANDLER, b Richmond, Va, July 29, 43; m 68; c 2. ORGANIC CHEMISTRY. Educ: Univ Richmond, BS, 65; Univ Fla, PhD(chem), 69. Prof Exp: Res assoc chem, State Univ NY Buffalo, 69-71; res assoc, Philip Morris, Inc, 71-73; ASST PROF CHEM, UNIV RICHMOND, 73- Mem: Am Chem Soc. Res: Thermal and photochemical rearrangements and reactions of small organic molecules with particular interest in the formation and characterization of reactive intermediates. Mailing Add: Dept of Chem Univ of Richmond Richmond VA 23173

CLOUTIER, ELMER JOSEPH, b New Richmond, Wis, Jan 22, 13; m 49; c 3. ENTOMOLOGY. Educ: Univ Notre Dame, BS, 36, MS, 48; Univ Wis, PhD, 63. Prof Exp: Pvt sch teacher, 35-48; from instr to asst prof, St Anselm's Col, 48-54; asst prof, Creighton Univ, 54-59; assoc prof biol, Elmira Col, 62-65, prof, 65-67; PROF BIOL, STATE UNIV NY COL BROCKPORT, 67-, MEM GRAD SCH FAC, 68- Mem: AAAS; Entom Soc Am; NY Acad Sci. Res: Photoperiod and hormonal control of insect morphogenesis; biological effects of magnetism; cornborer reproductive behavior. Mailing Add: Dept of Biol Sci State Univ NY Brockport NY 14420

CLOUTIER, GILLES GEORGES, b Quebec City, Que, June 27, 28; m 54; c 1. PHYSICS. Educ: Laval Univ, BA, 49, BASc, 53; McGill Univ, MSc, 56, PhD(physics), 59. Prof Exp: Tech officer, Defence Res Bd Can, 53-54; sr mem sci staff, plasma physics, res labs, RCA Victor Co, Ltd, Can, 59-64; assoc prof physics, Univ Montreal, 64-68; sci dir, 68-71, dir res, 71-74, ASST DIR, HYDRO-QUEBEC INST RES, 74- Concurrent Pos: Mem comn, Int Sci Radio Union, Can, 64-68; assoc comt space res, Nat Res Coun Can, 64-69, mem Coun, 73- Mem: Sr mem Inst Elec & Electronics Engrs; Am Phys Soc; Can Asn Physicists (pres, 72-73). Res: Plasma physics; microwave optics; electron impact phenomena; electric propulsion; electromagnetic waves and plasmas; arc physics. Mailing Add: Hydro-Quebec Inst of Res PO Box 1000 Varennes PQ Can

CLOUTIER, LOUIS, b Cap St-Ignace, Que, May 14, 05; m 31; c 4. CHEMISTRY. Educ: Laval Univ, BA, 24, ChD & LSc, 28; Super Sch Mines, Univ Paris, DSc(chem), 32. Hon Degrees: DSc, Univ Ottawa, 71. Prof Exp: PROF GEN CHEM, LAVAL UNIV, 34- Res: Precipitation of normal and basic compounds by means of a special apparatus. Mailing Add: Dept of Chem Laval Univ Quebec PQ Can

CLOUTIER, PAUL FREDERICK, b Keene, NH, Oct 13, 36; m 58; c 4. BIOCHEMISTRY. Educ: Univ NH, AB, 58; Univ Rochester, MS, 59, PhD(biochem), 65. Prof Exp: Instr, 65-68, ASST PROF DENT RES, UNIV ROCHESTER, 68- Res: Investigations of the organic matrix of calcified tissues, particularly in teeth. Mailing Add: Dept of Dent Res Univ of Rochester Sch of Med & Dent Rochester NY 14642

CLOUTIER, ROGER JOSEPH, b North Attleboro, Mass, July 25, 30; m 54; c 5. HEALTH PHYSICS. Educ: Univ Mass, BS, 56; Univ Rochester, MS, 57. Prof Exp: Assoc engr, Westinghouse Elec Corp, 57-59; scientist, Med Div, 59-74, CHMN, SPEC TRAINING DIV, OAK RIDGE ASSOC UNIVS, 74- Mem: Health Physics Soc; Am Asn Physicists in Med; Soc Nuclear Med. Res: Radiation safety and dosimetry; medical use of radioisotopes. Mailing Add: Spec Training Div Oak Ridge Assoc Univs Oak Ridge TN 37830

CLOUTMAN, LAWRENCE DEAN, b Pratt, Kans, Nov 5, 44; m 69; c 1. ASTROPHYSICS, FLUID DYNAMICS. Educ: Univ Kans, BS, 68; Ind Univ, MA, 71, PhD(astrophys), 72. Prof Exp: STAFF MEM NUMERICAL FLUID DYNAMICS, LOS ALAMOS SCI LAB, 72- Mem: Am Astron Soc; Royal Astron Soc; Sigma Xi. Res: Methodology development in numerical fluid dynamics; theoretical astrophysics. Mailing Add: Group T3 MS 216 Los Alamos Sci Lab Los Alamos NM 87545

CLOVER, RICHMOND BENNETT, b Johnson City, NY, Jan 30, 43; m 66; c 2. MAGNETISM. Educ: Cornell Univ, BS, 65; Yale Univ, MS, 67, PhD(appl physics), 69. Prof Exp: Mem tech staff, RCA Labs, 69-72; mem tech staff, 72-73, DEPT MGR, HEWLETT-PACKARD LAB, 72- Mem: Inst Elec & Electronic Engr. Res: Investigation of magnetic bubble domain materials, devices and memory systems. Mailing Add: Hewlett-Packard Co 3500 Deer Creek Rd Palo Alto CA 94304

CLOVIS, JAMES S, b Waynesburg, Pa, Aug 14, 37; m 70. PHYSICAL ORGANIC CHEMISTRY. Educ: Waynesburg Col, BS, 59; Calif Inst Technol, PhD(chem), 63. Prof Exp: Fel, Univ Munich, 62-63; mem staff, 63-70, lab head, process chem, 70-74, PROJ LEADER, POLLUTION CONTROL RES, ROHM AND HAAS CO, 74- Mem: Am Chem Soc. Res: Benzidine rearrangement; 1, 3-dipolar addition; phosphorus chemistry; plastics research; process research; pollution control research. Mailing Add: Rohm and Haas Co Spring House PA 19477

CLOVIS, JESSE FRANKLIN, b Clarksburg, WVa, Jan 31, 21; m 48; c 3. SYSTEMATIC BOTANY. Educ: WVa Univ, BSF, 47, MS, 52; Cornell Univ, PhD, 55. Prof Exp: Instr bot, Univ Conn, 55-57; from asst prof to assoc prof biol, 57-72, pre-med adv, 63-70, PROF BIOL, W VA UNIV, 72- Mem: Bot Soc Am; Am Soc Plant Taxonomists. Res: Aquatic plants; speciation. Mailing Add: Dept of Biol W Va Univ Morgantown WV 26506

CLOW, JAMES RODGERS, b Dayton, Ohio, Nov 7, 38. PHYSICS. Educ: Miami Univ, BA, 61, MA, 62; Yale Univ, MS, 64, PhD(physics), 67. Prof Exp: Asst prof physics, Tex A&M Univ, 67-68 & Mass Inst Technol, 68-74; MEM STAFF, C S DRAPER LAB, 74- Mem: Am Phys Soc. Res: Persistent current measurements of superfluid density and critical velocities in helium II; superfluid gyroscope. Mailing Add: C S Draper Lab D110 142 275 Massachusetts Ave Cambridge MA 02139

CLOWER, DAN FREDERIC, b Crystal Springs, Miss, Mar 9, 28; m 51; c 4. ENTOMOLOGY. Educ: La State Univ, BS, 49; Cornell Univ, PhD(econ entom), 55. Prof Exp: Asst entom, Cornell Univ, 50-55; asst entomologist, 55-58, assoc prof entom, 58-63, PROF ENTOM, LA STATE UNIV, BATON ROUGE, 63- Mem: Entom Soc Am. Res: Cotton entomology; ecological relationships; forest entomology. Mailing Add: Dept of Entom Life Sci Bldg La State Univ Baton Rouge LA 70803

CLOWER, EUGENE WESTON, physical chemistry, see 12th edition

CLOWERS, CHURBY CONRAD, JR, b Little Rock, Ark, Mar 23, 34; m 59; c 2. ANALYTICAL CHEMISTRY. Educ: Univ Mo, Kansas City, BS, 57; Univ Mo, Columbia, PhD(anal chem), 66. Prof Exp: Res chemist, Textile Fibers Dept, Nylon Tech Div, E I du Pont de Nemours & Co, Inc, 66-68; ANAL RES CHEMIST, ANAL CHEM DEPT, WM S MERRELL CO, 68- Mem: Am Chem Soc. Res: Pharmaceutical, biochemical and metabolic analytical research; chemical instrumentation; physicochemical studies of drug degradation; chromatographic separations; spectra-structure relationships. Mailing Add: Dept of Anal Chem Wm S Merrell Co Cincinnati OH 45215

CLOWES, RONALD MARTIN, b Calgary, Alta, Mar 18, 42; m 68. SEISMOLOGY. Educ: Univ Alta, BSc, 64, MSc, 66, PhD(geophys), 69. Prof Exp: Nat Res Coun Can postdoctoral & hon res fel geophys, Australian Nat Univ, 69-70; ASST PROF GEOPHYS, UNIV BC, 70- Concurrent Pos: Consult, Horton Maritime Explor Ltd, 75- Honors & Awards: Soc Explor Geophysicists Award, 68. Mem: Am Geophys Union; Seismol Soc Am; Can Soc Explor Geophysicists; Can Geophys Union. Res: Structure and properties of the earth's crust and upper mantle, at sea and on land, from detailed analysis of reflected and refracted seismic waves generated by chemical explosions. Mailing Add: Dept of Geophys & Astron Univ of BC Vancouver BC Can

CLOWES, ROYSTON COURTENAY, b Swansea, Wales, Sept 11, 21; m 52; c 3. MICROBIOLOGY, GENETICS. Educ: Univ Birmingham, BSc, 48, PhD(physiol), 51, DSc(genetics), 65. Prof Exp: Res assoc microbiol, Wright-Fleming Inst, St Mary's Hosp, London, Eng, 51-57; staff mem, Microbial Genetics Res Unit, Med Res Coun, Hammersmith Hosp, London, 57-65; prof biol, Grad Res Ctr of Southwest, 65-69, head div, 68-74, PROF BIOL, UNIV TEX, DALLAS, 69- Concurrent Pos: Damon Runyon Cancer Res fel, 55-56; vis prof, virus lab, Univ Calif, Berkeley, 61; mem, Microbial Chem Study Sect, NIH, 71-73, chmn, 73-75. Mem: Am Soc Microbiol; Genetics Soc Am; Brit Soc Gen Microbiol; Brit Genetical Soc. Res: Microbial genetics; molecular biology; biology of bacterial plasmids. Mailing Add: Div of Biol Univ Tex at Dallas PO Box 688 Richardson TX 75080

CLOYD, GROVER DAVID, b Mosheim, Tenn, Oct 25, 18; m 74; c 3. VETERINARY MEDICINE. Educ: Auburn Univ, DVM, 42. Prof Exp: Self employed, Vet Med Pract, 47-55; dir vet serv, Ky Chem Indust, 55-57; asst dir field res pharmaceut prod, Richardson-Merrell Inc, 57-59, dir field res, 59-62, asst dir res, 62-65, sci develop aide gen mgr, 65-68; vet med dir, 68-71, dir vet med, 71-72, DIR VET MED & CONSUMER PROD RES & DEVELOP, A H ROBINS CO, INC, 72- Mem: Indust Vets Asn (secy, 70-74, pres, 75-76). Res: Pharmaceutical products research and development, veterinary and human. Mailing Add: 2024 Floyd Ave Richmond VA 23220

CLUFF, CARWIN BRENT, b Central, Ariz, Feb 20, 35; m 68; c 5. HYDROLOGY. Educ: Univ Ariz, BS, 59, MS, 61. Prof Exp: Asst engr civil eng, Calif Dept Water Resources, 61-62; res assoc hydrol, Water Resources Res Ctr, Univ Ariz, 62-63, from asst hydrologist to assoc hydrologist, 63-75; expert in hydrol, Food & Agr Orgn, UN, 75-76; ASSOC HYDROLOGIST, WATER RESOURCES RES CTR, UNIV ARIZ, 76- Mem: Am Soc Civil Engrs; Am Water Resources Asn. Res: Evaporation and seepage control; water harvesting; reuse of municipal waste water. Mailing Add: Water Resources Res Ctr Univ of Ariz Tucson AZ 85721

CLUFF, EDWARD FULLER, b Dedham, Mass, Feb 14, 28; m 56; c 2. ORGANIC CHEMISTRY. Educ: Mass Inst Technol, BS, 49, PhD(org chem), 52. Prof Exp: Res chemist, 52-69, develop supvr, 69-70, res div head, 70-74, DEVELOP SUPT, E I DU PONT DE NEMOURS & CO, 74- Res: Polymer chemistry; elastomers, including polyurethanes and fluoroelastomers; hydrocarbon elastomers and olefin polymerization via transition metal catalysts. Mailing Add: 1616 Windybush Rd Wilmington DE 19810

CLUFF, LEIGHTON EGGERTSEN, b Salt Lake City, Utah, June 10, 23; m 44; c 2. ALLERGY, INFECTIOUS DISEASES. Educ: George Washington Univ, MD, 49. Prof Exp: House officer med, Johns Hopkins Hosp, Baltimore, 49-50; asst res physician, Hosp, Duke Univ, 50-51; asst res physician, Johns Hopkins Hosp, 51-52; asst physician & vis investr, Rockefeller Inst, 52-54; res physician, Johns Hopkins Hosp, 54-55, from instr to prof, 54-66; prof med & chmn dept, Col Med, Univ Fla, Gainesville, 66-76; VPRES, ROBERT WOOD JOHNSON FOUND, PRINCETON, 76- Concurrent Pos: Markle scholar, 55-62; consult, Food & Drug Admin; consult ed, Dermatol Dig; chmn training grant comn, Nat Inst Allergy & Infectious Dis; mem, Nat Res Coun-Nat Acad Sci Drug Res Bd; mem, Nat Comn Pharm & Pharm Educ; chmn comt biomed res & patient care in the Vet Admin, Nat Acad Sci-Nat Res Coun; expert adv panel on bact dis, WHO. Honors & Awards: Ordronaux Award, 49. Mem: Am Fedn Clin Res; Am Soc Clin Invest; Soc Exp Biol & Med; Am Acad Allergy; Am Phys Asn. Res: Elucidation of the role of psychologic factors in convalescence from acute infection; studies of the epidemiology of staphylococcal infections; pathogenesis of fever due to bacterial pyrogens; mechanism of host injury in infection; studies of the epidemiology of adverse drug reactions. Mailing Add: Robert Wood Johnson Found US Hwy No 1 Princeton NJ 08540

CLUFF, LLOYD STERLING, b Provo, Utah, Sept 29, 33; c 2. GEOLOGY. Educ: Univ Utah, BS, 60. Prof Exp: Geologist, Lottridge Thomas & Assoc, 60; staff geologist, 60-65, assoc & chief eng geologist, 65-71, VPRES, PRIN & DIR, WOODWARD-CLYDE CONSULT, 71- Concurrent Pos: Consult, Venezuelan Pres Earthquake Comn, 67-; mem consult bd, San Francisco Bay Conserv & Develop Comn, 68-73; mem consult panel to President & Secy Interior for Santa Barbara Oil Leak, 69; mem state Calif joint comt, Seismic Safety & Gov Earthquake Coun, 72-75; mem US Geol Surv earthquake adv panel, Nat Acad Sci comt seismol & Stanford Res Inst oversight comt earthquake probs, 75-; mem consult bd, Int Atomic Energy Agency, Vienna. Honors & Awards: Hogentogler Award, Am Soc Testing & Mat, 65. Mem: Asn Eng Geologists (pres, 68-69); Earthquake Eng Res Inst; Int Asn Eng Geologists (vpres, 70-74); Seismol Soc Am; Struct Engrs Asn. Res: Active faults, earthquake and geologic hazards. Mailing Add: Two Embarcadero Suite 700 San Francisco CA 94111

CLUM, FLOYD MYRON, b Thornville, Ohio, Feb 1, 19; m 41; c 2. PATHOLOGY, MYCOLOGY. Educ: Ohio State Univ, BSc, 42; Univ Wis, MSc, 50; Mich State Col, PhD(plant path), 54. Prof Exp: Asst bot, Univ Wis, 46-47; micro-technician, Triarch Prod, 47-50; asst bot & plant path, Mich State Col, 50-53, instr plant path, 53; plant scientist, US Govt, Washington, DC, 53-64, sci adminr, Off Resource Develop, Bur State Serv, 64-67, chief pub health training sect, Div Allied Health Manpower, Bur Health Manpower, NIH, 67-70, prog consult allied health prof br, 70-74; PROG CONSULT, EDUC DEVELOP BR, DIV ASSOC HEALTH PROF, HEALTH RESOURCES ADMIN, 74- Mem: Sigma Xi; Am Inst Biol Sci; Asn Schs Allied Health Professions. Res: Haustoria of Peronosporaceae; damping-off control; science administration; public health training; environmental health training; allied health training; health manpower. Mailing Add: 10527 Montrose Ave Apt 102 Bethesda MD 20014

CLUM, HAROLD HAYDN, b Cleveland, Ohio, Feb 16, 94; m 24; c 2. PLANT PHYSIOLOGY. Educ: Oberlin Col, AB, 17; Cornell Univ, PhD(plant physiol), 24. Prof Exp: Asst bot, Cornell Univ, 19-23; instr, Univ Mich, 23-24; prof, Univ PR, 24-26; asst prof, Syracuse Univ, 26-27; from instr to assoc prof, Hunter Col, 27-52, chmn dept, 44-56, prof biol sci, 53-63; hon plant physiologist, NY Bot Garden, New York, 63-67; HON RES ASSOC, KITCHAWAN RES LAB, BROOKLYN BOT GARDEN, 67- Mem: AAAS; Bot Soc Am; Am Phytopath Soc; Am Soc Plant Physiologists. Res: Effects of cytokinins in plant tissues. Mailing Add: 40 Smith St Chappaqua NY 10514

CLUNE, FRANCIS JOSEPH, JR, b San Antonio, Tex, Feb 23, 30; m 57; c 3. ARCHAEOLOGY, ANTHROPOLOGY. Educ: Univ Calif, Berkeley, BA, 57; Univ Calif, Los Angeles, PhD(anthrop), 63. Prof Exp: Teaching asst, Univ Calif, Los Angeles, 58-60; asst prof, San Fernando Valley State Col, 60-61; instr, Univ Ga, 61-63; asst prof anthrop, 63-67, ASSOC PROF ANTHROP, STATE UNIV NY COL BROCKPORT, 67- Concurrent Pos: Mem staff coop res proj, Anthrop Curric Proj, Univ Ga, 64-67. Res: Teaching anthropology in the elementary schools; mesoamerican archaeology; archaeology in the New World; technology. Mailing Add: Dept of Anthrop State Univ of NY Col Brockport NY 14420

CLUTTER, DALE R, physical chemistry, molecular spectroscopy, see 12th edition

CLUTTER, JEROME LEE, b Washington, Pa, Mar 19, 34; m 54; c 3. BIOMETRICS, OPERATIONS RESEARCH. Educ: Mich State Univ, BS, 56; Duke Univ, MF, 57, DF(biomet), 61. Prof Exp: Math statistician, Southeastern Forest Exp Sta, US Forest Serv, 57-58 & 59-63; prog analyst, Comput Ctr, Duke Univ, 58-59; opers analyst, Res Anal Corp, 63; ASSOC PROF BIOMET & OPERS RES, SCH FORESTRY, UNIV GA, 63- Mem: Biomet Soc; Opers Res Soc Am; Soc Am Foresters. Res: Prediction of growth and yield of forests; applications of operations research techniques in forest management. Mailing Add: Sch of Forestry Univ of Ga Athens GA 30601

CLUTTER, MARY ELIZABETH, b Charleroi, Pa; m. BOTANY. Educ: Allegheny Col, BS, 53; Univ Pittsburgh, MS, 57, PhD(bot), 60. Prof Exp: Res assoc, 61-73, LECTR BIOL, YALE UNIV, 65-, SR RES ASSOC, 73- Concurrent Pos: Prog dir, NSF, 76- Mem: AAAS; Am Soc Cell Biol; Am Soc Plant Physiologists; Scand Soc Plant Physiol; Soc Develop Biol. Res: Function of polytene chromosomes in plant embryo development. Mailing Add: OML-Yale Univ New Haven CT 06520

CLUTTERHAM, DAVID ROBERT, b Chicago, Ill, Feb 10, 22; m 45; c 4. MATHEMATICS. Educ: Cornell Col, BA, 45; Univ Ariz, MS, 48; Univ Ill, PhD(math), 53. Prof Exp: Asst Math, Univ Ariz, 46-48 & Univ Ill, 48-49, asst digital comput, 50-53; design specialist, Convair Div, Gen Dynamics Corp, 53-59, sect chief comput design, Martin-Orlando, 59-63, mgr, Info Sci Dept, 63-64; mem tech staff, Bunker-Ramo Corp, Calif, 64-65; sr scientist, Radiation Inc, Fla, 65-68; PROF MATH SCI & HEAD DEPT, FLA INST TECHNOL, 68- Mem: Am Math Soc; Asn Comput Mach; Inst Elec & Electronics Eng. Res: Digital computers and computing techniques. Mailing Add: Dept of Math Sci Fla Inst of Technol Melbourne FL 32901

CLUXTON, DAVID H, b Martinsville, Ohio, 1943. EXPERIMENTAL PHYSICS. Educ: Wilmington Col, AB, 65; Mich State Univ, MS, 67; Kent State Univ, PhD(physics), 72. Prof Exp: ASST PROF PHYSICS, RUSSELL SAGE COL, 72- Concurrent Pos: Consult, Gen Elec Res & Develop Ctr, Schenectady, 75- Mem: Am Phys Soc; AAAS. Res: Applications of physics techniques to biological systems; intensity fluctuation spectroscopy. Mailing Add: 23 Grissom Dr Clifton Park NY 12065

CLYDE, DAVID F, b Meerut, India, Jan 13, 25; US citizen; m 49; c 2. MEDICINE. Educ: Univ Kans, BA, 45; McGill Univ, MD & CM, 48; Univ London, dipl trop med & hyg, 52, PhD(med), 63. Prof Exp: Med officer, Govt Tanganyika, 49-60, malariologist, 60-64; epidemiologist & dep chief med officer, Govt Tanzania, 64-66; from assoc prof to prof int med, Univ Md, Baltimore, 66-75, dir inst int med, 67-75; PROF TROP MED & MED PARASITOL, LA STATE UNIV, NEW ORLEANS, 75- Concurrent Pos: Mem expert comt on malaria, WHO, 64- Mem: Am Soc Trop Med & Hyg; Royal Soc Trop Med & Hyg; Tanzania Soc. Res: Epidemiology and chemotherapy of malaria and associated tropical diseases. Mailing Add: La State Univ Med Ctr 1542 Tulane Ave New Orleans LA 70112

CLYDE, WALLACE ALEXANDER, JR, b Birmingham, Ala, Nov 7, 29; m 53; c 3. PEDIATRICS, INFECTIOUS DISEASES. Educ: Vanderbilt Univ, BA, 51, MD, 54. Prof Exp: Intern pediat, Univ Hosp, Vanderbilt Univ, 54-55; asst resident, NC Baptist Hosp, 55-56; resident, Univ Hosp, Vanderbilt Univ, 56-57; trainee infectious dis, 61-62, from instr to assoc prof pediat, 62-72, assoc prof bact, 68-72, PROF PEDIAT & BACT, UNIV NC, CHAPEL HILL, 72- Concurrent Pos: Nat Inst Allergy & Infectious Dis fel prev med, Case Western Reserve Univ, 59-61, career develop award, 63-73; assoc mem comn acute respiratory dis, Armed Forces Epidemiol Bd, 63-72; mem bact-mycol study sect, Div Res Grants, NIH, 66-70; vis assoc prof path,Yale Univ, 71-72. Mem: Am Soc Microbiol; Soc Pediat Res; Am Soc Clin Invest; Infectious Dis Soc Am; Am Pediat Soc. Res: Infectious diseases of children, especially nonbacterial respiratory infections; relationship of Mycoplasmataceae to human disease; respiratory disease pathogenesis. Mailing Add: Dept of Pediat Univ of NC Sch of Med Chapel Hill NC 27514

CLYDESDALE, FERGUS MACDONALD, b Toronto, Ont, Feb 19, 37; m 61; c 2. CHEMISTRY, FOOD SCIENCE. Educ: Univ Toronto, BA, 60, MA, 62; Univ Mass, PhD(food sci), 66. Prof Exp: Chemist, Can Industs Ltd, 60; physiol chemist, Can Defence Res Med Lab, 62; from fel to asst prof food sci, 66-72, ASSOC PROF FOOD SCI & NUTRIT, UNIV MASS, AMHERST, 72- Mem: AAAS; Inst Food Technologists; Am Chem Soc; Inter-Soc Color Coun. Res: Basic chemical changes in processed foods and their effect on quality; basic color measurement problems involved with foods. Mailing Add: Dept of Food Sci & Nutrit Univ of Mass Amherst MA 01002

CLYMER, HAROLD ARTHUR, b Philadelphia, Pa, May 19, 17; m 41. RESEARCH ADMINISTRATION. Educ: Philadelphia Col Pharm, BSc, 39. Prof Exp: Res pharmacist, Smith Kline Corp, 39-42, sr res pharmacist, 42-48, res lab adminr, 48-51, adminr develop, 51-60, assoc dir res & develop, 60-66, vpres, 66-71, vpres corp bus develop, 71-74; INDUST RES & DEVELOP CONSULT, 74- Concurrent Pos: Vpres, Menley & James Labs, 61-64. Mem: AAAS; Am Pharmaceut Asn; Am Chem Soc. Res: Development of new pharmaceuticals; regulatory economics and the impact of government regulation on innovation. Mailing Add: 1215 Gilvert Rd Meadowbrook PA 19046

CLYNE, ROBERT MARTIN, b Bridgeport, Conn, July 23, 18; m 43; c 3. INDUSTRIAL MEDICINE. Educ: Fordham Univ, BS, 39; Cornell Univ, MD, 43. Prof Exp: Intern med, Lenox Hill Hosp, 43-44; from asst resident to resident internal med, Lincoln Hosp, 44-45; resident, Bronx Vet Admin Hosp, 46-47; physician, 48-50, chief physician clin, 50-55, asst to the med dir, 55-56, asst med dir, 56-58, dir employee health, 58-67, CORP MED DIR, AM CYANAMID CO, 67- Concurrent Pos: Lectr alcoholism, Rutgers Univ, Univ Utah & Univ Miami; lectr occup health, Colby Col. Honors & Awards: Spec Award, Indust Med Asn, 72. Mem: Am Col

Physicians; Am Acad Occup Med; Am Occup Med Asn; Am Col Prev Med; AMA. Mailing Add: Am Cyanamid Co 859 Berdan Ave Wayne NJ 07470

CLYNES, MANFRED, b Vienna, Austria, Aug 14, 25; US citizen; div; c 3. NEUROPSYCHOLOGY, SENTICS. Educ: Univ Melbourne, BEngSci, 46, DSc(neurosci), 64; Juilliard Sch Music, MS, 49. Prof Exp: Chief study teacher pianoforte, Univ Melbourne, 50-52; instr music, Princeton Univ, 52-53; chief mathematician & comput specialist, Bogue Elec Mfg Co, 54-56; chief res scientist & dir biocybernetic labs, Rockland State Hosp, 56-73; DIR, BIOCYBERNETIC INST, 73- Concurrent Pos: Consult, Sonomedic Corp; staff consult, Feedback Syst Dynamics & Electronics, Bogue Elec Mfg Co, 56-; pres & chmn bd, Mnemotron Corp, 60-62; vpres, Tech Measurement Corp, 61-62. Mem: Soc Neurosci; Am Sentic Asn (pres, 74-); Inst Elec & Electronics Engrs; Am Phys Soc; Asn Hosp Psychologists. Res: Brain function; application of automatic control system theory to biological systems, neurophysiology, circulation, sentics; biologic basis of dynamic communication of emotions and qualities, neurophysiology of musical language, evolution of communication. Mailing Add: Biocybernetics Inst 8571 Villa La Jolla Dr La Jolla CA 92037

CMEJLA, HOWARD EDWARD, b Milwaukee, Wis, Dec 25, 26; m 48; c 6. PARASITOLOGY, ENTOMOLOGY. Educ: Univ Wis, BS, 50, MS, 51, PhD(entom, zool), 54. Prof Exp: Collabr, Div Apicult & Biol Control, USDA, Wis, 51-54; parasitologist, Parasitol Dept, Abbott Labs, 54-58, Food & Drug Admin liaison, 58-63; dir regulatory liaison, 63-70, VPRES PROD PLANNING & DEVELOP, AYERST LABS, 69- Concurrent Pos: Food & Drug Admin liaison. Mem: Am Soc Parasitologists; Entom Soc Am; Am Soc Trop Med & Hyg. Res: Drugs-product planning and development; amebiasis; apiculture; bee diseases and colony management. Mailing Add: Ayerst Labs 685 Third Ave New York NY 10017

COACHMAN, LAWRENCE KEYES, b Rochester, NY, Apr 25, 26; div; c 4. OCEANOGRAPHY. Educ: Dartmouth Col, AB, 48; Yale Univ, MF, 51; Univ Wash, PhD(oceanog), 62. Prof Exp: Hydrographer & oceanogr, Dartmouth Col Blue Dolphin Labrador Expeds, 50-55; sr scientist gases in glacier ice, Univ Oslo, 55-57; asst, 57-62, from asst prof to assoc prof, 62-72, PROF OCEANOG, UNIV WASH, 72- Concurrent Pos: Sr scientist, Arctic Inst NAm Expeds, Greenland, 58 & 63, Bering & Chukchi Seas, 64, 66 & 69-75, Cent Arctic Ocean, 70-72; chmn, US-USSR Oceanog Exchange Deleg, 64; Arctic Inst NAm/McGill Univ vis prof, 67. Mem: AAAS; Am Geophys Union; Am Soc Limnol & Oceanog; fel Arctic Inst NAm. Res: Physical oceanography of Arctic Ocean and peripheral seas; gas enclosures in glacier ice; dissolved gasses in sea water; oceanography of coastal, fjord and ice-covered waters. Mailing Add: Dept of Oceanog Univ of Wash Seattle WA 98105

COAD, L KEITH, b Glen Elder, Kans, Mar 31, 17; m 42; c 2. OPERATIONS RESEARCH, STATISTICS. Educ: Univ Kans, AB, 38; Univ Minn, PhD(phys chem), 42. Prof Exp: Teaching asst chem, Univ Minn, 38-42; res chemist, Gen Motors Res Lab, 42-46; phys res supvr, S C Johnson & Son, Inc, 47-52; sr res chemist, Pure Oil Co, 53-54; asst prof chem, Wis State Univ, 54-57; res assoc opers res, Case Western Reserve Univ, 57-58; STATIST ANALYST, 3 M Co, 59- Concurrent Pos: Consult tech writing, Erieside Assocs, Ohio, 59; lectr, Univ Minn, 66- Mem: Opers Res Soc Am. Res: Management information systems; statistics in engineering and research. Mailing Add: IS & DP Dept 0358 Bldg 224-4E 3M Ctr St Paul MN 55101

COAD, PETER, organic chemistry, see 12th edition

COAD, RAYLENE ADAMS, organic chemistry, see 12th edition

COAHRAN, DAVID RICHARD, biophysics, analytical chemistry, see 12th edition

COAKLEY, CHARLES SEYMOUR, b Washington, DC, July 4, 14; m; c 5. ANESTHESIOLOGY. Educ: George Washington Univ, MD, 37; Am Bd Anesthesiol, dipl, 48. Prof Exp: From instr to assoc prof, 39-49, PROF ANESTHESIOL & CHMN DEPT, SCH MED, GEORGE WASHINGTON UNIV, 49- Concurrent Pos: Consult, Walter Reed Army Med Ctr, Vet Admin Hosp, Washington, DC & NIH; med consult, CARE-Medico. Mem: AMA; Am Col Anesthesiol; Asn Am Med Cols; Am Soc Anesthesiol; Asn Univ Anesthetists. Res: Use of monitors for anesthetized and critically ill patients; applications of computer of continuous automated monitoring. Mailing Add: Dept of Anesthesiol George Washington Univ Hosp Washington DC 20037

COAKLEY, MARY PETER, b South Amboy, NJ, July 18, 15. PHYSICAL CHEMISTRY, INORGANIC CHEMISTRY. Educ: Georgian Court Col, AB, 47; Univ Notre Dame, MS, 53, PhD, 55. Prof Exp: Parochial high sch teacher, 42-47; TEACHER CHEM & CHMN DEPT, GEORGIAN COURT COL, 47- Concurrent Pos: Mem atomic energy proj, Univ Notre Dame, 54-55; res grants, AEC, 57-60 & 61-63, Petrol Res Fund, 66-68 & NSF, 68-70; res, Univ Calif, Berkeley, 68. Mem: Am Chem Soc; NY Acad Sci. Res: Nuclear magnetic resonance studies of tin complexes; spectroscopy; ultraviolet and infrared absorption; spectra of metal chelates. Mailing Add: Dept of Chem Georgian Court Col Lakewood NJ 08701

COALSON, JACQUELINE JONES, b Oklahoma City, Okla, Mar 12, 38; div; c 3. PATHOLOGY. Educ: Okla Baptist Univ, BS, 60; Univ Okla, MS, 63, PhD(path), 65. Prof Exp: Teaching asst histol & embryol, 60-63, from res asst to assoc prof, 63-75, PROF PATH, MED CTR, UNIV OKLA, 75- Mem: Am Thoracic Soc; Am Asn Path & Bact. Res: Electron microscopy and histochemistry of normal and diseased lungs of both human and experimental animals. Mailing Add: Dept of Path Univ of Okla Med Ctr Oklahoma City OK 73104

COALSON, JAMES ARTHUR, b Winters, Tex, Sept 6, 42; m 61; c 2. ANIMAL NUTRITION. Educ: Abilene Christian Col, BS, 66; Okla State Univ, MS, 69, PhD(animal nutrit), 71. Prof Exp: Res assoc, NC State Univ, 70-71, asst prof animal sci, 72-73; swine res specialist, 73-74, sr swine nutritionist, 74-75, SWINE FEEDS DIR, CENT SOYA, 75- Mem: Am Soc Animal Sci. Res: Effect of environment and management on growth and survival of neonatal pigs. Mailing Add: Cent Soya 1200 N Second St Decatur IN 46733

COALSON, RICHARD L, polymer physics, see 12th edition

COALSON, ROBERT ELLIS, b Hobart, Okla, Dec 7, 28; m 62; c 3. ANATOMY. Educ: Univ Okla, BS, 49, MS, 51, PhD(med sci), 55. Prof Exp: Instr anat, Sch Nursing, Univ Okla, 52-54; instr, Sch Med, Vanderbilt Univ, 57-60; from asst prof to assoc prof, 60-70, PROF ANAT, SCH MED, UNIV OKLA, 70-, ASST PROF PATH, 68- Concurrent Pos: Consult, Meharry Med Col, 58-60. Mem: Am Soc Zool; Tissue Cult Asn. Res: Comparative embryology, anatomy and histology; tissue transplantation. Mailing Add: Dept of Anat Univ of Okla Med Ctr Oklahoma City OK 73104

COAN, EUGENE VICTOR, b Los Angeles, Calif, Mar 26, 43. ENVIRONMENTAL SCIENCES, MALACOLOGY. Educ: Univ Calif, Santa Barbara, AB, 64; Stanford Univ, PhD(biol sci), 69. Prof Exp: Dir polit activ, Zero Pop Growth, 69-70; consult, 70-75, MAJOR ISSUES SPECIALIST, THE SIERRA CLUB, 75- Mem: Am Malacol Union. Res: United States energy policy; international aspects of environmental movement; protection of endangered species; taxonomy and distribution of northwest American bivalves. Mailing Add: San Francisco CA

COAN, STEPHEN B, b New York, NY, Apr 8, 21; m 42; c 2. ORGANIC CHEMISTRY. Educ: Univ Mich, BS, 41; Polytech Inst Brooklyn, MS, 50, PhD(chem), 54. Prof Exp: Anal chemist, Gen Chem Co, 41-42, TNT control chemist, 42-43, res chemist, 46; develop chemist, 46-51, res chemist, 51-54, patents-res liaison, 51-63, assoc dir patent dept, 63-65, DIR PATENT DEPT, SCHERING CORP, 65- Mem: Am Chem Soc. Res: Organic and medicinal research; organic synthesis; pharmaceuticals. Mailing Add: Schering-Plough Corp 2000 Galloping Hill Rd Kenilworth NJ 07033

COARTNEY, JAMES S, b Coles Co, Ill, Sept 3, 38; m 64; c 2. PLANT PHYSIOLOGY, WEED SCIENCE. Educ: Eastern Ill Univ, BSEd, 60; Purdue Univ, MS, 63, PhD(plant physiol), 67. Prof Exp: Res asst plant physiol, Purdue Univ, 60-66; asst prof, 66-75, ASSOC PROF PLANT PHYSIOL, VA POLYTECH INST & STATE UNIV, 75- Mem: Weed Sci Soc Am. Res: Plant growth regulation and mode of action of herbicides. Mailing Add: Dept of Plant Path & Physiol Va Polytech Inst & State Univ Blacksburg VA 24061

COASH, JOHN RUSSELL, b Denver, Colo, Sept 24, 22; m 48; c 3. GEOLOGY. Educ: Colo Col, BA, 47; Univ Colo, MA, 49; Yale Univ, PhD, 54. Prof Exp: From asst prof to assoc prof geol, Bowling Green State Univ, 49-61, chmn dept, 54-64, prof, 61-64, asst to provost & dir hon prog, 63-65, dir res, 65-66; assoc prog dir, N SF, 66-68; DEAN SCH NAT SCI, CALIF STATE COL, BAKERSFIELD, 68- Concurrent Pos: NSF vis lectr, 62, 63, 64 & 65; consult, geostudy & earth sci teacher training panel, Am Geol Inst; panelist, NSF, 68- & NIH, 72-73. Mem: AAAS; Geol Soc Am; Am Asn Petrol Geol; Nat Asn Geol Teachers; Nat Sci Teachers Asn. Res: Field geology of northern Nevada and central Rocky Mountains; stratigraphic and structural geology. Mailing Add: Sch of Nat Sci Calif State Col Bakersfield CA 93309

COATE, WILLIAM BLEECKER, b Pasadena, Calif, Sept 23, 21; m 41; c 2. INHALATION TOXICOLOGY. Educ: Cornell Univ, AB, 47, PhD(psychol), 50. Prof Exp: Asst prof psychol, Harpur Col, 50-55, chmn dept, 53-54; from asst prof to assoc prof, Wellesley Col, 55-63; assoc dir div pharmacol, 63-68, proj dir, 68-72, DIR INHALATION TOXICOL, HAZLETON LABS AM, INC, 72- Mem: Am Psychol Asn; Psychonomic Soc; Am Indust Hyg Asn. Res: Psychopharmacology; method development; impairment of pulmonary function by inhalants. Mailing Add: Hazleton Labs Am Inc 9200 Leesburg Turnpike Vienna VA 22180

COATES, ANTHONY GEORGE, b Staines, Eng, May 20, 36; m 61; c 1. PALEONTOLOGY, STRATIGRAPHY. Educ: Univ London, BSc, 59, PhD(geol), 63. Prof Exp: Geologist, Jamaican Geol Surv, 62-64; lectr geol, Univ West Indies, 64-67; assoc prof, 67-74, PROF GEOL, GEORGE WASHINGTON UNIV, 74- Concurrent Pos: Extra-mural lectr, Univ West Indies, 63-67, external examr, 70-; res assoc, Smithsonian Inst, 68-; NSF sci equip grant, 70; ed jour, Geol Soc Jamaica, 64-66. Mem: Paleont Soc Am; fel Geol Soc London; Geol Soc France; fel Geol Soc Jamaica. Res: Paleontology, stratigraphy and sedimentation of ordovician of Normandy; cretaceous Caribbean corals and biostratigraphy; Jamaican cretaceous stratigraphy. Mailing Add: Dept of Geol George Washington Univ Washington DC 20006

COATES, ARTHUR DONWELL, b Steubenville, Ohio, June 14, 28; m 57; c 3. PHYSICAL CHEMISTRY, FUEL SCIENCE. Educ: Col Steubenville, BS, 50; Univ Del, MS, 61. Prof Exp: Practice engr chem & metall, Wheeling Steel Corp, 50-51; chemist, 51-55, chief ignition sect, Combustion & Incendiary Effects Br, 55-57, Spec Prob Sect, Nuclear Physics Br, 57-59 & Radiation Damage Sect, 60-70, chief, Methodology Sect, Combustion & Incendiary Effects Br, 70-75, TECH ASST TO CHIEF DETONATION & DEFLAGRATION DYNAMICS LAB, BALLISTICS RES LABS, ABERDEEN PROVING GROUND, 75- Mem: AAAS; Am Chem Soc; fel Am Inst Chem; Combustion Inst; NY Acad Sci. Res: Fuels; combustion; powdered metals; pyrophoric materials; propellants; explosives; mass spectrometry; thin film physics; ignition; chemistry of exothermic reactions; thermal analysis. Mailing Add: Deton & Deflag Dynamics Lab Ballistic Res Labs Aberdeen Proving Ground MD 21005

COATES, DONALD ROBERT, b Grand Island, Nebr, July 23, 22; m 44; c 3. GEOLOGY. Educ: Col Wooster, BA, 44; Columbia Univ, MA, 48, PhD(geol), 56. Prof Exp: Asst geol, Columbia Univ, 46-48; asst prof & head dept, Earlham Col, 48-51, geologist, Ground Water Br, US Geol Surv, 51-54; chmn dept, 54-63, from instr to assoc prof, 54-63, PROF GEOL, STATE UNIV NY BINGHAMTON, 63- Concurrent Pos: Party chief, Ind Geol Surv, 49; lectr, Ind Univ, 50; res geologist, Off Naval Res, 54; consult, Chernin & Gold, NY, 55-; geologist, Gen Hydrol Br, US Geol Surv, 58-60; vis prof, Cornell Univ, 58, 60 & 61; State Univ NY Res Found fels, 61 & 66, grants-in-aid, 62-65; vis prof, Univ Ill, 63; assoc prof dir, NSF, 63-64, consult, 64-; vis geoscientist for Am Geol Inst, 63-65; consult, US Army Corps Engrs, 65-66, NY State Attorney Gen, 65-, US Dept Com, 72-75 & Consol Edison of New York, 75; proj dir, NY State Atomic & Space Develop Authority, 75. Honors & Awards: Award for Sustained Superior Performance, NSF, 64. Mem: Nat Asn Geol Teachers; Am Geophys Union; Geol Soc Am; Am Inst Prof Geologists. Res: Geomorphology; environmental geology and environmental lawsuits; glacial geology of eastern United States; analysis of man's changes of rivers and coasts; evaluation of water and earth surface resources. Mailing Add: Dept of Geol Sci State Univ of NY Binghamton NY 13901

COATES, GEOFFREY EDWARD, b London, Eng, May 14, 17; m 51; c 2. ORANOMETALLIC CHEMISTRY, INORGANIC CHEMISTRY. Educ: Oxford Univ, BA, 38, BSc, 39, MA, 42; DSc(chem), Bristol Univ, 54. Prof Exp: Res chemist, Magnesium Metal Corp, Eng, 40-45; lectr chem, Bristol Univ, 45-53; prof chem & head dept, Univ Durham, 53-68; PROF CHEM & HEAD DEPT, UNIV WYO, 68- Concurrent Pos: Consult, Rio Tinto Zinc Corp, 47-71; Imp Chem Indust, 55-68; Clarke, Chapman & Co, 55-69 & Ethyl Corp, 57-69. Mem: Am Chem Soc; The Chem Soc; Royal Inst Chem. Res: Beryllium chemistry; organometallic compounds. Mailing Add: 1801 Rainbow Ave Laramie WY 82070

COATES, JOHN STUART, organic chemistry, see 12th edition

COATES, ROBERT MERCER, b Evanston, Ill, May 21, 38; m 64. ORGANIC CHEMISTRY. Educ: Yale Univ, BS, 60; Univ Calif, Berkeley, PhD(chem), 64. Prof Exp: NIH fel, Stanford Univ, 63-65; ASST PROF CHEM, UNIV ILL, URBANA-CHAMPAIGN, 65- Concurrent Pos: A P Sloan Found fel, 71-73. Mem: Am Chem Soc; The Chem Soc. Res: Synthesis of natural products; synthetic methods; biogenetic-like rearrangement of terpenes; synthesis and reactions of polycyclic compounds. Mailing Add: Dept of Chem Univ of Ill Urbana IL 61801

COATS, ALFRED CORNELL, b Portland, Ore, Mar 12, 36; m 63; c 2.

NEUROPHYSIOLOGY. Educ: Stanford Univ, BA, 59; Baylor Univ, MD, 62, MS, 63. Prof Exp: From instr to assoc prof, 62-72, PROF PHYSIOL & OTOLARYNGOL, BAYLOR COL MED, 72- Concurrent Pos: Dir electronystagmography lab & mem consult staff, Methodist Hosp, 63-, St Luke's Hosp, 73- & Hermann Hosp, 75-; mem consult staff, Ben Taub Hosp, 65- Mem: Am Neuro-Otologic Asn; Soc Neurosci. Res: Physiology of peripheral auditory system; clinical vestibulometry; study of balance-and-equilibrium system in humans. Mailing Add: Dept of Physiol Baylor Col of Med Houston TX 77025

COATS, ALMA WINIFRED, b Edinburgh, Scotland; US citizen. POLYMER CHEMISTRY. Educ: Southwestern Univ, BS, 50. Prof Exp: Res chemist, Humble Oil & Refining Co, 50-57; res chemist biochem, Univ Tex Med Br, 58-65; res chemist, Milchem Inc, Tex, 66-68; SR CHEMIST RES & DEVELOP, HOOKER CHEM CORP, 69- Concurrent Pos: Res assoc, Ctr Fire Res, Nat Bur Stands, 75-77; mem working group 12, Int Stands Orgn, 76- Mem: Am Chem Soc. Res: Toxicology of combustion products. Mailing Add: Hooker Chem Corp Grand Island NY 14072

COATS, EUGENE ARTHUR, medicinal chemistry, organic chemistry, see 12th edition

COATS, RICHARD LEE, b Madill, Okla, Feb 14, 36; m 59; c 4. NUCLEAR PHYSICS, REACTOR PHYSICS. Educ: Univ Okla, BS, 59, MS, 63, PhD(nuclear eng), 66. Prof Exp: Mem tech staff, 66-69, DIV SUPVR, NUCLEAR ENG & SUPVR, REACTOR STUDIES DIV 5422, SANDIA LABS, 69- Mem: Am Nuclear Soc. Res: Experimental and theoretical nuclear reactor physics; coupled reactor dynamics; stochastic reactor kinetics; Monte Carlo reactor physics calculations. Mailing Add: Reactor Studies Div 5422 Sandia Labs Albuquerque NM 87115

COATS, ROBERT ROY, b Toronto, Ont, Nov 22, 10; US citizen; m 37; c 3. Educ: Univ Wash, BS, 31, MS, 32; Univ Calif, PhD(geol), 38. Prof Exp: Geologist, Storey County Mines, 37; asst prof geol, Univ Alaska, 37-39; from jr geologist to geologist, 39-64, RES GEOLOGIST, US GEOL SURV, 64- Mem: Fel Mineral Soc Am; fel Soc Econ Geologists; fel Geol Soc Am. Res: Tin deposits of Alaska; alteration by hydrothermal solutions; Aleutian volcanoes; geology of the northeastern Great Basin. Mailing Add: 345 Middlefield Rd Menlo Park CA 94025

COBB, ARTHUR LEE, applied statistics, analytical statistics, see 12th edition

COBB, BRYANT FRANKLIN, III, b Haskell Co, Tex, Nov 23, 35; m 58; c 4. FOOD SCIENCE, FISHERIES. Educ: Abilene Christian Col, BS, 61; Univ Tex, PhD(biochem), 67. Prof Exp: Res technician, Univ Tex M D Anderson Hosp & Tumor Inst, 61-62 & Med Br, Univ Tex, 62-63; asst res biochemist, Univ Calif, Berkeley, 67-68; asst prof, 68-70, ASSOC PROF ANIMAL SCI, TEX A&M UNIV, 70- Mem: Am Chem Soc; World Maricult Soc. Res: Protein chemistry; marine research. Mailing Add: Dept of Animal Sci Tex A&M Univ College Station TX 77843

COBB, CAROLUS M, b Lynn, Mass, Jan 22, 22; m 66; c 1. PHYSICAL CHEMISTRY. Educ: Mass Inst Technol, SB, 44, PhD(phys chem), 51. Prof Exp: Chemist, Tenn Eastman Corp, 44-46 & Ionics, Inc, 51-55; prin scientist, Allied Res Assocs, Inc, 55-60; CHIEF CHEMIST, AM SCI & ENG, INC, CAMBRIDGE, MASS, 60- Concurrent Pos: Affil speech & cancer res, Forsyth Dent Ctr, Harvard Sch Pub Health, 64- Mem: Am Chem Soc; Am Phys Soc. Res: Titanium and solution chemistry; ion exchange chromatography; atmospheric and physical optics; fluorescent materials; nuclear phenomena; high temperature and electronic materials; thermodynamics. Mailing Add: Am Sci & Eng Inc 955 Massachusetts Ave Cambridge MA 02139

COBB, CHARLES MADISON, b Kansas City, Mo, Sept 20, 40; m 64; c 1. DENTISTRY, PERIODONTICS. Educ: Univ Mo-Kansas City, DDS, cert periodont & MS, 64; Georgetown Univ, PhD(anat), 71. prof periodont, Sch Dent, La State Univ, 71-72; asst prof periodont, 73-74, ASSOC PROF ANAT & PERIODONT, SCH DENT, UNIV ALA, BIRMINGHAM, 74-, INVESTR PERIODONT, INST DENT RES, 73- Honors & Awards: Balant Orban Prize, Am Acad Periodont, 66. Mem: Int Asn Dent Res; Am Acad Periodont; Am Dent Asn. Res: Ultrastructure and histopathology of periodontal disease; ultrastructure and biochemistry of developing salivary glands; basic enzymology of mammalian type. Mailing Add: Dept of Periodont Sch of Dent Univ of Ala Univ Sta Birmingham AL 35294

COBB, EDWARD HUNTINGTON, b Great Barrington, Mass, Apr 23, 16; m 53. GEOLOGY. Educ: Yale Univ, BS, 38, MS, 41. GEOLOGIST, ALASKAN GEOL BR, US GEOL SURV, 46- Mem: AAAS; Geol Soc Am; Am Asn Petrol Geologists. Res: Mineral deposits geology; indices; bibliographies; collecting and synthesizing data on Alaskan mineral resources. Mailing Add: 1140 Cotton St Menlo Park CA 94025

COBB, EMERSON GILLMORE, b Slaughters, Ky, Nov 28, 07; m 29; c 2. ORGANIC CHEMISTRY. Educ: Union Col, Ky, AB, 28; Univ Ky, MS, 31; Univ NC, PhD(org chem), 41. Hon Degrees: LHD, Union Col, Ky, 61. Prof Exp: High sch instr, Ky, 28-29 & 32-40; asst prof chem, La Polytech Inst, 40-42; prof & head dept, Dakota Wesleyan Univ, 42-48; chmn dept, 48-74, PROF CHEM, UNIV OF THE PAC, 48- Concurrent Pos: Fulbright vis lectr, Univ Peshawar, 61-62; vis lectr, Univ Baja Calif & Univ Ciencias Marinas, 74-75. Mem: Am Chem Soc. Res: Natural plant products; protective coatings; constitution of tannins. Mailing Add: Dept of Chem Univ of the Pac Stockton CA 95211

COBB, ESTEL HERMAN, animal science, see 12th edition

COBB, FIELDS WHITE, JR, b Key West, Fla, Feb 16, 32; m 58; c 3. PLANT PATHOLOGY. Educ: NC State Univ, BS, 55; Yale Univ, MF, 56; Pa State Univ, PhD(plant path), 63. Prof Exp: Res forester, Southeastern Forest Exp Sta, US Forest Serv, 55-57, plant pathologist, Southern Forest Exp Sta, 57; statist clerk agr econ, Agr Mkt Serv, USDA, 57-58; instr plant path, Pa State Univ, 63; asst prof, 63-70, ASSOC PROF PLANT PATH, UNIV CALIF, BERKELEY, 70-, ASSOC PLANT PATHOLOGIST, EXP STA, 73- Mem: AAAS; Soc Am Foresters; Am Phytopath Soc; Mycol Soc Am. Res: Diseases of forest trees, particularly those of roots and the vascular system, their causes, epidemiology, development and control. Mailing Add: Dept of Plant Path Univ of Calif Berkeley CA 94720

COBB, GLENN WAYNE, b Jonesboro, La, Dec 1, 36; m 58; c 2. PLANT MORPHOGENESIS, PLANT PATHOLOGY. Educ: La Polytech Inst, BS, 58; Purdue Univ, MS, 61, PhD(plant morphol), 62. Prof Exp: Asst prof biol, Stephen F Austin State Univ, 62-65; assoc prof, 65-68, PROF BIOL, McNEESE STATE UNIV, 68- Concurrent Pos: Fac res grants, 63-64 & 65. Res: Plant disease research. Mailing Add: Dept of Biol McNeese State Univ Lake Charles LA 70601

COBB, GROVER CLEVELAND, JR, b Atlanta, Ga, Feb 6, 35; m 54; c 2. NUCLEAR PHYSICS. Educ: Univ Ga, BS, 56, MS, 57; Univ Va, PhD(physics), 60. Prof Exp: Asst prof, 60-69, ASSOC PROF PHYSICS, NC STATE UNIV, 69- Mem: Am Phys Soc. Res: Neutron scattering; optical spectroscopy; gaseous discharge experiments; nuclear cross sections; plasma oscillations. Mailing Add: Dept of Physics NC State Univ Raleigh NC 27607

COBB, HOWELL DEE, JR, b San Antonio, Tex, Sept 12, 30; m 52; c 5. CELL PHYSIOLOGY, MICROBIAL PHYSIOLOGY. Educ: Trinity Col, Tex, BS, 53, MS, 58; Univ Tex, PhD(nitrogen fixation), 63. Prof Exp: Asst prof physiol, Baylor Univ, 62-63; asst prof, 63-67, ASSOC PROF PHYSIOL, TRINITY UNIV, TEX, 67- Concurrent Pos: Res fel, Univ Tex, 63. Mem: AAAS; Am Soc Plant Physiologists; Am Soc Microbiol. Res: Relationships between photosynthesis and nitrogen metabolism in algae; study of microecological systems involved in the biodegradation of cresol. Mailing Add: Dept of Biol Trinity Univ San Antonio TX 78284

COBB, JEWEL PLUMMER, b Jan 17, 24; US citizen; div; c 1. CELL BIOLOGY. Educ: Talladega Col, AB, 44; NY Univ, MS, 47, PhD(cell biol), 50. Hon Degrees: LLD, Wheaton Col, Mass, 71; ScD, Lowell Technol Inst, 72. Prof Exp: Instr anat, Col Med, Univ Ill, 52-54; from instr to asst prof res surg, Post Grad Med Sch, NY Univ, 55-60; prof biol, Sarah Lawrence Col, 60-69; PROF ZOOL & DEAN, CONN COL, 69- Concurrent Pos: Nat Cancer Inst grant, 69-74 & 74-77; Am Cancer Soc grant, 71-73; vis lectr res assoc prog, Hunter Col, 56-57; mem bd, Nat Sci Bd, NSF, 74-; mem bd, Am Coun Educ, 74-; mem bd dirs, Nat Inst Med. Mem: Nat Inst Med; AAAS; fel NY Acad Sci; Tissue Cult Asn; Am Asn Cancer Res. Res: Research on mechanisms controlling differentiation and growth in malignant pigment cells. Mailing Add: Off of Dean Conn Col New London CT 06320

COBB, JOHN CANDLER, b Boston, Mass, July 8, 19; m 46; c 4. PREVENTIVE MEDICINE. Educ: Harvard Univ, BA, 41, MD, 48; Johns Hopkins Univ, MPH, 54. Prof Exp: Asst malaria control, Friends Serv Comt, 41-42; instr maternal & child health, Sch Hyg & Pub Health, Johns Hopkins Univ, 51-54, instr pediat, Sch Med, 51-56 & psychiat, 52-56, asst prof maternal & child health, Sch Hyg, 54-56; area consult, USPHS, Div Indian Health, NMex, 56-60; Johns Hopkins Univ & Ford Found dir med social res proj, Lahore, Pakistan, 60-64; chmn dept, 66-72, PROF PREV MED, SCH MED, UNIV COLO, DENVER, 65- Concurrent Pos: Mem comt peace educ & family planning, Am Friends Serv Comt, 64-74; WHO short term consult maternal & child health & family planning, Indonesia, 69-70 & family health educ, Western Pac Region, 71-72; mem gov sci adv coun, Colo; mem environ coun; mem air pollution control comn, Colo, 76-; mem gov's task force Nuclear Energy Plant, 75- Mem: AAAS; Int Solar Energy Soc; Asn Teachers of Prev Med. Res: Environmental health; food production from algae. Mailing Add: Dept of Prev Med Univ of Colo Sch of Med Denver CO 80220

COBB, JOHN IVERSON, b Marianna, Fla, Feb 9, 38. TOPOLOGY. Educ: Fla State Univ, BA, 60; Univ Wis, MA, 61, PhD(math), 66. Prof Exp: Instr math, Racine Ctr, Univ Wis, 66; asst prof, Rutgers Univ, 66-69; asst prof, 69-74, ASSOC PROF MATH, UNIV IDAHO, 74- Mem: Am Math Soc; Math Asn Am. Res: Point-set topology; piece-wise linear topology. Mailing Add: 500 Queen Rd Apt 44 Moscow ID 83843

COBB, R M KARAPETOFF, b Winthrop, Mass; m 36. PAPER CHEMISTRY. Educ: Tufts Col, BS, 22; Mass Inst Technol, MS, 23. Prof Exp: Asst, Mass Inst Technol, 23-24; res chemist, Larkin Co, Inc, 24 & Hunt-Rankin Leather Co, 24-26; res dir, 26-51, tech adv, 51-65, CONSULT, LOWE PAPER CO, RIDGEFIELD, NJ, 65- Concurrent Pos: Mem lithograph adv comt, Nat Bur Standards, 29-55, res assoc, Lithograph Tech Found, 31. Honors & Awards: Coating & Graphic Arts Div Award, Tech Asn Pulp & Paper Indust, 68. Mem: Am Chem Soc; Soc Rheol; fel Tech Asn Pulp & Paper Indust. Res: Paper sizing and coating; adhesives; lithography; emulsions. Mailing Add: 77 Grozier Rd Cambridge MA 02138

COBB, RAYMOND LYNN, b Ochelata, Okla, Dec 10, 29; m 66; c 1. ORGANIC CHEMISTRY. Educ: Ottawa Univ, BS, 51; Univ Kans, PhD(org chem), 55. Prof Exp: RES CHEMIST, PHILLIPS PETROL CO, 55- Mem: Am Chem Soc. Res: Organic nitrogen compounds; catalytic organic processes; thermal reactions; organosulfur compounds; reaction mechanisms. Mailing Add: Res & Develop Dept Phillips Petrol Co Bartlesville OK 74004

COBB, SIDNEY, b Cambridge, Mass, June 1, 16; m 41; c 4. EPIDEMIOLOGY. Educ: Harvard Univ, BS, 38, MD, 42, MPH, 51. Prof Exp: Intern med, Johns Hopkins Hosp, 42-43; chief resident, Sydenham Hosp Infectious Dis, 46-47; dir, Nashoba Assoc Bds Health, 48-52; from asst prof to assoc prof biostatist & epidemiol, Univ Pittsburgh, 52-61; assoc, Ment Health Res Inst, Univ Mich, Ann Arbor, 61-68, prog dir, Surv Res Ctr & lectr, Sch Pub Health, 61-73; PROF COMMUNITY HEALTH & PSYCHIAT, BROWN UNIV, 73- Concurrent Pos: Fel biochem, Harvard Med Sch, 47-48; NIH career develop award, 63-68; instr, Med Sch, Johns Hopkins Univ, 46-47 & Sch Pub Health, Harvard Univ, 48-52; res scientist, NIH, 68-, consult, NIMH, 69-, chief psychiat epidemiol, Butler Hosp, Providence, 73-74. Mem: Am Pub Health Asn; Am Rheumatism Asn; Am Psychosom Soc; Am Heart Asn; Am Epidemiol Soc. Res: Epidemiology of non-communicable diseases, especially psychosomatic disease and mental health. Mailing Add: Box G Brown Univ Providence RI 01912

COBB, THOMAS BERRY, b Atlanta, Ga, Nov 4, 39; m 64. PHYSICS, CHEMISTRY. Educ: Southern Missionary Col, BA, 60; Univ SC, MS, 63; NC State Univ, PhD(physics), 68. Prof Exp: Instr physics, Western Md Col, 63-65; res assoc chem, Univ NC, 68-69; asst prof, 69-74, ASSOC PROF PHYSICS, BOWLING GREEN STATE UNIV, 74-, ASST DEAN GRAD SCH, 74- Mem: AAAS; Am Asn Physics Teachers. Res: Nuclear magnetic resonance. Mailing Add: Dept of Physics Bowling Green State Univ Bowling Green OH 43402

COBB, WALTER R, b Canton, NJ, Jan 9, 15; m 37; c 1. AGRICULTURE, SCIENCE EDUCATION. Educ: Rutgers Univ, BS, 36, MEd, 48; Univ Thessalonika, DSc(agr exten educ), 56. Prof Exp: Instr high schs, 36-38, 42-51; sales work, Consolidated Prod Co, 38-40; chem operator, E I du Pont de Nemours & Co, 40-42; area dir, Near East Found, Iran & Greece, 51-56; farm mgr, 56-69, RES FARM MGR, MERCK & CO, 69- Res: Coccidiostats; anthelminias; hormones and their use in agricultural field. Mailing Add: Merck & Co 203 River Rd Somerville NJ 08876

COBB, WHITFIELD, b Winston-Salem, NC, Oct 1, 14; m 44; c 3. MATHEMATICAL STATISTICS. Educ: Univ NC, AB, 33, AM, 35, PhD(statist & math), 59. Prof Exp: Statist clerk, Div Social Res, Works Progress Admin, 35-36; teacher math & philos, Caney Jr Col, 37; actg assoc prof math & physics, Western Carolina Teachers Col, 37-38; instr math, Univ NC, 38-39; actg instr math, Agr & Mech Col Tex, 39-40; instr, Univ NC, 40-41, instr philos, 41-42; head dept math, Jr Col Div, Ward-Belmont, Nashville, Tenn, 46-47; assoc prof, Guilford Col, 47-57; asst prof, Womans Col, Univ NC, 58-61, assoc prof, 61-62; assoc prof statist, Hollins Col, 62-65; assoc prof, Va Polytech Inst & State Univ, 65-76; RETIRED. Mem: Am Statist Asn; Inst Math Statist; Soc Social Responsibility in Sci; Fedn Am Scientists. Res: Univariate and multivariate analysis of variance; philosophy of science and religion. Mailing Add: 800 Cupp St SE Blacksburg VA 24060

COBB, WILLIAM MONTAGUE, b Washington, DC, Oct 12, 04; m 29; c 2. ANATOMY. Educ: Amherst Col, AB, 25; Howard Univ, MD, 29; Western Reserve

Univ, PhD, 32. Hon Degrees: ScD, Amherst Col, 55; LLD, Morgan State Col, 64. Prof Exp: Instr embryol, 28-29, from asst prof to prof anat, 32-69, mem exec comt, Med Sch, 41 & 45-69, head dept anat, 47-69, DISTINGUISHED PROF ANAT, HOWARD UNIV, 69- Concurrent Pos: Fel, Western Reserve Univ, 33-39, Rosenwald fel, 41-42, assoc, 42-44; jr med officer, USDA, 35; chmn nat med comt, Nat Asn Advan Colored People, 44-; health specialist, Nat Urban League, 46 & 47; mem exec comt, White House Conf Health, 65; ed, J Nat Med Asn, 49-; assoc ed, J Am Asn Phys Anthrop, 44-48, ed, 49- Honors & Awards: Distinguished Serv Medal, Nat Med Asn, 55. Mem: AAAS (vpres, 55); Am Asn Anat; Am Soc Mammal; Nat Med Asn (pres, 64-65); Am Asn Phys Anthrop (vpres, 48-50 & 54-56, pres, 57-59). Res: Physical anthropology; collections of human materials; growth and development of the American Negro; aging in the adult skeleton; graphic method of anatomy. Mailing Add: Dept of Anat Howard Univ Washington DC 20001

COBB, WILLIAM Y, chemistry, see 12th edition

COBBAN, WILLIAM AUBREY, b Anaconda, Mont, Dec 31, 16; m 42; c 3. GEOLOGY. Educ: Univ Mont, BA, 40; Johns Hopkins Univ, PhD(geol), 49. Prof Exp: Plane table man, Carter Oil Co, Tulsa, 39-40, leader geol mapping field party, 40-41, dist geologist in chg field parties & for northeast Utah, 44-45; PALEONTOLOGIST & STRATIGRAPHER, US GEOL SURV, 46- Mem: Soc Econ Paleontologists & Mineralogists; Paleont Soc; Am Asn Petrol Geol; Soc Vert Paleont; Geol Soc Am. Res: Upper Cretaceous stratigraphy and paleontology of the Rocky Mountain area. Mailing Add: US Geol Surv Fed Ctr Bldg 25 Denver CO 80225

COBBE, THOMAS JAMES, b Cincinnati, Ohio, July 18, 18; m 44; c 2. FOREST ECOLOGY. Educ: Univ Cincinnati, BA, 40, MA, 41; Univ Mich, PhD(bot), 53. Prof Exp: Asst prof biol, Am Univ, 48-49; instr, Oberlin Col, 49-52; from instr to asst prof, Capital Univ, 52-57; asst prof, 57-71, ASSOC PROF BOT, MIAMI UNIV, 71- Mem: AAAS; Ecol Soc Am; Am Soc Photogram. Res: Secondary forest successions; Dutch elm disease. Mailing Add: RR 2 Juniper Hill Oxford OH 45056

COBBLE, JAMES WIKLE, b Kansas City, Mo, Mar 15, 26; m 49; c 2. PHYSICAL CHEMISTRY, INORGANIC CHEMISTRY. Educ: Ariz State Univ, AB, 46; Univ Southern Calif, MS, 49; Univ Tenn, PhD(phys chem), 52. Prof Exp: Chemist, Oak Ridge Nat Lab, 49-52 & Radiation Lab, Univ Calif, 52-54; from asst prof to prof chem, Purdue Univ, 55-73; PROF CHEM & DEAN GRAD DIV & RES, SAN DIEGO STATE UNIV, 73- Concurrent Pos: Lectr, Univ Calif, 53; Robert A Welch Found, 70-71. Honors & Awards: E O Lawrence Award, AEC, 70. Mem: Am Chem Soc; Am Phys Soc. Res: Radiochemistry; physical-inorganic chemistry; correlation of thermodynamic properties with structures; high temperature solutions; nuclear chemistry; mechanisms of nuclear reactions. Mailing Add: Dept of Chem San Diego State Univ San Diego CA 92115

COBBLE, JAMES WILLIAM, b Millersville, Mo, Apr 13, 20; m 40; c 2. DAIRY PRODUCTION. Educ: Univ Mo, BS, 47, MA, 48, PhD(dairy prod), 51. Prof Exp: Pub sch teacher, 37-40; social worker, Mo Social Security Comn, 41; from asst instr to instr, Dairy Dept, Univ Mo, 48-50; assoc dean col agr & assoc dir exp sta, 59-62, dean col resource develop, 62-72, PROF ANIMAL & DAIRY HUSB & HEAD DEPT, UNIV RI, 51-, PROF ANIMAL SCI, 73-, DIR, AGR EXP STA & AGR & HOME ECON EXTEN, 62- Concurrent Pos: Adv & chief party, AID, Korea, 71-73. Mem: Am Dairy Sci Asn; Am Soc Animal Sci. Res: Nutrition and physiology of dairy and animal husbandry. Mailing Add: Col of Resource Develop Univ of RI Kingston RI 02881

COBBOLD, R S C, b Worcester, Eng, Dec 10, 31; Can citizen; m 63; c 3. BIOMEDICAL ENGINEERING. Educ: Univ London, BSc, 56; Univ Sask, MSc, 61, PhD(elec eng), 65. Prof Exp: Asst exp officer electronics, Ministry of Supply, Eng, 49-53; sci officer, Defence Res Bd, Ottawa, Can, 56-59; from lectr to assoc prof elec eng, Univ Sask, 60-66; from assoc prof to prof elec eng, 66-75, DIR INST BIOMED ENG, UNIV TORONTO, 75- Mem: Inst Elec & Electronics Engrs; Int Fedn Med & Biol Eng. Res: Semiconductor electronics; biomedical transducers; physics of semiconductor devices. Mailing Add: Inst of Biomed Eng Univ of Toronto Toronto ON Can

COBBS, WALTER HERBERT, JR, physical chemistry, see 12th edition

COBLER, JOHN GEORGE, b Conneaut Lake, Pa, Sept 15, 18; m 41; c 2. ORGANIC POLYMER CHEMISTRY. Educ: Col Wooster, BA, 40. Prof Exp: Lab asst, Col Wooster, 38-39; control chemist, Ohio Exp Sta, 39-40; asst, Purdue Univ, 40-41; res chemist, Distillation Prod, Inc, NY, 42-45; lab dir, Bordon Co, NJ, 45-47; head spec prod develop, Bordens Soy Processing Co, 47-49; group leader, Anal Dept, Dow Chem Co, 49-59, tech expert, 59-69, tech consult, 69-70, assoc anal scientist, 70-75, ASSOC SCIENTIST, HEALTH & ENVIRON RES DEPT, DOW CHEM USA, 75- Concurrent Pos: Asst, Manhattan Proj, Univ Rochester, 44-45. Mem: Am Chem Soc; Sigma Xi; Am Soc Testing & Mat; NY Acad Sci. Res: Structure and composition of polymers; stability and degradation of polymers. Mailing Add: Health & Environ Res Dow Chem USA PO Box 1706 Midland MI 48640

COBURN, CORBETT BENJAMIN, JR, b Lake Providence, La, Dec 7, 40; m 61; c 1. ENVIRONMENTAL PHYSIOLOGY, PHYSIOLOGICAL ECOLOGY. Educ: La Polytech Inst, BS, 62, MS, 64; Univ Southern Miss, PhD(zool), 70. Prof Exp: Instr biol, Calif Baptist Col, 64-65 & East Central Jr Col, 66-68; prof, Wesleyan Col, 70-71; asst prof, West Liberty State Col, 71-72; ASST PROF BIOL, TENN TECHNOL UNIV, 72- Mem: Am Fisheries Soc. Res: Vertebrate hematology; metabolic responses of aquatic animals to pollutants; RNA/DNA ratio as affected by season and water quality; histological and physiological responses of fish to nitrogen supersaturation. Mailing Add: Dept of Biol Tenn Technol Univ Cookeville TN 38501

COBURN, EVERETT ROBERT, b Manchester, NH, Aug 10, 15; m 39; c 2. ORGANIC CHEMISTRY. Educ: Harvard Univ, SB, 38, AM, 40, PhD(org chem), 41. Prof Exp: Lilly fel polarog studies of quinones, Harvard Univ, 41-42; res chemist, Nat Defense Res Comt, 42-43; PROF CHEM, BENNINGTON COL, 43- Concurrent Pos: Lectr, Middlebury Col, 44; consult, Sprague Elec Co, 52- Mem: Am Chem Soc. Res: Diels-Alder reactions on quinones; polarographic work on quinones and related compounds; incendiary mixtures and design of apparatus for use. Mailing Add: Dept of Chem Bennington Col Bennington VT 05201

COBURN, FRANK EMERSON, b Toronto, Ont, Apr 25, 12; m 40; c 4. Educ: Univ Toronto, BA, 36, MD, 39; RCPS(C), cert psychiat. Prof Exp: Assoc prof psychiat, Univ Iowa, 50-55; assoc prof, 55-58, PROF PSYCHIAT, UNIV SASK, 58- Concurrent Pos: Baker lectr, Univ Mich, 54; mem, Am Bd Psychiat & Neurol, 47- Mem: Fel Am Psychiat Asn; Am Psychosom Soc; Can Psychiat Asn; Can Med Asn. Res: Teaching; therapy; community psychiatry. Mailing Add: Dept of Psychiat Univ Hosp Univ of Sask Saskatoon SK Can

COBURN, HORACE HUNTER, b Cambridge, Mass, May 10, 22; m 47; c 3. PHYSICS. Educ: Ohio State Univ, BS, 43; Univ Ill, MS, 47; Univ Pa, PhD(physics), 56. Prof Exp: Assoc prof physics, Moravian Col, 50-51; from asst prof to assoc prof, 54-69, PROF PHYSICS, NMEX STATE UNIV, 69- Concurrent Pos: Physicist, Manhattan Proj, Tenn, 44-46; consult, Los Alamos Sci Lab, 62-; mem fac, Inst Optics, Univ Rochester, 64-65. Mem: AAAS; Am Phys Soc; Am Asn Physics Teachers; Optical Soc Am. Res: Optics; biophysics. Mailing Add: NMex State Univ Box 3D Las Cruces NM 88001

COBURN, JACK WESLEY, b Fresno, Calif, Aug 6, 32; m 58; c 3. INTERNAL MEDICINE, NEPHROLOGY. Educ: Univ Redlands, BS, 53; Univ Calif, Los Angeles, MD, 57; Am Bd Internal Med, dipl, 65. Prof Exp: Intern med, Med Ctr, Univ Calif, Los Angeles, 57-58; asst res physician, Univ Wash Hosp Syst, Seattle, 58-60; assoc res physician, Med Ctr, Univ Calif, Los Angeles, 60-61; sect chief gen med, Wadsworth Hosp, Los Angeles, 68-69; from asst prof to assoc prof, 65-73, PROF MED, SCH MED, UNIV CALIF, LOS ANGELES, 73-; CHIEF NEPHROLOGY SECT, VET ADMIN WADSWORTH HOSP CTR, 70- Concurrent Pos: Nat Inst Arthritis & Metab res fel, Vet Admin Hosp & Univ Calif, Los Angeles, 61-63, clin investr award, Vet Admin Ctr, 65-67; chief metab res ward, Vet Admin Ctr, Los Angeles, 67-70. Mem: Am Fedn Clin Res; AMA; Am Soc Nephrology; fel Am Col Physicians; Am Physiol Soc. Res: Renal physiology, especially renal handling of divalent ions; pathophysiology of uremia; renal osteodystrophy; vitamin D and calcium metabolism. Mailing Add: Vet Admin Wadsworth Hosp Ctr Wilshire & Sawtelle Blvds Los Angeles CA 90073

COBURN, JOHN F, organic chemistry, see 12th edition

COBURN, JOHN WYLLIE, b Vancouver, BC, Nov 9, 33; m 66; c 1. PLASMA PHYSICS. Educ: Univ BC, BASc, 56, MASc, 58; Univ Minn, Minneapolis, PhD(elec eng), 67. Prof Exp: Fel physics, Simon Fraser Univ, 67-68; RES STAFF MEM, MAT SCI DEPT, RES LAB, INT BUS MACH CO, 68- Mem: Am Vacuum Soc. Res: Particle diagnostics in glow discharges; thin film formation by sputtering and by plasma polymerization. Mailing Add: IBM Res Lab K33/281 Monterey & Cottle Rds San Jose CA 95193

COBURN, KENNETH R, physiology, see 12th edition

COBURN, MICHAEL DOYLE, b Houston, Tex, Aug 6, 39; m 60; c 2. ORGANIC CHEMISTRY. Educ: Univ Tex, Austin, BS, 62, PhD(chem), 64. Prof Exp: STAFF MEM, LOS ALAMOS SCI LAB, UNIV CALIF, 64- Mem: AAA S; Am Chem Soc; Int Soc Heterocyclic Chem. Res: Synthesis of energetic organic compounds, predominantly in the heterocyclic field. Mailing Add: PO Box 1633 Los Alamos NM 87544

COBURN, NATHANIEL, mathematics, see 12th edition

COBURN, RICHARD KARL, b Salt Lake City, Utah, Feb 24, 20; m 42; c 16. APPLIED MATHEMATICS. Educ: Utah State Univ, BS, 42 & 43; Univ Wash, MS, 56; Univ Ill, MA, 62. Prof Exp: Instr math & physics, Pa State Univ, 46-48; prof chem & physics, Ricks Col, 48-58; PROF MATH & CHMN DEPT, THE BRIGHAM YOUNG UNIV, HAWAII CAMPUS, 58- Concurrent Pos: Math consult, Lockheed Airplane Co, 60-62; pres, Hawaii Coun Teachers Math; assoc dir, Hawaii Sci Fair; lectr, Nat Coun Teachers Math. Mem: Math Asn Am; Am Chem Soc. Res: Analytical methods of solving inequalities of order two and higher. Mailing Add: Dept of Math Box 47 Brigham Young Univ Laie Oahu HI 96762

COBURN, ROBERT A, b Akron, Ohio, Dec 31, 38; m 66; c 1. MEDICINAL CHEMISTRY, PHYSICAL ORGANIC CHEMISTRY. Educ: Univ Akron, BS, 60; Harvard Univ, AM, 62, PhD(org chem), 66. Prof Exp: Res chemist, US Army Natick Labs, Mass, 65-66; asst prof, 68-73, ASSOC PROF MED CHEM, SCH PHARM, STATE UNIV NY BUFFALO, 73- Mem: Am Chem Soc; The Chem Soc. Res: Synthesis and molecular structure studies of heterocyclic compounds of biological and/or pharmacological significance. Mailing Add: Dept Med Chem Health Sci Bldg State Univ of NY Buffalo NY 14214

COBURN, RONALD F, b Grand Rapids, Mich, Dec 10, 31; m 62; c 2. PHYSIOLOGY. Educ: Northwestern Univ, BS, 54, MD, 57. Prof Exp: Intern, Presby-St Luke's Hosp, Chicago, 57-58; resident internal med, Vet Admin Res Hosp, Chicago, 58-60; from instr to assoc med, 63-66 & asst prof, 67-68, from asst prof to assoc prof physiol, 66-75, PROF PHYSIOL & MED, SCH MED, UNIV PA, 75- Concurrent Pos: Fel physiol, Sch Med, Univ Pa, 60-63; mem, Vet Admin Respiratory Syst Res Eval Comt, 69-71. Mem: Am Physiol Soc; Am Soc Clin Invest; Am Fedn Clin Res. Res: Pulmonary and carbon monoxide physiology; heme catabolism; pulmonary gas exchange; tissue oxygenation; airway physiology. Mailing Add: Dept of Physiol Univ of Pa Sch of Med Philadelphia PA 19104

COBURN, STEPHEN PUTNAM, b Orange, NJ, Nov 10, 36. BIOCHEMISTRY. Educ: Rutgers Univ, BS, 58; Purdue Univ, MS, 61, PhD(biochem), 64. Prof Exp: DIR DEPT BIOCHEM, FT WAYNE STATE HOSP & TRAINING CTR, 63- Concurrent Pos: Vis asst prof, Purdue Univ; dipl, Am Bd Clin Chem. Mem: Am Chem Soc; Am Asn Ment Deficiency; Am Asn Clin Chem; Brit Biochem Soc; Am Inst Nutrit. Res: Biochemistry of mental retardation and other metabolic diseases. Mailing Add: Ft Wayne State Hosp & Training Ctr Dept Biochem 801 E State Blvd Ft Wayne IN 46805

COBURN, THEODORE JAMES, b Newton, Mass, June 11, 26; m 49; c 3. SOLID STATE PHYSICS. Educ: Ohio State Univ, BSc, 47, PhD(physics), 57. Prof Exp: Contract administr, Armaments Br, Off Naval Res, DC, 53-55; proj engr, Apparatus & Optical Div, 57-60, SR RES PHYSICIST, RES LABS, EASTMAN KODAK CO, NY, 60- Mem: Optical Soc Am; Am Phys Soc. Res: Surface state physics as applied to electrostatistics; infrared spectroscopy; military applications of infrared. Mailing Add: Res Labs Eastman Kodak Co Rochester NY 14650

COBURN, WILLIAM CARL, JR, b Duluth, Minn, Nov 2, 26; m 51; c 2. PHYSICAL ORGANIC CHEMISTRY. Educ: Univ Colo, BA, 48; Fla State Univ, MA, 51, PhD(phys chem), 54. Prof Exp: Phys chemist, 54-59, sr phys chemist, 59-67, HEAD MOLECULAR SPECTROS SECT, SOUTHERN RES INST, 67- Mem: AAAS; Coblentz Soc; Am Chem Soc. Res: Kinetics and mechanisms of organic reactions; molecular complexing; hydrogen bonding; theoretical and applied infrared spectroscopy; ultraviolet spectroscopy; nuclear magnetic resonance spectroscopy; mass spectroscopy. Mailing Add: Southern Res Inst 200 Ninth Ave S Birmingham AL 35205

COCCODRILLI, GUS D, JR, b Peckville, Pa, July 28, 45; m 67; c 2. NUTRITION. Educ: Pa State Univ, BS, 67, PhD(food sci), 71; Va Polytech Inst, MS, 69. Prof Exp: Sr chemist, 71-74, PROJ LEADER NUTRIT, GEN FOODS CORP, 74- Mem: Inst Food Technologists; Nutrit Today Soc. Res: Mineral nutrition research; trace element metabolism; vitamin nutrition; dental health research; cereal nutrition. Mailing Add: Tech Ctr 250 North St Gen Foods Corp White Plains NY 10625

COCEANI, FLAVIO, b Trieste, Italy, Jan 3, 37; m 69. NEUROPHYSIOLOGY. Educ: Univ Bologna, MD, 61, Docent(human physiol), 68. Prof Exp: Asst prof, Univ Bologna, 65-66; vis scientist neurochem, Montreal Neurol Inst, 66-68; asst prof, 68-70, ASSOC PROF PHYSIOL, UNIV TORONTO, 70-, ASST PROF PEDIAT, 69-; ASST SCIENTIST, RES INST, HOSP SICK CHILDREN, 68- Concurrent Pos: Nat Res Coun fel & res fel, Dept Physiol, Univ Bologna, 61-62; res fel neurophysiol, Montreal Neurol Inst, 62-64. Mem: AAAS; Am Soc Neurochem; Can Soc Clin Invest; NY Acad Sci; Can Physiol Soc. Res: Role of prostaglandins in brain function; effects of prostaglandins on smooth muscle contraction. Mailing Add: Res Inst Hosp for Sick Children Toronto ON Can

COCH, NICHOLAS KYROS, b New York, NY, Mar 30, 38; m 71. SEDIMENTOLOGY, MARINE GEOLOGY. Educ: City Col New York, BS, 59; Univ Rochester, MS, 61; Yale Univ, PhD(geol), 65. Prof Exp: Asst prof, Southampton Col, Long Island Univ, 65-67; from asst prof to assoc prof, 67-75, PROF GEOL, QUEENS COL, NY, 76- Concurrent Pos: Prin investr, NASA grant, 75- Mem: Nat Asn Geol Teachers; fel Geol Soc Am; Am Asn Petrol Geologists; Int Asn Sedimentol; Int Asn Gt Lakes Res. Res: Coastal and estuarine sedimentology; Atlantic Coastal Plain stratigraphy; environmental geology; pollution studies in Hudson Estuary and adjacent waters; determination of sedimentary structures and dispersal patterns in lunar cores. Mailing Add: Dept of Earth & Environ Sci Queens Col Flushing NY 11367

COCHIN, JOSEPH, b Winnipeg, Man, July 17, 16; nat US; m 51; c 3. PHARMACOLOGY. Educ: Wayne State Univ, BS, 37; Univ Mich, MD, 53, PhD(pharmacol), 55. Prof Exp: Chemist, Keystone Oil Refining Co, 37-39; teacher, Detroit Bd Educ, 39-41; asst pharmacol, Univ Mich, 47-53, intern, Univ Hosp, 53-54; pharmacologist, Sect Analgesics, Nat Inst Arthritis & Metab Dis, 54-62; assoc prof, 62-66, PROF PHARMACOL, SCH MED, BOSTON UNIV, 66-, PROF PSYCHIAT, 70- Concurrent Pos: Consult, Coun Drugs, AMA; mem sci rev comt, Ctr for Study Narcotic & Drug Abuse, NIMH, 67-70, chmn, 70-72; mem, Nat Acad Sci-Nat Res Coun Comt on Probs of Drug Dependence, 68-73; chmn sci consults, Alcohol & Drug Dependence Merit Rev Bd, Vet Admin, 72; mem ad hoc rev comt drug abuse training & educ, Nat Inst Drug Abuse, 72-; consult, Spec Action Off Drug Abuse Prev & Ctr Dis Control; specific field ed, J Pharmacol & Exp Therapeut. Mem: Fel AAAS; Am Soc Pharmacol & Exp Therapeut; Am Col Neuropsychopharmacol. Res: Tolerance, habituation and physical dependence to drugs; psychopharmacology; mode of action of analgesic drugs; drug metabolism. Mailing Add: Dept of Pharmacol Boston Univ Sch of Med Boston MA 02118

COCHIS, THOMAS, b Boston, Mass, June 24, 36; m 60; c 1. BOTANY, HORTICULTURE. Educ: McNeese State Col, BS, 60; La State Univ, MS, 62, PhD(hort, bot), 64. Prof Exp: Asst hort, La State Univ, 60-62, instr, 62-64; asst prof bot, Millsaps Col, 64-66; asst prof hort, Univ Fla, 66-68; PROF BOT, JACKSONVILLE STATE UNIV, 68- Mem: AAAS; Am Soc Hort Sci; Am Inst Biol Sci; Int Soc Hort Sci. Res: Horticultural and botanical research and teaching. Mailing Add: Dept of Biol Ayers Hall Jacksonville State Univ Jacksonville AL 36265

COCHOY, ROBERT EDMUND, organic chemistry, see 12th edition

COCHRAN, ALLAN CHESTER, b Long Beach, Calif, Jan 23, 42; m 62; c 2. TOPOLOGY. Educ: E Cent State Col, BS, 62; Univ Okla, MA, 64, PhD(math), 66. Prof Exp: Asst prof, 66-71, ASSOC PROF MATH, UNIV ARK, FAYETTEVILLE, 71- Mem: Am Math Soc; Math Asn Am. Res: Convergence space theory; theory of topological algebras. Mailing Add: Dept of Math Univ of Ark Fayetteville AR 72701

COCHRAN, ANDREW AARON, b West Plains, Mo, Sept 28, 19; m 42; c 3. BIOPHYSICS, QUANTUM PHYSICS. Educ: Univ Mo-Rolla, BS, 41, MS, 63. Prof Exp: Chem engr, Phillips Petrol Co, Kans, 41-46; org res chemist, Mallinckrodt Chem Works, Mo, 46-52; chemist, Metall Res, 52-60, supvry phys chemist, 60-62, SUPVRY RES CHEMIST, METALL RES, ROLLA METALL RES CTR, US BUR MINES, 62- Mem: Am Inst Mining, Metall & Petrol Eng. Res: The fundamental theory of biophysics; development of new processes for recovering titanium, tin, and manganese from their respective ores; development of processes for treating industrial wastes to recover metals and reduce pollution. Mailing Add: College Hills RFD 4 Box 20 Rolla MO 65401

COCHRAN, CHARLES NORMAN, b Pittsburgh, Pa, Mar 24, 25; m 46; c 3. PHYSICAL CHEMISTRY. Educ: Westminster Col, Pa, BS, 45; Ohio State Univ, MS, 47. Prof Exp: Scientist, Phys Chem Div, 47-58, sect head, 58-63, MGR, PHYS CHEM DIV, ALCOA LABS, 63- Honors & Awards: Dr Rene Wasserman Award, Am Welding Soc, 72. Mem: Am Chem Soc; Metall Soc; Am Soc Testing & Mat. Res: Oxidation of metals; gas in light metals; gas sorption; high temperature chemistry of aluminum; aluminum smelting; joining of aluminum; energy involvement of aluminum industry. Mailing Add: Phys Chem Div Alcoa Labs Alcoa Center PA 15069

COCHRAN, DONALD GORDON, b New Hampton, Iowa, July 5, 27; m 52; c 3. INSECT PHYSIOLOGY. Educ: Iowa State Univ, BS, 50; Va Polytech Inst, MS, 52; Rutgers Univ, PhD(entom), 55. Prof Exp: Entomologist, Chem Ctr, US Army, 55-57; assoc entomologist, Agr Exp Sta, 57-59, assoc prof, 59-64, PROF ENTOM, ENTOM INST, VA POLYTECH INST & STATE UNIV, 64- Mem: AAAS; Entom Soc Am. Res: Physiological and genetical aspects of insect resistance to insecticides; cockroach genetics and cytogenetics; biochemistry of insect excretion. Mailing Add: Dept of Entom Va Polytech Inst & State Univ Blacksburg VA 24061

COCHRAN, DONALD ROY FRANCIS, b San Francisco, Calif, Oct 6, 26; m 60; c 3. NUCLEAR SCIENCE. Educ: Univ Calif, Berkeley, BS, 48; Johns Hopkins Univ, PhD(nuclear chem), 54. Prof Exp: Instr chem, Johns Hopkins Univ, 52-54; mem staff physics, 54-68, group leader, 68-73, ASST DIV LEADER, MESON PHYSICS DIV, LOS ALAMOS SCI LAB, 73- Concurrent Pos: Mem adv panel on accelerator radiation safety, US AEC, 70-73. Mem: Am Phys Soc; Am Inst Physics; Am Nuclear Soc; Am Chem Soc. Res: Medium energy physics; nuclear chemistry and physics; accelerators. Mailing Add: PO Box 1663 MPDO-MS850 Los Alamos NM 87545

COCHRAN, EDWARD LEO, physical chemistry, see 12th edition

COCHRAN, FRED DERWARD, b Ware Shoals, SC, Nov 24, 10; m 37; c 2. HORTICULTURE. Educ: Clemson Col, BS, 32; La State Univ, MS, 33; Univ Calif, PhD, 42. Prof Exp: From asst horticulturist to assoc horticulturist, La State Univ, 36-47; head dept, 47-69, PROF HORT SCI & GENETICS, NC STATE UNIV, 47- Mem: Am Soc Hort Sci; Potato Asn Am. Res: Breeding and cytogenetics of vegetable crops. Mailing Add: Dept of Hort NC State Univ Raleigh NC 27607

COCHRAN, GEORGE THOMAS, b Washington, DC, Dec 28, 38; m 67; c 1. INORGANIC CHEMISTRY, ANALYTICAL CHEMISTRY. Educ: Univ Richmond, BS, 60; Univ Tenn, MS, 63; Clemson Univ, PhD(chem), 67. Prof Exp: Teacher high sch, Va, 60-61; asst prof, 67-71, ASSOC PROF CHEM, ROLLINS COL, 71- Mem: Am Chem Soc. Res: Coordination chemistry of univalent metal ions; extraction of

metal chelates; organic polarography. Mailing Add: Dept of Chem Rollins Col Winter Park FL 32789

COCHRAN, GEORGE WILSON, b Topeka, Kans, July 7, 19; m 43; c 2. PLANT VIROLOGY. Educ: Kans State Col, BS, 41, MS, 42; Cornell Univ, PhD(plant path), 46. Prof Exp: Asst, Rockefeller Inst, NJ, 46-47; PROF, DEPT BIOL, UTAH STATE UNIV, 48- Concurrent Pos: Fulbright fel, Neth, 57-58; Fulbright lectr, Italy, 58; partic, Int Tobacco Cong, Brussels, 58, Int Microbiol Cong, Stockholm, 58, Int Biophys Cong, Stockholm, 61 & Int Biochem Cong, Moscow, 61, New York, 64 & Tokyo, 67; guest scholar lectr, Kans State Univ, 64; partic, Ivanovsky Centennial Conf on Origin & Nature of Viruses, Moscow, 64, Int Conf Plant Viruses, Wageningen, Neth, 65 & Inf Conf Biochem Regulation in Diseased Plants or Injury, Tokyo, 67. Honors & Awards: First Prize, Am Urol Asn, 55. Mem: AAAS; Am Phytopath Soc; Electron Micros Soc Am; Am Chem Soc; Am Soc Microbiol. Res: Electron microscopy; chromatography; electrophoresis and dodder transmission of plant viruses; cell-free synthesis of viral nucleic acids and viruses. Mailing Add: Dept of Biol Utah State Univ Logan UT 84322

COCHRAN, HULON LILLEY, b Norman, Tex, Nov 11, 07; m. HORTICULTURE. Educ: Okla Agr & Mech Col, BS, 31; Cornell Univ, MS, 32, PhD(veg crops), 35. Prof Exp: Asst veg crops, Cornell Univ, 31-35; from asst horticulturist to assoc horticulturist, Exp Sta, Univ Ga, 35-44; horticulturist, Pomona Prod Co, 44-56, Mgr, Raw Prods Div, 56-74; RETIRED. Concurrent Pos: Agent, USDA, 38-44. Mem: Assoc Am Soc Plant Physiol; assoc Am Soc Hort Sci. Res: Fruit setting and seed germination in peppers; morphology of vegetable crop plants; pimiento breeding; soil reaction work; variety testing of green beans for processing and of peaches for pickling; factors affecting seed production in pimiento peppers. Mailing Add: 915 Maple Dr Griffin GA 30223

COCHRAN, JAMES ALAN, b San Francisco, Calif, May 12, 36; m 58; c 2. APPLIED MATHEMATICS. Educ: Stanford Univ, BS, 56, MS, 57, PhD(math), 62. Prof Exp: Res mathematician, Stanford Res Inst, 55-58; asst math, Stanford Univ, 58-61; mem tech staff, Bell Tel Labs, 62-65, supvr, Electromagnetic Res Dept, 65-68, supvr, Appl Math & Statist Dept, 69-72; PROF MATH, VA POLYTECH INST & STATE UNIV, 72- Concurrent Pos: Vis prof math, Stanford Univ, 68-69. Mem: Am Math Soc; Soc Indust & Appl Math; Math Asn Am; Sigma Xi. Res: Applied research in electromagnetic theory and microwave propagation; basic research in special functions, asymptotics, differential and integral equations, and operator theory. Mailing Add: Dept of Math Va Polytech Inst & State Univ Blacksburg VA 24061

COCHRAN, JAMES EDWARD, JR, organic chemistry, see 12th edition

COCHRAN, JOHN CHARLES, b Akron, Ohio, Feb 10, 35; m 58; c 3. ORGANIC CHEMISTRY. Educ: Col Wooster, BA, 57; Univ NC, MA, 60; Univ NH, PhD(org chem), 67. Prof Exp: Instr chem, Randolph-Macon Women's Col, 60-62; asst prof, 66-71, ASSOC PROF CHEM, COLGATE UNIV, 71- Mem: Am Chem Soc. Res: Preparation and reactions of organotin compounds; structure of allylic carbanions; reactions of sulfite esters. Mailing Add: Dept of Chem Colgate Univ Hamilton NY 13346

COCHRAN, JOHN FRANCIS, b Saskatoon, Sask, Jan 29, 30; m 57; c 2. SOLID STATE PHYSICS. Educ: Univ BC, BASc, 50, MASc, 51; Univ Ill, PhD(physics), 55. Prof Exp: Res assoc physics, Univ Ill, 55-56; Nat Res Coun Can fel, Clarendon Lab, Eng, 56-57; asst prof physics, Mass Inst Technol, 57-65; PROF PHYSICS, SIMON FRASER UNIV, 65- Concurrent Pos: Sloan fel, 58-60. Mem: Am Phys Soc; Can Asn Physicists. Res: Electronic properties of very pure metals. Mailing Add: Dept of Physics Simon Fraser Univ Burnaby BC Can

COCHRAN, JOHN RODNEY, b St Joseph, Mo, Feb 7, 20; m 41; c 2. SPEECH PATHOLOGY. Educ: Utah State Univ, BS, 49, MS, 50; Univ Utah, PhD(speech path), 59. Prof Exp: Asst speech path, Univ Utah, 54-59; assoc prof speech path & psychol, Southwest Tex State Col, 59-61; assoc prof speech & speech path, chmn dept speech path & audiol & dir speech & hearing clin, Eastern NMex Univ, 61-70; prof & dir speech & hearing clin, Univ Alaska, 70-71; assoc prof & dir speech & hearing clin, Leman Col, 71-73; ASSOC PROF SPEECH & DIR SPEECH & HEARING CLIN, KEARNEY STATE COL, 73- Mem: Am Psychol Asn; Am Speech & Hearing Asn. Res: Communication abilities of air force personnel. Mailing Add: Speech & Hearing Clin Kearney State Col Kearney NE 68847

COCHRAN, KENNETH WILLIAM, JR, b Chicago, Ill, Nov 2, 23; m 45; c 2. PHARMACOLOGY. Educ: Univ Chicago, SB, 47, PhD(pharmacol), 50; Am Bd Med Microbiol, dipl virol; Am Bd Indust Hyg, cert toxicol. Prof Exp: Res asst toxicity lab, Univ Chicago, 48-50, res assoc, US Air Force Radiation Lab & Dept Pharmacol, 50-52; instr pharmacol, Sch Med, 52-60, res assoc epidemiol, Sch Pub Health, 52-55, from asst prof to assoc prof, 55-68, secy fac, 70-73, ASST PROF PHARMACOL, SCH MED, UNIV MICH, ANN ARBOR, 60-, PROF EPIDEMIOL, SCH PUB HEALTH, 68- Concurrent Pos: Mem, pharmacol & endocrinol fel rev panel, NIH, 60-64. Mem: AAAS; Am Soc Microbiol; Am Acad Indust Hyg; Am Soc Pharmacol; Am Chem Soc. Res: Virus chemotherapy; toxicology. Mailing Add: 3556 Oakwood Ann Arbor MI 48104

COCHRAN, LEWIS WELLINGTON, b Perryville, Ky, Oct 12, 15; m 40; c 2. PHYSICS. Educ: Morehead State Col, BS, 36; Univ Ky, MS, 39 & 40, PhD(physics), 52. Prof Exp: Instr math & physics, Morehead State Col, 39-41 & Cumberland Univ, 41; from asst prof to assoc prof physics, 46-57, actg head dept, 56-58, assoc dean, 63-66, provost, 65-67, actg dean grad sch, 66-67, dean grad sch & vpres res, 67-70, PROF PHYSICS, UNIV KY, 57-, VPRES ACAD AFFAIRS, 70- Mem: Am Phys Soc; Am Asn Physics Teachers. Res: Proton induced nuclear reactions; interaction of radiation with matter; experimental nuclear physics; nuclear structure physics; gaseous electronics. Mailing Add: Univ of Ky 111 Admin Bldg Lexington KY 40506

COCHRAN, LLOYD CURTIS, horticulture, see 12th edition

COCHRAN, MICHAEL DAVID, geophysics, seismology, see 12th edition

COCHRAN, PATRICK HOLMES, b Guthrie Center, Iowa, Oct 14, 37; m 61; c 3. FOREST SOILS. Educ: Iowa State Univ, BSc, 59; Ore State Univ, MSc, 63, PhD(soils), 66. Prof Exp: Asst soils, Ore State Univ, 64-65; asst prof silvicult, State Univ NY Col Forestry, Syracuse Univ, 65-67; SOIL SCIENTIST, PAC NORTHWEST FOREST & RANGE EXP STA, US FOREST SERV, 67- Mem: Am Soc Agron; Soc Am Foresters. Res: Relationships of physical, thermal and chemical soil properties to the distribution of wild land vegetation and tree growth. Mailing Add: Silvicult Lab 1027 Trenton Ave Bend OR 97701

COCHRAN, PAUL TERRY, b Sullivan, Ind, Jan 27, 38; m 72; c 2. CARDIOLOGY, MEDICAL EDUCATION. Educ: DePauw Univ, BA, 60; Western Reserve Univ, MD, 64; Am Bd Internal Med, dipl; Am Bd Cardiovasc Dis, dipl. Prof Exp: Staff cardiologist, Malcolm Grow US Air Force Hosp, 69-71; clin cardiologist, Gallatin

Med Group, Downey, Calif, 71-72; ASST PROF MED, SCH MED, UNIV NMEX, 72- Concurrent Pos: Dir cardiac diag lab, Bernalillo County Med Ctr, 72-; staff cardiologist, Vet Admin Hosp, Albuquerque, 72-; fel coun clin cardiol, Am Heart Asn. Mem: Fel Am Col Cardiol. Res: Clinical investigations of the hemodynamics of heart disease; research in teaching methods in cardiovascular medicine. Mailing Add: Dept of Med Univ of NMex Sch of Med Albuquerque NM 87131

COCHRAN, ROBERT GLENN, b Indianapolis, Ind, July 12, 19; m 44; c 1. PHYSICS. Educ: Ind Univ, AB, 48, MS, 50; Pa State Univ, PhD(nuclear physics), 57. Prof Exp: Res asst cyclotron group, Ind Univ, 47-50; nuclear physicist & group leader chg swimming pool reactor fac, Physics Div, Oak Ridge Nat Lab, 50-54; assoc prof nuclear eng & dir res reactor facil, Pa State Univ, 54-59; PROF NUCLEAR ENG & HEAD DEPT & DIR NUCLEAR SCI CTR, TEX A&M UNIV, 59- Concurrent Pos: Consult, Nat Regulatory Comn; US Air Force; Sandia Corp & Watertown Arsenal. Mem: Am Nuclear Soc; Am Phys Soc; Am Soc Eng Educ; Nat Soc Prof Engrs. Res: Nuclear physics; decay schemes and isomeric states of short half-life nuclids; nuclear reactor physics and nuclear engineering; power reactors and their environmental effects. Mailing Add: Dept of Nuclear Eng Tex A&M Univ College Station TX 77843

COCHRAN, WILLIAM GEMMELL, b Rutherglen, Scotland, July 15, 09; nat; m 37; c 3. APPLIED STATISTICS. Educ: Glasgow Univ, MA, 31; Cambridge Univ, BA, 33, MA, 38. Hon Degrees: LLD, Glasgow Univ, 70; LLD, Johns Hopkins Univ, 75. Prof Exp: Statistician, Rothamsted Exp Sta, Eng, 34-39; prof math statistics, Iowa State Col, 39-46; assoc dir inst statist, NC State Col, 46-48; prof biostatist, Sch Hyg & Pub Health, Johns Hopkins Univ, 49-57; PROF STATIST, HARVARD UNIV, 57- Concurrent Pos: Res mathematician, Off Sci Res & Develop, Princeton Univ, 44; bombing res analyst, US War Dept, 45; mem Nat Acad Sci, 74; fel Am Acad Arts & Sci. Mem: Biomet Soc (pres, 54); fel Inst Math Statist (vpres, 44, pres, 46); fel Am Statist Asn (vpres, 43-45, ed, 45-50, pres, 53); hon fel Royal Statist Soc; Int Statist Inst (pres, 67-71). Res: Mathematical statistics and application to biology and social science. Mailing Add: Sci Ctr Rm 608 One Oxford St Cambridge MA 02138

COCHRAN, WILLIAM RONALD, b Kalamazoo, Mich, May 24, 40. PHYSICS. Educ: Univ Calif, Los Angeles, BA, 62, MS, 64, PhD(physics), 69. Prof Exp: Asst prof, 69-74, ASSOC PROF PHYSICS, YOUNGSTOWN STATE UNIV, 74- Mem: Am Phys Soc; Optical Soc Am. Res: Spectroscopy of ions in crystals; heterodyne spectroscopy of solids; optical physics. Mailing Add: Dept of Physics Youngstown State Univ Youngstown OH 44503

COCHRANE, CHAPPELLE CECIL, b Conway, SC, Oct 28, 13; m 40; c 3. ORGANIC CHEMISTRY. Educ: Howard Univ, BS, 38, MS, 40, Ohio State Univ, PhD(chem), 51. Prof Exp: Assoc, Howard Univ, 40-41; instr chem, Morgan State Col, 41-42 & Cent State Col, 46-47; res chemist, Glidden co, 51-58, Armour & Co, 58-60, US Army Chem Ctr, 60-62 & Nalco Chem Co, 62-74; ASST PROF CHEM, CHICAGO STATE UNIV, 74- Concurrent Pos: Assoc, Ohio State Univ, 51. Honors & Awards: Lloyd Hall Award, Howard Univ, 38. Mem: Am Chem Soc. Res: Polynuclear aromatic hydrocarbons; steroids; fatty acids. Mailing Add: 39 S 18th Ave Maywood IL 60153

COCHRANE, HECTOR, b Stowmarket, Eng, Mar 16, 40; m 65; c 2. PHYSICAL CHEMISTRY. Educ: Univ Nottingham, BSc, 61, PhD(chem), 64. Prof Exp: Fel fuel sci, Pa State Univ, 64-66; res chemist, 66-73, GROUP LEADER, CAB-O-SIL RES & DEVELOP, CABOT CORP, 73- Mem: Am Chem Soc. Res: Formation of fine particles in flames; study of the surface chemistry and aggregate morphology of particles and how these properties affect their theological properties in liquids and reinforcement properties in rubber. Mailing Add: 3 Conestoga Rd Chelmsford MA 01824

COCHRANE, ROBERT LOWE, b Morgantown, WVa, Feb 10, 31. REPRODUCTIVE PHYSIOLOGY, ENDOCRINOLOGY. Educ: WVa Univ, BA, 53; Univ Wis, MS, 54, PhD(genetics), 61. Prof Exp: Animal husb agent, USDA, 54-61; biologist, US Food & Drug Admin, 61-62; sr res fel primate reproduction, Med Sch, Univ Birmingham, 62-65; proj assoc, Sch Med, Univ Pittsburgh, 65-66; SR ENDOCRINOLOGIST, LILLY RES LABS, 66- Concurrent Pos: Res asst zool, Univ Wis, 57-60. Mem: AAAS; Am Soc Animal Sci; Am Inst Biol Sci; Soc Study Reproduction; Endocrine Soc. Res: Contraception; ovoimplantation; control of corpus luteum and ovary function; immunological tolerance, hormonal control of growth and effects of estrogens on reproduction in mustelids. Mailing Add: Lilly Res Labs Eli Lilly & Co Indianapolis IN 46206

COCHRANE, VINCENT WINNER, b Plainfield, NJ, Aug 21, 16; m 45; c 2. MICROBIOLOGY. Educ: Cornell Univ, BS, 39, PhD(plant path), 44. Hon Degrees: MA, Wesleyan Univ, 57. Prof Exp: Asst, Cornell Univ, 41-44; microbiologist, Lederle Labs, NY, 44-45; asst plant pathologist, Exp Sta, Univ Conn, 45-47; from asst prof to assoc prof, 47-57, PROF BIOL, WESLEYAN UNIV, 57- Mem: AAAS; fel Am Phytopath Soc; Bot Soc Am; Am Soc Microbiol; Mycol Soc Am. Res: Physiology of fungi and actinomycetes; ecology of soil microorganisms. Mailing Add: Dept of Biol Hall-Atwater Lab Wesleyan Univ Middletown CT 06457

COCHRANE, WILLIAM, b Toronto, Ont, Mar 18, 26; m 51; c 3. PEDIATRICS. Educ: Univ Toronto, MD, 49; FRCP(C), 56. Prof Exp: Clinician, Hosp Sick Children, Toronto, 56-58; assoc prof pediat, Fac Med, Dalhousie Univ, 58-63, prof & head dept, 63-67; dean fac med, 67-73, PRES & VCHANCELLOR, UNIV CALGARY, 74- Concurrent Pos: Dep minister health serv, Govt Alta, 73-74. Mem: Soc Pediat Res; fel Am Col Physicians; Can Pediat Soc; Can Soc Clin Invest. Res: Metabolic diseases of children; biochemical relationship of protein and amino acid metabolism to mental disease. Mailing Add: Off of the Pres Univ of Calgary Calgary AB Can

COCIVERA, MICHAEL, b Pittsburgh, Pa, Jan 21, 37; m 69. PHYSICAL CHEMISTRY. Educ: Carnegie-Mellon Univ, BSc, 59; Univ Calif, Los Angeles, PhD(phys chem), 63. Prof Exp: Mem tech staff chem, Bell Tel Labs, 63-69; assoc prof, 69-74, PROF CHEM, UNIV GUELPH, 74- Res: Nuclear magnetic resonance studies of fast thermal and photochemical reactions. Mailing Add: Dept of Chem Univ of Guelph Guelph ON Can

COCK, LORNE M, b Tatamagouche, NS, June 1, 32; m 57; c 2. ANIMAL NUTRITION. Educ: McGill Univ, BSc, 54; Univ Wis, MS, 60; Univ Maine, PhD(animal nutrit), 66. Prof Exp: From instr to asst prof animal sci, Univ Maine, 65-69; assoc prof, 69-74, PROF ANIMAL SCI, NS AGR COL, 74-, HEAD DEPT, 69- Mem: Am Soc Animal Sci; Am Dairy Sci Asn; Agr Inst Can. Res: Energy metabolism of ruminants; influence of dietary nitrogen on ruminant heat increment; fasting metabolism of sheep. Mailing Add: NS Agr Col Truro NS Can

COCKE, ALBERT KIRVEN, physics, see 12th edition

COCKE, ELTON CROMWELL, botany, deceased

COCKERELL, LEONE (DORIS), b Mabank, Tex, Mar 9, 09. ANALYTICAL CHEMISTRY, INORGANIC CHEMISTRY. Educ: NTex State Col, BS, 32, MS, 37; Univ Colo, PhD(chem), 54. Prof Exp: Teacher pub sch, Tex, 33-36 & 41-45 & NJ, 46-47; instr & supvr high sch sci, Henderson State Teachers Col, Ark, 37-41; instr chem, Univ Ill, 47-49; from assoc prof to prof chem, 49-74, EMER PROF CHEM, BAYLOR UNIV, 74- Res: Study of complex ions in solution and on exchange resins. Mailing Add: Dept of Chem Baylor Univ Waco TX 76706

COCKERHAM, COLUMBUS CLARK, b Mountain Park, NC, Dec 12, 21; m 44; c 3. POPULATION GENETICS, QUANTITATIVE GENETICS. Educ: NC State Col, BS, 43, MS, 49; Iowa State Univ, PhD(animal breeding & genetics), 52. Prof Exp: Asst prof biostatist, Sch Pub Health, Univ NC, 52-53; from assoc prof to prof exp statist, 53-72, WILLIAM NEAL REYNOLDS PROF STATIST & GENETICS, 72- Concurrent Pos: Prin investr, NIH Grant, 60-63, proj dir, 63-; mem Nat Acad Sci, 74; ed, Theoret Pop Biol, 75-77. Mem: AAAS; Biomet Soc; Am Soc Animal Sci; Genetics Soc Am; fel Am Soc Agron. Res: Population and quantitative genetic theory; estimation of genetic parameters in populations; development of mating and experimental designs for estimation; selection theory; applications to plant and animal breeding. Mailing Add: Dept of Statist NC State Univ Raleigh NC 27607

COCKERLINE, ALAN WESLEY, Toronto, Ont, Oct 2, 26; m 51; c 3. CYTOLOGY. Educ: Univ Mich, BS, 52, MS, 53; Mich State Univ, PhD(bot), 61. Prof Exp: Asst cancer & tissue cult, Univ Mich, 50-53; asst cytol, Mich State, 53-61; asst prof, 61-66, ASSOC PROF BIOL, TEX WOMAN'S UNIV, 66- Concurrent Pos: Proj dir undergrad res partic, NSF, 64-72. Mem: AAAS; Am Chem Soc; Am Asn Plant Physiologists; Mycol Soc Am; Tissue Cult Asn. Res: Mycotoxins; plant embryo culture; related callus formations. Mailing Add: Dept of Biol Tex Woman's Univ PO Box 22847 Denton TX 76204

COCKETT, ABRAHAM TIMOTHY K, b Maui, Hawaii, Sept 4, 28; m; c 4. UROLOGY, PHYSIOLOGY. Educ: Brigham Young Univ, BS, 50; Univ Utah, MD, 54. Prof Exp: Res fel, Dept Med, Univ Southern Calif, 57-58; assoc prof urol, Univ Calif, Los Angeles, 62-69; PROF UROL SURG, SCH MED, UNIV ROCHESTER, 69- Concurrent Pos: Chief urol, Harbor Gen Hosp, Torrance, Calif, 62-69; urologist in chief, Strong Mem Hosp, Rochester, NY, 69- Mem: AAAS; Soc Univ Surgeons; Am Urol Asn; Soc Univ Urologists; Undersea Med Soc. Res: Kidney physiology and transplantation; underwater physiology related to problems in decompression sickness and treatment of these alterations; support of man in outer space; studies related to renal, urinary and testicular physiology. Mailing Add: Div of Urol Univ of Rochester 601 Elmwood Ave Rochester NY 14642

COCKRELL, BEVERLY YVONNE, b Harrisonville, Mo, Feb 1, 37. VETERINARY PATHOLOGY, ELECTRON MICROSCOPY. Educ: Rockford Col, BA, 59; Univ Ill, Urbana-Champaign, BS, 63, DVM, 65, PhD(path), 69. Prof Exp: Res asst, Ben May Lab Cancer Res, Univ Chicago, 59-61; asst prof, Ctr Lab Animal Resources, Col Vet Med, Mich State Univ, 69-72; SR SCIENTIST, LITTON BIONETICS, INC, 72- Mem: AAAS; Am Vet Med Asn; Am Asn Lab Animal Sci; Int Acad Path. Res: Ultrastructure of drug-induced changes in the liver and kidney. Mailing Add: Litton Bionetics Inc 5510 Nicholson Lane Kensington MD 20795

COCKRELL, ROBERT ALEXANDER, b Yonkers, NY, Aug 11, 09; m 33; c 3. DENDROLOGY, WOOD SCIENCE & TECHNOLOGY. Educ: Syracuse Univ, BS, 30, MS, 31; Univ Mich, PhD(wood technol), 34. Prof Exp: Asst, NY State Col Forestry, Syracuse Univ, 30-32; asst, Sch Forestry & Conserv, Univ Mich, 32-34, jr forester, 34-35; assoc prof forestry, Clemson Col, 35-36; from asst prof to assoc prof, 36-50, assoc forester, Exp Sta, 45-62, assoc dean, Grad Div Univ, 56-67, PROF FORESTRY, UNIV CALIF, BERKELEY, 50-, SECY, ACAD SENATE, 68-, WOOD TECHNOLOGIST, EXP STA, 62- Concurrent Pos: Technologist, Forest Prod Lab, US Forest Serv, 42-45; forestry res specialist, US Army, Japan, 50; Orgn Europ Econ Coop sr vis fel, 61. Mem: AAAS; Soc Am Foresters; Forest Prod Res Soc. Res: Anatomy of tropical woods; mechanical properties of wood; gluing of wood; wood shrinkage; cell wall structure. Mailing Add: Forestry Dept Univ of Calif Berkeley CA 94720

COCKRELL, ROBERT GIFFORD, nuclear engineering, see 12th edition

COCKRELL, RONALD SPENCER, b Kansas City, Mo, June 26, 38; m 60; c 2. BIOCHEMISTRY. Educ: Univ Mo, BS, 63, BMedS, 64; Univ Pa, PhD(molecular biol), 68. Prof Exp: Asst prof, 69-74, ASSOC PROF BIOCHEM, SCH MED, ST LOUIS UNIV, 74- Concurrent Pos: USPHS fel, Cornell Univ, 68-69. Mem: AAAS; Am Soc Biol Chem. Res: Bioenergetics; ion transport and metabolism. Mailing Add: Dept of Biochem St Louis Univ Sch of Med St Louis MO 63104

COCKRUM, ELMER LENDELL, b Sesser, Ill, May 29, 20; m 43; c 3. VERTEBRATE ZOOLOGY. Educ: Univ Kans, PhD(zool), 51. Prof Exp: Asst cur mammal, Mus Natural Hist, Univ Kans, 46-48, res assoc, 51-52, fel embryol & gen zool, Univ, 48-49; from asst prof to assoc prof, 52-60, dir desert biol sta, 65-68, PROF ZOOL, UNIV ARIZ, 60-, CUR MAMMALS, 57-, MAMMALOGIST, AGR EXP STA, 74- Concurrent Pos: Assoc, Mus Natural Hist, 58- Mem: AAAS; Am Soc Zool; Am Soc Mammal. Res: Mammals of Kansas and Arizona; microtine rodents; life history studies of bats. Mailing Add: Dept of Biol Sci Univ of Ariz Tucson AZ 85721

COCKS, GEORGE GOSSON, b Sioux City, Iowa, Mar 22, 19; m 42; c 4. CHEMICAL MICROSCOPY. Educ: Iowa State Col, BS, 41; Cornell Univ, PhD(chem micros), 49. Prof Exp: Chemist, Allison Div, Gen Motors Corp, 41; asst chem micros, Cornell Univ, 46-49; asst chief physics solids div, Battelle Mem Inst, 49-64; ASSOC PROF CHEM MICROS, CORNELL UNIV, 64- Mem: AAAS; Am Chem Soc; Optical Soc Am; Electron Micros Soc Am (exec secy, 59-); Am Soc Metals. Res: Light and electron microscopy; chemical microscopy; optical crystallography; structure of polymer gels; growth of crystals in gels; crystallization of ice. Mailing Add: Sch of Chem Eng Cornell Univ Ithaca NY 14853

COCOLAS, GEORGE HARRY, b Flushing, NY, July 9, 29; m 53; c 4. MEDICINAL CHEMISTRY. Educ: Univ Conn, BS, 52; Univ NC, PhD(pharm), 56. Prof Exp: Org res chemist, Res Labs, Nat Drug Co, 56-58; from asst prof to assoc prof, 58-73, PROF MED CHEM, UNIV NC, CHAPEL HILL, 73-, HEAD DIV MED CHEM, SCH PHARM, 75- Mem: Am Chem Soc; Am Pharmaceut Asn. Res: Stereochemistry and biological activity; cholinergic mechanisms. Mailing Add: Sch of Pharm Univ of NC Chapel Hill NC 27514

CODD, JOHN EDWARD, b Spokane, Wash, Oct 13, 36; m 64; c 5. TRANSPLANTATION IMMUNOLOGY, THORACIC SURGERY. Educ: Gonzaga Univ, BA, 58; St Louis Univ, MD, 63. Prof Exp: Asst prof, 71-74, ASSOC PROF SURG, MED SCH, ST LOUIS UNIV, 75- Concurrent Pos: Chief, Unit II Surg, John Cochran Vet Admin Hosp, 71- Mem: Am Col Surgeons; Soc Transplant Surgeons; Asn Acad Surg. Res: Organ preservation for transplantation. Mailing Add: Dept Surg St Louis Univ Med Sch 1325 S Grand Blvd St Louis MO 63104

CODDING, EDWARD GEORGE, b Ionia, Mich, Jan 17, 42; m 68; c 1.

ANALYTICAL CHEMISTRY. Educ: Cent Mich Univ, BSc, 65; Mich State Univ, PhD(chem), 71. Prof Exp: ASST PROF CHEM, KENT STATE UNIV, 74- Mem: Am Chem Soc; Soc Appl Spectroscopy. Res: Application of solid state image sensors as spectrochemical detectors; digital data handling techniques; incorporation of digital and analog instrumentation techniques for the measurement of chemical information. Mailing Add: Dept of Chem Kent State Univ Kent OH 44242

CODDINGTON, EARL ALEXANDER, b Washington, DC, Dec 16, 20; m 45; c 3. MATHEMATICS. Educ: Johns Hopkins Univ, PhD(math), 48. Prof Exp: Physicist, Naval Ord Lab, Washington, DC, 42; mathematician, Navy Dept, 42-46; instr, Johns Hopkins Univ, 48-49; instr math, Mass Inst Technol, 49-52; from asst prof to assoc prof, 52-59, chmn dept, 68-71, PROF MATH, UNIV CALIF, LOS ANGELES, 59- Mem: AAAS; Math Asn Am; Am Math Soc. Res: Differential equations; analysis. Mailing Add: Dept of Math Univ of Calif Los Angeles CA 90024

CODE, ARTHUR DODD, b Brooklyn, NY, Aug 13, 23; m 43; c 4. ASTRONOMY, ASTROPHYSICS. Educ: Univ Chicago, MS, 47, PhD(astron & astrophysics), 50. Prof Exp: Asst, Yerkes Observ, Univ Chicago, 46-49; instr, Univ Va, 50; from instr to asst prof astron, Univ Wis, 51-56; assoc prof & mem staff, Mt Wilson & Palomar Observs, Calif Inst Technol, 56-58; prof, 58-69, JOEL STEBBINS PROF ASTRON, WASHBURN OBSERV, UNIV WIS-MADISON, 69-, DIR, 58- Honors & Awards: Pub Serv Award, Nat Aeronaut & Space Admin, 69; Prof Achievement Award, Univ Chicago Alumni Asn, 70. Mem: Nat Acad Sci; Am Astron Soc; Int Acad Astronaut. Res: Photoelectric photometry of stars and nebulae; stellar spectroscopy; development of instruments; satellite astronomy. Mailing Add: Washburn Observ Univ of Wis Madison WI 53706

CODE, CHARLES FREDERICK, b Can, Feb 1, 10; nat US; m 35; c 3. PHYSIOLOGY. Educ: Univ Man, MD & BSc, 34; Univ Minn, PhD, 39. Prof Exp: Lectr physiol, Univ London, 35-36; asst exp surg, Mayo Grad Sch Med, Univ Minn, 37, from instr to prof physiol, 38-75, dir med educ & res, 66-72; ASSOC DIR & SR RES SCIENTIST, CTR ULCER RES & EDUC, WADSWORTH VET ADMIN HOSP & UNIV CALIF, LOS ANGELES, 75- Concurrent Pos: Mem staff & consult, Mayo Clin, 40-75. Honors & Awards: Theobald Smith Award, AAAS, 38; Friedenwald Medal, Am Gastroenterol Asn, 74. Mem: AAAS; Am Physiol Soc; Brit Physiol Soc; Am Soc Pharmacol & Exp Therapeut; Am Soc Clin Invest. Res: Metabolism of histamine, relationship to gastric secretion; hypersensitive state; physiology of gastrointestinal tract; motor action of the alimentary canal, secretion and absorption from the stomach, small and large bowel. Mailing Add: CURE-Bldg 115 Vet Admin Wadsworth Hosp Ctr Los Angeles CA 90073

CODE, WILLIAM ROBERT, b Kemptville, Ont, Jan 2, 42; m 65; c 1. URBAN GEOGRAPHY, HISTORICAL GEOGRAPHY. Educ: Queen's Univ, BA, 65; Univ Calif, Berkeley, PhD(geog), 71. Prof Exp: ASST PROF GEOG, UNIV WESTERN ONT, 70- Mem: Can Asn Geog; Asn Am Geog. Res: Information theory and urban development; spatial organization of business decision-making; structural paradigms and urban planning. Mailing Add: Dept of Geog Univ of Western Ont London ON Can

CODELL, MAURICE, analytical chemistry, physical chemistry, see 12th edition

CODERE, HELEN, b Winnipeg, Man, Sept 10, 17; nat. ANTHROPOLOGY. Educ: Univ Minn, BA, 39; Univ Columbia, PhD, 50. Prof Exp: From instr to prof anthrop, Vassar Col, 46-63; mem fac, Bennington Col, 63-64; PROF ANTHROP, BRANDEIS UNIV, 64-, DEAN, GRAD SCH ARTS & SCI, 75- Concurrent Pos: Vis lectr, Univ BC, 54-55; Guggenheim fel, Africa, 59-60; lectr, Northwestern Univ, 63. Mem: Fel AAAS; fel Am Anthrop Asn; Soc Appl Anthrop; Am Ethnol Soc (pres-elect, 73). Res: Indians of North Pacific Coast; sociocultural change; economic anthropology; Rwanda, Central Africa. Mailing Add: Dept of Anthrop Brandeis Univ Waltham MA 02154

CODINGTON, JOHN F, b Macon, Ga, Feb 9, 20; m 52; c 3. BIOCHEMISTRY. Educ: Emory Univ, AB, 41, MA, 42; Univ Va, PhD(org chem), 45. Prof Exp: Res asst, Joint Off Sci Res & Develop & Comt Med Res Proj, Univ Va, 43-45; chemist, NIH, Md, 45-49; res biochem, Columbia Univ, 51-55; asst, Sloan-Kettering Inst Cancer Res, 55-59, assoc, 59-67; asst biochemist, 67-74, ASSOC BIOCHEMIST, DEPT MED, MASS GEN HOSP, 74-; PRIN RES ASSOC BIOL CHEM, HARVARD MED SCH, 67- Concurrent Pos: Asst prof, Cornell Med Col, 62-67. Res: Amino-alcohols; stilbenes; thiamine analogs; peptide antibiotics; nucleosides and sugars as potential anti-cancer agents; isolation, structures and immunological properties of glycoproteins of tumor cell surfaces. Mailing Add: 1725 Commonwealth Ave West Newton MA 02165

CODISPOTI, LOUIS ANTHONY, b Brooklyn, NY, June 6, 40; m 68. CHEMICAL OCEANOGRAPHY. Educ: Fordham Univ, BS, 62; Univ Wash, MS, 65, PhD(oceanog), 73. Prof Exp: Oceanogr, US Naval Hydrographic Off, 66-69; actg asst prof, 73-74, fel, 74-75, PRIN OCEANOGR CHEM OCEANOG, UNIV WASH, 75- Mem: Artic Inst NAm; Am Soc Limnol & Oceanog; Sigma Xi. Res: Chemical oceanography of the Arctic Ocean; upwelling; marine denitrification; global nutrient budgets, cycles and feedback mechanisms. Mailing Add: Dept of Oceanog Univ of Wash Seattle WA 98195

CODRINGTON, ROBERT SMITH, b Victoria, BC, Dec 11, 25; nat US; m 48; c 1. PHYSICS. Educ: Univ BC, BA, 46, MA, 48; Univ Notre Dame, PhD(physics), 51. Prof Exp: Asst physics, Univ BC, 45-47; res assoc, Univ Notre Dame, 48-51; asst res specialist, Rutgers Univ, 51-54; physicist, Schlumberger, Ltd, 54-62; mgr eng, 62-69, MGR NUCLEAR MAGNETIC RESONANCE, VARIAN ASSOCS, 69- Honors & Awards: Nat Telemetry Prize, 62. Mem: Am Phys Soc. Res: Magnetic instrumentation; geophysics; radar and telemetry synchronization; high resolution nuclear magnetic resonance. Mailing Add: Varian Assocs 611 Hansen Way Palo Alto CA 94304

CODY, D THANE, b St John, NB, June 23, 32; m 63; c 2. OTOLARYNGOLOGY, PHYSIOLOGY. Educ: Dalhousie Univ, MD & CM, 57; Univ Minn, PhD(otolaryngol), 66. Prof Exp: Asst to staff otorhinolaryngol, Mayo Clin, 62-63, consult, 63-68, from instr to assoc prof otolaryngol & rhinol, Mayo Grad Sch Med, 63-74, PROF OTOLARYNGOL, MAYO MED SCH, UNIV MINN, 74-, CHMN DEPT OTORHINOLARYNGOL, MAYO CLIN, 68- Concurrent Pos: Edward John Noble travel award, Mayo Found, 61; Am Acad Ophthal & Otolaryngol res award, 61. Honors & Awards: Dr John Black Award surg, Dalhousie Univ, 55. Mem: Fel Am Acad Ophthal & Otolaryngol; fel otolaryngol Am Col Surg; Am Laryngol, Rhinol & Otol Soc; Am Physiol Soc. Res: Histamine metabolism and anaphylaxis; averaged evoked responses to sound. Mailing Add: Mayo Clin Rochester MN 55901

CODY, GEORGE DEWEY, b New York, NY, May 16, 30. SOLID STATE PHYSICS. Educ: Harvard Univ, AB, 52, MA, 54, PhD(physics), 57. Prof Exp: Asst, Harvard Univ, 52-57; staff physicist, 57-69, DIR SOLID STATE LAB, RCA LABS, 69- Concurrent Pos: Harvard Univ Parker fel, Clarendon Lab, Oxford Univ, 57-58; regents prof, Univ Calif, San Diego, 69. Mem: Fel, Am Phys Soc. Res: Low temperature physics; superconductivity; high temperature thermal conductivity; thin films; magnetic properties of metals. Mailing Add: RCA Labs 201 Washington Rd Princeton NJ 08540

CODY, REYNOLDS M, b Asheville, NC, Apr 17, 29; m 58; c 2. MICROBIOLOGY. Educ: Univ Tenn, BA, 56; Miss State Univ, MS, 61, PhD(microbiol), 64. Prof Exp: Instr microbiol, Miss State Univ, 58-61, res asst, 61-64; asst prof bact, 64-65, ASSOC PROF MICROBIOL & BOT, SCH VET MED, AUBURN UNIV, 65-, RES FOUND RES GRANT, 64- Mem: Am Soc Microbiol; Soc Indust Microbiol. Res: Biochemistry and physiology of pathogenic microorganisms including viruses; elucidation of metabolic pathways in Serratia indica. Mailing Add: Sch of Vet Med Auburn Univ Auburn AL 36830

CODY, TERENCE EDWARD, b Orrville, Ohio, June 3, 38; m 61; c 2. ENVIRONMENTAL HEALTH. Educ: Mt Union Col, BS, 60; Case Western Reserve Univ, MS, 68; Ohio State Univ, PhD(bot), 72. Prof Exp: Asst ed biochem, Chem Abstracts Serv, 64-68; fel environ health, 72-73, ASST PROF ENVIRON HEALTH, UNIV CINCINNATI, 73- Mem: Am Inst Biol Sci; Am Soc Limnologists & Oceanogr; Ecol Soc Am; Int Asn Water Pollution Res; Sigma Xi. Res: The applicability of bioassays and bioindicators for evaluating environmental impacts and health effects of organic chemicals and heavy metals in aquatic systems. Mailing Add: Inst Environ Health 3223 Eden Ave Univ of Cincinnati Med Ctr Cincinnati OH 45267

CODY, VIVIAN, b San Diego, Calif, Jan 28, 43. CRYSTALLOGRAPHY. Educ: Univ Mich, BS, 65; Univ Cincinnati, PhD(chem), 69. Prof Exp: Res asst chem, Univ Mich, 63-65; teaching asst, Univ Cincinnati, 65-67, NSF fel, 67-69; fel, Univ Mo, St Louis, 69-70; fel endocrinol trainee, 70-72, RES SCIENTIST CRYSTALLOG, MED FOUND OF BUFFALO, 72- Mem: Endocrine Soc; Am Thyroid Asn; Am Crystallog Asn; Am Chem Soc; Biophys Soc. Res: Structure-function analysis of thyroid, steroid and polypeptide hormones using the techniques of x-ray crystallography. Mailing Add: Med Found of Buffalo 73 High St Buffalo NY 14203

CODY, WILLIAM JAMES, b Hamilton, Ont, Dec 2, 22; m 50; c 5. BOTANY. Educ: McMaster Univ, BA, 46. Prof Exp: ASSOC BOTANIST, CAN DEPT AGR, 46-, CUR, VASCULAR PLANT HERBARIUM, 59- Mem: Am Soc Plant Taxon; Int Asn Plant Taxon; Can Bot Asn. Res: Floristics of Mackenzie district, Northwest territory; Canadian ferns. Mailing Add: Dept of Agr Biosysts Res Inst Cent Exp Farm Ottawa ON Can

CODY, WILLIAM JAMES, JR, b Melrose Park, Ill, Nov 28, 29; m 53; c 5. NUMERICAL ANALYSIS. Educ: Elmhurst Col, BS, 51; Univ Okla, MA, 56. Prof Exp: Instr math, Univ Okla, 57-58 & Northwestern Univ, 58-59; asst mathematician, 59-66, ASSOC MATHEMATICIAN, ARGONNE NAT LAB, 66- Mem: Am Math Soc; Math Asn Am; Sigma Xi; Soc Indust & Appl Math; Asn Comput Mach. Res: Approximation of functions; computer arithmetic. Mailing Add: Appl Math Div Bldg 221 Argonne Nat Lab 9700 S Cass Ave Argonne IL 60439

COE, BERESFORD, b Philadelphia, Pa, May 4, 19; m 45; c 3. ORGANIC CHEMISTRY. Educ: Earlham Col, AB, 41. Prof Exp: Res chemist, Barrett Div, Allied Chem & Dye Corp, 41-46; RES CHEMIST, ROHM AND HAAS CO, 46- Mem: Am Chem Soc. Res: Adhesives technical service. Mailing Add: Rohm and Haas Co Spring House PA 19477

COE, EDWARD HAROLD, JR, b San Antonio, Tex, Dec 7, 26; m 49; c 2. GENETICS. Educ: Univ Minn, BS, 49, MS, 51; Univ Ill, PhD(bot), 54. Prof Exp: Res fel genetics, Calif Inst Technol, 54-55; res assoc field crops, 55-58, assoc prof, 59-63, chmn genetics comt, 62-64, GENETICIST, CROPS RES DIV, AGR RES SERV, USDA, 55-, PROF FIELD CROPS, UNIV MO, 64- Mem: AAAS; Genetics Soc Am; Am Soc Naturalists. Res: Genetics of maize; anthocyanin synthesis; unorthodox inheritance; fertilization. Mailing Add: Curtis Hall-Genetics Univ of Mo Columbia MO 65201

COE, ELMON LEE, b Phoenix, Ariz, Mar 6, 31; m 61. BIOCHEMISTRY. Educ: Harvard Univ, AB, 52; Univ Calif, Los Angeles, PhD(physiol chem), 61. Prof Exp: Res assoc biochem, Univ Ind, 60-61; from instr to asst prof, 61-68, ASSOC PROF BIOCHEM, MED SCH, NORTHWESTERN UNIV, CHICAGO, 68- Concurrent Pos: NIH grant, 64-67. Res: Interactions between metabolic pathways and metabolic control mechanisms; carbohydrate metabolism; biochemistry of tumors. Mailing Add: Dept of Biochem Northwestern Univ Med Sch Chicago IL 60611

COE, FREDRIC LAWRENCE, b Chicago, Ill, Dec 25, 36; m 65; c 2. INTERNAL MEDICINE, NEPHROLOGY. Educ: Univ Chicago, BA & BS, 57, MD, 61. Prof Exp: Asst prof, 69-72, ASSOC PROF MED, UNIV CHICAGO, 72- Concurrent Pos: USPHS fel, Univ Tex Southwestern Med Sch Dallas, 67-69; mem renal prog, Michael Reese Hosp, 69-72, chmn prog, 72- Mem: AAAS; Am Soc Nephrology; Am Fedn Clin Res; Cent Soc Clin Res. Res: Renal physiology; causes of renal calculi; computer medicine. Mailing Add: Michael Reese Hosp 2900 Ellis Ave Chicago IL 60616

COE, GERALD EDWIN, b Granville, Ill, Apr 21, 22; m 41; c 3. GENETICS, CYTOLOGY. Educ: Tex Col Arts & Indust, BS, 42; Univ Tex, PhD(bot), 52. Prof Exp: Asst, Univ Tex, 52; GENETICIST, AGR RES SERV, USDA, 52- Mem: Bot Soc Am. Res: Cytology and genetics in the genus Beta. Mailing Add: Bldg 009 Lab 6B Agr Res Ctr-W Beltsville MD 20705

COE, JOFFRE LANNING, b Greensboro, NC, July 5, 16; m 40; c 2. ANTHROPOLOGY. Educ: Univ NC, AB, 44; Univ Mich, MA, 48, PhD, 59. Prof Exp: Asst dir, Anthrop Lab, 38-42, from instr to assoc prof, 48-65, PROF ANTHROP, UNIV NC, CHAPEL HILL, 65-, DIR RES LAB ANTHROP, 48- Mem: AAAS; Soc Am Archaeol. Res: North American archaeology; photography; aerial mapping. Mailing Add: Dept of Anthrop Univ of NC Box 561 Chapel Hill NC 27514

COE, JOHN EMMONS, b Evanston, Ill, Sept 1, 31; m 54; c 3. IMMUNOLOGY. Educ: Oberlin Col, BA, 53; Hahnemann Med Col, MD, 57. Prof Exp: Intern, Univ Ill Res & Educ Hosp, 57-58; resident internal med, Med Ctr, Univ Colo, 58-60; surgeon, Rocky Mountain Lab, Nat Inst Allergy & Infectious Dis, NIH, 60-63; fel path, Scripps Clin & Res Found, 63-65; MED OFFICER, ROCKY MOUNTAIN LAB, NAT INST ALLERGY & INFECTIOUS DIS, NIH, 65- Concurrent Pos: Affil prof microbiol & zool, Univ Mont, 74- Mem: Am Asn Immunologists. Res: Selective induction of antibody formation in immunoglobulin classes of rodents, especially Syrian hamsters; immunity in lower vertebrates, especially amphibians and reptiles. Mailing Add: Rocky Mountain Lab Hamilton MT 59840

COE, JOHN IRA, b Chicago, Ill, Jan 21, 19; m 42. PATHOLOGY. Educ: Carleton Col, AB, 40; Univ Minn, MB, 44, MD, 45. Prof Exp: Fel, Univ Minn, 45-46 & 48-49; chief path, Vet Admin Hosp, Minneapolis, 49-50; from asst prof to assoc prof, 52-72, PROF PATH, UNIV MINN, MINNEAPOLIS, 72-; PATHOLOGIST & DIR LABS, HENNEPIN COUNTY GEN HOSP, 50-, MED EXAMR, 64- Concurrent Pos:

Pathologist, Drs Mem Hosp, 52-57. Mem: Am Soc Clin Path; Col Am Pathologists; AMA; Am Acad Forensic Sci; Int Acad Path. Res: Tumors; developmental anomalies forensic pathology. Mailing Add: 5108 Tifton Dr Minneapolis MN 55435

COE, KENNETH LOREN, b Omaha, Nebr, Apr 16, 27. CHEMISTRY, INFORMATION SCIENCE. Educ: Tarkio Col, BA, 49. Prof Exp: From asst ed to assoc ed chem abstr, 51-58, head ed dept, 59-61, managing ed, abstr issues, 62-69, MANAGING ED, PUBL, CHEM ABSTR, 69- Concurrent Pos: Mem, Nat Fedn Sci Abstracting & Indexing Serv. Mem: AAAS; Am Chem Soc; Coun Biol Ed; Asn Earth Sci Ed. Res: Chemical documentation. Mailing Add: 1631 Roxbury Rd Apt D1 Columbus OH 43212

COE, MICHAEL DOUGLAS, b New York, NY, May 14, 29; m 55; c 5. ANTHROPOLOGY, ARCHAEOLOGY. Educ: Harvard Univ, AB, 50, PhD(anthrop), 59. Prof Exp: Asst prof anthrop, Univ Tenn, 58-60; from instr to assoc prof, 60-68, PROF ANTHROP, YALE UNIV, 68- Concurrent Pos: Adv, Dumbarton Oaks Ctr Precolumbian Studies, 63- Mem: Soc Am Archaeol; fel Royal Anthrop Inst. Res: Archaeology of Mesoamerica; study of the Olmec civilization, Southern Veracruz, Mexico. Mailing Add: Dept of Anthrop Yale Univ New Haven CT 06520

COE, RICHARD HANSON, b Stamford, Conn, Jan 29, 20; m 55; c 6. CHEMISTRY. Educ: Wesleyan Univ, BA, 41, MA, 42; Stanford Univ, PhD(chem), 47. Prof Exp: Chemist, Conn State Water Comn, 41-42; actg instr chem, Stanford Univ, 43, asst, Nat Defense Res Comt, 44; asst res chemist, Calif Res Corp, 45-46; instr chem, Wesleyan Univ, 46-48; sr technologist, 49-64, sr res chemist, 64-67, supvr, 67-69, staff res chemist, 69-71, SR STAFF RES CHEMIST, SHELL OIL CO, 71- Honors & Awards: Indust Wastes Medal, Fed Sewage & Indust Wastes Asn, 52. Mem: AAAS; Am Chem Soc; Am Inst Chem. Res: Analytical chemical problems; oil refinery operating problems; catalysis. Mailing Add: Shell Oil Co PO Box 100 Deer Park TX 77536

COE, ROBERT STEPHEN, b Toronto, Ont, Feb 20, 39; m 69. GEOPHYSICS, GEOCHEMISTRY. Educ: Harvard Univ, BA, 61; Univ Calif, Berkeley, MS, 64, PhD(geophys), 66. Prof Exp: Asst prof, 68-74, ASSOC PROF EARTH SCI, UNIV CALIF, SANTA CRUZ, 74- Concurrent Pos: Fel, Australian Nat Univ, 66-68; starter grant, Petrol Res Fund, Am Chem Soc, 69-70; NSF grants, 70-72. Mem: AAAS; Am Geophys Union; Geol Soc Am; Soc Terrestrial Magnetism & Elec Japan. Res: Paleomagnetism, especially paleointensities of the geomagnetic field; effects of shear stress on polymorphic transitions in minerals. Mailing Add: Dept of Earth Sci Univ of Calif Santa Cruz CA 95060

COELHO, ANTHONY MENDES, JR, b Danbury, Conn, May 26, 47; m 74. PHYSICAL ANTHROPOLOGY, PRIMATOLOGY. Educ: Western Conn State Col, BS, 70; Univ Tex, Austin, MA, 73, PhD(anthrop), 75. Prof Exp: Asst prof anthrop, Tex Tech Univ, 74-75; ASSOC FOUND SCIENTIST & DIR PRIMATE ETHOLOGY LAB, SW FOUND RES & EDUC, 75- Mem: Am Asn Phys Anthrop; Int Primatological Soc; Am Anthrop Asn; AAAS. Res: Primate behavior; ecology and socio-bioenergetics; behavioral correlates of atherosclerosis; human and nonhuman primate growth, development, nutrition and evolution. Mailing Add: Primate Ethology Lab SW Found 8848 W Commerce St San Antonio TX 78284

COELHO, DAVID H, b New York, NY, Dec 25, 10; m 37; c 2. PROSTHODONTICS. Educ: City Col New York, AB, 31; NY Univ, DDS, 35; Am Acad Crown & Bridge Prosthodontics, dipl. Prof Exp: From instr to assoc prof, 36-56, chmn dept, 46-75, PROF FIXED PROSTHODONTICS, COL DENT, NY UNIV, 56- Concurrent Pos: USPHS dent auxiliary utilization grant, 60-73; consult, Vet Admin, Brooklyn, 66-73. Mem: Fel Am Col Dent; Am Dent Asn; Sigma Xi. Res: Materials and techniques involved in the fabrication of fixed prostheses for dental replacement. Mailing Add: Dept of Fixed Prosthodontics NY Univ Col of Dent New York NY 10016

COENSGEN, FREDERIC HARLEY, b Great Falls, Mont, Feb 10, 19; m 51; c 4. PHYSICS. Educ: Mont State Col, BS, 41; Calif Inst Technol, MS, 45; Univ Calif, PhD, 54. Prof Exp: Assoc physics, Mont State Col, 43-44; PHYSICIST, LAWRENCE LIVERMORE LAB, UNIV CALIF, 50- Mem: Fel Am Phys Soc. Res: High energy gamma rays; photonuclear reactions; controlled fusion research; plasma physics. Mailing Add: Lawrence Livermore Lab Box 808 Livermore CA 94550

COERVER, HELEN JOSEPH, b Prairie Du Rocher, Ill, Feb 16, 11. INORGANIC CHEMISTRY. Educ: Fontbonne Col, BS, 39; St Louis Univ, MS, 41; Univ Notre Dame, PhD(inorg chem), 52. Prof Exp: Instr chem, Fontbonne Col, 39-41; teacher high sch, Wis, 41-43; from asst prof to prof, 43-74, chmn dept, 52-63 & 69-74, REGISTRAR, FONTBONNE COL, 74- Concurrent Pos: Res Corp grant, 53-54; Petrol Res Corp grant, 61-63; mem fac, Dept Chem, Cardinal Glennon Col, 68- Mem: Am Chem Soc. Res: Preparation and determination of properties of coordination compounds of organic nitrogen compounds and inorganic ligands such as boron trichloride. Mailing Add: Dept of Chem Cardinal Glennon Col St Louis MO 63119

COESTER, FRITZ, b Berlin, Ger, Oct 16, 21; nat US; m 52; c 6. PHYSICS. Educ: Univ Zurich, PhD(theoret physics), 44. Prof Exp: Res, Sulzer Bros, Inc, Switz, 44-46; asst, Univ Geneva, 46-47; from asst prof to prof physics, Univ Iowa, 47-63; SR PHYSICIST, ARGONNE NAT LAB, 63- Concurrent Pos: Mem, Inst Advan Study, 53-54. Mem: Fel Am Phys Soc; Switz Phys Soc; Europ Phys Soc. Res: Quantum theory of fields; theoretical nuclear physics; theoretical physics. Mailing Add: Argonne Nat Lab Argonne IL 60439

COETZEE, JOHANNES FRANCOIS, b Bloemfontein, Union SAfrica, Nov 25, 24; m 54; c 1. ANALYTICAL CHEMISTRY. Educ: Univ Orange Free State, SAfrica, BSc, 44, MSc, 49; Univ Minn, PhD(chem), 55. Prof Exp: Instr chem, Univ Orange Free State, SAfrica, 45-48, lectr, 49-51 & Univ Witwatersrand, 56-57; from asst prof to assoc prof, 57-66, PROF CHEM, UNIV PITTSBURGH, 66- Concurrent Pos: Titular mem Int Union Pure & Appl Chem. Res: Non-aqueous solutions; electroanalytical chemistry; kinetics of fast ligand-substitution reactions. Mailing Add: Dept of Chem Univ of Pittsburgh Pittsburgh PA 15260

COFER, HARLAND E, JR, b Atlanta, Ga, Dec 28, 22; m 45; c 6. ECONOMIC GEOLOGY, MINERALOGY. Educ: Emory Univ, AB, 47, MS, 48; Univ Ill, Urbana-Champaign, PhD(geol), 57. Prof Exp: Asst prof geol, Emory Univ, 48-58; sr geologist, Indust Chem Div, Am Cyanamid Co, Ga, 58-66; PROF GEOL & CHMN DIV SCI & MATH, GA SOUTHWESTERN COL, 66- Concurrent Pos: Geologist, Groundwater Div, US Geol Surv, Ga, 53-54; consult, Consol Quarries Inc, Ga, 52-53 & Indust Chem Div, Am Cyanamid Co, 52-58. Mem: Geol Soc Am; Clay Minerals Soc. Res: Clay mineral research, especially utilization of kaolin group minerals as sources for alumina; mechanism of dissolution of clay minerals in natural and artificial environments. Mailing Add: Dept of Geol Ga Southwestern Col Americus GA 31709

COFFEN, DAVID LLEWELLYN, organic chemistry, see 12th edition

COFFER, HENRY FORD, b Phoenix, Ariz, Mar 9, 23; m 49; c 3. COLLOID CHEMISTRY. Educ: Univ Ariz, BA, 47, MS, 48; Univ Southern Calif, PhD, 51. Prof Exp: From asst dir to dir prod res div, Continental Oil Co, 51-66; VPRES, CER GEONUCLEAR CORP, 66- Mem: Am Inst Mining, Metall & Petrol Eng; Am Nuclear Soc. Res: Methods of producing oil; investigation of calcium stearate cetane gel system by means of x-ray diffraction, rheological properties and electron microscope. Mailing Add: CER Geonuclear Corp PO Box 15090 Las Vegas NV 89114

COFFEY, CHARLES EUGENE, b Bristol, Va, Feb 12, 31; m 54; c 3. PHYSICAL CHEMISTRY. Educ: King Col, AB, 52; Univ NC, PhD(phys chem), 56. Prof Exp: Fel phys chem, Univ Wash, 56-57; res chemist, 57-62, sr res chemist, 62-70, res supvr, 70-75, LAB DIR, E I DU PONT DE NEMOURS & CO, INC, FALLING WATERS, W VA, 75- Concurrent Pos: Fel inorg chem, Univ London, 61. Res: Extractive metallurgy; molecular structure; metal-organic and inorganic chemistry; explosives; roof support systems. Mailing Add: 26 Peachtree Lane Williamsport MD 21795

COFFEY, DEWITT, JR, b Gilmer, Tex, Apr 12, 35; m 68. CHEMICAL PHYSICS. Educ: Abilene Christian Col, BS(chem), 58; Univ Tex, BS(chem eng), 58, PhD(chem), 67. Prof Exp: Res asst drilling & well completion, Mobil Oil Field Res Lab, Tex, 58-61; McBean fel, Stanford Res Inst, 66-68; ASSOC PROF CHEM, SAN DIEGO STATE UNIV, 68- Concurrent Pos: Res Corp Frederick Gardner Cottrell grants-in-aid, 69-; consult, Stanford Res Inst, Calif, 69-; San Diego State Univ Found fac res grant, 69-; vis prof, Kyushu Univ, Fukuoka City, Japan, 75. Mem: AAAS; Am Chem Soc; Am Phys Soc. Res: infrared laser; microwave spectroscopy; stark-shifted spectroscopy. Mailing Add: Dept of Chem San Diego State Univ San Diego CA 92182

COFFEY, DONALD STRALEY, b Bristol, Va, Oct 10, 32; m 53; c 2. PHARMACOLOGY, BIOCHEMISTRY. Educ: ETenn State Univ, BS, 57; Johns Hopkins Univ, PhD(biochem), 64. Prof Exp: Chemist, NAm Rayon Corp, Tenn, 55-57; chem engr, Westinghouse Corp, Md, 57-59; actg dir James B Brady Urol Res Lab, Johns Hopkins Hosp, 59-60, instr physiol chem, 65-66, from asst prof to assoc prof pharmacol, 66-74, assoc prof, The Oncol Ctr, Sch Med, 73-74, dir, James Buchanan Brady Lab Reprod Biol, Johns Hopkins Hosp, 64-74, actg chmn, Dept Pharmacol & Exp Therapeut, Sch Med, 73-74, DIR, DEPT UROL RES LABS, JOHNS HOPKINS HOSP, 74-, PROF PHARMACOL & EXP THERAPEUT, SCH MED, JOHNS HOPKINS UNIV & PROF, ONCOL CTR, 74-, PROF UROL, 75- Concurrent Pos: USPHS res career award, 66-72; asst ed, J Molecular Pharmacol, 67-71. Mem: Am Soc Pharmacol & Exp Therapeut. Res: Control of cell replication and DNA synthesis; biochemistry of mammalian nuclei; growth and development of the prostate gland; control of reproduction; cancer chemotherapy. Mailing Add: Dept Pharmacol & Exp Therapeut Johns Hopkins Univ Sch of Med Baltimore MD 21205

COFFEY, HOWARD THOMAS, b Bristol, Va, Sept 23, 34; m 62; c 2. LOW TEMPERATURE PHYSICS. Educ: King Col, AB, 56; Univ NC, PhD(physics), 64. Prof Exp: Sr scientist, Res Div, Am Radiator & Stand Sanit Corp, 61-62 & Res Labs, Westinghouse Elec Corp, Pa, 63-67; physicist, 67-71, SR PHYSICIST, STANFORD RES INST, 71- Res: Fundamental and applied studies of superconductivity and superconducting devices; radiation effects on superconductors and insulators; magnetic suspension of high speed vehicles. Mailing Add: Stanford Res Inst 333 Ravenswood Ave Menlo Park CA 94025

COFFEY, JAMES CECIL, JR, b Salisbury, NC, July 2, 38; m 57; c 2. ENDOCRINOLOGY, PARASITOLOGY. Educ: Catawba Col, BA, 63; Univ NC, MSPH, 66, PhD(parasitol, biochem), 70. Prof Exp: Res assoc, Dent Res Ctr, 71-72, ASST PROF PEDIAT & ORAL BIOL, SCH MED & DENT RES CTR, UNIV NC, CHAPEL HILL, 72- Concurrent Pos: Fel pediat endocrinol, Sch Med, Univ NC, 70-71; NIH res career develop award. Res: Androgen metabolism; physiology of submaxillary glands. Mailing Add: Dent Res Ctr Univ of NC Chapel Hill NC 27514

COFFEY, JANICE CARLTON, b Lenoir, NC, July 8, 41. SYSTEMATIC BOTANY. Educ: Appalachian State Univ, BS, 62; Univ SC, MS, 64, PhD(biol), 66. Prof Exp: Assoc prof biol, Clemson Univ, 66-67; ASSOC PROF BIOL, QUEENS COL, NC, 67- Concurrent Pos: Presby Church Bd Christian Educ grant, 68; Piedmont Univ Ctr res grant, 70-71; Nat Acad Sci exchange scientist, USSR, 73 & 75. Mem: Am Soc Plant Taxonomists (secy, 75-78); Am Inst Biol Sci; Bot Soc Am; Int Asn Plant Taxon. Res: Systematics of the Juncaceae; biochemical, floristic and embryological investigations of Luzula. Mailing Add: Dept of Biol Queens Col Charlotte NC 28274

COFFEY, JERRY LEE, mathematical statistics, statistics, see 12th edition

COFFEY, JOHN JOSEPH, b Cambridge, Mass, Apr 24, 40. BIOCHEMISTRY, BIOCHEMICAL PHARMACOLOGY. Educ: Harvard Univ, AB, 61; Johns Hopkins Univ, PhD(biochem), 67. Prof Exp: NSF fel, Virus Lab, Univ Calif, Berkeley, 67-68; BIOCHEMIST, ARTHUR D LITTLE, INC, 68- Mem: Am Chem Soc; AAAS. Res: Kinetics of regulatory processes; enzyme induction; allosteric properties of proteins; drug distribution and metabolism; protein binding of drugs; pharmacokinetics; cancer chemotherapy. Mailing Add: Arthur D Little Inc Acorn Park Cambridge MA 02140

COFFEY, JOHN WILLIAM, b Sedalia, Mo, Jan 20, 37; m 63; c 2. BIOCHEMISTRY. Educ: Rockhurst Col, BS, 59; Tulane Univ, PhD(biochem), 63. Prof Exp: Res fel biochem, Touro Res Inst, 63-65 & Rockefeller Univ, 65-67; asst prof, Tulane Med Sch, 67-69; SR BIOCHEMIST, HOFFMANN-LA ROCHE INC, 69- Mem: AAAS; NY Acad Sci; Am Soc Biol Chemists. Res: Metabolism of L-fucose; lysosomal functions; vitamin B12 metabolism; collagen metabolism. Mailing Add: Dept of Pharmacol Hoffmann-La Roche Inc Nutley NJ 07110

COFFEY, JOSEPH FRANCIS, b East St Louis, Ill, June 6, 17; m 44; c 5. CHEMISTRY, CHEMICAL ENGINEERING. Educ: Mass Inst Technol, BS, 39, MS, 42; St Louis Univ, PhD(chem), 70. Prof Exp: Res staff mem, Standard Oil Co Ind, 39-40; asst, Mass Inst Technol, 40-42; res staff mem, Standard Oil Co Ind, 42-45, group leader process develop, 45-47, sect leader process design develop, 48-53; develop dir, Indoil Chem Co, 53-55; mgr commercial develop, Am Viscose Corp, 55-57; pvt consult, 57-62; asst prof chem, Westminster Col, 62-63; asst prof & actg chmn dept, Florrissant Valley Community Col, 63-65; PROF CHEM, FOREST PARK COMMUNITY COL, 65- Res: Petrochemicals; organic chemistry. Mailing Add: 7156 Kingsbury Blvd University City MO 63130

COFFEY, MARVIN DALE, b Midvale, Idaho, Apr 25, 30; m 52; c 4. ENTOMOLOGY, ZOOLOGY. Educ: Brigham Young Univ, AB, 52, MA, 53; Wash State Univ, PhD(entom), 57. Prof Exp: From instr to assoc prof zool & entom, 57-67, chmn dept biol, 65-70, PROF ZOOL & ENTOM, SOUTHERN ORE STATE COL, 67- Concurrent Pos: Asst prof, Fresno State Col, 64-65; vis prof, Tex A&M Univ, 69-70. Mem: AAAS; Entom Soc Am; Am Inst Biol Sci. Res: Taxonomy, ecology and medical importance of parasitic arthropods and of the Diptera. Mailing Add: Dept of Biol Southern Ore State Col Ashland OR 97520

COFFEY, MITCHAEL DEWAYNE, b Ada, Okla, Feb 2, 44; m 65; c 3. ORGANIC CHEMISTRY. Educ: ECent State Col, BS, 66; Purdue Univ, PhD(org chem), 72. Prof Exp: SR RES CHEMIST, DOW CHEM CO, 71- Mem: Am Chem Soc. Res: Development of products and services for the oilfield industry; synthetic-organosulfur, organonitrogen and organophosphorus; surface-scale inhibitors, corrosion inhibitors, electrophoretic mobilities and polarization studies. Mailing Add: Dowell Div of Dow Chem Co PO Box 21 Tulsa OK 74102

COFFEY, RONALD GIBSON, b Monte Vista, Colo, Dec 29, 36; m 59; c 3. BIOCHEMISTRY, PHARMACOLOGY. Educ: Colo State Univ, BS, 58; Ore State Univ, PhD(biochem), 63. Prof Exp: Res asst biochem, Ore State Univ, 58-63; chief div chem, Fifth US Army Med Lab, St Louis, Mo, 63-65; res assoc biochem, Univ Ore, 65-68; head div biochem, Children's Asthma Res Inst & Hosp, Denver, Colo, 68-73; ASSOC IMMUNOPHARMACOL, SLOAN-KETTERING INST CANCER RES, 73- Concurrent Pos: NIH grant, Children's Asthma Res Inst & Hosp, Denver, Colo, 70-73. Mem: Am Chem Soc; NY Acad Sci. Res: Regulation of cell membrane enzymes, especially adenylate cyclase, guanylate cyclase and ATPase by hormonal and pharmacologic agents and its application to immunology, allergy and cancer. Mailing Add: Dept of Immunopharmacol Sloan-Kettering Inst Cancer Res Rye NY 10580/

COFFIN, HAROLD GLEN, b Nanning, China, Apr 9, 26; US citizen; m 47; c 2. PALEONTOLOGY. Educ: Walla Walla Col, BA, 47, MA, 52; Univ Southern Calif, PhD, 55. Prof Exp: Head dept biol, Can Union Col, 47-52; res fel, Univ Southern Calif, 52-54; head div sci & math, Can Union Col, 54-56; assoc prof, Walla Walla Col, 56-58, head dept, 58-64; PROF ZOOL, GEOSCI RES INST, ANDREWS UNIV, 64-65, PALEONT, 65- Mem: AAAS; Geol Soc Am. Res: Marine invertebrates; science and religion, especially as related to geology and biology. Mailing Add: Geosci Res Inst Andrews Univ Berrien Springs MI 49104

COFFIN, KENNETH PUTNAM, physical chemistry, see 12th edition

COFFIN, LAURENCE HAINES, b Buenos Aires, Arg, June 4, 33; US citizen; m 58; c 3. THORACIC SURGERY, CARDIOVASCULAR SURGERY. Educ: Mass Inst Technol, BS, 55; Case Western Reserve Univ, MD, 59; Am Bd Surg & Bd Thoracic Surg, dipl, 68. Prof Exp: From instr to sr instr thoracic surg, Case Western Reserve Univ, 67-69; assoc prof surg, 69-75, PROF SURG, COL MED, UNIV VT, 75-, CHIEF SECT THORACIC & CARDIAC SURG, 69-, ATTEND & CHIEF THORACIC SERV, MED CTR HOSP VT, 70- Concurrent Pos: Surg intern, Univ Hosps Cleveland, 59-60, resident, 60-61 & 63-67, asst thoracic surgeon, 67-69; chief thoracic surg, Vet Admin Hosp Cleveland, 67-69; mem coun cardiovasc surg, Am Heart Asn, 70- Mem: Am Soc Clin Invest; As Am Acad Surg; fel Am Col Surg; AMA; Am Heart Asn. Res: Pathophysiology of burn shock; cardiovascular physiology. Mailing Add: Mary Fletcher Unit Med Ctr Hosp of Vt Burlington VT 05401

COFFIN, PERLEY ANDREWS, b Newburyport, Mass, Oct 8, 08; m 39; c 1. RUBBER CHEMISTRY, POLYMER CHEMISTRY. Educ: Northeastern Univ, BChE, 31; Mass Inst Technol, MS, 33. Prof Exp: Lab asst, Simplex Wire & Cable Co, Mass, 30-31; control chemist, Vultex Chem Co, 33-37; develop chemist, Gen Latex & Chem Corp, 37-42; lab mgr, Gen Tire & Rubber Co, Tex, 43-45; sect head qual control, 46-61, LAB MGR, GEN LATEX & CHEM CORP, 62- Mem: Am Chem Soc; Am Soc Testing & Mat. Res: Polymerization; rubber; synthetic resin and plastics; latex. Mailing Add: Gen Latex & Chem Corp 666 Main St Cambridge MA 02139

COFFMAN, CHARLES BENJAMIN, b Baltimore, Md, Dec 19, 41; m 71; c 1. AGRONOMY. Educ: Univ Md, BS, 66, MS, 69, PhD(soil mineral), 72. Prof Exp: Asst geol, Univ Md, 66-71, instr 72; RES AGRONOMIST, AGR ENVIRON QUAL INST, AGR RES SERV, USDA, 72- Concurrent Pos: Instr geol, Frederick Community Col, 73- Mem: Am Soc Agron; Soil Sci Soc Am; Clay Mineral Soc; AAAS. Res: Development of and control of narcotic plant species involving plant environment interrelationships. Mailing Add: AEQI Agr Res Serv USDA Range 4 Headhouse 2 Bldg 050 ARC-West Beltsville MD 20705

COFFMAN, CHARLES VERNON, b Hagerstown, Md, Oct 23, 35; m 63; c 3. MATHEMATICS. Educ: Johns Hopkins Univ, BES, 57, PhD(math), 62. Prof Exp: Assoc engr, Appl Physics Lab, Johns Hopkins Univ, 57; vis mem res staff math, Res Inst Advan Study, Martin Co, Md, 60; from asst prof to assoc prof, 62-71, PROF MATH, CARNEGIE-MELLON UNIV, 71- Mem: Am Math Soc. Res: Differential equations and functional analysis. Mailing Add: Dept of Math Carnegie-Mellon Univ Pittsburgh PA 15213

COFFMAN, HAROLD H, b Overbrook, Kans, Feb 16, 15; m 46; c 4. CHEMISTRY. Educ: Kans State Univ, BS, 40. Prof Exp: Res chemist, Bareco Oil Co, 40-42, 46-52, dir res, 52-55; dir, Barnsdall Res Group, 55-66, asst dir, Appln Res Lab, 66-70, MGR QUAL CONTROL, BARECO DIV, PETROLITE CORP, 70- Mem: Am Chem Soc; Am Soc Testing & Mat; Am Tech Asn Pulp & Paper Indust. Res: Microcrystalline petroleum waxes in regard to their properties; end use applications and processes of manufacturing. Mailing Add: Bareco Div Petrolite Corp PO Box 669 Barnsdall OK 74002

COFFMAN, JAMES PAUL, inorganic chemistry, physical chemistry, see 12th edition

COFFMAN, JAY D, b Quincy, Mass, Nov 17, 28; m 55; c 4. INTERNAL MEDICINE. Educ: Harvard Col, BA, 50; Boston Univ, MD, 54. Prof Exp: From assoc to asst prof, 60-70, PROF MED, MED CTR, BOSTON UNIV, 70- Mem: Am Physiol Soc; Am Fedn Clin Res; Am Heart Asn; Am Soc Clin Invests; NY Acad Sci. Res: Peripheral vascular physiology and disease. Mailing Add: Peripheral Vascular Lab Univ Hosp 75 E Newton St Boston MA 02118

COFFMAN, JOHN ROBERT, biochemistry, see 12th edition

COFFMAN, JOHN W, b El Dorado, Kans, Dec 19, 31; m 54; c 3. PHYSICS. Educ: Univ Kans, BS, 54, MS, 56. Prof Exp: Res asst, Univ Kans, 54-56; physicist atmospheric acoustics, Missile Geophys Div, White Sands Missile Range, 58-60, supvry physicist, Missile Meteorol Div, 60-63, res atmospheric physicist, Environ Sci Dept, 63-67; WITH GODDARD SPACE FLIGHT CTR, NASA, 67- Mem: Am Phys Soc. Res: Polarizable dielectrics; atmospheric infrared absorption and emission spectra; low frequency sound propagation in the atmosphere. Mailing Add: NASA Goddard Space Flight Ctr Greenbelt MD 20771

COFFMAN, MICHAEL S, b Anaheim, Calif, Sept 7, 43; m 64; c 2. FOREST ECOLOGY, PLANT PHYSIOLOGY. Educ: Northern Ariz Univ, BS, 66, MS, 67; Univ Idaho, PhD(forestry), 71. Prof Exp: ASST PROF FORESTRY, MICH TECHNOL UNIV, 70- Mem: AAAS; Soc Am Foresters; Am Soc Plant Physiol. Res: Growth characteristics of pinus ponderosa in the Southwest and root growth stimulation of pinus ponderosa seedlings; effects of timber species on forest hydrologic processes. Mailing Add: Dept of Forestry Mich Technol Univ Houghton MI 49931

COFFMAN, MOODY LEE, b Abilene, Tex, July 25, 25; m 47; c 4. THEORETICAL PHYSICS, ACOUSTICS. Educ: Abilene Christian Col, BA, 47; Univ Okla, MA & MS, 49; Agr & Mech Col Tex, PhD(physics), 54. Prof Exp: Instr physics, E Tex State Col, 49-51; instr, Agr & Mech Col Tex, 51-53, instr math, 53-54; sr nuclear engr, Convair, Tex, 54-55; from asst prof to assoc prof physics & math, Abilene Christian Col, 55-60, head dept, 56-60; sr physicist, Missile & Space Systs Dept, Hamilton Standard Div, United Aircraft Corp, Conn, 60-61; prof physics & head dept, Oklahoma City Univ, 61-69; PROF PHYSICS, CENT STATE UNIV, OKLA, 69- Concurrent Pos: Adj prof, Tex Christian Univ, 54-55; adj assoc prof physics, Hartford Grad Ctr, Rensselaer Polytech Inst, 60-61; consult, Convair, 55-57; consult physics, 69-; vpres res, Acoustic Controls, Inc, Tex, 69- Mem: Am Phys Soc; Am Math Soc; Am Asn Physics Teachers; Am Geophys Union; Sigma Xi. Res: Electromagnetic theory and quantum mechanics related to molecular and atomic structure; mechanics of charged particles; geomagnetism. Mailing Add: 3612 Ann Arbor Ave Oklahoma City OK 73122

COFFMAN, ROBERT EDGAR, b Grosse Pointe Farms, Mich, Jan 5, 31; m 59; c 3. CHEMICAL PHYSICS. Educ: Univ Ill, BS, 53; Univ Calif, Berkeley, MS, 55; Univ Minn, PhD(chem physics), 64. Prof Exp: Chemist, Hanford Atomic Prod Oper, Gen Elec Co, Wash, 55-56; chemist, Chemet Prog, NY, 56-57, phys chemist, Advan Semiconductor Lab, Semiconductor Prod Dept, 57-60; NSF fel physics, Nottingham Univ, 64-65; asst prof chem, Augsburg Col, 65-67; asst prof, 67-72, ASSOC PROF CHEM, UNIV IOWA, 72- Concurrent Pos: NATO sr fel, Cambridge Univ, 73. Mem: AAAS; Am Phys Soc; Am Chem Soc. Res: Quantum chemistry; electron paramagnetic resonance in inorganic, metal-ligand and biological molecules. Mailing Add: Dept of Chem Univ of Iowa Iowa City IA 52240

COFFMAN, WILLIAM PAGE, b Vandergrift, Pa, Jan 7, 42; m 65; c 2. ECOLOGY, LIMNOLOGY. Educ: Thiel Col, BS, 63; Univ Pittsburgh, PhD(biol), 67. Prof Exp: Hydrobiologist, Karlsruhe, Ger, 68-69; asst prof, 69-74, ASSOC PROF ECOL, UNIV PITTSBURGH, 74- Res: Energy flow in aquatic ecosystems; ecology and taxonomy of aquatic insects, particularly the Dipteran family Chironomidae. Mailing Add: Dept of Biol Univ of Pittsburgh Pittsburgh PA 15213

COFRANCESCO, ANTHONY J, b New Haven, Conn, Feb 24, 10; m 41; c 3. INDUSTRIAL ORGANIC CHEMISTRY. Educ: Wesleyan Univ, BA, 33, MA, 34; Yale Univ, PhD(org chem), 39. Prof Exp: Res chemist, Calco Chem Co, NJ, 39-44; chief chemist, Arnold & Hofmann, RI, 44-45 & Carwin Chem Co, Conn, 45-50; group leader, 50-54, sect mgr, TPM Dyes, 54-58 & Intermediates, 58-62, mgr, Chem Specialty Sect, GAF Corp, Rensselaer, 62-75; CONSULT INDUST ORG CHEM, 75- Res: Anthraquinone intermediates and dyes; industrial organic chemicals. Mailing Add: Windward Meadows Box 399-A Knight Rd Delanson NY 12053

COGAN, BRUCE CLAYTON, astrophysics, see 12th edition

COGAN, DAVID GLENDENNING, b Fall River, Mass, Feb 14, 08; m 34; c 3. OPHTHALMOLOGY. Educ: Dartmouth Col, AB, 29; Harvard Univ, MD, 32. Prof Exp: Asst ophthal, 34-40, from asst prof to prof ophthalmic res, 40-63, Henry Willard Williams prof ophthal, 63-70, prof, 70-74, actg dir lab, 40-42, dir, Howe Lab Ophthal, 43-74, EMER HENRY WILLARD WILLIAMS PROF OPHTHAL, HARVARD MED SCH, 74-; MED OFFICER, USPHS, 74- Concurrent Pos: Moseley traveling fel, Harvard Med Sch, 37-38; asst ophthal, Mass Eye & Ear Infirmary, 34-40, clin asst, 35-39, from asst surgeon to assoc surgeon, 39-54, surgeon, 54-, dir ophthal labs, 47-74, chief ophthal, 60-66; consult, Los Alamos Med Ctr; mem comt ophthalmic consults, Nat Res Coun; coun inst neurol & blindness, USPHS; mem coun, Nat Eye Inst, 70-73; ed-in-chief, AMA, Arch of Ophthal, 60-67. Mem: AAAS; AMA; Am Soc Clin Invests; Am Neurol Soc; Can Ophthal Soc. Res: Clinical physiology of the eye; neuro-ophthalmology. Mailing Add: Bldg 10 Rm 10-N-317 Nat Insts of Health Bethesda MD 22014

COGAN, EDWARD J, b Milwaukee, Wis, Jan 18, 25; m 47; c 1. MATHEMATICAL LOGIC. Educ: Univ Wis, BA, 46, MA, 48; Pa State Univ, PhD(math philos), 55. Prof Exp: Instr math, Pa State Univ, 48-50, 51-55 & Dartmouth Col, 55-57; MEM FAC, SARAH LAWRENCE COL, 57- Concurrent Pos: Dir, NSF Insts, 59-64, co-dir, Upward Bound Prog, 66-69; consult, Metrop Sch Study Coun, 59-60; co-dir, Spec Prog, Sarah Lawrence Col, 74- Mem: AAAS; Math Asn Am; Asn Symbolic Logic. Res: Foundations of mathematics; theory of sets; combinatory logic; automatic programming languages for computers. Mailing Add: Dept of Math Sarah Lawrence Col Bronxville NY 10708

COGAN, HAROLD LOUIS, b Framingham, Mass, May 30, 31; m 55; c 2. PHYSICAL CHEMISTRY. Educ: Univ Mass, BS, 54; Yale Univ, MS, 56, PhD(phys chem), 58. Prof Exp: Asst, Yale Univ, 54-56, NSF asst phys chem of electrolytes, 56-57; PRES, HAROLD L COGAN, INC, 57- Mem: Am Chem Soc. Res: Thermodynamic studies of the effect of pressure upon ionic equilibria; pressure and temperature dependence of the dielectric constant of water; use of coaxial cavity resonators for dielectric constant measurements. Mailing Add: 2600 Hampshire Rd SE Grand Rapids MI 49506

COGBURN, ROBERT RAY, b Weatherford, Tex, Mar 3, 35; m 58; c 4. ENTOMOLOGY. Educ: Univ Tex A&M Univ, BS, 58, MS, 61. Prof Exp: Entomologist, USDA, Agr Res Serv, Tex, 59-63, res entomologist, Stored Prod Insects Br, Calif, 63-66, Ga, 66-69, RES ENTOMOLOGIST, USDA, AGR RES SERV, STORED RICE INSECTS LAB, 69- Mem: Entom Soc Am. Res: Research with insects affecting stored rice. Mailing Add: Stored-Rice Insects Lab USDA Rte 5 Box 784 Beaumont TX 77706

COGDELL, THOMAS JAMES, b Quanah, Tex, Aug 19, 34; m 61; c 4. ORGANIC CHEMISTRY. Educ: Midwestern Univ, BA, 55; Univ Tex, MA, 62; Harvard Univ, PhD(chem), 65. Prof Exp: Chemist, Dow Chem Co, 55-58; mem tech staff, Bell Tel Labs, 65-66; asst prof, 66-74, ASSOC PROF CHEM, UNIV TEX, ARLINGTON, 74- Mem: AAAS; Am Chem Soc. Res: Organic reaction mechanisms; carbonium ion rearrangements; stable free radicals; benzyne intermediates. Mailing Add: Dept of Chem Univ of Tex Arlington TX 76010

COGEN, WILLIAM MAURICE, b Chicago, Ill, Mar 30, 09; m 41; c 2. GEOLOGY. Educ: Calif Inst Technol, BS, 31, MS, 33, PhD(geol), 37. Prof Exp: Petrol geologist, Superior Oil Co, Tex, 36-37 & Shell Oil Co, 37-62; consult, 63-67; supvr tech writers, Lockheed Electronics Co, 67-74; GEOL CONSULT, 74- Mem: Geol Soc Am; Am Asn Petrol Geologists. Res: Heavy minerals of Gulf Coast sediments; mechanics of landslides; geology of Texas Gulf coast; petroleum and engineering geology. Mailing Add: 4830 Imogene St Houston TX 77035

COGGESHALL, A DARLING, b Schenectady, NY, Mar 30, 17; m 46; c 2. POLYMER CHEMISTRY. Educ: Middlebury Col, AB, 40. Prof Exp: Anal chemist, Metal & Thermit Corp, NJ, 41-42; chemist develop elec insulation, 42-50, CHEMIST REINFORCED PLASTICS, MAT & PROCESSES LAB, GEN ELEC CO, 50- Mem: Am Chem Soc; Soc Plastics Indust. Res: Physical and electrical properties of structural electrical insulations useful in highly stressed applications in heavy electrical machinery. Mailing Add: Mat & Processes Lab Gen Elec Co Bldg 55-129 Schenectady NY 12345

COGGESHALL, LOWELL THELWELL, b Saratoga, Ind, May 7, 01; m 30; c 3. MEDICAL EDUCATION, TROPICAL MEDICINE. Educ: Ind Univ, AB, 22, AM, 23, MD, 28. Hon Degrees: LLD, Ind Univ, 47; LHD, Jefferson Med Col, 56, Lake Forest Col, 61 & Temple Univ, 65; ScD, Chicago Med Sch, 62 & Albany Med Col, 64. Prof Exp: Mem staff, Rockefeller Found, 24-26; instr anat, Sch Med, Ind Univ, 26-27; from instr to asst prof med, Univ Chicago, 31-35; mem staff, Rockefeller Found, 35-41; prof epidemiol, Sch Pub Health, Univ Mich, 41-43, chmn dept trop dis, 43; dean div biol sci, 47-60, chmn dept med, 47-66, vpres, 60-66, EMER VPRES, UNIV CHICAGO, 66- Concurrent Pos: Spec asst health & med affairs, Dept Health, Educ & Welfare, 56. Consult to Secy, US Dept Navy; chmn med consults, Secy of Defense; mem consult comt, Surv Med Res & Educ, 57-58; consult, Chicago Community Trust, Searle Fund; mem exec comt, Div Med Sci, Nat Res Coun. Trustee, Rockefeller Found, Josiah Macy, Jr Found, Crerar Libr, Mus Sci & Indust, LaRabida Sanitarium, Abbott Labs, Commonwealth Edison Co, Field Found, Ill, Inc, Univ Chicago, Mus Sci & Indust, Visual Info Systs, Inc & NY Ctr Continuing Med Educ. Mem, Int Develop Adv Bd, 59; mem panels, Rockefeller Bros Fund Spec Studies Proj Reports IV & V, 56-58; deleg, WHO, 59. Honors & Awards: Gorgas Medal, 45; Jesuit Centennial Citation, 57; Founders Day Award, Loyola Univ, 61; Abraham Flexner Award, 63. Mem: Nat Acad Sci; AAAS; Am Soc Clin Invest; Am Soc Trop Med & Hyg (vpres, 47); Am Col Physicians. Mailing Add: Rte 2 Foley AL 36535

COGGESHALL, NORMAN DAVID, b Ridgefarm, Ill, May 15, 16; m 40; c 4. PHYSICS. Educ: Univ Ill, BA, 37, MS, 39, PhD(physics), 42. Prof Exp: Asst physics, Univ Ill, 37-41, instr, 42-43; dir, ph Phys Sci Div, 43-66 & Proc Sci Dept, 66-70, dir, Explor & Prod Dept, 70-74, VPRES, EXPLOR & PROD, GULF RES & DEVELOP CO, 74- Concurrent Pos: Mem adv bd, Nat Bur Stand. Honors & Awards: Recipient of Resolution of Appreciation, Div Refining, Am Petrol Inst; Award, Am Chem Soc, 69. Mem: Am Chem Soc; Am Phys Soc. Res: Mass, infrared and ultraviolet spectroscopy; molecular physics; separation processes; process instrumentation. Mailing Add: Explor & Prod Dept PO Box 2038 Gulf Res & Develop Co Pittsburgh PA 15230

COGGESHALL, RICHARD E, b Chicago, Ill, May 29, 32; m 59; c 4. ANATOMY. Educ: Univ Chicago, BA, 51; Harvard Med Sch, MD, 56. Prof Exp: Instr anat, Harvard Med Sch, 64-65, assoc, 65-67, from asst prof to assoc prof, 67-71; PROF ANAT, UNIV TEX MED BR, GALVESTON, 71- Concurrent Pos: USPHS career develop award, 66-; NIH fels, Univ Tex Med Br, Galveston, 71-74 & 75-79, NIH grant, 74-77. Mem: Am Asn Anatomists; Am Soc Cell Biol; Am Soc Neurosci; Sigma Xi; AAAS. Res: Neurobiology and the structure of the nervous system. Mailing Add: Marine Biomed Inst 200 University Blvd Galveston TX 77550

COGGIN, JOSEPH HIRAM, b Birmingham, Ala, Feb 4, 38; m 57; c 4. MICROBIOLOGY, VIROLOGY. Educ: Vanderbilt Univ, BA, 59; Univ Tenn, MS, 61; Univ Chicago, PhD(midrobiol), 65. Prof Exp: Sr bacteriologist, Tenn Dept Pub Health, 59-60; sr res virologist virus & cell biol div, Merck Inst; therapeut res, 65-67; asst prof & virologist, 67-68, assoc prof microbiol, 68-73, PROF MICROBIOL, UNIV TENN, KNOXVILLE, 73- Concurrent Pos: Sect chief & consult, Molecular Anat Prog, Tumor transplanation Study, Oak Ridge Nat Labs, 68-; mem immunol sci study sect, Nat Cancer Inst, 75. Mem: AAAS; Tissue Culture Asn; Soc Exp Biol & Med; Am Soc Microbiol; Am Asn Immunol. Res: Drug resistance in microorganisms; tumor immunology; virology of cancer. Mailing Add: Dept of Microbiol Univ of Tenn Knoxville TN 37916

COGGINS, CHARLES WILLIAM, JR, b NC, Nov 17, 30; m 51; c 3. PLANT PHYSIOLOGY. Educ: NC State Col, BS, 52, MS, 54; Univ Calif, PhD(plant physiol), 58. Prof Exp: Asst plant physiologist, 57-64, assoc plant physiologist, 64-70, PLANT PHYSIOLOGIST, DEPT PLANT SCI, CITRUS RES CTR, UNIV CALIF, RIVERSIDE, 70-, PROF PLANT PHYSIOL & CHMN DEPT PLANT SCI, 75- Honors & Awards: Am Soc Hort Sci Award, 66. Mem: AAAS; Am Soc Hort Sci; Am Soc Plant Physiologists; Am Inst Biol Sci; Int Soc Citricult. Res: Evaluation of vegetative, reproductive and fruit quality responses of citrus, avocado and other subtropical fruit to plant regulators. Mailing Add: 5359 Brighton Dr Riverside CA 92504

COGGINS, LEROY, b Thomasville, NC, July 29, 32; m 56; c 5. VETERINARY VIROLOGY. Educ: NC State Col, BS, 55; Okla State Univ, DVM, 57; Cornell Univ, PhD(vet virol), 62. Prof Exp: Vet res officer virus res, Ft Detrick, Md, 57-59; asst, NY State Col Vet Med, Cornell Univ, 59-62; res assoc, Cornell Univ, 62-63; vet res officer, EAfrican Vet Res Orgn, USDA, 63-68; PROF VIROL, NY STATE COL VET MED, CORNELL UNIV, 68- Mem: Am Vet Med Asn; US Livestock Sanit Asn; Conf Res Workers Animal Dis. Res: Virus research; viruses of variola, hog cholera, bovine virus diarrhea and African swine fever and the host response to these agents; equine infectious anemia. Mailing Add: Dept of Path NY State Col Vet Med Cornell Univ Ithaca NY 14853

COGGON, PHILIP, b Kirkby, Eng, Mar 22, 42; US citizen; m 65; c 1. NATURAL PRODUCTS CHEMISTRY. Educ: Univ Nottingham, BS, 63, PhD(chem), 66. Prof Exp: Res fel chem, Univ Sussex, 66-68; res assoc crystallog, Duke Univ, 68-70; GROUP LEADER TEA RES, THOMAS J LIPTON, INC, 70- Mem: The Chem Soc; Am Chem Soc; Am Inst Chemists; Inst Food Technologists. Res: Organic chemistry of natural products, including structure elucidation and biochemical studies. Mailing Add: 20 Howard Ave Orangeburg NY 10962

COGGSHALL, WILLIAM LAMAR, apiculture, see 12th edition

COGHLAN, ANNE EVELINE, b Boston, Mass, Mar 29, 27. MICROBIOLOGY. Educ: Simmons Col, BS, 48; Boston Univ, MEd, 53; Univ Vt, MS, 57; Univ RI, PhD(biol sci), 65. Prof Exp: Instr bact, Colby Jr Col, 49-59; NSF fel, 59-61; PROF BIOL, SIMMONS COL, 62-, CHMN DEPT, 72- Mem: AAAS; Am Soc Microbiol; Am Soc Cell Biol. Res: General microbiology; basic bacteriology; immunology; host-parasite relationships. Mailing Add: Dept of Biol Simmons Col Boston MA 02115

COGLIANO, JOSEPH ALBERT, b Brooklyn, NY, Mar 4, 30. ORGANIC CHEMISTRY. Educ: Polytech Inst Brooklyn, BS, 51; Princeton Univ, MA, 56, PhD, 58; George Washington Univ, MS, 57. Prof Exp: Chemist, Nat Bur Stand, 51-53; res assoc, George Washington Univ, 53-54; asst, Princeton Univ, 54-57; RES ASSOC, W R GRACE & CO, 57- Concurrent Pos: Staff mem, Nat Acad Sci, 55. Mem: Am Chem Soc. Res: Organic research and synthesis; process development; organophosphorous chemistry; physico-chemical measurements; permeability; foams-preparation and properties; technical trouble shooting; technical program management; bound enzymes. Mailing Add: W R Grace & Co 7379 Rte 32 Columbia MD 21044

COGSWELL, GEORGE WALLACE, b New York, NY, Feb 8, 23; m 49; c 4. ORGANIC CHEMISTRY. Educ: City Col New York, BS, 53; Fordham Univ, MS, 55, PhD(org chem), 60. Prof Exp: Chemist indust detergents, Colgate-Palmolive Co, NJ, 50-53; lab asst, Fordham Univ, 54-57; res assoc, Dept Pharmacol, Med Col, Cornell Univ, 57-58; sr develop chemist-proj coord, A E Staley Mfg Co, 58-64; sect mgr, A-U Proj, Armour & Co, 64-65; mkt develop mgr, Hooker Chem Co, Inc, 65-68; vpres, Woodburn Anal Lab & Vanguard Chem Co, 68-72; PRES, ARLINGTON SERV CORP, 72- Concurrent Pos: Instr chem, Anderson Col, SC, 74- Mem: Am Chem Soc; Am Oil Chem Soc; Am Asn Textile Chem & Colorists; Am Pharmaceut Asn; Tech Asn Pulp & Paper Indust. Res: Synthetic and natural polymers; specialty chemicals; coatings; paper; detergents, ozonolysis; structure activity studies; instrumental and wet analyses. Mailing Add: 411 Ravenal Rd Anderson SC 29621

COGSWELL, HOWARD LYMAN, b Susquehanna Co, Pa, Jan 19, 15; m 38; c 1. ORNITHOLOGY, ECOLOGY. Educ: Whittier Col, BA, 48; Univ Calif, Berkeley, MA, 51, PhD(zool), 62. Prof Exp: Asst prof biol sci, Mills Col, 52-64; assoc prof, 64-69, PROF BIOL SCI, CALIF STATE COL HAYWARD, 69- Concurrent Pos: NSF sci fac fel, 63-64. Mem: Am Ornith Union; Cooper Ornith Soc; Ecol Soc Am; Am Soc Zool; Am Inst Biol Sci. Res: Habits, phenology and populations of birds of California; territory size in birds of chaparral; habitat distribution and selection in birds; place-related behavior among animals in general; solid waste disposal; bird hazard to aircraft. Mailing Add: Dept of Biol Sci Calif State Col Hayward CA 94542

COGSWELL, HOWARD WINWOOD, b Sherman, Tex, Apr 22, 23; m 47; c 1. ANALYTICAL CHEMISTRY. Educ: Austin Col, BS, 47. Prof Exp: Jr chemist, Cities Serv Refining Corp, 48-49, anal chemist, Cities Serv Res & Develop Co, 49-53, sr anal chemist, 53-56, sect leader, 56-60, sr res chemist, 58-60, res assoc, 60-61; HEAD ANAL RES GROUP, PETRO-TEX CHEM CORP, 61- Mem: Am Chem Soc. Res: Catalytic petroleum processing; catalyst reactivation and development; research, design, and construction in gas chromatography; analytical instrumentation and methods development. Mailing Add: 526 Shawnee Houston TX 77034

COHAN, LEONARD HECHT, b Baltimore, Md, Sept 23, 13; m 34; c 2. COLLOID CHEMISTRY. Educ: Johns Hopkins Univ, AB, 32, PhD(chem), 35. Prof Exp: Asst chem, Johns Hopkins Univ, 35-36; res chemist, Colloid Corp, 36-40 & Continental Carbon Co, 40-42, chief chemist, 42-44, dir res, 44-48; dir, Tech Serv Lab, Witco Chem Co, 46-48, tech dir, Carbon Black Div, 48-49; prod mgr, 49-61, PRES, HANLINE BROS, 61- Concurrent Pos: With War Prod Bd, 44. Honors & Awards: Herman H Shuger Mem Award, 59. Res: Paint; absorption; carbon black; rubber; colloid chemistry. Mailing Add: Hanline Bros Inc 1400 Warner St Baltimore MD 21230

COHART, EDWARD MAURICE, b New York, NY, Dec 8, 09; m 33; c 2. PUBLIC HEALTH. Educ: Columbia Univ, AB, 28, MD, 33, MPH, 47; Am Bd Prev Med & Pub Health, dipl. Hon Degrees: MA, Yale Univ, 56. Prof Exp: Assoc prof, 48-56, chmn dept epidemiol & pub health, 66-68, PROF PUB HEALTH, SCH MED, YALE UNIV, 56- Concurrent Pos: Cancer control consult, USPHS, 47-48; dep comnr, Dept Health, NY, 55-56, mem adv coun, Nat Inst Environ Health Sci, 69-73. Mem: AAAS; AMA; Am Cancer Asn; Am Col Prev Med (vpres pub health, 73-74); Am Pub Health Asn. Res: Epidemiology of chronic disease; public health practice. Mailing Add: Dept of Epidemiol & Pub Health Yale Univ New Haven CT 06510

COHEE, GEORGE VINCENT, b Indianapolis, Ind, Feb 4, 07; m 30. PETROLEUM GEOLOGY. Educ: Univ Ill, BS, 31-33, MS, 32, PhD(geol), 37. Prof Exp: Asst geologist, Oil & Gas Div, State Geol Surv, Ill, 36-42; asst state geologist, State Geol Surv, Ind, 42-43; petrol analyst, Petrol Admin for War, 43; geologist, Fuels Br, 43-47, sr geologist, 47-51, CHMN GEOL NAMES COMT, US GEOL SURV, 52- Concurrent Pos: Chmn geol dept, Univ Ark, 51-52. Mem: Fel Geol Soc Am; Soc Econ Paleontologists & Mineralogists; hon mem Am Asn Petrol Geologists (secy-treas, 60-62). Res: Stratigraphy and petroleum geology. Mailing Add: 5508 Namakagan Rd Washington DC 20016

COHEN, ABRAHAM BERNARD, b Philadelphia, Pa, July 19, 22; m 53; c 2. ORGANIC CHEMISTRY, PHOTOCHEMISTRY. Educ: Temple Univ, AB, 48; Cornell Univ, PhD(chem), 52. Prof Exp: Asst, Temple Univ, 47-48 & Cornell Univ, 48-50; sr res chemist, Photo Prod Dept, 51-55, res supvr, 55-65, res fel, 65-66, res mgr, 66-69, mgr new prod develop, 69-70, mgr photopolymer systs, 70-72, DIR RES, E I DU PONT DE NEMOURS & CO, INC, 72- Honors & Awards: Indust Res Mag IR 100 Award, 69, 74 & 75. Mem: Am Chem Soc; Soc Photog Sci & Eng. Res: Correlation of structure and properties of polymers; mechanism of polymer reactions; photopolymerization; nonconventional photographic systems; dimensionally stable film bases and coatings; photoresist films and equipment; photopolymer printing plates, color proofing systems and graphic arts systems; new venture management. Mailing Add: 33 Hemlock Terr Springfield NJ 07081

COHEN, ADOLPH IRVIN, b New York, NY, Apr 7, 24; m 55; c 2. NEUROBIOLOGY. Educ: City Col New York, BS, 48; Columbia Univ, MA, 50, PhD(zool), 54. Prof Exp: From instr to asst prof anat, 55-64, res assoc prof ophthal, 64-70, chmn univ comt neurobiol, 70-74, PROF ANAT IN OPHTHAL, SCH MED, WASH UNIV, 70-, PROF ANAT, 70- Concurrent Pos: USPHS fel, Univ Calif, Berkeley, 53-54 & Wash Univ, 54-55; mem bd trustees, Asn Res Vision & Ophthal, 73-, mem ed bd, Vision Res, Exp Eye Res & J Ultrastruct Res. Mem: AAAS; Soc Neurosci; Am Soc Cell Biol; Asn Res Vision & Ophthal; Am Asn Anat. Res: Vision; cell biology; receptor physiology. Mailing Add: 4550 Scott Ave St Louis MO 63110

COHEN, ALAN MATHEW, b Chicago, Ill, Mar 22, 43. BIOLOGY, EMBRYOLOGY. Educ: Univ Ill, BSc, 64; Univ Va, PhD(biol), 69. Prof Exp: ASST PROF ANAT, SCH MED, JOHNS HOPKINS UNIV, 71- Concurrent Pos: USPHS fel anat, Harvard Med Sch, 69-71 & grant, 74-77; Nat Found March Dimes Basil O'Conner grant, 74-76. Mem: AAAS; Am Soc Zool; Soc Develop Biol. Res: Regulation of cellular events during embryogenesis; neural crest as a model system for the study of development. Mailing Add: Dept of Anat Johns Hopkins Univ Sch of Med Baltimore MD 21205

COHEN, ALAN SEYMOUR, b Boston, Mass, Apr 9, 26; m 54; c 3. MEDICINE. Educ: Harvard Univ, AB, 47; Boston Univ, MD, 52. Prof Exp: Instr, Harvard Med Sch, 58-60; from asst prof to prof, 60-72, CONRAD WESSELHOEFT PROF MED, SCH MED, BOSTON UNIV, 72-, DIR ARTHRITIS & CONNECTIVE TISSUE DIS SECT, UNIV HOSP, 60-; CHIEF MED, HOSP & DIR THORNDIKE MEM LAB, BOSTON CITY HOSP, 73- Concurrent Pos: Fel med, Harvard Med Sch, 53, res fel, 56-58; consult, USPHS, 66-70; Bernadine Becker Mem lectr, 69; consult, Food & Drug Admin, 70-; mem gen med study sect A, Nat Inst Arthritis, Metab & Digestive Dis, 72-; Wallace-Graham Mem lectr, Queen's Univ, Ont, 73; chmn med & sci comt, Mass Arthritis Found. Honors & Awards: Maimonides Award, Boston Med Soc, 52. Mem: Am Soc Clin Invest; Am Fedn Clin Res; Asn Am Physicians; Am Soc Cell Biol; Am Soc Exp Path. Res: Internal medicine; rheumatology; electron microscopy. Mailing Add: Thorndike 314 Boston City Hosp 818 Harrison Ave Boston MA 02118

COHEN, ALEX, b New York, NY, Feb 7, 31. ORGANIC CHEMISTRY, CLINICAL CHEMISTRY. Educ: Brooklyn Col, 54. Prof Exp: ORG CHEMIST, NAT BUR STAND, 57- Mem: Am Chem Soc. Res: Synthesis and reaction mechanisms of carbohydrates and nitrogen mustards; clinical standard characterization-cholesterol; anticancer agents; fluorinated compounds; stable isotope-labeled clinical compounds; dilution studies of cholesterol, glucose, and uric acid. Mailing Add: 5811 14th st NW Washington DC 20011

COHEN, ALLEN IRVING, b New York, NY, May 26, 32; m 54; c 3. ANALYTICAL CHEMISTRY, STRUCTURAL CHEMISTRY. Educ: City Col New York, BS, 54; Syracuse Univ, PhD(chem), 58. Prof Exp: Asst instr, Syracuse Univ, 54, AEC asst, 54-57; res chemist anal chem, Gulf Res & Develop Co, 57-59; sr res chemist, 59-66, res assoc, 66-69, res group leader, 69-72, SECT HEAD MOLECULAR SPECTROS, SQUIBB INST MED RES, 72- Mem: Am Chem Soc; The Chem Soc; NY Acad Sci; Am Soc Mass Spectrometry. Res: Structure determination of organic compounds and natural products by physicochemical methods; nuclear magnetic resonance and ultraviolet spectroscopy; organic polarography; radio chemistry, coprecipitation phenomena; developmental analytical methods; mass spectrometry and development of computer data acquisition programs. Mailing Add: Squibb Inst for Med Res PO Box 4000 Princeton NJ 08540

COHEN, ALONZO CLIFFORD, JR, b Stone Co, Miss, Sept 4, 11; m 34; c 3. STATISTICS. Educ: Ala Polytech Univ, BS, 32, MS, 33; Univ Mich, MA, 40, PhD(statist), 41. Prof Exp: Student engr, Westinghouse Elec & Mfg Co, 33-34; instr math, Ala Polytech Univ, 34-40; from instr to asst prof, Mich State Univ, 40-47; assoc prof, 47-52, PROF MATH, UNIV GA, 52-, DIR INST STATIST, 59- Concurrent Pos: Consult, Opers Anal Off Hqs, US Air Force, 50- Honors & Awards: Michael Award, Univ Ga, 54. Mem: Fel AAAS; fel Am Soc Qual Control; Inst Math Statist; fel Am Statist Asn; Math Asn Am. Res: Truncated frequency distributions; statistical methods of quality control; mathematical statistics. Mailing Add: Dept of Statist Univ of Ga Athens GA 30601

COHEN, ALVIN JEROME, b Louisville, Ky, July 21, 18; m 43, 69; c 4. GEOCHEMISTRY. Educ: Univ Fla, BS, 40; Univ Ill, PhD(inorg chem), 49. Prof Exp: Anal chemist, Tenn Valley Authority, Wilson Dam, Ala, 41; physicist closed bomb ballistics, Ind Ord Works, 42; chemist war alcohol prod, Joseph Seagram & Sons, 43; asst, Purdue Univ, 46-47 & Univ Ill, 47-49; fel, Calif Inst Technol, 49-50; chemist phys chem, Naval Ord Test Sta, China Lake, 50-53; from fel to sr fel, Mellon Inst, 53-62; PROF GEOCHEM, UNIV PITTSBURGH, 63- Mem: Am Mineral Soc; Am Geochem Soc; Meteoritical Soc. Res: Radiation effects in silicate minerals and glasses; geochemistry of meteorites. Mailing Add: Dept of Earth & Planetary Sci Univ of Pittsburgh Pittsburgh PA 152160

COHEN, ANNA FONER, b Pittsburgh, Pa, July 4, 24; m 50; c 4. PHYSICS Educ: Univ Pittsburgh, BS, 46, PhD(higher educ), 71, environ systs eng, 75-; Carnegie Inst Technol, MS, 48, DSc(physics), 50. Prof Exp: Res engr physics, Westinghouse Res Labs, 44-46; asst, Carnegie Inst Technol, 46-48; sr scientist, Oak Ridge Inst Nuclear Studies, 50; physicist, Oak Ridge Nat Lab, 50-58; guest fel, Carnegie-Mellon Univ, 58-59, res physicist, 60-64, lectr physics, 64-70; prof math & physics & chmn dept, Point Park Col, 71-74; mem res staff, Oak Ridge Nat Lab, 74-75. Concurrent Pos: Consult, Gen Atomic, 59-60. Mem: Am Phys Soc; Am Soc Metals; Sigma Xi. Res: Solid state and low temperature physics; radiation damage; transport properties; magnetic transitions at low temperatures; physics teaching; environmental systems and technology. Mailing Add: 5414 Albemarle Ave Pittsburgh PA 15217

COHEN, ARON DAVID, inorganic chemistry, organometallic chemistry, see 12th edition

COHEN, ARTHUR, b Brooklyn, NY, May 23, 33; m 57; c 2. MATHEMATICAL STATISTICS. Educ: Brooklyn Col, BA, 55; Columbia Univ, MA, 58, PhD(math statist), 63. Prof Exp: Statistician, Epidemic Intel Serv, Communicable Dis Ctr, USPHS, 57-59; asst prof, 63-66, assoc prof & actg chmn dept, 68-70, PROF APPL & MATH STATIST & CHMN DEPT STATIST, RUTGERS UNIV, 70- Mem: Inst Math Statist; Am Statist Asn. Res: Decision theory approaches to statistical inference and regression analysis; application of statistics to problems arising in biological science. Mailing Add: Dept of Statist Rutgers Univ New Brunswick NJ 08903

COHEN, ARTHUR DAVID, b Wilmington, Del, Feb 26, 42; m 70. PETROLOGY, PALYNOLOGY. Educ: Univ Del, BS, 64; Pa State Univ, PhD(geol), 68. Prof Exp: Asst prof geol, Univ Ga, 68 & Southern Ill Univ, Carbondale, 69-74; geologist, Coal Resources Br, US Geol Surv, 74-75; ASSOC PROF GEOL, UNIV OF SC, 75- Concurrent Pos: Nat Sci Found grants, 69, 71, 73 & 75. Mem: AAAS; Geol Soc Am (vchmn, coal geol div, 74-75); Bot Soc Am; Am Asn Stratig Palynologists; Brit Palaeont Asn. Res: Petrologic investigation of the peats of southern Florida with special reference to the origin of coal; geologic history of the Okefenokee Swamp from study of its peat sediments; bermuda peats. Mailing Add: Dept of Geology Geol Univ of SC Columbia SC 29208

COHEN, ARTHUR ISAAC, b Malden, Mass, Mar 23, 23; m 53; c 1. PHYSIOLOGY, BIOCHEMISTRY. Educ: Yeshiva Univ, AB, 46; Univ Minn, MS, 49; Tufts Univ, PhD(physiol), 56. Prof Exp: Res assoc path, Children's Cancer Res Found, 55-59; res assoc, Boston Dispensary, 59-63; head, Cellular Physiol Sect, Abbott Labs, 63-68; asst prof pharm, Sch Med, Marquette Univ, 64-68; with Strasenburgh Labs, 68-74; VPRES RES & DEVELOP, UNION POLYMER APPLNS INC, 74- Concurrent Pos: Res career develop award, NIH, 61-63. Mem: Am Soc Exp Path; Am Asn Cancer Res; Biophys Soc; Endocrine Soc; Am Chem Soc. Res: Tissue culture; polypeptide and steroid hormones; hypothalamic-pituitary relationships. Mailing Add: Res & Develop Dept Union Polymer Applns Inc Scottsville NY 14546

COHEN, ARTHUR LEROY, b Newport News, Va, Jan 22, 16; m 43; c 4. ELECTRON MICROSCOPY, RESEARCH ADMINISTRATION. Educ: Stanford Univ, AB, 37; Harvard Univ, MA, 39, PhD(bot), 40. Prof Exp: Sheldon traveling fel from Harvard Univ, Hopkins Marine Sta, 40-41; res fel, Calif Inst Technol, 42-47; prof biol, Oglethorpe Univ, 47-62; ASSOC PROF BIOL & DIR ELECTRON MICROS CTR, WASH STATE UNIV, 62-, PROF BOT & BIOL SCI, 68- Concurrent Pos: Res assoc, Cedars of Lebanon Hosp, Los Angeles, 45-47; Guggenheim fel, Delft Univ Technol, 56-57; vis prof, Yale Sch Med, 71. Mem: AAAS; fel Am Soc Microbiol (pres, 62); Bot Soc Am; Soc Protozool; Mycol Soc Am. Res: Experimental morphogenesis of Myxomycetes; general biology, ultrastructure; critical point drying; enzymatic and chemical subcellular dissection. Mailing Add: Electron Micros Ctr Wash State Univ Pullman WA 99163

COHEN, BENNETT J, b Brooklyn, NY, Aug 2, 25; m 52; c 2. LABORATORY ANIMAL MEDICINE, COMPARATIVE MEDICINE. Educ: Cornell Univ, DVM, 49; Northwestern Univ, MS, 51, PhD(physiol), 53; Am Col Lab Animal Med, dipl, 58. Prof Exp: Veterinarian, Northwestern Univ, 49-53; statewide veterinarian, Univ Calif, Berkeley, 53-57; veterinarian, Univ Calif, Los Angeles, 54-62, from instr to asst prof, 56-62; assoc prof physiol & dir animal care unit, 62-67, PROF LAB ANIMAL MED & DIR UNIT FOR LAB ANIMAL MED, UNIV MICH, ANN ARBOR, 67- Concurrent Pos: Mem, Aging Rev Comt, Nat Insurance Asn, NIH, 74-78. Mem: Am Vet Med Asn; Am Asn Lab Animal Sci (pres, 58-60); Am Physiol Soc. Res: Pathology of aging in laboratory animals; animal models of inflammatory bowel disease; diseases of laboratory animals. Mailing Add: Unit for Lab Animal Med Univ of Mich Animal Res Facil Ann Arbor MI 48109

COHEN, BERNARD, b Wilmington, Del, Nov 2, 33; m 55; c 2. INORGANIC CHEMISTRY. Educ: Temple Univ, AB, 55; Univ Pa, MS, 57, PhD(chem), 63. Prof Exp: Chemist, Geol Surv, US Dept Interior, 55-57; res chemist, Pennsalt Chem Corp, 57-59 & Foote Mineral Co, 59-61; res fel chem, Univ Birmingham, 63-65; sr res chemist, 65-70, mgr process res, 70-74, DIR SPECIALTY CHEM, FMC CORP, 74- Mailing Add: Indust Chem Div FMC Corp PO Box 8 Princeton NJ 08540

COHEN, BERNARD, b Newark, NJ, Apr 30, 29; m 55; c 3. NEUROLOGY, NEUROPHYSIOLOGY. Educ: Middlebury Col, AB, 50; NY Univ, MD, 54; Am Bd Psychiat & Neurol, dipl, 61. Prof Exp: Asst attend neurologist, Mt Sinai Hosp, New York, 61-66; assoc prof physiol, 66-69, PROF NEUROL, MT SINAI SCH MED, 69- Concurrent Pos: Res fel neurophysiol, Col Physicians & Surgeons, Columbia Univ, 60-61; trainee, Nat Inst Neurol Dis & Blindness, 60-62; Nat Inst Neurol Dis & Stroke career res develop award, 67-73; assoc attend neurologist, Mt Sinai Hosp, New York, 66-69, attend neurologist, 69-; attend neurol, Elmhurst Gen Hosp; mem ed bd, Brain Res, 72-, Am J Physiol, 72 & J Appl Physiol, 72-; expert in neurol, FDA, 75- Mem: Am Physiol Soc; Am Acad Neurol; Am Neurol Asn; Barany Soc. Res: Physiology of oculomotor, vestibular and postural systems. Mailing Add: Mt Sinai Sch of Med Atran 710 One E 100th St New York NY 10029

COHEN, BERNARD LEONARD, b Pittsburgh, Pa, June 14, 24; m 50; c 4. NUCLEAR PHYSICS. Educ: Case Western Reserve Univ, BS, 44; Univ Pittsburgh, MS, 48; Carnegie Inst Technol, DSc(physics), 50. Prof Exp: Asst, Carnegie Inst Technol, 47-49; physicist & group leader, Oak Ridge Nat Lab, 50-58; assoc prof, 58-61, PROF PHYSICS, UNIV PITTSBURGH, 61-, DIR SCAIFE NUCLEAR PHYSICS LAB, 65- Concurrent Pos: Consult, Oak Ridge Nat Lab, 58-66, Nuclear Sci Eng Corp, 59-61, Gen Atomic, 59-60, NSF, 62, Inst Defense Anal, 62, Brookhaven Nat Lab, 65, Los Alamos Sci Lab, 68 & World Publ Co, 69-70; mem Nat Coun, Am Asn Physics Teachers, 73-76; mem exec comt, 71-73, chmn, Div of Nuclear Physics, Am Phys Soc, 74-75; vis staff, Inst Energy Anal, 74-75. Mem: Fel Am Phys Soc. Res: Nuclear structure; nuclear reactions and scattering; applied nuclear physics; environmental impacts of nuclear power. Mailing Add: 5414 Albemarle Ave Pittsburgh PA 15217

COHEN, BERNICE HIRSCHHORN, b Baltimore, Md, Apr 25, 24; m; c 2. HUMAN GENETICS, EPIDEMIOLOGY. Educ: Goucher Col, AB, 44; Johns Hopkins Univ, PhD(human genetics), 58, MPH, 59. Prof Exp: Nat Heart Inst fel, Div Med Genetics, Sch Med, 59-60, from asst prof to assoc prof, 60-70, PROF EPIDEMIOL, DEPT CHRONIC DIS, SCH HYG, JOHNS HOPKINS UNIV, 70- Concurrent Pos: Nat Inst Gen Med Sci res career develop award, Dept Chronic Dis, Johns Hopkins Univ, 60-70; assoc, Univ Sem in Genetics & Evolution of Man, Columbia Univ, 64-70; consult, Baltimore City Hosps, Md, 66-; asst prof, Sch Med, Johns Hopkins Univ, 70- Mem: AAAS; Am Soc Human Genetics; fel Am Pub Health Asn; Genetics Soc Am; Am Asn Phys Anthropologists. Res: Human epidemiological genetics, especially the role of genetic factors in chronic diseases, differential fertility, aging and mortality; congenital anomalies; genetically determined marker traits and disease; maternal-fetal blood group incompatibility. Mailing Add: Dept of Epidemiol Johns Hopkins Univ Sch of Hyg Baltimore MD 21205

COHEN, BERTRAM IRA, organic chemistry, biochemistry, see 12th edition

COHEN, BEVERLY SHAPIRO, b New York, NY, Mar 30, 44; m 63; c 1. MEDICAL BIOPHYSICS. Educ: Pa State Univ, BS, 63; Univ Pa, PhD(molecular biol), 68; Hahnemann Med Col, MD, 75. Prof Exp: Fel biochem, Univ Pa, 68-69; asst prof & grant-in-aid, Sch Dent, Temple Univ, 69-70, fel, Fels Inst, 70-75; intern, 75-76, RESIDENT INTERNAL MED, ALBERT EINSTEIN MED CTR, 76- Mem: AAAS; NY Acad Sci; Am Chem Soc. Res: Microsomal electron transport reactions; drug metabolism; cytochrome chemistry; induction of enzymes in mammalian systems; biological control; oxidation reduction reactions and bioenergetics; microsomes and metabolism; coordination of psychic and somatic energy systems; metabolism effect of mind on health. Mailing Add: 645 Lombard St Philadelphia PA 19147

COHEN, BRUCE IRA, b Los Angeles, Calif, Oct 26, 48; m 75. PLASMA PHYSICS. Educ: Harvey Mudd Col, BS, 70; Univ Calif, Berkeley, MA, 72, PhD(physics), 75. Prof Exp: RES ASSOC, PLASMA PHYSICS LAB, PRINCETON UNIV, 75- Concurrent Pos: Jr staff scientist, Phys Dynamics, Inc, Calif, 73- Mem: Am Phys Soc. Res: Theoretical plasma physics and physical oceanography; computational plasma physics and fluid mechanics. Mailing Add: Princeton Plasma Physics Lab Forrestal Campus PO Box 451 Princeton NJ 08540

COHEN, BURTON D, b Waterbury, Conn, Aug 10, 26; m 51; c 3. INTERNAL MEDICINE. Educ: Yale Univ, AB, 50; Columbia Univ, MD, 54; Am Bd Internal Med, dipl, 63. Prof Exp: Clin asst prof, 70-72, ASSOC PROF MED, ALBERT EINSTEIN COL MED, 72-; CHIEF METAB SECT, BRONX-LEBANON HOSP, 62- Concurrent Pos: Asst vis physician, Bellevue Hosp, 59-68; clin investr, Vet Admin, 59-62; career scientist, Health Res Coun New York, 65-73. Mem: Am Fedn Clin Res: Am Diabetes Asn; fel Am Col Physicians. Res: Metabolism and renal disease. Mailing Add: 1276 Fulton Ave Bronx NY 10456

COHEN, CARL, b Brooklyn, NY, Nov 15, 20; m 49; c 3. IMMUNOGENETICS. Educ: Ohio State Univ, BSc, 46, MSc, 48, PhD(bact), 51. Prof Exp: Asst bact, Ohio State Univ, 46-50; assoc immunogenetics, Jackson Mem Lab, 55-57; proj leader, Battelle Mem Inst, 57-62; prof biol & assoc prof exp path, Case Western Reserve Univ, 62-70; PROF GENETICS & DIR CTR GENETICS, UNIV ILL, MED CTR, 70- Concurrent Pos: Res fel immunogenetics, Jackson Mem Lab, 51-55. Mem: Fel AAAS; Am Asn Immunol; Genetics Soc Am; Am Soc Human Genetics. Res: Immunology; mammalian genetics. Mailing Add: Ctr Genetics Sch Basic Med Sci Univ of Ill Med Ctr Chicago IL 60612

COHEN, CAROLYN, b Long Island City, NY, June 18, 29. BIOPHYSICS. Educ: Bryn Mawr Col, AB, 50; Mass Inst Technol, PhD(biophys), 54. Prof Exp: Fulbright scholar, King's Col, Univ London, 54-55; res assoc, Children's Cancer Res Found, 55-56; instr biol, Mass Inst Technol, 57-58; res assoc path, Children's Cancer Res Found, Children's Hosp Med Ctr, 58-74; PROF BIOL & MEM ROSENSTIEL BASIC MED SCI RES CTR, BRANDEIS UNIV, 72- Concurrent Pos: Res assoc biol, Mass Inst Technol, 55-58; res assoc biochem, Harvard Med Sch, 58-64, Harvard Med Sch lectr biophys, Children's Hosp, 64-74. Mem: Am Crystallog Asn; Am Soc Biol Chem; Soc Gen Physiol; AAAS; Biophys Soc. Res: Structure of protein assemblies in the cell as determined by x-ray diffraction and electron micrographs; muscle structure and the contractile mechanism; structural aspects of cell division and blood coagulation. Mailing Add: Rosenstiel Ctr Brandeis Univ Waltham MA 02154

COHEN, DANIEL, b Brooklyn, NY, June 22, 24; m 47; c 3. PUBLIC HEALTH, EPIDEMIOLOGY. Educ: Univ Ill, BS, 53, DVM, 55; Univ Pittsburgh, MPH, 60.

Prof Exp: Vet officer, Commun Dis Ctr, USPHS, 55-60, assigned to Southern Health Dist, NJ, 55-56, NJ State Dept Health, 56-57,, Cancer Control Prog, 57-58 & fel, Grad Sch Pub Health, Univ Pittsburgh, 58-60; assoc mem, Lab Res Virol, Wistar Inst, Univ Pa, 60-61; dir, Vet Res Inst, Beit Dagan, Israel, 61-62; dir grad training prog epidemiol, Univ Pa, 62-71, from asst prof to prof vet pub health, Sch Vet Med, 63-71; pub health vet, Div Commun Dis, WHO, Switz, 71-73; PROF COMP MED, BEN GURION UNIV, ISRAEL, 73- Concurrent Pos: Consult, WHO & Pan Am Health Orgn, 65- & US AID, 66-; Am Pub Health Asn fel, Univ Pittsburgh, 66-; pres, Conf Pub Health Vets, 68. Mem: AAAS; Am Pub Health Asn; Am Vet Med Asn; Am Asn Cancer Res; Asn Teachers Vet Pub Health & Prev Med. Res: Epidemiology of acute and chronic disease of animals of comparative interest to man, zoonoses, and international aspects of veterinary public health; animal neoplasia; animal influenza; respiratory and enteric diseases of domestic animals. Mailing Add: Dept of Comp Med Ben Gurion Univ Beersheva Israel

COHEN, DANIEL MORRIS, b Chicago, Ill, July 6, 30; m 55; c 2. ICHTHYOLOGY. Educ: Stanford Univ, AB, 52, MA, 53, PhD, 58. Prof Exp: Asst gen biol, Natural Hist Mus, Stanford Univ, 53-55, actg instr, 55-57; asst prof, Univ Fla, 57-58; syst zoologist fishes, 58-60, lab dir ichthyol lab, 60-70, LAB DIR SYSTS LAB, US FISH & WILDLIFE SERV, NAT MARINE FISHERIES SERV, 70- Concurrent Pos: Vis researcher, Brit Mus Natural Hist, 64-65; mem, Nat Acad Sci Comn Ecol Res Interocean Canal, 69-70; res assoc, Smithsonian Inst, 69-; Ed-in-chief, Fishes of West NAtlantic, pt 6. Mem: AAAS; Am Soc Ichthyologists & Herpetologists (vpres, 69-70); Soc Study Evolution; Soc Syst Zool. Res: Biology of fishes, particularly systematics; deepsea fishes; general marine biology; museum collections; systematics and biology of deep benthic fishes. Mailing Add: Systs Lab US Nat Mus Nat Marine Fisheries Serv Washington DC 20560

COHEN, DAVID, b Winnipeg, Man, Dec 1, 27; US citizen. MAGNETISM. Educ: Univ Man, BA, 48; Univ Calif, Berkeley, PhD(exp nuclear physics), 55. Prof Exp: Assoc physicist, Defence Res Bd Can, 55-57; res assoc physics, Univ Rochester, 57-58; assoc physicist, Argonne Nat Lab, 58-65; assoc prof physics, Univ Ill, Chicago Circle, 65-68; SR SCIENTIST BIOPHYS, FRANCIS BITTER NAT MAGNET LAB, MASS INST TECHNOL, 68- Mem: AAAS; Am Phys Soc; sr mem Inst Elec & Electronics Engrs; Biophys Soc. Res: Measuring magnetic fields produced by the human body, including magnetocardiography and magnetoencephalography. Mailing Add: Nat Magnet Lab Mass Inst Technol Cambridge MA 02139

COHEN, DAVID HARRIS, b Springfield, Mass, Aug 26, 38; m 60; c 3. NEUROPHYSIOLOGY. Educ: Harvard Univ, AB, 60; Univ Calif, Berkeley, PhD(psychol), 63. Prof Exp: Asst prof physiol, Sch Med, Case Western Reserve Univ, 64-68; assoc prof, 68-71, PROF PHYSIOL, SCH MED, UNIV VA, 71-, DIR NEUROSCI PROG, 75- Concurrent Pos: NSF fel neurophysiol, Med Sch, Univ Calif, Los Angeles, 63-64; Nat Heart & Lung Inst career develop award, 69-74; mem adv panel neurobiol, NSF, 72-75; mem ed bd, Brain Res Bull, 75-; consult, Educ Resource Prog, Nat Heart & Lung Inst, 75-, Nat Sci Adv Comt, Brain Info Serv, 75- Mem: Am Asn Anat; Am Physiol Soc; Pavlovian Soc NAm; Soc Neurosci (secy, 75-). Res: Neural mechanisms of learning; comparative neurology. Mailing Add: Dept of Physiol Univ of Va Sch of Med Charlottesville VA 22903

COHEN, DAVID WALTER, b Philadelphia, Pa, Dec 15, 26; m 48; c 3. DENTISTRY. Educ: Univ Pa, DDS, 50. Hon Degrees: DSc, Boston Univ, 75. Prof Exp: From asst instr to assoc oral med & oral path, Sch Dent, 51-55, asst prof periodont, Dent Sch & Grad Sch Med & vchmn dept, Grad Sch Med, 55-59, assoc prof periodont, Dent Sch & Grad Sch Med & chmn dept, Grad Sch Med, 59-64, prof periodont & chmn depts, 69-72, DEAN, SCH DENT MED, UNIV PA, 72-; PROF DENT MED, MED COL PA, 73- Concurrent Pos: Res fel path & periodont, Beth Israel Hosp, Boston, 50-51; asst vis chief oral med, Philadelphia Gen Hosp, 51-; clin asst periodont, Albert Einstein Med Ctr, 53-; nat consult periodont, US Air Force, 65-69, nat emer consult, 69-; dir, Am Bd Periodont, 68, vchmn, 70-, chmn, 71; vis prof, Col Dent, Univ Ill; consult, Vet Admin Hosp, Philadelphia, Ft Dix Army Base & Walter Reed Army Med Ctr. Honors & Awards: Spec Citation Periodont, Boston Univ, 69; Israel Peace Award, 70; Gold Medal Award, Am Acad Periodont, 71; William J Gies Found Periodont Award, 75. Mem: Fel AAAS; fel Am Acad Oral Path; Int Asn Dent Res. Res: Vascular plexus of oral tissues; periodontal disease; treatment planning in dentistry; periodontal therapy. Mailing Add: Dept of Periodont Univ of Pa Sch of Dent Med Philadelphia PA 19104

COHEN, DAVID WARREN, b Hartford, Conn, Feb 28, 40; m 64; c 1. MATHEMATICS. Educ: Worcester Polytech Inst, BS, 62; Univ NH, MS, 64, PhD(math), 68. Prof Exp: Instr math, Exten Serv, Univ NH, 65-68 & Univ, 67-68; asst prof, 68-74, ASSOC PROF MATH, SMITH COL, 74- Mem: Math Asn Am. Res: Topological groups; Lie groups; Lie algebras; mathematical physics. Mailing Add: Dept of Math Smith Col Northampton MA 01060

COHEN, DONALD, solid state physics, space physics, see 12th edition

COHEN, DONALD, b Tom's River, NJ, Feb 9, 20; m 43; c 4. RADIOCHEMISTRY. Educ: Univ Buffalo, BS, 41; Purdue Univ, MS, 48, PhD(chem), 50. Prof Exp: Assoc chemist, 49-72, CHEMIST, ARGONNE NAT LAB, 72- Mem: Sigma Xi; Am Chem Soc. Res: Electrode potentials; chemistry of the transuranium elements. Mailing Add: Chem Div Argonne Nat Lab 9700 SCass Ave Argonne IL 60439

COHEN, DONALD SUSSMAN, b Providence, RI, Nov 30, 34; m 58; c 2. APPLIED MATHEMATICS. Educ: Brown Univ, ScB, 56; Cornell Univ, MS, 59; NY Univ, PhD(math), 62. Prof Exp: Preceptorship, Dept Eng Mech & Inst Flight Struct, Columbia Univ, 62-63; asst prof math, Rensselaer Polytech Inst, 63-65; asst prof math, 65-67, assoc prof appl math, 67-71, PROF APPL MATH, CALIF INST TECHNOL, 71- Mem: AAAS; Am Math Soc; Soc Indust & Appl Math (ed, SIAM Rev). Res: Wave propagation and vibration problems; partial differential equations; special functions; variational techniques; non-linear boundary value problems; bifurcation theory; perturbation and asymptotic methods; differential equations. Mailing Add: Dept of Appl Math Calif Inst of Technol Pasadena CA 91125

COHEN, E RICHARD, b Philadelphia, Pa, Dec 14, 22; m 53; c 1. MATHEMATICAL PHYSICS, REACTOR PHYSICS. Educ: Univ Pa, AB, 43; Calif Inst Technol, MS, 46, PhD(physics), 49. Prof Exp: Asst instr physics, Univ Pa, 43-44; jr physicist acoust-electronic res, Calif Inst Technol, 44-45; theoret physicist, NAm Rockwell Corp, 49-56, res adv, 56-61, assoc dir, Res Dept, 61-62 & Sci Ctr, 62-69, mem tech staff, 69-75, DISTINGUISHED FEL, SCI CTR, ROCKWELL INT CORP, 75- Concurrent Pos: Sr lectr, Calif Inst Technol, 62-63, res assoc, 63-72. Mem: AAAS; fel Am Phys Soc; fel Am Nuclear Soc; Asn Comput Mach; Sigma Xi. Res: Evaluation of the fundamental physical constants; nuclear reactor theory; molecular spectroscopy. Mailing Add: Sci Ctr Rockwell Int Corp 1049 Camino Dos Rios Thousand Oaks CA 91360

COHEN, ECKFORD, b Starkville, Miss, Mar 21, 20. ALGEBRA. Educ: Miss State Col, BS, 43; Duke Univ, MA, 45, PhD(math), 47. Prof Exp: Instr math, Miss State Col, 43-44 & Duke Univ, 47-48; from instr to asst prof, Syracuse Univ, 48-51; asst prof, Inst Advan Study, 51-53; asst prof math, Univ SC, 54-56; from asst prof to assoc prof, Univ Tenn, 56-65; prof, Univ Fla, 66-67 & Kans State Univ, 67-70; VIS PROF MATH, WESTERN MICH UNIV, 71- Res: Theory of numbers. Mailing Add: Dept of Math Western Mich Univ Kalamazoo MI 49001

COHEN, EDWARD DAVID, b Haverhill, Mass, Mar 12, 37; m 58; c 3. PHYSICAL CHEMISTRY, CHEMICAL ENGINEERING. Educ: Tufts Univ, BSChE, 58; Univ Del, PhD(phys chem), 64. Prof Exp: Jr engr, Elkton Div, Thiokol Chem Corp, 58-60; res chemist, 64-67, sr res chemist, 67-72, RES ASSOC, PHOTOPROD DEPT, E I DU PONT DE NEMOURS & CO, INC, 72- Mem: Am Inst Chem Eng; Am Chem Soc; Soc Photog Sci & Eng. Res: Reaction kinetics and mechanisms; radiation and polymer chemistry; analytical chemistry; gelatin and bio-polymers; emulsion chemistry. Mailing Add: Rt 4 107 Silver Creek Ct Greer SC 29651

COHEN, EDWARD HIRSCH, b Seattle, Wash, Aug 28, 47. CELL BIOLOGY. Educ: Univ Chicago, BS, 68; Yale Univ, MPhil, 70, PhD(biol), 73. Prof Exp: ASST PROF BIOL, PRINCETON UNIV, 74- Concurrent Pos: Fel zool, Univ Wash, 72-74. Mem: Am Soc Cell Biol; Genetics Soc Am. Res: Organization of DNA sequences in eukaryote chromosomes; structure and evolution of satellite DNAs of Drosophila. Mailing Add: Dept of Biol Guyot Hall Princeton Univ Princeton NJ 08540

COHEN, EDWARD MORTON, b New York, NY, May 12, 36; m 59; c 3. PHARMACEUTICAL CHEMISTRY, ANALYTICAL CHEMISTRY. Educ: Columbia Univ, BS, 57; Rutgers Univ, MS, 60, PhD(pharmaceut chem), 65. Prof Exp: Res chemist, Johnson & Johnson Res Ctr, 62-65; res assoc anal chem, 65-70, SR RES FEL ANAL CHEM, MERCK SHARP & DOHME RES LABS, 70- Mem: AAAS; Am Chem Soc; Am Pharmaceut Asn. Res: Development of analytical methods for pharmaceutical dosage forms; measurement of physical properties of compounds; polarography and other electroanalytical methods; thin layer chromatography; thermal analysis. Mailing Add: 3008 Eisenhower Dr Norristown PA 19401

COHEN, ELAINE, b NJ, July 17, 46; m 74. MATHEMATICAL ANALYSIS, SYSTEMS THEORY. Educ: Vassar Col, BA, 68; Syracuse Univ, MA, 70, PhD(math), 74. Prof Exp: Vis instr math, 74-75, RES ASST PROF COMPUT SCI, UNIV UTAH, 74-, ASSOC INSTR MATH, 75- Mem: Am Math Soc; Asn Comput Mach; Inst Elec & Electronics Engrs. Res: Mathematical structures in sensory information processing; Fourier analysis. Mailing Add: Comput Sci Univ of Utah Salt Lake City UT 84112

COHEN, ELIAS, b Baltimore, Md, Sept 17, 20; m 46; c 2. IMMUNOLOGY. Educ: Univ Md, BSc, 42; Johns Hopkins Univ, MA, 49; Rutgers Univ, PhD(immunol), 52. Prof Exp: Biologist, Sch Hyg & Pub Health, Johns Hopkins Univ, 46, jr instr biol & asst, 47-49; jr instr, Rutgers Univ, 49-50, asst genetics & physiol, 50-52; instr clin path & asst dir clin labs, Sch Med, Univ Okla, 52-54, res assoc clin path, 54-56; assoc cancer res scientist immunohemat, 56-69, prin cancer res scientist, 69-72, ASSOC CHIEF CANCER RES SCIENTIST, ROSWELL PARK MEM INST, 72-; RES ASSOC PROF MICROBIOL & IMMUNOL, SCH MED, STATE UNIV NY BUFFALO, 68- Concurrent Pos: Lectr, Okla City Univ, 53-54; prin investr, S R Noble Found, Inc, 54-56; lab consult, Erie County Lab, 57-64; lectr med genetics, Sch Med, State Univ NY Buffalo, 57-72; lab consult, Erie County Lab, 57-64; lectr, Queen's Univ, 58; mem acute leukemia task force & proj dir, Platelet Transfusion Eval Prog, Nat Cancer Inst, 64-; immunohemat consult, Children's Hosp, Buffalo, NY, 67-; US-USSR exchange scientist, 69, 73 & 75; pres, NY Publ Health Lab Asn, 74-75. Mem: Am Soc Exp Path; Am Soc Hemat; Am Asn Immunol; Int Soc Blood Transfusion; Int Soc Hemat. Res: Immunobiology and immunogenetics; comparative and mammalian immunohematology; physiology and biochemistry; biomedical applications of erythrocytes and serum proteins of human and other species; immunochemical nature of serum protein antigens. Mailing Add: Roswell Park Mem Inst 666 Elm St Buffalo NY 14263

COHEN, ELLIOTT, b New York, NY, Apr 14, 30; m 58; c 3. MEDICINAL CHEMISTRY. Educ: Syracuse Univ, BA, 51; Columbia Univ, PhD(chem), 56. Prof Exp: Chemist, NY State Psychiat Inst, 56-57; res chemist, 57-61, group leader org chem, 61-74, DEPT HEAD, CENT NERV SYST DIS THER SECT, LEDERLE LABS, DIV AM CYANAMID CO, 74- Mem: Am Chem Soc. Res: Organic synthetic work in heterocyclic and cardiovascular drugs, anti-inflammatory agents and central nervous systems agents; mechanism of action and structure activity relationships. Mailing Add: Lederle Labs Div of Am Cyanamid Co Pearl River NY 10965

COHEN, ELLIS N, b Des Moines, Iowa, June 5, 19; m 47; c 3. ANESTHESIOLOGY. Educ: Univ Minn, BS, 41, MD, 43, MS, 49. Prof Exp: From Clin instr to assoc clin prof anesthesiol, Univ Minn, 49-60; assoc prof, 60-65, PROF ANESTHESIOL, STANFORD UNIV, 65- Mem: AMA; Am Soc Anesthesiol. Res: Clinical pharmacology; toxicology; uptake and distribution and metabolism of drugs; analytic methods. Mailing Add: Dept of Anesthesia Stanford Univ Stanford CA 94305

COHEN, ERWIN (MORTON), physical chemistry, see 12th edition

COHEN, EZECHIEL GODERT DAVID, b Amsterdam, Neth, Jan 16, 23; m 50; c 2. PHYSICS. Educ: Univ Amsterdam, BS, 47, PhD(physics), 57. Prof Exp: First asst physics, Univ Amsterdam, 50-61, assoc prof, 61-63; PROF PHYSICS, ROCKEFELLER UNIV, 63- Concurrent Pos: Netherlands Orgn Pure Sci Res scholar, Univ Mich, 57-58 & Johns Hopkins Univ, 58-59; Van der Waals prof, Univ Amsterdam, 69; vis prof, Col France, Paris, 69-72. Mem: Fel Am Phys Soc; Neth Phys Soc. Res: Statistical mechanics, particularly applied to equilibrium and nonequilibrium properties of gases and liquids at normal and low temperatures. Mailing Add: Rockefeller Univ 1230 York Ave New York NY 10021

COHEN, FLOSSIE, b Calcutta, Brit India, May 10, 25; US citizen; m 58; c 1. IMMUNOLOGY, PEDIATRICS. Educ: Med Col, Calcutta, MB, 45; Univ Buffalo, MD, 50. Prof Exp: Hematologist, 56-58, from asst pediatrician & assoc hemat to assoc pediatrician, 58-73, DIR CLIN IMMUNOL, CHILDREN'S HOSP MICH, DETROIT, 69- Concurrent Pos: Res assoc, Child Res Ctr Mich, Detroit, 58-60, sr res assoc, 60-; univ assoc, Dept Affil-Pediat, Hutzel Hosp, Detroit, 64-; attend pediatrician, Dept Pediat Med, Children's Hosp Mich, 71-, attend pediatrician, Dept Lab Med, Immunol Sect, 74-; immunol consult, William Beaumont Hosp, Royal Oak, Mich, 73-; mem biomet & epidemiol contract rev comt, Nat Cancer Inst, 74-76. Mem: Am Pediat Soc; Am Soc Hemat; Am Soc Human Genetics; Soc Pediat Res; Am Asn Immunologists. Res: Immunodeficiency diseases; immunology of pregnancy; transplacental passage of cells. Mailing Add: 3901 Beaubien Blvd Detroit MI 48201

COHEN, FREDRIC SUMNER, b Boston, Mass, Dec 17, 35; m 62; c 2. POLYMER CHEMISTRY, OPTICS. Educ: Oberlin Col, AB, 57; Brandeis Univ, PhD, 63; Mass Inst Technol, MS, 76. Prof Exp: Res chemist, US Indust Chem Co, 59-60; sr res chemist, Diamond Alkali Co, 63-65; res chemist, Stauffer Chem Co, 65-67, sr res chemist, 67-69; scientist, 69-70, RES GROUP LEADER, POLAROID CORP, 71- Mem: Am Chem Soc; Soc Plastics Eng; Soc Rheol. Res: Physical polymer chemistry, especially related to optical systems and light polarizers; correlation of molecular and gross properties of synthetic polymers; polymer orientation and dyeing. Mailing Add: Res Div Polaroid Corp Cambridge MA 02139

COHEN, GARY H, b Brooklyn, NY, May 7, 34; m 59; c 2. MICROBIOLOGY, VIROLOGY. Educ: Brooklyn Col, BS, 56; Univ Vt, PhD(microbiol), 64. Prof Exp: ASSOC PROF MICROBIOL, SCH DENT MED, UNIV PA, 67- Concurrent Pos: USPHS fel virol, Univ Pa, 64-67; USPHS res career develop award, 69- Mem: AAAS; Am Soc Microbiol. Res: Chemical composition of the polysaccharides of Azotobacter; soluble antigens of vaccinia virus-infected mammalian cells; antigens of herpes simplex virus; DNA synthesis in herpes-infected mammalian cells; herpes infection in synchronized human cells. Mailing Add: Dept of Microbiol Univ of Pa Sch of Dent Med Philadelphia PA 19104

COHEN, GEORGE LESTER, b Brooklyn, NY, Dec 24, 39; m 62; c 3. PHYSICAL CHEMISTRY. Educ: Clarkson Col Technol, BS, 61, MS, 63; Univ Md, PhD(phys chem), 67. Prof Exp: Chemist, US Naval Ord Lab, Md, 62-67; proj supvr, Gillette Res Inst, 67-69; HEAD PHYS CHEM RES DEPT, PROD DIV, BRISTOL-MYERS, INC, 69- Concurrent Pos: Adj asst prof, Rutgers Univ, Newark, 70- Mem: Am Chem Soc; Soc Cosmetic Chemists. Res: Chemistry of silver oxides, pharmaceutical, cosmetic, polymer, surface and colloid chemistry. Mailing Add: Bristol-Myers Inc Prod Div 225 Long Ave Hillside NJ 07207

COHEN, GERALD, b New York, NY, Feb 1, 30. NEUROCHEMISTRY. Educ: City Col New York, BS, 50; Columbia Univ, MA, 52, PhD(chem), 55. Prof Exp: Assoc biochem, Col Physicians & Surgeons, Columbia Univ, 54-65, asst prof, 65-73; RES PROF NEUROL, MT SINAI SCH MED, 73- Honors & Awards: Claude Bernard Sci Jour Award, Nat Soc Med Res, 68. Mem: Am Soc Biol Chem; Am Soc Pharmacol & Exp Therapeut; Am Chem Soc. Res: Biochemical pharmacology. Mailing Add: Dept of Neurol Mt Sinai Sch of Med New York NY 10029

COHEN, GERALD STANLEY, b New York, NY, Nov 29, 26; m 51; c 2. BIOMEDICAL ENGINEERING. Educ: City Col New York, BS, 50; Univ Md, MS, 67; Univ NC, PhD, 76. Prof Exp: Electronic engr, Philco-Ford Corp, 50-58; electronic engr, Army Ord, Harry Diamond Labs, 58-60; chief electronic & elec eng sect, Div Res Servs, 60-68, ASSOC DIR HEALTH CARE TECHNOL DIV, NAT CTR HEALTH SERVS, RES & DEVELOP, USPHS, 68- Mem: Instrument Soc Am; Inst Elec & Electronics Eng; Opers Res Soc Am. Res: Design and development of electronic instrumentation for medical research; support applications of technology to the delivery of health services. Mailing Add: Ctr Health Serv Res Parklawn Bldg 15A38 Rockville MD 20852

COHEN, GERSON H, b Philadelphia, Pa, July 8, 39; m 63; c 3. PHYSICAL CHEMISTRY. Educ: Temple Univ, AB, 61; Cornell Univ, PhD(phys chem), 65. Prof Exp: RES CHEMIST, NIH, 65- Mem: AAAS; Am Chem Soc; Am Crystallog Asn. Res: Chemical crystallography with emphasis in protein structure. Mailing Add: NIH Bldg 2 Rm 312 Bethesda MD 20014

COHEN, GLORIA, b Leeds, Eng, Jan 2, 30; m 51; c 3. INFORMATION SCIENCE. Educ: Univ Birmingham, Eng, DDS, 52. Prof Exp: Dental res fel, London Hosp Med Col, 52-53, clin asst, 53-55; gen dent practr, London, 55-67; info scientist, Data & Info Ctr, Res & Develop Div, 68-69, supvr biomed doc, 69-70, from asst mgr to mgr, 70-72, DIR INFO SERV DEPT, G D SEARLE & CO, 72- Concurrent Pos: Prog evaluator, NSF, 74-; mem task force sci & technol, Indust Res Inst, 74- Mem: AAAS; Am Soc Info Sci; Drug Info Asn; Am Rec Mgt Asn; Int Asn Dent Res. Mailing Add: G D Searle & Co Box 1045 Skokie IL 60076

COHEN, GORDON MARK, b Chicago, Ill, Jan 7, 48; US & Can citizen. PHYSICAL ORGANIC CHEMISTRY. Educ: McGill Univ, BSc, 69; Harvard Univ, AM, 70, PhD(chem), 74. Prof Exp: RES CHEMIST, E I DU PONT DE NEMOURS & CO INC, 74- Mem: Am Chem Soc. Res: Preparative polymer chemistry and organic chemistry in polymer systems, mechanistic and synthetic aspects. Mailing Add: Elastomer Chem Dept Exp Sta E I du Pont de Nemours & Co Inc Wilmington DE 19898

COHEN, HAROLD KARL, b Trenton, NJ, Mar 12, 15; m 60; c 2. VETERINARY PATHOLOGY. Educ: City Col New York, BS, 38; Kans State Col, DVM, 47; Univ Wis, MS, 52, PhD(path), 55. Prof Exp: Instr vet path, Univ Wis, 50-55; vet pathologist, Ralph M Parsons Co, 55; vet, 55-73, VET PATHOLOGIST, ELI LILLY & CO, 73- Mem: Am Vet Med Asn; Am Col Vet Path. Res: Virology. Mailing Add: Eli Lilly & Co Greenfield IN 46140

COHEN, HAROLD P, b Brooklyn, NY, Sept 6, 24; m 57; c 3. BIOLOGICAL CHEMISTRY. Educ: City Col New York, BS, 48; Univ Iowa, MS, 51, PhD(biochem), 53. Prof Exp: Res assoc, Albert Einstein Med Ctr, 52-53; res assoc neurol, Med Sch, 53-55, asst prof, 55-63, ASSOC PROF NEUROL, COL MED SCI, UNIV MINN, MINNEAPOLIS, 65- Mem: AAAS; Am Chem Soc; Am Acad Neurol; Am Soc Biol Chem. Res: Central nervous system metabolism; amino acid metabolism; ultra micro techniques and the metabolism of single, functioning neurons. Mailing Add: Col of Med Sci Univ of Minn Minneapolis MN 55455

COHEN, HARRY, b Chicago, Ill, May 17, 16; m 44; c 1. CHEMISTRY. Educ: Univ Ill, BS, 38, MS, 39; Univ Wis, PhD(org chem), 41. Prof Exp: Fel, Univ Wis, 42-43; res chemist, Upjohn Co, Mich, 43-44; lectr org chem, 46-52, from asst prof to assoc prof chem, 52-65, PROF CHEM, ROOSEVELT UNIV, 65- Concurrent Pos: Res chemist, Armour & Co, 44-49. Mem: Am Chem Soc. Res: Organic synthesis; reactions of ketene diethylacetal; formation of heterocyclic compounds; formation of unsaturated compounds. Mailing Add: Dept of Chem Roosevelt Univ Chicago IL 60605

COHEN, HARVEY MARTIN, b Boston, Mass, Sept 8, 36; m 58; c 3. INORGANIC CHEMISTRY. Educ: Harvard Univ, AB, 58; Mass Inst Technol, PhD(inorg chem), 62. Prof Exp: Asst chemist, Res & Adv Develop Div, Avco Corp, Mass, 58-62; chemist, Nat Res Corp Div, Norton Co, Cambridge, 62-69; pres, 69-71, CHMN BD, TECHNOL ASSOCS, INC, 71- Concurrent Pos: Sr chemist, Norton Co & Nat Res Corp, 58-62. Mem: Am Electroplaters Soc; Am Chem Soc; The Chem Soc. Res: Chemical synthesis; reaction mechanism; metal-organic and organic chemistry; technology transfer and licensing. Mailing Add: Technol Assocs Inc 850 Providence Hwy Dedham MA 02026

COHEN, HASKELL, b Omaha, Nebr, Sept 12, 20; m 45; c 3. MATHEMATICS. Educ: Univ Omaha, AB, 42; Univ Chicago, SM, 47; Tulane Univ, PhD(math), 52. Prof Exp: Instr math, Univ Ala, 46-50; asst, Off Naval Res contract, Tulane Univ, 51-52; instr, Univ Tenn, 52-55; from asst prof to prof, La State Univ, 55-67; PROF MATH, UNIV MASS, AMHERST, 67- Concurrent Pos: Partic, US Air Force res contract, La State Univ, 56-57; consult, US Naval Ord Testing Sta, Calif, 58; mem, Inst Advan Study, 62. Mem: Am Math Soc; Math Asn Am. Res: Topology; fixed point theorems; dimension theory; topological semigroups. Mailing Add: Dept of Math Univ of Mass Amherst MS 01002

COHEN, HERBERT DANIEL, b New York, NY, Apr 27, 37; m 59. PHYSICS. Educ: Antioch Col, BS, 59; Stanford Univ, PhD(physics), 66. Prof Exp: Res assoc low temperature physics, Stanford Univ, 66-67; asst prof physics, Brandeis Univ, 67-72; ASSOC PROF PHYSICS, UNIV VT, 72- Concurrent Pos: NSF fel, 66-67. Mem: AAAS; Am Phys Soc. Res: Low temperature physics. Mailing Add: Cook Phys Sci Bldg Univ of Vt Burlington VT 05401

COHEN, HERMAN, b New York, NY, Mar 1, 15; m 42; c 2. BIOCHEMISTRY. Educ: City Col New York, BS, 40; NY Univ, MS, 42, PhD, 51. Prof Exp: Res assoc, E R Squibb & Sons, 46-53; dir prod develop, 53-71, VPRES & DIR RES, PRINCETON LABS, INC, 71- Mem: Fel AAAS; Soc Exp Biol & Med; Am Chem Soc; fel NY Acad Sci; Am Physiol Soc. Res: Endocrinology; estrogenic hormones, pituitary hormones; enzymes. Mailing Add: Princeton Labs Inc Princeton NJ 08540

COHEN, HERMAN JACOB, b New York, NY, Sept 18, 22. MATHEMATICS. Educ: City Col New York, BA, 43; Univ Wis, MA, 46, PhD(math), 49. Prof Exp: Jr physicist, Nat Bur Stand, 43-44; asst math, Univ Wis, 45-49; instr, Tulane Univ, 49-50; Fulbright scholar, Univ Paris, 50-51; from instr to assoc prof, 54-70, PROF MATH, CITY COL NEW YORK, 70- Mem: Am Math Soc; Math Asn Am. Res: General topology; plane continua; uniform spaces; theory of numbers; combinatorics. Mailing Add: 90 La Salle St Apt 11 H New York NY 10027

COHEN, HIRSH G, b St Paul, Minn, Oct 6, 25; m 52; c 3. APPLIED MATHEMATICS. Educ: Univ Wis, BS, 47; Brown Univ, MS, 48, PhD(appl math), 50. Prof Exp: Asst prof eng res, Pa State Univ, 50-51; res assoc aeronaut eng, Israel Inst Technol, 51-53; asst prof math, Carnegie Inst Technol, 53-55; sr engr, NAm Aviation Co, 55; from assoc prof to prof math, Rensselaer Polytech Inst, 55-59; mgr appl math, 61-65, from asst dir to dir math sci dept, IBM Res Ctr, 65-68, asst dir res, IBM Res Div, 69-72, consult to dir, 72-74, RES STAFF MEM, IBM CORP, 59-, CHMN RES REV BD, 74- Concurrent Pos: Fulbright vis res prof, Delft Univ Technol, 58-59; assoc scientist, Sloan-Kettering Inst Cancer Res, 64-70; consult biostatistician, Mem Hosp, 64-70; vis prof biomath, Med Col, Cornell Univ, 65-70; vis assoc appl math, Calif Inst Technol, 68-69; vis prof, Hebrew Univ, Jerusalem, 73-74; actg dir, IBM Res, Zurich, 74. Mem: AAAS; Am Math Soc; Biophys Soc; Soc Indust & Appl Math. Res: Applied mathematical investigations in acoustics, vibration theory, hydrodynamics, cavitation, nonlinear differential equations; superconductivity theory; mathematical biology. Mailing Add: IBM Corp PO Box 218 Yorktown Heights NY 10598

COHEN, HOWARD DAVID, b San Francisco, Calif, Jan 10, 40; m 65; c 1. CHEMICAL PHYSICS, ENERGY CONVERSION. Educ: Univ Calif, Berkeley, BS, 62; Univ Chicago, PhD(chem), 65. Prof Exp: Res scientist, Nat Bur Standards, 65-66; corp appointee chem, Harvard Univ, 66-67; mem tech staff, NAm Rockwell Corp, 67-69; res scientist, Systs Sci & Software Corp, 69-74; res scientist, STD Res Corp, 74-75; SR STAFF MEM, MASS INST TECHNOL, 75- Concurrent Pos: Instr chem, Univ Calif Exten, 68. Mem: AAAS; Am Phys Soc. Res: Hartree-Fock wave functions; polarizabilities; photoionization phenomena; molecular collisions; plasmas; numerical analysis; group theory; computers; magnetohydrodynamics power systems. Mailing Add: 26-153 Mass Inst of Technol Cambridge MA 02139

COHEN, HOWARD JOSEPH, b New York, NY, Jan 12, 28; m 52; c 2. CHEMISTRY. Educ: City Col New York, BS, 54; George Washington Univ, MSA, 75. Prof Exp: Technician powder metall & inorg synthesis, Sylvania Elec Prod Inc, 52-55; res chemist, Nat Lead Co, 55-56 & US Indust Chem Co, 56-61; proj leader corp res, Glidden Co, 61-64 & chem res ctr, 64, sr chemist pigments & color group, 64-69, mkt res, 69-71, prod, 72-73, SR CHEMIST PROCESS & PROD DEVELOP, GLIDDEN DURKEE DIV OF SCM CORP, 73- Concurrent Pos: Ed, The Chesapeake Chemist, 73-; consult organometallic compounds. Mem: Am Chem Soc. Res: Organometallics; polymers; inorganics; metal organics; catalysts; pigments; process and product development of silica gels as pigments, desiccants and catalysts. Mailing Add: Glidden-Durkee Div SCM Corp 3901 Hawkins Point Rd Baltimore MD 21226

COHEN, HOWARD LIONEL, b New York, NY, May 27, 40; m 62; c 1. ASTRONOMY. Educ: Univ Mich, BS, 62; Ind Univ, AM, 64, PhD(astron), 68. Prof Exp: Asst prof, 68-73, ASSOC PROF PHYS SCI & ASTRON, UNIV FLA, 73- Honors & Awards: Res Award, 70. Mem: AAAS; Am Astron Soc; Royal Astron Soc; Astron Soc Pac; Int Soc Planetarium Educr. Res: Photoelectric and photographic photometry of stars; spectroscopic and eclipsing binaries, variable stars; computer applications in astronomy; planetarium education. Mailing Add: Dept of Physics & Astron Univ of Fla Gainesville FL 32611

COHEN, HOWARD MELVIN, b Ft Wayne, Ind, May 22, 36; m 57; c 2. SOLID STATE CHEMISTRY. Educ: George Washington Univ, BS, 58; Pa State Univ, PhD(geochem), 62. Prof Exp: Mem tech staff, 62-64, SUPVR, BELL LABS, INC, 65- Mem: Am Ceramic Soc; fel Am Inst Chemists; Inst Elec & Electronic Engrs. Res: Glass technology; thermodynamics; defects in solids; phase equilibrium; high temperature chemistry of solids; magnetic oxides; thick film technology. Mailing Add: Bell Labs Inc 555 Union Blvd Allentown PA 18103

COHEN, HYMAN L, b New York, NY, Apr 11, 19; m 49; c 1. ORGANIC CHEMISTRY. Educ: City Col New York, BS, 39; Brooklyn Col, MA, 48; Univ Toronto, PhD(chem), 52. Prof Exp: Chemist, Felton Chem Co, NY, 46-47; res assoc, Jewish Hosp, Brooklyn, 47-48; chief chemist, Bell Craig, Ltd, Can, 48-49; res chemist, Glidden Co, Ill, 52-55; sr res chemist, 55-62, RES ASSOC, RES LABS, EASTMAN KODAK CO, 62- Mem: Am Chem Soc. Res: Organometallic compounds; reactions of polymers. Mailing Add: Res Labs Eastman Kodak Co Rochester NY 14650

COHEN, I BERNARD, b New York, NY, Mar 1, 14; m 44; c 1. HISTORY OF SCIENCE. Educ: Harvard Univ, SB, 37, PhD(hist sci), 47, LLD, 64. Prof Exp: Librn, Eliot House, 37-42, instr physics, 42-46 & hist sci, 46-49, from asst prof to assoc prof, 49-59, PROF HIST SCI, HARVARD UNIV, 59- Concurrent Pos: Guggenheim fel, 56; NSF sr fel, 60-61; vis fel, Clare Hall, Cambridge Univ, 65; spec lectr, Univ London, 59; Lowell lectr, Boston Univ, 61; chmn, US Nat Comt Hist & Philos Sci, 61-62; vpres, Int Union Hist & Philos Sci. Mem: Fel AAAS; Int Acad Hist Sci; Asn Hist Med; Am Acad Arts & Sci; Am Hist Sci Soc (pres, 61-62). Res: Newtonian science; science in America; effects of science on society. Mailing Add: Harvard Univ Cambridge MS 02138

COHEN, IRWIN, b Cleveland, Ohio, Feb 28, 24; m 45; c 3. ORGANIC CHEMISTRY, STRUCTURAL CHEMISTRY. Educ: Case Western Reserve Univ, AB, 44, MS, 48, PhD(chem), 50. Prof Exp: From asst prof to assoc prof, 49-58, PROF CHEM, YOUNGSTOWN STATE UNIV, 58- Mem: Am Chem Soc; Sci Res Soc Am. Res: Molecular structure through molecular orbital results by localization, density difference methods and population analysis. Mailing Add: Dept of Chem Youngstown State Univ Youngstown OH 44555

COHEN, IRWIN A, b New York, NY, Apr 28, 39; m 64; c 2. INORGANIC

CHEMISTRY. Educ: Boston Univ, BA, 60; Northwestern Univ, PhD(inorg chem), 64. Prof Exp: Fel, Med Sch, Johns Hopkins Univ, 64-66; from asst prof to assoc prof chem, Polytech Inst New York, 66-74; ASSOC PROF CHEM, BROOKLYN COL, 74- Mem: Am Chem Soc. Res: Biochemically significant reactions of inorganic metal complexes, metalloporphyrins, organometallics; mechanisms of inorganic and biochemical reactions. Mailing Add: Dept of Chem Brooklyn Col Brooklyn NY 11210

COHEN, ISADORE, b Haverhill, Mass, May 31, 11; m 49. BOTANY. Educ: Tufts Col, BS, 32, MS, 33; Univ Pa, PhD(bot), 36. Prof Exp: Asst, Philadelphia Col Pharm, 36-38; biologist, Philadelphia Inst Med Res, 38-39; res assoc, Belmont Labs, Philadelphia, 39-40; assoc prof bot, 46-51, PROF BIOL, AM INT COL, HEAD DEPT, 51- Mem: AAAS; Bot Soc Am; Soil Conserv Soc Am; Am Inst Biol Sci. Res: Chromosome structure; cell biology; plant ecology of western Massachusetts. Mailing Add: Dept of Biol Am Int Col Springfield MA 01109

COHEN, JACK, b New York, NY, Jan 31, 37; m 59; c 3. PHARMACY, ANALYTICAL CHEMISTRY. Educ: Columbia Univ, BS, 57; Univ Iowa, MS, 59, PhD(pharm, anal chem), 61. Prof Exp: Sr res chemist anal res, Lakeside Lab, Inc, 61-62; res chemist, Chas Pfizer & Co, Inc, Conn, 62-65; ASST DIR, INST PHARMACEUT SCI, SYNTEX CORP, 73-, HEAD PHARMACEUT ANAL DEPT, 65- Mem: Am Chem Soc; Am Pharmaceut Asn. Res: Pharmaceutical analysis; drug stability; physical chemistry and characterization of new compounds, kinetics and mechanisms of degradation. Mailing Add: Syntex Corp Stanford Indust Park Palo Alto CA 94304

COHEN, JACK SIDNEY, b London, Eng, Sept 6, 38; m 61; c 2. PHYSICAL BIOCHEMISTRY. Educ: Univ London, BSc, 61; Cambridge Univ, PhD(chem), 64. Prof Exp: Sci Res Coun UK fel, Weizman Inst, Israel, 64-66; fel, Harvard Med Sch, 66-67; sr res chemist, Merck Inst, 67-69; sr staff fel, Phys Sci Lab, Div Comput Res & Technol, 69-73, SR INVESTR, NAT INST CHILD HEALTH & HUMAN DEVELOP, NIH, 73- Mem: Am Chem Soc; Biophys Soc. Res: Studies of protein structure and function using nuclear magnetic resonance; biophysical applications of stable isotopes, such as carbon-13; development of computer methods; nucleotide and phosphorous chemistry; history of biochemistry. Mailing Add: Nat Inst Child Health & Human Dev NIH Bethesda MD 20014

COHEN, JACOB ISAAC, b Boston, Mass, Sept 8, 41; m 63; c 1. PHOTOGRAPHIC CHEMISTRY. Educ: Harvard Univ, BA, 63; Brandeis Univ, MA, 65, PhD(org chem), 67. Prof Exp: NIH fel, Dept Chem, Univ Chicago, 67-69; RES CHEMIST, EASTMAN KODAK CO, 69- Mem: Am Chem Soc; The Chem Soc; Soc Photog Sci & Eng. Res: Preparation and characterization of dispersions for photographic systems. Mailing Add: Res Labs Eastman Kodak Co Rochester NY 14650

COHEN, JACOB ORTLIEB, b Jacksonville, Fla, Feb 17, 30; m 52; c 2. MEDICAL MICROBIOLOGY. Educ: Univ Fla, BS, 52, MS, 55; Purdue Univ, PhD(bact), 59. Prof Exp: MICROBIOLOGIST COMMUN DIS CTR, USPHS, 59- Mem: Sigma Xi; Am Soc Microbiol; NY Acad Sci; fel Am Acad Microbiol. Res: Production and immunogenicity of M protein of group A Streptococci; natural antibodies for staphylococci in nonimmunized animals; serological relationships of strains of staphylococci; epidemiology of staphyloccus disease. Mailing Add: USPHS Ctr for Dis Control Atlanta GA 30333

COHEN, JEFFREY M, b Elizabeth, NJ, Aug 30, 40; m 64; c 1. ASTROPHYSICS, THEORETICAL PHYSICS. Educ: Newark Col Eng, BS, 62; Yale Univ, MS, 63, PhD(physics), 66. Prof Exp: Res staff physicist, Yale Univ, 65-66, vis fel physics, 66-67; resident res assoc gravitation & astrophys, Inst Space Studies, New York, 67-69; mem physics, Inst Advan Study, NJ, 69-71; ASSOC PROF PHYSICS, UNIV PA, 71- Concurrent Pos: Fel, Inst Space Studies, 66 & US AEC, 66-67; assoc, Nat Acad Sci-Nat Res Coun, 67-69; NSF grant, 70-; consult, Naval Res Lab, 74- Mem: AAAS; fel Am Phys Soc; Am Astron Soc; Int Astron Union; Int Soc Gen Relativity & Gravitation. Res: Theoretical astrophysics; rotating bodies in general relativity; neutron star models and pulsars; gravitational collapse; relativistic astrophysics; cosmology. Mailing Add: Dept of Physics Univ of Pa Philadelphia PA 19174

COHEN, JOEL M, b Worcester, Mass, Sept 27, 41. PURE MATHEMATICS. Educ: Brown Univ, ScB, 63; Mass Inst Technol, PhD(math), 66. Prof Exp: Instr math, Univ Chicago, 66-68; asst prof, Univ Pa, 68-75; ASSOC PROF MATH, UNIV MD, 75- Mem: Am Math Soc; Math Asn Am. Res: Algebraic topology, chiefly stable homotopy theory and low dimensional complexes; algebra and mathematical linguistics. Mailing Add: Dept of Math Univ of Md College Park MD 20742

COHEN, JOEL RALPH, b Chelsea, Mass, Oct 20, 26; m 47; c 3. CLINICAL MICROBIOLOGY. Educ: Univ Mass, BS, 49, MS, 50, PhD(microbiol), 75. Prof Exp: Microbiologist & supvr labs, Springfield Hosp, 50-66, chief clin labs, 66-68; ASSOC PROF BIOSCI, SPRINGFIELD COL, 68- Concurrent Pos: Lectr microbiol, Sch Nursing, 52-53; vis lectr, Univ Mass, 52- & Springfield Col, 54-68; consult, Ludlow Hosp, 52-68, Noble Hosp, 53-72, Wesson Mem Hosp, 54-, Wesson Maternity Hosp, 56-68, Springfield Munic Hosp, Vet Admin Hosp, Northampton & Springfield Health Dept; regist & specialist microbiologist, Am Bd Microbiol. Mem: Fel AAAS; Am Soc Microbiol; Sci Res Soc Am; fel Am Pub Health Asn; NY Acad Sci. Res: Methods in clinical microbiology in area of incidence of intrahospital infections and microbial susceptibility to antibiotics; rapid identification of bacterial agents. Mailing Add: Dept of Biol Springfield Col 263 Alden St Springfield MA 01109

COHEN, JOEL SEYMOUR, b Baltimore, Md, Aug 27, 41. MATHEMATICS. Educ: Univ Md, BSEE, 64, PhD(math), 70. Prof Exp: ASST PROF MATH, UNIV DENVER, 69- Mem: Am Math Soc; Math Asn Am. Res: Functional analysis; Banach spaces; absolutely p-summing operators. Mailing Add: Dept of Math Univ of Denver Denver CO 80210

COHEN, JONATHAN BREWER, b Akron, Ohio, Dec 17, 44. MOLECULAR PHARMACOLOGY. Educ: Harvard Col, BA, 66; Harvard Univ, MA, 67, PhD(chem), 72. Prof Exp: Res assoc neurobiol, Pasteur Inst, Paris, 71-74; ASST PROF PHARMACOL, HARVARD MED SCH, 75- Concurrent Pos: Lectr neurobiol, Ecole Normale Superieure, Paris, 73-74. Mem: AAAS. Res: Molecular basis of synaptic function; mechanism of permeability control by acetycholine receptors; mode of action of drugs acting at cholinergic synapses; structural and functional properties of synaptic membranes. Mailing Add: Dept of Pharmacol Harvard Med Sch 25 Shattuck St Boston MA 02115

COHEN, JORDAN J, b St Louis, Mo, June 18, 34; m 56; c 3. INTERNAL MEDICINE, NEPHROLOGY. Educ: Yale Univ, BA, 56; Harvard Univ, MD, 60. Prof Exp: Assoc physician, assoc div med res & dir div renal dis, RI Hosp, Providence, 65-71; ASSOC PROF MED, MED SCH, TUFTS UNIV & DIR RENAL DIV, TUFTS-NEW ENG MED CTR, 71- Concurrent Pos: From asst prof to assoc prof, Brown Univ, 65-71; instr med, Harvard Med Sch, 67-71. Mem: Fedn Clin Res; Am Soc Nephrology; Am Soc Internal Med. Res: Renal mechanisms involved in acid-base homeostasis; application of computer techniques to simulation analyses of body fluid; electrolyte physiology. Mailing Add: Tufts-New Eng Med Ctr 171 Harrison Ave Boston MA 02111

COHEN, JOSEPH, b Pittsburgh, Pa, Nov 11, 16; m 47; c 3. INORGANIC CHEMISTRY. Educ: Univ Pittsburgh, BSChem, 38. Prof Exp: Res fel chem & physics, Mellon Inst Indust Res, 45-47; salesman restaurant equip, Interstate Restaurant Supply, Calif, 47-48; res chemist, Naval Ord Test Sta, China Lake, Calif, 48-50, unit leader solid propellant res, combustion mech & heterogeneous interactions, 50-53; prin chemist, Solid Propellant Div, 53-58, dept mgr process control eng, 58-60, div mgr solid rocket prod, 60-62, PROG MGR SOLID PROPELLANT DEVELOP, SOLID PROPELLANT DIV, AEROJET-GEN CORP, 63- Mem: Am Chem Soc; Sigma Xi; fel Am Inst Chemists. Res: Synthesis of high energy tetrazole and guanidine derivatives; formulated low flame-temperature gun propellants; crystallization control of inorganic compounds; solid propellant research; internal ballistics of solid rockets. Mailing Add: 8129 La Riviera Dr Sacramento CA 95826

COHEN, JUDITH GAMORA, b NY, May 5, 46; m 73. ASTROPHYSICS. Educ: Radcliffe Col, BA, 69; Calif Inst Technol, MS, 71, PhD(astron), 73. Prof Exp: Miller fel astron, Univ Calif, Berkeley, 71-73; ASST ASTRONOMER, KITT PEAK NAT OBSERV, 73- Mem: Am Astron Soc; Astron Soc Pac. Res: Nucleosynthesis, interstellar medium, infrared sources. Mailing Add: Kitt Peak Nat Observ PO Box 26732 Tucson AZ 85726

COHEN, JULES, b Brooklyn, NY, Aug 26, 31; m 56; c 3. INTERNAL MEDICINE, CARDIOLOGY. Educ: Univ Rochester, AB, 53, MD, 57. Prof Exp: Intern, Beth Israel Hosp, 57-58; resident, Strong Mem Hosp, 58-59; res assoc, NIH, 60-62; res asst, Royal Postgrad Md Sch London, 62-63; sr instr med, 64-65, from asst prof to assoc prof, 66-73, PROF MED, UNIV ROCHESTER, 73- Concurrent Pos: USPHS trainee med, Med Ctr, Univ Rochester, 59-60, res fel, 64-65; USPHS res grants, 63-69 & 75-; Am Heart Asn res grant, 70-72. Mem: Am Physiol Soc; Am Fedn Clin Res; Royal Soc Med; Am Heart Asn; Am Col Physicians. Res: Cardiac hypertrophy; hemoglobin function and tissue oxygenation; cardiomyopathies. Mailing Add: Dept of Med Strong Mem Hosp Rochester NY 14642

COHEN, JULIUS, b Brooklyn, NY, Apr 16, 26. PHYSICS. Educ: NY Univ, AB, 50; Syracuse Univ, MS, 53. Prof Exp: Res assoc thin films, Syracuse Univ, 51-52; jr engr transistor physics, Gen Tel & Electronics Labs, 52-53, adv res engr emission & thin films, 54-67, eng specialist, 67-69; PHYSICIST, NAT BUR STAND, 69- Concurrent Pos: Assoc physicist thermistor bolometers, Bulova Res & Develop Labs, 53-54; Japanese Ministry Educ guest scholar tunnel emission, Japan & Osaka Univs, 64-65. Mem: Am Phys Soc; Sigma Xi. Res: Thin films; semiconductors; electrical conductivity; emission; vacuum techniques; photoconductivity; piezoeletricity and pyroelectricity in polymers; optical instrumentation; physics engineering. Mailing Add: Nat Bur of Stand Sect 446.02 Bldg 224 Rm A109 Washington DC 20234

COHEN, JULIUS JAY, b Newark, NJ, Apr 26, 23; m 52; c 2. PHYSIOLOGY. Educ: Rutgers Univ, BS, 45; NY Univ, MD, 48. Prof Exp: Intern, Cincinnati Gen Hosp, 48-49; from jr to sr resident, Dept Med, Col Med, Univ Cincinnati, 49-51, res fel dept clin physiol, 51 & dept physiol, 53-54, instr physiol, 54-55, asst prof physiol & instr med, 55-59; assoc prof physiol, 59-66, actg chmn dept physiol, 67-68, PROF PHYSIOL, SCH MED & DENT, UNIV ROCHESTER, 66- Concurrent Pos: Markle Found med sci scholar, 55-60; mem physiol study sect, NIH, 65-70. Mem: AAAS; Soc Exp Biol & Med; Am Heart Asn; Am Physiol Soc. Res: Renal physiology and metabolism; relationships between intermediary metabolism of the kidney and its excretory function; comparative physiology; pathological physiology; isotopes. Mailing Add: Dept of Physiol Box 642 Univ Rochester Sch Med & Dent Rochester NY 14642

COHEN, KARL (PALEY), b New York, NY, Feb 5, 13; m 38; c 3. NUCLEAR SCIENCE. Educ: Columbia Univ, AB, 33, AM, 34, PhD(phys chem), 37; Prof Exp: From asst to prof, Columbia Univ, 38-40, dir, Theoret Div SAM Labs, 40-44; head theoret physics group, Standard Oil Develop Co, 44-48; tech dir, H K Ferguson Co, 48-52; vpres, Walter Kidde Nuclear Labs, Inc, 52-55; mgr adv eng, Atomic Power Equip Dept, 55-65, gen mgr, Breeder Reactor Dept, 65-71, mgr, Oper Planning, 71-73, CHIEF SCIENTIST, NUCLEAR ENERGY DIV, GEN ELEC CO, 73- Concurrent Pos: Regents lectr, Univ Calif, Berkeley, 70; dir, US Nat Comt, World Energy Conf, 72- Mem: AAAS; Am Phys Soc; Am Nuclear Soc (treas, 55-57, pres, 68-69); Nat Acad Eng. Res: Applied nuclear energy; isotope separation; gaseous diffusion; gas centrifuges; fast breeder reactors. Mailing Add: 928 N California Ave Palo Alto CA 94303

COHEN, KENNETH JOEL, b Bridgeton, NJ, Sept 19, 41; m 68; c 1. EXPERIMENTAL HIGH ENERGY PHYSICS. Educ: Princeton Univ, AB, 64; Mass Inst Technol, PhD(physics), 67. Prof Exp: Res assoc physics, Lab Nuclear Sci, Mass Inst Technol, 67-68, Ger Electronic Synchrotron, 68-69 & Lab Nuclear Sci, Mass Inst Technol, 69-70; res assoc, 70-71, ASST PROF PHYSICS, RUTGERS UNIV, NEW BRUNSWICK, 71- Mem: Am Phys Soc. Res: Experimental high energy physics to study strong interactions of elementary particles; particle spectroscopy, proton-proton elastic scattering and inclusive particle production. Mailing Add: Dept of Physics Rutgers Univ New Brunswick NJ 08903

COHEN, KENNETH MICHAEL, organic chemistry, see 12th edition

COHEN, LARRY WILLIAM, b Winnipeg, Man, Dec 24, 36; US citizen; m 59; c 2. GENETICS, MICROBIOLOGY. Educ: Univ Calif, Los Angeles, BA, 60, MA, 62, PhD(zool), 64. Prof Exp: NIH fel, Med Sch Univ Mich, 63-65; asst prof, Dept Biol Sci, Douglass Col, Rutgers Univ, 65-67; asst prof zool, 67-70, ASSOC PROF ZOOL, POMONA COL, 70-, CHMN DEPT, 73- Concurrent Pos: NIH spec fel, Univ Calif, Riverside, 71; vis prof, Dept Molecular Virol, Hadassah Med Sch, Hebrew Univ Jerusalem, 74. Mem: AAAS; Am Soc Microbiol. Res: Microbial genetics; cell to cell agglutination phenomena; lysis in salmonella phage P22. Mailing Add: Dept of Zool Pomona Col Claremont CA 91711

COHEN, LAWRENCE, b Leeds, Eng, Nov 23, 26; m 51; c 3. ORAL MEDICINE, ORAL PATHOLOGY. Educ: Univ Leeds, BChD, 49, Univ London, MD, 56, PhD(histochem), 66. Prof Exp: Resident oral surg, Middlesex Hosp, London, Eng, 57-59; sr resident, Plastic & Jaw Unit, Stoke Mandeville Hosp, Buckinghamshire, 60-61; sr resident, Univ Col Hosp Dent Sch, 61-62; lectr oral path, Inst Dent Surg, London, 62-63, sr lectr oral med, 63-67; PROF ORAL DIAG & HEAD DEPT, UNIV ILL MED CTR, 67- Concurrent Pos: Consult, West Side Vet Admin Hosp, Chicago, Ill, 68-; dir dent educ, Ill Masonic Med Ctr, Chicago, 70-; chmn panel rev of oral cavity drug preparations, Food & Drug Admin, 74-76; vis prof med, Univ Chicago. Mem: Fel Royal Soc Med; Brit Bone & Tooth Soc; Int Asn Dent Res. Res: Keratinization and histochemistry of the oral mucosa. Mailing Add: Dept of Oral Diag Univ of Ill Med Ctr Chicago IL 60612

COHEN, LAWRENCE BARUCH, b Indianapolis, Ind, June 18, 39; c 2. NEUROPHYSIOLOGY. Educ: Univ Chicago, BS, 61; Columbia Univ, PhD(zool), 65.

Prof Exp: Asst prof, 68-71, ASSOC PROF PHYSIOL, SCH MED, YALE UNIV, 71- Concurrent Pos: NSF fel, Agr Res Coun Inst Physiol, Cambridge, Eng, 66-68. Honors & Awards: McMaster Award, Columbia Univ, 65. Mem: Soc Neurosci; Soc Gen Physiol; Biophys Soc. Res: Optical methods for measuring activity in invertebrate central nervous systems. Mailing Add: Dept of Physiol Yale Univ Sch of Med New Haven CT 06510

COHEN, LAWRENCE SOREL, b New York, NY, Mar 27, 33; m 61; c 2. INTERNAL MEDICINE, CARDIOLOGY. Educ: Harvard Col, AB, 54; NY Univ, MD, 58. Hon Degrees: MA, Yale Univ, 70. Prof Exp: Fel cardiol, Harvard Univ, 62-64; sr investr cardiol, NIH, 65-68; assoc prof med, Univ Tex Southwestern Med Sch Dallas, 68-70; PROF MED, SCH MED, YALE UNIV, 70- Concurrent Pos: Attend physician & chief cardiol, Yale-New Haven Med Ctr, 70- Honors & Awards: NY Univ Sch Med Award, 58. Mem: Am Fedn Clin Res; fel Am Col Cardiol; fel Am Col Physicians; Am Heart Asn. Res: Coronary artery disease; hemodynamics; radionuclide myocardial perfusion. Mailing Add: Dept of Med Yale Univ Sch of Med New Haven CT 06510

COHEN, LEON WARREN, b New York, NY, Apr 24, 03; m 27; c 1. MATHEMATICS. Educ: Columbia Univ, AB, 23, AM, 25; Univ Mich, PhD, 28. Prof Exp: Instr math, Univ Mich, 26-29; Nat Res fel, Princeton Univ, 29-31; from asst prof to prof, Univ Ky, 31-47; from asst prof to assoc prof, Queens Col, NY, 47-55; chmn dept, 58-68, prof, 58-73, EMER PROF MATH, UNIV MD, 73- Concurrent Pos: Lectr, Univ Wis, 42-44; res mathematician, Columbia Univ, 44-45 & Brown Univ, 45; prof, Ohio State Univ, Wright Field, 46; Ford fel, Inst Advan Study, 52-53; prog dir math sci, NSF, 53-58; mem, Sci Manpower Comn, 62-65 & exec bd, 64-65; exec secy, Conf Bd Math Sci, 62-65; exec secy, Div Math, Nat Acad Sci-Nat Res Coun, 64-72. Honors & Awards: Distinguished Serv to Math, Math Asn Am, 76. Mem: Am Math Soc; Math Asn Am. Res: Topology; analysis. Mailing Add: Dept of Math Univ of Md College Park MD 20742

COHEN, LEONARD, mathematics, deceased

COHEN, LEONARD A, b New York, NY, Mar 21, 39; m 67; c 1. CELL BIOLOGY. Educ: Univ Wis-Madison, BS, 60; City Univ New York, PhD(biol), 72. Prof Exp: Res asst microbiol, Kingsbrooke Jewish Med Ctr, 60-62; res asst cell biol, Albert Einstein Col Med, 63-66; instr biol, City Univ New York, 67-72; RES ASSOC CELL BIOL, AM HEALTH FOUND, 73- Mem: AAAS. Res: Effect of hormones on the cyclic adenosine monophosphate system in cultured normal and neoplastic mammary epithelial cells. Mailing Add: Naylor Dana Inst for Dis Prev Am Health Found Dana Rd Valhalla NY 10595

COHEN, LEONARD ARLIN, b Brooklyn, NY, May 4, 24; m 50; c 2. PHYSIOLOGY. Educ: Univ Conn, BA, 48; Yale Univ, PhD(physiol), 52. Prof Exp: From instr to asst prof physiol & pharmacol, Sch Med, Univ Pittsburgh, 51-61; head dept physiol, Albert Einstein Med Ctr, 61-67; prof phys med & rehab, Sch Med, Temple Univ, 67-70; PROF PHYSIOL, MICH STATE UNIV, 70-, DIR RES & DEVELOP, REHAB CTR, 73- Concurrent Pos: Fulbright fel, Oxford Univ, 56-57, NSF fel, 57-58. Mem: AAAS; Am Physiol Soc; Aerospace Med Asn; Asn Hosp Med Educ; fel Royal Soc Med. Res: Neurophysiology; proprioception; physiology of joint, muscle and tactile receptors; mechanisms for body orientation; proprioceptive mechanisms of extraocular muscles, vestibular apparatus, neck and foot; relationship between orientation and motor coordination; cardiac stress testing; disability evaluation; clinical physiology; hallucinogens and body-environment orientation. Mailing Add: 15951 Harden Circle Southfield MI 48075

COHEN, LEONARD DAVID, b Philadelphia, Pa, Aug 7, 32; m 57; c 3. NUCLEAR PHYSICS, REACTOR PHYSICS. Educ: Univ Pa, BA, 54, MS, 56, PhD(physics), 59. Prof Exp: Reactor physicist, Knolls Atomic Power Lab, Gen Elec Co, 59-62, geophysicist, Space Sci Lab, 62-64; asst prof, 64-73, ASSOC PROF PHYSICS, DREXEL UNIV, 73- Mem: Am Phys Soc; Am Nuclear Soc; Am Geophys Union. Res: Space radiation physics; atmospheric physics using radon gas as a tracer of atmospheric turbulence. Mailing Add: Dept of Physics Drexel Univ Philadelphia PA 19104

COHEN, LEONARD GEORGE, b Brooklyn, NY, Feb 28, 41. COMMUNICATIONS, PLASMA PHYSICS. Educ: City Col New York, BEE, 62; Brown Univ, ScM, 64, PhD(plasma physics), 68. Prof Exp: Res asst eng, Brown Univ, 64-66, plasma physics, 66-68; MEM TECH STAFF, GUIDED WAVE RES LAB, BELL TEL LABS, 68- Mem: Inst Elec & Electronics Eng. Res: Experimental and theoretical studies of the interaction between collisionless plasmas and electromagnetic fields; measured attenuation and depolarization of light transmitted along glass fibers; optical communications. Mailing Add: Dept 1264 Bell Tel Labs Box 400 Holmdel NJ 07733

COHEN, LEONARD HARVEY, b Winnipeg, Man, Mar 19, 25; m 49; c 2. BIOCHEMISTRY. Educ: Univ Man, BS, 48, MSc, 51; Univ Toronto, PhD(biochem), 54. Prof Exp: Asst pharmacol, Yale Univ, 54-55; res scientist, Roswell Mem Inst, 55-59; from asst prof pharmac- ol to prof biochem, Univ Man, 59-65; ASSOC MEM, INST CANCER RES, 65-; ASSOC PROF PHYS BIOCHEM, UNIV PA, 65- Mem: Am Soc Biol Chemists; Am Soc Cell Biol; Brit Biochem Soc; Can Biochem Soc. Res: Nucleic acid and nucleotide metabolism; enzymology; chromosomal proteins; histones, gene regulation. Mailing Add: Inst for Cancer Res 7701 Burholme Ave Philadelphia PA 19111

COHEN, LESLIE, b Baltimore, Md, Jan 14, 23; m 67; c 1. NUCLEAR PHYSICS. Educ: Johns Hopkins Univ, BA, 44, PhD(physics), 52. Prof Exp: Physicist, Bur Standards, 44-45; instr physics, Loyola Col, 47-48; jr instr, Johns Hopkins Univ, 48-51, asst, 51-52; res assoc, Knolls Atomic Power Lab, Gen Elec Co, 53-55; RES ASSOC, NUCLEAR PHYSICS DIV, NAVAL RES LAB, 55- Concurrent Pos: Vis scientist, Optical Sci Ctr, Univ of Ariz, 75-76. Mem: Am Phys Soc. Res: Nuclear spectroscopy; nuclear reactors; photonuclear and charged particle reactions; lasers. Mailing Add: Radiation Technol Div Naval Res Lab Washington DC 20375

COHEN, LESTER ALLAN, b Amsterdam, NY, June 7, 27; m 51; c 4. PHYSICAL ORGANIC CHEMISTRY. Educ: Union Col, BS, 50. Prof Exp: Chemist, Nat Dairies, 50-51, lab supvr, 51-55; res chemist, 55-60, res group leader, 61-66, tech mgr, 66-67, mgr res, 67-68, mgr tech dept, 68-72, MGR OPERS, MONSANTO CO, 72- Honors & Awards: Soc Plastics Eng Award, 64. Mem: Am Soc Testing & Mat. Res: Properties of polymeric materials, including thermal, dielectric, melt rheological and mechanical. Mailing Add: Tech Dept Monsanto Co 200 NSeventh St Kenilworth NJ 07033

COHEN, LEWIS H, b Dallas, Tex, Jan 2, 37. GEOCHEMISTRY. Educ: Mass Inst Technol, BS, 58; Univ Calif, Berkeley, MS, 61; Univ Calif, San Diego, PhD(earth sci), 65. Prof Exp: Asst prof geophys eng, Univ Calif, Berkeley, 65-66; asst prof geol, 66-69, ASSOC PROF GEOL, UNIV CALIF, RIVERSIDE, 69- Mem: Am Geophys Union; Geochem Soc; Mineral Soc Am. Res: Chemistry and physics at elevated temperatures and pressures. Mailing Add: Dept of Earth Sci Univ of Calif Riverside CA 92502

COHEN, LOUIS, b Chicago, Ill, Dec 5, 28; m 52; c 3. INTERNAL MEDICINE, CARDIOLOGY. Educ: Univ Chicago, BS, 48, MD, 53. Prof Exp: Res asst, 49-53, from intern to resident med, 53-55 & 57-58, instr, 59-61, asst prof, 61-68, assoc prof, 68-75, PROF MED UNIV CHICAGO, 75-, ATTEND PHYSICIAN, UNIV CHICAGO HOSPS & CLINS, 61- Concurrent Pos: Nat Heart Inst trainee, 55; Am Heart Asn res fel, 58-60, advan res fel, 60-62; vis prof, Shaare Zedek Hosp, Jerusalem, Israel, 66, Sacred Heart Hosp, Eugene, Ore, 68 & Univ Hawaii, 68; fel coun arteriosclerosis & coun clin cardiol, Am Heart Asn. Mem: Tissue Cult Asn; Am Soc Human Genetics; fel Am Col Physicians; fel Am Col Cardiol; fel Am Col Clin Pharmacol. Res: Lipoproteins; atherosclerosis; enzymology; tissue culture; muscular dystrophy; myocardial infarction. Mailing Add: Univ of Chicago Sch of Med 950 E 59th St Chicago IL 60637

COHEN, LOUIS ARTHUR, b Boston, Mass, July 12, 26; m 55; c 1. PHYSICAL ORGANIC CHEMISTRY, BIO-ORGANIC CHEMISTRY. Educ: Northeastern Univ, BS, 49; Mass Inst Technol, PhD(chem), 52. Prof Exp: Instr biochem, Med Sch, Yale Univ, 52-54; asst scientist, 54-57, from scientist to sr scientist, 57-64, CHIEF CHEM, SECT BIOCHEM MECHANISMS, NAT INST ARTHRITIS & METAB DIS, NIH, 65-, DIR GRAD SCH, FOUND ADVAN EDUC SCI, 70- Concurrent Pos: USPHS scientist, 54- Mem: Am Chem Soc. Res: Steroid synthesis; veratrum alkaloids; peptide synthesis; phosphorylation; oxidation of nitrogen compounds; amino acid metabolism and interconversion; protein structure; reaction kinetics and mechanism; nuclear magnetic resonance spectroscopy; biochemical mechanisms; drug design; antiviral antibiotics. Mailing Add: Lab of Chem Nat Inst Arthritis & Metab Dis NIH Bethesda MD 20014

COHEN, LUCY M, b San Jose, Costa Rica, May 9, 36; US citizen. ANTHROPOLOGY. Educ: Mt St Mary's Col, BA, 56; Cath Univ Am, MSW, 59, PhD(anthrop), 66. Prof Exp: Clin social worker, St Elizabeths Hosp, Washington, DC, 59-62; vis prof anthrop, Univ of the Andes, Colombia, 65; asst prof, Cath Univ Am, 66-67; chief, Prog Eval Community Ment Health, DC Dept Pub Health, Planning & Res, 67-69; ASSOC PROF ANTHROP, CATH UNIV AM, 69- Concurrent Pos: Vis prof, Pontif Univ, Javeriana, Colombia, 65; vis lectr, Cath Univ Am; USPHS biomed sci award, 70-71; Nat Inst Ment Health res award, Ctr Minority Ment Health Prog, 72- Mem: Fel Am Anthrop Asn; Nat Asn Social Workers; Soc Appl Anthrop. Res: Cultural anthropology; complex societies; applied anthropology; research methods; Latin America; culture and personality. Mailing Add: 2805 McKinley Pl NW Washington DC 20015

COHEN, MAIMON MOSES, b Baltimore, Md, Jan 24, 35; m 55; c 3. CYTOGENETICS. Educ: Johns Hopkins Univ, AB, 55; Univ Md, MS, 59, PhD(agron), 63. Prof Exp: Jr asst health serv officer, NIH, 60-62; NSF fel human genetics, Univ Mich, 62-64, instr, 64-65; from asst prof to assoc prof pediat & assoc res prof microbiol, Med Sch, State Univ NY Buffalo, 68-72; PROF HUMAN GENETICS & CHMN DEPT, HADASSAH-HEBREW UNIV MED CTR, 72- Concurrent Pos: Dir cytogenetics, Buffalo Children's Hosp, 67-72. Mem: AAAS; Genetics Soc Am; Soc Pediat Res; Am Soc Human Genetics; Tissue Cult Asn. Res: Structure and function of human chromosomes; effect of various mutagenic agents on chromosomes of cultured tissues. Mailing Add: Hadassah-Hebrew Univ Med Ctr PO Box 499 Jerusalem Israel

COHEN, MARGO NITA PANUSH, b Detroit, Mich, Oct 28, 40; m 61; c 3. INTERNAL MEDICINE, ENDOCRINOLOGY. Educ: Univ Mich, BS, 60, MD, 64; Univ Buenos Aires, PhD(biochem), 70. Prof Exp: Intern, Sinai Hosp, Detroit, 64-65; resident internal med, Henry Ford Hosp, 65-66; instr physiol, Univ Buenos Aires, 68-70; investr, Arg Nat Res Coun, 70-71; asst prof, 71-75, ASSOC PROF MED & PHYSIOL, SCH MED, WAYNE STATE UNIV, 75- Concurrent Pos: USPHS diabetes trainee, Wayne State Univ & Sinai Hosp, Detroit, 66-68; NIH spec fel, Wayne State Univ, 68-69. Mem: Am Fedn Clin Res; Biochem Soc; Am Diabetes Asn; Endocrine Soc; fel Am Col Physicians. Res: Diabetes, complications of; glomerular disease; basment membrane metabolism. Mailing Add: Dept of Med Wayne State Univ Sch of Med Detroit MI 48201

COHEN, MARION DEUTSCHE, b Perth Amboy, NJ, Jan 2, 43; m 64; c 2. MATHEMATICAL ANALYSIS. Educ: NY Univ, BA, 64; Wesleyan Univ, MA, 66, PhD(math), 70. Prof Exp: Asst prof, 69-73, ASSOC PROF MATH, NJ INST TECHNOL, 74- Mem: Am Math Soc. Res: Schwartz distribution theory, particularly the concepts of order and value of a distribution at a point. Mailing Add: Dept of Math NJ Inst of Technol Newark NJ 07102

COHEN, MARK EDGAR, inorganic chemistry, physical chemistry, see 12th edition

COHEN, MARLENE LOIS, b New Haven, Conn, May 5, 45. PHARMACOLOGY. Educ: Univ Conn, Storrs, BS, 68; Univ Calif, San Francisco, PhD(pharmacol), 73. Prof Exp: Fel, Roche Inst Molecular Biol, 73-75; SR PHARMACOLOGIST, LILLY RES LABS, ELI LILLY & CO, 75- Res: Pharmacology, physiology and biochemistry of vascular smooth muscle. Mailing Add: MC 304 Lilly Res Labs Div Eli Lilly & Co Indianapolis IN 46206

COHEN, MARSHALL HARRIS, b Manchester, NH, July 5, 26; m 48; c 3. ASTRONOMY. Educ: Ohio State Univ, BEE, 48, MSc, 49, PhD(physics), 52. Prof Exp: Res assoc, Antenna Lab, Ohio State Univ, 52-54; from asst prof to assoc prof, Sch Elec Eng, Cornell Univ, 54-64, assoc prof astron, 64-66; prof, Dept Appl Electrophys, Univ Calif, San Diego, 66-68; PROF RADIO-ASTRON, CALIF INST TECHNOL, 68- Concurrent Pos: Guggenheim fel, 60-61. Honors & Awards: Rumford Medal, Am Acad Arts & Sci, 71. Mem: Am Astron Soc; Int Union Radio Sci; Int Astron Union. Res: Radio astronomy. Mailing Add: Dept of Astron Calif Inst of Technol Pasadena CA 91125

COHEN, MARTIN ALLEN, b Butler, Pa, Aug 7, 49. INORGANIC CHEMISTRY, TEXTILE TECHNOLOGY. Educ: Univ Del, BS, 71; Univ Ill, PhD(chem), 76. Prof Exp: RES CHEMIST, E I DU PONT DE NEMOURS INC, 75- Mem: Sigma Xi; Am Chem Soc. Res: Textile fiber end-uses and textile technology. Mailing Add: E I du Pont de Nemours Inc Chestnut Run Wilmington DE 19898

COHEN, MARTIN GILBERT, b Brooklyn, NY, Jan 13, 38; m 61; c 3. LASERS. Educ: Columbia Col, AB, 57; Harvard Univ, MA, 58, PhD(appl physics), 64. Prof Exp: Mem tech staff, Bell Tel Labs, 64-69; DIR APPL RES, QUANTRONIX CORP, 69- Mem: Am Phys Soc. Res: Optical modulation and deflection; interaction of materials and laser light. Mailing Add: Quantronix Corp 225 Engineers Rd Smithtown NY 11787

COHEN, MARTIN JOSEPH, b Brooklyn, NY, May 6, 21; m 49; c 3. ELECTRONIC PHYSICS. Educ: Brooklyn Col, BA, 42; Princeton Univ, MA, 48, PhD(physics), 51. Prof Exp: Staff mem, Radiation Lab, Mass Inst Technol, 42-45; res assoc electronics,

Princeton Univ, 45-49; staff physicist, Princeton Lab, Radio Corp Am, 49-52; sr physicist, Radiation Res Corp, 52-55; consult physicist, 55-57; vpres, Franklin Systs, Inc, 57-63, VPRES, FRANKLIN GNO CORP, 63- Mem: Am Phys Soc; Inst Elec & Electronics Engrs; Am Chem Soc. Res: Microwave instrumentation; electron physics; experimental nuclear physics and cosmic rays; nuclear energy conversion; scintillation techniques; solid state electronics; ion dynamics applications. Mailing Add: Franklin GNO Corp Box 3206 West Palm Beach FL 33402

COHEN, MARTIN WILLIAM, b Brooklyn, NY, Feb 18, 35; m 62. IMMUNOLOGY, HUMAN PATHOLOGY. Educ: Columbia Col, AB, 56; State Univ NY, MD, 60. Prof Exp: Intern, NY Hosp, 60-61; instr, Albert Einstein Col Med, 67-68, asst prof, 69-70; ASSOC ATTEND PATHOLOGIST, BETH ISRAEL MED CTR, 70- Concurrent Pos: Fel, NY Univ Med Ctr, 61-65; spec fel, Nat Inst Allergy & Infectious Dis, 65; asst attend pathologist, Bronx Munic Hosp Ctr, 67-70; asst prof, Mt Sinai Med Sch, 70- Mem: Int Acad Path. Res: Cellular aspects of hormonal immunity; immunopathology. Mailing Add: Div of Labs Beth Israel Med Ctr New York NY 10003

COHEN, MARVIN, b New York, NY, Apr 26, 37; m 59; c 2. PHARMACOLOGY, PSYCHOPHARMACOLOGY. Educ: Fordham Univ, BS, 57; Rutgers Univ, MS, 59; Ohio State Univ, PhD(pharmacol), 62. Prof Exp: Proj leader pharmacol, IIT Res Inst, 62-66; GROUP LEADER PHARMACOL, ENDO LABS, INC, 66- Mem: Am Pharmaceut Asn. Res: Behavioral pharmacology, particularly the development of tests for screening compounds with activity on the central nervous system. Mailing Add: Endo Labs Inc 1000 Stewart Ave Garden City NY 11530

COHEN, MARVIN LOU, b Montreal, Que, Mar 3, 35; US citizen; m 58; c 2. PHYSICS. Educ: Univ Calif, Berkeley, AB, 57; Univ Chicago, MS, 58, PhD(theoret solid state physics), 64. Prof Exp: Mem tech staff, Bell Tel Labs, 63-64; from asst prof to assoc prof, 64-68, PROF PHYSICS, UNIV CALIF, BERKELEY, 69- Concurrent Pos: Prof, Miller Inst Basic Res Sci, Univ Calif, Berkeley, 69-70; exchange prof, Univ Paris, 72-73. Mem: Fel Am Phys Soc. Res: Theoretical solid state physics; superconductivity in semiconductors; optical properties and band structure of semiconductors; theory of superconductivity. Mailing Add: Dept of Physics Univ of Calif Berkeley CA 94720

COHEN, MARVIN MORRIS, b New York, NY, Apr 24, 40; m 61; c 4. SOLID STATE PHYSICS. Educ: Brooklyn Col, BA, 62; Am Univ, MS, 65, PhD(solid state physics), 67. Prof Exp: Res physicist, Harry Diamond Labs, 62-74; PHYSICIST, ENERGY RES & DEVELOP ADMIN, 74- Concurrent Pos: Adj prof, Am Univ, 67- Mem: AAAS; Am Phys Soc. Res: Impurity band conduction in solids; tunneling in solids; radiation damage in a fusion environment. Mailing Add: 14116 Bauer Dr Rockville MD 20853

COHEN, MAYNARD, b Regina, Sask, May 17, 20; US citizen; m 45; c 2. NEUROSCIENCES. Educ: Univ Mich, AB, 41; Wayne Univ, MD, 44; Univ Minn, PhD(path), 53. Prof Exp: Res assoc, Riks Hosp, Oslo, Norway, 51-52; from asst prof to prof neurol, Univ Minn, 53-63; prof & head div neurol & prof pharmacol, Col Med, Univ Ill, 63-71; PROF & HEAD DEPT NEUROL SCI & PROF BIOCHEM, RUSH MED COL, 71-, CHMN DEPT NEUROL, PRESBY-ST LUKE'S HOSP, 63- Concurrent Pos: NIH spec fel, Univ London, 57-58; consult, Nat Inst Neurol Dis & Blindness, 59-63 & NIMH, 68-72; mem prof adv bd, Epilepsy Founds. Honors & Awards: Distinguished Serv Award, Col Med, Wayne State Univ, 64. Mem: Biochem Soc; Am Acad Neurol (past vpres); Am Asn Neuropath; Int Soc Neurochem; Asn Univ Prof Neurol (past pres). Res: Nervous and mental diseases; biochemistry of the nervous system; cerebrovascular disease; neurotoxic agents; phosphorylated compounds in the brain; amino acid and carbohydrate interrelationships in the brain. Mailing Add: Dept of Neurol Sci Rush Med Col Chicago IL 60612

COHEN, MELVIN JOSEPH, b Los Angeles, Calif, Sept 28, 28; m 63; c 4. MICROBIOLOGY. Educ: Univ Calif, Los Angeles, BA, 49, MA, 52, PhD(zool), 54. Prof Exp: NSF fel, Stockholm, Sweden, 54-55; instr biol, Harvard Univ, 55-57; from asst prof to assoc prof, Univ Ore, 57-69; PROF BIOL, YALE UNIV, 69- Concurrent Pos: Guggenheim Found fel, 64-65. Mem: Am Soc Zool; Soc Gen Physiol. Res: Comparative neurophysiology; sensory physiology. Mailing Add: Dept of Biol Yale Univ New Haven CT 06520

COHEN, MERRILL, b Boston, Mass, Feb 5, 26; m 50; c 3. ORGANIC CHEMISTRY. Educ: Boston Univ, BA, 48; Univ Chicago, MS, 49, PhD(chem), 51. Prof Exp: Res assoc, Res Lab, Gen Elec Co, 51-52, Specialist org chem, Thomson Lab, 52-55, mgr chem & insulat- ion eng, Medium Steam Turbine Generator Dept, 55-70, MGR MAT & PROCESSING LAB, MEDIUM STEAM TURBINE GENERATOR DEPT, GEN ELEC CO, 70- Concurrent Pos: Jr chemist, Ionics Inc, 49, 50; mem tech adv comt, Nat Geothermal Info Resource Proj, Lawrence Berkeley Lab, Univ Calif, 74- Mem: Am Chem Soc; fel Am Inst Chem. Res: Organic resin and polymer chemistry; epoxy, polyester, silicone resins; electrical insulation; laminated plastics; adhesives; protective coatings, high temperature synthetic lubricants; gas chromatography, air, water pollution; geothermal energy and steam turbine materials. Mailing Add: 8 May St Marblehead MA 01945

COHEN, MICHAEL, b New York, NY, May 9, 30; m 58. THEORETICAL PHYSICS. Educ: Cornell Univ, AB,51; Calif Inst Technol, PhD(physics), 56. Prof Exp: Res fel theoret physics, Calif Inst Technol, 55-57; mem, Inst Advan Study, NJ, 57-58; asst prof, 58-60, ASSOC PROF PHYSICS, UNIV PA, 60-, ASSOC CHMN GRAD AFFAIRS, 69- Concurrent Pos: Consult, Los Alamos Sci Lab. Mem: Am Phys Soc. Res: Quantum and statistical mechanics; theory of liquid helium. Mailing Add: Dept of Physics Univ of Pa Philadelphia PA 19104

COHEN, MONROE W, b Montreal, Que, May 3, 40; m 65; c 2. NEUROSCIENCES. Educ: McGill Univ, BSc, 61, PhD(neurophys), 65. Prof Exp: Res fel neurobiol, Harvard Med Sch, 65-68; asst prof, 68-74, ASSOC PROF PHYSIOL, MCGILL UNIV, 74- Concurrent Pos: NATO sci fel, Nat Res Coun Can, 66-68; Med Res Coun Can scholar, 71-76; ed, The J Physiol, 75- Mem: Can Physiol Soc; Soc Neurosci; Physiol Soc; AAAS. Res: Formation and development of nerve-muscle synapses; regulation of the distribution of acetylcholine receptors in skeletal muscle. Mailing Add: Dept of Physiol McGill Univ Montreal PQ Can

COHEN, MONTAGUE, b London, Eng, July 24, 25; m 47; c 3. MEDICAL PHYSICS. Educ: Univ London, BSc, 46, PhD(physics), 58; Royal Col Sci, ARCS, 46; Inst Physics, Eng, F Inst P, 68. Prof Exp: Physicist, Royal Aircraft Estab, 46-47 & Gen Elec Co Res Labs, 47-48; physicist med physics, London Hosp, 48-61, chief physicist, 66-75; prof officer, Int Atomic Energy Agency, 61-66; PROF RADIO PHYSICS, McGILL UNIV, 75- Concurrent Pos: Consult radiother physics, Int Comn Radiation Units & Measurements, 64- & WHO, 70-; consult med physics, Int Atomic Energy Agency, 66-; dep ed, Brit J Radiol, 70-74. Honors & Awards: Roentgen Prize, Brit Inst Radiol, 74. Mem: Inst Physics; Brit Inst Radiol (secy, 73-75); Hosp Physicists Asn; Am Asn Physicists in Med; Can Asn Physicists. Res: Clinical dosimetry applied to radiotherapy, with special reference to the acquisition of patient data and the use of such data in computerized dosimetry systems. Mailing Add: Dept of Radiother Montreal Gen Hosp Montreal PQ Can

COHEN, MORREL HERMAN, b Boston, Mass, Sept 10, 27; m 50; c 4. THEORETICAL PHYSICS. Educ: Worcester Polytech Inst, BS, 47; Dartmouth Col, MA, 48; Univ Calif, Berkeley, PhD(physics), 52. Prof Exp: Res assoc physics, Dartmouth Col, 47-48; from instr to assoc prof theoret physics, 52-60, prof, Dept Physics, Univ Chicago & James Franck Inst, 60-72, actg dir, 65-66, dir, James Franck Inst, 68-71, prof theoret biol, 68-72, LOUIS BLOCK PROF PHYSICS & THEORET BIOL, UNIV CHICAGO, 72- Concurrent Pos: Guggenheim fel, Cambridge Univ, 57-58; vis scientist, Nat Res Coun Can, 60; NSF sr fel, Univ Rome, 64-65; consult, Westinghouse Res Labs, 53; Chicago Midway Labs, 54, 55; Gen Elec Co Res Labs, 57-65; Argonne Nat Lab, 59-70; Boeing Sci Res Labs, 60; Hughes Res Lab, 60, 62; NAm Aviation Co, Basic Sci Ctr, 62 & Energy Conversion Devices Inc, 67-; NASA mem, Adv Panel Electrophysics, 62-66; mem adv comt, Nat Magnet Lab, 63-66; review comt, Solid State Sci & Metall Div, Argonne Nat Lab, 64-66, chmn, 66; chmn, Gordon Conf Chem & Physics of Solids, NH, 68 & chmn, Fourth Int Conf Amorphous & Liquid Semiconductors, 71; assoc ed, J Chem Physics, 60-63; mem ed bd, Physics of Condensed Matter, 62-; McGraw-Hill Co, 63-70; publ bd, Univ Chicago, 69-70 & bd eds, J Statist Physics, 70-; NIH spec fel, 72-73; vis fel, Cambridge Univ, 72-73, assoc, 73-; Shrum lectr, Simon Fraser Univ, 73- Mem: AAAS; Am Inst Physics; fel Am Phys Soc. Res: Theoretical physics of condensed matter; developmental biology; quantum theory of solids; general physics of solids. Mailing Add: James Franck Inst Univ Chicago 5640 S Ellis Ave Chicago IL 60637

COHEN, MORRIS, b Santa Ana, Calif, July 20, 21. CELL BIOLOGY, MEDICAL RESEARCH. Educ: Univ Calif, BS, 47, PhD(plant path), 51. Prof Exp: Asst plant path, Univ Calif, Berkeley, 48-51; res assoc bot, Univ Calif, Los Angeles, 51-54, asst res botanist, 54-57; electron microscopist, St Joseph Hosp, Burbank, Calif, 57-62, res assoc, 62; RES BIOLOGIST, WADSWORTH VET ADMIN HOSP, 62- Concurrent Pos: Am Cancer Soc fel, 51-54. Mem: AAAS; Am Soc Cell Biol; Electron Micros Soc Am; NY Acad Sci; Am Soc Microbiol. Res: Submicroscopic morphology, pathogenesis of inflammatory diseases; bacteria; fungi; adipose tissue fine structure and function. Mailing Add: Electron Micros Res Lab Rm 209 Wadsworth Vet Admin Hosp Bldg 114 Los Angeles CA 90073

COHEN, MORRIS, b Regina, Can, July 10, 15; m 40; c 2. PHYSICAL CHEMISTRY. Educ: Brandon Col, BA, 34; Univ Toronto, MA, 35, PhD, 39. Prof Exp: Chemist, Monarch Battery Mfg Co, 40-43; PRIN RES OFFICER METALLIC CORROSION & OXIDATION & HEAD CORROSION LAB, NAT RES COUN CAN, 43- Concurrent Pos: Vis prof metall, Univ New South Wales, 66; chmn, Gordon Conf Corrosion, 61. Honors & Awards: Willis Rodney Whitney Award, Asn Corrosion Eng, 60. Mem: Fel Chem Inst Can. Res: Reactions of metal surfaces with their environment, oxidation of iron and its alloys; passivity; electrochemistry. Mailing Add: Div of Chem Nat Res Coun Can Montreal Rd Ottawa ON Can

COHEN, MORTIMER, b New York, NY, Mar 7, 16; m 43; c 2. PLANT PATHOLOGY. Educ: City Col New York, MSEd, 42; Univ Minn, Minneapolis, PhD(plant path), 51. Prof Exp: Plant pathologist, Div Plant Path, Calif Dept Agr, 51-52 & Fla State Plant Bd, 52-56; PROF & PLANT PATHOLOGIST, UNIV FLA, 56- Mem: Am Phytopath Soc. Res: Diseases of citrus trees. Mailing Add: Agr Res Ctr Univ of Fla PO Box 248 Ft Pierce FL 33450

COHEN, MORTON IRVING, b New York, NY, July 11, 23; m 58; c 2. NEUROPHYSIOLOGY. Educ: City Col New York, BS, 42; Columbia Univ, AM, 50, PhD(physiol), 57. Prof Exp: From instr to assoc prof, 57-73, PROF PHYSIOL, ALBERT EINSTEIN COL MED, 73- Mem: Am Physiol Soc; Biophys Soc; Soc Neurosci. Res: Neural regulation of respiration and circulation; spontaneous activity of neurons in central nervous system; patterns of synaptic excitation and inhibition; computer analysis of neuroelectric data. Mailing Add: Albert Einstein Col of Med Yeshiva Univ New York NY 10461

COHEN, MOSES E, Brit citizen. APPLIED MATHEMATICS. Educ: Univ London, BSc, 63; Univ Wales, PhD(theoret physics, appl math), 67. Prof Exp: Res fel astrophys, Fr Atomic Energy Comn, 68; asst prof math, Mich Technol Univ, 68-69; from asst prof to assoc prof, 69-74, PROF MATH, CALIF STATE UNIV, FRESNO, 74- Mem: Am Math Soc; assoc fel Brit Inst Math & Appln; assoc Brit Inst Physics & Phys Soc. Res: Astrophysics, especially cosmic radiation; applied mathematics, especially generating functions, combinatorial identities; solid state physics, especially solar cells. Mailing Add: Dept of Math Calif State Univ Fresno CA 93710

COHEN, MURRAY SAMUEL, b Brooklyn, NY, May 19, 25; m 48; c 2. ORGANIC CHEMISTRY. Educ: Univ Mo, BS, 48, MA, 50, PhD(chem), 52. Prof Exp: Res chemist, Schenley Labs, 52-53; res chemist reaction motors div, Thiokol Chem Corp, 53-54, proj leader synthetic org & inorg res, 54-56, chief propellant synthesis sect, 56-62, mgr chem dept, 62-67; staff adv advan planning, Esso Res & Eng Co, 67-68, dir new ventures & dir fuel additives labs, 68-73; VPRES RES, WESTON CHEM DIV & TECH DIR FOR CHEM, BORG WARNER CORP, 73- Mem: Am Chem Soc; Soc Plastics Indust. Res: Solid and liquid propellants; light weight metal hydrides and derivatives; organometallic polymers; additive research, antioxidants, heat and ultra violet stabilizers, copper deactivators. Mailing Add: Borg Warner Chem 103 Spring Valley Rd Montvale NJ 07645

COHEN, NATALIE SHULMAN, b New York, NY, Jan 16, 38; m 58; c 2. CELL PHYSIOLOGY, BIOCHEMISTRY. Educ: Cornell Univ, BA, 59; NY Univ, MS, 61, PhD(biol), 65. Prof Exp: Res fel biol, Calif Inst Technol, 66-70; RES ASSOC BIOCHEM, SCH MED, UNIV SOUTHERN CALIF, 70- Concurrent Pos: Res assoc biochem, Col Med, Univ Ariz, 73-74. Res: Physiological and biochemical aspects of cellular functions; relation to cellular structure; membrane structure and function. Mailing Add: 1725 Homet Rd Pasadena CA 91106

COHEN, NATHAN WOLF, b Richmond, Va, Oct 3, 19; m 46; c 2. HERPETOLOGY. Educ: Univ Calif, Los Angeles, AB, 44; Univ Calif, Berkeley, MA, 50; Ore State Univ, PhD(zool), 55. Prof Exp: Asst physiol & zool, Univ Calif, 48-50, res zoologist, San Joaquin Exp Range, 50-51; instr zool, Fresno State Col, 51-52; instr physiol & biol, Modesto Jr Col, 55-63; sci coordr, Lib Arts Dept, 63, head, Letters & Sci Exten, 64-70 & Continuing Educ in Sci & Math, 70-72, DIR CURRIC DEVELOP IN SCI, UNIV CALIF EXTEN, BERKELEY, 72- Concurrent Pos: Res assoc herpet, Los Angeles, County Mus Natural Hist & Mus Vert Zool, Univ Calif, Berkeley; mem steering comt, Ctr Biol Educ, Comn Undergrad Educ Biol Sci. Mem: Fel AAAS; Am Soc Ichthyologists & Herpetologists; fel Herpet League; Sigma Xi. Res: Environmental physiology of terrestrial cold-blooded vertebrates; color photography of amphibians and reptiles; environmental physiology and behavior of amphibians and reptiles. Mailing Add: Dept of Sci & Math Univ of Calif Exten 2223 Fulton St Berkeley CA 94720

COHEN, NICHOLAS, b New York, NY, Nov 20, 38; m 74; c 3. IMMUNOLOGY, DEVELOPMENTAL BIOLOGY. Educ: Princeton Univ, AB, 59; Univ Rochester,

PhD(biol), 66. Prof Exp: USPHS scholar med microbiol & immunol, Univ Calif, Los Angeles, 65-67; asst prof, 67-73, ASSOC PROF MICROBIOL-IMMUNOL, SCH MED & DENT, UNIV ROCHESTER, 73- Mem: AAAS; Transplantation Soc; Am Soc Zool; Am Asn Immunol. Res: Comparative and developmental immunology; transplantation and developmental biology. Mailing Add: Dept of Microbiol Univ of Rochester Sch Med & Dent Rochester NY 14642

COHEN, NOAL, b Rochester, NY, Dec 29, 37; m 60; c 2. SYNTHETIC ORGANIC CHEMISTRY, NATURAL PRODUCTS CHEMISTRY. Educ: Univ Rochester, BS, 59; Northwestern Univ, PhD(org chem), 65. Prof Exp: Chemist, Eastman Kodak Co, 59-61; sr chemist, Dept Chem Res, 67-75, RES FEL, HOFFMANN-LA ROCHE, INC, 75- Concurrent Pos: NSF fel, Stanford Univ, 65-67. Mem: Am Chem Soc; Sigma Xi; Int Soc Heterocyclic Chem. Res: Synthesis of natural products and other organic compounds possessing biological activity. Mailing Add: Hoffman-La Roche Inc Nutley NJ 07110

COHEN, NOEL LEE, b New York, NY, Sept 20, 30; m 57; c 1. OTOLARYNGOLOGY. Educ: NY Univ, BA, 51; State Univ Utrecht, MD, 57. Prof Exp: From instr to assoc prof, 62-72, PROF OTOLARYNGOL, POST-GRAD MED SCH, NY UNIV, 72- Concurrent Pos: Consult, USPHS Hosp, Staten Island, 67- & Manhattan Vet Admin Hosp, 74-; assoc attend, Univ Hosp, 68- & Bellevue Hosp, NY, 70- Mem: Am Neurotology Soc; Am Acad Ophthal & Otolaryngol; Am Col Surg; Soc Univ Otolaryngol; Am Laryngol, Rhinol & Otol Soc. Res: Neurotology. Mailing Add: Dept of Otolaryngol NY Univ Post-Grad Med Sch New York NY 10016

COHEN, NORMAN, b New York, NY, Dec 13, 36; m 59. PHYSICAL CHEMISTRY. Educ: Reed Col, AB, 58; Univ Calif, Berkeley, MA, 60, PhD(chem), 63. Prof Exp: Mem tech staff phys chem, 63-69, staff scientist, 69-73, HEAD CHEM KINETICS DEPT, AEROSPACE CORP, 73- Mem: Am Chem Soc; fel Am Inst Chemists; Am Phys Soc. Res: Gas phase chemical kinetics and photochemistry; reactions of free radicals and atoms; reactions in atmospheric systems and photochemical smog; kinetics of chemical lasers; decomposition of hydrogen halides; vibrational energy transfer. Mailing Add: Aerospace Corp PO Box 92957 Los Angeles CA 90009

COHEN, NORMAN, b Brooklyn, NY, Nov 6, 38; m 62; c 1. RADIOLOGICAL PHYSICS, RADIOBIOLOGY. Educ: Brooklyn Col, BS, 60; NY Univ, MS, 65, PhD(environ sci), 70. Prof Exp: Chemist, Columbia Presby Hosp, NY, 60-61; res assoc radiobiol & radiochem, 66-74, ASST PROF ENVIRON MED, MED CTR, NY UNIV, 74- Mem: AAAS; Am Indust Hyg Asn; Health Physics Soc; Radiation Res Soc. Res: Radiobiological research of the metabolism of various radionuclides and elements in man and other primates; research and evaluation of toxicological properties of heavy metals in man. Mailing Add: Inst Environ Med A J Lanza Lab NY Univ Med Ctr Tuxedo NY 10987

COHEN, OBADIAH PHILLIP, soil physics, see 12th edition

COHEN, OSCAR PAUL, chemistry, see 12th edition

COHEN, PAUL JOSEPH, b Long Branch, NJ, Apr 2, 34; m 63; c 2. MATHEMATICS. Educ: Univ Chicago, MS, 54, PhD(math), 58. Prof Exp: Instr math, Univ Rochester, 57-58 & Mass Inst Technol, 58-59; fel, Inst Advan Study, 59-61; from asst prof to assoc prof, 61-64, PROF MATH, STANFORD UNIV, 64- Honors & Awards: Bocher prize & Res Corp Award, 64; Fields Medal, Int Math Union, 66; Nat Medal Sci, 67. Mem: Am Math Soc. Res: Axiomatic set theory; harmonic analysis; partial differential equations. Mailing Add: Dept of Math Stanford Univ Palo Alto CA 94305

COHEN, PAUL SIDNEY, b Boston, Mass, Jan 20, 39; m 63. MOLECULAR BIOLOGY Educ: Brandeis Univ, AB, 60; Boston Univ, AM, 62, PhD(genetics), 64. Prof Exp: USPHS trainee, St Jude Hosp, 64-67; assoc prof bacteriol, 67-75, PROF MICROBIOL, UNIV RI, 75- Mem: Am Soc Biol Chemists; Am Soc Microbiol. Res: Regulation of enzyme synthesis in bacteriophage infected cells. Mailing Add: Dept of Microbiol Univ of RI Kingston RI 02881

COHEN, PETER, b New Haven, Conn, Aug 16, 06; m 33. PEDIATRICS. Educ: Yale Univ, BS, 27; Univ Rochester, MD, 33; Am Bd Pediat, dipl, 46. Prof Exp: Technician bact dept, Psychiat Inst, NY, 27-29; intern pediat, Univ Calif Hosp, 33-34; house off pediat & commun dis, San Francisco Gen Hosp, 34-35; resident pediat, Univ Calif Hosp, 35-36; dep health officer, County Health Dept, Santa Barbara, 36-42; lectr pediat, 42-51, assoc prof, 51-74, EMER PROF PEDIAT, UNIV CALIF, SAN FRANCISCO, 74- Concurrent Pos: Med Dir, Golden Gate Regional Ctr, 66-71. Mem: Am Acad Pediat; Am Pub Health Asn; AMA; Am Acad Cerebral Palsy; Am Acad Neurol. Res: Clinical research relating to handicapped children, particularly those suffering from cerebral palsy and mental retardation. Mailing Add: Dept of Pediat Univ of Calif San Francisco CA 94143

COHEN, PHILIP, solid state physics, see 12th edition

COHEN, PHILIP, b New York, NY, Dec 13, 31; m 54; c 1. HYDROGEOLOGY, GROUNDWATER GEOLOGY. Educ: Univ Rochester, MS, 56. Prof Exp: Asst, Univ Rochester, 54-56; geologist, 56-67, res hydrologist, 67-68, hydrologist in charge LI prog, 68-72, staff scientist, Off of the Dir, 72-74, ACTG ASSOC CHIEF RES & TECH COORD, LAND INFO & ANAL OFF, US GEOL SURV, 74- Honors & Awards: Ward Medal, 54; Meritorious Serv Award, Dept of the Interior, 75. Mem: Fel Geol Soc Am; Am Water Resources Asn; Am Inst Prof Geologists; Sigma Xi. Res: Artificial groundwater recharge; seawater encroachment; land-use planning implications of earth sciences. Mailing Add: US Geol Surv Nat Ctr MS 703 12201 Sunrise Valley Dr Reston VA 22092

COHEN, PHILIP IRA, b Baltimore, Md, Oct 27, 48. SURFACE PHYSICS. Educ: Johns Hopkins Univ, BA, 69; Univ Wis-Madison, PhD(physics), 75. Prof Exp: Teaching asst, Univ Wis, 70, res asst, 70-75; RES ASSOC PHYSICS & CHEM, CTR MAT RES, UNIV MD, COLLEGE PARK, 76- Concurrent Pos: Guest worker, Nat Bur Standards, 76- Mem: Sigma Xi; Am Inst Physics. Res: Surface physics and chemistry, physical adsorption, surface crystallography, low energy electron diffraction, catalysis and chemiluminescence. Mailing Add: Ctr Mat Res Univ of Md College Park MD 20742

COHEN, PHILIP PACY, b Derry, NH, Sept 26, 08; m 35; c 4. PHYSIOLOGICAL CHEMISTRY, ENZYMOLOGY. Educ: Tufts Col, BS, 30; Univ Wis, PhD(physiol chem), 37, MD, 38. Prof Exp: From asst prof to assoc prof, 43-47, prof physiol chem & chmn dept, 48-75, H C BRADLEY PROF PHYSIOL CHEM, UNIV WIS-MADISON, 68- Concurrent Pos: Nat Res Coun fel, Univ Sheffield, 38-39 & Yale Univ, 39-40; Commonwealth Fund fel, Oxford Univ, 58; mem comt growth, Nat Res Coun, 54-56; mem bd sci counr, Nat Cancer Inst, 57-61, mem adv cancer coun, 63-67; mem adv comt biol & med, US AEC, 63-71; mem adv coun to dir, NIH, 66-70, mem adv coun arthritis, metab & digestive dis, 70-74; mem adv comt med res, Pan-Am Health Orgn, 67-75. Honors & Awards: Hon mem fac, Univ Chile, 66. Mem: Nat Acad Sci; Am Soc Biol Chem (treas, 51-55); Brit Biochem Soc; hon mem Harvey Soc; hon mem Mex Nat Acad Med. Res: Intermediary nitrogen metabolism; action of thyroxine; differentiation and development; comparative biochemistry. Mailing Add: 694 Med Sci Bldg Univ of Wis Madison WI 53706

COHEN, PINYA, b Burlington, Vt, Dec 23, 35; m 60. BIOCHEMISTRY, IMMUNOLOGY. Educ: Del Valley Col, BS, 57; Univ Ga, MS, 59; Purdue Univ, PhD(microbiol), 64. Prof Exp: Res microbiologist, NIH, 64-68, chief plasma derivatives sect, Lab Blood & Blood Prod, 68-72; DIR PLASMA DERIVATIVES BR, BUR BIOLOGICS, FOOD & DRUG ADMIN, 72- Mem: AAAS; Am Soc Microbiol; Int Soc Blood Transfusion. Res: Immunology and biochemistry of plasma proteins. Mailing Add: Bur of Biologics Food & Drug Admin Bethesda MD 20014

COHEN, RICHARD LAWRENCE, b Philadelphia, Pa, Oct 6, 22; m 50; c 2. CHILD PSYCHIATRY. Educ: Univ Pa, AB, 43, MD, 47; Am Bd Psychiat & Neurol, cert psychiat, 53, cert child psychiat, 60. Prof Exp: Clin dir psychiat, Embreeville State Hosp, 51-52; dir child psychiat, Oakburne Hosp, 57-62; dir training, Philadelphia Child Guid Clin, 62-64; assoc prof child psychiat, Col Med, Univ Nebr, 64-67; assoc prof, 67-70, PROF CHILD PSYCHIAT, SCH MED, UNIV PITTSBURGH, 70- Concurrent Pos: Psychiat consult, Univ Settlement House, Jewish Family Serv Philadelphia, Asn Jewish Children, Nat Teacher Corps & Student Health Serv & Univ Nebr; mem deans comt psychiat residency, 48-51. Mem: AAAS; Am Acad Child Psychiat; Am Orthopsychiat Asn; Am Psychiat Asn; Am Pub Health Asn. Res: Prenatal prevention of developmental disorders in children; operations research into systems of delivery of medical service. Mailing Add: Child Psychiat Serv 201 De Soto St Pittsburgh PA 15213

COHEN, RICHARD LEWIS, b New York, NY, Sept 8, 36; m 59; c 1. SOLID STATE SCIENCE. Educ: Haverford Col, BS, 57; Calif Inst Technol, MS, 59, PhD(physics), 62. Prof Exp: MEM TECH STAFF PHYSICS, BELL LABS, 62- Concurrent Pos: Res fel, Inst Physics, Munich Tech Univ, 64-65; mem ed bd, Rev Sci Instruments, 75-78 Mem: AAAS; fel Am Phys Soc. Res: Use of Mössbauer effect to study nuclear and solid state physics, especially with rare-earth isotopes; x-ray photoelectric spectroscopy in solids; fiber optics; colloidal catalysts. Mailing Add: Bell Labs Murray Hill NJ 07974

COHEN, ROBERT, b Indianapolis, Ind, Oct 15, 24; m 63; c 2. ENERGY CONVERSION. Educ: Wayne Univ, BS, 47; Univ Mich, MS, 48; Cornell Univ, PhD(elec eng), 56. Prof Exp: Res asst, Univ Mich, 47-48, Purdue Univ, 48-51 & Cornell Univ, 51-56; physicist, Aeronomy Lab, Environ Res Labs, Nat Oceanic & Atmospheric Admin, 56-73, prog mgr ocean thermal energy conversion, NSF Res Appl to Nat Needs, 73-75; BR CHIEF OCEAN THERMAL ENERGY CONVERSION, DIV SOLAR ENERGY, ENERGY RES & DEVELOP ADMIN, 75- Honors & Awards: Boulder Scientist Award, 64. Mem: Sigma Xi; Inst Elec & Electronics Eng; Am Geophys Union (aeronomy secy); fel AAAS. Res: Ionospheric radio-wave propagation; irregularities in the ionosphere; equatorial ionosphere; ionospheric modification; aeronomy; program managment of the ocean thermal energy conversion program. Mailing Add: Div of Solar Energy Energy Res & Develop Admin Washington DC 20545

COHEN, ROBERT ABRAHAM, b Chicago, Ill, Nov 13, 09; m 33; c 2. PSYCHIATRY. Educ: Univ Chicago, SB, 30, PhD(physiol) & MD, 35. Prof Exp: Clin dir, Chestnut Lodge Sanatorium, 47-53; dir clin invests, 53-68, DIR DIV CLIN & BEHAV RES, NIMH, 68- Concurrent Pos: Training & supv analyst, Wash Psychoanal Inst, 50-, chmn educ comt, 54-59, dir, 59-62; dir, Found Fund Res Psychiat, 60-63, chmn, 62-63; pres, Wash Sch Psychiat, 73- Honors & Awards: Distinguished Serv Award, Dept Health, Educ & Welfare. Mem: Fel Am Psychiat Asn; Am Psychoanal Asn; Am Psychopathol Asn; Asn Res Nerv & Ment Dis. Res: Social psychiatry; nature of the psychotherapeutic process; research administration. Mailing Add: Div of Clin & Behav Res NIMH Bethesda MD 20014

COHEN, ROBERT JAY, b Milwaukee, Wis, May 31, 42; m 68; c 2. BIOPHYSICS, BIOCHEMISTRY. Educ: Univ Wis, BS, 64; Yale Univ, PhD(biophys chem), 69. Prof Exp: NIH fel, Calif Inst Technol, 69-71; ASST PROF BIOCHEM, COL MED, UNIV FLA, 71- Mem: AAAS. Res: Sensory and hormonal transduction in model system; nucleic acids; electrokinetics. Mailing Add: Dept of Biochem Univ of Fla Col of Med Gainesville FL 32610

COHEN, ROBERT LORING, organic chemistry, see 12th edition

COHEN, ROBERT ROY, b Duluth, Minn, June 3, 39. VERTEBRATE ZOOLOGY. Educ: Univ Minn, Duluth, BA, 61; Univ Colo, PhD(zool), 65. Prof Exp: Asst prof biol, Univ Sask, 65-67 & NMex Inst Mining & Technol, 67-69; assoc prof, 69-74, PROF BIOL, METROP STATE COL, DENVER, 74- Mem: AAAS; Am Inst Biol Sci; Am Ornith Union; Am Soc Zool. Res: Vertebrate zoology, especially with respect to birds and to physiological adaptations; ecological physiology; avian erythrocyte and hemoglobin kinetics; human population growth and its consequences. Mailing Add: Dept of Biol Metrop State Col Denver CO 80204

COHEN, ROBERT SONNE, b New York, NY, Feb 18, 23; m 44; c 3. THEORETICAL PHYSICS, PHILOSOPHY OF SCIENCE. Educ: Wesleyan Univ, AB, 43; Yale Univ, MS, 43, PhD(physics), 48. Prof Exp: Instr physics, Yale Univ, 43-44; mem sci staff, Div War Res, Columbia Univ, 44-46; Am Coun Learned Socs fel philos of sci, Yale Univ, 48-49, instr philos, 49-51; asst prof physics & philos, Wesleyan Univ, 49-57; assoc prof physics, 57-59, chmn dept, 58-73, actg dean, Col Lib Arts, 71-73, PROF PHYSICS, BOSTON UNIV, 59- Concurrent Pos: Mem tech staff, Joint Commun Bd, US Joint Chiefs of Staff, 44-46; consult, Fund Advan Educ, 51-53 & Nat Woodrow Wilson Found, 59-64; Ford fac fel, 55-56; vis prof, Mass Inst Technol, 58-, Brandeis Univ, 59-60 & Univ Calif, San Diego, 69; chmn, Boston Ctr Philos Sci, 61- & Ctr Philos & Hist of Sci, Boston Univ, 70-; vis lectr, Polish & Czech Acad Sci, 62, Yugoslav Philos Asn, 63 & Hungarian Acad Sci, 64; ed, Boston Studies in Philos of Sci, 63-; mem staff, Oak Ridge Conf Sci & Contemporary Social Probs, 64; chmn, Am Inst Marxist Studies, 64-; trustee, Inst Unity of Sci, 66- & Wesleyan Univ, 68-; chmn US nat comt, Int Union Hist & Philos of Sci, 67-75. Mem: Am Phys Soc; Am Asn Physics Teachers; Hist of Sci Soc; Am Philos Asn; Philos Sci Asn (vpres, 73-75). Res: Concept and theory formation in physical sciences; science and the social order; logical empiricism and natural science; dialectical materialism and science; general education in science; history of scientific concepts; comparative historical sociology of science. Mailing Add: Dept of Physics Boston Univ Boston MA 02215

COHEN, ROCHELLE SANDRA, b Brooklyn, NY, June 20, 45. ENDOCRINOLOGY, NEUROBIOLOGY. Educ: Rutgers Univ, AB, 67; Univ Conn, MS, 70, PhD(physiol, endocrinol), 73. Prof Exp: Res asst endocrinol, Univ Conn, 68, teaching asst biol, 68-72; asst prof, State Univ NY Col Purchase, 72-74; NIH fel cell biol, 74-75, RES ASSOC CELL BIOL, ROCKEFELLER UNIV, 75- Concurrent Pos: Asst prof biol, Empire State Col, 73-74; res assoc biochem genetics, Rockefeller Univ, 73-74. Mem: Am Soc Cell Biol. Res: Characterization of the synaptic junction by the use of biochemical, ultra structural and immunological techniques to find the relationship between its structure and function. Mailing Add: Dept of Cell Biol Rockefeller Univ New York NY 10021

COHEN, RONALD, b Toronto, Ont, Jan 22, 30; m 55; c 2. ANTHROPOLOGY, POLITICAL SCIENCE. Educ: Univ Toronto, BA, 51; Univ Wis, MSc, 55, PhD(anthrop), 60. Prof Exp: Lectr anthrop, Univ Toronto, 58-61; asst prof, McGill Univ, 61-63; assoc prof, 63-67, PROF ANTHROP & POLIT SCI, NORTHWESTERN UNIV, EVANSTON, 67- Concurrent Pos: NSF grants, 65-68, Evanston, 71-72, Nigeria, 72-74, Nat Endowment Humanities fel, 72-74; assoc ed, Anthropologica, 63- & Am Anthropologist, 71-; rev bd mem, NIMH small grants, 68-72; prof sociol & head dept, Ahmadu Bello Univ, Nigeria, 72-74. Mem: Fel Am Anthrop Asn; fel African Studies Asn; Brit Asn Social Anthrop; assoc Am Sociol Asn; Am Polit Sci Asn. Res: Social structure; political organization; methods of social research; Africa. Mailing Add: Dept of Anthrop Northwestern Univ Evanston IL 60201

COHEN, RONALD BRUCE, b New York, NY, Dec 18, 39; m 64; c 3. CHEMICAL PHYSICS. Educ: Brooklyn Col, BS, 60; Pa State Univ, PhD(phys chem), 66. Prof Exp: Res assoc mass spectros, Pa State Univ, 66; scientist, Nat Phys-Tech Inst, Braunshweig, Ger, 66-67; res assoc molecular beams, Space Res Coord Ctr, Pittsburgh, 67-69; asst prof chem, Ill Inst Technol, 69-75; MEM TECH STAFF, AEROSPACE CORP, 75- Mem: Am Phys Soc; Am Chem Soc. Res: Kinetics of elementary reactions in the gas phase using molecular beam and laser excitation techniques; the study of chemi-ionization reactions and reactions of electronically excited atoms and molecules. Mailing Add: Aerospace Corp PO Box 92957 Los Angeles CA 90009

COHEN, SAMUEL ALAN, b Brooklyn, NY, Feb 3, 47; m 73. PLASMA PHYSICS, SURFACE PHYSICS. Educ: Mass Inst Technol, BS, 68, PhD(physics), 72. Prof Exp: Teaching asst physics, Mass Inst Technol, 68-72; RES STAFF PHYSICS, PLASMA PHYSICS LAB, PRINCETON UNIV, 73- Concurrent Pos: Consult, Res Lab Electronics, Mass Inst Technol, 73- Mem: Am Phys Soc; Am Vacuum Soc. Res: Experimental work on surface physics and plasma physics related to controlled thermonuclear fusion. Mailing Add: Plasma Physics Lab Princeton Univ Forrestal Campus Princeton NJ 08540

COHEN, SANFORD I, b New York, NY, Sept 5, 28; m 52; c 3. PSYCHIATRY, PSYCHOPHYSIOLOGY. Educ: NY Univ, AB, 48; Chicago Med Sch, MD, 52; Am Bd Psychiat, dipl, 59. Prof Exp: Resident psychiat, Med Ctr, Univ Colo, 54 & Med Ctr, Duke Univ, 55, instr, 56-58, assoc, 58-59, from asst prof to assoc prof, 59-61, head div psychophysiol res, 60-65, prof & chmn exec comt inter-dept res training prog nerv syst sci, 64-68, head div psychosom med & psychophysiol res, 65-68; prof psychiat & biobehav sci & chmn dept, Med Ctr, La State Univ, New Orleans, 68-70; PROF PSYCHIAT & CHMN DIV, MED SCH, BOSTON UNIV, 70-, PSYCHIATRIST-IN-CHIEF, UNIV HOSP, 70- Concurrent Pos: Markle scholar, 57-62; fel psychoanal, Found Fund Res Psychiat, 58-63; instr, Wash Psychoanal Inst, 64-; mem ment health small grants comt, NIMH, 63-; supt, Dr Solomon Carter Fuller Ment Health Ctr, 70- Mem: AAAS; fel Am Psychiat Asn; Am Psychosom Soc; Am Fedn Clin Res; Soc Biol Psychiat. Res: Psychosomatic medicine; psychophysiology of emotions and behavior; neuroendocrinology and physiology of conditioned reflexes; effects of altered sensory environments; perceptual mode, personality and Pavlovian typology. Mailing Add: Div of Psychiat Boston Univ Med Sch Boston MA 02118

COHEN, SANFORD NED, b Bronx, NY, June 12, 35; m 58; c 1. PHARMACOLOGY, DEVELOPMENTAL BIOLOGY. Educ: Johns Hopkins Univ, AB, 56, MD, 60. Prof Exp: Res physician, Walter Reed Army Inst Res, 63-65; from instr to asst prof pharmacol, Sch Med, NY Univ, 65-71, asst prof pediat, 68-71, assoc prof pharmacol & pediat, 71-74; PROF PEDIAT & CHMN DEPT, SCH MED, WAYNE STATE UNIV, 74-; PEDIATRICIAN-IN-CHIEF, CHILDREN'S HOSP MICH, 74- Concurrent Pos: Nat Inst Child Health & Human Develop spec fel, 63-65; Markle Found scholar acad med, 68; physician-in-chg nurseries & assoc dir pediat, Bellevue Hosp, 69-74. Mem: AAAS; Am Pediat Soc; Am Acad Pediat; Soc Pediat Res; Am Soc Pharmacol & Exp Therapeut. Res: Developmental pharmacology; how developmental phenomena alter the rate of metabolism or the biological effects of drugs in the immature animal. Mailing Add: Dept of Pediat Wayne State Univ Sch Med Detroit MI 48202

COHEN, SAUL BERNARD, b Malden, Mass, July 28, 25; m 50; c 2. GEOGRAPHY. Educ: Harvard Univ, AB, 47, AM, 49, PhD(geog), 55. Prof Exp: From asst prof to prof geog, Boston Univ, 52-64; exec secy, Asn Am Geogr, 64-65; dean grad sch, 67-70, PROF GEOG & DIR GRAD SCH GEOG, CLARK UNIV, 65-, ADJ PROF GOVT & INT RELS, 74- Concurrent Pos: Vis lectr, Yale Univ, 55; vis prof, US Naval War Col, 57; vchmn, Consortium Prof Asn, 65-71; mem US comt, Int Geog Union, 65-69; chmn, Comn Col Geog, 65-67; mem comn geog, Nat Acad Sci-Nat Res Coun, 66-70. Mem: Fel AAAS; Am Geog Soc; Asn Am Geog. Res: Political geography; education. Mailing Add: Grad Sch of Geog Clark Univ Worcester MA 01610

COHEN, SAUL G, b Boston, Mass, May 10, 16; m 41; c 2. ORGANIC CHEMISTRY. Educ: Harvard Univ, AB, 37, MA, 38, PhD(org chem), 40. Prof Exp: Pvt asst, Harvard Univ, 39-40, instr chem, 40-41; res assoc, Nat Defense Res Comt, 41, res fel, 41-43; Nat Res fel & lectr chem, Univ Calif, Los Angeles, 43-44; res chemist, Pittsburgh Plate Glass Co, 44-45; sr chemist, Polaroid Corp, 45-50; assoc prof chem, 50-52, prof, 52-74, chmn dept, 59-72, dean fac, 55-58, UNIV PROF CHEM, BRANDEIS UNIV, 74- Concurrent Pos: Fulbright sr scholar & Guggenheim fel, UK, 58-59; consult, Polaroid Corp. Honors & Awards: James F Norris Award, Am Chem Soc, 72. Mem: Am Chem Soc; Am Acad Arts & Sci; The Chem Soc. Res: Mechanisms of organic reactions; free radicals; polymerization; stereochemistry; photography; enzyme reactions; photochemistry Mailing Add: Dept of Chem Brandeis Univ Waltham MA 02154

COHEN, SAUL ISRAEL, b Boston, Mass, Feb 15, 26; m 52; c 2. NUTRITION. Educ: Northeastern Univ, BS, 45; Boston Univ, MA, 46; Harvard Univ, MS, 51; Columbia Univ, PhD, 56. Prof Exp: Asst, Harvard Med Sch, 44-45; jr chemist & jr bacteriologist, State Dept Pub Health, Mass, 46-48; asst, Sch Pub Health, Harvard Univ, 49-52; instr, NY Med Col, 55; res assoc, Harvard Med Sch, 56-64; assoc prof, NEssex Community Col, 65-66; RES ASST, RETINA FOUND, 67- Concurrent Pos: Teacher high sch, Boston, 67-68. Mem: AAAS; Am Chem Soc. Res: Chemical interactions of the protein components of blood plasma; amino acids and proteins in liver repair. Mailing Add: 14 Corey Rd Brookline MA 02146

COHEN, SAUL LOUIS, b Regina, Sask, May 10, 13; m 38; c 3. BIOCHEMISTRY, ENDOCRINOLOGY. Educ: Brandon Col, BA, 32; Univ Toronto, PhD(biochem, physiol), 36. Prof Exp: Exhib of 1851 fel, Swiss Fed Inst Technol, 36-37; instr physiol, Med Sch, Ohio State Univ, 37-42; asst prof, Med Sch, Univ Mich, 42-46; from asst prof to assoc prof biochem, Univ Minn, 46-56; res asst, 62-67, ASST PROF PATH CHEM, OBSTET & GYNEC, MED SCH, UNIV TORONTO, 67- Honors & Awards: Reeve Prize, Toronto, 35. Mem: AAAS; Am Soc Biol Chem; Endocrine Soc; Can Fedn Biol Sci. Res: Biochemical endocrinology; hydrolysis, assay, concentration, isolation identification, metabolism and significance of conjugate steroids. Mailing Add: Rm 6366 Med Sci Bldg Univ of Toronto Toronto ON Can

COHEN, SAUL MARK, b Springfield, Mass, Oct 6, 24; m 53; c 1. ORGANIC CHEMISTRY. Educ: Univ Mass, BS, 48; Univ Ill, MS, 49, PhD(org chem), 52. Prof Exp: Asst, Univ Ill, 50-52; res chemist, Eastman Kodak Co, 52-55; res chemist, Shawinigan Resins Corp, 55-60, res group leader, 60-65; res specialist, 65-69, SR RES SPECIALIST, MONSANTO CO, 69- Honors & Awards: Arthur K Doolittle Award, Div Org Coatings & Plastic Chem, Am Chem Soc, 72. Mem: AAAS; NY Acad Sci; Am Chem Soc; fel Am Inst Chemists; Sigma Xi. Res: Organic synthesis; polymerization; polymer properties; mechanism studies. Mailing Add: 15 Lindsay Rd Springfield MA 01128

COHEN, SEYMOUR STANLEY, b New York, NY, Apr 30, 17; m 40; c 2. BIOCHEMISTRY. Educ: City Col New York, BS, 36; Columbia Univ, PhD(biochem), 41. Hon Degrees: Dr, Univ Louvain, 72. Prof Exp: Res assoc biochem, Columbia Univ, 42-43; instr pediat, Univ Pa, 45-47, assoc prof physiol chem, Dept Pediat, 47-54, prof, 54-57, Am Cancer Soc Charles Hayden prof biochem, 57-71, Hartzell prof therapeut res & chmn dept, 63-71; prof, 71-72, AM CANCER SOC PROF MICROBIOL, SCH MED, UNIV COLO, DENVER, 72- Concurrent Pos: Abbott Lab fel, Columbia Univ, 40-41; Nat Res Found fel plant viruses, Rockefeller Inst, 41-42; Johnson Found fel, Univ Pa, 43-45; Guggenheim fel, Pasteur Inst, Paris, 47-48; Lalor fel, Marine Biol Lab, 50-51; Fogarty scholar, Nat Cancer Inst, 73-74; ed, Virology, 54-59, J Biol Chem, 60-65 & J Bact Rev, 69-73; vis prof, Radium Inst, 67, Col France, 70, Hadassah Med Sch, 74 & Univ Tokyo, 74; Smithsonian scholar, 73-74; trustee, Marine Biol Lab; mem bd sci consult, Sloan-Kettering Inst; mem, Coun for Anal & Proj, Am Cancer Soc, 71- Honors & Awards: Eli Lilly Award, 51; Mead Johnson Award, 52; Cleveland Award, AAAS, 55; Borden Award, Am Asn Med Cols, 68; Passano Award, 74; French Soc Biol Chem Medal. Mem: Inst of Med of Nat Acad Sci; fel AAAS; Am Soc Biol Chem; fel Am Acad Arts & Sci; Soc Gen Physiol (pres, 68). Res: Chemistry of viruses and nucleoproteins; metabolism of bacteria and virus infected cells; nucleic acids and phosphate compounds; polyamines; cancer research. Mailing Add: Dept of Microbiol Univ of Colo Denver CO 80220

COHEN, SHELDON GILBERT, b Pittston, Pa, Sept 21, 18. IMMUNOLOGY, ALLERGY. Educ: Ohio State Univ, BA, 40; NY Univ, MD, 43. Prof Exp: Intern, Bellevue Hosp, 44; resident internal med, Vet Admin Hosp, Md, 47-48; resident allergy, Vet Admin Hosp & Univ Med Ctr, Univ Pittsburgh, 48-49; res fel, Addison H Gibson Lab of Appl Physiol, Univ Pittsburgh, 49-50, res assoc, 50-51; res assoc immunol, Dept Biol, Wilkes Col, 51-56, from assoc prof to prof biol res, 57-68, prof exp biol, 68-72; consult, Inst, 72-73, CHIEF EXTRAMURAL PROG, ALLERGY & IMMUNOL BR, NAT INST ALLERGY & INFECTIOUS DIS, 73- Concurrent Pos: Attend physician, Vet Admin Hosp, 51-60, consult internal med, 60-72, consult res, 61-72; chief allergy, Mercy Hosp, 51-72. Honors & Awards: Distinguished Serv Award, Am Acad Allergy, 71. Mem: Am Asn Immunol; Am Soc Exp Biol; fel Am Acad Allergy; fel Am Col Physicians; fel Am Col Allergists. Res: Immunologic basis of hypersensitivity reactions; experimental eosinophilia; histopathogenesis of allergic inflammation. Mailing Add: Allergy & Immunol Br Nat Inst Allergy & Infect Dis Bethesda MD 20014

COHEN, SHELDON H, b Milwaukee, Wis, May 21, 34; m 62; c 3. INORGANIC CHEMISTRY. Educ: Univ Wis, BS, 56; Univ Kans, PhD(chem), 62. Prof Exp: From asst prof to assoc prof, 60-70, PROF CHEM & CHMN DEPT, WASHBURN UNIV, 70- Mem: Am Chem Soc. Res: Polarography of inorganic systems; preparation of compounds with unusual oxidation states; preparation and stability of inorganic complexes. Mailing Add: Dept of Chem Washburn Univ Topeka KS 66621

COHEN, SIDNEY, b Philadelphia, Pa, Jan 31, 17; m 40; c 1. ORGANIC CHEMISTRY. Educ: Univ Pa, BS, 37, MS, 38, PhD(org chem), 42. Prof Exp: Assoc chemist, Frankford Arsenal, 41-43; res chemist, Off Sci Res & Develop Proj, Pa, 43-44 & De Paul Chem Co, 44-52; res chemist, Hart Prod Corp, NY, 52-55, tech dir, 55-61; dir textile res & develop, Onyx Chem Corp, NJ, 61-67, dir cent textile lab, Millmaster-Onyx Corp, 67-68; res & develop supvr, 68-70, prod mgr dyeing chem, 70-74, VPRES OPERS, TANATEX CHEM CORP, 74- Concurrent Pos: Mem, Textile Res Inst. Mem: Am Chem Soc; Am Asn Textile Chem & Colorists. Res: Textile chemistry; chemical specialties. Mailing Add: Tanatex Chem Corp Page & Schuyler Aves Lyndhurst NJ 07071

COHEN, SIDNEY, b Boston, Mass, June 3, 28; m 62; c 2. ORGANIC CHEMISTRY. Educ: Northeastern Univ, BS, 51; Tufts Univ, MS, 52; Univ Colo, PhD(org chem), 59. Prof Exp: Develop engr, Chem Div, Gen Elec Co, NY, 52-55; res assoc fluorine chem, Univ Colo, 60-61; from asst prof to assoc prof chem, Ft Lewis Col, 61-66; assoc prof, 66-69, PROF CHEM, STATE UNIV NY COL BUFFALO, 69- Mem: Am Chem Soc. Res: Small ring compounds; oxycarbons; fluorine chemistry; photochemistry. Mailing Add: Dept of Chem State Univ of NY Col Buffalo NY 14222

COHEN, SIDNEY, b Malden, Mass, Jan 29, 13; m 54; c 2. MEDICINE, MICROBIOLOGY. Educ: Harvard Univ, AB, 33; Harvard Med Sch, MD, 37; Am Bd Internal Med, dipl. Prof Exp: Intern, Mt Sinai Hosp, 37-40; resident med, Beth Israel Hosp, 40-41; res assoc, 41-42, assoc path & med & in chg bact lab, 46-51, dir clin labs, 51-56, from assoc vis physician to vis physician, 49-58; DIR DEPT MICROBIOL, MICHAEL REESE HOSP, 58-; PROF MED, PRITZKER SCH MED, UNIV CHICAGO, 71- Concurrent Pos: Res fel bact, Harvard Med Sch, 40; asst bact & med, Harvard Med Sch, 41-42, instr med, 45-51, assoc, 51-58. Mem: Infectious Dis Soc Am; Am Soc Microbiol; Am Col Physicians; Brit Soc Gen Microbiol. Res: Medical microbiology with relation to staphylococcal infection; action of antibiotics and mechanism of microbial resistance to antibiotics. Mailing Add: Dept of Microbiol Michael Reese Hosp Chicago IL 60616

COHEN, STANLEY, b Brooklyn, NY, Nov 17, 22; m 51; c 2. BIOCHEMISTRY. Educ: Brooklyn Col, BA, 43; Oberlin Col, MA, 45; Univ Mich, PhD(biochem), 48. Prof Exp: Instr pediat res, Univ Colo, 48-52; Am Cancer Soc fel, Wash Univ, 52-53, res assoc biochem, Dept Biol, 53-59; sr res fel, 59-62, assoc prof, 62-67, PROF BIOCHEM, VANDERBILT UNIV, 67- Mem: Am Chem Soc; Am Soc Biol Chem. Res: Metabolism of urea and creatine; carbon dioxide fixation; chemical embryology; protein hormones. Mailing Add: Dept of Biochem Vanderbilt Univ Sch of Med Nashville TN 37203

COHEN, STANLEY, b Los Angeles, Calif, Feb 5, 27. THEORETICAL PHYSICS. Educ: Univ Southern Calif, BS, 49; Cornell Univ, PhD(physics), 55. Prof Exp: AEC res assoc, Fermi Inst, Univ Chicago, 54-56; physicist, Lawrence Radiation Lab, Univ Calif, 56-59; assoc physicist, 59-68, SR PHYSICIST, ARGONNE NAT LAB, 68- Concurrent Pos: Consult, Rand Corp, 57-62. Mem: Am Phys Soc. Res: Theoretical nuclear physics; computational physics. Mailing Add: Physics Bldg 203 Argonne Nat Lab Argonne IL 60439

COHEN, STANLEY, b New York, NY, June 4, 37; m; c 3. IMMUNOLOGY, PATHOLOGY. Educ: Columbia Univ, BA, 57, MD, 61. Prof Exp: Asst attend pathologist, Med Ctr, NY Univ, 65-66; capt immunochem, Walter Reed Army Inst Res, 66-68; from assoc prof to prof path, State Univ NY Buffalo, 68-74; PROF PATH, UNIV CONN HEALTH CTR, 74- Concurrent Pos: Buswell fel, 68-; NIH grant, 69-72; instr, NY Univ, 66-68; lectr, Cath Univ, 68. Mem: Am Asn Immunol; Am Asn Path & Bact; Am Soc Exp Path; Reticuloendothelial Soc. Res: Biological activities of immunoglobulins; mechanism of delayed hypersensitivity; products of lymphatic secretion; biological activities of eosinophils. Mailing Add: Dept of Path Univ of Conn Health Ctr Farmington CT 06032

COHEN, STEPHEN ROBERT, b New York, NY, May 7, 28; m 54; c 2. NEUROCHEMISTRY, PHYSICAL CHEMISTRY. Educ: Cornell Univ, BChE, 51, PhD(phys chem), 56. Prof Exp: Asst phys chem, Cornell Univ, 51-55; instr chem, Brown Univ, 56-58; phys chemist, Itek Corp, 58-59; asst prof chem, Northeastern Univ, 59-61; asst prof, City Col New York, 61-64; res assoc, Col Physicians & Surgeons, Columbia Univ, 64-66; sr res scientist, 66-70, ASSOC RES SCIENTIST, NY STATE RES INST NEUROCHEM & DRUG ADDICTION, 70- Mem: AAAS; Am Chem Soc; Am Soc Neurochem; The Chem Soc; Fedn Am Socs Exp Biol. Res: Kinetics and reaction mechanisms; neurochemistry; active transport; extracellular spaces. Mailing Add: NY State Res Inst Neurochem & Drug Addiction Ward's Island New York NY 10035

COHEN, STEVEN DONALD, b Boston, Mass, Nov 22, 42; m 65; c 2. TOXICOLOGY, PHARMACOLOGY. Educ: Mass Col Pharm, BS, 65, MS, 67; Harvard Univ, DSc(toxicol), 70. Prof Exp: Res assoc toxicol, Harvard Univ, 70-72; ASST PROF TOXICOL, UNIV CONN, 72- Mem: AAAS; Soc Toxicol; Am Asn Cols Pharm; Am Conf Govt Indust Hyg. Res: Biochemical actions of toxicants; special emphasis on toxicologic interactions involving organophosphate insecticides. Mailing Add: Sect of Pharmacol & Toxicol Univ Conn Sch of Pharm Storrs CT 06268

COHEN, STUART COLIN, b London, Eng, Apr 13, 44; m 66; c 2. ORGANIC POLYMER CHEMISTRY. Educ: Univ London, BSc, 65, PhD(chem), 68. Prof Exp: Res assoc chem, Va Polytech Inst, 68-69; asst prof, Syracuse Univ, 69-73; sr chemist, Borg-Warner Chem, 73-75; PRIN SCIENTIST, LEEDS & NORTHRUP CO, 75- Concurrent Pos: Petrol Res Fund grant, 69-72. Mem: Am Chem Soc; The Chem Soc; Royal Inst Chem; Soc Plastics Engrs. Res: Reactions at surface of inorganic and organic polymers; immobilization of organic molecules onto inorganic and organic polymeric supports; infrared and ultraviolet/visible spectroscopy. Mailing Add: Tech Ctr Leeds & Northrup Co Dickerson Rd North Wales PA 19454

COHEN, STUART LYLE, organic chemistry, polymer chemistry, see 12th edition

COHEN, SYLVAN IRVING, plant pathology, see 12th edition

COHEN, THEODORE, b Arlington, Mass, May 11, 29; m 54; c 2. ORGANIC CHEMISTRY. Educ: Tufts Univ, BS, 51; Univ Southern Calif, PhD(chem), 55. Prof Exp: Asst chem, Univ Southern Calif, 52-55; asst lectr, Glasgow Univ, 55-56; from instr to assoc prof, 56-66, PROF CHEM, UNIV PITTSBURGH, 66- Concurrent Pos: Fulbright grant & Ramsay fel, Glasgow Univ, 55-56. Mem: Am Chem Soc; The Chem Soc. Res: New synthetic methods involving sulfur and/or copper; the biosynthesis of cyclopropane rings; mechanisms of reactions involving organocopper intermediates. Mailing Add: Dept of Chem Univ of Pittsburgh Pittsburgh PA 15260

COHEN, VICTOR DAVID, mechanical engineering, see 12th edition

COHEN, VICTOR WILLIAM, physics, deceased

COHEN, WERNER VICTOR, organic chemistry, polymer chemistry, see 12th edition

COHEN, WILLIAM b Brooklyn, NY, Apr 19, 31; m 63. BIOCHEMISTRY. Educ: Long Island Univ, BS, 51; Purdue Univ, MS, 54; Fordham Univ, PhD(biochem), 58. Prof Exp: Asst, Purdue Univ, 51-54; assoc dept microbiol, Col Physicians & Surgeons, Columbia, 58-63; from asst prof to assoc prof, 63-74, PROF BIOCHEM, SCH MED, TULANE UNIV, 74- Mem: Am Soc Biol Chem. Res: Enzyme therapy. Mailing Add: Dept of Biochem Tulane Univ New Orleans LA 70112

COHEN, WILLIAM DAVID, b Brooklyn, NY, Feb 24, 28; m 48; c 6. BIOCHEMISTRY. Educ: Univ Iowa, BS, 47; Univ Minn, MS, 50, PhD(biochem), 52. Prof Exp: Damon Runyan sr res fel, 52-53; chief biochemist, Mem Hosp, Worcester, Mass, 53-61; consult chemist, Minn State Dept Health, 62; asst prof obstet & gynec, Med Col, Univ Minn, 62-68; clin biochemist, 68-70, CHIEF BIOCHEMIST, GEN HOSP LAB, ST JOHN'S, 70- Concurrent Pos: Clin assoc path, Mem Univ Nfld, 71- Mem: Am Asn Clin Chem; Can Soc Clin Chemists. Res: Chemistry and metabolism of lipids and steroid hormones. Mailing Add: Gen Hosp Lab Forest Rd St John's NF Can

COHEN, YEHUDI ARYEH, b New York, NY, June 7, 28; m 62; c 1. ANTHROPOLOGY. Educ: Brooklyn Col, BA, 48; Yale Univ, PhD(anthrop), 53. Prof Exp: Instr anthrop, Albert Einstein Col Med, 55-59; res assoc, Social Res, Inc, 62-64; assoc prof, Univ Calif, Davis, 64-67; PROF ANTHROP, RUTGERS UNIV, NEW BRUNSWICK, 67- Concurrent Pos: Instr, Conn Col, 52-53; lectr, Columbia Univ, 56-62, Univ Chicago, 62-64 & Northwestern Univ, 62-64; Wenner-Gren Found grants anthrop res, Univ Calif, Davis, 66-67 & Rutgers Univ, 70-72; US Off Educ grant, Israel, 67-69. Honors & Awards: Socio-Psychol Award, AAAS, 55. Mem: Am Anthrop Asn. Res: Political organization; religion; education; evolution; cross-cultural research. Mailing Add: Dept of Anthrop Livingston Col Rutgers Univ New Brunswick NJ 08903

COHENOUR, FRANCIS D, b Muskogee, Okla, July 11, 17; m 44; c 3. ANIMAL PHYSIOLOGY, MICROBIOLOGY. Educ: Okla State Univ, BS, 39, BS & MS, 48; Miss State Univ, PhD(animal physiol), 66. Prof Exp: Trouble shooter, Armour & Co, 48-55; instr dairy & food technol, Iowa State Univ, 55-61; asst prof, Miss State Univ, 61-68; ASSOC PROF BIOL, UNION COL, KY, 68-, ACTG HEAD DEPT, 70- Mem: Am Dairy Sci Asn. Res: Zoology; embryology; pathogenic bacteriology and microbiology; epidemiology. Mailing Add: Dept of Biol Union Col Box 502 Barbourville KY 40906

COHICK, A DOYLE, JR, b Nevada, Mo, July 14, 39; m 61; c 2. ECONOMIC ENTOMOLOGY, INFORMATION SCIENCE. Educ: Cent Mo State Col, BSEd, 61, MA, 64; Cornell Univ, PhD(entom), 68. Prof Exp: Teacher pub schs, Raytown, Mo, 61-65; asst entom, Cornell Univ, 65-68; res biologist, Chemagro Corp, 68-72, MGR DATA DOCUMENTATION, CHEMAGRO AGR DIV, MOBAY CHEM CORP, 72- Mem: Entom Soc Am. Res: Odonate ethology; mammalian and insect ecology; stored products insect behavior and control; information storage and data processing systems. Mailing Add: Chemagro Agr Div Mobay Chem Corp Box 4913 Kansas City MO 64120

COHLAN, SIDNEY QUEX, b New York, NY, July 31, 15; m 51; c 2. PEDIATRICS. Educ: Brooklyn Col, AB, 34; NY Med Col, MD, 39; Am Bd Pediat, dipl, 52. Prof Exp: From asst prof to assoc prof clin pediat, 53-70, PROF PEDIAT, MED SCH, NY UNIV-BELLEVUE MED CTR, 70- Concurrent Pos: Adj pediatrician, Beth Israel Hosp, 50-56, assoc pediatrician, 56-; vis physician, Children's Med Serv, Bellevue Hosp, 53-58, vis physician, Hosp, 58-; asst vis physician, NY Univ Hosp, 53-, pediat in chg, Dept Pediat, 58-; mem med bd, Irvington House, 59-; study sect child health & human develop, NIH. Mem: AAAS; Am Pediat Soc; Am Acad Pediat; Soc Pediat Res. Res: Vitamin A metabolism; developmental malformations. Mailing Add: NY Univ 566 First Ave New York NY 10016

COHN, CHARLES ERWIN, b Chicago, Ill, Apr 25, 31. EXPERIMENTAL PHYSICS. Educ: Univ Chicago, PhD(physics), 57. Prof Exp: Asst physicist, 56-59, ASSOC PHYSICIST, ARGONNE NAT LAB, 59- Mem: Am Nuclear Soc; Am Phys Soc. Res: Nuclear reactor experimental physics; reactor kinetics, noise and computer-aided experimentation. Mailing Add: 445 Ridge Ave Clarendon Hills IL 60514

COHN, DANIEL ROSS, b Berkeley, Calif, Nov 28, 43. LASERS, PLASMA PHYSICS. Educ: Univ Calif, Berkeley, AB, 66; Mass Inst Technol, PhD(physics), 71. Prof Exp: Staff physicist, 71-74, tech asst to dir, 74-75, ASSOC GROUP LEADER QUANTUM OPTICS & PLASMA PHYSICS, FRANCIS BITTER NAT MAGNET LAB, MASS INST TECHNOL, 75- Res: Laser plasma interactions; infrared and submillimeter lasers; plasma diagnostics; controlled thermonuclear fusion. Mailing Add: Francis Bitter Nat Magnet Lab 170 Albany St Cambridge MA 02139

COHN, DAVID LIONEL, b Brooklyn, NY, Oct 3, 27; m 56; c 4. PHYSICS, THEORETICAL BIOLOGY. Educ: Harvard Univ, BA, 49; Univ Chicago, PhD(math biol), 55. Prof Exp: NIH fel, Univ Chicago, 55-56; mem tech staff, Bell Tel Labs, 56-58 & Space Tech Labs, 58-61; sr staff engr, Aerospace Corp, 61-66, prog dir, Prog 949, 66-68, assoc group dir, Group III Progs, 68-71, gen mgr, Eastern Tech Opers, 71-75; CONSULT, 75- Res: Structure of biological systems including the circulatory systems and musculatory systems; military systems research including communication systems and general system analysis; satellite systems and technology; civil system applications of satellite system for air traffic control, navigation and communications; energy systems. Mailing Add: 10400 Great Arbor Dr Potomac MD 20854

COHN, DAVID VALOR, b New York, NY, Nov 8, 26; m 47; c 2. BIOCHEMISTRY, ENDOCRINOLOGY. Educ: City Col New York, BS, 48; Duke Univ, PhD, 52. Prof Exp: USPHS fel, Western Reserve Univ, 52-53; prin scientist & actg chief radioisotope serv, Vet Admin Hosp, 53-68; from instr to assoc prof, 53-74, PROF BIOCHEM, SCH MED, & ASSOC DEAN, COL HEALTH SCI, UNIV KANS MED CTR, KANSAS CITY, 74-; ASSOC CHIEF STAFF & DIR CALCIUM RES LAB, VET ADMIN HOSP, 68- Concurrent Pos: From asst prof to assoc prof, Sch Dent, Univ Mo, 62-70, prof, 71-; mem gen med B study sect, USPHS, 71-75; chmn, Gordon Res Conf Chem, Physiol & Structure Bones & Teeth, 74. Mem: AAAS; Am Chem Soc; Am Soc Biol Chem; Endocrine Soc. Res: Chemistry and physiology of peptide hormones including parathyroid hormone and calcitonin; biology of bone growth and resorption; calcium binding proteins. Mailing Add: Vet Admin Hosp 4801 Linwood Blvd Kansas City MO 64128

COHN, ERNST M, b Mainz, Ger, Mar 31, 20; nat US; m 49. PHYSICAL CHEMISTRY. Educ: Univ Pittsburgh, BS, 42, MS, 52. Prof Exp: Chemist minimum ignition energies gas mixtures, US Bur Mines, 42-44, chemist Fischer-Tropsch synthesis & magnetochem, 46-53; chemist, Bituminous Coal Res Br, Dept Interior, 53-60; phys chemist, US Army Res Off, 60-62; MGR SOLAR & CHEM POWER, NASA, 62- Mem: Am Chem Soc; Electrochem Soc. Res: Kinetics of reactions in solids; carbides of iron, cobalt and nickel; Fischer-Tropsch synthesis; magnetochemistry; chemistry and physics of coal; batteries; fuel cells; bioelectrochemistry; electrocatalysis; photovoltaics and solar-cell arrays. Mailing Add: Apt 620-B 103 G St SW Washington DC 20024

COHN, GERALD EDWARD, b Buffalo, NY, June 10, 43; m 69; c 3. MOLECULAR BIOPHYSICS. Educ: Columbia Univ, AB, 65; Univ Wis-Madison, MA, 68, PhD(physics), 72. Prof Exp: Trainee biophys, Pa State Univ, 71-72, scholar, 73; ASST PROF PHYSICS, ILL INST TECHNOL, 73- Concurrent Pos: Cottrell Res Corp res grant, 74; fac res grant, Ill Inst Technol, 75. Mem: Am Phys Soc; Biophys Soc; Am Asn Physics Teachers. Res: Electron spin resonance studies of the photobiology of natural and model membrane systems; photodynamic damage in yeast. Mailing Add: Dept of Physics Siegel Hall Ill Inst of Technol Chicago IL 60616

COHN, HANS OTTO, b Berlin, Ger, Dec 27, 27; Nat US; m 60; c 3. PHYSICS. Educ: Ind Univ, BS, 49, MS, 50, PhD(physics), 54. Prof Exp: Asst physics, Ind Univ, 49-51, res asst, 52-54; PHYSICIST, OAK RIDGE NAT LAB, 54- Mem: Fel Am Phys Soc. Res: Strange particles; nuclear and high energy physics. Mailing Add: Oak Ridge Nat Lab Bldg 4500 PO Box X Oak Ridge TN 37830

COHN, HARVEY, b New York, NY, Dec 27, 23; m 51; c 2. MATHEMATICS. Educ: City Col New York, BS, 42; NY Univ, MS, 43; Harvard Univ, PhD(math), 48. Prof Exp: Asst prof math, Wayne Univ, 48-54; vis assoc prof, Stanford Univ, 54-55; assoc prof, Wayne Univ, 55-56; assoc prof, Wash Univ, 56-57, prof math & dir comput ctr, 57-58; prof math, Univ Ariz, 58-71, head dept, 58-67; DISTINGUISHED PROF MATH, CITY UNIV NEW YORK, 71- Concurrent Pos: Consult, Gen Motors Corp, 53, AEC, NY, 54, Nat Bur Standards, 56, Int Bus Mach Corp, 57 & Argonne Nat Lab, 58-68; mem comt regional develop for math, Nat Res Coun, 62-65; mem adv comt, Autonomous Univ Guadalajara, 63-65; vis mem, Inst Advan Study, Princeton, 70-71. Mem: Am Math Soc; Asn Comput Mach; Math Asn Am. Res: Number theory and modular functions, particularly use of computer techniques. Mailing Add: Dept of Math City Col of New York 138th St & Convent Ave New York NY 10031

COHN, ISIDORE, JR, b New Orleans, La, Sept 25, 21; m 44; c 2. SURGERY. Educ: Tulane Univ, BS, 42; Univ Pa, MD, 45, MSc, 52, DSc(med), 55; Am Bd Surg, dipl, 53. Prof Exp: From instr to assoc prof, 52-59, PROF SURG, SCH MED, LA STATE UNIV MED CTR, NEW ORLEANS, 59-, CHMN DEPT, 62- Concurrent Pos: Surgeon-in-chief, La State Univ Serv, Charity Hosp; mem Am Bd Surg, 69-75; chmn, clin invest adv comt, Am Cancer Soc, 69-73; mem ed staff, Am Surg, Rev Surg, Am J Surg & Surg Digest; dir, Nat Pancreatic Cancer Proj, 75- Mem: Am Surg Asn; Am Col Surgeons; Am Gastroenterol Asn; Soc Univ Surgeons; Int Soc Surg. Res: Gastrointestinal surgery, strangulation intestinal obstruction and secondary interest in problems involving biliary, pancreatic and tumor problems associated with antibacterial agents; surgical research in germ-free animals. Mailing Add: Dept of Surg La State Univ Med Ctr New Orleans LA 70112

COHN, JACK, b Rock Island, Ill, May 6, 32; m 53; c 4. THEORETICAL PHYSICS, ELECTRODYNAMICS. Educ: Univ Iowa, BA, 53, MS, 56, PhD(statist mech), 62. Prof Exp: From asst prof to assoc prof, 60-71, PROF PHYSICS, UNIV OKLA, 71- Res: Statistical mechanics; general relativity. Mailing Add: Dept of Physics Univ of Okla Norman OK 73069

COHN, JAY BINSWANGER, b Pelham, NY, Feb 22, 22; m 45; c 2. PSYCHIATRY. Educ: Amherst Col, BA, 42; Yale Univ, MD, 45; Univ Calif, Irvine, PhD(psychol), 74. Prof Exp: Intern, St Elizabeth's Hosp, Washington, DC, 45-46; resident, Crile Hosp, Cleveland, Ohio, 48-49 & Cleveland Receiving Hosp, 51-53; lectr psychiat, Western Reserve Univ, 54-59; instr, Univ Calif, Los Angeles, 59-65; asst clin prof, Univ Southern Calif, 65-68, assoc prof, 68-71; PROF PSYCHIAT & SOC SCI, UNIV CALIF, IRVINE, 71- Concurrent Pos: Consult, res projs, State Dept Ment Hyg, Ohio, 53-58, Vet Admin, Calif, 59-65, Indust Accident Comn, 60- & US Fed Court, Panel, 61- Mem: AAAS; fel Am Psychiat Asn; Am Psychol Asn; Am Acad Neurol; AMA; Acad Psychoanal. Res: Counter-transference and pharmacology; electrical measurement of counter-transference as measured in electrocardiogram. Mailing Add: Dept of Psychiat & Human Behav Univ of Calif Irvine Orange CA 92668

COHN, JAY NORMAN, b Schenectady, NY, July 6, 30; m 53; c 3. CARDIOVASCULAR DISEASES. Educ: Union Univ, NY, BS, 52; Cornell Univ, MD, 56; Am Bd Internal Med, dipl. Prof Exp: Intern med, Beth Israel Hosp, Boston, Mass, 56-57, asst resident, 57-58; chief resident, Vet Admin Hosp, 61-62, clin investr cardiovasc, 62-65; from instr to prof med, Georgetown Univ, 62-74; PROF MED & HEAD CARDIOVASC DIV, MED SCH, UNIV MINN, MINNEAPOLIS, 74- Concurrent Pos: Res fel cardiovasc, Georgetown Univ Hosp, 60-61; chief hypertension & clin hemodynamics, Vet Admin Hosp, DC, 65-74. Honors & Awards: Arthur S Flemming Award, 69. Mem: AAAS; Am Heart Asn; Am Fedn Clin Res; Am Soc Clin Invest; Am Soc Pharmacol & Exp Therapeut. Res: Hemodynamics of myocardial infarction; pathophysiology and treatment of hypertension in man; hemodynamic factors in clinical hypotension and shock; dynamics of regional blood flow in man. Mailing Add: Med Sch Univ of Minn Minneapolis MN 55455

COHN, JOHANN GUNTHER ERNST, b Berlin, Ger, May 6, 11; nat US; m 40; c 1. CHEMISTRY. Educ: Univ Berlin, PhD(chem), 34. Prof Exp: Asst, Nobel Inst, Sweden, 34-36; asst & instr, Chalmers Tech Univ, Sweden, 37-41; Carnegie fel, Univ Minn, 41-42, Welsh fel, 43; vpres, Engelhard Industs Div, 43-63, vpres & dir, Res & Develop Dept, 63-72, VPRES RES & TECHNOL, ENGELHARD INDUSTS DIV, ENGELHARD MINERALS & CHEM CORP, 72- Mem: Am Chem Soc; Electrochem Soc. Res: Corrosion of metals and reactions in solid state; catalytic properties of solids; photochemistry of solids; powder metallurgy; instrumentation of chemical processes; precious metals; electrochemistry. Mailing Add: Engelhard Industs Div Menlo Park Edison NJ 08817

COHN, KIM, b New York, NY, Jan 25, 39; m 62; c 1. INORGANIC CHEMISTRY. Educ: Queens Col, NY, BS, 60; Univ Mich, PhD(chem), 67. Prof Exp: Asst prof chem, Mich State Univ, 67-72; ASSOC PROF CHEM, CALIF STATE COL, BAKERSFIELD, 72- Mem: Am Chem Soc; The Chem Soc. Res: Synthesis, structure and bonding of non-transition elements. Mailing Add: Dept of Chem Calif State Col Bakersfield CA 93309

COHN, LESLIE, b Philadelphia, Pa, Feb 3, 43. MATHEMATICS. Educ: Univ Pa, BA, 65; Univ Chicago, MS, 67, PhD(math), 69. Prof Exp: Asst prof, Univ Ill, Chicago Circle, 69-70; vis mem, Inst Advan Study, 70-71; ASST PROF MATH, JOHNS HOPKINS UNIV, 71- Mem: Am Math Soc. Res: Number theory; automorphic forms; representations of live groups. Mailing Add: Dept of Math Johns Hopkins Univ Baltimore MD 21218

COHN, MARGARET FOREMAN, limnology, invertebrate zoology, see 12th edition

COHN, MARTIN, applied mathematics, see 12th edition

COHN, MAURICE LEON, b Chicago, Ill, Nov 1, 03; m 30; c 1. MEDICAL MICROBIOLOGY. Educ: Univ Chicago, BS, 25, MS, 27, PhD(org chem), 29. Prof Exp: Res chemist, Great Western Sugar Co, 29-30; asst dir res, 30-65, actg dir res & labs div, 65-66, assoc dir res, 66-70, ASSOC DIR RES EMER & CONSULT, NAT JEWISH HOSP, 70- Mem: Assoc fel AMA; assoc fel Am Soc Clin Path. Res: Chemistry and bacteriology of tubercle bacilli and tuberculosis. Mailing Add: Nat Jewish Hosp 3800 E Colfax Ave Denver CO 80206

COHN, MELVIN, b New York, NY, Mar 28, 22; m 62. IMMUNOCHEMISTRY, BIOCHEMISTRY. Educ: City Col New York, BS, 40; Columbia Univ, MA, 41; NY Univ, PhD(biochem), 49. Prof Exp: Merck Nat Res Coun fel, Pasteur Inst, 49-55; prof microbiol, Sch Med, Wash Univ, 55-58; prof biochem, Sch Med, Stanford Univ, 59-61; NSF fel cellular biochem, Pasteur Inst, 61-63; RESIDENT FEL PROTEIN SYNTHESIS, SALK INST BIOL STUDIES, 63- Honors & Awards: Eli Lilly Award, 57. Mem: AAAS; Am Soc Microbiol; Am Asn Immunol; Am Soc Biol Chem; French Soc Biol Chem. Res: Antibody-antigen reactions; antibody synthesis; induced enzyme synthesis. Mailing Add: Salk Inst for Biol Studies PO Box 1809 San Diego CA 92112

COHN, MILDRED, b New York, NY, July 12, 13; m 38; c 3. BIOCHEMISTRY, BIOPHYSICS. Educ: Hunter Col, BA, 31; Columbia Univ, PhD(phys chem), 38. Hon Degrees: ScD, Women's Med Col Pa, 66. Prof Exp: Res assoc biochem, George Washington Univ, 37-38, Cornell Univ, 38-46 & Harvard Med Sch, 50-51; res assoc, Wash Univ, 46-58, assoc prof, 58-60; PROF BIOPHYS & PHYS BIOCHEM, UNIV PA, 61- Concurrent Pos: Career investr, Am Heart Asn, 64-; vis prof biochem, Inst Biol Phys Chem, Paris, 66-67. Honors & Awards: Garvan Medal, Am Chem Soc, 63; Cresson Medal, Franklin Inst, 75. Mem: Nat Acad Sci; Am Philos Soc; Am Acad Arts & Sci; Am Soc Biol Chem; Am Chem Soc (chmn, Div Biol Chem, 73-74). Res: Metabolic studies with isotopes, stable and radioactive; mechanisms of enzymatic reactions; electron spin and nuclear magnetic resonance. Mailing Add: Univ of Pa Dept Biochem Biophys Richards Bldg Philadelphia PA 19174

COHN, NAOMI KENDA, b Syracuse, NY. CELL BIOLOGY, CANCER. Educ: Syracuse Univ, BS, 52; Univ Wis-Madison, MS, 54. Prof Exp: Microbiologist, Rocky Mountain Lab, NIH, USPHS, 57-61; res asst oncol & med genetics, MacArdle Lab, Univ Wis-Madison, 61-68, microbiologist III, Dept Radiobiol, 68-70; RES SCIENTIST CELL BIOL, BURROUGHS-WELLCOME CO, 70- Mem: Tissue Cult Asn; Am Soc Microbiol; AAAS; Sigma Xi. Res: Mouse mammary carcinoma; drug resistance. Mailing Add: Burroughs-Wellcome Co 3030 Cornwallis Rd Research Triangle Park NC 27709

COHN, NORMAN STANLEY, b Philadelphia, Pa, June 26, 30; m 56; c 3. CYTOLOGY, CYTOCHEMISTRY. Educ: Univ Pa, AB, 52; Univ Ky, MS, 53; Yale Univ, PhD(bot), 57. Prof Exp: Asst biol, Yale Univ, 53-56; NSF res fel, Johns Hopkins Univ, 57-59; from asst prof to assoc prof, 59-68, chmn dept, 69-70, DISTINGUISHED PROF BOT, OHIO UNIV, 68-, DEAN, GRAD COL & DIR RES, 70- Concurrent Pos: Fulbright res scholar, State Univ Leiden, 65-66 & 68-69; Nat Acad Sci vis lectr, Czech & Yugoslavia, 68 & 73. Mem: AAAS; Am Soc Cell Biol; Genetics Soc Am. Res: Radiation cytology; chromosome breakage and chemistry; radiomimetic agents and plant materials. Mailing Add: Grad Col Wilson Hall Ohio Univ Athens OH 45701

COHN, RICHARD MOSES, b New York, NY, Sept 2, 19. ALGEBRA. Educ: Columbia Univ, BA, 39, MA, 41, PhD(math), 47. Prof Exp: Engr, Inspection Agency, Signal Corps, US Army, 42-44; lectr elem math, Columbia Univ, 46-47; from instr to assoc prof, 47-59, PROF MATH, RUTGERS UNIV, 59- Mem: Am Math Soc. Res: Difference algebra; statistics; electrical networks. Mailing Add: Dept of Math Rutgers Univ New Brunswick NJ 08903

COHN, ROBERT, b Washington, DC, Feb 25, 09; m 37. NEUROLOGY. Educ: George Washington Univ, BS, 32, MD, 36. Prof Exp: Fel neurophysiol, St Elizabeth's Hosp, Washington, DC, 36-38; mem staff neurol & electroencephalog, St Elizabeth's Hosp, 38-43; dir neurol res, US Naval Hosp, Bethesda, 46-71; PROF NEUROL, HOWARD UNIV, 71- Concurrent Pos: Clin prof, Howard Univ, 60-65; prof lectr, Mt Sinai Med Sch, NY, 69-; vis prof, Boston Univ, 70- Mem: Am Neurol Asn; Am Electroencephalog Soc; Asn Res Nerv & Ment Dis. Res: Clinical and experimental neurology and electroencephalography; physiology of sensation. Mailing Add: 7221 Pyle Rd Bethesda MD 20034

COHN, ROBERT M, b Brooklyn, NY, Jan 18, 37; m 58; c 2. PHARMACEUTICAL CHEMISTRY. Educ: Long Island Univ, BS, 58; Columbia Univ, MS, 60; Univ Conn, PhD(pharm), 63. Prof Exp: ASST DIR, SQUIBB INST MED RES, 62- Mem: Am Pharmaceut Asn. Res: Evaluation and preparation of pharmaceutical dosage form. Mailing Add: 18 Stanford Ave Colonia NJ 07067

COHN, SIDNEY ARTHUR, b Toronto, Ont, May 8, 18; US citizen; m 46; c 2. BIOLOGY. Educ: Univ Conn, BS, 40, MS, 48; Brown Univ, PhD(biol), 51. Prof Exp: From instr to assoc prof, 51-66, PROF ANAT, UNIV TENN, MEMPHIS, 66- Mem: Am Asn Anat; Int Asn Dent Res. Res: Developmental and histological studies of teeth; long-term effects of non-function on supportive tissues of teeth; temporomandibular joint; attachment of fibers of periodontal ligament; study of the supportive tissues of the teeth. Mailing Add: Dept of Anat Univ of Tenn Memphis TN 38163

COHN, STANTON HARRY, b Chicago, Ill, Aug 25, 20; m 49; c 5. PHYSIOLOGY, RADIOBIOLOGY. Educ: Univ Chicago, SB, 46, SM, 49; Univ Calif, PhD(physiol, radiobiol), 52. Prof Exp: Chemist, Kankakee Ord Works, 41-42; chemist, Sherwin Williams, 42-43; jr scientist biochem, Argonne Nat Lab, Univ Chicago, 46-49; asst radiobiol, Crocker Radiation Lab, Univ Calif, 49-50; head internal toxicity br, Biomed Div, US Naval Radiation Lab, 50-58; scientist, 58-70, SR SCIENTIST, MED PHYSICS DIV, MED RES CTR, BROOKHAVEN NAT LAB, 70- Concurrent Pos: Mem subcomt inhalation hazards, Path Effects Atomic Radiation Comt, Nat Acad Sci; Nat Comt Radiation Protection subcomt II, Int Radiation Dose, 61- Mem: Radiation Res Soc; Am Physiol Soc. Res: Chemical dynamics of the mineral metabolism of bone; distribution and biological effects of internally deposited radioisotopes; whole-body neutron activation analysis and whole-body counting. Mailing Add: Med Res Ctr Brookhaven Nat Lab Upton NY 11973

COHN, VICTOR HUGO, b Reading, Pa, July 9, 30; m 53; c 3. BIOCHEMICAL PHARMACOLOGY. Educ: Lehigh Univ, BS, 52; Harvard Univ, AM, 54; George Washington Univ, PhD(biochem), 61. Prof Exp: Pharmacologist, Army Chem Ctr, 55-56; neuropharmacologist, Vet Admin Res Labs, 56-57; biochem pharmacologist, Nat Heart Inst, 57-61; from asst prof to assoc prof, 61-71, actg chmn dept, 70-71, PROF PHARMACOL, MED CTR, GEORGE WASHINGTON UNIV, 71- Concurrent Pos: Vis investr, Jackson Labs, 66. Mem: Fel AAAS; Am Soc Pharmacol & Exp Therapeut; Am Chem Soc; Int Soc Biochem Pharmacol. Res: Drug metabolism; drug abuse; fluorometric methods of analysis of biochemicals; histamine metabolism; pharmacogenetics. Mailing Add: Dept of Pharmacol George Washington Univ Med Ctr Washington DC 20037

COHN, WALDO E, b San Francisco, Calif, June 28, 10; m 38, 43; c 2. BIOCHEMISTRY. Educ: Univ Calif, BS, 31, MS, 32, PhD(biochem), 38. Prof Exp: Asst biochem, Huntington Labs, Harvard Med Sch, 39-42; tutor biochem sci, Harvard Univ, 39-42; biochem group leader, Plutonium Proj, Univ Chicago, 42-43; group leader, Manhattan Proj, Oak Ridge, 43-47; sr biochemist, Oak Ridge Nat Lab, 47-75; CONSULT, 75- Concurrent Pos: Fulbright scholar, Cambridge Univ, 55-56; Guggenheim fel, 55-56, 62-63; secy, Comn Biochem Nomenclature, Int Union Pure & Appl Chem-Int Union Biochem, 65-; dir off biochem nomenclature, Nat Acad Sci-Nat Res Coun, 65- Honors & Awards: Award, Am Chem Soc, 63. Mem: Fel AAAS; fel Am Acad Arts & Sci; Am Chem Soc; Am Soc Biol Chemists (treas, 59-64). Res: Ion-exchange separations of rare earth elements and fission products, also of nucleic acid constituents and related biochemical substances; chemistry and structure of nucleic acids; biochemical nomenclature. Mailing Add: Biol Div Oak Ridge Nat Lab Box Y Oak Ridge TN 37830

COHOON, DANIEL FRED, b Windsor, Ont, July 6, 29; m 52; c 3. PLANT PATHOLOGY. Educ: Univ Western Ont, BS, 52; Rutgers Univ, PhD(plant path), 56. Prof Exp: Res assoc, Rutgers Univ, 52-56; ASSOC PLANT PATHOLOGIST, EDISTO EXP STA, CLEMSON, 56-, SUPT, 62- Mem: Am Phytopath Soc. Res: Vegetable and fruit diseases; breeding for disease resistance; nematology. Mailing Add: Edisto Exp Sta Blackville SC 29817

COHOON, DAVID KENT, b Terre Haute, Ind, June 4, 40; m 64; c 5. APPLIED MATHEMATICS. Educ: Mass Inst Technol, SB, 62; Purdue Univ, MS, 64, PhD(math), 69. Prof Exp: Teaching asst, Purdue Univ, 62-63, res assoc, 63-65; res assoc, Inst Henri Poincare, Paris, 65-66 & Purdue Univ, 66-67; asst prof, Bucknell Univ, 67-68; res assoc math, Purdue Univ, 68; mem tech staff, Bell Tel Labs, 69-70; Off Naval Res res assoc partial differential equations, Univ Wis-Madison, 69-70; asst prof math, Univ Minn, Minneapolis, 70-74; MATHEMATICIAN, BIOMETRICS DIV, SCH AEROSPACE MED, BROOKS AFB, 74- Mem: Math Asn Am; Am Math Soc. Res: Propagation of zero sets of solutions of homogeneous partial differential equations; continuous right inverse problem for partial differential operators. Mailing Add: Biomet Div Sch Aerospace Med Brooks AFB San Antonio TX 78235

COIL, WILLIAM HERSCHELL, b Ft Wayne, Ind, July 6, 25; m 46; c 2. PARASITOLOGY. Educ: Purdue Univ, BS, 48, MS, 49; Ohio State Univ, PhD(zool), 53. Prof Exp: Asst zool, Univ Tenn, 49-50; instr, Purdue Univ, 53-54; asst prof, Univ Nebr, 56-64; assoc prof, 64-69, PROF ZOOL, UNIV KANS, 69-, ASSOC CHMN DEPT SYSTS & ECOL, 68- Concurrent Pos: Mem exped, Oaxaca & Chiapas, Mex, 54 & 55; Muellhaupt scholar, Ohio State Univ, 54-56; NSF fel, Ore Inst Marine Biol, 57; res fel, Biol Sta, Univ Okla, 58; fac fel, Univ Nebr, 59; res assoc, Duke Marine Lab, 62-65; vis prof, Kans State Univ, 75-76. Mem: Am Soc Zool; Soc Syst Zool; Am Micros Soc. Res: Histochemistry and electron microscopy of cestodes and trematodes, fascioliasis, life histories and bionomics. Mailing Add: Dept of Syst & Ecol Univ of Kans Lawrence KS 66045

COILE, RUSSELL CLEVEN, b Washington, DC, Mar 11, 17; m 51; c 3. OPERATIONS RESEARCH. Educ: Mass Inst Technol, SB, 38, SM, 39, EE, 50. Prof Exp: Asst, Mass Inst Technol, 38-39; magnetician, Carnegie Inst, 39-42; engr, Colton & Foss, Inc, 46-47; opers analyst, Opers Eval Group, Mass Inst Technol, 47-62; dir Marine Corps Opers anal group, Ctr Naval Anal, Franklin Inst, 62-67; OPERS

ANALYST, UNIV ROCHESTER, 67- Concurrent Pos: Dir res, Opers Res Group, Off Naval Res, 53-54 & 56-57; Am del, Int Fedn Oper Res Socs Conf, Oslo, Norway, 63; mem small arms adv comt, Advan Res Proj Agency, Dept Defense, 68-70. Mem: Inst Elec & Electronics Eng; Opers Res Soc Am; Brit Oper Res Soc. Res: Acoustics; electronics, radar; nomography; documentation; information retrieval; operational research; information science. Mailing Add: 4323 Rosedale Ave Bethesda MD 20014

COISH, HAROLD ROY, b Bronte, Ont, Aug 20, 18; m 42; c 4. THEORETICAL PHYSICS. Educ: Mt Allison Univ, BA, 38, BSc, 39; Dalhousie Univ, MSc, 42; Univ Toronto, MA, 46, PhD(appl math), 52. Prof Exp: Jr res scientist physics, Nat Res Coun Can, 40-42, jr engr theoret physics, 48-49; jr engr electronics, Can Marconi Co, 42-45; from asst prof to assoc prof, 49-55, PROF MATH PHYSICS, UNIV MAN, 55- Mem: Am Phys Soc; Can Asn Physicists; Can Math Cong. Res: Theory of internal conversion; physical geometry as a finite geometry and application to elementary particles. Mailing Add: Dept of Physics Univ of Man Winnipeg MB Can

COKAL, EDWARD J, analytical chemistry, see 12th edition

COKE, CHAUNCEY EUGENE, b Toronto, Ont, July 27, 05; m 41. POLYMER CHEMISTRY. Educ: Univ Man, BSc, 27, MSc, 29; Univ Toronto, MA, 30; Univ Leeds, PhD(polymer chem), 38. Prof Exp: Asst org chem, Yale Univ, 28-29; asst phys chem, Univ Toronto, 31-32; res fel, Ont Res Found, 32-36; in charge res, Courtaulds Can, Ltd, 39-42, mgr indust yarn sales & develop, 48-54, mgr develop, 54-59; dir res & develop, Hartford Fibres Co Div, Bigelow-Sanford Carpet Co, Inc, 59-62; tech dir, Drew Chem Corp, 62-63; mgr new prod, Fibers Div, Am Cyanamid Co, NJ, 63-70; PRES, COKE & ASSOC CONSULT, 70- Concurrent Pos: Chmn, Can Adv Comt on Int Orgn for Stand, 57-59, mem Can Nat Comt, 59; dir, Textile Technol Fedn Can, 58-59. Honors & Awards: Bronze Medal, Can Asn Textile Colorists & Chem, 63; Bronze Medal, Am Asn Textile Technol, 71. Mem: AAAS; NY Acad Sci; Am Asn Textile Technol (pres, 63-65); Can Inst Textile Sci (third pres); Can Asn Textile Colorists & Chem (vpres, 54, pres, 57-59). Res: Organic and physical chemistry; equilibrium; manufacture of cellulosic and acrylic fibers; physical properties; performance characteristics; processing into fabrics, end-use developments. Mailing Add: 26 Aqua Vista Dr Ormond Beach FL 32074

COKE, JAMES LOGAN, b Brownwood, Tex, Nov 1, 33; m 58; c 3. ORGANIC CHEMISTRY, BIOCHEMISTRY. Educ: Wash State Univ, BS, 56; Wayne State Univ, PhD(chem), 61 Prof Exp: Res assoc, Univ Wis, 60-61; from instr to asst prof, 61-67, ASSOC PROF ORG CHEM, UNIV NC, CHAPEL HILL, 67- Concurrent Pos: Res assoc, Univ Minn, 64. Mem: Am Chem Soc. Res: Physical organic chemistry; organic chemistry of natural products; plant biochemistry. Mailing Add: Dept of Chem Univ of NC Chapel Hill NC 27514

COKER, EARL HOWARD, JR, b Cottonwood, Calif, May 4, 34; m 60; c 4. PHYSICAL CHEMISTRY. Educ: Ore State Univ, PhD(phys chem), 62. Prof Exp: From asst prof to assoc prof, 61-71, PROF PHYS CHEM, UNIV SDAK, 71- Mem: Am Chem Soc; Am Sci Affil. Res: Spectroscopy of complex ions in dilute solid solution; internal forces in ionic solids; point defects in ionic solids. Mailing Add: Dept of Chem Univ of SDak Vermillion SD 57069

COKER, SAMUEL TERRY, b Evergreen, Ala, Nov 29, 26; m 54; c 4. PHARMACOLOGY. Educ: Auburn Univ, BS, 51; Purdue Univ, MS, 53, PhD(pharmacol), 55. Prof Exp: Instr pharmacol, Univ Pittsburgh, 53-54; assoc prof, Univ Miss, 55-56 & Univ Mo, Kansas City, 56-59; dean, Sch Pharm, 59-73, PROF PHARMACOL, AUBURN UNIV, 73- Mem: Am Pharmaceut Asn; Acad Pharmaceut Sci. Res: Toxicology, especially drug detoxification. Mailing Add: Rte 3 Box 99 Auburn AL 36830

COKER, WILLIAM RORY, b Athens, Ga, Dec 20, 39. THEORETICAL NUCLEAR PHYSICS, EXPERIMENTAL NUCLEAR PHYSICS. Educ: Univ Ga, BS, 61, MS, 64, PhD(physics), 66. Prof Exp: Res fel nuclear physics, Ctr Nuclear Studies, 66-68, asst prof physics, 68-72, ASSOC PROF PHYSICS, UNIV TEX, AUSTIN, 72- Mem: Am Phys Soc. Res: Mechanisms of direct nuclear reactions; reactions to particle-unstable states; several-nucleon transfer reactions; nuclear spectroscopy; low and medium energy nuclear physics. Mailing Add: Dept of Physics Univ of Tex Austin TX 78712

COLAIZZI, JOHN LOUIS, b Pittsburgh, Pa, May 10, 38. PHARMACEUTICS, PHARMACY. Educ: Univ Pittsburgh, BS, 60; Purdue Univ, MS, 62, PhD(pharm), 65. Prof Exp: Asst prof pharm, WVa Univ, 64-65; asst prof pharm, 65-68, assoc prof pharmaceut, 68-72, PROF PHARMACEUT & CHMN DEPT, SCH PHARM, UNIV PITTSBURGH, 72- Concurrent Pos: Mem comt revision, US Pharmacopeial Convention, 75- Mem: Am Pharmaceut Asn; Acad Pharmaceut Sci; Am Soc Hosp Pharmacists; Am Asn Cols Pharm. Res: Bioequivalency and biavailability of drugs; sterile dosage forms tech n technology; drug formulation factors related to pharmacological response. Mailing Add: Sch of Pharm Univ of Pittsburgh Pittsburgh PA 15261

COLAS, ANTONIO E, b Muel, Spain, June 22, 28; US citizen; m 55; c 4. BIOCHEMISTRY, REPRODUCTIVE PHYSIOLOGY. Educ: Univ Zaragoza, Lic, 51; Univ Madrid, MD, 53; Univ Edinburgh, PhD(biochem), 55. Prof Exp: Prof in chg physiol & biochem, Lit Univ Salamanca, 55-57; prof biochem, Univ Valle, Colombia, 57-62, dir grad div, 60-62; assoc prof biochem, obstet & gynec, Med Sch, Univ Ore, 62-66, prof obstet & gynec, 66-68, prof biochem, 68; PROF GYNEC & OBSTET & PHYSIOL CHEM, SCH MED, UNIV WIS-MADISON, 68- Concurrent Pos: Rockefeller Found grants, 57-62. Mem: AAAS; Brit Biochem Soc; Span Soc Phys Sci; Soc Gynec Invest; Span Soc Biochem. Res: Biochemistry and metabolism of steroid hormones; in vitro and in vivo studies during pregnancy; steroid hydroxylases; steroid hormone receptors. Mailing Add: Dept of Gynec & Obstet Univ of Wis Sch of Med Madison WI 53706

COLASANTI, BRENDA KAREN, b Charleston, WVa, Dec 5, 45; m 68. NEUROPHARMACOLOGY. Educ: WVa Univ, BS, 66, PhD(pharmacol), 70. Prof Exp: NIMH fel neurol & psychopharmacol, Mt Sinai Sch Med, 70-72; asst prof pharmacol & surg ophthal, 72-76, ASSOC PROF PHARMACOL & SURG OPHTHAL, W VA UNIV, 76- Mem: Sigma Xi; Asn Psychophys Study Sleep; Am Soc Pharmacol & Exp Therapeut; Am Soc Neurochem. Res: Effects of psychoactive drugs on brain neurochemistry and electrophysiology during the sleep-wakefulness cycle. Mailing Add: Dept of Pharmacol WVa Univ Med Ctr Morgantown WV 26506

COLASITO, DOMINIC JAMES, b Denver, Colo, July 10, 25; m 51; c 4. INDUSTRIAL MICROBIOLOGY. Educ: Purdue Univ, BS, 49, MS, 52, PhD(microbiol), 56. Prof Exp: Microbiologist, Swift & Co, 56-57; bact group leader, Chicago Div, Kendall Co, 57-61; actg chmn dept microbiol, Stanford Res Inst, 61-64; sr res microbiologist, Bioferm Div, Int Minerals & Chem Corp, 64-71; LAB DIR, BIO-CON, 72- Concurrent Pos: Lectr biol, Calif State Univ, Bakersfield, 73- Mem: Am Soc Microbiol; Am Chem Soc; AAAS. Res: Hydrocarbon microbiology; enzymatic attack on lignins; microbiological treatment of waste streams. Mailing Add: PO Box 628 Wasco CA 93280

COLBERT, EDWIN HARRIS, b Clarinda, Iowa, Sept 28, 05; m 33; c 5. VERTEBRATE PALEONTOLOGY. Educ: Univ Nebr, AB, 28; Columbia Univ, AM, 30, PhD(vert paleont), 35. Hon Degrees: DSc, Univ Nebr, 73. Prof Exp: Asst, Univ Nebr, 26-29; asst, Am Mus Natural Hist, 30-33, asst cur paleont, 33-42, cur fossil reptiles & amphibians, 43-70; prof, 45-69, EMER PROF VERT PALEONT, COLUMBIA UNIV, 70-; EMER CUR FOSSIL REPTILES & AMPHIBIANS, AM MUS NATURAL HIST, 70-; CUR VERT PALEONT, MUS NORTHERN ARIZ, 70- Concurrent Pos: Assoc cur, Acad Natural Sci, Philadelphia, 37-40, res assoc, 40-; lectr, Bryn Mawr Col, 39-42 & Univ Calif, 45; assoc cur, Mus Northern Ariz, 49-69. Honors & Awards: Elliot Medal, Nat Acad Sci, 35; Am Mus Natural Hist Medal, 69. Mem: Nat Acad Sci; fel Geol Soc Am; fel Am Paleont Soc (vpres, 63); Soc Vert Paleont (secy & treas, 44, pres, 46-47); Soc Study Evolution (pres, 58). Res: Evolution of fossil vertebrates, particularly fossil amphibians, reptiles and mammals; fossil reptiles of North and South America and Asia; fossil mammals of North America and Asia; past distribution and intercontinental migrations of land-living vertebrates; paleoecology as based upon study of fossil vertebrates; fossil amphibians and reptiles of Antarctica Mailing Add: Mus of Northern Ariz Rte 4 Box 720 Flagstaff AZ 86001

COLBERT, MARVIN J, b Spokane, Wash, Nov 6, 23; m 51; c 3. INTERNAL MEDICINE. Educ: Yale Univ, BS, 46; Boston Univ, MD, 49; Am Bd Internal Med, dipl, 58. Prof Exp: Intern, Presby Hosp, Chicago, Ill, 49-50, asst resident internal med, 50; indust physician, Pub Serv Co Northern Ill, 52; asst resident internal med, Vet Admin Hosp, Boston, Mass, 53-54; resident, Univ Ill Res & Educ Hosps, Chicago, 54-55, instr, Dept Med, 56-58; physician, Steele Mem Clin, Belmond, Iowa, 55-56; from instr to assoc prof, 56-65, PROF MED, COL MED, UNIV ILL MED CTR, 65-, DIR HEALTH SERV, 59- Concurrent Pos: Vis prof, Chiengmai Med Sch & Hosp, Thailand, 65-66; lectr, Ill Acad Gen Pract, 67-68; attend physician, West Side Vet Admin Hosp & Univ Ill Hosp, Chicago. Mem: Am Fed Clin Res; Am Asn Automotive Med; Am Col Physicians; Soc Adolescent Med. Res: Medical aspects of automotive safety; smallpox vaccination during pregnancy. Mailing Add: Univ of Ill at the Med Ctr PO Box 6998 Chicago IL 60680

COLBORN, GENE LOUIS, b Springfield, Ill, Nov 23, 35; m 56; c 3. ANATOMY. Educ: Ky Christian Col, BA, 57; Milligan Col, BS, 62; Bowman Gray Sch Med, MS, 64, PhD(anat), 67. Prof Exp: Asst prof, 68-71, assoc prof anat, Univ Tex Health Sci Ctr San Antonio, 71-75; ASSOC PROF ANAT, MED COL GA, 75- Concurrent Pos: Fel, Dept Anat, Sch Med, Univ NMex, 67-68; grants, Med Res Found Tex, 68-69 & Am Heart Asn, 70-; consult dept surg, Ft Sam Houston, 69-75. Mem: AAAS; Am Asn Anat; Am Soc Zool; Am Soc Cell Biol. Res: Morphology and cytochemistry of atrioventricular conduction system and autonomic ganglia; gross anatomy, especially primate. Mailing Add: Dept Anat Med Col of Ga Augusta GA 30904

COLBOURN, JOSEPH LEASON, b Baltimore, Md, Oct 29, 29; m 52; c 6. BIOCHEMISTRY. Educ: Loyola Col, Md, AB, 51; Univ Md, PhD(biochem), 63. Prof Exp: Instr biochem, Sch Med, Univ Md, 63; res assoc, Sch Hyg & Pub Health, Johns Hopkins Univ, 63-67, asst prof, 67-71; res scientist, 71-73, MGR PROD & PRODUCT DEVELOP, RES PRODUCTS, MILES LABS, 73- Mem: Am Soc Microbiol; Am Chem Soc. Res: Toxicology of military chemicals; polyamine stabilization of ribosomes; kinetics of M luteus DNA dependent RNA polymerase; purification of polynucleotide polymerizing enzymes; enzyme synthesis of polyribonucleotides and polydeoxyribonucleotides. Mailing Add: 3110 Cherry Tree Lane Elkhart IN 46514

COLBOW, KONRAD, b Bremen, Ger, May 23, 35; Can citizen; m 60; c 2. SOLID STATE PHYSICS. Educ: McMaster Univ, BSc, 59, MSc, 60; Univ BC, PhD(physics), 63. Prof Exp: Mem tech staff, Bell Tel Labs, 63-65; ASSOC PROF PHYSICS, SIMON FRASER UNIV, 65- Mem: Am Phys Soc. Res: Absorption and luminescent properties of semiconductors; light interaction with matter; liquid crystal physics. Mailing Add: Dept of Physics Simon Fraser Univ Burnaby BC Can

COLBURN, CHARLES BUFORD, b Harrisonville, Mo, July 6, 23; m 50; c 5. PHYSICAL INORGANIC CHEMISTRY. Educ: Kans State Col, BS, 44; Univ Utah, PhD(phys chem), 51. Prof Exp: Develop chemist, Lederle Labs, 44; res assoc, Univ Utah, 51-52; group leader phys chem, Restone Res Div, Rohm and Haas Co, 52-68; HEAD PROF CHEM, AUBURN UNIV, 68- Concurrent Pos: Centenary lectr, 65. Mem: Am Chem Soc; Am Phys Soc; fel Am Inst Chem; Brit Chem Soc. Res: Chemistry of nitrogen and fluorine; inorganic free radicals; physical chemistry of combustion and ignition. Mailing Add: 235 Cary Dr Auburn AL 36830

COLBURN, IVAN PAUL, b San Diego, Calif, June 5, 27; m 58; c 3. STRUCTURAL GEOLOGY, STRATIGRAPHY. Educ: Pomona Col, BA, 51; Claremont Grad Sch, MA, 53; Stanford Univ, PhD(geol), 61. Educ: Exploitation engr, Shell Oil Co, 53-56; Am Asn Petrol Geol res grant, 58-59; from asst prof to assoc prof geol, Calif State Col Hayward, 61-64, assoc prof, 64-70, PROF GEOL, CALIF STATE COL LOS ANGELES, 70- Concurrent Pos: NSF res grant, 63-65; Nat Sci grant, 69-71. Mem: Am Asn Petrol Geol; Soc Econ Paleont & Mineral; Geol Soc Am; Nat Asn Geol Teachers. Res: California Coast Range structure and stratigraphy; sedimentation and paleocurrent analysis of Jurassic-Cretaceous sediments in California Coast Ranges; statistical analysis of clastic rock fabrics. Mailing Add: Dept of Geol Calif State Col Los Angeles CA 90032

COLBURN, NANCY HALL, b Wilmington, Del, May 15, 41; div; c 2. BIOCHEMISTRY. Educ: Swarthmore Col, BA, 63; Univ Wis, PhD(oncol), 67. Prof Exp: Asst prof molecular biol, Univ Del, 68-72; spec res fel dermat/carcinogenesis, Univ Mich, 72-74, asst prof biochem, Depts Biol Chem & Dermat, 74-75, vis scientist carcinogenesis, Dept Environ & Indust Health, 75-76; EXPERT CHEM CARCINOGENESIS IN VITRO, EXP PATH BR, NAT CANCER INST, 76- Concurrent Pos: NIH spec res fel, 72-74. Mem: AAAS; Am Asn Cancer Res; NY Acad Sci; Sigma Xi. Res: Molecular and cellular mechanism of chemical carcinogenesis; use of epithelial cell culture model systems to test the somatic mutation theory and to investigate the role of differentiation in neoplastic transformation. Mailing Add: Exp Path Br Bldg 37 Nat Cancer Inst Bethesda MD 20014

COLBURN, ROBERT WARREN, b Rochester, NY, Aug 15, 27; m 52; c 4. BIOCHEMISTRY. Educ: Univ Wis, BS, 50, MS, 51; Stanford Univ, PhD(biol sci), 58. Prof Exp: Res biochemist, Merck Inst Therapeut Res, 51-53; res assoc biochem, Sch Med, Stanford Univ, 57-59; res assoc, Rockefeller Inst, 59-63; CHIEF, UNIT ANAL BIOCHEM, NIMH, 63- Mem: AAAS; Am Chem Soc. Res: Neurochemistry; behavior; gene structure and function. Mailing Add: Lab of Clin Sci NIMH Bldg 10 3N320 Bethesda MD 20014

COLBURN, WILLIAM, organic chemistry, see 12th edition

COLBY, BENJAMIN N, b Evanston, Ill, Sept 14, 31; m 56; c 2. ANTHROPOLOGY,

PSYCHOLOGY. Educ: Princeton Univ, BA, 53; Harvard Univ, PhD(social anthrop), 60. Prof Exp: Instr social anthrop, Harvard Univ, 60-63; cur, Lab Anthrop, Mus NMex, Sante Fe, 64-67; assoc prof, Univ NC, Chapel Hill, 68; assoc prof, 68-71, PROF ANTHROP, DEPT SOC SCI, UNIV CALIF, IRVINE, 71- Concurrent Pos: Res assoc anthrop, Cornell Univ, 67. Mem: Fel AAAS; fel Am Anthrop Asn. Res: Cognitive anthropology; experimental anthropology; cultural and psychological components of stress causation; study of narrative processes; analysis and interpretation of dream content; cultural parameters of individual life style; urban anthropology. Mailing Add: Sch of Soc Sci Univ of Calif Irvine CA 92664

COLBY, CLARENCE, b Memphis, Tenn, Sept 11, 39. BIOCHEMICAL GENETICS, VIROLOGY. Educ: Memphis State Univ, BS, 62; Univ Ky, PhD(biochem), 67. Prof Exp: Fel virol, Univ Calif, Berkeley, 67-69 & Univ Calif, San Diego, 69-70; asst prof microbiol, Univ Conn, 70-73; assoc prof genetics, Univ Calif, Davis, 73-76; ASSOC MEM, SLOAN-KETTERING CANCER RES CTR, 76- Concurrent Pos: Res career develop award, Nat Inst Allergy & Infectious Dis, 71-76. Mem: AAAS; Am Soc Microbiol; Fedn Am Scientists. Res: Biochemical and genetic characterizations of the molecular regulatory mechanisms involved in the induction of interferon and the establishment and maintenance of the interferon-induced antiviral states. Mailing Add: Mem Sloan-Kettering Cancer Ctr 1275 York Ave New York NY 10021

COLBY, FRANK GERHARDT, b Muhlhausen, Ger, Apr 10, 15; m 52; c 2. CHEMISTRY. Educ: Univ Geneva, ChemEng, 39, DSc, 51. Prof Exp: Consult chemist, Havana, Cuba, 42-46; res chemist, Indust Tape, NJ, 46-47; chem lit specialist, Com Solvents Corp, 47-51; dir res info, R J Reynolds Tobacco Co, 51-70, MGR SCI INFO, R J REYNOLDS INDUSTS, INC, 70- Mem: Fel AAAS; Am Chem Soc. Res: Chemical, bioscience and technological literature; tobacco; research analysis; report writing. Mailing Add: Res Dept R J Reynolds Indust Inc 115 Chestnut St SE Winston-Salem NC 27102

COLBY, HOWARD DAVID, b New York, NY, June 16, 44; m 70. ENDOCRINOLOGY. Educ: City Col New York, BS, 65; State Univ NY Buffalo, PhD(endocrinol), 70. Prof Exp: NIH fel physiol, Med Sch, Univ Va, 70-72; asst prof, 72-75, ASSOC PROF PHYSIOL, MED SCH, W VA UNIV, 75- Mem: Am Physiol Soc; Endocrine Soc; AAAS. Res: Hormonal regulation of adrenocortical secretion; hormonal control of hepatic steroid and drug metabolism. Mailing Add: Dept of Physiol WVa Univ Sch of Med Morgantown WV 26506

COLBY, PETER J, b Grand Rapids, Mich, Mar 26, 33; m; c 3. AQUATIC BIOLOGY, FISHERIES. Educ: Mich State Univ, BS, 55, MS, 58; Univ Minn, PhD(fishery biol), 66. Prof Exp: Food technologist, Gen Foods Corp, 57-62; res asst fishery biol, Univ Minn, 62-66; aquatic biologist & proj leader, US Bur Com Fisheries, 66-70, US Bur Sport Fisheries, 70-71; RES SCIENTIST, ONT MINISTRY NATURAL RESOURCES, 71- Concurrent Pos: Res assoc environ & indust health, Sch Pub Health, Univ Mich, 70-71. Mem: Am Soc Zool; Am Fisheries Soc; Int Asn Gt Lakes Res. Res: Physiological and behavioral responses of fish to environmental stress; response of fish communities to perturbations. Mailing Add: Fish & Wildlife Res Fisheries Sect Box 2089 Thunder Bay ON Can

COLBY, ROBERT WILLIAM, b Tallula, Ill, Nov 4, 20; m 46; c 1. ANIMAL NUTRITION, BIOCHEMISTRY. Educ: Univ Ill, BS, 41; Wash State Col, MS, 46, PhD(animal nutrit), 49. Prof Exp: Res asst, Univ Ill, 41-42; res asst, Wash State Col, 42-43 & 46-48, instr animal nutrit, 48-49; asst prof, Tex Agr & Mech Col, 49-51; animal nutritionist, Mich, 51-53, dir agr res dept, Tex, 53-64, tech specialist, Int Div, Japan, 64-66 & Nat Coord Nutrit Res, Bioprod Dept, 66-67, prod mgr, Agr Dept, Mich, 67-73, PROD TECH SPECIALIST, AGR DEPT, DOW CHEM CO, 73- Mem: Am Soc Animal Sci; Poultry Sci Asn. Res: Vitamin requirements and deficiency symptoms in swine and sheep; ration composition studies with beef cattle; effect of diet composition upon growth and blood constituents of cattle and sheep. Mailing Add: Agr Dept Dow Chem Co PO Box 1706 Midland MI 48640

COLCLOUGH, NORMA VESEY, b Melrose, Mass, Aug 31, 15. PHARMACOLOGY, TOXICOLOGY. Educ: Syracuse Univ, BS, 56; Univ Vt, MS, 61. Prof Exp: Teaching asst physiol, Green Mountain Col, 47-49; res asst hemat, State Univ NY Upstate Med Ctr, 49-55; pharmacol, Col Med, Univ Vermont, 55-61; sr biologist, Childrens Cancer Res Found, Boston, Mass, 61-63; res biologist, Burroughs Wellcome & Co, 63-65; pharmacologist, Vick Div Res & Develop, Richardson-Merrell Inc, 65-70; mgr pharmacol-toxicol, Beecham Pharmaceut, 70-73; CONSULT PHARMACOL-TOXICOL-MED COMMUN, COLCLOUGH & CO, 73- Mem: AAAS; Drug Info Asn; Environ Mutagen Soc; Int Primatol Soc. Res: Biochemical, pathological, hematological and pharmacologic changes induced by newly developed drugs; biochemical pharmacology and chemotherapeutics. Mailing Add: Putnam Green 14G Greenwich CT 06830

COLDWELL, BLAKE BURGESS, b NS, July 24, 19; m 45; c 2. TOXICOLOGY, FORENSIC SCIENCE. Educ: McGill Univ, BSc, 42, MSc, 48, PhD(agr chem), 51; Queen's Univ, Ont, MSc, 67. Prof Exp: Asst protective coatings, Nat Res Coun Can, 42-44; chemist, Soils & Fertilizers Div, NS Dept Agr & Mkt, 44-47; lectr chem, MacDonald Col, McGill Univ, 49-51; chemist, Res & Develop Sect, Crime Detection Lab, Can Royal Mounted Police, 51-65; toxicologist, 65-70, head toxicol sect, Pharmacol Div, Res Labs, Foods & Drug Directorate, 70-73, CHIEF DRUG TOXICOL DIV, DRUG RES LABS, DRUGS DIRECTORATE, CAN DEPT NAT HEALTH & WELFARE, 73- Concurrent Pos: Mem comt alcohol & drugs, Nat Safety Coun, US, 60- Mem: Fel Chem Inst Can; Can Soc Forensic Sci (pres, 64); Pharmacol Soc Can; Can Asn Res Toxicol. Res: Toxicity of drug interactions; analytical toxicology and alcohol-drug related traffic problems. Mailing Add: 1820 Botsford St Ottawa ON Can

COLDWELL, ROBERT LYNN, b Woodland, Wash, May 27, 41; m 62; c 2. STATISTICAL MECHANICS. Educ: Univ Wash, BS, 63, PhD(physics), 69. Prof Exp: Fel & teacher physics, Washington & Lee Univ, 69-71; res, Northwestern Univ, 71-72; RES PHYSICS, UNIV FLA, 72- Mem: Am Phys Soc. Res: Developing a biased selection Monte Carlo procedure for finding the correlation functions, slater sum and partition function of a system of nuclei and electrons at all temperatures. Mailing Add: Dept of Physics & Astron Univ of Fla Gainesville FL 32611

COLE, ALAN L, b Battle Creek, Mich, May 23, 22; m 46; c 2. METEOROLOGY, PHYSICS. Educ: Univ Mich, BSE, 48, MS, 49 & 60, PhD(physics), 61. Prof Exp: Res meteorologist, Univ Mich, 61-69; ASSOC PROF GEOL, NORTHERN ILL UNIV, 69- Mem: AAAS; Am Meteorol Soc; Air Pollution Control Asn. Res: Mesoscale winds in the Great Lakes; wave hindcasting and forecasting; air pollution. Mailing Add: Dept of Geog Northern Ill Univ DeKalb IL 60115

COLE, ALLEN THOMAS, chemistry, see 12th edition

COLE, ARTHUR, biophysics, see 12th edition

COLE, ARTHUR CHARLES, JR, b Ann Arbor, Mich, Mar 2, 08; m 39; c 1. ENTOMOLOGY. Educ: Ohio State Univ, AB, 29, MS, 30, PhD(entom), 33. Prof Exp: Asst bur entom, USDA, 27-28, field asst, 29-30; asst, Ohio State Univ, 30-31; field aide, Bur Entom, USDA, 32-34, agent, Bur Entom & Plant Quarantine, 35-37; from instr to prof, 37-75, EMER PROF ENTOM, UNIV TENN, KNOXVILLE, 75- Concurrent Pos: Res grants from Am Philos Soc & NSF. Mem: Fel Entom Soc Am; Soc Syst Zool; Int Union Study Soc Insects. Res: Taxonomy and ecology of ants. Mailing Add: Dept of Zool & Entom Univ of Tenn Knoxville TN 37916

COLE, AVEAN WAYNE, b Smithville, Miss, June 23, 34; m 60; c 2. WEED SCIENCE. Educ: Miss State Univ, BS, 62; Iowa State Univ, PhD(hort & plant physiol), 66. Prof Exp: Asst prof hort, Univ Wis, 66-68; PROF WEED SCI, MISS STATE UNIV, 68- Mem: Weed Sci Soc Am. Res: Weed control and related responses of both weed and crop plants to herbicides. Mailing Add: PO Drawer PG Miss State Univ Mississippi State MS 37962

COLE, BASIL CHAMBRUS, b Newton, Kans, Feb 23, 10; m 39; c 3. BACTERIOLOGY. Educ: Western Ky State Teachers Col, BS, 31; Univ Ky, MS, 32; Iowa State Col, PhD(bact), 38. Prof Exp: From instr to asst prof biol, Western Ky State Teachers Col, 30-32; teaching asst bact, Iowa State Col, 32-34, 36-38; assoc prof biol, Western Ky State Teachers Col, 38-45; tech dir, 45-60, vpres, 60-74, MEM BD DIRS, SEVEN-UP CO, 74- Mem: Inst Food Technol; Am Chem Soc; Soc Soft Drink Technol; Am Water Works Asn. Res: Carbonated beverage spoilage problems; sugar quality; water treatment and control. Mailing Add: Seven-Up Co 121 S Meramec St Louis MO 63105

COLE, BENJAMIN THEODORE, b New Brunswick, NJ, May 24, 21; m 43; c 2. PHYSIOLOGY. Educ: Duke Univ, BS, 49, MA, 51, PhD(physiol), 54. Prof Exp: Instr physiol, Sch Med, Duke Univ, 53-54; from asst prof to assoc prof zool, La State Univ, 54-58; res partic cell physiol & biol, Oak Ridge Nat Lab, 59-60; assoc prof, 60-63, head dept, 64-73, PROF BIOL, UNIV SC, 63- Concurrent Pos: Consult, Cell Physiol Sect, Biol Div, Oak Ridge Nat Lab. Mem: Fel AAAS; Soc Exp Biol & Med; Am Physiol Soc. Res: Digestion; absorption and metabolism of unsaturated fatty acids; in vitro autoxidation of unsaturated fatty acids; effects of sodium fluoroacetate on carbohydrate metabolism in cold-blooded animals. Mailing Add: Dept of Biol Univ of SC Columbia SC 29208

COLE, BERWIN ABBEY, parasitology, bacteriology, see 12th edition

COLE, CHARLES DANIEL, b Lockport, NY, Apr 14, 25; m 51; c 4. PHYSICS. Educ: Univ Buffalo, BA, 49, PhD(physics), 61. Prof Exp: Asst physics, Univ Buffalo, 48-54; mem appl math group, Bell Aerosysts Co, 54-56, group leader, 56-59, supvr math anal, 59-61; prof physics & chmn dept, Parsons Col, 61-62; PROF PHYSICS, LOWELL TECHNOL INST, 62- Concurrent Pos: Consult, Lowell Tech Inst Res Found, 62- Mem: AAAS; Am Phys Soc; Am Asn Physics Teachers; Optical Soc Am. Res: Atomic spectra; forbidden lines; mathematical modeling; applied mathematics; theoretical physics; philosophy of physics; biophysics. Mailing Add: Dept of Physics Lowell Technol Inst Lowell MA 01854

COLE, CHARLES FRANKLYN, b Beaver Falls, Pa, Aug 3, 28; m 52; c 3. FISH BIOLOGY. Educ: Cornell Univ, BA, 50, PhD(vert zool), 57. Prof Exp: Instr zool, Univ Ark, 57-60; asst prof biol, Univ SFla, 60-62, assoc prof zool & chmn prog, 62-64; assoc prof, 64-72, PROF FISHERY BIOL, UNIV MASS, AMHERST, 72- Mem: AAAS; Am Soc Ichthyol & Herpet; Am Fisheries Soc; Ecol Soc Am; Am Soc Limnol & Oceanog. Res: Percid and sciaenid biology; ecology of estuarine fishes. Mailing Add: 21 Columbia Circle Amherst MA 01002

COLE, CHESTER F, b Spokane, Wash, Mar 10, 16; div; c 2. GEOGRAPHY. Prof Exp: Eastern Wash Col Educ, 37; Univ Wash, MA, 41; Univ Nebr, PhD, 51. Prof Exp: Mem fac, Seattle Pub Schs, 38-42 & 46; instr geog, Univ Nebr, 46-47; from asst prof to assoc prof geog, 47-59, chmn dept, 60-70, actg dean sch soc sci, 70-71, PROF GEOG, CALIF STATE UNIV, FRESNO, 59-, CHMN DEPT, 72- Concurrent Pos: Fel & past mem exec bd, Nat Coun Geog Educ. Mem: AAAS; fel Am Geog Soc; fel Asn Am Geog; Regional Sci Asn. Res: Central Valley of California; irrigation agriculture; political geography. Mailing Add: Dept of Geog Calif State Univ Fresno CA 93710

COLE, CLARENCE LORRAINE, b Lowell, Mich, Apr 30, 05; m 29; c 4. ANIMAL HUSBANDRY. Educ: Mich State Col, BS, 29; Univ Minn, MS, 36, PhD, 56. Prof Exp: Instr animal sci, Univ Minn, 29-38; assoc prof, Mich State Col, 38-45; asst dir mkt, Willys Overland, Toledo, 45-46, farm coun & sales, 46-51; assoc prof & supt, NCent Sch & Exp Sta, 51-56, prof dairy husb & head dept, 56-70, EMER PROF DAIRY HUSB, UNIV MINN, 70-; AGR CONSULT, 73- Concurrent Pos: Collabr, Agr Res Serv, USDA, 59. Mem: Am Soc Animal Sci; Am Dairy Sci Asn. Res: Animal breeding; sheep parasites; nutrition. Mailing Add: 481 Lakeside Dr Stanton MI 48888

COLE, CLARENCE RUSSELL, b Crestline, Ohio, Nov 20, 18; m 45; c 3. VETERINARY PATHOLOGY, MEDICAL ADMINISTRATION. Educ: Ohio State Univ, DVM, 43, MSc, 44, PhD(comp path), 47. Prof Exp: Instr vet path, 44-46, prof, 46-66, chmn dept, 47-67, asst dean col vet med, 60-67, REGENTS PROF VET PATH, OHIO STATE UNIV, 66-, DEAN COL VET MED, 67- Concurrent Pos: Assoc vet res, Ohio Res & Develop Ctr, 47-; consult, US Air Force, 43-, Walter Reed Med Ctr, 51-, Armed Forces Inst Path, 53- & USPHS, 53-; mem spec bd, Am Col Vet Pathologists, 49; lectr, Auburn Univ, 53. Mem: AAAS; Am Vet Med Asn; Poultry Sci Asn; Am Col Vet Path (vpres, 56, pres, 57); Int Acad Path. Res: Animal infectious diseases: animal neoplasms; metabolic diseases, comparative pathology. Mailing Add: Col of Vet Med Ohio State Univ Columbus OH 43210

COLE, DALE WARREN, b Everett, Wash, May 28, 31; m 56; c 4. FOREST SOILS. Educ: Univ Wash, BSF, 55, PhD(forest soils), 63; Univ Wis, MS, 57. Prof Exp: Res assoc, 60-63, res instr, 63-64, from instr to assoc prof, 64-74, PROF FOREST SOILS, COL FOREST RESOURCES, UNIV WASH, 74-, DIR CTR ECOSYST STUDIES, 72- Mem: AAAS; Soil Sci Soc Am; Am Geophys Union; Ecol Soc Am. Res: Mineral cycling in a forest ecosystem; forest soil hydrology; factors influencing the movement of elements in a soil system; soils and land-use planning. Mailing Add: 6114 164th Ave SE Issaquah WA 98027

COLE, DAVID F, b Childress, Tex, Mar 3, 33. PHYSICAL CHEMISTRY. Educ: Univ Tex, Austin, BS, 57, PhD(phys chem), 64. Prof Exp: Jr chemist, Oak Ridge Nat Lab, 57; vis asst prof chem, La State Univ, 63-64; MEM TECH STAFF, CENT RES LABS, TEX INSTRUMENTS, 65- Mem: Electrochem Soc; Am Chem Soc. Res: Advanced manufacturing processes and yield improvement studies for the manufacture of semiconductor and other electronic devices. Mailing Add: PO Box 995 Richardson TX 75080

COLE, DAVID LE ROY, b Preston, Idaho, Aug 6, 39; m 65; c 2. PHYSICAL CHEMISTRY, CHEMICAL KINETICS Educ: Univ Utah, BS, 66, PhD(phys chem), 70 Prof Exp: SR RES CHEMIST, RES LABS, EASTMAN KODAK CO, 69- Mem:

Am Chem Soc. Res: Kinetic studies of group III metal ion hydrolysis using E-jump perturbation techniques; investigation of fundamental reaction kinetics between metal ions and organic and inorganic ligands. Mailing Add: Eastman Kodak Co Res Labs Bldg 59 343 State St Rochester NY 14650

COLE, EDMOND RAY, b Huntington, WVa, Dec 17, 28; m 55; c 3. BIOCHEMISTRY. Educ: WVa Univ, BS, 52; MS, 55; Purdue Univ, PhD(biochem), 61. Prof Exp: NIH grants, Purdue Univ, 60-61 & Wayne State Univ, 61-63; res assoc coagulation, 63-69, asst dir coagulation, 69-73, DIR COAGULATION, RUSH-PRESBY-ST LUKE'S MED CTR, 73-, ASSOC PROF BIOCHEM, RUSH MED COL, 73- Concurrent Pos: Assoc scientist, Presby-St Luke's Hosp, 73- Mem: AAAS; Int Soc Thrombosis & Haemostasis; Am Soc Hemat. Res: Coagulation; fibrinolysis; tissue activators of plasminogen; enzymology. Mailing Add: Sect of Hemat Dept of Med Rush-Presby-St Luke's Med Ctr Chicago IL 60612

COLE, EDWARD ANTHONY, b Boston, Mass, Oct 16, 32; m 64; c 3. MICROBIOLOGY, PHYSIOLOGY. Educ: Univ Notre Dame, BS, 56, PhD(microbiol), 67. Prof Exp: Teacher, St Charles High Sch, Wis, 54-56, Vincentian Inst, NY, 56-57, Boysville High Sch, Mich, 57-62 & Bradley High Sch, NH, 62-63; teaching asst biol, Univ Notre Dame, 63-64; from asst prof to assoc prof, 66-72, PROF BIOL, ANNA MARIA COL WOMEN, 72- Mem: AAAS; NY Acad Sci. Res: Experimental hematology; transplantation. Mailing Add: Dept of Biol Anna Maria Col Paxton MA 01612

COLE, EVELYN, b Aberdeen, Miss, June 10, 10. ZOOLOGY. Educ: Miss Univ Women, AB, 32; Duke Univ, MA, 43; Vanderbilt Univ, PhD(biol), 66. Prof Exp: From instr to asst prof biol, Greensboro Col, 45-57; from asst prof to prof, 60-75, EMER PROF BIOL, MURRAY STATE UNIV, 75- Mem: Am Soc Zoologists; Soc Syst Zool; Am Inst Biol Sci; Am Ornithologists Union. Res: Taxonomy and ecology of Ostracoda; taxonomy of invertebrates. Mailing Add: 1703 Ryan Ave Murray KY 42071

COLE, FRANCIS TALMAGE, b Lynbrook, NY, Oct 6, 25; m 55; c 4. PHYSICS. Educ: Oberlin Col, AB, 47; Cornell Univ, PhD(physics), 53. Prof Exp: Asst physics, Cornell Univ, 47-51; from instr to assoc prof, Univ Iowa, 51-64; physicist, Lawrence Radiation Lab, 64-67; PHYSICIST, NAT ACCELERATOR LAB, 67- Concurrent Pos: Physicist, Midwestern Univs Res Asn, 55-59, head theory sect, 59-60, head physics div, 60-64. Mem: AAAS; Am Phys Soc. Res High energy particle accelerators. Mailing Add: PO Box 500 Batavia IL 60510

COLE, FRANKLIN RUGGLES, b Newton, Mass, Aug 16, 25; m 47; c 2. PHARMACOGNOSY Educ: Mass Col Pharm, BS, 51, MS, 53; Univ Utah, PhD(pharmacog), 56. Prof Exp: From asst prof to assoc prof, 56-65, PROF PHARMACOG, IDAHO STATE UNIV, 65- Concurrent Pos: Fulbright lectr, Cairo Univ, 65 Mem: AAAS; Am Soc Pharmacog; Am Pharmaceut Asn; Sigma Xi. Res: Re-evaluation of ethnobotanical drugs of the Bannock-Shoshone Indians Mailing Add: Col of Pharm Idaho State Univ Pocatello ID 83201

COLE, GEORGE CHRISTOPHER, b Brooklyn, NY, Oct 12, 29; m 55; c 5. MICROBIOLOGY Educ: St John's Univ, NY, BS, 50, MS, 52; Univ Wis, PhD(bact), 57. Prof Exp: Lab asst parasitol, NY Univ-Bellevue Med Ctr, 51-52; asst virol, Univ Wis, 54-57; sr res scientist microbiol, E R Squibb & Sons, 57-66; SR RES MICROBIOLOGIST, PARKE, DAVIS & CO, DETROIT, 66- Mem: AAAS; Am Soc Microbiol. Res: Chemotherapy of viral, bacterial and parasitic infections; pleuropneumonia-like organisms of swine and chickens; colicines and phage; viral interference; Rubella vaccines; bacterial vaccines; viral and bacterial serology and immunology. Mailing Add: 3834 Quarton Rd Bloomfield Hills MI 48013

COLE, GEORGE DAVID, b Minden, La, June 23, 25; m 47; c 3. CHEMICAL PHYSICS. Educ: Northwestern State Col, La, BS, 50; Univ Ala, PhD(physics), 63. Prof Exp: Asst prof physics & math, Nicholls State Col, 54-60; from asst prof to assoc prof, 64-72, from actg head to head dept, 68-72, PROF PHYSICS & CHMN DEPT PHYSICS & ASTRON, UNIV ALA, 72- Concurrent Pos: Physics consult, Insts Int Educ, E & W Pakistan, 67-70. Mem: Am Phys Soc; Am Asn Physics Teachers. Res: Positron annihilation in liquid crystalline compounds; partial L- and M-shell fluorescence yields; polymorphic and mesomorphic behavior in organic compounds. Mailing Add: Dept of Physics & Astron Univ of Ala Box 1921 University AL 35486

COLE, GEORGE ROLLAND, b Lincoln, Kans, Dec 12, 25; m 50; c 4. SOLID STATE SCIENCE. Educ: Univ Kans, BS, 49, MA, 54, PhD(physics), 57. Prof Exp: Asst instr physics, Univ Kans, 51-56; res engr, Savannah River Lab, 56-65, res physicist, Electrochem Dept, Wilmington, Del, 65-69, qual control supvr, Electronic Prod Div, Niagara Falls, 69-73, SR RES PHYSICIST, PHOTO PROD DEPT, E I DU PONT DE NEMOURS & CO, INC, 73- Mem: Am Phys Soc. Res: Color centers in alkali halides; properties and irradiation behavior of uranium oxide; solid actinide compounds; thick film electronic components; photographic science; electron spectroscopy. Mailing Add: Exp Sta 352 E I du Pont de Nemours & Co Inc Wilmington DE 19898

COLE, GERALD AINSWORTH, b Hartford, Conn, Dec 25, 17; m 44; c 5. LIMNOLOGY. Educ: Middlebury Col, AB, 39; St Lawrence Univ, MS, 41; Univ Minn, PhD(zool), 49. Prof Exp: Teaching fel biol, St Lawrence Univ, 40-41; teacher chem, Milton Acad & instr biol, Phillips Acad, 42; asst zool, Univ Minn, 46-49; from asst prof to assoc prof biol, Univ Louisville, 49-58; from lectr to assoc prof zool, 58-63, PROF ZOOL, ARIZ STATE UNIV, 63- Concurrent Pos: Res, Douglas Lake State Biol Sta, Mich, 50; mem teaching staff, Lake Itasca Biol Sta, Minn, 64-66, 68 & 70; res, Coahuila, Mex, 67 & 68. Mem: Am Soc Limnol & Oceanog; Micros Soc Am; Ecol Soc Am; Int Asn Theoret & Appl Limnol. Res: Microcrustacea and regional limnology. Mailing Add: 2015 Sierra Vista Tempe AZ 85281

COLE, GERALD ALAN, b West New York, NJ, June 11, 31; m 58; c 3. VIROLOGY, IMMUNOLOGY. Educ: Wilson Teachers Col, BS, 52; Univ Md, PhD(microbiol), 66. Prof Exp: Med bacteriologist, Walter Reed Army Inst Res, 55-59; virologist, Sch Med, Univ Md, 59-60, res assoc microbiol, 66-67; from asst prof to assoc prof, 67-74, PROF EPIDEMIOL, SCH HYG & PUB HEALTH, JOHNS HOPKINS UNIV, 74- Concurrent Pos: USPHS res grant, Sch Hyg & Pub Health, Johns Hopkins Univ, 70-78, res career develop award, 71-76; Josiah Macy Jr Found fac scholar award, John Curtin Sch Med Res, Australian Nat Univ, 74-75; mem ed hoc study group virus & rickettsial dis, US Army Res & Develop Command, 73-; consult to Surgeon Gen, 73- Mem: AAAS; Am Soc Microbiol; Am Asn Immunol; Am Soc Trop Med & Hyg. Res: Role of the immune response in the outcome of viral infections. Mailing Add: Sch of Hyg & Pub Health Johns Hopkins Univ Baltimore MD 21205

COLE, HAROLD HARRISON, b Waterloo, Wis, Feb 11, 97; m 29, 55; c 2. REPRODUCTIVE ENDOCRINOLOGY. Educ: Univ Wis, BS, 20; Univ Calif, MS, 25; Univ Minn, PhD(physiol), 28. Hon Degrees: LLD, Univ Calif, Davis, 65. Prof Exp: Asst anat, Univ Calif, 26; asst dairy husb, Univ Minn, 27-28; from instr to assoc prof, 28-43, chmn dept, 52-60, PROF ANIMAL HUSB, UNIV CALIF, DAVIS, 43- Concurrent Pos: Fac res lectr, Univ Calif, Davis, 43. Honors & Awards: Morrison Award, 52; Endocrine & Physiol Award, Am Soc Animal Sci, 63. Mem: AAAS; Am Asn Anat; Soc Exp Biol & Med; Am Soc Animal Sci; Soc Study Reprod. Res: Endocrine control of reproductive processes with special reference to domestic animals; use of antigonadotropins in inhibiting fertility. Mailing Add: Dept of Animal Sci Univ of Calif Davis CA 95616

COLE, HAROLD S, b Brooklyn, NY, Apr 20, 16. MEDICINE. Educ: Univ Md, BS, 37; NY Univ, MD, 42; Am Bd Pediat, dipl, 49. Prof Exp: Assoc prof, 48-74, PROF PEDIAT & CHIEF, SECT METAB, DEPT PEDIAT, NY MED COL, 74- Concurrent Pos: Assoc attend pediatrician, Flower & Fifth Ave Hosps, New York, 48-74, attend pediatrician, 74-; assoc vis pediatrician, Metrop Hosp, 48-73, vis pediatrician, 73-, head, pediat diabetes clin. Mem: Fel Am Acad Pediat; Am Pediat Soc; Am Diabetes Asn. Res: Adolescent medicine; diabetes mellitus in children; the infant of the diabetic mother. Mailing Add: Dept of Pediat NY Med Col New York NY 10029

COLE, HARVEY E, b Otoe Co, Nebr, Dec 18, 10; m 38; c 2. ECOLOGY. Educ: Nebr State Teachers Col, Peru, AB, 32; Univ Denver, MA, 36. Prof Exp: Supt sch, Nebr, 33-35; teacher pub schs, 36-40, 41-46; asst prof, 47-59, ASSOC PROF BIOL, KEARNEY STATE COL, 59- Mem: AAAS; Am Bot Soc; Am Inst Biol Sci. Res: Plant ecology. Mailing Add: Dept of Biol Kearney State Col Kearney NE 68847

COLE, HENDERSON, b Wilmington, NC, Oct 2, 24; m 50; c 3. PHYSICS. Educ: Mass Inst Technol, BS, 50, PhD(physics), 52. Prof Exp: Asst physics, Mass Inst Technol, 49-52, instr, 53-55; Fulbright fel, Col France, 52-53; STAFF PHYSICIST, RES CTR, IBM CORP, 55- Mem: Am Phys Soc; Sigma Xi; Am Crystallog Asn. Res: Supervising service work; solid state physics; x-ray diffraction; computer control of instruments. Mailing Add: Old Shop Rd Cross River NY 10518

COLE, HERBERT, JR, b Long Island, NY, Mar 29, 33. PLANT PATHOLOGY, AGRICULTURAL CHEMISTRY. Educ: Pa State Univ, BS, 54, MS, 55, PhD(plant path, agr biochem), 57. Prof Exp: From asst prof to assoc prof plant path, 57-66, agr chem coord, Col Agr, 64-66, assoc prof plant path & chem pesticides, 66-70, PROF PLANT PATH & CHEM PESTICIDES, PA STATE UNIV, 70- Mem: Am Phytopath Soc; Potato Asn Am. Res: Side effects of pesticides; control of plant diseases; biological recycling and disposal of solid and liquid organic wastes. Mailing Add: 211 Buckhout Lab PA State Univ University Park PA 16802

COLE, JACK ROBERT, b Milwaukee, Wis, Dec 28, 29; m 52; c 3. MEDICINAL CHEMISTRY. Educ: Univ Ariz, BS, 53; Univ Minn, PhD(pharmaceut chem), 57. Prof Exp: From asst prof to assoc prof, 57-62, PROF MED CHEM, UNIV ARIZ, 62-, HEAD DEPT PHARMACEUT SCI, COL PHARM, 75- Res: Chemistry of natural medicinal products; chromatography; synthetic organic medicinals. Mailing Add: Col of Pharm Univ of Ariz Tucson AZ 85721

COLE, JACK WESTLEY, b Portland, Ore, Aug 28, 20; m 43; c 4. EXPERIMENTAL SURGERY. Educ: Univ Ore, AB, 41; Wash Univ, MD, 44; Am Bd Surg, dipl, 53. Prof Exp: From jr instr to sr instr surg, Western Reserve Univ, 52-54, asst surgeon, Univ Hosps, 54-56, from assoc prof to prof surg, Sch Med, 56-63; prof & chmn dept, Hahnemann Med Col & Hosp, 63-66; chmn dept surg, 66-74, ENSIGN PROF, DEPT SURG, YALE UNIV SCH MED, 66-; DIR, DIV ONCOL & YALE COMPREHENSIVE CANCER CTR, 75- Concurrent Pos: Chief, Hahnemann Serv, Div B, Dept Surg, Philadelphia Gen Hosp, 63-66; consult, Vet Admin Hosp, Philadelphia, 63-66; chief surg, Yale New Haven Hosp, 66-74; consult, West Haven Vet Hosp, 66- Mem: Soc Exp Biol & Med; Am Fedn Clin Res; fel Am Col Surg; Soc Univ Surg; Am Surg Asn. Res: Cellular kinetics of gastrointestinal epithelium, normal and neoplastic. Mailing Add: Dept of Surg Yale Univ Sch of Med New Haven CT 06520

COLE, JAMES A, b Albany, NY, Nov 18, 39; m 63; c 2. EXPERIMENTAL PHYSICS, SYSTEMS ENGINEERING. Educ: Union Col, NY, BS, 61; Johns Hopkins Univ, PhD(physics). Prof Exp: Instr physics, Johns Hopkins Univ, 65-66; res assoc, State Univ NY Stony Brook, 66-67, asst prof, 67-71; mgr systs eng, Elsytec Inc, NY, 71-74; MGR SYSTS ENG, MEGADATA INC, BOHEMIA, 74- Concurrent Pos: Guest res assoc, Brookhaven Nat Lab, 63-71. Mem: AAAS; Am Phys Soc; Inst Elec & Electronics Engrs. Res: Elementary particle physics; sound and vibration analysis. Mailing Add: 20 Upper Sheep Pasture Rd East Setauket NY 11733

COLE, JAMES EDWARD, b Detroit, Mich, Sept 10, 40; m 60; c 2. ETHOLOGY, VERTEBRATE ZOOLOGY. Educ: Western Mich Univ, BA, 62, MA, 63; Ill State Univ, PhD(zool), 68. Prof Exp: Instr biol, Highland Park Col, 63-64; fac asst zool, Ill State Univ, 65-67; assoc prof, 68-71, PROF BIOL, BLOOMSBURG STATE COL, 71-, PROG COORDR HEALTH SCI, 73- Mem: AAAS; Am Soc Zool; Animal Behav Soc; Am Inst Biol Sci. Res: Parent-young interactions of Cichlid fishes; behavior of lower vertebrates. Mailing Add: Dept of Biol Bloomsburg State Col Bloomsburg PA 17815

COLE, JAMES WEBB, JR, b Norfolk, Va, July 22, 10; m 36; c 2. CHEMISTRY. Educ: Univ Va, BS, 32, MS, 34, PhD(phys chem), 36. Prof Exp: Res chemist, Exp Sta, E I du Pont de Nemours & Co, 36-37; from asst prof to assoc prof, 37-56, PROF CHEM, UNIV VA, 56-, DEAN, 58- Concurrent Pos: Instr, USN, 44; prog dir, NSF, 52-53, mem adv panel, 53-; consult, Off Sci Res & Develop; investr, Nat Defense Res Comt. Mem: AAAS; fel Am Inst Chem; Am Chem Soc. Res: Thermal measurements; adsorption of gases on solids; vapor phase catalytic oxidation and hydrolysis; new analytical methods; gas analysis; synthesis of complex inorganic compounds and of organo metallic compounds; spectrophotometer used in analysis and in problems of structure; mechanisms of antioxidants and anticorrosion agents. Mailing Add: 900 Rosser Lane Charlottesville VA 22903

COLE, JEROME F, b Cincinnati, Ohio, Aug 8, 40; m 63; c 2. ENVIRONMENTAL HEALTH, PHARMACY. Educ: Univ Cincinnati, BS, 62, MS, 66, ScD(environ health), 68. Prof Exp: Pharmacist, Fidelity Prescriptions, Inc, 63-64; indust hygienist, Procter & Gamble Co, 68-69; mgr environ health res, 69-73, DEP DIR, INT LEAD ZINC RES ORGN, 73-, V PRES, 75- Mem: Am Indust Hyg Asn; Air Pollution Control Asn; Soc Environ Geochem & Health; Soc Toxicol. Res: Biochemistry of insecticides; toxicology and environmental health impact of the use of lead, zinc and cadmium in industry and by consumers. Mailing Add: Int Lead Zinc Res Orgn Inc 292 Madison Ave New York NY 10017

COLE, JERRY JOE, b Kansas City, Mo, May 22, 38; m 61; c 2. ANALYTICAL CHEMISTRY. Educ: Univ Kansas City, BS, 60; Univ Iowa, MS, 63, PhD(anal chem), 65. Prof Exp: Asst prof chem, Ft Hays Kans State Col, 64-67; assoc prof, 67-74, PROF CHEM, ASHLAND COL, 74- Mem: Am Chem Soc. Res: Complexing of metal ions with organic and inorganic moieties; organic precipitants as a means of gravimetric estimation; amperometric titrations of inorganic ions. Mailing Add: Dept of Chem Ashland Col Ashland OH 44805

COLE, JOHN OLIVER, b Jamestown, NY, Apr 15, 15; m 44; c 2. CHEMISTRY. Educ: Bethany Col, Kans, BS, 39; Univ Colo, MS, 41, PhD(org chem), 43. Prof Exp: Asst chem, Univ Colo, 40-43, asst chem eng, 43; res chemist, 43-57, head anal sect, 57-65, MGR ANAL CHEM, GOODYEAR TIRE & RUBBER CO, 65- Mem: Am Chem Soc. Res: Organic and analytical chemistry; reaction of phenyl glyoxal with aliphatic amidines; autoxidation of elastomers; rubber chemicals. Mailing Add: Res Div Goodyear Tire & Rubber Co Akron OH 44316

COLE, JOHN RUFUS, JR, b Baconton, Ga, Oct 27, 38; m 64; c 2. VETERINARY MICROBIOLOGY. Educ: Univ Ga, BS, 60, MS, 63, PhD(microbiol), 66. Prof Exp: Res asst microbiol, Poultry Dis Res Ctr, 61-63, MICROBIOLOGIST, DIAG & RES LABS, COL VET MED, UNIV GA, 66- Mem: Am Soc Microbiol. Res: Blood chemistry of chicks infected or endointoxicated with Escherichia coli; fat absorption from the small intestine of gnotobiotic chicks; efficacy of Pasteurella multocida bacterins in turkeys; fluorescent antibody tests for animal diseases; Leptospirosis. Mailing Add: Vet Diag & Investigational Labs Univ of Ga PO Box 1389 Tifton GA 31794

COLE, JOHN WALLACE, b Oshkosh, Wis, July 7, 34; m 56; c 2. CULTURAL ANTHROPOLOGY, ECONOMIC ANTHROPOLOGY. Educ: Univ Mich, AB, 57, MA, 63, PhD(anthrop), 69. Prof Exp: Instr anthrop, Wayne State Univ, 65; res asst, Univ Mich, 65-67; asst prof, Wayne State Univ, 67-71; asst prof, 71-73, ASSOC PROF ANTHROP, UNIV MASS, AMHERST, 73- Concurrent Pos: Fulbright-Hays sr res fel, 74; Ford Found res grant, 75. Mem: AAAS; fel Am Anthrop Asn; fel Royal Anthrop Inst Gt Brit & Ireland. Res: Peasant socio-economic organization and cultural ecology; political economy of underdeveloped societies; land tenure systems and inheritance; Socialist transformation in Southern Transylvania, Romania. Mailing Add: Dept of Anthrop Univ of Mass Amherst MA 01002

COLE, JOHNNETTA B, b Jacksonville, Fla, Oct 19, 36; m 60; c 2. ANTHROPOLOGY. Educ: Oberlin Col, BA, 57; Northwestern Univ, MA, 59, PhD(anthrop), 67. Prof Exp: Res assoc anthrop, Northwestern Univ Econ Surv Liberia, WAfrica, 60-62; instr, Wash State Univ, 63-65, collabr, 66-67; asst prof, 67-71, ASSOC PROF AFRO-AM STUDIES, UNIV MASS, AMHERST, 71- Concurrent Pos: Consult & lectr, Peace Corps Training Proj, San Francisco State Col, 63-65. Mem: Fel Am Anthrop Asn; Int African Inst. Res: African and Liberian ethnology. Mailing Add: Dept of Anthrop Machmer Hall Univ of Mass Amherst MA 01002

COLE, KENNETH STEWART, b Ithaca, NY, July 10, 00; m 32; c 2. BIOPHYSICS. Educ: Oberlin Col, AB, 22; Cornell Univ, PhD(exp physics), 26. Hon Degrees: ScD, Oberlin Col, 54 & Univ Chicago, 67; MD, Univ Uppsala, 67. Prof Exp: Instr, Cornell Univ, 22-26, Nat Res Coun fel, 26-29; from asst prof to assoc prof physiol, Col Physicians & Surgeons, Columbia Univ, 29-46; prin biophysicist, Metall Lab, Univ Chicago, 42-46, prof biophys, 46-49; dir, Naval Med Res Inst, 49-54; chief lab biophys, Nat Inst Neurol Dis & Blindness, NIH, 54-66, SR RES BIOPHYSICIST, NAT INST NEUROL & COMMUN DISORDERS & STROKE, NIH, 66- Concurrent Pos: Mem staff, Cold Spring Harbor, 34-37, mem bd, 40-45; Guggenheim fel, Inst Advan Study, 42; trustee, Marine Biol Lab, Woods Hole, 47-55, 56-64, emer trustee, 66-, mem exec comt, 56-59, 60-64; Regents prof, Univ Calif, Berkeley, 63-64, prof, 65- Priestley lectr, Pa State Col, 39; Tennant lectr, Bryn Mawr Col, 41; guest lectr, Pa State Univ, 61 & Yale Univ, 62; Whitehead lectr, Nat Acad Sci-Nat Res Coun Conf Elec Insulation, 64; vis prof, Univ Tex Med Br Galveston, 74. Honors & Awards: Order of the Southern Cross, Brazil, 66; Nat Medal Sci, 67; Bicentennial Medal, Col Physicians & Surgeons, Columbia Univ, 67. Mem: Nat Acad Sci; fel Am Phys Soc; Biophys Soc (pres elect, 62-63, pres, 63-64); hon mem Brazilian Biol Soc; for mem Royal Soc London. Res: Photographic action of electrons; heat production and surface force of Arbacia eggs; electrical impedance of tissues, cells, natural and artificial membrane; electrical analysis of nerve impulse; ionic conductances of squid axon membrane. Mailing Add: Nat Inst of Neurol & Commun Disorders & Stroke NIH Bethesda MD 20014

COLE, LAMONT COOK, b Chicago, Ill, July 15, 16; m 40; c 2. ZOOLOGY. Educ: Univ Chicago, SB, 38, PhD(zool), 44; Univ Utah, MS, 40. Hon Degrees: ScD, Univ Vt, 69 & Ripon Col, 71. Prof Exp: Asst zool, Univ Chicago, 41-44; asst sanitarian, USPHS, Washington, DC, 44-45, sr asst sanitarian, 45-46; from instr to asst prof zool, Ind Univ, 46-48; from asst prof to assoc prof, 48-53, chmn dept, 64-66, PROF ZOOL, CORNELL UNIV, 53- Concurrent Pos: Consult, NIH, USPHS & Off Naval Res; asst, US Army specialized training prog, 43-44; mem panel environ biol, NSF Comt Prog, AAAS, 62-64; rev ed zool, Ecol; assoc ed, Ecol Monographs. Mem: AAAS; Am Soc Naturalists; Am Soc Zool; Ecol Soc Am (vpres, 63, pres, 68); Am Inst Biol Sci (vpres, 68, pres, 69). Res: Autecology of reptiles; field study invertebrate populations; cryptozoic fauna of an Illinois woodland; laboratory populations; mathematical analysis of populations; population cycles. Mailing Add: 241 Langmuir Lab Cornell Univ Ithaca NY 14850

COLE, LEONARD JAY, biological chemistry, deceased

COLE, MADISON BROOKS, JR, b Worcester, Mass, Aug 30, 40; m 67; c 1. CELL BIOLOGY, ORTHOPEDICS. Educ: Colgate Univ, AB, 62; Univ Tenn, PhD(zool), 69. Prof Exp: Res asst biol, Brown Univ, 62-63 & Univ Tenn, 64-68; from instr to asst prof biol sci, Oakland Univ, 68-74; spec trainee biophys, NY Univ, 74-75; ASST PROF ORTHOPEDIC SURG, LOYOLA UNIV MED CTR, 75- Mem: AAAS; Am Inst Biol Sci; Electron Micros Soc Am; Am Soc Cell Biol. Res: Electron microscopy; cytochemistry; nuclear-cytoplasmic relationships; cell division; growth and differentiation of bone; electro-magnetic effects on living systems. Mailing Add: Dept of Orthopedic Surg Loyola Univ Med Ctr 2160 S1st Ave Maywood IL 60153

COLE, MICHAEL ALLEN, b Denver, Colo, Dec 15, 43. MICROBIAL GENETICS, SOIL MICROBIOLOGY. Educ: Cornell Univ, BS, 67; NC State Univ, Raleigh, MS, 71, PhD(microbiol), 72. Prof Exp: Asst prof microbiol, Southern Ill Univ, Edwardsville, 72-73; ASST PROF SOIL MICROBIOL, UNIV ILL, URBANA, 74- Mem: Sigma Xi; Am Soc Microbiol. Res: Genetics of Rhizobium, particularly plasmid genetics; effects of agricultural chemicals and pollutants on soil micro-organisms. Mailing Add: Dept of Agron Univ of Ill Urbana IL 61801

COLE, MILTON WALTER, b Washington, DC, Dec 14, 42. SOLID STATE PHYSICS, LOW TEMPERATURE PHYSICS. Educ: Johns Hopkins Univ, BA, 64; Univ Chicago, MS, 65, PhD(physics), 70. Prof Exp: Fel physics, Univ Toronto, 70-72; res assoc, Univ Wash, 72-74; asst prof, Pa State Univ, 74-75; ASSOC PROF PHYSICS, BROOKLYN COL, 75- Concurrent Pos: Consult, Jet Propulsion Lab, Calif Inst Technol, 74- Mem: Am Phys Soc. Res: Surface physics, especially thin films, liquid helium and electronic properties; inhomogenous quantum systems. Mailing Add: Dept of Physics Brooklyn Col Brooklyn NY 11210

COLE, MONROE, b New York, NY, Mar 21, 33; m 58; c 4. NEUROLOGY. Educ: Amherst Col, AB, 53; Georgetown Univ, MD, 57. Prof Exp: From intern to asst resident, Seton Hall Col Med, 57-59; asst resident & fel, Mass Gen Hosp, 59-60; neurologist & neuropathologist, Walter Reed Army Inst Res, 62-65; from asst prof to assoc prof neurol, Bowman Gray Sch Med, 65-70; assoc prof, 70-72, ASSOC CLIN PROF NEUROL, SCH MED, CASE WESTERN RESERVE UNIV, 72-; CHIEF NEUROL, HIGHLAND VIEW HOSP, 70- Concurrent Pos: Teaching fel, Harvard Univ, 59-62; fels, Mass Gen Hosp, 60-62. Mem: Acad Aphasia; fel Am Acad Neurol; fel Am Col Physicians; Asn Res Nerv & Ment Dis; Am Fedn Clin Res. Res: Clinical neurology; experimental neuropathology. Mailing Add: River Rd Gates Mills OH 44040

COLE, MORTON S, b Chicago, Ill, June 16, 29; m 52; c 3. FOOD TECHNOLOGY. Educ: Univ Ill, BS, 51, MS, 53; Iowa State Univ, PhD(food technol), 61. Prof Exp: Scientist grocery prod develop, Pillsbury Co, 59-61, dried prod develop, 61-63; food technologist, Archer-Daniels-Midland Co, 63-64, group leader edible prod, 64-68; dir res, Paniplus Co, Div ITT Continental Baking Co, 68-74; ASSOC DIR RES, ARCHER-DANIELS-MIDLAND CO, 74- Mem: Inst Food Technologists; Am Asn Cereal Chem; Am Oil Chem Soc. Res: Porphyrin pigments in fresh and processed meats; lipid oxidation; pectic enzymes; vegetable dehydration; edible oxygen and moisture barriers; fabricated foods from vegetable proteins; food color, flavor and texture; bakery ingredients; chemical surfactants. Mailing Add: Archer-Daniels-Midland Co PO Box 1470 Decatur IL 62525

COLE, NANCY, b Boston, Mass, Oct 15, 02. MATHEMATICS. Educ: Vassar Col, AB, 24; Radcliffe Col, AM, 29, PhD(math), 34. Prof Exp: Instr, Oxford Sch, 24-26 & Vassar Col, 27-28; tutor, Radcliffe Col, 28-29; instr, Wells Col, 31-32; instr math, Sweet Briar Col, 33-42; from vis asst prof to vis assoc prof, Kenyon Col, 43-44; asst prof, Conn Col, 44-47; from asst prof to assoc prof, 47-71, EMER ASSOC PROF MATH, SYRACUSE UNIV, 71- Concurrent Pos: Actg head dept, Sweet Briar Col, 34-35 & 41-42. Mem: AAAS; Am Math Soc; Math Asn Am. Res: Calculus of variations; index form associated with an extremaloid. Mailing Add: 1214 Westcott St Syracuse NY 13210

COLE, NYLA J, b Wesco, Calif, Dec 5, 25; m 55. PSYCHIATRY. Educ: Univ Calif, Berkeley, BA, 47; Univ Rochester, MD, 51. Prof Exp: Intern, 52, resident psychiat, 55, resident instr, 56-60, dir outpatient div, 56-62, asst prof, 60-68, dir adult psychiat, 62-65, ASSOC PROF PSYCHIAT, COL MED, UNIV UTAH, 68-, DIR PSYCHIAT OUTPATIENT DEPT, 70- Concurrent Pos: Lectr, Dept Social Work, 61-; mem, President's Comt Employ Handicapped, 67-72; chmn med subcomt, Gov Comt Employ Handicapped, 68-72. Mem: Fel Am Psychiat Asn; AMA. Res: Natural history of psychiatric disease; mental health care delivery systems; social survival of discharged patients. Mailing Add: Dept of Psychiat Univ of Utah Col of Med Salt Lake City UT 84112

COLE, QUINTIN PERRY, b Ballston Spa, NY, Aug 31, 17; m 46; c 3. CHEMISTRY. Educ: Union Univ, NY, BS, 40; Yale Univ, PhD(org chem), 43. Prof Exp: Res chemist, Am Cyanamid Co, Conn, 43-46; instr chem, Conn, 46-47; instr, Trinity Col, Conn, 47-51; res chemist, 51-61, mgr chem & insulation, 61-69, MGR ADVAN DEVELOP LAB, GEN ELEC CO, 69- Mem: AAAS; Am Chem Soc; Am Inst Chem. Res: Synthesis of organic compounds; silicones; irradiation of polymers; production of hyperpure silicon; thermally stable polymers. Mailing Add: Gen Elec Co Bldg 9 3001 E Lake Rd Erie PA 16501

COLE, RANDAL HUDIE, b Clinton, Ont, Nov 6, 08; m 40; c 3. MATHEMATICS. Educ: Univ Western Ont, BA, 36; Univ Wis, AM, 37, PhD(math), 40. Prof Exp: Asst math, Univ Wis, 38-40; from instr to prof, 40-74, EMER PROF MATH, UNIV WESTERN ONT, 74- Concurrent Pos: Res assoc, Princeton Univ, 48-49. Mem: Am Math Soc; Inst Math Statist; Math Asn Am. Res: Analysis; statistics. Mailing Add: RR1 Arva ON Can

COLE, RANDALL KNIGHT, b Putnam, Conn, Sept 21, 12; m 39; c 3. ANIMAL GENETICS. Educ: Mass State Col, BS, 34; Cornell Univ, MS, 37, PhD(animal breeding), 39. Prof Exp: Asst animal dis, Univ Conn, 34-35; instr poultry husb, 35-40, asst prof poultry sci & animal genetics, 40-48, from assoc prof to prof, 48-73, EMER PROF POULTRY SCI & ANIMAL GENETICS, CORNELL UNIV, 73- Concurrent Pos: Animal geneticist, Exp Sta, NY State Col Agr, Cornell Univ, 40-73; consult, Shaver Poultry Breeding Farms, Ltd, Cambridge, Ont, 56-; hon mem, 1st World Cong Genetics Appl to Livestock Prod, Madrid, 74. Honors & Awards: Poultry Sci Asn Res Award, 49; Tom Newman Mem Int Award, 69. Mem: AAAS; Am Genetic Asn; Poultry Sci Asn; World Poultry Sci Asn; Am Inst Biol Sci. Res: Genetics of disease resistance in poultry, especially to neoplasms; avian genetics; animal models of specific diseases. Mailing Add: Dept of Poultry Sci Cornell Univ Ithaca NY 14853

COLE, RAYMOND THOMAS, physical organic chemistry, see 12th edition

COLE, RICHARD, b New York, NY, Apr 16, 24; m 47; c 2. PHYSICAL CHEMISTRY, ENVIRONMENTAL MANAGEMENT. Educ: City Col New York, BChE, 44; Univ Ill, MS, 48, PhD(phys chem), 52. Prof Exp: Asst phys chem, Univ Ill, 47-49, 50-51; radiol chemist, US Naval Radiol Defense Lab, 52-61, head countermeasures eval br, 61-64, chem tech div, 64-66 & nuclear tech div, 66-69; sr res assoc, 69-74, VPRES, ENVIRON SCI ASSOCS, FOSTER CITY, 74- Mem: Am Chem Soc; Opers Res Soc Am; Asn Environ Prof. Res: Formation, transport, deposition and removal of radioactive and other contamination; on-site inspection for clandestine nuclear operations; radioactivity in the oceans; environmental impact analysis and mitigation Mailing Add: 1431 Tarrytown St San Mateo CA 94402

COLE, RICHARD ALLEN, b Suffern, NY, Oct 27, 42. AQUATIC ECOLOGY. Educ: State Univ NY Col Forestry, Syracuse Univ, BS, 64, MS, 66; Pa State Univ, PhD(zool), 69. Prof Exp: Res assoc, 69-73, ASST PROF AQUATIC ECOL, MICH STATE UNIV, 73- Mem: Am Soc Limnologists & Oceanogr; Am Fisheries Soc; Int Soc Limnol; Ecol Soc Am; NAm Benthological Soc. Res: Aquatic community ecology, particularly as influenced by eutrophication, thermal discharge and other watershed disturbances. Mailing Add: Dept of Fisheries & Wildlife Mich State Univ East Lansing MI 48824

COLE, RICHARD H, b Woodstock, Maine, Mar 2, 30; m 53; c 4. AGRONOMY. Educ: Univ Maine, BS, 52; Pa State Univ, PhD, 60. Prof Exp: Assoc county agent crops, Mass Exten Serv, 54-55; asst prof agron & agr eng, Univ Del, 60-65, chmn dept, 65-68, assoc prof plant sci, 68-70; ASSOC PROF INT AGRON, PA STATE UNIV, 70- Concurrent Pos: Asst coordr, Int Agr Prog, 73- Mem: Am Soc Agron; Crop Sci Soc Am; Weed Sci Soc Am. Res: Crop management; seed quality; weed control Mailing Add: 119 Tyson Bldg Pa State Univ University Park PA 16802

COLE, RICHARD WALLACE, physics, see 12th edition

COLE, ROBERT HUGH, b Oberlin, Ohio, Oct 26, 14; m 43. CHEMICAL PHYSICS. Educ: Oberlin Col, AB, 35; Harvard Univ, AM, 36, PhD(physics), 40. Prof Exp: Instr & tutor physics, Harvard Univ, 39-41; res supvr underwater explosives, Oceanog Inst, Woods Hole, 41-46; asst prof physics, Univ Mo, 46-47; from assoc prof to prof, 47-60, chmn dept, 49-61, JESSE H & LOUISA D SHARPE METCALF PROF CHEM, BROWN UNIV, 60- Concurrent Pos: Fulbright lectr & Guggenheim fel, State Univ

Leiden, 55-56; NSF sr fel & Guggenheim fel, Oxford Univ, 61-62; vis prof, Univ Paris, Orsay, 69-70. Honors & Awards: Langmuir Prize Chem Physics, Am Phys Soc, 75. Mem: Fel Am Phys Soc; fel Am Acad Arts & Sci; Am Chem Soc. Res: Dielectric properties of matter; intermolecular forces. Mailing Add: Dept of Chem Brown Univ Providence RI 02912

COLE, ROBERT KLEIV, b San Francisco, Calif, Dec 19, 28; m 53; c 4. PHYSICS. Educ: Univ Calif, Berkeley, AB, 52; Univ Wash, PhD(physics), 59. Prof Exp: Physicist, Hanford Works, Gen Elec Corp, Wash, 52-54; vis asst prof, Univ Calif, Los Angeles, 59-62; asst prof, 62-66, ASSOC PROF PHYSICS, UNIV SOUTHERN CALIF, 66- Mem: Am Phys Soc. Res: Nuclear scattering; nuclear reactions studies; nuclear physics instrumentation. Mailing Add: Dept of Physics Univ of Southern Calif Los Angeles CA 90007

COLE, ROBERT STEPHEN, b Los Angeles, Calif, Apr 10, 43; m 65; c 2. SOLAR PHYSICS, QUANTUM PHYSICS. Educ: Univ Calif, Berkeley, AB, 65; Univ Wash, MS, 67; Mich State Univ, PhD(physics), 72. Prof Exp: Instr physics, St Martins Col, 67-69; ASST PROF PHYSICS, UNIV NC, ASHEVILLE, 72- Concurrent Pos: Consult solar energy, Asheville Orthop Hosp & Rehab Ctr, 75- Mem: Am Asn Physics Teachers; Nat Asn Environ Educ; Int Solar Energy Soc. Res: Solar energy research, especially passive systems for residential heating and cooling. Mailing Add: Dept of Physics Univ of NC Asheville NC 28804

COLE, ROGER DAVID, b Berkeley, Calif, Nov 17, 24; m 44; c 3. PROTEIN CHEMISTRY. Educ: Univ Calif, BS, 48, PhD(biochem), 54. Prof Exp: Asst phys chem, Atomic Res Inst, Univ Iowa, 48-49; chemist, Tidewater Oil Co, Calif, 49-51; jr res biochemist, Univ Calif, 54-55; fel, Nat Inst Med Res, London & Nat Found Infantile Paralysis, 55-56; res assoc biochem, Rockefeller Inst, 56-58; from asst prof to assoc prof, 58-65, chmn dept, 68-73, PROF BIOCHEM, UNIV CALIF, BERKELEY, 65- Concurrent Pos: Guggenheim fel, Cambridge Univ, 66-67. Mem: AAAS; Am Soc Biol Chem; Am Chem Soc. Res: Protein and peptide isolation and structural determination; relation of structure and biological activity of enzymes and hormones; protein biosynthesis; histones Mailing Add: Dept of Biochem Univ of Calif Berkeley CA 94720

COLE, ROGER M, b Akron, Ohio, Nov 5, 34; m 58; c 2. INORGANIC CHEMISTRY. Educ: Kent State Univ, BS, 56; Univ Minn, MS, 58. Prof Exp: From res chemist to sr res chemist, 58-69, HEAD PROCESSING RES LAB, EASTMAN KODAK CO, 69- Mem: Soc Photog Sci & Eng. Res: Mechanism of photographic development; photographic chemistry of silver ion complexes; mechanism and application of physical development; methods to simplify photographic processing. Mailing Add: Eastman Kodak Co Res Labs 1669 Lake Ave Rochester NY 14650

COLE, RONALD SINCLAIR, b San Pedro, Calif, Mar 8, 40. MOLECULAR BIOLOGY, RADIOBIOLOGY. Educ: Univ Calif, AB, 63; Calif Inst Technol, PhD(chem), 67. Prof Exp: Res assoc radiobiol, Sch Med, Yale Univ, 67-72, res assoc molecular biol, Dept Molecular Biophys & Biochem, 68-72; asst prof, 72-74, ASSOC PROF BIOCHEM & MICROBIOL, UNIV GA, 74- Concurrent Pos: NIH Cancer Agency career develop res award, 75-80. Mem: Am Soc Biol Chemists; Biophys Soc; Am Soc Microbiol; Am Chem Soc. Res: Molecular mechanisms of DNA repair and genetic recombination; drug interactions with DNA; photobiology of nucleic acids. Mailing Add: Dept of Biochem Univ of Ga Athens GA 30602

COLE, TERRY, b Albion, NY, Mar 28, 31; m 55; c 3. CHEMICAL PHYSICS. Educ: Univ Minn, BS, 54; Calif Inst Technol, PhD(chem), 58. Prof Exp: Asst, Calif Inst Technol, 54-57; mgr chem dept, 59-75, MGR CHEM ENG DEPT, FORD MOTOR CO, 75- Concurrent Pos: Adj prof, Univ Mich. Mem: Am Phys Soc; Biophys Soc. Res: Electron spin and nuclear magnetic resonance; radiation damage; free radical chemistry; quasielastic light scattering; thermo-electric energy conversion. Mailing Add: Ford Res Staff Box 2053 Dearborn MI 48121

COLE, THOMAS A, b Harrisburg, Ill, Jan 9, 36; m 67; c 2. BIOCHEMISTRY, GENETICS. Educ: Wabash Col, BA, 58; Calif Inst Technol, PhD(biochem), 63. Prof Exp: Asst prof, 62-67, ASSOC PROF BIOL, WABASH COL, 67-, CHMN DEPT, 68- Concurrent Pos: Comnr, Comn Undergrad Educ Biol Sci, 68-71; consult-examr, NCent Asn, 72-; panelist, Educ Directorate, NSF, 75- Mem: Fel AAAS; NY Acad Sci. Res: Biochemistry of metamorphosing Drosophila; protein synthesis in Drosophila; centrifugation techniques; indentification of hydrolytic enzymes in substrate-included gels for electrophoresis. Mailing Add: Dept of Biol Wabash Col Crawfordsville IN 47933

COLE, THOMAS WINSTON, JR, b Vernon, Tex, Jan 11, 41; m 64. ORGANIC CHEMISTRY. Educ: Wiley Col, BS, 61; Univ Chicago, PhD(chem), 66. Prof Exp: Asst prof, 66-69, FULLER E CALLAWAY PROF CHEM, ATLANTA UNIV, 69-, CHMN DEPT, 71- Concurrent Pos: Vis prof, Dept Chem, Mass Inst Technol, 73-74. Mem: AAAS; Am Chem Soc; Nat Inst Sci. Res: Chemistry of cubane; small ring compounds; photochemistry; application of gas chromatography-mass spectrometry to problems in clinical chemistry; bio-organic chemistry. Mailing Add: Dept of Chem Atlanta Univ Atlanta GA 30314

COLE, VERNON C, b Wenatchee, Wash, Nov 12, 22; m 48; c 3. SOIL SCIENCE. Educ: Univ Mass, BS, 47, MS, 48; PhD(soil sci), 50. Prof Exp: RES SOIL SCIENTIST, SOIL & WATER CONSERV RES DIV, AGR RES SERV, USDA, 50- Mem: Am Soc Agron; Soil Sci Soc Am; Am Soc Plant Physiol. Res: Soil physical chemistry; plant mineral nutrition. Mailing Add: Plant Sci Bldg Colo State Univ Ft Collins CO 80521

COLE, W STORRS, b Albany, NY, July 16, 02; m 26. GEOLOGY. Educ: Cornell Univ, BS, 25, MS, 28, PhD(micropaleont), 30. Prof Exp: Asst instr hist geol, Cornell Univ, 26; paleontologist, Huasteca Petrol Co, Mex, 26-27; instr physiog & com geog, Cornell Univ, 28-30; paleontologist, Sun Oil Co, Tex, 30-31; from instr to prof geomorphol, Ohio State Univ, 31-45; prof paleont & stratig, 46-68, chmn dept geol, 47-62, EMER PROF PALEONT & STRATIG, CORNELL UNIV, 68- Concurrent Pos: Mem div geol & geog, Nat Res Coun, 44-47; geologist, US Geol Surv, Wash, 47-75; mem, NY State Mus Coun, 58-63. Mem: AAAS; fel Asn Am Geog; fel Am Geol Soc (3rd vpres, 54); fel Paleont Soc (pres, 53); Paleont Res Inst (vpres, 54). Res: Micropaleontology; geomorphology; fossil larger Foraminifera of Florida, Mexico, Trinidad, Panama, Cuba, Guam, Saipan, Bikini and Fiji Islands; stratigraphy and micropaleontology of the Tampico Embayment, Mexico and of Florida; erosion surfaces of the Appalachians. Mailing Add: 310 Fall Creek Dr Ithaca NY 14850

COLE, WALTER EARL, b Hunter, Ark, Aug 5, 21; m 47; c 2. ORGANIC CHEMISTRY. Educ: Ark State Teachers Col, BS, 43; Vanderbilt Univ, MS, 48, PhD(chem), 57. Prof Exp: Instr chem, Vanderbilt Univ, 48-51, asst, 52-53; res chemist tech sect, Film Dept, E I du Pont de Nemours & Co, Inc, Tenn, 53-55, develop group leader, 55-56, develop supvr, Iowa, 56-57, coating area supvr, 57-59, mfg supt, Spruance Film Plant, 59-62, tech supt, 62-64, acetate supt, 64-73; MGR, TEE-PAK INC, 73- Mem: Am Chem Soc; Am Inst Chemists. Res: Synthesis of quinolines; halogenation of heterocyclic compounds; preparation of oximes; Beckman rearrangement; cellulose chemistry; manufacture of packaging films. Mailing Add: Tee-Pak Inc 915 N Michigan Ave Danville IL 61832

COLE, WALTER ECKLE, b Muskogee, Okla, Sept 2, 28; m 55; c 3. ENTOMOLOGY, METHEMATICAL BIOLOGY. Educ: Colo State Univ, BSc, 50, MSc, 55; NC State Univ, PhD, 72. Prof Exp: Entomologist forest insect res, 54-60, actg proj leader pop dynamics res, 60-65, PROJ LEADER POP DYNAMICS RES, US FOREST SERV, 65-, PRIN ENTOMOLOGIST, 70- Concurrent Pos: Counr, Western Forest Insect Work Conf, 70-73; chmn working party, Int Union Forest Res Orgns. Honors & Awards: Superior Serv Award, Forest Serv, US Dept Agr, 55. Mem: Am Statist Asn; Biomet Soc; Entom Soc Am; Japanese Soc Pop Ecol. Res: Population dynamics of forest insects; mensurational aspects of behavioral sampling; analysis and modeling of populations. Mailing Add: Forest Serv Bldg 507 25th St Ogden UT 84401

COLE, WARREN HENRY, b Clay Center, Kans, July 24, 98; m 42. SURGERY. Educ: Univ Kans, BS, 18; Wash Univ, MD, 20. Hon Degrees: DSc, Wash Univ, 67 & Univ Ill, 70. Prof Exp: Assoc prof surg, Sch Med, Wash Univ; prof & head dept, 36-67, EMER PROF SURG & EMER HEAD DEPT, COL MED, UNIV ILL, CHICAGO, 67- Honors & Awards: Distinguished Serv Award, AMA, 66; Nat Award, Am Cancer Soc, 67. Mem: Am Surg Asn (vpres, 47, pres, 59); Soc Univ Surgeons (pres, 40); fel Am Col Surgeons (pres, 56); Am Cancer Soc (pres, 59); hon fel Royal Col Surgeons. Res: Gall bladder; thyroid; bile ducts; development of cholecystography; first aid; operative technique; cancer dissemination; chemotherapy of cancer; textbook of surgery. Mailing Add: 8 W Kensington Rd Asheville NC 28804

COLE, WAYNE, b Indianapolis, Ind, Nov 5, 13; m 36; c 5. NATURAL PRODUCTS CHEMISTRY. Educ: DePauw Univ, AB, 35; Univ Ill, AM, 36, PhD(org chem), 38. Prof Exp: Asst chem, Univ Mich, 38-39; res chemist, Glidden Co, 39-46, from asst dir to dir res, 46-58; proj leader steroid res, 58-72, RES FEL, ABBOTT LABS, 72- Mem: Fel AAAS; Am Chem Soc; Am Oil Chem Soc; Soc Chem Indust; Swiss Chem Soc. Res: Synthesis of carcinogenic hydrocarbons, protein derivatives; sterols and hormones; steroid and lipid research; peptide synthesis. Mailing Add: Org Res Dept Abbott Labs North Chicago IL 60064

COLE, WILBUR VOSE, b Waterville, Maine, Jan 19, 13; m 36; c 3. NEUROANATOMY. Educ: Univ NH, BS, 35; Kirksville Col Osteop, DO, 43; Northeast Mo State Teachers Col, MA, 54. Prof Exp: Lab asst histol, Kirksville Col Osteop, 43-44, instr histol & embryol, Dept Anat, 44, asst prof histol & neuroanat, 44-51; ASSOC PROF ANAT, KANSAS CITY COL OSTEOP MED, 51-, PROF CLIN NEUROL, 69-, DEAN OF THE COL, 73- Mem: AAAS; Photog Biol Asn; Nat Asn Biol Teachers; Sci Res Soc Am; Am Micros Soc. Res: Histopathology; polarized light in photomicroscopy; gallocyanin as a nuclear stain; comparative anatomy and physiology of motor and sensory endings in straited muscle. Mailing Add: Dept of Clin Neurol Kansas City Col of Osteop Med Kansas City MO 64124

COLE, WILFRED Q, b Jackson, Miss, Nov 28, 24; m 49; c 3. PEDIATRICS. Educ: Univ Va, MD, 51; Am Bd Allergy & Immunol, dipl, 74. Prof Exp: From intern to resident pediat, Univ Hosp, Birmingham, Ala, 51-54; from clin instr to clin asst prof, 55-68, CLIN ASSOC PROF PEDIAT, SCH MED, UNIV MISS, 68-, DIR PEDIAT ALLERGY CLIN, 62-, DIR CYSTIC FIBROSIS RES CTR, 62- Concurrent Pos: Pediat allergy fel, Duke Med Ctr, 61; co-chmn clin comt, Nat Cystic Fibrosis Res Found, 66-70. Mem: Am Acad Pediat; Am Col Allergists; Am Acad Allergy. Res: Cystic fibrosis; pediatric allergy. Mailing Add: Cystic Fibrosis Res Ctr Univ of Miss Jackson MS 39201

COLEBROOK, LAWRENCE DAVID, b Helensville, NZ, Dec 29, 30; m 59. PHYSICAL ORGANIC CHEMISTRY. Educ: Univ NZ, BSc, 54, MSc, 55, PhD(chem), 61. Prof Exp: From jr lectr to lectr chem, Univ Auckland, 57-60; fel, Univ Rochester, 61-63, asst prof, 63-69; ASSOC PROF CHEM, SIR GEORGE WILLIAMS UNIV, 69- Mem: Brit Chem Soc; assoc NZ Inst Chem. Res: Nuclear magnetic resonance spectroscopy; infrared spectroscopy; natural products. Mailing Add: Dept of Chem Sir George Williams Univ Montreal PQ Can

COLEGROVE, FORREST DONALD, b Madeira, Ohio, Nov 21, 29; m 56; c 2. APPLIED PHYSICS. Educ: Purdue Univ, BS, 51; Univ Mich, MS, 54, PhD(physics), 60. Prof Exp: MEM TECH STAFF PHYSICS, TEX INSTRUMENTS, INC, 59- Concurrent Pos: Vis scientist, Southwest Ctr Advan Studies, 65-69. Mem: AAAS; Am Phys Soc. Res: Superconducting device research; magnetic resonance; infrared systems; physics of the upper atmosphere. Mailing Add: 15022 N Lakes Dr Dallas TX 75240

COLELLA, DONALD FRANCIS, b Utica, NY, Mar 22, 38; m 69; c 1. PHARMACOLOGY. Educ: Rensselaer Polytech Inst, BS, 61; Col St Rose, MS, 68; Drexel Univ, MBA, 75. Prof Exp: Assoc res biologist pharmacol, Sterling-Winthrop Res Inst, 61-68; assoc pharmacologist, 68-69, pharmacologist, 69-72, SR PHARMACOLOGIST, SMITH KLINE & FRENCH LABS, 72- Mem: NY Acad Sci; Am Chem Soc. Res: Cardiovascular and respiratory pharmacology; pharmacology of cardiac and smooth muscle; adrenergic mechanisms; theory of drug-receptor interactions; structure-activity relationships of medicinal agents. Mailing Add: Smith Kline & French Labs 1500 Spring Garden St PO Box 7929 Philadelphia PA 19101

COLELLA, ROBERTO, b Milano, Italy, May 22, 35; m 60; c 3. SOLID STATE PHYSICS. Educ: Univ Milano, Laurea(physics), 58. Prof Exp: Trainee physics, Nat Nuclear Energy Comt, Casaccia, Rome, Italy, 60-61; staff scientist, Europe Atomic Energy Comn, Common Ctr Res, Ispra, Italy, 61-67; res assoc physics, Cornell Univ, 67-70 & Catholic Univ, Washington, DC, 70-71; ASSOC PROF PHYSICS, PURDUE UNIV, WEST LAFAYETTE, 71- Mem: Am Phys Soc; Am Crystallog Asn; Italian Phys Soc; Europ Phys Soc; Italian Crystallog Soc. Res: Diffraction physics in perfect and imperfect crystals; phonons, charge densities and interferometry. Mailing Add: Dept of Physics Purdue Univ West Lafayette IN 47907

COLEMAN, ALBERT JOHN, b Toronto, Ont, Can, May 20, 18; m 53; c 2. APPLIED MATHEMATICS. Educ: Univ Toronto, BA, 39, PhD, 43; Princeton Univ, MA, 43. Prof Exp: Lectr math, Queen's Col, Can, 43-45; travelling secy, World's Student Christian Fedn, 45-49; from lectr to assoc prof math, Univ Toronto, 49-60; HEAD DEPT MATH, QUEEN'S UNIV, ONT, 60- Mem: Am Math Soc; Math Asn Am. Res: Eddington's fundamental theory; group theory; quantum mechanics. Mailing Add: 108 Albert St Kingston ON Can

COLEMAN, ANNA M, b New Concord, Ohio, Jan 5, 13. INFORMATION SCIENCE. Educ: Geneva Col, BS, 33; Univ Pa, MS, 34; Univ Pittsburgh, PhD(chem), 58. Prof Exp: Teacher high sch, Pa, 34-42; Koppers Co res asst, Mellon Inst, 42-44, jr fel, 44-50; chem librn, 50-60, supvr res info serv, 60-67, MGR TECH INFO SERV, DOW CORNING CORP, 67- Mem: Am Chem Soc; Sigma Xi. Res: Chemical documentation; dipole moments. Mailing Add: Tech Info Servs Dow Corning Corp Midland MI 48640

COLEMAN, ANNETTE WILBOIS, b Des Moines, Iowa, Feb 28, 34; m 58; c 3. CELL BIOLOGY. Educ: Barnard Col, AB, 55; Univ Ind, PhD(bot), 58. Prof Exp: NSF res fel, Johns Hopkins Univ, 58-61; res assoc, Univ Conn, 62-63; res assoc, 63-72, ASST PROF BIOL, BROWN UNIV, 72- Mem: Bot Soc Am; Soc Protozool; Phycol Soc Am; Soc Gen Physiol; NY Acad Sci. Res: Physiological control of mating in algae, inheritance of mating type; geographical distribution and speciation in algae; genetics; vertebrate muscle tissue culture; problems of cell fusion. Mailing Add: Div of Biol & Med Sci Brown Univ Dept of Biol Providence RI 02912

COLEMAN, BABETTE BROWN, b Bellevue, Pa, Mar 4, 08; m 55. BOTANY. Educ: Wilson Col, AB, 30; Cornell Univ, MA, 40, PhD(bot), 46. Prof Exp: Teacher sci, Grier Sch, 30-36, Milwaukee Downer Sem, 36-38 & Northrop Collegiate Sch, 42-43; asst bot, Cornell Univ, 43-45; instr physiol, Miami Univ, 45-46; from instr to assoc prof, 46-70, EMER ASSOC PROF BOT, UNIV ROCHESTER, 71- Mem: Am Bryol & Lichenol Soc; Ecol Soc Am; Bot Soc Am. Res: Biology and distribution of cryptogams; distribution of higher plants; ecology of local areas; arctic botany; preservation of natural areas; flora of Genesee country. Mailing Add: 3699 Lake Ave Rochester NY 14612

COLEMAN, BERNARD DAVID, b New York, NY, July 5, 30; m 65; c 2. CONTINUUM MECHANICS, MATHEMATICAL ANALYSIS. Educ: Ind Univ, BS, 51; Yale Univ, MS, 53, PhD(chem), 54. Prof Exp: Res chemist, E I du Pont de Nemours & Co, 54-57; SR FEL, MELLON INST SCI, CARNEGIE-MELLON UNIV, 57-, PROF MATH & BIOL, UNIV, 67- Concurrent Pos: Visitor, Inst Math, Univ Bologna, 60-61; vis prof, Johns Hopkins Univ, 62-63; adj prof, Univ Pittsburgh, 64-65; ed-in-chief, Springer Tracts in Natural Philos, 67-; lectr, State Univ Col Pisa, 69 & 70 & Int Ctr Mech Sci, Udine, Italy, 71-73. Mem: Soc Natural Philos (treas, 67-68); Soc Rheol; Soc Neurosci; Asn Res Vision & Ophthalmol. Res: Viscoelasticity; foundations of thermodynamics; functional analysis; differential equations; mathematical biology; theories of visual perception. Mailing Add: Mellon Inst of Sci Carnegie-Mellon Univ 440 5th Ave Pittsburgh PA 15213

COLEMAN, BERNELL, b Lorman, Miss, Apr 26, 29; m 62; c 2. PHYSIOLOGY. Educ: Alcorn Agr & Mech Col, BS, 52; Loyola Univ, Ill, PhD(physiol), 64. Prof Exp: Asst path, Med Ctr, Univ Kans, 52-53; asst biochem, Univ Chicago, 56-57; asst cancer res, Hines Vet Admin Hosp, Ill, 57-59; instr physiol, Sch Med St Louis Univ, 63-65, asst prof, 65-69; assoc prof physiol & cardiovasc res, 69-76, PROF PHYSIOL, CHICAGO MED SCH, 76- Concurrent Pos: Mem coun basic sci, Am Heart Asn. Mem: AAAS; Am Heart Asn; assoc Am Physiol Soc. Res: Myocardial catecholamines in hemorrhagic hypotension; electrolytes and water metabolism of the heart in hemorrhagic shock; effects of norepinephrine on electrolytes and water content of cardiac muscle; cardiodynamics in irreversible hemorrhagic shock; cardiodynamic and circulatory responses to heat; angiotensin II and cardiac function; carotid sinus reflex control of the heart. Mailing Add: Dept of Physiol Chicago Med Sch Chicago IL 60612

COLEMAN, CHARLES CLYDE, b York, Eng, July 31, 37; US citizen. SOLID STATE PHYSICS. Educ: Univ Calif, Los Angeles, BA, 59, MA, 61, PhD(physics), 68. ASSOC PROF PHYSICS, CALIF STATE UNIV, LOS ANGELES, 68- Concurrent Pos: Res fel, Cavendish Lab, Cambridge Univ, 75-76. Mem: Am Phys Soc; fel Brit Interplanetary Soc. Res: Semiconductors; superconductors; crystal growth; thin films; cryogenics; simulated bioluminescence; modulation spectroscopy; ion implantation. Mailing Add: Dept of Physics Calif State Univ Los Angeles CA 90032

COLEMAN, CHARLES FRANKLIN, b Burley, Idaho, Dec 30, 17; m 52; c 3. PHYSICAL CHEMISTRY. Educ: Univ Utah, BS, 41; Purdue Univ, MS, 43, PhD(phys chem), 48. Prof Exp: Asst chem, Univ Utah, 39-41 & Purdue Univ, 42-44; chemist, Substitute Alloy Mat Labs, Columbia Univ, 44, Tenn Eastman Corp & Clinton Eng Works, Oak Ridge, 44-46; asst chem, Purdue Univ, 46-47; chemist, Y-12 Plant, Carbide & Carbon Chem Corp, 48-51, chemist, Union Carbide Corp, 51-67, ASST SECT CHIEF, UNION CARBIDE CORP, OAK RIDGE NAT LAB, 67- Mem: AAAS; Am Chem Soc; Sigma Xi Res: Calorimetry; phase equilibria; solution chemistry; separations chemistry; solvent extraction reagents; equilibria, kinetics, applications; sol-gel processes; actinide-lanthanide chemistry. Mailing Add: Union Carbide Corp-ORNL PO Box X Oak Ridge TN 37830

COLEMAN, CHARLES MOSBY, b New York, NY, Oct 14, 25; m 51; c 5. CLINICAL CHEMISTRY, MICROBIOLOGY. Educ: Univ Mich, BS, 49; Univ Chicago, MS, 54; Univ Colo, PhD(microbiol), 56. Prof Exp: Asst med bact & virol, Univ Chicago, 52; biochemist, Nat Jewish Hosp, Denver, Colo, 56-58, chief clin labs, 58-60, res biochemist, 60-62; biochemist, Warner-Lambert Res Inst NJ, 63-64; chief clin chem, Vet Admin Hosp, Pittsburgh, 64-69; INVENTOR & CONSULT, 69- Mem: AAAS; Am Chem Soc; Am Asn Clin Chem; NY Acad Sci. Res: Clinical laboratory diagnostic devices; analytical and organic chemistry; instrumentation and automation; biochemistry of trace metals and chelates; chemistry of mycobacteria. Mailing Add: 958 Washington Rd Pittsburgh PA 15228

COLEMAN, CHARLES R, b New York, NY, May 30, 12; m 69. OCEANOGRAPHY. Educ: City Col New York, BS, 34, MS, 36. Prof Exp: Head ed sect, Div Oceanog, US Navy Hydrog Off, 51-53, tech serv br, 53-54 & marine geog sect, 54-60; head environ br, Oceanog Anal Div, US Navy Oceanog Off, 60-70, prod mgr, 70-71, head phys oceanog br, 71-75; RETIRED. Res: Production supervision and administration of oceanographic publications, technical editing and review, descriptive oceanography. Mailing Add: PO Box 9252 Suitland MD 20023

COLEMAN, COURTNEY (STAFFORD), b Ventura, Calif, July 19, 30; m 54; c 3. MATHEMATICAL ANALYSIS. Educ: Univ Calif, BA, 51; Princeton Univ, MA, 53, PhD, 55. Prof Exp: Instr math, Princeton Univ, 54-55; from instr to asst prof, Wesleyan Univ, 55-59; from asst prof to assoc prof, 59-66, PROF MATH, HARVEY MUDD COL, 66- Concurrent Pos: Vis scientist, Res Inst Advan Studies, 58-59 & 63-64; mem fac, Claremont Grad Sch, 68- Mem: Am Math Soc; Math Asn Am. Res: Ordinary differential equations. Mailing Add: 675 Northwestern Dr Claremont CA 91711

COLEMAN, CURTIS BURGER, b Alliance, Ohio, Nov 17, 21; m 44; c 5. ENVIRONMENTAL CHEMISTRY. Educ: Mt Union Col, BS, 43; Western Reserve Univ, MS, 49, PhD(chem), 50. Prof Exp: Asst chem, Syracuse Univ, 43-44 & Western Reserve Univ, 46-50; asst prof, Knox Col, 50-56, assoc prof & chmn dept, 56-59; from assoc prof to prof, NMex State Univ, 59-71, chmn dept, 60-63; prof, Univ Mo-Columbia, 71-73; DIR CHEM DIV, SCI LAB SYST OF N MEX, 73- Mem: Am Chem Soc; Sigma Xi. Res: Free radical aromatic substitution; quantitative separation of isomeric diphenyls; derivative formation of isomeric diphenyls; molecular compounds; vinylation reactions; optical resolution; elimination reactions. Mailing Add: Chem Div Sci Lab Syst of NMex Albuquerque NM 87125

COLEMAN, DAVID COWAN, b Bennington, Vt, Nov 7, 38; m 65; c 2. ECOLOGY. Educ: Reed Col, BA, 60; Univ Ore, MA, 62, PhD(biol), 64. Prof Exp: Demonstr, Univ Col Swansea, Wales, 64-65; res assoc & fel, Inst Radiation Ecol, Savannah River Ecol Lab, Univ Ga, 65-72, asst prof zool, 67-72; res assoc agron, Natural Res Ecol Lab, 72-75, ASST PROF ZOOL, COLO STATE UNIV, 75- Mem: AAAS; Ecol Soc Am; Am Soc Microbiol; Brit Ecol Soc; Sigma Xi. Res: Decomposition and nutrient cycles; terrestrial ecosystems; flora-fauna interactions in the soil. Mailing Add: Natural Resource Ecol Lab Colo State Univ Ft Collins CO 80523

COLEMAN, DENIS, b Manchester, Eng, Dec 2, 15; nat US; m 51; c 5. ORGANIC CHEMISTRY. Educ: Univ London, BSc, 39, PhD(protein chem), 44. Prof Exp: Res officer natural silk, Shirley Inst, Brit Cotton Indust Res Asn, 39-47; sr res chemist, Imp Chem Industs, 47-57; head dept polymers, Monsanto Can, Ltd, 57-61; SR RES CHEMIST, FIBERS DIV, AM CYANAMID CO, 61- Mem: Sr mem Am Chem Soc. Res: High polymers; chemistry of silk fibroin; synthesis of polypeptides and block polymers; new polyamides, polyesters and acid-dyeable acrylic fibers; novel flame-retardants. Mailing Add: 223 Glenbrook Rd Stamford CT 06906

COLEMAN, DONALD BROOKS, b Russellville, Ky, June 18, 34; m 54; c 4. MATHEMATICS. Educ: Union Univ, BA, 56; Purdue Univ, MS, 58, PhD(math), 61. Prof Exp: Asst instr math, Mich State Univ, 60-61; from asst prof to assoc prof, Vanderbilt Univ, 61-66; assoc prof, 66-75, PROF MATH, UNIV KY, 75- Mem: Am Math Soc; Math Asn Am; Nat Coun Teachers Math. Res: Algebra. Mailing Add: 1146 Athenia Dr Lexington KY 40504

COLEMAN, DONALD GEORGE, b Denver, Colo, Dec 17, 96; m 24; c 2. FORESTRY. Educ: George Washington Univ, AB, 21. Prof Exp: Magnetic observer, Carnegie Inst, Washington, DC, 21-23, magnetician, 23-24; asst engr, Nat Adv Comt Aeronaut, 25-28; tech ed, Forest Prod Lab, US Forest Serv, 28-45, chief ed & tech reviewer, 45-50, chief res publ & info, 50-65; ed-in-chief, Forest Prod J, 66-68; cartog ed, World Bk Encycl, 68-70; COMPILER, COMPUTERIZED INFO RETRIEVAL SYST, WOOD SCI & FOREST PROD J, 71- Concurrent Pos: Consult, US Army, US Army Air Force, US Navy & Nat Adv Comt Aeronaut; consult, Nat Bur Standards, 44-46. Mem: Forest Prod Res Soc. Res: Magnetics; aircraft instruments; forest products. Mailing Add: 3401 Viburnum Dr Madison WI 53705

COLEMAN, DONALD JAMES, JR, b Tampa, Fla, Apr 20, 39; m 60; c 2. PHYSICS, SOLID STATE ELECTRONICS. Educ: Fla State Univ, BS, 61, PhD(physics), 67. Prof Exp: Engr space track, NASA, 61-63; mem tech staff, Bell Tel Labs, 67-73; MEM TECH STAFF, SOLID STATE DEVICE, TEX INSTRUMENTS, INC, 73- Mem: Am Phys Soc. Res: Solid state microwave device physics; surface properties of solids and device fabrication. Mailing Add: Box 5936 Mail Sta 118 Tex Instruments Inc Dallas TX 75222

COLEMAN, ERNEST, b Detroit, Mich, Aug 31, 43. NUCLEAR PHYSICS, HIGH ENERGY PHYSICS. Educ: Univ Mich, BS & MS, 63, PhD(physics), 66. Prof Exp: Instr physics, Univ Mich, 66-67; asst prof, 68-71, spec asst to actg vpres, 73-74, ASSOC PROF PHYSICS, UNIV MINN, MINNEAPOLIS, 71-; HEAD, CENT LAB RES, DIV PHYS RES, ENERGY RES & DEVELOP ADMIN, 75- Concurrent Pos: Deutsches Elektronen-Synchrotron fel, Hamburg, Ger, 67-68; vis assoc prof, Stanford Univ, 71-72; vis prof, Univ Heidelberg, Ger, 72 & Univ Tel-Aviv, 73; head, Cent Lab Res, High Energy Phys Prog, AEC, 74-75; mem, Coun Int Exchange Scholars, Grad Records Exam Comt & High Energy Phys Adv Panel. Mem: Am Phys Soc; Am Asn Phys Teachers; Ger Phys Soc; Math Asn Am; AAAS. Res: Experimental particle physics; theoretical nuclear physics; quantum electrodynamics; boundary layer problems; education of gifted; research management; complex analysis; fluid flow. Mailing Add: Div of Phys Res Energy Res & Develop Admin Washington DC 20545

COLEMAN, ERNEST ALBERT, organic chemistry, see 12th edition

COLEMAN, EUGENE ALFRED, b Stamford, Tex, May 5, 35; m 56; c 4. PLANT PHYSIOLOGY. Educ: Tex Tech Col, BS, 60; Purdue Univ, MS, 62, PhD(plant physiol & breeding), 64. Prof Exp: Asst prof agron, Tex Tech Univ, 64 & 65; agronomist, Int Minerals & Chem Co, 65-67; ASSOC PROF AGRON, TEX TECH UNIV, 67- Mem: AAAS; Am Soc Agron; Crop Sci Soc Am ; Soil Sci Soc Am. Res: Protein physiology of plants and inorganic nutrition of plants. Mailing Add: Dept of Agron Tex Tech Univ Lubbock TX 79410

COLEMAN, GEORGE HUNT, b San Gabriel, Calif, Oct 15, 28; m 53; c 3. NUCLEAR CHEMISTRY. Educ: Univ Calif, Berkeley, AB, 50; Univ Calif, Los Angeles, PhD(chem), 58. Prof Exp: Chemist, Calif Res & Develop Co, Livermore, 51-53; sr chemist, Lawrence Radiation Lab, 57-69; ASSOC PROF CHEM, NEBR WESLEYAN UNIV, 69- Mem: Am Chem Soc. Res: Radiocarbon dating; activation analysis. Mailing Add: Dept of Chem Nebr Wesleyan Univ Lincoln NE 68504

COLEMAN, HAROLD MITCHELL, b Chicago, Ill, 13; m 37; c 2. PHYSICAL CHEMISTRY. Educ: Univ Chicago, BS, 34, MS, 36, PhD(org chem), 42. Prof Exp: Asst org chem, Univ Chicago, 36-37; instr chem, Northern Ill Col Optom, 37-39; instr phys scis, Evening City Cols, Chicago, 40-48; sr chemist, Cook County Coroner's Toxicol & Chem Labs, 42-44; res scientist, Armour & Co, 44-53, supvr, Armour Res Found, 53-56; asst mgr, Res & Develop Labs, Gen Am Transp Corp, 56-60; mem fac, Univ Ill & Chicago City Cols, 60-72; ASSOC PROF CHEM & PHYS SCI, WRIGHT COL, 72- Concurrent Pos: TV lectr, 63 & 65; tech consult. Mem: AAAS; Am Chem Soc; Am Inst Chemists; Am Soc Metals; Nat Asn Corrosion Eng. Res: Toxicology; reaction kinetics of organic compounds; carbohydrates; microtoxicology; tissue pigments; enzymes; corrosion; metallurgy; research administration; teaching. Mailing Add: Dept Chem & Phys Sci Wright Col 3400 N Austin Ave Chicago IL 60634

COLEMAN, HOWARD S, b Everett, Pa, Jan 10, 17; m 41; c 4. ELECTROOPTICS, ELECTRICAL ENGINEERING. Educ: Pa State Univ, BS, 38, MS, 39, PhD(physics), 42. Prof Exp: Asst, Pa State Univ, 39-40, instr phys sci, 40-42; dir optical inspection lab, 42-47; assoc prof physics & tech dir optical res lab, Univ Tex, 47-51; dir, Sci Bur, Bausch & Lomb Optical Co, NY, 51-56, vpres in charge res & eng, 56-62; head physics res & tech asst to vpres for res, Melpar, Inc, Va, 62-64; dean, Col Eng, Univ Ariz, 64-68; dir spec projs, Ctr Schellenger Res Labs, Univ Tex, El Paso, 68-75; DIR, HOWARD S COLEMAN & ASSOCS, 75- Concurrent Pos: Consult, Xerox Corp, Burr-Brown Co, Kollsman Instrument Co, Melpar, Inc, Singer Co, NSF, NASA & Univ Calif, San Diego. Mem: Nat Soc Prof Eng; Am Inst Aeronaut & Astronaut; Am Phys Soc; Am Meteorol Soc; Optical Soc Am. Res: Properties of optical instruments; guided missiles; atmospheric physics; solar energy; lasers; hydrology; optical radiometry; night vision; countermeasures; camouflage; topography; simulation and consulting assistance in test and evaluation of weapon systems. Mailing Add: PO Box 26368 El Paso TX 79926

COLEMAN, JAMES ANDREW, b Niagara Falls, NY, Mar 11, 21; m 47; c 2. THEORETICAL PHYSICS. Educ: NY Univ, BA, 46; Columbia Univ, MA, 47. Prof Exp: Assoc physicist, Appl Physics Lab, Johns Hopkins Univ, 47-50; instr physics & astron, Conn Col Women, 50-57; PROF PHYSICS & CHMN DEPT, AM INT COL, 57- Concurrent Pos: Mem guided missile subcomt, Res & Develop Bd, 48-50; NSF fac fel, 58- Mem: AAAS; Am Phys Soc; Am Astron Soc; Am Asn Physics Teachers; Nat

Asn Sci Writers. Res: Astronomy; relativity; cosmology. Mailing Add: Dept of Physics Am Int Col Springfield MA 01109

COLEMAN, JAMES EDWARD, b Newport, Ark, Oct 6, 28; m 54; c 3. PHYSICAL CHEMISTRY. Educ: La State Univ, BS, 50, MS, 52; Ohio State Univ, PhD(chem), 59. Prof Exp: Lab asst, La State Univ, 50-52; asst boron chem, Ohio State Univ, 52-58; chemist long range fuels, Esso Stand Oil Co, La, 59-60 & spec proj unit, Esso Res & Eng Co, NJ, 60-63, sr chemist, 63-64, sr chemist, Process Res Div, 64-68; assoc prof, 68-72, PROF CHEM, FAIRMONT STATE COL, 72- Mem: Am Chem Soc. Res: Boron chemistry; rocket propellants; synthesis and thermal stability of high energy fuels. Mailing Add: Dept of Chem Fairmont State Col Fairmont WV 26554

COLEMAN, JAMES MOLCOLM, b Vinton, La, Nov 19, 35; m 58. GEOMORPHOLOGY. Educ: La State Univ, BS, 58, MS, 62, PhD(geol), 66. Prof Exp: Assoc researcher geol, Coastal Studies Inst, 60-66, from asst prof to assoc prof, 66-74, asst dir inst, 70-75, DIR COASTAL STUDIES INST, LA STATE UNIV, BATON ROUGE, 75-, PROF SEDIMENTOL, 74- Concurrent Pos: Leader of numerous clastic sandstone seminars for industry; dir recent & ancient deltaic deposits seminar, NSF grants, 67, 69 & 71; mem sedimentary processes panel, Gulf Univs Res Corp, 67; mem ad hoc comt on EPakistan, Nat Acad Sci. Honors & Awards: A I Levorsen Award, Am Asn Petrol Geologists, 74. Res: Relationships between process and form and sedimentary characteristics of recent environments, especially in deltaic regions. Mailing Add: Coastal Studies Inst La State Univ Baton Rouge LA 70803

COLEMAN, JAMES R, b New York, NY, Nov 24, 37; m 59; c 2. CELL PHYSIOLOGY. Educ: St Peter's Col, BS, 59; NY Univ, MS, 61; Duke Univ, PhD(cytol), 64. Prof Exp: Asst cytol, Inst Muscle Dis, 60; asst zool, Duke Univ, 61-62, res assoc cell biol, Dept Anat, 64; NIH fel, 64-65; asst prof, 65-71, ASSOC PROF, DEPT RADIATION BIOL & BIOPHYS, SCH MED & DENT, UNIV ROCHESTER, 71- Mem: Am Soc Cell Biol; Histochem Soc; Electron Probe Anal Soc Am. Res: Cytology; histochemistry; light and electron microscopy; electron probe analysis. Mailing Add: Dept of Radiation Biol & Biophys Univ of Rochester Med Ctr Rochester NY 14620

COLEMAN, JAMES THOMAS, plasma physics, electrical engineering, see 12th edition

COLEMAN, JOHN FRANKLIN, b Akron, Ohio, July 15, 39; m 61; c 2. ORGANIC CHEMISTRY, POLYMER CHEMISTRY. Educ: Univ Akron, BS, 61; Univ Ill, MS, 63, PhD(org chem), 66. Prof Exp: Asst gen chemist inorg chem, Univ Ill, 61-62, res asst org chem, 62-65; polymer res chemist, 65-67, 69-70, sr adhesives res chemist, 70-73, group leader, 73-74, SECT LEADER MAT, B F GOODRICH RES & DEVELOP CTR, 74- Mem: Am Chem Soc. Res: Polymerization of vinyl monomers; polymer modifications; condensation polymerization; adhesives; fire retardant additives; thermally stable polymers; materials technology. Mailing Add: 2465 Olentangy Dr Akron OH 44313

COLEMAN, JOHN RUSSELL, b Medford, Ore, Nov 4, 33; m 58; c 3 DEVELOPMENTAL BIOLOGY. Educ: Univ Minn, AB, 55; Ind Univ, MA, 57; Johns Hopkins Univ, PhD(biol), 61. Prof Exp: Res assoc develop biol, Univ Conn, 62-63; asst prof, 63-69, ASSOC PROF BIOL, BROWN UNIV, 69- Concurrent Pos: Nat Inst Child Health & Human Develop grant, 64-; spec res fel, Univ Calif, San Diego, 69-70; mem cell biol study sect, NIH, 74-78. Mem: AAAS; Am Soc Cell Biol; Am Soc Zool; Soc Develop Biol; Int Soc Differentiation. Res: Differentiation of chicken embryo cells in culture—myogenesis, protein and RNA synthesis, genome organization and expression, physiological maturation, hormone function. Mailing Add: Div of Biol & Med Sci Brown Univ Providence RI 02912

COLEMAN, JOHN SHERRARD, b Honolulu, Hawaii, Jan 15, 14; m 44; c 2. PHYSICS. Educ: Col William & Mary, BS, 35; Mass Inst Technol, SM, 40. Prof Exp: Construct supt, Washington, DC, 36-37; tech aide, Nat Defense Res Comt, Mass, 40-43; res assoc, Harvard Univ, 43-44; tech aide, Off Sci Res & Develop, London, 44-45; asst dir summary reports group, Columbia Univ, 45-46; exec secy, Comt Undersea Warfare, 47-53, exec secy, Div Phys Sci, 53-65, EXEC OFFICER, DIV PHYS SCI, NAT ACAD SCI-NAT RES COUN, 65- Concurrent Pos: Prof, Pa State Univ, 53. Honors & Awards: Meritorious Pub Serv Award, US Navy, 58. Res: Systems analysis; research and development administration. Mailing Add: Nat Acad of Sci 2101 Constitution Ave Washington DC 20418

COLEMAN, JOSEPH EMORY, b Iowa City, Iowa, Oct 11, 30; m 61; c 1. BIOCHEMISTRY, BIOPHYSICS. Educ: Univ Va, BA, 53, MD, 57; Mass Inst Technol, PhD(biophys), 63. Prof Exp: Intern med, Peter Bent Brigham Hosp, Harvard Sch Med, 57-58, Nat Acad Sci fel biophys, 48-59, NIH fel, 59-62, univ res fel, Biophys Res Lab, 58-63, sr resident med, Peter Bent Brigham Hosp, 63-64; from asst prof to assoc prof biochem, 64-74, PROF MOLECULAR BIOPHYS & BIOCHEM, YALE UNIV, 74- Mem: Am Chem Soc; Am Soc Biol Chemists. Res: Physical chemistry of proteins; mechanisms of enzyme action; metalloenzymes. Mailing Add: Dept of Molecular Biophys & Biochem Yale Univ New Haven CT 06510

COLEMAN, JOSEPH JOHNSTON, b Boulder, Colo, Sept 14, 07; m 47. PHYSICS Educ: Univ Colo, AB, 31, AM, 34, PhD(chem), 36. Prof Exp: Res chemist & physicist, Burgess Battery Co Div, Servel, Inc, 36-40, chief engr, 40-55, vpres eng & res, 55-68; CONSULT, 68- Res: Reaction kinetics of dry cells; administration of engineering and research. Mailing Add: 1547 W Harrison St Freeport IL 61032

COLEMAN, JULES VICTOR, b Brooklyn, NY, Nov 2, 07; m 32; c 2. PSYCHIATRY. Educ: Cornell Univ, AB, 28; Univ Vienna, MD, 34; Am Bd Psychiat & Neurol, dipl, 46. Prof Exp: Dir, East Harlem Unit, Delinquency Proj, Bur Child Guid, 41-42; head ment hyg div, Med Ctr, Univ Colo, 46-50; psychiatrist, Dept Univ Health, 50-52, clin prof psychiat, Sch Med, 52-62, chief ment health sect, Dept Epidemiol & Pub Health, 62-74, dir social & community psychiat training prog, 70-73, CLIN PROF PUB HEALTH & PSYCHIAT, SCH MED, YALE UNIV, 62-; CHIEF MENT HEALTH & PSYCHIAT, COMMUNITY HEALTH CARE CTR PLAN, NEW HAVEN, 71- Concurrent Pos: Lectr, Sch Social Work, Univ Denver, 46-50; consult, Community Serv Comn, USPHS, 47-49, State Dept Health, Colo, 47-50 & Health Dept, New York, 49; mem nat tech fact finding comt, Mid-Century White House Conf Children & Youth, 50; assoc dir, Bur Ment Hyg, State Dept Health, Conn, 50-52; physician-in-chief, Psychiat Clin, New Haven Hosp, 52-56; consult, Vet Admin Hosp, West Haven, 53-70; chmn psychiat training rev comt, NIH, 65-69; Conn State Bd Ment Health chmn, 69-71; consult, US Vet Admin, Fitzsimons Gen Hosp, Nat Jewish Hosp & Winter Gen Hosp, Topeka. Mem: Am Psychiat Asn; Am Orthopsychiat Asn; Asn Psychiat Clins Children (pres, 49-51); Am Psychoanal Asn. Res: Psychotherapy; social psychiatry. Mailing Add: 135 Whitney Ave New Haven CT 06510

COLEMAN, LESLIE CHARLES, b Toronto, Ont, Oct 22, 26; m 52; c 2. GEOLOGY. Educ: Queen's Univ, Can, BA, 50, MA, 52; Princeton Univ, PhD(geol), 55. Prof Exp: Instr geol, Tulane Univ, 55-56; vis asst prof, Lafayette Col, 56-57; asst prof mineral, Ohio State Univ, 57-60; from asst prof to assoc prof, 60-70, PROF GEOL, UNIV SASK, 70- Mem: Geol Soc Am; Mineral Soc Am; Geochem Soc; Geol Asn Can; Mineral Asn Can. Res: Distribution of trace metals in bedrock and relationship to the geology of the Hanson Lake area in Saskatchewan; mineralogy, petrology and geochemistry of meteorites and volcanic rocks. Mailing Add: Dept of Geol Sci Univ of Sask Saskatoon SK Can

COLEMAN, LESTER EARL, (JR), b Akron, Ohio, Nov 6, 30; m 51; c 2. ORGANIC CHEMISTRY, POLYMER CHEMISTRY. Educ: Univ Akron, BS, 52; Univ Ill, MS, 53, PhD(chem), 55. Prof Exp: Chemist, Polymer Res Div, Goodyear Tire & Rubber Co, 51-52; asst gen chem, Univ Ill, 52-53, chemist res, Govt Synthetic Rubber Prog, 53-55; chemist, Lubrizol Corp, 55; proj engr, Polymer Sect, Mat Lab, Wright Air Develop Ctr, US Air Force, 55-57; proj leader additive res dept, 59-64, dir org res, 64-68, asst div head res & develop, 68-72, asst to the pres, 72-73, EXEC VPRES & DIR, LUBRIZOL CORP, 73- Mem: Am Chem Soc. Res: Synthesis and polymerization of vinyl monomers; synthetic organic chemistry; lubricant additives. Mailing Add: 35850 Eddy Rd Willoughby Hills OH 44094

COLEMAN, LESTER LYMAN, b Maricopa, Calif, Nov 6, 12; m 39; c 2. BIOCHEMISTRY. Educ: Cornell Col, AB, 35; NY Univ, MS, 38; Univ Iowa, PhD(biochem), 43. Prof Exp: Res chemist, Masonite Corp, Miss, 38-41; asst path chem, Univ Iowa, 41-43; patent chemist, Hercules Powder Co, Del, 43-44; res chemist, Upjohn Co, 44-54, sect head dept biochem, 54-58, HEAD PROD DEVELOP, UPJOHN INT INC, 58- Mem: Am Chem Soc; fel Am Inst Chem. Res: Fat metabolism; fermentation of wood sugars; metabolism of branched chain alcohols; partial synthesis and metabolism of steroid hormones; blood anticoagulants; anti-lipemic agents. Mailing Add: Upjohn Int Inc 320 Portage St Kalamazoo MI 49001

COLEMAN, MARCIA LEPRI, b New Haven, Conn. CHEMICAL PHYSICS, POLYMER PHYSICS. Educ: Mt Holyoke Col, BA, 69; Mass Inst Technol, PhD(chem physics), 73. Prof Exp: RES CHEMIST POLYMER PHYSICS & CHEM, TEXTILE FIBERS DEPT, E I DU PONT DE NEMOURS & CO, INC, 73- Mem: AAAS; Am Chem Soc; Am Inst Physics. Res: Polymer and synthetic fiber chemistry and physics. Mailing Add: Exp Sta Bldg 262 E I du Pont de Nemours & Co Inc Wilmington DE 19803

COLEMAN, MICHAEL MURRAY, b Herne Bay, Eng, Jan 24, 38. POLYMER SCIENCE. Educ: Borough Polytech, Eng, BSc, 68; Case Western Reserve Univ, MS, 71, PhD(polymer sci), 73. Prof Exp: Assayer chem, Rhokana Corp Ltd, Zambia, 55-61; anal chemist, Johnson Mathey Ltd, Eng, 63-64; res chemist polymers, Revertex Ltd, Eng, 68-69 & E I du Pont de Nemours & Co, 73-75; ASST PROF POLYMERS, PA STATE UNIV, UNIVERSITY PARK, 75- Mem: Am Chem Soc; Royal Inst Chem; Am Phys Soc. Res: Polymer physical chemistry; polymer characterization; infrared, Raman and NMR spectroscopy as applied to polymers; vulcanization elastomers. Mailing Add: Dept of Mat Sci Pa State Univ University Park PA 16802

COLEMAN, MORTON, b Norfolk, Va, Sept 15, 39; m 68; c 2. HEMATOLOGY, ONCOLOGY. Educ: Johns Hopkins Univ, BA, 59; Med Col Va, MD, 63. Prof Exp: Asst prof, 68-75, ASSOC PROF MED, MED COL, CORNELL UNIV, 75- Concurrent Pos: Asst attend physician, New York Hosp, 68-75, assoc attend physician, 75-; assoc dir oncol serv, New York Hosp-Cornell Med Ctr, 68-; consult, Doctors Hosp, New York, 69- & Manhattan Eye, Ear, Nose & Throat Hosp, 70-; assoc dir clin chemother prog cancer control, Nat Cancer Inst-New York Hosp, 74-; chmn new agents comt, Cancer & Leukemia Group B, 75- Mem: Int Soc Hemat; Am Soc Hemat; Am Soc Clin Oncol; Soc Study Blood; Harvey Soc. Res: Clinical research in new chemotherapeutic agents for blood and lymphatic malignancies. Mailing Add: 525 E 68th St New York NY 10021

COLEMAN, NATHANIEL T, soil chemistry, deceased

COLEMAN, NEIL LLOYD, b Belvidere, Ill, Sept 3, 30; m 52. GEOLOGY, FLUID MECHANICS. Educ: Cornell Univ, BA, 52; Univ Chicago, MS, 57, PhD(geol), 60. Prof Exp: GEOLOGIST, SEDIMENTATION LAB, AGR RES SERV, USDA, 59-, RES LEADER, SEDIMENT TRANSPORT RES UNIT, 70- Concurrent Pos: Assoc prof civil eng, Univ Miss, 61- Mem: Am Geophys Union; Int Asn Hydraul Res. Res: Soil erosion; sediment transportation and depostion; mechanics of flow in natural and artificial streams or channels. Mailing Add: Sedimentation Lab USDA PO Box 1157 Oxford MI 38655

COLEMAN, NORMAN P, JR, b Richmond, Va, Mar 13, 42. APPLIED MATHEMATICS. Educ: Univ Va, BA, 65; Vanderbilt Univ, MA, 68, PhD(math), 69. Prof Exp: Mathematician, E I du Pont de Nemours & Co, Inc, 65; instr math, Vanderbilt Univ, 68-69; MATHEMATICIAN, HQ, US ARMY WEAPONS COMMAND, 69- Concurrent Pos: Adj prof, Col Eng, Univ Iowa, 71- Mem: Soc Indust & Appl Math; Am Math Soc; Sigma Xi. Res: Function algebras; necessary conditions for existence of complemented subspaces; operator theory; application of the theory of perturbation for linear operations to development of algorithms in optimal design and optimal control theory. Mailing Add: 3435 60th St Apt 1-D Moline IL 61265

COLEMAN, OTTO HARVEY, b Denver, Colo, June 26, 05; m 35; c 3. AGRONOMY. Educ: Colo State Univ, BS, 34, MS, 37. Prof Exp: Asst agronomist, State Agr Exp Sta, Univ Colo, 35-42; asst agronomist, Sugar Plant Field Sta, Miss, 42-44 & Fla, 45-46, assoc agronomist in charge sorgo breeding, 47-54, sta supt & res agronomist, Miss, 54-70, COLLABR SUGAR CROPS FIELD STA, AGR RES SERV, USDA, 70- Mem: AAAS; Am Soc Agron; Am Soc Sugar Cane Technol; Am Genetic Asn. Res: Genetics of barley and of HCN in Sudan grass; design of agricultural experiments; sugar cane breeding for sirup; breeding sorgo for sirup and sugar; sorghum genetics. Mailing Add: PO Box 3188 Meridian MS 39301

COLEMAN, PAUL DAVID, b New York, NY, Dec 2, 27; m 55; c 2. NEUROBIOLOGY. Educ: Tufts Univ, AB, 48; Univ Rochester, PhD(psychol), 53. Prof Exp: Asst auditory psychophysiol, Univ Rochester, 48-51, asst statist, 51-52; res psychologist, Army Med Res Lab, 54-56; asst prof & res assoc, Inst Appl Exp Psychol, Tufts Univ, 56-59; assoc prof physiol, Sch Med, Univ Md, Baltimore County, 62-67; PROF ANAT, SCH MED, UNIV ROCHESTER, 67- Concurrent Pos: USPHS fel, Sch Med, Johns Hopkins Univ, 59-62; instr, Rochester Gen Hosp, 52-53. Mem: AAAS; Am Psychol Asn; Am Statist Asn; Soc Neurosci. Res: Neuroanatomy; senosry physiology; brain-behavior relations; effects of early environment on quantitative aspects of brain. Mailing Add: Dept of Anat Univ of Rochester Sch of Med Rochester NY 14620

COLEMAN, PAUL JEROME, JR, b Evanston, Ill, Mar 7, 32; m 64; c 1. SPACE PHYSICS. Educ: Univ Mich, BS(eng math) & BS(eng physics), 54, MS, 58; Univ Calif, Los Angeles, PhD(space physics), 66. Prof Exp: Mem tech staff, Space Tech Labs, Inc, Calif, 58-61; head interplanetary sci prog, NASA, DC, 61-62; res scientist, Inst Geophys, 62-66, assoc prof, 66-71, PROF PLANETARY & SPACE SCI, 71- Concurrent Pos: Mem comt on planetary surfaces & atmospheres, Space Sci Bd, Nat Acad Sci; working groups 3 & 5, Inter-Union Comn Solar-Terrestrial Physics; Comn

V, Int Asn Geomag & Aeronomy; Comn IV, Int Sci Radio Union. Honors & Awards: NASA Exceptional Sci Achievement Medal, 70. Mem: AAAS; Am Geophys Union; Am Phys Soc. Res: Experimental space physics, including measurements of radiation-belt particles and magnetic fields of the earth and the planets and in interplanetary space. Mailing Add: Dept of Planetary & Space Sci Inst of Geophys Univ of Calif Los Angeles CA 90024

COLEMAN, PETER STEPHEN, b New York, NY, Feb 10, 38; m 69; c 1. BIOCHEMISTRY, BIOPHYSICS. Educ: Columbia Univ, AB, 59, PhD(biophys, biol), 66. Prof Exp: Res fel mechanochem, Weizmann Inst Sci, Israel, 67-68; res fel biochem & biophys, Yale Univ, 68-70, Nat Cancer Inst fel, 68-69; asst prof, 70-75, ASSOC PROF BIOCHEM, NY UNIV, 75- Mem: AAAS; Am Chem Soc; Am Soc Cell Biol; Biophys Soc. Res: Oxidative phosphorylation, mechanisms of coupling; free-radical research in uncoupling oxidative phosphorylation; bioenergetics of early embryogenesis; membrane biochemistry. Mailing Add: Dept of Biol NY Univ New York NY 10003

COLEMAN, PHILIP HOXIE, b Fredericksburg, Va, May 11, 33; m 53. VIROLOGY. Educ: Univ Ga, DVM, 56; Univ Wis, MS, 57, PhD(vet microbiol), 59. Prof Exp: Asst chief southeast rabies lab, Nat Commun Dis Ctr, USPHS, 50-60, asst chief zoonosis res unit, 60-61, in chg arbovirus lab, 61-66, asst chief biol reagents sect, 66-68, chief arbovirus infectious unit, 68-69; PROF MICROBIOL & DIR CENT ANIMAL FACIL, MED COL VA, VA COMMONWEALTH UNIV, 69- Mem: Am Vet Med Asn; Sci Res Soc Am; Am Soc Microbiol; Soc Exp Biol & Med; Am Soc Trop Med & Hyg. Res: Laboratory aspects of arthropod-borne viruses in relation to their public health significance. Mailing Add: Med Col of Va Health Sci Div Va Commonwealth Univ Richmond VA 23219

COLEMAN, PHILIP LYNN, b Denver, Colo, Dec 25, 44; m 68; c 1. APPLIED PHYSICS. Educ: Calif Inst Technol, BS, 66; Univ Wis-Madison, PhD(physics), 71. Prof Exp: Res assoc space physics, Univ Wis-Madison, 71-72 & Rice Univ, 72-73; SCI STAFF MEM SHOCK PHYSICS, SYSTS, SCI & SOFTWARE, 73- Mem: AAAS; Am Astron Soc. Res: Shock-wave physics; explosively induced ground motion; instrumentation for transient, high pressure shocks; applied physics. Mailing Add: Systs Sci & Software PO Box 1620 La Jolla CA 92038

COLEMAN, RALPH H, b Winslow, Ind, Jan 20, 12; m 46; c 3. MATHEMATICS, STATISTICS. Educ: Oakland City Col, AB, 31; Ind Univ, AM, 32, PhD(math, educ), 56. Prof Exp: Teacher & prin high sch, 32-43; army training specialist math, Ind Univ, 43-44; teacher & prin high sch, 44-46; INSTR MATH, UNIV EVANSVILLE, 46-, CHMN DEPT, 56- Res: Components of success in college mathematics. Mailing Add: Dept of Math Univ of Evansville Evansville IN 47704

COLEMAN, RALPH ORVAL, JR, b Corvallis, Ore, Dec 9, 31; m 64; c 2. SPEECH PATHOLOGY. Educ: Ore State Univ, BS, 54; Univ Ore, MS, 60; Northwestern Univ, PhD(speech path), 63. Prof Exp: Asst prof speech path, Univ Nebr, 63-65; res speech pathologist, Lancaster Cleft Palate Clin, 65-66; ASSOC PROF SPEECH PATH, UNIV ORE HEALTH SCI CTR, PORTLAND, 66- Concurrent Pos: Mem summer fac, Eastern Ore Col, 69 & 71; guest researcher, Speech Transmission Lab, Royal Inst Technol, Stockholm, Sweden, 73. Mem: Am Asn Univ Prof; AAAS; Am Speech & Hearing Asn. Res: Disorders of human communication; the development of language in humans. Mailing Add: 2923 SE Tolman Portland OR 97202

COLEMAN, RICHARD J, b West New York, NJ, July 23, 24; m 49; c 4. ORGANIC CHEMISTRY, FOOD TECHNOLOGY. Educ: City Col New York, BS, 48; Fordham Univ, MS, 50, PhD(org chem), 53. Prof Exp: Res chemist proteins, Fleichmann Labs, Standard Brands, Inc, 52-53, group leader food prod, 53-57, from asst div head to div head, 57-60; mgr prod develop, Thomas J Lipton, Inc, 60-63; dir tech develop flavors, 63-73, SR VPRES, FLORASYNTH INC, 73- Mem: Inst Food Tech; Soc Cosmetic Chem; Am Asn Cereal Chem. Res: Development of flavors and aromatic chemicals for use in the food and cosmetic fields. Mailing Add: 249 Oak Knoll Terr Highland Park IL 60035

COLEMAN, RICHARD WALTER, b San Francisco, Calif, Sept 10, 22; div; c 1. ECOLOGY, BIOLOGY. Educ: Univ Calif, Berkeley, BA, 45, PhD(parasitol), 51. Prof Exp: Asst med entom, Univ Calif, Berkeley, 46, res asst med entom & helminth, 46-47 & 49-50; pvt supported res study, 51-61; prof biol & chmn dept, Curry Col, 61-63; chmn sci & math div, Monticello Col & Preparatory Sch, Ill, 63-64; vis prof biol, Wilberforce, 64-65; PROF SCI, UPPER IOWA COL, 65-, HEAD DEPT BIOL, 75- Concurrent Pos: Collabr nat hist div, Nat Park Serv, 52-53; spec consult, Arctic Health Res Ctr, USPHS Alaska, 54-64. Mem: AAAS; Am Inst Biol Sci; Ecol Soc Am; Nat Sci Teachers Asn; Human Ecol Soc. Res: Epidemiology; zoology; botany; parasitology. Mailing Add: Dept of Biol Upper Iowa Col Fayette IA 52142

COLEMAN, ROBERT E, b Trenton, NJ, Apr 22, 21; m 44; c 5. PLANT PHYSIOLOGY. Educ: Swarthmore Col, AB, 43; Univ Pa, MS, 53; Univ Hawaii, PhD(bot), 56. Prof Exp: Plant physiologist, Field Crops Res Br, Agr Res Serv, USDA, La, 48-54, Co-op Proj Exp Sta, Hawaiian Sugar Planters Asn, Hawaii, 54-56, Field Crops Res Br, Agr Res Serv, USDA, 56-60 & Co-op Proj Exp Sta, Hawaiian Sugar Planters Asn, 60-67; invest leader sugarcane & sweet sorghum invests, Tobacco & Sugar Res Br, Plant Sci Res Div, 67-72, STAFF SCIENTIST SUGAR CROPS, NAT PROG STAFF, AGR RES SERV, USDA, 72- Mem: Fel AAAS; Am Soc Plant Physiol; Soc Sugar Cane Technol; Sigma Xi; Am Inst Biol Sci. Res: Sugar cane physiology; post harvest deterioration; freeze injury; germination; flowering. Mailing Add: Nat Prog Staff Rm 337 Bldg 005 Agr Res Ctr-W USDA Beltsville MD 20705

COLEMAN, ROBERT GRIFFIN, b Twin Falls, Idaho, Jan 5, 23; m 48; c 3. GEOLOGY. Educ: Ore State Col, BS, 48, MS, 50; Stanford Univ, PhD(geol), 57. Prof Exp: Instr geol, Ore State Col, 49 & La State Univ, 51-52; mineralogist, US AEC, 52-55; br chief isotope geol, 64-68, GEOLOGIST, US GEOL SURV, 55- Concurrent Pos: Vis lectr, Stanford Univ, 60; vis geologist, NZ Geol Surv, 62; tech adv, Saudi Arabia, 70-71; consult geol, Sultanate Oman, 73-74. Mem: Geol Soc Am; Mineral Soc Am. Res: Mineralogy and geochemistry of uranium ore deposits; geology and mineralogy of silicate and sulfide minerals as related to their origin; glaucophane schists and ultramafic rocks of California and New Zealand; new plate tectonic theories as applied to formation of ophiolites and glaucophane schists. Mailing Add: US Geol Surv 345 Middlefield Rd Menlo Park CA 94025

COLEMAN, ROBERT MARSHALL, b Bridgton, Maine, Sept 27, 25; m 47; c 2. PARASITOLOGY, IMMUNOLOGY. Educ: Bates Col, BS, 50; Univ NH, MS, 51; Univ Notre Dame, PhD(parasitol), 54. Prof Exp: From instr to assoc prof microbiol, Russell Sage Col, 54-62; assoc prof, Boston Col, 62-68; PROF MICROBIOL & HEAD DEPT BIOL SCI, UNIV LOWELL, 68- Concurrent Pos: Consult, AID, India, 65 & 68. Mem: NY Acad Sci; Am Soc Parasitol; Am Soc Microbiol; Am Soc Trop Med & Hyg. Res: Immunology of animal parasites; helminth antigens; fluorescent antibodies; cell-mediated immunity; cytotoxic systems malaria. Mailing Add: Dept of Biol Sci Univ of Lowell Lowell MA 01854

COLEMAN, ROBERT VINCENT, b Iowa City, Iowa, Oct 11, 30. PHYSICS. Educ: Univ Va, BA, 53, PhD(physics), 56. Prof Exp: Mem tech staff physics, Res Labs, Gen Elec Co, 55; sr res physicist, Res Labs, Gen Motors Corp, 56-58; asst prof physics, Univ Ill, 58-60; assoc prof, 60-64, PROF PHYSICS, UNIV VA, 64- Mem: Am Phys Soc. Res: Solid state physics; growth and properties of crystals; magnetism and low temperature physics; metals. Mailing Add: Dept of Physics Univ of Va McCormick Rd Charlottesville VA 22903

COLEMAN, RONALD LEON, b Wellington, Tex, Aug 20, 34; m 56; c 3. BIOCHEMISTRY. Educ: Abilene Christian Col, BS, 56; Univ Okla, PhD(biochem), 63. Prof Exp: From instr to assoc prof biochem, 63-75, PROF BIOCHEM & ENVIRON HEALTH, SCH MED, UNIV OKLA, 75- Mem: Am Chem Soc; Am Pub Health Asn; Am Indust Hyg Asn; Am Conf Govt Indust Hygienists. Res: Trace metal metabolism; toxicology of carbon monoxide, cadmium, nickel, chromium, silver and mercury; biochemical function of zinc, manganese, magnesium, iron, cadmium, nickel and chromium. Mailing Add: Dept of Environ Health Univ of Okla Health Sci Ctr Oklahoma City OK 73104

COLEMAN, RUSSELL, b Montpelier, Miss, Dec 23, 13; m; c 3. SOILS. Educ: Miss State Col, BS, 36, MS, 37; Univ Wis, PhD(soil chem), 41. Prof Exp: Asst, Miss Agr Exp Sta, 36-37, asst agronomist, 37-39; asst, Univ Wis, 39-40; assoc prof soils, Miss State Col, 40-48, assoc agronomist, Miss Agr Exp Sta, 40-46, from assoc dir to dir, 46-48; pres, Nat Fertilizer Asn, 48-55; exec vpres, Nat Plant Food Inst, 55-60; PRES, SULPHUR INST, 60- Mem: Fel Am Soc Agron; Soil Sci Soc Am. Res: Phosphate studies in soils; mineral composition of the colloidal fraction of coastal plain soils. Mailing Add: Sulphur Inst 1725 K St NW Washington DC 20006

COLEMAN, SAMUEL HENRY, mathematics, deceased

COLEMAN, SIDNEY RICHARD, b Chicago, Ill, Mar 7, 37. THEORETICAL PHYSICS. Educ: Ill Inst Technol, BS, 57; Calif Inst Technol, PhD(physics), 62. Prof Exp: Corning Glass Works res fel, 61-63, from asst prof to assoc prof, 63-69, PROF PHYSICS, HARVARD UNIV, 69- Concurrent Pos: Sloan Found fel, 64-66. Mem: Am Phys Soc. Res: Theoretical high-energy physics; symmetry principles. Mailing Add: Lyman Lab Dept of Physics Harvard Univ Cambridge MA 02138

COLEMAN, SYLVIA ETHEL, microbiology, see 12th edition

COLEMAN, THEO HOUGHTON, b Millport, Ala, Oct 25, 21; m 49. POULTRY SCIENCE. Educ: Ala Polytech Inst, BS, 43, MS, 48; Ohio State Univ, PhD(poultry genetics), 53. Prof Exp: Instr, Dept Poultry Sci, Ohio State Univ, 51-52; with com poultry farm, 53-54; from asst prof to assoc prof, 55-63, PROF POULTRY SCI, MICH STATE UNIV, 63- Mem: AAAS; Genetics Soc Am; Am Genetic Asn; Poultry Sci Asn Am; Am Soc Study Reprod. Res: Fertility and hatchability in turkeys; effective use of visual aids in teaching; genetics, reproduction and behavior of quail. Mailing Add: Dept of Poultry Sci Mich State Univ East Lansing MI 48823

COLEMAN, WILLIAM EARL, b Greenup, Ill, July 22, 33; m 54; c 3. ORGANIC CHEMISTRY. Educ: Univ Ill, BS, 56; Univ Calif, Berkeley, PhD(chem), 59. Prof Exp: Res chemist, Arapahoe Chem, Inc, 59-62, mgr prod develop, 62-63, sales mgr, 63-65, vpres mkt, 65 & Arapahoe Chem Div, Syntex Corp, 65-66, gen mgr, Bahamas Chem Div, 66-74; PRES, ARAPAHOE CHEM, INC, 74- Mem: AAAS; Am Chem Soc. Res: Organic synthesis of small ring compounds and steroids. Mailing Add: Arapahoe Chem Inc PO Box 511 Boulder CO 80302

COLEMAN, WILLIAM FLETCHER, b Montgomery, WVa, Sept 15, 45; m 65. PHYSICAL INORGANIC CHEMISTRY, MOLECULAR SPECTROSCOPY. Educ: Eckerd Col, BS, 66; Ind Univ, PhD(chem), 70. Prof Exp: Fel, Univ Ariz, 70-71; asst prof, 71-75, ASSOC PROF CHEM, UNIV NMEX, 75- Concurrent Pos: Vis staff mem, Los Alamos Sci Lab, 75- Mem: Am Chem Soc; AAAS. Res: Photochemistry and spectroscopy of metal complexes; ligand field theory; matrix isolated luminescence spectroscopy; inorganic complexes of biological interest; dye laser development; excited state electron spin resonance spectroscopy; nonradiative transitions. Mailing Add: Dept of Chem Univ of NMex Albuquerque NM 87131

COLEMAN, WILLIAM GILMORE, JR, b Birmingham, Ala, May 6, 42. MOLECULAR BIOLOGY. Educ: Talladega Col, BS, 64; Atlanta Univ, MS, 70; Purdue Univ, PhD(molecular biol), 73. Prof Exp: Lab instr physiol & cytol, Atlanta Univ, 64-66; biol teacher, David T Howard Community Adult Sch & elem sch teacher, Whiteford Elem Sch, 66-69; teaching asst cell biol, Purdue Univ, 70-73, vis prof, 73-74; RES STAFF FEL, LAB BIOCHEM PHARMACOL, NAT INST ARTHRITIS METAB & DIGESTIVE DIS, NIH, 74- Mem: Am Soc Microbiol; Genetics Soc Am. Res: Mode of assembly and function of the outer membrane of Escherichia coli. Mailing Add: Nat Inst of Arthritis Metab & Digestive Dis NIH Bldg 4 Rm 116 Bethesda MD 20014

COLEMAN, WILLIAM H, b Chestertown, Md, Mar 6, 37. MICROBIOLOGY. Educ: Wash Col, BS, 59; Univ Chicago, MS, 62, PhD(microbiol), 67. Prof Exp: Instr biol, Univ Chicago, 65-66, 67-68, sr res technologist microbiol, 66-67; fel, Univ Colo, 68-71; ASST PROF BIOL, UNIV HARTFORD, 71- Concurrent Pos: Lilly Found fel, Yale Univ, 75. Mem: AAAS; Am Soc Microbiol; Sigma Xi. Res: RNA, protein purifications; protein synthesis in mammalian cells. Mailing Add: Dept of Biol Univ of Hartford West Hartford CT 06117

COLEN, ALAN HUGH, b Brooklyn, NY, Jan 29, 39; m 60; c 3. BIOPHYSICAL CHEMISTRY. Educ: Cornell Univ, BA, 60, PhD(phys chem), 67. Prof Exp: Great Lakes Cols Asn-Kettering Found teaching intern chem, Kalamazoo Col, 66-67, asst prof, 67-70; res biochemist, Med Sch, Univ Kans, 70-72; RES CHEMIST, VET ADMIN HOSP, KANSAS CITY, MO, 72- Concurrent Pos: Consult, Upjohn Co, Mich, 67-68; Res Corp grant, 70. Mem: AAAS; Am Chem Soc; Am Phys Soc; NY Acad Sci. Res: Kinetics of fast reactions in solution; relaxation kinetics; theory of fluids; critical phenomena; enzyme kinetics. Mailing Add: Lab Molec Biochem Vet Admin Hosp 4801 Linwood Blvd Kansas City MO 64128

COLENBRANDER, AUGUST, b Leiden, Netherlands, Aug 30, 31; m 63; c 2. OPHTHALMOLOGY. Educ: State Univ Leiden, MD, 59; State Univ Utrecht, PhD(ophthal), 63. Prof Exp: From resident to chief resident ophthal, State Univ Utrecht, 60-64; assoc prof, State Univ Leiden, 64-69; vis assoc prof, Univ Iowa, 69-71; ASSOC PROF OPHTHAL, SCH MED SCI, UNIV OF THE PAC, 71- Concurrent Pos: Consult terminology, Am Acad Ophthal & Otolaryngol, 70- & AMA, Calif Med Asn & Col Am Pathologists, 71-; secy, Comt Info, Int Coun Ophthal, 71-74, chmn, 74-; co-dir proj self instr mat ophthal, Sloan Found & US Off Educ, 70-71, Nat Med AV Ctr & Bur Health Resources Develop, 73-74; Nat Libr Med prin investr, Int Nomenclature Ophthal, 73-76. Mem: Am Acad Ophthal & Otolaryngol. Res: Medical education; medical audit; nomenclature and terminology. Mailing Add: Sch of Med Sci Pac Med Ctr Clay & Webster San Francisco CA 94120

COLER, ROBERT A, b Hartford, Conn, July 23, 28; m 53; c 5. LIMNOLOGY. Educ:

Champlain Col, BA, 52; Col Educ, Albany, MA, 54; State Univ NY Col Forestry, Syracuse, PhD(biol), 61. Prof Exp: Asst prof physics, NY State Col Educ, Cortland, 55-56; res biologist, Univ Md, 60-61; asst prof comp anat, Mass State Col, Bridgewater, 61-67; fel, 67-69, sr res assoc, 69-70, lectr, 70-71, ASST PROF ENVIRON SCI, UNIV MASS, 71-, DIR, ENVIRON TECHNOL TRAINING PROG, 70- Concurrent Pos: Dir, NSF res grant, 63-65. Mem: AAAS; Am Soc Zoologists. Res: Pollution biology and ecology; rhizosphere population responses to stress. Mailing Add: Dept of Environ Sci Univ Mass Flint Lab Amherst MA 01002

COLES, DONALD KENNEDY, physics, see 12th edition

COLES, EMBERT HARVEY, JR, b Garden City, Kans, Oct 12, 23; m 46; c 2. CLINICAL PATHOLOGY, MICROBIOLOGY. Educ: Kans State Univ, DVM, 45, PhD(bact, path), 58; Iowa State Univ, MS, 46. Prof Exp: Instr vet hyg, Iowa State Univ, 46-48; pvt pract, Kans, 48-54; from asst prof to prof path & head dept, 54-68, PROF INFECTIOUS DIS & HEAD DEPT, KANS STATE UNIV, 68- Concurrent Pos: Mem, Nat Conf Vet Lab Diagnosticians; dean fac vet med, Ahmadu Bello Univ, Zaria, Nigeria, 70-72; chief of party, Kans State Univ-Agency Int Develop Proj, Abu, Zaria, Nigeria, 70-72; consult, Develop Planning & Res Assoc, Ethiopia, 75. Mem: Am Vet Med Asn; Conf Res Workers Animal Dis; Am Soc Vet Clin Path (pres, 65-66); Am Soc Microbiologists. Res: Animal staphylococci; anemia of trypanosomiasis; bovine leukemia. Mailing Add: Dept of Infectious Dis Kans State Univ Manhattan KS 66506

COLES, JAMES STACY, b Mansfield, Pa, June 3, 13; m 38; c 3. PHYSICAL CHEMISTRY. Educ: Mansfield State Col, BS, 34; Columbia Univ, AB, 36, AM, 39, PhD(phys chem), 41. Hon Degrees: LLD, Brown Univ, 55, Univ Maine, 56, Colby Col, 59, Middlebury Col & Columbia Univ, 62 & Bowdoin Col, 68; DSc, Univ NB, 58; ScD, Merrimac Col, 64. Prof Exp: Instr chem, City Col, 36-41; from instr to asst prof, Middlebury Col, 41-43; res group leader, Underwater Explosives Res Lab, Woods Hole Oceanog Inst, Mass, 43-45, res supvr, 45-46, res consult, 46-68; from asst prof to assoc prof chem, Brown Univ, 46-52, exec off dept chem, 48-52; pres, Bowdoin Col, 52-68; dir, 58-68, PRES, RES CORP, 68- Concurrent Pos: Mem adv comt, NSF, 53-55; trustee, Woods Hole Oceanog Inst, 53-; dir, Dixon Corp, 53-; civilian aide to Secy US Dept Army, 54-57; mem adv comt educ, Int Geophys Year, 58-; dir, Coun Libr Resources, 60-; Am Coun Educ, 62; Chem Fund, Inc, 68-; chmn mine adv comt, Nat Res Coun-Nat Acad Sci, 68-; dir, Res-Cottrell, Inc, 68-; mem adv comt, US Coast Guard Acad, 69-; trustee-at-large, Independent Col Funds of Am, 70-; dir, Pennwalt Corp, 71-; Edo Corp, 71- Mem: AAAS; Am Chem Soc; Am Acad Arts & Sci; assoc NY Acad Sci; hon mem Am Inst Chem. Res: Physical properties of natural high polymers; ultracentrifuge; underwater explosives and measurement of shock waves. Mailing Add: Res Corp 405 Lexington Ave New York NY 10017

COLES, LESLIE STEPHEN, b New York, NY, Jan 19, 41; m 67; c 1. INFORMATION SCIENCE. Educ: Rensselaer Polytech Inst, BS, 62; Carnegie Inst Technol, MS, 64, PhD(systs & commun sci), 67. Prof Exp: Engr assoc, Syst Develop Corp, 62-63; res mathematician, Info Sci Lab, Stanford Res Inst, 67-75; CONSULT, IBM FRANCE, 75- Concurrent Pos: Lectr, Dept Comput Sci, Stanford Univ & Dept Elec Eng & Comput Sci, Univ Calif, Berkeley, 67-75; res grant study info retrieval, NSF, 70. Mem: AAAS; sr mem Inst Elec & Electronics Eng; Systs Man & Cybernet Soc; Asn Comput Mach; Asn Comput Ling. Res: Natural language input to a computer for question-answering and medical information retrieval systems; sematics of English query languages; language understanding for robots; computational linguistics; artificial intelligence; computers in medicine. Mailing Add: IBM France 36 Ave Raymond Poincare 75116 Paris France

COLES, RICHARD WARREN, b Philadelphia, Pa, Sept 16, 39; m 62; c 2. PHYSIOLOGICAL ECOLOGY, ANIMAL BEHAVIOR. Educ: Swarthmore Col, BA, 61; Harvard Univ, MA & PhD(biol), 67. Prof Exp: Asst prof biol, Claremont Cols, 66-70; adj asst prof biol, 70-73, DIR, TYSON RES CTR, WASHINGTON UNIV, 70-, FAC ASSOC BIOL, 73- Concurrent Pos: Consult, Ealing Corp, Mass, 65; Claremont Cols grant, 67-69; dir res field exped, Colo Rockies, 67-70; dir NSF undergrad instrnl equip grant, Claremont Cols, 69-71; prin investr, NIH Animal Care Facil improvement grant, 72-73; ed consult, Nat Geog Mag, 73-74. Mem: AAAS; Am Soc Mammal; Am Ornith Union; Am Soc Zool; Ecol Soc Am. Res: Thermoregulation of the beaver with special emphasis on the role of the beaver's expanded tail in thermoregulation; diving behavior and physiology of the dipper, or water ouzel. Mailing Add: Tyson Res Ctr Washington Univ PO Box 258 Eureka MO 63025

COLES, WILLIAM JEFFREY, b Marquette, Mich, Oct 31, 29; m 55; c 3. MATHEMATICS. Educ: Northern Mich Col, BA, 50; Duke Univ, MA, 52, PhD(math), 54. Prof Exp: Asst, Duke Univ, 50-51; instr math, Univ Wis, 54-55; analyst, Dept Defense, Washington, DC, 55-56; from asst prof to assoc prof math, Univ Utah, 56-63; vis assoc prof, Math Res Ctr, US Army, Univ Wis, 63-64; PROF MATH, UNIV UTAH, 64- Mem: Am Math Soc; Math Asn Am; Soc Indust & Appl Math. Res: Differential equations. Mailing Add: Dept of Math Univ of Utah Salt Lake City UT 84112

COLEY, RONALD FRANK, b Chicago, Ill, Dec 27, 41; m 62; c 2. PHYSICAL CHEMISTRY, INORGANIC CHEMISTRY. Educ: St Procopius Col, 63; Iowa State Univ, PhD(inorg & phys chem), 69. Prof Exp: Asst chem, Iowa State Univ, 63-69; assoc, Biol & Med Res Div, Argonne Nat Lab, 69-70; CHEMIST, COMMONWEALTH EDISON CO, 70- Concurrent Pos: Res assoc, Biol & Med Res Div, Argonne Nat Lab, 70- Mem: Am Chem Soc. Res: Neutron therapy of cancer; Monte Carlo computations of neutron and gamma interactions with matter; gamma and neutron spectrometry; chemical and radionuclide analysis of nuclear power plant systems and effluents. Mailing Add: Commonwealth Edison Co PO Box 767 Chicago IL 60690

COLGATE, SAMUEL ORAN, b Amarillo, Tex, Oct 5, 33; m 55; c 2. PHYSICAL CHEMISTRY. Educ: West Tex State Univ, BS, 55; Okla State Univ, MS, 56; Mass Inst Technol, PhD(phys chem), 59. Prof Exp: Asst prof, 59-66, ASSOC PROF CHEM, UNIV FLA, 66- Concurrent Pos: Vis prof, Harvard & Boston Univs, 69-70. Mem: Am Phys Soc. Res: Derivation of the intermolecular potential from scattering of molecular beams. Mailing Add: Dept of Chem Univ of Fla Gainesville FL 32601

COLGATE, STIRLING AUCHINCLOSS, b New York, NY, Nov 14, 25; m 47; c 3. PHYSICS. Educ: Cornell Univ, BA, 48, PhD(physics), 52. Prof Exp: Physicist, Radiation Lab, Univ Calif, 51-64; pres, 65-74, ADJ PROF PHYSICS, N MEX INST MINING & TECHNOL, 75- Concurrent Pos: Mem sci adv bd, US Air Force, 58-61, gas centrifuge comt, Atomic Energy Comn, 61-69 & fluid dynamics comt, NASA, 61-63; consult, Conf Cessation Nuclear Weapons Tests, US State Dept, 59; lectr, Univ Calif, 59-64; ed, Nuclear Fusion; trustee-at-large, Assoc Univs, Inc, 70-73; chmn subfield surv plasma physics & physics of fluids panel, Nat Acad Sci Physics Surv Comt, 70- Mem: Fel Am Phys Soc; Am Astron Soc; NY Acad Sci; Am Geophys Union. Res: Gamma ray absorption; positron annihilation; linear accelerators; nuclear weapon physics; controlled thermonuclear fusion; plasma physics; linear and toroidal pinches; origin of cosmic rays; astrophysics. Mailing Add: NMex Inst of Mining & Technol Campus Sta Socorro NM 87801

COLGLAZIER, MERLE LEE, b Holyoke, Colo, Aug 6, 20; m 48; c 3. VETERINARY PARASITOLOGY. Educ: Univ Colo, BA, 48. Prof Exp: From parasitologist to sr parasitologist, 48-68, ZOOLOGIST VET CHEMOTHER, ANIMAL PARASITOL INST, AGR RES SERV, BELTSVILLE AGR RES CTR, USDA, 68- Mem: Am Soc Parasitologists. Res: Antiparasitic investigations dealing with chemotherapy and chemical control of helminthic diseases and parasites that affect domestic animals; poultry and fur-bearing animals raised in captivity. Mailing Add: 2712 Philben Dr Adelphi MD 20783

COLICHMAN, EUGENE LOUIS, b Colorado City, Tex, July 11, 18; m 47; c 2. PHYSICAL CHEMISTRY. Educ: Univ Calif, Los Angeles, BA, 40, MA, 42, PhD(phys chem), 44. Prof Exp: Asst chem, Univ Calif, Los Angeles, 41-44; res chemist, Turco Prod, Inc, Calif, 44-47; instr chem, Yale Univ, 48-49; asst prof, Univ Portland, 49-52; res assoc, Stanford Res Inst, 52; supvr anal chem, Calif Res & Develop Co, 52-53; supvr radiation chem, NAm Aviation, 53-57; mem tech staff, Propulsion Res, Space Tech Labs, Inc, 57-60, staff res consult, Marquardt Corp, 60-64; INDUST CONSULT, 64- Concurrent Pos: Instr chem, Univ Southern Calif, 46-48. Mem: Am Chem Soc. Res: Analytical chemistry; physical chemistry of non-aqueous systems; reaction kinetics; surface active compounds; onium salt chemistry; polarography; radiation, propulsion and electrochemistry. Mailing Add: 509 Greencraig Rd Los Angeles CA 90049

COLING, FORREST L, b Lafayette, Ind, Jan 8, 24; m 47; c 3. COMPUTER SCIENCES. Educ: Purdue Univ, BS, 50; Ind State Univ, MS, 51. Prof Exp: Teacher high schs, 51-54; sr engr, NAm Aviation, Inc, 54-59; dept mgr, Raytheon Co, 59-61; group engr, Douglas Aircraft Co, 61-64; indust & govt consult, 64-66; res specialist, Lockheed Calif Co, 66-67; res engr, TRW Syst Div, 67-68; instr comput prog & technol, Hughes Aircraft Co, 68-70; CONSULT COMPUT SOFTWARE, BUDCO DATA SYSTS, 64-67 & 70- Concurrent Pos: Mem, Nat Comt Ceramic Mat. Mem: Inst Elec & Electronics Engrs; Am Inst Aeronaut & Astronaut; Math Asn Am. Res: Microwave devices; propagation; electromagnetic windows; testing, manufacturing and marketing; computer technology; software techniques; medical applications. Mailing Add: 2650 Dalemead St Torrance CA 90505

COLINGSWORTH, DONALD RUDOLPH, b Beaver Dam, Wis, June 20, 12; m 38; c 2. CHEMISTRY, BACTERIOLOGY. Educ: Univ Wis, BS, 34, MS, 36, PhD(bact), 38. Prof Exp: Biochemist, Red Star Yeast & Prod Co, 38-43; res chemist, Heyden Chem Corp, NJ, 43-44; sr res scientist, 44-56, mgr fermentation res & develop, 56-67, GROUP MGR FERMENTATION PROD, UPJOHN CO, 67- Mem: Am Chem Soc; Am Soc Microbiol. Res: Fermentation chemistry; antibiotics. Mailing Add: 1215 Miles Ave Kalamazoo MI 49001

COLINVAUX, LLEWELLYA WILLIAMS, see Hillis-Colinvaux, Llewellya Williams, see 12th edition

COLINVAUX, PAUL ALFRED, b St Albans, Eng, Sept 22, 30; m 60; c 2. ECOLOGY, PALEOECOLOGY. Educ: Cambridge Univ, BA, 56, MA, 60; Duke Univ, PhD(zool), 62. Prof Exp: Res officer pedology, Can Dept Agr, NB, 56-59; NATO fel biol, Queen's Univ, N Ireland, 62-63; res biologist, Yale Univ, 63-64; from asst prof to assoc prof, 64-71, PROF ZOOL, OHIO STATE UNIV, 71-, MEM INST POLAR STUDIES, 64- Concurrent Pos: Guggenheim fel, 71-72; counr for Galapagos Islands, Charles Darwin Found, 75- Mem: Am Soc Limnol & Oceanog; Arctic Inst NAm; Ecol Soc Am; Am Soc Nat. Res: Environmental history of Bering land bridge and Galapagos Archipelago; Galapagos limnology and ecology; pollen analysis of Galapagos, Andean, and Arctic vegetation histories; chronology of Quaternary; ecological models of human history. Mailing Add: Dept of Zool Ohio State Univ 484 W12th Ave Columbus OH 43210

COLL, HANS, b Graz, Austria, June 8, 29. CHEMISTRY. Educ: La State Univ, MS, 55, PhD(chem), 57. Prof Exp: Res assoc phys chem, Cornell Univ, 57-60; res assoc, Mass Inst Technol, 60-62; res chemist, Shell Develop Co, Calif, 62-75; RES CHEMIST, RES LABS, EASTMAN KODAK CO, 75- Concurrent Pos: Grants, Off of Naval Res, 57-60 & NIH, 60-62. Mem: Am Chem Soc. Res: Chemistry of metal complexes; physical chemistry of detergents and high polymers. Mailing Add: Res Labs Eastman Kodak Co Kodak Park Bldg 82C Rochester NY 14650

COLLAGAN, ROBERT BRUCE, b Quincy, Mass, Jan 15, 33; m 54; c 3. GEOLOGY, ASTRONOMY. Educ: Univ Mass, BS, 55; Un Reno, MS, 58; Cath Univ Am, PhD(geol, higher educ), 6 Prof Exp: Resident engr, Commonwealth Mass, 55-56; teacher h Md, 58-61; instr phys sci, 61-62, from asst prof to a prof astron & geol, 62-69, PROF ASTRON & GEOL, MORG COL, 70- Concurrent Pos: Vis lectr, Community Col Baltimore, 62-66 Prince George Community Col, 66-68 & Catonsville V Community Col, 68-; researcher earth & plant sci, Space Flight Ctr, Nat Aeronaut & Space Admin, 70-7 Fulbright lectr, Coun Int Exchange of Scholars-Nat Re Univ of Philippines, Manila, 76-77. Mem: AAAS; Nat Sc Teachers Asn; Nat Asn Res Sci Teaching; Nat Asn Advan Colored People; Am Asn Physics Teachers. Res: Stratig differentiation of selected Nevada black shales by analyses; construction and evaluation of corridor-typ stration experiments in phyiscs; construction and eva of programmed course in mathematics; developing ial modules using multi-media techniques. Mailing Add: Apt B One ship Court Towson MD 21204

COLLARD, WILLIAM DAVID, b Poplar Bluff, Mo, July 6, 36; m 71; c 1. ONCOLOGY, VIROLOGY. Educ: Ark State Univ, BS, 58; Univ Okla, MS, 61; Univ Mo-Columbia, PhD, 68. Prof Exp: Teaching asst microbiol, Univ Okla, 58-61; head tetanus vaccine unit, Mich Dept Health, Lansing, 61-63; teaching asst microbiol, Univ Mo-Columbia, 64-66; RES ASSOC MOLECULAR VIROL, MED SCH, ST LOUIS UNIV, 71- Concurrent Pos: NIH fel, Molecular Virol Inst, Sch Med, St Louis Univ, 68-71. Mem: Am Soc Microbiol. Res: Viral oncology; molecular virology; biochemistry. Mailing Add: Inst Molecular Virol St Louis Univ Sch of Med St Louis MO 63110

COLLAT, JUSTIN WHITE, b New York, NY, Sept 29, 28; m 60; c 3. ANALYTICAL CHEMISTRY. Educ: Harvard Univ, AB, 49, AM, 51, PhD(chem), 53. Prof Exp: Fel chem, Mass Inst Technol, 54; from asst prof to assoc prof chem, Ohio State Univ, 54-66; asst prog adminr, 66-70, PROG ADMINR, PETROL RES FUND, AM CHEM SOC, 70- Mem: Am Chem Soc. Res: Electroanalytical chemistry. Mailing Add: Petrol Res Fund Am Chem Soc 1155 16th St NW Washington DC 20036

COLLEN, MORRIS F, b St Paul, Minn, Nov 12, 13; m 37; c 4. MEDICINE. Educ: Univ Minn, BEE, 34, MB, 38, MD, 39; Am Bd Internal Med, dipl, 46. Prof Exp: Intern Med, Michael Reese Hosp, Chicago, 38-40; resident internal med, Los Angeles County Hosp, 40-42; chief med serv, Kaiser Found Hosp, Oakland, 42-52, med dir, 52-53; MED DIR WEST BAY DIV & CHMN EXEC COMT, PERMANENTE MED GROUP, 53-, DIR MED METHODS RES, 61- Concurrent Pos: Chief staff, Kaiser Found Hosp, San Francisco, 53-61; lectr, Sch Pub Health, Univ Calif, Berkeley,

65-; consult, USPHS, 65-, mem adv comt on demonstration grants, 66-69; chmn health care systs study sect, US Dept Health, Educ & Welfare, 69-72; lectr, Med Sch, Univ Calif, San Francisco; consult, WHO, Europe & Pan-Am Health Orgn. Mem: Inst Med of Nat Acad Sci; fel Am Col Physicians; fel Am Col Cardiol; fel Am Col Chest Physicians; Am Pub Health Asn. Res: Medical research and administration; internal medicine. Mailing Add: Permanente Med Group 3779 Piedmont Ave Oakland CA 94611

COLLERSON, KENNETH DAVID, b Sydney, Australia, Aug 30, 45. PETROLOGY, GEOLOGY. Educ: Univ New Eng, Australia, BSc, 67; Univ Adelaide, PhD(geol), 73. Prof Exp: Asst prof, 72-75, ASSOC PROF GEOL, MEM UNIV NFLD, 75- Concurrent Pos: Mem Can working group archean geochem, Int Union Geol Sci Proj, 75-76. Mem: Mineral Soc; assoc Australian Inst Mining & Metall; fel Geol Asn Can; Mineral Asn Can. Res: Early Archaean gneisses in Northern Labrador which provide data regarding the formation and subsequent tectonic modification of the early terrestrial continental crust. Mailing Add: Dept of Geol Mem Univ of Nfld St John's NF Can

COLLETT, LEONARD STANIER, b Brantford, Ont, Sept 19, 22; m 53. GEOPHYSICS. Educ: McMaster Univ, BSc, 45; Univ Toronto, MA, 48. Prof Exp: Geophysicist induced polarization res, Newmont Explor, Ltd, Ariz, 49-52; EXPLOR GEOPHYSICIST & HEAD TERRAIN GEOPHYS PROG & ELEC METHODS SECT, RESOURCE GEOPHYS & GEOCHEM DIV, GEOL SURV CAN, 53- Mem: Soc Explor Geophys; Am Geophys Union; Can Asn Physicists; Can Soc Explor Geophys; Europ Asn Explor Geophys. Res: Electrical properties of rocks and minerals; electrical methods applied to mineral exploration; electrical and seismic methods applied to engineering geology and geological mapping; former PI lunar materials. Mailing Add: Dept of Energy Mines & Resources Geol Surv of Can 601 Booth St Ottawa ON Can

COLLETTE, ALFRED THOMAS, b Syracuse, NY, Sept 10, 22. BOTANY, HUMAN GENETICS. Educ: Syracuse Univ, AB, 47, MS, 48, PhD, 52. Prof Exp: Asst sci educ, 49-50, instr genetics, 50-53; from asst prof to assoc prof genetics & sci educ, 54-60, PROF GENETICS & SCI EDUC & CHMN DIV SCI TEACHING, 60- Mem: Am Soc Human Genetics; Am Genetic Asn; Am Eugenics Soc; Nat Sci Teachers Asn; Nat Asn Res Sci Teaching. Res: Science education. Mailing Add: Ctr for Sci Educ 410 Lyman Hall Syracuse Univ Syracuse NY 13210

COLLETTE, BRUCE BADEN, b Brooklyn, NY, Mar 14, 34; m 56; c 3. ICHTHYOLOGY Educ: Cornell Univ, BS, 56, PhD(vert zool), 60. Prof Exp: SYST ZOOLOGIST, NAT MARINE FISHERIES SERV SYST LAB, NAT MUS NATURAL HIST, 60-, ASST LAB DIR, 63-, RES ASSOC, DEPT VERT ZOOL, SMITHSONIAN INST, 67- Concurrent Pos: Ichthyol ed, Copeia, Am Soc Ichthyologists & Herpetologists, 64-69; sci ed, Nat Marine Fisheries Serv, 74- Mem: Fel AAAS; Am Soc Ichthyologists & Herpetologists (secy, 74-); Am Fisheries Soc; Soc Study Evolution; Ecol Soc Am. Res: Systematics, distribution and evolution of fishes, especially Hemiramphidae, Belonidae, Percidae and Scombridae. Mailing Add: Nat Marine Fish Serv Syst Lab Nat Mus of Natural Hist Washington DC 20560

COLLETTE, JOHN WILFRED, b Calgary, Alta, Can, July 20, 33; nat US; m 52; c 3. ORGANIC CHEMISTRY. Educ: Univ Alta, BSc, 55; Univ Calif, PhD(chem), 58. Prof Exp: Res chemist, 58-64, develop chemist, 64-66, res supvr, 66-69, SUPVR CENT RES DEPT, E I DU PONT DE NEMOURS & CO, INC, 69- Mem: Am Chem Soc; The Chem Soc; AAAS. Res: Olefin polymerization; chemistry of organo metallic compounds; effect of micro structure on elastomer properties; chemistry of higher acetylenics. Mailing Add: Cent Res Dept du Pont Exp Sta E I du Pont de Nemours & Co Inc Wilmington DE 19898

COLLEY, DANIEL GEORGE, b Buffalo, NY, Jan 21, 43; m 65. IMMUNOLOGY, MICROBIOLOGY. Educ: Cent Col Ky, BA, 64; Tulane Univ, PhD(microbiol), 68. Prof Exp: Fel, Yale Univ, 68-70; vis prof immunol, Fed Univ Pernambuco, 70-71; asst prof, 71-74, ASSOC PROF IMMUNOL, VANDERBILT UNIV, 74- Concurrent Pos: Res immunologist, Vet Admin Hosp, 71-; vis prof immunol, Fed Univ Minas Gerais, 74; assoc ed, J Immunol, 74- Mem: AAAS; Am Asn Immunologists; Am Soc Trop Med & Hyg; Am Soc Microbiol. Res: Immunobiology with specific activity in the areas of cellular hypersensitivity and thymic development also extending into immunopathology and resistance with specific reference to schistosomiasis. Mailing Add: Vet Admin Hosp 1310 24th Ave S Nashville TN 37203

COLLEY, FREDRICK CHRISTENSEN, b Indio, Calif, Jan 5, 38; m 65. PARASITOLOGY, PUBLIC HEALTH. Educ: Univ Calif, Riverside, BA, 59; San Diego State Col, MA, 62; Ariz State Univ, PhD(zool), 65. Prof Exp: Asst biol, San Diego State Col, 61-62; asst zool, Ariz State Univ, 62-64, fac assoc, 64-65; clin instr trop med, Sch Med, La State Univ, New Orleans, 65-67; asst res parasitologist, 67-75, RES ASSOC, GEORGE WILLIAMS HOOPER FOUND, MED CTR, UNIV CALIF, SAN FRANCISCO, 75- Concurrent Pos: USPHS res fel, 65-67; fel, La State Univ Inter-Am Prog Trop Med, 66; MPH, Univ Calif, Berkeley, 73; overseas assignment with Univ Calif Int Ctr for Med Res, Inst Med Res, Kuala Lumpur, Malaysia, 70-72; dep resident coordr, Univ Calif Int Ctr for Med Res, Malaysia, 73-75. Mem: Nat Audubon Soc; Am Soc Parasitol; Soc Protozool; Malaysian Soc Parasitol & Trop Med. Res: Biology, taxonomy and fine structure of parasitic protozoa; biological control of parasitic helminths; community health. Mailing Add: George Williams Hooper Found Univ of Calif Med Ctr San Francisco CA 94122

COLLIAS, ELSIE COLE (MRS NICHOLAS E COLLIAS), b Tiffin, Ohio, Mar 24, 20; m 48; c 1. ZOOLOGY. Educ: Heidelberg Col, BA, 42; Univ Wis, MS, 44, PhD(zool), 48. Prof Exp: Asst zool, Univ Wis, 42-46, asst econ entom, 47-48; asst prof biol, Heidelberg Col, 48-49; instr, Univ Wis, 49-50; assoc prof, Ill Col, 53-57; res assoc entom, Univ Calif, Los Angeles, 59-62; RES ASSOC, LOS ANGELES COUNTY MUS, 62-; RES ASSOC ZOOL, UNIV CALIF, 64- Concurrent Pos: Entomologist, USPHS, Ga, 46-47. Mem: Animal Behav Soc; Am Ornithologist Union. Res: Bird behavior; invertebrate zoology; insect physiology and behavior. Mailing Add: Dept of Zool Univ of Calif Los Angeles CA 90024

COLLIAS, EUGENE EVANS, b Cumberland, Wash, Feb 3, 25; m 49; c 2. OCEANOGRAPHY. Educ: Univ Wash, MS, 51 & Scripps Inst, Univ Calif, 59. Prof Exp: Res instr oceanog, 54-56, sr oceanogr, 59-70, PRIN OCEANOGR, UNIV WASH, 70- Mem: Marine Technol Soc; Am Geophys Union; Sigma Xi; Am Soc Limnol & Oceanog. Res: Descriptive and chemical oceanography; field methods of oceanography. Mailing Add: Mailstop WB-10 Dept of Oceanog Univ of Wash Seattle WA 98195

COLLIAS, NICHOLAS ELIAS, b Chicago Heights, Ill, July 19, 14; m 48; c 1. ZOOLOGY. Educ: Univ Chicago, BS, 37, PhD(zool), 42. Prof Exp: Asst zool, Univ Chicago, 37-42; instr biol, Chicago City Jr Col, 46 & Amherst Col, 46-47; inst zool, Univ Wis, 47-51; conserv biologist, State Conserv Dept, Wis, 51-52; USPHS spec res fel, Cornell Univ, 52-53; prof biol, Ill Col, 53-58; from asst prof to assoc prof, 58-65, PROF ZOOL, UNIV CALIF, LOS ANGELES, 66- Concurrent Pos: Guggenheim fel, 62-63; hon res assoc, Los Angeles County Mus, 62-; hon res assoc, Percy Fitzpatrick Inst African Ornith, Univ Capetown, 69-70 & Nat Mus Kenya, Nairobi, 73. Mem: Am Soc Zool; Ecol Soc Am; fel Am Ornith Union; Cooper Ornith Soc; fel Animal Behav Soc. Res: Animal sociology and ecology; field studies of behavior and populations in birds and mammals; hormones and behavior; analysis of vocal communication in animals; nest-building in birds; ornithology; behavioral energetics. Mailing Add: Dept of Biol Univ of Calif Los Angeles CA 90024

COLLIER, ALBERT WALKER, b Nowata, Okla, Dec 12, 10; m 37, 63; c 2. MARINE BIOLOGY, OCEANOGRAPHY. Educ: Rice Inst, BA, 33. Prof Exp: Marine biologist & dir marine lab, Tex Game, Fish & Oyster Comn, 35-39; jr aquatic biologist, US Fish & Wildlife Serv, Washington, DC, 39-42; instrument mechanic, Graphic Arts Lab, US Navy, 42-45; mgr coastal prod, Wicker Fish & Poultry Co, Tex, 46; consult marine biologist, 46-50; chief, Gulf Fishery Invests, 50-57; chief scientist, Galveston Marine Lab, Agr & Mech Col, Tex, 58-62; dir oceanog inst, 62-65, PROF BIOL, FLA STATE UNIV, 62- Mem: Fel AAAS; Biol Photog Asn; Am Geophys Union; Am Chem Soc; hon fel Am Inst Fishery Res Biologists. Res: Marsh ecology; marine productivity; mechanisms of plankton blooming; natural organic compounds in seawater; diatom taxonomy; oyster biology. Mailing Add: Dept of Biol Sci Fla State Univ Tallahassee FL 32306

COLLIER, BOYD DAVID, b Sacramento, Calif, Aug 14, 37; m 58; c 2. POPULATION ECOLOGY. Educ: Univ Calif, Berkeley, BA, 60; Cornell Univ, MST, 64, PhD(evolutionary biol), 66. Prof Exp: From asst prof to assoc prof, 66-72, PROF BIOL, SAN DIEGO STATE UNIV, 72- Mem: Ecol Soc Am; Brit Ecol Soc; Japanese Soc Pop Ecol. Res: Population ecology of terrestrial animals, particularly insects; computer simulation of ecological systems. Mailing Add: Dept of Biol San Diego State Univ San Diego CA 92182

COLLIER, BRIAN, b York, Eng, June 8, 40; m 65; c 3. PHARMACOLOGY, PHYSIOLOGY. Educ: Univ Leeds, BSc, 62, PhD(pharmacol), 65. Prof Exp: Brit Med Res Coun fel, Cambridge Univ, 65-66; lectr physiol, 66-67, asst prof, 67-72, ASSOC PROF PHARMACOL, McGILL UNIV, 72- Concurrent Pos: Med Res Coun Can scholar, 68- Mem: AAAS; Am Soc Pharmacol & Exp Therapeut; Can Physiol Soc; Pharmacol Soc Can. Res: Identity, synthesis, storage, release and fate of neurotransmitter substances in the central and peripheral nervous system of mammals; physiology and pharmacology of cholinergic synapses. Mailing Add: Dept of Pharmacol & Therapeut McGill Univ Montreal PQ Can

COLLIER, CLARENCE ROBERT, b Freeport, Ill, Mar 25, 19; m 42; c 3. PHYSIOLOGY. Educ: Andrews Univ, BA, 40; Loma Linda Univ, MD, 49. Prof Exp: Instr sci, Hylandale Acad, 40-41; from instr to asst prof med, Loma Linda Univ, 52-57, assoc prof physiol, 57-64, prof physiol & biophys & chmn dept, 64-70; assoc prof med, 70-71, PROF MED & PHYSIOL, SCH MED, UNIV SOUTHERN CALIF, 71- Concurrent Pos: Nat Found Infantile Paralysis res fel, Sch Pub Health, Harvard Univ, 55-56; sr res fel, NIH, 59-62; res assoc, Rancho Los Amigos Hosp, 52-55, dir res, 56-57, chief med sci serv, 62-64, consult, 58-; consult, Rand Corp, 63-65; consult physiol, Christian Med Col, Vellore, India, 72- Mem: AAAS; Am Physiol Soc; Am Fedn Clin Res. Res: Pulmonary physiology; pulmonary exchange and circulation; mechanics of breathing; biological models. Mailing Add: Sch of Med Univ of Southern Calif Los Angeles CA 90033

COLLIER, DONALD, b Sparkill, NY, May 1, 11; m 39; c 2. ANTHROPOLOGY, ARCHAEOLOGY. Educ: Univ Calif, Berkeley, BA, 33; Univ Chicago, PhD(anthrop), 55. Prof Exp: Instr anthrop, Wash State Col, 40-41; from asst cur to cur SAm archaeol & ethnol, 41-64, chief cur anthrop, 64-70, CUR MID & S AM ARCHAEOL & ETHNOL, FIELD MUS NATURAL HIST, 71- Concurrent Pos: Inst Andean Res grant archaeol fieldwork, Ecuador, 41-42; Wenner-Gren Found grant archaeol res, Viru Valley, Peru, 46; rev ed, Am Anthrop Asn, 49-50; lectr anthrop, Univ Chicago, 49-; NSF grants, Peru, 56 & Int Cong, 58, 60, 64 & 68; rev ed, Soc Am Archaeol, 58-62. Mem: Fel Am Anthrop Asn; Soc Am Archaeol; Inst Andean Res. Res: Indians of North, Middle and South America; the rise of urbanism and civilization in Mexico and Peru. Mailing Add: Field Mus of Natural Hist Chicago IL 60605

COLLIER, FRANCIS NASH, JR, b New York, NY, Feb 11, 17; m 47; c 3. INORGANIC CHEMISTRY. Educ: Howard Col, BS, 42; Ohio State Univ, MS, 49, PhD, 57. Prof Exp: Asst prof chem & physics, Howard Col, 43-46, assoc prof chem, 49-53, assoc prof chem & physics & chmn dept physics, 56-57; assoc prof, 57-70, PROF CHEM, UNIV NC, 70- Mem: Am Chem Soc. Res: Kinetic studies of halomines in liquid ammonia; molecular addition compounds; anhydrous metalhalides; isotope exchange studies. Mailing Add: Dept of Chem Venable Hall Univ of NC Chapel Hill NC 27515

COLLIER, GERALD, b Monterey Park, Calif, Nov 16, 30; m; c 1. VERTEBRATE ZOOLOGY, ANIMAL BEHAVIOR. Educ: Univ CAlif, Los Angeles, BA, 53, MA, 58, PhD(zool), 64. Prof Exp: Asst prof, 61-69, ASSOC PROF ZOOL, SAN DIEGO STATE UNIV, 69- Concurrent Pos: Grant, Mex, 65; Res Found grant, Mex & Costa Rica, 66 & 69, Sigma Xi grant, Costa Rica, 70; ecol consult, Dillingham Environ Co, 70- Mem: Cooper Ornith Soc; Am Ornith Union; Soc Study Evolution; Wilson Ornith Soc; Soc Syst Zool. Res: Avian behavior and ecology; functional vertebrate anatomy. Mailing Add: Dept of Zool San Diego State Univ San Diego CA 92115

COLLIER, GERALD LOYD, b Farmersville, Tex, Aug 20, 28; m 52; c 2. ECONOMIC GEOGRAPHY. Educ: Southern Methodist Univ, BA, 52; Univ Nebr, MA, 54, PhD(geog), 64. Prof Exp: Intel officer, US Govt, 56-58; from asst prof to assoc prof geog, Univ Southern Miss, 58-64; from asst prof to assoc prof, 64-70, head dept, 67-74, PROF GEOG, STEPHEN F AUSTIN STATE UNIV, 70- Mem: Asn Am Geog; Nat Coun Geog Educ; Am Geog Soc. Res: Geography of Anglo-America. Mailing Add: Dept Geog S F Austin State Univ SFA Sta Box 3012 Nacogdoches TX 75961

COLLIER, HERBERT BRUCE, b Toronto, Ont, Oct 10, 05; m 30; c 2. BIOCHEMISTRY. Educ: Univ Toronto, BA, 27, MA, 29, PhD(biochem), 30. Prof Exp: From asst prof to assoc prof biochem, Col Med & Dent, WChina Union, 32-39; biochemist, Inst Parasitol, Macdonald Col, McGill Univ, 39-42; from asst prof to assoc prof biochem, Dalhousie Univ, 42-46; prof & head dept, Univ Sask, 46-49; head dept biochem, 49-61, prof, 49-64, prof clin biochem, 64-71, EMER PROF CLIN BIOCHEM, UNIV ALTA, 71- Concurrent Pos: Consult biochem, Path Assocs, Edmonton, 71- Mem: Am Chem Soc; Am Soc Biol Chem; Royal Soc Can; fel Int Soc Hemat. Res: Clinical biochemistry; chemistry of enzymes; drug action; biochemistry of erythrocytes. Mailing Add: Dept of Path Clin Sci Bldg Univ of Alta Edmonton AB Can

COLLIER, HERMAN EDWARD, JR, b St Louis, Mo, Aug 8, 27; m 48; c 3. ANALYTICAL CHEMISTRY, INORGANIC CHEMISTRY. Educ: Randolph-Macon Col, BS, 50; Lehigh Univ, MS, 52, PhD(anal chem), 55. Hon Degrees: LLD, Lehigh Univ, 71. Prof Exp: Prof chem & chmn dept, Moravian Col, 55-57; res chemist, E I du Pont de Nemours & Co, 57-63; prof chem & chmn div natural sci & math, 63-69, PRES, MORAVIAN COL, 69- Mem: Am Chem Soc; fel Am Inst

Chemists; Sigma Xi. Res: New developments in flame photometry; the hydrogenfluorine flame; quantitative analytical infrared analysis; determination of enol content; differential reaction rates as an analytical tool. Mailing Add: Moravian Col Main St & Elizabeth Ave Bethlehem PA 18018

COLLIER, JACK REED, b Louisville, Ky, Aug 19, 26. EMBRYOLOGY. Educ: Univ Ky, BS, 48, MS, 50, PhD(zool), Univ NC, 54. Prof Exp: USPHS fel, Tokyo Metrop Univ, 54-55 & Calif Inst Technol, 55-56; instr biol, Univ Vt, 56-57; asst prof, La State Univ, 57-59; independent investr, Marine Biol Lab, Woods Hole, 59-63; assoc prof, Rensselaer Polytech Inst, 63-66; PROF BIOL, BROOKLYN COL, 66- Concurrent Pos: Mem corp, Marine Biol Lab. Mem: Am Soc Zool; Soc Study Develop Biol; Biophys Soc; Soc Gen Physiol; Am Soc Cell Biol. Res: Invertebrate chemical embryology; molluscan and echinoderm fertilization; nucleic acid and protein metabolism of the gastropod embryo. Mailing Add: Dept of Biol Brooklyn Col New York NY 11210

COLLIER, JAMES BRYAN, b Portland, Maine, Apr 9, 44; m 72. GEOMETRY. Educ: Carleton Col, BA, 66; Univ Wash, PhD(math), 72. Prof Exp: Fel math, Dalhousie Univ, 72-73; ASST PROF MATH, UNIV SOUTHERN CALIF, 73- Mem: Am Math Soc. Res: Convex functions; convex sets; polytopes; graphs. Mailing Add: Dept of Math Univ of Southern Calif Los Angeles CA 90007

COLLIER, JAMES ELI, b Mt Sharon, Ky, Jan 17, 13; m 36; c 2. GEOGRAPHY. Educ: Western Ky State Teachers Col, BS, 36; Univ Cincinnati, MA, 38; Univ Nebr, PhD(geog), 51. Prof Exp: Instr geog, Henderson State Teachers Col, 40-41, Univ Mo, 41-45 & Univ Nebr, 45-46; asst prof, Fresno State Col, 46-47; from asst prof to assoc prof, Univ Mo, 47-64; PROF GEOG, SOUTHERN ILL UNIV, 64- Mem: Asn Am Geog; Nat Coun Geog Educ. Res: Regional economic geography, including regional development of the Ozark-Ouachita region of Missouri-Arkansas-Oklahoma; also cartographic symbolism. Mailing Add: Dept of Earth Sci Southern Ill Univ Edwardsville IL 62025

COLLIER, JESSE WILTON, b Killeen, Tex, Dec 20, 14; m 40; c 2. AGRONOMY, GENETICS. Educ: Tex A&M Univ, BS, 38, MS, 52; Rutgers Univ, PhD(agron), 57. Prof Exp: Jr soil surveyor, Soil Conserv Serv, USDA, 38-41; instr agr, Tex A&M Univ, 41-43; asst agr, Tex Agr Exp Sta, 43-53; res assoc, Rutgers Univ, 53-55; asst agronomist, 55-57, assoc prof, 57-72, PROF AGRON, TEX AGR EXP STA, 72-; ASSOC PROF AGRON, TEX A&M UNIV, 64- Mem: Am Soc Agron; Crop Sci Soc Am. Res: Plant breeding methods in corn and grain sorghum; foundation seed production of several crops including small grains, corn, grain sorghum and vegetables. Mailing Add: Dept of Soil & Crop Sci Tex A&M Univ College Station TX 77843

COLLIER, JOHN RAYMOND, b Pratt, Kans, Dec 21, 17; m 43; c 3. VETERINARY MICROBIOLOGY. Educ: Ohio State Univ, DVM, 41; Iowa State Univ, MS, 51, PhD(vet bact), 55; Am Col Vet Microbiol, dipl. Prof Exp: Asst prof vet hyg, Iowa State Col, 49-56; from assoc prof to prof vet path & bact, 56-60, prof vet med microbiol & head dept path & microbiol, 60-66, PROF VET MICROBIOL, COLO STATE UNIV, 66- Mem: Am Vet Med Asn; Am Pub Health Asn; Am Soc Microbiol. Res: Infectious diseases of domestic animals; etiology; pathogenesis; diagnosis; immunology; streptoccal infections of swine; cattle respiratory disease; enteric disease of calves. Mailing Add: Dept of Microbiol Col of Vet Med Colo State Univ Ft Collins CO 80521

COLLIER, ROBERT EUGENE, b Douglas, Ariz, Sept 7, 26; m 51; c 3. MICROBIOLOGY. Educ: Univ Okla, BS, 49, MS, 50; Univ Ill, PhD(bact), 58. Prof Exp: Asst prof biol & chem, Phillips Univ, 50-53; instr bact, Univ Ill, 57; res bacteriologist & asst dir labs, Swift & Co, 58-59; assoc prof bact & asst dean col arts & sci, Univ Okla, 59-67; prof biol & vpres acad affairs, ETex State Univ, 67-70; PRES, NORTHEASTERN OKLA STATE UNIV, 70- Mem: Am Soc Microbiol. Res: Bacterial spore formation and germination; industrial microbiology. Mailing Add: Northeastern Okla State Univ Tahlequah OK 74464

COLLIER, ROBERT JACOB, b Springfield, Mass, May 27, 26; m 56; c 2. ELECTRON OPTICS. Educ: Yale Univ, BS, 50, MS, 51, PhD(physics), 54. Prof Exp: Mem tech staff res & develop, 54-59, supvr, 59-71, SUPVR ELECTRON OPTICS GROUP, BELL LABS, 71- Mem: Am Phys Soc; fel Optical Soc Am. Res: High power traveling wave tube amplifiers; optical memory devices; holography; scanning electron beam devices; design of electron optical systems for electron lithography. Mailing Add: Bell Labs Murray Hill NJ 07974

COLLIER, ROBERT JOHN, b Wichita Falls, Tex, Aug 6, 38; m 62; c 3. PATHOBIOLOGY, BIOCHEMISTRY. Educ: Rice Univ, BA, 59; Harvard Univ, MS, 61, PhD(biol), 64. Prof Exp: NIH fel Harvard Univ, 64; NSF fel, Inst Molecular Biol, Geneva, Switz, 64-66; from asst prof to assoc prof, 66-74, PROF BACT, UNIV CALIF, LOS ANGELES, 74- Concurrent Pos: Guggenheim Found fel, Pasteur Inst, Paris, France, 73-74. Honors & Awards: Eli Lilly & Co Award Microbiol & Immunol, Am Soc Microbiol, 72. Mem: AAAS; Am Soc Microbiol; Am Soc Biol Chem. Res: Structure and activity of bacterial toxins; protein synthesis in mammalian systems. Mailing Add: Dept of Bact Univ of Calif Los Angeles CA 90024

COLLIER, SUSAN S, b Washington, DC, Nov 5, 39. PHOTOCHEMISTRY. Educ: Cornell Univ, BA, 61; Univ Rochester, PhD(spectros), 66. Prof Exp: Fel, Univ Rochester, 66; D J Wilson vis res assoc, Ohio State Univ, 66-69, J G Calvert spec fel air pollution, 69; sr chemist, 69-74, RES ASSOC, RES LABS, EASTMAN KODAK CO, 74- Mem: Am Chem Soc; Soc Photog Sci & Eng. Res: Ultraviolet spectroscopy; photochemistry of gaseous air pollutants; spectral sensitization; low temperature luminescence; microwave photoconductivity. Mailing Add: Res Lab Eastman Kodak Co Rochester NY 14650

COLLIGAN, JOHN JOSEPH, b Watertown, NY, Feb 8, 37; m 58; c 3. ELECTROMAGNETICS. Educ: Le Moyne Col, NY, BS, 58; Univ Notre Dame, MS, 60; State Univ NY Buffalo, PhD(policy sci), 72. Prof Exp: Jr engr, Int Bus Mach Corp, 60-61, from assoc engr to sr assoc engr, 61-64; asst prof physics, Broome Tech Commun Col, 64-67; admin asst, 67-68, asst dean, 68-69, ASSOC DEAN, SCH ADVAN TECHNOL, STATE UNIV NY BINGHAMTON, 69-, DIR STATE TECH SERV, 67- Concurrent Pos: Consult, Comn Col Physics Teachers, 69- Mem: Am Phys Soc; Am Soc Eng Educ. Mailing Add: Sch of Advan Technol State Univ of NY Binghamton NY 13901

COLLIN, PIERRE-PAUL, b Montreal, Que, July 23, 20; m 49; c 5. PEDIATRIC SURGERY. Educ: Univ Montreal, BA, 41, MD, 48. Prof Exp: Asst prof, 66-70, ASSOC PROF SURG, FAC MED, UNIV MONTREAL, 70-; DIR SURG, STE JUSTINE HOSP, 66- Concurrent Pos: Consult, Hopital Misericorde, 65-, Hopital Marie-Enfant, 70- Mem: Fel Am Col Surg; fel Int Col Surg; fel Am Acad Pediat; foreign mem Brit Asn Pediat Surg; Can Asn Pediat Surg. Mailing Add: Dept of Surg Univ of Montreal Montreal PQ Can

COLLIN, ROBERT LOUIS, physical chemistry, see 12th edition

COLLIN, WILLIAM KENT, b Los Angeles, Calif, Mar 31, 38; m 64; c 2. MEDICAL MICROBIOLOGY. Educ: Univ Calif, Davis, BA, 61; Univ Calif, Los Angeles, PhD(med microbiol & immunol), 68. Prof Exp: Fel electron micros virol, parasitol, Sch Pub Health, Univ Calif, Los Angeles, 68-70, from asst res virologist to asst res pediatrician, Dept Pediat, Sch Med, 70-74; biol res coordr dermatol, Redken Labs Inc, 74-75; LECTR MICROBIOL, CALIF STATE UNIV, FRESNO, 75- Concurrent Pos: NIH fel, 68. Mem: Am Soc Parasitologists; AAAS; Sigma Xi. Res: Cytodifferentiation and histogenesis of cestode larval forms; lymphocytic responsiveness in viropathology. Mailing Add: Dept of Biol Calif State Univ Fresno CA 93740

COLLINGS, CHARLES KENNETH, b Princeton, Mo, Apr 18, 05; m 26; c 2. DENTISTRY. Educ: Univ Mo, BS, 27, MA, 32; Baylor Univ, DDS, 49; Am Bd Periodont, dipl, 54. Prof Exp: High sch teacher, Mo, 27-31; asst zool, Univ Mo, 31-32, instr, 32-33; high sch teacher, Mo, 33-44; asst prof anat & physiol, 44-48, assoc prof, 48-49, prof, 49-51, chmn dept periodont, 51-73, EMER PROF PERIODONT, COL DENT, BAYLOR UNIV, 73- Concurrent Pos: Consult, William Beaumont Army Hosp, El Paso, Tex, USPHS Hosp & Carswell AFB, Ft Worth, Vet Admin Hosp, Dallas & Brooke Army Hosp, San Antonio. Honors & Awards: Piper Prof Tex, Minnie Stevens Piper Found, 62; Distinguished Serv Award, Tex State Dental Asn, 75. Mem: AAAS; Am Acad Periodont; Am Dent Asn; Soc Exp Biol & Med; fel Am Col Dent. Res: Physiological effects of terramycin mouthwash on oral flora; rate of alveolar bone resorption in dentulous and edentulous mouths. Mailing Add: 2020 W Five Mile Pkwy Dallas TX 75224

COLLINGS, EDWARD WILLIAM, b New Plymouth, NZ, Jan 22, 30; m 53; c 2. EXPERIMENTAL SOLID STATE PHYSICS. Educ: Univ NZ, BSc, 51, MSc, 52, PhD(physics), 58; Victoria Univ Wellington, DSc(physics), 70. Prof Exp: From jr lectr to lectr physics, Victoria Univ Wellington, 52-61; sr res physicist, Franklin Inst Res Labs, Philadelphia, Pa, 62-66; sr physicist, 66-67, assoc fel, 67-68, fel, 68-72, PRIN PHYSICIST, COLUMBUS LABS, BATTELLE MEM INST, 72- Concurrent Pos: Nat Res Coun Can fel, 58-60. Mem: AAAS; Am Phys Soc; Metall Soc; Am Inst Mining, Metall & Petrol Eng; Mat Res Soc. Res: Magnetic susceptibility, low-temperature specific heat and other electronic properties; rapidly quenched and glassy metals; superconducting materials—properties, fabrication and applications. Mailing Add: Battelle Columbus Labs 505 King Ave Columbus OH 43201

COLLINGS, WILLIAM DOYNE, b Rockville, Ind, Oct 11, 13; m 41; c 2. BIOLOGY. Educ: DePauw Univ, AB, 35; Princeton Univ, AM, 37, PhD(biol), 38. Prof Exp: Proctor fel, Princeton Univ, 38-39, E R Squibb & Sons fel biol, 39-40; asst prof physiol, Univ Okla, 40-42; from instr to asst prof, Med Col, Univ Tex, 43-46; asst prof, Univ Iowa, 46-49; assoc prof, 49-57, PROF PHYSIOL, MICH STATE UNIV, 57-, ASSOC CHMN DEPT, 67- Mem: Am Physiol Soc; Soc Exp Biol & Med; Am Soc Nephrol. Res: Physiology of the adrenal cortex; experimental hypertension; renal function. Mailing Add: Dept of Physiol Mich State Univ East Lansing MI 48823

COLLINS, ALFRED PATTERSON, b Anamosa, Iowa, Apr 5, 27; m 56; c 4. PHARMACEUTICAL CHEMISTRY, PHARMACOLOGY. Educ: Univ Iowa, BS, 52, MD, 54, PhD(pharm), 56. Prof Exp: Dept head, Vet Pharmaceut Labs, Chas Pfizer & Co, Inc, 55-60; sect head phys & anal chem, Strasenburgh Labs, 60-63, head, Pharmaceut Chem Dept, 63-68; MGR PHARMACEUT PROD CONTROL, PFIZER, INC, 68- Mem: AAAS; Am Chem Soc; Am Pharmaceut Asn. Res: Analytical chemistry; absorption spectroscopy; gas chromatography; formulation and development of new pharmaceutical products; stability and degradation of new drugs, vitamins, antibiotics and hormones. Mailing Add: Pfizer Inc Eastern Point Rd Groton CT 06355

COLLINS, ALLAN CLIFFORD, b Milwaukee, Wis, June 14, 42; m 62; c 2. BIOCHEMICAL PHARMACOLOGY, BEHAVIORAL GENETICS. Educ: Univ Wis-Milwaukee, BS, 65, MS, 67, PhD(pharmacol), 69. Prof Exp: Instr pharmacol, Univ Wis-Milwaukee, 68-69; NIH fel, Med Sch, Univ Colo, Denver, 69-71; res pharmacologist, Vet Admin Hosp, Houston, Tex, 71-72; ASST PROF PHARMACOL, UNIV COLO, BOULDER, 72-, FEL, INST BEHAV GENETICS, 73- Mem: Am Soc Neurochem. Res: Biochemical bases of tolerance to and dependence upon alcohol, barbiturates and nicotine; use of behavior genetic techniques to test hypotheses concerning mechanisms by which these drugs exert their behavioral effects. Mailing Add: Sch of Pharm Univ of Colo Boulder CO 80309

COLLINS, ALVA LEROY, JR, b Sanford, Fla, May 15, 40; m 67. INORGANIC CHEMISTRY, ORGANOMETALLIC CHEMISTRY. Educ: Oberlin Col, AB, 62; Duke Univ, MA, 66, PhD(chem), 67. Prof Exp: Res chemist, Am Cyanamid Co, 66-67; res assoc inorg chem, Ind Univ, 67-68; ASST PROF CHEM, SAM HOUSTON STATE UNIV, 68- Mem: Am Chem Soc. Res: Reactions of silicon-nitrogen and boron-nitrogen compounds; chemistry of boranes. Mailing Add: Dept of Chem Sam Houston State Univ Huntsville TX 77340

COLLINS, AMY L TSUI, b Shanghai, China, Apr 30, 48; Hong Kong citizen; m 74. RADIOTHERAPY, IMMUNOLOGY. Educ: Univ Minn, Minneapolis, BA, 70; Stanford Univ, PhD(med microbiol), 75. Prof Exp: Res specialist genetics & cell biol, 74-75, RES ASSOC RADIOTHER & IMMUNOTHER, UNIV MINN, MINNEAPOLIS, 75- Mem: Am Soc Microbiol. Res: Investigation of the combined effects of immunotherapy and radiotherapy. Mailing Add: Dept of Therapeut Radiol Box 494 Mayo Bldg Univ of Minn Minneapolis MN 55455

COLLINS, ARLEE GENE, b Forest City, Iowa, Dec 20, 27; m 53; c 2. GEOCHEMISTRY. Educ: Kletzing Col, BA, 51; Kans State Col Pittsburg, MS, 55; Univ Tulsa, MS, 72. Prof Exp: Chemist, H C Maffitt, Consult Chemist, Iowa, 51-53 & Spencer Chem Co, Kans, 53-56; asst chief chemist, Consumers Coop Refinery, 56; proj leader geochem of petrol reservoirs, US Bur Mines, 56-75, PROJ LEADER, BARTLESVILLE ENERGY RES CTR, ENERGY RES & DEVELOP ADMIN, 75- Concurrent Pos: Mem task group, Fed Adv Comt on Water Data. Mem: Am Chem Soc; Geochem Soc Am; Am Soc Testing & Mat; Int Asn Geochem & Cosmochem; AAAS. Res: Geochemistry of oil and gas reservoirs for the characterization of reservoirs for enhanced recovery. Mailing Add: Bartlesville Energy Res Ctr Energy Res & Develop Admin Bartlesville OK 74003

COLLINS, BARBARA JANE, b Passaic, NJ, Apr 29, 29; m 55; c 5. BOTANY. Educ: Bates Col, BS, 51; Smith Col, MS, 53; Univ Ill, PhD(geol), 55, MS, 59. Prof Exp: Instr geol, Univ Ill, 57, teaching asst bot, 57-59; ASSOC PROF BIOL, CALIF LUTHERAN COL, 63- Mem: Am Soc Microbiol. Res: Taxonomy of gymnosperms and angiosperms; keys to the flora of regional areas; electron microscopy of clay minerals; replication. Mailing Add: Dept of Bot Calif Lutheran Col Thousand Oaks CA 91360

COLLINS, BILL MARTIN, b Tiptonville, Tenn, July 23, 37; div; c 1. SPEECH PATHOLOGY. Educ: Southeast Mo State Univ, BS, 61; Univ Mo, Columbia, AM, 67, PhD(speech path), 70. Prof Exp: Speech clinician, Charleston Pub Schs, Mo, 61-62; ins adjuster, Gen Adjust Bur, Hannibal, 62-64; speech clinician, Lee County Bd Educ, Iowa, 64-66; CHIEF SPEECH PATH & AUDIO, HEALTH SCI, CHILD

DEVELOP CTR, UNIV TENN, 70- Mem: Am Speech & Hearing Asn; Am Cleft Palate Asn; Am Asn Ment Deficiency; Am Asn Univ Prof. Res: Language development of small-for-date and true premature children and language characteristics of overgrown infants above eight pounds; velopharyngeal closure in infants with palatal clefts. Mailing Add: 711 Jefferson Ave Memphis TN 38105

COLLINS, CARL BAXTER, JR, b San Antonio, Tex, Mar 4, 40; m 60, 73; c 3. LASERS. Educ: Univ Tex, BS, 60, MA, 61, PhD, 63. Prof Exp: Instr physics, Univ Tex, 62-64; from asst prof to assoc prof, Southwest Ctr Advan Studies, 64-69, assoc prof, Univ, 69-74, head grad physics prog, 72-75, PROF PHYSICS UNIV TEX, DALLAS, 74- Concurrent Pos: Res asst, Univ Tex, 63-64. Mem: Fel Am Phys Soc. Res: Ion-electron recombination processes; gaseous electronics; low energy plasmas; atomic and molecular collision processes; high energy lasers; multiphoton spectroscopy; isotope separation. Mailing Add: Dept of Physics Univ of Tex Dallas PO Box 668 Richardson TX 75080

COLLINS, CAROL HOLLINGWORTH, radiation chemistry, see 12th edition

COLLINS, CAROLYN JANE, b White Plains, NY, Sept 18, 42. MOLECULAR BIOLOGY, VIROLOGY. Educ: Skidmore Col, BA, 64; Duke Univ, PhD(biochem), 70. Prof Exp: Fel tumor viruses, Ger Ctr Cancer Res, 70-72; FEL, DEPT MICROBIOL, UNIV MICH, ANN ARBOR, 73- Concurrent Pos: F G Novy fel, Univ Mich, Ann Arbor, 73-74, USPHS fel, 75-76. Mem: Am Soc Microbiol; Sigma Xi. Res: Molecular biology of tumor virus-animal cell interactions; isolation of specific mammalian genes. Mailing Add: Dept of Microbiol Univ of Mich Ann Arbor MI 48104

COLLINS, CARTER COMPTON, b San Francisco, Calif, Aug 3, 25; m 61; c 2. BIOPHYSICS. Educ: Univ Calif, BS, 49, MS, 53, PhD(biophys), 66. Prof Exp: Consult biomech group, Med Ctr, Univ Calif, San Francisco, 50-54, assoc engr, Res & Develop Labs, 55, dir, 59-63; res mem, Inst Visual Sci, Pac Med Ctr, San Francisco, 63-69; assoc prof visual sci, Univ Pac, 69-71; SR SCIENTIST, SMITH-KETTLEWELL INST VISUAL SCI, INST MED SCI, 71- Concurrent Pos: Res & develop engr, Donner Sci Co, 53; pres, Sutter Instruments, Inc, 53-; Consult, Neurosurg Inst, Mt Zion Hosp, 54 & Dept Physiol, Univ Pa, 57-58; lectr, Cardiovasc Res Inst, 59; mem armed forces comt vision, Nat Acad Sci-Nat Res Coun. Honors & Awards: Hektoen Silver Medal Award, AMA, 72. Mem: AAAS; NY Acad Sci; Asn Res Ophthal; Inst Elec & Electronics Eng. Res: Conception and design of instrumentation for extracting information from physical and biological systems; basic correlates of recognition; neural information transfer codes; control of eye movements; vision substitution by tactile TV image projection. Mailing Add: 8 Ridge Ave Mill Valley CA 94941

COLLINS, CHARLES THOMPSON, b Long Branch, NJ, Mar 9, 38. ZOOLOGY, ORNITHOLOGY. Educ: Amherst Col, AB, 60; Univ Mich, MS, 62; Univ Fla, PhD(zool), 66. Prof Exp: Am Mus Natural Hist Chapman res fel, 66-67; asst prof biol, Fairleigh Dickinson Univ, Florham-Madison Campus, 67-68; asst prof, 68-72, ASSOC PROF BIOL, CALIF STATE UNIV, LONG BEACH, 72- Concurrent Pos: Fulbright res scholar, India, 74-75. Mem: Am Ornith Union; Wilson Ornith Soc; Cooper Ornith Soc; Am Inst Biol Sci; Asn Trop Biol. Res: Ornithology, particularly biology and ecology of swifts. Mailing Add: Dept of Biol Calif State Univ Long Beach CA 90840

COLLINS, CLAIR JOSEPH, b Austin, Minn, Aug 16, 15; m 49; c 3. PHYSICAL ORGANIC CHEMISTRY. Educ: Univ Minn, BChemEng, 37, MS, 39; Northwestern Univ, PhD(org chem), 44. Prof Exp: Org chemist, Eli Lilly & Co, Ind, 39-41; asst, Northwestern Univ, 41-44; GROUP LEADER, OAK RIDGE NAT LAB, 47-; PROF CHEM, UNIV TENN, KNOXVILLE, 64- Concurrent Pos: Fulbright lectr, Univ Tübingen, 68-69. Mem: Am Chem Soc. Res: Reaction mechanisms with deuterium, tritium and carbon-14; molecular rearrangements; isotope effects. Mailing Add: Chem Div Oak Ridge Nat Lab Oak Ridge TN 37830

COLLINS, CLIFFORD B, b Can, Nov 12, 16; nat US; m 42; c 3. PHYSICS. Educ: Univ Western Ont, BSc, 47; McMaster Univ, MA, 48; Univ Toronto, PhD(physics), 51. Prof Exp: Res assoc physics, Univ Toronto, 51-52; res assoc, Res Labs, 52-56, physicist lamp develop, 56-66, GROUP LEADER LAMP DEPT, GEN ELEC CO, 66- Mem: Am Phys Soc; Electrochem Soc. Res: Lamp development; materials liaison; technical planning. Mailing Add: Lamp Dept Gen Elec Co Nela Park Cleveland OH 44112

COLLINS, CONRAD GREEN, gynecology, deceased

COLLINS, CURTIS ALLAN, b Des Moines, Iowa, Sept m 62; c 1. PHYSICAL OCEANOGRAPHY. Educ: US Merchant Mari BS, 62; Ore State Univ, PhD(oceanog), 67. Prof Exp: Res scientist, Pac Oceanog Group, Nanaimo, BC, 68-70; sr tech adv oceanog, Cities Serv Oil Co, Okla, 70-72; PROG MGR, ENVIRON FORECASTING, OFF INT DECADE OCEAN EXPLOR, NSF, 72- Mem: Am Geophys Union. Res: Descriptive physical oceanography of the Pacific Ocean. Mailing Add: Int Decade Ocean Explor Rm 605 NSF Washington DC 20550

COLLINS, DELWOOD C, b Cairo, Ga, Oct 7, 37; m 62; c 3. BIOCHEMISTRY, REPRODUCTIVE PHYSIOLOGY. Educ: Emory Univ, AB, 59; Univ Ga, MS, 63, PhD(endocrinol, physiol), 66. Prof Exp: Res scientist, Div Pharmacol, Food & Drug Directorate, Ottawa, Ont, 67-68; res assoc steroid biochem, Univ Ottawa, 68-69; asst prof med & instr biochem, 69-72, ASSOC PROF MED & ASST PROF BIOCHEM, MED SCH, EMORY UNIV, 72-; COLLAB SCIENTIST, YERKES PRIMATE CTR, 71- Concurrent Pos: Squibb Ayrest traveling fel, 68; Nat Inst Arthritis, Metab & Digestive Dis career develop award, 72-77. Mem: AAAS; Soc Study Reproduction; Endocrine Soc; Can Biochem Soc. Res: Metabolism and conjugation of steroids; characterization of steroid glucuronyl transferase and steroid n-acetyl glucosaminyl transferase; radioimmunoassay of steroids; gas chromatography of steroids; reproductive physiology of lower primates. Mailing Add: Dept of Med Emory Univ Med Sch Atlanta GA 30303

COLLINS, DESMOND H, b Daylesford, Australia, July 15, 38; m 64; c 3. PALEONTOLOGY. Educ: Univ Western Australia, BSc, 60; Univ Iowa, PhD(geol), 66. Prof Exp: Tech officer, Geol Surv Can, 64-65; sr res fel paleont, Brit Mus, London, 66-68; CUR INVERT PALEONT, ROYAL ONT MUS, 68- Concurrent Pos: Assoc prof zool, Univ Toronto, 70-; res assoc, McMaster Univ, 71- Mem: Am Paleont Soc; Brit Palaeont Asn; Australian Geol Soc; Int Paleont Union; Geol Soc Can. Res: Paleozoic nautiloid cephalopod systematics and shell function. Mailing Add: Invert Palaeont Royal Ont Mus 100 Queen's Park Toronto ON Can

COLLINS, DON DESMOND, b Cardwell, Mont, Jan 10, 34; m 56; c 1. PLANT ECOLOGY, PLANT GENETICS. Educ: Mont State Univ, BS, 61, PhD(genetics), 65. Prof Exp: Asst prof, 65-69, ASSOC PROF BOT, MONT STATE UNIV, 69- Concurrent Pos: Nat Park Serv grant, 67-68; NSF grant, 68-70; Int Biol Prog grant, 69-70; consult, Grassland Info Synthesis Proj, Int Biol Prog, Grassland Biome, 68-69. Mem: Ecol Soc Am; Bot Soc Am; Nat Parks Asn. Res: Ecological research on effects of weather modification and vegetational response to campground usage; mountain grasslands. Mailing Add: Dept of Bot Mont State Univ Bozeman MT 59715

COLLINS, DONALD LOUIS, entomology, see 12th edition

COLLINS, EDWARD A, b Winnipeg, Man, Can, May 22, 28; US citizen; m 52; c 3. PHYSICAL CHEMISTRY, POLYMER CHEMISTRY. Educ: Univ Man, BS, 50, MS, 52, PhD(phys chem), 67. Prof Exp: Lectr phys chem, Royal Mil Col, Ont, 51-52; develop chemist, 52-54, assoc develop scientist, 56-59, develop scientist, 59-62, sr scientist, 62-65, DEVELOP CONSULT, B F GOODRICH CHEM CO, 65- Concurrent Pos: Adj prof, Rensselaer Polytech Inst, 68-76, Cleveland State Univ, 69-76 & Case Western Reserve Univ, 75-76. Mem: Am Chem Soc; Soc Rheol; Brit Soc Rheol. Res: Polymer characterization and relation of molecular structure to physical and mechanical properties; rheology. Mailing Add: 255 Glen View Dr Avon Lake OH 44012

COLLINS, EDWIN BRUCE, b Conway, SC, Aug 5, 21; m 53. FOOD MICROBIOLOGY. Educ: Clemson Col, BS, 43; Iowa State Col, MS, 48, PhD(dairy bact), 49. Prof Exp: From instr dairy indust & jr dairy bacteriologist to assoc prof dairy indust & assoc dairy bacteriologist, 49-64, PROF FOOD SCI, UNIV CALIF, DAVIS & DAIRY BACTERIOLOGIST, EXP STA, 64- Mem: Am Soc Microbiol; Am Dairy Sci Asn. Mailing Add: 808 Cherry Lane Davis CA 95616

COLLINS, ELLIOTT JOEL, b New York, NY, June 12, 19; m 47; c 2. ENDOCRINOLOGY. Educ: Col of Charleston, BS, 49; Princeton Univ, MA, 51, PhD(biol), 52. Prof Exp: Res scientist, Upjohn Co, 52-62; PRIN SCIENTIST & GROUP LEADER, DEPT PHYSIOL, SCHERING CORP, 62- Mem: AAAS; Soc Exp Biol & Med; Endocrine Soc; Orthopaedic Res Soc; affil Royal Soc Med. Res: Endocrine physiology; metabolic bone disease; connective tissue disease. Mailing Add: Schering Corp Bloomfield NJ 07003

COLLINS, ERNEST HOBART, b Hopkins, Mo, Nov 18, 96; m 26; c 2. EXPERIMENTAL PHYSICS. Educ: William Jewell Col, AB, 19; Univ Iowa, MS, 23, PhD(physics), 28. Prof Exp: Prof physics, Parsons Col, 24-39 & Albany Col, 39-41; from instr to asst prof, Univ Ore, 41-46; physicist, Develop Lab, Weyerhaeuser Timber Co, Wash, 46-61; assoc prof, 61-72, EMER ASSOC PROF PHYSICS, UNIV PORTLAND, 72- Concurrent Pos: Consult, Weyerhaeuser Co, 61- Mem: Am Phys Soc; Am Asn Physics Teachers; Inst Elec & Electronics Engrs; Forest Prod Res Soc. Res: Fundamental properties of wood fibers; plywood; spectroscopy; nondestructive testing of lumber; hyperfine structure of mercury resonance. Mailing Add: Dept of Physics Univ of Portland Portland OR 97203

COLLINS, FRANCES WILMOTH, b Georgetown, Ky, Dec 19, 24; m 47; c 5. ORGANIC CHEMISTRY. Educ: Georgetown Col, AB, 45; Mass Inst Technol, PhD(chem), 48. Prof Exp: Instr, Mt Holyoke Col, 48-49 & 55-58, asst prof, 58-68; ASST PROF CHEM, UNIV MASS, 68- Res: General organic chemistry. Mailing Add: Dept of Chem Univ of Mass Amherst MA 01003

COLLINS, FRANCIS ALLEN, b Wichita, Kans, July 24, 31. SOLID STATE PHYSICS, CHEMICAL PHYSICS. Educ: Univ Tex, BS, 54, MA, 57; Harvard Univ, PhD(appl physics), 64. Prof Exp: Physicist, Defense Res Labs, Univ Tex, 53-58; asst prop optics, Univ Rochester, 64-69; WITH APPL RES LABS, 69- Mem: Am Phys Soc. Res: Properties of insulating solids; airborne sensors. Mailing Add: Appl Res Labs PO Drawer 8029 Austin TX 78712

COLLINS, FRANK CHARLES, b Marton, NZ, Sept 18, 11; nat US; m 41; c 1. PHYSICAL CHEMISTRY. Educ: Univ Calif, AB, 39; Columbia Univ, AM, 47, PhD(chem), 49. Prof Exp: Chemist, Shell Develop Co, Calif, 39-46; lectr anal chem, Columbia Univ, 47; asst prof, 48-51, PROF PHYS & ENVIRON CHEM, POLYTECH INST NY, 51- Mem: Am Chem Soc; Am Phys Soc; Electrochem Soc. Res: Diffusion-controlled reactions; solid surfaces; thin films; semiconductors. Mailing Add: Polytech Inst of NY 333 Jay St Brooklyn NY 11201

COLLINS, FRANK E, JR, organic chemistry, biochemistry, see 12th edition

COLLINS, FRANKLYN, b Toronto, Ont, Mar 13, 29; nat US; m 51; c 3. SOLID STATE PHYSICS. Educ: Mich State Col, BS, 50; Univ Buffalo, PhD(physics), 58. Prof Exp: MGR ELECTRONICS RES, RES & DEVELOP LABS, AIRCO SPEER, 57- Mem: Am Phys Soc; Inst Elec & Electronics Engrs. Res: Carbon and graphite research; electronic components. Mailing Add: Airco Speer Res & Develop Labs Packard Ave & 47th St Niagara Falls NY 14302

COLLINS, FREDERICK CLINTON, b Prairie Grove, Ark, May 31, 41; m 61. PLANT BREEDING, PLANT GENETICS. Educ: Univ Ark, Fayetteville, BSA, 63, MS, 65; Purdue Univ, PhD(plant breeding, genetics), 69. Prof Exp: Res asst agron, Univ Ark, 63-64; geneticist sugarbeet invests, Agr Res Serv, USDA, 64-66; asst prof, 69-73, ASSOC PROF AGRON, UNIV ARK, FAYETTEVILLE, 73- Mem: Am Soc Agron; Crop Sci Soc Am. Res: Plant breeding; genetic control and physiology of yield and chemical characteristics, particularly those involved with nutritional quality of plants. Mailing Add: Dept of Agron Univ of Ark Fayetteville AR 72701

COLLINS, GALEN FRANKLIN, b Winona Lake, Ind, Dec 29, 27; m 56; c 4. CLINICAL BIOCHEMISTRY. Educ: Purdue Univ, BS, 49, MS, 52, PhD(pharmaceut chem), 54. Prof Exp: Asst pharm, Purdue Univ, 49-52; pharmaceut chemist, Miles Labs, Inc, 53-58, asst to dir, Miles-Ames Pharmaceut Res Lab, 58-59, head, Ames Pharmaceut Res Sect, 59-60; chief prod develop, Norwich Pharmacal Co, 60-63; mgr res div, S E Massengill Co, 63-67, dir res, 67-71; vpres res & develop, 71-75, VPRES & SCI DIR, DADE DIV, AM HOSP SUPPLY CORP, 75- Mem: Fel AAAS; Am Pharmaceut Asn; fel Am Inst Chem; Am Asn Clin Chemists; Asn Clin Scientists. Res: Clinical diagnostics which covers clinical instruments, hematology, blood coagulation, clinical chemistry, immunochemistry, immunohematology and immunology. Mailing Add: Dade Div Am Hosp Supply Co PO Box 5-0672 Miami FL 33152

COLLINS, GARY BRENT, b Clare, Mich, Sept 23, 40; m 62; c 2. PHYCOLOGY, AQUATIC ECOLOGY. Educ: Cent Mich Univ, BA, 62, MA, 64; Iowa State Univ, PhD(bot), 68. Prof Exp: Asst prof bot, Ohio State Univ, 68-72; RES AQUATIC BIOLOGIST, US ENVIRON PROTECTION AGENCY, 72- Mem: AAAS; Am Micros Soc; Int Phycol Soc; Bot Soc Am; Phycol Soc Am. Res: Algal ecology; freshwater diatom ecology and taxonomy; diatom paleoecology; plankton and periphyton methods development. Mailing Add: 8109 Woodruff Rd Cincinnati OH 45230

COLLINS, GEORGE BRIGGS, b Washington, DC, Jan 3, 06; m 34; c 3. PHYSICS. Educ: Johns Hopkins Univ, PhD(physics), 31. Prof Exp: Instr physics, Johns Hopkins Univ, 27-30; instr, Univ Notre Dame, 33-41; mem radiation lab, Mass Inst Technol, 41-46; prof physics & chmn dept, Univ Rochester, 46-50; chmn cosmotron dept, Brookhaven Nat Lab, 50-62, sr physicist, 62-71; PROF PHYSICS & SR PHYSICIST,

VA POLYTECH INST & STATE UNIV, 71- Concurrent Pos: Coun mem, State Univ NY Stony Brook, 53-70; Fulbright fel, Belg, 57; with Deutsches Elektronen-Synchrotron, 65-66. Mem: Fel Am Phys Soc. Res: Raman effect; hyperfine structure of iodine; far ultraviolet spectroscope; excitation of nuclei by electrons; nuclear physics; high energy electrons and protons; particle physics; development of wire chamber spectrometers; high energy multiparticle production. Mailing Add: Dept of Physics Va Polytech Inst & State Univ Blacksburg VA 24061

COLLINS, GEORGE EDWIN, b Stuart, Iowa, Jan 10, 28; m 54; c 3. COMPUTER SCIENCE, MATHEMATICS. Educ: Univ Iowa, BA, 51, MS, 52; Cornell Univ, PhD(math), 55. Prof Exp: Mathematician, Int Bus Mach Corp, 55-66; assoc prof, 66-68, chmn dept, 70-72, PROF COMPUT SCI, UNIV WIS-MADISON, 68- Concurrent Pos: Res fel math, Calif Inst Technol, 63-64; vis prof, Stanford Univ, 72-73; guest prof, Univ Kaiserslautern, 74-75. Mem: AAAS; Asn Comput Mach; Am Math Soc; Math Asn Am; Soc Indust & Appl Math. Res: Algebraic algorithms; algorithm analysis; computations with polynomials and rational functions; decision methods in mathematical logic; computer algebra systems. Mailing Add: Dept of Comput Sci Univ of Wis 1210 WDayton St Madison WI 53706

COLLINS, GEORGE H, b Albany, NY, Sept 11, 27; m 51; c 6. PATHOLOGY, NEUROLOGY. Educ: Univ Vt, AB, 49, MD, 53. Prof Exp: From asst prof to prof path & med, Col Med, Univ Fla, 62-73; PROF PATH, COL MED, STATE UNIV NY UPSTATE MED CTR, 73- Concurrent Pos: Teaching fel neuropath, Harvard Med Sch, 58-59, res fel, 59-62; clin & res fel, Mass Gen Hosp, 59-62; asst resident path, Mass Gen Hosp, 58-59. Mem: AAAS; Am Asn Neuropath. Res: Effect of nutritional deficiency upon the ultra structure of the nervous system; glial cell function; extracellular space of the central nervous system. Mailing Add: Dept of Path State Univ NY Upstate Med Ctr Syracuse NY 13210

COLLINS, GEORGE W, II, b Waukegan, Ill, July 18, 37; m 61; c 1. ASTRONOMY. Educ: Princeton Univ, AB, 59; Univ Wis, PhD(astron), 62. Prof Exp: Asst prof numerical anal, Univ Wis, 62-63; from asst prof to assoc prof, 63-69, PROF ASTRON, OHIO STATE UNIV, 69- Mem: AAAS; Am Astron Soc; fel Royal Astron Soc. Res: Eclipsing binary stars; stellar atmospheres; general problems in radiative transfer; numerical analysis. Mailing Add: Dept of Astron Smith Physics Lab Ohio State Univ 174 W 18th Ave Columbus OH 43085

COLLINS, GERALD BRADFORD, biology, see 12th edition

COLLINS, GLENN BURTON, b Folsom, Ky, Aug 7, 39; m 59; c 2. PLANT GENETICS, PLANT BREEDING. Educ: Univ Ky, BS, 61, MS, 63; NC State Univ, PhD(genetics), 67. Prof Exp: Res asst plant genetics, Univ Ky, 61-63; plant cytogenetics, NC State Univ, 63-66; from asst prof to assoc prof, 66-75, PROF CYTOGENETICS, UNIV KY, 75- Concurrent Pos: Sabbatical, John Innes Inst, Norwich, Eng, 73. Honors & Awards: Cooper Res Award, Col Agr, Univ Ky, 75. Mem: Am Soc Agron; Am Genetics Soc; Genetics Soc Am; Can Genetics Soc. Res: Genetics and breeding of tobacco; development and use of haploid procedures for breeding and genetic studies; use of tissue and cell culture methods in plant improvement. Mailing Add: Dept of Agron Univ of Ky Lexington KY 40506

COLLINS, HENRY A, b Machipongo, Va, Sept 21, 32; m 62; c 3. AGRONOMY, PLANT PHYSIOLOGY. Educ: Md State Col, BS, 55; Rutgers Univ, MS, 57, PhD(farm crops), 62. Prof Exp: Asst farm crops, Rutgers Univ, 59-62; from asst prof to assoc prof biol, Tuskegee Inst, 62-68; res specialist, 68-73, MKT PLANNING SPECIALIST, DIV AGR CHEM, CIBA-GEIGY CORP, 73- Mem: Am Soc Agron; Crop Sci Soc Am; Weed Sci Soc Am; Am Inst Biol Sci. Res: Selective absorption of strontium by plants; factors affecting weed seed germination. Mailing Add: Agr Div Ciba-Geigy Corp PO Box 11422 Greensboro NC 27409

COLLINS, HENRY B, b Geneva, Ala, Apr 9, 99; m 31; c 1. ANTHROPOLOGY, ARCHAEOLOGY. Educ: Millsaps Col, AB, 22; George Washington Univ, AM, 25. Hon Degrees: ScD, Millsaps Col, 40. Prof Exp: Asst, Miss Dept Arch & Hist, 23; aide, Div Ethnol, US Nat Mus, 24-25, asst cur, 25-37, assoc cur, 38-39; sr ethnologist, Bur Am Ethnol, 39-51, sr anthropologist, 51-63, actg dir, 63-65, sr scientist, Inst, 65-66, ARCHAEOLOGIST EMER, SMITHSONIAN INST, 66- Concurrent Pos: Vpres, Int Cong Ethnol & Anthrop Sci, Copenhagen, 38, Moscow, 64 & mem permanent coun, 52-;asst dir, Ethnogeog Bd, 43-44, dir, 44-46; mem coun int rels, Nat Res Coun, 45-56; chmn, Dir Comt Arctic Bibliog, Arctic Inst NAm, 47-67, bd gov, 48; hon vpres, Int Cong Americanists, Copenhagen, 56. Honors & Awards: Gold Medal, Royal Danish Acad Sci & Lett, 36. Mem: AAAS; Am Anthrop Asn; Soc Am Archaeol (vpres, 42-52); Am Asn Phys Anthrop; Arctic Inst NAm. Res: Archaeology and anthropology of the Arctic; physical anthropology, Indians and Eskimos. Mailing Add: Smithsonian Inst Washington DC 20560

COLLINS, HERON SHERWOOD, b Charlotte, NC, Nov 17, 22; m 52; c 4. MATHEMATICS. Educ: Wofford Col, BS, 48; Tulane Univ, MS, 50, PhD, 52. Prof Exp: Asst prof math, Univ Md, 52-53; asst prof, Univ SC, 53-54; from asst prof to assoc prof, 54-62, PROF MATH, LA STATE UNIV, 62- Mem: Am Math Soc. Res: Banach spaces and algebras; measure and integration theory; analytic function theory. Mailing Add: Dept of Math La State Univ Univ Sta Baton Rouge LA 70803

COLLINS, HOLLIE L, b Laona, Wis, May 20, 38; m 57; c 2. ZOOLOGY. Educ: Wis State Univ, BS, 60; Mich State Univ, MS, 62, PhD(zool), 65. Prof Exp: ASSOC PROF BIOL, UNIV MINN, DULUTH, 64-, DIR GRAD STUDIES BIOL, 70- Mem: Animal Behavior Soc; Soc Ichthyol & Herpet. Res: Animal behavior of aquatic organisms, especially social behavior of fishes. Mailing Add: Dept of Biol Univ of Minn Duluth MN 55812

COLLINS, HORACE RUTTER, b Shawnee, Okla, Feb 4, 30. GEOLOGY. Educ: Ohio Univ, BS, 54; WVa Univ, MS, 59. Prof Exp: Instr geol, Ohio Univ, 58-59; geologist, Coal Sect, 59-60, head, 60-63, head regional sect, 63-66, asst state geologist, 66-68, asst chief Ohio Div, 67-68, STATE GEOLOGIST & CHIEF OHIO DIV, OHIO GEOL SURV, 68- Mem: Geol Soc Am; Am Inst Prof Geol; Asn Am State Geol. Res: Geology of Ohio; Pennsylvania stratigraphy; paleobotany; coal geology. Mailing Add: Ohio Div Natural Resources Div of Geol Survey Fountain Sq Columbus OH 43224

COLLINS, JAMES FRANCIS, b Baltimore, Md, Jan 26, 42; m 69; c 2. BIOCHEMISTRY. Educ: Loyola Col, Md, BS, 63; Univ NC, Chapel Hill, PhD(genetics), 68. Prof Exp: Fel, Nat Inst Arthritis, Metab & Digestive Dis, NIH, 68-70, staff fel, 70-72, sr staff fel, Nat Heart & Lung Inst, 73-75; RES CHEMIST, AUDIE L MURPHY VET ADMIN HOSP & ASST PROF MED & BIOCHEM, UNIV TEX HEALTH SCI CTR, SAN ANTONIO, 75- Concurrent Pos: Consult, Biol & Chem Sci Degree Prog, Upward Mobility Col, Fed City Col Exp Prog, Washington, DC, 72. Mem: AAAS; Am Thoracic Soc. Res: Protein biosynthesis; connective tissue of the lung in experimental lung disease. Mailing Add: Dept of Med Univ of Tex Health Sci Ctr San Antonio TX 78284

COLLINS, JAMES MALCOLM, b Atlanta, Ga, Mar 21, 38; m 59; c 2. BIOCHEMISTRY, DEVELOPMENTAL BIOLOGY. Educ: Univ Southern Miss, BS, 62; Univ Tenn, PhD(biochem), 68. Prof Exp: USPHS fel develop biol, Oak Ridge Nat Lab, 68-69; asst prof, 69-75, ASSOC PROF BIOCHEM, HEALTH SCI DIV, VA COMMONWEALTH UNIV, 75- Mem: AAAS; Am Soc Biol Chemists; Am Chem Soc. Res: DNA synthesis; cell cycle; cancer. Mailing Add: Dept of Biochem Health Sci Div Va Commonwealth Univ Richmond VA 23219

COLLINS, JAMES PAUL, b New York, NY, July 3, 47; m 70; c 2. ECOLOGY. Educ: Manhattan Col, BS, 69; Univ Mich, MS, 71, PhD(zool), 75. Prof Exp: ASST PROF ZOOL, ARIZ STATE UNIV, 75- Mem: Ecol Soc Am; Am Soc Study Evolution; Am Soc Ichthyologists & Herpetologists; Sigma Xi. Res: Investigation of the selective advantage of life history characters, especially in relation to the predictability of the organism's breeding habitat; the effect of competition and predation on life history characters. Mailing Add: Dept of Zool Ariz State Univ Tempe AZ 85281

COLLINS, JANET VALERIE, b Riversdale, Jamaica, Oct 11, 39. CELL BIOLOGY, INSECT PHYSIOLOGY. Educ: Univ London (Univ WIndies), BSc, 63; Western Reserve Univ, PhD(biol), 67. Prof Exp: Res asst, Western Reserve Univ, 63-67, fel, 67-68; fel, Queen's Univ, Ont, 68-70; ASST PROF BIOL, DALHOUSIE UNIV, 70- Mem: Am Soc Cell Biol; Soc Develop Biol. Res: Effects of hormones at the cellular level, particularly on control of pinocytosis, storage and release of haemolymph proteins before and during molting; selective uptake of proteins by fat body cells. Mailing Add: Dept of Biol Dalhousie Univ Halifax NS Can

COLLINS, JASON HAYDEL, b New Orleans, La, Aug 13, 18; m 49; c 6. OBSTETRICS & GYNECOLOGY. Educ: Tulane Univ, BS, 38, MD, 41; Am Bd Obstet & Gynec, dipl, 50. Prof Exp: From instr to assoc prof obstet & gynec, 48-59, prof clin obstet & gynec, 59-72, actg chmn dept obstet & gynec, 71-72, C J MILLER PROF OBSTET & GYNEC & CHMN DEPT, SCH MED, TULANE UNIV, 72- Concurrent Pos: Sr vis surgeon, Hosps; dir med ed, Southern Baptist Hosp, New Orleans; sr consult, USPHS Hosp. Mem: Am Asn Obstet & Gynec; Am Col Obstet & Gynec; Am Gynec Soc; Am Col Surg. Mailing Add: Sch of Med Tulane Univ New Orleans LA 70112

COLLINS, JEFFERY ALLEN, b Oakland, Calif, Feb 16, 44. ANIMAL PARASITOLOGY. Educ: Purdue Univ, DVM, 67. Prof Exp: Vet lab animal med, Shell Develop Co, Div Shell Oil Co, Calif, 67-68; prod develop technologist, Shell Chem Co, New York, 68-70; staff vet anthelmintic prod, 70-72, SUPVR VET THERAPEUT, SHELL DEVELOP CO, DIV SHELL OIL CO, 72- Concurrent Pos: Consult parasitol, Am Asn Zoo Animal Vet, 73-; mkt res technologist, Shell Int Chem Co, Ltd, London, 73-74. Mem: Am Vet Med Asn; Am Soc Vet Parasitologists; Am Asn Lab Animal Sci; Indust Vet Asn; Am Asn Animal Welfare Vet. Res: Veterinary parasitology; administrative role re all aspects of parasiticide research and product development. Mailing Add: Shell Develop Co PO Box 4248 Modesto CA 95352

COLLINS, JERRY DALE, b Greeneville, Tenn, June 3, 43; m 61; c 2. ORGANIC CHEMISTRY, POLYMER CHEMISTRY. Educ: Tusculum Col, BS, 65; Univ Ark, Fayetteville, PhD(carbene chem), 70. Prof Exp: Res chemist, 70-71, spec process engr, 71-73, TECH SUPVR NYLON DEVELOP, AM ENKA DIV, AKZONA CORP, LOWLAND, 73- Mem: Am Chem Soc; NY Acad Sci. Res: Steric effects associated with insertion of carbenes into carbon-hydrogen bonds; determination of insertion reaction mechanism; modification of fiber surface by carbene insertion reactions; new process development concerning nylon textile yarns. Mailing Add: Rte 5 Greeneville TN 37743

COLLINS, JIMMIE LEE, b Vicksburg, Miss, Nov 24, 34; m 59; c 3. FOOD SCIENCE. Educ: La State Univ, BS, 61; Univ Md, MS, 63, PhD(food tech), 65. Prof Exp: Asst prof, 65-69, ASSOC PROF FOOD TECHNOL, UNIV TENN, KNOXVILLE, 69- Mem: Am Inst Food Technol. Res: Textural characteristics of fruits and vegetables; enzymes of fresh and processed vegetables. Mailing Add: Dept of Food Technol Univ of Tenn Knoxville TN 37916

COLLINS, JOHN W, b Rhinebeck, NY, Aug 19, 35; m 60; c 4. BIOCHEMISTRY, ORGANIC CHEMISTRY. Educ: Bowdoin Col, AB, 57; Middlebury Col, MS, 59; Univ Vt, PhD(biochem), 63. Prof Exp: Biochemist, WVa Pulp & Paper Co, 63-67; biochemist, Pulp Mfrs Res League, 67-70; res assoc, 70-74, CHIEF EFFLUENT PROCESSES, INST PAPER CHEM, 74- Mem: Am Chem Soc; Tech Asn Pulp & Paper Indust. Res: Leaf protein biochemistry; biodegradability of lignin and model ethers; pulp and paper mill byproduct fractionation and chemistry; ultra filtration and reverse osmosis, development and application to pulp and paper mill effluents; lignin hydrolysis, pyrolysis and utilization. Mailing Add: Inst of Paper Chem Div of Environ Systs Box 1048 Appleton WI 54911

COLLINS, JON GERALD, physical chemistry, analytical chemistry, see 12th edition

COLLINS, JOSEPH CHARLES, JR, b Pontiac,Mich, May 7, 31; m 53; c 4. ORGANIC CHEMISTRY. Educ: Wayne State Univ, BS, 53; Univ Wis, PhD(chem), 58. Prof Exp: Asst, Gen Motors Res, 53 & Chem Dept, E I du Pont de Nemours & Co, 56; res assoc, Univ Wis, 58 & Sterling-Winthrop Res Inst, 58-62; assoc prof chem, Ill Wesleyan Univ, 62-67; res assoc, 67-68, assoc dir chem div, 68-69, DIR CHEM DIV, STERLING-WINTHROP RES INST, 69- Mem: AAAS; Am Chem Soc; The Chem Soc. Res: Organic synthesis and conformational analysis of biomolecular systems. Mailing Add: Sterling-Winthrop Res Inst Columbia Turnpike Rensselaer NY 12144

COLLINS, JOSEPH WILFRED, organic chemistry, see 12th edition

COLLINS, JUNE MCCORMICK, b Seattle, Wash, Jan 28, 20; m 42; c 3. ANTHROPOLOGY, ETHNOLOGY. Educ: Univ Wash, BA, 41; Univ Chicago, MA, 46, PhD(anthrop), 49. Prof Exp: Res asst human rels in indust, Univ Chicago, 45-49; teaching assoc, Northwestern Univ, 49-50; asst prof, Univ Mo, 51-52; lectr, Seattle Univ, 54-56; assoc prof, Pa State Univ, 56-58; asst prof psychol, Chicago State Univ, 58-59; asst prof anthrop, Oakland Univ, 59-62; assoc prof, Southern Ill Univ, 65-68; chmn dept, 68-74, PROF ANTHROP, STATE UNIV NY COL BUFFALO, 68- Concurrent Pos: Carnegie grant social struct Am labor unions, Oakland Univ, 60-62; NIMH Small grant, Yucatan, 66; fac res grant, State Univ NY Col Buffalo, 70. Mem: AAAS; Am Anthrop Asn; Am Ethnol Soc (secy, 66-). Res: Social or cultural anthropology; interaction, social structure, theory, linguistics; geographical areas—North America, Middle America. Mailing Add: Dept of Anthrop State Univ of NY Col Buffalo NY 14222

COLLINS, LIMONE C, physiology, see 12th edition

COLLINS, LLOYD RAYMOND, b Oakland, Calif, Nov 7, 20; m 49; c 3. ANTHROPOLOGY. Educ: Univ Ore, BS, 49, MS, 51; Univ Ariz, PhD, 62. Prof Exp: Res assoc anthrop, Univ Ore, 51-52; phys anthropologist human factors, US Navy Electronics Lab, 58-61; reliability engr, Human Eng Group, Gen Dynamics/Astronaut, 61-62; staff eng anthropologist, McDonnell Aircraft Corp, 62-

69; assoc prof sociol & anthrop, Eastern Ill Univ, 69-71; ASSOC PROF ANTHROP, UNIV MO-ST LOUIS, 71- Concurrent Pos: Asst prof, Southern Ill Univ, 63-65 & Univ Col, Washington Univ, 65-67. Mem: Fel Am Anthrop Asn; Am Asn Phys Anthrop. Res: Applied anthropology; cultural and physical anthropology; systems research; electronics. Mailing Add: PO Box 160 Godfrey IL 62035

COLLINS, LOIS COWAN, b McKeesport, Pa, July 1, 14; m 42; c 3. RADIOLOGY. Educ: Univ Pittsburgh, BS & MD, 37. Prof Exp: Instr radiol, Boston Univ, 39-42; from assoc to assoc prof, Col Physicians & Surg, Columbia Univ, 42-52; ASSOC PROF RADIOL, POSTGRAD SCH, UNIV TEX, 52- Concurrent Pos: Asst radiologist, Presby Hosp & Psychiat Inst, NY, 42-52; consult, US Marine Hosp, Staten Island, 47-52; assoc prof, Col Med Baylor Univ, 52-58; asst radiologist, M D Anderson Hosp, Tex, 53-62; radiologist, Methodist Hosp, 63-66. Mem: Fel Am Col Radiol; Am Roentgen Ray Soc; Radiol Soc NAm. Mailing Add: 9200 Westheimer Ave Houston TX 77042

COLLINS, LORENCE GENE, b Vernon, Kans, Nov 19, 31; m 55; c 5. PHOTOGEOLOGY. Educ: Univ Ill, BS, 53, MS, 57, PhD(geol), 59. Prof Exp: Instr phys sci, Univ Ill, 58-59; from asst prof to prof geol, San Fernando Valley State Col, 59-74, fac res coordr, 70-74; PROF GEOSCI & RES COORDR, CALIF STATE UNIV, 74- Mem: Fel Geol Soc Am; Am Asn Geol Teachers. Res: Mineral deposits in metamorphic terrains, particularly magnetite in granitic gneisses; refractive index studies of ferromagnesian silicates. Mailing Add: Dept of Geol Calif State Univ Northridge CA 91324

COLLINS, MALCOLM FRANK, b Crewe, Eng, Dec 15, 35; m 61; c 3. PHYSICS. Educ: Cambridge Univ, BA, 57, MA, 61, PhD(physics), 62. Prof Exp: Physicist, Solid State Physics Div, Atomic Energy Res Estab, Harwell, Eng, 61-69; assoc prof, 69-73, PROF PHYSICS, MCMASTER UNIV, 73-, ASSOC CHMN DEPT, 74- Concurrent Pos: Guest, Brookhaven Nat Lab, 67-68; fel, Alfred P Sloan Found, 70-72. Mem: Fel Can Asn Physicists; Brit Inst Physics and Phys Soc. Res: Various aspects of slow neutron scattering, especially applications to magnetism, critical phenomena, transition metals, lattice vibrations, crystallography and molecular vibrations; theory of paramagnetism and theory of neutron spectrometer operation. Mailing Add: Dept of Physics McMaster Univ Hamilton ON Can

COLLINS, MARY JANE, b Miami, Fla, Oct 17, 40. PSYCHOACOUSTICS, AUDIOLOGY. Educ: Vanderbilt Univ, BA, 61, MS, 63; Univ Iowa, PhD(hearing sci), 70. Prof Exp: Audiologist, Bristol Speech & Hearing Ctr, 64-65; res asst audiol, Vet Admin Hosp, Coral Gables, Fla, 65-66; AUDIOLOGIST, VET ADMIN HOSP, NASHVILLE, 70- Concurrent Pos: Asst prof audiol, Vanderbilt Univ, 70-72. Mem: Acoust Soc Am; Am Speech & Hearing Asn. Res: Psychophysical phenomena in the normal and pathological human auditory system. Mailing Add: Ear Nose & Throat Clin Vet Admin Hosp Nashville TN 37212

COLLINS, MICHAEL FREDERICK, b Detroit, Mich, July 27, 37; m 60; c 1. DEVELOPMENTAL BIOLOGY. Educ: Brown Univ, AB, 59; Johns Hopkins Univ, PhD(biol), 64. Prof Exp: NIH fel physiol chem, Sch Med, Johns Hopkins Univ, 64-66; asst prof zool, Univ Tex, Austin, 66-71; ASST PROF ANAT, SCH MED, UNIV CONN, 71- Mem: AAAS; Am Asn Anat. Res: Cellular surface properties and morphogenesis. Mailing Add: Dept of Anat Univ Conn Sch of Med Hartford CT 06105

COLLINS, NICHOLAS CLARK, b Ogden, Utah, Mar 2, 46; m 75. POPULATION ECOLOGY. Educ: Pomona Col, BA, 68; Univ Ga, PhD(zool), 72. Prof Exp: Fel acarology, Ohio State Univ, 72-73; ASST PROF ZOOL, ERINDALE COL, UNIV TORONTO, 73- Mem: Ecol Soc Am; AAAS; Am Soc Naturalists; Entom Soc Can. Res: Comparative functional analyses of simple ecosystems like hypersaline lakes and thermal springs; analyses of life history variations among brine shrimp and brine fly populations from simple environments. Mailing Add: Dept of Zool Erindale Col Mississauga ON Can

COLLINS, O'NEIL RAY, b Opelousas, La, Mar 9, 31; m 59; c 2. MYCOLOGY. Educ: Southern Univ, BS, 57; Univ Iowa, MS, 59, PhD(bot), 61. Prof Exp: Instr biol, Queens Col, 61-63; assoc prof, Southern Univ, 63-65; assoc prof, Wayne State Univ, 65-69; assoc prof, 69-73, PROF BOT, UNIV CALIF, BERKELEY, 73- Mem: Bot Soc Am; Mycol Soc Am. Res: Mating types in the Myxomycetes. Mailing Add: Dept of Bot Univ of Calif Berkeley CA 94720

COLLINS, PAUL EVERETT, b White Rock, Minn, Feb 22, 17; m 48; c 2. SILVICULTURE. Educ: Gustavus Adolphus Col, BA, 39; Univ Minn, BS, 48, MS, 49, PhD, 67. Prof Exp: Asst, Univ Minn, 48-49; asst prof & exten forester, Kans State Col, 49-51; from asst prof & asst forester to assoc prof & assoc forester, 52-74, PROF HORT & FORESTRY, S DAK STATE UNIV, 74- Mem: Soc Am Foresters; Sigma Xi. Res: Plains windbreak and shelter belt research on matters of cultural practices, design and tree breeding. Mailing Add: Dept of Hort & Forestry SDak State Univ Brookings SD 57006

COLLINS, PAUL WADDELL, b Greenville, SC, Feb 26, 40; m 63; c 3. MEDICINAL CHEMISTRY. Educ: Univ SC, BS, 62; Med Col Va, PhD(med chem), 66. Prof Exp: Res fel org synthesis, Univ Va, 66-67; RES SCIENTIST, G D SEARLE & CO, 67- Mem: Am Chem Soc. Res: Synthesis of heterocyclic spiro compounds; synthesis of cyclopropylogs of naturally occurring amines and amino acids; synthesis of prostaglandins. Mailing Add: G D Searle & Co Chem Res Dept PO Box 5110 Chicago IL 60680

COLLINS, PETER FAY, analytical chemistry, see 12th edition

COLLINS, RALPH PORTER, b Alpena, Mich, Nov 26, 27; m 55; c 3. BOTANY. Educ: Mich State Univ, BA, 50, MS, 52, PhD(bot), 57. Prof Exp: Asst bot, Mich State Univ, 52-54 & 55-57; from instr to assoc prof, 57-69, PROF BOT, UNIV CONN, 69- Concurrent Pos: NIH spec res fel, 64-65; Smithsonian Inst sr res fel, 72. Mem: AAAS; Mycol Soc Am; Am Chem Soc; Am Soc Pharmacog. Res: Fungal physiology; natural products chemistry. Mailing Add: Dept of Biol-Bot Sect Univ of Conn Storrs CT 06268

COLLINS, RICHARD ANDREW, b Norristown, Pa, Oct 27, 24; m 55; c 3. PATHOLOGY. Educ: Pa State Univ, BS, 48; Univ Wis, MS, 50, PhD(biochem), 52; Marquette Univ, MD, 62. Prof Exp: Asst biochem, Marquette Univ, 58-60; intern, Mary Fletcher Hosp, Burlington, Vt, 62-63; asst prof path, Med Col Wis, 65-67; PATHOLOGIST, REGIONAL MED LABS, 72- Concurrent Pos: Nat Heart Inst spec fel path, Univ Vt, 63-65. Mem: Am Soc Exp Path; Am Asn Clin Chem; Am Soc Clin Path; Col Am Path; Int Acad Path. Mailing Add: Regional Med Labs 175 College St Battle Creek MI 49017

COLLINS, RICHARD ARLEN, b Shaw, Miss, Aug 4, 30; m; c 5. FISHERIES BIOLOGY. Educ: Delta State Col, BSE, 53; Univ Southern Miss, MA, 57; Univ Southern Ill, PhD(zool), 68. Prof Exp: Marine biologist, Gulf Coast Res Lab, Miss, 57-59; assoc prof biol, State Col Ark, 59-65; instr zool, Univ Southern Ill, 65-68; PROF BIOL, STATE COL ARK, 68- Mem: AAAS; Am Fisheries Soc. Res: Fish culture; aquatic ecology. Mailing Add: Dept of Biol State Col of Ark Conway AR 72032

COLLINS, RICHARD CORNELIUS, b Phoenix, Ariz, Feb 16, 41; m 61; c 1. PARASITOLOGY, ENTOMOLOGY. Educ: Ariz State Univ, BS, 63; Univ Ariz, MS, 66, PhD(entom, parasitol), 70. Prof Exp: Res asst entom, Dept Animal Path, Univ Ariz, 64-66, parasitol, 68-70; self-employed, 66-68; fel entom & parasitol, Dept Parasitol, Tulane Univ, 70-74; RES ENTOMOLOGIST PARASITOL, CTR DIS CONTROL, BUR TROP DIS, CENT AM RES STA, USPHS, 74- Mem: Am Soc Parasitol; Entom Soc Am; Am Soc Trop Med & Hyg; AAAS. Res: Bionomics of parasites and insects; chemotherapy of parasitic diseases; transmission dynamics of vector-borne diseases of man and animals. Mailing Add: Cent Am Res Sta c/o Am Embassy APO New York NY 09889

COLLINS, RICHARD LAPOINTE, b New York, NY, May 10, 38; m 65. INVERTEBRATE ZOOLOGY, PROTOZOOLOGY. Educ: Boston Univ, AB, 61, MA, 62; Univ Calif, Berkeley, PhD(zool), 69. Prof Exp: Actg asst prof zool, Univ Calif, Berkeley, 68-69; ASST PROF ZOOL & PHYSIOL, LA STATE UNIV, BATON ROUGE, 69- Mem: AAAS; Am Inst Biol Sci; Soc Protozool; Am Micros Soc. Mailing Add: Dept of Zool & Physiol La State Univ Baton Rouge LA 70803

COLLINS, ROBERT JAMES, organic chemistry, see 12th edition

COLLINS, ROBERT JAMES, b Hazel Park, Mich, July 15, 28; m 52; c 2. PHYSIOLOGY, PHARMACOLOGY. Educ: Alma Col, BS, 51; Mich State Univ, 54, PhD, 58. Prof Exp: Bacteriologist, Mich Dept Health, 51-52; lab technician, E W Sparrow Hosp, 54-56; RES HEAD, UPJOHN CO, 58- Mem: AAAS; Am Pharmacol Soc; Soc Exp Biol & Med. Res: Central nervous system. Mailing Add: CNS Unit Upjohn Co Kalamazoo MI 49001

COLLINS, ROBERT JOSEPH, b Philadelphia, Pa, July 23, 23; m 45; c 4. PHYSICS. Educ: Univ Mich, AB, 47, MS, 48; Purdue Univ, PhD(physics), 53. Prof Exp: Asst prof physics, Rose Polytech Inst, 49-50; asst, Purdue Univ, 50-53; mem staff, Bell Tel Lab, 53-62; Inst Defense Anal, DC, 62-63; PROF ELEC ENG, UNIV MINN, MINNEAPOLIS, 63-, HEAD DEPT, 64- Mem: Am Phys Soc. Res: Optical properties of solid state; radiation effects in solids; quantum electronics. Mailing Add: Dept of Elec Eng Univ of Minn Minneapolis MN 55455

COLLINS, ROBERT LOUIS, genetics, see 12th edition

COLLINS, ROBERT MATTHEW, b West Bend, Iowa, Dec 28, 20; m 46; c 3. AGRICULTURAL EDUCATION. Educ: Iowa State Univ, BS, 43, MS, 51, PhD(voc educ), 53. Prof Exp: Training specialist agr, US Vet Admin, 46-49; soil conservstionist, US Soil Conserv Serv, 49-50; instr sociol, Iowa State Univ, 51-52, farm oper, 52-54, asst prof farm oper & voc educ, 54-58, asst prof voc educ, 58-60; from asst prof to assoc prof, 60-69, PROF VOC EDUC, UNIV MINN, 69-, DIR ACAD AFFAIRS, 74-, PRIN, SOUTHERN SCH & EXP STA, 60- Res: Vocational education; agronomy; animal husbandry. Mailing Add: 423 NE Fourth St Waseca MN 56093

COLLINS, RONALD WILLIAM, b Dayton, Ohio, Feb 5, 36; m 60; c 2. INORGANIC CHEMISTRY. Educ: Univ Dayton, BS, 57; Ind Univ, PhD(inorg chem), 62. Prof Exp: Inorg res chemist, Wyandotte Chem Corp, 62-65; from asst prof to assoc prof, 65-71, PROF CHEM, EASTERN MICH UNIV, 71- Concurrent Pos: Vis prof, Mich State Univ, 70-71. Mem: Am Chem Soc; The Chem Soc; Am Crystallog Asn; AAAS; Am Soc Eng Educ. Res: Inorganic compounds of group IV metals; synthetic apatites; x-ray crystallography; instructional uses of digital computers; crystal lattice energies. Mailing Add: Dept of Chem Eastern Mich Univ Ypsilanti MI 48197

COLLINS, ROYAL EUGENE, b Corsicana, Tex, Feb 25, 25; m 62; c 2. THEORETICAL PHYSICS. Educ: Univ Houston, BS, 49; Tex A&M Univ, MS, 50, PhD(physics), 54. Prof Exp: Res engr, Stanolind Oil Co, 54-55; sr res engr, Humble Oil Co, 55-59; assoc prof, 59-71, PROF PHYSICS, UNIV HOUSTON, 71- Concurrent Pos: Consult, Vet Admin Hosp, Houston, 58-62, Exxon Prod Res Corp, 59-68, M D Anderson Tumor Inst, 62, Col Med, Baylor Univ, 62- & US Bur Mines, 65-67. Mem: Am Phys Soc; Am Asn Physics Teachers. Res: Statistical mechanics; quantum theory; mathematical biophysics; applied mathematics; porous materials. Mailing Add: Dept of Physics Univ of Houston Houston TX 77004

COLLINS, RUSSELL LEWIS, b Coffeyville, Kans, Sept 21, 28; m 48; c 2. CHEMICAL PHYSICS, PLANT EMBRYOLOGY. Educ: Univ Tulsa, BS, 48; Univ Okla, MS, 50, PhD(physics), 53. Prof Exp: Res physicist, Phillips Petrol Co, Okla, 53-58, group leader, 58-62; asst prof, 62-66, ASSOC PROF PHYSICS, UNIV TEX, AUSTIN, 66- Concurrent Pos: Petrol Res Fund grant, 63-66; founder & pres, Austin Sci Assocs, Inc, 64-; Robert A Welch Found grant, 65-; Off Naval Res contract, 69-; US Army Mobile Equip Res & Develop Lab contract, 70-75. Mem: Am Phys Soc; Weed Sci Soc Am; Am Chem Soc. Res: Mössbauer effect as applied to iron organometallic complexes and stresses in ferrous metals; water-degradable polymers for controlled release of pesticides and medicines; thermodynamics of kinetic temperature. Mailing Add: 2000 Westlake Dr Austin TX 78746

COLLINS, STEPHEN, b Chicago, Ill, May 14, 27; m 54; c 3. ECOLOGY. Educ: Cornell Univ, BS, 49; Rutgers Univ, PhD(ecol), 56. Prof Exp: Resident naturalist, Palisades Nature Asn, NJ, 49-51; asst forester, Conn Agr Exp Sta, 57-62; from asst prof to assoc prof, 62-71, PROF BIOL, SOUTHERN CONN STATE COL, 71- Concurrent Pos: Consult, McGraw-Hill Bk Co, Inc, 57. Mem: Ecol Soc Am; Am Soc Mammal; Wildlife Soc; Wilderness Soc; Am Nature Study Soc. Res: Relation of land use, climatic, biotic and fire factors on biotic communities and their successions; silvics, vertebrate ecology, biogeography and general natural history; environmental problems; natural areas; pesticides; biological photography. Mailing Add: Dept of Biol Southern Conn State Col New Haven CT 06515

COLLINS, THOMAS C, solid state physics, see 12th edition

COLLINS, TIMOTHY LEO, JR, b Troy, NY, Jan 22, 25; m 49; c 4. PHYSICAL CHEMISTRY. Educ: St Bernardine of Siena Col, BS, 48, MS, 55. Prof Exp: Physicist, Knolls Atomic Power Lab, 48-56, mgr mass spectrometry, 57-73; MGR MASS SPECTROMETRY, GEN ELEC CO, 73- Mem: Am Nuclear Soc; Am Soc Testing & Mat; fel Am Inst Chemists. Res: Mass spectrometry in nuclear research. Mailing Add: Gen Elec Co PO Box 1072 Schenectady NY 12301

COLLINS, VERNON KIRKPATRICK, b Advocate, NS, May 14, 17; m 39; c 2. BIOCHEMISTRY. Educ: Acadia Univ, BSc, 37; McGill Univ, MS, 46. Prof Exp: Chemist, Can Packers, Montreal, 41-43; res chemist, Ogilvie Flour Mills, 43-46; CHIEF CHEMIST, VIO BIN CORP, MONTICELLO, 46- Mem: Am Chem Soc; Am

Oil Chem Soc. Res: Food spoilage; fat deterioration. Mailing Add: 201 S New St Champaign IL 61820

COLLINS, VINCENT J, b Haverstraw, NY, Nov 24, 14; m 44; c 8. MEDICINE, ANESTHESIOLOGY. Educ: Marietta Col, BS, 36; Brown Univ, MS, 38; Yale Univ, MD, 42. Prof Exp: Assoc, Doctors Hosp, New York, 46-49; dir anesthesiol, St Vincent's Hosp, 49-57; asst dir, Bellevue Hosp, 57-61; assoc prof surg, 61-66, PROF ANESTHESIOL, SCH MED, NORTHWESTERN UNIV, 66-; DIR DIV ANESTHESIOL, COOK COUNTY HOSP, CHICAGO, 61- Mem: Am Soc Anesthesiol; AMA (secy, 57-62), AAAS. Res: Investigation of drugs, shock and endocrine problems. Mailing Add: Div of Anesthesiol Cook County Hosp Chicago IL 60612

COLLINS, WALTER MARSHALL, b Enfield, Conn, Nov 6, 17; m 43; c 4. ANIMAL GENETICS. Educ: Univ Conn, BS, 40, MS, 49; Iowa State Univ, PhD(poultry breeding), 60. Prof Exp: Instr poultry husb, Univ Conn, 47-49; from asst prof to assoc prof, 51-63, chmn genetics prog, 65-66, PROF POULTRY SCI, UNIV NH, 63- Concurrent Pos: NIH spec fel, Univ Calif, Davis, 64-65. Mem: AAAS; Am Poultry Sci Asn; Genetics Soc Am; Am Genetic Asn. Res: Applied population genetics; experimental verification of quantitative genetics theory. Mailing Add: Dept Animal Sci Kendall Hall Univ of NH Durham NH 03824

COLLINS, WARREN EUGENE, b Memphis, Tenn, Jan 26, 47; m 71. NUCLEAR PHYSICS. Educ: Christian Bros Col, BS, 68; Vanderbilt Univ, MS, 70, PhD(physics), 72. Prof Exp: Assoc prof physics, Southern Univ, Baton Rouge, 72-73; asst prof, 73-74, ASSOC PROF PHYSICS, FISK UNIV, 74-; RES ASSOC, VANDERBILT UNIV, 75- Mem: Am Asn Physics Teachers; Am Phys Soc. Res: Level structure studies of selenium and arsenic isotopes with mass numbers between 68 and 76, using various methods; singles analysis, gamma-gamma angular correlation, gamma-gamma coincidences, and life-time studies. Mailing Add: Dept of Physics Fisk Univ Nashville TN 37203

COLLINS, WILLIAM BECK, b Port Williams, NS, Dec 5, 26; m 50; c 1. PLANT PHYSIOLOGY, HORTICULTURE. Educ: McGill Univ, BSc, 48, MSc, 54; Rutgers Univ, PhD(hort, plant physiol), 61. Prof Exp: Res scientist, 48-73, POTATO PHYSIOLOGIST, RES STA, CAN DEPT AGR, 73- Mem: Int Soc Hort Sci; Am Soc Hort Sci; Can Soc Hort Sci. Res: Relationships of growth regulators, endogenous and applied, to growth and development in potato; growth analysis and management studies with potato. Mailing Add: Can Dept of Agr Res Sta PO Box 280 Fredericton NB Can

COLLINS, WILLIAM CARRIDINE, b Henderson, NC, Jan 4, 41; m 60; c 2. SOLID STATE PHYSICS. Educ: NC State Univ, BS, 65; Univ NC, PhD(physics), 71. Prof Exp: Res assoc physics, Nat Res Coun, 71-72; RES PHYSICIST, NAVAL RES LAB, 72- Mem: Sigma Xi. Res: Optical properties of solids and optical data processing. Mailing Add: Code 6440 Naval Res Lab Washington DC 20375

COLLINS, WILLIAM E, b Beachburg, Ont, Aug 2, 15; m 42; c 1. UROLOGY. Educ: Queen's Univ, Ont, MD & CM, 38; FRCPS(C), 49. Prof Exp: PROF SURG & HEAD SUB-DEPT UROL, UNIV OTTAWA, 59-; CHIEF DEPT UROL, OTTAWA CIVIC HOSP, 60- Concurrent Pos: Consult, Royal Ottawa Sanatorium, 50; examr urol, Royal Col Physicians & Surgeons Can, 60-64, chmn urol nucleus comt, 74-76; rep bd gov, Am Col Surgeons, 62-65, mem urol comt, 64-66; chmn urol working party, Nat Med Manpower Comn Can, 73-75. Mem: Fel Am Col Surg; Am Urol Asn; Can Urol Asn (pres, 62-63); Int Soc Urol; Can Acad Urol Surg (pres, 73-74). Res: Bladder neck obstruction; storage and transplant of urine; etiology and treatment of bladder carcinoma. Mailing Add: 1105 Carling Ave Suite 207 Ottawa ON Can

COLLINS, WILLIAM EDGAR, b Terra Alta, WVa, Mar 31, 35; m 59; c 2. ENDOCRINOLOGY. Educ: Univ WVa, BS, 57; Univ Wis, MS, 61, PhD(endocrinol), 65. Prof Exp: Prog specialist, Reprod Physiol for Ford Found, Inst Agr, Gujarat, India, 65-67; from asst prof to assoc prof, 67-74, actg dean, 74-75, PROF BIOL, WVA UNIV, 74-, DEAN COL ARTS & SCI, 75- Mem: Brit Soc Study Fertil; Endocrine Soc; Am Soc Zoologists. Res: Function of corpus luteum and control of the estrous cycle. Mailing Add: 104 Woodburn Hall WVa Univ Morgantown WV 26506

COLLINS, WILLIAM ERLE, b Lansing, Mich, July 9, 29; m 56. MEDICAL ENTOMOLOGY, ECONOMIC ENTOMOLOGY. Educ: Mich State Univ, BS, 51, MS, 52; Rutgers Univ, PhD(entom), 54. Prof Exp: Asst entom, Rutgers Univ, 53-54; entomologist, Diamond Alkali Chem Co, 54; med entomologist, Biol Warfare Labs, Ft Detrick, Md, 55-58; exten specialist entom, Rutgers Univ, 58-59; med biologist, Lab Parasitic Dis, 59-74, MED BIOLOGIST, BUR TROP DIR, CTR DIS CONTROL, USPHS, 74- Mem: Entom Soc Am; Am Soc Trop Med & Hyg; Am Mosquito Control Asn. Res: Medical entomology in field of virus and malaria transmission by arthropod vectors. Mailing Add: Bur Trop Dis Ctr Dis Control 1600 Clifton Rd Atlanta GA 30333

COLLINS, WILLIAM F, b Laceyville, Pa, May 9, 18; m 45; c 3. FOOD SCIENCE. Educ: Pa State Col, BS, 42, MS, 48, PhD(dairy sci), 49. Prof Exp: Dairy res chemist, Swift & Co, Chicago, 49-53, head, Ice Cream & Stabilizer Res Div, Res Labs, 53-56, mem staff, Mkt Develop, Hammond, Ind, 56-57, tech sales rep US & Can, Gen Gelatin Dept, Kearny, NJ, 57-67, mgr tech sci, Swift Chem Co, Oak Brook, Ill, 67-68; pres & gen mgr, Nutriprod Ltd & Topping Co, Can, 68-73; ASST PROF FOOD SCI & TECHNOL & EXTEN SPECIALIST, VA POLYTECH INST & STATE UNIV, 73- Honors & Awards: Award of Merit, Nat Confectioners Asn, 68. Mem: Am Dairy Sci Asn; Inst Food Technologists; Int Asn Milk Food & Environ Sanitarians; Am Asn Candy Technologists. Res: Milk quality and flavor as affected by microbial flora and ultra high pasteurization; temperatures and asceptic packaging practices. Mailing Add: Dept of Food Sci & Technol Va Polytech Inst & State Univ Blacksburg VA 24061

COLLINS, WILLIAM FRANCIS, JR, b New Haven, Conn, Jan 20, 24; m 51; c 3. NEUROSURGERY. Educ: Yale Univ, BS, 44, MD, 47; Am Bd Neurol Surg, dipl, 56. Prof Exp: Resident neurosurg, Barnes Hosp, St Louis, Mo, 47-49, from asst res neurosurgeon to res neurosurgeon, 51-53; from instr to assoc prof neurosurg, Sch Med, Western Reserve Univ, 54-63; prof neurol surg, dir div & neurosurgeon-in-chief, Med Col Va, 63-67; prof neurol surg & chmn dept, 67-70, CUSHING PROF SURG, SCH MED, YALE UNIV, 70- Concurrent Pos: Nat Found Infantile Paralysis fel, Wash Univ, 53-54; consult, West Haven Vet Admin Hosp, 67-; neurosurgeon in chief, Yale New Haven Med Ctr, 67- Mem: Acad Neurol Surg; Am Col Surg; Asn Res Nerv & Ment Dis; Harvey Cushing Soc; Neurosurg Soc Am. Res: Neurophysiology of afferent pathway systems. Mailing Add: Yale Univ Sch of Med 333 Cedar St New Haven CT 06510

COLLINS, WILLIAM JOHN, b Mt Union, Pa, July 5, 34; m 58; c 3. ENTOMOLOGY. Educ: Juniata Col, BS, 56; Rutgers Univ, PhD(entom), 65. Prof Exp: Asst prof entom, Rutgers Univ, 64-66; ASST PROF ENTOM, OHIO STATE UNIV, 67- Mem: Entom Soc Am; AAAS; Am Chem Soc. Res: Toxicology; insecticide resistance, especially in cockroaches; mechanisms of resistance; environmental toxicology. Mailing Add: Dept of Entom Ohio State Univ Columbus OH 43210

COLLINS, WILLIAM KERR, b Vance Co, NC, June 21, 31; m 54; c 3. PLANT BREEDING, GENETICS. Educ: NC State Col, BS, 54, MS, 61; Iowa State Univ, PhD(plant breeding), 63. Prof Exp: Res instr, Crops Dept, NC State Col, 56-60; asst agron, Iowa State Univ, 60-63; agronomist, Res Dept, R J Reynolds Tobacco Co, 63-66; assoc prof, 66-70, PROF AGRON, NC STATE UNIV, 70-, TOBACCO EXTEN SPECIALIST, 66- Mem: Am Soc Agron. Res: Fluecured tobacco variety evaluation for agronomic, chemical, physical and pathological properties; tobacco herbicides and sucker control chemicals; tobacco fertility; tobacco production in Greece, Iran, Afghanistan, British Honduras, Costa Rica, Uruguay, Italy and Venezuela. Mailing Add: Dept Crop Sci 454 Williams Hall NC State Univ Raleigh NC 27607

COLLINSON, CHARLES WILLIAM, b Wichita, Kans, Dec 15, 23; m 44; c 2. GEOLOGY. Educ: Augustana Col, AB, 49; Univ Iowa, MS, 50-51, PhD(geol), 52. Prof Exp: Asst, Univ Iowa, 48 & 51-52; from asst geologist to geologist, 52-69, HEAD STRATIG & AREAL GEOL, STATE GEOL SURV, ILL, 69-; PROF GEOL, UNIV ILL, URBANA-CHAMPAIGN, 68- Concurrent Pos: Lectr, Univ Ill, 56-57 & Stanford Univ, 60; ed, Jour Paleont, 58-64; Guggenheim fel, Gt Brit & Ger, 63-64; mem Paleont Res Inst. Mem: Paleont Soc; fel Geol Soc Am; Am Asn Petrol Geol; Soc Econ Paleontologists & Mineralogists; Ger Paleont Soc. Res: Mississippian stratigraphy of the Mississippi Valley; Devonian-Mississippian conodonts; sedimentation and paleolimnology of Lake Michigan. Mailing Add: Dept of Geol Univ of Ill Urbana IL 61801

COLLINSON, JAMES W, b Moline, Ill, June 24, 38; m 61; c 2. GEOLOGY. Educ: Augustana Col, Ill, AB, 60; Stanford Univ, PhD(geol), 66. Prof Exp: Asst prof, 66-71, ASSOC PROF GEOL, OHIO STATE UNIV, 71- Mem: AAAS; Geol Soc Am; Paleont Soc;Am Asn Petrol Geol; Soc Econ Paleont & Mineral. Res: Permian and Triassic stratigraphy and paleontology; Antarctic geology. Mailing Add: Dept of Geol Ohio State Univ Columbus OH 43210

COLLINS-WILLIAMS, CECIL, b Toronto, Ont, Dec 31, 18; m 44; c 2. PEDIATRICS, CLINICAL IMMUNOLOGY. Educ: Univ Toronto, BA, 41, MD, 44. Prof Exp: Assoc prof, Univ Toronto, 68-74; SR STAFF PHYSICIAN, HOSP FOR SICK CHILDREN, 65-; PROF PEDIAT, UNIV TORONTO, 74- Concurrent Pos: Consult, Ont Crippled Children's Ctr, Toronto, 66- Mem: Am Col Allergists; Am Acad Allergy; Can Soc Allergy & Clin Immunol (pres, 62); Can Soc Immunol; Can Pediat Soc. Mailing Add: Hosp for Sick Children 555 University Ave Toronto ON Can

COLLIPP, PLATON JACK, b Niagara Falls, NY, Nov 4, 32; m 56; c 3. PEDIATRICS. Educ: Univ Rochester, AB, 54, MD, 57. Prof Exp: UPSHS trainee biochem, Univ Wash, 57-59; from intern to resident pediat, Univ Southern Calif, 59-61, asst prof, 63-65; clin prof, State Univ NY Downstate Med Ctr, 65-69; PROF PEDIAT, HEALTH SCI CTR, STATE UNIV NY STONYBROOK, 69- Concurrent Pos: Assoc chmn pediat, Maimonides Med Ctr, 65-67; chmn pediat, Nassau County Med Ctr, 67- Mem: Am Acad Pediat; Am Physiol Soc; Soc Pediat Res. Res: Pediatric endocrinology and metabolism, especially growth hormone, growth disorders and childhood obesity. Mailing Add: Nassau County Med Ctr East Meadow NY 11553

COLLIS, RONALD THOMAS, b London, Eng, July 22, 20; m 51; c 1. ATMOSPHERIC PHYSICS. Educ: Oxford Univ, MA, 51. Prof Exp: Meteorologist, Decca Radar Ltd, Eng, 55-58; head radar aerophys group, 58-67, DIR ATMOSPHERIC SCI LAB, STANFORD RES INST, 67- Concurrent Pos: Consult, NASA, 74- Mem: Assoc fel Am Inst Aeronaut & Astronaut; fel Am Meteorol Soc; Am Geophys Union; fel Royal Meteorol Soc. Res: Atmospheric factors in propagation of electromagnetic energy; weather radar; lidar; instrumental and data processing aspects of air pollution, aviation and general meteorology. Mailing Add: Atmospheric Sci Lab Stanford Res Inst Menlo Park CA 94025

COLLISTER, EARL HAROLD, b Galva, Ill, Mar 25, 23; m 45; c 3. PLANT BREEDING. Educ: Purdue Univ, BS, 47, MS, 48, PhD(plant genetics), 50. Prof Exp: Assoc agronomist oilseed res, Tex Res Found, 50-51, agronomist & chmn field crops dept, 51-56, from sr agronomist to prin agronomist & chmn plant sci dept, 56-59; from asst dir & chief agronomist to dir & chief agronomist, High Plains Res Found, 59-67; exec vpres res sales mkt in US & overseas, World Seeds, Inc, 67-68; PRES, INT GRAIN, INC, 68-; PRES, TRANSERA RES, INC, 71- Concurrent Pos: Consult, Francisco Sugar Co, Cuba, 56-58; hon trustee, Int Sesanum Found. Mem: Am Soc Agron; Am Genetic Asn; AAAS; NY Acad Sci. Res: Sesame; sunflowers; soybeans; triticale; wheat; safflower; corn; grain sorghum; millet. Mailing Add: 530 Park Lane Richardson TX 75080

COLLITON, MARGARET ANNETTE, psychiatric nursing, see 12th edition

COLLIVER, GARY WINSTON, biostatistics, plant nutrition, see 12th edition

COLLMAN, JAMES PADDOCK, b Beatrice, Nebr, Oct 31, 32; m 55. INORGANIC CHEMISTRY. Educ: Univ Nebr, BSc, 54, MS, 56; Univ Ill, PhD, 58. Prof Exp: From instr to prof chem, Univ NC, 58-67; PROF CHEM, STANFORD UNIV, 67- Concurrent Pos: A P Sloan Found fel, 64-66; Frontiers in Chem lectr, 64. Honors & Awards: Calif Sect Award, Am Chem Soc, 72, Inorg Chem Award, 75. Mem: Nat Acad Sci; Am Acad Sci; Am Chem Soc; The Chem Soc. Res: Synthesis and electron transport properties of metal-metal bonds, reactions of coordinated dioxygen, homogeneous catalysis; reactions of coordinated ligands and homogeneous catalysis; mixed-functions oxygenase models. Mailing Add: Dept of Chem Stanford Univ Stanford CA 94305

COLLORD, JAMES, b Buffalo, NY, May 23, 25; m 50; c 2. PEDODONTICS. Educ: Ohio Wesleyan Univ, BA, 50; Univ Buffalo, DDS, 58. Prof Exp: Supvr com factoring, William Iselin & Co, Inc, Mich, 50-54; teaching fel, Sch Dent, Univ Buffalo, 58-59, instr clin dent, 59-60, instr clin prosthetics, 60-61, instr clin dent, 61-62; asst prof clin dent, 62-68, ASSOC PROF PEDODONTICS, SCH DENT, STATE UNIV NY BUFFALO, 68-; DIR DENT AUXILIARY UTILIZATION, USPHS RES FOUND, 61- Concurrent Pos: Prog dir, Div Dent Health, Undergrad Dent Auxiliary Utilization Training, USPHS, 61- Res: Dental auxiliary utilization and expanded duties research. Mailing Add: Sch of Dent State Univ of NY Buffalo NY 14214

COLLVER, MICHAEL MOORE, b Los Angeles, Calif, Feb 9, 41; m 62; c 3. SOLID STATE PHYSICS. Educ: Univ Calif, Los Angeles, BA, 63; Univ Calif, Berkeley, MA, 70, PhD(mat sci), 71. Prof Exp: Res physicist, US Naval Radiol Lab, 68, asst superconductivity, 68-72; assoc sr res physicist, Gen Motors Res Labs, Mich, 72-76; MEM FAC PHYSICS, UNIV CAMPINAS, BRAZIL, 76- Mem: Am Phys Soc; Am Vacuum Soc. Res: Superconductivity; magnetic and semiconductor materials research; thin film physics. Mailing Add: Inst of Physics Univ of Campinas Campinas Brazil

COLMAN, ARTHUR DAVID, b New York, NY, m 63; c 3. PSYCHIATRY. Educ: Harvard Univ, AB, 58, MD, 62. Prof Exp: Intern med, Beth Israel Hosp, Boston,

Mass, 62-63; res psychiat, Langley Porter Neuropsychiat Inst, Med Ctr, Univ Calif, San Francisco, 63-66; res psychiatrist, Walter Reed Army Inst res & staff psychiatrist & mem teaching fac psychiat, Walter Reed Gen Hosp, 67-70; career teacher & resident asst prof psychiat, Med Ctr, 70-72, ASSOC CLIN PROF PSYCHIAT, MED SCH, UNIV CALIF, SAN FRANCISCO, 72- Concurrent Pos: State of Calif Dept Ment Hyg grant, 65-66; assoc fac prog in community psychiat, Washington Sch Psychiat, DC, 66-68; US Army Res & Develop Command grant, 67-69; lectr, Hebrew Univ Jerusalem, 69-70; lectr behav sci, Sch Environ Design, Univ Calif, Berkeley, 72- Mem: Asn Adv Psychother; Am Psychiat Asn. Res: Psychology of pregnancy; design of treatment environments; family therapy; experimental analysis of behavior. Mailing Add: Dept of Psychiat Sch of Med Univ of Calif San Francisco CA 94122

COLMAN, BRIAN, b Stockport, Eng, Oct 19, 33; m 64; c 3. BIOCHEMISTRY. Educ: Univ Keele, BA, 58; Univ Wales, PhD(plant physiol), 61. Prof Exp: Nat Res Coun Can fel biol, Queen's Univ, Ont, 61-62; res assoc, Univ Rochester, 62-65; asst prof, 65-70, ASSOC PROF BIOL, YORK UNIV, ONT, 70- Mem: Am Soc Plant Physiologists; Phycol Soc Am. Res: Photosynthesis and photorespiration in algae and higher plants; carbon metabolism in blue-green algae. Mailing Add: Dept of Biol York Univ 4700 Keele St Downsview ON Can

COLMAN, ROBERTA F, b New York, NY, July 21, 38; m 57; c 2. BIOCHEMISTRY, PROTEIN CHEMISTRY. Educ: Radcliffe Col, AB, 59, AM, 60, PhD(biochem), 62. Prof Exp: NIH fel enzym, 62-64; USPHS fel, Sch Med, Wash Univ, 64-66, asst prof enzym & protein chem, 66-67; assoc enzym, Harvard Med Sch, 67-69, from asst prof to assoc prof biochem, 69-73; PROF BIOCHEM, UNIV DEL, 73- Concurrent Pos: USPHS career develop award & res grant, Sch Med, Wash Univ, 66-67; Med Found Boston fel, Harvard Med Sch, 67-68, USPHS res grant, 67-, career develop award, 68- , Med Found res grant, 70-; Am Soc Biol Chem travel award, Int Cong Biochem, 67 & 70. Mem: AAAS; Am Soc Biol Chem; Am Chem Soc. Res: Mechanism of enzyme action; active sites of dehydrogenases; chemical basis of regulation of allosteric enzymes; investigation of specific enzymes such as acetoacetate decarboxylose, glutathione reductase, glutamate dehydrogenase, isocitrate dehydrogenase. Mailing Add: Dept of Biochem Univ of Del Newark DE 19711

COLMANO, GERMILLE, b Pola, Italy, Aug 22, 21; nat US; m 47; c 3. VETERINARY PHYSIOLOGY, MEDICAL BIOPHYSICS. Educ: Univ Bologna, DVM, 49, PhD(physiol, biochem), 50. Prof Exp: Asst physiol, Univ Bologna, 47-49, from instr to asst prof, 49-51; asst vet, Phillips Vet Hosp, Colo, 51-52; asst vet sci, Univ Wis, 52-53; proj asst, Inst Enzyme Res, Wis, 54-56; scientist biophys, Res Inst Advan Study, 56-61; Stoner fel biophys res lab, Eye & Ear Hosp, Sch Med, Univ Pittsburgh, 61-62; PROF VET SCI, VA POLYTECH INST & STATE UNIV, 62- Mem: Am Asn Vet Clin; Am Soc Vet Physiol & Pharmacol; Asn Am Vet Med Cols; Biophys Soc; fel Royal Soc Health. Res: Lamellar function-structure; monomolecular films; absorption spectra of chromophores; trace elements; steroids and stress in health and disease; bacteriostasis of silver in intramedullary orthopedic pins. Mailing Add: Dept of Vet Sci Va Polytech Inst & State Univ Blacksburg VA 24061

COLMENARES, CARLOS ADOLFO, b Ocana, Colombia, June 17, 32; US citizen; m 56; c 2. PHYSICAL CHEMISTRY, CHEMICAL ENGINEERING. Educ: Univ Calif, Berkeley, BS, 53; Wash State Univ, MS, 56; Rensselaer Polytech Inst, PhD(chem eng), 60. Prof Exp: Actg instr chem eng, Wash State Univ, 54-56; instr, Rensselaer Polytech Inst, 56-60; CHEMIST, LAWRENCE LIVERMORE LAB, 60- Mem: Am Vacuum Soc. Res: Gas-solid interactions; surface chemistry and physics; radiation chemistry; catalysis. Mailing Add: 2211 Granite Dr Alamo CA 94507

COLMER, ARTHUR RUSSELL, b Moss Point, Miss, Apr 12, 04. BACTERIOLOGY. Educ: Miss State Col, BSc, 25; Univ Wis, MS, 26, PhD(bact), 43. Prof Exp: Teacher high sch, Miss, 30-35, prin, 35-40; asst, Univ Wis, 40-41, bact, 41-42, instr physiol, 42-43; instr, Univ Tex, 43-45; assoc prof, WVa Univ, 45-47; from assoc prof to prof bact, 47-74; EMER ALUMNI PROF MICROBIOL, 74- Mem: AAAS; Am Soc Microbiol. Res: Bacillus cereus lecithinase; microbiology acid mine drainage; effects of herbicides on microorganisms. Mailing Add: Dept of Microbiol La State Univ Baton Rouge LA 70803

COLMEY, JOHN C, b St Louis, Mo, May 28, 30; m 53; c 2. FOOD SCIENCE. Educ: Colo State Univ, BS, 53, MS, 58; Univ Ill, PhD(dairy tech), 62. Prof Exp: Prod develop scientist, Res & Develop Lab, Pillsbury Co, Minn, 62-64, tech mgr new prod develop, 64-68; group leader, Foremost Foods Co, 68-71; dir res admin, 71-74, DIR RES & PROD DEVELOP, ITT CONTINENTAL BAKING CO, 74- Mem: Am Chem Soc; Inst Food Technologists; Am Asn Cereal Chemists; NY Acad Sci; Sigma Xi. Res: Chromatographic identification and quantitative measurement of flavor components in cheese; identification of changes in milk during manufacture of cottage cheese; product development; development of food and bakery products; research on dairy products and functionality of protein fractions. Mailing Add: ITT Continental Baking Co Box 731 Rye NY 10580

COLODNY, PAUL CHARLES, b Springfield, Mass, Feb 17, 30. PHYSICAL CHEMISTRY. Educ: Univ Mass, BS, 51; Princeton Univ, MA, 53, PhD(chem), 57. Prof Exp: Fiber physicist, Dow Chem Co, 57-60; from res chemist to sr res chemist, Aerojet-Gen Corp, 60-63, res chem specialist, 63-65; sr res chemist, Gen Tire & Rubber Co, Ohio, 65-66; sr staff mem, Raychem Corp, 66-70; res scientist, Lockheed Missile & Space Co, 70-73; GROUP LEADER, MCA DISCO-VISION, 74- Concurrent Pos: Consult, Raychem Corp, 71-73. Mem: Am Chem Soc; Soc Rheol; AAAS. Res: Fiber physics; textile fiber spinning; elastomers; solid rocket propellants; crystalline polymers; graphite, photo-polymerization. Mailing Add: MCA Labs 1640 West 228th St Torrance CA 90501

COLOM, JUAN, soil science, agricultural chemistry, see 12th edition

COLOMB, HENRY OCTAVE, JR, organic chemistry, see 12th edition

COLOMBINI, VICTOR DOMENIC, b Boston, Mass, Feb 2, 24. GEOLOGY. Educ: Boston Univ, AB, 50, AM, 52 & 53, PhD, 61. Prof Exp: Geologist, USAEC, 55; asst prof geol, La Polytech Inst, 56-61; actg asst prof, Univ Miss, 61; assoc prof earth sci & head dept, Findlay Col, 61-66; mem fac, 66-68, asst prof, 68-73, ASSOC PROF GEOG, OHIO STATE UNIV, LIMA BR, 73- Mem: Geol Soc Am; Mineral Soc Am; Nat Asn Geol Teachers; Asn Am Geog; Nat Coun Geog Educ. Res: Physical geography; introductory economic geography; economic geology; crystallography; mineralogy; geomorphology. Mailing Add: Dept of Geog Ohio State Univ 4300 Campus Dr Lima OH 48504

COLON, JOSE A, b Coamo, PR, Nov 24, 21; m 51; c 3. METEOROLOGY. Educ: Univ PR, BA, 44; Univ Chicago, MS, 50, PhD(air-sea interactions), 60. Prof Exp: Asst meteorol, Inst Trop Meteorol, PR, 45-46; instr math & meteorol, Univ PR, 46-49; instr meteorol, Univ Chicago, 51-52, res assoc, 52-54; res forecaster, US Weather Bur, PR, 54-48; supvry res meteorologist, Nat Hurricane Res Lab, Fla, 49-62, Int Meteorol Ctr, Int Indian Ocean Exped, Bombay, India, 62-64 & Nat Hurricane Res Lab, Fla, 64; METEOROLOGIST CHG, US WEATHER BUR, SAN JUAN, PR, 64- Mem: Am Meteorol Soc; Am Geophys Union; Royal Meteorol Soc. Res: Tropical cyclones, their motion and forecast problems, development, structure and evolution; air-sea interactions and atmospheric energy sources; cyclogenesis over Indian Ocean and Indian monsoon circulations. Mailing Add: US Weather Bur Isla Verde Int Airport San Juan PR 00913

COLON, JULIO ISMAEL, b Coamo, PR, June 19, 28; US citizen; m 55; c 2. VIROLOGY. Educ: Univ PR, BS, 50; Univ Chicago, PhD(microbiol), 59. Prof Exp: Fel, Nat Acad Sci-Nat Res Coun, 59-60; microbiologist, Ft Detrick, Md, 60-63; microbiologist, US Trop Res Med Lab, 63-65; ASSOC PROF VIROL, SCH MED, UNIV PA, SAN JUAN, 64- Concurrent Pos: Dir grad studies, Univ PR, San Juan, 72-74. Mem: Am Soc Microbiol; fel Am Acad Microbiologist; Tissue Culture Asn; Sigma Xi. Res: Intermediary metabolism; biochemistry and genetics of psittacosis group of microorganism; biochemistry and genetics of the arthropod-borne viruses; origin and source of viral infectious nucleic acid; radiobiology of viruses. Mailing Add: Dept of Microbiol Univ of PR Sch of Med San Juan PR 00936

COLONNIER, MARC, b Quebec, Que, May 12, 30; m 59; c 1. ANATOMY. Educ: Univ Ottawa, BA, 51, MD, 59, MSc, 60; Univ London, PhD(neurobiol), 63. Prof Exp: Med Res Coun Can fel, Univ Ottawa & Univ London, 59-63; asst prof anat, Univ Ottawa, 63-65; from asst prof to assoc prof neuroanat, Neurol Sci Lab, Univ Montreal, 65-69; PROF ANAT & HEAD DEPT, UNIV OTTAWA, 69- Honors & Awards: Lederle Med Fac Award, 66; Charles Judson Herrick Award, Am Asn Anat, 67. Mem: Can Asn Anat; Am Asn Anat; fel Royal Soc Can; Soc Neurosci. Res: Cerebral cortex; synapses; visual system. Mailing Add: Dept of Anat Univ of Ottawa Fac of Med Ottawa ON Can

COLOSI, NATALE, b Messina, Italy, Mar 5, 02. BACTERIOLOGY. Educ: St Francis Col, BS, 28; NY Univ, MS, 30, PhD, 35. Prof Exp: Instr bact, Col Med, NY Univ, 32-36; prof, 34-74, EMER PROF BACT & PUB HEALTH, WAGNER COL, 74-, PROF, NY POLYCLIN MED SCH & HOSP, 59-, DEAN, 65- Concurrent Pos: Dir & vpres bd trustees, Ital Hosp, 36-, NY State comnr, Inter-State Sanit Comn, 45. Mem: AAAS; fel Am Pub Health Asn; Am Soc Microbiol; Air Pollution Control Asn; Acad Polit Sci. Res: Serology. Mailing Add: Wagner Col 631 Howard Ave Staten Island NY 10301

COLOWICK, SIDNEY PAUL, b St Louis, Mo, Jan 12, 16; m 43, 51; c 3. BIOCHEMISTRY. Educ: Wash Univ, BS, 36, MS, 39, PhD(biochem), 42. Prof Exp: Asst biol chem, Med Sch, Wash Univ, 36-42, from instr to asst prof pharmacol, 42-46; assoc, Div Nutrit & Physiol, Pub Health Res Inst, NY, 46-48; assoc prof biochem, Col Med, Univ Ill, 48-50; prof biol, Johns Hopkins Univ, 50-59; PROF MICROBIOL & AM CANCER SOC PROF, VANDERBILT UNIV, 59- Honors & Awards: Eli Lilly Award, 47. Mem: Nat Acad Sci; Am Soc Biol Chem; Am Acad Arts & Sci. Res: Intermediary metabolism of carbohydrates; action of hormones. Mailing Add: Dept of Microbiol Vanderbilt Univ Sch of Med Nashville TN 37232

COLPA, JOHANNES PIETER, b Arnhem, Netherlands, Jan 26, 26; m 51; c 1. THEORETICAL CHEMISTRY, MAGNETIC RESONANCE. Educ: Univ Amsterdam, PhD(chem), 57. Prof Exp: Res chemist, Shell Res Labs, Amsterdam, 57-66; assoc prof chem, Univ Amsterdam, 63-69; PROF CHEM, QUEEN'S UNIV, ONT, 69- Concurrent Pos: Fel, Cambridge Univ, 61-62; consult, Max Planck Inst, Ger, 67-69; assoc ed, Molecular Physics, 64-69. Mem: Am Phys Soc; Royal Netherlands Chem Soc; Netherlands Phys Soc. Res: High pressure spectroscopy; pressure induced spectroscopic transitions; molecular orbital theory and magnetic resonance. Mailing Add: Dept of Chem Queen's Univ Kingston ON Can

COLQUHOUN, DONALD JOHN, b Toronto, Ont, Mar 29, 32; m 56; c 2. GEOLOGY. Educ: Univ Toronto, BA, 53, MA, 56; Univ Ill, PhD(geol), 60. Prof Exp: Asst geol, Univ Ill, 58-59; from asst prof to assoc prof, 60-70, PROF GEOL & MARINE BIOL & CHMN DEPT, 70- Concurrent Pos: Proj geologist, SC State Develop Bd, 60-; NSF grant, 63- Mem: Geol Soc Am; Am Asn Petrol Geol; Soc Econ Paleont & Mineral; Paleont Soc. Res: Application of sedimentalogical studies and techniques to geomorphology and stratigraphy; interpretation of coastal plain terraces; regional stratigraphy. Mailing Add: Dept of Geol Univ of SC Columbia SC 29208

COLQUITT, LANDON AUGUSTUS, b Ft Worth, Tex, Jan 25, 19; m 54. MATHEMATICS. Educ: Tex Christian Univ, BA, 39; Ohio State Univ, MA, 41, PhD(math), 48. Prof Exp: Instr math, Ohio State Univ, 46-48; from asst prof to assoc prof, 48-54, PROF MATH, TEX CHRISTIAN UNIV, 55- Mem: AAAS; Am Math Soc; Math Asn Am; Am Meteorol Soc; Soc Indust & Appl Math. Res: Mathematical analysis; applied mathematics. Mailing Add: 2601 McPherson Ft Worth TX 76109

COLQUITT, LEROY, JR, solid state physics, see 12th edition

COLSKY, JACOB, b Memphis, Tenn, Dec 5, 21; m 53; c 3. INTERNAL MEDICINE, ONCOLOGY. Educ: Memphis State Col, 38-40; Univ Tenn, MD, 44; Am Bd Internal Med, dipl. Prof Exp: Intern med, Jackson Mem Hosp, Miami, Fla, 44-45; asst chief clin res, Nat Cancer Inst, USPHS Hosp, Baltimore, Md, 50-51, actg chief, 51-52; from instr to assoc prof med, Col Med, State Univ NY, 52-57; assoc prof, 57-66, clin assoc prof, 66-74, CLIN PROF MED, SCH MED, UNIV MIAMI, 74-, PROF ONCOL, 75- Concurrent Pos: Fel prev med, Sch Med, Johns Hopkins Univ, 47-50; asst physician, Johns Hopkins Hosp, 47-52; instr med, Sch Med, Johns Hopkins Univ, 50-51; assoc dir med, Maimonides Hosp, Brooklyn, NY, 52-57; dir med oncol sect, Dept Med, Univ Miami & Jackson Mem Hosp, 60-70; sr investr, Eastern Co-op Group Solid Tumor Chemother, 60-; attend physician, Cedars of Lebanon Hosp & Jackson Mem Hosp, Miami, 61-; dir med oncol sect, Cedars of Lebanon Hosp, 72-; consult, Baptist Hosp, Mt Sinai Hosp & Vet Admin Hosp. Mem: AAAS; fel Am Col Physicians; AMA; Am Soc Clin Oncol; Am Asn Cancer Res. Res: Biology and chemotherapy of animal and human malignant neoplasms. Mailing Add: Prof Arts Ctr 1150 NW 14th St Miami FL 33136

COLSON, STEVEN DOUGLAS, b Idaho Falls, Idaho, Aug 16, 41; m 62; c 5. CHEMICAL PHYSICS. Educ: Utah State Univ, BS, 63; Calif Inst Technol, PhD(chem), 68. Prof Exp: Asst prof chem, 68-73, ASSOC PROF CHEM, YALE UNIV, 73- Concurrent Pos: Jr fac fel, Yale Univ, 72-73; mem, Nat Res Coun Adv Bd to US Army Res Off, 72-75. Res: Intermolecular interactions and energy transfer in molecular crystals supplemented by high resolution gas phase spectroscopic studies; a variety of spectroscopic and crystallographic techniques are used. Mailing Add: Dept of Chem Yale Univ 225 Prospect St New Haven CT 06520

COLTEN, HARVEY RADIN, b Houston, Tex, Jan 11, 39; m 59; c 3. IMMUNOLOGY, PEDIATRICS. Educ: Cornell Univ, BA, 59; Western Reserve Univ, MD, 63. Prof Exp: From intern to resident pediat, Univ Cleveland Hosps, 63-65; res assoc immunol, NIH, 65-67, sr scientist, 67-69, head molecular separations unit, 69-70; asst prof pediat, 70-73, ASSOC PROF PEDIAT, HARVARD MED SCH, 73- Concurrent Pos: Chief div allergy, Childrens Hosp Med Ctr, Boston. Mem: AAAS; Am Soc Clin Invest; Am Acad Allergy; Soc Pediat Res; Am Asn Immunol.

Res: Immunochemistry; allergy. Mailing Add: Div of Allergy Dept of Pediat Harvard Med Sch Boston MA 02115

COLTER, ALLAN KENNEDY, b Edmonton, Alta, Mar 14, 29; US citizen; m 55; c 3. ORGANIC CHEMISTRY. Educ: Univ Alta, BSc, 51; Univ Calif, Los Angeles, PhD(chem), 56. Prof Exp: Fel, Harvard Univ, 56-57; from instr to prof chem, Carnegie Inst Technol, 57-68; chmn dept, 68-75, PROF CHEM, UNIV GUELPH, 68- Mem: Am Chem Soc; The Chem Soc. Res: Mechanisms of organic reactions. Mailing Add: Dept of Chem Univ of Guelph Guelph ON Can

COLTER, JOHN SPARBY, b Bawlf, Alta, July 23, 22; m 50; c 2. VIROLOGY. Educ: Univ Alta, BS, 45; McGill Univ, PhD(biochem), 51. Prof Exp: Mem staff, Virus & Rickettsial Res Div, Lederle Labs, 51-57 & Wistar Inst, Pa, 57-61; PROF BIOCHEM & HEAD DEPT, UNIV ALTA, 61- Mem: Am Soc Biol Chem; Am Soc Cell Biol; Am Asn Cancer Res; Can Biochem Soc; fel Royal Soc Can. Res: Biochemistry of virus infection; mechanisms of viral replication and oncogenesis. Mailing Add: Dept of Biochem Univ of Alta Edmonton AB Can

COLTHARP, FORREST LEE, b Caney, Kans, Oct 30, 33; m 52; c 3. MATHEMATICS EDUCATION. Educ: Okla State Univ, BS, 57, MS, 60, EdD(higher educ), 68. Prof Exp: Teacher & math consult, pub schs, Okla, 57-64; asst prof math educ, Kans State Col, 64-66; instr, Okla State Univ, 66-67; PROF MATH EDUC, KANS STATE COL, 67- Mem: Nat Coun Teachers Math. Mailing Add: 1402 S Homer Pittsburg KS 66762

COLTHARP, GEORGE B, b Maringouin, La, Nov 28, 28; m 53; c 2. FOREST HYDROLOGY, WATER POLLUTION. Educ: La State Univ, BS, 51; Colo State Univ, MS, 55; Mich State Univ, PhD(forest hydrol), 58. Prof Exp: Proj leader watershed mgt, Rocky Mountain Forest & Range Exp Sta, US Forest Serv, NMex, 58-61; asst mgr, Coltharp's Livestock Mkt, La, 61-64; from asst prof to assoc prof range watershed mgt, Utah State Univ, 64-74; ASSOC PROF FORESTRY, UNIV KY, 75- Mem: Am Water Resources Asn; Soc Am Foresters; Am Soc Range Mgt; Soil Conserv Soc Am. Res: Forest hydrology studies; wildland water quality and watershed management. Mailing Add: Dept of Forestry Univ of Ky Lexington KY 40506

COLTHUP, NORMAN BERTRAM, b Paris, France, July 6, 24; nat US; m 57; c 3. SPECTROSCOPY. Educ: Antioch Col, BS, 49. Hon Degrees: DSc, Fisk Univ, 74. Prof Exp: INFRARED SPECTROSCOPIST, AM CYANAMID CO, 44- Concurrent Pos: Infrared course lectr, Fisk Univ. Mem: Coblentz Soc. Res: Vibrational spectroscopy; molecular structure studies using infrared spectroscopy; infrared spectra-structure correlations. Mailing Add: Am Cyanamid Co Chem Res Div 1937 W Main St Stamford CT 06904

COLTMAN, CHARLES ARTHUR, JR, b Pittsburgh, Pa, Nov 7, 30; m 51; c 4. HEMATOLOGY, ONCOLOGY. Educ: Univ Pittsburgh, BS, 52, MD, 56; Ohio State Univ, MMS, 63; Am Bd Internal Med, dipl, 63, cert hemat, 72, cert med oncol, 73. Prof Exp: From intern to resident path, Del Hosp, Wilmington, 56-57; Med Corps, US Air Force, 57-, flight surgeon, Walker AFB, NMex, 57-59, resident, Univ Hosp, Ohio State Univ, 59-63, staff hematologist, 63-66, CHIEF HEMAT-ON-COL SERV, WILFORD HALL AIR FORCE MED CTR, 66-, CHMN DEPT MED, 75- Concurrent Pos: Asst med, Ohio State Univ Hosp, 59-61, from jr asst resident to sr asst resident, 59-62, asst instr med, 61-62, chief med resident & demonstr med, 62-63, attend physician & instr med, 63; mil consult to Surg Gen, US Air Force, 64-; clin assoc prof physiol & med, Univ Tex Med Sch, San Antonio, 70- Honors & Awards: Stitt Award, Asn Mil Surg US, 70; cert of achievement in internal med, Surg Gen. Mem: AAAS; fel Am Col Physicians; Am Asn Cancer Res; Am Fedn Clin Res; Am Soc Hemat. Res: Research in clinical cancer chemotherapy. Mailing Add: 10411 Moonglow San Antonio TX 78216

COLTMAN, JOHN WESLEY, b Cleveland, Ohio, July 19, 15; m 41; c 2. MUSICAL ACOUSTICS, ELECTRON OPTICS. Educ: Case Inst Technol, BS, 37; Univ Ill, MS, 39, PhD(physics), 41. Prof Exp: Res engr, 41-44, sect mgr, 44-49, mgr electronics & nuclear physics dept, 49-60, assoc dir, 60-64, dir math & radiation, 64-69, res dir indust & defense prod, 69-73, DIR RES PLANNING, RES LABS, WESTINGHOUSE ELEC CORP, 73- Concurrent Pos: Mem adv group electron devices, US Dept Defense, 62-66; mem adv comt, NASA, 64-66; mem numerical data adv bd, Nat Acad Sci, 68-71. Honors & Awards: Longstreth Medal, Franklin Inst, 60; Westinghouse Order of Merit, 68; Roentgen Medal, 70. Mem: Fel Am Phys Soc; fel Inst Elec & Electronics Engrs. Res: Slow neutrons; microwave tubes; x-ray fluorescence; scintillation counters; image amplifier tubes; energy conversion; control mechanisms; lasers. Mailing Add: Westinghouse Elec Corp Res Labs Beulah Rd Pittsburgh PA 15235

COLTMAN, RALPH READ, JR, b Pittsburgh, Pa, Nov 15, 24; m 44; c 3. SOLID STATE PHYSICS. Educ: Carnegie Inst Technol, BS, 50. Prof Exp: PHYSICIST, OAK RIDGE NAT LAB, 50- Mem: Am Phys Soc. Res: Irradiation damage in metals; cryogenics. Mailing Add: Solid State Div Oak Ridge Nat Lab Oak Ridge TN 37830

COLTON, DAVID L, b San Francisco, Calif, Mar 14, 43; m 68. MATHEMATICS. Educ: Calif Inst Technol, BS, 64; Univ Wis, MS, 65; Univ Edinburgh, PhD(math), 67. Prof Exp: From asst prof to assoc prof math, Ind Univ, Bloomington, 67-75; PROF MATH, UNIV STRATHCLYDE, 75- Concurrent Pos: Asst prof, McGill Univ, 68-69; assoc ed, Applicable Anal; vis res fel, Univ Glasgow, 71-72; guest prof, Univ Konstanz, 74-75. Mem: Edinburgh Math Soc; Soc Indust & Appl Math. Res: Analytic theory of partial differential equations; improperly posed problems; applied mathematics. Mailing Add: Dept of Math Univ of Strathclyde Glasgow Scotland

COLTON, ERVIN, b Omaha, Nebr, June 25, 27; m 55; c 4. INORGANIC CHEMISTRY. Educ: Ga Inst Technol, BS, 50; Univ Kans, MS, 52; Univ Ill, PhD(inorg chem), 54. Prof Exp: Asst prof chem, Ga Inst Technol, 54-56; prin res chemist, Int Minerals & Chem Corp, 56-58; res chemist, Allis-Chalmers Mfg Co, 58-62, mgr, Cerac Sect, New Prod Dept, 62-64, PRES, CERAC, INC, 64-, CERAC HOT-PRESSING, INC, 67- & CERAC/PURE, INC, 70- Mem: Am Chem Soc; Am Ceramic Soc. Res: Liquid ammonia and inorganic fluorine chemistry; hydrazine chemistry and synthesis; high temperature refractories; potassium chemicals. Mailing Add: Cerac Inc Box 1178 Milwaukee WI 53201

COLTON, FRANK BENJAMIN, b Poland, Mar 3, 23; nat US; m; c 4. CHEMISTRY. Educ: Northwestern Univ, BS, 45, MS, 46; Univ Chicago, PhD(chem), 49. Prof Exp: Fel biochem, Mayo Clin, 49-51; asst dir chem res, 51-70, RES ADV, G D SEARLE & CO, 70- Mem: Am Chem Soc; The Chem Soc. Res: Organic chemistry; biochemistry. Mailing Add: 3901 Lyons St Evanston IL 60203

COLTON, RAYMOND H, b Springfield, Mass, Mar 17, 42; m 66; c 3. OTOLARYNGOLOGY, COMMUNICATION SCIENCE. Educ: Cent Conn State Col, BSc, 63; Univ Conn, MA, 67; Univ Fla, PhD(commun sci), 69. Prof Exp: Social worker child welfare, State of Conn, 63-64; res assoc speech, Commun Sci Lab, Univ Fla, 68-69; ASST PROF OTOLARYNGOL, STATE UNIV NY UPSTATE MED CTR, 69- Concurrent Pos: Lectr, Syracuse Univ, 70- Mem: AAAS; Am Speech & Hearing Asn; Acoust Soc Am; Int Soc Phonetic Sci. Res: Normal operation of the human larynx; experiments related to the acoustical, physiological and perceptual correlates of human speech. Mailing Add: Dept Otolaryngol & Commun Sci State Univ NY Upstate Med Ctr Syracuse NY 13210

COLTON, ROGER BURNHAM, b Windsor Locks, Conn, Jan 1, 24; m 47, 73; c 5. ENVIRONMENTAL GEOLOGY. Educ: Yale Univ, BS, 47, MS, 49. Prof Exp: GEOLOGIST, US GEOL SURV, 49- Mem: Geol Soc Am; Am Asn Petrol Geol; Am Inst Prof Geologists; Asn Eng Geologists. Res: Landslides, geomorphology and photogeology. Mailing Add: US Geol Surv Box 25046 Denver Fed Ctr Denver CO 80225

COLUCCI, ANTHONY VITO, b Chicago, Ill, Sept 24, 38. ENVIRONMENTAL HEALTH, TOXICOLOGY. Educ: Loyola Univ, Ill, BS, 61; Johns Hopkins Univ, ScD(pathobiol), 66. Prof Exp: Res asst bact res, Lutheran Gen Hosp, Park Ridge, Ill, 60-61; res assoc parasitol res, Loyola Univ, Ill, 61; NSF fel biol res, Univ Wis, 66-67; chemist III, R J Reynolds Tobacco Co, 67-70; chief biochem & physiol br, Human Studies Lab, US Environ Protection Agency, 70-74; VPRES HEALTH PROGS, GREENFIELD, ATTAWAY & TYLER INC, 74- Concurrent Pos: Mem adv comt coord res coun subcomt health effect mobile source emissions, 70-; NSF consult, Dir Joint US Environ Protection Agency & Nsf Spanish Am Mercury Health Effects Prog, 70-74. Mem: Soc Toxicol; Sigma Xi. Res: Parasite biochemistry; fate of smoke constituents in animals; intermediary metabolism of parasitic animals; pharmacology of helminth infections; health effects of pollutants in both occupational and environmental setting. Mailing Add: Greenfield Attaway & Tyler Inc 91 Larkspur St San Rafael CA 94901

COLVARD, DEAN WALLACE, b Ashe Co, NC, July 10, 13; m 39; c 3. ANIMAL SCIENCE, ANIMAL ECONOMICS. Educ: Berea Col, BS, 35; Univ Mo, MA, 38; Purdue Univ, PhD, 50. Hon Degrees: Dr, Purdue Univ, 60. Prof Exp: Instr agr, Brevard Col, 35-37; res asst, Univ Mo, 37-38; supt, NC Agr Res Sta, 38-46; prof animal sci, NC State Col, 47-53, head dept, 48-53, dean agr, 53-60; pres, Miss State Univ, 60-66; CHANCELLOR, UNIV NC, CHARLOTTE, 66- Concurrent Pos: Trustee, Berea Col, 56- & St Andrews Presby Col, 69- Honors & Awards: Distinguished Serv Award, US Army. Res: Research, teaching and writing; animal physiology; economic geography and education. Mailing Add: Off of the Chancellor Univ of NC Charlotte NC 28223

COLVILLE, WILLIAM LYTLE, agronomy, see 12th edition

COLVIN, BURTON HOUSTON, b West Warwick, RI, July 12, 16; m 47; c 3. MATHEMATICS. Educ: Brown Univ, AB, 38, AM, 39; Univ Wis, PhD(math), 43. Prof Exp: Instr math, Univ Wis, 40-42, 43-44, 45-46, asst prof, 47-51; tech aid appl math panel, Nat Defense Res Comt & Off Sci Res & Develop, 44-45; mathematician, Phys Res Staff, Boeing Co, 51-55, supvr math anal, 55-58, assoc head math res lab, Sci Res Labs, 58-59, head, 59-70, head math & Info Sci Lab, 70-72; CHIEF APPLIED MATH DIV, NAT BUR STANDS, 72- Concurrent Pos: Vis sci lectr, Soc Indust & Appl Math, 59-60 & Math Asn Am, 63-65; chmn, Conf Bd Math Scis, 75-76; consult, NSF, 76- Mem: Fel AAAS; Soc Indust & Appl Math (pres, 71-72); Am Math Soc; Math Asn Am; Inst Math Statist. Res: Differential equations; applied mathematics. Mailing Add: Appl Math Div Nat Bur Stands Washington DC 20234

COLVIN, CLAIR IVAN, b Clyde, Ohio, Sept 16, 27. PHYSICAL CHEMISTRY. Educ: Ohio Univ, BS, 49; Univ Miami, MS, 61, PhD(phys chem), 63. Prof Exp: Anal chemist, Aluminum & Magnesium, Inc, 51-52 & Nat Carbon Co, 52-53; instr chem, Racine Exten Ctr, Univ Wis, 53-56; teacher high sch, 56-57; univ fel chem & USPHS grant, Sch Med, Univ Miami, 63-64; from asst prof phys chem to assoc prof chem, 64-70, PROF CHEM & CHMN DEPT, GA SOUTHERN COL, 70- Mem: Am Chem Soc; The Chem Soc. Res: Kinetics of the thermal decomposition of ammonium nitrate in the presence of catalysts; molecular orbital calculations for conjugated organic compounds of biological interest. Mailing Add: Dept of Chem Ga Southern Col Statesboro GA 30458

COLVIN, CURTIS A, b Provo, Utah, May 14, 28; m 54; c 4. NUCLEAR CHEMISTRY, ANALYTICAL CHEMISTRY. Educ: Brigham Young Univ, BS, 51. Prof Exp: Jr chemist, Gen Elec Co, Hanford, 51-54, chemist, 54-65, sr chemist, Isochem Inc, 65-67, sr chemist, 67-72, MGR ANAL CHEM, ATLANTIC RICHFIELD HANFORD CO, 72- Mem: Am Chem Soc; Inst Nuclear Mat Mgt. Res: Chemistry involved in the separation of actinide elements; analytical chemistry of the actinide elements. Mailing Add: 1410 Sunset Richland WA 99352

COLVIN, DALLAS VERNE, b Westport, Ore, May 30, 37; m 64. ANIMAL BEHAVIOR, ANIMAL ECOLOGY. Educ: Portland State Univ, BS, 63; Univ Colo, PhD(zool), 70. Prof Exp: Teaching asst biol, Univ Colo, Boulder, 67-68; ASST PROF BIOL SCI, CALIF STATE COL, DOMINGUEZ HILLS, 70- Mem: AAAS; Am Inst Biol Sci; Am Soc Mammal. Res: Behavioral studies of small mammals; analysis and biological bases of ultrasounds used by Microtus as a form of communication between neonates and adults. Mailing Add: Dept of Biol Calif State Col Dominguez Hills CA 90246

COLVIN, HARRY WALTER, JR, b Schellsburg, Pa, Dec 5, 21; m 50; c 2. PHYSIOLOGY, BIOCHEMISTRY. Educ: Pa State Univ, BS, 50; Univ Calif, Davis, PhD(comp physiol), 57. Prof Exp: Instr physiol, Okla State Univ, 56-57; from asst prof to assoc prof, Univ Ark, 57-65; from asst prof to assoc prof, 65-75, PROF PHYSIOL, UNIV CALIF, DAVIS, 75- Concurrent Pos: Coun Int Exchange Scholars Fulbright-Hays Award, 72-73. Mem: Sigma Xi; AAAS; Am Dairy Sci Asn; Am Soc Animal Sci; Am Inst Biol Sci. Res: Rumen physiology; carbohydrate metabolism in calves; blood coagulation. Mailing Add: Dept of Animal Physiol Univ of Calif Davis CA 95616

COLVIN, JOHN ROSS, b Regina, Sask, Jan 6, 21; m; c 2. BIOCHEMISTRY, BIOPHYSICS. Educ: Univ Sask, BSA, 46; Univ Alta, MSc, 48; Univ Minn, PhD(biochem), 51. Prof Exp: From asst res officer to sr res officer, 51-66, HEAD BIOPHYS SECT, DIV BIOL SCI, NAT RES COUN CAN, 57-, PRIN RES OFFICER, 67- Concurrent Pos: Spec lectr, Univ Ottawa, 61-; vis prof, Laval Univ, 69-70. Mem, Nat Comt Biophys Can, 61-71, mem assoc comt, 62-71. Mem: Biophys Soc; Chem Inst Can; fel Can Physiol Soc; Can Biochem Soc. Res: Physical chemistry of proteins; biological fibrogenesis; fine cell structure and function. Mailing Add: Div of Biol Sci Nat Res Coun of Can Ottawa ON Can

COLWELL, CHARLES E, organic chemistry, see 12th edition

COLWELL, JACK HAROLD, b Wooster, Ohio, Dec 29, 31; m 56; c 2. EXPERIMENTAL SOLID STATE PHYSICS. Educ: Mt Union Col, BS, 53; Purdue Univ, MS, 58; Univ Wash, PhD(chem), 61. Prof Exp: Nat Res Coun Can fel, 61-63; chemist, Cryophysics Sect, 63-73, RES CHEMIST, PRESSURE & VACUUM SECT,

NAT BUR STAND, 73- Mem: Am Chem Soc; Am Phys Soc. Res: Low temperature calorimetry; properties of molecular crystals, superconductors and magnetic insulators at low temperatures; properties of materials at high pressures. Mailing Add: Heat Div Nat Bur of Stand Washington DC 20234

COLWELL, JOHN AMORY, b Boston, Mass, Nov 4, 28; m 54; c 4. INTERNAL MEDICINE, PHYSIOLOGY. Educ: Princeton Univ, AB, 50; Northwestern Univ, MD, 54, MS, 57, PhD(physiol), 68; Am Bd Internal Med, dipl, 62. Prof Exp: Intern med, Med Sch, Western Reserve Univ, 54-55; resident, Northwestern Univ, 54-56, 59-60; instr, Med Sch, 60-62, assoc, 62-65, from asst prof to assoc prof, 65-71; PROF MED, MED UNIV SC, 71-, DIR ENDOCRINOL & METAB, NUTRIT DIV, 72-, UNIV RES COORDR, 73- Concurrent Pos: NIH fel, Northwestern Univ, 56-57, Am Diabetes Asn & univ fels, 60-61, NIH res grant, 62-71; clin investr, Vet Admin Res Hosp, Chicago, 61-63, chief sect, 63-71; res, Vet Admin Hosp, Charleston, 71-74. Mem: AAAS; fel Am Col Physicians; Am Diabetes Asn; Am Fedn Clin Res; Endocrine Soc. Res: Insulin secretion, degradation and action in animals and man; selected clinical studies in subjects with disorder of metabolism and endocrinology; platelet function in diabetes. Mailing Add: Dept of Med Med Univ of SC Charleston SC 29401

COLWELL, JOHN EDWIN, physical chemistry, see 12th edition

COLWELL, JOSEPH F, b Brush, Colo, Mar 16, 29; m 53; c 3. SOLID STATE PHYSICS. Educ: Colo State Univ, BS, 51; Cornell Univ, PhD(physics), 60. Prof Exp: Physicist, Navy Electronics Lab, 51-53; asst physics, Cornell Univ, 53-54 & 56-60; res staff mem, Gen Atomic Div, Gen Dynamics Corp, 60-67; staff mem, Gulf Radiation Technol Div, Gulf Energy & Environ Systs, Inc, 67-74; MEM STAFF, INTELCOM/RADIATION TECHNOL, 74- Mem: AAAS; Am Phys Soc; Soc Explor Geophys. Res: Solid state theory, especially as related to direct energy conversion devices and radiation damage in solids; seismic methods of geophysical exploration. Mailing Add: Intelcom/Radiation Technol Box 80817 San Diego CA 92138

COLWELL, PRISCILLA J, b Boston, Mass, Feb 6, 44. EXPERIMENTAL SOLID STATE PHYSICS, QUANTUM ELECTRONICS. Educ: Emmanuel Col, Boston, AB, 65; Univ Ill, Urbana, MS, 67, PhD(physics), 71. Prof Exp: Res assoc physics, Northwestern Univ, 71-72 & Univ Chicago, 72-73; ASST PROF PHYSICS, MICH STATE UNIV, 73- Concurrent Pos: Louis Bloch fel, Univ Chicago, 72-73. Mem: Am Phys Soc. Res: Effects of radiation damage in ion-bombarded semiconductors and the electronic properties of doped semiconductors using laser light scattering techniques. Mailing Add: Dept of Physics Mich State Univ East Lansing MI 48824

COLWELL, RITA R, b Beverly, Mass, Nov 23, 34; m 56; c 2. MICROBIOLOGY. Educ: Purdue Univ, BS, 56, MS, 58; Univ Wash, PhD(marine microbiol), 61. Prof Exp: Res asst prof, Univ Wash, 61-64; from asst prof to assoc prof biol, Georgetown Univ, 64-72, vis asst prof, 63-64; PROF BIOL, UNIV MD, 72- Concurrent Pos: Guest scientist, Nat Res Coun Can, 61-63; mem classification res group, London; mem bd trustees, Am Type Cult Collection; consult, Bur Higher Educ, Dept Health, Educ & Welfare, 68-70; consult div res grants, NIH, 70; consult adv comt sci educ & biolog oceanog, NSF, 70-75; consult, Environ Protection Agency, 75- Honors & Awards: Phi Sigma Serv Award, Am Chem Soc, 75. Mem: AAAS; Am Soc Microbiol; Soc Invert Path; Soc Indust Microbiol; Am Acad Microbiol. Res: Marine microbiology; numerical taxonomy; uses of high-speed electronic computers in biology and medicine. Mailing Add: Dept of Microbiol Univ of Md College Park MD 20742

COLWELL, ROBERT KNIGHT, b Denver, Colo, Oct 9, 43; m 65. ECOLOGY, ENTOMOLOGY. Educ: Harvard Univ, AB, 65; Univ Mich, Ann Arbor, PhD(zool), 69. Prof Exp: Asst to cur econ & ethnobot, Bot Mus, Harvard Univ, 66; Ford Found fel math biol, Univ Chicago, 69-70; ASST PROF ZOOL, UNIV CALIF, BERKELEY, 70- Concurrent Pos: Coordr, Org Trop Studies grad course in trop biol, 71. Mem: AAAS; Ecol Soc Am; Soc Study Evolution. Res: Ecology and evolution of biological communities; tropical biology; behavioral and theoretical ecology. Mailing Add: Dept of Zool Univ of Calif Berkeley CA 94720

COLWELL, ROBERT NEIL, b Star, Idaho, Feb 4, 18; m 42; c 4. FOREST MENSURATION. Educ: Univ Calif, BS, 38, PhD(plant physiol), 42. Prof Exp: Asst bot, 38-42, from asst prof to assoc prof, 47-57, PROF FORESTRY, UNIV CALIF, BERKELEY, 57-, ASSOC DIR SPACE SCI LAB, 69-, DIR BERKELEY OFF, EARTH SATELLITE CORP, 70- Concurrent Pos: Chmn comt crop geog & veg anal, Nat Res Coun, 53-54. Honors & Awards: Abrams Award, 54; Fairchild Photogram Award, 57; Photo Interpretation Award, Am Soc Photogram, 64. Mem: Soc Am Foresters; Am Soc Photogram (vpres, 54); Int Soc Photogram. Res: Identification and mapping of vegetation types from aerial photographs; use of radioactive tracers in biological studies; applications of remote sensing to the space sciences; aerospace and earth sciences. Mailing Add: 145 Mulford Hall Univ of Calif Berkeley CA 94720

COLWELL, WILLIAM MAXWELL, b Blairsville, Ga, May 28, 31; m 56; c 4. VETERINARY MICROBIOLOGY. Educ: Berry Col, BS, 52; Univ Ga, DVM, 59, MS, 68, PhD(microbiol), 69; Am Col Vet Microbiologists, dipl. Prof Exp: Vet, Vanderbilt Vet Hosp, Durham, NC, 59-60 & diag lab, Ga Poultry Lab, Oakwood, 60-64; area vet, Elanco Prod Co, Ind, 64-66; Campbell fel, Avian Dis Res Ctr, Univ Ga, 66-69; asst prof vet sci, Univ Fla, 69-70; assoc prof poultry sci, 70-74, PROF VET SCI, NC STATE UNIV, 74- Mem: Am Asn Avian Pathologists; Am Vet Med Asn. Res: Avian disease research; epidemiology of avian tumor viruses; avian respiratory viruses; oncogenic viruses of poultry; organ culture techniques; bioassay of aflatoxins. Mailing Add: Dept of Vet Sci NC State Univ 1111 Grinnells Lab Raleigh NC 27607

COLWELL, WILLIAM TRACY, b Joliet, Ill, Oct 18, 34. PHARMACEUTICAL CHEMISTRY, SYNTHETIC ORGANIC CHEMISTRY. Educ: Occidental Col, BA, 56; Univ Calif, Los Angeles, PhD(ort chem), 62. Prof Exp: SR ORG CHEMIST PHARMACEUT CHEM, STANFORD RES INST, 62- Mem: Am Chem Soc. Res: Synthesis of pteridines and related heterocycles as antifolates; antiparasitic compounds; synthesis of prostaglandin metabolites and prostaglandin synthetase inhibitors. Mailing Add: Stanford Res Inst Menlo Park CA 94025

COLWILL, JACK M, b Cleveland, Ohio, June 15, 32; m 54; c 3. INTERNAL MEDICINE. Educ: Oberlin Col, BA, 55; Univ Rochester, MD, 57; Am Bd Internal Med, dipl, 64. Prof Exp: Intern, Barnes Hosp, Washington Univ, 57-58; res, Univ Wash Affiliated Hosps, 58-60, chief res, Univ Hosp, 60-61; from instr to sr instr med & dir med outpatient dept, Med Sch, Univ Rochester, 61-64; asst prof, 64-70, asst dean, Sch Med, 64-67, assoc dean, 67-69, ASSOC DEAN ACAD AFFAIRS, 69-, ASSOC PROF MED & COMMUNITY HEALTH MED PRACT, SCH MED, UNIV MO-COLUMBIA, 70-, DIR FAMILY MED PROG, MED CTR, 74- Mem: AMA; Asn Am Med Cols. Mailing Add: Dean's Off Univ of Mo Sch of Med Columbia MO 65201

COLWIN, ARTHUR LENTZ, b Sydney, Australia, Jan 26, 11; nat US; m 40. ZOOLOGY. Educ: McGill Univ, BSc, 33, MSc, 34, PhD(embryol), 36. Prof Exp: Moyse traveling fel, Sir William Dunn Inst Biochem & Dept Exp Zool, Univ Cambridge, 34-35; Seessel fel, Osborn Zool Lab, Yale Univ, 36-37, Royal Soc Can fel, 37-38; instr biol, NY Univ, 38-39; from instr to prof biol, Queens Col, NY, 40-73; ADJ PROF, ROSENSTEIL SCH MARINE & ATMOSPHERIC SCI, UNIV MIAMI, 73- Concurrent Pos: Instr embryol, Marine Biol Lab, Woods Hole, 48-50; Fulbright res scholar, Misaki Marine Biol Sta, Tokyo, 53-54; mem corp, Marine Biol Lab, Woods Hole & trustee, 62-75. Honors & Awards: Am Mills Gold Medal. Mem: Am Soc Zoologists; fel NY Acad Sci; Am Soc Cell Biol; Soc Develop Biol; Int Soc Develop Biol. Res: Normal and experimental embryology; cell division and differentiation; fertilization; sperm-egg association; egg cortical changes. Mailing Add: 320 Woodcrest Rd Key Biscayne FL 33149

COLWIN, LAURA HUNTER, b Philadelphia, Pa, July 5, 11; m 40. ZOOLOGY. Educ: Bryn Mawr Col, AB, 32, Univ Pa, MA, 34, PhD(protozool), 38. Prof Exp: Instr biol, Pa Col Women, 36-37, asst prof, 37-40; instr zool, Vassar Col, 40-43; instr biol, Pa Col Women, 45-46; lectr biol, Queens Col, NY, 47-66, prof, 66-73; ADJ PROF, ROSENSTEIL SCH MARINE & ATMOSPHERIC SCI, UNIV MIAMI, 73- Concurrent Pos: Morrison fel, Am Asn Univ Women, Misaki Marine Biol Sta, Tokyo, 53-54; trustee, Marine Biol Lab, Woods Hole, 71-75. Mem: Am Soc Zoologists; Am Soc Cell Biol; Soc Develop Biol; Int Soc Develop Biol; Int Soc Cell Biol. Res: Normal and experimental embryology; cell division and differentiation; fertilization; sperm-egg association; egg cortical changes. Mailing Add: 320 Woodcrest Rd Key Biscayne FL 33149

COMAN, DALE REX, b Hartford, Conn, Feb 22, 06; m 37; c 2. PATHOLOGY. Educ: Univ Mich, AB, 28; McGill Univ, MD, 33. Prof Exp: Asst, Inst Path, McGill Univ, 33-34; resident, Univ Pa Hosp, 34-35; resident, Mass State Tumor Hosp, Pondville, 35-36; instr, Sch Med, NY Univ, 36-37; instr, 37-41, assoc, 41-42, from asst prof to prof exp path, 42-54, prof path, 54-72, chmn dept, 54-67, EMER PROF PATH, SCH MED, UNIV PA, 72- Mem: AAAS; NY Acad Sci; Am Soc Cell Biol; Int Soc Cell Biol; Soc Exp Path & Med. Res: Cancer. Mailing Add: Dept of Path Univ of Pa Sch of Med Philadelphia PA 19104

COMAR, CYRIL LEWIS, b Dudley, Eng, Mar 28, 14; nat US; m 39; c 3. ENVIRONMENTAL HEALTH. Educ: Univ Calif, BS, 36; Purdue Univ, PhD(biochem), 41. Prof Exp: Asst, Univ Calif, 37-38; asst chemist, Exp Sta, Purdue Univ, 40-41 & Mich State Univ, 41-43; biochemist, Univ Fla, 43-48; lab dir, AEC Agr Res Prog, Univ Tenn, 48-54; chief biomed res, Oak Ridge Inst Nuclear Studies, 54-57; prof radiation biol, head dept phys biol & dir lab radiation biol, 57-75, EMER PROF RADIATION BIOL, STATE UNIV NY VET COL, CORNELL UNIV, 75-; DIR ENVIRON ASSESSMENT DEPT, ELEC POWER RES INST, 75- Concurrent Pos: Mem, US deleg to Conf Peaceful Uses Atomic Energy, Geneva, 55; mem comt effects radiation on agr & food supplies, Nat Acad Sci, 56, mem food protection comt, 61-; mem, UN Sci Comt Effects Atomic Radiation, 56-58; consult, Div Biol Med, US AEC, Oak Ridge Inst Nuclear Studies & Food & Agr Orgn, US, 57-; consult, USPHS, 58-, mem nat adv comt radiation, Surgeon Gen, 62-; mem ad hoc comt, Fed Radiation Coun, 64-; chmn adv comt of Nat Acad Sci to Fed Radiation Coun, 64-; mem, Nat Coun Radiation Protection & Measurements, 68. Honors & Awards: Borden Award, Am Inst Nutrit, 68. Mem: AAAS; Am Chem Soc; Am Vet Med Asn; Soc Exp Biol & Med; Radiation Res Soc. Res: Biological and environmental effects of energy production. Mailing Add: Elec Power Res Inst 3412 Hillview Ave Palo Alto CA 94303

COMARR, AVROM ESTIN, b Chicago, Ill, July 15, 15; m 38; c 1. UROLOGY. Educ: Univ Southern Calif, AB, 37; Chicago Med Sch, MB, 40, MD, 41; Am Bd Urol, dipl, 53. Prof Exp: Asst chief, Spinal Cord Injury Serv, Vet Admin Hosp, Long Beach, 42-70; clin instr, Med Sch, Univ Calif, Los Angeles, 53-55; asst clin prof, 55-58, asst prof urol, 58-59, assoc clin prof, 59-67, CLIN PROF UROL, SCH MED, LOMA LINDA UNIV, 67-, DIR SPINAL CORD INJURY CLIN, 73- Concurrent Pos: Attend head urol, Rancho Los Amigos Hosp, Downey, 55-73, mem staff, 73-; attend physician, Los Angeles County Hosp, 55-70; clin prof surg, Univ Southern Calif, 67-; sr attend physician, Los Angeles County Hosp-Univ Southern Calif Med Ctr, 70-; chief spinal cord injury serv, Vet Admin Hosp, Long Beach, 70-73; neurol & urol consult, 73- Honors & Awards: Paralyzed Vet Am Award, 57-58; Meritorious Award, Chicago Med Sch, 63. Mem: Am Paraplegia Soc (pres, 54-); Am Urol Asn; AMA; fel Am Col Surgeons; fel Int Col Surgeons. Res: Neurological urology; spinal cord injuries. Mailing Add: Rancho Los Amigos Hosp 7601 E Imperial Hwy Downey CA 90242

COMAS, JUAN, anthropology, see 12th edition

COMBA, PAUL GUSTAVO, b Tunis, Tunisia, Mar 6, 26; nat US. COMPUTER SCIENCE. Educ: Bluffton Col, AB, 47; Calif Inst Technol, PhD(math), 52. Prof Exp: Asst math, Calif Inst Technol, 47-51; from asst prof to assoc prof, Univ Hawaii, 51-60; math systs analyst, 60-63, mgr advan comput technol, 63, mgr prog lang eval, 63-64, SCI STAFF MEM, IBM CORP, 65- Concurrent Pos: Adj prof, NY Univ, 70. Mem: AAAS; Am Math Soc; Math Asn Am; Am Asn Comput Mach. Res: Computer programming languages; programming applications; digital simulation; heuristics; computer graphics; computer aided design; computer applications development; design and management of data bases. Mailing Add: IBM Corp Cambridge Sci Ctr 545 Technol Sq Cambridge MA 02139

COMBELLACK, WILFRED JAMES, b New Gloucester, Maine, June 27, 15; m 37; c 2. MATHEMATICS. Educ: Colby Col, AB, 37, MA, 38; Boston Univ, PhD(physics), 44. Prof Exp: From instr to assoc prof math, Northeastern Univ, 38-48; head dept, 48-70, PROF MATH, COLBY COL, 48- Mem: Math Asn Am; Am Math Soc. Res: Summation of series; table of summations. Mailing Add: Dept of Math Colby Col Waterville ME 04901

COMBES, BURTON, b New York, NY, June 30, 27; m 48; c 3. INTERNAL MEDICINE. Educ: Columbia Univ, AB, 47, MD, 51; Am Bd Internal Med, dipl, 59. Prof Exp: Intern med, Columbia-Presby Med Ctr, 51-52, asst resident, 52-53, asst physician, 53-56; from instr to assoc prof, 57-67, PROF INTERNAL MED, UNIV TEX HEALTH SCI CTR DALLAS, 67- Concurrent Pos: Res fel, Col Physicians & Surgeons, Columbia Univ, 53-55, Am Heart Asn res fel, 55-56; Am Heart Asn res fel, Univ Col Hosp, Med Sch, Univ London, 56-57; USPHS res career develop award, 62-72; estab investr, Am Heart Asn, 57-62; consult, Dallas Vet Admin Hosp, Tex, 65-; mem adv coun, Nat Inst Arthritis, Metab & Digestive Dis, 75. Mem: Am Fedn Clin Res; AMA; Am Soc Clin Invest; Am Asn Study Liver Dis; Am Gastroenterol Asn. Res: Hepatic excretory function; sulfobromophthalein metabolism; liver function during pregnancy. Mailing Add: Dept of Internal Med Univ of Tex Health Sci Ctr Dallas TX 75235

COMBS, ALAN B, b Boulder, Colo, July 4, 39; m 61; c 2. PHARMACOLOGY. Educ: Univ of the Pac, BSc, 62, MSc, 64; Univ Calif, Davis, PhD(comp pharmacol), 70. Prof Exp: ASST PROF PHARMACOL, SCH PHARM, UNIV TEX, AUSTIN, 70- Mem: AAAS; Am Pharmaceut Asn; Am Asn Col Pharm. Res: Pharmacology of compounds that cause pulmonary edema; cardiovascular pharmacology. Mailing Add: Dept of Pharmacol Univ of Tex Sch of Pharm Austin TX 78712

COMBS, CLARENCE MURPHY, b Louisville, Ky, Apr 13, 25; m 46; c 3. ANATOMY. Educ: Transylvania Col, AB, 46; Northwestern Univ, MS, 48, PhD(anat), 50. Prof Exp: Instr neuroanat, Med Sch, Univ WVa, 48; from instr to prof, Northwestern Univ, 50-66; PROF ANAT & CHMN DEPT, CHICAGO MED SCH, 66- Concurrent Pos: Nat Inst Neurol Dis & Blindness career res develop award, 59-64; assoc prof, Med Sch, Univ PR, 58-60; sect chief, Perinatal Physiol Lab, Nat Inst Neurol Dis & Blindness, PR, 58-60, spec consult & trainee, 58; mem, Int Brain Res Orgn; actg dean, Sch Grad & Postdoctoral Studies, Chicago Med Sch, Univ Health Sci, 75-76. Mem: Am Asn Anat; Biol Stain Comn; Soc Neurosci. Res: Electroanatomical studies of cerebellar connections; thalamocortical connections; spinal cord structure; neurophysiological regulation of lingual movement. Mailing Add: Dept of Anat Chicago Med Sch Chicago IL 60612

COMBS, GEORGE ERNEST, b Arcadia, Fla, Feb 21, 27; m 48; c 3. ANIMAL NUTRITION. Educ: Univ Fla, BSA, 51, MSA, 53; Iowa State Univ, PhD(animal nutrit), 55. Prof Exp: Asst animal husb, Univ Fla, 51-52, instr, 52-53; asst animal nutrit, Iowa State Univ, 53-55; from asst prof to assoc prof, 55-67, PROF ANIMAL NUTRIT, UNIV FLA, 67- Mem: Am Inst Nutrit; Am Soc Animal Sci. Res: Mineral, energy and amino acid metabolism with swine. Mailing Add: Dept of Animal Sci Univ of Fla Gainesville FL 32601

COMBS, GERALD FUSON, b Olney, Ill, Feb 23, 20; m 43; c 1. ANIMAL NUTRITION. Educ: Univ Ill, BS, 40; Cornell Univ, PhD(animal nutrit), 48. Prof Exp: Asst animal nutrit, Cornell Univ, 40-41; prof poultry nutrit, Univ Md, College Park, 68-69; dep chief nutrit prog, US Dept Health, Educ & Welfare, 69-71; nutrit & food safety coordr, USDA, 71-73; prof foods & nutrit & head dept, Univ Ga, 73-75; NUTRIT PROG DIR, NAT INST ARTHRITIS, METAB & DIGESTIVE DIS, NIH, 75- Honors & Awards: Poultry Nutrit Res Award, Am Feed Mfg, 53; Man of the Year in Md Agr, 67; Cert of Appreciation, Delmarva Poultry Ind, 69; Meterious Nutritionist Award, Distillers Res Inst, 70. Mem: AAAS; Poultry Sci Asn; Soc Exp Biol & Med; Am Inst Nutrit; Brit Nutrit Soc. Res: Poultry nutrition; factors concerned in bone formation; unidentified factors; energy-protein balance; antibiotics; amino acid requirements; human and international nutrition; basic and clinical human and animal nutrition research. Mailing Add: Nutrit Prog NIH Nat Inst Arthritis Metab Dig Dis Bethesda MD 20014

COMBS, GERALD FUSON, JR, b Ithaca, NY, June 10, 47; m 69; c 2. NUTRITION. Educ: Univ Md, College Park, BS, 69; Cornell Univ, MS, 71, PhD(nutrit), 74. Prof Exp: Asst prof biochem & nutrit, Auburn Univ, 74-75; ASST PROF NUTRIT, CORNELL UNIV, 75- Mem: AAAS; Poultry Sci Asn; World's Poultry Sci Asn; Sigma Xi. Res: Nutrient interrelationships and mechanisms of action; influences of foreign compounds on nutrient function. Mailing Add: Dept of Poultry Sci Rice Hall Cornell Univ Ithaca NY 14853

COMBS, LEON LAMAR, III, b Meridian, Miss, Sept 19, 38; m 62; c 1. CHEMICAL PHYSICS. Educ: Miss State Univ, BS, 61; La State Univ, PhD(chem physics), 68. Prof Exp: Res chemist, Devoe & Reynolds Co, Inc, Ky, 61-64; from asst prof to assoc prof, 67-75, PROF CHEM & PHYSICS, MISS STATE UNIV, 75- Mem: Am Phys Soc; Am Chem Soc; Sigma Xi. Res: Quantum chemistry of small molecules; application of statistical mechanics to study of phase transitions; theoretical conformational analysis; quantum mechanical studies in pharmacology. Mailing Add: Box CH Miss State Univ Mississippi State MS 39762

COMBS, OVA BEETEM, b Emmalena, Ky, Nov 12, 08; m 33; c 4. HORTICULTURE. Educ: Purdue Univ, BS, 30; Univ Wis, MS, 32. Prof Exp: Asst hort, 30-39, from instr to assoc prof, 39-49, chmn dept, 49-65, PROF HORT, UNIV WIS-MADISON, 49- Concurrent Pos: Chief party, USAID-Univ Wis contract, Fac Agr, Univ Ife, Nigeria, 65-68. Mem: Am Soc Hort Sci. Res: Culture and improvement of vegetable crops; eggplant; lima beans; peppers; vegetable-type soybeans; tomatoes; squash; olericulture. Mailing Add: Dept of Hort Univ of Wis Madison WI 53705

COMBS, ROBERT L, JR, b Fayetteville, Ark, Nov 6, 28; m 50; c 3. ENTOMOLOGY. Educ: Univ Ark, BS, 61, MS, 63; Miss State Univ, PhD(entom), 67. Prof Exp: Asst entom, Univ Ark, 63-64; asst entom, 64-66, asst entomologist, 66-70, ASSOC ENTOMOLOGIST, MISS STATE UNIV, 70- Mem: Entom Soc Am. Res: Veterinary entomology; applied and basic entomological problems. Mailing Add: Dept of Entom Miss State Univ Mississippi State MS 39762

COMBS, ROBERT LEONARD, b Elizabethton, Tenn, Nov 10, 29; m 54; c 1. PHYSICAL CHEMISTRY, POLYMER CHEMISTRY. Educ: Eastern Tenn State Univ, BS, 51; Univ Tenn, MS, 52, PhD(chem), 55. Prof Exp: Chemist, Union Carbide Corp, Tenn, 52; res chemist, 55-59, sr res chemist, 60-69, RES ASSOC, TENN EASTMAN CO DIV, EASTMAN KODAK CO, 69- Mem: Am Chem Soc; Soc Plastics Eng. Res: Characterization of polymers; rheology; kinetics and mechanisms of polymerization; polymers application requirements; adhesives; moldings; coatings. Mailing Add: 4509 Chickasaw Rd Kingsport TN 37664

COMBS, WESLIE, animal genetics, see 12th edition

COME, THOMAS V, b Titusville, Pa, Sept 3, 27; m 50; c 2. SCIENCE EDUCATION. Educ: Pa State Teachers Col, Edinboro, BS, 50; Pa State Univ, MEd, 58, DEd(genetics), 63. Prof Exp: Teacher pub schs, NY & Pa, 50-59; assoc prof, 59-70, PROF SCI, EDINBORO STATE COL, 70- Mem: AAAS; Nat Asn Biol Teachers. Res: Genetics. Mailing Add: Dept of Sci Edinboro State Col Edinboro PA 16444

COMEAU, ROGER WILLIAM, b Quincy, Mass, Apr 22, 33; m 61; c 3. MAMMALIAN PHYSIOLOGY. Educ: Boston Univ, AB, 55; State Univ NY Buffalo, PhD(physiol), 67. Prof Exp: Mem staff, Arthur D Little, Inc, 59-61; teaching asst physiol, State Univ NY Buffalo, 61-63, from asst instr to asst prof, 63-68; assoc dir sci info & regulatory affairs, Mead Johnson & Co, 68-69; assoc prof, 70-75, PROF BIOL, MID GA COL, 75-, CHMN DEPT BIOL SCI, 73- Mem: AAAS; Assoc Am Physiol Soc. Res: Membrane transport; teaching mammalian physiology; pharmacology and toxicology of cancer chemotherapeutic agents. Mailing Add: Dept of Biol Mid Ga Col Cochran GA 31014

COMEAUX, MALCOLM LOUIS, b Lafayette, La, Apr 19, 38; m 67; c 2. CULTURAL GEOGRAPHY. Educ: Univ Southwestern La, BA, 63; Southern Ill Univ, MA, 66; La State Univ, PhD(geog), 69. Prof Exp: Instr geog, Univ Southwestern La, 65-66; asst prof, 69-75, fac res grant, 70-71, ASSOC PROF GEOG, ARIZ STATE UNIV, 75- Mem: Am Geog Soc; Asn Am Geog. Res: Acadians of south Louisiana; fishing in the Mississippi River. Mailing Add: Dept of Geog Ariz State Univ Tempe AZ 85281

COMEFORD, JOHN J, b Schenectady, NY, Apr 30, 28. ANALYTICAL CHEMISTRY. Educ: Colo State Univ, BS, 50; Wash State Univ, MS, 53; Georgetown Univ, PhD(molecular spectros), 66. Prof Exp: Phys chemist, 67-68, RES CHEMIST, NAT BUR STANDARDS, 68- Concurrent Pos: Secy comt on spectral absorption data, Nat Res Coun-Nat Acad Sci, 59-62; vis assoc prof, Dept Mat Sci & Eng, Univ Utah, 75. Mem: Am Chem Soc; Combustion Inst; Am Soc Mass Spectrometry; AAAS. Res: Low temperature matrix-isolation spectroscopy; infrared spectra of unstable molecules; mass spectrometry of combustion products. Mailing Add: 10430 Haywood Dr Silver Spring MD 20902

COMEN, ALAN LEE, b Bridgeport, Conn, Aug 14, 37; m 59; c 2. ORGANIC CHEMISTRY. Educ: Bates Col, BS, 59; Purdue Univ, PhD(org chem), 64. Prof Exp: Res chemist, Pine & Paper Chem Div, Res Ctr, 63-73, SR MKT DEVELOP REP, HERCULES INC, 73- Mem: Am Chem Soc. Res: Physical organic chemistry; fatty acids; textile chemicals. Mailing Add: 421C RD 1 Hockessin DE 19707

COMER, FREDERICK WILLIAM, organic chemistry, see 12th edition

COMER, JACK PAYNE, b Indianapolis, Ind, Dec 22, 22; m 43; c 1. ANALYTICAL CHEMISTRY. Educ: Purdue Univ, BS, 46, MS, 48, PhD(pharmaceut chem), 50. Prof Exp: Pharmaceut chemist, Ind State Bd Health, 50-52; control assoc, 52-60, HEAD ANAL DEVELOP, ELI LILLY & CO, 60- Mem: Am Chem Soc; Am Pharmaceut Asn; Acad Pharmaceut Sci. Res: Analytical methods for pharmaceutical and clinical samples; diagnostic aid for enzymatic paper test for glucose in diabetic urine. Mailing Add: 732 Gettysburg Ct Indianapolis IN 46217

COMER, JAMES PIERPONT, b East Chicago, Ind, Sept 25, 34; m 59; c 2. CHILD PSYCHIATRY, PUBLIC HEALTH ADMINISTRATION. Educ: Ind Univ, AB, 56; Howard Univ, MD, 60; Univ Mich, MPH, 64. Prof Exp: Intern, St Catherine's Hosp, 60-61; staff physician, NIMH, 67-68; asst prof, 68-70, assoc prof, 70-75, PROF PSYCHIAT, CHILD STUDY CTR, YALE UNIV, 75-, ASSOC DEAN STUDENT AFFAIRS, MED SCH, 69- Concurrent Pos: Fel psychiat, Med Sch, Yale Univ, 64-66 & Child Study Ctr, 66-67; NIMH fel, Hillcrest Children's Ctr, Washington, DC, 67-68; Markle scholar, 69; mem ed bd, Am J Orthopsychiat, 69-, J Youth & Adolescence, 71- & J Negro Educ, 73-; adv & consult, Children's Television Workshop, 70-; mem prof adv coun, Nat Asn Ment Health, 71-; mem comn, Joint Inst Judicial Admin-Am Bar Asn Juv Justice Standards Proj, 73-75. Mem: Am Psychiat Asn; Am Orthopsychiat Asn; Am Acad Child Psychiat. Res: Race relations; elementary school education and mental health. Mailing Add: Child Study Ctr Med Sch Yale Univ 333 Cedar St New Haven CT 06510

COMER, JOSEPH JOHN, b Brooklyn, NY, Dec 8, 20; m 47; c 4. INORGANIC CHEMISTRY, ELECTRON MICROSCOPY. Educ: Pa State Univ, BS, 44, MS, 47. Prof Exp: Chemist, Naval Res Lab, 44-45; electron microscopist, Cent Res Lab, Gen Aniline & Film Corp, 46-52; res assoc, Col Mineral Indust, Pa State Univ, 52-55, from asst prof to assoc prof mineral sci, 55-62, head mineral const labs, 55-62; scientist, Res Ctr, Sperry Rand Corp, 62-67; RES CHEMIST, AIR FORCE CAMBRIDGE RES LABS, 67- Mem: AAAS; Am Chem Soc; Electron Micros Soc Am. Res: Electron microscope studies of electronic, electrooptic materials and thin films. Mailing Add: Air Force Cambridge Res Labs Hanscom AFB Bedford MA 01731

COMER, RALPH DUDLEY, b Kansas City, Mo, Sept 28, 27; m 50; c 4. PREVENTIVE MEDICINE, TROPICAL MEDICINE. Educ: Univ Kans, BA, 50, MA, 52; Med Col SC, PhD(anat), 55, MD, 57; Johns Hopkins Univ, MPH, 65; Am Bd Prev Med, dipl gen prev med, 69. Prof Exp: Instr gross anat, Med Col SC, 52-55; Med Corps, US Navy, 57-, med officer, Rodman Naval Sta, CZ, 59-61, resident internal med, Naval Hosp, Oakland, 61-62, resident gen surg, 62-63, head tuberc & venereal dis control, Prev Med Div, Bur Med & Surg, DC, 63-64, officer-in-chg, Naval Med Sci Unit, Gorgas Mem Lab, 65-68, officer-in-chg, Naval Prev Med Unit 2, Va, 68-70, head community health br, Prev Med Div, Bur Med & Surg, 70-73, STAFF MEM, REGIONAL DISPENSARY, NAVY REGIONAL MED CTR, SAN DIEGO, 73- Mem: Am Pub Health Asn; Am Soc Trop Med & Hyg; Royal Soc Trop Med & Hyg. Res: Malaria. Mailing Add: Regional Dispensary MCRD Navy Regional Med Ctr San Diego CA 92440

COMER, STEPHEN DANIEL, b Covington, Ky, May 2, 41; m 63. MATHEMATICAL LOGIC, ALGEBRA. Educ: Ohio State Univ, BSc, 62; Univ Calif, Berkeley, MA, 64; Univ Colo, Boulder, PhD(math), 67. Prof Exp: Asst prof math, Vanderbilt Univ, 67-74; vis asst prof, Clemson Univ, 74-75; ASSOC PROF MATH, THE CITADEL, 75- Mem: Am Math Soc; Asn Symbolic Logic. Res: Algebra and logic; algebraic logic; universal algebra; model theory; decision problems; sheaf theory. Mailing Add: Dept of Math The Citadel Charleston SC 29409

COMER, WILLIAM TIMMEY, b Ottumwa, Iowa, Jan 11, 36; m 63; c 2. ORGANIC CHEMISTRY. Educ: Carleton Col, BA, 57; Univ Iowa, PhD(org chem), 62. Prof Exp: Sr scientist, 61-67, res group leader, 67-68, sect leader chem res, 68-70, from prin investr to sr prin investr, 70-74, DIR PHARMACEUT RES, MEAD JOHNSON & CO, 75- Mem: AAAS; Am Chem Soc; Sigma Xi. Res: Medicinal chemistry; adrenergic agents; catecholamines; antihypertensive agents; sulfonamides; medium ring heterocycles; phosphamides; mercaptans. Mailing Add: 8234 Larch Lane Evansville IN 47710

COMERFORD, JOHN RICHARD, JR, inorganic chemistry, nuclear chemistry, see 12th edition

COMERFORD, JOHN ROGER, physiology, see 12th edition

COMES, RICHARD DURWARD, b Nisland, SDak, Nov 16, 31; m 54; c 3. WEED SCIENCE. Educ: Univ Wyo, BS, 58, MS, 60; Ore State Univ, PhD(weed sci), 71. Prof Exp: Res agronomist, 60-65, PLANT PHYSIOLOGIST, AGR RES SERV, USDA, 65- Mem: Weed Sci Soc Am. Res: Management of vegetation in aquatic and marginal areas; biology and ecology of aquatic and ditchbank vegetation; fate of herbicides in water; effect of herbicides in irrigation water on crops. Mailing Add: Irrigated Agr Res & Exten Ctr Agr Res Serv USDA Prosser WA 99350

COMFORT, JOSEPH ROBERT, b Fayetteville, Ark, July 18, 40. NUCLEAR PHYSICS. Educ: Ripon Col, AB, 62; Yale Univ, MS, 63, PhD(nuclear physics), 68. Prof Exp: Res physicist, Nuclear Struct Lab, Yale Univ, 67-68; fel nuclear physics, Argonne Nat Lab, 68-70; instr physics, Princeton Univ, 70-72; ASST PROF, OHIO UNIV, ATHENS, 72- Concurrent Pos: Vis scientist, Univ Groningen, Netherlands, 74-75 & Ind Univ, Bloomington, 76- Mem: Am Phys Soc; Am Asn Physics Teachers. Res: Penetration of charged particles in matter; nuclear structure physics; nuclear reaction mechanisms. Mailing Add: Cyclotron Lab Ind Univ Bloomington IN 47401

COMFORT, WILLIAM WISTAR, b Bryn Mawr, Pa, Apr 19, 33; m 57; c 2. MATHEMATICS. Educ: Haverford Col, BA, 54; Univ Wash, MSc, 57, PhD(math), 58. Hon Degrees: MA, Wesleyan Univ, 69. Prof Exp: Asst math, Univ Wash, 56-58; Benjamin Peirce instr, Harvard Univ, 58-61; asst prof, Univ Rochester, 61-65; assoc prof, Univ Mass, Amherst, 65-67; chmn dept, 69-70, PROF MATH, WESLEYAN UNIV, 67- Concurrent Pos: Managing ed proc, Am Math Soc, 74-75. Mem: Am Math Soc; Math Asn Am. Res: General topology; topological analysis; Stone-Cech compactification; the theory of ultrafilters. Mailing Add: Dept of Math Wesleyan Univ Middletown CT 06457

COMINGS, DAVID EDWARD, b Beacon, NY, Mar 8, 35; m 58; c 3. MEDICAL GENETICS, CELL BIOLOGY. Educ: Northwestern Univ, BS, 55, MD, 58; Am Bd Internal Med, dipl. Prof Exp: From intern to resident internal med, Cook County Hosp, Chicago, 58-61; chief hemat, Madigan Gen Hosp, Tacoma, Wash, 62-64; DIR DEPT MED GENETICS, CITY OF HOPE MED CTR, 66- Concurrent Pos: Fel hemat, Cook County Hosp, Chicago, 61-62; fel med genetics, Univ Wash, 64-66. Mem: AAAS; Am Soc Human Genetics; Am Soc Cell Biol; Am Fedn Clin Res; Am Soc Clin Invest. Res: Human genetics; biochemistry and physiology of chromosomes; hemoglobinopathies and thalassemia; mechanisms of DNA replication; molecular aging; differentiation. Mailing Add: Dept of Med Genetics City of Hope Med Ctr Duarte CA 91010

COMINSKY, CATHERINE, b Las Animas, Colo, May 11, 20; m 49; c 2. ZOOLOGY, HISTOLOGY. Educ: Univ Colo, BA, 42, MA, 44, PhD, 46. Prof Exp: From instr to assoc prof, 46-54, actg head dept, 46-49, PROF BIOL, UNIV HOUSTON, 55- Res: Vitamin-B deficiencies; histologic effects of doses of morphine and demerol and injections of bacillus anthracis on mouse tissue. Mailing Add: Dept of Biol Univ of Houston Houston TX 77004

COMITA, GABRIEL WILLIAM, b Minneapolis, Minn, July 27, 15; m 51; c 2. ZOOLOGY. Educ: Col St Thomas, BS, 37; Univ Minn, MA, 49; Univ Wash, PhD(zool), 53. Prof Exp: Instr water purification, Ft Belvoir, Va, 41-43; asst chemist, Sanit Dist, Minneapolis & St Paul, 46-47; asst, Univ Minn, 48-49; jr res zoologist, Univ Wash, 51-53; from asst prof to assoc prof, 53-60, PROF ZOOL, NDAK STATE UNIV, 60- Concurrent Pos: Mem staff, Arctic Res Lab, Point Barrow, Alaska, 51 & 52. Mem: Fel AAAS; Soc Syst Zool; Am Soc Limnol & Oceanog; Am Micros Soc; Ecol Soc Am. Res: Limnology and invertebrate zoology; copepods, their biology, energy transformations. Mailing Add: Dept of Zool NDak State Univ Fargo ND 58102

COMIZZOLI, ROBERT BENEDICT, b Union City, NJ, Apr 22, 40; m 65; c 2. SEMICONDUCTORS, SOLID STATE ELECTRONICS. Educ: Boston Col, BS, 62; Princeton Univ, MA, 64, PhD(physics), 67. Prof Exp: MEM TECH STAFF RES, RCA LABS, RCA CORP, 66- Mem: Inst Elec & Electronics Engrs; Electrochem Soc. Res: Semiconductor devices; integrated circuits; power devices; reliability; passivation; semiconductor processing; electrophotography. Mailing Add: Knickerbocker Dr RD 1 Belle Mead NJ 08502

COMLY, HUNTER HALL, b Denver, Colo, July 21, 19; m 41; c 5. PSYCHIATRY. Educ: Yale Univ, BS, 41, MD, 43. Prof Exp: Intern pediat, Mass Gen Hosp, Boston, 44; asst resident, Children's Hosp, Iowa City, Iowa, 44-46, resident psychiat, Psychopathic Hosp, Iowa City, 46-47; asst prof pediat in psychiat, Univ Iowa, 48-56; staff psychiatrist, Children's Div, Lafayette Clin, Detroit, Mich, 56-58; dir, Children's Ctr Wayne County, Detroit, 58-67; assoc prof child psychiat, 67-71, PROF CHILD PSYCHIAT & HEAD DIV, UNIV IOWA, 71- Concurrent Pos: Fel child psychiat, Univ Minn Hosps, Minneapolis, 47-48; fac mem, Continuation Courses Child Psychiat, Univ Minn, 48, 54 & 61; ed, Presch study course, Nat Parent-Teacher Mag, 49-51; consult, Iowa Child Welfare Res Sta, 50-51; psychiatrist, Univ Iowa, 51-52, child psychiatrist, 52-56; workshop chmn, Psychopharmacol in Children's Learning & Behav Disorders, 63-66 & 69-70. Mem: AMA; fel Am Psychiat Asn; fel Am Orthopsychiat Asn; Am Acad Child Psychiat. Res: Learning and behavior disorders of children; effects of psychoactive drugs on learning and behavior. Mailing Add: Child Psychiat Serv 500 Newton Rd Iowa City IA 52242

COMLY, JAMES B, b New York, NY, Nov 28, 36; m 59; c 1. APPLIED PHYSICS. Educ: Cornell Univ, BEE, 59; Harvard Univ, MA, 60, PhD(appl physics), 65. Prof Exp: NSF fel, Atomic Energy Res Estab, Eng, 65-66; res physicist, 66-69, mgr planning & resources, 69-72, MGR THERMAL BR, GEN ELEC RES & DEVELOP CTR, 72- Concurrent Pos: Chmn rev comt, Toms River Energy Proj, Princeton Univ, 73-; assoc ed, Int J Energy, 75- Mem: Am Phys Soc; Inst Elec & Electronics Engrs; AAAS; Am Soc Mech Engrs. Res: Energy utilization in power plants, buildings, industry, including solar energy, heat pumps, industrial processes and power plant cycles. Mailing Add: Gen Elec Res & Develop Ctr PO Box 43 Schenectady NY 12301

COMMARATO, MICHAEL A, b Montclair, NJ, Apr 13, 40; m 67; c 1. PHARMACOLOGY. Educ: Rutgers Univ, BS, 62; Marquette Univ, PhD(pharmacol), 68. Prof Exp: USPHS fel pharmacol, Mich State Univ, 68-69; sr pharmacologist, William H Rorer, Inc, 69-71; SR SCIENTIST PHARMACOL, WARNER LAMBERT RES INST, 71- Mem: Am Soc Pharmacol & Exp Therapeut; Am Heart Asn. Res: Primary detection and secondary cardiovascular and autonomic evaluation of antihypertensive drugs. Mailing Add: Warner Lambert Res Inst Morris Plains NJ 07950

COMMERFORD, JOHN D, b Deadwood, SDak, Aug 23, 29; m 53; c 5. ORGANIC CHEMISTRY. Educ: Carroll Col, Mont, AB, 50; St Louis Univ, PhD(chem), 55. Prof Exp: Res chemist, Callery Chem Co, 54-57; sr res scientist, Anheuser-Busch, 57-67, mgr com develop, 67-69; DIR TECH DEVELOP, CORN REFINERS ASN, INC, 69- Mem: Am Chem Soc; Am Asn Cereal Chem; Inst Food Technol; Tech Inst Pulp & Paper Indust. Res: Carbohydrates; corn products; boron hydrides; medicinal chemistry. Mailing Add: Corn Refiners Asn Inc 1001 Connecticut Ave NW Washington DC 20036

COMMERFORD, SPENCER LEWIS, b Toledo, Ohio, May 23, 30. BIOCHEMISTRY. Educ: Mass Inst Technol, BS & MS, 52; Harvard Univ, PhD(biochem), 59. Prof Exp: Res collabr biochem, 59-61, from asst scientist to assoc scientist, 61-68, SCIENTIST, BROOKHAVEN NAT LAB, 68- Concurrent Pos: NIH res fel, 59-61. Mem: Am Soc Biol Chem; Biophys Soc; Harvey Soc. Res: Structure and function of DNA and deoxyribonucleohistone; kinetics of cell proliferation and death; cell differentiation. Mailing Add: Med Dept Brookhaven Nat Lab Upton NY 11973

COMMON, ROBERT HADDON, b Larne, Northern Ireland, Feb 25, 07; m 35; c 6. AGRICULTURAL CHEMISTRY. Educ: Queen's Univ Belfast, BSc, 28, BAgr, 29, MAgr, 31, DSc, 57; Univ London, BSc, 30, PhD(biochem), 35, DSc, 44. Hon Degrees: LLD, Queen's Univ Belfast, 74. Prof Exp: Asst, Chem Res Div, Ministry Agr & instr agr chem, Queen's Univ Belfast, 29-47; prof, 47-74, chmn dept, 47-72, EMER PROF AGR CHEM, MACDONALD COL, McGILL UNIV, 75- Honors & Awards: E W McHenry Award, Nutrit Soc Can, 75. Mem: Fel Royal Soc Can; fel Can Inst Chem; fel Agr Inst Can; fel Royal Inst Chem Gt Brit. Res: Mineral metabolism in the domestic fowl; biochemical effects of gonadal hormones in the fowl; composition and digestibility of feedstuffs; metabolism of estrogens in the fowl. Mailing Add: Box 223 Macdonald Col Quebec PQ Can

COMMONER, BARRY, b New York, NY, May 28, 17; m; c 2. BIOLOGY. Educ: Columbia Univ, AB, 37; Harvard Univ, MA, 38, PhD(biol), 41. Hon Degrees: DSc, Hahnemann Med Col, 63, Colgate Univ, 72 & Clark Univ, 74; LLD, Univ Calif, 67. Prof Exp: Asst biol, Harvard Univ, 38-40; instr, Queens Col, NY, 40-42; assoc ed, Sci Illustrated, NY, 46-47; assoc prof, 47-53, dept bot, 65-69, PROF PLANT PHYSIOL, WASH UNIV, 53-, DIR CTR BIOL NATURAL SYSTS, 65- . Concurrent Pos: Naval liaison off, US Senate Comt Mil Affairs, 46; mem bd dirs, Scientists Inst Pub Info, 63-, co-chmn bd, 67-69, chmn, 69-; pres, St Louis Comt Nuclear Info, 65-66; mem bd dirs & exec comt sci div, St Louis Comt Environ Info, 66-; mem spec study group on sonic boom, US Dept Interior, 67-68; mem bd consult experts, Rachel Carson Trust for Living Environ, 67- & law ctr comn, Univ Okla, 69-70; mem bd, Univs Nat Anti-War Fund; mem adv comt, Coalition for Health of Communities, 75. Honors & Awards: Newcomb Cleveland Prize, AAAS, 53; First Int Humanist Award, Int Humanist & Ethical Union, 70. Mem: Fel AAAS; Soc Gen Physiol; Am Inst Biol Sci; Am Chem Soc; Am Soc Plant Physiol. Res: Alterations in the environment in relation to modern technology; current status of the nitrogen cycle; roles of free radicals in biological processes; the origins and significance of the environmental and energy crises; environmental carcinogenesis; development of strategies to reduce the vulnerability of United States agriculture to disruptions from energy shortages. Mailing Add: Ctr Biol Natural Systs Wash Univ St Louis MO 63130

COMPAAN, ALVIN DELL, b Hull, NDak, June 11, 43; m 69; c 3. SOLID STATE PHYSICS. Educ: Calvin Col, AB, 65; Univ Chicago, MS, 66, PhD(physics), 71. Prof Exp: Res assoc physics, NY Univ, 71-73; ASST PROF PHYSICS, KANS STATE UNIV, 73- Mem: Am Phys Soc. Res: Raman acattering and photoluminescence studies of semiconductors; dynamics of excitons and electronphonon interactions in semiconductors; ion implantation effects in semiconductors; solar cell development. Mailing Add: Dept of Physics Kans State Univ Manhattan KS 66506

COMPANION, AUDREY (LEE), b Tarentum, Pa, Aug 19, 32. QUANTUM CHEMISTRY. Educ: Carnegie Inst Technol, BS, 54, MS, 56, PhD(phys chem), 58. Prof Exp: From instr to assoc prof chem, Ill Inst Technol, 58-75; ASSOC PROF CHEM, UNIV KY, 75- Mem: Am Chem Soc; Am Phys Soc; AAAS. Res: Molecular orbital theories; electronic spectroscopy; crystal field theory; theories of chemisorption. Mailing Add: Dept of Chem Univ of Ky Lexington KY 40506

COMPANS, RICHARD W, b Syracuse, NY, Sept 15, 40; m 65. VIROLOGY. Educ: Kalamazoo Col, BA, 63; Rockefeller Univ, PhD(virol), 68. Prof Exp: Guest investr electron micros, Inst Sci Res Cancer, Villejuif, France, 68; Am Cancer Soc hon fel microbiol, John Curtin Sch Med Res, Australian Nat Univ, 68-69; from asst prof to assoc prof virol, Rockefeller Univ, 69-75; PROF MICROBIOL, UNIV ALA, BIRMINGHAM & SR SCIENTIST, CANCER RES & TRAINING CTR, 75- Mem: Am Soc Cell Biol; Soc Gen Microbiol; Am Asn Immunol; Am Chem Soc; Am Soc Microbiol. Res: Cell biology; biochemistry; structure and assembly of viruses. Mailing Add: Dept of Microbiol Univ of Ala Birmingham AL 35294

COMPERE, CLINTON LEE, b Greenville, Tex, Feb 17, 11; m 32; c 2. ORTHOPEDIC SURGERY. Educ: Univ Chicago, BS, 36, MD, 37. Prof Exp: From asst prof to assoc prof, 46-65, PROF ORTHOP SURG, MED SCH, NORTHWESTERN UNIV, 65-, ACAD DIR PROSTHETIC-ORTHOTIC ED & PROSTHETIC RES CTR, 55-, DIR, REHAB ENG CTR, 72- Concurrent Pos: Pvt pract, 46-; sr attend staff, Northwestern Mem Hosp, 46-64, chief staff, 64-; sr consult, US Vet Admin, 46-; consult, Henrotin Hosp, Div Handicapped & Crippled Children; vchmn med coun, Rehab Inst Chicago; mem med adv comt, Div Voc Ed & Rehab, Ill; mem comt prosthetic-orthotic ed, Nat Res Coun-Nat Acad Sci, 60- Mem: AMA; Am Col Surg; Am Acad Orthop Surg (pres, 63-64); Am Orthop Asn. Res: Pathology of bone; prosthetic research for upper and lower extremity amputees; neoplasms; joint implants; myoelectric assistive devices. Mailing Add: 233 E Erie St Chicago IL 60611

COMPERE, EDGAR LATTIMORE, b Hamburg, Ark, Jan 23, 17; m 45; c 3. PHYSICAL CHEMISTRY. Educ: Ouachita Col, AB, 38; La Tech Univ, MS, 40, PhD(phys chem), 43. Prof Exp: Chemist, La Div, Stand Oil Co, NJ, 42-46; from asst prof to assoc prof chem, La Tech Univ, 46-51; sr chemist, Chem Div, 51-56, group leader corrosion sect, Reactor Exp Eng Div, 53-55, asst sect chief, 55-58, chief slurry mat sect, 58-61, sr chemist, Reactor Chem Div, 61-73, SR CHEMIST, CHEM TECHNOL DIV, OAK RIDGE NAT LAB, 73- Mem: Fel AAAS; Am Chem Soc; Sigma Xi; Am Nuclear Soc; fel Am Inst Chemists. Res: Reaction kinetics; vacuum evaporation of metals; fractional liquid extraction; nuclear reactor chemistry; corrosion; reactor materials; fission product transport. Mailing Add: Chem Technol Div Oak Ridge Nat Lab Oak Ridge TN 37830

COMPERE, EDWARD L, JR, b Detroit, Mich, June 22, 27; m 54; c 3. ORGANIC CHEMISTRY. Educ: Beloit Col, BS, 50; Univ Chicago, MS, 54; Univ Md, PhD, 58. Prof Exp: Asst prof chem, Univ WVa, 58-59 & Kans State Teachers Col, 59-60; from asst prof to assoc prof, Mich Tech Univ, 60-64; assoc prof, 64-67, dir state tech serv prog, 66-67, PROF CHEM, EASTERN MICH UNIV, 67- Mem: AAAS; Am Chem Soc; Sigma Xi. Res: Physical-organic chemistry; inorganic chemistry; chemistry in aquatic biology; lattice-salt structure. Mailing Add: Dept of Chem Eastern Mich Univ Ypsilanti MI 48197

COMPHER, MARVIN KEEN, JR, b Clifton Forge, Va, May 17, 42. DEVELOPMENTAL BIOLOGY, ENDOCRINOLOGY. Educ: Wake Forest Univ, BS, 64; Univ Va, PhD(biol), 68. Prof Exp: Asst prof biol, Col Wooster, 68-72; ASST PROF BIOL, CHATHAM COL, 72- Res: Regulation of the newt thyroid gland; effects of hypothalamic lesions and pituitary autotransplantations on thyroid activity. Mailing Add: Dept of Biol Chatham Col Woodland Rd Pittsburgh PA 15232

COMPTON, CHARLES CHALMER, b Brookline, Vt, Oct 25, 98; m 24; c 3. ENTOMOLOGY. Educ: Univ Conn, BS, 21; Univ Ill, MS, 34, PhD(entom), 40. Prof Exp: Entomologist, Ill Natural Hist Surv, 21-44 & Julius Hyman & Co, 44-52; mgr sales develop dept, Shell Chem Corp, DC, 52-59, div rep, 59-64; PROF ENTOM & COORDR INTERREGIONAL PROJ 4, RUTGERS UNIV, NEW BRUNSWICK, 64- Mem: AAAS; Entom Soc Am; Am Inst Biol Sci. Res: Truck crop and field insect control; control of greenhouse and ornamental insects; administration of sales development activities. Mailing Add: Dept of Entom & Econ Zool Rutgers Univ New Brunswick NJ 08903

COMPTON, CHARLES (DANIEL), b Elizabeth, NJ, Jan 8, 15; m 53. CHEMISTRY. Educ: Princeton Univ, AB, 40; Yale Univ, PhD(org chem), 43. Prof Exp: Asst chemist, Calco Chem Co, NJ, 34-38, res chemist, 43; instr chem, Princeton Univ, 43-44, res assoc, Manhattan Dist proj, 44-45; instr org &gen chem, 46, from asst prof to prof chem, 46-74, chmn dept, 64-74, EBENEZER FITCH PROF CHEM, WILLIAMS COL, 74- Mem: Am Chem Soc. Res: Correlation of spectra and structure of organic compounds. Mailing Add: Dept of Chem Williams Col Williamstown MA 01267

COMPTON, ELL DEE, b Wilmington, Ohio, Mar 16, 16; m 44; c 1. ORGANIC CHEMISTRY, ENVIRONMENTAL MANAGEMENT. Educ: Univ Cincinnati, ChE, 39, MS, 40, PhD(tanning res), 42. Prof Exp: Monsanto Chem Co fel tanned calf skin, Univ Cincinnati, 42-43; chemist, Merrimac Div, Monsanto Chem Co, 43-46, group leader, 47-52; dir res, Eagle-Ottawa Leather Co, 52-60; res group leader, Maumee Chem Co, 60-61, appl res dir, 61-63, chem res dir, 63-69; lab dir, Sherwin-

Williams Chem Div, 69-73, GROUP DIR ENVIRON CONTROL-CHEMS, SHERWIN-WILLIAMS CO, 73- Mem: AAAS; Am Chem Soc; Am Inst Chem. Res: Applications of organic chemicals; statistics. Mailing Add: 8457 Whitewood Rd Brecksville OH 44141

COMPTON, JACK, b Myrtlewood, Ala, July 19, 09; m 36; c 3. TEXTILE TECHNOLOGY. Educ: Howard Col, BSc, 30; Ohio State Univ, MSc, 31, PhD(org chem), 33. Prof Exp: Asst chem, Ohio State Univ, 30-33; Kendall Co fel & demonstr chem, McGill Univ, 33-34; asst res chemist, Dept Chem, Rockefeller Inst, 34-36; res chemist, Div Cellulose Chem, Boyce Thompson Inst, 36-40; textile res dept, B F Goodrich Co, 40-44; group leader & res chemist, Tubize Rayon Corp, 44-46, group leader, Celanese Corp Am, 46-49; head chem div, Inst Textile Technol, 49-50, tech dir, 50-59, vpres & dir res, 59-64, exec vpres, 64-73; RETIRED. Concurrent Pos: Pvt consult chem & textile technol, 73- Honors & Awards: Harold DeWitt Smith Mem Award, Am Soc Testing & Mat, 72. Mem: AAAS; Am Chem Soc; fel Am Inst Chemists; Am Oil Chemists Soc; Fiber Soc. Res: Carbohydrates; cellulose; lignin; nucleosides and nucleic acids; textiles; adhesives; resins; natural and synthetic rubber. Mailing Add: Drawer W Linden AL 36748

COMPTON, KENNETH GORDON, b Seattle, Wash, Sept 27, 04; m 29, 61; c 5. ELECTROCHEMISTRY. Educ: Wash State Univ, BS(elec eng) & BS(chem eng), 26, MS, 29. Prof Exp: Res scientist, Bell Tel Labs, 29-37, supvr phys chem mat & processes, 37-43, head dept mat & processes & electrochem, 43-64; PROF OCEAN ENG, UNIV MIAMI, 66-, DIR CTR FOR STUDY MAT IN THE SEA, 68- Concurrent Pos: Consult, Nat Defense Res Coun, Washington, DC, 42-46, Ord Dept, US Army, 44-45 & Bur Ships & Bur Ord, US Navy, 44-47. Honors & Awards: Bronze Medal, Am Electroplaters Soc, 52; Sam Tour Award, Am Soc Testing & Mat, 57, Merit Award, 64; Speller Award, Nat Asn Corrosion Eng, 62. Mem: Nat Asn Corrosion Eng; fel Am Soc Testing & Mat; Electrochem Soc. Res: Electrochemistry of the corrosion process; polarization behavior of metal surfaces in electrolytes; collection and concentration of metal ions and gases at metallic surfaces by electric current. Mailing Add: Dept Ocean Eng Sch Mar & Atmos Sci Univ of Miami Coral Gables FL 33124

COMPTON, LESLIE ELLWYN, b San Diego, Calif, Mar 24, 43; m 69; c 1. PHYSICAL CHEMISTRY. Educ: Stanford Univ, BS, 66; Univ Calif, Santa Barbara, PhD(chem), 70. Prof Exp: Sr res chemist, Garrett Res & Develop Co, Occidental Petrol, 70-71; staff scientist, Sci Applications Inc, 72-75; SR RES CHEMIST, OCCIDENTAL RES CORP, OCCIDENTAL PETROL, 75- Concurrent Pos: Fel, Univ Calif, Santa Barbara, 71-72; consult, Radiation & Environ Mat, Inc, 71-72. Mem: Am Phys Soc. Res: Gas phase and heterogeneous reactions of chemically high energy ions, atoms and small molecules; phase kinetics and heterogeneous catalysis of reactions occurring in fuels synthesis; instrumental analysis and instrumentation research. Mailing Add: Occidental Res Corp 1855 Carrion Rd La Verne CA 91750

COMPTON, OLIVER CECIL, b Seattle, Wash, Mar 1, 03; m 50. POMOLOGY. Educ: Univ Calif, BS, 31, MS, 32; Cornell Univ, PhD(pomol), 47. Prof Exp: Assoc, Exp Sta, Univ Calif, 32-40; from asst to instr pomol, Cornell Univ, 40-47; from asst horticulturist to assoc horticulturist, 48-60, from assoc prof to prof, 49-72, EMER PROF HORT, AGR EXP STA, ORE STATE UNIV, 72- Mem: Am Soc Hort Sci; Am Soc Plant Physiol. Res: Use of water by citrus and avocado trees; effect of aeration on absorption of nutritients by apple trees; physiological effects of fluorine on plants; response of tree and fruit to climate and nutrient level. Mailing Add: Dept of Hort Ore State Univ Corvallis OR 97331

COMPTON, ROBERT NORMAN, b Metropolis, Ill, Nov 28, 38; m 61. ATOMIC PHYSICS, MOLECULAR PHYSICS. Educ: Berea Col, BA, 60; Univ Fla, MS, 62; Univ Tenn, PhD(physics), 66. Prof Exp: Consult, 63-64, PHYSICIST, HEALTH PHYSICS DIV, OAK RIDGE NAT LAB, 66- Concurrent Pos: Ford Found fel, Univ Tenn, 68, lectr, 69- Mem: Am Phys Soc. Res: Negative ion-molecule reactions; interaction of electrons with atoms and molecules. Mailing Add: Health Physics Div Oak Ridge Nat Lab Oak Ridge TN 37830

COMPTON, ROBERT ROSS, b Los Angeles, Calif, July 21, 22; m 48; c 5. GEOLOGY. Educ: Stanford Univ, BA, 43, PhD(geol), 49. Prof Exp: Geologist, P-1, US Geol Surv, 43-44; from instr to assoc prof, 47-61, PROF GEOL, STANFORD UNIV, 61-, VCHMN DEPT, 69- Concurrent Pos: Geologist, P-3, US Geol Surv, 51-52; NSF fel, 55-56; Guggenheim fel, 63-64. Mem: Geol Soc Am; Brit Geol Soc. Res: Igneous and metamorphic petrology and structure. Mailing Add: Sch of Earth Sci Stanford Univ Stanford CA 94305

COMPTON, WALTER AMES, b Elkhart, Ind, Apr 22, 11; m 35; c 5. MEDICINE, PHARMACEUTICS. Educ: Princeton Univ, AB, 33; Harvard Univ, MD, 37. Prof Exp: Intern, Billings Hosp, Chicago, 37-38; med & res dir, 38-46, vpres res & med affairs, 46-60, exec vpres, 60-64, pres, 64-73, CHMN & CHIEF EXEC OFFICER, MILES LABS, INC, 73-, MEM BD DIRS, 36- Concurrent Pos: Mem bd dirs various subsidiaries of Miles Labs, Inc, Oaklawn Found Ment Health, Royal Soc Med Found, Weizmann Inst Sci & First Nat Bank, Elkhart; mem adv bd, Goshen Col, Col Sci Notre Dame Univ & South Bend Ctr Med Educ. Mem: AAAS; AMA; Royal Soc Health; Royal Soc Med; NY Acad Sci. Mailing Add: Miles Labs Inc Myrtle & McNaughton Sts Elkhart IN 46514

COMPTON, WALTER DALE, b Chrisman, Ill, Jan 7, 29; m 51; c 3. PHYSICS. Educ: Wabash Col, BA, 49; Univ Okla, MS, 51; Univ Ill, PhD, 55. Prof Exp: Physicist, US Naval Ord Test Sta, Inyokern, 51-52; physicist, US Naval Res Lab, 55-61; prof physics, Univ Ill, Urbana, 61-70, dir coord sci lab, 65-70; dir chem & phys sci, 70-75, VPRES, FORD MOTOR CO, 75- Concurrent Pos: Fel, US Naval Ord Test Sta, 55-56; US Naval Res Lab award, 58; mem adv bd, Naval Weapons Ctr. Mem: Fel Am Phys Soc; Sigma Xi. Res: Solid state physics; radiation effects in solids; color centers in insulating crystals; luminescence; metal semiconductor junction. Mailing Add: Ford Motor Co Sci Res Staff PO Box 2053 Dearborn MI 48121

COMPTON, WILLIAM A, b Richmond, Va, Aug 2, 27; m 55. GENETICS, STATISTICS. Educ: NC State Col, BS, 58, MS, 60; Univ Nebr, PhD(agron), 63. Prof Exp: Res asst genetics, NC State Col, 58-60; res asst, Univ Nebr, 60-62, consult statistician, 62-63; asst prof, NC State Univ, 63-69; ASSOC PROF AGRON, UNIV NEBR, LINCOLN, 67- Concurrent Pos: Consult, Agrarian Univ, Peru, 63- Mem: Biomet Soc; Am Soc Agron; Crop Sci Soc Am. Res: Applied quantitative genetics research with corn; statistical consulting work in experimental design and analysis in agronomy; computer programming. Mailing Add: Dept of Agron Univ of Nebr Lincoln NE 68503

COMPTON, WILLIAM DAVID, b DeLeon, Tex, Oct 21, 27; m 50; c 3. ORGANIC CHEMISTRY. Educ: NTex State Col, BS, 48, MS, 49; Univ Tex, PhD(chem), 56. Prof Exp: Instr chem, Arlington State Col, 49-50; from assoc prof to prof chem, West Tex State Col, 55-59, head dept, 58-59; from asst prof to assoc prof, Colo Sch Mines, 49-67; PROF CHEM, PRESCOTT COL, 67- Concurrent Pos: Vis prof, Imp Col Sci Technol, 71-72. Mem: Am Chem Soc. Res: Organometallic compounds; Grignard reagents and alkylcadmiums; synthesis and properties of heterocyclic nitrogen compounds; specifically ethylenimines. Mailing Add: Prescott Col Prescott AZ 86301

COMROE, JULIUS HIRAM, JR, b York, Pa, Mar 13, 11; m 36; c 1. PHYSIOLOGY. Educ: Univ Pa, AB, 31, MD, 34. Hon Degrees: MD, Karolinska Inst, Sweden, 68; DSc, Univ Chicago, 68. Prof Exp: Instr pharmacol, Sch Med, Univ Pa, 36-40, assoc, 40-42, asst prof, 42-46, prof physiol & pharmacol, Grad Sch Med & clin physiologist, Hosp, 46-57; PROF PHYSIOL, 57-, DIR CARDIOVASCULAR RES INST, 73-, MORRIS HERZSTEIN PROF BIOL, UNIV CALIF, SAN FRANCISCO, 73- Concurrent Pos: Commonwealth Fund fel, Nat Inst Med Res, London, 39; chmn physiol sect, USPHS, 55-58; mem bd sci counsellors, Nat Heart Inst, 57-61; mem, Nat Adv Ment Health Coun, 58-62 & Nat Adv Heart Coun, 63-67; ed, Circulation Res, 66-70; mem, Nat Adv Heart & Lung Coun, 70-74; ed, Annual Rev Physiol, 71-74; mem, President's Panel on Heart Dis, 72. Honors & Awards: Res Achievement Award, Am Heart Asn, 68; Carl J Wiggers Award, 74; Trudeau Medal, Am Lung Asn, 74; Gold Heart Award, Am Heart Asn, 75. Mem: Nat Acad Sci; Am Physiol Soc (pres, 60-61); Am Acad Arts & Sci; Am Soc Clin Invest. Res: Carotid and aortic bodies; autonomic drugs; regulation of respiration circulation; pulmonary function; neuromuscular transmission. Mailing Add: Cardiovasc Res Inst Univ of Calif San Francisco CA 94143

COMSTOCK, CRAIG, b Long Beach, Calif, June 11, 34; m 57; c 3. APPLIED MATHEMATICS. Educ: Cornell Univ, BEngPhys, 56; US Naval Postgrad Sch, MS, 61; Harvard Univ, PhD(appl math), 65. Prof Exp: Teaching fel appl math, Harvard Univ, 63-64; asst prof math, Pa State Univ, 64-68 & Univ Mich, 68-70; assoc prof, 70-73, PROF MATH, NAVAL POSTGRAD SCH, 73- Concurrent Pos: Consult, HRB-Singer, Inc, 65 & US Naval Ord Res Lab, 65-68. Mem: Math Asn Am; Soc Indust & Appl Math. Res: Asymptotic expansion of differential equations; wave propagation; plasmas in the geomagnetic field; finite element calculations in meteorology. Mailing Add: 3096 Sloat Rd Pebble Beach CA 93953

COMSTOCK, DALE ROBERT, b Frederic, Wis, Jan 18, 34; m 56; c 2. MATHEMATICS. Educ: Cent Wash State Col, BA, 55; Ore State Univ, MS, 62, PhD(algebra), 66. Prof Exp: Instr math, Columbia Basin Col, 56-57 & 59-60; teaching asst, Ore State Univ, 61-64; from asst prof to assoc prof, 64-70, PROF MATH & DEAN GRAD STUDIES, CENT WASH STATE COL, 70- Mem: Math Asn Am; Am Math Soc; Soc Indust & Appl Math; Asn Comput Mach. Res: Algebra; computability. Mailing Add: Dept of Math Cent Wash State Col Ellensburg WA 98926

COMSTOCK, GEORGE MILTON, b Charleston, SC, May 5, 40; m 63; c 1. PHYSICS. Educ: Univ Chicago, SB, 61, SM, 62, PhD(physics), 68. Prof Exp: Res assoc, Univ Chicago, 68-69; PHYSICIST, GEN ELEC RES LAB, 70- Mem: Am Phys Soc; Am Astron Soc; Am Geophys Union. Res: Cosmic rays; astrophysics; space science; lunar studies. Mailing Add: Phys Sci Br Gen Elec Co Res & Develop Ctr Schenectady NY 12301

COMSTOCK, GEORGE WILLS, b Niagara Falls, NY, Jan 7, 15; m 39; c 3. EPIDEMIOLOGY. Educ: Antioch Col, BS, 37; Harvard Univ, MD, 41; Univ Mich, MPH, 51; Johns Hopkins Univ, DrPH, 56. Prof Exp: Dir, Muscogee County Tuberc Study, USPHS, 46-55; chief epidemiol studies, Tuberc Prog, 56-62; assoc prof, 62-65, PROF EPIDEMIOL, JOHNS HOPKINS UNIV, Concurrent Pos: Consult tuberc prog, USPHS, 62-; dir, Training Ctr for Pub Health Res, Hagerstown, Md, 63- Mem: Am Pub Health Asn; Am Thoracic Soc; AMA; Am Epidemiol Soc. Res: Epidemiology of chronic diseases, especially tuberculosis. Mailing Add: Johns Hopkins Sch of Hyg 615 N Wolfe St Baltimore MD 21205

COMSTOCK, GILBERT LEROY, b Albia, Iowa, Aug 20, 36; m 58; c 4. FOREST PRODUCTS. Educ: Iowa State Univ, BS, 58; NC State Univ, MS, 62; NY State Col Forestry, PhD(wood prod eng), 68. Prof Exp: Res asst wood technol, NC State Univ, 59-62; forest prod technologist, Forest Prod Lab, US Forest Serv, 62-68; prof specialist wood drying res & develop, 68-72, sect mgr lumber processing res & develop, 72-76, DEPT MGR PROCESS CONTROL RES & DEVELOP, WEYERHAEUSER CO, 76- Honors & Awards: Wood Structure Design Award, Nat Forest Prod Asn, 62. Mem: Forest Prod Res Soc; Soc Wood Sci & Technol (treas & pres, 69-72). Res: Development and application of measuring and control systems for wood processing operations using the latest electronics and computer technology. Mailing Add: 327 Cedar Lane Longview WA 98632

COMSTOCK, JACK CHARLES, b Detroit, Mich, June 13, 43. PLANT PATHOLOGY. Educ: Mich State Univ, BS, 65, PhD(plant path), 71. Prof Exp: Res assoc corn path, Iowa State Univ, 71-74; asst pathologist, 74-75, ASSOC PATHOLOGIST, HAWAIIAN SUGAR PLANTERS ASN, 75- Concurrent Pos: Affil fac mem, Univ Hawaii, 75- Mem: Am Phytopath Soc; Am Soc Plant Physiologists; AAAS; Sigma Xi. Res: Sugarcane pathology; disease control, screening resistance. Mailing Add: Hawaiian Sugar Planters Asn 99-193 Aiea Heights Dr Aiea HI 96701

COMSTOCK, RALPH ERNEST, b Spring Valley, Minn, July 19, 12; m 36; c 3. GENETICS, ANIMAL BREEDING. Educ: Univ Minn, BS, 34, MS, 36, PhD(animal genetics), 38. Prof Exp: From instr to asst prof animal breeding, Univ Minn, 37-43; from assoc prof to prof, NC State Col, 43-46, consult animal sci statist, 43-46; animal husbandman & head animal husb dept, Exp Sta, Univ PR, 46-47; prof, Inst Statist, State Col Agr & Eng, Univ NC, 47-57; prof animal husb, 57-65, prof genetics & head dept, 65-68, REGENT'S PROF GENETICS, UNIV MINN, ST PAUL, 68- Honors & Awards: Animal Breeding & Genetics Award, Am Soc Animal Sci, 66. Mem: Am Soc Naturalists; Am Soc Animal Sci; Genetics Soc Am; Biomet Soc; Am Soc Agron. Res: Population genetics, especially mathematical theory and methods of investigation with reference to problems in animal and plant breeding. Mailing Add: Dept of Genetics & Cell Biol Univ of Minn St Paul MN 55101

COMSTOCK, VERNE EDWARD, b Kildeer, NDak, July 18, 19; m 41; c 6. PLANT BREEDING, PLANT GENETICS. Educ: State Col Wash, BS, 41, MS, 47; Univ Minn, PhD, 59. Prof Exp: Asst agronomist forage invest, Div Forage Crops & Dis, Bur Plant Indust, USDA, State Col Wash, 47-50, res agronomist flax breeding & invest, Southwestern Irrig Field Sta, Brawley, Calif, 50-53, flax qual invests, cereal crop sect, Crops Res Div, Agr Res Serv, Univ Minn, St Paul, 53-57, leader seedflax invests, Indust Crop Sect, 57-73; PROF AGRON & PLANT GENETICS, UNIV MINN, ST PAUL, 74- Mem: Am Soc Agron. Res: Range grass breeding; flax cultural studies under irrigation and quality investigations, flax breeding for disease resistance. Mailing Add: Dept of Agron Univ of Minn St Paul MN 55101

COMUNALE, GIUSEPPE VINCENT, analytical chemistry, polymer chemistry, see 12th edition

CONAN, NEAL JOSEPH, JR, b Syracuse, NY, Aug 7, 18; m 42; c 4. MEDICINE. Educ: Col of the Holy Cross, AB, 40; Columbia Univ, MD, 43, ScD(med), 49. Prof Exp: Intern, Presby Hosp, New York, 44; fel antimalarial res, Off Sci Res & Develop Proj, Goldwater Mem Hosp, 44-45; asst resident, Presby Hosp, 45-47; fel med, Col

Med, NY Univ, 47-48, instr med & asst dir cancer teaching prog, 48-49, clin asst attend physician, Med Clin, 48-49; prof internal med & chmn dept, Sch Med, Am Univ Beirut & chief med serv, Hosp, 49-52, clin prof, 52-54; head dept internal med, Arabian Am Oil Co, Dhahran, Saudi Arabia, 52-54; asst med, 54-70, INSTR MED, COL PHYSICIANS & SURGEONS, COLUMBIA UNIV, 70- Concurrent Pos: Asst physician, Vanderbilt Clin, Presby Med Ctr, 47-49; clin asst vis physician, Bellevue Hosp, 48-49. Mem: AAAS; Am Soc Trop Med & Hyg; NY Acad Sci; Royal Soc Trop Med & Hyg. Res: Chemotherapy of infectious diseases; physiology of the kidney and liver; pharmacology of antimalarial drugs in man; hepatic dysfunction during fever in man; antimebic drugs; renal mechanisms for electrolytes; metabolic studies in cancer patients. Mailing Add: Col of Physicians & Surgeons Columbia Univ New York NY 10032

CONAN, ROBERT JAMES, JR, b Syracuse, NY, Oct 30, 24. PHYSICAL CHEMISTRY. Educ: Syracuse Univ, BS, 45, MS, 47; Fordham Univ, PhD(phys chem), 50. Prof Exp: Instr gen chem, Fordham Univ, 48; asst prof phys chem, 49-52, assoc prof, 53-56, PROF PHYS CHEM, LE MOYNE COL, NY, 57-, CHMN DEPT CHEM, 59-67 & 73- Concurrent Pos: Researcher, Stockholm, Sweden, 53 & Res Lab Phys Chem, Swiss Fed Inst Technol, 66-67. Mem: Am Chem Soc; Am Phys Soc. Res: Theory of liquids and solutions; surface phenomena; thermodynamics. Mailing Add: Dept of Chem Le Moyne Col Syracuse NY 13214

CONANT, DALE HOLDREGE, b Casper, Wyo, July 5, 39. PHOTOGRAPHIC CHEMISTRY. Educ: Col Idaho, BS, 61; Ore State Univ, MS, 64; Ohio State Univ, PhD(phys chem), 69. Prof Exp: Engr, Kaiser Refractories, 63-65; sr chemist, Rochester, 69-74, res assoc, Photog Res Div, 74-75, DEVELOP ENGR, PLATE MFG DIV, EASTMAN KODAK CO, 75- Mem: Am Chem Soc; Soc Photog Sci & Eng. Res: Interaction of chlorophyll and its derivatives with II-aromatic electron acceptor systems involving visible spectrum measurements, fluorescence quenching measurements and nuclear magnetic resonance measurements; research and development of photolithography. Mailing Add: 508 Canadian Parkway Fort Collins CO 80521

CONANT, DONALD ROBERTSON, JR, physical chemistry, see 12th edition

CONANT, FLOYD SANFORD, b Leroy, WVa, Nov 27, 14; m 38; c 2. POLYMER PHYSICS. Educ: Morris Harvey Col, BS, 34; WVa Univ, MS, 35. Prof Exp: Teacher high sch, 35-42; sr res scientist, 42-75, RES ASSOC, FIRESTONE TIRE & RUBBER CO, 75- Concurrent Pos: Assoc ed, J Rubber Chem & Technol, Rubber Div Am Chem Soc, 74-; mem US deleg, Tech Comt 45, Int Stand Orgn, 75-79. Mem: Am Chem Soc; Am Soc Testing & Mat. Res: Low temperature properties of elastomers; vibration properties of pneumatic tires; coefficient of friction of rubber; tire dynamics. Mailing Add: Firestone Tire & Rubber Co Cent Res 1200 Firestone Pkwy Akron OH 44317

CONANT, FRANCIS PAINE, b New York, NY, Feb 27, 26; m 51; c 2. CULTURAL ANTHROPOLOGY, ETHNOLOGY. Educ: Cornell Univ, BA, 50; Columbia Univ, PhD, 60. Prof Exp: Lectr anthrop, Columbia Univ, 56-57; asst prof, Univ Mass, Amherst, 60-61; assoc prof, 62-67, PROF ANTHROP, HUNTER COL, 67- Concurrent Pos: Consult, Am Mus Natural Hist, 60-; NIMH-NSF-Univ Calif, Los Angeles grant cult ecol, Africa, 61-64; Fulbright res fel, Oxford Univ, 68-69. Honors & Awards: NSF Award Anthrop Use of Satellite Data, 75. Res: Cultural ecology; settlement pattern; religion; kinship; anthropological use of satellite data in studying human ecology. Mailing Add: Dept of Anthrop Hunter Col New York NY 10021

CONANT, JAMES BRYANT, b Boston, Mass, Mar 26, 93; m 21; c 2. CHEMISTRY. Educ: Harvard Univ, AB, 13, PhD(chem), 16. Hon Degrees: Numerous from US and foreign univs, 33-56. Prof Exp: From instr to prof chem, 16-29, Emory prof org chem, 29-33, pres, 33-53, EMER PRES, HARVARD UNIV, 53- Concurrent Pos: US High Cmnr for Ger, 53-55; US Ambassador Fed Repub Ger, 55-57; dir, A Study of the Am High Sch, 57-62 & Study of Educ of Am Teachers, 62-63. Ed adv, Ford Found, Berlin, 63-64. Res assoc, Calif Inst Technol, 27. Mem comt sci aids to learning, Nat Res Coun, 37-42; mem bd sci dirs, Rockefeller Inst, 30-49; ed Policies Comn, 41-46 & 47-; NSF, 50-53, chmn Nat Defense Res Comt, 41-46. With AEC, 47-52. Honors & Awards: Chandler Med, Columbia Univ, 32; Medal, Am Inst Chem, 34; Priestley Medal, 44; Freedom House Award, 52; Comdr, Legion of Hon; Hon Comdr, Most Excellent Order of Brit Empire; Medal for Merit, Oak Leaf Cluster, 48; Nichols Medal, Am Chem Soc. Mem: Nat Acad Sci; AAAS; Am Chem Soc; Am Soc Biol Chem; Am Philos Soc. Res: Organic chemistry; reduction and oxidation; hemoglobin; free radicals; quantitative study of organic reactions; superacid solutions; chlorophyll; chemistry of organic compounds. Mailing Add: 200 E 66th St New York NY 10021

CONANT, JOHN WESLEY, physical chemistry, see 12th edition

CONANT, LOUIS COWLES, b Orford, NH, Sept 14, 02; m 30; c 2. GEOLOGY. Educ: Dartmouth Col, AB, 26; Cornell Univ, AM, 29, PhD(geol), 34. Prof Exp: Instr geol, Dartmouth Col, 26-27; asst, Cornell Univ, 27-28, instr, 28-29, 30-37; field geologist, NRhodesia, 29-30; asst geologist, Miss Geol Surv, 37-38, 40-42; supvr mineral surv, Works Progress Admin, 38-39; lectr, Smith Col, 39-40; geologist, US Geol Surv, 42-72. Concurrent Pos: Assoc prof, Univ Miss, 37-39, 40-42; geol map compiler, Agency Int Develop, Libya, 60-62. Mem: Fel AAAS; fel Geol Soc Am; Am Asn Petrol Geol. Res: Non-metallic economic geology; stratigraphy; Chattanooga shale; east Gulf Coastal Plain; fuels; geology of Libya. Mailing Add: 3070 Porter St NW Washington DC 20008

CONANT, NORMAN FRANCIS, b Walpole, Mass, Mar 9, 08; m 29; c 7. MYCOLOGY, BACTERIOLOGY. Educ: Bates Col, BS, 30; Harvard Univ, MA, 31, PhD(mycol), 33. Prof Exp: Asst, Harvard Med Sch, 34-35; instr bact, Sch Med, Duke Univ, 35-36, assoc, 36-38, from asst prof to assoc prof bact & mycol, 38-46, prof mycol & assoc prof bact, 46-58, prof microbiol, 58-74, chmn dept, 58-68; RETIRED. Concurrent Pos: Sheldon traveling fel med mycol from Harvard, Sch Med, Paris, 33-34; Markle fel, Army Med Sch, Inst Oswaldo Cruz & Univ Sao Paulo, 44; lectr, Army Med Sch, 43-46; exp consult, Secy of War, 43-46; consult, Commun Dis Ctr, USPHS, Ga, 47-49; mem adv panel microbiol, Off Naval Res, 52-57; grad training prog study sect, Nat Inst Allergy & Infectious Dis, 57-60, chmn, Infectious Dis & Trop Med Training Grant Comt, 60-61, mem nat adv allergy & infectious dis coun, 62-65; mem & chmn stands & exam comt, Am Bd Microbiol, 62-63. Mem: Soc Exp Biol & Med; Med Mycol Soc of the Americas; Mycol Soc Am; Int Soc Human & Animal Mycol (pres, 62). Res: Medical mycology. Mailing Add: 5622 Garrett Rd Durham NC 27710

CONANT, ROBERT HENRY, b Rockland, Mass, Oct 5, 16; m 46; c 5. PHOTOGRAPHIC CHEMISTRY. Educ: Loyola Col, Md, BS, 37; Georgetown Univ, MS, 39, PhD(biochem), 42. Prof Exp: Instr chem, Georgetown Univ, 37-42; res chemist, Photo Repro Div, Gen Aniline & Film Corp, 45-47, sr res chemist & res group leader, 47-51, tech asst to film plant mgr, 51-54, sr opers supvr film emulsions dept, 54-56, sr emulsion specialist, 56-60, mgr film qual control dept, 60-63 & sensitometry dept, 63-64, res chemist, 64-67, qual control specialist gelatin, 67, prod qual specialist gelatin, 67-74, SUPVR EMULSION MFG, PHOTO-REPRO DIV, GAF CORP, 74- Res: Photographic emulsion; gelatin; tests for sugars; S-amino acids in proteins; cystine and methionine distribution in proteins of egg whites. Mailing Add: Emulsion Mfg 37-4 GAF Corp Charles St Binghamton NY 13902

CONANT, ROBERT M, b Binghamton, NY, Oct 19, 27; m 51; c 3. VIROLOGY, IMMUNOLOGY. Educ: Harpur Col, BA, 58; State Univ NY Buffalo, MA, 63, PhD(virol, immunol), 66. Prof Exp: Instr pub schs, NY, 59-60; from instr pediat to asst prof pediat & med microbiol, Col Med, Ohio State Univ, 65-70; scientist & adminr, Div Res Grants, NIH, 70-71, asst chief, Div Allied Health Manpower, Bur Health Manpower Educ, 71-73, chief off spec studies, 73-74, CHIEF MANPOWER UTILIZATION BR, BUR HEALTH MANPOWER, HEALTH RESOURCES ADMIN, DEPT HEALTH, EDUC & WELFARE, 74- Concurrent Pos: Mem bd vaccine develop, Rhinovirus Ref Ctr, Nat Inst Allergy & Infectious Dis, 65-70; NIH res grants, Children's Hosp Res Found & Res Found, Ohio State Univ, 66-67; regist, Registry Am Type Cult Collection, 67- Mem: AAAS; Am Soc Microbiol; NY Acad Sci. Res: Characterization of antigens distributed among enteroviruses by immunodiffusion methods; relationships betweeen variant strains of same enterovirus serotype by immunodiffusion; identification and classification of rhinoviruses; application of plaque and immunodiffusion techniques to the study of rhinoviruses. Mailing Add: Div of Assoc Health Professions Bur of Health Manpower Bethesda MD 20014

CONANT, ROGER, b Mamaroneck, NY, May 6, 09; m 47; c 2. HERPETOLOGY. Hon Degrees: ScD, Univ Colo, 71. Prof Exp: Cur reptiles, Toledo Zoo, Ohio, 29-33, educ dir, 31-33, cur, 33-35; cur reptiles, Philadelphia Zool Garden, 35-73, dir, 67-73; ADJ PROF BIOL, UNIV NMEX, 73- Concurrent Pos: Res assoc, Am Mus Natural Hist & Acad Natural Sci Philadelphia; consult, Am Philos Soc Proj for Adult Educ & Partic in Sci, 40-42 & Nat Res Coun, 59-62. Mem: Am Soc Ichthyol & Herpet (first vpres, 46, 56, secy, 58-60, pres, 62); Soc Systs Zool; Zool Soc London. Res: Distribution, natural history and speciation in reptiles and amphibians of the United States and Mexico. Mailing Add: Dept of Biol Univ of NMex Albuquerque NM 87131

CONARD, ELVERNE CLYDE, b Mowequa, Ill, Oct 14, 09; m 33; c 2. AGRONOMY. Educ: Colo Agr & Mech Col, BS, 32; Univ Nebr, MS, 38; Agr & Mech Col Tex, PhD(range mgt), 53. Prof Exp: Agronomist, Soil Conserv Serv, USDA, Nebr, 35-39, agronomist & nursery mgr, 39-45; ASSOC AGRONOMIST, EXP STA, UNIV NEBR, LINCOLN, 45-, ASSOC PROF AGRON, UNIV, 56- Mem: Am Soc Agron; Am Soc Range Mgt. Res: Range and pasture management; grass seed production; ecology. Mailing Add: Dept of Agron Univ of Nebr Lincoln NE 68506

CONARD, GORDON JOSEPH b Milwaukee, Wis, Sept 22, 39; m 62; c 3. BIOCHEMICAL PHARMACOLOGY. Educ: Univ Wis-Madison, BS, 61, MS, 67, PhD(pharmacol), 69. Prof Exp: Instr, Sch Pharm, Univ Wis-Milwaukee, 68; res assoc biochem, Res Inst, Am Dent Asn, 68-72; sr biochem pharmacologist, Drug Metab, 73-74, RES SPECIALIST DRUG METAB, RIKER LAB, INC, 3M CO, 75- Concurrent Pos: Reviewer, J Pharmaceut Sci, 72-; asst prof, Col Pharm, Univ Minn, Minneapolis, 73-; consult, Div Biochem, Res Inst, Am Dent Asn, 73- Mem: Int Asn Dent Res; AAAS; Am Pharmaceut Asn; Acad Pharmaceut Sci; Sigma Xi. Res: Metabolic disposition of new drug molecules in laboratory animals and in man with emphasis on relationships to pharmacological and toxicological activity. Mailing Add: Riker Lab Inc Bldg 218-2 3M Ctr St Paul MN 55101

CONARD, ROBERT ALLEN, b Jacksonville, Fla, July 29, 13; m 48; c 4. MEDICAL RESEARCH. Educ: Univ SC, BS, 36, MD, 41. Prof Exp: Med officer, US Navy, 41-47, proj officer, Radiol Defense Lab, 47-50, with med res dist, 50-56; scientist & chief, Marshall Island Med Survs, 56-68, SR SCIENTIST, BROOKHAVEN NAT LAB, 68- Concurrent Pos: Prof path, State Univ NY Stony Brook. Mem: Am Soc Hemat; Tissue Cult Soc; Radiation Res Soc. Res: Radiation effects; medical surveys of Marshallese people exposed to radioactive fallout. Mailing Add: Brookhaven Nat Lab Upton NY 11973

CONARY, ROBERT EKVALL, b Minneapolis, Minn, Aug 14, 13; m 39; c 3. PHYSICAL CHEMISTRY. Educ: Univ Minn, BChE, 34; Univ Wis, PhD(chem), 38. Prof Exp: Asst chem, Univ Wis, 34-38; res chemist, 38-42, asst supvr, Fuels Res Dept, 42-43, chems res dept, 43-45, supvr, 45-54, res admin, 54-56, dir res, 56-57, asst mgr res & develop, 57-60, mgr, Beacon Res Labs, 60-68, Europ Res Ctr, Texaco Belgium, 68-69, DIR RES, TEXACO SERV EUROPE LTD, TEXACO, INC, 69- Mem: Fel AAAS; Am Inst Chem Eng; Am Chem Soc; fel Am Inst Chem; NY Acad Sci. Res: Development of fuels and lubricants for internal combustion engines; production of chemicals from petroleum; chemistry of thiophene; plastics. Mailing Add: Texaco Servs Europe Ltd 149 Ave Louise B-1050 Brussels Belgium

CONAWAY, CLINTON HARPER, zoology, see 12th edition

CONAWAY, HOWARD HERSCHEL, b Fairmont, WVa, Oct 2, 40; m 69; c 1. PHYSIOLOGY. Educ: Fairmont State Col, BS, 63; WVa Univ, MS, 67; Univ Mo-Columbia, PhD, 70. Prof Exp: NIH fel, Dept Pediat, Univ Mo-Columbia, 70-71; ASST PROF PHYSIOL, SCH MED, UNIV ARK, LITTLE ROCK, 71- Mem: AAAS. Res: Endocrinology and metabolism, especially calcium metabolism and pancreas interrelationships. Mailing Add: Dept of Physiol Univ of Ark Sch of Med Little Rock AR 72201

CONBERE, JOHN PHILIP, b Shamokin, Pa, May 23, 25; m 46; c 4. INDUSTRIAL CHEMISTRY. Educ: Pa State Univ, BS, 47; Univ Notre Dame, MS, 49, PhD(org chem), 50. Prof Exp: Res chemist synthetic org med, Merck & Co, 50-53; res chemist org chem, Arnold Hoffman & Co, Inc, 53-59; head, Exp Dept, ICI Organics, Inc, 59-64, develop mgr, 64-67, com develop mgr, ICI United States, Inc, 67-70, TECH MGR, ICI UNITED STATES, 71- Mem: Am Chem Soc. Res: Antimalarials; amoebicides; steroids; amino acids; textile auxiliaries; dyestuffs; fluoropolymers; chlorinated organics; condensation polymers. Mailing Add: ICI United States Inc East 22nd St & Ave J Bayonne NJ 07002

CONCA, ROMEO JOHN, b New Haven, Conn, May 11, 26; m 46; c 2. ORGANIC CHEMISTRY. Educ: Yale Univ, BS, 49, PhD(org chem), 53. Prof Exp: Asst, Princeton Univ, 52-53; res chemist, G D Searle & Co, 53-55; res chemist, 55-59, group leader, 59-62, sect leader, 62-64, RES SUPVR, OLYMPIC RES DIV, ITT RAYONIER, INC, 64- Mem: Tech Asn Pulp & Paper Indust; Am Chem Soc. Res: Cellulose and wood chemistry; carbohydrates. Mailing Add: Olympic Res Div ITT Rayonier Inc 409 E Harvard Ave Shelton WA 98584

CONCANNON, JOSEPH N, b New York, NY, Sept 25, 20; m 64; c 3. PARASITOLOGY, RADIOBIOLOGY. Educ: Univ Dayton, BS, 42; Ohio State Univ, MS, 51; St John's Univ, MS, 56, PhD(parasitol), 59. Prof Exp: Teacher high sch, 42-51; instr biol, Cath Univ PR, 52-54; asst prof, Univ Dayton, 59-62; asst prof, 62-64, ASSOC PROF BIOL, ST JOHN'S UNIV, NY, 64-, ACTG CHAIRPERSON, DEPT BIOL SCI, 74- Concurrent Pos: NSF res fel, Univ Rochester, 62. Mem: AAAS; Soc Protozool; Am Soc Parasitol; Nat Sci Teachers Asn. Res: Protozoan parasitology, especially the trichomonads; thyroid physiology. Mailing Add: Dept of Biol St John's Univ Jamaica NY 11432

CONCIATORI, ANTHONY BERNARD, b New York, NY, Mar 4, 16; m 51; c 1. ORGANIC CHEMISTRY. Educ: Fordham Univ, BSc, 38; Univ Cincinnati, PhD(chem), 49. Prof Exp: Res chemist, Interchem Res Labs, 39-44; sr res chemist, 49-60, res assoc, 60-68, sect head, 68-71, MGR, CELANESE CORP, 71- Mem: Am Chem Soc; Sigma Xi (pres, Sci Res Soc Am, 63-64). Res: Polymers; catalysis; coatings; fibers. Mailing Add: 27 Orchard St Chatham NJ 07928

CONCUS, PAUL, b Los Angeles, Calif, June 18, 33; m 59; c 2. APPLIED MATHEMATICS, NUMERICAL ANALYSIS. Educ: Calif Inst Technol, BS, 54; Harvard Univ, AM, 55, PhD(appl math), 59. Prof Exp: Appl mathematician, Int Bus Mach Corp, 59-60; MATHEMATICIAN, LAWRENCE BERKELEY LAB, UNIV CALIF, BERKELEY, 60- Concurrent Pos: Lectr, Univ Calif, Berkeley, 63-65; consult, Lockheed Res Labs, 61-70 & Gen Elec, 73-; sr vis fel, Sci Res Coun Gt Brit, 70-71. Mem: Soc Indust & Appl Math. Res: Capillary fluid mechanics; computation. Mailing Add: Lawrence Berkeley Lab Univ of Calif Berkeley CA 94720

CONDELL, WILLIAM JOHN, JR, b Melrose, Mass, Mar 29, 27; m 52; c 2. RESEARCH ADMINISTRATION, OPTICAL PHYSICS. Educ: Cath Univ Am, BChemE, 49, MS, 52, PhD(physics), 59. Prof Exp: Physicist, Naval Ord Lab, 51-52, Eng Res & Develop Lab, US Army, 52-58 & Lab Phys Sci, 58-66; physicist, 66-74, DIR PHYSICS PROGS, OFF NAVAL RES, 74- Concurrent Pos: Asst prof lectr, George Washington Univ, 57-66. Mem: Am Phys Soc; Optical Soc Am; Am Asn Physics Teachers. Res: Atomic spectroscopy; optics; lasers. Mailing Add: Physics Prog Off Code 421 Off of Naval Res Arlington VA 22217

CONDELL, YVONNE C, b Quitman, Ga, Aug 29, 31; m 52. HUMAN GENETICS. Educ: Fla Agr & Mech Col, BS, 52; Univ Conn, MA, 58, PhD(cellular biol), 65. Prof Exp: Teacher high sch, Fla, 55-57 & Minn, 58-60; instr biol, Fergus Falls Jr Col, 60-65; asst prof, 65-67, ASSOC PROF BIOL, MOORHEAD STATE UNIV, 67- Concurrent Pos: Lectr, Univ Conn, 63. Mem: AAAS; Soc Study Social Biol; Am Inst Biol Sci; Nat Asn Biol Teachers. Res: Cellular biology; biology education; biochemical and genetic bases of human defects. Mailing Add: Dept of Biol Moorhead State Univ Moorhead MN 56560

CONDER, HAROLD LEE, b Salem, Ohio, Nov 26, 45; m 63; c 1. ORGANOMETALLIC CHEMISTRY. Educ: Youngstown Univ, BS, 67; Purdue Univ, PhD(inorg chem), 71. Prof Exp: Res asst chem, Tulane Univ, 71-73; ASST PROF CHEM, GROVE CITY COL, 73- Mem: Am Chem Soc. Res: Photochemical substitution reactions of transition metal phosphites and phosphines. Mailing Add: Grove City Col Grove City PA 16127

CONDIE, KENT CARL, b Salt Lake City, Utah, Nov 28, 36; m 63; c 1. PETROLOGY, GEOCHEMISTRY. Educ: Univ Utah, BS, 59, MA, 62; Univ Calif, San Diego, PhD(geochem), 65. Prof Exp: From asst prof to assoc prof geochem & petrol, Wash Univ, 64-70; ASSOC PROF GEOCHEM, NMEX INST MINING & TECHNOL, 70- Mem: Geol Soc Am; Geochem Soc; Am Geophys Union. Res: Trace element geochemistry; origin and growth of continents. Mailing Add: Dept of Geosci NMex Inst of Mining & Technol Socorro NM 87801

CONDIKE, GEORGE FRANCIS, b Brockton, Mass, Dec 1, 16; m 41; c 2. INORGANIC CHEMISTRY. Educ: DePauw Univ, AB, 40; Cornell Univ, PhD(inorg chem), 43. Prof Exp: Fel, Mellon Inst Indust Res, 43-44; sr engr, Sylvania Elec Prod, Inc, 44; tech rep, Rohm and Haas Co, 44-47; assoc prof chem, 47-53, dean col, 54-56, PROF CHEM, FITCHBURG STATE COL, 56- Concurrent Pos: NSF grant, 64-66, mem, NSF Equip Comt, 68-69. Mem: Am Chem Soc. Res: Chelate compounds; donor-acceptor bonding. Mailing Add: Dept of Chem Fitchburg State Col Fitchburg MA 01420

CONDIT, CARLTON, b Oakland, Calif, May 28, 06; m 35; c 3. GEOLOGY. Educ: Univ Calif, AB, 35, PhD(paleobot), 39. Prof Exp: Asst paleobot, Univ Calif, 35-39; asst, R W Chaney, 35-40; teacher, Williams Col, Calif, 40-41; instr, Exten Div, Univ Calif, 41-42 & San Bernardino Valley Union Jr Col, 44-45; asst prof geol, Univ Iowa, 45-50; chmn geol sect, Ill State Mus, 50-61; prof geol, Curry Col, 63-73; RETIRED. Mem: AAAS; Paleont Soc; Geol Soc Am. Res: Pleistocene paleontology; Pennsylvanian paleobotany. Mailing Add: 4 Amherst Rd Stoughton MA 02072

CONDIT, PAUL BRAINARD, b Berkeley, Calif, Mar 12, 43; m 66; c 1. ORGANIC CHEMISTRY. Educ: Univ Calif, Riverside, BA, 65; Univ Mich, PhD(org chem), 70. Prof Exp: Res fel, Calif Inst Technol, 70-71; RES CHEMIST, EASTMAN KODAK CO, 71- Mem: AAAS; Am Chem Soc; Soc Photog Sci & Eng. Res: Stereochemistry and mechanism of organic reactions; organic chemistry of color photography. Mailing Add: Eastman Kodak Co Res Labs B-59 Rochester NY 14650

CONDIT, PAUL CARR, b Cleveland, Ohio, Sept 19, 14; m 39; c 2. PETROLEUM CHEMISTRY. Educ: Yale Univ, BS, 36, PhD(org chem), 39. Prof Exp: Res chemist, Calif Res Corp, 39-46, sr res chemist, 46-52, sect supvr, 52-66; sect supvr, 66-67, mgr polymer div, 67-69, SR RES ASSOC, PATENT DEPT, CHEVRON RES CO, 69- Mem: Am Chem Soc; AAAS. Res: Petrochemical research and development; polymers; patents. Mailing Add: 720 Butterfield Rd San Anselmo CA 94960

CONDIT, RALPH HOWELL, b Hollywood, Calif, May 12, 29; m 66; c 2. SOLID STATE CHEMISTRY. Educ: Princeton Univ, BA, 51, PhD(chem), 60. Prof Exp: Res adminr, Air Force Off Sci Res, 58-60; CHEMIST, LAWRENCE LIVERMORE LAB, UNIV CALIF, 60- Concurrent Pos: Consult, Air Force Off Sci Res; dir, Geos Corp, Calif. Mem: Am Chem Soc; Am Phys Soc; Am Inst Aeronaut & Astronaut; Am Inst Mining, Metall & Petrol Eng; Am Ceramic Soc. Res: Tracer techniques and diffusion in solids; nuclear explosives security; hydrogen fuel economy; ceramics for turbines; laser isotope separation. Mailing Add: 4602 Almond Circle Livermore CA 94550

CONDLIFFE, PETER GEORGE, b Christchurch, NZ, June 30, 22; nat US; m 42; c 3. BIOCHEMISTRY, RESEARCH ADMINISTRATION. Educ: Univ Calif, BA, 47, PhD(biochem), 52. Prof Exp: Asst biochem, Univ Calif, 50-52; res assoc, Med Col, Cornell Univ, 52-54; chemist, Nat Inst Arthritis & Metab Dis, 54-66, chief, Europ Off, Paris, 66-68, chief, Conf & Sem Prog Br, John E Fogerty Int Ctr Advan Study Health Sci, 68-73, RES BIOCHEMIST & HORMONE DISTRIB OFFICER, LAB NUTRIT & ENDOCRINOL, NAT INST ARTHRITIS, METAB & DIGESTIVE DIS, NIH, BETHESDA, 73- Concurrent Pos: Lectr, USDA Grad Sch, Washington, DC, 55-; fel, Nat Found Carlsberg Lab, Copenhagen, 59-60; vis res assoc comp physiol lab, Mus Natural Hist, Paris, 66-68. Mem: AAAS; Am Soc Biol Chemists; Endocrine Soc. Res: Biochemical endocrinology; pituitary biochemistry; chemistry of pituitary hormones; reproductive biology; social implications of biomedical research; ethical issues in biology and medicine; international aspects of research in biomedicine. Mailing Add: Lab Nutrit & Endocrinol Nat Inst Arthrit Metab & Digest Dis NIH Bethesda MD 20014

CONDO, ALBERT CARMAN, JR, b Hackensack, NJ, May 25, 24; m 47; c 3. PETROLEUM CHEMISTRY. Educ: Cornell Univ, BS, 49, MS, 51. Prof Exp: Instr anal chem, Cornell Univ, 49-51; from res chemist to sr chemist, 51-64, dir plastics develop, 64-69, mgr protective eng systs, 69-72, SR TASK MGR ARCTIC-CIVIL ENG, ATLANTIC RICHFIELD CO, 72- Concurrent Pos: From instr to asst prof, eve col, Drexel Univ, 54-72. Mem: Am Chem Soc; Soc Plastics Eng; Nat Asn Corrosion Eng; Sigma Xi; Soil Conserv Soc Am. Res: Petrochemicals; polymerization; plastics; coatings and corrosion control; protective environmental systems; insulated roads on permafrost; hydraulic and thermal erosion control; restoration; revegetation; arctic thermal regimes; oil spill cleanup. Mailing Add: 3424 Ivy Lane Newton Square PA 19073

CONDO, GEORGE T, b East St Louis, Ill, May 14, 34; m 58; c 2. HIGH ENERGY PHYSICS. Educ: Univ Ill, BS, 56, MS, 57, PhD(physics), 62. Prof Exp: Res assoc physics, Univ Ill, 62-63; asst prof, 63-70, ASSOC PROF PHYSICS, UNIV TENN, 70- Concurrent Pos: Consult, Oak Ridge Nat Lab, 63- Mem: Am Phys Soc. Res: Elementary particles using nuclear research emulsions and bubble chamber technique. Mailing Add: Dept of Physics Univ of Tenn Knoxville TN 37916

CONDON, EDWARD UHLER, physics, deceased

CONDON, FRANCIS EDWARD, b Abington, Mass, Oct 12, 19; m 43; c 7. CHEMISTRY. Educ: Harvard Univ, AB, 41, AM, 43, PhD(org chem), 44. Prof Exp: Res chemist, Phillips Petrol Co, 44-52; from asst prof to assoc prof, 52-67, PROF CHEM, CITY COL NEW YORK, 67- Concurrent Pos: NSF fac fel, Univ Southern Calif, 64-65. Res: Hydrocarbon chemistry; structure-reactivity correlations; hydration and base strength; hydrazines; field effects in electrophilic aromatic substitution. Mailing Add: 471 Larch Ave Bogota NJ 07603

CONDON, JAMES BENTON, b Buffalo, NY, Aug 20, 40; c 1. PHYSICAL CHEMISTRY, SURFACE CHEMISTRY. Educ: State Univ NY Binghamton, AB, 62; Iowa State Univ, PhD(phys chem), 68. Prof Exp: DEVELOP CHEMIST, NUCLEAR DIV, UNION CARBIDE CO, 68- Mem: Am Chem Soc. Res: Catalysis; electrochemistry; corrosion. Mailing Add: 511 Robertsville Rd Oak Ridge TN 37830

CONDON, ROBERT EDWARD, b Albany, NY, Aug 13, 29; m 51; c 2. SURGERY, PHYSIOLOGY. Educ: Univ Rochester, AB, 51, MD, 57; Univ Wash, MS, 65; Am Bd Surg, dipl, 66. Prof Exp: Asst resident & intern, Univ Wash Hosps, Seattle, 57-63, chief resident surg, 63-65, res assoc, Sch Med, Univ Wash, 59-61, instr, 61-65; asst prof, Col Med, Baylor Univ, 65-67; from assoc prof to prof, Col Med, Univ Ill, Chicago, 67-71; prof surg & head dept, Col Med, Univ Iowa, 71-72; PROF SURG, MED COL WIS, 72- Concurrent Pos: Hon clin asst, Royal Free Hosp, London, 63-64; asst chief surg, Vet Admin Hosp, Houston, Tex, 65-67; attend surgeon, Univ Ill Hosp, 67-71; chief surg serv, Univ Iowa Hosp, 71-72 & Vet Admin Hosp, Wood, Wis, 72- Mem: Am Col Surg; Soc Univ Surg; Am Surg Asn; Soc Surg Alimentary Tract; Soc Clin Surg. Res: Gastric and intestinal physiology; hernia. Mailing Add: 8700 W Wisconsin Ave Milwaukee WI 53226

CONDOULIS, WILLIAM V, b Brooklyn, NY, Oct 20, 39. IMMUNOLOGY, DEVELOPMENTAL BIOLOGY. Educ: Univ Va, BS, 61; Western Reserve Univ, MS, 64; Univ Notre Dame, PhD(biol), 67. Prof Exp: Res technician histol, Univ Va, 59-61; res asst, Univ Notre Dame, 65-67; res assoc allergy, Michael Reese Hosp, 67-68; RES ASSOC IMMUNOL, AM DENT ASN, 68- Mem: AAAS; NY Acad Sci. Res: Delayed hypersensitivity; arthropod development; insect endocrinology. Mailing Add: Am Dent Asn 211 E Chicago Ave Chicago IL 60611

CONDOURIS, GEORGE ANTHONY, b Passaic, NJ, Dec 9, 25; m 49; c 5. PHARMACOLOGY. Educ: Rutgers Univ, BS, 49; Yale Univ, MS, 53; Cornell Univ, PhD(pharmacol), 55. Prof Exp: Instr, Med Col, Cornell Univ, 56-57; vis investr, Rockefeller Inst, 57; from asst prof to assoc prof, 57-67, PROF PHARMACOL, COL MED & DENT NJ, 67-, CHMN DEPT, 72- Concurrent Pos: NSF fel, Med Col, Cornell Univ, 54-55, res fel, 55-56. Mem: AAAS; Am Soc Pharmacol & Exp Therapeut. Res: Pharmacology of the nervous system; biometrics. Mailing Add: NJ Med Sch Col of Med & Dent of NJ Newark NJ 07103

CONDRATE, ROBERT ADAM, b Worcester, Mass, Jan 19, 38; m 60; c 2. SOLID STATE CHEMISTRY. Educ: Worcester Polytech Inst, BS, 60; Ill Inst Technol, PhD(chem), 66. Prof Exp: NSF fel & res assoc chem, Univ Ariz, 66-67; asst prof, 67-71, ASSOC PROF SPECTROS, ALFRED UNIV, 71- Concurrent Pos: Finger Lakes grant-in-aid, 69; vis prof, Los Alamos Sci Lab, 72-73; Corning Glass Works Found grant-in-aid, 75-76. Honors & Awards: Award, Soc Appl Spectros, 64. Mem: AAAS; Am Chem Soc; Soc Appl Spectros; Am Phys Soc; Am Ceramic Soc. Res: Application of spectroscopy to elucidate the structure of molecules in solids. Mailing Add: Dept of Ceramic Sci Alfred Univ Alfred NY 14802

CONDRAY, BEN ROGERS, b Waco, Tex, July 4, 25; m 51; c 3. ORGANIC CHEMISTRY. Educ: Baylor Univ, BS, 48, PhD(org chem), 64; Purdue Univ, MS, 50. Prof Exp: From asst prof to assoc prof chem, ETex Baptist Col, 50-55; instr, Baylor Univ, 55-57; assoc prof, 58-65, PROF CHEM, E TEX BAPTIST COL, 65- Mem: AAAS; Am Chem Soc. Res: Synthesis of organosilicon compounds; separation and identification of natural products; organic-ozone chemistry and stabilities of organic ozonides; charge-transfer complexes. Mailing Add: Dept of Chem ETex Baptist Col Marshall TX 75670

CONE, CLARENCE DONALD, JR, b Savannah, Ga, Apr 17, 31; m 54; c 2. CELL BIOLOGY, ONCOLOGY. Educ: Ga Inst Technol, BChE, 54; Univ Va, MAeronE, 59; Med Col Va, PhD(biophys), 65. Chem engr, Buckeye Cellulose Corp, 54-55; res chemist, Herty Found Lab, 55-57; res aerodynamicist, Nat Adv Comt Aeronaut, 57-59; head hypersonic res, NASA, 59-61, head subsonic theory res, 61-64, res biophysicist, 64-65, head molecular biophys res, 65-72; dir molecular biol lab, Eastern Va Med Authority, 72-74; DIR CELLULAR & MOLECULAR BIOL LAB, VET ADMIN, 75- Concurrent Pos: Res assoc, Va Inst Marine Sci, 62-68 & Smithsonian Inst, 64-66; mem, Third Int Cong, Int Tech & Sci Orgn Soaring Flight, 63. Honors & Awards: Medal, Except Sci Achievement, NASA, 69. Mem: AAAS; Sigma Xi; Tissue Cult Asn; Biophys Soc; Am Soc Microbiol. Res: Molecular mechanisms of mitogenesis regulation; cytogenetic regulation; carcinogenesis mechanisms. Mailing Add: 104 Harbour Dr Yorktown VA 23690

CONE, CONRAD, b Wellington, NZ, Dec 3, 39; US citizen. ORGANIC CHEMISTRY, MASS SPECTROMETRY. Educ: Bristol Univ, BSc, 62, PhD(org chem), 66. Prof Exp: Fel, Mass Inst Technol, 65-68; fac assoc, 68-70, assoc prof, 70-73, RES SCIENTIST ORG CHEM, UNIV TEX, AUSTIN, 73- Mem: Am Chem Soc; Am Soc Mass Spectrometry; The Chem Soc. Res: Mass spectrometry of organic compounds; negative ion mass spectrometry; laboratory computers. Mailing Add: Dept of Chem Univ of Tex Austin TX 78712

CONE, EDWARD JACKSON, b Mobile, Ala, Sept 17, 42; m 63; c 2. DRUG METABOLISM. Educ: Mobile Col, BS, 67; Univ Ala, PhD(org chem), 71. Prof Exp:

Lab instr chem, Mobile Col, 65-69; chemist, Shell Oil Co, 69-71; instr chem, Univ Ala, 71; fel tobacco chem, Univ Ky, 71-72; CHEMIST, NAT INST DRUG ABUSE ADDICTION RES CTR, 72- Concurrent Pos: Asst prof, Sch Pharm, Univ Ky, 75- Mem: Am Chem Soc. Res: Studies on biotransformation of drugs of abuse in man and other animal species; concomitant development of analytical procedures useful for detection, isolation and quantification of drugs and their metabolites. Mailing Add: NIDA Addiction Res Ctr PO Box 12390 Lexington KY 40511

CONE, RICHARD ALLEN, b St Paul, Minn, May 23, 36. BIOPHYSICS. Educ: Mass Inst Technol, SB, 58; Univ Chicago, SM, 59, PhD(physics), 63. Prof Exp: Res assoc biophys, Univ Chicago, 63-64; from instr to asst prof biol, Harvard Univ, 64-69; assoc prof, 69-73, PROF BIOPHYS, JOHNS HOPKINS UNIV, 73- Concurrent Pos: USPHS fel, Univ Chicago, 64. Res: Mechanism of visual excitation; neural organization of the retina; membrane structure and function. Mailing Add: Dept of Biophys Johns Hopkins Univ Baltimore MD 21218

CONE, ROBERT EDWARD, b Brooklyn, NY, Aug 18, 43; m 66; c 2. IMMUNOLOGY. Educ: Brooklyn Col, BS, 64; Fla State Univ, MS, 67; Univ Mich, Ann Arbor, PhD(microbiol), 70. Prof Exp: Fel immunol, Walter & Eliza Hall Inst Med Res, Melbourne, Australia, 71-73; immunol & path mem, Basel Inst Immunol, Switz, 73-74; ASST PROF PATH & SURG, SCH MED, YALE UNIV, 74- Concurrent Pos: Damon Runyon Cancer Fund fel, 70. Mem: Sigma Xi; AAAS; Am Soc Microbiol. Res: Structural and physiological properties of lymphocyte membranes; relationship of membrane structure to lymphocyte function. Mailing Add: Dept of Surg Yale Univ Sch of Med New Haven CT 06510

CONE, THOMAS E, JR, b Brooklyn, NY, Aug 15, 15; m 39; c 3. PEDIATRICS. Educ: Columbia Univ, BA, 36, MD, 39. Prof Exp: Chief pediat serv, US Naval Hosp, Philadelphia, Pa, 49-53; chief pediat serv, Nat Naval Med Ctr, Bethesda, Md, 53-63; chief med out-patient serv & asst dir adolescent unit, 63-73, chief med ambulatory serv & sr assoc med, 67-73, SR ASSOC CLIN GENETICS, CHILDREN'S HOSP MED CTR, 71- Concurrent Pos: Sr consult, Children's Hosp, Washington, DC, 53-63; consult, Nat Inst Neurol Dis & Blindness, 58-63; mem nat adv coun, Nat Inst Child Health & Human Develop, 63-64; assoc clin prof pediat, Harvard Med Sch, 63-67, clin prof, 67-; prof pediat, Southwest Med Sch, Univ Tex, Dallas; assoc ed, Pediatrics, 65-74. Honors & Awards: Officer, Order Naval Merit, Repub Brazil. Mem: AMA; fel Am Acad Pediat; Am Pediat Soc; NY Acad Sci; Am Asn Hist Med. Res: Physical growth and development of children; history of pediatrics; clinical pediatrics; adolescent medicine. Mailing Add: 300 Longwood Ave Boston MA 02115

CONE, WYATT WAYNE, b Plains, Mont, Mar 16, 34; m 56; c 3. ENTOMOLOGY. Educ: San Diego State Col, AB, 56; Wash State Univ, PhD(entom), 62. Prof Exp: From jr entomologist to assoc entomologist, 61-72, ENTOMOLOGIST, WASH STATE UNIV, 72- Mem: Entom Soc Am; Entom Soc Can. Res: Insecticide effect on predator-prey relationship in mite populations; biology and control of Brachyrhinus weevils on grapes; ecology of arthropods in native grasslands; biology and control of two-spotted spider mites, two-spotted spider mite sex attractant and pheromones. Mailing Add: Irrigated Agr Res & Exten Ctr Prosser WA 99350

CONEN, PATRICK E, b London, Eng, Mar 5, 28. PATHOLOGY. Educ: Univ London, MB, BS, 51, MD, 55; Royal Col Physicians & Surgeons, Can, cert, 60. Prof Exp: House physician, Southern Hosp, Dartford, Eng, 51-52; house surgeon, Cent Middlesex Hosp, 52, sr house officer, 52-53; sr house officer, St Margaret's Hosp, Essex, 53-54, St George's Hosp, London, 54-55 & St Anne's & Prince of Wales Gen Hosps, 55-56; registr, St George's Hosp, London, 56-57, asst pathologist, 57-59; res fel path, 59-61, RES ASSOC PATH, HOSP SICK CHILDREN, TORONTO, 61- Concurrent Pos: Res assoc, Univ Toronto, 61-, asst prof, 65- Mem: Am Asn Pathologists & Bacteriologists; Am Soc Exp Path; Am Asn Cancer Res; Electron Micros Soc; Am Soc Human Genetics. Res: Human chromosome studies of tumors, leukemias and congenital malformations; electron microscopy of human biopsies, especially liver, muscle and kidney and the ultrastructure of developing human tissues. Mailing Add: Dept of Path Hosp for Sick Children Toronto ON Can

CONFER, JOHN L, b Dayton, Ohio, Sept 15, 40. AQUATIC ECOLOGY, POPULATION ECOLOGY. Educ: Earlham Col, BA, 62; Wash State Univ, MS, 64; Univ Toronto, PhD(zool), 69. Prof Exp: Asst prof zool, Univ Fla, 69-70; ASST PROF BIOL, ITHACA COL, 70- Res: Phosphorus circulation in lakes; regulation of aquatic populations. Mailing Add: Dept of Biol Ithaca Col Ithaca NY 14850

CONGDON, CHARLES C, b Dunkirk, NY, Dec 13, 20; m 47; c 5. PATHOLOGY. Educ: Univ Mich, AB, 42, MD, 44, MS, 50. Prof Exp: Instr path, Univ Mich, 49-51; vis scientist, Nat Cancer Inst, NIH, 51-52, med officer path, 52-55; sr biologist, Biol Div, Oak Ridge Nat Lab, 55-73; part-time prof, Univ Tenn, Knoxville, 66-73, RES PROF & ASST DIR, UNIV TENN MEM RES CTR, 73- Mem: AAAS; Am Soc Exp Path; Am Asn Cancer Res; Am Asn Path & Bact; Soc Exp Biol & Med. Res: Biomedical research administration. Mailing Add: Univ of Tenn Mem Res Ctr 1924 Alcoa Hwy Knoxville TN 37920

CONGEL, FRANK JOSEPH, b Syracuse, NY, Mar 6, 43; m 65; c 3. NUCLEAR PHYSICS. Educ: LeMoyne Col, BS, 64; Clarkson Col Technol, MS, 67, PhD(physics), 68. Prof Exp: Asst, Argonne Nat Lab, 68-69; asst prof physics, Macalester Col, 69-72; ENVIRON PHYSICIST, NUCLEAR REGULATORY COMN, 72- Mem; Am Phys Soc. Res: Measurement of cosmic ray fluences; calculation of individual and integrated population doses due to radionuclides dispersed in the environment. Mailing Add: 10515 Sweepstakes Rd Damascus MD 20750

CONGER, ALAN DOUGLAS, b Muskegon, Mich, Mar 23, 17; m 44; c 4. RADIOBIOLOGY, BIOPHYSICS. Educ: Harvard Univ, AB, 40, MA & PhD(biol), 47. Prof Exp: Sr res scientist biol, Oak Ridge Nat Lab, USAEC, 47-58; res prof radiobiol, Univ Fla, 58-65; PROF RADIOBIOL & HEAD DEPT, SCH MED, TEMPLE UNIV, 65- Concurrent Pos: Fulbright sr res scholar, Radiobiol Res Unit, Med Res Coun, London, Eng, 52-53; USAEC grant nuclear sci, Univ Fla, 58-65; Nat Cancer Inst fel radiobiol, Sch Med, Temple Univ, 65-; sci proj officer, Oper Greenhouse Atomic Bomb Tests, Marshall Islands, 51-52; assoc ed, Radiation Res, 58-62, Radiation Bot, 61- & Mutation Res, 62-; consult biol, Oak Ridge & Brookhaven Nat Labs, USAEC, 58-; mem radiation study sect, NIH, 63-66; chmn panel low dose report, Nat Acad Sci-Nat Res Coun, 72-73, mem subcomt radiobiol, 72- Mem: Genetics Soc Am; Am Soc Nat; Radiation Res Soc (vpres, pres, 71-73). Res: Effects of radiation on cells; genetics effects of radiation. Mailing Add: Dept of Radiation Biol Temple Univ Sch of Med Philadelphia PA 19140

CONGER, BOB VERNON, b Greeley, Colo, July 2, 38; m 60; c 4. PLANT GENETICS. Educ: Colo State Univ, BS, 63; Wash State Univ, PhD(genetics), 67. Prof Exp: Asst prof genetics, Wash State Univ, 67-68; asst prof agron, Univ Tenn, 68-73, ASSOC PROF, UNIV TENN-ENERGY RES DEVELOP ADMIN COMP ANIMAL RES LAB, 73- Mem: AAAS; Am Soc Agron; Am Genetic Asn; Genetics Soc Am; Radiation Res Soc. Res: Mutagenic effectiveness, efficiency and specificity; nature of induced mutations in higher plants. Mailing Add: 723 Robertsville Rd Oak Ridge TN 37830

CONGER, KYRIL BAILEY, b Berlin, Ger, Apr 11, 13; US citizen; m 46; c 4. SURGERY. Educ: Univ Mich, AB, 33, MD, 36; Am Bd Urol, dipl, 47. Prof Exp: Consult urol, US Army Mid Pac Area & Tripler Gen Hosp, Honolulu, 46; PROF UROL & HEAD DEPT, HOSP & MED SCH, TEMPLE UNIV, 47- Concurrent Pos: Consult, Vet Admin Hosp, Philadelphia & Mid-Atlantic Area, Vet Admin; attend urologist, St Christopher's Hosp for Children. Mem: Am Urol Asn; fel Am Col Surg. Res: Urological and prostatic surgery. Mailing Add: 3401 N Broad St Philadelphia PA 19148

CONGER, ROBERT LYNN, physics, see 12th edition

CONGER, ROBERT PERRIGO, b Youngstown, Ohio, Dec 1, 22; m 48; c 2. PLASTICS CHEMISTRY. Educ: Cornell Univ, AB, 43, PhD(chem), 50. Prof Exp: Lab asst, Cornell Univ, 43-44, 46-50; instr chem, Col Med, Univ Tenn, 50-51; res chemist, Gen Labs, US Rubber Co, 51-61; MGR RES, CONGOLEUM INDUSTS, INC, 61- Mem: AAAS; Am Chem Soc; Soc Plastics Eng. Res: Foamed plastics, especially polyvinyl chloride; chemical embossing; clear films of elastomers such as polyvinyl chloride and polyurethane; compounding and modification of various elastomers. Mailing Add: 87 Oak Ave Park Ridge NJ 07656

CONGER, THEODORE WILLIAM, biochemistry, see 12th edition

CONGLETON, JAMES LEE, b Lexington, Ky, Dec 20, 42; c 1. PHYSIOLOGICAL ECOLOGY, FISHERIES. Educ: Univ Ky, BS, 64; Univ Calif, San Diego, PhD(marine biol), 70. Prof Exp: NIH fel, Univ Wash, 70-71; biologist aquaculture, Kramer, Chin & Mayo, Inc, 71-75; ASST PROF FISHERIES, UNIV WASH, 75- Mem: Am Fisheries Soc. Res: Fish respiratory physiology and bioenergetics. Mailing Add: Coop Fisheries Res Unit Univ of Wash Seattle WA 98195

CONIGLIARO, PETER JAMES, b Milwaukee, Wis, Jan 27, 42; m 65; c 3. ORGANIC CHEMISTRY, CHEMICAL INSTRUMENTATION. Educ: Marquette Univ, BS, 63, MS, 65; Ohio State Univ, PhD(org chem), 67. Prof Exp: Res chemist, 68-70, SR RES CHEMIST, S C JOHNSON & SON, 70- Mem: Am Chem Soc. Res: Specialized organic analysis; analysis of chemical specialty consumer products. Mailing Add: S C Johnson & Son 1525 Howe St Racine WI 53403

CONIGLIO, JOHN GIGLIO, b Tampa, Fla, July 21, 19; m 42; c 3. BIOCHEMISTRY. Educ: Furman Univ, BS, 40; Vanderbilt Univ, PhD(biochem), 49. Prof Exp: From instr to assoc prof, 51-63, PROF BIOCHEM, SCH MED, VANDERBILT UNIV, 63- Concurrent Pos: AEC fel biophys, Colo Med Ctr, 49-50; AEC fel biochem, Sch Med, Vanderbilt Univ, 50-51; assoc ed, Lipids, 69- Mem: AAAS; Am Inst Nutrit; Am Oil Chem Soc; Am Soc Biol Chem; Soc Exp Biol & Med. Res: Fat absorption and distribution acetate utilization and fatty acid synthesis; effects of x-irradiation on fat absorption and on fatty acid synthesis; determination and metabolism of essential fatty acids; use of tracers in metabolism; interconversion of fatty acids; lipids in reproductive tissue. Mailing Add: Dept of Biochem Vanderbilt Med Sch Nashville TN 37232

CONINE, JAMES WILLIAM, b Newton, Iowa, May 13, 26; m 56; c 4. PHARMACEUTICAL CHEMISTRY. Educ: Univ Iowa, BS, 50, MS, 52, PhD(pharmaceut chem), 54. Prof Exp: Instr pharm, Univ Iowa, 52-54; PHARMACEUT CHEMIST, ELI LILLY & CO, 54- Mem: AAAS; Am Chem Soc; Am Pharmaceut Asn. Res: Stability of pharmaceutical products; product development of tablets and capsules. Mailing Add: Eli Lilly & Co 307 E McCarty St Indianapolis IN 46206

CONKIE, WILLIAM R, b Ayr, Scotland, Jan 10, 32; m 54; c 2. PHYSICS. Educ: Univ Toronto, BASc, 53; McGill Univ, MSc, 54; Univ Sask, PhD, 56. Prof Exp: Asst res officer physics, Atomic Energy Can, Ltd, 56-60; PROF PHYSICS, QUEEN'S UNIV, ONT, 60- Mem: Am Phys Soc; Can Asn Physicists. Res: Theoretical physics. Mailing Add: Dept of Physics Queen's Univ Kingston ON Can

CONKIN, JAMES E, b Glasgow, Ky, Oct 14, 24; m 51; c 4. GEOLOGY, PALEONTOLOGY. Educ: Univ Ky, BS, 50; Univ Kans, MS, 53; Univ Cincinnati, PhD(geol), 60. Prof Exp: Paleontologist, Union Producing Co Div, United Gas Corp, 53-56; instr geol, Univ Cincinnati, 56-57; from instr to asst prof natural sci, 57-63, assoc prof geol, 63-67, PROF GEOL, UNIV LOUISVILLE, 67-, CHMN DEPT, 63- Concurrent Pos: Fulbright res fel micropaleont, Tasmania, 64-65; mem, Paleont Res Inst. Mem: Paleont Soc; fel Geol Soc Am. Res: Paleozoic Foraminifera; paleozoic stratigraphy, the Mississippian system and Devonian bone beds. Mailing Add: Dept of Geol Univ of Louisville Louisville KY 40208

CONKIN, ROBERT A, b Green City, Mo, Dec 31, 20; m 46; c 1. ANALYTICAL CHEMISTRY, ORGANIC CHEMISTRY. Educ: Northeast Mo State Teachers Col, BS, 42, AB, 46. Prof Exp: Analyst, Ill, 46-47, lab supvr, 47-50, chief chemist, Ala, 50-53, res chemist, MO, 53-62, RES GROUP LEADER, MONSANTO CO, 62- Mem: Am Chem Soc. Res: Development and application of micro-analytical methods and techniques in the field of agricultural pesticide residues in or on fram produce and products. Mailing Add: Monsanto Co 800 N Lindbergh Blvd St Louis MO 63166

CONKLIN, EDWARD KIRKHAM, b Hartford, Conn, Sept 25, 41. RADIO ASTRONOMY. Educ: Yale Univ, BSEE, 64; Stanford Univ, MSEE, 65, PhD(radio astron), 69. Prof Exp: Asst scientist & head, Tucson Div, Nat Radio Astron Observ, 69-73; RES ASSOC, NAT ASTRON & IONOSPHERE CTR, ARECIBO OBSERV, 73- Mem: AAAS; Inst Elec & Electronics Eng; Am Astron Soc. Res: Cosmic microwave background; spectra and variability of millimeter-wave sources; radio emission from elliptical galaxies; low-frequency variable radio sources. Mailing Add: Nat Astron & Ionosphere Ctr Arecibo Observ PO Box 995 Arecibo PR 00612

CONKLIN, GLENN ERNEST, b Lyndon, Kans, June 2, 29; m 57; c 2. PHYSICS. Educ: Univ Wichita, BA, 51, MS, 53; Univ Kans, PhD(physics), 61. Prof Exp: Mem tech staff physics, Bell Tel Labs, 60-66; res scientist, Singer-Gen Precision, Inc, 66-70; radiation physicist, Off Criteria & Stand, 70-74, STAND CONSULT, BUR RADIOL HEALTH, FOOD & DRUG ADMIN, 74- Mem: Am Inst Physics; Inst Elec & Electronics Eng. Res: Dielectric behavior of plastics at millimeter wavelengths; application of optically pumped nuclear magnetic resonance to communications; nonionizing radiation. Mailing Add: Bur of Radiol Health Food & Drug Admin 12721 Twinbrook Pkwy Rockville MD 20852

CONKLIN, HAROLD COLYER, b Easton, Pa, Apr 27, 26; m 54; c 2. ANTHROPOLOGY, ETHNOGRAPHY. Educ: Univ Calif, Berkeley, AB, 50; Yale Univ, PhD(anthrop), 55. Prof Exp: From instr to assoc prof anthrop, Columbia Univ, 54-62; chmn dept, 64-68, PROF ANTHROP, YALE UNIV, 62- Concurrent Pos: NSF grant, Columbia Univ grant & Yale Univ grant, Philippines, Malaysia, Columbia Univ & Yale Univ, 56-; Sigma Xi Coun res grant, 62; lectr, Rockefeller Inst, 61-62;

mem, Pac Sci Bd, Nat Acad Sci-Nat Res Coun, 62-66 & Probs & Policy Comt & bd dirs, Sigma Xi Coun, 64-70. Mem: Nat Acad Sci; Am Anthrop Asn; Ling Soc Am; NY Acad Sci; Soc Am Archaeol. Res: Ethnography; linguistics; cultural ecology; folk classification; semantics. Mailing Add: Dept of Anthrop Yale Univ New Haven CT 06520

CONKLIN, JAMES BYRON, JR, b Charlotte, NC, July 29, 37; m 62; c 2. SOLID STATE PHYSICS, ACADEMIC ADMINISTRATION. Educ: Mass Inst Technol, SB, 59, SM, 61, ScD(solid state physics), 64. Prof Exp: Asst prof, 64-71, ASSOC PROF PHYSICS, UNIV FLA, 71-, DIR CTR INSTRNL & RES COMPUT ACTIV, 75- Concurrent Pos: Assoc dir, Northeast Regional Data Ctr, State Univ Syst Fla, 73-74. Mem: AAAS; Asn Comput Mach; Am Asn Physics Teachers; Am Phys Soc. Res: Computers as instructional tools; numerical mathematics; calculation of energy band structure and related electronic properties of semiconductors. Mailing Add: Ctr Instrnl & Res Comput Activ Univ of Fla Gainesville FL 32611

CONKLIN, JAMES L, b Owosso, Mich, Oct 30, 28; m 50; c 4. ANATOMY. Educ: Albion Col, AB, 57; Univ Mich, MS, 59, PhD(anat), 61. Prof Exp: Lectr embryol, Med Sch, Univ Mich, 60, from instr to assoc prof, 67-71, PROF ANAT & ASSOC DEAN STUDENT AFFAIRS, MICH STATE UNIV, 71- Concurrent Pos: Res grants, Instnl Cancer Comt & NIH, 61- Mem: AAAS; Histochem Soc; Am Asn Anatomists; Health Sci Commun Asn. Res: Developmental anatomy; histochemical and biochemical studies of enzyme development; instructional technology; medical education. Mailing Add: Off of Student Affairs Mich State Univ East Lansing MI 48823

CONKLIN, JOHN DOUGLAS, b Middletown, NY, Mar 1, 33; m 59; c 2. BIOPHARMACEUTICS. Educ: Col Holy Cross, BSc, 56. Prof Exp: From jr res biochemist to res biochemist, 59-65, SR RES BIOCHEMIST, DRUG DISTRIB UNIT, PHARMACOMETRICS DIV, NORWICH PHARMACAL CO, 65- Mem: Acad Pharmaceut Sci; Am Pharmaceut Asn. Res: The study of the bioavailability of drugs in man and animals by investigating drug absorption, distribution and excretion to optimize drug pharmacologic or therapeutic activity for clinical application. Mailing Add: Drug Distrb Unit Pharmacomet Div Norwich Pharm Co Norwich NY 13815

CONKLIN, MARIE ECKHARDT, b Derby, Conn, Sept 30, 08; m 31; c 2. BOTANY, GENETICS. Educ: Wellesley Col, AB, 29; Univ Wis, MS, 30; Columbia Univ, PhD(bot), 36. Prof Exp: Asst bact, Wellesley Col, 30-31; vol res, Brooklyn Bot Garden, 35-36; asst, Carnegie Inst, 37-41; instr bact, genetics & bot, 43-46, from asst prof to prof biol, 46-74, head dept, 53-67, EMER PROF BIOL, ADELPHI UNIV, 74- Concurrent Pos: Dir serv prog for high sch teachers sci, 58-64; res collabr, Brookhaven Nat Labs. Mem: AAAS; Sigma Xi; Genetics Soc Am; Bot Soc Am; Radiation Res Soc. Res: Dir, Serv Prog for High Sch Teachers Sci, 58-64; res collabr, Brookhaven Nat Labs. Res: Parthenogenesis and embryo culture of Datura; radiation effects, isozme production and plant tumorization and cytogenetics. Mailing Add: Dept of Biol Adelphi Univ Garden City NY 11530

CONKLIN, RICHARD LOUIS, b Rockford, Ill, Dec 9, 23; m 50; c 4. PHYSICAL OPTICS. Educ: Univ Ill, BS, 44, MS, 48; Univ Colo, PhD(solid state physics), 57. Prof Exp: Jr scientist, Los Alamos Lab, Univ Calif, 44-46; asst physics, Univ Ill, 46-49; asst prof, Huron Col, 49-53; instr, Univ Colo, 53-57; assoc prof, 57-58, PROF PHYSICS, HANOVER COL, 58- Concurrent Pos: Vis colleague, Univ Hawaii, 67-68. Mem: Am Asn Physics Teachers; Am Phys Soc. Res: Electron accelerators and their application to nuclear physics; luminescence of solids. Mailing Add: Dept of Physics Hanover Col Hanover IN 47243

CONKLING, EDGAR CLARK, b Marion, Ind, Mar 29, 21; m 52. GEOGRAPHY. Educ: Morehead State Col, BA, 43; Univ Chicago, MA, 57; Northwestern Univ, MS, 60, PhD(geog), 62. Prof Exp: Export traffic mgr, Signode Int, Ill, 52-58; asst prof econ geog, Kent State Univ, 61-64; assoc prof, Queen's Univ, Ont, 64-68; PROF GEOG, STATE UNIV NY BUFFALO, 68-, CHMN DEPT, 74- Mem: Asn Am Geog; Regional Sci Asn; Am Geog Soc; Coun Latin Am Geogrs; AAAS. Res: Diversification of industry; international trade and economic development; land use theory. Mailing Add: Dept of Geog 4224 Ridge Lea Rd State Univ of NY Buffalo Amherst NY 14226

CONKLING, RANDALL MURRAY, mathematics, see 12th edition

CONLAN, JAMES, b San Francisco, Calif, Apr 15, 23; m 46. MATHEMATICS. Educ: Univ Calif, BA, 45; Univ Md, PhD(math), 58. Prof Exp: Mathematician, Aberdeen Proving Ground, 50-53 & Naval Ord Lab, 53-67; assoc prof math, Howard Univ, 67-68; PROF MATH, UNIV REGINA, 68- Mem: Am Math Soc. Res: Partial differential equations; fluid mechanics. Mailing Add: Dept of Math Univ of Regina Regina SK Can

CONLEY, BERNARD EDWARD, b Medina, NY, Jan 28, 19; m 45; c 4. PHARMACOLOGY, TOXICOLOGY. Educ: Duquesne Univ, BS, 42; Univ Chicago, SM, 51, PhD(pharmacol), 56. Prof Exp: Regional chief pharm serv, Vet Admin, Ohio, Mich & Ky, 46-48; admin assoc, AMA, 48-50, secy comt pesticides, 50-59, secy comt toxicol, 55-59, dir toxicol, 59-60; pharmacologist, US Pharmacopeia, 61-64; pharmacologist, Nat Inst Neurol Dis & Blindness, 64-66; chief commun studies prog, Off Pesticides, 66-67; chief air qual criteria, Nat Ctr Air Pollution Control, 67-69, dir grants & contracts rev br, 69-71, CHIEF DRUG UTILIZATION STUDIES, NAT CTR HEALTH SERV RES, USPHS, 71- Concurrent Pos: Tech ed, New & Nonofficial Remedies, 48-50; AMA rep, Food Protection Comt & Chem Comt, Nat Res Coun, 50-60; contrib ed, Am J Hosp Pharm, 50-60; mem comt adverse reactions drugs, US Food & Drug Admin, 56-60; sci consult, Nat Asn Practical Nurse Educ Serv, 62- Mem: AAAS; fel Am Pub Health Asn; Am Pharmaceut Asn; Am Soc Hosp Pharmacists; Am Pub Health Asn. Res: Biomedical and legal problems of hazardous substances; socio-economic aspects of drug utilization research. Mailing Add: Nat Ctr for Health Serv Res Parklawn Bldg Rockville MD 29852

CONLEY, CARROLL LOCKARD, b Baltimore, Md, May 14, 15; m 43; c 2. MEDICINE. Educ: Johns Hopkins Univ, AB, 35; Columbia Univ, MD, 40. Prof Exp: Fel physiol, Univ Md, 36-37; intern & asst med, Presby Hosp, New York, 40-42; asst, 46-47, from instr to assoc prof, 47-56, PROF MED, SCH MED, JOHNS HOPKINS UNIV, 56-, PHYSICIAN IN CHARGE HEMAT DIV, 47- Concurrent Pos: Mem hematol study sect, NIH, 52-56, chmn, 62-65; mem comt on blood coagulation, Nat Res Coun, 54-59, mem comt blood related probs, 59-63, mem comt thrombosis & hemorrhage, 61-64, chmn, 62-64; secy anti-anemia preparations adv bd, US Pharmacopoeia, 54-59; hon assoc prof, Guys Hosp Med Sch, London, 57; consult, US Food & Drug Admin, 67-; mem NIH Arthritis & Metab Dis Prog Proj Comt, 67-71; mem, Comt to Rev Life Sci Prog of NASA, Nat Acad Sci, 70; mem sickle cell anemia adv comt, Dept Health, Educ & Welfare, 71-73; consult, US Army, US Vet Admin & USPHS. Mem: Am Soc Hemat (pres, 75-76); AMA; Am Soc Clin Invest; Soc Exp Biol & Med; fel Am Col Physicians. Res: Hematology; blood coagulation; hemoglobin. Mailing Add: Dept of Med Johns Hopkins Hosp Baltimore MD 21205

CONLEY, CECIL, b Tomahawk, Ky, June 1, 22; m 64. BIOCHEMISTRY, AGRICULTURE. Educ: Univ Ky, BS, 48, MS, 50; NC State Col, PhD(biochem), 54. Prof Exp: Exten agent, Ky, 55-56; asst res scientist, Clemson Col, 56-63, statistician, Exp Sta, 60-62; chmn dept sci & math, Pembroke State Col, 63-68; prof chem & chmn div sci & math, Livingston Univ, 68-70; prof chem, 70-72, exec dean, 72-75, VPRES, PALM BEACH JR COL, 75- Res: Protein bound iodine; iron and copper requirements; irrigation. Mailing Add: 1041 S E Second St Belle Glade FL 33430

CONLEY, CHARLES CAMERON, b Royal Oak, Mich, Sept 26, 33; m 63; c 1. MATHEMATICS. Educ: Mass Inst Technol, PhD(math), 61. Prof Exp: Asst res scientist, Courant Inst Math Sci, NY Univ, 61-63, temp mem, 63; from asst prof to assoc prof, 63-68, PROF MATH, UNIV WIS-MADISON, 68- Concurrent Pos: Consult, NASA, 63-66; prof, Math Res Ctr, Univ Wis, 73-75. Mem: Am Math Soc. Res: Ordinary differential equations with emphasis on dynamical systems and topological dynamics. Mailing Add: Dept of Math Univ of Wis Madison WI 53706

CONLEY, FRANCIS RAYMOND, b Donora, Pa, Feb 11, 16; m 44; c 5. INORGANIC CHEMISTRY. Educ: St Bonaventure Univ, BS, 41. Prof Exp: Chemist, Ryder Scott Co, 41-43, lab foreman, 43, waterflood engr, 44-47, dir res, 47-51; res engr, 51-55, res group leader, 55, supvr res chemist, 55-59, asst dir prod res, 59-65, dir prod res, 65-67, MGR PROD RES, CONTINENTAL OIL CO, 67- Mem: Marine Technol Soc; Am Inst Mining, Metall & Petrol Eng. Res: Petroleum secondary recovery; petrophysics reservoir mechanics; fluid flow in porous media; electric and radiation logging; coring; water flooding; reservoir geology. Mailing Add: Res & Develop Dept Continental Oil Co PO Box 1267 Ponca City OK 74601

CONLEY, HARRY LEE, JR, b Somerset, Ky, June 8, 35; m 58; c 3. PHYSICAL CHEMISTRY. Educ: Univ Ky, BS, 57; Univ Calif, Berkeley, MS, 60; Univ Va, PhD(chem), 64. Prof Exp: Sr chemist, Res Ctr, Sprague Elec Co, 63-68; asst prof, 68-71, ASSOC PROF CHEM, MURRAY STATE UNIV, 71- Mem: AAAS; Am Chem Soc. Res: Kinetics; electrochemistry; thermodynamics. Mailing Add: Dept of Chem Murray State Univ Murray KY 42071

CONLEY, JACK MICHAEL, b Wichita, Kans, Aug 9, 43; m 75. ANALYTICAL CHEMISTRY. Educ: Colo Sch Mines, MEC, 69; Univ Ill, MS, 72, PhD(anal chem), 75. Prof Exp: Teaching asst chem, Univ Ill, 69-72, res asst, 72-74; RES SCIENTIST CHEM, RES & DEVELOP DIV, UNION CAMP CORP, 74- Mem: AAAS; Am Chem Soc. Res: Applied research in analysis of products from pulp and paper industry. Mailing Add: Union Camp Corp PO Box 412 Princeton NJ 08540

CONLEY, JAMES FRANKLIN, b Forest City, NC, Dec 28, 31; m 54; c 3. GEOLOGY. Educ: Berea Col, AB, 54; Ohio State Univ, MS, 56. Prof Exp: Geologist, NC Div Mineral Resources, 56-65; GEOLOGIST, VA DIV MINERAL RESOURCES, 65- Mem: Fel Geol Soc Am; Am Inst Mining, Metall & Petrol Engrs. Res: Geologic mapping in central and southern Virginia Piedmont. Mailing Add: 1614 Trailridge Rd Charlottesville VA 22903

CONLEY, PATRICK, b Roby, Tex, Oct 10, 21; m 42; c 3. ELECTROMAGNETICS. Educ: Rice Inst, BS, 42; Harvard Univ, ScM, 46, PhD(appl physics), 48, MBA, 55. Prof Exp: Mgr underwater sound sect, Electronics Dept, Res Labs, Westinghouse Elec Co, 49-50, mgr radar & acoustics sect, 51-52, asst mgr electronics & nuclear physics dept, 53, spec assignment to mgr develop, 54, staff vpres eng, 55-56, exec engr, Baltimore Divs, 57-58, tech dir to defense prod vpres, 58, mgr air arm div, 59-61, vpres indust systs, 61-64; mem staff, Off Sci & Tech, Exec Off of the President, 64; prof eng, Carnegie Inst Technol, 64-65; assoc sci adv to gov Pa & mem gov sci adv comt, 65-67; VPRES, BOSTON CONSULT GROUP, INC, 67- Concurrent Pos: Adj sr fel, Mellon Inst, 64. 64; consult, Off Sci & Technol, DC & Regional Indust Develop Corp, Pa, 64. Mem: Acoustical Soc Am; Inst Elec & Electronics Engrs. Res: Electronics; physics, research management; business administration. Mailing Add: Boston Consult Group Inc One Boston Pl Boston MA 02106

CONLEY, ROBERT F, b Kokomo, Ind, Nov 23, 28; m 57; c 3. INORGANIC CHEMISTRY. Educ: Ind Univ, BS, 53, PhD, 58. Prof Exp: Spectrographer, Ind Geol Surv, 54-57; chief chemist, Ga Kaolin Co, 58-62, asst dir res labs, 62-69, mgr basic res, 69-72; dir res, Tech Assocs, 72-74; VPRES, MINERAL RESOURCE ASSOCS, 74- Concurrent Pos: Prof chem, Rutgers Univ, 62-; US del, World Mining Cong, 74. Mem: Am Ceramic Soc; Clay Minerals Soc; Am Chem Soc. Res: Physical inorganic chemistry; surface chemistry; geochemistry; industrial process design; mineral recovery; pollution abatement; electrochemical recovery. Mailing Add: Box 1218 Mountainside NJ 07092

CONLEY, ROBERT T, b Summit, NJ, Dec 27, 31; m 55; c 3. ORGANIC CHEMISTRY, POLYMER CHEMISTRY. Educ: Seton Hall Univ, BS, 53; Princeton Univ, MA, 55, PhD(chem), 57. Prof Exp: Asst prof chem, Canisius Col, 56-61; from asst prof to prof, Seton Hall Univ, 61-67; chmn dept chem, 67-68, dean col sci & eng, 68-, PROF CHEM, WRIGHT STATE UNIV, 67-, VPRES PLANNING & DEVELOP, 74- Concurrent Pos: Consult, Carborundum Co. Mem: AAAS; Am Chem Soc. Res: Thermal stability of polymers; molecular rearrangements; reaction mechanisms; synthesis of compounds of pharmacological activity; synthetic methods in alicyclic systems; infrared spectroscopy. Mailing Add: Col of Sci & Eng Wright State Univ Dayton OH 45431

CONLEY, THOMAS DANIEL, b Somerville, Mass, Apr 26, 28; m 55; c 3. PHYSICS. Educ: Boston Univ, BS, 49, MA, 51. Prof Exp: Asst, Boston Col, 49 & Boston Univ, 49-50; physicist, Navy Underwater Sound Lab, 51; electronic scientist, 51-54 & 56-64, RES PHYSICIST, OPTICAL PHYSICS LAB, AIR FORCE CAMBRIDGE RES CTR, 64- Concurrent Pos: Res engr, Ballistic Res Labs, 54-56. Mem: Am Inst Aeronaut & Astronaut; Int Union Radio Sci. Res: Radio wave propagation; ionospheric physics; plasma probe; gas dynamics. Mailing Add: Hq AFCRL Optical Physics Lab OPR Hanscom AFB Bedford MA 01730

CONLEY, VERONICA LUCEY, b Taunton, Mass, July 13, 19; m 45; c 4. PUBLIC HEALTH. Educ: Boston Univ, AB, 40; Yale Univ, MSN, 43; Univ Chicago, MA, 53, PhD(sci educ), 59. Prof Exp: Instr nursing, Ohio State Univ, 46-48; secy comt cosmetics, AMA, 48-60, dir dept nursing, 60-62; exec dir, Nat Asn Practical Nurse Educ & Serv, Inc, 62-65; chief allied health sect, Div Regional Med Progs, Pub Health Serv, 67-73, CHIEF, OFF OF COMT & REV ACTIVITIES, NAT CANCER INST, 73- Concurrent Pos: Ed monthly series, Today's Health Mag, 52-58; exec ed, J Practical Nursing, 62- Mem: AAAS; affil mem AMA. Res: Special problems of health education through mass media in field of biochemistry and physiology; medical aspects of the skin; nursing and administration. Mailing Add: 14706 Crossway Rd Rockville MD 20853

CONLIN, BERNARD JOSEPH, b Columbus, Wis, Mar 15, 35; m 65; c 1. DAIRY SCIENCE, ANIMAL BREEDING. Educ: Univ Wis, BS, 57, MS, 63; Univ Minn, PhD(dairy sci), 66. Prof Exp: 4-H Club exten agent, Agr Exten Serv, Univ Wis, 57-58; prom dir, E Cent Breeders Coop, 58-60; res asst dairy sci, Univ Wis, 60-61, exten dairy specialist, 61-62, res asst dairy sci, 62; res asst dairy husb, 62-66, from asst prof to assoc prof animal sci, 66-71, PROF ANIMAL SCI, UNIV MINN, ST PAUL, 71-EXTEN DAIRYMAN, 66- Mem: Am Dairy Sci Soc; Am Soc Animal Sci. Res: Relative merits of inbred and non-inbred sires for use in artificial insemination and uniformity of their sire families. Mailing Add: Univ of Minn Haecker Hall St Paul MN 55101

CONLON, DANIEL RUPERT, b Brockton, Mass, Sept 20, 12; m 40; c 5. CHEMISTRY, INSTRUMENTATION. Educ: Union Col, BS, 33. Prof Exp: Res chemist phys chem, Atlantic Refining Co, 35-45; sr scientist, Rohm and Haas Co, 45-52, sr scientist instrumentation, 52-57; PRES, INSTRUMENTS FOR RES & INDUST, 57- Mem: Am Chem Soc; Instrument Soc Am. Res: Development of instruments to detect toxic gases, and for research and plant operations; design and manufacture of safety products for laboratory sciences and instruments for automatizing tedious laboratory tasks; design and manufacture of other devices that facilitate laboratory work in fields of chemistry and biochemistry. Mailing Add: Instruments for Res & Indust 108 Franklin Ave Cheltenham PA 19012

CONLY, JAMES CARROLL, organic chemistry, see 12th edition

CONN, ERIC EDWARD, b Berthoud, Colo, Jan 6, 23; m 59; c 2. BIOCHEMISTRY. Educ: Univ Colo, AB, 44; Univ Chicago, PhD(biochem), 50. Prof Exp: Chemist, Manhattan Dist, 44-46; instr biochem, Univ Chicago, 50-52; plant nutrit & jr plant physiologist, Exp Sta, Univ Calif, Berkeley, 52-53; lectr & asst plant physiologist, 53-54, asst prof plant biochem & asst plant biochemist, 54-58; assoc prof & assoc biochemist, 58-63, head dept biochem & biophys, 63-66, PROF BIOCHEM, UNIV CALIF, DAVIS, 63- Concurrent Pos: NIH sr res fel, 60, mem, USPHS Fel Rev Panel for Biochem Nutrit, 63-66; Fulbright scholar, NZ, 65-66. Mem: Am Soc Biol Chem; Am Soc Plant Physiol; Brit Biochem Soc; Phytochem Soc NAm. Res: Plant enzymes; intermediary metabolism of secondary plant products; cyanogenic glycosides. Mailing Add: Dept of Biochem Univ of Calif Davis CA 95616

CONN, HAROLD O, b Newark, NJ, Nov 16, 25; m 51. INTERNAL MEDICINE. Educ: Univ Mich, BS, 46, MD, 50. Prof Exp: Intern, Johns Hopkins Hosp, 50-51; asst resident physician, Grace New Haven Community Hosp, 51-52, chief resident physician, 55-56; from instr to assoc prof, 55-71, PROF INTERNAL MED, SCH MED, YALE UNIV, 71- Concurrent Pos: Brown res fel, Sch Med, Yale Univ, 52-53; dir med ed, Middlesex Mem Hosp, 56-57; clin investr, Vet Admin Hosp, West Haven, 57-60, actg chief med serv, 59-60, chief, Hepatic Res Lab, 60-; counr, Am Asn Study Liver Dis, 67-, vpres elect, 71, vpres, 72, pres, 73; vis prof, Sch Med, Univ Wash Univ, 68-69; mem med sch coun, Yale Univ Sch Med, 70. Mem: Am Fedn Clin Res; Am Soc Clin Invest; Asn Am Physicians; Int Asn Study Liver; Am Gastroenterol Asn. Res: Clinical management of liver disease; treatment of hepatic coma, esophageal varices and ammonia metabolism; abnormalities of protein metabolism. Mailing Add: Yale Univ Sch of Med 333 Cedar St New Haven CT 06510

CONN, JAMES FREDERICK, b Osborne, Kans, July 2, 24; m 48; c 2. CEREAL CHEMISTRY. Educ: Kans State Univ, BS, 48, MS, 49. Prof Exp: Wheat qual chemist, Int Multifoods Co, 49-57; sr res chemist, 57-66, RES SPECIALIST CHEM LEAVENING CHEESE EMULSIFICATION RES, MONSANTO CO, 66- Mem: Am Asn Cereal Chemists; Inst Food Technologists; Sigma Xi. Res: Determining the functions of ortho-, pyro- and polyphosphates in chemical leavening and process cheese with particular interest in the interactions with other ingredients and the rheological effects. Mailing Add: Monsanto Co 800 N Lindbergh Blvd St Louis MO 63166

CONN, JEROME W, b New York, NY, Sept 24, 07; m; c 2. MEDICINE. Educ: Univ Mich, MD, 32. Hon Degrees: DSc, Rutgers Univ, 64; Univ Turin, MD, 75. Prof Exp: From instr to prof internal med, 35-68, Louis Harry Newburgh Univ Prof med, 68-74, EMER DISTINGUISHED UNIV PROF INTERNAL MED, MED SCH, UNIV MICH, ANN ARBOR, 74-, CONSULT CLIN INVEST, DEPT INTERNAL MED, 74- Concurrent Pos: Distinguished physician, Vet Admin, 73- Honors & Awards: Mod Med Mag Award, 57; Bernard Medal, Univ Montreal, 57; Banting Medal, Am Diabetes Asn, 58 & Banting Mem Award, 63; Henry Russell Lectr Award, Univ Mich, 61; Wilson Medal, Am Clin & Climat Asn, 62; Gairdner Found Int Prize, 65; Phillips Mem Award, Am Col Physicians, 65; Howard Taylor Award, Am Therapeut Soc, 67; Ruth Gray Mem Medal, Evanston Hosp, 68; Stouffer Int Prize, 69; Gold Medal, Int Soc Progress Internal Med, 69; Heath Med Award & Medal, Univ Tex, Houston, 71; Award, Am Col Nutrit, 73. Mem: Nat Acad Sci; Nat Inst Med; Asn Am Physicians; Am Diabetes Asn (pres, 62-63, first vpres, 61, 2nd vpres, 60); hon fel Am Col Surgeons. Res: Human nutrition; normal metabolism; disorders of metabolism; endocrinology. Mailing Add: Univ Hosp Univ of Mich Med Ctr Ann Arbor MI 48104

CONN, PAUL JOSEPH, b Kalispell, Mont, May 17, 40; m 61; c 1. PHYSICAL CHEMISTRY. Educ: Univ Calif, Davis, BS, 62; Univ Ore, MS, 64, PhD(phys chem), 66. Prof Exp: STAFF RES CHEMIST, SHELL DEVELOP CO, 64- Mem: Am Chem Soc; Catalysis Soc. Res: Heterogeneous catalysis, preparation and characterization of catalysts, catalytic kinetics and mechanism. Mailing Add: 1414 Scenic Ridge Houston TX 77043

CONN, PAUL KOHLER, b Akron, Ohio, July 25, 29; m 54; c 3. MATERIALS SCIENCE. Educ: Kenyon Col, AB, 51; Kans State Univ, MS, 53, PhD(phys chem), 56. Prof Exp: Sr engr & prin engr, Aircraft Nuclear Propulsion Dept, Gen Elec Co, 55-61, supvr chem res, Nuclear Mat & Propulsion Oper, 61-64, unit mgr, Solid State Chem Res, 64-69; PRIN INVESTR & ASSOC DIR ADVAN MAT RES, BELL AEROSPACE CO, TEXTRON INC, 69- Mem: Am Chem Soc. Res: High temperature materials and coatings; solid state chemistry, including kinetics, diffusion, fission product transport processes; gas-solid reactions; hot atom chemistry; radiation effects in materials; viscoelastic composite materials; organic coatings. Mailing Add: 4682 W Park Dr Lewiston NY 14092

CONN, REX BOLAND, b Marengo, Iowa, Aug 3, 27; m 50; c 3. MEDICINE, CLINICAL PATHOLOGY. Educ: Iowa State Univ, BS, 49; Yale Univ, MD, 53; Oxford Univ, BSc, 55; Univ Minn, Minneapolis, MS, 60. Prof Exp: Instr lab med, Univ Minn, Minneapolis, 60; asst prof med, WVa Univ, 60-61, from asst prof to prof, 61-68; PROF LAB MED & DIR SUBDEPT LAB MED, JOHNS HOPKINS UNIV, 68- Concurrent Pos: Mem path study sect, NIH, 68-72, mem path training comt, 72-73; mem comt chem path, Am Bd Path, 74-; mem bd dirs, Am Soc Clin Path, 75- Mem: Fel Am Soc Clin Path; fel Col Am Path; Am Asn Path & Bact; Am Fedn Clin Res; Acad Clin Lab Physicians & Scientists (pres, 72-73). Res: Creatine metabolism; methods in clinical chemistry. Mailing Add: 592 Carnegie Bldg Johns Hopkins Hosp Baltimore MD 21205

CONNALLY, GEORGE GORDON, b Passaic, NJ, Aug 11, 34; m 55; c 3. GEOLOGY, PETROLOGY. Educ: Lafayette Col, AB, 56; Univ Rochester, MS, 59; Mich State Univ, PhD(geol), 64. Prof Exp: Asst prof geol, State Univ NY Col New Paltz, 61-67; asst prof, Lafayette Col, 67-71; vis assoc prof, State Univ NY Buffalo, 71-74; MEM FAC, DEPT GEOL, STATEN ISLAND COMMUNITY COL, 74- Concurrent Pos: New York State res grants, 62- Mem: Geol Soc Am; Soc Econ Paleontologists &

Mineralogists; Glaciol Soc; Nat Asn Geol Teachers; Am Soc Photogram. Res: Glacial geology and petrology of glacial deposits; heavy mineral stability. Mailing Add: Dept of Geol Staten Island Commun Col Staten Island NY 10301

CONNAMACHER, ROBERT HENLE, b Newark, NJ, Dec 20, 33; m 66; c 3. PHARMACOLOGY. Educ: Oberlin Col, AB, 55; NY Univ, MS, 59; George Washington Univ, PhD(pharmacol), 66. Prof Exp: From instr to asst prof, 67-73, ASSOC PROF PHARMACOL, SCH MED, UNIV PITTSBURGH, 73- Concurrent Pos: Fel molecular biol, Sch Med, Univ Pittsburgh, 66-67. Mem: Am Soc Microbiol; Biophys Soc; Int Soc Chemother. Res: Mechanism of action of antibiotics and biochemical mechanisms of adaptation, especially protein synthesis inhibitors, using tetracycline as the model drug. Mailing Add: Dept of Pharmacol Univ of Pittsburgh Sch of Med Pittsburgh PA 15261

CONNAR, RICHARD GRIGSBY, b Zanesville, Ohio, Jan 11, 20; m 46; c 3. THORACIC SURGERY, CARDIOVASCULAR SURGERY. Educ: Duke Univ, AB, 41, MD, 44. Prof Exp: Intern & asst resident internal med, Duke Univ Hosp, 44-46, asst resident & resident surg, 48-53, from instr to asst prof, Sch Med, 50-55; CHMN DEPT SURG, TAMPA GEN HOSP, 55-68, 70-; PROF SURG, COL MED, UNIV S FLA, 68-, CHIEF SECT THORACIC & CARDIOVASC SURG, 74- Concurrent Pos: Chief thoracic surg sect, Vet Admin Hosp, 53-55; thoracic & cardiovasc surgeon, St Josephs Hosp, 55; consult, MacDill AFB Hosp; chmn, Duke Univ Nat Coun, 70-71; mem med adv comt, Col Med, Univ SFla, 70-; deleg, AMA, 71-, mem coun med educ, 74-; mem liaison comt grad med educ, 74- Mem: Am Col Surg; Am Asn Thoracic Surg; Soc Thoracic Surg; Am Col Chest Physicians. Res: Vascular and esophageal reconstruction; dermal transplants; veratrum derivatives and hypertension; cardiac surgery. Mailing Add: Suite 703 One Davis Blvd Tampa FL 33606

CONNELL, ALASTAIR MCCRAE, b Glasgow, Scotland, Dec 21, 29; m 55; c 5. GASTROENTEROLOGY. Educ: Univ Glasgow, BSc, 51, MB, ChB, 54, MD, 69; FRCP, 72; FACP, 73. Prof Exp: House physician, Stobhill Gen Hosp & house surgeon, Univ Dept Surg, Western Infirmary, Glasgow, 54-55; clin & res asst, Cent Middlesex Hosp, London, 57-60; mem sci staff, Med Res Coun Gastroenterol, Cen Middlesex Hosp & St Mark's Hosp London, 61-64; sr lectr clin sci, Queen's Univ, Belfast, 64-70; MARK BROWN PROF MED, PROF PHYSIOL & DIR DIV DIGESTIVE DIS, COL MED, UNIV CINCINNATI, 70-, ASSOC DEAN, 75- Concurrent Pos: Consult physician, Northern Ireland Hosp Authority, Royal Victoria Hosp & SBelfast Hosp, Belfast, 64-70; attend physician, Cincinnati Gen Hosp, Ohio, 70-; chief clinician, Div Digestive Dis, Med Ctr, Univ Cincinnati, 70-; consult, Vet Admin Hosp, Cincinnati, 70-, Jewish & Drake Hosps, Cincinnati, 73- Mem: Brit Soc Gastroenterol; Am Gastroenterol Asn; Am Fedn Clin Res; Am Soc Digestive Endoscopy; Am Asn Study Liver Dis. Res: Motility of gastrointestinal tract; role of gastrointestinal hormones in control of motor activity; assessment of therapy of gastrointestinal disease; nutritional factors in pathogenesis of gastrointestinal disease. Mailing Add: Dept of Med Div Digestive Dis Univ of Cincinnati Col of Med Cincinnati OH 45267

CONNELL, BALFOUR, b Lexington, Ky, July 11, 14; m 39; c 3. RUBBER CHEMISTRY. Educ: Univ Ky, BS, 38. Prof Exp: Group head tire compounding & develop, B F Goodrich Co, 38-52; asst tech dir sponge rubber prod, Dryden Rubber Div, 52-57; TECH DIR MECH GOODS, GEN TIRE & RUBBER CO, 57- Mem: Am Chem Soc; Am Soc Testing & Mat; Am Ord Asn; Soc Automotive Engrs. Res: Rubber; synthetic rubber; plastics; urethane; molded, extruded and cellular products. Mailing Add: Gen Tire & Rubber Co PO Box 507 Wabash IN 46992

CONNELL, ELIZABETH BISHOP, b Springfield, Mass, Oct 17, 25; m 49; c 6. OBSTETRICS & GYNECOLOGY. Educ: Univ Pa, AB, 47, MD, 51; Am Bd Obstet & Gynec, dipl, 65. Prof Exp: Intern, Lankenau Hosp, Philadelphia, Pa, 51-52, resident path & anesthesia, 52-53; gen pract & anesthetist, Maine, 53-58; resident gynec, Grad Hosp, Univ Pa, 58-60; resident obstet, Mt Sinai Hosp, New York, 60-61; assoc prof obstet & gynec, NY Med Col, 62-69; assoc prof obstet & gynec, Col Physicians & Surgeons, Columbia Univ, 70-73, dir res & develop, Family Planning Serv, Int Inst Study Human Reproduction, 70-73; ASSOC DIR HEALTH SCI, ROCKEFELLER FOUND, 73- Concurrent Pos: Am CAncer fel, Kings County Hosp, State Univ NY, 61-62; dir, Family Planning Ctr, NY Med Col-Metrop Hosp Med Ctr, 64-69; mem med adv bd, Planned Parenthood, New York, 64-; mem nat adv coun, Alan Guttmacher Inst, Planned Parenthood World Pop, 68-; chmn nat med comt, 74-; mem exec comt, Comt Med & Pub Health Asn Voluntary Sterilization, Inc; mem obstet & gynec adv comt, Food & Drug Admin, chmn over-the-counter rev panel; consult, Family Planning, New York Dept Health; family planning proj consult, Human Resources Admin; mem res adv comt, Agency Int Develop. Mem: Am Col Obstet & Gynec; Am Col Surg; Am Pub Health Asn; Am Fertil Soc; AMA. Res: Medicine; contraception. Mailing Add: Rockefeller Found 1133 Ave of the Americas New York NY 10036

CONNELL, FRANK HERMAN, b Hudson, NH, 1905; m 31; c 1. PARASITOLOGY. Educ: Dartmouth Col, BS, 28; Univ Calif, AM, 29, PhD(zool), 31. Prof Exp: From instr to prof, Dartmouth Col, 31-55, parasitol, Med Sch, 35-41, prof, 41-55; chief labs, Hiroshima Atomic Bomb Casualty Comn, Nat Res Coun, 50-51, assoc dir, Nagasaki, 52-54, actg dir, 54, exec dir, Comt on Atomic Casualties, 54-55; head lab servs, M D Anderson Hosp & Tumor Inst & prof parasitol, Postgrad Med Sch, Tex, 55-60; asst dir, China Med Bd New York, 60-68, assoc dir, 68-72; RETIRED. Concurrent Pos: Protozoologist, Harvard Med Sch & Carnegie Inst Med Exped, Yucatan, 31; consult parasitologist, State Dept Fish & Game, NH, 36-42 & US Vet Admin, 46-50, 52; clin prof, Baylor Univ, 55-60. Mem: AAAS; Am Soc Trop Med & Hyg. Res: Amebiasis; medical education. Mailing Add: Fairview Plantation PO Box 305 Berwick LA 70342

CONNELL, GEORGE EDWARD, b Saskatoon, Sask, June 20, 30; m 55; c 4. BIOCHEMISTRY. Educ: Univ Toronto, BA, 51, PhD(biochem), 55. Prof Exp: Nat Res Coun Can fel, 55-56; Nat Acad Sci-Nat Res Coun fel biochem, Sch Med, NY Univ, 56-57; asst prof biochem, 57-62, chmn dept, 65-70, assoc dean fac med, 72-74, ASSOC PROF BIOCHEM, UNIV TORONTO, 62-, VPRES, 74- Concurrent Pos: Mem, Med Res Coun Can, 66-70. Mem: Am Soc Human Genetics; Can Biochem Soc (pres, 73-74); fel Chem Inst Can; Royal Soc Can; Can Arthritis & Rheumatism Soc. Res: Protein chemistry; enzymology; immunochemistry; chemistry of human plasma proteins. Mailing Add: Dept of Biochem Univ of Toronto Toronto ON Can

CONNELL, JAMES FREDERICK LOUIS, b Baltimore, Md, June 25, 20; m 43; c 1. GEOLOGY. Educ: La State Univ, BS, 49; Univ Okla, MS, 51, PhD, 55. Prof Exp: Mus preparator, Univ Okla, 49-51, instr geol, 51-53; asst prof, La Polytech Inst, 53-56; assoc prof, Univ Southern Miss, 56-57, State Univ NY, 57-58 & Univ Southwestern La, 58-62; PROF GEOL, UNIV MONTEVALLO, 62-, ACAD MARSHAL, 65- Concurrent Pos: Consult geologist, 62- Mem: AAAS; Paleont Res Inst; Am Inst Prof Geol; Am Asn Petrol Geol; Paleont Soc. Res: Appalachian and Gulf Coastal Plain stratigraphy and paleontology; Pennsylvanian flora of Alabama. Mailing Add: Drawer B Univ of Montevallo Montevallo AL 35115

CONNELL, JOSEPH H, b Gary, Ind, Oct 5, 23; m 54; c 4. POPULATION BIOLOGY. Educ: Univ Chicago, BS, 46; Univ Calif, MA, 53; Glasgow Univ, PhD(zool), 56. Prof Exp: Res assoc marine biol, Woods Hole Oceanog Inst, 55-56; PROF ZOOL, UNIV CALIF, SANTA BARBARA, 56- Concurrent Pos: Guggenheim fel, 62-63, 71-72. Honors & Awards: Mercer Award, Ecol Soc Am, 63. Mem: AAAS; Am Soc Limnol & Oceanog; Brit Ecol Soc; Australian Ecol Soc; Am Soc Naturalists. Res: Species diversity of tropical communities, especially rain forests and coral reefs; population ecology of marine intertidal organisms, particularly predation, competition, spatial distribution and territoriality. Mailing Add: Dept of Biol Sci Univ of Calif Santa Barbara CA 93106

CONNELL, LOUIS FRED, JR, b Honey Grove, Tex, June 25, 14; m 38; c 3. PHYSICS. Educ: Tex Col Arts & Indust, BA, 34; Univ Tex, MA, 36, PhD(physics), 48. Prof Exp: Teacher high schs, Tex, 34-37; from instr to asst prof physics, N Tex State Univ, 37-42; asst prof physics, Univ Tex, 47-51, mathematician, Mil Physics Lab, 48-51; prof, 51-75, dir dept, 51-69, EMER PROF PHYSICS, N TEX STATE UNIV, 75- Mem: AAAS; Am Asn Physics Teachers; Acoustical Soc Am; Am Phys Soc. Res: Electron diffraction; x-rays and crystal structure; electrical properties of semiconductors; acoustics and noise. Mailing Add: 924 Ridgecrest Circle Denton TX 76201

CONNELL, RICHARD ALLEN, b Lincoln, Nebr, Oct 31, 29; m 56; c 2. PHYSICS. Educ: Nebr Wesleyan Univ, BA, 51; Northwestern Univ, PhD, 57. Prof Exp: Asst gaseous discharges, Northwestern Univ, 51-54, low temperature solid state, 54-57; assoc physicist superconductivity, Res Lab, Int Bus Mach Corp, 57-59, proj physicist, 59-62, prof engr, 62-63; sr physicist, Midwest Res Lab, 62-67; sr physicist, 67-70, MGR MAT SCI GROUP, PITNEY-BOWES, INC, 70- Mem: Am Phys Soc; Sigma Xi. Res: Superconductivity; thin film physics; biophysics; photoconductivity; electrophotography. Mailing Add: Pitney-Bowes Inc Walnut & Pacific St Stamford CT 06903

CONNELL, ROSEMARY, b St Louis, Mo, Jan 31, 22. PHYSIOLOGY. Educ: Fontbonne Col, BS, 53; Univ Notre Dame, MS, 61, PhD(biol), 65. Prof Exp: Secondary teacher, St Joseph's Acad, 53-61 & Little Flower High Sch, 61-62; from instr to asst prof, 65-71, ASSOC PROF BIOL, FONTBONNE COL, 71-, CHMN DEPT, 66- Concurrent Pos: Assoc prof, Maryville Col, 74-75. Mem: Radiation Res Soc; Transplantation Soc; Exp Hemat Soc. Res: Radiation treatment and physiology; hematology. Mailing Add: 6800 Wydown St Louis MO 63105

CONNELL, TERRENCE LEE, mathematical statistics, see 12th edition

CONNELL, WALTER ANTHONY, b New London Township, Pa, Aug 9, 09; m 43; c 1. INSECT TAXONOMY. Educ: Univ Md, BS, 33, PhD, 57; Univ Minn, MS, 41. Prof Exp: Jr entomologist, US Forest Serv, Baltimore, 35-37; asst entomologist, Bur Entom & Plant Quarantine, USDA, 37-38; asst entomologist, 39-41, from assoc prof to prof, 46-75, EMER PROF ENTOM, UNIV DEL, 75- Concurrent Pos: Taxonomist-Consult, Insect Identification Lab, Agr Res Serv, USDA, 61- Mem: AAAS; Entom Soc Am; Royal Entom Soc London; Soc Syst Zool; Coleopterist's Soc. Res: Biology and control of insects; life history and control of the oak lace bug; biology and taxonomy of Nitidulidae, the sap beetles; diseases of Acarina. Mailing Add: Dept Entom & Appl Ecol Univ of Del Newark DE 19711

CONNELL, WALTER FORD, b Kingston, Ont, Aug 24, 06; m 33; c 3. INTERNAL MEDICINE, CARDIOLOGY. Educ: Queen's Univ, Ont, MD & CM, 29; FRCP; FRCPS(C). Hon Degrees: LLD, Queen's Univ, Ont, 73. Prof Exp: Dir cardiol, Kingston Gen Hosp, 33-68; prof, 43-76, EMER PROF MED, QUEEN'S UNIV, ONT, 76- Concurrent Pos: Mem coun arteriosclerosis, Am Heart Asn. Mem: Am Heart Asn; Can Heart Asn; Royal Soc Med. Res: Coronary heart disease. Mailing Add: 11 Arch St Kingston ON Can

CONNELLY, CLARENCE MORLEY, b Jamestown, NY, Nov 4, 16; m 46; c 1. BIOPHYSICS. Educ: Cornell Univ, AB, 38; Univ Pa, PhD(biophys), 49. Prof Exp: Asst physics, Cornell Univ, 38-42; staff mem, Radiation Lab, Mass Inst Technol, 42-46; asst biophys, Univ Pa, 46-49; from instr to asst prof, Johns Hopkins Univ, 49-54; asst prof, 54-60, ASSOC PROF BIOPHYS, ROCKEFELLER UNIV, 60-, ASSOC DEAN GRAD STUDIES, 62- Concurrent Pos: Ed, J Gen Physiol, 61-64. Mem: Soc Gen Physiol; Am Phys Soc; Biophys Soc. Res: Physiology of nerve and muscle. Mailing Add: Rockefeller Univ 66th St & York Ave New York NY 10021

CONNELLY, DAMIAN, b Pittsburgh, Pa, Aug 9, 18. MATHEMATICS. Educ: Catholic Univ, BA, 39, MS, 41, PhD(math), 48; Univ Notre Dame, MA, 53. Prof Exp: Teacher high sch, Pa, 39-40; instr math, Catholic Univ, 40-41, teacher high sch, Pa, 41-46; instr, De La Salle Col, 40-41 & 46-48; from asst prof to assoc prof, 48-69, PROF MATH, LA SALLE COL, 69- Mem: AAAS; Am Math Soc; Math Asn Am. Res: Algebraic geometry; analytic additive theory of numbers; investigation of conditions for reality of certain characteristics of the bicircular cuspidal quartic. Mailing Add: Dept of Math La Salle Col Philadelphia PA 19141

CONNELLY, JERALD LEONARD, b Rochester, NY, Jan 8, 28; m 53; c 3. BIOCHEMISTRY. Educ: Univ Rochester, BS, 50, PhD(biochem), 58. Prof Exp: Asst prof biochem, Med Ctr, WVa Univ, 61-63; from asst prof to assoc prof, 63-70, PROF BIOCHEM, GUY & BERTHA IRELAND RES LAB, SCH MED, UNIV N DAK, 70- Concurrent Pos: McEchern fel, Enzyme Inst, Univ Wis, 58-61; NIH career develop fel, 66. Mem: Am Soc Biol Chem. Res: Metabolic diseases. Mailing Add: Dept of Biochem Univ of NDak Sch of Med Grand Forks ND 58201

CONNELLY, JOHN JOSEPH, JR, b Syracuse, NY, Apr 14, 25; m 47; c 3. EXPERIMENTAL SOLID STATE PHYSICS. Educ: Rensselaer Polytech Inst, BAeroE, 45; Univ Va, MS, 55, PhD(physics), 56. Prof Exp: Sci adv to chief aircraft nuclear propulsion off & mat technologist, US AEC, 56-60; prog officer energy conversion, Power Br, Off Naval Res, 60-64; assoc prof, 64-68, PROF PHYSICS, STATE UNIV NY COL FREDONIA, 68- Concurrent Pos: Mem comt elec eng systs, NASA. Mem: Assoc fel Am Inst Aeronaut & Astronaut; Am Phys Soc. Res: Very high speed ultra-centrifuges; high-temperature materials; advance type propulsion systems; direct energy conversion; crystal growth of halides, calcite and tellurates. Mailing Add: Dept of Physics State Univ of NY Col Fredonia NY 14063

CONNELLY, THOMAS GEORGE, b Oak Park, Ill, Sept 22, 45; m 66; c 2. ANATOMY, ZOOLOGY. Educ: Monmouth Col, Ill, AB, 66; Mich State Univ, PhD(zool), 70. Prof Exp: ASST PROF ANAT, MED SCH, UNIV MICH, ANN ARBOR, 72-, ASST RES SCIENTIST, CTR HUMAN GROWTH & DEVELOP, 75- Concurrent Pos: NIH res fel, Biol Div, Oak Ridge Nat Lab, 70-72. Mem: Am Soc Zool; Soc Develop Biol. Res: Amphibian lens and limb regeneration; amphibian pituitary cytology. Mailing Add: Dept of Anat Univ of Mich Med Sch Ann Arbor MI 48104

CONNER, ALBERT Z, b Philadelphia, Pa, Dec 21, 21. ANALYTICAL CHEMISTRY. Educ: Drexel Inst Technol, BS, 47; Univ Del, MS, 52. Prof Exp: Chemist, Publicker Industs, 46-47; res chemist, Hercules Powder Co, 47-65, sr res chemist, Hercules Inc, 65-70, RES SCIENTIST, HERCULES INC, 70- Mem: Am Chem Soc; Am Inst Chemists. Res: Chromatography; organic analysis; analysis of rocket propellants. Mailing Add: Anal Div Hercules Res Ctr Wilmington DE 19899

CONNER, GABEL HENRY, b Uniontown, Wash, Jan 1, 19; m 43; c 3. VETERINARY MEDICINE, PHYSIOLOGY. Educ: Wash State Univ, BS, 40, DVM, 41; Iowa State Univ, MS, 43; Univ Minn, PhD(physiol), 59. Prof Exp: Asst prof surg & med, Iowa State Univ, 41-43; pvt pract, Wash, 43-45; ass prof surg & med, Wash State Univ, 45-47; instr surg, Univ Minn, 47-50; assoc prof, 50-58, PROF VET SURG & MED, MICH STATE UNIV, 58- Concurrent Pos: Res grants, 62- Mem: Vet Med Asn. Res: Oncological research in animals, primarily cattle and dogs. Mailing Add: Dept of Large Animal Surg & Med Mich State Univ East Lansing MI 48823

CONNER, GEORGE W, genetics, see 12th edition

CONNER, HOWARD EMMETT, b Madison, Wis, Sept 26, 30; m 54; c 3. MATHEMATICS. Educ: Univ Wis, BS, 56; Mass Inst Technol, PhD(math), 61. Prof Exp: Staff assoc math, Lincoln Lab, Mass Inst Technol, 57-61; mem staff, US Army Math Res Ctr, 61-62, asst prof, univ, 62-67, assoc prof, 67-73, PROF MATH, UNIV WIS-MADISON, 73- Concurrent Pos: Vis assoc prof, Rockefeller Univ, 67-68. Mem: Am Math Soc. Res: Systematic study of the properties of the solutions of the integral-differential equations used in gas dynamics, plasma theory and kinetic theory. Mailing Add: Dept of Math Van Vleck Hall Univ of Wis Madison WI 53706

CONNER, JACK MICHAEL, b Jackson, Miss, Nov 2, 35. INORGANIC CHEMISTRY, PHYSICAL CHEMISTRY. Educ: Millsaps Col, BS, 56; Univ Wyo, PhD(chem), 66. Prof Exp: Chemist, Baxter Labs, Inc, Miss, 57-58; res chemist, Sch Med, Univ Miss, 58-59; res assoc inorg chem, Univ Kans, 66-67; asst prof chem, 67-75, ASSOC PROF CHEM, REGIS COLL, 75- Mem: AAAS; Am Inst Chem; Am Chem Soc. Res: Preparation and chemistry of transition-metal coordination compounds and oxides; aqueous solution equilibria. Mailing Add: Dept of Chem Regis Col Denver CO 80221

CONNER, JERRY POWER, b Sherman, Tex, Mar 20, 27; m 47; c 3. NUCLEAR PHYSICS. Educ: Rice Inst, PhD(physics), 52. Prof Exp: MEM STAFF & PHYSICIST, LOS ALAMOS SCI LAB, 52- Mem: Am Phys Soc; Am Geophys Union; Am Astron Soc. Res: Space radiations. Mailing Add: 2848 Walnut St Los Alamos NM 87544

CONNER, MARK HALE, physiology, see 12th edition

CONNER, PIERRE EUCLIDE, JR, b Houston, Tex, June 27, 32; m 58; c 2. MATHEMATICS. Educ: Tulane Univ, BS, 52, MS, 53; Princeton Univ, PhD(math), 55. Concurrent Pos: Vis fel math, Inst Adv Study, 55-57; asst prof, Univ Mich, 57-58; from asst prof to prof, Univ Va, 48-71; NICHOLSON PROF MATH, LA STATE UNIV, 71- Concurrent Pos: Sloan fel, 60-64; vis mem, Inst Adv Study, 61-62; mem comt sci confs, Div Math, Nat Acad Sci, Nat Res Coun, 64-67. Res: Applications of the methods of differential and algebraic topology to the study of transformation groups, especially periodic maps. Mailing Add: Dept of Math La State Univ Baton Rouge LA 70803

CONNER, RAY M, b Ft Sam Houston, Tex, Aug 28, 25; m 54. BACTERIOLOGY. Educ: Univ NC, BS, 49, MS, 51; Univ Md, PhD(bact), 66. Prof Exp: Asst, Sch Pub Health, Univ NC, 48; bacteriologist & dir shellfish lab, NC State Lab Hyg, 48-56; asst, Dept Bact, Univ NC, 57-59; bacteriologist, Dairy Cattle Res Br, Agr Res Serv, USDA, 59-61, res bacteriologist, Insect Path Lab, 61-71; MICROBIOLOGIST, PESTICIDE REGULATION DIV, ENVIRON PROTECTION AGENCY, 71- Mem: Am Soc Microbiol. Res: Detection, identification; bacterial physiology; spore forming bacteria; insect pathogens; germicides; disinfectants; sporocides; fungicides. Mailing Add: Pesticide Regulation Div Agr Res Ctr Beltsville MD 20705

CONNER, ROBERT LOUIS, b Wabash, Ind, Feb 3, 27; m 49; c 5. ZOOLOGY. Educ: Wash Univ, AB, 49; Ind Univ, PhD(zool), 54. Prof Exp: Res assoc, Ind Univ, 54; from asst prof to assoc prof, 54-66, actg chmn dept, 67 & 69-70, PROF BIOL, BRYN MAWR COL, 66- Concurrent Pos: Lalor fel, 55; USPHS fels, 61-62 & 68-69. Mem: AAAS; Soc Protozool (pres, 74-76); Brit Soc Gen Microbiol. Res: Biological chemistry; mode of action of steroids; nutritional requirements and physiology of protozoa; membrane chemistry; membrane biochemistry of protozoa. Mailing Add: Dept of Biol Bryn Mawr Col Bryn Mawr PA 19010

CONNER, ROBERT THOMAS, b Lisbon, NH, Oct 2, 10; m 40; c 1. BIOCHEMISTRY. Educ: Univ Vt, BS, 32; Columbia Univ, PhD(biochem), 36. Prof Exp: Asst chem, Columbia Univ, 32-36, instr, 36-39; chemist in chg div biochem, Cent Res Labs, Gen Foods Corp, 39-45; tech dir, William R Warner Co & actg dir Warner Inst Therapeut Res, New York, 45-46; vpres res & develop, Harrower Lab, Inc, 46-50, dir develop labs, 50-51; dir res & develop labs, Smith Kline & French Labs, 51-61 & Strasenburgh Labs, 61-63; DIR LABS, MAX FACTOR & CO, 63- Mem: Fel AAAS; Am Chem Soc; fel Am Inst Chemists; fel NY Acad Sci. Res: Organic, nutritional dnd medicinal chemistry; bacteriology.

CONNER, WILLARD PRESTON, (JR), physical chemistry, see 12th edition

CONNERS, GARY HAMILTON, b Rochester, NY, Feb 15, 36; m 59; c 4. APPLIED MECHANICS. Educ: St Lawrence Univ, BS, 57; Mich State Univ, PhD(appl mech), 63. Prof Exp: Physicist, Delco Appliance Div Gen Motors Corp, 57-59, sr physicist, 62-64; instr appl mech, Mich State Univ, 61-62; assoc lectr mech eng, Univ Rochester, 63-64, asst prof mech & aerospace sci, 64-67; res supvr, Apparatus Div, 67-70, SUPVR APPL MATH RES, HAWK-EYE WORKS, EASTMAN KODAK CO, 70- Concurrent Pos: Lectr mech & aerospace sci, Univ Rochester. Mem: Am Phys Soc; Am Soc Mech Eng; Soc Indust & Appl Math. Res: Solid mechanics including elasticity, plasticity and thermal mechanics; dislocation mechanics; applied solid state physics; optical and mechanical properties of continuous media; liquid crystals; solid mechanics and mechanics of structures. Mailing Add: Apparatus Div Eastman Kodak Co 901 Elmgrove Rd Rochester NY 14650

CONNETT, WILLIAM C, b Mexico City, Mex, Feb 22, 39; m 69. MATHEMATICAL ANALYSIS. Educ: Georgetown Univ, BS, 61; Univ Chicago, MS, 63, PhD(math), 69. Prof Exp: ASST PROF MATH, UNIV MO-ST LOUIS, 69- Mem: Am Math Soc. Res: Multiple Fourier series; multiplier theory; singular integrals. Mailing Add: Dept of Math Univ of Mo St Louis MO 63121

CONNEY, ALLAN HOWARD, b Chicago, Ill, Mar 23, 30; m 54; c 2. BIOCHEMISTRY, PHARMACOLOGY. Educ: Univ Wis, BS, 52, MS, 54, PhD(oncol), 56. Prof Exp: Asst, Univ Wis, 52-56; pharmacologist, Nat Heart Inst, 57-60; head biochem pharmacol sect, Pharmacodynamics Div, Burroughs Wellcome & Co, 60-70; DIR DEPT BIOCHEM & DRUG METAB, HOFFMANN-LA ROCHE INC, 70- Mem: Am Soc Pharmacol & Exp Therapeut; Am Soc Biol Chem; Am Asn

Cancer Res; Am Soc Toxicol; Acad Pharmaceut Sci. Res: Induced enzyme synthesis in mammals; metabolism of drugs, carcinogens, and steroid hormones; mechanism of drug action; ascorbic acid biosynthesis; carcinogenesis. Mailing Add: Dept of Biochem & Drug Metab Hoffmann-La Roche Inc Nutley NJ 07110

CONNICK, ROBERT ELWELL, b Eureka, Calif, July 29, 17; m 52; c 6. INORGANIC CHEMISTRY. Educ: Univ Calif, BS, 39, PhD(phys chem), 42. Prof Exp: Asst chem, 39-42, instr, 42-43, res, Manhattan Proj, 43-46, from asst prof to assoc prof chem, 45-52, chmn dept, 58-60, dean col chem, 60-65, vchancellor acad affairs, 65-67, vchancellor, 69-71, PROF CHEM, UNIV CALIF, BERKELEY, 52- Concurrent Pos: Guggenheim fels, 49 & 59. Mem: Am Acad Sci; Am Chem Soc. Res: Radio chemistry; mechanisms of reactions; complex ions; aqueous solution chemistry of chromium and ruthenium; nuclear magnetic resonance studies of inorganic systems. Mailing Add: Dept of Chem Univ of Calif Berkeley CA 94720

CONNOLA, DONALD PASCAL, b Dobbs Ferry, NY, Nov 15, 13; m 40; c 2. FOREST ENTOMOLOGY, FOREST GENETICS. Educ: Rutgers Univ, BS, 50. Prof Exp: Res assoc, Carbide & Carbon Chem Corp, 50-51; SR SCIENTIST, NY STATE MUS & SCI SERV, NY STATE EDUC DEPT, 51- Mem: Entom Soc Am; Soc Am Foresters. Res: Study of white pine weevil resistance in white pine. Mailing Add: NY State Mus & Sci Serv NY State Educ Dept Albany NY 12234

CONNOLLY, JAMES DONALD, biochemistry, animal nutrition, see 12th edition

CONNOLLY, JOHN E, b Omaha, Nebr, May 21, 23. SURGERY. Educ: Harvard Univ, AB, 45, MD, 48. Prof Exp: Giannini Found fel, Stanford Univ, 57, Markle scholar, 57-62, from instr to assoc prof surg, 57-65; PROF SURG & CHMN DEPT, UNIV CALIF, IRVINE, 65- Concurrent Pos: Chief consult, Vet Admin Hosp, Long Beach, 65-; chief surg, Orange County Med Ctr; staff mem, St Joseph's Hosp & Children's Hosp, Orange. Mem: Fel Am Col Surgeons; Am Surg Asn; Soc Univ Surgeons; Asn Thoracic Surg; Int Cardiovasc Soc. Res: Cardiovascular surgery. Mailing Add: Dept of Surg Univ of Calif Irvine CA 92664

CONNOLLY, JOHN FRANCIS, physical chemistry, see 12th edition

CONNOLLY, JOHN FRANCIS, b Teaneck, NJ, Jan 22, 36; m 63; c 6. ORTHOPEDICS. Educ: St Peter's Col, NJ, BA, 57; NJ Col Med, MD, 61. Prof Exp: Asst prof orthop surg, Vanderbilt Univ, 68-73, asst prof biomed eng, 72-73; PROF ORTHOP SURG & REHAB, MED CTR, UNIV NEBR, OMAHA, 74- Concurrent Pos: Dir amputee clin & dir cerebral palsy clin, Med Ctr, Vanderbilt Univ, 68-73; chief orthop surg, Nashville Vet Admin Hosp, 68-73, Vet Admin Fund grant biomech fractures, 69-73; chief orthop surg, Omaha Vet Admin Hosp, 74- Mem: Orthop Res Soc. Res: Pathophysiology and biomechanics of trauma and fracture healing, including analysis of fracture union by external nondestructive measurement techniques. Mailing Add: Dept of Orthop Univ of Nebr Med Ctr Omaha NE 68105

CONNOLLY, JOHN IRVING, JR, b Boston, Mass, June 23, 36. LOW TEMPERATURE PHYSICS, OPTICS. Educ: Mass Inst Technol, BS, 58; Univ Ill, Urbana, MS, 59, PhD(physics), 65; Northeastern Univ, MS, 71. Prof Exp: Fulbright lectr physics, Coun Sci Invests, 65-66; tech staff mem, Mitre Corp, 66-74; MEM STAFF, SCI APPLNS, INC, 74- Mem: Am Phys Soc; Soc Indust & Appl Math; Inst Elec & Electronics Engrs. Res: Laser technology with emphasis on operational devices. Mailing Add: Sci Applns Inc 1911 N H Myer Dr Rosslyn VA 22209

CONNOLLY, JOHN JOSEPH, b Council Bluffs, Iowa, Oct 23, 34; m 57; c 4. PHYSIOLOGY, IMMUNOLOGY. Educ: Creighton Univ, MD, 59. Prof Exp: Instr, 67-68, ASST PROF MED, SCH MED, YALE UNIV, 68-; DIR EMPHYSEMA UNIT, VET ADMIN HOSP, WEST HAVEN, 67- Mem: AMA; Am Thoracic Soc. Res: Pulmonary physiology and immunology. Mailing Add: Dept of Internal Med Yale Univ Sch of Med New Haven CT 06510

CONNOLLY, JOHN W, b Cincinnati, Ohio, Apr 4, 36; m 60; c 2. INORGANIC CHEMISTRY, BIOLOGICAL CHEMISTRY. Educ: Xavier Univ, Ohio, BS, 58; Purdue Univ, PhD(chem), 63. Prof Exp: Asst chem, Yale Univ, 62-64; asst prof, Marietta Col, 64-65; asst prof, 65-74, ASSOC PROF CHEM, UNIV MO-KANSAS CITY, 74- Mem: AAAS; Am Chem Soc. Res: Organometallic chemistry and enzyme model chemistry. Mailing Add: Dept of Chem Univ of Mo Kansas City MO 62214

CONNOLLY, JOHN WILLIAM DOMVILLE, b South Porcupine, Ont, July 18, 38; m 62; c 2. THEORETICAL PHYSICS, SOLID STATE PHYSICS. Educ: Univ Toronto, BA, 60; Univ Fla, PhD(physics), 66. Prof Exp: Sr res assoc, Pratt & Whitney Aircraft Div, United Aircraft Corp, Conn, 67-70; ASSOC PROF PHYSICS & MEM STAFF, QUANTUM THEORY PROJ, UNIV FLA, 70- Concurrent Pos: Res affil, Mass Inst Technol, 69. Mem: AAAS; Am Phys Soc. Res: Electronic structure of solids. Mailing Add: Dept of Physics Univ of Fla Gainesville FL 32601

CONNOLLY, LEWIS TIMOTHY, b Indiana, Pa, Dec 21, 30; m 52; c 5. PHOTOGRAPHIC CHEMISTRY. Educ: Univ Rochester, BS, 63. Prof Exp: Res chemist, 63-67, sr chemist, 67-71, res assoc, 71-73, LAB HEAD PHOTOG CHEM, EASTMAN KODAK RES LABS, 73- Mem: Soc Photog Scientists & Engrs; Tech Asn Graphic Arts. Res: Photographic materials used in the graphic arts; processing of photographic materials. Mailing Add: 250 Rhea Crescent Rochester NY 14615

CONNOLLY, PHILIP LOUIS, b New Glasgow, NS, Can, Jan 9, 30; m 62. PHYSICS. Educ: St Francis Xavier, BA, 50; Catholic Univ, MSc, 52; Cornell Univ, PhD(physics), 59. Prof Exp: Res assoc physics, Cornell Univ, 59-60; asst physicist, 60-62, ASSOC PHYSICIST, BROOKHAVEN NAT LAB, 62- Concurrent Pos: Ford Found vis scientist, Europ Orgn Nuclear Res, Switz, 64-65. Mem: Am Phys Soc. Res: High energy physics; bubble chambers; data processing. Mailing Add: Dept of Physics Bldg 510 Brookhaven Nat Lab Upton NY 11973

CONNOLLY, THOMAS WORTHINGTON, b Washington, DC, June 17, 23; m 49; c 4. NUCLEAR PHYSICS. Educ: Ga Inst Technol, BChE, 49; Columbia Univ, MA, 53. Prof Exp: Res officer nuclear weapons effects, Chem Corps Sch, US Army, 50-52, Res & Eng Command, 52-53, cryogenic eng, Nat Bur Stand, 53-55 & Armed Forces Spec Weapons Proj, 55-57, from instr to asst prof chem, US Mil Acad, 58-61, res officer, Army Mat Command, 64-65; RES SCIENTIST, KAMAN SCI CORP, 65- Res: Scientific applications of nuclear weapons, their employment and effects. Mailing Add: 12830 Falcon Dr Colorado Springs CO 80908

CONNOLLY, WALTER CURTIS, b Marysville, Ohio, May 1, 22; m 44; c 2. PHYSICS. Educ: Miami Univ, AB, 44; Univ Ill, MS, 46; Cath Univ Am, PhD, 54. Prof Exp: Asst physics, Univ Ill, 44-46; assoc prof, US Naval Acad, 46-55; sr scientist, Westinghouse Atomic Power, 55-56; assoc prof physics, Ala Polytech Inst, 56-58; sr physicist, Res Lab Eng Sci, Univ Va, 58-63; PROF PHYSICS, APPALACHIAN STATE UNIV, 63- Mem: Am Phys Soc; Am Asn Physics Teachers. Res: Astronomy. Mailing Add: Dept of Physics Appalachian State Univ Boone NC 28608

CONNON, NEIL WILLIAM, organic chemistry, physical chemistry, see 12th edition

CONNOR, CHARLES ASHLEY RICHARD, cardiovascular diseases, deceased

CONNOR, DANIEL HENRY, b Aylmer, Ont, Mar 26, 28; US citizen; m 53; c 3. PATHOLOGY. Educ: Queen's Univ, MD, CM, 53. Prof Exp: From intern med to jr resident path, Emergency Hosp, Washington, DC, 53-55; from resident to chief resident path, Med Sch Hosp, George Washington Univ, 55-57; chief lab serv, Irwin Army Hosp, Ft Riley, Kans, 57-59; assoc pathologist, Liver & Pediat Br, Armed Forces Inst Path, 59-60, assoc pathologist, Skin & Gastrointestinal Br, 60-61; prin investr study path of endomyocardial fibrosis, WHO, 62-64; assoc pathologist, Infectious Dis Br, 64-67, chmn dept infectious & parasitic dis path, 74, CHIEF INFECTIOUS DIS BR, ARMED FORCES INST PATH, 67-, CHIEF GEOG PATH DIV, 70- Concurrent Pos: Dir path, US Med Res & Develop Proj, Kampala, Uganda, 62-64; hon lectr path, Makerere Col Med Sch, Kampala, 62-64; adv African cardiopathies, WHO, 64; assoc mem, Comn Parasitic Dis, Armed Forces Epidemiol Bd, 69; consult med parasitol, Am Bd Path, 72, mem test comt med microbiol-med parasitol, 74-; consult onchocerciasis, WHO, 72, mem sci adv panel, Onchocerciasis Control Prog, Volta River Basin, 74-79, consult, Spec Comt Parasitic Dis, 76-81. Honors & Awards: Official Commendations, Dept of Army, 64 & 71, Decoration for Meritorious Civilian Serv, 71. Mem: Int Acad Path; Am Asn Pathologists & Bacteriologists; Am Soc Exp Path; Col Am Pathologists; Am Soc Trop Med & Hyg. Res: Pathogenesis of tropical and exotic infectious diseases, especially those of tropical Africa, including onchocerciasis, Mycobacterium ulcerans infection, endomyocardial fibrosis, streptocerciasis and others. Mailing Add: Infectious & Parasitic Dis Path Armed Forces Inst of Path Washington DC 20306

CONNOR, DANIEL S, b Cleveland, Ohio, Feb 25, 38. SYNTHETIC ORGANIC CHEMISTRY, APPLIED CHEMISTRY. Educ: Brown Univ, ScB, 60; Yale Univ, PhD(chem), 65. Prof Exp: RES CHEMIST, MIAMI VALLEY LABS RES DIV, PROCTER & GAMBLE CO, 65- Mem: Am Chem Soc; Sigma Xi. Res: Analytical and polymer chemistry; immunochemistry. Mailing Add: Miami Valley Labs Procter & Gamble Co Cincinnati OH 45239

CONNOR, DAVID THOMAS, b Batley, Eng, Nov 6, 39; m 67; c 2. ORGANIC CHEMISTRY. Educ: Univ Manchester, BSc, 62, MSc, 64, PhD(org chem), 65. Prof Exp: Res assoc org chem, Univ Chicago, 65-68; SCIENTIST, WARNER-LAMBERT RES INST, 69- Mem: Am Chem Soc. Res: Steroid chemistry; organic reaction mechanisms; synthetic methods for heterocyclic chemistry; biological activity of organic molecules. Mailing Add: Warner-Lambert Res Inst Morris Plains NJ 07950

CONNOR, DONALD W, b Chicago, Ill, Jan 2, 23; m 43; c 4. SOLID STATE PHYSICS. Educ: Univ Chicago, SB, 43, SM, 48, PhD, 60. Prof Exp: Jr physicist, Argonne Nat Lab, 46-49; elec engr, Univ Chicago, 49-50; assoc physicist, Brookhaven Nat Lab, 51; scientist, Chicago Midway Labs, 51-52; assoc physicist, 52-69, physicist, Solid State Div, 69-74, SR PHYSICIST, ENVIRON STATEMENT PROJ, ARGONNE NAT LAB, 74- Mem: AAAS; Inst Elec & Electronic Eng; Am Nuclear Soc; Sigma Xi. Res: Nuclear moments; lattice vibrations; neutron scattering. Mailing Add: Environ Statement Proj Argonne Nat Lab 9700 S Cass Ave Argonne IL 60439

CONNOR, FRANK FIELD, b Chicago, Ill, June 15, 32; m 58; c 2. MATHEMATICS. Educ: Ill Inst Technol, BS, 54, MS, 56, PhD(math), 59. Prof Exp: Instr math, Ill Inst Technol, 58-60; res instr, La State Univ, 60-61, from asst prof to assoc prof, 61-70; chmn dept, 70-75, PROF MATH, N TEX STATE UNIV, 70- Mem: Am Math Soc; Math Asn Am. Res: Measure and integration; linear operators. Mailing Add: 1813 Stonegate Dr Denton TX 76201

CONNOR, JAMES D, b Collenton, SC, Nov 19, 26; m; c 3. PEDIATRICS, MICROBIOLOGY. Educ: Clemson Col, BS, 53; Med Col SC, MD, 53; Am Bd Pediat, dipl, 59. Prof Exp: Intern, Wayne County Gen Hosp, Eloise, Mich, 53-54; resident pediat, Children's Hosp, Mich, 54-55; instr, Sch Med, Wayne State Univ, 55-57; instr, Sch Med, Univ Miami, 57-59, clin asst prof, 59-60, from asst prof to assoc prof, 61-69; ASSOC PROF PEDIAT, UNIV CALIF, SAN DIEGO, 70- Concurrent Pos: Chief resident pediat, Children's Hosp, Mich, 55-57; med coordr, Variety Children's Hosp, Miami, 58-60, res assoc virol, Variety Children's Res Found, 60-; USPHS fel virol, 60-62; res asst prof microbiol, Univ Miami, 61-70. Mem: AAAS; AMA; Am Acad Pediat; Am Pub Health Asn; Am Soc Microbiol. Res: Parasitic infections of the intestinal tract; infectious diarrheas; influenza in infants; pertussis. Mailing Add: Dept of Pediat & Microbiol Univ of Calif San Diego PO Box 109 La Jolla CA 92037

CONNOR, JAMES EDWARD, JR, b New Haven, Conn, Feb 14, 24; m 51; c 6. PHYSICAL ORGANIC CHEMISTRY. Educ: Harvard Univ, BA, 44, MS, 48, PhD(phys org chem), 49. Prof Exp: Sr res chemist, Atlantic Richfield Co, Glenolden, 49-54, supv chemist, 54-60, asst mgr basic res div, 60-61, mgr chem res, 61-62, mgr res div, Res & Develop Dept, Arco Chem Co Div, 62-71, MGR RES & DEVELOP DIV, RES & ENG DEPT, ARCO CHEM CO DIV, ATLANTIC RICHFIELD CO, GLENOLDEN, 71- Mem: Am Chem Soc. Res: Petroleum refining methods, especially hydrocarbon reactions and catalytic reactions; petrochemicals. Mailing Add: 1421 Hillside Rd Wynnewood PA 19096

CONNOR, JOHN D, b Coatesville, Pa, Jan 15, 33; m 63; c 2. PHARMACOLOGY. Educ: Philadelphia Col Pharm, BS, 60, MS, 62, PhD(pharmacol), 66. Prof Exp: Lab asst zool, Philadelphia Col Pharm, 60-62, lab asst pharmacol, 62-63; res assoc neuropharmacol, East Pa Psychiat Inst, 63-66; from asst prof to assoc prof, 69-75, PROF PHARMACOL, HERSHEY MED CTR, PA STATE UNIV, 75- Concurrent Pos: NIMH lab fel, 66-69. Mem: Am Soc Pharmacol & Exp Therapeut; Soc Neurosci; Sigma Xi. Res: Pharmacology and physiology of involuntary motor activity; temperature regulation; synaptic transmission in the central nervous system. Mailing Add: Dept of Pharmacol Hershey Med Ctr Pa State Univ Hershey PA 17033

CONNOR, JON JAMES, b Columbus, Ohio, Dec 12, 32; m 63. GEOLOGY. Educ: Ohio State Univ, BSc, 55; Univ Colo, PhD(geol), 63. Prof Exp: Geologist, 55-72, CHIEF BR REGIONAL GEOCHEM, US GEOL SURV, 72- Concurrent Pos: Liaison mem, US Nat Comt Geochem, Nat Acad Sci-Nat Res Coun, 72- Mem: Soc Environ Geochem & Health; Geol Soc Am; Int Asn Math Geol. Res: Geology of Colorado Plateau and Black Hills uranium deposits; groundwater investigations around Carlsbad, New Mexico; regional geochemistry of sedimentary rocks; environmental geochemistry and its relation to human and animal health. Mailing Add: US Geol Surv Denver Fed Ctr Lakewood CO 80225

CONNOR, JOSEPH GERARD, JR, b West Chester, Pa, Aug 15, 36; m 61; c 3. ENGINEERING PHYSICS. Educ: Georgetown Univ, BS, 57; Pa State Univ, MS, 61, PhD(physics), 63. Prof Exp: RES PHYSICIST, NAVAL SURFACE WEAPONS CTR, WHITE OAK, 63- Concurrent Pos: Lectr, Montgomery Jr Col, 64-65 & Trinity Col, DC, 65-72. Mem: AAAS; Am Asn Physics Teachers; Am Phys Soc. Res: Structural response; explosion effects. Mailing Add: 17805 Dominion Dr Sandy Spring MD 20860

CONNOR, LAWRENCE JOHN, b Kalamazoo, Mich, Aug 15, 45; m 68; c 2. EMTOMOLOGY, APICULTURE. Educ: Mich State Univ, BS, 67, MS, 69, PhD(entom), 72. Prof Exp: ASST PROF ENTOM, OHIO STATE UNIV, 72-; EXTEN ENTOMOLOGIST, OHIO COOP EXTEN SERV, 72-; ASST PROF ENTOM, OHIO AGR RES & DEVELOP CTR, 75- Mem: Sigma Xi; Bee Res Asn. Res: Crop pollination requirements and mechanisms. Mailing Add: 1735 Neil Ave Columbus OH 43210

CONNOR, NOLEN DUNCAN, b Montgomery, Ala, Oct 13, 21; m 44; c 4. PHARMACOLOGY, PATHOLOGY. Educ: Ala Polytech Inst, DVM, 48; Ohio State Univ, MS, 50. Prof Exp: From instr to asst prof anat & hist, Ala Polytech Inst, 47-51; res scientist, Upjohn Co, 51-66, vet chg, Vet Exp Sta, 56-64, sr scientist, 58-71; ASSOC PROF COMP MED & ASSOC DIR DIV LAB ANIMAL RESOURCES, SCH MED, WAYNE STATE UNIV, 71- Concurrent Pos: Mem, Nat Pest Control Asn. Mem: AAAS; Am Vet Med Asn; Am Soc Vet Physiol & Pharmacol; Indust Vet Asn; Royal Soc Health. Res: Veterinary and microscopic anatomy; embryology; tuberculosis; toxicopharmacology; endocrinology; parasitology; clinical drug testing; vertebrate pest control research. Mailing Add: Div of Lab Animal Resource Sch of Med Wayne State Univ Detroit MI 48207

CONNOR, RALPH (ALEXANDER), b Newton, Ill, July 12, 07; m 31; c 1. CHEMISTRY. Educ: Univ Ill, BS, 29; Univ Wis, PhD(org chem), 32. Hon Degrees: DSc, Phila Col Pharm, 54, Univ Pa, 59, Polytech Inst Brooklyn, 67; LLD, Lehigh Univ, 66. Prof Exp: Asst chem, Univ Wis, 29-31; instr org chem, Cornell Univ, 32-35; from asst prof to assoc prof, Univ Pa, 35-44; assoc dir res, Rohm and Haas Co & Resinous Prod & Chem Co, 45-48, vpres res, Rohm and Haas Co, 48-70, dir & mem exec comt, 49-73, chmn bd, 60-70, vpres & chmn exec comt, 70-73; RETIRED. Concurrent Pos: Res chemist, du Pont Co, 34; tech aide, sect chief & div chief, Nat Defense Res Comn, 41-46; mem div chem & chem tech, Nat Res Coun, 53-58; mem, Tech Adv Panel Biol & Chem Warfare, 54-60; chmn, US Nat Comt on Int Union Pure & Appl Chem; bd dirs, Ursinus Col, 71- Honors & Awards: Naval Ord Develop Award; Medal for Serv in Cause of Freedom (Brit); Medal for Merit; Gold Medal, Am Inst Chemists, 63, Chem Pioneer Award, 68; Chem Indust Medal, Soc Chem Indust, 65; Priestley Medal, Am Chem Soc, 67; Outstanding Civilian Serv Award, US Dept Army, 70; Achievement Award, Univ Ill, 71. Mem: Am Chem Soc; Am Inst Chemists; Soc Chem Indust. Res: Organic chemistry; catalysis; synthesis; explosives; mechanisms. Mailing Add: 234 N Bent Rd Wyncote PA 19095

CONNOR, ROBERT DICKSON, b Edinburgh, Scotland, May 15, 22; m 48; c 2. NUCLEAR PHYSICS. Educ: Univ Edinburgh, BSc, 42, PhD, 49. Prof Exp: From asst lectr to lectr physics, Univ Edinburgh, 48-57; assoc prof, 57-60, assoc dean arts & sci, 63-70, PROF PHYSICS, UNIV MAN, 60-, DEAN SCI, 70- Concurrent Pos: Mem Fisheries Res Bd Can, 67-; mem univ grants comn, Prov Man, 68- Mem: Can Asn Physicists; Sigma Xi; fel Brit Inst Physics. Res: Alpha, beta and gamma ray spectroscopy. Mailing Add: 239 N E M P Univ of Man Winnipeg MB Can

CONNOR, ROBERT SHERMAN, b Ann Arbor, Mich, Sept 6, 18; m 44; c 2. PARASITOLOGY. Educ: Mich State Univ, BS, 49, MS, 51; Purdue Univ, PhD(zool, parasitol), 56. Prof Exp: From instr to asst prof, 56-69, ASSOC PROF BIOL, UTICA COL, SYRACUSE UNIV, 69- Mem: Am Soc Parasitol. Res: Taxonomy of digenetic trematodes; comparative histology. Mailing Add: Dept of Biol Utica Col of Syracuse Univ Utica NY 13502

CONNOR, STEPHEN R, b Philadelphia, Pa, Mar 22, 40; m 63; c 2. PLANT PATHOLOGY, PLANT PHYSIOLOGY. Educ: Ursinus Col, BS, 63; Univ Del, MS, 65, PhD(plant path & physiol), 68. Prof Exp: Res asst, Univ Del, 65-67; scientist, 68, plant pathologist, 68-72, new prod develop mgr, 72-75, RES & DEVELOP FIELD SUPVR, ROHM AND HAAS CO, 75- Mem: AAAS; Am Phytopath Soc. Res: Eradicant fungicides; epiphytology of plant pathogens; moisture stress and diseases; physiology of microorganisms. Mailing Add: Rohm and Haas Co Spring House PA 19095

CONNOR, THOMAS BYRNE, b Baltimore, Md; m 57; c 2. MEDICINE. Educ: Loyola Col, Md, BA, 43; Univ Md, MD, 46. Prof Exp: Intern, Mercy Hosp, Baltimore, 46-47, resident, 49-51; from asst prof to assoc prof, 56-67, PROF MED, SCH MED, UNIV MD, BALTIMORE, 67-, DIR DIV ENDOCRINOL & METAB, 56-, DIR CLIN RES CTR, 62- Concurrent Pos: Fel endocrine & metab dis, Johns Hopkins Hosp, 51-56; asst physician outpatient dept, Diabetic Clin & Endocrine Clin, Johns Hopkins Hosp, 51-59; staff physician, Univ Md Hosp, 56-; consult med, Mercy Hosp, Baltimore, 60- & Baltimore Vet Admin Hosp, 65- Mem: AAAS; Endocrine Soc; Am Diabetes Asn; Am Fedn Clin Res; fel Am Col Physicians. Res: Clinical research in calcium and bone metabolism, parathyroid disorders and hypertension. Mailing Add: Dept of Med Sch of Med Univ of Md Baltimore MD 21201

CONNOR, WILLIAM ELLIOTT, b Pittsburgh, Pa, Sept 14, 21; m 46; c 3. INTERNAL MEDICINE. Educ: Univ Iowa, BA, 42, MD, 50; Am Bd Internal Med, dipl, 57; Am Bd Nutrit, dipl, 67. Prof Exp: Intern, USPHS Hosp, Calif, 50-51; asst med resident, San Joaquin Gen Hosp, 51-52; mem med staff, Enloe Hosp, 52-54; med resident, Vet Admin Hosp, Iowa, 54-56; from instr to prof internal med, Univ Iowa, 56-75, dir clin res ctr, 68-75; PROF MED, MED SCH, UNIV ORE, 75- Concurrent Pos: Am Heart Asn res fel, Univ Iowa, 56-58; Am Col Physicians traveling fel, Oxford Univ, 60; vis prof, Med Cols & Basic Sci Med Inst, Karachi, Pakistan, 61; ed, J Lab & Clin Med, 70-73; mem & chmn, Coun Arteriosclerosis, Am Heart Asn. Mem: AAAS; Am Inst Nutrit; Am Soc Clin Invest; Asn Am Physicians; Am Fedn Clin Res. Res: Atherosclerosis; lipid metabolism; nutrition. Mailing Add: Dept of Med Univ of Ore Med Sch Portland OR 97201

CONNOR, WILLIAM GORDEN, b El Paso, Tex, Nov 1, 36. MEDICAL PHYSICS. Educ: Tex Western Col, BSc, 62; Vanderbilt Univ, MSc, 64; Univ Calif, Los Angeles, PhD(med physics), 70. Prof Exp: Physicist, Michael Reese Hosp, Chicago, Ill, 64-66; asst prof radiol, Univ Wis-Madison, 70-72; ASST PROF RADIOL, UNIV ARIZ, 72- Mem: Am Asn Physicists in Med; Health Physics Soc. Res: Application of ionizing radiations to the treatment of malignant diseases; cellular repair of damage due to ionizing radiations. Mailing Add: Dept of Radiol Univ of Ariz Med Ctr Tucson AZ 85724

CONNOR, WILLIAM KEITH, b Houston, Tex, Dec 19, 31; m 59; c 3. ACOUSTICS. Educ: Rice Inst, BA, 55; Southern Methodist Univ, MS, 61. Prof Exp: Engr trainee, Gen Motors Proving Ground, 55-56; proj engr, Electro-Mech Labs, White Sands Proving Ground, 57-58, Gen Motors Proving Ground, 58-59 & Noise & Vibration Lab, 61-62; consult acoust, Rudmose Assocs Inc, 62-63; consult acoust, 63-69, dir acoust res dept, 69-70, DIR ACOUST & SOCIOMETRIC RES DEPT, TRACOR, INC, 70- Concurrent Pos: Vis lectr, Sch Archit & Planning, Univ Tex, Austin, 73-75. Mem: Inst Noise Control Eng; Sigma Xi; Acoust Soc Am; Audio Eng Soc. Res: Community noise, psychoacoustics; architectural acoustics; noise and vibration control; electroacoustics. Mailing Add: Tracor Inc 6500 Tracor Lane Austin TX 78721

CONNORS, KENNETH A, b Torrington, Conn, Feb 19, 32. PHARMACEUTICS, ANALYTICAL CHEMISTRY. Educ: Univ Conn, BS, 54; Univ Wis, MS, 57, PhD(pharm), 59. Prof Exp: Res assoc phys-org chem, Ill Inst Technol, 59-60 & Northwestern Univ, 60-62; from asst prof to assoc prof pharmaceut anal, 62-70, asst dean grad studies, 68-72, PROF PHARMACEUT ANAL, UNIV WIS-MADISON, 70- Concurrent Pos: NIH fel, 60-62. Mem: AAAS; Am Chem Soc; Am Pharmaceut Asn. Res: Pharmaceutical analysis; mechanisms of organic reactions; molecular complexes. Mailing Add: Sch of Pharm Univ of Wis Madison WI 53706

CONNORS, NATALIE ANN, b St Louis, Mo. HISTOLOGY, HISTOCHEMISTRY. Educ: St Louis Univ, BS, 50, PhD(anat), 68; Univ Ill, Urbana-Champaign, MS, 52. Prof Exp: Technician, Monsanto Chem Co, Ill, 52-56; res assoc anat, 56-61, teaching asst, 61-68, instr, 68-70, ASST PROF ANAT, SCH MED, ST LOUIS UNIV, 70- Mem: Sigma Xi; Am Asn Univ Profs; AAAS. Res: Histochemical study of the developing chick retina; effects of antiviral drugs on laboratory animals; teratogenic effects of chemical substances on the developing chick. Mailing Add: Dept of Anat St Louis Univ Sch of Med St Louis MO 63104

CONNORS, PHILIP IRVING, b Norfolk, Va, Oct 7, 37; m 59; c 3. ACADEMIC ADMINISTRATION, EXPERIMENTAL NUCLEAR PHYSICS. Educ: Univ Notre Dame, BS, 59; Pa State Univ, MS, 62, PhD(physics), 66. Prof Exp: Asst physics, Pa State Univ, 59-63; jr res assoc, Brookhaven Nat Lab, 63-65; res assoc, Dept Physics & Astron, Univ Md, 65-69, asst prof physics & dir col sci improv prog, 69-75; ASSOC PROF & CHMN, DIV ENVIRON & NATURAL SCI, NORTHERN VA COMMUNITY COL, WOODBRIDGE CAMPUS, 75- Concurrent Pos: Field Ctr Coordr, NSF Chautauqua Short Courses, 71-75. Mem: Fel AAAS; Am Phys Soc; Am Asn Physics Teachers; Nat Sci Teachers Asn. Res: Education and teacher training in physics on all levels; low energy nuclear experimental physics. Mailing Add: Northern Va Commun Col 15200 Smoketown Rd Woodbridge VA 22191

CONNORS, THEODORE THOMAS, b San Diego, Calif, Dec 11, 25; m 57; c 2. PETROLEUM, GEOGRAPHY. Educ: San Diego State Col, BA, 50; Univ Calif, Los Angeles, MA, 63, CPhil, 69. Prof Exp: Asst social scientist, Rand Corp, 54; geog analyst, Cent Intel Agency, 55-56; asst social scientist, Rand Corp, 56; guest lectr Eng, Asia Found, Okayama Univ, 56-58; assoc systs anal, Planning Res Corp, 58-64; PHYS SCIENTIST SYSTS ANAL, RAND CORP, 64- Mem: AAAS; Asn Am Geogr; Asn Pac Coast Geogr; Am Defense Preparedness Asn. Res: Analysis of problems concerned with international production, distribution and consumption of petroleum; environmental context for military systems; cross-country mobility; systems analysis of tactical weapons systems. Mailing Add: Rand Corp 1700 Main St Santa Monica CA 90406

CONNORS, WILLIAM MATTHEW, b Canandaigua, NY, Sept 16, 21. BIOCHEMISTRY, INDUSTRIAL CHEMISTRY. Educ: St Bonaventure Col, BS, 42; Univ Southern Calif, MS, 47. Hon Degrees: PhD, James Martin Col, 69. Prof Exp: Chemist, Pillsbury Mills, 45; group leader, Nat Dairy Res Labs, Inc, 49-56; anal supvr, Gen Cigar Res & Develop Ctr, 56-73; MGR, CONNORS RES ASSOCS, 73- Concurrent Pos: Mem Manhattan Proj, 44-45. Honors & Awards: Meritorious Serv Award, Am Inst Chemists, 72. Mem: Fel AAAS; Am Chem Soc; fel Am Inst Chemists; Am Soc Microbiol. Res: Biochemistry and analytical chemistry of tobacco products; commercial production of enzymes; nutrition of dairy products; energetics of ATPase; pharmacology and toxicology of uranium compounds; industrial chemical consulting; research and development. Mailing Add: 314 College Ave Lancaster PA 17603

CONOLLY, JOHN R, b Sydney, Australia, July 23, 36; m 70. GEOLOGY, NATURAL RESOURCES. Educ: Univ Sydney, BSc, 58; Univ New South Wales, MSc, 60, PhD(geol), 63. Prof Exp: Sr demonstr geol, Univ New South Wales, 60-63; Ford Found fel, Lamont Geol Observ, NY, 63-65; vis prof, La State Univ, 65-66; Queen Elizabeth fel, Univ Sydney, 66-68; from assoc prof to prof, Univ SC, 69-72; explor geologist, B P Alaska Explor Inc, 72-74; PRES, ERA NORTH AM INC, 75- Concurrent Pos: Fulbright travel award, 63-66; consult, Scripps Inst Oceanog, 66; consult geologist, Univ Sydney, 68; consult, Oceanog Off, US Navy, 69-; consult, John R Conolly & Assocs Inc, 74-; adj assoc prof, Columbia Univ, 73-75; adj prof, City Col New York, 75 & C W Post Col, Long Island Univ, 75-76. Honors & Awards: Olle Prize, Royal Soc NSW, 67. Mem: Fel Geol Soc Am; Am Asn Petrol Geol; Geol Soc Australia; Australian Inst Mining & Metall; Petrol Explor Soc NY (vpres, 75-76). Res: Sedimentology of recent and ancient rocks; marine geology; petrology; glacial marine geology; origin of continental margins and geosyndines; exploration geology. Mailing Add: ERA NAm Inc 200 Railroad Ave Greenwich CT 06830

CONOMOS, TASSO JOHN, b New Kensington, Pa, Sept 11, 38; m 69; c 3. OCEANOGRAPHY, GEOCHEMISTRY. Educ: San Jose State Univ, BS, 61, MS, 63; Univ Wash, PhD(oceanog), 68. Prof Exp: Tech asst geol, San Jose State Univ, 60-62; phys sci technician paleont & stratig, US Geol Surv, 62-63; res assoc oceanog, Univ Wash, 63-64 & 68-69, teaching assoc, 64-65; intern, Smithsonian Inst, 66-68; fel, 69-70, RES OCEANOGR, WATER RESOURCES DIV, US GEOL SURV, 70- Mem: AAAS; Soc Econ Paleontologists & Mineralogists; Am Soc Limnol & Oceanog. Res: Geochemistry and distribution of suspended particulate matter in river-ocean mixing systems; descriptive chemical oceanography of near-shore and in-shore waters; sedimentological—geochemical studies of biogenic sediments. Mailing Add: Water Resources Div US Geol Surv 345 Middlefield Rd Menlo Park CA 94025

CONOMY, JOHN PAUL, b Cleveland, Ohio, July 31, 38; m 63; c 3. NEUROLOGY. Educ: John Carroll Univ, BS, 60; St Louis Univ, MD, 64. Prof Exp: Intern med, St Louis Univ Hosps, 64-65; resident neurol, Univ Hosps Cleveland, Case Western Reserve Univ, 65-68, fel neuropath, Cleveland Metrop Gen Hosp, 68; neurologist, US Air Force, 69 & 70; res fel neuroanat, Univ Pa, 70-71; ASST PROF MED, CASE WESTERN RESERVE UNIV MED SCH, 72-; CHMN DEPT NEUROL, CLEVELAND CLIN FOUND, 75- Concurrent Pos: Career teaching fel award, Case Western Reserve Univ, 70; consult, Vet Admin Hosps, 72-; grants in aid, Vet Admin, 72 & 74, Mary B Lee Fund, 74, Mellon Fund & Reinberger Found, 76. Mem: Am Acad Neurol; Asn Res Nerv & Ment Dis; Soc Neurosci; fel Am Col Physicians; Asn Univ Profs Neurol. Res: Behavioral aspects of neurology, especially correlative studies of unit peripheral nerve and neuronal activity and behavior in animals and man; neurophysiologic action of central neurotransmitters. Mailing Add: Cleveland Clin Found 9500 Euclid Ave Cleveland OH 44106

CONOVER, CHARLES ALBERT, b Elizabeth, NJ, Apr 22, 34; m 57; c 1. ORNAMENTAL HORTICULTURE. Educ: Univ Fla, BSA, 62, MSA, 63; Univ Ga, PhD(plant sci), 70. Prof Exp: Asst ornamental horticulturist, Agr Exten Serv, 63-70, ORNAMENTAL HORTICULTURIST & CTR DIR, AGR RES CTR, UNIV FLA, 71- Concurrent Pos: Ornamental hort consult, United Brands Co, 69-74 & Rainbird, 74- Mem: Int Soc Hort Sci; Am Soc Hort Sci. Res: Tropical ornamental plant nutrition; acclimatization of tropical ornamental plants for interior use; development of synthetic soil media; propagation of tropical ornamentals; commercial production of tropical ornamental foliage crops. Mailing Add: Agr Res Ctr Rt 3 Box 580 Apopka FL 32703

CONOVER, JOHN HOAGLAND, b McKeesport, Pa, Oct 26, 16; m 40; c 2. METEOROLOGY. Prof Exp: Weather observer, Blue Hill Meteorol Observ, Harvard Univ, 36-40, chief observer, 40-47, res asst & tech mgr, 47-52, meteorologist, 52-59, tech mgr, 52-57, actg dir, 57-58; METEOROLOGIST, AIR FORCE CAMBRIDGE RES LAB, 59-; RES ASSOC, MT WASHINGTON OBSERV, 69- Concurrent Pos: Lab instr, Harvard Univ, 41-43; instr, US Weather Bur, 44; consult meteorologist, 47-56. Mem: Am Meteorol Soc. Res: Climatic change; microclimates in the arctic; cloud studies; satellite meteorology. Mailing Add: 15 Nobel Rd Dedham MA 02026

CONOVER, LLOYD HILLYARD, b Orange, NJ, June 13, 23; m 44; c 4. ORGANIC CHEMISTRY. Educ: Amherst Col, AB, 47; Univ Rochester, PhD(chem), 50. Prof Exp: Res chemist, Pfizer, Inc, 50-58, res supvr, 58-61, res mgr, 61-68, dir chem res-chemother, 68-73, MEM STAFF, PFIZER LTD, ENG, 73- Mem: Am Chem Soc; NY Acad Sci; Am Soc Microbiol; Marine Technol Soc; Int Col Trop Med. Res: Synthesis of heterocycles; hydrogenolysis of oxygen functions; structure and synthesis of tetracycline antibiotics; synthesis of antiparasitic agents; drugs of microbiological origin. Mailing Add: Pfizer Ltd Ramsgate Rd Sandwich Kent England

CONOVER, ROBERT ARMINE, b Lima, Ill, Nov 5, 16; m 46; c 3. BOTANY. Educ: Culver-Stockton Col, BS, 39; Univ Iowa, MS, 41; Univ Ill, PhD(plant path), 47. Prof Exp: Asst bot, Univ Iowa, 39-41; asst, Univ Ill, 41-42 & 46-47; from assoc plant pathologist to plant pathologist chg, Res & Educ Ctr, Univ Fla, Homestead, 47-65, plant pathologist & dir ctr, 65-74, PROF PLANT PATH, UNIV FLA, GAINESVILLE, 74- Mem: AAAS; Am Phytopath Soc; Am Soc Hort Sci. Res: Nature and control of vegetable diseases. Mailing Add: Dept of Plant Path Univ of Fla Gainesville FL 32611

CONOVER, ROBERT JAMES, biological oceanography, see 12th edition

CONOVER, THOMAS ELLSWORTH, b Plainfield, NJ, Nov 20, 31; m 66; c 3. BIOCHEMISTRY. Educ: Oberlin Col, BA, 53; Univ Rochester, PhD(biochem), 59. Prof Exp: Res assoc, Johnson Found, Univ Pa, 62-64; asst mem, Inst Muscle Dis, 64-69; ASSOC PROF BIOL CHEM, HAHNEMANN MED COL, 70- Concurrent Pos: Nat Found fel, Wenner-Gren Inst, Univ Stockholm, 58-60; USPHS fel, Pub Health Res Inst, New York, 60-62; Muscular Dystrophy Asn fel, Inst Gen Path, Univ Padua, 69-70. Mem: Am Soc Biol Chem. Res: Oxidative phosphorylation; mitochondrial structure and function; respiration and phosphorylation in cell nuclei; metabolic significance of cellular structures. Mailing Add: Dept of Biol Chem Hahnemann Med Col & Hosp Philadelphia PA 19102

CONOVER, WILLIAM JAY, b Hays, Kans, Dec 6, 36; m 60; c 5. STATISTICS. Educ: Iowa State Univ, BS, 58; Cath Univ, MA, 62, PhD(math statist), 64. Prof Exp: Asst prof statist, Kans State Univ, 64-67, assoc prof statist & comput sci, 67-73, PROF MATH & STATIST, TEX TECH UNIV, 73- Concurrent Pos: Consult, Water Resources Div, US Geol Surv, 62-68; NSF res grant, 67-68; NIH career develop award, 69-73; prof, Univ Zurich, 70-71; consult, Upjohn Co, 75; vis staff mem, Los Alamos Sci Labs, 75- Mem: Inst Math Statist; Biomet Soc; Am Statist Asn. Res: Nonparametric statistics; stochastic models in hydrology and hydraulics. Mailing Add: Dept of Math Tex Tech Univ Lubbock TX 79406

CONOVER, WOODROW WILSON, b Terre Haute, Ind, July 30, 47; m 72. PHYSICAL BIOCHEMISTRY. Educ: Rose Hulman Polytech Inst, BS, 69; Ind Univ, Bloomington, PhD(chem), 73. Prof Exp: Fel biochem, Univ Chicago, 73-75; SR RES ASSOC BIOCHEM & OPERS MGR, STANFORD MAGNETIC RESONANCE LAB, STANFORD UNIV, 75- Mem: Am Chem Soc. Res: Exploitation of nuclear magnetic resonance techniques in the study of fundamental biochemical mechanisms. Mailing Add: Stanford Magnetic Resonance Lab Stanford Univ Stanford CA 94305

CONOYER, JOHN WEEDON, b St Charles, Mo, Jan 6, 05; m 38; c 3. GEOGRAPHY. Educ: Culver-Stockton Col, AB, 36; Wash Univ, MS, 41. Prof Exp: Chmn dept geog, 43-71, prof classroom on wheels annual bus tour filed geog, 61-71, prof, 43-74, EMER PROF GEOG, ST LOUIS UNIV, 74- Concurrent Pos: TV prof, Series on mod world in perspective & Russia, Latin Am & Anglo-Am in perspective; consult, William H Sadlier Col. Mem: Asn Am Geog; Nat Coun Geog Educ. Res: Human geography-regional geography of Africa; geography of commerce. Mailing Add: Dept of Geog St Louis Univ St Louis MO 63103

CONRAD, BRUCE, b Ann Arbor, Mich, July 2, 43; m 64. TOPOLOGY. Educ: Harvey Mudd Col, BS, 64; Univ Calif, Berkeley, PhD(math), 69. Prof Exp: Asst prof, 69-74, ASSOC PROF MATH, TEMPLE UNIV, 74- Mem: AAAS; Am Math Soc; Math Asn Am. Res: Homology of groups; algebraic K-theory. Mailing Add: Dept of Math Temple Univ Philadelphia PA 19122

CONRAD, EDWARD EZRA, b Richmond, Calif, June 11, 27; m 51; c 3 5LOLID STATE PHYSICS, RADIATION CHEMISTRY. Educ: Univ Calif, BA, 50; Univ Md, MS, 55, PhD(radiation chem), 70. Prof Exp: Physicist solid state, Nat Bur Standards, 51-52; physicist, 52-70, CHIEF NUCLEAR RADIATION EFFECTS LAB, HARRY DIAMOND LABS, 70-, ACTG ASSOC TECH DIR, 75- Concurrent Pos: Secy Army res & develop fel, 59; lectr, Univ Md, 70-; rep, US Nat Comt Int Electrotech Comn, 58- Mem: Am Phys Soc; Inst Elec & Electronics Engrs. Res: Semiconductors; radiation effects; magnetics and dielectric measurements; radiation chemistry; pulse radiolysis. Mailing Add: Harry Diamond Labs 2800 Powder Mill Rd Adelphi MD 20783

CONRAD, EUGENE ANTHONY, b Clinton, Mass, Aug 15, 27; m 49; c 2. PHARMACOLOGY. Educ: Col of the Holy Cross, BS, 50; Univ NH, MS, 52; Vanderbilt Univ, PhD(pharmacol), 56. Prof Exp: Res bacteriologist, Charles Pfizer & Co, 52; asst pharmacol, Vanderbilt Univ, 55-56; instr physiol & pharmacol, Bowman Gray Sch Med, 56-58; res assoc pharmacol, Sterling-Winthrop Res Inst, 58-60, asst dir coord sect, 60-62, clin pharmacologist, 62-63; admin asst dept drugs, AMA, Ill, 63-65, dir drug doc sect, 64-66; dir res admin, Denver Chem Mfg Co, 66-67, dir clin res, 67-70; ASSOC DIR MED, PURDUE FREDERICK CO, NORWALK, 70- Concurrent Pos: Mem behav pharmacol comt, NIMH, 65-66. Mem: AAAS; Int Asn Dent Res; AMA; Drug Info Asn; Soc Pharmacol & Exp Therapeut. Res: Drug research. Mailing Add: 154 Cold Spring Rd 76 Stamford CT 06905

CONRAD, FRANKLIN, b Smithville, Ohio, Sept 27, 21; m 49; c 4. INDUSTRIAL CHEMISTRY. Educ: Col Wooster, BA, 43; Ohio State Univ, MS, 43, PhD(chem), 52. Prof Exp: Chemist, 52-55, supvr, 55-58, proj mgr chlorinated hydrocarbons, 58-60, supvr, 60-63, asst dir contract res, 63-66, DIR INDUST CHEM RES, ETHYL CORP, 66- Mem: Am Chem Soc. Res: Organometallics; metal hydrides; propellant chemicals; chlorinated hydrocarbons; all aspects of industrial chemicals research and development. Mailing Add: 1881 Madras Dr Baton Rouge LA 70815

CONRAD, GARY WARREN, b Amsterdam, NY, Mar 24, 41. DEVELOPMENTAL BIOLOGY. Educ: Union Col, BS, 63; Yale Univ, MS, 65, PhD(biol), 68. Prof Exp: NIH fel polysaccharide biochem, Univ Chicago, 68-70; asst prof, 71-75, ASSOC PROF DEVELOP BIOL, KANS STATE UNIV, 75- Concurrent Pos: Dahlgren res fel, Mt Desert Island Biol Lab, 71, ed bulletin & trustee, 75- Mem: Am Soc Cell Biol; Int Soc Develop Biologists; Sigma Xi; AAAS. Res: Differentiation of connective tissue; control of synthesis, polymerization and degradation of extracellular matrices; mosaic development; mechanisms of cell movement, cell adhesion and cell shape change. Mailing Add: Div of Biol Kans State Univ Manhattan KS 66506

CONRAD, GEOFFREY WENTWORTH, b Boston, Mass, Dec 24, 47; m 71. ANTHROPOLOGY. Educ: Harvard Univ, AB, 69, PhD(anthrop), 74. Prof Exp: Exhibits res archaeologist, 74-75, RES ASSOC ANTHROP, NAT MUS NATURAL HIST, SMITHSONIAN INST, 75- Concurrent Pos: Archaeologist, Md Geol Surv, 75- Mem: Soc Am Archaeol; Am Anthrop Asn; AAAS. Res: Archaeology of the Andean area, with emphasis on prehistoric urbanism of the North Coast of Peru. Mailing Add: 4021 Davis Pl NW Washington DC 20007

CONRAD, HARRY EDWARD, b Washington, DC, Jan 21, 29; m 52; c 2. BIOCHEMISTRY. Educ: La State Univ, BS, 49; Purdue Univ, MS, 52, PhD(biochem), 54. Prof Exp: Res chemist, Mead Johnson & Co, 54-58; res assoc, 58-60, from instr to assoc prof chem, 60-72, PROF BIOCHEM, UNIV ILL, URBANA-CHAMPAIGN, 72- Mem: Am Chem Soc; Am Soc Biol Chemists. Res: Chemistry and biochemistry of mucopolysaccharides; changes in metabolism of mucopolysaccharides and complex cell surface carbohydrates during embryonic development. Mailing Add: Dept of Biochem Univ of Ill Urbana IL 61801

CONRAD, HARRY RUSSELL, b Burlington, Ky, Oct 3, 25; c 2. NUTRITION. Educ: Univ Ky, BSc, 48; Ohio State Univ, MSc, 49, PhD(dairy sci), 52. Prof Exp: From instr to assoc prof, 52-64, PROF DAIRY SCI, OHIO AGR RES & DEVELOP CTR, 64- Res: Rumen physiology; digestion; nitrogen metabolism and growth in cattle. Mailing Add: Dept of Dairy Sci Ohio Agr Res & Develop Ctr Wooster OH 44691

CONRAD, HERBERT M, b New York, NY, Feb 20, 27; m 51; c 3. BIOCHEMISTRY, NUTRITION. Educ: Cornell Univ, BS, 49; Univ Southern Calif, MS, 59, PhD(biochem), 65. Prof Exp: Chemist, Calif Grape Prod Corp, 50-52; chemist, Star-Kist Foods, Inc, 52-54; lab dir, Long Beach Water Dept, 54-59; proj engr, NAm Aviation, Inc, 62-67; dir biochem, RPC Corp, 67-71; PRES, ECOL SYSTS CORP, 71- Concurrent Pos: Consult munic water dist. Mem: Fel AAAS; Am Chem Soc; NY Acad Sci. Res: The effect of weightlessness upon physiological processes; mechanisms of plant hormones; biodegradation of hazardous materials. Mailing Add: Ecol Systs Corp 2200 Colorado Ave Santa Monica CA 90404

CONRAD, JACK RANDOLPH, b Atlanta, Ga, July 25, 23; m 48; c 3. ANTHROPOLOGY. Educ: Emory Univ, AB, 49, MA, 51; Duke Univ, PhD(anthrop, 54. Prof Exp: From asst prof to assoc prof anthrop, 55-65, PROF ANTHROP & HEAD DEPT, SOUTHWESTERN AT MEMPHIS, 65- Concurrent Pos: NSF course improv grant, Southwestern at Memphis, 60-61; NIMH spec fel, Univ Calif, Berkeley, 63-64 & Yale Univ, 65-66; Rockefeller Found grant, Southwestern at Memphis, 67-70. Mem: Fel AAAS; fel Am Anthrop Asn; fel Royal Anthrop Inst. Res: Ethnopsychology; creativity; ethnoaesthetics; operant vs analytic ethnography. Mailing Add: 1751 Forrest Ave Memphis TN 38112

CONRAD, JOHN RUDOLPH, b San Antonio, Tex, Mar 21, 47; m 71. PLASMA PHYSICS. Educ: St Mary's Univ, Tex, BS, 68; Dartmouth Col, PhD(physics), 73. Prof Exp: Res assoc plasma physics, Inst Fluid Dynamics & Appl Math, Univ Md, 73-75; ASST PROF PLASMA PHYSICS, UNIV WIS, MADISON, 75- Mem: Am Phys Soc; Sigma Xi. Res: Experimental research in the areas of beam-plasma interaction; plasma transport properties and ion source technology. Mailing Add: Dept of Nuclear Eng Univ of Wis Madison WI 53706

CONRAD, JOHN TERRY, b New York, NY, Feb 13, 28; m 53; c 3. PHYSIOLOGY. Educ: NY Univ, AB, 51, MS, 55, PhD(muscle physiol), 61. Prof Exp: Res asst physiol, Wash Sq Col, NY Univ, 52-53; res asst radiobiol, Sloan-Kettering Inst, 53-54; res asst physiol, Sch Med, Yale Univ, 57-60, instr, 60-62; asst prof physiol, 62-67, ASSOC PROF PHYSIOL, BIOPHYS, OBSTET & GYNEC, SCH MED, UNIV WASH, 67-, CHMN PERINATAL BIOL, 75- Mem: AAAS; Am Physiol Soc. Res: Uterine muscle physiology; muscular dystrophy; radiation induced skin changes; bioelectrical phenomena in muscle. Mailing Add: Dept of Obstet & Gynec Sch of Med Univ of Wash Seattle WA 98105

CONRAD, JOSEPH H, b Cass Co, Ind, Dec 6, 26; m 50; c 4. ANIMAL NUTRITION, BIOCHEMISTRY. Educ: Purdue Univ, BSA, 50, MS, 54, PhD(animal nutrit), 58. Prof Exp: From instr to prof animal sci, Purdue Univ, 53-71; PROF ANIMAL NUTRIT & COORDR TROP ANIMAL SCI PROGS, UNIV FLA, 71- Concurrent Pos: Animal nutritionist from Purdue Univ, Brazil Tech Asst Prog, Agr Univ Minas Gerais, USAID, 61-65, hon prof, 65. Honors & Awards: Distinguished Nutrit Award, Distillers Feed Res Coun, 64. Mem: AAAS; Am Soc Animal Sci; Latin Am Soc Animal Sci; Brazilian Soc Animal Sci. Res: Swine and poultry nutrition; amino acids; proteins; minerals; trace minerals; unidentified factors and antibiotics; ruminant nutrition in tropical forage utilization; phosphorus and cobalt deficiencies. Mailing Add: Dept of Animal Sci Univ of Fla Gainesville FL 32611

CONRAD, LESTER I, b New York, NY, Mar 12, 12; m 38; c 2. COSMETIC CHEMISTRY. Educ: Brooklyn Col, BS, 33. Prof Exp: Res chemist immunochem, Jewish Hosp, Brooklyn, 34-39; tech dir & vpres, Am Cholesterol Prod, Inc, 39-70, PRES, AMERCHOL, CPC INT INC, EDISON, 70- Concurrent Pos: Mem praesidium, Int Fedn Socs Cosmetic Chem, 66-69, pres, 67-68; mem bd trustees, Col Pharmaceut Sci, Columbia Univ, 67-, secy, 69- Honors & Awards: Medal Award, Soc Cosmetic Chem, 68. Mem: Fel AAAS; fel Am Inst Chemists; Am Chem Soc; Soc Cosmetic Chem (treas, 59-61, pres, 63); Mex Soc Cosmetic Chemists (hon pres, 73). Res: Sterol chemistry as applied to surface activity; emulsifiers; lanolin derivatives. Mailing Add: 240 S Adelaide Ave Highland Park NJ 08904

CONRAD, LOUIS JOHNSON, b Harrisburg, Pa, Sept 28, 17; m 42; c 1. CHEMISTRY. Educ: Lebanon Valley Col, BS, 39; Univ NC, MA, 41; La State Univ, PhD, 50. Prof Exp: Sr res chemist, Manhattan Proj, E I du Pont de Nemours & Co, Inc, NJ, 41-50; asst chief chemist, Kind & Knox Gelatin Co, 50-55, vpres photog div, 55-69, vpres res, 69-74, TECH MGR, CAMDEN GELATIN DIV, PETER COOPER CORP, 74- Concurrent Pos: Asst, La State Univ, 48. Mem: AAAS; Am Chem Soc; Am Soc Photog Sci & Eng; Tech Asn Pulp & Paper Indust; Royal Photog Soc Gt Brit. Res: Photographic gelatin; photographic emulsions; analytical chemistry; emission spectroscopy; organic synthesis. Mailing Add: Camden Gel Div Peter Cooper Corp 1000 N Fifth St Camden NJ 08102

CONRAD, MALCOLM ALVIN, b Chicago, Ill, Apr 2, 27; m 53; c 2. MINERALOGY, CRYSTALLOGRAPHY. Educ: Mich Tech Univ, BS & MS, 52; Univ Mich, PhD(mineral), 60. Prof Exp: Geologist, Kennecott Copper Corp, 52-53; res assoc mineral, Univ Mich, 59-60; res engr, 60-62, res scientist, 62-66, CHIEF EXP

MINERAL, OWENS-ILL INC, 66- Mem: Mineral Soc Am; Am Ceramic Soc; Sci Res Soc Am. Res: Industrial mineralogy; glass-ceramics; ceramic raw materials. Mailing Add: Tech Ctr Owens-Ill Inc 1700 N Westwood Toledo OH 43607

CONRAD, MARCEL E, b New York, NY, Aug 15, 28; m 48; c 3. INTERNAL MEDICINE, HEMATOLOGY. Educ: Georgetown Univ, BS, 49, MD, 53; Nat Bd Internal Med, dipl, 61. Prof Exp: Intern med, Med Corps, US Army, 53-54, resident internal med, 55-58, chief resident, 58-59, asst chief hemat, 58-60 & 61-65, chief hemat, Walter Reed Army Med Ctr, 65-74, dir div med, 69-71, dir clin invest serv, 71-74; PROF MED & DIR DIV HEMAT & ONCOL, UNIV ALA, BIRMINGHAM, 74- Concurrent Pos: From clin asst prof to clin assoc prof med, Sch Med, Georgetown Univ, 64-74. Mem: AAAS; fel Am Col Physicians; Am Soc Clin Invest; fel Int Soc Hemat; Asn Am Physicians. Res: Gastroenterology; iron metabolism; hemolytic disorders; intestinal transport; infectious hepatitis. Mailing Add: Div of Hemat & Oncol Univ of Ala Birmingham AL 35294

CONRAD, MARGARET C, b Burlington, NC, Apr 4, 30; div; c 3. PHYSIOLOGY. Educ: Catawba Col, AB, 52; Univ NC, PhD(physiol, biochem), 55. Prof Exp: Asst physiol, Univ NC, 52-55; res asst path & oncol, Univ Kans Med Ctr, Kansas City, 55-56; from res asst to res assoc physiol & pharmacol, Bowman Gray Sch Med, 59-63; from instr to prof physiol, Med Univ SC, 63-73; PROF PHYSIOL, EASTERN VA MED SCH, 73- Mem: Am Physiol Soc; Am Heart Asn. Res: Physiology of peripheral vascular circulation. Mailing Add: Dept of Physiol Eastern Va Med Sch Norfolk VA 23507

CONRAD, MICHAEL, b New York, NY, Apr 30, 41. BIOPHYSICS, BIOMATHEMATICS. Educ: Harvard Univ, AB, 63; Stanford Univ, PhD(biophys), 69. Prof Exp: Fel biophys, Ctr Theoret Studies, Univ Miami, 69-70; fel math, Univ Calif, Berkeley, 72; asst prof, Inst Info Sci, Univ Tübingen, 72-74; assoc prof biol, City Col, 74-75; ASSOC PROF COMPUT & COMMUN SCI, UNIV MICH, ANN ARBOR, 75- Concurrent Pos: Res assoc, Inst Info Sci, Univ Tübingen, 74-76. Mem: Biophys Soc; Ecol Soc; Soc Math Biol; Am Inst Biol Sci; Neth Soc Theoret Biol. Res: Biological information processing; brain models; biological adaptability; experimental ecology; physics of enzymes. Mailing Add: Dept of Comput & Commun Sci Univ of Mich Ann Arbor MI 48104

CONRAD, PAUL, b Hempstead, NY, Oct 7, 21; m 43; c 1. MATHEMATICS. Educ: Univ Ill, PhD(math), 51. Prof Exp: From asst prof to prof math, Newcomb Col, Tulane Univ, 51-70; PROF MATH, UNIV KANS, 70- Concurrent Pos: Fulbright lectr, Univ Ceylon, 56-57; NSF sr fel, Australian Nat Univ, 64-65; vis prof, Univ Paris, 67. Mem: Am Math Soc. Res: Ordered algebraic systems; group theory. Mailing Add: Dept of Math Univ of Kans Lawrence KS 66044

CONRAD, ROBERT DEAN, b El Reno, Okla, Sept 20, 23; m 47; c 3. LABORATORY ANIMAL MEDICINE. Educ: Univ Okla, BS, 49; Okla State Univ, MS & DVM, 53; Univ Calif, Davis, PhD(comp path), 70. Prof Exp: Asst prof vet microbiol, Wash State Univ, 53-59; pathologist, Heisdorf & Nelson Farms, Inc, 59-65; NIH fel epidemiol & prev med, Univ Calif, Davis, 65-67, specialist, Div Exp Animal Resources, Sch Vet Med, 67-71; ASSOC PROF MED MICROBIOL, COL MED, DIR ANIMAL RESOURCE FACILITIES, UNIV NEBR, OMAHA, 71- Concurrent Pos: Dr Salsbury fel, 57-58. Mem: Am Vet Med Asn; Am Asn Avian Path; Am Soc Microbiol. Res: Bacterial and viral diseases of domestic and wild fowl. Mailing Add: Univ of Nebr Col of Med 42nd St & Dewey Ave Omaha NE 68105

CONRAD, RODDY MERL, physical chemistry, polymer chemistry, see 12th edition

CONRAD, WALTER EDMUND, b Forward, Pa, Nov 16, 20; m 49; c 2. ORGANIC CHEMISTRY. Educ: Wayne Univ, BS, 44, MS, 46; Univ Kans, PhD(org chem), 51. Prof Exp: Chemist, Armour Labs, 45-47; chemist, Sterling-Winthrop Res Inst, 47-48; res assoc, Med Sch, Tufts Univ, 51-53; instr chem, Univ Mass, 53-55; res chemist, Celanese Corp, 55-57, assoc prof & chmn dept, Ohio Northern Univ, 57-59; PROF CHEM, SOUTHEAST MASS UNIV, 59- Mem: Am Chem Soc. Res: Schmidt reaction; effect of ultraviolet irradiation on pyrimidines; catalytic debenzylation; lubricants; chemistry of phosphoranes. Mailing Add: Dept of Chem Southeastern Mass Univ North Dartmouth MA 02747

CONRADI, JAN, b Greenville, SC, Mar 13, 39; Can citizen; m 66; c 2. SOLID STATE PHYSICS. Educ: Queen's Univ, Ont, BSc, 62; Imp Col, Univ London, dipl mat sci, 63; Univ Birmingham, MSc, 64; Simon Fraser Univ, PhD(solid state physics), 68. Prof Exp: MEM SCI STAFF SEMICONDUCTOR ELECTRONICS, RES LABS, RCA LTD, 69- Mem: Can Asn Physicists. Res: Luminescence studies in single crystal and evaporated films of cadmium sulphide; device physics, especially semiconductor photosensors. Mailing Add: RCA Ltd Head Off Res Labs Ste Anne de Bellevue PQ Can

CONREY, BERT L, b Glendale, Calif, Sept 9, 20; m 47. GEOLOGY. Educ: Univ Calif, Berkeley, AB, 47, MA, 48; Univ Southern Calif, PhD(geol), 59. Prof Exp: Field party chief petrol geol, Stanolind Oil & Gas Co, 48-50, div photogeologist, 50-51; from asst prof to assoc prof, 55-63, chmn dept, 60-64, PROF GEOL, CALIF STATE UNIV, LONG BEACH, 63- Concurrent Pos: NSF res grant, 63-64. Mem: Am Asn Petrol Geol; Geol Soc Am; Soc Econ Paleontologists & Mineralogists; Int Asn Sedimentol. Res: Marine geology and sedimentology. Mailing Add: Dept of Geol Calif State Univ Long Beach CA 90804

CONROW, KENNETH, b Philadelphia, Pa, Jan 22, 33; m 55; c 3. COMPUTER SCIENCE. Educ: Swarthmore Col, BA, 54; Univ Ill, PhD(org chem), 57. Prof Exp: From instr to asst prof chem, Univ Calif, Los Angeles, 57-61; from asst prof to assoc prof org chem, 61-71, ASSOC PROF COMPUT SCI, KANS STATE UNIV, 71-, ASST DIR, COMPUT CTR, 74- Mem: Am Chem Soc; Asn Comput Mach. Res: Programming language/1 preprocessors; non-numeric programming. Mailing Add: Comput Ctr Kans State Univ Manhattan KS 66506

CONROY, CHARLES WILLIAM, b Neodesha, Kans, Dec 29, 27; m 49; c 5. PERIODONTICS, ORAL PATHOLOGY. Educ: Univ Kans, AB, 50; Univ Mo, DDS, 54, MSD, 57; Ohio State Univ, MSc, 58. Prof Exp: Asst prof med periodont, Univ Tex, 57-61; from asst prof to assoc prof, 61-70, PROF DENT, COL DENT, OHIO STATE UNIV, 70- Concurrent Pos: Consult, Procter & Gamble Co, 61- & US Air Force, Wright-Patterson AFB, 63- Mem: Int Asn Dent Res; Am Acad Periodont; Am Acad Dent Electrosurg (pres, 73-74). Res: Studies relating to the prevention of dental deposits on the teeth which cause periodontal disease. Mailing Add: Ohio State Univ Col of Dent 305 W 12th Ave Columbus OH 43210

CONROY, HAROLD, b Brooklyn, NY, Apr 1, 28; m 48; c 3. THEORETICAL CHEMISTRY. Educ: Mass Inst Technol, BS, 48; Harvard Univ, PhD(org chem), 50. Prof Exp: NIH fel, Harvard Univ, 50-51; instr chem, Columbia Univ, 51-52; sr chemist, Merck & Co, Inc, 52-55; asst prof chem, Brandeis Univ, 55-58 & Yale Univ, 58-61; prof biol & chem, 67-72, SR FEL, MELLON INST, 61-, PROF CHEM, CARNEGIE-MELLON UNIV, 72- Concurrent Pos: NIH spec res fel, 70-71; vis prof, Mass Inst Technol, 70-71; exchange visitorship for Eastern Europe, Nat Acad Sci, 74. Mem: AAAS; Sigma Xi. Res: Mathematical properties of molecular wavejunctions and density matrices; quantum chemistry; structure of natural products; physical organic chemistry; organic reaction mechanisms. Mailing Add: Dept of Chem Carnegie-Mellon Univ 4400 Fifth Ave Pittsburgh PA 15213

CONROY, JAMES D, b Dayton, Ohio, Dec 15, 33; m 62; c 3. VETERINARY PATHOLOGY. Educ: Ohio State Univ, DVM, 60; Univ Ill, Urbana-Champaign, PhD(vet path), 68. Prof Exp: Fel clin res internal med, Animal Med Ctr, NY, 60-61, fel comp dermat, 61-62, staff vet, 62-64, sr resident vet path, 64; USPHS fel vet med sci, 64-68, from instr to asst prof vet path & hyg, 68-70, ASSOC PROF VET PATH & HYG, COL VET MED, UNIV ILL, URBANA-CHAMPAIGN, 70- Honors & Awards: Clin Proficiency Award, Upjohn Co, Kalamazoo, Mich, 60. Mem: Am Vet Med Asn; Am Col Vet Path; Int Acad Path; Am Acad Vet Dermat (secy-treas, 64-68, vpres, 68-70, pres, 71-73); Am Col Vet Internal Med. Res: Development of cutaneous prenatal pigmentation; melanocyticmelanocytic tumors in domestic animals; cytopathology of infectious diseases; comparative dermatology; dermal pathology. Mailing Add: Dept of Vet Path & Hyg Col of Vet Med Univ of Ill Urbana IL 61801

CONROY, JAMES STRICKLER, b Philadelphia, Pa, Aug 24, 31. FUEL CHEMISTRY. Educ: Univ Pa, AB, 53; Pa State Univ, MS, 56, PhD(fuel tech), 59. Prof Exp: Asst prof, 57-62, ASSOC PROF CHEM, WIDENER COL, 62-, CHMN SCI, 73- Mem: Am Chem Soc; Am Inst Chem. Res: Condensed aromatic hydrocarbons; fuels and combustion; reaction mechanisms. Mailing Add: 1307 Maryland Ave Havertown PA 19083

CONROY, LAWRENCE EDWARD, b Providence, RI, Aug 29, 26; m 62; c 2. INORGANIC CHEMISTRY. Educ: Univ RI, BS, 49; Cornell Univ, MS, 52, PhD, 55. Prof Exp: Proj chemist phys chem, Colgate-Palmolive Co, 51-53; asst prof chem, Temple Univ, 55-59; asst prof, 59-63, ASSOC PROF INORG CHEM, UNIV MINN, MINNEAPOLIS, 63- Concurrent Pos: Consult, Minn Mining & Mfg Co, 62-65. Mem: AAAS; Am Chem Soc; Brit Chem Soc. Res: Solid state and high temperature inorganic chemistry; nonstoichiometric and metallic compounds; transport properties. Mailing Add: 1515 East River Rd Minneapolis MN 55414

CONROY, MARGARET FRANCES, b Boston, Mass, Jan 1, 23. APPLIED MATHEMATICS. Educ: Regis Col, AB, 44; Boston Col, AM, 46; Brown Univ, PhD, 53. Prof Exp: Instr physics, Womans Col, NC, 46-48; mathematician, Watertown Arsenal Res Lab, Mass, 52-53; instr math, Purdue Univ, 53-54; asst prof, Washington Univ, 54-55; asst, Boston Col, 55-58; mathematician, Parke Math Lab, 58-59 & Raytheon Corp, 59-61; res scientist, Sperry Rand Res Ctr, Sperry Rand Corp, 61-65; asst prof math biol, Harvard Med Sch, 65-72; PROF MATH, COLO WOMEN'S COL, 72- Mem: Am Math Soc; NY Acad Sci. Res: Applied mechanics; numerical analysis; biomathematics. Mailing Add: Dept of Math Colo Women's Col Denver CO 80220

CONS, JEAN MARIE ABELE, b Lancaster, Pa; m; c 2. DEVELOPMENTAL PHYSIOLOGY. Educ: Calif State Univ, San Francisco, BA, 60; Univ Calif, San Francisco, MS, 63, PhD(endocrinol), 72. Prof Exp: Lab technician develop anat, Univ Calif, San Francisco, 60-64, res anat, 64-68; fel, 72-74, ASST RES DEVELOP PHYSIOL, UNIV CALIF, BERKELEY, 74- Concurrent Pos: Consult, Med Sci Prog, Univ Calif, Berkeley, 72-73; lectr, Col Notre Dame, Calif, 75- Mem: Soc Study Reproduction. Res: Developmental patterns of pituitary and plasma glycoproteins and the actions of these hormones during development. Mailing Add: Col of Notre Dame Ralston Ave Belmont CA 94002

CONSELMAN, FRANK BUCKLEY, b New York, NY, Oct 1, 10; m 34; c 3. PETROLEUM GEOLOGY. Educ: NY Univ, BSc, 30, ScM, 31; Univ Mo, PhD(geol), 34. Prof Exp: Asst geol, NY Univ, 30-31; asst geol, Univ Mo, 31-32 & 33-34; geologist, Mo Geol Surv, 34-35; geologist, Gulf Oil Corp, 35-41; dist geologist, Great Lakes Carbon Corp, Tex, 45-46; dist mgr, Am Trading & Prod Corp, 47; CONSULT GEOLOGIST, 47-; PROF GEOSCI & DIR INT CTR ARID & SEMI-ARID LAND STUDIES, TEX TECH UNIV, 69- Concurrent Pos: Vis lectr, Univ Tex, 57 & 68; mem US Nat Comn Geol, 69- Mem: Hon mem Am Asn Petrol Geologists (vpres, 60-61, pres, 68-69); Am Inst Mining, Metall & Petrol Engrs; Am Inst Prof Geol (pres, 74); Am Geol Inst (vpres, 74, pres, 75); fel Geol Soc Am. Res: Subsurface stratigraphy; oil and gas exploration; water supplies for arid lands; mineral resources. Mailing Add: Lake Ransom Canyon Route 2 Slaton TX 79364

CONSIDINE, JUDITH MAYBERRY, b Albany, NY, Sept 9, 38; m 64. BIOLOGICAL CHEMISTRY. Educ: St Lawrence Univ, BS, 59; Pa State Univ, MS, 61, PhD(biochem), 63. Prof Exp: Sr res staff, Merck Sharp & Dohme Div, 63-64; instr chem, Ind Univ, South Bend, 64-67; asst prof, Trinity Col, DC, 67-69; asst prof, 69-75, ASSOC PROF BIOL, SIMPSON COL, 75- Concurrent Pos: Asst dir, Higher Educ Resource Serv, 74- Mem: NY Acad Sci; Tissue Cult Asn. Res: Enzymology; tissue culture. Mailing Add: Dept of Biol Simpson Col Indianola IA 50125

CONSIDINE, RICHARD GEORGE, b Erie, Pa, June 20, 36; m 64. VIROLOGY, IMMUNOLOGY. Educ: Univ Notre Dame, BS, 58, PhD(microbiol), 67; Pa State Univ, MS, 63; Col Osteop Med & Surg, DO, 73. Prof Exp: Res assoc microbiol, Naval Med Res Inst, 67-68; asst prof microbiol, Col Osteop Med & Surg, 68-70; resident med, Miriam Hosp, Providence, 74-75; RESIDENT FAMILY PRACT, SOMERSET HOSP, SOMERVILLE, NJ, 75- Concurrent Pos: Nat Acad Sci-Nat Res Coun fel, 67-68. Mem: Am Soc Microbiol; Am Pub Health Asn; NY Acad Sci; AMA. Res: Gnobiotics; enzymology. Mailing Add: 381 Gemini Dr 9 Hillsborough NJ 08876

CONSIDINE, WILLIAM JAMES, organic chemistry, see 12th edition

CONSIGLI, RICHARD ALBERT, b Brooklyn, NY, Mar 2, 31; m 60; c 2. VIROLOGY, CANCER. Educ: Brooklyn Col, BS, 54; Univ Kans, MA, 56, PhD(bact), 60. Prof Exp: Asst bact, Univ Kans, 54-59, instr, 59-60; USPHS fel virol, Univ Pa, 60-62; from asst prof to assoc prof bact, 62-69, PROF BACT, KANS STATE UNIV, 69-, SR SCIENTIST MID-AM CANCER CTR, UNIV MED CTR, 75- Concurrent Pos: Pa Plan scholar, 61-62; NIH res grants, 62-70; career develop award, USPHS, 67-; consult, NSF; assoc ed, Appl Microbiol. Mem: Am Soc Microbiol; Tissue Cult Asn; Soc Exp Biol & Med; NY Acad Sci. Res: Investigation of the biochemical events during the animal virus infection of tissue culture cells; investigation of the host-parasite interrelationship during rickettsial infections; cancer research, biochemistry of tumor virus infection of cultured cells. Mailing Add: Subdiv Moleclr Biol & Genetics Div Biol Kans State Univ Manhattan KS 66504

CONSOLAZIO, CARLO FRANK, b Cambridge, Mass, Mar 9, 13; m 38; c 2. NUTRITION, PHYSIOLOGY. Educ: Am Bd Clin Nutrit, cert human nutrit, 67. Prof Exp: Chief lab technician, Fatigue Lab, Harvard Univ, 29-42, asst physiol, 43-47; chief biochem br, Med Nutrit Lab, 47-51, bioenergetics div, Med Res & Nutrit Lab, 52-75, CHIEF BIOENERGETICS DIV, US ARMY, LETTERMAN ARMY INST RES, 75- Concurrent Pos: Affil prof, Colo State Univ, 64-65; spec consult, Interdept Comt

Nutrit for Nat Defense, 56- & WHO, 62; mem subcomt calories, fat, carbohydrates and alcohol, Nat Acad Sci-Nat Res Coun, 65-68, mem comt population biol of altitude, Int Biol Prog with WHO, Washington, 67; sr lectr physical educ, Univ of Colo, 70-; consult, State Dept Agency Int Develop, 71- & Nat Inst Child Health & Develop, 71; monitor, Nutrition Res Info Syst & Int Biol Prog Sect on Energy Requirements, 71. Honors & Awards: McLester Award, 63. Mem: AAAS; Am Chem Soc; Am Inst Nutrit; Am Soc Clin Nutrit. Res: Environmental nutrition; nutritional requirements; performance evaluation; energy metabolism; evaluation of nutritional status; nutritional surveys; mineral metabolisms; body composition. Mailing Add: Dept of the Army Letterman Army Inst of Res Presidio of San Francisco CA 94129

CONSOLAZIO, WILLIAM VENERANDO, chemistry, see 12th edition

CONSROE, PAUL F, b Cortland, NY, Oct 18, 42; m 71. PHARMACOLOGY. Educ: Albany Col Pharm, BS, 66; Univ Tenn, Memphis, MS, 69, PhD(pharmacol), 71. Prof Exp: ASSOC PROF PHARMACOL, COL PHARM, UNIV ARIZ, 71- Concurrent Pos: NIMH grant, Univ Ariz, 72-; consult, Ariz Poison Control Inform Serv, 71- Mem: Soc Neurosci; Am Pharmaceut Asn; Am Soc Pharmacol & Exp Therapeut. Res: Neuropsychopharmacological investigations of hallucinogens, marijuana and other psychotropic drugs. Mailing Add: Dept of Pharmacol & Toxicol Univ of Ariz Col of Pharm Tucson AZ 85721

CONSTABLE, JAMES HARRIS, b Dayton, Ohio, Mar 9, 42; m 68. SOLID STATE PHYSICS, LOW TEMPERATURE PHYSICS. Educ: Ohio State Univ, BSc, 66, MSc, 67, PhD(physics), 69. Prof Exp: Res assoc physics, Ohio State Univ, 69-72, vis asst prof, 72-74; ASST PROF PHYSICS, STATE UNIV NY BINGHAMTON, 74- Mem: Am Phys Soc. Res: Experimental studies of three phonon processes in solids and the magnetic and thermal properties of the hydrogen solids down to very low temperatures. Mailing Add: Dept of Physics State Univ of NY Binghamton NY 13901

CONSTABLE, ROBERT L, b Detroit, Mich, Jan 20, 42; m 64; c 1. COMPUTER SCIENCE, MATHEMATICS. Educ: Princeton Univ, AB, 64; Univ Wis, MA, 65, PhD(math), 68. Prof Exp: Instr comput sci, Univ Wis, 68; asst prof, 68-72, ASSOC PROF COMPUT SCI, CORNELL UNIV, 72- Mem: Asn Symbolic Logic; Asn Comput Mach; Am Math Soc; Soc Indust & Appl Math. Res: Theory of computation, especially semantics of programming languages and computational complexity. Mailing Add: Dept of Comput Sci Cornell Univ Ithaca NY 14850

CONSTANCE, LINCOLN, b Eugene, Ore, Feb 16, 09; m 36; c 1. BOTANY. Educ: Univ Calif, AM, 32, PhD(bot), 34. Prof Exp: Instr bot & cur herbarium, State Col Wash, 34-36, asst prof, 36-37; from asst prof to prof bot, 37-76, cur seed plant collections, 47-63, chmn dept bot, 54-55, dean col letters & sci, 55-62, vchancellor, 62-65, dir herbarium, 63-75, EMER PROF BOT, UNIV CALIF, BERKELEY, 76- Concurrent Pos: Vis lectr & acting dir, Gray Herbarium, Harvard Univ, 47-48; Guggenheim fel, 53-54. Mem: Bot Soc Am (pres, 70); Torrey Bot Club; Am Soc Plant Taxonomists (pres, 53); Soc Study Evolution; fel Am Acad Arts & Sci. Res: Systematic botany of Umbelliferae; cytotaxonomy of Hydrophyllaceae. Mailing Add: Dept of Botany Univ of Calif Berkeley CA 94720

CONSTANT, CLINTON, b Nelson, BC, Can, Mar 20, 12; US citizen; m 50. INORGANIC CHEMISTRY, CHEMICAL ENGINEERING. Educ: Univ Alta, BSc, 35. Prof Exp: Develop engr, Harshaw Chem Co, Ohio, 36-38, foreman & engr, Acid Plant, 38-43; plant supt, Nyotex Chem, Inc, Tex, 43-47, chief develop engr, 47-48; sr chem engr, Harshaw Chem Co, Ohio, 48-50; mgr eng, Ferro Chem Corp, 50-52; tech asst mfg dept, Plant Develop & Eng, Armour Agr Chem Co, Fla, 52-61, mgr, Fla Res Div, 61-63, proj mgr eng & design & mgr spec proj, Ga, 63-70; CHEM ENGR, ROBERT & CO ASSOC, 70- Mem: Fel AAAS; fel Am Inst Chem; Am Inst Chem Eng; Am Inst Aeronaut & Astronaut; Water Pollution Control Fedn. Res: Chemical engineering design; phosphates; chemistry and production of anhydrous hydrofluoric acid; slide rule for complex chemical formulation. Mailing Add: PO Box 1221 Atlanta GA 30301

CONSTANT, FRANK WOODBRIDGE, b Minneapolis, Minn, June 1, 04; m 40; c 3. ENVIRONMENTAL PHYSICS. Educ: Princeton Univ, BS, 25; Yale Univ, PhD(physics), 28. Prof Exp: Nat res fel physics, Calif Inst Technol, 28-30; from instr to assoc prof, Duke Univ, 30-46; JARVIS PROF PHYSICS, TRINITY COL, CONN, 46- Concurrent Pos: Mem & off investr, Nat Defense Res Comt, Duke Univ, 42-46; instr, Dorr-Loomis Pre-Col Sci Ctr, 57, dir, 58 & 59. Mem: Fel Am Phys Soc; Am Asn Physics Teachers. Res: Ferromagnetism; theoretical physics; mechanics; electromagnetism; fundamental laws of physics and their environmental applications. Mailing Add: Dept of Physics Trinity Col Hartford CT 06106

CONSTANT, MARC DUNCAN, b Aledo, Ill, July 17, 41. ANALYTICAL CHEMISTRY. Educ: Monmouth Col, BA, 63; Southern Ill Univ, MA, 65; Kans State Univ, PhD(anal chem), 70. Prof Exp: SUPVR ANAL CHEM, AM MAIZE PROD CO, 69- Mem: Am Chem Soc. Res: Applications of instrumental analysis to the food processing industries, including gas and liquid chromatography, infrared, and general applications of automated wet methods. Mailing Add: Am Maize Prod Co 113th & Indianapolis Blvd Hammond IN 46326

CONSTANTIN, JAMES MICHAEL, b Chicago, Ill, Sept 13, 22; m 48; c 3. LEATHER CHEMISTRY. Educ: Univ Notre Dame, BS, 43, PhD(org chem), 49. Prof Exp: Res chemist, Merck & Co, Inc, 49-53, sr res chemist, 53-54, purchasing engr, 54-55, purchasing res analyst, 55-57, prod develop specialist, 57-59, sr technologist, 59-63, mgr sales & mkt, 63-65; res assoc, Albert Trostel & Sons Co, 65-67; dir, 67-70, VPRES RES & DEVELOP, PFISTER & VOGEL TANNING CORP, INC, 70- Concurrent Pos: Sci adv comt, Tanners Coun Am, 68-74. Mem: AAAS; Am Chem Soc; Am Leather Chem Asn; Brit Soc Leather Trades Chem. Res: Steroids and steroid total synthesis; medicinal chemicals; enzymatic processes in leather processing; purchasing research and value analysis; market development in leather chemicals; water pollution. Mailing Add: Pfister & Vogel Tanning Corp 1531 N Water St Milwaukee WI 53201

CONSTANTIN, MILTON J, b Duson, La, Mar 25, 34; m 58; c 4. PLANT BREEDING, BOTANY. Educ: Southwestern La Inst, BS, 56; La State Univ, MS, 58, PhD(plant breeding), 60. Prof Exp: From asst prof to assoc prof, 60-71, PROF RADIOBOT, UNIV TENN-ENERGY RES & DEVELOP ADMIN COMP ANIMAL RES LAB, OAK RIDGE, 71- Mem: Radiation Res Soc; Genetic Soc Am; Am Genetic Asn. Res: Radiobotany; factors that modify radiation response of plants; effects of nuclear energy on genetics of agronomic plants; induction, recovery, characterization and utilization of mutations in higher plants. Mailing Add: U Tenn-ERDA Comp Animal Res Lab 1299 Bethel Valley Rd Oak Ridge TN 37830

CONSTANTINE, ANTHONY BENEDICT, b Buffalo, NY, Jan 30, 16; m 43; c 2. PATHOLOGY. Educ: Univ Buffalo, BA, 38, MD, 43. Prof Exp: Asst anat, 39-41, instr path, 43-44, assoc, 46-56, ASST PROF PATH, SCH MED, STATE UNIV NY BUFFALO, 56- Concurrent Pos: Res fel, Buffalo Gen Hosp, 46-47, asst pathology, 47-51; pathologist, Vet Admin Hosp, NY, 47 & Buffalo Mercy Hosp, 47-; consult pathologist, Buffalo Vet Admin Hosp, 56- Res: Pathology of lymphoid diseases. Mailing Add: 51 Ruskin Rd East Aurora NY 14052

CONSTANTINE, DENNY G, b San Jose, Calif, May 5, 25; m 52; c 2. VETERINARY MEDICINE, EPIDEMIOLOGY. Educ: Univ Calif, Davis, BS, 53, DVM, 55; Univ Calif, Berkeley, MPH, 65. Prof Exp: Head wildlife ecol care prog, Arctic Health Res Ctr, Alaska, 50-51, chief wildlife rabies res, Communicable Dis Ctr, Ga, 55-56, rabies field unit, NMex, 56-58, southwest rabies invest sta, 58-66, CHIEF NAVAL BIOMED RES LAB, CTR DIS CONTROL ACTIVITIES, USPHS, 66- Concurrent Pos: Asst mammalogist, Los Angeles County Mus, 42-46; pvt res, 46-50; zool mus cur, Univ Calif, Davis, 51-55, mammalogist & lab technician, Sch Vet Med, 54-55; consult, Armed Forces Bat Bomb Proj, 43-44. Mem: Am Pub Health Asn; Am Vet Med Asn; Am Soc Mammal. Res: Public health, veterinary medicine; ecology, virology; physiology; mammalogy; wildlife diseases. Mailing Add: USPHS Ctr for Dis Control Naval Biomed Res Lab Naval Supply Ctr Oakland CA 94625

CONSTANTINE, GEORGE HARMON, JR, b San Francisco, Calif, Sept 30, 36; m 62; c 3. PHARMACY. Educ: Univ Utah, BS, 60, MS, 62, PhD(pharmacog), 66. Prof Exp: Asst chem, 62-63, asst pharm, 63-64, asst pharmacog, 64-66, asst prof, 66-71, ASSOC PROF PHARMACOG, ORE STATE UNIV, 71- Mem: Am Pharmaceut Asn; Acad Pharmaceut Sci; Am Soc Pharmacog. Res: Natural products; steroid and triterpenoid constituents of higher plants; plant alkaloids; marine biomedicinals. Mailing Add: Dept of Pharmacog Ore State Univ Corvallis OR 97331

CONSTANTINE, HERBERT PATRICK, b Buffalo, NY, May 10, 29; m 54; c 2. PHYSIOLOGY, MEDICINE. Educ: Univ Buffalo, MD, 53. Hon Degrees: MA, Brown Univ, 67. Prof Exp: Nat Tuberc Asn fel, 57-59; Am Heart Asn res fel & instr med, Univ Rochester, 59-60; instr physiol, Univ Pa, 60-63; asst prof med, Boston Univ, 63-66; ASSOC PROF MED SCI, BROWN UNIV, 66- Concurrent Pos: NIMH grants, 64-68; consult, Vet Admin Hosp, Providence, RI, 66- Mem: Am Fedn Clin Res; Am Thoracic Soc; Am Physiol Soc. Res: Respiratory physiology and mechanics; carbon dioxide reaction rates; laboratory automation. Mailing Add: Div of Biol & Med Brown Univ Providence RI 02912

CONSTANTINE, JAY WINFRED, b New York, NY, Feb 8, 26; m 52; c 2. PHYSIOLOGY. Educ: McGill Univ, BS, 51; Ohio State Univ, PhD(physiol), 59. Prof Exp: From asst to instr physiol, Ohio State Univ, 56-59; asst prof, NDak State Univ, 59-61; sr pharmacologist, 61-67, mgr gen pharmacol, 67-72, ASST DIR DEPT PHARMACOL, MED RES LABS, PFIZER, INC, 72- Mem: Am Soc Pharmacol & Exp Therapeut; Am Physiol Soc. Res: Cardiovascular; platelet aggregation; general pharmacology. Mailing Add: Dept of Pharmacol Pfizer Inc Med Res Labs Groton CT 06340

CONSTANTINIDES, EUSTRATIOS, atomic physics, see 12th edition

CONSTANTINIDES, PARIS, b Smyrna, Asia Minor, Dec 21, 19; Can citizen; m 50; c 1. ANATOMY, ELECTRON MICROSCOPY. Educ: Univ Vienna, MD, 43; Univ Montreal, PhD(exp med), 53. Prof Exp: Asst exp med, Inst Exp Med, Univ Montreal, 47-50; from asst prof to prof anat, 50-64, hon prof path, 64-65, prof anat, 65-67, PROF PATH, MED SCH, UNIV BC, 67- Concurrent Pos: Vis prof, Wash Univ, 63-64; ed, J Atherosclerosis Res, 61; mem adv comt artificial heart, USPHS, 66; ed, Can J Physiol & Pharmacol, 69; consult, Nat Heart & Lung Inst. Mem: Soc Exp Biol & Med; Am Heart Asn; Am Asn Anatomists; Can Physiol Soc; Can Asn Anatomists. Res: Experimental pathology; aging degenerative diseases; endocrine. Mailing Add: Dept of Path Univ of BC Med Sch Vancouver BC Can

CONSTANTINIDES, SPIROS MINAS, b Thessaloniki, Greece, Nov 4, 32; m 59; c 2. FOOD SCIENCE, BIOCHEMISTRY. Educ: Univ Thessaloniki, BS, 57; Mich State Univ, MS, 63, PhD(food sci), 66. Prof Exp: Res assoc food technol, Univ Thessaloniki, 57-61; NIH fel biochem, Mich State Univ, 66-68; from asst prof to assoc prof, 68-74, PROF FOOD & NUTRIT SCI & BIOCHEM, UNIV RI, 74- Concurrent Pos: Vis prof, Cath Univ Valparaiso, Chile, 72 & Univ Campinas, Brazil, 75-76. Mem: Am Chem Soc; Inst Food Technologists; Am Soc Biol Chemists. Res: Structure and function of enzymes; multiple molecular forms and control mechanisms of enzymes; lysosomes and proteolytic enzymes in marine animals; biochemical aspects of preservation of marine foods; utilization of unconventional food resources; food and nutritional science for developing nations. Mailing Add: Dept of Food & Nutrit Sci 211 Quinn Hall Univ of RI Kingston RI 02881

CONSTANTOPOULOS, GEORGE, b Greece, Feb 21, 23; US citizen; m 54; c 1. BIOCHEMISTRY. Educ: Nat Univ Athens, BSc, 48; Wayne State Univ, PhD(biochem), 62. Prof Exp: Clin chemist, Hotel Dieu Hosp, Windsor, Ont, 55-58; res assoc biochem, Wayne State Univ, 62-63; RES BIOCHEMIST, NAT INST NEUROL DIS & STROKE, 66- Concurrent Pos: Univ res fel chem, Harvard Univ, 63-66, NIH fel, 64-65. Mem: AAAS; Am Chem Soc; Soc Biol Chem; Soc Microbiol; Soc Plant Physiol. Res: Neurochemistry, mucopolysaccharidoses. Mailing Add: Nat Inst of Neurol Dis & Stroke Bethesda MD 20014

CONSUL, PREM CHANDRA, b Meerut, India, Aug 10, 23; m 51; c 5. STATISTICS, MATHEMATICS. Educ: Agra Univ, BSc, 43, MSc, 46 & 47, PhD(math, statist), 57. Prof Exp: Lectr math, SD Col, Muzaffarnagar, India, 46-47; sr lectr, SM Col, Chandausi, 47-48; asst prof, NREC Col, Khurja, 48-50, assoc prof, 50-53, prof math & statist & head dept, 53-57; head dept statis & prin, MS Col, Saharanpur, 57-61; prof math & statist, Libya, Tripoli, 61-67; assoc prof, 67-71, PROF MATH & STATIST, UNIV CALGARY, 71- Concurrent Pos: Mem, Indian Statist Inst, 54-; convenor bd studies in statist, mem acad coun & fac sci, Agra Univ, 55-58, senate, 58-61, panel univ inspectros, 59-61. Mem: Am Math Soc; Inst Math Statist; Can Math Cong; fel Royal Statist Soc; Indian Math Soc. Res: Multivariate distribution theory; properties of multivariate distributions; some special functions and integral transforms; new approximations in gamma functions; generalizations of probability distributions. Mailing Add: Dept of Math Univ of Calgary Calgary AB Can

CONTACOS, PETER GEORGE, b Springfield, Mass, Mar 19, 26. MEDICAL PARASITOLOGY, TROPICAL MEDICINE. Educ: Harvard Univ, AB, 47; Boston Univ, MA, 49; Tulane Univ, PhD(med parasitol), 54, MD, 57. Prof Exp: Instr parasitol, Sch Med, Tulane Univ, 51-53; intern, Methodist Hosp, Brooklyn, 57-58; intern, USPHS, 58-, pub health physician, USPHS-US Tech Coop Mission, India, 58-60; med officer in charge, NIH Malaria Proj, Lab Parasite Chemother, 60-65, head sect cytol, 65-66, head sect primate malaria, Lab Parasitic Dis, Nat Inst Allergy & Infectious Dis, 66-72, asst to dir res, Malaria Prog, Ctr Dis Control, 72-74, CHIEF, HOST-PARASITE STUDIES BR, VECTOR BIOL & CONTROL DIV, BUR TROP DIS, CTR DIS CONTROL, 74- Concurrent Pos: Mem expert adv panel malaria, WHO, 66-, adv sci group on immunol, Geneva, 67, consult to visit malaria res labs in Europe & Mid East, 67; mem, Expert Comt on Malaria, Geneva, 67; assoc mem comn on malaria, Armed Forces Epidemiol Bd, 67-71; consult past & present studies, Cent Am Malaria Res Sta, San Salvador, El Salvador, 70. Honors & Awards: Walter Reed Mem Medal, La State Med Soc, 57. Mem: Am Acad Microbiol; Am Soc Trop Med & Hyg; AMA; fel Royal Soc Trop Med & Hyg; AMA. Res: Chemotherapy of malaria and other parasitic diseases; immunology of parasitic diseases; epidemiology. Mailing Add: Bur of Trop Dis Ctr for Dis Control Atlanta GA 30333

CONTE, FRANK PHILIP, b South Gate, Calif, Feb 2, 29; m 54; c 3. COMPARATIVE PHYSIOLOGY, CELL PHYSIOLOGY. Educ: Univ Calif, Berkeley, AB, 51, PhD(physiol, biochem), 61. Prof Exp: Asst physiol, Univ Calif, Berkeley, 51-52 & aero-med lab, Wright Air Develop Ctr, Ohio, 52-53; asst biol div, Oak Ridge Nat Lab, Tenn, 53-56; asst chem, Wash State Univ, 56-57; asst prof, Cent Ore Col, 57-59; asst radiation physiol, Donner Lab, Univ Calif, Berkeley, 59-60; from asst prof to assoc prof, 61-71, PROF ZOOL, ORE STATE UNIV, 71- Concurrent Pos: Sr fel, NIH, 68-69; vis prof, Dept Zool, Duke Univ, 68-69; prog dir regulatory biol, NSF, 72-73. Mem: AAAS; Am Soc Zoologists; Am Physiol Soc; Soc Gen Physiologists; Am Zool Soc. Res: Cellular regeneration in aquatic vertebrate; regulation of internal body fluids in aquatic vertebrates; biogenesis of cell membranes; exocrine glands in invertebrates. Mailing Add: Dept of Zool Ore State Univ Corvallis OR 97331

CONTE, JOHN SALVATORE, b Philadelphia, Pa, June 12, 32; m 60. ORGANIC CHEMISTRY. Educ: La Salle Col, AB, 54; La State Univ, MS, 56; Univ Pa, PhD(chem), 59. Prof Exp: Asst instr gen chem, La State Univ, 54-56; asst instr org chem, Univ Pa, 56-57; res group leader, Org Synthetic Sect, Scott Paper Co, 59-69; PROD MGR, QUAKER CHEM CO, 69- Mem: Am Chem Soc; The Chem Soc. Res: Kinetics and mechanism of organic reactions; mechanism of polymer reactions; resin synthesis and free radical polymerization; specialty chemicals for the pulp and paper industries. Mailing Add: 908 Pierce Rd Norristown PA 19403

CONTE, SAMUEL D, b Lackawanna, NY, June 5, 17; m 48; c 5. NUMERICAL ANALYSIS. Educ: Buffalo State Teachers Col, BS, 39; Univ Buffalo, MS, 43; Univ Mich, MA, 48, PhD(math), 50. Prof Exp: From instr to assoc prof math, Wayne Univ, 46-56; mgr math anal dept, Space Tech Labs, 56-61; mgr math dept, Aerospace Corp, 61-62; dir comput sci ctr, 62-66, HEAD COMPUT SCI DEPT, PURDUE UNIV, 66- Concurrent Pos: Consult, Aerospace Corp. Mem: Soc Indust & Appl Math; Am Math Soc; Math Asn Am; Asn Comput Mach. Res: Numerical analysis and computation; mathematical analysis. Mailing Add: 2241 Indian Trails Dr West Lafayette IN 47906

CONTENTO, ISOBEL CORNEIL, b Wuwei, China, Sept 15, 40; US citizen; m 66. IMMUNOLOGY, NUTRITION. Educ: Univ Edinburgh, BSc, 62; Univ Calif, Berkeley, MA, 64, PhD(immunochem), 69. Prof Exp: Asst prof biol, Merritt Col, 64-65; FAC FEL, JOHNSTON COL, UNIV REDLANDS, 69- Res: Affinity labeling studies on antibodies to a polysaccharide antigen and to a carbohydrate hapten; environmental aspects of gene expression; food consumption patterns of contemporary United States college students. Mailing Add: Johnston Col Univ of Redlands Redlands CA 92373

CONTI, PETER SELBY, astrophysics, see 12th edition

CONTI, PIERRE ANDRE, b Williamsport, Pa, Dec 30, 34; m 60; c 2. LABORATORY ANIMAL MEDICINE. Educ: Pa State Univ, BS, 56; Univ Pa, VMD, 60; Am Col Lab Animal Med, dipl. Prof Exp: SR STAFF VET, MERCK, SHARP & DOHME, 63- Mem: Am Vet Med Asn; Am Asn Lab Animal Sci; Am Soc Lab Animal Practitioners; Am Col Lab Animal Med. Res: Laboratory animal medicine; colony management; primatology. Mailing Add: Merck Sharp & Dohme Res Labs West Point PA 19486

CONTI, SAMUEL FRANCIS, b Brooklyn, NY, Dec 24, 31; m 54; c 3. MICROBIOLOGY. Educ: Brooklyn Col, BS, 52; Univ Conn, MS, 56; Cornell Univ, PhD(bact), 59. Prof Exp: Asst, Univ Conn, 52-53 & 55-56; asst, Cornell Univ, 56-58; res assoc, Brookhaven Nat Lab, 59-61; from instr to assoc prof microbiol, Dartmouth Med Sch, 61-66; PROF MICROBIOL & CHMN DEPT, DIR T H MORGAN SCH BIOL SCI & ASSOC DEAN COL ARTS & S SCI, UNIV KY, 66- Concurrent Pos: Mem ed bd, J Bact, 63-; NIH career develop award, 64-66; lectr, Nat Found Microbiol, 67-68. Mem: AAAS; Am Acad Microbiol; Am Soc Microbiologists; Am Soc Biol Chemists; Am Soc Plant Physiol. Res: Microbial ultrastructure; development of microbial organelles; biology of bdellovibrio; germination and formation of endospores; fungal and algal development and structure. Mailing Add: T H Morgan Sch of Biol Sci Univ of Ky Lexington KY 40506

CONTOGOURIS, ANDREAS P, b Athens, Greece, Oct 25, 31; m 57; c 2. THEORETICAL PHYSICS. Educ: Nat Tech Univ Athens, dipl elec eng, 54; Cornell Univ, PhD(physics), 61. Prof Exp: Res assoc thepret particle physics, Cornell Univ, 61-62; prof theoret physics, Democritus Nuclear Res Ctr, Greece, 62-64; res assoc theoret particle physics, Europ Orgn Nuclear Res, Geneva, Switz, 64-66; assoc lectr theoret physics, Univ Paris, Orsay, 66-68; ASSOC PROF THEORET PARTICLE PHYSICS, McGILL UNIV, 68- Concurrent Pos: Greek rep ro gov coun, Europ Orgn Nuclear Res, 65-66; Nat Res Coun Can grantee, 68- Mem: Am Phys Soc. Res: THeoretical problems and phenomenological applications related to the interactions of elementary particles at high energies. Mailing Add: Dept of Physics McGill Univ Montreal PQ Can

CONTOIS, DAVID ELY, b Battle Creek, Mich, Jan 18, 28; m 52; c 2. MICROBIOLOGY. Educ: Univ Calif, Los Angeles, BA, 50, PhD(microbiol), 57; Univ Hawaii, MS, 52. Prof Exp: Res microbiologist, Scripps Inst Oceanog, Univ Calif, 53-58; from asst prof to assoc prof, 58-68, asst dean, 64-69, PROF MICROBIOL, UNIV HAWAII, 68-, DEAN, 69- Mem: Am Soc Microbiologists; Brit Soc Gen Microbiol; AAAS. Res: Population dynamics; growth kinetics; marine microbiology; geomicrobiology; theory of steady-state microbial growth. Mailing Add: Dept of Microbiol Univ of Hawaii Honolulu HI 96822

CONTOS, GEORGE A, physical chemistry, see 12th edition

CONTRERA, JOSEPH FABIAN, b New York, NY, Nov 18, 38; m 62; c 1. VERTEBRATE PHYSIOLOGY, NEUROPHARMACOLOGY. Educ: NY Univ, BA, 60, MS, 61, PhD(endocrine physiol), 66. Prof Exp: Res asst neurochem, Sch Med, NY Univ, 60-62, res assoc physiol, Lab Exp Hemat, 63-66; asst prof, 67-70, ASSOC PROF PHYSIOL, UNIV MD, COLLEGE PARK, 70- Concurrent Pos: Lectr biol, Hunter Col, 63-64; trainee pharmacol & psychiat, Sch Med, Yale Univ, 66-67; vis instr, Dept Pharmacol & Exp Therapeut, Johns Hopkins Univ Sch Med, 75. Mem: AAAS; Soc Exp Biol & Med; Am Physiol Soc. Res: Physiology and pharmacology of the autonomic nervous system; experimental hematology; endocrinology. Mailing Add: Dept of Zool Univ of Md College Park MD 20742

CONTROULIS, JOHN, b Chicago, Ill, Mar 31, 19; m 52; c 2. ORGANIC CHEMISTRY. Educ: Transylvania Col, AB, 40; Univ Cincinnati, MA, 41; Univ Mich, PhD(chem), 50. Prof Exp: Sr res chemist, 41-47, patent chemist, 49-50, asst dir prods develop, 50-62, mgr mkt res & develop, 62-71, SUPT CAPSULE DEVELOP DIV, PARKE DAVIS & CO, 71- Mem: Am Chem Soc; AAAS; Sigma Xi. Res: Synthesis of organic chemicals for medicinal uses; organometallic compounds; nitrogen heterocyclic compounds; chloromycetin and intermediates. Mailing Add: 968 Westchester Grosse Pointe Park MI 48230

CONTU, PAOLO, b Orani, Italy, Aug 17, 21; m 59; c 2. ANATOMY, NEUROANATOMY. Educ: Univ Cagliari, Sardinia, MD, 47; Univ Recife, Physician lic, 57; Univ Rio Grande do Sul, Brazil, MD, 59. Prof Exp: Asst prof anat, Sch Med, Univ Bologna, 47-53, prof, 53-54; prof, Sch Med, Univ Recife, 54-58; HEAD RESEARCHER NEUROANAT, SCH MED, UNIV RIO GRANDE DO SUL, BRAZIL, 58-, PROF ANAT, 64- Concurrent Pos: CAPES Brazil grants, 61-62 & 66-67; notorio saber, Sch Dent, Univ Recife, 64; res ccun grant, Univ Rio Grande do Sul, Brazil, 66, res found grant, 69; grant & res lectr, Res Coun Brazil, 69. Mem: AAAS; Ital Soc Anat; Brazilian Soc Anat; Am Asn Anatomists; Brazilian Genetics Soc. Res: Glial architecture of man and vertebrates; electrolytic decalcification; coclear annervation in cats; nervous cells on peripheral arteriopathies; regeneration of nervous fibers in central and peripheral nervous system; reticular cells of human central nervous system. Mailing Add: Med Sch Port Alegre Inst Anat Univ Rio Grande do Sul Porto Alegre Brazil

CONVERSE, GLENN, seismology, applied mathematics, see 12th edition

CONVERSE, JIMMY G, b Scotts Bluff, Nebr, Aug 28, 38. CHEMICAL INSTRUMENTATION. Educ: San Diego State Col, BS, 63; Iowa State Univ, PhD(magnetic susceptibilities), 68. Prof Exp: Tool maker, Gen Dynamics-Convair, 57-61; SR RES CHEMIST, MONSANTO CO, 68- Mem: Am Chem Soc; Instrument Soc Am; Sigma Xi. Res: Molecular structure of molecules in the gaseous, liquid and solid state; electronic configuration; physical properties; instrumentation. Mailing Add: Monsanto Co 800 N Lindbergh Blvd T2D St Louis MO 63166

CONVERSE, JOHN MARQUIS, b San Francisco, Calif, Sept 29, 09. PLASTIC SURGERY, TRANSPLANTATION BIOLOGY. Educ: Univ Paris, MD, 35; Am Bd Plastic Surg, dipl, 51. Prof Exp: LAWRENCE D BELL PROF PLASTIC SURG, SCH MED & DIR INST RECONSTRUCTIVE PLASTIC SURG, MED CTR, NY UNIV, 57- Concurrent Pos: Chmn dept plastic surg, Manhattan Eye, Ear & Throat Hosp, 52-; dir plastic surg serv, Bellevue Hosp, 60- & NY Vet Admin Hosp, 62-; mem adv panel, Med & Dent Br, Off Naval Res, 64-; mem nat adv dent res coun, NIH, 65-; pres, Found Res Med & Biol; gov, Am Hosp Paris. Mem: Am Asn Plastic Surg; Am Soc Plastic & Reconstruct Surg; fel Am Col Surg; Am Cleft Palate Asn; Transplantation Soc (pres). Res: Vascularization of skin grafts; transplantation; burns research; clinical studies of congenital facial deformities, with emphasis on craniofacial malformations, hemifacial microsomia and second arch syndromes; reconstruction of the auricle. Mailing Add: Inst of Reconstructive Plastic Surg NY Univ Med Ctr New York NY 10003

CONVERSE, RICHARD HUGO, b Greenwich, Conn, Sept 18, 25; m 47; c 3. PHYTOPATHOLOGY. Educ: Univ Calif, BS, 47, MS, 48, PhD, 51. Prof Exp: Asst prof plant path, SDak State Col, 50-52; asst prof bot & plant path, Okla State Univ & asst plant pathologist & agent, USDA, 52-57; plant pathologist, Plant Indust Sta, USDA, 57-67; plant pathologist, USDA, 67-72, RES LEADER HORT CROPS, AGR RES SERV, USDA, 72-, PROF BOT, ORE STATE UNIV, 67- Mem: Am Phytopath Soc. Res: Small fruit diseases, particularly virus and viruslike diseases of Rubus and Fragaria. Mailing Add: Dept Bot & Plant Path Ore State Univ Corvallis OR 97331

CONVERY, F RICHARD, b Olympia, Wash, June 12, 32; m 55; c 3. ORTHOPEDIC SURGERY. Educ: Univ Wash, BA, 54, MD, 58; Am Bd Orthop Surg, dipl, 69. Prof Exp: Resident, 61-66, from instr to assoc prof orthop surg, Sch Med, Univ Wash, 67-72; ASSOC PROF ORTHOP SURG & DIR REHAB, DIV ORTHOP & REHAB, SCH MED, UNIV CALIF, SAN DIEGO, 72- Concurrent Pos: Sr fel orthop surg, Sch Med, Univ Wash, 63-64; Southern Calif Arthritis Found clin fel, Ranchos Los Amigos Hosp, 66-67. Honors & Awards: Kappa Delta Award, Am Acad Orthop Surg, 73. Mem: Am Rheumatism Asn. Res: Degeneration and repair of articular cartilage as related to joint reconstruction; surgical management of rheumatoid arthritis. Mailing Add: Div of Orthop & Rehab Univ of Calif Sch of Med San Diego CA 92103

CONVERY, ROBERT JAMES, b Philadelphia, Pa, Jan 17, 31; m 52; c 3. ORGANIC CHEMISTRY. Educ: St Joseph's Col, Philadelphia, BS, 52; Univ Notre Dame, MS, 56; Univ Pa, PhD(chem), 59. Prof Exp: Res chemist, Rohm and Haas Co, Pa, 54, Atlantic Refining Co, 55 & E I du Pont de Nemours & Co, 56-59; res chemist, Res & Develop Div, Sun Oil Co, 59-62 & Res Group, 62-63, asst sect chief, 63-64; from asst prof to assoc prof chem, 64-70, actg head dept, 64-66, assoc dean sci, 71-74, PROF CHEM, COL STEUBENVILLE, 70-, HEAD DEPT, 66-71 & 74- Concurrent Pos: Vis lectr, St Mary's Col, Ind, 54-55; instr, Evening Div, St Joseph's Col, Pa, 60-62. Mem: AAAS; Am Chem Soc; Am Inst Chem. Res: Organic mechanisms and synthesis; free radicals. Mailing Add: Dept of Chem Col of Steubenville Steubenville OH 43952

CONVEY, EDWARD MICHAEL, b Hicksville, NY, Oct 12, 39; m 62; c 2. PHYSIOLOGY, ENDOCRINOLOGY. Educ: Mich State Univ, BS, 63, MS, 65; Rutgers Univ, PhD(physiol), 68. Prof Exp: Asst prof, 68-74, ASSOC PROF DAIRY PHYSIOL, MICH STATE UNIV, 74- Mem: Am Dairy Sci Asn; Am Soc Animal Sci. Res: Lactational physiology; anterior pituitary. Mailing Add: Dept of Dairy Mich State Univ East Lansing MI 48823

CONWAY, ALVIN CHARLES, b Chicago, Ill, Jan 5, 22; m 48; c 2. PHARMACOLOGY. Educ: Univ Chicago, BS, 47, MS, 48. Prof Exp: Asst anesthesiol & pharmacol, Univ Chicago, 47-48; res chemist, Swift & Co, 48-49; res pharmacologist, Irwin, Neisler & Co, 49-51; sr res pharmacologist, Lakeside Labs, 51-60, chief gen pharmacol sect, 60-65; sr pharmacologist, 65-67, SR RES SPECIALIST, MINN MINING & MFG CO, 67- Mem: AAAS; NY Acad Sci; Am Chem Soc; Sigma Xi. Res: Behavioral screening; psychotropics. Mailing Add: Riker Res & Dev Div 218-2 Minn Mining & Mfg Co 3M Ctr St Paul MN 55101

CONWAY, BRIAN EVANS, b London, Eng, Jan 26, 27; m 54; c 1. PHYSICAL CHEMISTRY. Educ: Imp Col, London, BSc, 46, PhD & dipl, 49, DSc, 61. Prof Exp: Res assoc chem, Inst Cancer Res, London, 49-54; asst prof, Univ Pa, 54-56; from asst prof to assoc prof, 56-60, chmn dept, 66-69, PROF CHEM, UNIV OTTAWA, 60- Concurrent Pos: Consult tech ctr, Gen Motors; Commonwealth vis prof, Univs Southampton & Newcastle, 69-70. Honors & Awards: Novanda lectr award, Chem Inst Can, 64. Mem: The Chem Soc; Electrochem Soc; fel Brit Chem Soc; fel Royal Inst Chem; fel Royal Soc Can. Res: Electrochemistry; kinetics of electrode processes; adsorption at electrodes; isotopic effects in electrode reactions; polyelectrolytes; thermodynamics of polymer solutions. Mailing Add: Dept of Chem Univ of Ottawa 365 Nicholas St Ottawa ON Can

CONWAY, DWIGHT COLBUR, b Long Beach, Calif, Nov 14, 30; m 62; c 4. PHYSICAL CHEMISTRY. Educ: Univ Calif, BS, 52; Univ Chicago, PhD(chem), 56. Prof Exp: From instr to asst prof chem, Purdue Univ, 56-63; assoc prof, 63-67, PROF CHEM, TEX A&M UNIV, 67- Mem: Am Chem Soc; Am Phys Soc; Am Soc Mass Spectros. Res: Ion-molecule reactions; self consistent field-molecular orbit calculations;

mass spectroscopy. Mailing Add: Dept of Chem Tex A&M Univ College Station TX 77843

CONWAY, EDWARD DAIRE, III, b New Orleans, La, Feb 7, 37; m 61; c 2. MATHEMATICAL ANALYSIS. Educ: Loyola Univ, BS, 59; Ind Univ, MA, 63, PhD(math), 64. Prof Exp: Vis mem, Courant Inst Math Sci, NY Univ, 64-65; asst prof math, Univ Calif, San Diego, 65-67; from asst prof to assoc prof, 67-74, PROF MATH, TULANE UNIV, 74- Mem: Am Math Soc; Soc Indust & Appl Math; Math Asn Am; AAAS. Res: Nonlinear partial differential equations. Mailing Add: Dept of Math Tulane Univ New Orleans LA 70118

CONWAY, GENE FARRIS, b Cynthiana, Ky, Aug 24, 28; m 50; c 3. INTERNAL MEDICINE, CARDIOVASCULAR DISEASES. Educ: Univ Ky, BS, 49; Univ Cincinnati, MD, 52. Prof Exp: Intern med, Philadelphia Gen Hosp, 52-53; resident internal med, Louisville Gen Hosp, 53-54; chief med serv, US Air Force Hosp, Topeka, Kans, 54-56; from resident to chief resident internal med, Cincinnati Gen Hosp, 56-59, USPHS res fel cardiol, 59-61, from asst prof to assoc prof, 61-70, PROF MED, COL MED, UNIV CINCINNATI, 70-, ASST DIR DEPT INTERNAL MED, 73-; CHIEF MED SERV, VET ADMIN HOSP, CINCINNATI, 72- Concurrent Pos: Clin investr, Vet Admin Hosp, Cincinnati, 61-63, chief of cardiol, 63-, assoc chief of staff for res, 72-74; fel, Coun Clin Cardiol, Am Heart Asn, 69-; prog adv cardiovasc res, Res Serv, Vet Admin, 74- Mem: Am Col Cardiol. Res: Chemistry of myocardial contractile proteins; myocardial biology; congestive heart failure; arrhythmias; ischemic heart disease. Mailing Add: Vet Admin Hosp Med Serv Univ of Cincinnati Med Ctr Cincinnati OH 45220

CONWAY, HERTSELL S, b Peoria, Ill, Sept 30, 14; m 40; c 2. INFORMATION SCIENCE, CHEMICAL LITERATURE. Educ: Univ Chicago, SB, 32, PhD(org chem), 37. Prof Exp: Asst, Emulsol Corp, Ill, 37-39; chemist, Food & Drug Admin, USDA, 39-41 & Lambert Pharmacal Co, Mo, 41-46; org chemist, US Govt Rubber Labs, 46-47; chemist, Standard Oil Co, Ind, 47-49, group leader, 49-55, sect leader, 55-60; res supvr, Am Oil Co, 60-65; SR INFO SCIENTIST, STANDARD OIL CO, IND, 65- Mem: Am Chem Soc; Am Soc Info Sci. Res: Analysis of petroleum products and petrochemicals; indexing and searching; technical writing and editing. Mailing Add: Standard Oil Co Ind Box 400 Naperville IL 60540

CONWAY, JOHN BLIGH, b New Orleans, La, Sept 22, 39; m 64. MATHEMATICAL ANALYSIS. Educ: Loyola Univ, BS, 61; La State Univ, PhD(math), 65 Prof Exp: Asst prof, 65-70, ASSOC PROF MATH, IND UNIV, BLOOMINGTON, 70- Mem: Am Math Soc; Math Asn Am. Res: Functional analysis. Mailing Add: Dept of Math Ind Univ Bloomington IN 47401

CONWAY, JOHN GEORGE, JR, b Pittsburgh, Pa, May 16, 22; m 47; c 7. ATOMIC SPECTROSCOPY. Educ: Univ Pittsburgh, BS, 44. Prof Exp: Chemist, Los Alamos Sci Lab, NMex, 44-46; asst physics, Univ Pittsburgh, 46; PHYSICIST, LAWRENCE BERKELEY LAB, UNIV CALIF, BERKELEY, 46- Concurrent Pos: Mem comt, Line Spectra of the Elements, Nat Acad Sci, 67; res assoc, Nat Ctr Sci Res, Orsay, France, 73-74; chmn comt line spectra of elements, Nat Res Coun, 74-75. Mem: Soc Appl Spectros; assoc Am Phys Soc; fel Optical Soc Am. Res: Spectroscopy; chemical analysis and absorption and emission of transuranium elements; the spectra of higher ionized atoms. Mailing Add: Lawrence Berkeley Lab Univ of Calif Berkeley CA 94720

CONWAY, KENNETH EDWARD, b Philadelphia, Pa, June 7, 43; m 68; c 1. MYCOLOGY, PLANT PATHOLOGY. Educ: State Univ NY Potsdam, BS, 66; State Univ NY Col of Forestry, Syracuse, MS, 68; Univ Fla, PhD(bot), 73. Prof Exp: Teacher biol & ecol, Alachua County Bd Pub Instr, 70-73; res asst, 73-75, ASST RES SCIENTIST PLANT PATH, UNIV FLA, 75- Mem: Mycol Soc Am; Am Photopath Soc; Hyacinth Control Soc; Sigma Xi. Res: Biological control of aquatic plants utilizing plant pathogens. Mailing Add: Dept of Plant Path Univ of Fla Gainsville FL 32611

CONWAY, LYNN ANN, b Mt Vernon, NY, Jan 2, 38. COMPUTER SCIENCE, ELECTRICAL ENGINEERING. Educ: Columbia Univ, BS, 62, MSEE, 63. Prof Exp: Mem res staff comput archit, Int Bus Mach Corp, 64-69; sr staff engr, Memorex Corp, 69-73; MEM RES STAFF, DIGITAL SYST ARCHIT, XEROX PALO ALTO RES CTR, 73- Concurrent Pos: Consult, Syst Industs, 73-74. Mem: Inst Elec & Electronics Engrs; Asn Comput Mach. Res: Computer architecture; digital system architecture; digital image processing; optical character recognition; digital system design automation. Mailing Add: Xerox Palo Alto Res Ctr 3333 Coyote Hill Rd Palo Alto CA 94304

CONWAY, RICHARD WALTER, b Milwaukee, Wis, Dec 12, 31; m 53; c 2. OPERATIONS RESEARCH. Educ: Cornell Univ, BME, 54, PhD(opers res), 58. Prof Exp: Assoc prof indust eng & opers res, 58-67, PROF COMPUT SCI & OPERS RES, CORNELL UNIV, 67- Concurrent Pos: Consult, Gen Elec Co, Int Bus Mach Corp, Western Elec Co & Rand Corp. Mem: Inst Mgt Sci; Asn Comput Mach; Opers Res Soc Am. Res: Production control; computer sciences. Mailing Add: Upson Hall Cornell Univ Ithaca NY 14850

CONWAY, THOMAS WILLIAM, b Aberdeen, SDak, June 6, 31; m 57; c 2. BIOCHEMISTRY. Educ: Col St Thomas, BS, 53; Univ Tex, MA, 55, PhD(biochem), 62. Prof Exp: Fel biochem, Rockefeller Inst, 62-64, res assoc, 64; from asst prof to assoc prof, 64-73, PROF BIOCHEM, UNIV IOWA, 73- Concurrent Pos: Vis prof, Univ Chile, 68. Mem: AAAS; Am Chem Soc; Am Soc Microbiol; Am Soc Biol Chem; hon mem Biol Soc Chile. Res: Mechanism and control of protein biosynthesis in phage-infected bacteria. Mailing Add: Dept of Biochem Univ of Iowa Iowa City IA 52242

CONWAY, WALTER DONALD, b Troy, NY, Feb 4, 31; m 58; c 4. ORGANIC CHEMISTRY Educ: Rensselaer Polytech Inst, BS, 52; Univ Rochester, PhD(org chem), 56. Prof Exp: Res chemist, Esso Res & Eng Co, 56-57, Nat Cancer Inst, US Dept Health, Educ & Welfare, 57-62 & Sterling-Winthrop Res Inst, 62-65; res assoc, Lab Chem Pharmacol, Nat Heart Inst, 65-67; asst prof, 67-71, ASSOC PROF PHARMACEUT, SCH PHARM, STATE UNIV NY BUFFALO, 71- Concurrent Pos: Mem US pharmacopeia comt rev, US Pharmacopeial Conv, 75-80. Mem: Am Chem Soc; Am Pharmaceut Asn; Acad Pharmaceut Sci; Am Soc Pharmacol & Exp Therapeut. Res: Analytical methodology; identification of drug metabolites; effect of species differences and route of administration on the metabolic fate of drugs. Mailing Add: Sch of Pharm State Univ of NY Buffalo NY 14214

CONWELL, ESTHER MARLY, b New York, NY, May 23, 22; m 45; c 1. SOLID STATE PHYSICS. Educ: Brooklyn Col, BA, 42; Univ Rochester, MS, 45; Univ Chicago, PhD(physics), 48. Prof Exp: Instr physics, Brooklyn Col, 45-51; mem tech staff, Bell Tel Labs, 51-52; eng specialist, GTE Labs, 52-63, mgr physics dept, 63-72; PRIN SCIENTIST, XEROX, 72- Concurrent Pos: Vis prof, Univ Paris, 62-63; Abbie Rockefeller Mauze prof, Mass Inst Technol, 72. Honors & Awards: Annual Award, Soc Women Eng, 60. Mem: Fel AAAS; fel Am Phys Soc; Inst Elec & Electronics Eng. Res: Integrated optics. Mailing Add: Xerox Xerox Square W114 Rochester NY 14644

CONYERS, EMERY SWINFORD, b Cynthiana, Ky, Aug 16, 39; m 64; c 2. SOIL SCIENCE. Educ: Univ Ky, BS, 61; Ohio State Univ, MS, 63, PhD(soil chem), 66. Prof Exp: Res asst soils, Ohio State Univ, 61-66, teaching assoc, 66; chemist, US Army Aviation Mat Labs, 66-68; res specialist, 68-74, proj mgr construct mat, 74-75, GROUP LEADER MEMBRANE SYSTS, DOW CHEM CO, 75- Mem: Am Soc Agron; Soil Sci Soc Am; Sigma Xi. Res: Fixation and release of potassium by soils and clays, metal ion-clay interactions, chemical control of soil erosion, land treatment of wastewater and membrane systems for water treatment. Mailing Add: Dow Chem Co 2800 Mitchell Dr Walnut Creek CA 94598

CONYNE, RICHARD FRANCIS, b Canandaigua, NY, June 26, 19; m 47; c 4. POLYMER CHEMISTRY. Educ: Univ Rochester, AB, 41. Prof Exp: Chemist & group leader, 41-56, head lab, 56-68, res supvr, 68-73, PLASTICS MGR, INT MKT DEPT, ROHM AND HAAS CO, 73- Mem: Am Chem Soc; Soc Plastics Indust; Soc Plastics Eng. Res: Plastics; plasticizers; coatings. Mailing Add: Rohm & Haas Co Independence Mall Philadelphia PA 19105

CONZELMAN, GAYLORD MAURICE, JR, b Republic, Kans, Mar 30, 23; m 53; c 3. PHARMACOLOGY. Educ: Idaho State Col, BS, 49; George Washington Univ, PhD, 53. Prof Exp: Lilly Res Labs fel, Christ Hosp Inst Med Res, Cincinnati, Ohio, 53-55, res assoc, 55-63; res pharmacologist, 63-68, lectr, 68-69, asst prof, 69-74, ASSOC PROF PHARMACOL, DEPT PHYSIOL SCI, SCH VET MED, UNIV CALIF, DAVIS, 74- Mem: AAAS; Am Soc Pharmacol & Exp Therapeut; Am Asn Cancer Res; Soc Toxicol. Res: Cancer biology; biosynthesis of purines; chemotherapy; drug metabolism; carcinogenesis; drug induced nephropathy. Mailing Add: Dept of Physiol Sci Sch Vet Med Univ of Calif Davis CA 95616

CONZETT, HOMER EUGENE, b Dubuque, Iowa, Oct 16, 20; m 60; c 2. EXPERIMENTAL NUCLEAR PHYSICS. Educ: Univ Dubuque, BS, 42; Univ Calif, PhD(physics), 56. Prof Exp: Degaussing physicist, Bur Ord, Dept Navy, 42-44; res physicist, Radiation Lab, Univ Calif, 56-57; vis Fulbright lectr, Univ Tokyo, 57-58; res physicist, 58-64, SR RES PHYSICIST IN CHG OPER & DEVELOP 88 INCH CYCLOTRON, LAWRENCE BERKELEY LAB, UNIV CALIF, 64- Concurrent Pos: Vis res physicist, Inst Nuclear Sci, Univ Grenoble, 66-67. Mem: Am Phys Soc. Res: Nuclear reactions and scattering below 100 mev; spin-polarization phenomena in nuclear physics. Mailing Add: 318 Vassar Ave Kensington CA 94708

COOCH, FREDERICK GRAHAM, b Winnipeg, Man, May 4, 28; m 58; c 3. ECOLOGY, ORNITHOLOGY. Educ: Queen's Univ, Ont, BA, 51; Cornell Univ, MS, 53, PhD(wildlife mgt), 58. Prof Exp: Arctic ornithologist, Dept Environ, Can Wildlife Serv, 54-62, head biocide invests, 62-64, STAFF SPECIALIST, MIGRATORY BIRDS INVESTS, HEAD MIGRATORY BIRD POP SECT, HEAD CAN BANDING OFF & MIGRATORY BIRD COORDR, CAN WILDLIFE SERV, 64- Mem: Fel AAAS; fel Arctic Inst NAm; Am Ornith Union; Wildlife Soc. Res: Wildlife ecology, especially Arctic; insecticides; vertebrate systematics. Mailing Add: 685 Echo Dr Ottawa ON Can

COODLEY, EUGENE LEON, b Los Angeles, Calif, Jan 14, 20; m 47; c 3. INTERNAL MEDICINE, CARDIOLOGY. Educ: Univ Calif, Berkeley, BA, 40; Univ Calif, San Francisco, MD, 43. Prof Exp: Consult cardiac dis & rehab, Calif Dept Rehab, 55-61; dir dept med, Lidcombe Hosp, Australia, 61-62; dir dept med, Kern County Hosp, 65-67; DIR DEPT MED, PHILADELPHIA GEN HOSP, 67-; PROF MED, HAHNEMANN MED COL, 67-, ASSOC CHMN DEPT, DIR DIV INTERNAL MED & DIR RESIDENCY TRAINING PROG, 74- Concurrent Pos: Guest lectr, Europ Cong Rheumatism, 59 & Int Cong Surg, 65; consult, Dept Rehab, Sydney, Australia, 61-62. Mem: Fel Am Col Physicians; Am Col Angiol; fel Am Col Cardiol; fel Am Col Gastroenterol. Res: Enzyme research, development of new enzyme procedures for diagnosis in medicine. Mailing Add: Dept of Med Hahnemann Med Col Philadelphia PA 19102

COOGAN, ALAN H, b Brooklyn, NY, Dec 19, 29. PALEONTOLOGY, GEOLOGY. Educ: Univ Calif, Berkeley, BA, 56, MA, 57; Univ Ill, PhD, 62. Prof Exp: Instr geol, Cornell Col, 57-60; geologist, Humble Oil & Refining Co, 62-65, geologist, Esso Prod Res Co, 65-67; PROF GEOL, KENT STATE UNIV, 67-, ASSOC DEAN RES, 69- Mem: Am Asn Petrol Geologists; Am Paleont Soc; Soc Econ Paleontologists & Mineralogists; Soc Res Adminr. Res: Paleontology and stratigraphy; carbonate petrology; Rudist paleontology; environmental geology. Mailing Add: 102 Kent Hall Kent State Univ Kent OH 44242

COOGAN, JOHN MICHAEL, b Seattle, Wash, Apr 23, 31; m 58; c 3. RESEARCH ADMINISTRATION, SCIENCE POLICY. Educ: Univ Wash, BS, 53; Standford Univ, MS, 66. Prof Exp: Physicist systems anal, US Air Force, 57-59; asst proj mgr, Explorer XI Satellite, Goddard Space Flight Ctr, NASA, 59-62, physicist, Off Adv Technol, Ames Res Ctr, 62-76; staff consult space sci, US Air Force, 67-73; DEP TO COUNR SCI AFFAIRS, US EMBASSY, PARIS, 73- Mem: French Phys Soc; Europ Phys Soc. Res: Physiology of nutrition; molecular biology. Mailing Add: US Embassy-Sci Att APO New York NY 09777

COOHILL, THOMAS PATRICK, b Brooklyn, NY, Aug 25, 41; m 62; c 3. BIOPHYSICS. Educ: Univ Toronto, BSc, 62; Univ Toledo, MSc, 64; Pa State Univ, PhD(biophys), 68. Prof Exp: Asst prof cell biol, Med Sch, Univ Pittsburgh, 68-72; res physicist, Vet Admin Hosp, Leech Farm, 68-72; ASSOC PROF BIOPHYS, WESTERN KY UNIV, 72- Mem: AAAS; Biophys Soc; Am Soc Photobiol. Res: Effects of ultraviolet light on fungi, on mammalian tissue culture cells, on human viruses and crustaceans. Mailing Add: Biophys Prog Western Ky Univ Bowling Green KY 42101

COOIL, BRUCE JAMES, b Colfax, Wash, Aug 21, 14; m 44; c 2. PLANT PHYSIOLOGY. Educ: State Col Wash, BSc, 36; Univ Hawaii, MS, 39; Univ Calif, PhD(plant physiol), 47. Prof Exp: Asst physiologist guayule res proj, Bur Plant Indust, Soils & Agr Eng, USDA, Calif, 42-45; asst plant physiologist, US Regional Salinity Lab, 45-47; assoc plant physiologist, 47-54, PROF BOT & PLANT PHYSIOLOGIST, AGR EXP STA, UNIV HAWAII, 54- Mem: Bot Soc Am; Am Soc Plant Physiol. Res: Translocation of organic materials in plants; mineral nutrition of plants; salt absorbtion and transport in plant roots. Mailing Add: Dept of Bot Univ Hawaii 3190 Maile Way Honolulu HI 96822

COOK, ADDISON GILBERT, b Caracas, Venezuela, Apr 1, 33; US citizen; m 56; c 3. ORGANIC CHEMISTRY. Educ: Wheaton Col, BS, 55; Univ Ill, PhD(org chem), 59. Prof Exp: Asst chem, Univ Ill, 55-56; fel, Cornell Univ, 59-60; from asst prof to assoc prof, 60-70, PROF CHEM & CHMN DEPT, VALPARAISO UNIV, 70- Concurrent Pos: Consult, Argonne Nat Labs, 62-73. Mem: Am Chem Soc. Res: Amines, heterocyclic compounds, bicyclic compounds, organophosphorus compounds and small ring compounds. Mailing Add: Dept of Chem Valparaiso Univ Valparaiso IN 46383

COOK, ALAN, b Harrow, Eng, July 15, 39; US citizen; m 64; c 1. CHEMISTRY. Educ: Univ Birmingham, BSc, 61; Univ London, PhD(chem), 64. Prof Exp: Fel, Syntex Corp, Calif, 64-66; CHEMIST, HOFFMANN-LA ROCHE INC, 66- Mem: Am Chem Soc. Res: Chemistry of nucleosides and nucleotides Mailing Add: Hoffmann-La Roche Inc 340 Kingsland St Nutley NJ 07110

COOK, ALBERT WILLIAM, b Brooklyn, NY, July 23, 22; m 47; c 2. NEUROSURGERY. Educ: Dartmouth Col, AB, 44; Long Island Col Med, MD, 46; Am Bd Neurol Surg, dipl, 56. Prof Exp: Assoc prof surg & head div neurosurg, 59-71, PROF NEUROSURG & CHMN DEPT, STATE UNIV NY DOWNSTATE MED CTR, 71- Concurrent Pos: Consult regional hosps, Brooklyn, NY & Vet Admin Hosps, 59-; mem coop study treatment intracranial aneurysms & subarachnoid hemorrhages, 63- Mem: AMA; Am Col Surgeons; Am Asn Neurol Surgeons; Soc Neurosci; Asn Res Nerv & Ment Dis. Res: Craniocerebral trauma; intracranial blood clots; cerebrovascular disease; surgical treatment of vascular anomalies; cerebral hemodynamics; respiratory and metabolic aspects of cerebral lesions; cerebral neoplasms; pain control; electrical stimulation of the spinal cord. Mailing Add: 200 Hicks St Brooklyn NY 11201

COOK, ALLAN FAIRCHILD, II, b New York, NY, May 9, 22; m 59; c 2. ASTROPHYSICS. Educ: Princeton Univ, BSE, 47, AM, 50, PhD(astron), 52. Prof Exp: Asst astron, Princeton Univ, 48-49; instr astron & physics, Carleton Col, 50-51; asst, Col Observ, Harvard Univ, 51-57, res assoc, 57-59, res fel, 59-61, res assoc, 61-62, lectr, 61-74; physicist, 61-64, ASTROPHYSICIST, SMITHSONIAN ASTROPHYS OBSERV, 64- Mem: Am Astron Soc; Meteoritical Soc. Res: Asteroids; meteors; meteor spectra; Saturn's rings. Mailing Add: Smithsonian Astrophys Observ 60 Garden St Cambridge MA 02138

COOK, ALLYN AUSTIN, b Grandview, Ill, Feb 14, 27; m 48; c 2. PLANT PATHOLOGY. Educ: Univ Wis, PhD(plant path), 51. Prof Exp: Asst pathologist, SDak State Col, 52-54; agent pathologist, Spec Crops Sect, USDA, 54-56; from assoc plant pathologist to plant pathologist, Agr Exp Sta, 56-74, PROF PLANT PATH, UNIV FLA, 74- Concurrent Pos: Fulbright lectr, Ain Shams Univ, Cairo, 63-64; vis specialist, Univ Hawaii, 70-71. Mem: Am Phytopath Soc; Am Hort Soc. Mailing Add: Dept Plant Path Univ of Fla Gainesville FL 32601

COOK, ANCEL EUGENE, b Sadieville, Ky, Sept 15, 09; m 37; c 3. PHYSICAL OPTICS, MICROELECTRONICS. Educ: Georgetown Col, BA, 35; Univ Ky, MS, 48. Prof Exp: Teacher pub sch, Ky, 35-42; physicist, Indust Mgr Off, US Navy, La, 42-43; asst prof physics, Georgetown Col, 47-49; instr, Univ Ky, 49-51; physicist, Bur Ships, Navy Dept, 54-59; PHYSICIST, OFF NAVAL RES, 59- Concurrent Pos: Sci secy laser res & explor develop panel, US Navy, 63-, mem microelectronics panel, 66-, coordr adv groups, industry & independent res & develop progs; US Navy mem, Dept Defense Adv Group Electron Devices; asst to US Navy mem, Armed Serv Res Specialist Comt; secy, Govt Microelectronics Applns Confs. Mem: Am Phys Soc; Am Soc Naval Eng; sr mem Inst Elec & Electronic Engrs. Res: Lasers and fluidic research and exploratory development; physical sciences. Mailing Add: 3021 Park Dr S E Washington DC 20020

COOK, BARNETT C, b Chicago, Ill, Nov 25, 23. PHYSICS. Educ: Northwestern Univ, BS, 46; Univ Chicago, PhD(physics), 56. Prof Exp: Physicist, Inst Nuclear Res, Univ Chicago, 52-56; asst prof physics, Univ Pa, 56-59; physicist, Midwest Univs Res Asn, 59-60; physicist, 60-61, asst prof physics, 61-75, ASSOC PHYSICIST, AMES LAB, IOWA STATE UNIV, 61-, ASSOC PROF PHYSICS, 74- Mem: AAAS; Am Phys Soc. Res: Photonuclear reactions; accelerator design. Mailing Add: Dept of Physics Iowa State Univ Ames IA 50010

COOK, BENJAMIN JACOB, b Upper Darby, Pa, Sept 26, 30; m 52; c 2. INSECT PHYSIOLOGY, BIOCHEMISTRY. Educ: Providence Col, AB, 58; Rutgers Univ, MS, 61, PhD(insect physiol), 63. Prof Exp: Asst prof biol, Col Holy Cross, 62-64; ADJ PROF ZOOL & ASST PROF ENTOM, N DAK STATE UNIV & INSECT PHYSIOLOGIST, METAB & RADIATION LAB, USDA, 64-, ADJ PROF AGRON, 74- Mem: Am Soc Zool; Entom Soc Am. Res: Insect neurophysiology and neurochemistry; peripheral and central mechanisms of nerve impulse transmission; the isolation and pharmacodynamics of natural neurotransmitters and neurohormones. Mailing Add: Metab & Radiation Lab USDA NDak State Univ Univ Sta Fargo ND 58103

COOK, BILLY DEAN, b Oklahoma City, Okla, July 28, 35; m 60; c 2. PHYSICS. Educ: Okla State Univ, BS, 57; Mich State Univ, MS, 59, PhD(physics), 62. Prof Exp: Res instr, Mich State Univ, 62-63, asst prof res physics, 65-68; res assoc, 68-69, ASSOC PROF MECH & ELEC ENG, UNIV HOUSTON, 69- Res: Ultrasonic light diffraction; non-linear propagation of acoustical waves; optics and noise control. Mailing Add: Dept of Mech & Elec Eng Univ of Houston Houston TX 77004

COOK, CHARLES DAVENPORT, b, Minneapolis, Minn, Nov 30, 19; m 45; c 4. MEDICINE. Educ: Princeton Univ, AB, 41; Harvard Univ, MD, 44. Prof Exp: Asst prof pediat, Harvard Med Sch, 57-63, assoc clin prof, 63-64; prof pediat, Yale Univ, 64-75; PROF PEDIAT, STATE UNIV NY DOWNSTATE MED CTR, 75- Concurrent Pos: Physician, Boston Children's Hosp, 58-64; chmn, Joint Coun Nat Pediat Socs, 70-73. Mem: Soc Pediat Res; Am Acad Pediat; Am Pediat Soc (secy-treas, 64-75); Am Physiol Soc. Res: Respiratory physiology in children; health care. Mailing Add: 450 Clarkson Ave Brooklyn NY 11203

COOK, CHARLES FALK, b Jonesboro, Ark, Dec 21, 28; m 49; c 2. NUCLEAR PHYSICS, CHEMICAL PHYSICS. Educ: Tex Christian Univ, BA, 48, MA, 50; Rice Inst, PhD(physics), 53. Prof Exp: Fel physics, Rice Inst, 53-54; asst prof, Univ Fla, 54-55; sr nuclear eng, Convair Div, Gen Dynamics Corp, Ft Worth, Tex, 55-56, nuclear test lab engr, 56-58; nuclear physics sect mgr, 58-62, physics br mgr, 62-70, PHYSICS & ANAL BR MGR, RES DIV, PHILLIPS PETROL CO, 70- Mem: Am Phys Soc. Res: Nuclear reaction cross sections; electron spin resonance; molecular energy levels in hydrocarbon molecules using spectroscopic techniques; analytical chemistry; molecular structure. Mailing Add: Phillips Petrol Res & Develop Co 248 Res Bldg 1 Res Ctr Bartlesville OK 74004

COOK, CHARLES J, b West Point, Nebr, Oct 2, 23; m 45; c 2. ATOMIC PHYSICS, ELECTRICAL ENGINEERING. Educ: Univ Nebr, BS, 48, MA, 50, PhD(physics), 53. Prof Exp: Res assoc, Univ Nebr, 53-54; physicist, 54-56, mgr, Molecular Physics Sect, 56-62, dir chem physics div, 62-69, exec dir phys sci, 69-76, VPRES OFF RES OPERS, STANFORD RES INST, 76- Concurrent Pos: Instr, San Jose City Col, 57-58 & Foothill Col, 59-62; sr res assoc, Queen's Univ, Belfast, 62-63; partic, Advan Mgt Prog, Harvard Grad Sch Bus, 68. Mem: Am Phys Soc; Am Asn Physics Teachers; Am Inst Aeronaut & Astronaut; Am Inst Physics; Am Defense Preparedness Asn. Res: Ionic and atomic impact phenomena; technology utilization and transfer. Mailing Add: Stanford Res Inst Menlo Park CA 94025

COOK, CHARLES MARSHALL, JR, physical chemistry, see 12th edition

COOK, CHARLES WAYNE, b Gove, Kans, Oct 28, 14; m 40; c 1. RANGE SCIENCE. Educ: Ft Hays Kans State Univ, BS, 40; Utah State Univ, MS, 42; Agr & Mech Univ, Tex, PhD, 50. Prof Exp: Range conservationist, Soil Conserv Serv, USDA, 42-43; res prof range mgt, Utah State Univ, 43-46, prof range sci, 46-67; PROF RANGE SCI & HEAD DEPT, COLO STATE UNIV, 67- Honors & Awards: Hoblitzelle Award, 53. Mem: AAAS; Am Inst Biol Sci; assoc Am Soc Animal Sci; assoc Soc Range Mgt; Soil Conserv Soc Am. Res: Range seeding and forage nutrition; bioecology of cactus on the central great plains; utilization of range plants and forage by herbivores; physiological responses of plants; energy flow in the range ecosystem. Mailing Add: Dept of Range Sci Colo State Univ Ft Collins CO 80521

COOK, CHARLES WILLIAM, b Yankton, SDak, Sept 27, 27; m 50; c 3. EXPERIMENTAL NUCLEAR PHYSICS. Educ: Univ SDak, AB, 51; Calif Inst Technol, MS, 54, PhD(physics), 57. Prof Exp: Sr res engr, Convair, 57-58, design specialist & head nuclear physics, 58-60; chief res ballistic missile defense, Inst Defense Anal, 60-61; adv res proj agency, Dept Defense, 61-62; res & develop specialist, NAm Aviation, Inc, Calif, 62-63, corp dir electronics, 63-67; independent consult, 67-71; asst dir, Dept Res & Eng, Dept Defense, 71-74; DEP UNDER SECY, AIR FORCE SPACE SYSTS, WASHINGTON, DC, 74- Concurrent Pos: Consult, Inst Defense Anal, 62-63 & McGraw-Hill Book Co, Inc, 62- Honors & Awards: Secy Defense Meritorious Civil Serv Award, Dept Defense, 74. Mem: Am Phys Soc; Am Inst Phys; Am Inst Aeronaut & Astronaut; Inst Elec & Electronics Engrs. Res: Energy generation and element synthesis reactions occurring in stellar interiors; electronics research and development. Mailing Add: 1180 Daleview Dr McLean VA 22101

COOK, CLARENCE EDGAR, b Jefferson City, Tenn, Apr 27, 36; m 57; c 3. ORGANIC CHEMISTRY. Educ: Carson-Newman Col, BS, 57; Univ NC, PhD(org chem), 61. Prof Exp: Am Chem Soc Petrol Res Fund fel, Cambridge Univ, 61-62; sr chemist, 62-68, group leader, 68-71, ASST DIR, CHEM & LIFE SCI DIV, RES TRIANGLE INST, 71- Mem: AAAS; Am Chem Soc; Phytochem Soc NAm; NY Acad Sci; Am Soc Pharmacol & Exp Therapeut. Res: Drug metabolism; oral contraceptives; synthesis of medicinal compounds; oxygen and nitrogen heterocycles; steroid chemistry; natural products; agricultural chemistry; immunoassay development. Mailing Add: Chem & Life Sci Div Res Triangle Inst PO Box 12194 Research Triangle Park NC 27709

COOK, CLARENCE HARLAN, b Winthrop, Iowa, Jan 16, 25; m 45; c 4. MATHEMATICS. Educ: Univ Iowa, BA, 48, MS, 50; Univ Colo, PhD(math), 62. Prof Exp: Asst prof math, Western State Col, Colo, 50-52; asst, Univ Tex, 52-53; engr, Convair, Tex, 53-54; sr mathematician, Martin Co, Md, 54-55, prin engr, Colo, 55-59, sragg engr opers res, 60-62; asst prof math, Univ Okla, 62-65; asst prof, 65-69, ASSOC PROF MATH, UNIV MD, COLLEGE PARK, 69- Mem: Am Math Soc; Math Asn Am; Math Soc France; Math Soc Belg. Res: Functional analysis; topology. Mailing Add: Dept of Math Univ of Md College Park MD 20742

COOK, CLARENCE SHARP, b St Louis Crossing, Ind, Aug 18, 18; m 43; c 2. PHYSICS. Educ: DePauw Univ, AB, 40; Ind Univ, MA, 42, PhD(physics), 48. Prof Exp: Asst physics, Ind Univ, 40-42 & 46-48; asst prof, Wash Univ, St Louis, 48-53; head nuclear radiation br, US Naval Radiol Defense Lab, 53-59, head radiation effects br, 59-60, head nucleonics div, 60-61, physics consult to sci dir, 62-65, head radiation physics div, 65-69; lectr, Univ Santa Clara, 69-70; chmn dept, 70-72, PROF PHYSICS, UNIV TEX, EL PASO, 70- Concurrent Pos: Mem bd exam scientists & engrs, US Civil Serv Comn, 55-58, chmn bd, 57-58, mem prof coun scientists & engrs, Calif-Nev area, 67-69; Fulbright res scholar, Aarhus Univ, 61-62. Mem: Am Phys Soc; Am Asn Physics Teachers; Am Geophys Union; Health Physics Soc. Res: Energy education; effects of ionizing radiations. Mailing Add: Box 204 Univ of Tex El Paso TX 79968

COOK, COURTNEY FREDERICK, physical chemistry, see 12th edition

COOK, DAVID ALLAN, b Colby, Wis, Mar 11, 40; m 62; c 4. NUTORITIONAL BIOCHEMISTRY. Educ: Wis State Univ-River Falls, BS, 62; Iowa State Univ, PhD(animal nutrit), 67. Prof Exp: NIH res fel biochem, Univ Minn, St Paul, 67-68; sr investr, Dept Nutrit Res, 69-73, prin investr, 73-74, clin investr, Dept Clin Invest, 74-75, NUTRIT INVESTR, DEPT CLIN NUTRIT, MEAD JOHNSON RES CTR, 75- Mem: AAAS; Am Soc Animal Sci; Am Inst Nutrit; Sigma Xi; Nutrit Today Soc. Res: Ruminant fatty acid metabolism and absorption; oxalic acid formation from aromatic amino acids; infant nutrition; bile acid metabolism; atherosclerosis; mineral metabolism and bioavailability; clinical nutrition; gastroenterology. Mailing Add: Dept of Clin Nutrit Mead Johnson Res Ctr Evansville IN 47721

COOK, DAVID EDGAR, b Corpus Christi, Tex, Dec 13, 40; m 63; c 2. BIOCHEMISTRY. Educ: Southwest Tex State Col, BS, 62; Okla State Univ, MS, 65; Univ Tex, Austin, PhD(chem), 70. Prof Exp: From instr to asst prof chem, Northeastern State Col, Okla, 64-70; NIH res fel & res assoc biochem, Inst Enzyme Res, Univ Wis-Madison, 70-73; ASST PROF BIOCHEM, COL MED, UNIV NEBR MED CTR, OMAHA, 73- Concurrent Pos: NSF sci fac fel, Univ Tex, Austin, 68-69. Mem: Am Chem Soc; AAAS; Sigma Xi. Mailing Add: Dept of Biochem Univ of Nebr Med Ctr Omaha NE 68105

COOK, DAVID EDWIN, b Houston, Tex, Dec 3, 35; m 58; c 2. TOPOLOGY. Educ: Univ Tex, BA, 56, MA, 60, PhD(math), 67. Prof Exp: Spec instr math, Univ Tex, 61-64, teaching assoc, 64-67; asst prof, 67-69, ASSOC PROF MATH, UNIV MISS, 69- Mem: Am Math Soc; Math Asn Am. Res: Point set topology. Mailing Add: Dept of Math Univ of Miss University MS 38677

COOK, DAVID LEWIS, b Grand Junction, Colo, July 8, 16; m 41; c 4. DATA PROCESSING. Educ: Munic Univ Wichita, BA, 39; Univ NC, PhD(chem), 43. Prof Exp: Lab asst chem, Munic Univ Wichita, 36-37; asst chemist, Naval Res Lab, DC, 37-38; chemist, Kans State Grain Inspection, 39-41; asst instr chem, Univ NC, 39-42; chemist, Shell Develop Co, 43-56, technologist, Shell Oil Soc, 56-64, chemist, Shell Develop Co, 64-70; OWNER, TECH-SERVAS TIME-SHARING COMPUT, 70- Res: Spectroscopy; computing; physical chemistry; reaction kinetics; magnetochemistry; electronics; mass spectrometry. Mailing Add: 602 Colusa Ave El Cerrito CA 94530

COOK, DAVID MARSDEN, b Troy, NY, Apr 3, 38; m 65; c 2. THEORETICAL PHYSICS, MATHEMATICAL PHYSICS. Educ: Rernsselaer Polytech Inst, BS, 59; Harvard Univ, AM, 60, PhD(physics), 65. Prof Exp: Asst prof, 65-71, ASSOC PROF PHYSICS, LAWRENCE UNIV, 71- Concurrent Pos: NSF sci fac fel, Dartmouth Col, 71-72. Mem: Am Phys Soc; Am Asn Physics Teachers. Res: Plasma physics; applied mathematics; computers in physics education. Mailing Add: Dept of Physics Lawrence Univ Appleton WI 54911

COOK, DAVID RUSSELL, b Hastings, Mich, Aug 9, 22; m 52; c 2. INVERTEBRATE ZOOLOGY, ENTOMOLOGY. Educ: Univ Mich, BS, 48, MA, 51, PhD, 52. Prof Exp: USDA entomologist, US Nat Mus, 52-53; from asst prof to assoc prof, 53-65,

PROF BIOL, WAYNE STATE UNIV, 65- Concurrent Pos: Entomologist & malariologist, Int, Co-op Admin Malaria Control Prog, US Opers Mission to Liberia, 56-58; Fulbright res fel, India, 62-63. Res: Acarina; Hydracarina. Mailing Add: Dept of Biol Wayne State Univ Detroit MI 48202

COOK, DAVID WILSON, b Wilkinson County, Miss, Nov 3, 39; m 61; c 2. MICROBIOLOGY. Educ: Miss State Univ, BS, 61, MS, 63, PhD(microbiol), 66. Prof Exp: HEAD MICROBIOL SECT, GULF COAST RES LAB, 66-, REGISTR, 71-, ASST DIR ADMIN & ACAD AFFAIRS, 72- Concurrent Pos: Assoc mem grad fac, Miss State Univ, 68-; asst prof biol, Univ Miss, 71-; mem grad fac, Univ Southern Miss, 74- Mem: World Mariculture Soc; Am Soc Microbiol; Gulf Estuarine Soc; Sigma Xi. Res: Microbiology of the estuarine environment including pollution; nutrient turnover; diseases of fish and shellfish and spoilage of seafoods. Mailing Add: PO Box AG Gulf Coast Res Lab Ocean Springs MS 39564

COOK, DONALD BOWKER, b Easthampton, Mass, Jan 14, 17; m 43; c 6. PHYSICS. Educ: Princeton Univ, AB, 38; Columbia Univ, MA, 39, PhD(physics), 50. Prof Exp: Physicist, Manhattan Proj, Columbia Univ & Carbon & Carbide Chem Corp, NY, 42-44, sect leader, 44-45, group leader, 45-46, group leader, Div Govt Aid Res, Columbia Univ, 46-50; group leader, Nevis Cyclotron Lab, 47; res physicist, 50-57, SR RES PHYSICIST, E I DU PONT DE NEMOURS & CO, INC, 57- Mem: Am Phys Soc; Am Chem Soc; Sigma Xi. Res: Multilayer films; gas diffusion; cryogenics; structure of polyfibers. Mailing Add: RD1 Box 126 Hockessin DE 19707

COOK, DONALD JACK, b Rock Island, Ill, Feb 12, 15; m 39; c 2. ORGANIC CHEMISTRY. Educ: Augustana Col, AB, 37; Univ Ill, MA, 38; Ind Univ, PhD(org chem, 44. Prof Exp: City chemist, Rock Island, Ill, 38-39; Am Container Corp, 39-40; instr chem, Augustana Col, 40-41; chemist, Texas Co, NY, 41-42; asst sci, Ind Univ, 42-44; res chemist, Lubri-Zol Corp, 44-45; from asst prof to assoc prof, 45-54, PROF CHEM, DePAUW UNIV, 54-, HEAD DEPT, 64- Concurrent Pos: Assoc prof dir, Div Sci Personnel & Ed, NSF, 61-62. Mem: Am Chem Soc. Res: N-substituted carbostyrils; preparation and properties of substituted lepidones; SeO2 oxidations. Mailing Add: Dept of Chem DePauw Univ Greencastle IN 46135

COOK, DONALD LATIMER, b Arena, Wis, July 31, 16; m 43; c 2. PHARMACOLOGY. Educ: Univ Wis, BS, 38, PhD(pharmaceut chem), 43. Prof Exp: Instr pharmacol & physiol, Sch Pharm, Western Reserve Univ, 42-43; pharmacologist, G D Searle & Co, 46-70, head dept pharmacol, 70-74, ASSOC DIR DEPT BIOL RES, SEARLE LABS, G D SEARLE & CO, 74- Mem: Am Soc Pharmacol & Exp Therapeut; Soc Exp Biol & Med. Res: Phytochemical study of the leaves of Celastrus scandens Linne. Mailing Add: Searle Labs PO Box 5110 Chicago IL 60680

COOK, EARL FERGUSON, b Bellingham, Wash, May 24, 20; m 47, 69; c 3. GEOLOGY, GEOGRAPHY. Educ: Univ Wash, Seattle, BS, 43, MS, 47, PhD(geol), 54. Prof Exp: Instr, Univ Wash, Seattle, 47-48; instr, Stanford Univ, 48; photogeologist, Geophoto Servs, 49-51; asst prof geol & geog & actg head, Univ Idaho, 51-52, assoc prof & head, 52-57, prof & dean, Col Mines, 57-65; exec secy, Div Earth Sci, Nat Acad Sci-Nat Res Coun, 64-66; prof geol & assoc dean col geosci, 66-71, actg dean col geosci, 68-69, PROF GEOG & GEOL & DEAN COL GEOSCI, TEX A&M UNIV, 71- Concurrent Pos: Dir, State Bur Mines & Geol, Idaho, 57-65; consult, Nat Acad Sci, 66-67, mem comt Alaska earthquake, 67-; mem US del, Orgn Econ Coop & Develop Adv Conf Tunnelling, 70. Mem: Geol Soc Am; Soc Econ Geologists; Am Inst Mining, Metall & Petrol Engrs; Am Geophys Union; Asn Am State Geol (pres, 63-64). Res: Environmental decision-making related to resource development and use; ecoethics; environmental attitudes and behavior; volcanic geology and structures related to volcanism. Mailing Add: Col of Geosci Tex A&M Univ College Station TX 77843

COOK, EDWARD HOOPES, JR, b Harrisburg, Pa, May 21, 29; m 51; c 2. PHYSICAL CHEMISTRY, INORGANIC CHEMISTRY. Educ: Elizabethtown Col, BS, 50; Pa State Col, PhD(chem), 53. Prof Exp: SUPVR, HOOKER CHEM CORP, 63- Mem: Am Chem Soc; Electrochem Soc. Res: Electrochemical development; industrial electrolytic; theoretical electrochemistry; mechanism of corrosion processes. Mailing Add: 4698 Fifth St Lewistown NY 14092

COOK, EDWARD WERNER, organic chemistry, see 12th edition

COOK, EDWIN AUBREY, b Honolulu, Hawaii, Sept 7, 32; m 55; c 2. ANTHROPOLOGY. Educ: Univ Ariz, AB, 59; Yale Univ, PhD(anthrop), 67. Prof Exp: Asst prof anthrop, Univ Hawaii, 65-67, mem Res Coun, 65-66; asst prof, Univ Calif, Davis, 67-71; ASSOC PROF ANTHROP, SOUTHERN ILL UNIV, CARBONDALE, 71- Mem: Am Anthrop Asn; Am Ethnol Soc; Polynesian Soc; Royal Anthrop Inst Gt Brit & Ireland. Res: Kinship; social structure; psychological anthropology; language and culture. Mailing Add: Dept of Anthrop Southern Ill Univ Carbondale IL 62901

COOK, EDWIN FRANCIS, b San Francisco, Calif, Sept 11, 18; m 49; c 5. ENTOMOLOGY. Educ: Stanford Univ, AB, 43, AM, 44, PhD(biol sci), 48. Prof Exp: Actg instr gen biol, Stanford Univ, 46-48, actg asst prof entom, 48-49; PROF ENTOM, UNIV MINN, ST PAUL, 49- Mem: AAAS; Am Soc Syst Zool; Entom Soc Am. Res: Systematic entomology. Mailing Add: 2750 Sheldon St St Paul MN 55113

COOK, ELBERT GARY, JR, b Birmingham, Ala, Dec 13, 44; m 66; c 1. RESEARCH ADMINISTRATION. Educ: Univ Va, BS, 66; Va Polytech Inst, PhD(chem), 70. Prof Exp: Chemist, 69-73, res supvr, 73-75, SR RES SUPVR PLASTIC PROD & RESINS DEPT, E I DU PONT DE NEMOURS & CO, INC, 75- Mem: Am Chem Soc. Res: Product and process research in plastic filament and strap products. Mailing Add: Plastic Prod & Resins Dept E I du Pont de Nemours & Co Inc Parkersburg WV 26101

COOK, ELIZABETH ANNE, b Colorado Springs, Colo, Sept 19, 26. MICROBIOLOGY. Educ: Univ Colo, BA, 48; Ind Univ, MA, 56, PhD(bact), 58. Prof Exp: Instr bact, Mankato State Col, 57-58; from instr to asst prof, 62-70, ASSOC PROF BACT, DOUGLASS COL, RUTGERS UNIV, 70- Mem: AAAS; Am Soc Microbiol. Res: Microbiological assays for vitamin B12 and folic acid; bacterial nutrition. Mailing Add: Dept Bact Douglass Col Rutgers the State Univ New Brunswick NJ 08903

COOK, ELLSWORTH BARRETT, b Springfield, Mass, Jan 29, 16; m 41. PHARMACOLOGY. Educ: Springfield Col, BS, 38; Duke Univ, 38-39; Tufts Univ, PhD(med sci), 51. Prof Exp: Res assoc, Fatigue Labs, Harvard, 41-45; head visual screening & statist facil, Med Res Lab Submarine Base, Conn, 45-49; environ physiologist, Cold Injury Res Team, Korea, 51-52; consult biometrician, Army Med Res Lab, Ft Knox, Ky, 52; res pharmacologist, Nat Naval Med Res Inst, 52-57; head dept exp biol, Naval Med Field Res Lab, Camp Lejeune, NC, 57-61; EXEC OFFICER, AM SOC PHARMACOL & EXP THERAPEUT, 61- Concurrent Pos: Lectr, Med Sch, Howard Univ, 62- Mem: AAAS; Am Soc Pharmacol & Exp Therapeut; Biomet Soc; Psychomet Soc; Coun Biol Ed. Res: Acetylcholine analogues; standardization of bioassay techniques; biometrics and psychometrics. Mailing Add: Am Soc Pharmacol & Exp Therapeut 9650 Rockville Pike Bethesda MD 20014

COOK, ELTON DAVIS, b Hale County, Tex, Apr 9, 04; m 29. AGRONOMY. Educ: Tex Tech Col, BS, 35; Kans State Col, MS, 48; Univ Nebr, PhD, 50. Prof Exp: Jr agronomist, Soil Conserv Serv, USDA, Tex, 35-36, asst soil conservationist, 36-44; supt agr res sta, Tex Res Found, 44-47; supt sub-sta, Tex Agr Sta, 49-51, agronomist, 52-69; chmn dept , 69-74. LUBBOCK CHRISTIAN COL, 74- Mem: Am Soc Agron; Ecol Soc Am. Res: Use og green manure crops and commercial fertilizers in field crop production; root development; defoliation and mechanical harvesting of cotton. Mailing Add: Dept of Agr Lubbock Christian Col 5601 W 19th St Lubbock TX 79407

COOK, ELTON STRAUS, b Oberlin, Ohio, Dec 24, 09; m 35; c 2. MEDICINAL CHEMISTRY. Educ: Oberlin Col, AB, 30; Yale Univ, PhD(org chem), 33. Prof Exp: Asst chem, Yale Univ, 30-33, hon fel, 33-34; res chemist, Wm S Merrell Co, 34-37; asst dir res activities, 43-45, vpres, 55-70, PROF & HEAD DIV CHEM & BIOCHEM, ST THOMAS INST, 37-, DEAN, 46-, MEM CORP, 70- Concurrent Pos: Consult, War Prod Bd, 42; chmn & organizer Gibson Island Res Conf on Chem Growth promoters, AAAS, 42. Honors & Awards: Dipl Honor, Pan Am Cancer Cytol Cong, 57; Cert Award, Gordon Res Conf, 59; Am Chem Soc Award, 64. Mem: AAAS; Am Chem Soc; hon mem Am Inst Chem; Am Pharmaceut Asn; Am Asn Cancer Res. Res: Synthesis and pharmacology of local anesthetics and other drugs; cellular metabolism; cancer; mechanism of drug action approached through metabolism and enzymes; anti-infectious drugs. Mailing Add: St Thomas Inst 1842 Madison Rd Cincinnati OH 45206

COOK, EMORY, theoretical physics, see 12th edition

COOK, ERNEST EWART, b Stratton St Margaret, Eng, Mar 23, 26; m 53; c 1. EXPLORATION GEOPHYSICS. Educ: Cambridge Univ, BA, 46, MA, 50. Prof Exp: Geophysicist, Cia Shell de Venezuela, 47-56, chief geophysicist, Pakistan Shell Oil Co, 56-57; chief geophysicist, Signal Oil & Gas Co, Venezuela, 57-60, geophysicist, Tex, 60-62, chief geophysicist, 62-65, asst mgr int explor, Calif, 65-68; vpres, Seismic Comput Corp, 68-71; PRES, INVENT INC, 71- Mem: Am Asn Petrol Geologists; Soc Explor Geophys; Am Geophys Union; Geol Soc London; Europ Asn Explor Geophys. Res: Seismic refraction methods and interpretation techniques; application of seismic velocities to geologic interpretation. Mailing Add: 9235 Katy Fwy Houston TX 77024

COOK, EUGENE WILBUR, JR, b Danville, Ky, Jan 1, 02. BIOLOGY. Educ: Centre Col, AB, 23; Ohio State Univ, AM, 25, PhD(bact), 44. Prof Exp: Asst serol, Dept Health, Columbus, Ohio, 25-26; from asst prof to prof, 26-72, head dept, 36-72, actg dean fac, 48-49, 52-53, EMER PROF BIOL, CENTRE COL, 72- Concurrent Pos: Mem, Boyle County Health Bd, 56-; bd comnr, Ky Sch Deaf, vpres, 59-60, chmn adv bd, 60-; mem, Ky State Fulbright Comt; mem, Govs Comn Adult Educ, Ky. Mem: Fel AAAS; Am Soc Microbiol. Res: 8acterial mutation; physiology of anaerobes; lepidoptera; nature photography. Mailing Add: Dept of Biol Centre Col Danville KY 40422

COOK, EULA BELLE MALEY, b Oakville, Tex, Nov 13, 11; m 32. IMMUNOLOGY. Educ: Univ Tex, BA & MA, 34. Prof Exp: Bacteriologist, 35-42, IMMUNOLOGIST, TEX STATE DEPT HEALTH, 42- Mem: Am Soc Microbiol; Am Pub Health Asn; Am Acad Microbiol. Res: Production and standardization of biologicals; developmental research on production methods; immunological techniques and viral multiplication in embryonic tissues and tissue culture; laboratory safety; training of laboratory personnel. Mailing Add: 1810 San Gabriel St Austin TX 78701

COOK, EVIN LEE, b Waco, Tex, July 11, 18; m 41; c 1. PHYSICAL CHEMISTRY. Educ: Baylor Univ, AB, 39; Rice Inst, MA, 41; Univ Tex, PhD(phys chem), 49. Prof Exp: Res chemist petrol ref, Humble Oil & Refining Co, 41-43; res chemist petrol ref, Pan Am Ref Corp, 44-47; MGR OIL RECOVERY RES, MOBIL RES & DEVELOP CORP, 49- Mem: Am Chem Soc; Am Inst Mining, Metall & Petrol Engrs; Sigma Xi. Res: Hydrous oxides; aviation gasoline research; unit processes; surface chemistry; fluid flow through porous media. Mailing Add: 1315 Boca Chica Dr Dallas TX 75232

COOK, FRANKLAND SHAW, b Toronto, Nov 30, 21; m 45; c 3. BOTANY. Educ: Univ Toronto, BA, 50, PhD(bot), 56. Prof Exp: From asst prof to assoc prof, 52-69, PROF BOT, UNIV WESTERN ONT, 69- Mem: Can Bot Asn; Am Bryol & Lichenological Soc; Can Soc Plant Physiol. Res: Growth and physiology of pollen and spores; physiology of mosses; emzyme kinetics. Mailing Add: Dept Bot Univ of Western Ont London ON Can

COOK, FRED D, b Ottawa, Ont, Oct 30, 21; m 48; c 3. SOIL MICROBIOLOGY. Educ: Univ BC, BSA, 45, MSA, 47; Univ Edinburgh, PhD(microbiol), 60. Prof Exp: Res officer soil microbiol, Exp Farm, Swift Current, Sask, 50-57; officer soil microbiol, Microbiol Res Inst, Ottawa, 57-64; mem fac, 64-69, PROF SOIL SCI, UNIV ALTA, 69- Mem: Am Soc Microbiol; Can Soc Microbiol. Res: Soil organic matter chemistry; soil genesis; microbial ecology. Mailing Add: Dept of Soil Science Univ of Alta Edmonton AB Can

COOK, FREDERICK LEE, b Baltimore, Md, Mar 15, 40; m 62; c 2. MATHEMATICS. Educ: Ga Inst Technol, BS, 61, MS, 63, PhD(math), 67. Prof Exp: Res mathematician, George C Marshall Space Flight Ctr, NASA, 67-69; asst prof, 69-73, part time, 67-69, ASSOC PROF & CHMN DEPT MATH, UNIV ALA, HUNTSVILLE, 73- Mem: Am Math Soc; Math Asn Am. Res: Characterizations and applications of recursively generated Sturm-Liouville polynomial sequences; solutions of countably infinite systems of differential equations; determination of weight functions for given polynomial sequences. Mailing Add: Dept of Math Univ of Ala Huntsville AL 35807

COOK, GERHARD ALBERT, b Berea, Ky, May 2, 07; m 35; c 4. PHYSICAL CHEMISTRY. Educ: Univ Mich, BS, 28, MS, 33, PhD(phys chem), 35. Prof Exp: Teacher chem, High Sch, Mich, 28-31; asst, Johns Hopkins Univ, 31-32; asst, Univ Mich, 32-35, instr, Case Inst Technol, 35-37; res chemist, Linde Div, Union Carbide Corp, 37-57, asst mgr res lab, 57-67; consult, Nat Univ Asuncion, Paraguay, 67-69; EXEC OFF & ADJ PROF CHEM, STATE UNIV NY BUFFALO, 69- Mem: Am Chem Soc. Res: Reactions of oxygen; production of hydrogen peroxide, ozone and propylene oxide; properties of ozone, hydrogen and the inert gases. Mailing Add: Dept of Chem Acheson Hall State Univ of NY Buffalo NY 14214

COOK, GILBERT R, b Washington, DC, Apr 7, 16; m 64; c 4. PHYSICS Educ: Carnegie-Mellon 0niv, BS, 40; Univ Southern Calif, MS, 54, PhD(physics), 59. Prof Exp: Elec engr, Western Elec Co, 40-41; elec engr, Pac Tel & Tel Co, 46-50; lab assoc, Univ Southern Calif, 54-56; staff physicist, TRW Systs, Inc, 59-60; mem tech staff, Aerospace Corp, 60-75. Mem: Fel Am Phys Soc; Optical Soc Am; NY Acad Sci; Am Geophys Union. Res: Experimental physics; interaction of radiation of the vacuum ultraviolet region with atmospheric gases; cross sections for absorption and photoionization; ion-molecule reaction rates. Mailing Add: PO Box 425 El Segundo CA 90245

COOK, GLENN MELVIN, b Los Angeles, Calif, Sept 26, 35; m 60; c 2. PHYSICAL CHEMISTRY. Educ: Univ Calif, Berkeley, BS, 57; Univ Ill, PhD(phys chem), 61. Prof Exp: Res asst, Radiation Lab, Univ Calif, 57; teaching asst, Univ Ill, 57-60; from assoc res scientist to res scientist, Lockheed Missiles & Space Co, Calif, 61-66; mem tech staff, Sprague Elec Co, 66-70; sr electro chemist, 70-73, SR PROCESS SCIENTIST, LEDGEMONT LAB, KENNECOTT COPPER CO, 73- Mem: AAAS; Am Chem Soc; Am Phys Soc; Electrochem Soc; Am Inst Mining, Metall & Petrol Engrs. Res: Visible, fluorescent and phosphorescent spectra; molecular and ionic interactions in non-aqueous electrolytes; non-aqueous electrochemical cells; electroplating and corrosion; electrochemical processing of metals; porous electrode; dilute solution technology. Mailing Add: Kennecott Copper Co Ledgemont Lab 128 Spring St Lexington MA 02173

COOK, GORDON SMITH, b Newark, NY, Mar 5, 14; m 39; c 3. ORGANIC CHEMISTRY. Educ: Hope Col, AB, 37; Syracuse Univ, MS, 39. Prof Exp: Asst gen chem, Syracuse Univ, 37-39; control & develop chemist, Chambers Works, 40-49, tech rep org chem dept, Rubber Chem Div, 50-53, head div prod control & specifications, Elastomers Dept, 53-57, head foam-elastomers chems dept, 57-58, head prod control & specifications, 58-70, HEAD TECH DEVELOP DIV, E I DU PONT DE NEMOURS & CO, 70- Mem: Am Chem Soc; Am Soc Qual Control. Res: Synthetic elastomers; cellulose chemistry. Mailing Add: 21 Briar Rd Wilmington DE 19803

COOK, HAROLD ANDREW, b Wheeling, WVa, July 10, 41; m 64; c 1. AGRICULTURAL MICROBIOLOGY. Educ: West Liberty State Col, BS, 64; WVa Univ, MS, 66, PhD(microbiol), 69. Prof Exp: Asst prof, 69-73, ASSOC PROF BIOL, WEST LIBERTY STATE COL, 73-, CHMN DEPT MED TECHNOL, 70-, CHMN DEPT BIOL, 71- Concurrent Pos: Reviewer, Am Biol Teacher Today, 74-; adj course dir, Chautauqua-Type Short Course Prog, NSF, 75-76; consult, AAAS. Mem: Am Soc Microbiol. Res: Pollution; sanitary landfills; acid mine drainage; developing biology laboratory manual for non-science majors. Mailing Add: Dept of Biol West Liberty State Col West Liberty WV 26074

COOK, HAROLD DALE, b Rio Caribe, Venezuela, Oct 7, 30; US citizen; m 56; c 4. PHYSICS. Educ: Wheaton Col, Ill, BA, 52. Prof Exp: Engr, 56-58, res & develop engr, 58-61, proj supvr, 61-69, PROJ DIR, TELETYPE CORP, SKOKIE, 69- Res: Design of digital logic circuits for data transmission equipment and magnetic tape data recording equipment; data error detection and correction system; integrated circuit design; visual display devices for data communication. Mailing Add: 725 E Elm St Wheaton IL 60187

COOK, HAROLD THURSTON, plant pathology, deceased

COOK, HARRY, b Neth, May 29, 37; Can citizen; m 64; c 2. ENDOCRINOLOGY, PHYSIOLOGY. Educ: Univ BC, BSA, 60, MSA, 62; Free Univ, Amsterdam, PhD(zool), 66. Prof Exp: Acad asst zool, Free Univ, Amsterdam, 63-66; fel, Simon Fraser Univ, 66-69; ASST PROF ZOOL, TRINITY CHRISTIAN COL, 69- Concurrent Pos: Nat Marine Fisheries Serv grant, 69-71. Mem: Am Soc Zoologists; Can Soc Zoologists; Can Soc Cell Biol; Am Sci Affil; Neth Royal Zool Soc. Res: Pituitary gland of fishes, especially the prolactin producing cells; investigation of sockeye salmon and alewife, specifically salinity effects. Mailing Add: Dept of Biol Trinity Christian Col Palos Heights IL 60463

COOK, HARRY E, III, b Fresno, Calif, June 11, 35; m 61; c 2. GEOLOGY. Educ: Univ Calif, Santa Barbara, BS, 61, Univ Calif, Berkeley, PhD(geol), 66. Prof Exp: Consult petrog, Hales Labs, Calif, 63-65; res geologist, Denver Res Ctr, Marathon Oil Co, 65-74; RES GEOLOGIST, US GEOL SURV, 74- Mem: AAAS; Geol Soc Am; Am Asn Petrol Geologists; Soc Econ Paleontologists & Mineralogists. Res: Sedimentary carbonate bank and reef facies; carbon and oxygen isotopes in carbonates; diagenesis; petroleum reservoirs; turbidity currents; organic geochemistry of lime muds; stratigraphy of Western Canada and Texas; ignimbrites. Mailing Add: US Geol Surv 345 Middlefield Rd Menlo Park CA 94025

COOK, HARRY LEE, b Terre Haute, Ind, Nov 13, 11; m 37; c 1. AGRONOMY. Educ: Purdue Univ, BS, 34, MS, 36. Prof Exp: Res assoc agron, Purdue Univ, 37-41; agronomist, Farm Bur Co-op Asn, Inc, 46-53, mgr res & control dept, 53-57, dir res & tech serv, 57-68, DIR RES & TECH SERV, LANDMARK, INC, 69- Concurrent Pos: Bd mem, F F Redich Assoc, Lafayette, Inc & Fertilizer Inst Round Table, Baltimore, Md. Mem: Am Soc Agron; Soil Sci Soc Am; Am Soc Agr Eng; Am Chem Soc. Res: Soils; fertilizer manufacturing chemistry and processes. Mailing Add: Landmark Inc 245 N High St Columbus OH 43216

COOK, HOLLIS LEE, b Calera, Ala, Jan 30, 13; m 35; c 1. APPLIED MATHEMATICS. Educ: STephen F Austin State Univ, BS, 33; Tex A&M Univ, MS, 35; George Peabody Col, PhD(math), 51. Prof Exp: Prin high sch, Tex, 35-37; instr math, Pan Am Col, 37-39; asst prof, Auburn Univ, 39-42; indust engr, Consol Vultee, Ga, 42-44; assoc prof math, Ga Inst Technol, 44-49; head dept, Little Rock Univ, 51-54; res mathematician, NASA, Ala, 54-55; PROF & HEAD DEPT MATH, W TEX STATE UNIV, 55- Concurrent Pos: Consult educ math, Manned Spacecraft Ctr, NASA, Tex, 65-67; consult, Panhandle Educ Serv Orgn, 65-68. Mem: Math Asn Am. Res: Techniques in operational calculus; statistical studies; applications of mathematics. Mailing Add: Dept of Math Box 417 WTex State Univ Canyon TX 79015

COOK, HOWARD, b Spartanburg, SC, June 13, 33; m 60; c 3. TOPOLOGY. Educ: Clemson Col, BS, 56; Univ Tex, PhD(math), 62. Prof Exp: Spec instr math, Univ Tex, 58-62; asst prof, Auburn Univ, 62-64, res asst prof, 64; asst prof, Univ NC, Chapel Hill, 64-66; assoc prof, 66-71, PROF MATH, UNIV HOUSTON, 71- Concurrent Pos: Vis prof pure math, Univ Tasmania, 72-73. Mem: Am Math Soc. Res: Point set theory. Mailing Add: Dept of Math Univ of Houston Houston TX 77004

COOK, JACK E, b Ind, Feb 3, 31; m 54; c 2. ORGANIC POLYMER CHEMISTRY. Educ: DePauw Univ, BA, 53; Northwestern Univ, PhD, 57. Prof Exp: Res chemist exploratory plastics, Phillips Petrol Co, 57-64; sr res chemist, 64-66, RES SPECIALIST NEW PROD DEVELOP, MINN MINING & MFG CO, 66- Mem: Am Chem Soc. Res: Acrylics; electroplating; elastomers; polymer characterization; injection molding; surface topography; polymer synthesis; personal identification systems; retroreflective products; adhesion Mailing Add: 56 Michael St St Paul MN 55119

COOK, JAMES ALLISON, JR, organic chemistry, polymer chemistry, see 12th edition

COOK, JAMES ARTHUR, b Ga, May 19, 20; m 54; c 5. PLANT NUTRITION, SOIL FERTILITY. Educ: Cornell Univ, PhD(pomol), 51. Prof Exp: From asst plant physiologist to assoc plant physiologist, USDA, 48-52; asst viticulturist, 53-58, assoc viticulturist & assoc prof viticulture, 58-64, chmn dept viticult & enol, 62-66, VITICULTURIST, EXP STA, UNIV CALIF, DAVIS, & PROF VITICULTURE, UNIV, 64- Mem: Am Soc Plant Physiol; Am Soc Hort Sci; Am Soc Enol & Viticulture. Res: Fertilizer requirements of dates, apples and grapes; foliage nutrient sprays; plant nutrition; mineral nutrition of grapevines; deficiencies as well as toxicities, as determined by visual symptoms, tissue analysis and field trial responses. Mailing Add: Dept Viticult & Enol Univ of Calif Davis CA 95616

COOK, JAMES ELLSWORTH, b Eureka, Kans, Oct 20, 23; m 50; c 4. VETERINARY PATHOLOGY, COMPARATIVE PATHOLOGY. Educ: Okla State Univ, DVM, 51; Am Col Vet Path, dipl, 56; Kans State Univ, PhD(path), 70. Prof Exp: Asst prof path, Okla State Univ, 51-52; vet pathologist, Jensen Salsburys Labs, 52-53; resident comp path, Armed Forces Inst Path, 53-56, assoc pathologist, Aerospace Path Br, 56-57, pathologist, Vet Path Div, 65-69; chief vivarium br primate med, 6571st Aerospace Med Lab, Holloman AFB, NMex, 57-63; instr path, 69-70, actg head dept, 72-75, PROF PATH, KANS STATE UNIV, 70-, DIR ANIMAL RESOURCE FACILITY, COL VET MED, 74-, HEAD DEPT PATH, 75- Mem: Am Vet Med Asn; Am Soc Vet Clin Path. Res: Infectious and neoplastic diseases of primates and other laboratory animals; leptospirosis in domestic animals. Mailing Add: Dept of Path Col of Vet Med Kans State Univ Manhattan KS 66506

COOK, JAMES MARION, b Franklin, Ky, Aug 16, 41; m 64; c 2. THEORETICAL PHYSICS, MATHEMATICAL ANALYSIS. Educ: Western Ky Univ, BS, 62; Vanderbilt Univ, PhD(physics), 67. Prof Exp: ASSOC PROF PHYSICS, MID TENN STATE UNIV, 66- Mem: Am Asn Physics Teachers. Res: Differential equations. Mailing Add: Dept of Chem & Physics Mid Tenn State Univ Murfreesboro TN 37130

COOK, JAMES MINTON, b Bluefield, WVa, Aug 6, 45; c 1. NATURAL PRODUCTS CHEMISTRY, MEDICINAL CHEMISTRY. Educ: WVa Univ, BS, 67; Univ Mich, PhD(org chem), 71. Prof Exp: NIH fel natural prod, Univ BC, 72-73; ASST PROF CHEM, UNIV WIS, MILWAUKEE, 73- Mem: Am Chem Soc; The Chem Soc; Am Soc Pharamacog. Res: Synthesis of beta adrenergic antagonists for antihypertensive drug studies; synthesis of thiocarborhydrates for enzyme induction studies; studies on the reactions of dicarbonyl compounds with dimethyl-3-ketoglutarate; structure determinations. Mailing Add: Dept of Chem Univ of Wis Milwaukee WI 53201

COOK, JAMES RICHARD, b Maben, WVa, Nov 22, 29; m 59; c 2. CELL PHYSIOLOGY. Educ: Concord Col, BS, 50; WVa Univ, MS, 55; Univ Calif, Los Angeles, PhD(zool), 60. Prof Exp: NIH fel zool, Misaki Marine Biol Sta, Japan, 60-61; asst biophysicist, Lab Nuclear Med & Radiation Biol, Univ Calif, Los Angeles, 61-63; from asst prof to assoc prof zool, 63-74, PROF ZOOL & BOT, UNIV MAINE, ORONO, 74- Concurrent Pos: NIH res grant, 64-67, res career develop award, 68- Mem: Am Soc Cell Biol; Soc Protozool; Am Soc Plant Physiol. Res: Cell growth and division; adaptations of cells. Mailing Add: Murray Hall Univ of Maine Orono ME 04473

COOK, JOHN CALL, b Afton, Wyo, Apr 7, 18; m 49, 62; c 2. GEOPHYSICS. Educ: Univ Utah, BS, 42; Pa State Univ, MS, 47, PhD(geophys), 51. Prof Exp: Electronics res staff, Radiation Lab, Mass Inst Technol, 42-45; asst physics, Pa State Univ, 45-47, asst geophys, 47-49, res assoc, 49-51; sr physicist, Southwest Res Inst, 51-55, mgr geophys sect, 55-64; chief geophysicist, Geotech Corp, 64-66; PRIN GEOPHYSICIST, TELEDYNE GEOTECH CO, 66- Concurrent Pos: Instr math, Eve Div, San Antonio Col, 53-54; Int Geophys Year seismologist, Antarctica, Arctic Inst NAm, 57-58; asst prof physics, Trinity Univ, Tex, 58-59; assoc geophysicist, Scripps Inst Oceanog, 61; vis investr, Woods Hole Oceanog Inst Int Indian Ocean Exped, 64. Mem: Am Geophys Union; Soc Explor Geophys; Sigma Xi. Res: Detection and sensors; seismic, electromagnetic, infrared, magnetic, radioactivity and thermal; unorthodox prospecting methods; ground-probing radar; electronic circuits; properties of soils, ice and rock; rock probing radar. Mailing Add: Teledyne Geotech Co 3401 Shiloh Rd Garland TX 75040

COOK, JOHN SAMUEL, b Wilmington, Del, Sept 7, 27; m 65. CELL PHYSIOLOGY. Educ: Princeton Univ, AB, 50, MA, 53, PhD(biol), 55. Prof Exp: US Pub Health fel, Bern Univ, 55-56; from instr to assoc prof physiol, Sch Med, NY Univ, 56-67; prof biol, Univ Tenn-Oak Ridge Grad Sch Biomed Sci, 67-70, assoc dir sch, 68-70; STAFF MEM BIOL DIV, OAK RIDGE NAT LAB, 70- Mem: AAAS; Soc Gen Physiol; Biophys Soc; Am Physiol Soc. Res: Physiology of cell division; effects of ultraviolet and visible radiations on living cells; membrane physiology; biogenesis and turnover of cell membranes. Mailing Add: Biol Div Oak Ridge Nat Lab PO Box Y Oak Ridge TN 37830

COOK, JOSEPH MARION, b Oak Park, Ill, Feb 18, 24; m 56; c 3. APPLIED MATHEMATICS. Educ: Univ Ill, BS, 47, MS, 48; Univ Chicago, PhD(math), 51. Prof Exp: Instr math, Johns Hopkins Univ, 51-52; NSF fel quantum mech, Harvard Univ, 52-53; SR MATHEMATICIAN, ARGONNE NAT LAB, 53- Mem: Soc Indust & Appl Math. Mailing Add: Argonne Nat Lab Bldg 221 9700 S Cass Ave Argonne IL 60439

COOK, KENNETH EMERY, b Nebr, June 23, 28; m 52; c 3. ORGANIC CHEMISTRY. Educ: Hastings Col, BA, 53; Univ Nebr, MSc, 55, PhD(chem), 57. Prof Exp: From asst prof to assoc prof, 57-68, chmn dept, 62-74, PROF CHEM, ANDERSON COL, 68- Mem: Am Chem Soc; AAAS. Res: Synthetic organic chemistry; heterocyclic compounds. Mailing Add: 705 Maplewood Ave Anderson IN 46012

COOK, KENNETH LORIMER, b Middleton, NH, June 8, 15; m 46; c 3. GEOPHYSICS. Educ: Mass Inst Technol, BS, 39; Univ Chicago, PhD(geol & phyiscs), 43. Prof Exp: Part-time instr phys scis, Univ Chicago, 41-43, geophysicist, US Bur Mines, Reno, Nev, 43-46; geophysicist, US Geol Surv, Nev, 46-49, Utah, 49-56; head dept, 52-68, PROF GEOPHYS, UNIV UTAH, 52-, DIR UNIV SEISMOG STAS, 68- Mem: AAAS; Am Inst Mining, Metall & Petrol Eng; Soc Explor Geophys; Geol Soc Am; Am Geophys Union. Res: Mass spectroscopy; magnetic, gravitational and electrical geophysical interpretation; relative abundance of isotopes of potassium in Pacific kelps and rocks; vertical magnetic intensity over veins; resitivity data over filled sinks; regional megnetic and gravity surveys; gravity and magnetics of Utah; crustal structure of earth; seismic recording of large blasts. Mailing Add: Dept of Geol & Geophys Scis Univ of Utah Mines Bldg Rm 319 Salt Lake City UT 84112

COOK, KENNETH MARLIN, b Braddock, Pa, Aug 5, 20; m 44; c 2. ZOOLOGY. Educ: Univ Pittsburgh, BS, 43, MS, 48, PhD(biol sci), 53. Prof Exp: Asst instr biol, Univ Pittsburgh, 46-50, res assoc, Grad Sch Pub Health, 52-54; from asst prof to assocprof, 54-67, PROF BIOL, COE COL, 67- Honors & Awards: Co-winner of award, Indust Med Asn, 55. Mem: AAAS. Res: Retention of particule matter in the human lung; measurement of pulmonary functional capacity; effects of antithyroid drugs on reproduction; oxygen consumption of small mammals. Mailing Add: Dept of Biol Coe Col Cedar Rapids IA 52402

COOK, LAWRENCE C, b July 5, 25; US citizen; m 47; c 4. ORGANIC CHEMISTRY. Educ: East Tenn State Univ, BS, 57. Prof Exp: Clin chemist, Mem Hosp, Johnson

City, Tenn, 53-57; res org chemist, 57-67, develop chemist, 67-70, GROUP LEADER, R J REYNOLDS TOBACCO CO, 70- Mem: Am Chem Soc. Res: Clinical chemistry; isolation and identification of naturally occurring compounds. Mailing Add: Develop Ctr R J Reynolds Tobac Co Shorefair Dr Winston-Salem NC 27102

COOK, LAWRENCE HARVEY, b Paterson, NJ, Sept 23, 96; m 19, 34; c 3. CHEMISTRY. Educ: Stanford Univ, AB, 21, AM, 22; Univ Santa Clara, PhD(chem eng), 31. Prof Exp: From asst prof to prof chem eng, Univ Santa Clara, 27-36; consult sanit engr, Cook Res Labs, 36-73, pres & chmn bd, 66-73. Concurrent Pos: With water works sanit control, Menlo Park, Calif, 26-73; secy, Menlo Park Sanit Dist, 26-74. Mem: Am Chem Soc; Am Soc Civil Engrs; Am Water Works Asn; Water Pollution Control Fedn; Int Asn Water Pollution Res. Res: Organic chemistry; water works chemistry. Mailing Add: Cook Labs of NUS Corp Edison Way 11th Ave PO Box 2266 Menlo Park CA 94025

COOK, LEONARD, b Newark, NJ, June 27, 24; m 46; c 3. PHARMACOLOGY. Educ: Rutgers Univ, BA, 48; Yale Univ, PhD, 51. Prof Exp: Sr pharmacologist neuropharmacol, Smith, Kline & French Labs, 51-56, dir psychopharmacol res, 56-61, asst head pharmacol in charge res, 58-61, head psychopharmacol sect, 61-67, assoc dir pharmacol, 67-69; assoc dir pharmacol, 69-75, DIR PSYCHOTHERAPEUT RES, HOFFMANN-LaROCHE INC, 69-, DIR PHARMACOL, 75- Concurrent Pos: Lectr, Woman's Med Col Pa, 59-; comt mem, Psychopharmacol Serv Ctr, NIH. Honors & Awards: AMA Sci Exhibit Awards, 54. Mem: Am Soc Pharmacol & Exp Therapeut; Biomet Soc; Am Psychol Asn (pres, Psychopharmacol Div, 73); fel Am Col Neuropsychopharmacol; fel Int Col Neuropsychopharmacol. Res: Neuropharmacology; central nervous system stimulants and depressants; drug potentiators; psychopharmacology; operant and classical conditioning techniques; neurobiochemistry; physiological conditioning; analgesics; gastrointestinal drugs; drugs affecting memory learning. Mailing Add: Hoffman-La Roche Inc Nutley NJ 07110

COOK, LEROY FRANKLIN, (JR), b Ashland, Ky, Dec 12, 31; m 57; c 3. PHYSICS. Educ: Univ Calif, Berkeley, AB, 53, MA, 57, PhD(physics), 59. Prof Exp: From instr to asst prof physics, Princeton Univ, 59-65; assoc prof, 65-68, actg head dept, 69-71, PROF PHYSICS, UNIV MASS, AMHERST, 68-, HEAD DEPT PHYSICS & ASTRON, 71- Concurrent Pos: Vis Fel, Clare Hall, Cambridge Univ, 71-72. Mem: Fel Am Phys Soc. Res: Dispersion relations; applications of gauge theories to weak and electromagnetic interactions. Mailing Add: Dept of Physics Univ of Mass Amherst MA 01002

COOK, MARGARET MARY, b Mandan, NDak, Oct 15, 16. MICROBIOLOGY. Educ: Univ Wash, BS, 39; Univ Mich, MPH, 52, PhD(epidemiol sci), 62. Prof Exp: From jr bacteriologist to sr bacteriologist, Pub Health Lab, Wash State Dept Health, 39-43; officer in chg serol lab, US Navy Hosp, St Albans, NY, 44-46; prin bacteriologist, Pub Health Lab, Wash State Dept Health, 47-51; asst & teacher sch pub health, Univ Mich, 53-60, trainee epidemiol & sci, 61-62; actg chief serol training, Nat Communicable Dis Ctr, USPHS, 62-67, serology consult, 67-69; SR RES VIROLOGIST, DIV VIRUS & CELL BIOL RES, MERCK, SHARP & DOHME RES LABS, 69- Concurrent Pos: Asst dir, Div Labs, Utah State Health Dept, 64-67. Mem: AAAS; Am Soc Microbiol. Res: Epidemiologic science; interaction between staphylococci and influenza virus; study of reaction of bacteria and their products in tissue culture and chick embryo; thymus factors in cellular and humoral immunity. Mailing Add: Div Virus & Cell Biol Res Merck Sharp & Dohme Res Labs West Point PA 19486

COOK, MARIE MILDRED, b Bridgeport, Conn, Nov 22, 39. ZOOLOGY, DEVELOPMENTAL BIOLOGY. Educ: Georgian Ct Col, AB, 64; Rutgers Univ, MS, 70, PhD(zool), 74. Prof Exp: Teacher, Camden Cath High Sch, 66-69; instr & asst, 64-66, ASST PROF BIOL & CHMN DEPT, GEORGIAN CT COL, 69- Mem: Am Soc Zoologists; Am Inst Biol Sci; Nat Asn Biol Teachers; Nat Sci Teachers Asn. Res: Developmental patterns and mechanisms of expression of esterase isozymes in hybrid species of the teleost genus Brachydanio. Mailing Add: Georgian Ct Col Lakewood NJ 08701

COOK, MAURICE GAYLE, b Frankfort, Ky, Dec 26, 32; m 66; c 1. SOIL SCIENCE, AGRONOMY. Educ: Univ Ky, BS, 57, MS, 59; Va Polytech Inst & State Univ, PhD(agron), 61. Prof Exp: From asst prof to assoc prof, 61-70, PROF SOIL SCI, NC STATE UNIV, 70- Res: Soil mineralogy and its applications to soil genesis, morphology and classification; chemical and clay mineralogical interactions in soils. Mailing Add: Dept of Soil Sci NC State Univ Raleigh NC 27605

COOK, MICHAEL ARNOLD, b London, Eng, Dec 22y 44; m 68; c 2. PHARMACOLOGY, GASTROENTEROLOGY. Educ: Univ London, BSc, 67; Univ BC, PhD(physiol), 72. Prof Exp: Biochemist, Renal Unit, Royal Free Hosp, London, 67-68; Med Res Coun Can fel pharmacol, Univ Alta, 72-74; ASST PROF PHARMACOL, UNIV WESTERN ONT, 76- Mem: Can Physiol Soc; NY Acad Sci. Res: Studies on the neural and hormonal control of gastrointestinal motor activity; pharmacology of the gastrointestinal polypeptide hormones. Mailing Add: Dept of Pharmacol Univ of Western Ont London ON Can

COOK, NATHAN HOWARD, b Winston-Salem, NC, Apr 26, 39; m 61; c 2. CYTOLOGY, ZOOLOGY. Educ: NC Cent Univ, BS, 61, MA, 63; Okla State Univ, PhD(zool), 72. Prof Exp: Asst prof biol, Barber-Scotia Col, 62-68; teaching asst zool, Okla State Univ, 68-69; PROF BIOL, LINCOLN UNIV, 71-, HEAD DEPT, 74- Concurrent Pos: Proj dir, NSF Sci Equip Prog, 72-74 & Minority Biomed Support Prog Biol, NIH, Dept Health, Educ & Welfare, 72-; chmn, Mo Sickle Cell Anemia Adv Comt, 72- Mem: AAAS; Tissue Cult Asn; Sigma Xi. Res: In vitro effects of certain chemical carcinogens on the growth and chromosomes of mammalian cells. Mailing Add: Dept of Biol Lincoln Univ Jefferson City MO 65101

COOK, NEWELL CHOICE, b Chattahoochee, Fla, Dec 15, 16; m 39; c 2. ORGANIC CHEMISTRY. Educ: Fla Southern Col, BS, 39; Pa State Univ, PhD(chem), 43. Prof Exp: Asst prof chem, Pa State Univ, 45-50; STAFF SCIENTIST, R & D CTR, GEN ELEC CORP, 50- Mem: Am Chem Soc. Res: Hydrocarbon chemistry; high temperature reactions; electrolyses in fused salts; surface alloying of metals; dihydropyridine chemistry; nitration studies and processes Mailing Add: Gen Elec Corp Res & Develop Ctr Schenectady NY 12301

COOK, PAUL LAVERNE, b Holland, Mich, Mar 2, 25; m 51; c 4. ORGANIC CHEMISTRY. Educ: Hope Col, BS, 50; Univ Ill, MS, 52, PhD, 54 Prof Exp: From instr to assoc prof, 54-64, PROF CHEM, ALBION COL, 65- Mem: Am Chem Soc. Res: Organic synthesis; products of pharmaceutical interest. Mailing Add: Dept of Chem Albion Col Albion MI 49224

COOK, PAUL PAKES, JR, b Topeka, Kans, Nov 25, 27; m 49; c 2. ECOLOGY, EVOLUTION. Educ: Univ Kans, AB, 51, MA, 52; Univ Calif, Berkeley, PhD(entom), 62. Prof Exp: Field rep insect control, Calif Spray-Chem Corp, 54-59; asst entom, Univ Calif, Berkeley, 61-62; asst prof, 62-68, ASSOC PROF BIOL, SEATTLE UNIV, 68- Mem: Entom Soc Am; Soc Syst Zool; Soc Study Evolution; Ecol Soc Am; Brit Ecol Soc. Res: Population biology, ecology and evolution; animal behavior; systematics. Mailing Add: Dept of Biol Seattle Univ Seattle WA 98122

COOK, PHILIP W, b Underhill, Vt, Oct 6, 36. BOTANY. Educ: Univ Vt, BSc, 57, MSc, 59; Ind Univ, PhD(bot), 62. Prof Exp: NSF fel, 62-63; asst prof, 63-67, ASSOC PROF BOT, UNIV VT, 67- Mem: Bot Soc Am; Phycological Soc Am (pres, 71-72); Mycological Soc Am. Res: Fresh water algae; fungal parasites of algae. Mailing Add: Dept of Bot Univ of Vt Burlington VT 05401

COOK, RAY LEWIS, b Okemos, Mich, Mar 10, 04; m 20; c 1. SOILS. Educ: Mich State Col, BS, 27, MS, 29; Univ Wis, PhD(soils), 34. Prof Exp: Asst soils, 27-29, res asst, 29-38, from asst prof to prof, 38-73, head dept soil sci, 53-69, EMER PROF SOILS, MICH STATE UNIV, 73- Concurrent Pos: Vis prof, Chung Hsing Univ, Taiwan, 69-70; consult, UN Develop Prog. Mem: Am Soc Agron; Soil Sci Soc Am; Int Soc Soil Sci; Soil Conserv Soc Am; hon mem Soil Sci Soc Taiwan. Res: Field experimental work, tillage nutrient levels; symptoms of nutritional disorders; soil and plant analyses. Mailing Add: 830 Newton Ave Lansing MI 48912

COOK, RICHARD JAMES, b Alpena, Mich, Oct 20, 47; m 73. PHYSICAL ORGANIC CHEMISTRY. Educ: Univ Mich, BS, 69; Princeton Univ, MA, 71, PhD(chem), 73. Prof Exp: Res fel chem, Princeton Univ, 70-73; ASST PROF CHEM, KALAMAZOO COL, 73- Concurrent Pos: Res grant, Res Corp, 75-77. Res: Electronic effects upon the barriers to inversion of substituted imines. Mailing Add: Dept of Chem Kalamazoo Col Kalamazoo MI 49007

COOK, RICHARD KAUFMAN, b Chicago, Ill, June 30, 10; m 38; c 1. ACOUSTICS, GEOPHYSICS. Educ: Univ Ill, BS, 31, MS, 32, PhD(physics), 35. Prof Exp: Asst physics, Univ Ill, 30-35; physicist, Nat Bur Standards, 35-42, chief sound sect, 42-66; chief geoacoustics group, Nat Oceanic & Atmospheric Admin, 66-71; CONSULT PHYSICIST, NAT BUR STANDARDS, 71- Concurrent Pos: Mem tech staff, Bell Tel Labs, 55-56; adj prof elec eng, Brooklyn Polytech Inst, 56. Honors & Awards: Wash Acad Eng Sci Award, 49; US Dept Com Except Serv Award, 64. Mem: Fel AAAS; fel Am Phys Soc; fel Acoust Soc Am (pres, 57-58); Am Geophys Union. Res: Solid state physics; applied mathematics; geophysics; physical acoustics; atmospheric sound propagation; acoustical measurements; mathematical acoustics. Mailing Add: Nat Bur of Standards Washington DC 20234

COOK, RICHARD SHERRARD, b Philadelphia, Pa, Apr 11, 21. ORGANIC CHEMISTRY. Educ: Philadelphia Col Pharm, BS, 43; Temple Univ, MA, 56. Prof Exp: Control chemist, Barrett Div, Allied Chem & Dye Corp, 43-45; chemist, 47-56, GROUP LEADER CHEM, ROHM AND HAAS CO, 56- Mem: Am Chem Soc; fel Am Inst Chemists Res: Synthesis of new organic agricultural pesticides, including herbicides, fungicides and insecticides. Mailing Add: Res Labs Rohm and Haas Co Spring House PA 19477

COOK, ROBERT CARTER, b Washington, DC, Apr 9, 98; m; c 3. GENETICS, DEMOGRAPHY. Prof Exp: Sci aide aeronaut, Nat Bur Stands, 16-19; disciplinarian, Tucson Indian Sch, 20-21; managing ed, J Heredity, 22-52, ed, 52-62; dir & ed, Pop Reference Bur, 52-58, pres & ed, 59-68, sr consult, 68; consult pop & genetics, Nat Parks & Conserv Asn, 68-75; RETIRED. Concurrent Pos: Lectr, George Washington Univ, 44-63; mem adv coun, Conserv Found, 49-; bd mem, Nat Parks Asn, 59- Honors & Awards: Lasker Award, Am Pub Health Asn, 57. Mem: Fel AAAS; Nat Asn Sci Writers; Am Eugenics Soc; Pop Asn Am; Am Genetic Asn. Res: Lucidity of the genetic written word; analysis and interpretation of population trends. Mailing Add: PO Box 6025 Washington DC 20005

COOK, ROBERT CROSSLAND, b New Haven, Conn, June 5, 47; m 65; c 2. CHEMICAL PHYSICS. Educ: Lafayette Col, BS, 69; Yale Univ, MPh, 71, PhD(phys chem), 73. Prof Exp: ASST PROF CHEM, LAFAYETTE COL, 73- Mem: Am Phys Soc; AAAS. Res: Solid and liquid state theory; phase transitions; theory of brittle fracture; statistical mechanics; melting. Mailing Add: Dept of Chem Lafayette Col Easton PA 18042

COOK, ROBERT DOUGLAS, physical organic chemistry, see 12th edition

COOK, ROBERT EDWARD, b Springhill, WVa, Aug 26, 27; m 50; c 2. GENETICS. Educ: WVa Univ, BS, 49, MS, 56; NC State Col, PhD(genetics), 58. Prof Exp: Instr poultry, WVa Univ, 54-56; asst prof, Univ Fla, 58-61; coordinator genetics, Agr Res Serv, USDA, 61-64, leader, Genetics Invests, 64-65; head dept poultry sci, Ohio State Univ, 65-69; HEAD DEPT POULTRY SCI, NC STATE UNIV, 69- Mem: Poultry Sci Asn; World Poultry Sci Asn; Am Genetics Asn. Res: Basic genetics of the domestic fowl and systems of breeding for the improvement of poultry. Mailing Add: Dept of Poultry Sci NC State Univ PO Box 5307 Raleigh NC 27607

COOK, ROBERT JAMES, b Moorhead, Minn, Jan 14, 37; m 58; c 4. PLANT PATHOLOGY. Educ: NDak State Univ, BS, 58, MS, 61; Univ Calif, Berkeley, PhD(phytopath), 64. Prof Exp: NATO fel, Waite Inst, Australia, 64-65; asst, 65-68, PROJ LEADER, REGIONAL CEREAL DIS RES LAB, AGR RES SERV, USDA, WASH STATE UNIV, 68- Concurrent Pos: Guggenheim Mem Found fel, 73-74. Mem: Am Phytopath Soc; AAAS; Soil Sci Soc Am; Australian Soc Plant Path; Am Inst Biol Sci. Res: Biological control of soil born plant pathogens; water relations of soil microorganisms; cereal root rots. Mailing Add: Reg Cereal Dis Res Lab USDA Wash State Univ Pullman WA 99163

COOK, ROBERT LEE, b Hollywood, Fla, Aug 30, 36; m 65; c 2. MOLECULAR SPECTROSCOPY. Educ: Univ Miami, BS, 58, MS, 60; Univ Notre Dame, PhD(phys chem), 63. Prof Exp: Res assoc microwave spectros, Duke Univ, 63-65; from instr to asst prof physics, 65-71; assoc prof physics, 71-74, PROF PHYSICS & CHEM, MISS STATE UNIV, 74- Mem: Am Phys Soc; Am Asn Physics Teachers; Am Chem Soc. Res: Microwave spectroscopy; centrifugal distortion effects in asymmetric rotors; determination of molecular force constants; molecular structure; hyperfine interactions; ring conformations; spectrochemical analysis. Mailing Add: Dept of Physics Miss State Univ Mississippi State MS 39762

COOK, ROBERT MEROLD, b Bethany, Ill, Aug 5, 30; m 49; c 5. ANIMAL NUTRITION, BIOCHEMISTRY. Educ: Univ Ill, BS, 57, MS, 61, PhD(dairy sci), 62. Prof Exp: Res asst dairy sci, Univ Ill, 57-62; asst prof, Univ Idaho, 62-66; asst prof, 66-70, ASSOC PROF DAIRY SCI, MICH STATE UNIV, 70- Mem: Am Chem Soc; Am Dairy Sci Asn; Am Inst Nutrit; Am Inst Chemists; Biomet Soc. Res: Ruminant nutrition, control mechanisms regulating volatile fatty acid metabolism in ruminants; protein metabolism in rumen; glucose metabolism in cows; xenobiotic metabolism in ruminants. Mailing Add: Dept of Dairy Sci Mich State Univ East Lansing MI 48823

COOK, ROBERT SEWELL, b Unity, Wis, Nov 25, 29; m 53; c 2. VERTEBRATE ECOLOGY. Educ: Wis State Univ-Stevens Point, BS, 51; Univ Wis-Madison, MS, 58, PhD(vet sci), 66. Prof Exp: Res asst wildlife ecol, Univ Wis-Madison, 54-57; biologist, Wis Conserv Dept, 57-59; gen secy, Appleton YMCA, 59-61; biol instr high sch, Wis, 61-63; res asst vet sci, Univ Wis-Madison, 63-66; res participation grant, 66-69, asst prof physiol, 66-70, Alumni Res Found grant, 67-68, asst prof environ control, 70-71, ASSOC PROF ENVIRON CONTROL, UNIV WIS-GREEN BAY, 71- Mem: AAAS; Wildlife Soc; Wildlife Dis Asn. Res: Ecology of diseases in populations of wildlife that are transmissible to domestic animals and man; biological aspects of land-use planning; environmental impact analysis. Mailing Add: Col of Environ Sci Univ of Wis Green Bay WI 54302

COOK, ROBERT THOMAS, b Nebraska City, Nebr, Apr 27, 37; m 61; c 3. PATHOLOGY, BIOCHEMISTRY. Educ: Univ Kans, AB, 58, MD, 62, PhD(biochem), 67. Prof Exp: Assoc pathologist, Walter Reed Army Inst Res, 67-69; ASST PROF PATH, INST PATH, CASE WESTERN RESERVE UNIV, 69- Mem: Am Chem Soc. Res: Cell control mechanisms in neoplasia; chemical carcinogenesis. Mailing Add: Inst of Path Case Western Reserve Univ Cleveland OH 44106

COOK, RONALD FRANK, b Buffalo, NY, July 22, 39; m 64; c 2. ANALYTICAL CHEMISTRY. Educ: Univ Buffalo, BA, 61. Prof Exp: From chemist to res chemist, 62-72, SUPVR, RESIDUE LAB, NIAGARA CHEM DIV, FMC CORP, 72- Concurrent Pos: Assoc referee, Asn Off Anal Chemists, 65- Mem: Am Chem Soc. Res: Development and application of analytical methods to the determination of pesticide content of formulations and the residue content of agricultural commodities. Mailing Add: Niagara Chem Div FMC Corp 100 Niagara St Middleport NY 14105

COOK, SHERBURNE FRIEND, physiology, deceased

COOK, SHIRL ELDON, b Paris, Idaho, Mar 15, 18; m 44; c 1. CHEMISTRY Educ: Brigham Young Univ, BS, 39; La State Univ, MS, 41. Prof Exp: Chemist, US Army, La, 41-42; chemist, La Ord Plant, 42-45; CHEMIST, ETHYL CORP, 45- Mem: AAAS; Am Chem Soc. Res: Alkyl metal compounds; organometallics. Mailing Add: Ethyl Corp PO Box 341 Baton Rouge LA 70821

COOK, STANTON ARNOLD, b Oakland, Calif, Dec 10, 29; m 59; c 2. ECOLOGY, EVOLUTION. Educ: Harvard Univ, AB, 51; Univ Calif, Berkeley, PhD(bot), 61. Prof Exp: Asst prof, 60-67, ASSOC PROF BIOL, UNIV ORE, 67- Mem: AAAS; Ecol Soc Am; Soc Study Evolution; Am Soc Naturalists. Res: Plant genetical ecology; vascular plant population and genetical ecology; terrestrial ecosystem analysis. Mailing Add: Dept of Biol Univ of Ore Eugene OR 97403

COOK, STUART D, b Boston, Mass, Oct 23, 36; m 60; c 3. NEUROLOGY, NEUROSCIENCES. Educ: Brandeis Univ, AB, 57; Univ Vt, MS, 59, MD, 62. Prof Exp: Intern med & surg, State Univ NY Upstate Med Ctr, 62-63; resident neurol, Albert Einstein Col Med, 65-68, instr, 68-69; asst prof, Col Physicians & Surgeons, Columbia Univ, 69-71; PROF MED, NJ MED SCH, COL MED & DENT NJ, 71-, PROF NEUROSCI & CHMN, 72- Honors & Awards: S Weir Mitchell Award, Am Acad Neurol, 69. Mem: Am Acad Neurol; Am Asn Neuropath; Am Fedn Clin Res; Harvey Soc; Reticuloendothelial Soc. Res: Neuroimmunology; demyelinating diseases. Mailing Add: NJ Med Sch 100 Bergen St Newark NJ 07103

COOK, THEODORE DAVIS, b Kentfield, Calif, Jan 23, 24; m 48; c 4. GEOLOGY. Educ: Univ Utah, BS, 48; Univ Calif, MA, 50. Prof Exp: SR STAFF GEOLOGIST, SHELL OIL CO, 50- Mem: Geol Soc Am. Res: Tertiary micropaleontology and stratigraphy; North and Central America stratigraphy. Mailing Add: Shell Oil Co Box 481 Houston TX 77001

COOK, THEODORE WARREN, b Eckford, Mich, Aug 20, 10; m 31; c 2. CHEMISTRY. Educ: Battle Creek Col, AB, 30, BS, 32; Albion Col, AM, 39. Prof Exp: Anal chemist, Kellogg Co, Mich, 30 & 31-36; anal chemist, Pontiac Motor Co, 37-38; from asst prof to assoc prof, 38-75, EMER ASSOC PROF CHEM, CENT MICH UNIV, 75- Mem: Am Chem Soc. Res: Metallurgy; electronics; radar. Mailing Add: Dept of Chem Cent Mich Univ Mt Pleasant MI 48858

COOK, THOMAS BRATTON, JR, b Rich Pond, Ky, Aug 28, 26; m 47; c 2. PHYSICS. Educ: Western Ky Univ, BS, 47; Vanderbilt Univ, MS, 49, PhD(physics), 51. Prof Exp: Mem staff, Weapons Effects Dept, Sandia Lab, 51-55, supvr, Vulnerability Studies Sect, 55-56, supvr nuclear burst studies div, 56-59, mgr nuclear burst dept, 59-62, dir nuclear burst physics & math res, 62-67, VPRES, SANDIA LABS, 67- Concurrent Pos: Consult, Dir Defense Res & Eng & Defense Atomic Support Agency, 61-; consult, Aerospace Corp, 61-; mem, US Air Force Sci Adv Bd, 64- Mailing Add: Sandia Labs Livermore CA 94550

COOK, THOMAS HENRY, organic chemistry, see 12th edition

COOK, THOMAS M, b Miami, Fla, Mar 12, 31; m 50; c 3. MICROBIOLOGY. Educ: Univ Md, BS, 55, MS, 57; Rutgers Univ, PhD(bact), 63. Prof Exp: Microbiologist, Merck Sharp & Dohme Res Labs, 57-61; assoc res biologist, Sterling-Winthrop Res Inst, 63-66; asst prof, 66-70, ASSOC PROF MICROBIOL, UNIV MD, COL PARK, 70- Mem: Am Soc Microbiol. Res: Microbial physiology and biochemistry; oxidative metabolism; fermentations; action of antimicrobial agents. Mailing Add: Dept of Microbiol Univ of Md College Park MD 20742

COOK, THOMAS TRAGER, physics, see 12th edition

COOK, THURLOW ADREAN, b Utica, NY, June 2, 39; m 61. MATHEMATICS. Educ: Univ Rochester, BA, 61; State Univ NY Buffalo, MA, 65; Fla State Univ, PhD(math), 67. Prof Exp: asst prof, 67, ASSOC PROF MATH, UNIV MASS, AMHERST, 67- Mem: Am Math Soc; Math Asn Am. Res: Functional analysis; general theory of Schauder bases in locally convex topological vector spaces, particularly Schauder bases in Banach spaces; foundations of quantum mechanics. Mailing Add: Dept of Math Univ of Mass Amherst MA 01002

COOK, VICTOR, b Palenville, NY, July 13, 29; m 57; c 2. PHYSICS. Educ: Univ Calif, Berkeley, AB, 57, PhD(physics), 62. Prof Exp: Res physicist, Lawrence Radiation Lab, Univ Calif, 62-63; asst prof, 63-67, ASSOC PROF PHYSICS, UNIV WASH, 67- Mem: Am Phys Soc; Am Asn Physics Teachers; Fedn Am Sci. Res: Properties and interactions of elementary particles. Mailing Add: Dept of Physics Univ of Wash Seattle WA 98105

COOK, WARREN AYER, b Conway, Mass, July 22, 00; m 28; c 1. INDUSTRIAL HYGIENE. Educ: Dartmouth Col, AB, 23. Prof Exp: Head chem unit, Eng & Inspection Div, Travelers Ins Co, 25-28; chief indust hygienist, Bur Indust Hyg, State Dept Health, Conn, 28-37; dir div indust hyg & eng res, Zurich-Am Ins Co, 37-53; res assoc & from assoc prof to prof, 53-71, EMER PROF INDUST HEALTH, INST INDUST HEALTH, UNIV MICH, 71- Concurrent Pos: Mem comt Z37 acceptable concentration toxic dusts & gases, Am Nat Standards Inst; consult subcomt 3 on nitrogen gases & oxidants, Intersoc Comt Manual for Air Sampling & Anal; adj prof indust health, Univ NC, 51- Honors & Awards: Cummings Award, Am Indust Hyg Asn, 52; Meritorious Achievement Award, Am Conf Govt Indust Hygienists, 73. Mem: Hon mem Am Indust Hyg Asn (pres, 40); hon mem Am Acad Occup Med; mem emer Am Soc Safety Eng; Am Pub Health Asn; Am Chem Soc. Res: Methods of determination of atmospheric contaminants; administrative phases of industrial hygiene; educational and training programs for occupational safety and health personnel. Mailing Add: 713 Emory Dr Chapel Hill NC 27514

COOK, WENDELL SHERWOOD, b Youngstown, Ohio, June 14, 16; m 47. ORGANIC CHEMISTRY. Educ: Miami Univ, Ohio, BA, 38; Mich Technol Univ, MS, 48. Prof Exp: Chemist, Mineral Aggregates, France Co Labs, 36-42; res scientist, Cent Res Labs, 42-46; instr org biochem & gen chem, Mich Col Mining & Technol, 47-51; RES SCIENTIST, CHEM & PHYS RES LABS, FIRESTONE TIRE & RUBBER CO, 50- Mem: Am Chem Soc. Res: Rubber chemistry; x-ray fluorescence of trace elements in polymers and vulcanizates; atomic absorption analysis of polymers and vulcanizates; use of ultraviolet, infrared, NMR and mass spectrometric analysis of rubber and rubber chemicals research; organic chemicals as accelerators and oxidation inhibitors in synthetic polymers and vulcanizates. Mailing Add: 1200 Firestone Pkwy Akron OH 44317

COOK, WILLIAM BOYD, b Dallas, Tex, July. 20, 18; m 42; c 1. ORGANIC CHEMISTRY. Educ: Univ Tex, AB, 40; Univ Colo, MS, 42; Univ Wyo, PhD(chem), 50. Prof Exp: Asst, Univ Colo, 40-42; chemist-analyst, Monsanto Chem Co, 42-43, res assoc, 43-47; from instr to asst prof chem, Univ Wyo, 47-53; assoc prof, Baylor Univ, 53-57; prof & head dept, Mont State Col, 57-65; vis prof, Stanford Univ, 65-67; dean col sci & arts, 67-68, PROF CHEM, COLO STATE UNIV, 67-, DEAN COL NATURAL SCI, 68- Concurrent Pos: Fund Adv Educ fel, 52-53; NSF grant, Cambridge Univ, 62-63; exec dir, Adv Coun Chem, 65-67, mem, 66-69; mem adv comt grants of Res Corp, 69-75; US rep comt teaching chem, Int Union Pure & Appl Chem, 70- Honors & Awards: Gold Medal, Am Chem Soc, 73. Mem: Fel AAAS; fel Am Inst Chemists; Am Chem Soc. Res: Isolation and structural studies of alkaloids; nitrogen heterocyclic compounds; science education curricula. Mailing Add: Col of Natural Sci Colo State Univ Ft Collins CO 80521

COOK, WILLIAM H, b Brooklyn, NY, Mar 12, 24; m 52; c 5. OPERATIONS RESEARCH. Educ: Hofstra Univ, AB, 49. Prof Exp: Mathematician, Navy Dept, DC, 49-50; math statistician, US Bur Census, 50-61; sr staff math statistician, Opers Res Inc, Silver Spring, Md, 61-63, sr scientist, 63-72; CHMN BD, COBRO CORP, SILVER SPRING, MD, 72- Honors & Awards: Meritorious Serv Award, US Govt, 59. Mem: Fel Am Statist Asn; Inst Math Statist. Res: Probability theory and statistics, application to system cost-effectiveness studies. Mailing Add: 936 Dead Run Dr McLean VA 22101

COOK, WILLIAM HARRISON, b Alnwick, Northumberland, Eng, 03; m 32; c 3. BIOCHEMISTRY. Educ: Univ Alta, BSc, 26, MSc, 28; Stanford Univ, PhD(chem), 31; Univ Sask, LLD, 48. Hon Degrees: DSc, Laval Univ, 63. Prof Exp: Asst plant biochem, Univ Alta, 26-30; assoc res biologist, 30-41, dir div biosci, 41-68, exec dir, 68-69, dir, Gen Can Comt Int Biol Prog, Nat Res Coun Can, 69-74. Concurrent Pos: Pres, Biol Coun Can, 68-69. Honors & Awards: Officer, Order of Brit Empire, 46; Medal of Serv, Order of Can, 69. Mem: Am Inst Food Technol; fel Royal Soc Can (pres, 62-63); fel Agr Inst Can; Can Biochem Soc (pres, 65-66). Res: Food preservation; proteins; physical biochemistry. Mailing Add: 201 Maple Lane Ottawa ON Can

COOK, WILLIAM R, JR, b Boston, Mass, Nov 28, 27; m 50; c 4 SOLID STATE CHEMISTRY, MINERALOGY. Educ: Oberlin Col, BA, 49; Columbia Univ, MA, 50; Case Western Reserve Univ, PhD, 71. Prof Exp: Crystallographer, Brush Develop Co, 51-53; head crystallog sect, Electronic Res Div, Clevite Corp, Gould, Inc, 53-74; CONSULT, 74-; SECY, CLEVELAND CRYSTALS, INC, 73- Concurrent Pos: Mem, Int Conf for Thermal Anal. Mem: Am Crystallog Asn; Mineral Soc Am; Am Ceramic Soc; Am Chem Soc. Res: Ferroelectricity and piezoelectricity; nonlinear optical materials; mineral chemistry. Mailing Add: 684 Quilliams Rd Cleveland OH 44121

COOKE, ANSON RICHARD, b Lawrence, Mass, Jan 12, 26; m 48; c 4. PLANT PHYSIOLOGY. Educ: Univ Mass, BS, 49, MS, 50; Univ Mich, PhD(bot), 53. Prof Exp: Asst bot, Univ Mich, 50-53; asst prof plant biochem, Univ Hawaii, 53-54; asst prof plant physiol, Okla State Univ, 55-56; res plant physiologist, E I du Pont de Nemours & Co, 56-63; DIR BIOL RES, AMCHEM PROD, INC, 63- Mem: Am Chem Soc; Am Soc Plant Physiol; Weed Sci Soc Am. Res: Plant growth regulators; herbicides; flowering; stress physiology. Mailing Add: Amchem Prod Inc Ambler PA 19002

COOKE, CHARLES ROBERT, b Oak Hill, WVa, June 12, 29; m 51; c 4. MEDICINE. Educ: WVa Univ, AB, 50, BS, 52; Johns Hopkins Univ, MD, 54. Prof Exp: From instr to asst prof, 63-70, ASSOC PROF MED, SCH MED, JOHNS HOPKINS UNIV, 70- Concurrent Pos: Am Heart Asn fel, 59-60; USPHS fel, 60-61; asst chief med, Baltimore City Hosps, 63-67; consult, Vet Admin, 67- Mem: Fel Am Col Physicians; Am Fedn Clin Res; Am Soc Nephrology. Res: Renal physiology and electrolyte metabolism. Mailing Add: Sch of Med Dept of Med Johns Hopkins Univ Baltimore MD 21218

COOKE, DEAN WILLIAM, b Uniontown, Pa, Mar 12, 31; m 56; c 4. INORGANIC CHEMISTRY. Educ: Ohio State Univ, BS, 55, PhD(chem), 59. Prof Exp: From instr to assoc prof, 59-72, PROF CHEM, WESTERN MICH UNIV, 72- Mem: Am Chem Soc; Sigma Xi; The Chem Soc. Res: Coordination chemistry; stereochemistry; reactions of coordinated ligands; homogeneous catalysis; optical activity; mechanisms of substitution reactions. Mailing Add: Dept of Chem Western Mich Univ Kalamazoo MI 49008

COOKE, DERRY DOUGLAS, b Schenectady, NY, Jan 29, 37. PHYSICAL CHEMISTRY. Educ: Parsons Col, BS, 63; Clarkson Col Technol, PhD(phys chem), 69. Prof Exp: Fel phys chem, 68-73, RES ASST PROF CHEM, CLARKSON COL TECHNOL, 73- Mem: Am Chem Soc. Res: Aerosols; light scattering; submicron cylinders. Mailing Add: Fac of Arts & Sci Clarkson Col of Technol Potsdam NY 13676

COOKE, FRED, b Darlington, Eng, Oct 13, 36; m 63; c 2. GENETICS. Educ: Cambridge Univ, BA, 60, MA, 63, PhD(biol), 65. Prof Exp: Asst prof, 64-69, ASSOC PROF BIOL, QUEEN'S UNIV, ONT, 69- Mem: Genetics Soc Am; Genetics Soc Can. Res: Genetic recombination in fungi; population genetics in the snow goose. Mailing Add: Dept of Biol Queen's Univ Kingston ON Can

COOKE, GEORGE DENNIS, b Ravenna, Ohio, June 29, 37; m 62; c 2. AQUATIC ECOLOGY. Educ: Kent State Univ, BS, 59; Univ Iowa, MS, 63, PhD(zool), 65. Prof Exp: Asst biol, Kent State Univ, 60; asst zool & ecol, Univ Iowa, 60-65; USPHS fel ecol, Univ Ga, 65-67; asst prof, 67-71, ASSOC PROF, DEPT BIOL SCI, KENT STATE UNIV, 71-, RES ASSOC, CTR URBAN REGIONALISM & ENVIRON SYSTS, 72- Mem: AAAS; Ecol Soc Am; Am Soc Limnol & Oceanog; Int Asn Gt Lakes Res; Int Asn Theoret & Appl Limnol. Res: Plankton productivity; factors regulating plankton population size; effects of pesticides on metabolism and species diversity of aquatic ecosystems; microecosystems; eutrophication; human ecology; life support systems; lake restoration. Mailing Add: Dept of Biol Sci Kent State Univ Kent OH 44242

COOKE, GILES BUCKNER, organic chemistry, deceased

COOKE, HELEN JOAN, b Greenfield, Mass, May 21, 43; m; c 2. DEVELOPMENTAL PHYSIOLOGY. Educ: Univ Mass, BS, 65; Univ Calif, Los Angeles, MS, 67; Univ Sydney, PhD(physiol), 71. Prof Exp: Instr, 71-73, ASST PROF PHYSIOL, UNIV IOWA, 73- Mem: Am Fedn Clin Res; Sigma Xi. Res: Development of amino acid transport systems in the newborn rabbit. Mailing Add: Dept of Physiol Univ of Iowa Iowa City IA 52242

COOKE, HENRY CHARLES, b Poughkeepsie, NY, June 24, 13; m 39; c 2. MATHEMATICS. Educ: NC State Col, BS, 37, MS, 51. Prof Exp: Teacher high sch, NC, 37-40; from instr to asst prof, 40-50, ASSOC PROF MATH, NC STATE UNIV, 51-, TV INSTR, 55- Mem: Sigma Xi. Res: Methods and techniques of audiovisual television presentation of instructional material. Mailing Add: Dept of Math NC State Univ Raleigh NC 27607

COOKE, HERBERT BASIL SUTTON, b Johannesburg, SAfrica, Oct 17, 15; m 43; c 2. GEOLOGY. Educ: Cambridge Univ, BA, 36, MA, 41; Univ Witwatersrand, MSc, 41, DSc(geol), 47. Prof Exp: Geologist, Cent Mining & Investment Co, Ltd, SAfrica, 36-38; lectr geol, Univ Witwatersrand, 38-47; sr lectr, 53-58, reader, 58-61; private consult, 47-52; assoc prof geol, 61-63, dean arts & sci, 63-68, PROF GEOL, DALHOUSIE UNIV, 63- Concurrent Pos: Ed jour, SAfrican Asn Advan Sci, 45-57; Nuffield Found bursary, 55-58; Du Toit Mem Lectr, 57; vis res assoc, Univ Calif, Berkeley, 57-58; chmn, Bernard Price Inst Paleont Res, 58-61. Mem: Fel Geol Soc Am; fel Royal Soc SAfrica; SAfrica Archaeol Soc (pres, 51); SAfrican Geog Soc (pres, 46); S African Asn Advan Sci (vpres, 60). Res: Pleistocene geology and fossil mammals, particularly African. Mailing Add: Dept of Geol Dalhousie Univ Halifax NS Can

COOKE, HERMAN GLENN, b Petersburg, Va, Nov 28, 18. ZOOLOGY, ENTOMOLOGY. Educ: Va State Col, BS, 36; Univ Pa, MS, 39; Univ Wis, PhD(zool), 62. Prof Exp: Taxonomist entom, Agr Res Ctr, USDA, 44-45; asst prof biol, Hampton Inst, 46-53; PROF BIOL, ELIZABETH CITY STATE UNIV, 61- Concurrent Pos: Researcher, Max Planck Inst Limnol, 67. Mem: AAAS; Am Entom Soc; Int Asn Theoret & Appl Limnol. Res: Mating flights of mayflies; bob-tailed rats after twenty-two successive generations; ecology and biology of immature Tendipedidae; Aklabesmyia cookei. Mailing Add: Dept of Biol Elizabeth State Univ Elizabeth City NC 27909

COOKE, HERMON RICHARD, JR, b Tonopah, Nev, Jan 3, 14; m 43; c 3. MINING GEOLOGY. Educ: Univ Nev, BA, 34, BS, 35; Harvard Univ, MA, 39, PhD(mining geol), 45. Prof Exp: Geologist, Black Mammoth Mining Co, Nev, 36; topog survr, US Nat Mus, Mo, 38; geologist, Original Sixteen to One Mine, Inc, Calif, 39-40; geologist, Am Metal Co, Ltd, Nev, 42; geologist, Basic Refractories, Inc, Nev, 42-43; geologist, US Geol Surv, Mont, 46; geologist, Chile Explor Co, 46-49; geologist, Patino Co, Bolivia & Volcan Mines, Peru, 49; consult geol, Graff & Kruger, Peru, 50, Am Smelting & Refining Co, Peru, 51-54; consult, Martin Sykes & Assocs, Venezuela, 54-57; geologist, Cooke, Everett & Assocs, Nev, 58-68; chief geologist, Oper Hardrock, Parsons Corp, 68-72; MEM STAFF, GRONLANDS GEOLOGISKE, DENMARK, 72- Mem: AAAS; Nat Soc Prof Engrs; Am Inst Mining, Metall & Petrol Engrs; Am Soc Photogram; Mineral Soc Am. Res: Mining exploration; applied geology. Mailing Add: Gronlands Geologiske Undersea Observ Copenhagen Denmark

COOKE, IAIN, b Glasgow, Scotland, Aug 1, 32; Can citizen; m 57; c 4. SOLID STATE PHYSICS. Educ: Glasgow Univ, BSc, 55; Univ Birmingham, PhD(physics), 58. Prof Exp: Asst prof, 58-64, ASSOC PROF PHYSICS, UNIV MAN, 64-, ASSOC DEAN SCI, 70- Mem: Can Asn Physicists; Brit Inst Physics & Phys Soc; Am Phys Soc. Res: Electrical and optical properties of insulating solids. Mailing Add: Dept of Physics Univ Man Winnipeg MB Can

COOKE, IAN MCLEAN, b Honolulu, Hawaii, Feb 6, 33; m 59; c 3. NEUROPHYSIOLOGY, COMPARATIVE PHYSIOLOGY. Educ: Harvard Univ, AB, 55, AM, 59, PhD(biol), 62. Prof Exp: Instr biol, Harvard Univ, 62; res assoc biophys, Univ Col London, 62-63; from instr to asst prof biol, Harvard Univ, 63-70; res assoc, Lab Cellular Neurophysiol, Nat Ctr Sci Res, Paris, 70-72; PROF ZOOL, UNIV HAWAII, MANOA, 72-, PROG DIR, LAB SENSORY SCI, 75- Concurrent Pos: NATO res fel, 62-63. Mem: Am Soc Zool; Soc Gen Physiol; Am Physiol Soc. Res: Control and mechanisms of release of neurosecretory material; mechanisms of synaptic transmission; cellular neurophysiology. Mailing Add: Lab Sensory Sci Univ of Hawaii-Manoa Honolulu HI 96822

COOKE, JAMES HORTON, b Ft Worth, Tex, Apr 26, 40. THEORETICAL PHYSICS. Educ: North Tex State Univ, BA, 62; Univ NC, PhD(physics), 67. Prof Exp: Fel physics, Univ Man, 66-68; asst prof, 68-74, ASSOC PROF PHYSICS, UNIV TEX, ARLINGTON, 74- Res: Investigation of various theories of relativistic interacting particles. Mailing Add: Dept of Physics Univ of Tex Arlington TX 76010

COOKE, JOHN COOPER, b Lawrence, Mass, May 12, 39; m 63; c 1. MYCOLOGY. Educ: Univ Mass, BS, 61, MA, 63; Univ Ga, PhD(mycol), 67. Prof Exp: Teacher high sch, RI, 64; asst prof biol, Elizabethtown Col, 67-69; res assoc, 69-70, asst prof, 70-75, ASSOC PROF BIOL, UNIV CONN, 75- Mem: Mycol Soc Am; Bot Soc Am; Am Inst Biol Sci. Res: Morphology of fungi; ecology of soil fungi. Mailing Add: Dept of Biol Univ of Conn Avery Pt Groton CT 06340

COOKE, KENNETH LLOYD, b Kansas City, Mo, Aug 13, 25; m 50; c 3. APPLIED MATHEMATICS. Educ: Pomona Col, BA, 47; Stanford Univ, MS, 49, PhD(math), 52. Prof Exp: From instr to asst prof math, State Col Wash, 50-57; asst prof, 57-62, chmn dept, 61-71, PROF MATH, POMONA COL, 62- Concurrent Pos: Consult, Rand Corp, Calif, 56-65; assoc ed, J Math Anal & Appln; researcher, Res Inst Adv Study, 63-64; NSF sci fac fel, Stanford Univ, 66-67; assoc ed, Utilitas Mathematica, 71-; Fulbright res scholar, Univ Florence, 72; assoc ed, J Computational & Applied Math, 74-; assoc, Ctr Study Dem Inst, 74-75. Mem: Am Math Soc; Math Asn Am; Soc Indust & Appl Math; Soc Comput Simulation; Ital Math Union. Res: Ordinary, partial and functional differential equations; integral equations; difference equations; dynamic programming; mathematical models in the biological and social sciences Mailing Add: Dept of Math Pomona Col Claremont CA 91711

COOKE, LLOYD MILLER, b La Salle, Ill, June 7, 16; m 57. CHEMISTRY. Educ: Univ Wis, BS, 37; McGill Univ, PhD(org chem), 41. Prof Exp: Lectr org chem, McGill Univ, 41-42; res chemist & sect leader, Corn Prod Refining Co, Ill, 42-46; group leader res, Food Prod Div, 46-49; mgr cellulose & casing res dept, 50-54, asst to mgr tech div, 54-59, asst dir res, 59-64, mgr mkt res, 64-67, mgr planning, 65-70, dir urban affairs, 70-73, CORP DIR UNIV RELS, UNION CARBIDE CORP, 73- Concurrent Pos: Trustee, Chicago Chem Libr Found; mem, Nat Sci Bd, 70-76; trustee, Carver Res Found, Tuskegee Inst, 71- & McCormick Theol Sem, Chicago, 73;

consult, Off Technol Assessment, US Cong, 74- Honors & Awards: Proctor Prize in Sci, Sci Res Asn Am, 70. Mem: Am Inst Chem; Am Chem Soc. Res: Structure of lignin; starch modifications and derivatives; cellulose derivatives; viscose chemistry; carbohydrate and polymer chemistry. Mailing Add: Union Carbide Corp 270 Park Ave New York NY 10017

COOKE, MANNING PATRICK, JR, b Suffolk, Va, July 24, 41; m 63; c 3. ORGANIC CHEMISTRY. Educ: Univ NC, AB, 63, MS, 66, PhD(chem), 67. Prof Exp: Fel org chem, Harvard Univ, 68-70 & Stanford Univ, 70-71; ASST PROF ORG CHEM, WASH STATE UNIV, 71- Mem: Am Chem Soc. Res: Synthesis and new synthetic methods in organic chemistry. Mailing Add: Dept of Chem Wash State Univ Pullman WA 99163

COOKE, PATRICIA M, b Vancouver, BC, Apr 10, 26; m 47; c 1. BACTERIOLOGY, VIROLOGY. Educ: Univ BC, BA, 47; McGill Univ, MSc, 58, PhD(bact), 61. Prof Exp: Lectr, 60-62, asst prof, 62-66, ASSOC PROF BACT, McGILL UNIV, 66-, MICROBIOL & IMMUNOL, 73- Mem: Am Soc Microbiol; Can Soc Microbiol; Brit Soc Gen Microbiol. Res: Study of antiviral agent using influenza and other viruses. Mailing Add: Dept of Microbiol & Immunol McGill Univ Box 6070 Sta A Montreal PQ Can

COOKE, PETER HAYMAN, b Beverly, Mass, Feb 4, 43; c 2. CYTOLOGY. Educ: Springfield Col, BS, 64; Univ NH, PhD(zool), 67. Prof Exp: Res fel cell biol, Harvard Univ, 67-69; res fel muscle, Boston Biomed Res Inst, 69-71; asst prof physiol & cell biol, Univ Kans, 71-75; ASSOC PROF PHYSIOL, HEALTH CTR, UNIV CONN, 75- Concurrent Pos: Res fel, Muscular Dystrophy Asn Am, 69 & Am Heart Asn-Brit Heart Found, 74; vis res fel physics, The Open Univ, UK, 74-75. Mem: Am Soc Cell Biol; Sigma Xi. Res: Contractile mechanism of muscle. Mailing Add: Dept of Physiol Health Ctr Univ of Conn Farmington CT 06032

COOKE, QUINTON EDWIN, JR, organic chemistry, see 12th edition

COOKE, ROBERT E, b Attleboro, Mass, Nov 13, 20; m 42; c 5. PEDIATRICS. Educ: Yale Univ, BS, 41, MD, 44. Hon Degrees: ScD, Univ Miami, 71. Prof Exp: Intern pediat, New Haven Hosp, 44-45, asst resident, 45-46; instr, Sch Med, Yale Univ, 50-51, from asst prof to assoc prof pediat & physiol, 51-56; Given Found prof pediat, Johns Hopkins Univ, 56-73, pediatrician-in-chief, Johns Hopkins Hosp, 56-73; PROF PEDIAT & V CHANCELLOR HEALTH SCI, UNIV WIS-MADISON, 73- Concurrent Pos: NIH fel, 48-50; Markle scholar, 51-55; resident, Grace New Haven Hosps, 50-51; mem, President's Panel Ment Retardation, 61-62 & President's Comt Ment Retardation, 66-69; consult div hosps & med facil & mem comt areawide planning of facil ment retarded, USPHS, 63-65; mem, White House Adv Comt Ment Retardation, 63-65; mem res & demonstration panel, Off Educ, Dept Health, Educ & Welfare, 63-66; consult, Nat Found-March of Dimes, 68-70; mem, Nat Comn Protection Human Subjects of Biomed & Behav Res, Off Asst Secy Health, 74-76; mem, Health Manpower Training Assistance Rev Comt, Vet Admin, Washington, DC, 74-76; consult, Off Technol Assessment, Cong of US, 74-; mem, Nat Asn Retarded Children & Joseph P Kennedy Jr Mem Found; mem adv coun, Nat Inst Child Health & Human Develop; chmn steering comt, Oper Head Start, Off Econ Opportunity. Honors & Awards: Johnson Award, 54; St Coletta Award, Caritas Soc, 67; Kennedy Int Award, 68. Mem: AAAS; AMA; Am Asn Med Cols; Am Fedn Clin Res; Am Pub Health Asn. Res: Mental retardation; water and electrolyte physiology. Mailing Add: 1007 WARF Bldg 610 N Walnut St Madison WI 53705

COOKE, ROGER, b Ann Arbor, Mich, Feb 22, 40. BIOPHYSICS. Educ: Mass Inst Technol, BS, 62; Univ Ill, MS, 64, PhD(physics), 68. Prof Exp: Res assoc, 70-71, ASST PROF BIOPHYS, UNIV CALIF, SAN FRANCISCO, 71- Concurrent Pos: USPHS fel, Univ Calif, San Francisco, 68-70; estab investr, Am Heart Asn, 71-; mem res coun, Am Heart Asn. Mem: Biophys Soc; Am Heart Asn. Res: Muscle biochemistry and biophysics; protein interactions; fluorescence; electron paramagnetic resonance; nuclear magnetic resonance; structure of intracellular water. Mailing Add: Dept of Biochem & Biophys Univ of Calif San Francisco CA 94143

COOKE, ROGER LEE, b Alton, Ill, July 31, 42; m 68; c 1. MATHEMATICS. Educ: Northwestern Univ, BA, 63; Princeton Univ, MA & PhD(math), 66. Prof Exp: Asst prof math, Vanderbilt Univ, 66-68; asst prof, 68-72, ASSOC PROF MATH, UNIV VT, 72- Concurrent Pos: NSF res grant, 69-71. Mem: Am Math Soc; Math Asn Am. Res: Trigonometric series in several variables. Mailing Add: Dept of Math Univ of Vt Burlington VT 05401

COOKE, RON CHARLES, b Chico, Calif, Dec 31, 47; c 1. PLANT SCIENCE. Educ: Calif State Univ, Chico, BS, 70; Univ of the Pacific, MS, 73. Prof Exp: Chemist org chem, Calif State Univ, Chico, 70-71; instr pharmaceut sci, Univ of the Pacific, 73-75; LAB DIR PLANT TISSUE CULT, BAILEY'S NURSERY, INC, 75- Mem: Am Soc Pharmacog; Electron Micros Soc; Tissue Cult Asn; AAAS. Res: Plant morphogenesis and asexual propagation of plants by tissue culture; production of drugs by plants in tissue culture. Mailing Add: PO Box 268 Lodi CA 95240

COOKE, SAMUEL LEONARD, JR, b Atlanta, Ga, Nov 30, 31; m 54; c 3. PHYSICAL CHEMISTRY, ANALYTICAL CHEMISTRY. Educ: Univ Richmond, BS, 52, MS, 54; Baylor Univ, PhD(phys chem), 57. Prof Exp: Res chemist, E I du Pont de Nemours & Co, 57-58; instrument designer, Intersci, Inc, 58-61, assoc prof chem, Ala Col, 61-63; from asst prof to assoc prof, 63-70, PROF CHEM, UNIV LOUISVILLE, 70- Mem: Elctrochem Soc; Am Chem Soc; Asn Comput Mach. Res: Instrumental methods of analysis; electrochemistry; chemical application of computers. Mailing Add: Dept of Chem Univ of Louisville 2301 S Third St Louisville KY 40208

COOKE, THEODORE FREDERIC, b Pittsfield, Mass, Jan 28, 13; m 40, 73; c 4. TEXTILE CHEMISTRY. Educ: Univ Mass, BS, 34; Yale Univ, PhD(phys chem), 37. Prof Exp: Res chemist, Standard Oil Develop Co, NJ, 37-40; res chemist, Org Chem Div, 40-42, asst dir phys chem res, 45-48, asst dir appln res dept, 48-52, mgr textile resin lab, 52-54, asst to mgr textile resin dept, 54-58, mgr commercial develop, 58-60, dir chem res, 60-62, asst dir res & develop, 62-72, DIR SCI SERVS DEPT, CHEM RES DIV, AM CYANAMID CO, 72- Concurrent Pos: Consult, Southern Regional Res Labs, USDA, 54-60; chmn, Gordon Res Conf Textiles; chmn comt textile finishing, Nat Res Coun Adv Bd Qm Res & Develop. Mem: Am Chem Soc; Am Asn Textile Chem & Color; fel Am Inst Chem; Asn Res Dirs; NY Acad Sci. Res: Analytical chemistry; physical chemistry; cosmetic chemistry; polymer chemistry; cellulose chemistry. Mailing Add: Stamford Res Lab Am Cyanamid Co Stamford CT 06904

COOKE, WILLIAM BRIDGE, b Foster, Ohio, July 16, 08; m 42. MYCOLOGY. Educ: Univ Cincinnati, BA, 37; Ore State Col, MS, 39; State Col Wash, PhD(bot), 50. Prof Exp: Mycologist, Trop Deterioration Res Lab, US Qm Corps, 45-46; res assoc, Dept Plant Path, State Col Wash, 50-51; mycologist, Bact Sect, Environ Health Ctr, US Pub Health Serv, 52-53, prin mycologist, Robert A Taft Sanit Eng Ctr, 53-56, sr mycologist, Microbiol Activities, Cincinnati Water Res Lab, Fed Water Pollution Control Admin, Dept Health, Educ & Welfare, 56-66; mycologist biol treatment activities, Advan Waste Treatment Prog, US Dept Interior, 66-69; res assoc dept bot, Miami Univ, 69-70; SR RES ASSOC, DEPT BIOL SCI, UNIV CINCINNATI, 70- Honors & Awards: Superior Service Award, Dept Health, Educ & Welfare, 59; Fed Water Pollution Control Admin Award, US Dept Interior. Mem: Fel AAAS; Mycol Soc Am; Soc Indust Microbiol; Am Soc Agron; Bot Soc Am. Res: Fungi of polluted water and sewage; taxonomy of Polyporaceae; flora and fungi of Mt Shasta and fungi of national parks; fungi of Ohio. Mailing Add: 1135 Wilshire Ct Cincinnati OH 45230

COOKE, WILLIAM DONALD, b Philadelphia, Pa, May 15, 18; m 46; c 6. ANALYTICAL CHEMISTRY. Educ: St Joseph's Col, Philadelphia, BS, 40; Univ Pa, MS & PhD, 49. Prof Exp: Chemist, Harshaw Chem Co, Pa, 40-42; Nat Res Coun fel, Princeton Univ, 49-51; from asst prof to assoc prof, 51-59, assoc dean col arts & sci, 62-64, dean grad sch, 64-69, PROF CHEM, CORNELL UNIV, 59-, VPRES RES, 69- Concurrent Pos: Pres, Asn Grad Schs, 70-71; mem bd trustees, Fordham Univ, 70- & Assoc Univs, Inc, 74-; mem, Nat Bd Grad Educ, 72-75. Mem: Am Chem Soc. Res: Electrochemical methods; absorption spectra; flame spectroscopy; nuclear magnetic resonance; gas chromatography. Mailing Add: Day Hall Cornell Univ Ithaca NY 14853

COOKE, WILLIAM JOSEPH, b Newark, NJ, Sept 15, 40; c 4. BIOCHEMICAL PHARMACOLOGY. Educ: St Bernadine Siena Col, BS, 62; Col St Rose, MS, 66; State Univ NY, PhD(pharmacol), 70. Prof Exp: Fel, Darmouth Med Sch, 70-71, res assoc pharmacol, 71-72; asst prof physiol & biochem, 72-74, ASSOC PROF PHARMACOL, MED SCH, UNIV MASS, 74- Mem: Am Soc Pharmacol & Exp Therapeut; Sigma Xi; AAAS; Am Heart Asn. Res: Hepatic accumulation, metabolism and excretion of organic anions; ion movement across excitable membranes. Mailing Add: Dept of Pharmacol Univ of Mass Med Sch Worcester MA 01605

COOKE, WILLIAM PEYTON, JR, b Hobart, Okla, Jan 4, 34; m 61; c 3. MATHEMATICS, STATISTICS. Educ: West Tex State Univ, BS, 59; Tex Tech Col, MS, 61; Tex A&M Univ, PhD(statist), 68. Prof Exp: Instr math, Amarillo Col, 60-61; instr, Tex Tech Col, 61-62; from asst prof to assoc prof, WTex State Univ, 64-69; ASSOC PROF STATIST, UNIV WYO, 69- Mem: Math Asn Am; Am Statist Asn. Res: Mathematical programming; statistical reliability. Mailing Add: Dept of Statist Univ of Wyo Laramie WY 82070

COOK-IOANNIDIS, LESLIE PAMELA, b Kingston, Ont, Aug 23, 46; US citizen; m 72. APPLIED MATHEMATICS. Educ: Univ Rochester, BA, 67; Cornell Univ, MS, 69, PhD(appl math), 71. Prof Exp: NATO fel appl math, 71-72; instr & res assoc, Dept Theoret & Appl Mech & Dept Math, Cornell Univ, 72-73; adj asst prof, 73-75, ASST PROF, DEPT MATH, UNIV CALIF, LOS ANGELES, 75- Mem: Sigma Xi; Soc Indust & Appl Math; Am Math Soc. Res: Aerodynamics, biomathematics. Mailing Add: Dept of Math Univ of Calif Los Angeles CA 90024

COOKSEY, DONALD ERNEST, b Duncan, Okla, Dec 3, 15; m 40; c 1. ORAL SURGERY. Educ: Univ Southern Calif, DDS, 40; Georgetown Univ, MS, 54. Prof Exp: Chief oral surg sect, US Naval Dent Sch, US Dept Navy, 54-62, cmndg officer, US Naval Dent Clin, Yokosuka, Japan, 62-66, dist dent officer, Sixth Naval Dist, Charleston, SC, 66-67; ORAL SURGEON, 67-; PROF ORAL SURG, SCH DENT, UNIV SOUTHERN CALIF, 71- Concurrent Pos: Instr, Georgetown Univ, 53-54; spec lectr, Univ Pa, 55; lectr, Col Dent, Univ Calif, Los Angeles; consult, Nat Bd Dent Examrs & Coun Dent Educ, Am Dent Asn; past pres, Am Bd Oral Surg. Mem: Fel Am Col Dentists; Am Dent Asn; Am Soc Oral Surg. Res: Transplantation of freeze dried tissues to defects of the jaws. Mailing Add: Dept of Oral Surg Sch of Dent Univ of Southern Calif Los Angeles CA 90007

COOKSON, FRANCIS BERNARD, b Preston, Eng, Oct 30, 28; m 53; c 2. NEUROANATOMY, HISTOLOGY. Educ: Univ Manchester, BSc, 53, MB & ChB, 56; Royal Col Obstetricians & Gynaecologists, Eng, dipl obstet, 57. Prof Exp: Demonstr anat, Univ Manchester, 57-58; pvt pract med, Eng, 58-64; asst prof anat, Univ Sask, 64-66; assoc prof, 66-71, PROF ANAT, UNIV ALTA, 71-, HON LECTR MED, 66-, ASST DEAN MED, 74- Concurrent Pos: Med Res Coun res grants, 65-69; Alta Heart Found res grant, 67-68. Mem: Fel Am Heart Asn; Anat Soc Gt Brit & Ireland; Can Med Asn; Brit Med Asn. Res: Histopathology; experimental atherosclerosis, etiology pathogenesis and preventions; hypertension incidence in university students. Mailing Add: Dept of Anat Univ of Alta Edmonton AB Can

COOL, BINGHAM MERCUR, b Marion, Ill, Dec 21, 18; m 43; c 3. FORESTRY. Educ: La State Univ, BS, 40; Iowa State Univ, MS, 41; Mich State Univ, PhD(forestry), 57. Prof Exp: Asst, Iowa State Univ, 40-41; asst agr aide, Soil Conserv Serv, 41; asst forestry aide, Tenn Valley Authority, 42; asst county agent forestry, Ala Exten Serv, 45-47; timber mkt specialist, Miss Exten Serv, 47-48, state forest prods marketing specialist, 48-49; asst prof forestry, Ala Polytech Inst, 49-54, 56-58; asst, Mich State Univ, 54-56; assoc prof, 58-66, PROF FORESTRY, CLEMSON UNIV, 66- Mem: Soc Am Foresters. Res: Siviculture. Mailing Add: Dept of Forestry Clemson Univ Clemson SC 29631

COOL, RAYMOND DEAN, b Winchester, Va, Mar 14, 02. CHEMISTRY. Educ: Bridgewater Col, BS, 22; Univ Va, MS, 26, PhD(chem), 28. Prof Exp: Instr high sch, Va, 22-24; instr chem, Univ Nev, 28-29; instr, Univ Ore, 29-30; res assoc dept pharmacol, Sch Med, Univ Pa, 30-34; from instr to asst prof, Univ Akron, 34-41; asst prof, Univ Okla, 41-46; asst prof, WVa Univ, 46; prof, 46-72, EMER PROF CHEM, MADISON COL, 72- Concurrent Pos: Chemist, US Naval Ord Lab, 49-52 & 54-57. Mem: AAAS; Am Chem Soc; Am Microchem Soc; Am Soc Testing & Mat. Res: Microanalytical methods for iodides and rarer elements; volumetric methods for nitrites; polymorphic transitions; distribution coefficients in gas-liquid systems; chemical microscopy; metallic complexes of diketones and picolines. Mailing Add: 405 E College St Bridgewater VA 22812

COOL, RODNEY LEE, b Platte, SDak, Mar 8, 20; m 49; c 4. PHYSICS. Educ: Univ SDak, BA, 42; Harvard Univ, MA, 47, PhD(physics), 49. Prof Exp: Res physicist, Brookhaven Nat Lab, 49-59, dep chmn high energy physics, 60-64, from asst dir to assoc dir, 64-70; PROF EXP HIGH ENERGY PHYSICS, ROCKEFELLER UNIV, 70- Concurrent Pos: Mem policy comt, Stanford Linear Acceleration Ctr, 62-67; mem high energy panel, Assoc Univs Inc, 63-70; Walker Panel Comt on Sci & Pub Policy, Nat Acad Sci, 64; high energy physics adv panel, AEC, 67-70; chmn high energy adv comt, Brookhaven Nat Lab, 67-70; chmn physics adv comt, Nat Accelerator Lab, 67-70; mem adv panel physics, NSF, 70-73. Mem: Nat Acad Sci; fel Am Phys Soc. Res: Experimental high energy physics. Mailing Add: Rockefeller Univ New York NY 10021

COOLBAUGH, JAMES CAMERON, b Philadelphia, Pa, Sept 21, 42; m 75. MICROBIOLOGY. Educ: Calif State Univ, Long Beach, BS, 65; Baylor Col Med, PhD(microbiol), 74. Prof Exp: Virologist, Merck Inst Therapeutic Res, 65-66; MICROBIOLOGIST, US NAVY MED SERV CORPS, 67- Mem: Am Soc Microbiol; Sigma Xi. Res: Isolation, growth, purification and physiology of typhus and scrub typhus rickettsiae. Mailing Add: Naval Med Res Inst Bethesda MD 20014

COOLER, FREDERICK WILLIAM, b Knoxville, Tenn, Dec 7, 30; m 52. FOOD TECHNOLOGY. Educ: Univ Tenn, BS, 52, MS, 58; Univ Md, PhD(food technol), 62. Prof Exp: Asst horticulturist, Univ Tenn, 56-58; asst food technol, Univ Md, 58-62; ASSOC PROF FOOD SCI & TECHNOL, VA POLYTECH INST & STATE UNIV, 62- Mem: Inst Food Technologists; Am Soc Hort Sci. Res: Rheological properties of foods; quality evaluation and control of foods; transportation, storage and quality maintenance of fruits and vegetables; food product development. Mailing Add: Dept of Food Sci & Technol Va Polytech Inst & State Univ Blacksburg VA 24061

COOLEY, ADRIAN B, JR, b Amelia, Tex, Sept 28, 28; m 48; c 3. PHYSICS, MATHEMATICS. Educ: Sam Houston State Univ, BS, 50, MA, 57; Univ Tex, Austin, PhD(sci educ), 70. Prof Exp: Teacher high sch, Tex, 50-56; asst prof physics, Sam Houston State Univ, 56-66; asst prof, Southwestern Univ, 66-67; teaching assoc educ, Univ Tex, Austin, 67-68; from asst prof to assoc prof, 68-72, PROF PHYSICS, SAM HOUSTON STATE UNIV, 72- Mem: Am Asn Physics Teachers; Nat Sci Teachers Asn. Res: Physical sciences. Mailing Add: Dept of Physics Sam Houston State Univ Huntsville TX 77340

COOLEY, DENTON ARTHUR, b Houston, Tex, Aug 22, 20; m 49; c 5. SURGERY. Educ: Univ Tex, BA, 41; Johns Hopkins Univ, MD, 44. Prof Exp: Resident surg, Johns Hopkins Hosp, 44-50; sr surg registr, Brompton Hosp Chest Dis, London, 50-51; from assoc prof to prof surg, Baylor Col Med, 51-69; SURGEON-IN-CHIEF, TEX HEART INST, 69-; PROF CLIN SURG, UNIV TEX MED SCH, HOUSTON, 75- Concurrent Pos: Consult surg serv, Tex Children's & St Lukes Episcopal Hosps, Houston, 63- Mem: AMA; Am Asn Thoracic Surg; Soc Vascular Surg; Am Col Surg; Soc Clin Surg. Res: Cardio-vascular surgery and diseases of the chest. Mailing Add: Tex Heart Inst 6621 Fannin St Houston TX 77025

COOLEY, DUANE STUART, b Batavia, NY, May 9, 23; m 49; c 2. METEOROLOGY. Educ: Mass Inst Technol, BS, 48, MS, 49, PhD(meteorol), 59. Prof Exp: Instr meteorol, Weather Sch, Chanute AFB, Dept Air Force, 49-51, meteorologist, Air Force Cambridge Res Ctr, 51-57, supvry atmospheric physicist, 57-60, supvry physicist, Electronic Systs Div, 496L Syst Proj Off, Air Force Syst Command, 60-61; sr res scientist, Travelers Res Ctr, Inc, 61-69; actg exec scientist, US Comt Global Atmospheric Res Prog, Nat Acad Sci, 69-70; CHIEF TECH PROCEDURES BR, NAT WEATHER SERV, NAT OCEANIC & ATMOSPHERIC ADMIN, 70- Mem: Am Meteorol Soc; Am Geophys Union. Res: General atmospheric circulation; statistical and dynamical weather forecasting; atmospheric radiation; analysis and interpretation of meteorological satellite data. Mailing Add: 4503 Libbey Dr Fairfax VA 22030

COOLEY, JAMES AVAS, mathematics, deceased

COOLEY, JAMES HOLLIS, b New York, NY, March 25, 30; m 55; c 2. ORGANIC CHEMISTRY. Educ: Middlebury Col, AB, 52, MS, 54; Univ Minn, PhD(org chem), 58. Prof Exp: Asst, Univ Minn, 54-57; from asst prof to assoc prof, 57-68, PROF CHEM, UNIV IDAHO, 68- Concurrent Pos: Res assoc, Columbia Univ, 63-66. Mem: Am Chem Soc. Res: Chemistry of the hydroxylamine compounds. Mailing Add: Dept of Chem Univ of Idaho Moscow ID 83844

COOLEY, JAMES WILLIAM, b New York, NY, Sept 18, 26; m 57; c 3. APPLIED MATHEMATICS. Educ: Manhattan Col, BA, 49; Columbia Univ, MA, 51, PhD(math), 61. Prof Exp: Res asst math, Courant Inst, NY Univ, 56-62; RES STAFF MATH, IBM WATSON RES CTR, 62- Concurrent Pos: Prof comput sci, Royal Inst Technol, Sweden, 73-74. Mem: Inst Elec & Electronics Engrs; Sigma Xi. Res: Numerical methods; solution of partial and ordinary differential equations; digital signal processing; discrete Fourier methods; mathematical modeling of nerve membranes. Mailing Add: IBM Watson Res Ctr Box 218 Yorktown Heights NY 10598

COOLEY, MAXWELL LOUIS, b Toledo, Ohio, Sept 13, 11; m 36; c 2. NUTRITION. Educ: Univ Toledo, BS, 35, MSc, 40. Prof Exp: Chemist, Larrowe Div, Gen Mills, 36-39, chief chemist, 39-50, dir prod control, Feed Div, 50-55; tech dir, Hoffman-Taff, Inc, 55-70, MGR TECH SALES SERV, SYNTEX AGRIBUS, INC, 70- Mem: Am Chem Soc; Am Asn Cereal Chem; Asn Off Anal Chem; Am Asn Feed Micros; Animal Nutrit Res Coun. Res: Animal nutrition; biochemistry; analytical chemistry. Mailing Add: Syntex Agribus Inc Box 1246 SSS Springfield MO 65805

COOLEY, NELSON REEDE, b Mobile, Ala, Nov 30, 20; m 51; c 2. PROTOZOOLOGY. Educ: Spring Hill Col, BS, 42; Univ Ala, MS, 47; Univ Ill, PhD(zool), 54. Prof Exp: From asst to instr biol, 46-48; instr anat, Druid City Hosp, Ala, 47; asst zool, Univ Ill, 48-53; instr zool & physiol, Okla Agr & Mech Col, 53-54, asst prof zool, 54-56; fishery res biologist, US Fish & Wildlife Serv, 56-70; res biologist, Gulf Breeze Lab, 70-74, MICROBIOLOGIST, US ENVIRON PROTECTION AGENCY, 74- Honors & Awards: Bronze Medal, US Environ Protection Agency, 75. Mem: Soc Protozoologists. Res: Physiology of ciliate protozoa; blood protozoa of bats; biological control of oyster predators; estuarine faunal biology; pesticides vs ciliate protozoan population growth; pesticide bioaccumulation by ciliates. Mailing Add: US Environ Protection Agency Environ Res Lab Sabine Island Gulf Breeze FL 32561

COOLEY, RICHARD LEWIS, b Akron, Colo, Jan 11, 40; m 60; c 1. HYDROGEOLOGY, GEOMORPHOLOGY. Educ: Ariz State Univ, BS, 62; Pa State Univ, PhD(geol), 68. Prof Exp: Hydrologist, Hydrol Eng Ctr, US Corps Engrs, 68-70; res assoc, 70-72, ASSOC RES PROF, CTR WATER RESOURCES RES, DESERT RES INST, UNIV NEV SYST, RENO, 72- Concurrent Pos: Inst Res Land & Water Resources scholar appointee, Pa State Univ, 68. Mem: Soc Econ Paleontologists & Mineralogists; Geol Soc Am; Am Geophys Union; AAAS. Res: Physics of water movement through porous media and its application in inferring the influence of geological conditions on ground water movement and recharge; analysis of ground water flow systems. Mailing Add: Ctr for Water Resources Res Desert Res Inst Univ of Nev Reno NV 89507

COOLEY, ROBERT LEE, b Birmingham, Ala, Feb 20, 27; m 52; c 4. MATHEMATICS. Educ: Univ Ala, BS, 48; Univ Va, LLB, 51; Purdue Univ, MS, 57, PhD(math), 64. Prof Exp: From asst to asst prof, 57-67, ASSOC PROF MATH, WABASH COL, 67- Mem: Math Asn Am. Res: Topological algebra. Mailing Add: Dept of Math Wabash Col Crawfordsville IN 47933

COOLEY, ROBERT NELSON, b Woodlawn, Va, Mar 12, 11; m 48; c 3. RADIOLOGY. Educ: Univ Va, MD, 34. Prof Exp: From asst prof to assoc prof radiol, Med Sch, Johns Hopkins Univ, 48-53; PROF RADIOL & CHMN DEPT, UNIV TEX MED BR GALVESTON, 53- Concurrent Pos: Pres & trustee, Am Bd Radiol. Mem: Am Col Radiol; Radiol Soc NAm; Am Roengen Ray Soc; AMA; Asn Univ Radiol. Res: Radiological investigations in congenital and acquired heart disease. Mailing Add: Dept of Radiol Univ of Tex Med Br Galveston TX 77550

COOLEY, STONE DEAVOURS, b Laurel, Miss, Jan 13, 22; m 48; c 3. PHYSICAL

CHEMISTRY. Educ: Univ Tex, BSChE, 49, PhD(chem), 53. Prof Exp: Res chemist, Celanese Corp Am, 53-55, group leader, 55-59, sect head, 59-62, asst to tech dir, Celanese Chem Co, 62-63, facil mgr Summit Res Lab, 64-68; asst dir res, 68-74, MGR TECH SERV, PETRO-TEX CHEM CORP, 74- Mem: Am Chem Soc. Res: Hydrocarbon oxidation. Mailing Add: 7719 Glenheath Houston TX 77017

COOLEY, WILLIAM EDWARD, b St Louis, Mo, March 7, 30; m 52; c 4. DENTAL RESEARCH. Educ: Cent Col, Mo, AB, 51; Univ Ill, PhD(chem), 54. Prof Exp: Res chemist, 54-71, SECT HEAD, PROD DEVELOP, PROCTER & GAMBLE CO, 71- Mem: Am Chem Soc; Int Asn Dent Res; Am Asn Dent Res. Res: Chemistry of dental systems; fluorides; dentifrice abrasives; professional and regulatory affairs in oral health. Mailing Add: 531 Chisholm Trail Wyoming OH 45215

COOLIDGE, ARDATH ANDERS, b Chicago, Ill, July 22, 19; m 49; c 5. NUTRITION. Educ: Earlham Col, AB, 41; Iowa State Col, PhD(nutrit), 46. Prof Exp: Teacher pub sch, Ill, 41-42; asst prof foods & nutrit, Western Reserve Univ, 46-47; asst prof, Berea Col, 47-49; home economist, Sensory Testing Food Res, Armour & Co, 58-62; lit scientist, Nutrit Res Div, Nat Dairy Coun, 62-66; asst prof, 66-69, ASSOC PROF HOME ECON, PURDUE UNIV, CALUMET CAMPUS, 69- Mem: Am Dietetic Asn; Am Home Econ Asn. Res: Biological utilization of ascorbic acid in apples. Mailing Add: Dept of Home Econ Purdue Univ Calumet Campus Hammond IN 46323

COOLIDGE, EDWIN CHANNING, b Gambier, Ohio, Jan 30, 25; m 53; c 1. ANALYTICAL CHEMISTRY, ORGANIC CHEMISTRY. Educ: Kenyon Col, AB, 44; Johns Hopkins Univ, PhD(chem), 49. Prof Exp: Res org chemist, Procter & Gamble Co, 49-50 & 53-54; asst prof, Univ Utah, 51-52; res org chemist, Dugway Proving Ground, 52-53; asst prof chem, Hamilton Col, 54-58; asst prof, NMex Inst Mining & Technol, 58-61; assoc prof, 61-64, PROF CHEM, STETSON UNIV, 64- Concurrent Pos: Dir, Assoc Mid-Fla Cols Year Abroad Prog, 68, Freiburg, WGer, 69-70; consult, Tech Adv Serv for Attorneys, 75. Mem: Am Chem Soc; The Chem Soc; AAAS. Res: Organometallic chemistry; pyrrole and porphyrin synthesis; organophosphorus compounds; metal chelate stability and structure; ecological analysis. Mailing Add: Dept of Chem Stetson Univ De Land FL 32720

COOLIDGE, HAROLD JEFFERSON, b Boston, Mass, Jan 15, 04; m 31, 72; c 3. MAMMALOGY, CONSERVATION. Educ: Harvard Univ, SB, 27. Hon Degrees: DSc, George Washington Univ, 59, Seoul Nat Univ, 65 & Brandeis Univ, 70. Prof Exp: Asst mammalogist, Harvard African Exped, Liberia, Belgian Congo, 26-27; leader Indo-China div, Kelley-Roosevelt's Field Mus Exped, 28-29; asst cur mammals, Mus Comp Zool, Harvard Univ, 29-46, assoc mammal, 46-70; exec dir Pac Sci Bd, Nat Acad Sci-Nat Res Coun, 46-70; pres, 66-72, HON PRES, INT UNION CONSERV NATURE & NATURAL RESOURCES, 72- Concurrent Pos: Secy Am comt, Int Wildlife Protection, 30-51, chmn, 51-71, hon chmn, Am Comt Int Conserv, 71-; collabr, US Nat Park Serv, 48-; hon consult, Bernice P Bishop Mus, 53-; adv Pac studies, Peabody Mus of Salem, 74-; mem adv bd, Cult Survival, Inc, 75- Vpres, Int Union Conserv Nature & Natural Resources, 49-55, chmn int comn nat parks, 58-66, chmn, Survival Serv Comn, 49-58, mem & hon pres, 72-; secy gen, Tenth Pac Sci Cong, Honolulu, 61; mem orgn comt, XVI Int Zool Cong, 61-63; chmn, First World Conf Nat Parks, Seattle, 62; mem bd dirs, African Wildlife Leadership Found, Charles Darwin Found for Galapagos Islands, Chocorua Island Chapel Asn, hon dir, Nat Inst Conserv Nature, Zaire, vchmn, Island Resources Found, L S B Leakey Found; emer mem, Pac Trop Bot Garden Found, 74-; mem bd, Threshold, Inc, US & Int World Wildlife Fund; mem corp, Boston Mus Sci. Honors & Awards: Frances K Hutchinson Medal, Garden Clubs of Am, 63; Horace Marden Albright Scenic Preserv Medal, Am Scenic & Hist Preserv Soc, 69; Silver Medal, US Nat Parks Centennial Comn, 72. Mem: hon life mem Pac Sci Asn; Am Soc Mammalogists; Int Inst Differing Civilizations; Nat Parks & Conserv Asn (secy, 46-59); corresp mem Zool Soc London. Res: Classification of gorillas, chimpanzees and gibbons; classification of bovids; curating and comparative sociology of mammals; international conservation and preservation of vanishing species; world extension of national parks and primative areas, marine parks, international cooperation in Pacific science. Mailing Add: 38 Standley St Beverly MA 01915

COOLIDGE, THOMAS BUCKINGHAM, b Concord, Mass, July 2, 01; m 27, 44; c 5. BIOCHEMISTRY. Educ: Harvard Univ, AB, 23, MD, 27; Columbia Univ, PhD(biochem), 36. Prof Exp: Intern, Mass Gen Hosp; tutor biochem, Harvard Univ, 29-32, asst med, 32; asst biochem, Col Physicians & Surgeons, Columbia Univ, 33-35; from instr to asst prof, Sch Med, Duke Univ, 34-35; from assoc prof to prof, 45-75, EMER PROF BIOCHEM, UNIV CHICAGO, 75- Concurrent Pos: Vis prof, Meharry Med Col, 68-74. Mem: Am Soc Biol Chem. Res: Urinary pigments; bacterialmetabolism; dental caries; calcium metabolism. Mailing Add: 5755 S Dorchester Chicago IL 60637

COOLIDGE, WILLIAM DAVID, physical chemistry, deceased

COOMBES, CHARLES ALLAN, b Nevada City, Calif, Feb 25, 34; m 56; c 3. PHYSICS. Educ: Univ Calif, AB, 55, PhD(physics), 60. Prof Exp: Asst prof physics, Idaho State Univ, 59-62 & San Jose State Col, 62-63; asst prof, 63-66, actg vdean fac arts & sci, 70-72, ASSOC PROF PHYSICS, UNIV CALGARY, 66- Mem: AAAS; Am Asn Physics Teachers. Res: Field theory; properties of elementary particles; foundations of quantum mechanics. Mailing Add: Dept of Physics Univ of Calgary Calgary AB Can

COOMBS, HOWARD ABBOTT, b Dallas, Tex, Apr 10, 06; m 36; c 1. GEOLOGY. Educ: Univ Wash, Seattle, BS, 29, MS, 31, PhD(geol), 34. Prof Exp: With Mt Rainier Nat Park, 30-33; from instr to prof, 34-52, HEAD DEPT GEOL, UNIV WASH, 52- Concurrent Pos: Consult geologist numerous cities, industs, state & fed orgn & foreign countries, 35-; collabr, Environ Sci Serv Admin, State Wash; mem, US Comn Large Dams. Mem: Fel Geol Soc Am; Seismol Soc Am; Soc Econ Paleontologists & Mineralogists. Res: Petrology of igneous rocks; engineering geology, especially to dam and nuclear sites. Mailing Add: Dept of Geol Univ of Wash Seattle WA 98105

COOMBS, MARGERY CHALIFOUX, b Nashua, NH, Aug 12, 45; m 69; c 1. VERTEBRATE PALEONTOLOGY. Educ: Oberlin Col, BA, 67; Columbia Univ, MA, 68, PhD(biol sci), 73. Prof Exp: ASST PROF ZOOL, UNIV MASS, AMHERST, 73- Mem: Soc Vert Paleont. Res: Chalicothere systematics and function; early Miocene biostratigraphy. Mailing Add: Dept of Zool Univ of Mass Amherst MA 01002

COOMBS, RENATE BANGERT, b Erlangen, Ger, June 1, 37; m 63. ORGANIC CHEMISTRY. Educ: Brunswick Tech Univ, dipl, 63; Univ Wis, PhD(org chem), 64. Prof Exp: Res chemist, Brit Drug Houses Ltd, Eng, 64-66; fel org photochem, Synvar Res Inst, Calif, 66-67; res chemist, Org Chem Div, Am Cyanamid Co, 67-69; sr scientist, Pharmaceut Div, Sandoz-Wander, Inc, 69-74, GROUP LEADER, PHARMACEUT DIV, SANDOZ, INC, 75- Mem: Am Chem Soc; The Chem Soc; Soc Ger Chem; Am Soc Mass Spectrometry. Res: Synthetic organic chemistry related to natural products; organic mass spectrometry. Mailing Add: Sandoz Inc Rt 10 East Hanover NJ 07936

COOMBS, ROBERT E, ecology, phycology, see 12th edition

COOMBS, ROBERT VICTOR, b Brighton, Eng, June 24, 37; m 63. ORGANIC CHEMISTRY. Educ: Univ London, BSc, 58, PhD(org chem), 61. Prof Exp: Fel, Univ Wis, 61-62 & 63-64; fel, Columbia Univ, 62-63; res chemist, Brit Drug Houses, Ltd, Eng, 64-66; fel, Synvar Res Inst, Calif, 66-67; sr scientist, Chem Dept, Sandoz-Wander, Inc, 67-71, GROUP LEADER, SANDOZ, INC, 71- Mem: Am Chem Soc; The Chem Soc. Res: Synthetic organic and medicinal chemistry. Mailing Add: Chem Dept Sandoz Inc Rt 10 East Hanover NJ 07936

COOMBS, WILLIAM, JR, b Brooklyn, NY, Apr 30, 24; m 45; c 2. BIOMEDICAL ENGINEERING. Educ: Mass Inst Technol, BS, 47, MS, 48; Univ Rochester, MS, 56. Prof Exp: Res assoc biol, Mass Inst Technol, 47-48; develop engr, Eastman Kodak Co, 48-51; chief elec engr, Dept Physics, Univ Rochester, 51-59, head dept electronics res & develop, 59-63, dir electronics & biophys res & develop, 63-67, dir cent res labs, 67-68, gen mgr soflens contact lens, 68-71, VPRES & DIR TECH OPERS, SOFLENS CONTACT LENS DIV, BAUSCH & LOMB, INC, 71- Concurrent Pos: Consult, Xerox Corp & Taylor Instrument Co, 54-59; lectr univ sch, Univ Rochester, 55-60; indust rep, Ophthalmic Prosthetic Devices Subcomt, Food & Drug Admin, 76. Mem: Inst Elec & Electronics Eng. Res: Biomedical instruments and ophthalmic prosthetics. Mailing Add: Bausch & Lomb Inc 1400 N Goodman St Rochester NY 14602

COOMBS, WILLIAM CHRISTOPHER, radio physics, see 12th edition

COOMES, EDWARD ARTHUR, b Louisville, Ky, June 27, 09; m 40; c 5. PHYSICS. Educ: Univ Notre Dame, BS, 31, MS, 33; Mass Inst Technol, ScD(physics), 38. Prof Exp: From instr to assoc prof math & physics, 33-42, prof physics, 45-74, EMER PROF PHYSICS, UNIV NOTRE DAME, 74- Concurrent Pos: Mem staff radiation lab, Mass Inst Technol, 42-45; consult various industs & labs, 52- Honors & Awards: US Army-Navy Res Citation, 47. Mem: Fel Am Phys Soc; sr mem Inst Elec & Electronics Engrs. Res: Solid state electronics; oxide cathodes; surface physics and chemistry; thermionics; energy conversion. Mailing Add: 1036 N Johnson St South Bend IN 46628

COON, BECKFORD FEDDERSEN, b Fruitland, Idaho, Sept 25, 15; m 40; c 2. ENTOMOLOGY. Educ: Univ Idaho, BS, 37, MS, 38; Ohio State Univ, PhD(entom), 42. Prof Exp: Asst sugar beet leafhopper control, Univ Idaho, 37-38; asst zool & entom, Ohio State Univ, 38-42; jr entomologist malaria control in war areas, USPHS, 42; asst prof cotton insect control & asst exten entomologist, NC State Col, 42-43; asst, Pa State Col, 39-41, from asst prof to assoc prof, 43-53, actg head, 63-65, PROF ECON ENTOM, PA STATE UNIV, 53-, HEAD DEPT, 65- Mem: AAAS; Entom Soc Am. Res: Potato and tobacco insect control; cotton insect control; breeding for plants resistant to insects; insect transmission of plant pathogens. Mailing Add: Dept of Entom Pa State Univ 106 Patterson University Park PA 16802

COON, CARLETON STEVENS, b Wakefield, Mass, June 23, 04; m 26, 45; c 2. ANTHROPOLOGY. Educ: Harvard Univ, AB, 25, MA & PhD(anthrop), 28. Prof Exp: From tutor to prof anthrop, Harvard Univ, 27-48, asst cur, Peabody Mus, 36-48; cur ethnol, Univ Mus, Univ Pa, 48-63, RES CUR ANTHROP, UNIV MUS, UNIV PA, 63- Concurrent Pos: Sheldon traveling fel, Morocco, 25; partic, Bur Int Res Exped, Morocco, 26-27, pvt exped, Senhaja & Ghomara, 28; fel, Exped Northern Albania, 29-30; partic, Peabody Mus Exped, Ethiopia & Southern Arabia, 33-34, Morocco, 39; spec asst to Am Legation, Tangier, Morocco, 42-43; partic, Am Sch Prehist Res Exped, Morocco, 47-; partic, Univ Pa Exped, Iraq & Iran & mem, Soc Sci Res Coun, 48-49; partic, Univ Pa Exped, Iran, 51; consult, Arabian-Am Oil Co, Saudi Arabia, 52; partic, Univ Pa Exped, Afghanistan & Australia, 54; consult, Life Mag, 54-55 & Pan-African Prehist Cong, Rhodesia, 55; partic, Univ Pa Exped, Syria, 55; consult, Scott Foresman Co, Ill, 56- & comt sci personnel, NSF, 61-64; partic, Univ Pa Exped, Jebel Ighoud Man No 2 Exped, Morocco, 62-63; consult, Smithsonian Foreign Currency Prog, 64-; partic, Univ Pa Exped, Yengema Cave Excavation Exped, Sierra Leone, 65; consult, Archaeol TV Show, 66-67; partic, Saharan Rock Paintings Exped, Chad & Libya, 66-67; hon cur anthrop, Peabody Mus, Salem, Mass, 66-; mem, Nat Res Coun. Honors & Awards: Viking Medal, Nat Acad Sci & Am Acad Arts & Sci, 51; Gold Medal, Philadelphia Athenaeum, 62. Mem: Am Anthrop Asn; Am Asn Phys Anthrop (pres, 62-64). Res: Physical anthropology; ethnography of the Middle East; archaeology of North Africa and the Middle East. Mailing Add: 207 Concord St Gloucester MA 01930

COON, CRAIG NELSON, b Big Springs, Tex, May 17, 44; m 62; c 1. NUTRITION, METABOLISM. Educ: Tex A&M Univ, BS, 66, MS, 70, PhD(biochem & nutrit), 73. Prof Exp: Asst prof nutrit & poultry sci, Univ Md, College Park, 73-75; ASST PROF NUTRIT & ANIMAL SCI, WASH STATE UNIV, 75- Mem: AAAS; Poultry Sci Asn. Res: Regulation of nitrogen metabolism; hormone-enzyme relationships as influenced by nutrition; amino acid availability and carbohydrate availability in feedstuffs. Mailing Add: Dept of Animal Sci Wash State Univ Pullman WA 99163

COON, GERALDINE ALMA, b North Stonington, Conn, Sept 13, 13. MATHEMATICS. Educ: Conn Col Women, BA, 35; Brown Univ, MS, 37; Univ Rochester, PhD(math), 50. Prof Exp: Instr shop math, Scovill Mfg Co, 39-44; res mathematician, Taylor Instrument Cos, 44-58; from asst prof to assoc math, Univ Conn, 58-64; PROF MATH, GOUCHER COL, 64- Concurrent Pos: Mem, Courant Inst Math Sci, 59-60. Mem: Am Math Soc; Math Asn Am; Soc Indust & Appl Math. Res: Numerical analysis; applied mathematics. Mailing Add: Dept of Math Goucher Col Towson MD 21204

COON, JAMES HUNTINGTON, b Liberty, Mo, Nov 9, 14; m 55; c 3. SPACE PHYSICS. Educ: Ind Univ, AB, 37; Univ Chicago, PhD(physics), 42. Prof Exp: Res staff, Metall Lab, Univ Chicago, 42-43 & Los Alamos Sci Labs, 43-46; res assoc, Univ Wis, 47; res staff, 48-50, GROUP LEADER, LOS ALAMOS SCI LABS, 50- Mem: Fel AAAS; fel Am Phys Soc; Am Geophys Union; Am Astron Soc. Res: Thermal neutron diffusion and capture; nuclear interactions between light particles; scattering of fast neutrons; space physics. Mailing Add: Los Alamos Sci Labs Los Alamos NM 87545

COON, JESSE BRYAN, b Liberty, Mo, Oct 29, 10; m 35; c 2. PHYSICS. Educ: Ind Univ, AB, 32, AM, 35; Univ Chicago, PhD(physics), 48. Prof Exp: Instr physics, Colo State Univ, 37-40; instr, Ind Univ, 42-46; PROF PHYSICS, TEX A&M UNIV, 46- Mem: Am Phys Soc. Res: Vibrational and rotational structure of visible and ultraviolet absorption spectra of simple polyatomic molecules. Mailing Add: Dept of Physics Tex A&M Univ College Station TX 77843

COON, JULIUS MOSHER, b Liberty, Mo, Oct 29, 10; m 47; c 2. PHARMACOLOGY. Educ: Ind Univ, AB, 32; Univ Chicago, PhD(pharmacol), 38; Univ Ill, MD, 45. Prof Exp: Asst pharmacol, Univ Chicago, 35-39, instr, 39-45, pharmacologist, Toxicity Lab, 41-45; pharmacologist, US Food & Drug Admin, Washington, DC, 46; from asst prof to assoc prof pharmacol, Univ Chicago, 46-53, dir, Toxicity Lab, 48-51, dir, US Air Force Radiation Lab, 51-53; PROF PHARMACOL & CHMN DEPT, THOMAS JEFFERSON UNIV, 53- Concurrent Pos: Mem pharmacol test comt, Nat Bd Med Exam, 54-57; mem food protection comt, Nat Acad Sci-Nat Res Coun, 54-, mem comt radiation preservation of food, 69-74; mem toxicol study sect, NIH, 58-62, chmn sect, 62-64; mem pharmacol adv comt, Walter Reed Army Inst Res, 66-70; mem adv comt protocols for safety eval, Food & Drug Admin, 66-, mem panel rev of internal analgesic agents, 72-; mem expert adv panel food additives, WHO, 66-; chmn panel food safety, White House Conf Food, Nutrit & Health, 69; mem nominating comt gen comt revision, US Pharmacopoeia, 70; mem comt admissions, Nat Formulary, 70-75; mem subcomt interpretation of relevant human experience versus newly acquired exp data, Citizens' Comn Sci, Law & Food Supply, 73-; mem expert panel on food safety & nutrit, Inst Food Technol, 74-; mem select comt on flavor evaluation criteria, Life Sci Res Off, Fedn Am Socs for Exp Biol, 75- Mem: Am Soc Pharmacol & Exp Therapeut (treas, 64-66); Soc Toxicol; Inst Food Technol; Soc Exp Biol & Med; fel NY Acad Sci. Res: Toxicology of insecticides; food toxicology; autonomic pharmacology. Mailing Add: Dept of Pharmacol Jefferson Med Col Jefferson Univ Philadelphia PA 19107

COON, LEWIS HULBERT, b Oklahoma City, Okla, Feb 26, 25; m 48; c 3. MATHEMATICS. Educ: Okla Agr & Mech Col, BS, 50; Ind Univ, MS, 51; Okla State Univ, MS, 58, EdD(math), 63. Prof Exp: Teacher Okla City Pub Schs, 50-57; asst prof math, Southwestern State Col, Okla, 58-61; staff asst, Okla State Univ, 62; asst prof educ res, Ohio State Univ, 63-65; PROF MATH, EASTERN ILL UNIV, 65- Concurrent Pos: Mem, NSF Col Conf, Carleton Col, 63. Mem: Math Asn Am; Am Math Asn; Nat Coun Teachers Math; Am Nat Metric Coun. Res: Hawthorne effect in mathematics education. Mailing Add: Dept of Math Eastern Ill Univ Charleston IL 61920

COON, MARVIN DUAIN, organic chemistry, see 12th edition

COON, MINOR J, b Englewood, Colo, July 29, 21; m 48; c 2. BIOCHEMISTRY, PHARMACOLOGY. Educ: Univ Colo, BA, 43; Univ Ill, PhD(biochem), 46. Prof Exp: Res asst, Univ Ill, Urbana, 46-47; from instr to assoc prof physiol chem, Univ Pa, 47-55; PROF BIOL CHEM, UNIV MICH, ANN ARBOR, 55-, CHMN DEPT, 70- Concurrent Pos: USPHS spec fel, NY Univ, 52-53; res fel, Swiss Fed Polytech Inst, Zurich, Switz, 61-62; NSF travel award, 69; mem biochem study sect, NIH, 63-66; mem res career award comt, Nat Inst Gen Med Sci, 66-70. Honors & Awards: Paul Lewis Award, Am Chem Soc, 59. Mem: Am Chem Soc; Am Soc Biol Chem; Am Soc Pharmacol & Exp Therapeut; Am Soc Microbiol. Res: Enzyme reaction mechanisms; amino acid and lipid metabolism; cytochrome P-450; drug metabolism; detoxication. Mailing Add: Dept of Biol Chem Med Sch Univ of Mich Ann Arbor MI 48104

COON, ROBERT WILLIAM, b Billings, Mont, July 13, 20; m 47; c 3. PATHOLOGY. Educ: NDak Agr Col, BS, 42; Univ Rochester, MD, 44; Am Bd Path, dipl, 51. Prof Exp: Intern path, Stong Mem Hosp, 44-45; asst, Sch Med, Emory Univ, 45-46; lab officer, US Naval Hosp, RI, 46-47; fel path, Univ Rochester, 48-49; from assoc to assoc prof, Col Physicians & Surgeons, Columbia Univ, 49-55; chmn dept, 55-72, PROF PATH, SCH MED, UNIV VT, 55-, ASSOC DEAN DIV HEALTH SCI, 68- Concurrent Pos: Trustee, Am Bd Path, 60- Mem: Am Soc Clin Path (pres, 63-64); Am Soc Exp Path; Am Asn Pathologists & Bacteriologists; Col Am Pathologists; Int Acad Path. Res: Coagulation of blood. Mailing Add: Div of Health Sci Univ of Vt Burlington VT 05401

COON, WILLIAM WARNER, b Saginaw, Mich, Aug 10, 25; m 49; c 3. SURGERY. Educ: Johns Hopkins Univ, MD, 49; Am Bd Surg, dipl, 57. Prof Exp: Dir blood bank, Univ Hosp, 56-64, from instr to assoc prof, 56-67, PROF SURG, UNIV MICH, ANN ARBOR, 67- Concurrent Pos: Attend surgeon, Ann Arbor Vet admin Hosp, 56-62, consult surgeon, 62- Mem: AAAS; Soc Univ Surg; Soc Exp Biol & Med. Res: Blood coagulation; thromboembolism; metabolism. Mailing Add: 1405 E Ann St Ann Arbor MI 48104

COONCE, HARRY B, b Independence, Mo, Mar 19, 38; m 65; c 2. MATHEMATICS. Educ: Iowa State Univ, BS, 59. Univ Del, PhD(math), 69. Prof Exp: Asst prof math, Wichita State Univ, 64-66, Del State Col, 66-67 & US Naval Acad, 67-69; asst prof, 69-70, ASSOC PROF MATH, MANKATO STATE COL, 70- Concurrent Pos: Consult, Boeing Co, Kans, 65-67. Mem: Am Math Soc. Res: Geometric function theory; univalent functions; functions with boundary rotation; functions with bounded radius rotation; variational methods. Mailing Add: Dept of Math Mankato State Col Mankato MN 56001

COONEY, DONALD GEORGE, b Virginia City, Nev, June 13, 18; m 39; c 4. MICROBIOLOGY. Educ: Univ Nev, BS, 47; Univ Calif, PhD(bot), 52. Prof Exp: Asst, Univ Calif, 47-48, 50-52; instr, 48-49, from asst prof to assoc prof, 52-65, chmn dept, 63-69, PROF BIOL, UNIV NEV, RENO, 65- Concurrent Pos: Chmn, Nev Basic Sci Bd, 70- Mem: AAAS; Bot Soc Am; Am Inst Biol Sci; Am Soc Microbiol. Res: Environmental bacteriology and mycology; physiology of thermophilic fungi. Mailing Add: Dept of Biol Univ of Nev Reno NV 89507

COONEY, JOHN ANTHONY, b Jersey City, NJ, Oct 31, 22; m 50; c 3. ATMOSPHERIC PHYSICS. Educ: Fordham Univ, BS, 49, MS, 50; NY Univ, PhD(physics), 67. Prof Exp: Res engr syst eng, Vitro Labs, Pullman Corp, 50-54; res engr microwave eng, Sperry Corp, 55-60; res scientist plasma & atmospherics, RCA Labs, 60-70; ASSOC PROF PHYSICS, DREXEL UNIV, 70- Res: Laser radar probing of the atmosphere. Mailing Add: Dept Physics & Atmospheric Sci Drexel Univ Philadelphia PA 19104

COONEY, JOHN LEO, b Washington, DC, June 26, 28; m 56; c 1. PHYSICAL CHEMISTRY. Educ: Loyola Col, Md, BS, 52; Fordham Univ, MS, 54, PhD(phys chem), 57. Prof Exp: Lectr phys chem, Notre Dame Col, Staten Island, 54-55; SR PHYS CHEMIST, E I DU PONT DE NEMOURS & CO, INC, 57- Mem: Am Phys Soc; Am Chem Soc. Res: Physical properties of polymers; thermodynamics of irreversible processes; polymer morphology; vapor phase reactions. Mailing Add: Exp Station E I du Pont de Nemours & Co Wilmington DE 19898

COONEY, JOSEPH CHARLES, medical entomology, wildlife management, see 12th edition

COONEY, JOSEPH JUDE, b Syracuse, NY, Jan 16, 34; m 57; c 4. MICROBIOLOGY. Educ: Le Moyne Col, NY, BS, 56; Syracuse Univ, MS, 58, PhD(microbiol), 61. Prof Exp: From asst prof to assoc prof bact, Loyola Univ, 61-65; assoc prof, 65-70, PROF BIOL, UNIV DAYTON, 70- Concurrent Pos: Res grants, NIH, 63-64 & La Div, Am Cancer Asn, 64-; sabbatical leave, dept biochem, Med Sch, Tufts Univ, 71; grants, Firestone Coated Fabrics Div, Monsanto Corp, US Dept Interior. Mem: AAAS; Am Soc Microbiol; Soc Gen Microbiol; Soc Indust Microbiol; fel Am Acad Microbiol. Res: Microbial carotenoid pigments; metabolism of hydrocarbons; ecology and physiology of hydrocarbon-using organisms; oil pollution. Mailing Add: 62 Bizzell Ave Dayton OH 45459

COONEY, MARION KATHLEEN, b Mercedes, Tex, Feb 2, 20. MICROBIOLOGY. Educ: Col St Benedict, BS, 39; Univ Minn, Minneapolis, MS, 53, PhD(microbiol), 62; Am Bd Microbiol, dipl. Prof Exp: Med technologist, Fairview Hosp, Minn, 40-43; bacteriologist, Minn Dept Health, 43-46, bacteriologist & supvr sect virus & rickettsia, 46-53, bacteriologist, virologist & chief sect, 53-66; asst prof, 66-72, ASSOC PROF PATH, SCH PUB HEALTH & COMMUNITY MED, UNIV WASH, 72- Mem: AAAS; Am Soc Microbiol; Am Pub Health Asn; Am Asn Immunol; fel Am Acad Microbiol. Res: Cell culture, virology and immunology as related to the epidemiology of infectious disease; human rhinoviruses; rhinovirus infections; adeno-associated viruses; family studies of influenza. Mailing Add: Dept Pathobiol F262 Health Sci Univ Wash Sch Pub Hlth & Com Med Seattle WA 98105

COONEY, MIRIAM PATRICK, b South Bend, Ind, May 6, 25 MATHEMATICS. Educ: St Mary's Col, Ind, BS, 51; Univ Notre Dame, MS, 53; Univ Chicago, SM, 63, PhD(math), 69. Prof Exp: Assoc prof, 63-72, PROF MATH, ST MARY'S COL, IND, 72- Mem: Am Math Soc; Math Asn Am. Res: Algebra, especially finite groups; mathematical education on all levels. Mailing Add: Dept of Math St Mary's Col Notre Dame IN 46556

COONEY, ROBERT CLAIR, b Ashtabula, Ohio, Feb 8, 13; m 38; c 2. ORGANIC CHEMISTRY. Educ: Union Univ, NY, BS, 34; Polytech Inst Brooklyn, MS, 38, PhD(org chem), 43. Prof Exp: Res chemist, H Kohnstamm & Co, Inc, 34-44, supvr org res, 44-53, dir res, 53-58, asst vpres, 58-63, mem, Bd Dirs, 62-74, vpres res & develop, 63-74; RETIRED. Concurrent Pos: Instr, Polytech Inst Brooklyn, 44-48. Mem: AAAS; Am Chem Soc; fel Am Inst Chemists; Am Soc Testing & Mat; Am Asn Textile Chemists & Colorists. Res: Synthesis of organic intermediates and dyestuffs; organic and inorganic pigments; synthetic detergents and aromatic chemicals; laundry products; textile coatings; chemical modification cotton; emulsion colloids; food processing; flavoring materials. Mailing Add: 937 Andover Terr Ridgewood NJ 07540

COONRADT, HARRY LYNN, chemistry, see 12th edition

COONS, ALBERT HEWETT, b Gloversville, NY, June 28, 12; m 47; c 5. IMMUNOLOGY. Educ: Williams Col, AB, 33, Harvard Univ, MD, 37. Hon Degrees: ScD, Williams Col, 60, Yale Univ, 61, Emory Univ, 69. Prof Exp: House officer med, Mass Gen Hosp, 37-39; instr, 47-48, assoc, 48-50, Houghton asst prof, 50-53, vis prof, 53-70, PROF BACT & IMMUNOL, HARVARD MED SCH, 70-; CAREER INVESTR, AM HEART ASN, 53- Concurrent Pos: Res fel, Harvard Med Sch, 39-40, univ res fel bact & immunol & Nat Res Coun fel med sci, 40-42; asst resident, Thorndike Lab & Boston City Hosp, 39-40; responsible investr comn immunization, Armed Forces Epidemiol Bd, 49-59, comn mem, 59-, dep dir, 61-63; consult, USPHS, 56-, spec consult, Commun Dis Ctr, 59-62, mem bd sci coun, Nat Inst Allergy & Infectious Dis, 59-62; Harvey lectr, 57; Gordon Wilson lectr, 63. Honors & Awards: Kimble Methodol Award, 58; Lasker Award, 59; Paul Ehrlich Award, 61; Passano Award, 62; T Duckett Jones Mem Award, 62; Gairdner Found Annual Award, 62; Emil von Behring Prize, 66; Albion O Bernstein MD Award, 73; Boerhaave Medal, Univ Leiden, 74. Mem: Nat Acad Sci; AAAS; Am Asn Immunol (pres, 60-61); Histochem Soc (vpres, 63-64, pres, 64-65); Am Acad Arts & Sci. Res: Fate of injected antigens; site of antibody formation; development of methods for study of immune reactions in vivo; immunohistochemistry. Mailing Add: Dept of Path Harvard Med Sch Boston MA 02115

COONS, LEWIS BENNION, b Salt Lake City, Utah, July 28, 38; m 60; c 4. ELECTRON MICROSCOPY, HISTOLOGY. Educ: Utah State Univ, BS, 64, MS, 66; NC State Univ, PhD, 70. Prof Exp: Fel electron micros, Molecular Toxicol Prog, NC State Univ, 70-73; head, Electron Micros Ctr, Miss State Univ, 73-76; HEAD, ELECTRON MICROS CTR, MEMPHIS STATE UNIV, 76- Mem: AAAS; Electron Micros Soc Am. Res: Scanning and transmission electron microscopy of animal cell and organ systems. Mailing Add: Dept of Biol Electron Micros Ctr Memphis State Univ Memphis TN 38152

COONS, RICHARD DANIEL, atmospheric physics, see 12th edition

COOPE, JOHN ARTHUR ROBERT, b Liverpool, Eng, June 9, 31; Can citizen; m 54. MOLECULAR PHYSICS. Educ: Univ BC, BA, 50, MSc, 52; Univ Oxford, DPhil(theoret chem), 56. Prof Exp: Fel molecular physics, 56-57, lectr, 57-59, from instr to assoc prof, 59-73, PROF CHEM, UNIV BC, 73- Mem: Am Phys Soc. Res: Quantum theory; physics of atoms and small molecules; theoretical chemistry; solid state physics. Mailing Add: Dept of Chem Univ of BC Vancouver BC Can

COOPER, AARON DAVID, b Phildelphia, Pa, Nov 17, 28; m 56; c 2. ANALYTICAL CHEMISTRY. Educ: Temple Univ, BS, 50; Univ Wis, MS, 52, PhD(pharmaceut), 54. Prof Exp: Res scientist, Anal Chem, Upjohn Co, 54-55; sr asst scientist, NIH, 55-57; res chemist, Atlas Powder Co, 57-61; head anal chem sect, 61-69, DIR ANAL CHEM, VICK DIV RES & DEVELOP, RICHARDSON-MERRELL, INC, 69- Concurrent Pos: Mem gen comn of revision US Pharmacopeia, 70-75. Mem: Am Chem Soc; Am Pharmaceut Asn. Res: Chromatography; pharmaceutical analyses; instrumental methods; quality control; biological analyses. Mailing Add: Vick Div Res & Develop Richardson-Merrell Inc Mt Vernon NY 10553

COOPER, ALFRED WILLIAM MADISON, b Dublin, Ireland, June 12, 32; US citizen; m 61; c 1. PHYSICS. Educ: Trinity Col, Dublin, BA, 55, MA, 58; Queen's Univ, Ireland, PhD(physics), 61. Prof Exp: Res asst physics, Queen's Univ, Belfast, 55-56, asst lectr, 56-57; asst prof, 57-64, ASSOC PROF PHYSICS, NAVAL POSTGRAD SCH, 64- Concurrent Pos: Mem tech staff, Aerospace Corp, 64. Mem: Am Phys Soc. Res: Low frequency oscillations in gas discharges; particle transport and instabilities in plasma. Mailing Add: 1087 Trappers Trail Pebble Beach CA 93953

COOPER, ARTHUR WELLS, b Washington, DC, Aug 15, 31; m 53; ECOLOGY. Educ: Colgate Univ, BA, 53, MA, 55; Univ Mich, PhD(bot), 58. Prof Exp: Preceptor, Colgate Univ, 53-55, instr biol, 54-55; from asst prof to prof bot, 58-71; dep dir, NC Dept Conserv & Develop 71, dir, 72-74; ASST SECY NC DEPT NATURAL & ECON RESOURCES, 71- Concurrent Pos: Bot ed, Ecol Monogr, 69-72, mem bd ed, 72-; prof bot & forestry, NC State Univ, 71- Honors & Awards: Conservation Award, Am Motors Corp, 73. Mem: AAAS (secy, sect G, 67-71); Bot Soc Am; Ecol Soc Am. Res: Plant ecology; general plant sociology; microenvironments; resource management; forest productivity. Mailing Add: NC Dept Natural & Econ Resources 217 W Jones St Raleigh NC 27611

COOPER, BENJAMIN FRANKLIN, (JR), pharmacy, see 12th edition

COOPER, BENJAMIN STUBBS, b Schenectady, NY, Apr 12, 41; m 62; c 4. NUCLEAR PHYSICS. Educ: Swarthmore Col, BA, 63; Univ Va, PhD(physics), 68. Prof Exp: Asst prof physics, Iowa State Univ, 67-73; cong sci fel, 73-74, MEM PROF STAFF, SENATE COMT INTERIOR & INSULAR AFFAIRS, 74- Concurrent Pos:

Mem, Panel Physics & Pub Affairs, Am Phys Soc, 74-, mem exec comt, Forum on Physics & Soc, 74- Mem: Am Phys Soc; AAAS. Res: Theoretical nuclear physics; science and public policy. Mailing Add: 3106 Dirksen Senate Off Bldg Washington DC 20510

COOPER, BERNARD A, b Plainfield, NJ, July 2, 28; Can citizen; m 55; c 3. HEMATOLOGY. Educ: McGill Univ, BSc, 49, MD, CM, 53; FRCPS(C), 58. Prof Exp: Demonstr med, 60-62, lectr, 62-63, asst prof, 63-65, assoc prof med & clin med, 65-70, PROF EXP MED, McGILL UNIV, 70-; DIR HEMAT DIV, ROYAL VICTORIA HOSP, 68- Concurrent Pos: Am Col Physicians res fel med, Harvard Univ, 56-57, Markle scholar med sci, 57-58; Med Res Coun Can career investr, 60-; asst physician, Royal Victoria Hosp, 63-68, physician, 69- Mem: Am Fedn Clin Res; Am Physiol Soc; Am Soc Clin Invest; fel Am Col Physicians; Can Soc Clin Invest. Res: Transport of vitamin B-twelve across the intestine and into other cells, and its interrelationship with folate in human metabolism. Mailing Add: Royal Victoria Hosp Montreal PQ Can

COOPER, BERNARD RICHARD, b Everett, Mass, Apr 15, 36; m 62; c 3. SOLID STATE PHYSICS. Educ: Mass Inst Technol, BS, 57; Univ Calif, Berkeley, PhD(physics), 61. Prof Exp: Res assoc theoret physics, UK Atomic Energy Res Estab, Harwell, 62-63; res fel theoret solid state physics, Harvard Univ, 63-64; physicist, Gen Elec Res Lab, 64-68, Gen Elec Res & Develop Ctr, 68-74; CLAUDE WORTHINGTON BENEDUM PROF PHYSICS, W VA UNIV, 74- Mem: Fel Am Phys Soc. Res: Theory of magnetic properties of solids; physics of rare earth metals and compounds; theory of elecronic and optical properties of solids; theory of electronic properties of surface and interfaces. Mailing Add: Dept of Physics WVa Univ Morgantown WV 26506

COOPER, BILLY HOWARD, b Tyler, Tex, Mar 15, 36; m 62; c 2. MEDICAL MYCOLOGY, MICROBIOLOGY. Educ: NTex State Univ, BA, 61, MA, 64; Tulane Univ, PhD(microbiol), 68. Prof Exp: NSF Univ Sci Develop Prog fac assoc microbiol, Univ Tex, Austin, 68-70; asst prof microbiol, Sch Med, Temple Univ, 70-75; DIR MYCOL, DEPT PATH, BAYLOR UNIV MED CTR, DALLAS, 75- Concurrent Pos: Vis assoc prof biol, Southern Methodist Univ. Mem: Am Soc Microbiol; Med Mycol Soc of the Americas; Int Soc Human & Animal Mycol. Res: Study of serological and biochemical activities of pathogenic dematiaceous fungi; host-parasite interactions in Candida albicans infections; morphogenetic control mechanisms in human pathogenic fungi. Mailing Add: Dept of Path Baylor Univ Med Ctr Dallas TX 75246

COOPER, CARY WAYNE, b Camden, Maine, Sept 1, 39; m 62; c 2. BIOLOGY, PHARMACOLOGY. Educ: Bowdoin Col, AB, 61; Rice Univ, PhD(biol), 65. Prof Exp: From instr to asst prof, 68-73, ASSOC PROF PHARMACOL, SCH MED, UNIV NC, CHAPEL HILL, 73- Concurrent Pos: Nat Inst Dent Res fel pharmacol, Sch Dent Med, Harvard Univ, 65-66; Nat Inst Dent Res fel, Sch Med, Univ NC, Chapel Hill, 66-67, Nat Inst Arthritis & Metab Dis spec fel, 67-69, career develop award, 72-76, res grant, 74-77; Merck Co Found grant fac develop, 69-70. Mem: Endocrine Soc; Am Soc Pharmacol & Exp Therapeut; AAAS. Res: Endocrine physiology and pharmacology, especially calcium and bone metabolism; hormonal control of calcium homeostasis and bone metabolism; parathyroid hormone and thyrocalcitonin; radioimmunoassay; peptide hormones. Mailing Add: Dept of Pharmacol Univ of NC Sch of Med Chapel Hill NC 27514

COOPER, CECIL, b Philadelphia, Pa, Dec 24, 22; m 46; c 2. BIOCHEMISTRY. Educ: George Washington Univ, BS, 49; Univ Pa, PhD(biochem), 54. Prof Exp: Instr physiol chem, Johns Hopkins Univ, 55-56; from asst prof to assoc prof biochem, 56-68, PROF BIOCHEM, CASE WESTERN RESERVE UNIV, 68- Mem: Am Soc Biol Chemists; Am Chem Soc; Brit Biochem Soc. Res: Oxidative phosphorylation; metal-nucleotide complexes; whole body lipid metabolism; effect of ethanol on liver. Mailing Add: Dept of Biochem Case Western Reserve Univ Cleveland OH 44106

COOPER, CHARLES BURLEIGH, b Parkesburg, Pa, Apr 18, 20; m 46; c 3. PHYSICS. Educ: Franklin & Marshall Col, BS, 41; Cornell Univ, MS, 43; Univ Md, PhD(physics), 51. Prof Exp: Asst physics, Cornell Univ, 41-44; tech supvr, Tenn Eastman Corp, Tenn, 44-46; instr physics, Univ Del, 46-49; asst physics, Univ Md, 49-51, asst prof, 51-52; vpres, Tagcraft Corp, Pa, 52-58; assoc prof, 58-65, PROF PHYSICS, UNIV DEL, 65- Mem: Am Phys Soc; Am Asn Physics Teachers. Res: Mass spectroscopy; calutron; surface physics; sputtering; high temperature specific heat measurements; optical properties. Mailing Add: Dept of Physics Univ of Del Newark DE 19711

COOPER, CHARLES DEWEY, b Whittier, NC, Jan 11, 24; m 46; c 3. PHYSICS. Educ: Berry Col, BS, 44; Duke Univ, AM, 48, PhD(physics), 50. Prof Exp: From asst prof to assoc prof, 50-61, PROF PHYSICS, UNIV GA, 61-, DIR ARTS & SCI SELF STUDY, 70- Concurrent Pos: Res fel, Harvard Univ, 54-55. Mem: AAAS; Am Phys Soc. Res: Visible and ultraviolet spectroscopy; atmospheric physics; electron and negative ion phyiscs. Mailing Add: Dept of Physics Univ of Ga Athens GA 30601

COOPER, CHARLES F, b Kenosha, Wis, Sept 26, 24. ECOLOGY. Educ: Univ Minn, BS, 50; Univ Ariz, MS, 57; Duke Univ, PhD(bot), 58. Prof Exp: Forester, Bur Land Mgt, US Dept Interior, 50-55, forester, Watershed Prog, Ariz, 56; asst prof natural resources, Humboldt State Col, 58-60; ecologist, Agr Res Serv, USDA, Idaho, 60-64; lectr, Sch Natural Resources, Univ Mich, 64-65, from assoc prof to prof natural resources ecol, 65-71; PROF BIOL & DIR CTR REGIONAL ENVIRON STUDIES, SAN DIEGO STATE UNIV, 71- Concurrent Pos: Fulbright res fel, Australia, 62-63; hydrologist, USPHS, 64-65; prog dir, Ecosystem Anal, NSF, 69-71; mem US deleg, UNESCO, Man and the Biosphere, Intergovt Coord Coun, 71; chmn US deleg, US-Taiwan Coop Sci Prog Sem Forest Ecol, 72. Mem: Fel AAAS; Soc Am Foresters; Ecol Soc Am; Am Geophys Union; Soc Range Mgt. Res: Ecological hydrology; systems ecology; ecology and public policy. Mailing Add: Ctr for Regional Environ Studies San Diego State Univ San Diego CA 92182

COOPER, CLEE S, b Willow Creek, Mont, July 11, 22; m 47; c 2. AGRONOMY, PLANT PHYSIOLOGY. Educ: Mont State Col, BS, 48, MS, 50; Ore State Univ, PhD(agron), 64. Prof Exp: AGRONOMIST FORAGE CROPS, AGR RES SERV, US DEPT AGR, 51- Mem: Am Soc Agron. Res: Irrigated pasture establishment and management; forage crop physiology; pasture ecology. Mailing Add: 621 Johnson Hall Mont State Univ Bozeman MT 59715

COOPER, DAVID CLAYTON, aquatic ecology, see 12th edition

COOPER, DAVID JOHN, b Slough, Eng, May 19, 40; m 61; c 3. ORGANIC CHEMISTRY, BIO-ORGANIC CHEMISTRY. Educ: Univ London, BSc, 62, DIC & PhD(org chem), 65. Prof Exp: Res assoc pharmacol, Col Med, Baylor Univ, 65-66; scientist, Schering Corp, 66-68, sr scientist, 68-70, prin scientist, 70; SR MED CHEMIST, SMITH KLINE & FRENCH LABS, 70- Mem: Am Chem Soc; assoc Royal Inst Chem; fel The Chem Soc. Res: Chemistry of the aminoglycoside antibiotics and their semisynthetic derivatives; chemistry of antibiotics. Mailing Add: Smith Kline & French Labs 709 Swedeland Rd Swedeland PA 19479

COOPER, DAVID YOUNG, b Henderson, NC, Aug 14, 24; m 55; c 2. MEDICINE. Educ: Univ Pa, MD, 48. Prof Exp: Intern, 48-49, asst instr pharmacol, 49-50, asst resident, 50, asst instr surg, 53-57, assoc, 57-59, from asst prof to assoc prof surg res, 59-68, PROF SURG RES, HOSP UNIV PA, 68- Concurrent Pos: Fel surg, Harrison Dept Surg Res, Univ Pa, 53-56, Kirby fel, 56-57, Finley fel, Col Surgeons, 57-60; estab investr, Am Heart Asn, 60- Mem: Am Physiol Soc; Endocrine Soc; Am Fedn Clin Res; Am Soc Biol Chem. Res: Adrenal physiology; relation of adrenal to hypertension; pulmonary physiology; oxygenases. Mailing Add: 1507 Ravdin Bldg Hosp Univ of Pa 34th & Spruce St Philadelphia PA 19104

COOPER, DONALD RUSSELL, b Kalamazoo, Mich, Sept 8, 17; m 42; c 2. SURGERY. Educ: Univ Mich, AB, 39, MD, 42. Prof Exp: Intern, Univ Mich Hosp, 42-43, resident surg & instr, 46-50; assoc, 50-51, from asst prof to assoc prof, 51-59, PROF SURG, MED COL PA, 59- Concurrent Pos: Asst instr, Grad Sch Med, Univ Pa, 50-51, assoc, 51-52, from asst prof to assoc prof, 52-65; consult, US Navy Hosp, 55-; chief surgeon, Philadelphia Gen Hosp, 56-65; consult surgeon, Vet Admin Hosp, 57- Mem: Am Col Surg; AMA; Pan-Pac Surg Soc; Int Soc Surg; Soc Surg Alimentary Tract. Res: Gastrointestinal surgery; surgical research in gastrointestinal and vascular fields. Mailing Add: Dept of Surg Med Col Pa 3300 Henry Ave Philadelphia PA 19129

COOPER, DOUGLAS ELHOFF, b New Boston, Ohio, May 21, 12; m 59. ORGANIC CHEMISTRY. Educ: Eastern Ky Univ, BS, 39; Univ Tenn, MS, 40; Purdue Univ, PhD(org chem), 43. Prof Exp: Asst chem, Univ Tenn, 39-40; asst, Purdue Univ, 40-41; res chemist, Bristol Labs, 43-52; res chemist, Res Labs, Ethyl Corp, Detroit, 52-66, res assoc, 66-69 & mkt anal, 69-74; RES ASSOC TOXICOL SUPPORT, ETHYL CORP, BATON ROUGE, 74- Mem: Am Chem Soc. Res: Organic synthesis; antimalarial synthesis; penicillin isolation, purification and production; dosage forms for penicillin; derivatives of 4-chloroquinoline; chemical aspects of internal combustion engine problems; radioisotopic tracer applications; steroids; chemical and engineering support of toxicology and medical activities. Mailing Add: Ethyl Corp 451 Florida Baton Rouge LA 70801

COOPER, EDWIN LAVERN, b Utica, Mich, Aug 31, 19; m 41; c 2. FISH BIOLOGY. Educ: Univ Mich, BS, 40, MS, 47, PhD(zool), 49. Prof Exp: Res assoc, State Dept Conserv, Mich, 49-52; chief fishery biologist, Wis, 52-56; assoc prof zool, 56-62, PROF ZOOL, PA STATE UNIV, 62- Mem: AAAS; Am Inst Fishery Res Biologists (pres, 69-70); Am Fisheries Soc (pres, 71); Am Soc Ichthyologists & Herpetologists; Am Soc Limnol & Oceanog. Res: Ecology and zoogeography of fishes. Mailing Add: 315 Life Sci Bldg Pa State Univ University Park PA 16802

COOPER, EDWIN LOWELL, b Oakland, Tex, Dec 23, 36; m 69. IMMUNOLOGY, BIOLOGY. Educ: Tex Southern Univ, BS, 57; Atlanta Univ, MS, 59; Brown Univ, PhD(biol), 63. Prof Exp: From asst prof to assoc prof, 64-73, PROF ANAT, UNIV CALIF, LOS ANGELES, 73- Concurrent Pos: Nat Cancer Inst fel, Univ Calif, Los Angeles, 62-64; Guggenheim fel & hon Fulbright scholar, Bact Inst, Karolinska Inst, Sweden, 70-71; vis asst prof, Nat Polytech Inst, Mex, 66; mem, Pan Am Cong Anat; mem corp, Mt Desert Island Biol Lab, Maine; mem adv comt, Nat Res Coun, 72-73, mem comn human resources, 73-74; mem bd sci counr, Nat Inst Dent Res, 74-78. Mem: Fel AAAS; Am Soc Zool; Am Asn Immunol; Soc Invert Path; Am Asn Anat. Res: Transplantation immunology; developmental immuno-biology; cellular immunology. Mailing Add: Dept of Anat Univ of Calif Sch of Med Los Angeles CA 90024

COOPER, ELMER JAMES b Milwaukee, Wis, Mar 31, 20; m 43; c 2. CEREAL CHEMISTRY. Educ: Marquette Univ, BS, 41, MS, 49. Prof Exp: Res chemist, Carnation Milk Co, 49-51 & Milwaukee County Hosp, 51; TECH MGR, CEREAL TECHNOL, UNIVERSAL FOODS CORP, 51- Mem: AAAS; Am Chem Soc; Am Asn Cereal Chemists; Am Soc Bakery Engrs. Res: Fermentation; baking; cereal technology. Mailing Add: 1633 N 57th St Milwaukee WI 53208

COOPER, EMERSON AMENHOTEP, b Panama, Jan 15, 24; US citizen; m 49; c 3. ORGANIC CHEMISTRY. Educ: Oakwood Col, BA, 49; Polytech Inst Brooklyn, MS, 54; Mich State Univ, PhD(org chem), 59. Prof Exp: Instr sci, Oakwood Acad, 48-50; chemist, Metric Chem Co, 50-51; from asst prof to assoc prof, 51-59, PROF CHEM, OAKWOOD COL, 59- Mem: Am Chem Soc. Res: Cyanine dyes; synthesis of ninhydrin analogs. Mailing Add: Dept of Chem Oakwood Col Huntsville AL 35806

COOPER, EUGENE PERRY, b Somerville, Mass, Aug 15, 15; m 42; c 4. PHYSICS. Educ: Mass Inst Technol, BS, 37; Univ Calif, PhD(theoret physics), 42. Prof Exp: Asst prof physics, Univ NC, 41-43; res physicist, Franklin Inst, Philadelphia, 43-45 & US Naval Ord Test Sta, Inyokern, Calif, 45-47; assoc prof physics, Univ Ore, 47-48; res physicist, US Naval Ord Test Sta, Calif, 48-51; assoc sci dir, US Naval Radiol Defense Lab, 52-60, sci dir, 60-70, CONSULT TO TECH DIR, NAVAL UNDERSEA CTR, 70- Concurrent Pos: Instr math & theoret physics, Univ Exten, Univ Calif; consult physicist, Off Sci Res & Develop, 44. Mem: AAAS; fel Am Phys Soc. Res: Theoretical atomic and nuclear physics; interactions of matter and radiation; beta disintegration; nuclear isomerism; hydrodynamics; electromechanical instrument theory; radioactivity; nuclear weapon effects; physical and biological radiation effects. Mailing Add: Naval Undersea Ctr San Diego CA 92132

COOPER, FRANKLIN SEANEY, b Robinson, Ill, Apr 29, 08; m 35; c 2. SPEECH, COMMUNICATION SCIENCE. Educ: Univ Ill, BS, 31; Mass Inst Technol, PhD(physics), 36. Prof Exp: Res engr, Gen Elec Res Labs, 36-39, assoc res dir, 39-55, pres & res dir, 55-75, ASSOC RES DIR, HASKINS LABS, 75- Concurrent Pos: Sci consult, Atomic Energy Comn Group, UN Secretariat, 46-47; consult, Off Secy of Defense, 49-50; mem vis comt, Modern Language Dept, Mass Inst Technol, 49-65; mem adv comt, Res Div, Col Eng, NY Univ, 49-65; adj prof phonetics, Columbia Univ, 55-65; mem bd dirs, Ctr Appl Linguistics, 68-74; chmn, Communicative Sci Interdisciplinary Cluster, President's Biomed Res Panel, 75; adj prof linguistics, Univ Conn, 69-; sr res assoc linguistics, Yale Univ, 70- Honors & Awards: President's Certificate of Merit, 48; Pioneer Award in Speech Communication, Inst Elec & Electronics Eng, 72; Warren Medal, Soc Exp Psychol, 75; Silver Medal in Speech Communication, Acoust Soc Am, 75. Mem: AAAS; Am Phys Soc; fel Acoust Soc Am; fel Inst Elec & Electronics Eng; fel Am Speech & Hearing Asn. Res: Perception and production of speech; voice communications systems; prosthetic aids for the blind. Mailing Add: Haskins Labs 270 Crown St New Haven CT 06510

COOPER, FREDERICK MICHAEL, b New York, NY, Apr 1, 44; m 72; c 1. ELEMENTARY PARTICLE PHYSICS. Educ: City Col New York, BS, 64; Harvard Univ, MA, 65, PhD(physics), 68. Prof Exp: Instr physics, Cornell Univ, 68-70; asst prof, Belfer Grad Sch Sci, Yeshiva Univ, 70-75; STAFF MEM PHYSICS, LOS ALAMOS SCI LAB, 75- Concurrent Pos: Frederick Cottrell Res Corp res grant, 71. Mem: Am Phys Soc. Res: Develop a transport formalism for understanding multiparticle production in high energy collisions and to understand the formation of bound states and their interactions. Mailing Add: Theory Div Los Alamos Sci Lab Los Alamos NM 87545

COOPER, GARRETT, b Watertown, Wis, July 24, 04; m 30; c 3. DERMATOLOGY. Educ: Univ Wis, BA, 32, MA, 33, MD, 35; Am Bd Dermat & Syphil, dipl. Prof Exp: CLIN PROF MED & DERMAT, MED SCH, UNIV WIS-MADISON. Concurrent Pos: Chief consult dermat, Madison Vet Admin Hosp; mem staff, Univ Hosps & Madison Gen Hosp; mem bd dirs, Health Planning Coun & Wis State Bd Health. Mem: Am Acad Dermat; AMA. Res: Histochemistry and lipoids of skin; histochemistry of psoriasis. Mailing Add: 110 E Main St Madison WI 53703

COOPER, GARY PETTUS, b York, Ala, Aug 30, 33; m 61; c 2. NEUROPHYSIOLOGY. Educ: Univ Ala, BA, 56; Tulane Univ, MS, 59, PhD(physiol), 63. Prof Exp: Res physiologist, US Naval Radiol Defense Lab, 63-66; asst prof, 66-72, ASSOC PROF PHYSIOL, COL MED, UNIV CINCINNATI, 72- Mem: AAAS; Am Physiol Soc; Soc Neurosci; Radiation Res Soc. Res: Chemical senses; effects of ionizing radiation on the nervous system; neural regulation of food intake. Mailing Add: Dept of Environ Health Univ of Cincinnati Col of Med Cincinnati OH 45219

COOPER, GEORGE RAYMOND, b Denver, Colo, May 3, 16; m 43; c 1. PLANT ECOLOGY, PLANT PHYSIOLOGY. Educ: Univ Northern Colo, BS, 42; Iowa State Univ, MS, 48, PhD(plant ecol), 50. Prof Exp: Assoc prof, 50-59, PROF BOT, PLANT PHYSIOL & ECOL, UNIV MAINE, 59- Mem: AAAS; Am Soc Plant Physiol: Ecol Soc Am. Res: Remote sensing of the environment by aerial infrared photography; physiological changes in higher plants caused by fungicidal sprays. Mailing Add: Dept of Bot Deering Hall Univ of Maine Orono ME 04473

COOPER, GEORGE S, b Medicine Hat, Alta, Oct 18, 14; m 35; c 2. CROP SCIENCE, SOIL SCIENCE. Educ: Univ Alta, BSc, 49, MSc, 51; Univ Ill, PhD(crops, soils), Univ Ill, 53. Prof Exp: MGR TECH SERV & DEVELOP, CYANAMID OF CAN LTD, 53- Concurrent Pos: Exec mem, Can Comt Pesticide Use in Agr, 65-; mem, Can Assoc Comt Agr & Forestry Aviation, 65-; dir, Fourth Int Agr Aviation Cong, 69; chmn subcomt on pesticides & related compounds, Nat Res Coun Assoc Comt on Sci Criteria for Environ Qual, 72- Mem: Entom Soc Can; Can Agr Chem Asn (exec pres, 68-69 & 71); Agr Inst Can; Can Soc Agron (pres, 75-76); Can Soc Soil Sci. Res: Crop dessication; soil fertility; pesticide use patterns; pollution control measures for water and soils. Mailing Add: Cyanamid of Can Ltd Plaza One 2000 Argentina Rd Mississauga ON Can

COOPER, GEORGE WALLACE, JR, b Warrensburg, Mo, June 7, 36. REPRODUCTIVE BIOLOGY. Educ: Brown Univ, AB, 58; Stanford Univ, PhD(embryol), 64. Prof Exp: Res fel, Dept Anat, Sch Med, Univ Pa, 63-66; res assoc, Clin Res Ctr, Phila Gen Hosp, Pa, 66-67; asst prof anat, Col Physicians & Surgeons, Columbia Univ, 67-73; ASST PROF REPRODUCTIVE BIOL, DEPT OBSTET & GYNEC, MED COL, CORNELL UNIV, 73- Concurrent Pos: Vis lectr, Dept Biol, Swarthmore Col, 66-67; vis scientist, Lab Physiol of Reproduction, Europ Orgn Nuclear Res, Hospital Bicetre, Le Kremlin Bicetre, France, 72-73. Mem: AAAS; Am Asn Anat; Am Soc Nat; Am Soc Zool. Res: Surface properties of mammalian gametes; sperm maturation and transport in mammals. Mailing Add: Dept of Obster & Gynec Cornell Univ Med Col 1300 York Ave New York NY 10021

COOPER, GEORGE WILLIAM, b Mt Vernon, NY, Dec 16, 28; m 56; c 2. PHYSIOLOGY, HEMATOLOGY. Educ: NY Univ, AB, 49, PhD(physiol), 60; Columbia Univ, MA, 51. Prof Exp: Res asst radiobiol, Sloan-Kettering Inst, 57-59; USPHS fel physiol, Columbia Univ, 60-62; res assoc biol, NY Univ, 62-68; lectr, 62, from instr to asst prof, 62-72, ASSOC PROF BIOL, CITY COL NEW YORK, 73- Honors & Awards: Founders Day Award, NY Univ, 60. Mem: AAAS; Am Soc Hemat; Am Soc Physiol; Soc Exp Biol & Med; Harvey Soc. Res: Endocrine regulation of blood cell formation and release; effects of ionizing radiations on hematopoietic system and fetal development; action of hormones on enzyme systems; blood coagulation. Mailing Add: Dept of Biol City Col of New York New York NY 10031

COOPER, GERALD PAUL, b Alma, Mich, June 26, 10; m 35; c 3. FISH BIOLOGY. Educ: Mich State Norm Col, BS, 31; Univ Mich, AM, 32, PhD(fisheries), 38. Prof Exp: From instr to asst prof zool, Univ Maine, 36-44; from assoc fisheries biologist to fisheries biologist, Mich State Dept Conserv, 45-55, dir, 56-65, SUPVR FISHERIES RES, MICH DEPT NATURAL RESOURCES, INST FISHERIES RES, 65- Concurrent Pos: Dir fish serv, Maine Dept Inland Fish & Game, 36-44; ichthyol ed, Copeia, 47-49, ed-in-chief, 50-55. Mem: AAAS; Am Fisheries Soc; Am Soc Ichthyol & Herpet; Am Soc Limnol & Oceanog; Wildlife Soc. Res: Lake and stream biological surveys; fish life histories; sport fish management. Mailing Add: Inst for Fisheries Res Univ Mus Annex Ann Arbor MI 48104

COOPER, GERALD RICE, b Scranton, SC, Nov 19, 14; m 46; c 3. PHYSICAL CHEMISTRY, MEDICINE. Educ: Duke Univ, AB, 36, AM, 38, PhD(phys chem), 39, MD, 50. Prof Exp: Res assoc, Med Sch, Duke Univ, 39-50; intern & resident, US Vet Admin Hosp, Atlanta, 50-52; chief phys chem lab, 52-54, chief hemat & biochem sect, 54-63, chief med lab sect, 63-70, chief clin chem & hemat br, 70-72, CHIEF, METAB BIOCHEM BR, CTR DIS CONTROL, USPHS, 72- Concurrent Pos: Fel, Med Sch, Duke Univ, 39-50. Honors & Awards: Hektoen Silver Medal, AMA, 54, Billings Silver Medal, 56; Commendation Medal, USPHS, 64; Am Asn Clin Chem Fisher Award, 75. Mem: AAAS; Am Asn Clin Chem; Am Soc Clin Path; Am Chem Soc; Soc Exp Biol & Med. Res: Electrophoresis; ultracentrifuge; diffusion; viscosity and electron microscopy of proteins and viruses; liver diseases and protein metabolism; lipids; standardization. Mailing Add: Metab Biochem Br Ctr for Dis Control Atlanta GA 30333

COOPER, GILBERT E, JR, nuclear physics, see 12th edition

COOPER, GLENN ADAIR, JR, b Chicago, Ill, Aug 17, 31; m 56; c 3. WOOD SCIENCE, WOOD TECHNOLOGY. Educ: Iowa State Univ, BS, 53, MS, 59; Univ Minn, PhD, 70. Prof Exp: Prin wood scientist proj leader, Forestry Sci Lab, Southern Ill Univ, Carbondale, 59-74, STAFF RES FOREST PRODS TECHNOLOGIST, US FOREST SERV, WASHINGTON, DC, 74- Concurrent Pos: Adj asst prof, Southern Ill Univ. Mem: Forest Prod Res Soc; Soc Wood Sci & Technol; Soc Am Foresters. Res: Wood-moisture relations; physical properties of wood; treatments and processes which alter wood properties; utilization of hardwoods; wood residues for energy. Mailing Add: Forest Prods & Eng Res US Dept Agr 14th & Independence Washington DC 20250

COOPER, GLENN DALE, b Hoxie, Ark, May 4, 19; m 49; c 2. ORGANIC CHEMISTRY. Educ: Ark State Col, BS, 40; Purdue Univ, PhD(chem), 49. Prof Exp: Res assoc, Northwestern Univ, 49-51; res assoc res lab, Gen Elec Co, 51-62; res prof chem, NMex State Univ, 62-67; SR SCIENTIST, CHEM DEVELOP OPER, GEN ELEC CO, 67- Mem: Am Chem Soc. Res: Organosilicon compounds; oxidative coupling; polymer blends. Mailing Add: Chem Develop Opers Gen Elec Co Selkirk NY 12158

COOPER, GUSTAV ARTHUR, b College Point, NY, Feb 9, 02; m 30; c 2. PALEOBIOLOGY. Educ: Colgate Univ, BS, 24, MS, 26, DSc(geol), 53; Yale Univ, PhD(geol), 29. Prof Exp: Asst, Peabody Mus, Yale Univ, 28-29, res assoc, 29-30; from asst cur to cur invert fossils, US Nat Mus, 30-56, head cur dept geol, 56-63, chmn dept paleobiol, 63-67, sr paleobiologist, 67-72, EMER PALEOBIOLOGIST, US NAT MUS, SMITHSONIAN INST, 72- Mem: Fel Geol Soc Am; Paleont Soc (pres, 56-57). Res: Stratigraphy; paleontology; stratigraphy of the Hamilton group of New York; invertebrate fossils; investigations on modern and fossil Brachiopoda. Mailing Add: US Nat Mus Smithsonian Inst Washington DC 20560

COOPER, HAROLD EUGENE, b Warren, Ark, July 14, 28; m 44; c 3. GENETICS, PHYSICAL ANTHROPOLOGY. Educ: State Col Ark, BSE, 59, MSE, 61; Univ Minn, PhD(genetics), 65. Prof Exp: Chmn dept sci, 59-70, ACAD DEAN, CENT BAPTIST COL, 70-; ASSOC PROF BIOL, STATE COL ARK, 65- Mem: AAAS; Am Soc Human Genetics; Am Sci Affiliation. Res: Hereditary factors in the Stein-Leventhal syndrome; sex chromatin in the premature. Mailing Add: Dept of Sci Cent Baptist Col Conway AR 72032

COOPER, HERBERT ASEL, b Grand Junction, Colo, Feb 21, 38; m 63; c 2. EXPERIMENTAL PATHOLOGY, PEDIATRICS. Educ: Univ Kans, BA, 60, MD, 64. Prof Exp: Intern, Charles T Miller Hosp, St Paul, Minn, 64-65; resident pediat, Mayo Grad Sch Med, 65-67, assoc consult, Mayo Clin, 67, resident pediat hemat, Mayo Grad Sch Med, 69-71; fel exp path, 71-74, ASST PROF PATH & PEDIAT, SCH MED, UNIV NC, CHAPEL HILL, 74- Concurrent Pos: NIH res career develop award, 75; mem coun thrombosis, Am Heart Asn. Mem: Am Soc Hemat; Soc Exp Biol & Med; Am Soc Exp Pathologists; Int Soc Hemostasis & Thrombosis. Res: Hemostasis and thrombosis; the biochemistry of factor VIII and von Willebrand factor; interaction of platelets with bovine platelet aggregation factor and other macromolecules. Mailing Add: Dept of Path Univ of NC Sch of Med Chapel Hill NC 27514

COOPER, HOWARD GORDON, b Joliet, Ill, Feb 16, 27; m 53; c 2. RESEARCH ADMINISTRATION, ELECTROOPTICS. Educ: Univ Ill, BS, 49, MS, 50, PhD(physics), 54. Prof Exp: Mem tech staff, Bell Tel Labs, Inc, 54-70; DIR RES, RECOGNITION EQUIP INC, 70- Mem: Am Phys Soc; Optical Soc Am; Inst Elec & Electronics Eng; Pattern Recognition Soc. Res: Technology forecasting; optical scanning; pattern recognition; image processing and information processing systems. Mailing Add: Recognition Equip Inc, PO Box 22307 Dallas TX 75222

COOPER, HUBERT B, JR, genetics, plant breeding, see 12th edition

COOPER, IRVING S, b Atlantic City, NJ, July 15, 22; m 44; c 3. NEUROSURGERY. Educ: George Washington Univ, AB, 42, MD, 45; Univ Minn, MS & PhD, 51; Am Bd Psychiat & Neurol, dipl, 51; Am Bd Neurol Surg, dipl, 53. Prof Exp: Asst prof neurosurg, Postgrad Med Sch, NY Univ, 51-57, dir neuromuscular dis, 57-66, prof res surg, Bellevue Med Ctr, 57-66; DIR DEPT NEUROSURG, ST BARNABAS HOSP, 54-; RES PROF NEUROANAT, NY MED COL, 66- Concurrent Pos: Eliza Savage fel, Australia; assoc attend neurosurgeon, NY Univ Hosp, 51-66; assoc vis neurosurgeon & consult neurosurgeon, Inst Rehab, Bellevue Hosp, 51-; asst attend neurosurgeon, Hosp, Spec Surg & St Joseph's Hops, 51-; foreign acad corresp, Royal Acad Med, Madrid, 63. Honors & Awards: Hektoen Bronze Medal, AMA, 57 & 58, Cert of Merit, 61; Taylor Award, Am Therapeut Soc, 57; Award, St Barnabas Hosp, 59; Mod Med Award, 60; Alumni Award, George Washington Univ, 60, Merit Award, 67; Nat Cystic Fibrosis Found Award, 62; Merit Award, Chicago Nat Parkinson Found, 62; Outstanding Achievement Award, Univ Minn, 64; Merit Award, United Parkinson Found, 65; Henderson Lect Award, 67; Gold Medal, Worshipful Soc Apothecaries London, 67; Bronze Award, Am Cong Rehab Med, 67; Medal, Comenius Univ, 71. Mem: Am Asn Neurol Surg; AMA; Neurosurg Soc Am; Am Acad Neurol; fel Am Col Surg. Res: Neurosurgery in Parkinsonism; development of cryogenic surgery in humans. Mailing Add: St Barnabas Hosp E 183rd St & Third Ave New York NY 10457

COOPER, JACK LORING, b Steubenville, Ohio, Oct 26, 25; m 46; c 4. POLYMER CHEMISTRY. Educ: Univ Akron, BS, 50. Prof Exp: Res chemist, Gen Tire & Rubber Co, 50-56, group leader, Pilot Plant Opers, 56-62, sect head aqueous polymerization, 62-66, resident mgr res, Chem Pilot Plant, 66-68, MGR CHEM PILOT PLANT, 68- Mem: Am Chem Soc. Res: Rubber and polymer chemistry; pilot plant operations and administration; rubber and plastics development. Mailing Add: Chem Pilot Plant Gen Tire & Rubber Co Mogadore OH 44260

COOPER, JACK ROSS, b Ottawa, Can, July 26, 24; nat US; m 51; c 3. PHARMACOLOGY. Educ: Queen's Univ, Ont, BA, 48; George Washington Univ, MA, 52, PhD(biochem), 54. Prof Exp: Asst, NY Univ Res Serv, Goldwater Mem Hosp, 48-50; from instr to assoc prof, 56-71, PROF PHARMACOL, YALE UNIV, 71- Concurrent Pos: Fel, Pub Health Res Inst, NY, 54-56; Wellcome travel grant, Eng, 59; NIH spec fel & vis scientist, Maudsley Hosp, London, 65-66. Mem: Am Soc Pharmacol & Exp Therapeut; Int Soc Neurochem; Am Soc Neurochem; Soc Neurosci. Res: Thiamine; acetylcholine; neurochemistry; neuropharmacology. Mailing Add: Dept of Pharmacol Yale Univ Sch of Med New Haven CT 06510

COOPER, JAMES ERWIN, b Waxahachie, Tex, Aug 30, 33; m 54; c 2. ORGANIC GEOCHEMISTRY. Educ: NTex State Univ, BS, 54, MS, 55; Rice Univ, PhD(chem), 59. Prof Exp: Chemist, Core Labs, Inc, 55-56; sr res technologist, Field Res Labs, Socony Mobil Oil Co, Inc, 59-68, res assoc explor-prod res div, Mobil Res & Develop Corp, 68-73; ADJ PROF GEOL, UNIV TEX, ARLINGTON, 74- Concurrent Pos: Lectr, Dallas Baptist Col, 69, 74-75. Mem: Am Chem Soc; Org Geochem Soc; Sigma Xi. Res: Organic reaction mechanisms; organic geochemistry and synthesis. Mailing Add: 2423 Bonnywood Lane Dallas TX 75233

COOPER, JAMES WILLIAM, b Buffalo, NY, Feb 7, 43; m 69; c 2. CHEMICAL INSTRUMENTATION, ORGANOMETALLIC CHEMISTRY. Educ: Oberlin Col, AB, 64; Ohio State Univ, MS, 67, PhD(chem), 69. Prof Exp: Instr chem, State Univ NY Buffalo, 69-70; Nmr appln programmer, Digital Equip Co, 70-71; anal appln mgr, Nicolet Instrument Corp, 71-74; ASST PROF CHEM, TUFTS UNIV, 74- Concurrent Pos: Consult, Bruker Instruments, Inc, 74- Mem: Am Chem Soc. Res: Fourier transform Nmr; computers in chemistry; organometallic chemistry; non-benzenoid aromatic and pseudoaromatic systems. Mailing Add: Dept of Chem Tufts Univ Medford MA 02155

COOPER, JANE ELIZABETH, b Bethlehem, Pa, June 2, 37. GENETICS. Educ: Lindenwood Col, BA, 59; Univ Pa, PhD(zool), 65. Prof Exp: From instr to asst prof biol, Drexel Inst, 64-67; asst prof, 67-73, ASSOC PROF BIOL, PA STATE UNIV, 73- Mem: AAAS; Genetics Soc Am; Asn Adv Health Professions; Sigma Xi. Res: Swimming rates in Paramecium aurelia; control of protein system in Paramecium aurelia; host-endosymbiont interactions in Paramecium biaurelia. Mailing Add: Dept of Biol Pa State Univ Media PA 19063

COOPER, JOHN ALLEN DICKS, b El Paso, Tex, Dec 22, 18; m 44; c 4.

BIOCHEMISTRY. Educ: NMex State Univ, BS, 39; Northwestern Univ, PhD(biochem), 43, MB, 50, MD, 51. Hon Degrees: Dr, Univ Brazil, 58; DSc, Northwestern Univ, 72, Duke Univ, 73 & Med Col Ohio, 74; DMedSc, Med Col Pa, 73. Prof Exp: From instr to prof biochem, Med Sch, Northwestern Univ, 43-69; PRES, ASN AM MED COLS, 69- Concurrent Pos: Markle scholar acad med, 51-56; intern, Passavant Mem Hosp, Chicago, Ill, 50-51; consult, Vet Admin Res Hosp, Chicago, 54-69, dir radioisotope serv, 54-65; mem grad sch fac, Northwestern Univ, 55-69, assoc dean med sch, 59-63, dir integrated prog med educ, 60-68, dean sci & assoc dean fac, 63-69, mem pres admin coun, 63-69, chmn res comt, 64-69, mem patents & inventions comt, 65-69, chmn int progs comt, 66-69; vis prof, Univ Brazil, 56 & Univ Buenos Aires, 58; mem comt licensure, US AEC, 56-68, mem adv comt educ & training, Div Biol & Med, 57-63; mem policy adv bd, Argonne Nat Lab, 57-63, chmn rev comt, Div Biol & Med Res & Radiol Physics, 58-62; chmn comt sci exhibs, Second World Conf Med Educ, 58-59; mem med adv comt, PR Nuclear Ctr, 59-60; mem bd pub health adv, State of Ill, 62-69, mem, Ill Legis Comn on Atomic Energy, 63-69, mem bd higher educ, 64-69, mem, Gov Sci Adv Coun, 65-69, chmn, Sci Adv Coun of Ill, 67-69; ed, J Med Educ, 62-71; mem adv comt on investigational drugs, Food & Drug Admin, 63-65, consult to adminr, 65-70; mem coun, Assoc Midwest Univs, 63-68, vpres bd dirs, 64-65, pres, 65-66; mem med adv comt, W K Kellogg Found, 63-68, mem Latin Am adv comt, 64-68; treas admin comt, Pan Am Fedn Asns Med Schs, 63-; mem adv comt on health sci, Eng & Biotechnol & Int Fel Rev Panel, NIH, 64, adv coun health res facil, 65-69, spec consult to dir, 68-70, consult to div physician & health professions educ, Bur Health Manpower Educ, 70-73; US specialist, Brazil, 65; chmn extramural educ surv comt, Mayo Found, 65-66; vpres, Argonne Univs Asn, 65-68, mem bd trustees, 65-69; adv to adminr int health manpower, Agency Int Develop, 66-71; chmn adv comt on study of training progs in gen med sci, Nat Acad Sci -Nat Res Coun, 66-71, mem adv comt off sci personnel, 67-70; mem adv comt for instnl rels, NSF, 67-71; mem bd dirs, Evanston Hosp, 68-69; mem US health Planning del to USSR, 70; prof lectr, Georgetown Univ, 70-; mem, Inst Med, Nat Acad Sci, 72-; prof, Inst Policy Sci & Pub Health, Duke Univ, 73-; mem bd dir, Nat Med Fels Inc, 73-; treas, World Fedn Med Educ, 74. Mem: AAAS; AMA; Am Pub Health Asn; Am Soc Biol Chem; Asn Am Med Cols. Res: Educational administration; radiobiology. Mailing Add: Asn of Am Med Col One DuPont Circle Washington DC 20036

COOPER, JOHN ARTHUR, nuclear chemistry, see 12th edition

COOPER, JOHN C, JR, b Fullerton, Calif, Jan 16, 36; m 58; c 2. AUDIOLOGY. Educ: Auburn Univ, BS, 57; Wayne State Univ, MA, 65, PhD(audiol), 68. Prof Exp: Res asst audiol, Wayne State Univ, 65-67; asst prof, Vanderbilt Univ, 67-69; ASSOC PROF AUDIOL, UNIV TEX MED SCH, SAN ANTONIO, 69- Concurrent Pos: Assoc prof, Trinity Univ & Univ Tex, Austin, 70- Mem: Am Speech & Hearing Asn; Acad Rehab Audiol. Res: Verbal learning in normal, hard of hearing and deaf children; discrimination of speech through hearing aids and in noise; hearing screening. Mailing Add: 123 Tall Oak San Antonio TX 78232

COOPER, JOHN (HANWELL), b Tynemouth, Eng, Mar 15, 22; Can citizen; m 53; c 2. HUMAN PATHOLOGY, HISTOCHEMISTRY. Educ: Glasgow Univ, MB & ChB, 45; FRCPath, 69. Prof Exp: Registr path, Victoria Infirmary, Scotland, 49-56; assoc, Gen Hosp, St John's Nfld, 56-57; dir, Glace Bay Hosps, NS, 57-62; assoc, Path Inst, Dept Pub Health, Prov NS, 62-63, dir anat path, 63-65; assoc prof, 62-66, PROF PATH, DALHOUSIE UNIV, 66-; MED DIR, PATH INST, DEPT PUB HEALTH, PROV NS, 65- Mem: AAAS; Histochem Soc; Int Acad Path; Can Med Asn; Can Asn Path. Res: Connective tissue histochemistry and pathology, especially elastic sheath-elastofibril system, amyloid, splenic follicular hyaline, actinic elastosis, aldehyde fuchsin and scleroderma. Mailing Add: Dept of Path Univ Dalhousie Halifax NS Can

COOPER, JOHN (JINX), b Norwich, Eng, Nov 30, 37; m 62; c 3. ATOMIC PHYSICS, PLASMA PHYSICS. Educ: Cambridge Univ, BA, 59, MA, 63; Imp Col, London, dipl, 61, PhD(physics), 62. Prof Exp: Asst lectr physics, Imp Col, London, 62-63, lectr, 63-65; from asst prof to assoc prof, 65-70, mem, 65-67, PROF PHYSICS & ASTROPHYS, UNIV COLO, BOULDER, 70-, FEL, JOINT INST LAB ASTROPHYS, 67- Concurrent Pos: Consult, Atomic Energy Auth, UK, 63-65 & Radio Stands Lab, Nat Bur Stands, Colo, 68- Mem: Fel Am Inst Physics. Res: Experimental and theoretical interests in the radiation from hot plasmas of laboratory and astrophysical importance; line broadening and other aspects of plasma spectroscopy; optical diagnostic techniques. Mailing Add: Joint Inst for Lab Astrophys Univ of Colo Boulder CO 80302

COOPER, JOHN NEALE, b San Antonio, Tex, May 25, 38; m 60; c 1. PHYSICAL CHEMISTRY, INORGANIC CHEMISTRY. Educ: Calif Inst Technol, BS, 60; Univ Calif, Berkeley, PhD(chem), 64. Prof Exp: Lectr inorg chem, Makerere Univ Col, Uganda, 64-66; asst prof chem, Carleton Col, 66-67; asst prof, 67-74, ASSOC PROF CHEM, BUCKNELL UNIV, 74- Concurrent Pos: Vis asst prof, Univ Ill, Urbana, 71-72. Mem: Am Chem Soc. Res: Kinetics and mechanisms of inorganic reactions. Mailing Add: Dept of Chem Bucknell Univ Lewisburg PA 17837

COOPER, JOHN NIESSINK, b Kalamazoo, Mich, Feb 4, 14; m 36. PHYSICS. Educ: Kalamazoo Col, AB, 35; Cornell Univ, PhD(physics), 40. Prof Exp: Asst physics, Cornell Univ, 35-40; instr, Univ Southern Calif, 40-43; asst prof, Univ Okla, 43-46; from asst prof to prof, Ohio State Univ, 46-56; PROF PHYSICS, NAVAL POSTGRAD SCH, 56- Concurrent Pos: Res physicist, Radiation Lab, Univ Calif, 44-45; staff mem, Sandia Corp, 51-54; consult, Ramo-Woolridge Corp, 55-58, Space Tech Labs, 58-66 & Kaman Nuclear, 65. Mem: Fel Am Phys Soc; Am Asn Physics Teachers; sr mem Inst Elec & Electronics Eng. Res: X-rays; nuclear spectroscopy; stopping of protons; superconductivity. Mailing Add: Dept of Physics Naval Postgrad Sch Monterey CA 93940

COOPER, JOHN R, organic chemistry, polymer chemistry, see 12th edition

COOPER, JOHN RAYMOND, b Lafayette, ALA, Jan 2, 31; m 52; c 2. PHYSICS. Educ: Auburn Univ, BS, 52, PhD, 70; Ohio State Univ, MS, 55. Prof Exp: Sr physicist, Aircraft Nuclear Propulsion Dept, Gen Elec Co, 58-60; res physicist, Southern Res Inst, 60-65, sr physicist, 65-66; mem staff dept physics, 66-71, DIR NUCLEAR SCI CTR & ASST PROF PHYSICS, AUBURN UNIV, 71- Mem: Am Asn Physics Teachers. Res: Neutron scattering; scattering; neutron induced reactions; nuclear instrumentation; reactor physics; electro optics. Mailing Add: Dept of Physics Auburn Univ Auburn AL 36830

COOPER, JOHN WESLEY, b Delta, Colo, Sept 7, 46; m 73. EXPERIMENTAL HIGH ENERGY PHYSICS. Educ: Univ Colo, Boulder, BA, 68; Univ Mich, Ann Arbor, MA, 69, PhD(physics), 75. Prof Exp: Res asst, Bubble Chamber Group, Univ Mich, 71-75; RES ASSOC EXP HIGH ENERGY PHYSICS, UNIV ILL, 75- Mem: Sigma Xi; Am Phys Soc. Res: Experimental elementary particle research using the Fermi National Accelerator Lab 30 inch bubble chamber hybrid spectrometer and the Fermilab Chicago Cyclotron Magnet Spectrometer. Mailing Add: Dept of Physics Univ of Ill Urbana IL 61801

COOPER, KEITH EDWARD, b Frome, Eng, Aug 7, 22; m 46; c 2. PHYSIOLOGY. Educ: Univ London, MB, BS, 45, BSc, 48, MSc, 50; Oxford Univ, MA, 60, DSc(physiol), 70. Prof Exp: Resident, St Mary's Hosp, Univ London, 45-46, lectr physiol, 46-48; mem div human physiol, Med Res Coun Labs, Eng, 50-54; from founding mem to dir, Med Res Coun Body Temperature Res Unit, London & Oxford, 54-69; PROF MED PHYSIOL & HEAD DIV, FAC MED, UNIV CALGARY, 69- Concurrent Pos: Vis lectr, Vat Admin Hosp, Cincinnati, Ohio, 66; vis lectr clin invest, Yale Univ, 66; vis lectr biophys, Univ Western Ont & Rutgers Univ, 66; consult, Cowley Rd Hosp, Oxford, Eng, 66; examr, Oxford Univ, Univ WI & Queen's Univ, Belfast, 66; mem panel arctic med & climatic physiol, Can Defense Res Bd, 70- Mem: Brit Physiol Soc; Can Physiol Soc (vpres, 74, pres, 75-76); Am Physiol Soc; Brit Med Res Soc. Res: Body temperature regulation; mechanism of fever. Mailing Add: Fac of Med Univ of Calgary Calgary AB Can

COOPER, KENNETH WILLARD, b Flushing, NY, Nov 29, 12; m 37, c 2. CYTOGENETICS, ENTOMOLOGY. Educ: Columbia Univ, AB, 34, AM, 35, PhD(cytol), 39. Hon Degrees: MA, Dartmouth Col, 62. Prof Exp: Asst zool, Columbia Univ, 34-37; instr, Univ Rochester, 38-39; from instr to assoc prof biol, Princeton Univ, 39-53; prof & chmn dept, Univ Rochester, 53-57; grad res prof, Univ Fla, 57-59; prof cytol, Dartmouth Med Sch, 59-62, cytol & genetics, 62-67; PROF BIOL, UNIV CALIF, RIVERSIDE, 67- Concurrent Pos: Guggenheim fel, Calif Inst Technol, 44, 45; vis lectr, Univ Colo, 50; mem vis comt, Brookhaven Nat Lab, 58-61; mem gen training panel, NIH, 58-66, health res facilities, 66-69; mem FCP Coun, Smithsonian Inst, 72-76; consult, Energy Res & Develop Agency, 76- Mem: Fel AAAS; fel Am Acad Arts & Sci; fel Royal Entom Soc; Genetics Soc Am; Soc Study Evolution. Res: Experimental ecology; insect communication; biology and systematics of moss insects, beetles, wasps; fossil insects; chromosome structure and segregation; interchromosomal effects; non-gametic functions of gametes; mechanisms of mitosis and meiosis; evolution. Mailing Add: Dept of Biol Univ of Calif Riverside CA 92502

COOPER, LARRY RUSSELL, b Los Angeles, Calif, Sept 19, 34; m 65; c 3. NUCLEAR PHYSICS. Educ: Univ Calif, Los Angeles, BS, 56; Ore State Univ, PhD(physics), 67. Prof Exp: Jr scientist physics, Edwards AFB, US Air Force, 56-57; res assoc, Ore State Univ, 63-67; Nat Acad Sci-Nat Res Coun res assoc, Cyclotron Lab, Naval Res Lab, 67-69; physicist, Nuclear Physics Br, 69-74, PHYSICIST, ELECTRONICS & SOLID STATE SCI PROG, OFF NAVAL RES, 74- Mem: Am Phys Soc. Res: Electronics and solid state sciences; physics of electron and electrooptic devices; solid state surfaces and interfaces; radiation effects in electronic materials; ion implantation; surface analysis techniques; solid state physics. Mailing Add: Elec & Solid State Sci Prog Off of Naval Res Code 427 Arlington VA 22217

COOPER, LEON N, b New York, NY, Feb 28, 30; m 69; c 2. THEORETICAL PHYSICS. Educ: Columbia Univ, AB, 51, AM, 53, PhD(physics), 54. Hon Degrees: DSc, Columbia Univ, 73, Univ Sussex, 73, Univ Ill, 74, Brown Univ, 74 & Gustavus Adolphus Col, 75. Prof Exp: NSF fel, Inst Advan Study, 54-55; res assoc physics, Univ Ill, 55-57; asst prof, Ohio State Univ, 57-58; from assoc prof to prof physics, 58-66, Henry Ledyard Goddard univ prof, 66-74, THOMAS J WATSON, SR PROF SCI, BROWN UNIV, 74- Concurrent Pos: Vis prof, var univs & schs; consult, var govt agencies, indust & educ orgn; var pub lectr, int confs; Sloan Found res fel, 59-66; Guggenheim fel, 65-66. Honors & Awards: Comstock Prize, Nat Acad Sci, 68; Nobel Prize, 72. Mem: Nat Acad Sci; fel Am Phys Soc; Am Acad Arts & Sci; Am Philos Soc; Fedn Am Scientists. Res: Nuclear, low temperature and elementary particle physics; field theory; superconductivity; many body problems. Mailing Add: Dept of Physics Brown Univ Providence RI 02912

COOPER, LOUIS ZUCKER, b Albany, Ga, Dec 25, 31; m 55; c 4. INFECTIOUS DISEASES, PEDIATRICS. Educ: Yale Univ, BS, 54, MD, 57. Prof Exp: USPHS fel, 61-63; instr med, Sch Med, Tufts Univ, 63-64; from instr to assoc prof pediat, Sch Med, NY Univ, 64-73; PROF PEDIAT, COL PHYSICIANS & SURGEONS, COLUMBIA UNIV, 73-; DIR PEDIAT SERV, ROOSEVELT HOSP, 73- Concurrent Pos: Career scientist, Health Res Coun, New York, 67-73; dir, Rubella Proj; consult, Bur Educ Handicapped, US Off Educ, 68-72; consult, President's Comt Ment Retardation, 70-; mem, NY State Comt Children, 70-74; mem, Nat Adv Coun Develop Disabilities, Rehab Servs Admin. Mem: AAAS; Infectious Dis Soc Am; Am Pub Health Asn; Soc Pediat Res. Res: Rubella; handicapped children; viral immunology and vaccines; chemotherapy of infectious diseases. Mailing Add: Roosevelt Hosp Pediat Serv 428 W 59th St New York NY 10019

COOPER, MARGARET HARDESTY, b St Louis, Mo, June 7, 44. ANATOMY. Educ: Drury Col, AB, 66; St Louis Univ, MS, 69, PhD(anat), 71. Prof Exp: ASST PROF ANAT, SCH MED, WAYNE STATE UNIV, 71- Res: Neuroanatomy. Mailing Add: Dept of Anat Wayne State Univ Sch of Med Detroit MI 48201

COOPER, MARGARET MOORE, b Villisca, Iowa, Feb 17, 09. TEXTILE CHEMISTRY. Educ: Univ Iowa, AB, 30, MS, 32, PhD(org chem), 34. Prof Exp: Chemist, Child Welfare Res Sta, Iowa, 36-37; head dept sci, Fairfax Hall, 37-38; asst prof chem, Meredith Col, 38-41; asst prof textile chem, Pa State Univ, 41-42; asst prof chem, Tex State Col Women, 42-45; from assoc prof to prof home econ, 45-74, EMER PROF HOME ECON, UNIV WIS, MADISON, 74- Mem: AAAS; Am Asn Textile Chemists & Colorists; Am Chem Soc; Am Soc Testing & Mat; Am Home Econ Asn. Res: Colorfastness and chemical structure; surface active agents-structure and effectiveness; textile finishes; identification of dyes on textiles. Mailing Add: 4626 Gregg Rd Madison WI 53705

COOPER, MARTIN JACOB, b Detroit, Mich, June 27, 39; m 65. STATISTICAL MECHANICS, SCIENCE ADMINISTRATION. Educ: Univ Mich, BSE, 61, MS, 63; Brandeis Univ, PhD(physics), 67. Prof Exp: Nat Res Coun res assoc, Nat Bur Standards, 66-68, res assoc statist physics, 68-72, prog analyst, 72-74; Presidential interchange exec, Corp Res & Develop, Gen Elec Co, 74-75; SCI ADMINR, NAT BUR STANDARDS, 75- Mem: AAAS; Am Phys Soc. Res: Many-body physics, theory of cooperative phenomena, phase transitions; science policy. Mailing Add: 9536 Whetstone Dr Gaithersburg MD 20760

COOPER, MATTHEW OWEN, b Brooklyn, NY, June 4, 43; m 71. ANTHROPOLOGY. Educ: Brooklyn Col, BA, 64; Yale Univ, MPhil, 68, PhD, 70. Prof Exp: From instr to asst prof anthrop, Univ NH, 69-71; asst prof, 71-75, ASSOC PROF ANTHROP, McMASTER UNIV, 75- Concurrent Pos: Vis prof anthrop, Univ Bergen, 72. Mem: Fel Am Anthrop Asn; Asn Social Anthrop Oceania; Can Asn Latin Am Studies; Latin Am Studies Asn. Res: Economic and social development; Brazil; Melanesia. Mailing Add: Dept of Anthrop McMaster Univ Hamilton ON Can

COOPER, MAURICE D, b Jackson, Mich, Jan 1, 11; m 37; c 2. INORGANIC CHEMISTRY, ANALYTICAL CHEMISTRY. Educ: Albion Col, AB, 33, MA, 42. Prof Exp: Payroll acct, Sparks-Withington Co, Mich, 33-35; gen acct asst to dept head, Consumers Power Co, 35-42; anal chemist, Buick Motor Div, 42-44 & Res Labs 44-53, from asst head to head chem dept, 53-72, HEAD ANAL CHEM DEPT, GEN MOTORS CORP, 72- Honors & Awards: Lundell-Bright Mem Award, Am Soc Testing & Mat, 67, Award of Merit, 74. Mem: Am Chem Soc; fel Am Soc Testing &

Mat. Res: Instrumental methods of inorganic analytical chemistry. Mailing Add: Gen Motors Corp Res Labs 12 Mile & Mound Rd Warren MI 48090

COOPER, MAURICE ZEALOT, b Brooklyn, NY, Feb 19, 08; m 41; c 1. MEDICINE. Educ: Long Island Col Med, MD, 31. Prof Exp: Intern, Beth-el Hosp, Brooklyn, NY, 31-32; physician, Health Dept, 33-36; physician, Vet Admin, 36-40, chief outpatient & reception serv, Vet Admin Hosp, Togus, Maine, 40-42, chief med officer, Vet Admin Regional Off, RI, 42-46, chief outpatient serv, New Eng Br, 46-49, chief plans & policy develop, Med Serv, Washington, DC, 49-56, dir med criteria ed bd, 56-59, asst dir, Med Serv, 59-62, chief-of-staff, Vet Admin Hosp, Seattle, Wash, 62-68, dir, Vet Admin Hosp, Tex, 68-69, dir, Domiciliary, Los Angeles Ctr, 69-70, DIR, VET ADMIN OUTPATIENT CLIN, 70- Mem: AMA; Am Col Physicians. Res: Residuals and evaluations of medical diseases; hospital and medical system administration, effects of malnutrition and other hardships on the morality and morbidity of former United States prisoners of war and civilian internees of World War II. Mailing Add: PO Box 49771 Los Angeles CA 90049

COOPER, MAX DALE, b Hazlehurst, Miss, Aug 31, 33; m 60; c 4. PEDIATRICS, IMMUNOLOGY. Educ: Tulane Univ, MD, 57; Am Bd Pediat, dipl, 62. Prof Exp: Intern med, Saginaw Gen Hosp, 57-58; resident, Sch Med, Tulane Univ, 58-60; house officer, Hosp Sick Children, London, Eng, 60, res asst neurophysiol, 61; instr pediat, Sch Med, Tulane Univ, 62-63; from instr to asst prof, Univ Minn, Minneapolis, 64-67; assoc prof microbiol, 67-73, PROF PEDIAT, SCH MED, UNIV ALA, BIRMINGHAM, 67-, PROF MICROBIOL, 73- Concurrent Pos: Fel pediat, Univ Calif, San Francisco, 61-62; Nat Tuberc Asn teaching traineeship award, Univ Minn, Minneapolis, 63-64; USPHS spec res fel, 64-66. Honors & Awards: Samuel J Melzer Award, Soc Exp Biol, 66. Mem: AAAS; Soc Pediat Res; Am Soc Exp Path; Am Asn Immunol; Am Soc Clin Invest. Res: Development and function of the lymphoid system; immunologic deficiency diseases; lymphoid malignancies. Mailing Add: Dept of Pediat Univ of Ala Med Ctr Birmingham AL 35294

COOPER, MILES ROBERT, b Elizabeth City, NC, Oct 21, 33; m 55; c 2. ONCOLOGY, HEMATOLOGY. Educ: Univ NC, Raleigh, BS, 55; Bowman Gray Sch Med, MD, 62. Prof Exp: From instr to assoc prof med, 67-75, PROF MED, BOWMAN GRAY SCH MED, 75- Mem: Am Col Physicians; Am Soc Hemat; Am Fedn Clin Res; Soc Clin Oncol. Res: Laboratory and clinical studies of the pathophysiology of leukocytes and platelets; evaluation of supportive therapies consisting of leukocyte and platelet transfusions and definitive therapy with various therapeutic programs in neoplastic disease. Mailing Add: Bowman Gray Sch of Med 300 S Hawthorne Rd Winston-Salem NC 27103

COOPER, MILTON, organic chemistry, see 12th edition

COOPER, MURRAY IRVING, b New York, NY, Aug 8, 19; m 43; c 3. ENTOMOLOGY. Educ: Cornell Univ, BS, 47; Univ Ill, MS, 49, PhD(entom), 51. Prof Exp: Asst, Univ Ill, 47-51; entomologist, Union Starch & Ref Co, 51; sanit consult, Orkin Inst Indust Sanit, 52-53; DIR, PHILADELPHIA OFF, INSECT CONTROL & RES, INC, 53- Mem: Entom Soc Am; Inst Food Technol. Res: Insect ecology and nutrition; control of stored products and household insects. Mailing Add: Pa Insect Control & Res Inc 2641 Mt Carmel Ave Glenside PA 19038

COOPER, MURRAY SAM, b Poland, Jan 14, 23; nat US; m 48; c 3. MICROBIOLOGY. Educ: City Col New York, BS, 47; Univ Ill, MS, 48, PhD(bact), 50. Prof Exp: With biol develop, 50-68, MGR BIOL QUALITY CONTROL, LEDERLE LABS DIV, AM CYANAMID CO, 68- Mem: Am Soc Microbiol; Am Acad Microbiol. Res: Development of bacterial and virus vaccines; immune response; clostridial anaerobes. Mailing Add: Lederle Labs Div Am Cyanamid Co Pearl River NY 10965

COOPER, NORMAN S, b Brooklyn, NY, Dec 23, 20; m 45; c 1. PATHOLOGY. Educ: Columbia Univ, AB, 40; Univ Rochester, MD, 43; Am Bd Path, dipl, 52. Prof Exp: Asst path, Med Col, Cornell Univ, 45-46; asst microbiol, 49-50, instr microbiol, 50-51, from instr to assoc prof path, 51-67, PROF PATH, SCH MED, NY UNIV, 67-; CHIEF LAB SERV, NY VET ADMIN HOSP, 67- Concurrent Pos: Hoskins fel, 51-53; Polachek Found med res fel, 54-59; intern, NY Hosp, 44 & 48-49, asst res, 44-45, res pathologist, 45-46. Mem: Am Asn Path & Bact; Reticuloendothelial Soc; Am Soc Exp Path; Am Asn Immunol; Int Acad Path. Res: Immunopathology; pathology of rheumatic diseases. Mailing Add: Dept of Path NY Univ Sch of Med New York NY 10016

COOPER, OWEN, b New York, NY, July 30, 20; m 42; c 2. FOOD TECHNOLOGY. Educ: Brooklyn Col, AB, 42. Prof Exp: Supv chemist, Explosives Lab, Hercules Powder Co, 42-43; res chemist, Thomas J Lipton, Inc, 46-48; res chemist, Ann Page Div, Great Atlantic & Pac Tea Co, 48-50, asst chief chemist, 50-54, head prod improv div, 55-57, chief chemist, 57-59; tech dir, Food Div, Carter-Wallace, Inc, 59-67, vpres, Res & Develop, 67-71; DIR PROD DEVELOP, AM HOME FOODS DIV, AM HOME PROD CORP, 71- Concurrent Pos: Tech chmn, Mayonnaise & Salad Dressings Inst, 66-71; mem sci comt, Calorie Control Coun, 67-71; tech comt for food protection, Grocery Mfrs Am, 70-71. Mem: AAAS; Inst Food Technol. Res: Low calorie food product development and process development design; packaging development and shelf life studies; microbiological stability; psychometrics; rheological studies; plant sanitation; food law and regulatory agencies. Mailing Add: 63 Knoll Dr Princeton NJ 08540

COOPER, PASCAL WILSON, organic chemistry, see 12th edition

COOPER, PAUL DAVID, b Ashland, Ky, Mar 17, 36; m 66. GEOGRAPHY. Educ: Univ Ky, AB, 60; Univ Ga, MA, 63, PhD(geog), 65. Prof Exp: Econ res analyst, US Govt, 65-67; asst prof geog, Univ Ky, 67-68; GEOG RES ANALYST, US GOVT, 69- Mem: Asn Am Geog. Res: Regional geography of China and East Asia; climatology. Mailing Add: 1225 N Pierce St Apt 503 Arlington VA 22209

COOPER, PAUL DAVID, b Winnipeg, Man, May 10, 35; m 58; c 3. PHARMACOLOGY, ORGANIC CHEMISTRY. Educ: Univ Toronto, BSc, 58, MSc, 59; Univ Ottawa, PhD(chem), 62. Prof Exp: Res assoc chem, Banting Inst, Univ Toronto, 62-64; res dir pharmaceut, Penick Can Ltd, Ont, 64-65; res assoc, Pharmacol, Univ Toronto, 65-66, from asst prof to assoc prof, 66-70; ASSOC PROF PHARMACOL, FAC PHARM, UNIV MONTREAL, 70- Concurrent Pos: Lectr, Univ Toronto, 67-68; consult, Astra Pharmaceut Ltd, Ont, 67-69; vis scientist, Govt France, 68; fel, Int Cong Phatmacol, Switz, 69. Mem: Pharmacol Soc Can. Res: Structure-action relationships among cholinergic and adrenergic drugs, especially stereochemical and electronic aspects of molecules representing hallucinogenic drugs; receptor theory and the events leading from excitation to muscle contraction. Mailing Add: Dept of Biochem Pharmacol Fac of Pharm Univ of Montreal Montreal PQ Can

COOPER, PHILIP HARLAN, b Charleston, WVa, Feb 20, 35; m 58; c 3. RADIATION BIOLOGY, RADIOLGOICAL PHYSICS. Educ: Vanderbilt Univ, BA, 59; Univ Rochester, MS, 60, PhD(radiation hemat), 66. Prof Exp: USAEC fel, Pac Northwest Labs, Battelle Mem Inst, 66-68; RES PHYSICIST, VET ADMIN HOSP, 68-, ASST CHIEF, 74-; RES INSTR EXP BIOL, COL MED, BAYLOR UNIV, 68- Mem: Am Asn Physicists in Med; Soc Nuclear Med. Res: Quantitative scintiscanning of isotopes used in nuclear medicine; hematopoietic cell proliferation following acute X-irradiation of dogs and chronic strontium-90 ingestion in swine. Mailing Add: Vet Admin Hosp 2002 Holcombe Blvd Houston TX 77031

COOPER, RALPH SHERMAN, b Newark, NJ, June 25, 31; m 56; c 2. THEORETICAL PHYSICS. Educ: Cooper Union, BChE, 53; Univ Ill, MS, 55, PhD(physics), 57. Prof Exp: Mem staff theoret physics, Los Alamos Sci Lab, Univ Calif, 57-65; chief scientist, Nuclear Lab, Donald W Douglas Labs, Douglas Aircraft Co, 65-69; staff mem, 69-70, alt group leader, 70-71, group leader 71-72, ASST LASER DIV LEADER, LOS ALAMOS SCI LAB, UNIV CALIF, 72- Concurrent Pos: Consult, Inst Defense Anal, Washington, DC, 62-65; consult, Douglas Aircraft Co, Calif, 63- Mem: Am Phys Soc; Am Nuclear Soc; Res: Laser fusion and laser isotope enrichment; solid state physics; reactor physics; nuclear rocket propulsion; mission analysis; radiation effects; ion exchange column theory. Mailing Add: Los Alamos Sci Lab PO Box 1663 Los Alamos NM 87544

COOPER, RAYMOND B, b Thomasville, Ga, July 21, 39; m 63; c 3. WEED SCIENCE. Educ: Univ Ga, BS, 62, MS, 64; Va Polytech Inst, PhD(agron, physiol), 66. Prof Exp: Plant sci rep, 66-70, REGIONAL COORD, ELI LILLY & CO, 70- Mem: Am Soc Agron; Crop Sci Soc Am; Weed Sci Soc Am. Res: Physiology of Bermuda grass forage species; alfalfa growth and morphology as influenced by potassium nutrition; research and development of agricultural chemicals with emphasis on herbicides but with effort in the areas of plant pathology and entomology. Mailing Add: Eli Lilly & Co PO Box 3008 Omaha NE 68103

COOPER, RAYMOND DAVID, b Kansas City, Kans, Dec 13, 27; m 54; c 3. PHYSICS, RESEARCH ADMINISTRATION. Educ: Univ Ill, BS, 51; Iowa State Col, MS, 54; Mass Inst Technol, PhD, 67. Prof Exp: Asst physics, Ames Lab, AEC, 53-54; nuclear physicist, Pioneering Res Div, Quartermaster Res & Eng Ctr, 54-62; chief linear accelerator sect, US Army Natick Labs, 62-70; radiation physicist, USAEC, 70-75; PHYSICIST, US ENERGY RES & DEVELOP ADMIN, 75- Concurrent Pos: Lectr, Tufts Univ, 56-61. Mem: AAAS; Am Phys Soc; Sci Res Soc Am. Res: Physical mechanisms in radiation biology; dosimetry; accelerators; radiation interactions; environmental systems analysis. Mailing Add: DBER US Energy Res & Develop Admin Washington DC 20545

COOPER, REGINALD RUDYARD, b Elkins, WVa, Jan 6, 32; m 54; c 4. ORTHOPEDIC SURGERY. Educ: Univ WVa, BA, 52, BS, 53; Med Col Va, MD, 55; Univ Iowa, MS, 60. Prof Exp: Resident surg, 56-57, resident orthop, 57-60, assoc, 62-65, from asst prof to assoc prof orthop surg, 65-71, PROF ORTHOP SURG, UNIV IOWA, 71-, CHMN DEPT ORTHOP, 73- Concurrent Pos: Res fel orthop surg & anat, Johns Hopkins Hosp, 64-65; Am, Brit & Can Orthop Asns exchange fel, 69. Honors & Awards: Outstanding Orthop Res Award, Kappa Delta, 71. Mem: AMA; fel Am Col Surg; Am Acad Orthop Surg; Orthop Res Soc; Am Asn Mil Surg. Res: Electron microscopy of bone and skeletal muscle as related to disuse atrophy and regeneration. Mailing Add: Dept of Orthop Surg Univ of Iowa Iowa City IA 52242

COOPER, RICHARD GRANT, b New York, NY, Mar 8, 34. PHARMACOLOGY, PHYSIOLOGY. Educ: Univ Ky, BS, 56, MS, 60; Univ Tex, PhD(physiol), 64. Prof Exp: From instr to asst prof physiol & agr chem, Univ Mo, 64-71, res assoc space sci res ctr, 65-71; ASSOC PROF PHYSIOL, COL OSTEOP MED & SURG, 72- Concurrent Pos: Res assoc, Mo Regional Med Program, 67-68; mem coun thrombosis, Am Heart Asn. Mem: AAAS. Res: Mammalian physiology; blood coagulation and fibrinolysis; hemorrhagic diseases; hemodynamics; depressed metabolism. Mailing Add: 1136 33rd St West Des Moines IA 50265

COOPER, RICHARD KENT, b Detroit, Mich, Apr 13, 37; m 57; c 2. ELECTROMAGNETICS, THEORETICAL PHYSICS. Educ: Calif Inst Technol, BS, 58, MS, 59; Univ Ariz, MS, 62, PhD(physics), 64. Prof Exp: Fulbright Scholar, Niels Bohr Inst, 64; PROF PHYSICS & CHMN DEPT, CALIF STATE UNIV, HAYWARD, 65- Concurrent Pos: Vis staff mem, Los Alamos Sci Lab, Univ Calif, 73-74, consult, 75-76; physicist, Lawrence Livermore Lab, Univ Calif, 74-75, consult, 75-76. Mem: Sigma Xi. Res: Electromagnetic phenomena in charged particle beams; accelerator theory. Mailing Add: Dept of Physics Calif State Univ 25800 Hillary St Hayward CA 94542

COOPER, RICHARD LEE, b Rensselear, Ind, Feb 28, 32; m 52; c 4. PLANT BREEDING, PLANT GENETICS. Educ: Purdue Univ, BS, 57; Mich State Univ, MS, 58, PhD(plant breeding & genetics), 62. Prof Exp: Res assoc soybean breeding & genetics, Dept Agron & Plant Genetics, Univ Minn, 61-67; assoc prof agron, 69-73, RES LEADER, US REGIONAL SOYBEAN LAB, UNIV ILL, URBANA-CHAMPAIGN, 67-, PROF AGRON, 73- Mem: Am Soc Agron; Crop Sci Soc Am. Res: Soybean breeding, genetics and cultural practices. Mailing Add: US Regional Soybean Lab Univ of Ill Urbana IL 61801

COOPER, ROBERT ARTHUR, JR, b St Paul, Minn, Aug 27, 32; m 59; c 3. MEDICINE, ONCOLOGY. Educ: Univ Pa, AB, 54; Jefferson Med Col, MD, 58; Am Bd Path, cert path anat, 63. Prof Exp: From instr to prof path, Med Sch, Univ Ore, 62-69; from assoc prof to prof, 69-74, assoc dean curricular affairs, 69-73, head div surg path, 72-75, PROF ONCOL IN PATH & DIR CANCER CTR, SCH MED, UNIV ROCHESTER, 74- Concurrent Pos: Am Cancer Soc fel, 60-61; teaching fel path, Harvard Med Sch, 62-63; Cancer Res Ctr grant, 75-; mem treatment comt, Breast Cancer Force, Nat Cancer Inst. Mem: AAAS; Am Soc Cancer Educ; Sigma Xi. Res: Experimental radiation pathology and toxicology; radiation carcinogenesis; mammary tumor biology. Mailing Add: Cancer Ctr Univ of Rochester Rochester NY 14642

COOPER, ROBERT CHAUNCEY, b San Francisco, Calif, July 4, 28; m 56; c 4. MICROBIOLOGY, PUBLIC HEALTH. Educ: Univ Calif, Berkeley, BS, 52; Mich State Univ, MS, 53, PhD(microbiol), 58. Prof Exp: Asst prof, 58-66, ASSOC PROF PUB HEALTH, SCH PUB HEALTH, UNIV CALIF, BERKELEY, 66- Mem: AAAS; Am Soc Microbiol; Water Pollution Control Fedn; Int Asn Water Pollution Res; Am Inst Biol Sci. Res: Microbiological aspects of water quality. Mailing Add: Sch of Pub Health Univ of Calif Berkeley CA 94720

COOPER, ROBERT MICHAEL, b Oakland, Calif, Oct 21, 39. PHARMACY. Educ: Univ Calif, San Francisco, PharmD. Prof Exp: Mem staff pharmaceut develop serv, Clin Ctr, NIH, 64-66; clin pharmacist, Long Beach Mem Hosp, Calif, 66; mem staff dose formulation unit, Cancer Chemother Serv Ctr, Nat Cancer Inst, 66-67; asst prof pharm, 67-71, actg chmn dept, 71-72, ASSOC PROF PHARM, SCH PHARM, STATE UNIV NY BUFFALO, 71-, ASST DEAN, 72- Mem: Am Pharmaceut Asn; Am Soc Hosp Pharmacists. Res: Clinical pharmacy; sterile manufacturing. Mailing Add: Rm 114 Cary Hall State Univ of NY Sch of Pharm Buffalo NY 14214

COOPER, ROBERT WARREN, b New York NY, Aug 18, 21; m 72; c 2.

FORESTRY. Educ: NY State Col Forestry, BS, 42, MS, 53. Prof Exp: Res forester, US Forest Serv, 46-70, asst dir, Southeastern Forest Exp Sta, Ga, 70-72, prog mgr, 72-75, ASST DIR, SOUTHEASTERN FOREST EXP STA, GAINESVILLE, FLA, 75- Honors & Awards: Cert of Merit, USDA, 56. Mem: Soc Am Foresters. Res: Silviculture; sand pine; slash pine; loblolly pine; longleaf pine; fire research; forest fire smoke research; forest recreation research; watershed management; range and wildlife habitat research. Mailing Add: Univ of Fla Sch of Forest Resources & Conserv Gainesville FL 32601

COOPER, ROBERT WOODROW, b Scotia, Calif, Dec 9, 38; m 60; c 3. PRIMATE BIOLOGY. Educ: Univ Calif, Davis, BS, 60, DVM, 62. Prof Exp: Assoc res biologist & dir primate res colony, Inst Comp Biol, Zool Soc San Diego, 62-74; ASSOC PROF ZOOL, SAN DIEGO STATE UNIV, 74- Concurrent Pos: Mem, Subcomt Primate Stand, Inst Lab Animal Resources, Div Biol & Agr, Nat Acad Sci-Nat Res Coun, 64-67. Mem: AAAS; Am Asn Lab Animal Sci; Am Vet Med Asn. Res: Comparative biology of small primate species, especially reproduction, social behavior, and development of baseline biometric parameters prerequisite to effective biomedical research utilization. Mailing Add: Dept of Zool San Diego State Univ San Diego CA 92115

COOPER, ROBIN D G, b Eastbourne, Eng, Sept 26, 38; m 70. ORGANIC CHEMISTRY. Educ: Univ London, BSc, 59, PhD(org chem), 62, DIC, 60; FRIC. Prof Exp: Glaxo res fel org chem, Imp Col, Univ London, 62-63; Nat Acad Sci vis res assoc, US Army Natick Labs, 63-65; sr res chemist, 65-70, res scientist, 70-75, RES ASSOC, ELI LILLY & CO, 75- Mem: Am Chem Soc; fel The Chem Soc. Res: Structural determination and synthesis of antibiotics, especially penicillin and cephalosporins. Mailing Add: Eli Lilly & Co 231 E McCarty Indianapolis IN 46225

COOPER, RONDA FERN, b Schenectady, NY, June 20, 43. MEDICAL MICROBIOLOGY. Educ: Okla State Univ, BS, 64; Kans State Univ, MS, 66, PhD(microbiol), 71. Prof Exp: Res assoc & instr path, Kans State Univ, 67-73; asst prof biol, Univ NMex, 73-74; ASST PROF MICROBIOL, LA STATE UNIV, BATON ROUGE, 74- Concurrent Pos: Lalor Found fel, 75- Mem: Am Soc Microbiol; Conf Pub Health Lab Dirs. Res: Interaction of microbial toxins with host tissues and other aspects of host-parasite relationships. Mailing Add: Dept of Microbiol La State Univ Baton Rouge LA 70803

COOPER, STEPHEN, b Brooklyn, NY, Aug 6, 37; m 60; c 2. MICROBIOLOGY, GENETICS. Educ: Union Col, NY, BA, 59; Rockefeller Inst, PhD(microbiol), 63. Prof Exp: NSF res fels, Univ Inst Microbiol, Copenhagen Univ, 63-64 & Med Res Coun Microbial Genetics Res Unit, Univ London, 64-65; res assoc biol chem, Med Sch, Tufts Univ, 65-66; asst res prof pediat, State Univ NY, Buffalo, 66-70, asst res prof biochem, 67-70, lectr biol, 69-70; ASSOC PROF MICROBIOL, SCH MED, UNIV MICH, 70- Mem: AAAS; Am Soc Microbiol. Res: Biochemistry and genetics of viruses; protein synthesis; control of DNA replication and cell division in bacteria; microbial genetics. Mailing Add: Dept of Microbiol Sch of Med Univ of Mich Ann Arbor MI 48104

COOPER, TERENCE ALFRED, b Oxford Eng, Feb 8, 41; m 66; c 1. POLYMER CHEMISTRY. Educ: Oxford Univ, BA, 62, BSc, 64, DPhil(phys org chem) & MA, 66; Drexel Univ, MBA, 73. Prof Exp: Res staff chemist, Sterling Chem Lab, Yale Univ, 66-68; RES CHEMIST, ELASTOMERS RES LAB, EXP STA, E I DU PONT DE NEMOURS & CO, INC, 68- Mem: Am Chem Soc; The Chem Soc; Am Inst Chem Engrs. Res: Polymer engineering; colloid chemistry; elastomers technology. Mailing Add: Elastomer Chem Dept Exp Sta E I du Pont de Nemours & Co Inc Wilmington DE 19898

COOPER, THEODORE, b Trenton, NJ, Dec 28, 28; m 56; c 4. PHYSIOLOGY, PHARMACOLOGY. Educ: Georgetown Univ, BS, 49; St Louis Univ, MD, 54, PhD(physiol), 56. Prof Exp: Intern, St Louis Univ Hosps, 54-55; sr asst surgeon, Nat Heart Inst, 56-58, asst res cardiovascular surg, 58-59, mem staff clin surg, 59-60; from asst prof to prof surg, St Louis Univ, 60-66, mem bd grad studies, 62-66; prof pharmacol & surg & chmn dept pharmacol, Sch Med, Univ NMex, 66-68; dir, Nat Heart & Lung Inst, 68-74; dep asst secy health, 74-75, ASST SECY HEALTH, US DEPT HEALTH, EDUC & WELFARE, 75- Concurrent Pos: USPHS career develop award, 62-; mem pharmacol & exp therapeut study sect, USPHS, 64-68. Honors & Awards: Borden Award, 54. Mem: AAAS; Am Physiol Soc; Soc Exp Biol & Med; Am Soc Pharmacol; Am Col Chest Physicians. Res: Experimental and clinical cardiovascular physiology and pharmacology. Mailing Add: US Dept of Health Educ & Welfare 330 Independence Ave SW Washington DC 20201

COOPER, TOMMYE, b Bandana, Ky, May 17, 38; m 61; c 1. AGRICULTURE, STATISTICS. Educ: Murray State Univ, BS, 60; Univ Ky, MS, 62, PhD(dairy sci, statist), 66. Prof Exp: Anal statistician, 66-67, chief methods & opers unit, Comput Lab, 67-68, actg dir, Data Systs Appln Div, 68-70, dir, 70-73, DIR, FT COLLINS COMPUT CTR, AGR RES SERV, US DEPT AGR, 73- Mem: Am Dairy Sci Asn; Am Soc Animal Sci; Am Asn Comput Mach. Res: Dairy cattle reproductive and genetic research using digital computer and statistical techniques to achieve these goals. Mailing Add: Ft Collins Comput Ctr 3825 E Mulberry St Ft Collins CO 80521

COOPER, WALTER, b Clairton, Pa, July 18, 28; m 53; c 2. PHYSICAL CHEMISTRY. Educ: Washington & Jefferson Col, BA, 50; Univ Rochester, PhD(phys chem), 57. Prof Exp: From res chemist to sr res chemist, 56-66, RES ASSOC, EASTMAN KODAK CO, 66- Mem: AAAS; Am Chem Soc; Am Phys Soc. Res: Gas-phase kinetics; photographic theory; solid state chemistry of silver halides; luminescence properties of dyes. Mailing Add: 68 Skyview Lane Rochester NY 14625

COOPER, WILLIAM ANDERSON, b Archer City, Tex, Feb 4, 27; m 52; c 4. ZOOLOGY, PHYSIOLOGY. Educ: NTex State Univ, BS, 48, MS, 50; Tex A&M Univ, PhD(zool), 57. Prof Exp: Instr natural sci, Paris Jr Col, 53-54; asst biol, Tex A&M Univ, 54-55, instr, 55-57; assoc prof, 57-65, PROF BIOL, WTEX STATE UNIV, 65- Concurrent Pos: Res assoc, Agr & Mech Res Found, NIH, 55-57. Mem: Am Micros Soc; Am Soc Zoologists; Sigma Xi. Res: Effects of vitamin deficiencies on embryo development in white rats; limnology; water quality and fisheries biology research. Mailing Add: Dept of Biol WTex State Univ Canyon TX 79016

COOPER, WILLIAM CECIL, b Salisbury, Md, Apr 6, 09; m 32; c 2. PLANT PHYSIOLOGY. Educ: Univ Md, BS, 29; Calif Inst Technol, MS, 36, PhD, 38. Prof Exp: Jr pomologist, Bur Plant Indust, 29-38, assoc plant physiologist, Subtrop Fruit Prod, 38-42, plant physiologist, Trop Tree Crops Propagation, Office For Agr Relations, 43-44, citrus rootstock invest, Bur Plant Indust, 44-54, sr plant physiologist, Hort Crops Br, 55-59, LEADER CITRUS INVESTS, PLANT SCI RES DIV, AGR RES SERV, USDA, 59- Mem: Am Soc Plant Physiol; Am Soc Hort Sci. Res: Plant hormones and root and flower formation; salt tolerance; cold hardiness; citrus rootstocks and citrus pheonology; fruit abscission; ethylene physiology. Mailing Add: US Hort Sta 2120 Camden Rd Orlando FL 32803

COOPER, WILLIAM CHARLES, chemistry, see 12th edition

COOPER, WILLIAM CLARK, b Manila, Philippines, June 22, 12; m 37; c 4. MEDICINE. Educ: Univ Va, MD, 34; Harvard Univ, MPH, 58; Am Bd Internal Med, Am Bd Prev Med & Am Bd Indust Hyg, dipl. Prof Exp: Intern & asst resident, Univ Hosp, Cleveland, 34-37; instr bact, Sch Hyg & Pub Health, Johns Hopkins Univ, 40; from asst surgeon to surgeon, NIH, 41-51, chief & med dir occup health field hqs, 52-57, chief epidemiol serv, Occup Health Prog, Bur State Serv, 57-61, dep chief, 61-62, chief div occup health, USPHS, 62-63; res physician, Sch Pub Health, Univ Calif, Berkeley, 63-65, prof in residence occup health, 65-72; VPRES, EQUITABLE ENVIRON HEALTH, INC, 73- Concurrent Pos: Med consult, AEC, 64-72. Mem: AAAS; fel AMA; fel Am Col Chest Physicians; fel Am Acad Occup Med; fel Am Acad Indust Hyg. Res: Nutrition; malaria; occupational health. Mailing Add: Equitable Environ Health Inc 2180 Milvia St Berkeley CA 94704

COOPER, WILLIAM E, b Orono, Maine, May 8, 38; m 60; c 1. ZOOLOGY. Educ: Mich State Univ, BS, 60; Univ Mich, MS, 62, PhD(zool), 64. Prof Exp: Asst prof zool, Univ Mass, 64-65; from asst prof to assoc prof, 65-72, PROF ZOOL, MICH STATE UNIV, 72- Mem: Ecol Soc Am; Am Soc Zoologists. Res: Population dynamics and regulation of fresh-water invertebrate populations. Mailing Add: Dept of Zool Mich State Univ Col Natural Sci East Lansing MI 48823

COOPER, WILLIAM GREGORY, b Cincinnati, Ohio, Sept 12, 26; m 70; c 4. MEDICAL EDUCATION. Educ: Univ Cincinnati, BS, 49; Cornell Univ, MS, 50; Columbia Univ, PhD(anat), 54. Prof Exp: Res asst anat, Med Col, Cornell Univ, 49-50; instr, Sch Med, Univ PR, 50-52; asst, Col Physicians & Surgeons, Columbia Univ, 52-54, instr, 54-55; from instr to assoc prof, 55-66, dir unit teaching labs basic sci, 64-70, dir off educ resources, 69-71, PROF ANAT, UNIV COLO MED CTR, DENVER, 66-, ASST VPRES HEALTH AFFAIRS, 71- Concurrent Pos: USPHS res career develop award, 59-64; vis res fel, Univ Calif, Berkeley, 61-62; mem biomed libr rev comt, NIH, 71-76. Mem: Asn Am Med Cols (dir, Div Educ Resources & dep dir, Dept Acad Affairs, 73-). Res: Myogenesis; cardiogenesis; cell attachment to substrates; establishment, maintenance and alterations of cell lines and cell strains; determinations of generation times of cells in vitro; cell, tissue and organ culture techniques; planning, development and operation of multi-disciplinary student laboratories and health science television facilities and networks. Mailing Add: Off of VPres Health Affairs Univ of Colo Med Ctr Denver CO 80220

COOPER, WILLIAM S, b Winnipeg, Man, Nov 7, 35; m 64; c 3. INFORMATION SCIENCE. Educ: Principia Col, BSc, 56; Mass Inst Technol, MSc, 59; Univ Calif, Berkeley, PhD(logic & methodology), 64. Prof Exp: Alexander von Humboldt scholar, Univ Erlangen, 64-65; asst prof info sci, Univ Chicago, 66-70; actg assoc prof info sci & actg dir inst libr res, 70-71, ASSOC PROF INFO SCI, UNIV CALIF, BERKELEY, 71- Concurrent Pos: Miller prof, Miller Inst, Berkeley, Calif, 75-76. Mem: AAAS; Asn Symbolic Logic; Soc Info Sci. Res: Symbolic logic; descriptive linguistics; foundation of language; information storage and retrieval; theory of indexing. Mailing Add: Sch of Librarianship Univ of Calif Berkeley CA 94720

COOPER, WILSON WAYNE, b Checotah, Okla, July 28, 42; m 63; c 3. INDUSTRIAL CHEMISTRY, CHEMICAL ENGINEERING. Educ: Northeastern State Col, BS, 64; Univ Ark, Fayetteville, MS, 67, PhD(chem), 69. Prof Exp: Instr chem, Northwestern State Col, 66-67 & 69-70; RES CHEM ENGR, PILOT PLANT CTR, US BORAX RES CORP, 70- Mem: Am Chem Soc. Res: Process development for both organic and inorganic systems and analytical techniques to support these studies. Mailing Add: US Borax Res Corp Pilot Plant Ctr Boron CA 93516

COOPERBAND, SIDNEY R, b Boston, Mass, June 3, 31; m 55; c 3. IMMUNOLOGY, CELL BIOLOGY. Educ: Harvard Univ, AB, 53; Univ Pa, MD, 57. Prof Exp: Intern, Hosp Univ Pa, 57-58; instr med, 62-65, assoc, 65-67, asst prof med & microbiol, 67-69, assoc prof, 69-75, PROF MED & MICROBIOL, SCH MED, BOSTON UNIV, 75- Concurrent Pos: Clin fel internal med, Hosp Univ Pa, 58-59; res fel med, Mass Gen Hosp, 59-61; res fel bacteriol & immunol, Harvard Med Sch, 61-62; Nat Inst Allergy & Infectious Dis res career develop award, 67; assisting physician med serv, Boston City Hosp, 62-; asst vis physician med, Boston Univ Hosp, 67-; mem, Fulbright-Hays Comt Int Exchange Persons, 70; vis scientist bacteriol, Karolinska Inst, Sweden, 70-71. Honors & Awards: Frances Stone Burns Award, Am Cancer Soc, 70. Mem: AAAS; Am Fedn Clin Res; Am Soc Clin Invest; Am Asn Immunol. Res: Cellular immunology and immunobiology, basic areas of investigation being applied to clinical problems involving derangements of the immune system in man and attempts at immunotherapy of cancer. Mailing Add: Dept of Med Sch of Med Boston Univ Boston MA 02118

COOPERMAN, BARRY S, b New York, NY, Dec 11, 41; m 63; c 1. PHYSICAL ORGANIC CHEMISTRY, BIOCHEMISTRY. Educ: Columbia Univ, BA, 62; Harvard Univ, PhD(chem), 68. Prof Exp: NATO fel biochem, Pasteur Inst, 67-68; asst prof, 68-72, ASSOC PROF BIOORG CHEM, UNIV PA, 72- Mem: Am Chem Soc. Res: Mechanism of phosphoryl transfer; allosteric properties of fructose diphosphatase; affinity labels for adenylic acid reception. Mailing Add: Dept of Chem 16 E F Smith Lab Univ of Pa Philadelphia PA 19104

COOPERMAN, EDWARD LEE, b Sept 4, 36; US citizen; m 63; c 2. NUCLEAR PHYSICS. Educ: Lehigh Univ, BS, 58; Pa State Univ, PhD(physics), 63. Prof Exp: Asst prof physics, Ariz State Univ, 63-64; physicist, Los Alamos Sci Lab, 64; res physicist, Nat Res Ctr, Univ Strasbourg, 64-67; from asst prof to assoc prof, 67-74, actg chmn dept, 69-71, PROF PHYSICS, CALIF STATE UNIV, FULLERTON, 74-, CHMN DEPT, 74- Res: Fission process, especially U-235 and Cf-252; angular correlations and distributions of medium nuclei with the help of the Litherland method. Mailing Add: Dept of Physics Calif State Univ 800 N State College Blvd Fullerton CA 92631

COOPERMAN, JACK M, b New York, NY, Jan 13, 21; m 49; c 1. NUTRITIONAL BIOCHEMISTRY. Educ: City Col New York, BS, 41; Univ Wis, MS, 43, PhD(biochem), 45. Prof Exp: Sr res biochemist, Hoffman-La Roche Inc, 46-56; from asst prof to assoc prof, 57-71, PROF PEDIAT, NY MED COL, 71-, PROF COMMUNITY & PREV MED, 75- Concurrent Pos: Nat Found Infantile Paralysis fel, Univ Wis, 45-46; mem med adv bd, Cooley's Anemia & Res Found Children, Inc; staff mem, Off Sci Res & Develop. Mem: AAAS; Am Chem Soc; Geront Soc; Soc Exp Biol & Med; Am Soc Clin Nutrit. Res: Nutrition of hamster; nutrition of monkey; relation of nutrition of prevention of infectious diseases; microbiological vitamin assays; vitamins; unknown growth factors for animals and bacteria; enzymes; amino acids and anemias; vitamin B-twelve, folic acid and riboflavin metabolism; placental transfer; purine and pyrimidine metabolism and synthesis. Mailing Add: Dept of Pediat NY Med Col 1249 Fifth Ave New York NY 10029

COOPERMAN, PHILIP, b US, Dec 3, 18; m 50; c 2. MATHEMATICS. Educ: City Col New York, 38; NY Univ, MS, 48, PhD(math), 51. Prof Exp: Teacher high sch, NY, 38-39; statist clerk, Dept Welfare, New York, 39-41; jr physicist, US Navy, 42-43; physicist, Fed Tel & Radio Corp, 43; physicist ultrasonics, Balco Labs, 46; asst physics, NY Univ, 48-49; physicist, Gaseous Electronics, Res Corp, 51-56; sr mathematician, Gulf Res & Develop Co, 56-58; asst prof math, Univ Pittsburgh, 58-

61; assoc prof, 61-63, actg chmn dept, 63-64, PROF MATH, FARLEIGH DICKINSON UNIV, 63- Concurrent Pos: Dir res & develop, Res-Cottrell, Inc, 60, consult, 60-62; consult appl res labs, US Steel Corp; consult, Precipitair Pollution Control Work, 67- Mem: Am Phys Soc; Am Math Soc; Air Pollution Control Asn; Math Asn Am. Res: Electricity; electronics; corona discharge id gases; calculus of variations; partial differential equations. Mailing Add: Dept of Math FArleigh Dickinson Univ 1000 River Rd Teaneck NJ 07666

COOPERRIDER, DONALD ELMER, b Thornville, Ohio, Sept 21, 14; m 36. VETERINARY MEDICINE. Educ: Ohio State Univ, DVM, 36, MS, 42. Prof Exp: Jr veterinarian, USDA, 36-40; veterinarian, Civilian Conserv Corps, 40-41; asst vet parasitol, Ohio State Univ, 41-42; vet diagnostician, State Dept Agr, Ohio, 46; assoc veterinarian vet parasitol, Okla Agr & Mech Col, 46-48; assoc parasitologist, Univ Tenn, 48-49; assoc prof vet parasitol & path, Univ Ga, 49-54; chief diag labs, State Dept Agr, NC, 54-59; parasitologist, 59-68, CHIEF, BUR DIAG LABS, FLA DEPT AGR, 68- Mem: Am Vet Med Asn; US Livestock Sanit Asn; Poultry Sci Asn; Am Asn Vet Parasitol (secy-treas, 61-69, pres, 71-73); Am Asn Vet Lab Diagnosticians (pres, 71-73). Res: Veterinary parasitology; poultry diseases. Mailing Add: Bur of Diag Labs Fla Dept of Agr Box 460 Kissimmee FL 32741

COOPERRIDER, TOM SMITH, b Newark, Ohio, Apr 15, 27; m 53; c 2. PLANT TAXONOMY. Educ: Denison Univ, BA, 50; Univ Iowa, MS, 55, PhD(bot), 58. Prof Exp: Asst bot, Univ Iowa, 53-57; from instr to asst prof biol sci, Kent State Univ, 58-62; asst prof bot, Univ Hawaii, 62-63; from asst prof to assoc prof, 63-69, PROF BIOL SCI, KENT STATE UNIV, 69-, CUR HERBARIUM, 66-, DIR BOT GARDENS & ARBORETUM, 72- Mem: Bot Soc Am; Am Soc Plant Taxon; Int Asn Plant Taxon. Res: Angiosperm taxonomy; floristics; interspecific hybridization; biosystematics. Mailing Add: Dept of Biol Sci Kent State Univ Kent OH 44242

COOPERSMITH, ALAN, radiation biology, hematology, deceased

COOPERSMITH, MICHAEL HENRY, b Brooklyn, NY, Aug 11, 36; m 59; c 3. THEORETICAL PHYSICS. Educ: Swarthmore Col, BA, 57; Cornell Univ, PhD(physics), 62. Prof Exp: NSF fel, Univ Paris, 61-62; res assoc physics, Univ Chicago, 62-64; asst prof, Case Inst Technol, 64-69; ASSOC PROF PHYSICS, UNIV VA, 69- Res: Statistical mechanics; phase transitions; homogeneity properties of thermodynamic systems; many-body theory; mobility of ions in helium. Mailing Add: Dept of Physics Univ of Va Charlottesville VA 22901

COOPERSTEIN, RAYMOND, b New York, NY, Nov 19, 24; m 52; c 2. INORGANIC CHEMISTRY. Educ: City Col New York, BS, 47; Syracuse Univ, MS, 49; Pa State Univ, PhD(inorg chem), 52. Prof Exp: Res assoc ceramics, Exp Sta, Sch Mineral Indust, Pa State Univ, 52-53; sr chemist inorg chem, Navy Ord Div, Eastman Kodak Co, 53-57; prin engr, Dept Aircraft Nuclear propulsion, Gen Elec Co, 57-59; sect chief, Beryllium Corp, 59-61; chemist refractory mat, Lawrence Radiation Lab, Univ Calif, Berkeley, 61-65; mgr new prod develop, Wood Ridge Chem Corp, 65; sr engr, Nuclear Reactor Dept, Gen Elec Co, 65-67; sr engr, Douglas United Nuclear, Inc, 67-72; MEM STAFF, JRB ASSOCS, McLEAN, VA, 72- Mem: Am Chem Soc; Am Ceramic Soc; Am Soc Testing & Mat. Res: Coordination compounds; ferrites; photoconductors; nuclear ceramics; inorganic synthesis; silicate and molten salt chemistry. Mailing Add: JRB Assocs 1600 Anderson Rd McLean VA 22101

COOPERSTEIN, SHERWIN JEROME, b New York, NY, Sept 14, 23; m 47; c 2. ANATOMY, CELL PHYSIOLOGY. Educ: City Col New York, BS, 43; NY Univ, DDS, 48; Western Reserve Univ, PhD(anat), 51. Prof Exp: Instr biol, City Col New York, 43 & 46-48; instr anat, Western Reserve Univ, 48-49, sr instr, 51-52, from asst prof to assoc prof, 52-64, asst dean, 57-64; prof, Sch Dent Med, 64-65, PROF ANAT & HEAD DEPT, SCHS MED & DENT MED, UNIV CONN HEALTH CTR, 65- Concurrent Pos: Res assoc physiol, Col Dent, NY Univ, 46-48; mem anat sci training comt, Nat Inst Gen Med Sci, 66-70; mem spec study sect diabetes ctrs, Nat Inst Arthritis, Metab & Digestive Dis, 73-75. Mem: AAAS; Am Diabetes Asn; Am Chem Soc; Asn Anat; Am Soc Biol Chem. Res: Metabolism of the islets of Langerhans; insulin secretion; mechanism of the diabetogenic action of alloxan. Mailing Add: Dept of Anat Univ of Conn Sch of Med Farmington CT 06032

COOPERSTOCK, FRED ISAAC, b Winnipeg, Man, Aug 20, 40; m 62; c 2. THEORETICAL PHYSICS. Educ: Univ Man, BSc, 62; Brown Univ, PhD(physics), 66. Prof Exp: Res scholar theoret physics, Dublin Inst Adv Studies, 66-67; asst prof physics, 67-71, ASSOC PROF PHYSICS, UNIV VICTORIA, 71- Concurrent Pos: Can-France sci exchange visitor, Inst Henri Poincare, Paris, 73-74. Mem: Am Phys Soc. Res: General relativity; gravitational waves; relativistic astrophysics and cosmology; two-body problem; singularities; alternative theories of gravitation. Mailing Add: Dept of Physics Univ of Victoria Victoria BC Can

COOR, THOMAS, JR, physics, see 12th edition

COORTS, GERALD DUANE, b Emden, Ill, Feb 3, 32; m 57; c 3. ORNAMENTAL HORTICULTURE. Educ: Univ Mo, BS, 54, MS, 58; Univ Ill, PhD(hort), 64. Prof Exp: Asst prof hort, Univ RI, 64-68; assoc prof, 68-72, PROF PLANT & SOIL SCI, SOUTHERN ILL UNIV, CARBONDALE, 72-, CHMN DEPT, 73- Mem: Am Hort Soc; Am Soc Hort Sci; Int Plant Propagators Soc; Am Inst Biol Sci; Am Soc Agron. Res: Plant growth regulators; growing ornamental plants in artificial media; grades and standards on cut roses; post-harvest physiology of roses; mineral nutrition of ornamental crops. Mailing Add: Dept of Plant & Soil Sci Southern Ill Univ Carbondale IL 62901

COOTS, ALONZO FREEMAN, b Little Rock, Ark, May 6, 27; m 47. PHYSICAL CHEMISTRY. Educ: Vanderbilt Univ, BE, 49, PhD(chem), 54. Prof Exp: Instr phys chem, Fisk Univ, 52-53; res chemist, Anal Sta, Del, 56-58; ASSOC PROF CHEM, NC STATE UNIV, 58- Mem: Am Chem Soc. Res: Radioisotope principles and techniques; general and physical chemistry; radiochemistry. Mailing Add: Dept of Chem NC State Univ Raleigh NC 27607

COOTS, ROBERT HERMAN, b Kansas City, Mo, Feb 24, 28; m 57; c 1. BIOCHEMISTRY. Educ: Univ Mo, BS, 54; Univ Wis, MS, 56, PhD(biochem), 58. Prof Exp: Res biochemist, Procter & Gamble Co, 58-65, head physiol chem sect, 65-67, dent res sect, Miami Valley Labs, 67-69, dent & toxicol res sect, 69-70 & pharmacol & metab res sect, 70-71, ASSOC DIR RES & DEVELOP DEPT, MIAMI VALLEY LABS, PROCTER & GAMBLE CO, 71- Mem: Am Soc Biol Chem. Res: Intermediary metabolism; lipid metabolism; pharmacology and toxicology. Mailing Add: PO Box 39175 Cincinnati OH 45239

COOVER, HARRY WESLEY, JR, b Newark, Del, Mar 6, 18, m 41; c 3. ORGANOMETALLIC CHEMISTRY, POLYMER CHEMISTRY. Educ: Hobart Col, BS, 41; Cornell Univ, MS, 42, PhD, 44. Prof Exp: Res chemist, Eastman Kodak Co, 44-50; res assoc, 51-62, div head, 63-65, vpres, 70-73, DIR RES, TENN EASTMAN CO, 65-, EXEC VPRES, 73- Honors & Awards: Southern Chemist Award, Am Chem Soc, 60. Mem: Am Chem Soc; Textile Res Inst; Am Asn Textile Technologists; Indust Res Inst; NY Acad Sci. Res: Adhesives; insecticides; fungicides; high polymer chemistry; organophosphorus chemistry; synthetic fiber research. Mailing Add: Res Labs PO Box 511 Tenn Eastman Co Kingsport TN 37662

COPE, DAVID FRANKLIN, b Crumpler, WVa, June 28, 12; m 36; c 2. NUCLEAR SCIENCE, ENERGY CONVERSION. Educ: WVa Univ, AB, 33, MS, 34; Univ Va, PhD(physics), 52. Prof Exp: Instr phsyics, WVa Univ, 35-36; instr math, Agr & Mech Col, Texas, 37-38, physics, 46-47; asst prof, NMex Agr & Mech Col, 47-52; physicist, AEC, 52, chief res br, 53-56, dep dir res & develop div, Oak Ridge Opers, 56-59, dir reactor div, 59-66, sr site rep reactor develop at Oak Ridge Nat Lab, 66-74; NUCLEAR ENERGY EXPERT, ENERGY RES & DEVELOP OFF, OAK RIDGE 74- Concurrent Pos: US team leader, US-Mexico Study of Dual Purpose Nuclear Plants, 68-; chmn tech subcontm NSF Energy Fac Sifting Comn, 74-; energy consult, 75- Mem: AAAS; Am Nuclear Soc; Am Phys Soc; Am Asn Physics Teachers. Res: Magnetic properties of iron and nickel at high temperatures; analysis of rocket trajectories; nuclear science and technology; the nation's future energy requirement and assessamnt of various options for meeting these needs. Mailing Add: 113 Orange Lane Oak Ridge TN 37830

COPE, FREEMAN WIDENER, b Peekskill, NY, Aug 4, 30. PHYSICAL BIOCHEMISTRY, PHYSIOLOGY. Educ: Harvard Col, AB, 51; Johns Hopkins Univ, MD, 55. Prof Exp: Intern, Church Home Hosp, Md, 55-56; proj officer physiol & aviation med, 57-59, BIOCHEMIST & PHYSIOLOGIST, BIOCHEM LAB, US NAVAL AIR DEVELOP CTR, 59- Concurrent Pos: Dir, Soc Math Biol, assoc ed, Bull Math Biol. Mem: Am Chem Soc; Am Physiol Soc; Am Phys Soc; Biophys Soc; Soc Math Biol. Res: Applications of solid state physics to biochemistry and biology; sodium and potassium ion complexing in tissues by nuclear magnetic resonance; physics of human arterial system; aerospace medicine. Mailing Add: Biochem Lab US Naval Air Develop Ctr Warminster PA 18974

COPE, JAMES FRANCIS, b Charleston, SC, Sept 10, 43; m 68; c 1. ORGANIC POLYMER CHEMISTRY, ORGANOMETALLIC CHEMISTRY. Educ: Hampden-Sydney Col, BS, 65; Clemson Univ, MS, 70, PhD(org chem), 71. Prof Exp: Res chemist textiles, West Point-Pepperell Inc, 71-72; RES CHEMIST SYNTHETIC FIBERS, PHILLIPS FIBERS CORP, DIV PHILLIPS PETROL CO, 72- Mem: Am Chem Soc; The Chem Soc. Res: Modification and development of synthetic fibers. Mailing Add: Phillips Fibers Corp PO Box 66 Greenville SC 29602

COPE, JOHN THOMAS, JR, b Akron, Ohio, July 24, 21; m 45; c 3. SOIL FERTILITY. Educ: Auburn Univ, BS, 42, MS, 46; Cornell Univ, PhD(soil sci), 50. Prof Exp: Farm mgr, Fla, 46-47; assoc soil chemist, Nitrogen Res, 50-52, assoc agronomist in charge exp fields, 52-59, agronomist, 59-66, SOIL TESTER, AUBURN UNIV, 66- Mem: Am Soc Agron; Soil Sci Soc Am. Res: Soil fertility; nitrogen fertilization; soil organic matter; crop rotations; soil and crop management; soil test fertilizer recommendations and computer programs; soil test calibration; fertility index. Mailing Add: Dept of Agron Auburn Univ Auburn AL 36830

COPE, OLIVER BREWERN, b San Francisco, Calif, June 16, 16; m 42; c 2. FISHERY BIOLOGY. Educ: Stanford Univ, AB, 38, AM, 40, PhD(entomol), 42. Prof Exp: Agent, Bur Entomol & Plant Quarantine, US Dept Agr, Calif, 39; asst, Stanford Univ, 39-42; aquatic biologist, US Fish & Wildlife Serv, 46-50, chief cent valley invest, 50-52, chief Rocky Mt invests, 52-59, fish-pesticide res lab, 59-69, chief br fish husb res, US Bur Sport Fisheries & Wildlife, 69-71; phys scientist, Off of Water Resources Res, 71-74; CONSULT-ED, 74- Concurrent Pos: Jr quarantine inspector, Calif State Dept Agr, 41. Mem: AAAS; Am Entom Soc; Am Fisheries Soc; Wildlife Soc; Am Inst Fisheries Res Biol. Res: Insect morphology; fresh water fisheries; economic poisons and fish; fish husbandry. Mailing Add: 3145 Zinnia St Golden CO 80401

COPE, OSWALD JAMES, b Hanley, Eng, June 8, 34; m 58; c 2. POLYMER CHEMISTRY. Educ: Univ London, BSc, 55; Univ Alta, PhD(org chem), 62. Prof Exp: Asst lectr chem, Western Col, 56-57; chemist, Paint Res Labs, Can Indust Ltd, 57-58; asst chem, Univ Alta, 58-62; res assoc org chem, Purdue Univ, 62-63; chemist, Plastics Dept, Exp Sta, E I du Pont de Nemours & Co, Del, 63-69; sr chemist, Memorex Corp, Calif, 69-70; mgr, 70-74, VPRES RES & DEVELOP, XIDEX CORP, 74- Mem: Am Chem Soc; Soc Photog Sci & Eng. Res: Polymer synthesis and structure property relationships; non-silver duplication film processes; vesicular and diazo film. Mailing Add: Xidex Corp 305 Soquel Way Sunnyvale CA 94086

COPE, VIRGIL W, b Storm Lake, Iowa, Feb 4, 43; m 65. INORGANIC CHEMISTRY. Educ: Iowa State Teachers Col, BA, 65; Univ Kans, PhD(inorg chem), 68. Prof Exp: Asst prof, 68-74, ASSOC PROF CHEM, UNIV MICH-FLINT, 74- Concurrent Pos: Vis prof chem, Boston Univ, 74-75. Mem: Am Chem Soc. Res: Mechanisms of reactions of coordination compounds, especially electron transfer reaction and photosubstitution. Mailing Add: Dept of Chem Univ of Mich Flint MI 48503

COPE, WILL ALLEN, b Inverness, Ala, June 23, 22; m 50; c 1. GENETICS, PLANT BREEDING. Educ: Ala Polytech Inst, BS, 48, MS, 49; NC State Col, PhD(field crops), 56. Prof Exp: Asst prof field crops, 55-64, res assoc prof crop sci, 64-69, assoc prof crop sci & genetics, 69-71, PROF CROP SCI, NC STATE UNIV, 71-, RES AGRONOMIST, AGR RES SERV, US DEPT AGR, 55- Mem: Am Soc Agron; Genetics Soc Am. Res: Breeding and genetics of Trifolium repens. Mailing Add: Dept of Crop Sci NC State Univ Raleigh NC 27607

COPELAND, ARTHUR HERBERT, JR, b Columbus, Tex, June 11, 26; m 47; c 3. MATHEMATICS. Educ: Univ Mich, BS, 49, MA, 50; Mass Inst Technol, PhD(math), 54. Prof Exp: Instr, Mass Inst Technol, 51-54; from instr to assoc prof math, Purdue Univ, 54-61; assoc prof, Northwestern Univ, 61-68; assoc prof, 68-74, PROF MATH, UNIV N H, 74- Mem: Am Math Soc; Math Asn Am. Res: Algebraic topology; homotopy theory; Hopf and fibre spaces. Mailing Add: Dept of Math Univ of NH Durham NH 03824

COPELAND, BILLY JOE, b Mannsville, Okla, Nov 20, 36; m 63; c 2. ECOLOGY. Educ: Okla State Univ, BS, 59, MS, 61, PhD(zool), 63. Prof Exp: Asst limnol, Okla State Univ, 59-62; res scientist marine ecol, Inst Marine Sci, Univ Tex, 62-65, asst prof, Univ, 65-68, assoc prof, 68-70; ASSOC PROF ZOOL & BOT & DIR PAMLICO MARINE LAB, N C STATE UNIV, 70- Concurrent Pos: Grants estuarine ecol & pollution control; consult, indust; chmn ecol comt, Univ Coun Water Resources. Mem: Am Soc Limnol & Oceanog; Ecol Soc Am; Water Pollution Control Fedn. Res: Estuarine ecology; effects of water resources development and pollution on fresh and saltwater ecology; systems analysis of estuarine ecosystems. Mailing Add: Dept of Zool NC State Univ Raleigh NC 27607

COPELAND, BRADLEY ELLSWORTH, b Wilkinsburg, Pa, Sept 8, 21; m 43; c 2. MEDICINE, CLINICAL PATHOLOGY. Educ: Dartmouth Col, AB, 43; Univ Pa, MD, 45. Prof Exp: Assoc path, 50-59, clin pathologist, 59-72, chmn dept path, 72-75, CHIEF, DIV CLIN PATH, NEW ENG DEACONESS HOSP & NEW ENG

BAPTIST HOSP, 75- Concurrent Pos: Chmn comt world stands, World Asn Socs Anat & Clin Path, 60-; chmn subcomt documentation, Comt Path, Nat Acad Sci-Nat Res Coun, 69-71. Mem: Am Chem Soc; AMA; Am Soc Clin Path; Col Am Path. Res: Electrolytes; magnesium; hemoglobin; statistics and quality control; instrumentation. Mailing Add: New Eng Deaconess Hosp 185 Pilgrim Rd Boston MA 02215

COPELAND, CHARLES WESLEY, JR, b Hueytown, Ala, Oct 1, 32; m 57; c 2. STRATIGRAPHY, INVERTEBRATE PALEONTOLOGY. Educ: Birmingham-Southern Col, BS, 54; Univ NC, Chapel Hill, MS, 61. Prof Exp: Paleontologist & core libr supvr, 61-68, geol mapping supvr, 64-68, chief paleont & stratig div, 68-74, CHIEF GEOL DIV, GEOL SURV ALA, 74- Mem: Geol Soc Am; Sigma Xi; Brit Paleont Soc. Res: Subsurface stratigraphy, paleontology and structure; Pleistocene geology; coastal plain faults; solution collapse phenomena; ground water aquifer studies. Mailing Add: Geol Surv of Ala PO Drawer O University AL 35486

COPELAND, DAVID ANTHONY, b Jasper, Ala, Dec 4, 42; m 67. CHEMICAL PHYSICS, INORGANIC CHEMISTRY. Educ: Univ Ala, BSChem, 65, MA, 66; La State Univ, PhD(chem physics), 70. Prof Exp: Res assoc inorg chem, La State Univ, 69-70; asst prof, 70-74, ASSOC PROF INORG PHYS CHEM, UNIV TENN, MARTIN, 74- Mem: Am Chem Soc; Am Phys Soc. Res: Theoretical electronic structures of transition metal tetrahedral complexes and the theories of electrons in polar solvents. Mailing Add: Dept of Chem Univ of Tenn Martin TN 38237

COPELAND, DONALD EUGENE, b Mendon, Ohio, Feb 6, 12; m 41; c 3. BIOLOGY. Educ: Univ Rochester, AB, 35; Amherst Col, MA, 37; Harvard Univ, PhD(zool), 41. Prof Exp: Instr biol, Univ NC, 41-42; from asst prof to assoc prof, Brown Univ, 46-54; prof assoc, Med Sci Div, Nat Res Coun, 53-55; exec secy, Div Res Grants, NIH, 56-59; prof zool, 59-70, PROF BIOL, TULANE UNIV, 70- Concurrent Pos: Asst, Marine Biol Lab, Woods Hole, 37-40; Porter fel, Bermuda Biol Sta, 39; fel, Atkins Trop Sta, Cuba, 41; mem physiol study sect & morph & gen study sect, NIH-USPHS, 51-53; mem physiol panel, Res & Develop Bd, US Dept Defense, 52-53. Mem: Am Soc Zool; Am Asn Anat; Soc Develop Biol; Am Physiol Soc. Res: Electron microscopy of inorganic ion transport; salt secretion in fish gills; functioning of swimbladder; pseudobranch; salt absorption in crab gill, brine shrimp and mosquito larvae; gas secretion in Portuguese man-of-war. Mailing Add: Dept of Biol Tulane Univ New Orleans LA 70118

COPELAND, EDMUND SARGENT, b Lancaster, Pa, Mar 2, 36; m 65; c 2. BIOPHYSICAL CHEMISTRY, MAGNETIC RESONANCE. Educ: Cornell Univ, AB, 58; Univ Rochester, MS, 61, PhD(radiation biol), 64. Prof Exp: Fel biophys, Roswell Park Mem Inst, 64-65; USPHS fel biochem, Norsk Hydro's Inst Cancer Res, 65-67; CHEMIST, WALTER REED ARMY INST RES, 67- Mem: AAAS; Am Chem Soc; Radiation Res Soc; Biophys Soc. Res: Mechanism of action of phototoxic drugs; chemical protection against phototoxicity; the narcotic receptor, mechanism of action of drugs of abuse. Mailing Add: Div of Biochem Walter Reed Army Inst of Res Washington DC 20012

COPELAND, FREDERICK CLEVELAND, b Brunswick, Maine, Oct 9, 12; m 39; c 3. BIOLOGY. Educ: Williams Col, AB, 35; Harvard Univ, AM, 37, PhD(biol), 40. Prof Exp: Asst biol, Harvard Univ, 37-40; instr, Trinity Col, Conn, 40-46, dir admis, 45-46; from asst prof to assoc prof, 46-58, PROF BIOL, WILLIAMS COL, 58-, DEAN ADMIS, 46- Concurrent Pos: Trustee, Hotchkiss Sch, Lenox Sch & Educ Rec Bur. Honors & Awards: Rogerson Cup Award, 69. Mem: Genetics Soc Am. Res: Cytogenetics; plant growth rates. Mailing Add: Admis Off Williams Col Williamstown MA 01267

COPELAND, GARY EARL, b Maud, Okla, Aug 15, 40. CHEMICAL PHYSICS. Educ: Univ Okla, BS, 64, MS, 65, PhD(physics), 70. Prof Exp: Engr plasma physics, Tex Instruments, 64-65; instr physics, Univ Okla, 69-70; fel laser physics, Langley Res Ctr, NASA, 70-71; PROF PHYSICS, OLD DOMINION UNIV, 71- Concurrent Pos: Consult, Va State Air Pollution Control, 72- & Earth Resources Observ Bd, Goddard Space Flight Ctr, NASA, 74. Mem: Am Phys Soc; Am Chem Soc; Am Geophys Union; Air Pollution Control Asn; AAAS. Res: Chemical physics, laser physics, air pollution monitoring, atmospheric physics, environmental modelling. Mailing Add: Dept of Physics & Geophys Sci Old Dominion Univ Norfolk VA 23508

COPELAND, JAMES CLINTON, b Chicago, Ill, Nov 15, 37; m 60; c 5. MICROBIAL GENETICS, MOLECULAR GENETICS. Educ: Univ Ill, Urbana-Champaign, BS, 59; Univ Tenn, MS, 61; Rutgers Univ, PhD(microbiol), 65. Prof Exp: Am Cancer Soc fel molecular genetics, Albert Einstein Col Med, 65-67; from asst geneticist to assoc geneticist, Argonne Nat Lab, 67-72; ASSOC PROF MICROBIOL, OHIO STATE UNIV, 72- Concurrent Pos: Vis lectr, Northern Ill Univ, 70-71, adj assoc prof, 71-72; NIH res career develop award, 72-; ed Microbial Genetics Bull, 74- Mem: AAAS; Am Soc Microbiol; Genetics Soc Am; Fedn Am Scientists. Res: Structure and function of microbial chromosomes, especially their organization, regulation, replication, evolution, information storage and retrieval; genetic engineering. Mailing Add: Dept of Microbiol Ohio State Univ Columbus OH 43210

COPELAND, JAMES LEWIS, b Champaign, Ill, Apr 7, 31; m 55; c 2. PHYSICAL CHEMISTRY. Educ: Univ Ill, BS, 52; Ind Univ, PhD(phys chem), 62. Prof Exp: Res assoc & fel phys chem, Inst Atomic Res, Iowa State Univ, 61-62; from asst prof to assoc prof chem, 62-74, PROF CHEM, KANS STATE UNIV, 74- Concurrent Pos: Grants, Bur of Gen Res, 63, Res Coord Coun, 64-65 & NSF, 65-74. Mem: Am Chem Soc; Am Phys Soc; Sigma Xi. Res: Physical chemistry of fused salt systems, including properties, electrochemistry, and theories of behavior; physical chemistry of high-temperature systems in general; high temperature fuel cell research; high temperature reaction kinetics. Mailing Add: Dept of Chem Kans State Univ Manhattan KS 66506

COPELAND, JOHN ALEXANDER, b Atlanta, Ga, Feb 6, 41; m 60; c 2. SOLID STATE PHYSICS. Educ: Ga Inst Technol, BS, 62, MA, 63, PhD(physics), 65. Prof Exp: Res assoc, Jet Propulsion Lab, 62-65; res physicist, Eng Exp Sta, Ga Inst Technol, 65; MEM TECH STAFF, BELL TEL LABS, 65- Concurrent Pos: Ed, IEEE Trans Electron Devices, 71- Honors & Awards: Morris N Liebmann Award, Inst Elec & Electronics Engrs, 70. Mem: Am Phys Soc; Inst Elec & Electronics Engrs. Res: Gallium arsenide microwave devices; electronic properties of semiconducting material; magnetic domain device theory and technology. Mailing Add: Bell Tel Labs 600 Mountain Ave Murray Hill NJ 07974

COPELAND, LAWRENCE O, b Battiest, Okla, Mar 1, 36; m 64. CROP SCIENCE. Educ: Ore State Univ, BS, 63, MS, 66, PhD(farm crops), 68. Prof Exp: Asst farm crops, Ore State Univ, 64-65; asst prof crop sci, 68-74, ASSOC PROF CROP & SOIL SCI, MICH STATE UNIV, 74- Mem: Am Soc Agron. Res: Seed production, testing, crop improvement. Mailing Add: Dept of Crop & Soil Sci Mich State Univ East Lansing MI 48823

COPELAND, MURRAY JOHN, b Toronto, Ont, Apr 23, 28; m 54; c 1. MICROPALEONTOLOGY, INVERTEBRATE PALEONTOLOGY. Educ: Univ Toronto, BA, 49, MA, 51; Univ Mich, PhD(geol), 55. Prof Exp: Field officer geol, 49-55, geologist, 55-65, RES SCIENTIST, GEOL SURV, CAN, 65- Mem: Fel Geol Soc Am; Am Paleont Soc; Am Asn Petrol Geologists; Geol Asn Can. Res: Micropaleontology; Paleozoic and Mesozoic Ostracoda; Paleozoic and Mesozoic Arthropoda. Mailing Add: Geol Surv Can Bldg 601 Booth St Ottawa ON Can

COPELAND, MURRAY MARCUS, b McDonough, Ga, June 23, 02; m 31. SURGERY. Educ: Oglethorpe Univ, AB, 23; Johns Hopkins Univ, MD, 27; Am Bd Surg, dipl. Hon Degrees: DSc, Oglethorpe Univ, 53. Prof Exp: Intern, City Hosp, Baltimore, 27-28; resident surg, Union Mem Hosp, Baltimore, 33-37; instr, Univ Md, 37-44; instr, Johns Hopkins Hosp, 44-45; chief, Kennedy Vet Admin Hosp, Memphis, 46-47; prof & dir dept, 47-60, EMER PROF ONCOL, MED CTR, GEORGETOWN UNIV, 60-; PROF SURG, UNIV TEX MED SCH & GRAD SCH BIOMED SCI, HOUSTON, 68-, VPRES, UNIV CANCER FOUND, 67- Concurrent Pos: Fel, Mayo Clin, 29-30; fel, Mem Hosp, New York, 30-33; chmn cancer control comt, Nat Cancer Inst, 56-58, mem, Nat Adv Cancer Coun, 58-61 & 66-69, prog dir, Nat Large Bowel Cancer Proj, Nat Cancer Inst, Div Cancer Res, Resources & Ctrs, 71-; chmn, USA comt, Int Union Against Cancer, 65-71, vpres for NAm, 70-74. Mem: Am Cancer Soc; Am Radium Soc; Am Orthop Asn; Am Acad Orthop Surg; Am Col Surg. Res: Cancer control; neoplastic diseases; professional education. Mailing Add: Univ of Tex M D Anderson Hosp & Tumor Inst Houston TX 77030

COPELAND, OTIS LEE, b Ehrhardt, SC, Feb 10, 19; c 2. ENVIRONMENTAL MANAGEMENT, RESOURCE MANAGEMENT. Educ: Clemson Col, BS, 39; Pa State Univ, MS, 41, PhD, 55. Prof Exp: Asst, Pa State Univ, 39-41; jr soil scientist, US Dept Agr, SC, 41-42; asst prof agron, Univ Ga, 46-47; soil scientist, US Dept Agr, 47-53, inland empire res ctr, US Forest Serv, 53-58, chief, Div Watershed Mgt Res, 58-65, asst dir, 65-74, group leader, Northern Gt Plains Resources Prog, 73-74; CONSULT, 75- Mem: So Am Foresters; Soil Conserv Soc Am. Mailing Add: 590 Ben Lomond Av Ogden UT 84403

COPELAND, RICHARD FRANKLIN, b Tyler, Tex, Oct 10, 38; m 68. PHYSICAL CHEMISTRY, ORGANOMETALLIC CHEMISTRY. Educ: Tex A&M Univ, BS, 61, MS, 63, PhD(chem), 65. Prof Exp: Robert A Welch fels, Tex A&M Univ, 65-66 & Univ Tex, Austin, 66-67; asst prof chem, Ball State Univ, 67-69 & Univ Mich, 69-71; ASSOC PROF CHEM, BETHUNE-COOKMAN COL, 71- Concurrent Pos: NSF grants, 68-69, 74- & NIH, 74- Mem: Am Chem Soc; Am Crystallog Asn. Res: Crystal and molecular structure of organometallic complexes; iridium-isonitrile complexes; electronic properties of materials; educational uses of computers. Mailing Add: Dept of Chem Bethune-Cookman Col Daytona Beach FL 32015

COPELAND, THOMPSON PRESTON, b Ark, March 11, 21; m 50; c 2. ENTOMOLOGY. Educ: Ouachita Baptist Col, BS, 47; George Peabody Col, MA, 50; Univ Tenn, PhD, 62. Prof Exp: Instr biol, Ouachita Baptist Col, 47-49; asst prof, Union Univ, Tenn, 51-52; assoc prof, 54-64, PROF & CHMN DEPT BIOL, E TENN STATE UNIV, 64- Concurrent Pos: Danforth assoc, 64. Mem: AAAS; Am Entom Soc. Res: Taxonomy and ecology of Protura, Collembola and Zoraptera. Mailing Add: Dept of Biol ETenn State Univ Johnson City TN 37601

COPELIN, EDWARD CASIMERE, b Philadelphia, Pa, Feb 16, 31; m 56; c 2. GEOCHEMISTRY. Educ: Pa State Univ, BS, 53; Purdue Univ, MS, 55, PhD(chem), 58. Prof Exp: RES CHEMIST, UNION OIL CO CALIF, 58- Mem: Am Chem Soc; Soc Appl Spectros. Res: Petroleum geochemistry. Mailing Add: Union Oil Co of Calif Dept of Res PO Box 76 Brea CA 92621

COPELIN, HARRY B, b Staten Island, NY, Aug 9, 18; m 43; c 3. ORGANIC CHEMISTRY. Educ: Cornell Univ, BS, 40, MS, 41. Prof Exp: Chemist, 41-58, tech assoc, 58-63, res assoc, 63-71, RES FEL, E I DU PONT DE NEMOURS & CO, INC, 71- Res: Organic chlorine compounds; coordination chemistry; catalysis; heterocyclic compounds; hydroformylation processes. Mailing Add: 2019 Longcome Dr Wilmington DE 19810

COPENHAVER, JOHN HARRISON, JR, b Ralston, Nebr, Dec 21, 22; m 46; c 5. BIOCHEMISTRY. Educ: Dartmouth Col, BA, 46; Univ Wis, MS, 49, PhD(zool), 50. Prof Exp: Nat Heart Inst fel enzyme res, Univ Wis, 50-51; asst prof pharmacol, Southwestern Med Sch, Univ Tex, 51-53; from asst prof to assoc prof zool, 53-60, PROF BIOL SCI, DARTMOUTH COL, 60- Mem: Am Soc Biol Chem. Res: Oxidative phosphorylation; enzyme chemistry of renal function; cell transport mechanisms. Mailing Add: Dept of Biol Sci Dartmouth Col Hanover NH 03755

COPENHAVER, JOHN WILLIAM, b Tazewell, Va, Oct 27, 10; m 40; c 4. ORGANIC CHEMISTRY, RESEARCH ADMINISTRATION. Educ: Emory & Henry Col, BS, 29; WVa Univ, MS, 31; Univ Ill, PhD(chem), 34. Hon Degrees: DSc, Emory & Henry Col, 70. Prof Exp: Res chemist, Socony-Vacuum Oil Co, 34-36 & Rohm and Haas Co, 36-43; sect leader, Cent Res Lab, Gen Aniline & Film Corp, 43-49; assoc dir res, M W Kellogg Co, 49-57; from assoc dir to dir res, 57-63, Cent Res Labs, Minn Mining & Mfg Co, exec dir, 63-70, vpres res & develop, Riker Labs, Inc, 70-73, consult, 3M Co, 73-75; RETIRED. Res: Lube oil additives; leather chemistry; synthetic tanning agents; synthetic organic; acetylene chemistry; fluorocarbon polymers. Mailing Add: 3830 Bow Sprit Circle Westlake Village CA 91361

COPENHAVER, WILFRED MONROE, b Westminster, Md, Dec 26, 98; m 27; c 2. ANATOMY. Educ: Western Md Col, AB, 21; Yale Univ, PhD(zool), 25. Prof Exp: Asst instr biol, Yale Univ, 21-24, asst instr anat, Sch Med, 24-25; from instr to asst prof, Univ Rochester, 25-28; from asst prof to prof, 28-67, chmn dept, 57-66, EMER PROF ANAT, COL PHYSICIANS & SURGEONS, COLUMBIA UNIV, 67-; PROF ANAT, SCH MED, UNIV MIAMI, 67- Concurrent Pos: Interim chmn dept biol struct, Sch Med, Univ Miami, 67-69. Mem: AAAS; Am Asn Anat; Am Soc Zool; Am Soc Exp Biol & Med; Soc Develop Biol. Res: Experimental embryology, especially of heart and blood; developmental anomalies; histology of heart conduction systems. Mailing Add: Dept Biol Struct Univ of Miami Biscayne Annex Box 520875 Miami FL 33152

COPES, DONALD LOUIS, forest genetics, forest physiology, see 12th edition

COPES, FREDERICK ALBERT, b Tomahawk, Wis, Dec 25, 37; m 59; c 5. ECOLOGY, FISHERIES. Educ: Wis State Univ, Stevens Point, BS, 60; Univ NDak, MS, 65; Univ Wyo, PhD(zool), 70. Prof Exp: Teacher high sch, Wis, 60-63; from instr to asst prof, 64-67, ASSOC PROF BIOL & ZOOL, UNIV WIS-STEVENS POINT, 70- Concurrent Pos: Secy, Wis Environ Pract Comn, 74-76. Mem: Am Fisheries Soc; Sigma Xi. Res: Fishes of the Red River tributaries of North Dakota; ecology of the native fishes of east Wyoming; ecology of fishes; fishery production; vital statistics of the Lake Michigan Whitefish Fishery. Mailing Add: Dept of Biol Univ of Wis Stevens Point WI 54481

COPES, JOSEPH PAUL, b Savannah, Ga, Apr 13, 17; m 42; c 6. PHYSICAL CHEMISTRY, ORGANIC CHEMISTRY. Educ: Lafayette Col, cert, 42, 46. Prof Exp: Lab control, Riegel Paper, NJ, 39-40; res asst, C K Williams & Co, Pa, 41-43; technician & apprentice, 45-46, CHEMIST, CENT RES LABS, GAF CORP, 47-

Mem: AAAS; Am Chem Soc. Res: Vapor phase catalysis; physical-chemical constants; distillation; corrosion; organic synthesis; literature and nomenclature. Mailing Add: RFD 4 Easton PA 18042

COPLAN, MICHAEL ALAN, b Cleveland, Ohio, Apr 26, 38. PHYSICAL CHEMISTRY. Educ: Williams Col, BA, 60; Yale Univ, MS, 61, PhD(electrolytic conductance), 63. Prof Exp: NIH res fel electrochem, Univ Paris, 63-64, NATO fel, 64-65; res assoc, Univ Chicago, 65-67; res asst prof, 67-72, RES ASSOC PROF ELECTROCHEM, INST FLUID DYNAMICS, UNIV MD, 72- Mem: Am Phys Soc. Res: Chemical physics; atomic and molecular collisions; charged particle optics. Mailing Add: Inst Fluid Dynamics & Appl Math Univ of Md College Park MD 20742

COPLAN, MYRON JULIUS, b Chicago, Ill, Jan 5, 22; m 52; c 2. CHEMICAL PHYSICS. Educ: Brooklyn Col, AB, 43. Prof Exp: Prod mgr, Montrose Chem Co, 43-48; head chem eng sect, Inst Textile Technol, 48-51; sr res assoc, Fabric Res Labs, Inc, 51-55, from asst dir res to assoc dir res, 55-63, vpres, 63-74, DIR, FRL, ALBANY INT CO, 74- Concurrent Pos: Lectr, Northeastern Univ; mem, Textile Res Inst. Mem: Am Chem Soc; Fiber Soc; fel Brit Textile Inst. Res: Chemical physics of textiles. Mailing Add: 47 Speen St Natick MA 01760

COPLEY, ALFRED LEWIN, b Ger, June 19, 10; nat US; wid; c 1. PHYSIOLOGY, EXPERIMENTAL MEDICINE. Educ: Univ Heidelberg, MD, 35; Univ Basel, MD, 36. Hon Degrees: Dr Med, Univ Heidelberg, 72. Prof Exp: Asst med, Univ Basel, 36-37; intern, Trinity Lutheran Hosp, Mo, 39; res assoc, Hixon Lab Med Res, Univ Kans, 40-42; asst resident med, Goldwater Mem Hosp, New York, 42-43; head lab exp surg, Univ Va, 43-44, res assoc prev med & bact, 44-45; res assoc lab cellular physiol, NY Univ, 45-49, res fel hemat, Mt Sinai Hosp, 48-49; res assoc med & asst clin prof, NY Med Col, 49-52; sr researcher & head lab blood & vascular physiol, Res Labs, Int Children's Ctr, Paris, 52-55; sr researcher, Nat Inst Hyg & head res lab microcirculation, Nat Ctr Blood Transfusion, 55-57; dir exp res vascular dis, Med Res Labs, Charing Cross Hosp, London, 57-59; vis prof path, Royal Col Surgeons Eng, 59; assoc prof physiol, 60-61, head hemorrheol lab, 60-65, prof physiol, 62-64, RES PROF PHARMACOL, NEW YORK MED COL, 65-, RES PROF MED, 72-; RES PROF LIFE SCI & BIOENG & DIR LAB BIORHEOL, POLYTECH INST NEW YORK, 74- Concurrent Pos: Celler fel, Univ Kans, 40; fel surg, Univ Va, 43-44; vis res assoc, Grad Sch Arts & Sci, NY Univ, 47-48; gen lectr, Int Rheol Cong, Neth, 48; dir, AEC Proj, 49-52; chief investr, Off Naval Res Proj, 50-52; mem med adv bd, Nat Blood Res Found, 54-; mem, Marine Biol Lab Corp, Woods Hole, 54-; gen lectr, Int Cong Blood Transfusion, Japan, 60; co-ed-in-chief & co-founder, Biorheol, 62-; co-ed, Rheologica Acta; assoc chief of staff res & educ, Vet Admin Hosp, East Orange, NJ, 65-67, chief hemorrhage & thrombosis res labs, 65-71; chmn sci organizing comt & conf chmn, Int Conf Hemorheol, Iceland, 66; adj res prof bioeng, NJ Inst Technol, 67-; invited lectr, Int Cong Rheol, Kyoto, Japan, 68, St Louis, France, 72; co-ed-in-chief & founder, Thrombosis Res, 72-; co-founder & adv, Acupuncture & Electrotherapeutic Res, 75- Honors & Awards: Poiseuille Gold Medal, Int Cong Biorheol, 72. Mem: Fel AAAS; Am Physiol Soc; Soc Study Blood; Int Soc Hemat (pres, 66-69); Int Soc Biorheol (pres, 69-72, asst pres, 72-78). Res: Blood clotting; mechanisms of thrombosis, hemorrhage and hemostasis; blood vessel wall; surface rheology and adsorption of proteins; comparative hematology; blood platelets; physiology of the spleen; immunology; brain and liver metabolism; cholinesterase; radiobiology; biorheology of hair; experimental tuberculosis; microcirculation; hemorheology; survey and studies of snake venoms; biorheology. Mailing Add: Lab of Biorheol Polytech Inst of New York Brooklyn NY 11201

COPLEY, LAWRENCE GORDON, b Reading, Eng, July 17, 39. APPLIED PHYSICS. Educ: Univ Queensland, BMechEng, 60; Harvard Univ, SM, 62, PhD(appl physics), 65. Prof Exp: Mech engr, Queensland Railways, Australia, 60-61, res engr appl physics, 65-66; sr scientist, Cambridge Acoust Assocs, Inc, 66-70; INDEPENDENT CONSULT, 70- Mem: Acoust Soc Am; Am Inst Aeronaut & Astronaut; Inst Elec & Electronics Engrs. Res: Theoretical applied mechanics with emphasis on elasticity, structural dynamics and acoustics. Mailing Add: 10 Bowers St Newton MA 02160

COPLIN, DAVID LOUIS, b Albuquerque, NMex, July 7, 45; m 68. PLANT PATHOLOGY. Educ: Univ Calif, Davis, BS, 67; Univ Wis, Madison, MS, 71, PhD(plant path & bact), 72. Prof Exp: Res assoc plant path, Univ Nebr, 72-74; ASST PROF PLANT PATH, OHIO AGR RES & DEVELOP CTR & OHIO STATE UNIV, 74- Mem: Am Phytopath Soc; Am Soc Microbiol. Res: Molecular genetics and physiology of plant pathogenic bacteria. Mailing Add: Dept Plant Path Ohio Agr Res & Develop Ctr Wooster OH 44691

COPP, DOUGLAS HAROLD, b Toronto, Ont, Jan 16, 15; m 39; c 3. PHYSIOLOGY. Educ: Univ Toronto, BA, 36, MD, 39; Univ Calif, PhD(biochem), 43. Hon Degrees: LLD, Univ Toronto, 70, Queen's Univ, Ont, 60; DSc, Univ Ottawa, 73, Acadia Univ, NS, 75; FRCP(C), 74. Prof Exp: Lectr biochem, Univ Calif, 42-43, from instr to asst prof physiol, 44-50; PROF PHYSIOL & HEAD DEPT, UNIV BC, 50- Concurrent Pos: Consult, Nat Res Coun, DC, 46-49; mem subcomt human applns, Adv Comt Isotope Distribution, US AEC, 49-50, mem, Adv Comt Clin Uses Radioisotopes, 52-; mem, Panel Radiation Protection, Defense Res Bd Can, 51-, chmn, 57-59; mem assoc comt dent res, Nat Res Coun Can, 53-59, chmn, 57-59, mem adv comt med res, 54-57; chmn, Gordon Res Conf Bones & Teeth, 57; mem sci secretariat, UN Int Conf Peaceful Uses Atomic Energy, 58; pres, Nat Cancer Inst Can, 68-70; Beaumont lectr, 70; Jacobaeus Mem lectr, Scandinavia, 71. Honors & Awards: Gairdner Found Award, 67; Nicolas Andry Award, Asn Bone & Joint Surgeons, 68; Officer, Order of Can, 71; Flavelle Medal, Royal Soc Can, 72; Steindler Award, Orthop Res Soc, 74. Mem: Am Physiol Soc; Soc Exp Biol & Med; Endocrine Soc; Can Physiol Soc (pres, 63-64); Royal Soc Can. Res: Iron and bone metabolism; fission product metabolism; severe phosphorus deficiency; regulation of blood calcium; parathyroid function; calcitonin; bone blood flow; ultimobranchial function. Mailing Add: 4755 Belmont Ave Vancouver BC Can

COPPAGE, WILLIAM EUGENE, b Geary, Okla, Nov 24, 34; m 58; c 2. ALGEBRA. Educ: Tex A&M Univ, BA, 55, MS, 56; Ohio State Univ, PhD(math), 63. Prof Exp: Assoc prof math, Ind State Col, 63-64; asst prof, 64-67, coord dept math, 65-66, ASSOC PROF MATH, WRIGHT STATE UNIV, 67- Concurrent Pos: Consult, Inst Defense Anal, 63-69; consult, Aerospace Med Res Labs, 66- Mem: Am Math Soc; Math Asn Am. Res: Non-associative algebra and applications to geometry. Mailing Add: Dept of Math Wright State Univ Colonel Glenn Hwy Dayton OH 45431

COPPEL, CLAUDE PETER, b Zweibrucken, Ger, Aug 14, 32; nat US; m 60; c 2. PHYSICAL CHEMISTRY. Educ: Univ Denver, BS, 54; Univ Calif, PhD(chem), 58. Prof Exp: Asst, Univ Calif, 54-55, asst radiation lab, 55-58; res scientist, Marathon Oil Res Ctr, Colo, 58-63; res chemist, Arthur D Little, Inc, Mass, 63-66; from res chemist to sr res chemist, 66-74, SR RES ASSOC, CHEVRON OIL FIELD RES CO, 74- Mem: AAAS; Am Chem Soc; Soc Petrol Engrs. Res: Colloidal and surface chemistry; high temperature physical measurements; fluid flow in porous media; well stimulation technology. Mailing Add: Chevron Oil Field Res Co PO Box 446 La Habra CA 90631

COPPEL, HARRY CHARLES, b Galt, Ont, Jan 2, 18; nat US; m 50; c 2. ENTOMOLOGY. Educ: Ont Agr Col, BSA, 43; Univ Wis, MSc, 46; NY State Col Forestry, Syracuse Univ, PhD(forest entom), 49. Prof Exp: Agr res officer, Entom Lab Sci Serv, Can Dept Agr, 43-57; from asst prof to assoc prof, 57-65, PROF ENTOM, UNIV WIS-MADISON, 65- Mem: Entom Soc Am; Entom Soc Can; Int Orgn Biol Control. Res: Biological control of forest insect pests. Mailing Add: Dept of Entom Univ of Wis Madison WI 53706

COPPENGER, CLAUDE JACKSON, b Beaumont, Tex, Oct 12, 27. PHYSIOLOGY. Educ: Stephen F Austin State Col, BS, 51; Tex A&M Univ, MS, 53, PhD(physiol), 64. Prof Exp: Asst physiol, Tex A&M Univ, 51-53, biochem, 54-55; instr, Pub Sch, Tex, 57-58; assoc prof zool, Amarillo Jr Col, 58-61; res assoc radiation biol, Tex A&M Univ, 61-63; from asst prof to assoc prof physiol, 63-74, PROF BIOL, SAN FRANCISCO STATE UNIV, 74-, CHMN DEPT PHYSIOL & BEHAV BIOL, 68- Mem: AAAS. Res: Mammalian physiology; effects of low intensity continuous gamma irradiation on the prenatal development of the albino rat. Mailing Add: Dept of Physiol & Behav Biol San Francisco State Col San Francisco CA 94132

COPPENS, ALAN BERCHARD, b Los Angeles, Calif, June 26, 36; c 2. PHYSICS, ACOUSTICS. Educ: Cornell Univ, BEngPhys, 59; Brown Univ, MS, 62, PhD(physics), 65. Prof Exp: Asst prof, 64-69, ASSOC PROF PHYSICS, US NAVAL POST GRAD SCH, 69- Mem: Acoust Soc Am; Am Asn Physics Teachers. Res: Finite-amplitude acoustic processes; properties of liquids; propagation of acoustic transients. Mailing Add: 3H 1155 Monarch Lane Pacific Grove CA 93950

COPPENS, PHILIP, b Amersfoort, Holland, Oct 24, 30; m 56; c 3. CRYSTALLOGRAPHY, CHEMISTRY. Educ: Univ Amsterdam, Drs, 57, PhD(crystallog), 60. Prof Exp: Res asst, Weizmann Inst, 57-60, res assoc, 62-65; res assoc, Brookhaven Nat Lab, 60-62, scientist 65-68; assoc prof, 68-71, PROF CHEM, STATE UNIV NY BUFFALO, 71- Concurrent Pos: Vis prof, Fordham Univ, 66-67; res grants, NSF, Petrol Res Fund & AEC, 69; mem nat comt crystallog, Nat Res Coun-Nat Acad Sci, 72-75; adj prof, Univ Grenoble, 74-75; mem comn charge spin & momentum densities, Int Union Crystallog, 72- & comn neutron diffraction, 75- Mem: AAAS; Am Crystallog Asn; Am Chem Soc. Res: Crystal structure determination; crystallographic computing; neutron diffraction; electron density determination by accurate diffraction methods; crystallography at liquid helium temperatures. Mailing Add: 90 Scamridge Curve Williamsville NY 14221

COPPER, PAUL, b Surabaya, Indonesia, May 6, 40; Can citizen; m 67. PALEONTOLOGY. Educ: Univ Sask, BA, 60, MA, 62; Univ London, DIC & PhD(paleontol), 65. Prof Exp: Nat Res Coun fel paleontol, Queen's Univ, Ont, 65-67; asst prof geol, 67-70, ASSOC PROF GEOL, LAURENTIAN UNIV, 70-, CHMN DEPT, 74- Honors & Awards: Huxley Prize, London, 68. Res: Brachiopod morphology, ecology and evolution, especially atrypida; evolution and paleoecology of early Paleozoic reef ecosystems. Mailing Add: Dept of Geol Laurentian Univ Sudbury ON Can

COPPI, BRUNO, b Gonzaga, Italy, Nov 19, 35; m 63; c 3. PHYSICS. Educ: Milan Polytech Inst, Dr Nuclear Eng, 59. Prof Exp: Res assoc physics, Univ Milan, 60-61; mem plasma physics lab, Princeton Univ, 61-63; asst prof & res asst physics, Univ Calif, San Diego, 64-67; mem, Inst Advan Study, 67-69; PROF PHYSICS, MASS INST TECHNOL, 69- Concurrent Pos: Ital Acad Sci grant, 62; mem, Int Ctr Theoret Physics, Trieste, 66; consult, Princeton Univ & Am Sci & Eng, Cambridge, Mass, 69-; sci counr to bd dirs, Nat Comt Nuclear Energy Italy, 71. Mem: Am Phys Soc; Ital Physics Soc. Res: Basic plasma physics; controlled thermonuclear fusion research; astrophysics; space physics; neutron transport theory. Mailing Add: Dept of Physics Mass Inst of Technol Cambridge MA 02139

COPPIN, CHARLES ARTHUR, b Belton, Tex, July 18, 41; m 64; c 2. MATHEMATICAL ANALYSIS. Educ: Southwestern Univ, Tex, BS, 63; Univ Tex, Austin, MS, 65, PhD(math), 68. Prof Exp: Instr math, Univ Tex, Austin, 68; asst prof, 68-75, ASSOC PROF MATH, UNIV DALLAS, 75-, CHMN DEPT, 73- Mem: Am Math Soc; Math Asn Am. Res: Integration theory; primitive dispersion sets; relations on topological spaces. Mailing Add: Dept of Math Univ of Dallas Irving TX 75061

COPPINGER, RAYMOND PARKE, b Boston, Mass, Feb 7, 37; m 58; c 2. BIOLOGY, ECOLOGY. Educ: Boston Univ, AB, 59; Univ Mass, Amherst, MA, 64; Amherst Col via Univ Mass, PhD(biol), 68. Prof Exp: ASSOC PROF BIOL, HAMPSHIRE COL, 69- Concurrent Pos: Res assoc, Dept Biol, Amherst Col, 68-70; mem sci adv bd, Behav Sci Found, Mass, 70- Mem: AAAS; Am Ornith Union; Brit Ornith Union. Res: Feeding behavior of birds exposed to novel food items; evolutionary and adaptive significance of acceptance or rejection of novelty; comparative biology and behavior of canids. Mailing Add: Sch of Natural Sci & Math Hampshire Col Amherst MA 01002

COPPOC, GORDON LLOYD, b Larned, Kans, Nov 11, 39; m 62; c 2. PHARMACOLOGY, VETERINARY PHARMACOLOGY. Educ: Kans State Univ, BS, 61, DVM, 63; Harvard Univ PhD(pharmacol), 68. Prof Exp: Fel, Harvard Univ, 63-66; instr pharmacol, Sch Med, Univ NC, 66-67; res pharmacologist, US Air Force Sch Aerospace Med, 67-69; res fel, Ben May Lab Cancer Res, Univ Chicago, 69-71; asst prof, 71-73, ASSOC PROF PHARMACOL, SCH VET SCI & MED, PURDUE UNIV, 73- Mem: AAAS; Am Vet Med Asn; NY Acad Sci; Asn Am Vet Med Col; Am Asn Vet Physiol & Pharmacol. Res: Steroid endocrinology; reproduction; biochemical pharmacology; tumor cell growth regulation and polyamines; pharmacology of shock. Mailing Add: Sch of Vet Med Purdue Univ W Lafayette IN 47907

COPPOC, WILLIAM JOSEPH, b Cumberland, Iowa, July 14, 13; m 39; c 2. CHEMISTRY. Educ: Ottawa Univ, Kans, BSc, 35; Rice Inst Technol, MA, 37, PhD(chem), 39. Hon Degrees: DSc, Ottawa Univ, 55. Prof Exp: Chemist, Port Arthur, Tex, 39-44, asst to asst chief chemist, 44-47, 48-49, actg asst supvr, Grease Res, Beacon, NY, 47-48, asst dir, Res, 49-51, assoc dir, New York, 51-53, dir, 53-54, mgr res, Beacon, 54-57, res & develop, 57-60, sci planning & Info, 60-65, gen mgr res & tech dept, 65-68, vpres res & tech dept, 68-71, VPRES ENVIRON PROTECTION, TEXACO, INC, 71- Concurrent Pos: Mem, Gordon Res Conf Coun, 49-, chmn; mem environ studies bd, Nat Res Coun-Nat Acad Sci, 74- Mem: AAAS; Am Chem Soc; fel Am Inst Chem; Soc Automotive Eng; Sigma Xi. Res: Coagulation of colloidal solutions; composition of hydrates; development of block type greases; ball and roller bearing greases; various industrial lubricants; soluble and non-soluble cutting oils; research administration; atmospheric chemistry. Mailing Add: Texaco Inc PO Box 509 Beacon NY 12508

COPPOCK, CARL EDWARD, b Dayton, Ohio, Dec 1, 32; m 59; c 3. ANIMAL NUTRITION. Educ: Ohio State Univ, BS, 54; Tex A&M Univ, MS, 55; Univ Md, PhD(dairy sci), 64. Prof Exp: Dairy husbandman, Agr Res Serv, USDA, 58-64; asst prof dairy cattle nutrit, 64-69, ASSOC PROF ANIMAL SCI, CORNELL UNIV, 69- Concurrent Pos: Animal husbandman, Int Voluntary Serv, Laos, 56-58. Mem: Am Dairy Sci Asn; Am Soc Animal Sci. Res: Energy metabolism of lactating cows;

nonprotein nitrogen utilization by lactating cows. Mailing Add: Rm 130 Morrison Hall Dept of Animal Sci Cornell Univ Ithaca NY 14850

COPPOCK, HENRY AARON, b Iowa City, Iowa, Jan 6, 37; m 64; c 2. HISTORICAL GEOGRAPHY, CULTURAL GEOGRAPHY. Educ: St Cloud State Col, BS, 64; Mich State Univ, MA, 66, PhD(geog), 70. Prof Exp: Asst prof geog, Univ Wyo, 69-72; ASST PROF GEOG, ST CLOUD STATE UNIV, 72- Concurrent Pos: Asst prof, Col St Benedict, Minn, 72, 74 & St John's Univ, Minn, 73, 75; consult, Apache Oro, Ideas Inc, Wyo, 72- Mem: Asn Am Geogr. Res: Process of pioneer settlement. Mailing Add: Dept of Geog St Cloud State Univ St Cloud MN 56301

COPPOCK, WILLIAM HOMER, b Lincoln, Nebr, June 8, 11; m 37; c 4. ORGANIC CHEMISTRY. Educ: Monmouth Col, BS, 33; Univ Iowa, MS, 35, PhD(org chem), 39. Prof Exp: Asst prof, Eastern Ill State Col, 39-41; asst chemist, Sangamon Ord Plant, 41-43; asst prof, Minn State Teachers Col, 44-46; sci instr, Univ Notre Dame, 43-44; head dept, 46-75, PROF CHEM, DRAKE UNIV, 75- Mem: Am Chem Soc. Res: Friedel-Crafts reaction. Mailing Add: Dept of Chem Drake Univ Des Moines IA 50311

COPPOLA, EDWARD DANTE, b Providence, RI, Dec 19, 30; m 56; c 4. SURGERY, MEDICAL EDUCATION. Educ: Amherst Col, AB, 51; Yale Univ, MD, 55; Am Bd Surg, dipl, 64. Prof Exp: Intern surg, Yale Univ Med Ctr, 55-56, asst resident, 56-61; from resident to chief resident, Hahnemann Med Col & Hosp, 61-63, from instr to sr instr, 63-66, from asst prof to assoc prof, 66-71; PROF SURG & CHMN DEPT, COL HUMAN MED, MICH STATE UNIV, 71-, CHIEF OF SURG, UNIV HEALTH CTR, 71- Concurrent Pos: Clin fel, Am Cancer Soc, 61-63, advan clin fel, 63-66; attend surgeon, Hahnemann Med Col & Hosp, 63-71 & Philadelphia Vet Admin Hosp, 63-69; Markle scholar acad med, 66-71; asst attend physician, Philadelphia Gen Hosp, 63-67, attend physician, 67-71; attend staff, Edward W Sparrow Hosp, St Lawrence Hosp & Ingham Med Ctr, 71-; consult, Depts Surg, Wayne County Gen Hosp, Gratiot Community Hosp & Saginaw Vet Admin Hosp, 72- Mem: Fel Am Col Surgeons; Transplantation Soc; Am Asn Immunologists; Asn Acad Surg; Soc Exp Biol & Med. Res: Transplantation immunology. Mailing Add: Dept of Surg Col of Med Mich State Univ East Lansing MI 48824

COPPOLA, ELIA DOMENICO, b S Salvatore Telesino, Italy, Aug 2, 41; m 72; c 1. ANALYTICAL CHEMISTRY. Educ: Southern Conn State Col, BA, 67; Rensselaer Polytech Inst, PhD(anal chem), 72. Prof Exp: Chem technician, Upjohn Co, Conn, 64-67; RES CHEMIST ANAL CHEM, CONN AGR EXP STA, 71- Mem: Am Chem Soc; Asn Official Anal Chemists. Res: The development of new analytical methods in food analysis, especially food additives; fluorometric applications in analytical chemistry. Mailing Add: 547 Evergreen Ave Hamden CT 06518

COPPOLA, JOHN ANTHONY, b Philadelphia, Pa, Jan 28, 38; m 60; c 2. ENDOCRINOLOGY, PHARMACOLOGY. Educ: La Salle Col, AB, 59; Hahnemann Med Col, MS, 61; Jefferson Med Col, PhD(pharmacol), 63. Prof Exp: Group leader endocrine pharmacol, Lederle Labs, Am Cyanamid Co, 63-72; DIR MED AFFAIRS, SCHERING CORP, 72- Concurrent Pos: Lederle fel, Univ Wis, 63-64. Mem: Endocrine Soc; Am Soc Pharmacol & Exp Therapeut; Soc Study Reproduction; Soc Study Fertil; Int Soc Res Reproduction. Mailing Add: Schering Corp Kenilworth NJ 07033

COPPOLA, PATRICK PAUL, b Buffalo, NY, June 30, 17; m 44; c 5. PHYSICAL CHEMISTRY, ELECTRONICS. Educ: Canisius Col, BS, 41, MS, 51. Prof Exp: Chemist, Anal Lab, Union Carbide & Carbon Corp, 41-42, chemist group leader, Metals Lab, 42-43, chemist, Res Lab, Manhattan Dist Contract, 43-46; res chemist, Philips Res Lab, 46-55; sr res engr, Radio Corp Am, 55-56; physicist & leader res group, 56-60, mgr chem & phys electronics adv develop, 60-66, mgr mat develop, 66-68, mgr eng visual commun prod dept, 68-72, MGR ENG VIDEO DISPLAY EQUIP OPER, GEN ELEC CO, 72- Honors & Awards: Award, US War Dept, 45. Mem: Am Chem Soc; NY Acad Sci; fel Am Inst Chem. Res: Electronic emission; vacuum physics; semiconductors; materials and processes; new business planning and organization. Mailing Add: Video Display Equip Oper Gen Elec Co Electronics Park Syracuse NY 13201

COPPOLILLO, HENRY, b Cervicati, Italy, July 27, 26; US citizen; m 62; c 3. PSYCHIATRY. Educ: Univ Rome, MD, 55. Prof Exp: Intern, Cook County Hosp, Ill, 55-56; res psychiat, Univ Chicago, 56-59; asst attend physician, Michael Reese Hosp, 59-61, pediat liaison psychiatrist, 61-65, assoc physician, 61-66, asst chief child psychiat, 63-65; asst prof, Med Sch, Univ Mich, 66-68, dir day care serv, Children's Psychiat Hosp, 66-68, assoc prof psychiat, Univ, 68-71, prof, 71, asst to chmn dept, 68-69, assoc chmn, 69-70 & 70-71, actg chmn, 70; PROF PSYCHIAT & DIR DIV CHILD PSYCHIAT, VANDERBILT UNIV, 71- Concurrent Pos: Fel, Michael Reese Hosp, Chicago, 61-63; res asst psychiat, Northwestern Univ, 59-61, instr, 61-63; child psychiat consult, McLean County Health Clin, Ill, 59-66 & Med Sch, Univ Wis, 64; lectr, Col Lit, Sci & Arts, Univ Mich, 67; fac asst, Mich Psychoanal Inst, 69-70, lectr, 70-71. Mem: Fel Am Psychiat Asn; Am Acad Child Psychiat; fel Am Col Psychiat; Sigma Xi. Res: Child and adult psychiatry and psychoanalysis. Mailing Add: Div of Child Psychiat Vanderbilt Univ Med Ctr South 2100 Pierce Ave Nashville TN 37212

COPSON, DAVID ARTHUR, b Boston, Mass, June 16, 18; m 60; c 6. BIOPHYSICS, MICROBIOLOGY. Educ: Univ Mass, BS, 40; Mass Inst Technol, PhD(tech), 53. Prof Exp: Group mgr res div, Raytheon Co, 53-58, consult microwave heating, 58-60; PROF BIOL, UNIV PR, MAYAGUEZ, 60- Concurrent Pos: Consult, IBM Corp, Whirlpool Corp, Sunbeam & FMC Corp, 58-67. Mem: AAAS; Int Microwave Power Inst; Biophys Soc. Res: Microwave biophysics; athermic microwave effects; radiation and theoretical biology; information theory and biological communication; radiation hazards; automation of telecommunications analysis; electromagnetic spectrum characteristics, documentation and classification; integration of new technology in science education. Mailing Add: Box 3661 Mayaguez PR 00708

COPSON, HARRY ROLLASON, b Easthampton, Mass, July 8, 08; m 39, 73; c 3. CORROSION. Educ: Univ Mass, BS, 29; Yale Univ, PhD(phys chem), 32. Prof Exp: Res chemist, State Water Comn, Conn, 29-30; chief chemist, Apothecaries Hall Co, 33-34; res chemist, Int Nickel Co, 34-48, supvr corrosion sect, Res Lab, 48-62, chem res mgr, 62-72, res fel, 72-73; CONSULT, INT NICKEL CO, 73- Honors & Awards: Whitney Award, Nat Asn Corrosion Eng, 60; Dudley Medal, Am Soc Test & Mat, 46. Mem: Am Chem Soc; Electrochem Soc; Am Inst Chem Eng; Nat Asn Corrosion Eng. Res: Aqueous corrosion; high temperature dry corrosion; corrosion resisting alloys; coatings; electroplating; chemical engineering; nuclear engineering; electrochemistry; metallurgy. Mailing Add: 1 Deerfield Terrace Mahwah NJ 07430

COPULSKY, WILLIAM, b Zhitomir, Russia, Apr 4, 22; US citizen; m 48; c 3. INDUSTRIAL CHEMISTRY. Educ: NY Univ, BA, 42, PhD(econ), 57; City Col New York, MBA, 48. Prof Exp: Asst res dir, J J Beruner & Staff, 46-48; asst to pres, R S Aries & Assocs, 47-51; MGR COM DEVELOP, W R GRACE & CO, 51, DIR ELECTRONUCLEONICS LABS, 69-, VPRES OPER SERV GROUP, 74- Concurrent Pos: Adj assoc prof, Baruch Col, 67- Mem: Am Chem Soc; Am Statist Asn; Am Mkt Asn. Res: Long range planning; forecasting of new technology; correlation techniques in forecasting; industrial chemical development. Mailing Add: W R Grace & Co 1114 Ave of the Americas New York NY 10036

CORAK, WILLIAM SYDNEY, b Philadelphia, Pa, Mar 10, 22; m 46; c 5. MATERIALS DEVELOPMENT, LOW TEMPERATURE PHYSICS. Educ: Univ Pa, BS, 43; Ohio State Univ, MS, 47; Univ Pittsburgh, PhD, 54. Prof Exp: Chem control engr, Davison Chem Corp, 43-44; res assoc, Johns Hopkins Univ, 44-45 & Ohio State Univ, 45-47; res engr, Westinghouse Res Labs, 47-55, supvr engr, Westinghouse Semi-Conductor Dept, 55-64, MGR SOLID STATE TECHNOL, SCI & TECHNOL DEPT, SYSTS DEVELOP DIV, WESTINGHOUSE DEFENSE CTR, 64- Mem: Am Phys Soc; AAAS; Inst Elec & Electronics Eng. Res: Low temperature gas thermodynamics; low temperature techniques; superconductivity; low temperature specific heats; materials and process development; integrated circuit technology; solid state physics. Mailing Add: Systs Develop Div Westinghouse Defense Ctr B1521 Baltimore MD 21203

CORAN, AUBERT Y, b St Louis, Mo, Mar 24, 32; m 58; c 2. PHARMACY, ORGANIC CHEMISTRY. Educ: St Louis Col Pharm, BS, 53, MS, 55. Prof Exp: Instr chem, St Louis Col Pharm, 53-55; res chemist, Mo, 55-62, res group leader, WVa, 62-64, scientist, Mo, 64-70, sect mgr, Rubber Chem Res, Ohio, 70-75, DISTINGUISHED SCI FEL, MONSANTO CO, 75- Mem: AAAS; Am Chem Soc; Sigma Xi. Res: Polymer and rubber chemistry; vulcanization kinetics and chemistry. Mailing Add: Monsanto Co Akron OH 44313

CORAOR, GEORGE ROBERT, b Jacksonville, Ill, May 10, 24; m 47; c 3. ORGANIC CHEMISTRY. Educ: Ill Col, AB, 47; Univ Ill, PhD(org chem), 50. Prof Exp: Res assoc org chem, Mass Inst Technol, 50-51; sr chemist, 51-53, group leader, 53-55, sect head, 55-60, DIV HEAD ORG CHEM, E I DU PONT DE NEMOURS & CO, INC, 60- Mem: AAAS; Am Chem Soc. Res: Free radical reaction; coordination chemistry; photochemistry. Mailing Add: 29 Paxon Dr Penarth Wilmington DE 19803

CORBASCIO, ALDO NICOLA, b Castellana, Italy, Mar 21, 28; nat US; m 55; c 1. PHARMACOLOGY. Educ: Q Orazio Flacco, Bari, Italy, MA, 47; Univ Bari, MD, 53; Univ Pa, DSc(pharmacol), 58. Prof Exp: Intern, Med Clin, Univ Bari, 53-54; resident med, Univ Hosp, Univ Pa, 54-56, instr pharmacol, Sch Med, 56-59; asst res pharmacologist, Med Ctr, Univ Calif, San Francisco, 59-65; assoc prof, 63-67, PROF PHARMACOL & CHMN DEPT, SCH DENT, UNIV OF THE PAC, 68- Concurrent Pos: Lectr, Med Sch, Univ Ore; consult, Surgeon Gen, US Army Dept Anesthesiol, Med Sch, Stanford Univ. Mem: Am Soc Pharmacol & Exp Therapeut; Am Therapeut Soc. Res: Cardiovascular pharmacology. Mailing Add: Dept of Pharmacol Univ of the Pac Sch of Dent San Francisco CA 94115

CORBATO, CHARLES EDWARD, b Los Angeles, Calif, July 12, 32; m 57; c 3. GEOLOGY, GEOPHYSICS. Educ: Univ Calif, Los Angeles, BA, 54, PhD(geol), 60. Prof Exp: Instr geol, Univ Calif, Riverside, 59; from instr to asst prof, Univ Calif, Los Angeles, 59-66; assoc prof, 66-69, PROF GEOL, OHIO STATE UNIV, 69-, CHMN DEPT GEOL & MINERAL, 72- Mem: Geol Soc Am; Soc Explor Geophys; Am Geophys Union; Int Asn Math Geol. Res: Exploration geophysics; structural geology; computer applications to geological problems. Mailing Add: Dept of Geol & Mineral Ohio State Univ Columbus OH 43210

CORBATO, FERNANDO JOSE, b Oakland, Calif, July 1, 26; m 62. PHYSICS. Educ: Calif Inst Technol, BS, 50; Mass Inst Technol, PhD(physics), 56. Prof Exp: Res assoc, 56-59, asst dir in charge programming res, Comput Ctr, 59-60, assoc dir, 60-63, from assoc prof to prof elec eng, 62-74, DEP DIR COMPUT CTR, MASS INST TECHNOL, 63-, ASSOC HEAD DEPT COMPUT SCI & ENG, MASS INST TECHNOL, 74- Concurrent Pos: Mem comput sci & eng bd, Nat Acad Sci, 71-73. Honors & Awards: W W McDowall Award, Inst Elec & Electronics Eng, 66. Mem: Am Phys Soc; Asn Comput Mach; fel Inst Elec & Electronics Eng; fel Am Acad Arts & Sci. Res: Computer operating systems; time-sharing systems; automatic programming and knowledge-based application systems. Mailing Add: Dept of Elec Eng & Comput Sci Mass Inst of Technol Cambridge MA 02139

CORBEELS, ROGER, b Kessel-Loo, Belgium, Apr 16, 36; m 62. FUEL SCIENCE, HIGH TEMPERATURE CHEMISTRY. Educ: Cath Univ Louvain, BS, 56, MS, 58, PhD(chem), 60. Prof Exp: Vis res assoc flame kinetics, Aerospace Res Labs, Wright-Patterson Air Force Base, Ohio, 62-64; sr chemist, 64-70, res chemist, 70-73, SR RES CHEMIST, TEXACO INC, 73- Mem: Am Chem Soc; Combustion Inst. Res: Kinetics of combustion reactions; coal chemistry. Mailing Add: Pine Ridge Dr Wappingers Falls NY 12590

CORBEN, HERBERT CHARLES, b Dorset, Eng, Apr 18, 14; US citizen; m 41, 57; c 3. THEORETICAL PHYSICS. Educ: Univ Melbourne, MA & MSc, 36; Cambridge Univ, PhD(theoret physics), 39. Prof Exp: Lectr math & physics, Univ Col Armidale, NSW, 41; dean, Trinity Col & lectr math & physics, Univ Melbourne, 42-44, sr lectr, 45-46; from asst prof to prof physics, Carnegie Inst Technol, 46-56; mem tech staff, Ramo-Wooldridge Corp, 54-55, 56-57, assoc dir res lab, 57-61; dir quantum physics lab, TRW, Inc, 61-68, chief scientist, Phys Res Ctr, 66-68; prof physics, Cleveland State Univ, 68-72, dean faculties & grad studies, 68-70, vpres acad affairs, 70-72; PROF PHYSICS & CHMN PHYS SCI DEPT, SCARBOROUGH COL, UNIV TORONTO, 72- Concurrent Pos: Rouse Ball res student, Trinity Col, Cambridge Univ, 36-39; Fulbright vis prof, Univs Genoa, Milan & Bologna, 51-53; consult, Ramo-Wooldridge Corp, 55-56. Mem: Fel Am Phys Soc; Can Asn Physicists. Res: Quantum theory; relativity theory; theory of nuclear forces; electromagnetic propagation; nuclear reactor theory; coherent radiation; theory of elementary particles. Mailing Add: Phys Sci Group Scarborough Col Univ of Toronto Toronto ON Can

CORBET, JOHN HARRY, b Memphis, Tenn, Jan 10, 31; m 53; c 1. GEOGRAPHY. Educ: Memphis State Univ, BS, 53, MA, 54; Univ Fla, PhD(geog), 66. Prof Exp: From instr to asst prof, 58-67, ASSOC PROF GEOG, MEMPHIS STATE UNIV, 67- Mem: Asn Am Geog. Res: Physical and agricultural geography; geomorphology; Union of Soviet Socialist Republic; air photo interpretation. Mailing Add: Dept of Geog Memphis State Univ Memphis TN 38111

CORBET, PHILIP STEVEN, b Kuala Lumpur, Malaya, May 21, 29; m 57. ZOOLOGY, ENTOMOLOGY. Educ: Univ Reading, BSc, 49, BSc, 50, DSc(zool), 62; Cambridge Univ, PhD(entom), 53. Prof Exp: Res officer, EAfrican Fisheries Res Orgn, Uganda, 54-57; entomologist, EAfrican Virus Res Inst, 57-62; entomologist, Entom Res Inst, Can Dept Agr, 62-67, dir, Can Agr Res Inst, 67-71; prof biol & chmn dept, Univ Waterloo, 71-74, adj prof, 74-75; PROF BIOL & DIR, JOINT CTR ENVIRON SCI, UNIV CANTERBURY, 75- Mem: Brit Ecol Soc; Royal Entom Soc London; Brit Inst Biol; Entom Soc Can. Res: Ecological and physiological aspects of seasonal regulation and rhythmic behavior of insects, particularly Odonata and Diptera of medical importance; pest management. Mailing Add: Joint Ctr for Environ Sci Univ of Canterbury Christchurch New Zealand

CORBETT, EDWARD GEORGE, horticulture, botany, see 12th edition

CORBETT, GAIL RUSHFORD, b Rapid River, Mich, May 23, 36; m 59; c 2. PLANT ECOLOGY, TAXONOMIC BOTANY. Educ: Univ Mich, BA, 58, MS, 60, PhD(bot), 67. Prof Exp: Instr bot, 63, RES CONSULT FLORA, W VA UNIV, 68- Mem: Sigma Xi; Ecol Soc Am; Bot Soc Am; AAAS. Res: Phytosociological study of disturbed habitats; taxonomic investigation of the goldenrods of West Virginia. Mailing Add: 220 Atterbury Blvd Hudson OH 44236

CORBETT, JAMES MURRAY, b Welland, Ont, Can, Jan 29, 38; m 58; c 3. PHYSICS. Educ: Univ Toronto, BASc, 60; Univ Waterloo, MSc, 61, PhD(physics), 66. Prof Exp: Lectr, 62-66, asst prof, 66-74, ASSOC PROF PHYSICS, UNIV WATERLOO, 74- Concurrent Pos: Nat Res Coun Can res fel, Dept Metall, Oxford Univ, 67-68. Mem: Electron Micros Soc Am; Can Asn Physicists. Res: Nucleation, growth and structure of vacuum-deposited films; electron microscopy of thin crystals. Mailing Add: Dept of Physics Univ of Waterloo Waterloo ON Can

CORBETT, JAMES WILLIAM, b New York, NY, Aug 25, 28; m 54, 72; c 2. SOLID STATE PHYSICS. Educ: Univ Mo, BS, 51, MA, 52; Yale Univ, PhD(physics), 55. Prof Exp: Res assoc chem, Yale Univ, 55; res assoc, Gen Elec Res Lab, 55-68; chmn physics dept, 69- 70, PROF PHYSICS, STATE UNIV NY ALBANY, 68-, DIR, INST STUDY DEFECTS IN SOLIDS, 74- Concurrent Pos: Adj prof, Rensselaer Polytech Inst, 64-68; chmn, Int Conf Radiation Effects in Semiconductors, 70 & Int Conf Radiation-Induced Voids in Metals, 71; exchange scholar, US-USSR Exchange Prog, Nat Acad Sci, 73; Guggenheim fel, 75; lectr oenology, State Univ NY Albany, 72- Mem: Fel Am Phys Soc; Am Asn Physics Teachers; Inst Elec & Electronics Engrs; NY Acad Sci. Res: Point defects in semiconductors and metals; radiation damage; reaction kinetics; nucleation theory. Mailing Add: Dept of Physics State Univ of NY Albany NY 12222

CORBETT, JOHN DUDLEY, b Yakima, Wash, Mar 23, 26; m 48; c 3. INORGANIC CHEMISTRY. Educ: Univ Wash, BS, 48, PhD(chem), 52. Prof Exp: From assoc chemist to sr chemist, Ames Lab, US AEC, 52-68, div chief, 68-73; from asst prof to assoc prof chem, 52-63, chmn dept, 68-73, PROF CHEM, IOWA STATE UNIV, 63-, ASST PROG DIR, AMES LAB, ENERGY RES & DEVELOP ADMIN, 74- Mem: Am Chem Soc. Res: Inorganic and physical chemistry; unfamiliar oxidation states; solid state chemistry. Mailing Add: Dept of Chem Iowa State Univ Ames IA 50011

CORBETT, JOHN FRANK, b Doncaster, Eng, May 8, 35; m 59; c 3. ORGANIC CHEMISTRY, COSMETIC CHEMISTRY. Educ: Royal Inst Chem, London, ARIC, 57; Univ Reading, PhD(org chem), 61. Prof Exp: Res scientist chem, Gillette Res Labs, Eng, 61-62, head dept org chem, 63-69; sr res scientist chem, Gillette Co, Chicago, 70-71; mgr special proj, Gillette Res Labs, Eng, 72; dir res chem, 72-74, VPRES RES, CLAIROL RES LAB, STAMFORD, CONN, 74- Concurrent Pos: Chmn hair color tech comt, Cosmetics, Toiletry & Frangrance Asn, 73- Mem: Fel Royal Inst Chem; The Chem Soc; Am Chem Soc; Soc Cosmetic Chemists; Fel Soc Dyers & Colorists. Res: Synthesis and properties of dyestuffs, mechanism of oxidative dyeing processes. Mailing Add: Clairol Res Lab 2 Blachley Rd Stamford CT 06902

CORBETT, JOHN JOSEPH, biology, see 12th edition

CORBETT, JULES JOHN, b Natrona, Pa, Apr 12, 19; m 50; c 3. MICROBIOLOGY. Educ: Univ Chicago, BS, 50; Ill Inst Technol, MS, 57. Prof Exp: Instr microbiol & chem, Sch Nursing & bacteriologist, Englewood Hosp, 50-54; dir labs, Beverly Med Arts Bldg, 54-55; from instr to assoc prof, 56-72, chmn biol dept, 74-76, PROF BIOL, ROOSEVELT UNIV, 72- Concurrent Pos: Bacteriologist, Borden Co, Ill, 55-64. Mem: AAAS; Am Soc Microbiol. Res: Immunology and pathogenicity of chromobacterium violaceum. Mailing Add: Dept of Biol Roosevelt Univ Chicago IL 60605

CORBETT, M KENNETH, b Port Lorne, NS, Sept 12, 27; m 50; c 3. PLANT PATHOLOGY. Educ: McGill Univ, BSc, 50; Cornell Univ, PhD(plant path), 54. Prof Exp: Asst plant path, Cornell Univ, 50-54; asst plant pathologist & asst prof, Univ Fla, 54-60, assoc virologist, Agr Exp Sta, 60-66; assoc prof, 66-69, PROF BOT, UNIV MD, 69- Concurrent Pos: Guggenheim fel, Netherlands, 64-65. Mem: Am Phytopath Soc. Res: Virology; biochemistry; plant physiology; plant breeding and pathology. Mailing Add: Dept of Bot Univ of Md Col of Agr College Park MD 20742

CORBETT, ROBERT G, b Chicago, Ill, Mar 13, 35; m 59; c 1. ECONOMIC GEOLOGY, GEOCHEMISTRY. Educ: Univ Mich, BS, 58, MS, 59, PhD(geol), 64. Prof Exp: From asst prof to assoc prof geol, WVa Univ, 62-69, prin investr, Water Res Inst, 67-69; assoc prof, 69-75, PROF GEOL, UNIV AKRON, 75-, COORDR RES, 72- Mem: Mineral Soc Am; Geochem Soc; Nat Asn Geol Teachers; Geol Soc Am; Soc Res Adminr. Res: Formation of hydroxylapatite; geology and mineralogy of uranium deposits; chemical characteristics of natural waters; mineral deposits. Mailing Add: Dept of Geol Univ of Akron Akron OH 44304

CORBIN, ALAN, b New York, NY, Sept 3, 34; m 59. NEUROENDOCRINOLOGY, REPRODUCTIVE PHYSIOLOGY. Educ: City Col New York, BS, 56; Univ Iowa, MS, 60, PhD(physiol), 61. Prof Exp: Instr anat, Albert Einstein Col Med, 61-63; instr, Inst Pharmacol, Milan, Italy, 63-64; sr investr endocrinol, Sci Div, Abbott Labs, 64-66; sr res scientist, Squibb Inst Med Res, 66-71; HEAD ENDOCRINOL SECT, WYETH LABS, 71- Concurrent Pos: Fel neuroanat, Albert Einstein Col Med, 61-63 & Inst Pharmacol, Milan, Italy, 63-64; NIH fel, 61-64; adj prof physiol, Rutgers Univ, New Brunswick; mem res grant review adv panel, NSF. Mem: Endocrine Soc; Int Soc Res Reproduction; Int Soc Neuroendocrinol; Am Physiol Soc. Res: Neuroendocrinology of mammalian reproduction and pituitary-adrenal axis; general endocrinology; neuroanatomy; pharmacology of hypothalamic releasing factors; neuropharmacology. Mailing Add: Endocrinol Sect Wyeth Labs Box 8299 Philadelphia PA 19101

CORBIN, FREDERICK THOMAS, b Franklin, NC, Dec 2, 29; m 52; c 2. WEED SCIENCE, PLANT PHYSIOLOGY. Educ: Wake Forest Col, BS, 51; Univ NC, MEd, 56; NC State Univ, PhD(physiol, microbiol), 65. Prof Exp: Teacher high sch, NC, 53-56; instr chem & physics, Mars Hill Col, 56-60; asst prof physics, ECarolina Col, 60-62; res asst, 62-65, ASST PROF WEED SCI, N C STATE UNIV, 65- Concurrent Pos: Agr Res Serv grant, 65-68. Mem: AAAS; Weed Sci Soc Am; Am Phys Soc. Res: Interactions between major classes of chemical pesticides; biotransformation of herbicides. Mailing Add: Dept Crop Sci 457 Williams Hall N C State Univ Raleigh NC 27607

CORBIN, JACK DAVID, b Franklin, NC, Feb 8, 41; m 65; c 1. PHYSIOLOGY, BIOCHEMISTRY. Educ: Tenn Technol Univ, BS, 63; Vanderbilt Univ, PhD(physiol), 68. Prof Exp: ASST PROF PHYSIOL, VANDERBILT UNIV, 71- Concurrent Pos: NIH fel, Univ Calif, Davis, 68-70; Am Diabetes Asn fel, Univ Calif, Davis & Vanderbilt Univ, 70-71 & Vanderbilt Univ, 71-72; NIH res grant, 72, Diabetes Ctr grant, 74. Mem: AAAS. Res: Molecular endocrinology; hormone regulation; cyclic nucleotide regulation; protein kinase. Mailing Add: Dept of Physiol Vanderbilt Univ Sch of Med Nashville TN 37232

CORBIN, JAMES EDWARD, b Providence, Ky, July 14, 21; m 50; c 4. ANIMAL NUTRITION. Educ: Univ Ky, BS, 43, MS, 47; Univ Ill, PhD(animal nutrit), 51. Prof Exp: Dir res, Nat Oats Co, 51-54; mgr special chows res, Ralston Purina Co, 54-59, mgr dog res, 59-67; dir pet care ctr, 67-73; PROF ANIMAL SCI, UNIV ILL, URBANA, 73- Concurrent Pos: Chmn dog nutrit subcomt, Nat Res Coun, 68-; chmn dog & cat standards, Inst Lab Animal Resources, Nat Acad Sci. Mem: Am Soc Animal Sci; Am Asn Lab Animal Sci (pres, 72-73); Am Inst Nutrit; Am Asn Lab Animal Sci; Brit Small Animal Vet Med Asn. Res: Nutrition of dogs and laboratory animals. Mailing Add: 160 Animal Sci Bldg Dept of Animal Sci Univ of Ill Urbana IL 61801

CORBIN, JAMES LEE, b Coshocton, Ohio, Oct 20, 35; m 60. ORGANIC CHEMISTRY. Educ: Bowling Green State Univ, BA, 57; Mich State Univ, PhD, 62. Prof Exp: Proj chemist, Am Oil Co, Ind, 62-63; INVESTR, KETTERING RES LAB, 63- Mem: Am Chem Soc; The Chem Soc. Res: Structure of natural products; synthesis; design and synthesis of novel ligand systems; metal complexes. Mailing Add: Charles F Kettering Res Lab 150 E Colleg St Yellow Springs OH 45387

CORBIN, KENDALL BROOKS, b Oak Park, Ill, Dec 31, 07; m 32; c 2. NEUROLOGY, MEDICAL ADMINISTRATION. Educ: Stanford Univ, AB, 31, MD, 35. Prof Exp: Instr anat, Stanford Univ, 34-38; from assoc prof to prof, Col Med, Univ Tenn, 38-46, chief div, 41-46, in charge neurol, 43-46; prof neuroanat, 46-54, prof neurol, 54-72, EMER PROF NEUROL, MAYO GRAD SCH, 72- Concurrent Pos: Nat Res Coun fel, Neurol Inst, Northwestern Univ, 37-38; assoc dir, Mayo Found, 50-54, pres staff, 68; head sect neurol, Mayo Clin, 57-63, sr consult, 63-72, emer consult, Mayo Clin & Mayo Found, 72-; pres, Friends of Gardens, Mario Selby Bot Gardens, Fla, 75- Mem: Am Physiol Soc; Soc Exp Biol & Med; fel AMA; Am Asn Anat; fel Am Acad Neurol. Res: Neuroanatomy; neurophysiology; clinical neurology; medical administration. Mailing Add: 7323 Pine Needle Rd Sarasota FL 33581

CORBIN, KENDALL WALLACE, b Memphis, Tenn, Apr 5, 39; m 61; c 2. EVOLUTIONARY BIOLOGY, POPULATION ECOLOGY. Educ: Carleton Col, BA, 61; Cornell Univ, PhD(vert biol), 65. Prof Exp: NIH res fel, 65-67; res assoc & lectr biol, Yale Univ, 67-70; asst prof ecol & behav biol, 70-73, ASSOC PROF ECOL & BEHAV BIOL, UNIV MINN, MINNEAPOLIS, 73-, CUR SYSTS, 70-, CHMN PROG EVOLUTIONARY & SYST BIOL, 74- Mem: AAAS; Soc Study Evolution; Am Ornith Union; Cooper Ornith Soc. Res: Evolutionary relationships among bird species using comparative biochemical data on the structure of protein molecules; studies of protein polymorphisms; gene flow and natural selection in animal populations. Mailing Add: Dept of Ecol & Behav Biol Univ of Minn Bell Mus Natural Hist Minneapolis MN 55455

CORBIN, LUDLOW, b Croton, Ohio, Feb 21, 15; m 40; c 3. PHYCOLOGY. Educ: Marion Col, Ind, ThB, 39, AB, 44; Ball State Teachers Col, AM, 46. Prof Exp: Prof biol & chmn div sci & math, Cascade Col, 46-69; PROF BIOL & CHMN DEPT SCI & MATH, WARNER PAC COL, 69- Mem: AAAS; Bot Soc Am; Am Inst Biol Sci. Res: Preservation of biological specimens; plant physiology; fossils; fresh water and marine algae. Mailing Add: Warner Pac Col Dept Math & Sci 2219 SE 68th Ave Portland OR 97215

CORBIN, THOMAS ELBERT, b Orange, NJ, Sept 6, 40; m 66. ASTRONOMY. Educ: Harvard Univ, AB, 62; Georgetown Univ, MA, 69; Univ Va, PhD(astron), 76. Prof Exp: ASTRONOMER, US NAVAL OBSERV, 64- Mem: Am Astron Soc. Res: Positional astronomy; fundamental astrometry; proper motion systems. Mailing Add: US Naval Observ Washington DC 20390

CORBIN, THOMAS F, physical organic chemistry, see 12th edition

CORBIT, JOHN DARLINGTON, JR, obstetrics, gynecology, deceased

CORBY, DONALD G, b Jamestown, NDak, Jan 13, 34; m 59; c 6. PEDIATRICS, HEMATOLOGY. Educ: Univ NDak, BSc, 57; Northwestern Univ, MD, 59; Am Bd Pediat, dipl, 65. Prof Exp: Intern, Evanston Hosp Asn, Ill, 59-60; resident pediat, Children's Mem Hosp, Chicago, 60-61; resident pediat, US Army, 61-, chief pediat, 225th Sta Hosp, Europe, 61-63, resident pediat, Brooke Gen Hosp, Ft Sam Houston, Tex, 63-65, asst chief pediat, William Beaumont Gen Hosp, El Paso, 65-68, dir spec educ & clin res pediat, 70-71, CHIEF CLIN RES SERV, FITZSIMONS GEN HOSP, 71-; ASST CLIN PROF PEDIAT, SCH MED, UNIV COLO, DENVER, 70- Concurrent Pos: Fel pediat hemat, Univ Ill, 68-70. Honors & Awards: Cert of Achievement & Army Commendation Medal, William Beaumont Gen Hosp, 68. Mem: AAAS; fel Am Acad Pediat; Am Acad Clin Toxicol; Am Soc Hemat; Soc Pediat Res. Res: Toxicology; specifically prevention and treatment of accidental poisoning in childhood; coagulation physiology. Mailing Add: Clin Res Serv Fitzsimons Gen Hosp Denver CO 80240

CORCHARY, GEORGE SUTTER, b Concord, NH, Apr 14, 23; m 59; c 1. GEOLOGY. Educ: Univ NH, BS, 48; Univ Ill, MS, 54. Prof Exp: GEOLOGIST, US GEOL SURV, 48- Mem: AAAS; Am Quaternary Asn. Res: Engineering and glacial geology; geomorphology. Mailing Add: US Geol Surv Fed Ctr Denver CO 80225

CORCINO, JOSE JUAN, b Humacao, PR, May 22, 38; c 2. HEMATOLOGY, GASTROENTEROLOGY. Educ: Univ PR, BS, 58, MD, 62; Tulane Univ, MS, 64. Prof Exp: Fel hemat, Sch Med, Univ PR, 65-66; USPHS fel, Mt Sinai Sch Med, 68-70; trainee & instr gastroenterol, Sch Med, Univ Rochester, 70-72; DIR TROP MALABSORPTION UNIT, SCH MED, UNIV PR, 72-, DIR GEN CLIN RES CTR, 73- Concurrent Pos: NIH acad career award nutrit, Sch Med, Univ PR, 73- Mem: Am Soc Hemat; Am Soc Clin Nutrit; Am Fedn Clin Res; AAAS; Am Col Nutrit. Res: Nutritional anemias, especially megaloblastic anemias; vitamin B12 and folate metabolism; intestinal absorption and malabsorption; tropical sprue, etiology, pathogenism and therapy. Mailing Add: Gen Clin Res Ctr Univ of PR Sch of Med GPO Box 5067 San Juan PR 00936

CORCORAN, EUGENE FRANCIS, b Arthur, NDak, Nov 28, 16; m 40; c 2. OCEANOGRAPHY. Educ: NDak Agr Col, BS, 40; Univ Calif, PhD(oceanog), 58. Prof Exp: Instr math, San Diego State Col, 46-49; asst marine biochem, Scripps Inst, Univ Calif, 49-50, res biochemist, 52-57; from asst prof to assoc prof, 57-72, PROF MARINE BIOCHEM, DOROTHY H & LEWIS ROSENSTIEL SCH MARINE & ATMOSPHERIC SCI, UNIV MIAMI, 72- Concurrent Pos: Assoc prog dir facils & spec progs, NSF, 67-68. Mem: AAAS; Am Chem Soc; Geochem Soc; Am Soc Limnol & Oceanog; Marine Biol Asn UK. Res: Organic productivity of sea water; photosynthesis in marine plants; organic constituents of sea water and marine sediments. Mailing Add: Sch of Marine & Atmospheric Sci Univ of Miami 4600 Rickenbacker Miami FL 33149

CORCORAN, JOHN W, b Dayton, Ohio, Sept 23, 24; m 49; c 2. PEDODONTICS. Educ: Miami Univ, BA, 48; Western Reserve Univ, DDS, 53; Am Bd Pedodontics, dipl, 62. Prof Exp: Intern, Univ Hosps, Cleveland, Ohio, 53-54, resident, 54-55; pvt pract pedodontics, 55-65; from asst prof to assoc prof pedodontics, Case Western Reserve Univ, 65-68; PROF PEDODONTICS & CHMN DEPT, MED UNIV SC, 68-, STAFF MEM, DEPT DENT, UNIV HOSP, 69- Concurrent Pos: Supvr, Dent Fillings Clin, City Cleveland Babies & Children's Hosp, 58-65; assoc dent surgeon & dent surgeon, Univ Hosps, Cleveland, 65-68; dir dent serv, Coastal Habilitation Ctr, 68- Mem: Am Dent Asn; Am Acad Pedodontics; Am Soc Dent for Children. Res: Clinical research for the handicapped child. Mailing Add: Dept of Pediat Dent Med Univ of SC Charleston SC 29401

CORCORAN, JOHN WILLIAM, b Des Moines, Iowa, June 12, 27; m 48; c 2. BIOCHEMISTRY. Educ: Iowa State Univ, BS, 49; Western Reserve Univ, PhD, 55. Prof Exp: Instr biochem, Columbia Univ, 56-57; from asst prof to assoc prof, Western Reserve Univ, 57-68; PROF BIOCHEM & CHMN DEPT, MED SCH, NORTHWESTERN UNIV, CHICAGO, 68- Concurrent Pos: Vis fel, Columbia Univ, 55-56; Am Heart Asn res fel, 55-58; USPHS career develop award, 64-; estab investr, Am Heart Asn, 58-63; acad guest, Lab Org Chem, Swiss Fed Inst Technol, 64-65; pharmaceut consult, 68- Mem: AAAS; Am Soc Biol Chem; Am Soc Microbiol; Am Chem Soc. Res: Mechanisms of sensitivity and resistance to antibiotics; antibiotic chemistry; natural product chemistry; carbohydrate chemistry; biosynthesis of natural products; carbohydrate metabolism; actinomycete metabolism. Mailing Add: Dept of Biochem Northwestern Univ Med Sch Chicago IL 60611

CORCORAN, MARY RITZEL, b Los Angeles, Calif, July 3, 28; m 57; c 2. GENETICS, PLANT PHYSIOLOGY. Educ: Univ Calif, Los Angeles, BA, 53, PhD(bot), 59. Prof Exp: Asst bot, Univ Calif, Los Angeles, 54-59, jr res botanist, 59-62; from asst prof to assoc prof, 62-71, PROF BIOL, CALIF STATE UNIV, NORTHRIDGE, 71- Mem: AAAS; Bot Soc Am; Am Soc Plant Physiol. Res: Plant growth hormones and inhibitors. Mailing Add: Dept of Biol Calif State Univ Northridge CA 91324

CORCORAN, VINCENT JOHN, b Chicago, Ill, Oct 7, 34; m 57; c 5. LASERS. Educ: Univ Notre Dame, BSEE, 57; Univ Ill, MSEE, 58; Univ Fla, PhD(elec eng), 68. Prof Exp: Staff mem infrared, Lab Appl Sci, Univ Chicago, 58-62; vpres lasers, Astromarine Prod Corp, 62-63; sr res scientist, Martin Marietta Aerospace, 63-73; RES STAFF MEM LASERS, INST DEFENSE ANAL, 73- Mem: Optical Soc Am; Inst Elec & Electronic Engrs. Res: Laser technology needed for technology base for future Department of Defense systems including blue-green laser, search radar, laser aided imaging, coherent receivers, optical phased arrays and atmospheric transmission properties in the far infrared region. Mailing Add: Inst for Defense Anal 400 Army-Navy Dr Arlington VA 22202

CORCOS, ALAIN FRANCOIS, b Paris, France, June 7, 25; US citizen; m 50; c 2. GENETICS. Educ: Mich State Univ, BS, 51, MS, 52, PhD(plant breeding), 60. Prof Exp: Plant breeder, Grant Merrill Orchards, Inc, 58-60; assoc biol, Univ Calif, Santa Barbara, 61-63; instr, Ore Col Educ, 63-64; res assoc virol, Inst Cancer Res, Philadelphia, 64-65; from asst prof to assoc prof natural sci, 65-73, NATURAL SCI, MICH STATE UNIV, 73- Concurrent Pos: Co-ed, Arabidopsis Inf Serv Newslett, 74- Mem: Genetics Soc Am; Am Genetic Asn; NY Acad Sci. Res: Genetics of bacteriophages; genetics and molecular biology of Arabidopsis thaliana. Mailing Add: 129 Highland East Lansing MI 48823

CORDARO, J CHRISTOPHER, molecular biology, genetics, see 12th edition

CORDELL, RICHARD WILLIAM, b Brooklyn, NY, Oct 4, 39; m 66. ANALYTICAL CHEMISTRY. Educ: Villanova Univ, BS, 61; Ohio Univ, PhD(anal chem), 66. Prof Exp: Asst prof, 65-69, ASSOC PROF CHEM, HEIDELBERG COL, 69- Mem: Am Chem Soc. Res: Platinum group metals; organic analytical reagents. Mailing Add: Dept of Chem Heidelberg Col Tiffin OH 44883

CORDELL, ROBERT JAMES, b Quincy, Ill, Jan 7, 17; m 42; c 3. PETROLEUM GEOLOGY. Educ: Univ Ill, BS, 39, MS, 40; Univ Mo, PhD(geol), 49. Prof Exp: Asst, Univ Ill, 39-40; asst instr geol, Univ Mo, 41-42, instr 46-47; instr Colgate Univ, 47-51; res geologist, Sun Oil Co, 51-53, dir res, Abilene Labs, 53-55, mgr geol res, 55-63, supvr basic res, 63-66, sr sect mgr basic res, 66-68 & paleontology, 68-70, res scientist, 70-75, SR RES SCIENTIST, SUN OIL CO, 75- Concurrent Pos: Mem US Potential Gas Comt, 75- Mem: Fel AAAS; Paleont Soc; Soc Econ Paleontologists & Mineralogists; Am Asn Petrol Geologists; fel Geol Soc Am. Res: Origin, migration and accumulation of oil and natural gas; determination of geological parameters-categories preferentially associated with major petroleum; prediction of amount and distribution of undiscovered oil and gas. Mailing Add: Sun Oil Co Prod Res Lab 503 N Central Expressway Richardson TX 75080

CORDEN, BRIAN JOSEPH, b Chattanooga, Tenn, Oct 11, 43. MEDICINE, BIOINORGANIC CHEMISTRY. Educ: Georgetown Univ, BSc, 65; Brown Univ, PhD(inorg chem), 70; Georgetown Univ, med, 72- Prof Exp: Kettering Found fel bioinorg chem, Kettering Res Lab, Ohio, 69-70; Nat Sci Coun fel, Univ Col, Dublin, 70-72; fel, Brown Univ, 72 & Georgetown Univ, 72-73. Res: Electron spin resonance of transition metal complexes; eight coordination; nitrogen fixation; trace metal incorporation and transport; pharmacological action of sodium nitroprusside. Mailing Add: 407 Brookfield Ave Chattanooga TN 37411

CORDEN, MALCOLM ERNEST, b Portland, Ore, Nov 8, 27; m 48; c 6. PLANT PATHOLOGY. Educ: Ore State Univ, BS, 52, PhD(plant path), 55. Prof Exp: Asst plant pathologist, Crop Protection Inst, Conn Agr Exp Sta, 55-58; assoc prof, 58, PROF PLANT PATH, ORE STATE UNIV, 58- Mem: AAAS; Am Phytopath Soc. Res: Physiology of parasitism in fungal diseases of plants and the mechanism of fungicidal action. Mailing Add: Dept of Bot & Plant Path Ore State Univ Corvallis OR 97331

CORDEN, PIERCE STEPHEN, b Chattanooga, Tenn, June 26, 41; m 75. PHYSICS. Educ: Georgetown Univ, BS, 63; Univ Pa, MS, 66, PhD(physics), 71. Prof Exp: PHYS SCI OFFICER WEAPONS TECHNOL & ARMS CONTROL, US ARMS CONTROL & DISARMAMENT AGENCY, WASHINGTON, DC, 71- Mem: Am Phys Soc. Res: Science and technology for arms control and disarmament, especially verification of nuclear testing and aspects of environmental warfare. Mailing Add: 4914 Crescent St Chevy Chase MD 20016

CORDER, CLINTON NICHOLAS, b Oberlin, Kans, Aug 1, 41; m 61; c 5. PHARMACOLOGY, MEDICINE. Educ: Univ Kans, BS, 64; Marquette Univ, PhD(pharmacol), 68; Wash Univ, MD, 71. Prof Exp: Res asst pharmacol, Sch Med, Wash Univ, 71-72; ASST PROF PHARMACOL & MED, SCH MED, UNIV PITTSBURGH, 72- Concurrent Pos: Nat Inst Neurol Dis & Blindness fel, 68-69; Am Cancer Soc scholar, 69-71; NIH trainee clin pharm, Univ Pittsburgh, 72-74. Mem: Am Pharmaceut Asn; Am Chem Soc; AMA. Res: Biochemical pharmacology, quantitative microbiochemical analysis, and drug biotransformation in clinical pharmacology. with emphasis on kidney, hepatic, and neoplastic tissues. Mailing Add: Dept of Pharmacol Univ of Pittsburgh Sch of Med Pittsburgh PA 15261

CORDES, ARTHUR WALLACE, b Freeport, Ill, June 29, 34; m 56; c 4. INORGANIC CHEMISTRY. Educ: Northern Ill Univ, BS, 56; Univ Ill, MS, 58, PhD(inorg chem), 60. Prof Exp: From asst prof to assoc prof, 59-66, PROF INORG CHEM, UNIV ARK, FAYETTEVILLE, 66- Mem: Am Chem Soc; Am Crystallog Asn. Res: Inorganic chemistry of the elements phosphorus, nitrogen, sulfur and arsenic; crystal structure determinations. Mailing Add: Dept of Chem Univ of Ark Fayetteville AR 72701

CORDES, DAVIES MARCIA A, b Omaha, Nebr, Nov 24, 38; m 71. MOLECULAR SPECTROSCOPY. Educ: Duchesne Univ, BA, 61; Univ Notre Dame, PhD(inorg chem), 66. Prof Exp: Res assoc, Univ Ariz, 66-67; asst prof, 67-74, ASSOC PROF CHEM, CREIGHTON UNIV, 74- Mem: Am Chem Soc; Soc Appl Spectros. Res: Metal isotope effects in far infrared spectra of metal complexes; magnetic susceptibilities of metal complexes. Mailing Add: 3031 Lincoln Blvd Omaha NE 68131

CORDES, EUGENE H, b York, Nebr, Apr 7, 36; m 57; c 2. BIOCHEMISTRY, ORGANIC CHEMISTRY. Educ: Calif Inst Tech, BS, 58; Brandeis Univ, PhD(biochem), 62. Prof Exp: From instr to assoc prof chem, 62-70, PROF CHEM, IND UNIV, BLOOMINGTON, 70-, CHMN DEPT, 72- Concurrent Pos: Grants, NSF, 62-75 & NIH, 64-75. Mem: Am Chem Soc; Am Soc Biol Chem. Res: Mechanism of enzyme-catalyzed reactions; mechanism and catalysis of carbonyl addition reactions. Mailing Add: Dept of Chem Ind Univ Col of Arts & Sci Bloomington IN 47401

CORDES, HERMAN FREDRICK, b Upland, Calif, May 30, 27. PHYSICAL CHEMISTRY. Educ: Pomona Col, BA, 50; Stanford Univ, PhD(chem), 54. Prof Exp: RES CHEMIST, NAVAL WEAPONS CTR, 54- Mem: AAAS; Am Chem Soc. Res: Chemical kinetics; physical chemistry of propellant and explosive ingredients. Mailing Add: Michelson Lab Code 6054 Naval Weapons Ctr China Lake CA 93555

CORDES, JAMES MARTIN, b Conneaut, Ohio, Dec 3, 49. RADIO ASTRONOMY. Educ: Univ Calif, Santa Barbara, BS, 71; Univ Calif, San Diego, PhD(appl physics), 75. Prof Exp: RES ASSOC ASTRON, UNIV MASS, AMHERST, 75- Res: Analyses of signals from pulsars aim at understanding the many observed time scales of polarization fluctuations, the emission mechanism and how the objects are coupled to the interstellar medium. Mailing Add: Dept of Physics & Astron Univ of Mass Amherst MA 01002

CORDES, WILLIAM CHARLES, b St Louis, Mo, Aug 17, 29; m 57; c 6. PLANT PHYSIOLOGY, CYTOCHEMISTRY. Educ: Univ Mo, BS, 55, MA, 57, PhD(plant physiol), 60. Prof Exp: From instr to asst prof biol, Creighton Univ, 60-65; actg chmn dept, 70-72, ASST PROF BIOL, LOYOLA UNIV CHICAGO, 65- Mem: AAAS; Bot Soc Am; Japanese Soc Plant Physiol. Res: Physiology of differentiation in plants; appearance of enzyme systems in time; plant wound enzymes. Mailing Add: Dept of Biol Loyola Univ Chicago IL 60626

CORDINGLY, RICHARD HENRY, b Denver, Colo, Aug 9, 31; m 56; c 2. PULP & PAPER TECHNOLOGY. Educ: Univ Colo, BS, 53; Inst Paper Chem, MS, 55, PhD(chem), 58. Prof Exp: Proj chemist papermaking, 58-60, asst tech dir, 61-62, res engr, 63-70, MGR RES & DEVELOP, WEYERHAEUSER CO, 71- Mem: Tech Asn Pulp & Paper Indust. Res: Wood pulping and bleaching; pulp refining; paper forming, pressing and drying; surface treating; paper product technology and new paper product development. Mailing Add: 7505 90th Ave SW Tacoma WA 98498

CORDON, MARTIN, b West New York, NJ, Aug 10, 28; m 53; c 3. ORGANIC CHEMISTRY. Educ: Rutgers Univ, BS, 50; Univ Calif, Los Angeles, PhD(chem), 55. Prof Exp: Sr res chemist, 55-70, RES ASSOC, COLGATE-PALMOLIVE CO, 70- Mem: Am Chem Soc. Res: Biochemistry of saliva; chemistry of hair; organic synthesis. Mailing Add: 55 Grant Ave Highland Park NJ 08904

CORDON, THEONE CHANDLER, microbiology, see 12th edition

CORDS, CARL ERNEST, JR, b South Bend, Ind, Aug 24, 33; m 58; c 2. VIROLOGY, MEDICAL MICROBIOLOGY. Educ: Ariz State Univ, BS, 58; Univ Wash, PhD(microbiol), 64. Prof Exp: NIH fel, 64-65, from instr to asst prof microbiol, 65-72, ASSOC PROF MICROBIOL, SCH MED, UNIV NMEX, 72- Mem: AAAS; Am Soc Microbiol; Tissue Cult Asn. Res: Genetics and mechanisms of replication of animal viruses and bacteriophage; tissue culture. Mailing Add: Dept of Microbiol Univ of NMex Sch of Med Albuquerque NM 87131

CORDS, DONALD PHILIP, b Evanston, Ill, Sept 18, 40; m 64; c 2. ORGANIC CHEMISTRY. Educ: Northwestern Univ, BA, 62; Ind Univ, PhD(org chem), 66. Prof Exp: Res chemist, 66-73, SR SUPVR, E I DU PONT DE NEMOURS & CO, INC, 73- Mem: Am Chem Soc. Res: Applications of organometallics in organic synthesis; free radical rearrangement and elimination reactions; organic reaction mechanisms. Mailing Add: E I du Pont de Nemours & Co Org Chem Dept PO Box 525 Wilmington DE 19898

CORDS, HELMUTH, organic chemistry, deceased

CORDS, HOWARD PAUL, b Glendale, Ariz, Mar 6, 19; m 46. AGRONOMY, WEED SCIENCE. Educ: Univ Ariz, BSA, 41, MS, 42; Ohio State Univ, PhD(agron), 54. Prof Exp: From instr to asst prof agron, Univ Ariz, 46-54; from asst prof to assoc prof, 54-64, PROF AGRON, UNIV NEV, RENO, 64-, CHMN DIV PLANT, SOIL & WATER SCI, 75- Mem: AAAS; Am Soc Agron; Weed Sci Soc; Crop Sci Soc. Res: Weed control; crop production. Mailing Add: Div of Plant Soil & Water Sci Univ of Nev Reno NV 89507

CORDTS, RICHARD HENRY, JR, b Teaneck, NJ, May 1, 34; m 66; c 1. NUTRITION. Educ: Rutgers Univ, BS, 56, MS, 61, PhD(nutrit), 64. Prof Exp: Asst, Rutgers Univ, 61-64; dir nutrit res, Whitmoyer Labs, Inc, Rohm and Haas Co, 64-65, dir nutrit, 65-68; assoc dir res & develop, Nat Molasses Co, 68-70; nutritionist, 70-73, assoc mgr animal health tech serv, 73-75, MGR MKT RES & PLANNING, HOFFMANN LaROCHE, INC, NUTLEY, 75- Mem: AAAS; Animal Nutrit Res Coun; NY Acad Sci; fel Am Inst Chemists; Am Mkt Asn. Res: Roughage utilization by ruminants; protein metabolism in monogastrics. Mailing Add: 1 Sylvan Pl Nutley NJ 07110

CORDY, DONALD R, b Fall River, Wis, Feb 17, 13. VETERINARY PATHOLOGY. Educ: Univ Calif, Los Angeles, BA, 34; Iowa State Col, DVM, 37; Cornell Univ, MS, 38, PhD(vet path), 40. Prof Exp: Instr vet path, State Col Wash, 40-42, assoc prof, 46-50; from asst prof to assoc prof, 50-58, head dept, 58-69, PROF VET PATH, UNIV CALIF, DAVIS, 58- Mem: Am Col Vet Pathologists; Am Vet Med Asn; Am Soc Exp Path. Res: Pathology of communicable animal diseases; animal neuropathology. Mailing Add: Dept of Vet Path Univ of Calif Davis CA 95616

CORE, EARL LEMLEY, b Core, WVa, Jan 20, 02; m 25; c 4. BOTANY. Educ: WVa

Univ, AB, 26, AM, 28; Columbia Univ, PhD(bot), 36. Hon Degrees: DSc, Waynesburg Col, 57; DSc, WVa Univ, 74. Prof Exp: From instr to assoc prof bot, 28-42, prof biol, 42-72, chmn dept bot, 48-66, cur herbarium, 66-72, EMER PROF BIOL, WVA UNIV, 72- Concurrent Pos: Botanist, Foreign Econ Admin, Colombia, 43-45; ed, Castanea, 36-70. Mem: AAAS; Bot Soc Am; Am Soc Plant Taxonomists. Res: Taxonomy of vascular plants of eastern United States; American species of Scleria; flora of West Virginia. Mailing Add: Dept of Biol WVa Univ Morgantown, WV 26506

CORE, HAROLD ADDISON, b Cassville, WVa, Nov 4, 20; m 43; c 2. FORESTRY. Educ: WVa Univ, BSF, 42; State Univ NY, MS, 49, PhD, 62. Prof Exp: From asst prof to assoc prof wood prod eng, State Univ NY Col Forestry, Syracuse, 46-66; PROF FORESTRY, COL AGR, UNIV TENN, KNOXVILLE, 66- Mem: Soc Am Foresters; Forest Prod Res Soc. Res: Wood and fiber anatomy; foreign woods. Mailing Add: Col of Agr Univ of Tenn Knoxville TN 37916

CORE, SUE KARICKHOFF, biochemistry, organic chemistry, see 12th edition

CORELLI, JOHN CHARLES, b Providence, RI, Aug 6, 30; m 59; c 1. RADIATION BIOPHYSICS, SOLID STATE SCIENCE. Educ: Providence Col, BSc, 52; Brown Univ, MSc, 54; Purdue Univ, PhD(physics), 58. Prof Exp: Physicist, Knolls Atomic Power Lab, Gen Elec Co, 58-62; assoc prof, 62-65, PROF, DEPT NUCLEAR ENG & SCI, RENSSELAER POLYTECH INST, 65- Concurrent Pos: NIH fel, Univ Rochester, 71. Mem: Am Phys Soc; Radiation Res Soc. Res: Radiation damage studies in silicon; low energy charged particle nuclear scattering; properties of thermoelectric materials; effects of ionizing radiation in nucleic acids and cells studied by electron spin resonance and infrared spectroscopy. Mailing Add: 33 Belle Ave Troy NY 12180

CORET, IRVING ALLEN, b Salt Lake City, Utah, Apr 28, 20; m 46; c 1. PHARMACOLOGY. Educ: Emory Univ, AB, 40, MD, 43. Prof Exp: Intern, Piedmont Hosp, Ga, 44; instr pharmacol, Col Physicians & Surgeons, Columbia Univ, 47-48; from instr to assoc prof, 50-73, PROF PHARMACOL, SCH MED, ST LOUIS UNIV, 73- Concurrent Pos: Rockefeller fel, Col Physicians & Surgeons, Columbia Univ, 46-47; NIH fel, Univ Pa, 48-50. Mem: AAAS; Am Soc Pharmacol & Exp Therapeut. Res: Antibiotics; autonomic and cellular pharmacology. Mailing Add: Dept of Pharmacol Sch of Med St Louis Univ St Louis MO 63103

COREY, ALBERT EUGENE, b Gardner, Mass, July 4, 28; m 50; c 2. ORGANIC CHEMISTRY. Educ: Rensselaer Polytech, BS, 50. Prof Exp: Assoc chemist, Allied Chem & Dye Corp, 51-54; res chemist, Plymouth Cordage Co, 54-55; res specialist, Shaw- inigan Resins Corp, 55-69 & Plastics Prod & Resins Div, Monsanto Co, 69-73, SR RES SPECIALIST, MONSANTO POLYMERS & PETROCHEMICALS CO, 73- Mem: Am Chem Soc. Res: Glycerol; hydroxylation; non-electrolytic hydrogen peroxide; polymers; emulsion polymers; paper and surface coatings; textile applications; pressure sensitive adhesive applications. Mailing Add: 185 Mountainview Rd East Longmeadow MA 01028

COREY, ALFRED J, physical chemistry, see 12th edition

COREY, ELIAS JAMES, b Methuen, Mass, July 12, 28; m 61; c 3. ORGANIC CHEMISTRY. Educ: Mass Inst Technol, BS, 48, PhD(chem), 51. Hon Degrees: DSc, Univ Chicago, 68, Hofstra Univ, 74. Prof Exp: Res chemist, A D Little Co, Inc, 48; from instr to prof chem, Univ Ill, 51-59; PROF CHEM, HARVARD UNIV, 59- Concurrent Pos: Consult, Chas Pfizer Co, Honors & Awards: Chem Award, Am Chem Soc, 60, Fritzsche Award, 68, Synthetic Chem & Harrison Howe Awards, 70, Linus Pauling Award, 73, Remsen Award, 74; Ciba Found Medal & Evans Award, Ohio State Univ, 72; Dickson Prize Sci, Carnegie Mellon Univ, 73; George Ledlie Prize Sci, Harvard Univ, 73. Mem: Am Chem Soc. Res: Stereochemistry; structural, synthetic and theoretical organic chemistry. Mailing Add: Dept of Chem Harvard Univ Cambridge MA 02138

COREY, EUGENE R, b Oregon City, Ore, Nov 18, 35; m 62. INORGANIC CHEMISTRY, CRYSTALLOGRAPHY. Educ: Willamette Univ, BS, 58; Univ Wis, PhD(inorg chem), 63. Prof Exp: Instr chem, Univ Wis, 63-64; assoc prof, Univ Cincinnati, 64-69; ASSOC PROF CHEM, UNIV MO-ST LOUIS, 69- Mem: Am Chem Soc; Am Crystallog Asn; The Chem Soc; NY Acad Sci. Res: Chemical crystallography; single crystal structure determinations of organometallic compounds by the method of x-ray diffraction. Mailing Add: Dept of Chem Univ of Mo St Louis MO 63121

COREY, JAMES LAURENCE, physical chemistry, analytical chemistry, see 12th edition

COREY, JOHN CHARLES, b Toronto, Ont, May 7, 38; US citizen; m 66; c 2. SOIL PHYSICS. Educ: Univ Toronto, BSA, 60; Univ Calif, MS, 62; Iowa State Univ, PhD(soil physics), 66. Prof Exp: Res physicist, 66-73, TECH SUPVR, SAVANNAH RIVER LAB, E I DU PONT DE NEMOURS & CO, INC, 73- Mem: AAAS; Am Soc Agron; Am Geophys Soc; Int Soc Soil Sci. Res: Management of radioactive waste; ion transport; nuclear techniques for soil water movement studies. Mailing Add: Savannah River Lab E I du Pont de Nemours & Co Inc Aiken SC 29801

COREY, JOYCE YAGLA, b Waverly, Iowa, May 26, 38; m 62. INORGANIC CHEMISTRY. Educ: Univ NDak, BS, 60, MS, 61; Univ Wis, PhD(inorg chem), 64. Prof Exp: From instr to asst prof chem, Villa Madonna Col, 64-68; asst prof, 68-71, ASSOC PROF CHEM, UNIV MO-ST LOUIS, 71- Mem: Am Chem Soc; The Chem Soc; Sigma Xi. Res: Synthetic organometallic chemistry of group IV, specifically analogs of antidepressants. Mailing Add: Dept of Chem Univ of Mo St Louis MO 63121

COREY, RICHARD BOARDMAN, b Wisconsin Rapids, Wis, Dec 25, 27. SOIL CHEMISTRY, SOIL FERTILITY. Educ: Univ Wis-Madison, BS, 49, MS, 51, PhD(soil Chem), 53. Prof Exp: From asst prof to assoc prof soil sci, 54-65, PROF SOIL SCI, UNIV WIS-MADISON, 65- Concurrent Pos: Vis prof, Postgrad Col, Nat Sch Agr, Mex, 64; consult, US AID-Univ Wis Proj, Porto Alegre, Brazil, 65, prof, Univ Ife, Nigeria, 67-71, dean fac agr, 70-71; consult, US AID-Midwest Univ Consortium Int Activ Proj, Bogor, Indonesia, 72. Mem: AAAS; Am Soc Agron; Soil Sci Soc Am; Int Soc Soil Sci. Res: Reactions of phosphorus and potassium in soils; development of methods for determining available nutrients in soils. Mailing Add: Dept of Soil Sci Univ of Wis Madison WI 53706

COREY, ROBERT ARDEN, b Wheeling, WVa, Aug 31, 20; m 44; c 3. ENTOMOLOGY. Educ: WVa Univ, BS, 47; Univ Calif, MS, 49. Prof Exp: Field aide forest insect div, Bur Entom & Plant Quar, USDA, 48; ENTOMOLOGIST, SHELL DEVELOP CO, 49- Mem: Entom Soc Am. Res: Testing new compounds as potential insecticides. Mailing Add: Shell Develop Co Box 4248 Modesto CA 95353

COREY, ROLAND REECE, JR, b Easton, Md, Oct 25, 26; m 49; c 3. MICROBIOLOGY. Educ: Washington Col, BS, 48; Univ Md, MS, 51; Univ Calif, Davis, PhD(microbiol), 55. Prof Exp: From instr to assoc prof bot & bact, Univ Ark, 55-64; prof biol & chmn dept, Anne Arundel Col, 64-69; part-time assoc prof, 65-69, PROF BIOSCI, DEPT CHEM, US NAVAL ACAD, 69- Concurrent Pos: Classified res, US Chem Corp, 55-58; fel, Inst Col Teachers Bot, Ind Univ, 59; US Fish & Wildlife res grant, 61-64. Mem: Bot Soc Am; Am Soc Microbiol. Res: Fungicides and fungicidal activity; transformation of Xanthomonas; microbial genetics; aquatic microbiology; disease of wildlife. Mailing Add: Div of Sci US Naval Acad Annapolis MD 21402

COREY, VICTOR BREWER, b Bynumville, Mo, Feb 9, 15; m 42; c 2. PHYSICS. Educ: Cent Col, Mo, AB, 37; Univ Iowa, MS, 39, PhD(physics), 42. Prof Exp: Asst physics, Univ Iowa, 37-42; res physicist, Sylvania Elec Prods, Inc, Pa & NY, 42-45; head electro-acoustics sect, Curtiss-Wright-Cornell Res Lab, 45-46; head electro-acoustics dept & actg head nuclear physics dept, Fredric Flader, Inc, 46-48, res coordr, 48-49; mgr eng physics div, 49-50, dir res, 50-51; exec engr, Electronics Div, Willys Motors, Inc, Toledo, 51-53; tech dir, Donner Sci Co, 53-59; pres, Palomar Sci Corp, Calif, 60-63; vpres & bd dir, United Control Corp, 63-67, gen mgr transducer div, 66, int vpres, 66-71, VPRES NEW BUS DEVELOP, SUNDSTRAND DATA CONTROL, INC, 71- Mem: AAAS; Am Phys Soc; Acoust Soc Am; Inst Elec & Electronics Engrs; Am Inst Aeronaut & Astronaut. Res: Simulators and analog computers; servomechanisms; telemetering; missile guidance, control components; trainers; servo accelerometers, electromechanical amplifiers, integrators; applied physics instrumentation; electronic test and measurement equipment; digital transducers; analog-digital converters; microelectronics; radio altimeters. Mailing Add: Sundstrand Data Control Inc Overlake Indust Park Redmond WA 98052

CORFIELD, PETER WILLIAM REGINALD, b Manchester, Eng, Sept 14, 37; m 63; c 2. INORGANIC CHEMISTRY, X-RAY CRYSTALLOGRAPHY. Educ: Univ Durham, BSc, 59, PhD(x-ray), 63. Prof Exp: Res assoc, Crystallog Lab, Univ Pittsburgh, 63-65; res instr chem, Northwestern Univ, 65-66, instr, 66-67; asst prof, Ohio State Univ, 67-73; ASSOC PROF CHEM, KING'S COL, 73- Mem: AAAS; Am Crystallog Asn; Am Chem Soc; The Chem Soc. Res: Crystal and molecular structure by x-ray methods; development of computer programs in crystallography. Mailing Add: Dept of Chem The Kings Col Briarcliff Manor NY 10510

CORFMAN, PHILIP ALBERT, b Berea, Ohio, July 19, 26; m 50; c 4. OBSTETRICS & GYNECOLOGY. Educ: Oberlin Col, BA, 50; Harvard Univ, MD, 54; Am Bd Obstet & Gynec, dipl. Prof Exp: Mem staff clin obstet & gynec, Rip Van Winkle Clin, Hudson, NY, 59-63; prog assoc population res, 65-67, asst to dir population res, 67-68, DIR CTR POPULATION RES, NAT INST CHILD HEALTH & HUMAN DEVELOP, 68- Concurrent Pos: Josiah Macy, Jr Found res fel cervical carcinogenesis & population res, Col Physicians & Surgeons, Columbia Univ, 63-65; adv, WHO. Mem: Am Col Obstet & Gynec. Res: Population research, including biological studies in animals and humans; research administration of this field. Mailing Add: Landow Bldg A721 Inst Child Health Human Develop Bethesda MD 20014

CORI, CARL FERDINAND, b Prague, Austria, Dec 5, 96; nat US; m 20; c 1. PHARMACOLOGY, BIOCHEMISTRY. Educ: Ger Univ, Prague, MD, 20. Hon Degrees: ScD, Western Reserve Univ & Yale Univ, 47, Boston Univ, 48, Cambridge Univ, 49, Gustavus Adolphus Col, 64, St Louis Univ & Brandeis Univ, 65, Wash Univ & Monash Univ, 66 & Univ Granada, 67; MD, Univ Trieste, 72. Prof Exp: Instr, 2nd Med Clin, Prague Univ, 19-20; asst, 1st Med Clin, Univ Vienna, 20-21; asst pharmacol, Graz Univ, 21; biochemist, State Inst Study Malignant Dis, NY, 22-31; prof pharmacol & biochem, Sch Med, Washington Univ, 31-66; DIR ENZYME RES LAB, MASS GEN HOSP, HARVARD MED SCH, 66- Concurrent Pos: Asst prof physiol, Univ Buffalo, 30-31; pres, Int Cong Biochem, Vienna, 58; cis prof biochem, Harvard Med Sch, 67- Honors & Awards: Nobel Prize, 47. Mem: Nat Acad Sci; AAAS; Am Soc Pharmacol & Exp Therapeut; Am Soc Biol Chem (pres, 50); Am Chem Soc. Res: Influence of ovariectomy on tumor incidence; fate of sugar in the animal body; intestinal absorption; action of epinephrin on metabolism; phosphate changes in muscle; isolation of glucose-1 phosphoric acid; enzymatic synthesis of glycogen; aerobic phosphorylation; isolation of crystalline enzymes; mechanism of action of insulin; regualtions; inborn errors of metabolism. Mailing Add: Enzyme Res Lab Mass Gen Hosp Boston MA 02114

CORIELL, KATHLEEN PATRICIA, b Cumberland, Md, Feb 19, 35; m 63. COMPUTER SCIENCES. Educ: Univ Md, BS, 59, MS, 66; Howard Univ, PhD(physics), 69. Prof Exp: Physicist, Nat Bur Standards, 59-63; asst prof physics, Hood Col, 69-70; instr, Montgomery Col, Md, 72; programmer analyst, Greenwich Data Systs, 73; MEM TECH STAFF, COMPUT SCI CORP, 74- Mem: Am Phys Soc. Res: Computer programming and analysis; satellite attitude determination and control. Mailing Add: 18948 Whetstone Circle Gaithersburg MD 20760

CORIELL, LEWIS L, b Sciotoville, Ohio, June 19, 11; m 36; c 3. BACTERIOLOGY. Educ: Univ Mont, BA, 34; Univ Kans, MA, 36, PhD(bact), 40, MD, 42. Prof Exp: Asst instr bot, Univ Mont, 34; from asst instr to instr bact, Univ Kans, 34-40; instr pediat, 46-49, assoc prof immunol pediat, 49-63, PROF PEDIAT, UNIV PA, 63-; DIR, INST MED RES, 55- Concurrent Pos: Nat Res sr fel virus dis, Inst Med Res, 47-49; med dir, Camden Munic Hosp for Contagious Dis, 49-61; pediatrist, Cooper Hosp, 49-; sr physician, Children's Hosp, 54-; consult, Philadelphia Naval Hosp, 56-66. Mem: AAAS; assoc AMA; assoc Am Soc Microbiol; assoc Asn Mil Surg US; assoc Soc Pediat Res. Res: Preservation of bacteria lymphilization; natural immunity and streptomycin therapy of tularemia; botulism; herpes simplex; herpes zoster; natural immunity of cats; poliomyelitis; antibiotics; tissue culture; cancer; pediatrics. Mailing Add: Inst for Med Res Copewood & Davis Sts Camden NJ 08103

CORIELL, SAM RAY, b Greenfield, Ohio, Dec 21, 35; m 63. PHYSICAL CHEMISTRY, PHYSICAL METALLURGY. Educ: Ohio State Univ, BSc, 56, PhD(phys chem), 61. Prof Exp: Phys chemist, Statist Physics Sect, 61-63, PHYS CHEMIST, METALL DIV, NAT BUR STANDARDS, 63- Concurrent Pos: Nat Res Coun-Nat Bur Standards res assoc, 61-63. Mem: AAAS; Am Chem Soc; Am Phys Soc; Am Inst Chemists; Metall Soc. Res: Crystal growth; solidification; heat flow and diffusion. Mailing Add: Nat Bur Standards Washington DC 20234

CORINALDESI, ERNESTO, b Italy, Aug 20, 23. QUANTUM MECHANICS. Educ: Univ Rome, BS, 44; Univ Manchester, PhD(theoret physics), 51. Prof Exp: Asst, Univ Rome, 45-47; res fel, Nat Res Coun Can, 51-52; Higgins vis fel & instr, Palmer Phys Lab, Princeton Univ, 52-53; asst, Dublin Inst Adv Studies, 53-55; Imp Chem Indust res fel, Univ Glasgow, 55-57; lectr math, Univ Col North Staffordshire, 57-58; dir grad sch nuclear studies & in charge of theoret physics course, Nat Inst Nuclear Physics, Pisa, 59-61; mem, Inst Adv Study, 61-62; assoc vis prof, Univ Iowa, 62-63; vis prof, Univ Toronto, 63-64; vis prof, Boston Univ, 65; fel scientist, Westinghouse Res & Develop Ctr, Pa, 65-66; PROF PHYSICS, BOSTON UNIV, 66- Mem: Italian Phys Soc. Res: Quantum mechanics. Mailing Add: Dept of Physics Boston Univ 700 Commonwealth Ave Boston MA 02215

CORK, BRUCE, b Peck, Mich, Oct 21, 15; m 46; c 4. PARTICLE PHYSICS. Educ:

Univ Mich, BS, 37; Polytech Inst New York, MS, 41; Univ Calif, Berkeley, PhD, 60. Prof Exp: Physicist, Tube Lab, Radio Corp Am, 37-40; physicist, Radiation Lab, Mass Inst Technol, 41-45 & Dept Physics, 46; physicist, Los Alamos Sci Lab, Univ Calif, 45-46 & Lawrence Radiation Lab, 46-68; assoc lab dir high energy physics, Argonne Nat Lab, 68-73; PHYSICIST, LAWRENCE BERKELEY LAB, UNIV CALIF, BERKELEY, 73- Concurrent Pos: Chmn LAMPF policy comt, Los Alamos Sci Lab, 73- Mem: Am Phys Soc. Res: Nuclear scattering; antiproton; antineutron; parity nonconservation; resonant states of nucleons. Mailing Add: Lawrence Berkeley Lab Univ of Calif Berkeley CA 94720

CORKE, CHARLES THOMAS, b Stratford, Ont, Mar 19, 21; m 45; c 4. SOIL MICROBIOLOGY. Educ: Univ Western Ont, BSc, 50, MSc, 51; Rutgers Univ, PhD, 54. Prof Exp: Assoc prof, 54-68, PROF MICROBIOL, UNIV GUELPH, 68- Mem: Can Soc Microbiol; Agr Inst Can. Res: Forest microbiology; nitrogen transformations in acid forest soils of Ontario; thermophilic actinomycetes, growth, nutrition and phage relationships; importance and ecology of soil actinomycete viruses; effects of pesticides on microbial activities in agricultural soils. Mailing Add: Dept of Microbiol Univ of Guelph Guelph ON Can

CORKER, GERALD ALOYSIUS, physical organic chemistry, biochemistry, see 12th edition

CORKERN, WALTER HAROLD, b Washington Parish, La, Mar 28, 39; m 59; c 3. ORGANIC CHEMISTRY. Educ: La State Univ, BS, 61; Univ Ark, PhD(org chem), 66. Prof Exp: Res chemist, E I du Pont de Nemours & Co, Tex, 65-66; asst prof chem, 66-72, ASSOC PROF CHEM, SOUTHEASTERN LA UNIV, 72- Mem: AAAS. Res: Acid catalized ketone rearrangements; mechanistic studies; organic reaction mechanisms. Mailing Add: 205 Alexander Hammond LA 70401

CORKINS, JACK PHILIPS, b Hamilton, Mont, Feb 13, 21; m 48; c 5. ENTOMOLOGY, APPLIED PHYSIOLOGY. Educ: Mont State Col, BS, 50. Prof Exp: Head farm sci dept, Occident Elevators, 46-49; asst state entomologist, Mont State Col, 49-54; field entomologist, Skyway Flying Serv, 54; area supvr res & develop, Naugatuck Chem Div, US Rubber Co, 55-65; SR RES BIOLOGIST, UNIROYAL CHEM DIV, UNIROYAL INC, 65- Mem: Am Entom Soc; Am Phytopath Soc; Weed Sci Soc Am. Res: Foliar absorption, translocation and plant responses of certain growth regulators; acaricide and insecticide research; cuticular penetration and translocation of systemic pesticides in plants. Mailing Add: 1696 S Leggett Ave Porterville CA 93257

CORKUM, KENNETH C, b Aurora, Ill, Aug 9, 30. PARASITOLOGY, INVERTEBRATE ZOOLOGY. Educ: Aurora Col BS, 58; La State Univ, MS, 60, PhD(zool), 63. Prof Exp: NIH fel parasitol, Sch Med, Tulane Univ, 63-65; asst prof , 65-69, ASSOC PROF ZOOL & PHYSIOL, LA STATE UNIV, BATON ROUGE, 69- Mem: Am Soc Parasitol; Am Micros Soc; Soc Syst Zool; Am Soc Zool. Res: Marine trematode taxonomy; cestode life cycles. Mailing Add: Dept of Zool & Physiol La State Univ Baton Rouge LA 70803

CORLESS, JOSEPH MICHAEL JAMES, b Orlando, Fla, July 28, 44; m 74; c 1. BIOPHYSICS, VISION. Educ: Georgetown Univ, BS, 66; Duke Univ, PhD(anat), 71, MD, 72. Prof Exp: Teaching asst bot & zool, Georgetown Univ, 65-66; instr micros anat, 68 & 69, teaching asst phys anthrop, 69, instr micros anat, 70, assoc photoreceptor struct, 72-73, ASST PROF ANAT & ASSOC OPHTHAL, MED CTR, DUKE UNIV, 74- Concurrent Pos: Fel, Med Res Coun Lab Molecular Biol, Cambridge, Eng, 73-74. Mem: Biophys Soc; AAAS; Asn Am Med Cols; NY Acad Sci; Fedn Am Scientists. Res: Structure of visual photoreceptors. Mailing Add: Lab of Biophys Cytol Dept Anat Box 3011 Duke Univ Med Ctr Durham NC 27710

CORLETT, MABEL ISOBEL, b Noranda, Que, Feb 7, 39. MINERALOGY. Educ: Queen's Univ, Ont, BSc, 60; Univ Chicago, SM, 62, PhD(mineral), 64. Prof Exp: Res asst mineral, Inst Crystallog & Petrog, Swiss Fed Inst Technol, 65-69; res assoc, 69-71, asst prof geol sci, 71-75, ASSOC PROF GEOL SCI, QUEEN'S UNIV, ONT, 75- Concurrent Pos: Asst ed minerals, Joint Comt Powder Diffraction Stand. Mem: Mineral Soc Am; Mineral Asn Can; Swiss Soc Mineral & Petrog. Res: Electron probe microanalysis; relation between chemistry of minerals and order/disorder. Mailing Add: Dept of Geol Sci Queen's Univ Kingston ON Can

CORLETT, MICHAEL PHILIP, b Toronto, Ont, Apr 28, 37. MYCOLOGY. Educ: Univ Toronto, BA, 59, MA, 62, PhD(mycol), 65. Prof Exp: RES MYCOLOGIST, PLANT RES INST, CAN DEPT AGR, 65- Mem: Mycol Soc Am; Can Bot Asn; Can Phytopath Soc. Res: Histology, morphology and development of fungi using the standard histological and cytological technics as well as electron microscopy. Mailing Add: Can Dept Agr Plant Res Inst Ottawa ON Can

CORLEY, CHARLES CALHOUN, JR, b Charlotte, NC, June 30, 27; m 52; c 4. HEMATOLOGY, INTERNAL MEDICINE. Educ: Clemson Univ, BS, 53; Emory Univ, MD, 53. Prof Exp: From instr to assoc prof, 56-70, PROF HEMAT, EMORY UNIV, 70- Concurrent Pos: USPHS fel hemat, Emory Univ, 56-57. Mem: Am Col Physicians; Am Soc Hemat; Am Fed Clin Res; AMA. Res: Chemotherapy of hematologic malignancies. Mailing Add: Emory Univ Clin Sect Hemat 1365 Clifton Rd NE Atlanta GA 30322

CORLEY, ERNEST L, dairy science, see 12th edition

CORLEY, GLYN JACKSON, b Carson, La, Jan 23, 16; m 39; c 5. MATHEMATICS. Educ: Northwestern State Univ, AB, 38; Columbia Univ, MA, 40; George Peabody Col, PhD(math), 59. Prof Exp: Instr math & physics, Springfield Col, 40; from instr to assoc prof math, Northwestern State Univ, 41-62; prof, ETex State Univ, 62-67, head dept, 66-67; PROF MATH & HEAD DEPT, LA STATE UNIV, SHREVEPORT, 67- Concurrent Pos: Instr, Univ Tex, 46-47 & Vanderbilt Univ, 54-55. Mem: Math Asn Am. Res: Analysis; statistics. Mailing Add: Dept of Math La State Univ Shreveport LA 71105

CORLEY, JOHN BRYSON, b Calgary, Alta, Aug 29, 13; m 47; c 2. FAMILY MEDICINE. Educ: Univ Alta, BA, 36, MD, 42. Prof Exp: Sr partner, Chinook Med Clin, Calgary, 58-73; PROF FAMILY PRACT, MED UNIV SC, 73- Concurrent Pos: Prin investr, Nat Health Res Grant, Develop Grad Training Family Physicians, 66-69; vis prof dept community med, Univ Conn, 72; consult educ adv comt, Col Family Physicians Can, 73-; assoc ed self-assessment, Continuing Educ for Family Physician, 73-; mem clin prob solving skills comt, Nat Bd Med Examr, 74-75. Mem: Royal Col Med; fel Am Soc Clin Hypnosis; fel Can Col Family Physicians; Int Soc Res Med Educ; Am Educ Res Asn. Res: Development of prototype models of formative evaluation of post-graduate programs of specialty medical education. Mailing Add: 377 Grove St Charleston SC 29403

CORLEY, KARL C, JR, psychology, see 12th edition

CORLEY, RICHARD STANCLIFFE, analytical chemistry, see 12th edition

CORLISS, CHARLES HOWARD, b Medford, Mass, Oct 30, 19; m. ATOMIC SPECTROSCOPY. Educ: Mass Inst Technol, BS, 41. Prof Exp: PHYSICIST, NAT BUR STANDARDS, 42- Honors & Awards: Silver Medal, US Dept Com, 63. Mem: AAAS; Optical Soc Am; Am Astron Soc; Royal Astron Soc. Res: Description and analysis of atomic spectra; measurement of spectral intensity; transition probabilities; light sources; astronomical spectroscopy. Mailing Add: Nat Bur Standards Washington DC 20234

CORLISS, CLARK EDWARD, b Coats, Kans, Nov 11, 19; m 50; c 3. ANATOMY. Educ: Univ Vt, BS, 42; Univ Mass, MS, 49; Brown Univ, PhD(biol), 52. Prof Exp: Asst prof microanat, Med Sch, Dalhousie Univ, 51-53; from isntr to asst prof, Med Units, 53-63, ASSOC PROF ANAT, CTR HEALTH SCI, UNIV TENN, MEMPHIS, 63- Mem: Am Asn Anat; Soc Develop Biol. Res: Study of normal and abnormal development of the central nervous system in vertebrate embryos. Mailing Add: Dept of Anat Univ of Tenn Ctr for Health Sci Memphis TN 38163

CORLISS, EDITH LOU ROVNER, b Cleveland, Ohio, Sept 8, 20; m 43; c 2. PHYSICS. Educ: Mass Inst Technol, BS & MS, 41. Prof Exp: Jr physicist, Nat Bur Standards, 41-42 & US Weather Bur, 42-43; jr astronomer, US Naval Observ, 43-44; PHYSICIST, NAT BUR STANDARDS, 44- Mem: Am Phys Soc; Acoust Soc Am; Am Hort Soc; Inst Elec & Electronics Eng. Res: Analysis of transients; speech communication; physical problems in measurement of hearing. Mailing Add: 2955 Albemarle St NW Washington DC 20008

CORLISS, GLENN ARTHUR, b Albion, Mich, Oct 16, 38; m 63; c 2. FOOD SCIENCE. Educ: Albion Col, BA, 61; Mich State Univ, MS, 63, PhD(food sci), 68. Prof Exp: Res chemist, Carnation Res Labs, Calif, 63-65; RES SCIENTIST & SECT HEAD, MILES LABS, 68- Mem: Am Oil Chem Soc; Inst Food Tech. Res: Phospholipid oxidation in emulsions; development of new food products; development and improvement of meat analogs and other vegetable protein based foods. Mailing Add: Miles Labs 601 East Algonquin Rd Schaumburg IL 60195

CORLISS, JOHN FRANKLIN, b Yakima, Wash, Mar 26, 31; m 56; c 2. SOIL SCIENCE. Educ: Wash State Univ, BS, 53; Iowa State Univ, MS, 55, PhD, 58. Prof Exp: Asst prof soils, asst soil scientist, Exp Sta & party chief, Alsea Basin Soil Veg Surv, Ore State Univ, 58-61; soil scientist, Div Watershed Mgt, 61-67, chief soil mgt br, 67-74, DIR WATERSHED & MINERALS UNIT, US FOREST SERV, US DEPT AGR, 74- Mem: Am Soc Agron; Soil Sci Soc Am; Am Soc Photogram; Soil Conserv Soc Am. Res: Land use planning; soil interpretation for forest land management; soil classification, survey, genesis and morphology; soil vegetation survey. Mailing Add: Watershed & Minerals Unit US Forest Serv 1720 Peachtree St NW Atlanta GA 30309

CORLISS, JOHN OZRO, b Coats, Kans, Feb 23, 22; m 68; c 5. PROTOZOOLOGY. Educ: Univ Chicago, BS, 44; Univ Vt, AB, 47; NY Univ, PhD(biol), 51. Prof Exp: US Atomic Energy Comn res fel biol sci, Col France, 51-52; instr zool, Yale Univ, 52-54; from asst prof to prof, 54-64, head dept biol sci, Univ Ill, 64-69; dir syst biol, NSF, 69-70; PROF ZOOL & HEAD DEPT, UNIV MD, COLLEGE PARK, 70- Concurrent Pos: Vis prof, Univs London & Exeter, 60-62; chmn US Nat comt, Int Union Biol Sci, 71-73; mem & comnr, Int Comn Zool Nomenclature, 72-; mem, Corp of Marine Biol Lab, Woods Hole, Mass, 74- Mem: Soc Protozool (secy, 58-61, pres, 64-65); Am Soc Zool (pres, 71-72); Soc Syst Zool (pres, 69-70); Am Micros Soc (pres, 65-66); Coun Biol Eds. Res: Comparative morphology, systematics, evolution, and phylogeny of ciliate protozoa; anatomy of the infraciliature; morphogenesis; nomenclature; international collection of ciliate type-specimens. Mailing Add: Dept of Zool Univ of Md College Park MD 20742

CORLISS, LESTER MYRON, b NJ, Mar 29, 19; m 41; c 2. SOLID STATE PHYSICS. Educ: City Col New York, BS, 40; Harvard Univ, MA, 48, PhD(chem physics), 49. Prof Exp: Chemist, Manhattan Proj, 43-46; SR CHEMIST, BROOKHAVEN NAT LAB, 49- Concurrent Pos: NSF sr fel, Univ Grenoble, 59-60; mem comn neutron diffraction, Int Union Crystallog, 66-75, chmn 69-75. Mem: Fel Am Phys Soc. Res: Neutron diffraction; magnetism; phase transformations; critical phenomena. Mailing Add: Brookhaven Nat Lab Upton NY 11973

CORMACK, DOUGLAS VILLY, b Lacombe, Alta, Aug 30, 26; m; c 3. BIOPHYSICS. Educ: Univ Alta, BSc, 49, MSc, 50; Univ Sask, PhD(physics), 53. Prof Exp: Res assoc physics, Univ Sask, 54-56, from asst prof to assoc prof, 56-67; ASST PROF MED MICROBIOL, UNIV MAN & BIOPHYSICIST, MAN CANCER FOUND, 67- Concurrent Pos: Consult, Int Comn Radiol Units & Measurements. Mem: Radiation Res Soc; Can Asn Physicists; Can Asn Radiol; Brit Inst Radiol. Res: Radiation physics applied to radiology and radiobiology; cancer research; virology. Mailing Add: Dept of Med Microbiol Univ of Man Winnipeg MB Can

CORMACK, JAMES FREDERICK, b Portland, Ore, Mar 13, 27; m 52; c 5. ENVIRONMENTAL CHEMISTRY. Educ: Reed Col, BA, 48; Ore State Col, PhD(biochem), 53. Prof Exp: Asst, Ore State Col, 48-49; res chemist, 53-59, SUPVR, CROWN-ZELLERBACH, 59- Mem: AAAS; Am Chem Soc; Tech Asn Pulp & Paper Indust. Res: Industrial fermentations; stream improvement; waste treatment; industrial slimicides; continuous analysis. Mailing Add: Environ Serv Div Crown-Zellerbach Camas WA 98607

CORMACK, MELVILLE WALLACE, b Rossburn, Man, July 29, 08; m 33; c 2. PLANT PATHOLOGY. Educ: Univ Man, BSA, 30; Univ Alta, MSc, 31; Univ Minn, PhD(plant path), 37. Prof Exp: Plant dis investr, Dom Lab Plant Path, 28-31, asst plant pathologist, 31-47, head plant path sect, Sci Serv Lab, 48-57; dir, Can Agr Res Sta, 57-69; DIR PLANT BREEDING STA, NJORO, KENYA, 69- Mem; Am Phytopath Soc; fel Royal Soc Can; Agr Inst Can. Res: Diseases of forage crop legumes and grasses; low temperature fungi; wheat improvement. Mailing Add: Plant Breeding Sta PO Njoro Kenya

CORMACK, ROBERT GEORGE HALL, b Cedar Rapids, Iowa, Feb 2, 04; m 39; c 2. BOTANY. Educ: Univ Toronto, BA, 29, MA, 31, PhD(biol), 34. Prof Exp: Demonstr bot, Univ Toronto, 29-36; lectr, 36-45, from asst prof to prof, 45-69, EMER PROF BOT, UNIV ALTA, 69- Concurrent Pos: Bot consult, Northwest Proj Study Group, MacKenzie Valley, NWT, Can, 70-71. Mem: Can Inst Forestry; Royal Soc Can. Res: Developmental anatomy; forest and wildlife conservation. Mailing Add: 9737 112th St Apt 703 Edmonton AB Can

CORMAN, BLAINE G, organic chemistry, rubber chemistry, see 12th edition

CORMAN, EMMETT GARY, b Kansas City, Kans, Aug 2, 30; m 57; c 4. THEORETICAL PHYSICS, NUCLEAR PHYSICS. Educ: Univ Kans, BS, 52, MS, 54, PhD(physics), 60. Prof Exp: Assoc physicist, Vitro Corp Am, 52-53; physicist, Los Alamos Sci Lab, 54-56; RES PHYSICIST, LAWRENCE RADIATION LAB, UNIV CALIF, LIVERMORE, 60- Res: Optical model analysis of high energy nucleons; Monte Carlo photon-matter interactions; physics of nuclear weapons; numerical solutions on high speed computers; physics of plasma-magnetic field interactions.

Mailing Add: Theoret Div Bldg 113 Lawrence Radiation Lab PO Box 808 Livermore CA 94550

CORMAN, LEW ANDRE, b Balti, Rumania, May 14, 15; US citizen; m 41. CLINICAL PHARMACOLOGY, INTERNAL MEDICINE. Educ: Univ Toulouse, PCN, 34; Univ Iasi, MD, 44. Prof Exp: Prin investr, Geriat Inst, Univ Bucharest, 52-58; head res, Ctr Studies Pharmaceut Indust, Univ Toulouse, 59-63; ASST PROF MED, HAHNEMANN MED COL & HOSP, 63- Concurrent Pos: Fel isotope methodology, Hahnemann Med Col, 64, fel clin pharmacol, 64-65. Mem: Fel Am Soc Clin Pharmacol & Therapeut; Phlebology Soc Am; Fr Soc Therapeut & Pharmacodynamics; Fedn Europ Biochem Socs. Res: Peripheral vascular disease; lipid metabolism. Mailing Add: Hahnemann Med Col & Hosp 230 N Broad St Philadelphia PA 19102

CORMIER, BRUNO M, b Laurierville, Que, Nov 14, 19; m. PSYCHIATRY. Educ: Univ Montreal, BA, 42, MD, 47; McGill Univ, dipl psychiat, 52. Prof Exp: From asst prof to assoc prof psychiat, 53-67, dir criminal res, 67-74, lectr psychiat, 63-74, ASSOC PROF PSYCHIAT, McGILL UNIV, 74- Concurrent Pos: Clin asst psychiat, Royal Victoria Hosp, Montreal, 53-55, assoc psychiatrist, 56-63; psychiatrist-in-chief, St Vincent de Paul Penitentiary, 55-70; consult, NY State Correctional Serv, 66-72. Mem: Am Psychiat Asn; Can Psychiat Asn; Can Corrections Asn; Can Psychoanal Soc. Res: Clinical criminology; psychopathology of deprivation of liberty; persistent criminality. Mailing Add: Dept of Psychiat McGill Univ Montreal PQ Can

CORMIER, MILTON JOSEPH, b DeRidder, La, Nov 29, 26; m 51; c 3. BIOCHEMISTRY. Educ: Southwestern La Inst, BS, 48; Univ Tex, MA, 51; Univ Tenn, PhD(microbiol), 56. Prof Exp: Assoc biochemist, Biol Div, Oak Ridge Nat Lab, 56-58; from asst prof to prof bioluminescence, 58-66, RES PROF BIOLUMINESCENCE, UNIV GA, 66- Mem: Am Chem Soc; Am Soc Microbiol; Am Soc Biol Chem. Res: Mechanisms of bioluminescent reactions. Mailing Add: Dept of Biochem Univ of Ga Athens GA 30601

CORMIER, RANDAL, b Truro, NS, Mar 9, 30; m 55; c 6. GEOLOGY. Educ: St Francis Xavier Univ, Can. BSc, 51; Mass Inst Technol, PhD(geol), 56. Prof Exp: Assoc prof, 57-74, PROF GEOL, ST FRANCIS XAVIER UNIV, 74- Mem: Can Geol Soc; Can Inst Min & Metall. Res: Geochronology; geochemical prospecting. Mailing Add: Dept of Geol St Francis Xavier Univ Antigonish NS Can

CORN, HERMAN, b Philadelphia, Pa, Oct 7, 21; c 3. PERIODONTOLOGY. Educ: Temple Univ, DDS, 44; Am Bd Periodont, dipl, 63. Prof Exp: Instr periodont, Grad Sch Med, 58-63, assoc, Sch Dent Med & Grad Sch Med, 63-65, asst prof, Sch Dent Med, 65-68, assoc prof, 68-74, PROF PERIODONT, SCH DENT MED, UNIV PA, 74- Concurrent Pos: Oral surgeon, Lower Bucks County Hosp, 54-; lectr, Dent Schs, Wash Univ & Univ Ky, 68, Temple Univ & Case Western Reserve Univ, 69 & Boston Univ, 69-; lectr post grad educ, Sch Dent Med, Univ Pa; clin assoc prof & dir prev dent, Med Col Pa, 74- Mem: Fel Am Col Dentists; Am Dent Asn; Am Acad Periodont; Am Soc Prev Dent (pres, 73-74). Res: Clinical periodontics; periodontal therapy; mucogingival surgery.

CORNATZER, WILLIAM EUGENE, b Mocksville, NC, Sept 28, 18; m 46; c 2. BIOCHEMISTRY. Educ: Wake Forest Col, BS, 39; Bowman Gray Sch Med, MD, 51; Univ NC, MS, 41, PhD(biochem), 44; Am Bd Clin Chem, dipl. Prof Exp: Asst zool, Wake Forest Col, 37-38, asst phys chem, 38-39; asst biol chem, Univ NC, 39-41; asst prof biochem, Bowman Gray Sch Med, 46-51; CHESTER FRITZ DISTINGUISHED PROF, SCH MED, UNIV N DAK, 73-, PROF BIOCHEM & HEAD DEPT, 51-, DIR IRELAND RES LAB, 53- Concurrent Pos: NSF travel award, Int Cong Biochem, Paris, 52, Tokyo, 67; USAEC travel award, Int Cancer Cong, London, 58; Int Union Physiol Sci travel award, Int Cong Pharmacol, Stockholm, 61; Am Inst Nutrit travel award, Int Cong Nutrit, Prague, 69 & Mexico City, 72; consult med div, Oak Ridge Inst Nuclear Studies, 51-; mem, Am Bd Clin Chem; mem biochem test comt, Nat Bd Med Exam. Honors & Awards: Billing Award, AMA, 51. Mem: AAAS; Am Chem Soc; Am Soc Biol Chem; Am Fedn Clin Res; Soc Exp Biol & Med. Res: Properties of proteins; quinine metabolism and absorption; antimalarial testing; phospholipid metabolism; liver function tests and disease; radiation of effects and toxicity of isotopes; lipotropic agents. Mailing Add: Dept of Biochem Univ of NDak Med SCh Grand Forks ND 58201

CORNBLATH, MARVIN, b St Louis, Mo, June 18, 25; m 48; c 3. PEDIATRICS, BIOCHEMISTRY. Educ: Wash Univ, MD, 47. Prof Exp: Asst pediat, Wash Univ, 49-50; from instr to asst prof, Johns Hopkins Univ, 53-59; from asst prof to assoc prof, Northwestern Univ, 59-61; from assoc prof to prof, Univ Ill, 61-68; PROF PEDIAT & CHMN DEPT, SCH MED, UNIV MD, BALTIMORE CITY, 68- Concurrent Pos: USPHS fel biochem, Wash Univ, 50-51; res assoc, Sinai Hosp, Md, 53-59; asst chmn div pediat, Michael Reese Hosp, Ill, 59-61. Mem: Soc Pediat Res; Am Physiol Soc; Brit Biochem Soc. Res: Physiological and biochemical maturation of newborn and premature infants; carbohydrate metabolism and enzymology. Mailing Add: 3809 St Paul St Baltimore MD 21201

CORNEIL, PAUL HAMPTON, b Baytown, Tex, Dec 7, 41; m 66. PHYSICAL CHEMISTRY. Educ: Rice Univ, BA, 63; Univ Calif, Berkeley, PhD(chem), 67. Prof Exp: Jr res chemist, Esso Res & Eng, NJ, 63; US Atomic Energy Comn fel & res chemist, Univ Calif, Los Angeles, 67-69; FAC FEL CHEM, JOHNSTON COL, UNIV REDLANDS, 69- Concurrent Pos: Vis fac mem, Thomas Jefferson Col, Grand Valley State Cols, Allendale, Mich, 74-75. Mem: AAAS; Nat Geog Soc. Res: Hydrogen chloride chemical laser; atmosphere of Venus; worldwide environmental quality; mathematical analysis of chemical lasers. Mailing Add: Johnston Col Univ of Redlands Redlands CA 92373

CORNELIA, RICHARD HARTMAN, organic chemistry, see 12th edition

CORNELISON, FLOYD S, JR, b San Angelo, Tex, Apr 30, 18; m 40, 66; c 1. PSYCHIATRY. Educ: Baylor Univ, AB, 39; Cornell Univ, MD, 50; Boston Univ, MS, 58. Prof Exp: Lectr psychol, Tufts Univ, 54-56; from asst to instr psychiat, Sch Med, Boston Univ, 51-58; from asst prof to assoc prof, Sch Med, Univ Okla, 58-62; chmn dept, 62-74, PROF PSYCHIAT, JEFFERSON MED COL, 62- Concurrent Pos: Attend physician, Thomas Jefferson Univ Hosp, 62-; mem, Ment Health Film Bd, NY; mem adv comt psychiat, Neurol & Psychol Serv, Dept Med & Surg, Vet Admin, Washington, DC. Mem: Am Med Soc Alcoholism; fel Am Psychiat Asn; Am Sci Film Asn; Asn Am Med Cols; fel Am Col Psychiat. Res: Psychiatric education; self-image experience; research involving motion picture films and television in study of alcoholism, obesity, anorexia nervosa and hemophilia. Mailing Add: Dept of Psychiat & Human Behav Jefferson Med Col Philadelphia PA 19107

CORNELIUS, CHARLES EDWARD, b Walnut Park, Calif, Dec 19, 27; m 48; c 4. Educ: Univ Calif, BS, 49 & 51, DVM, 53, PhD(comp path), 58. Prof Exp: Lectr clin path, Univ Calif, Davis, 54-58, from asst prof to assoc prof, 58-66, assoc dean, 62-64; prof physiol & dean col vet med, Kans State Univ, 66-71; DEAN COL VET MED, UNIV FLA, 71-, PROF PATH, 74- Concurrent Pos: Mem res grants study sect gen med, NIH, 65-69, nat adv coun health res facilities, 69- Mem: Am Physiol Soc; Soc Exp Biol & Med; Am Gastroenterol Asn; Am Vet Med Asn. Res: Hepatic physiology; bile pigment and acion transport; comparative hepatic dysfunction. Mailing Add: Col of Vet Med Univ of Fla Gainesville FL 32603

CORNELIUS, DONALD RISDON, agronomy, see 12th edition

CORNELIUS, LARRY MAX, b Washington, Ind, Apr 30, 43; m 65; c 1. VETERINARY MEDICINE. Educ: Purdue Univ, DVM, 67; Univ Mo-Columbia, PhD(clin path), 71. Prof Exp: Intern, Angell Mem Animal Hosp, Boston, Mass, 67-68; res assoc vet med, Univ Mo-Columbia, 68-74; MEM FAC ANIMAL MED, UNIV GA, 74- Concurrent Pos: Speaker, Technicon Int Cong, Chicago, 69; Nat Inst Arthritis & Metab Dis fel, 69-72. Mem: Am Vet Med Asn; Comp Gastroenterol Soc. Res: Fluid and electrolyte balance in veterinary medicine. Mailing Add: Dept Animal Med Univ of Ga Athens, GA 30602

CORNELL, ALAN, b Fall River, Mass, May 4, 29; m 52; c 3. FOOD SCIENCE. Educ: Univ Mass, BS, 51; Mass Inst Technol, PhD(food sci), 60. Prof Exp: Res chemist, Res Ctr, Philip Morris, Inc, 60-65, sr scientist, 65-67; mgr food sci div, Cent Labs, Ralston Purina Co, 67-69; asst vpres res & develop, 69-73, VPRES RES & PROD DEVELOP, CONSOL CIGAR CORP, 73- Mem: Am Chem Soc; Inst Food Technol; NY Acad Sci. Res: Subjective effects on constituents in tobacco and resultant smoke; subjective evaluation; product development of consumer products; research management; tobacco and smoke chemistry; quality control; product safety and health. Mailing Add: Consol Cigar Corp One Gulf & Western Plaza New York NY 10023

CORNELL, CREIGHTON N, b Rolfe, Iowa, Mar 20, 33; m 58; c 2. PHYSIOLOGY, BIOCHEMISTRY. Educ: Univ Mo, BS, 54, DVM, 62, MS, 69. Prof Exp: Instr agr chem & vet physiol, 62-63, instr agr chem, 63-65, ASST PROF AGR CHEM, UNIV MO-COLUMBIA, 65- Res: Biochemistry and physiology of the intoxication occurring in cattle feeding on toxic fescue grass pastures or hay, causing gangrene and sloughing of the distal portions of the extremities. Mailing Add: 105 Schweitzer Hall Univ of Mo Columbia MO 65201

CORNELL, DAVID ALLAN, b St Paul, Minn, Dec 29, 37; m 62; c 2. PHYSICS. Educ: Principia Col, BS, 59; Univ Calif, Berkeley, PhD(physics), 64. Prof Exp: From instr to asst prof, 64-71, ASSOC PROF PHYSICS, PRINCIPIA COL, 71- Concurrent Pos: Researcher-demonstr, Dept Physics, Univ Warwick, Eng, 71-72. Mem: Am Asn Physics Teachers. Res: Nuclear magnetic resonance phenomena in metals and alloys; hydrogen diffusion in metallic crystals. Mailing Add: Dept of Physics Principia Col Elsah IL 62028

CORNELL, HOWARD VERNON, b Berwyn, Ill, Apr 13, 47; m 71. BIOGEOGRAPHY, ECOLOGY. Educ: Tufts Univ, BS, 69; Cornell Univ, PhD(ecol), 75. Prof Exp: ASST PROF BIOL SCI, UNIV DEL, 75- Mem: AAAS; Ecol Soc Am. Res: Ecology and host finding characteristics of cynipid oak galls on oak trees. Mailing Add: Dept of Biol Sci Univ of Del Newark DE 19711

CORNELL, JAMES MORRIS, b Bismarck, NDak, Sept 8, 37; m 57; c 2. ANIMAL BEHAVIOR. Educ: Univ Wash, BS, 60, MS, 62, PhD(psychol), 63. Prof Exp: Asst prof psychol, Mont State Col, 63-64; asst prof, 64-68, ASSOC PROF PSYCHOL, UNIV WATERLOO, 68- Concurrent Pos: Nat Res Coun Can res grants, 64- Mem: AAAS; Am Psychol Asn; Animal Behavior Soc. Res: Conidtioning in animals; imprinting; memory. Mailing Add: Dept of Psychol Univ of Waterloo Waterloo ON Can

CORNELL, JAMES S, b Harrisburg, Pa, Apr 9, 47. BIOLOGICAL CHEMISTRY, ENDOCRINOLOGY. Educ: Mich State Univ, BS, 69; Univ Calif, Los Angeles, PhD(biol chem), 73. Prof Exp: Fel, Univ Calif, Los Angeles, 73; res assoc, 73-75, ASST PROF BIOCHEM & ASST DIR LAB CLIN BIOCHEM, MED COL, CORNELL UNIV, 75- Mem: Am Chem Soc; AAAS; NY Acad Sci. Res: Reproductive biochemistry—the protein chemistry of the placenta; factors influencing the induction of labour. Mailing Add: Dept of Biochem Cornell Univ Med Col New York NY 10021

CORNELL, JOHN, polymer chemistry, see 12th edition

CORNELL, JOHN ANDREW, b Westerly, RI, Apr 29, 41; m 63; c 2. APPLIED STATISTICS. Educ: Univ Fla, BSEd, 62, MStat, 66; Va Polytech Inst, PhD(statist), 69. Prof Exp: Appl statistician, Tenn Eastman Co, 65-66; asst prof statist, Univ Fla, 68-72; lectr statist, Birkbeck Col, Univ London, 72-73; ASSOC PROF STATIST, UNIV FLA, 73- Concurrent Pos: Statist consult, Tenn Eastman Co, 65-66 & Agr Exp Sta, Univ Fla, 68-72. Mem: Am Statist Asn; Inst Math Statist. Res: Design and analysis of statistical experiments in agricultural, social, physical and biological sciences, design and analysis of experiments with mixtures. Mailing Add: Dept of Statist 219 Rolfs Hall Univ of Fla Gainesville FL 32601

CORNELL, JOHN B, b East Chicago, Ind, Nov 25, 21; m 53; c 1. CULTURAL ANTHROPOLOGY, SOCIAL ANTHROPOLOGY. Educ: Univ Mich, BA, 46, MA, 47, PhD(anthrop), 53. Prof Exp: Lectr anthrop, Univ Mich, 53-54; asst prof, Wayne State Univ, 54-55; prog dir anthrop, NSF, 71-74; from asst prof to assoc prof, 55-65, PROF ANTHROP, UNIV TEX, AUSTIN, 65-, CHMN DEPT, 74- Mem: Am Anthrop Asn; Am Ethnol Soc (vpres, 73, pres elect, 74, pres, 74-75); Soc Appl Anthrop; Asn Asian Studies; Japanese Soc Ethnol. Res: Rural social organization, Japan; outcaste status and mobility, Japan; bicultural behavior of immigrants and immigrant decendants, Brazil. Mailing Add: Dept of Anthrop Univ of Texas Austin TX 78712

CORNELL, NEAL WILLIAM, b Savannah, Ga, July 29, 37; m 67. BIOCHEMISTRY. Educ: Univ Redlands, BS, 60; Univ Calif, Los Angeles, PhD(biochem), 64. Prof Exp: Res fel biol chem, Harvard Med Sch, 63-66; from asst prof to assoc prof chem, Pomona Col, 66-72; Med Res Coun investr, Oxford Univ, 72-74; RES CHEMIST, LAB ALCOHOL RES, NAT INST ALCOHOL ABUSE & ALCOHOLISM, 74- Concurrent Pos: Investr, Marine Biol Lab, Woods Hole, 70, 71 & 72, mem corp, 71- Mem: Brit Biochem Soc; Am Soc Biol Chem. Res: Metabolic control; chemistry and function of peptide antibiotics; comparative biochemistry of regulatory enzymes; computer simulation. Mailing Add: NIAAA/WAW Bldg Lab of Alcohol Res St Elizabeths Hosp Washington DC 20032

CORNELL, RICHARD GARTH, b Cleveland, Ohio, Nov 18, 30; m 61; c 3. BIOSTATISTICS. Educ: Univ Rochester, AB, 52; Va Polytech Inst, MS, 54, PhD(statist), 56. Prof Exp: Asst, Va Polytech Inst, 52-54, Oak Ridge Inst Nuclear Studies fel, 55-56; statistician, Commun Dis Ctr, USPHS, Ga, 56-58, chief lab & field sta, Statist Unit, 58-60; from assoc prof to prof statist, Fla State Univ, 60-71; PROF BIOSTATIST & CHMN DEPT, UNIV MICH, ANN ARBOR, 71- Mem: AAAS; fel Am Statist Asn; Biomet Soc; Am Sci Affil; Am Pub Health Asn. Res: Biometry; nonlinear estimation; sampling. Mailing Add: Dept of Biostatist Univ of Mich Ann Arbor MI 48104

CORNELL, RICHARD HENRY, b Port Washington, NY, June 18, 31; m 59; c 2. PAPER CHEMISTRY. Educ: Colgate Univ, AB, 53; Lawrence Inst Technol, MS, 55, PhD(paper chem). 60. Prof Exp: Res chemist, Riegel Paper Corp, NJ, 56-57, 60-62, supvr process res, 62-64; mgr res & develop, Northwest Paper Co, 64-65; mgr pulp & paper res, 65-67, DIR PAPER & PAPERBOARD RES, POTLATCH FORESTS, INC, 67- Concurrent Pos: Indust res fel, Inst Paper Chem, 60-61. Mem: Am Chem Soc. Am Tech Asn Pulp & Paper Indust. Res: Cellulose chemistry, reactions and modification; graft polymers of cellulose; cellulose xanthate chemistry. Mailing Add: Dept of Paper & Paperboard Res Potlach Forests, Inc East End Cloquet MN 55720

CORNELL, ROBERT JOSEPH, b Westerly, RI, Oct 1, 40; m 67; c 2. POLYMER CHEMISTRY. Educ: Clarkson Col Technol, BS, 62, MS, 64; Worcester Polytech Inst, PhD(org chem), 67. Prof Exp: Res chemist polymer chem, 67-70, res scientist, Polymer Res Group, 70-73, SR GROUP LEADER PARACRIL RES, CHEM DIV, UNIROYAL INC, 73- Mem: Am Chem Soc; Sigma Xi. Res: The development of new high heat-oil resistant elastomers; the development of unique new rubber blends for enhanced vulcanizate properties. Mailing Add: Uniroyal Inc Chem Div Elm St Naugatuck CT 06770

CORNELL, SAMUEL DOUGLAS, b Buffalo, NY, Apr 16, 15; m 39; c 4. SCIENCE ADMINISTRATION, ACADEMIC ADMINISTRATION. Educ: Yale Univ, BA, 35, PhD(physics), 38. Prof Exp: Develop physicist, Eastman Kodak Co, NY, 38-42; sci warfare adv, Res & Develop Bd, Washington, DC, 46-52; exec officer, Nat Acad Sci-Nat Res Coun, 52-65; pres, Mackinac Col, 65-70; consult sci admin, 70-72, ASST TO THE PRES, NAT ACAD SCI, 72- Mem: AAAS; Am Phys Soc. Res: Molecular spectroscopy; design of motion picture equipment. Mailing Add: Nat Acad of Sci 2101 Constitution Ave NW Washington DC 20418

CORNELL, WILLIAM CROWNSHIELD, b Attleboro, Mass, May 5, 41; m 69; c 1. MICROPALEONTOLOGY, PALYNOLOGY. Educ: Univ RI, BS, 65, MS, 67; Univ Calif, Los Angeles, PhD(geol), 72. Prof Exp: Asst prof, 71-75, ASSOC PROF GEOL, UNIV TEX, EL PASO, 75- Concurrent Pos: Actg chmn dept geol sci, Univ Tex, El Paso, 7S-76. Mem: Soc Econ Paleontologists & Mineralogists; AAAS; Am Asn Stratig Palynologists. Res: Mesozoic and Cenozoic siliceous; organic walled phytoplankton. Mailing Add: Dept of Geol Sci Univ of Tex El Paso TX 79968

CORNELY, PAUL BERTAU, b French West Indies, Mar 9, 06; nat US; m 34; c 1. PUBLIC HEALTH. Educ: Univ Mich, AB, 28, MD, 31, DrPH, 34; Am Bd Prev Med, dipl. Hon Degrees: DSc, Univ Mich, 68; DPS, Univ of the Pac, 73. Prof Exp: Asst prof, 34-35, assoc prof, 35-47, dir health serv, 37-47, head dept community health pract, 55-70, prof, 47-73, EMER PROF PREV MED & PUB HEALTH, COL MED, HOWARD UNIV, 73-; MEM STAFF, HEALTH SERVS EVAL SYST SCI, INC, 73- Concurrent Pos: Med dir, Freedmen's Hosp, 47-58; consult, Nat Urban League, 44-47 & Am Health Educ African Develop, US AID; pres, Community Group Health Found, 68-; mem, President's Comn Population Growth & Am Future, 70-72. Mem: Fel Am Pub Health Asn (pres, 69-70); fel Am Col Prev Med; hon fel Am Col Hosp Adminr. Res: Medical education; distribution and supply of professional personnel; Negro health problems; student health program; health motivation among low income families. Mailing Add: 1338 Geranium St NW Washington DC 20012

CORNER, GEORGE WASHINGTON, b Baltimore, Md, Dec 12, 89; m 15; c 2. ANATOMY. Educ: Johns Hopkins Univ, AB, 09, MD, 13. Hon Degrees: Dr, Cath Univ Chile, 42; DSc, Univ Rochester, 44, Boston Univ, 48, Oxford Univ, 50, Univ Chicago, 58 & Thomas Jefferson Univ, 71; LLD, Tulane Univ, 55 & Temple Univ, 56; Dr Med Sci, Woman's Med Col Philadelphia, 58; DLitt, Univ Pa, 65. Prof Exp: Asst anat, Med Dept, Johns Hopkins Univ, 13-14, assoc prof, 19-23, res house officer, Johns Hopkins Hosp, 14-15; asst prof anat, Univ Calif, 15-19; prof, Sch Med, Univ Rochester, 23-40, cur med libr, 38-40; dir dept embryol, Carnegie Inst Technol, 40-55; historian & affil mem, Rockefeller Inst, 56-60; EXEC OFFICER, AM PHILOS SOC, 60- Concurrent Pos: Vicary lectr, Royal Col Surg, London, 36; Vanuxem lectr, Princeton Univ, 42; Terry lectr, Yale Univ, 44; Eastman vis prof, Oxford Univ, 52-53; hon prof, Univ Santiago, Chile, 59; trustee, Samuel Ready Sch. Honors & Awards: Passano Found Award, 58; Dale Medal, Brit Soc Endocrinol, 64; Squibb Award, Endocrine Soc, 40; Schering Award, Am Gynec Soc, 51. Mem: Emer mem Nat Acad Sci (vpres, 53-57); Am Asn Anat (secy-treas, 30-38, pres, 46-48); Am Philos Soc (vpres, 53-56); hon mem Am Gynec Soc; hon mem Anat Soc Gt Brit & Ireland. Res: Anatomy and physiology of mammalian reproductive system; history of science. Mailing Add: Am Philos Soc 104 S Fifth St Philadelphia PA 19106

CORNER, JAMES OLIVER, b Toronto, Ont, July 19, 17; nat US; m 45; c 3. ORGANIC POLYMER CHEMISTRY. Educ: Dartmouth Col, AB, 39; Univ Ill, AM, 40, PhD(org chem), 42. Prof Exp: Chemist, 42-52, res supvr, Exp Sta, 52-57, sr supvr, 57-61, res mgr, 61, res mgr, Textile Res Lab, 62-63, lab dir, Dacron Res Lab, 63-64, actg tech supt, Kinston Plant, 64-65, res dir, Dacron Tech Div, 65-68, Nylon Tech Div, 68-69 & Indust Fibers Div, 70-71, TECH MGR NYLON TECH DIV, E I DU PONT DE NEMOURS & CO, INC, 71- Mem: Am Chem Soc. Res: Synthetic organic chemistry; polymer chemistry; textile fibers. Mailing Add: Textile Fibers Dept E I du Pont de Nemours & Co Inc Wilmington DE 19898

CORNER, THOMAS RICHARD, b Waterloo, NY, Jan 6, 40; m 61; c 3. MICROBIOLOGY. Educ: Cornell Univ, BS, 62, MS, 64; Univ Rochester, PhD(microbiol), 68. Prof Exp: NIH fel, 68-70, ASST PROF MICROBIOL, MICH STATE UNIV, 70- Mem: Am Soc Microbiol; Soc Gen Microbiol; Int Soc Biorheology. Res: Structure and function of bacterial membranes and cell walls; action of organochlorine compounds on bacteria; biomechanical and biorheological properties of cells. Mailing Add: Dept of Microbiol Mich State Univ East Lansing MI 48824

CORNESKY, ROBERT ANDREW, b Aliquippa, Pa, Jan 21, 39; m 60; c 2. HEALTH SCIENCES. Educ: Geneva Col, BS, 60; George Washington Univ, MS, 64; Univ Pittsburgh, ScD(epidemiol), 71. Prof Exp: Chief serologist microbiol, Bionetics Res Lab, Inc, 64-65; instr biol sci, Carnegie Inst Technol, 65-68; asst prof, Carnegie-Mellon Univ, 68-71; assoc prof biol, 71-73, PROF HEALTH SCI & CHMN DEPT, CALIF STATE COL, BAKERSFIELD, 73- Concurrent Pos: Chmn bd dirs, Ctr Allied Health Sci Studies, 74-75; campus coordr, Health Educ Manpower Educ Proj, Calif, 75; chmn health careers comt, Kern Health Manpower Consortium, 74-; proj evaluator, External Degree Prog in Nursing, Bakersfield, 74- Mem: Am Soc Allied Health Prof; AAAS; Am Soc Microbiol. Res: Education and rural health care delivery models. Mailing Add: Dept of Health Sci Calif State Col 9001 Stockdale Hwy Bakersfield CA 93309

CORNETT, RICHARD ORIN, b Driftwood, Okla, Nov 14, 13; m 43; c 3. PHYSICS, COMMUNICATIONS SCIENCE. Educ: Okla Baptist Univ, BS, 34; Okla Univ, MS, 37; Univ Tex, PhD(physics), 40. Hon Degrees: DSc, Hardin Simmons Univ, 54; LittD, Jacksonville Univ, 64; LLD, Belknap Col, 67. Prof Exp: From instr to assoc prof physics, Okla Baptist Univ, 35-41; asst supvr physics, eng, sci & mgt defense training prog, Pa State Univ, 41-42; lectr electronics, Harvard Univ, 42-45; prof physics & asst to the pres, Okla Baptist Univ, 45-46, head dept physics & vpres, 46-47, exec vpres, 47-51; exec secy educ comn, Southern Baptist Convention & ed, Southern Baptist Educr, 51-58; specialist col & univ orgn, Off Educ, Dept Health, Educ & Welfare, Washington, DC, 59, exec asst to dir div higher educ, 60-61, actg asst comnr & dir div, 61-63, dir div educ admin, 64-65; vpres, 65-75, RES PROF & DIR CUED SPEECH PROGS, GALLAUDET COL, 75- Concurrent Pos: Originator of Cued Speech, new method of commun for deaf, 66; consult, Am Optometric Asn, 68- Mem: Acoustical Soc Am. Res: Acoustics; theory of hearing; diplacusis; communication for the hearing impaired; development of electronic lipreading aid for the deaf; design of hearing aids for improved auditory processing. Mailing Add: 8702 Royal Ridge Lane Laurel MD 20811

CORNETTÉ, JAMES L, b Bowling Green, Ky, May 8, 35; m 62; c 2. MATHEMATICS. Educ: WTex State Col, BS, 55; Univ Tex, MA, 59, PhD(math), 62. Prof Exp: From asst prof to assoc prof, 63-70, PROF MATH, IOWA STATE UNIV, 70- Concurrent Pos: Fulbright lectr, Nat Univ Malaysia, 73-74. Mem: Am Math Soc; Math Asn Am (chmn elect, 75-76); Soc Indust & Appl Math. Res: Biomathematics; point set topology; mathematical models in population genetics. Mailing Add: Dept of Math Iowa State Univ Ames IA 50010

CORNFELD, DAVID, b Philadelphia, Pa, Apr 5, 26; m 56; c 3. PEDIATRICS. Educ: Univ Pa, MD, 48, MSc, 66. Prof Exp: Asst instr, 51-52, assoc, 54-57, from asst prof to assoc prof, 57-72, PROF PEDIAT & ACTG CHMN DEPT, SCH MED, UNIV PA, 72- Concurrent Pos: Asst chief univ serv, Philadelphia Gen Hosp, 55-68; dir pediat clin, Hosp Univ Pa, 56-62; dir outpatient dept & sr physician, Children's Hosp Philadelphia, 62-, co-dir nephrology serv, 63- Mem: Am Pediat Soc. Res: Nephrology; patient care. Mailing Add: Children's Hosp of Philadelphia 1740 Bainbridge St Philadelphia PA 19146

CORNGOLD, NOEL ROBERT DAVID, b New York, NY, Jan 20, 29; m 52; c 2. REACTOR PHYSICS, STATISTICAL MECHANICS. Educ: Columbia Univ, AB, 49; Harvard Univ, AM, 50, PhD(physics), 54. Prof Exp: Asst phys sci, Harvard Univ, 50-51; res assoc, Brookhaven Nat Labs, 51-54, from assoc physicist to physicist, 54-66; PROF APPL SCI, DIV ENG & APPL SCI, CALIF INST TECHNOL, 66- Concurrent Pos: Consult, Los Alamos Sci Lab, 67-73 & Brookhaven Nat Lab, 74- Honors & Awards: Certificate of Merit, Am Nuclear Soc, 66. Mem: Fel Am Nuclear Soc; Am Phys Soc. Res: Theory of neutron scattering and transport; reactor physics; statistical physics. Mailing Add: Dept of Appl Sci Calif Inst of Technol Pasadena CA 91125

CORNING, MARY ELIZABETH, b Norwich, Conn, Oct 19, 25. PHYSICAL CHEMISTRY Educ: Conn Col, BA, 47; Mt Holyoke Col, MA, 49. Prof Exp: Asst chemist, Mt Holyoke Col, 47-49; chemist, Nat Bur Standards, 49-58; spec asst to sci adv, US Dept State, 58-60; proj dir, Ed & Int Activities Planning Group, NSF, 60-61, spec asst to head, Off Int Sci Activities, 61-62, assoc proj dir, 62-64; chief publ & trans div, 64-66, spec asst to dep dir, 66-67, spec asst to dir, 67-72, ASST DIR INT PROGS, NAT LIBR MED, DEPT HEALTH, EDUC & WELFARE, 72- Concurrent Pos: Mem chem panel, US Civil Serv Bd Examr, 53- & tech ed & writer panel, 55-60; US Nat Liaison Officer to Orgn Econ Coop & Develop, 62; mem, US Nat Comt, Int Comn Optics, 61- & Int Fedn Doc, 64-; mem bd dirs & exec comt, Gorgas Mem Inst Trop & Prev Med, Inc, 72-, secy, 74-; consult on biomed commun to nat & int bodies. Honors & Awards: Silver Medal for Superior Serv, Dept Health, Educ & welfare, 71. Mem: AAAS; Med Libr Asn; Am Chem Soc; fel Optical Soc Am. Res: Far ultraviolet absorption spectra of organic compounds; color and chemical constitution; spectrophotomgtry; organization of science; science information and documentation; development of national and regional biomedical communication programs. Mailing Add: Nat Libr of Med 8600 Rockville Pike Bethesda MD 20014

CORNISH, ALBERT JOHN, physical chemistry, see 12th edition

CORNISH, HERBERT HARRY, b Fremont, Ohio, Sept 22, 16; m 43; c 4. TOXICOLOGY. Educ: Bowling Green State Univ, BS, 39; Univ Mich, MS, 52, PhD(biol chem), 56. Prof Exp: Clin chemist, 46-50; instr biol chem, 54-59, from asst prof to assoc prof indust health, 59-67, PROF ENVIRON & INDUST HEALTH, UNIV MICH, ANN ARBOR, 67- Mem: Am Chem Soc; Soc Toxicol; Am Indust Hyg Asn. Res: Metabolism of caffeine, chlorinated aromatics and aromatic nitro compounds; potentiation of toxicity and organic solvents; toxicity of combustion products of plastics. Mailing Add: Sch of Pub Health Univ of Mich Ann Arbor MI 48104

CORNISH, JOHN HENRY, b Arkansas City, Kans, July 14, 12; m 37; c 1. HYDROLOGY. Educ: Southwestern Col, Kans, BA, 37; Univ Tulsa, MS, 56. Prof Exp: Meteorologist, US Weather Bur, 40-51, hydrologist, 51-71; RETIRED. Res: Computer programming of river forecasting; mathematical analysis of unit hydrograph of runoff. Mailing Add: 1569 N Joplin Ave Tulsa OK 74115

CORNMAN, IVOR, b Cleveland, Ohio, May 22, 14; m 47. ZOOLOGY. Educ: Oberlin Col, AB, 36; NY Univ, MS, 39; Univ Mich, PhD, 49. Prof Exp: Asst zool, Univ Mich, 41-42; fel cytol, Sloan-Kettering Inst, 46-49; asst prof anat, Med Sch, George Washington Univ, 49-55; head dept cellular physiol, Hazleton Labs, Va, 55-59, asst dir res, 59-64; independent biol consult, 64-68; vpres, Environ Develop, Inc, 68-69; BIOL CONSULT, 69- Concurrent Pos: Mem corp, Marine Biol Lab, Woods Hole, 47-; guest, Univ West Indies, 64- Mem: Am Soc Zool. Res: Cancer chemotherapy; carcinogenesis; experimental alteration of cell division in normal and malignant plant and animal cells; marine biology; biological systems adapted to development of new drugs. Mailing Add: 10-A Orchard St Woods Hole MA 02543

CORNMAN, JOHN FARNSWORTH, floriculture, see 12th edition

CORNS, JOSEPH BARR, horticulture, see 12th edition

CORNS, WILLIAM GEORGE, b Taber, Alta, Oct 27, 16; m 43; c 2. PLANT SCIENCE, WEED SCIENCE. Educ: Univ Alta, BSc, 42, MSc, 44; Univ Toronto, PhD (plant physiol & ecol), 46. Prof Exp: Teacher high sch, 33-38; asst plant biochem, 41-44, asst prof plant sci, 46-50, assoc prof crop ecol, 50-56, head dept, 61-71, PROF CROP ECOL, UNIV ALTA, 56- Mem: Can Soc Agron; Agr Inst Can. Res: Crop ecology. Mailing Add: Dept of Plant Sci Univ of Alta Edmonton AB Can

CORNSWEET, TOM NORMAN, b Cleveland, Ohio, Apr 29, 29. VISION. Educ: Cornell Univ, AB, 51; Brown Univ, MSc, 53, PhD(exp psychol), 55. Prof Exp: From instr to asst prof psychol, Yale Univ, 55-59; from asst prof to prof, Univ Calif, Berkeley, 59-65; staff scientist biomed res, Stanford Res Inst, 65-71; chief scientist, Acuity Systs, Inc, Va, 71-73; ASSOC PROF OPHTHAL, BAYLOR COL MED, 74- Concurrent Pos: NIH grants, Yale Univ, Univ Calif, Berkeley & Stanford Res Inst, 56-71; NASA grants, Stanford Res Inst, 66-71; Retina Res Found grant, Baylor Col Med, 74- Mem: AAAS; fel Optical Soc Am; Asn Res Vision & Ophthal. Res: Glaucoma; oculomotor system; ophthalmic instrumentation; visual aids for visually handicapped. Mailing Add: Dept of Ophthal Baylor Col of Med Tex Med Ctr Houston TX 77025

CORNWALL, HENRY ROWLAND, b Middlebury, Vt, Apr 14, 13; m 37; c 3.

ECONOMIC GEOLOGY, PETROLOGY Educ: Princeton Univ, AB, 35, PhD(geol), 47. Prof Exp: Geologist, Anaconda Copper Co, Mex, 35, engr, Ariz, 36; geologist, Dana & Co, NY, 36; GEOLOGIST, US GEOL SURV, 43- Concurrent Pos: Instr, Mich Col Mining & Metall, 48. Mem: Geol Soc Am; Soc Econ Geol; Mineral Soc Am. Res: Chemical petrogeny; differentiation in doleritic lavas of the Michigan Keweenawan and the origin of copper deposits; nickel deposits of Canada, United States and Cuba; native copper deposits of the world; Tertiary volcanic rocks, calderas and associated ore deposits in Nevada; porphyry copper deposits in Arizona. Mailing Add: Pac Mineral Res Br US Geol Surv 345 Middlefield Rd Menlo Park CA 94025

CORNWALL, JOHN MICHAEL, b Denver, Colo, Aug 19, 34; m 65. THEORETICAL PHYSICS. Educ: Harvard Univ, AB, 56; Univ Denver, MS, 59; Univ Calif, Berkeley, PhD(physics), 62. Prof Exp: NSF fel physics, Calif Inst Technol, 62-63; mem, Inst Adv Study, 63-65; from asst prof to assoc prof, 65-74, PROF PHYSICS, UNIV CALIF, LOS ANGELES, 74- Concurrent Pos: Consult, Aerospace Corp, El Segundo, 62-; Alfred P Sloan Found fel, 67-69; consult, NASA, 75- Mem: Am Phys Soc; Am Geophys Union. Res: Theoretical elementary particle and high energy physics; emphasis on group symmetries; theoretical space and plasma physics; Van Allen belt studies. Mailing Add: Dept of Physics Univ of Calif Los Angeles CA 90024

CORNWELL, CHARLES DANIEL, b Williamsport, Pa, Dec 27, 24; m 51. CHEMICAL PHYSICS. Educ: Cornell Univ, AB, 47; Harvard Univ, MS, 49, PhD(chem physics), 51. Prof Exp: Res assoc chem, Univ Iowa, 50-52; from instr to assoc prof, 52-60, PROF CHEM, UNIV WIS, MADISON, 60- Mem: AAAS; Am Phys Soc; Am Chem Soc. Res: Nuclear magnetic resonance, nuclear quadrupole resonance and microwave molecular rotational spectra; relationship of spectroscopic parameters to molecular structure and chemical binding; spin relaxation in gases. Mailing Add: Dept of Chem Univ of Wis Madison WI 53706

CORNWELL, DAVID GEORGE, b San Rafael, Calif, Oct 8, 27; m 59; c 2. BIOCHEMISTRY. Educ: Col of Wooster, BA, 50; Ohio State Univ, MA, 52; Stanford Univ, PhD(chem), 55. Prof Exp: Asst physiol chem, Ohio State Univ, 50-52; asst chem, Stanford Univ, 52-53; fel, Nat Res Coun, Harvard Univ, 54-56; from asst prof to assoc prof, 56-63, PROF PHYSIOL CHEM, OHIO STATE UNIV, 63-, CHMN DEPT, 65- Concurrent Pos: Mem nutrit study sect, NIH, 66-70, nutrit sci training comt, 70-73; mem adv bd, J Lipid Res, 73- Honors & Awards: Hon Mention, Int Cong Hematol, 56. Mem: Am Chem Soc; Biophys Soc; Am Soc Biol Chem; Am Inst Nutrit; Am Oil Chem Soc. Res: Chemistry and metabolism of lipids, lipoproteins and membranes. Mailing Add: Dept of Physiol Chem Ohio State Univ 410 W 10th Ave Columbus OH 43210

CORNWELL, GEORGE WILLIAM, b Benton Harbor, Mich, Dec 4, 29; m 52. WILDLIFE ECOLOGY, ENVIRONMENTAL SCIENCES Educ: Mich State Univ, BS, 55; Univ Utah, MScE, 59; Univ Mich, PhD(wildlife path), 66. Prof Exp: Conserv aide, Mich Dept Conserv, 53-55; wildlife biol aide, US Fish & Wildlife Serv, 56; teacher high schs, Mich, 56-59; sta pathologist, Delta Waterfowl Res Sta, Man, Can, 60-62; asst prof wildlife biol & ornith, Va Polytech Inst & State Univ, 63-67; assoc prof wildlife ecol, Sch Forestry, Univ Fla, 67-73; PRES, ECOIMPACT, INC, 73- Honors & Awards: Conserv Award, Gov of Fla, 70. Mem: Wildlife Soc; Wildlife Dis Asn; Am Ornith Union; Wilson Ornith Soc; Am Inst Planners. Res: Avian parasites and diseases; waterfowl biology; resource management and conservation; regional planning; transportation ecology; environmental impact assessment; ecological planning. Mailing Add: EcoImpact Inc 6615 SW 13th St Gainesville FL 32608

CORONITI, FERDINAND VINCENT, b Boston, Mass, June 14, 43; m 69; c 2. SPACE PHYSICS, PLASMA PHYSICS. Educ: Harvard Univ, BA, 65; Univ Calif, Berkeley, PhD(physics), 69. Prof Exp: Res asst physics, Univ Calif, Berkeley, 65-69; asst res physicist, 69-70, asst prof, 70-74, ASSOC PROF PHYSICS & SPACE PHYSICS, UNIV CALIF, LOS ANGELES, 74- Concurrent Pos: Consult, TRW Syst, 69- & Los Alamos Sci Lab, 73- Mem: Am Geophys Union; Int Union Radio Sci. Res: Magnetospheric dynamics; Jupiter radiation belts and magnetospheric structure; magnetic field reconnection; nonlinear plasma theory. Mailing Add: Dept oQ of Physics Univ of Calif Los Angeles CA 90024

CORREIA, JOHN ARTHUR, b Brookline, Mass, June 8, 45; m 67; c 1. MEDICAL PHYSICS. Educ: Lowell Technol Inst, BS, 67, PhD(nuclear physics), 73. Prof Exp: Res fel physics, 72-74, ASST PHYSICIST, MASS GEN HOSP, 74- Concurrent Pos: Res fel, Sch Med, Harvard Univ, 72-74, res assoc, 74- Mem: Am Phys Soc; Am Nuclear Soc; Am Asn Physicists Med. Res: Application of computers to nuclear medicine; development of instrumentation and computational schemes for the measurement of cerebral blood flow and for three dimensional reconstruction using radioisotopes. Mailing Add: Physics Res Lab Dept of Radiol Mass Gen Hosp Boston MA 02114

CORRICK, JAMES ADAM, JR, b Parsons, WVa, Sept 21, 13; m 37; c 2. ANIMAL SCIENCE. Educ: WVa Univ, MS, 39; Univ Tenn, BS, 35, PhD(animal sci), 68. Prof Exp: County agent agr, WVa Univ, 35-41; ASST PROF ANIMAL SCI, UNIV TENN, KNOXVILLE, 68- Mem: AAAS; Am Soc Animal Sci; Inst Food Technologists; Am Soc Mil Engrs. Res: Interaction, deficiency excess of selenium and vitamin E in the diet; non-protein-nitrogen in ruminant diet; correlation of body size and composition with the ruminant performance. Mailing Add: 2116 Lake Ave Knoxville TN 37916

CORRIERE, JOSEPH N, JR, b Easton, Pa, Apr 3, 37; m 60; c 4. UROLOGY. Educ: Univ Pa, BA, 59; Seton Hall Col Med, MD, 63. Prof Exp: Resident urol, Hosp Univ Pa, 64-69; asst prof, Univ Pa, 71-74; PROF UROL & DIR PROG, UNIV TEX MED SCH HOUSTON, 74- Concurrent Pos: USPHS res trainee urol, Univ Pa, 67-68. Mem: Asn Acad Surg; Soc Nuclear Med; Acad Pediat; Am Col Surg; Am Urol Asn. Res: Urinary tract infection and the use of isotopes in the urinary tract. Mailing Add: Prog in Urol Univ of Tex Med Sch Houston TX 77025

CORRIGAN, JAMES JOHN, JR, b Pittsburgh, Pa, Aug 28, 35; m 60; c 2. HEMATOLOGY, PEDIATRICS. Educ: Juniata Col, BS, 57; Univ Pittsburgh, MD, 61. Prof Exp: Intern med, Univ Colo, 61-62, resident pediat, 62-64; assoc, Sch Med, Emory Univ, 66-67, asst prof, 67-70; assoc prof, 71-74, PROF PEDIAT, COL MED, UNIV ARIZ, 74- Concurrent Pos: Fel pediat hemat, Col Med, Univ Ill, 64-66; NIH res grants, 67-; Ga Heart Inc res grant, 68-70. Mem: AAAS; Fel Am Acad Pediat; Am Soc Hemat; Soc Pediat Res. Res: Disorders of blood coagulation mechanisms, particularly those conditions associated with disseminated intravascular coagulation and hyperfibrinolysis. Mailing Add: Dept of Pediat Univ of Ariz Col of Med Tucson AZ 85724

CORRIGAN, JOHN JOSEPH, b Chicago, Ill, Jan 17, 29; m 54; c 4. BIOCHEMISTRY. Educ: Carleton Col, AB, 50; Univ Ill, MS, 57, PhD(entom), 59. Prof Exp: Physiologist, Baxter Labs, 50-52; asst entom, Ill Natural Hist Surv, 52-53; res fel biochem, Sch Med, Tufts Univ, 59-62, sr instr, 62-63, asst prof, 63-70; ASSOC PROF LIFE SCI & BIOCHEM & ASSOC DEAN COL ARTS & SCI, IND STATE UNIV, 70- Mem: AAAS; Am Chem Soc; Entom Soc Am. Res: Metabolism of amino acids and proteins in insects; biochemistry of insect metamorphosis; silk fibroin biosynthesis in lepidoptera; biochemical basis of insecticide action; stereospecific metabolism of amino acids; biochemistry of development. Mailing Add: Dept of Life Sci Ind State Univ Terre Haute IN 47809

CORRIGAN, JOHN RAYMOND, b Fargo, NDak, Apr 28, 19; m 51; c 4. ORGANIC CHEMISTRY. Educ: Univ Portland, BS, 40; Univ Notre Dame, MS, 42, PhD(chem), 49. Prof Exp: Asst, Univ Notre Dame, 41-43; res assoc, Frederick Stearns & Co, 43-47; res fel, Sterling-Winthrop Res Inst, 47-49, res assoc, 49-51; res assoc, Sharp & Dohme, 51-55; sr chemist, 55-57, group leader, 57-59, from asst dir to dir dept, 59-68, DIR PROCESS ENG, MEAD JOHNSON & CO, 68- Mem: AAAS; Am Chem Soc; Am Inst Chemists; Am Inst Chem Engrs. Res: Synthetic organic medicinals. Mailing Add: Mead Johnson & Co Evansville IN 47721

CORRIGAN, KENNETH EDWIN, b New York, NY, June 10, 05; m 50. PHYSICS. Educ: Knox Col, BS, 28; Univ Ill, MS, 30, PhD(physics), 32. Prof Exp: Dir diffraction lab, Barrett Co, NJ, 30; physicist, Harper Hosp, 32-34, chief radiol res lab, 35-37; dir res div, 47-49, special instr flight sci, aircraft & engine res, 42-45; ASSOC PROF RADIOL, SCH MED, WAYNE STATE UNIV, 57-, DIR DIV RADIOL SCI, 67- Concurrent Pos: Dir dept med physics, William Beaumont Hosp, 57-67; past sci assoc, Sloan-Kettering Inst. Mem: Am Phys Soc; Radiol Soc NAm; assoc mem Am Roentgen Ray Soc; fel Am Col Radiol; Soc Nuclear Med. Res: Radiation measurements; thermal neutrons; production and utilization of radioisotopes; activation analysis; cryogenic trapping systems for radioactive gases. Mailing Add: Dept of Radiol 1400 Chrysler Freeway Wayne State Univ Sch of Med Detroit MI 48207

CORRIGAN, SAMUEL WALTER, b St Boniface, Man, Oct 19, 39; c 1. APPLIED ANTHROPOLOGY, SOCIAL ANTHROPOLOGY. Educ: Univ Man, BA, 62; Univ BC, MA, 64; Cambridge Univ, PhD(anthrop), 70. Prof Exp: Asst prof anthrop, Univ Man, 69-70; coordr educ admin, Spec Mature Student Prog, Three Univs of Man, 70-71; ASST PROF ANTHROP, BRANDON UNIV, 71- Concurrent Pos: Sessional lectr, Brandon Univ, 70-71, adv native studies, 71-; Brandon Univ & Secy of State res grants, Univ Alta & Brandon Univ, 71; consult regional & nat native orgn. Mem: Fel Royal Anthrop Inst Gt Brit & Ireland; foreign fel Am Anthrop Asn; Can Ethnol Soc. Res: Contemporary Canadian native people, especially changes in social structure, availability and manipulation of resources, and native organizations; patterns of urbanization of native peoples; Canadian Dakota Indians. Mailing Add: Dept of Anthrop Brandon Univ Brandon MB Can

CORRIN, MYRON LEE, b Waupaca, Wis, Sept 6, 14. PHYSICAL CHEMISTRY. Educ: Marquette Univ, BS, 39, MS, 40; Univ Chicago, PhD(chem), 46. Prof Exp: Res assoc, Univ Chicago, 46-47; instr chem, Univ Calif, 47-48; from res assoc to asst prof, Univ Chicago, 48-51; res assoc, Gen Elec Res Labs, 51-55; prof chem, Univ Ariz, 59-67; PROF ATMOSPHERIC SCI, COLO STATE UNIV, 67- Mem: Am Chem Soc. Res: Atmospheric aerosols; air pollution; solution of long-chain electrolytes; heterogeneous nucleation; surface chemistry. Mailing Add: Dept of Atmospheric Sci Colo State Univ Ft Collins CO 80521

CORRIVAULT, GEORGES WILFRID, histology, see 12th edition

CORRUCCINI, LINTON REID, b Corvallis, Ore, Jan 1, 44; m 72. LOW TEMPERATURE PHYSICS. Educ: Swarthmore Col, BA, 66; Cornell Univ, PhD(physics), 72. Prof Exp: NSF fel physics, Cornell Univ, 67-71; mem tech staff, Aerospace Corp, 71-73; ASST PROF PHYSICS, UNIV CALIF, DAVIS, 73- Mem: Am Phys Soc. Res: Liquid helium three-helium four solutions; superfluid helium 3; nuclear magnetic resonance. Mailing Add: Dept of Physics Univ of Calif Davis CA 95616

CORRUCCINI, ROBERT JOE, physical chemistry, see 12th edition

CORRY, MARTHA LUCILLE, b Springfield, Ohio, Feb 27, 19. GEOGRAPHY. Educ: Ohio State Univ, BA & BScEd, 41; Univ Iowa, MA, 47, PhD(geog), 53; McGill Univ, cert, 48. Prof Exp: Instr soc sci & geog, Univ Iowa, 49-51; from asst prof to assoc prof, 51-59, PROF GEOG, STATE UNIV NY COL ONEONTA, 59-, CHMN DEPT, 70- Concurrent Pos: Partic, Field Study, Soviet Union, 58 & 67; NY State Educ Dept Ctr Int Progs & Comp Studies grants, Seminars Yugoslavian Lit, State Univ NY Albany, 70 & Univ Belgrade, 72. Mem: Asn Am Geogr; Am Geog Soc; Nat Coun Geog Educ. Res: Location of economic activity; spatial patterns of consumer behavior; economic regions of the Union of Soviet Socialist Republics. Mailing Add: Dept of Geog 317 Milne Libr State Univ of NY Col Oneonta NY 13820

CORSARO, ROBERT DOMINIC, b Elizabeth, NJ, Nov 5, 44; m 67; c 1. PHYSICAL CHEMISTRY, ACOUSTICS. Educ: Lebanon Valley Col, BS, 66; Univ Md, PhD(chem), 71. Prof Exp: RES CHEMIST ACOUST, NAVAL RES LAB, 70- Mem: Am Chem Soc; Am Phys Soc; Acoust Soc Am; Am Ceramic Soc. Res: Rheology of liquids and glasses; acoustic techniques in material science; fast reaction kinetics; underwater sound and parametric sonar. Mailing Add: Code 8131 Naval Res Lab Washington DC 20375

CORSE, JOSEPH WALTERS, b Denver, Colo, Sept 7, 13; m 36; c 4. PHYTOCHEMISTRY. Educ: Univ Calif, Los Angeles, BA, 36; Univ Ill, PhD(org chem), 40. Prof Exp: Asst, Univ Ill, 37-39; res chemist, Eli Lilly Co, 40-46; asst, Calif Inst Technol, 46-47; res assoc, Univ Calif, Los Angeles, 47-48; CHEMIST, WESTERN REGIONAL RES LAB, AGR RES SERV, US DEPT AGR, 48- Mem: Am Chem Soc; Phytochem Soc NAm. Res: Synthetic drugs; biosynthesis of penicillin; enzyme reactions and synthesis of metabolites; chemical interactions in plant growth. Mailing Add: Western Regional Res Lab US Dept of Agr Agr Res Serv Albany CA 94706

CORSINI, A, b Hamilton, Ont, Apr 28, 34; m 61; c 1. ANALYTICAL CHEMISTRY. Educ: McMaster Univ, BSc, 56, PhD(anal chem), 61. Prof Exp: Atomic Energy Comn fel chelate chem, Univ Ariz, 61-63; from asst prof to assoc prof anal & inorg chem, 63-72, PROF CHEM, McMASTER UNIV, 72- Concurrent Pos: Nat Res Coun Can grant, 63- Honors & Awards: Louis Gordon Mem Award, Pergamon Press, 73. Mem: Chem Inst Can. Res: Instrumental analysis; chemistry of metal chelates, heteropolytungstates, metalloporphyrins and analytical applications. Mailing Add: Dept of Chem McMaster Univ Hamilton ON Can

CORSINI, DENNIS LEE, b Los Angeles, Calif, Nov 8, 42; m 63; c 3. AGRICULTURAL BIOCHEMISTRY. Educ: Univ Calif, Los Angeles, BA, 65; Univ Idaho, PhD(agr biochem), 71. Prof Exp: Fel plant path, Univ Calif, Riverside, 71-72; RES ASSOC PLANT SCI, UNIV IDAHO, 73- Mem: Potato Asn Am. Res: Chemical determinants of potato quality. Mailing Add: Univ of Idaho Res & Exten Ctr Aberdeen ID 83210

CORSON, DALE RAYMOND, b Pittsburg, Kans, Apr 5, 14; m 38; c 4. NUCLEAR PHYSICS. Educ: Col Emporia, AB, 34; Univ Kans, AM, 35; Univ Calif,

PhD(physics), 38. Hon Degrees: LHD, Emporia Col, 70; LLD, Columbia Univ, 72 & Hamilton Col, 73; DSc, Univ Rochester, 75. Prof Exp: Asst, Univ Calif, 36-38, res fel, 38-39, instr physics, 39-40; from asst prof to assoc prof, Univ Mo, 40-45; staff mem, Los Alamos Sci Lab, 45-46; from asst prof to prof physics, 46-63, chmn dept, 56-59, dean col eng, 59-63, provost, 63-69, PRES, CORNELL UNIV, 69- Concurrent Pos: Mem staff, Radiation Lab, Mass Inst Technol, 41-43; radar consult, US Army Air Force, 43-45. Mem: Am Phys Soc; fel Am Acad Arts & Sci. Res: Cosmic rays; nuclear physics; engineering. Mailing Add: 300 Day Hall Cornell Univ Ithaca NY 14853

CORSON, GEORGE EDWIN, JR, b Sebastopol, Calif, Aug 2, 40; m 64; c 2. PLANT MORPHOGENETICS. Educ: Univ of the Pac, MS, 64; Univ Calif, Davis, PhD(bot), 68. Prof Exp: NSF trainee, Univ Calif, Davis, 66-68; ASSOC PROF, CALIF STATE UNIV, CHICO, 68- Mem: Bot Soc Am; Int Soc Plant Morphol. Res: Cell division rates of the root apical cell and its derivatives in Equisetum hyemale; callus and protoplast formation and differentiation in Equisetum hyemale. Mailing Add: Dept of Biol Sci Calif State Univ Chico CA 95929

CORSON, HARRY HERBERT, b Nashville, Tenn, Feb 2, 31; m 53; c 2. MATHEMATICS. Educ: Vanderbilt Univ, AB, 52; Duke Univ, MA, 54, PhD(math), 57. Prof Exp: Res instr math, Tulane Univ, 56-58; res asst prof, 58-62, assoc prof, 62-65, PROF MATH, UNIV WASH, 65- Concurrent Pos: Fel, Off Naval Res, 58-59. Mem: Am Math Soc. Res: Topology; linear spaces. Mailing Add: Dept of Math Univ of Wash Seattle WA 98105

CORSON, SAMUEL ABRAHAM, b Odessa, Russia, Dec 31, 09; nat US; m 47; c 3. PHYSIOLOGY, PHARMACOLOGY. Educ: NY Univ, BS, 30; Univ Pa, MS, 31; Univ Tex, PhD, 42. Prof Exp: Asst physiol, Sch Med, Univ Pa, 32-35; res assoc cell physiol, NY Univ, 35-37; instr physiol, Div Gen Educ, NY, 37-39; consult physiologist, New York, 38-40; instr zool & physiol, Univ Tex, 40-42; asst prof physiol, Sch Med, Univ Okla, 42-43; instr pharmacol, Sch Med, Georgetown Univ, 43-44; from instr to asst prof physiol, Med Col, Univ Minn, 44-47, instr sci Russian & coordr contemp Russian, Exten Div, 45-47; assoc prof, Col Med, Howard Univ, 47-56; chief dept physiol, Toledo Hosp Inst Med Res, 50-51; prof pharmacol & head dept, Kirksville Col Osteop & Surg, 51-54; assoc prof pharmacol & physiol, Sch Med, Univ Ark, 54-59; res assoc hist med, Sch Med, Yale Univ, 59-60; assoc prof psychiat, 60-67, PROF PSYCHIAT, COL MED, OHIO STATE UNIV, 67-, PROF BIOPHYS, COL BIOL SCI, 69-, DIR LAB CEREBROVISCERAL PHYSIOL, COL MED, 60- Concurrent Pos: Consult, hosp staff, Univ Ark, 58-59; ed in chief, Int J Psychobiol, 70-73. Mem: Fel AAAS; Am Physiol Soc; fel Am Col Cardiol; Am Psychosom Soc. Res: Cerebrovisceral and renal physiology; psychopharmacology; cybernetics and systems approach in psychophysiology; individual differences in reactions to psychologic stress; Pavlovian and operant conditioned reflexes; interaction of pharmacologic and psychosocial factors in the control of aggression and hyperkinesis; minimal brain dysfunction; paradoxical effects of amphetamines. Mailing Add: Dept Psychiat Div of Behav Sci Ohio State Univ Col of Med Columbus OH 43210

CORSTVET, RICHARD E, b Big Bend, Wis, Oct 12, 28; m 54; c 3. VETERINARY MICROBIOLOGY, VETERINARY PATHOLOGY. Educ: Univ Wis, BS, 51, MS, 55; Univ Calif, Davis, PhD(comp path), 65. Prof Exp: Lab technician avian med, Sch Vet Med, Univ Calif, Davis, 55-59, pub health, 59-65; asst res microbiologist, 65; from asst prof to assoc prof microbiol, 65-72, PROF PARASITOL & PUB HEALTH, COL VET MED, OKLA STATE UNIV, 72- Mem: Poultry Sci Asn; US Animal Health Asn; NY Acad Sci; Am Asn Avian Path. Res: Infection and immunity in Newcastle disease; relationship of various disease syndromes to the wholesomeness of market poultry; carrier states in animals; multiple infections in animals. Mailing Add: Col of Vet Med Okla State Univ Stillwater OK 74074

CORT, WINIFRED MITCHELL, b Cleveland, Ohio, Dec 9, 17; m 47; c 2. MICROBIAL BIOCHEMISTRY. Educ: Univ Ill, BS, 41, MS, 43, PhD(bact, biochem), 46. Prof Exp: Res microbiologist, Commercial Solvents Corp, 43-45; res asst biochem, Univ Ill, 45; res biochemist, Chas Pfizer & Co, 45-47; res microbiologist, Nat Dairy Res Labs, Inc, 51-54; sect leader, Res & Develop Div, 54-58, sect leader enzymes, Evans Res & Develop Corp, 58-64; group leader, Res & Develop, Beechnut Life Savers, 64-65; sr biochemist, 65-73, GROUP LEADER, HOFFMANN-LA ROCHE INC, 73- Concurrent Pos: Lectr, Adelphi Univ, 59-62; consult, Evans Res & Develop Corp, 61-62 & Food & Drug Res, Inc, 62. Honors & Awards: Co-author Bond Award, Am Oil Chem Soc, 74. Mem: Am Soc Microbiol; Am Chem Soc; Soc Indust Microbiol; AAAS; Inst Food Technol. Res: Mechanism of action of antioxidants; DeV. test systs for antioxidants and antibiotics, chemical and microbiological. Mailing Add: 120 Francisco Ave Falls NJ 07424

CORTEZ, HENRY VALDEZ, organic chemistry, see 12th edition

CORTH, RICHARD, b New York, NY, Apr 14, 25; m 44; c 3. PHYSICAL CHEMISTRY, ANALYTICAL CHEMISTRY. Educ: Brooklyn Col, BS, 48; Polytech Inst Brooklyn, PhD(radiation chem), 63. Prof Exp: Res engr, 55-56, sr res scientist, 56-73, FEL RES SCIENTIST, WESTINGHOUSE ELEC CORP, 73- Mem: AAAS; Am Inst Biol Sci; Am Chem Soc. Res: Photobiology; radiochemistry; radiation chemistry; solid state diffusion; gas-metal reactions; gas chromatography; activation analysis. Mailing Add: Lamp Div Westinghouse Elec Corp Bloomfield NJ 07003

CORTNER, JEAN A, b Nashville, Tenn, Nov 11, 30; m; c 3. BIOCHEMICAL GENETICS. Educ: Vanderbilt Univ, BA, 52, MD, 55. Prof Exp: Guest investr human genetics & asst physician, Rockefeller Inst, 62-63; chief dept pediat, Roswell Park Mem Inst, 63-67; prof pediat & chmn dept, State Univ NY, Buffalo, 67-74, physician-in-chief, Children's Hosp, 70-74, pediatrician-in-chief, 67-70; PHYSICIAN-IN-CHIEF, CHILDREN'S HOSP PHILADELPHIA, 74-, PROF & CHMN DEPT, SCH MED, UNIV PA, 74- Concurrent Pos: NIH vis fel dept pediat & biochem, Babies Hosp & Columbia Univ, 61-63; NIH fel, Galton Lab Human Genetics, Univ London, 72-73. Mem: AAAS; Am Soc Human Genetics; Am Fedn Clin Res; Soc Pediat Res; Am Acad Pediat. Res: Human genetics; pediatrics. Mailing Add: Dept of Pediat Children's Hosp 34th St & Civic Ctr Blvd Philadelphia PA 19104

CORTNEY, MARSHALL ALLEN, b Detroit, Mich, Oct 17, 33. PHYSIOLOGY. Educ: Univ Mich, BA, 56; Univ Calif, San Francisco, PhD(physiol), 62. Prof Exp: NIH fel physiol, Univ NC, 62-64 & Inst Physiol, Univ Göttingen, 64-65; asst prof, 65-70, ASSOC PROF PHYSIOL, COL MED, UNIV IOWA, 70- Mem: Am Physiol Soc; Soc Exp Biol & Med; Am Soc Nephrology. Res: Renal physiology; regulation of electrolyte reabsorption by the renal tubules. Mailing Add: Dept of Physiol & Biophys Univ of Iowa Col of Med Iowa City IA 52240

CORUM, CYRIL JOSEPH, b Louisville, Ky, Dec 21, 14; m 38. MYCOLOGY. Educ: Miami Univ, AB, 36; Univ Wis, PhD(mycol, plant physiol), 40. Prof Exp: Instr biol, Western Reserve Univ, 40-42, asst prof, 46-47; microbiologist, Res & Develop Div, Com Solvents Corp, Ind, 47-52; biochemist, 52-53, head fermentation develop & assay labs, 53-63, sr microbiologist, 63-65, RES SCIENTIST, BIOTICS DEVELOP DIV, ELI LILLY & CO, 65- Mem: AAAS; Am Chem Soc; Am Soc Microbiol; Soc Indust Microbiol; NY Acad Sci. Res: Fermentation; physiology of microorganisms. Mailing Add: Dept K-418 Eli Lilly & Co 740 S Alabama St Indianapolis IN 46206

CORWIN, ALSOPH HENRY, b Marietta, Ohio, Jan 11, 08; m 38. ORGANIC CHEMISTRY. Educ: Marietta Col, AB, 28, DSc, 53; Harvard Univ, PhD(chem), 32. Prof Exp: Assoc, 32-39, from assoc prof to prof, 39-74, chmn dept, 44-47, EMER PROF CHEM, JOHNS HOPKINS UNIV, 74- Mem: AAAS; Am Chem Soc. Res: Organic synthesis and structure; reaction mechanisms; reaction kinetics; microchemistry; construction of balances and precision weighing. Mailing Add: Dept of Chem Johns Hopkins Univ Baltimore MD 21218

CORWIN, HARRY O, b Los Angeles, Calif, Apr 30, 38; m 61; c 1. GENETICS. Educ: Univ Calif, Santa Barbara, BA, 61; Univ Calif, Los Angeles, MA, 62, PhD(zool), 66. Prof Exp: Teaching asst biol, embryol, anat & genetics, Univ Calif, Los Angeles, 62-66; asst prof genetics, 66-71, ASSOC PROF BIOL, UNIV PITTSBURGH, 71- Mem: AAAS. Res: Genetic analysis of mutagens effect on the dumpy locus of the fruit fly, Drosophila melanogaster. Mailing Add: Dept of Biol Univ of Pittsburgh Pittsburgh PA 15213

CORWIN, JAMES FAY, b Blanchester, Ohio, Oct 31, 07; m 31; c 1. CHEMISTRY. Educ: Ohio Univ, BA, 32, MA, 34; Ohio State Univ, PhD(chem), 44. Prof Exp: Teacher high sch, Ohio, 34-35; res chemist, Wheeling Steel Corp, WVa, 35-37; teacher high sch, Ohio, 37-40; asst anal chem, Ohio State Univ, 40-41; from instr to prof gen & anal chem & metall, 41-73, chmn dept chem, 48-61, chmn phys sci area, 64-70, EMER PROF GEN & ANAL CHEM & METALL, ANTIOCH COL, 73- Concurrent Pos: Res dir & asst to pres, Vernay Labs, Inc, Ohio, 44-74, mem bd dirs, 74-77, consult, 74-; dir teachers progs, NSF, 57-73, dir hydrothermal res, 54-70; vis res prof, Inst Marine Sci, Univ Miami, 60-68, adj prof, 67-72; consult, US AID, India, 65. Mem: Fel AAAS; Am Chem Soc; Soc Plastics Engrs. Res: Quantitative analytical chemistry; physical organic chemistry; conductometric titrations with organic reagents; quartz crystal growth; reactions in water solutions in the super-critical state; oceanography and environmental studies; consulting in industrial and academic science services. Mailing Add: 528 Palo Verde Dr Hawthorne at Leesburg Leesburg FL 32748

CORWIN, LAURENCE MARTIN, b Rochester, NY, Aug 26, 29; m 52; c 3. BIOCHEMISTRY. Educ: Univ Chicago, PhB, 48; Syracuse Univ, BA, 50; Wayne State Univ, PhD(biochem), 56. Prof Exp: Res fel biol, Calif Inst Technol, 57-58; biochemist, NIH, 58-61; biochemist, Walter Reed Army Inst Res, 61-67; assoc prof, 67-73, PROF MICROBIOL & NUTRIT SCI, SCH MED, BOSTON UNIV, 73- Mem: Am Soc Biol Chemists; Am Inst Nutrit; Am Soc Microbiol. Res: Biochemical function of vitamin E; biochemical genetics in microorganisms. Mailing Add: Dept of Microbiol Boston Univ Sch of Med Boston MA 02118

CORWIN, ROBERT MARVIN, zoology, microbiology, see 12th edition

CORWIN, THOMAS LEWIS, b Newburgh, NY, Oct 9, 47; m 71; c 1. MATHEMATICAL STATISTICS. Educ: Villanova Univ, BS, 69; Princeton Univ, MS, 71, PhD(statist), 73. Prof Exp: ASSOC STATIST, DANIEL H WAGNER ASSOC, 73- Mem: Inst Math Statist; Am Statist Asn; Soc Appl & Indust Math. Res: Development of analytical tools to perform real-time, operational analysis of naval search problems; developing Bayesian information processing techniques for use in search problems. Mailing Add: Daniel H Wagner Assoc Station Square One Paoli PA 19301

CORY, JOSEPH G, b Tampa, Fla, Jan 27, 37; m 63; c 1. BIOCHEMISTRY. Educ: Univ Tampa, BS, 58; Fla State Univ, PhD(chem), 63. Prof Exp: Fel chem, Fla State Univ, 63; fel biochem, Albert Einstein Med Ctr, 63-64, asst mem, 64-65; asst prof, Fla State Univ, 65-69; assoc prof biochem, 69-73, prof med microbiol, Col Med, 73-74, PROF BIOCHEM & CHMN DEPT, COL MED, UNIV S FLA, 74- Mem: Am Chem Soc; Am Soc Biol Chem; Am Asn Cancer Res; Soc Exp Biol & Med. Res: Enzymology of nucleotide interconversions and nucleic acid synthesis; cancer biochemistry. Mailing Add: Dept of Biochem Univ of SFla Col of Med Tampa FL 33620

CORY, MICHAEL, b New York, NY, Dec 5, 41; m 66; c 2. MEDICINAL CHEMISTRY. Educ: San Jose State Col, BS, 64; Univ Calif, Santa Barbara, PhD(org chem), 71. Prof Exp: CHEMIST ORG CHEM, STANFORD RES INST, 64-68, 71- Mem: Am Chem Soc. Res: Drug design; enzyme inhibitor design; interaction of drugs with macromolecules. Mailing Add: Stanford Res Inst Menlo Park CA 94025

CORY, ROBERT MACKENZIE, b Washington, DC, Feb 22, 43; m 66; c 1. SYNTHETIC ORGANIC CHEMISTRY. Educ: Harvey Mudd Col, BS, 65; Univ Wis, PhD(org chem), 71. Prof Exp: Fel, Univ Colo, 71-72 & Rice Univ, 72-73; asst prof chem, Ohio Univ, 73-74; ASST PROF CHEM, UNIV WESTERN ONT, 74- Mem: Am Chem Soc; The Chem Soc; Chem Inst Can. Res: Organic synthesis; bicycloannulation, carbon atom equivalents; diazo compounds, vinyl phosphonium and sulfonium salts, terpenes, strained polycyclic systems, cryptate complexes. Mailing Add: Dept of Chem Univ of Western Ont London ON Can

CORY, ROBERT PAUL, biochemistry, organic chemistry, deceased

CORYELL, MARGARET E, b Great Falls, Mont, Jan 25, 13; m 41; c 3. BIOLOGICAL CHEMISTRY. Educ: Univ Colo, AB, 35; Univ Mich, MA, 37, PhD, 41. Prof Exp: Asst, Childrens Fund of Mich, 40-48; INSTR CELL BIOL, MED COL GA, 56- Res: Biochemical defects in mentally retarded children. Mailing Add: Dept of Biochem Med Col of Ga Augusta GA 30902

COSBY, JOHN NORMAN, b Mayfield, Ky, July 4, 15; m 40; c 4. ORGANIC CHEMISTRY, ENGINEERING. Educ: Univ Ill, BS, 37; Pa State Univ, MS, 39, PhD(org chem), 41. Prof Exp: Proj leader, Pa State Univ, 40-42; res chemist, Gen Aniline & Film Corp, 42-43; res chemist, Allied Chem & Dye Corp, 43-44, proj leader, 44-52, asst lab dir, 52-58, dir res & develop, Barrett Div, 58-60, tech dir, 60-62; assoc dir res, Int Minerals & Chem Corp, 62-67, vpres com develop, Lavino Div, 67-71; MGR BUS DEVELOP, UNITED ENGRS & CONSTRUCTORS, INC, 72- Res: Synthesis and properties of high molecular weight hydrocarbons in the lubricating oil range; discovery of the ammoxidation process; liquid phase oxidation of organic compounds; structure of triisobutylenes and tetraisobutylenes; process development for adipic acid, cyclohexanone, hydrogen peroxide, emulsifiable polyethylene wax, chlorinated polyethylene and polyethers. Mailing Add: 813 Milmar Rd Newton Square PA 19073

COSBY, LUCILLE A, biophysics, radiation biology, see 12th edition

COSBY, LYNWOOD ANTHONY, b Richmond, Va, June 11, 28; m 51; c 7. ELECTRONICS. Educ: Univ Richmond, BS, 49; Va Polytech Inst, MS, 51. Prof Exp: Instr physics, Va Mil Inst, 50-51; from physicist to head br advan tech, 51-71, HEAD DIV ELECTRONIC WARFARE, US NAVAL RES LAB, 71- Concurrent Pos: Chmn advan tech objectives working group, Electronic Warfare, 73-; mem threat environ steering comt, Airborne Warning & Control Syst, 75- Honors & Awards: US Navy Distinguished Civilian Serv Award, 58; Am Soc Naval Engrs Gold Medal Award, 68; Dept of Defense Distinguished Civilian Serv Award, 74. Mem: Fel Inst Elec & Electronic Engrs (pres, 70-71); Sigma Xi; Am Soc Naval Engrs. Res: Administrative and technical leadership for advancement of technology and systems for all Navy electronic warfare applications; development of circuits, electron devices, antennas and the design of special research development technical and engineering facilities. Mailing Add: US Naval Res Lab Code 5700 Washington DC 20375

COSBY, RICHARD SHERIDAN, b New York, NY, July 6, 13; m 48. MEDICINE. Educ: Harvard Univ, AB, 34, MD, 38; Am Bd Internal Med, dipl, 48; Am Bd Cardiovasc Dis, dipl, 49. Prof Exp: Fel physiol, Sch Med, Western Reserve Univ, 39; resident cardiol, Mass Gen Hosp, 40-41; instr med, Harvard Med Sch, 42; assoc clin prof, 54-67, CLIN PROF MED, SCH MED, UNIV SOUTHERN CALIF, 67-; DIR, FOUND CARDIOVASC RES, 66- Concurrent Pos: Consult, Huntington Mem Hosp, Pasadena, Good Samaritan Hosp, La Vina Sanitarium & Greater El Monte Community Hosp. Mem: AMA; Am Heart Asn; fel Am Col Physicians; Am Col Chest Physicians; Am Col Cardiol. Res: Electrocardiography; cardiopulmonary physiology. Mailing Add: 1127 Wilshire Blvd Los Angeles CA 90017

COSCARELLI, WALDIMERO, b Brooklyn, NY, Nov 5, 26; m 48; c 3. MICROBIOLOGY, BIOCHEMISTRY. Educ: Wagner Col, BS, 56; Rutgers Univ, PhD(microbiol), 61. Prof Exp: Bacteriologist, Chase Chem Co, NJ, 56; microbiologist, Schering Corp, 56-58; mem tech staff microbiol deterioration, Bell Tel Labs, Inc, 61-68; sr scientist, 68-69, MGR MICROBIOL, DIV CENT RES, SHULTON, INC, 69- Mem: Am Soc Microbiol. Res: Effects of microorganisms on plastics, including bacteria and fungi; isolation of organisms to study their metabolic requirements from plastics; transformations of steroid compound by microorganisms. Mailing Add: Div of Cent Res Shulton Inc 697 Route 46 Clifton NJ 07015

COSCIA, ANTHONY THOMAS, b New York, NY, Nov 1, 28; m 51; c 4. POLYMER CHEMISTRY. Educ: Fordham Col, BS, 50; NY Univ, PhD(chem), 56. Prof Exp: Asst, NY Univ, 50-52; res chemist, Fleischmann Labs, 52-55; res chemist, 55-60, group leader org chem, 60-65, group leader polymer res, 65-66, mgr org & polymer res, Indust Chem Div, 66-71, MGR FLOCCULANTS, INDUST CHEM & PLASTICS DIV, AM CYANAMID CO, 71- Mem: Am Chem Soc. Res: Polysaccharide isolation and identification; synthesis of nitrogen heterocycles; studies of alkylation reactions; synthesis and polymerization of new monomers; water soluble polymers; resins and coatings for paper; direct new product research on flocculants and water treating chemicals. Mailing Add: Indust Chem & Plastics Div Am Cyanamid Co 1937 W Main St Stamford CT 06904

COSCIA, CARMINE JAMES, b New York, NY, July 26, 35; m 56; c 3. BIOCHEMISTRY. Educ: Manhattan Col, BS, 57; Fordham Univ, MS, 60, PhD(org chem), 62. Prof Exp: NATO res fel org chem, Swiss Fed Inst Technol, 62-63, USPHS fel, 63-64; res fel biochem, Univ Pittsburgh, 64-65; from asst prof to assoc prof, 65-73, PROF BIOCHEM, SCH MED, ST LOUIS UNIV, 73- Mem: Am Soc Biol Chemists. Res: Secondary metabolism in plants and animals. Mailing Add: Dept of Biochem St Louis Univ Sch of Med St Louis MO 63104

COSENZA, BENJAMIN JOHN, b New Haven, Conn, May 15, 26; m 51; c 4. BACTERIOLOGY. Educ: Univ Conn, BA, 51, PhD(bact), 59; Univ Vt, MS, 54. Prof Exp: Asst bact & bot, Univ Vt, 51-54; asst instr bact, Univ Conn, 54-57, instr, 57-59; asst prof bot, Univ Vt, 60; from asst prof to assoc prof bact, Univ Conn, 61-74; ASSOC PROF BIOL, SOUTHERN CONN STATE COL, 74- Mem: Am Soc Microbiol; NY Acad Sci. Res: Bacterial cytology and histochemistry; insect and marine microbiology. Mailing Add: Dept of Biol Southern Conn State Col New Haven CT 06515

COSGRIFF, JOHN W, JR, b Denver, Colo, Nov 10, 31; m 57; c 2. VERTEBRATE PALEONTOLOGY. Educ: Univ Ariz, BA, 53; Univ Calif, Berkeley, MA, 60, PhD(paleont), 63. Prof Exp: Sr res fel paleont, Univ Tasmania, 64-67; ASSOC PROF BIOL, WAYNE STATE UNIV, 67- Concurrent Pos: Res assoc, Mus Northern Ariz, 69-; res assoc geol, Univ Tasmania, 75- Mem: AAAS; Soc Vert Paleont; Australian Geol Soc; Paleont Soc. Res: Mesozoic vertebrate faunas of Australia and North America. Mailing Add: Dept of Biol Col of Lib Arts Wayne State Univ Detroit MI 48202

COSGROVE, CLIFFORD JAMES, b Torrington, Conn, Jan 31, 27; m 52; c 3. DAIRY SCIENCE, FOOD SCIENCE. Educ: Univ Conn, BS, 51; Southern Conn State Col, BS, 53; Univ RI, MS, 57. Prof Exp: Assoc prof animal sci, 53-76, PROF FOOD & RESOURCE CHEM, UNIV R I, 76- Concurrent Pos: Expert dairy technol, Food & Agr Orgn of UN, 76- Res: Food technology, especially dairy products and their substitution by imitations or synthetics. Mailing Add: Dept of Food & Resource Chem 237 Woodward Hall Univ of RI Kingston RI 02881

COSGROVE, FRANK P, b Peekskill, NY, June 28, 14; m 51. PHARMACY. Educ: Univ Notre Dame, BS, 38; NY Univ, MA, 41; Univ Colo, MS, 49; Ohio State Univ, PhD(pharmacy), 53. Prof Exp: Asst prof pharm, Loyola Univ, La, 47-50; instr, Ohio State Univ, 50-53; assoc prof, Univ Nebr, 53-55; prof, Loyola Univ, 55-58; from asst prof to assoc prof, Univ Tex, 58-68; PROF PHARM & DEAN, IDAHO STATE UNIV, 68- Res: Cardiac glycosides; particles size in pharmaceutical preparations; biopharmaceutics. Mailing Add: Col of Pharm Idaho State Univ Pocatello ID 83201

COSGROVE, GERALD EDWARD, b Dubuque, Iowa, July 13, 20; m 43; c 7. PATHOLOGY. Educ: Univ Notre Dame, BS, 45; Univ Mich, MD, 44; Am Bd Path, dipl, 50. Prof Exp: Hosp pathologist, Klamath Falls, Ore, 50-52; hosp pathologist, Rapid City, SDak, 52-55; hosp pathologist, Gorgas Hosp, Ancon, CZ, 55-57; BIOLOGIST, BIOL DIV, OAK RIDGE NAT LAB, 57- Mem: Radiation Res Soc; Am Soc Exp Path; Am Soc Parasitol; Am Soc Ichthyologists & Herpetologists. Res: Delayed effects of radiation; radiation protective treatments; parasitology; experimental parasitology. Mailing Add: Div of Biol Oak Ridge Nat Lab Oak Ridge TN 37830

COSGROVE, LEE ALBERT, physical chemistry, see 12th edition

COSGROVE, WILLIAM BURNHAM, b New York, NY, June 11, 20; m 49; c 2. PHYSIOLOGY. Educ: Cornell Univ, AB, 41; NY Univ, MS, 47, PhD(zool), 49. Prof Exp: Assoc zool, Univ Iowa, 49-51, from asst prof to assoc prof, 51-57; from assoc prof to prof, 57-69, head dept, 64-69, ALUMNI FOUND PROF ZOOL, UNIV GA, 69- Honors & Awards: Michael Award, 67. Mem: AAAS; Am Soc Cell Biol; Am Soc Zool; Soc Protozool. Res: Comparative physiology of respiratory pigments; cell biology of trypanosomatids; physiology of protozoa; membrane transport. Mailing Add: Dept of Zool Univ of Ga Athens GA 30602

COSMATOS, ALEXANDROS, b Port Sudan, Sudan, June 26, 27; m 67. ORGANIC CHEMISTRY. Educ: Nat Univ Athens, dipl, 54, PhD(chem), 59. Prof Exp: Asst org chem, Nat Univ Athens, 54-60, sr instr, 60-67; assoc, Med Ctr, Duke Univ, 67-68; supvr peptide chem, Cyclo Chem Corp, Calif, 68-69; assoc, 69-71, RES ASST PROF BIOCHEM, MT SINAI SCH OF MED, 71- Concurrent Pos: Fel, Tex A&M Univ & Univ of the Pac, 60-61. Mem: Am Chem Soc; Hellenic Chem Soc. Res: Peptides and amino sugars. Mailing Add: Dept of Biochem Mt Sinai Sch of Med New York NY 10029

COSMIDES, GEORGE JAMES, b Pittsburgh, Pa, July 23, 26; m 48; c 1. PHARMACOLOGY. Educ: Univ Pittsburgh, BS, 52; Purdue Univ, MS, 54, PhD(pharmacol), 56. Prof Exp: Pharmacist, Pittsburgh, 52; pharmacist, Purdue Univ, 52-54; sr scientist, Smith, Kline & French Labs, 56-57; asst prof pharmacol, Univ RI, 57-59; sr res pharmacologist, Psychopharmacol Serv Ctr, NIMH, 59-63, prog adminr, Pharmacol Training Prog, Nat Inst Gen Med Sci, 63-64, exec secy pharmacol-toxicol prog comt, 64-74, prog dir pharmacol-toxicol prog & mem pharmacol res assoc training comt, NIH, 64-74, DEP DIR TOXICOL INFO PROG & DEP ASSOC DIR, NAT LIBR MED, NIH, 74- Concurrent Pos: Consult, Sci Criminal Invest, RI, 57-59; mem reference panel, Am Hosp Formulary Serv, 63-; adj prof pharmacol, Univ Pittsburgh, 63-; mem comt probs of drug safety, Nat Acad Sci-Nat Res Coun, 64-69; consult, WHO, 65; consult biomed commun for radio & TV, NIH, 70- & Surgeon Gen, USPHS, 71- Honors & Awards: Vavro Mem Award & Bristol Award, 51-52; Distinguished Scientist Award Pharmaceut Sci, AAAS, 71. Mem: AAAS; Am Chem Soc; Am Soc Pharmacol & Exp Therapeut; Environ Mutagen Soc; Soc Toxicol. Res: Psychopharmacology; effects of drugs on human behavior; drug metabolism; research training in pharmacology and toxicology; environmental toxicology; toxicology information systems; drug use, misuse and abuse; drug safety and efficacy; rational pharmacotherapy. Mailing Add: 639 Crocus Dr Rockville MD 20850

COSPER, DAVID RUSSELL, b Ypsilanti, Mich, Oct 3, 42; m 65; c 1. ORGANIC CHEMISTRY, POLYMER CHEMISTRY. Educ: Purdue Univ, BS, 64; Univ Wis, Madison, PhD(chem), 69. Prof Exp: Chemist, 69-73, GROUP LEADER, NALCO CHEM CO, 73- Mem: Am Chem Soc; Tech Asn Pulp & Paper Indust. Res: Synthesis of organic compounds and polymers for water treatment applications; study of chemical additives for paper manufacturing. Mailing Add: 6824 Valley View Dr Downers Grove IL 60515

COSPER, SAMMIE WAYNE, b Greggton, Tex, Oct 8, 33; m 54; c 3. NUCLEAR PHYSICS. Educ: Southwestern La Univ, BS, 60; Purdue Univ, PhD(nuclear physics), 65. Prof Exp: Res appointee nuclear chem, Lawrence Radiation Lab, Univ Calif, 65-67; assoc prof, 67-70, head dept, 67-72, PROF PHYSICS, UNIV SOUTHWESTERN LA, 70- Mem: Am Asn Physics Teachers; Am Phys Soc. Res: Low-energy charged particle nuclear reactions and scattering; charged particle identification techniques; charged particles from spontaneous fission; nuclear physics; instrumentation. Mailing Add: Dept of Physics USL Box 1210 Univ of Southwestern La Lafayette LA 70501

COSSABOOM, ROBERT T, b Boston, Mass, Jan 12, 14; m 42; c 2. GEOGRAPHY, GEOLOGY. Educ: Clark Univ, AB, 41; Ohio State Univ, MS, 43. Prof Exp: Asst zool, Ohio State Univ, 42-43; from instr to assoc prof biol, 44-51, assoc prof geog & geol, 51-57, PROF GEOG & GEOL & HEAD DEPT, BALDWIN-WALLACE COL, 57- Honors & Awards: NSF Award, 59. Mem: Am Geog Soc; Paleont Soc; Nat Coun Geog Educ; Am Polar Soc; Asn Am Geog. Res: Taxonomy of Odonata; physical geography; geomorphology; climatic classification. Mailing Add: Dept of Earth Sci Baldwin-Wallace Col Berea OH 44017

COSSINS, EDWIN ALBERT, b Romford, Eng, Feb 28, 37; m 62; c 2. PLANT PHYSIOLOGY, BIOCHEMISTRY. Educ: Univ London, BSc, 58, PhD(plant biochem), 61. Prof Exp: Res assoc biol sci, Purdue Univ, 61-62; from asst prof to assoc prof, 62-69, actg head dept, 65-66, PROF BOT, UNIV ALTA, 69-, COORDR, INTROD BIOL PROG, 74- Concurrent Pos: Vis prof, Univ Geneva, Switz, 72-73. Honors & Awards: Centennial Medal, Govt Can, 67. Mem: Am Soc Plant Physiol; Can Soc Plant Physiol; Brit Biochem Soc; fel Royal Soc Can; NY Acad Sci. Res: Pteroylglutamates in plants; amino acid biosynthesis; biochemistry of germinating tissues; intermediary metabolism of plant tissues. Mailing Add: Dept of Bot Univ of Alta Edmonton AB Can

COST, KONSTANTINE, plant physiology, bioenergetics, see 12th edition

COSTA, ERMINIO, b Cagliari, Italy, Mar 9, 24; US citizen; m 50; c 3. PHARMACOLOGY. Educ: Univ Cagliari, Sardinia, MD, 47. Prof Exp: Asst prof, Univ Cagliari, Sardinia, 47-54, assoc prof pharmacol, 54-55; Fulbright res fel, Dept Physiol & Pharmacol, Chicago Med Sch, 55; med res assoc, Galesburg State Res Hosp, 56-61; vis scientist, Nat Heart Inst, 60-61, head sect clin pharmacol, Lab Chem Pharmacol, 61-65; assoc prof pharmacol, Col Physicians & Surgeons, Columbia Univ, 65-68; CHIEF LAB PRECLIN PHARMACOL, NIMH, 68- Honors & Awards: Bennett Found Award, 60. Mem: Fel AAAS; Am Soc Pharmacol & Exp Therapeut; Am Acad Neurol; fel Am Col Neuropsychopharmacol; Am Physiol Soc. Res: Antirheumatic drugs; anticoagulants; curari; cholinesterase inhibitors; chemotherapy of cancer; chemotherapy of tuberculosis; glucagon; effect of drugs and physiological control of the turnover rate of neuronal catecholamines, indolealkylamines, acetylcholine and cyclic adenosine monophosphate; mechanism of habituation to amphetamines and morphine. Mailing Add: William A White Bldg St Elizabeth's Hosp Washington DC 20032

COSTA, JOHN EMIL, b Ithaca, NY, Aug 14, 47; m 73. GEOMORPHOLOGY, ENVIRONMENTAL GEOLOGY. Educ: State Univ NY Col Oneonta, BS, 69; Johns Hopkins Univ, PhD(geog & environ eng), 73. Prof Exp: Instr geog, Towson State Col, 71-72; geologist, Md Geol Surv, 70-73; hydrologist, US Geol Surv, 74; ASST PROF GEOG, UNIV DENVER, 73- Mem: Geol Soc Am; Asn Am Geogrs; Sigma Xi. Res: Fluvial processes in urban streams; geomorphic responses to extreme events. Mailing Add: Dept of Geog Univ of Denver Denver CO 80210

COSTA, LORENZO F, b Genova, Italy, Feb 10, 31; m 66; c 4. ELECTRONIC SPECTROSCOPY. Educ: Birmingham-Southern Col, BS, 61; State Univ NY Buffalo, MS, 68. Prof Exp: Cancer scientist, Roswell Park Mem Inst, 61-67; res chemist, 67-74, SR RES CHEMIST, EASTMAN KODAK CO, 74- Mem: Soc Appl Spectros; Am Chem Soc; Soc Photog Scientists & Engrs; Am Optical Soc. Res: Photophysics; absolute luminescence methodology; optical spectroscopy; spectrophotometry of optically turbid media; photosensitivity mechanisms. Mailing Add: Res Labs Bldg 82A Eastman Kodak Co Rochester NY 14650

COSTA, ROBERT RICHARD, ecology, invertebrate zoology, see 12th edition

COSTAIN, CARMAN HUDSON, radio astronomy, see 12th edition

COSTAIN, CECIL CLIFFORD, spectroscopy, see 12th edition

COSTAIN, JOHN KENDALL, b Boston, Mass, Nov 18, 29; m 56; c 3. GEOPHYSICS. Educ: Boston Univ, BA, 51; Univ Utah, PhD(geol), 60. Prof Exp: Geophysicist, Socony Vacuum Oil Co Venezuela, 51-54; asst prof physics, San Jose State Col, 59-

60; from asst prof to assoc prof geophys, Univ Utah, 60-67; assoc prof, 67-69, PROF GEOPHYS, VA POLYTECH INST & STATE UNIV, 69- Concurrent Pos: NSF grants, 64- Mem: Am Geophys Union; Am Asn Petrol Geol; Geol Soc Am; Soc Explor Geophys; Seismol Soc Am. Res: Exploration seismology; measurement of terrestrial heat flow. Mailing Add: Dept of Geol Sci Va Polytech Inst & State Univ Blacksburg VA 24060

COSTAIN, ROBERT ANTHONY, b Bromborough, Eng, May 5, 35; Can citizen; m 69. NUTRITION. Educ: Univ Nottingham, BSc, 57, PhD(animal nutrit), 60. Prof Exp: Res assoc nutrit, Macdonald Col, McGill Univ, 60-63; dir nutrit & tech serv, Robin Hood Flour Mills, Ltd, Can, 63-67; mgr nutrit res, Int Milling Co, 67-70; mgr res & nutrit, United Coops Ont, 70-74; MEM STAFF, ROTHSAY CONCENTRATES, 74- Concurrent Pos: Nat Res Coun Can fel, 61-63. Mem: Am Soc Animal Sci; Poultry Sci Asn. Res: Animal nutrition, especially swine, poultry and cattle. Mailing Add: Rothsay Concentrates Elmira ON Can

COSTANTIN, LEROY LIBERAL, physiology, deceased

COSTANTINO, MARC SHAW, b Oakland, Calif, Nov 8, 45; m 66; c 3. HIGH PRESSURE PHYSICS, GEOPHYSICS. Educ: Rensselaer Polytech Inst, BS, 67; Princeton Univ, PhD(solid state physics), 72. Prof Exp: Physicist, US Mil Acad, West Point, 71-74; PHYSICIST GEOPHYS, LAWRENCE LIVERMORE LAB, 74- Mem: Am Phys Soc. Res: High pressure thermodynamics; physics of geological materials. Mailing Add: Lawrence Livermore Lab L-437 Livermore CA 94550

COSTANTINO, ROBERT FRANCIS, b Everett, Mass, Mar 21, 41; m 63; c 4. POPULATION BIOLOGY. Educ: Univ NH, BS, 63; Purdue Univ, MS, 65, PhD(genetics), 67. Prof Exp: Asst prof genetics, Pa State Univ, 67-72; ASSOC PROF ZOOL, UNIV RI, 72- Mem: Genetics Soc Am; AAAS; Sigma Xi. Res: Population genetics of tribolium; genetics of competing species. Mailing Add: Dept of Zool Univ of RI Kingston RI 02881

COSTANTINO-CECCARINI, ELVIRA, b Italy, May 16, 42; m 67; c 1. NEUROCHEMISTRY. Educ: Palermo Univ, Italy, Dr Biol, 65. Prof Exp: Fel molecular biol, Inst Comp Anat, Palermo, 65-68, asst prof, 68-69; assoc, 70-75, ASST PROF NEUROCHEM, ALBERT EINSTEIN COL MED, 75- Mem: Ital Soc Biologists; Am Soc Neurochem; Int Soc Neurochem; Soc Neurosci. Res: Neurochemistry of developing brain. Mailing Add: Dept of Neurol Albert Einstein Col of Med Bronx NY 10461

COSTANZA, ALBERT JAMES, b Pittsburgh, Pa, Dec 3, 17; m 43; c 3. POLYMER CHEMISTRY, RUBBER CHEMISTRY. Educ: Univ Pittsburgh, BS, 40; Univ Akron, MS, 50. Prof Exp: Jr chemist, US Govt, 40-43; proj chemist, Govt Synthetic Rubber Lab, Univ Akron, 46-50; sr res chemist, 50-71, sect head specialty rubbers, 71-74, RES SCIENTIST, GOODYEAR TIRE & RUBBER CO, 74- Mem: Am Chem Soc REs: Synthetic latices and polymers; emulsion and emulsifier free polymerizations; liquid polymers; functional end group polymers; reactive copolymers; specialty rubbers. Mailing Add: Res Div Goodyear Tire & Rubber Co Akron OH 44316

COSTANZA, JAMES L, education, engineering, see 12th edition

COSTANZI, FRANK ALBERT, elementary particle physics, see 12th edition

COSTEA, NICOLAS V, b Bucharest, Romania, Nov 10, 27; US citizen. IMMUNOHEMATOLOGY. Educ: Univ Paris, MD, 56. Prof Exp: Instr med, Sch Med, Tufts Univ, 59-63; Lederle fac award, 63-66; from asst prof to prof med, Sch Med, Univ Ill, Chicago Circle, 70-72; chief hemat sect, Univ Calif, Los Angeles-San Fernando Valley Prog, 72; PROF MED, SCH MED, UNIV CALIF, LOS ANGELES, 72- Concurrent Pos: Dir hemat, Cook County Hosp, 70-72; NIH grants. Mem: Am Soc Hemat; Am Fedn Clin Res; Am Asn Immunologists; Am Rheumatism Asn. Res: Hematology. Mailing Add: Hemat Dept Vet Admin Sepulveda Hosp Los Angeles CA 91343

COSTELLO, CATHERINE E, b Medford, Mass, May 11, 43. MASS SPECTROMETRY, ORGANIC BIOCHEMISTRY. Educ: Emmanuel Col, AB, 64; Georgetown Univ, MS, 67, PhD(org chem), 70. Prof Exp: Chemist, Div Food Chem, US Food & Drug Admin, Dept Health Educ & Welfare, 66-67 & USDA, 69; RES ASSOC MASS SPECTROMETRY, MASS INST TECHNOL, 70- Mem: Am Chem Soc; Am Soc Mass Spectrometry; AAAS. Res: Applications of advanced methods of organic chemical analysis, particularly mass spectrometry to biomedical problems including metabolic disorders, drug metabolism and toxicology. Mailing Add: Rm 56-012 Mass Inst of Technol Cambridge MA 02139

COSTELLO, CHRISTOPHER HOLLET, b St John's, Nfld, Jan 19, 13; nat US; m 46; c 5. PHARMACOLOGY. Educ: Mass Col Pharm, BS, 36, MS, 41, PhD(chem & pharmacol), 50. Prof Exp: Pharmacist, Mass Gen Hosp, 35-36; chief chemist, Pinkham Co, 38-41; instr chem, Mass Col Pharm, 46-49; vpres & sci dir, Columbus Pharmacol Co, 49-60; dir res, Pharmaceut Labs, 60-62, dir res biol prod, 62-69, assoc dir corp res, 69-75, ADMINR REGULATORY AFFAIRS, COLGATE-PALMOLIVE CO, 76- Concurrent Pos: Drug indust liaison, Rev Panel Oral Cavity Drugs, Food & Drug Admin, 74- Mem: Fel AAAS; Am Chem Soc; Am Pharm Asn; fel Am Inst Chemists; Soc Clin Pharmacol & Therapeut. Res: Pharmaceutical and cosmetic chemistry; chemistry and pharmacology of botanical drugs; isolated estrone from licorice; sustained release medication; antilipemic agents; proprietary drugs; dermatological, dental and hair products. Mailing Add: Colgate-Palmolive Co 909 River Rd Piscataway NJ 08854

COSTELLO, DAVID FRANCIS, b Nebr, Sept 1, 04; m 29; c 3. ENVIRONMENTAL BIOLOGY. Educ: Nebr State Teachers Col, AB, 25; Univ Chicago, MS, 26, PhD(plant ecol), 34. Prof Exp: Instr bot, Marquette Univ, 26-32; forest ecologist, Rocky Mountain Forest & Range Exp Sta, US Forest Serv, 34-37; chief div range res, 37-53, chief div range res, Pac NW Forest & Range Exp Sta, 53-61, chief div range, wildlife habitat & recreation res, 61-65; WRITER NATURAL SCI BOOKS, 65-; CONSULT ENVIRON BIOL, 71- Concurrent Pos: Spec lectr, Colo State Univ, 42-53; guest prof, State Col Wash, 57; guest speaker, Univ Col NWales, 62; mem range comt, Western Agr Res Coun, 55-64. Honors & Awards: Citation for Outstanding Achievement & Serv, Am Soc Range Mgt, 70. Mem: Fel AAAS; Ecol Soc Am. Res: Plant ecology; environmental relations of plants and animals; social background of outdoor recreationists; psychology of recreation. Mailing Add: 4965 Hogan Dr Ft Collins CO 80521

COSTELLO, DONALD F, b New York, NY, Mar 12, 34; m 61; c 5. STATISTICS, COMPUTER SCIENCE. Educ: Manhattan Col, BS, 54; Univ Notre Dame, MS, 59. Prof Exp: Instr math, Univ Alaska, 56-57 & La Salle Col, 59-60; teaching asst, Univ Nebr, 60-63; asst prof & dir comput ctr, Wis State Univ, 63-64; res assoc, Univ Wis, 64-65; asst dir comput ctr, 65-70, ASSOC PROF COMPUT SCI, UNIV NEBR, 70-, DIR COMPUT CTR, 72- Concurrent Pos: Asst prof, Colo State Univ, 67-68; proj dir, Mo Valley Planning Info Ctr. Mem: Asn Comput Mach; Inst Math Statist. Res: Multivariate statistical computing algorithms in interactive systems environment. Mailing Add: Comput Ctr 225 Nebraska Hall Univ of Nebr Lincoln NE 68508

COSTELLO, DONALD PAUL, b Detroit, Mich, Sept 27, 09; m 36; c 2. ZOOLOGY, CELL BIOLOGY. Educ: Wayne State Univ, AB, 30; Univ Pa, PhD(zool), 34. Prof Exp: Instr zool, Univ Pa, 30-34; Nat Res Coun fel, Stanford Univ, 34-35; from asst prof to prof, 35-49, chmn dept, 47-57, Kenan prof, 49-75, KENAN EMER PROF ZOOL, UNIV N C, CHAPEL HILL, 75- Concurrent Pos: Rockefeller fel, Stanford Univ, 41-42; mem corp, Marine Biol Lab, Woods Hole, 34-, trustee, 46-54, 55-64, 64-75, emer trustee, 75-, mem exec comt, 51-54, 55-58, dir, embryol staff, 46-50; mem exec comt, NC Inst Fisheries Res, 49-53, chmn, 53-57, adminr, 55-57; mem div comt biol, NSF, 51-53; managing ed, Biol Bul, 51-68; mem study sect morphol & genetics, NIH, 56-58, study sect for cell biol, 59-60, 63-68, chmn sect B, 66-68. Mem: Am Soc Nat (treas, 48-50); fel Int Inst Embryol; Int Soc Cell Biol; Am Soc Cell Biol. Res: Experimental cytology; experimental embryology and breeding habits of marine invertebrates; microtubule structure and function; comparative spermatology; electron microscopy. Mailing Add: Dept of Zool Univ of N C Chapel Hill NC 27514

COSTELLO, ERNEST F, JR, b Fall River, Mass, June 9, 23; m 48; c 4. PHYSICS. Educ: Boston Univ, AB, 49; Lehigh Univ, MS, 51, PhD(physics), 59. Prof Exp: Asst physics, Lehigh Univ, 49-52, instr, 52-59; assoc prof, 59-67, chmn dept, 68-69, dir div lib arts & sci, 69-74, asst dir, 63-69, PROF PHYSICS, MERRIMACK COL, 67-, DEAN SCI & ENG, 74- Concurrent Pos: Sr engr physics, Raytheon Co, 62- Mem: Am Phys Soc; Am Asn Physics Teachers. Res: Magnet thin films; magnetic powders; thin film circuits. Mailing Add: Div of Lib Arts & Sci Merrimack Col North Andover MA 01845

COSTELLO, LESLIE CARL, b Brooklyn, NY, May 11, 30; m 59; c 2. CELL PHYSIOLOGY, ENDOCRINOLOGY. Educ: Univ Md, BS, 52, MS, 54, PhD(zool, biochem), 57. Prof Exp: Asst zool, Univ Md, 52-56, instr, 56, asst prof zool & physiol, 57-60, assoc prof anat & physiol & head dept, 60-67; PROF PHYSIOL, COL MED, HOWARD UNIV, 67- Concurrent Pos: Parasitologist, Livestock Lab, 53 & Animal Dis & Parasite Res Br, 54-55; Lederle fac award, 63; evaluator, NSF grants; reviewer res articles, Sci Mag. Mem: AAAS; Am Physiol Soc; Endocrine Soc. Res: Cellular physiology; physiological chemistry; developmental and environmental biochemistry; metabolism; endocrine physiology. Mailing Add: Dept of Physiol Howard Univ Col of Med Washington DC 20001

COSTELLO, RICHARD LYNN, microbiology, see 12th edition

COSTELLO, WILLIAM JAMES, b Cavalier, NDak, Aug 2, 32; m 58; c 2. ANIMAL SCIENCE, MEAT SCIENCE. Educ: NDak State Univ, BS, 54; Okla State Univ, MS, 60, PhD(meat sci), 63. Prof Exp: Res nutritionist, Res Lab, John Morrell & Co, 62-65; asst prof, 65-75, ASSOC PROF ANIMAL SCI, S DAK STATE UNIV, 75- Mem: Am Meat Sci Asn; Am Soc Animal Sci; Inst Food Technol. Res: Quality studies of beef and pork; quantity aspects of beef, pork and lamb carcasses; fast freezing and packaging beef; tranquilizers associated with beef cattle marketing. Mailing Add: Dept of Animal Sci S Dak State Univ Brookings SD 57006

COSTER, HENDRIK PAULUS, bHague, Neth, Nov 22, 19; US citizen; m 50; c 2. GEOPHYSICS. Educ: State Univ Groningen, BS, 40; State Univ Utrecht, PhD(geol & geophys), 45. Prof Exp: Geophysicist & party chief seismic explor, Seismog Serv Ltd, Eng, 47-52, party chief seismic & gravity interpretation, Seismog Serv Corp, Okla, 52-54; geophysicist, Shell Oil Co, 54-64, SR GEOPHYSICIST, SHELL DEVELOP CO, 64- Mem: Soc Explor Geophys; European Asn Explor Geophys; Am Geophys Union. Res: Exploration geophysics; seismology; digital processing; crustal studies of the earth. Mailing Add: Shell Develop Co PO Box 481 Houston TX 77001

COSTERO, ISAAC (TUDANCA), b Burgos, Spain, Dec 9, 03; m 28; c 4. PATHOLOGY. Educ: Univ Zaragosa, Bachiller, 21, lic med & surg, 29; Univ Madrid, Dr(med, surg), 31. Hon Degrees: Hon Dr, Univ Guadalajara, 51. Prof Exp: Prof histol & path anat, Univ Valladolid, 31-37; PROF PATH, NAT UNIV MEX, 39-; HEAD DEPT, NAT INST CARDIOL, 44- Concurrent Pos: Hon prof, Univ Michoacan, 39, Univ San Carlos, Guatemala, 47, Univ El Salvador, 48 & Univ Puebla, 50; prof path, Nat Polytech Inst, 44-; lectr path, Med Br, Univ Tex, 50; mem, WHO. Mem: Col Am Pathologists; Am Soc Clin Pathologists; Am Asn Pathologists & Bacteriologists; Int Acad Path; Mex Nat Acad Med. Res: Neuropathology, particularly brain tumors; connective tissue, especially collagen disease; tissue culture of nerve and connective tissue; normal carotid body; carotid body tumors; arteriovenous glomic anastomoses. Mailing Add: Nat Inst of Cardiol Avenida Cuauhtemoc 300 Mexico City Mexico

COSTERTON, J WILLIAM F, b Vernon, BC, July 21, 34; m 55; c 4. MICROBIOLOGY Educ: Univ BC, BA, 55, MA, 56; Univ Western Ont, PhD(microbiol), 60. Prof Exp: Prof biol, Baring Union Christian Col, Punjab, India, 60-62, dean sci, 63-64; fel bot, Cambridge Univ, 65; prof assoc microbiol, McGill Univ, 66-67, asst prof, 68-70; assoc prof, 70-75, PROF MICROBIOL, UNIV CALGARY, 75- Mem: Can Soc Microbiol; Am Soc Microbiol. Res: Architecture of bacterial cell walls, including extracellular carbohydrate coats, especially as it relates to physiological processes and to the presence of periplasmic enzymes. Mailing Add: Dept of Biol Univ of Calgary Calgary AB Can

COSTICH, EMMETT RAND, b Rochester, NY, July 15, 21; m 45; c 6. ORAL SURGERY. Educ: Univ Pa, DDS, 45; Univ Rochester, MS, 49, PhD(path), 54; Colgate Univ, BA, 56. Prof Exp: Instr dent surg, Sch Med & Dent, Univ Rochester, 47-55; asst prof dent, Sch Dent, Univ Mich, 55-58, assoc prof, 58-62; chmn dept oral surg, 62-69, assoc dean col dent, 69-72, PROF ORAL SURG, COL DENT, UNIV KY, 62-, ASST FOR EXTRAMURAL EDUC PROG COORD TO VPRES FOR MED CTR, 72- Concurrent Pos: Res assoc & supvr periodont & oral path, Eastman Dent Dispensary, 49-55; consult, US Vet Hosp, Lexington, Ky, 62- & NIMH Hosp, 63-73. Mem: Am Dent Asn; Int Asn Dent Res. Res: Transplantation; bone physiology. Mailing Add: Rm 205 Med Ctr Annex 2 Univ of Ky Lexington KY 40506

COSTICH, VERNE ROBERT, physics, optics, see 12th edition

COSTILL, DAVID LEE, b Feb 7, 36; US citizen; m 60; c 2. HUMAN PHYSIOLOGY. Educ: Ohio Univ, BSEd, 59; Miami Univ, MA, 61; Ohio State Univ, PhD(physiol), 65. Prof Exp: Instr physiol, Ohio State Univ, 63-64; asst prof, State Univ NY Cortland, 64-66; PROF & DIR, HUMAN PERFORMANCE LAB, BALL STATE UNIV, 66- Concurrent Pos: Vis fac, Desert Res Inst, 68; hon lectr, Univ Sulford, Eng, 72; res assoc, Gymnastik-och idrottshögskolan, Stockholm, 72-73. Honors & Awards: McClintock Award, Ball State Univ Found, 72. Mem: Am Physiol Soc; Am Soc Zoologists. Res: Alterations in skeletal muscle water and electrolytes following prolonged exercise and dehydration in man; glycolytic-oxidative enzyme and fiber composition in human skeletal muscle. Mailing Add: Human Performance Lab Ball State Univ Muncie IN 47306

COSTILOW, RALPH NORMAN, b Oxford, WVa, Oct 23, 22; c 2.

MICROBIOLOGY. Educ: WVa Univ, BS, 48; NC State Col, MS, 50; Mich State Col, PhD(bact), 53. Prof Exp: Instr bact, NC State Col, 49-51; from asst prof to assoc prof, 53-60, PROF MICROBIOL & PUB HEALTH, MICH STATE UNIV, 60- Concurrent Pos: NIH spec fel biochem, Univ Calif, Berkeley, 64-65. Mem: AAAS; Am Soc Microbiol; Am Soc Biol Chemists. Res: Microbial physiology; amino acid metabolism; industrial and foods microbiology. Mailing Add: Dept of Microbiol & Pub Health Mich State Univ East Lansing MI 48824

COSTIN, ANATOL, b Bucharest, Rumania, Aug 26, 26; m 55; c 2. NEUROSCIENCES. Educ: Inst Med & Pharm, Bucharest, 52; Rumanian Acad Sci, PhD(physiol, physiopath), 56. Prof Exp: Asst prof physiol, Inst Med & Pharm, Bucharest, 49-57; sr res fel, Inst Physiol, Rumanian Acad Sci, 56-61; sr lectr pharmacol, Hadassah Med Sch, Hebrew Univ, Israel, 61-70; res assoc, 70-72, ASSOC PROF ANAT, UNIV CALIF, LOS ANGELES, 72- Concurrent Pos: Res asst anat, Univ Calif, Los Angeles, 67-68. Mem: Am Physiol Soc; Soc Neurosci; Am Asn Anatomists; Royal Soc Med; Int Brain Res Orgn. Res: Neurophysiology; neurochemistry. Mailing Add: Brain Res Inst Univ of Calif Los Angeles CA 90024

COSTLOW, JOHN DEFOREST, b Brookville, Pa, Jan 28, 27; m 52; c 2. MARINE INVERTEBRATE ZOOLOGY. Educ: Western Md Col, BS, 50; Duke Univ, PhD(zool), 55. Prof Exp: Asst biol, Western Md Col, 50; asst zool, 50-51, 53-54, marine zool, Off Naval Res, 51-53, res assoc, NSF contracts, 54-59, from asst prof to assoc prof, 59-68, PROF ZOOL, DUKE UNIV, 68-, DIR MARINE LAB, 68- Mem: Am Soc Limnol & Oceanog; Am Soc Zool; Brit Marine Biol Asn. Res: Organogenesis; larval development; molting, growth and physiology of Cirripedia; larval development of Crustacea in relation to environmental factors; endocrine mechanisms in larvae of marine Crustacea. Mailing Add: Marine Lab Duke Univ Beaufort NC 28516

COSTLOW, RICHARD DALE, b Johnstown, Pa, July 19, 25; m 45; c 4. MICROBIOLOGY, BIOCHEMISTRY. Educ: Pa State Univ, BS, 49, MS, 52, PhD(bact), 54. Prof Exp: Res microbiologist, US Army Biolabs, Ft Detrick, 54-69, chief virus & rickettsia div, 69-71; dir cancer chemother dept, Microbiol Assocs, Inc, 71-75; PROG DIR CANCER PREVENTION BR, DIV CANCER CONTROL & REHABILITATION, NAT CANCER INST, 75- Mem: Am Soc Microbiol; AAAS; Sigma Xi; NY Acad Sci. Res: Bacterial physiology; enzymology; microbiological chemistry; virology; research administration. Mailing Add: Nat Cancer Inst Nat Insts of Health Rm 614 Bethesda MD 20014

COSTOFF, ALLEN, b Milwaukee, Wis, Sept 26, 35. ZOOLOGY, ENDOCRINOLOGY. Educ: Marquette Univ, BS, 57, MS, 59; Univ Wis-Madison, PhD(zool, biochem), 69. Prof Exp: Res technician life sci, Marquette Univ, 59-60; head biol res, Aldrich Chem Co, 60-61; teaching asst zool, Univ Wis-Madison, 61-63, NIH-NSF res asst, 63-65, NIH-Ford Found trainee, 65-69, trainee reproductive physiol, 69-70; instr, 70-73, ASST PROF ENDOCRINOL, MED COL GA, 73- Mem: AAAS; Am Inst Biol Sci; Am Soc Zoologists; Am Fertil Soc; NY Acad Sci. Res: Ultrastructure of anterior pituitary gland; fractionation and bioassay of pituitary secretory granules; steroid and gonadotropin levels in monkey blood and urine; experimentally produced and natural occurring tumors in rats; control of gonadotropins. Mailing Add: Dept of Endocrinol Med Col of Ga Augusta GA 30902

COSTON, TULLOS OSWELL, b Dixie, La, Sept 17, 05; m 30; c 3. OPHTHALMOLOGY. Educ: Univ Tex, AB, 26; Johns Hopkins Univ, MD, 30; Am Bd Ophthal, dipl, 39. Prof Exp: Intern med, Johns Hopkins Hosp, 31-32, resident ophthal, 34-35, mem fac, Sch Med, Univ, 35-36; instr, 36-54, clin prof, 54-70, PROF OPHTHAL, HEALTH SCI CTR, UNIV OKLA, 70-, CHMN DEPT, 62- Mem: Am Soc Prev Blindness; Am Ophthal Soc; Am Acad Ophthal & Otolaryngol; Pan-Am Asn Ophthal; Nat Med Found Eye Care. Mailing Add: Dean McGee Eye Inst Univ of Okla Health Sci Ctr Oklahoma City OK 73069

COSTRELL, LOUIS, b Bangor, Maine, June 26, 15; m 42; c 3. NUCLEAR PHYSICS, NUCLEAR SCIENCE. Educ: Univ Maine, BS, 39; Univ Md, MS, 49. Prof Exp: Elec designer, Westinghouse Elec Corp, Pa, 40-41; elec engr, Bur Ships, US Dept Navy, 41-46; PHYSICIST, NAT BUR STANDARDS, 46- Concurrent Pos: Chmn comt N42 on nuclear instrumentation, Am Nat Standards Inst, 62-; tech adv to US Nat Comt of Int Electrotech Comn, 62- Honors & Awards: Outstanding Achievement Award, US Dept Com, 63, Gold Medal Award, 67; Standards Citation, Inst Elec & Electronics Engrs, 73, Harry Diamond Mem Award, 75 Mem: Am Phys Soc; fel Inst Elec & Electronics Engrs Res: Nucleonic instrumentation; electronics. Mailing Add: Nat Bur of Standards Washington DC 20234

COSULICH, DONNA BERNICE, b Albuquerque, NMex, Dec 2, 18. DRUG METABOLISM. Educ: Univ Ariz, BS, 39, MS, 40; Stanford Univ, PhD(org chem), 43. Prof Exp: Res chemist, Calco Chem Div, 43-55, RES CHEMIST, LEDERLE LABS, AM CYANAMID CO, 55- Honors & Awards: Sr Res Award, Am Cyanamid Co, Geneva, 58; Iota Sigma Pi Res Award, 54. Mem: Am Chem Soc. Res: Crystal and molecular structures of pharmaceutical compounds by x-ray crystallographic analysis; pteroylglutamic acid and related compounds; proof of structure work by degradation of hemicellulose, echinocystaic acid and antibiotics; pharmaceutical chemistry; synthetic organic chemistry. Mailing Add: Lederle Labs Am Cyanamid Co Pearl River NY 10965

COTA-ROBLES, EUGENE H, b Nogales, Ariz, July 13, 26; m 57; c 3. MICROBIOLOGY, CYTOLOGY. Educ: Univ Ariz, BS, 50; Univ Calif, MA, 54, PhD(microbiol), 56. Prof Exp: Instr bact, Univ Calif, Riverside, 56-58, from asst prof to prof microbiol, 58-70, asst dean col letters & sci, 68-69, spec asst to chancellor, 69-70; prof & head dept, Pa State Univ, 70-73; vchancellor-acad admin & prof biol, 73-75, VCHANCELLOR & PROF BIOL, UNIV CALIF, SANTA CRUZ, 75- Concurrent Pos: USPHS fel, Sabbatsberg Hosp, Stockholm, Sweden, 57-58; fel, Biochem Inst, Univ Uppsala, 63-64; consult, Mex-Am & Puerto Rican predoctoral fel selection comt, Ford Found, 69-71. Mem: AAAS; Am Soc Microbiol; Electron Micros Soc Am. Res: Biochemical organization of microbial cells; chemical structure of microbial membranes; membrane morphogenesis in lipid containing bacterophages. Mailing Add: 192 Cent Serv Univ of Calif Santa Cruz CA 95064

COTE, LOUIS J, b Detroit, Mich, July 18, 21; m 48; c 4. MATHEMATICAL STATISTICS. Educ: Univ Mich, AB, 43, AM, 47; Columbia Univ, PhD(math statist), 54. Prof Exp: Asst prof math & statist, Purdue Univ, 54-56; asst prof math, Syracuse Univ, 56-59; ASSOC PROF MATH & STATIST, PURDUE UNIV, 59- Concurrent Pos: Consult, Midwest Appl Sci Corp, 61- Mem: Am Math Soc; Math Asn Am. Res: Sums of random variables; statistical estimation. Mailing Add: Dept of Statist Purdue Univ West Lafayette IN 47907

COTE, LUCIEN JOSEPH, b Angers, PQ, Jan 4, 28; m 60; c 1. BIOCHEMISTRY, NEUROLOGY. Educ: Univ Vt, BS, 51, MD, 54. Prof Exp: Physician, Buffalo Gen Hosp, NY, 54-56; resident neurol, Neurol Inst, New York, 58-61; NIH fel, NY State Psychiat Inst, 61-68, asst prof, 68-70, ASSOC PROF NEUROL, MED CTR, COLUMBIA UNIV, 70- Mem: AAAS. Res: Clinical neurochemistry. Mailing Add: Dept of Neurol Columbia Univ Med Ctr New York NY 10032

COTE, PHILIP NORMAN, b Norwich, Conn, Oct 1, 42; m 67. ORGANIC CHEMISTRY. Educ: Univ Conn, BA, 64; Univ RI, PhD(chem), 70. Prof Exp: Spec instr chem, Univ RI, 70-71; chemist res & develop, Toms River Chem Corp, 72-76; CHEMIST RES & DEVELOP, SODYECO DIV, MARTIN MARIETTA CHEMS, 76- Concurrent Pos: Res assoc, Univ RI, 71-72. Mem: Am Chem Soc; Sigma Xi. Res: Electronic effects in free radical rearrangement reactions; synthesis of glycosides; synthesis of disperse dyes for synthetic fibers. Mailing Add: Martin Marietta Chem Sodyeco Div PO Box 10098 Charlotte NC 28237

COTE, RAYMOND-HENRI, b Quebec, Que, June 25, 23; m 49; c 6. IMMUNOCHEMISTRY, BIOPHYSICAL CHEMISTRY. Educ: Laval Univ, BA, 44, BSc, 49, PhD(org chem), 52. Prof Exp: Researcher immunochem, Pasteur Inst, Univ Paris, 52-54; Beit Mem res fel, Lister Inst Prev Med, Univ London, 54-58; res asst microbiol, Sch Med, 58-63, assoc prof, 63-67, PROF BIOCHEM, FAC SCI, LAVAL UNIV, 67- Concurrent Pos: Nat Res Coun Can grant, 59- Mem: The Chem Soc; Brit Biochem Soc; Can Soc Immunol; NY Acad Sci; Nutrit Today Soc. Res: Structural studies on blood-group substances; physico-chemical properties and fine chemical composition in relation with serological activity. Mailing Add: Dept of Biochem Fac of Sci Laval Univ Quebec PQ Can

COTE, ROGER ALBERT, b Manchester, NH, Aug 28, 28; m 52; c 5. PATHOLOGY. Educ: Assumption Col, Mass, BA, 50; Univ Montreal, MD, 55; Marquette Univ, MS, 64; FRCP(C). Prof Exp: Chief immunohemat sect, Vet Admin Hosp, Milwaukee, Wis, 60-61, asst chief lab serv, 61-62, actg chief, 62-63, chief anat path, 63-64, chief lab serv, Boston, 64-69, chief, Boston Area Reference Lab Syst, 67-69; PROF PATH & CHMN DEPT, MED SCH, UNIV SHERBROOKE, 69- Concurrent Pos: Asst prof, Sch Med, Marquette Univ, 62-64; instr, Sch Med, Harvard Univ, 64-69; asst prof, Sch Med, Tufts Univ, 64-68, assoc prof, 68-69, lectr, Sch Dent Med, 67-69; co-dir, Lab Comput Proj, Boston Vet Admin Hosp, 67-69; mem path & lab med res eval comt, Vet Admin Cent Off, Washington, DC, 67-69; ed-in-chief, Systematized Nomenclature of Med; Can rep sci adv bd, Institut de Recherch d'Informatique et d'Automatique. Mem: Int Acad Path; Am Soc Clin Path; fel Col Am Path; Am Fedn Clin Res; Am Thoracic Soc. Res: Pulmonary pathology, emphysema; clinical pathology, methodology; medical nomenclature; computer technology; laboratory medicine. Mailing Add: Dept of Path Univ of Sherbrooke Sherbrooke PQ Can

COTE, WILFRED ARTHUR, JR, b Willimantic, Conn, May 27, 24; m 47; c 5. ELECTRON MICROSCOPY, WOOD SCIENCE. Educ: Univ Maine, BS, 49; Duke Univ, MF, 50; State Univ NY Col Forestry, Syracuse, PhD(wood prod eng), 58. Prof Exp: Instr wood technol, 50-58, from asst prof to assoc prof wood prod eng, 58-65, PROF WOOD PROD ENG, STATE UNIV NY COL ENVIRON SCI & FORESTRY, 65-, DIR N C BROWN CTR ULTRASTRUCTURE STUDIES, 70- Concurrent Pos: Fulbright res fel, Univ Munich, 59-60; Walker-Ames vis prof, Univ Wash, 66; vis prof, Tech Univ Denmark, 72. Mem: Electron Micros Soc Am; Soc Wood Sci & Technol; fel Int Acad Wood Sci; Int Asn Wood Anat (secy, 70-). Res: Ultrastructure of the cells of wood as revealed by light and electron microscopy; interaction of wood ultrastructure with adhesives, coatings and processing chemicals. Mailing Add: 207 Brookford Rd Syracuse NY 13224

COTERA, AUGUSTUS S, JR, b Houston, Tex, Jan 2, 31; m 55; c 2. GEOLOGY. Educ: Univ Tex, BS, 52, MA, 56, PhD(geol), 62. Prof Exp: Res geologist, Tex Pac Coal & Oil Co, 52 & 54; explor geologist, Carter Oil Co, 56-57; from asst prof to assoc prof geol, Allegheny Col, 61-67; assoc prof, 67-70, PROF GEOL, NORTHERN ARIZ UNIV, 70-, CHMN DEPT, 68- Concurrent Pos: NSF res grant, 64-68. Mem: Geol Soc Am; Soc Econ Paleont & Mineral. Res: Sedimentary petrology, paleogeography and environments of deposition during the Mississippian, Pennsylvanian and Cretaceous Periods. Mailing Add: Dept of Geol Northern Ariz Univ Flagstaff AZ 86001

COTHERN, CHARLES RICHARD, b Indianapolis, Ind, Sept 6, 37; div; c 2. NUCLEAR PHYSICS. Educ: Miami Univ, BA, 59; Yale Univ, MS, 60; Univ Man, PhD(physics), 65. Prof Exp: Lectr physics, Univ Man, 61-65; asst prof, 65-70, ASSOC PROF PHYSICS, UNIV DAYTON, 70- Mem: Am Asn Physics Teachers; Am Phys Soc; Am Chem Soc; Sigma Xi (pres, 75-76). Res: Low energy nuclear spectroscopy; environmental physics—heavy metals in air, water and sludge; surface physics—photoelectron and Auger spectroscopies. Mailing Add: Dept of Physics Univ of Dayton Dayton OH 45469

COTHRAN, WARREN RODERIC, b San Francisco, Calif, Mar 14, 38. INSECT ECOLOGY. Educ: San Jose State Col, BA, 61, gen sec teaching credential, 62; Cornell Univ, PhD(entom), 67. Prof Exp: Instr ecol, Cornell Univ, 67-68; asst prof, 68-74, ASSOC PROF ENTOM, UNIV CALIF, DAVIS, 74- Mem: AAAS; Entom Soc Am; Ecol Soc Am. Res: Biology of legume weevils of the genus Hypera; general ecology of certain scavenging insects. Mailing Add: Dept of Entom Univ of Calif Davis CA 95616

COTRAN, RAMZI S, b Haifa, Palestine, Dec 7, 32; m 56; c 4. MEDICINE, PATHOLOGY. Educ: Am Univ Beirut, BA, 52, MD, 56; Am Bd Path, dipl. Prof Exp: From intern to chief resident path, Mallory Inst, Boston, 56-59; from instr to assoc prof, 60-72; FRANK B MALLORY PROF PATH, HARVARD MED SCH, 72-; PATHOLOGIST-IN-CHIEF, PETER BENT BRIGHAM HOSP, 74- Concurrent Pos: Fel, Mem Ctr Cancer & Allied Dis, NY, 59-60; vis res fel, Sloan-Kettering Inst Cancer Res, NY, 59-60; assoc dir, Mallory Inst Path, 69-74; mem coun circ & renal sect, Am Heart Asn. Mem: Am Asn Path & Bact; Am Soc Nephrol; Am Soc Cell Biologists; Int Acad Path; Am Soc Exp Path. Res: Pathology and pathogenesis of renal infections; electron microscopy; pathogenesis of inflammation; endothelial injury. Mailing Add: Peter Bent Brigham Hosp 721 Huntington Ave Boston MA 02115

COTRUFO, COSIMO (GUS), b New York, NY, May 7, 24; m 56; c 5. PLANT PHYSIOLOGY. Educ: Univ Mo, BS, 52, MS, 53, PhD(bot), 58. Prof Exp: Asst prof hort, NDak State Univ, 57-61; RES PLANT PHYSIOLOGIST, SOUTHEASTERN FOREST EXP STA, US FOREST SERV, 61- Mem: Am Soc Hort Sci; Am Soc Plant Physiol. Res: Physiology of forest trees; hardwoods. Mailing Add: Southeastern Forest Exp Sta Rt 3 Box 1249 Asheville NC 28806

COTRUVO, JOSEPH ALFRED, b Toledo, Ohio, Aug 3, 42. ORGANIC CHEMISTRY. Educ: Toledo Univ, BS, 63; Ohio State Univ, PhD(chem), 68. Prof Exp: Ital Nat Res Coun res fel heterocyclic chem, Bologna, 69-70; chemist, Chem Samples Co, 70-73; tech policy analyst, US Environ Protection Agency, 73-75, SCI ADV, WATER SUPPLY OFF, US ENVIRON PROTECTION AGENCY, 75- Concurrent Pos: Prof lectr, Prince Georges Community Col, 74-75. Mem: Am Chem Soc. Res: Synthesis and application of insect sex attractants; heterocyclic chemistry; diazo compounds and electronic properties of carbenes; chemical products of chlorine and ozone disinfection of drinking water and waste water. Mailing Add: 5015 46th St NW Washington DC 20016

COTSONAS, NICHOLAS JOHN, JR, b Boston, Mass, Jan 28, 19; m 70; c 3. MEDICINE. Educ: Harvard Col, AB, 40; Georgetown Univ, MD, 43. Prof Exp:

Rotating intern, Gallinger Munic Hosp, Washington, DC, 44, res clin dir, Tuberc Div, 46-47, asst med res, Med Div, 47-48, chief med res, 48-49, chief med officer, 51-53; from instr to prof, 49-70, PROF MED & DEAN, PEORIA SCH MED, UNIV ILL COL MED, 70- Concurrent Pos: Chief med serv, Res & Educ Hosp, Univ Ill Med Ctr, 68-70; fel coun clin cardiol, Am Heart Asn. Mem: AMA; fel Am Col Cardiol; Am Fedn Clin Res; fel Am Col Physicians. Res: Internal medicine; cardiology; medical education. Mailing Add: Peoria Sch of Med Univ of Ill Col of Med Peoria IL 61606

COTSORADIS, BRILLE R PERRY, inorganic chemistry, see 12th edition

COTTAM, GENE LARRY, b Coffeyville, Kans, Nov 3, 40; m 63; c 2. BIOCHEMISTRY. Educ: Univ Kans, AB, 62; Univ Mich, MS, 63 & 65, PhD(biochem), 67. Prof Exp: USPHS trainee biochem, Univ Tex & Vet Admin Hosp, Dallas, 67-68; ASSOC PROF BIOCHEM, UNIV TEX HEALTH SCI CTR DALLAS, 68- Concurrent Pos: Consult, Lawrence Radiation Lab, Univ Calif, Livermore, 67- Mem: Am Chem Soc. Res: Elucidation of mechanism of chemical reactions catalyzed by enzymes; active sites. Mailing Add: Dept of Biochem Univ of Tex Health Sci Ctr Dallas TX 75235

COTTAM, GRANT, b Sandy, Utah, Aug 26, 18; m 42; c 5. ECOLOGY, ACADEMIC ADMINISTRATION. Educ: Univ Utah, BA, 39; Univ Wis, PhD(bot), 48. Prof Exp: Teacher high sch, Utah, 39-40; asst prof bot, Univ Hawaii, 48-49; from asst prof to assoc prof, 49-60, chmn dept, 70-73, PROF BOT, UNIV WIS-MADISON, 60-, CHMN INSTRNL PROG, INST ENVIRON STUDIES, 74- Concurrent Pos: Guggenheim fel, 54-55. Mem: AAAS; Ecol Soc Am; Brit Ecol Soc; Inst Ecol. Res: Methods of phytosociology; forest phytosociology; interdisciplinary environmental research. Mailing Add: Dept of Bot Birge Hall Univ of Wis Madison WI 53706

COTTEN, GEORGE RICHARD, b Warsaw, Poland, Mar 21, 29; US citizen; m 57; c 3. POLYMER CHEMISTRY, PHYSICAL CHEMISTRY. Educ: Univ London, BS, 52, PhD(chem), 56. Prof Exp: Nat Res Coun Can fel, 56-57; res chemist, Visking Co Div, Union Carbide Corp, 58-60 & Fibers Div, Am Cyanamid Co, 60-63; group leader carbon black res dept, 63-69, res assoc, Res & Develop Div, 69-72, SR RES ASSOC, RES & DEVELOP DIV, CABOT CORP, 72- Mem: Am Chem Soc; assoc Brit Inst Rubber Indust. Res: Reaction kinetics; characterization and viscoelastic properties of polymers; rubber reinforcement; statistics; polymer characterization. Mailing Add: Res & Develop Div Cabot Corp Concord Rd Billerica MA 01821

COTTEN, MARION DEVEAUX, b Charleston, SC, Nov 11, 27. PHARMACOLOGY. Educ: Col Charleston, BS, 48; Med Col SC, MS, 51, PhD(pharmacol), 52; Augusta Law Sch, JD, 74. Prof Exp: Res asst pharmacol, Med Col SC, 48-52, assoc, 52-53; asst prof, Sch Med, Tulane Univ, 54-55; sr asst scientist, Nat Heart Inst, 55-56, scientist & head sect physiol, 56-57; assoc prof pharmacol & physiol, Sch Med, Emory Univ, 57-61; prof pharmacol & chmn dept, Sch Med, Univ Okla, 61-69; prof & chmn dept, Univ Nebr, 69-70; prof pharmacol & med, Med Col Ga, 70-73. Concurrent Pos: Nat Heart Inst fel, 53-54, sr res fel, 57-61; assoc ed, J Pharmacol, 57-60, asst ed, 60-61, cardiovasc field ed, 61-62, ed-in-chief, J Pharmacol & Exp Therapeut, 68-; mem pharmacol & exp therapeut study sect, NIH, 61-65; mem pharmacol test comt, Nat Bd Med Examrs, 64-68; ed-in-chief, Pharmacol Rev, 70- Mem: Am Soc Pharmacol & Exp Therapeut; Am Bar Asn. Res: Cardiovascular pharmacology and physiology. Mailing Add: Rte 3 Box 229 Sylvania GA 30467

COTTER, DAVID JAMES, b Glens Falls, NY, July 24, 32; m 53; c 3. PLANT ECOLOGY. Educ: Univ Ala, BS, 52, AB, 53, MS, 55; Emory Univ, PhD(biol), 58. Prof Exp: Res assoc, Atomic Energy Comn, Emory Univ, 57-58; from asst prof to assoc prof biol, Ala Col, 58-66; dir ctr regional resources, 70-72, PROF BIOL, & CHMN DEPT, GA COL MILLEDGEVILLE, 66- Concurrent Pos: Res assoc, Atomic Energy Comn, Emory Univ, 63; consult, Bd Educ & dir summer insts, Talladega, Shelby & Montgomery Counties, 64 & 65; NSF Undergrad Res Participation, 67, 70, 71, 73, 74, & 75; fel, Inst Radiation Ecol, Savannah River Plant, 65-67; staff biologist, Comn Undergrad Educ Biol Sci, Am Inst Biol Sci; consult, Outdoor Educ Inst, Ga Col, 69- Mem: Ecol Soc Am. Res: Radiation biology. Mailing Add: 1652 Pine Valley Rd Milledgeville GA 31061

COTTER, DONALD JAMES, b Providence, RI, Jan 13, 30; m 52; c 3. HORTICULTURE, PLANT PHYSIOLOGY. Educ: Univ RI, BS, 52; Cornell Univ, MS, 54, PhD(veg crops), 56. Prof Exp: From asst prof to assoc prof hort, Univ Ky, 56-69; PROF HORT, N MEX STATE UNIV, 69- Mem: Soc Hort Sci; Am Soc Plant Physiol; Am Soc Hort Sci; Sigma Xi; Int Soc Hort Sci. Res: Physiological and environmental studies on vegetable plants; media for crop culture; water use on urban landscapes. Mailing Add: Dept of Hort N Mex State Univ Las Cruces NM 88001

COTTER, EDWARD, b Everett, Mass, May 30, 36. SEDIMENTOLOGY. Educ: Tufts Univ, BS, 58; Princeton Univ, MA, 61, PhD(geol), 63. Prof Exp: Res geologist, Jersey Prod Res Co, Okla, 62-64; asst prof geol, Tufts Univ, 64-65; ASST PROF GEOL, BUCKNELL UNIV, 65- Mem: Geol Soc Am; Soc Econ Paleont & Mineral; Int Asn Sedimentol. Res: Sedimentary petrology of carbonate and clastic rocks. Mailing Add: Dept of Geol & Geog Bucknell Univ Lewisburg PA 17837

COTTER, EDWARD F, b Baltimore, Md, Feb 15, 10; m 45; c 1. INTERNAL MEDICINE. Educ: Univ Md, MD, 35. Prof Exp: Asst med, 39-40, instr path, 40-42, asst neurol, 40-47, instr, 47-48, asst prof med, 47-53, ASSOC PROF MED, SCH MED, UNIV MD, BALTIMORE CITY, 53-, ASSOC NEUROL, 48- Concurrent Pos: Hitchcock fel neurosurg, Sch Med, Univ Md, 40-42; pvt pract, 46-63 & 69-; consult, Vet Admin, 50; dir demyelinating dis clin, Univ Md Hosp, 54-63; chief dept med, Md Gen Hosp, 52-63, dir educ, Dept Med, 63-69. Mem: AMA; Am Col Physicians. Res: Clinical medicine. Mailing Add: 216 Med Arts Bldg Univ of Md Baltimore MD 21201

COTTER, MARY VIRGINIA, b New York, NY, May 12, 19. GENETICS. Educ: Notre Dame Col, NY, BA, 38; Fordham Univ, MS, 50; NY Univ, PhD, 57. Prof Exp: From asst to prof biol, Notre Dame Col, NY, 47-70, chmn natural sci div, 67-70; assoc prof, 70-74, PROF BIOL, ST FRANCIS COL, NY, 74- Concurrent Pos: Secy, Cath Round Table Sci, 57-61. Mem: NY Acad Sci. Res: Cytology and genetics; developmental biology. Mailing Add: 329 W 25th St New York NY 10001

COTTER, MAURICE JOSEPH, b New York, NY, Apr 20, 33. PHYSICS. Educ: Fordham Univ, AB, 54, MS, 59, PhD(physics), 62. Prof Exp: Mathematician, US Navy Bur Ships, Washington, DC, 54-55; jr res assoc, Brookhaven Nat Lab, 60-62; asst prof, 62-71, ASSOC PROF PHYSICS, QUEENS COL, NY, 71- Concurrent Pos: Res assoc, Chem Dept, Brookhaven Nat Lab, 69-; guest scientist, Dept Physics, State Univ NY Stony Brook, 75-76. Mem: Am Asn Physics Teachers. Res: Nuclear reactor physics; neutron spectroscopy; application of neutron activation analysis techniques to oil paintings. Mailing Add: Dept of Physics Queens Col Flushing NY 11367

COTTER, ROBERT JAMES, b New Bedford, Mass, Apr 15, 30; m 59; c 3. POLYMER CHEMISTRY. Educ: Brown Univ, ScB, 51; Mass Inst Technol, PhD(org chem), 54. Prof Exp: Chemist, Res Dept, Bakelite Div, 54-56, group leader, Res & Develop Dept, Plastics Div, 56-61, sr group leader, Chem & Plastics Opers Div, 61-71, res assoc, 71-75, RES FEL, UNION CARBIDE CORP, 75- Mem: AAAS; Am Chem Soc; Sigma Xi. Res: Product and process development; polystyrenes; high performance engineering plastics; polymer additives; organic chemicals; pollution control polymers. Mailing Add: Chem & Plastics Union Carbode Corp PO Box 670 Bound Brook NJ 08805

COTTER, WILLIAM BRYAN, JR, b Hartford, Conn, May 8, 26; m 48; c 6. GENETICS, ANATOMY. Educ: Wesleyan Univ, BA, 49, MA, 51; Yale Univ, PhD(zool), 56. Prof Exp: Asst prof biol, Col Charleston, 55-56; asst prof, Wesleyan Univ, 56-57; asst prof biol, Col Charleston, 57-59; teaching fel anat, Med Col SC, 59-60; from asst prof to assoc prof, 60-74, actg chmn dept, 63-64, PROF ANAT, MED CTR, UNIV KY, 74- Mem: AAAS; Am Soc Human Genetics; Soc Study Evolution; Genetics Soc Am; Am Asn Anat. Res: Physiological genetics and evolution; effects of genes on behavior. Mailing Add: Dept of Anat Univ of Ky Med Ctr Lexington KY 40506

COTTERILL, OWEN JAY, food science, see 12th edition

COTTINGHAM, ROBERT WARREN, organic chemistry, see 12th edition

COTTIS, STEVE GUST, polymer chemistry, organic chemistry, see 12th edition

COTTLE, MERVA KATHRYN WARREN, b Calgary, Alta, Oct 8, 28; m 50; c 3. PHYSIOLOGY, PHARMACOLOGY. Educ: Univ BC, BA, 49, MA, 51; Univ Wash, PhD(physiol), 56. Prof Exp: Lectr physiol, Univ Alta, 56-67; Wellcome res fel, Inst Physiol, Agr Res Coun, Cambridge, Eng, 67-68; RES ASSOC & LECTR PHARMACOL, UNIV ALTA, 69- Mem: Can Physiol Soc. Res: Histological studies of central nervous system and peripheral nervous system, including afferent and efferent innervation of the heart. Mailing Add: Dept of Pharmacol Univ of Alta Edmonton AB Can

COTTLE, RICHARD W, b Chicago, Ill, June 29, 34; m 59; c 2. MATHEMATICS, OPERATIONS RESEARCH. Educ: Harvard Univ, AB, 57, AM, 58; Univ Calif, Berkeley, PhD(math), 64. Prof Exp: Instr math, Middlesex Sch, Mass, 58-60; mem tech staff, Bell Tel Labs, Inc, NJ, 64-69; ASSOC PROF OPERS RES, STANFORD UNIV, 69- Mem: Am Math Soc; Math Asn Am; Opers Res Soc Am; Asn Comput Mach; Soc Indust & Appl Math. Res: Mathematical programming and combinatorial mathematics. Mailing Add: Dept of Opers Res Stanford Univ Stanford CA 94305

COTTLE, WALTER HENRY, b Edmonton, June 9, 21; m 50; c 3. PHYSIOLOGY. Educ: Univ BC, BA, 49, MA, 51; Univ Wash, Seattle, PhD, 56. Prof Exp: Asst prof, 56-60, ASSOC PROF PHYSIOL, UNIV ALTA, 60- Mem: Can Physiol Soc; Am Physiol Soc. Res: Thermal responses and adaptations of homeotherms to environmental temperature; thyroid physiology. Mailing Add: Dept of Physiol Univ of Alta Edmonton AB Can

COTTON, DONALD J, surface chemistry, physical chemistry, see 12th edition

COTTON, FRANK ALBERT, b Philadelphia, Pa, Apr 9, 30; m 59; c 2. CHEMISTRY. Educ: Temple Univ, AB, 52; Harvard Univ, PhD(chem), 55. Hon Degrees: DSc, Temple Univ. Prof Exp: From instr to prof chem, Mass Inst Technol, 55-72; PROF CHEM, TEX A&M UNIV, 72- Concurrent Pos: Guggenheim fel, 56; Sloan fel, 61-64; Gouch lectr, Yale Univ, 62; Reilly lectr, Univ Notre Dame, 63. Honors & Awards: Am Chem Soc Award, 62; Baekkeland Medal, 64. Mem: Nat Acad Sci; AAAS; Am Acad Arts & Sci; Am Chem Soc; Am Crystallog Asn. Res: Application of valence theory and physical and preparative studies to elucidate molecular structures and bonding in inorganic compounds; molecular structure of enzymes and proteins. Mailing Add: Dept of Chem Mass Inst of Technol Cambridge MA 02139

COTTON, JAMES V, b Jamestown, NY, Sept 4, 30; m 58; c 1. GEOGRAPHY. Educ: Pa State Teachers Col, Slippery Rock, BS, 56; Pa State Univ, DEd, 58. Prof Exp: Assoc prof, 58-71, PROF GEOG, FROSTBURG STATE COL, 71-, CHMN DEPT, 58- Mem: Asn Am Geog; Nat Coun Geog Educ; Am Geog Soc. Res: Status of public high school geography in Maryland. Mailing Add: Dept of Geog Frostburg State Col Frostburg MD 21532

COTTON, JOHN EDWARD, b Minneapolis, Minn, Dec 21, 24; m 51; c 7. PHYSICAL CHEMISTRY. Educ: San Francisco State Col, AB, 53; Univ Ore, MS, 55, PhD(phys chem), 59. Prof Exp: Asst, Univ Ore, 54-57; PHYS CHEMIST, BOEING CO, 59- Mem: AAAS; Am Chem Soc. Res: Transport properties of gases; polargraphy; gas chromatography; mass spectroscopy. Mailing Add: 2512 102nd Ave NE Bellevue WA 98804

COTTON, ROBERT HENRY, b Newton, Mass, Nov 17, 14; m 48; c 2. FOOD SCIENCE, NUTRITION. Educ: Bowdoin Col, BS, 37; Mass Inst Technol, SM, 39; Pa State Univ, PhD(plant nutrit), 44. Prof Exp: Chemist, Gen Elec Co, Mass, 39-40; asst plant nutrit, Pa State Univ, 40-43, instr & asst prof human nutrit res, 43-45; dir, Plymouth, Fla Div, Nat Res Corp, 45-47; prof & supvry chemist, Citrus Exp Sta, Fla, 47-48; dir res, Holly Sugar Corp, 48-53; dir res, Huron Milling Co, 54-58; dir res, Continental Baking Co, 58-65, VPRES, ITT CONTINENTAL BAKING CO, INC, INT TEL & TEL CORP, 65-, CHIEF SCIENTIST, ITT FOOD GROUP, 74- Concurrent Pos: Mem indust adv comt, Sugar Res Found, 48-53 & sci adv comt, Am Inst Baking, 58-; chmn tech liaison comt, Am Bakers Asn-US Dept Agr, 59-; mem vis comt, Dept Nutrit & Food Sci, Mass Inst Technol, 65-69; mem panel V-3, White House Conf Food, Nutrit & Health, 69; chmn comt cereals & gen prod adv bd mil personnel supplies, Nat Res Coun-Nat Acad Sci, 69-, adv comt nutrit guidelines for foods, 70; dir gen, Found Chile, Santiago, Chile, 75- Mem: Fel AAAS; Am Chem Soc; Inst Food Technol; Asn Res Dir (pres, 64-65); Am Asn Cereal Chem (pres, 65-66). Res: Plant nutrition; food technology; analytical chemistry; new technics to increase storage life of orange juice powder; cattle nutrition; sugar beet technology; by-product development; application food science to combat malnutrition. Mailing Add: ITT Continental Baking Co PO Box 731 Rye NY 10580

COTTON, WILLIAM ROBERT, b Miami, Fla, Nov 29, 31; m 74; c 3. MICROSCOPIC ANATOMY, ORAL MICROANATOMY. Educ: Univ Md, DDS, 55; Northwestern Univ, MS, 63; Roosevelt Univ, MA, 73. Prof Exp: Asst dent officer, Naval Training Ctr, Bainbridge, Md & Camp Lejeune, NC, 55-57 & Mobilization Team, Miami, Fla, 57-58, asst dent officer & clin supvr, Dent Detachment, Marine Corps Sch, Quantico, Va, 58-59, postgrad officer, Naval Dent Sch, Bethesda, Md, 59-60, asst dent officer, USS F D Roosevelt (CVA-42), Mayport, Fla, 60-61, head exp path div, Dent Res Dept, Naval Med Res Inst, Nat Naval Med Ctr, 63-67, dent officer, USS Fulton (AS-11), New London, Conn, 67-69, chief histopath div, Naval Dent Res Inst, Great Lakes, Ill, 69-76, exec officer, 72-73, dep cmndg officer, 73-76, CHMN DENT SCI DEPT, NAVAL MED RES INST, 76- Concurrent Pos: Mem tech comt 106, Joint Working Group 6 & secy, Int Dent Fedn-Int Standards Orgn, 74-; mem adv comt, Dent Lab Technol Prog, Sch Tech Careers, Southern Ill Univ,

Carbondale, 75-; mem, Am Asn Den Schs. Mem: Am Dent Asn; Int Asn Dent Res; Am Asn Dent Res. Res: Biological and toxicity testing of dental restorative materials in animals and human clinical trials; bone resorption research—introduced toothless (tl) rat which lacks bone resorption; dental caries research. Mailing Add: Dent Sci Dept Naval Med Res Inst Bethesda MD 20014

COTTON, WYATT DANIEL, b Mexia, Tex, Feb 2, 43; m 68; c 1. ORGANIC CHEMISTRY. Educ: Calif State Univ, Los Angeles, BS, 69; Univ Calif, Los Angeles, PhD(org chem), 74. Prof Exp: RES CHEMIST PHYS & SYNTHETIC ORG CHEM, PROCTOR & GAMBLE CO, 74- Mem: Am Chem Soc. Res: Amino acid synthesis and sultone chemistry. Mailing Add: Miami Valley Lab PO Box 39175 Proctor & Gamble Co Cincinnati OH 45239

COTTRAL, GEORGE EDWARD, veterinary medicine, see 12th edition

COTTRELL, IAN WILLIAM, b York, Eng, June 18, 43; m 69. CARBOHYDRATE CHEMISTRY. Educ: Univ Edinburgh, BSc, 65, PhD(polysaccharide chem), 68. Prof Exp: Fel carbohydrate chem, Trent Univ, 68-70; res chem guar gum, Stein Hall & Co, Celanese Corp, 70-72; sr res chemist, 72-73, SECT HEAD, KELCO CO, MERCK, INC, 73- Mem: Am Chem Soc; AAAS; Brit Biophys Soc. Res: Preparation of industrially useful polymers with special emphasis on polysaccharides. Mailing Add: 3268 Caminito Ameca La Jolla CA 92037

COTTRELL, STEPHEN FRANCIS, cell biology, biochemistry, see 12th edition

COTTRELL, THOMAS H E, viticulture, enology, see 12th edition

COTTRELL, THOMAS S, b Chicago, Ill, Feb 2, 34; m 59; c 3. PATHOLOGY. Educ: Brown Univ, AB, 55; Columbia Univ, MD, 65. Prof Exp: Fel path, Columbia Univ-Presby Hosp, 65-68, instr, Columbia Univ, 68-69; ASSOC PROF PATH, NEW YORK MED COL, 68- Concurrent Pos: Asst attend pathologist, Flower & Fifth Ave Hosps, 68-; asst vis pathologist, Metrop Hosp, NY, 68-; Markle scholar acad med, 69- Res: Ultrastructural interpretation of normal and abnormal cardio-respiratory physiology. Mailing Add: Dept of Path New York Med Col Valhalla NY 10595

COTTS, ROBERT MILO, b Green Bay, Wis, Aug 22, 27; m 50; c 4. NUCLEAR MAGNETIC RESONANCE. Educ: Univ Wis, BS, 50; Univ Calif, PhD(physics), 54. Prof Exp: Instr physics, Stanford Univ, 54-57; from asst prof to assoc prof, 57-67, PROF PHYSICS, CORNELL UNIV, 67- Concurrent Pos: Physicist, Nat Bur Standards, 63-64; vis prof physics, Univ BC, 70-71. Mem: Am Phys Soc; Am Asn Physics Teachers. Res: Solid state physics. Mailing Add: Clark Hall Cornell Univ Ithaca NY 14853

COTTS, RONALD FRANCIS, physical chemistry, see 12th edition

COTTY, VAL FRANCIS, b New York, NY, July 11, 26; m 51; c 3. PHARMACOLOGY, TOXICOLOGY. Educ: St John's Col, BS, 48, MSc, 50; NY Univ, PhD(biol), 55. Prof Exp: From instr to asst prof biol, St John's Col, 50-55; staff mem, Boyce Thompson Inst, 55-57; head dept biochem, 57-70, DIR BIOL RES, BRISTOL-MYERS CO, 70- Concurrent Pos: Res assoc, Sch Med, NY Univ, 61-65. Mem: NY Acad Sci; Am Chem Soc; Am Soc Microbiol; Int Asn Dent Res; AAAS. Res: Pharmacology and toxicology of drugs and personal products. Mailing Add: 236 Avon Rd Westfield NJ 07090

COTY, VERNON FRANK, b St Paul, Minn, Sept 18, 22; m 48; c 6. MICROBIOLOGY. Educ: St Thomas Col, BS, 48; Marquette Univ, MS, 50; Purdue Univ, PhD(bact), 54. Prof Exp: Teaching asst biol, Marquette Univ, 48-50; teaching asst bact, Purdue Univ, 51-52, res fel, 50-51, 52-53; sr res technologist, Socony Mobil Oil Co, 53-68, RES ASSOC, MOBIL OIL RES & DEVELOP CORP, 68- Mem: Am Soc Microbiol; Sigma Xi. Res: Physiological studies of streptomyces griseus; microbiology related to petroleum industry; environmental pollution control. Mailing Add: Mobil Oil Res & Develop Corp PO Box 1025 Princeton NJ 08540

COTZIAS, GEORGE CONSTANTIN, b Greece, June 16, 18; nat US; m 50; c 1. MEDICINE. Educ: Harvard Univ, MD, 43. Hon Degrees: DSc, Cath Univ Santiago, 69, Med Col Pa, 70 & St John's Univ, NY, 71. Prof Exp: Intern path, Peter Bent Brigham Hosp, 43; intern med, Mass Gen Hosp, 44, resident neurol, 44-45, asst resident med, 45-46; asst physician, Rockefeller Inst Hosp, 46-50; fel, Nat Res Coun, 51-52; asst, Rockefeller Inst, 52-53; exec officer, Physiol Div, physician, Med Res Ctr & scientist, Brookhaven Nat Lab, 53-55, head physiol div, sr physician, Med Res Ctr & sr scientist, 55-75, actg head, Hosp, Med Res Ctr, 66-67; PROF NEUROL, MED COL, CORNELL UNIV, 74-; SPEC ASST TO PRES & ATTEND PHYSICIAN, NEUROPSYCHIAT SERV, DEPT MED, MEM SLOAN-KETTERING CANCER CTR, 75- Concurrent Pos: Asst neurol, Harvard Med Sch, 44-45; attend physician, Brookhaven Nat Lab Hosp & consult, Surgeon Gen, US Army, 58-; prof neurol, Mt Sinai Sch Med, 69-74; prof med, State Univ NY Stony Brook, 69-74; attend physician, New York Hosp, 74-; res collabr med dept, Brookhaven Nat Lab, 75-76, vis clinician to courtesy staff div med staff, 75-76 & mem gov body of med res ctr, 76-77; adj prof, Rockefeller Univ, 75- Honors & Awards: Albert Lasker Award Clin Med Res, 69; Citation, US AEC, 72; Harvey Lect, 72; Oscar B Hunter Award, Am Soc Clin Pharmacol & Therapeut, 73; Award, Am Col Physicians, 74. Mem: Nat Acad Sci; Am Soc Clin Invest; fel Am Acad Arts & Sci; World Fedn Neurol; Am Neurol Asn. Res: Enzymes from biochemical and physiological viewpoint; isotopes; irradiation; central nervous system disease; trace elements; Parkinson's disease. Mailing Add: Mem Sloan-Kettering Cancer Ctr 1275 York Ave New York NY 10021

COUCH, ELTON LEROY, geology, geochemistry, see 12th edition

COUCH, HOUSTON BROWN, b Estill Springs, Tenn, July 1, 24; m 45; c 3. PLANT PATHOLOGY. Educ: Tenn Polytech Inst, BS, 50; Univ Calif, PhD(plant path), 54. Prof Exp: From asst prof to assoc prof plant path, Pa State Univ, 54-65; PROF PLANT PHYSIOL & PATH & HEAD DEPT, VA POLYTECH INST, 65- Mem: Am Phytopath Soc; Am Agron Soc; Soil Sci Soc Am; Crop Sci Soc Am. Res: Diseases of turfgrasses and forage crops; physiology of parasitism; role of physical environment in plant disease development. Mailing Add: 608 Lansdowne Blacksburg VA 24060

COUCH, JACK GARY, b Pocatello, Idaho, Apr 5, 36; m 55; c 5. NUCLEAR PHYSICS. Educ: Utah State Univ, BS, 58; Vanderbilt Univ, MS, 59; Tex A&M Univ, PhD(physics), 66. Prof Exp: Chmn dept phys sci, Church Col Hawaii, 59-61; asst prof physics, Wash State Univ, 66-67; chmn dept, 67-74, ASSOC PROF PHYSICS, SOUTHERN ORE COL, 67- Mem: Am Asn Physics Teachers; Am Phys Soc. Res: Neutron transfer between heavy ions; neutron scattering by metal hydride systems; bose condensate in liquid helium; molecular infrared spectroscopy; physics education. Mailing Add: Dept of Physics Southern Ore Col Ashland OR 97520

COUCH, JAMES RUSSELL, b Grandview, Tex, June 10, 09; m 34; c 4 BIOCHEMISTRY, NUTRITION Educ: Agr & Mech Col Tex, BS, 31, MS, 34; Univ Wis, PhD(biochem), 48. Prof Exp: From asst poultry husbandman to poultry husbandman, Exp Sta, Univ Tex, 31-41; PROF POULTRY NUTRIT & BIOCHEM, TEX A&M UNIV, 48- Honors & Awards: Am Feed Mfrs Award, Poultry Sci Asn, 51. Mem: AAAS; fel Poultry Sci Asn; Am Chem Soc; Fedn Am Socs Exp Biol; Am Soc Biol Chem. Res: Poultry nutrition; embryology; physiology; nutritional significance and metabolic functions of vitamins and trace elements in the domestic fowl; vitamin B12; folic acid; antibiotics; unidentified growth factors; trace elements; proteins and amino acids; fats and fatty acids. Mailing Add: Dept of Poultry Sci Tex A&M Univ College Station TX 77841

COUCH, JAMES RUSSELL, JR, b Bryan, Tex, Oct 25, 39; m 64; c 1. NEUROLOGY, NEUROPHARMACOLOGY. Educ: Tex A&M Univ, BS, 61; Baylor Col Med, MD, 65, PhD(physiol), 66. Prof Exp: Nat Heart Inst fel, Baylor Col Med, 65-66; NIH staff fel, Lab Neuropharmacol, NIMH, 67-69; Nat Inst Neurol Dis & Stroke spec trainee, Washington Univ, 69-72; ASST PROF NEUROL, UNIV KANS MED CTR, 72- Concurrent Pos: Consult, Kansas City Vet Admin Hosp & Kansas City Gen Hosp, Kans. Mem: Am Acad Neurol; Soc Neurosci; Am Geriatrics Soc. Res: Neuropharmacology and neurotransmitters; movement disorders; headache; stroke. Mailing Add: Dept of Neurol Univ of Kans Med Ctr Kansas City KS 66103

COUCH, JOHN ALEXANDER, b Washington, DC, Feb 12, 38; m 63; c 2. PATHOBIOLOGY, PROTOZOOLOGY. Educ: Univ Ala, BS, 61; Fla State Univ, MS, 64, PhD(morphogenesis, cell biol), 71. Prof Exp: Teaching asst zool, Fla State Univ, 61-64; parasitologist, Biol Lab, Nat Marine Fisheries Serv, Nat Oceanic & Atmospheric Admin, 64-71; PARASITOLOGIST, BIOL LAB, ENVIRON PROTECTION AGENCY, 71- Concurrent Pos: US Dept Interior training assignment, Fla State Univ, 67-68; fac assoc, Univ WFla, 75- Mem: Am Soc Parasitol; Soc Invert Path; Soc Protozoologists. Res: Evolution of commensal-host relationships of marine protozoa; aquatic animal pathology; interaction of pollutants and natural disease; neoplasia. Mailing Add: Biol Lab Environ Protect Agency Sabine Island Gulf Breeze FL 32561

COUCH, JOHN NATHANIEL, b Prince Edward Co, Va, Oct 12, 96; m 27; c 2. BOTANY. Educ: Univ NC, AB, 19, AM, 22, PhD(bot), 24. Hon Degrees: ScD, Catawba Col, 46, Duke Univ, 65 & Univ NC, 72. Prof Exp: Teacher high schs, NC, 19-21; instr, 17-18, 22-25, from asst prof to prof, 25-45, Kenan prof, 45-68, chmn dept, 44-60, EMER KENAN PROF BOT, UNIV N C, CHAPEL HILL, 68- Concurrent Pos: Nat Res fel, Sta Exp Evolution, Carnegie Inst, 25-26 & Mo Bot Garden, 26-27; vis assoc prof, Johns Hopkins Univ, 33, vis prof, 34, 35; vis researcher, Exp Sta, NC State Col, 43; mem adv panel syst biol, NSF, 55-58. Honors & Awards: Walker Grand Prize, Boston Soc Natural Hist, 38; Jefferson Medal, NC Acad, 37; Cert of Merit, Mycol Soc Am, 56. Mem: Nat Acad Sci; AAAS (vpres, 62); Am Soc Nat; Bot Soc Am; Mycol Soc Am (secy-treas, 39-41, vpres, 42, pres, 43). Res: Culture, sexuality and ciliary structure of water fungi; fungi parasitic in mosquito larvae, symbiosis between fungi and scale insects; Actinomy cetales; fungi which parasitize mosquitoes. Mailing Add: Dept of Bot Univ of NC PO Box 443 Chapel Hill NC 27514

COUCH, MARGARET WHELAND, b Chicago, Ill, Aug 27, 41; m 64; c 2. ORGANIC CHEMISTRY. Educ: Duke Univ, BS, 63; Univ Fla, MS, 66, PhD(org chem), 69 Prof Exp: RES CHEMIST, VET ADMIN HOSP, GAINESVILLE, 71- Concurrent Pos: Res asst, Dept Radiol, Univ Fla, 69-70, asst res prof, 71- Mem: Am Chem Soc; Am Soc Mass Spectrometry. Res: Identification by means of mass spectrometry-gas chromatography of aromatic acids and amines present in biological fluids of patients with neurological disorders; radiopharmaceuticals for adrenal scanning. Mailing Add: 3524 NW 51st Ave Gainesville FL 32601

COUCH, RICHARD W, b Dayton, Ohio, June 9, 31; m 62; c 2. GEOPHYSICS, SEISMOLOGY. Educ: Mich State Univ, BS, 58; Ore State Univ, MS, 63, PhD(geophys), 69. Prof Exp: Electronics engr, Gen Dynamics/Electronics, NY, 58-60, mem res staff solid state design, 60-62; res asst marine geophys, 63-65, res fel geophys, 65-66, from instr to asst prof, 66-73, ASSOC PROF GEOPHYS, SCH OCEANOG, ORE STATE UNIV, 73- Mem: Am Geophys Union; Seismol Soc Am. Res: Structure and tectonics of continental margins of the Eastern Pacific Ocean; geophysical exploration for geothermal resources; earthquake seismology and ground response. Mailing Add: Sch of Oceanog Ore State Univ Corvallis OR 97331

COUCH, RICHARD WESLEY, b Pryor, Okla, Mar 30, 37; m 60; c 2. PLANT PHYSIOLOGY. Educ: Okla State Univ, BS, 59; Univ Tenn, MS, 61; Auburn Univ, PhD(bot biochem), 66. Prof Exp: Asst county agent, Exten Serv, Univ Tenn, 61-63; prof biol, Athens Col, 65-73, chmn dept, 66-73; ASSOC PROF BIOL, ORAL ROBERTS UNIV, 73- Concurrent Pos: US Corps Engrs, contract, 68- Mem: Am Asn Plant Physiol; Weed Sci Soc Am; Am Inst Biol Sci; Nat Asn Biol Teachers. Res:Herbicidal plant physiology; laser effects on plants; aquatic biology; aquatic weed control; microbiological photobiology. Mailing Add: Dept of Natural Sci Oral Roberts Univ Tulsa OK 74102

COUCH, ROBERT BARNARD, b Guntersville, Ala, Sept 25, 30; m 55; c 4. INTERNAL MEDICINE, INFECTIOUS DISEASES. Educ: Vanderbilt Univ, BA, 52, MD, 56. Prof Exp: From intern to chief resident, Vanderbilt Univ Hosp, 56-61; clin assoc surg, Nat Cancer Inst, 57-59, sr investr, Lab Clin Invest, Nat Inst Allergy & Infectious Dis, 61-65, head clin virol, 65-66; assoc prof, 66-71, PROF MICROBIOL & MED, BAYLOR COL MED, 71- Mem: Am Fedn Clin Res; Soc Exp Biol & Med; Am Soc Microbiol. Res: Clinical and general virology; immunology. Mailing Add: Dept of Microbiol Baylor Col of Med Houston TX 77025

COUCH, TERRY LEE, b Middletown, Pa, Jan 8, 44; m 65; c 2. ENTOMOLOGY. Educ: Franklin & Marshall Col, AB, 65; Pa State Univ, MS, 68, PhD(entom), 70. Prof Exp: Res entomologist, 70-75, GROUP LEADER ENTOM RES, AGR & VET PROD DIV, ABBOTT LABS, 75- Mem: Entom Soc Am; Soc Invert Path. Res: Discovery and development of microbial insecticides and narrow spectrum and chemical insecticides. Mailing Add: Abbott Labs D-911 Bldg T-9 14th St & Sheridan Rd North Chicago IL 60064

COUCHELL, GUS PERRY, b Henderson, NC, Apr 14, 39; m 68. NUCLEAR PHYSICS. Educ: NC State Univ, BS, 61, MS, 63; Columbia Univ, PhD(physics), 68. Prof Exp: Asst prof physics & appl physics, 68-73, ASSOC PROF PHYSICS, LOWELL TECHNOL INST, 73- Res: Isobaric analog states; nuclear spectroscopy; simulation of neutron sources using pulsed alpha beams; nuclear resonance fluorescence. Mailing Add: Dept of Physics Lowell Technol Inst Lowell MA 01854

COUCHMAN, JAMES C, b Cincinnati, Ohio, Aug 28, 29; m 48; c 3. PHYSICS. Educ: Cent Col, Iowa, BA, 53; Vanderbilt Univ, MA, 55; Tex Christian Univ, PhD(math physics), 65. Prof Exp: Radiation safety area rep, Argonne Nat Lab, 55-56; nuclear engr, Convair, Tex, 56-58; sr nuclear engr, Gen Dynamics/Ft Worth, 58-61, 63-65; sci specialist, Edgerton, Germeshausen & Grier, Calif, 65-67; PROJ NUCLEAR PHYSICIST, GEN DYNAMICS CORP, 67- Mem: Health Physics Soc; Am Nuclear Soc. Res: Nuclear weapons effects research; aerospace nuclear safeguards; health physics; use of digital computer techniques in studying nuclear reactor, criticality; evaluation of potential environmental hazards associated with the uses of nuclear energy. Mailing Add: 8112 Rush St Ft Worth TX 76116

COUCOUVANIS, DIMITRI N, b Athens, Greece, Nov 20, 40; US citizen; m 63; c 2. INORGANIC CHEMISTRY, CRYSTALLOGRAPHY. Educ: Allegheny Col, BS, 63; Case Inst Technol, PhD(chem), 67. Prof Exp: Res assoc, Case Inst Technol, 67 & Columbia Univ, 67-68; assoc prof, 68-75, PROF CHEM, UNIV IOWA, 75- Mem: Am Chem Soc. Res: Synthesis and structure of polynuclear coordination complexes and their use as models for metal containing enzymes. Mailing Add: Dept of Chem Univ of Iowa Iowa City IA 52240

COUEY, H MELVIN, b Shedd, Ore, May 22, 26; m 55; c 2. PLANT PHYSIOLOGY. Educ: Ore State Col, BS, 51, MS, 54; Iowa State Col, PhD(plant physiol), 56. Prof Exp: From assoc plant physiologist to plant physiologist, 56-63, sr plant pathologist, 63-68, invests leader, Northwest Fruit Invests, 68-73, LOCATION LEADER, PROD, HARVESTING & HANDLING TREE FRUITS, MKT QUAL RES DIV, AGR RES SERV, US DEPT AGR, 73- Mem: Am Soc Hort Sci; Am Soc Plant Physiol; Am Phytopath Soc; Scand Soc Plant Physiol. Res: Post-harvest physiology of fruits; fruit storage and storage disorders; physiology of fungus spores. Mailing Add: Box 99 Annex 111 US Dept of Agr Wenatchee WA 98801

COUGHANOUR, LESLIE WARREN, chemistry, see 12th edition

COUGHLIN, JOHN W, b Chicago, Ill, Sept 26, 32; m 61; c 4. DENTISTRY. Educ: Loyola Univ Ill, DDS, 63; Univ Wash, MSD, 65. Prof Exp: Instr oper dent, Univ Ky, 65-66, instr restorative dent, 66-67, asst prof, 67-68; PROF CROWN & BRIDGE DENT, SCH DENT, LA STATE UNIV MED CTR, NEW ORLEANS, 68-, ASST DEAN, 70- Concurrent Pos: Assoc prof dent, Loyola Univ La, 69-71; mem, Am Asn Dent Schs. Honors & Awards: C N Johnson Mem Award, Loyola Univ Ill, 63. Mem: Acad Gold Foil Opers; Am Dent Asn. Res: Oral embryology; sterilization of instruments; clinical dental restorations; cariostatic effect of fluoride. Mailing Add: Sch of Dent La State Univ Med Ctr New Orleans LA 70119

COUGHLIN, RAYMOND FRANCIS, b Chicago, Ill, Oct 6, 43; m 70; c 2. ALGEBRA. Educ: Lewis Col, BA, 65; Loyola Univ Chicago, MA, 67; Ill Inst Technol, PhD(math), 69. Prof Exp: From instr to asst prof math, Loyola Univ, Chicago, 67-70; ASST PROF MATH, TEMPLE UNIV, 70- Mem: Am Math Soc; Math Asn Am. Res: Non-associative algebras satisfying the associo-symmetric identity and non-associative rings satisfying the m-associative ring identity. Mailing Add: Dept of Math Temple Univ Philadelphia PA 19122

COUILLARD, PIERRE, b Montmagny, Que, Mar 19, 28; m 55; c 3. CELL PHYSIOLOGY. Educ: Laval Univ, BA, 47, BSc, 51; Univ Pa, PhD(zool), 55. Prof Exp: Fel, Belgium, 55-56; from asst prof to assoc prof, 56-70, head dept, 63-67, PROF BIOL, UNIV MONTREAL, 70- Mem: AAAS; Soc Protozool; Fr Asn Physiol; Can Biochem Soc; Can Soc Cell Biol. Res: Physiology of contractility in protozoa. Mailing Add: 631 Davarr Outremont PQ Can

COULEHAN, ROBERT E, physics, see 12th edition

COULL, BRUCE CHARLES, b New York, NY, Sept 16, 42; m 67. ECOLOGY, BIOLOGICAL OCEANOGRAPHY. Educ: Moravian Col, BS, 64; Lehigh Univ, MS, 66, PhD(biol), 68. Prof Exp: Res asst biol, Lehigh Univ, 64-68; NSF award biol oceanog, Marine Lab, Duke Univ, 68-70; asst prof zool, Clark Univ, 70-73; ASSOC PROF BIOL & MARINE SCI, UNIV SC, 73- Concurrent Pos: Ed, Psammonalia, 73-75. Mem: Am Soc Limnol & Oceanog; Am Soc Zoologists; Soc Syst Zool; Sigma Xi; Int Asn Meiobenthologists (chmn, 73-75). Res: Meiobenthic ecology; harpacticoid copepod systematics; benthic metabolism; population dynamics; zoogeography. Mailing Add: Baruch Inst Mar Biol & Coast Res Univ of SC Columbia SC 29208

COULOMBE, HARRY N, b Long Beach, Calif, Oct 7, 39; m 75. VERTEBRATE BIOLOGY, ENVIRONMENTAL MANAGEMENT. Educ: Univ Calif, Los Angeles, 62, MA, 65, PhD(zool), 68. Prof Exp: Asst prof zoophysiol, Inst Arctic Biol, Univ Alaska, 68-69; asst prof ecol, San Diego State Col, 69-74, dir bur ecol & mem exec comt, Ctr Regional Environ Studies, 70-74, proj mgr, 74-75; OIL SHALE RES MGR, WESTERN ENERGY & LAND USE TEAM, US FISH & WILDLIFE SERV, 75- Concurrent Pos: Anal & modeling coordr, Tundra Biome, Anal of Ecosyst Sect, US Int Biol Prog, 68-; chmn ad hoc comt rabies control, County of San Diego, 70- Mem: AAAS; Ecol Soc Am; Am Soc Mammal; Cooper Ornith Soc; Wildlife Soc. Res: Project development and management in regional environmental planning; ecosystem modeling and simulation; predator population ecology; ecology of cetaceans; adaptive physiology of vertebrates. Mailing Add: Western Energy & Land Use Team US Fish & Wildlife Serv Ft Collins CO 80521

COULOMBE, LOUIS JOSEPH, b Lac-St-Jean, Que, Dec 16, 20; m 47; c 7. AGRICULTURE. Educ: Laval Univ, BA, 43, BSA, 47; McGill Univ, MSc, 49, PhD, 56. Prof Exp: Res officer, Can Dept Agr, 48-61; mem tech dept & sales, Niagara Brand Chem Co, 61-64; RES SCIENTIST, CAN DEPT AGR, 64- Mem: Can Phytopath Soc. Res: Creation of apple cultivars resistant to scab, mildew and fire blight; ecological aspect of fruit pesticides; new programs for scab and mildew control. Mailing Add: 1415 DuVallon Beloeil PQ Can

COULOMBRE, ALFRED JOSEPH, b Boston, Mass, Aug 15, 22; m 48, 75. EMBRYOLOGY. Educ: Catholic Univ, BS, 47, MS, 49; Johns Hopkins Univ, PhD(embryol), 53. Prof Exp: Instr biol, Wabash Col, 48; from instr to asst prof anat, Sch Med, Yale Univ, 53-61; head, Sect Exp Embryol, Nat Inst Neurol Dis & Blindness, 61-67, chief lab neuroanat sci, 62-67; assoc dir intramural res, Nat Inst Child Health & Human Develop, 67-68; HEAD SECT EXP EMBRYOL, NAT EYE INST, 68- Concurrent Pos: Mem develop biol panel, NSF, 58-62; training comt mem, Nat Inst Child Health & Human Develop, 62-67; asst ed, J Exp Zool, 63-66, 69-73; comnr & vpres, Sci Manpower Comn, 63-68; asst ed, Develop Biol, 64-68; mem panel develop biol, Subcomt Life Sci, Nat Acad Sci-Nat Res Coun, 66-67. Honors & Awards: Jonas Friedenwald Award, 69. Mem: AAAS; Asn Res Ophthal; Soc Develop Biol (treas, 65-68); Int Soc Develop Biol. Res: Morphogenesis and developmental physiology of the vertebrate eye. Mailing Add: Nat Eye Inst Bldg 6 Rm 203 Bethesda MD 20014

COULSON, DALE ROBERT, b Monessen, Pa, Oct 26, 38; m 67; c 2. ORGANIC CHEMISTRY, ORGANOMETALLIC CHEMISTRY. Educ: Carnegie Inst Technol, BS, 60; Columbia Univ, MA, 61, PhD(chem), 64. Prof Exp: NSF fel photochem, Univ Chicago, 64-65, NIH fel, 65-66; res chemist, 66-74, RES SUPVR, E I DU PONT DE NEMOURS & CO, INC, 74- Mem: Am Chem Soc. Res: Homogeneous catalysis; organometallic synthesis. Mailing Add: 2314 Empire Dr Wilmington DE 19810

COULSON, JACK RICHARD, b Manhattan, Kans, Jan 31, 31; m 64; c 2. ENTOMOLOGY Educ: Iowa State Univ, BS, 52. Prof Exp: Biologist, Insect Identification & Parasite Introd Res Br, Washington, DC, 56-61, biologist, Plant Indust Sta, Md, 61-63, entomologist, 63-64, entomologist, Introduced Beneficial Insects Lab, NJ, 64-65 & Europ Parasite Lab, Paris, France, 65-67, asst to br chief taxon & biol control, Plant Indust Sta, Beltsville, Md, 67-72, CHIEF BENEFICIAL INSECT INTRODUCTION LAB, INSECT IDENTIFICATION & PARASITE INTROD RES BR, ENTOM RES DIV, AGR RES SERV, USDA, BELTSVILLE, MD, 72- Mem: Entom Soc Am. Res: Taxonomic entomology, especially bibliographic; biological control of insect pests. Mailing Add: Rte 1 Box 96-D Leesburg VA 22075

COULSON, KINSELL LEROY, b Hatfield, Mo, Oct 7, 16; m 47. METEOROLOGY, ATMOSPHERIC PHYSICS. Educ: Northwest Mo State Col, BS, 42; Univ Calif, Los Angeles, MA, 52, PhD(meteorol), 59. Prof Exp: Meteorologist, Univ Chicago, 42-43 & US Naval Ord Test Sta, Calif, 49-51; res meteorologist, Univ Calif, Los Angeles, 51-59 & Stanford Res Inst, 59-60; mgr geophys, Space Sci Lab, Gen Elec Co, 60-65; prof agr eng, 65-66, PROF METEOROL, UNIV CALIF, DAVIS, 67- Concurrent Pos: USSR exchange fel, Nat Acad Sci, 72; consult, Lawrence Livermore Lab, 74- Mem: AAAS; Am Meteorol Soc; Am Geophys Union; Solar Energy Soc. Res: Atmospheric radiation, especially solar radiation regime; molecular and aerosol scattering of radiation in planetary atmospheres; reflection from planetary surfaces; planetary albedo; space environment. Mailing Add: Dept Land Air & Water Resources Univ of Calif Davis CA 95616

COULSON, LARRY VERNON, b LaFollette, Tenn, Oct 15, 43; m 66; c 3. PARTICLE PHYSICS, RADIATION PHYSICS. Educ: Kans State Univ, BS, 65; Univ Va, PhD(physics), 70. Prof Exp: Fel particle physics, Rice Univ, 70-72; STAFF PHYSICIST, FERMI NAT ACCELERATOR LAB, 72- Res: Radiation related problems and dosimetry of accelerator produced radiation; isotope production cross sections at high energies. Mailing Add: Fermi Nat Accelerator Lab PO Box 500 Batavia IL 60510

COULSON, MICHAEL ROBERT CUMMINS, b Gloucester, Eng, Nov 29, 35; Can citizen; m 63; c 4. GEOGRAPHY, CARTOGRAPHY. Educ: Univ Durham, BA, 59; Univ Kans, MA, 63, PhD(geog), 66. Prof Exp: Asst geog, Univ Kans, 60-63; asst prof, 63-68, admin officer, Dept Geog, 67-69, from actg head to head dept, 69-73, ASSOC PROF GEOG, UNIV CALGARY, 68- Concurrent Pos: Mem planning adv comt, City of Calgary, 67-69; mem, Nat Adv Comt Geog Res, 70-71, chmn awards subcomt, 71; field prog coordr, Int Geog Cong Orgn Comt, 70-73; mem, Can Nat Comt Geog, 72- Mem: Can Asn Geogr; Can Inst Surv; Asn Am Geogr; Am Geog Soc; Regional Sci Asn. Res: Population age structures and federal and provincial electoral boundaries. Mailing Add: Dept of Geog Univ of Calgary Calgary AB Can

COULSON, PATRICIA BUNKER, b Kankakee, Ill, Apr 27, 42; m 65; c 2. REPRODUCTIVE ENDOCRINOLOGY, CELL PHYSIOLOGY. Educ: Univ Ill, BS, 64, MS, 65, PhD(reproductive endocrinol), 70. Prof Exp: Lab asst reproductive endocrinol, Univ Ill, 65-66; res reproduction, 70-72, ASST PROF ENDOCRINOL, UNIV TENN, KNOXVILLE, 72- Mem: AAAS; Sigma Xi; Soc Study Reproduction; Am Tissue Cult Asn; Endocrine Soc. Res: Endocrine control of the female reproductive tract emphasizing modulation of receptors for estrogen, progesterone and the gonadotropic hormones in the uterus, vagina, pituitary, hypothalamus, ovary and mammary cells. Mailing Add: Dept of Zool Life Sci Div 223 Hesler Biol Bldg Univ Tenn Knoxville TN 37916

COULSON, ROBERT N, b Dallas, Tex, Mar 1, 43. INSECT ECOLOGY, FOREST ENTOMOLOGY. Educ: Furman Univ, BS, 65; Univ Ga, MS, 67, PhD(entom), 69. Prof Exp: Prin entomologist pest control sect, Tex Forest Serv, 69-73; ASST PROF ENTOM, TEX A&M UNIV, 73- Concurrent Pos: Res assoc entom, Univ Ga, 67-70. Mem: Entom Soc Am; Entom Soc Can; Ecol Soc Am. Res: Forest insect community and population ecology in relation to pest management. Mailing Add: Dept of Entom Tex A&M Univ College Station TX 77845

COULSON, ROLAND ARMSTRONG, b Rolla, Kans, Dec 20, 15; m 44; c 2. BIOCHEMISTRY. Educ: Univ Wichita, BA, 37; La State Univ, MS, 39; Univ London, PhD(biochem), 44. Prof Exp: From instr to assoc prof, 44-53, PROF BIOCHEM, SCH MED, LA STATE UNIV, NEW ORLEANS, 53- Mem: Soc Exp Biol & Med; Am Soc Biol Chem; Am Asn Clin Chem; NY Acad Sci. Res: Nutrition; biochemical studies on Alligator mississippiensis. Mailing Add: Dept of Biochem La State Univ Sch of Med New Orleans LA 70112

COULSON, WALTER F, b Harrogate, Eng, Dec 17, 26; nat US; m 59; c 4. PATHOLOGY. Educ: Univ Edinburgh, MB, ChB, 49, BSc, 54, MD, 67; FRCPath, 72. Prof Exp: Resident med, Royal Infirmary, Edinburgh, Scotland, 49-50; registr path, Western Gen Hosp, 54-55; lectr, Univ Edinburgh, 55-60; asst prof, Univ Utah, 60-64, assoc prof, 64-68; assoc prof, 68-70, PROF PATH, CTR HEALTH SCI, SCH MED, UNIV CALIF, LOS ANGELES, 70-, VCHMN DEPT, 71- Concurrent Pos: Brown res fel, Yale Univ, 57-58; consult, Salt Lake Gen Hosp, 60-68; chief lab serv, Vet Admin Hosp, 62-68; res assoc, Univ Col, Univ London, 66-67; head div surg path, Univ Calif, Los Angeles, 68- Mem: Am Asn Path & Bact; Path Soc Gt Brit & Ireland; Am Soc Exp Path; Am Inst Nutrit; Am Soc Clin Path. Res: Morphology and biochemistry of connective tissue, including bone, particularly mechanical properties and the changes induced by copper deficiency. Mailing Add: Dept of Path Ctr for Health Sci Univ of Calif Los Angeles CA 90024

COULSTON, FREDERICK, b New York, NY, Dec 4, 14; m 40; c 2. PATHOLOGY, PARASITOLOGY. Educ: Syracuse Univ, AB, 36, MA, 39, PhD(zool), 42. Prof Exp: Asst zool, Syracuse Univ, 37-40; res fel bact & parasitol, Univ Chicago, 40-41, res asst parasitol, 41-43, res assoc & instr parasitol & bact, 43-46; res group leader, E I du Pont de Nemours & Co, Inc, 46-48; asst dir, Inst Med Res, Christ Hosp, Cincinnati, 48-52; dir exp path & toxicol & sr mem, Sterling-Winthrop Res Inst, 52-63, asst dir, Biol Div, 59-63; assoc path & bact, 52-62, res assoc path, 62-63, PROF PHARMACOL & PATH, ALBANY MED COL, 62-, PROF TOXICOL & DIR INST COMP & HUMAN TOXICOL, 63-, DIR ANIMAL FACIL, 63- Concurrent Pos: Consult human malaria proj, Comt Med Res, Off Sci Res & Develop, 41-46; vis lectr, State Univ NY Upstate Med Ctr, 52-54; ed, J Toxicol & Appl Pharmacol, 59- & J Exp & Molecular Path, 61-; mem, Animal Health Inst, Sci Adv Comt, 60-63; mem drug safety eval comt, Pharmaceut Mfrs Asn, 60-63; vchmn & chmn, Gordon Res Conf Toxicol & Safety Eval, 61-63, mem coun, Gordon Res Conf, 62-64. Mem: Fel NY Acad Sci; Am Soc Toxicol (pres elect, 63); fel Royal Soc Trop Med & Hyg; Am Soc Parasitol; Am Soc Clin Path. Res: Avian, simian and human malaria; cellular immunology and inflammation; chemotherapy; experimental pathology related to chemotherapy and pharmacology; toxicology; drug and vaccine therapy of poultry and large animals. Mailing Add: Inst of Comp & Human Toxicol Albany Med Col Albany NY 12207

COULTER, BYRON LEONARD, b Phenix City, Ala, Aug 16, 41; m 62; c 1. THEORETICAL PHYSICS. Educ: 0niv Ala, BS, 62, PhD(physics), 66. Prof Exp: Asst prof, 66-72, ASSOC PROF PHYSICS, E CAROLINA UNIV, 72- Mem: Am Phys Soc; Am Asn Physics Teachers; Nat Sci Teachers Asn; Sigma Xi. Res: Computer simulation of physics problems. Mailing Add: Dept of Physics E Carolina Univ Greenville NC 27834

COULTER, CHARLES L, b Akron, Ohio, Jan 10, 33; m 55; c 3. STRUCTURAL CHEMISTRY, BIOPHYSICAL CHEMISTRY. Educ: Miami Univ, AB, 54, MA, 56; Univ Calif, Los Angeles, PhD(phys chem), 60. Prof Exp: USPHS fel, Med Res Coun Unit for Molecular Biol, Cambridge, Eng, 60-62; fel, Lab Molecular Biol, NIH, 62-64, prog dir anal biochem, Div Res Facil & Resources, 65-66; asst prof, 66-72, ASSOC PROF ANAT, UNIV CHICAGO, 72- Mem: AAAS; Am Crystallog Asn; NY Acad Sci; fel Am Inst Chemists. Res: Protein crystallography; crystal structures of biologically important compounds; structural biochemistry. Mailing Add: Dept of Anat Univ of Chicago Chicago IL 60637

COULTER, CLAUDE ALTON, b Phenix City, Ala, Mar 30, 36; m 60; c 2. RADIATION PHYSICS, QUANTUM OPTICS. Educ: Samford Univ, BA, 56; Univ Ala, MS, 59; Harvard Univ, MA, 63, PhD(physics), 64. Prof Exp: Asst prof physics, Univ Ala, 63-66; from asst prof to assoc prof, Clark Univ, 66-71; ASSOC PROF PHYSICS, UNIV ALA, TUSCALOOSA, 71- Concurrent Pos: Consult, Phys Sci Directorate, US Army Missile Command, Redstone Arsenal, Ala, 63-; vis staff mem, Los Alamos Sci Lab, 74-76. Mem: Am Phys Soc. Res: Radiation damage in metals; quantum optics. Mailing Add: Dept of Physics Univ of Ala University AL 35486

COULTER, DWIGHT BERNARD, b Iowa City, Iowa, Jan 8, 35; m 61; c 2. VETERINARY PHYSIOLOGY. Educ: Iowa State Univ, DVM, 60, MS, 65, PhD(physiol), 69. Prof Exp: From instr to assoc prof physiol, Col Vet Med, Iowa State Univ, 62-72; ASSOC PROF PHYSIOL, COL VET MED, UNIV GA, 72- Mem: Am Vet Med Asn; Conf Res Workers Animal Dis. Res: Plasmatic and erythrocytic electrolyte concentrations in animals; effects of plasmatic electrolytes on electrocardiograms; comparative neurophysiology. Mailing Add: Dept of Physiol Univ of Ga Col of Vet Med Athens GA 30602

COULTER, HERBERT DAVID, JR, b Enid, Okla, Dec 31, 39; m 64; c 2. ANATOMY. Educ: Westminster Col, Mo, BA, 61; Univ Tenn, PhD(anat), 68. Prof Exp: Instr, 68-70, ASST PROF ANAT, SCH MED, UNIV MINN, MINNEAPOLIS, 70- Mem: AAAS; Am Soc Cell Biol; Electron Micros Soc Am; Am Asn Anatomists. Res: Biological membranes; neurocytology. Mailing Add: Dept of Anat Univ of Minn Sch of Med Minneapolis MN 55455

COULTER, JOE DAN, b Victoria, Tex, July 25, 44; m 67; c 2. NEUROPHYSIOLOGY, PSYCHOLOGY. Educ: Univ Okla, BA, 66, PhD(biol psychol), 71. Prof Exp: NIH grant, Marine Biomed Inst, Univ Tex Med Br, Galveston, 71-73; Found Fund Res Psychiat grant, Inst Physiol, Univ Pisa, 73-74 & Univ Edinburgh, 74-75; ASST PROF PHYSIOL & PSYCHOL, MARINE BIOMED INST, UNIV TEX MED BR, GALVESTON, 75- Mem: AAAS; Asn Psychophysiol Study Sleep; Soc Neurosci. Res: Neural basis of perception; motor control; states of consciousness and sleep. Mailing Add: Marine Biomed Inst Univ of Tex Med Br Galveston TX 77550

COULTER, LLEWELLYN LEGRANDE, b Rochester, Mich, Jan 29, 21; m 42; c 5. WEED SCIENCE, WILDLIFE MANAGEMENT. Educ: Mich State Univ, BS, 47, MS, 48. Prof Exp: Field res specialist herbicides, Dow Chem Co, 46-53, proj leader, Bioprod Dept, 53-59, proj leader & sect supvr, 59-64, mgr, Dow Int Bioprod Bus Develop, 64-66, mgr bioprod, Europ Area, Dow Chem Int, 66-71, INTER-AREA TECH MGR AGR PROD, DOW CHEM CO, 71- Mem: Soc Am Foresters; Weed Sci Soc Am. Res: Woody plant control with herbicides; industrial vegetation control; wildlife management; weed control in crops. Mailing Add: Agr Chem Res & Develop Dow Chem Co Agr Dept Midland MI 48640

COULTER, LOWELL VERNON, b Marion, Ohio, July 3, 13; m 37; c 2. PHYSICAL CHEMISTRY, SOLID STATE CHEMISTRY. Educ: Heidelberg Col, BS, 35; Colo Col, AM, 37; Univ Calif, Berkeley, PhD(chem), 40. Prof Exp: Instr chem, Colo Col, 35-37; asst, Univ Calif, 37-40; instr, Univ Idaho, 40-42; instr, 42-44, from asst prof to assoc prof, 46-55, chmn dept, 61-72, PROF CHEM, BOSTON UNIV, 55- Concurrent Pos: Group leader, Manhattan Proj, Monsanto Chem Co, Ohio, 44-45. Mem: Fel AAAS; Am Chem Soc; Am Phys Soc; NY Acad Sci; Sigma Xi. Res: Application of the third law of thermodynamics; low temperature calorimetry; solution calorimetry; properties of liquid ammonia solutions of metals; thermodynamic properties of clathrates. Mailing Add: Dept of Chem Boston Univ 675 Commonwealth Ave Boston MA 02215

COULTER, MALCOLM WILFORD, b Suffield, Conn, Dec 30, 20; m 48; c 5. WILDLIFE ECOLOGY. Educ: Univ Conn, BS, 42; Univ Maine, MS, 48; Syracuse Univ, PhD, 66. Prof Exp: Proj leader furbearers, Vt State Fish & Game Serv, 48-49; instr wildlife resources & asst leader res unit, 49-52, from asst prof to assoc prof, 52-62, PROF WILDLIFE RESOURCES, UNIV MAINE, ORONO, 62-, ASSOC DIR SCH FOREST RESOURCES, 68- Mem: AAAS; Wildlife Soc; Am Soc Mammal. Res: Ecology and behavior of furbearing animals; waterfowl breeding biology, ecology, behavior and population dynamics; marsh ecology and management. Mailing Add: Sch of Forest Resources Univ of Maine Orono ME 04473

COULTER, MURRAY WHITFIELD, b El Dorado, Ark, May 2, 32; m 59; c 2. GENETICS, PLANT PHYSIOLOGY. Educ: Emory Univ, BA, 54; Univ Ariz, MS, 56; Univ Calif, Los Angeles, PhD(plant sci), 63. Prof Exp: Teaching fel bot, Univ Ariz, 55-56; asst, Univ Calif, Los Angeles, 56-58, teaching & res assoc, 58-59; asst prof biol, San Fernando Valley State Col, 59-62; res botanist, Inst Geophys & Planetary Physics & fel bot & plant biochem, Univ Calif, Los Angeles, 62-64; asst prof, 64-67, ASSOC PROF BIOL SCI, TEX TECH UNIV, 67- Concurrent Pos: Calif Res Found grant, 58-59; consult, NAm Aviation, Inc, 62-63. Mem: Fel AAAS; Genetics Soc Am; Am Soc Plant Physiol; Bot Soc Am; Am Inst Biol Sci. Res: Physiological genetics; Gibberellin studies with microorganisms and genetic mutants of maize; photoperiod and endogenous rhythms as related to biological clocks; environmental control of plant growth and development; hormones and plant growth regulators. Mailing Add: Dept of Biol Tex Tech Univ Lubbock TX 79409

COULTER, NEAL STANLEY, b Columbus, Ga, May 3, 44; m 66; c 2. COMPUTER SCIENCES, INFORMATION SCIENCE. Educ: Univ Ala, BS, 65, MA, 66; Ga Inst Technol, MS, 72, PhD(info & comput sci), 74. Prof Exp: Assoc res engr appl math, Boeing Co, Ala, 66-67; asst prof math & comput sci, Columbus Col, 67-75; ASSOC PROF COMPUT SYSTS, FLA ATLANTIC UNIV, 75- Mem: Asn Comput Mach. Res: Application of semantic information measures based on logical probability in the simulation of inductive and deductive processes. Mailing Add: Dept of Comput Systs Fla Atlantic Univ Boca Raton FL 33431

COULTER, NORMAN ARTHUR, JR, b Atlanta, Ga, Jan 9, 20; m 51; c 1. PHYSIOLOGY, BIOPHYSICS. Educ: Va Polytech Inst, BS, 41; Harvard Univ, MD, 50. Prof Exp: Instr math, Va Polytech Inst, 46; Nat Res Coun fel, Johns Hopkins Univ, 50-52; asst prof physiol, Ohio State Univ, 52-55, from asst prof to assoc prof physiol & biophys, 55-65; assoc prof, 65-67, PROF BIOENG & BIOMATH, UNIV NC, CHAPEL HILL, 67-, CHMN BIOMED ENG & MATH CURRICULUM, 69-, SCI DIR, A F FORTUNE BIOMED COMPUT CTR, 70- Concurrent Pos: Consult, NIH. Mem: Am Soc Cybernet; Biophys Soc; Am Physiol Soc; Biomed Eng Soc; Inst Elec & Electronics Engrs. Res: Hemodynamics; neural net theory; biological cybernetics; mathematical biophysics. biomedical computing. Mailing Add: Dept of Surg Univ of NC Chapel Hill NC 27514

COULTER, PAUL (DAVID TODD), b Dayton, Ohio, Apr 4, 38; m 58; c 3. ANALYTICAL CHEMISTRY, APPLIED MATHEMATICS. Educ: Univ Dayton, BS, 60; Univ Kans, PhD(chem), 65. Prof Exp: Res scientist, 65-66, head anal res & develop, 66-69, mgr anal & phys test, 70-73, PROG MGR CONTROLLED LINE FACIL, CARBON PROD DIV, UNION CARBIDE CORP, 73- Mem: AAAS; Am Inst Chemists; Am Chem Soc; Soc Appl Spectros. Res: Electrochemistry in nonaqueous solvents; emission spectroscopy; porosity and surface area determinations using both sorption and liquid intrusion techniques; applications of computers in analytical chemistry and process control. Mailing Add: Union Carbide Corp 12900 Snow Rd Parma OH 44130

COULTER, PHILIP W, b Phenix City, Ala, Apr 19, 38; m 60; c 2. THEORETICAL NUCLEAR PHYSICS. Educ: Univ Ala, BS, 59, MS, 61; Stanford Univ, PhD(physics), 65. Prof Exp: Res assoc physics, Univ Mich, 65-67; asst res physicist, Univ Calif, Irvine, 67-68, asst prof physics, 67-71; ASSOC PROF PHYSICS, UNIV ALA, 71- Mem: Am Phys Soc; Sigma Xi. Res: Direct nuclear interactions. Mailing Add: Dept of Physics & Astron Univ of Ala University AL 35486

COULTER, SAMUEL TODD, b Weiser, Idaho, Sept 15, 03; m 28; c 3. DAIRY MANUFACTURING. Educ: Ore State Col, BS, 25; Univ Minn, MS, 30, PhD(dairy husb), 33. Prof Exp: Asst, Dairy Div, Univ Minn, 25-28; mgr, State Exp Creamery, Minn, 28-30; from instr to prof dairy husb, 30-72, head dept dairy indust, 59-66 & dept food sci & indust, 66-72, EMER PROF FOOD SCI & NUTRIT, UNIV MINN, ST PAUL, 72- Concurrent Pos: Sect chmn, Int Dairy Cong, Australia, 70. Honors & Awards: Borden Award. Mem: Fel AAAS; Inst Food Technol; Am Dairy Sci Asn (vpres, 62-63, pres, 63-64). Res: Quality and deterioration of dry milk products; design of spray driers; manufacture of foreign type cheeses; food dehydration; processes for cheese manufacture. Mailing Add: Dept of Food Sci & Indust Univ of Minn St Paul MN 55101

COULTER, WILSON H, b San Jose, Calif, Jan 10, 36; Can citizen; m 62; c 2. MICROBIOLOGY. Educ: Univ Toronto, BSA, 62, MSA, 64; Mich State Univ, PhD(microbiol), 68. Prof Exp: Res assoc microbiol, Mich State Univ, 68-69; asst prof microbiol, 69-74, ASSOC PROF BIOL, UNIV NB, 74- Mem: AAAS; Am Soc Microbiol; Can Soc Microbiol. Res: Bacterial physiology; problems associated with water pollution; industrial microbiology. Mailing Add: Dept of Biol Univ of NB Fredericton NB Can

COUMBIS, RICHARD J, toxicology, biology, see 12th edition

COUNCE, SHEILA JEAN, b Hayes Center, Nebr, Mar 18, 27; m 60. GENETICS, EMBRYOLOGY. Educ: Univ Colo, BA, 48, MA, 50; Univ Edinburgh, PhD(genetics), 54. Prof Exp: Lab asst biol, embryol & genetics, Univ Colo, 48-50, instr, 50; demonstr gentics, Univ Edinburgh, 51; lab assoc, Jackson Mem Lab, 54-55; Macauley fel, Univ Edinburgh, 55-56; NSF fel, Zurich, Switz, 56-57; from res asst to res assoc biol, Yale Univ, 57-65; assoc anat, 67-68, asst prof, 69-72, ASSOC PROF ANAT, SCH MED, DUKE UNIV, 72-, RES ASSOC ZOOL, 65- Mem: AAAS; Soc Develop Biol; Am Soc Zool; Am Soc Naturalists; Genetics Soc Am. Res: Developmental genetics; experimental embryology, especially with insects. Mailing Add: Dept of Anat Duke Univ Med Ctr Durham NC 27710

COUNCELL, CLARA ELIZABETH, b Baltimore, Md, July 4, 06. PUBLIC HEALTH, STATISTICS. Educ: Goucher Col, AB, 27; Johns Hopkins Univ, MS, 41; Yale Univ, PhD(pub health), 47. Prof Exp: Technician, Dept Epidemiol, Sch Hyg & Pub Health, Johns Hopkins Univ, 27-29; statistician, Md State Dept Health, 29-36 & Div Pub Health Methods, USPHS, 36-43; pub health officer, Health & Sanit Div, Inst Inter-Am Affairs, 43-48; dir clearinghouse for res, Children's Bur, Fed Security Agency, 48-51; consult, US Govt, 51-62; dep chief, Off Health Statist Anal, Nat Ctr Health Statist, 62-67 & chief, Int Training Br, Off Int Statist Progs, 67-70; consult, 70-74; RETIRED. Concurrent Pos: Instr statist, Dept Econ & Sociol, Goucher Col, 31; instr, Sch Med, Yale Univ, 42-45. Mem: Fel Am Pub Health Asn; Am Pop Asn; Am Statist Asn. Res: Morbidity and mortality studies. Mailing Add: 3036 O St NW Washington DC 20007

COUNCILL, RICHARD J, b Greenville, SC, May 26, 23; m 49; c 5. PETROGRAPHY, SEDIMENTOLOGY. Educ: Univ NC, BS, 48, MS, 56. Prof Exp: Geologist, US Geol Surv, 48-51; econ geologist, NC Dept Conserv & Develop, 52-55; indust geologist, Atlantic Coast Line RR, 55-60, gen indust geologist, 60-67; gen indust geologist, Seaboard Coast Line RR, 67-73, MGR INDUST DEVELOP & CHIEF GEOLOGIST, SEABOARD COAST LINE RR & LOUISVILLE & NASHVILLE RR, 73-; DIR, INT RESOURCES DEVELOP CORP, 68- Mem: Am Inst Prof Geol; Am Inst Mining & Metall Eng; Geol Soc Am. Res: Resources development. Mailing Add: 3126 Bridlewood Lane Jacksonville FL 32217

COUNSELL, RAYMOND ERNEST, b Vancouver, BC, Aug 20, 30; US citizen; c 3. MEDICINAL CHEMISTRY, PHARMACOLOGY. Educ: Univ BC, BSP, 53; Univ Minn, PhD(pharmaceut & org chem), 57. Prof Exp: Lectr, Univ BC, 53-54; sr res chemist, G D Searle & Co, 57-64; assoc prof, 64-69, PROF MED CHEM, UNIV MICH, ANN ARBOR, 69-, PROF PHARMACOL, 72- Concurrent Pos: Res assoc, Am Cancer Soc, 64-71; mem med chem study sect A, NIH, 68-72; E Roosevelt Inst fel cancer res, Univ Milan & Univ Uppsala, 72-73; mem prog comt pharmacol & toxicol, Nat Inst Gen Med Sci; consult, Nat Inst Child Health & Develop & G D Searle & Co; sect ed, Ann Reports in Med Chem. Assoc Award, Am Honors & Awards: Czerniak Prize, Ahavat Zion Found, Israel, 74. Mem: AAAS; Am Chem Soc; Am Soc Pharmacol & Exp Therapeut; fel Acad Pharmaceut Sci; Soc Nuclear Med. Res: Synthesis and molecular mode of action of chemical regulators of biological processes, especially adrenal hormone biogenesis; radiopharmaceuticals for diagnosis and treatment of cancer. Mailing Add: Dept of Pharmacol Univ of Mich Med Sch Ann Arbor MI 48109

COUNSELMAN, C J, b West Palm Beach, Fla, July 4, 25; m 49; c 4. ENTOMOLOGY, ICHTHYOLOGY. Educ: Auburn Univ, BS, 52, MS, 53. Prof Exp: Prod mgr, Big Springs Minnow Farm, 53-55; proj leader, Biol Surv Unit, State of Fla, 55-57; entom & plant path, Vero Beach Labs, Inc, 57-63; asst dir, 63-67, DIR AGR CHEM RES & DEVELOP, CIBA AGROCHEM CO, 67- Mem: Entom Soc Am; Am Phytopath Soc; Soc Nematol; Am Inst Biol Sci; NY Acad Sci. Res: Research, development and registration of chemicals for agricultural purposes. Mailing Add: PO Box 1090 Vero Beach FL 32960

COUNSELMAN, CHARLES CLAUDE, III, b Baltimore, Md, Apr 27, 43; m 66. PLANETARY SCIENCES, RADIO ASTRONOMY. Educ: Mass Inst Technol, BS, 64, MS, 65, PhD(instrumentation), 69. Prof Exp: Asst prof, 69-74, ASSOC PROF PLANETARY SCI, MASS INST TECHNOL, 74- Mem: Int Astron Union; Int Sci Radio Union; Am Geophys Union; Am Astron Soc; Inst Elec & Electronics Eng. Res:

Long baseline radio interferometry applications to astronomy, geodesy, geophysics, lunar libration and orbit, solar system dynamics and testing general relativity. Mailing Add: Dept Earth & Planetary Sci Mass Inst Technol Cambridge MA 02139

COUNTER, FREDERICK T, JR, b Lowell, Mass, Dec 16, 34; m 63. VIROLOGY. Educ: Mass Col Pharm, BS, 56, MS, 58; Univ Mass, PhD(microbiol), 63. Prof Exp: Sr bacteriologist, 63-68, mgr biol develop, 68-69, HEAD IMMUNIZATION BIOL RES & DEVELOP, ELI LILLY & CO, 69- Mem: AAAS; Am Soc Microbiol; NY Acad Sci. Res: Biological research and development in bacterial and virus vaccines. Mailing Add: Eli Lilly & Co W National Rd Greenfield IN 46140

COUNTRYMAN, DAVID WAYNE, b Ottumwa, Iowa, May 21, 43; m 65; c 2. FOREST MANAGEMENT. Educ: Iowa State Univ, BS, 66, MF, 68; Univ Mich, PhD(forest mgt & planning), 73. Prof Exp: Actg exten forester, Iowa State Univ Exten Serv, USDA, 67-68, forester, Poplar Bluff Ranger Dist, Forest Serv, 68-70, Region 9 Off, 70-73 & Washington, DC Off, 73-74; ASSOC PROF FOREST MGT, IOWA STATE UNIV, 75- Mem: Soc Am Foresters. Res: Forest administration and planning systems and processes; policy analysis and land use planning. Mailing Add: 251 Bessey Hall Iowa State Univ Ames IA 50011

COUNTRYMAN, JAMES JOSEPH, b Trego, Wis, June 18, 34; m 54; c 2. ALGEBRA. Educ: Wis State Col, Superior, BS, 59; Univ Notre Dame, MS, 63, PhD(math), 70. Prof Exp: Instr high schs, Minn, 59-62; instr math, asst to chmn dept & asst dir NSF Insts, Univ Notre Dame, 63-69; ASSOC PROF MATH, PURDUE UNIV, N CENT CAMPUS, 69-, DEAN ACAD SERV, 70- Mem: Am Math Soc. Res: Commutation semigroups of pq groups. Mailing Add: Purdue Univ N Cent Campus Westville IN 46391

COUNTS, JON MILTON, b Richlands, Va, July 31, 37. PUBLIC HEALTH ADMINSTRATION. Educ: Univ Ariz, BS, 59; Tulane Univ, MPH, 63; Univ NC, DrPH(lab pract), 66. Prof Exp: Bacteriologist pub health, 60-62, asst dir, 66-73, dir lab licensure, 70-75, actg chief, 73-75, CHIEF PUB HEALTH, STATE LAB, ARIZ DEPT HEALTH SERV, 75- Concurrent Pos: Chmn comt lab mgt, Nat Commun Dis Ctr Task Force, 71-73; qual assurance coordr, Adv Comt, Region IX, Environ Protection Agency, 74- Mem: Am Pub Health Asn; Am Soc Microbiologists; Conf State & Prov Pub Health Lab Dirs. Res: Laboratory management; development of clinical laboratories in quality assurance. Mailing Add: 1716 W Adams Phoenix AZ 85007

COUNTS, WAYNE BOYD, b Prosperity, SC, Oct 27, 36; m 62; c 1. ORGANIC CHEMISTRY. Educ: Furman Univ, BS, 58; Univ NC, PhD(org chem), 64. Prof Exp: Staff fel, Nat Cancer Inst, 63-66; prof chem, Lincoln Mem Univ, 66-69, head dept, 68-69; ASSOC PROF CHEM, GA SOUTHWESTERN COL, 69- Mem: Am Chem Soc. Res: Organic synthesis; heterocyclic chemistry. Mailing Add: Dept of Chem Ga Southwestern Col Americus GA 31709

COUPER, MONROE, organic chemistry, see 12th edition

COUPERUS, MOLLEURUS, b Essen, Ger, Jan 27, 06; nat US; m 39; c 4. DERMATOLOGY. Educ: Andrews Univ, BA, 27; Loma Linda Univ, MD, 34. Prof Exp: PROF DERMAT & SYPHILOL, LOMA LINDA UNIV, 47-, CHMN DEPT DERMAT, 64- Mem: Am Acad Dermat; Soc Invest Dermat. Res: Tumors of the skin; physical anthropology. Mailing Add: 24887 Taylor St Loma Linda CA 92354

COUPERUS, PIERCE GERARD, physics, see 12th edition

COUPLAND, ROBERT THOMAS, b Winnipeg, Man, Jan 24, 20; m 45; c 1. PLANT ECOLOGY, SYSTEMS ECOLOGY. Educ: Univ Man, BSA, 46; Univ Nebr, PhD(bot), 49. Prof Exp: Student asst pasture studies, Dom Dept Agr, Sask, 41-46; officer in chg, Dom Forest Serv, Man, 46; asst bot, Univ Nebr, 46-47; from asst prof to assoc prof, 48-57, PROF PLANT ECOL, UNIV SASK, 57-, HEAD DEPT, 48- Concurrent Pos: Dir Matador Proj, Int Ctr Grasslands Studies, Int Biol Prog, 67-, chmn int coord comt grasslands & mem productivity terrestrial sect comt. Honors & Awards: Can Centennial Medal. Mem: Fel AAAS; Am Inst Biol Sci; Int Asn Ecol; Ecol Soc Am. Res: Function of grassland ecosystems; primary production of grasslands; classification of grasslands; autecology of native plants, weeds and crop plants; studies of succession resulting from grazing and land abandonment. Mailing Add: Dept of Plant Ecol Univ of Sask Saskatoon SK Can

COURANT, ERNEST DAVID, b Goettingen, Ger, May 26, 20; nat US; m 44; c 2. THEORETICAL PHYSICS. Educ: Swarthmore Col, BA, 40; Univ Rochester, MS, 42, PhD(physics), 43. Prof Exp: Asst physics, Univ Rochester, 40-43; sci officer, Nat Res Coun Can, 43-46; res assoc theoret physics, Cornell Univ, 46-48; consult, 47-48, from assoc physicist to physicist, 48-60, SR PHYSICIST, BROOKHAVEN NAT LAB, 60- Concurrent Pos: Vis asst prof, Princeton Univ, 50-51; Fulbright res grant, Cambridge Univ, 56; consult, Gen Atomic Div, Gen Dynamics Corp, 58; vis prof, Yale Univ, 61-62, Brookhaven prof, 62-67; prof, State Univ NY Stony Brook, 67-; vis physicist, Nat Accelerator Lab, 68-69; vis scientist, Europ Orgn Nuclear Res, 74. Mem: Nat Acad Sci; AAAS; Am Phys Soc; NY Acad Sci. Res: Theory of solids; chain reactors, particle accelerators and nuclear reactions. Mailing Add: Brookhaven Nat Lab Upton NY 11973

COURANT, HANS WOLFGANG JULIUS, b Ger, Oct 30, 24; nat US; m 49; c 3. HIGH ENERGY PHYSICS. Educ: Mass Inst Technol, BS, 49, PhD(physics), 54. Prof Exp: Asst, Mass Inst Technol, 49-54, res assoc physics, 54; Fulbright grant, Polytech Sch, Paris, 54-55; asst res physicist, Univ Calif, 55-56; from instr to asst prof, Yale Univ, 56-61; Ford Found fel, 61-62; NSF sr fel, Europ Orgn Nuclear Res, Geneva, 62-63; assoc prof, 63-68, PROF PHYSICS, UNIV MINN, MINNEAPOLIS, 68- Concurrent Pos: Vis assoc physicist, Brookhaven Nat Lab, 56-57, 58, 59-, guest physicist, 57, 58-; physicist, Radiation Lab, Univ Calif, 57; consult, Argonne Nat Lab, 64-70; vis prof, Univ Heidelberg, 69-70. Mem: Am Phys Soc. Res: Experimental high energy physics; cosmic rays. Mailing Add: Sch of Physics Univ of Minn Minneapolis MN 55455

COURANT, RICHARD, mathematics, deceased

COURCHENE, WILLIAM LEON, b Springfield, Mass, Nov 15, 26; m 48; c 3. PHYSICAL CHEMISTRY. Educ: Univ Mass, BS, 48; Cornell Univ, PhD(chem), 52. Prof Exp: Asst, Cornell Univ, 48-52; res chemist, 52-67, sect head, 67-74, ASSOC DIR, PROCTER & GAMBLE, 74- Mem: Am Chem Soc. Res: Infrared and mass spectroscopy; solution thermodynamics; proteins. Mailing Add: 8678 Elmtree Ave Cincinnati OH 45231

COURNAND, ANDRE FREDERIC, b Paris, France, Sept 24, 95; nat US; m 24; c 3. MEDICINE. Educ: Univ Paris, BA, 13, PCN, 14, MD, 30; Univ Nancy, DHC, 69. Hon Degrees: DUniv, Univ Strasbourg, 57, Univ Lyon, 58, Free Univ Brussels, 59, Univ Pisa, 60; DSc, Univ Birmingham, 61, Gustavus Adolphus Col, 63, Columbia Univ, 65. Prof Exp: From instr to assoc, 34-42, from asst prof to prof, 42-64, EMER PROF MED, COL PHYSICIANS & SURGEONS, COLUMBIA UNIV, 64- Honors & Awards: Nobel Prize Med & Physiol, 56; Retzius Silver Medal, Swedish Soc Internal Med, 46; Lasker Award, Am Pub Health Asn, 49; Phillips Award, Am Col Physicians, 52; Gold Medal, Royal Acad Med, Brussels, 56; Officer, Legion d'honneur, 57, Commandeur, 70; Jiminez Diaz Award, Madrid, 70. Mem: Nat Acad Sci; Am Physiol Soc; Asn Thoracic Surg; Am Thoracic Soc; hon fel Royal Soc Med. Res: Applied physiology of respiration and circulation. Mailing Add: 1361 Madison Ave New York NY 10010

COURSEN, BRADNER WOOD, b Roselle Park, NJ, Feb 10, 29; m 54; c 3. PHYSIOLOGY. Educ: Drew Univ, BA, 52; Univ Md, MS, 57, PhD(fungus physiol), 59. Prof Exp: Jr chemist, Res Dept, Am Can Co, 52-55; instr biol & plant physiol, Lawrence Univ, 59-61, from asst prof to assoc prof biol, 61-68; assoc prof, 68-69, PROF BIOL, COL WILLIAM & MARY, 69- Concurrent Pos: Pres, Wis State Bd Exam in Basic Sci, 65-69. Mem: Bot Soc Am. Res: Mechanism of action of antibiotics; fungus physiology and metabolism; biology and biochemistry of cellular aging. Mailing Add: Dept of Biol Col of William & Mary Williamsburg VA 23185

COURSEN, DAVID LINN, b Newark, NJ, May 7, 23; m 45; c 4. ROCK MECHANICS. Educ: Columbia Univ, BA, 43; Cornell Univ, PhD(chem), 51. Prof Exp: Res chemist, 50-58, sr res chemist, 58-59, tech assoc, 59-61, res assoc, 61-66, prod mgr, 66-67, res & develop mgr, 67-70, RES FEL, E I DU PONT DE NEMOURS & CO, INC, 70- Concurrent Pos: Assoc, Woods Hole Oceanog Inst. Mem: Am Chem Soc; Int Soc Rock Mech; Soc Explor Geophysicists. Res: Crystal structure; x-ray diffraction; explosives; radiography; underwater acoustics; mining and quarring; oil and gas well stimulation; working of mineral deposits in place. Mailing Add: RD 2 Mercersburg PA 17236

COURSEY, BERT MARCEL, physical chemistry, radiochemistry, see 12th edition

COURSIN, DAVID BAIRD, b Steubenville, Ohio, Jan 14, 19; m 45; c 3. PEDIATRICS. Educ: Haverford Col, BS, 40; Univ Pa, MD, 43; Am Bd Pediat, dipl. Prof Exp: DIR RES, ST JOSEPH HOSP, LANCASTER, PA, 48- Concurrent Pos: Consult, NIH; mem, Food & Nutrit Bd, Nat Res Coun. Mem: AMA; fel Am Acad Pediat; Am Col Physicians; Soc Res Child Develop; Asn Res Nerv & Ment Dis. Res: Fundamental biochemistry and functional physiology of cellular metabolism as they relate to central nervous system activity and to congenital anomalies, especially oral-facial communicative disorders. Mailing Add: Res Inst St Joseph Hosp Lancaster PA 17604

COURT, ANITA, b Chicago, Ill, Aug 15, 30. REACTOR PHYSICS, PHYSICAL INORGANIC CHEMISTRY. Educ: Ill Inst Technol, BS, 52; Mich State Univ, PhD(chem), 56. Prof Exp: Asst chemist, M W Kellogg Co Div, Pullman, Inc, 56-59; chemist, Brookhaven Nat Lab, 59-70; scientist, 70-75, PRIN ENGR, ADVAN REACTORS DIV, WESTINGHOUSE ELEC CORP, 75- Mem: Am Chem Soc; Am Nuclear Soc. Res: Shielding, safety analysis. Mailing Add: Advan Reactor Div Westinghouse Elec Corp Madison PA 15663

COURT, ARNOLD, b Seattle, Wash, June 20, 14; m 41; c 3. CLIMATOLOGY. Educ: Univ Okla, BA, 34, Univ Wash, Seattle, MS, 49; Univ Calif, PhD(geog), 56. Prof Exp: Meteorologist, US Weather Bur, 38-43; climatologist & head climat unit, Res & Develop Br, Off Qm Gen, US Dept Army, 46-51; res meteorologist, Statist Lab, Univ Calif, 52-54, lectr meteorol & climat, 56-57 & 58; meteorologist, Pac Southwest Forest & Range Exp Sta, US Forest Serv, 56-60; chief appl climat br, Geophys Res Directorate, Air Force Cambridge Res Labs, 60-62; res scientist, Lockheed-Calif Co, 62-64; chmn dept geog, 70-72, PROF CLIMAT, CALIF STATE UNIV, NORTHRIDGE, 62- Concurrent Pos: Chief meteorologist, US Antarctic Serv, Little Am, 40-41; consult, Lockheed-Calif Co, 64- Mem: AAAS; Am Geophys Union; Asn Am Geog; Am Meteorol Soc; Am Statist Asn. Res: Climates of polar regions; statistical analysis of climate; cloud seeding evaluation; wind energy computations. Mailing Add: 17168 Septo St Northridge CA 91324

COURT, DONALD LEE, molecular biology, see 12th edition

COURTENAY, WALTER ROWE, JR, b Neenah, Wis, Nov 6, 33; m 60; c 2. MARINE BIOLOGY, ICHTHYOLOGY. Educ: Vanderbilt Univ, BA, 56; Univ Miami, Fla, MS, 60, PhD(marine biol), 65. Prof Exp: Temp instr zool, Duke Univ, 63-64, vis asst prof, 64-65; asst prof biol, Univ Boston, 65-67; assoc ichthyol, Mus Comp Zool, Harvard Univ, 65-68; from asst prof to assoc prof, 67-72, PROF ZOOL, FLA ATLANTIC UNIV, 72- Concurrent Pos: NSF res grants, 65-69. Mem: Am Soc Ichthyol & Herpet; Am Fisheries Soc; Soc Syst Zool. Res: Systematics, functional morphology and ecology of marine fishes, especially within the families of Ophidiidae, Brotulidae, Serranidae, Pomadasyidae and related groups; sonic mechanisms in fishes; introduced exotic freshwater fishes. Mailing Add: Dept of Biol Sci Fla Atlantic Univ Boca Raton FL 33432

COURTER, RICHARD CARSON, mathematics, see 12th edition

COURTER, ROBERT DAVID, veterinary medicine, see 12th edition

COURTNEY, CAMERON B, b Ottawa, Ont, Apr 24, 15; US citizen; m 44; c 2. MATHEMATICS. Educ: Queen's Univ, Ont, BA, 48; Univ Mich, AM, 57, EdD(math), 65. Prof Exp: Teacher elem sch, 34-47, asst prin, 48-49; teacher com subj & math & head dept, Forster Collegiate Inst, 49-57; jr lectr math, Univ Mich, 57-61, lectr, Univ Mich, Dearborn, 61-66, from asst prof to assoc prof, Univ Mich, Flint, 66-73, PROF MATH & ASSOC DEAN ACAD AFFAIRS, UNIV MICH-FLINT, 73- Mem: Math Asn Am. Res: History of elementary transcendental functions; history of mathematics; mathematics education. Mailing Add: Dept of Math Univ of Mich Flint MI 48503

COURTNEY, DALE ELLIOTT, b Ashton, Idaho, Mar 3, 18; m 42. GEOGRAPHY. Educ: Western Wash Col Educ, BA, 40; Univ Wash, MA, 50, PhD(geog), 59. Prof Exp: Instr geog, Bowling Green State Univ, 52-56; from instr to assoc prof, 56-67, PROF GEOG, PORTLAND STATE UNIV, 67- Mem: Asn Am Geogr; Am Geog Soc; Nat Coun Geog Educ. Res: Low altitude agriculture, land use, settlement and energy resources; historical geography of western United States. Mailing Add: Dept of Geog Portland State Univ PO Box 751 Portland OR 97207

COURTNEY, GLADYS A, b Erwin, Tenn, June 10, 30; m 67; c 2. MAMMALIAN PHYSIOLOGY, ENDOCRINOLOGY. Educ: La Col, BS, 56; La State Univ, MS, 58; Univ Ill Med Ctr, USPHS fel, 63-64, PhD(physiol), 64. Prof Exp: Gen duty nurse, Baptist Hosp, La, 52-56; asst zool, La State Univ, 56-58; asst physiol, Univ Ill, Urbana, 58-60; asst, Univ Ill Med Ctr, 60-65; asst prof biol, Malone Col, 65-68; assoc prof nursing & physiol, Univ Ill Med Ctr, 69-71, proj dir nurse scientist prog, Col Nursing, 68-76, prof nursing & physiol & head dept gen nursing, 71-76; DEAN SCH NURSING, UNIV MO-COLUMBIA, 76- Res: Relationship of adrenal cortex to ovarian function; prolongation of pseudopregnancy in hamsters; adrenal cortical function during environmental stress. Mailing Add: Sch of Nursing Univ of Mo Columbia MO 65201

COURTNEY, KATHERINE DIANE, b Boston, Mass, Nov 13, 33. PHARMACOLOGY. Educ: Univ Mass, BS, 55; Univ Calif, San Francisco, MS, 61; State Univ NY, PhD(pharmacol), 64. Prof Exp: Scientist, Bionetics Res Labs Inc, 64-68; scientist, Nat Inst Environ Health Sci, 68-71; scientist, Perrine Primate Lab, Fla, 71-73, SCIENTIST, PESTICIDE & TOXIC SUBSTANCES EFFECTS LAB, NAT ENVIRON RES CTR, ENVIRON PROTECTION AGENCY, 73- Res: Perinatal pharmacology; influence of drugs and environmental agagents with regards to congenital defects, drug metabolism and placental transport. Mailing Add: Environ Biol Div EPA Environ Res Ctr Research Triangle Park NC 27711

COURTNEY, KENNETH OLIVER, b St Paul, Minn, Nov 13, 06; m 41; c 3. MEDICAL RESEARCH. Educ: Ore State Col, BS, 29; McGill Univ, MD, 35; London Sch Trop Med, DMT & H, 48; Am Bd Prev Med, dipl, 49. Prof Exp: Technician, Mediter Fruitfly Campaign, USDA, 29-30; instr bact, San Diego State Col, 30-31, assoc prof, 32-33; intern, Gorgas Hosp, Panama CZ, 35-36, resident, 36-37, med officer, outpatient clins & leper colony, 37-39, health officer, Panama City, 39-40, asst chief health officer, 41-51, yellow fever campaign coordr for Armed Forces, Panama Canal & Repub of Panama, 49-50; zone rep, Pan-Am Serv Bur & WHO, Brazil, 51-58; mem staff, Dept Clin Invest, Parke, Davis & Co, 58-67, assoc dir clin therapeut, 67-69; INDEPENDENT CONSULT TROP MED, 69- Mem: AMA; Am Pub Health Asn; Am Soc Trop Med & Hyg; Royal Soc Trop Med & Hyg; Brazilian Soc Hyg & Pub Health. Res: Tropical diseases; malaria; yellow fever; leprosy; gastroenteritis; public health administration. Mailing Add: 12338 Filera Rd San Diego CA 92128

COURTNEY, RICHARD JAMES, b Greenville, Pa, July 2, 41; m 66; c 2. VIROLOGY. Educ: Grove City Col, BS, 63; Syracuse Univ, MS, 66, PhD(microbiol), 68. Prof Exp: NIH fel, 68-70, ASST PROF VIROL, BAYLOR COL MED, 70- Mem: Am Soc Microbiol; Sigma Xi. Res: The biochemical and immunological characterization of the proteins and glycoproteins of herpes simplex viruses and to identify their functional role in virus-infected and transformed cells. Mailing Add: Dept of Virol & Epidemiol Baylor Col of Med Houston TX 77030

COURTNEY, WELBY GILLETTE, b Hamilton, Ohio, Sept 17, 25. PHYSICAL CHEMISTRY. Educ: Oberlin Col, BA, 49; Iowa State Col, PhD(phys chem), 51. Prof Exp: Fel, Inst Atomic Res, Iowa, 51-52; sr chemist, Chem Construct Corp, NY, 52-55; res chemist, Freeport Sulphur Co, La, 55-56; sr scientist, Exp, Inc, 56-62; sect supvr heterogeneous combustion, Thiokol Chem Corp, 62-65, actg mgr physics dept, 65-68; SUPVRY RES CHEMIST, US BUR MINES, 68- Concurrent Pos: Consult phase transformation, Switz, 58. Res: Kinetics and thermodynamics of phase transformation; coupled kinetics-gas dynamics; reduction of respirable dust in coal and noncoal mines. Mailing Add: US Bur of Mines 4800 Forbes Ave Pittsburgh PA 15213

COURTNEY-PRATT, JEOFRY STUART, b Hobart, Australia, Jan 31, 20; m 57. PHYSICS, ENGINEERING. Educ: Univ Tasmania, BE, 42; Cambridge Univ, PhD, 49, ScD, 58. Prof Exp: From asst res officer to res officer, Lubricants & Bearings Div, Coun Sci & Indust Res, Australia, 41-44; res officer ballistic instrumentation, Brit Admiralty, Eng, 44-45; asst dir res, Depts Phys Chem & Physics, Cambridge Univ, 52-57; head, Dept Mech & Optics, 58-69, HEAD APPL PHYSICS DEPT, BELL TEL LABS, 69- Concurrent Pos: Consult, Tube Investments, Ltd, 53-55, Gen Elec Co, 55, Bell Tel Labs, 56-58 & Brit Ministry Supply; lectr, US, 52 & German Phys Soc, 55. Honors & Awards: Stewart Prize, 46; Civic Medal, High Speed Photog Cong, Paris, 54; Boys Prize, 54; Gold Medal, Photog Soc Vienna, 61; Dupont Gold Medal, Soc Motion Picture & TV Eng, 61, Progress Medal, 69; Alan Gordon Mem Award, Soc Photo-Optical Instrument Engrs, 74. Mem: Hon mem Soc Motion Picture & TV Eng; fel Optical Soc Am; fel Royal Photog Soc Gt Brit; assoc Brit Inst Mech Eng; assoc Brit Inst Physics. Res: Applied physics; instrumentation; high speed photography; optics; ballistics; physics of contact of solids friction; adhesion; electrical contacts; optical measurements on satellites; acoustics; electro-optics; recording systems. Mailing Add: Bell Tel Labs Holmdel NJ 07733

COURVILLE, JACQUES, b Montreal, PQ, Mar 17, 35; m 58; c 2. NEUROANATOMY. Educ: Univ Montreal, BA, 54, MD, 60, MSc, 62; Univ Oslo, Dr Med, 68. Prof Exp: Asst prof neurol & neurosurg, McGill Univ, 66-70; ASSOC PROF NEUROANAT, UNIV MONTREAL, 70- Concurrent Pos: Can Med Res Coun scholar, 67; mem res group neurol sci, Univ Montreal, 72. Mem: Am Asn Anatomists; Can Asn Anatomists; Soc Neurosci. Res: Experimental neuroanatomy with silver impregnation methods and injections of labelled amino acids; electron microscopy; red nucleus, facial nucleus, intracerebellar nuclei, cerebellar cortex, thalamic nuclei, inferior olive and pontine nuclei. Mailing Add: Dept of Physiol Univ of Montreal CP 6128 Montreal PQ Can

COURY, ARTHUR JOSEPH, b Coaldale, Pa, Dec 5, 40; m 67; c 2. ORGANIC CHEMISTRY, POLYMER CHEMISTRY. Educ: Univ Del, BS, 62; Univ Minn, PhD(org chem), 65. Prof Exp: SR RES CHEMIST II, GEN MILLS CHEM, INC, 65- Mem: Am Chem Soc. Res: Synthesis of monomers and polymers. Mailing Add: Gen Mills Chem E Hennepin Lab 2010 E Hennepin Ave Minneapolis MN 55413

COUSER, RAYMOND DOWELL, b Oklahoma City, Okla, Mar 11, 31; m 54; c 2. ENTOMOLOGY, INVERTEBRATE ZOOLOGY Educ: Northeastern State Col, BSEd, 56; Univ Okla, MS, 59, PhD(zool), 67. Prof Exp: Instr biol, Baker Univ, 58-61; ASSOC PROF ZOOL, ARK POLYTECH COL, 65- Res: Insect physiology; cold tolerance. Mailing Add: Dept of Biol Ark Polytech Col Russellville AR 72801

COUSINEAU, GILLES H, b Montreal, Que, Sept 19, 32; m 59; c 2. MOLECULAR BIOLOGY, BIOCHEMISTRY. Educ: Col Ste-Marie de Montreal, BA, 54; Univ Montreal, BS, 58; NY Univ, MS, 61; Brown Uni , PhD(biochem), 64. Prof Exp: Res asst biol, Haskins Labs, 58-61; teaching fel, Brown Univ, 61-65; asst prof molecular biol, 65-71, ASSOC PROF BIOL SCI, UNIV MONTREAL, 71- Concurrent Pos: Res asst, Marine Biol Lab, Woods Hole, 58-63, independent investr, 66-; fel biol, Zool Sta, Naples, Italy, 64-65; grants, Nat Res Coun Can, 65-69, Montreal, 65-69, Damon Runyon Mem Fund, 66-68, Childs Mem Fund, 66-67, Donner Can Found, 66-67, Soc Sigma Xi, 66-67 & Defense Res Bd Can, 67-69. Mem: Electron Micros Soc Am; Can Soc Cell Biol; Am Soc Cell Biol; NY Acad Sci; Int Soc Develop Biol. Res: Synthesis of macromolecules in developing sea urchin eggs; cell division and differentiation. Mailing Add: Lab of Molecular Biol Univ of Montreal Montreal PQ Can

COUSINEAU, LEO, b Widewater, Alta, Mar 26, 32; m 55; c 2. HEMATOLOGY, COMPUTER SCIENCE. Educ: Univ Montreal, BA, 52, MD, 58. Prof Exp: Intern med, Univ Montreal, 57-58; resident & chief resident, Grade Hosp, Detroit, 58-61; jr clin instr radioisotope hemat, Univ Mich & Vet Admin, 62-63, instr med & res assoc hemat, Univ, 63-64; mem staff hemat, Sacre-Coeur Hosp, Montreal, 64-65 & Verdun Gen Hosp, 65-66; lectr path & clin hemat, Univ Montreal, 66-68; asst prof, 68-69, ASSOC PROF MED, FAC MED, UNIV SHERBROOKE, 69-, CHIEF DEPT MED BIOL, 73- Concurrent Pos: Consult, Sherbrooke Gen Hosp, D'Youville Hosp, Sherbrooke, St-Vincent de Paul Hosp, Sherbrooke & Prov Hosp, Magog, 68-; mem coun med adv, Am Red Cross. Mem: Can Soc Hemat (vpres, 69); Can Fedn Res; Can Nat Defense Med Asn; Am Soc Hemat; Am Fedn Clin Res. Res: Neoplastic diseases and cryobiology; medical computer sciences and biomedical engineering. Mailing Add: Fac of Med Univ of Sherbrooke Sherbrooke PQ Can

COUSINS, ROBERT JOHN, b New York, NY, Apr 5, 41; m 69; c 2. NUTRITIONAL BIOCHEMISTRY, NUTRITION. Educ: Univ Vt, BA, 63; Univ Conn, MS, 65, PhD(exp nutrit), 68. Prof Exp: Res assoc biochem, Univ Wis, 68-69, NIH fel, 69-70; asst prof, 71-74, ASSOC PROF NUTRIT, RUTGERS UNIV, 74- Mem: Am Chem Soc; Am Inst Nutrit. Res: Function of metals in mammalian systems, emphasizing nutrition and disease; regulatory aspects of zinc, copper and cadmium metabolism; control of metalloprotein biosynthesis; mineral nutrition, deficiency diseases and toxicology. Mailing Add: Dept of Nutrit Cook Col Rutgers Univ PO Box 231 New Brunswick NJ 08903

COUSMINER, HAROLD L, b New York, NY, Mar 9, 25; m 47; c 3. MICROPALEONTOLOGY, GEOLOGY. Educ: NY Univ, BA, 49, MS, 56, PhD(micropaleont, geol), 64. Prof Exp: Paleontologist, Gulf Oil Corp, NY, 58-61, dir geol lab, Western Hemisphere Extraterritorial Div, Fla, 61-64, res paleontologist, Gulf Res & Develop Co, Pa, 64-66; res assoc, Dept Microbiol, Am Mus Natural Hist, 66-69; asst prof geol, Newark Col Arts & Sci, Rutgers Univ, 69-74; MEM STAFF, AM MUS NATURAL HIST, 74- Mem: AAAS; Am Paleont Soc; Am Asn Petrol Geologists; Geol Soc Am. Res: Biostratigraphic and paleoecologic applications of microfossil groups in local and regional stratigraphic syntheses. Mailing Add: Am Mus of Natural Hist Central Park W at 79th St New York NY 10024

COUTANT, CHARLES COE, aquatic ecology, see 12th edition

COUTINHO, CLAUDE BERNARD, b Bombay, India, Aug 19, 31; m 57; c 5. BIOCHEMISTRY, EXPERIMENTAL MEDICINE. Educ: Univ Bombay, BS, 52; Univ Bristol, dipl, 53; Inst Divi Thomae, PhD(exp med, biochem), 63. Prof Exp: Res technologist, Inst Laryngol & Otol, London, 55-57; asst chief chemother & microbiol, Aspro-Nicholas Pharmaceut Co, Ltd, Eng, 57-58, dept chief, 58-60; res scientist, Warner-Lambert Res Inst, 63-65; sr res assoc, Arthur D Little, Inc, 65-67; RES GROUP CHIEF, HOFFMANN-LA ROCHE, INC, 67- Mem: AAAS; Am Soc Microbiol; Am Soc Pharmacol & Exp Therapeut; Am Soc Toxicol; NY Acad Sci. Res: Mechanism of pharmacological and toxicological drug action as related to biochemical changes and drug metabolism; pharmacokinetics; pharmacodynamics; tissue immunity; cancer; cellular metabolism; pharmaceuticals and general chemotherapy. Mailing Add: Hoffmann-La Roche Inc Nutley NJ 07110

COUTTS, JOHN WALLACE, b Neepawa, Man, Feb 2, 23; nat US; m 59; c 1. PHYSICAL CHEMISTRY. Educ: Univ Man, BSc, 45, MSc, 47; Purdue Univ, PhD(chem), 50. Prof Exp: Asst, Purdue Univ, 46-48; asst prof chem, Mt Union Col, 50-55; assoc prof, 55-62, PROF CHEM, LAKE FORES COL, 62-, CHMN DEPT, 61- Concurrent Pos: NSF fac fel, Univ Calif, Berkeley, 67-68, vis prof, 68; vis prof, Rensselaer Polytech Inst, 74-75. Mem: AAAS; Am Chem Soc; Sigma Xi. Res: Molecular structure; thermodynamics of non-aqueous solutions. Mailing Add: 106 E Sheridan Rd Lake Bluff IL 60044

COUTTS, RONALD THOMSON, b Glasgow, Scotland, June 19, 31; m 57; c 3. ORGANIC CHEMISTRY, MEDICINAL CHEMISTRY. Educ: Univ Glasgow, BSc, 55, PhD(pharmaceut chem), 60. Prof Exp: Asst lectr pharmaceut chem, Royal Col Sci & Technol, Univ Glasgow, 56-59; lectr, Sunderland Tech Col, Eng, 59-63; from asst prof to assoc prof, Univ Sask, 63-66; assoc prof, 66-69, PROF PHARMACEUT CHEM, UNIV ALTA, 69- Concurrent Pos: Asst sci ed, Can J Pharmaceut Sci, Can Pharmaceut Asn, 66-69, sci ed, 69-72; fel, Chem Inst Can, 71; vis res prof, Dept Pharm, Chelsea Col, Univ London, Eng, 72-73; fel, Royal Inst Chem, 75. Mem: Chem Inst Can; Brit Pharmaceut Soc; fel Royal Inst Chem; The Chem Soc. Res: Synthesis, pharmacology, toxicology and metabolism of physiologically-active N-hydroxy compounds and amines; mass spectrometry of medicinal compounds and metabolites. Mailing Add: Fac of Pharm Univ of Alta Edmonton AB Can

COUTURE, ROGER, b Hull, PQ, May 7, 30; m 58; c 6. INTERNAL MEDICINE, NEPHROLOGY. Educ: Univ Ottawa, BA & BSc, 52; McGill Univ, MD, 57; FRCP(C), 64. Prof Exp: Lectr, 65-70, ASST PROF MED, FAC MED, UNIV OTTAWA, 70- Mem: Can Med Asn; Can Soc Nephrology. Mailing Add: Dept of Med Ottawa Gen Hosp Ottawa ON Can

COUVILLION, JOHN LEE, b Jackson, Miss, Oct 20, 41; m 63; c 2. ORGANIC CHEMISTRY. Educ: La State Univ, New Orleans, BS, 63, Baton Rouge, MS, 65, PhD(chem), 67. Prof Exp: Teaching asst org chem, La State Univ, Baton Rouge, 63-65; res chemist, Tech Ctr, Celanese Chem Co, 67-72; res chemist, Food & Drug Admin, Washington, DC, 72-74; ASSOC PROF CHEM, LA STATE UNIV, EUNICE, 74- Mem: Am Chem Soc. Res: Organic chemical synthesis; heterogeneous catalysis; toxicology; biochemical toxicology. Mailing Add: 110 S Tanglewood Dr Eunice LA 70535

COVALT-DUNNING, DOROTHY, b Washington, DC, Jan 1, 37; m 60. ANIMAL PHYSIOLOGY, ANIMAL BEHAVIOR. Educ: Mt Holyoke Col, MA, 60; Tufts Univ, PhD(biol), 66. Prof Exp: Asst zool, Yale Univ, 60-61; asst biol, Harvard Univ, 61-62; fel, Max-Planck Inst Behav Physiol, 67; temporary instr zool, Duke Univ, 68-69; asst prof, 69-74, ASSOC PROF BIOL, WVA UNIV, 74- Mem: AAAS; Am Soc Zool; Am Soc Mammal. Res: Sensory physiology and behavior of bats; prey-predator interactions between bats and moths; hibernation and social behavior in bats; auditory neurophysiology and behavior in insects. Mailing Add: Dept of Biol WVa Univ Morgantown WV 26506

COVE, JOHN JAMES, b Beverly, Mass, Sept 19, 41; Can citizen; m 70; c 1. ANTHROPOLOGY, SOCIOLOGY. Educ: Dalhousie Univ, BA, 67, MA, 68; UnivBC, PhD(anthrop), 71. Prof Exp: ASST PROF SOCIOL & ANTHROP, CARLETON UNIV, 71- Concurrent Pos: Lectr, Mt St Vincent Univ, 67-68, Univ BC, 70 & Univ Victoria, BC, 72; consult epidemiol & community med, Fac Med, Univ Ottawa, 72-73. Mem: Can Sociol & Anthrop Asn; Am Anthrop Asn. Res: Symbolic systems, especially myth and taboo; fishing technology and labor organization. Mailing Add: Dept of Sociol & Anthrop Carleton Univ Ottawa ON Can

COVEL, MITCHEL DALE, b Oakland, Calif, July 10, 17; m 71. MEDICINE. Educ: Univ Calif, AB, 39, MD, 42. Prof Exp: Assoc clin prof med, Loma Linda Univ, 51-60; asst clin prof, Ctr Health Sci, 60-65, assoc clin prof, 65-73, dep dir regional med progs, 67-73, CLIN PROF MED, MED CTR, UNIV CALIF, LOS ANGELES, 65-, DEP DIR REGIONAL MED PROGS, INST CHRONIC DIS & REHAB, 68- Concurrent Pos: Chief outpatient serv, Los Angeles County Hosp, 49-52; mem staff, Hosp Good Samaritan, 51-; fel coun clin cardiol, Am Heart Asn. Mem: Am Heart Asn; fel Am Col Physicians; fel Am Col Cardiol; fel NY Acad Sci; Am Heart Asn. Mailing Add: 9730 Wilshire Blvd Beverly Hills CA 90212

COVELL, CHARLES VAN ORDEN, JR, b Washington, DC, Dec 10, 35; m 58; c 1. ENTOMOLOGY. Educ: Univ NC, BA, 58; Va Polytech Inst, MS, 62, PhD(entom), 65. Prof Exp: Teacher, Norfolk Acad Sch Boys, 58-60; asst entom, Va Polytech Inst, 60-64; from instr to assoc prof, 64-74, PROF BIOL, UNIV LOUISVILLE, 74- Concurrent Pos: Consult, Health Dept, Louisville-Jefferson County, Ky, 65- Mem: Entom Soc Am; Am Entom Soc; Lepidop Soc; Am Mosquito Control Asn; Soc Syst Zool. Res: Taxonomy, distribution, ecology and life history of lepidopterous insects, especially the moth family Geometridae; faunistics of Kentucky insects, especially lepidoptera and mosquitoes. Mailing Add: Dept of Biol Univ of Louisville Louisville KY 40208

COVELL, JAMES WACHOB, b San Francisco, Calif, Aug 13, 36; m 63; c 2. CARDIOVASCULAR PHYSIOLOGY, BIOMEDICAL ENGINEERING. Educ: Carleton Col, BA, 58; Univ Chicago, MD, 62. Prof Exp: Intern surg, Univ Chicago Hosps & Clin, 62-63, resident, 63-64; res assoc cardiol, Nat Heart & Lung Inst, 64-66, sr investr, 66-68; asst prof, 68-70, ASSOC PROF MED, SCH MED, UNIV CALIF, SAN DIEGO, 70- Concurrent Pos: USPHS fel, 63-64, career develop award, 70-75. Mem: Am Fedn Clin Res; Am Heart Asn; Am Physiol Soc; Cardiac Muscle Soc. Res: Factors influencing cardiac performance in the normal and diseased heart; mechanics of muscle contraction; on-line data analysis and control of hemodynamic parameters. Mailing Add: Dept of Med M-013 Univ of Calif San Diego La Jolla CA 92093

COVENEY, RAYMOND MARTIN, JR, b Marlboro, Mass, Oct 15, 42; m 65; c 3. GEOLOGY. Educ: Tufts Univ, BS, 64; Univ Mich, MS, 68, PhD(geol), 71. Prof Exp: Mine geologist, Dickey Explor Co, 69-71; ASST PROF GEOL, UNIV MO, KANSAS CITY, 71- Mem: Geol Soc Am; Mineral Soc Am; Sigma Xi. Res: Geology and mineralogy of ore deposits; fluid inclusion research; determinative mineralogy with emphasis on x-ray and optical techniques. Mailing Add: Dept of Geosci Univ of Mo Kansas City MO 64110

COVENTRY, MARK BINGHAM, b Duluth, Minn, Mar 30, 13; m 37; c 3. ORTHOPEDIC SURGERY. Educ: Univ Mich, MD, 37; Univ Minn, MS, 42. Prof Exp: PROF ORTHOP SURG, MAYO MED SCH, 58-; CONSULT, SECT ORTHOP SURG, MAYO CLIN, 42- Concurrent Pos: Head dept orthop, Mayo Clin, 68-74. Mem: Am Acad Orthop Surg; Clin Orthop Soc; Am Orthop Asn; Am Col Surg; hon fel Brit Orthop Asn. Mailing Add: Mayo Clin 200 First St SW Rochester MN 55901

COVER, HERBERT LEE, b Elkton, Va, Dec 28, 21; m 45; c 1. PHYSICAL CHEMISTRY. Educ: Univ Va, BS, 45, MS, 46, PhD(chem), 49. Prof Exp: From instr to prof qual & quant anal, 49-74, PROF CHEM, MARY WASHINGTON COL, 74- Res: Catalytic oxidation of carbon monoxide; catalytic hydration of ethylene; analytical chemistry and electronics. Mailing Add: 1607 Franklin St Fredericksburg VA 22401

COVER, MORRIS SEIFERT, b Harrisburg, Pa, July 25, 16; m 38; c 2. VETERINARY PATHOLOGY. Educ: Univ Pa, VMD, 38; Kans State Col, MS, 43; Univ Ill, PhD, 52. Prof Exp: Asst poultry pathologist, Univ NH, 38-40; assoc prof anat & physiol, Kans State Col, 40-46; asst prof anat & histol, Univ Ill, 47-52; assoc prof poultry path, Univ Del, 52-59, prof & head dept, 59-67; dir agr exp sta, 62-67; mgr vet serv & vet labs, 67-73, DIR VET SERV & REGULATORY DEPT, RALSTON PURINA CO, 73- Concurrent Pos: Inspector, Sheffield Farms, Pa, 46. Mem: Am Vet Med Asn. Res: Pathology; anatomy; histology; embryology; cytology; veterinary surgery; mycotoxicosis. Mailing Add: Vet Serv & Regulatory Dept Ralston Purina Co Checkerboard Sq St Louis MO 63188

COVER, RICHARD EDWARD, b Youngsville, Pa, Nov 7, 26; m 48; c 3. ANALYTICAL CHEMISTRY. Educ: Pa State Univ, BS, 47; Polytech Inst Brooklyn, MS, 60, PhD(anal chem), 62. Prof Exp: Chemist, US Steel Corp, 48-50 & Fairchild Camera & Instrument Co, 50-57; sr chemist, Socony Mobil Oil Co, 57-64; PROF ANAL CHEM, ST JOHN'S UNIV, NY, 64- Concurrent Pos: Mem, Simulation Coun, 65- Mem: Am Chem Soc; Soc Indust & Appl Math. Res: Electroanalytical chemistry; gas chromatography; chemical kinetics; computer applications to chemistry; applied mathematics. Mailing Add: Dept of Chem St John's Univ Grad Sch Jamaica NY 11439

COVERT, SCOTT VEASEY, b Camden, NJ, Apr 30, 08; m 38; c 1. MEDICAL MICROBIOLOGY. Educ: Ursinus Col, BS, 32; Univ Pa, MA, 47, PhD, 49; Am Bd Med Microbiol, dipl. Prof Exp: Instr microbiol, Univ Pa, 47-49; from assoc prof to prof, 49-73, chmn dept, 56-73, EMER PROF MICROBIOL, ALBANY MED COL, 73- Mem: Fel Am Acad Microbiol; NY Acad Sci; Royal Soc Health. Res: Bacterial toxins. Mailing Add: Dept of Microbiol Albany Med Col Albany NY 12208

COVEY, RONALD PERRIN, JR, b Jamestown, NY, Aug 19, 29; m 52; c 3. PLANT PATHOLOGY. Educ: Univ Minn, BS, 56, MS, 59, PhD(plant path), 62. Prof Exp: Asst plant pathologist, 62-73, ASSOC PLANT PATHOLOGIST, TREE FRUIT RES CTR, UNIV WASH, 73- Mem: Am Phytopath Soc. Res: Bacterial and fungal disease of tree fruits; phytophthora cactorum. Mailing Add: Tree Fruit Res Ctr Univ of Wash 1100 N Western Ave Wenatchee WA 98801

COVEY, RUPERT ALDEN, b Manchester, NH, July 24, 29; m 56; c 3. ORGANIC CHEMISTRY. Educ: Middlebury Col, AB, 51; Univ NH, MS, 53; Univ Mich, PhD(pharmaceut chem), 58. Prof Exp: Sr res chemist agr chem, Naugatuck Chem Div, US Rubber Co, 57-67; sr res chemist, 67-70, RES SCIENTIST, UNIROYAL CHEM DIV, UNIROYAL, INC, 70- Mem: Am Chem Soc. Res: Agricultural chemicals; synthesis of organic compounds as pesticides; seven-membered rings; bicyclic compounds; organic sulfite esters. Mailing Add: Uniroyal Chem Div Uniroyal Inc Naugatuck CT 06770

COVEY, WINTON GUY, JR, b Glen Daniel, WVa, Feb 3, 29; m 52; c 2. MICROMETEOROLOGY. Educ: Johns Hopkins Univ, AB, 49; Tex A&M Univ, MS, 59, PhD(soil physics), 65. Prof Exp: Micrometeorologist, Tex A&M Univ, 55-59; soil physicist, Tex Agr Exp Sta, 59-60; res soil scientist, Agr Res Serv, USDA, 60-65; assoc prof meteorol, Cornell Univ, 65-68; assoc prof, 68-74, PROF NATURAL SCI, CONCORD COL, 74- Mailing Add: Dept of Phys Sic Concord Col Athens WV 24712

COVINO, BENJAMIN GENE, b Lawrence, Mass, Sept 12, 30; m 53; c 2. PHYSIOLOGY. Educ: Col of the Holy Cross, AB, 51; Boston Col, MS, 53; Boston Univ, PhD(physiol), 55; State Univ NY Buffalo, MD, 61. Prof Exp: Res scientist, Arctic Aeromed Lab, 55-57; asst prof pharmacol, Sch Med, Tufts Univ, 57-59; asst prof physiol, Sch Med, State Univ NY Buffalo, 59-63; MED DIR, ASTRA PHARMACEUT PROD, INC, 63-, VPRES RES & DEVELOP, 67-; PROF ANESTHESIOL, MED SCH, UNIV MASS, WORCESTER, 75- Concurrent Pos: Consult physiologist & clin pharmacologist, St Vincent's Hosp, Worcester. Mem: Am Physiol Soc; Am Fedn Clin Res; Int Col Angiol; Am Col Clin Pharmacol & Chemother. Res: Cardiovascular physiology and pharmacology of local anesthetics; physiology of temperature regulation. Mailing Add: Astra Pharmaceut Prod Inc PO Box 1089 Framingham MA 01701

COWAN, ARCHIBALD B, b Ambia, Ind, July 14, 15; m 42; c 3. WILDLIFE DISEASES, WILDLIFE MANAGEMENT. Educ: Univ Mich, BSF, 40, PhD(wildlife mgt), 54. Prof Exp: Asst leader pheasant res proj, State Natural Hist Surv, Ill, 46-49; wildlife res biologist, Patuxent Res Refuge, 54-56; asst prof, 56-62, ASSOC PROF WILDLIFE MGT, UNIV MICH, ANN ARBOR, 62- Mem: Wildlife Soc; Am Soc Parasitol; Wildlife Dis Asn (treas, 58-63, vpres, 65-67, pres, 67-69); Am Inst Biol Sci. Res: Parasites and diseases of wildlife. Mailing Add: Sch of Natural Resources Univ of Mich Ann Arbor MI 48104

COWAN, C MICHAEL, comparative physiology, environmental physiology, see 12th edition

COWAN, CLYDE LORRAIN, JR, nuclear physics, deceased

COWAN, DANIEL FRANCIS, b Mineola, NY, Aug 7, 34; m 60; c 3. PATHOLOGY. Educ: Antioch Col, BA, 56; McGill Univ, MDCM, 60. Prof Exp: NIH spec fel comp path, 66-67; from asst prof to assoc prof path, Mich State Univ, 67-73; CHMN DEPT PATH, EASTERN VA MED SCH, 73- Mem: NY Acad Sci; AMA; Am Fedn Clin Res; Wildlife Dis Asn. Res: Research in anatomic pathology; comparative pathology; stress diseases; wildlife diseases. Mailing Add: Dept of Path Eastern Va Med Sch Norfolk VA 23501

COWAN, DAVID J, b San Antonio, Tex, Aug 5, 36; m 65. PHYSICS. Educ: Univ Tex, BS, 58, MA, 60, PhD(physics), 65. Prof Exp: Assoc physicist, Int Bus Mach Corp, 60-61; NIH traineeship, 64-65; asst prof, 65-74, ASSOC PROF PHYSICS & CHMN DEPT, GETTYSBURG COL, 74- Mem: Am Phys Soc; Am Asn Physics Teachers. Res: Investigation of biological ultrastructure by x-ray diffraction. Mailing Add: Dept of Physics Gettysburg Col Gettysburg PA 17325

COWAN, DAVID LAWRENCE, b Havre, Mont, Oct 18, 36; m 60. SOLID STATE PHYSICS. Educ: Univ Wis, BS, 56, MS, 58, PhD(physics), 64. Prof Exp: Res engr, NAm Aviation, Inc, Calif, 58-60; res assoc physics, Cornell Univ, 64-67; mem tech staff, Sandia Corp, NMex, 67-68; asst prof, 68-74, ASSOC PROF PHYSICS, UNIV MO-COLUMBIA, 74- Mem: Am Phys Soc. Res: Ferromagnetic proper- ties of metals; electron spin resonance; nuclear magnetic resonance; optical properties of solids. Mailing Add: Dept of Physics Univ of Mo Columbia MO 65201

COWAN, DONALD, b Ft Worth, Tex, May 26, 14; m 39; c 1. PHYSICS. Educ: Tex Christian Univ, AB, 47; Vanderbilt Univ, PhD(physics), 51. Prof Exp: Asst chief engr, Am Type Founders, 43-45; from instr to asst prof physics, Vanderbilt Univ, 48-53; lab dir, Convair, 53-55, mgr atomic indust div, 55-56; assoc prof physics, Tex Christian Univ, 56-59; chmn dept, 59-62, PROF PHYSICS, UNIV DALLAS, 59-, PRES, 62- Res: Fluid dynamics; history of physics. Mailing Add: Univ of Dallas Irving TX 75061

COWAN, DONALD WILLIAM, b Rochester, Minn, Mar 26, 07; m 29; c 1. MEDICINE. Educ: Univ Minn, BS, MB, MS & MD, 31; Am Bd Prev Med, dipl. Prof Exp: Lab technician, Mayo Clin, 23-27; asst physiol, Univ Minn, 27-31; assoc, Univ Iowa, 31-36; med intern, USPHS, Ill, 36-37; prof pub health & dir univ health serv, 37-72, EMER PROF PUB HEALTH & EMER DIR UNIV HEALTH SERV, UNIV MINN, MINNEAPOLIS, 72- Mem: AMA; Am Pub Health Asn; Am Col Health Asn. Res: Creatin metabolism of heart muscle; cold prevention and treatment; drugs used in treatment of allergy; air pollution and allergy. Mailing Add: Univ Health Serv Univ of Minn Minneapolis MN 55455

COWAN, DWAINE O, b Fresno, Calif, Nov 25, 35; m 63. ORGANIC CHEMISTRY. Educ: Fresno State Col, BS, 58; Stanford Univ, PhD(org chem), 62. Prof Exp: Res fel chem, Calif Inst Technol, 62-63; from asst prof to assoc prof, 63-72, PROF CHEM, JOHNS HOPKINS UNIV, 72- Concurrent Pos: Sloan res fel, 68-70; Guggenheim fel, Phys Chem Inst, Univ Basel, 70-71. Mem: Am Chem Soc; Brit Chem Soc. Res: Physical organic chemistry; mixed valence organometallic compounds, electron transfer reactions, organic solid state chemistry, synthesis and study of organic metals, organic photochemistry. Mailing Add: Dept of Chem Johns Hopkins Univ Baltimore MD 21218

COWAN, EUGENE WOODVILLE, b Ree Heights, SDak, Sept 30, 20; m 56; c 2. PHYSICS. Educ: Univ Mo, BS, 41; Mass Inst Technol, SM, 43; Calif Inst Technol, PhD(physics), 48. Prof Exp: Instr elec commun, Mass Inst Technol, 43-44, staff mem, Radiation Lab, 44-45; res fel, 48-50, from asst prof to assoc prof, 50-61, PROF PHYSICS, CALIF INST TECHNOL, 61- Mem: Am Phys Soc. Res: Radar; electrical circuits; electrical measurements; cosmic rays. Mailing Add: Dept of Physics Calif Inst of Technol Pasadena CA 91109

COWAN, F BRIAN M, b Chatham, Ont, Apr 15, 38; m 62; c 2. ZOOLOGY, PHYSIOLOGY. Educ: Queen's Univ, BA, 62; Univ Toronto, MA & dipl electron micros, 65, PhD(zool), 70. Prof Exp: Instr zool, 69-70, ASST PROF BIOL, UNIV NB, 70- Mem: AAAS; Am Soc Zoologists; Int Soc Stereology; Can Soc Zoologists. Res: Activation of salt secreting epithelia during times of osmotic stress using euryhaline reptilian species as test system and studying ultrastrucure, stereology, histochemistry and biochemistry; electrical events in cardiac cycle in animals with apparent physiological alterations in plasma ion concentrations; effect of pesticides on blood cell structure. Mailing Add: Dept of Biol Univ of NB Fredericton NB Can

COWAN, FREDERICK FLETCHER, JR, b Washington, DC, Jan 17, 33; m 60; c 1. PHARMACOLOGY. Educ: George Washington Univ, BS, 55; Georgetown Univ, PhD(pharmacol), 59. Prof Exp: Pharmacist, Bretler Pharm, 55-56; instr pharmacol, Schs Med & Dent, Georgetown Univ, 60-62, asst prof, 62-66; assoc prof, 66-73, PROF PHARMACOL & CHMN DEPT, DENT SCH, UNIV ORE, 73- Concurrent Pos: Fel, Nat Heart Inst, 59; prin co-investr, USPHS res grants, 61- & prin investr, 64-; prin investr, NSF res grant, 62. Mem: AAAS; Am Pharmaceut Asn. Res: Pharmacology of the autonomic ganglia and the peripheral sympathetic nervous system; tachyphylaxis of sympathomimetic amines; pharmacology of carotid body chemoreceptors. Mailing Add: 7027 SW Eighth Ave Portland OR 97219

COWAN, FREDERICK PIERCE, b Bar Harbor, Maine, July 3, 06; m 34. HEALTH PHYSICS. Educ: Bowdoin Col, AB, 28; Harvard Univ, AM, 31, PhD(physics), 35. Prof Exp: Instr physics, Bowdoin Col, 28-29; asst, Harvard Univ, 29-34, instr, 34-35; from instr to asst prof, Rensselaer Polytech Inst, 35-43; res assoc, Radio Res Lab, Harvard Univ, 43-45; res assoc, Eng Div, Chrysler Corp, Detroit, 45-47; head health physics div, Brookhaven Nat Lab, 47-71; MEM ATOMIC SAFETY & LICENSING BD, NUCLEAR REGULATORY COMN, 72- Concurrent Pos: Asst, Radcliffe Col, 34-35; chmn, Am Bd Health Physics, 63-64; mem comt 3, Int Comn Radiation Protection, 65-73; mem, Int Comn Radiation Units & Measurements, 65-73; mem, Nat Coun Radiation Protection, 67-73; bd dirs, 69-73; mem adv comt radiobiol aspects of supersonic transport, Fed Aviation Admin, 67-74. Mem: AAAS; Am Phys Soc; Radiation Res Soc; Health Physics Soc (pres elect, 56-57, pres, 57-58). Res: Thermal measurements on electron tubes; electronic temperature control; radiation dosimetry. Mailing Add: 1800 SE St Lucie Blvd Apt 2-104 Stuart FL 33494

COWAN, GARRY IAN MCTAGGART, b Victoria, BC, July 9, 40; m 62; c 2. SYSTEMATICS. Educ: Univ BC, BSc, 63, MSc, 66, PhD(zool), 68. Prof Exp: Asst prof, 68-71, ASSOC PROF BIOL, MEM UNIV NFLD, 71- Concurrent Pos: Nat Res Coun Can res grant in aid, 68-; res scientist, Marine Sci Res Lab, 70- Mem: Am Soc Ichthyol & Herpet; Soc Study Evolution; Soc Syst Zool. Res: Systematics of marine organisms based on their morphological and biochemical characteristics. Mailing Add: Marine Sci Res Lab Mem Univ of Nfld St John's NF Can

COWAN, GEORGE A, b Worcester, Mass, Feb 15, 20; m 46. RADIOCHEMISTRY. Educ: Worcester Polytech Inst, BS, 41; Carnegie Inst Technol, DSc(chem), 50. Prof Exp: Res scientist, Metall Lab, Univ Chicago, 42-45; res scientist, 45-46, assoc div leader, Test Div, 56-70, MEM STAFF RADIOCHEM, LOS ALAMOS SCI LAB, UNIV CALIF, 49-, GROUP LEADER, 55-, DIV LEADER CHEM, 71- Honors & Awards: E O Lawrence Award, 65. Mem: Fel AAAS; Am Chem Soc; fel Am Phys Soc; Am Nuclear Soc. Res: Radiochemical diagnostics; nuclear reactions. Mailing Add: PO Box 1663 Los Alamos NM 87544

COWAN, GEORGE ROBERT, chemistry, see 12th edition

COWAN, IAN MCTAGGART, b Edinburgh, Scotland, June 25, 10; m 36; c 2. MAMMALOGY, WILDLIFE ECOLOGY. Educ: Univ BC, BA, 32; Univ Calif, PhD(vert zool), 35. Hon Degrees: LLD, Univ Alta, 71. Prof Exp: Insect pest investr, Dom Govt, Can, 29; field asst, Nat Mus Can, 30-31; asst biologist, BC Prov Mus, 35-38, asst dir, 38-40; from asst prof to prof zool, 40-75, head dept, 53-64, dean fac grad studies, 64-75, EMER PROF ZOOL, UNIV BC, 75- Concurrent Pos: Carnegie traveling fel, Am Mus & US Nat Mus, 37; Nuffield fel, 52; Am Inst Biol Sci foreign vis lectr, 63; Erskine fel, Univ Canterbury, 69; mem, Environ Protection Bd, 69-75; chmn, Acad Bd BC, 69-75; mem, Can Environ Adv Coun, 72-75, chmn, 75-78; with Dom Parks Bur & Northwest Territories Order of Can, 72; Leopold Medal & Arthur Einarsen Award, Wildlife Soc, 70. Mem: Am Ornith Union; Wildlife Soc (pres, 49); Am Soc Mammalogists; Soc Syst Zool; fel Royal Soc Can. Res: Ecology and environmental physiology of native ungulates; problems of insularity in vertebrates; vertebrate distribution and speciation in Canada. Mailing Add: 2088 Acadia Rd Vancouver BC Can

COWAN, JACK DAVID, b Leeds, Eng, Aug 24, 33; m 58; c 2. BIOPHYSICS, THEORETICAL BIOLOGY. Educ: Univ Edinburgh, BSc, 55; Mass Inst Technol, SM, 60; Imp Col, dipl elec eng, 59, Univ London, PhD(elec eng), 67. Prof Exp: Res engr & mathematician, Instrument & Fire Control Sect, Ferranti Ltd, Scotland, 55-58; staff mem neurophysiol group, Res Lab Electronics, Mass Inst Technol, 60-62; acad vis math biol, Imp Col, 62-67; chmn dept & prof theoret biol, 67-73, PROF BIOPHYS & THEORET BIOL, UNIV CHICAGO, 73- Concurrent Pos: Res assoc & consult, Northeastern Univ & US Air Force Cambridge Res Lab, 60; vis prof, Univ Naples, 65; guest worker, Nat Phys Lab, Eng, 66-67; mem, NIH comn training grants in epidemiol & biomet, 69-73; chmn adv coun, Lab Cybernet, Nat Res Coun, Italy, 69-73; mem vis comt, Dept Psychol, Mass Inst Technol, 72-75; mem, Am Math Soc comt math in life sci, 70-; mem comn on communication & control, Int Union Pure & Appl Biophys, 75-; mem ed bd, J Theoret Biol, Acta Biotheoret, Biol Cybernet, Lect Notes Biomath, Brain Theory Newsletter & Soc Indust & Appl Math J Appl Math. Mem: AAAS; Am Math Soc; Biophys Soc; Soc Am Neurosci; Brain Res Asn UK. Res: Theory of nervous activity; theoretical biology; vision research; theoretical immunology. Mailing Add: Dept Biophys & Theoret Biol Univ of Chicago 920 E 58th St Chicago IL 60637

COWAN, JAMES W, b Beaver Falls, Pa, Aug 23, 30; m 60. NUTRITION, BIOCHEMISTRY. Educ: Pa State Univ, BS, 55, MS, 59, PhD(biochem), 61. Prof Exp: Assoc prof food tech & nutrit, 61-73, assoc dean sch agr sci, 70-73, PROF NUTRIT & DEAN SCH AGR SCI, AM UNIV BEIRUT, 73- Mem: AAAS; Am Nutrit Soc; Am Soc Clin Nutrit. Res: Iodine metabolism; iron utilization. Mailing Add: Dept of Nutrit American Univ of Beirut Beirut Lebanon

COWAN, JOHN ARTHUR, b Winnipeg, Man, July 8, 21; m 46; c 4. FLUID PHYSICS. Educ: Univ Man, BSc, 43; Univ Toronto, MA, 47, PhD(physics), 50. Prof Exp: Lectr, Dept Math, Univ Man, 46; sci officer, Can Defense Res Bd, 50-57; chmn dept, 57-68, PROF PHYSICS, UNIV WATERLOO, 57- Mem: Can Acoustical Asn; Can Asn Physicists. Res: Speech communication; experimental studies of transport properties of liquids. Mailing Add: Dept of Physics Univ of Waterloo Waterloo ON Can

COWAN, JOHN C, b Danville, Ill, Oct 25, 11; m 38; c 2. ORGANIC CHEMISTRY. Educ: Univ Ill, AB, 34, PhD(chem), 38. Prof Exp: Asst, Univ Ill, 35-38, du Pont fel, 38-39; instr, DePauw Univ, 39-40; assoc chemist, Northern Regional Res Lab, Agr Res Serv, USDA, 40-41, from chemist to head oil & protein sect, Agr Res Serv, 53-57, chief oilseed crops lab, 57-73; CONSULT, 73- Concurrent Pos: Mem, Soybean Res Coun, 45-; adj prof chem, Bradley Univ, 73- Honors & Awards: Superior Serv Award, USDA, 48, Superior Serv Team Awards, 52, 63; A E Bailey Medal, Am Oil Chem Soc, 61; Meritorious Serv Award, Am Soybean Asn, 70; Cherreal Medal, French Asn for Study of Oil Substances, 75. Mem: Am Chem Soc; Am Oil Chem Soc (pres, 68-69); Inst Food Technol; Am Asn Cereal Chemists. Res: Condensation and vinyl polymers; edible oil spreads; flavor stability of soybean oil; metal-inactivation agents; cyclic fatty acids; aldehydic esters; amino acid derivatives; uses for soybeans and linseed oil. Mailing Add: 225 Olin Hall Bradley Univ Peoria IL 61625

COWAN, JOHN RITCHIE, b Leamington, Ont, Feb 3, 16; nat US; m 47; c 2. PLANT BREEDING. Educ: Univ Toronto, BSA, 39; Univ Minn, MS, 42, PhD, 52. Prof Exp: Asst corn & soybean breeding, Dom Exp Farm, Harrow, Ont, 37-42, in charge cereal & forage res, NS, 42-45; in charge grass breeding, Eastern Can Cent Exp Farm, Ont, 45-47; asst agronomist & asst prof forage breeding, Macdonald Col, McGill Univ, 47-48; tall fescue & alfalfa breeding, 48-52, assoc agronomist & assoc prof, 52-55, AGRONOMIST & PROF FARM CROPS, ORE STATE UNIV, 55-, HEAD DEPT, 59- Concurrent Pos: Sr vpres, Am Forage & Grassland Coun, 71, pres, 73; consult, Food & Agr Orgn, 72-76 & Res Appl Nat Needs, 75-76; pres, League Int Food Educ, 75-76. Honors & Awards: Distinguished Serv Award, Asn Seed Certifying Agencies, 72. Mem: Fel AAAS; fel Am Soc Agron (pres-elect, 71, pres, 72); Coun Agr Sci & Technol (pres, 74); Crops Sci Soc Am (pres, 60). Res: Forage crop plant breeding. Mailing Add: Dept of Crop Sci Ore State Univ Corvallis OR 97331

COWAN, KEITH MORRIS, b Salt Lake City, Utah, Sept 7, 21; m 44; c 3. IMMUNOCHEMISTRY. Educ: Southern Calif Univ, AB, 48, MS, 50; Johns Hopkins Univ, ScD(immunochem), 55. Prof Exp: Res assoc, Johns Hopkins Univ, 52-55; dir biol control, Cutter Labs, Calif, 55-56; investr for Howard Hughes Med Inst & res assoc, Dept Microbiol, Yale Univ, 56-58; MICROBIOLOGIST, ANIMAL DIS & PARASITE RES DIV, AGR RES SERV, USDA, 58- Res: Immunochemical characterization of antibodies and virus antigens pertaining primarily to foot-and-mouth disease. Mailing Add: Plum Island Animal Dis Ctr Box 848 Greenport NY 11944

COWAN, MAYNARD, JR, b Independence, Mo, Dec 15, 25; m 45; c 2. PHYSICS. Educ: William Jewel Col, BA, 48; Univ NMex, MS, 51. Prof Exp: Mem staff, Weapons Effects Studies, 51-60, div supvr, Magneto Physics Res Div, 60-69, mgr, shock simulation dept, 69-75, MGR, SIMULATION RES DEPT, SANDIA CORP, 75- Res: High magnetic fields; plasma physics; nuclear burst effects. Mailing Add: 1107 Stagecoach SE Albuquerque NM 87123

COWAN, RAYMOND, b Marion, Ind, Nov 14, 14; m 40; c 2. PHYSICS. Educ: Butler Univ, BS, 54, MS, 55. Prof Exp: From asst prof to assoc prof, 55-67, PROF PHYSICS, FRANKLIN COL, 67-, CHMN DEPT, 57-, CHMN SCI DIV, 60- Mem: Am Asn Physics Teachers. Res: Electronics. Mailing Add: Dept of Physics Franklin Col Franklin IN 46131

COWAN, RICHARD SUMNER, b Crawfordsville, Ind, Jan 23, 21; m 41; c 3. PLANT TAXONOMY. Educ: Wabash Col, AB, 42; Univ Hawaii, MS, 48; Columbia Univ, PhD, 52. Prof Exp: Tech asst, NY Bot Garden, 48-52, asst cur, 52-57; assoc cur, Smithsonian Inst, 57-62, asst dir, Nat Mus Natural Hist, 62-65, dir, 65-73, SR BOTANIST, DEPT BOT, SMITHSONIAN INST, NAT MUS NATURAL HIST, 73- Concurrent Pos: NSF fel, 52-53. Mem: AAAS; Am Inst Biol Sci; Am Soc Plant Taxon; Asn Trop Biol; Biol Soc Am. Res: Phanerogamic taxonomy; Leguminosae of northern South America; Rutaceae of Guayana Highland; flora of Venezuela. Mailing Add: Nat Mus Natural Hist Smithsonian Inst Washington DC 20560

COWAN, ROBERT DUANE, b Lincoln, Nebr, Nov 24, 19; m 44; c 4. ATOMIC PHYSICS. Educ: Friends Univ, AB, 42; Johns Hopkins Univ, PhD(physics), 46. Prof Exp: Jr instr physics, Johns Hopkins Univ, 43-46, lab asst, 44-46; Nat Res fel, Univ Chicago, 46-47, res assoc spectros, 47-48; prof physics & head dept, Friends Univ, 48-51; PHYSICIST, LOS ALAMOS SCI LAB, 51- Concurrent Pos: Fulbright lectr, Peru, 58-59; vis prof, Purdue Univ, 71. Mem: Fel Am Phys Soc; Optical Soc Am. Res: Visible and ultraviolet spectroscopy; selfabsorption of spectral lines; equations of state at extreme pressures and temperatures; theoretical atomic spectroscopy. Mailing Add: Los Alamos Sci Lab Los Alamos NM 87544

COWAN, ROBERT LEE, b Beaver County, Pa, Nov 20, 20; m 45; c 5. ANIMAL NUTRITION. Educ: Pa State Col, BS, 43, MS, 49, PhD, 52. Prof Exp: Chemist food processing, Gen Foods, Inc, 43-44; asst agr biol chem, 46-48, from instr to assoc prof, 48-65, PROF ANIMAL NUTRIT, PA STATE UNIV, 65- Concurrent Pos: Fulbright res scholar, Massey Col, NZ, 56-57. Mem: Am Soc Animal Sci; NY Acad Sci; Am Inst Nutrit; Wildlife Soc. Res: Nutritive values of forages; preservation of grass silage; nutrition of deer and mink. Mailing Add: Dept of Animal Sci 305 Animal Industs Bldg Pa State Univ University Park PA 16802

COWAN, RUSSELL (WALTER), b Oakland, Calif, Feb 26, 12; m 48. APPLIED MATHEMATICS. Educ: Univ Calif, AB, 32, MA, 33, PhD(math), 35. Prof Exp: Instr math & astron, Col of St Scholastica, 35-38; from instr to asst prof math, Univ Ala, 38-47; assoc prof, Univ Fla, 47-66; PROF MATH, LAMAR UNIV, 66- Mem: Am Math Soc; Math Asn Am. Res: Analysis; differential equations; difference equations; solution of a linear difference equation of the second order with quadratic coefficients; differential equations, special functions and gamma functions. Mailing Add: Dept of Math Lamar Univ Beaumont TX 77710

COWAN, W MAXWELL, b Johannesburg, SAfrica, Sept 27, 31; m 56; c 3. NEUROANATOMY. Educ: Univ Witwatersrand, BSc, 52; Oxford Univ, DPhil(neuroanat), 56, BM & BCh, 58. Hon Degrees: MA, Oxford Univ, 58. Prof Exp: Asst anat, Univ Witwatersrand, 51-53; dept demonstr, Oxford Univ, 53-58, lectr, 59-66, tutor, Pembroke Col, 56-66; assoc prof, Univ Wis-Madison, 66-68; prof & head dept, Sch Med, Wash Univ, 68-74; CHMN DEPTS ANAT & PHYSIOL, SCH MED, STANFORD UNIV, 74- Concurrent Pos: Fel, Pembroke Col, 56-66, lectr, Balliol Col, 62-66; vis assoc prof, Wash Univ, 64-65; managing ed, J Comp Neurol, 69- Mem: Am Asn Anat; Anat Soc Gt Brit & Ireland; fel Royal Micros Soc; Soc Neurosci. Res: Neuroembryology, especially the structure and development of the mammalian forebrain and the avian visual system. Mailing Add: Depts of Anat & Physiol Stanford Univ Sch of Med Stanford CA 94305

COWAN, WILLIAM ALLEN, b Pittsfield, Mass, Oct 4, 20; m 46. ANIMAL BREEDING. Educ: Mass State Col, BS, 42; Univ Minn, MS, 48, PhD(animal breeding, dairy prod), 52. Prof Exp: Asst farm mgr, Grafton State Hosp, Mass, 42-43; instr & farm supt, RI State Col, 43-45; asst & assoc prof animal husb, Univ Mass, 46-52; head dept, 52-74, PROF ANIMAL INDUSTS, RATCLIFFE HICKS SCH AGR, UNIV CONN, 52- Mem: Am Soc Animal Sci; Am Dairy Sci Asn; Am Genetic Asn. Res: Animal production; dairy herd and livestock management. Mailing Add: Dept of Animal Industs Univ of Conn Storrs CT 06268

COWARD, JAMES KENDERDINE, b Buffalo, NY, Oct 13, 38. BIO-ORGANIC CHEMISTRY, MEDICINAL CHEMISTRY. Educ: Middlebury Col, AB, 60; Duke Univ, MA, 64; State Univ NY Buffalo, PhD(med chem), 67. Prof Exp: NIH fel, 66-68; asst prof, 69-74, ASSOC PROF PHARMACOL, SCH MED, YALE UNIV, 75- Mem: Am Chem Soc; The Chem Soc; Am Soc Biol Chem. Res: Physical-organic and biochemical investigations of enzyme mechanisms as a basis for design of new drugs. Mailing Add: Dept of Pharmacol Yale Univ Sch Med 333 Cedar St New Haven CT 06510

COWARD, JOE EDWIN, b Searcy, Ark, Mar 27, 38; m 59; c 2. VIROLOGY, MICROBIOLOGY. Educ: State Col Ark, BSE, 59; Univ Ark, Fayetteville, MS, 62; Univ Miss, PhD(microbiol), 68. Prof Exp: Res assoc, Col Physicians & Surgeons, Columbia Univ, 68-70, asst prof microbiol, 70-75; ASSOC PROF MICROBIOL, SCH MED, LA STATE UNIV MED CTR, NEW ORLEANS, 75- Mem: Am Soc Microbiol. Res: Viral replication as studied by electron microscopy; development of the slow viruses and their relationship to human disease. Mailing Add: Dept of Microbiol Sch of Med La State Univ Med Ctr New Orleans LA 70112

COWARD, NATHAN A, b Belton, SC, Jan 7, 27; m 53; c 4. PHYSICAL CHEMISTRY. Educ: The Citadel, BS, 50; Univ Rochester, PhD(phys chem), 54. Prof Exp: Res chemist, F H Levey Co Div, Columbian Carbon Co, 54-57 & Am Viscose Corp, 57-59; from instr to assoc prof, 59-69, actg chmn dept, 64-65, 68-69, PROF CHEM, UNIV WIS-SUPERIOR, 69- Concurrent Pos: Vis prof, Univ Wis-Madison, 67-68. Mem: Am Chem Soc; Sigma Xi. Res: Fluorescence; photochemistry; printing inks; analytical methods and instrumentation; liquid phase weak complexes of lanthanide ions; trace metals in fuels and natural waters. Mailing Add: Dept of Chem Univ of Wis Superior WI 54880

COWARD, STUART JESS, b New York, NY, June 7, 36; m 65. DEVELOPMENTAL BIOLOGY. Educ: Univ Miami, Fla, BS, 58; Univ Iowa, MS, 61; Univ Calif, Davis, PhD(zool), 64. Prof Exp: Res assoc, State Univ NY Buffalo, 64-65; asst prof, 65-69, ASSOC PROF ZOOL, UNIV GA, 69- Concurrent Pos: NIH fel, 64-65; NIH spec fel, Harvard Univ Med Sch, 70-71. Mem: AAAS; Soc Develop Biol; Am Soc Cell Biol. Res: Regeneration in planaria; biochemical aspects of amphibian development. Mailing Add: Dept of Zool Univ of Ga Athens GA 30602

COWDEN, RONALD REED, b Memphis, Tenn, July 9, 31; m 56. CYTOLOGY, EMBRYOLOGY. Educ: La State Univ, Baton Rouge, BS, 53; Univ Vienna, DrPhil(zool), 56. Prof Exp: USPHS fel biol, Oak Ridge Nat Lab, 56-57; asst prof biol, Johns Hopkins Univ, 57-60; asst mem cell biol, Inst Muscle Dis, 60-61; asst prof path, Col Med, Univ Fla, 61-66, USPHS career develop award, 62-66; assoc prof anat, La State Univ Med Ctr, New Orleans, 66-68; prof biol sci & chmn dept, Univ Denver, 68-71; prof anat & chmn dept, Albany Med Col, 72-75; ASSOC DEAN BASIC SCI, COL MED, E TENN STATE UNIV, 75- Mem: Am Soc Cell Biologists; Soc Develop Biol; Am Asn Anatomists; Am Zool Soc; Royal Micros Soc. Res: Quantitative cytochemistry; cytochemistry of oogenesis and development; comparative hematology and immunity. Mailing Add: Col of Med E Tenn State Univ Johnson City TN 37601

COWELL, BRUCE CRAIG, b Buffalo, NY, Oct 20, 37; m 65; c 2. LIMNOLOGY. Educ: Bowling Green State Univ, BA, 58, MA, 59; Cornell Univ, PhD(limnol ecol), 63. Prof Exp: Fishery res biologist limnol, NCent Reservoir Invest, US Fish & Wildlife Serv, 63-75; asst prof zool, 67-75, ASSOC PROF BIOL, UNIV S FLA, 75- Mem: Am Fisheries Soc; Am Soc Limnol & Oceanog; Ecol Soc Am; Int Soc Limnol; NAm Benthological Soc. Res: Production and population dynamics of plankton and benthic invertebrate communities; trophic levels of freshwater environments; reservoir limnology; pollution ecology. Mailing Add: Dept of Biol Univ of SFla Tampa FL 33620

COWELL, JAMES LEO, b Wilkes-Barre, Pa, May 4, 44; m 69. MICROBIOLOGY. Educ: King's Col, Pa, BS, 66; St John's Univ, NY, MS, 68; Univ Ill, Urbana, PhD(microbiol), 72. Prof Exp: Res asst biochem, Cornell Univ, 72-74; RES ASST MICROBIOL, SCH MED, NY UNIV, 74- Concurrent Pos: NIH fel, 72. Mem: Am Soc Microbiol; Sigma Xi; Am Inst Biol Sci. Res: Mechanism of action of cytolytic bacterial toxins. Mailing Add: Dept Microbiol NY Univ Med Ctr 550 First Ave New York NY 10016

COWELL, WAYNE RUSSELL, b Wakefield, Kans, June 27, 26; m 53; c 3. APPLIED MATHEMATICS. Educ: Kans State Univ, BS, 48, MS, 50; Univ Wis, PhD(math), 54. Prof Exp: Asst math, Kans State Univ, 48-50 & Univ Wis, 50-54; asst prof, Mont State Univ, 54-59; mem tech staff, Bell Tel Labs, 59-61; assoc mathematician, 61-72, COMPUTER SCIENTIST, ARGONNE NAT LAB, 72- Concurrent Pos: Asst to pres, Argonne Univs Asn, 68-71. Mem: Asn Comput Mach; Soc Indust & Appl Math; Math Asn Am. Res: Mathematical software. Mailing Add: Appl Math Div Argonne Nat Lab Argonne IL 60439

COWEN, DAVID, b New York, NY, July 29, 07. NEUROPATHOLOGY. Educ: Columbia Col, AB, 28; Columbia Univ, MD, 32. Prof Exp: Asst pathologist, Neurol Inst, NY, 37-46; instr neurol, 37-39, from instr neuropath to assoc prof, 39-63, PROF NEUROPATH, COL PHYSICIANS & SURGEONS, COLUMBIA UNIV, 63- Concurrent Pos: From asst attend neuropathologist to attend neuropathologist, Columbia-Presby Med Ctr, 45-73, consult neuropathologist, 73-; consult neuropathologist, Vet Admin Hosp, East Orange, NJ, 53- & Lenox Hill Hosp, NY; ed, J Neuropath & Exp Neurol. Mem: Am Asn Neuropath; Am Neurol Asn; Harvey Soc; Am Soc Exp Path; NY Acad Med. Res: Infections of the nervous system; neurological diseases of prenatal origin. Mailing Add: Columbia Univ Div Neuropath Col of Physicians & Surgeons New York NY 10032

COWEN, JERRY ARNOLD, b Toledo, Ohio, July 17, 24; m 46; c 5. PHYSICS. Educ: Harvard Col, BS, 48; Mich State Col, MS, 50, PhD, 54. Prof Exp: Asst physics, Mich State Col, 49-53; asst prof, Colo Agr & Mech Col, 53-55; from asst prof to assoc prof, 55-69, PROF PHYSICS, COL NATURAL SCI, MICH STATE UNIV, 69- Concurrent Pos: NSF fel, Saclay Nuclear Res Ctr, France, 64-65; res physicist, Nat Bur Standards, 55 & Lockheed Res Lab, 60-61. Mem: Am Phys Soc; Am Asn Physics Teachers. Res: Microwave resonance. Mailing Add: Dept of Physics Mich State Univ East Lansing MI 48823

COWEN, RICHARD, b Workington, Eng, Jan 24, 40. PALEONTOLOGY. Educ: Cambridge Univ, BA, 62, PhD(geol), 66. Prof Exp: Res asst geol, Cambridge Univ, 65-67; asst prof, 67-73, ASSOC PROF GEOL, UNIV CALIF, DAVIS, 73- Mem: Paleontol Soc; Brit Palaeontol Asn; AAAS. Res: Anatomical and functional studies in living and fossil invertebrates, especially brachiopods, their implications for evolution and paleobiogeography. Mailing Add: Dept of Geol Univ of Calif Davis CA 95616

COWEN, WILLIAM FRANK, b Oshkosh, Wis, Aug 20, 45; m 67; c 3. WATER CHEMISTRY. Educ: Univ Wis, Madison, BS, 67, MS, 69, PhD(water chem), 74. Prof Exp: RES CHEMIST ENVIRON CHEM, US ARMY BIOENG RES & DEVELOP LAB, ENVIRON PROTECTION DIV, 73- Mem: Am Chem Soc; Water Pollution Control Fedn. Res: Development of methods for analysis of trace organic compounds in water supplies and in wastewaters. Mailing Add: USAMBDL Bldg 459 EPA Ft Detrick Frederick MD 21701

COWETT, EVERETT R, b Ashland, Maine, Mar 6, 35; m 60; c 5. AGRONOMY, PLANT PHYSIOLOGY. Educ: Univ Maine, BS, 57, MS, 58; Rutgers Univ, PhD(farm crops), 61. Prof Exp: Asst prof agron, Univ NH, 61-63; rep field res plant protection, Geigy Chem Corp, 63-65, herbicide specialist, 65-67, field res mgr plant protection, 67-70, mgr plant sci res, Ciba-Geigy Ltd, Ardsley, 70-73, DIR TECH SERVS, AGR DIV, CIBA-GEIGY CORP, 73- Mem: Am Soc Agron; Weed Sci Soc Am; Entom Soc Am. Res: Plant physiology; forage crop management; weed control; insect and disease control; plant growth regulators. Mailing Add: Ciba-Geigy Corp PO Box 11422 Greensboro NC 27409

COWGER, MARILYN L, b Douglas, Nebr, June 7, 31. PEDIATRICS, BIOCHEMISTRY. Educ: Univ Omaha, BA, 53; Univ Nebr, MD, 56; Am Bd Pediat, dipl, 62. Prof Exp: Intern med, Bryan Mem Hosp, Lincoln, Nebr, 56-57; fel pediat, Mayo Clin, Minn, 57-60; from asst to assoc prof pediat, Sch Med, Univ Wash, 60-70; ASSOC PROF PEDIAT, ALBANY MED COL, 70-; RES ASSOC PROF CHEM, STATE UNIV NY ALBANY, 70- Concurrent Pos: NIH trainee, 60-63, career develop award, 65-74; Am Heart Asn advan res fel, 63-65. Mem: Am Acad Pediat; Soc Pediat Res. Res: Mechanism of bilirubin toxicity; electron transport inhibitors; clinical and biochemical studies of the porphyrias; mechanism of mental retardation in inborn errors of metabolism. Mailing Add: Dept of Chem State Univ of NY Albany NY 12203

COWGILL, GEORGE L, b Grangeville, Idaho, Dec 19, 29; m 72. ANTHROPOLOGY. Educ: Stanford Univ, BS, 52; Iowa State Univ, MS, 54; Univ Chicago, MA, 56; Harvard Univ, PhD(anthrop), 63. Prof Exp: From instr to asst prof, 60-67, ASSOC PROF ANTHROP, BRANDEIS UNIV, 67- Concurrent Pos: Assoc, Rene Millon, 64-; res assoc, Univ Rochester, 67; vis prof anthrop, Harvard Univ, 71 & 73; NSF res grant, 73-75. Mem: Fel AAAS; fel Am Anthrop Asn; Soc Am Archaeol. Res: Middle American prehistory; computer and mathematical methods in anthropology; cultural ecology; archaeological theory. Mailing Add: Dept of Anthrop Brandeis Univ Waltham MA 02154

COWGILL, JAMES JOSEPH, b Ronan, Mont, Mar 30, 14. PHYSICAL CHEMISTRY. Educ: Gonzaga Univ, BS, 38, MS, 39, MA, 40; Alma Col, STL, 46; Univ Notre Dame, PhD(chem), 57. Prof Exp: Instr chem, Gonzaga Univ, 39-42; asst phys chem, Univ Notre Dame, 48-50; asst prof physics & math, 50-51, 53-57, assoc prof physics, 57-60, head dept physics, 56-57, mem bd trustees, 57-65, assoc dean grad sch, 67-70, dir res & instnl develop, 68-72, PROF PHYSICS, SEATTLE UNIV, 60-, DEAN GRAD SCH, 70-, DIR ACAD RES, 72- Concurrent Pos: Mem Task Force Master's Degree Progs, Coun Grad Schs, 74 & 75. Mem: AAAS; Am Chem Soc; Am Asn Physics Teachers; NY Acad Sci; Sigma Xi. Res: Ion-exchange resins; radioactive tracers. Mailing Add: Dept of Physics Seattle Univ Grad Sch Seattle WA 98122

COWGILL, ROBERT WARREN, b Topeka, Kans, Jan 31, 20; m 44; c 2. BIOCHEMISTRY. Educ: Univ Kans, BA, 41; Rensselaer Polytech Inst, MS, 42; Johns Hopkins Univ, PhD(biochem), 50. Prof Exp: Res chemist, Hercules Powder Co, 42-45; instr biochem, Wash Univ, 50-52; instr, Univ Calif, 53-56; asst prof, Univ Colo, 56-62; PROF BIOCHEM, BOWMAN GRAY SCH MED, 62- Mem: Am Chem Soc. Res: Mechanism of enzyme action; protein structure. Mailing Add: Dept of Biochem Bowman Gray Sch of Med Winston-Salem NC 27103

COWGILL, URSULA MOSER, b Bern, Switz, Nov 9, 27; US citizen; m 54. GEOCHEMISTRY. Educ: Hunter Col, AB, 48; Kans State Univ, MS, 52; Iowa State Univ, PhD(soil chem), 56. Prof Exp: Mem staff, Lincoln Lab, Mass Inst Technol, 57-58 & Doherty Charitable Found Inc, Guatemala, 58-60; res assoc ecol & anthrop geochem, Yale Univ, 60-68; PROF BIOL, UNIV PITTSBURGH, 68-, PROF ANTHROP, 72- Mem: Geochem Soc; Mineral Soc Am; Clay Minerals Soc; Mineral Soc Gt Brit & Ireland; Clay Minerals Group Gt Brit. Res: Ecological and anthropological geochemistry; primitive agriculture; phosphate mineralogy; history of lake basins; exotic demography; prosimian behavior. Mailing Add: Dept of Biol Univ of Pittsburgh Pittsburgh PA 15260

COWIN, STEPHEN CORTEEN, b Elmira, NY, Oct 26, 34; m 56; c 2. APPLIED MECHANICS, ENGINEERING MECHANICS. Educ: Johns Hopkins Univ, BS, 56, MS, 58; Pa State Univ, PhD(eng mech), 62. Prof Exp: Asst dynamics, Johns Hopkins Univ, 56-58; struct engr, Aircraft Armaments, Inc, Md, 58, sr struct engr, 58-59; from instr to asst prof mech, Pa State Univ, 59-63; from asst prof to assoc prof mech eng, 63-69, prof in charge mech & mat sci, 69-75, PROF MECH ENG, TULANE UNIV, 69- Concurrent Pos: Instr, Loyola Col, Md, 58-59; sr vis fel, Res Coun Gt Brit, 74; prof-in-charge, Tulane Univ-Newcomb Jr Yr Abroad Prog, Gt Brit, 74-75; chmn, Master Sci Appl Math Prog & Eng Curriculum Prog, 75- Mem: Soc Rheol; Soc Natural Philos; Soc Eng Sci; Math Asn Am; Sigma Xi. Res: Continuum mechanics, rheology, mechanics of granular media, continuum theories representing the microstructure of materials; theory of constitutive relations; biomechanics. Mailing Add: Sch of Eng Tulane Univ New Orleans LA 70118

COWLES, CRAIG SCHUYLER, b Omaha, Nebr, May 13, 44; m 64; c 2. INDUSTRIAL ORGANIC CHEMISTRY. Educ: Univ Nebr, BS, 66; Univ Kans, PhD(org chem), 71. Prof Exp: Teaching asst org chem, Univ Kans, 66-67, res asst, 67-71; fel, Univ Nebr, 72-73; from process chemist to sr process chemist, 73-75, PROD SUPT DYES, AM CYANAMID CO, 75- Mem: Am Chem Soc. Res: Development and production of dyes and chemical intermediates. Mailing Add: 1394 Mt Vernon Rd Bridgewater NJ 08807

COWLES, EDWARD J, b Careywood, Idaho, July 15, 18; m 48; c 2. CHEMISTRY. Educ: Univ Wash, Seattle, BS, 40, PhD(chem), 53. Prof Exp: Jr chemist, Western Regional Res Lab, Bur Agr & Indust Chem, USDA, 41-42; shift chemist, Rayonier, Inc, 46; instr chem & Ger, Grays Harbor Col, 46-47; chemist, Philippine Fishery Prog, 47-48; assoc prof chem, Whitworth Col, 52-54; from asst prof to assoc prof, 54-69, PROF CHEM, UNIV MINN, DULUTH, 69- Concurrent Pos: Consult, Hilding Res Fund, US Dept Health, Educ & Welfare, 62-72. Mem: AAAS; Am Chem Soc; Sigma Xi. Res: Electrophilic substitution reactions of azulene; identification of the pigments of maize; determination of ammonia; spectra of azulene and related compounds; solubility of inorganic compounds; empirical applications of the computer to chemical problems. Mailing Add: Dept of Chem Univ of Minn Duluth MN 55812

COWLES, JOE RICHARD, b Edmonson County, Ky, Oct 29, 41; m 65; c 2. PLANT PHYSIOLOGY. Educ: Western Ky Univ, BS, 63; Univ Ky, MS, 65; Ore State Univ, PhD(plant physiol), 68. Prof Exp: Res assoc plant physiol, Purdue Univ, 68-69, Univ Ga, 69-70; ASST PROF BIOL, UNIV HOUSTON, 70- Mem: Am Soc Plant Physiol; AAAS; Sigma Xi. Res: Physiological and biochemical parameters associated with the establishment of symbiosis in plants capable of biological nitrogen fixation and developmental changes associated with the regulation of aromatic biosynthesis in plant tissues. Mailing Add: Dept of Biol Univ of Houston Houston TX 77004

COWLEY, A RONALD, b Birmingham, Ala, Apr 30, 44; m 67. NEUROANATOMY, NEUROSCIENCES. Educ: Ala Col, BS, 66; Univ Ala, MS, 68, PhD(neuroanat), 70. Prof Exp: CLIN ASST PROF ANAT, SCH MED, LA STATE UNIV, SHREVEPORT, 70- Res: Limbic lobe areas pertaining to respiration in primates and other mammals; ultrastructure effects of vasoactive amines on cerebral blood vessels; comparative aspects of the forebrain in mammalian forms. Mailing Add: Dept of Anat Sch of Med La State Univ Shreveport LA 71101

COWLEY, ALAN H, b Manchester, Eng, Jan 29, 34; m 60; c 3. INORGANIC CHEMISTRY. Educ: Univ Manchester, BS, 55, MS, 56, PhD(chem), 58. Prof Exp: Fel chem, Univ Fla, 58-60; tech officer, Billingham Div, Imp Chem Industs Gt Brit, 60-62; from asst prof to assoc prof, 62-70, PROF CHEM, UNIV TEX, AUSTIN, 70- Concurrent Pos: Welch res grant, 64-; NIH res grant, 64-67; NSF grant, 69- Mem: Am Chem Soc; Brit Chem Soc. Res: Chemistry of non-metals; inorganic free radicals; nuclear magnetic resonance; vibrational spectroscopy. Mailing Add: Dept of Chem Univ of Tex Austin TX 78712

COWLEY, ALLEN WILSON, JR, b Harrisburg, Pa, Jan 21, 40; m 65. PHYSIOLOGY, CARDIOVASCULAR PHYSIOLOGY. Educ: Trinity Col, BA, 61; Hahnemann Med Col, MS, 65, PhD(physiol & biophys), 68. Prof Exp: NIH grant, Med Ctr, 68-69, from instr to assoc prof, 68-74, PROF PHYSIOL & BIOPHYS, MED CTR, UNIV MISS, 74- Concurrent Pos: NIH res grant hypertension res, 71-75; Am Heart Asn estab investr, 73. Mem: AAAS; Am Physiol Soc; Am Heart Asn. Res: Quantitative analysis of interacting neural and hormonal control systems in the overall regulation of arterial blood pressure. Mailing Add: Dept of Physiol & Biophys Univ of Miss Med Ctr Jackson MS 39216

COWLEY, ANNE PYNE, b Boston, Mass, Feb 25, 38; m 60; c 2. ASTRONOMY. Educ: Wellesley Col, BA, 59; Univ Mich, MA, 61, PhD(astron), 63. Prof Exp: Res assoc astron, Yerkes Observ, Univ Chicago, 63-68; res assoc astron, 68-73, ASSOC RES SCIENTIST, UNIV MICH, ANN ARBOR, 73- Concurrent Pos: Guest worker, Dom Astrophys Observ, 74-75. Mem: Am Astron Soc; Int Astron Union. Res: Stellar spectroscopy; astrophysics of stellar atmospheres. Mailing Add: Dept of Astron Physics-Astron Bldg Univ of Mich Ann Arbor MI 48104

COWLEY, CHARLES RAMSAY, b Aguana, Guam, Sept 13, 34; US citizen; m 60; c 2. ASTRONOMY. Educ: Univ Va, BA, 55, MA, 58; Univ Mich, PhD(astron), 63. Prof Exp: From instr to asst prof astron, Univ Chicago, 63-67; asst prof, 67-70, ASSOC PROF ASTRON, UNIV MICH, 70- Mem: Am Astron Soc; Royal Astron Soc; Int Astron Union. Res: Stellar atmospheres. Mailing Add: Dept of Astron Univ of Mich Ann Arbor MI 48104

COWLEY, GERALD TAYLOR, b Barron, Wis, Aug 1, 31; m 57; c 4. MYCOLOGY Educ: Univ Wis, BS, 53, MS, 57, PhD(bot), 62. Prof Exp: Teacher high sch, Wis, 55-56; asst prof biol, 62-67, asst dean col arts & sci, 69-72, asst vprovost, Div Lib & Cult Disciplines, 72-73, ASSOC PROF BIOL, UNIV SC, 67-, ASST HEAD DEPT, 73- Mem: Mycol Soc Am; Bot Soc Am; Am Soc Microbiol. Res: Ecology and physiology of soil and litter microfungi. Mailing Add: Dept of Biol Univ of SC Columbia SC 29208

COWLEY, MILFORD A, b Rio, Wis, Nov 15, 04; m 34; c 3. FOOD CHEMISTRY. Educ: Univ Wis, PhD(food chem), 33. Prof Exp: Teacher chem, Univ Wis, La Crosse, 33-38, chmn dept chem, 39-44 & 64-68, phys sci & math, 45-56 & phys sci, 57-63, prof chem, 66-74; RETIRED. Mem: AAAS; Am Chem Soc. Res: Sugar as a chemical raw material; fat and oil analysis. Mailing Add: Dept of Chem Univ of Wis La Crosse WI 54601

COWLEY, THOMAS GLADMAN, b Clifton Springs, NY, Aug 20, 38; m 60; c 2. ANALYTICAL CHEMISTRY, SPECTROCHEMISTRY. Educ: Rochester Inst Technol, BS, 61; Iowa State Univ, PhD(anal chem), 66. Prof Exp: Assoc, AEC Ames Lab, 66-67; RES SCIENTIST, CONTINENTAL OIL CO, 67- Mem: Am Chem Soc; Soc Appl Spectros. Res: Chemical and physical behavior of spectroscopic sources and their uses in analytical chemistry; development of flame emission spectroscopy as an analytical tool; application of spectrochemistry to environmental sciences; analysis of petroleum and petroleum products. Mailing Add: 2316 El Camino Ponca City OK 74601

COWLEY, WALTER RAYMOND, agriculture, see 12th edition

COWLING, ELLIS BREVIER, b Waukegan, Ill, Dec 11, 32; m 56; c 2. PLANT PATHOLOGY. Educ: State Univ NY Col Forestry, Syracuse, BS, 54, MS, 56; Univ Wis, PhD(plant path & biochem), 59; Univ Uppsala, FilDr(physiol bot), 70. Prof Exp: Chemist wood properties res, Dow Chem Co, 55-56; wood pathologist, Forest Prod Lab, USDA, 56-59; asst prof forest path, Sch Forestry, Yale Univ, 60-68; assoc prof, 65-69, PROF PLANT PATH, FORESTRY & WOOD & PAPER SCI, NC STATE UNIV, 68- Concurrent Pos: USPHS fel, Royal Pharmaceut Inst Stockholm, Sweden, 59-60; vis prof, Inst Physiol Bot, Univ Uppsala, Sweden, 70-71; assoc ed, Ann Rev Phytopath, Ann Revs, Inc, Palo Alto, Calif, 71-76. Honors & Awards: Res Award, Sigma Xi, 68. Mem: Nat Acad Sci; AAAS; fel Int Acad Wood Sci; Am Phytopath Soc; Soc Wood Sci & Technol. Res: Forest and wood products pathology; physiology of trees and of tree diseases; enzymatic degradation of plant cell walls; air pollution effects on vegetation. Mailing Add: Dept of Plant Path NC State Univ Raleigh NC 27607

COWLING, HALE, chemistry, see 12th edition

COWLING, VINCENT FREDERICK, b St Louis, Mo, Dec 15, 18; m 44; c 1. MATHEMATICS. Educ: Rice Inst, BA, 41, MA, 43, PhD(math), 44. Prof Exp: Instr math, Ohio State Univ, 45-46; asst prof, Lehigh Univ, 46-49; from assoc prof to prof, Univ Ky, 49-61; prof, Rutgers Univ, 61-67; PROF MATH, STATE UNIV NY ALBANY, 66- Concurrent Pos: Ford Found fel, Yale Univ, 52-53. Mem: Am Math Soc; Math Asn Am. Res: Complex variable theory; applications of functional analysis. Mailing Add: Dept of Math State Univ of NY Albany NY 12203

COWLISHAW, JOHN DAVID, b Grand Rapids, Mich, Sept 10, 38; m 61; c 3. BIOPHYSICS. Educ: Univ Mich, BS, 60, MS, 61; Pa State Univ, PhD(biophys), 68. Prof Exp: Asst engr, Air Arm Div, Westinghouse Elec Corp, 61-63; instr physics, Westminster Col, Pa, 63-65; asst prof, 68-74, ASSOC PROF BIOL SCI, OAKLAND UNIV, 74- Mem: AAAS; Biophys Soc. Res: Cell biophysics; macromolecular dynamics; viruses of bacteria and algae; radiation effects on cells; bioelectrochemistry; systems theory and biology. Mailing Add: Dept of Biol Sci Oakland Univ Rochester MN 48063

COWMAN, RICHARD AMMON, b Brainerd, Minn, Apr 24, 38; m 65; c 1. MICROBIOLOGY, BIOCHEMISTRY. Educ: Univ Minn, BS, 61; NC State Univ, MS, 63, PhD(food sci), 66. Prof Exp: Instr food microbiol, NC State Univ, 66-67, asst prof, 67-70; asst prof microbiol & biochem, Inst Oral Biol, Univ Miami, 70-74; RES MICROBIOLOGIST, US VET ADMIN HOSP, MIAMI, 70- Mem: Am Chem Soc; Am Soc Microbiol; Soc Cryobiol; NY Acad Sci. Res: Freeze-damage of microorganisms; nutritional aspects of microorganisms involved in dental caries. Mailing Add: Vet Admin Hosp Dent Res Unit 151 1201 NW 16th St Miami FL 33125

COWPER, GEORGE, b Newcastle-upon-Tyne, Eng, Sept 6, 21; Can citizen; m 51; c 3. PHYSICS. Educ: Durham Univ, BSc, 43. Prof Exp: Sci officer, Telecommun Res Estab, Eng, 43-48; from jr res officer to assoc res officer, 48-58, sr res officer & head radiation dosimetry br, 58-67, HEAD HEALTH PHYSICS BR, ATOMIC ENERGY CAN, LTD, 67- Concurrent Pos: Chmn subcomt electronic mat res, Defence Res Bd Can, 57-63, mem, 63-65, mem panel radiation protection & treatment, 58-66; chmn panel adequate stand personnel dosimetry, Int Atomic Energy Agency, 63-64; WHO vis prof, Bhabha Atomic Res Ctr, Bombay, India, 71; news ed, Health Physics, Health Phys Soc, 71-75; health physics adv comt, Oak Ridge Nat Lab, 71-75; assoc comt environ quality, Nat Res Coun Can, 75- Mem: Health Phys Soc; Brit Inst Physics & Phys Soc; Brit Inst Elec Eng; Int Radiation Protection Asn. Res: Instrumentation for health physics; radiation dosimetry; detection of ionizing radiation as applied to nuclear chemistry and geophysical prospecting. Mailing Add: Health Physics Br Atomic Energy Can Ltd Chalk River ON Can

COWSAR, DONALD ROY, b Baton Rouge, La, Dec 12, 42; m 66; c 3. BIOMATERIALS, POLYMER CHEMISTRY. Educ: La State Univ, BS, 64; Rice Univ, PhD(org chem), 69. Prof Exp: Res chemist, 68-70, sr chemist, 70-73, HEAD BIOMAT SECT, SOUTHERN RES INST, 74- Concurrent Pos: Consult, NIH, 73- Mem: Am Chem Soc. Res: New biomaterials and biomedical devices including controlled-release drug-delivery systems, biodegradable implants, hemoperfusion devices, prosthetic polymers, and hydrogels; in vivo evaluation of materials and devices. Mailing Add: Southern Res Inst 2000 Ninth Ave S Birmingham AL 35205

COWSER, KENNETH EMERY, b Chicago, Ill, Apr 12, 26; m 47; c 3. HEALTH PHYSICS, ENVIRONMENTAL ENGINEERING. Educ: Univ Ill, Champaign-Urbana, BS, 47; Univ Tenn, Knoxville, MS, 59; Oak Ridge Sch Reactor Technol, grad, 63. Prof Exp: Eng aide, Ill Dept Hwys, 46; dist engr, Ill Dept Pub Health, 47-51, actg regional engr, 51-53; engr leader res & develop, Oak Ridge Nat Lab, 53-67, sect chief res & develop, Radioactive Waste Disposal Sect, Health Physics Div, 67-71, asst to assoc dir biomed & environ sci, 71-75; ASST TO ASST ADMINR ENVIRON & SAFETY, ENERGY RES & DEVELOP ADMIN, 75- Concurrent Pos: Mem subcomt radioactive waste disposal, Am Standards Asn, 60-69; mem subcomt radioactive waste mgt, Am Nat Standards, 69-71; mem sci comt, task group krypton-85 & sci comt radiation nuclear power generation, Nat Coun Radiation Protection & Measurements; vis lectr health physics prog, Vanderbilt Univ. Mem: Health Physics Soc; Sigma Xi. Res: Radioactive waste management, including treatment and disposal of solid, liquid and gaseous wastes; siting and operation of nuclear facilities and activities; environmental monitoring; radiological hazards evaluations. Mailing Add: Off Asst Adminr Environ & Safety Energy Res & Develop Admin Washington DC 20545

COX, ALLAN CLAYTON, b Eriksdale, Man, May 23, 39; m 64; c 2. NUTRITION, PHYSIOLOGY. Educ: Univ Man, BSA, 64, MSc, 66; Iowa State Univ, PhD(nutrit), 69. Prof Exp: RES SCIENTIST NUTRIT PHYSIOL OF POULTRY, CAN AGR RES STA, 69- Res: Mineral metabolism; development and maintenance of the skeleton; nutrition-genetic interactions; strain evaluation. Mailing Add: Res Sta Can Dept of Agr Kentville NS Can

COX, ALLAN VERNE, b Santa Ana, Calif, Dec 17, 26. GEOPHYSICS Educ: Univ Calif, BA & MA, 57, PhD, 59. Prof Exp: Geol field asst, US Geol Surv, 50-51, 54; asst geol, Univ Calif, 56-57; geophysicist, US Geol Surv, 57-68; CECIL & IDA GREEN PROF GEOPHYS, STANFORD UNIV, 68- Honors & Awards: John A Fleming Medal, Am Geophys Union; Vetlesen Prize, 71; Day Medal, Geol Soc Am, 75. Mem: Nat Acad Sci; Am Geophys Union; Geol Soc Am; Soc Terrestrial Magnetism & Elec of Japan; Am Acad Arts & Sci. Res: Pleistocene geology; paleomagnetism; geomagnetism; seismology; geomorphology; history of earth's magnetic field determined from paleomagnetism of rocks. Mailing Add: Dept of Geophys Stanford Univ Stanford CA 94305

COX, ALVIN JOSEPH, JR, b Manila, PR, Mar 6, 07; m 46; c 3. PATHOLOGY. Educ: Stanford Univ, AB, 27, MD, 31. Prof Exp: From asst to prof path, 31-64, PROF PATH DERMAT, SCH MED, STANFORD UNIV, 64- Concurrent Pos: Fel Harvard Univ, 57-58; exchange asst, Path Inst, Freiburg, 35-36; consult pathologist, Western Utilization Res Br, USDA, 37; consult pathologist, San Francisco Hosp, 41-59. Mem: Fel AMA; Am Asn Path & Bact; Soc Exp Biol & Med; Soc Invest Dermat; Am Soc Dermatopath. Res Coccidioidal infection; pathology of stomach; arteriosclerosis; experimental tumor production; dermatopathology. Mailing Add: Rm R166 Stanford Univ Sch of Med Stanford CA 94305

COX, ANDREW CHADWICK, b Hattiesburg, Miss, July 20, 36; m 61; c 2. BIOCHEMISTRY, BIOPHYSICS. Educ: Univ Tex, BS, 59; Univ Houston, MS, 63; Duke Univ, PhD(biochem), 67. Prof Exp: Fel, Am Cancer Soc, 66-68; asst prof biochem, 69-72, ASSOC PROF BIOCHEM, SCH MED, UNIV OKLA, 72- Mem: Am Chem Soc. Res: Elucidation of lipid-protein interactions and their relations to mechanisms and structures, particularly plasma alpha-lipoprotein, prothrombin and cellular membranes. Mailing Add: Dept of Biochem Univ of Okla Health Sci Ctr Oklahoma City OK 73190

COX, ANNA LUCILE, b Birmingham, Ala, Aug 4, 22. PHYSICAL CHEMISTRY. Educ: Birmingham Southern Col, BS, 43; Univ Tenn, MS, 44; Purdue Univ, PhD(chem), 48. Prof Exp: Asst chemist, Birmingham Southern Col, Swann & Co, Univ Tenn, Univ Southern Calif & Purdue Univ, 41-48; instr, Univ Hawaii, 48-49; biochemist, Raukura Animal Res Sta, NZ, 51, head dept chem, 53-54 & 55-56; asst prof, Bridgeport Col, 54-55; res chemist, Lab Gordon Alles, 56; chemist, Aerojet-Gen Corp, 56-58; res specialist, aircraft specialist, Aircraft Div, Hughes Tool Co, 58-59; sr chemist, Rocketdyne div, NAm Aviation, Inc, 59-61; sr physicist, Ion Physics Corp, 61-63; prin res physicist, Brown Eng Co, Inc, 64-68; RES PHYSICIST, NAVAL RES LAB, 68- Mem: AAAS; Am Inst Aeronaut & Astronaut; Am Phys Soc. Res: Electrostatic propulsion and power generation; colloidal ion sources; condensation in rarefied gases; high intensity molecular beams; skinner nozzle applications; gaseous ion-molecule reactions. Mailing Add: 2904 Blooming Ct Oxon Hill MD 20022

COX, ARTHUR NELSON, b Van Nuys, Calif, Oct 12, 27; m 73; c 5. ASTROPHYSICS, HYDRODYNAMICS. Educ: Calif Inst Technol, BA, 48; Ind Univ, AM, 52, PhD(astron), 53. Hon Degrees: DSc, Ind Univ, 73. Prof Exp: Staff mem physics, 48-49, staff mem field testing, 53-57, group leader, 57-74, STAFF MEM THEORET DIV, LOS ALAMOS SCI LAB, 75- Concurrent Pos: Vis prof, Univ Calif, Los Angeles, 66; NATO sr fel sci, Univ Liege, Belgium, 68, Fulbright res scholar, 68-69; NSF prog dir, 73-74. Mem: AAAS; Am Astron Soc; Int Astron Union. Res: Calculations of stellar stability and pulsation; compilation of equations of state and opacities for astrophysics; studies of stellar atmosphere, interior structure and stellar evolution; hydrodynamical problems in astrophysics; total solar eclipses. Mailing Add: PO Box 1663 Los Alamos Sci Lab Los Alamos NM 87545

COX, BEVERLEY LENORE, b Huntington, Pa, Jan 11, 29. ZOOLOGY, PHYSIOLOGY. Educ: Pa State Univ, BS, 51, MS, 53; Univ Okla, PhD(invert physiol), 60. Prof Exp: Instr physiol, Univ Okla, 59-60; res assoc & NIH fel, Univ Ore, 60-61; from asst prof to assoc prof, 61-70, PROF BIOL, CENT STATE UNIV, OKLA, 70- Concurrent Pos: NIH fel cardiovasc physiol, Sch Med, Univ Okla, 63-, vis assoc prof, 67 & 69, NSF fel isotope & nuclear reactor technol. Mem: AAAS; Am Soc Zool. Res: Physiology of insect tarsal chemoreception; gull food-finding behavior; recorded gull calls as repellants and attractants; endocrinology of mammalian salivary glands, molting in arachnids and electrolyte balance in lower vertebrates. Mailing Add: Dept of Biol Cent State Univ Edmond OK 73034

COX, BILLY JOE, b Carroll Co, Mo, May 2, 41; m 61; c 2. ENVIRONMENTAL SCIENCES. Educ: Univ Mo-Columbia, BA, 68, MA, 70, PhD(bot), 72. Prof Exp: Instr ecol, Univ Mo-Columbia, 68-73; plant sci group leader, Indust Bio-Test Labs, 73-74, sect head terrestrial ecol, 74-75, SECT HEAD TERRESTRIAL ECOL, NALCO ENVIRON SCI, NALCO CHEM CO, 75- Mem: Ecol Soc Am; Am Forestry Asn; Am Inst Biol Sci; Am Soc Plant Taxonomists; Torrey Bot Club. Res: Lupinus and Catalpa plant taxonomy, ecology and chemosystematics; physical activities, technical design and quality of ecological studies relative to environmental assessments of areas considered for industrial or commercial development. Mailing Add: NALCO Environ Sci 1500 Frontage Rd Northbrook IL 60062

COX, BRADLEY BURTON, b Danville, Ky, Oct 29, 41; m 64; c 1. ELEMENTARY PARTICLE PHYSICS. Educ: Duke Univ, PhD(physics), 67. Prof Exp: Res assoc high energy physics, Johns Hopkins Univ, 67, consult, 67-69, asst prof high energy physics, 69-73; group leader proton lab, 73-74, assoc head, 74-76, HEAD PROTON LAB, FERMI NAT ACCELERATOR LAB, 76- Concurrent Pos: Guest appointment, Dept Physics, Brookhaven Nat Lab, 67-, mem high energy discussion group, 69-; consult, Nuclear Effects Lab, 69-70. Mem: AAAS; Am Phys Soc. Res: Electromagnetic and strong interactions of elementary particles as they occur in high energy production and decay processes; low energy fusion reactions and identical target-projectile heavy ion reaction such as lithium-lithium or nitrogen-nitrogen scattering. Mailing Add: Fermi Nat Accelerator Lab PO Box 500 Batavia IL 60510

COX, BRUCE ALDEN, b Santa Rosa, Calif, June 29, 34; m 61; c 3. ANTHROPOLOGY, APPLIED ANTHROPOLOGY. Educ: Reed Col, BA, 56; Univ Ore, MA, 59; Univ Calif, Berkeley, PhD(anthrop), 68. Prof Exp: Vis instr anthrop, Lewis & Clark Col, 64-65; asst prof, Uni Fla, 66; asst prof, Univ Alta, 67-69; asst prof, 69-74, ASSOC PROF ANTHROP, CARLETON UNIV, 74- Concurrent Pos: Can Coun fel, Univ Alta & Carleton Univ, 69-71; mem ed bd, Western Can J Anthrop, 69-70 & Newstatements, 71-72; assoc, Clare Hall, Univ of Cambridge, 75-76; vis, Scott Polar Res Inst, Univ Cambridge, 75-76. Mem: Fel Royal Anthrop Inst Gt Brit & Ireland; Am Anthrop Asn; Soc Appl Anthrop. Res: Historical studies of human ecology of Subarctic Indians; cultural ecology of American indigenous peoples; law and indigenous peoples. Mailing Add: Dept of Sociol & Anthrop Carleton Univ Colonel By Dr Ottawa ON Can

COX, CHARLES DONALD, b Danville, Ill, Sept 10, 18; m 42; c 2. MEDICAL MICROBIOLOGY. Educ: Univ Ill, BS, 40, MS, 41, PhD(bact), 47. Prof Exp: Asst bact, Univ Ill, 40-42, 46-47; asst prof, Med Col Va, 47-49; assoc prof, Pa State Col, 49-51; prof microbiol & pub health & head dept, Sch Med, Univ SDak, 51-60; head microbiol br, Off Naval Res, Washington, DC, 60-62; head dept, 62-72, PROF MICROBIOL, UNIV MASS, AMHERST, 62- Concurrent Pos: Mem ed bd, J Bact, 57-70; mem adv panel environ biol, NASA, 65-72; mem comt naval med res, Nat Res Coun, 66-73; mem exobiol panel, Space Sci Bd, Nat Acad Sci, 74- Mem: AAAS; Am Soc Microbiol; Am Acad Microbiol; Soc Exp Biol & Med; NY Acad Sci. Res: Medical bacteriology and immunology; physiology and virulence of spirochetes. Mailing Add: Dept of Microbiol Univ of Mass Amherst MA 01002

COX, CHARLES PHILIP, b Eng, Dec 15, 19; m 62. STATISTICS, BIOMETRICS. Educ: Oxford Univ, BA, 40, MA, 47. Prof Exp: Head sect 4a, Army Oper Res Group, Eng, 45-46; head statist, Nat Inst Res Dairying, 48-61; PROF STATIST, IOWA STATE UNIV, 61- Concurrent Pos: Ministry of Agr res scholar, Exp Sta, Rothamsted, 47-48. Mem: Am Statist Asn; Biomet Soc; Royal Statist Soc. Res: Design and analysis of experiments; biological assay; biomathematics. Mailing Add: Statist Lab Iowa State Univ Ames IA 50010

COX, CHARLES SHIPLEY, b Hawaii, Sept 11, 22; m 51; c 5. OCEANOGRAPHY. Educ: Calif Inst Technol, BS, 44; Univ Calif, PhD, 54. Prof Exp: Asst oceanogr, 54-60, assoc prof, 60-66, PROF OCEANOG, SCRIPPS INST OCEANOG, UNIV CALIF, SAN DIEGO, 66-, CHMN OCEAN RES DIV, 73- Concurrent Pos: NSF fel, 57; Fulbright res fel, 57-58; vis prof, Mass Inst Technol, 69-70. Mem: Am Geophys Union; Royal Astron Soc. Res: Physical oceanography; geophysics relating to oceanic microstructure and magnetism. Mailing Add: Scripps Inst of Oceanog Univ of Calif at San Diego La Jolla CA 92093

COX, CLAIR EDWARD, II, b Lawrence Co, Ill, Sept 2, 33; m 58; c 4. SURGERY, UROLOGY. Educ: Univ Mich, MD, 58; Am Bd Urol, dipl, 66. Prof Exp: Intern med, Med Ctr, Univ Colo, 58-59, resident surg, 59-60; resident urol, Med Ctr, Univ Calif, San Francisco, 60-63; from instr to prof, Bowman Gray Sch Med, 63-73; CHMN DEPT UROL, COL MED, UNIV TENN, MEMPHIS, 73- Concurrent Pos: Nat Cancer Inst res grant, 69- Mem: Am Urol Asn; Am Col Surg; Am Asn Genitourinary Surg; Infectious Dis Soc Am; Soc Univ Urol. Res: Epidemiology of urinary tract infection; urological cancer. Mailing Add: Dept of Urol Univ of Tenn Col of Med Memphis TN 38163

COX, CLARENCE DONALD, solid state physics, see 12th edition

COX, DAVID BUCHTEL, b Denver, Colo, Jan 25, 27; m 53; c 2. RHEOLOGY, ORGANIC CHEMISTRY. Educ: DePauw Univ, AB, 48; Stanford Univ, MS, 50; Univ NMex, PhD(org chem), 53. Prof Exp: From technologist to sr technologist, Socony Mobil Oil Co, Inc, 53-58, res assoc, 58-67; sr res chemist, Battelle Mem Inst, 67-69, assoc chief lubrication mech div, 69-70; chief chemist, 70-73, TECH DIR, CHEM-TREND, INC, 73- Mem: Am Chem Soc; Soc Rheol. Res: Rheology of lubricants, fuels and other petroleum products; polymer solutions; colloid chemistry of lubricants; mold release agents for urethane foams; die-casting lubricants; fire-resistant hydraulic fluids. Mailing Add: 1620 Sheridan Dr Ann Arbor MI 48104

COX, DAVID ERNEST, b Rochford, Eng, Dec 12, 34; m 57; c 3. CRYSTALLOGRAPHY, SOLID STATE CHEMISTRY. Educ: Univ London, BSc, 55, PhD(inorg chem), 59. Prof Exp: Tech officer ceramics, Steatite & Porcelain Prod Div, Imp Chem Industs, Ltd, 58-59; from chemist to sr chemist, Res Labs, Westinghouse Elec Corp, 59-63; assoc physicist, 63-66, PHYSICIST, BROOKHAVEN NAT LAB, 66- Concurrent Pos: Asst ed, J Physics & Chem of Solids, 69; mem neutron diffraction comn, Int Union Crystallog, 72- Mem: Fel Am Phys Soc; Am Asn Crystal Growth. Res: Neutron diffraction; magnetic and crystal structures; synthesis and characterization of inorganic materials; crystal growth; magnetic measurements. Mailing Add: Physics Dept Brookhaven Nat Lab Upton NY 11973

COX, DAVID FRAME, b New York, NY, Feb 19, 31; m 54; c 1. STATISTICS, ANIMAL SCIENCE. Educ: Cornell Univ, BS, 53; NC State Col, MS, 57; Iowa State Univ, PhD(animal breeding, genetics), 59. Prof Exp: Assoc prof animal sci, 59-66, PROF STATIST, IOWA STATE UNIV, 66- Mem: Biomet Soc; Am Soc Animal Sci. Res: Design and analysis of experiments; quantitative inheritance; genetic statistics. Mailing Add: Dept of Statist Beardshear Hall Iowa State Univ Ames IA 50010

COX, DAVID JACKSON, b New York, NY, Dec 22, 34; m 58; c 3. PHYSICAL BIOCHEMISTRY. Educ: Wesleyan Univ, BA, 56; Univ Pa, PhD(biochem), 60. Prof Exp: Instr biochem, Univ Wash, Seattle, 60-63; from asst prof to assoc prof chem, Univ Tex, Austin, 63-73; PROF BIOCHEM & HEAD DEPT, KANS STATE UNIV, 73- Concurrent Pos: Investr, Howard Hughes Med Inst, 60-63; NSF fel & vis prof, Univ Va, 70-71. Mem: AAAS; Am Chem Soc; Am Soc Biol Chem; NY Acad Sci. Res: Physical chemistry of macromolecules; protein chemistry. Mailing Add: Dept of Biochem Kans State Univ Manhattan KS 66506

COX, DELANO KIMBERLING, plant genetics, plant physiology, see 12th edition

COX, DENNIS HENRY, b St Paul, Minn, Aug 12, 25; m 50; c 1. ANIMAL NUTRITION. Educ: Univ Minn, BS, 50, MS, 53; Univ Fla, PhD(animal nutrit), 55. Prof Exp: Assoc toxicologist, Ga Coastal Plain Exp Sta, 55-62; asst prof, Univ Iowa, 62-63 & Coker Col, 63-64; asst prof foods & nutrit, Pa State Univ, 64-68, assoc prof nutrit, 68-71; RES CHEMIST, DEPT HEALTH, EDUC & WELFARE, 71- Mem: AAAS; Am Chem Soc; Am Inst Nutrit; Soc Environ Geochem & Health. Res: Various aspects of animal and human nutrition, especially trace and macro minerals and their various interrelationships. Mailing Add: 3278 Lansbury Village Dr NE Atlanta GA 30341

COX, DENNIS PURVER, b Seattle, Wash, Sept 12, 29; m 56; c 3. ECONOMIC GEOLOGY. Educ: Stanford Univ, BS, 51, MS, 54, PhD(geol), 56. Prof Exp: Geologist mineral deposits explor, Anaconda Co, 56-59; vis prof geol, Univ Bahia, Brazil, 59-61; geologist mil geol, 61-65, geologist mineral resources, 65-72, copper resources specialist, 72-76, MGR MINERAL RESOURCE SPECIALIST PROG, US GEOL SURV, 76- Concurrent Pos: Staff consult, Off Technol Assessment, US Cong. Mem: Geol Soc Am; Soc Econ Geol. Res: Primary dipersion geochemistry in mineral exploration; geology of porphyry copper deposits; mineral resource modeling and resource estimation. Mailing Add: US Geol Surv Nat Ctr 913 Reston VA 22092

COX, DIANE WILSON, b Belleville, Ont, May 18, 37; m 61; c 3. HUMAN GENETICS. Educ: Univ Western Ont, BSc, 59; Univ Toronto, MA, 61; McGill Univ, PhD(genetics), 68. Prof Exp: Res asst genetics, McGill Univ, 63-64; fel, 67-70, INVESTR, HOSP FOR SICK CHILDREN, 70-; ASST PROF, DEPTS PEDIAT & MED GENETICS, UNIV TORONTO, 75- Mem: Genetics Soc Can; Am Soc Human Genetics. Res: Human serum protein polymorphisms; ceruloplasmin, alpha-1-antitrypsin; population studies; copper metabolism; Wilson's disease; metabolic diseases; role of proteases and protease inhibitors in disease. Mailing Add: Pediat Dept & Res Inst Hosp for Sick Children Toronto ON Can

COX, DOAK CAREY, b Wailuku, Maui, Hawaii, Jan 16, 17; m 41; c 5. ENVIRONMENTAL MANAGEMENT, ENVIRONMENTAL GEOLOGY. Educ: Univ Hawaii, BS, 38; Harvard Univ, AM, 41, PhD, 65. Prof Exp: Geologist, Metals Sect, US Geol Surv, 41-46; geophysicist, Exp Sta, Hawaiian Sugar Planters Asn, 46-60; geophysicist, Inst Geophys, 60-64, dir, Water Resources Res Ctr, 64-70, PROF GEOL, UNIV HAWAII, 60-, DIR ENVIRON CTR, 70- Concurrent Pos: Consult, Pac Islands, 46; hydrologist, Arno Exped, Pac Sci Bd, 50; secy, Tsunami Comn Int Union Geod & Geophys, 60-67; Tsunami adv, Hawaii State Civil Defense Div, 60-; chmn oceanog panel, Comn on Alaska Earthquake, Nat Acad Sci-Nat Res Coun, 64-73. Mem: Fel AAAS; fel Geol Soc Am; Seismol Soc Am; Am Geophys Union; Ecol Soc Am Res: Hawaii and Pacific geology and hydrology; tsunamis. Mailing Add: Environ Ctr Univ of Hawaii 2550 Campus Rd Honolulu HI 96822

COX, DONALD CODY, b Peoria, Ill, Mar 31, 36; m 63; c 2. VIROLOGY, MOLECULAR BIOLOGY. Educ: Northwestern Univ, BA, 58; Univ Mich, PhD(epidemiol sci), 65. Prof Exp: From asst prof to assoc prof microbiol, 65-72, ASSOC PROF BOT & MICROBIOL, UNIV OKLA, 72- Mem: AAAS; Am Soc Microbiol. Res: Studies concerning biochemical alterations occurring in virus-infected cells. Mailing Add: Dept of Bot & Microbiol Univ of Okla Norman OK 73069

COX, DONALD DAVID, b Maben, WVa, Aug 2, 26; m 46; c 3. PLANT SCIENCE. Educ: Marshall Univ, AB, 49, MA, 50; Syracuse Univ, PhD(plant sci), 58. Prof Exp: Teacher pub sch, WVa, 49-50; prof biol, Marshall Univ, 50-62; consult biol sci curric study, Univ Colo, Boulder, 62-63; PROF BIOL, STATE UNIV NY COL OSWEGO, 63-, ACTG CHMN DEPT, 74- Concurrent Pos: Chmn dept biol, Marshall Univ, 57-62; res botanist, Corps Engrs, US Army, Miss Waterways Exp Sta, 62; staff biologist, Comn Undergrad Educ Biol Sci, NSF, 70-71. Res: Post glacial forests in New York State as determined by the method of pollen analysis. Mailing Add: Dept of Biol State Univ of NY Col Oswego NY 13126

COX, DUDLEY, b Brooklyn, NY, Mar 3, 29; m 55; c 2. MICROBIOLOGY, CELL PHYSIOLOGY. Educ: Howard Univ, BS, 56; Long Island Univ, MS, 58; NY Univ, PhD(microbiol), 66. Prof Exp: RES ASSOC MICROBIOL, HASKINS LABS, 58-; ASST PROF BIOL, PACE COL, 66- Mem: AAAS; Soc Protozool; Brit Soc Gen Microbiol. Res: Nutrition and metabolic activity of microorganisms; use of physical and chemical variations to determine membrane integrity. Mailing Add: Pace Univ Pace Plaza New York NY 10038

COX, EDMOND RUDOLPH, JR, b Pascagoula, Miss, Nov 15, 32; m 59; c 1. PHYCOLOGY, MICROBIOLOGY. Educ: Univ Ala, BS, 57, MS, 60, PhD(bot), 67. Prof Exp: From instr to assoc prof biol, Middle Tenn State Univ, 61-68; PROF BIOL & CHMN DEPT, TENN WESLEYAN COL, 68- Mem: Phycol Soc Am; Int Phycol Soc. Res: Soil algae. Mailing Add: Dept of Biol Tenn Wesleyan Col Athens TN 37303

COX, EDWARD CHARLES, b Alberni, Can, June 28, 37; US citizen; m 60; c 3. GENETICS. Educ: Univ BC, BSc, 59; Univ Pa, PhD(biochem), 64. Prof Exp: Instr biochem, Univ Pa, 61-62; fel genetics, Stanford Univ, 64-67; asst prof biol & biochem, 67-72, ASSOC PROF BIOL & ASSOC DEAN OF COL, PRINCETON UNIV, 72- Mem: Genetics Soc Am. Res: The genetic control of mutation rates; developmental genetics of the cellular slime molds. Mailing Add: Dept of Biol Princeton Univ Princeton NJ 08540

COX, EDWIN, b Richmond, Va, Sept 20, 02; m 27; c 1. CHEMISTRY, CHEMICAL ENGINEERING. Educ: Va Mil Inst, BS & MS. Prof Exp: Chemist, Va-Carolina Chem Corp, Richmond, 20-21; res chemist, Tobacco By-Prod & Chem Corp, 22-26; vpres, Phosphate Prod Corp, 27-34; mgr chem div, Va-Carolina Chem Corp, 34-40, 46-48, dir res & develop, 46-52, vpres res, 48-52, opers, 52-58; chemist, chem engr & partner, Cox & Gillespie, 58-66, CHEMIST, CHEM ENGR & PARTNER, EDWIN COX ASSOCS, 66- Concurrent Pos: Pres, Tobacco By-Prod & Chem Corp; chmn, Sci Adv Comt, Va Selective Serv, 56-; trustee, Va Inst Sci Res: former mem bd dirs, Agr Res Inst & Indust Res Inst; US govt adv air pollution, water archaeol & state hwy markers. Honors & Awards: Order Knighthood St John; Distinguished Serv Award, Am Chem Soc, 52; Gold Medal Award, Am Inst Chemists, 65. Mem: AAAS; Am Chem Soc (secy, 24); Am Inst Chem Engrs; fel Am Inst Chemists. Res: Agricultural chemicals; phosphorus chemistry; nicotine; sulphur; proteins; biological phosphate chemistry; air and surface pollution; nonmetallic mining and metallurgy; process engineering. Mailing Add: Edwin Cox Assocs 2209 E Broadway Richmond VA 23223

COX, EDWIN LORY, b Moncton, NB, Feb 26, 14; nat US; m 56. MATHEMATICAL STATISTICS. Educ: Mt Allison Univ, BA, 33; Acadia Univ, MA, 40; Va Polytech Inst, MS, 49; NC State Univ, PhD(statist), 52. Prof Exp: Instr math, Va Polytech Inst, 46-47 & 48-49; asst biologist, Va Fisheries Lab, 47-48; asst, NC State Univ, 49-51; chief statistician, Dugway Proving Ground, Tooele, Utah, 51-53; res assoc, Case Western Reserve Univ, 53-54; chief prog res br, Biol Warfare Labs, US Army, Md, 54-58; MATH STATISTICIAN, AGR RES SERV, USDA, 58- Concurrent Pos: Lectr, Univ NB, 41-42 & Univ Md, 59-; anal statistician, US Fish & Wildlife Serv, 50-51; consult, USPHS, 57-58. Mem: AAAS; Am Soc Limnol & Oceanog; Inst Math Statist; Biomet Soc. Res: Application of statistics and mathematics to biological problems; model building for dynamic situations such as growth, population change, meteorology and ecology. Mailing Add: Agr Res Serv DSAD Nat Agr Libr Rm 013 Beltsville MD 20705

COX, EUGENE FLOYD, b Tifton, Ga, Dec 4, 28; m 52; c 3. ORGANIC CHEMISTRY, PLANT PHYSIOLOGY. Educ: Ga Inst Technol, BS, 50, MS, 51; Calif Inst Technol, PhD, 54. Prof Exp: NSF fel, Harvard Univ, 54-55; res chemist res & develop dept, Chem Div, Union Carbide Corp, 55-66, group leader res & develop dept, Chems & Plastics, 61-66, from asst dir to dir, 66-74; DIR CHEM HYG LAB, CARNEGIE-MELLON INST RES, 74- Mem: Am Chem Soc. Res: Plastic foams; small ring compounds; carbonium ion rearrangements; sulfur chemistry; reaction kinetics; addition and condensation polymerization; polyurethanes; isocyanate reactions; catalysis; plastics technology; toxicology, industrial hygiene, safety, environmental protection. Mailing Add: Carnegie-Mellon Inst of Res 4400 Fifth Ave Pittsburgh PA 15213

COX, EVERETT FRANKLIN, b Eaton, Ohio, Sept 8, 08; m 34, 67; c 2. ENGINEERING PHYSICS. Educ: Miami Univ, AB, 30; Calif Inst Technol, PhD(physics), 33. Prof Exp: From instr to asst prof physics, Colgate Univ, 33-39; asst dir, Buhl Planetarium & Inst Popular Sci, Univ Pittsburgh, 39-40; chief degaussing engr, Bur Ord, US Navy Dept, Pearl Harbor, 40-42, contract physicist, Naval Ord Lab, 42-48; mgr weapons effects dept, Sandia Corp, 48-56; res scientist, Whirlpool Corp, Mich, 56-68; CONSULT, 70- Concurrent Pos: Assoc, George Washington Univ, 46; consult, Sandia Corp, 46-51, div biol & med, US AEC, 49-71, Atlantic-Pac Interocean Canal Study Group, 65-70, Test Organ, 68-71; mem comt supersonic transport sonic boom, Nat Acad Sci, 64-71; vis prof, Fla Atlantic Univ, 70. Mem: Fel Am Phys Soc; Sigma Xi. Res: Effects of atomic weapons; sonic booms from supersonic aircraft; thermoelectric heat pumping. Mailing Add: 2001 Leisure Dr NW Winter Haven FL 33880

COX, FRED WARD, JR, b Atlanta, Ga, Dec 10, 14; m 39; c 3. INDUSTRIAL CHEMISTRY, RESEARCH ADMINISTRATION. Educ: Ga Inst Technol, BS, 36; Univ Wis, PhD(org chem), 39. Prof Exp: Res chemist, Rohm and Haas Co, Philadelphia, 38; res chemist, Goodyear Tire & Rubber Co, Akron, 39-42, group leader, 42-45; group leader, Southern Res Inst, 45-49, head appl chem div, 46-49; asst dir, Eng Exp Sta, Ga Inst Technol, 49-53; chief chemist, Deering Milliken Res Trust, 53-56; MGR ATLAS RES & DEVELOP LAB, ATLAS POWDER CO, TYLER CORP, 56- Concurrent Pos: Mem, Tamaqua Area Sch Bd, 63-73, pres, 70-73; mem, Schuylkill County Bd Educ, 65-73 & Schuylkill County Area Voc Schs Oper Bd. Mem: Am Chem Soc; Am Soc Testing & Mat; Franklin Inst; Soc Rheol. Res: High polymer technology; organic process development; explosives; research administration; explosive slurries; blasting initiators. Mailing Add: Atlas Res & Develop Lab Atlas Powder Co PO Box 271 Tamaqua PA 18252

COX, FREDERICK EUGENE, b Quincy, Mass, Nov 4, 38; m 67; c 2. PEDIATRICS, INFECTIOUS DISEASES. Educ: Boston Univ, BA, 60; Boston Univ, MD, 64. Prof Exp: Surg intern, Univ Hosp, Boston, 64-65; gen med officer, US Coast Guard Acad Hosp, USPHS, New London, Conn, 65-67; resident path, Boston City Hosp, Mass, 67-68; resident pediat, St Elizabeth's Hosp, Brighton, Mass, 68-69 & Boston Floating Hosp, 69-70; fel pediat infectious dis, Cleveland Metrop Gen Hosp, Ohio, 70-72; fel, 72-74, RES ASSOC INFECTIOUS DIS SERV, ST JUDE CHILDREN'S RES HOSP, MEMPHIS, TENN, 74- Mem: Fel Am Acad Pediat; Sigma Xi. Res: Cytomegalovirus infections and hospital acquired infections in children with cancer. Mailing Add: St Jude Children's Res Hosp 332 N Lauderdale Memphis TN 38101

COX, FREDERICK RUSSELL, b Sutherland, Nebr, Mar 11, 32; m 59; c 2. SOIL SCIENCE. Educ: Univ Nebr, BS, 53, MS, 58; NC State Univ, PhD(soils), 61. Prof Exp: ASSOC PROF SOIL SCI, NC STATE UNIV, 61- Mem: Am Soc Agron; Soil Sci Soc Am. Res: Soil fertility; micronutrient research. Mailing Add: Dept of Soil Sci NC State Univ Raleigh NC 27607

COX, GENE SPRACHER, b Norton, Va, Mar 21, 21; m 46; c 2. FOREST SOILS. Educ: Duke Univ, BS, 47, MF, 48, PhD(forestry), 53. Prof Exp: Asst prof forestry, Stephen F Austin State Univ, 51-53; from asst prof to assoc prof, Univ Mont, 53-60; assoc prof, 60-63, PROF FORESTRY, UNIV MO, COLUMBIA, 63- Mem: Am Soc Agron; Soil Sci Soc Am; Soc Am Foresters; Ecol Soc Am. Res: Forest ecology. Mailing Add: Sch of Forestry, Fish & Wildlife Univ of Mo Columbia MO 65201

COX, GEORGE ELTON, b Ayden, NC, July 22, 31; m 56; c 2. EXPERIMENTAL PATHOLOGY. Educ: Univ NC, BA, 51, MS, 54; Univ Ill, MD, 56. Prof Exp: Intern path, Presby Hosp, Chicago, 56-57; resident, Presby-St Luke's Hosp, Chicago, 57-60; dir exp path, Evanston Hosp, 61-62; res assoc biol chem, Col Med, Univ Ill, 62-64; res assoc & asst prof path, Med Units, Univ Tenn, 64-65; res assoc, May Inst Med Res, Cincinnati, 65-69; asst prof path, Col Med, Univ Cincinnati, 66-69; PATHOLOGIST, FOOD & DRUG RES LABS, INC, 69- Mem: Am Soc Exp Path. Res: Experimental study of etiology of atherosclerosis; spontaneous neoplasia in laboratory animals; histopathologic effects of environmental chemicals. Mailing Add: PO Box 107 Rte 17 Waverly NY 14892

COX, GEORGE STANLEY, b Roswell, NMex, Jan 26, 46; m 67; c 2. BIOCHEMISTRY, MOLECULAR BIOLOGY. Educ: NMex State Univ, BS, 68; Univ Iowa, PhD(biochem), 73. Prof Exp: Fel, Roche Inst Molecular Biol, 72-74; staff fel, Lab Molecular Biol, Nat Inst Arthritis, Metab & Digestive Dis, NIH, 74-76; ASST PROF BIOCHEM & BIOPHYS, IOWA STATE UNIV, 76- Mem: Am Soc Microbiol; Sigma Xi. Res: Regulation of gene expression in eukaryotes including transcription, translation and nuclear-cytoplasmic transport of macromolecules. Mailing Add: Dept of Biochem & Biophys Iowa State Univ Ames IA 50011

COX, GEORGE W, b Williamson, WVa, Feb 10, 35; m 57; c 2. ECOLOGY, ORNITHOLOGY. Educ: Ohio Wesleyan Univ, AB, 56; Univ Ill, MS, 58, PhD(zool), 60. Prof Exp: Asst prof biol, Univ Alaska, 60-61 & Calif Western Univ, 61-62; from asst prof to assoc prof, 62-69, PROF BIOL, SAN DIEGO STATE UNIV, 69- Mem: AAAS; Ecol Soc Am; Am Ornith Union; Wilson Ornith Soc. Res: Physiological ecology of birds; evolution and speciation in birds; ecology of chaparral communities; agricultural and conservation ecology. Mailing Add: Dept of Biol San Diego State Univ San Diego CA 92182

COX, H C, b Melrose, NMex, May 18, 27; m 54; c 2. ENTOMOLOGY. Educ: Univ NMex, BS, 50; Iowa State Univ, MS, 52, PhD(entom), 55. Prof Exp: Asst, Iowa State Univ, 50-53; entomologist, Europ Corn Borer Res Lab, Iowa, 53-56, small grain insects, Utah, 56-58, corn insects invests, Miss, 58-60, dir & invest leader, Southern Grain Insects Res Lab, Ga, 60-67, asst dir entom res div, Md, 67-71 & dir, 71-72, assoc dep adminr, La, 72-76, DEP ADMINR, AGR RES SERV, USDA, CALIF, 76- Concurrent Pos: Fel, Princeton Univ, 66-67. Mem: Entom Soc Am. Res: Animal and veterinary sciences; agricultural engineering; soil and water conservation; plant and entomological sciences; market quality; agricultural product transportation and storage; agricultural products utilization and industrial processing research. Mailing Add: 2850 Telegraph Ave Berkeley CA 94705

COX, HENRY MIOT, b Stephens Co, Ga, 07; m 35; c 1. MATHEMATICS, PSYCHOMETRICS. Educ: Emory Univ, BS, 28; Duke Univ, AM, 31. Prof Exp: Instr math & asst exam, Univ Syst Ga, 30-39; from asst dir to dir, 39-73, EMER DIR, BUR INSTR RES & EXAM SERV, UNIV NEBR, LINCOLN, 73- Concurrent Pos: Exec dir, Ann High Sch Math Exam, Math Asn Am, Soc Actuaries, Mu Alpha Theta & Nat Coun Teachers Math, 70- Mem: AAAS; Math Asn Am; Psychomet Soc; Nat Coun Measurement in Educ. Res: Measurement; instructional research; student guidance. Mailing Add: 1145 N 44th St Lincoln NE 68503

COX, HERALD REA, b Rosedale, Ind, Feb 28, 07; m 32; c 3. VIROLOGY. Educ: Ind State Col, AB, 28; Johns Hopkins Univ, ScD(filterable viruses), 31. Hon Degrees: ScD, Univ Mont, 42, Ind State Col, 64 & Roswell Park Mem Inst, 72. Prof Exp: Instr immunol, Johns Hopkins Univ, 31-32; asst path & bact, Rockefeller Inst, 32-36; assoc bacteriologist, USPHS, 36-40, prin bacteriologist, 40-42; from assoc dir to dir viral res, Lederle Labs, 42-68; dir cancer res, Viral Oncol Sect, Roswell Park Mem Inst, 68-72; RETIRED. Honors & Awards: Theobald Smith Award, 41; Typhus Comn Medal, Secy of War, 46; Ricketts Award, Univ Chicago, 51. Mem: AAAS; hon mem Am Soc Microbiol (vpres, Am Soc Bact, 59-60, pres, 60-61); Am Soc Immunologists; Am Soc Trop Med & Hyg; Am Pub Health Asn. Res: Neurotropic viral diseases; rickettsial diseases; viral infections of man and animal; living trivalent polio vaccine; living virus vaccines, human and veterinary. Mailing Add: PO Box 937 Hamilton MT 59840

COX, HERBERT WALTON, b Clarkton, NC, Mar 15, 18; m 45. MALARIOLOGY. Educ: Univ NC, AB, 41, MPH, 48, PhD(parasitol), 52. Prof Exp: Instr prev med, Bowman Gray Sch Med, 48-49; instr bact, Med Col, Cornell Univ, 52-54; from instr to asst prof microbiol, State Univ NY Downstate Med Ctr, 54-64; asst prof path & hyg, Col Vet Med, Univ Ill, 64-66; assoc prof microbiol & pub health, 66-74, PROF MICROBIOL & PUB HEALTH, MICH STATE UNIV, 75- Mem: Am Soc Trop Med & Hyg; Am Soc Parasitol; Royal Soc Trop Med & Hyg; Soc Protozool; NY Acad Sci. Res: Immunopathology of malaria and infectious anemias. Mailing Add: Dept of Microbiol & Pub Health Mich State Univ East Lansing MI 48823

COX, HIDEN TOY, b Greenville, SC, Mar 3, 17; m 43; c 1. PLANT MORPHOLOGY, PLANT ANATOMY. Educ: Furman Univ, BS, 36, BA, 37; Univ NC, MA, 39, PhD(bot), 47. Prof Exp: Asst prof biol, Howard Col, 41-46; assoc prof bot, Agnes Scott Col, 46-49; from assoc prof to prof, Va Polytech Inst, 49-55; dep exec dir, Am Inst Biol Sci, 53-54, exec dir, 55-63; dean sch letters & sci, 67-72, PROF BIOL & COORDR RES, CALIF STATE UNIV, LONG BEACH, 63- Concurrent Pos: Asst adminr pub affairs, NASA, 61-62; Beattie lectr, 64; mem, Nat Coun Univ Res Adminr. Honors & Awards: Distinguished Serv Citation, NASA, 62. Mem: Fel AAAS; Bot Soc Am; Asn Advan Biomed Educ. Res: Comparative wood anatomy. Mailing Add: Dept of Biol Calif State Univ 6101 E Seventh St Long Beach CA 90804

COX, HOLLACE LAWTON, JR, b Oak Park, Ill, Nov 17, 35; m 59. MEDICAL PHYSICS, ATOMIC PHYSICS. Educ: Univ Rochester, AB, 59; Ind Univ, Bloomington, PhD(chem physics), 67. Prof Exp: Mem tech staff, Tex Instruments, Inc, 67-69; Robert A Welch res fel, Baylor Univ, 70-73; Robert A Welch res fel path, 73-75, RES FEL PHYSICS, UNIV TEX SYST CANCER CTR, M D ANDERSON HOSP & TUMOR INST HOUSTON, 75- Mem: Sigma Xi; Am Phys Soc; Optical Soc Am. Res: X-ray spectroscopy; x-ray and electron scattering; linear energy transfer measurements; stopping power of alpha particles. Mailing Add: Dept of Physics Univ of Tex Syst Cancer Ctr M D Anderson Hosp & Tumor Inst Houston TX 77025

COX, JAMES ALLAN, b Chisholm, Minn, Sept 19, 41; m 65; c 2. ANALYTICAL CHEMISTRY, ELECTROCHEMISTRY. Educ: Univ Minn, BS, 63; Univ Ill, PhD(chem), 67. Prof Exp: Lectr chem, Univ Wis, 67-69; asst prof, 69-74, ASSOC PROF CHEM, SOUTHERN ILL UNIV, CARBONDALE, 74- Mem: Am Chem Soc; Soc Appl Spectroscopists; Sigma Xi. Res: Trace methods for anions; ion exchange membrane applications to trace analysis, rates and mechanisms of heterogeneous electron transfer reactions. Mailing Add: Dept of Chem Neckers Bldg Southern Ill Univ Carbondale IL 62901

COX, JAMES CARL, JR, b Wolf Summit, WVa, June 17, 19; m 45; c 5. PHYSICAL CHEMISTRY, ORGANIC CHEMISTRY Educ: WVa Wesleyan Col, BS, 40; Univ Del, MS, 47, PhD(org chem), 49; Univ Md, LLB, 55, JD(constitutional law), 65. Prof Exp: Lab instr chem & physics, WVa Wesleyan Col, 37-40; res chemist, Ammonia Dept, E I du Pont de Nemours & Co, 40-43; instr chem, Univ Del, 46-49; prof chem & physics & head dept, Wesleyan Col, 49-51; civilian prof chem, Dept Elec Eng, US Naval Acad, 51-55; assoc prof, Lamar State Col, 55-56; prof, Univ Baghdad, 57-58 & Lamar State Col, 58-65; dir div sci & math, Oral Roberts Univ, 65-68; HEAD DEPT CHEM, WAYLAND BAPTIST COL, 68- Concurrent Pos: Transl, Chem Abstr, 47- & Trans-Chem, Inc, Tenn; Carnegie res fel, 49-51; prof, Med Tenn State Col, 50; ed, Condenser, 59-; consult, E I du Pont de Nemours & Co, 62-, Continental Oil Co, 62- & Texaco Inc, 63-; Nat Sci Found vis scientist, 63-64. Mem: AAAS; Am Chem Soc; Chem Soc Iraq; fel Acad Sci. Res: Synthesis and dehydration of secondary alcohols; intramolecular rearrangements; Grignard reactions; oxidation of organic compounds; aldol condensations; organometallic compounds; organic components of industrial water; Leuckart reaction; environmental protection; hydrocarbon synthesis; alicyclic compounds; carbohydrates; natural compounds. Mailing Add: Dept of Chem Wayland Baptist Col Plainview TX 79072

COX, JAMES LEE, b Wayne Co, Ind, Oct 5, 38; m 71. ANIMAL NUTRITION. Educ: Purdue Univ, BS, 61; Univ Ill, MS, 63, PhD(animal nutrit), 67. Prof Exp: Sr res physiologist, 67-71, animal sci data analyst, 71-74, RES FEL, MERCK SHARP & DOHME RES LABS, 74- Mem: AAAS; Am Soc Animal Sci; Poultry Sci Asn; Sigma Xi; World Poultry Sci Asn. Res: Design and analysis of animal experiments; environmental effects on pregnant animals; mineral nutrition and physiological functions; influence of drugs on growth of animals. Mailing Add: Merck Sharp & Dohme Res Labs Rahway NJ 07065

COX, JAMES LESTER, JR, b Reidsville, NC, Nov 12, 42; m 63; c 2. PLASMA PHYSICS. Educ: NC State Univ, BS, 63, MS, 65, PhD(physics), 69. Prof Exp: ASST PROF PHYSICS, OLD DOMINION UNIV, 69- Mem: AAAS; Am Phys Soc. Res: Relativistic elecfron beams; laser induced electron emission. Mailing Add: 600 Downing Crescent Virginia Beach VA 23452

COX, JAMES REED, JR, b Nashville, Tenn, Mar 25, 32. ORGANIC CHEMISTRY. Educ: Vanderbilt Univ, BA, 54, MA, 55; Harvard Univ, PhD(chem), 59. Prof Exp: NSF fel chem, Univ Munich, 58-59; from asst prof to assoc prof, Ga Inst Technol, 59-66; ASSOC PROF CHEM, UNIV HOUSTON, 66- Mem: Am Chem Soc. Res: Structure-reactivity relationships; mechanisms of reaction of organic and bio-organic chemical systems. Mailing Add: Dept of Chem Univ of Houston Houston TX 77004

COX, JOHN PAUL, b Ft Myers, Fla, Nov 4, 26. THEORETICAL ASTROPHYSICS. Educ: Ind Univ, AB, 49, MS, 50, PhD(theoret astrophys), 54. Prof Exp: Asst, Ind Univ, 49-54; from instr to asst prof astron, Cornell Univ, 54-62; vis scientist, Courant Inst Math Sci, NY Univ, 62-63; vis fel, Joint Inst Lab Astrophys, 63, assoc prof, 63-65, PROF ASTROPHYS, UNIV COLO, BOULDER, 65- Concurrent Pos: Cotterell res grant, 56; consult, Los Alamos Sci Lab, 60- Mem: Int Astron Union; Am Astron Soc; Am Phys Soc; NY Acad Sci. Res: Theory of stellar variability; stellar interiors and evolution; cosmology and cosmogony. Mailing Add: Joint Inst for Lab Astrophys Univ of Colo Boulder CO 80304

COX, JOHN WESLEY, JR, inorganic chemistry, x-ray crystallography, see 12th edition

COX, JOHN WILLIAM, b St Louis, Mo, Aug 31, 26; m 49; c 1. MEDICAL EDUCATION, CARDIOLOGY. Educ: St Louis Univ, MD, 51, PhD(physiol), 53; Am Bd Internal Med, dipl. Prof Exp: Chief res, Labs, Vet Admin Hosp, St Louis, 53-54; US Navy, 54-, chief pulmonary lab, Dept Internal Med, US Naval Hosp, San Diego, 54-61, chief med & dir clin serv, US Naval Hosp, Subic Bay, Philippines, 61-63, head cardio-respiratory dis br, US Naval Hosp, Philadelphia, 63-66, chief med & dir res, 66-69, head training & clin serv br, 69-72, dir med educ & training, Navy

Dept, 72-75, CMNDG OFFICER, NAVAL HEALTH SCI EDUC & TRAINING COMMAND, 75- Concurrent Pos: Assoc prof med, Jefferson Med Col, 64-72. Mem: Fel Am Col Physicians; Am Col Chest Physicians; Am Col Cardiol; Am Heart Asn; Asn Mil Surg US. Res: Clinical investigations; cardiopulmonary physiology. Mailing Add: Naval Health Sci Educ & Training Command Bethesda MD 20014

COX, JOSEPH ROBERT, b Lafayette, Ind, Mar 15, 34; m 56; c 3. PHYSICS. Educ: Harvard Univ, BA, 56; Ind Univ, MS, 57, PhD(physics), 62. Prof Exp: Asst prof physics, Univ Miami, 62-64; asst prof, 64-67, ASSOC PROF PHYSICS, FLA ATLANTIC UNIV, 67- Concurrent Pos: Res assoc, Yale Univ, 67-68. Mem: Am Phys Soc. Res: Scattering theory. Mailing Add: Dept of Physics Fla Atlantic Univ Boca Raton FL 33432

COX, KENT WALTER, physical chemistry, see 12th edition

COX, KEVIN ROBERT, b Warwick, Eng, Mar 22, 39; m 65. GEOGRAPHY Educ: Cambridge Univ, BA, 61; Univ Ill, MA, 63, PhD(geog), 66. Prof Exp: From asst prof to assoc prof, 65-71, PROF GEOG, OHIO STATE UNIV, 71- Mem: Asn Am Geogr; Regional Sci Asn; Regional Studies Asn; Peace Sci Soc. Res: Behavior in a spatial context. Mailing Add: Dept of Geog Ohio State Univ Columbus OH 43210

COX, LAWRENCE EDWARD, b Salina, Kans, May 4, 44; m 67; c 1. SPECTROCHEMISTRY. Educ: Kans State Univ, BS, 66; Ind Univ, PhD(chem), 70. Prof Exp: Fel chem, Univ Ga, 70-72; STAFF MEM CHEM, LOS ALAMOS SCI LAB, 72- Mem: Am Chem Soc. Res: Spectrochemical multielement analysis by means of flameless atomic absorption and inductively coupled plasma emission; surface analysis utilizing electron spectroscopic chemical analysis and auger spectroscopy. Mailing Add: Los Alamos Sci Lab CMB-1 MS 740 Los Alamos NM 87545

COX, LIONEL AUDLEY, b Winnipeg, Man, Sept 18, 16; m 42; c 3. ORGANIC CHEMISTRY, PHYSICAL CHEMISTRY. Educ: Univ BC, BA, 41, MA, 43; McGill Univ, PhD(chem), 46, PEng, 75. Prof Exp: Teacher sci, Univ Sch, Victoria Univ, 35-40; chief chemist & consult, Sidney Roofing & Paper Co, BC, 41-44; res chemist, Am Viscose Corp, Pa, 46-51, sr res chemist, 51-53; vpres & dir res, Johnson & Johnson, Ltd, Can, 53-61, vpres & dir res & eng, Personal Prod Co, div Johnson & Johnson, NJ, 61-65; dir res, 65-73, DIR TECHNOL ASSESSMENT, MacMILLAN BLOEDEL LTD, 73- Concurrent Pos: Lectr, Univ BC, 43-44; mem Sci Coun Can; mem bd dir, Educ Inst BC. Mem: Fel AAAS; chem Inst Can; The Chem Soc; Tech Asn Pulp & Paper Indust; Am Chem Soc. Res: Natural and man-made fibers; wood products; pulp and paper; non-woven fibers; pollution control technology; management and administration of research and development engineering. Mailing Add: MacMillan Bloedel Ltd 1075 W Georgia St Vancouver BC Can

COX, LOUIS THOMAS, JR, b Elizabeth City, NC, Oct 17, 15; m 39; c 2. SCIENCE EDUCATION. Educ: Towson State Col, BS, 39; Columbia Univ, MA, 47, EdD, 59. pub schs, Md, 39-43; chmn dept, 64-67, PROF SCI, TOWSON STATE COL, 47- Concurrent Pos: Consult var schs & teacher groups, Md, 50-; NSF lectr, 60-64. Mem: Nat Sci Teachers Asn. Res: Physical and earth science; science teacher education; working with kindergarten children in science; energy in waves. Mailing Add: Dept of Physics Towson State Col Towson MD 21204

COX, MARTHA, b Chappaqua, NY, Oct 23, 08. PHYSICS Educ: Cornell Univ, AB, 29; Bryn Mawr Col, AM, 36, PhD(physics), 42. Prof Exp: Asst to res physicist, Taylor Instrument Co, 29-30; lectr physics, Huguenot Univ Col, SAfrica, 31-33; demonstr, Bryn Mawr Col, 34-36; teacher, Shipley Sch, 36-38; instr, Bryn Mawr Col, 39-43; asst prof, Newcomb Col, Tulane Univ, 43-44; res physicist, Lukas Harold Corp, Ind, 44-45; physicist, US Naval Avionics Facil, 45-46, from actg head to head physics div, 46-58, tech consult, Appl Res Dept, 58-73, mgr mat lab & consult div, 73-75; RETIRED. Mem: Am Phys Soc. Res: Design of avionics equipment; guidance and control; mathematical and numerical analysis. Mailing Add: 3042-A Lake Shore Dr Indianapolis IN 46205

COX, MARY GRIFFITH, organic chemistry, see 12th edition

COX, MILTON D, b Indianapolis, Ind, Jan 13, 39; m 66; c 2. ALGEBRA, GEOMETRY. Educ: DePauw Univ, BA, 61; Ind Univ, MA, 64, PhD(quasi-finite fields), 66. Prof Exp: From asst prof to ASSOC PROF MATH, MIAMI UNIV, 66- Mem: Am Math Soc; Math Asn Am. Res: Class field theory; quasi-finite fields. Mailing Add: Dept of Math & Statist Miami Univ Oxford OH 45056

COX, NELSON ANTHONY, b New Orleans, La, Jan 6, 43; m 63; c 3. MICROBIOLOGY. Educ: La State Univ, Baton Rouge, BS, 66, MS, 68, PhD(poultry sci), 71. Prof Exp: Microbiol consult, Supreme Sugar Refinery, La, 69-70; MICROBIOLOGIST, RICHARD B RUSSELL AGR RES CTR, AGR RES SERV, USDA, 71- Honors & Awards: Ralston Purina Res Award, Southern Asn Agr Scientists, 72. Mem: Am Soc Microbiologists; Inst Food Technologists; Poultry Sci Asn; Soc Appl Bact; World's Poultry Sci Asn. Res: Destruction of salmonellae on poultry carcasses; development of improved sampling and cultural methods for detection of salmonellae in poultry; microbiological evaluation of immersion versus air chilling of broilers. Mailing Add: Animal Prod Lab Russell Res Ctr Athens GA 30604

COX, NORMAN LOUIS, chemistry, see 12th edition

COX, PRENTISS GWENDOLYN, b New Augusta, Miss, June 9, 32; m 53; c 4. DEVELOPMENTAL BIOLOGY. Educ: Southern Miss Univ, BS, 57; Univ Miss, MS, 61; Case Western Reserve Univ, PhD(regeneration), 68. Prof Exp: Instr sci, Clark Mem Col, 57-64; NIH res fel, Case Western Reserve Univ, 68-69; ASSOC PROF BIOL, MISS COL, 69- Mem: AAAS; Am Inst Biol Sci; Am Soc Zool; Soc Develop Biol. Res: Vertebrate regeneration, especially lizard tail regeneration and in vitro culture of myogeneic cells from the regenerating tail. Mailing Add: Dept of Biol Sci Miss Col Clinton MS 39058

COX, RAY, b Donalsonville, Ga, Dec 2, 43; m 65; c 2. BIOCHEMISTRY. Educ: Berry Col, BS, 65; Auburn Univ, MS, 67, PhD(biochem), 70. Prof Exp: ASST PROF BIOCHEM, VET ADMIN HOSP & UNIV TENN, 72- Concurrent Pos: Fel, Fels Res Inst, Sch Med, Temple Univ, 71-72. Mem: Am Asn Cancer Res. Res: Biochemistry of chemical carcinogenesis. Mailing Add: Vet Admin Hosp Res Serv 1030 Jefferson Memphis TN 38104

COX, RAYMOND H, b Meadville, Pa, Mar 26, 36. MATHEMATICS. Educ: Allegheny Col, BS, 58; Univ NC, MA, 61, PhD(math), 63. Prof Exp: Asst prof, 63-69, ASSOC PROF MATH, UNIV KY, 69- Mem: Am Math Soc. Res: Hilbert space theory; linear space theory. Mailing Add: Dept of Math Univ of Ky Lexington KY 40506

COX, RAYMOND H, JR, b Mt Vernon, Ind, Aug 17, 43; m 62; c 2. PSYCHOPHARMACOLOGY. Educ: Ind Univ, BA, 66, MS, 68, PhD(pharmacol), 70. Prof Exp: Res asst psychopharmacol, Ind Univ, 66-70, sr scientist, 70-73, SR INVESTR, MEAD JOHNSON & CO, 73- Concurrent Pos: Adj lectr, Ind State Univ, Evansville, 74-; adj lectr, Univ Evansville, 75- Mem: AAAS; Sigma Xi; Soc Neurosci. Res: Investigations into the mechanism action and behavioral effects of psychotropic drugs and the development of new behavioral methods based on a criterion of analysis of aberrant human behavior. Mailing Add: Mead Johnson Res Ctr Evansville IN 47712

COX, RICHARD HARVEY, b Oakland, Ky, May 21, 43; m 67. NUCLEAR MAGNETIC RESONANCE, ORGANIC CHEMISTRY. Educ: Univ Western Ky, BS, 63; Univ Ky, PhD(org chem), 66. Prof Exp: Res chemist, Northern Regional Lab, USDA, summer 63; res fel chem, Mellon Inst, 67; asst prof, 68-73, ASSOC PROF CHEM, UNIV GA, 73- Mem: Am Chem Soc; The Chem Soc. Res: Applications of nuclear magnetic resonance in chemistry; conformational analysis; alkai metal reductions of hydrocarbons; theoretical aspects of nuclear magnetic resonance spectroscopy; drug binding to proteins; organometallic chemistry. Mailing Add: Dept of Chem Univ of Ga Athens GA 30601

COX, ROBERT HAMES, pharmaceutical chemistry, organic chemistry, see 12th edition

COX, ROBERT HAROLD, b Philadelphia, Pa, Sept 10, 37; m 62; c 2. PHYSIOLOGY, BIOENGINEERING. Educ: Drexel Inst Technol, BS, 61, MS, 62; Univ Pa, PhD(biomed eng), 67. Prof Exp: Assoc physiol, 67-69, asst prof, 69-72, ASSOC PROF PHYSIOL, UNIV PA, 72-, ASSOC PROF BIOMECH, 73-, ASSOC DIR BOCKUS RES INST, 70- Concurrent Pos: Nat Heart & Lung Inst grant, Bockus Res Inst, Univ Pa, 75-78. Mem: Inst Elec & Electronic Engrs; AAAS; Sigma Xi; Am Physiol Soc. Res: Vascular smooth muscle mechanics; arterial wall physiology; hypertension; carotid sinus reflex. Mailing Add: Univ of Pa Bockus Res Inst 19th & Lombard Sts Philadelphia PA 19146

COX, RODY POWELL, b New Brighton, Pa, June 24, 26; m 53; c 3. MEDICAL GENETICS. Educ: Univ Pa, MD, 52. Prof Exp: Asst clin instr med, Univ Mich, 53-54; instr, Univ Pa, 54-56, assoc, 56-68, asst prof med & res med, 58-60; res assoc genetics, Glasgow Univ, 60-61; from asst prof to assoc prof med, 61-71, assoc prof pharmacol, 70-71, PROF MED & PHARMACOL, NY UNIV, 71- Concurrent Pos: Fel, Arthritis & Rheumatism Found, 57-59; USPHS res fel, 60-61; career scientist, Health Res Coun, New York, 61-; dir summer res inst, Will Rogers Hosp, 62-66; dir coun, Asn Career Scientists, 66-68; dir med scientist training prog, NY Univ, 67-; mem metab study sect, NIH, 70-73; dir div human genetics, NY Univ. Mem: Am Soc Clin Invest; Am Soc Human Genetics; Asn Am Physicians; Am Col Physicians; Harvey Soc. Res: Biochemical genetics; somatic cell genetics; mammalian cell regulatory mechanism; mechanisms of hormone action; pharmacology; tissue culture. Mailing Add: NY Univ Med Ctr 550 First Ave New York NY 10016

COX, SAMSON ARTHUR, nuclear physics, see 12th edition

COX, STEPHEN KENT, b Galesburg, Ill, Sept 2, 40; m 61; c 3. ATMOSPHERIC PHYSICS. Educ: Knox Col, Ill, BA, 62; Univ Wis, Madison, MS, 64, PhD(meteorol), 67. Prof Exp: Res meteorologist, Atmospheric Physics & Chem Lab, Environ Sci Serv Admin, 64-66; scientist, Space Sci & Eng Ctr, Univ Wis, 66-69; asst prof, 69-72, ASSOC PROF ATMOSPHERIC SCI, COLO STATE UNIV, 72- Concurrent Pos: Scientist, Dept Meteorol, Univ Wis, 67-69, prin investr, Environ Sci Serv Admin grant, Univ Wis, 68-69 & Colo State Univ, 69-70; NSF grant, Colo State Univ, 69-; chmn flight facil adv panel, Nat Ctr Atmospheric Res, 70-73; GATE Radiation Subprog scientist, 74, mem, Nat Acad Sci GATE adv panel, 74-, US GATE Radiation Subprog coordr, 75- Mem: Am Meteorol Soc. Res: Atmospheric heat budget; radiation parameterization for numerical models; meteorological field experiments; radiative transfer. Mailing Add: Dept of Atmospheric Sci Colo State Univ Ft Collins CO 80521

COX, WILLIAM LESTER, b Youngstown, Ohio, Sept 18, 24; m 47, 67; c 9. PHYSICAL CHEMISTRY, ORGANIC CHEMISTRY. Educ: Muskingum Col, BS, 48; Case Western Reserve Univ, MS, 50, PhD(org chem), 53. Prof Exp: Instr chem, Case Western Reserve Univ, 52-54; res chemist, Universal Oil Prod Co, 54-59, res coordr, 59-62, asst dir chem prod res, 62-64; res scientist, 64-66, mgr rubber chem res, 66-75, mgr sci liaison, 73-75, MGR ORG CHEM RES, GOODYEAR TIRE & RUBBER CO, 75- Mem: Am Chem Soc; Am Mgt Asn; AAAS. Res: Rubber chemistry; organic synthesis; polymer degradation. Mailing Add: 285 Pheasant Dr Mogadore OH 44260

COXETER, HAROLD SCOTT MACDONALD, b London, Eng, Feb 9, 07; m 36; c 2. MATHEMATICS. Educ: Cambridge Univ, BA, 29, PhD(geom), 31. Honors & Awards: LLD, Univ Alta, 56, Trent Univ, 73; DMath, Univ Waterloo, 69; DSc, Acadia Univ, 71. Prof Exp: Fel, Trinity Col, Cambridge Univ, 31-35; from asst prof to assoc prof, 36-48, PROF MATH, UNIV TORONTO, 48- Concurrent Pos: Rockefeller Found fel, Princeton Univ, 32-33, Procter fel, 34-35; vis prof, Univ Notre Dame, 47, Columbia Univ, 49, Dartmouth Col, 64, Fla Atlantic Univ, 65, Univ Amsterdam, 66, Univ Edinburgh, 67, Univ E Anglia, 68, Australian Nat Univ, 69, Univ Sussex, 72 & Univ Warwick, 76; ed in chief, Can J Math, 49-58; pres, Int Cong Mathematicians, 74. Honors & Awards: Tory Medal, 49. Mem: Am Math Soc; Math Asn Am; fel Royal Soc; fel Royal Soc Can; for mem Royal Netherlands Acad Arts & Sci. Res: Regular and semi-regular polytopes; abstract groups; non-Euclidean geometry; configurations. Mailing Add: 67 Roxborough Dr Toronto ON Can

COY, DAVID HOWARD, b Manchester, Eng, Sept 15, 44; m 69; c 1. ENDOCRINOLOGY. Educ: Univ Manchester, BSc, 66, PhD(chem), 69. Prof Exp: Res assoc chem, Univ Toledo, 69-70; teaching assoc biochem, Med Col Ohio, 70-72; ASST PROF MED, SCH MED, TULANE UNIV, 72- Concurrent Pos: Res assoc, Vet Admin Hosp, New Orleans, 72- Mem: Am Chem Soc. Res: Chemistry and biological properties of peptide hormones and their analogs. Mailing Add: Dept of Med Tulane Univ Sch of Med New Orleans LA 70112

COY, NETTIE HELENA, physics, see 12th edition

COY, RICHARD EUGENE, b New Kensington, Pa, Oct 28, 25; m 57; c 2. DENTISTRY. Educ: Univ Pittsburgh, BS, 49, DDS, 51, MS, 59. Prof Exp: Asst prof prosthodontics, Univ Pittsburgh, 60-70; PROF PROSTHODONTICS, SCH DENT MED, SOUTHERN ILL UNIV, EDWARDSVILLE, 70- Concurrent Pos: Consult, Vet Admin Hosps, 63-; assoc prof, Med Ctr, St Louis Univ, 71-; chmn continuing educ, Am Asn Dent Schs, 74, chmn prosthodont, 76. Mem: Fel Am Col Dentists; fel Am Col Prosthodontists; Int Asn Dent Res; Int Asn Dento-Facial Abnormalities; Am Equilibration Soc (secy, 68-). Res: Study of craniofacial abnormalities; study of pain involved in temporomandibular dysfunction syndrome. Mailing Add: Sch of Dent Med Southern Ill Univ Edwardsville IL 62025

COYE, ROBERT DUDLEY, b Los Angeles, Calif, Dec 17, 24; m 46; c 3. ANATOMIC PATHOLOGY. Educ: Williams Col, BA, 48; Univ Rochester, MD, 52. Prof Exp: Resident path, Univ Rochester, 52-55; from asst prof to prof, Univ Wis-Madison, 55-72; DEAN SCH MED, WAYNE STATE UNIV, 72- Res: Pathology of kidney. Mailing Add: 540 E Canfield Ave Detroit MI 48201

COYIER, DUANE L, b Aurora, Ill, Mar 14, 26; m 47; c 3. HORTICULTURE. Educ: Univ Wis, BS, 50, PhD(plant path), 61. Prof Exp: Plant pathologist, Tree Fruit Dis Invests, 61-73, PLANT PATHOLOGIST, FUNGUS & BACT DIS TREE FRUITS, FOLIAR DIS ORNAMENTAL PLANTS, AGR RES SERV, USDA, 73- Mem: Am Phytopath Soc. Mailing Add: Ornamental Res Lab Agr Res Serv USDA 3420 SW Orchard St Corvallis OR 97330

COYKENDALL, ALAN LITTLEFIELD, b Hartford, Conn, Jan 11, 37; m 59; c 1. ORAL MICROBIOLOGY. Educ: Bates Col, BS, 59; Tufts Univ, DMD, 63; George Washington Univ, MS, 70. Prof Exp: Fel microbiol, US Naval Dent Res Inst, 64-65, microbiologist, US Naval Med Res Inst, 67-71, microbiologist, US Naval Dent Res Inst, 71-72; res assoc dent, 72-75, CLIN INVESTR, VET ADMIN HOSP, NEWINGTON, 75- Mem: AAAS; Int Asn Dent Res; Am Soc Microbiol. Res: Tooth replantation; taxonomy of oral bacteria; retention of sugar in the mouth; nucleic acids of cariogenic streptococci; guanine-cytosine content and homologies. Mailing Add: Dent Res Vet Admin Hosp Newington CT 06111

COYLE, BERNARD ANDREW, b Pukekohe, NZ, May 2, 34; m 60; c 3. INORGANIC CHEMISTRY, CRYSTALLOGRAPHY. Educ: Univ NZ, BSc, 55, MSc, 56; Northwestern Univ, PhD(inorg chem), 69. Prof Exp: Instr chem, City Col San Francisco, 60-66; asst prof, NCent Col, Ill, 69-71; PROF CHEM, CITY COL SAN FRANCISCO, 71- Concurrent Pos: Guest scientist, Argonne Nat Lab, 67-73; vis chemist, Brookhaven Nat Lab, 70-71; lectr chem, Univ of San Francisco, 74- Mem: Am Chem Soc; Am Crystallog Asn; Asn Educ Data Systs. Res: Crystallography of inorganic and hydrogen bonded compounds. Mailing Add: Dept of Chem City Col Ocean Ave-Phelan San Francisco CA 94112

COYLE, ELIZABETH ELEANOR, b Galion, Ohio, Aug 26, 04. BIOLOGY, BOTANY. Educ: Col Wooster, BS, 26; Ohio State Univ, MS, 29, PhD(bot), 35. Prof Exp: From instr to prof, 26-72, chmn dept, 63-72, EMER DANFORTH PROF BIOL, COL WOOSTER, 72- Mem: AAAS; Bot Soc Am. Res: Algal food of fishes; soil algae; marine algae. Mailing Add: Dept of Biol Col of Wooster Wooster OH 44691

COYLE, FREDERICK ALEXANDER, b Port Jefferson, NY, May 31, 42; m 65; c 2. BIOLOGY. Educ: Col Wooster, BA, 64; Harvard Univ, MA, 66, PhD(biol), 70. Prof Exp: ASST PROF BIOL, WESTERN CAROLINA UNIV, 69- Mem: AAAS; Am Soc Zoologists. Res: Systematics, evolution and comparative behavior of arthropods, particularly arachnids. Mailing Add: Dept of Biol Western Carolina Univ Cullowhee NC 28723

COYLE, MARIE BRIDGET, b Chicago, Ill, May 13, 35. MEDICAL MICROBIOLOGY, MICROBIAL GENETICS. Educ: St Louis Univ, MS, 63; Kans State Univ, PhD(genetics), 65. Prof Exp: Instr sci, Columbus Hosp Sch Nursing, Chicago, 57-59; fel microbiol, Univ Chicago, 64-67, res assoc molecular genetics, 67-70; instr microbiol, Med Ctr, Univ Ill, Chicago Circle, 70-71; fel microbiol, Temple Univ, 71-72; ASST PROF MICROBIOL, SCH MED, UNIV WASH, 72- Mem: Am Soc Microbiol; Sigma Xi; Am Asn Univ Prof; Acad Clin Lab Physicians & Scientists. Res: Mitotic recombination in Neurospora crassa; DNA repair mechanisms in mammalian cells after treatment with alkylating agents; genetics of bacterial virulence. Mailing Add: Dept of Microbiol Univ of Wash Sch of Med Seattle WA 98195

COYLE, MARY ANN, organic chemistry, see 12th edition

COYLE, PETER, b Hanover, NH, Mar 4, 39; m 67. NEUROANATOMY, NEUROPHYSIOLOGY. Educ: Univ Vt, BA, 62; Univ Mich, MS, 64, PhD(anat), 67. Prof Exp: From instr to asst prof, 67-75, ASSOC PROF ANAT, MED SCH, UNIV MICH, ANN ARBOR, 76- Concurrent Pos: Rackham fac res grant, Med Sch, Univ Mich, Ann Arbor, 68-70, USPHS grant, 70-74. Mem: AAAS; Sigma Xi; assoc Am Acad Neurol. Res: Nervous system information coding; limbic lobe anatomy and physiology; cerebrovasculature. Mailing Add: Dept of Anat Univ of Mich Med Sch Ann Arbor MI 48104

COYLE, THOMAS DAVIDSON, b Glen Cove, NY, Sept 25, 31; m 54. INORGANIC CHEMISTRY, ORGANOMETALLIC CHEMISTRY. Educ: Univ Rochester, BS, 52; Harvard Univ, AM, 59, PhD(chem), 61. Prof Exp: NATO fel, 61-62; chemist, 62-64, CHIEF INORG CHEM SECT, NAT BUR STANDARDS, 64- Mem: AAAS; Am Chem Soc; The Chem Soc. Res: Synthetic inorganic and organometallic chemistry; boron chemistry; nuclear magnetic resonance and applications to inorganic chemistry; coordination chemistry of main group elements; inorganic halides. Mailing Add: Inorg Chem Sect Nat Bur of Standards Washington DC 20234

COYNE, DERMOT P, b Dublin, Ireland, July 4, 29; US citizen; m 57; c 6. PLANT BREEDING. Educ: Univ Col, Dublin, BAgrSc, 53, MAgrSc, 54; Cornell Univ, PhD(plant breeding), 58. Prof Exp: Asst mgr agr develop, Campbell Soups, Ltd, Eng, 58-60; asst prof, 61-68, PROF PLANT BREEDING, UNIV NEBR, LINCOLN, 68- Concurrent Pos: Chmn, Nat Bean Improv Coop, 67-; rev ed, Am Soc Hort Sci, 73-; assoc ed, Hort Sci, 74- Honors & Awards: Nat Canner's Asn Award, Am Soc Hort Sci. Mem: Am Soc Hort Sci; Am Genetic Asn; Int Soc Hort Sci. Res: Germplasm identification; breeding and genetics of tolerance to bacterial root rot and white mold pathogens in beans; genetics and breeding investigation in the following areas in beans, interspecific hybridization, photoperiodium, adaptation, physiological and morpho genetical components of yields and seed quality; breeding improved types of winter squash and effects of genotype X environmental interactions on fruit shape. Mailing Add: Dept of Hort & Forestry Univ of Nebr Lincoln NE 68503

COYNE, DONALD MANLEY, organic chemistry, see 12th edition

COYNE, GEORGE VINCENT, b Baltimore, Md, Jan 19, 33. ASTRONOMY. Educ: Fordham Univ, AB, 57, PhilosL, 58; Georgetown Univ, PhD(astron), 62. Prof Exp: Res assoc astron & investr NASA grant, Res Inst Natural Sci, Woodstock Col, 63-70; ASST PROF ASTRON, LUNAR & PLANETARY LAB, UNIV ARIZ, 70- Concurrent Pos: Asst astronr, Vatican Observ, Italy, 70- Mem: Assoc mem Am Astron Soc; assoc mem Soc Photog Sci & Eng; Int Astron Union. Res: Evolution in young stellar associations; polarimetry; interstellar material; stars with extended atmospheres. Mailing Add: Lunar & Planetary Lab Univ of Ariz Tucson AZ 85721

COYNE, MARY FRANCES D, b Lynn, Mass, Jan 17, 38; m 61; c 2. ENDOCRINOLOGY. Educ: Emmanuel Col, Mass, AB, 59; Wellesley Col, MA, 61; Univ Va, PhD(physiol), 64. Prof Exp: Instr physiol, Sch Med, Univ Va, 66-67, asst prof, 67-68; asst prof physiol, La State Univ Med Ctr, New Orleans, 68-73; asst prof, 70-74, ASSOC PROF BIOL SCI & CHMN DEPT, WELLESLEY COL, 74- Concurrent Pos: USPHS fel, 64-66 & res grant, 67-72. Mem: Endocrine Soc; Am Soc Zoologists. Res: Pituitary and adrenal gland interrelationships particularly mechanisms regulating pituitary secretion of corticotropin. Mailing Add: Dept of Biol Sci Wellesley Col Wellesley MA 02181

COYNE, PATRICK IVAN, b Wichita, Kans, Feb 26, 44; m 64; c 2. PHYSIOLOGICAL ECOLOGY. Educ: Kans State Univ, BS, 66; Utah State Univ, PhD(range sci), 70. Prof Exp: Plant physiologist arctic tundra, US Army Cold Regions Res & Eng Lab, 70-72; asst prof forest tree physiol, Soil Lab, Univ Alaska, 73-74; PLANT PHYSIOLOGIST AIR POLLUTION EFFECTS ON PLANTS, LAWRENCE LIVERMORE LAB, UNIV CALIF, 75- Concurrent Pos: Consult, US Army Cold Regions & Eng Lab, 73-74. Mem: AAAS; Soc Range Mgt; Am Soc Agron; Crop Sci Soc Am; Soil Sci Soc Am. Res: Effects of interactions of gaseous air pollutants with other environmental stresses on plant physiology and plant community ecology. Mailing Add: Lawrence Livermore Lab Univ of Calif PO Box 808 Livermore CA 94550

COYNE, VERONICA E, b Quincy, Mass, July 2, 37. INTERNAL MEDICINE, IMMUNOLOGY. Educ: Trinity Col, DC, AB, 59; Woman's Med Col Pa, MD, 63. Prof Exp: Intern med, Woman's Med Col Hosp, 63-64, resident, 64-65; house physician, Holy Redeemer Hosp, Meadowbrook, Pa, 65-66; res fel med, Inst Cancer Res, 66-68, res physician, 68-71; MEM STAFF, DOYLESTOWN HOSP, 71- Concurrent Pos: Damon Runyan cancer res fel, 67-69. Res: Relationship of immunologic function to the development of malignancy; Australia antigen and its relationship to hepatitis and leukemia. Mailing Add: Landisville Rd RD 2 Doylestown PA 18901

COYNE, WILLIAM E, organic chemistry, pharmaceutical chemistry, see 12th edition

COYNER, EUGENE CASPER, b Conover, NC, Dec 25, 18; m 43; c 3. CHEMISTRY. Educ: Univ Ill, BS, 40; Univ Minn, PhD(org chem), 44. Prof Exp: From asst to instr chem, Univ Minn, 40-44; res chemist, chem dept, E I du Pont de Nemours & Co, 44-46; asst prof chem, Univ Tenn, 46-48; group leader, res dept, Mallinckrodt Chem Works, 49-54; res supvr, 55, tech asst sales, 55-66, tech assoc, freon prod div, 66-69, SR BUS ANALYST, ORG CHEM DEPT, E I DU PONT DE NEMOURS & CO, 69- Mem: AAAS; Am Chem Soc. Res: General organic chemistry; fluoro-organic compounds; product development. Mailing Add: 1225 Evergreen Rd Wilmington DE 19803

COZAD, GEORGE CARMON, b Corning, Kans, Mar 5, 27; m 51; c 4. MEDICAL MYCOLOGY, IMMUNOLOGY. Educ: Univ Kans, AB, 50; Univ Okla, MS, 54; Duke Univ, PhD(microbiol), 57. Prof Exp: From res assoc med mycol to assoc prof microbiol, 57-70, PROF MICROBIOL, UNIV OKLA, 70- Concurrent Pos: La State Univ trainee, Costa Rica & Colombia, 69. Mem: Am Soc Microbiol; Mycol Soc Am; Int Soc Human & Animal Mycol; Am Thoracic Soc; Am Soc Trop Med & Hyg. Res: Pathogenic mechanisms of systemic fungal agents; immunology of systemic fungus diseases; effects of fungus infection on basic immune mechanisms of host. Mailing Add: Dept of Microbiol Univ of Okla Norman OK 73069

COZZARELLI, NICHOLAS ROBERT, b Jersey City, NJ, Mar 26, 38; m 67. BIOCHEMISTRY. Educ: Princeton Univ, AB, 60; Harvard Univ, PhD(biochem), 66. Prof Exp: NSF fel biochem, Stanford Univ, 66-68; asst prof, 68-74, ASSOC PROF BIOCHEM, UNIV CHICAGO, 74- Res: Regulation of enzyme synthesis in bacteria; synthesis of DNA, in vitro and in vivo. Mailing Add: Dept of Biochem & Biophys Univ of Chicago Chicago IL 60637

COZZENS, ROBERT F, b Alexandria, Va, Sept 6, 41. PHYSICAL CHEMISTRY. Educ: Univ Va, BS, 63, PhD(chem), 66. Prof Exp: Nat Res Coun-Nat Acad Sci fel, US Naval Res Lab, 66-67; ASSOC PROF CHEM, GEORGE MASON UNIV, 67- Concurrent Pos: Consult, chem div, US Naval Res Lab, 67- Mem: AAAS; Am Chem Soc; Sigma Xi. Res: Photochemistry and energy transfer processes, especially polymer intramolecular energy transfer and free radicals. Mailing Add: Dept of Chem George Mason Univ Fairfax VA 22030

COZZINI, BRUCE O, physical chemistry, see 12th edition

CRABB, DAVID WENDELL, b North Adams, Mass, Sept 10, 25. ANTHROPOLOGY. Educ: Colo Col, BA, 51; Columbia Univ, PhD(anthrop), 62. Prof Exp: ASSOC PROF ANTHROP, PRINCETON UNIV, 63- Mem: African Studies Asn. Res: Symbolics and language; the new ethnography. Mailing Add: Dept of Anthrop 103 Green Annex Princeton Univ Princeton NJ 08540

CRABILL, EDWARD VAUGHN, b Winamac, Ind, May 30, 30. ANATOMY. Educ: DePauw Univ, BA, 52; NY Univ, PhD(anat), 57. Prof Exp: From instr to asst prof, 56-68, ASSOC PROF ANAT, ALBANY MED COL, 68- Mem: Sigma Xi. Res: Pituitary cytology; hair growth; skin physiology. Mailing Add: Dept of Anat Albany Med Col Albany NY 12208

CRABLE, GEORGE FRANCIS, b New Castle, Pa, June 10, 22; m 45; c 3. PHYSICS. Educ: Geneva Col, BS, 43; Univ Mich, MS, 47; Duke Univ, PhD(physics), 51. Prof Exp: Asst, Carnegie Inst Technol, 43-44; chemist, Koppers Co, 44-45; physicist, Gulf Res & Develop Co, Pa, 51-61; chmn dept physics, Geneva Col, 61-63; physicist, 63-70, SR RES PHYSICIST, DOW CHEM CO, 70- Mem: Am Phys Soc; Am Chem Soc; NY Acad Sci; Sigma Xi. Res: Microwave; mass and infrared spectroscopy; nuclear magnetic resonance; x-ray photoelectron spectroscopy. Mailing Add: Dow Chem Co Res & Develop 2020 Bldg Midland MI 48640

CRABTREE, DAVID MELVIN, b Upper Lake, Calif, Aug 29, 45; m 72; c 1. MARINE BIOLOGY. Educ: Pac Union Col, BS, 68, MA, 70; Loma Linda Univ, PhD(biol), 75. Prof Exp: CHMN DEPT BIOL, ANTILLIAN COL, 75- Mem: AAAS. Res: Invertebrate growth lines, ecological and paleoecological applications; effects of environmental conditions on coral growth. Mailing Add: Dept of Biol Antillian Col Mayaguez PR 00708

CRABTREE, DOUGLAS EVERETT, b Boston, Mass, June 14, 38; m 59; c 3. MATHEMATICS. Educ: Bowdoin Col, BA, 60; Harvard Univ, MA, 61; Univ NC, PhD(amth), 65. Prof Exp: Fel math, Univ NC, 61-64; asst prof, Univ Mass, 64-66; asst prof, Amherst Col, 66-72; INSTR MATH, PHILLIPS ACAD, MASS, 72- Mem: Am Math Soc; Math Asn Am. Res: Theory of matrices; theory of rings. Mailing Add: Dept of Math Phillips Acad Andover MA 01810

CRABTREE, ELEANOR VOORHEES, b Freehold, NJ, Aug 31, 18; m 45; c 3. ORGANIC CHEMISTRY, ANALYTICAL CHEMISTRY. Educ: Rutgers Univ, AB, 40. Prof Exp: Org chemist, Plastics Div, Celanese Corp Am, NJ, 41-43; jr chemist, Res & Develop Labs, Merck & Co Inc, 43-47; biochemist, Vet Admin Hosp, 56-60; org chemist, US Naval Propellant Plant, Md, 60-62; RES CHEMIST, US ARMY CHEM RES & DEVELOP LABS, ARMY CHEM CTR, 62- Honors & Awards: Res & Develop Achievement Award, US Army, 69. Mem: AAAS; Am Chem Soc; Sigma Xi. Res: Development of analytical methods; pollution control; synthesis of new analytical reagents; ultramicro detection and identification of toxic chemicals. Mailing Add: Environ Res Div US Army Chem Lab Edgewood Arsenal MD 21010

CRABTREE, GARVIN (DUDLEY), b Eugene, Ore, Nov 29, 29; M 65; c 2. HORTICULTURE, WEED SCIENCE. Educ: Ore State Univ, BS, 51; Cornell Univ,

MS, 55, PhD, 58. Prof Exp: From asst prof & asst horticulturist to ASSOC PROF HORT, ORE STATE UNIV, 58- Concurrent Pos: Res assoc, Mich State Univ, 75-76. Mem: Weed Sci Soc Am; Am Soc Hort Sci. Res: Weed control in horticultural crops. Mailing Add: Dept of Hort Ore State Univ Corvallis OR 97331

CRABTREE, GERALD WINSTON, b Manchester, Eng, June 29, 41; Can citizen; m 65; c 2. BIOCHEMISTRY, PHARMACOLOGY. Educ: Univ Guelph, BSA, 63, MS, 65; Univ Alta, PhD(purine metab), 70. Prof Exp: Res assoc purine metab, 70-72, instr, 72-74, ASST PROF PURINE METAB, BROWN UNIV, 74- Mem: AAAS; Can Biochem Soc. Res: Purine nucleotide metabolism in intact mammalian cells; effects of purine analogues on normal purine metabolic pathways; metabolism of purine analogues; purine metabolism of schistosomes. Mailing Add: Div Biol & Med Sci Brown Univ Providence RI 02912

CRABTREE, JAMES BRUCE, b Wichita, Kans, Dec 11, 18. MATHEMATICS. Educ: Univ Kans, AB, 41, MA, 42; Harvard Univ, PhD(math), 50. Prof Exp: Instr math, Univ Chicago, 47-48; from asst prof to assoc prof, Univ NH, 50-56; ASSOC PROF MATH, STEVENS INST TECHNOL, 56- Mem: Am Math Soc. Res: Functional analysis; analytic functions on algebras. Mailing Add: Dept of Math Stevens Inst of Technol Hoboken NJ 07030

CRABTREE, KOBY TAKAYASHI, b Tokyo, Japan, Apr 29, 34; US citizen; m 58; c 2. MICROBIOLOGY, CIVIL ENGINEERING. Educ: Ohio Wesleyan Univ, BA, 58; Univ Wis, MS, 63, PhD(bact, civil eng), 65. Prof Exp: Fel civil eng, Univ Wis, 65-66, asst prof, 66-70, ASSOC PROF BACT, UNIV WIS, MARATHON CAMPUS, 70- Concurrent Pos: USDA Off Solid Wastes grant, 66-68; consult, Dept Mineral Res, Mich Technol Univ, 67-69; Wis State Dept Natural Resources grants, 67-70; consult, Zimpro, Wis, 67-68. Mem: AAAS; Am Soc Microbiol; Am Che Chem Soc. Res: Microbial ecology; water and solid waste pollution; nitrogen cycle. Mailing Add: Dept of Bact Sci Hall Univ of Wis Wausau WI 54401

CRABTREE, ROSS EDWARD, b Arkansas City, Kans, Mar 20, 32; m 54; c 2. DRUG METABOLISM. Educ: Southwestern State Col, Okla, BS, 54; Purdue Univ, MS, 56, PhD, 57. Prof Exp: Sr phys chemist, 57-63, sr anal res chemist, 63-70, RES SCIENTIST, LILLY CLIN, ELI LILLY & CO, 70- Mem: Am Chem Soc; Health Phys Soc. Res: Use of radioisotopes in diagnostic tests and in metabolism studies. Mailing Add: 5344 Daniel Dr Indianapolis IN 46226

CRACRAFT, JOEL LESTER, b Wichita, Kans, July 31, 42. VERTEBRATE MORPHOLOGY, BIOSYSTEMATICS. Educ: Univ Okla, BS, 64; La State Univ, MS, 66; Columbia Univ, PhD(biol), 69. Prof Exp: Res fel, Am Mus Natural Hist, 69-70; ASST PROF ANAT, UNIV ILL MED CTR, 70- Mem: Am Soc Naturalists; Soc Study Evolution; Soc Syst Zool; Am Soc Zoologists; Am Ornithologists Union. Res: Functional morphology of birds; multivariate morphometric analysis of size and shape; avian evolution; systematic theory; vertebrate biogeography. Mailing Add: Dept of Anat Univ of Ill Med Ctr Chicago IL 60680

CRADDOCK, ELYSSE MARGARET, b Sydney, Australia, Sept 7, 44; m 73. EVOLUTIONARY BIOLOGY, CYTOGENETICS. Educ: Univ Sydney, BSc, 65, PhD(cytoevolution), 71. Prof Exp: Fel evolution, Univ Hawaii, 71-72 & Yale Univ, 72; res fel pop biol, Australian Nat Univ, 73; RES SCIENTIST, DEPT BIOL, NY UNIV, 74- Mem: Genetics Soc Am; Soc Study Evolution; AAAS; NY Acad Sci. Res: Chromosomal and genetic aspects of the speciation process, in particular, chromosome rearrangements, evolutionary changes in the satellite DNA sequences and the role of regulatory genes in evolution. Mailing Add: Dept of Biol 952 Brown Bldg NY Univ New York NY 10003

CRADDOCK, GARNET ROY, b Chatham, Va, May 7, 26; m 49; c 2. AGRONOMY. Educ: Va Polytech Inst, BS, 52; Univ Wis, PhD(soils), 55. Prof Exp: From asst prof to assoc prof, 55-67, PROF AGRON, CLEMSON UNIV, 67-, HEAD DEPT AGRON & SOILS, 72- Concurrent Pos: Asst soil scientist, Exp Sta, Clemson Univ, 55-57, assoc soil scientist, 57-66. Mem: Am Soc Agron; Soil Sci Soc Am; Soil Conserv Soc Am. Res: Potassium and its relationship to soil mineralogy; pediological investigations relative to southeastern soils. Mailing Add: Dept of Agron & Soils Col of Agr Sci Clemson Univ Clemson SC 29631

CRADDOCK, (JOHN) CAMPBELL, b Chicago, Ill, Apr 3, 30; m 53; c 3. STRUCTURAL GEOLOGY, TECTONICS. Educ: DePauw Univ, BA, 51; Columbia Univ, MA, 53, PhD(geol), 54. Prof Exp: Asst geol, Columbia Univ, 53-54; geologist, Shell Oil Co, 54-56; from asst prof to assoc prof geol, Univ Minn, Minneapolis, 56-67; PROF GEOL, UNIV WIS, MADISON, 67- Concurrent Pos: Dir, Antarctic Res Expeds, 59-69; geologist, Minn Geol Surv, 59,64; vis scientist, NZ Geol Surv, 62-63; consult, Corps Engrs, US Army; NStar Res Inst & Am Geog Soc; US mem & chmn, Working Group on Geol & Int Union Geol Sci del, Sci Comt on Antarctic Res; mem comn struct geol, Int Union Geol Sci. Honors & Awards: US Antarctic Serv Medal, 68; Bellingshausen-Lazarev Medal, Soviet Acad Sci, 70. Mem: AAAS; fel Geol Soc Am; Am Asn Petrol Geol; Am Geophys Union; Seismol Soc Am. Res: Overthrusts; folds; transcurrent faults; Antarctic and Alaskan geology; gravity; precambrian geology. Mailing Add: Dept of Geol & Geophys Univ of Wis Madison WI 53706

CRADDOCK, JOHN HARVEY, b Memphis, Tenn, May 30, 36; m 67; c 2. INORGANIC CHEMISTRY. Educ: Memphis State Univ, BS, 58; Vanderbilt Univ, PhD(inorg chem), 61. Prof Exp: Res chem, M W Kellogg Co Div, Pullman, Inc, 61-65; res specialist, Cent Res Dept, 65-68, res group leader, 68-73, supvr com develop, Polymers & Petrochem Co, 73-74, MGR COM DEVELOP, INDUST CHEM CO, MONSANTO CO, 74- Concurrent Pos: Res assoc, Princeton Univ, 64. Mem: Am Chem Soc; The Chem Soc; Inst Food Technol. Res: Homogeneous catalysis; petrochemical reactions and processes; coordination chemistry; chemical preservations and antimicrobials. Mailing Add: Monsanto Co 800 N Lindbergh Blvd St Louis MO 63166

CRADDOCK, MICHAEL KEVIN, b Portsmouth, Eng, Apr 15, 36; m 70. NUCLEAR PHYSICS. Educ: Oxford Univ, MA, 61, DPhil(nuclear physics), 64. Prof Exp: Sci officer nuclear physics, Rutherford High Energy Lab, Nat Inst Res Nuclear Sci, Chilton, Eng, 61-64; asst prof, 64-68, ASSOC PROF PHYSICS, UNIV BC, 68- Concurrent Pos: Group leader beam dynamics, Triumf, BC, 68- Mem: Brit Inst Physics & Phys Soc (fel Phys Soc); Can Asn Physicists. Res: Medium energy proton scattering; polarized ion sources; cyclotrons. Mailing Add: Dept of Physics Univ of BC Vancouver BC Can

CRADDUCK, TREVOR DAVID, b London, Eng; m 59; c 3. MEDICAL PHYSICS. Educ: Bristol Univ, Eng, BSc, 58; Univ Sask, MSc, 63, Dr Phil (physics), 66. Prof Exp: Apprentice elec eng, Gen Elec Co, Coventry, Eng, 58-60; res fel, Nat Cancer Inst Can, 61-65; physicist & tech dir nuclear med, Manitoba Cancer Found, Winnipeg, 65-67; consult physicist, Foothills Hosp, Calgary, Alta, 67-70; PHYSICIST, TORONTO GEN HOSP, 70- Concurrent Pos: Assoc prof, Fac Med, Univ Calgary, 68-70; asst prof, Fac Med, Univ Toronto, 73- Mem: Soc Nuclear Med (secy, 74-77); Can Asn Physicists; Brit Hosp Physicists Asn; assoc Brit Inst Elec Eng; Am Col Nuclear Physicians. Res: Physics in nuclear medicine; scintillation cameras; computers applied to nuclear medicine and radiology. Mailing Add: Div of Nuclear Med Toronto Gen Hosp Toronto ON Can

CRAFT, GEORGE ARTHUR, b Youngstown, Ohio, Nov 16, 16; m 46; c 1. MATHEMATICS. Educ: Miami Univ, BS, 39; Ind Univ, MA, 50; Ohio State Univ, PhD(math), 57. Prof Exp: Instr math, Mont State Univ, 50-53 & Ohio State Univ, 57-58; asst prof, Denison Univ, 58-61; asst prof, 61-63, ASSOC PROF MATH, HARPUR COL, STATE UNIV NY BINGHAMTON, 63- Mem: Am Math Soc. Res: Continuous transformations in Euclidean n-space; group spaces and topological vector spaces. Mailing Add: Dept of Math Harpur Col State Univ of NY Binghamton NY 13901

CRAFT, HAROLD DUMONT, JR, b Newark, NJ, May 28, 38; m 62; c 2. RADIO ASTRONOMY, RADIOPHYSICS. Educ: Cornell Univ, BEE, 61; NY Univ, MEE, 63; Cornell Univ, PhD(radio astron), 70. Prof Exp: Staff mem tech commun syst, Bell Tel Lab, 61-65; mem tech radio propagation, COMSAT Lab, 69-71; tech coordr astron, Nat Astron & Ionosphere Ctr, Cornell Univ, 71-73; DIR OPERS, ARECIBO OBSERV, NAT ASTRON & IONOSPHERE CTR, CORNELL UNIV, 73- Mem: AAAS; Am Astron Soc; Int Union Radio Sci. Res: Pulsar physics and radio emission mechanisms; radio propagation studies particularly with respect to atmospheric and ionospheric effects. Mailing Add: Arecibo Observ PO Box 995 Arecibo PR 00612

CRAFT, JAMES HARVEY, b Princeton, WVa, Jan 21, 14; m; c 1. BOTANY. Educ: Concord Col, BA, 35; WVa Univ, MS, 39; Univ Iowa, PhD(plant histol), 43. Prof Exp: Teacher high sch, WVa, 35-38; asst biol, WVa Univ, 38-39; asst bot, Univ Iowa, 39-43; from assoc prof to prof biol, 43-75, chmn div sci & math, 46-75, EMER PROF BIOL, ADAMS STATE COL, 75- Mem: Fel AAAS; Am Bryol & Lichenological Soc. Res: Physiology and morphology of Bryophyllum calycinum; effective and practicable methods of mosquito abatement in the towns of the San Luis Valley, Colorado; bryophytes, lichens and mosses of southern Colorado. Mailing Add: 220 Alamosa Ave Alamosa CO 81101

CRAFT, THOMAS FISHER, b Macon, Ga, 1924, m 48; c 1. ENVIRONMENTAL SCIENCES, NUCLEAR SCIENCE. Educ: Mercer Univ, AB, 45; Emory Univ, MA, 47; Ga Inst Technol, MS, 65, PhD(nuclear eng), 69. Prof Exp: Instr chem, Mercer Univ, 45-46; tech salesman chem, Dow Chem Co, 49-62; SR RES SCIENTIST ENVIRON & NUCLEAR SCI, ENG EXP STA, GA INST TECHNOL, 62- Mem: Water Pollution Control Asn; Soc Environ Geochem & Health; Am Water Works Asn; Am Nuclear Soc; AAAS. Res: Radiation processing of industrial wastewater; fixation and mobilization of radionuclides in soil; water and wastewater treatment processes; neutron activation analysis; trace elements in water and elsewhere; radioactive tracer techniques. Mailing Add: 116 Ridley Circle Decatur GA 30030

CRAFT, THOMAS JACOB, SR, b Monticello, Ky, Dec 27, 24; m 48; c 2. CELL BIOLOGY, DEVELOPMENTAL BIOLOGY. Educ: Cent State Univ, BS, 48; Kent State Univ, MA, 50; Ohio State Univ, PhD(develop biol), 63. Prof Exp: Lab asst biol, Cent State Col, Ohio, 47-48; asst lab instr vert anat, Kent State Univ, 49-50; from instr to assoc prof, 50-67, dir summer session, 64-65, PROF BIOL, CENT STATE UNIV, 67- Concurrent Pos: Consult, NSF, US AID, India, 67-69; Eli Lilly grant; mem, Nat Adv Res Resources Coun, NIH; adv, Ohio Health Manpower Linkage Syst Proj, Ohio Dept Health; mem exec comt, Ohio Acad Sci; adj prof anat, Sch Med, Wright State Univ. Mem: AAAS; NY Acad Sci; Am Inst Biol Sci; Sigma Xi. Res: Experimental morphology and embryology; pigment cell biology; swimming in common brown bats; melanogenesis and the homograft reaction; effect of atmospheric pollutants. Mailing Add: Dept of Biol Cent State Univ Wilberforce OH 45384

CRAFT, WILLARD LEAHMAN, JR, b Benton Harbor, Mich, Nov 1, 38; m 61. CHEMISTRY. Educ: Univ Mich, BS, 61; Univ Wash, PhD(chem), 66. Prof Exp: Great Lakes Cols Asn Kettering intern, Col Wooster, 66-67; asst prof, 67-74, ASSOC PROF CHEM, ADRIAN COL, 74- Mem: Am Chem Soc; Am Phys Soc. Res: Lambda phase transitions in solids and helix coil transitions in biologically interesting polymers. Mailing Add: Dept of Chem Adrian Col Adrian MI 49221

CRAFTS, ALDEN SPRINGER, b Ft Collins, Colo, June 25, 97; m 26; c 2. BOTANY, WEED SCIENCE. Educ: Univ Calif, Berkeley, BS, 27, PhD(plant physiol), 30. Hon Degrees: MS, Oxford Univ, 57; LLD, Univ Calif, Davis, 66. Prof Exp: Agent, Blister Rust Control, Bur Plant Indust, USDA, 28-29; Nat Res Coun fel biol, Cornell Univ, 30-31; from asst prof bot & asst botanist to prof & botanist, Exp Sta, 31-64, EMER PROF & WRITER, UNIV CALIF, DAVIS, 64- Concurrent Pos: Guggenheim fels, Harvard Univ, 38 & Oxford Univ, 57-58; grant-in-aid, PR Exp Sta, 47; deleg, Univ Calif & Am Soc Plant Physiol Bot Cong, Paris, 47; mem, Panel on Arsenic, Nat Acad Sci, 72- Mem: AAAS; Bot Soc Am; Am Soc Plant Physiol (secy, 51-53, vpres, 53, pres, 55); Am Soc Range Mgt; hon mem Weed Sci Soc Am (vpres, 56-58, pres, 58-60). Res: Weed control; structure and function of the phloem; plant cell water relations; radiation biology. Mailing Add: Dept of Bot Univ of Calif Davis CA 95616

CRAFTS, ROGER CONANT, b Lewiston, Maine, Jan 26, 11; m 38; c 2. ANATOMY. Educ: Bates Col, BS, 33; Columbia Univ, PhD(anat), 41. Prof Exp: Instr, Col Physicians & Surgeons, Columbia Univ, 39-40; from instr to assoc prof anat, Sch Med, Boston Univ, 43-50; PROF ANAT & CHMN DEPT, COL MED, UNIV CINCINNATI, 50- Concurrent Pos: Instr, Boston City Hosp, 42-49; consult, Div Fels, NIH, 60-63. Mem: Fel AAAS; Am Asn Anatomists; Asn Am Med Cols; Soc Exp Biol & Med. Res: Endocrinology; reproduction; pituitary gland; endocrines and hemopoiesis. Mailing Add: Dept of Anat Univ of Cincinnati Col of Med Cincinnati OH 45267

CRAGG, HOYT J, b Rabun Co, Ga, Dec 12, 19; m 42; c 2. ORGANIC CHEMISTRY. Educ: Berry Col, BS, 41; Emory Univ, MS, 42. Prof Exp: Chemist, Chem Warfare Serv, 42-43 & Tenn Eastman, AEC, 43-47; CHEMIST, ETHYL CORP, 47- Mem: AAAS; fel Am Inst Chem; Am Chem Soc. Res: Chlorination; hydrochlorination; organometallics; separation and purification. Mailing Add: 6045 Hibiscus Dr Baton Rouge LA 70808

CRAGG, JAMES BIRKETT, b North Shields, Eng, Nov 8, 10. ECOLOGY, ENVIRONMENTAL SCIENCES. Educ: Durham Univ, BSc, 33, dipl, 34, MSc, 37; Univ Newcastle, DSc, 65. Prof Exp: Demonstr comp physiol, Manchester Univ, 35-37; asst lectr zool, Univ Col NWales, 37-40, on leave to, Sch Agr, 40-42, lectr zool, 42, on leave to, Agr Res Coun, UK, 42-44, sci officer, Unit Insect Physiol, 44-46; reader zool, Durham Col, Durham Univ, 46-50, prof, 50-61, head dept, 46-61; dir, Merlewood & Moor House Res Sta, Nature Conserv, UK, 61-66; head dept biol, 66-68, DIR ENVIRON SCI CTR, UNIV CALGARY, 66-, ACAD VPRES, 70- Concurrent Pos: Prog consult, Ford Found, 65; convener, Int Comt Terrestrial Prod, Can Comt Sec Prod & Can Comt Conserv Ecosysts, Int Biol Prog; chmn comn ecol, Int Union Conserv Nature & Natural Resources; mem sect ecol, Int Union Biol Sci; mem nat comt wildlife lands, Can Wildlife Serv. Honors & Awards: Killam Mem Chair Award, 67. Mem: Ecol Soc Am; Can Soc Zool (2nd vpres, 68-69); Brit Ecol Soc (pres, 60-61); Brit Soc Exp Biol; fel Brit Inst Biol. Res: Studies on moorland and forest ecosystems, particularly the decomposer system. Mailing Add: Fac Environ Design Univ of Calgary Calgary AB Can

CRAGGS, ROBERT F, b South Charleston, Ohio, June 9, 37. TOPOLOGY. Educ: Ohio Univ, Athens, AB, 59; Univ Wis-Madison, MS, 60, PhD(math), 66. Prof Exp: Instr math, Ohio Univ, Athens, 61-63; fel math, Inst Adv Study, Princeton, 66-68; asst prof, 68-74, ASSOC PROF MATH, UNIV ILL, URBANA, 74- Concurrent Pos: Sabbatical vis, Sci Inst, Univ Iceland, 74-75. Mem: Am Math Soc. Res: Topology of 3-and 4-manifolds; Heegaard theory; Poincare conjecture; mapping class groups. Mailing Add: Dept of Math Univ of Ill Urbana IL 61801

CRAGLE, RAYMOND GEORGE, b Orangeville, Pa, Feb 28, 26; m 50; c 3. ANIMAL PHYSIOLOGY, NUTRITION. Educ: NC State Col, BS, 51, MS, 54; Univ Ill, PhD(dairy sci), 57. Prof Exp: From asst prof to prof physiol & nutrit, Agr Res Lab, AEC, Univ Tenn, 57-68; vis prof, Lab Genetics, Univ Wis, Madison, 68-69; PROF DAIRY SCI & HEAD DEPT, VA POLYTECH INST & STATE UNIV, 70- Mem: Am Dairy Sci Asn; Am Soc Animal Sci; Am Inst Nutrit; Soc Study Reproduction. Res: Physiology and nutrition of the dairy cow; metabolism of fission products; mineral metabolism; gastrointestinal absorption and secretion of mineral elements; chimerism in large animals; tissue transplantation in cattle. Mailing Add: Dept of Dairy Sci Va Polytech Inst & State Univ Blacksburg VA 24061

CRAGOE, EDWARD JETHRO, JR, b Tulsa, Okla, July 16, 17; c 3. MEDICINAL CHEMISTRY, ORGANIC CHEMISTRY. Educ: Baker Univ, BA, 39; Univ Nebr-Lincoln, MA, 41, PhD(org chem), 44. Prof Exp: Res assoc, Sharp and Dohme, Inc, 44-56; instr chem, Pa State Col, 47-48; res assoc, 56-60, from asst dir to dir, 60-74, SR DIR MED CHEM, MERCK SHARP & DOHME RES LAB, 75- Mem: Am Chem Soc; Sigma Xi; fel NY Acad Sci; fel Am Inst Chemists; Soc Chem Indust. Res: Medicinal chemistry; synthetic drugs; renal agents; diuretics; antikaliuretics; carbonic anhydrase inhibitors; prostaglandin analogs; chemotherapy; monoamine oxidase inhibitors; antidiabetic agents; antihypertensive agents; mental health drugs; gastrointestinal drugs; organic synthesis; heterocyclic compounds. Mailing Add: Dept of Med Chem Merck Sharp & Dohme Res Lab West Point PA 19486

CRAIG, ALAN DANIEL, b Hempstead, NY, Feb 11, 35; m 60; c 4. INDUSTRIAL CHEMISTRY, PHARMACEUTICAL CHEMISTRY. Educ: Hofstra Univ, BA, 56; Univ Pa, PhD(inorg chem), 61. Prof Exp: Asst instr chem, Univ Pa, 56-59; res chemist res ctr, Hercules Inc, 61-67, mgr high energy res div, 67-70, sr venture analyst, New Enterprise Dept, 70-74; VPRES, ADRIA LABS, INC, 74- Mem: Am Chem Soc; Acad Pharmaceut Sci; Sigma Xi. Res: Nitrogen-fluorine chemistry; silicon and other light metal hydride chemistry; infra-red spectroscopy of these compounds; pharmaceutical development. Mailing Add: Adria Labs, Inc 1105 Market St Wilmington DE 19899

CRAIG, ALAN KNOWLTON, b Ft Sill, Okla, Mar 7, 30; m 59. GEOGRAPHY, ANTHROPOLOGY. Educ: La State Univ, BS, 58, PhD(geog), 66. Prof Exp: Asst prof geog, DePaul Univ, 66; asst prof, 66-69, ASSOC PROF GEOG, FLA ATLANTIC UNIV, 69- Concurrent Pos: Off Naval Res grant, 67-68. Mem: AAAS; Asn Am Geog; Am Geog Soc. Res: Origin and dispersal of folk fishing techniques; physical ecology of the marine littoral. Mailing Add: Dept of Geog Fla Atlantic Univ Boca Raton FL 33432

CRAIG, ALBERT BURCHFIELD, JR, b Sewickley, Pa, April 19, 24; m 47; c 4. PHYSIOLOGY, MEDICINE. Educ: Cornell Univ, MD, 48. Prof Exp: Intern, Dept Med, 48-49, asst resident, 50-51, instr, 53-55, instr physiol & med, 55-59, from asst prof to assoc prof physiol, 59-72, PROF PHYSIOL, SCH MED & DENT, UNIV ROCHESTER, 72- Concurrent Pos: Fel med, Sch Med & Dent, Univ Rochester, 49-50, USPHS res fel, 53-55; estab investr, Am Heart Asn, 61-66. Mem: Am Col Sports Med; Am Physiol Soc. Res: Man in water; respiration. Mailing Add: Dept of Physiol Univ of Rochester Sch of Med & Dent Rochester NY 14642

CRAIG, ALBERT MORRISON, b San Francisco, Calif, Oct 3, 42; m 74. BIOPHYSICS. Educ: Ore State Univ, BS, 65, PhD(biochem, biophys), 70. Prof Exp: Am Cancer Inst res assoc, Ore State Univ, 70-73; NIH res assoc, Univ Ore, 73-75; ASST PROF BIOPHYS, ORE STATE UNIV, 75- Mem: Biochem-Biophys Soc; Sigma Xi; AAAS. Mailing Add: Dept of Vet Med Ore State Univ Corvallis OR 97331

CRAIG, ARNOLD CHARLES, b Johnstown, NY, Sept 5, 33; m 59; c 1. PHYSICAL ORGANIC CHEMISTRY. Educ: Syracuse Univ, BA, 54; Cornell Univ, PhD(org chem), 59. Prof Exp: Res chemist, Eastman Kodak Co, 59-67; asst prof, 67-73, ASSOC PROF CHEM, MONT STATE UNIV, 73- Mem: Am Chem Soc. Res: Synthesis of natural products and heterocycles; color and constitution relation of sensitizing dyes; effect of environment on radiative transitions. Mailing Add: Dept of Chem Mont State Univ Bozeman MT 59715

CRAIG, BRUCE GORDON, b London, Ont, Mar 30, 22; m 48; c 2. GLACIAL GEOLOGY. Educ: Univ Western Ont, BSc, 49; Univ Mich, MS, 50, PhD, 56. Prof Exp: GEOLOGIST, DIV QUATERNARY RES & GEOMORPHOL, GEOL SURV CAN, 49- Mem: Geol Soc Am. Res: Pleistocene geology, especially of Arctic Canada. Mailing Add: 1125 Sherman Dr Ottawa ON Can

CRAIG, BURTON MACKAY, b Vermilion, Alta, May 29, 18; m 45; c 2. AGRICULTURAL BIOCHEMISTRY. Educ: Univ Sask, BScA, 44, MSc, 46; Univ Minn, PhD, 50. Prof Exp: Lab asst, 41-44, asst, 44-46, from asst res officer to prin res officer, 50-69, assoc dir, 69-70, DIR FATS & OILS LAB, PRAIRIE REGIONAL LAB, UNIV SASK, 70- Mem: Am Oil Chem Soc; Chem Inst Can. Res: Nutrition of fats; biosynthesis of fatty acids; fatty acid composition of oils and fats; gas liquid chromatography in fats and oils. Mailing Add: Prairie Regional Lab Nat Res Coun Saskatoon SK Can

CRAIG, CECIL CALVERT, b Otwell, Ind, Apr 14, 98; m 27; c 1. MATHEMATICS. Educ: Ind Univ, AB, 20, AM, 22; Univ Mich, PhD(math), 27. Prof Exp: Instr math, Ind Univ, 20-22 & Univ Mich, 22-24, 25-29; Nat Res fel, Princeton Univ, 29-30 & Stanford Univ, 30-31; from asst prof to prof, 31-68, dir, Statist Lab, 46-68, EMER PROF MATH, UNIV MICH, 68- Concurrent Pos: Rockefeller Found fel, Univ London, 37-38; vis prof, Mich Technol Univ, 68. Honors & Awards: Shewhart Medal, Am Soc Qual Control, 57. Mem: Am Math Soc; Math Asn Am; fel Inst Math Statist (pres, 42-43); fel Am Statist Asn; fel Am Soc Qual Control. Res: Mathematical statistics. Mailing Add: Dept of Math Univ of Mich Ann Arbor MI 48104

CRAIG, CHARLES ROBERT, b Buckhannon, WVa, Jan 24, 36; m 60; c 1. PHARMACOLOGY. Educ: Univ Wis, PhD(pharmacol), 64. Prof Exp: Sr investr neuropharmacol, G D Searle & Co, 64-66; from asst prof to assoc prof pharmacol, 66-75, PROF PHARMACOL, MED CTR, WVA UNIV, 75- Mem: Am Soc Pharmacol & Exp Therapeut; Am Soc Neurochem; Sigma Xi. Res: Pharmacological changes in the chronically epileptic rat. Mailing Add: Dept of Pharmacol WVa Univ Med Ctr Morgantown WV 26506

CRAIG, DEXTER HILDRETH, b Pontiac, Mich, Feb 12, 24; m 49; c 2. GEOLOGY. Educ: Univ Mich, BS, 50; Univ Tex, MA, 52. Prof Exp: From asst geologist to geologist, Tex, 52-69, SR GEOLOGIST, DENVER RES CTR, MARATHON OIL CO, COLO, 69- Mem: Fel AAAS; Am Asn Petrol Geol; Soc Econ Paleont & Mineral; Am Inst Mining, Metall & Petrol Eng; Sigma Xi. Res: Geology of petroleum deposits; stratigraphy; sedimentary petrology and petrography. Mailing Add: Marathon Oil Co PO Box 269 Littleton CO 80120

CRAIG, DONALD LAIRD, b Kentville, NS, Dec 18, 23; m 48; c 4. HORTICULTURE. Educ: McGill Univ, BSc, 47; Univ NH, MSc, 55, PhD(plant breeding), 59. Prof Exp: AGR RES SCIENTIST, RES STA, CAN DEPT AGR, 47-, HEAD, BERRY CROPS SECT, 66- Mem: Am Soc Hort Sci; Can Soc Hort Sci. Res: Culture and breeding of berry crops. Mailing Add: Res Sta Can Dept of Agr Kentville NS Can

CRAIG, DONALD SPENCE, b Ridgetown, Ont, Aug 13, 23; m 52; c 3. NUCLEAR PHYSICS. Educ: Queen's Univ, Ont, BSc, 45; Univ Wis, PhD(physics), 52. Prof Exp: Jr res physicist, Nat Res Coun Can, 45-47; asst res physicist, 52-57, assoc res officer, 57-67, SR RES OFFICER, ATOMIC ENERGY LTD, CAN, 67- Mem: Am Nuclear Soc; Can Asn Physicists. Res: Reactor physics. Mailing Add: 18 Cabot Pl Deep River ON Can

CRAIG, DOUGLAS ABERCROMBIE M, b Nelson, NZ, Oct 24, 39; m 62; c 1. INSECT MORPHOLOGY, INVERTEBRATE LIMNOLOGY. Educ: Univ Canterbury, BSc, 62, PhD(zool), 66. Prof Exp: Asst lectr zool, Univ Canterbury, 62-66; session lectr, 66-67, lectr, 67-68, asst prof, 68-74, ASSOC PROF ENTOM, UNIV ALTA, 74- Mem: Entom Soc Can; Royal Soc NZ; Ecol Soc NZ; Entom Soc NZ; NZ Limnol Soc. Res: Biology of Blepharoceridae; embryogenesis of larval simuliidae heads; larval sensory organs. Mailing Add: Dept of Entom Univ of Alta Edmonton AB Can

CRAIG, DOUGLAS KENNETH, b Stutterheim, SAfrica, Apr 23, 32; m 60; c 2. AIR POLLUTION, RADIOBIOLOGY. Educ: Univ Witwatersrand, BS, 53, MS, 60; Janeesburg Col Educ, Transvaal higher teachers dipl, 53; Univ Potchefstroom, BS, 57; Univ Rochester, MS, 61, PhD(radiation biol), 64. Prof Exp: High sch teacher, SAfrica, 54-55; physicist, Dust & Vent Res Lab, Transvaal & Orange Free State Chamber of Mines, 56-60; res officer health physics, isotopes & radiation div, Atomic Energy Bd, 60-64, sr res officer & head health physics subdiv, 64-65, prin res officer, 65-66, prin scientist, 66-69; sr scientist, 69-72, res assoc, 72-73, MGR INHALATION TOXICOL SECT, DEPT BIOL, PAC NORTHWEST LABS, BATTELLE MEM INST, 73- Mem: Am Indust Hyg Asn; Health Physics Soc. Res: Aerosol physics. Mailing Add: Pac Northwest Labs Biol Dept PO Box 999 Richland WA 99352

CRAIG, FRANCIS NORTHROP, b Englewood, NJ, June 2, 11; m 40; c 3. PHYSIOLOGY. Educ: Rutgers Univ, BS, 32; Harvard Univ, AM, 33, PhD(biol), 37. Prof Exp: Asst physiol, Harvard Univ, 33-35; asst, Radcliffe Col, 34-35; asst bot, Columbia Univ, 36; res asst, Harvard Med Sch, 36-39; instr physiol, Col Med, NY Univ, 43-46; physiologist, 46-49, chief appl physiol br, 49-73, CHIEF MED PHYSIOL BR, BIOMED LAB, EDGEWOOD ARSENAL, 73- Concurrent Pos: Teaching fel physiol, Harvard Med Sch, 38-39, res fel med, 39-41, res fel anesthesia, 41-43; vis scientist, Chem Defense Exp Estab, UK, 54. Mem: Am Physiol Soc; Soc Exp Biol & Med; Am Col Sports Med. Res: Cell metabolism; kidney physiology; environmental physiology; respiration; physiology of exercise. Mailing Add: Biomed Lab Edgewood Arsenal MD 21010

CRAIG, FRANK RANKIN, b Mt Holly, NC, Apr 6, 21; m 47; c 1. VETERINARY MEDICINE. Educ: NC State Univ, BS, 46, MS, 52; Univ Ga, DVM, 56. Prof Exp: Assoc poultry pathologist, NC State Univ, 48-52, from exten poultry pathologist & assoc prof to prof poultry sci, 56-70; DIR HEALTH & HEALTH SERV, A W PERDUE & SONS, INC, 70- Mem: Am Vet Med Asn; Poultry Sci Asn. Res: Microbiology, hematology and pathology of fowl. Mailing Add: A W Perdue & Sons Inc Salisbury MD 21801

CRAIG, GEORGE BROWNLEE, JR, b Chicago, Ill, July 8, 30; m 54; c 4. ENTOMOLOGY. Educ: Ind Univ, BA, 51; Univ Ill, MS, 52, PhD(entom), 56. Prof Exp: Asst, Dept Entom, Univ Ill, 51-53; from asst prof to prof biol, 57-74, CLARK DISTINGUISHED PROF BIOL, UNIV NOTRE DAME, 74- Concurrent Pos: WHO travel fel, 60, consult, 63; entomologist, Prev Med Univ, Md, 54 & Chem Corps Med Labs, Army Chem Ctr, 54-57; dir, WHO Int Ref Ctr for Aëdes, 66- & N D Vector Biol Lab, 60-; res dir, Int Ctr Insect Physiol & Ecol, Nairobi, Kenya, 70-; NIH study sect trop med, 69-74. Mem: Fel AAAS; Entom Soc Am; Am Soc Trop Med & Hyg; Genetics Soc Am; fel Am Acad Arts & Sci. Res: Culicidae; genetics, systematics bionomics and physiology of Aëdes; systematics of aedine eggs; evolutionary mechanisms; zoogeography; Arctic insects; vector genetics. Mailing Add: Dept of Biol Univ of Notre Dame Notre Dame IN 46556

CRAIG, HARMON, b New York, NY, Mar 15, 26; m 47; c 3. GEOCHEMISTRY, OCEANOGRAPHY. Educ: Univ Chicago, PhD(geol), 51. Prof Exp: Res assoc geochem, Inst Nuclear Studies, Univ Chicago, 51-55; assoc prof geochem, 59-64, RES GEOCHEMIST, SCRIPPS INST, UNIV CALIF, 55-, PROF GEOCHEM, LA JOLLA, 64- Concurrent Pos: Guggenheim fel, Univ Pisa, 62-63; mem oceanog expeds, Monsoon, 61, Zephyrus, 63 & Carrousel, 64. Mem: Am Geophys Union. Res: Isotopic geochemistry; geothermal areas; thermodynamics; physical oceanography; atmospheric chemistry; origin and history of ocean and atmosphere; Lake Tanganyika studies; Ethiopian Rift Valley. Mailing Add: Scripps Inst Oceanog Univ of Calif La Jolla CA 92093

CRAIG, JAMES MORRISON, b Drayton, NDak, June 21, 16; m 39; c 2. MICROBIOLOGY. Educ: San Jose State Col, AB, 38; Stanford Univ, MA, 48; Ore State Univ, PhD(microbiol), 69. Prof Exp: Chemist, B Cribari & Sons, 39-40; chemist & bacteriologist, Eng-Skell Co, 40-42; chief chemist & bacteriologist, Goldfield Consol Mines, 42-45; chief chemist & asst supt, L De-Martini Co, 45-48; instr zool & biol, 48-51, asst prof bact, 51-57, assoc prof, 57-68, PROF MICROBIOL, SAN JOSE STATE UNIV, 68- Concurrent Pos: NSF grant, Ore State Univ, 58 & Univ PR, 63; AEC grant, Univ Wash; NIH spec grant, Ore State Univ, 64. Mem: AAAS; Am Inst Biol Sci; Am Soc Microbiol; Soc Indust Microbiol; Nat Sci Teachers Asn. Res: General bacteriology; aquatic, sanitary and industrial microbiology. Mailing Add: 2201 Gundersen Dr San Jose CA 95125

CRAIG, JAMES PORTER, JR, b Mobile, Ala, Oct 4, 26; m 50; c 2. PHYSICAL CHEMISTRY. Educ: La State Univ, BS, 48, MS, 50, PhD(chem), 53. Prof Exp: Chemist, Bound Brook Labs, Am Cyanamid Co, 50-51; finish chemist, Nylon plant, Chemstrand Corp, 53-54, res chemist, 54-60, group leader phys chem, Res Ctr, NC, 60-70, GROUP LEADER PHYS CHEM, TECH CTR, MONSANTO TEXTILE DIV, 70- Mem: AAAS; Am Chem Soc. Res: Colloid and surface chemistry of polymers; polymer physical chemistry. Mailing Add: Tech Ctr Monsanto Textile Div Decatur AL 35601

CRAIG, JAMES ROLAND, b Philadelphia, Pa, Feb 16, 40; m 62; c 2. GEOCHEMISTRY. Educ: Univ Pa, BA, 62; Lehigh Univ, MS, 64, PhD(geol), 65. Prof Exp: Fel, Carnegie Inst, 65-67; asst prof geochem, Tex Tech Univ, 67-70; ASST PROF GEOCHEM & ASSOC PROF GEOL, VA POLYTECH INST & STATE UNIV, 70- Mem: Mineral Soc Am; Geol Soc Am; Soc Econ Geologists; Mineral Asn Can. Res: Phase relations of ore minerals; biogeochemistry; thermochemistry and phase equilebria of ore minerals; evaporite minerals. Mailing Add: Dept of Geol Sci Va Polytech Inst & State Univ Blacksburg VA 24061

CRAIG, JAMES VERNE, b Bonner Springs, Kans, Feb 7, 24; m 48; c 3. GENETICS. Educ: Univ Ill, BS, 48, MS, 49; Univ Wis, PhD(genetics), 52. Prof Exp: First asst animal sci, Univ Ill, 52-54, asst prof, 54-55; assoc prof poultry husb, 55-60, PROF POULTRY GENETICS, KANS STATE UNIV, 60- Concurrent Pos: NIH spec res fel, Poultry Res Ctr, Scotland, 61-62. Honors & Awards: Poultry Sci Res Award, 61. Mem: AAAS; Poultry Sci Asn; Animal Behav Soc. Res: Animal behavior; population genetics. Mailing Add: Dept of Poultry Sci Kans State Univ Manhattan KS 66506

CRAIG, JAMES WILLIAM, b West Liberty, Ohio, Jan 23, 21; m; c 8. MEDICINE. Educ: Case Western Reserve Univ, BS, 43, MD, 45. Prof Exp: From instr to assoc prof med, Sch Med, Case Western Reserve Univ, 52-72; PROF MED & ASSOC DEAN SCH MED, UNIV VA, 72- Mem: Soc Exp Biol & Med; Am Fedn Clin Res; Am Diabetes Asn; Am Inst Nutrit. Res: Internal medicine; clinical research in intermediary metabolism, particularly carbohydrate metabolism and diabetes mellitus. Mailing Add: Sch of Med Univ of Va Charlottesville VA 22903

CRAIG, JOHN CYMERMAN, b Berlin, Ger, Jan 23, 20; m 45; c 2. ORGANIC CHEMISTRY. Educ: Univ London, BSc, 42, PhD(org chem), 45; Univ Sydney, DSc, 61. Prof Exp: Res chemist, Boots Pure Drug Co, Eng, 45-47; lectr org chem, Univ London, 47-48; from lectr to sr lectr, Univ Sydney, 48-60; from vchmn to chmn dept, 63-70, PROF CHEM & PHARMACEUT CHEM, UNIV CALIF, SAN FRANCISCO, 60- Concurrent Pos: Nuffield Found Dom traveling fel, Dyson Perrins Lab, Oxford Univ, 56-57; vis scientist, Lab Chem Natural Prod, NIH, 59 mem panel, psychopharmacol chem NIMH, 63-68 & mem Preclin psychopharmacol res rev comt, NIMH, 68-72; mem panel, Polycyclic Org Matter, Nat Acad Sci, 70-72 & Vapor-Phase Org Pollutants, 72-75; fac res lectr, Acad Senate, Univ Calif, San Francisco, 74-75. Honors & Awards: Res Achievement Award, Am Pharmaceut Asn Res Found, 67. Mem: Am Chem Soc; The Chem Soc; Swiss Chem Soc; fel Acad Pharmaceut Sci. Res: Biomedical mass spectrometry; deuterium labeling in clinical research; acetylene chemistry and biosynthesis; mechanisms of biological reactions; chemistry of sulfur compounds. Mailing Add: Dept of Pharmaceut Chem Univ of Calif San Francisco CA 94143

CRAIG, JOHN HORACE, b Macon, Ga, Dec 25, 42; m 68. ORGANIC CHEMISTRY. Educ: George Washington Univ, BS, 64; Georgetown Univ, PhD(org chem), 69. Prof Exp: NIH Fel chem, Univ Ill, Urbana-Champaign, 69-70, res assoc org chem, 70-71; asst prof, 71-75, ASSOC PROF ORG CHEM, CALIF STATE COL, SAN BERNARDINO, 75- Mem: Am Chem Soc. Res: Reaction mechanisms; strained sigma and pi bonds; bridged small and medium ring compounds; aliphatic nitrogen heterocycles; conformational analysis; stereo-chemistry; molecular rearrangements; physiologically active compounds. Mailing Add: Dept of Chem Calif State Col San Bernardino CA 92407

CRAIG, JOHN MERRILL, b Pasadena, Calif, Oct 14, 13; m 49; c 3. PATHOLOGY. Educ: Univ Calif, AB, 36, MA, 38; Harvard Univ, MD, 41. Prof Exp: Asst prof path, Children's Hosp, Harvard Med Sch, 55-59; prof, Sch Med, Univ Pittsburgh, 59-60; clin prof, 60-70, PROF PATH, HARVARD MED SCH, 70-; PATHOLOGIST-IN-CHIEF, BOSTON HOSP WOMEN, 60- Concurrent Pos: Res assoc, Children's Cancer Res Found, 50-59; pathologist, Children's Hosp, 59; dir labs, E S Magee Hosp, 59-60; mem path study sect, USPHS, 59-64; assoc ed, Am J Path, 64- Mem: Am Soc Exp Path; Am Asn Path & Bact; Soc Pediat Res; Int Acad Path. Res: Experimental and morphological pediatric, gynecologic and obstetric pathology. Mailing Add: 41 Sargent-Beechwood Brookline MA 02146

CRAIG, JOHN PHILIP, b West Liberty, Ohio, Nov 29, 23. MICROBIOLOGY, INFECTIOUS DISEASES. Educ: Western Reserve Univ, MD, 47; Harvard Univ, MPH, 53. Prof Exp: Epidemiologist, 406th Med Gen Lab, Tokyo, Japan, 51-52; fel virol, Div Med & Pub Health, Rockefeller Found, 53-54; from asst prof to assoc prof, 54-72, PROF MICROBIOL & IMMUNOL, STATE UNIV NY DOWNSTATE MED CTR, 72- Concurrent Pos: NIH spec res fel, Lister Inst Prev Med, London, Eng, 58-59 & Pakistan-SEATO Cholera Res Lab, Dacca, EPakistan, 64; Nat Inst Allergy & Infectious Dis res grants, 60-; mem, NIH Cholera Adv Comt, 70-73; mem cholera panel, US-Japan Coop Med Sci Prog, 70-, chmn, 72- Mem: Infectious Dis Soc Am; Harvey Soc; Am Soc Trop Med & Hyg; fel Am Pub Health Asn; Am Acad Microbiol. Res: Pathogenesis and epidemiology of infectious diseases; bacterial toxins; cholera and other enteric diseases. Mailing Add: Dept of Microbiol & Immunol State Univ NY Downstate Med Ctr Brooklyn NY 11203

CRAIG, JOHN R, b Wichita, Kans, Apr 3, 28; m 48; c 4. RESEARCH ADMINISTRATION. Educ: Univ Chicago, PhB, 49, MS, 50. Prof Exp: Scientist, US Govt, 51-62; dir res grants, NASA, 62-67; ASST DIR, DENVER RES INST, 67- Concurrent Pos: Consult, NSF, 69-70 & Environ Protection Agency, 70-75. Mem: AAAS. Res: International science and technology policy and technology transfer; technology management; research and development management; related curricula development; environmental management. Mailing Add: Denver Res Inst Univ Denver Denver CO 80210

CRAIG, KENNETH ALEXANDER, b Keosauqua, Iowa, Apr 4, 08; m 39, 63. ANALYTICAL CHEMISTRY. Educ: Iowa Wesleyan Col, BS, 30; Pa State Col, MS & PhD(anal chem), 34. Prof Exp: Asst, Iowa Wesleyan Col, 28-30 & Pa State Col, 30-34, instr chem, exten serv, 34-36; asst prof, Lawrence Col, 36-39; res chemist, Kimberly Clark Corp, 39-44, coordr for plastics, 44-48, res chemist in charge pulp res, 48-50, supt pulp paper & newsprint lab, 50-58, chief paper & newsprint dept, 58-63, mgr paper dept, 63-65, sr res assoc, 65-73; CONSULT, PULP & PAPER INDUSTS, 73- Mem: Am Chem Soc; Am Tech Asn Pulp & Paper Indust; assoc Soc Plastics Engrs. Res: Starch modification; paper machine efficiency; secondary fiber utilization; groundwood pulp bleaching; groundwood pitch control. Mailing Add: 329 Ninth St Neenah WI 54956

CRAIG, LAWRENCE CAREY, b Plainview, Tex, May 14, 18; m 41; c 3. STRATIGRAPHY, ECONOMIC GEOLOGY. Educ: Swarthmore Col, AB, 39; Columbia Univ, MA, 41, PhD(geol), 49. Prof Exp: Geologist, Tenn, 42-43, Calif & Nev, 43-45, Colo, 45-61, chief paleotectonic map sect, 61-74, dep chief br oil & gas resources, 74, GEOLOGIST, BR URANIUM & THORIUM RESOURCES, US GEOL SURV, 74- Mem: AAAS; Geol Soc Am; Am Asn Petrol Geol; Paleont Soc; Soc Econ Paleont & Mineral. Res: Frontier uranium studies in Lower Cretaceous rocks of Colorado and Utah; Middle Ordovician stratigraphy of New York and Pennsylvania; Mesozoic stratigraphy of the Colorado Plateau; manganese deposits of eastern Tennessee; lead-zinc deposits of California and Nevada; paleotectonic maps of United States. Mailing Add: US Geol Surv Stop 916 Box 25046 Denver Fed Ctr Denver CO 80225

CRAIG, LOUIS ELWOOD, b Clifton Hill, Mo, Dec 10, 21; m 43; c 4. ORGANIC CHEMISTRY. Educ: Cent Col, Mo, BA, 43; Univ Rochester, PhD(org chem), 48. Prof Exp: Chemist, Am Cyanamid Co, Conn, 43-46 & Gen Aniline & Film Corp, 48-54; supv res chemist, Grand River Chem Div, Deere & Co, 54-56, dir res, 56-58, dir res & tech serv, John Deere Chem Co, 58-65; mgr mkt res & develop, Kerr-McGee Chem Corp, 65-68, mgr mfg, 68-69, vpres mfg, 69-70, vpres info ser, Kerr-McGee Corp, 70-72, VPRES CHEM MFG DIV, KERR-McGEE CHEM CORP, 72- Mem: AAAS; Am Chem Soc; Am Mgt Asn. Res: Polymerization; pharmaceuticals; heterocyclics; inorganic chemicals; fertilizers; marketing; manufacturing. Mailing Add: Kerr-McGee Chem Corp Kerr-McGee Ctr Oklahoma City OK 73125

CRAIG, LYMAN CREIGHTON, chemistry, deceased

CRAIG, NESSLY COILE, b Honolulu, Hawaii, Nov 20, 42; m 67. CELL BIOLOGY. Educ: Reed Col, BA, 63; Univ Pa, PhD(biol), 67. Prof Exp: Fel molecular biol, Inst Cancer Res, 67-70; asst prof, 70-75, ASSOC PROF BIOL SCI, UNIV MD, BALTIMORE COUNTY, 75- Concurrent Pos: USPHS fel, NIH, 68-70. Mem: AAAS; Am Soc Cell Biol; Soc Develop Biol. Res: Origin of ribsomes; synthesis and function of nucleic acids; control of growth and development in higher organisms. Mailing Add: Div of Biol Sci Univ of Md Baltimore County Baltimore MD 21228

CRAIG, NORMAN CASTLEMAN, b Washington, DC, Nov 12, 31; m 55; c 3. PHYSICAL CHEMISTRY. Educ: Oberlin Col, BA, 53; Harvard Univ, MA, 55, PhD(chem), 57. Prof Exp: From asst prof to assoc prof, 57-65, assoc dean, 67-68, chmn dept, 73-74, PROF CHEM, OBERLIN COL, 65-, CHMN DEPT, 75- Concurrent Pos: Hon fel chem, Univ Minn, 63-64; NSF sci fac fel, Univ Calif, Berkeley, 70-71; vis prof chem, Princeton Univ, 74-75. Mem: AAAS; Am Chem Soc. Res: Infrared and Raman spectroscopy; normal coordinate analysis; laser applications. Mailing Add: Dept of Chem Oberlin Col Oberlin OH 44074

CRAIG, PAUL NORMAN, b Minneapolis, Minn, Jan 2, 21; m 43; c 4. MEDICINAL CHEMISTRY, INFORMATION SCIENCE. Educ: Hamline Univ, BS, 41; Univ Minn, PhD(org chem), 48. Prof Exp: Res chemist, US Naval Res Lab, Washington, DC, 42-45 & Smith, Kline & French Labs, 48-71; PRES, CRAIG CHEM CONSULT SERV, 71-; ASST DIR SCI INFO SERV DEPT, FRANKLIN INST RES LABS, 73- Concurrent Pos: Mem, Comt Mod Methods Handling Chem Info, Nat Res Coun, 59-65; chmn, Gordon Conf on Structure-Activity Relationships in Biol, 75. Mem: AAAS; Am Chem Soc; Am Soc Info Sci. Res: Synthesis of new drugs; correlation of chemical structure and biological activities; biomedical information storage, retrieval and evaluation. Mailing Add: 120 Stout Rd Ambler PA 19002

CRAIG, PAUL PALMER, b Reading, Pa, July 29, 33; div; c 2. PHYSICS, SCIENCE. Educ: Haverford Col, BS, 54; Calif Inst Technol, PhD(physics), 59. Concurrent Pos: Mem staff cryogenics, Los Alamos Sci Lab, 58-62; assoc physicist, 62-66, GROUP LEADER, BROOKHAVEN NAT LAB, 62-, PHYSICIST, 66- Concurrent Pos: Guggenheim Found fel, 65-66; assoc prof, State Univ NY Stony Brook, 67-71; mem bd trustees, Environ Defense Fund; mem staff, NSF, 71-74, dep dir & actg dir, Off Energy Res & Develop Policy, 74-75; dir, Energy Resources Coun, Off of the Pres, Univ of Calif, 75- Mem: Fel Am Phys Soc. Res: Energy policy; energy conservation; cryogenics; Mössbauer effect; critical phenomena. Mailing Add: 5 Acacia Ave Berkeley CA 94708

CRAIG, PETER HARRY, b Pittsburgh, Pa, Dec 25, 29; m 54; c 3. PATHOLOGY. Educ: Pa State Univ, BS, 52; Univ Pa, VMD, 55, MS, 58. Prof Exp: Instr vet path, Univ Pa, 55-58; mem staff, Armed Forces Inst Path, 58-59 & 60-61; mem staff, Aviation Med Acceleration Lab, 59-60; asst prof vet path, Univ Pa, 61-65; a assoc prof path, State Univ NY Vet Col, Cornell Univ, 65-73; SECT HEAD PATH, TOXICOL & SURG, ETHICON, INC, 73- Concurrent Pos: Consult, LaWall & Harrisson Res Labs, 62-; asst prof, Grad Sch Arts & Sci, Univ Pa, 63-; lectr, Philadelphia Col Pharm, 63-64. Mem: Am Vet Med Asn; NY Acad Sci; Am Col Vet Path. Res: Experimental pathology, transplantation, cancer, radiation; laboratory animal, aviation, and veterinary pathology; bone and mineral metabolism. Mailing Add: Sect Path Toxicol & Surg Ethicon Inc Somerville NJ 08876

CRAIG, RAYMOND ALLEN, b Mansfield, Ohio, Apr 19, 20; m 48; c 3. CHEMISTRY. Educ: Muskingum Col, BS, 43; Ohio State Univ, MS & PhD(phys & org chem), 48. Prof Exp: Chemist, Am Petrol Inst, Ohio, 43-44; asst chem, Ohio State Univ, 44-47; res chemist nylon res lab, 48-51, res supvr, 52-54, dacron res lab, 54-57, sr res supvr, 57-59, res mgr, 59-63, textile res lab, Del, 63-66, dir, Benger Lab, Va, 66-68, tech dir, Du Pont Int, SA, Switz, 68-74, TECH MGR, E I DU PONT DE NEMOURS & CO, INC, 74- Mem: Am Chem Soc. Res: High polymer chemistry and polymer chemistry; synthetic fibers. Mailing Add: Du Pont Co Nemours Bldg Wilmington DE 19898

CRAIG, RAYMOND S, b DeLand, Fla, Jan 11, 17; m 50; c 2. PHYSICAL CHEMISTRY. Educ: Stetson Univ, BS, 39; Univ Pittsburgh, PhD(chem), 44. Prof Exp: Res assoc, Allegany Ballistics Lab, 44-46; sr res fel, 46-49, from asst res prof to assoc res prof, 49-59, PROF CHEM, UNIV PITTSBURGH, 59- Mem: Am Chem Soc. Res: Magnetic and thermal properties of metals and intermetallic compounds; low temperature calorimetry. Mailing Add: Dept of Chem Univ of Pittsburgh Pittsburgh PA 15213

CRAIG, RICHARD, b Carnegie, Pa, July 14, 37; m 61; c 3. PLANT GENETICS, PLANT BREEDING. Educ: Pa State Univ, BS, 59, MS, 60, PhD(genetics & breeding), 63. Prof Exp: Asst hort, 59-61 & comput sci, 61-62, instr plant breeding & asst prof, 63-71, ASSOC PROF PLANT BREEDING, PA STATE UNIV, 71- Honors & Awards: Spec Recommendation Hort Achievement, 64. Mem: Am Soc Hort Sci. Res: Genetics, cytology and breeding of floricultural plants, including geraniums, zinnias, and holly. Mailing Add: Dept of Hort 106 Tyson Bldg Pa State Univ University Park PA 16802

CRAIG, RICHARD ANDERSON, b New York, NY, Aug 9, 36. THEORETICAL PHYSICS, SOLID STATE PHYSICS. Educ: Univ Ill, BS, 59, MS, 60, PhD(physics), 66. Prof Exp: Res assoc physics, Univ Ore, 66-67; lectr physics, Univ Calif, 67-69; asst prof physics, Ore State Univ, 69-71; mem staff, Columbus Ctr, 71-74, PHYSICIST, COLUMBUS LABS, BATTELLE MEM INST, 74- Concurrent Pos: Consult, Naval Weapons Ctr, Corona, 69. Mem: Am Phys Soc; Sigma Xi. Res: Solid state theory, especially many-particle contributions to solid state phenomena; technical and techno-economic aspects of energy conversion, with particular emphasis on solar energy questions. Mailing Add: Battelle Mem Inst Columbus Labs 505 King Ave Columbus OH 43201

CRAIG, RICHARD ANSEL, b Abington, Mass, Mar 23, 22; m 44; c 5. METEOROLOGY. Educ: Harvard Univ, AB, 42; Mass Inst Technol, SM, 44, ScD(meteorol), 48. Prof Exp: Asst astron, Harvard Observ, 40-42; asst, Mass Inst

Technol, 44, res assoc meteorol, 47; fel, Harvard Observ, 48-51; meteorologist, Air Force Cambridge Res Ctr, 51-58; PROF METEOROL, FLA STATE UNIV, 58- Concurrent Pos: Mem staff, Oceanog Inst, Woods Hole, 46. Mem: Fel Am Meteorol Soc; Am Geophys Union. Res: Spherical harmonics applied to atmosphere; atmospheric ozone; stratospheric meteorology. Mailing Add: 2326 Amelia Circle Tallahassee FL 32304

CRAIG, ROBERT GEORGE, b Charlevoix, Mich, Sept 8, 23; m 45; c 3. DENTAL RESEARCH. Educ: Univ Mich, BS, 44, MS, 51, PhD(chem), 55. Prof Exp: Chemist, anal res, Linde Air Prods, 44-50 & friction & lubrication, Tex Co, 54-55; assoc res chemist high polymers, Eng Res Inst, 55-57, from asst prof to assoc prof, 57-65, PROF DENT, UNIV MICH, 65-, CHMN DEPT DENT MAT, 69- Honors & Awards: Thomas Young Award, 69; Wilmer Souder Award in Dent Mat, Dent Mat Group, Int Asn Dent Res, 75. Mem: Am Chem Soc; Am Soc Testing & Mat; Soc Exp Stress Anal; Int Asn Dent Res; Soc Biomat. Res: Colloid and surface chemistry; polymer chemistry; dental materials; bio-engineering; stress analysis. Mailing Add: 1503 Wells St Ann Arbor MI 48104

CRAIG, ROY PHILLIP, b Durango, Colo, May 10, 24. ENVIRONMENTAL MANAGEMENT, SCIENCE WRITING. Educ: Univ Colo, BA, 48; Calif Inst Technol, MS, 50; Iowa State Col, PhD(chem), 52. Prof Exp: Asst, Iowa State Col, 49-52; prof chemist, Dow Chem Co, 52-56, group leader, 56-60; lectr phys chem, Univ Colo, 61-65, assoc prof phys chem & coordr phys sci, div integrated studies, 65-68; vis prof, Univ Hawaii, 69; TECH & EDUC CONSULT & SCI WRITER, 69-; PRES, FOUR CORNERS ENVIRON RES INST, 74- Mem: AAAS. Res: Impact of science on society; relation between science and the humanities; physical sciences for nonscience majors; solution adsorption; environmental management; solar energy utilization. Mailing Add: Rte 2 Box B-10 Durango CO 81301

CRAIG, STANLEY HAROLD, b New York, NY, Oct 24, 09. RADIOLOGY. Educ: NY Univ, BS, 30; Univ Basel, MD, 35. Prof Exp: Resident radiol, Beth Israel Hosp, NY, 35-38; PROF RADIOL, NEW YORK MED COL, 48- Concurrent Pos: Assoc prof, Med Col, New York Univ, 47-54; consult radiol, Med Dept, NY Times, 48- Mem: Radiol Soc NAm; Int Skeletal Soc; AMA; Fel Am Col Gastroenterol; Asn Am Med Cols. Mailing Add: Metrop Hosp 1901 First Ave New York NY 10029

CRAIG, SUSAN WALKER, b New York, NY; m 68. IMMUNOBIOLOGY. Educ: Univ Pa, BA, 67; Johns Hopkins Univ, PhD(biol), 73. Prof Exp: Fel molecular pharmacol, Dept Pharmacol, 73-75, ASST PROF PHYSIOL CHEM, SCH MED, JOHNS HOPKINS UNIV, 75- Concurrent Pos: Fel, Jane Coffin Childs Mem Fund Med Res, 73-75. Res: Cell surface control of lymphocyte physiology. Mailing Add: Sch of Med Johns Hopkins Univ Dept Physiol Chem 725 N Wolfe St Baltimore MD 21205

CRAIG, THEODORE WARREN, b San Francisco, Calif, May 21, 40; m 67; c 1. FOOD SCIENCE. Educ: Univ Calif, Berkeley, BS, 62, MBA, 75; Mass Inst Technol, PhD(org chem), 66. Prof Exp: Sr res chemist, Cent Res Labs, Gen Mills, Inc, Minn, 65-68; proj leader, Foremost Res Ctr, 68-71, group leader, 71-73, tech mgr, Foods Prod Div, Foremost Tech Ctr, 73-74, mgr prod/process develop, 74-76, DIR RES & DEVELOP CTR, FOREMOST FOODS CO, 76- Mem: Inst Food Technol; Am Dairy Sci Asn. Res: Food and protein chemistry; cereal and dairy by-products technology; food product development. Mailing Add: Foremost Res & Develop Ctr Foremost Foods Co 6363 Clark Ave Dublin CA 94566

CRAIG, W E (JACK), chemistry, deceased

CRAIG, WILFRED STUART, b New Hartford, Mo, Aug 11, 16; m 44; c 3. ENTOMOLOGY. Educ: Univ Mo, BS & AB, 47; Iowa State Col, PhD(entom), 53. Prof Exp: Asst state entomologist, Mo, 47-49; asst entomologist, Iowa State Col, 49-53, asst state entomologist, Iowa State Univ, 53-60, state entomologist, Univ & entomologist, Exp Sta, 60-65; EXTEN ENTOMOLOGIST, UNIV MO, COLUMBIA, 65- Mem: Entom Soc Am. Res: Biology and control of nursery insects; taxonomy of Anthicidae and other Coleoptera. Mailing Add: 1-68 Agr Bldg Univ of Mo Columbia MO 65201

CRAIG, WILLIAM F, b Olean, NY, Sept 12, 29; m 55; c 2. AGRONOMY, PLANT GENETICS. Educ: Pa State Univ, BS, 51, MS, 63, PhD(genetics), 66. Prof Exp: Teacher pub schs, Pa, 54-60; from asst to instr, Pa State Univ, 60-66; corn breeder, Funk Bros Seed Co, 66-70, mgr tech serv & wheat opers, 70-73, MGR TECH SERV, FUNK SEEDS INT, INC, 73- Mem: Am Soc Agron; Crop Sci Soc Am; Genetics Soc Am; Am Inst Biol Sci. Res: Corn breeding. Mailing Add: Funk Seeds Int Inc 1300 W Washington St Bloomington IL 61701

CRAIG, WILLIAM GEORGE, organic chemistry, medicinal chemistry, see 12th edition

CRAIG, WILLIAM WARREN, b Kansas City, Mo, Apr 1, 35; m 61; c 4. GEOLOGY. Educ: Univ Mo, Columbia, BA, 57, MA, 61; Univ Tex, Austin, PhD(geol), 68. Prof Exp: Assoc prof geol, Northeastern Mo State Col, 65-68; asst prof, 68-71, ASSOC PROF EARTH SCI, LA STATE UNIV, NEW ORLEANS, 71- Mem: Geol Soc Am; Soc Econ Paleont & Mineral. Res: Cretaceous nonmarine Ostracodes; midcontinent Ordovician and Silurian stratigraphy and conodont biostratigraphy. Mailing Add: Dept of Earth Sci La State Univ Lakefront New Orleans LA 70122

CRAIGE, ERNEST, b El Paso, Tex, 1918; m 46; c 4. MEDICINE. Educ: Univ NC, BA, 39; Harvard Univ, MD, 43. Prof Exp: From intern to resident med, Mass Gen Hosp, 43-48, resident cardiol, 49-50, asst med, 50-52; from asst prof to assoc prof, 52-62, PROF MED, SCH MED, UNIV NC, CHAPEL HILL, 62- Concurrent Pos: Clin & res fel med, Mass Gen Hosp, 49-S0; fel med, Harvard Med Sch, 49-50; resident, House of Good Samaritan, Mass, 48-49, asst vis physician, 50-52; chief cardiol, NC Mem Hosp, 52- Mem: AMA; Am Heart Asn; fel Am Col Physicians; fel Am Col Cardiol; Am Clin & Climat Asn. Res: Phonocardiography, echo cardiography and other non-invasive methods of studying cardiac function. Mailing Add: Dept of Med Univ of NC Sch of Med Chapel Hill NC 27514

CRAIGHEAD, JOHN EDWARD, b Pittsburgh, Pa, Aug 14, 30; m 57; c 2. PATHOLOGY, VIROLOGY. Educ: Univ Utah, BS, 52, MD, 56. Prof Exp: Intern med, Barnes Hosp, St Louis, Mo, 56-57; jr asst resident, Peter Bent Brigham Hosp, 60-61, sr asst resident, 61-62, chief resident path, 62-63, assoc, 63-68; PROF PATH, COL MED, UNIV VT, 68-, CHMN DEPT, 74- Concurrent Pos: Teaching fel, Harvard Med Sch, 61-63; assoc, 63-66, asst prof, 66-68; assoc mem comn viral infections, Armed Forces Epidemiol Bd, 66; mem infectious dis adv comt, Nat Inst Allergy & Infectious Dis, 71-75. Mem: Am Soc Exp path; Am Asn Path & Bact; Int Acad Path; Soc Exp Biol & Med; Am Soc Clin Path. Res: Biology, pathogenesis and pathology of virus disease in man and animals. Mailing Add: Dept of Path Univ of Vt Col of Med Burlington VT 05401

CRAIGHEAD, JOHN J, b Washington, DC, Aug 14, 16; m 44; c 3. ECOLOGY. Educ: Pa State Col, BA, 39; Univ Mich, MS, 40, PhD, 50. Prof Exp: Biologist, NY Zool Soc, 47-49; dir survival training for armed forces, US Dept Defense, 50-52; WILDLIFE BIOLOGIST & LEADER MONT COOP WILDLIFE RES UNIT, BUR SPORT FISHERIES & WILDLIFE, 52-; PROF ZOOL & FORESTRY, UNIV MONT, 52- Concurrent Pos: Grants, Wildlife Mgt Inst, raptor predation; NSF, ecol of grizzly bear & radiotracking grizzly bears; AEC, radiotracking & telemetering systs large western mammals; NASA, satellite tracking large mammals & habitat mapping; Bur Sport Fisheries & Wildlife, surv peregrine eyries & studies ecol grizzly bear; US Forest Serv & Mont Fish & Game Dept, eval grizzly bear habitat. Honors & Awards: Bur Sport Fisheries & Wildlife Superior Performance Award, 65; Citation for Organizing & Admin Navy's Land Survival Training Prog, Secy Defense. Mem: Wildlife Soc (vpres, 62-63); Wilson Ornith Soc; Ecol Soc Am. Res: Raptor predation; waterfowl; population dynamics. Mailing Add: Mont Coop Wildlife Res Unit Bur Spt Fish & Wildlf Univ Mont Missoula MT 59801

CRAIGMILES, JULIAN PRYOR, b Thomasville, Ga, Jan 17, 21; m 48; c 3. AGRONOMY. Educ: Univ Ga, BSA, 42, MSA, 48; Cornell Univ, PhD, 52. Prof Exp: Asst, Coker Pedigreed Seed Co, SC, 41; asst agronomist, Ga Crop Improv Asn, 46-48; agronomist, Ga Exp Sta, 48-64; PROF & RESIDENT DIR, AGR RES & EXTEN CTR, TEX A&M UNIV, 64- Concurrent Pos: Res award, Ga Plant Food Educ Soc, 59- Honors & Awards: Outstanding Researcher, Sears Roebuck & Co, 58. Mem: Am Soc Agron; Weed Sci Soc Am; Am Genetic Asn; Crop Sci Soc Am. Res: Soybean breeding; rice and forage management; seed production of rice, soybeans and forages. Mailing Add: Tex A&M Agr Res & Exten Ctr Rte 5 Box 784 Beaumont TX 77706

CRAIK, EVA LEE, b Gatesville, Tex, Aug 12, 19; m 41; c 4. BIOLOGY, SCIENCE EDUCATION. Educ: Tex Women's Univ, BS, 40; Hardin-Simmons Univ, MEd, 60; N Tex State Univ, EdD(biol), 66. Prof Exp: Teacher, Ranger High Sch & Jr Col, Tex, 40-43, pub sch, Kans, 43-44 & pub schs, Tex, 59-62; from instr to assoc prof, 62-73, coordr sci educ, 67-73, PROF BIOL, HARDIN-SIMMONS UNIV, 73- Mem: Nat Sci Teachers Asn; Am Inst Biol Sci. Res: Relative effectiveness of inductive-deductive and deductive-descriptive methods of teaching college zoology. Mailing Add: 1802 N 11th Abilene TX 79603

CRAIN, ALFRED V R, b Rochester, NY, Apr 29, 24; m 51; c 3. ANALYTICAL CHEMISTRY. Educ: Brooklyn Polytech Inst, BS, 57. Prof Exp: Chemist, Charles Pfizer Co, 53-58; from asst res chemist to sr res chemist, 58-75, GROUP LEADER, STERLING WINTHROP RES INST, 75- Mem: Sigma Xi. Res: Thin layer chromatography and liquid chromatographic separation of pharmaceutical mixtures and drug metabolites. Mailing Add: Crarer Rd West Sand Lake NY 12156

CRAIN, DONALD LEE, organic chemistry, see 12th edition

CRAIN, JAMES LARRY, b Franklinton, La, July 16, 35; m 55; c 3. ZOOLOGY. Educ: Southern Miss Univ, BS, 57, PhD(zool), 66; Stephen F Austin State Univ, MA, 63. Prof Exp: Teacher high sch, Wash, 57-64; ASSOC PROF BIOL SCI, SOUTHEASTERN LA UNIV, 66-; PRES, LA ENVIRON CONSULTS, INC, 75- Mem: Am Soc Mammal; Am Fisheries Soc. Res: Vertebrate taxonomy and ecology; environmental impact. Mailing Add: Dept Biol Southeastern La Univ Box 745 Univ Sta Hammond LA 70401

CRAIN, JAY BOUTON, b Paso Robles, Calif, May 10, 41; m 63; c 1. ANTHROPOLOGY, PSYCHIATRY. Educ: Calif State Univ, Sacramento, BA, 63; Cornell Univ, PhD(social anthrop), 70. Prof Exp: Asst prof, 67-71, ASSOC PROF ANTHROP, CALIF STATE UNIV, SACRAMENTO, 72-; ASST PROF MED ANTHROP, DEPT PSYCHIAT, SCH MED, UNIV CALIF, DAVIS, 71- Concurrent Pos: Consult, Sacramento County Health Dept, 72-73. Mem: Fel Royal Anthrop Inst Gt Brit & Ireland; fel Am Anthrop Asn. Res: Ritual communication in medical and psychiatric settings; social and medical systems of Borneo; palaeopathology and palaeoepidemiology; communication of emotions in focused interaction. Mailing Add: Dept of Psychiat Univ of Calif Sch of Med Davis CA 95616

CRAIN, RONALD DEE, organic chemistry, see 12th edition

CRAIN, STANLEY M, b New York, NY, Feb 5, 23; m 46; c 2. NEUROPHYSIOLOGY. Educ: Brooklyn Col, AB, 43; Columbia Univ, PhD(biophys), 54. Prof Exp: Group leader radiol res, Health Physics Div, Argonne Nat Lab, 45-47; res electrophysiologist, Dept Neurol, Columbia Univ, 50-57; res cell physiologist & sect head nerve tissue cult lab, Abbott Labs, Ill, 57-61; asst prof anat, Dept Neurol, Col Physicians & Surgeons, Columbia Univ, 61-65; from assoc prof to prof physiol, 65-74, PROF NEUROSCI & PHYSIOL, ALBERT EINSTEIN COL MED, 69- Concurrent Pos: Grass res fel, Marine Biol Lab, Woods Hole, Mass, 57; NIH res career develop fel, 61-65; Kennedy scholar, Rose F Kennedy Ctr Ment Retardation & Human Develop, 65-75; assoc ed, J Neurobiol, 70- Honors & Awards: Lucy Moses Prize Neurol Award, Col Physicians & Surgeons, Columbia Univ, 66. Mem: Am Physiol Soc; Am Asn Anatomists; Am Soc Cell Biologists; Soc Neurosci; Biophys Soc. Res: Electrophysiologic and cytologic research on tissue cultures of mammalian brain, spinal cord and neuromuscular systems, especially during development of synaptic and other organotypic relations after isolation of fetal neural tissues in vitro. Mailing Add: Rose F Kennedy Ctr 1410 Pelham Pkwy Bronx NY 10461

CRAINE, ELLIOTT MAURICE, b Burlington, Iowa, Oct 5, 24; m 50; c 2. AGRICULTURAL BIOCHEMISTRY. Educ: Univ Ill, AB, 49, BS, 51, PhD(biochem), 52. Prof Exp: Chemist, Agr Res Serv, USDA, 54-58; res chemist, Upjohn Co, 58-63; res biochemist, 63-65, HEAD BIOCHEM, HESS & CLARK DIV, RHODIA, INC, 65- Mem: AAAS; Am Chem Soc. Res: Chemistry of seed proteins; isolation of antibiotics; drug metabolism; biochemistry of ruminants; metabolism of foreign compounds in animals, plants and soil; drugs and pesticides; analytical biochemistry; isolation of natural chemicals. Mailing Add: 827 Ridge Rd Ashland OH 44805

CRAKER, LYLE E, b Reedsburg, Wis, Feb 3, 41; m 63; c 3. PLANT PHYSIOLOGY. Educ: Univ Wis, Madison, BS, 63; Univ Minn, St Paul, PhD(agron), 67. Prof Exp: From asst prof to assoc prof, 69-75, PROF PLANT PHYSIOL, UNIV MASS, 75- Mem: Am Soc Plant Physiol; Am Soc Agron; Crop Sci Soc Am. Res: Utilization of lights and plant hormones for regulated crop production and storage. Mailing Add: Dept of Plant & Soil Sci Univ of Mass Amherst MA 01002

CRALL, HOWARD WILLIAM, b Akron, Ohio, Sept 30, 15; m 42; c 1. BIOLOGY. Educ: Ohio State Univ, BS, 38, MSc, 41, PhD(biol, educ), 50. Prof Exp: Teacher high sch, Ohio, 38-39; asst zool, Purdue Univ, 41-42; asst prof biol & head dept natural sci, Culver-Stockton Col, 46-47; PROF BIOL SCI & SUPVR OFF CAMPUS STUDENT TEACHING BIOL, WESTERN ILL UNIV, 47-, COORDR GEN BIOL, 69- Concurrent Pos: Dir, NSF Biol Teachers Inst, 63-64. Mem: Fel AAAS; Nat Sci Teachers Asn; Ecol Soc Am; Nat Asn Biol Teachers; Entom Soc Am. Res: Investigation of arthropod hibernation; teaching and evaluation of achievement in applying principles in high school biology; evaluation techniques in high school teaching; moral curriculum for biology teachers; improving teaching of biology in the secondary school. Mailing Add: Dept of Biol Sci Western Ill Univ Macomb IL 61455

CRALL, JAMES MONROE, b Monongahela, Pa, July 13, 14; m 43; c 2. PLANT BREEDING, PLANT PATHOLOGY. Educ: Purdue Univ, BS, 39; Univ Mo, AM, 41, PhD(plant path), 48. Prof Exp: Tomato blight agent, Exp Sta, Purdue Univ & Bur Plant Indust, USDA, 38-39; asst bot, Univ Mo, 39-42, jr plant pathologist, Exp Sta, 46-48; res asst prof plant path, Iowa State Univ & assoc path, USDA, 48-52, DIR, AGR RES CTR, LEESBURG, AGR EXP STA, INST FOOD & AGR SCI, USDA, UNIV FLA & PROF PLANT PATH, UNIV, 52- Mem: AAAS; Am Soc Hort Sci; Am Phytopath Soc; Mycol Soc Am; Coun Agr Sci & Technol. Res: Diseases of watermelon and grape; watermelon breeding. Mailing Add: Agr Res Ctr Univ of Fla PO Box 388 Leesburg FL 32748

CRALLEY, JOHN CLEMENT, b Carmi, Ill, Oct 16, 32; m 63; c 3. HUMAN ANATOMY, PHYSIOLOGY. Educ: Univ Ill, BS, 56, MS, 60, PhD(zool), 65. Prof Exp: ASST PROF ZOOL, ILL STATE UNIV, 63- Mem: AAAS; Am Soc Zoologists; Am Inst Biol Sci. Res: Fish physiology and acclimation; blood vascular systems; functional anatomy of the human foot. Mailing Add: Dept of Biol Ill State Univ Normal IL 61761

CRAM, DONALD JAMES, b Chester, Vt, Apr 22, 19; m 41. ORGANIC CHEMISTRY. Educ: Rollins Col, BS, 41; Univ Nebr, MS, 42; Harvard Univ, PhD(org chem), 47. Prof Exp: Res chemist, Merck & Co, NJ, 42-45; instr org chem & Am Chem Soc fel mold metabolites, 47-48, from asst prof to assoc prof chem, 48-56, PROF CHEM, UNIV CALIF, LOS ANGELES, 56- Concurrent Pos: Guggenheim fel, 54-55. Honors & Awards: Herbert Newby McCoy Award; Soc Chem Mfg Asn Award; Award Creative Work Synthetic Org Chem, Am Chem Soc, 65, Arthur C Cope Award, 74; Calif Scientist of Year, Calif Mus Sci & Indust, 74. Mem: Nat Acad Sci; Am Chem Soc; Swiss Chem Soc; Am Acad Arts & Sci. Res: Stereochemistry, especially of carbanions and substitutions at sulfur; macro ring chemistry; synthetic multiheteromacrocycles that model enzyme systems in complexation and catalysis; highly structured molecular complexes; host-guest chemistry. Mailing Add: Dept of Chem Univ of Calif Los Angeles CA 90024

CRAM, LEIGHTON SCOTT, b Emporia, Kans, Oct 21, 42; m 65. BIOPHYSICS. Educ: Kans State Teachers Col, BA, 64; Vanderbilt Univ, MS, 66; Pa State Univ, PhD(biophys), 69. Prof Exp: Fel, 69-71, STAFF MEM BIOPHYS, BIOMED RES GROUP, LOS ALAMOS SCI LAB, 71- Concurrent Pos: Consult, Particle Technol Inc, 72- Mem: AAAS; Biophys Soc. Res: Photon emission from thin films bombarded by electrons and DNA conformation in bacterial viruses; development and application of high speed methods of obtaining data on cellular properties and intracellular components. Mailing Add: Biomed Res Group H-10 Los Alamos Sci Lab Los Alamos NM 87544

CRAM, SHELDON LEWIS, b Maxbass, NDak, Nov 21, 19; m 43; c 5. PHYSICS. Educ: NDak State Teacher's Col, BA, 40; Ore State Col, BS, 41; Colo State Univ, AM, 50. Prof Exp: Prof physics, NDak State Teacher's Col, 46-49; PROF PHYSICS, WESTMAR COL, 50- Mem: AAAS; Am Asn Physics Teachers; Am Inst Physics. Res: Science education; electronics. Mailing Add: Dept of Physics Westmar Col Le Mars IA 51031

CRAM, STANFORD WINSTON, b Waterville, Minn, Aug 28, 07; m 30; c 3. PHYSICS. Educ: St Olaf Col, AB, 29; Univ Wis, PhD(physics), 35. Prof Exp: Instr physics, St Olaf Col, 29-30; asst, Univ Wis, 31-35; Kosciuszko fel, Univ Warsaw, 35-36; instr physics, Univ Wis, 36; prof physics, 37-72, head dept, 37-44, head dept phys sci, 45-72, EMER PROF PHYSICS, EMPORIA KANS STATE COL, 72- Concurrent Pos: Supvr electronics res, Sylvania Elec Prods, Inc, NY, 44-45. Honors & Awards: Distinguished Serv Citation, Am Asn Physics Teachers, 68. Mem: Assoc Am Phys Soc; assoc Am Asn Physics Teachers. Res; Spectrographic analysis band spectra; construction of high pressure manometer; gaseous discharge in mercury plus gas mixture; decay-time of fluorescent solutions. Mailing Add: 1620 Sherwood Way Emporia KS 66801

CRAM, STUART PROUD, b Emporia, Kans, Apr 23, 39; m 62; c 2. ANALYTICAL CHEMISTRY. Educ: Kans State Teachers Col, BA, 61; Univ Wis, MS, 63; Univ Ill, PhD(chem), 66. Prof Exp: Asst prof chem, Univ Fla, 66-74; MEM STAFF, VARIAN INSTRUMENT DIV, 74- Mem: Am Chem Soc; Soc Appl Spectros. Res: Analytical chemistry, including gas chromatography, neutron activation analysis, and the development of analytical instrumentation and methodology. Mailing Add: Varian Instrument Div 2700 Mitchell Dr Walnut Creek CA 94598

CRAM, WILLIAM HUGH, plant breeding, genetics, see 12th edition

CRAM, WILLIAM THOMAS, b South Burnaby, BC, Oct 20, 27; m 51; c 2. ENTOMOLOGY. Educ: Univ BC, BSA, 50; Ore State Univ, MS, 55, PhD(entom), 64. Prof Exp: SCIENTIST, RES STA, CAN DEPT AGR, 50- Mem: Entom Soc Am; Entom Soc Can. Res: Ecology of root weevils as pests of berry crops; nutrition, host selection and fecundity of root weevils, especially Otiorhynchus sulcatus; pests of berry crops. Mailing Add: Res Sta Can Dept of Agr 6660 NW Marine Dr Vancouver BC Can

CRAMBLET, WILBUR HAVERFIELD, mathematics, deceased

CRAMBLETT, HENRY G, b Scio, Ohio, Feb 8, 29; m 60; c 2. PEDIATRICS, VIROLOGY. Educ: Mt Union Col, BS, 50; Univ Cincinnati, MD, 53; Am Bd Pediat, dipl; Am Bd Med Microbiol, dipl. Prof Exp: Intern med, Harvard Med Serv, Boston City Hosp, 53-54; resident pediat, Children's Hosp, Cincinnati, Ohio, 54-55; res assoc, Clin Ctr, NIH, 55-57; resident & instr pediat, Univ Iowa, 57-58, assoc prof, 58-60; dir virol lab, Bowman Gray Sch Med, 60-64, assoc prof pediat, 60-63, prof, 63-64; chmn dept med microbiol, 66-73, PROF PEDIAT, OHIO STATE UNIV, 64-, PROF MED MICROBIOL, 66-, DEAN COL MED, 73-, ACTG VPRES MED AFFAIRS, 74- Concurrent Pos: Leukemia chemother grant, 58-60; career develop award, 62; Carey & Hofheimer scholastic awards, Univ Cincinnati; dir res & div microbiol, Children's Hosp, Columbus, 64-66, exec dir, Res Found, 66-73. Mem: Soc Pediat Res; Soc Exp Biol & Med; Am Acad Pediat; Infectious Dis Soc Am; fel Am Acad Microbiol. Res: Pediatric virology; infectious diseases; antibiotic research and immunology of the newborn and infant. Mailing Add: Ohio State Univ Col of Med 370 W Ninth Ave Columbus OH 43210

CRAMER, ARCHIE BARRETT, b Winnipeg, Man, Oct 23, 09; m 35; c 2. FOOD CHEMISTRY. Educ: Univ Man, BSc, 34, MSc, 35; McGill Univ, PhD(wood chem), 39. Prof Exp: Asst, Carnegie Inst Technol, 39-41; res chemist, Miner Labs, 41-46; chief chemist, 46-63, VPRES RES & TECH DIR, F & F LABS, 63- Honors & Awards: Stroud Jordan Award, 63. Mem: AAAS; Am Chem Soc; Am Inst Chem; Am Asn Candy Technol; Nat Confectioners Asn US. Res: Wood and lignin chemistry; chemistry and preparation of lecithin; general carbohydrate chemistry; general chemistry of candy and food manufacture; preparation of certain oils and derivatives from lignin; preparation of glycerine derivatives; carbohydrate food flavor. Mailing Add: F & F Labs Inc 3501 W 48th Pl Chicago IL 60632

CRAMER, ARDIS LAHANN, b Chicago, Ill, Jan 22, 27; m 50. PARASITOLOGY, INVERTEBRATE PHYSIOLOGY. Educ: Northwestern Univ, BSE, 48; Emory Univ, MS, 63, PhD(biol), 67. Prof Exp: Teacher pub schs, Ill & Ga, 48-53 & 58-61; teaching asst biol, Emory Univ, 61-62; vis instr, Agnes Scott Col, 68-69, vis asst prof, 69-70, asst prof biol, 70-73; vis asst prof, Ga Inst Technol, 73 & summer 74; VIS ASST PROF BIOL, EMORY UNIV, 75- Concurrent Pos: Consult, Westminster schs, 73-75. Mem: AAAS; Am Soc Parasitol. Res: Protein nutrition in Hymenolepis diminuta and rat host; lipid nutrition in Hymenolepis diminuta and rat host; use of bile components by Hymenolepis diminuta. Mailing Add: 1431 Cornell Rd NE Atlanta GA 30306

CRAMER, CARL FREDERICK, b Raton, NMex, June 7, 22; m 49; c 2. PHYSIOLOGY. Educ: Univ NMex, BS, 44, MS, 47; Univ Calif, PhD(physiol), 53. Prof Exp: Res assoc physiol, Med Sch, Tulane Univ, 54; ASSOC PROF PHYSIOL, UNIV BC, 55- Mem: Am Physiol Soc; Am Inst Nutrit; Can Physiol Soc. Res: Bone and gastrointestinal mineral physiology; turnover of radioactive tracers and removal of fission products. Mailing Add: Dept of Physiol Univ of BC Vancouver BC Can

CRAMER, DAVID ALAN, b Ann Arbor, Mich, Aug 7, 25; m 45; c 3. ANIMAL NUTRITION. Educ: Colo State Univ, BS, 49, MS, 55; Ore State Univ, PhD(animal nutrit & biochem), 60. Prof Exp: Res chemist, Arapahoe Chem, Inc, 55-56; asst prof animal nutrit & vet sci, Calif State Polytech Col, 56-58; from asst prof to assoc prof, 60-68, PROF BIOCHEM OF ANIMAL PRODS, COLO STATE UNIV, 68- Concurrent Pos: Fulbright scholar, Dept Sci & Indust Res, NZ, 65-66. Mem: Am Soc Animal Sci; Am Meat Sci Asn; Am Oil Chem Soc; Inst Food Technol; Am Dairy Sci Asn. Res: Lipid metabolism in domestic animals, especially in the ruminant; meat flavor analysis; rumen metabolism. Mailing Add: Dept of Animal Sci Colo State Univ Ft Collins CO 80521

CRAMER, EVA BROWN, b New York, NY, June 6, 44; m 68; c 2. ENDOCRINOLOGY, CELL BIOLOGY. Educ: Cornell Univ, BS, 65; Jefferson Med Col, MS, 67, PhD(anat), 69. Prof Exp: ASST PROF ANAT, STATE UNIV NY DOWNSTATE MED CTR, 73- Concurrent Pos: NIH fel, Col Physicians & Surgeons, Columbia Univ, 69-70; NH Heart Asn fel, Univ NH, 72-73. Mem: Am Asn Anatomists. Res: The ultrastructural localization of cations in the anterior pituitary gland; ultrastructural effects of hypothyroidism on the development of the endocrine system. Mailing Add: Dept of Anat State Univ NY Downstate Med Ctr 450 Clarkson Ave Brooklyn NY 11203

CRAMER, FRANCIS BARNARD, b Oneida, NY, Feb 3, 12; m 37; c 2. ORGANIC CHEMISTRY. Educ: Am Univ, AB, 33; Princeton Univ, AM, 34, PhD(carbohydrates), 37. Prof Exp: Asst, Princeton Univ, 34-37; res assoc, Mass Inst Technol, 38-39; res chemist, rayon tech div, E I du Pont de Nemours & Co, Del, 39-43 & Va, 43-45, res supvr, 45-48; org chemist, Franklin Inst Res Found, 48-53; res chemist, pioneer res div, Textile Fibers Dept, E I du Pont de Nemours & Co, Inc, 53-73; RETIRED. Mem: AAAS; Am Chem Soc. Res: Ketone sugar derivatives; rayon and cellulose derivatives; organic biochemistry; concentration of viscose by extraction; spinning process for surface-delustered rayon; alkoxymethyl derivatives of cellulose acetate for cross-linkage; enzyme inhibitors in carbohydrate metabolism; condensation polymers. Mailing Add: 420 Townsend Rd Newark DE 19711

CRAMER, GEORGE FRANKLIN, mathematics, see 12th edition

CRAMER, HARRISON EMERY, b Johnstown, Pa, May 27, 19; m 42; c 4. AIR POLLUTION, METEOROLOGY. Educ: Amherst Col, AB, 41; Mass Inst Technol, SM, 43, ScD(meteorol), 48. Prof Exp: Instr meteorol, Mass Inst Technol, 42-44; meteorologist, Am Export Air Lines, 44; res assoc, Mass Inst Technol, 46-48, res meteorologist, 48-65; sr staff scientist, GCA Technol Div, 65-66, dir environ sci lab, 67-72, vpres, 69-72; PRES, H E CRAMAR CO, INC, 72- Mem: Fel AAAS; NY Acad Sci; Am Meteorol Soc; Am Geophys Union; Royal Meteorol Soc. Res: Development of mathematical models for air pollution applications, meteorological instrumentation and air pollution data systems; quantitative assessment of air pollution problems; analysis of air pollution measurements. Mailing Add: H E Cramar Co Inc 540 Arapeen Dr Salt Lake City UT 84108

CRAMER, HOWARD ROSS, b Chicago, Ill, Sept 17, 25; m 50. STRATIGRAPHY, INVERTEBRATE PALEONTOLOGY. Educ: Univ Ill, BS, 49, MS, 50; Northwestern Univ, PhD(geol), 54. Prof Exp: From instr to asst prof geol, Franklin & Marshall Col, 53-58; asst prof, 58-62, ASSOC PROF GEOL, EMORY UNIV, 62- Mem: Geol Soc Am; Am Asn Petrol Geol; Paleont Soc; Nat Asn Geol Teachers. Res: Bibliography; geology and history; petroleum and stratigraphy of Georgia. Mailing Add: Dept of Geol Emory Univ Atlanta GA 30322

CRAMER, JAMES D, b Canton, Ohio, Aug 4, 37; m 57; c 2. NUCLEAR PHYSICS. Educ: Fresno State Col, BS, 60; Univ Ore, MS, 62; Univ NMex, PhD(physics), 69. Prof Exp: Staff mem, Los Alamos Sci Lab, 62-70; staff scientist, 70-75, DIR, SCI APPLNS INC, 72-, VPRES, 75- Concurrent Pos: Consult, Los Alamos Sci Lab, 70-72. Mem: Am Phys Soc; Inst Elec & Electronics Engrs. Res: Physics of gases and nuclear physics. Mailing Add: Sci Applns Inc PO Box 3507 Albuquerque NM 87110

CRAMER, JANE HARRIS, b Chicago, Ill, Dec 1, 42; m 66; c 1. MOLECULAR BIOLOGY. Educ: Carleton Col, BA, 64; Northwestern Univ, PhD(microbiol), 70. Prof Exp: Fel, 70-74, ASST SCIENTIST MOLECULAR BIOL, UNIV WIS-MADISON, 74- Mem: AAAS. Res: Structure and organization of ribosomal RNA cistrons in yeast, their arrangement on different chromosomes and their replication during the cell cycle. Mailing Add: Lab of Molecular Biol Univ of Wis Madison WI 53706

CRAMER, JOHN GLEASON, b Houston, Tex, Oct 24, 34; m 61; c 3. NUCLEAR PHYSICS. Educ: Rice Univ, BA, 57, MA, 59, PhD(physics), 61. Prof Exp: Res assoc nuclear physics, Univ Ind, 61-63, asst prof physics, 63-64; from asst prof to assoc prof, 64-73, PROF PHYSICS, UNIV WASH, 73- Mem: AAAS; fel Am Phys Soc. Res: Heavy ion reactions and scattering; reactions with polarized particles; nuclear reactions and reaction mechanisms through measurements of reaction cross sections and angular correlations; applications of computers to nuclear research; isospin reactions. Mailing Add: Nuclear Physics Lab GL-10 Univ of Wash Seattle WA 98195

CRAMER, JOHN JOSEPH, b Mo, Aug 14, 14; m 38; c 4. TEXTILE CHEMISTRY. Educ: Park Col, AB, 36. Prof Exp: Lab technician, Procter & Gamble Co, 37-41; res chemist, 42-54, SUPVR LAUNDRY RES, WYANDOTTE CHEM CORP, 55- Mem: Am Chem Soc; Am Oil Chemists Soc; Am Asn Textile Chemists & Colorists; Am Soc Testing & Mat. Res: Laundering of textiles; detergency, bleaching and sizing; physical and colloidal chemistry; chemistry of textile processing and finishing. Mailing Add: 2666 22nd St Wyandotte MI 48192

CRAMER, JOHN WESLEY, b Freeport, Ill, Oct 15, 28; m 67; c 3. BIOCHEMISTRY. Educ: Beloit Col, BS, 50; Univ Wis, MS, 52, PhD(biochem), 55. Prof Exp: Instr oncol, Univ Wis, 59; instr pharmacol, Sch Med, Yale Univ, 59-61, asst prof, 61-67; assoc prof, Sch Med, Univ Fla, 67-73; PROF PHARMACOL, SCH MED SCI, UNIV

NEV, RENO, 74- Concurrent Pos: Fel biochem, Univ Wis, 55-57, fel oncol, 57-59; NIH & Am Cancer Soc grants. Mem: AAAS; Am Soc Pharmacol & Exp Therapeut; Am Asn Cancer Res; Am Soc Microbiol; Tissue Cult Asn. Res: Nutrition; vitamin D; cancer aromatic hydrocarbons; cell and viral nucleic acid synthesis and inhibition; antimetabolites. Mailing Add: Div of Biomed Sci Univ of Nev Sch of Med Sci Reno NV 89507

CRAMER, JOSEPH BENJAMIN, b Rochester, NY, Aug 24, 14; m 46; c 2. PSYCHIATRY, PSYCHOANALYSIS. Educ: Univ Rochester, BA, 36; NY Med Col, MD, 41; Inst Psychoanal, Chicago, cert psychoanal, 55. Prof Exp: Scottish Rite res fel schizophrenia, Sch Med, NY Univ, 47-49; assoc prof child psychiat, Univ Pittsburgh, 51-55; assoc prof & dir child psychiat, 55-68, PROF PSYCHIAT & PEDIAT, ALBERT EINSTEIN COL MED, 68- Concurrent Pos: Dir, Pittsburgh Guid Ctr, 51-53; consult, Scranton Guid Ctr, Pa, 55-57, Jewish Child Care Asn NY, 55-65 & Wiltwyck Sch, 73-74; mem steering comt, Conf Training in Child Psychiat, 63; mem, Conf Psychiat & Med Educ, 67. Mem: Am Acad Child Psychiat; Am Psychoanal Asn; Am Psychiat Asn; Am Orthopsychiat Asn. Res: Childhood schizophrenia; psychotherapy; medical education. Mailing Add: Dept of Psychiat Albert Einstein Col of Med New York NY 10461

CRAMER, MICHAEL BROWN, b Dayton, Ohio, July 4, 38; m 62; c 3. PHARMACOLOGY, TOXICOLOGY. Educ: Purdue Univ, BS, 61, MS, 53, PhD(pharmacol), 65. Prof Exp: Prin investr, Armed Forces Radiobiol Res Inst, Nat Naval Med Ctr, Md, 65-67; res pharmacologist, Miami Valley Labs, Procter & Gamble Co, Ohio, 67-70; asst prof pharmacol, 70-75, ASSOC PROF PHARMACOL, UNIV HOUSTON, 75- Honors & Awards: Cert of Achievement, Defense Atomic Suport Agency, 67. Mem: AAAS; Am Pharmaceut Asn. Res: Effects of heavy metals on drug action and results of clinical laboratory tests; chronic toxicity of heavy metals in mammals. Mailing Add: Col of Pharm Univ of Houston Houston TX 77004

CRAMER, ONEIDA MORNINGSTAR, b Staunton, Va, May 2, 45; m 66. NEUROENDOCRINOLOGY. Educ: Univ Md, BS, 67, PhD(physiol), 71. Prof Exp: Fel physiol, Southwestern Med Sch, Univ Tex, 71-72; res assoc physiol, Sch Med, Univ Md, 72-73, asst prof, 73-76; ASST PROF OBSTET & GYNEC, HEALTH SCI CTR, UNIV TEX, DALLAS, 76- Concurrent Pos: NIH fel, 71-72, res grant, 73-76. Mem: AAAS; NY Acad Sci; Soc Study Reproduction. Res: Hypothalamus to determine the necessary mechanisms for neuroendocrine regulation of the anterior pituitary. Mailing Add: Dept of Obstet & Gynec Health Sci Ctr Univ of Tex Dallas TX 75235

CRAMER, RICHARD (DAVID), b Mifflin, Pa, Aug 12, 13; m 37; c 4. ORGANOMETALLIC CHEMISTRY. Educ: Juniata Col, BS, 35; Harvard Univ, MA, 36, PhD(chem), 40. Prof Exp: Asst prof org chem, Carnegie Inst Technol, 40-41; RES CHEMIST, E I DU PONT DE NEMOURS & CO, INC, 41- Honors & Awards: Del Sect Awards, Am Chem Soc, 65, 67. Mem: Am Chem Soc. Res: Synthesis of organics from radioactive carbon; chemistry of organic fluorine compounds; polymer synthesis; modified polyethylenes and synthesis of fluorine containing monomers; radioactive lactic acid in biological studies; chemistry of triphenylfuryl ketones; organo-transition metal chemistry. Mailing Add: Cent Res Dept E I du Pont de Nemours & Co, Inc Wilmington DE 19898

CRAMER, RICHARD DAVID, III, b Wilmington, Del, Feb 26, 42; m 69; c 1. MEDICINAL CHEMISTRY, PHYSICAL ORGANIC CHEMISTRY. Educ: Harvard Univ, AB, 63; Mass Inst Technol, PhD(org chem), 67. Prof Exp: Scientist org chem, Polaroid Corp, 67-69; res fel, Harvard Univ, 69-71; sr scientist technol assessment, 71-74, SR INVESTR RES CHEM, SMITH KLINE & FRENCH LABS, 74- Mem: AAAS; Am Chem Soc; Chem Inst Can. Res: Quantitative structure/activity relationships and other applications of computer science to organic chemistry. Mailing Add: Smith Kline & French Labs 1500 Spring Garden St Philadelphia PA 19101

CRAMER, ROBERT ELI, b Washington, DC, June 19, 19; m 41; c 3. CARTOGRAPHY, GEOGRAPHY OF NORTH AMERICA. Educ: Ohio Univ, AB, 42; Univ Chicago, SM, 47, PhD(geog), 52. Prof Exp: Cartogr, Aeronaut Chart & Info Ctr, US Air Force, 42 & 45-46; asst prof geog, Memphis State Univ, 50-51; sr res analyst, Air Force Intel, 51-54; PROF GEOG, E CAROLINA UNIV, 54-, CHMN DEPT, 62- Concurrent Pos: Consult, Aeronaut Chart & Info Ctr, US Air Force, 56; mem, NC Tech Adv Bd, 60-64. Mem: Asn Am Geog; Nat Coun Geog Educ. Res: Geographical research on North Carolina; application of Kodachrome slides to the study of geography, especially North Carolina, earth science, Mexico and Europe. Mailing Add: Dept of Geog East Carolina Univ Box 2723 Greenville NC 27834

CRAMER, ROGER EARL, b Findlay, Ohio, Sept 14, 43; m 67; c 1. INORGANIC CHEMISTRY. Educ: Bowling Green State Univ, BS, 65; Univ Ill, MS, 67, PhD(inorg chem), 69. Prof Exp: Asst prof, 69-73, ASSOC PROF CHEM, UNIV HAWAII, 73- Mem: Am Chem Soc. Res: Nuclear magnetic resonance of paramagnetic molecules; coordination chemistry; semi-empirical molecular orbital calculations; hydrogen bonding; free radicals; interaction of metal ions with nucleotides. Mailing Add: Dept of Chem 2524 The Mall Univ of Hawaii Honolulu HI 96822

CRAMER, WILLIAM ANTHONY, b New York, NY, June 11, 38; m 64; c 4. BIOPHYSICS. Educ: Mass Inst Technol, BS, 59; Univ Chicago, MS, 60, PhD(biophys), 65. Prof Exp: NSF fel photosynthesis, Univ Calif, San Diego, 65-67, res assoc, 67-68; asst prof, 68-73, ASSOC PROF BIOL SCI, PURDUE UNIV, 73- Concurrent Pos: Nat Inst Gen Med Sci res career develop award, 70- Mem: Am Soc Biol Chem; Biophys Soc. Res: Photosynthetic electron transport and energy coupling; electron transport and phosphorylation in bacteria; mechanism of action of colicin. Mailing Add: Dept of Biol Sci Purdue Univ West Lafayette IN 47907

CRAMER, WILLIAM HERBERT, b Chambersburg, Pa, Dec 4, 21. PHYSICAL CHEMISTRY. Educ: Pa State Univ, BS, 42, PhD(phys chem), 50. Prof Exp: Supvr, Control Lab, Photo Prod Dept, E I du Pont de Nemours & Co, 43-44; asst phys chem, Pa State Univ, 46-49; fel, Mellon Inst, 50-51; res assoc, Forrestal Res Ctr, 51-53; asst prof, Univ Fla, 53-62; asst prog dir chem, 62-63, assoc prog dir phys chem, 64-68, PROG DIR QUANTUM CHEM, NSF, 68- Mem: AAAS; Am Chem Soc; Am Phys Soc. Res: Atomic and molecular structure and properties; positive ion scattering; gaseous electrical discharges. Mailing Add: Chem Sect NSF Washington DC 20550

CRAMER, WILLIAM SMITH, b Frederick, Md, Aug 25, 14; m 47; c 2. PHYSICS. Educ: Ursinus Col, BS, 37; Brown Univ, MS, 38, PhD(physics), 48. Prof Exp: Asst math, Univ Md, 38-39; teacher math & sci, Pikeville Col, 39-40; physicist, US Naval Ord Lab, 42-56 & Off Naval Res, 56-66, PHYSICIST, NAVAL SHIP RES & DEVELOP CTR, WASHINGTON, DC, 66- Mem: Am Phys Soc; Acoust Soc Am (vpres, 75-76). Res: Acoustic properties of plastics; underwater and physical acoustics. Mailing Add: 11512 Colt Terr Silver Spring Md 20902

CRAMPTON, DAVID, b Eng, Jan 27, 42; Can citizen; m 64; c 2. ASTRONOMY. Educ: Univ Toronto, BSc, 63, MA, 64, PhD(astron), 67. Prof Exp: RES SCIENTIST ASTRON, DOMINION ASTROPHYS OBSERV, 67- Res: Galactic structure, spectroscopic observations of x-ray binaries. Mailing Add: Dominion Astrophys Observ 5071 W Saanich Rd Victoria BC Can

CRAMPTON, EARLE WILCOX, b Middletown, Conn, Aug 15, 95; m 20. NUTRITION. Educ: Conn Agr Col, BS, 19; Iowa State Col, MS, 22; Cornell Univ, PhD(nutrit), 37. Hon Degrees: DSc, Univ Reading, 60. Prof Exp: Lectr animal husb, Macdonald Col, McGill Univ, 22-26, from asst prof to prof, 26-65, EMER PROF NUTRIT, MCGILL UNIV, 65- Concurrent Pos: Mem, comt animal nutrit, Nat Acad Sci-Nat Res Coun, 55-70. Honors & Awards: Earle Willard McHenry Award, Nutrit Soc Can, 74. Mem: Soc Animal Sci; Agr Inst Can; Chem Inst Can; fel Royal Soc Can. Res: Requirements of weaning pigs; effect of food on bacon quality; coorelation of bacon carcass measurements and feed requirements; nutritional value of pasture herbage; feed nomenclature; nutrient composition. Mailing Add: 4 Governors Grove Wesleyan Hills Middletown CT 06457

CRAMPTON, JAMES MYLAN, b Mitchell, SDak, Nov 26, 23; m 50; c 6. PHARMACOLOGY. Educ: Creighton Univ, BS, 50; Univ Fla, MS, 51, PhD(pharmacol), 53. Prof Exp: Asst prof pharmacol, Xavier Univ, 55-58; asst prof biol sci, Sch Pharm, 58 & 67, prof, 67-70, dir dept biol sci, 58-71, dir clin studies, 71-74, PROF BIOPHARM, SCH PHARM & PROF PHYSIOL & PHARMACOL, SCH MED, CREIGHTON UNIV, 70-, ASSOC DEAN SCH PHARM, 75- Mem: Am Pharmaceut Asn. Res: Castrix toxicity; salicyclamide. Mailing Add: Sch of Pharm Creighton Univ Omaha NE 68102

CRAMPTON, STUART J B, b New York, NY, Nov 3, 36; m 61; c 3. ATOMIC PHYSICS. Educ: Williams Col, BA, 58; Oxford Univ, BA, 60; Harvard Univ, PhD(physics), 64. Hon Degrees: MA, Oxford Univ, 65. Prof Exp: From asst prof to assoc prof, 65-75, PROF PHYSICS, WILLIAMS COL, 75- Concurrent Pos: NSF fel, Harvard Univ, 64-65; Alfred P Sloan Found res fel, 67-69; NATO sr fel, 74. Mem: Fel Am Phys Soc; Am Asn Physics Teachers. Res: Atom-atom interactions, particularly electron spin exchange collisions and atom-surface interactions. Mailing Add: Dept of Physics Williams Col Williamstown MA 01267

CRAMPTON, THEODORE HENRY MILLER, b Patchogue, NY, Apr 4, 26; m 55. MATHEMATICS, RADIOBIOLOGY. Educ: Hamilton Col, AB, 49; Ind Univ, MA, 54, PhD(math), 55. Prof Exp: Instr math, Mt Holyoke Col, 55-57, asst prof, 57-58; instr nuclear weapons, US Army Engr Sch, Va, 58-61; mathematician, Armed Forces Radiobiol Res Inst, Nat Naval Med Ctr, Md, 61-64; instr math, US Mil Acad, 64-66, asst prof, 66-69; chief weapons sect, Instnl Div, Joint Strategic Target Planning Staff, US Army, 69-72; SYSTS ANALYST, INT SECURITY AFFAIRS, ENERGY RES & DEVELOP ADMIN, 72- Mem: AAAS; Am Math Soc; Math Asn Am; Am Nuclear Soc. Res: Class field theory; radiation transport theory; radiation dosimetry; radiation shielding. Mailing Add: Int Security Affairs ERDA Washington DC 20545

CRAMTON, THOMAS JAMES, b Chadron, Nebr, Oct 23, 38; m 60; c 2. MATHEMATICS. Educ: Harvard Univ, AB, 60; Clark Univ, AM, 62; Dartmouth Col, PhD(math), 70. Prof Exp: Instr math, US Naval Nuclear Power Sch, Md, 62-66, dir off dept, 63-66; ASST PROF MATH, PURDUE UNIV, FT WAYNE, 69- Mem: Am Math Soc; Math Asn Am. Res: Inverse eigenvalue problems in dimension 2 or higher. Mailing Add: Dept of Math Purdue Univ Ft Wayne IN 46805

CRANBERG, LAWRENCE, physics, see 12th edition

CRANDALL, ARTHUR JARED, b Syracuse, NY, Mar 14, 39; m 64; c 3. PHYSICS. Educ: St Lawrence Univ, BS, 61; Mich State Univ, MS, 64, PhD(physics), 67. Prof Exp: Asst prof, 67-74, ASSOC PROF PHYSICS, BOWLING GREEN STATE UNIV, 74- Mem: Acoust Soc Am. Res: Physics teaching; ultrasonics; diffraction effects and interaction of light with sound. Mailing Add: Dept of Physics Bowling Green State Univ Bowling Green OH 43402

CRANDALL, DANA IRVING, b New York, NY, Sept 4, 15; m 42; c 3. BIOCHEMISTRY. Educ: Columbia Univ, AB, 36; Univ Pa, PhD(biochem), 45. Prof Exp: From instr to asst prof physiol chem, Univ Pa, 46-50; assoc prof, 50-62, PROF BIOL CHEM, UNIV CINCINNATI, 62- Mem: Am Soc Biol Chemists. Res: Metabolism of tyrosine in mammalian tissues; properties and mechanism of action of homogentisate dioxygenase; role of oxygenases in mammalian metabolism. Mailing Add: Dept of Biol Chem Univ of Cincinnati Col of Med Cincinnati OH 45267

CRANDALL, DAVID HUGH, b Chicago, Ill, July 9, 42; c 2. ATOMIC PHYSICS, ELECTRON PHYSICS. Educ: Sioux Falls Col, BS, 64; Univ Nebr-Lincoln, MS, 67, PhD(physics), 70. Prof Exp: Vis asst prof physics, Univ Mo-Rolla, 70-71; res assoc, Joint Inst Lab Astrophys, Nat Bur Standards & Univ Colo, Boulder, 71-74; RES PHYSICIST, OAK RIDGE NAT LAB, 74- Mem: Am Phys Soc; Sigma Xi. Res: Atomic collisions between ions and atoms or electrons using ion beams of a few keV energy, important processes of charge exchange, impact excitation and ionization. Mailing Add: Physics Div Bldg 6000 Box X Oak Ridge Nat Lab Oak Ridge TN 37830

CRANDALL, EDWARD D, b Brooklyn, NY, Aug 10, 38; m 63; c 2. PULMONARY PHYSIOLOGY. Educ: Cooper Union Univ, BChE, 60; Northwestern Univ, MS, 62, PhD(chem eng), 64; Univ Pa, MD, 72. Prof Exp: From asst prof to assoc prof chem eng, Univ Notre Dame, 64-68; lectr, Dept Physiol, 68-74, ASST PROF PHYSIOL, SCH MED, UNIV PA, 74- Concurrent Pos: USPHS spec fel, 68-70; med internship & residency, Hosp, Univ Pa, 72-74, pulmonary fel, Hosp, 74-76. Honors & Awards: Res Career Develop Award, NIH, 75. Mem: Am Col Physicians; Am Thoracic Soc; Am Col Chest Physicians; Am Chem Soc; Am Inst Chem Eng. Res: Biomedical engineering; pulmonary physiology and disease; pulmonary gas exchange; red cell membrane structure and function. Mailing Add: Dept of Physiol Univ of Pa Sch of Med Philadelphia PA 19174

CRANDALL, ELBERT WILLIAMS, b Normal, Ill, Nov 4, 20; m 51; c 4. ORGANIC CHEMISTRY. Educ: Ill State Norm Univ, BEd, 42; Univ Mo, MA, 48, PhD(org chem), 50. Prof Exp: Prof chem, Ky Wesleyan Col, 50-51; res & develop, US Rubber Co, 51-52; PROF CHEM, KANS STATE COL PITTSBURG, 52- Mem: Am Chem Soc. Res: Nitric acid oxidations; near infrared spectra of polymers; organic compounds in natural waters. Mailing Add: Dept of Chem Kans State Col of Pittsburg Pittsburg KS 66762

CRANDALL, HAROLD FRANCIS, physical chemistry, see 12th edition

CRANDALL, JACK KENNETH, b Fillmore, Calif, June 8, 37; m 60; c 1. ORGANIC CHEMISTRY. Educ: Univ Calif, Berkeley, BS, 60; Cornell Univ, PhD(org chem), 63. Prof Exp: NIH fel, 64; from instr to assoc prof, 64-72, PROF CHEM, IND UNIV, BLOOMINGTON, 72- Concurrent Pos: Alfred P Sloan res fel, 68-70; John Simon Guggenheim fel, 70-71. Mem: Am Chem Soc; The Chem Soc. Res: Synthetic organic chemistry; small ring compounds; epoxides; allenes; organometallic chemistry; C-13 nuclear magnetic resonance. Mailing Add: Dept of Chem Ind Univ Bloomington IN 47401

CRANDALL, JOHN LOU, b Hart, Mich, Sept 18, 20; m 43; c 1. CHEMISTRY. Educ: Mass Inst Technol, BS, 42, PhD(phys chem), 48. Prof Exp: Res chemist, Chem Dept, 48-52, res mgr exp phys, Savannah River Lab, 53-67, from asst dir advan oper planning to dir advan oper planning, Atomic Energy Div, 67-75, DIR ENVIRON SCI SECT, SAVANNAH RIVER LAB, E I DU PONT DE NEMOURS & CO, INC, 75- Mem: Am Chem Soc; fel Am Nuclear Soc. Res: High polymers; photochemistry; reactor physics; operations analysis; environmental management. Mailing Add: E I du Pont de Nemours & Co Inc Aiken SC 29801

CRANDALL, PAUL HERBERT, b Essex, Vt, Feb 15, 23; m 51; c 4. MEDICINE. Educ: Univ Vt, BS, 43, MD, 47; Am Bd Neurol Surg, dipl, 56. Prof Exp: Asst chief neurosurg serv, Wadsworth Vet Admin Hosp, Calif, 54-56; from instr to assoc prof surg & neurosurg, 54-72, PROF NEUROSURG & NEUROL, SCH MED, UNIV CALIF, LOS ANGELES, 72- Concurrent Pos: Consult, Vet Admin Hosp & Wadsworth Gen Hosp, Calif. Mem: AMA; Am Asn Neurol Surg; NY Acad Sci; fel Am Col Surg. Res: Stereotaxic surgery; depth electrode studies in epilepsy; radioisotopic brain scanning. Mailing Add: Dept of Surg Univ of Calif Med Sch Los Angeles CA 90024

CRANDALL, PERRY CLARENCE, b Chillicothe, Mo, May 9, 15; m 41; c 4. HORTICULTURE. Educ: Iowa State Univ, BS, 40, PhD(hort, plant physiol), 51; Ohio State Univ, MS, 41. Prof Exp: Supt, Bluffs Exp Fruit Farm, Iowa State Univ, 46-51; asst prof hort, 51-58, SUPT & HORTICULTURIST, SOUTHWESTERN WASH RES UNIT, WASH STATE UNIV, 58- Mem: Am Soc Hort Sci; Am Pomol Soc; Int Soc Hort Sci. Res: Small fruits; tree fruits, and vegetables physiology and culture. Mailing Add: Southwestern Wash Res Unit 1918 NE 78th St Vancouver WA 98665

CRANDALL, RICHARD B, b Greencastle, Ind, Sept 8, 28; m 58. PARASITOLOGY, IMMUNOLOGY. Educ: DePauw Univ, AB, 49; Univ Mass, MA, 53, Purdue Univ, PhD(zool), 59. Prof Exp: NIH trainee, Sch Pub Health, Univ NC, 59-60; instr med microbiol, 60-64, from asst prof to assoc prof, 65-73, PROF PARASITOL, COL MED, UNIV FLA, 74- Mem: Am Soc Parasitol; Am Soc Trop Med & Hyg; Am Asn Immunol. Res: Immunology of parasitic infections; medical parasitology and microbiology. Mailing Add: Dept of Immunol & Med Microbiol Univ of Fla Col of Med Gainesville FL 32601

CRANDALL, TERRY GENE, organic chemistry, see 12th edition

CRANDALL, WALTER ELLIS, b Norwich, Conn, Dec 18, 16; m 44; c 2. NUCLEAR PHYSICS. Educ: Worcester Polytech Inst, BS, 40; Univ Calif, PhD(physics), 52. Prof Exp: Physicist, US Naval Ord Lab, 41-42; physicist, Radiation Lab, Univ Calif, 48-52 & 54-62; physicist, Calif Res & Develop Co, 53; mgr sci & technol dept, Northrop Corp, 62-70, VPRES & MGR, NORTHROP CORP LABS, HAWTHORNE, 70- Concurrent Pos: Consult, Boeing Airplane Co. Mem: Am Phys Soc. Res: High energy nuclear physics; neutral meson; nuclear radii; nuclear weapons; magnetics; electrooptics; biophysics. Mailing Add: 21930 Carbon Mesa Rd Malibu CA 90265

CRANDELL, CLIFTON E, US citizen. DENTISTRY. Educ: ECarolina Col, BS, 49; Med Col Va, DDS, 53; Univ Pa, MS, 61; Duke Univ, MEd, 61. Prof Exp: Pvt pract, Southport, NC, 53-55; PROF ORAL DIAG & TREATMENT, SCH DENT, UNIV NC, CHAPEL HILL, 55- Concurrent Pos: Consult, Womack Army Hosp, Ft Bragg, NC, 59- Mem: Am Dent Asn; Int Asn Dent Res; Am Acad Dent Radiol (pres, 69-70). Res: Dental radiology; oral diagnosis; half-value layer relation to film quality; radiation protection; computers in dental education; linear programming. Mailing Add: Univ of NC Sch of Dent Chapel Hill NC 27514

CRANDELL, DWIGHT RAYMOND, b Galesburg, Ill, Jan 25, 23; m 43; c 3. GEOLOGY. Educ: Knox Col, BA, 46; Yale Univ, MS, 48, PhD(geol), 51. Prof Exp: Field asst, 47, GEOLOGIST, US GEOL SURV, 48- Concurrent Pos: Asst, Yale Univ, 48-51. Mem: Fel Geol Soc Am. Res: Stratigraphy of unconsolidated deposits; glacial geology; volcanic mudflows; volcanic hazards in Western United States. Mailing Add: Eng Geol Br US Geol Surv Fed Ctr Denver CO 80225

CRANDELL, GEORGE FRANK, b Astoria, Ore, Dec 5, 32; m 56; c 5. BIOLOGICAL OCEANOGRAPHY. Educ: Ore State Univ, BS, 60, MS, 63, PhD(oceanog), 66. Prof Exp: From asst prof to assoc prof, 66-75, PROF OCEANOG, HUMBOLDT STATE UNIV, 75- Mem: Am Soc Limnol & Oceanog. Res: Zooplankton and benthic ecology. Mailing Add: Dept of Oceanog Humboldt State Univ Arcata CA 95521

CRANDELL, MERRELL EDWARD, b Clearfield, Pa, Mar 19, 38; m 60; c 2. PHYSICS. Educ: Hobart Col, BS, 59; Syracuse Univ, MS, 61, PhD(physics), 67. Prof Exp: Asst physics, Syracuse Univ, 59-63; instr physics, Le Moyne Col, NY, 63-67; asst prof, 67-70, ASSOC PROF PHYSICS, MUSKINGUM COL, 70- Concurrent Pos: Vis res fel, Kammerlingh Onnes Lab, State Univ Leiden, 74-75. Mem: Am Phys Soc; Am Asn Physics Teachers. Res: Optical and transport properties of metals and alloys. Mailing Add: Dept of Physics Muskingum Col New Concord OH 43762

CRANDELL, ROBERT ALLEN, b Three Rivers, Mich, July 30, 24; m 50; c 4. VETERINARY VIROLOGY. Educ: Mich State Col, BS, 47, DVM, 49; Univ Calif, MPH, 55; Am Bd Vet Pub Health, dipl; Am Col Vet Microbiologists, dipl. Prof Exp: Practitioner vet med, 49-51; base vet, Vet Corps, US Air Force, Selfridge AFB, Mich, 51-52, asst chief animal farm, Ft Detrick, Md, 52-54, lab officer, Naval Biol Lab, 54-56, asst chief virol br, Armed Forces Inst Path, 56-60, chief virol br, US Air Force Epidemiol Lab, Lackland AFB, Tex, 60-67, res epidemiologist, Pan Am Foot-and-Mouth Dis Ctr, Brazil, 67-69, chief biosci div, US Air Force Sch Aerospace Med, 69-70, chief epidemiol div, 70-71; SR MICROBIOLOGIST, COL VET MED, VET DIAG MED, UNIV ILL, URBANA, 71-, PROF, DEPT PATH & HYG, 75- Concurrent Pos: Mem, Western Hemisphere Animal Virus Characterization Comt; mem, Animal Resources Adv Comt, NIH, 75-79. Mem: Am Vet Med Asn; Am Soc Microbiol; US Animal Health Asn; Conf Pub Health Vets; Am Asn Vet Lab Daiagnosticians. Res: Isolation and charcaterization of new respiratory viruses of the domestic feline and bovine species; virologic studies and control of zoonoses and diseases of laboratory animals; basic studies with rabies virus; diagnostic virology; research administration. Mailing Add: Univ of Ill Col of Vet Med Vet Diag Med Urbana IL 61801

CRANDELL, WALTER BAIN, b New York, NY, July 26, 11; m 35; c 4. SURGERY, METABOLISM. Educ: Dartmouth Col, AB, 34; NY Univ, MD, 37. Prof Exp: Intern, Mary Hitchcock Mem Hosp, 37-38; clin asst chest surg, Hitchcock Clin, Hanover, NH, 39; asst resident surg, Mass Gen Hosp, Boston, 39-41; from asst resident to resident chest surg, Bellevue Hosp, New York, 41-42, asst vis consult surg & chest surg, 46-47; attend thoracic surg, Vet Admin Hosp, Bronx, NY, 47; CHIEF SURG SERV, VET ADMIN HOSP, WHITE RIVER JUNCTION, 47-; PROF SURG, DARTMOUTH MED SCH, 75- Concurrent Pos: Instr anat, Dartmouth Med Sch, 39; from asst clin prof to assoc clin prof surg, Dartmouth Med Sch, 47-68, clin prof surg, 68-75; adj surgeon, Lenox Hill Hosp, New York, 47; instr anat & oper surg, Postgrad Sch, NY Univ, 46, instr anat, Col Med, 46-47; consult surg, Mary Hitchcock Hosp, Hanover, NH, 70- Mem: Am Col Surg; Am Asn Thoracic Surg. Res: Metabolic disorders in surgical patients, especially disorders of acid-base and osmolality; pulmonary and renal insufficiency. Mailing Add: Vet Admin Ctr Surg Serv N Hartland Rd White River Junction VT 05001

CRANDLEMERE, ROBERT WAYNE, b South Weymouth, Mass, Mar 5, 47; m 66; c 2. APPLIED CHEMISTRY, FORENSIC SCIENCE. Educ: Suffolk Univ, BS, 70, MS, 75; Am Inst Chemists, cert, 76. Prof Exp: Assoc res scientist, Factory Mutual Res Corp, 67-70; chemist, NE Indust Chem Corp, 72-73; CHIEF CHEMIST & VPRES, BRIGGS ENG & TESTING CO, INC, 73- Concurrent Pos: Mem, Chem Week Adv Bd, 76. Mem: Am Chem Soc; Am Inst Chemists; AAAS; Nat Fire Protection Asn. Res: Investigation of weathering and other effects on bituminous materials used in roofing and development of test procedures in analysis of cement concrete. Mailing Add: Briggs Eng & Testing Co Inc 164 Washington St Norwell MA 02061

CRANE, ANATOLE, b New Brunswick, NJ, Feb 23, 33; m 56; c 2. MICROBIOLOGY. Educ: Univ Ill, BS, 54, MS, 56; Univ Ind, PhD(bact), 60. Prof Exp: Proj leader microbiol, 59-68, MGR MICROBIOL, QUAKER OATS CO, 68- Mem: Am Soc Microbiol; Am Asn Cereal Chemists; Inst Food Technologists. Res: Food microbiology and preservation. Mailing Add: Quaker Oats Co 617 W Main St Barrington IL 60010

CRANE, AUGUST REYNOLDS, b Brooklyn, NY, Dec 16, 08; m 35. MEDICINE, PATHOLOGY. Educ: Hamilton Col, AB, 29; Cornell Univ, MD, 33. Prof Exp: Asst prof path, Jefferson Med Col, 47-57; prof, 57-75, EMER PROF PATH, SCH MED, UNIV PA, 74- 57-74, USPHS, 44-46 & US Army Med Corps, 47-73; dir, Ayer Clin Lab, Pa Hosp, 46-74, consult, 74-; trustee, Hamilton Col. Mem: Fel Am Soc Clin Path; Am Asn Path & Bact; AMA; fel Col Am Path; fel Am Col Physicians. Res: Cancer research; teaching. Mailing Add: Pa Hosp Philadelphia PA 19107

CRANE, CHARLES RUSSELL, b Mangum, Okla, Jan 19, 28; m 54; c 4. BIOCHEMISTRY, TOXICOLOGY. Educ: Univ Okla, BS, 51, MS, 52; Fla State Univ, PhD(biochem), 56. Prof Exp: Asst prof biochem, Okla State Univ, 56-61; chief biochem labs, 61-68, CHIEF BIOCHEM RES, CIVIL AEROMED INST, FED AVIATION ADMIN, DEPT TRANSP, 68- Mem: AAAS; Am Chem Soc. Res: Aviation toxicology; drug metabolism; pesticides; enzymology; composition and inhalation toxicology of combustion/pyrolysis products. Mailing Add: Civil Aeromed Inst Fed Aviation Admin AAC-114 PO Box 25082 Oklahoma City OK 73125

CRANE, EDWARD MASTIN, b Brattleboro, Vt, Mar 23, 20; m 44; c 3. PHOTOGRAPHIC SCIENCE. Educ: Dartmouth Col, AB, 42; Harvard Univ, AM, 43, PhD(chem), 49. Prof Exp: Res chemist, 49-56, RES ASSOC, EASTMAN KODAK CO, 57- Res: Communication theory; photographic sharpness, graininess, detail rendition, etc; computer programming. Mailing Add: Eastman Kodak Res Labs 343 State St Rochester NY 14650

CRANE, FRANK A, pharmacognosy, plant physiology, see 12th edition

CRANE, FREDERICK LORING, b Mass, Dec 3, 25; m 50; c 4. BIOCHEMISTRY. Educ: Univ Mich, BS, 50, MS, 51, PhD(bot), 53. Prof Exp: Trainee, Inst Enzyme Res, Univ Wis, 53-57, asst prof, 57-59; asst prof chem, Univ Tex, 59-60; assoc prof, 60-62, PROF BIOL, PURDUE UNIV, WEST LAFAYETTE, 62- Concurrent Pos: Newcombe fel, Univ Mich, 52-53; NSF sr fel, Univ Stockholm, 63-64; NIH career investr, 64; Fulbright vis fel, Australian Nat Univ, 71-72. Honors & Awards: Eli Lilly Award, Am Chem Soc, 61. Mem: Am Soc Plant Physiol; Am Soc Biol Chem; Am Chem Soc; Am Soc Cell Biol; Am Inst Biol Sci. Res: Vitamin biosynthesis in plants; fatty acid metabolism; biological oxidations and energy coupling; coenzyme Q and plastoquinones; ultrastructure of subcellular particles. Mailing Add: Dept of Biol Sci Purdue Univ West Lafayette IN 47907

CRANE, GEORGE THOMAS, b Nephi, Utah, Nov 6, 28; m 54; c 4. MICROBIOLOGY, VIROLOGY. Educ: Utah State Univ, BS, 58; Colo State Univ, 62-63; Brigham Young Univ, MSc, 69. Prof Exp: Microbiologist, Dis Ecol Sect, USPHS, Colo, 58-66; MICROBIOLOGIST, ENVIRON & ECOL BR, DUGWAY, 66- Mem: Am Soc Trop Med & Hyg; Am Mosquito Control Asn. Res: Arbovirus and rickettsial surveillance of the fauna of North America. Mailing Add: 698 N Nelson Tooele UT 84074

CRANE, GRANT, b Columbus, Ohio, Sept 1, 15; m 47; c 3. ORGANIC CHEMISTRY. Educ: Dartmouth Col, AB, 37; Ohio State Univ, PhD(chem), 40. Prof Exp: Res chemist, Pittsburgh Plate Glass Co, 40-41; RES CHEMIST, FIRESTONE TIRE & RUBBER CO, 46- Mem: Am Chem Soc. Res: Preparation of pure hydrocarbons and of monomers for plastics; drying oils; synthetic rubbers; petrochemicals. Mailing Add: Firestone Tire & Rubber Co 1200 Firestone Pkwy Akron OH 44317

CRANE, HORACE RICHARD, b Turlock, Calif, Nov 4, 07; m 34; c 2. PHYSICS. Educ: Calif Inst Technol, BS, 30, PhD(physics), 34. Prof Exp: Res fel physics, Calif Inst Technol, 34-35; instr & res physicist, 35-38, from asst prof to assoc prof, 38-46, chmn dept, 65-72, PROF PHYSICS, UNIV MICH, ANN ARBOR, 46- Concurrent Pos: Res assoc, Mass Inst Technol, 40 & dept terrestrial magnetism, Carnegie Inst Technol, 41; dir proximity fuze proj, Univ Mich, 42-45; pres, Midwestern Univs Res Asn, 57-60; mem policy adv bd, Argonne Nat Lab, 57-67; mem, Comn Col Physics, 62-70, vpres, 68-70; mem standing comt controlled thermonuclear res, Atomic Energy Comn, 69-72; chmn bd gov, Am Inst Physics, 71-75. Honors & Awards: Davisson-Germer Prize, Am Phys Soc, 68. Mem: Nat Acad Sci; fel AAAS; fel Am Phys Soc; Am Asn Physics Teachers (pres, 65); fel Am Acad Arts & Sci. Res: Nuclear physics; high energy accelerators; g-factor of electron; physics teaching methods; geomagnetism. Mailing Add: Dept of Physics Univ of Mich Ann Arbor MI 48104

CRANE, JOSEPH LELAND, b Wilmot, NH, Jan 9, 35. MYCOLOGY. Educ: Univ Maine, BS, 61; Univ Del, MS, 64; Univ Md, PhD(mycol), 67. Prof Exp: Asst mycologist, 67-72, ASSOC MYCOLOGIST, ILL NATURAL HIST SURV, 72- Mem: Bot Soc Am; Mycol Soc Am; Am Inst Biol Sci; Brit Mycol Soc. Res: Aquatic hyphomycetes; marine fungi; taxonomy of fungi imperfecti. Mailing Add: 218 Natural Resources Annex Ill Natural Hist Surv Urbana IL 61801

CRANE, JULES M, JR, b New York, NY, Sept 5, 28; m 55; c 4. ICHTHYOLOGY, MARINE BIOLOGY. Educ: NY Univ, AB, 54; Calif State Col Los Angeles, MA, 63. Prof Exp: Chmn dept biol, 64-70, PROF BIOL, CERRITOS COL, 62-, CHMN DEPT, 74- Concurrent Pos: Lectr, Calif State Col Long Beach, 65- Mem: AAAS; Am Soc Ichthyologists & Herpetologists; Marine Biol Soc UK. Res: Behavioral and biochemical aspects of bioluminescence in fishes; evolution of deep sea fishes; marine ecology. Mailing Add: Dept of Biol Cerritos Col 11110 E Alondra Blvd Norwalk CA 90650

CRANE, JULIA GORHAM, b Mt Kisco, NY, Nov 8, 25. CULTURAL ANTHROPOLOGY. Educ: Columbia Univ, BS, 59, PhD(anthrop), 66. Prof Exp: Res asst anthrop with Dr Margaret Mead, Am Mus Natural Hist, 56-59; asst, Columbia

Univ, 59-61; asst prof, 66-72, ASSOC PROF ANTHROP, UNIV NC, CHAPEL HILL, 72- Concurrent Pos: Inst Res Soc Sci res grant, Saba, Netherlands Antilles, 70; Prince Bernhard Fund Award, 70-71. Mem: Am Anthrop Asn; Am Ethnol Soc. Res: Social anthropology; Caribbean studies. Mailing Add: Dept of Anthrop Univ of NC Chapel Hill NC 27514

CRANE, JULIAN COBURN, b Morgantown, WVa, Mar 7, 18; m 42. POMOLOGY. Educ: Univ Md, BS, 39, PhD(hort), 42. Prof Exp: Asst, Univ Md, 39-42, asst horticulturist, 42; assoc agronomist, Off for Agr Rels, USDA, 43-45, horticulturist, 46; from asst prof to assoc prof, 46-58, PROF POMOL, COL AGR, UNIV CALIF, DAVIS, 58- Concurrent Pos: NSF fel, 57. Mem: AAAS; Am Soc Plant Physiol; fel Am Soc Hort Sci; Am Inst Biol Sci. Res: Plant physiology; hormones in fruit set, growth and maturation; fig and pistachio culture. Mailing Add: Dept of Pomol Univ of Calif Davis CA 95616

CRANE, LANGDON TEACHOUT, JR, b Detroit, Mich, Feb 23, 30; div; c 2. SOLID STATE PHYSICS, RESEARCH ADMINISTRATION. Educ: Amherst Col, BA, 52; Univ Md, PhD(physics), 60. Prof Exp: Res physicist, Sci Lab, Ford Motor Co, 59-63; from asst prog dir to assoc prog dir, Physics Sect, NSF, 63-68, prog dir atomic & molecular physics, 68-69; res prof & dir inst fluid dynamics & appl math, 69-74, PROF INST FLUID DYNAMICS & APPL MATH, UNIV MD, 74-; SPECIALIST SCI & TECHNOL & HEAD MGT & POLICY SCI SECT, SCI POLICY RES DIV, CONG RES SERV, LIBR CONG, WASHINGTON, DC, 74- Concurrent Pos: Mem comt atomic & nuclear physics, Nat Acad Sci-Nat Res Coun, 71-73. Mem: AAAS; fel Am Phys Soc. Res: Low temperature, atomic, molecular and plasma physics; areas of superconductivity; low temperature specific heats and magnetic properties of metals and alloys; nuclear magnetic resonance; analysis of national policies and legislation affecting science and effectiveness of federally sponsored research programs; technology transfer. Mailing Add: 7103 Oakridge Ave Chevy Chase MD 20015

CRANE, LAURA JANE, b Middletown, Ohio, Nov 2, 41; m 72. CLINICAL BIOCHEMISTRY. Educ: Carnegie-Mellon Univ, BS, 63; Harvard Univ, MA, 64; Rutgers Univ, PhD(biochem), 73. Prof Exp: Asst scientist clin biochem, Warner-Lambert Pharmaceut Co, 67-68; assoc scientist polymer chem, W R Grace, Inc, 68-69; from fel to res assoc biochem, Roche Inst Molecular Biol, Hoffmann-La Roche Inc, 72-75; SCIENTIST CLIN BIOCHEM, WARNER-LAMBERT CO, 75- Mem: Am Chem Soc. Res: Development of immunological assays for clinically significant blood constituents and constituents of other body fluids. Mailing Add: 70 W Valley Brook Rd Long Valley NJ 07853

CRANE, PAUL LEVI, b Clayton, NMex, Oct 17, 25; m 51; c 3. PLANT BREEDING, AGRONOMY. Educ: NMex State Col, BS, 50; Iowa State Col, MS, 51; Purdue Univ, PhD, 56. Prof Exp: Asst agronomist, 51-60, assoc prof, 60-71, PROF AGRON, PURDUE UNIV, 71- Res: Genetics and breeding of maize. Mailing Add: Dept of Agron Purdue Univ West Lafayette IN 47907

CRANE, ROBERT ANTHONY, b Toronto, Ont, July 30, 34; m 60; c 4. SPECTROSCOPY, LASER PHYSICS. Educ: Univ Toronto, BASc, 57, MA, 61, PhD(molecular physics), 65. Prof Exp: Lectr math, Univ Waterloo, 58-60; mem sci staff, 65-74, SR MEM SCI STAFF, RCA, LTD, UNIV MONTREAL, 74- Mem: Can Asn Physicists; Inst Elec & Electronics Eng. Res: Induced infrared absorption in molecular solids; laser beam interactions in plasmas; infrared laser studies; data processing. Mailing Add: RCA Ltd Ste Anne de Bellevue PQ Can

CRANE, ROBERT KELLOGG, b Palmyra, NJ, Dec 20, 19; m 41; c 2. BIOLOGICAL CHEMISTRY, PHYSIOLOGY. Educ: Wash Col, BS, 42; Harvard Univ, PhD(biol chem), 50. Prof Exp: Res assoc, Reynolds Exp Lab, Atlas Powder Co, 42-43; instr chem, Northeast Mo State Teachers Col, 43-44; asst biochemist, Mass Gen Hosp, 49-50; from instr to assoc prof biol chem, Sch Med, Washington Univ, 50-61; prof biochem & head dept, Chicago Med Sch, 61-66; PROF PHYSIOL & CHMN DEPT, COL MED & DENT NJ, RUTGERS MED SCH, PISCATAWAY, 66- Mem: Am Physiol Soc; Am Soc Cell Biologists; Biophys Soc; Am Chem Soc; Am Soc Biol Chemists. Res: Intermediary metabolism; transport of carbohydrates. Mailing Add: Col Med & Dent NJ Dept Physiol Rutgers Med Sch Piscataway NJ 08854

CRANE, ROBERT KENDALL, b Worcester, Mass, Dec 9, 35; m 57; c 4. RADIOPHYSICS, METEOROLOGY. Educ: Worcester Polytech Inst, BS, 57, MS, 59, PhD(elec eng), 70. Prof Exp: Mem tech staff, Mitre Corp, 59-64; MEM TECH STAFF, LINCOLN LAB, MASS INST TECHNOL, 64- Concurrent Pos: Mem, Comn II, Int Sci Radio Union, 67-; mem US study groups 5 & 6, Int Radio Consultative Comt, 68-; asst ed, Trans Antennas & Propagation, Inst Elec & Electronic Engrs, 72-74. Mem: Inst Elec & Electronic Engrs; Am Meteorol Soc; Am Geophys Union. Res: Radio propagation; underwater acoustic propagation; turbulence theory. Mailing Add: 21 Deacon Hunt Dr Acton MA 01720

CRANE, ROGER L, b Monroe, Iowa, June 27, 33; m 59; c 1. MATHEMATICS. Educ: Iowa State Univ, BS, 56, MS, 61, PhD(math), 62. Prof Exp: Engr, Univac Remington-Rand Corp, Minn, 56-59; instr math, Iowa State Univ, 62-63; mem tech staff, Radio Corp Am Labs, 63-69, MEM TECH STAFF, RCA CORP, 69- Mem: Soc Indust & Appl Math; Inst Elec & Electronics Eng; Asn Comput Mach. Res: Numerical analysis and scientific applications of digital computers. Mailing Add: RCA Corp Princeton NJ 08540

CRANE, SHELDON CYR, b Long Beach, Calif, Dec 22, 18; m 56; c 1. ENVIRONMENTAL SCIENCES. Educ: Calif Inst Technol, BS, 40; Univ Calif, Los Angeles, PhD(phys biol sci), 49. Prof Exp: Res fel chem, Calif Inst Technol, 49-55; proj engr, Gyro Div, Giannini Controls Corp, 56-59; specialist engr inertial navig, Monterey Eng Lab, Dalmo Victor Corp, 59-60; res opers engr, Combat Develop Exp Ctr, Stanford Res Inst, 60-61; partner, Del Monte Tech Assocs, 61-62; opers res analyst, Combat Develop Exp Ctr, Stanford Res Inst, Ft Ord, 62-63, US Army Concept Team, Viet Nam, 63-64, combat develop exp ctr, 64, sr res engr marine tech, Systs Sci Dept, 64-65, bioengr med res, Life Sci, 65-72; sr res assoc, Haile Sellassie I Univ, Addis Ababa, Ethiopia, 72-74; HEAD, MONITORING DIV, GUAM ENVIRON PROTECTION AGENCY, 74- Mem: AAAS; Nat Environ Health Asn. Res: Design of electromechanical instruments for research or industry; operations research for United States Army; marine technology and biochemistry; respiratory physiology; biomedical instrumentation; studies of respiratory function and effects of artificially induced emphysema in rats and monkeys; development of pilot manufacturing facility for concentrating the molluscicidal principle in the soap berry Endod for schistosomiasis control; instrumentation for air pollution monitoring; noise pollution studies; chemical analysis for water pollution. Mailing Add: PO Box 7143 Tamuning GU 96911

CRANEFIELD, PAUL FREDERIC, b Madison, Wis, Apr 28, 25. PHYSIOLOGY. Educ: Univ Wis, PhB, 46, PhD(physiol), 51; Albert Einstein Col Med, MD, 64. Prof Exp: Rockefeller Found-Nat Res Coun fel, Dept Biophys, Johns Hopkins Univ, 51-53; sr res fel, Dept Psychiat, Albert Einstein Col Med, 60-64. Prof Exp: From instr to assoc prof, Dept Physiol, State Univ NY Downstate Med Ctr, 53-62; exec secy, Comt Pub & Med Info & ed bull, NY Acad Med, 63-66; assoc prof physiol, 66-75; PROF PHYSIOL, ROCKEFELLER UNIV, 75-; ED, J GEN PHYSIOL, 66- Concurrent Pos: Rockefeller Found-Nat Res Coun fel, Dept Biophys, Johns Hopkins Univ, 51-53; sr res fel, Dept Psychiat, Albert Einstein Col Med, 60-64; assoc prof pharm, Col Physicians & Surgeons, Columbia Univ, 64-66, adj assoc prof, 66- Honors & Awards: Schumann Prize, 56. Mem: Am Physiol Soc; Cardiac Muscle Soc; Soc Neurosci; Am Asn Hist Med; fel Int Acad Hist Med. Res: Cardiac electrophysiology; history of medicine. Mailing Add: Rockefeller Univ New York NY 10021

CRANFORD, JERRY L, b Heber Springs, Ark, Mar 12, 42; m 67. NEUROPSYCHOLOGY, PSYCHOPHYSIOLOGY. Educ: Wichita State Univ, BA, 64; Vanderbilt Univ, PhD(exp psychol), 68. Prof Exp: USPHS trainee, Med Ctr, Duke Univ, 68-70; res assoc neuropsychol, Ind Univ, Bloomington, 70-73, asst prof physiol psychol, 72-73; ASST PROF AUDITORY PSYCHOPHYSIOL, BAYLOR COL MED, 73- Mem: Soc Neurosci; Int Brain Res Orgn. Res: Neuroanatomical and neurophysiological substrates of auditory nervous system function. Mailing Add: Otorhinolaryngol & Commun Sci Baylor Col of Med Houston TX 77025

CRANFORD, ROBERT HENRY, b Columbia, Tenn, Sept 10, 35; m 60; c 3. MATHEMATICS. Educ: Mid Tenn State Col, BS, 57; La State Univ, MS, 59, PhD(algebra), 64. Prof Exp: Asst prof math, NTex State Univ, 64-73; ASSOC PROF MATH, TEX EASTERN UNIV, 73- Mem: Math Asn Am; Nat Coun Teachers Math. Res: Ideal theory in commutative rings with unity. Mailing Add: Dept of Math Tex Eastern Univ Tyler TX 75701

CRANG, RICHARD EARL, b Clinton, Ill, Dec 2, 36; m 58; c 2. PLANT CYTOLOGY, ELECTRON MICROSCOPY. Educ: Eastern Ill Univ, BS, 58; Univ SDak, MA, 62; Univ Iowa, PhD(bot), 65. Prof Exp: Teacher high sch, Ill, 58-61; asst prof biol, Wittenberg Univ, 65-69; assoc prof biol, 69-74, PROF BIOL SCI, BOWLING GREEN STATE UNIV, 74- Concurrent Pos: Mem educ comt, Electron Microscopy Soc Am, 72-; adj prof anat, Med Col Ohio, 74- Mem: AAAS; Am Soc Cell Biol; Bot Soc Am; Electron Micros Soc Am; Sigma Xi. Res: Plant ultrastructure, particularly fine structure of mutant chloroplasts and pollen tube growth; analyses with transmission and scanning electron microscopy and x-ray microanalysis. Mailing Add: Dept of Biol Sci Bowling Green State Univ Bowling Green OH 43403

CRANKSHAW, WILLIAM BLISS, b Ft Wayne, Ind, Dec 2, 25; m 54; c 4. FOREST ECOLOGY. Educ: Purdue Univ, BS, 50, PhD(plant ecol), 64; Ind Univ, MS, 55. Prof Exp: Forester, US Forest Serv, 50-54; teacher high sch, Calif, 55-59; teacher biol & bot, Joliet Jr Col, 59-61; from asst prof to assoc prof ecol, 64-75, PROF BIOL, BALL STATE UNIV, 75- Mem: Ecol Soc Am; Soc Am Foresters; Am Forestry Asn. Res: Edaphology of forest trees, especially the deciduous species; vegetation mapping; edaphic relationships of mycorrhiza. Mailing Add: Dept of Biol Ball State Univ Muncie IN 47306

CRANMER, MORRIS F, b Columbus, Ohio, June 1, 39. PHARMACOLOGY, BIOCHEMISTRY. Educ: Sul Ross State Univ, BS, 60; Univ Tex, Austin, PhD(cell biol), 65. Prof Exp: Leader metab group, Pesticide Res Lab, USPHS, Fla, 65-68; leader pesticide metab & biochem groups, Perrine Primate Res Lab, 68-69, chief pharmacol sect, 69-72; DIR NAT CTR TOXICOL RES, FOOD & DRUG ADMIN, JEFFERSON, ARK, 72-; PROF BIOCHEM, SCH MED, UNIV ARK, LITTLE ROCK, 72- Concurrent Pos: Adj prof biol, Univ Miami, 68-70; proj reviewer, Div Pesticide Community Studies, Food & Drug Admin, 69-70, asst ed, Manual Anal Methods, 70- Mem: AAAS; Am Chem Soc; Am Inst Biol Sci. Res: Biochemical systematic application of natural products; studies on consideration of environmental contaminating residues persistance; distribution magnification in biological systems and resulting alterations of the metabolism in man and experimental animals especially primates. Mailing Add: Dept of Biochem Univ of Ark Sch of Med Little Rock AR 72201

CRANNELL, CAROL JO ARGUS, b Columbus, Ohio, Nov 15, 38; m 61; c 3. ASTROPHYSICS. Educ: Miami Univ, Ohio, BA, 60; Stanford Univ, PhD(physics), 67. Prof Exp: Res assoc high energy physics lab, Stanford Univ, summer 67; fel exp nuclear physics, Cath Univ, 67-70; res assoc high energy cosmic ray physics, Fed City Col & Goddard Space Flight Ctr, NASA, 70-73; res scientist, Imp Col, Univ London, 73-74; vis scientist, Mass Inst Technol, 74; ASTROPHYSICIST, GODDARD SPACE FLIGHT CTR, NASA, 74- Concurrent Pos: Asst prof, Howard Univ, 68-69. Mem: Am Phys Soc. Res: Solar physics. Mailing Add: Code 682 NASA Goddard Space Flight Ctr Greenbelt MD 20771

CRANNELL, HALL L, b Berkeley, Calif, Feb 23, 36; m 61; c 3. NUCLEAR PHYSICS. Educ: Miami Univ, BA, 56, MA, 58; Stanford Univ, PhD(physics), 64. Prof Exp: Res assoc high energy physics, Stanford Univ, 64-67; assoc prof, 67-72, PROF PHYSICS, CATH UNIV AM, 72- Concurrent Pos: Vis sr scientist & vis prof, Westfield Col, Univ London, 73-74; vis prof, Mass Inst Technol, 74. Mem: AAAS; Am Phys Soc. Res: Electron scattering at medium energies; nuclear instrumentation. Mailing Add: 10000 Branch View Ct Silver Spring MD 20903

CRANO, JOHN CARL, b Akron, Ohio, Nov 16, 35; m 58; c 3. ORGANIC CHEMISTRY. Educ: Univ Notre Dame, BS, 57; Case Inst Technol, MS, 59, PhD(org chem), 62. Prof Exp: Sr res chemist, Chem Div, Pittsburgh Plate Glass Co, 61-63, res supvr, 63-65, RES ASSOC, CHEM DIV, PPG INDUSTS, INC, 65- Mem: AAAS; Am Chem Soc. Res: Mechanism of peroxide decomposition; fluorine chemistry; heterogeneous catalysis; halogen exchange reactions; polymerization; bromine chemistry. Mailing Add: PPG Industs Inc PO Box 31 Barberton OH 44203

CRANSTON, FREDERICK PITKIN, JR, b Denver, Colo, Aug 28, 22; m 46; c 4. PHYSICS. Educ: Colgate Univ, BA, 43; Stanford Univ, MS, 50, PhD(physics), 59. Prof Exp: Asst physics, Stanford Univ, 47-53; mem staff nuclear physics, Los Alamos Sci Lab, 53-62; assoc prof, 62-66, PROF PHYSICS, HUMBOLDT STATE UNIV, 66- Concurrent Pos: Consult, Lawrence Livermore Lab, Univ Calif, 64-70. Mem: Am Phys Soc; Am Asn Physics Teachers. Res: Nuclear spectroscopy; radiation hazards; nuclear weapons effects; x-ray fluorescence. Mailing Add: Dept of Physics Humboldt State Univ Arcata CA 95521

CRAPANZANO, VINCENT BERNARD, b Glen Ridge, NJ, Apr 15, 39; m 68; c 1. ANTHROPOLOGY. Educ: Harvard Univ, AB, 60; Columbia Univ, PhD(anthrop), 70. Prof Exp: Asst prof anthrop, Princeton Univ, 70-74; ASSOC PROF ANTHROP, QUEENS COL, 74- Concurrent Pos: Princeton Univ Res Fund grant, Aix-en-Province, Madrid & Paris, 71; Comn Int & Regional Studies grant, Princeton Univ, 72;, Lehrmann Inst, 76-78. Mem: Fel Am Anthrop Asn; fel Royal Anthrop Inst. Res: Transcultural psychiatry; rituals of curing; the individual in culture; relationship between individual and cultural symbols; articulation of psychic reality; ethnopsychology; structuralism; hermeneutics; anthropology of literature. Mailing Add: 333 Central Park W New York NY 10035

CRAPO, HENRY HOWLAND, b Detroit, Mich, Aug 12, 32; m 62; c 1. MATHEMATICS. Educ: Univ Mich, AB, 54; Mass Inst Technol, PhD(math), 64.

Prof Exp: PROF MATH, UNIV WATERLOO, 65- Concurrent Pos: Mem coun, Can Math Cong, 75-77; mem, Ctr Math Res, Montreal. Mem: Am Math Soc; Math Asn Am; NY Acad Sci. Res: Combinatorial geometry; abstract linear dependence; lattices and ordered sets. Mailing Add: Dept of Pure Math Univ of Waterloo Waterloo ON Can

CRAPO, RICHLEY H, b La Habra, Calif, Apr 15, 43; m 65; c 1. ANTHROPOLOGICAL LINGUISTICS, CULTURAL ANTHROPOLOGY. Educ: Calif State Col, BA, 67; Univ Utah, MA, 68, PhD(anthrop), 70. Prof Exp: ASST PROF SOCIOL, SOCIAL WORK & ANTHROP, UTAH STATE UNIV, 70- Mem: Am Anthrop Asn; AAAS. Res: Uto-Aztecan linguistics, especially Numic languages and classical Aztec; new world ethnography; culturology; cultural evolution; psycholinguistics; psychiatric anthropology; religion. Mailing Add: Dept of Sociol Social Work & Anthrop Utah State Univ Logan UT 84322

CRAPPLE, GEORGE A, chemistry, see 12th edition

CRARY, ALBERT PADDOCK, b Pierrepont, NY, July 25, 11; m 68; c 1. GEOPHYSICS. Educ: St Lawrence Univ, BS, 31; Lehigh Univ, MS, 33. Hon Degrees: PhD, St Lawrence Univ, 59. Prof Exp: Explor geophysicist, Independent Explor Co, 35-41; proj scientist, Woods Hole Oceanog Inst, 41-42; explor geophysicist, United Geophys Co, 42-46; proj scientist, US Air Force Cambridge Res Ctr, 46-60; chief scientist, Off Antarctic Prog, 60-67, dep dir div environ sci, 67-69, DIR DIV ENVIRON SCI, NSF, 69- Concurrent Pos: Dep chief scientist, US Nat Comt, Int Geophys Year, Antarctic Prog, Nat Acad Sci, 56-58. Honors & Awards: Distinguished Pub Serv Award, US Dept Navy, 59; Cullum Medal, Am Geog Soc, 59; Distinguished Civilian Serv Award, US Dept Defense, 60; Patrons Medal, Royal Geog Soc, 63. Mem: AAAS; Am Geophys Union; Am Geog Soc; Geol Soc Am. Res: Seismic exploration for oil; atmospheric acoustic studies; geophysical work in Arctic Ocean Basin and Antarctica. Mailing Add: NSF 1800 G St NW Washington DC 20550

CRARY, DOUGLAS DUNHAM, b Warren, Pa, Sept 1, 10; m 34; c 4. GEOGRAPHY. Educ: Univ Mich, AB, 33, MA, 34, PhD, 47. Prof Exp: From instr to prof geog, Univ Mich, Ann Arbor, 43-75; RETIRED. Concurrent Pos: Consult res & develop bd, US Dept Defense, 49-51; mem comt Near East, Soc Sci Res Coun, 51-53; US Dept State lectr, 60-61. Mem: Asn Am Geog; Am Geog Soc; African Studies Asn; Mid East Inst. Res: Geography of the Near East and Africa; applied urban geography; land use, settlement and mainland relationships in the Far East, Near East and Africa. Mailing Add: 1842 Cambridge Rd Ann Arbor MI 48104

CRARY, JAMES WALTER, b Lexington, Ky, June 21, 30; m 54; c 2. ORGANIC CHEMISTRY. Educ: Univ Ky, BS, 52; Emory Univ, PhD(chem), 55. Prof Exp: RES CHEMIST, EXP STA, E I DU PONT DE NEMOURS & CO, 55- Res: Polymer chemistry, particularly elastomers. Mailing Add: 3017 Ridgevale Rd Wilmington DE 19808

CRASEMANN, BERND, b Hamburg, Ger, Jan 23, 22; nat; m 52. ATOMIC PHYSICS. Educ: Univ Calif, Los Angeles, AB, 48; Univ Calif, Berkeley, PhD(physics), 53. Prof Exp: From asst prof to assoc prof, 53-63, PROF PHYSICS, UNIV ORE, 63- Concurrent Pos: Vis prof, physics dept, Univ Calif, Berkeley, 68-69; chmn ad hoc panel on accelerator-related atomic physics res, Comt Atomic & Molecular Physics, Nat Acad Sci-Nat Res Coun, 74-; vis scientist, Ames Res Ctr, NASA, 75-76. Mem: Am Phys Soc; Am Asn Physics Teachers; AAAS. Res: Atomic inner-shell processes; transition probabilities; interface of atomic and nuclear physics. Mailing Add: Dept of Physics Univ of Ore Eugene OR 97403

CRASEMANN, JEAN M, b Saskatoon, Sask, Oct 21, 21; nat; m 52. MOLECULAR GENETICS. Educ: Univ Sask, BA, 42, MA, 46; Univ Calif, PhD(bot), 52. Prof Exp: Res assoc, Univ Calif, 52-53; instr biol, 55, 58-59, RES ASSOC MOLECULAR BIOL, UNIV ORE, 59- Res: Bacteriophage genetics; recombination in bacteriophage lambda. Mailing Add: Inst of Molecular Biol Univ of Ore Eugene OR 97403

CRASS, GWENDOLYN, b Ada, Okla, Sept 13, 12. PATHOLOGY. Educ: Southern Methodist Univ, BA, 34; Univ Tex, MD, 44. Prof Exp: Instr path, Med Br, Univ Tex, 52-54, asst prof, 54-59; assoc pathologist, Med Ctr, Baylor Univ, 59-75; RETIRED. Concurrent Pos: Fel, Med Br, Univ Tex, 51-52; clin prof, Univ Tex Health Sci Ctr, Dallas, 59-74. Mem: Col Am Path; Am Soc Hemat; Am Soc Cytol; Int Acad Path; Int Soc Hemat. Res: Hematology; electron microscopy. Mailing Add: 6520 Linden Dallas TX 75230

CRASS, MAURICE FREDERICK, III, b Akron, Ohio, Nov 15, 34; m 60; c 3. PHYSIOLOGY, BIOCHEMISTRY. Educ: Univ Md, BS, 57, MS, 59; Vanderbilt Univ, PhD(physiol), 65. Prof Exp: Instr physiol, Sch Med, Vanderbilt Univ, 65-66; asst prof med & physiol, Col Med, Univ Fla, 69-70; from asst prof to assoc prof biochem & med, Col Med, Univ Nebr, 70-73; ASSOC PROF PHYSIOL, SCH MED, TEX TECH UNIV, 73- Concurrent Pos: Res fel, Tenn Heart Asn, 65-66; res fel med, Col Med, Univ Fla, 66-69. Mem: AAAS; Am Physiol Soc; Soc Exp Biol & Med; NY Acad Sci. Res: Cardiovascular physiology and biochemistry; carbohydrate and lipid metabolism; hormonal regulation of metabolism. Mailing Add: Dept of Physiol Tex Tech Univ Sch of Med Lubbock TX 79406

CRASWELL, KEITH J, b Kent, Wash, Dec 17, 36; m 58; c 2. MATHEMATICS. Educ: Univ Wash, BS, 59, MSc, 61, PhD(math), 63. Prof Exp: Actg instr math, Univ Wash, 62-63; staff mem, Statist Res Div, Sandia Corp, 63-65; asst prof math & statist, Colo State Univ, 65-67; ASSOC PROF MATH, WESTERN WASH STATE COL, 67- Mem: Inst Math Statist; Am Math Soc; Math Asn Am. Res: Stochastic processes. Mailing Add: Dept of Math Western Wash State Col Bellingham WA 98225

CRATER, HORACE WILLIAM, b Washington, DC, May 26, 42; m 63; c 2. PARTICLE PHYSICS, QUANTUM MECHANICS. Educ: Col William & Mary, BS, 64; Yale Univ, MS, 65, MPhil & PhD(physics), 68. Prof Exp: Vis mem, Inst Advan Study, 68-70; ASST PROF PHYSICS, VANDERBILT UNIV, 70- Mem: Am Phys Soc. Res: Theoretical particle physics, current algebra; meson-meson scattering; phenomenological quantum field theory; non-perturbative techniques in quantum field theory; Pade approximants. Mailing Add: Dept of Physics & Astron Vanderbilt Univ Nashville TN 37203

CRATIN, PAUL DAVID, b Moline, Ill, Feb 26, 29; m 57; c 5. PHYSICAL CHEMISTRY. Educ: Spring Hill Col, BS, 51; St Louis Univ, MS, 54; Tex A&M Univ, PhD(thermodyn), 62. Prof Exp: Anal chemist, Chemstrand Corp, 55-58; instr chem, Ga Inst Technol, 58-59 & Tex A&M Univ, 59-62; res chemist, Jersey Prod Res Co, 62-64; assoc prof, Spring Hill Col, 64-67; res assoc, St Regis Paper Co, 67-68; assoc prof, Univ Miami, 68-70; chem dept chem, 71-73, PROF CHEM, CENT MICH UNIV, 71- Mem: Am Chem Soc. Res: Excess thermodynamic properties of binary liquid solutions; application of thermodynamics to surface and interfacial phenomena; kinetics and mechanisms of homogeneous chemical reactions. Mailing Add: Dept of Chem Cent Mich Univ Mt Pleasant MI 48859

CRATTY, LELAND EARL, JR, b Oregon, Ill, June 3, 30; m 56; c 3. PHYSICAL CHEMISTRY, SURFACE CHEMISTRY. Educ: Beloit Col, BS, 52; Brown Univ, PhD(chem), 57. Prof Exp: Res chemist, Linde Co, 56-58; from asst prof to assoc prof, 58-73, PROF CHEM, HAMILTON COL, 73- Concurrent Pos: Vis fel, Mellon Inst, 65-66; chemist, Ames Lab, US Atomic Energy Comn, 69-70. Mem: AAAS; Am Chem Soc. Res: Ultra high vacuum techniques; adsorption state in clean surfaces; surface chemistry and catalysis, particularly at metal and alloy surfaces. Mailing Add: Dept of Chem Hamilton Col Clinton NY 13323

CRAUL, PHILLIP JAN, forest soils, soil classification, see 12th edition

CRAVEN, BRYAN MAXWELL, b Wellington, NZ, Feb 12, 32; m 56; c 1. STRUCTURAL CHEMISTRY. Educ: Univ Auckland, BSc, 53, MSc, 54, PhD(chem), 58. Prof Exp: Jr lectr, Univ Auckland, 56; res assoc & instr, 57-59, from asst res prof to assoc prof, 59-71, PROF CHEM CRYSTALLOG, UNIV PITTSBURGH, 71-, CHMN DEPT CRYSTALLOG, 74- Concurrent Pos: Rothmans sr fel, Univ Sydney, 62-64. Res: Crystal structure determination of drugs and biomolecules, including lipids, serum albumin. Mailing Add: Dept of Crystallog Univ of Pittsburgh Pittsburgh PA 15260

CRAVEN, CHARLES WALLER, b Waurika, Okla, Jan 5, 20; m 43; c 2. CHEMISTRY, BIOCHEMISTRY. Educ: Okla Agr & Mech Col, BS, 47, MS, 49; Ohio State Univ, PhD(biochem), 53. Prof Exp: Chief biochem br, Air Force Armament Ctr, US Air Force, Elgin Air Force Base, Fla, 53-55, dir biosci, Air Force Off Sci Res, Washington, DC, 55-57, dept dir adv studies, Pasadena, Calif, 57-59, chief, Tech Div, Regional Off, Air Force Res & Develop Command, Hollywood, 59-60, dept chief bioastronaut, Space Systs Div, Los Angeles, 60-62, manned military vehicle directorate, 62-63, dir, Manned Environ Systs Directorate, 63-65, dept dir space prog, 65-67; planetary quarantine mgr, Voyager Proj, Jet Propulsion Lab, Calif Inst Technol, 67-69, mem tech staff long range planning & exec asst to dir, 69-73, MGR PLANETARY QUARANTINE PROJ OFF, CALIF INST TECHNOL, 73- Mem: AAAS; Sigma Xi. Mailing Add: 4535 Alveo Rd La Canada CA 91011

CRAVEN, CLAUDE JACKSON, b Concord, NC, Jan 13, 08. PHYSICS. Educ: Univ NC, AB, 31, MA, 33, PhD(physics), 35. Prof Exp: Asst prof physics & math, Delta State Teachers Col, 35-36; assoc prof physics, Furman Univ, 36-39; asst prof physics & math, Emory Univ, 39-42; scientist, Columbia Univ, 42-45; physicist, Carbide & Carbon Chem Corp, NY & Tenn, 45-46; physicist, Fiber Res Lab, 46-52, from assoc physicist agr exp sta to assoc prof univ, 52-69, PROF, UNIV TENN, 69- Concurrent Pos: Consult, Oak Ridge Nat Lab, 46-; prin scientist, Oak Ridge Inst Nuclear Studies, 60-66. Mem: AAAS; Am Phys Soc. Res: Infrared spectroscopy; gaseous diffusion through porous media; physical properties of fibers. Mailing Add: Dept of Physics Univ of Tenn Knoxville TN 37916

CRAVEN, DONALD ALLEN, analytical biochemistry, see 12th edition

CRAVEN, JAMES MILTON, b Los Angeles, Calif, Dec 12, 30; m 52; c 2. POLYMER CHEMISTRY. Educ: Univ Calif, Berkeley, BA, 53; Univ Wash, Seattle, PhD(org chem), 59. Prof Exp: Control chemist, Cutter Labs, 53-55; NSF fel, Inst Org Chem, Munich, Ger, 59-60; res chemist, Fabrics & Finishes Dept, 60-63, staff chemist, 64-68, RES ASSOC, E I DU PONT DE NEMOURS & CO, INC, 68- Mem: AAAS; Am Chem Soc. Res: Electronic spectra of aromatic compounds; 1, 3-dipolar additions; physical properties of polymers; synthesis of aromatic-hetero-cyclic polymers and elastomers; chemistry and physics of polymers, adhesives and surface coatings. Mailing Add: E I du Pont de Nemours & Co Inc Wilmington DE 19898

CRAVEN, ROBERT ALAN, b Holyoke, Mass, Nov 12, 45; m 69; c 1. SOLID STATE PHYSICS. Educ: Trinity Col, BS, 67; Univ Rochester, MA, 69, PhD(physics), 73. Prof Exp: Res assoc physics, Univ Ill, 73-75; MEM STAFF PHYSICS, THOMAS J WATSON RES CTR, IBM, 75- Mem: Am Phys Soc. Res: Thermal and transport properties of organic conducting materials and magnetic properties of amorphous metallic alloys. Mailing Add: IBM Thomas J Watson Res Ctr Yorktown Heights NY 10598

CRAVEN, ROBERT LEE, b Detroit, Mich, Apr 9, 23; m 49; c 2. ORGANIC CHEMISTRY. Educ: Amherst Col, BA, 44; Univ Mich, MS, 48, PhD(chem), 54. Prof Exp: Res chemist, Jackson Lab, E I du Pont de Nemours & Co, 54-59; from asst prof to assoc prof, 59-68, dept head, 68-70, staff chmn dept, 63-68, PROF CHEM, ROCHESTER INST TECHNOL, 68- Mem: Am Chem Soc. Res: Mechanisms; organic reactions; elastomers. Mailing Add: Dept of Chem Rochester Inst of Technol One Lomb Memorial Dr Rochester NY 14623

CRAVEN, WILLIAM JAMES, organic chemistry, see 12th edition

CRAVENS, WILLIAM WINDSOR, b Daviess Co, Ky, Oct 24, 14; m 39; c 5. NUTRITION, RESEARCH ADMINISTRATION. Educ: Univ Ky, BS, 35; Iowa State Col, MS, 37; Univ Wis, PhD(biochem), 40. Prof Exp: Asst animal chem & nutrit, Iowa State Col, 35-37; from asst to instr poultry, Univ Wis, 37-41, from asst prof to prof poultry husb, 41-53; dir feed res, 53-68, VPRES RES, CENT SOYA CO, 68- Honors & Awards: Am Feed Mfrs Award, 50. Mem: Am Chem Soc; Poultry Sci Asn; Soc Exp Biol & Med; Am Soc Animal Sci; Am Inst Nutrit. Res: Nutrition of livestock and poultry; animal production; feed formulation; soybean processing; food technology. Mailing Add: Cent Soya Co Inc Ft Wayne Nat Bank Bldg Ft Wayne IN 46802

CRAVER, BRADFORD NORTH, pharmacology, see 12th edition

CRAVER, CLARA DIDDLE (SMITH), b Portsmouth, Ohio, Dec 3, 24; m 46, 70; c 4. SPECTROCHEMISTRY, PETROLEUM CHEMISTRY. Educ: Ohio State Univ, BSc, 45. Hon Degrees: DSc, Fisk Univ, 74. Prof Exp: Tech man spectros, Esso Res Labs, 45-48; res engr, Battelle Mem Inst, 49-55, group leader, 55-57, consult, 57-58; SPECTROS CONSULT & OWNER CHEMIR LABS, 58- Concurrent Pos: Carbide & carbon chem award, Am Chem Soc, 55-56; ed, Coblentz Soc Spectral Data, 56-; instr chem, Ohio State Univ, 57-58; guest lectr, Infrared Inst, Fisk Univ, 59-69 & Fisk Infrared Inst, Sao Paulo, Brazil, 65 & fac mem, 65; guest lectr, Infrared Inst, Univ Minn, 61-62, 64-65 & 67-68 & Canisius Col, 61; consult, Nat Stand Ref Data Prog, 65-; consult, US Coast Guard, 75. Honors & Awards: A K Doolittle Award, Union Carbide & Am Chem Soc, 56. Mem: Am Chem Soc; Optical Soc Am; Coblentz Soc; Soc Appl Spectros; Am Soc Test & Mat. Res: Application of chemical spectroscopy to research and analysis in polymers, coatings, marine environment, air pollution, petroleum products, asphalts, paper and cellulose chemistry; standard data compilations and computer retrieval of spectroscopic data; absorption spectroscopy; applications of infrared spectroscopy. Mailing Add: 761 W Kirkham Ave Glendale MO 63122

CRAVER, JOHN KENNETH, b Jonesboro, Ill, May 1, 15; m 39, 70; c 4. CHEMISTRY, TECHNOLOGICAL FORECASTING. Educ: Southern Ill Univ, BEd, 37; Syracuse Univ, MS, 38. Prof Exp: Res chemist, Monsanto Co, 38-46, coordr plasticizers, 46-51, develop mgr resin mat & functional fluids, 51-54; develop dir, Gen Mills, Inc, 55-56; develop mgr, Res & Eng Div, 56-61, res assoc org div, 61-67, sr res assoc, Cent Res Dept, 67-70, MGR FUTURES RES, MONSANTO CO, 70- Concurrent Pos: Chmn, Gordon Res Conf, 66; chmn chem mktg & econ div, Am Chem Soc, 75. Mem: AAAS; Am Chem Soc; Commercial Chem Develop Asn; Int Soc Technol Assessment. Res: Long range planning; decision analysis; paper and polymer chemistry; applied research; commercial development. Mailing Add: Monsanto Co 800 N Lindbergh St St Louis MO 63166

CRAVIOTO, HUMBERTO, b Pachuca, Mex, Oct 4, 24. NEUROPATHOLOGY, NEUROLOGY. Educ: Sci & Lit Inst, Mex, BS, 45; Nat Univ Mex, MD, 52; State Univ NY, MD, 64. Prof Exp: Jr & sr resident path, Univ Vt, 52-54; resident neurol, Bellevue Med Ctr, NY Univ, 54-56, instr neurol & neuropath, 56-58, asst prof neuropath, 58-64; assoc prof path & neurol, Sch Med, Univ Southern Calif, 64-68; ASSOC PROF NEUROPATH, SCH MED, NY UNIV, 68- Concurrent Pos: Alexander von Humboldt Soc res fel electron micros, Berlin, 61-62. Mem: Am Acad Neurol; Am Asn Neuropath (asst secy-treas, 61-64); Histochem Soc; NY Acad Sci; AAAS. Res: Electron microscopy; brain tumor immunology. Mailing Add: NY Univ Sch of Med 550 First Ave New York NY 10016

CRAVITZ, LEO, b Chelsea, Mass, Nov 26, 18; m 44; c 3. BACTERIOLOGY, IMMUNOLOGY. Educ: Boston Univ, BS, 41; Mass Inst Technol, DrPH, 44. Prof Exp: Fel pub health lab methods, Mass Inst Technol, 42-44; MED MICROBIOLOGIST, ROCHESTER GEN HOSP, 46- Concurrent Pos: Res assoc, Boston Health Dept, 42-44; instr, Wilson Sch, Mass, 43-44; consult, Strasenburgh Labs, Pennwalt Corp, 50-, Castle Co, Sybron Corp, 58-, Park-Ridge Hosp, Rochester, 63- & Myers Community Hosp, Sodus, NY, 67- Mem: Inst Environ Sci; Sigma Xi; Am Soc Microbiol; Am Pub Health Asn. Res: Pertussis immunization; immunology of glanders and meliodosis; clostridium perfringens food poisoning; laboratory diagnosis of rheumatic diseases; immunologic factors in allergic disease; antibiotic synergism; sterilization with chemical agents; surgical antisepsis and hospital infection. Mailing Add: Rochester Gen Hosp 1425 Portland Ave Rochester NY 14621

CRAW, ALEXANDER R, b Chicago, Ill, July 29, 19; m 42; c 4. MATHEMATICS, OPERATIONS RESEARCH. Educ: DePaul Univ, BS, 39; Univ Notre Dame, MS, 41; Univ Chicago, cert meteorol, 42; American Univ, PhD(math), 62. Prof Exp: Instr math, DePaul Univ, 41-42; asst prof, US Naval Acad, 46-51; mathematician, Dept Army, Ft Detrick, Md, 55-67; MATHEMATICIAN, NAT BUR STANDARDS, 67- Mem: Soc Indust & Appl Math; Opers Res Soc Am; Am Meteorol Soc; Sigma Xi. Res: Stochastic processes; atmospheric diffusion; transportation systems; environmental systems. Mailing Add: 330 W College Terr Frederick MD 21701

CRAWFORD, BRYCE (LOW), JR, b New Orleans, La, Nov 27, 14; m 40; c 3. PHYSICAL CHEMISTRY. Educ: Stanford Univ, AB, 34, AM, 35, PhD(chem), 37. Prof Exp: Asst chem, Stanford Univ, 33-35; Nat Res Found fel, Harvard Univ, 37-39; instr chem, Yale Univ, 39-40; from asst prof to assoc prof, 40-46, chmn dept, 55-60, dean grad sch, 60-72, mem grad record exam bd, 68-72, PROF PHYS CHEM, UNIV MINN, MINNEAPOLIS, 46- Concurrent Pos: Guggenheim Found fel, Calif Inst Technol & Oxford Univ, 50-51; Fulbright fel, Oxford Univ, 51; chmn, Coun Grad Schs US, 62-63; mem bd dirs, NStar Res & Develop Inst, 63-; Fulbright prof, Univ Tokyo, 66; mem adv comt, Off Sci Personnel, Nat Res Coun, 67-71; ed, J Phys Chem, 70- Honors & Awards: Presidential Cert Merit; Minn Award, Am Chem Soc, 68. Mem: Nat Acad Sci; Am Chem Soc; Asn Grad Schs (pres, 70); Optical Soc Am; Am Philos Soc. Res: Molecular spectra; statistical thermodynamics; kinetics; molecular dynamics. Mailing Add: Grad Sch 13 Smith Hall Univ of Minn Minneapolis MN 55455

CRAWFORD, CHARLES KIMBALL, atomic physics, see 12th edition

CRAWFORD, CLIFFORD SMEED, b Beirut, Lebanon, July 30, 32; US citizen; m 58; c 3. INVERTEBRATE ECOLOGY, INVERTEBRATE PHYSIOLOGY. Educ: Whitman Col, BA, 54; Wash State Univ, MS, 58, PhD(entom), 61. Prof Exp: Instr biol, Portland State Col, 61-64; from asst prof to assoc prof, 64-73, PROF BIOL, UNIV NMEX, 73-, CHMN DEPT, 75- Mem: AAAS; Entom Soc Am; Ecol Soc Am; Int Asn Ecol. Res: Ecology and physiology of arid-land invertebrates. Mailing Add: Dept of Biol Univ of NMex Albuquerque NM 87131

CRAWFORD, CRAYTON MCCANTS, b Greenville, SC, Sept 20, 26; m 55; c 3. PHYSICAL CHEMISTRY. Educ: Clemson Col, BS, 49; Univ NC, PhD(phys chem), 57. Prof Exp: Mem staff, Los Alamos Sci Lab, 55-59; asst prof, 59, ASSOC PROF PHYS CHEM, MISS STATE UNIV, 59- Concurrent Pos: Ed, J Miss Acad Sci, 68- Mem: Am Chem Soc; Sigma Xi. Res: Precision absorptiometry; photochemical kinetics. Mailing Add: Dept of Chem Miss State Univ Drawer CH Mississippi State MS 39762

CRAWFORD, DANIEL JOHN, b Columbus Junction, Iowa, May 27, 42; m 61; c 2. PLANT TAXONOMY. Educ: Univ Iowa, BA, 64, MS, 66, PhD(bot), 69. Prof Exp: Asst prof, 69-74, ASSOC PROF BOT, UNIV WYO, 74- Mem: Bot Soc Am; Int Asn Plant Taxon; Am Soc Plant Taxon. Res: Systematics and evolution of flowering plants; use of flavonoid compounds in elucidating plant relationships. Mailing Add: Dept of Bot Univ of Wyo Laramie WY 82070

CRAWFORD, DAVID LEE, b Hays, Kans, Nov 30, 35; m 57; c 2. FOOD SCIENCE, FOOD BIOCHEMISTRY. Educ: Ore State Col, BS, 58, Ore State Univ, MS, 61, PhD(food sci), 66. Prof Exp: Chemist I, 58-60, from asst to asst prof food sci, 64-70, ASSOC PROF FOOD SCI & TECHNOL, ORE STATE UNIV, 70-, PROG DIR SEAFOODS LAB, 66- Concurrent Pos: Grants, Sea Grant Prog, Bur Com Fisheries, 66- & Fish Comn Ore, 66- Mem: Inst Food Technologists. Res: Basic and applied research of food science, especially investigation of basic chemistry of biological systems as related to food preservation, quality and physiological response. Mailing Add: Seafoods Lab Ore State Univ 250 36th St Astoria OR 97103

CRAWFORD, DAVID LIVINGSTONE, b Tarentum, Pa, Mar 2, 31; m 63; c 3. ASTRONOMY. Educ: Univ Chicago, PhD(astron), 58. Prof Exp: Asst, Yerkes Observ, Univ Chicago, 53-57; asst prof physics & astron, Vanderbilt Univ, 58-60; assoc dir, 70-74, ASTRONR, KITT PEAK NAT OBSERV, 60- Mem: Am Astron Soc; Int Astron Union. Res: Galactic structure; stellar photometry; observational instruments and techniques. Mailing Add: Kitt Peak Nat Observ Box 26732 Tucson AZ 85726

CRAWFORD, DONALD LEE, b Santa Ana, Tex, Sept 28, 47; m 70; c 1. MICROBIOLOGY. Educ: Oklahoma City Univ, BA, 70; Univ Wis, MS, 72, PhD(bact), 73. Prof Exp: ASST PROF, DEPT BIOL, GEORGE MASON UNIV, 73- Mem: Am Soc Microbiol; Sigma Xi. Res: General microbiology of soil and water microorganisms with emphasis of microbial recycling of organic nutrients in nature. Mailing Add: Dept of Biol George Mason Univ Fairfax VA 22030

CRAWFORD, DONALD W, b St Louis, Mo, Mar 9, 28; m 57; c 3. CARDIOVASCULAR DISEASES. Educ: Washington Univ, BA, 50, MD, 54. Prof Exp: Intern, Grady Mem Hosp, Emory Univ, 54-55; resident med, Med Ctr, Stanford Univ, 58-60; NIH trainee clin cardiol, Univ Calif, San Francisco, 60-62, fel cardiopulmonary physiol, Cardiovasc Res Inst, 62-63; chief cardiol sect & cardiopulmonary lab, Long Beach Vet Admin Hosp, 63-66; asst clin prof med, Col Med, Univ Calif, 65-66; asst prof, 66-70, ASSOC PROF MED, SCH MED, UNIV SOUTHERN CALIF, 70- Concurrent Pos: Fel coun clin cardiol & coun arteriosclerosis, Am Heart Asn. Honors & Awards: Long Beach Vet Admin Hosp Award, 65. Mem: AMA; Am Fedn Clin Res; fel Am Col Cardiol. Res: Atherosclerosis and clinical physiology. Mailing Add: Dept of Med Univ of Southern Calif Med Ctr Los Angeles CA 90033

CRAWFORD, EUGENE CARSON, JR, b Mt Gilead, NC, Nov 13, 31; m 55; c 2. PHYSIOLOGY, ZOOLOGY. Educ: ECarolina Col, AB, 57; Duke Univ, MA, 60, PhD(physiol), 65. Prof Exp: From asst prof to assoc prof, 65-75, PROF ZOOL, PHYSIOL & BIOPHYS, UNIV KY, 75- Concurrent Pos: NIH fel zool, Duke Univ, 65; NSF res grants, 68-77; res fel, Max Planck Inst Exp Med, 73. Mem: AAAS; Am Soc Zool; Am Physiol Soc. Res: Comparative physiology; temperature regulation; respiration metabolism. Mailing Add: Sch of Biol Sci Univ of Ky Lexington KY 40506

CRAWFORD, FRANCIS WELDON, b Soldier, Kans, Dec 19, 06; m 29; c 2. PHYSICS. Educ: Phillips Univ, AB, 28; Univ Okla, MS, 29, PhD(physics), 34. Prof Exp: Instr physics, Univ Okla, 31-37; res physicist, Phillips Petrol Co, 37-42, sect chief, 42-51, tech dir, atomic energy div, 51-53, mgr prod res, 53-60; assoc prof, 60-72, EMER PROF PHYSICS, KANS STATE UNIV, 72- Mem: AAAS; Am Phys Soc; Am Geophys Union; Soc Explor Geophysicists; Am Asn Physics Teachers. Res: Raman effect; spectrophotometry; infrared spectroscopy; electron microscopy; physical methods of analysis of hydrocarbons; instrumentation for plant process control; physical properties of cereal grains. Mailing Add: Dept of Physics Kans State Univ Manhattan KS 66052

CRAWFORD, FRANK STEVENS, JR, b Scranton, Pa, Oct 25, 23; m 62. PHYSICS. Educ: Univ Calif, AB, 48, PhD(physics), 53. Prof Exp: Res assoc, Radiation Lab, 53-58, from asst prof to assoc prof, 58-65, PROF PHYSICS, UNIV CALIF, BERKELEY, 65- Mem: Am Phys Soc; Am Asn Physics Teachers. Res: Experimental nuclear physics. Mailing Add: 2826 Garber St Berkeley CA 94705

CRAWFORD, FREDERICK WILLIAM, b Birmingham, Eng, July 28, 31; m 63; c 2. PLASMA PHYSICS. Educ: Univ London, BSc, 52 & 54, MSc, 58; Univ Liverpool, PhD(electronics), 55, dipl ed, 56, DEng, 65. Prof Exp: Res trainee, J Lucas Ltd, Birmingham, Eng, 48-52; scientist, Mining Res Estab, Nat Coal Bd, Isleworth, 56-57; sr lectr elec eng, Col Advan Technol, Univ Birmingham, 58-59; res assoc, microwave lab, Stanford Univ, 59-61; vis scientist, Fr Atomic Energy Comn, Saclay, 61-62; res physicist, microwave lab, 62-64, sr res assoc, 64-67, assoc prof, 67-69, PROF ELEC ENG, INST PLASMA RES, STANFORD UNIV, 69- Concurrent Pos: Consult, Compagnie Francaise Thomson-Houston, Paris, 61-62; mem comn III & IV, Int Sci Radio Union. Mem: Fel AAAS; fel Inst Elec & Electronics Engrs; fel Am Phys Soc; fel Brit Inst Elec Engrs; fel Brit Inst Physics. Res: Plasma wave propagation phenomena and diagnostic techniques. Mailing Add: Inst Plasma Res Stanford Univ Via Crespi Stanford CA 94305

CRAWFORD, GEORGE HOMER, polymer chemistry, see 12th edition

CRAWFORD, GEORGE WILLIAM, b Statesville, NC, Oct 21, 06; m 34. PHYSICS. Educ: Davidson Col, BS, 29; Univ NC, MS, 49; Ohio State Univ, PhD, 59. Prof Exp: Teacher, pub schs, NC, 29-40; instr physics, NC State Col, 46-50; asst prof, Davidson Col, 51-60; from assoc prof to prof, 60-73, EMER PROF PHYSICS, COL WILLIAM & MARY, 73- Mem: AAAS; Am Asn Physics Teachers; Optical Soc Am. Mailing Add: Dept of Physics Col of William & Mary Williamsburg VA 23185

CRAWFORD, GEORGE WOLF, b San Antonio, Tex, May 7, 22; m 46; c 3. ENVIRONMENTAL PHYSICS, NUCLEAR PHYSICS. Educ: Univ Tex, BS, 47, MA, 49, PhD(physics), 51. Prof Exp: Res physicist, Optical Res Lab, Univ Tex, 49-51; assoc prof physics, Clemson Col, 51-55; asst dir, Tex Petrol Res Comt & asst prof petrol eng, Univ Tex, 55-59; nuclear physicist & chief physics br, Sch Aerospace Med, Brooks AFB, 59-63; PROF PHYSICS, SOUTHERN METHODIST UNIV, 63-, CHMN DEPT, 75- Concurrent Pos: Res physicist, Deering Millikin Res Trust, 52-54 & Grad Res Ctr Southwest, 63-64. Mem: AAAS; Am Phys Soc; Am Asn Physics Teachers; Sigma Xi. Res: Solar and wind energy storage and conversion; solid state physics. Mailing Add: Dept of Physics Southern Methodist Univ Dallas TX 75275

CRAWFORD, HEWLETTE SPENCER, JR, b Syracuse, NY, June 4, 31; m 52; c 4. WILDLIFE ECOLOGY, FOREST ECOLOGY. Educ: Univ Mich, BS, 54, MS, 57; Univ Mo, Columbia, PhD(ecol, forest wildlife), 67. Prof Exp: Asst wildlife mgt, Univ Mich, 53-54 & 56-57; res forester, Southern Exp Sta, 57-64, proj leader, Cent States Exp Sta, 64-68, prin wildlife ecologist & proj leader wildlife habitat res, Southeastern Forest Exp Sta, 68-74, prin ecologist, Northeastern Forest Exp Sta, 74-75, PRIN RES WILDLIF ECOLOGIST, NORTHEASTERN FOREST EXP STA, ORONO, MAINE, FOREST SERV, USDA, 75- Concurrent Pos: Mem grad fac, Col Life Sci & Agr, Univ Maine, 75- Mem: Wildlife Soc; Ecol Soc Am. Res: Wildlife habitat research, especially as influenced by forest management practices. Mailing Add: USDA Bldg Univ of Maine Orono ME 04473

CRAWFORD, IRVING POPE, b Cleveland, Ohio, Nov 20, 30; m 55; c 2. MICROBIAL GENETICS. Educ: Stanford Univ, AB, 51, MD, 55. Prof Exp: Med officer, virus diag sect, Walter Reed Army Inst Res, 55-57; res assoc biol sci, Stanford Univ, 58-59; assoc prof microbiol, sch med, Western Reserve Univ, 59-65; MEM DEPT MICROBIOL, SCRIPPS CLIN & RES FOUND, 65- Concurrent Pos: Fel, Nat Found Infantile Paralysis, 58-59; sr fel, USPHS, 59, career develop award, 61-65. Honors & Awards: Borden Res Award, 54. Mem: AAAS; Am Soc Microbiol; Genetics Soc Am; Am Soc Biol Chemists; Genetic Soc Japan. Res: Genetics of microorganisms; human genetics. Mailing Add: Dept of Microbiol Scripps Clin & Res Found La Jolla CA 92037

CRAWFORD, JAMES DALTON, b Clyde, Tex, Aug 2, 19; m 43; c 1. ORGANIC CHEMISTRY. Educ: Hardin-Simmons Univ, BA, 52, MA, 53. Prof Exp: Chemist, Cardinal Chem Inc, 53-54, dir res, 54-60; TECH DIR, CONTINENTAL PROD TEX, 60- Mem: AAAS; Am Chem Soc; Nat Asn Corrosion Engrs. Res: Oil field chemicals for prevention of corrosion in production of petroleum; water treatment; demulsifiers; surfactants. Mailing Add: PO Box 3627 Odessa TX 79760

CRAWFORD, JAMES GILMORE, petroleum chemistry, deceased

CRAWFORD, JAMES GORDON, b Alma, Mich, Sept 12, 29; m 53; c 3. MICROBIOLOGY. Educ: Alma Col, BS, 51; Univ Mich, MS, 53, PhD(bact), 55. Prof Exp: Sr asst scientist, Div Biol Standards, NIH, 55-58; head biol control labs, 58-61, proj dir infectious hepatitis res, 62-65, dir biol develop, 63-68, DIR FERMENTATION DEVELOP, PFIZER, INC, 68- Concurrent Pos: Vis prof, Ind

State Col, 63- Mem: Am Soc Microbiol; NY Acad Sci. Res: Etiology and prophylaxis of infectious hepatitis; development of human and veterinary vaccines. Mailing Add: Fermentation Devel Dept Pfizer Inc Terre Haute IN 47808

CRAWFORD, JAMES HOMER, JR, b Union, SC, May 19, 22; m 44; c 1. SOLID STATE PHYSICS. Educ: Wofford Col, AB, 43; Univ NC, PhD(chem), 49. Hon Degrees: DSc, Wofford Col, 68. Prof Exp: Instr chem, Univ NC, 48-49; res physicist solid state physics, Oak Ridge Nat Lab, 49-52, assoc dir solid state div, 52-66; PROF PHYSICS & CHMN DEPT PHYSICS & ASTRON, UNIV NC, CHAPEL HILL, 66- Mem: AAAS; Am Chem Soc; Am Phys Soc. Res: Properties of semiconductors; radiation effects in solids; electronic processes in ionic crystals. Mailing Add: Dept of Physics Univ of NC Chapel Hill NC 27514

CRAWFORD, JAMES JOSEPH L, b June 23, 31; US citizen; m 51; c 6. MICROBIOLOGY. Educ: Univ Mo, BA, 53, MA, 54; Univ NC, PhD(microbiol), 62. Prof Exp: Asst instr microbiol, Sch Med, Univ Mo, 53-54; teaching asst, Med Sch, Univ Minn, 54-56; from res asst to res assoc, Sch Med, 57-60, instr oral microbiol, Sch Dent, 60-65, trainee, Dept Bact, Sch Med, 62-63, from asst prof to assoc prof oral microbiol, Sch Dent, 65-74, ASSOC PROF ENDODONTICS, SCH DENT & LECTR BACTERIOL, SCH MED, UNIV NC, CHAPEL HILL, 74- Concurrent Pos: Consult, Microbiol Labs, NC Mem Hosp, 59-; prin investr, USPHS, 65-68; mem, Am Asn Dent Schs; consult, Womack Army Hosp, Ft Bragg, 67- Mem: Am Soc Microbiol; Am Asn Endodontics. Res: Mechanisms of host resistance; clinical and oral microbiology. Mailing Add: Dent Res Ctr Sch of Dent Univ of NC Chapel Hill NC 27514

CRAWFORD, JAMES WELDON, b Napoleon, Ohio, Oct 27, 27; m 55; c 1. PSYCHIATRY, EXPERIMENTAL PSYCHOLOGY. Educ: Oberlin Col, AB, 50; Univ Chicago, MD, 54, PhD, 61. Prof Exp: Clin instr psychiat, Chicago Med Sch, 62-63, clin assoc, 63-65, clin asst prof, 65-69, clin assoc prof & assoc dir undergrad educ, Dept Psychiat & Behav Sci, 69-70; CHMN & ORGANIZER OF DEPT PSYCHIAT, RAVENSWOOD HOSP MED CTR, 70-; ATTEND STAFF, FOX RIVER HOSP, 71- Concurrent Pos: Nat Inst Neurol Dis & Blindness res grants, 62-64 & 67; staff psychiatrist, Field Clin, Chicago, Ill, 62-65, partner, 65-; assoc staff, Mt Sinai Hosp, 66-; courtesy staff, Louis A Weiss Mem Hosp, 71- Mem: AAAS; fel Am Psychiat Asn; Asn Am Med Cols; Am Med Asn; Am Asn Univ Profs. Res: Changes in attitudes of medical students during psychiatric training and differences between individual and couples or family therapy; visual behavior and thyroid function in cats; various organic therapies in psychiatry; empathy in psychotherapy. Mailing Add: 2418 Lincoln St Evanston IL 60201

CRAWFORD, JAMES WORTHINGTON, b Newport, RI, Feb 25, 44; m 65; c 1. ORGANIC CHEMISTRY, PHOTOCHEMISTRY. Educ: The Citadel, BS, 65; Univ SC, PhD(org photochem), 69. Prof Exp: From staff scientist to DEVELOP ENGR, OFF PROD DIV, IBM CORP, 69- Mem: Am Chem Soc. Res: Development of organic photoconducting polymers and small molecule photoconductors for use in electrophotography. Mailing Add: Off Prod Div IBM Corp Boulder CO 80302

CRAWFORD, JEAN VEGHTE, b Buffalo, NY, Mar 13, 19. ORGANIC CHEMISTRY. Educ: Mt Holyoke Col, AB, 40; Oberlin Col, AM, 42; Univ Ill, PhD(chem), 50. Prof Exp: Instr chem, Mt Holyoke Col, 42-45; chemist, Eastman Kodak Co, 45-47; adj prof chem, Randolph-Macon Woman's Col, 50-51; from asst prof to prof, 51-74, CHARLES FITCH ROBERTS PROF CHEM, WELLESLEY COL, 74- Mem: Am Chem Soc. Res: Heterocyclic nitrogen compounds; mechanism of organic reactions. Mailing Add: Dept of Chem Wellesley Col Wellesley MA 02181

CRAWFORD, JOHN CLARK, b Liberty, Tex, Sept 27, 35; m 55; c 2. PHYSICS. Educ: Phillips Univ, BA, 57; Kans State Univ, MS, 59, PhD(physics), 62. Prof Exp: Mem staff physics, 62-67, div supvr, Solid State Electronics Res, 67-71, MGR ELECTRON TUBE DEVELOP DEPT, SANDIA LAB, 71- Mem: Am Phys Soc; sr mem Inst Elec & Electronics Engrs. Res: X-ray diffraction as applied to ferroelectric whiskers; pulsed high magnetic fields; thin films semiconductors; piezoelectric devices; neutron sources. Mailing Add: Sandia Labs Div 1410 Albuquerque NM 87115

CRAWFORD, JOHN DOUGLAS, b Boston, Mass, Apr 16, 20; m 49; c 3. ENDOCRINOLOGY, METABOLISM. Educ: Harvard Med Sch, MD, 44. Prof Exp: Asst prof pediat, 54-63, ASSOC PROF PEDIAT, HARVARD MED SCH, 63- Concurrent Pos: Chief pediat, Burns Inst, Boston Univ, Shriners Hosps Crippled Children; chief endocrine-metab unit, Children's Serv, Mass Gen Hosp, Boston. Mem: Am Pediat Soc; Soc Pediat Res; Am Soc Clin Invest; Endocrine Soc. Res: Clinical endocrinology; renal physiology. Mailing Add: Dept of Pediat Harvard Med Sch Boston MA 02115

CRAWFORD, JOHN S, b Toronto, Ont, Dec 5, 21; m 44; c 3. INTERNAL MEDICINE, REHABILITATION MEDICINE. Educ: Univ Toronto, MD, 44; FRCP(C), 52. Prof Exp: J J McKenzie fel path, Banting Inst, Univ Toronto, 48-49, McLaughlin traveling fel med, Fac Med, 53-54, clin teacher med & rehab med, 54-60, assoc rehab med, 60-63, assoc med, 60-64, asst prof rehab med, 63-68, asst prof med, 64-68, assoc prof med & rehab med, 68-75, PROF MED, FAC MED, UNIV TORONTO, 68-, PROF REHAB MED & CHMN DEPT, 73-; DIR REHAB MED, TORONTO WESTERN HOSP, 54- Concurrent Pos: Mem med adv bd, Rehab Found Disabled, 61-; med dir, Hillcrest Hosp, 64-75; chmn med adv bd, Toronto Rehab Ctr, 66-; mem div rehab, Coun Health, Prov Ont, 69- Mem: Can Asn Phys Med & Rehab (pres, 65). Res: Electrodiagnosis; electromyegraphy. Mailing Add: Dept of Rehab Med Toronto Western Hosp Toronto ON Can

CRAWFORD, MARIA LUISA BUSE b Beverly, Mass, July 18, 39; m 63. PETROLOGY, MINERALOGY. Educ: Bryn Mawr Col, BA, 60; Univ Calif, Berkeley, PhD(geol), 65. Prof Exp: Asst prof, 65-73, ASSOC PROF GEOL, BRYN MAWR COL, 73- Mem: Geol Soc Am; Norweg Geol Soc; Am Geophys Union; Microbeam Anal Soc. Res: Petrology, mineralogy and geochemistry of metamorphic and igneous rocks; petrologic study of lunar samples. Mailing Add: Dept of Geol Bryn Mawr Col Bryn Mawr PA 19010

CRAWFORD, MARVIN PATRICK, cardiovascular physiology, see 12th edition

CRAWFORD, MICHAEL H, b Shanghai, China, July 25, 39; US citizen; m 61; c 1. ANTHROPOLOGY. Educ: Univ Wash, BA, 60, MA, 65, PhD(anthrop), 67. Prof Exp: Res asst surg, Univ Wash, 61-63, asst biol anthrop, 63-65, mem res staff primate genetics, Regional Primate Ctr, 65-67; asst prof anthrop, Univ Pittsburgh, 67-71; ASSOC PROF ANTHROP, UNIV KANS, 71- Concurrent Pos: Assoc ed, J Ethnol, 67-69; Wenner-Gren Found grant study pop genetics, Mex, 69, grant study Irish tinkers, 70; US Off Educ grant study nutrit in Afro-Am community, 72; consult, Pastoral Inst, Pittsburgh, 72, forensic med, Kans Bur Invest, 73 & assoc prog, Am Anthrop Asn, 73. Mem: Am Anthrop Asn; Am Asn Phys Anthropologists; Brit Soc Study Human Biol; Int Asn Human Biol; Am Soc Human Genetics. Res: Anthropological genetics of Mexican transplanted populations; genetics of mental retardation; primate biochemical genetics. Mailing Add: Dept of Anthrop Univ of Kans Lawrence KS 66044

CRAWFORD, MORRIS LEE JACKSON, b Ellijay, Ga, Jan 20, 33; m 60; c 1. PHYSIOLOGICAL PSYCHOLOGY. Educ: Univ Ga, BS, 59, MS, 60, PhD(psychol), 62. Prof Exp: Asst prof res & psychol, Univ Ga, 62-63; Nat Inst Neurol Dis & Blindness fel, 63-64; instr res, Sch Med, Univ Miss, 65; asst prof psychol, Col Med, Baylor Univ, 66-70; ASSOC PROF NEURAL SCI, UNIV TEX HEALTH CTR HOUSTON, 70- Mem: Am Psychol Asn. Res: Behavioral science; central nervous system control of behavior; visual system encoding. Mailing Add: Grad Sch Biomed Sci Univ Tex Health Ctr 6414 Fannin Houston TX 77025

CRAWFORD, OAKLEY H, b Bridgeton, NJ, Sept 29, 38; m 63; c 2. PHYSICAL CHEMISTRY. Educ: Carson-Newman Col, BS, 59; Univ Ill, PhD(phys chem), 66. Prof Exp: Sr res fel appl math, Queen's Univ, Belfast, 66-67; asst prof chem, Pa State Univ, 67-74; ASSOC PROF CHEM, BARNARD COL, COLUMBIA UNIV, 74- Mem: AAAS; Am Phys Soc. Res: Reaction dynamics in crossed molecular beams; electron-molecule scattering; calculation of chemical reaction rates from first principles; molecule scattering and intramolecule forces. Mailing Add: Dept of Chem Barnard Col Columbia Univ New York NY 10027

CRAWFORD, PAUL VINCENT, b Concord, NH, Jan 9, 33; m 65; c 2. CARTOGRAPHY, PHYSICAL GEOGRAPHY. Educ: Univ Okla, BA, 56, MA, 58; Univ Kans, PhD(geog), 69. Prof Exp: Instr geog, RI Col, 60-65; asst prof, 69-71, ASSOC PROF GEOG, BOWLING GREEN STATE UNIV, 71- Mem: Am Geog Soc; Asn Am Geog; Am Cong Surv & Mapping. Res: Cartographic perception; three dimensional mapping; interrelationships of soil genesis and geomorphic processes. Mailing Add: Dept of Geog Bowling Green State Univ Bowling Green OH 43403

CRAWFORD, RAYMOND BARTLETT, organic chemistry, see 12th edition

CRAWFORD, RAYMOND BERTRAM, b Chipman, NB, Jan 10, 20; US citizen; m 49; c 2. MEDICINE. Educ: Andrews Univ, BA, 45; Loma Linda Univ, MD, 49. Prof Exp: Resident internal med, White Mem Hosp, Calif, 49-52; asst prof med, 54-62, dir cardiac diag lab, 62-69, ASSOC PROF MED, LOMA LINDA UNIV, 62- Concurrent Pos: Nat Heart Inst fel, 60-61. Mem: Am Col Cardiol; Am Col Physicians. Res: Cardiovascular dynamics. Mailing Add: Loma Linda Univ Sch of Med Loma Linda CA 92354

CRAWFORD, RICHARD BRADWAY, b Kalamazoo, Mich, Feb 16, 33; m 54; c 4. BIOCHEMISTRY. Educ: Kalamazoo Col, AB, 54; Univ Rochester, PhD(biochem), 59. Prof Exp: Fel biochem, Univ Rochester, 59; instr microbiol, Sch Dent, Univ Pa, 59-60, assoc, 60-61, from asst prof to assoc prof microbiol, 61-67, instr biochem, Sch Med, 59-61, assoc, 61-67; assoc prof, 67-74, PROF BIOL, TRINITY COL, CONN, 74- Mem: AAAS; Am Chem Soc; Am Soc Zoologists. Res: Role of lipids in bioenergetics; biochemistry of fertilization and embryogenesis; xenobiotic effects on development. Mailing Add: Dept of Biol Trinity Col Hartford CT 06106

CRAWFORD, RICHARD DWIGHT, b Kirksville, Mo, Nov 16, 47; m 66; c 1. WILDLIFE ECOLOGY. Educ: Northeast Mo State Univ, BS, 68, MS, 69; Iowa State Univ, PhD(wildlife biol), 75. Prof Exp: Sec sch teacher math-biol, Adair County RII Sch Dist, 68-69; instr wildlife biol, Iowa State Univ, 73-75; ASST PROF WILDLIFE COL, UNIV NDAK, 75- Mem: Sigma Xi; Wildlife Soc; Am Ornithologists Union; Wilson Ornith Soc; Cooper Ornith Soc. Res: Population ecology; habitat management and conservation of waterfowl; upland game and nongame birds. Mailing Add: Dept of Biol Univ of NDak Grand Forks ND 58201

CRAWFORD, ROBERT FIELD, b Martinez, Calif, Feb 18, 30; m 55; c 2. AGRONOMY, SOIL SCIENCE. Educ: Calif State Polytech Col, BS, 56; Cornell Univ, MS, 58, PhD(agron), 60. Prof Exp: Lab asst, Union Oil Co, 52; agr res scientist, US Borax Res Corp, 60-64, mgr agr res & develop dept, 64-67; ASSOC PROF CHEM & DEAN COL, BIOLA COL, 67- Mem: Am Soc Agron; Weed Sci Soc Am. Res: Crop production and physiology; plant-animal relations; chemistry. Mailing Add: Biola Col La Mirada CA 90638

CRAWFORD, ROBERT JAMES, b Edmonton, Alta, July 8, 29; m 56; c 4. ORGANIC CHEMISTRY. Educ: Univ Alta, BSc, 52, MSc, 54; Univ Ill, PhD, 56. Prof Exp: From asst prof to assoc prof, 56-67, PROF CHEM, UNIV ALTA, 67- Mem: Am Chem Soc; Chem Inst Can; The Chem Soc. Res: Reaction mechanisms; azo and diazo chemistry; racemization of cycloalkanes and epoxides. Mailing Add: Dept of Chem Univ of Alta Edmonton AB Can

CRAWFORD, ROGER ALLEN, physical chemistry, see 12th edition

CRAWFORD, ROGER JAMES, JR, geography, see 12th edition

CRAWFORD, RONALD LYLE, b Santa Ana, Tex, Sept 28, 47; m 67; c 1. MICROBIOLOGY, ECOLOGY. Educ: Okla City Univ, BA, 70; Univ Wis, MS, 72, PhD(bact), 73. Prof Exp: Assoc, Univ Minn, St Paul, 73-74; res scientist, Div Labs & Res, NY State Health Dept, 74-75; ASST PROF MICROBIOL, FRESHWATER BIOL INST, UNIV MINN, TWIN CITIES, 75- Mem: Am Soc Microbiol. Res: Degradation of aromatic compounds in natural environments; microbial ecology. Mailing Add: Freshwater Biol Inst PO Box 100 Navarre MN 55392

CRAWFORD, RONALD WARD, b Sheldon, Iowa, July 2, 23; m 46; c 1. ICHTHYOLOGY, LIMNOLOGY. Educ: San Diego State Col, AB, 48; Cornell Univ, PhD(vert zool), 53. Prof Exp: Asst instr zool, San Diego State Col, 48-50, from instr to prof, 53-75, chmn dept, 60-64. Concurrent Pos: Res assoc, Cornell Univ, 51-53; fishery consult, 60- Mem: Am Soc Ichthyol & Herpet; Am Fisheries Soc; Am Inst Fishery Res Biologists. Res: Biology and systematics of North American fishes. Mailing Add: Dept of Zool San Diego State Col San Diego CA 92115

CRAWFORD, ROY DOUGLAS, b Vancouver, BC, June 6, 33. POULTRY GENETICS. Educ: Univ Sask, BSA, 55; Cornell Univ, MS, 57; Univ Mass, PhD(poultry genetics), 63. Prof Exp: Poultry geneticist, Res Br, Can Dept Agr, Ottawa, 57-58, Charlottetown, 58-63 & Kentville, 63-64; from asst prof to assoc prof, 64-74, PROF POULTRY GENETICS, UNIV SASK, 74- Mem: Poultry Sci Asn; Am Genetic Asn; Agr Inst Can; World Poultry Sci Asn. Res: Physiological and behavioral genetics of domestic fowl; effects of environment and management on domestic fowl. Mailing Add: Dept of Poultry Sci Univ of Sask Saskatoon SK Can

CRAWFORD, ROY KENT, b Wilmington, NC, Oct 21, 41. MATERIALS SCIENCE, FLUID PHYSICS. Educ: Kans State Univ, BS, 63; Princeton Univ, PhD(physics), 68. Prof Exp: Sloan fel, Princeton Univ, 68-69, res assoc mat sci, 69-70, mem res staff & lectr, 70-72; ASST PROF PHYSICS, UNIV ILL, URBANA, 72- Mem: Am Phys Soc. Res: Application of high pressure as a tool for research into the equilibrium and dynamical properties of fluids and solids; study of melting at high pressures. Mailing Add: Dept of Physics 372 Physics Bldg Univ of Ill Urbana IL 61801

CRAWFORD, STANLEY EVERETT, b Dallas, Tex, Nov 9, 24; m 48; c 3. PEDIATRICS, MEDICAL ADMINISTRATION. Educ: Univ Tex, BA, 45, MD, 48; Am Bd Pediat, dipl, 54. Prof Exp: Intern, Univ Chicago Clins & Albert Merritt Billings Hosp, 48-49; resident, Univ Tex, Galveston, 50-51; resident child psychiat, Univ Minn Hosps, 52-53; pvt pract, Children's Clin, Jackson, Tenn, 54-61; asst prof pediat, LeBonheur Children's Hosp, Col Med, Univ Tenn, 63-68; prof pediat & chmn dept, 68-73, DEAN, UNIV TEX MED SCH, SAN ANTONIO, 73- Concurrent Pos: Mem staff, Jackson-Madison County Gen Hosp, Tenn, 54-61; mem adv bd, Educ Film Prod, 67; off examr, Am Bd Pediat, 67-, mem residency rev comt, 70; mem exec comt, Children's Hosp Found, San Antonio, 68, bd dirs, 68-; pediatrician-in-chief, Bexar County Hosp Dist Teaching Hosps, 68-73; civilian regional consult, Wilford Hall, US Air Force Hosp, 69-; mem vis staff, Brooke Army Hosp, 70; mem adv comt, Foster Grandparent Proj, 70; mem, Gov Conf Children & Youth, Austin, Tex, 70; mem, Abraham Jacobi Award Nominating Comt, 70-73; mem adv bd, Child Develop Ctr, San Antonio, 71. Mem: AMA; fel Am Acad Pediat; Asn Med Schs; Am Soc Pediat Nephrology. Res: Clinical pediatrics. Mailing Add: Dept of Pediat Univ of Tex Med Sch San Antonio TX 78284

CRAWFORD, SUSAN N, b Vancouver, BC, May 11, 27; m 56; c 1. INFORMATION SCIENCE. Educ: Univ BC, BA, 48; Univ Toronto, BLS, 50; Univ Chicago, MA, 56, PhD(info sci), 70. Prof Exp: Res assoc, 56-59, DIR INFO SCI, AM MED ASN, 60- Concurrent Pos: Prin investr, Nat Libr Med-NIH grant, 68-; chmn, Comn Surv & Statist, Med Libr Asn, 67-75; assoc prof, Columbia Univ, 71-75; mem bd regents, US Nat Libr Med, 71-75. Honors & Awards: Cert Achievement, USPHS, 75. Mem: AAAS; Am Soc Info Sci; Med Libr Asn; Am Libr Asn; Spec Libr Asn. Res: Social organization of scientists in communication; statistical survey of health sciences libraries in the United States. Mailing Add: 2418 Lincoln St Evanston IL 60201

CRAWFORD, THOMAS CHARLES, b Muskegon, Mich, July 31, 45; m 72. ORGANIC CHEMISTRY. Educ: Kalamazoo Col, BA, 67; Univ Calif, Los Angeles, MS, 69, PhD(chem), 74. Prof Exp: RESEARCHER SYNTHETIC ORG CHEM, PFIZER, INC, 74- Mem: Am Chem Soc; The Chem Soc; Sigma Xi. Res: Synthesis of biologically important molecules, development of new synthetic methods and utilization of organometallics in synthesis. Mailing Add: Cent Res Pfizer Inc Groton CT 06340

CRAWFORD, THOMAS H, b Oct 22, 31; US citizen; m 54; c 2. INORGANIC CHEMISTRY. Educ: Univ Louisville, BS, 58, PhD(phys chem), 61. Prof Exp: From asst prof to assoc prof, 61-70, chmn dept, 71-75, actg dean Col Arts & Sci, 73-74, PROF CHEM, UNIV LOUISVILLE, 70-, FAC ASSOC, OFF OF PRES, 75- Concurrent Pos: Vis assoc prof, Calif Inst Technol, 68-69. Mem: AAAS; Am Chem Soc. Res: Preparation and characterization of transition metal coordination compounds and the study of molecular complexes of antimony trichloride and organic substrates. Mailing Add: Off of Pres Univ of Louisville Louisville KY 40208

CRAWFORD, THOMAS MICHAEL, b Cleveland, Ohio, Aug 13, 28; m 51; c 3. FOOD TECHNOLOGY. Educ: Ohio State Univ, BA, 50, MSc, 54, PhD(food tech), 57. Prof Exp: Technician, Cleveland Clin Found, 50; chemist, Walter Reed Army Med Serv Grad Sch, 53; asst, Ohio Agr Exp Sta, 54-57; food technologist, Res & Develop Div, Nat Dairy Prod Corp, 57-62; sect chief res & develop ctr, Pet Milk Co, Ill, 62-66, group mgr, 66-68; tech mgr, 68-71, SR SCIENTIST, FOOD TECHNOL GROUP, RES & DEVELOP, PILLSBURY CO, MINNEAPOLIS, 71- Mem: Inst Food Technol; Am Soc Hort Sci; Am Asn Cereal Chem. Res: Research and product development in the areas of refrigerated, frozen and shelf cereal, fruit and vegetable food products. Mailing Add: 213 Spring Valley Dr Bloomington MN 55420

CRAWFORD, TIMOTHY B, b Shreveport, La, July 31, 42; m 60; c 2. VETERINARY PATHOLOGY. Educ: Tex A&M Univ, BS, 65, DVM, 66; Wash State Univ, PhD, 73. Prof Exp: NIH fel vet path, 66-69; ASST PROF VET PATH, WASH STATE UNIV, 69- Concurrent Pos: NIH res grant, 69-72. Res: Mechanisms of immunological diseases, particularly those which involve viruses; mechanisms of peripheral vascular disease. Mailing Add: Dept of Vet Path Wash State Univ Pullman WA 99163

CRAWFORD, TODD V, b Los Angeles, Calif, Aug 9, 31; m 59; c 3. METEOROLOGY. Educ: Calif Polytech State Col, BS, 53; Univ Calif, Los Angeles, MA, 58, PhD(meteorol), 65. Prof Exp: Res asst numerical meteorol, Univ Calif, Los Angeles, 57-58, asst agr engr & lectr, Davis, 58-65, physicist, Lawrence Livermore Lab, 65-72; RES MGR, SAVANNAH RIVER LAB, E I DU PONT DE NEMOURS & CO, 72- Mem: Am Meteorol Soc; Am Soc Agr Eng; Am Geophys Union; Royal Meteorol Soc. Res: Energy balance, turbulent properties of the lower atmosphere and atmospheric diffusion from small scale to continental scale; managing a multi disciplined environmental research group. Mailing Add: Savannah River Lab E I du Pont de Nemours & Co Aiken SC 29801

CRAWFORD, VERNON, b Amherst, NS, Feb 13, 19; nat US; m 43; c 2. PHYSICS. Educ: Mt Allison Univ, BA, 39; Dalhousie Univ, MSc, 44; Univ Va, PhD(physics), 49. Prof Exp: Lectr physics, Dalhousie Univ, 44-47; assoc prof, 49-56, dir physics, 64-68, PROF PHYSICS, GA INST TECHNOL, 56-, VPRES ACAD AFFAIRS, 68- Res: Optics; electromagnetics. Mailing Add: Ga Inst of Technol Atlanta GA 30332

CRAWFORD, WHEELER CONRAD, organic chemistry, see 12th edition

CRAWFORD, WILLIAM ARTHUR, b Norman, Okla, Mar 25, 35; m 63. GEOCHEMISTRY. Educ: Kans State Univ, BS, 57; Univ Kans, MS, 60; Univ Calif, Berkeley, PhD(geol, geochem), 65. Prof Exp: Teaching asst geol, Univ Kans, 59-60; instr geol, Univ Calif, Berkeley, 65; asst prof, 65-73, ASSOC PROF GEOL, BRYN MAWR COL, 73- Mem: Mineral Soc Am; Nat Asn Geol Teachers; fel Geol Soc Am; Geochem Soc. Res: High pressure-high temperature experimental petrology; chemical analysis of rocks and minerals; studying field relations and mapping of igneous and metamorphic rock bodies. Mailing Add: Dept of Geol Bryn Mawr Col Bryn Mawr PA 19010

CRAWFORD, WILLIAM HOWARD, JR, b Montclair, NJ, Apr 14, 37; m 58; c 2. PATHOLOGY, ORAL PATHOLOGY. Educ: Univ Southern Calif, BA, 58, DDS, 62, MS, 64. Prof Exp: Asst prof, 66-68, asst dean, 69-71, ASSOC PROF PATH & ORAL PATH, SCH DENT, UNIV SOUTHERN CALIF, 68-, ASSOC DEAN ACAD AFFAIRS, 71-, INTERIM DEAN SCH DENT, 72- Concurrent Pos: Nat Inst Dent Res res fel, 62-64. Mem: AAAS; fel Am Acad Oral Path; Am Dent Asn; Am Asn Dent Schs. Res: Detailed histology and pathogenesis of keratinizing cysts of the oral cavity. Mailing Add: Sch of Dent Univ of Southern Calif Los Angeles CA 90007

CRAWFORD, WILLIAM STANLEY HAYES, b St John, NB, Apr 17, 18; m 43; c 2. MATHEMATICS. Educ: Mt Allison Univ, BA, 39; Univ Minn, MA, 42, PhD(math), 50. Prof Exp: Instr math, Univ Minn, 42-43; from asst prof to prof & head dept, 43-73, dean sci, 56-62, dean fac, 62-65, vpres fac, 62-69, OBED EDMUND SMITH PROF MATH, MT ALLISON UNIV, 73- Concurrent Pos: Mem, NB Higher Educ Comn, 67- Mem: AAAS; Am Math Soc; Math Asn Am; Can Math Cong (vpres, 67-69). Res: Analysis; integration in function space; history of mathematics. Mailing Add: Dept of Math Mt Allison Univ Sackville NB Can

CRAWHALL, JOHN C, b Harrow, Eng, Oct 15, 28; m 56; c 3. BIOCHEMISTRY, MEDICINE. Educ: Univ London, BSc, 48, PhD(org chem), 52, MBBS, 61, MD, 67. Prof Exp: Res scientist, Nat Inst Med Res, London, 49-53; lectr biochem, St Bartholomew's Hosp, 54-58, lectr med, 63-68; DIR CLIN BIOCHEM, ROYAL VICTORIA HOSP, 68- Concurrent Pos: Vis scientist, Nat Inst Arthritis & Metab Dis, 64-66; assoc prof, McGill Univ, 68-; vis prof pediat, Univ Calif, San Diego, 73-74. Mem: Brit Biochem Soc. Res: Biochemistry of amino acids; inherited metabolic disorders, particularly those related to sulfur amino acid metabolism; application of mass spectrometry to structural determinations of natural products. Mailing Add: Div of Clin Biochem Royal Victoria Hosp Montreal PQ Can

CRAWLEY, GERARD MARCUS, b Airdrie, Scotland, Apr 10, 38; m 61; c 4. NUCLEAR PHYSICS. Educ: Univ Melbourne, BSc, 59, MSc, 61; Princeton Univ, PhD(physics), 65. Prof Exp: Res assoc, Cyclotron Lab, Mich State Univ, 65-66; Queen Elizabeth fel, Dept Nuclear Physics, Australian Nat Univ, 66-68; from asst prof to assoc prof, 68-74, PROF PHYSICS, MICH STATE UNIV, 74- Concurrent Pos: Fulbright scholar & Ford Int fel, 61; vis fel, Australian Nat Univ, 74-75; prog officer nuclear physics, NSF, 75- Mem: Am Phys Soc. Res: Nuclear reactions, particularly inelastic scattering and multinuclear transfer reactions. Mailing Add: Cyclotron Lab Mich State Univ East Lansing MI 48824

CRAWLEY, PETER L, mathematics, see 12th edition

CRAWSHAW, LARRY INGRAM, b Los Angeles, Calif, Nov 5, 42; m 68; c 1. COMPARATIVE PHYSIOLOGY. Educ: Univ Calif, Los Angeles, BA, 64; Univ Calif, Santa Barbara, PhD(physiol psychol), 70. Prof Exp: Res asst psychol, Univ Calif, Los Angeles, 63-64; res & teaching asst, Univ Calif, Santa Barbara, 68-70, NSF res assoc, 68-70; NSF res assoc physiol, Scripps Inst Oceanog, Univ Calif, San Diego, 70-71, NIMH fel, 71-72; ASST FEL, JOHN B PIERCE FOUND, 72-; ASST PROF, SCH MED, YALE UNIV, 73- Concurrent Pos: NSF res grant, 74. Mem: Am Physiol Soc; AAAS; Am Psychol Asn; Psychonomic Soc. Res: Temperature regulation in vertebrates. Mailing Add: John B Pierce Found 290 Congress Ave New Haven CT 06519

CRAYTHORNE, N W BRAIN, b Belfast, Northern Ireland, Jan 1, 31; m 57; c 2. ANESTHESIOLOGY. Educ: Queen's Univ, Belfast, MB, BCh, 54; Am Bd Anesthesiol, dipl, 63. Prof Exp: Asst instr anesthesiol, Univ Pa, 57-58, instr, 58-59; clin asst, Royal Victoria Hosp, 59-60; consult anesthetist, Queen Mary Vet Hosp, 60-61; prof anesthesiol & chmn dept, WVa Univ Med Ctr, 61-70; prof anaesthesia & chmn dept, Univ Cincinnati Med Ctr, 70-74; PROF ANESTHESIOL & CHMN DEPT, SCH MED, UNIV MIAMI, 75- Concurrent Pos: Demonstr anesthesiol, McGill Univ, 60-61; anesthetist-in-chief, Cincinnati Gen Hosp, 70-75. Mem: Am Soc Anesthesiologists; Int Anesthesia Res Soc; Am Med Asn; Fel Am Col Anesthesiologists. Mailing Add: Dept of Anesthesiol Sch of Med Univ of Miami Biscayne Annex Miami FL 33152

CRAYTON, PHILIP HASTINGS, b Seneca Falls, NY, Jan 22, 28; m 51; c 2. INORGANIC CHEMISTRY. Educ: Alfred Univ, BA, 49; Univ Buffalo, MA, 51, PhD(chem), 56. Prof Exp: Res chemist, Metals Res Labs, Union Carbide Metals Co, 52-59, sect leader, 59-61; sr res chemist res & develop div, Carborundum Co, 61-63; from asst prof to assoc prof, 63-74, PROF CHEM & HEAD DIV ENG & SCI, NY STATE COL CERAMICS, ALFRED UNIV, 74- Concurrent Pos: Consult, Carborundum Co, 63-66. Mem: AAAS; Am Chem Soc. Res: Synthesis and properties of the transition metal carbides, borides and nitrides; description and mechanism of ceramic hot pressing; chemistry of process metallurgy; synthesis of oxides, carbides, nitrides and borides. Mailing Add: Div of Eng & Sci NY State Col Ceramic Alfred Univ Alfred NY 14802

CREAGAN, ROBERT JOSEPH, b Rockford, Ill, Aug 24, 19; m 48; c 4. APPLIED PHYSICS. Educ: Ill Inst Technol, BS, 42; Yale Univ, MS, 43, PhD(physics), 49. Prof Exp: Physicist, Argonne Nat Lab, 46-47; physicist, Atomic Power Div, Westinghouse Elec Co, 49-57; dir nuclear prog, Bendix Aviation Corp, 57-60; eng mgr, Atomic Power Div, 60-69, proj mgr liquid metal fast breeder reactor, Adv Reactor Div, 69-75, DIR TECHNOL ASSESSMENT, WESTINGHOUSE ELEC CO, 75- Mem: Am Phys Soc; fel Am Nuclear Soc. Res: Nuclear power reactors; instrumentation. Mailing Add: 2305 Haymaker Rd Monroeville PA 15146

CREAGER, CHARLES BICKNELL, b Bicknell, Ind, Oct 5, 24; m 51; c 3. NUCLEAR PHYSICS. Educ: Western Reserve Univ, BS, 51, MS, 53; Ind Univ, PhD, 59. Prof Exp: From asst prof to assoc prof physics, Kans Wesleyan Univ, 53-69, prof physics & fac dir res & grants, 69-71; PROF PHYSICS & CHMN DIV PHYS SCI, EMPORIA KANS STATE COL, 71- Concurrent Pos: Res assoc, Ind Univ, 56-59. Mem: AAAS; Am Phys Soc; Am Asn Physics Teachers. Res: Beta-gamma ray spectroscopy. Mailing Add: Div of Phys Sci Emporia Kans State Col Emporia KS 66801

CREAGER, JOAN GUYNN, b Austin, Ind, Dec 8, 32; m 52; c 4. EMBRYOLOGY. Educ: Trinity Univ, Tex, BS, 55, MS, 58; George Washington Univ, PhD(zool), 64. Prof Exp: Part-time res analyst, Bionetics Res Labs, Va, 63-67; res assoc, off sci personnel, Nat Acad Sci, Washington, DC, 67-69; ASSOC PROF BIOL, NORTHERN VA COMMUNITY COL, EASTERN CAMPUS, 69-72, 74- Concurrent Pos: Staff biologist, Comn Undergrad Educ Biol Sci, 69; consult, AAAS, 70-; ed, Am Biol Teacher, 74- Mem: AAAS; Nat Asn Biol Teachers; Nat Sci Teachers Asn. Res: Automated processing and analysis of biomedical research data; personnel research, especially two-year college biologists; curriculum development in undergraduate education in biology. Mailing Add: 1101 N Potomac St Arlington VA 22205

CREAGER, JOE SCOTT, b Vernon, Tex, Aug 30, 29; m 51; c 2. GEOLOGICAL OCEANOGRAPHY. Educ: Colo Col, BS, 51; Agr & Mech Col Tex, MS, 53, PhD(oceanog), 58. Prof Exp: Asst geol & eng oceanog, Agr & Mech Col Tex, 51-52; phys sci asst, US Army Beach Erosion Bd, 53-55; from asst prof to assoc prof geol oceanog, 58-66, PROF GEOL OCEANOG & ASSOC DEAN COL ARTS & SCI, UNIV WASH, 66- Concurrent Pos: Asst geol & eng oceanog, Agr & Mech Col, Tex, 55-58; prog dir oceanog, NSF, 65-66. Mem: Geol Soc Am; Soc Econ Paleont & Mineral; Am Asn Quaternary Environ; Am Geophys Union; Int Asn Sedimentol. Res: Submarine geology relating to continental shelves, slopes and shorelines; sedimentation and micropaleontology; bottom sediment transport. Mailing Add: Dept of Oceanog Univ of Wash Seattle WA 98105

CREAGH, LINDA TRUITT, b Denton, Tex, May 25, 41; m 61; c 2. PHYSICAL ORGANIC CHEMISTRY. Educ: NTex State Univ, BS, 62, MS, 63, PhD(chem), 67. Prof Exp: Res technician org synthesis, NTex State Univ, 58-63, teaching fel, 62-66; mem tech staff, cent res & eng, Tex Instruments Inc, 66-68; asst prof chem, Tex Woman's Univ, 68-69; mem staff res & develop, Tex Instruments Inc, 69-73; sr scientist, 73-75, MGR, XEROX INC, 75- Mem: AAAS; Am Inst Chemists; Am

Chem Soc; Sigma Xi. Res: Laser development and applications; photochemistry; liquid crystals chemistry, characterization for displays; development and evaluation of materials for marking technologies. Mailing Add: Xerox Inc 1341 W Mockingbird Dallas TX 75247

CREAMER, GEORGE BERNARD, organic chemistry, see 12th edition

CREAMER, ROBERT M, b Baltimore, Md, Nov 17, 17; m 46; c 2. ANALYTICAL CHEMISTRY, PHYSICAL ORGANIC CHEMISTRY. Educ: Univ Md, BS, 38, PhD(inorg chem), 52. Prof Exp: Chemist, Celanese Corp Am, 38-45; res chemist, US Bur Mines, 51-54; res assoc, USDA, 54-60; from res chemist to sr res chemist, 60-68, SR SCIENTIST, PHILIP MORRIS RES CTR, 68- Concurrent Pos: Vis sr fel, City Univ, London, Eng, 73-74. Mem: Am Chem Soc; Am Inst Chem. Res: Electroplating; fertilizers and insecticides; razor blade lubricants; tobacco and smoke chemistry; aerosol science. Mailing Add: Philip Morris Res Ctr Box 26583 Richmond VA 23261

CREAN, JOSEPH GAYLORD, b Chicago, Ill, Oct 6, 19; m 41; c 2. PLANT PHYSIOLOGY, PLANT BIOCHEMISTRY. Educ: Chicago Teachers Col, BEd, 42; De Paul Univ, MS, 49; Univ Chicago, PhD(bot), 66. Prof Exp: Teacher, Northwestern Mil Acad, 45-48; teacher pub schs, Ill, 48-56, Wilson Br, Chicago Jr Col Syst, 56-62 & Loop Br, 62-66; PROF BIOL, NORTHEASTERN ILL UNIV, 66- Mem: AAAS; Am Inst Biol Sci; Bot Soc Am; Phycol Soc Am. Res: Systematics, biochemical genetics and elucidation of synthetic pathways of enzymes and pigments; general physiological study of the Lemnaceae; research in water relations and translocation. Mailing Add: 10331 S Avers Ave Chicago IL 60655

CREAN, PATRICK J, b Brooklyn, NY, July 15, 37; m 60; c 3. ORGANIC CHEMISTRY. Educ: State Univ NY Stony Brook, BS, 61; Purdue Univ, PhD(org chem), 66. Prof Exp: CHEMIST, JACKSON LAB, E I DU PONT DE NEMOURS & CO, 66- Mem: Am Chem Soc. Res: Mechanistic organic and petroleum chemistry. Mailing Add: E I du Pont de Nemours & Co Org Chem Dept Jackson Lab Wilmington DE 19898

CREANGE, JOHN ELLYSON, b Amityville, NY, Feb 25, 35; m 61; c 2. ENDOCRINOLOGY. Educ: Univ Calif, Los Angeles, AB, 56, MA, 58, PhD(zool), 64. Prof Exp: Res biol chemist, Univ Calif, 64-66; from res biol chemist to assoc res biologist, 66-70, RES BIOLOGIST & GROUP LEADER, STERLING-WINTHROP RES INST, 70- Mem: Assoc Am Physiol Soc; Am Soc Zoologists; Endocrine Soc. Res: Control of endocrine disorders and reproductive processes in man. Mailing Add: Sterling-Winthrop Res Inst Rensselaer NY 12144

CREASEY, SAVILL CYRUS, b Portland, Ore, July 17, 17; m 43; c 2. GEOLOGY. Educ: Univ Calif, Los Angeles, AB, 39, AM, 41, PhD(geol), 49. Prof Exp: GEOLOGIST, US GEOL SURV, 41- Honors & Awards: Meritorious Serv Award, US Geol Surv, 71. Mem: Am Asn Petrol Geol; Geol Soc Am; Soc Econ Geol; Geochem Soc. Res: Geology of base metal deposits, especially porphyry coppers; structural geology; petrology of igneous and metamorphic rocks. Mailing Add: US Geol Surv 345 Middlefield Rd Menlo Park CA 94025

CREASEY, WILLIAM ALFRED, b London, Eng, May 12, 33; US citizen; m 57; c 1. BIOCHEMISTRY, PHARMACOLOGY. Educ: Oxford Univ, BA, 55, MA & DPhil(radiobiol), 59. Prof Exp: Asst tutor org chem, St Catherine's Col, Oxford Univ, 58-59; from res asst to asst prof, 61-68, ASSOC PROF PHARMACOL, YALE UNIV, 68- Concurrent Pos: USPHS fel, Yale Univ, 59-61; mem cancer clin invest rev comt, Nat Cancer Inst, 72-76; mem working cadre, Nat Bladder Cancer Proj, 74-78. Mem: AAAS; fel The Chem Soc; Brit Biochem Soc; Am Soc Biol Chem; Am Asn Cancer Res. Res: Biochemical effects of ionizing radiations; influence of dietary pyrimidines on pyrimidine and lipid metabolism; enzymatic studies with agents that influence nucleotide metabolism; metabolic studies with antineoplastic and antimitotic agents; studies with plant derivatives; clinical pharmacology. Mailing Add: Dept of Internal Med Yale Univ Sch of Med New Haven CT 06510

CREASY, LEROY L, b White Plains, NY, Feb 21, 38; m 60; c 2. PLANT PHYSIOLOGY. Educ: Cornell Univ, BS, 60, MS, 61; Univ Calif, Davis, PhD(plant physiol), 64. Prof Exp: NSF fel, Low Temperature Res Sta, Cambridge Univ, Eng, 64-65; asst prof, 65-69, ASSOC PROF POMOL, CORNELL UNIV, 69- Mem: Phytochem Soc NAm; Am Soc Hort Sci. Res: Physiology and biochemistry of secondary plant products derived from phenylalanine. Mailing Add: Dept of Pomol 117 Plant Sci Bldg Cornell Univ Ithaca NY 14850

CREAVEN, PATRICK JOSEPH, b London, Eng, Jan 31, 33; m 63; c 4. CLINICAL PHARMACOLOGY. Educ: Univ London, MB, BS, 56, PhD(biochem), 64. Prof Exp: Asst lectr biochem, St Mary's Hosp Med Sch, Univ London, 63-64, lectr, 64-66; chief biochem sect, Tex Res Inst Ment Sci, 66-69; chief oncol pharmacol, Med Oncol Br, Nat Cancer Inst, Vet Admin Hosp, 69-75; CHIEF CANCER RES CLINICIAN, DEPT EXP THERAPEUT & DEPT MED A, ROSWELL PARK MEM INST, 75- Concurrent Pos: Asst prof, Col Med, Baylor Univ, 66-69; asst prof, Grad Sch Biomed Sci, Univ Tex, 67-69, assoc prof, 69; assoc res prof, Grad Sch, State Univ NY Buffalo. Mem: Fel Royal Soc Health; Am Soc Pharmacol & Exp Therapeut; NY Acad Sci; Acad Pharm Sci; Am Asn Cancer Res. Res: Clinical pharmacology, pharmacokinetics, metabolism and biochemical pharmacology of anti-cancer agents; initial clinical testing o of anticancer agents; induction of drug metabolizing enzymes. Mailing Add: Grace Cancer Drug Ctr 666 Elm St Buffalo NY 14263

CRECELIUS, HARRY GILBERT, b Hope, NMex, Jan 5, 12; m 39; c 2. MEDICAL MICROBIOLOGY. Educ: Univ SDak, BA, 34, MA, 37; Yale Univ, PhD(bact), 41. Prof Exp: Instr bact, Univ NH, 41-42; lab officer, Camp Edwards, Mass, 42-44, chief bur labs, US Army Mil Govt, Korea, Dept Health & Welfare, 45-46; asst dir labs, State Bd Health, Kans, 46-48; dir labs, Ariz State Dept Health, 48-76; RETIRED. Concurrent Pos: Chmn, Conf State & Prov Pub Health Lab Dirs, 54. Mem: Am Soc Microbiol; Am Pub Health Asn. Mailing Add: 1830 W Coolidge Phoenix AZ 85015

CRECELIUS, ROBERT LEE, b Volin, SDak, Dec 8, 22; m 45; c 4. CHEMISTRY. Educ: Mont State Col, BSChem E, 47, MS, 49; Univ Wyo, PhD(chem), 54. Prof Exp: Chem engr, US Bur Mines, 49-52, 53-54; CHEMIST, SHELL DEVELOP CO, 54- Res: Catalytic polyforming of shale oil; para-Claisen rearrangement; high pressure hydrogenation of shale oil; oil reaction processes; hydrotreating; hydrocracking; catalytic cracking; catalysis; catalyst formulation and development. Mailing Add: 12527 Blackstone Ct Houston TX 77077

CRECELIUS, SAMUEL BROWN, b Louisville, Ky, Apr 12, 12; m 41; c 2. BIOCHEMISTRY. Educ: Univ Louisville, BS, 34; Georgetown Univ, MS, 54, PhD(biochem), 56. Prof Exp: Org res chemist, DeVoe Raynolds Co, 36-42, sr org chemist, 45-46; polymer chemist, Chem Div, Res Lab, US Navy, 43-45; head varnish & resin sect, Res & Develop Dept, S C Johnson & Sons, 46-49; develop chemist org coatings, Am Cyanamid Co, 49-50; head resins sect, Chem Div, Naval Res Lab, 50-56; head org group, Res & Develop Div, Econ Lab, Inc, 56-59; dir develop, Ariz Chem Co, 59-60; res mgr, Econ Lab, Inc, 60-62, asst dir, Res & Develop Dept, 63-70, asst vpres corp technol, 70-71, CONSULT, ECON LAB, INC, 71- Mem: Am Chem Soc; Am Oil Chemists' Soc. Res: Basic organic chemistry of drying oils; coatings; polymers; biochemistry of bacteria and fungi; enzymes; electrophoresis; organic surfactants. Mailing Add: 8837 N Lagoon Dr Panama City FL 32401

CRECELY, ROGER WILLIAM, b Rochester, NY, Jan 12, 42; m 67. ANALYTICAL CHEMISTRY. Educ: Univ Rochester, BS, 64; Emory Univ, PhD(chem), 69. Prof Exp: LAB DIR, BRANDYWINE RES LAB, INC, 70- Mem: Am Chem Soc; Coblentz Soc. Res: Proton and carbon nuclear magnetic resonance studies; high pressure liquid chromatography; infrared spectroscopic analysis of polymers. Mailing Add: PO Box 89 Claymont DE 19703

CREDE, ROBERT H, b Chicago, Ill, Aug 11, 15; m 47; c 3. INTERNAL MEDICINE. Educ: Univ Calif, Berkeley, AB, 37; Univ Calif, San Francisco, MD, 41. Prof Exp: Commonwealth fel med & instr med, Col Med, Univ Cincinnati, 47-49; from asst prof to assoc prof, 49-60, asst dean sch med, 56-60, PROF MED, SCH MED, UNIV CALIF, SAN FRANCISCO, 60-, CHMN DIV AMBULATORY & COMMUNITY MED, 67- Mem: Am Psychosom Soc; Am Fedn Clin Res; AMA; Asn Teachers Prev Med; Am Pub Health Asn. Res: Psychosomatic medicine; delivery of health services; community medicine. Mailing Add: Div Ambulatory & Community Med Univ of Calif Med Ctr San Francisco CA 94143

CREDITOR, MORTON C, b Brooklyn, NY, Oct 4, 23; c 4. MEDICAL EDUCATION. Educ: Purdue Univ, BS, 42; Columbia Univ, MD, 47. Prof Exp: Assoc prof med, Sch Med, Univ Kans & dir med educ, Menorah Med Ctr, 57-61; assoc prof med, Med Sch, Univ Chicago, 61-67, dir prof affairs, Michael Reese Hosp & Med Ctr, 61-69, assoc dir community & social med, Hosp, 69-70; exec dir, Ill Regional Med Prog, 70-74; PROF MED & ASSOC DEAN SCH BASIC MED SCI, UNIV ILL, URBANA-CHAMPAIGN, 74- Mem: Am Fedn Clin Res; fel Am Col Physicians. Res: Renal physiology and disease; humeral factors in human hypertension. Mailing Add: Sch of Basic Med Sci Univ of Ill Med Sci Bldg Urbana IL 61801

CREE, ALLAN, b Congress, Ariz, July 10, 19; m 37; c 2. PETROLEUM GEOLOGY. Educ: Northern Ariz Univ, AB, 33; Ohio Univ, Athens, MA, 35; Univ Colo, PhD(geol), 48. Prof Exp: Tech asst, Lowell Observ, 28-33; teacher high sch, Ariz, 35-36; supv critic phys sci & math, Ohio Univ, Athens, 36-41; asst geol, Univ Colo, 41-43; geologist, Shell Oil Co, Inc, Wyo, 43-46; asst prof geol, Univ Nev, 46-48; geologist, Cities Serv Oil Co, 48-50, dist geologist, 50-53, actg div geologist, 53-54, chief geologist, Cities Serv Petrol, Inc, 54-59, Cities Serv Co, 59-63 & Int Cities Serv Oil Co, 63-67, mgr explor, Int Div, Cities Serv Oil Co, 67-70, explor coordr, Cities Serv Int, 70-72; PETROL CONSULT, ULSTER PETROLS LTD, 72-, MEM BD DIRS, 74- Mem: Am Asn Petrol Geol; Am Inst Prof Geol. Res: Subsurface geology; structural and stratigraphic geology and photogeologic mapping; petrographic and petrologic study of igneous rocks; metasomatism and replacement phenomena resulting in igneous rocks; exploration for petroleum on concessions in the Middle East, Europe, Africa, South America, Indonesia, Canada, Australia, Southeast Asia and India. Mailing Add: Box 3197 West Sedona AZ 86340

CREECH, HENRY BRYANT, b Roanoke Rapids, NC, Nov 20, 31; m 54; c 3. AUDIOLOGY, SPEECH PATHOLOGY. Educ: ECarolina Col, BS, 58; Ohio State Univ, MA, 60, PhD(speech & hearing sci), 62. Prof Exp: Assoc prof audiol, Univ Southern Miss, 62-65; CHIEF AUDIOL-SPEECH PATH, VET ADMIN HOSP & CLIN ASSOC OTOLARYNGOL, MED COL VA, 65- Concurrent Pos: Consult, Indust Hearing Conserv Prog, 63-; Am Speech & Hearing Asn int travel grant, Int Cong Audiol, 64. Mem: Am Speech & Hearing Asn. Res: Clinical behavior during the process of audition; communication problems after major maxillofacial surgery. Mailing Add: 9 Swanage Rd Richmond VA 23235

CREECH, HUGH JOHN, b Exeter, Ont, June 27, 10; nat US; m 37; c 2. IMMUNOCHEMISTRY, CANCER. Educ: Univ Western Ont, BA, 33, MA, 35; Univ Toronto, PhD(biochem), 38. Prof Exp: Asst chem, Harvard Univ, 38-41; from asst prof to assoc prof, Univ Md, 41-45; immunochemist, Lankenau Hosp Res Inst, 45-47; chmn admin comt, Inst Cancer Res, 47-54, head dept chemother & immunochem, 47-57, chmn div chemother, 57-70, SR MEM, INST CANCER RES, 49- Concurrent Pos: Lectr, Bryn Mawr Col, 45-47; mem US comt, Int Union Against Cancer, 57-60. Mem: Sigma Xi; Am Asn Cancer Res (secy-treas, 52-). Res: Synthetic organic chemistry; chemotherapy and immunology of polysaccharides and protein complexes in cancer research; chemoantigens; carcinogenesis; alkylating agents; antimalarials; fluorescent antibodies; frameshift mutagens. Mailing Add: Inst for Cancer Res Fox Chase Philadelphia PA 19111

CREECH, ROY G, b Center, Tex, Jan 24, 35; m 57; c 3. GENETICS, PLANT BREEDING. Educ: Stephen F Austin State Col, BS, 56; Purdue Univ, MS, 58, PhD(genetics), 60. Prof Exp: Asst genetics, Purdue Univ, 56-60; from asst prof to prof plant genetics, Pa State Univ, 60-72; PROF AGRON & HEAD DEPT, MISS STATE UNIV, 72- Concurrent Pos: Am Soc Agron rep, Am Inst Biol Sci-Campbell Award Comt, 63-72, NE-66 Regional Proj Comt, 63-; vis prof, Univ Ill, 69-70; secy, S-9 Tech Comt Germplasm Resources & New Crops, 73-; chmn, Southern Region Task Force New Crops, 74- Mem: Genetics Soc Am; Crop Sci Soc Am; Am Soc Agron; Am Soc Hort Sci; Nat Sweet Corn Breeders Asn (pres, 73). Res: Genetic regulation of metabolic pathways in plants, especially carbohydate metabolism in maize. Mailing Add: Dept of Agron Miss State Univ Mississippi State MS 39762

CREED, DAVID, b Colchester, Eng, Sept 22, 43. PHOTOCHEMISTRY. Educ: Univ Manchester, BS, 65, MS, 66, PhD(chem), 68. Prof Exp: Res assoc biol, Southwest Ctr, Advan Studies, 68-69; res assoc biol, Univ Tex, Dallas, 69-71; SRC res fel, Royal Inst, Univ London, 71-72; temp lectr, Univ Warwick, 72-73; ROBERT A WELCH FOUND FEL, UNIV TEX, DALLAS, 73- Mem: Am Soc Photobiol; The Chem Soc; Am Chem Soc. Res: Reaction mechanisms in organic photochemistry; the molecular basis of some biological effects of ultraviolet light. Mailing Add: Inst Chem Sci Univ of Tex Dallas Box 688 Richardson TX 75080

CREEDON, JOHN E, b Quincy, Mass, May 25, 25; m 55; c 2. PLASMA PHYSICS. Educ: Boston Col, BS, 50; Monmouth Col, MS, 74. Prof Exp: Electronic scientist, Nat Bur Standards, 50-55; PHYSICIST, US ARMY ELECTRONICS COMMAND, 55- Mem: Am Phys Soc; Inst Elec & Electronics Engrs. Res: Gaseous electronics, especially infrared sources; optical pumps; lasers; hydrogen plasmas; hydrogen thyratrons; high power switching—modulator component development; IR source and materials. Mailing Add: US Army Electronics Labs Ft Monmouth NJ 07703

CREEK, ROBERT OMER, b Harrisburg, Ill, June 30, 28; m 58; c 2. ENDOCRINOLOGY. Educ: Univ Ill, BS, 50; Southern Ill Univ, MS, 55; Univ Ind, PhD(zool), 60. Prof Exp: NIH fel, Sloan-Kettering Inst, NY, 60-61; trainee endocrinol, Univ Wis, 61-64; from asst prof to assoc prof physiol, 64-72, PROF PHYSIOL, SCH MED, CREIGHTON UNIV, 72- Mem: Endocrine Soc. Res: Mechanisms of hormone action; effect of thyroid stimulating hormone on the thyroid. Mailing Add: Dept of Physiol Sch of Med Creighton Univ Omaha NE 68178

CREEL, DONNELL JOSEPH, b Kansas City, Mo, June 17, 42. NEUROPSYCHOLOGY. Educ: Univ Mo, Kansas City, BA, 64, MA, 66; Univ Utah, PhD(neuropsychol), 69. Prof Exp: Res assoc, Vet Admin Hosp, Kansas City, 69-71; asst prof psychol, Univ Mo-Kansas City, 71; chief psychol res, Vet Admin Hosp, Phoenix, 71-76; RES NEUROPSYCHOLOGIST, VET ADMIN HOSP, SALT LAKE CITY, 76- Concurrent Pos: Adj assoc prof, Ariz State Univ, 71-76 & Univ Utah, 76- Mem: Soc Neurosci; Sigma Xi; AAAS; NY Acad Sci. Res: Functional anatomy of the visual system; inherited anomalies of sensory systems specifically those correlated with genes controlling pigmentation, albinism; the scalp-recorded evoked potential as a diagnostic tool. Mailing Add: Neuropsychol Res Vet Admin Hosp Salt Lake City UT 84113

CREEL, GORDON C, b Daisetta, Tex, Oct 14, 26; m 49; c 3. GENETICS. Educ: Howard Payne Col, BA, 49; Univ Tex, MA, 58; Mont State Univ, PhD(genetics), 64. Prof Exp: Teacher high sch, 50-57; asst prof biol, Wayland Baptist Col, 58-62; NSF res grant, 62-63; prof biol & chmn div, Howard Payne Col, 64-65; PROF BIOL, ANGELO STATE UNIV, 65- Mem: AAAS; Am Genetic Asn. Res: Genetics of vertebrates. Mailing Add: Dept of Biol Angelo State Univ ASU Sta Box 10890 San Angelo TX 76901

CREELY, ROBERT SCOTT, b Kentfield, Calif, Aug 29, 26; m 52; c 4. GEOLOGY. Educ: Univ Calif, Berkeley, BS, 50, PhD(geol), 55. Prof Exp: Geologist, Wm Ross Cabeen & Assoc, Colo, 55-60; from instr to assoc prof geol, Colo State Univ, 60-68, chmn dept, 65-68; PROF GEOL & CHMN DEPT, SAN JOSE STATE UNIV, 68- Mem: AAAS; Geol Soc Am; Nat Asn Geol Teachers. Res: Structural geology; petrology. Mailing Add: Dept of Geol San Jose State Univ San Jose CA 95192

CREESE, THOMAS MORTON, b New York, NY, June 19, 34. MATHEMATICS. Educ: Mass Inst Technol, BS, 56; Univ Calif, Berkeley, MA, 63, PhD(math), 64. Prof Exp: Res mathematician, Univ Calif, Berkeley, 64; ASST PROF MATH, UNIV KANS, 64- Concurrent Pos: Lectr, Univ Oslo, 67-68. Mem: Am Math Soc; Math Asn Am. Res: Complex and functional analysis; several complex variables. Mailing Add: Dept of Math Univ of Kans Lawrence KS 66044

CREGER, CLARENCE R, b Carona, Kans, June 8, 34. BIOCHEMISTRY. Educ: Kans State Univ, BS, 56, MS, 57; Tex A&M Univ, PhD(biochem, nutrit), 61. Prof Exp: Res asst, Kans State Univ, 56-57 & Tex A&M Univ, 58-61; res assoc, Allied Mills Inc, 61-62; asst prof, 62-74, PROF POULTRY SCI, BIOCHEM & NUTRIT, TEX A&M UNIV, 74- Mem: Am Chem Soc; Am Inst Chem; Poultry Sci Asn. Res: Metabolism of various radio elements; intermediary metabolism in biosynthetic mechanism. Mailing Add: Dept of Poultry Sci & Biochem Tex A&M Univ College Station TX 77843

CREGER, PAUL LEROY, b Cresco, Iowa, Oct 26, 30; m 60; c 3. ORGANIC CHEMISTRY. Educ: Univ Ill, BS, 52; Univ Nebr, MS, 55, PhD(chem), 57. Prof Exp: From assoc res chemist to res chemist, 58-71, SR RES CHEMIST, PARKE, DAVIS & CO, 71- Concurrent Pos: Fel, Columbia Univ, 57-58. Honors & Awards: Indust Res Award, Am Chem Soc, 71. Mem: Am Chem Soc. Res: Syntheses and reactions of simple nitrogenous heterocycles; qualitative structure; spectral relationships; reaction mechanisms; medicinal chemistry. Mailing Add: Park Davis & Co Res Labs 2800 Plymouth Rd Ann Arbor MI 48106

CREIDER, CHESTER ARTHUR, III, b Chicago, Ill, Oct 30, 42. ANTHROPOLOGY. Educ: Reed Col, BA, 65; Univ Minn, MA, 68, PhD(anthrop), 73. Prof Exp: ASST PROF ANTHROP, UNIV WESTERN ONT, 73- Mem: Can Ethnol Soc; Ling Soc Am; Am Anthrop Asn. Res: Discourse, conversational analysis; semantic and syntactic theory; non-verbal communication; African languages. Mailing Add: Dept of Anthrop Univ of Western Ont London ON Can

CREIGHTON, CHARLIE SCATTERGOOD, b Orlando, Fla, Aug 23, 26; m 52; c 3. ENTOMOLOGY. Educ: Clemson Univ, BS, 50. Prof Exp: Biol aide tobacco insect invests, Entom Res Serv, 52-56, entomologist, 56-61, ENTOMOLOGIST, VEG INSECT INVESTS, USDA, 61- Mem: Entom Soc Am. Res: Biology and control of insects affecting tobacco; investigation of the biology, ecology and control of insects affecting vegetable crops in the South. Mailing Add: Veg Insect Invests PO Box 3187 Charleston SC 29407

CREIGHTON, HARRIET BALDWIN, b Delavan, Ill, June 27, 09. BOTANY. Educ: Wellesley Col, AB, 29; Cornell Univ, PhD(bot), 33. Prof Exp: Asst bot, Cornell Univ, 29-32, instr cytol & microtechnique, 32-34; asst prof bot, Conn Col, 34-40; from assoc prof to prof, 40-74, EMER PROF BOT, WELLESLEY COL, 74- Concurrent Pos: Fulbright lectr, Australia, 52-53 & Peru, 59-60. Mem: Fel AAAS (secy, 60-63, vpres, 64); Genetics Soc Am; Am Soc Naturalists; Soc Develop Biol; Bot Soc Am (secy, 50-54, vpres, 55, pres, 56). Res: Cell physiology; plant growth hormones; plant morphogenesis. Mailing Add: Dept of Biol Sci Wellesley Col Wellesley MA 02181

CREIGHTON, JOHN THOMAS, b Orangeburg, SC, Aug 18, 05; m 35. ENTOMOLOGY. Educ: Univ Fla, BS, 26, MS, 29; Ohio State Univ, PhD(entom), 35. Prof Exp: Prof chem, Gainesville High Sch, Fla, 26-27; from instr to prof entom & head dept, 27-65, head div pest control, 32-65, EMER PROF ENTOM, UNIV FLA, 69-; PRES, CREIGHTON INT SCI CONSULT, 69- Concurrent Pos: Field agt, Bur Entom, USDA, 30, tech adv, Bur Entom & Plant Quarantine, 38, consult & adv, 39; entom res consult, Stanco Inc, 45-48, Velsicol Corp, 46-49, Micronizer Corp, 47, Dow Chem Co, 47-48, Hercules Powder Co, 46-47, Carbide & Carbon Chem Co, 46-47, Max Weedels Tobacco Co, Am Sumatra Tobacco Co, 48, Inter Am Inst Agr Sci, Costa Rica, 49-50, Stand Oil Co, NJ, Esso Stand Oil Co, Cent Am, 49-51; attache, US State Dept; dir res, AID, spec econ & tech mission to Thailand, Bangkok & Bangken, 51-52; chmn, Struct Pest Control Comn, 47-51; organizer & dir int conf use radioisotopes & radiation in entom res, AEC, Vienna, Austria & Food & Agr Div UN, Rome, Italy, 63-64; vol, Int Exec Serv Corps, 72-76; honored consult entom, Nat Asn Stand Med Vocab. Mem: Hon fel Entom Soc Am (vpres, Am Asn Econ Entomologists, 40-49); AAAS; Sigma Xi; Siam Sci Soc. Mailing Add: Devils Millhopper Rd Gainesville FL 32605

CREIGHTON, ROBERT HERVEY JERMAIN, b Swarthmore, Pa, Feb 2, 18; m 46; c 2. ORGANIC CHEMISTRY. Educ: Swarthmore Col, AB, 39; McGill Univ, PhD(chem), 43. Prof Exp: Res chemist, Shawinigan Chem Ltd, 43-48, res supvr, 48-54; sr res supvr, Macmillan, Bloedel & Powell River, Ltd, 54-64, asst dir res wood prod, 64-66, mgr bldg mat div, Corp Res & Develop Dept, 66-68, ASST DIR RES, MacMILLAN BLOEDEL RES LTD, 68- Mem: Am Chem Soc; Forest Prod Res Soc; NY Acad Sci; fel Chem Inst Can; Can Pulp & Paper Asn. Res: Oxidation studies on wood and lignin; reactions of acetylene extractives from wood and bark; coatings; adhesives. Mailing Add: MacMillan Bloedel Res Ltd 3350 E Broadway Vancouver BC Can

CREIGHTON, STEPHEN MARK, b Sask, Aug 2, 20; m 46; c 1. ORGANIC CHEMISTRY, PHYSICAL CHEMISTRY. Educ: Queen's Univ, Ont, BA, 48, MA, 49; Yale Univ, PhD(org chem), 54. Prof Exp: Chemist, Can Packers Ltd, 49-51; chemist, Hooker Chem Corp, 53-56, res assoc polyolefins, 56-59, res supvr polymers, 59-63; HEAD PROD RES & DEVELOP, RES COUN ALTA, 63- Concurrent Pos: Lectr & adv, Niagara Univ, 57-58. Mem: Am Chem Soc; fel Chem Inst Can. Res: Steroids; fat and fatty acids; chlorinated hydrocarbons; pesticides; light stability and fire-retardance of plastics; polyurethane foams; polyolefins; polythers; polyesters; expoxies; sugar and sugar cane; building products. Mailing Add: Alta Res Coun 11315 87th Ave Edmonton AB Can

CREIGHTON, THOMAS EDWIN, b St Louis, Mo, Apr 20, 40; m 63; c 3. MOLECULAR BIOLOGY. Educ: Calif Inst Technol, BS, 62; Stanford Univ, PhD(biol), 66. Prof Exp: Air Force Off Sci Res-Nat Acad Sci res fel, Med Res Coun Lab Molecular Biol, Cambridge Univ, 66-67; asst prof biol, Yale Univ, 67-69; MEM SCI STAFF, MED RES COUN LAB MOLECULAR BIOL, CAMBRIDGE, 69- Res: Protein structure and chemistry; enzyme structure and function. Mailing Add: Med Res Coun Lab Molecular Biol Hills Rd Cambridge England

CREININ, HOWARD LEE, b Chicago, Ill, Sept 5, 42; m 63; c 2. FOOD SCIENCE. Educ: Univ Ill, Urbana, BS, 63, PhD(food sci), 71. Prof Exp: Prod develop chemist food sci, Res & Develop Div, Lever Brothers Co, NJ, 70-74; PROD DEVELOP SCIENTIST FOOD SCI, MARSCHALL DIV, MILES LABS, 74- Mailing Add: Marschall Div Miles Labs Worthington OH 43085

CREITZ, JOSEPH REUEL, b Beloit, Kans, Dec 29, 13; m 40; c 6. MICROBIOLOGY. Educ: Univ Chicago, SB, 39. Prof Exp: Bacteriologist, Jefferson County Bd Health, Ala, 38-40, in chg, 40-43; prin serologist, Ala State Bd Health, 46; microbiologist, Doctors Hosp, 46-47; bacteriologist, Tacoma Gen Hosp, 48-49; bacteriologist, Norwood Clin, 49-50; chief lab, US Army Hosp, Ft Campbell, Ky, 50-51, chief bact, parasitol & serol, Fitzsimons Gen Hosp, 51-56, microbiologist, Ryukyas Hosp, 56-59, chief virol & serol, Second Army Med Lab, 59-61, chief microbiol, 61-68, exec officer, 64-68, EXEC OFFICER & CHIEF MICROBIOL, THIRD US ARMY MED LAB, FT McPHERSON, GA, US ARMY, 68- Res: Medical microbiology; spherule development and safe curing of Coccidioides immitis. Mailing Add: Microbiol Dis Third USA Med Lab Ft McPherson GA 30336

CRELIN, EDMUND SLOCUM, b Red Bank, NJ, Apr 26, 23; m 48; c 4. HUMAN ANATOMY, HUMAN DEVELOPMENT. Educ: Cent Col, Iowa, BA, 47; Yale Univ, PhD(anat), 51. Hon Degrees: DSc, Cent Col, Iowa, 69. Prof Exp: From instr to assoc prof human anat, 51-68, PROF ANAT, SCH MED, YALE UNIV, 68-; CHMN HUMAN GROWTH & DEVELOP STUDY UNIT, YALE-NEW HAVEN MED CTR, 72- Concurrent Pos: Consult, Ciba-Geigy Pharmaceut Co Inc, 61-; assoc ed, Anat Record, 68-74. Honors & Awards: Award, Sch Med, Yale Univ, 61; Kappa Delta Res Award, Am Acad Orthop Surgeons, 76. Mem: AAAS; AMA; Am Asn Anatomists. Res: Structure and physiology of connective tissues; developmental biology; anthropology. Mailing Add: Dept of Surg Yale Univ Sch of Med New Haven CT 06510

CRELLING, JOHN CRAWFORD, b Philadelphia, Pa, June 13, 41; m 67; c 1. GEOLOGY. Educ: Univ Del, BA, 64; Pa State Univ, MS, 67, PhD(geol), 73. Prof Exp: RES GEOLOGIST, BETHLEHEM STEEL CORP, 72- Mem: Geol Soc Am; Am Inst Mining, Metall & Petrol Engrs; AAAS; Am Soc Photogram. Res: Applied coal petrology and coal geology. Mailing Add: Homer Res Labs Bethlehem Steel Corp Bethlehem PA 18015

CREMER, NATALIE E, b Minot, NDak, Sept 13, 19. IMMUNOLOGY, VIROLOGY. Educ: Univ Minn, BS, 44, MS, 56, PhD, 60. Prof Exp: RES SPECIALIST IMMUNOL, CALIF STATE DEPT HEALTH, 62- Concurrent Pos: NIH res fel immunochem, Calif Inst Technol, 60-62. Mem: AAAS; Am Soc Microbiol; Am Asn Immunologists; NY Acad Sci; Soc Exp Biol & Med. Res: Immunology of viral infection; tumor immunology; latent viral infections; chronic degenerative disease. Mailing Add: Viral & Rickettsial Dis Lab Calif State Dept Health Berkeley CA 97404

CREMER, SHELDON E, b Parkersburg, WVa, Oct 24, 35; m 65; c 2. ORGANIC CHEMISTRY. Educ: Carnegie Inst Technol, BS, 57; Univ Rochester, PhD(org chem), 61. Prof Exp: Summer res scientist, E I du Pont de Nemours & Co, Inc, 62; asst prof chem, Ill Inst Technol, 63-69; assoc prof, 69-74, PROF CHEM, MARQUETTE UNIV, 74- Concurrent Pos: Fels, Ohio State Univ, 61-62 & Univ Ill, 62-63; sr res associateship, Nat Res Coun, Wright-Patterson Air Force Base, Dayton, Ohio, 74-75. Mem: Fel Am Inst Chem; Brit Chem Soc; Am Chem Soc. Res: Organic photochemistry; organo-phosphorus chemistry; organosilicon and organoarsenic chemistry. Mailing Add: Dept of Chem Marquette Univ 535 N 14th St Milwaukee WI 53233

CRENSHAW, CRAIG MOFFETT, b Chinkyiang, China, Sept 24, 16; US citizen; m 42; c 4. PHYSICS. Educ: Southwestern Univ, Tenn, BS, 37; NY Univ, PhD(physics), 42. Prof Exp: Asst, NY Univ, 37-42; physicist, Res & Develop Labs, Ft Monmouth, NJ, 42-56, div dir, 56-57, chief scientist res & develop for chief signal officer, 57-62, CHIEF SCIENTIST, MATERIEL COMMAND, US ARMY, 62- Honors & Awards: US Army Awards, 58 & 59. Mem: AAAS; Am Phys Soc; Inst Elec & Electronics Eng; NY Acad Sci. Res: Sound ranging and propagation; wave motion; upper atmospheric research; research and development management. Mailing Add: US Army Materiel Command Hq 5001 Eisenhower Ave Alexandria VA 22304

CRENSHAW, DAVID BROOKS, b Columbia, Mo, May 15, 45; m 66; c 2. ANIMAL BREEDING. Educ: Univ Mo-Columbia, BS, 68, MS, 69, PhD(animal husb & cytogenetics), 72. Prof Exp: ASST PROF ANIMAL SCI, TEX A&I UNIV, 72- Mem: Am Soc Animal Sci. Res: Genetics of fertility in Texas Longhorn cattle; cytogenetics of early embryonic mortality in beef cattle; hormonal manipulation of postpartum estrus in range beef cattle. Mailing Add: Col of Agr Box 156 Tex A&I Univ Kingsville TX 78363

CRENSHAW, JACK WESTCOTT, b Montgomery, Ala, Aug 27, 34; m 56; c 3. COMPUTER SCIENCES. Educ: Auburn Univ, BS, 58, MS, 59, PhD(physics), 68. Prof Exp: Aeronaut res engr, Langley Res Ctr, NASA, Va, 59-60; physicist, Missile & Space Vehicle Dept, Gen Elec Co, Pa, 60-61, engr, 61-62, engr, Apollo Support Dept, Fla, 62-63; staff engr, fed systs div, Int Bus Mach Corp, Ala, 67-68; mem tech staff, Northrop Corp, 68-75; mgr eng, Comp-Sultants, Inc, 75-76; ASSOC PROF, ALA A&M UNIV, 76- Mem: Am Inst Aeronaut & Astronaut; Am Astronaut Soc. Res: Hardware and software for microprocessor applications. Mailing Add: 1409 Blevins Gap Rd Huntsville AL 35802

CRENSHAW, JOHN WALDEN, JR, b Atlanta, Ga, May 17, 23; m 46; c 2. POPULATION GENETICS. Educ: Emory Univ, AB, 48; Univ Ga, MS, 51; Univ Fla, PhD(zool), 55. Prof Exp: Instr zool, Univ Mo, 55-56; asst prof biol, Antioch Col, 56-60; asst prof zool, Southern Ill Univ, 60-62; assoc prof zool, Univ Md, 62-65 & prof, 65-67; prof zool, Univ RI, 67-72; DIR & PROF SCH BIOL, GA INST TECHNOL, ATLANTA, 72- Concurrent Pos: Am Philos Soc res grant, 57-58; NSF res grants, 58-64 & sci fac fel, Univ Calif, Berkeley, 59-60; USPHS res grants, 62-67 & res contracts, 68-70 & 75; consult, NSF Int Sci Activities, 63-67, Biol Sci

Curriculum Study, 60-68 & US AID, Latin Am, 69. Mem: Fel AAAS; Am Soc Nat; Genetics Soc Am; Am Genetic Asn; Am Inst Biol Sci. Res: Ascertainment of and effects on fitness of mutagen induced polygenic mutations in mice and fish; origin and maintenance of polygenic variation. Mailing Add: Sch of Biol Ga Inst Technol Atlanta GA 30332

CRENSHAW, MILES AUBREY, b Earlysville, Va, Mar 22, 32; div; c 3. PHYSIOLOGY. Educ: Univ Va, BA, 59; Duke Univ, MA, 62, PhD(zool), 64. Prof Exp: Asst physiol, Med Sch, Univ Va, 54-59; asst zool, Duke Univ, 60, instr, 64; instr physiol, Sch Dent Med, Harvard Univ & staff assoc histol, Forsyth Dent Ctr, 64-67; asst prof dent sci, 67-71, ASSOC PROF ORAL BIOL, DENT RES CTR, UNIV NC CHAPEL HILL, 71-, ADJ ASSOC PROF ZOOL, 73- Mem: AAAS; assoc Am Dent Asn; Am Chem Soc; Am Soc Zoologists. Res: Comparative biology of mineralizing tissues. Mailing Add: Dent Res Ctr Univ of NC Chapel Hill NC 27514

CRENSHAW, RONNIE RAY, b Earlington, Ky, Dec 24, 36; m 59; c 2. MEDICINAL CHEMISTRY. Educ: Vanderbilt Univ, BA, 58, PhD(org chem), 63. Prof Exp: Assoc chemist, Mead Johnson Labs, Ind, 58-59; SR CHEMIST, BRISTOL LABS, 63- Mem: Am Chem Soc. Res: Synthetic organic chemistry; organic sulfur chemistry; pharmaceuticals. Mailing Add: 107 Charing Rd DeWitt NY 13214

CRENTZ, WILLIAM LUTHER, b Baltimore, Md, May 1, 10; m 57. CHEMISTRY. Educ: Univ Md, BS, 32, MS, 33. Prof Exp: Statistician, Smokeless Coal Code Authority, Washington, DC, 34-35; researcher, US Procurement Div, 35-37; econ analyst, Coal Div, US Dept Interior, 37-43; coal technologist, US Bur Mines, 43-52, chem engr, 52-60, asst to chief, Div Bituminous Coal, 60-63, from asst dir to dir coal res, 63-70, asst dir energy, 70-73; consult, Off Secy, US Dept Interior, 74-75; CONSULT, ENERGY RES DEVELOP ADMIN, 75- Honors & Awards: Chevalier de l'Ordre de la Courounne, Belg, 59; Distinguished Serv Award & Gold Medal, US Dept Interior, 68. Mem: Am Inst Mining, Metall & Petrol Eng. Res: Coal preparation and utilization; petroleum and natural gas research; shale oil research; production and conservation of helium. Mailing Add: 3850 Tunlaw Rd NW Washington DC 20007

CREPEA, SEYMOUR B, b New York, NY, Oct 25, 18; m 53; c 2. MEDICINE, ALLERGY. Educ: Tulane Univ, BS, 39, MD, 42; Columbia Univ, DMedSc, 48. Prof Exp: Assoc prof med & head dept allergy, Sch Med, Univ Wis, 49-60; resident med dir & resident physician, Dept Res Eng, Sahuaro Sch Asthmatic Children, Univ Ariz, 60-70; ASSOC DIR MED, SYNTEX LABS, INC, 70- Concurrent Pos: Consult, Vet Admin Hosp. Mem: AMA; Trudeau Soc; fel Am Acad Allergy; fel Am Col Physicians. Res: Immunology. Mailing Add: Syntex Labs Inc 3401 Hillview Ave Palo Alto CA 94304

CREPEAU, RICHARD HANES, b Schenectady, NY, May 26, 44; m 65; c 2. BIOPHYSICS. Educ: State Univ NY, Albany, BS, 66; Cornell Univ, MS, 69, PhD(physics), 71. Prof Exp: RES ASSOC BIOCHEM, CORNELL UNIV, 71- Res: Application of electron microscopy, image reconstruction, computer studies and ultracentrifugation to problems of sickle cell hemoglobin and virus assembly. Mailing Add: Dept of Biochem Wing Hall Cornell Univ Ithaca NY 14853

CREPET, WILLIAM LOUIS, b New York, NY, Aug 10, 46; m 72. PALEOBOTANY, EVOLUTIONARY BIOLOGY. Educ: State Univ NY Binghamton, BA, 69; Yale Univ, MPh, 71, PhD(biol), 73. Prof Exp: Lectr bot, Ind Univ, 73-75; ASST PROF BIOL, UNIV CONN, 75- Mem: Am Inst Biol Sci; AAAS; Bot Soc Am; Int Asn Angiosperm Paleobot. Res: Evolution of the angiosperms with emphasis on the evolution of floral structure and the evolution of pollination mechanisms; cycadophyte evolution. Mailing Add: Dept of Biol Sci Univ of Conn U-42 Storrs CT 06268

CREPS, ELAINE SUE, b Toledo, Ohio, July 14, 46; m 70; c 1. NEUROANATOMY, NEUROENDOCRINOLOGY. Educ: Ohio Univ, BS, 68; Univ Mich, MS, 69; Univ Ariz, PhD(anat), 73. Prof Exp: Fel physiol, Univ Ariz, 73-74; ASST PROF ANAT, UNIV N MEX, 74- Mem: Soc Neurosci; Am Asn Anatomists. Res: Hypothalamic regulation of the endocrine system, especially in response to stress, in the genetically obese rat. Mailing Add: Dept of Anat Sch of Med Univ of N Mex Albuquerque NM 87131

CRESCITELLI, FREDERICK, b Providence, RI, June 23, 09; m 41; c 3. VISUAL PHYSIOLOGY. Educ: Brown Univ, PhB, 30, MS, 32, PhD(physiol), 34. Hon Degrees: MD, Univ Linköping, Sweden, 75. Prof Exp: Keen fel, Brown Univ, 34-35; instr, Colby Jr Col, 35-36; res assoc zool, Univ Iowa, 36-40; instr, Univ Wash, 40-42; actg asst prof, Stanford Univ, 42-43; vis asst prof, Univ Southern Calif, 43-44; physiologist, Chem Warfare Serv, US Army, Edgewood Arsenal, Md, 44-46; Mellon fel, Johns Hopkins Hosp, 46; assoc prof zool, 46-51, PROF CELL BIOL, UNIV CALIF, LOS ANGELES, 51- Concurrent Pos: Nat Res Coun fel, 35-36; vis prof, Univ Cologne, 66-67; ed, Vision Res; mem adv bd, J Comparable Physiol; mem adv bd, Nat Eye Inst, 74-75. Mem: Am Physiol Soc; Soc Gen Physiol; fel Optical Soc Am. Res: Neurophysiology; effects of drugs; visual physiology; aviation physiology; physiology and biochemistry of retina. Mailing Add: Dept of Biol Univ of Calif Los Angeles CA 90024

CRESPI, HENRY LEWIS, b Joliet, Ill, Mar 18, 26; m 55; c 6. PHYSICAL CHEMISTRY. Educ: Univ Ill, BS, 52, PhD(chem), 55. Prof Exp: Asst chemist, 55-59, ASSOC CHEMIST, ARGONNE NAT LAB, 59- Concurrent Pos: Vis lectr, St Procopius Col, 57-62. Mem: Am Chem Soc. Res: Biological effects of deuterium, with special reference to algae and other microorganisms; physical chemistry of proteins and lipoproteins. Mailing Add: Argonne Nat Lab 9700 Cass Ave Argonne IL 60439

CRESPI, MURIEL K, cultural anthropology, social anthropology, see 12th edition

CRESS, CHARLES EDWIN, b Rowan Co, NC, Aug 17, 34; m 57. STATISTICS, QUANTITATIVE GENETICS. Educ: NC State Univ, BS, 56, MS, 61; Iowa State Univ, PhD(statist), 65. Prof Exp: Asst prof statist, Rutgers Univ, 65-66; from asst prof to assoc prof, 66-75, PROF CROP & SOIL SCI, MICH STATE UNIV, 75- Mem: Biomet Soc; Am Statist Asn; Am Soc Agron. Res: Methods of statistical data analysis. Mailing Add: Dept of Crop & Soil Sci Mich State Univ East Lansing MI 48824

CRESS, DONALD CHAUNCEY, b Canon City, Colo, July 6, 41; m 65. ENTOMOLOGY. Educ: Colo State Univ, BS, 64; Univ Wyo, MS, 67; Okla State Univ, PhD(entom), 70. Prof Exp: EXTEN SPECIALIST ENTOM & VEG INSECT RES, MICH STATE UNIV, 70- Mem: Entom Soc Am. Res: Mating behavior of Pasimachus elongatus; life history and habits of P elongatus; nutritional requirements of Schizaphis graminum, I, II, III. Mailing Add: Dept of Entom Mich State Univ East Lansing MI 48823

CRESSEY, ROGER F, b Stoughton, Mass, June 9, 30. PARASITOLOGY. Educ: Boston Univ, AB, 56, AM, 58, PhD(biol), 65. Prof Exp: Instr biol, Boston Univ, 64-65; CUR CRUSTACEA, SMITHSONIAN INST, 65- Concurrent Pos: Res assoc, Mote Marine Lab, Fla; adj lectr biol sci, George Washington Univ; ed, Biol Soc Wash; adj prof, Dunbarton Col, 74; panel mem, Food & Agr Orgn, UN, 74- Mem: Am Soc Zool. Res: Taxonomy, evolution and systematics of copepods parasitic on fishes. Mailing Add: Dept of Crustacea Smithsonian Inst Washington DC 20560

CRESSMAN, GEORGE PARMLEY, b West Chester, Pa, Oct 7, 19; m 42; c 4. METEOROLOGY. Educ: Univ Chicago, PhD(meteorol), 49. Prof Exp: Asst meteorol, Univ Chicago, 45-49; consult, Air Weather Serv, 49-54; dir Joint Numerical Weather Prediction Unit, 54-58, Nat Meteorol Serv, US Weather Bur, 58-65, Environ Sci Serv Admin, 65-70, DIR NAT OCEANIC & ATMOSPHERIC AGENCY, NAT WEATHER SERV, 70- Honors & Awards: Losey Award, Am Inst Astronaut & Aeronaut, 67; Appl Meteorol Award, Am Meteorol Soc, 72 & Cleveland Abbe Award, 75. Mem: Am Meteorol Soc; Am Geophys Union. Res: Synoptic meteorology. Mailing Add: 9 Old Stage Ct Rockville MD 20852

CRESSMAN, HARRY KEITH, b New Hamburg, Ont, July 6, 25; m 54; c 2. SOIL SCIENCE, PLANT PHYSIOLOGY. Educ: Goshen Col, BA, 48; Purdue Univ, MA, 54; Mich State Univ, PhD(soil sci), 61. Prof Exp: Asst control chemist anal chem, Naugatuck Chem Ltd, 48-50; control chemist, Snyder's Potato Chips, 50-52 & Smith Douglass Inc, 54-55; res chemist, Int Minerals & Chem Corp, 55-58; res officer soil fertility, Can Dept Agr, 61-62; res agronomist, Int Minerals & Chem Corp, Ill, 62-67; sr res scientist, Continental Oil Co, Okla, 67-69; ASST PROF AGR SCI, COL AGR, UNIV WIS-PLATTESVILLE, 69- Concurrent Pos: Proj dir environ awareness ctr, Univ Wis-Madison, 72- Mem: AAAS; Soil Conserv Soc Am; Am Soc Agron; Soil Sci Soc Am. Res: Soil fertility; plant nutrition; investigation of new phosphorus compounds as possible fertilizer materials; fertilizer formulations for optimum yields. Mailing Add: Dept of Agr Univ of Wis Platteville WI 53818

CRESSMAN, HOMER WILLIAM JOHN, chemistry, deceased

CRESSMAN, RICHARD MORRIS, plant physiology, see 12th edition

CRESSMAN, WILLIAM ARTHUR, b Philadelphia, Pa, Feb 1, 41; m 63; c 2. PHARMACOLOGY. Educ: Philadelphia Col Pharm, BSc, 63, MSc, 65, PhD(biopharmaceut), 67. Prof Exp: Group leader, 67-73, SR PROJ COORDR, McNEIL LABS, INC, 73- Mem: Am Pharmaceut Asn; Acad Pharmaceut Sci; Am Chem Soc; Am Soc Pharmacol & Exp Therapeut; Am Soc Clin Pharmacol. Res: Physical pharmacy; drug dosage form design and evaluation; drug metabolism and kinetics; bioavailability and new drug development. Mailing Add: 1014 Stevens Dr Ft Washington PA 19034

CRESSWELL, ARTHUR, b Leicester, Eng, Oct 3, 03; nat US; m 33; c 2. ORGANIC CHEMISTRY. Educ: Tufts Col, BS, 25, MS, 26; Mass Inst Technol, PhD(org chem), 32. Prof Exp: Instr chem, Mass Inst Technol, 26-32; control lab foreman in chg routine & spec testing, NAm Rayon Corp, Tenn, 33-36, asst res dir in chg cellulose & spinning, 36-42; group leader in chg synthetic fiber develop, Am Cyanamid Co, 43-54, tech asst to mgr, 54-56; head cellulose & viscose res, NAm Rayon Corp, 56-66; assoc prof chem, ETenn State Univ, 66-71; RETIRED. Mem: Fel Am Chem Soc. Res: Cellulose chemistry; protein fibers; synthetic resin fibers; treatment of rayon velvet for crush-proofing; addition of polyethylene oxides to spin bath to reduce incrustation of spinnerettes; thermal decomposition ethers of triphenylcarbinol and esters of triphenylacetic acid; acrylic polymerization and acrylic fiber spinning and aftertreatment. Mailing Add: Dept of Chem ETenn State Univ Johnson City TN 37601

CRESSWELL, MICHAEL WILLIAM b Gloucester, Eng, Apr 26, 37; m 63; c 2. ELECTRONICS. Educ: Univ London, BSc, 58; Pa State Univ, MS, 61, PhD(physics), 65; Univ Pittsburgh, MBA, 75. Prof Exp: SR ENGR, POWER DEVICES LAB, WESTINGHOUSE RES & DEVELOP CTR, 65- Mem: Am Phys Soc; Inst Elec & Electronics Engrs. Res: Nuclear physics, including observation and analysis of reactions with photographic emulsions; semiconductors, including influence of mechanical stress on electrical properties of semiconductor materials; computer process control in power semiconductor device manufacture; effects of nuclear radiation on silicon devices. Mailing Add: Westinghouse Res & Develop Ctr Bldg 501-2Y59 Beulah Rd Pittsburgh PA 15235

CRESSY, NORMAN J, quantum chemistry, see 12th edition

CRESTFIELD, ARTHUR MEADE, biochemistry, see 12th edition

CREUTZ, CAROL ANN, b Washington, DC, Oct 20, 44; m 65; c 1. INORGANIC CHEMISTRY. Educ: Univ Calif, Los Angeles, BS, 66; Stanford Univ, PhD(chem), 70. Prof Exp: Asst prof chem, Georgetown Univ, 70-72; res assoc, 72-75, ASSOC CHEMIST, BROOKHAVEN NAT LAB, 75- Mem: Am Chem Soc. Res: Synthesis and properties of unusual transition metal complexes; dynamics of inorganic reactions in solution. Mailing Add: Dept of Chem Brookhaven Nat Lab Upton NY 11973

CREUTZ, EDWARD (CHESTER), b Beaver Dam, Wis, Jan 23, 13; m 37; c 3. NUCLEAR PHYSICS. Educ: Univ Wis, BS, 36, PhD(physics), 39. Prof Exp: Res assoc, Princeton Univ, 38-39, instr physics, 40-41; physicist, Nat Defense Res Coun, 41-42, metall lab, Univ Chicago, 42-44 & Manhattan Proj, Univ Calif, Los Alamos, 44-45; assoc prof physics, Carnegie Inst Technol, 46-49, prof physics, head dept & dir nuclear res ctr, 49-55; dir res, Gen Atomic Div, Gen Dynamics Corp, 55-59, vpres res & develop, 59-67; vpres res, Gulf Gen Atomic, Inc, 67-70; asst dir res, 70-75, ASST DIR MATH & PHYS SCI & ENG, NSF, 75- Concurrent Pos: Consult, Manhattan Proj, 46, Oak Ridge Nat Lab, 46-58, Lawrence Radiation Lab, Univ Calif, 46, 56 & NSF, 50-; mem coun exec bd, Argonne Nat Lab, 46-51; appointments comt, Am Inst Physics, 55-58, vis scientists prog comt, 58-, col physics comt, 59, adv comt corp assocs, 64 & dir-at-large bd gov, 65-; scientist-at-large, Proj Sherwood Div Res, US AEC, 55-56; dir, John Jay Hopkins Lab Pure & Appl Sci, 55-67; mem adv coun & seawater conversion tech adv comt, Water Resources Ctr, Univ Calif, 58-; adv panel gen sci, US Dept Defense, 59-63; adv comt to off sci personnel, Nat Res Coun, 60-; res adv comt electrophys, NASA, 64-; mem comt sr reviewers, Energy Res & Develop Admin, 73-, fusion power coord comt, 74- Mem: Nat Acad Sci; fel Am Phys Soc; fel Am Nuclear Soc; NY Acad Sci; Am Soc Eng Educ. Res: Proton-proton and proton-lithium scattering; artificial radioactivity; metallurgy of uranium and beryllium; deuteron-neutron reactions; neutron absorption in uranium; synchrocyclotron design; meson reactions; nuclear reactors; thermonuclear reactions; gas flow in porous media. Mailing Add: Nat Sci Found 1800 G St Washington DC 20550

CREUTZ, MICHAEL JOHN, b Los Alamos, NMex, Nov 24, 44; m 66; c 1. THEORETICAL PHYSICS. Educ: Calif Inst Technol, BS, 66; Stanford Univ, MS, 68, PhD(physics), 70. Prof Exp: Res assoc, Stanford Linear Accelerator Ctr, 70; fel physics, Univ Md, 70-72; asst physicist, 72-74, ASSOC PHYSICIST, BROOKHAVEN NAT LAB, 74- Res: Theoretical applications of quantum field theory to elementary particle physics. Mailing Add: Dept of Physics Brookhaven Nat Lab Upton NY 11973

CREVASSE, GARY A, b Cedar Key, Fla, Oct 16, 34; m 59. FOOD SCIENCE. Educ: Univ Fla, BS, 61, MS, 63; Mich State Univ, PhD(food sci), 67. Prof Exp: Asst prof food sci, Univ Ariz, 67-68; DIR RES COLLAGEN SAUSAGE CASINGS, BRECHTEEN CO DIV, HYGRADE FOOD PROD CORP, 68- Mem: Inst Food Technol. Res: Meat science, specifically collagen and its application to meat industry as a packaging medium; proteolytic enzyme effects on acid soluble collagen and isolation of the resulting components. Mailing Add: Brechteen Co 50750 E Russell Schmidt Mt Clemens MI 48043

CREVELING, CYRUS ROBBINS, b Washington, DC, May 30, 30; m 54; c 2. BIOCHEMISTRY, PHARMACOLOGY. Educ: George Washington Univ, BS, 53, MS, 55, PhD(pharmacol), 62. Prof Exp: Chemist, Naval Ord Res Lab, Washington, DC, 52-53; asst biochem, George Washington Univ, 54-55; chemist, Hunter Mem Labs, DC, 56-58; biochemist, Nat Heart Inst, 58-62; from asst to assoc biochem, Sch Med, Harvard Univ, 62-64; PHARMACOLOGIST, NAT INST ARTHRITIS, METAB & DIGESTIVE DIS, 64- Concurrent Pos: Mem staff, Pharmacol Study Sect, Dis Res Grants, NIH, 66-70; assoc, Med Sch, Howard Univ, 70- Mem: AAAS; Am Soc Pharmacol & Exp Therapeut; Am Chem Soc. Res: Biosynthesis and metabolism of biogenic amines; drug metabolism; central nervous system pharmacology; cyclic adenosine monophosphate. Mailing Add: Sect Pharmacodynamics Nat Inst Arth Metab & Dig Dis Bethesda MD 20014

CREVELING, LOUIS, JR, physics, see 12th edition

CREVIER, WILLIAM FRANCIS, b Los Angeles, Calif, Oct 29, 41; m 65; c 2. PLASMA PHYSICS, ELECTRODYNAMICS. Educ: Univ Santa Clara, BS, 63; Univ Southern Calif, MS, 65; Univ Md, PhD(physics), 70. Prof Exp: Res engr, Fairchild Semiconductor, 65-66; physicist, Gen Elec Co, 70-71; PHYSICIST, MISSION RES CORP, 71- Res: Effects of electromagnetic fields produced by nuclear explosions interacting with the atmosphere or directly with a particular system such as a satellite. Mailing Add: Mission Res Corp Drawer 719 Santa Barbara CA 93102

CREW, HENRY, physical oceanography, see 12th edition

CREW, JOHN EDWIN, b Chicago, Ill, July 10, 30; m 58; c 4. NUCLEAR PHYSICS. Educ: Univ Chicago, BS, 52, MS, 53; Univ Ill, PhD(physics), 57. Prof Exp: Physicist, X-ray Sect, Radiation Physics Lab, Nat Bur Standards, 57-59; res asst prof physics, Univ Ill, 59-61; assoc prof, Millikin Univ, 61-63; assoc prof, 63-69, PROF PHYSICS, ILL STATE UNIV, 69- Mem: Am Phys Soc; Am Asn Physics Teachers. Res: High energy nuclear physics; penetration of matter by fast electrons. Mailing Add: Dept of Physics Ill State Univ Normal IL 61761

CREW, MALCOLM CHARLES, b Columbus, Ohio, May 11, 27; m 52; c 3. DRUG METABOLISM. Educ: Ohio Wesleyan Univ, BA, 48; Columbia Univ, MA, 50, PhD(chem), 54. Prof Exp: Asst chem, Columbia Univ, 49-51; pharmaceut res chemist, Wallace & Tiernan, Inc, 52-56; sr proj leader, Fleischmann Labs, Stand Brands, Inc, Conn, 56-65; scientist, Biochem Dept, 65-68, sr scientist, 68-71, SR RES ASSOC, DEPT DRUG METAB, WARNER-LAMBERT RES INST, 71- Mem: Am Chem Soc; The Chem Soc; Am Soc Pharmacol & Exp Therapeut. Res: Pharmaceuticals; foods; flavor; analytical instrumentation. Mailing Add: Dept of Drug Metab Warner-Lambert Res Inst Morris Plains NJ 07950

CREWDSON, RICHARD CLARK, solid state physics, see 12th edition

CREWE, ALBERT VICTOR, b Bradford, Eng, Feb 18, 27; m 49; c 4. PHYSICS. Educ: Univ Liverpool, BSc, 47, PhD(cosmic rays), 50. Prof Exp: Lectr physics, Univ Liverpool, 50-55; res assoc, 55-56, from asst prof to assoc prof, 56-63, PROF PHYSICS, UNIV CHICAGO, 63-, DEAN PHYS SCI DIV, 71-; MEM STAFF, ENRICO FERMI INST, 67- Concurrent Pos: Dir, Argonne Nat Lab, 61-67, dir particle accelerator div, 58-61; tech dir cyclotron, Univ Chicago, 56-58; consult, Sweden, 55, Switz, 56 & Arg, 57. Mem: Am Phys Soc; Am Nuclear Soc; Electron Micros Soc Am. Res: High energy physics; electron microscopes. Mailing Add: Enrico Fermi Inst 5630 Ellis Ave Chicago IL 60637

CREWS, LOWELL THOMAS, b University City, Mo, Oct 16, 15; m 43; c 2. PETROLEUM CHEMISTRY. Educ: Southern Ill Univ, BEd, 37; Okla Agr & Mech Col, MS, 39. Prof Exp: Asst chem, Okla Agr & Mech Col, 37-39; res chemist, Armour & Co, 43-48; group leader, Toni Co, 48-51; res chemist, Standard Oil Co Ind, 51-52, group leader, 52-61; group leader, Am Oil Co, 62-63, PROJ MGR, AM OIL CO, 64- Mem: Am Chem Soc; Am Tech Asn Pulp & Paper Indust; Am Soc Testing & Mat. Res: Petroleum waxes; asphalts. Mailing Add: 17851 Gladville Ave Homewood IL 60430

CREWS, PHILLIP O, b Urbana, Ill, Aug 15, 43; m 67. ORGANIC CHEMISTRY. Educ: Univ Calif, Los Angeles, BS, 66; Univ Calif, Santa Barbara, PhD(org chem), 69. Prof Exp: From asst to assoc chem, Univ Calif, Santa Barbara, 66-68; NSF fel, Princeton Univ, 69-70; ASST PROF CHEM, UNIV CALIF, SANTA CRUZ, 70- Mem: Am Chem Soc; The Chem Soc. Res: Application of nuclear magnetic resonance to problems of organic structure and stereochemistry; marine natural products chemistry; synthesis and study of organometalloids. Mailing Add: Thimann Labs Univ of Calif Santa Cruz CA 95064

CREWS, ROBERT WAYNE, b Pendleton, Ore, Feb 11, 19; m 45; c 4. PHYSICS. Educ: Ore State Univ, BS, 47, MA, 48, PhD(physics), 52. Prof Exp: Sr res physicist eng, Stanford Res Inst, 52-64; mem fac, Col San Mateo, 64-65; MEM STAFF, CHABOT COL, 65- Mem: Am Asn Physics Teachers. Res: Charged particle interactions; electron devices. Mailing Add: Dept of Math & Sci Chabot Col 25555 Hesperian Blvd Hayward CA 94545

CRIBBEN, LARRY DEAN, b Jackson, Ohio, July 3, 40; m 72. PLANT ECOLOGY. Educ: Rio Grande Col, BS, 62; Univ Okla, MNS, 68; Ohio Univ, PhD(bot), 72. Prof Exp: ASST PROF BIOL, MONTCLAIR STATE COL, 72- Mem: Sigma Xi. Res: The effects of acid mine drainage on river bottom plant communities. Mailing Add: Dept of Biol Montclair State Col Upper Montclair NJ 07043

CRIBBS, RICHARD MADISON, b Derry, Pa, July 1, 32; m 55; c 1. GENETICS. Educ: Univ Pittsburgh, BS, 55, MS, 58, PhD(genetics), 62. Prof Exp: Asst prof genetics, 62-68, ASSOC PROF GENETICS & BIOL, MED COL VA, 68- Mem: Am Soc Microbiol; Genetics Soc Am. Res: Microbial genetics; gene-enzyme relationships in microorganisms and tissue cells. Mailing Add: Dept of Genetics & Biol Med Col of Va Richmond VA 23219

CRICHTON, DAVID, b Central Falls, RI, Mar 7, 31. ANALYTICAL CHEMISTRY. Educ: Hope Col, AB, 52; Purdue Univ, MS, 55; Univ Iowa, PhD(spectrophotom), 62. Prof Exp: From instr to assoc prof, 55-71, PROF CHEM, CENT COL, IOWA, 71- Mem: Am Chem Soc. Res: Spectrophotometry; stability constants. Mailing Add: Dept of Chem Cent Col Pella IA 50219

CRICKMAY, COLIN HAYTER, b Vancouver, BC, Apr 6, 99; m 27; c 5. GEOLOGY. Educ: Univ BC, BA, 22; Stanford Univ, PhD(geol), 25. Prof Exp: Asst prof geol, Univ Calif, Los Angeles, 26-31 & Univ Ill, 31-33; consult, 33-45; geologist, Imperial Oil Ltd, 45-69; RETIRED. Mem: Hon mem Can Soc Petrol Geologists. Res: Paleontology and geologic history; geomorphology, particularly the geologic work of rivers. Mailing Add: 525 Salem Ave Calgary AB Can

CRIDDLE, RICHARD S, b Logan, Utah, Sept 20, 36. BIOCHEMISTRY. Educ: Utah State Univ, BS, 58; Univ Wis, MS, 60, PhD(biochem), 62. Prof Exp: Asst prof biophys, 62-73, PROF BIOPHYS & BIOCHEM & BIOPHYSICIST EXP STA, UNIV CALIF, DAVIS, 73- Res: Protein structure and protein-protein interactions in relation to enzyme activity; biosynthesis of organelles. Mailing Add: Dept of Biochem Univ of Calif Davis CA 95616

CRIDER, FRETWELL GOER, b Centerville, Ala, June 8, 23; m 47; c 4. PHYSICAL CHEMISTRY. Educ: Univ NC, BS, 45, PhD(phys chem), 53. Prof Exp: Instr chem, Armstrong Col, Ga, 47-48; asst, Univ NC, 48-52 & Off Naval Res, 52-53; sr res technologist, Field Res Labs, Socony Mobil Oil Co, Inc, 53-64; chmn dept chem & physics, Armstrong State Col, 64-72; dean, Gordon Jr Col, 72-73; DEAN ADMIN, MID GA COL, 73- Concurrent Pos: Exec dir, Armstrong Res Inst, 66-72. Mem: Am Chem Soc; Combustion Inst; fel Am Inst Chemists. Res: Kinetics of photochemical decomposition; combustion kinetics; oxidation of carbon and hydrocarbons; diffusion flames; mechanics and chemistry of geological formations; corrosion; pollution chemistry; water analysis. Mailing Add: Off of Admin Mid Ga Col Cochran GA 31014

CRIDLAND, ARTHUR A, b London, Eng, Mar 29, 36; m 56; c 2. PALEOBOTANY, BRYOLOGY. Educ: Univ Reading, BSc, 57; Univ Kans, PhD(bot), 61. Prof Exp: Res assoc, Inst Polar Studies, Ohio State Univ, 61-62; asst prof bot & biol sci, 62-66, ASSOC PROF BOT, WASH STATE UNIV, 66- Mem: Bot Soc Am; Bryol & Lichenol Soc; Geol Soc Am; Brit Bryol Soc; Int Orgn Paleobot. Mailing Add: Dept of Bot Wash State Univ Pullman WA 99163

CRIGLER, JOHN F, JR, b Charlotte, NC, Sept 11, 19; m 44; c 4. PEDIATRICS. Educ: Duke Univ, AB, 39; Johns Hopkins Univ, MD, 43. Prof Exp: Intern med, Mass Mem Hosp, 43-44; intern pediat, Johns Hopkins Hosp, 46-47, asst resident, 47-48, physician-in-chg outpatient dept, 48-49, resident, 49-50; instr pediat, 55-56, assoc, 56-62, asst prof, 62-68, assoc clin prof, 68-69, ASSOC PROF PEDIAT, HARVARD MED SCH, CHILDREN'S HOSP, BOSTON, 69- Concurrent Pos: Fel pediat endocrinol, Johns Hopkins Hosp, 50-51; Nat Found fel biol, Mass Inst Technol, 51-55; instr, Sch Med, Johns Hopkins Univ, 48-50; assoc physician, Children's Hosp Med Ctr, 56-61, physician, 61-62, sr assoc med, 62-, dir gen clin res ctr for children, 64-, chief endocrine div, Dept Med, 66- Mem: Am Acad Pediat; Endocrine Soc; Soc Pediat Res; Am Pediat Soc; Am Fedn Clin Res. Res: Pediatric endocrinology and metabolism; biological effects of pituitary hormones in human beings; steroid hormone metabolism. Mailing Add: 300 Longwood Ave Boston MA 02115

CRIKELAIR, GEORGE F, b Wis, July 15, 20; m; c 7. SURGERY. Educ: Univ Wis, BA, 42, MD, 44. Prof Exp: Intern, US Naval Hosp, San Diego, Calif, 44-45; preceptor, Dr M G Rice, Wis, 46-47; asst instr surg, Col Med, Wayne State Univ, 49-50; from instr surg to assoc prof clin surg, 52-60, PROF CLIN SURG, COL PHYSICIANS & SURGEONS, COLUMBIA UNIV, 60-; DIR PLASTIC SURG SERV, PRESBY HOSP, 59- Concurrent Pos: Resident, US Vet Hosp, Dearborn, Mich 47-50; resident plastic surg, Presby Hosp, New York, 50-52, from asst attend surgeon to attending surgeon, 52-60; asst vis surgeon, Francis Delafield Hosp, 52-57, assoc vis surgeon, 58; assoc attend, Vanderbilt Clin, 58; mem staff, St Elizabeth's Hosp, New York & Valley Hosp, Ridgewood, NJ. Mem: Am Soc Plastic & Reconstruct Surg (pres-elect, 70); AMA; Am Asn Plastic Surg; NY Acad Med; Am Col Surg. Res: Plastic and reconstructive surgery. Mailing Add: Col of Physicians & Surgeons Columbia Univ New York NY 10032

CRILEY, BRUCE, b Chicago, Ill, Apr 13, 39; m 67; c 2. EXPERIMENTAL EMBRYOLOGY. Educ: Univ Ill, BS, 60, MS, 62, PhD(zool), 67. Prof Exp: Instr embryol, Univ Ill, 65-66; asst prof, Univ Colo, 66-70, assoc prof embryol & chmn organismic biol div, 70-71; CHMN DEPT BIOL, ILL WESLEYAN UNIV, 71-, PROF BIOL, 73- Mem: AAAS; Am Zool Soc. Res: Neuroembryology; vertebrate morphogenesis. Mailing Add: Dept of Biol Ill Wesleyan Univ Bloomington IL 61701

CRILL, PAT, b Blaine, Colo, June 29, 39; m 61; c 2. PLANT BREEDING, PLANT PATHOLOGY. Educ: Panhandle State Col, BS, 61; Okla State Univ, MS, 63; Univ Wis, PhD(plant path), 68. Prof Exp: Soil conservationist, Soil Conserv Serv, USDA, 61; teaching asst agron, Okla State Univ, 61-63; res asst, Kans State Univ, 63-64; teaching asst plant path, Univ Wis, 66-68; from asst prof to assoc prof, Univ Fla, 68-74; SR PLANT BREEDER, PETOSEED RES CTR, 74- Concurrent Pos: Consult, Univ Rio Grande do Sul, Brazil, 68; assoc ed, Plant Dis Reporter, 74- Mem: Am Phytopath Soc; Am Soc Agron; Crop Sci Soc Am; Am Genetics Asn; Tomato Genetics Coop. Res: Evaluation of pathogen variation and identification of host resistance genes in normal-exotic germplasms; mechanical harvesting of fresh market tomatoes, development and evaluation of plant breeding methods and systems. Mailing Add: Petoseed Res Ctr Rt 4 Box 1255 Woodland CA 95695

CRIM, STERLING CROMWELL, b Corsicana, Tex, Jan 5, 27; m 54; c 2. MATHEMATICS. Educ: Baylor Univ, BS, 50; NTex Univ, MEd, 53; George Peabody Col, MA, 58; Univ Tex, Austin, PhD(math), 68. Prof Exp: Teacher high sch, Tex, 51-53 & jr high sch, 53-57; teacher math, WGa Col, 58-59; spec instr, Univ Tex, Austin, 59-64; PROF MATH, LAMAR UNIV, 64- Mem: Math Asn Am. Res: Integral transforms. Mailing Add: Dept of Math Lamar Univ Beaumont TX 77704

CRIMINALE, WILLIAM OLIVER, JR, b Mobile, Ala, Nov 29, 33; m 62; c 2. PHYSICAL OCEANOGRAPHY, APPLIED MATHEMATICS. Educ: Univ Ala, BS, 55; John Hopkins Univ, PhD(aeronaut), 60. Prof Exp: Asst fluid mech & appl math, Johns Hopkins Univ, 56-60; asst prof, Dept Aerospace & Mech Sci, Princeton Univ, 62-68; assoc prof oceanog & geophys, 69-73, PROF OCEANOG & GEOPHYS, UNIV WASH, 73- Concurrent Pos: NATO fel, Inst Appl Math, Ger, 60-61; sr vis, Cambridge Univ, 61-; consult, Aerospace Corp, Calif, 63, 65; guest prof, Can Armament Res & Develop Estab, Que, 65; guest prof, Inst Mech Statist of Turbulence, France, 67-68; consult adv group aeronaut res & develop, NATO, 67-68; Nat Acad Sci exchange scientist, USSR, 69; consult, Boeing Sci Res Labs, 69-70, Math Sci Northwest Inc, 70-73 & Appl Physics Lab, 73; sr res award, Alexander von Humboldt Found, Ger, 73-74; guest prof, Royal Inst Technol, Stockholm & Inst Oceanog, Gothenburg, 73-74. Mem: AAAS; Am Geophys Union; NY Acad Sci; Am Phys Soc; Soc Natural Philos. Res: Non-linear mechanics; geophysical fluid dynamics, especially stability and turbulence. Mailing Add: Dept of Oceanog & Geophys Prog Univ of Wash Seattle WA 98195

CRIMMINS, TIMOTHY FRANCIS, b Hempstead, NY, Oct 14, 39; m 66. ORGANIC CHEMISTRY. Educ: St Johns Univ, NY, BS, 61; Purdue Univ, MS, 63, PhD(org chem), 65. Prof Exp: Res assoc of Dr C R Hauser & dir res, Duke Univ, 65-66; ASSOC PROF CHEM, WIS STATE UNIV, OSHKOSH, 66- Mem: Am Chem Soc. Res: Organosodium metalation reactions; alkylation of acetyl-acetone by means of boron trifluoride; equilibrium studies in liquid ammonia; phenyl migrations through carbanion intermediates. Mailing Add: Dept of Chem Univ of Wis Oshkosh WI 54901

CRIPPEN, GORDON MARVIN, b Cheyenne, Wyo, Apr 2, 45; m 70; c 1. BIOPHYSICAL CHEMISTRY, THEORETICAL CHEMISTRY. Educ: Univ Wash, BS, 67; Cornell Univ, PhD(biophys chem), 72. Prof Exp: Res assoc phys chem, Cardiovasc Res Inst, Univ Calif, San Francisco, 72-73; instr chem & physics, Bur Schs, Hamburg, Ger, 73-75; ASST PROF PHYS CHEM, SCH PHARM, UNIV CALIF, SAN FRANCISCO, 75- Res: Theoretical studies on the conformation of proteins and related subjects. Mailing Add: Dept of Pharmaceut Chem Sch of Pharm Univ Calif San Francisco CA 94143

CRIPPS, DEREK J, b London, Eng, Sept 17, 28; m 63; c 4. DERMATOLOGY, PHOTOBIOLOGY. Educ: Univ London, MB & BS, 53, MD, 65; Univ Mich, MS, 61; Am Bd Dermat, dipl, 69. Prof Exp: Intern med, London Hosps, 53-54; resident dermat, Med Ctr, Univ Mich, 59-62; sr registr, Inst Dermat, Eng, 62-65; asst prof med, 65-68, assoc prof dermat, 68-72, PROF DERMAT & CHMN DEPT, MED CTR, UNIV WIS-MADISON, 72- Concurrent Pos: NIH grant, 66-; mem, Study Comt, NIH, 69- Mem: Brit Dermat Asn; Am Acad Dermat; Am Fedn Clin Res; Soc Invest Dermat; fel Am Col Physicians. Res: Investigation of persons sensitive to sunlight and diseases of prophyrin metabolism. Mailing Add: Med Ctr Univ of Wis 1300 University Ave Madison WI 53706

CRIPPS, HARRY NORMAN, b Webster, NY, May 14, 25; m 48; c 4. ORGANIC CHEMISTRY. Educ: Ga Inst Technol, BS, 46; Univ Rochester, BS, 48; Univ Ill, PhD(chem), 51. Prof Exp: Chemist, Eastman Kodak Co, 48; CHEMIST, ORG CHEM DEPT, E I DU PONT DE NEMOURS & CO, 51- Mem: Sigma Xi. Res: Organic synthesis; polymers. Mailing Add: E I du Pont de Nemours & Co Wilmington DE 19898

CRISAN, ELI VICTOR, b Sharon, Pa, July 14, 34. MYCOLOGY. Educ: Youngstown Univ, AB, 56; Purdue Univ, MS, 59, PhD(plant sci), 62. Prof Exp: Asst plant pathologist, Boyce Thompson Inst Plant Res, NY, 61-68; ASST PROF FOOD SCI & ASST MICROBIOLOGIST, AGR EXP STA, UNIV CALIF, DAVIS, 68- Concurrent Pos: Lectr, Sch Gen Studies, Hunter Col, 62-64 & Grad Sch, 65-68; NSF grant, 63-65; NIH grant, 64-66; AID contract, 69-; Nutrit Found grant, 71-; prin investr, Sea Grant Prog, Nat Oceanic & Atmospheric Agency, 71-75. Mem: AAAS; Am Inst Biol Sci; Mycol Soc Am; Soc Indust Microbiol; Inst Food Technologists. Res: Thermophilic fungi; mycotoxins; fish protein concentrate; food fermentations; food from microbial sources. Mailing Add: Dept of Food Sci & Technol Univ of Calif Davis CA 95616

CRISCUOLO, DOMINIC, b New Haven, Conn, June 14, 08; m 46; c 2. BIOCHEMISTRY. Educ: Univ Pittsburgh, BS, 32; Trinity Univ, MS, 53. Prof Exp: Biochemist, Food Res Labs, 39-41, Sch Aviation Med, Randolph Field, 46-58 & Surg Res Lab, Lackland AFB Hosp, 58-68; chemist, Regional Environ Lab, Kelly AFB, 68-73; ADMINR, ST ANTHONY'S SPEC EDUC SCH, 73- Mem: Fel AAAS; Am Chem Soc; fel Am Inst Chemists; NY Acad Sci; Sigma Xi. Res: Enzymes; hematology; blood volumes; acclimatization to adverse environments; anesthesiology. Mailing Add: 321 Concord Ave San Antonio TX 78201

CRISLER, JOSEPH PRESLEY, b Hedley, Tex, Sept 12, 22; m 60; c 4. MICROCHEMISTRY. Educ: WTex State Univ, BS, 42, MA, 47; Univ Colo, PhD(chem micros), 62; Drexel Inst Technol, dipl elec micros, 63. Prof Exp: Head dept chem, Buena Vista Col, 47-54; instr inorg chem, Univ Colo, 54-55; asst prof anal chem, Tex Col Arts & Industs, 55-56; instr, Colo Sch Mines, 56-59; chemist micros, Univ Colo, 59-62; res chemist, US Naval Ord Sta, 62-69; CHIEF CHEM BR, DEPT HUMAN RESOURCES, GOVT DC, 69- Mem: Am Chem Soc; fel Royal Micros Soc. Res: Micro methods for the determination of trace elements, erythrocyte protoporphyrin and drugs of abuse in body fluids; chemical microscopy. Mailing Add: Chem Br Div of Labs Rm 6128 Munic Ctr 300 Indiana Ave NW Washington DC 20001

CRISLER, ROBERT MORRIS, b Columbia, Mo, Jan 5, 21; m 43; c 2. GEOGRAPHY. Educ: Univ Mo, AB, 41; Northwestern Univ, MS, 47, PhD(geog), 49. Prof Exp: Asst prof geog, Wash Univ, 48-54; assoc prof geog, Univ Southwestern La, 54-56, head dept & from assoc prof to prof social studies, 56-72, prof geog, 72-74; RETIRED. Mem: Asn Am Geogr; Nat Coun Geog Educ. Res: Electoral geography; resources and government of Louisiana; trade area delimitation. Mailing Add: 154 Ronald Blvd Lafayette LA 70501

CRISLEY, FRANCIS DANIEL, b Braddock, Pa, Aug 19, 26; m 60; c 3. MICROBIOLOGY, BACTERIOLOGY. Educ: Univ Pittsburgh, BS, 50, MS, 52, PhD(microbiol), 59. Prof Exp: Asst bact, Univ Pittsburgh, 50-52; instr, Miami Univ, 52-54 & Univ Pittsburgh, 54-58, res assoc microbiol, Dept Biol Sci, 59-61; microbiologist, Robert A Taft Sanit Eng Ctr, Ohio, 61-67; chmn dept biol, 67-75, PROF BIOL, NORTHEASTERN UNIV, 67- Honors & Awards: PHS Employee Award, 64. Mem: Am Soc Microbiol; Brit Soc Gen Microbiol; Soc Indust Microbiol. Res: Bacterial physiology; microbial ecology; public health microbiology; microbial toxins; food and industrial microbiology. Mailing Add: Dept Biol 360 Huntington Ave Northeastern Univ Boston MA 02115

CRISMON, JEFFERSON MARTINEAU, b Philadelphia, Pa, Feb 4, 08; m 37; c 2. PHYSIOLOGY. Educ: Stanford Univ, AB, 31, MD, 38. Prof Exp: Asst physiol, 32-33, asst pharmacol, 36-37, from instr to prof physiol, 37-73, from actg exec to exec dept, 49-63, EMER PROF PHYSIOL, STANFORD UNIV, 73- Concurrent Pos: Hon fel, Yale Univ, 40-41; Guggenheim fel, 57-58; consult, Off Naval Res, 50-51 & Surgeon-Gen, US Army, 52-53, 58-70. Mem: AAAS; Am Physiol Soc; Microcirculatory Soc. Res: Cardiac actions of vasconstrictor amines; isolated tissue metabolism; metabolism of water and electrolytes; effects of low body temperature on mammals; effects of regional ischemia; frostbite; control capillary blood-flow; human skin blood flow. Mailing Add: 1805 Guinda St Palo Alto CA 94303

CRISP, CARL EUGENE, b Buhl, Idaho, Aug 1, 31; m 55. PLANT PHYSIOLOGY. Educ: Univ Idaho, BS, 55, MS, 59; Univ Calif, Davis, PhD(plant physiol), 65. Prof Exp: Jr flight test engr, Lockheed Missile Systs Div, 55; teaching asst plant physiol, Univ Calif, Davis, 59-64; PLANT PHYSIOLOGIST, INSECTICIDE EVAL PROJ, US FOREST SERV, 65- Concurrent Pos: Consult, Diversified Sci Instruments, Inc, 67- Mem: AAAS; Am Chem Soc; Am Inst Biol Sci; Am Soc Plant Physiol. Res: Investigations into the anatomy and chemical composition of plant surfaces, the biopolymer cutin, the design of systemic insecticides and the mechanisms of phloem transport in plants. Mailing Add: Insecticide Eval Proj US Forest Serv PO Box 245 Berkeley CA 94701

CRISP, EDWARD LEE, b Elliottville, Ky, Dec 11, 46. PETROLEUM GEOLOGY, GEOCHEMISTRY. Educ: Morehead State Univ, BS, 69; Univ Ky, MS, 73; Ind Univ, PhD(geol), 75. Prof Exp: PETROL GEOLOGIST, TEXACO, INC, 75- Mem: Geol Soc Am; Am Asn Petrol Geologists; Nat Audubon Soc. Res: Carbonate, aquatic, marine and trace element geochemistry; petroleum exploration; paleoecology and paleobiology. Mailing Add: Texaco Inc Box 36650 Houston TX 77036

CRISP, MICHAEL DENNIS, b Elmhurst, Ill, Apr 27, 42; m 65. QUANTUM OPTICS, LASERS. Educ: Bradley Univ, AB, 64; Washington Univ, MS, 66, PhD(physics), 68. Prof Exp: Res assoc physics, Columbia Univ, 68-70; scientist, 70-72, SR SCIENTIST, OWENS-ILL INC, TOLEDO, 72- Concurrent Pos: Res assoc, Argonne Nat Lab, 65; instr physics, Univ Toledo, 72; Cong sci & eng fel, AAAS/Optical Soc Am, 74-76. Mem: Am Phys Soc; Optical Soc Am; Inst Elec & Electronics Engrs; Soc Info Display; AAAS. Res: Interaction of coherent light with matter, including laser physics, nonlinear optics, coherent pulse propagation and coherent spectroscopy; neoclassical radiation theory, laser-induced damage and gas-discharge display devices. Mailing Add: 6758 Gettysburg Dr Sylvania OH 43560

CRISP, THOMAS MITCHELL, JR, b San Antonio, Tex, Sept 29, 39; m 65; c 2. ANATOMY, ENDOCRINOLOGY. Educ: Univ St Thomas, Tex, 61; Rice Univ, MA, 64; Univ Tex, PhD(anat), 66. Prof Exp: Res asst endocrinol, Dent Br, Univ Tex, 60-61; teaching asst biol, Rice Univ, 61-63; from instr to asst prof, 66-72, ASSOC PROF ANAT, SCH MED & DENT, GEORGETOWN UNIV, 72- Concurrent Pos: NIH teaching fel anat, Med Br, Univ Tex, 63-66; consult, Army Oral Biol Prog, 66-70 & Navy Dent Sch, Bethesda Naval Hosp, 67-69; hon res fel, Dept Clin Endocrinol, Women's Hosp, Univ Birmingham, 74-75; spec res fel, Rockefeller Found, 74-75. Mem: AAAS; Am Soc Zoologists; Am Asn Anatomists; Soc Study Reprod; Endocrine Soc. Res: Electron microscopy of mammalian corpora lutea; ovarian enzymology. Mailing Add: Dept of Anat Georgetown Univ Washington DC 20007

CRISPELL, KENNETH RAYMOND, b Ithaca, NY, Oct 30, 16; m 42; c 6. MEDICINE. Educ: Philadelphia Col Pharm & Sci, BS, 38; Univ Mich, MD, 43; Am Bd Internal Med, dipl, 50. Prof Exp: From instr to assoc prof internal med, Univ Va, 49-58; prof med & dir dept, NY Med Col, 58-60; asst dean, Sch Med, 62, actg dean, 62-64, dean, 64-71, PROF MED, SCH MED, UNIV VA, 60-, VPRES HEALTH SCI, 71- Concurrent Pos: Fel internal med, Ochsner Clin, La, 47-48; fel biophys, Tulane Univ, 48-49; Commonwealth fel, Univ Va, 49-51; physician, Univ Hosp, Va, 49-58 & 60. Mem: Endocrine Soc; Am Soc Clin Invest; Am Thyroid Asn; AMA; fel Am Col Physicians. Res: Endocrinology. Mailing Add: Med Ctr Box 333 Univ of Va Charlottesville VA 22901

CRISPENS, CHARLES GANGLOFF, JR, b Bellevue, Pa, Aug 3, 30; m 53; c 1. ONCOLOGY. Educ: Pa State Univ, BS, 53; Ohio State Univ, MS, 55; Wash State Univ, PhD(zool), 59. Prof Exp: Fel, Jackson Lab, Bar Harbor, Maine, 59-60; from instr to assoc prof anat, Sch Med, Univ Md, 60-68; fac med, Univ Sherbrooke, 68-69; PROF BIOL, UNIV ALA, BIRMINGHAM, 69- Honors & Awards: Lederle Med Fac Award, 64. Mem: AAAS; Am Asn Anat; Am Med Asn; Am Asn Cancer Res; Am Soc Microbiol. Res: Chemically and virus-induced neoplasms in chickens; serum enzymes and diagnosis of cancer; lactate dehydrogenase virus and murine tumorigenesis; reticulum cell neoplasms of mice. Mailing Add: Dept of Biol Univ of Ala Univ Sta Birmingham AL 35294

CRISS, CECIL M, b Wheeling, WVa, Apr 22, 34; m 58; c 2. PHYSICAL CHEMISTRY. Educ: Kenyon Col, AB, 56; Purdue Univ, PhD(phys chem), 61. Prof Exp: Asst prof chem, Univ Vt, 61-65; asst prof, 65, ASSOC PROF CHEM, UNIV MIAMI, 65- Mem: AAAS; Am Chem Soc; The Chem Soc. Res: Thermodynamic properties of aqueous ionic solutions at higher temperatures; ionic heat capacities; ionic entropies and oxidation-reduction potentials in nonaqueous solutions. Mailing Add: Dept of Chem Univ of Miami Coral Gables FL 33124

CRISS, WAYNE ELDON, b Washington, Iowa, Mar 7, 40. CANCER, ENDOCRINOLOGY. Educ: William Penn Col, BS, 62; Univ Fla, MS, 64, PhD(biochem), 68. Prof Exp: Asst prof obstet & gynec, Med Sch, Univ Fla, 70-75, biochem, 74-75; DIR CANCER RES, COMPREHENSIVE CANCER CTR, HOWARD UNIV, 75- Concurrent Pos: Am Cancer Soc fel, Fels Inst, Temple Univ, 68-70; Nat Cancer Inst res career develop award, 74-79; consult, NSF, NIH, Cancer Res, Biochimica Biophysica Acta & Infection & Immunity. Mem: Endocrine Soc; Am Asn Univ Profs; Nat Tissue Cult Asn; Am Asn Cancer Res; Am Soc Biol Chemists. Res: Cancer metabolism, modified regulatory control mechanisms in cancer tissues. Mailing Add: Comprehensive Cancer Ctr Howard Univ Washington DC 20060

CRISSEY, LAVERNE WILLIAM, chemistry, see 12th edition

CRISSEY, WALTER FORD, b Ithaca, NY, Nov 4, 15; m 39; c 3. BIOLOGY. Educ: Cornell Univ, BS, 37. Prof Exp: Res proj leader game res, NY State Conserv Dept, 37-42 & 46-49; from asst chief to chief, Waterfowl Mgt Invest Sect, 49-57, staff specialist migratory game birds, 57-61, DIR MIGRATORY BIRD POP STA, BUR SPORT FISHERIES & WILDLIFE, 61- Mem: Wildlife Soc. Res: Population dynamics and life history research on North American waterfowl and other migratory birds; mourning doves, woodcock, snipe, rails and song birds. Mailing Add: Migratory Bird Pop Sta Bur Sport Fisheries & Wildlife Laurel MD 20810

CRISSMAN, JOHN MATTHEWS, b Evanston, Ill, Oct 21, 35; m 68. POLYMER SCIENCE. Educ: Pa State Univ, PhD(physics), 63. Prof Exp: PHYSICIST, NAT BUR STANDARDS, 63- Concurrent Pos: Nat Acad Sci-Nat Res Coun physicist, Nat Bur Standards, 63-65. Mem: Am Phys Soc. Res: Mechanical and other bulk properties of polymeric materials. Mailing Add: 7708 Maryknoll Ave Bethesda MD 20034

CRISSMAN, JUDITH ANNE, b Clarion, Pa, July 11, 42. INORGANIC CHEMISTRY. Educ: Thiel Col, BA, 64; Univ NC, Chapel Hill, PhD(inorg chem), 70. Prof Exp: Asst prof, 68-73, ASSOC PROF CHEM, MARY WASHINGTON COL, 73- Mem: Am Chem Soc. Res: Schiff base complexes of transition metals. Mailing Add: Dept of Chem Mary Washington Col Fredericksburg VA 22401

CRIST, BUCKLEY, JR, b Plainfield, NJ, Jan 12, 41; m 66; c 2. POLYMER SCIENCE. Educ: Williams Col, BA, 62; Duke Univ, PhD(chem), 66. Prof Exp: Phys chemist, Camille Dreyfus Lab, Res Triangle Inst, 66-73; ASST PROF MAT SCI & ENG & CHEM ENG, NORTHWESTERN UNIV, EVANSTON, 73- Mem: Am Chem Soc; Am Phys Soc. Res: Morphology and mechanical properties of semicrystalline polymers; scattering of light and x-rays by polymer solids; molecular motion and relaxation effects. Mailing Add: Dept of Mat Sci & Eng Northwestern Univ Evanston IL 60201

CRIST, DELANSON ROSS, b New York, NY, July 16, 40; m 62; c 3. PHYSICAL ORGANIC CHEMISTRY. Educ: Swarthmore Col, AB, 62; Mass Inst Technol, PhD(org chem), 67. Concurrent Pos: NSF fel, 67-68; asst prof org chem, 68-72, ASSOC PROF ORG CHEM, GEORGETOWN UNIV, 72- Mem: AAAS; Am Chem Soc. Res: Mechanisms of organic reactions of synthetic or biological importance; investigations concerning the stability and reactions of species containing positive-charged nitrogen atoms. Mailing Add: Dept of Chem Georgetown Univ Washington DC 20057

CRIST, RAY HENRY, b Mechanicsburg, Pa, Mar 8, 00; m; c 3. CHEMISTRY. Educ: Dickinson Col, AB, 20; Columbia Univ, AM, 22, PhD(chem), 26. Hon Degrees: DSc, Dickinson Col, 60. Prof Exp: Instr chem, Williamsport-Dickinson Sem, 20-21; asst, Columbia Univ, 22-26, from instr to assoc prof, 27-45, dir res substitute alloy mat labs, Manhattan Dist, 45-46; sr res chemist, Union Carbide Corp, 46-53, dir res coal utilization, 53-57; dir res, Union Carbide Olefins Co, 57-59, dir, Union Carbide Res Inst, 59-63; prof chem, 63-74, EMER PROF CHEM, DICKINSON COL, 74- Concurrent Pos: Cutting traveling fel, Inst Phys Chem, Berlin, 28-29. Mem: Am Chem Soc. Res: Photochemistry; reaction kinetics; reactions and properties of deuterium and its compounds. Mailing Add: Dept of Chem Dickinson Col Carlisle PA 17013

CRISTOFALO, VINCENT JOSEPH, b Philadelphia, Pa, Mar, 19, 33; m 64; c 3. PHYSIOLOGY, BIOCHEMISTRY. Educ: St Joseph's Col, BS, 55; Temple Univ, MA, 58; Univ Del, PhD, 62. Prof Exp: Asst instr gen biol, Temple Univ, 57-58; res asst, Univ Del, 58-60, NSF fel, 60-61; USPHS fel, Temple Univ, 61-62; mem, Wistar, Inst Anat & Biol, 63-69; from asst prof to assoc prof biochem, 67-74, PROF BIOCHEM, DIV ANIMAL BIOL, SCH VET MED, UNIV PA, 74-; INSTR DEPT CHEM, TEMPLE UNIV, 62-; ASSOC MEM, WISTAR INST ANAT & BIOL, 69- Concurrent Pos: Res asst, Oak Ridge Nat lab, 59; res assoc, Temple Univ, 63; reviewer, NIH, Study Sect, 74- & Molecular Cytol Study Sect, 75-78. Mem: AAAS; Tissue Culture Asn; Am Cell Biol Geront Soc; Soc Exp Biol & Med. Res: Effects of oxygen and radiation on development; intermediary metabolism; neoplastic tissues; tissue culture cells; aging in cell and tissue culture. Mailing Add: Wistar Inst of Anat & Biol 36th St & Spruce Philadelphia PA 19104

CRISTOL, STANLEY JEROME, b Chicago, Ill, June 14, 16; m 57; c 2. CHEMISTRY. Educ: Northwestern Univ, BS, 37; Univ Calif, Los Angeles, MA, 39, PhD(org chem), 43. Prof Exp: Asst chem, Univ Calif, Los Angeles, 37-38; res chemist, Stand Oil Co, Calif, 38-41; from asst to instr chem, Univ Calif, Los Angeles, 41-43; res fel, Univ Ill, 43-44; res chemist, USDA, Md, 44-46; from asst prof to assoc prof chem, 46-55, chmn dept, 60-62, fac res lectr, 60, PROF CHEM, UNIV COLO, BOULDER, 55- Concurrent Pos: Guggenheim fel, 55-56; assoc ed, Chem Rev, 57-59; consult, E I du Pont de Nemours & Co, Inc; NSF Adv Comt, 57-63, 69- & NIH Adv Comt, 68- Honors & Awards: Stearns Award, 71; James Flack Norris Award, Am Chem Soc, 72. Mem: Nat Acad Sci; AAAS; Am Chem Soc; The Chem Soc. Res: Organic chemistry; mechanisms of organic reactions. Mailing Add: Dept of Chem Univ of Colo Boulder CO 80302

CRISWELL, BENNIE SUE, b Huntsville, Tex, Nov 17, 42; m 64; c 2. IMMUNOLOGY, INFECTIOUS DISEASES. Educ: NTex State Univ, BS, 64; Registry Med Technol, cert, 64; Baylor Col Med, MS, 68, PhD(immunol), 69. Prof Exp: Instr, 72-74, ASST PROF MICROBIOL, BAYLOR COL MED, 74- Concurrent Pos: NASA fel, Manned Spacecraft Ctr, Houston, 69-72; vis scientist, NASA Johnson Space Ctr, 72- Res: Early detection of viral respiratory illness. Mailing Add: 15710 Buccaneer Houston TX 77058

CRISWELL, JEROME GLENN, b Oshkosh, Nebr, June 27, 44. CROP PHYSIOLOGY. Educ: Univ Nebr, Lincoln, BS, 66; Iowa State Univ, MS, 68, PhD(agron), 70. Prof Exp: Asst agron, Iowa State Univ, 66-67; res assoc, Univ Guelph, 70-74; RES SCIENTIST, E I DUPONT DE NEMOURS & CO, 74- Mem: Am Soc Agron; Crop Sci Soc Am. Res: Crop production; physiological basis for genotypic variation in net photosynthesis; carbohydrate distribution pattern in cereals; photo-periodic insensitiveness in soybeans; nitrogen and carbon nutrition in legumes. Mailing Add: Cent Res & Develop Dept Exp Sta E I duPont de Nemours & Co Wilmington DE 19898

CRITCHFIELD, CHARLES LOUIS, b Shreve, Ohio, June 7, 10; m 35; c 4. MATHEMATICAL PHYSICS. Educ: George Washington Univ, BS, 34, MA, 36, PhD(physics), 39. Prof Exp: Instr, Univ Rochester, 39-40 & Harvard Univ, 41-42; physicist, Geophys Lab, 42-43 & Monsanto Chem Co, 46-47; assoc prof physics, George Washington Univ, 46; from assoc prof to prof, Univ Minn, 47-55; dir sci res, Convair Div, Gen Dynamics, 55-60; vpres res, Telecomput Corp, 60-61; physicist, 43-46, ASSOC DIV & GROUP LEADER NUCLEAR PHYSICS, LOS ALAMOS SCI LAB, 61- Concurrent Pos: Mem planetology subcomt, NASA, 64-69. Mem: Fel Am Phys Soc. Res: Scalar potentials in the Dirac equation for nuclear and particle physics. Mailing Add: 391 El Conejo Los Alamos NM 87544

CRITCHFIELD, FRANK EDWARD, polymer chemistry, see 12th edition

CRITCHFIELD, HOWARD JOHN, b Vernon, Colo, Sept 24, 20; wid; c 2. GEOGRAPHY, CLIMATOLOGY. Educ: Univ Wash, BA, 46, MA, 47, PhD(geog), 52. Prof Exp: Elem teacher, Boundary County Sch Dist 2, Idaho, 40-42; instr geog, Wash State Univ, 47-48; vis lectr, Univ Canterbury, 48-50; from asst prof to prof geog, 51-75, chmn dept, 60-74, PROF GEOG & REGIONAL PLANNING, WESTERN WASH STATE COL, 75- Concurrent Pos: Assoc geog, Univ Wash, 51; vis lectr, Univ Canterbury, 59, res prof, 71; Univ NZ travel grant, Univ Canterbury, 59; travel grant, Nat Acad Sci 21st Int Geog Cong, New Delhi, 68; hon res assoc, Univ London, 64; vis prof, Univ Stellenbosch, 72. Mem: AAAS; Asn Am Geog; Am Meteorol Soc; Can Asn Geog; NZ Geog Soc. Res: Climatic perception; climate of Puget Sound Lowland. Mailing Add: Dept of Geog & Regional Planning Western Wash State Col Bellingham WA 98225

CRITCHFIELD, WILLIAM BURKE, b Minneapolis, Minn, Nov 21, 23. FOREST GENETICS. Educ: Univ Calif, BS, 49, PhD(bot), 56. Prof Exp: Forest geneticist, Cabot Found, Harvard Univ, 56-59; GENETICIST, PAC SOUTHWEST FOREST & RANGE EXP STA, US FOREST SERV, 59- Mem: Bot Soc Am. Res: Geographic variation and evolution in forest trees; leaf variation and shoot development in trees. Mailing Add: Pac Southwest Forest & Range Exp Sta PO Box 245 Berkeley CA 94701

CRITCHLOW, BURTIS VAUGHN, b Hotchkiss, Colo, Mar 5, 27; m 48; c 4. ANATOMY. Educ: Occidental Col, BA, 51; Univ Calif, Los Angeles, PhD(neuroendocrinol), 57. Prof Exp: Instr anat, Baylor Col Med, 58-63, from assoc prof to prof, 63-72; PROF ANAT & CHMN DEPT, HEALTH SCI CTR, UNIV ORE, 72- Concurrent Pos: Sr res fel, 59-64; USPHS res career develop award, 64-69; mem, Reprod Biol Study Sect, NIH, 69-73, consult. Mem: Am Asn Anatomists; Endocrine Soc; Int Soc Res Reprod; Am Physiol Soc; Int Brain Res Orgn. Res: Brain and endocrine interrelations; neural control of pituitary functions. Mailing Add: Med Sch Dept of Anat Univ of Ore Health Sci Ctr Portland OR 97201

CRITES, JOHN L, b Wilmington, Ohio, July 10, 23; m 46; c 2. ZOOLOGY. Educ: Univ Idaho, BS, 49, MSc, 51; Ohio State Univ, PhD(zool), 56. Prof Exp: Asst zool, Univ Idaho, 49-51; asst, 51-54, asst instr, 54-55, from instr to assoc prof, 55-67, PROF ZOOL, OHIO STATE UNIV, 67- Concurrent Pos: Consult, US AID, Sci Inst, India, 63- Mem: Am Soc Parasitol; Wildlife Dis Asn; Int Soc Nematol. Res: Nematode parasites of animals; nematode parasites of plants; free-living nematodes, both marine and fresh water. Mailing Add: Ohio State Univ Dept of Zool 1735 Neil Ave Columbus OH 43210

CRITTENDEN, ALDEN LA RUE, b Wichita, Kans, Nov 27, 20; m 58. CHEMISTRY. Educ: Univ Ill, BS, 42, PhD(chem), 47. Prof Exp: From instr to asst prof, 47-60, ASSOC PROF CHEM, UNIV WASH, 60- Mem: Am Chem Soc. Res: Polarography. Mailing Add: Dept of Chem Univ of Wash Seattle WA 98105

CRITTENDEN, EUGENE CASSON, JR, b Washington, DC, Dec 25, 14; m 42; c 2. SOLID STATE PHYSICS, OPTICS. Educ: Cornell Univ, AB, 34, PhD(physics), 39. Prof Exp: Asst physics, Cornell Univ, 34-38; from instr to prof, Case Inst Technol, 38-53; prof physics, 53-71, chmn dept, 64-67, DISTINGUISHED PROF PHYSICS, NAVAL POSTGRAD SCH, 71- Concurrent Pos: Res physicist, Radiation Lab, Univ Calif, 44-45 & Atomic Energy Res Dept, NAm Aviation, 52-53. Mem: Fel Am Phys Soc. Res: Ferromagnetism; superconductivity; optical transmission in the marine boundary layer. Mailing Add: Dept of Physics Naval Postgrad Sch Monterey CA 93940

CRITTENDEN, HENRY WILLIAM, b Burton, Ohio, July 7, 16; m 45; c 2. PLANT PATHOLOGY. Educ: Kent State Univ, BS, 39; Ohio State Univ, MS, 40, PhD(bot, plant path), 58. Prof Exp: Asst bot, Ohio State Univ, 41-42 & 46-47; from asst prof to assoc prof plant path, 48-73, PROF PLANT SCI, UNIV DEL, 73- Mem: Bot Soc Am; Am Phytopath Soc. Res: Control of plant pathogenic organisms in the soil; soybean diseases. Mailing Add: Dept of Plant Sci Univ of Del Newark DE 19711

CRITTENDEN, LYMAN BUTLER, b New Haven, Conn, May 27, 26; m 59; c 2. GENETICS. Educ: Calif Polytech Inst, BS, 51; Purdue Univ, MS, 55, PhD, 58. Prof Exp: Geneticist, Nedlar Farms, 51-53; instr genetics, Purdue Univ, 55-57; geneticist, Creighton Bros, 57-58 & Nat Cancer Inst, 58-61; geneticist, Regional Poultry Res Lab, 61-67 & Animal Physiol & Genetics Inst, Beltsville, Md, 67-75, RES GENETICIST, REGIONAL POULTRY RES LAB, MICH, USDA, 75- Mem: AAAS; Poultry Sci Asn; Genetics Soc Am. Res: Quantitative genetics; genetics of disease resistance; virology. Mailing Add: Regional Poultry Res Lab 3606 E Mt Hope Rd E Lansing MI 48823

CRITTENDEN, MAX DERMONT, JR, b Seattle, Wash, May 12, 17; m 42; c 4. ENVIRONMENTAL GEOLOGY. Educ: San Jose State Col, BA, 39; Univ Calif, PhD, 49. Prof Exp: Asst geol, Univ Calif, 40-42; geologist, 42-65, chief southwestern br, 65-70, RES GEOLOGIST, US GEOL SURV, 65- Concurrent Pos: Vis prof, Univ Calif, Santa Cruz, 69; chmn, Geothermal Environ Adv Panel, 74- Mem: Fel Geol Soc Am. Res: Geology of manganese deposits especially origin and occurence in the Western states; geologic history of Wasatch Mountains, Utah, especially Pre-Cambrian rocks and Laramide structural history; isostatic recovery of Lake Bonneville, Utah and viscosity of earth. Mailing Add: US Geol Surv 345 Middlefield Rd Menlo Park CA 94025

CRITTENDEN, RAY RYLAND, b Galesburg, Mich, Mar 19, 31; m 62; c 3. PHYSICS. Educ: Willamette Univ, BA, 54; Univ Wis, MS, 56, PhD(physics), 60. Prof Exp: Assoc scientist, Brookhaven Nat Lab, 60-63; from asst prof to assoc prof, 63-74, PROF PHYSICS, COL ARTS & SCI, GRAD SCH, IND UNIV, BLOOMINGTON, 74- Mem: Am Phys Soc. Res: High energy nuclear physics. Mailing Add: Dept of Physics Ind Univ Bloomington IN 47405

CRITTENDEN, REBECCA SLOVER, b Lake City, Tenn, July 10, 36; m 66. ALGEBRA. Educ: Georgetown Col, BS, 58; Univ NC, Chapel Hill, MA, 62, PhD(math), 63. Prof Exp: Asst prof math, Georgetown Col, 63-64, Va Polytech, 64-66 & Vanderbilt Univ, 66-67; asst prof, 67-70, ASSOC PROF MATH, VA POLYTECH INST & STATE UNIV, 70- Concurrent Pos: Woodrow Wilson fel. Mem: Am Math Soc; Math Asn Am. Res: Ring theory. Mailing Add: Dept of Math Va Polytech Inst & State Univ Blacksburg VA 24061

CRITTENDEN, RICHARD JAMES, b Milwaukee, Wis, Feb 28, 30; m 53; c 3. MATHEMATICS. Educ: Williams Col, AB, 52; Oxford Univ, BA, 54; Mass Inst Technol, PhD(math), 60. Prof Exp: Asst prof math, Northwestern Univ, 60-64; assoc ed, Math Rev, Am Math Soc, 64-68, exec ed, 68-71; chmn dept, 71-74, PROF MATH, COL GEN STUDIES, UNIV ALA, BIRMINGHAM, 71- Mem: Am Math Soc. Res: Differential geometry; Riemannian geometry; G-structures. Mailing Add: Dept of Math Col Gen Studies Univ of Ala Birmingham AL 35233

CRITZ, JERRY B, b Boonville, Mo, Apr 9, 34; m 55; c 2. PHYSIOLOGY. Educ: Univ Mo, BS, 56, MA, 58, PhD(physiol), 61. Prof Exp: From asst prof to assoc prof physiol, Univ SDak, 61-67; assoc prof, Univ Western Ont, 67-73; ASSOC PROF PHYSIOL, CTR MED EDUC, IND UNIV SOUTH BEND, 73- Concurrent Pos: NIH fel, 62-64. Mem: AAAS; assoc Am Physiol Soc. Res: Cardiovascular tolerance to exercise. Mailing Add: Ind Univ Ctr for Med Educ 1825 Northside Blvd South Bend IN 46615

CRIVELLO, JAMES V, b Grand Rapids, Mich, July 30, 40. ORGANIC CHEMISTRY. Educ: Aquinas Col, BS, 62; Univ Notre Dame, PhD(org chem), 66. Prof Exp: RES CHEMIST, RES & DEVELOP CTR, GEN ELEC CO, 66- Mem: Am Chem Soc. Res: Polymer science; thermally stable polymers; synthesis and characterization; oxidation and nitration chemistry; organic photochemistry. Mailing Add: Gen Elec Res & Develop Ctr Bldg K-1, PO Box 8 Schenectady NY 12301

CROACH, JESSE WILLIAM, JR, b Dyersburg, Tenn, Dec 16, 18; m 44; c 1. PHYSICS, RESEARCH ADMINISTRATION. Educ: Harvard Univ, AB, 40. Prof Exp: Physicist, Explosives Dept, Burnside Lab, E I du Pont de Nemours & Co, 40-41, ballistic supvr Ala ord works, 42-43, Burnside Lab, 43-44, asst tech supt Ind ord works plant II, 44-45; arms res physicist, Remington Arms, 45-47; supvr chem eng, Burnside Lab, 47-50, supvr atomic energy div, Chicago Liaison Off, 51-53, mgr theoret physics div, Savannah River Lab, 53-54, dir pile eng & mat sect, 54-59, gen supt works tech dept, Savannah River Plant, 60-61, asst dir, Savannah River Lab, 62-63, asst tech dir, Atomic Energy Div, Wilmington, 64-67, TECH DIR, ATOMIC ENERGY DIV, E I DU PONT DE NEMOURS & CO, WILMINGTON, 67- Mem: AAAS; Fedn Am Scientists. Res: Management science; nuclear reactor physics; reactor engineering; fuel element development; reactor safety; large-scale isotope production; orientation of materials; applied mathematics; computer applications. Mailing Add: 1007 Market St Wilmington DE 19898

CROASDALE, HANNAH THOMPSON, b Daylesford, Pa, Nov 18, 05. BOTANY. Educ: Univ Pa, BS, 28, MS, 31, PhD(bot), 35. Prof Exp: Mem staff, Biol Abstracts, 28-32; prepateur bot, Univ Pa, 32-33; res asst, 35-46, assoc zool, 46-59, from asst prof to prof biol, 59-71, EMER PROF BIOL, DARTMOUTH COL, 71- Mem: Am Phycol Soc; Am Micros Soc. Res: Systematics of freshwater algae; arctic and tropical desmids; desmid flora of North America. Mailing Add: Dept of Biol Dartmouth Col Hanover NH 03755

CROCE, LOUIS J, b New York, NY, Sept 17, 21; m 48; c 2. PHYSICAL

CHEMISTRY, ORGANIC CHEMISTRY. Educ: St John's Univ, NY, BS, 48; NY Univ, PhD(phys org chem), 52. Prof Exp: Group leader org res, Evans Res & Develop Co, 51-52; res chemist, Socony-Mobile Oil Co, Inc, 52-55; res chemist, 55-58, group leader, 58-60, res supvr, 60-61, res mgr, 61-66, assoc dir res, 66-74, DIR RES, PETRO-TEX CHEM CORP, 74- Mem: AAAS; Am Chem Soc. Res: Petrochemicals, including monomer synthesis, catalytic dehydrogenation, dehydrocyclization, ammoxidation; reactions of olefins and dienes; polyolefins; petroleum research in hydrocracking, hydrocarbon separations, alkylation. Mailing Add: 135 Driftwood Dr Seabrook TX 77586

CROCKER, ALLEN CARROL, b Boston, Mass, Dec 25, 25; m 53; c 3. PEDIATRICS. Educ: Mass Inst Technol, 42-44; Harvard Univ, MD, 48. Prof Exp: Lab house officer, 48-49, jr asst resident med, 49-51, from asst physician to assoc physician, 56-62, res assoc path, 56-68, assoc med, 62-66, SR ASSOC MED, CHILDREN'S HOSP MED CTR, 66- Concurrent Pos: Fel path, Children's Hosp Med Ctr, 53-56; res assoc path, Harvard Med Sch, 56-60, res assoc pediat, 60-66, asst clin prof, 66-69, assoc prof pediat, 69-, tutor med sci, 64- Res: Clinical investigation; pediatric metabolic diseases; biochemistry of the lipids; mental retardation. Mailing Add: Children's Hosp Med Ctr 300 Longwood Ave Boston MA 02115

CROCKER, DENTON WINSLOW, b Salem, Mass, May 1, 19; m 46; c 4. INVERTEBRATE ZOOLOGY. Educ: Northeastern Univ, BA, 42; Cornell Univ, MA, 48, PhD(zool), 52. Prof Exp: Instr biol, Amherst Col, 51-53; from instr to assoc prof, Colby Col, 53-60; PROF BIOL & CHMN DEPT, SKIDMORE COL, 60- Mem: AAAS; Soc Syst Zool; Am Soc Zoologists; Soc Study Evolution. Res: Systematics and physiological ecology of crayfishes. Mailing Add: Dept of Biol Skidmore Col Saratoga Springs NY 12866

CROCKER, DIANE WINSTON, b Cambridge, Mass, Nov 23, 26; m 49, 74; c 3. PATHOLOGY, DATA PROCESSING. Educ: Wellesley Col, BA, 46; Brown Univ, MS, 48; Boston Univ, MD, 52. Prof Exp: Asst path, Univ Southern Calif, 55-56; asst, Col Physicians & Surgeons, Columbia Univ, 57-58; instr, Harvard Med Sch, 58-65, assoc, 65-68, asst clin prof, 68-69; from asst prof to prof path & chief dept, Health Sci Ctr Hosp, Temple Univ, 69-73; PROF PATH, UNIV SOUTHERN CALIF, 73-, PROF & CHIEF ANAT PATH DATA PROCESSING, LOS ANGELES COUNTY-UNIV SOUTHERN CALIF MED CTR, 73- Concurrent Pos: Los Angeles County Heart Asn res grant, Los Angeles Children's Hosp, 55-56; Am Cancer Soc fel, Francis Delafield Hosp, New York, 56-57 & Presby Hosp, 58; surg pathologist & chief cytol, Peter Bent Brigham Hosp, Boston, 58-70, attend & consult, 70-71; attend, Vet Admin Hosp, West Roxbury, Mass, 60-70. Mem: AAAS; fel Royal Soc Med; Am Soc Nephrology; Int Acad Path; Int Soc Nephrology. Res: Renal disease and hypertension. Mailing Add: Box 2111 LA Co-Univ So Calif Med Ctr Los Angeles CA 90033

CROCKER, IAIN HAY, b Hamilton, Ont, July 28, 28; m 55; c 3. ANALYTICAL CHEMISTRY. Educ: McMaster Univ, BSc, 50. Prof Exp: Anal chemist, NAm Cyanamid Co, 50-52; chemist, NRX Reactor Opers, 52-53, Chem Process Develop, 53-55 & Develop Chem, 55-62, assoc res officer, 62-72, SECT HEAD, MASS SPECTROMETRY & FUEL ANAL, CHALK RIVER NUCLEAR LABS, ATOMIC ENERGY, CAN, LTD, 72- Mem: Chem Inst Can; Am Soc Mass Spectrometry. Res: Mass spectrometry-spark source and thermionic; nuclear fuel analyses; burnup; trace analyses; environmental analyses. Mailing Add: Chalk River Nuclear Labs Atomic Energy of Can Ltd Chalk River ON Can

CROCKER, THOMAS TIMOTHY, b Barranquilla, Colombia, May 9, 20; m 45; c 4. INFECTIOUS DISEASES, VIROLOGY. Educ: Univ Calif, AB, 42, MD, 44. Prof Exp: Asst med, Hosp & Sch Med, Univ Calif, 45-46; asst med, Grace-New Haven Community Hosp, 48-49; from asst prof to prof med, Univ Calif, San Francisco, 50-71; PROF MED & CHMN DEPT COMMUN & ENVIRON MED, UNIV CALIF, IRVINE, 71- Concurrent Pos: Nat Res Coun fel virol, Sch Med, Yale Univ, 49-50; Markle Found scholar, 50-55; Guggenheim fel, Clare Col, Cambridge Univ, 57-58; res assoc, Cancer Res Inst, Univ Calif, San Francisco, 57-71; consult, Calif State Dept Pub Health, 58-, Nat Cancer Inst, 69-70 & Biol Div, Oak Ridge Nat Lab, 69-70; mem extramural grants adv comt, Air Pollution Control Off, Environ Protection Agency, 69- Mem: AAAS; Am Tissue Cult Asn; Am Soc Cell Biol; Am Thoracic Soc; Am Inst Biol Sci. Res: Cellular physiology of mammalian respiratory epithelia and tumors; physiology of virus-infected cells; chemical and viral carcinogenesis with correlation between rodent and primate susceptibility to epithelial versus fibroblastic transformation. Mailing Add: Col of Med Univ of Calif Irvine CA 92664

CROCKER, WILLIAM HENRY, b San Francisco, Calif, Aug 20, 24; div. CULTURAL ANTHROPOLOGY. Educ: Yale Univ, BA, 50; Stanford Univ, MA, 53; Univ Wis, PhD, 62. Prof Exp: Assoc cur SAm Ethnol, 62-74; CUR ETHNOL, SMITHSONIAN INST, 74- Concurrent Pos: NSF grant, 64; Wenner-Gren Found Anthrop Res grant, 64. Mem: AAAS; Am Anthrop Asn; Latin Am Studies Asn. Res: Ethnology, especially of the Ramkokamekra-Canela Indians of Maranhao, Brazil; process of cultural change and conservatism over a long period of time. Mailing Add: Div Lat Am Anthrop Dept Anthrop Smithsonian Inst Washington DC 20560

CROCKET, JAMES HARVIE, b Fredericton, NB, June 27, 32; m 58; c 2. GEOCHEMISTRY, GEOLOGY. Educ: Univ NB, BSc, 55; Oxford Univ, BSc, 57; Mass Inst Technol, PhD(geochem), 61. Prof Exp: From asst prof to assoc prof, 61-74, PROF GEOL, McMASTER UNIV, 74- Mem: Geochem Soc; Am Geophys Union. Res: Neutron activation studies of the geochemistry of precious metals in basic rocks; genesis of ore deposits; geochronology. Mailing Add: Dept of Geol McMaster Univ Hamilton ON Can

CROCKETT, DAVID SCOTT, b Cranford, NJ, Mar 6, 31; m 53; c 5. INORGANIC CHEMISTRY. Educ: Colby Col, AB, 52; Univ NH, MS, 54, PhD(inorg chem), 60. Prof Exp: Trainee, Owens-Corning Fiberglass, 53-54; asst prof inorg chem, 59-65, asst dean acad affairs, 65-67, assoc dean col, 67-68, dean spec progs, 68-73, ASSOC PROF CHEM, LAFAYETTE COL, 65-, ASSOC PROVOST, 73- Mem: Am Chem Soc. Res: Study of complex fluorides, particularly in the solid state by means of x-ray diffraction; infrared spectro-photometry. Mailing Add: 109 Markle Hall Lafayette Col Easton PA 18042

CROCKETT, JERRY J, b Chickasha, Okla, May 25, 28; m 51; c 3. PLANT ECOLOGY. Educ: Northwestern State Col, Okla, BS, 51; Ft Hays Kans State Col, MS, 60; Univ Okla, PhD(bot), 62. Prof Exp: Teacher high school, Tex, 51-52; develop & res chemist, Continental Oil Co, Okla, 52-55; corrosion engr, Tech Serv Dept, Petrolite Corp, 55-59; asst bot & plant ecol, Ft Hays Kans State Col, 59-60; asst & instr bot, Univ Okla, 60-62; from asst prof to assoc prof, Okla State Univ, 62-67; prof & assoc dean col letters & sci, Univ Idaho, 67-68; PROF BOT & DIR ARTS & SCI EXTEN, OKLA STATE UNIV & DIR OKLA JR ACAD SCI, 68- Mem: Ecol Soc Am; Am Soc Range Mgt; Grassland Res Found (secy, 62-64). Res: Mechanisms of secondary plant succession; productivity of grasslands; soil-geology-plant relationships; terrestrial pollution. Mailing Add: Dept of Bot Okla State Univ Stillwater OK 74074

CROCKETT, JOE RICHARD, b Stamford, Tex, Aug 7, 26; m 56; c 2. ANIMAL

GENETICS. Educ: NMex State Univ, BSA, 53; Univ Fla, MSA, 58, PhD(animal breeding), 62. Prof Exp: Asst prof, Univ & asst animal geneticist, Exp Sta, 62-69, ASSOC PROF ANIMAL SCI, UNIV FLA & ASSOC ANIMAL GENETICIST, EVERGLADES EXP STA, 69- Mem: Am Soc Animal Sci; Am Genetic Asn. Res: Breeding and management of beef cattle adapted to subtropical conditions. Mailing Add: IFAS Agr Res & Educ Ctr 579 SE Fifth St Belle Glade FL 33430

CROCKFORD, HORACE DOWNS, chemistry, see 12th edition

CROCKFORD, JACK ALFRED, wildlife management, see 12th edition

CROFFORD, OSCAR BLEDSOE, b Chickasha, Okla, Mar 29, 30; m 57; c 3. MEDICAL RESEARCH, INTERNAL MEDICINE. Educ: Vanderbilt Univ, AB, 52, MD, 55. Prof Exp: Intern med, Hosp, Vanderbilt Univ, 55-56, asst resident, 56-57, USPHS res fel clin physiol, Univ, 59-62, resident med, Hosp, 62-63; USPHS fel clin biochem, Univ Geneva, 63-65; from asst prof to assoc prof, 65-74, PROF MED, SCH MED, VANDERBILT UNIV, 74-, ASSOC PROF PHYSIOL, 70- Concurrent Pos: Investr, Howard Hughes Med Inst, 65-71; mem, Metab Study Sect, NIH, 70-74, chmn, 72-74; Addison B Scoville, Jr Chair Diabetes & Metab, Sch Med, Vanderbilt Univ, 73-, div head diabetes & metab, Dept Med, 73-, dir, Diabetes-Endocrinol Ctr, 73-; mem biomed adv bd, Gen Foods Corp, 74-; chmn, Nat Comn Diabetes, 75-76. Honors & Awards: Lilly Award, Am Diabetes Asn, 70, Charles H Best Award, 76; Humanitarian Award, Juv Diabetes Found, 76. Mem: Am Diabetes Asn; Am Physiol Soc; Endocrine Soc; Am Soc Clin Invest; Asn Am Physicians. Res: Hormone control of metabolism in adipocytes; mechanism of action of insulin; sugar transport; pathophysiology and treatment of Diabetes Mellitus; pathophysiology and treatment of obesity. Mailing Add: A-5119 Med Ctr Vanderbilt Univ Sch of Med Nashville TN 37232

CROFT, ALFRED RUSSELL, b Ogden, Utah, July 29, 96; m 21; c 4. ECOLOGY. Educ: Utah State Univ, BS, 20, MS, 25. Prof Exp: Teacher high sch, Utah, 20-24 & Weber Col, 26-34; forest ecologist, Intermountain Forest & Range Exp Sta, US Forest Serv, 34-50, forest hydrologist, Nat Forest Admin, 50-62; consult forest hydrologist, 62-64; prof, Univ Ariz, 64-67; CONSULT HYDROLOGIST, 68- Honors & Awards: Super Serv Award, USDA, 57; Distinguished Serv Award, Ariz Water Resources Comt, 66. Mem: Soc Am Foresters; hon mem Soil Conserv Soc Am; Am Geophys Union. Res: Forest hydrology; consumptive use of water; erosion control. Mailing Add: 3921 S 895 E Ogden UT 84403

CROFT, CHARLES CLAYTON, b Washington, DC, June 16, 14; m 44; c 2. MICROBIOLOGY, PUBLIC HEALTH. Educ: Univ Md, BS, 36, MS, 37; Johns Hopkins Univ, ScD(bact), 49; Am Bd Microbiol, dipl. Prof Exp: Lab technician, Md State Health Dept, 37; sr bacteriologist, Ariz State Lab, 38-43; asst chief labs, 49-61, CHIEF BUR LABS, OHIO DEPT HEALTH, 61- Concurrent Pos: Consult, WHO, 60; chmn, Intersoc Comt Lab Serv Related to Health, 61-64. Mem: Am Soc Microbiol; fel Am Pub Health Asn; fel Am Acad Microbiol. Res: Public health laboratory administration; virology; syphilis serology; enteric bacteriology; hemolytic antibody. Mailing Add: Ohio Dept of Health Lab 1571 Perry St Columbus OH 43210

CROFT, GEORGE THOMAS, b Washington, DC, Sept 29, 26; m 48; c 3. PHYSICS. Educ: Western Md Col, BS, 48; Univ Pa, PhD(physics), 53. Prof Exp: Physicist, Sound Div, Naval Res Lab, 47-48; physicist, Frankford Arsenal, 49-50; res physicist, Edison Lab, McGraw Edison, Inc, 53-58; supvr appl res, Pitney-Bowes Inc, 58-61, mgr, Appl Res Lab, 61-68, dir res & develop, 68, dir corp res, develop & eng, 68-74; V PRES CORP RES & DEVELOP, ADDRESSOGRAPH MULTIGRAPH CORP, 74- Mem: Am Phys Soc; Sigma Xi; Inst Elec & Electronics Engrs; Electrochem Soc; NY Acad Sci. Res: Low temperature physics; semiconductors; mechanisms of electrochemical reactions; mathematical analysis. Mailing Add: Addressograph Multigraph Corp 20600 Chagrin Blvd Cleveland OH 44122

CROFT, PAUL DOUGLAS, b Ft Erie, Ont, Oct 16, 37; US citizen; m 61; c 3. NUCLEAR CHEMISTRY. Educ: Univ Western Ont, BSc, 59; Univ Calif, Berkeley, PhD(chem), 64. Prof Exp: Res assoc & instr chem, State Univ NY Stony Brook, 64-66, lectr chem & dir chem labs, 66-72; dean for admin, Univ Mich, Flint, 72-75; EXEC OFFICER CHEM, UNIV CALIF, SAN DIEGO, 75- Concurrent Pos: Res collab, Brookhaven Nat Labs, 64-66. Mem: AAAS; Am Chem Soc; Chem Inst Can. Mailing Add: Univ of Calif at San Diego B-014 La Jolla CA 92093

CROFT, THOMAS STONE, b Marfa, Tex, May 9, 38; m. ORGANIC CHEMISTRY. Educ: Univ Fla, BS, 61; Univ Colo, PhD(org chem), 66. Prof Exp: RES SPECIALIST, 3M CO, 66- Mem: Am Chem Soc. Res: Organic fluorine chemistry; photochemistry. Mailing Add: Cent Res Lab 3M Co St Paul MN 55119

CROFT, WALTER LAWRENCE, b Longview, Miss, June 28, 35; m 68. PHYSICS. Educ: Miss State Univ, BS, 57; Vanderbilt Univ, PhD(physics), 64. Prof Exp: From asst prof to assoc prof, 62-71, PROF PHYSICS, MISS STATE UNIV, 71- Mem: Am Asn Physics Teachers. Res: Nuclear spectroscopy; Mössbauer effect; microwave spectroscopy. Mailing Add: 120 Hilburn Hall Miss State Univ State College MS 39762

CROFT, WILLIAM JOSEPH, b New York, NY, Nov 29, 26; m 49; c 2. CRYSTALLOGRAPHY. Educ: Columbia Univ, BS, 50, MA, 52, PhD(mineral, crystallog), 54. Prof Exp: Asst mineral, Columbia Univ, 52-54; instr, Hofstra Col, 54-55; staff mem crystallog, Lincoln Lab, Mass Inst Technol, 55-57; sr staff mem, Radio Corp Am, 57-61; staff mem, Sperry Rand Res Ctr, 61-69; RES CHEMIST, ARMY MAT & MECH RES CTR, 69- Concurrent Pos: Vis prof, Brown Univ, 70-72. Mem: AAAS; fel Mineral Soc Am; Am Crystallog Asn; Mineral Soc Gt Brit & Ireland. Res: X-ray crystallographic studies of oxides; high and low temperature phase transformations; lattice distortion and crystallite size in multiple oxides. Mailing Add: Army Mat & Mech Res Ctr Watertown MA 02172

CROFT, WILMA JANICE, b Brooksville, Miss, Oct 16, 40; div. CHEMISTRY, INFORMATION SCIENCE. Educ: Miss State Col Women, BS, 62; Fla State Univ, MS, 67. Prof Exp: Teacher, Pub Schs, Fla, 62-65; res librn, Tenneco Chem, Tenneco Co, 65-68, anal chemist, 68-69; asst res librn, Monsanto Textiles Co, 69-73; SR RES INFO SPECIALIST, CIBA-GEIGY CORP, 73- Mem: Am Chem Soc; Am Asn Textile Technologists; Am Soc Info Sci. Res: Pesticide products, providing ongoing assistance to research and development departments to facilitate labeling, registration of divisional compounds by classifying, processing, storing, retrieving and distributing data. Mailing Add: Agr Div Ciba-Geigy Corp PO Box 11422 Greensboro NC 27409

CROG, RICHARD STANLEY, b Minneapolis, Minn, Aug 4, 15; m 40; c 4. CHEMISTRY, RESEARCH ADMINISTRATION. Educ: Col of St Thomas, BS, 37; Purdue Univ, MS, 40, PhD(phys chem), 42. Prof Exp: Res chemist, 41-42, 46-52, res supvr, 52-62, mgr, Chem Res Div, 62-65, assoc dir res, Petrochem Res Div, 65-66, DIR IN-CHG EXPLOR & PROD RES, 66- Mem: Am Inst Mining, Metall & Petrol Eng; Am Chem Soc. Res: Heats of combustion of methylethyl ketone and ethylene oxide; lubricating oils; aviation gasoline; gas analysis; production research problems;

petrochemical research. Mailing Add: Union Oil Co of Calif PO Box 76 Brea CA 92621

CROKER, ROBERT ARTHUR, b New York, NY, Sept 4, 32; m; c 1. MARINE ECOLOGY. Educ: Adelphi Univ, AB, 58; Univ Miami, MS, 60; Emory Univ, PhD(biol), 66. Prof Exp: Asst fisheries, Univ Miami, 58-59; marine biologist, Bur Sport Fisheries & Wildlife, US Dept Interior, 60-62 & Bur Commercial Fisheries, 62-64; res asst marine benthic communities, Marine Inst, Univ Ga, 64; asst prof, 66-71, ASSOC PROF ZOOL, UNIV NH, 71- Concurrent Pos: Res asst Oyster Lab, Rutgers Univ, 60; mem marine biol working group, Int Biol Prog, 67; panel mem, Nat Res Coun, 73-; consult, Battelle-Columbus, 75- Mem: Fel AAAS; Soc Syst Zool; Am Acad Arts & Sci; Ecol Soc Am; Marine Biol Asn UK. Res: Systematics-ecology of Gammaridean Amphipods; marine sand communities; littoral and estuarine ecology; coral reef ecology; coastal land-use. Mailing Add: Jackson Estuarine Lab RFD Adams Pt Durham NH 03824

CROLEY, THOMAS EDGAR, b Gladewater, Tex, Apr 30, 40; m 64; c 2. ANATOMY, MICROBIOLOGY. Educ: ETex State Univ, BS, 63, MS, 64; Southwestern Univ, cytotechnol degree, 66; Baylor Univ, PhD(anat), 71. Prof Exp: Adj prof, NTex State Univ, 71-72; ASST PROF ANAT, LA STATE UNIV MED CTR, 72- Concurrent Pos: NASA-Manned Spacecraft Ctr fel, 71-72. Mem: Am Soc Clin Path; Int Asn Dent Res. Res: Calcified tissue research, particularly developing teeth; hormone relationship to epithelial development; placenta morphology. Mailing Add: Dept of Anat La State Univ Med Ctr New Orleans LA 70119

CROLL, IAN MURRAY, b Regina, Sask, Dec 7, 29; m 55; c 2. PHYSICAL CHEMISTRY. Educ: Univ Man, BSc, 50, MSc, 58; Univ Calif, Los Angeles, PhD(chem), 58. Prof Exp: Chemist, Defence Res Bd Can, 51-52; chemist fed systs div, Int Bus Mach Corp, NY, 58-61, mem res staff, Watson Res Ctr, 61-63, mgr chem res, 63-67, mgr mem & storage res, 67-74, MGR MAT TECHNOL, IBM CORP, 74- Mem: Am Chem Soc; Inst Elec & Electronics Eng; Electrochem Soc. Res: Thermodynamics; solution chemistry; electrochemistry; electrodeposition; structure and magnetic properties of metal films. Mailing Add: IBM Corp Monterey & Cottle Rds San Jose CA 95123

CROLL, NEIL ARGO, b Capetown, SAfrica, Oct 25, 41; Brit citizen; m 64; c 1. PARASITOLOGY, BEHAVIORAL PHYSIOLOGY. Educ: London Univ, BSc, & ARCS, 63, DIC, DCC & PhD(parasitol), 66. Prof Exp: Fulbright scholar nematol, Univ Calif, Davis, 67-68; Churchill fel parasitol, Univ Nairobi, 69; vis prof, Univ Colombo, Sri Lanka, 72; DIR & PROF, INST PARASITOL, MACDONALD COL, McGILL UNIV, 73- Concurrent Pos: Lectr zool, Imperial Col, London Univ, 66-74; dir, Nat Ref Ctr Parasitol, Can, 74-; mem, various govt adv comts. Mem: Brit Soc Parasitol; Soc Nematol; Can Soc Zool; Inst Biol. Res: Behavior physiology of parasites, epidemiology and medical parasitology in Canada; physiology of the nematode level of organization. Mailing Add: Inst of Parasitol McGill Univ Montreal PQ Can

CROMARTIE, THOMAS HOUSTON, b Raleigh, NC, Aug 9, 46. BIOCHEMISTRY. Educ: Duke Univ, BS, 68; Mass Inst Technol, PhD(chem), 73. Prof Exp: Res assoc, Mass Inst Technol, 73-75; ASST PROF CHEM, UNIV VA, 75- Mem: Am Chem Soc; Sigma Xi; AAAS. Res: Reaction mechanisms of enzymes, especially flavoenzymes and isomerases; synthesis of coenzyme and amino acid analogs; heavy atom and hydrogen kinetic isotope effects. Mailing Add: Dept of Chem Univ of Va Charlottesville VA 22901

CROMARTIE, WILLIAM JAMES, b Garland, NC, May 19, 13; m 45; c 5. MICROBIOLOGY, INFECTIOUS DISEASES. Educ: Emory Univ, MD, 37; Am Bd Path, dipl, 48, Am Bd Internal Med, dipl, 51. Prof Exp: Intern, Grady Hosp, Emory Univ, 37-38; asst resident path, Vanderbilt Univ Hosp, 38-39, instr, Sch Med, 39-41; asst resident med, Bowman-Gray Sch Med, 42; dir res lab & asst chief med serv, Vet Admin Hosp, 46-49; assoc prof med & bact, Med Sch, Univ Minn, 49-51; assoc prof, 51-59, PROF MED & BACT, SCH MED, UNIV NC, CHAPEL HILL, 59-, ASSOC DEAN CLIN SCI, 69- Concurrent Pos: Instr, Med Col, Southwestern Univ, 47-49; mem, Adv Panel Microbiol, Off Naval Res, 50-55, Bact Test Comt, Nat Bd Med Examrs, 61-69; chief staff, NC Mem Hosp, 69-74; mem, Infectious Dis Adv Comt, Nat Inst Allergy & Infectious Dis, 71-75. Mem: Am Acad Microbiol; fel Am Col Physicians; Am Asn Path & Bact; Am Soc Exp Path; Am Soc Microbiol. Res: Pathogenesis of rheumatoid arthritis and rheumatic fever. Mailing Add: Univ of NC Sch of Med Chapel Hill NC 27514

CROMARTIE, WILLIAM JAMES, JR, b Dallas, Tex, Oct 14, 47. ECOLOGY, NATURAL HISTORY. Educ: St Johns Col, AB, 69; Cornell Univ, PhD(ecol), 74. Prof Exp: ASST PROF ENVIRON STUDIES, STOCKTON STATE COL, 74- Mem: AAAS; Ecol Soc Am; Entom Soc Am; Brit Ecol Soc. Res: Biogeography and evolution of plant-arthropod associations; coastal plain biogeography, especially pine barrens and pocosins; insects associated with pines. Mailing Add: Dept of Environ Studies Stockton State Col Pomona NJ 08240

CROMBIE, LANCE BRIAN, microbiology, biochemistry, see 12th edition

CROMER, ALAN H, b Chicago, Ill, Aug 15, 35; m 61. SCIENCE WRITING, PHYSICS. Educ: Univ Wis, BS, 54; Cornell Univ, PHD(theoret physics), 60. Prof Exp: Res fel physics, Harvard Univ, 59-61; from asst prof to assoc prof, 61-70, PROF PHYSICS, NORTHEASTERN UNIV, 70- Mem: AAAS; Am Phys Soc; Am Asn Physics Teachers. Res: Physics for the life sciences; theoretical mechanics. Mailing Add: Dept of Physics Northeastern Univ Boston MA 02115

CROMER, JERRY HALTIWANGER, b Anderson, SC, Apr 4, 35; m 62. DEVELOPMENTAL BIOLOGY, PHYSIOLOGY. Educ: Wofford Col, BS, 57; Univ SC, MS, 65; Vanderbilt Univ, PhD(develop biol), 68. Prof Exp: Asst prof, 68-74, ASSOC PROF BIOL, CONVERSE COL, 74- Concurrent Pos: Fac res grant, 68-69. Mem: AAAS; Am Soc Zool; NY Acad Sci. Res: Quantitative analysis of impulse transmission in the aortic depressor nerve of rabbits; development of hepatic xanthine dehydrogenase in hatching chick embryos; histochemistry; enzyme development; developmental genetics and physiology. Mailing Add: Dept of Biol Converse Col Spartanburg SC 29301

CROMPTON, ALFRED W, b Durban, SAfrica, Feb 21, 27; m 54; c 3. VERTEBRATE PALEONTOLOGY. Educ: Univ Stellenbosch, BSc, 47, MSc, 49, PhD(zool), 51; Cambridge Univ, PhD(paleont), 53. Prof Exp: Cur fossil vertebrates, Nat Mus, Bloemfontein, SAfrica, 54-56; dir, SAfrican Mus, Cape Town, 56-64; dir, Peabody Mus Natural Hist & prof biol & geol, Yale Univ, 64-70; DIR, MUS COMP ZOOL, HARVARD UNIV, 70- Concurrent Pos: Lectr, Univ Cape Town, 58-64. Mem: Fel Zool Soc London; SAfrican Asn Adv Sci; fel Am Acad Arts & Sci; Soc Vertebrate Paleont; Am Soc Zool. Res: Evolution of mammals and dinosaurs; functional anatomy. Mailing Add: Mus of Comp Zool Harvard Univ Cambridge MA 02138

CROMPTON, CHARLES EDWARD, b St George Island, Alaska, Oct 8, 22; m 45; c 2. PHYSICAL CHEMISTRY. Educ: Univ Calif, Berkeley, BS, 43; Univ Tenn, PhD(chem), 49. Prof Exp: Res chemist, Manhattan Proj, Berkeley, Calif & Oak Ridge, Tenn, 43-46; tech reviewer, Isotopes Div, USAEC, 49-51; dep dir, 53-56; dir radioactivity div, US Testing Co, 51-53; assoc tech dir, Nat Lead Co Ohio, 56-60; dir nuclear chem, Martin-Nuclear Div, Martin Marietta Corp, 60-62; tech dir pigments div, Chem Group, Glidden Co, 62-63, dir res, Consol Inorg Res, 63-66; dir res inorg div, 66-74, DIR DEVELOP, CHEM GROUP, CHEMETRON CORP, 74- Concurrent Pos: Consult, USAEC, 51-52 & Am Med Asn, 54-55; mem sci adv bd, New Eng Nuclear Corp, 56-; consult, Martin Co, Md, 62-63; mem comt on depleted uranium, Nat Mat Adv Bd, 70-71. Mem: Fel Am Inst Chem; Am Chem Soc; Asn Res Dirs; Com Develop Asn; Sigma Xi. Res: Nuclear chemistry; homogeneous and heterogeneous catalysis; kinetics surface chemistry of inorganic solids, metals and ceramics; catalytic hydrogenation; physical chemistry and metallurgy of uranium and plutonium; pigments; fire retardants; plastic additives. Mailing Add: Chemetron Corp 12555 W Higgins Rd Chicago IL 60666

CROMROY, HARVEY LEONARD, b Lawrence, Mass, Jan 5, 30; m 57, 70; c 4. RADIOBIOLOGY, ENTOMOLOGY. Educ: Northeastern Univ, BSc, 51; NC State Col, PhD(entom), 58. Prof Exp: Entomologist, NC State Bd Health, 56-57; assoc prof entom & biophys, Col Agr, Univ PR, 57-60; res scientist, Oak Ridge Nat Lab, 60-61; sci adminr radiol health, USPHS, 61-64; assoc prof nuclear sci, 64-70, PROF ENTOM & NEMATOL, UNIV FLA, 70- Concurrent Pos: Adv mem on agr, Southern Interstate Nuclear Bd, 73- Mem: Health Physics Soc; Enom Soc Am; Am Acarological Soc. Res: Taxonomy of plant feeding mites; chemical and physical modification of the effects of ionizing radiation; genetic effects ozone; uses of isotopes in entomology. Mailing Add: Newell 102 Univ of Fla Gainesville FL 32611

CROMWELL, GARY LEON, b Salina, Kans, Oct 6, 38; m 60; c 3. ANIMAL NUTRITION. Educ: Kans State Univ, BS, 60; Purdue Univ, MS, 65, PhD(animal nutrit), 67. Prof Exp: Teacher high sch, Kans, 60-64; res asst animal nutrit, Purdue Univ, 64-67; asst prof, 67-70, ASSOC PROF ANIMAL SCI, UNIV KY, 70- Mem: Am Soc Animal Sci. Res: Swine nutrition. Mailing Add: Dept of Animal Sci Univ of Ky Lexington KY 40506

CROMWELL, NORMAN HENRY, b Terre Haute, Ind, Nov 22, 13; m 55; c 2. ORGANIC CHEMISTRY, CHEMOTHERAPY. Educ: Rose Polytech Inst, BS, 35; Univ Minn, PhD(org chem), 39. Prof Exp: Asst chem, Univ Minn, 35-39; from instr to prof org chem, 39-60, mem res coun, 47-49, mem grad coun, 51-60, Howard S Wilson prof chem, 60-70, chmn dept, 64-70, vpres grad studies & res, Univ Nebr Syst, 72-73, REGENTS PROF CHEM, UNIV NEBR-LINCOLN, 73- Concurrent Pos: Consult, Parke, Davis & Co, 44-45, Smith Kline & French, Pa, 46-49, USPHS, 53-, Philip Morris, Inc & Nat Cancer Inst, 54-65; Fulbright scholar & Guggenheim fel, 50 & 58; hon res assoc, Univ Col, Univ London, 50-51, 58-59 & Calif Inst Technol, 58; guest, Mass Inst Technol, 67; pres, Int Heterocyclic Chem Cong, 69; asst ed, Int J Heterocyclic Chem, 73- Mem: Am Chem Soc; The Chem Soc; Sigma Xi; Int Soc Heterocyclic Chem. Res: Reaction mechanisms; aziridines and azetidines; amino ketones; beta chloro amines; benzacridines; synthetic and physical organic chemistry; small-ring compounds; heterocyclics of biological interest in carcinogenesis and carcinostasis; pyrrols; activated allylsystems. Mailing Add: Dept of Chem Univ of Nebr Lincoln NE 68508

CRON, MARTIN JOHN, organic chemistry, see 12th edition

CRONAN, JOHN EMERSON, JR, b Long Beach, Calif, Dec 2, 42; m 73. BIOCHEMISTRY, MOLECULAR BIOLOGY. Educ: San Fernando Valley State Col, BA, 65; Univ Calif, Irvine, PhD(molecular biol), 68. Prof Exp: Instr biol chem, Sch Med, Washington Univ, 68-70; asst prof, 71-74, ASSOC PROF MOLECULAR BIOPHYS & BIOCHEM, YALE UNIV, 74- Concurrent Pos: NIH fel, Sch Med, Washington Univ, 68-70; NIH res grant & NSF res grant, Yale Univ, 71, NIH career development award, 72-; mem, Metab Biol Panel, NSF, 75- Mem: Am Soc Microbiol; Am Soc Biol Chemists. Res: Microbial lipid metabolism and chemistry; biogenesis of biological membranes; bacteriophage infection physiology; biochemical genetics. Mailing Add: Dept of Molec Biophys & Biochem Yale Univ Med Sch New Haven CT 06510

CRONE, LAWRENCE JOHN, b Orangeville, Ill, May 18, 35. PLANT PATHOLOGY. Educ: Carthage Col, AB, 56; Rutgers Univ, PhD(plant path), 62. Prof Exp: Asst prof, 62-67, ASSOC PROF BIOL, UNIV WIS-WHITEWATER, 67- Mem: Mycol Soc Am. Res: Diseases of trees and ornamental plants. Mailing Add: Dept of Biol Univ of Wis Whitewater WI 53190

CRONEIS, CAREY, geology, paleontology, deceased

CRONEMEYER, DONALD CHARLES, b Chanute, Kans, Nov 10, 25; m 53; c 5. PHYSICS. Educ: Mass Inst Technol, ScD(physics), 51. Prof Exp: Res physicist, Gen Elec Co, 51-57 & Bendix Aviation Res Lab, Mich, 57-67; RES PHYSICIST, T J WATSON RES CTR, IBM CORP, 68- Concurrent Pos: Instr physics, Wayne State Univ, 58- Mem: Am Sci Affil; Electrochem Soc; Am Phys Soc. Res: Electrical and otpical properties of semi-conductors; lasers. Mailing Add: T J Watson Res Ctr IBM PO Box 218 Yorktown Heights NY 10596

CRONHEIM, GEORG ERICH, b Berlin, Ger, Jan 15, 06; nat US; m 32. BIOCHEMISTRY, PHARMACOLOGY. Educ: Univ Berlin, PhD(phys chem), 30. Prof Exp: Asst & instr, Inst Phys Chem, Univ Berlin, 30-32; head biochem lab, Inst Phys Agron & Res Chemist, Inst Exp Med, Univ Leningrad, 32-36; res chemist, Phys Inst, Univ Stockholm, 37-39; res chemist, NY State Res Inst, 40; chief chemist, G F Harvey Co, NY, 41-45; res dir, S E Massengill Co, 46-52; dir biol sci, 52-67, tech liaison exec, 67-73, CONSULT, RIKER LABS, INC, 73- Mem: AAAS; Am Chem Soc; Am Soc Pharmacol; Soc Exp Biol & Med; NY Acad Sci. Res: Biochemistry; pharmacology; physical and pharmaceutical chemistry. Mailing Add: Riker Labs Inc 19901 Nordhoff St Northridge CA 91324

CRONHOLM, LOIS S, b St Louis, Mo, Aug 15, 30; div; c 2. MICROBIOLOGY. Educ: Univ Louisville, BA, 62, PhD(bot), 67. Prof Exp: NIH fel dept microbiol, 66-69, ASST PROF MICROBIOL DEPT BIOL, UNIV LOUISVILLE, 72-, RES ASSOC, WATER RESOURCES LAB, 73-, ASSOC DEPT MICROBIOL, SCH MED, 75- Mem: Am Soc Microbiol; AAAS; Am Water Resources Asn. Res: Histamine sensitivity in mice; mechanism pathogenicity; crown gall disease; identification, isolation and removal of bacteria and viruses in water and wastewater. Mailing Add: Dept of Biol Univ of Louisville Louisville KY 40208

CRONIN, GEORGE RICHARD, inorganic chemistry, see 12th edition

CRONIN, JAMES WATSON, b Chicago, Ill, Sept 29, 21; m 54; c 2. NUCLEAR PHYSICS. Educ: Southern Methodist Univ, BS, 51; Univ Chicago, MS, 53, PhD(physics), 55. Prof Exp: Asst physicist, Brookhaven Nat Lab, 55-58; from asst prof to prof, Princeton Univ, 58-71, PROF PHYSICS, UNIV CHICAGO, 71- Concurrent Pos: Physicist, Nat Accelerator, Ill, 70-71; mem panel elem particle physics, Div Phys Sci, Nat Acad Sci. Mem: Nat Acad Sci; Am Phys Soc. Res: Elementary particles; experiments on pion-proton total cross sections; hyperon decay asymmetries; development of improved detection techniques. Mailing Add: Dept of Physics Univ of Chicago Chicago IL 60637

CRONIN, JANE SMILEY (MRS JOSEPH C SCANLON), b New York, NY, July 17, 22; m 53; c 4. BIOMATHEMATICS. Educ: Univ Mich, PhD(math), 49. Prof Exp: Mathematician, US Air Force Cambridge Res Ctr, 51-54; instr, Wheaton Col, 54-55; mathematician, Am Optical Co, 56; from asst prof to prof math, Polytech Inst Brooklyn, 57-65, PROF MATH, RUTGERS UNIV, 65- Mem: Soc Indust & Appl Math; Am Math Soc. Res: Topological degree; functional analysis; ordinary differential equations; applications of analysis in medicine and biology. Mailing Add: 110 Valentine St Highland Park NJ 08904

CRONIN, JOHN READ, b Marietta, Ohio, Mar 5, 37; m 63; c 3. BIOCHEMISTRY. Educ: Col Wooster, BA, 59; Univ Colo, PhD(biochem), 64. Prof Exp: Res assoc biochem, Sch Med, Yale Univ, 64-66; asst prof, 66-72, ASSOC PROF, DEPT CHEM, ARIZ STATE UNIV, 72- Concurrent Pos: NIH fel, 64-65; temp staff mem org geochem, Carnegie Inst Wash, 74-75. Mem: AAAS; Am Chem Soc. Res: Organic chemistry of carbonaceous meteorites; chemical evolution; analytical biochemistry. Mailing Add: Dept of Chem Ariz State Univ Tempe AZ 85281

CRONIN, LEWIS EUGENE, b Aberdeen, Md, May 11, 17; m 45; c 3. MARINE ECOLOGY, ZOOLOGY. Educ: Western Md Col, AB, 38; Univ Md, MS, 42, PhD(zool), 46. Hon Degrees: DSc, Western Md Col, 66. Prof Exp: Teacher high sch, Md, 38-40; biologist, State Dept Res & Educ, Md, 43-50; dir marine lab & assoc prof biol sci, Univ Del, 50-55; dir & biologist, Dept Res & Educ, 55-61; dir, Natural Resources Inst & Chesapeake Biol Lab, 61-75, RES PROF, UNIV MD, 61-, ASSOC DIR RES, CTR ENVIRON & ESTUARINE STUDIES, 75- Concurrent Pos: Consult coastal ecol off chief engr, CEngr, US Army; consult pesticides, USDA & Environ Protection Agency & Mem Md Water Pollution Comn, 55-63; mem, Md Bd Nat Resources, 55-69, Chesapeake Res Coun, chmn, 66-67, Md Comn Pesticides, chmn, 67-69 & Md Comn Environ Educ, 70-; mem, Interstate Comn Potomac River Basin, 63-, Secys Comn Pesticides & Relationship Environ Health, Dept Health, Educ & Welfare, 69, Adv Comt Water Resources Sci Info Ctr, Dept Interior, 69-73 & Panel Oceanog & Marine Res, Comn Space Prog Earth Observation, Adv to Dept Interior, Nat Acad Sci; mem, Nat Marine Fish Adv Comn, Dept Com, 71-73; mem, Law of the Sea Adv Comn, US State Dept, 73- Honors & Awards: Award, Oyster Inst NAm, Isaac Walton League & Chesapeake Bay Seafood Indust Asn; Am Motors consult Award, 71. Mem: AAAS; Am Soc Zool; Am Inst Fishery Res Biol; Am Fisheries Soc; Nat Shell Fisheries Asn (pres, 60); Estuarine Res Fedn (pres, 71-73). Res: Fisheries and pollution; resource management; environmental education; estuarine ecology and biology; research administration. Mailing Add: Univ Md Ctr for Environ & Estuarine Studies Cambridge MD 21613

CRONIN, MICHAEL THOMAS IGNATIUS, b Glasgow, Scotland, Feb 1, 24; nat US; m 50; c 8. PATHOLOGY. Educ: Vet Col Ireland, MRCVS, 45; Univ Dublin, MSc, 46, PhD(path, bact), 48; Georgetown Univ, MD, 65. Prof Exp: Res officer, Equine Res Sta, Eng, 50-52; dir, Regional Diag Lab, Va, 52-53; bacteriologist, Dept Animal Path, Univ Ky, 53-55; assoc pathologist, Penrose Res Lab & asst prof vet path, Univ Pa, 55-57; head dept path & toxicol, Schering Corp, 57-61; pathologist, Woodard Res Corp, 61-65; intern, Grace-New Haven Community Hosp, Med Ctr, Yale Univ, 65-66; asst resident, Yale Univ-New Haven Hosp, 66-67 & Vet Admin Hosp, 67-68; PATHOLOGIST, MEM HOSP, MERIDEN, 68-72, 74- Concurrent Pos: Consult pathologist, Woodard Res Corp, 65-71; pathologist, Masonic Home & Hosp, Wallingford, 68-70, 72-; assoc pathologist, Hosp of St Raphael, 72- Mem: Am Asn Path & Bact; Am Col Vet Path; Int Acad Path; Col Am Pathologists. Mailing Add: 67 Edgehill Rd New Haven CT 06511

CRONIN, ROBERT FRANCIS PATRICK, b London, Eng, Sept 1, 26; Can citizen; m 54; c 3. PHYSIOLOGY. Educ: McGill Univ, MD & CM, 53, MSc, 60; FRCPS. Prof Exp: Lectr med, 59-64, asst prof med & physiol, 64-68, assoc prof med & assoc dean med fac, 68-72, PROF MED & DEAN MED FAC, McGILL UNIV, 72- Concurrent Pos: Can Life Ins res fel, 59-62; consult cardiol, Can Dept Vet Affairs, 61-70; res assoc, Can Heart Found, 62-66. Mem: Fel Am Col Physicians; Can Soc Clin Invest; Can Physiol Soc; Am Physiol Soc. Res: Exercise physiology; physiology of the coronary circulation; myocardial metabolism. Mailing Add: Fac of Med McGill Univ Montreal PQ Can

CRONIN, TIMOTHY H, b Boston, Mass, June 4, 39; m 63; c 2. ORGANIC CHEMISTRY. Educ: Boston Col, BS, 60; Mass Inst Technol, PhD(org chem), 64. Prof Exp: Res chemist, Chas Pfizer & Co, Inc, 65-70, proj leader, Pfizer, Inc, 70-71, mgr, 71-75, DIR MED CHEM, INFECTIOUS DIS, PFIZER, INC, 75- Mem: Am Chem Soc. Res: Intramolecular Diels-Alder reactions; synthetic organic chemistry. Mailing Add: Pfizer Inc Eastern Point Rd Groton CT 06340

CRONK, CASPAR, b West Newton, Mass, Apr 21, 35; m 60; c 3. EXPLORATION GEOPHYSICS, GEOLOGY. Educ: Harvard Univ, AB, 57; Ohio State Univ, PhD(geol), 68. Prof Exp: Glaciologist Antarctic res, Arctic Inst NAm, 57-59; res geophysicist explor processing & interpreting, Gulf Res & Develop Co, Gulf Oil Corp, 67-69, geophysicist, Gulf Oil Korea, 69-72, Houston Tech Serv Ctr, 72-73; ASST PROF GEOL GEOPHYS, WESTERN MICH UNIV, 73- Mem: AAAS; Soc Explor Geologists; Int Glaciol Soc; Am Geophys Union. Res: Gravity anomalies related to reef structures; gravity studies for gravel and ground water location and to delineate muck deposits; seismic studies for engineering; plate tectonics of the Indonesia area. Mailing Add: Dept of Geol Western Mich Univ Kalamazoo MI 49001

CRONK, GARY ARNOLD, b Syracuse, NY, July 15, 14; m 39; c 3. PHARMACOLOGY. Educ: Syracuse Univ, BA, 36, MD, 39. Prof Exp: Assoc prof health & prev med, State Univ NY Upstate Med Ctr, 46-60; assoc med dir, Wallace Labs, Carter Prod, 60-61; dir clin res, 62-65, DIR RES PHARMACEUT, ORTHO PHARMACEUT CORP, 65-, VPRES, 66- Mem: Fel Am Col Physicians; AMA; Am Col Pharmacol & Chemother; NY Acad Sci. Res: Drug metabolism; clinical chemotherapy. Mailing Add: Ortho Pharmaceut Corp Raritan NJ 08869

CRONK, TED CLIFFORD, b Ridgway, Pa, May 25, 46. FOOD SCIENCE. Educ: Pa State Univ, BS, 68, MS, 71; Cornell Univ, PhD(food sci), 75. Prof Exp: Asst prof food & nutrit, Univ Ga, 74-75; SCIENTIST, PILLSBURY CO, 75- Mem: Am Soc Microbiol; Mycol Soc Am; Inst Food Technologist. Mailing Add: Pillsbury Co 311 Second St SE Minneapolis MN 55414

CRONKITE, EUGENE PITCHER, b Los Angeles, Calif, Dec 11, 14; m 40; c 1. MEDICINE. Educ: Stanford Univ, AB, 36, MD, 41. Hon Degrees: DSc, Univ Long Island, 62. Prof Exp: Intern, Stanford Univ Hosp, 40-41, asst resident, 41-42; hematologist, Naval Med Res Inst, 46-54; HEMATOLOGIST, MED RES CTR, BROOKHAVEN NAT LAB, 54-, PHYSICIAN & CHMN MED DEPT, 67- Concurrent Pos: Mem, Naval Studies Bd, Nat Acad Sci; prof med & dean, Brookhaven Clin Campus, State Univ NY Stony Brook; hematologist, Atomic Bomb Tests, 46, 51-54. Honors & Awards: Wellcome Prize, 48; Alfred Benzon Prize, Govt Denmark; Ludwig Heilmeyer Gold Medal, Govt WGer, 74; Semmelweis Medal, Govt Hungary, 75. Mem: Am Asn Physicians; Am Soc Hemat; Soc Exp Biol & Med; Am Physiol Soc; Am Soc Clin Invest. Res: Control of hemopoiesis in health and disease. Mailing Add: Med Res Ctr Brookhaven Nat Lab Upton NY 11973

CRONKRIGHT, WALTER ALLYN, JR, b East Orange, NJ, July 23, 31; m 54; c 3. POLLUTION CHEMISTRY, ANALYTICAL CHEMISTRY. Educ: Rutgers Univ, BS, 52; Drexel Inst Technol, MS, 60; Polytech Inst Brooklyn, PhD(anal chem), 68. Prof Exp: Anal chemist, M W Kellogg Co, Lab, Pullman Inc, 59-61, res chemist, 61-65, supvr, 65-68, sect head, Pullman-Kellog Div, Pullman Inc, 68-71, MGR ANAL SERVS, PULLMAN-KELLOGG DIV, PULLMAN INC, 71- Mem: Am Chem Soc; Am Soc Test & Mat; Sigma Xi. Res: Development of improved processes for removal of SO2 from waste gases; standardization of analytical methods for atmospheric pollutants. Mailing Add: Pullman-Kellogg Res & Develop Ctr 16200 Industrial Park Terr Houston TX 77084

CRONQUIST, ARTHUR JOHN, b San Jose, Calif, Mar 19, 19; m 40; c 2. SYSTEMATIC BOTANY. Educ: Utah State Col, BS, 38, MS, 40; Univ Minn, PhD(bot), 44. Prof Exp: Tech asst, NY Bot Garden, 43-44, asst cur, 44-46; asst prof bot, Univ Ga, 46-48; asst prof, State Col Wash, 48-51; assoc cur, 52-57, cur, 57-65, sr cur, 65-71, dir bot, 71-74, SR SCIENTIST, NEW YORK BOT GARDEN, 74- Concurrent Pos: Tech consult, Econ Co-op Admin, Brussels, 51-52; res assoc, State Col Wash, 53- Mem: Am Soc Plant Taxon (pres, 62); Bot Soc Am; Asn Trop Biol; Ecol Soc Am; Torrey Bot Club. Res: Taxonomy of American species of Compositae; flora of western United States; systems of angiosperms; general system of plants. Mailing Add: New York Bot Garden Bronx NY 10458

CRONSHAW, JAMES, b Oswaldtwistle, Eng, Mar 11, 33; m 56; c 1. BOTANY, CELL BIOLOGY. Educ: Univ Leeds, BSc, 54, PhD(bot), 57. Prof Exp: Res officer, Commonwealth Sci & Indust Res Orgn, Australia, 57-62; asst prof biol, Yale Univ, 62-65; assoc prof, 65-70, PROF BIOL, UNIV CALIF, SANTA BARBARA, 70- Mem: AAAS; Bot Soc Am; Electron Micros Soc Am; Am Soc Cell Biol; Soc Exp Biol. Res: Fine structure of plant cells; structural aspects of growth and differentiation. Mailing Add: Dept of Biol Sci Univ of Calif Santa Barbara CA 93106

CRONYN, MARSHALL WILLIAM, b Oakland, Calif, June 22, 19; m 42; c 3. ORGANIC CHEMISTRY. Educ: Reed Col, BA, 40; Univ Mich, PhD(org chem), 44. Prof Exp: Res assoc comt med res, Univ Mich, 44-46; lectr & Am Chem Soc fel, Univ Calif, 46-48, from instr to asst prof org chem, 48-52; from asst prof to assoc prof, 52-60, chmn dept, 66-73, PROF ORG CHEM, REED COL, 60- Concurrent Pos: NIH res fel, Cambridge Univ, 60-61; consult, USPHS, 61-66. Mem: AAAS; Am Chem Soc. Res: Organic sulfur chemistry; lignin; chemotherapy. Mailing Add: Dept of Chem Reed Col Portland OR 97202

CROOK, GEORGE H, physical chemistry, see 12th edition

CROOK, JAMES RICHARD, b Ft Worth, Tex, Dec 20, 35; m 69. PARASITOLOGY, EPIDEMIOLOGY. Educ: Univ Utah, BS, 58, MS, 59, PhD(parasitol), 64; Univ Med Sci, Bankgog, DTM & Sch Trop Med, UK, DSc, 67; Am Bd Path, dipl, 50; Med Res Prog, Manila, P.I., dipl epidemiol. Prof Exp: Crew chief, Glen Canyon Ecol Res, Utah, 58; path res asst tumor res, 60-61, NSF grant & res asst parasitol, 61-64; res parasitologist, Walter Reed Inst Res, 64-65; chief med zool div, US Army Med Component-SEATO & vis prof helminth, Fac Trop Med, Univ Med Sci, Bangkok, 65-67; mem staff, 67-69, PROF BIOL, LINFIELD COL, 69- Concurrent Pos: Consult, Upper Mekong Delta Eval Bd, WHO & Org Am States. Mem: AAAS; Am Soc Parasitol; Int Col Trop Med; Royal Soc Trop Med & Hyg; Am Soc Trop Med & Hyg. Res: Endoparasite life cycles and natural ecology and pathology. Mailing Add: Dept of Microbiol Linfield Col McMinnville OR 97128

CROOK, JAMES WASHINGTON, b Baltimore, Md, July 23, 20; m 46; c 5. PHARMACOLOGY. Educ: Loyola Col, Md, BS, 43; Univ Md, BS, 49. Prof Exp: Aircraft assemblyman, Glenn L Martin Co, 46; technician med biol, 50, RES PHARMACOLOGIST, EDGEWOOD ARSENAL, 51- Res: Chemical warfare; vapor toxicology; respiration; pharmacology of toxic agents; prophylaxis and treatment of casualties; research and development on new drugs and protective devices; air pollution. Mailing Add: Res Labs Med Res Lab Toxicology Dept Edgewood Arsenal MD 21010

CROOK, JOSEPH RAYMOND, b Reno, Nev, Oct 16, 36; m 58; c 2. INORGANIC CHEMISTRY. Educ: Univ Nev, BS, 58; Ill Inst Technol, PhD(chem), 63. Prof Exp: NSF fel, Univ Colo, 63-64; asst prof inorg chem, San Jose State Col, 64-66; from asst to assoc prof chem, Cleveland State Univ, 66-70; ASSOC PROF CHEM, WESTERN WASH STATE COL, 70- Mem: Am Chem Soc; The Chem Soc. Res: Synthetic and mechanistic chemistry of transition elements. Mailing Add: Dept of Chem Western Wash State Col Bellingham WA 98225

CROOK, LYNN, biochemistry, see 12th edition

CROOK, PHILIP GEORGE, b Washington, DC, Oct 21, 25. MICROBIOLOGY. Educ: Univ Md, BS, 49; Univ NMex, MS, 51; Pa State Univ, PhD(bact), 55. Prof Exp: Asst instr bact, Pa State Univ, 53-55; assoc prof biol, Hope Col, 55-69, chmn dept, 62-69; CHAS A DANA PROF BIOL, COLGATE UNIV, 69- Concurrent Pos: NSF grant, Am Cancer Soc grant & Nat Cancer Inst grant, Hope Col, 58-60. Mem: AAAS; Am Soc Parasitol; Am Soc Microbiol; Soc Protozool. Res: Intermediate metabolism of carbohydrates in bacteria and protozoa; axenic culture of protozoa; biochemical action of hormones. Mailing Add: Dept of Biol Colgate Univ Hamilton NY 13346

CROOKE, PHILIP SCHUYLER, b Summit, NJ, Mar 10, 44; m 68. APPLIED MATHEMATICS. Educ: Stevens Inst Technol, BS, 66; Cornell Univ, PhD(applied math), 70. Prof Exp: ASST PROF MATH, VANDERBILT UNIV, 70- Mem: Soc Indust & Appl Math. Res: Partial differential equations; isoperimetric inequalities; dusty gas equations. Mailing Add: Box 6205 Sta B Vanderbilt Univ Nashville TN 37235

CROOKER, ARTHUR MERVYN, b Cayuga, Ont, Sept 19, 09; m 39; c 1. PHYSICS. Educ: McMaster Univ, BA, 30; Univ Toronto, MA & PhD(physics), 35. Prof Exp: Demonstr, Univ London, 36-37; asst prof physics, Univ BC, 37-41; optical computer, Res Enterprises, Ltd, 41-45; PROF PHYSICS, UNIV BC, 45- Mem: Am Phys Soc. Res: Multiplet structure in zinc, arsenic, selenium, antimony, tellurium and bismuth spark spectra; identification of hydrogenic terms. Mailing Add: Dept of Physics Univ of BC Vancouver BC Can

CROOKER, NANCY USS, b Chicago, Ill, Apr 1, 44. SPACE PHYSICS. Educ: Knox Col, Ill, AB, 66; Univ Calif, Los Angeles, MS, 68, PhD(meteorol), 72. Prof Exp: Engr scientist atmospheric sci, McDonnell Douglas Astronaut Co, 69; res assoc space physics, Cornell Univ, 73 & Mass Inst Technol, 73-75; ASST RES

METEOROLOGIST SPACE PHYSICS, UNIV CALIF, LOS ANGELES, 75- Concurrent Pos: Consult, Physics Dept, Boston Col, 75- Mem: Am Geophys Union. Res: The coupling of the solar wind and the earth's magnetosphere. Mailing Add: Dept of Meteorol Univ of Calif Los Angeles CA 90024

CROOKER, PETER PEIRCE, b Westerly, RI, Apr 4, 37; m 67; c 2. OPTICAL PHYSICS. Educ: Ore State Univ, BS, 59; Naval Postgrad Sch, PhD(physics), 67. Prof Exp: Proj officer, David Taylor Model Basin, 59-60; instr physics, Naval Postgrad Sch, 60-67; fel, Lincoln Lab, Mass Inst Technol, 68-70; asst prof physics, Calif State Polytech Col, 70; ASST PROF PHYSICS, UNIV HAWAII, 70- Mem: Am Phys Soc. Res: Electron paramagnetic resonance; acoustic paramagnetic resonance; raman spectroscopy; Rayleigh and Brillouin spectroscopy in liquid crystals. Mailing Add: Dept of Physics & Astron Univ of Hawaii Honolulu HI 96822

CROOKS, GEORGE CHAPMAN, b North Brookfield, Mass, Jan 21, 05; m 34; c 2. PHYSIOLOGICAL CHEMISTRY. Educ: Amherst Col, AB, 28; Mass State Col, PhD(biochem), 37. Prof Exp: Asst chem, Mass State Col, 28-30; instr, 30-35, 36-39; asst prof, 39-40, 46-48; assoc prof, 48-50, 52-69, PROF CHEM, UNIV VT, 69- Concurrent Pos: Consult, Univ Vt, 71- Mem: Am Chem Soc. Res: Blood proteins; chemical and nutritive studies on fish muscle; carotene studies in blood, milk and plant materials. Mailing Add: Cook Phys Sci Hall Univ of Vt Burlington VT 05401

CROOKS, HARRY MEANS, JR, b Albany, Ore, Oct 14, 12; m 38; c 3. ORGANIC CHEMISTRY. Educ: Alma Col, AB, 32; Pa State Univ, PhD(chem), 38. Prof Exp: Res chemist, 37-52, asst dir chem res, 52-63, group dir org chem res, 63-70, SECT DIR ORG CHEM RES, PARKE DAVIS & CO, 70- Mem: NY Acad Sci; Am Chem Soc. Res: Steroid degradation and synthesis; chemistry of natural products; antibiotics, degradation and synthesis. Mailing Add: Parke Davis & Co 2800 Plymouth Rd Ann Arbor MI 48106

CROOKS, MICHAEL JOHN CHAMBERLAIN, b Victoria, BC, Oct 25, 30; m 59; c 3. LOW TEMPERATURE PHYSICS. Educ: Reed Col, BA, 53; Univ BC, MA, 57; Yale Univ, PhD(physics), 63. Prof Exp: Res assoc physics, Duke Univ, 62-63; asst prof, 63-72, ASSOC PROF PHYSICS, UNIV BC, 72- Concurrent Pos: Sr vis res fel, Univ of Sussex, Eng, 71-72. Mem: Am Phys Soc; Am Asn Physics Teachers; Can Asn Physicists; Sigma Xi. Res: Properties of liquid and solid states of helium-three and helium-four. Mailing Add: Dept of Physics Univ of BC Vancouver BC Can

CROOKSHANK, HERMAN ROBERT, b Linneus, Mo, June 7, 16; m 45. CLINICAL CHEMISTRY. Educ: Northeast Mo State Teachers Col, BS, 38; Univ Iowa, MS, 40, PhD(biochem), 42. Prof Exp: Asst biochem, Univ Iowa, 38-42; instr, Med Col, Univ Ala, 46-49; asst secy, Am Inst Biol Sci, Washington, DC, 49; animal nutritionist, Animal Husb Res Div, 49-70, RES CHEMIST, AGR RES SERV, USDA, 70- Concurrent Pos: Assoc prof animal husb, Tex Technol Col, 54-59; asst exec secy, Div Biol & Agr, Nat Res Coun, 49-; dipl, Am Bd Nutrit & Am Bd Clin Chem. Mem: Fel AAAS; Am Chem Soc; fel Am Inst Chem; Am Soc Animal Sci; fel Am Asn Clin Chemists. Res: Grass tetany and urinary caluli; pesticide residues in livestock; physiological and biochemical interactions of mineral in livestock. Mailing Add: Vet Toxicol & Entom Res Lab USDA College Station TX 77840

CROOKSTON, JOHN HAMILL, b Fernie, BC, July 30, 22; m 57; c 2. HEMATOLOGY. Educ: Univ Toronto, MD, 47, BSc, 48; Cambridge Univ, PhD(biol), 54; FRCP(C), 68. Prof Exp: PROF MED & PATH, FAC MED, UNIV TORONTO, 68-; HEMATOLOGIST-IN-CHIEF, TORONTO GEN HOSP, 70- Mem: Am Soc Hemat; Int Soc Blood Transfusion. Res: Clinical and laboratory hematology, especially hereditary and acquired hemolytic anemias. Mailing Add: Dept of Hematol Toronto Gen Hosp Toronto ON Can

CROOKSTON, MARIE CUTBUSH, b Dean, Victoria, Australia, Aug 15, 20; m 57; c 2. IMMUNOHEMATOLOGY. Educ: Univ Melbourne, BSc, 40. Prof Exp: Head lab clin path, St Andrew's Hosp, 41-46; res scientist, Blood Transfusion Res Unit, Med Res Coun, Eng, 47-57; RES ASSOC IMMUNOHEMAT, UNIV TORONTO, 64- Concurrent Pos: Nuffield grant, India, 52. Mem: Am Soc Human Genetics; Can Soc Immunol; Brit Soc Immunol; Int Soc Blood Transfusion. Res: Hemolytic disease of the newborn; Duffy blood group system; survival of red cells after low temperature storage; effect of incompatible antibodies on red cell survival; specificity and behavior of auto-antibodies. Mailing Add: Dept of Path Toronto Gen Hosp Univ of Toronto Toronto ON Can

CROOM, FREDERICK HAILEY, b Lumberton, NC, Aug 6, 41; m 63; c 2. MATHEMATICS. Educ: Univ NC, BS, 63, PhD(math), 67. Prof Exp: Asst prof math, Univ Ky, 67-71; asst prof, 71-74, ASSOC PROF MATH, UNIV OF THE SOUTH, 74- Mem: Am Math Soc; Math Asn Am; Sigma Xi. Res: Algebraic topology; point set topology. Mailing Add: Dept of Math Univ of the South Sewanee TN 37375

CROOM, HENRIETTA BROWN, b Burlington, NC, Sept 23, 40; m 63; c 2. BIOCHEMISTRY. Educ: Univ NC, AB, 62, PhD(biochem), 68. Prof Exp: Res assoc anat, Univ Ky, 69-70; ASST PROF BIOL, UNIV OF THE SOUTH, 72- Mem: Sigma Xi. Res: Mode of action of antibiotics which inhibit protein synthesis in Escherichia coli at the ribosome level. Mailing Add: Box 1247 SPO Univ of the South Sewanee TN 37375

CROOM, HERBERT GEORGE, animal science, dairy science, deceased

CROOM, HERMAN LEE, b Wilmington, NC, Nov 21, 09; m 35; c 1. SCIENCE ADMINISTRATION, METEOROLOGY. Educ: NY Univ, BS, 50; Am Univ, MA, 55. Prof Exp: Meteorologist, US Weather Bur, 36-53; res adminr phys sci, US Govt, 53-70; consult, 70-74; RETIRED. Honors & Awards: Sustained Outstanding Performance Award, US Govt, 59, Merit Award, 70. Mem: Am Meteorol Soc. Res: Physical sciences; synoptic meteorology and weather forecasting; upper air and space physica. Mailing Add: 610 N Ford Rd Stuart FL 33494

CROPP, FREDERICK WILLIAM, III, b Wheeling, WVa, Dec 9, 32; m 55. GEOLOGY, PALYNOLOGY. Educ: Col Wooster, BA, 54; Univ Ill, MS, 56, PhD(geol), 58. Prof Exp: Asst, Univ Ill, 54-58, from instr to asst prof geol, 58-64; assoc prof & assoc dean, 64-68, PROF GEOL, COL WOOSTER, DEAN & VPRES ACAD AFFAIRS, 68- Concurrent Pos: Ellis L Phillips Found intern acad admin, 62-63. Mem: Geol Soc Am; Soc Econ Paleontologists & Mineralogists; Nat Asn Geol Teachers; Am Asn Petrol Geol. Res: Use of Pennsylvania spores for stratigraphic purposes. Mailing Add: Dept of Geol Col of Wooster Wooster OH 44691

CROPP, GERD J A, b Delmenhorst, WGer, July 2, 30; Can citizen; m 57; c 3. PHYSIOLOGY, PEDIATRICS. Educ: Univ Western Ont, MD, 58, PhD(biophys), 65. Prof Exp: From asst prof to assoc prof, Sch Med, Univ Colo, 65-72, assoc clin prof pediat, Sch Med, 72-76; PROF PEDIAT, STATE UNIV NY BUFFALO & DIR, CHILDREN'S LUNG CTR, CHILDREN'S HOSP, BUFFALO, 76- Concurrent Pos: NIH career develop award, 67-72; dir dept clin physiol, Nat Asthma Ctr, Denver, Colo, 72-76. Mem: Soc Pediat Res; Am Physiol Soc. Res: Pathophysiology of anemia; control of respiration; exercise; pathophysiology of asthma. Mailing Add: Children's Lung Ctr Children's Hosp 219 Bryant St Buffalo NY 14222

CROPPER, WALTER V, b Xenia, Ohio, July 31, 17; m 41, 55; c 5. INSTRUMENTATION, RESEARCH ADMINISTRATION. Educ: Ky Wesleyan Col, BS, 38; Univ Ky, MS, 40. Prof Exp: Materials engr, State Hwy Dept, Ky, 38-39; res engr, Servel, Inc, 40-42; chemist, Stand Oil Co, Ind, 42-46, group leader, 46-52, asst chief chemist, 52-53, chief chemist, 54-60; gen mgr, Precision Sci Develop Co, 60-65, vpres, 65-69; consult, 69-70; DIR, AM SOC TEST & MAT, 70- Mem: Instrument Soc Am; Air Pollution Control Asn; Am Chem Soc. Res: Catalyzed photochemical reactions; corrosion in saline systems; petroleum technology; automatic analyzers; quality sensitive processing monitors; instrument development; air pollution measurement methods and instruments; energy conservation; solar heating and cooling; biomass conversion processes. Mailing Add: Am Soc Test & Mat 1916 Race St Philadelphia PA 19103

CROSBIE, EDWIN ALEXANDER, b Washington, Pa, July 23, 21; m 46; c 2. PHYSICS. Educ: Washington & Jefferson Col, BA, 42, MA, 48; Univ Pittsburgh, PhD(physics), 51. Prof Exp: ASSOC PHYSICIST, ARGONNE NAT LAB, 52- Mem: Am Phys Soc. Res: Theoretical nuclear physics; high energy particle physics, particle accelerators; field theory. Mailing Add: Argonne Nat Lab 360 9700 S Cass Ave Argonne IL 60439

CROSBY, ALAN HUBERT, b Columbia, Pa, Oct 17, 22; m 53; c 3. ORGANIC CHEMISTRY. Educ: Ursinus Col, BS, 43; Univ Va, MS, 50, PhD, 51. Prof Exp: From asst prof to prof chem, Northwestern State Col La, 50-69, head dept phys sci, 57-69; PROF CHEM, DEPT CHEM & PHYSICS, LOCK HAVEN STATE COL, 69- Mem: Am Chem Soc. Res: Chemistry of quinones. Mailing Add: Dept of Chem & Physics Lock Haven State Col Lock Haven PA 17745

CROSBY, DAVID S, b St George, Utah, June 4, 38; m 62; c 1. APPLIED STATISTICS, MATHEMATICAL STATISTICS. Educ: Am Univ, BA, 62; Univ Ariz, MA, 64, PhD(math), 66. Prof Exp: Mathematician, Harry Diamond Labs, 62-64; asst prof math & statist, 66-75, PROF MATH, STATIST & COMPUT SCI, AM UNIV, 75-, CHMN DEPT, 75- Concurrent Pos: Statist consult, Nat Environ Satellite Ctr, 68- Mem: Math Asn Am; Inst Math Statist; Am Statist Asn. Res: Multidimensional probability distributions; statistical problems of inversion techniques. Mailing Add: Dept of Math & Statist Am Univ Massachusetts & Nebraska Ave Washington DC 20016

CROSBY, DONALD GIBSON, b Portland, Ore, Sept 11, 28; m 53; c 2. ENVIRONMENTAL CHEMISTRY, PESTICIDE CHEMISTRY. Educ: Pomona Col, BA, 50; Calif Inst Technol, PhD(chem), 54. Prof Exp: Chemist, Union Carbide Chem Co, 54-55, group leader biol chem, 56-61; assoc toxicologist, Exp Sta, 61-62, toxicologist & lectr food sci & technol, 62-69, PROF ENVIRON TOXICOL, UNIV CALIF, DAVIS, 69- Prof Exp: Chmn, Agr Toxicol & Residue Res Lab, 62-66 & Regional Res Proj W-45, 68-70; mem comn terminal pesticide residues, Int Union Pure Appl Chem, 74- & hazard mat adv comt, Environ Protection Agency, 75- Mem: AAAS; fel Am Chem Soc; Oceanic Soc. Res: Chemistry of natural products; nutritional and food chemistry; pesticide chemistry and metabolism; chemical ecology; environmental chemistry. Mailing Add: Dept of Environ Toxicol Univ of Calif Davis CA 95616

CROSBY, EMORY SPEAR, b Georgetown, SC, Jan 17, 28; m 53; c 2. FOREST PATHOLOGY, PLANT PHYSIOLOGY. Educ: Western Ky Univ, BS, 60, MA, 61; Clemson Univ, PhD(plant path), 65. Prof Exp: Assoc prof biol, Armstrong State Col, 66-68, ASSOC PROF BIOL, THE CITADEL, 68- Res: Fungus morphology and physiology. Mailing Add: Dept of Biol The Citadel Charleston SC 29409

CROSBY, GARY WAYNE, b Vidor, Tex, June 13, 31; m 56; c 3. STRUCTURAL GEOLOGY, GEOPHYSICS. Educ: Brigham Young Univ, BS, 58, MS, 59; Columbia Univ, PhD(struct geol, geophys), 63. Prof Exp: Res geologist, Atlantic Ref Co, 62-65; asst prof struct geol & geophys, ETex State Univ, 65-66; ASSOC PROF GEOPHYS, UNIV MONT, 66- Mem: Geol Soc Am; Am Geophys Union; Am Asn Petrol Geol; Soc Explor Geophys. Res: Gravity interpretation; isostasy; mechanical principles of rock deformation. Mailing Add: Dept of Geol Univ of Mont Missoula MT 59801

CROSBY, GAYLE MARCELLA, b Battle Creek, Mich. DEVELOPMENTAL BIOLOGY. Educ: Albion Col, BA, 55; Univ Mich, MS, 57; Brandeis Univ, PhD(develop biol), 67. Prof Exp: Asst prof biol, St Mary's Col, 69-73; res asst anat, Univ Wis, 73-75; ASST PROF ANAT, HOWARD UNIV, 75- Mem: Soc Develop Biol. Res: Determination and analysis of the morphogenetic factors responsible for normal development of the vertebrate limb. Mailing Add: Dept of Anat Howard Univ 520 W St NW Washington DC 20059

CROSBY, GLENN ARTHUR, b Hempfield Township, Pa, July 30, 28; m 50; c 3. PHYSICAL CHEMISTRY. Educ: Waynesburg Col, BS, 50; Univ Wash, PhD(phys chem), 54. Prof Exp: Res assoc chem, Fla State Univ, 55-57, vis asst prof physics, 57; from asst prof to assoc prof chem, Univ NMex, 57-67; PROF CHEM, WASH STATE UNIV, 67- Concurrent Pos: Vis prof physics, Univ Canterbury, Christchurch, NZ, 74; consult, Indust & Govt Agencies. Mem: Fel AAAS; Am Chem Soc; Am Phys Soc. Res: Perturbations of ions and molecules by chemical environments; luminescence of transition-metal and rare-earth complexes; optical, magnetic and electrical properties of complexes and solids with extended interactions; design of electro-optical materials. Mailing Add: Dept of Chem Wash State Univ Pullman WA 99163

CROSBY, HARVEY JOE, organic chemistry, inorganic chemistry, see 12th edition

CROSBY, JOHN ALBERT, b Chicago, Ill, Feb 2, 21; m 51; c 4. URBAN GEOGRAPHY, URBAN SOCIOLOGY. Educ: Univ Chicago, BS, 43; Univ Wash, MA, 51, PhD(geog), 59. Prof Exp: Instr geog & geol, Eastern Wash Col Educ, 51; lectr geog, Univ Toronto, 53-56; from asst prof geog to assoc prof, 56-65, PROF GEOL, CALIF STATE UNIV, FRESNO, 65- Concurrent Pos: Consult, Bur Bus Res, Calif State Univ, Fresno, 64-66, mem, Bd Dirs, Ctr Urban & Regional Studies, 70-72. Honors & Awards: State Merit Award, Nat Coun Geog Educ, 72. Mem: Asn Am Geog; Asn Pac Coast Geog; Nat Sci Teachers Asn. Res: Cartography; photo interpretation. Mailing Add: Dept of Geog Calif State Univ Fresno CA 93710

CROSBY, LON OWEN, b Webster City, Iowa, Aug 6, 45; m 67. NUTRITION, NUTRITIONAL BIOCHEMISTRY. Educ: Iowa State Univ, BS, 67; Purdue Univ, PhD(nutrit), 71. Prof Exp: Res assoc biochem nutrit, Cornell Univ, 71-73; head monogastric nutrit, Syntex Corp, 73-75; SR PROG SCIENTIST NUTRIT, ENVIRO CONTROL, INC, 75- Mem: Am Soc Animal Sci; Poultry Sci Asn; AAAS; Inst Food Technologists. Res: Interrelationship between diet, nutrition and cancer, including nutritional and biochemical interactions and the use of nutrition in prevention and treatment. Mailing Add: 13814 Drake Dr Rockville MD 20853

CROSBY, PAUL FALJEAN, b New York, NY, Sept 18, 28; m 52; c 5, CHEMISTRY.

Educ: Queen's Col, BS, 50; Univ Colo, PhD(chem), 56. Prof Exp: Control chemist, Bymart, Inc, 50-51; sr biochemist, Grain Processing Corp, 56-59; res biochemist, US Army Trop Res Med Labs, PR, 60-62; sr res scientist, E R Squibb & Sons, 63-66; ASST CHIEF RADIOISOTOPES SERV, VET ADMIN HOSP, SAN JUAN, PR, 66- Concurrent Pos: Lectr, Sch Med, Univ PR. Mem: Am Chem Soc. Res: Enzymes; fermentation; synthesis of fermentation precursors; nutrition of fungi; steroid metabolism; mental retardation due to faulty amino acid metabolism; enzyme activity alterations caused by Spirometra mansoni infection. Mailing Add: Vet Admin Hosp GPO Box 4867 San Juan PR 00936

CROSBY, PERCY, b Washington, DC, Jan 25, 30. PETROLOGY. Educ: Dartmouth Col, AB, 52; Harvard Univ, AM, 55, PhD(geol), 60. Prof Exp: Geologist, US Geol Surv, 53-54; asst prof geol, George Washington Univ, 59-61; assoc prof, State Univ NY Col Plattsburgh, 61-62, Northeastern Ill State Col, 63-67, ECarolina Univ, 67-69 & Miami Univ, 69-70; RES & WRITING, 70- Concurrent Pos: Grants, State Univ NY, 61-62, NSF, 64-66, 67-69; vis prof, Pahlavi Univ, Shiraz, Iran, 72-73. Mem: Fel Geol Soc Am; Mineral Soc Am. Res: Precambrian geology for history of northeastern Adirondacks; geology and structural and metamorphic evolution of southeastern British Columbia. Mailing Add: 35 W Lenox St Chevy Chase MD 20015

CROSBY, WARREN MELVILLE, b Topeka, Kans, Mar 19, 31; m 54; c 1. OBSTETRICS & GYNECOLOGY. Educ: Washburn Univ, Topeka, BS, 53; Univ Kans, MD, 57; Am Bd Obstet & Gynec, dipl, 65. Prof Exp: Intern, St Luke's Hosp, Kansas City, Mo, 57-58; resident obstet & gynec, Univ Calif, San Francisco, 58-62; from instr to assoc prof, 62-70, PROF OBSTET & GYNEC & VCHMN DEPT, UNIV OKLA, 70- Mem: AMA; fel Am Col Obstet & Gynec. Res: Erythroblastosis Fetalis; intrauterine fetal transfusion; studies of impact and decompression in pregnant women and animals. Mailing Add: Dept of Gynec & Obstet Univ of Okla Oklahoma City OK 73104

CROSBY, WILLIAM HOLMES, JR, b Wheeling, WVa, Dec 1, 14; m 40, 59; c 7. HEMATOLOGY. Educ: Univ Pa, AB, 36, MD, 40. Prof Exp: Intern, Walter Reed Gen Hosp, US Army, 40-41, instr, Off Cand Sch, Med Field Serv, 41-43, 45-46, regimental surgeon, 85th Infantry Div, Italy, 43-45, resident internal med, Brooke Gen Hosp, 46-48, res fel hemat, Pratt Diagnostic Hosp, 49-50, med specialist, Queen Alexandria Mil Hosp, London, 50-51, chief dept hematol, Walter Reed Army Inst Res, 51-65; chief hemat, dir blood res & sr physician, NEng Med Ctr Hosps & prof med, Sch Med, Tufts Univ, 65-72; HEAD DIV HEMATOL & ONCOL & DIR L C JACOBSON BLOOD CTR, SCRIPPS CLIN & RES FOUND, 72- Concurrent Pos: Dir surg res team, Korea, 52-53; chief hematol serv, Walter Reed Gen Hosp, 53-65, cancer chemother prog, 60-65, dir div med, 59-65; spec lectr hemat, Dept Med, George Washington Univ, 53-65; consult, Surgeon Gen, US Army, Vet Admin Hosp, La Jolla, El Centro Hosp & San Diego Naval Hosp; mem med adv comt, Blood Res Found; mem hematol test comt, Am Bd Internal Med, 70-; mem bd trustees, Nat Hemophilia Found, 71-; adj prof med, Univ Calif, San Diego, 72- Mem: Honors & Awards: Order of Carlos J Finlay, Cuba, 55; Stitt Award, 64; McCollum Award, 70. Mem: Am Soc Clin Invest; fel Am Col Physicians; hon mem Italian Soc Hematol; hon mem Dutch Soc Hematol; hon mem Europ Soc Hematol. Res: Clinical hematology; iron metabolism; marrow; spleen. Mailing Add: Scripps Clin & Res Found La Jolla CA 92037

CROSEN, ROBERT GLENN, b Maitland, Mo, Mar 12, 00; m 25; c 3. ANALYTICAL CHEMISTRY, SCIENCE EDUCATION. Educ: Tarkio Col, BS, 23; Univ SDak, AM, 25; Columbia Univ, PhD(chem), 33. Prof Exp: Instr chem, Univ SDak, 23-28; asst instr, Exten Div, Columbia Univ, 28-29; from instr to assoc prof, Lafayette Col, 31-46, dean col, 41-46, dean fac, 46-57; prof chem & chmn dept, Abadan Inst Technol, Iran, 57-62; prof, 62-70, chmn dept, 63-70, EMER PROF CHEM, STERLING COL, 70- Mem: Am Chem Soc; Sigma Xi. Res: Saponification of oils, fats and waxes; quantitative spectrographic analysis by emission and absorption. Mailing Add: 718 N Sixth St Sterling KS 67579

CROSLEY, DAVID RISDON, b Webster City, Iowa, Mar 4, 41; m 63. PHYSICAL CHEMISTRY. Educ: Iowa State Univ, BS, 62; Columbia Univ, MA, 63, PhD(phys chem), 66. Prof Exp: Res assoc physics, Joint Inst Lab Astrophys, 66-68; asst prof phys chem, Univ Wis-Madison, 68-75; PROJ LEADER, BALLISTIC RES LABS, ABERDEEN PROVING GROUND, MD, 75- Concurrent Pos: Joint Inst Lab Astrophys fel, 66-68. Mem: Am Phys Soc; AAAS. Res: Spectroscopy and chemical dynamics, reaction and energy transfer, of small state-selected molecules; level crossing and double resonance studies. Mailing Add: Ballistic Res Labs Aberdeen Proving Ground MD 21005

CROSMAN, ARTHUR MARSTON, b Monmouth, Maine, Sept 18, 02; m 47; c 4. ZOOLOGY. Educ: Dartmouth Col, BS, 24; Columbia Univ, AM, 27; Cornell Univ, PhD(morphol), 35. Prof Exp: From asst to prof, 25-72, asst dean, 62-70, assoc dean, 70-72, EMER PROF BIOL, NY UNIV, 72- Mem: AAAS; assoc Am Soc Zool; NY Acad Sci. Res: Experimental embryology; educational evaluation; dissolution and absorption of retained dead fetuses. Mailing Add: 270 Hardenburg Ave Demarest NJ 07627

CROSS, ALEXANDER DENNIS, b Leicester, Eng, Mar 29, 32. ORGANIC CHEMISTRY. Educ: Univ Nottingham, BSc, 52, PhD(org chem), 55; FRIC. Hon Degrees: DSc, Univ Nottingham, 66. Prof Exp: Fulbright travel scholar & fel, Univ Rochester, 55-57; sr fel & res asst, Dept Sci & Indust Res, Imp Col, London, 57-58, from asst lectr to lectr, 58-60; sr chemist, Syntex SAm, 61-62, asst dir chem res, 62-64, from assoc dir to dir, Syntex Inst Steroid Chem, 64-67, vpres, Syntex Res, 66-67, vpres, Com Rels, 67-70 & Chem Group 67-72, pres, Syntex Sci Systs, 70-74, PRES, INT PHARM DIV, SYNTEX CORP, 74- Concurrent Pos: Vis lectr, Stanford Univ, 67. Mem: Am Chem Soc; The Chem Soc. Res: Elucidation of structure and synthesis of natural products; synthesis of biologically active steroids; applications of spectroscopic methods to problems in stereochemistry and structure. Mailing Add: Syntex Corp 3401 Hillview Ave Palo Alto CA 94304

CROSS, CHARLES KENNETH, b Toronto, Ont, June 25, 27; m 54; c 2. ANALYTICAL CHEMISTRY. Educ: Univ Toronto, BA, 52, MA, 53. Prof Exp: RES CHEMIST, CAN PACKERS LTD, 53- Mem: Chem Inst Can. Res: Trace contaminants in food, particularly nitrosamines in cured meats; analytical chemistry in the field of fats and oils. Mailing Add: Can Packers Ltd 2211 St Clair Ave W Toronto ON Can

CROSS, CHESTER ELLSWORTH, b Boston, Mass, May 5, 13; m 41; c 3. BOTANY. Educ: Mass State Col, BS, 35, MS, 37; Harvard Univ, PhD(paleobot), 40. Prof Exp: Res specialist, Mass State Col, 37-41; from asst prof to assoc prof, 41-51; res prof in chg, 52-56, HEAD, CRANBERRY EXP STA, UNIV MASS, 56- Concurrent Pos: Assoc, Northeastern States Weed Control Conf, 47 & 48. Mem: Bot Soc Am. Res: Weed control and weather in cranberry culture. Mailing Add: Cranberry Exp Sta Univ of Mass East Wareham MA 02538

CROSS, CLARK IRWIN, b Olds, Alta, Sept 20, 13; US citizen; m 46; c 2. GEOGRAPHY, PHYSICAL SCIENCE. Educ: Univ Wash, MS, 47, PhD(geog), 51. Prof Exp: Instr geog, Southern Methodist Univ, 46-47; from instr to asst prof, 49-54, ASSOC PROF PHYS SCI, UNIV FLA, 54- Concurrent Pos: Ranger naturalist, Mt Rainier Nat Park, 54-62. Mem: Soc Am Foresters; Asn Am Geog. Res: Geography of USSR; air photo interpretation. Mailing Add: Dept of Geog 102 Bryan Hall Univ of Fla Gainesville FL 32601

CROSS, DAVID RALSTON, b Lawrence, Mass, Mar 16, 28; m 58; c 2. ORGANIC CHEMISTRY, PHYSICAL CHEMISTRY. Educ: Wesleyan Univ, AB, 49, MA, 51; Syracuse Univ, 52-53, PhD(phys & org chem), 60. Prof Exp: Sr res chemist, Res Labs, Eastman Kodak Co, NY, 57-64; from asst prof to assoc prof phys chem, 64-74, PROF CHEM, WESTERN MD COL, 74- Concurrent Pos: NSF res grant, 68-; vis prof, Case Western Reserve Univ, 70-71. Mem: Am Chem Soc. Res: Chromatography; chemiluminescence; photochemistry; chemical and electrode kinetics; electrochemistry; solvated electrons; photochemistry of phytochrome. Mailing Add: Dept of Chem Western Md Col Westminster MD 21157

CROSS, EARLE ALBRIGHT, JR, b Memphis, Tenn, Nov 23, 25; m 48; c 4. INSECT TAXONOMY, INSECT ECOLOGY. Educ: Utah State Univ, BS, 51; Univ Kans, MA, 55, PhD(entom), 62. Prof Exp: Instr entom, Purdue Univ, 57-58; vis instr, Univ Kans, 58-59, NIH res assoc, 59-60; from asst prof to prof biol sci, Northwestern State Col La, 60-70; assoc prof biol, 70-73, PROF BIOL, UNIV ALA, 73- Concurrent Pos: NIH res grant, 64-67; US Forest Serv coop res matching grant, 66-67; NSF teaching grant, 72 & res grant, 72-74. Mem: Soc Syst Zool; Entom Soc Am; Ecol Soc Am. Res: Systematics and ecology of insects and acarines; systematics and ecology of mite family Repnotidae; ecological energetics; strip mine succession. Mailing Add: Dept of Biol Univ of Ala Tuscaloosa AL 35486

CROSS, FRANK BERNARD, b Kansas City, Mo, Sept 17, 25; m 54; c 3. ZOOLOGY. Educ: Okla Agr & Mech Col, BS, 47, MS, 49, PhD(zool), 51. Prof Exp: Instr biol, 51-53, asst prof zool, 53-59, assoc prof & assoc dir, State Biol Surv, 59-67, PROF SYSTS & ECOL & DIR STATE BIOL SURV, 67-, CUR FISHES, MUS NATURAL HIST, 67- Mem: AAAS; Am Fisheries Soc; Am Soc Ichthyologists & Herpetologists; Wildlife Soc. Res: Ichthyology and fishery biology. Mailing Add: Dyche Hall Univ of Kans Lawrence KS 66044

CROSS, GEORGE ELLIOT, b Auburndale, NS, Apr 17, 28; m 52; c 5. PURE MATHEMATICS. Educ: Dalhousie Univ, BA, 52, MA, 54; Univ BC, PhD(math), 58. Prof Exp: From instr to asst prof math, Victoria Col, BC, 56-59; from asst prof to assoc prof, Univ Western Ont, 59-63; assoc prof, 63-65, dean grad studies, 67-72, PROF MATH, UNIV WATERLOO, 65- Mem: Can Math Cong. Res: General theories of integration and summability of series as they relate to problems in Fourier analysis. Mailing Add: Dept of Pure Math Univ of Waterloo Waterloo ON Can

CROSS, HANSELL FLYNN, b Wilson, La, Oct 30, 13; m 43; c 2. ACAROLOGY. Educ: La State Univ, BA, 36, MS, 41; Univ Md, PhD(zool), 55. Prof Exp: Teacher pub sch, La, 36-42; asst entomologist, Exp Sta, La State Univ, 42-43 & Orlando Lab, Dept of Agr, 46-49; asst prof biol, Univ Ga, 49-53; asst, Univ Md, 53-55; asst prof, Northeast La State Col, 55-59, assoc prof biol & dir Northeast La Res Lab for Host Reaction Studies, 59-60; prof biol & head dept, Huntingdon Col, 60-62; prof, 62-74, EMER PROF BIOL, GA STATE UNIV, 74- Concurrent Pos: NIH awards, 57 & 59; partic, Int Cong Acarology, Colo. Res: Host specificity of mites; effect of arthropods on host tissue. Mailing Add: Dept of Biol Ga State Univ Atlanta GA 30303

CROSS, HIRAM RUSSELL, b Dade City, Fla, Feb 11, 44; m 63; c 2. FOOD SCIENCE, ANIMAL SCIENCE. Educ: Univ Fla, BSA, 66, MSA, 69; Tex A&M Univ, PhD(meat sci), 72. Prof Exp: Instr meat sci, Univ Fla, 67-69 & Tex A&M Univ, 69-72; US meat grader, Agr Marketing Serv, 66-67, livestock & meat marketing specialist, 72-73, RES FOOD TECHNOLOGIST, MEAT SCI RES LAB, AGR RES SERV, USDA, 73- Concurrent Pos: Mem, Agr Marketing Serv, USDA Task Force Develop US Feeder Grades, 72-74. Mem: Am Meat Sci Asn; Am Soc Animal Sci; Inst Food Technologists. Res: Develop improved objective methods for evaluating meat and livestock products and investigate factors associated with quality loss of these products during marketing. Mailing Add: Bldg 201 Agr Res Ctr E Beltsville MD 20705

CROSS, JOHN HENRY, b Lynn, Mass, Sept 25, 25; m 52; c 2. MEDICAL PARASITOLOGY. Educ: Miami Univ, AB, 53, MA, 55; Univ Tex, PhD(parasitol), 58. Prof Exp: Med supply technician, UN Relief & Rehab Admin, China Mission, 46-47 & US Econ Coop Admin, 48-50; asst zool, Miami Univ, 53-55; asst parasitol, Med Br, Univ Tex, 55-58, res assoc, 58-60; from instr to assoc prof microbiol, Sch Med, Univ Ark, 60-66; HEAD MED ECOL DEPT, US NAVAL MED RES UNIT, 66-, SCI DIR, 74- Concurrent Pos: Fel trop med, Sch Med, La State Univ, 60; training fel, Inst Acarology, Ohio State Univ, 64; vis prof, Nat Taiwan Univ Med Sch, 67 & Chinese Nat Defense Med Ctr, 67-; clin assoc prof, Univ Wash Med Sch, 68-70 & Wash Sch Pub Health, 70-; dipl, Am Bd Microbiol. Mem: Am Soc Parasitol; Am Soc Trop Med & Hyg; Soc Exp Biol & Med; Wildlife Dis Asn. Res: Immunity to helminthic infections; epidemiology of parasitic diseases; experimental and diagnostic parasitology; ecology of Arthropod-borne diseases. Mailing Add: Med Ecol Dept Box 14 US Naval Med Res Unit No 2 APO San Francisco CA 96263

CROSS, JOHN MILTON, b Little Falls, NJ, Jan 2, 15; m 45; c 4. PHARMACEUTICAL CHEMISTRY. Educ: Rutgers Univ, BS, 36; Univ Md, MS, 39, PhD(pharm), 43. Prof Exp: Res chemist, Merck & Co, Inc, NJ, 42-46; asst prof pharm, 46-49, assoc prof chem, 49-52, PROF PHARMACEUT CHEM & CHMN DEPT, COL PHARM, RUTGERS UNIV, 52- Mem: Am Pharmaceut Asn; Am Chem Soc. Res: Preparation of iodine compounds for roentgenology; fats and oils; analysis of foods and drugs; synthesis of fungicides; photo-decomposition of foods and drugs. Mailing Add: Col of Pharm Rutgers Univ New Brunswick NJ 08903

CROSS, JON BYRON, b New York, NY, July 16, 37; m 60; c 3. CHEMICAL PHYSICS, PHYSICAL CHEMISTRY. Educ: Univ Colo, BS, 60; Univ Ill, PhD(chem physics), 67. Prof Exp: MEM STAFF, LOS ALAMOS SCI LAB, UNIV CALIF, 66- Mem: Am Phys Soc. Res: Dynamics of chemical reactions; measurement of reactive scattering differential cross sections and product translational energy distributions using crossed molecular beam apparatus and mass spectrometer detector; photoionization of large organic molecules. Mailing Add: Los Alamos Sci Lab CNC-3 Los Alamos NM 87544

CROSS, LESLIE ERIC, b Leeds, Eng, Aug 14, 23; m 50; c 5. PHYSICS. Educ: Univ Leeds, BSc, 48, PhD(physics), 52. Prof Exp: Exp off electronics, Brit Admiralty, 43-46; asst lectr physics, Univ Leeds, 49-51; Imp Chem Indust fel, 51-54; sr res assoc, Brit Elec Res Assoc, 54-61; sr res assoc, 61-63, assoc prof physics, 63-68, PROF ELEC ENG, PA STATE UNIV, 68-, ASSOC DIR MAT RES LAB, 69- Mem: Assoc Am Ceramic Soc; Phys Soc Japan; fel Am Inst Phys. Res: Material science; ferroelectric and antiferroelectric properties of titanates and niobates; thermodynamics of ferroelectricity; high permittivity materials; dielectric measuring techniques;

dielectric properties of glass systems. Mailing Add: Mat Res Lab Rm 251-A Eng Sci Bldg Pa State Univ University Park PA 16802

CROSS, MORTON H, reproductive physiology, deceased

CROSS, PATRICIA CATHERINE, reproductive physiology, see 12th edition

CROSS, RALPH DONALD, b Quincy, Ill, Dec 31, 31; m 55; c 4. CLIMATOLOGY, WATER RESOURCES GEOGRAPHY. Educ: Eastern Mich Univ, AB, 60; Univ Okla, MA, 61; Mich State Univ, PhD(geog), 68. Prof Exp: Instr geog, Southeast Mo State Col, 61-63; asst prof, Okla State Univ, 66-68 & Boston Univ, 68-71; ASSOC PROF GEOG, UNIV SOUTHERN MISS, 71- Concurrent Pos: Boston Univ Grad Sch grant soil moisture, Payne County, Okla, 68-70; Univ Southern Miss grant, Miss State Atlas, 72-74; res consult, Miss Marine Resources Coun, 75& Miss Water Resources Inst, 75-76, Mem: Am Water Resources Asn; Soil Conserv Soc Am; Asn Am Geog; Am Meteorol Soc. Res: Hydroclimatology; water resources; a regional interest in the Union of Soviet Socialist Republics; air quality analysis. Mailing Add: Southern Sta Box 352 Univ of Southern Miss Hattiesburg MS 39401

CROSS, RICHARD JAMES, b New York, NY, Mar 31, 15; m 39; c 5. INTERNAL MEDICINE, COMMUNITY HEALTH. Educ: Yale Univ, BA, 37; Columbia Univ, MD, 41, ScD(med), 49. Prof Exp: From instr to asst prof med, Columbia Univ, 47-59, asst dean, 57-59; assoc dean, Univ Pittsburgh, 59-63; assoc prof, Sch Med, Temple Univ, 63-64; adminr, Asn Am Med Cols, 64-65; prof med & assoc dean, 65-70, PROF COMMUNITY MED & CHMN DEPT, COL MED & DENT NJ-RUTGERS MED SCH, PISCATAWAY, NJ, 70- Concurrent Pos: Dean, Fac Med, Univ Ghana, 63-64. Res: Treatment of thrombo-embolic disease; aerobic phosphorylation; renal tubular transport in rabbit kidney slices; medical school administration. Mailing Add: Col of Med & Dent of NJ-Rutgers Med Sch PO Box 101 Piscataway NJ 08854

CROSS, ROBERT FRANKLIN, b Columbia, Ohio, May 6, 24; m 46; c 4. VETERINARY PATHOLOGY. Educ: Ohio State Univ, DVM, 46, MSc, 50; Purdue Univ, PhD(vet path), 61. Prof Exp: Instr vet path, Ohio State Univ, 48-50; vet pathologist, Territory of Hawaii, 50-58; from instr to assoc prof, Purdue Univ, 58-66; VET PATHOLOGIST, OHIO AGR RES & DEVELOP CTR, 66- Concurrent Pos: Lectr, Univ Hawaii, 55-58. Mem: Am Vet Med Asn; Am Col Vet Path; Conf Res Workers Animal Dis. Res: Veterinary clinical pathology. Mailing Add: Dept of Vet Sci Ohio Agr Res & Develop Ctr Wooster OH 44691

CROSS, RONALD ALLAN, b Chicago, Ill, Sept 19, 31. MICROBIAL GENETICS. Educ: Univ Calif, Los Angeles, AB, 54, MA, 62, PhD(genetics), 65. Prof Exp: Res assoc microbial genetics, 65-67, vis asst prof microbiol, 67-70, ASST PROF MICROBIOL & CURRICULUM COORD, SCH MED, UNIV SOUTHERN CALIF, 70- Concurrent Pos: NIH fel, 65-67. Mem: AAAS; Genetics Soc Am; Am Soc Microbiol. Res: Mutagenesis in microorganisms; mechanisms of gene action and control in bacteriophage lambda; genetics of the genus Neisseria and neisseriaphage. Mailing Add: Dept of Microbiol Sch Med Univ of Southern Calif Los Angeles CA 90033

CROSS, STEPHEN P, b Santa Monica, Calif, Apr 10, 38; m 59; c 4. MAMMALOGY. Educ: Calif State Polytech Col, BS, 60; Univ Ariz, MS, 62, PhD(zool), 69. Prof Exp: From instr to asst prof biol, 63-67, asst prof, 69-71, ASSOC PROF BIOL, SOUTHERN ORE STATE COL, 71- Concurrent Pos: Consult, Bur Reclamation, 73-74. Mem: Am Inst Biol Sci; Am Soc Mammal. Res: Vertebrate natural history; behavioral ecology of the Sciuridae; general ecology of North American Chiroptera. Mailing Add: Dept of Biol Southern Ore State Col Ashland OR 97520

CROSS, TIMOTHY AUREAL, b Pittsburgh, Pa, Jan 22, 46; m 70. SEDIMENTOLOGY, TECTONICS. Educ: Oberlin Col, BA, 67; Univ Mich, Ann Arbor, MS, 69; Univ Southern Calif, PhD(geol), 76. Prof Exp: Explor geologist, Texaco, Inc, 69-75; ASST PROF SEDIMENTOLOGY, TECTONICS, UNIV N DAK, GRAND FORKS, 75- Mem: Am Asn Petrol Geologists; Soc Econ Paleontologists & Mineralogists; Geol Soc Am. Res: History of igneous activity, western United States with relation to plate tectonics; Mississippian bioherms; Precambrian sedimentation, Canada carbonate sedimentology, Williston Basin. Mailing Add: Dept of Geol Univ of NDak Grand Forks ND 58201

CROSS, WILLIAM GUNN, b Detroit, Mich, Oct 1, 22; Can citizen; m 45; c 5. NUCLEAR PHYSICS. Educ: Univ Toronto, BA, 43, MA, 46; Harvard Univ, AM, 48, PhD(physics), 50. Prof Exp: Actg assoc prof physics, Washington Univ, 57-58; RES PHYSICIST, CHALK RIVER NUCLEAR LABS, ATOMIC ENERGY CAN, LTD, 50-57, 59- Mem: Am Phys Soc; Can Asn Physicists. Res: Compton scattering; elastic and inelastic scattering of fast neutrons; nuclear reactions and cross sections; dosimetry of beta x-rays and neutrons; particle optics; fission. Mailing Add: Chalk River Nuclear Lab Atomic Energy of Can Ltd Chalk River ON Can

CROSS, WILLIAM HENLEY, b Baker Co, Ga, Dec 10, 28; m 52; c 5. ENTOMOLOGY, ECOLOGY. Educ: Fla State Univ, BS, 49, MS, 51; Univ Ga, PhD(zool), 56. Prof Exp: Asst, Fla State Univ, 49-51 & Univ Ga, 51-55; ENTOMOLOGIST, AGR RES SERV, USDA, 57-; PROF ZOOL, MISS STATE UNIV, 63- Mem: Entom Soc Am; Ecol Soc Am. Res: Taxonomy and ecology in insect orders, especially Odonata, Orthoptera and Coleoptera; behavior and population ecology, especially insects. Mailing Add: Boll Weevil Res Lab PO Box 5367 Mississippi State MS 39762

CROSSAN, DONALD FRANKLIN, b Wilmington, Del, Apr 8, 26; m 48; c 3. PLANT PATHOLOGY. Educ: Univ Del, BS, 50; NC State Col, MS, 52, PhD(plant path), 54. Prof Exp: Res asst veg dis, NC State Col, 50-54; assoc prof, 54-66, asst dean col agr sci & asst dir agr exp sta, 66-69, ASSOC DEAN COL AGR SCI & ASSOC DIR, DEL AGR EXP STA, UNIV DEL, 69-, VPRES UNIV RES, 72-, PROF VEG DIS, 66- Mem: AAAS; Am Phytopath Soc. Res: Vegetable plant diseases; disease resistance; fungus physiology. Mailing Add: 155 Woodshade Dr Newark DE 19702

CROSSLEY, DAVID JOHN, b Salisbury, Eng, May 10, 44. GEOPHYSICS. Educ: Univ Newcastle-upon-Tyne, BSc, 66; Univ BC, MSc, 69, PhD(geophys), 73. Prof Exp: Res scientist, ICI Fibers Div, 66-67; fel, 73-74, RES ASSOC PHYSICS, MEM UNIV NFLD, 74- Mem: Can Asn Physicists; Am Geophys Union. Res: Solid earth geophysics, especially dynamics of the earth's core and free oscillation theory; planetary geophysics, crustal seismology and instrument design. Mailing Add: Dept of Physics Mem Univ of Nfld St John's NF Can

CROSSLEY, DERYEE ASHTON, JR, b Kingsville, Tex, Nov 6, 27; m 50, 61; c 1. RADIATION ECOLOGY. Educ: Tex Technol Col, BA, 49, MS, 51; Univ Kans, PhD(entom), 57. Prof Exp: Instr biol, Tex Technol Col, 49-51; asst, Univ Kans, 51-56; biologist, Oak Ridge Nat Lab, 56-67; PROF ENTOM, UNIV GA, 67- Mem: Ecol Soc Am; Entom Soc Am; Am Soc Naturalists. Res: Radioisotope movement in food chains; mineral cycling in ecosystems; role of soil arthropods in ecosystems; taxonomy of soil mites. Mailing Add: Dept of Entom Univ of Ga Athens GA 30602

CROSSMAN, EDWIN JOHN, b Niagara Falls, Ont, Sept 21, 29; m 52; c 2. ICHTHYOLOGY. Educ: Queen's Univ, Can, BA, 52; Univ Toronto, MA, 54; Univ BC, PhD, 57. Prof Exp: Fishery biologist, Biol Sta, Queen's Univ, Can, 50; biologist, Toronto Anglers & Hunters Conserv Proj, 51-54; fishery biologist, BC Game Comn, 54-57; asst cur, Dept Fishes, 57-64, assoc cur, Dept Ichthyol & Herpet, 64-68, CUR, DEPT ICHTHYOL & HERPET, ROYAL ONT MUS, 68-; PROF ZOOL, UNIV TORONTO, 68- Mem: Am Fisheries Soc Am; Soc Ichthyologists & Herpetologists; Am Soc Syst Zool; Am Inst Fishery Res Biol; Can Soc Zool. Res: Biology, distributions and systematics of freshwater fishes, particularly esocoid fishes. Mailing Add: Dept of Ichthyol Royal Ont Mus 100 Queens Park Toronto ON Can

CROSSMON, GERMAIN CHARLES, b Prattsburg, NY, May 9, 05; m 31. MICROSCOPY. Educ: Alfred Univ, BS, 28. Prof Exp: Teacher high sch, NJ, 29-31; bacteriologist, Sch Med & Dent, Univ Rochester, 35-42; biol & chem microscopist & indust hyg chemist, Bausch & Lomb, Inc, 42-74; CONSULT INDUST HYG & MICROS, 74- Mem: AAAS; Am Indust Hyg Asn; Am Chem Soc; Am Micros Soc; Am soc Microbiol. Res: Biological and chemical microscopy; industrial hygiene chemistry. Mailing Add: 23 Esternay Lane Pittsford NY 14534

CROSSON, JOHN WILLIAM, b Philadelphia, Pa, Nov 17, 07; m 39; c 2. MEDICINE. Educ: Univ Pa, AB, 30; Temple Univ, MD, 34. Prof Exp: Intern, Germantown Hosp, Philadelphia, 34-36, resident, 36-38; asst surgeon, Div Indust Hyg, NIH, New York & Salt Lake City, 38-40; dir, Bur Indust Hyg, WVa State Dept Health, 40-43; assoc med dir, Med Res Div, Sharp & Dohme, Inc, 43-46, med dir, 46-50; asst dir clin res, G D Searle & Co, 50-72; RETIRED. Honors & Awards: Cert Merit, AMA, 48. Mem: AAAS; fel AMA; fel Am Col Physicians; Am Therapeut Soc; NY Acad Sci. Res: Experimental therapeutics. Mailing Add: 127 Church Rd Winnetka IL 60093

CROSSON, ROBERT SCOTT, b Fairbanks, Alaska, Oct 19, 38; m 59; c 2. GEOPHYSICS, SEISMOLOGY. Educ: Univ Wash, BS, 61; Univ Utah, MS, 63; Stanford Univ, PhD(geophys), 66. Prof Exp: Asst prof, 66-72, ASSOC PROF GEOPHYS & GEOL, UNIV WASH, 72- Mem: Seismol Soc Am; Soc Explor Geophys; Am Geophys Union. Res: Physical properties and structure of earth's crust and upper mantle; characteristics and distribution of earthquakes and their tectonic implications; elastic characteristics of rocks. Mailing Add: Geophys Prog Univ of Wash Seattle WA 98105

CROSSWHITE, CAROL D, b Perth Amboy, NJ, Aug 1, 40; m 61; c 3. DESERT ECOLOGY, ENTOMOLOGY. Educ: Univ Calif, Riverside, BS, 61; Univ Wis, Madison, MS, 64, PhD(entom), 68. Prof Exp: Instr entom, Univ Wis, Madison, 68-70; CUR ZOOL, DESERT BIOL STA, BOYCE THOMPSON SOUTHWESTERN ARBORETUM, UNIV ARIZ, 71- Mem: Ecol Soc Am; Soc Syst Zool. Res: Classification and ecology of Hymenoptera; pollination ecology; desert biology. Mailing Add: Univ Ariz Desert Biol Sta PO Box AB Boyce Thompson Southwestern Arboretum Superior AZ 85273

CROSSWHITE, F JOE, b Springfield, Mo, Oct 13, 29; m 49; c 3. MATHEMATICS EDUCATION. Educ: Univ Mo, BSEd, 53, MEd, 58; Ohio State Univ, PhD(math educ), 64. Prof Exp: Teacher math, Salem High Sch, Mo, 53-57; instr, Keokuk Community Col, Iowa, 57-61; from instr to assoc prof, 62-70, PROF MATH EDUC, OHIO STATE UNIV, 70- Concurrent Pos: US Off Educ fel, Stanford Univ, 68-69; prog mgr, NSF, 75- Mem: Nat Coun Teachers Math; Math Asn Am; Am Educ Res Asn. Res: Teaching and learning of mathematics at the precollege level including the education of teachers for this level. Mailing Add: 283 Arps Hall Ohio State Univ Columbus OH 43210

CROSSWHITE, FRANK SAMUEL, b Atchison, Kans, Sept 23, 40; m 61; c 3. BOTANY, BIOGEOGRAPHY. Educ: Ariz State Univ, BS, 62; Univ Wis, Madison, MS, 65, PhD, 71. Prof Exp: Res asst bot, Univ Wis, Madison, 69-70; asst prof, Univ Wis Ctr, Waukesha, 70-71; CUR BOT, BOYCE THOMPSON SOUTHWESTERN ARBORETUM, UNIV ARIZ, 71- Mem: Am Soc Plant Taxon; Int Asn Plant Taxon; Bot Soc Am. Res: Classification of Scrophulariaceae; pollination ecology; history of botany; geography of North American communities. Mailing Add: Univ of Ariz PO Box AB Boyce Thompson Southwestern Arboretum Superior AZ 85273

CROSSWHITE, HENRY MILTON, JR, b Riverdale, Md, March 26, 19; m 44; c 3. PHYSICS. Educ: Western Md Col, AB, 40; Johns Hopkins Univ, PhD(physics), 46. Prof Exp: Asst, Nat Res Coun war contract, Johns Hopkins Univ, 42-46, instr physics, 46-47, from asst prof to adj prof, 47-72, from res scientist to prin res scientist, 58-72; MEM STAFF, CHEM DIV, ARGONNE NAT LAB, 72- Mem: Am Phys Soc. Res: Visible and ultraviolet spectroscopy, Mailing Add: Chem Div Argonne Nat Lab Argonne IL 60439

CROSTON, CLARENCE BRADFORD, biochemistry, see 12th edition

CROTEAU, RODNEY, b Springfield, Mass, Dec 6, 45; m 67; c 2. BIOCHEMISTRY. Educ: Univ Mass, Amherst, BS, 67, PhD(food sci), 70. Prof Exp: NIH res assoc biochem, Ore State Univ, 70-72; res assoc, 73-75, ASST AGR CHEMIST, WASH STATE UNIV, 75- Mem: Am Chem Soc; Am Soc Plant Physiologists; Phytochem Soc NAm; Inst Food Technologists. Res: Biosynthesis and catabolism of terpenoid natural products by plants; metabolism of monoterpenes and sesquiterpenes by animals. Mailing Add: Dept of Agr Chem Wash State Univ Pullman WA 99163

CROTHERS, DONALD M, b Fatehgarh, India, Jan 28, 37; US citizen; m 60; c 2. BIOPHYSICAL CHEMISTRY. Educ: Yale Univ, BS, 58; Cambridge Univ, BA, 60; Univ Calif, San Diego, PhD(chem), 63. Prof Exp: NSF fel, 63-64; from asst prof to assoc prof, 64-71, PROF CHEM & MOLECULAR BIOPHYS, YALE UNIV, 71- Res: Physical chemistry of biological macromolecules. Mailing Add: Surrey Dr Northford CT 06472

CROTTY, WILLIAM JOSEPH, b NJ, Sept 25, 22; m 46; c 4. PLANT MORPHOGENESIS. Educ: City Col New York, BS, 48; Rutgers Univ, PhD(bot), 52. Prof Exp: Teaching fel bot, Rutgers Univ, 49-52; from asst prof to assoc prof, 52-73, chmn dept, 52-73, PROF BIOL & DIR UNDERGRAD STUDIES, WASH SQ COL, NY UNIV, 73- Mem: Torrey Bot Club; Bot Soc Am; Am Soc Cell Biol. Res: Plant development. Mailing Add: Dept of Biol Wash Sq Col NY Univ New York NY 10003

CROUCH, BILLY G, b Port Lavaca, Tex, May 14, 30. PHYSIOLOGY, RADIATION BIOLOGY. Educ: Baylor Univ, BS, 54, MS, 55; Univ Tenn, PhD(clin physiol), 58. Prof Exp: Instr biol, Baylor Univ, 54-55; asst clin physiol, Col Med, Univ Tenn, 55-58, lectr, 56-58; radiation biologist, US Naval Radiol Defense Lab, Calif, 60-63; clin res pharmacologist, Sterling-Winthrop Res Inst, 63-64, sr clin pharmacologist, 64-65; dir med commun, Winthrop Labs, 65-68; DIR DRUG REGULATORY AFFAIRS, STERLING DRUG, INC, 68- Concurrent Pos: Nat Acad Sci-Nat Res Coun Donner fel, Radiobiol Sect, Nat Defense Res Coun, Netherlands, 58-59; Nat Heart Inst res fel, 59-60; vis lectr, Fac Med, State Univ Leiden, 59-60. Mem: Am Physiol Soc; Am

Fedn Clin Res; Radiation Res Soc; Royal Soc Med; Netherlands Soc Radiobiol. Res: Clinical physiology; bone marrow transplantation in irradiation sickness; clinical evaluation of new drug substances. Mailing Add: Sterling Drug Inc 90 Park Ave New York NY 10016

CROUCH, GLENN LEROY, wildlife management, range management, see 12th edition

CROUCH, HARRY R, JR, physics, see 12th edition

CROUCH, HUBERT BRANCH, b Jacksonville, Tex, Dec 1, 06; m 30; c 2. ZOOLOGY. Educ: Tex Col, AB, 27; Iowa State Univ, MS, 30, PhD(parasitol), 36. Prof Exp: Assoc prof, Dept Nat Sci, Ky State Indust Col, 31-34, prof biol, 35-44, chmn div arts & sci, 37-44; head dept, 44-70, PROF BIOL, TENN STATE UNIV, 44-, DIR DIV SCI & DEAN GRAD SCH, 58- Concurrent Pos: Dir, Ky Syphilis Servs, 39-43; State Ky fel, US Marine Hosp, 39, Gen Educ Bd fel, Columbia Univ, 43. Mem: Nat Inst Sci (exec secy-treas, 43-53, pres, 58-59); Nat Asn Res Sci Teaching; Nat Asn Biol Teachers. Res: Parasitic protozoa and nematodes. Mailing Add: Dept of Biol Tenn State Univ Nashville TN 37203

CROUCH, JAMES ENSIGN, b Urbana, Ill, Jan 28, 08; m 31; c 2. ORNITHOLOGY. Educ: Cornell Univ, MS, 31; Univ Southern Calif, PhD(vert zool), 39. Prof Exp: Mem dept zool, 32-42, prof, 42-73, chmn div life sci, 62-69, EMER PROF ZOOL, SAN DIEGO STATE UNIV, 73- Mem: AAAS; assoc Am Ornithologists Union; assoc Cooper Ornith Soc. Res: Bird behavior; life history of Phainopepla nitens lepida. Mailing Add: Dept of Zool San Diego State Univ San Diego CA 92182

CROUCH, MADGE LOUISE, b Winston-Salem, NC, Sept 21, 19. PUBLIC HEALTH ADMINISTRATION. Educ: Methodist Hosp Sch Nursing, Brooklyn, dipl, 41; Columbia Univ, BS, 47; George Washington Univ, MA, 61. Prof Exp: Instr basic nursing & microbiol, Methodist Sch Nursing, Brooklyn, 41-43; nat dir, Am Nat Res Cross, 48-65; asst to chief blood bank prod lab, Div Biol Studies, NIH, 65-71; DIR BLOOD BANK PROD BR, DIV BLOOD & BLOOD PROD, BUR BIOL, FOOD & DRUG ADMIN, 72- Concurrent Pos: Mem comt plasma & plasma substitutes, Div Med Sci, Nat Acad Sci, 69-70; mem secy task force nat blood policy, Dept of Health, Educ & Welfare, 72-74. Honors & Awards: Pfizer Merit Award, Pfizer Lab, 61; Food & Drug Admin Commendation Award, 74. Mem: Int Soc Blood Transfusion. Res: Development of improved blood banking procedures for prolonged storage of red blood cells; greater efficacy of components; training of personnel; techniques and equipment. Mailing Add: Bur of Biol 8800 Rockville Pike Bethesda MD 20014

CROUCH, MARSHALL FOX, b St Louis, Mo, Nov 22, 20; m 49; c 4. PHYSICS. Educ: Univ Mich, BS, 41; Univ Wash, PhD(physics), 50. Prof Exp: Res assoc, Nat Defense Res Comt, Univ Mich, 41-42; mem staff, Radiation Lab, Mass Inst Technol, 42-43 & Los Alamos Lab, 43; PROF PHYSICS, CASE WESTERN RESERVE UNIV, 50- Concurrent Pos: Fulbright res prof, Univ Tokyo, 56-57; dep sci attache, Am Embassy, Tokyo, 59-61. Mem: Am Phys Soc. Res: Nuetrino physics; cosmic rays. Mailing Add: Dept of Physics Case Western Reserve Univ Cleveland OH 44106

CROUCH, NORMAN ALBERT, b Monroe, Wis, June 7, 40; m 63; c 2. VIROLOGY. Educ: Univ Wis, BS, 62, MS, 66, PhD(med microbiol), 69. Prof Exp: Fel, Col Med, Baylor Univ, 69-70 & Pa State Univ, 70-72; ASST PROF MICROBIOL, UNIV IOWA, 72- Mem: Am Soc Microbiol. Res: The mechanism of alteration of plasma membranes induced by tumor viruses and the role of such alterations in the abnormal proliferation of tumor cells. Mailing Add: Dept of Microbiol Col of Med Univ of Iowa Iowa City IA 52242

CROUCH, RALPH BOYETT, b Lynn Grove, Ky, Aug 10, 22; m 44; c 4. MATHEMATICS. Educ: Murray State Col, BS, 42; Univ Ill, MS, 47; Univ Kans, PhD(math), 54. Prof Exp: Teacher pub schs, Ill, 42-47; from instr to prof math, NMex State Univ, 47-66; dean col sci, 68-69, PROF MATH & HEAD DEPT, DREXEL UNIV, 66-, VPRES ACAD AFFAIRS, 69- Mem: Am Math Soc; Math Asn Am. Res: Algebra; group theory. Mailing Add: Drexel Univ 32nd & Chestnut St Philadelphia PA 19104

CROUCH, ROBERT THOMAS, b Rock Hall, Md, Oct 23, 25; m 47; c 4. PLASTICS CHEMISTRY. Educ: Franklin & Marshall Col, BS, 48; Carnegie Inst Technol, MS, 51, DSc, 52. Prof Exp: Proj leader, Shell Chem Corp, Tex, 52-54; res chemist, Res Ctr, Johns-Manville Prod Corp, 54-56, chief plastics & resin prod sect, 56-59, mgr mfg & tech serv, New Bus Develop Dept, 59-62, coordr rigid vinyl activities, 62-64; mgr mkt develop, Richmond Res Ctr, Stauffer Chem Co, Calif, 64-66; dir res & develop, Avery Label Co, 66-69; vpres develop div, Compac Corp, NJ, 69-73; PRES, HAMILTON MFG CO, INC, 73- Mem: Am Chem Soc; Soc Plastics Engrs; Am Soc Testing & Mat. Res: Asbestos reinforced plastics; rigid vinyl compounds; pressure sensitive adhesives; lacquers and coatings; hot melt laminates. Mailing Add: 7400 Ranco Rd Richmond VA 23228

CROUCH, ROBERT WHEELER, b Detroit, Mich, Feb 27, 21; m 48. MICROPALEONTOLOGY, PALEOECOLOGY. Educ: Univ Southern Calif, BS, 47, MS, 48. Prof Exp: Paleontologist foraminifera, Richfield Oil Corp, 48-51; chief paleontologist, John W Mecom-Oil Independent, 53-65; sr paleontologist, 65-73, CONSULT PALEONTOLOGIST FORAMINIFERA, AMOCO PROD CO, STANDARD OIL, IND, 73- Mem: Am Asn Petrol Geologists; fel Geol Soc Am; Sigma Xi. Res: World wide planctonic foraminiferal research. Mailing Add: 633 Ramona Ave Space 96 Los Osos CA 93402

CROUCH, STANLEY ROSS, b Turlock, Calif, Sept 23, 40. ANALYTICAL CHEMISTRY. Educ: Stanford Univ, MS, 63; Univ Ill, PhD(chem), 67. Prof Exp: Instr chem, Univ Ill, 67, vis asst prof, 67-68; asst prof, 68-74, ASSOC PROF CHEM, MICH STATE UNIV, 74- Mem: AAAS; Am Chem Soc; Optical Soc Am; Soc Appl Spectros. Res: Kinetics and mechanisms of analytical reactions; fast reaction kinetics; chemical instrumentation; analytical spectroscopy. Mailing Add: Dept of Chem Mich State Univ East Lansing MI 48823

CROUNSE, NATHAN NORMAN, b Omaha, Nebr, May 25, 17; m 39; c 3. ORGANIC CHEMISTRY. Educ: Iowa State Col, BS, 38; Univ Iowa, PhD(chem), 42. Prof Exp: Chemist, C M Bundy Co, 42-43; sr chemist org res, Hilton-Davis Chem Co, 43-47 & Inst Med Res, Christ Hosp, 47-51; assoc dir res, 51-62, DIR CHEM RES, HILTON-DAVIS CHEM CO, 62- Mem: Am Chem Soc. Res: pigments, fluorescent compounds; colorless duplicating papers; structure studies in dyes and drug metabolism; germicides; aliphatic synthesis; dyes. Mailing Add: Hilton-Davis Chem Co 2235 Langdon Farm Rd Cincinnati OH 45237

CROUNSE, ROBERT GRIFFITH, b Albany, NY, Mar 23, 31; m 55; c 2. DERMATOLOGY, BIOCHEMISTRY. Educ: Yale Univ, BS, 52, MD, 55; Am Bd Dermat, dipl, 63. Prof Exp: Intern, Grace-New Haven Hosp, Conn, 55-56; clin assoc, Nat Cancer Inst, 57-58; res asst prof, Sch Med, Univ Miami, 61, from asst prof to assoc prof, 63-64; assoc prof & chmn, sub-dept, Sch Med, Johns Hopkins Univ, 64-67; prof dermat & biochem & res dir, Med Col Ga, 67-73, assoc dean, Sch Med, & dir, Off Instrnl Systs, 71-73; prof dermat, 73-75, ASSOC DEAN, SCH MED, UNIV NC, CHAPEL HILL, 73- Concurrent Pos: Fel dermat, Sch Med, Yale Univ, 56-57; Nat Cancer Inst spec res fel biochem, Sch Med, Univ Miami, 61-62; Nat Inst Arthritis & Metab Dis res career develop award, 62-64. Mem: Am Chem Soc; AMA; Soc Invest Dermat; Am Fedn Clin Res. Res: Biochemistry of epidermal keratinization; protein chemistry of epidermal structures; skin physiology and pharmacology. Mailing Add: Dept of Med Allied Hlth Profs Univ of NC Sch of Med Chapel Hill NC 27514

CROUSE, DALE MCCLISH, b Los Angeles, Calif, Aug 27, 41; m 66. ORGANIC CHEMISTRY. Educ: Stanford Univ, BS, 64; Ore State Univ, MS, 67; Univ Del, PhD(org chem), 70. Prof Exp: RES CHEMIST, JACKSON LABS, E I DU PONT DE NEMOURS & CO, INC, 69- Concurrent Pos: Instr, Univ Del, 69-71. Mem: Am Chem Soc. Res: Dialkyl carbene chemistry; Wittig reaction; organophosphorus; nuclear magnetic resonance spectroscopy; fluorocarbons; aromatic intermediates. Mailing Add: Jackson Labs E I du Pont de Nemours & Co Inc Deepwater NJ 08023

CROUSE, DAVID AUSTIN, b Canton, Ill, Aug 29, 44; m 68; c 2. RADIOBIOLOGY. Educ: Western Ill Univ, BS, 66, MS, 68; Univ Iowa, PhD(radiobiol), 74. Prof Exp: NDEA IV fel radiobiol, Univ Iowa, 71-74; APP, ARGONNE NAT LAB, 75- Mem: Radiation Res Soc. Res: Comparative studies of late effects resulting from single or fractionated doses of gamma or fast neutron exposures of mammalian systems with special emphasis on the cell mediated immune response. Mailing Add: Div of Biol & Med Res Argonne Nat Lab Argonne IL 60439

CROUSE, GAIL, b Connellsville, Pa, May 10, 23; m 51; c 3. ANATOMY. Educ: Heidelberg Col, BS, 50; Univ Mich, AM, 52, PhD(zool), 56. Prof Exp: From instr to asst prof anat, Hahnemann Med Col, 55-61; asst prof, 61-67, ASSOC PROF ANAT, SCH MED, TEMPLE UNIV, 67- Mem: Am Asn Anatomists. Res: Experimental mammalian tissue transplantation, especially differentiation of embryonic tissue after transplantation to the brain; reactivity to prostheses in animals; transplantation site; progressive development of the autonomic nervous system in human fetuses. Mailing Add: Dept of Anat Temple Univ Sch of Med Philadelphia PA 19140

CROUT, JOHN RICHARD, b Portland, Ore, Dec 30, 29; m 54; c 3. CLINICAL PHARMACOLOGY, PUBLIC HEALTH ADMINISTRATION. Educ: Oberlin Col, AB, 51; Northwestern Univ, MD, 55, MS, 56; Am Bd Internal Med, dipl, 62. Prof Exp: Intern, Passavant Mem Hosp, Chicago, Ill, 55-56; asst resident med, Vet Admin Res Hosp, Chicago, 56-57; clin assoc, Nat Heart Inst, 57-60; asst resident, NY Univ-Bellevue Hosp Med Ctr, 60-61; instr pharmacol, Harvard Med Sch, 61-63; from asst prof to assoc prof pharmacol & internal med, Univ Tex Southwestern Med Sch, 63-70; prof pharmacol & med, Col Human Med, Mich State Univ, 70-71; dep dir, Bur Drugs, 71-72, dir, off Sci Eval, 72-73, DIR, BUR DRUGS, FOOD & DRUG ADMIN, 73- Concurrent Pos: Res fel pharmacol, Harvard Med Sch, 61-62; USPHS fel, 61-63, Burroughs-Wellcome scholar clin pharmacol, 65-70; mem, Coun High Blood Pressure Res, Am Heart Asn, Anesthesiol Training Grant Comt, Nat Inst Gen Med Sci, 66-68, Pharmacol, Toxicol Prog Comt, 69-71, Comt Myocardial Info Study Ctrs, NIH, 67 & Ad Hoc Sci Adv Comt, Food & Drug Admin, 70-71; field ed, J Am Soc Pharmacol & Exp Therapeut, 68-71. Mem: Am Fedn Clin Res; Am Soc Pharmacol & Exp Therapeut; Am Soc Clin Pharmacol & Therapeut; Am Soc Clin Invest; fel Am Col Physicians. Res: Catecholamine metabolism; pheochromocytoma; cardiovascular pharmacology; hypertension; clinical pharmacology; drug regulation policy. Mailing Add: HFD-1 Food & Drug Admin 5600 Fishers Lane Rockville MD 20852

CROUT, PRESCOTT DURAND, b Columbus, Ohio, July 28, 07; m 33; c 4. APPLIED MATHEMATICS. Educ: Mass Inst Technol, PhD(math), 30. Prof Exp: Elec engr, Gen Elec Co, NY, 30-32; elec & mech engr, Raytheon Mfg Co, Mass, 32-34; from assoc prof to prof, 34-74, EMER PROF MATH, MASS INST TECHNOL, 74- Concurrent Pos: Mem staff, Radiation Lab, Mass Inst Technol, 41-45; consult, US Naval Ord Test Sta, Calif, 55- Mem: Soc Indust & Appl Math; Am Math Soc; Inst Elec & Electronics Eng. Res: Mechanics; electromagnetic theory; microwaves. Mailing Add: Dept of Math Mass Inst Technol Cambridge MA 02139

CROUTHAMEL, CARL EUGENE, b Lansdale, Pa, Dec 25, 20; m 44; c 3. INORGANIC CHEMISTRY. Educ: Eastern Nazarene Col, BS, 42; Boston Univ, MA, 43; Iowa State Col, PhD(chem), 50. Prof Exp: Inorg chem res, Manhattan Proj, 43-44; res assoc, Inst Atomic Res, Iowa State Col, 46-50; sr chemist, Argonne Lab, 50-66, 67-73; SR ENGR, EXXON NUCLEAR CORP, 73- Mem: AAAS; NY Acad Sci; Am Chem Soc; Sci Res Soc Am; Am Nuclear Soc. Res: Inorganic chemistry of transuranium elements; rare earths; fused salt-liquid metals chemistry; development of high temperature fast breeder reactor fuel and materials; fuel performance of LWR reactor fuel. Mailing Add: 71 Park St Richland WA 99352

CROUTHAMEL, WILLIAM GUY, b Sellersville, Pa, Feb 11, 42; m 66; c 1. PHARMACY, PHARMACEUTICS. Educ: Philadelphia Col Pharm & Sci, BS, 65, MS, 67; Univ Ky, PhD(pharmaceut), 70. Prof Exp: From asst prof to assoc prof pharmaceut, WVa Univ, 70-75; ASSOC PROF PHARM, UNIV MD, BALTIMORE CITY, 75- Mem: Am Pharmaceut Asn; Am Asn Cols Pharm. Res: Effect of various factors such as pH and intestinal blood flow on the kinetics of gastrointestinal drug absorption; effect of disease states on drug kinetics; lipid pharmacology; gastrointestinal absorption of drugs. Mailing Add: Sch of Pharm Univ of Md 636 W Lombard St Baltimore MD 21201

CROVELLO, THEODORE JOHN, b Brooklyn, NY, Nov 20, 40; m 62; c 2. SYSTEMATICS. Educ: State Univ NY Col Forestry, Syracuse, BS, 62; Univ Calif, Berkeley, PhD(bot), 66. Prof Exp: Res assoc bot & entom, Univ Kans, 66-67; from asst prof to assoc prof biol, 66-75, PROF BIOL, UNIV NOTRE DAME, 75-, CHMN DEPT, 75-, CUR HERBARIUM, 66- Concurrent Pos: Chmn, Int Register Comput Proj Systs, 74-; mem adv bd, Plant Systs & Evolution, Int J, 74- Mem: Am Inst Biol Sci; Bot Soc Am; Soc Syst Zool; Am Soc Plant Taxon; Soc Study Evolution. Res: Computerized information retrieval of herbarium and floristic data; evolutionary strategies of the mustard family; computers in biological teaching. Mailing Add: Dept of Biol Univ of Notre Dame Notre Dame IN 46556

CROVETTI, ALDO JOSEPH, b Lake Forest, Ill, Apr 2, 30; m 59. ORGANIC CHEMISTRY. Educ: Lake Forest Col, BA, 51; Univ Ill, MS, 52, PhD(org chem), 55. Prof Exp: Res chemist, 57-65, group leader, 65-67, res org chemist, 67-70, DEPT MGR, AGR & INDUST PROD RES & DEVELOP, ABBOTT LABS, 70- Mem: Am Chem Soc; Sigma Xi; Am Soc Hort Sci; Weed Soc Am; Int Soc Heterocyclic Chem. Res: Synthesis in the fields of heterocyclic compounds, mainly nitrogen; agricultural chemistry and biological research, including field research. Mailing Add: Agr Prod Res & Develop Dept Abbott Labs Chicago IL 60064

CROW, ALONZO BIGLER, b Warren, Pa, Aug 27, 10; m 35; c 1. FOREST ECOLOGY. Educ: NC State Col, BS, 34; Yale Univ, MF, 41. Prof Exp: Jr forester, US Forest Serv, Mo, 34-35 & Pa, 35-40; asst agr aide & jr soil conservationist, Soil Conserv Serv, USDA, Md, 41-42, asst soil conservationist, 42-45; from asst to regional consult, Am Forestry Asn, Washington, DC, 45-46; from asst prof to assoc prof, 46-64, PROF FORESTRY, LA STATE UNIV, BATON ROUGE, 64- Mem:

Soc Am Foresters. Res: Forest ecology; southern pines; geographic seed sources loblolly pine; forest fire control and use. Mailing Add: Sch of Forestry & Wildlife Mgt La State Univ Baton Rouge LA 70803

CROW, EDWIN LEE, b Clinton, Mo, Apr 26, 39; m 62. ORGANIC CHEMISTRY. Educ: Univ Miss, PhD(chem), 66. Prof Exp: Chemist, Electrochem Dept, Del, 65-68, develop chemist, Tex, 68-70, res supvr, Wilmington, Del, 70-74, TECH SUPT, E I DU PONT DE NEMOURS & CO, GIBBSTOWN, NJ & MEMPHIS, TENN, 74- Mem: AAAS; The Chem Soc; Am Chem Soc. Res: Catalytic reactions; organometallic chemistry; chemistry of carbenes; chemistry of olefins. Mailing Add: 45 S Rose Rd Memphis TN 38117

CROW, EDWIN LOUIS, b Cadiz, Wis, Sept 15, 16; m 42; c 2. STATISTICS, MATHEMATICS. Educ: Beloit Col, BS, 37; Univ Wis, PhM, 38, PhD(math), 41. Prof Exp: Instr math, Case Univ, 41-42; mathematician res & develop div, Bur Ord, US Navy Dept, 42-46; mathematician, US Naval Ord Test Sta, 46-54, head statist br, 50-54; consult statist, Boulder Labs, Nat Bur Stand, 54-65, Environ Sci Serv Admin, 65-70 & Off Telecommun, US Dept Com, 70-73; STATISTICIAN, OFF TELECOMMUN, US DEPT COM, 74-, NAT CTR ATMOSPHERIC RES, 75- Concurrent Pos: Instr exten div, Univ Calif, Los Angeles, 47-54; Govt Employees Training Act trainee, London, 61-62; adj prof appl math, Univ Colo, 63- Honors & Awards: Bronze Medal, US Dept Com, 70. Mem: Fel AAAS; Soc Indust & Appl Math; Inst Math Statist; fel Am Statist Asn; Royal Statist Soc. Res: Expansion problems associated with ordinary differential equations; mathematical statistics with applications in ordance, radio propagation, radio standards and weather modification. Mailing Add: 605 20th St Boulder CO 80302

CROW, GARRETT EUGENE, b Phoenix, Ariz, Dec 11, 42; m 72. SYSTEMATIC BOTANY. Educ: Taylor Univ, AB, 65; Mich State Univ, MS, 68, PhD(bot), 74. Prof Exp: Instr, Mich State Univ, 73-74, fel syst bot & man & biosphere, 74-75; ASST PROF SYST BOT & CUR, HERBARIUM, UNIV NH, 75- Mem: Int Asn Plant Taxonomists; Am Soc Plant Taxonomists; Sigma Xi; Bot Soc Am; Am Inst Biol Sci. Res: Plant systematics, particularly Caryophyllaceae sagina; scanning electron microscope studies in seed morphology in Caryophyllaceae floristics, ecology and phytogeography of arctic, subantarctic, alpine plants, bogs; flora of New Hampshire. Mailing Add: Dept of Bot & Plant Path Univ of NH Durham NH 03824

CROW, JAMES FRANKLIN, b Phoenixville, Pa, Jan 18, 16; m 41; c 3. GENETICS. Educ: Friends Univ, AB, 37; Univ Tex, PhD(genetics), 41. Prof Exp: Tutor zool, Univ Tex, 37-40; from instr zool to asst prof zool & prev med, Med Sch, Dartmouth Col, 41-48; from aasasst prof to prof zool & genetics, 48-58, actg dean med sch, 63-65, chmn dept genetics & med genetics, 65-70, PROF MED GENETICS, UNIV WIS-MADISON, 58- Mem: AAAS; Genetics Soc Am (pres, 60); Am Soc Human Genetics (pres, 63); Am Statist Asn; Soc Study Evolution. Res: Genetics of Drosophila; population genetics. Mailing Add: Dept of Genetics Univ of Wis Madison WI 53706

CROW, JOHN H, b San Pedro, Calif, Nov 18, 42; m 66. PLANT ECOLOGY. Educ: Whittier Col, BA, 64; Wash State Univ, PhD(bot), 68. Prof Exp: Asst prof bot, 68-71, ASSOC PROF BOT, RUTGERS UNIV, 71-, CHMN DEPT, 72- Concurrent Pos: Ecol consult, State of Wash, 68-69, US Dept Interior, 74- & State of NJ Dept Pub Advocate, 75. Mem: AAAS; Am Inst Biol Sci; Ecol Soc Am. Res: Ecological investigations of the salt marshes of Pacific Coastal Alaska; wetland vegetation of New Jersey; upland forest composition and environmental studies; salt marsh pollution. Mailing Add: Dept of Bot Rutgers Univ Newark NJ 07102

CROW, TERRY TOM, b Sapulpa, Okla, Sept 16, 31; m 54; c 3. ELECTROMAGNETISM. Educ: Miss State Univ, BS, 53; Vanderbilt Univ, MA, 57, PhD(physics), 60. Prof Exp: Asst prof physics, Miss State Univ, 60-62; physicist, Lawrence Radiation Lab, Univ Calif, 62-64; assoc prof, 64-67, PROF PHYSICS, MISS STATE UNIV, 67- Mem: Inst Elec & Electronics Engrs. Res: Electromagnetic theory. Mailing Add: Dept of Physics Miss State Univ State College MS 39762

CROW-BASTE, CLAUDIA ADKISON, b Montgomery, Ala, Dec 9, 41; m 72. ANATOMY. Educ: Huntingdon Col, BA, 64; Tulane Univ, PhD(anat), 69. Prof Exp: Instr myocardial biol, Baylor Col Med & anat, Sch Med, Tulane Univ, 70-71; ASST PROF ANAT, SCH MED, EMORY UNIV, 71- Concurrent Pos: NIH training grant pharmacol, Baylor Col Med, 69-70. Mem: Am Soc Cell Biol; Electron Micros Soc Am; Am Heart Asn. Res: Heart disease; heart ultrastructure; heart organelle function. Mailing Add: Dept of Anat Emory Univ Sch of Med Atlanta GA 30322

CROWDER, ADELE A, b Dublin, Ireland, Jan 14, 26; m 57; c 3. PLANT ECOLOGY. Educ: Univ Dublin, BA, 47, MA & PhD(bot), 55. Prof Exp: Asst to prof bot, Trinity Col, Dublin, 49-51; res assoc, Queen's Univ, Belfast, 63-66; CUR HERBARIUM, QUEEN'S UNIV, ONT, 69-, ASST PROF BIOL, 71- Concurrent Pos: Dept Univ Affairs grant, 70. Honors & Awards: Hackett Mem Prize, Univ Dublin, 68. Mem: Brit Ecol Soc. Res: Chemistry of bog soils; British Droseraceae; Neolithic forest clearances in North Ireland; Quaternary ecology in British Isles; vegetational history of Ontario; bryophytes, pollen-rain and vegetation of the Kingston region; pollen analysis. Mailing Add: Dept of Biol Queen's Univ Kingston ON Can

CROWDER, GENE AUTRY, b Wichita Falls, Tex, Oct 25, 36; m 65; c 2. PHYSICAL CHEMISTRY. Educ: Cent State Col, Okla, BS, 58; Univ Fla, MS, 61; Okla State Univ, PhD(phys chem), 64. Prof Exp: Asst res chemist, Petrol Chems, Inc & Cities Serv-Continental Oil Co, 58-59; from asst prof to assoc prof, 64-68, PROF CHEM, WEST TEX STATE UNIV, 68-, HEAD DEPT, 70- Mem: Am Chem Soc; Soc Appl Spectros; Coblentz Soc; Sigma Xi. Res: Molecular spectroscopy; vibrational assignments and normal coordinate calculations; rotational isomerism. Mailing Add: Dept of Chem West Tex State Univ Canyon TX 79015

CROWDER, LARRY A, b Mattoon, Ill, Mar 12, 42; m 64; c 2. INSECT PHYSIOLOGY, TOXICOLOGY. Educ: Eastern Ill Univ, BS, 64; Purdue Univ, MS, 66, PhD(entom), 70. Prof Exp: ASSOC PROF INSECT PHYSIOL, UNIV ARIZ, 69- Mem: AAAS; Entom Soc Am; Am Chem Soc; Am Registry Prof Entomologists; Sigma Xi. Res: Significance of 5-hydroxytryptamine in insects; mode of action of cyclodiene insecticides; regulation of insect diapause. Mailing Add: Dept of Entom Univ of Ariz Tucson AZ 85721

CROWDER, LOY VAN, b Cleveland Co, NC, Feb 5, 20; m 43; c 3. PLANT BREEDING. Educ: Berry Col, BSA, 42; Univ Ga, MSA, 47; Cornell Univ, PhD(plant breeding), 52. Prof Exp: Agronomist, Exp Sta, Univ Ga, 47-49 & 52-55; agronomist, Rockefeller Found, 55-63; assoc prof, 63-70, PROF PLANT BREEDING, CORNELL UNIV, 70- Concurrent Pos: On leave with dept agron, Univ Ibadan, 71-72. Mem: Am Soc Agron; Int Crop Improv Asn. Res: International agricultural development; graduate student training in plant breeding; tropical foliage crop breeding and improvement; management and fertility requirements of forage and pasture crops; cytological behavior of tall fescue grass and fescue-ryegrass hybrids. Mailing Add: Dept of Plant Breeding Cornell Univ Ithaca NY 14850

CROWE, ARLENE JOYCE, b Wakaw, Sask, Can, Oct 8, 31. BIOCHEMISTRY. Educ: Univ Alta, BSc, 50; McGill Univ, MSc, 56, PhD(biochem), 62. Prof Exp: Res asst liver dis, Hammersmith Hosp, Postgrad Sch, London, 56-57; chief technologist biochem, Montreal Children's Hosp, Que, 57-59; BIOCHEMIST, HOTEL DIEU HOSP, 62- Concurrent Pos: Lectr, Queen's Univ, Ont, 62- Mem: Can Biochem Soc; Can Soc Clin Chem; Am Asn Clin Chem; NY Acad Sci. Res: Liver metabolism; acid-base balance; bioenergetics. Mailing Add: Dept of Biochem Hotel Dieu Hosp Kingston ON Can

CROWE, BERNARD FRANCIS, b New York, NY, Nov 2, 21; m 52; c 3. ORGANIC CHEMISTRY. Educ: Fordham Univ, BS, 43, MS, 48, PhD(chem), 50; NY Law Sch, LLB, 62. Prof Exp: Res chemist org chem, Tex Co, 50-52; proj leader polymers, Olin Mathieson Chem Corp, 52-55; sect head, Air Reduction Co, Inc, 55-63; atty, 63-64, PATENT ATTY, UNION CARBIDE CORP, 64- Res: Thiophene chemistry; oxo reaction; lube oil additives; polymers; patent law. Mailing Add: 15 Kendall Ave Maplewood NJ 07040

CROWE, CHRISTOPHER, b London, Eng, Dec 4, 28; m 52; c 3. GEOPHYSICS. Educ: Western Ont Univ, BSc, 52, PhD(physics), 56. Prof Exp: Teacher pub schs, Ont, 46-47; physics demonstr, Western Ont Univ, 52-55; 1851 Exhib Overseas scholar, Dept Geog & Geophys, Cambridge Univ, 56-58; asst prof geophys, Pa State Univ, 58-63; staff geophysicist, Tex Instruments, Inc, 64-66, scientist, 66-69; sr geophysicist, 69-70, RES SCIENTIST, EXPLOR & PROD RES LAB, SUN OIL CO, 70- Concurrent Pos: Consult, Earth Sci Curric Proj, Am Geol Inst, 63-65. Mem: AAAS; fel Brit Inst Physics & Phys Soc; Am Geophys Union; Soc Explor Geophys; Can Asn Physicists. Res: Heat flow; geophysical exploration; physical properties of materials; seismic absorption and dispersion; direct hydrocarbon detection. Mailing Add: Explor & Prod Res Lab Sun Oil Co 503 N Cent Expressway Richardson TX 75080

CROWE, DAVID BURNS, b New Brighton, Pa, Oct 6, 30; m 56; c 2. ZOOLOGY. Educ: Washington & Jefferson Col, BA, 52; Univ Mich, MS, 57; Univ Louisville, PhD(biol), 64. Prof Exp: Instr biol, State Univ NY, 58-61; PROF BIOL, UNIV WIS-EAU CLAIRE, 64- Mem: Am Soc Ichthyologists & Herpetologists. Res: Orientation of social piscine groups with emphasis on integrative role of the sense organs. Mailing Add: Dept of Biol Univ of Wis Eau Claire WI 54701

CROWE, DONALD WARREN, b Lincoln, Nebr, Oct 28, 27; m 53; c 3. MATHEMATICS. Educ: Univ Nebr, BS, 49, MA, 51; Univ Mich, PhD(math), 59. Prof Exp: Instr math, Univ BC, 55-57; lectr, Univ Toronto, 57-59 & Univ Col, Ibadan, 59-62; from asst prof to assoc prof, 62-68, PROF MATH, UNIV WIS, MADISON, 68- Mem: Math Asn Am; Can Math Cong. Res: Geometry, especially finite planes and combinatorial problems. Mailing Add: Dept of Math Univ of Wis Madison WI 53706

CROWE, GEORGE A, b Jackson, Mo, Aug 11, 17; m 40; c 3. PHYSICAL CHEMISTRY. Educ: Univ Mo, AB, 38, MA, 40; Univ Del, PhD(chem), 50. Prof Exp: Chemist explosives, Western Cartridge Co, 40-41; inorg chem, Gen Chem Co, 41-42 & plastics, Hercules Powder Co, 42-50; sr chemist plastics, Bakelite Co, 50-53; sr res scientist, Johnson & Johnson, 53-66; SR CHEMIST & MGR RUBBER & PLASTICS RES DEPT, THOMPSON, WIENMAN & CO, 66- Mem: Am Chem Soc; Soc Plastics Engrs. Res: Plastics; polymers; electrochemistry; kinetics; surgical dressings; compounding synthetic rubbers and plastics. Mailing Add: Thompson, Weinman & Co Res Labs PO Box 125 Montclair NJ 07042

CROWE, GEORGE JOSEPH, b Brooklyn, NY, Oct 1, 21, m 52; c 11. SOLID STATE PHYSICS. Educ: Manhattan Col, BS, 43; Columbia Univ, MA, 47; Carnegie Inst Technol, MS, 61, PhD(physics), 66. Prof Exp: Instr physics, Manhattan Col, 46-48; assoc prof, Seton Hill Col, 48-64; proj physicist, Carnegie Inst Technol, 64-65, instr physics, 65; assoc prof, 65-74, PROF PHYSICS, MANHATTAN COL, 74-, CHMN DEPT, 68- Mem: AAAS; Am Phys Soc; Am Asn Physics Teachers. Res: Point defects in alkali halides; electron irradiated cadmium sulfide x-ray spectroscopy. Mailing Add: Dept of Physics Manhattan Col Riverdale Bronx NY 10471

CROWE, JOHN H, b Columbia, SC, Apr 12, 43. COMPARATIVE PHYSIOLOGY. Educ: Wake Forest Univ, BS, 65, MA, 67; Univ Calif, Riverside, PhD(biol), 70. Prof Exp: Asst prof, 70-75, ASSOC PROF ZOOL, UNIV CALIF, DAVIS, 75- Concurrent Pos: Consult, Review Panelist, Nat Sci Found, 75- Mem: Sigma Xi; Am Soc Zoologists; AAAS; Am Micros Soc; Biophys Soc Am. Res: Physiology of the induction of cryptobiotic states; fine structure, chemical composition, formation and physiology of invertebrate cuticles, particularly those of tardigrades and mites. Mailing Add: Dept of Zool Univ of Calif Davis CA 95616

CROWE, KENNETH MORSE, b Boston, Mass, Oct 6, 26; m 63; c 3. PHYSICS. Educ: Brown Univ, BS, 48, PhD(physics), 53. Prof Exp: Physicist, Radiation Lab, Univ Calif, 49-51; res assoc, High-Energy Physics Lab, Stanford Univ, 51-56; from asst prof to assoc prof, 58-69, PHYSICIST, LAWRENCE RADIATION LAB, UNIV CALIF, BERKELEY, 56-, PROF PHYSICS, UNIV, 69- Mem: Am Phys Soc. Res: High-energy particle physics. Mailing Add: Lawrence Radiation Lab Univ of Calif Berkeley CA 94720

CROWELL, ALBERT DARY, b Dover, NH, Feb 12, 25; m 47; c 3. SURFACE PHYSICS. Educ: Brown Univ, BEE, 46, PhD(physics), 50; Harvard Univ, MS, 47. Prof Exp: From instr to asst prof, Amherst Col, 50-55; from asst prof to assoc prof, 55-61, PROF PHYSICS & CHMN DEPT, UNIV VT, 61- Concurrent Pos: Vis prof, Bristol Univ, 68. Mem: AAAS; Am Asn Physics Teachers; Am Phys Soc; Am Vacuum Soc. Res: Adsorption of gases. Mailing Add: Dept of Physics Univ of Vt Burlington VT 05401

CROWELL, EDWIN PATRICK, b Elizabeth, NJ, Feb 27, 34; m 58; c 5. CHEMISTRY. Educ: Seton Hall Univ, BS, 56; Univ Richmond, MS, 62. Prof Exp: Control anal chemist, Ethylene Oxide Unit, Gen Aniline & Film Corp, 56-58; assoc chemist, Anal Res, Phillip Morris, Inc, 58-62; res chemist, Agr Div, Am Cyanamid Co, 62-63; res scientist, Union Bag-Camp Paper Corp, 63-70, group leader, 70-74, SECT LEADER, RES & DEVELOP DIV, UNION CAMP CORP, 74- Mem: Am Chem Soc; Tech Asn Pulp & Paper Indust; Am Soc Appl Spectros. Res: Development of instrumental and chemical analytical procedures in support of research program in paper pulp and chemical biproducts; tobacco and agricultural chemicals. Mailing Add: Res Div Union Camp Corp Box 412 Princeton NJ 08540

CROWELL, HAMBLIN HOWES, b Portland, Ore, Aug 23, 13; m 39; c 1. ENTOMOLOGY. Educ: Ore State Col, BS, 35, MS, 37; Ohio State Univ, PhD(entom), 40. Prof Exp: Asst biol sci surv, Ore State Col, 35-37 & zool & entom, Ohio State Univ, 37-40; sanit inspector, Health Dept, CZ, 40-44; asst entomologist, Exp Sta, 46-51, assoc prof, 51-66, actg head entom dept, 58-59, PROF ENTOM, ORE STATE UNIV, 66- Mem: AAAS; Entom Soc Am. Res: Economic entomology and malacology. Mailing Add: Dept of Entom Ore State Univ Corvallis OR 97331

CROWELL, JACK WESLEY, b Itta Bena, Miss, May 2, 26; m 50; c 2. PHYSIOLOGY, BIOPHYSICS. Educ: Univ Miss, BA, 50, MS, 51, med cert, 54, PhD(physiol), 57. Prof Exp: Lab asst physics, 48, res assoc physiol, 49, lab instr physiol & biophys, 50-54, from asst prof to assoc prof, 54-66, PROF PHYSIOL & BIOPHYS, UNIV MISS, 66- Mem: AAAS; Am Physiol Soc. Res: Cardiovascular research. Mailing Add: Depts of Physiol & Biophys Univ of Miss Med Ctr Jackson MS 39216

CROWELL, JOHN CHAMBERS, b State College, Pa, May 12, 17; m 46; c 1. GEOLOGY. Educ: Univ Tex, BS, 39; Univ Calif, Los Angeles, MA, 46, PhD(geol), 47. Hon Degrees: DSc, Cath Univ Louvain, 66. Prof Exp: Asst geologist, Shell Oil Co, Inc, 41-43; from instr to prof geol, Univ Calif, Los Angeles, 47-67, chmn dept, 57-60, 63-67; PROF GEOL, UNIV CALIF, SANTA BARBARA, 67- Concurrent Pos: Guggenheim Found fel, 53-54; Fulbright res prof, Austria, 53-54; NSF sr res fel, Scotland, 60-61. Honors & Awards: Chrestien Mica Gondwanaland Medal, Mining, Geol & Metall Inst India, 72. Mem: Geol Soc Am; Am Asn Petrol Geol; Am Geophys Union; Soc Econ Paleont & Mineral. Res: Structural and general geology; tectonics; paleoclimatology of ancient ice ages. Mailing Add: Dept of Geol Univ of Calif Santa Barbara CA 93106

CROWELL, JOHN MARSHALL, b Mobile, Ala, June 30, 42. BIOPHYSICS. Educ: Ga Inst Technol, BS, 64; Johns Hopkins Univ, PhD(physics), 73. Prof Exp: PHYSICIST, LOS ALAMOS SCI LAB, UNIV CALIF, 73- Mem: Am Phys Soc; Inst Elec & Electronic Engrs; Biophys Soc. Res: Applications of modern technologies to cytology, biological cell analysis and sorting. Mailing Add: Biophys Group H-10 MS 888 Los Alamos Sci Lab PO Box 1663 Los Alamos NM 87545

CROWELL, JULIAN, b Shelbyville, Tenn, Jan 24, 34; m 58; c 3. MATHEMATICS. Educ: Univ Tenn, BS, 56; Vanderbilt Univ, MS, 59, PhD(physics), 66. Prof Exp: Lectr physics, Gordon Col, Rawalpindi, Pakistan, 58-61; asst prof physics, Roanoke Col, 65-67; assoc prof physics & math, St Andrews Presby Col, 67-69; ASSOC PROF MATH, BOSPHORUS UNIV, ISTANBUL, TURKEY, 69-, CHMN DEPT, 71- Concurrent Pos: Consult- 66-70. Mem: Math Asn Am; Am Phys Soc. Res: Positron annihilation in metals; collective phenomena in metals. Mailing Add: Bosphorus Univ Dept of Math PK2 Bebek Istanbul Turkey

CROWELL, KENNETH L, b Glen Ridge, NJ, July 19, 33; m 62; c 1. VERTEBRATE ECOLOGY, ZOOGEOGRAPHY. Educ: Yale Univ, BS, 55; Univ Pa, PhD(zool), 61. Prof Exp: Instr zool, Duke Univ, 61-62; mem fac biol, Marlboro Col, 62-66; fel zool, Calgary Univ, 66-67, instr, 67; asst prof, 67-74, ASSOC PROF BIOL, ST LAWRENCE UNIV, 74- Concurrent Pos: Chapman Mem Fund grant, Am Mus Natural Hist, 59-60, 66; Jessup Fund grant, Acad Natural Sci Philadelphia, 61; Res Soc grant, Soc Sigma Xi, 62; NSF grants, 62-72; dir, Planned Parenthood Northern NY, 69-71; trustee, Adirondack Conservancy Comt, 75- Mem: Fel AAAS; Ecol Soc Am; Am Ornith Union; Am Soc Mammal. Res: Population dynamics and niche segregation through studies of competition and habitat selection in mammals of islands of Gulf of Maine and Adirondacks and birds of Bermuda and West Indies. Mailing Add: Dept of Biol St Lawrence Univ Canton NY 13617

CROWELL, (PRINCE) SEARS, (JR), b Natick, Mass, May 2, 09; m 38; c 3. Educ: Bowdoin Col, AB, 30; Harvard Univ, AM, 31, PhD(biol), 35. Prof Exp: Instr, Brooklyn Col, 35-36; from instr to assoc prof zool, Miami Univ, Ohio, 36-48; from asst prof to assoc prof, 48-62, PROF ZOOL, IND UNIV, BLOOMINGTON, 62- Concurrent Pos: Mem corp, Marine Biol Lab, Woods Hole, trustee, 58-66, 67-75, mem exec comt trustees, 63-66, 67-70 & secy bd trustees, 72-75; manag- ing ed, Am Zoologist, Am Soc Zoologists, 61-65. Mem: AAAS; Am Soc Zoologists; Soc Develop Biol; Int Soc Develop Biol. Res: Morphogenesis and natural history of hydroids; regeneration in worms. Mailing Add: Dept of Zool Indiana Univ Bloomington IN 47401

CROWELL, RICHARD HENRY, b Northeast, Pa, Apr 6, 28; m 55; c 2. MATHEMATICS. Educ: Harvard Univ, AB, 49; Princeton Univ, MA, 53, PhD(math), 55. Hon Degrees: MA, Dartmouth Col, 67. Prof Exp: Asst anal res group, Forrestal Res Ctr, Princeton Univ, 55-57; lectr, Mass Inst Technol, 57-58; from asst prof to assoc prof, 58-67, PROF MATH, DARTMOUTH COL, 67-, CHMN DEPT, 73- Mem: Am Math Soc; Math Asn Am. Res: Topology, knot theory and algebraic topology. Mailing Add: Dept of Math Dartmouth Col Hanover NH 03755

CROWELL, RICHARD LANE, b Springfield, Mo, Sept 27, 30; m 53; c 4. VIROLOGY. Educ: Univ Buffalo, BA, 52; Univ Minn, MS, 54, PhD(microbiol), 58; Am Bd Med Microbiol dipl. Mailing Add: Webb-Waring Inst Div Immunol Univ of Colo Med Ctr Box 122 Denver CO 80220

CROWELL, ROBERT MERRILL, b Sandusky, Ohio, May 29, 21; m 46; c 3. ZOOLOGY, ENTOMOLOGY. Educ: Bowling Green State Univ, AB, 45, MA, 47; Ohio State Univ, PhD(biol), 57. Prof Exp: Asst biol, Bowling Green State Univ, 45-47; instr, Kent State Univ, 47-48; asst zool, Duke Univ, 48-51; instr biol, Col Wooster, 51-55; from asst prof to assoc prof, 56-67, PROF BIOL, ST LAWRENCE UNIV, 67- Concurrent Pos: NSF res grants, 61-63 & 69-71; trainee, Ohio State Univ, 67-68. Mem: Entom Soc Am; Am Soc Parasitol; Entom Soc Can. Res: Systematics; developmental cycles; host-parasite relationships of the Hydracarina. Mailing Add: Dept of Biol St Lawrence Univ Canton NY 13617

CROWELL, THOMAS IRVING, b Glen Ridge, NJ, July 9, 21; m 50; c 2. PHYSICAL ORGANIC CHEMISTRY. Educ: Harvard Univ, BS, 43; Columbia Univ, AM, 47, PhD(chem), 48. Prof Exp: Asst, Manhattan Proj, 43-45; asst chem, Columbia Univ, 45-46; instr math, Bard Col, 46-47; from asst prof to assoc prof, 48-61, chmn dept, 57-62, PROF CHEM, UNIV VA, 61- Mem: Am Chem Soc. Res: Organic reaction kinetics; fused salts. Mailing Add: Dept of Chem Univ of Va Charlottesville VA 22901

CROWELL, WILFRED J, b Can, Jan 23, 18; US citizen; m 53; c 3. PHARMACEUTICAL CHEMISTRY. Educ: Univ Sask, BA, 48; Univ Ill, MS, 52; Univ Calif, PhD(pharmaceut chem), 58. Prof Exp: Asst prof, 57-61, ASSOC PROF PHARM, UNIV SOUTHERN CALIF, 61- Res: Factors influencing absorption of medicaments from suppository and suspension dosage forms. Mailing Add: Sch of Pharm Univ of Southern Calif Los Angeles CA 90007

CROWL, GEORGE HENRY, b Wooster, Ohio, Apr 10, 10; m 35; c 3. GLACIAL GEOLOGY, GEOMORPHOLOGY. Educ: Col Wooster, BA, 32; Harvard Univ, MA, 34; Princeton Univ, PhD, 50. Prof Exp: Topographer, US Engrs, Ohio, 34-35; jr geologist, Gulf Oil Corp, Arabia & Venezuela, 35-37; geologist, Shell Oil Co, Ill, 38-39; instr geol, Rutgers Univ, 39-41 & Vanderbilt Univ, 42-43; asst prof, Hamilton Col, 43-44; geologist, Carter Oil Co, Wyo, 44-46; asst prof geol, Pa State Col, 46-47; from asst prof to prof, 47-75, head dept, 47-62, EMER PROF GEOL, OHIO WESLEYAN UNIV, 75- Concurrent Pos: Vis prof, US State Dept, Univ Rangoon, 52-54; with exped, Greenland, 63-65 & US Educ Found, India, 66; exped glacial geol, Del Valley, Pa, 68, 69, 71. Mem: Geol Soc Am; Nat Asn Geol Teachers. Res: Geomorphology; structural geology; glacial geology of Greenland; erosion surfaces of Adirondacks; glacial geology of Northern Pennsylvania. Mailing Add: Dept of Geol Ohio Wesleyan Univ Delaware OH 43015

CROWL, ROBERT HAROLD, b Wellsville, Ohio, Apr 17, 25; m 47; c 3. ANIMAL GENETICS. Educ: Harvard Univ, SB, 49; Miami Univ, MS, 50; Ohio State Univ, PhD, 64. Prof Exp: Instr zool, Miami Univ, 50-51; sales rep pharmaceut, Bowman Bros Drug Co, 51-54, Winthrop Labs, 54-59 & Columbus Pharmacal Co, 59-60; teaching asst zool, Ohio State Univ, 60-64; from asst prof to assoc prof biol, 64-71, PROF BIOL, PFEIFFER COL, 71- Mem: NY Acad Sci; AAAS. Res: Pysiological genetics. Mailing Add: Dept of Biol Pfeiffer Col Misenheimer NC 28109

CROWLE, ALFRED JOHN, b Mexico, DF, Mex, Apr 15, 30; US citizen; m 54; c 2. IMMUNOLOGY, MICROBIOLOGY. Educ: San Jose State Univ, AB, 51; Stanford Univ, PhD(microbiol), 54. Prof Exp: Researcher, Webb-Waring Lung Inst, 56-59, from instr to assoc prof, Sch Med, 56-74, PROF MICROBIOL, SCH MED, UNIV COLO MED CTR, DENVER, 74-, HEAD DIV IMMUNOL, WEBB-WARING LUNG INST, 59- Concurrent Pos: Nat Tuberc Asn med res fel, Stanford Univ, 53-55, Nat Acad Sci-Nat Res Coun fel, 55-56; NY Tuberc & Health Asn fel, Sch Med, Univ Colo Med Ctr, Denver, 59-61. Mem: Am Acad Allergy; Am Asn Immunol; Am Soc Microbiol; Soc Exp Biol & Med; Am Col Allergists. Res: Immunochemistry, immunodiffusion and applications of immunodiffusion techniques; delayed hypersensitivity, its induction and control; specific acquired immunity and tuberculoimmunity. Mailing Add: Webb-Waring Inst Div Immunol Univ of Colo Med Ctr Box 122 Denver CO 80220

CROWLEY, DANIEL JOHN, b Peoria, Ill, Nov 27, 21; m 58; c 3. CULTURAL ANTHROPOLOGY. Educ: Northwestern Univ, AB, 43, PhD(anthrop), 56; Bradley Univ, MA, 48. Prof Exp: Instr art hist, Bradley Univ, 48-50; tutor anthrop, Trinidad Extramural Dept, Univ WI, 53-56; asst prof, Univ Notre Dame, 58-59; from asst prof to assoc prof anthrop & art, 61-67, PROF ANTHROP & ART, UNIV CALIF, DAVIS, 67- Concurrent Pos: Ford Found foreign area training fel, Africa, 59-60; vis res prof, Inst African Studies, Univ Ghana, 69-71; vis lectr, Univ WI, Trinidad, 73-74; del of Am Folklore Soc, US Nat Comn UNESCO, 75-77. Honors & Awards: Stafford Prize, Am Folklore Soc, 52; Centennial Citation, Univ Calif, Santa Cruz, 68. Mem: Am Folklore Soc (pres, 69-71); fel Am Anthrop Asn; Soc Ethnomusicol; Am Soc Aesthetics; fel African Studies Asn. Res: Anthropological approach to the study of art, music and folklore, especially of African and Afroamerican peoples. Mailing Add: Dept of Anthrop Univ of Calif Davis CA 95616

CROWLEY, JAMES WILLIAM, dairy husbandry, see 12th edition

CROWLEY, JOHN JAMES, b San Diego, Calif, Feb 20, 46; m 69. BIOSTATISTICS. Educ: Pomona Col, BA, 68; Univ Wash, MS, 70, PhD(biomath), 73. Prof Exp: Fel biostatist, Stanford Univ, 73-74; ASST PROF BIOSTATIST, UNIV WIS-MADISON, 74- Mem: Am Statist Asn; Biometric Soc; Inst Math Statist; AAAS. Res: Methods for analyzing censored survival data; biostatistical methods in cancer research. Mailing Add: Dept of Statist Univ of Wis 1210 W Dayton St Madison WI 53706

CROWLEY, JOHN MAX, b Mound City, Mo, May 27, 33. GEOGRAPHY. Educ: Univ Idaho, BS, 57; Univ Minn, Minneapolis, MS, 60, PhD(geog), 64. Prof Exp: Asst prof geog, Laval Univ, 62-69; vis assoc prof geog & planning, Univ Waterloo, 69-70; assoc prof geog, 70-74, chmn dept, 72-75, PROF GEOG, UNIV MONT, 74- Mem: Am Geog Soc; Asn Am Geog; Can Asn Geog; Fr Soc Biogeog. Res: Mountain geography; biogeography, western North America; methodology of regional geography; Rocky Mountains; Montana; ecosystem mapping in western Montana valleys. Mailing Add: Dept of Geog Univ of Mont Missoula MT 59801

CROWLEY, LAWRENCE GRANDJEAN, b Newark, NJ, July 2, 19; m 45; c 3. SURGERY. Educ: Yale Univ, BA, 41, MD, 44. Prof Exp: From instr to asst prof surg, Sch Med, Yale Univ, 51-53; attend, Southern Calif Permanente Med Group, 53-55; clin asst prof, Sch Med, Univ Southern Calif, 55-64; from assoc prof to prof, Sch Med, Stanford Univ, 64-74, assoc dean, 72-74; PROF SURG & DEAN, SCH MED, UNIV WIS-MADISON, 74- Concurrent Pos: Am Cancer Soc clin fel, Med Ctr, Yale Univ, 49-51; mem, Bd Dirs, Casa Colina Rehab Hosp, Pomona, Calif, 59-74, pres, 63-66; mem spec grants comt, Calif Div, Am Cancer Soc, 67-70; mem, Cancer Categorical Comt, Calif Regional Med Prog, 68-71; mem, Cancer Adv Coun, State Calif, 70-74. Honors & Awards: Cert Merit, Conn Div, Am Cancer Soc, 53. Mem: Am Col Surgeons. Res: Endocrine relationships with mammary carcinoma. Mailing Add: Off of the Dean Univ of Wis Sch of Med Madison WI 53706

CROWLEY, LEONARD VINCENT, b Binghamton, NY, Jan 12, 26; m 51; c 5. MEDICINE, PATHOLOGY. Educ: Univ Vt, MD, 49; Ohio State Univ, MSc, 56. Prof Exp: Instr path, Col Physicians & Surgeons, Columbia Univ, 52-54; instr, Med Sch, Ohio State Univ, 54-56; from asst prof to assoc prof, Col Med, Univ Vt, 56-60; ASST PROF PATH, MED COL, UNIV MINN, MINNEAPOLIS, 62- Mem: AMA; Am Soc Clin Path; Col Am Pathologists; Am Fedn Clin Res; Am Soc Exp Path. Res: Immunohematology; clinical chemistry. Mailing Add: 5337 Kellogg Ave Minneapolis MN 55424

CROWLEY, PATRICK ARTHUR, b Titusville, Pa, June 6, 41; m 64; c 1. NUCLEAR PHYSICS, FLUID PHYSICS. Educ: Carnegie Inst Technol, BS, 63; Univ Pittsburgh, PhD(physics), 68. Prof Exp: Staff physicist, Columbia Res Corp, 68-70; PHYSICIST, US ARMY FOREIGN SCI & TECHNOL CTR, 70- Mem: Am Phys Soc; Am Inst Physics; Am Geophys Union. Res: Nuclear structures and reactions; ionospheric physics; radiation effects; atmopheric phenomena. Mailing Add: US Army Foreign Sci & Technol Ctr Charlottesville VA 22901

CROWLEY, PAUL JOSEPH, inorganic chemistry, see 12th edition

CROWLEY, THOMAS HENRY, b Bowling Green, Ohio, June 7, 24; m 47; c 14. MATHEMATICS. Educ: Ohio State Univ, BEE, 48, MA, 50, PhD(math), 54. Prof Exp: Res assoc antenna lab, Ohio State Univ, 48-54; mem tech staff, Bell Tel Labs, Inc, 54-64, head comput res dept, 64-65, dir comput sci res ctr, 65-68, EXEC DIR SAFEGUARD DESIGN DIV, BELL LABS, 68- Mem: Soc Indust & Appl Math; Inst Elec & Electronics Eng. Res: Computers; switching and control. Mailing Add: Safeguard Design Div Bell Labs 175 Park Ave Madison NJ 07940

CROWLEY, WALTER A, chemistry, see 12th edition

CROWLEY, WILLIAM PATRICK, b Ft Edward, NY, Aug 20, 36; m 58; c 2. GEOLOGY. Educ: Union Col, BS, 62; Yale Univ, MS, 64, PhD(geol), 67. Prof Exp: GEOLOGIST IV, MD GEOL SURV, 67- Mem: Geol Soc Am. Res: Stratigraphy, structure, petrology, economic geology and engineering geology of the crystalline rocks of Maryland. Mailing Add: Md Geol Surv Johns Hopkins Univ Baltimore MD 21218

CROWNFIELD, FREDERIC RUDOLPH, JR, b Boston, Mass, Feb 25, 27; m 56, 71; c 2. PLASMA PHYSICS. Educ: Harvard Univ, AB, 48; Lehigh Univ, MS, 49,

PhD(physics), 53. Prof Exp: Asst physics, Lehigh Univ, 48-50; instr, Univ Akron, 51-53 & NC State Col, 53-56; from asst prof to assoc prof, 56-68, PROF PHYSICS, COL WILLIAM & MARY, 68- Concurrent Pos: Consult, Guilford Tel Co, 50-56. Mem: Am Phys Soc; Am Asn Physics Teachers; assoc Inst Elec & Electronic Engrs; Nuclear & Plasma Physics Soc. Res: Ionized gases; quantum theory; field theory; plasma kinetic theory; stability and wave properties of spatially non-uniform plasmas. Mailing Add: Dept of Physics Col of William & Mary Williamsburg VA 23185

CROWNOVER, RICHARD MCCRANIE, b Quincy, Fla, Oct 11, 36; m 58; c 3. MATHEMATICAL ANALYSIS. Educ: Ga Inst Technol, BS, 58, MS, 60; La State Univ, PhD(math), 64. Prof Exp: Instr math, Ga Inst Technol, 59-61 & La State Univ, 63-64; asst prof, 64-72, ASSOC PROF MATH, UNIV MO-COLUMBIA, 72- Mem: Am Math Soc; Math Asn Am; Sigma Xi. Res: Spaces and algebras of functions and operators. Mailing Add: Dept of Math Univ of Mo Columbia MO 65201

CROWSHAW, KEITH, b Liverpool, Eng, Jan 11, 39; m 62; c 3. BIOCHEMISTRY, ORGANIC CHEMISTRY. Educ: Univ Birmingham, BSc, 60, PhD(org chem), 64. Prof Exp: Med Res Coun staff scientist, Dept Physiol, Med Sch, Univ Birmingham, 63-64; staff scientist, Worcester Found Exp Biol, 64-66; res chemist, St Vincent Hosp, Worcester, Mass, 66-68; asst prof internal med & biochem, Sch Med, St Louis Univ, 68-70; ASST PROF PHARMACOL, MED COL WIS, 70- Concurrent Pos: Res fel chem, Harvard Univ, 67-68; estab investr, Am Heart Asn, 70-75. Mem: Am Chem Soc; Brit Chem Soc; Brit Biochem Soc. Res: Chromatographic methods for isolation and identification of biological compounds; distribution, physiology and biochemistry of prostaglandins present in mammalian tissues and fluids. Mailing Add: Dept of Pharmacol Med Col of Wis Milwaukee WI 53233

CROWSON, CHARLES NEVILLE, b Ottawa, Ont, Sept 21, 19; m 51; c 2. PATHOLOGY. Educ: Queen's Univ, Ont, BA, 41, MA, 43; McGill Univ, MD, 49; Univ Edinburgh, PhD(path), 54; FRCPS(C). Prof Exp: Biochemist, Nat Res Coun, Can, 42-43; lectr path, Queen's Univ, Ont, 54-55; dir labs, Deer Lodge Hosp, 55-60; DIR LABS, MISERICORDIA GEN HOSP, 60- Concurrent Pos: Asst prof, Univ Man, 55-67; dir, Cent Med Lab, 59-; consult pathologist aviation crash path, Can Armed Forces, 55- & traumatics, Can Forces Med Coun, 71- Mem: Am Soc Clin Path; Am Asn Path & Bact; Can Asn Pathologists; Int Acad Path. Res: Hepato-renal syndrome and renal pathology; pathogenesis of nephrotoxic chemical lesions. Mailing Add: Misericordia Gen Hosp Winnipeg MB Can

CROWSON, HENRY L, b Okeechobee, Fla, Apr 16, 27; m 51; c 3. MATHEMATICS. Educ: Univ Fla, BChE, 53, MS, 55, PhD(amth), 59. Prof Exp: Instr math, Univ Fla, 58-59. asst prof, 59-60; staff mathematician, 60-65, ADV MATHEMATICIAN, IBM CORP, 65- Mem: Am Math Soc; Math Asn Am; Soc Indust & Appl Math. Res: Classical mathematical analysis, especially ordinary differential equations; signal analysis. Mailing Add: PO Box 30166 Bethesda MD 20014

CROWTHER, C RICHARD, b Waterloo, Iowa, July 16, 24; m 49; c 3. FORESTRY. Educ: Iowa State Col, BS, 47, MS, 56; Univ Mich, PhD, 71. Prof Exp: Forester, Cent States Exp Sta, US Forest Serv, 47; assoc ed, Naval Stores Rev, 47-48; ed, Lake Mills Graphic, Iowa, 48-53; from instr to assoc prof, 56-72, PROF FORESTRY, MICH TECHNOL UNIV, 72- Concurrent Pos: Alumni Found res grant, 64. Mem: Soc Am Foresters. Res: Forest recreation. Mailing Add: Dept of Forestry Mich Technol Univ Houghton MI 49931

CROXALL, WILLARD (JOSEPH), b Aberdeen, Wash, Nov 11, 10; m 32; c 5. CHEMISTRY. Educ: Univ Notre Dame, BS, 32, MS, 33, PhD(org chem), 35. Prof Exp: Garyin fel, Univ Notre Dame, 35; res chemist, Gen Chem Co, NY, 35-36; res chemist, Rohm & Haas Co, 36-46, res chemist & lab head hydrocarbon explor, 46-48, res chemist & lab head insecticides, 48-50, from asst dir res to dir res, Sumner Chem Co, Inc, 50-56, mem bd dirs, 53-36, gen mgr, 56-59, asst to mgr & coordr res & develop, Miles Chem Co, 59-62, dir, 62-64, vpres, 64-69, VPRES & SR SCI OFFICER, PROCESS INDUST GROUP, MILES CHEM CO, 69- Mem: Am Chem Soc. Res: Emulsion polymerization; synthetic monomers; acetylene chemistry; organic insecticides and fungicides; organic chemical development. Mailing Add: Process Indust Group Miles Labs Inc 1127 Myrtle St Elkhart IN 46514

CROXDALE, JUDITH GEROW, b Oakland, Calif, Aug 27, 41; div; c 1. BOTANY. Educ: Univ Calif, Berkeley, AB, 71, PhD(bot), 75. Prof Exp: ASST PROF BOT, VA POLYTECH INST & STATE UNIV, 75- Mem: Bot Soc Am; Am Soc Plant Physiologists; Sigma Xi. Res: Structural and functional aspects of vegetative plant development. Mailing Add: Dept of Biol Va Polytech Inst & State Univ Blacksburg VA 24061

CROXTON, FRANK CUTSHAW, b Washington, DC, June 26, 07; m 30; c 2. CHEMISTRY, RESEARCH ADMINISTRATION. Educ: Ohio State Univ, AB, 27, AM, 28, PhD(phys chem), 30. Hon Degrees: DSc, Denison Univ, 65. Prof Exp: Res chemist, Standard Oil Co of Ind, 30-39; res chemist, Battelle Mem Inst, 39-42, from asst supvr to supvr, 42-47, asst dir, 47-53, tech dir, 53-64, asst dir, 64-72; CONSULTANT, 72- Concurrent Pos: Mem bd trustees, Denison Univ Res Found & Children's Hosp Res Found. Mem: AAAS; Am Chem Soc; Am Inst Chem Eng; Com Develop Asn; Sigma Xi (pres, 71-72). Res: Petroleum technology; lubrication oil refining; lubricant additive; structure of aluminum carbide. Mailing Add: 1921 Collingswood Rd Columbus OH 43221

CROY, LAVOY I, b Pauls Valley, Okla, Dec 4, 30; m 54; c 5. CROP PHYSIOLOGY, PLANT BIOCHEMISTRY. Educ: Okla State Univ, BS, 55, MS, 59; Univ Ill, Urbana-Champaign, PhD(crop physiol), 67. Prof Exp: Instr, 55-70, ASSOC PROF AGRON, OKLA STATE UNIV, 70- Concurrent Pos: Asst to head dept agron, Univ Ill, Urbana-Champaign, 64-66. Mem: AAAS; Am Soc Agron; Crop Sci Soc Am; Am Soc Plant Physiol. Res: Protein production in cereal grains; physiology of nitrate reduction in wheat; protease system in vegetative portion of wheat plant. Mailing Add: Dept of Agron Okla State Univ Stillwater OK 74074

CROZIER, DAN, b Matoaka, WVa, Dec 25, 14; m 44. INTERNAL MEDICINE. Educ: Univ WVa, AB, 35; Harvard Med Sch, MD, 39; Am Bd Internal Med, dipl, 53. Prof Exp: Chief med serv, US Army Hosp, Camp Rucker, Ala, 50-53, med consult, Eighth US Army, Korea, 53-54, chief med serv, Ryukyus Army Hosp, Okinawa, 54-56, asst chief med consult, Off Surgeon Gen, Dept Army, DC, 56-58, chief med consult, 58-61, comdg officer, US Army Med Res Inst Infectious Dis, Ft Detrick, Md, 61-73; CONSULT MED & HOSP ADMIN, 73- Concurrent Pos: Fel, Vanderbilt Univ Hosp, Nashville, Tenn, 47-48, St Thomas Hosp, 48-49 & Oliver Gen Hosp, Augusta, Ga, 49-50. Honors & Awards: Gorgas Medal, Asn Mil Surgeons US, 70; Superior Serv Award, USDA, 71; Distinguished Serv Medal, US Army. Mem: Fel Am Col Physicians; Asn Mil Surgeons US; AMA. Mailing Add: Cherry Hill Dr Rte 7 Frederick MD 21701

CROZIER, EDGAR DARYL, b Montreal, Que, Apr 9, 39; m 64; c 2. METAL PHYSICS, SEMICONDUCTORS. Educ: Univ Toronto, BSc, 61; Queen's Univ, Ont, PhD(phys chem), 65. Prof Exp: Nat Res Coun Can overseas fel, 65-67; asst prof, 67-74, ASSOC PROF PHYSICS, SIMON FRASER UNIV, 74- Mem: Can Asn Physicists. Res: Fluid and electronic properties of liquid metals and liquid semiconductors being investigated by linear and non-linear optical measurements, ultrasonics, extreme x-ray absorption of fine structure. Mailing Add: Dept of Physics Simon Fraser Univ Burnaby BC Can

CROZIER, GEORGE FREDERICK, marine biology, comparative biochemistry, see 12th edition

CRUDDACE, RAYMOND GIBSON, b Richmond, Eng, June 3, 36; US citizen. XRAY ASTRONOMY. Educ: Imp Col, Univ London, BSc, 58; Linacre Col, Oxford Univ, DPhil(physics), 68; Univ Calif, Berkeley, MA, 73. Prof Exp: Sci officer, Rocket Propulsion Estab, UK Ministry Aviation, 59-62; asst prof mech eng, Sacramento State Col, 65-66; physicist, Nuclear Rocket Opers, Aerojet-Gen Corp, 65-69; res physicist, Space Sci Lab, Univ Calif, Berkeley, 71-74; ASTROPHYSICIST, SPACE SCI DIV, NAVAL RES LAB, 74- Mem: Am Astron Soc; Brit Interplanetary Soc; Am Inst Aeronaut & Astronaut. Mailing Add: Code 7129.5 Space Sci Div Naval Res Lab Washington DC 20375

CRUDEN, ROBERT WILLIAM, b Cleveland, Ohio, Mar 18, 36; m 67; c 2. ECOLOGY, EVOLUTION. Educ: Hiram Col, BA, 58; Ohio State Univ, MSc, 60; Univ Calif, Berkeley, PhD(bot), 67. Prof Exp: Asst prof, 67-71, ASSOC PROF BOT, UNIV IOWA, 71- Mem: Ecol Soc Am; Brit Ecol Soc; Plant Taxonomists; Am Bot Soc. Res: Breeding systems, pollination biology, other animal-plant interactions and other life history parameters of plants and their evolution. Mailing Add: Dept of Bot Univ of Iowa Iowa City IA 52240

CRUICKSHANK, ALEXANDER MIDDLETON, b Marlboro, NH, Dec 13, 19; m 45; c 2. INORGANIC CHEMISTRY, ANALYTICAL CHEMISTRY. Educ: RI State Col, BS, 43, MS, 45; Univ Mass, PhD(chem), 54. Prof Exp: Asst chem, RI State Col, 43-45, instr, 45-48; instr, Univ Mass, 48-53; from asst prof to assoc prof, 53-69, PROF CHEM, UNIV RI, 69- Concurrent Pos: Instr, Holyoke Jr Col, 50-53; asst to dir, Gordon Res Confs, AAAS, 47-68, dir, 68- Mem: Am Chem Soc; Am Asn Textile Chem & Colorists. Res: Polarography; metal complexes of amino and hydroxyl acids; alkyd resins; textiles. Mailing Add: Dept of Chem Univ of RI Kingston RI 02879

CRUICKSHANK, BRUCE, b Edinburgh, Scotland, May 27, 20; m 43; c 2. ANATOMIC PATHOLOGY, CYTOLOGY. Educ: Univ Edinburgh, MB & ChB, 43, PhD(rheumatic dis), 52, MD, 58; FRCPS(G), 65; Royal Col Path, fel, 68; FRCPS(C), 69. Prof Exp: Lectr path, Univ Edinburgh, 52-56; sr lectr, Glasgow Univ, 56-59 & 60-63; assoc dir, Mallory Inst Path, 59-60; prof & head dept, Univ Col Rhodesia & Nyasaland, 64-68; PROF PATH, UNIV TORONTO & HEAD DEPT, SUNNYBROOK HOSP, 68- Concurrent Pos: Vis prof, Boston Univ & vis lectr, Harvard Univ & Tufts Univ, 59 60; hon consult pathologist, Glasgow Royal Infirmary, 60-63; mem, Panel, Can Tumor Ref Ctr, 68- Mem: Am Asn Path & Bact; Int Acad Path; Path Soc Gt Brit & Ireland; Brit Soc Immunol; Heberden Soc. Res: Histopathology and pathogenesis of rheumatic diseases; immunofluorescence of tissue antigens. Mailing Add: Dept of Path Sunnybrook Hosp Univ of Toronto Toronto ON Can

CRUICKSHANK, P A, b Bremerton, Wash, Oct 11, 29; m 53; c 3. PESTICIDE CHEMISTRY. Educ: Univ Wash, BSc, 51; Mass Inst Technol, PhD(org chem), 55. Prof Exp: Chemist, E I du Pont de Nemours Co, 55-58; group leader peptides, Res Inst Med & Chem, 58-63; interdisciplinary scientist, 63-66, RES MGR, FMC CORP, 66- Mem: Am Chem Soc; NY Acad Sci. Res: Isolation and characterization of natural products with physiological activity; synthesis of peptides and steroids; chromatographic methods of analysis; insect hormones; pesticides. Mailing Add: Agr Chem Div FMC CorpPO Box 8 Princeton NJ 08540

CRUIKSHANK, DONALD BURGOYNE, JR, b Boise, Idaho, Aug 2, 39; m 65; c 1. ACOUSTICS. Educ: Kalamazoo Col, BA, 64; Univ Rochester, MS, 66, PhD(acoustics), 70. Prof Exp: Asst prof physics, Cornell Col, 69-71; asst prof, 71-74, ASSOC PROF PHYSICS, ANDERSON COL, 74- Mem: Acoust Soc Am; Am Asn Physics Teachers. Res: Finite-amplitude acoustic resonance oscillations in closed, rigid-walled tubes. Mailing Add: Dept of Physics Anderson Col Anderson IN 46011

CRUISE, DONALD RICHARD, b Los Angeles, Calif, Feb 17, 34; m 65; c 2. APPLIED MATHEMATICS. Educ: Fresno State Col, AB, 56. Prof Exp: MATHEMATICIAN, NAVAL WEAPONS CTR, 56- Mem: Sigma Xi. Res: Mathematical models for chemical equilibrium, cavities and far fields of cylindrical lasers; digital twirling. Mailing Add: PO Box 5453 China Lake CA 93555

CRUISE, JAMES E, b Port Dover, Ont, June 26, 25. PLANT TAXONOMY. Educ: Univ Toronto, BA, 50; Cornell Univ, MS, 51, PhD, 54. Prof Exp: Lab instr, Cornell Univ, 51-56; from asst prof to prof biol, Trenton State Col, 56-63; assoc prof, Univ Toronto, 63-69, prof biol & cur, 69-75; DIR, ROYAL ONT MUS, 75- Mem: Am Soc Plant Taxon; Can Bot Asn; Int Soc Plant Taxon. Res: Taxonomy of vascular plants; flora of Ontario. Mailing Add: 100 Queen's Park Toronto ON Can

CRULL, HARRY EDWARD, mathematics, astronomy, see 12th edition

CRUM, CHARLES WILLIAM, plant breeding, see 12th edition

CRUM, HOWARD ALVIN, b Mishawaka, Ind, July 14, 22; m 60; c 2. BRYOLOGY. Educ: Western Mich Col, BS, 47; Univ Mich, MS, 49, PhD(bot), 51. Prof Exp: Instr, Western Mich Col, 46-47; teaching fel, Univ Mich, 48-49; res biol & actg asst prof, Stanford Univ, 51-53; asst prof, Univ Louisville, 53-54; biologist, Nat Mus Can, 54-65; PROF BOT & CUR BRYOPHYTES & LICHENS, UNIV MICH, 65- Concurrent Pos: Assoc ed, Am Bryological & Lichenological Soc, 53,62-, ed, 54-62; assoc ed, Brit Bryological Soc, 72-; pres, Mich Bot Club, 73-74, assoc ed, 75. Mem: Am Bryological & Lichenological Soc (pres, 62-63); Bot Soc Am; Brit Bryological Soc. Res: Taxonomy of bryophytes; mosses; monographic studies in the genus Sphagnum. Mailing Add: Herbarium 2002 N Univ Bldg Univ of Mich Ann Arbor MI 48104

CRUM, JAMES DAVIDSON, b Ironton, Ohio, July 29,30; m 56; c 3. ORGANIC CHEMISTRY. Educ: Ohio State Univ, BSc, 52, PhD(org chem), 58; Marshall Col, MSc, 53. Prof Exp: Asst, Marshall Col, 52-53; from asst to asst instr, Ohio State Univ, 53-57; NIH res fel, Harvard Univ, 58-59; from instr to asst prof chem, Case Western Reserve Univ, 59-66; assoc prof, 66-69, chmn dept, 70-72, PROF CHEM, CALIF STATE COL, SAN BERNARDINO, 69-, DEAN, SCH NATURAL SCI, 72- Concurrent Pos: Consult, Nat Sci Found-US Agency Int Develop, 65-68. Mem: Fel AAAS; fel Am Inst Chem; Am Chem Soc; fel India Inst Chem; The Chem Soc. Res: Natural products; carbohydrates and steroids; reaction mechanisms; synthesis. Mailing Add: Sch of Natural Sci Calif State Col San Bernardino CA 92407

CRUM, JOHN KISTLER, b Brownsville, Tex, July 28, 36. INORGANIC CHEMISTRY, SCIENCE ADMINISTRATION. Educ: Univ Tex, BSChE, 60, PhD(chem), 64; Harvard Univ, Advan Mgt Prog dipl, 75. Prof Exp: From asst ed to assoc ed anal chem, 64-68, managing ed publ, 68-70, group mgr jour, 70-71, dir books & jour div, 71-75, TREAS & CHIEF FINANCIAL OFFICER, AM CHEM SOC, 75- Concurrent Pos: Consult, Joint Comt Atomic & Molecular Phys Data. Mem: Am Chem Soc; The Chem Soc; Coun Eng & Sci Soc Execs. Res: Chemistry of reactions in liquid ammonia; design of scientific publications; chemical literature; information retrieval. Mailing Add: Am Chem Soc 1155 16th St NW Washington DC 20036

CRUM, LAWRENCE ARTHUR, b Caldwell, Ohio, Oct 25, 41; m 63; c 3. ACOUSTICS. Educ: Ohio Univ, BS, 63, MS, 65, PhD(physics), 67. Prof Exp: Res fel, Harvard Univ, 67-68; asst prof, 68-72, ASSOC PROF PHYSICS, US NAVAL ACAD, 72- Concurrent Pos: Consult, Planning Syst Inc, 72- & Naval Ord Sta, 73-; pres, Acad Assoc, Inc, 75- Mem: Acoust Soc Am; Am Asn Physics Teachers. Res: Acoustic cavitation and effects of sound fields on bubbles. Mailing Add: Dept of Physics US Naval Acad Annapolis MD 21402

CRUMB, GLENN HOWARD, b Burlingame, Kans, Dec 21, 27; m 50; c 4. ACADEMIC ADMINISTRATION. Educ: Kans State Teachers Col, BS, 51, MS, 56; Univ Nebr, PhD(sci educ), 64. Prof Exp: Teacher high schs, Kans, 51-56; instr phys sci, Kans State Teachers Col, 56-59; physics, Univ Nebr High Sch, 59-63; assoc prof phys sci, Kans State Teachers Col, 63-65, prof, 65-71, head dept, 64-71, dir res & grants ctr, 69-71; PROF EDUC & DIR GRANT & CONTRACT SERVS, WESTERN KY UNIV, 71- Concurrent Pos: Shell Oil Co Merit fac fel, 69. Mem: Am Asn Physics Teachers; Nat Sci Teachers Asn; Nat Coun Univ Res Adminrs. Res: Atomic and nuclear physics; optics. Mailing Add: Dir Grant & Contract Servs Western Ky Univ Bowling Green KY 42101

CRUMLEY, CAROLE LINDA, b Ann Arbor, Mich, June 2, 44. ANTHROPOLOGY. Educ: Univ Mich, BA, 66; Univ Calgary, MA, 67; Univ Wis-Madison, PhD(anthrop), 72. Prof Exp: Asst prof anthrop, Western Mich Univ, 72 & Carleton Col, 73-74; ASST PROF ANTHROP, UNIV MO-COLUMBIA, 74- Concurrent Pos: Vis asst prof, Washington Univ, 72-73; NSF res grant, Univ Mo, 75. Mem: Soc Am Archaeol; Am Anthrop Asn. Res: Emergence of the state; anthropological method in use of classical and historical data; network and locational analysis; social structure, class, role; Celtic France; European Iron Age. Mailing Add: Dept of Anthrop 210 Switzler Hall Univ of Mo Columbia MO 65201

CRUMLEY, RICHARD D, b Lancaster, Ohio, Nov 24, 21; m 47; c 2. MATHEMATICS. Educ: Univ Ohio, BS, 42; Univ Chicago, MS, 50, PhD(educ), 56. Prof Exp: Actg instr math, Ohio Univ, 46-48; asst prof educ, Univ SC, 53-56; assoc prof math, State Col Iowa, 56-62; ASSOC PROF MATH, ILL STATE UNIV, 62- Concurrent Pos: Adv for Ohio Univ, US Agency Int Develop Proj, Northern Nigeria, 64-66. Mem: Math Asn Am. Res: Mathematics education; foundations of arithmetic. Mailing Add: Dept of Math Ill State Univ Normal IL 61761

CRUMMETT, WARREN B, b Moyers, WVa, Apr 4, 22; m 48; c 2. ANALYTICAL CHEMISTRY. Educ: Bridgewater Col, BA, 43; Ohio State Univ, PhD(chem), 51. Prof Exp: Control chemist, Solvay Process Co, 43-46; anal chemist, 51-55, group leader, 55-61, asst dir anal labs, 61-71, ANAL SCIENTIST, DOW CHEM CO, 71- Concurrent Pos: Mem adv bd, Anal Chem, 74-76. Mem: AAAS; Am Inst chem; Am Chem Soc; Sigma Xi. Res: Ion exchange; platinum metals; ultraviolet and near infrared spectrophotometry; purity of organic compounds; liquid chromatography; analytical systems; environmental analysis. Mailing Add: 808 Crescent Dr Midland MI 48640

CRUMMY, ANDREW B, b Newark, NJ, Jan 30, 30; m 58; c 3. RADIOLOGY. Educ: Bowdoin Col, AB, 51; Boston Univ, MD, 55. Prof Exp: Intern, Univ Wis Hosps, 55-56, resident radiol, 58-61; asst prof, Univ Colo, 63-64; from asst prof to assoc prof, 64-70, PROF RADIOL, UNIV WIS-MADISON, 70- Concurrent Pos: Fel radiol, Mt Auburn Hosp, Cambridge, Mass, 61-62; fel cardiovasc radiol, Yale Univ, 62-63. Mem: Fel Am Col Angiol; fel Am Col Radiol; fel Am Heart Asn; AMA. Res: Diagnostic radiology; cardiovascular radiology. Mailing Add: Dept of Radiol 203 Bradley Univ Hosps 1300 University Ave Madison WI 53706

CRUMMY, PRESSLEY LEE, b Glade Mills, Pa, Oct 1, 06; m 34. HISTOLOGY, HUMAN ANATOMY. Educ: Grove City Col, BS, 29; Univ Pittsburgh, MS, 32, PhD(zool), 34. Prof Exp: Teacher rural sch, Pa, 25-26, high sch, 29-30; asst biol, Grove City Col, 27-29 & zool, Univ Pittsburgh, 30-34; head dept sci, High Sch, Pa, 34-35; from instr to asst prof biol, Juniata Col, 35-47, prof, 47-49, from actg registr to registr, 42-49; assoc prof, 49-60, PROF ANAT, KIRKSVILLE COL OSTEOP MED, 60- Mem: Fel AAAS. Res: Histology; gross human anatomy; morphological anomalies; biological effects of x-rays. Mailing Add: Dept of Anat Kirksville Col of Osteop Med Kirksville MO 63501

CRUMP, JESSE FRANKLIN, b Pine Bluff, Ark, July 20, 27; m 59. BIOMEDICAL ENGINEERING, INTERNAL MEDICINE. Educ: Univ Nebr, BSEE, 50, MD, 56. Prof Exp: Intern, Methodist Hosp, Brooklyn, 56-57, asst resident, 57-58; resident internal med, Long Island Col Hosp, 61-62; from asst prof to assoc prof elec eng, 62-68, ASSOC PROF BIOENG, POLYTECH INST NEW YORK, 68- Concurrent Pos: NIH res fel psychosom med, State Univ NY, 58-59 & cardiol, Long Island Col Hosp, Brooklyn, 59-61; consult, Dept Psychol, Princeton Univ, 57-62; Biomed Eng Rehab Dept, Montefiore Hosp, New York, 65-; vis res physiologist, Princeton Univ, 62-75; assoc attend, Long Island Col Hosp, 62-; prin investr, USPHS grant, 66-68. Mem: AAAS; NY Acad Sci; Asn Advan Med Instrumentation; Inst Elec & Electronics Engrs. Res: Auditory, cardiology and cardiovascular physiology; biomedical instrumentation. Mailing Add: Dept of Bioeng Polytech Inst of NY Brooklyn NY 11201

CRUMP, JOHN C, III, b Richmond, Va, Feb 21, 40. SOLID STATE PHYSICS. Educ: Hampden-Sydney Col, BS, 60; Univ Va, MS, 62, PhD(physics), 64. Prof Exp: Res assoc physics, Univ Va, 64-66; res physicist, Oak Ridge Nat Lab, 66-69; RES SCIENTIST, PHILIP MORRIS INC, 69- Mem: Am Phys Soc; Electron Micros Soc Am. Res: Dislocation phenomena and irradiation effects in solid materials. Mailing Add: Philip Morris Inc Res Ctr Po Box 3 D Richmond VA 23206

CRUMP, JOHN WILLIAM, b Santa Rosa, Calif, Jan 18, 32; m 55; c 4. ORGANIC CHEMISTRY. Educ: Univ Calif, BA, 53; Univ Ill, PhD(org chem), 57. Prof Exp: Res chemist, Dow Chem Co, 57-62; from asst prof to assoc prof, 62-69, PROF CHEM, ALBION COL, 69-, CHAIRPERSON DEPT, 75- Mem: Am Chem Soc. Res: Organic reaction mechanisms and their application to synthetic control. Mailing Add: Dept of Chem Albion Col Albion MI 49224

CRUMP, KENNY SHERMAN, b Haynesville, La, Oct 13, 39; m 61; c 3. MATHEMATICS, STATISTICS. Educ: La Tech Univ, BS, 61; Univ Denver, MA, 63; Mont State Univ, PhD(math), 68. Prof Exp: Instr, Mont State Univ, 63-66; ASSOC PROF MATH, LA TECH UNIV, 66- Concurrent Pos: Res assoc, State Univ NY Buffalo, 67-68. Mem: Inst Math Statist. Res: Probability theory; branching processes; renewal theory; stochastic models. Mailing Add: Dept of Math La Tech Univ Ruston LA 72170

CRUMP, MALCOLM HART, b Culpeper, Va, Aug 10, 26; m 52; c 3. GASTROENTEROLOGY. Educ: Va Polytech Inst, BS, 51; Univ Ga, DVM, 58; Univ Wis, MS, 61, PhD(physiol), 65. Prof Exp: Res assoc physiol, Univ Wis, 60-65; asst prof, 64-66, ASSOC PROF PHYSIOL, IOWA STATE UNIV, 66- Mem: Am Vet Med Asn; assoc Am Physiol Soc. Res: Pharmacological action of mycotoxins; drug metabolism; comparative gastroenterology; control mechanisms in the large intestine. Mailing Add: Dept of Physiol Iowa State Univ Col of Vet Med Ames IA 50010

CRUMP, PHELPS PUTNAM, statistics, see 12th edition

CRUMP, ROBERT MYERS, b Brazil, Ind, June 5, 15; m 45; c 4. GEOLOGY. Educ: Univ Wis, BS, 37, MA, 39, PhD(geol), 48. Prof Exp: Geologist, Philippine Bur Mines, 39-41; from res geologist to staff geologist, 48-62, CHIEF MINING & DEVELOP ENG, JONES & LAUGHLIN STEEL CORP, 62- Mem: Geol Soc Am; Soc Econ Geologists. Res: Iron ore deposits; gold deposits; feldspars. Mailing Add: Jones & Laughlin Steel Corp Geol Div 401 Liberty Ave Pittsburgh PA 15222

CRUMP, STUART FAULKNER, b Boston, Mass, Feb 20, 21; m 44; c 4. PHYSICS, RESEARCH ADMINISTRATION. Educ: Brown Univ, BA, 43. Prof Exp: Physicist, David W Taylor Model Basin, 43-45, 56-59, naval architect, 45-49, 52-56, hydraul engr, 49-52, phys sci administr, 59-60; physicist, Navy Bur of Ships, 61-62, phys sci adminstr, 62-67, CONTRACT RES ADMINSTR, DAVID W TAYLOR NAVAL SHIP RES & DEVELOP CTR, 67- Concurrent Pos: Consult, 61-62. Honors & Awards: Super Accomplishment Award, US Navy, 52 & 60. Mem: Acoust Soc Am; Marine Technol Soc; Philos Soc. Res: Hydromechanics of naval architecture; underwater acoustics; cavitation research; engineering physics; contract research administration. Mailing Add: David W Taylor Naval Ship Res & Develop Ctr Code 1505 Bethesda MD 20084

CRUMPACKER, DAVID WILSON, b Enid, Okla, Mar 29, 29; m 55; c 4. GENETICS, AGRONOMY. Educ: Okla Agr & Mech Col, BS, 51; Univ Calif, Davis, PhD(genetics), 59. Prof Exp: Asst agron, Univ Calif, Davis, 55-59; assoc prof, Colo State Univ, 59-70; PROF ENVIRON, POP & ORGANISMIC BIOL, 70- Concurrent Pos: US Pub Health Serv spec fel, Rockefeller Univ, 65-66. Mem: AAAS; Am Soc Agron; Crop Sci Soc Am; fel Am Soc Naturalists; Genetics Soc Am. Res: Evolution and population genetics of maize and Drosophila; biometrical genetics of wheat; maize breeding. Mailing Add: Dept of Biol Univ of Colo Boulder CO 80302

CRUMPLER, THOMAS BIGELOW, b Louisa, Ky, July 20, 09; m 35; c 2. ANALYTICAL CHEMISTRY. Educ: Va Polytech Inst, BS, 31, MS, 32; Univ Va, PhD(chem), 36. Prof Exp: Fel, Univ Va, 36-37; from instr to assoc prof, 37-43, head dept, 42-62, prof, 43-74, EMER PROF CHEM, TULANE UNIV, 74- Concurrent Pos: Ford fac fel, 51-52. Mem: Fel Am Inst Chem; Am Chem Soc. Res: Photoelectric colorimetry and polarimetry; colorimetric reagents; paper chromatography. Mailing Add: PO Box 831 Highlands NC 28741

CRUMRINE, ANN LOUISE, b Ashland, Ohio, Apr 17, 48. SYNTHETIC ORGANIC CHEMISTRY. Educ: Ashland Col, BA, 70; Univ Ill, PhD(org chem), 75. Prof Exp: SR RES CHEMIST, MONSANTO AGR PROD CO, 75- Mem: Am Chem Soc. Res: Phosphonate systems in agriculture chemicals; correlation of structure and activity via computers. Mailing Add: Monsanto Agr Prod Co T4D 800 N Lindbergh St Louis MO 63166

CRUMRINE, DAVID SHAFER, b Memphis, Tenn, Aug, 12, 44; m 67; c 2. PHYSICAL ORGANIC CHEMISTRY. Educ: Ashland Col, AB, 66; Univ Wis-Madison, PhD(org chem), 71. Prof Exp: Fel org chem, Mass Inst Technol & Ga Inst Technol, 71-72; ASST PROF ORG PHYS CHEM, LOYOLA UNIV, CHICAGO, 72- Mem: Am Chem Soc; Sigma Xi. Res: Carbene and carbenoid reactions; mechanistic and exploratory organic photochemistry; synthesis of novel systems; organic electrochemistry. Mailing Add: Dept of Chem Loyola Univ Chicago IL 60626

CRUMRINE, NORMAN ROSS, II, b Beaver, Pa, May 22, 34; m 71; c 3. ANTHROPOLOGY. Educ: Northwestern Univ, BA, 57; Univ Ariz, MA, 62, PhD(anthrop), 68. Prof Exp: Lectr anthrop, Univ Ariz, 64; asst prof, Calif State Col, Hayward, 65-68; asst prof, 68-71, ASSOC PROF ANTHROP, UNIV VICTORIA, 71- Concurrent Pos: Res fel, Can Coun, 72. Mem: Fel Am Anthrop Asn; Am Folklore Soc; Am Ethnol Soc; Soc Sci Study Relig. Res: Symbolic and structural anthropology; myth, ritual symbolism and folk and ritual drama; culture change; ethnography of Northwest Mexico; Middle America and South America; Spanish contact in the Pacific, in the Marianas and in the Philippines. Mailing Add: 1670 Earlston Ave Victoria BC Can

CRUSBERG, THEODORE CLIFFORD, b Meriden, Conn, Feb 23, 41; m 66; c 3. BIOCHEMISTRY. Educ: Univ Conn, BA, 63; Yale Univ, MS, 64; Clark Univ, PhD(chem), 68. Prof Exp: NIH trainee biochem, Sch Med, Tufts Univ, 68-69; ASSOC PROF LIFE SCI, WORCESTER POLYTECH INST, 69- Mem: NY Acad Sci; AAAS; Am Chem Soc. Res: Cell surface characteristics determined by scanning electron microscopy and biochemical methods; aging; erythrocyte membrane proteins and erythrocyte morphology and function. Mailing Add: Dept of Life Sci Worcester Polytech Inst Worcester MA 01609

CRUSE, JULIUS MAJOR, JR, b New Albany, Miss, Feb 15, 37. IMMUNOLOGY, PATHOLOGY. Educ: Univ Miss, BA, 58, BS, 59; Graz Univ, MS, 60; Univ Tenn, MD, 64, PhD(immunol & path), 66. Prof Exp: Prof biol & res prof immunol, Grad Sch, 67-74, asst prof microbiol, Sch Med, 68-74, ASSOC PROF MICROBIOL, PROF PATH & DIR GRAD STUDIES PATH, SCH MED, UNIV MISS, 74- Concurrent Pos: USPHS fel path, Inst Path, Tenn Med Units, 64-67; lectr, Col Med, Univ Tenn, 67- Honors & Awards: Physician's Recognition Award in Continuing Med Educ, AMA, 69 & 75. Mem: Fel AAAS; Am Asn Path & Bact; Am Soc Exp Path; Am Chem Soc; Am Soc Microbiol. Res: Immunopathology; tumor immunology; histocompatability antigens; cellular and humoral immune responses to tissue allografts; immunological enhancement. Mailing Add: Dept of Path Univ of Miss Med Ctr Jackson MI 39216

CRUSE, ROBERT RIDGELY, b Tucson, Ariz, Aug 20, 20; m 47; c 1. APPLIED CHEMISTRY. Educ: Antioch Col, BS, 42. Prof Exp: Res engr, Battelle Mem Inst, 42-47; anal chemist, US Bur Mines, Ariz, 53-55, extractive metallurgist, 55; assoc indust chemist, Southwest Res Inst, 55-61; res chemist, Nitrogen Div, Allied Chem Corp, 61-68; RES CHEMIST, FOOD CROPS UTILIZATION RES LAB, USDA, 68- Concurrent Pos: Instr org chem, Trinity Univ, 56-57; assoc, Southwest Agr Inst, Tex, 59-60. Mem: Fel AAAS; fel Am Inst Chem; Am Chem Soc; Sigma Xi. Res: Chemurgy; organic synthesis; pharmaceuticals; fuels; insecticides; fertilizers; foods. Mailing Add: 1106 W Third St Weslaco TX 78596

CRUSER, STEPHEN ALAN, b Greensburg, Ind, Dec 12, 42; m 67; c 2. ANALYTICAL CHEMISTRY, PHYSICAL CHEMISTRY. Educ: Ind Univ, Bloomington, BS, 64; Univ Tex, Austin, PhD(chem), 68. Prof Exp: Chemist, 68-69, sr chemist, 70-72, RES CHEMIST, TEXACO INC, BELLAIRE, 72- Mem: Nat Asn

Corrosion Engrs. Res: Electroanalytical chemistry; corrosion control; inhibitor mechanisms. Mailing Add: 8026 Belle Glen Dr Houston TX 77072

CRUTCHER, HAROLD L, b Cheraw, Colo, Nov 18, 13; m 43; c 2. METEOROLOGY, CHEMISTRY. Educ: Southeastern State Col, BA, 33, BS, 34; NY Univ, MS, 51, PhD(meteorol), 60. Prof Exp: Teacher pub schs, Okla, 35-39; from jr observer to sr observer, 39-42, meteorologist, 42-45, forecaster, 45-47, meteorologist, Opers Div, 47-50, Climat & Hydrol Serv Div, 51-52, Climat Serv Div, 52-53 & Nat Climatic Ctr, 53-63, RES METEOROLOGIST, NAT CLIMATIC CTR, NAT OCEANIC & ATMOSPHERIC ADMIN, 63- Concurrent Pos: Mem opers anal standby unit, Univ NC, 59-69. Honors & Awards: Silver Award, US Dept Com, 62, Gold Medal, 71. Mem: Am Meteorol Soc; Am Geophys Union; Am Soc Qual Control; Am Statist Asn; fel Am Inst Chem. Res: Upper air climatology; stochastic and dynamic models. Mailing Add: Nat Climatic Ctr Nat Oceanic & Atmospheric Admin Asheville NC 28801

CRUTCHFIELD, CHARLIE, b Norwood, Pa, Dec 29, 28; m 62; c 2. ANALYTICAL CHEMISTRY, INORGANIC CHEMISTRY. Educ: Univ Pa, BA, 50, MS, 53, PhD(chem), 60. Prof Exp: Anal chemist, Dalare Assocs, Pa, 50-55; instr gen & anal chem, Flint Jr Col, 59-61; electrochemist, Stanford Res Inst, 61-66, anal chemist, 66-69; TECH DIR, TRUESDAIL LABS, INC, 69- Mem: Am Chem Soc. Res: Complex compounds. Mailing Add: Truesdail Labs Inc 4101 N Figuerosa Los Angeles CA 90065

CRUTCHFIELD, FLOY LOVE, b Branch, La, Dec 22, 22; m 43; c 2. ENDOCRINOLOGY. Educ: La Col, BS, 43; Med Col Pa, MS, 69, PhD(anat), 72. Prof Exp: RES ASSOC ENDOCRINOL, VET ADMIN HOSP & MED COL PA, 72- Concurrent Pos: Instr histol neuroanat, Med Col Pa, 72-73. Mem: Am Soc Zoologists; Soc Develop Biol. Res: Investigating the metabolism of thyroxine in the adrenergic nervous system. Mailing Add: Dept of Med Med Col of Pa 3300 Henry Ave Philadelphia PA 19129

CRUTCHFIELD, MARVIN MACK, b Oxford, NC, Sept 15, 34; m 56; c 2. PHYSICAL CHEMISTRY. Educ: Duke Univ, BS, 56; Brown Univ, PhD(phys chem), 60. Prof Exp: Sr res chemist, 60-63, res specialist, 63-65, scientist, 65-70, SCI FEL, MONSANTO CO, 70- Mem: AAAS; Am Chem Soc; Sigma Xi. Res: Peroxoanions; nuclear magnetic resonance, metal ion complexing; chemistry of phosphorus compounds; surface active agents; detergents; polyelectrolytes; biodegradation; molecular structure-physical property relationships; calcium ion equilibria. Mailing Add: MICC R&D NIA Monsanto Co 800 N Lindbergh Blvd St Louis MO 63166

CRUTHERS, LARRY RANDALL, b Kenosha, Wis, Mar 15, 45; m 67; c 2. VETERINARY PARASITOLOGY. Educ: Univ Wis, Stevens Point, BS, 67; Kans State Univ, MS, 71, PhD(parasitol), 73. Prof Exp: Instr biol, Kans State Univ, 70-73; RES INVESTR PARASITOL, E R SQUIBB & SONS, INC, 74- Mem: Am Soc Parasitol; Sigma Xi. Res: Chemotherapy in relation to veterinary parasitology. Mailing Add: Squibb Agr Res Ctr Three Bridges NJ 08887

CRUTY, MICHAEL ROBERT, b Johnson City, NY, June 11, 41. RADIOLOGICAL PHYSICS. Educ: State Univ NY Stony Brook, BS, 63; Clarkson Col Technol, MSEE, 71, PhD(physics), 72. Prof Exp: Res physicist & lectr physics, Univ San Francisco, 72-74; BIOPHYSICIST, LAWRENCE BERKELEY LAB, 74- Res: Diagnostic radiography with accelerated heavy particles using plastic nuclear track detectors for image registration. Mailing Add: Lawrence Berkeley Lab Univ of Calif Berkeley CA 94720

CRUZ, ALEXANDER, b New York, NY, July 12, 41; m 63; c 2. ECOLOGY, VERTEBRATE ZOOLOGY. Educ: City Univ New York, BS, 64; Univ Fla, PhD(ecol & zool), 73. Prof Exp: Microbiologist, New York Dept Health, Bur Labs, 64-68; ASST PROF BIOL, UNIV COLO, BOULDER, 73- Concurrent Pos: Consult, Biol & Health Sci Educ Opportunities Prog, Univ Colo, 73- & Ecol Analysts Inc, 74-; fac res initiation fel, Univ Colo, 74. Mem: Ecol Soc Am; Am Ornithologists Union; Wilson Ornith Soc; Sigma Xi. Res: Ecology and behavior of vertebrates with a special interest in avian ecology and behavior, ornithology, community analysis, tropical and insular biology. Mailing Add: Dept of Environ Pop & Organismic Biol Univ of Colo Boulder CO 80302

CRUZ, CALVIN J, analytical chemistry, see 12th edition

CRUZ, CARLOS, b Aguadilla, PR, Dec 24, 40; US citizen; m 71; c 2. ENTOMOLOGY. Educ: Univ PR, Mayagüez, BSA, 63; Rutgers Univ, MS, 68, PhD(entom), 72. Prof Exp: Agr agt, Agr Exten Serv, 63-65; res asst sugar cane, Agr Exp Sta, PR, 65-66; res asst entom, Rutgers Univ, 66-68 & Agr Exp Sta, PR, 68-69; ASST ENTOMOLOGIST, AGR EXP STA, UNIV PR, 72- Mem: Entom Soc Am; PR Soc Agr Sci; Sigma Xi; Caribbean Food Crop Soc. Res: Insect pest management on legumes and vegetables; study of all measures of insect control, particularly the use of resistant varieties, parasites, predators and selective insecticides. Mailing Add: Agr Exp Sta Univ of PR Isabela Substa Box 506 Isabela PR 00662

CRUZ, MAMERTO MANAHAN, JR, b Manila, Philippines, July 15, 18; US citizen; m 50; c 2. CHEMISTRY. Educ: Univ Philippines, BS, 39; Mass Inst Technol, SM, 41; State Univ NY Col Forestry, Syracuse, PhD(chem), 54. Prof Exp: Chemist, Am Viscose Corp, Pa, 42-47; lectr chem processes, Mapua Inst Technol, Philippines, 48; chem engr, Philippines Mission to Japan, 48, tech investr, 48-49; sr res chemist, Am Viscose Corp, Pa, 53-55; sect chief colloid chem, Olin Mathieson Chem Corp, Conn, 55-57; group leader pulp & bleaching, Ketchikan Res, Am Viscose Corp, 57-59, paper fibers, Com Develop Dept, 59-60, head avicel pilot opers, Corp Res Dept, 61-62, avicel plant mgr, 62, sect leader prod develop, 63, & spec prod, Am Viscose Div, 63, sr scientist new prod explor res, 63-64 & chem div, 64-66, mgr probing res, Cent Res Dept, 66-70, mgr avicon proj, 71-74, SR RES SCIENTIST, CHEM & DEVELOP CTR, FMC CORP, 74- Mem: Fel AAAS; Am Chem Soc; Tech Asn Pulp & Paper Indust. Res: Textile engineering; pulping and bleaching of wood pulps; rayon fiber and cellophane film technology; cellulose reactions; cellulose derivatives; synthetic polymeric coatings; wet strength paper and water laid non-wovens; wood chemistry; microcrystalline cellulose. Mailing Add: Chem & Develop Ctr FMC Corp PO Box 8 Princeton NJ 08540

CRUZAN, CHARLES GRANT, b Cushing, Okla, Feb 23, 12; m 37; c 4. ENGINEERING PHYSICS. Educ: Okla State Univ, BS, 34, MS, 38. Prof Exp: Asst math, Okla State Univ, 37-38; instr math & physics, Woodward Jr Col, 38-41; instr, Army Air Forces Tech Schs, Chanute Field, 41-43; CHIEF PHYSICIST, PATENT DIV, PHILLIPS PETROL CO, 47- Mem: Am Phys Soc. Res: Oxide protective coatings of metals; computing devices; geophysics; electronics; oil production. Mailing Add: 1950 Dewey St Bartlesville OK 74003

CRUZAN, GEORGE, biochemistry, see 12th edition

CRUZAN, JOHN, b Bridgeton, NJ, Jan 6, 42; m 64; c 2. ECOLOGY. Educ: King's Col, BA, 65; Univ Colo, PhD(zool), 68. Prof Exp: Asst prof, 68-73, ASSOC PROF BIOL, GENEVA COL, 73- Mem: Am Soc Mammal; Am Sci Affiliation; Ecol Soc Am; Am Inst Biol Sci. Res: Ecology, behavior and geography of mammals and other vertebrates. Mailing Add: Dept of Biol Geneva Col Beaver Falls PA 15010

CRUZ-VIDAL, BALTASAR AUGUSTO, solid state physics, radiation physics, see 12th edition

CRYBERG, RICHARD LEE, b Los Angeles, Calif, Nov 2, 41; m 63; c 2. ORGANIC CHEMISTRY. Educ: Iowa State Univ, BS, 63; Ohio State Univ, PhD(org chem), 69. Prof Exp: Chief chemist, G F Smith Chem Soc, 63-68; sr res chemist, 69-74, GROUP LEADER CHEM, T R EVANS RES CTR, DIAMOND SHAMROCK CORP, 74- Mem: Am Chem Soc. Res: Agricultural process development and optimization. Mailing Add: T R Evans Res Ctr Box 348 Painesville OH 44077

CRYER, COLIN WALKER, b Leeds, Eng, Aug 28, 35; m 68; c 1. COMPUTER SCIENCE. Educ: Univ Pretoria, BS, 54, MS, 58; Cambridge Univ, PhD(comput sci), 62. Prof Exp: Asst res officer, Nat Phys Lab, SAfrica, 55-58; res officer, Nat Res Inst Math Sci SAfrica, 62-63; res fel comput sci, Calif Inst Technol, 63-65; mem res staff, Math Res Ctr, 65-66, asst prof, 66-73, ASSOC PROF COMPUT SCI, UNIV WIS-MADISON, 73- Mem: Am Math Soc; Asn Comput Mach. Res: Numerical solution of integral and partial differential equations. Mailing Add: Dept of Comput Sci Univ of Wis Madison WI 53706

CRYER, JONATHAN D, b Toledo, Ohio, Feb 10, 39; m 61; c 3. STATISTICS. Educ: DePauw Univ, BA, 61; Univ NC, PhD(statist), 66. Prof Exp: Statistician, Res Triangle Inst, 63-66; asst prof, 66-70, ASSOC PROF STATIST, UNIV IOWA, 70-, STATISTICIAN, 66- Concurrent Pos: Consult, Res Triangle Inst, 66- Mem: Inst Math Statist; Am Statist Asn. Res: Stochastic processes and mathematical statistics. Mailing Add: Dept of Statist Univ of Iowa Iowa City IA 52242

CRYSTAL, MAXWELL MELVIN, b New York, NY, Oct 9, 24; m 49; c 1. VETERINARY ENTOMOLOGY. Educ: Brooklyn Col, AB, 47; Ohio State Univ, MSc, 48; Univ Calif, PhD(parasitol), 56. Prof Exp: Asst helminth & med entom, Univ Calif, 51-55; asst prof biol, NY Teachers Col, New Paltz, 56-57; instr parasitol, Albert Einstein Col Med, 57-58; Nat Cancer Inst res grant, 58-59; asst med physics, Montefiore Hosp, 59-60; RES ENTOMOLOGIST, SCREWWORM RES LAB, AGR RES SERV, USDA, 61- Mem: AAAS; Soc Parasitol; Inst Biol Sci; Entom Soc Am. Res: Insect sterility and biology. Mailing Add: Screwworm Res Lab Agr Res Serv USDA PO Box 986 Mission TX 78572

CSAKY, TIHAMER ZOLTAN, b Hungary, Aug 12, 15; nat US; m 53; c 2. PHYSIOLOGY, PHARMACOLOGY. Educ: Univ Budapest, MD, 39. Prof Exp: Asst prof physiol, Univ Budapest, 40-45; res adj, Hungarian Biol Res Inst, 46-47; res assoc, Duke Univ, 49-51; from asst prof to assoc prof pharmacol, Sch Med, Univ NC, 51-61; PROF PHARMACOL & CHMN DEPT, SCH MED, UNIV KY, LEXINGTON, 61- Concurrent Pos: Res fel, Biochem Inst, Helsinki, 47-48; res fel, Microbiol Inst, Uppsala, 48-49; vis prof & USPHS spec fel, Univ Milan, 68-69. Mem: Int Soc Biochem Pharmacol; Soc Exp Biol & Med; Am Soc Pharmacol & Exp Therapeut; Am Physiol Soc; Soc Gen Physiol. Res: Biological transport. Mailing Add: Dept of Pharmacol Univ of Ky Sch of Med Lexington KY 40506

CSALLANY, AGNES SAARI, b Budapest, Hungary, Apr 20, 32; US citizen; m 54. ORGANIC CHEMISTRY. Educ: Budapest Tech Univ, BS, 54, MS, 55, ScD(lipid chem), 70. Prof Exp: Head qual control lab, Duna Canning Co, Budapest, 55-56; from res asst to res assoc animal sci, Univ Ill, Urbana, 57-65, res asst prof, 65-69, res asst prof food sci, 69-71; ASSOC PROF FOOD CHEM, UNIV MINN, ST PAUL, 72- Concurrent Pos: Deleg, Int Nutrit Cong, 60, 69 & 75 & Int Biochem Cong, 64; Am Inst Nutrit travel grants, 69 & 75. Mem: Am Chem Soc; Am Inst Nutrit; Sigma Xi. Res: Basic research in the field of vitamin E; determination of chemical structures; air and water pollution; effect of nitrate, nitrogen dioxide and ozone on rats and mice. Mailing Add: Dept of Food Sci & Nutrit Univ of Minn St Paul MN 55101

CSALLANY, SANDOR CSERGO, b Szentes, Hungary, Apr 15, 28; m 54. HYDROLOGY, WATER RESOURCES. Educ: Col Com & Bus Admin, Budapest, BA, 48; Univ Tech Sci, Budapest, BSc, 54, MSc, 55, ScD(civil eng), 70. Prof Exp: Design engr hydrol, Melyepterv, Budapest, 55-56; prof scientist, Ill State Water Surv, Urbana, 57-74; PROF HYDROL, WESTERN KY UNIV, 74- Concurrent Pos: Co-founder, gen secy & mem bd dir, Am Water Resources Asn, 64-; consult, Comm Water Resources Coun, Rome, 70-73, UN, New York, 72-73 & BRADD, Ky, 75; founding mem, dep secy & gen mem exec bd, Int Water Resources Asn, 72-; del, Univ Coun Water Resources, 75; dir hydrol prog & chmn fac adv comt hydrol, Western Ky Univ, 75- Mem: Am Water Resources Asn; Int Water Resources Asn. Res: Water demands forecasting in water resources planning; mathematical models; education in hydrology. Mailing Add: Dept of Geog & Geol Western Ky Univ Bowling Green KY 42101

CSAPILLA, JOSEPH, b Budapest, Hungary, Feb 26, 34; US citizen. ORGANIC CHEMISTRY, TEXTILE CHEMISTRY. Educ: Univ Basel, PhD(org chem), 61. Prof Exp: USPHS fel reaction mechanism, Univ Mich, 61-63; res chemist, 63-70, SR RES CHEMIST, AM CYANAMID CO, 70- Mem: Am Chem Soc. Res: Synthetic, physical-organic and photochemistry; acrylic fibers. Mailing Add: Res Labs Am Cyanamid Co Stamford CT 06904

CSAPO, ARPAD ISTVAN, b Szeged, Hungary, Jan 15, 18; m 45; c 2. PHYSIOLOGY. Educ: Univ Szeged, MD, 43. Prof Exp: Instr bact & path, Univ Szeged, 40-41; asst prof pharmacol, 41-43; asst obstet & gynec, Midwifery Sch, Nagyvarad Univ & Univ Budapest, 43-48; mem staff, Dept Embryol, Carnegie Inst, 51-56; assoc prof, Rockefeller Inst, 56-61, head lab physiol reprod, 61-63; PROF OBSTET & GYNEC, WASHINGTON UNIV, 63- Concurrent Pos: Mannheimer fel, Univ Uppsala, 48-49, fel, Carnegie Inst, 49-51; Guggenheim fel, 54-55; lectr, Dept Obstet, Johns Hopkins Univ, 51-; CIBA lectr, Univ London, 52; vis prof, 62; Holmes lectr, 54; hon prof, Bahia Univ, Brz Brazil, 58. Honors & Awards: De Snoovan't Found prize, 64. Res: Reproductive physiology: biochemistry of the contractile system; biophysics and physiology of uterine muscle; mechanism of uterine muscle function and its endocrine regulation; endocrine control of pregnancy and functional myometrial disorders; induction of labor; mechanism of action and therapeutic application of fertility controlling agents. Mailing Add: Dept of Obstet & Gynec Washington Univ St Louis MO 63110

CSAVINSZKY, PETER JOHN, b Budapest, Hungary, July 10, 31; US citizen. THEORETICAL SOLID STATE PHYSICS. Educ: Tech Univ Budapest, diplom ing chem, 54; Univ Ottawa, PhD(theoret physics), 59. Prof Exp: Mem tech staff, Semiconductor Div, Hughes Aircraft Co, 59-60; sr physicist, Electronics Div, Gen Dynamics Corp, 60-62; mem tech staff, Tex Instruments Inc, 62-65 & TRW Systems Inc, 65-70; assoc prof, 70-75, PROF PHYSICS, UNIV MAINE, ORONO, 75- Mem: Fel Am Phys Soc; Am Asn Univ Prof; AAAS; NY Acad Sci. Res: Quantum theoretical studies of charge transport in solids; the structure of impurities and defects in solids. Mailing Add: Dept of Physics Univ of Maine Orono ME 04473

CSEJKA, DAVID ANDREW, b Passaic, NJ, June 9, 35; m 65; c 5. PHYSICAL CHEMISTRY, ANALYTICAL CHEMISTRY. Educ: Fordham Univ, BS, 56; Iowa State Univ, PhD(phys chem), 61. Prof Exp: Sr res chemist, 61-74, RES ASSOC CHEM, OLIN CORP, 74- Mem: Am Chem Soc. Res: Chemical and physical properties of aqueous and non-aqueous solutions; electroanalytical chemistry, polarography, coulometry, voltammetry. Mailing Add: Olin Corp 275 Winchester Ave New Haven CT 06504

CSENDES, ERNEST, b Satu-Mare, Rumania, Mar 2, 26; nat US; m 53; c 2. ORGANIC CHEMISTRY. Educ: Univ Hungary, BA, 44; Univ Heidelberg, BS, 48, MS & PhD, 51. Prof Exp: Asst, Univ Heidelberg, 51; res assoc biochem, Tulane Univ, 52; fel, Harvard Univ, 52-53; res chemist, Org Chem Dept, E I du Pont de Nemours & Co, Del, 53-56, Elastomer Chem Dept, 56-61; dir res, Armour & Co, 61-62, dir res & develop, Agr Chem Div, 62-63; vpres corp develop, Occidental Petrol Corp, 63-64, exec vpres res eng & develop & mem exec comt, 64-68, exec vpres, Occidental Res Eng Corp, 63-68, dir, Occidental Res & Eng, Ltd, UK, 64-68; PRES & CHIEF EXEC, TEX REPUB INDUSTS, 68- Concurrent Pos: Pres & chief exec, TRI Ltd, Bermuda, 71-, chmn & dir, TRI Int Ltd, Bermuda, 71-, managing dir, TRI Holdings SA, Luxembourg, 71- & TRI Capital NV, Neth, 71- Mem: Fel AAAS; fel Am Inst Chemists; Am Chem Soc; Am Defense Preparedness Asn; Am Inst Aeronaut & Astronaut. Res: International projects related to energy resources and agriculture; international finance related to leasing; banking, trusts, insurance; administration of research, engineering and industrial development. Mailing Add: 1651 San Onofre Dr Pacific Palisades CA 90272

CSERNA, EUGENE GEORGE, b Budapest, Hungary, Jan 2, 20; nat US; m 50; c 3. STRUCTURAL GEOLOGY. Educ: Univ Budapest, Dr Polit Sci, 43; Columbia Univ, MA, 50, PhD(geol), 56. Prof Exp: Asst geol, Columbia Univ, 50-51; geologist & explorer, Gulf Oil Corp, 51-54; instr geol, Hunter Col, 55; field geologist, US Geol Surv, 55-57; asst prof, Idaho State Col, 57-59; PROF GEOL, CALIF STATE UNIV, FRESNO, 59- Concurrent Pos: NSF fac fel, Swiss Fed Inst Technol, 66-67. Mem: Am Petrol Inst; Geol Soc Am. Res: structure and stratigraphy of Fra Cristobal Area, New Mexico, Poison Spider, Oil Mountain, Pine Mountain in Wyoming; revision of portions of the New Tectonic Map of the United States. Mailing Add: Dept of Geol Calif State Univ Fresno CA 93740

CSERR, HELEN F, b Boston, Mass, June 23, 37; m 62; c 1. PHYSIOLOGY. Educ: Middlebury Col, BA, 59; Harvard Univ, PhD(physiol), 65. Prof Exp: Instr neurol, Harvard Med Sch, 68-70; lectr, 70-71, ASST PROF MED SCI, BROWN UNIV, 71- Concurrent Pos: United Cerebral Palsy Res & Educ Found fel physiol, Harvard Med Sch, 65-68; Nat Inst Neurol Dis & Stroke career develop award, 73-; trustee, Mt Desert Island Biol Lab, 71-, mem, Exec Comt, 75-; mem, Physiol Study Sect, NIH, 75- Mem: AAAS; Soc Gen Physiol; Soc Neurosci; Am Physiol Soc. Res: Physiology of cerebrospinal fluid. Mailing Add: Div of Biomed Sci Brown Univ Providence RI 02912

CSICSERY, SIGMUND MARIA, b Budapest, Hungary, Feb 3, 29; US citizen; m 56. SURFACE CHEMISTRY, PETROLEUM CHEMISTRY. Educ: Budapest Tech Univ, MS, 51; Northwestern Univ, PhD(org chem), 61. Prof Exp: Res chemist, Res & Eng Div, Monsanto Chem Co, Ohio, 57-59 & Calif Res Corp, 61-66; sr res chemist, 66-70, SR RES ASSOC, CHEVRON RES CO, 70- Concurrent Pos: Exten instr, Univ Calif, Berkeley, 65- Mem: Am Chem Soc. Res: Catalysis and chemistry of petroleum hydrocarbons; structure, activation, and preparation of acidic and metal-containing heterogeneous catalysts; application of molecular sieves as catalysts. Mailing Add: Chevron Res Co PO Box 1627 Richmond CA 94802

CSIZMADIA, IMRE GYULA, physical organic chemistry, see 12th edition

CSONKA, PAUL L, b Budapest, Hungary, Aug 10, 38; US citizen. ELEMENTARY PARTICLE PHYSICS. Educ: Johns Hopkins Univ, PhD(theoret physics), 64. Prof Exp: Fel theoret physics, Lawrence Radiation Lab, Univ Calif, 64-66; NSF fel, Europ Org Nuclear Res Labs, 66-67; asst prof, 68-71, ASSOC PROF THEORET PHYSICS, UNIV ORE, 71-, RES ASSOC, INST THEORET SCI, 68- Concurrent Pos: Alfred P Sloan fel, 70-72. Mem: Am Phys Soc. Res: Invariance principles; statistical mechanics; causality; lasers. Mailing Add: Inst of Theoret Sci Univ of Ore Eugene OR 97403

CSORGO, MIKLOS, b Egerfarmos, Hungary, Mar 12, 32; Can citizen; m 57; c 1. MATHEMATICS. Educ: Eötvös Lorand, Budapest, BA, 55; McGill Univ, MA, 61, PhD(math), 63. Prof Exp: Lectr statist, Sch Econ, Eötvös Lorand, Budapest, 55-56; instr math, Princeton Univ, 63-65; from asst prof to assoc prof, McGill Univ, 65-74; PROF MATH, CARLETON UNIV, 74- Concurrent Pos: Vis prof, Math Inst, Univ Vienna, 69-70. Mem: Can Math Cong; Am Math Soc; Inst Math Statist. Res: Probability theory; mathematical statistics. Mailing Add: Dept of Math McGill Univ Montreal PQ Can

CUANY, ROBIN LOUIS, b Glasgow, Scotland, Oct 17, 26; US citizen; m 51; c 6. PLANT GENETICS. Educ: Cambridge Univ, BA, 47, MA, 51; Iowa State Univ, PhD(crop breeding, genetics), 58. Prof Exp: Cytogeneticist cotton breeding sect, Empire Cotton Growing Corp & Sudan Govt, 48-53; asst dept agron, Iowa State Univ, 54-56; res assoc biol dept, Brookhaven Nat Lab, 56-58, asst geneticist, 58-59; geneticist nuclear energy prog, Inter-Am Inst Agr Sci, 59-62; asst prof bot, Univ Iowa, 62-68; asst prof agron, crops, 68-75, ASSOC PROF AGRON, CROPS, COLO STATE UNIV, 75- Mem: Am Soc Agron; Am Genetic Asn; Asn Trop Biol. Res: Genetics and breeding of maize; grasses for forage and revegetation; photoperiodism and climatic adaptation; genecology; use of marker genes in detecting outcrossing and somatic mutation; cytogenetics. Mailing Add: Dept of Agron Colo State Univ Ft Collins CO 80523

CUATRECASAS, PEDRO, b Madrid, Spain, Sept 27, 36; US citizen; m 59; c 4. BIOCHEMISTRY, MEDICINE. Educ: Washington Univ, AB, 58, MD, 62. Prof Exp: Intern & resident internal med, Johns Hopkins Univ Hosp, 62-64; clin assoc clin endocrinol, Nat Inst Arthritis & Metab Dis, 64-66, med officer, Lab Chem Biol, 68-70; assoc prof med & pharmacol, Burroughs Wellcome prof clin pharmacol & dir div, 70-72, prof pharmacol & assoc prof med, Sch Med, Johns Hopkins Univ, 72-75; VPRES RES, DEVELOP & MED, WELLCOME RES LABS & DIR, BURROUGHS WELLCOME CO, 75- Concurrent Pos: USPHS spec fel, NIH Lab Chem Biol, 67-68; prof lectr biochem, George Washington Univ, 67-70; mem, Adv Comt Personnel for Res, Am Cancer Soc, 73-75; dir, Burroughs Wellcome Fund, 75-; adj prof, Dept Med & Depts Pharmacol & Physiol, Duke Univ, 75-; adj prof, Dept Med & Dept Pharmacol, Univ NC, 75-; ed, J Solid-Phase Biochem; mem, Adv Comt Cancer Res Prog, Univ NC. Honors & Awards: John Jacob Abel Prize Pharmacol, 72; Lilly Award, Am Diabetes Asn, 75; Laude Prize, Pharmaceut World, 74 & 75. Mem: Am Chem Soc; Sigma Xi; Am Soc Biol Chemists; Am Soc Pharmacol & Exp Therapeut; Am Soc Clin Invest. Res: Cell membranes; protein and glycolipid chemistry; mechanism of action of hormones; membrane receptors; cell growth; affinity chromatography; pharmacology. Mailing Add: Wellcome Res Labs Burroughs Wellcome 3030 Cornwallis Rd Research Triangle Park NC 27709

CUBBERLEY, ADRIAN H, b Westfield, NJ, Feb 1, 18; m 41; c 2. RESEARCH ADMINISTRATION, ORGANIC CHEMISTRY. Educ: Columbia Univ, AB, 39. Prof Exp: Res chemist, Norda Essential Oil & Chem Co, 39-42; res chemist, Barrett Div, Allied Chem Corp, 42-54, admin asst res & develop, 54-58 & plastics div, 58-62, proj mgr com develop, 62-65, dep sci dir, Allied Chem SA, Belg, 66-69, sr technico-econ planner, Allied Chem Corp, NJ, 69-70; res planner, Corp Res Ctr, 70-74, MGR RES INTEL, CORP RES & DEVELOP DIV, INT PAPER CO, TUXEDO PARK, 74- Mem: Am Chem Soc; fel Am Inst Chem; Tech Asn Pulp & Paper Indust; Coun Agr & Chemurgic Res; Soc Indust Chem. Res: Research management, strategy and planning; technical liaison; commercial development. Mailing Add: Sands Point Rd Washingtonville NY 10992

CUBICCIOTTI, DANIEL DAVID, JR, b Philadelphia, Pa, June 28, 21; m 48; c 3. HIGH TEMPERATURE CHEMISTRY, PHYSICAL INORGANIC CHEMISTRY. Educ: Univ Calif, BS, 42, PhD(chem), 46. Prof Exp: Asst chem, Manhattan Proj, Univ Calif, 44-46; instr, NY Univ, 46-47 & Univ Calif, 47-48; res asst prof, Ill Inst Technol, 48-51; res chemist, Atomic Energy Res Lab, NAm Aviation, Inc, 51-55; res chemist, Stanford Res Inst, 55-63, sci fel, 63-72; tech specialist, Nuclear Energy Div, Gen Elec Co, 72-74; SR SCIENTIST, STANFORD RES INST, 74- Concurrent Pos: Mem, Nat Res Coun-Nat Acad Sci Comt High Temperature Sci & Technol, 67-74, chmn, 71-74; US rep, Int Union Pure & Appl Chem Comn High Temperatures & Refractories, 70-; div ed jour, Electrochem Soc, 75- Mem: Am Chem Soc; Electrochem Soc. Res: Reactions of metals with molten salts and with gases at high temperatures; evaporation of materials; chemistry of nuclear reactor fuels; thermodynamics of inorganic systems. Mailing Add: Stanford Res Inst Menlo Park CA 94025

CUBITT, JOHN MALCOLM, b London, Eng, June 15, 49; m 72. GEOLOGY, STATISTICAL ANALYSIS. Educ: Leicester Univ, BSc, 70, PhD(geol), 75. Prof Exp: Res asst geol, Kans Geol Surv, 70-71; syst analyst comput unit, Inst Geol Sci, London, 73-75; ASST PROF GEOL, SYRACUSE UNIV, 75- Concurrent Pos: Free lance ed x-ray diffraction, Sci & Technol Agency, London, 72-75. Mem: Fel Geol Soc London; Int Asn Math Geol; Soc Econ Paleontologists & Mineralogists; Inst Brit Geog. Res: Applications of statistical, mathematical and computer techniques in geology; instrumental analysis in geology; mineral resources; geochemistry, mineralogy and petrology of shales; regional tectonics of the midcontinent, USA; Welsh gold mining. Mailing Add: Dept of Geol Syracuse Univ Syracuse NY 13210

CUCCI, CESARE ELEUTERIO, b Italy, Dec 22, 25; US citizen; m 49, 66; c 3. MEDICINE. Educ: Univ Perugia, MD, 49; Univ Rome, dipl cardiol, 53; Am Bd Pediat, dipl, 58, cert cardiol, 63. Prof Exp: ASSOC PROF CLIN PEDIAT, NY UNIV, 60-; CHIEF CHILDREN'S CARDIAC SERV, LENOX HILL HOSP, 63- Concurrent Pos: Vis pediatrician, Bellevue Hosp, New York, 65-; consult pediat cardiol, Booth Mem Hosp, 60- & Flushing Hosp & Methodist Hosp, 63- Mem: AAAS; Am Col Physicians; Am Acad Pediat; Am Col Chest Physicians; Am Col Cardiol. Res: Cardio-pulmonary physiology; pediatric cardiology. Mailing Add: 45 E 62nd St New York NY 10021

CUCKA, PAUL, crystallography, chemistry, see 12th edition

CUCKLER, ASHTON CLINTON, b Wilsonville, Nebr, Mar 16, 10; m 41; c 2. ANIMAL PARASITOLOGY. Educ: Univ Nebr, AB, 35, AM, 36; Univ Minn, PhD(parasitol), 41; Army Med Sch, dipl, 44. Prof Exp: Asst zool, Univ Nebr, 34-36, asst instr, 36-38; teaching asst, Univ Minn, 38-41; asst parasitologist, Univ Hawaii, 41-42; from instr to asst prof zool, Univ Minn, 42-46; head parasitol lab, Merck Inst Therapeut Res, 47-51, asst dir parasitol, 51-56, dir, 56-63, assoc dir, 63-69, dir infectious dis res, 66-69, exec dir, Quinton Res Labs, Merck & Co, Inc, 69-72, exec dir, Animal Sci Res, 72-75; RETIRED. Concurrent Pos: Asn Am Med Cols traveling fel, Cent Am, 44; adj assoc prof parasitol, Col Physicians & Parasit Surg, Columbia Univ, 65-; in chg labstudies med, Mission, UN Relief & Rehab Admin, Rome, Italy, 45. Mem: AAAS; Am Soc Parasitol; Am Soc Trop Med & Hyg; Am Micros Soc; NY Acad Sci. Res: Nematode morphology and taxonomy; trematode life cycles; medical parasitology and entomology; parasites and diseases of fresh water fish; helminthology and parasitology; chemotherapy of parasitic infections. Mailing Add: 31 Hawthorn Dr Westfield NJ 07090

CUDABACK, DAVID DILL, b Napa, Calif, Jan 18, 29; m 53; c 1. RADIO ASTRONOMY. Educ: Univ Calif, Berkeley, BA, 51, PhD(astron), 62. Prof Exp: Physicist, Lawrence Radiation Lab, Univ Calif, 50-57; astronr, Radio Astron Inst, Stanford Univ, 58-62; ASTRONR, UNIV CALIF, BERKELEY, 63-, LECTR ASTRON, 64-, ASSOC ASTRONR, RADIO ASTRON LAB, 71-, ASSOC DIR ASTRON, WHITE MOUNTAIN RES STA, 73- Concurrent Pos: Physicist, Los Alamos Sci Lab, 53-54. Mem: AAAS; Am Astron Soc; Int Astron Union; Int Sci Radio Union. Res: Radio and infrared studies of interstellar material and star formation; instrumentation for same; high altitude observatory development for same. Mailing Add: Dept of Astron Univ of Calif Berkeley CA 94720

CUDD, HERSCHEL HERBERT, b Memphis, Tex, July 29, 12; m. CHEMISTRY. Educ: Tex Col Arts & Indust, AB, 33; Univ Tex, AM, 36, PhD(phys chem), 41. Prof Exp: Teacher schs, Tex, 33-34; tutor anal chem, Univ Tex, 34-37, instr, 39-40; chemist, State Liquor Control Bd, Tex, 37-39, 40-41; res chemist, Rayon Tech Div, E I du Pont de Nemours & Co, 41-42; res chemist, Int Minerals & Chem Corp, 42-46; res engr, West Point Mfg Co, Ga, 46-50; head chem div, Eng Exp Sta, Ga Inst Technol, 50-52, dir, 52-54; vpres, Am Viscose Corp, 54-60; pres, Avisun Corp, 60-62; pres, Amoco Chem Corp, 63-72; vpres chem, 63, 67-69, DIR, STAND OIL CO, IND, 72- Concurrent Pos: Dir, R J Reynolds Indust, Inc. Mem: Am Chem Soc; Am Inst Chem Engrs. Res: General business management. Mailing Add: Stand Oil Co of Ind 200 E Randolph Dr Chicago IL 60601

CUDDIHY, RICHARD GEORGE, radiation biology, biophysics, see 12th edition

CUDDY, THOMAS FOSTER, b Winnipeg, Man, May 11, 10; m 45. PLANT PHYSIOLOGY. Educ: Univ Man, BSA, 41; Univ London, PhD, 53; Imp Col, London, dipl, 53. Prof Exp: Asst barley breeding, Univ Man, 40-41; asst, Plant Prods Div, 46-49, plant physiologist seed germination, Seed Res Lab, 54-63, off-in-chg germination & seed physiol sect, 63-66, HEAD SEED BIOL LAB, ANAL SERV SECT, AGR CAN, 66- Mem: Asn Off Seed Anal; Agr Inst Can; Can Soc Plant Physiol; Brit Soc Exp Biol. Res: Seed dormancy and related problems in seed germination. Mailing Add: Plant Prod Div Anal Serv Sect Agr Can Carling Ave Ottawa ON Can

CUDE, JOE E, b Austin, Tex, Feb 23, 39; m 68. TOPOLOGY, ALGEBRA. Educ: Southwest Tex State Univ, BS, 60; Univ Tex, MA, 62, PhD(math), 66. Prof Exp: Asst prof math, Wash State Univ, 66-69; assoc prof & chmn dept, Dallas Baptist Col, 69-72; PROF MATH & HEAD DEPT, TARLETON STATE COL, 72- Mem: Am Math

Soc; Math Asn Am. Res: Topological rings. Mailing Add: Dept of Math Tarleton State Col Stephenville TX 76401

CUDE, WILLIS AUGUSTUS, JR, b Luling, Tex, Jan 2, 22; m 44; c 4. INORGANIC CHEMISTRY, CRYSTALLOGRAPHY. Educ: Univ Tex, Austin, BS, 42, PhD(chem), 68; Ohio State Univ, MS, 53. Prof Exp: Meterologist, US Army Air Corps, US, 43-44 & Okinawa, 45-49; instr meteorol, Chanute Air Force Base, Ill, 49-51; sci liaison nuclear sci, Los Alamos Sci Labs, 53-56; res scientist, Aeronaut Res Labs, Wright Patterson AFB, Ohio, 56-59; from instr to assoc prof chem, US Air Force Acad, 59-63; asst prof, 63-66, ASSOC PROF CHEM, SOUTHWEST TEX STATE UNIV, 67- Concurrent Pos: Partic, NSF Col Teacher Res Prog, Univ Ark, Fayetteville, 68-69. Mem: Am Crystallog Asn; Am Chem Soc. Res: Coordination compounds; structure of mu-bridged complexes. Mailing Add: Dept of Chem Southwest Tex State Univ San Marcos TX 78666

CUDERMAN, JERRY FERDINAND, b Crosby, Minn, Aug 1, 35; m 61; c 2. EXPERIMENTAL ATOMIC PHYSICS. Educ: Univ Minn, BME, 58; Ore State Univ, PhD(physics), 66. Prof Exp: Exp engr, United Aircraft Corp, 58-59; STAFF MEM, SANDIA LAB, 65- Mem: Am Phys Soc. Res: Atomic physics; fast alkali metal atom interactions with other atomic species and surfaces; laser plasma physics; x-ray spectroscopy. Mailing Add: 4300 Andrew Dr NE Albuquerque NM 87109

CUDKOWICZ, GUSTAVO, b Zurich, Switz, July 27, 27; m 57; c 3. IMMUNOBIOLOGY, TRANSPLANTATION IMMUNOLOGY. Educ: Univ Milan, MD, 52, cert med radiol, 55. Prof Exp: Asst prof exp path, Inst Gen Path, Univ Milan, 52-53, 54-55; vis fel biochem, Univ Uppsala, 53-54; asst prof path, Nat Cancer Inst, Milan, Italy, 56-59; vis investr, Biol Div, Oak Ridge Nat Lab, 60, biologist, 61-65; assoc cancer res scientist, Roswell Park Mem Inst, 65-67, prin cancer res scientist, 67-69; actg chmn dept path, 73-74, PROF PATH & MICROBIOL, STATE UNIV NY BUFFALO, 69- Concurrent Pos: Nat Res Coun Italy fel, 53-54, Ital League Against Cancer fel, 59; Int Atomic Energy Agency, Austria fel, 60. Mem: Am Soc Exp Path; Soc Exp Biol & Med; Transplantation Soc; Am Asn Cancer Res; Am Asn Immunologists. Res: Functional characterization of progenitor and effector cells of the immune system; immunobiology and genetics of blood-forming allografts. Mailing Add: 232 Farber Hall State Univ of NY Buffalo NY 14214

CUDWORTH, KYLE MCCABE, b Minneapolis, Minn, June 7, 47. ASTRONOMY. Educ: Univ Minn, BPhys, 69; Univ Calif, Santa Cruz, PhD(astron), 74. Prof Exp: ASST PROF ASTRON, YERKES OBSERV, UNIV CHICAGO, 74- Mem: Am Astron Soc; Am Sci Affil. Res: Measuring proper motions and parallaxes of stars; low luminosity stars, planetary nebulae and star clusters. Mailing Add: Yerkes Observ Williams Bay WI 53191

CUE, NELSON, b Cavite City, Philippines, Aug 10, 41. EXPERIMENTAL NUCLEAR PHYSICS. Educ: Feati Univ, Philippines, BS, 61; Univ Wash, PhD, 67. Prof Exp: Res assoc, Univ Wash, 67 & State Univ NY Stony Brook, 67-70; ASST PROF PHYSICS, STATE UNIV NY ALBANY, 70- Mem: AAAS; Am Phys Soc; Am Asn Physics Teachers. Res: Nuclear reaction and spectroscopy; atomic collisions. Mailing Add: Dept of Physics State Univ of NY Albany NY 12222

CUETO, CIPRIANO, JR, b Tampa, Fla, July 15, 23; m 41; c 3. PHARMACOLOGY, TOXICOLOGY. Educ: Univ Tampa, BS, 49; Emory Univ, MS, 57, PhD(pharmacol), 64. Prof Exp: Chemist, Bur Agr & Indust Chem, USDA, Fla, 50-51, chemist, Bur Entom & Plant Quarantine, Ga, 51-53; chemist, Tech Develop Labs, Dept Health, Educ & Welfare, USPHS, Ga, 53-58, chemist, Phoenix Field Sta, Ariz, 58-59, chemist, Toxicol Sect, Commun Dis Ctr, 59-66, supvry res pharmacologist & chief pharmacol sect, Fla, 66-69; chief staff officer, Human Safety Eval of Pesticides, USDA & Environ Protection Agency, Washington, DC, 69-71; chief chronic studies div, Nat Ctr Toxicol Res, Ark, 71-74; PHARMACOLOGIST-TOXICOLOGIST, NAT CANCER INST, 74- Mem: Am Chem Soc. Res: Toxicology of pesticides; adrenocortical inhibitions and its effects on the response of the cardiovascular system to catecholamines; chemical carcinogen bioassay and research. Mailing Add: Carcinogen Bioassay & Res Br Nat Cancer Res Inst Landow Bldg Bethesda MD 20014

CUFF, DAVID J, b Edmonton, Alta, May 11, 33; m 61; c 1. CARTOGRAPHY, PHYSICAL GEOGRAPHY. Educ: Univ Alta, BSc, 54; Pa State Univ, MSc, 68, PhD(geog), 72. Prof Exp: Geologist, Texaco Explor Co, Calgary, 54-56, Gallup, Buckland & Farney Co, 56-63 & Can Indust Gas Corp, 63-64; ASST PROF GEOG, TEMPLE UNIV, 68- Mem: Asn Am Geog; Soc Univ Cartogr. Res: The role of color in map design. Mailing Add: Dept of Geog Temple Univ 1822 Park Ave Philadelphia PA 19122

CUFFEY, JAMES, b Chicago, Ill, Oct 8, 11; m 38; c 4. ASTRONOMY. Educ: Northwestern Univ, BS, 34; Harvard Univ, AM, 36, PhD(astron), 38. Prof Exp: Res fel, Ind Univ, 39-41; instr navig, US Naval Acad, 41-46; from asst prof to assoc prof, Kirkwood Observ, Ind Univ, Bloomington, 46-66; PROF ASTRON, NMEX STATE UNIV, 66- Mem: Am Astron Soc. Res: Photometric work on the colors of stars. Mailing Add: Dept of Astron NMex State Univ Box 4500 Las Cruces NM 88001

CUFFEY, ROGER JAMES, b Indianapolis, Ind, May 2, 39; m 64; c 2. INVERTEBRATE PALEONTOLOGY. Educ: Ind Univ, Bloomington, BA, 61, MA, 65, PhD(paleont), 66. Prof Exp: ASSOC & ASST PROF PALEONT, PA STATE UNIV, UNIVERSITY PARK, 67- Mem: AAAS; Paleont Soc; Soc Vert Paleont; Geol Soc Am; Am Sci Affiliation. Res: Taxonomy, morphology, evolution and paleoecology of fossil and living bryozoans; role of bryozoans in fossil and living reefs. Mailing Add: Dept of Geosci Pa State Univ University Park PA 16802

CUGELL, DAVID WOLF, b New Haven, Conn, Sept 19, 23. PULMONARY DISEASES. Educ: Yale Univ, BS, 45; LI Col Med, MD, 47. Prof Exp: Asst med, Albany Med Col, 49-50; assoc, 55-57, from asst prof to assoc prof, 57-69, BAZLEY PROF MED, SCH MED, NORTHWESTERN UNIV, 69- Concurrent Pos: Res fel, Harvard Med Sch, 50-51, 53-55, Am Heart Asn fel, 53-55; USPHS res career develop award, 62-68; dir pulmonary labs, Northwestern Univ; attend physician, Vet Admin Res Hosp, Cook County Hosp, Northwestern Mem Hosp; mem pulmonary dis subspeciality bd, Am Bd Internal Med, 66-72. Mem: Am Col Physicians; Am Thoracic Soc; Am Col Chest Physicians; Am Fedn Clin Res. Res: Cardiopulmonary disease. Mailing Add: Northwestern Univ Med Sch Chicago IL 60611

CUJEC, BIBIJANA DOBOVISEK, b Ljubljana, Yugoslavia, Dec 25, 26; Can citizen; m 56; c 4. NUCLEAR PHYSICS. Educ: Univ Ljubljana, MSc, 51, PhD(physics), 59. Prof Exp: Res asst physics, Inst J Stefan, Ljubljana, 50-55, res officer, 55-61; res assoc, Univ Pittsburgh, 61-63 & Univ Alta, 63-64; from asst prof to assoc prof, 64-70, PROF PHYSICS, LAVAL UNIV, 70- Concurrent Pos: Vis assoc, Calif Inst Technol, 71-72. Mem: Am Phys Soc; Can Asn Physicists. Res: Nuclear structure; nuclear reactions and scattering. Mailing Add: Dept of Physics Laval Univ Quebec PQ Can

CUKIER, ROBERT ISAAC, b New York, NY, Oct 10, 44. THEORETICAL CHEMISTRY. Educ: Harpur Col, BA, 65; Princeton Univ, MS, 67, PhD(chem), 69. Prof Exp: NATO fel, Lorentz Inst Theoret Physics, 69-71; res chemist, Univ Calif, San Diego, 71-72; ASST PROF CHEM, MICH STATE UNIV, 72- Mem: Am Phys Soc. Res: Applications of nonequilibrium statistical mechanics to physical chemistry problems. Mailing Add: Dept of Chem Mich State Univ East Lansing MI 48824

CUKOR, PETER, b Szolnok, Hungary, Aug 29, 36; US citizen; m 64; c 3. ANALYTICAL CHEMISTRY. Educ: City Col New York, BChEng, 61; St John's Univ, NY, MS, 63, PhD(phys & anal chem), 66. Prof Exp: Res technologist anal chem, Customer Serv Dept, Mobil Oil Co, 64-66, eng specialist anal chem & metal-org, 67-69, head absorption spectros & chromatog sect, 69-74, HEAD ORG ANAL & ORG MAT SECT, GEN TEL & ELECTRONICS RES LABS, INC, 74- Mem: AAAS; Am Chem Soc; NY Acad Sci. Res: Absorption spectroscopy; thermal analysis; gas and liquid chromatography; fluorimetry and metalorganic compounds; radioimmunoassay; photopolymers; photoresist technology; metalorganic compounds; fire retardant polymers. Mailing Add: Gen Tel & Electronics Res Labs 40 Sylvan Rd Waltham MA 02154

CULBERSON, CHICITA FRANCES, b Philadelphia, Pa, Nov 1, 31; m 53. ORGANIC CHEMISTRY. Educ: Univ Cincinnati, BS, 53; Univ Wis, MS, 54; Duke Univ, PhD(chem), 59. Prof Exp: Res assoc org chem, 59-61, SR RES ASSOC BOT, DUKE UNIV, 61-, LECTR, 71- Mem: AAAS; Am Chem Soc. Res: Chemistry of lichen substances. Mailing Add: Dept of Bot Duke Univ Durham NC 27706

CULBERSON, JAMES LEE, b Pana, Ill, Sept 18, 41; m 62; c 2. ANATOMY. Educ: Ill Wesleyan Univ, AB, 63; Tulane Univ, PhD(anat), 68. Prof Exp: Instr anat, Tulane Univ, 67-68; from instr to asst prof, 68-72, ASSOC PROF ANAT, MED CTR, W VA UNIV, 72- Res: Comparative neuroanatomy and neurophysiology; relations of limbic system and hypothalamus. Mailing Add: Dept of Anat WVa Univ Med Ctr Morgantown WV 26506

CULBERSON, WILLIAM LOUIS, b Indianapolis, Ind, Apr 5, 29; m 53. BOTANY. Educ: Univ Cincinnati, BS, 51; Sorbonne, France, dipl, 52; Univ Wis, PhD(bot), 54. Prof Exp: From instr to assoc prof, 55-70, PROF BOT, DUKE UNIV, 70- Concurrent Pos: Grants, NSF & Lalor Found, 57-75; ed, Bryologist, Am Bryol & Lichenological Soc, 63-70 ; ed, Brittonia, Am Soc Plant Taxon, 75- Mem: Bot Soc Am; Am Soc Plant Taxon; Am Bryol & Lichenological Soc; Mycol Soc Am. Res: Taxonomy, ecology, morphology and biochemistry of lichens. Mailing Add: Dept of Bot Duke Univ Durham NC 27706

CULBERT, HARVEY V, solid state physics, see 12th edition

CULBERT, JOHN ROBERT, b Rossville, Ill, Dec 18, 14; m 45; c 2. FLORICULTURE. Educ: Univ Ill, BS, 37; Ohio State Univ, MS, 39. Prof Exp: Instr floricult, Pa State Col, 39-43 & 45-46; from asst prof to assoc prof, 46-65, PROF FLORICULT, UNIV ILL, URBANA-CHAMPAIGN, 65- Concurrent Pos: Consult, Framptons Nurseries, Eng, 59-60. Honors & Awards: Res award, Soc Am Florists, 56. Mem: Am Soc Hort Sci. Res: Breeding of chrysanthemums. Mailing Add: 100 Floricult Bldg Univ of Ill Urbana IL 61801

CULBERT, THOMAS PATRICK, b Minneapolis, Minn, June 13, 30; m 58; c 4. ANTHROPOLOGY, ARCHAEOLOGY. Educ: Univ Minn, BA, 51; Univ Chicago, MA, 57, PhD(anthrop), 62. Prof Exp: Actg asst prof anthrop, Univ Miss, 60; proj ceramicist, Tikal Proj, Univ Pa, 60-62; vis prof anthrop, San Carlos Univ Guatemala, 62; asst prof, Southern Ill Univ, Edwardsville, 62-64; from asst prof to assoc prof, 64-74, PROF ANTHROP, UNIV ARIZ, 75- Mem: Soc Am Archaeol (secy, 68-72); Am Anthrop Asn. Res: Development of complex societies; archaeology of Mesoamerica. Mailing Add: Dept of Anthrop Univ of Ariz Tucson AZ 85721

CULBERTSON, BILLY MURIEL, b Hillside, Ky, Aug 23, 29; m 49; c 3. ORGANIC CHEMISTRY, PHYSICAL CHEMISTRY. Educ: Augustana Col, Ill, BA, 59; Univ Iowa, MS, 62, PhD(org chem), 63. Prof Exp: Sr res chemist, Archer Daniels Midland Co, Minn, 62-67; sr res chemist, 67-69, RES ASSOC, ASHLAND CHEM CO, 69- Concurrent Pos: Ed, Minn Chemist. Mem: Indust Res Inst; Am Chem Soc; Mfg Chemists Asn; The Chem Soc; Sigma Xi. Res: Organic syntheses, monomer syntheses and polymerization studies; polymerization mechanisms and kinetics studies; cyclopolymerization studies; synthesis of plastics for space applications; synthesis and physical studies of polymers having high thermal stability; amine acylimide monomers and polymers. Mailing Add: Ashland Chem Co PO Box 2219 Columbus OH 43216

CULBERTSON, CLYDE GRAY, b Vevay, Ind, July 27, 06; m 31. PATHOLOGY. Educ: Ind Univ, BS, 28, MD, 31. Prof Exp: From instr to asst prof path, Sch Med, Ind Univ, 31-42, prof clin path & chmn dept, 43-63; from asst dir to dir biol div, 46-63, res adv, 63-70, Lilly res consult, Lilly Res Labs, 70-71, CONSULT, ELI LILLY & CO, 71-; PROF PATH, SCH MED, IND UNIV, INDIANAPOLIS, 63- Concurrent Pos: Dir lab, Sch Med, Ind Univ, 31-46, dir labs, State Bd Health, Ind, 33-46. Mem: AMA; Am Asn Immunologists; Soc Protozoologists; Tissue Cult Asn; Am Soc Clin Path (secy-treas, 49-58, vpres, 59-68, pres, 70). Res: Clinical pathology; medical microbiology and parasitology. Mailing Add: Lilly Lab for Clin Res Wishard Mem Hosp Indianapolis IN 46202

CULBERTSON, GEORGE EDWARD, b Cranes Nest, Va, Oct 23, 37; m 59; c 3. MATHEMATICS. Educ: Va Polytech Inst, BS, 59, MS, 62, PhD(math), 70. Prof Exp: Engr, Radford Army Ammunition Plant, Hercules, Inc, Va, 62-63, sr engr, 63-64, asst area supvr, 64; asst prof math, US Naval Acad, 64-70; assoc prof, 70-71, REGISTR & ASST DEAN, CLINCH VALLEY COL, UNIV VA, 71-, PROF MATH, 75- Concurrent Pos: Pres, Systech Corp, Md, 69-70. Mem: Math Asn Am. Res: A basic variable study of the propellant used in the Minuteman Missile; geometry of paths of particles traveling in force fields under certain constraints. Mailing Add: Dept of Math Clinch Valley Col Univ of Va Wise VA 24293

CULBERTSON, TOWNLEY PAYNE, organic chemistry, see 12th edition

CULBERTSON, WILLIAM RICHARDSON, b Coeburn, Va, May 16, 16; m 50; c 2. MEDICINE. Educ: Transylvania Col, AB, 37; Vanderbilt Univ, MD, 41; Am Bd Surg, dipl, 55. Prof Exp: Intern, St Joseph's Hosp, Lexington, Ky, 41-42; asst resident surg, Cincinnati Gen Hosp, 46-53; from asst prof to assoc prof, 57-70, PROF SURG, UNIV CINCINNATI, 70- Mem: Fel Am Col Surg; Am Surg Asn; Soc Surg Alimentary Tract; AMA. Res: Bacteriology of surgical infections; hemorrhage shock; shock due to sepsis. Mailing Add: Dept of Surg Univ of Cincinnati Med Ctr Cincinnati OH 45267

CULKOWSKI, WALTER MARTIN, b Cleveland, Ohio, Sept 26, 28; m 53. METEOROLOGY. Educ: Western Reserve Univ, BS, 50. Prof Exp: Engr, Nat Tool Co, 51-52; RES METEOROLOGIST, ENVIRON SCI SERV ADMIN, 56- Mem: AAAS; Am Meteorol Soc; Prof Photogr Asn. Res: Problems and instrumentation in the fields of atmospheric transport, diffusion and depletion of material, especially associated with nuclear industry, considering various environments and source geometries; problems in atmospheric optics and laser applications; numerical modeling techniques for various scales of atmospheric pollution. Mailing Add: Atomic Energy Comn Box E Oak Ridge TN 37830

CULL, NEVILLE, b Victoria, BC, Can, July 24, 20; nat US; m 44; c 3. INORGANIC CHEMISTRY. Educ: Tulane Univ, BS, 47, MS, 48, PhD(chem), 50. Prof Exp: Asst prof, Inst Sci & Technol, Univ Ark, 50-51; govt res, Ordark,, 50-51; RES ASSOC, ESSO RES LABS, STANDARD OIL CO, NJ, 51- Mem: Am Chem Soc. Res: Petrochemicals; catalyst research petroleum processes; development of catalysts and processes for NOx and SOx abatement. Mailing Add: Rt 2 Box 262 Baker LA 70714

CULLEN, ABBEY BOYD, JR, b Oxford, Miss, Dec 25, 15; m 43; c 1. APPLIED PHYSICS. Educ: Univ Miss, BA, 37, MS, 42; Univ Va, PhD(physics), 47. Prof Exp: Technician develop physiol res apparatus, Sch Med, Univ Miss, 37-40; engr & physicist, Naval Res Lab, 42-45; assoc prof, 47-70, PROF PHYSICS, UNIV MISS, 70-, CHMN DEPT PHYSICS & ASTRON, 57- Mem: Am Phys Soc. Res: Microwave electron accelerator; electronic instruments. Mailing Add: 108 Physics Bldg Univ of Miss University MS 38677

CULLEN, BRUCE F, b Iowa City, Iowa, May 6, 40; m 60; c 3. ANESTHESIOLOGY. Educ: Stanford Univ, BS, 62; Univ Calif, Los Angeles, MD, 66. Prof Exp: Staff anesthesiologist, NIH, 70-72; asst prof, 72-75, ASSOC PROF ANESTHESIOL, UNIV WASH, 75- Mem: Am Soc Anesthesiologists; Int Anesthesia Res Soc; Asn Univ Anesthetists; Soc Neuroanesthesia & Neurol Intensive Care. Res: Immunologic and cellular effects of anesthesia. Mailing Add: Dept of Anesthesiol Univ of Wash Sch of Med Seattle WA 98195

CULLEN, CHARLES G, b Elmira, NY, Nov 6, 32; m 54; c 2. MATHEMATICS. Educ: State Univ NY Albany, BA, 54; Univ NH, MA, 56; Case Inst Technol, PhD(math), 62. Prof Exp: Instr math, Worcester Polytech Inst, 56-59 & Case Inst Technol, 59-61; asst prof, 62-66, ASSOC PROF MATH, UNIV PITTSBURGH, 66- Mem: Am Math Soc; Math Asn Am. Res: Linear algebra; matrix analysis; functions of matrices and matrix algebras; numerical analysis. Mailing Add: Dept of Math Univ of Pittsburgh Pittsburgh PA 15213

CULLEN, DANIEL EDWARD, b Oak Park, Ill, Feb 16, 42; m 63; c 2. OPERATIONS RESEARCH. Educ: Stanford Univ, BS, 63; Univ Ill, MS, 64; Wash Univ, ScD(appl math), 67. Prof Exp: Mem tech staff appl math, Bell Tel Labs, 67-68; ASST DIR OPERS RES, MATHEMATICA, 68- Mem: Am Math Soc; Soc Indust & Appl Math; Opers Res Soc Am; Inst Mgt Sci; Financial Mgt Asn. Res: Development of analytic and scientific methods for the solution of business and government problems. Mailing Add: 3 Turner Ct Princeton NJ 08540

CULLEN, DERMOTT EDWARD, b Brooklyn, NY, Nov 22, 39; m 65; c 1. NUCLEAR PHYSICS, COMPUTER SCIENCE. Educ: US Merchant Marine Acad, BS, 61; Columbia Univ, MS, 64, PhD(nuclear eng), 68. Prof Exp: Asst prof elec eng, US Merchant Marine Acad, 63-64; nuclear consult, Nat Lead Co, 66-67; from asst scientist to scientist, Nat Neutron Cross Sect Ctr, Brookhaven Nat Lab, 67-72; PHYSICIST, LAWRENCE LIVERMORE LAB, 72- Concurrent Pos: Nuclear consult, 61-71, Lawrence Radiation Lab, 67-71. Mem: Am Nuclear Soc. Res: Neutron transport theory, theory and application to high speed computers; organization of nuclear data files; interactive graphics. Mailing Add: Lawrence Livermore Lab PO Box 808 Livermore CA 94550

CULLEN, GLENN WHERRY, b Nashville, Tenn, June 27, 31; m 56. SOLID STATE CHEMISTRY. Educ: Univ Cincinnati, BS, 53; Univ Ill, MS, 54, PhD(chem), 56. Prof Exp: HEAD MAT APPL SYNTHESIS GROUP, DAVID SARNOFF LABS, RCA CORP, 58- Concurrent Pos: Assoc ed, J Crystal Growth, 74-; div ed jour, Electrochem Soc, 75- Mem: Am Asn Crystal Growth; Am Chem Soc; Electrochem Soc. Res: Chemistry of rare earth metals; semiconductor materials chemistry; superconductor materials chemistry; heteroepitaxial thin film growth; single crystal growth. Mailing Add: David Sarnoff Labs RCA Corp Princeton NJ 08540

CULLEN, HELEN FRANCES, b Boston, Mass, Jan 4, 19. MATHEMATICS. Educ: Radcliffe Col, AB, 40; Univ Mich, AM, 44, PhD(math), 50. Prof Exp: From asst prof to assoc prof, 49-71, PROF MATH, UNIV MASS, AMHERST, 71- Mem: Am Math Soc; Math Asn Am; Sigma Xi. Res: Topology; algebraic geometry. Mailing Add: Dept of Math Univ of Mass Amherst MA 01002

CULLEN, JAMES HENRY, b New York, NY, Jan 23, 11. MEDICINE. Educ: NY Univ, BS, 33, MD, 36. Prof Exp: Assoc physician, House of Rest, Yonkers, 40-52; from instr to assoc prof med, 52-62, PROF MED, ALBANY MED COL, 62- Concurrent Pos: Chief pulmonary dis sect, Vet Admin Hosp, Albany, NY, 52-56; chief med serv, 56-72, consult pulmonary dis, 72-74. Mem: Am Thoracic Soc; Am Col Physicians; Am Col Chest Physicians. Res: Disturbed physiology and response to treatment in pulmonary disease; direct pulmonary function laboratory. Mailing Add: Dept of Med Albany Med Col Albany NY 12208

CULLEN, JAMES ROBERT, b Brooklyn, NY, Jan 28, 36; m; c 1. SOLID STATE PHYSICS. Educ: St John's Univ, NY, BS, 58; Univ Md, PhD(physics), 65. Prof Exp: Physicist, US Army Night Vision Lab, Ft Belvoir, Va, 65-66; prof,xDept Physics,xUniv PHYSICIST, US NAVAL ORD LAB, 66- Concurrent Pos: Lectr, George Washington Univ, 66; vis asst prof, Dept Physics, Univ Wis-Milwaukee, 68-69. Mem: Am Phys Soc. Res: Superconductivity, fluctuation phenomena, theory of ultrasonic waves in superconductors; semiconductors, interactions between impurities; magnetism, theory of electron-electron interactions and their effects on susceptibility and neutron scattering experiments; Hall effect in rare-earth metals; metal-insulator transition in transition metal-oxides, especially in magnetite. Mailing Add: US Naval Ord Lab 212 Silver Spring MD 20910

CULLEN, JOHN KNOX, b Denver, Colo, Jan 2, 36; m 57; c 2. PHYSIOLOGY, ELECTRICAL ENGINEERING. Educ: Univ Md, BS, 60, MS, 69; La State Med Ctr, New Orleans, PhD, 75. Prof Exp: Engr, NIH, 60-62, res engr, Psychopharmacol Res Ctr, 62-64; res engr, Neurocommun Lab, Sch Med, Johns Hopkins Univ, 64-68; RES ENGR, KRESGE HEARING LAB, LA STATE UNIV MED CTR, NEW ORLEANS, 68- Mem: AAAS; NY Acad Sci; Inst Elec & Electronics Engrs. Res: Biomedical research; speech, hearing physiology; animal communications. Mailing Add: Bldg 164 La State Univ Med Ctr New Orleans LA 70119

CULLEN, MARION PERMILLA, b Pittsburgh, Pa. BIOCHEMISTRY, NUTRITION. Educ: Pa State Univ, BS, 54, MS, 56; Univ Calif, Berkeley, PhD(nutrit, biochem), 68. Prof Exp: Asst animal nutrit, Am Meat Inst Found, Univ Chicago, 56-57, asst biochemist, 57-60; res asst nutrit, Univ Calif, Berkeley, 62-68, res biochemist vitamin deficiency, 68-69; SR RES SCIENTIST BLOOD RESOURCES, DIV LABS & RES, NY STATE DEPT HEALTH, 71- Concurrent Pos: Fel surg, Sch Med, Johns Hopkins Univ, 69-71; assoc prof, Albany Med Col, 72- Mem: Am Chem Soc; Animal Nutrit Res Coun. Res: Energy metabolism of animal by-products; correlation of iron deficiency anemia in adolescent population with food consumption; pantothenic acid deficiency; serum protein isolation, in vitro labeling of serum proteins, and turnover rate studies; isolation of serum born virus; antigen and antibody assay techniques. Mailing Add: 103 Patroon Dr Guilderland NY 12084

CULLEN, MARY URBAN, b Marietta, Ohio, Apr 5, 07. GENETICS. Educ: Albertus Magnus Col, AB, 30; Ohio State Univ, MA, 42; Yale Univ, PhD, 48. Prof Exp: Prof biol, Col of St Mary of the Springs, 47-51; PROF BIOL, ALBERTUS MAGNUS COL, 51-, CHMN DEPT, 55- Mem: AAAS; Bot Soc Am; Genetics Soc Am; Am Soc Zoologists. Res: Developmental genetics. Mailing Add: Dept of Biol Albertus Magnus Col New Haven CT 06511

CULLEN, MICHAEL ROBERT, mathematics, see 12th edition

CULLEN, STUART CHESTER, b Milton Junction, Wis, Jan 31, 09; m 32; c 2. MEDICINE. Educ: Univ Wis, BS, 31, MD, 33. Prof Exp: Rotating intern, Multnomah Hosp, Ore, 33-34; surg intern, State Wis Gen Hosp, 34-35; res anesthesiol, Bellevue Hosp, New York, 36-38; from asst prof & anesthesiol to prof surg, Univ Iowa, 38-58, chmn div anesthesiol, 38-58; chmn dept, 58-66, prof, 58-73, assoc dean, Sch Med, 63-66, dean, 66-70, EMER PROF ANESTHESIA, SCH MED, UNIV CALIF, SAN FRANCISCO, 73- Concurrent Pos: Mem, Med Mission, Unitarian Serv Comn & WHO, Austrian Med Schs, 47, India, 53; sr instr, WHO, Denmark, 50, 52, 54; assoc ed, J Am Soc Anesthesiol. Honors & Awards: Distinguished Serv Award, Am Soc Anesthesiol, 64; Fel, Fac Anesthetists, Royal Col Surgeons, 75. Mem: Am Soc Anesthesiol (vpres, 47-49); Soc Exp Biol & Med; Am Soc Pharmacol & Exp Therapeut; AMA; hon mem, Danish Soc Anesthesiol. Res: Pharmacology and physiology as they apply to anesthesia. Mailing Add: 73 W Shore Rd Belvedere CA 94920

CULLEN, THEODORE JOHN, b St Louis, Mo, Dec 19, 28; m 47; c 4. MATHEMATICS. Educ: DePaul Univ, BS, 55, MS, 56. Prof Exp: Asst math, Univ Ill, 56-57; asst prof, Ariz State Col, 57-59 & Los Angeles State Col, 59-68; ASSOC PROF CALIF STATE POLYTECH UNIV, 68- Concurrent Pos: Engr, Jet Propulsion Lab, Pasadena, 62- Mem: Math Asn Am. Res: Abstract algebra; hilbert and banach spaces; set theory and point set toplogy; functional analysis; numerical analysis; computers. Mailing Add: Dept of Math Calif State Polytech Univ Pomona CA 91766

CULLEN, THOMAS L, physics, see 12th edition

CULLEN, WILLIAM CHARLES, b Buffalo, NY, Nov 6, 19; m 52; c 3. ORGANIC CHEMISTRY. Educ: Canisius Col, BS, 48. Prof Exp: Chemist, Bldg Res Div, Inst Appl Tech, 48-67, sect chief, 67-73, DEP CHIEF STRUCT, MAT & SAFETY DIV, NAT BUR STAND, 73- Concurrent Pos: US Dept Com sci fel, 65-66. Honors & Awards: J A Piper Award, Nat Roofing Contractors Asn, 74; Award of Merit, Am Soc Testing & Mat, 74. Mem: Am Soc Testing & Mat. Res: Mechanisms of chemical degradation and physical deterioration of organic roofing materials; methods to determine engineering properties of roof systems; durability and performance of building materials. Mailing Add: B368 Bldg Res Nat Bur of Stand Washington DC 20234

CULLEN, WILLIAM ROBERT, b Dunedin, NZ, May 4, 33; m 56; c 3. INORGANIC CHEMISTRY. Educ: Univ Otago, NZ, BSc, 55, MSc, 57; Cambridge Univ, PhD(chem), 59. Prof Exp: From instr to assoc prof, 58-69, PROF CHEM, UNIV BC, 69- Concurrent Pos: Nat Res Coun Can fel, 66-67. Mem: Am Chem Soc; Chem Inst Can. Res: Organometallic chemistry of the main group elements, especially fluorocarbon derivatives; coordination chemistry of arsines and phosphines; conformational problems in coordination chemistry. Mailing Add: Dept of Chem Univ of BC Vancouver BC Can

CULLER, VAUGHN EDGAR, b Martinsburg, WVa, Oct 7, 27; m 49; c 2. PHYSICS. Educ: Univ WVa, AB & MS, 49; Harvard Univ, PhD(physics), 57. Prof Exp: Jr physicist, Argonne Nat Lab, 50; asst, Harvard Univ, 51-55; res assoc, Corning Glass Works, 55-67; RES ASSOC, LAWRENCE LIVERMORE LAB, 67- Concurrent Pos: Staff mem, Lincoln Lab, Mass Inst Technol, 54; mem oper comt, Ind Reactor Labs, 63- Mem: Am Phys Soc; Am Nuclear Soc. Res: Effects of radiation on glass and ceramics; effects of radiation on electrical properties of dielectric materials; radiation shielding. Mailing Add: Lawrence Livermore Lab Radiation-L 531 Box 808 Livermore CA 94550

CULLERS, ROBERT LEE, b North Manchester, Ind, May 19, 37; m 70; c 2. GEOCHEMISTRY. Educ: Ind Univ, Bloomington, BS, 59, MA, 62; Univ Wis-Madison, PhD(geochem), 71. Prof Exp: Teacher math, Vevay Town Schs, Ind, 60-61; teacher math-chem, Goshen City Schs, 61-63; teacher chem, Park Ridge, Ill, 64-67; res asst, Univ Wis-Madison, 67-69, NSF fel geochem, 69-71; ASST PROF GEOCHEM, KANS STATE UNIV, MANHATTAN, 71- Mem: Geochem Soc Am; Am Chem Soc; Geol Soc Am. Res: Trace element geochemistry; experimental igneous and metamorphic trace element partitioning; neutron activation analysis of igneous, metamorphic and sedimentary rocks. Mailing Add: Dept of Geol Kans State Univ Manhattan KS 66506

CULLEY, BENJAMIN HAYS, b Hollywood, Calif, Oct 16, 13. MATHEMATICS. Educ: Univ Southern Calif, AB, 34, MS, 36, EdD, 49. Prof Exp: Teacher high sch, Calif, 36-43; from instr to asst prof math, 43-59, assoc prof, 60-66, dean men, 44-69, PROF MATH, OCCIDENTAL COL, 66-, ASSOC DEAN STUDENTS, 69- Res: Statistics in physical and social sciences. Mailing Add: Dept of Math Occidental Col Los Angeles CA 90041

CULLEY, DUDLEY DEAN, JR, b Jackson, Miss, May 14, 37. ZOOLOGY. Educ: Millsaps Col, BS, 59; Univ Miss, MEd, 61; Miss State Univ, MS, 64, PhD(zool), 68. Prof Exp: Teacher, Univ High Sch, Miss, 59-61; teacher-counsr pub schs, 61-63; asst prof fisheries, pollution biol & aquatic ecol, 68-74, ASSOC PROF FORESTRY & WILDLIFE MGT, LA STATE UNIV, BATON ROUGE, 74- Mem: Am Fisheries Soc; Am Chem Soc; Ecol Soc Am; Wildlife Soc. Res: Pesticide effects on wildlife; productivity in lakes; pollution surveys; waterfowl restoration; aquatic plant studies; development of laboratory amphibians. Mailing Add: Dept of Forestry & Wildlife Mgt La State Univ Baton Rouge LA 70803

CULLEY, WILLIAM JAMES, b Peoria, Ill, Nov 13, 28; m 59; c 4. BIOCHEMISTRY. Educ: Bradley Univ, BS, 53; Purdue Univ, MS, 57, PhD(biochem), 59. Prof Exp: DIR MENT RETARDATION RES LAB, MUSCATATUCK STATE HOSP, 59- Mem: AAAS; Am Chem Soc; Am Asn Ment Deficiency; Fedn Am Soc Exp Biol. Res: Biochemical aspects of mental retardation; neurochemistry; nutrition. Mailing Add: Ment Retardation Res Lab Muscatatuck State Hosp Butlerville IN 47223

CULLIMORE, DENIS ROY, b Oxford, Eng, Apr 7, 36; m 62. MICROBIAL ECOLOGY, BACTERIOLOGY. Educ: Univ Nottingham, BSc, 59, PhD(agr microbiol), 62. Prof Exp: Lectr microbiol, Univ Surrey, 62-68; from asst prof to assoc prof, 68-74, PROF MICROBIOL, UNIV REGINA, 74- Mem: Can Soc Microbiol;

Brit Soc Appl Bact. Res: Bioassay systems using algae; effects of pollution on soil and water microflora; simplified classification of bacteria; novel uses of microorganisms. Mailing Add: Dept of Biol Univ of Regina Regina SK Can

CULLISON, ARTHUR EDISON, b Lawrence Co, Ill, Oct 30, 14; m 39; c 3. ANIMAL NUTRITION. Educ: Univ Ill, BS, 36, MS, 37, PhD(animal husb), 48. Prof Exp: Asst animal husb, Univ Ill, 36-39; asst prof, Miss State Col, 39-43; prof, Ala Polytech Inst, 46-48; head dept, 48-58, PROF ANIMAL HUSB, UNIV GA, 58- Mem: Fel AAAS; Am Soc Animal Sci. Res: Ruminant nutrition; silage preservation; by-products for livestock feeding; recycling animal wastes. Mailing Add: Dept of Animal & Dairy Sci Univ of Ga Athens GA 30601

CULLISON, DAVID ARTHUR, b Auburn, Ala, Sept 29, 46; m 67; c 2. MEDICINAL CHEMISTRY. Educ: Univ Ga, BS, 68, PhD(org chem), 72. Prof Exp: NIH fel chem, Univ Chicago, 72-74; RES INVESTR CHEM, E R SQUIBB & SONS INC, 74- Mem: Am Chem Soc; The Chem Soc. Res: Synthesis and study of structure-activity relationships of heterocyclic and carbocyclic psychopharmacological agents. Mailing Add: Org Chem Dept PO Box 4000 E R Squibb & Sons Inc Princeton NJ 08540

CULLISON, JAMES SHELLEY, b Lawrence Co, Ill, July 22, 06; m 31; c 2. GEOLOGY. Educ: Univ Ill, AB, 28; Univ Mo, MS, 30; Yale Univ, PhD(stratig), 42. Prof Exp: Instr geol, Mo Sch Mines, 30-35, asst prof, 35-45; sr geologist, asst chief paleontologist & asst supt lab, Creole Petrol Corp, 45-50; prof geol & head dept, Fla State Univ, 50-54; chief subsurface & res geologist, Sahara Petrol Co, Continental Oil Co, 54-57, chief geologist, 57-62, sr geologist, Tex, 62-68, NJ, 68; consult, paleont & stratig, Oasis Oil Co, Libya, 58-; assoc geologist, Tenn Valley Authority, 34-35; assoc & geologist, US Geol Surv, 41-45. Mem: Fel Geol Soc Am; fel Paleont Soc; Am Asn Petrol Geologists. Res: Non-metallic ore deposits; micropaleontology; field geologic mapping; marine geology. Mailing Add: RFD 1 Box 254 Monticello FL 32344

CULLMANN, RALPH E, b West Salem, Wis, July 9, 17. INORGANIC CHEMISTRY. Educ: Wis State Col, La Crosse, BS, 40; Columbia Univ, MA, 49, EdD(sci educ), 61. Prof Exp: Instr chem, Springfield Jr Col, 46-48; part-time instr sci educ, Teachers Col, Columbia Univ, 48-50; asst prof chem & phys sci, Western Wash Col Educ, 50-53; asst chem, Robert Col, Istanbul, 53-56; part-time instr sci educ, Teachers Col, Columbia Univ, 56-57; asst phys sci, Wis State Col, Eau Claire, 57-59; prof, Goddard Col, 59-63; assoc prof chem, 63-70, PROF CHEM-PHYSICS, KEAN COL NJ, 70-, CHMN DEPT, 73- Mem: AAAS; Nat Sci Teachers Asn; Brit Asn Sci Educ. Res: General and analytical chemistry; quantitative analysis. Mailing Add: Dept of Chem-Physics Kean Col of NJ Union NJ 07083

CULLUM, JANE KEHOE, b Norfolk, Va, Sept 17, 38; m 59; c 2. APPLIED MATHEMATICS. Educ: Va Polytech Inst, BS, 60, MS, 62; Univ Calif, Berkeley, PhD(appl math), 66. Prof Exp: Res staff mem math sci, 66-70, tech asst to dir res, 75, RES STAFF MEM MATH SCI, T J WATSON RES CTR, IBM CORP, 72- Concurrent Pos: Mem adv panel elec sci & anal sect engr div, NSF, 75-76. Mem: Soc Indust & Appl Math (secy, 72-); Am Math Soc; Math Asn Am; AAAS. Res: Analysis and development of algorithms for optimization including mathematical programming and optimal control, and for numerical algebra including numerical differentiation and the solution of large, eigenelement computations. Mailing Add: T J Watson Res Ctr IBM Corp Yorktown Heights NY 10598

CULLUMBINE, HARRY, b Eng, Dec 29, 12; m 59; c 1. PHARMACOLOGY. Educ: Univ Sheffield, BSc, 33, MSc, 34, MB & ChB, 37, MD, 45. Prof Exp: Prof & head dept physiol & pharmacol, Univ Ceylon, 47-51; chief med officer res, Chem Defense Exp Estab, Eng, 52-56; prof & head dept pharmacol, Univ Toronto, 56-58; pres, Air Shields, Inc, 58-67; VPRES & CORP MED DIR, NARCO SCI INDUST, 67- Concurrent Pos: Lectr, Grad Sch Med, Univ Pa, 58- Mem: AAAS; Am Soc Pharmacol & Exp Therapeut; Can Physiol Soc; Can Pharmacol Soc; Brit Physiol Soc. Res: Anticholinergic; anticholinesterase agents; toxicology; atmospheric pollution; physiology of respiratory mechanisms. Mailing Add: NARCO Sci Indust Ft Wash Industrial Park Ft Washington PA 19034

CULNAN, ROBERT NEVILLE, b Riverside, Calif, July 10, 15; m 41; c 3. METEOROLOGY. Educ: Univ Calif, AB, 37; NY Univ, MS, 41. Prof Exp: Meteorologist, US Weather Bur, 39-40; instr, NY Univ, 40-42, asst prof & exec secy, Dept Meteorol, 43-46; meteorologist, US Weather Bur, 45-62, exec officer, Off Meteorol Res, 62-65, liaison officer, 65-71, DEP DIR OFF PROGS, ENVIRON RES LABS, NAT OCEANIC & ATMOSPHERIC ADMIN, 71- Mem: Am Geophys Union; Am Meteorol Soc. Res: Administration of research. Mailing Add: Nat Oceanic & Atmospheric Admin Boulder CO 80302

CULO, DAVID ALBERT, b Sunbury, Pa, Oct 19, 19; m 43. MEDICINE. Educ: Bucknell Univ, BS, 41; Jefferson Med Col, MD, 44; Am Bd Urol, dipl. Prof Exp: Intern, Geiginger Mem Hosp, 44-45, resident urol, 45-46; resident, Watts Hosp, 48-50; asst, 50-51, from instr to assoc prof, 51-61, PROF UROL, HOSP & COL MED, UNIV IOWA, 61-, CHMN DEPT, 70- Mem: AMA; fel Am Col Surgeons; Am Urol Asn; Am Asn Genito-Urinary Surgeons; Am Clin Soc Genito-Urinary Surg. Res: Clinical use of radioactive gold in treatment of carcinoma of the prostate gland. Mailing Add: Dept of Urol Univ of Iowa Col of Med Iowa City IA 52240

CULP, FREDERICK LYNN, b Duquesne, Pa, May 12, 27; m 53; c 2. PHYSICS. Educ: Carnegie Inst Technol, BS, 49, MS, 60; Vanderbilt Univ, PhD, 66. Prof Exp: Asst, US Steel Res Lab, 50-51; electronics engr, Pratt & Whitney Air Craft Corp, 52-54; res physicist, Stand Piezo Co, Pa, 54-55 & Westinghouse Lab, 55-56; asst elec & magnetism, Carnegie Inst Technol, 57-59, asst shaped charges & hyperballistics, 57-59; from asst prof to assoc prof, 59-64, PROF PHYSICS & CHMN DEPT, TENN TECHNOL UNIV, 64- Mem: AAAS; Am Asn Physics Teachers; Am Phys Soc. Res: Physics of fluids; electronic instrumentation; piezoelectricity; photoconductivity; shaped charges and hyperballistics; exploding wires; relaxation times in gases. Mailing Add: Dept of Physics Tenn Technol Univ Cookeville TN 38501

CULP, LLOYD ANTHONY, b Elkhart, Ind, Dec 23, 42; m 65; c 2. VIROLOGY, CELL BIOLOGY. Educ: Case Inst Technol, BS, 64; Mass Inst Technol, PhD(biochem), 69. Prof Exp: Fel virol, Harvard Med Sch & Mass Gen Hosp, 69-71; ASST PROF MICROBIOL, SCH MED, CASE WESTERN RESERVE UNIV, 72- Mem: AAAS; Am Soc Microbiol; Sigma Xi. Res: Study of the molecular mechanism of substrate adhesion of normal growth-controlled mammalian cells and possible alteration after transformation to a malignant state by oncogenic viruses. Mailing Add: Dept of Microbiol Sch of Med Case Western Reserve Univ Cleveland OH 44106

CULP, ORMOND S, b Toronto, Ohio, Nov 18, 10; m 38; c 1. UROLOGY. Educ: Ohio Wesleyan Univ, AB, 31; Johns Hopkins Univ, MD, 35; Am Bd Urol, dipl. Hon Degrees: DSc, Ohio Wesleyan Univ, 63. Prof Exp: Intern, Johns Hopkins Univ Hosp, 35-36, from asst resident to resident urol, 38-42, instr, Univ, 38-42; instr path, McGill Univ, 36-37; resident physician, St Mary's Hosp, Pierre, SDak, 37-38 & urol, Ancker Hosp, St Paul, Minn, 40-41; assoc surgeon-in-chg, Henry Ford Hosp, Detroit, 42-50; from assoc prof to prof, 50-72, head sect, 62-72, EMER PROF UROL, UNIV MINN, 72- Concurrent Pos: Consult, US Vet Admin, 46-52 & Mayo Clin, 50-75; mem emer staff, Mayo Found, 76-; mem, Urol Forum Clin Invest. Mem: Am Urol Asn; Am Asn Genito-Urinary Surg; Am Clin Soc Genito-Urinary Surg; fel AMA; fel Am Col Surgeons. Mailing Add: Mayo Clinic 200 First St SW Rochester MN 55901

CULP, TOM W, biochemistry, see 12th edition

CULPEPPER, GIDEON ALSTON, b Denver, Colo, Dec 12, 18; m 45; c 2. MATHEMATICS, STATISTICS. Educ: Univ Colo, BA, 47, MA, 48. Prof Exp: Mathematician, 50-59, math statistician, Methodology & Instrumentation Div, 59-66, MATH STATISTICIAN, RELIABILITY, AVAILABILITY & MAINTAINABILITY DIV, ARMY MISSILE TEST & EVAL DIRECTORATE, WHITE SANDS MISSILE RANGE, 66- Res: Reliability; failure time analysis; safety analysis. Mailing Add: Box 14 Mesilla Park NM 88047

CULPEPPER, THOMAS JAMES, biological oceanography, marine ecology, see 12th edition

CULSHAW, WILLIAM, physics, mathematics, see 12th edition

CULVAHOUSE, JACK WAYNE, b Mt Park, Okla, Sept 15, 29; m 52; c 3. MAGNETISM. Educ: Univ Okla, BS, 51, AM, 54; Harvard Univ, PhD(physics), 58. Prof Exp: Physicist, Gen Elec Co, 51-53; asst prof physics, Univ Okla, 57-58; from asst prof to assoc prof, 58-64, PROF PHYSICS, UNIV KANS, 64- Concurrent Pos: Consult, Hycon Eastern Inc, 58-59; Guggenheim fel, 68-69. Mem: Fel Am Phys Soc. Res: Low temperature physics; magnetic resonance; solid state physics. Mailing Add: Dept of Physics & Astron Univ of Kans Lawrence KS 66045

CULVER, DAVID CLAIR, b Waverly, Iowa, Sept 23, 44; m 68. ECOLOGY. Educ: Grinnell Col, BA, 66; Yale Univ, PhD(biol), 70. Prof Exp: Fel pop biol, Univ Chicago, 70-71; asst prof, 71-76, ASSOC PROF BIOL, NORTHWESTERN UNIV, EVANSTON, 76- Concurrent Pos: Vis assoc prof human ecol, Harvard Sch Pub Health, 75. Mem: Nat Speleol Soc. Res: Cave biogeography; competition in discontinuous environments; control of species diversity; competition in ants; theoretical ecology. Mailing Add: Dept of Biol Sci Northwestern Univ Evanston IL 60201

CULVER, JAMES F, b Macon, Ga, June 10, 21; m 47. OPHTHALMOLOGY. Educ: Univ Ga, MD, 45; Am Bd Ophthal, dipl, 53. Prof Exp: Clin asst, Med Sch, Northwestern Univ, 52; pvt pract, 53-59; chief, Ophthal Br, US Air Force Sch Aerospace Med, 59-66, asst dir res & develop, Aerospace Med Div, Brooks AFB, Tex, 66-69, CHIEF MED RES GROUP, OFF SURGEON GEN, US AIR FORCE, 69- Concurrent Pos: Fel ophthal, Wesley Mem & Passavant Mem Hosps, Chicago, 51-52; mem, Consult Group & Med Debriefing Team, Projs Mercury & Gemini, 61-, Exec Coun, Armed Forces-Nat Res Coun Comt Vision, 62- & Vision Comt, Adv Group Aerospace Res & Develop, Aerospace Med Panel, NATO, 64- Honors & Awards: Tuttle Award, Aerospace Med Asn, 66. Mem: AMA; Aerospace Med Asn; Am Acad Ophthal & Otolaryngol. Res: Medicine and surgery; aerospace medicine and ophthalmology. Mailing Add: Command Surgeon Hq Pac Air Forces APO San Francisco CA 96553

CULVER, ROGER BRUCE, b Brigham City, Utah, Sept 6, 40; m 65; c 3. ASTRONOMY. Educ: Univ Calif, Riverside, BA, 62; Ohio State Univ, MSc, 66, PhD(astron), 71. Prof Exp: Instr astron & math, 66-70; ASSOC PROF ASTRON, COLO STATE UNIV, 70- Concurrent Pos: Mem user's comt, Kitt Peak Nat Observ, 74-76. Mem: Am Astron Soc. Res: Physical properties of cool stars having unusual chemical compositions. Mailing Add: Dept of Physics Colo State Univ Ft Collins CO 80523

CULVER, WILLIAM HOWARD, b Eau Claire, Wis, Feb 17, 27; m 59; c 3. PHYSICS. Educ: Mass Inst Technol, BS, 50; Univ Calif, Los Angeles, MS, 54, PhD, 61. Prof Exp: Res asst, Mass Inst Technol, 50-52 & Scripps Inst Oceanog, Univ Calif, 52-54; physicist, Rand Corp, 55-61 & Inst Defense Anal, 61-66; mgr, Quantum Electronics Dept, IBM Corp, 66-72, corp laser strategist, 67-69; PRES, OPTELECOM INC, GAITHERSBURG, MD, 72- Concurrent Pos: Mem, Spec Group Optical Masers, Dept Defense, 62-66; mem interdept comt atmospheric sci, Dept Com, 64-65; Nat Acad Sci adv comt to Nat Bur Standards Cent Radio Propagation Lab, 65; mem, NASA Res & Technol Adv Comt on Commun & Tracking, 67-69. Mem: Optical Soc Am; Am Phys Soc; Inst Elec & Electronics Engrs. Res: Quantum electronics; optical and microwave spectroscopy; applications of lasers to communication systems. Mailing Add: 2841 Chesapeake St NW Washington DC 20008

CUMBERBATCH, ELLIS, b Eng, Apr 19, 34; m 57; c 4. APPLIED MATHEMATICS. Educ: Univ Manchester, BSc, 55, PhD(appl math), 58. Prof Exp: Res fel appl mech, Calif Inst Technol, 58-60; res assoc, Courant Inst Math Sci, NY Univ, 60-61; lectr, Univ Leeds, 61-64; assoc prof, 64-68, PROF MATH, PURDUE UNIV, WEST LAFAYETTE, 68- Res: Fluid dynamics; mathematics. Mailing Add: Dept of Math Purdue Univ West Lafayette IN 47907

CUMBIE, BILLY GLENN, b Dickens, Tex, Mar 21, 30; m 51; c 3. BOTANY. Educ: Tex Tech Col, BS, 51, MS, 52; Univ Tex, PhD(bot), 60. Prof Exp: Spec instr bot, Univ Tex, 57-58; from instr to asst prof, Tex Tech Col, 58-61; ASST PROF BOT, UNIV MO-COLUMBIA, 61- Mem: Bot Soc Am; Int Asn Wood Anat. Res: Development of the vascular cambium and xylem in dicotyledons. Mailing Add: Div of Biol Sci Univ of Mo Columbia MO 65201

CUMMEROW, ROBERT LEGGETT, b Toledo, Ohio, Jan 7, 15; m 48; c 2. PHYSICS. Educ: Univ Toledo, BE, 38; Univ Pittsburgh, MS, 40, PhD(physics), 47. Prof Exp: Asst physics, Univ Pittsburgh, 38-42; spec res assoc underwater sound lab, Harvard Univ, 42-45; mem staff, US Navy Underwater Sound Lab, Conn, 45-46; res assoc, Knolls Atomic Power Lab, Gen Elec Co, 48-55; group leader solid state physics, Nat Carbon Res Labs, 55-63; group leader, Union Carbide Res Inst, 63-70, SR SCIENTIST, TARRYTOWN TECH CTR, UNION CARBIDE CORP, 70- Mem: Am Phys Soc. Res: Elastic and damping constants of metals and alloys; underwater sound detection equipment; paramagnetic resonance absorption in salts of the iron group; radiation-induced property changes in semi-conductors and metals; photo-effects in semiconductors; plastic behavior of semiconductors and refractory materials; superconductivity. Mailing Add: Tarrytown Tech Ctr Union Carbide Corp Tarrytown NY 10591

CUMMIN, ALFRED SAMUEL, b London, Eng, Sept 5, 24; nat US; m 45; c 1. PHYSICAL CHEMISTRY. Educ: Polytech Inst Brooklyn, BS, 43, PhD, 46; Univ Buffalo, MBA, 59. Prof Exp: Res chemist, Substitute Alloy Mats Lab, Manhattan Proj, Columbia Univ, 43-44; plant supvr & head res, Metal & Plastic Processing Co, 46-51; res chemist gen chem div, Allied Chem & Dye Corp, 51-53; sr chemist, Congoleum Nairn, 53-54; prof math & sci, US Merchant Marine Acad, 54; capacitor div, Gen Elec Co, 54-56, supvr dielectric adv develop, 54-56; mgr indust prof res dept, Spencer Kellog & Sons, Inc, 56-59; mgr plastics div, Trancoa Chem Corp, 59-62;

assoc dir, Prod Develop & Serv Labs, Chem Div, Merck & Co, Inc, NJ, 62-69; dir prod develop, Chem Div, Borden Co, 69-72, tech dir, Borden Chem Div, Borden Inc, NY, 72-73, CORPORATE TECH DIR, BORDEN INC, NY, 73- Concurrent Pos: Instr, Polytech Inst Brooklyn, 46-47; asst prof, Adelphi Col, 52-54; adj prof mgt, Mgt Inst, NY Univ, 68- Honors & Awards: Roon Cert Award, Fedn Socs Paint Technol, 65. Mem: Am Chem Soc; Inst Food Technol; Fedn Coatings Technol; Am Soc Tes Mat; Nutrit Found. Res: Polymers; electrochemistry; food packaging; colloid chemistry; preservatives; agricultural chemicals; dielectrics and insulating materials; paints; adhesives; surgical adhesives; nutrition; industrial hygiene; occupational health. Mailing Add: Borden Inc 277 Park Ave New York NY 10017

CUMMING, BRUCE GORDON, b London, Eng, Oct 12, 25; Can citizen; m 69. PLANT PHYSIOLOGY. Educ: Univ Reading, BS, 52; McGill Univ, PhD(plant physiol, agron), 56. Prof Exp: Res off, Plant Res Inst, Can Dept Agr, 57-65; prof bot, Western Ont Univ, 65-71; chmn dept, 71-74, PROF BIOL, UNIV NB, 71- Concurrent Pos: Mem exec comt, Can Photobiol Group; mem Can Nat Comt, Int Union Biol Sci; del gen assembly, Int Union Biol Sci, 70 & 73. Mem: Can Soc Plant Physiol (pres, 69-70). Res: Physiology; photobiology; photoperiodism; endogenous rhythms and phytochrome, particularly in germination and flowering; tissue culture and morphogenesis. Mailing Add: Dept of Biol Univ of NB Fredericton NB Can

CUMMING, CAMERON, experimental physics, see 12th edition

CUMMING, JAMES BURTON, b Jamaica, NY, June 6, 28; m 53; c 3. NUCLEAR CHEMISTRY. Educ: Yale Univ, BS, 49; Columbia Univ, MA, 51, PhD(nuclear chem), 54. Prof Exp: Res assoc, 54-55, from assoc chemist to chemist, 55-69, SR CHEMIST, BROOKHAVEN NAT LAB, 69- Mem: Am Phys Soc; Am Chem Soc. Res: Nuclear reactions at high energies; nuclear decay schemes. Mailing Add: Chem Dept Brookhaven Nat Lab Upton NY 11973

CUMMING, LESLIE MERRILL, b Joggins, NS, Nov 5, 25; m 58; c 3. GEOLOGY. Educ: Univ NB, BSc, 48, MSc, 51; Univ Wis, PhD(geol), 55. Prof Exp: Asst to prov geologist, NB Dept Mines, 44-46; field asst, 47-49, tech officer, 49-54, geologist, 55-67, RES SCIENTIST, GEOL SURV CAN, 67- Mem: Am Paleont Soc; fel Geol Asn Can; fel Royal Can Geog Soc; Brit Palaeontolograph Soc. Res: Regional Paleozoic geology of the Appalachians; regional geology of the Hudson Bay lowlands; Paleozoic faunas; graptolite morphology. Mailing Add: Geol Surv of Can 601 Booth St Ottawa ON Can

CUMMINGS, CHARLES SUMNER, II, b Rochester, NY, Aug 28, 14; m 44; c 3. PHYSICS. Educ: Princeton Univ, AB, 36, MA, 38, PhD(physics), 40. Prof Exp: Res physicist, Remington Arms Co, 40-41, asst chief ballistic engr, 41, chief ballistic engr, 41-45, supvr ballistics res, 45-48, supvr physics & ballistics res, 48-55, supvr fundamental res, 55-61; mgr systs eval, Missile Test Proj, Radio Corp Am, 61-62, systs anal, 62-66, sr mem tech staff, Astroelectronics Div, RCA Corp, 66-74; RETIRED Concurrent Pos: Instr, Bridgeport Eng Inst, 49-51, lectr, 52-55; dean grad sch, Brevard Eng Col, 61- Mem: Inst Elec & Electronics Eng; Inst Mgt Sci. Res: Technical management; aerospace engineering; system analysis; ballistics; statistics; electrochemistry; internal combustion engines; operations research; welding; computers; high speed photography; academic administration. Mailing Add: PO Box 376 Enfield NH 03748

CUMMINGS, DAVID, b New York, NY, Feb 11, 32. GEOLOGY. Educ: City Col New York, BS, 57; Univ Tenn, MS, 59; Mich State Univ, PhD(geol), 62. Prof Exp: Mineralogist & petrologist, Electrotech Res Lab, US Bur Mines, 59; geologist, US Geol Surv, 62-68; PROF GEOL, OCCIDENTAL COL, 68- Concurrent Pos: Consult, US Geol Surv, 68-75; independent consult, 70- Mem: Am Geophys Union; Geol Soc Am. Res: Theoretical and applied mechanics used to solve structural geologic problems, engineering geologic problems. Mailing Add: Dept of Geol Occidental Col Los Angeles CA 90041

CUMMINGS, DENNIS PAUL, b Yonkers, NY, Apr 19, 40. MICROBIOLOGY. Educ: Manhattan Col, BS, 61; St John's Univ, NY, MS, 63, PhD(microbiol), 68. Prof Exp: Asst microbiol, St John's Univ, NY, 62-63; res microbiologist, 68-72, SR RES SCIENTIST, MILES LABS, INC, 72- Mem: Am Soc Microbiol; Soc Indust Microbiol. Res: Medical microbiology; antifungal and antibacterial agents; analytical microbiology; microbiological aspects of pharmaceutical development. Mailing Add: Dome Labs Div of Miles Labs Inc 400 Morgan Lane West Haven CT 06516

CUMMINGS, DONALD JOSEPH, b Staten Island, NY, Mar 4, 30; m 58; c 3. BIOPHYSICS. Educ: George Washington Univ, BS, 55; Univ Chicago, MS, 57, PhD(biophys), 59. Prof Exp: Biophysicist, NIH, 53-55; USPHS fel, Copenhagen Univ, 59-60; res physicist, Nat Inst Neurol Dis & Blindness, 60-64; from asst prof to assoc prof, 64-71, PROF MICROBIOL, MED SCH, UNIV COLO, DENVER, 71- Res: Effect of canavanine on head morphogenesis in T-even bacteriophages of Entameba coli; replication and function of mitochondria from Paramecium; biochemistry and morphology of aging in Paramecium. Mailing Add: Dept of Microbiol Univ of Colo Med Sch Denver CO 80220

CUMMINGS, EDMUND GEORGE, b Albany, NY, Aug 2, 28; m 55; c 2. PHYSIOLOGY. Educ: Union Col, BS, 50; NC State Col, MS, 53, PhD(ecol), 55. Prof Exp: Asst zool, NC State Col, 51-55, instr, Exten Sch, 54; instr, Duke Univ, 55-56; PHYSIOLOGIST & CHIEF RESPIRATORY SECT, EDGEWOOD ARSENAL, 56- Concurrent Pos: Asst prof, Harford Col, 59-62, assoc prof, 67-69; asst prof, Univ Md Exten Sch, 64- Mem: AAAS; Am Soc Zool; Am Physiol Soc. Res: Culture of slime molds; food habits of game birds; osmoregulation in fish; respiratory and exercise physiology; thermoregulation and anticholinergics; environment and skin penetration. Mailing Add: Biomed Labs Med Physiol Edgewood Arsenal MD 21010

CUMMINGS, FREDERICK W, b New Orleans, La, Nov 21, 31. THEORETICAL PHYSICS, MATHEMATICS. Educ: La State Univ, BS, 55; Stanford Univ, PhD(physics), 60. Prof Exp: Res scientist theoret physics, Aeronutronic Div, Philco Corp, Calif, 60-63; from asst prof to assoc prof, 63-74, PROF PHYSICS, UNIV CALIF, RIVERSIDE, 74- Concurrent Pos: Consult, Aeronutronic Div, Philco Corp, 64-66; ed, Coop Phenomena. Mem: AAAS; Am Phys Soc; Am Sci Teachers Asn. Res: Coherence in radiation; solid state; many particles. Mailing Add: Dept of Physics Univ of Calif Riverside CA 92507

CUMMINGS, GEORGE AUGUST, b Cortland, Ind, Dec 17, 27; m 53; c 3. SOIL SCIENCE. Educ: Purdue Univ, BS, 51, MS, 57, PhD(agron), 61. Prof Exp: Teacher high sch, Ind, 51-58; asst prof, 61-65, ASSOC PROF SOIL FERTIL, NC STATE UNIV, 65- Mem: Am Soc Hort Sci; Am Soc Agron; Soil Sci Soc Am. Res: Determination of effects of plant nutrition upon biochemical constituents of plants; influence of nutrition upon yield and quality of fruits; influence of animal waste upon soil plant systems; runoff and ground water. Mailing Add: Dept of Soil Sci NC State Univ Raleigh NC 27607

CUMMINGS, JEAN MARIE, b Cleveland, Ohio, Apr 1, 19. BIOLOGY. Educ: Western Reserve Univ, BA, 42, MA, 43; Smith Col, PhD(cytol), 47. Prof Exp: Lab asst bot & bact, Western Reserve Univ, 41-42, from instr to asst prof, 46-57; assoc prof, 57-62, PROF BOT, JOHN CARROLL UNIV, 63- Concurrent Pos: Prof, Smith Col, 55. Mem: Bot Soc Am. Res: Cytology of Datura; effects of radiation on Datura; cytology of Chaetomium. Mailing Add: Dept of Biol John Carroll Univ University Heights OH 44118

CUMMINGS, JOHN ALBERT, b Evanston, Ill, May 3, 31; m 51; c 2. RADIOBIOLOGY. Educ: Wis State Univ-Whitewater, BS, 53; Univ Wis, MS, 59; Univ Northern Colo, EdD, 66. Prof Exp: Teacher high sch, Wis, 53-61; PROF BIOL, UNIV WIS-WHITEWATER, 61- Concurrent Pos: Dir, NSF Inserv Inst Molecular Biol, 64-65 & 68-69, dir, NSF Inst Environ Sci, 71-72; AEC grant, Nuclear Sci Instrumentation for Radiation Biol Lab, Univ Wis-Whitewater, 64-65. Mem: Nat Asn Biol Teachers; Am Inst Biol Sci; Soc Syst Zool. Res: Concentration of various radioisotopes in Orconectes virilis in various intermolt stages; radioisotope accumulation by radioautographic methods in crustaceans and reptiles; crayfish population of Wisconsin. Mailing Add: 1264 Satinwood Lane Whitewater WI 53190

CUMMINGS, JOHN FRANCIS, b Newark, NJ, Sept 3, 36; m 61; c 4. VETERINARY ANATOMY, COMPARATIVE NEUROLOGY. Educ: Cornell Univ, BS, 58, DVM, 62, MS, 63, PhD(vet anat), 66. Prof Exp: Asst vet anat, 63-65, asst prof anat, 67-71, ASSOC PROF ANAT, NY STATE VET COL, CORNELL UNIV, 71- Concurrent Pos: Consult, Div Neuropsychiat, Walter Reed Army Inst Res, 75- Mem: Am Asn Anatomists; Am Asn Vet Anat; World Asn Vet Anat. Res: Neuroanatomy and neuropathology. Mailing Add: Dept of Anat NY State Vet Col Cornell Univ Ithaca NY 14850

CUMMINGS, JOHN (NELSON), animal physiology, animal breeding, see 12th edition

CUMMINGS, JOHN RHODES, b Detroit, Mich, Feb 4, 26; m 53; c 2. PHARMACOLOGY. Educ: Kalamazoo Col, BA, 50; Wayne State Univ, MS, 52, PhD(pharmacol), 54. Prof Exp: Asst prof pharmacol, Med Sch, Tufts Univ, 54-57; group leader cardiovasc pharmacol, Lederle Labs Div, Am Cyanamid Co, 57-69, head dept cardiovasc-renal pharmacol, 69-73; DIR DEPT PHARMACOL, AYERST RES LABS DIV, AM HOME PROD CORP, 73- Mem: Am Soc Pharmacol & Exp Therapeut; Soc Exp Biol & Med; Am Heart Asn; Am Chem Soc. Res: Cardiovascular pharmacology, particularly in the fields of arrhythmia, hypertension and diuretic research. Mailing Add: Dept of Pharmacol Ayerst Labs Montreal PQ Can

CUMMINGS, JON CLARK, b Saranac Lake, NY, Apr 30, 30; m 56; c 3. GEOLOGY. Educ: Stanford Univ, BS, 52, MS, 56, PhD(geol), 60. Prof Exp: Asst prof geol, Ore State Univ, 58-64; assoc prof, 64-71, PROF GEOL, CALIF STATE UNIV, HAYWARD, 71-, CHMN DEPT EARTH SCI, 69- Concurrent Pos: Geologist, WAE, US Geol Surv, 67-71 & Portola Valley & Woodside, Calif, 71- Mem: AAAS; Geol Soc Am; Am Asn Petrol Geol; Soc Econ Paleont & Mineral; Nat Asn Geol Teachers. Res: Stratigraphy; sedimentation; California coast range geology; environmental geology. Mailing Add: Dept of Earth Sci Calif State Univ Hayward CA 94542

CUMMINGS, JOSEPH GERARD, b Dunmore, Pa, May 23, 23; m 51; c 5. PESTICIDE CHEMISTRY. Educ: Univ Scranton, BS, 50. Prof Exp: Anal chemist, Bur Mines, 51-52; research chemist, USDA, 52-55, chief staff officer, 55-61; chief chem br, Food & Drug Admin, 61-70; CHIEF CHEM BR, ENVIRON PROTECTION AGENCY, 70- Concurrent Pos: Consult, Secretariat, Comt Pesticide Residues, Food & Agr Orgn & WHO, 73-; adv pesticide chem, UN Develop Prog, Govt Thailand, 74-75. Mem: Am Chem Soc; Asn Off Anal Chemists. Mailing Add: EPA Waterside Mall 401 M St SW Washington DC 20460

CUMMINGS, KENNETH ROSS, b Leslie, Ark, Jan 8, 40; m 62. ANIMAL NUTRITION. Educ: Okla State Univ, BS, 61; Purdue Univ, MS, 65, PhD(animal nutrit), 67. Prof Exp: Asst prof, 67-70, ASSOC PROF DAIRY PROD, MISS STATE UNIV, 70- Mem: Am Soc Animal Sci; Am Dairy Sci Asn. Res: Nutrition research with dairy animals; forage utilization and preservation; utilization of moist grain; complete rations for lactating cows. Mailing Add: Drawer DD Dept of Dairy Sci Miss State Univ Mississippi State MS 39762

CUMMINGS, LARRY JEAN, b Chicago, Ill, Oct 1, 37; m 63; c 2. ALGEBRA. Educ: Roosevelt Univ, BS, 61; DePaul Univ, MS, 63; Univ BC, PhD(math), 67. Prof Exp: Teaching asst math, Univ BC, 63-67; ASST PROF MATH, UNIV WATERLOO, 67- Res: Multilinear algebra; generalized matrix functions and matrix inequalities. Mailing Add: Dept of Math Univ of Waterloo Waterloo ON Can

CUMMINGS, MARTIN MARC, b Camden, NJ, Sept 7, 20; m 42; c 3. MEDICINE. Educ: Bucknell Univ, BS, 41; Duke Univ, MD, 44. Hon Degrees: DSc, Bucknell Univ, 68; ScD, Univ Nebr, Emory Univ & Georgetown Univ, 71. Prof Exp: Med intern, Boston Marine Hosp, USPHS, 44, asst resident med, 45, med officer, Tuberc Div, 46-47, dir, Tuberc Eval Lab, Commun Dis Ctr, 47-49; dir, Tuberc Res Lab, Vet Admin Hosp, Atlanta, Ga, 49-53; dir res serv, US Vet Admin, 53-59; prof microbiol & chief dept, Univ Okla, 59-61; chief off int res, NIH, 61-63, assoc dir res grant, 63; DIR, NAT LIBR MED, USPHS, 64- Concurrent Pos: Asst prof & assoc prof, Sch Med, Emory Univ, 51-53; prof lectr, Sch Med, George Washington Univ, 53-54. Mem: Am Fedn Clin Res; Am Soc Clin Invest; Am Clin & Climat Asn; Am Acad Microbiol; Asn Am Med Cols. Res: Laboratory diagnosis of tuberculosis and experimental methods; epidemiology of sarcoidosis; library and information science. Mailing Add: Nat Libr of Med 8600 Rockville Pike Bethesda MD 20014

CUMMINGS, MICHAEL R, b Chicago, Ill, July 7, 41; m 66; c 2. CYTOLOGY, DEVELOPMENTAL GENETICS. Educ: St Mary's Col, Minn, BA, 63; Northwestern Univ, MS, 65, PhD(biol), 68. Prof Exp: Instr genetics, Northwestern Univ, 68-69; asst prof genetics, 69-74, ASSOC PROF BIOL SCI, UNIV ILL, CHICAGO CIRCLE, 74- Mem: AAAS; Genetics Soc Am. Res: Genetic control of development and maturation in female insect reproductive systems; biochemical and ultrastructural processes associated with vitellogenesis; analysis of determination and differentiation in imaginal discs. Mailing Add: Dept of Biol Sci Univ of Ill Chicago Cir Chicago IL 60680

CUMMINGS, NANCY BOUCOT, b Phildelphia, Pa, Feb 21, 27; m 59; c 3. NEPHROLOGY, INTERNAL MEDICINE. Educ: Oberlin Col, BA, 47; Univ Pa, MD, 51. Prof Exp: Rotating intern, Pa Hosp, 51-52; resident internal med, Univ Pa Hosp, 52-54; res & clin asst med, Royal Hosp St Bartholomew, London, 54-55; res & clin asst med, Manchester Royal Infirmary, Eng, 55; asst med, Peter Bent Brigham Hosp, 55-58; guest worker, Nat Inst Arthritis '& Metab Dis, 59-62; res med officer, Walter Reed Army Inst Res, 62-66; res med officer, Div Exp Med, Naval Med Res Inst & consult nephrol, Dept Med, Naval Hosp, Md, 66-72; prog officer, 72-73, spec asst to dir renal & urol dir, 73-74, ACTG ASSOC DIR RENAL & UROL DIS, NAT INST ARTHRITIS, METAB & DIGESTIVE DIS, 74- Concurrent Pos: Res fel med, Harvard Med Sch, 55-58, res fel biol chem, 58-59; Am Heart Asn fel, 55-57, advan res fel, 57-62; Nat Found res fel, Royal Hosp St Bartholomew & Manchester Royal Infirmary, 54-55; mem, Res Comt, Washington Heart Asn & med adv bd, Washington

Chap, Nat Kidney Found, 69-; clin instr med, Georgetown Univ, 60-70, clin asst prof, 70-; co-chmn, Adv Comt Epidemiol & Statist Kidney Dis, Nat Kidney Found, mem, Sci Adv Bd & trustee at large; consult, Coord Comt Coun Urol; mem, Res Comt, Am Urol Asn. Mem: AAAS; Am Soc Hemat; Am Soc Nephrology; Am Soc Artificial Internal Organs; Am Fedn Clin Res. Res: Epidemiology and statistics of renal disease; biochemistry of uremia; renal physiology and pathophysiology. Mailing Add: 2811 35th St NW Washington DC 20007

CUMMINGS, NORMAN ALLEN, b New York, NY, Mar 26, 35; m 60; c 2. INTERNAL MEDICINE. Educ: NY Univ, AB, 55; State Univ NY, MD, 59; Am Bd Internal Med, dipl, 72. Prof Exp: Intern med, Jewish Hosp, Brooklyn, 59-60, resident, 61; resident med, Med Sch, Univ Mich, 61-62; res assoc protein chem, Nat Cancer Inst, 64-66; asst prof internal med, Col Med, Baylor Univ, 66-67; med officer & head connective tissue dis prog, Oral Med & Surg Br, Nat Inst Dent Res, 67-74; ASSOC PROF MED, CHIEF CLIN IMMUNOL & CONNECTIVE TISSUE DIS SECT & DIR ARTHRITIS CTR, SCH MED, UNIV LOUISVILLE, 74- Concurrent Pos: Vis fel rheumatic dis, NY Univ-Bellevue Hosp Med Ctr, 62; USPHS physician trainee grant rheumatic dis, 62-64; fel biophys & arthritis, Med Sch, Univ Mich, 62-64; Kayser Found sci grant, 66-67; asst attend physician, Ben Taub Gen Hosp & Vet Admin Hosp, Houston, 66-67; attend physician, Louisville Gen Hosp; consult, Louisville Vet Admin Hosp, Jewish Hosp & Nat Inst Dent Res, NIH, 67-74; consult, Norton's Children's Hosp. Honors & Awards: Res Award, Jewish Hosp Brooklyn, 61. Mem: Am Rheumatism Asn. Res: Protein-mucopolysaccharide interactions; metallo-proteins of serum; joint pH; protein solubility and conformation; cryoprecipitation of cryoglobulins; oral-mucosal manifestations of connective tissue diseases; immunochemistry and cellular immunology in rheumatic diseases. Mailing Add: Col of Med Univ Louisville Health Sci Ctr Louisville KY 40201

CUMMINGS, RALPH WALDO, JR, b Ithaca, NY, July 20, 38; m 61; c 2. AGRICULTURAL ECONOMICS. Educ: Univ NC, AB, 60; Univ Mich, PhD(econ), 65. Prof Exp: Asst prof econ, Univ Ill, 65-70; consult, Bur Near East & S Asia, AID, 70; adv agr econ, Harvard Adv Group & Nat Develop Planning Agency Indonesia, 70-72; AGR ECONOMIST, ROCKEFELLER FOUND, 72- Concurrent Pos: Asst dir, Midwestern Univs Consortium Int Activ, 66-67; chief agr econ div, Off Agr Develop, AID, India, 67-69. Mem: Am Agr Econ Asn; Am Econ Asn. Res: General problems of agricultural development, particularly improving standards of living in rural areas. Mailing Add: Rockefeller Found 1133 Ave of the Americas New York NY 10036

CUMMINGS, ROBERT HOPKINS, plant pathology, see 12th edition

CUMMINGS, SUE CAROL, b Dayton, Ohio, Apr 24, 41. INORGANIC CHEMISTRY, BIOINORGANIC CHEMISTRY. Educ: Northwestern Univ, BA, 63; Ohio State Univ, MSc, 65, PhD(inorg chem), 68. Prof Exp: Vis res assoc inorg chem, Aerospace Res Labs, Wright-Patterson AFB, 68-69; asst prof, 69-73, ASSOC PROF CHEM, WRIGHT STATE UNIV, 73- Concurrent Pos: Petrol Res Fund grant, 69-72; Cottrell res grant, 73; Nat Heart & Lung Inst res grant, 73-75. Mem: AAAS; Am Chem Soc. Res: Coordination chemistry; synthesis, characterization, stereochemistry and reactions of metal complexes containing multidentate, macrocyclic or cage-type ligands; synthetic oxygen carriers; study of metal complexes as models for biologically important molecules. Mailing Add: Dept of Chem Wright State Univ Dayton OH 45431

CUMMINGS, THOMAS FULTON, b Taxila, India, Oct 25, 25; US citizen; m 48; c 4. PHYSICAL CHEMISTRY, ORGANIC CHEMISTRY. Educ: Mass Inst Technol, BSc, 47; Case Inst Technol, MSc, 52, PhD(chem), 55. Prof Exp: Chem engr, Res Ctr, B F Goodrich Co, 48-49; asst chem, Case Inst Technol, 49-52; instr, Westminster Col, 52-55; from asst prof to assoc prof, 55-67, PROF CHEM, BRADLEY UNIV, 67- Concurrent Pos: Vis prof chem, Univ Birmingham, Eng, 73-74. Mem: Am Chem Soc; Am Sci Affiliation; Soc Appl Spectros. Res: Kinetics; instrumental analysis; water and air pollution analysis, government support. Mailing Add: Dept of Chem Bradley Univ Peoria IL 61625

CUMMINGS, WILLIAM CHARLES, b Boston, Mass, Apr 6, 32; m 55; c 2. MARINE BIOLOGY, BIOACOUSTICS. Educ: Bates Col, BS, 54; Univ Miami, MS, 60, PhD(marine biol sci), 68. Prof Exp: Biol oceanogr, Univ RI, 58-60; instr bioacoustics, Univ Miami, 60-65; oceanogr, 65-70, HEAD UNDERWATER BIOACOUSTICS BR, NAVAL UNDERSEA RES & DEVELOP CTR, 70- Concurrent Pos: Mem, Int Comt Bioacoustics, 62-; consult, 63-; prof writer & ed, 68-; mem sci adv bd, Am Cetacean Soc, 68-; dir, Oceanog Consult, 70-; mem marine mammal coun, NSF, 71- Mem: Asn Island Marine Labs Caribbean; Am Fisheries Soc; Acoustical Soc Am. Res: Underwater bioacoustics; tropical and subtropical marine biology and ecology; research methodology; fisheries biology; mammalogy and Antarctic research. Mailing Add: Code 5054 Naval Undersea Res & Develop Ctr San Diego CA 92132

CUMMINGS, WILLIAM HAWKE, forestry, see 12th edition

CUMMINS, ALVIN J, b Wheeling, WVa, Apr 26, 19; m 47; c 3. INTERNAL MEDICINE. Educ: Georgetown Univ, BS, 41; Johns Hopkins Univ, MD, 44. Prof Exp: Asst instr med, Med Sch, Univ Pa, 51-52, instr, 52-53, assoc, 53-54, from asst prof to assoc prof, 57-63, prof med & chief sect gastroenterol, 63-71, CLIN PROF MED, MED COL, UNIV TENN, 71- Concurrent Pos: Fel med, Cornell Med Ctr, 50-51; hon consult, Blytheville AFB, Ark, 61; chmn, Gastroenterol Res Group, 61-62; consult, Vet Admin Hosp, Memphis, 62-71; pvt pract gastroenterol. Mem: AMA; Am Col Physicians; Am Gastroenterol Asn; Am Fedn Clin Res. Res: Clinical and investigative gastroenterology; intestinal absorption; intestinal blood flow; pharmacology of gastrointestinal tract. Mailing Add: 1324 Peabody Memphis TN 38104

CUMMINS, CECIL STRATFORD, b Monkstown, Ireland, Nov 20, 18; m 59; c 2. MICROBIOLOGY. Educ: Univ Dublin, BA, 41, MB, BCh, BAO, 43, ScD, 64. Prof Exp: House physician, Sir Patrick Dun's Hosp, Dublin, Ireland, 43; asst to prof bact, Trinity Col, Dublin, 44; lectr bact, London Hosp Med Col, London, 48-64, reader, 64-67; PROF MICROBIOL, ANAEROBE LAB, DIV BASIC SCI, COL AGR, VA POLYTECH INST & STATE UNIV, 67- Mem: Am Soc Microbiol; Brit Soc Gen Microbiol; NY Acad Sci; Path Soc Gt Brit. Res: Chemical morphology; taxonomy. Mailing Add: Anaerobe Lab Div of Basic Sci Va Polytech Inst & State Univ Blacksburg VA 24061

CUMMINS, DAVID GRAY, b Cookeville, Tenn, June 29, 36; m 59; c 3. AGRONOMY. Educ: Tenn Polytech Inst, BS, 57; Univ Tenn, MS, 59; Univ Ga, PhD(agron), 62. Prof Exp: Soil scientist, US Forest Serv, Ky, 62-63; asst agronomist, 63-68, ASSOC AGRONOMIST, GA EXP STA, UNIV GA, 68- Mem: Am Soc Agron; Crop Sci Soc Am. Res: Determining more efficient practices of fertilization and management of annual silage crops for maximum production and quality. Mailing Add: Ga Exp Sta Univ of Ga Experiment GA 30212

CUMMINS, EARL WESLEY, b Woodbine, Ky, Dec 12, 23; m 50; c 3. ORGANIC CHEMISTRY. Educ: Detroit Inst Technol, BSc, 43; Wayne Univ, PhD(chem), 51. Prof Exp: Res chemist, R P Scherer Co, 45-47; RES CHEMIST, E I DU PONT DE NEMOURS & CO, INC, 51- Mem: Am Chem Soc. Res: Biological chemicals in general; sulfones. Mailing Add: 2410 Shellpot Dr Oaklane Manor Wilmington DE 19803

CUMMINS, ERNIE LEE, b Warrenton, Ore, July 13, 21; m 43; c 1. SCIENCE EDUCATION. Educ: Ore State Univ, BS, 43, MS, 52, EdD(sci educ), 60. Prof Exp: Chemist, Scott Paper Co, 47-48 & Evans Prod Co, 49-50; teacher gen sci, Jr High Sch, Ore, 51-53 & High Sch, 53-57; asst prof phys sci & sci educ, 57-61, assoc prof phys sci, 61-66, PROF PHYS SCI & SCI EDUC, ORE COL EDUC, 66- Concurrent Pos: Dir in-serv insts, NSF, 62-69. Mem: Fel AAAS; Nat Asn Res Sci Teaching; Am Asn Physics Teachers; Nat Sci Teachers Asn. Res: Physics; general science. Mailing Add: Dept of Sci Ore Col of Educ Monmouth OR 97361

CUMMINS, HERMAN Z, b Rochester, NY, Apr 23, 33; m 63. QUANTUM OPTICS, SOLID STATE PHYSICS. Educ: Ohio State Univ, BS & MS, 56; Columbia Univ, PhD(physics), 63. Prof Exp: Res physicist, Radiation Lab, Columbia Univ, 62-64; from asst prof to prof physics, Johns Hopkins Univ, 64-71; prof, NY Univ, 71-73; DISTINGUISHED PROF PHYSICS, CITY COL NEW YORK, 73- Mem: Fel Am Phys Soc. Res: Phase transitions in liquids and crystals, ferroelectrics and light scattering spectroscopy; critical phenomena. Mailing Add: Dept of Physics City Col of New York New York NY 10031

CUMMINS, JACK D, b Shreveport, La, Dec 28, 39; m 60; c 2. INORGANIC CHEMISTRY. Educ: Western State Col Colo, BA, 61; Univ MNex, MS, 66, PhD(inorg chem), 67. Prof Exp: asst prof, 66-74, PROF CHEM, METROP STATE COL, 74-, CHMN DEPT, 71- Concurrent Pos: Res grant, Educ Media Inst, Univ Colo, 67-68; fel, Univ Mo, St Louis, 69-70. Mem: Am Chem Soc. Res: Study of bonding by characterization and preparation of boron hydrides and boronium cations; x-ray crystallography of amalgams. Mailing Add: Dept of Chem Metrop State Col Denver CO 80204

CUMMINS, JAMES NELSON, b Dix, Ill, Jan 22, 25; m 48; c 5. POMOLOGY, PLANT MORPHOGENESIS. Educ: Univ Ill, BS, 48; Univ Southern Ill, MS, 60, PhD(bot), 66; Univ Wis, MS, 61. Prof Exp: Exec secy, Ill Fruit Coun, 48-50; self employed orchardist, 53-55; instr, Anna-Jonesboro High Sch, Ill, 55-57 & Mt Vernon High Sch & Mt Vernon Community Col, 60; asst prof sci educ, Southern Ill Univ, 61-67; ASSOC PROF POMOL, NY STATE AGR EXP STA, CORNELL UNIV, 67- Mem: Am Soc Hort Sci; Am Pomol Soc; Int Soc Hort Sci; Int Asn Plant Tissue Cult. Res: Breeding, testing and development of rootstocks for deciduous fruit trees; rhizogenesis in stem tissues of woody plants; asexual plant propagation of woody plants. Mailing Add: Hedrick Hall New York State Agr Exp Sta Geneva NY 14456

CUMMINS, JOSEPH E, b Whitefish, Mont, Feb 5, 33; m 62; c 1. GENETICS, CELL BIOLOGY. Educ: Washington State Univ, BS, 55; Univ Wis, PhD(bot), 62. Prof Exp: NIH res fel zool, Univ Edinburgh, 62-64; cancer res, McArdle Lab, Univ Wis, 64-66; asst prof biol sci, Rutgers Univ, 66-67; asst prof zool, Univ Wash, 67-72; ASST PROF PLANT SCI, UNIV WESTERN ONT, 72- Concurrent Pos: Fel, Karolinska Inst, Stockholm, Sweden, 69. Mem: Am Soc Cell Biol; Brit Biochem Soc; Brit Soc Cell Biol; Can Genetics Soc. Res: Cell cycle; morphogenesis; environmental gene damage. Mailing Add: Dept of Plant Sci Univ of Western Ont London ON Can

CUMMINS, KENNETH BURDETTE, b New Washington, Ohio, July 27, 11. MATHEMATICS. Educ: Ohio Wesleyan Univ, AB, 33; Bowling Green State Univ, MA, 39; Ohio State Univ, PhD(math educ), 58. Prof Exp: Teacher pub schs, Ohio, 33-40, 41-55, 56-57; from asst prof to assoc prof math, 57-64, chmn dept, 64-65, PROF MATH, KENT STATE UNIV, 64- Honors & Awards: G O Higley Award, Ohio Wesleyan Univ, 33. Mem: Am Math Asn. Res: Methodology in the teaching of collegiate and high school mathematics; an experiential approach to mathematics education. Mailing Add: Dept of Math Kent State Univ Kent OH 44240

CUMMINS, KENNETH WILLIAM, b Chicago, Ill, Mar 28, 33; m 55; c 2. LIMNOLOGY. Educ: Lawrence Col, BA, 55; Univ Mich, MS, 57, PhD(zool), 61. Prof Exp: Instr zool, Univ Mich, 60-61; asst prof biol sci, Northwestern Univ, 61-62 & biol, Univ Pittsburgh, 62-68; PROF BIOL, KELLOGG BIOL LAB, MICH STATE UNIV, 68- Concurrent Pos: Prin investr, USPHS res grant, 63-, AEC res grant, 66- & NSF res grants, 70-; chmn water ecosysts, Inst Ecol, 73-; aquatics ed, Ecol Soc Am, 74- Mem: Am Inst Biol Sci; Ecol Soc Am; Am Soc Limnol & Oceanog; Brit Freshwater Biol Asn; Int Asn Theoret & Appl Limnol. Res: Structure and function of stream ecosystems. Mailing Add: Kellogg Biol Lab Mich State Univ Hickory Corners MI 49060

CUMMINS, LAURENCE MARK, analytical chemistry, biochemistry, see 12th edition

CUMMINS, RICHARD WILLIAMSON, b Allen, Mich, May 10, 20; m 43; c 2. SYNTHETIC ORGANIC CHEMISTRY. Educ: Univ Mich, BS, 42; Polytech Inst Brooklyn, MS, 46, PhD(chem), 53. Prof Exp: Res chemist, Westvaco Mineral Prod Div, Food Mach & Chem Corp, 42-64, sr res chemist, FMC Corp, 64-68, res assoc, 68-73, SR RES ASSOC, FMC CORP, 73- Mem: Am Chem Soc. Res: Organic synthesis reaction mechanisms; organic phosphorus compounds; triazine chemistry; detergent builders. Mailing Add: FMC Corp PO Box 8 Princeton NJ 08540

CUMMISKEY, CHARLES, b St Louis, Mo, Feb 12, 24. INORGANIC CHEMISTRY. Educ: Dayton Univ, BS, 43; Northwestern Univ, MS, 52; Univ Notre Dame, PhD(chem), 56. Prof Exp: Teacher sec schs, 43-52; assoc, Univ Notre Dame, 53-55; from asst prof to assoc prof chem, 55-65, head dept, 57-66, vpres & dean faculties, 66-75, PROF CHEM, ST MARY'S UNIV, TEX, 65- Concurrent Pos: Am Chem Soc vis scientist to sec schs, 62-66. Mem: Am Chem Soc; Sigma Xi. Res: Chemical effects of radioactive decay; ion exchange methodology, especially as applied to inorganic analytical separations and complex ions. Mailing Add: Dept of Chem St Mary's Univ San Antonio TX 78284

CUNDIFF, LARRY VERL, b Abilene, Kans, Dec 9, 39; m 60; c 3. ANIMAL BREEDING, POPULATION GENETICS. Educ: Kans State Univ, BS, 61; Okla State Univ, MS, 64, PhD(animal breeding), 66. Prof Exp: Asst prof animal sci, Univ Ky, 65-67; REGIONAL COORDR & GENETICIST, US MEAT ANIMAL RES CTR, NCENT REGION, AGR RES SERV, USDA, 67- Concurrent Pos: Assoc prof animal sci, Univ Nebr, Lincoln, 75-76. Mem: Am Soc Animal Sci. Res: Beef cattle breeding. Mailing Add: US Meat Animal Res Ctr PO Box 166 Clay Center NE 68933

CUNDIFF, MILFORD FIELDS, b Baker, Ore, Dec 7, 36; div; c 2. VERTEBRATE PHYSIOLOGY. Educ: Univ Colo, BS, 60, PhD(zool), 66. Prof Exp: Teaching assoc biol, Univ Colo, 61-62; from instr to assoc prof, Austin Col, 64-70; ASSOC PROF BIOL & COORDR BIOL SCI, UNIV COLO, BOULDER, 70- Mem: AAAS; Am Inst Biol Sci. Res: Altitude physiology of blood; nuclear activation analysis of trace metals in biological systems. Mailing Add: Dept of Integrated Studies Univ of Colo Boulder CO 80302

CUNDIFF, ROBERT HALL, b Winchester, Ky, Apr 10, 22; m 44; c 2. RESEARCH ADMINISTRATION, ANALYTICAL CHEMISTRY. Educ: Univ Ky, BS, 48, MS, 49. Prof Exp: Res chemist, Com Solvents Corp, Ind, 49-52; res chemist, 52-66, head anal serv sect & blends & filter develop sect, 66-70, MGR TOBACCO PROD DEVELOP DIV, R J REYNOLDS TOBACCO CO, 70- Honors & Awards: Philip Morris Award, 67. Mem: Am Chem Soc. Res: Tobacco products development, including chewing and smoking tobacco, cigarettes and filter technology; tobacco processing, blending and flavoring; analytical chemistry of tobacco, tobacco additives and tobacco smoke. Mailing Add: R J Reynolds Tobacco Co Winston-Salem NC 27102

CUNDY, KENNETH RAYMOND, b Spearfish, SDak, Dec 22, 29; m 57. MEDICAL MICROBIOLOGY, CLINICAL MICROBIOLOGY. Educ: Stanford Univ, BA, 50; Univ Wash, MS, 53; Univ Calif, Davis, PhD(comp path), 65; Am Bd Med Microbiol, dipl, 69. Prof Exp: Res microbiologist, Univ Calif, Berkeley, 57-60; bacteriologist, Gerber Prod Co, Calif, 60-61; res microbiologist, Univ Calif, Davis, 61-65; from instr to assoc prof, 67-71, ASSOC PROF MICROBIOL, SCH MED, TEMPLE UNIV, 71-, DIR CLIN MICROBIOL LABS, UNIV HOSP & HEALTH SCI CTR, 70- Concurrent Pos: Fel microbiol, Sch Med, Temple Univ, 65-67 & Nat Cystic Fibrosis Res Found, 68-69; assoc dir microbiol, Lab, St Christopher's Hosp Children, 67-68, dir diag microbiol, 68-70. Mem: Fel Am Acad Microbiol; AAAS; Am Soc Microbiol; NY Acad Sci. Res: Mechanisms of pathogenicity as related to microorganisms, particularly with reference to Pseudomonas and anaerobes; infectious diseases and their laboratory diagnosis. Mailing Add: Dept of Microbiol Temple Univ Sch of Med Philadelphia PA 19140

CUNDY, PAUL FRANKLIN, b Hibbing, Minn, June 17, 10; m 34; c 2. INORGANIC CHEMISTRY. Educ: Univ Ill, AB, 30, MS, 32, PhD(inorg chem), 39. Prof Exp: Instr chem, Virginia Jr Col, Minn, 32-38, 39-43; from res asst to res assoc, Inst Paper Chem, Lawrence Col, 43-47, assoc prof chem, 47-51; res chemist, Am Can Co, 51-57, mgr pkg mat res, 57-67, asst to dir res & develop, 67, adminr regulatory compliance, res & develop, 67-75; RETIRED. Mem: AAAS; Am Chem Soc; Tech Asn Pulp & Paper Indust. Res: Formation of complex compounds in solution; analysis of cellulose products and related substances; surface chemistry; graphic arts. Mailing Add: 1515 S Mason St Appleton WI 54911

CUNHA, GERALD R, anatomy, embryology, see 12th edition

CUNHA, TONY JOSEPH, b Los Banos, Calif, Aug 22, 16; m 41; c 3. ANIMAL HUSBANDRY, NUTRITION. Educ: Utah State Agr Col, BS, 40, MS, 41; Univ Wis, PhD(animal nutrit), 44. Prof Exp: From instr to assoc prof animal husb, State Col Wash, 44-50; prof animal sci & head dept, Univ Fla, 50-75; DEAN SCH AGR, CALIF POLYTECH UNIV, 75- Concurrent Pos: Am Soc Animal Sci res award, 68-; chmn swine nutrient requirements comt & chmn animal nutrit comt, Nat Res Coun, 72; mem, White House Conf on Food, Nutrit & Health, 69; consult, Univ Tenn-AEC, Oak Ridge & USDA Meat Animal Res Ctr, Nebr. Mem: AAAS; Am Soc Animal Sci (vpres, 61, pres, 62); Soc Exp Biol & Med; Am Inst Nutrit. Res: Nutrition and feeding of swine, beef, cattle, sheep, horses and small animals. Mailing Add: Dean Sch of Agr Calif Polytech Univ Pomona CA 91769

CUNIA, TIBERIUS, b Edessa, Greece, Jan 10, 26; Can citizen; m 57; c 4. STATISTICS. Educ: McGill Univ, MSc, 57. Prof Exp: Forester mensuration, Can Inst Paper Co, 51-52, forest opers, 52-54 & comput appln, 54-58, forest statistician, 58-68; prof forest mensuration & statist, 68-70, PROF MENSURATION & OPERS RES, STATE UNIV NY COL FORESTRY, SYRACUSE UNIV, 70- Mem: Am Statist Asn; Biomet Soc; Can Pulp & Paper Asn; Can Inst Forestry; Soc Am Foresters. Res: Forest mensuration and inventory; applications of the statistical methodology, operations research methods and computers to forestry problems; experimental designs in pulp and paper mills. Mailing Add: Sch Envir Resources Mgt SUNY Col Forestry Syracuse Univ Syracuse NY 13210

CUNICO, ROBERT FREDERICK, b Detroit, Mich, Feb 25, 41; m 66. ORGANIC CHEMISTRY. Educ: Univ Detroit, BS, 62; Purdue Univ, PhD(organosilicon chem), 66. Prof Exp: Res assoc with Dr Melvin S Newman, Ohio State Univ, 66-68; asst prof, 68-73, ASSOC PROF CHEM, NORTHERN ILL UNIV, 73- Mem: Am Chem Soc. Res: Organosilicon chemistry; synthetic methods in organic chemistry. Mailing Add: Dept of Chem Northern Ill Univ De Kalb IL 60115

CUNKLE, CHARLES HENRY, b Ft Smith, Ark, May 28, 15; m 49; c 2. MATHEMATICS. Educ: Ind Univ, AB, 38; La State Univ, MA, 41; Univ Mo, PhD(math), 55. Prof Exp: Instr math, Agr & Mech Col, Univ Tex, 41-42; adv, US Mil Govt, Ger, 47-49; instr math, La State Univ, 49-51 & Univ Mo, 51-55; asst prof, Univ Idaho, 55-56; sr res engr, Convair, Calif, 56-57; asst prof math, Colo State Univ, 57-58; res mathematician, Aeronaut Lab, Cornell Univ, 58-59; assoc prof math, Utah State Univ, 59-63; prof & chmn dept, Clarkson Col Technol, 63-65; prof, Kans State Univ, 65-67; chmn dept, 67-71, PROF MATH, SLIPPERY ROCK STATE COL, 71- Concurrent Pos: Mathematician, Ballistics Res Labs, 58. Mem: Am Math Soc; Math Asn Am; Nat Coun Teachers Math. Res: Topology; Boolean algebras. Mailing Add: Dept of Math Slippery Rock State Col Slippery Rock PA 16057

CUNLIFFE, HARRY R, b Stanford, Conn, Aug 28, 29; m 59; c 2. VETERINARY IMMUNOLOGY, ANIMAL VIROLOGY. Educ: Univ Toronto, DVM, 57; Iowa State Univ, MS, 67. Prof Exp: Res assoc vet immunol, Plum Island Animal Dis Lab, 58-60, res vet, 60-63, res vet, Nat Animal Dis Lab, 63-67, vet immunol, Plum Island Animal Dis Lab, 67-74, BIOHAZARD/SAFETY RES VET MED OFFICER, PLUM ISLAND ANIMAL DIS CTR, USDA, 74- Mem: AAAS; Am Soc Microbiol; US Animal Health Asn; Int Asn Milk, Food & Environ Sanitarians. Res: Tissue cultures of foot-and-mouth and hog cholera viruses; chemical inactivation of virus, quantitation of antigens, vaccine adjuvants and potency assays; survival of exotic viruses in food products. Mailing Add: Plum Island Animal Dis Ctr PO Box 848 Greenport NY 11944

CUNNEA, WILLIAM M, b Chicago, Ill, Oct 31, 27; m 71. MATHEMATICS. Educ: Univ Chicago, PhB, 46, MS, 56; Univ Calif, Berkeley, PhD(math), 62. Prof Exp: Asst prof, 61-66, ASSOC PROF MATH, WASH STATE UNIV, 66- Concurrent Pos: Vis scholar, Univ Victoria, 68-69. Mem: AAAS; Math Asn Am; Am Math Soc. Res: Ideal and valuation theory; algebraic geometry. Mailing Add: Dept of Math Wash State Univ Pullman WA 99163

CUNNIFF, PATRICIA A, b Washington, DC, Dec 6, 38; m 60; c 4. PHYSICAL CHEMISTRY. Educ: Dunbarton Col Holy Cross, BA, 59; Univ Md, College Park, MS, 62, PhD(phys chem), 72. Prof Exp: Ed asst, Anal Chem, Am Chem Soc, 61-62; asst prof, 72-75, ASSOC PROF CHEM, PRINCE GEORGE'S COMMUNITY COL, 75- Mem: AAAS; Am Chem Soc; Sigma Xi. Res: Academic education in chemistry; instrumental analysis. Mailing Add: Prince George's Community Col Largo MD 20870

CUNNINGHAM, ALICE JEANNE, b Walnut Ridge, Ark, Sept 23, 37. ELECTROCHEMISTRY, POLAROGRAPHY. Educ: Univ Ark, BA, 59; Emory Univ, PhD(chem), 66. Prof Exp: Chemist, Layne Res, Tenn, 59; instr chem, Brenau Acad, 59-61 & Atlanta Pub Schs, 61-62; vis asst prof, Agnes Scott Col, 66-67; res assoc, Univ Tex, 67-68; asst prof, 68-72, ASSOC PROF CHEM, AGNES SCOTT COL, 72- Mem: AAAS; Am Chem Soc; Electrochem Soc. Res: Electrochemistry of pyridine nucleotides, ribonuclease, indophenols and ferredoxin; mechanisms of biological oxidation-reduction reactions; free radical intermediates in electro-organic chemistry. Mailing Add: Dept of Chem Agnes Scott Col Decatur GA 30030

CUNNINGHAM, ALLEN BYRON, b Smithfield, WVa, Mar 24, 12. MATHEMATICS. Educ: Fairmont State Col, AB, 34; Univ WVa, MS, 36, PhD(math), 40. Prof Exp: Instr math, physics & geog, Davis & Elkins Col, 38-39; instr math, Pa State Univ, 40-46, instr math & mech, Eng Defense Training Prog, 41; asst to chief engr, Northwest Mining & Mach Co, Pa, 43; instr math, Pa State Univ, 43-46; from asst prof to assoc prof, 46-60, PROF MATH, WVA UNIV, 60- Mem: Am Math Soc; Math Asn Am. Res: Algebraic geometry; analysis. Mailing Add: Dept of Math WVa Univ Morgantown WV 26506

CUNNINGHAM, BRUCE ARTHUR, b Winnebago, Ill, Jan 18, 40; m 65; c 1. BIOCHEMISTRY. Educ: Univ Dubuque, BS, 62; Yale Univ, PhD(biochem), 66. Prof Exp: NSF fel biochem, 66-68, asst prof, 68-71, ASSOC PROF BIOCHEM, ROCKEFELLER UNIV, 71- Concurrent Pos: Camille & Henry Dreyfus Found grant, 70-75. Honors & Awards: Career Scientist, Irma T Hirschl Trust, 75. Mem: AAAS; Am Chem Soc; Am Asn Immunologists; Am Soc Biol Chemists; Harvey Soc. Res: Structure and function of cell-surface proteins; structure of antibodies; primary structure of proteins. Mailing Add: Dept of Biochem Rockefeller Univ New York NY 10021

CUNNINGHAM, BRYCE A, b Brainerd, Minn, June 21, 32; m 56; c 4. BIOCHEMISTRY. Educ: Univ Minn, BA, 55, BS, 58, PhD(biochem), 63. Prof Exp: Asst prof, 63-72, ASSOC PROF BIOCHEM, KANS STATE UNIV, 72- Mem: AAAS; Am Chem Soc. Res: Enzyme chemistry; peroxidases. Mailing Add: Dept of Biochem Kans State Univ Manhattan KS 66506

CUNNINGHAM, CHARLES EVERETT, b Washburn, Maine, Dec 21, 24; m 45; c 3. PLANT GENETICS, AGRONOMY. Educ: Univ Maine, BS, 48, MS, 52; Univ Wis, PhD(genetics), 62. Prof Exp: Asst agronomist, Univ Maine, 49-55, asst prof agron, 55-57; plant geneticist, Red Dot Foods, Inc, Wis, 57-61; GENETICIST, CAMPBELL SOUP CO, 62- Mem: Potato Asn Am (secy, 63-65, vpres, 66, pres, 68). Res: Nutrition and culture of the potato; development of new potato varieties for processing. Mailing Add: Campbell Inst for Agr Res Riverton NJ 08077

CUNNINGHAM, CHARLES HENRY, b Washington, DC, Apr 12, 13; m 40; c 4. VETERINARY MICROBIOLOGY. Educ: Univ Md, BS, 34; Iowa State Univ, MS, 37, DVM, 38; Mich State Univ, PhD, 53; Am Bd Med Microbiol, dipl; Am Col Vet Microbiol, dipl. Prof Exp: Asst prof & vet inspector, Univ Md, 38-42; assoc prof poultry husb, Univ RI, 42-45; assoc prof microbiol & pub health, 45-54, PROF MICROBIOL & PUB HEALTH, MICH STATE UNIV, 54- Concurrent Pos: Consult vet microbiol, Univ Tenn Int Agr Prog. Honors & Awards: Distinguished Fac Award, Mich State Univ, 73. Mem: AAAS; Am Asn Avian Path; NY Acad Sci; Am Soc Microbiol; Am Vet Med Asn. Res: Virology; veterinary virology; animal diseases. Mailing Add: Dept of Microbiol & Pub Health Mich State Univ East Lansing MI 48824

CUNNINGHAM, CLARENCE MARION, b Cooper, Tex, July 24, 20; m 51; c 4. PHYSICAL CHEMISTRY. Educ: Agr & Mech Col Tex, BS, 42; Univ Calif, MS, 48; Ohio State Univ, PhD(chem), 54. Prof Exp: Instr chem, Calif State Polytech Col, 48-49; cryogenic engr, AEC Proj, 52; consult, Herrick L Johnston, Inc, 53; ASSOC PROF CHEM, OKLA STATE UNIV, 54- Mem: AAAS; Am Chem Soc; Am Phys Soc; Am Asn Univ Prof; Nat Sci Teachers Asn. Res: Theoretical and experimental investigation of physical absorption of gases on solid surfaces at low temperatures; electrochemical properties on nonaqueous solution of electrolytes. Mailing Add: Dept of Chem Okla State Univ Stillwater OK 74074

CUNNINGHAM, DAVID A, b Toronto, Ont, Mar 18, 37; m 61; c 3. PHYSIOLOGY. Educ: Univ Western Ont, BA, 60; Univ Alta, MSc, 63; Univ Mich, PhD(phys educ, physiol), 66. Prof Exp: Res assoc epidemiol, Univ Mich, 66-69; ASSOC PROF PHYSIOL & PHYS EDUC, UNIV WESTERN ONT, 69- Concurrent Pos: Res assoc, Dept Nat Health & Welfare, Can, 69-72. Mem: Am Physiol Soc; Can Physiol Soc; Can Asn Sports Sci. Res: Physiology of exercise with special interest in the relationship of heart disease and physical activity. Mailing Add: Dept of Physiol Univ of Western Ont London ON Can

CUNNINGHAM, DAVID KENNETH, b Victoria, BC, Feb 28, 20; m 49. AGRICULTURAL BIOCHEMISTRY. Educ: Univ BC, BA, 41; Univ Minn, PhD(biochem), 53. Prof Exp: Chemist, Defense Industs, 41-43 & Grain Res Lab, 45-58; tech mgr refrig res & develop, 58-64, res assoc methods & develop, 64-68, RES ASSOC CORP RES, PILLSBURY CO, 69- Mem: Am Chem Soc; Am Asn Cereal Chem; NY Acad Sci. Res: Cereal and analytical chemistry; flour quality and improving agents; physical properties of doughs; chemistry of leavening agents. Mailing Add: Pillsbury Co 311 Second St SE Minneapolis MN 55414

CUNNINGHAM, DENNIS DEAN, b Des Moines, Iowa, Aug 16, 39. CELL BIOLOGY, BIOCHEMISTRY. Educ: State Univ Iowa, BA, 61; Univ Chicago, PhD(biochem), 67. Prof Exp: Asst prof, 70-74, ASSOC PROF MED MICROBIOL, COL MED, UNIV CALIF, IRVINE, 74- Concurrent Pos: NSF res fel, Princeton Univ, 67-70; USPHS res grant, 71. Res: Control of cell division in cultured fibroblasts. Mailing Add: Dept of Med Microbiol Univ of Calif Col of Med Irvine CA 92717

CUNNINGHAM, DONALD EUGENE, b Providence, RI, May 18, 30; m 57; c 2. PHYSICS. Educ: Brown Univ, AB, 51; Case Univ, MA, 54, PhD(physics), 59. Prof Exp: Instr physics, Case Univ, 53-57; plasma physics group leader, Thompson Ramo Wooldridge, Inc, 57-59; dir vis scientist progs, Am Inst Physics, 59-62; asst prof physics & asst dir inst sci & math, Adelphi Univ, 62-63, assoc prof physics & dir progs space related sci, 63-66; prof physics & dean res, Miami Univ, 66-70; spec asst to dir, NSF, Washington, DC, 70-74; SR RES SCIENTIST, UNIV DENVER RES INST, 74- Concurrent Pos: Res engr, Westinghouse Elec Co, 54; consult, Thompson Ramo Wooldridge, Inc, 59-61, Mayer & Sklar, Inc, 61 & Am Inst Physics, 62-; lectr, Manhattan Col, 61-62; dir, NSF High Sch Teacher Inst, 62; chmn vis scientist prog high schs, NY, 63-66; prin investr, NASA Grant, 63-67. Mem: Am Phys Soc; Am Asn Physics Teachers; fel AAAS. Res: Atomic collision processes; optical pumping techniques; plasma and low temperature physics; magnetism; impacts of resource development on a regional bases. Mailing Add: Univ of Denver Res Inst Denver CO 80210

CUNNINGHAM, DOROTHY J, b Jersey City, NJ, Nov 7, 27. PHYSIOLOGY. Educ: Caldwell Col Women, BA, 49; Cath Univ, MS, 51; Yale Univ, PhD(physiol), 66. Prof Exp: Res asst physiol, Sch Med, Univ Pa, 57-58; asst prof biol sci, Montclair State

Col, 58-62; fel epidemiol & pub health & environ physiol, Sch Med, Yale Univ, 66-67, lectr, 67-69, asst prof, 69-70, asst fel, John B Pierce Found Lab, 66-70; assoc prof, Sch Health Sci, 70-74, PROF PHYSIOL, SCH HEALTH SCI, HUNTER COL, 75- Concurrent Pos: Lectr dept community med, Div Environ Med, Mt Sinae Sch Med, 71- Mem: AAAS; Am Physiol Soc; fel NY Acad Sci; assoc fel NY Acad Med. Res: Temperature regulation; environmental physiology. Mailing Add: Inst of Health Sci Hunter Col 695 Park Ave New York NY 10021

CUNNINGHAM, EARLENE BROWN, b Cleveland, Ohio, Aug 27, 30. BIOCHEMISTRY, ORGANIC CHEMISTRY. Educ: Univ Ill, BS, 49; Univ Calif, Los Angeles, MS, 51; Univ Southern Calif, PhD(org chem), 54. Prof Exp: Asst clin path, Sch Med, Ind Univ, 54-56, res assoc med, 56-59; from asst prof to assoc prof, Col Med, Howard Univ, 59-64, res physiologist, Univ Calif, Berkeley, 66-68; res assoc chem, Univ SC, 68-69, asst prof, 69-70, lectr, 70-71; ASSOC PROF BIOCHEM, MED UNIV SC, 71- Honors & Awards: Lederle Med Fac Award, 61-64. Mem: Am Chem Soc. Res: Kinetic and enzymatic mechanisms of biochemical reactions; mechanisms of reactions influenced by 3', 5' adenosine monophosphate; mechanisms of reactions involving phosphoryl transfer. Mailing Add: Dept of Biochem Med Univ of SC Charleston SC 29401

CUNNINGHAM, ELLEN M, b Chicago, Ill, June 20, 40. MATHEMATICAL LOGIC. Educ: St Mary of the Woods Col, BA, 63; Cath Univ Am, MA, 70; Univ Md, PhD(math), 74. Prof Exp: ASST PROF MATH, ST MARY OF THE WOODS COL, 74- Mem: Math Asn Am; Am Math Soc; Asn Symbolic Logic. Res: Chain models and infinite-quantifier languages. Mailing Add: St Mary of the Woods Col St Mary of the Woods IN 47876

CUNNINGHAM, FRANCES, b New York, NY. CELL BIOLOGY. Educ: Manhattanville Col, BA, 30; Villanova Col, MS, 50; Cath Univ Am, PhD, 56. Prof Exp: Teacher, Convents Sacred Heart, 32-52; prof, 52-74, EMER PROF BIOL, NEWTON COL SACRED HEART, 74-; RESEARCHER CYTOL, SIAS MED RES LAB, LAHEY CLIN FOUND-BROOKS HOSP, 72- Concurrent Pos: Adj asst mem, Sci Resources Found, Mass, 65- Mem: AAAS; Am Soc Cell Biol; Am Soc Cytol. Res: Biological sciences; exfoliative cytology; respiratory cells; histochemistry of oxidative enzymes. Mailing Add: Convent of the Sacred Heart 785 Centre St Newton MA 02158

CUNNINGHAM, FRANK FIRMAN, b Wallsend, Eng, Nov 5, 16; m 44; c 2. GEOGRAPHY. Educ: Univ Durham, BA, 37, dipl educ, 38, MA, 40. Prof Exp: Teacher grammar sch, Eng, 38-40; head dept geog, Royal Acad, Scotland, 46-60; prin lectr educ & geog, Col Educ, Univ Nottingham, 60-65; assoc prof geog, 65-69, PROF GEOG, SIMON FRASER UNIV, 69- Concurrent Pos: Mem, Everest Found Expeds to Andes of Colombia, 56-57 & 60-61; mem, Int Geog Cong, Mexico City, 66; Nat Res Coun Can grant, 67 & 68. Mem: Fel Royal Geog Soc; Can Asn Geographers. Res: Geomorphology of Bornhardts & Tors; geomorphological mapping; models of landscape development; education; Latin America. Mailing Add: Dept of Geog Simon Fraser Univ Burnaby BC Can

CUNNINGHAM, FRANK W, b LeCompton, Kans, Nov 19, 96; m 21; c 2. CHEMISTRY. Educ: Kans Wesleyan, AB, 21; Ft Hays Kans State Col, MS, 36. Hon Degrees: DSc, Col Ozarks, 67. Prof Exp: Teacher pub schs, Kans, 31-42; instr chem & physics, Northeastern Okla Agr & Mech Col, 42-50; prof chem, Col Ozarks, 57-71, chmn div sci, 64-71; RETIRED. Concurrent Pos: Danforth fel, Univ Southern Calif, 55; NSF fel, Ind Univ, 56. Mem: AAAS; Am Chem Soc; fel Am Inst Chem. Res: Analytical research in lead and zinc. Mailing Add: 502 Park St Miami OK 74354

CUNNINGHAM, FRANKLIN E, b Huntington, WVa, Mar 2, 27; m 51; c 5. FOOD SCIENCE, BIOCHEMISTRY. Educ: Kans State Univ, BS, 57; Univ Mo, MS, 59, PhD(poultry prod), 63. Prof Exp: Res chemist biochem, Western Regional Res Lab, USDA, 63-69; ASSOC PROF DAIRY & POULTRY SCI, KANS STATE UNIV, 69- Mem: Poultry Sci Asn; Inst Food Technol. Res: Constituents of poultry and eggs; new and improved food items or processes. Mailing Add: Dept of Dairy & Poultry Sci Call Hall Kans State Univ Manhattan KS 66502

CUNNINGHAM, FREDERIC, JR, b Cooperstown, NY, Sept 6, 21; m 47; c 3. MATHEMATICS. Educ: Harvard Univ, BS, 43, MA, 47, PhD(math), 53. Prof Exp: Teaching fel math, Harvard Univ, 47-50; from instr to asst prof, Univ NH, 51-56; lectr, Bryn Mawr Col, 56-57; asst prof, Wesleyan Univ, 57-59; assoc prof, 59-69, PROF MATH, BRYN MAWR COL, 69- Mem: AAAS; Am Math Soc; Math Asn Am; Math Soc France. Res: L-structure and related structures on Banach spaces. Mailing Add: Dept of Math Bryn Mawr Col Bryn Mawr PA 19010

CUNNINGHAM, FREDERICK WILLIAM, b Stamford, Conn, Mar 23, 02; m 41; c 3. PHYSICS. Educ: Mass Inst Technol, SB, 24, ScD(physics), 36. Prof Exp: Asst physics, Mass Inst Technol, 25-26, res assoc, 26-28; develop engr, Arma Eng Co, Inc, 34-38; develop eng, Arma Corp, 36-44, asst develop engr in chg servomech, 44-48, gen consult, 48-60; PRES, CUNNINGHAM INDUSTS, INC, 53- Honors & Awards: Naval Ord Develop Award, 45. Mem: Am Phys Soc; Optical Soc Am. Res: Automatic spectrophotometers; servomechanisms; illumination for dark adaptation; high intensity searchlights; electrical computers; alternating current potentiometers; non-circular gears; color reproduction. Mailing Add: 56 Hubbard Ave Stamford CT 06905

CUNNINGHAM, GEOFFREY EVERETT, b Hazleton, Ind, Jan 15, 99; m 50; c 1. PHYSICAL CHEMISTRY. Educ: Tulane Univ, BS, 23, MS, 25; Rice Inst, PhD(chem), 28. Prof Exp: Instr chem, Tulane Univ, 23-25; asst prof, Shorter Col, 26, Macalester Col, 28-29 & ceramic eng, Iowa State Col, 29-30; from asst prof to assoc prof chem, Clarkson Tech, 30-42; dir res & develop & later tech dir, Dollinger Corp, 42-56; head struct physics sect, Stromberg-Carlson Co, 56-60; head mat eng, Gen Dynamics-Electtronics, 60-64; consult chemist & engr, 64-75; RETIRED. Concurrent Pos: Heckscher fel, Cornell Univ, 29. Mem: Am Chem Soc; Inst Elec & Electronics Engrs; NY Acad Sci. Res: adsorptive power of lignite char; colors imparted to glass by oxides of iron; sedimentation; colloidal properties of talc; theory of plastic flow and of adsorption; effect of adsorbed ions on the physical character of precipitates; design of oil separators and gas and liquid filters; fluid flow through textiles; detergent mixtures; fabrication of transistors and ferrite cores. Mailing Add: 11 Creek Side Rd Fairport NY 14450

CUNNINGHAM, GEORGE J, b Belfast, N Ireland, Sept 7, 06; m 57. MEDICINE, PATHOLOGY. Educ: Univ London, MBBS, 33, MD, 37; FRCP, 63. Prof Exp: Sr asst pathologist, Royal Sussex County Hosp, Brighton, 36-42; sr lectr path, St Bartholomew's Hosp, Univ London, 46-55, prof, Royal Col Surgeons, Eng, 55-68; PROF PATH, MED COL VA, VA COMMONWEALTH UNIV, 68-, CHMN DEPT ACAD PATH, 73- Concurrent Pos: Med Res Coun traveling fel, 51; vis prof path, State Univ NY Downstate Med Ctr, 62, Cairo Univ, 63 & Univ Ala, Birmingham, 67; mem, Ministry Health Comt Radiation & Cancer Ther, 58-66 & Brit Govt Safety-in-Drugs Comt, 65-68. Honors & Awards: Mem mil div, Order of Brit Empire, 45. Mem: Int Acad Path (pres, 66-); fel NY Acad Sci; Brit Asn Clin Path (pres, 65-66); Path Soc Gt Brit & Ireland. Res: General pathology, especially cellular damage; cancer research, especially cancer of liver and lung. Mailing Add: Dept of Path Va Commonwealth Univ Richmond VA 23219

CUNNINGHAM, GEORGE LEWIS, JR, b New Orleans, La, Nov 20, 23; m 53; c 2. PHYSICAL CHEMISTRY. Educ: Tulane Univ, BE, 44; Univ Calif, Berkeley, MS, 47, PhD(phys chem), 50. Prof Exp: Asst prof chem, La State Univ, 50-53; from assoc prof to prof, Southwest State Col, Okla, 53-61; assoc prof, 61-66, PROF CHEM, EASTERN ILL UNIV, 66- Mem: AAAS; Am Chem Soc. Res: Microwave spectra; molecular structure; digital computer applications to chemistry. Mailing Add: Dept of Chem Eastern Ill Univ Charleston IL 61920

CUNNINGHAM, GLENN N, b Spring City, Tenn, Sept 13, 40; m 62; c 1. BIOCHEMISTRY, NUTRITION. Educ: Univ Tenn, Knoxville, BS, 61; NC State Univ, MS, 64, PhD(nutrit, biochem), 66. Prof Exp: Fel biochem, Clayton Found Biochem Inst, Univ Tex, Austin, 66-68; asst prof chem, Northeast La State Col, 68-69; asst prof, 69-71, ASSOC PROF CHEM, FLA TECHNOL UNIV, 71- Mem: AAAS; Am Chem Soc; Am Inst Biol Sci. Res: Control mechanisms in biosynthesis; eucaryotic developmental biochemistry. Mailing Add: Dept of Chem Fla Technol Univ Orlando FL 32816

CUNNINGHAM, GORDON ROWE, b Oswego, Kans, May 11, 22; m 46; c 3. FORESTRY. Educ: Mich State Univ, BS, 48, PhD(forest econ), 66; Pa State Univ, MS, 50. Prof Exp: Asst exten forester, Univ Ill, 49-54; from asst prof to assoc prof forestry & exten forester, Cornell Univ, 54-61; asst prof, 63-66, ASSOC PROF FORESTRY, UNIV WIS-MADISON, 66-, EXTEN FORESTER, 63- Mem: Soc Am Foresters. Res: Management of small woodlands. Mailing Add: Dept of Forestry Univ of Wis Madison WI 53706

CUNNINGHAM, HARRY N, JR, b Imperial, Pa, Mar 7, 35; m 57; c 2. VERTEBRATE BIOLOGY. Educ: Univ Pittsburgh, BS, 55, MS, 60, PhD(ecol), 66. Prof Exp: Instr biol, Mt Union Col, 59-61; asst prof, Thiel Col, 63-67, chmn dept, 65-67; asst prof, 67-71, ASSOC PROF BIOL, BEHREND COL, PA STATE UNIV, 71- Concurrent Pos: NSF instnl grant, Pa State Univ, 67-; consult, Aquatic Ecol Assocs, 73- Mem: Ecol Soc Am; Am Soc Mammal; Am Inst Biol Sci. Res: Energy relationships of small mammal populations; aspects of ecology of short-tailed shrew, Blarina brevicauda; food studies of stream fishes and salamanders; lipid changes in small mammals. Mailing Add: Dept of Biol Behrend Col of Pa State Univ Erie PA 16510

CUNNINGHAM, HOWARD CHARLES, b Philadelphia, Pa, Aug 22, 42. ORGANIC CHEMISTRY. Educ: Univ Pa, BS, 64, PhD(org chem), 69. Prof Exp: Sr develop chemist, Res Ctr, Lever Bros Co, NJ, 69-71; tech serv rep, Chem Div, Pfizer, Inc, 71-74; MKT RES ANALYST, WITCO CHEM CORP, 74- Mem: Am Chem Soc. Res: Organic synthesis-pharmaceuticals; preparation and evaluation of new actives in detergent formulations; organic specialty chemicals. Mailing Add: 323 W Champlost Ave Philadelphia PA 19120

CUNNINGHAM, HUGH BENSON, b Collinsville, Ala, Apr 8, 18; m 46; c 4. ENTOMOLOGY. Educ: Ala Polytech Inst, BS, 50, MS, 53; Univ Ill, PhD(entom), 62. Prof Exp: Asst zool, Ala Polytech Inst, 50-52, asst entom, 52-53, tech counr, USDA, 53; asst entomologist, Ala Polytech Inst, 53-54; asst entomol, Univ Ill, 54-56; res assoc, Ill Natural Hist Surv, 56-58, asst taxonomist, 58-64, assoc prof scientist, 64-66; ASSOC PROF ZOOL & ENTOM, AUBURN UNIV, 66- Concurrent Pos: Rockefeller Found fel, 61-62. Mem: Entom Soc Am; Soc Study Evolution; Am Inst Biol Sci. Res: Phylogeny and classification of the leafhopper genus Empoasca and the subfamily typhlocybina. Mailing Add: Dept of Zool-Entom Auburn Univ Auburn AL 36830

CUNNINGHAM, HUGH MEREDITH, b Brandon, Man, Dec 28, 27; m 50; c 4. FOOD CHEMISTRY. Educ: Univ Man, BSA, 49, MSc, 50; Cornell Univ, PhD(animal nutrit), 53. Prof Exp: Res scientist, Can Dept Agr, 53-70; HEAD FOOD ADDITIVES & CONTAMINANTS, FOOD DIV, FOOD & DRUG DIRECTORATE, DEPT NAT HEALTH & WELFARE, 70- Concurrent Pos: Res scientist, Inst Animal Physiol, Cambridge, Eng, 67-68. Mem: Am Soc Animal Sci; Nutrit Soc Can; Agr Inst Can; Prof Inst Pub Serv Can; Can Soc Animal Prod. Res: Lipid biochemistry; physiology; nutrition. Mailing Add: Food Div Food & Drug Dir Dept of Nat Health & Welfare Ottawa ON Can

CUNNINGHAM, JAMES A, inorganic chemistry, see 12th edition

CUNNINGHAM, JAMES GORDON, b Pittsburgh, Pa, Dec 13, 40; m; c 3. NEUROPHYSIOLOGY, NEUROLOGY. Educ: Purdue Univ, DVM, 64; Univ Calif, Davis, PhD(physiol), 71. Prof Exp: Lectr physiol, Univ Ibadan, 64-66; ASSOC PROF PHYSIOL & NEUROL, MICH STATE UNIV, 72- Concurrent Pos: Consult, Peace Corps. Mem: AAAS; Am Vet Neurol Asn; Am Soc Vet Physiol & Pharmacol. Res: Neurophysiologic basis of seizures; animal models for epilepsy. Mailing Add: Dept of Physiol Mich State Univ East Lansing MI 48824

CUNNINGHAM, JOHN CASTEL, b Aberdeen, Scotland, Jan 28, 42; m 68; c 1. INSECT PATHOLOGY. Educ: Glasgow Univ, BSc, 64; Oxford Univ, DPhil(insect virol), 67. Prof Exp: RES SCIENTIST, INSECT PATH RES INST, 67- Mem: Soc Invert Path; Brit Soc Gen Microbiol; Entom Soc Can. Res: Various aspects of insect virology, including use of viruses in insect control, electron microscopy and serological relationships of insect viruses. Mailing Add: Insect Path Res Inst PO Box 490 Sault Ste Marie ON Can

CUNNINGHAM, JOHN E, b Malone, NY, Apr 18, 31; m 59; c 4. GEOLOGY. Educ: Dartmouth Col, AB, 53; Univ Ariz, PhD(geol), 65. Prof Exp: Asst geologist, Bear Creek Mining Co, 56 & 57; asst prof geol & anthrop, Eastern NMex Univ, 62-63; geologist, Minerals Br, Superior Oil Co, 63; asst prof, 64-70, ASSOC PROF GEOL & ANTHROP, WESTERN NMEX UNIV, 70- Res: Economic and structural geology; geologic map of Circle Mesa Quadrangle, New Mexico. Mailing Add: Dept of Phys Sci Western NMex Univ Silver City NM 88061

CUNNINGHAM, JOHN L, b Tifton, Ga, Sept 2, 35; m 58; c 2. MYCOLOGY, PLANT PATHOLOGY. Educ: Bowling Green State Univ, BS, 57; Univ Wis, PhD(plant path), 61. Prof Exp: Instr bot, Univ Toledo, 61-62, asst prof, 62-65; res mycologist, Crops Protection Res Br, USDA, 65-69; sr mycologist, Am Type Cult Collection, 70-73; asst prof, 73-74, ASSOC PROF BIOL, NORTHWESTERN COL, 74-, CHMN ARTS & SCI DIV, 73- Concurrent Pos: NSF lectr, 62-63, res grant, 63-65; hon cur fungi, Smithsonian Inst, 66-69; partic, Bredin-Archbold-Smithsonian Biol Surv of Dominica, BWI, 66; mem adv sci comt, Life Sci B, Coun Int Exchange Scholars, 73- Mem: Am Phytopath Soc; Mycol Soc Am. Res: Rusts and other parasitic fungi; taxonomy and evolution of fungi; microfossils. Mailing Add: Dept of Biol Northwestern Col Roseville MN 55113

CUNNINGHAM, JOHN (ROBERT), b Regina, Sask, Jan 5, 27; m 51; c 5. MEDICAL

PHYSICS. Educ: Univ Sask, BEng, 50, MSc, 51; Univ Toronto, PhD(physics), 55. Prof Exp: Physicist, Grain Res Lab, 51-53 & Can Defense Res Bd, 55-58; PHYSICIST, ONT CANCER INST, TORONTO, 58- Concurrent Pos: Tech adv to Govt of Ceylon, Int Atomic Energy Agency, 64-65; assoc ed, Brit Inst Radiol, 67- Mem: Can Asn Physicists; Brit Hosp Physicists Asn; Am Asn Physicists in Med; Brit Inst Radiol. Res: Scattering absorption and measurement of x-rays, gamma rays; application to medical physics and radiobiology; computer applications in radiotherapy. Mailing Add: 6 Marshfield Ct Don Mills ON Can

CUNNINGHAM, LELAND E, b Wiscasset, Maine, Feb 10, 04. CELESTIAL MECHANICS. Educ: Harvard Univ, PhD(astron), 46. Prof Exp: Jr engr, Ballistic Res Labs, Aberdeen Proving Ground, 42, from asst mathematician to mathematician, 42-46; from asst prof to prof, 47-71, EMER PROF ASTRON, UNIV CALIF, BERKELEY, 71- Mem: Am Astron Soc; fel Royal Astron Soc; Brit Astron Asn; Int Astron Union. Res: Orbits of comets, minor planets; satellites and planets; astronometry; motions of artificial satellites and space probes; applications of electronic calculators to problems in astronomy and celestial mechanics. Mailing Add: Dept of Astron Univ of Calif Berkeley CA 94720

CUNNINGHAM, LEON WILLIAM, b Columbus, Ga, June 9, 27; m 48; c 3. BIOCHEMISTRY. Educ: Auburn Univ, BS, 47; Univ Ill, MS, 49, PhD(biochem), 51. Prof Exp: From asst prof to assoc prof, 53-65, assoc dean, 67-72, PROF BIOCHEM, SCH MED, VANDERBILT UNIV, 65-, CHMN DEPT, 72- Concurrent Pos: Fel, Dept Biochem, Univ Wash, 51-53; USPHS spec fel, Netherlands Nat Defense Orgn, 61-62. Mem: AAAS; NY Acad Sci; Am Chem Soc; Am Soc Biol Chemists. Res: Protein structure and function; glycoproteins; enzyme mechanisms; collagens. Mailing Add: Dept of Biochem Vanderbilt Univ Sch of Med Nashville TN 37232

CUNNINGHAM, MARY ELIZABETH, b Newark, NJ, Apr 21, 31. FLUID DYNAMICS. Educ: Mt Holyoke Col, AB, 53; Univ Ill, MS, 55; Univ Ore, PhD(physics), 64. Prof Exp: SR PHYSICIST, LAWRENCE LIVERMORE LAB, 64- Concurrent Pos: Reviewer, Am J Physics, Am Phys Soc, 75- Mem: Am Phys Soc; NY Acad Sci. Res: High temperatures fluid dynamics, 1, 2, and 3 dimensional computer codes for work in this field. Mailing Add: Lawrence Livermore Lab PO Box 808 Livermore CA 94550

CUNNINGHAM, MICHAEL PAUL, b St John, NB, Mar 5, 43; m 66; c 1. PHOTOCHEMISTRY, PHOTOGRAPHIC CHEMISTRY. Educ: St Francis Xavier Univ, BSc, 66; McGill Univ, MSc, 69, PhD(chem), 71. Prof Exp: Indust fel, Bristol Meyers Corp, 70-71; CHEMIST, EASTMAN KODAK CO RES LABS, 71- Res: Organic photochemistry; chemistry of light-sensitive polymers; photo-imaging processes; photographic sensitizing dyes and dye forming processes. Mailing Add: Eastman Kodak Co Res Labs 1669 Lake Ave Rochester NY 14650

CUNNINGHAM, NEWLIN BUCHANAN, b Boone's Path, Va, July 15, 17; m 47; c 4. INDUSTRIAL CHEMISTRY. Educ: Emory & Henry Col, BS, 38; WVa Univ, MS, 63. Prof Exp: Chemist, 38-43, asst chief chemist, 43-70, SR CHEMIST, BELLE WORKS, E I DU PONT DE NEMOURS & CO, INC, 70- Mem: AAAS; Am Chem Soc; Am Inst Chem. Res: Nylon intermediates; antifreezes; methacrylates; urea products; methyl amines; fungicides; waste water treatment. Mailing Add: 1531 Bedford Rd Charleston WV 25314

CUNNINGHAM, NICHOLAS, b Cooperstown, NY, July 30, 28; m; c 3. PEDIATRICS, PUBLIC HEALTH. Educ: Harvard Col, AB, 50; Johns Hopkins Univ, MD, 55; London Sch Hyg & Trop Med, dipl, 65. Prof Exp: Med consult, US Peace Corps, 62-64; lectr, Johns Hopkins Univ, 69; ASST PROF PEDIAT & COMMUNITY MED, MT SINAI SCH MED, 69- Concurrent Pos: NIH fel neonatal physiol, Sloane Hosp, Columbia-Presby Med Ctr, 61-62; spec fel int health, Sch Hyg, Johns Hopkins Univ, 66-69. Mem: AMA; fel Am Acad Pediat; Am Pub Health Asn. Res: Growth and development; nutrition; tropical pediatrics; child care systems. Mailing Add: Dept of Community Med Mt Sinai Sch of Med New York NY 10029

CUNNINGHAM, PAUL THOMAS, b Newton, Iowa, Oct 16, 36; m 61; c 3. PHYSICAL CHEMISTRY. Educ: Univ Idaho, BS, 58; San Diego State Col, MS, 65; Univ Calif, Berkeley, PhD(phys chem), 68. Prof Exp: Mem staff, 68-75, GROUP LEADER ENVIRON CHEM & MGR ANAL CHEM LAB, ARGONNE NAT LAB, 75- Mem: Optical Soc Am; AAAS; Am Chem Soc. Res: Atmospheric chemistry; chemical analysis of airborne particulates; sulfur emission control chemistry; molecular spectroscopy; chemical kinetics of reactions related to environmental pollution. Mailing Add: Chem Eng Div Argonne Nat Lab Argonne IL 60439

CUNNINGHAM, PETER JOHN, b Winterset, Iowa, Oct 27, 41; m 64; c 3. ANIMAL BREEDING. Educ: Iowa State Univ, BS, 64; Okla State Univ, MS, 67, PhD(animal breeding), 69. Prof Exp: Asst animal sci, Okla State Univ, 64-68; asst prof, 68-73, ASSOC PROF ANIMAL SCI, UNIV NEBR-LINCOLN, 73- Concurrent Pos: Mem, Rec & Stand Comt, Nat Swine Improv Fedn, 75- Mem: Am Soc Animal Sci. Res: Swine breeding, with emphasis in the areas of selection for ovulation rate and selection for lean growth. Mailing Add: 211 Marvel Baker Hall Univ of Nebr Lincoln NE 68583

CUNNINGHAM, ROBERT ELWIN, b Parkersburg, WVa, Mar 16, 29; m 54; c 2. POLYMER CHEMISTRY. Educ: WVa Univ, BS, 51, PhD(chem), 55. Prof Exp: Instr WVa Univ, 54; sr res chemist, 55-75, RES SCIENTIST, GOODYEAR TIRE & RUBBER CO, 75- Mem: Am Chem Soc. Res: Stereospecific polymerizations of olefins and diolefins; anionic polymerizations initiated by lithium alkyls; preparation and properties of block polymers using anionic techniques. Mailing Add: 342 Kenilworth Dr Akron OH 44313

CUNNINGHAM, ROBERT ERNEST, physical chemistry, see 12th edition

CUNNINGHAM, ROBERT GAIL, b Manton, Mich, July 2, 28; m 47; c 2. SURFACE PHYSICS, SOLID STATE PHYSICS. Educ: Mich State Univ, BS, 50, PhD(physics), 57. Prof Exp: Res physicist, Eastman Kodak Res Labs, 57-69, SR ENGR, EASTMAN KODAK ENG DIV, 69- Mem: Am Phys Soc; Optical Soc Am; Soc Motion Picture & TV Eng; Sigma Xi. Res: Static electrification of photographic materials. Mailing Add: Bldg 23 Eastman Kodak Co Kodak Park Rochester NY 14650

CUNNINGHAM, ROBERT LESTER, b Fullerton, Nebr, July 6, 29; m; c 4. SOIL GENESIS & MORPHOLOGY. Educ: Univ Nebr, BS, 59, MS, 61; Wash State Univ, PhD(soil genesis & morphol), 64. Prof Exp: Res asst soil genesis, Wash State Univ, 60-64; asst prof soil technol, 64-70, ASSOC PROF SOIL GENESIS & MORPHOL, PA STATE UNIV, 70- Mem: Soil Sci Soc Am. Res: Movement of soil materials; arrangement of genetic soil parts; relationships of soils to the ecosystem of which they are a part; characteristics, interpretations and use of Pennsylvania soils; micromorphology examination of certain soil features; application of remote sensing techniques in differentiating soils in the field. Mailing Add: 311 Tyson Bldg Pa State Univ University Park PA 16802

CUNNINGHAM, ROBERT M, b Boston, Mass, July 1, 19; m 45; c 3. METEOROLOGY. Educ: Mass Inst Technol, SB, 42, ScD(meteorol), 52. Prof Exp: Instr meteorol instruments, Mass Inst Technol, 42-43, asst meteorol, 43-46, res assoc, 46-52; proj scientist, 52-61, br chief, 61-74, SR SCIENTIST METEOROL, AIR FORCE CAMBRIDGE RES LAB, 74- Concurrent Pos: Mem, Nat Adv Comt Aeronaut Subcomt Icing, 52-57; mem, Weather Modification Panel, World Meteorol Orgn, 66- Mem: Am Meteorol Soc; Am Geophys Union; Royal Meteorol Soc. Res: Physics of the atmosphere; cloud physics; particle size; water content; cloud dynamics; airborne meteorological instrumentation; weather erosion effects on high speed flight. Mailing Add: Meteorol Lab Air Force Cambridge Res Labs Lawrence G Hanscom Field Bedford MA 01730

CUNNINGHAM, ROBERT STEPHEN, b Springfield, Mo, June 28, 42; m 64. MATHEMATICS. Educ: Drury Col, BA, 64; Univ Ore, MA, 66, PhD(math), 69. Prof Exp: ASST PROF MATH, UNIV KANS, 69- Mem: Am Math Soc; Math Asn Am. Res: Noncommutative ring theory. Mailing Add: Dept of Math Univ of Kans Lawrence KS 66044

CUNNINGHAM, ROBERT W, solid state physics, see 12th edition

CUNNINGHAM, ROY THOMAS, b Wexford, Pa, Oct 14, 31; m 56; c 7. ENTOMOLOGY. Educ: Bucknell Univ, BS, 53; Cornell Univ, PhD(entom), 60. Prof Exp: Entomologist, Birds Eye Div, Gen Foods Corp, 60-63; SUPVRY RES ENTOMOLGIST, AGR RES SERV, USDA, 63- Mem: AAAS; Entom Soc Am. Res: Development of methods for control or eradication of tephritid subtropical fruit flies. Mailing Add: US Dept of Agr PO Box 1365 Hilo HI 96720

CUNNINGHAM, RUSSELL D, b Mobile, Ala, Oct 29, 32. ENDOCRINOLOGY, METABOLISM. Educ: Miami Univ, BA, 54; Vanderbilt Univ, MD, 58. Prof Exp: Intern, Vanderbilt Univ Hosp, 59; resident, Los Angeles Children's Hosp, 60, chief resident, 61; asst prof, 64-67, ASSOC PROF PEDIAT, MED CTR, UNIV ALA, 67- Concurrent Pos: USPHS res trainee, Harvard Univ, 62. Res: Athyreotic cretinism; mitochondrial protein synthesis; cardiovascular malformations in Turner's syndrome; long-term effects of adrenalectomy and gonadectomy on thyroid function in the rat. Mailing Add: Dept of Pediat Univ of Ala Med Ctr Birmingham AL 35233

CUNNINGHAM, SAMUEL PRESTON, b Beaumont, Tex, Apr 4, 24; m 54; c 1. PHYSICS. Educ: La State Univ, BS, 43, MS, 46; Johns Hopkins Univ, PhD(physics), 51. Prof Exp: Instr physics, La State Univ, 43-44; instr & res assoc, Johns Hopkins Univ, 51-53, asst prof, 53-54; res assoc, Proj Matterhorn, Princeton Univ, 54-56; sr staff scientist, Gen Atomic Div, Gen Dynamics Corp, 56-60; res scientist, Rand Corp, 60-68, ASSOC PROF PHYSICS, CALIF STATE UNIV, NORTHRIDGE, 68- Res: Plasma physics; optical spectra. Mailing Add: Dept of Physics & Astron Calif State Univ Northridge CA 91324

CUNNINGHAM, VIRGIL DWAYNE, b Artesian, SDak, Nov 29, 30; m 56; c 1. ENTOMOLOGY, PLANT PATHOLOGY. Educ: NDak State Univ, BS, 58, MS, 60; Iowa State Univ, PhD(entom), 63. Prof Exp: Entomologist, 63-69, supvr entom, 69-70, DEPT HEAD ENTOM, BIOL SCI RES CTR, SHELL DEVELOP CO, 70- Mem: Entom Soc Am; Sigma Xi. Res: Agricultural chemicals; insecticides, herbicides and plant growth regulators. Mailing Add: 1114 Durant Modesto CA 95350

CUNNINGHAM, WILLIAM GLENN, b West Liberty, WVa, Feb 11, 08. GEOGRAPHY. Educ: Univ Calif, Los Angeles, BA, 31; Univ Pa, MA, 40, PhD(econ & geog), 50. Prof Exp: Instr geog, Univ Calif, Los Angeles, 31-35; instr geog & indust, Univ Pa, 37-42; lectr geog, Univ Calif, Los Angeles, 46-51; rep econ res, Am Can Co, 51-53; asst field prod mgr documentary films, Herbert Knapp Prod, 55-56; from instr to assoc prof earth sci, 58-67, PROF EARTH SCI, LOS ANGELES CITY COL, 67- Concurrent Pos: Ed, Publ Los Angeles Geog Soc, 61-; prof geog, Seven Seas Div, Chapman Col, 66. Mem: Asn Am Geog; Royal Geog Soc. Res: Economic geography; geography of manufacturing; geography of California and Western United States; cultural geography of Polynesia and Melanesia. Mailing Add: Dept of Earth Sci Los Angeles City Col Los Angeles CA 90029

CUNNINGHAM, WILLIAM JOHN, b Ubly, Mich, Dec 8, 14; m 38; c 2. DENTISTRY, PREVENTIVE DENTISTRY. Educ: Univ Western Ont, AB, 34; Univ Detroit, DDS, 37; Mich State Univ, MBA, 65. Prof Exp: Pvt pract, 38-69; assoc prof bus admin, Manchester Col, 69-70; PROF DENT, SCH DENT, MEHARRY MED COL, 71- Mem: Fel Acad Gen Dent; Am Dent Asn. Mailing Add: Sch of Dent Meharry Med Col Nashville TN 37208

CUNNINGHAM, WILLIAM PEYTON, b Wilkes-Barre, Pa, Dec 26, 06; m 31; c 1. PHYSICS. Educ: Sheffield Sci Sch, BS, 28; Yale Univ, PhD(spectros), 32. Prof Exp: Asst, Yale Univ, 31; teacher high sch, Pa, 33-34; head sci dept, Calif State Teachers Col, 34-38 & Tower Hill Sch, Del, 38-43; asst supvr lectr staff, Radar Sch, Mass Inst Technol, 43-45; prof physics, Naval Postgrad Sch, Annapolis, 46-50, prof physics, 50-73, DISTINGUISHED PROF PHYSICS & OPER RES, NAVAL POSTGRAD SCH, MONTEREY, 73- Concurrent Pos: Instr, Temple Univ, 42-43. Mem: Am Phys Soc; Am Asn Physics Teachers; Opers Res Soc Am. Res: Optics; operations research. Mailing Add: Dept of Physics Naval Postgrad Sch Monterey CA 93940

CUNNOLD, DEREK M, b Reading, Eng, July 10, 40; US citizen; m 65; c 1. METEOROLOGY. Educ: Cambridge Univ, BA, 62, MA, 66; Cornell Univ, PhD(elec eng), 65. Prof Exp: Adv res engr, Appl Res Lab, Sylvania Electronics Systs Div, Gen Tel & Electronics Corp, 65-67; res fel ionospheric physics, Harvard Col Observ, 67-68; res physicist, Smithsonian Astrophys Observ, Cambridge, 68-70; RES ASSOC METEOROL, MASS INST TECHNOL, 70- Concurrent Pos: Consult, Appl Res Lab, Sylvania Electronic Systs, 67-69 & Lincoln Lab, Mass Inst Technol, 69-; hon res fel, Div Eng & Appl Physics, Harvard Univ, 68-70; mem, Int Comn Upper Atmosphere, Int Asn Meteorol & Atmospheric Physics, 75- Mem: Am Meteorol Soc; Am Geophys Union. Res: Chemical-dynamical modeling of the stratosphere; stratospheric ozone and pollution; dynamics of the stratosphere and mesosphere; radar observations of the mesosphere. Mailing Add: 54-1517 Dept of Meteorol Mass Inst of Technol Cambridge MA 02139

CUNOV, CARL HENRY, b Detroit, Mich, Feb 19, 18; m 42; c 3. MEDICINAL CHEMISTRY, ORGANIC CHEMISTRY. Educ: Wayne State Univ, BS, 41, PhD(chem), 51. Prof Exp: Chief chemist drug mfg, Jamieson Pharamcal, 39-46 & McKay-Davis, 46-47; asst plant tech dir drug mfg, Smith, Kline & French Labs, 50-54, plant tech dir, 54-58, process dept mgr, 58-60, mfr proj dir, 60-61; assoc prof chem, Cent Mo State Col, 62-65; asst prof Keuka Col, 65-66, prof & head dept, 66-68; CHEM RES SUPVR, PENNWALT CORP, 68- Concurrent Pos: Consult, 61- Mem: AAAS; Am Chem Soc; Am Pharmaceut Asn; Pharmaceut Mfrs Asn. Res: Mills-Nixon effect; cyclobutane derivatives; medicinal chemistry; analytical methods. Mailing Add: Pharmaceut Div Pennwalt Corp 755 Jefferson Rd Rochester NY 14623

CUPAS, CHRIS ANGELO, b Boston, Mass, June 9, 37; m 66. ORGANIC CHEMISTRY. Educ: Harvard Col, AB, 59; Princeton Univ, MA, 62, PhD(chem), 65. Prof Exp: Asst prof, 65-69, ASSOC PROF CHEM, CASE WESTERN RESERVE UNIV, 69- Concurrent Pos: Consult, Diamond Shamrock Corp. Mem: Am Chem Soc; The Chem Soc. Res: Synthetic organic and physical organic chemistry; nuclear magnetic resonance studies of transient species; new synthetic reactions; rearrangements of polycyclic molecules. Mailing Add: Dept of Chem Case Western Reserve Univ Cleveland OH 44106

CUPERY, KENNETH N, b Wilmington, Del, May 27, 37; m 60; c 2. OPTICS. Educ: Oberlin Col, BA, 59; Johns Hopkins Univ, MA, 64. Prof Exp: Develop engr, 65-66, sr develop engr, 66-69, sr res physicist, 69-72, SUPVR OPTICS RES, KODAK APPARATUS DIV, RES LAB, 72- Mem: Optical Soc Am. Res: Modulation transfer function and relation to subjective quality on optical images; MTF based optical test systems; micrographies systems development. Mailing Add: 400 Westminster Rd Rochester NY 14607

CUPERY, WILLIS ELI, b Wilmington, Del, July 1, 32; m 55; c 3. PESTICIDE CHEMISTRY. Educ: Oberlin Col, AB, 54; Univ Ill, PhD(org chem), 58. Prof Exp: Res chemist, 58-66, SR RES CHEMIST, E I DU PONT DE NEMOURS & CO, INC, 66- Mem: Am Chem Soc. Res: Pesticide formulation development. Mailing Add: 13 Crestfield Rd Wilmington DE 19810

CUPP, EDDIE WAYNE, b Highsplint, Ky, Apr 17, 41; m 64; c 1. ENTOMOLOGY, INVERTEBRATE ZOOLOGY. Educ: Murray State Univ, BA, 64; Univ Ill, Urbana, PhD(entom), 69. Prof Exp: Res fel, Sch Pub Health & Trop Med, Tulane Univ, 69-70; res biologist, Gulf South Res Inst, La, 70-71; asst prof biol, Univ Southern Miss, 71-75; ASST PROF ENTOM, CORNELL UNIV, 75- Concurrent Pos: Adj asst prof, Sch Pub Health & Trop Med, Tulane Univ, 70-71. Mem: Entom Soc Am; Am Soc Trop Med & Hyg; Am Soc Parasitol; Am Mosquito Control Asn. Res: Bionomics of medically important insects; developmental biology of mosquitos; invertebrate tissue culture; in vitro cultivation of filariae; dynamics of transmission of filariae. Mailing Add: Dept of Entom Cornell Univ Ithaca NY 14853

CUPP, PAUL VERNON, JR, b Corbin, Ky, Oct 18, 42; m 69. PHYSIOLOGICAL ECOLOGY, HERPETOLOGY. Educ: Eastern Ky Univ, BS, 65, MS, 70; Clemson Univ, PhD(zool), 74. Prof Exp: Asst prof biol, Ga Southern Col, 73-74; ASST PROF BIOL, EASTERN KY UNIV, 74- Mem: Sigma Xi; Am Soc Zoologists; Am Inst Biol Sci; Ecol Soc Am; Soc Study Amphibians & Reptiles. Res: Thermal tolerance and acclimation and physiological responses to temperature in amphibians and reptiles; aggressive and courtship behavior of amphibians. Mailing Add: Dept of Biol Sci Eastern Ky Univ Richmond KY 40475

CUPPAGE, FRANCIS EDWARD, b Cleveland, Ohio, Aug 17, 32; m 56; c 3. PATHOLOGY. Educ: Case Western Univ, BS, 54; Ohio State Univ, MS, 59, MD, 59. Prof Exp: Rotating intern med, Univ Hosps, Cleveland, 59-60; resident path, 60-63; instr, Case Western Reserve Univ, 64-65; asst prof, Ohio State Univ, 65-67; from asst prof to assoc prof, 67-73, PROF PATH, UNIV KANS MED CTR, KANSAS CITY, 73- Concurrent Pos: NIH fel path, Case Western Univ, 63-64. Mem: Int Soc Nephrol; Am Asn Path & Bact; Int Acad Path; Am Soc Exp Path. Res: Kidney response of nephron to injury and repair. Mailing Add: Dept of Path Univ of Kans Med Ctr Kansas City KS 66103

CUPPER, ROBERT ALTON, b Tyrone, Pa, Jan 7, 18; m 41; c 4. ORGANIC CHEMISTRY, PETROLEUM CHEMISTRY. Educ: Juniata Col, BS, 40. Prof Exp: Chemist chem petrol prods, United Ref Co, 41-46; indust fel org chem, Mellon Inst, 46-59; group leader, Chem & Plastic Div, 60-69, RES SCIENTIST, RES & DEVELOP DEPT, UNION CARBIDE CORP, 69- Mem: AAAS; Am Chem Soc; Soc Automotive Eng; Am Soc Lubrication Eng; NY Acad Sci. Res: Fuels and lubricants; additives and chemical inter-synthetic lubricants; synthesis and formulation. Mailing Add: Res & Develop Dept Union Carbide Corp Box 65 Tarrytown NY 10592

CUPPETT, CHARLES CECIL, biochemistry, physical chemistry, see 12th edition

CUPPS, PERRY THOMAS, b Granby, Mo, June 18, 16; m 44; c 4. ANIMAL PHYSIOLOGY. Educ: Univ Mo, BS, 39; Cornell Univ, PhD(animal physiol), 43. Prof Exp: Asst, Cornell Univ, 39-43; from asst prof to assoc prof, 46-59, PROF ANIMAL HUSB, UNIV CALIF, DAVIS, 59- Concurrent Pos: NSF fel, 61-62. Res: Endocrinology; thyrotrophic hormone assays; normal change of reproductive tract during estrous cycle; joint lesions in horses; estrous cycle in cattle; adrenal physiology. Mailing Add: Dept of Animal Sci Univ of Calif Davis CA 95616

CUPRAK, LUCIAN J, biochemistry, see 12th edition

CURBY, WILLIAM ADOLPH, b Evanston, Ill, Sept 25, 27; m 53; c 3. PHYSICAL BIOLOGY. Educ: Tufts Col, BS, 50, MS, 53. Prof Exp: Asst dent sch, Tufts Col, 50-51, res coordr, Cleft Palate Inst, 53-57, asst prod dent, 56-58; dir res, Int Dynamics Corp, 59-60; mem staff, Charles D Sias Res Labs, Brooks Hosp, 61-70, DIR, SIAS MED RES LABS, LAHEY CLIN FOUND, 71- Concurrent Pos: Res consult, US Naval Med Ctr, Md, 51-52; lectr, Northeastern Univ, 63-; res assoc, Lahey Clin Found, 70- Mem: AAAS; Am Soc Microbiol; NY Acad Sci. Res: Organization and development of specialized diagnostic techniques in the fields of pressure, electrophysiology, sound, x-ray, optics, enzymology, statistics, biochemistry, hematology, and microbiology. Mailing Add: Sias Med Res Labs 211 Summitt Ave Brookline MA 02146

CURD, MILTON RAYBURN, b Tulsa, Okla, July 18, 28; m 52; c 4. ZOOLOGY. Educ: Okla State Univ, BS, 50, MS, 51, PhD(zool), 66. Prof Exp: Asst prof, 64-68, ASSOC PROF ZOOL, OKLA STATE UNIV, 68- Mem: Am Soc Ichthyol & Herpet. Res: Anatomy and taxonomy of freshwater fishes; structure and phylogeny of sense organs, especially the eye. Mailing Add: Sch of Biol Sci Okla State Univ Stillwater OK 74074

CURD, RUDY LEROY, b Watauga, Tenn, May 20, 43; m 65; c 1. MATHEMATICS. Educ: Lincoln Mem Univ, BS, 65; Univ Ky, MA, 66, PhD(math), 69. Prof Exp: Instr, Ky State Col, 69; ASST PROF MATH, APPALACHIAN STATE UNIV, 69- Mem: Am Math Soc; Math Asn Am. Res: Topologic algebra, especially topological semigroups. Mailing Add: Dept of Math Appalachian State Univ Boone NC 28607

CURET, JUAN DANIEL, b Arecibo, PR, Dec 24, 14; m 39, 49; c 2. PHYSICAL CHEMISTRY. Educ: Univ PR, BS, 36; Univ Mich, MS, 45, PhD(chem), 48. Prof Exp: Teacher high sch, PR, 36-40, prin, 40-42; from instr to assoc prof chem, 42-53, extramural ctr dir, 48-54, chmn dept chem, 59-60, dean col natural sci, 60-67, PROF CHEM, UNIV PR, RIO PIEDRAS, 53-, COORDR GRAD PROGS & RES, 75- Concurrent Pos: Vis fel, Ohio State Univ, 55-56; lectr, PR Nuclear Ctr, 58-59; res scientist, NY Univ, 67-68; consult, PR Develop Admin, 67-70. Mem: Am Chem Soc. Res: Magnetic susceptibility; radiochemistry; x-ray crystallography; atmospheric chemistry. Mailing Add: Col of Natural Sci Univ of PR Rio Piedras PR 00931

CURETON, GLEN LEE, b Santa Cruz, Calif, Mar 29, 38; m 64; c 1. PHARMACEUTICS. Educ: Univ Calif, San Francisco, PharmD, 62; Harvard Univ, MBA, 64. Prof Exp: Asst dir life sci res, Stanford Res Inst, 64-65, indust economist, 65-66, health economist, 66-67; dir new prod develop, Chattem Drug & Chem Co, 67-68, dir res & new prod develop, 68-72; DIR RES & DEVELOP, BARNES-HIND PHARMACEUTICALS, INC, 72- Mem: AAAS; Am Chem Soc; Am Pharmaceut Asn; Am Acad Dermat; Asn Res in Vision & Ophthalmol. Res: Antacids, anitperspirants and other chemicals, inculding high purity aluminas; development of pharmaceuticals; ophthalmologic contact lens solutions; dermatologicals; technoeconomic analyses of life sciences research, pharmaceutical industry and cosmetics. Mailing Add: 714 Raymundo Ave Los Altos CA 94022

CURETON, THOMAS (KIRK), JR, b Fernandina, Fla, Aug 4, 01; m; c 3. APPLIED PHYSIOLOGY. Educ: Yale Univ, BS, 25; Int YMCA Col, BPE, 29, MPE, 30; Columbia Univ, AM, 36, PhD(educ res), 39. Hon Degrees: DSc, Univ Ottawa, 68, Univ Southern Ill, 69. Prof Exp: Phys dir & teacher biol & gen sci, Suffield Sch, Conn, 25-29; prof appl physics & animal mech, Springfield Col, 29-41; dir res nat sci & phys educ, 35-42; from assoc prof to prof, 42-74, EMER PROF PHYS EDUC, UNIV ILL, URBANA-CHAMPAIGN, 74-, DIR PHYS FITNESS INST, 70- Concurrent Pos: Supvr res, Col Phys Educ, Univ Ill, Urbana-Champaign, 42-74. Honors & Awards: Robert-Gulick Award, Phys Educ Soc, 45; Am Col Sports Med Award, 68; Am Dent Asn Citation, 68. Mem: Am Physiol Soc; fel Soc Res Child Develop; fel Am Pub Health Asn; fel Am Asn Health, Phys Educ & Recreation; NY Acad Sci. Res: Body mechanics; anthropometry; physical fitness; physiology of exercise; aquatics; athletics; gymnastics; sport medicine; physical education. Mailing Add: Col of Phys Educ Univ of Ill Urbana IL 61820

CURJEL, CASPAR ROBERT, b Switz, Nov 21, 31; m 55; c 2. MATHEMATICS. Educ: Swiss Fed Inst Technol, dipl, 54, DrScMath, 60. Prof Exp: From instr to asst prof, Cornell Univ, 60-64; vis assoc prof, 64-65, assoc prof, 65-69, PROF MATH, UNIV WASH, 69- Concurrent Pos: Mem, Inst Advan Study, 63-64. Res: Algebraic topology; algebra. Mailing Add: Dept of Math Univ of Wash Seattle WA 98105

CURL, ELROY ARVEL, b Marmaduke, Ark, Dec 1, 21. PLANT PATHOLOGY. Educ: La Polytech Inst, BS, 49; Univ Ark, MS, 50; Univ Ill, PhD(plant path), 54. Prof Exp: Spec res asst plant path, State Natural Hist Surv, Ill, 50-54, asst plant pathologist, 54; asst plant pathologist, 54-57, assoc prof, 57-67, PROF BOT & MICROBIOL, AUBURN UNIV, 67- Mem: Am Phytopath Soc; Mycol Soc Am. Res: Soil microbiology; soil fungus ecology; soil microbe-pesticide interactions; biological control; root diseases of plants. Mailing Add: Dept of Bot & Microbiol Auburn Univ Auburn AL 36830

CURL, HERBERT (CHARLES), JR, b New York, NY, Feb 26, 28; m 73; c 2. BIOLOGICAL OCEANOGRAPHY. Educ: Wagner Col, BS, 50; Ohio State Univ, MS, 51; Fla State Univ, PhD(biol oceanog), 56. Prof Exp: Asst, Oceanog Inst, 54-56; res assoc, Woods Hole Oceanog Inst, 56-61; from assoc prof to prof oceanog, Ore State Univ, 61-74; fisheries oceanog coordr, Nat Marine Fisheries Serv, 74-75; ECOLOGIST, ENVIRON RES LABS, NAT OCEANIC & ATMOSPHERIC ADMIN, 75- Concurrent Pos: Fulbright grant, Colombia, SAm, 71. Mem: Fel AAAS; Ecol Soc Am; Am Soc Naturalists; Am Soc Limnol & Oceanog(secy, 64-66); Int Asn Limnol. Res: Physiological ecology of marine organisms; ecosystem computer simulation; snow algae and snow field ecology. Mailing Add: RX4 OCSEP Off Environ Res Labs Nat Oceanic & Atmospheric Admin Boulder CO 80302

CURL, ROBERT FLOYD, JR, b Alice, Tex, Aug 23, 33; m 55; c 2. PHYSICAL CHEMISTRY. Educ: Rice Inst, BA, 54; Univ Calif, PhD, 57. Prof Exp: Res fel, Harvard Univ, 57-58; from asst prof to assoc prof, 58-67, PROF CHEM, RICE UNIV, 67- Concurrent Pos: NATO fel, Oxford Univ, 64-65; vis res officer, Nat Res Coun Can, 72-73. Honors & Awards: Clayton Prize, Inst Mech Eng, 58. Mem: Am Chem Soc. Res: Microwave and laser spectroscopy. Mailing Add: Dept of Chem Rice Univ Houston TX 77001

CURL, SAMUEL EVERETT, b Tolar, Tex, Dec 26, 37; m 57; c 3. ANIMAL PHYSIOLOGY, ENDOCRINOLOGY. Educ: Sam Houston State Teachers Col, BS, 59; Univ Mo, MS, 61; Tex A&M Univ, PhD(physiol reprod), 63. Prof Exp: Asst, Univ Mo, 60-61; instr endocrinol, Tex Tech Col, 61; asst, Tex A&M Univ, 61-63; from asst prof to assoc prof physiol reprod & endocrinol, 63-70, asst dean res & interim dean agr, 68-71, assoc dean & dir res col agr sci, 71-73, PROF ANIMAL SCI, TEX TECH UNIV, 70-, ASSOC VPRES ACAD AFFAIRS, 73- Concurrent Pos: Fel, Am Coun Educ Acad Admin Internship Prog, Off Pres, Okla State Univ, 72-73. Mem: Am Soc Animal Sci. Res: Physiology of dwarfism in beef cattle; synchronization of estrus and hormone-induced multiple birth in cattle and sheep; environmental influence on reproductive efficiency of livestock; physiology of growth. Mailing Add: 104 Admin Bldg Tex Tech Univ Lubbock TX 79409

CURLESS, WILLIAM TOOLE, b Quincy, Ill, Mar 11, 28; m 55; c 3. ENVIRONMENTAL SCIENCES, PESTICIDE CHEMISTRY. Educ: Univ Ill, BS, 49; Univ Kans, MS, 56. Prof Exp: Anal chemist, Victor Chem Works, 49-51, inorg res chemist, 51-53; res chemist, Columbia-Southern Chem Corp, 55-61; res chemist, Spencer Chem Co, 62-66; SR RES CHEMIST, GULF OIL CORP, 66- Mem: Am Chem Soc. Res: Nonaqueous solvents; condensed phosphates; phase diagrams; fertilizer technology; pesticide environmental studies in soil; pesticide residue analytical method development; packaging of pesticides. Mailing Add: Gulf Oil Chem Co 9009 W 67th St Merriam KS 66202

CURLEY, JAMES EDWARD, b Winchester, Mass, Apr 10, 44; m 66; c 1. ANALYTICAL CHEMISTRY. Educ: Univ Mass, Amherst, BS, 66, MS, 68, PhD(chem), 70. Prof Exp: Staff chemist, Med Res Lab, 70-73, sr res scientist, Pfizer Cent Res, 73-75, SECT SUPVR, PFIZER QUAL CONTROL, PFIZER, INC, 75- Mem: Am Chem Soc; Sigma Xi. Res: Analytical methodology for penicillin products; analysis of pharmaceuticals by high pressure liquid chromatography. Mailing Add: Pfizer Inc Groton CT 06340

CURLIN, LEMUEL CALVERT, b Waxahachie, Tex, Feb 1, 13; m 40; c 3. PHARMACEUTICAL CHEMISTRY. Educ: Trinity Univ, Tex, BS, 34; Univ Chicago, MS, 36, PhD(biochem), 40. Prof Exp: Chemist, Ideal Lab, Univ Tex, 35-36; lab asst, Univ Chicago, 37-40, Armour & Co fel, 40; asst dir develop lab, Armour & Co, 40-41, res chemist, 41-43; secy, 64-69, CHIEF CHEMIST, L PERRIGO CO, 43-, EXEC VPRES, 69- Res: Separation plasma proteins; synthetic thyroid; organic synthesis. Mailing Add: L Perrigo Co 100 Brady St Allegan MI 49010

CURME, GEORGE OLIVER, JR, b Mt Vernon, Iowa, Dec 24, 88; m 16; c 5. CHEMISTRY. Educ: Northwestern Univ, BS, 09; Univ Chicago, PhD(org chem), 13. Hon Degrees: ScD, Northwestern Univ, 33; DSc, Univ Chicago, 54. Prof Exp: Fel, Mellon Inst, 14-20; chief chemist, Carbide & Carbon Chem Co, 20-29, vpres, 29-51, dir res, 29-44; vpres, Res Labs, Union Carbide Corp, 38-52, chmn bd, 52-54, vpres, res, 51-55, dir, 52-61; RETIRED. Concurrent Pos: Vpres, Bakelite Co, 39-51; vpres & dir, Carbide & Carbon Chem Ltd, 44-55. Honors & Awards: Chandler Medal,

Columbia Univ, 33; Perkins Medal, 35; Cresson Medal, Franklin Inst, 36; Nat Mod Pioneer Award, Nat Mfrs Asn, 40; Gibbs Medal, 44. Mem: Nat Acad Sci; Am Chem Soc; Am Inst Chem Eng; Am Soc Chem Indust; NY Acad Sci. Res: Organic chemical reactions; hydrocarbon gases; plastics. Mailing Add: 9 Ednam Village Rte 3 Charlottesville VA 22901

CURME, HENRY GARRETT, b Niagara Falls, NY, Dec 31, 23; m 48; c 3. PHYSICAL CHEMISTRY. Educ: Northwestern Univ, BS, 45, MS, 47; Univ Calif, PhD(chem), 50. Prof Exp: Chemist, Manhattan Proj, Tenn Eastman Co, 44-46; asst, Northwestern Univ, 46-47 & Univ Calif, 47-50; res chemist & res assoc, 50-65, lab head, 65-70, SR LAB HEAD, EASTMAN KODAK CO, 70- Mem: Am Chem Soc. Res: Interactions between proteins and small molecules; photochemistry; polymer solutions; builk properties and morphology of hydrophilic polymers; polymer networks; adsorption of polymers; surface properties of polymers. Mailing Add: Res Lab Kodak Park Works Eastman Kodak Co Rochester NY 14604

CURME, JOHN HENRY, b White Plains, NY, Apr 20, 29; m 55; c 3. PLANT GENETICS. Educ: Harvard Univ, AB, 49; Kans State Univ, MS, 51; Iowa State Col, PhD(agron), 55. Prof Exp: Plant pathologist & geneticist, Campbell Soup Co, 57-68; GENETICIST, AGR RES DEPT, GREEN GIANT CO, 68- Concurrent Pos: Fel biol, Calif Inst Technol, 58-60. Mem: Am Soc Agron; NY Acad Sci. Res: Mushroom genetics and strain improvement. Mailing Add: 1332 N Court LeSueur MN 56058

CURNEN, EDWARD CHARLES, JR, b Yonkers, NY, Jan 5, 09; m 42, 68; c 7. PEDIATRICS, MICROBIOLOGY. Educ: Yale Univ, AB, 31; Harvard Univ, MD, 35. Prof Exp: Res bacteriologist, Children's Hosp, Boston, Mass, 35-36, med intern, 36-38, med resident, 38-39; asst, Rockefeller Inst, NY, 39-46; from asst prof prev med to assoc prof prev med & pediat, Sch Med, Yale Univ, 46-52; prof pediat & chmn dept, Univ NC, 52-60; chmn dept, 60-70, Carpentier prof, 60-74, EMER CARPENTIER PROF PEDIAT, COL PHYSICIANS & SURGEONS, COLUMBIA UNIV, 74-, SPEC LECTR, 74- Concurrent Pos: Harvard Med Sch fel, Thorndike Mem Lab, Boston City Hosp, 38; asst, Harvard Med Sch, 38-39; asst resident physician, Rockefeller Inst Hosp, 39-46, mem, US Naval Res Unit, 41-46; res worker, Off Sci Res, 41-46; assoc physician, New Haven Community Hosp, Conn, 46-48, assoc pediatrician, 48-52; chief pediat serv, NC Mem Hosp, 52-60, chief staff, 59-60; consult pediatrician, US Army Hosp, Ft Bragg, NC & NC State Bd Health, 52-60; dir pediat serv, Babies Hosp, Columbia Presby Med Ctr, 60-70, attend pediatrician, 60-74, consult, 74-; Schick lectr, Mt Sinai Hosp, New York, 62; consult pediatrician, St Albans Naval Hosp, NY, 63-73 & St Luke's Hosp, New York, 66-; attend pediatrician, Harlem Hosp, 74-; assoc, Comn Influenza, US Armed Forces Epidemiol Bd, 47-69; mem, Planning Comt Studies Cardiovasc Effects Oriental Influenza, Nat Heart Inst & Influenza Res Comt, NIH, 57, Bd Sci counr, Div Biol Stand, 59-63, Infectious Dis & Trop Med Training Grant Comt, Nat Int Allergy & Infectious Dis, 60-63; chmn, Allergy & Infectious Dis Panel & mem, Rev Panel, Health Res Coun New York, 62-68, mem, Exec Comt, 68-72. Honors & Awards: Presidential Award, Int Poliomyelitis Cong, 58. Mem: Fel AAAS; Am Acad Pediat; Am Pediat Soc (vpres, 69); fel Am Pub Health Asn; fel NY Acad Sci. Res: Infectious diseases; bacteriology and virology; hemagglutination phenomena. Mailing Add: Dept of Pediat Columbia Presby Med Ctr New York NY 10032

CURNEN, MARY G MCCREA, b Belgium; m 55, 68; c 7. EPIDEMIOLOGY, PEDIATRICS Educ: Cath Univ Louvain, MD, 48; Inst Trop Med, Belg, DTM, 49; Columbia Univ, DrPH(epidemiol), 73. Prof Exp: Intern pediat, Cath Univ Louvain Hosp, 47-48; resident, Bellevue Med Ctr, Col Med, NY Univ, 52-53; instr prev med, Sch Med, Yale Univ, 53-54, instr prev med & pediat, 54-56, clin physician child health and pub schs, New Haven City Health Dept, 59-64; res assoc pediat, Yale Univ, 64-66, res assoc med, Yale Univ Virol Lab, Vet Admin Hosp, West Haven & Yale Univ, 67-69; res assoc epidemiol, 73-74, ASST PROF EPIDEMIOL, SCH PUB HEALTH, COLUMBIA UNIV, 74-; ASSOC MED & PEDIAT, YALE-NEW HAVEN HOSP, 68- Concurrent Pos: Clin res fel, Sch Med, Yale Univ, 49-52; exec dir & pediatrician in chg, New Haven Live Poliomyelitis Vaccine Trial, 60. Mem: Am Pub Health Asn; Harvey Soc; Soc Epidemiol Res; hon mem Peruvian Pediat Soc; Sigma Xi. Res: Cancer epidemiology; virology. Mailing Add: Fac of Med Dept of Epidemiol Columbia Univ Sch Pub Health New York NY 10032

CURNUTT, JERRY LEE, b Elwood, Ind, Nov 15, 42; m 63; c 1. PHYSICAL CHEMISTRY. Educ: Franklin Col, BA, 64; Univ Nebr, MS, 66, PhD(chem), 68. Prof Exp: NASA res traineeship, 66-68; res chemist, Thermal Res Lab, 68-76, RES SPECIALIST, DOW CHEM CO, 76- Mem: Am Chem Soc. Res: Thermodynamic properties of electrolytes; use of gas adsorption techniques to characterize bare structures in solid materials. Mailing Add: 2800 Mt Vernon Dr Midland MI 48640

CURNUTTE, BASIL, JR, b Portsmouth, Ohio, Mar 1, 23; m 45; c 2. PHYSICS. Educ: US Naval Acad, BS, 45; Ohio State Univ, PhD(physics), 53. Prof Exp: Res assoc physics, Ohio State Univ, 53-54; from asst prof to assoc prof, 54-64, PROF PHYSICS, KANS STATE UNIV, 64- Mem: AAAS; Am Phys Soc; Am Asn Physics Teachers; Optical Soc Am. Res: Chemical physics; molecular spectra and structure; atomic spectroscopy. Mailing Add: Dept of Physics Kans State Univ Manhattan KS 66504

CUROTT, DAVID RICHARD, b Passaic, NJ, June 3, 37; m 64; c 1. PHYSICS, ATROPHYSICS. Educ: Stevens Inst Technol, BSc, 59; Princeton Univ, MA, 62, PhD(physics), 65. Prof Exp: Res assoc astrophys, Princeton Univ, 65-66, instr physics, 66-67; asst prof, Wesleyan Univ, 67-75; ASSOC PROF PHYSICS, UNIV N ALA, 75- Mem: Am Asn Physics Teachers; Am Phys Soc; Am Astron Soc; Am Optical Soc. Res: Relativity; experimental and observational cosmology and astrophysics; astronomical electronic instrumentation. Mailing Add: Box 5050 Univ NAla Florence AL 35630

CURPHEY, THOMAS JOHN, b New York, NY, Oct 7, 34; m 59; c 2. ORGANIC CHEMISTRY. Educ: Harvard Univ, BA, 56, PhD(chem), 60. Prof Exp: Res assoc chem, Univ Wis, 60-62; instr, Yale Univ, 62-64; asst prof, 64-68, ASSOC PROF CHEM, ST LOUIS UNIV, 68- Mem: Am Chem Soc; The Chem Soc. Res: Synthetic organic chemistry; chemical and biochemical reaction mechanisms; natural products chemistry. Mailing Add: Dept of Chem St Louis Univ Box 8089 St Louis MO 63156

CURRAH, JACK ELLWOOD, b Toronto, Ont, Sept 12, 20; m 44; c 3. ANALYTICAL CHEMISTRY. Educ: Univ Toronto, BA, 41, MA, 42, PhD(anal & inorg chem), 44. Prof Exp: Asst chem, Univ Toronto, 41-43; res chemist, Atomic Energy Proj, Nat Res Coun Can, 44-46; res chemist, Cent Res Lab, Can Industs, Ltd, 46-51, res group leader, 51-56, sect leader, Res Dept, Textile Fibers Div, 56-57 & Process & Prod Res Sect, Tech Dept, 57-62, res sect mgr, Tech Dept, 62-65; tech mgr, Millhaven Fibres Ltd, 65-67; res & tech mgr plastics group, 67-70, res leader, Cent Res Lab, 70-73, SR RES CHEMIST, CHEM RES LAB, CAN INDUSTS LTD, McMASTERVILLE, 74- Mem: Chem Inst Can; Can Soc Chem Eng; Am Soc Mass Spectrometry. Res: Synthetic textile fibers; polyester fibers; mass spectrometry. Mailing Add: 122 Bathurst Ave Pointe Claire PQ Can

CURRAN, COLUMBA, b Cincinnati, Ohio, Apr 8, 11. INORGANIC CHEMISTRY. Educ: Univ Notre Dame, BS, 33, MS, 35, PhD(phys chem), 37. Prof Exp: Instr high sch, Mass, 33-34; from instr to assoc prof chem, 37-57, PROF CHEM, UNIV NOTRE DAME, 57- Concurrent Pos: Fel, Calif Inst Technol, 45-46. Mem: Am Chem Soc. Res: Dielectric constants of solutions; absorption spectra of solids and solutions; hydrogen bonding; coordinate bonding; solid state reactions, Mossbauer effect in metal complexes. Mailing Add: Dept of Chem Univ of Notre Dame Notre Dame IN 46556

CURRAN, DANIEL R, b Homestead, Pa, Mar 6, 25; m 45; c 3. PHYSICS, MATHEMATICS. Educ: Pa State Univ, BS, 49, MS, 50. Prof Exp: Asst acoust, Pa State Univ, 48-50 & ultrasonics, Brown Univ, 50-53; mem res staff, Res Div, Raytheon Mfg Co, 53-55; proj engr, Pioneer Cent Div, Bendix Aviation Corp, 55-56; sr physicist, Electronic Res Div, Clevite Corp, 56-57, head electronic device develop sect, 57-65, mgr filter prod, Piezoelec Div, 65-68, opers mgr magnetic prod, 68-70; mgr mkt develop, 70-75, PROD MGR TEST & MEASUREMENT, BRUSH DIV, GOULD INC, 75- Concurrent Pos: Consult ultrasonics, 73- Honors & Awards: C B Sawyer Mem Award, 68. Mem: Sr mem inst Elec & Electronics Engrs; Acoust Soc Am; Asn Comput Mach; Instrument Soc Am. Res: Physical acoustics; high intensity ultrasonics; solid state physics; elastic wave theory; electric network theory; piezoelectric resonators and filters; ferroelectric devices. Mailing Add: 145 Timberlane Dr Aurora OH 44202

CURRAN, DAVID JAMES, b Kitchener, Ont, Sept 19, 32; US citizen; m 57; c 6. ANALYTICAL CHEMISTRY. Educ: Univ Mass, BS, 53; Boston Col, MS, 58; Univ Ill, PhD(polarography), Univ Ill, 61. Prof Exp: Asst prof anal chem, Seton Hall Univ, 61-63; asst prof, 63-69, ASSOC PROF ANAL CHEM, UNIV MASS, AMHERST, 69- Mem: Am Chem Soc; fel Am Inst Chem. Res: Electroanalytical chemistry; solid electrodes in potentiometry, coulometry, amperometry and voltammetry-polarography; chemical instrumentation; pressure transducers in chemical analysis. Mailing Add: Dept of Chem Univ of Mass Amherst MA 01002

CURRAN, DONALD ROBERT, b Aurora, Ill, Mar 10, 32; m 59. PHYSICS. Educ: Iowa State Univ, BS, 53; Wash State Univ, MS, 56, PhD(physics), 60. Prof Exp: Physicist, Stanford Res Inst, 56-62; res physicist, Norweg Defense Res Estab, 62-67; physicist, Ernst Mach Inst, WGer, 67-70; PHYSICIST, STANFORD RES INST, 70- Res: Shock wave and high pressure physics; fluid dynamics; mechanical behavior of solids; fracture mechanics. Mailing Add: Stanford Res Inst Menlo Park CA 94025

CURRAN, HAROLD ALLEN, b Washington, DC, Dec 2, 40; m 63; c 2. PALEONTOLOGY, PALEOECOLOGY. Educ: Washington & Lee Univ, BS, 62; Univ NC, MS, 65, PhD(geol), 68. Prof Exp: Asst dept geol, Univ NC, 63-68; asst prof dept earth, space & graphic sci, US Mil Acad, 68-70; asst prof geol, 70-75, ASSOC PROF GEOL, SMITH COL, 75- Mem: Geol Soc Am; Soc Econ Paleont & Mineral; AAAS. Res: Upper Cretaceous and Cenozoic Foraminifera; trace fossils; Atlantic Coastal Plain stratigraphy; problems in paleoecology and marine geology. Mailing Add: Dept of Geol Clark Sci Ctr Smith Col Northampton MA 01060

CURRAN, JOHN PHINEAS, b New Brunswick, NJ, Sept 21, 28; m 53; c 4 PEDIATRICS, MEDICAL GENETICS Educ: Rutgers Univ, BS, 51; New York Med Col, MD, 55. Prof Exp: Intern, Mountainside Hosp, Montclair, NJ, 55-56; resident, Children's Hosp, Philadelphia, 59-60, resident, US Naval Hosp, Philadelphia, 59-61; assoc prof pediat, New York Med Col, 63-66; pediatrician in chief, Margaret Hague Maternity Hosp, 66-75; PROF GEN & ORAL PATH, NJ DENT SCH, 71-, ASSOC PROF PEDIAT, NJ MED SCH, 74-; MED DIR, CHILDREN'S SPECIALIZED HOSP, MOUNTAINSIDE-WESTFIELD, NJ, 75- Concurrent Pos: Consult pediat, Pollack Hosp, Jersey City, 69-; dir pediat, Jersey City Med Ctr, 66-75; clin assoc prof, NJ Med Sch, 73-74. Mem: AAAS; fel Am Acad Pediat; Am Soc Human Genetics; Am Col Clin Pharmacol. Res: Malformation syndromes; genetics. Mailing Add: Children's Spec Hosp New Providence Rd Westfield-Mountainside NJ 07091

CURRAN, PETER FERGUSON, biophysics, deceased

CURRAN, ROBERT KYRAN, b Kingston, Pa, Aug 8, 31; m 55; c 4. PHYSICS. Educ: Univ Scranton, BS, 53; Univ Pittsburgh, PhD(physics), 59. Prof Exp: Physicist, Westinghouse Elec Corp, 59-62 & Air Prod & Chem Inc, Pa, 62-64; PHYSICIST, BELL LABS, INC, 64- Mem: Am Phys Soc. Res: Atomic collisions; mass spectrometry; vacuum technique. Mailing Add: Bell Labs Inc Murray Hill NJ 07974

CURRAN, WILLIAM VINCENT, b Easton, Pa, Nov 30, 29; m 52; c 5. ORGANIC CHEMISTRY. Educ: Lafayette Col, BS, 52; Col Holy Cross, MS, 53; NY Univ, PhD, 68. Prof Exp: Res worker org chem, Columbia Univ, 54-56; res chemist, Lederle Labs, Am Cyanamid Co, 56-65; asst prof chem, Fairleigh Dickinson Univ, 65-67; res chemist, 68-69, SR RES CHEMIST, AM CYANAMID CO, 69- Mem: Am Chem Soc. Mailing Add: Am Cyanamid Co Lederle Labs Div Pearl River NY 10965

CURRARINO, GUIDO, b Levanto, Italy, Dec 17, 20. MEDICINE. Educ: Univ Genoa, MD, 45. Prof Exp: Asst prof radiol & pediat, Med Sch, Univ Cincinnati, 56-60; assoc prof, Med Sch, Cornell Univ, 61-64; PROF RADIOL & PEDIAT, UNIV TEX HEALTH SCI CTR, DALLAS, 65- Res: Pediatric radiology. Mailing Add: Children's Med Ctr 1935 Amelia St Dallas TX 75235

CURRAY, JOSEPH ROSS, b Cedar Rapids, Iowa, Jan 19, 27; m 49; c 3. MARINE GEOLOGY. Educ: Calif Inst Technol, BS, 49; Pa State Univ, MS, 51; Univ Calif, PhD(oceanog), 59. Prof Exp: Asst & instr geol & mineral, Pa State Univ, 49-51; res geologist, Res Lab, Carter Oil Co, 51-53; asst marine geol & res staff, 53-67, assoc prof oceanog, 67-70, chmn grad dept, 73-75, PROF OCEANOG & RES GEOLOGIST, SCRIPPS INST OCEANOG, UNIV CALIF, 70-, CHMN GEOL RES DIV, 76- Concurrent Pos: Dir, Gen Oceanographics, Inc, 58-; ed, Marine Geol, 63-66; trustee, Found Ocean Res, 71- Honors & Awards: Shepard Medal, Soc Econ Paleontologists & Mineralogists, 70. Mem: AAAS; Geol Soc Am; Am Asn Petrol Geologists; Soc Econ Paleontologists & Mineralogists; Am Geophys Union. Res: Sediments; history of continental margin. Mailing Add: Scripps Inst of Oceanog A-015 La Jolla CA 92093

CURRELL, DOUGLAS LEO, b Tulsa, Okla, Feb 5, 27. ORGANIC CHEMISTRY. Educ: Univ Colo, BA, 50, MA, 54; Univ Ark, PhD(chem), 56. Prof Exp: Res fel chem, Calif Inst Technol, 56-57; from asst prof to assoc prof, 57-65, chmn dept, 65-68, PROF CHEM, CALIF STATE UNIV, LOS ANGELES, 65- Mem: Am Chem Soc; Sigma Xi. Res: Mechanisms of organic reactions; hemoglobin chemistry. Mailing Add: Dept of Chem Calif State Univ Los Angeles CA 90032

CURREN, THOMAS, b St John's, Nfld, Aug 2, 39; m 62; c 2. PLANT PATHOLOGY. Educ: Mem Univ, BSc, 61; Univ Toronto, MA, 63, PhD(plant path), 66. Prof Exp: Res scientist, Res Sta, Can Dept Agr, 66-69; lectr biol, Carleton Univ, 69-70; HERBICIDE EVAL OFF, PLANT PROD DIV, CONTROL PROD SECT, AGR CAN, 70- Concurrent Pos: Mem, Can Weed Comt, Can Phytopath Soc. Res: Physiology of parasitism of storage diseases of carrot and squash, especially the role of pectic enzymes in disease process; herbicides. Mailing Add: Plant Prod Div Control Prod Sect Agr Can Ottawa ON Can

CURRENT, DAVID HARLAN, b Connerville, Ind, July 26, 41; m 66; c 2. EXPERIMENTAL SOLID STATE PHYSICS, NUCLEAR MAGNETIC RESONANCE. Educ: Carleton Col, BA, 63; Northwestern Univ, MS, 66; Mich State Univ, PhD(physics), 71. Prof Exp: From instr to asst prof, 66-75, ASSOC PROF PHYSICS, CENT MICH UNIV, 75- Mem: Am Phys Soc; Am Asn Physics Teachers; Am Asn Univ Prof. Res: Experimental chlorine nuclear quadrupole resonance in insulators, including temperature dependence and phase transitions; theoretical models of electric field gradients, especially in ionic solids. Mailing Add: Dept of Physics Cent Mich Univ Mt Pleasant MI 48859

CURRENT, JERRY H, b Anderson, Ind, Mar 2, 35; m 61; c 3. PHYSICAL CHEMISTRY, ANALYTICAL CHEMISTRY. Educ: Ind Univ, BS, 57; Univ Wash, PhD(phys chem), 61. Prof Exp: NSF fel, Univ Calif, Berkeley, 62-64; asst prof chem, Univ Mich, 64-70; RES CHEMIST, GULF RES & DEVELOP CO, 70- Mem: Am Chem Soc; Am Phys Soc. Res: Gas phase kinetics; x-ray photoelectron spectroscopy; process control. Mailing Add: Gulf Res & Develop Co PO Drawer 2038 Pittsburgh PA 15230

CURRENT, MICHAEL IRA, b Washington, DC, Dec 14, 45; m 74. SURFACE PHYSICS. Educ: Rensselaer Polytech Inst, BS, 67, MS, 73, PhD(physics), 74. Prof Exp: FEL, MAT DIV, RENSSELAER POLYTECH INST, 74- Mem: Am Asn Physics Teachers. Res: Use of surface analytical techniques and modulated molecular beams to follow the kinetics of catalytic oxidation of sulfur oxides by stack particulate materials. Mailing Add: Surface Studies Lab Mat Res Ctr Rensselaer Polytech Inst Troy NY 12181

CURRERI, ANTHONY RUDOLPH, b New York, NY, Sept 18, 09; m 35; c 3. THORACIC SURGERY. Educ: Univ Wis, BA, 30, MA, 31, MD, 33; Am Bd Surg, dipl, 42; Am Bd Thoracic Surg, dipl, 48. Prof Exp: Intern, Columbia & Milwaukee Children's Hosp, 33-35; resident surg, Univ Hosps, 36-39, from instr to assoc prof, 39-54, dir cancer res hosp, 48-61, chief of staff, 60-61, dir div clin oncol, 63-70, PROF SURG, UNIV WISMADISON, 54-, CHMN DEPT, 68- Concurrent Pos: Consult, USPHS, Nat Cancer Inst, 66-, Surgeon Gen, US Army, Am Col Surg & Am Cancer Soc; sci adv bd consult, Armed Forces Inst Path, 67- & Defense Sci Bd, 70-; mem, Nat Adv Cancer Coun, 59-62, Army Sci Adv Panel, 65-, Panel Anti-Neoplastic Surg, Nat Acad Sci-Nat Res Coun, 66-69, Spec Study Sect Clin Cancer Progs, Cancer Ctr & Spec Studies, 67-, Nat Adv Coun Regional Med Progs, 69-70 & Nat Adv Gen Med Sci Coun, 70-; chmn, Chemother Rev Bd, 60-62, Adv Comt to Int Res, 60-62, Spec Study Panel, Nat Cancer Inst, 63-65 & Cent Clin Drug Eval Prog. Honors & Awards: Bronze Medal, Am Cancer Soc, 50; Shahbanou Award & Gold Medal, Lila Motley Found. Mem: AAAS; Am Asn Cancer Res; Am Surg Asn; Int Surg Asn; Am Asn Thoracic Surg. Res: General and thoracic surgery and cancer. Mailing Add: Dept of Surg Univ Hosp 1300 University Ave Madison WI 53706

CURREY, JOHN E, physics, see 12th edition

CURRIE, ALLAN, organic chemistry, see 12th edition

CURRIE, BALFOUR WATSON, b Helena, Mont, Nov 1, 03; m 34; c 3. PHYSICS. Educ: Univ Sask, BSc, 25, MSc, 27; McGill Univ, PhD(physics), 30. Hon Degrees: LLD, Univ Sask, 75. Prof Exp: From instr to prof physics, Univ Sask, 28-73, head dept, 52-61, dean grad studies, 59-73, dean faculties, 64-73, vpres res, 67-73; RETIRED. Concurrent Pos: Meteorologist, Meteorol Serv Can, 32-34. Honors & Awards: Companion, Order of Can, 72. Mem: AAAS; Am Meteorol Soc; fel Royal Soc Can; Royal Meteorol Soc; Can Asn Physicists. Res: Absorption of ions at liquid gas surfaces; electro-endosmosis; atmospheric electricity; earth currents and aurorae; climate of prairie provinces and northwest territories. Mailing Add: 416 Bate Crescent Saskatoon SK Can

CURRIE, BRUCE LAMONTE, b Pasadena, Calif, Mar 1, 45; m 65; c 3. MEDICINAL CHEMISTRY, ORGANIC CHEMISTRY. Educ: Ariz State Univ, BS, 66; Univ Utah, PhD(org chem), 70. Prof Exp: Robert A Welch fel, Inst Biomed Res, Univ Tex, Austin, 69-72, res assoc biochem & endocrinol, 72-74; ASST PROF MEDICINAL CHEM, COL PHARM, UNIV ILL MED CTR, 74- Mem: Am Chem Soc. Res: Design and synthesis of small peptide hormone agonists and antagonists; peptide sequence and structure determination; anticancer nucleoside analogs. Mailing Add: Dept of Medicinal Chem Col Pharm PO Box 6998 Univ of Ill Med Ctr Chicago IL 60680

CURRIE, GUSTAVUS NOEL, II, b Presque Isle, Maine, May 31, 38; m 61; c 2. PHARMACOLOGY, ENDOCRINOLOGY. Educ: Univ Maine, BS, 61, MS, 62; WVa Univ, PhD(pharmacol), 65. Prof Exp: Trainee, Endocrine Res Lab, Dent Sch, Harvard Univ, 65, res assoc anat, 65-67; trainee reproductive endocrinol, Lab Reproductive Physiol, Univ Mass, 67; SR RES TOXICOLOGIST, NORWICH PHARMACAL CO, 69- Mem: AAAS. Res: Thyroparathyroid axis in calcium hameostasis; special involvements of autonomic pharmacology and sex differences; adrenal and ovarian steroidogenesis. Mailing Add: PO Box 139 Oxford NY 13830

CURRIE, JOHN BICKELL, b Guelph Township, Ont, May 29, 22; m 48; c 2. GEOLOGY. Educ: McMaster Univ, BA, 46; Univ Toronto, MA, 47, PhD, 50. Prof Exp: Res geologist & sect head, Geol & Geochem Div, Gulf Res & Develop Co, 50-59; assoc prof, 59-64, PROF GEOL SCI, UNIV TORONTO, 64- Mem: Geol Soc Am; Am Asn Petrol Geol; Am Geophys Union. Res: Structural geology; mechanical development of geological structures; regional tectonics; petroleum geology. Mailing Add: Dept of Geol Univ of Toronto Toronto ON Can

CURRIE, JULIA RUTH, b Freeport, Tex, Dec 13, 44; m 71; c 2. NEUROSCIENCES. Educ: Radcliffe Col, AB, 67; Washington Univ, PhD(neurobiol), 74. Prof Exp: Res asst neurobiol, Harvard Med Sch, 67-69; RES ASSOC PHYSIOL, MED COL, CORNELL UNIV, 74-, NIH FEL, 75- Mem: AAAS; Soc Neurosci; Fedn Am Scientists. Res: Neuroanatomy including classical cell staining, experimental fiber staining, histochemical and autoradiographic technique; experimental embryology, neuroembryology, vertebrate visual system, nerve regeneration axonal transport, plasticity in the nervous system. Mailing Add: Dept of Physiol Cornell Univ Med Col New York NY 10021

CURRIE, KENNETH LYELL, petrology, see 12th edition

CURRIE, LLOYD ARTHUR, b Portland, Ore, Mar 14, 30; m 59; c 4. PHYSICAL CHEMISTRY, RADIOCHEMISTRY. Educ: Mass Inst Technol, BS, 52; Univ Chicago, PhD(phys chem), 55. Prof Exp: Asst prof chem, Pa State Univ, 55-62; NUCLEAR CHEMIST, NAT BUR STANDARDS, 62- Concurrent Pos: Lectr, Am Univ, 62-66; mem, Int Comn on Radiation Units & Measurements Task Group, 65-; vis prof, Inst Nuclear Sci, Univ Ghent & phys inst, Univ Bern, 70-71. Mem: Fel Am Inst Chemists; NY Acad Sci. Res: Nuclear reactions; electromagnetic isotope-separation of reaction products; low-level and environmental radioactivity; statistical aspects of nuclear decay and data analysis; application of nuclear methods to trace analysis and physical chemistry. Mailing Add: Nat Bur of Standards Washington DC 20234

CURRIE, WILLIAM DEEMS, b Wallace, NC, Sept 4, 35; m 61; c 2. BIOCHEMISTRY. Educ: Davidson Col, BS, 57; Univ NC, MS, 62, PhD(biochem), 64. Prof Exp: Res asst biochem, Univ NC, 60-62; appointee, Biol & Med Res Group, Los Alamos Sci Lab, Univ Calif, 64-66; asst prof biol, East Carolina Univ, 66-67; asst prof radiobiol, 67-73, ASSOC PROF RADIOBIOL, MED CTR, DUKE UNIV, 73- Mem: AAAS; Am Chem Soc; Biophys Soc. Res: Control of energy metabolism; nuclear metabolism; effects of hyperbaric oxygen; oxidative phosphorylation; metabolism of cancer cells; seizure activity and energy metabolism. Mailing Add: Box 3224 Duke Univ Med Ctr Durham NC 27706

CURRIER, ALBERT (ELDRED), b Brooklyn, NY, Feb 8, 05; m 47; c 1. MATHEMATICS. Educ: Harvard Univ, BS, 27, AM, 28, PhD(math), 30. Prof Exp: Instr math, Harvard Univ, 29-30, Sheldon fel, 30-31, instr, 31-35; from instr to prof, US Naval Acad, 35-75; RETIRED. Mem: Am Math Soc; Math Asn Am. Res: Fundamental theorem on second order cross partial derivatives; variable end point of calculus of variations; mathematical analysis. Mailing Add: 1925 Harwood Rd Annapolis MD 21401

CURRIER, HERBERT BASHFORD, b Richwood, Ohio, Oct 16, 11; m 34; c 2. PLANT PHYSIOLOGY. Educ: Ohio State Univ, BS, 32; Utah State Col, MS, 38; Univ Calif, PhD(plant physiol), 43. Prof Exp: Asst, Utah State Col, 37-38; asst col agr, Univ Calif, 38-39, assoc bot, 39-42; res chemist, Basic Veg Prod Co, 43-46; asst prof bot & asst botanist, 46-52, assoc prof bot & assoc botanist, 52-58, PROF BOT, UNIV CALIF, DAVIS, 58-, BOTANIST, EXP STA, 58- Concurrent Pos: Guggenheim fel, 54-55 & 61-62; guest prof, Univ Göttingen, 74. Mem: Am Soc Plant Physiol; Bot Soc Am; Scand Soc Plant Physiol; Japanese Soc Plant Physiol. Res: Cellular physiology; phloem translocation; plant water relations; toxicity; structural-functional studies of the phloem tissue of higher plants. Mailing Add: Dept of Bot Univ of Calif Davis CA 95616

CURRIER, ROBERT DAVID, b Grand Rapids, Mich, Feb 19, 25; m 51; c 2. NEUROLOGY. Educ: Univ Mich, AB, 48, MD, 52, MS, 55. Prof Exp: From asst prof to assoc prof neurol, Univ Mich, 57-61; assoc prof, 61-69, PROF MED, MED CTR, UNIV MISS, 69-, CHIEF DIV NEUROL, 61- Concurrent Pos: Consult, Wayne County Gen Hosp & Ann Arbor Vet Admin Hosp, Mich, 59-61 & Jackson Vet Admin Hosp, Miss, 61- Mem: Fel Am Acad Neurol. Res: Clinical, academic and research neurology. Mailing Add: Div of Neurol Univ of Miss Med Ctr Jackson MS 39216

CURRIER, VERNON ARTHUR, b Sheboygan, Wis, Dec 13, 25; m 48; c 4. PETROLEUM CHEMISTRY. Educ: Univ Wis, BS, 50. Prof Exp: From chemist to sr res chemist, 50-62, res specialist tech serv, 62-63, sr proj chemist, 63-65, sect supvr tech serv, 65-71, MGR TECH SERV, JEFFERSON CHEM CO, INC, 71- Mem: Am Chem Soc; Soc Plastics Indust. Res: Petrochemical research in ethylene and propylene based chemicals. Mailing Add: 11311 Hunters Lane Austin TX 78753

CURRO, JOHN GILLETTE, b Detroit, Mich, Oct 5, 42; m 67; c 2. POLYMER PHYSICS. Educ: Univ Detroit, BChE, 65; Calif Inst Technol, PhD(mat sci), 69. Prof Exp: Res assoc theoret physics, Inst Appl Math & Fluid Dynamics, Univ Md, College Park, 69-70; staff mem, 70-75, DIV SUPVR POLYMER PHYSICS, SANDIA LABS, 75- Mem: Am Phys Soc; Am Inst Chem Engrs. Res: Conformation of polymer chains and its relationship to physical properties; viscoelasticity of polymers undergoing chemical reaction. Mailing Add: Div 5813 Sandia Labs Albuquerque NM 87115

CURRY, DONALD LAWRENCE, b Bridgeport, Conn, Mar 26, 36; m 69; c 3. MEDICAL PHYSIOLOGY. Educ: Sacramento State Col, AB, 63; Univ Calif, San Francisco, PhD(physiol), 67. Prof Exp: Asst res physiologist, Med Ctr, Univ Calif, San Francisco, 67-69, lectr, 68-69; asst prof, 69-75, ASSOC PROF PHYSIOL SCI, SCH VET MED, UNIV CALIF, DAVIS, 75- Concurrent Pos: NIH-Am Diabetes Asn res grants, 74- Mem: Am Physiol Soc; Am Diabetes Asn; Endocrine Soc. Res: Study of factors which control insulin secretion by the isolated perfused rat pancreas. Mailing Add: Dept of Physiol Sci Univ Calif Sch of Vet Med Davis CA 95616

CURRY, FRANCIS J, b San Francisco, Calif, July 19, 11; m 48; c 8. INTERNAL MEDICINE. Educ: San Francisco Univ, BSc, 36; Stanford Univ, MD, 46; Univ Calif, MPH, 64. Prof Exp: Asst med bact, Stanford Univ, 41-42, asst path, 42-43; chief tuberc & pulmonary dis sect, Army Specialized Treatment Ctr, Ft Carson, Colo, 53-54, chief tuberc sect, Army & Air Force Specialized Treatment Ctr & coordr clin res, Fitzsimons Army Hosp, Denver, 54-56; chief tuberc control div, 56-60, asst dir pub health, 60-70, DIR PUB HEALTH, HOSPS & MENT HEALTH, SAN FRANCISCO HEALTH DEPT, 70- Concurrent Pos: Dir chest clin, San Francisco Gen Hosp, 56-; prof med, Univ Calif, 58-, lectr, Sch Pub Health, 60-, prof ambulatory & community med, Sch Med, 66-; assoc prof, Stanford Univ, 62; proj dir, USPHS clin res grant, 62-; consult, Nat Jewish Hosp, Denver, Colo, 64-; consult, Tuberc Commun Dis Div & mem, Adv Comt Tuberc Control, USPHS, Ga, 65-; consult, Nat Commun Dis Ctr Mem: Fel Am Col Chest Physicians; fel Am Col Prev Med; Am Thoracic Soc; fel Am Col Physicians; fel Am Trauma Soc. Res: Pulmonary fungus diseases, particularly histoplasmosis and coccidioidomycosis; atypical mycobacteria infections of man; social and behavioral problems of staff and patients and their effect upon each other and upon the maintenance of treatment by patients; tuberculin skin testing on large population groups. Mailing Add: 101 Grove St San Francisco CA 94102

CURRY, GEORGE MONTGOMERY, b Halifax, NS, Nov 7, 26; nat US; m 50; c 2. PLANT PHYSIOLOGY. Educ: Acadia Univ, BA, 46; Yale Univ, MS, 47; Harvard Univ, AM, 54, PhD, 58. Prof Exp: Teacher pub sch, Mass, 47-52; from instr to assoc prof biol, Tufts Univ, 58-70; PROF BIOL & DEPT HEAD, ACADIA UNIV, 70- Mem: AAAS; Am Soc Plant Physiol. Res: Effects of light on unicellular and multicellular plants. Mailing Add: RR 1 Port Williams Kings Co NS Can

CURRY, HASKELL BROOKS, b Millis, Mass, Sept 12, 00; m 28; c 2. MATHEMATICAL LOGIC. Educ: Harvard Univ, AB, 20, AM, 24; Univ Göttingen, PhD(math anal), 30. Hon Degrees: DSc, Curry Col, 66. Prof Exp: Asst to Prof Bridgman, Harvard Univ, 22-23, instr math, 26-27; instr, Princeton Univ, 27-28; from asst prof to prof, 29-60, Evan Pugh res prof, 60-66, EMER PROF MATH, PA STATE UNIV, 66- Concurrent Pos: Nat Res Coun fel, Univ Chicago, 31-32; mem, Inst Advan Study, 38-39; mem corp, Curry Col, 40-51; mathematician, Frankford Arsenal, 42-44, appl physics lab, Johns Hopkins Univ, 44-45 & Aberdeen Proving Grounds, Md, 45-46; lectr, Univ Notre Dame, 48; Fulbright res scholar & vis prof, Cath Univ Louvain, 50-51, hon vis prof, 51; prof math, Inst voor Grondslagenonderzoek & Filosofie der Exacte Wetenschappen, Univ Amsterdam, 66-70, emer prof, 70-; vis Andrew Mellon prof, Univ Pittsburgh, 71-72. Mem math panel, Nat Defense Res Comt, 43-44, mem policy comt math, 56-57; mem, Conf Bd Math Sci, 58-63; mem US Nat Comt, Int Union Hist & Philos Sci, 59-63. Honors &

Awards: Medal, Cath Univ Louvain, 51. Mem: AAAS; Am Math Soc; Math Asn Am; Am Philos Asn; Asn Symbolic Logic (vpres, 36-37, pres, 38-40). Res: Combinatory logic. Mailing Add: 228 E Prospect Ave State College PA 16801

CURRY, HIRAM BENJAMIN, b Midville, Ga, Sept 19, 27; m 51; c 4. NEUROLOGY. Educ: Col Charleston, BS, 47; Med Col SC, MD, 50. Prof Exp: Gen practr, 51-57; intern internal med, Vet Admin Hosp, Lake City, Fla, 58; assoc, 63-64, asst prof, 64-68, ASSOC PROF NEUROL, MED UNIV SC, 68- PROF FAMILY PRACT & CHMN SECT, 70- Concurrent Pos: Nat Inst Neurol Dis & Blindness spec fel, 60-63; prin investr, USPHS grant, 63-70, co-prin investr, 67-69; chief neurol serv, Vet Admin Hosp, Charleston, SC, 66-68; pres prof staff & mem exec comt, Med Univ Hosp, Charleston, SC, 70-; med dir outpatient clins, dir family pract residency prog & mem behav sci comt, Med Univ SC, 70- Mem: Am Acad Neurol; Asn Res Nerv & Ment Dis; Am Aacd Gen Pract. Res: Cerebral circulation; cerebral vascular disease; epidemiology of vascular disease; health care delivery; medical education. Mailing Add: Dept of Family Pract Med Univ of SC Charleston SC 29401

CURRY, HOWARD MILLARD, b Haverhill, Mass, Mar 8, 24; m 47; c 2. ORGANIC CHEMISTRY. Educ: Northeastern Univ, BS, 45; Boston Univ, AM, 47, PhD(org chem), 50. Prof Exp: Instr chem, Bates Col, 49-50; assoc prof, 51-60, PROF CHEM, WITTENBERG UNIV, 60-, CHMN DEPT, 74- Concurrent Pos: Consult, C F Kettering Found, 57- Mem: Am Chem Soc. Res: Nitrogen organics. Mailing Add: 1328 N Lowry Ave Springfield OH 45504

CURRY, JAMES KENNETH, b Amarillo, Tex, Nov 30, 20; m 45; c 3. GEOLOGY. Educ: Tex Tech Col, BS, 42; Univ Calif, Los Angeles, cert meteorol. Prof Exp: Geologist, 46-48, asst dist geologist, 48-52, dist geologist, 52-54, div geologist, 54-55, mgr geol, 55-56, geol 60-64, Geol Div, 64-69, MGR EXPLOR, PANHANDLE EASTERN PIPELINE CO, 69- Mem: Am Asn Prof Geologists; Am Inst Mining, Metall & Petrol Engrs. Res: Exploratory geology in search of oil and gas. Mailing Add: 13418 Pinerock Houston TX 77079

CURRY, JOHN D, b Xenia, Ohio, Nov 16, 38; m 60; c 2. INORGANIC CHEMISTRY. Educ: Wilmington Col, Ohio, BA, 60; Ohio State Univ, PhD(inorg chem), 64. Prof Exp: William Ramsay Mem & NATO fels, Oxford Univ, 64-65; RES CHEMIST, MIAMI VALLEY LABS, PROCTER & GAMBLE CO, CINCINNATI, 65- Mem: Am Chem Soc; The Chem Soc; Sigma Xi. Res: Homogeneous catalysis; reactions of coordinated ligands; phosphorus chemistry. Mailing Add: 709 Marcia Dr Oxford OH 45056

CURRY, JOHN JOSEPH, III, b Brooklyn, NY, Nov 1, 40; m 66; c 2. NEUROENDOCRINOLOGY. Educ: State Univ NY Col Plattsburgh, BS, 62; Adelphi Univ, MS, 64; Univ Calif, Berkeley, PhD(physiol), 69. Prof Exp: From instr to asst prof physiol, Sch Med, Boston Univ, 68-73; ASST PROF PHYSIOL, OHIO STATE UNIV, 73- Mem: Am Physiol Soc; Int Soc Psychoneuroendocrinol; AAAS. Res: Central nervous system regulation of reproductive behavior and gonadotrophin secretion; sensory and hormonal factors controlling ovulation. Mailing Add: Dept of Physiol Ohio State Univ 1645 Neil Ave Columbus OH 43210

CURRY, LA VERNE LEON, b Stony Creek, Mich, Apr 17, 14; m 39; c 6. INSECT TAXONOMY. Educ: Eastern Mich Univ, AB, 38; Univ Mich, MA, 49; Mich State Univ, PhD(zool), 52. Prof Exp: Teacher pub schs, Mich, 38-46; head dept, 60-65, PROF BIOL, CENT MICH UNIV, 46- Concurrent Pos: Consult, AEC, 54, 59-, NIH, 59, Fisheries Div, State Dept Conserv, Mich, 53-54, 55-58 & Great Lakes Ill River Basin Proj, 62-; mem, Mich Bd Basic Sci, 62-, pres, 70-71. Mem: AAAS. Res: Biology of freshwater midges; their role in the uptake of radioisotopes from the hydrosols of fresh-water lakes and streams; taxonomy of chironomids, Chironomidae, Diptera; morphological study of all stages for use in taxonomic keys. Mailing Add: Dept of Biol Cent Mich Univ Mt Pleasant MI 48859

CURRY, LESLIE, b Newcastle, Eng, Nov 23, 22; m 52; c 3. GEOGRAPHY. Educ: Univ Durham, BA, 49; Johns Hopkins Univ, MA, 51; Univ NZ, PhD(geog), 59. Prof Exp: Res officer, Northern Indust Group, 49-50; econ affairs officer, UN Secretariat, 51; asst, Climat Lab, Johns Hopkins Univ, 52; instr geog, Univ Wash, 52; lectr, Univ NZ, 53-60; asst prof, Univ Md, 60-63; assoc prof, Ariz State Univ, 63-64; PROF GEOG, UNIV TORONTO, 64- Concurrent Pos: Vis Commonwealth prof, UK, 69-70; vis fel, Australian Nat Univ, 74. Honors & Awards: Citation for Meritorious Contrib, Asn Am Geogr, 69. Mem: Asn Am Geogr; Regional Sci Asn; Can Asn Geogr. Res: Economic geography, especially theoretical; statistical and mathematical methods; climatology. Mailing Add: Dept of Geog Univ of Toronto 150 St George St Toronto ON Can

CURRY, MICHAEL JOSEPH, b Brooklyn, NY, Aug 15, 20; m 58; c 4. PLASTICS CHEMISTRY. Educ: St John's Col, NY, BS, 41; Univ Wis-Madison, PhD(org chem), 48. Prof Exp: Jr chemist chem warfare, US Naval Res Lab, 42-45; asst prof org chem, Col St Thomas, 48-50; develop chemist, Calco Div, Am Cyanamid Co, 50-51; prod develop engr, Celanese Corp, 51-54, tech serv mgr, 54-59, lab mgr, 60-61, plastics develop dir, 61-67; assoc, Heidrick & Struggles, 67-71; PRES, MICHAEL J CURRY ASSOCS, INC, 71- Mem: Am Chem Soc; Asn Consult Chemists & Chem Engrs. Res: Plastics research; chemical marketing; laboratory administration; technical personnel administration. Mailing Add: 941 St Marks Ave Westfield NJ 07090

CURRY, ROBERT PATRICK, analytical chemistry, see 12th edition

CURRY, ROBERT RODNEY, b Santa Monica, Calif, Mar 21, 37; m 60; c 1. GEOMORPHOLOGY, ECOLOGY. Educ: Univ Colo, BA, 60, MS, 62; Univ Calif, Berkeley, PhD(geol, geophys), 68. Prof Exp: Asst prof environ sci, Univ Calif, Santa Barbara, 67-69; ASSOC PROF ENVIRON SCI, UNIV MONT, 69- Concurrent Pos: Res hydrologist, US Geol Surv, 67-; NSF environ sci grant for instnl prog soil sci, 68-69; consult environ panel, US Senate Comt Pub Works; dir res, Sierra Club, 74-75; adv panelist, NSF Div Advan Energy Res & Technol, 75-76. Mem: Ecol Soc Am; Geol Soc Am; Glaciol Soc; Int Asn Quaternary Res; Int Ecol Soc. Res: Dynamics of slope profile development; quaternary paleoclimatic fluctuations; causes of ice ages; effects of land use upon watersheds and rivers; floods; channel geometry; land use planning; hazard analysis. Mailing Add: Dept of Geol Univ of Mont Missoula MT 59801

CURRY, RONALD HOWARD, physical chemistry, analytical chemistry, see 12th edition

CURRY, THOMAS HARVEY, b Sullivan Co, Ind, Oct 7, 21; m 45; c 3. ORGANIC CHEMISTRY, CHEMICAL ENGINEERING. Educ: Purdue Univ, BChE, 42; Ohio State Univ, PhD(org chem), 53. Prof Exp: Tech supvr, Holston Ord Works, Tenn Eastman Corp, 43-45; instr chem, Antioch Col, 50-52; from asst prof to assoc prof, Univ Ohio, Athens, 53-61, chmn dept chem eng, 56-61; dean, Col Technol, Univ Maine, 61-67; dir resident res associateships, 67-69, DIR ASSOCIATESHIPS, COMN HUMAN RESOURCES, NAT ACAD SCI, 69- Concurrent Pos: NSF sci fac fel, 57-60. Mem: Am Chem Soc. Res: Chemistry of pyrroles and porphyrins. Mailing Add: Nat Acad of Sci 2101 Constitution Ave Washington DC 20418

CURRY, WILLIAM HIRST, III, b San Antonio, Tex, Feb 3, 32; m 54; c 4. GEOLOGY. Educ: Cornell Univ, BA, 54; Princeton Univ, PhD(geol), 59. Prof Exp: Geologist, Marathon Oil Co, 59-65; PRES, CURRY OIL CO, 66- Concurrent Pos: Consult geologist, 65- Mem: Sigma Xi; Am Asn Petrol Geologists. Res: Search for more effective oil and gas exploration tools and combinations of concepts and tools. Mailing Add: PO Box 3001 Casper WY 82602

CURTI, GABRIEL PHILIP, b San Francisco, Calif, Jan 28, 14; m. GEOGRAPHY. Educ: Univ London, dipl teaching, 38; Univ Calif, Los Angeles, BA, 53, MA, 55, PhD, 60. Prof Exp: Asst prof geog, Calif Lutheran Col, 63-66; assoc prof, Calif State Univ, Hayward, 66-71; PROF GEOG, N TEX STATE UNIV, 71- Mem: Royal Geog Soc; Asn Am Geogr; Nat Coun Geog Educ. Res: Albania, especially origins of people and present distribution and movement; Sahara; especially subsistence nomadism and incipient sedentarism; impact of Europeanisation on Saharan peoples with particular reference to the Ahagarr Tuareg. Mailing Add: Dept of Geog NTex State Univ Denton TX 76201

CURTICE, JAY STEPHEN, b Dallas, Tex, Jan 19, 28; m 57; c 2. PHYSICAL ORGANIC CHEMISTRY. Educ: Southern Methodist Univ, BS, 48; Iowa State Univ, PhD(chem), 54. Prof Exp: Res chemist, Sinclair Res, Inc, 54-62; assoc prof, 62-68, PROF CHEM, ROOSEVELT UNIV, 68- Mem: AAAS; Am Chem Soc. Res: Mechanisms of free radical reactions; organometallics. Mailing Add: 3135 Priscilla Ave Highland Park IL 60035

CURTIN, CHARLES BYRON, b Cohoes, NY, Aug 2, 17; m 45; c 2. BIOLOGY. Educ: George Washington Univ, BS, 45; Catholic Univ, MS, 47; Univ Pittsburgh, PhD, 56. Prof Exp: Asst, Catholic Univ, 46-47; prof biol, Mt St Mary's Col, Md, 47-57; staff ed life sci, McGraw-Hill Encycl Sci & Technol, 57-62; ASSOC PROF BIOL, CREIGHTON UNIV, 62- Concurrent Pos: Asst, Univ Pittsburgh, 52-53; consult ed, McGraw-Hill, 71. Mem: AAAS; Am Soc Syst Zool; Ecol Soc Am; Sigma Xi; Am Micros Soc. Res: Taxonomy and ecology of Tardigrada; parasitology; microtechnique. Mailing Add: 6218 Florence Blvd Omaha NE 68110

CURTIN, DAVID YARROW, b Philadelphia, Pa, Aug 22, 20; m 50; c 4. ORGANIC CHEMISTRY. Educ: Swarthmore Col, AB, 43; Univ Ill, PhD(chem), 45. Prof Exp: Rockefeller Inst grant, Harvard Univ, 45-46; from instr to asst prof chem, Columbia Univ, 46-51; from asst prof to assoc prof, 51-54, PROF CHEM, UNIV ILL, URBANA, 54- Mem: Am Chem Soc; fel The Chem Soc; Swiss Chem Soc; Am Crystallog Asn. Res: Reaction mechanisms and exploratory synthetic organic chemistry; solid state organic chemistry. Mailing Add: Dept of Chem Univ of Ill Urbana IL 61801

CURTIN, LEO VINCENT, b Christian Co, Ill, Mar 2, 23; m 47; c 2. ANIMAL NUTRITION. Educ: Univ Ill, BS, 47, MS, 48; Cornell Univ, PhD(nutrit), 52. Prof Exp: Animal nutritionist, Buckeye Cotton Oil Co, 48-50, head prod res dept, 52-55; asst dir feed res & nutrit, McMillen Food Mills, Ind, 55-66; dir, 66-69, vpres, 69-75, SR VPRES, RES & DEVELOP, NAT MOLASSES CO, 75- Mem: Am Soc Animal Sci; Poultry Sci Asn; Am Dairy Sci Asn. Res: Nutritional requirements of farm animals; soybean processing; feed manufacturing. Mailing Add: Nat Molasses Co Willow Grove PA 19090

CURTIN, TERRENCE M, b Spencer, SDak, June 9, 26; m 53; c 4. VETERINARY MEDICINE, PHYSIOLOGY. Educ: Univ Minn, BS & DVM, 54; Purdue Univ, MS, 63, PhD(vet physiol), 64. Prof Exp: Pvt vet practice, SDak, 54-58; exten vet, Purdue Univ, 58-61, NIH fel, 61-64 & grant, 61-65, asst prof vet physiol, 64-66; prof vet physiol & dir continuing educ, Univ Mo-Columbia, 66-68, prof & chmn dept vet physiol & pharmacol, Sch Vet Med, 68-73; PROF VET SCI & HEAD DEPT, NC STATE UNIV, 73- Concurrent Pos: USDA res grant, 65-66. Mem: AAAS; Am Vet Med Asn; US Animal Health Asn; Am Phys Soc; NY Acad Sci. Res: Esophagogastric ulcers of swine; mycotoxin induced hepatitis. Mailing Add: Dept of Vet Sci Box 5658 NC State Univ Raleigh NC 27607

CURTIN, THOMAS J, b New York, NY; c 4. ENTOMOLOGY, RESEARCH ADMINISTRATION. Educ: Manhattan Col, BS, 40; Univ Fla, MS, 51; Univ Md, College Park, PhD(entom), 59. Prof Exp: Consult drugs, Sandoz Pharmaceut, 46-51; consult entom, Strategic Air Command, USAF, 54-59, sect chief entom, USAF Epidemial Lab, Turkey, 59-62, br chief, Tex, 62-68; assoc dir life sci res, Ohio State Univ Res Found, 68-73; DIR, OFF RES & SPONSORED PROGS SERV, WAYNE STATE UNIV, 73- Concurrent Pos: Consult, US Army, 50-51; adv, UN Korean Relief Admin, 52-53 & Turkist Govt, 61-62; del, Int Cong Entom, USAF, 60; Dept Air Force del, US Dept Defense Pesticide Bd, 62-68. Mem: AAAS; Am Inst Biol Sci; Entom Soc Am; Am Mosquito Control Asn. Res: Physiology of insect reproductive systems; insect susceptibility to pesticides; plant hormonal activity. Mailing Add: Off of Res & Sponsored Progs Serv Wayne State Univ 1064 MacKenzie Hall Detroit MI 48202

CURTIS, ALLEN JAMES, applied mechanics, see 12th edition

CURTIS, BRIAN ALBERT, b New York, NY, Nov 25, 36; m 60; c 1. PHYSIOLOGY. Educ: Univ Rochester, BA, 58; Rockefeller Inst, PhD(physiol), 63. Prof Exp: Guest investr, Physiol Lab, Cambridge Univ, 60-61; from instr to asst prof, 65-71, ASSOC PROF PHYSIOL, SCH MED & SCH DENT, TUFTS UNIV, 71-, ASST DEAN EDUC PLANNING, 70- Mem: Soc Gen Physiol (secy, 67-69); Biophys Soc; Am Physiol Soc. Res: Muscle physiology. Mailing Add: Dept of Physiol Tufts Univ 136 Harrison Ave Boston MA 02111

CURTIS, BRUCE FRANKLIN, b Denver, Colo, Dec 16, 18; m 58. GEOLOGY, HYDROLOGY. Educ: Oberlin Col, AB, 41; Univ Colo, MA, 42; Harvard Univ, PhD(geol), 49. Prof Exp: Asst geol, Univ Colo, 41-42; field asst, US Geol Surv, 42; geologist to regional geologist, Continental Oil Co, 49-57; assoc prof, 57-61, chmn dept, 61-68, PROF GEOL, UNIV COLO, BOULDER, 61- Mem: Fel Geol Soc Am; Am Asn Petrol Geologists; Geochem Soc. Res: Subsurface fluids; petroleum; sedimentation. Mailing Add: 375 Harvard Lane Boulder CO 80303

CURTIS, BYRD COLLINS, b Roosevelt, Okla, Feb 25, 26; m 45; c 6. PLANT BREEDING, PLANT GENETICS. Educ: Okla State Univ, BS, 50, PhD(plant breeding, plant genetics), 59; Kans State Univ, MS, 51. Prof Exp: Vet instr agr pub sch, Okla, 52-53; from asst prof to assoc prof agron, Okla State Univ, 53-62; prof, Colo State Univ, 63-67; mgr wheat res, 67-75, MGR, HYBRID SMALL GRAINS BREEDING, CARGILL, INC, 75- Mem: Am Soc Agron. Res: Wheat breeding, particularly hybrid wheat. Mailing Add: Cargill Inc 2540 E Drake Rd Ft Collins CO 80521

CURTIS, CASSIUS W, b Indianapolis, Ind, Mar 9, 06; m 42; c 3. PHYSICS. Educ: Williams Col, AB, 28; Hamilton Col, MA, 30; Princeton Univ, PhD(physics), 36. Prof

Exp: Instr physics, Hamilton Col, 28-30; asst prof, Western Reserve Univ, 36-42; mem Nat Defense Res Comt, 42-46; from assoc prof to prof, 46-74, EMER PROF PHYSICS, LEHIGH UNIV, 74- Concurrent Pos: Consult, Frankford Arsenal, US Army, 46-66. Mem: Fel Am Phys Soc; Am Asn Physics Teachers. Res: Spectroscopy, ballistics and armor plate; solid state physics; dynamic behavior of metals. Mailing Add: Dept of Physics Lehigh Univ Bethlehem PA 18015

CURTIS, CHARLES ELLIOTT, b Bentonville, Ark, Mar 16, 37; m 60; c 3. ENTOMOLOGY. Educ: Univ Tex, BA, 59; Univ Ark, MS, 62; Purdue Univ, PhD(entom), 66. Prof Exp: Inspector, Food & Drug Admin, Dept Health, Educ & Welfare, 59-60; res asst, Univ Ark, 60-62 & Purdue Univ, 62-66; RES ENTOMOLOGIST, AGR RES SERV, USDA, 66- Mem: Entom Soc Am. Res: Pest management of insects in tree nuts and dried fruits with emphasis on area-wide suppression of insect populations with non-insecticidal approaches in almond orchards. Mailing Add: Agr Res Serv USDA Air Terminal Dr Fresno CA 93727

CURTIS, CHARLES R, b Ault, Colo, Oct 6, 38; m 66; c 2. PLANT PATHOLOGY. Educ: Colo State Univ, BS, 61, MS, 63, PhD(bot sci), 65. Prof Exp: Asst prof plant path, 67-72, ASSOC PROF PLANT PATH, UNIV MD, 72- Mem: Am Phytopath Soc; Am Soc Plant Physiol; Am Bot Soc. Res: Host-parasite relationships; physiology of disease; environmental pollution. Mailing Add: 4700 Yates Rd Beltsville MD 20705

CURTIS, CHARLES WHITTLESEY, b Providence, RI, Oct 13, 26; m 50; c 3. MATHEMATICS. Educ: Bowdoin Col, BA, 47; Yale Univ, MA, 48, PhD(math), 51. Prof Exp: Asst instr math, Yale Univ, 49-51; from instr to prof, Univ Wis, 51-63; PROF MATH, UNIV ORE, 63- Concurrent Pos: Nat Res Coun fel, 54-55; NSF sr fel, 63-64. Mem: Am Math Soc; Math Asn Am; London Math Soc. Res: Representation theory. Mailing Add: Dept of Math Univ of Ore Eugene OR 97403

CURTIS, CYRIL DEAN, b Albion, Ill, Sept 18, 20; m 48; c 2. PHYSICS. Educ: McKendree Col, BS, 43; Univ Ill, MS, 47, PhD(physics), 51. Prof Exp: Asst, McKendree Col, 41-43 & Univ Ill, 46-50; assoc physicist, Argonne Nat Lab, 51-53; asst prof physics, Vanderbilt Univ, 53-59; scientist, Midwest Univs Res Asn, 59-67; PHYSICIST, NAT ACCELERATOR LAB, 67- Mem: Am Phys Soc; Sigma Xi. Res: Experimental nuclear reactions with charged particles and neutrons; electronic instrumentation; reactor physics; bremsstrahlung production; accelerator physics. Mailing Add: 230 Woodland Hills Rd Batavia IL 60510

CURTIS, DAVID WILLIAM, b Kalamazoo, Mich, Oct 2, 24; m 55; c 4. APPLIED PHYSICS. Educ: Western Mich Univ, BS, 48; Iowa State Univ, PhD(physics), 55. Prof Exp: Asst, Ames Lab, AEC, 49-55; sr develop engr, Goodyear Aerospace Corp, 55-57, head theoret group, Aerophys Dept, 57-63, head physics group, Res & Develop Sect, Electronics Eng Div, 63-66; mem tech staff, Advan Develop Oper, Aeronaut Div, Philco-Ford Corp, 66-71; TECH CONSULT RES & DEVELOP, ARIZ ENG DIV, GOODYEAR AEROSPACE CORP, 71- Mem: Am Phys Soc. Res: Engineering analysis; information theory; probability theory in physics and engineering; synthetic array radar; microwave halography; physical optics; data processing systems. Mailing Add: 10042 N 26th St Phoenix AZ 85028

CURTIS, DORIS MALKIN, b Brooklyn, NY, Jan 12, 14. GEOLOGY. Educ: Brooklyn Col, BA, 33; Columbia Univ, MA, 34, PhD(geol), 49. Prof Exp: Paleontologist, Shell Oil Co, 41-45, stratigrapher, 45-48, geologist, 48-50; from asst to assoc prof geol, Univ Houston, 50-52; asst res geologist, Scripps Inst, Calif, 52-54; from asst to assoc prof geol, Univ Okla, 54-59; sr geologist, 59-66, STAFF GEOLOGIST, SHELL OIL CO, 66- Mem: Fel AAAS; Am Asn Petrol Geol; hon mem Soc Econ Paleont & Mineral (secy-treas, 64-66); fel Geol Soc Am; Int Asn Sedimentol. Res: Sedimentology; petroleum geology. Mailing Add: Shell Develop Co PO Box 481 Houston TX 77001

CURTIS, DWAYNE H, b Caldwell, Idaho, May 9, 30; m 54; c 4. PHYSIOLOGY. Educ: Idaho State Col, BS, 53; Univ Utah, MA, 60, PhD(exp biol), 63. Prof Exp: From asst prof to assoc prof, 63-73, PROF BIOL SCI, CALIF STATE UNIV, CHICO, 73- Concurrent Pos: Pulmonary function testing technician & consult, N T Enloe Mem Hosp, 69-72. Mem: Am Chem Soc; Mycol Soc Am. Res: Effect of chemical crosslinking agents on the mechanical properties of rat-tail tendon; taxonomy of Myxomycetes in the states of California, Idaho and Oregon; pulmonary physiology function testing. Mailing Add: Dept of Biol Calif State Univ Chico CA 95926

CURTIS, EARL CLIFTON, JR, b Vt, Oct 15, 32; m 61; c 2. PHYSICAL CHEMISTRY. Educ: Univ Vt, BS, 54; Univ Minn, PhD(phys chem), 59. Prof Exp: MEM TECH STAFF, ROCKETDYNE DIV, ROCKWELL INT CORP, 60- Mem: Am Phys Soc. Res: Molecular vibrations; infrared spectroscopy; lasers. Mailing Add: Rocketdyne Div Rockwell Int Corp Canoga Park CA 91304

CURTIS, GARNISS HEARFIELD, b San Rafael, Calif, May 27, 19; m 42; c 3. GEOLOGY. Educ: Univ Calif, PhD(geol), 51. Prof Exp: Mining engr, Christmas Copper Corp, 42-45; geologist, Shell Oil Co, 45-46; asst, lectr & instr, 46-51, from asst prof to assoc prof geol, 52-64, PROF GEOL & GEOPHYS, UNIV CALIF, BERKELEY, 64- Concurrent Pos: Fel, Miller Inst Basic Res in Sci, 58-61. Honors & Awards: Newcomb Cleveland Award, AAAS, 62. Mem: Am Geol Soc; Am Geophys Union. Res: Geology of the Sierra Nevada; 1912 eruption of Mt Katmai, Alaska; potassium argon dating. Mailing Add: Dept of Geol & Geophys Univ of Calif Berkeley CA 94720

CURTIS, GARY LYNN, b Belleville, Ill, Jan 21, 44; m 70 BIOCHEMISTRY, IMMUNOLOGY. Educ: Nebr Wesleyan Univ, BA, 66; Univ Nebr, PhD(biochem), 71. Prof Exp: Instr biochem, Med Ctr, 73-76, oncol, Eppley Cancer Inst, 73-75, ASST PROF BIOCHEM, MED CTR & RES ASST PROF OBSTET & GYNEC, UNIV NEBR, 76- Concurrent Pos: USPHS trainee, Eppley Cancer Inst, Univ Nebr, 72-75. Res: Cyclic nucleotides; immunological aspects of chemical carcinogenesis. Mailing Add: Dept of Biochem Univ of Nebr Med Ctr Omaha NE 68105

CURTIS, GEORGE CLIFTON, b St Petersburg, Fla, Dec 10, 26; m 57; c 3. PSYCHIATRY, PSYCHOSOMATIC MEDICINE. Educ: Lambuth Col, BA, 50; Vanderbilt Univ, MD, 53; McGill Univ, MSc, 59. Prof Exp: Demonstr psychiat, McGill Univ, 57-59; assoc, Univ Pa, 59-60, from asst prof to assoc prof, 60-72; PROF PSYCHIAT, UNIV MICH, ANN ARBOR, 72-, RES SCIENTIST, 73- Concurrent Pos: USPHS res fel psychiat, McGill Univ, 58-59; NIMH career investr, Univ Pa, 61-66; clin asst psychiat, Royal Victoria Hosp, Montreal, Que, 57-59; med res scientist, Eastern Pa Psychiat Inst, 59-72; consult clin res, Mercy Douglass Hosp, Philadelphia, Pa, 60-62; assoc mem, Albert Einstein Med Ctr, 62-68; actg dir clin res, Philadelphia Gen Hosp, 68-72. Mem: Am Psychiat Asn; Am Psychosom Soc; Int Soc Chronobiol. Res: Psychobiology; psychosomatics; chronobiology; psychotherapy. Mailing Add: Neuropsychiat Inst Univ of Mich Ann Arbor MI 48104

CURTIS, GEORGE GRAYDON, organic chemistry, see 12th edition

CURTIS, GEORGE WILLIAM, b Brussels, Belg, Mar 9, 25; US citizen; m 51; c 4. ASTROPHYSICS, SOLAR PHYSICS. Educ: Univ Colo, BS, 49, MS, 52, PhD(astrophys), 63. Prof Exp: Physicist, Nat Bur Standards, 52-55; Fulbright fel, Inst Astrophys, Paris, 55-56; observer in charge, High Altitude Observ, 56-59, fel solar physics, Joint Inst Lab Astrophys, 63-64; res scientist, Sacramento Peak Observ, Air Force Cambridge Res Labs, 64-67; res scientist, 67-73, DEP DIR HIGH ALTITUDE OBSERV, NAT CTR ATMOSPHERIC RES, 73- Concurrent Pos: Sci coordr & exped leader, Solar Eclipse Exped, Nat Ctr Atmospheric Res, 71-73. Mem: Am Astron Soc; Royal Astron Soc; Am Inst Physics. Res: Low temperature properties of metal and insulating materials; design and development of low temperature instrumentation; acquisition, reduction and analysis of solar observational data on the corona, chromosphere and photosphere, especially spectral data. Mailing Add: High Altitude Observ Nat Ctr for Atmospheric Res Boulder CO 80302

CURTIS, HAROLD ORMAND, b Orono, Maine, July 15, 23; m 46; c 5. PHYSICS. Educ: Bowdoin Col, AB, 47; Harvard Univ, MA, 48, PhD(physics), 51. Prof Exp: Instr physics, Hamilton Col, 50-51; physicist, Air Force Cambridge Res Ctr, 51-59 & electronic systs div, 59-61; group leader, Lincoln Lab, Mass Inst Technol, 61-68; PRES, HARRINGTON, DAVENPORT & CURTIS, INC, 68- Mem: Am Phys Soc; Am Meteorol Soc; NY Acad Sci. Res: Atmospheric electricity; space vehicle tracking and orbit computations; space surveillance; reentry physics; instrumentation; space vehicle dynamics. Mailing Add: Harrington Davenport & Curtis Inc 74 Loomis St Bedford MA 02154

CURTIS, HENRY L, b Tucson, Ariz, Jan 2, 36; m 49; c 2. PLANT PHYSIOLOGY. Educ: Brigham Young Univ, BS, 61, MS, 65; Ohio State Univ, PhD(bot), 70. Prof Exp: Plant physiologist, Crops Res Div, Agr Res Serv, USDA, 69-70; ASST PROF BIOL, GA SOUTHERN COL, 70- Mem: AAAS; Bot Soc Am; Am Soc Plant Physiol. Res: Photoperiodic and hormonal control of flowering in lemna; mechanisms governing establishment and breaking of dormancy in higher plant organs and germination of spores and seeds. Mailing Add: Dept of Biol Ga Southern Col Statesboro GA 30458

CURTIS, HERBERT JOHN, b Oak Park, Ill, Aug 18, 18; m 50; c 2. MATHEMATICS. Educ: Yale Univ, BA, 39; Ill Inst Technol, MS, 48, PhD(math), 54. Prof Exp: Instr math, Ill Inst Technol, 48-54; from asst prof to assoc prof, 54-65, from actg head to head div, 60-64, exec secy dept, 64-73, PROF MATH, UNIV ILL, CHICAGO CIRCLE, 65-, ACTG HEAD DEPT, 74- Mem: Am Math Soc; Math Asn Am. Mailing Add: Dept of Math Univ of Ill at Chicago Circle Chicago IL 60680

CURTIS, HOWARD BENTON, JR, b McKinney, Tex, July 14, 24. MATHEMATICS. Educ: Univ Okla, BSCE, 46; Univ Ark, MA, 48; Rice Inst, PhD(math), 58. Prof Exp: Actg instr math, Univ Ark, 47; from instr to asst prof, Agr & Mech Col, Tex, 50-55; asst, Rice Inst, 54-58; from asst prof to assoc prof, 58-73, PROF MATH, UNIV TEX, AUSTIN, 73- Mem: Am Math Soc; Math Asn Am. Res: Theory of functions; Riemann surfaces. Mailing Add: Dept of Math Univ of Tex Austin TX 78712

CURTIS, HOWARD JAMES, physics, physiology, deceased

CURTIS, JERRY LEON, b Mooresville, NC, June 27, 41; m 68. BIOCHEMISTRY. Educ: Pfeiffer Col, AB, 63; Fla State Univ, PhD(biochem), 68. Prof Exp: Res assoc biochem, Fla State Univ, 68-69; RES SCIENTIST, CENT STATE HOSP, 69- Mem: Am Chem Soc; NY Acad Sci. Res: Enzyme control mechanisms, especially as related to human mental retardation; biochemical basis of mental disease. Mailing Add: Res Dept Cent State Hosp Milledgeville GA 31061

CURTIS, JOHN RUSSELL, b Bessemer City, NC, Nov 7, 34; m 58; c 3. PSYCHIATRY. Educ: Univ NC, AB, 56, MD, 60. Prof Exp: Clin instr psychiat, Sch Med, Univ Ky, 64-66, consult psychiatrist, Student Health Serv, 64-55, dir psychiat sect, 66-68, asst prof psychiat, 66-68; assoc prof psychiat, Dept Psychol, 68-74, CHIEF PSYCHIATRIST & DIR, UNIV HEALTH SERV, UNIV GA, 68-; ASSOC CLIN PROF NEUROL, MED COL GA, 69- Concurrent Pos: Staff psychiatrist, Clin Res Ctr, NIMH, 64-65, chief ment addiction serv, 65-66; consult psychiatrist, Berea Col, 66-69, Clin Ctr Addiction Res, Ky, 67-69 & Stephens County Pub Health Dept, Ga, 69- Mem: Fel Am Psychiat Asn; Am Col Health Asn. Res: College health with emphasis on college mental health; health care delivery systems; community health and mental health. Mailing Add: Off of the Dir Univ Health Serv Univ of Ga Athens GA 30601

CURTIS, JOSEPH C, b Manchester, NH, Feb 14, 30; m 59; c 4. CELL BIOLOGY, BIOPHYSICS. Educ: Cornell Univ, BA, 51; Brown Univ, PhD(biol), 60. Prof Exp: Res assoc biol, Brown Univ, 60-63; from asst prof to prof biol, 63-74, PROF ZOOL, CLARK UNIV, 74- Concurrent Pos: USPHS fel, 60-62. Mem: Am Soc Zoologists; Am Soc Cell Biol; Am Inst Biol Sci; Biophys Soc; NY Acad Sci. Res: Biological effects of ultrasound; cytology; cytochemistry; cell fine structure and biochemistry of steroid-producing tissues; effects of ultrasound on cell ultrastructure; experimental pathology. Mailing Add: Dept of Zool Clark Univ Worcester MA 01610

CURTIS, LORENZO JAN, b St Johns, Mich, Nov 4, 35; div. ATOMIC PHYSICS. Educ: Univ Toledo, BS, 58; Univ Mich, MS, 61, PhD(high energy physics), 63. Prof Exp: Asst prof physics, 63-68, assoc prof physics & astron, 68-72, ASSOC PROF PHYSICS, UNIV TOLEDO, 72- Concurrent Pos: NSF res grant, 64-65, fel, Latin Am Sch Physics, Mexico City, 65; vis scientist, Woods Hole Oceanog Inst, 67 & Nobel Inst, Sweden, 70-71. Mem: Am Phys Soc. Res: Atomic and ionic spectra; transition probabilities; collision processes, utilizing beam-foil and pulsed electron beam methods. Mailing Add: Dept of Physics Univ of Toledo Toledo OH 43606

CURTIS, MORTON LANDERS, b Tex, Nov 11, 21; m 44; c 2. MATHEMATICS. Educ: Tex Col Arts & Indust, BS, 43; Univ Mich, PhD(math), 51. Prof Exp: Instr math, Northwestern Univ, 50-51; mem, Inst Adv Study, 51-53; asst prof math, Northwestern Univ, 53-56; prof, Univ Ga, 56-59 & Fla State Univ, 59-64; chmn dept, 64-69, W L MOODY JR PROF MATH, RICE UNIV, 67- Concurrent Pos: NSF fel, Cauis Col, Cambidge, 59-60. Mem: Am Math Soc. Res: Topology. Mailing Add: Dept of Math Rice Univ Houston TX 77001

CURTIS, MYRON DAVID, b Mt Vernon, Ind, Oct 17, 38; m 66; c 2. INORGANIC CHEMISTRY, ORGANOMETALLIC CHEMISTRY. Educ: Wabash Col, AB, 60; Northwestern Univ, PhD, 65. Prof Exp: Res chemist, Hooker Chem Corp, 64-66; instr chem, Northwestern Univ, 66-67; asst prof, 67-72, ASSOC PROF CHEM, UNIV MICH, ANN ARBOR, 72- Concurrent Pos: Petrol Res Fund grant, 67-69; fac res fel, Univ Mich, 69. Mem: AAAS; Am Chem Soc. Res: Nature of bonding in organometallic compounds; synthesis of organometallic and inorganic systems to test bonding theories; structural studies in homogeneous catalysis. Mailing Add: Dept of Chem 4019 Chem Bldg Univ of Mich Ann Arbor MI 48104

CURTIS, ORLIE LINDSEY, JR, b Hutchinson, Kans, Feb 27, 34; m 55; c 2. SOLID STATE PHYSICS. Educ: Union Col, Nebr, BA, 54; Purdue Univ, MS, 56; Univ Tenn, PhD, 61. Prof Exp: Chief semiconductors, Oak Ridge Nat Lab, 56-63; chief, Solid State Physics, Ventura Div, 63-67, asst dir, Solid State Electronics Lab, 67-68, DIR, SOLID STATE ELECTRONICS LAB, NORTHROP CORP, 68- Concurrent

Pos: Vis lectr, Univ Calif, Berkeley, 70-71. Mem: Fel Am Phys Soc; fel Inst Elec & Electronics Eng. Res: Radiation effects on materials, devices, circuits and systems; electronic properties of solids; device physics. Mailing Add: Northrop Corp 3401 W Broadway Hawthorne CA 90250

CURTIS, OTIS FREEMAN, JR, b Ithaca, NY, Jan 28, 15; m 39; c 2. PLANT PHYSIOLOGY. Educ: Oberlin Col, AB, 36; Cornell Univ, PhD(plant physiol), 40. Prof Exp: Asst bot, Cornell Univ, 36-40; instr floricult & jr plant physiologist, Univ Calif, Los Angeles, 40-43; assoc plant physiologist, Bur Plant Indust, USDA, Calif, 43-45, assoc plant physiologist, Indio, 45-46; asst prof, 46-70, ASSOC PROF POMOL, NY AGR EXP STA, CORNELL UNIV, 70- Mem: Am Soc Plant Physiologists; Bot Soc Am; Am Soc Hort Sci. Res: Chemical weed killers; plant growth regulation; dormancy; flowering; factors influencing plant constituents; plant propagation; carotene; carbohydrates; rubber. Mailing Add: NY Agr Exp Sta Cornell Univ Geneva NY 14456

CURTIS, PAUL ROBINSON, b Hanumkonda, S India, Aug 14, 31; US citizen; m 61; c 2. MICROBIOLOGY. Educ: Col Wooster, BA, 52; Ohio State Univ, MS, 57, PhD(bact), 61. Prof Exp: Asst, Ohio State Univ, 55-60; from asst prof to assoc prof biol, Am Univ, 60-74; MEM FAC, DEPT BIOL, EISENHOWER COL, 74- Concurrent Pos: NSF grant, Am Univ, 63-64. Mem: AAAS; Am Soc Microbiol; Am Inst Biol Sci. Res: Bacteriophages; bacterial and viral genetics. Mailing Add: Dept of Biol Eisenhower Col Seneca Falls NY 13148

CURTIS, PHILIP CHADSEY, JR, b Providence, RI, Mar 6, 28; m 50; c 5. MATHEMATICS. Educ: Brown Univ, AB, 50; Yale Univ, MA, 52, PhD(math), 55. Prof Exp: From instr to assoc prof, 55-67, chmn dept, 71-75, PROF MATH, UNIV CALIF, LOS ANGELES, 67- Concurrent Pos: Fulbright travel fel, Aarhus Univ, 69-70; consult, Space Tech Labs, TRW Inc, 56-; vis prof, Univ Copenhagen, 75-76. Mem: Am Math Soc. Res: Functional analysis; Banach algebras; harmonic analysis. Mailing Add: Dept of Math Univ of Calif Los Angeles CA 90024

CURTIS, RALPH WENDELL, b Cuba City, Wis, Oct 20, 36; m 57; c 3. METALLURGICAL CHEMISTRY. Educ: Univ Wis-Platteville, BS, 58; Iowa State Univ, PhD(metall), 62. Prof Exp: AEC fel metall, Iowa State Univ, 62-63; teacher chem, 63-69, HEAD DEPT CHEM, UNIV WIS-PLATTEVILLE, 69- Mem: Am Chem Soc. Mailing Add: Dept of Chem Univ of Wis Platteville WI 53818

CURTIS, RICHARD BERTRAM, theoretical physics, see 12th edition

CURTIS, RICHARD MILTON, physical chemistry, see 12th edition

CURTIS, ROBERT ORIN, b Portland, Maine, Oct 27, 27; m 52; c 3. FOREST MENSURATION. Educ: Yale Univ, BS, 50, MF, 51; Univ Wash, PhD(forestry), 65. Prof Exp: Forester, Northeastern Forest Exp Sta, 51-54, res forester, 54-62, PRIN MENSURATIONIST, PAC NORTHWEST FOREST & RANGE EXP STA, US FOREST SERV, 65- Mem: Soc Am Foresters. Res: Mensuration research in forest yield studies, site-growth relationships and related measurement problems. Mailing Add: Pac Northwest Forest & Range Exp Sta Box 3141 Portland OR 97208

CURTIS, ROBIN LIVINGSTONE, b Bellingham, Wash, Jan, 16, 26; m 50, 60; c 1. NEUROANATOMY, PHYSIOLOGICAL PSYCHOLOGY. Educ: Wesleyan Univ, BA, 48, MA, 50; Brown Univ, PhD(exp psychol), 54. Prof Exp: Chem analyst, E I du Pont de Nemours & Co, NY, 46-47; psychiat aide, Middletown State Hosp, Conn, 47-48; asst biol, Wesleyan Univ, 49-50; spec asst, Brown Univ, 52, asst psychol, 52-53; from instr to assoc prof, NJ Col Med & Dent, 56-67; ASSOC PROF ANAT, MED COL WIS, 67- Concurrent Pos: USPHS fel, 53-55; Nat Multiple Sclerosis Soc fel anat, Col Med, NY Univ, 55-56. Mem: AAAS; Am Psychol Asn; Tissue Cult Asn; Am Asn Anat; Am Genetic Asn. Res: Genetics, behavior and neuroanatomy of hereditary neuromuscular abnormalities in rodents; comparative neurology and animal behavior. Mailing Add: Dept of Anat Med Col of Wis Milwaukee WI 53233

CURTIS, ROGER WILLIAM, b East Lansing, Mich, Oct 4, 04; m 27; c 3. EXPERIMENTAL PHYSICS. Educ: Univ Mich, AB, 26; Johns Hopkins Univ, PhD(physics), 34. Prof Exp: Jr physicist, Nat Bur Stand, 26-27, 28-30, asst physicist, 30-36, assoc physicist, 36-41, physicist, 41-42; physicist, Naval Ord Lab, 42-43, Dept Navy, 43-44, Underwater Sound Lab, Harvard Univ, 44, Nat Bur Stand, 44-53 & Diamond Ord Fuze Labs, 53-57; res & develop adv to mil assistance adv group, Am Embassy, Bonn, Ger, 57-61; consult, 61-63; staff mem, Stand Lab, Nortronics, Mass, 63-65; prod reliability res & develop mgr, Westinghouse Elec Corp, Baltimore, 65-66; phys sci adminstr mat test lab, Nat Bur Stand, 66-68; dep chief, Mat Eval Develop Lab, Fed Supply Serv, Gen Serv Admin, 68-73, actg dir, 73-74; CONSULT, 74- Mem: Am Phys Soc; Inst Elec & Electronics Eng. Res: Absolute measurement of electric current; ultrasonic absorption in gases; electronics. Mailing Add: 6308 Valley Rd Bethesda MD 20034

CURTIS, ROY WALTER, b Chicago, Ill, Mar 17, 27; m 51; c 3. PLANT PHYSIOLOGY. Educ: Northern Ill State Teachers Col, BS, 50; Univ Wis, MS, 51, PhD(bot), 54. Prof Exp: Asst, Univ Wis, 50-54, proj assoc, 54; asst microbiologist, 54-57, assoc prof, 57-70, PROF MICROBIOL, PURDUE UNIV, WEST LAFAYETTE, 70- Res: Effects of ultraviolet radiation of fungus spores; acenaphthene-requiring fungi; effects of microbial products on plants. Mailing Add: Dept of Bot & Plant Path Purdue Univ West Lafayette IN 47906

CURTIS, STANLEY BARTLETT, b Evanston, Ill, Feb 16, 32; m 57; c 3. RADIATION BIOPHYSICS. Educ: Carleton Col, BA, 54; Wash Univ, PhD(physics), 62. Prof Exp: Res scientist, Biophys Res Div, Lockheed-Calif Co, 62-63; res specialist, Space Physics Group, Boeing Co, 63-66; BIOPHYSICIST, DONNER LAB, UNIV CALIF, BERKELEY, 66- Concurrent Pos: Mem radiation biol panel, NASA, 60; Eleanor Roosevelt fel, 70-71; physics counr, Coun Rehab Res Soc, 72-75. Mem: Am Phys Soc; Radiation Res Soc. Res: Biological effects of ionizing radiation in space; high linear energy transfer effects on biological systems; tumor cell kinetics. Mailing Add: Lawrence Berkeley Lab Univ of Calif Berkeley CA 94720

CURTIS, STANLEY EVAN, b Plymouth, Ind, Apr 3, 42; m 67; c 3. ANIMAL SCIENCE, ENVIRONMENTAL PHYSIOLOGY. Educ: Purdue Univ, BS, 64, MS, 66, PhD(environ physiol), 68. Prof Exp: Vis fel biomet, Univ Wis, 66-67; asst prof dairy husb & physiologist, Univ Mo-Columbia, 68-70; asst prof animal sci, 70-75, ASSOC PROF ANIMAL SCI, UNIV ILL, URBANA, 75- Mem: Soc Animal Sci; Int Soc Biometeorol. Res: Environmental adaptability of livestock, especially newborns; metabolic responses to cold; biometeorology; air environment. Mailing Add: 126 Animal Sci Lab Univ of Ill Urbana IL 61801

CURTIS, THOMAS EDWIN, b Miami, Okla, Oct 2, 27; m 46; c 4. PSYCHIATRY. Educ: Duke Univ, MD, 50; Univ NC-Duke Univ Training Prog, cand, 60. Prof Exp: Intern, St John's Hosp, Okla, 50-51; resident psychiat, Fairfield State Hosp, Conn, 52; resident, Dorthea Dix State Hosp, NC, 52-54; from instr to assoc prof, 54-69, PROF PSYCHIAT, SCH MED, UNIC NC, CHAPEL HILL, 69-, CHMN DEPT, 73- Concurrent Pos: Resident, NC Mem Hosp, 54-55. Mem: AMA; fel Am Psychiat Asn; Am Asn Med Cols. Res: Experimental teaching; group analytic psychotherapy; family study and treatment; clinical research in psychotherapy; psychiatric nursing, teaching and administration. Mailing Add: Dept of Psychiat Univ of NC Sch of Med Chapel Hill NC 27514

CURTIS, THOMAS HASBROOK, b Detroit, Mich, July 26, 41; div; c 2. APPLIED PHYSICS, ACOUSTICS. Educ: Kenyon Col, BA, 63; Yale Univ, MS, 65, PhD(physics), 68. Prof Exp: Res physicist, Lawrence Radiation Lab, Univ Calif, 68-70; MEM TECH STAFF, BELL LABS, 70- Concurrent Pos: Consult, Lycoming Div, Avco Corp, Conn, 67 & Automation Res Mechanisms, Inc, Calif, 68-69. Mem: Inst Elec & Electronics Engrs; AAAS; Am Phys Soc; Acoust Soc Am. Res: Physics of speech production, transmission and hearing; digital processing of acoustic signals; applications of computers to system control; nuclear physics. Mailing Add: Bell Labs 4F513 Holmdel NJ 07733

CURTIS, WILLIAM EDGAR, animal physiology, see 12th edition

CURTISS, CHARLES FRANCIS, b Chicago, Ill, Apr 4, 21; m 46; c 3. THEORETICAL CHEMISTRY. Educ: Univ Wis, BS, 42, PhD(chem), 48. Prof Exp: Chemist, Geophys Lab, Carnegie Inst, 42-45; assoc, Allegany Ballistics Lab, George Washington Univ, 45; proj assoc, 48-49, from asst prof to assoc prof, 49-60, PROF CHEM, UNIV WIS-MADISON, 60- Mem: AAAS; Am Chem Soc; Am Phys Soc. Res: Statistical mechanics; kinetic theory of gases. Mailing Add: Dept of Chem Univ of Wis Madison WI 53706

CURTISS, DERRY HASBROUCK, physical chemistry, see 12th edition

CURTISS, JOHN HAMILTON, b Evanston, Ill, Dec 23, 09. MATHEMATICAL ANALYSIS, ANALYTICAL STATISTICS. Educ: Northwestern Univ, AB, 30; Univ Iowa, SM, 31; Harvard Univ, PhD(math), 35. Prof Exp: Instr math, Harvard Univ, 32-33; asst, Johns Hopkins Univ, 35; from instr to asst prof, Cornell Univ, 35-43; asst to dir, Nat Bur Standards, 46-47, chief appl math div, 47-53; sr scientist, Inst Math Sci & adj prof math, NY Univ, 53-54; exec dir, Am Math Soc, 54-59; prof, 59-75, chmn dept, 59-61, EMER PROF MATH, UNIV MIAMI, 75- Concurrent Pos: Vis lectr, Univ Calif, Los Angeles, 51 & Harvard Univ, 53. Honors & Awards: Meritorious Serv Medal, US Dept Commerce, 49. Mem: AAAS; Am Math Soc; Math Asn Am; fel Am Statist Asn; fel Inst Math Statist (vpres, 47). Res: Complex function theory; probability and statistics. Mailing Add: 9888 N Kendall Dr Miami FL 33176

CURTISS, ROY, III, b New York, NY, May 27, 34; c 4. MICROBIAL GENETICS, MOLECULAR GENETICS. Educ: Cornell Univ, BS, 56; Univ Chicago, PhD(microbiol), 62. Prof Exp: Jr technical specialist, Biol Div, Brookhaven Nat Lab, 56-58; biologist, Oak Ridge Nat Lab, 63-72, group leader microbial genetics & radiation microbiol group, Biol Div, 69-72; prof microbiol, Univ Tenn-Oak Ridge Grad Sch Biomed Sci, 69-72, assoc dir, 70-71, interim dir, 71-72; PROF MICROBIOL, UNIV ALA, BIRMINGHAM, SR SCIENTIST, INST DENT RES & CANCER RES & TRAINING CTR, 72-, DIR MOLECULAR CELL BIOL GRAD TRAINING PROG, 73- Concurrent Pos: Lectr microbiol, Univ Tenn, 65-72 & Oak Ridge Grad Sch Biomed Sci, 67-69; vis prof, Venezuelan Inst Sci Res, 69, Univ PR, 72 & Cath Univ Chile, 73; lectr, Am Found Microbiol, 69-70; ed, J Bact, 70-; mem, NIH Recombinant DNA Molecule Prog Adv Comt, 74-; mem, NSF Genetic Biol Study Sect, 75-; dir, Nat Inst Gen Med Sci Predoctoral & Postdoctoral Training Grants, 75-; mem bd dirs, Coun Advan Sci Writing, 75- Mem: Fel Am Acad Microbiol; hon mem Microbiol Asn Chile; Am Soc Microbiol; Genetics Soc Am; Brit Soc Gen Microbiol. Res: Mechanisms of conjugation and recombination in bacteria; plasmid evolution, transmission, replication and functions. Mailing Add: Dept of Microbiol Box 11 DB Univ of Ala Birmingham AL 35294

CURTS, KEPHART MAYNARD, b Kansas City, Mo, Feb 29, 16; m 39; c 2. VETERINARY PHARMACOLOGY. Educ: Tex A&M Univ, BS, 40, DVM, 41. Prof Exp: Asst bact, Tex A&M Univ, 39, asst vet physiol, pharmacol & toxicol, 39-41, instr vet anat, 41-42; asst prof, Univ Ga, 46-47; staff vet, Curts-Folse Labs, 47-53; secy, 53-72, vet dir, 53-63, vpres, 63-72, PRES, CURTS LAB, INC, 72- Mem: Am Vet Med Asn; US Animal Health Asn; fel Am Col Vet Toxicol; Indust Vet Asn. Res: Veterinary pharmacology as regards to chemotherapy of bacterial and fungus diseases; toxicology. Mailing Add: 5236 Del Mar Ave Shawnee Mission KS 66205

CURTZ, THADDEUS BANKSON, b New York, NY, Mar 16, 22; m 42; c 4. MATHEMATICS. Educ: Baldwin-Wallace Col, BA, 48; Yale Univ, MA, 49, PhD, 60. Prof Exp: Res assoc & proj mgr, Willow Run Labs, Univ Mich, 51-58, head comput dept, 58-61; dir comput ctr, Conductron Corp, 61-67; vpres, KMS Industs, Inc, 67-68; vis prof comput sci, 69-70, assoc prof, 70-72, PROF COMPUT SCI, UNIV KY, 72-, CHMN DEPT, 70- Mem: Am Math Soc. Res: Data processing; differential equations; computation. Mailing Add: Dept of Comput Sci Univ of Ky Lexington KY 40506

CURWEN, DAVID, b Ridgewood, NJ, May 25, 35; m 57; c 2. HORTICULTURE, FOOD TECHNOLOGY. Educ: Univ Vt, BS, 57; Pa State Univ, MS, 60, PhD(hort), 64. Prof Exp: Asst prof & area hort agent veg exten & res, 63-71, ASSOC PROF AGR EXTEN & HORT & AREA EXTEN HORTICULTURIST, UNIV WIS-MADISON, 71- Mem: Am Soc Hort Sci; Am Inst Biol Sci. Mailing Add: Dept of Hort Univ of Wis Madison WI 58706

CUSACHS, LOUIS CHOPIN, b Orange, NJ, Sept 9, 33; m 64; c 3. COMPUTER SCIENCE, QUANTUM CHEMISTRY. Educ: US Naval Acad, BS, 56; Univ Paris, dipl, 61; Northwestern Univ, PhD, 61. Prof Exp: Du Pont teaching asst, Northwestern Univ, 58-59; Fulbright exchange prof, Univ Valencia, 61-62; from asst prof to assoc prof phys chem, Tulane Univ, 62-72; PROF COMPUT SCI, LOYOLA UNIV, LA, 72- Concurrent Pos: Lectr, Fulbright Prog, Nat Univ Buenos Aires & Univ La Plata, Arg, 74. Mem: Am Chem Soc; Inst Elec & Electronics Eng; Am Phys Soc; Int Soc Quantum Biol (vpres, 73-75, pres, 75-77). Res: Scientific computation; electronic structure of molecules. Mailing Add: 7711 Willow St New Orleans LA 70118

CUSANOVICH, MICHAEL A, b Los Angeles, Calif, Mar 2, 42; m 63; c 2. BIOCHEMISTRY. Educ: Univ of the Pac, BS, 63; Univ Calif, San Diego, PhD(chem), 67. Prof Exp: NIH fel biochem, Univ Calif, San Diego, 67-68 & Cornell Univ, 68-69; asst prof, 69-74, ASSOC PROF CHEM, UNIV ARIZ, 74- Concurrent Pos: NIH Career Develop Award, 75. Mem: Am Chem Soc; Biophys Soc; NY Acad Sci. Res: Mechanism of electron transfer as catalyzed by cytochromes and coupled energy conservation; mechanism of action of the visual pigment rhodopsin. Mailing Add: Dept of Chem Univ of Ariz Tucson AZ 85721

CUSCURIDA, MICHAEL, b Waco, Tex, Nov 16, 26. ORGANIC CHEMISTRY. Educ: Tex Agr & Mech Col, BS, 48; Baylor Univ, MS, 51, PhD, 55. Prof Exp: Assoc res chemist, Midwest Res Inst, 55-57; sr res chemist, 57-65, proj chemist, 65-69, SR PROJ CHEMIST, JEFFERSON CHEM CO, INC, 69- Mem: Am Chem Soc; Soc Plastics Engrs. Res: Organic chemistry; polyurethane chemistry; polymers of alkylene oxides; inorganic chemistry; chemistry of metal hydrides; high energy fuels based on boron compounds. Mailing Add: Jefferson Chem Co Inc Box 4128 N Austin Sta Austin TX 78765

CUSHEN, WALTER EDWARD, b Hagerstown, Md, Mar 21, 25; m 49; c 2. OPERATIONS RESEARCH. Educ: Western Md Col, BA, 48, DSc, 66; Univ Edinburgh, PhD(metaphys), 51. Prof Exp: Mathematician, Ballistic Res Labs, 48 & 51-52; sr programmer, Remington Rand, Inc, 52; chmn res group, Opers Res Off, 52-61; assoc prof opers res, Case Inst Technol, 61-63; staff mem, Inst Defense Anal, 63-64; chief, Tech Anal Div, Nat Bur Standards, 64-74; actg asst dir, Off Energy Systs, FPC, 74-75; VPRES & DIR, MATHTECH WASHINGTON, MATHEMATICA, INC, 75- Concurrent Pos: Consult, FAA, 63-64; mem, Md Water Sci Adv Bd, 68-76, Gov Sci Adv Coun, 67-73 & Bd Trustees, Western Md Col, 71-75. Mem: Fel AAAS; Opers Res Soc Am (pres, 70-71); Inst Mgt Sci; Am Judicature Soc. Res: Systems analysis; operations research. Mailing Add: 6910 Maple Ave Chevy Chase MD 20015

CUSHING, COLBERT ELLIS, JR, b Ft Collins, Colo, Jan 9, 31; m 59; c 3. FRESH WATER ECOLOGY. Educ: Colo State Univ, BSc, 52, MSc, 56; Univ Sask, PhD(biol), 61. Prof Exp: Fishery biologist, Mont Fish & Game Dept, 56-58; biol scientist, Hanford Labs, Gen Elec Co, 61-65; res scientist, 65-67, SR RES SCIENTIST, PAC NORTHWEST LAB, BATTELLE MEM INST, 67- Mem: Am Soc Limnol & Oceanog; Ecol Soc Am; Am Fisheries Soc; Int Asn Theoret & Appl Limnol; Am Inst Biol Sci. Res: Stream ecology; radioecological and nutrient cycling studies of aquatic ecosystems; primary productivity. Mailing Add: Ecosystems Dept Pac Northwest Lab Battelle Mem Inst Richland WA 99352

CUSHING, EDWARD JOHN, b St Louis, Mo, Nov 15, 33; m 55; c 5. ECOLOGY. Educ: Wash Univ, AB, 54; Univ Minn, PhD(geol), 63. Prof Exp: Res fel geol, Sch Earth Sci, Univ Minn, Minneapolis, 63-64, asst prof, 64-66, from asst prof to prof bot, 66-74; PROF ECOL & BEHAV BIOL, UNIV MINN, ST PAUL, 75- Concurrent Pos: NATO fel, Univ NWales, Bangor, 66-67. Mem: AAAS; Geol Soc Am; Ecol Soc Am; Bot Soc Am; Brit Ecol Soc. Res: Quaternary pollen analysis and paleoecology; plant ecology; glacial geology. Mailing Add: Dept of Ecol & Behav Biol Univ of Minn St Paul MN 55108

CUSHING, JAMES THOMAS, b Long Beach, Calif, Feb 4, 37; m 63. THEORETICAL PHYSICS. Educ: Loyola Univ, Ill, BS, 59; Northwestern Univ, MS, 60; Univ Iowa, PhD(physics), 63. Prof Exp: Res assoc theoret physics, Univ Iowa, 63-64; NSF fel, Imp Col, Univ London, 64-65; resident res assoc, Argonne Nat Lab, 65-66; asst prof, 66-69, ASSOC PROF PHYSICS, UNIV NOTRE DAME, 69- Mem: Am Phys Soc. Res: Mathematical aspects of strong interactions; origin of internal symmetries. Mailing Add: Dept of Physics Univ of Notre Dame Notre Dame IN 46556

CUSHING, JIM MICHAEL, b North Platte, Nebr, Mar 20, 42. APPLIED MATHEMATICS. Educ: Univ Colo, BA, 64; Univ Md, PhD(math), 68. Prof Exp: Asst prof, 68-74, ASSOC PROF MATH, UNIV ARIZ, 74- Concurrent Pos: IBM fel, 71-72. Mem: Am Math Soc; Math Asn Am; Soc Indust & Appl Math. Res: Qualitative theory of integral, differential and integrodifferential equations and applications. Mailing Add: Dept of Math Univ of Ariz Tucson AZ 85721

CUSHING, JOHN (ELDRIDGE), JR, b San Francisco, Calif, Aug 25, 16; m 41; c 2. BIOLOGY. Educ: Univ Calif, AB, 38; Calif Inst Technol, PhD(genetics & immunol), 43. Prof Exp: Asst immunochem, Calif Inst Technol, 43-45; instr biol, Johns Hopkins Univ, 45-48; from asst prof to prof biol, 48-69, PROF IMMUNOL, UNIV CALIF, SANTA BARBARA, 69- Concurrent Pos: Consult, US Fish & Wildlife Serv, 56-; Guggenheim fel, 58-59. Mem: Am Asn Immunologists; Am Soc Naturalists. Res: Comparative immunology; genetics of Neurospora; biochemical mutations by chemical means; evolution studies; blood groups of marine animals; immune reactions of invertebrates. Mailing Add: Dept of Immunol Univ of Calif Santa Barbara CA 93106

CUSHING, MERCHANT LEROY, b Haverhill, Mass, Jan 31, 10; m 39; c 4. CARBOHYDRATE CHEMISTRY. Educ: Univ NH, BS, 31, MS, 33; Columbia Univ, PhD(Org chem), 41. Prof Exp: Res chemist, Stein, Hall & Co, NY, 37-42, chief chemist, Starch Explor Lab, 46-48, dir, NY Labs, 48-50, chief chemist, Paper Lab, 52-55; head paper lab, A E Staley Mfg Co, Ill, 55-62; mgr paper res, Fiber Prods Res Ctr, Inc, 62-65; group leader, Paper Appln, CIBA Corp, NJ, 65-69; group leader paper res & develop, Am Maize Prods Co, 69-74; RES ASSOC, BERGSTROM PAPER CO, 74- Mem: Tech Asn Pulp & Paper Indust. Res: Carbohydrate chemistry oriented to new industrial products; modifications of natural raw materials for use in papermaking; wet end additives for improving interfiber bonding and filler retention; enzyme conversion of starches; deinking recycled waste papers; new product development. Mailing Add: 225 W Wisconsin Ave Neenah WI 54956

CUSHING, ROBERT LEAVITT, b Ord, Nebr, Apr 12, 14; m 38; c 3. AGRONOMY. Educ: Univ Nebr, BSc, 36, MSc, 38. Hon Degrees: DSc, Univ Hawaii, 62. Prof Exp: Asst agronomist, Nebr Agr Exp Sta, 38-42 & USDA, 42-43; from asst prof to assoc prof plant breeding, Cornell Univ, 43-47; agronomist, Hawaiian Pineapple Co, 47-49; from assoc prof to prof plant breeding, Cornell Univ, 49-51; asst dir, Pineapple Res Inst Hawaii, 51-52, dir, 52-63, pres, 58-63; DIR EXP STA, HAWAIIAN SUGAR PLANTERS ASN, 63-, VPRES & SECY, 66- Concurrent Pos: Mem bd regents, Univ Hawaii, 67, chmn, 68. Mem: AAAS; Am Soc Agron; Am Genetic Asn; Inst Food Technol. Res: Plant breeding; plant genetics. Mailing Add: 99-193 Aied Hts Dr Aied HI 96701

CUSHING, VINCENT JEROME, b Evanston, Ill, Apr 17, 24; m 45; c 7. APPLIED PHYSICS. Educ: Univ Notre Dame, BS, 45, MS, 46; Ill Inst Technol, PhD(physics), 56. Prof Exp: Develop engr, Eugene Mittelman, Consults, Ill, 49-50; scientist, Propulsion & Fluids Res Div, Armour Res Found, Ill Inst Technol, 50-55, dir, 55-59; dir appl sci div, Fairchild Engine & Airplane Corp, 59-60; pres, Eng-Physics Co, 60-72; PRES, CUSHING ENG, INC, 73- Mem: Am Phys Soc; Inst Elec & Electronics Eng; Am Soc Test & Mat. Res: Induction flow meter; magneto-fluid dynamics; shock hydrodynamics; time series; error analysis applied to physical systems; electromagnetic flowmeasurement in pipes and open channels. Mailing Add: Cushing Eng Inc 3364 Commercial Ave Northbrook IL 60062

CUSHLEY, ROBERT JOHN, b Edmonton, Alta, July 12, 36; m 65; c 2. SPECTROSCOPY, ORGANIC CHEMISTRY. Educ: Univ Alta, BSc, 57, MSc, 59, PhD(org chem), 65. Prof Exp: Res assoc chem, Sloan-Kettering Inst Cancer Res, 65-68; asst prof, Med Sch, Yale Univ, 68-74; ASSOC PROF CHEM, SIMON FRASER UNIV, 74- Concurrent Pos: Lectr, Sloan-Kettering Div, Cornell Univ, 67-68. Mem: Am Chem Soc. Res: Fourier transform spectroscopy; nuclear magnetic resonance studies of compounds of biological significance; computer applications to chemistry. Mailing Add: Dept of Chem Simon Fraser Univ Burnaby BC Can

CUSHMAC, GEORGE EDWARD, organic chemistry, see 12th edition

CUSHMAN, DAVID WAYNE, b Indianapolis, Ind, Nov 15, 39; m 64; c 2. BIOCHEMICAL PHARMACOLOGY. Educ: Wabash Col, AB, 61; Univ Ill, Urbana, PhD(biochem), 66. Prof Exp: Sr res scientist, 66-70, sr res investr, 70-74, RES FEL, SQUIBB INST MED RES, 74- Mem: AAAS; Am Chem Soc; Am Soc Pharmacol & Exp Therapeut. Res: Bacterial hydroxylases and related electron transport systems; non-heme iron proteins; enzymic halogenations; enzymology and pharmacology of vasoactive polypeptides and Prostaglandins. Mailing Add: Squibb Inst for Med Res Princeton NJ 08540

CUSHMAN, MARK, b Fresno, Calif, Aug 20, 45. ORGANIC CHEMISTRY, MEDICINAL CHEMISTRY. Educ: Univ Calif, San Francisco, Pharm D, 69, PhD(med chem), 73. Prof Exp: NIH fel chem, Mass Inst Technol, 73-75; ASST PROF MED CHEM, PURDUE UNIV, 75- Mem: Am Chem Soc. Res: New synthetic methods; organic reaction mechanisms; natural products synthesis; structure-activity relationships in pharmacology and the isolation and structure elucidation of natural products. Mailing Add: Dept of Med Chem & Pharmacog Purdue Univ West Lafayette IN 47907

CUSHMAN, PAUL, JR, b New York, NY, Feb 4, 30; m 59; c 2. ENDOCRINOLOGY, DRUG ABUSE. Educ: Yale Univ, BA, 51; Columbia Univ, MD, 55. Prof Exp: Instr med, 62-68, assoc, 68-69, asst clin prof, 69-71, NY Heart Asn sr investr, 71-72, ASST PROF MED, COLUMBIA UNIV, 71-, ASSOC ATTEND PHYSICIAN, ST LUKE'S HOSP, 65-, DIR CLIN PHARMACOL, 73- Concurrent Pos: NIH res fel endocrinol, Univ Rochester, 59-61, St Luke's Hosp, 62-64 & 67-75; consult, Vet Admin Hosp, Batavia, NY, 60-62; from assoc dir to dir endocrinol, St Luke's Hosp, 62-73. Mem: AAAS; Endocrine Soc; Am Fedn Clin Res; Am Therapeut Soc; fel Am Col Physicians; NY Acad Med. Res: Narcotics and marijuana. Mailing Add: St Luke's Hosp Morningside Heights Amsterdam Ave & 113th St New York NY 10025

CUSHMAN, ROBERT VITTUM, b Middlebury, Vt, Dec 14, 16; m 42; c 3. HYDROGEOLOGY. Educ: Middlebury Col, AB, 39; Northwestern Univ, MS, 41. Prof Exp: Recorder, US Geol Surv, Alaska, 41; mining geologist, Slide Mines, Inc, Colo, 41-42; mineral economist, US Bur Mines, 42-45; geologist, Water Resources Div, US Geol Surv, NY, 46-47, Conn, 47-61, Ky, 61-71, assoc dist chief, 67-71, dist chief, 71-75; CONSULT GROUND WATER, GEOL & HYDROLOGY, 75- Mem: Geol Soc Am; Am Geophys Union. Mailing Add: 20 Court St Middlebury VT 05753

CUSHMAN, SAMUEL WRIGHT, b Bryn Mawr, Pa, Oct 2, 41; m 64; c 3. CELL PHYSIOLOGY, BIOCHEMISTRY. Educ: Bowdoin Col, AB, 63; Rockefeller Univ, PhD(physiol chem), 69. Prof Exp: Res asst, Inst Clin Biochem, Sch Med, Univ Geneva, 69-71; res asst prof med, 71-73, ASST PROF MED & ADJ ASST PROF BIOCHEM, DARTMOUTH MED SCH, 73- Concurrent Pos: Am Cancer Soc fel, Inst Clin Biochem, Geneva, Switz, 69-71; Am Diabetes Asn career develop award, Dartmouth Med Sch, 72-74. Mem: Am Diabetes Asn; AAAS; Am Fedn Clin Res; Europ Asn Study Diabetes. Res: Adipose tissue and cell structure and function; obesity; mechanism of hormone action, especially insulin and epinephrine. Mailing Add: Endocrine-Metab Div Dept Med Dartmouth Med Sch Hanover NH 03755

CUSHWA, CHARLES T, wildlife biology, see 12th edition

CUSICK, THOMAS WILLIAM, b Joliet, Ill, Sept 18, 43; m 65; c 2. NUMBER THEORY, COMBINATORICS. Educ: Univ Ill, Urbana, BS, 64; Cambridge Univ, PhD(math), 67. Prof Exp: Asst prof, 68-71, ASSOC PROF MATH, STATE UNIV NY, BUFFALO, 71- Mem: Am Math Soc. Res: Diophantine approximation. Mailing Add: Dept of Math 4246 Ridge Lea Rd State Univ of NY Buffalo Amherst NY 14226

CUSSON, RONALD YVON, b Drummondville, Que, July 3, 38; US citizen. THEORETICAL NUCLEAR PHYSICS. Educ: Univ Montreal, BSc, 60; Calif Inst Technol, PhD(physics, math), 65. Prof Exp: Res fel physics, Calif Inst Technol, 65-66; Alexander von Humboldt fel, Inst Theoret Physics, Univ Heidelberg, 66-67; physicist, Chalk River Nuclear Lab, Atomic Energy Can Ltd, 67-70; ASSOC PROF PHYSICS, DUKE UNIV, 70- Concurrent Pos: Consult, Oak Ridge Assoc Univs, 71- Mem: Am Phys Soc. Res: Group theory of nuclear collective motion; parastatistics and its applications; nuclear structure calculations. Mailing Add: Dept of Physics Duke Univ Durham NC 27706

CUSTARD, HERMAN CECIL, b Cleburne, Tex, Aug 19, 29; m 55; c 3. PHYSICAL CHEMISTRY. Educ: Baylor Univ, BS, 57, PhD(chem), 62. Prof Exp: RES ASSOC & ACTIV LEADER, FIELD RES LAB, MOBIL RES & DEVELOP CORP, 62- Mem: Am Chem Soc; Electrochem Soc. Res: Dipole moments; electrochemistry of membrane systems; physical chemistry of electrolytic solutions. Mailing Add: Mobil Res & Develop Corp PO Box 900 Dallas TX 75221

CUSTER, FREDERIC, b Newton Falls, Ohio, Feb 7, 32; m 55; c 2. DENTISTRY. Educ: Kent State Univ, BS, 58; Univ Mich, DDS, 60, MS, 61; Temple Univ, MEd, 65. Prof Exp: Researcher & instr, Univ Mich, 61; pvt pract, 61-62; asst prof dent mat, Sch Dent, Temple Univ, 62-66; assoc prof, chmn oper dent & dir phys res, Sch Dent, Wash Univ, 66-68; dir dent progs, Forest Park Community Col, 68-70; PROF CLIN SCI, SCH DENT MED, SOUTHERN ILL UNIV, EDWARDSVILLE, 70- Concurrent Pos: Consult, Vet Admin Hosp, Philadelphia & Kensington Hosp, 63-65; Fulbright vis prof, Damascus Univ, 64-65; partic, Proj Vietnam, 70; lectr, Forest Park Community Col. Mem: AAAS; Int Asn Dent Res; Am Dent Asn; Am Acad Gold Foil Opers; Asn Schs Allied Health Professions. Res: Dental materials; prosthodontics. Mailing Add: Sch of Dent Med Southern Ill Univ Edwardsville IL 62025

CUSTER, HUBERT MINTER, b Johnstown, Pa, Sept 19, 22; m 51; c 1. PHYSICS. Educ: Carnegie Inst Technol, BSEE, 43; Franklin & Marshall Col, MS, 59. Prof Exp: Asst prof, 52-61, ASSOC PROF PHYSICS, ELIZABETHTOWN COL, 61- Concurrent Pos: Asst prof physics, Johnstown Col, 61-63, assoc prof, 63- Mem: Inst Elec & Electronics Engrs; Am Asn Physics Teachers. Res: Trapping levels in zinc sulfide. Mailing Add: Dept of Physics Elizabethtown Col Elizabethtown PA 17022

CUSTER, MICHAEL, b Lowell, Mass, Oct 29, 10. CHEMISTRY. Educ: St Anselm's Col, AB, 35; Cath Univ Am, MS, 44. Prof Exp: CHMN DEPT CHEM, ST ANSELM'S COL, 48- Mem: Am Chem Soc. Mailing Add: Dept of Chem St Anselm's Col Manchester NH 03102

CUSTER, RICHARD PHILIP, b Munhall, Pa, July 14, 03; m 39; c 2. ONCOLOGY. Educ: Jefferson Med Col, MD, 28. Prof Exp: From asst instr to prof, 29-70, EMER PROF PATH, SCH MED, UNIV PA, 70-; SR MEM, INST CANCER RES, FOX CHASE, 71- Concurrent Pos: Fel, Path Anat Inst, Innsbruck, 30 & 32; chief path, Philadelphia Gen Hosp, 31-38 & consult, 38-; dir labs, Presby Hosp, Philadelphia, 38-59, chmn chief hemat & chemother, 59-68; trustee, Am Bd Path, 44-52, life trustee, 61; consult, Children's Heart Hosp, Philadelphia, 46-, Armed Forces Inst Path, 46-56 & US Naval Hospital, Philadelphia, 47-70. Honors & Awards: Gerhard Gold Medalist, Philadelphia Path Soc, 55. Mem: AAAS; AMA; Am Asn Path & Bact. Res: Hematology, particularly etiology, interrelationships and treatment of the lymphomas and leukemias; basic cancer research. Mailing Add: Inst for Cancer Res Fox Chase Philadelphia PA 19111

CUSTER, ROBERT LOUIS, b Manchester, Conn, Aug 31, 21; m 44; c 5. SCIENCE EDUCATION. Educ: Univ Conn, BS, 47, PhD(chem), 52. Prof Exp: Asst, Univ Conn, 50-51; res chemist, E I du Pont de Nemours & Co, Inc, 51-73; INSTR CHEM, FISHBURNE SCH, VA, 75- Mailing Add: 705 Walnut Ave Waynesboro VA 22980

CUSUMANO, CHARLES LOUIS, b Brooklyn, NY, Jan 16, 36; m 58; c 4. VIROLOGY, IMMUNOLOGY. Educ: Univ Notre Dame, BA, 57; Georgetown Univ, MD, 61. Prof Exp: Med intern, Georgetown Univ Hosp, DC, 61-62; resident physician internal med, Vet Admin Hosp, DC, 62-63; res assoc virol, Nat Inst Neurol Dis & Blindness, 63-65; resident physician, Georgetown Univ Hosp, 65-66; asst prof microbiol & med, 68-74, ASSOC PROF IMMUNOL & MED, COL MED, UNIV FLA, 74- Concurrent Pos: Fel protein chem, Nat Inst Arthritis & Metab Dis, 66-68; Am Cancer Soc Instnl grant, 68-; clin investr, Vet Admin Hosp, Gainesville, Fla, 70-; attend physician med & oncol, Shands Teaching Hosp & Gainesville Vet Admin Hosp, 72- Mem: Am Fedn Clin Res: NY Acad Sci. Res: Clinical and basic studies on rubella and rubella virus; circulating immune complexes in cancer patients; cancer immunochemotherapy; sequence and structure of staphylococcal nuclease; immunochemistry of tumor transplantation. antigens. Mailing Add: Col of Med Univ of Fla Gainesville FL 32601

CUSUMANO, JAMES A, b Elizabeth, NJ, Apr 14, 42; m 64; c 1. CATALYSIS. Educ: Rutgers Univ, BA, 64, PhD(chem physics), 68. Prof Exp: Res chemist heterogeneous catalysis, Esso Res & Eng Co, 67-74; PRES, CATALYTICA ASSOC, INC, 74- Honors & Awards: Colloid & Surface Chem Award, Continental Oil Co, 64. Mem: Am Chem Soc; Catalysis Soc; Am Phys Soc. Res: Catalysis, chemistry and physics of small metal particle systems; infrared spectroscopic studies of adsorbed species; chemical kinetics of catalytic reactions; theory and measurement of surface thermal transients. Mailing Add: Catalytica Assoc Inc 5 Palo Alto Square Palo Alto CA 94304

CUTCHINS, ERNEST CHARLES, b Newsoms, Va, Aug 19, 22. BACTERIOLOGY, VIROLOGY. Educ: Va Polytech Inst, BS, 43; Univ Md, MS, 51, PhD(bact), 55; Am Bd Med Microbiol, dipl. Prof Exp: Bacteriologist, Walter Reed Army Inst Res, 55-56 & NIH, 56-59; res assoc, Chas Pfizer & Co, Inc, 60-61; ASSOC PROF VIROL, DEPT BIOL, CATH UNIV AM, 61- Mem: Am Asn Immunologists; Am Soc Microbiol. Res: Viral inactivation; virus-host interaction; immunology and serology of viruses. Mailing Add: Dept of Biol Cath Univ of Am Washington DC 20017

CUTCLIFFE, JACK ALEXANDER, b Quincy, Mass, Aug 10, 29; Can citizen; m 54; c 6. OLERICULTURE. Educ: McGill Univ, BSc, 52, MSc, 54. Prof Exp: Exten hort crops, Ont Dept Agr, 54-62; RES OLERICULT, AGR CAN RES BR, 62-, SECT HEAD HORT, AGR CAN RES STA, 70- Mem: Agr Inst Can; Can Soc Hort Sci; Am Soc Hort Sci; Int Soc Hort Sci. Res: Conduct research on macro- and micro-nutrient requirements and on management practices to improve efficiency of production of broccoli, cauliflower, brussel sprouts, rutabagas and peas. Mailing Add: Agr Can Res Sta PO Box 1210 Charlottetown PE Can

CUTFORTH, HOWARD GLEN, b Topeka, Kans, May 6, 20; m 44; c 3. CHEMISTRY, AEROSPACE SCIENCES. Educ: Univ Wichita, BA, 42; Northwestern Univ, PhD(phys chem), 48. Prof Exp: Res chemist, Rohm & Haas Co, 47-48 & Phillips Petrol Co, 48-59; SECT CHIEF, CHEM SYSTS DIV, UNITED TECHNOL CORP, 59- Mem: Am Chem Soc. Res: Magnetic study of free radicals; chemistry of rare earths; colloid problems in crude oil production; magnetic susceptibility of substituted hexanyl ethanes and related compounds; direct tailoring of standard and development of new solid, composite, rocket fuels. Mailing Add: Sunnyvale CA

CUTHBERT, FREDERICK LEICESTER, geology, see 12th edition

CUTHBERT, NICHOLAS LE HURAY, b Westmont, NJ, Mar 25, 14; m 38; c 1. ORNITHOLOGY. Educ: Univ Iowa, AB, 36; Univ Wis, AM, 37, PhD(genetics), 41. Prof Exp: Instr biol, Iowa Wesleyan Col, 41-44; assoc prof anat, Chicago Col Osteop, 44-46; lectr genetics & zool, Univ Wis, 46-47; from asst prof to assoc prof, 47-55, PROF BIOL, CENT MICH UNIV, 55- Mem: Genetics Soc Am; Wilson Ornith Soc; Am Ornithologists Union; Cooper Ornith Soc. Res: Field ornithology studies. Mailing Add: Dept of Biol Cent Mich Univ Mt Pleasant MI 48858

CUTHBERTSON, GEORGE RAYMOND, b Liberty, Mo, June 23, 10; m 31; c 3. PHYSICAL CHEMISTRY. Educ: William Jewell Col, AB, 31; Harvard Univ, AM, 33, PhD(phys chem), 35. Hon Degrees: LLD, William Jewell Col, 64. Prof Exp: Res assoc, Harvard Univ, 35-36; res chemist, US Rubber Co, 36-43, sr scientist, 43-45, admin asst to dir develop, Tire Div, 45-48, asst dir develop, 48-52, mgr Los Angeles plant, 52-53, from asst prod mgr to prod mgr, Tire Div, 53-55, from asst gen mgr to vpres & gen mgr, 55-61, vpres res & develop, 61-62, tech dir, Fibers & Textile Div, Uniroyal, Inc, 62-70; assoc prof, 70-75, PROF ENG MGT, UNIV MO-ROLLA, 75- Mem: Am Chem Soc; Am Soc Eng Educ. Res: Resonance fluorescence of benzene; thermal equilibrium of ethylene iodide; heats of reaction of free radicals; reaction rates; sublimation pressures; rubber; synthetic and natural fibers; engineering management. Mailing Add: 1106 Sycamore Dr Rolla MO 65401

CUTHILL, ELIZABETH, b Southington, Conn, Oct 16, 23; m 50; c 1. MATHEMATICS. Educ: Univ Buffalo, BA, 44; Brown Univ, MA, 46; Univ Minn, PhD, 51. Prof Exp: Chemist, Res & Develop Labs, Socony Vacuum Oil Co, 44-45; asst math, Univ Minn, 46-48 & Conn Col, 48-49; instr, Purdue Univ, 50-52 & Univ Md, 52-53; supvry mathematician, Appl Math Lab, David Taylor Model Basin, 53-70, SUPVRY MATHEMATICIAN, APPL MATH LAB, DAVID W TAYLOR NAVAL SHIP RES & DEVELOP CTR, 70- Concurrent Pos: Prof lectr, Am Univ, 62- Mem: Fel AAAS; Am Math Soc; Soc Indust & Appl Math; Math Asn Am; Asn Comput Mach. Res: Development of mathematical methods and digital computer programs for solution of problems arising in the areas of nuclear reactor design; structural mechanics; hydrodynamics. Mailing Add: David W Taylor Naval Ship Ctr Washington DC 20084

CUTHRELL, ROBERT EUGENE, b Houston, Tex, Nov 6, 33; m 54; c 4. SURFACE CHEMISTRY, SURFACE PHYSICS. Educ: Univ Tex, Austin, BA, 54, BS, 61, PhD(inorg chem), 64. Prof Exp: STAFF MEM CHEM, SANDIA LABS, 63-68 & 69- Concurrent Pos: Res scientist, Air Force Weapons Lab, Kirtland AFB, 70-76; counr, NMex Inst Chemists, 74-75. Mem: Fel Chem Soc London; fel Am Inst Chemists; Sigma Xi; Am Chem Soc. Res: Ultrahigh vacuum study of the chemistry, physics and mechanics of the surface and the near-surface region of metals; applied research in metal-metal bonding, design of electrical contacts, arc theory, hydrogen embrittlement, friction and wear, and atmospheric pollution. Mailing Add: Sandia Labs PO Box 5834 Albuquerque NM 87115

CUTKOMP, LAURENCE KREMER, b Wapello, Iowa, Jan 24, 16; m 39; c 4. INSECT TOXICOLOGY, ECONOMIC ENTOMOLOGY. Educ: Iowa Wesleyan Col, BA, 36; Cornell Univ, PhD(econ entom), 42. Prof Exp: Asst entom, Cornell Univ, 42-43; fel zool, Univ Pa, 43-45; res assoc entom, Univ Minn, 45-46; assoc entomologist, TVA, 46-47; from asst prof to assoc prof, 47-60, PROF ENTOM, UNIV MINN, ST PAUL, 60- Concurrent Pos: Consult, Eli Lilly & Co, 59-64 & Onamia Corp, 63-64; entomologist, Int Atomic Energy Agency, Vienna, 65-67 & consult entom & insect biochem, India, 74. Mem: AAAS; Entom Soc Am; Sigma Xi. Res: Insecticidal action; toxicity of insecticides; insect rhythms as related to insecticide effectiveness. Mailing Add: Dept of Entom, Fisheries & Wildlife Univ of Minn St Paul MN 55108

CUTKOSKY, RICHARD EDWIN, b Minneapolis, Minn, July 29, 28; m 52; c 2. THEORETICAL HIGH ENERGY PHYSICS. Educ: Carnegie Inst Technol, BS & MS, 50, PhD(physics), 53. Prof Exp: Res physicist, Carnegie Inst Technol, 53-54; NSF fel, Inst Theoret Physics, Denmark, 54-55; from asst prof to assoc prof physics, 55-61, PROF PHYSICS, CARNEGIE-MELLON UNIV, 61-, BUHL PROF THEORET PHYSICS, 63- Concurrent Pos: Sloan Found res fel, 56-61; NSF res fel, Nordic Inst Theoret Atomic Physics, Denmark, 61-62. Mem: AAAS; Am Phys Soc. Res: Field theory; high energy physics. Mailing Add: Dept of Physics Carnegie-Mellon Univ Pittsburgh PA 15213

CUTLER, CASSIUS CHAPIN, b Springfield, Mass, Dec 16, 14; m 41; c 3. PHYSICS. Educ: Worcester Polytech Inst, 37. Prof Exp: DIR ELECTRONIC SYSTS RES, BELL TEL LABS, 37- Concurrent Pos: Ed, Spectrum, 66-68; mem bd eng manpower & educ policy, Nat Acad Eng. Mem: Nat Acad Sci; Nat Acad Eng; fel Inst Elec & Electronics Engrs. Res: Radio transmitter design; design of radar antennas; microwave electron tube research. Mailing Add: Bell Tel Labs Rm 4C-502 Holmdel NJ 07733

CUTLER, DOYLE O, b Corpus Christi, Tex, Jan 20, 36; m 60; c 2. MATHEMATICS. Educ: Tex Christian Univ, BA, 59, MA, 61; NMex State Univ, PhD(math), 64. Prof Exp: Asst prof math, Tex Christian Univ, 64-65; asst prof, 65-71, ASSOC PROF MATH, UNIV CALIF, DAVIS, 71- Mem: Am Math Soc; Math Asn Am. Res: Infinite Abelian groups; algebra. Mailing Add: Dept of Math Univ of Calif Davis CA 95616

CUTLER, EDWARD BAYLER, b Plymouth, Mich, May 28, 35; m 61; c 2. MARINE BIOLOGY. Educ: Wayne State Univ, BS, 61; Univ Mich, MS, 62; Univ RI, PhD(zool), 67. Prof Exp: Asst prof biol, Lynchburg Col, 62-64; oceanog trainee, Coop Oceanog Prog, Marine Lab, Duke Univ, 66-67; asst prof, 67-71, ASSOC PROF BIOL, UTICA COL, 71- Mem: AAAS; Am Soc Zoologists; Soc Syst Zool; Am Soc Limnol & Oceanog; Marine Biol Asn UK. Res: Systematics; zoogeography; ecology; evolution of benthic marine invertebrates, particularly the Sipuncula and Pogonophora. Mailing Add: Dept of Biol Utica Col Utica NY 13502

CUTLER, FRANK ALLEN, JR, b Pana, Ill, Nov 14, 20; m 46; c 3. CHEMISTRY, RESEARCH ADMINISTRATION. Educ: Univ Ill, BS, 42; Univ Minn, PhD(org chem), 48. Prof Exp: Jr chemist synthetic rubber res, B F Goodrich Co, 44-46; sr chemist develop res, 48-59, tech adv to patent dept, 59-61, coordr compound eval, 61-65, MGR RES COMPOUNDS, MERCK & CO, INC, 65- Mem: Am Chem Soc. Res: Quinones; cortical steroids; pharmaceuticals; information retrieval; substructure searching by computer. Mailing Add: 41 Faulkner Dr Westfield NJ 07090

CUTLER, HORACE GARNETT, b London, Eng, Nov 21, 32; US citizen; m 55; c 7. PLANT PHYSIOLOGY, PLANT NEMATOLOGY. Educ: Univ Md, BS, 64, MS, 66, PhD(plant path & nematol), 67. Prof Exp: Res asst plant physiol, Boyce Thompson Inst Plant Res, 54-59; plant physiologist, Cent Agr Res Sta, WI, 59-62; asst bot, Univ Md, 63-67; PLANT PHYSIOLOGIST, GA COASTAL PLAIN EXP STA, USDA, 67- Honors & Awards: Carroll E Cox Scholarship Award, 66. Mem: NY Acad Sci; Japanese Soc Plant Physiol; Sigma Xi. Res: Herbicides; defoliants; growth substances from sugarcane; nematodes and micro-organisms; growth inhibitors from tobacco; mode of action and biochemistry of auxins and other regulators; biologically active natural products from microorganisms. Mailing Add: Ga Coastal Plain Exp Sta US Dept of Agr Tifton GA 31794

CUTLER, HUGH CARSON, b Milwaukee, Wis, Sept 8, 12; m 40; c 1. BOTANY. Educ: Univ Wis, BA, 35, MA, 36; Univ Wash, PhD(bot), 39. Prof Exp: Fel, Univ Wash, 29-40; res assoc, Harvard Univ, 40-43; field technician, Rubber Develop Corp, Brazil, 43-45; res assoc, Harvard Univ, 46-47; cur econ bot, Chicago Natural Hist Mus, 47-52; cur econ bot, Mus Useful Plants, 53-54, from asst dir to exec dir, 54-64, CUR USEFUL PLANTS, MO BOT GARDEN, 64- Concurrent Pos: Guggenheim fel, 42-43 & 46-47; assoc prof bot, Wash Univ, 53-72, adj prof anthrop & biol, 72- Res: Useful cultivated plants of the New World and their wild relatives; evolution of maize and cucurbits; ethnobotany; economic botany. Mailing Add: Mo Bot Garden 2315 Tower Grove Ave St Louis MO 63110

CUTLER, IRVING HERBERT, b Chicago, Ill, Apr 11, 23; m 51; c 2. GEOGRAPHY. Educ: Univ Chicago, MA, 48; Northwestern Univ, PhD(geog), 64. Prof Exp: Housing investr, US Dept Labor, 46-47; teacher geog, Mich & Ill High Schs, 48-57; transp economist, US Corps Army Engrs, 57-58; instr social sci, Crane Jr Col, 58-60; asst geog, Northwestern Univ, 60-61; PROF GEOG, CHICAGO STATE UNIV, 61-, CHMN DEPT, 74- Concurrent Pos: Film script writer, Ginn Co, 64-65; consult, OEO, 65-66 & Jour Films, 67-69; lectureship, De Paul Univ, 67 & 72; writer, Follett Co, 70- Honors & Awards: Award, Haas Res Fund, 64. Mem: Asn Am Geogr; Nat Coun Geog Educ; Int Geog Union. Res: Geography of Chicago metropolitan area; urban transportation; air transportation; industrial geography. Mailing Add: Dept of Geog Chicago State Univ 95th St at King Dr Chicago IL 60628

CUTLER, JANICE ANN, b Perth Amboy, NJ, May 25, 18; m 47. PHYSICAL CHEMISTRY. Educ: NY Univ, BA, 37, PhD(chem), 46. Prof Exp: Phys chemist, Brookhaven Nat Lab, 46-47 & Delner Corp, 47; res assoc, Polytech Inst Brooklyn, 47-49; sr scientist, Markite Co, 49-66; asst dir, 66-72, ASSOC DIR, COPES PROJ, NY UNIV 72- Mem: AAAS; Am Chem Soc; NY Acad Sci; Nat Sci Teachers Asn. Res: Vacuum spectroscopy; permeation and sorption by polymers; absorption spectra of alkyl nitriles and allyl halides in the vacuum ultraviolet; solid state physics chemistry as related to interatomic and intermolecular forces; development of elementary science curriculum. Mailing Add: NY Univ Four Washington Pl New York NY 10003

CUTLER, JANICE ZEMANEK, b Chicago, Ill, May 3, 42; m 73; c 1. ALGEBRA. Educ: Univ Wis-Madison, BS, 64; Univ Ill, Urbana-Champaign, MA, 65, PhD(math), 70. Prof Exp: Instr math, Univ Ill, Urbana-Champaign, 70; ASST PROF MATH, LA STATE UNIV, BATON ROUGE, 70- Mem: Am Math Soc; Math Asn Am. Res: Integral representation theory. Mailing Add: Dept of Math La State Univ Baton Rouge LA 70803

CUTLER, JIMMY EDWARD, b Los Angeles, Calif, Nov 28, 44; m 65; c 2. MICROBIOLOGY. Educ: Calif State Univ, Long Beach, BS, 67, MS, 69; Sch Med, Tulane Univ, PhD(microbiol), 72. Prof Exp: NIH fel microbiol, Rocky Mountain Lab, Nat Inst Allergy & Infectious Disease, 73-74; ASST PROF MICROBIOL, MONT STATE UNIV, 74- Mem: Am Soc Microbiol. Res: Host parasite relationships; host innate and acquired immune responses; parasite virulence factors. Mailing Add: Dept of Microbiol Mont State Univ Bozeman MT 59715

CUTLER, JOHN CHARLES, b Cleveland, Ohio, June 29, 15; m 42. PUBLIC HEALTH. Educ: Western Reserve Univ, BA, 37, MD, 41; Johns Hopkins Univ, MPH, 51; Am Bd Prev Med, dipl, 51. Prof Exp: Mem staff, Venereal Dis Res Lab, USPHS, 43-46 & Pan-Am Sanit Bur, 46-48; head, Venereal Dis Demonstration Team, WHO, Southeast Asia, 48-50; head, Venereal Dis Div, Bur State Serv, 50-58; asst dir & asst surgeon gen, Nat Inst Allergy & Infectious Dis, 58-59; dist health officer, Allegheny County Health Dept, Pa, 59-60; dep dir, Pan-Am Sanit Bur, 60-67; PROF INT HEALTH & DIR POP PROG, GRAD SCH PUBLIC HEALTH, UNIV PITTSBURGH, 67- Concurrent Pos: Chmn, Int Comt & mem, Bd Dirs, Asn Voluntary Sterilization, 71-; mem, Bd Dirs, Family Planning Coun Southwestern Pa, Inc, 71- Mem: Fel Am Pub Health Asn; Am Venereal Dis Asn. Res: Combined contraceptive venereal disease prophylactic preparation. Mailing Add: Pop Prog Grad Sch Pub Health Univ of Pittsburgh Pittsburgh PA 15261

CUTLER, JOHN FREDRICK, paleontology, geology, see 12th edition

CUTLER, LEONARD HARRY, physical chemistry, see 12th edition

CUTLER, LEONARD SAMUEL, b Los Angeles, Calif, Jan 10, 28; m 54; c 4. PHYSICS, MATHEMATICS. Educ: Stanford Univ, BS, 58, MS, 60, PhD(physics), 66. Prof Exp: Chief engr, Gertsch Prod Inc, 49-55, vpres, 55-57; res engr, 57-62, eng sect leader, 62-64, dir res, 64-67, mgr frequency & time div, 67-69, DIR PHYS RES LAB, HEWLETT-PACKARD CO, 69- Concurrent Pos: Mem adv panel, Nat Bur Standards under Nat Acad Sci contract, 68- Mem: Fel Inst Elec & Electronics Engrs; AAAS. Res: Frequency measuring devices; quartz and atomic frequency standards; noise theory; frequency stability theory; quantum electronics; atomic beam work; lasers; precision alternating current transformers; magnetic bubble memories. Mailing Add: Phys Res Lab Hewlett-Packard Co 3500 Deer Creek Rd Palo Alto CA 94304

CUTLER, LESLIE STUART, b New Brunswick, NJ, Jan 20, 43; m 66; c 2. DEVELOPMENTAL BIOLOGY, PATHOBIOLOGY. Educ: Washington Univ, DDS, 68; State Univ NY Buffalo, PhD(path), 73. Prof Exp: Fel path, Sch Med, State Univ NY Buffalo, 68-73; ASST PROF ORAL BIOL, SCH DENT MED, HEALTH CTR, UNIV CONN, 73- Mem: AAAS; Am Soc Cell Biol; Soc Develop Biol; Tissue Cult Asn; Int Asn Dent Res. Res: Epithelial-mesenchymal interactions; adenylate cyclase and other enzymes in cyclic nucleotide metabolism; biochemical and electron microscopic cytochemical studies-salivary glands-developmental aspects of neoplasia. Mailing Add: Dept of Oral Biol Sch of Dent Med Univ of Conn Health Ctr Farmington CT 06032

CUTLER, LOUISE MARIE, b Troy, NY, Oct 1, 21. CHEMISTRY. Educ: Good Counsel Col, BS, 43; Fordham Univ, MS, 52; Columbia Univ, PhD(chem), 62. Prof Exp: Instr chem, Good Counsel Acad, 43-44; from instr to assoc prof, 45-62, PROF CHEM, GOOD COUNSEL COL, 62- Concurrent Pos: Trustee, Good Counsel Col, 63-; NSF grant, NY Univ, 65. Mem: AAAS; Am Chem Soc; Hist Sci Soc. Res: Chelation equilibria and kinetics; study of complexing agents by gas chromatography procedures; history and philosophy of science. Mailing Add: Dept of Chem Good Counsel Col White Plains NY 10603

CUTLER, MELVIN, b Beaumont, Tex, Dec 18, 23; m 50; c 2. SOLID STATE PHYSICS. Educ: City Col New York, BS, 43; Columbia Univ, AM, 47, PhD(chem), 51. Prof Exp: Mem tech staff, Hughes Aircraft Co, 51-58; mem staff, John Jay Hopkins Lab Pure & Appl Sci, Gen Atomic Div, Gen Dynamics Corp, 58-63; assoc prof, 63-69, PROF PHYSICS, ORE STATE UNIV, 69- Mem: Am Phys Soc. Res: Physical chemistry; electronic properties of disordered systems. Mailing Add: 835 Merrie Dr Corvallis OR 97330

CUTLER, PAUL, b Philadelphia, Pa, Mar 27, 20; m 47; c 2. INTERNAL MEDICINE. Educ: Univ Pa, BA, 40; Jefferson Med Col, MD, 44. Prof Exp: Pvt pract, 50-65; team capt, Care-Medico, Afghanistan, 65-67; prof med & physiol, 68-74, ASSOC DEAN CLIN AFFAIRS, UNIV TEX HEALTH SCI CTR SAN ANTONIO, 71- Mem: Fel Am Col Physicians. Mailing Add: Univ Tex Health Sci Ctr 7703 Floyd Curl Dr San Antonio TX 78229

CUTLER, RICHARD GAIL, b Lovell, Wyo, Aug 6, 35; m 62; c 3. CELL BIOLOGY. Educ: Long Beach State Col, BS, 61; Univ Houston, MS & PhD(biophys), 66. Prof Exp: Fel geront, Brookhaven Nat Lab, 66-69; ASST PROF BIOL, UNIV TEX, DALLAS, 69- Mem: AAAS; Geront Soc; Biophys Soc. Res: Gerontology; developmental biology; control of DNA, RNA and protein synthesis; DNA replication; RNA transcription patterns; bacteria conjugation; synchronous bacteria cultures; DNA-RNA hybridization techniques; aging at cellular level in mammals. Mailing Add: Inst for Molecular Biol Univ of Tex at Dallas PO Box 688 Richardson TX 75080

CUTLER, RICHARD OSCAR, b Delavan, Ill, Mar 8, 30; m 56; c 4. GEOGRAPHY. Educ: Fla State Univ, BA, 55; Univ Mich, MA, 57; Univ Fla, PhD(geog), 65. Prof Exp: Instr phys sci, Univ Fla, 57-62; asst prof geog, Ind State Univ, 62-66; assoc prof, Adams State Col, 66-70; ASSOC PROF GEOG & CHMN DEPT, STATE UNIV NY COL OSWEGO, 70- Mem: Asn Am Geogr; Am Geog Soc; Nat Coun Geog Educ. Res: Recreational potential of northeast Florida, especially outdoor recreation and tourism; geography of the San Luis Valley, Colorado, especially outdoor recreation. Mailing Add: Dept of Geog State Univ of NY Oswego NY 13126

CUTLER, ROBERT W P, b New York, NY, Aug 3, 33; m 61; c 3. NEUROLOGY, NEUROCHEMISTRY. Educ: Tufts Univ, MD, 57. Prof Exp: Assoc neurol, Med Sch, Harvard Univ, 64-68; ASSOC PROF NEUROL, UNIV CHICAGO, 68- Concurrent Pos: Asst, Children's Hosp Med Ctr, Boston, Mass, 64-68; prin investr, USPHS grant, 69. Mem: AAAS; Am Soc Neurochem; Am Acad Neurol; Asn Res Nerv & Ment Dis. Res: Physiology of cerebrospinal fluid and bloodbrain barrier mechanisms. Mailing Add: Dept of Med Pritzker Sch of Med Univ of Chicago Chicago IL 60637

CUTLER, ROYAL ANZLY, JR, b Spokane, Wash, March 10, 18; m 43; c 3. ORGANIC CHEMISTRY. Educ: Whitman Col, AB, 41; Rensselaer Polytech Inst, MS, 42, PhD(org chem), 47. Prof Exp: Lab asst chem, Whitman Col, 40-41; asst, Rensselaer Polytech Inst, 41-42, instr chem, 43-44; jr res chemist, Winthrop Chem Co, 44-47, sr res chemist, 47-60, sr res chemist & tech liaison, 60-66, ASST DIR PROD DEVELOP, STERLING-WINTHROP RES INST, 66- Mem: Am Chem Soc. Res: Product development; pharmaceutical chemicals; pharmaceutical dosage forms; antimicrobials and chemical specialties. Mailing Add: Sterling-Winthrop Res Inst Rensselaer NY 12144

CUTLER, SIDNEY JOSHUA, b Russia, Apr 13, 17; nat US; m 47; c 3. MEDICAL STATISTICS, EPIDEMIOLOGY. Educ: City Col New York, BS, 38; Columbia Univ, MA, 41; Univ Pittsburgh, ScD, 61. Prof Exp: Statistician, United Serv for New Am, 39-42; sr statistician, City Dept Health, New York, 46-47; med res statistician, Sch Aviation Med, 47-48; anal statistician, Nat Cancer Inst, 48-75; EPIDEMIOLOGIST, SCH MED, WAYNE STATE UNIV & MICH CANCER FOUND, 75- Mem: Fel Am Statist Asn; Soc Epidemiol Res. Res: Epidemiology of cancer. Mailing Add: 23775 Riverside Dr Southfield MI 48075

CUTLER, WARREN GALE, b Avon, Ill, Jan 31, 22; m 49; c 4. FLUID PHYSICS. Educ: Monmouth Col, BA, 47; Pa State Univ, MS, 53, PhD(physics), 55. Prof Exp: Instr physics & math, Monmouth Col, 47-51; asst physics, Pa State Univ, 51-53; assoc prof, Mankato State Col, 55-57, head dept, 56-57; assoc res physicist, 57-59, mgr mech eng res, 59-60, res dir, 60-68, DIR CORP RES, WHIRLPOOL CORP RES LABS, 68- Concurrent Pos: Sr exec course, Sloan Sch Mgt, Mass Inst Technol, 66. Mem: Am Phys Soc; Am Chem Soc; Am Oil Chemists' Soc; Sigma Xi. Res: Physics of high pressure; physical properties of liquid; surface properties of solutions containing surface-active agents. Mailing Add: 218 Crofton Circle St Joseph MI 49085

CUTLIP, RANDALL CURRY, b Hillsbora, WVa, Sept 30, 34; m 57; c 2. VETERINARY PATHOLOGY. Educ: Ohio State Univ, DVM, 61; Iowa State Univ, MS, 65, PhD(vet path), 71. Prof Exp: RES VET PATH, NAT ANIMAL DIS CTR, AGR RES SERV, USDA 61- Mem: Am Col Vet Pathologists; Am Vet Med Asn; Conf Res Workers in Animal Dis; Am Asn Sheep & Goat Practr. Res: Respiratory diseases of sheep-identify individual disease entities, establish their etiologies and evaluate their pathogenesis with emphasis on chronic diseases of the sheep lung. Mailing Add: Nat Animal Dis Ctr PO Box 70 Agr Res Serv USDA Ames IA 50010

CUTLIP, WILLIAM FREDERICK, b Lincoln, Ill, Oct 20, 36; m 57; c 3. MATHEMATICS, COMPUTER SCIENCE. Educ: Eastern Ill Univ, BSEd, 58; Univ Ill, MA, 61; Mich State Univ, PhD(math), 68. Prof Exp: Pub sch prin & teacher, Ill, 55-56, teacher, 58-60; from instr to assoc prof math, Northern Mich Univ, 61-64; asst prof, 68-70, ASSOC PROF MATH, CENT WASH STATE COL, 70- Mem: Asn Comput Mach; Math Asn Am. Res: Decomposition and synthesis of finite automata. Mailing Add: Dept of Math Cent Wash State Col Ellensburg WA 98926

CUTNELL, JOHN DANIEL, b Pittsburgh, Pa, Aug 30, 40; m 68. PHYSICAL CHEMISTRY. Educ: Lehigh Univ, AB, 62; Univ Wis-Madison, PhD(phys chem), 67. Prof Exp: Monsanto Chem Co fel, 67-68; ASST PROF PHYSICS, SOUTHERN ILL UNIV, CARBONDALE, 68- Mem: Am Phys Soc. Res: Molecular motions by way of relaxation and pulsed nuclear magnetic resonance. Mailing Add: Dept of Physics Southern Ill Univ Carbondale IL 62901

CUTRESS, CHARLES ERNEST, b Calgary, Can, Mar 8, 21; nat US; m 46. INVERTEBRATE ZOOLOGY. Educ: Ore State Col, BS, 48, MS, 49. Prof Exp: USPHS fel, Univ Hawaii, 53-55; assoc cur, Div Marine Invert, US Nat Mus, Smithsonian Inst, 56-65; assoc prof marine biol, 65-70, PROF MARINE SCI, UNIV PR, MAYAGUEZ, 70- Mem: AAAS; Soc Syst Zool; Am Soc Zoologists. Res: Systematics; biology; behavior; marine invertebrates, particularly coelenterates. Mailing Add: Dept of Marine Sci Univ of PR Mayaguez PR 00708

CUTRIGHT, DUANE EDWIN, b Bliss, Idaho, Oct 1, 26; m 44; c 2. PATHOLOGY. Educ: St Louis Univ, DDS, 56; Georgetown Univ, PhD(nucleic acids), 65. Prof Exp: Resident, 65-66, chief div oral path, 70-74, CHIEF, US ARMY INST DENT RES, 74- ASST PROF PATH & ASSOC PROF CLIN PATH, MED SCH, GEORGETOWN UNIV, 66- Mem: Am Dent Asn; Am Acad Oral Path; Asn Mil Surg US. Res: Uses for tissue adhesives; turnover rates of oral epithelium; nucleic acid reutilization and fate in oral epithelium; geriatric oral diseases; microcirculation; cyanocrylates; cryobiology; biodegradable sutures and bone plates; tissue levels in mercury exposure; long tern subcutaneous drug administration; implants; tendon gliding devices. Mailing Add: 6601 Glenbrook Rd Chevy Chase MD 20015

CUTSHALL, ALDEN DENZEL, b Olney, Ill, Apr 12, 11; m 33; c 2. GEOGRAPHY. Educ: Eastern Ill State Univ, BEd, 32; Univ Ill, MA, 35; Ohio State Univ, PhD(geog), 40. Prof Exp: Teacher, Ill Pub Schs, 32-34 & 37-39; from instr to asst prof, Univ Ill, Urbana, 40-47; asst prof in charge geog, Undergrad Div, Univ Ill, Chicago, 47-49, from assoc prof to prof, 49-64; prof geog, 64-72, head dept, 64-69, EMER PROF GEOG, UNIV ILL, CHICAGO CIRCLE, 72- Concurrent Pos: From res analyst to dep sect chief, Off Strategic Serv, Washington, DC, 44-45; prin res analyst, US State Dept, Philippines, 45-46; Fulbright res grant, Philippines, 50-51; Fulbright lectr, Univ Philippines, 57-58; assoc mem, Ctr Advan Study, Univ Ill, 62-63; prof geog, Aurora Col, 73-74; mem, Nat Coun Geog Educ. Mem: AAAS; Asn Am Geogr; Philippine Geog Soc. Res: East and Southeast Asia, especially the Philippines and Thailand; midwestern United States, especially Illinois; resource analysis and regional development, especially the humid tropics; historical geography of Illinois. Mailing Add: Dept of Geog Univ of Ill at Chicago Circle Chicago IL 60680

CUTSHALL, NORMAN HOLLIS, b Glens Falls, NY, Mar 21, 38; m 61; c 2. CHEMICAL OCEANOGRAPHY. Educ: Ore State Univ, BS, 61, MS, 63, PhD(oceanog), 67. Prof Exp: Instr oceanog, Ore State Univ, 65-66; chemist, Oak Ridge Nat Lab, 67-68; marine scientist, AEC, 68-69; RES ASSOC OCEANOG, ORE STATE UNIV, 69- Mem: AAAS; Am Chem Soc; Am Soc Limnol & Oceanog; Ecol Soc Am. Res: Environmental radioactivity; geochemistry; economic geology. Mailing Add: Dept of Oceanog Ore State Univ Corvallis OR 97331

CUTSHALL, THEODORE WAYNE, b Lafayette, Ind, Mar 22, 28; m 52; c 2. ORGANIC CHEMISTRY. Educ: Purdue Univ, BS, 49; Northwestern Univ, MS, 59, PhD(org chem), 64. Prof Exp: Res engr, Am Can Co, 49-50 & 53-58; from asst prof to assoc prof chem, Purdue Univ, Indianapolis, 61-69; ASSOC PROF CHEM, IND UNIV-PURDUE UNIV, INDIANAPOLIS, 69- Concurrent Pos: Vis lectr, Franklin Col, 62-63 & Butler Univ, 65-66. Mem: Am Chem Soc. Res: Synthesis of heteroaromatic systems; reaction mechanisms; stereochemistry. Mailing Add: Dept of Chem Ind Univ-Purdue Univ Indianapolis IN 46205

CUTT, ROGER ALAN, b Rochester, NY, Aug 4, 36; m 59; c 2. PHYSIOLOGICAL PSYCHOLOGY, AUDIOLOGY. Educ: Franklin & Marshall Col, BA, 57; Univ Del, MA, 59, PhD(physiol psychol), 62. Prof Exp: Res assoc, Otol Res Lab, Presby-Univ Pa Med Ctr, 62-65, dir res, 65-70; med serv specialist, Social & Rehab Serv, Dept Health, Educ & Welfare, Philadelphia, 70-73; COMNR MED PROGS, PA DEPT PUB WELFARE, HARRISBURG, 73- Concurrent Pos: NIH & John A Hartford Found res grants, 63-70; instr, Dept Otolaryngol, Sch Med, Univ Pa, 62-65, assoc, 65-70, asst prof, 70-73; asst prof community med, Med Col Pa, 73- Mem: AAAS; Acoust Soc Am; Aerospace Med Asn. Res: Basic and medical research in auditory and vestibular physiology, including behavioral hearing tests on animals, temporal bone histology, electrophysiology, electron microscopy and electronystagmography. Mailing Add: 1016 Westwood Dr Springfield PA 19064

CUTTER, LOIS JOTTER, b Weaverville, Calif, Mar 11, 14; m 42; c 2. SYSTEMATIC BOTANY, PLANT MORPHOLOGY. Educ: Univ Mich, BA, 35, MS, 36, PhD, 43. Prof Exp: Res asst plant breeding, Parke, Davis & Co, Mich, 42; lectr biol, 63-65,

ASST PROF BIOL, UNIV NC, GREENSBORO, 65- Mem: Sigma Xi. Mailing Add: Dept of Biol Univ of NC Greensboro NC 27412

CUTTER, PAUL RAMEY, b Hugoton, Kans, Aug 2, 15. PHYSICAL CHEMISTRY. Educ: Panhandle State Col, BS, 39; Okla State Univ, MS, 42; Univ Okla, PhD(phys chem), 50. Prof Exp: Asst chief chemist, Consol Vultee Aircraft Corp, 42-45; electrochemist, Hanson Van Winkle Munning Co, 45-46; instr chem, Univ Okla, 46-50; prof chem & head dept sci, Panhandle State Col, 50-53; tech dir & chief chemist, Texas City Chem Co, 53-55; res assoc, Diamond Shamrock Co, 55-66; res scientist, Armour Agr Chem Co, 66-68; ASSOC PROF CHEM & PHYSICS, PANHANDLE STATE COL, 68-, HEAD DEPT PHYSICS, 73- Mem: Am Chem Soc. Res: Anodic finishes on light metals; electron diffraction of molecular vapors; recovery of sodium salts from natural alkali deposits; liquid detergents; substrate coatings on metals; manufacturing of dicalcium phosphate. Mailing Add: Dept of Chem Panhandle State Col Goodwell OK 73939

CUTTING, WINDSOR COOPER, pharmacology, therapeutics, see 12th edition

CUTTITA, JOSEPH ANTHONY, b New York, NY, Oct 25, 10; m 40; c 6. DENTISTRY. Educ: Fordham Univ, AB, 32, MS, 35; Columbia Univ, DDS, 39. Prof Exp: From asst prof to assoc prof, 45-65, MEM FAC, SCH DENT & ORAL SURG, COLUMBIA UNIV, 39-, PROF DENT, 65-, ASST DEAN ADMIS, 74- Concurrent Pos: Consult, Vet Admin Hosp, New York, 50- Mem: AAAS; Am Dent Asn; NY Acad Sci; fel Am Col Dent; Int Asn Dent Res. Res: Oral diagnosis and oral medicine; mouth diseases. Mailing Add: 630 W 168th St New York NY 10032

CUTTITTA, FRANK, analytical chemistry, deceased

CUTTS, DAVID, b Providence, RI, Dec 12, 40; c 2. EXPERIMENTAL PARTICLE PHYSICS. Educ: Harvard Univ, AB, 62; Univ Calif, Berkeley, PhD(physics), 68. Prof Exp: Res assoc physics, State Univ NY Stony Brook, 68-71 & Rutherford High Energy Lab, Eng, 71-73; ASST PROF PHYSICS, BROWN UNIV, 73- Mem: Am Phys Soc. Res: Experimental studies of particle interactions and decays, using electronic, counter and wire chamber techniques. Mailing Add: Dept of Physics Brown Univ Providence RI 02912

CUTTS, JAMES HENRY, b Barnsley, Eng, June 12, 26; Can citizen; m; c 2. HISTOLOGY, HEMATOLOGY. Educ: Univ Sask, BA, 48; Dalhousie Univ, MSc, 56; Univ Western Ont, PhD(med res), 58. Prof Exp: Chief technician hemat, Gray Nuns Hosp, Regina, Sask, 48-50; technician & head dept, St Paul's Hosp, Vancouver, BC, 50-53; prof hematologist, NS Dept Pub Health, 53-56; from asst prof to assoc prof anat, Univ Western Ont, 60-68; assoc prof, 68-73, chmn dept, 72-73, PROF ANAT, SCH MED, UNIV MO-COLUMBIA, 73- Concurrent Pos: Nat Cancer Inst Can fel med res, Univ Western Ont, 56-68, Ross Mem fel, 58-60; lectr, Dalhousie Univ, 53-56; adv fac, Belknap Col, 64- Honors & Awards: SAMA Golden Apple Award, 72. Mem: Am Asn Cancer Res; NY Acad Sci; Am Asn Anatomists; Am Soc Zoologists; Can Asn Anatomists. Res: Vascular patterns in bone marrow and relations to cell types and hemopoiesis; oncology; postnatal development of the opossum; comparative hematology. Mailing Add: Dept of Anat Sch of Med Univ of Mo Columbia MO 65201

CUTTS, ROBERT IRVING, b Chicago, Ill, Jan 16, 15; m 48; c 2. PSYCHIATRY. Educ: Lewis Inst Technol, BS, 36; Univ Ill, MD, 40. Prof Exp: Clin asst internal med, Res & Educ Hosp, Univ Ill, 46-47; staff psychiatrist, Vet Admin Hosp, Ill, 49-53 chief continued treatment serv, 53-57; chief serv, Vet Admin Hosp, Tucson, Ariz, 57-59; pvt practr, 59-66; dir, Southern Ariz Ment Health Ctr, 66-68; chief psychiat serv, Vet Admin Hosp, 68-72; ASST PROF PSYCHIAT, MED COL, UNIV ARIZ, 72- Concurrent Pos: Clin asst psychiat, Med Sch, Northwestern Univ, 53-55; consult, Southern Ariz Med Health Ctr & Vet Admin Hosp; adj assoc prof, Med Col, Univ Ariz. Mem: Fel AAAS; AMA; fel Am Psychiat Asn. Res: Outpatient psychiatry. Mailing Add: 5232 E Rosewood Tucson AZ 85711

CUYKENDALL, TREVOR RHYS, b Denver, Colo, Nov 30, 05; m 31; c 2. APPLIED PHYSICS. Educ: Univ Denver, BSc, 26, MSc, 27; Cornell Univ, PhD(physics), 35. Prof Exp: Instr math, Univ Denver, 27-29; instr physics, 29-36, res assoc, 37-39, asst prof appl mech, 40-41, prof eng physics, 46-66, Spencer T Olin prof eng, 66-74, dir, Sch Eng Physics, 67-71, SPENCER T OLIN EMER PROF ENG, 74- Concurrent Pos: Consult, US Eng Dept, NY, 39-41 & Oak Ridge Nat Lab, 50-51; physicist, Naval Ord Lab, Washington, DC, 41-43; scientist, Manhattan Dist Proj, Los Alamos Sci Lab, 43-46; from asst dir to dir dept eng physics, Cornell Univ, 51-62, mem educ policy comt eng, 51-53, chmn, 53, mem gen comt, Grad Sch, 52-56 & 63, chmn univ fel bd, 54-57, mem univ admin bd, Div Unclassified Studies, 60-63, assoc dir eng physics & mat sci, 62-67; mem, Oak Ridge Inst Nuclear Studies-AEC Nuclear Eng Fel Bd, 63-65, chmn, 65. Honors & Awards: Naval Res & Develop Award, US Navy. Mem: Fel Am Phys Soc; Am Soc Eng Educ; Am Nuclear Soc. Res: Physics of solids; properties of materials; x-rays; electricity and magnetism; photoelasticity; nuclear reactor physics. Mailing Add: Sch of Eng Physics Cornell Univ Ithaca NY 14850

CVANCARA, ALAN MILTON, b Ross, NDak, Mar 7, 33; m 59; c 2. PALEOBIOLOGY. Educ: Univ NDak, BS, 55, MS, 57; Univ Mich, PhD(geol), 65. Prof Exp: From asst prof to assoc prof, 63-72, PROF GEOL, UNIV N DAK, 72- Mem: Am Paleont Soc; Geol Soc Am; Brit Palaeont Asn; Sigma Xi; Am Malacol Union. Res: Upper Cretaceous and Lower Tertiary stratigraphy and mollusks in north central United States; Quaternary freshwater and terrestrial mollusks. Mailing Add: Dept of Geol Univ of NDak Grand Forks ND 58202

CVANCARA, VICTOR ALAN, b Ross, NDak, Aug 29, 37; m 58; c 4. PHYSIOLOGY. Educ: Minot State Col, BS, 59; Univ SDak, MS, 64, PhD(physiol), 68. Prof Exp: Sci instr, Seyyida Maatuka Col, Zanzibar, EAfrica, 61-63; teaching asst, Univ SDak, 64-66, res assoc physiol, 67-69; fel, Ore State Univ, 69-71; asst prof, Beloit Col, 71-72; ASSOC PROF PHYSIOL, UNIV WIS-EAU CLAIRE, 72- Mem: Am Soc Zoologists; Sigma Xi; Am Fisheries Soc. Res: Temperature adaptation, protein synthesis and enzyme induction in teleost fishes. Mailing Add: Dept of Biol Univ of Wis Eau Claire WI 54701

CVETANOVIC, RATIMIR J, b Belgrade, Yugoslavia, May 19, 13; nat US; m 55; c 3. PHYSICAL CHEMISTRY. Educ: Univ Edinburgh, BSc, 36; Univ Belgrade, BASc, 42; Univ Toronto, MA, 50, PhD(chem), 51. Prof Exp: With lead-zinc flotation plant, Trepca Mines, Ltd, 37-38; with agr res div, Ministry Agr, Yugoslavia, 38-42; supt, Coop Veg Oil Factory, 42-45; asst lectr phys chem, Univ Belgrade, 45-47; fel, 51-52, res officer, 52-59, PRIN RES OFFICER, DIV CHEM, NAT RES COUN CAN, 59-, HEAD KINETICS, PHOTOCHEM & CATALYSIS SECT, 55- Concurrent Pos: Vis prof, Cornell Univ, 63 & Univ Calif, Davis, 64; overseas fel, Churchhill Col, Cambridge, 67. Mem: Fel Royal Soc Can. Res: Chemical kinetics; photochemistry; heterogeneous catalysis. Mailing Add: Div of Chem Nat Res Coun Ottawa ON Can

CVIJANOVICH, GEORGE, b Maradik, Yugoslavia, Oct 29, 21; nat US; m 50. PHYSICS. Educ: Univ Vienna, BA, 43; Columbia Univ, MA, 56; Univ Bern, PhD(physics), 61. Prof Exp: Asst, Nat Ctr Sci Res, France, 47-51; head res & develop, Comet A G Bern, Switz, 51-52; asst, Radiation Lab, Columbia Univ, 52-54; asst prof physics, Upsala Col, 54-57; res asst, Lamont Geol Observ, Columbia Univ, 57-64; privat docent theoret physics, Phys Inst, Univ Bern, 64-67; prof, 67-73, WAHLSTROM FOUND PROF PHYSICS, UPSALA COL, 73- Concurrent Pos: Chief scientist, Proj Ice Skate, Drift Sta Alpha, Arctic, 58. Mem: Am Phys Soc; Am Asn Physics Teachers. Res: Geophysics; magnetohydrodynamics; nuclear physics as applied to earth science; physics of elementary particles. Mailing Add: Dept of Physics Upsala Col East Orange NJ 07017

CWALINA, GUSTAV EDWARD, b Baltimore, Md, Feb 6, 09; m 43; c 2. MEDICINAL CHEMISTRY. Educ: Univ Md, BS, 31, MS, 33, PhD(pharmaceut chem), 37. Prof Exp: Asst chem, Univ Md, 31-37; from asst prof to assoc prof, Creighton Univ, 37-42; from asst prof to prof, 46-75, EMER PROF PHARMACEUT SCI, COL PHARM, PURDUE UNIV, 75-, ASSOC DEAN, 63- Mem: Am Chem Soc; Am Pharmaceut Asn. Res: Phytochemical investigation of Ipomea pes-caprae; plant analysis. Mailing Add: 1634 Northwestern Ave West Lafayette IN 47906

CYBRIWSKY, ALEX, b Pidsosniw, Ukraine, Mar 26, 14; US citizen; m 45; c 4. PHYSICS. Educ: Univ Vienna, PhD(physics), 45. Prof Exp: Asst physics, Univ Vienna, 43-46; chief chemist, Archer Chem Co, Ky, 50-51; sect head, Reynolds Metals Co, 51-58, proj dir, 58-60; res physicist, Gen Elec Co, 60-62; RES PHYSICIST, ALLIS-CHALMERS MFG CO, 62- Mem: Am Phys Soc; Am Soc Metals. Res: Nuclear and cosmic rays studies by nuclear photographic emulsions; recovery, finishing and alloying of aluminum; alumina catalyst; thermoelectric and galvano-thermomagnetic phenomena; energy conversion. Mailing Add: Allis-Chalmers Mfg Co PO Box 512 Milwaukee WI 53201

CYGAN, NORBERT EVERETT, b Chicago, Ill, June 5, 30; m 57; c 2. GEOLOGY, PALEONTOLOGY. Educ: Univ Ill, PhD(geol), 62. Prof Exp: Instr geol, Ohio Wesleyan Univ, 56-59; asst, Univ Ill, 59-62; div stratigrapher, 62-71, MINERALS STAFF GEOLOGIST, CHEVRON OIL CO, 71- Concurrent Pos: Lectr, Univ Houston. Mem: Am Asn Petrol Geol; fel Geol Soc Am. Res: Cambrian and Ordovician conodonts; tertiary micro fauna; stratigraphy; uranium geology; paleoecology. Mailing Add: Chevron Oil Co Minerals Staff PO Box 599 Denver CO 80201

CYMERMAN, ALLEN, b New York, NY, Feb 4, 42; m 66; c 1. PHYSIOLOGY, BIOCHEMISTRY. Educ: Rutgers Univ, BA, 63; Jefferson Med Col, MS, 66, PhD(physiol), 68. Prof Exp: Res technician clin endocrinol, Temple Univ Hosp & Med Sch, 63-64; sr investr biochem & pharmacol, 68-70, SR INVEST ALTITUDE RES DIV, US ARMY RES INST ENVIRON MED, 70- Res: Effects of environmental extreme on normal and drug altered physiological and biochemical mechanisms. Mailing Add: Altitude Res Div US Army Inst of Environ Med Natick MA 01760

CYNKIN, MORRIS ABRAHAM, b Brooklyn, NY, Nov 23, 30; m; c 2. BIOCHEMISTRY, MICROBIOLOGY. Educ: City Col New York, BS, 52; Cornell Univ, PhD(bact), 57. Prof Exp: Vis scientist, NIH, 58-60; from asst prof to assoc prof, 60-71, PROF BIOCHEM, SCH MED, TUFTS UNIV, 71- Concurrent Pos: Vis fel biochem, Cornell Univ, 56-58; USPHS fel, 56-58; Arthritis & Rheumatism Soc fel, 58-59; Am Cancer Soc fel, 59-60; USPHS res career develop award, 61-71. Mem: Am Soc Microbiol; Am Chem Soc; Am Soc Biol Chem. Res: Carbohydrate metabolism; structure and biosynthesis of bacterial and mammalian polysaccharides and glycoproteins. Mailing Add: Dept of Biochem & Pharmacol Tufts Univ Med Sch Boston MA 02111

CYR, GILMAN NORMAN, b Van Buren, Maine, Feb 14, 23; m 45; c 6. PHARMACY. Educ: Mass Col Pharm, BS, 43, MS, 47; Purdue Univ, PhD(pharm), 49. Prof Exp: Asst pharm, Mass Col Pharm, 46, asst prof, 48-52; head dept pharm res, 52-66, asst dir pharmaceut res, 66-70, DIR PHARMACEUT RES & DEVELOP, SQUIBB INST MED RES, 70- Mem: AAAS; Am Pharmaceut Asn; Soc Cosmetic Chem. Res: Pharmaceutical formulations, especially oral and dermatological products. Mailing Add: Squibb Inst for Med Res Pharmaceut Res & Develop Dept New Brunswick NJ 08903

CYR, WILBUR HOWARD, b New Haven, Vt, Mar 16, 43. RADIATION GENETICS. Educ: Univ Vt, BA, 65; Pa State Univ, MS, 66, PhD(biophys), 72. Prof Exp: Res biophys, Armed Force Inst Path, 70-74; RES RADIATION GENETICS, DIV BIOL EFFECTS, BUR RADIOL HEALTH, FOOD & DRUG ADMIN, DEPT HEALTH, EDUC & WELFARE, 74- Mem: Biophys Soc; Sigma Xi. Res: Electron microscopic studies of radiation induced chromosomal aberrations. Mailing Add: Bur Radiol Health HFX-120 5600 Fishers Lane Rockville MD 20852

CYWINSKI, NORBERT FRANCIS, b Brown Co, Wis, Aug 24, 29; m 61; c 1. ORGANIC CHEMISTRY. Educ: Univ Wis, BS, 54; Northwestern Univ, PhD(org chem), 62. Prof Exp: Res chemist, Phillips Petrol Co, 59-65, SR RES CHEMIST, EL PASO PROD CO, 65- Concurrent Pos: Part-time instr org chem, Odessa Col. Mem: Am Chem Soc. Res: Bicyclic systems; cyclization and dehydrocyclization reactions; free radical reactions; diolefin synthesis; hydrocarbon oxidation; petrochemicals; nylon intermediates. Mailing Add: El Paso Prod Co PO Box 3986 Odessa TX 79760

CZAMANSKE, GERALD KENT, b Chicago, Ill, Jan 17, 34; m 57; c 2. GEOCHEMISTRY. Educ: Univ Chicago, BA, 53, BS, 55; Stanford Univ, PhD(geol), 61. Prof Exp: NSF fel, Univ Oslo, 60-61; res assoc geochem, Mass Inst Technol, 61-62; asst prof geol, Univ Wash, 62-65; GEOLOGIST, US GEOL SURV, 65- Mem: Mineral Soc Am; Geochem Soc; Soc Econ Geologists; Mineral Asn Can. Res: Inorganic geochemistry; chemistry of hydrothermal ore metal transport and deposition; sulfide mineralogy and phase relationships; geochemical problems relating to igneous and metamorphic petrology. Mailing Add: 750 W Greenwich Pl Palo Alto CA 94303

CZANDERNA, ALVIN WARREN, b LaPorte, Ind, May 27, 30; m 53; c 3. SURFACE PHYSICS, SURFACE CHEMISTRY. Educ: Purdue Univ, BS, 51, PhD(phys chem), 57. Prof Exp: Jr res metallurgist, Aluminum Co Am, Pa, 53; sr res physicist, Parma Res Lab, Union Carbide Corp, 57-63, res scientist, Chem Div, 63-65; assoc prof physics & charter mem inst colloid & surface sci, 65-68, assoc dir inst colloid & surface sci, 68-74, COUN MEM, INST COLLOID & SURFACE SCI, CLARKSON COL TECHNOL, 66-, PROF PHYSICS, COL, 68- Concurrent Pos: Consult, Los Alamos Sci Lab, 68- & Owens-Ill Inc, 69-70; vis scientist, Fritz-Haber-Inst, Max Planck Soc, Berlin, Ger, 71-72. Mem: Am Chem Soc; fel Am Phys Soc; sr mem Am Vacuum Soc; fel NY Acad Sci; Catalysis Soc. Res: Surface science; desorption; chemisorption; reactions at interfaces; properties of metal and metal oxide thin films; catalysis; oxidation; applications of vacuum ultramicrogravimetry and ion scattering spectrometry. Mailing Add: 3 Berkley Dr Potsdam NY 13676

CZAPSKI, ULRICH HANS, b Munich, Ger, July 15, 25; m 54. METEOROLOGY, GEOPHYSICS. Educ: Univ Hamburg, PhD(meteorol), 53. Prof Exp: Guest staff mem,

Int Inst Meteorol, Stockholm, 52; staff mem geophys instruments, Askania-Werke, AG, Berlin, Ger, 53-55; geophysicist, Seismos GmbH, Hannover, 55-61; UN tech expert meteorol & geophys, Mission to Pakistan, World Meteorol Orgn, UN Tech Assistance Admin, 61-62; lectr meteorol, Imp Col, Univ London, 62-64; ASSOC PROF ATMOSPHERIC SCI, STATE UNIV NY ALBANY, 64- Mem: Am Meteorol Soc; Am Geophys Union. Res: General circulation of the atmosphere; numerical forecasting; turbulent transfer and convection; hydrometeorology. Mailing Add: Dept of Atmospheric Sci State Univ of NY Albany NY 12222

CZARNECKI, CAROLINE MARY ANNE, b Detroit, Mich, Aug 3, 29. VETERINARY ANATOMY. Educ: Bemidji State Col, BS, 50; State Col Iowa, MA, 60; Univ Minn, Minneapolis, PhD(anat), 67. Prof Exp: Instr high schs, 50-62; teaching asst anat, 62-67, instr, 67, asst prof vet anat, 67-71, ASSOC PROF VET ANAT, UNIV MINN, ST PAUL, 71- Mem: AAAS; Am Asn Vet Anat; World Asn Vet Anat; Am Asn Anatomists. Res: Histochemistry of glycogen; Purkinje system of heart; roundheart disease in turkeys; glycogen storage diseases. Mailing Add: Dept of Vet Biol Univ of Minn St Paul MN 55108

CZARNECKI, REYNOLD BERNARD, b Erie, Pa, Oct 28, 13. MICROBIOLOGY. Educ: Pa State Univ, BS, 39; Univ Ill, MS, 42, PhD(bact), 48. Prof Exp: Instr bact, Mich State Univ, 48-51; ASST PROF BACT, UNIV MASS, 51- Mem: Am Soc Microbiologists; fel Am Pub Health Asn; Royal Soc Health. Res: General and medical microbiology. Mailing Add: Dept of Microbiol Univ of Mass Amherst MA 01002

CZARNETZKY, EDWARD JOHN, b La Crosse, Wis, May 20, 05; m 43; c 2. AGRICULTURE. Educ: Univ Minn, BS, 30; Univ Calif, PhD(biochem), 33. Prof Exp: Assoc, Mt Zion Hosp, 34; Merck fel, Univ Pa, 35, Merck fel bact, Sch Med & instr, Grad Sch Med, 36-39; chief biologist, Wilson & Co, Inc, 39-55; dean agr educ, 55-66, vpres, 66-71, EMER PROF AGR, WILLIAM H MINER AGR RES INST, 71-; DIR RES, INT STOCK FOOD CORP, 73- Concurrent Pos: Consult, Almatroud Enterprises, Saudi Arabia, 74- Res: Phase rule applied to biology; biochemistry of proteins; enzymes of leucocytes; antigens of hemolytic streptococci; intravenous antisepsis; mercurial protein compounds; electro-ultrafiltration; nutrition of animals; bacteriology of foods. Mailing Add: Miner Inst Chazy NY 12921

CZECH, MICHAEL PAUL, b Pawtucket, RI, June 24, 45; m 69; c 1. BIOCHEMISTRY. Educ: Brown Univ, BA, 67; Duke Univ, PhD(biochem), 72; Duke Univ, MA, 69. Prof Exp: Fel biochem, Duke Univ, 72-74; ASST PROF PHYSIOL CHEM, DIV BIOL & MED SCI, BROWN UNIV, 74- Honors & Awards: Res & Develop Award, Am Diabetes Asn, 74; Elliot P Joslin Award, 75. Mem: Endocrine Soc; Am Diabetes Asn. Res: Biochemistry of the modulation of hexose transport activity by insulin and other agents in plasma membranes from isolated fat cells. Mailing Add: Div of Biol & Med Sci Brown Univ Providence RI 02912

CZEISLER, JEFFREY LANCE, b New York, NY, July 20, 42; m 73. BIOPHYSICAL CHEMISTRY. Educ: Harpur Col, BA, 64; State Univ NY Binghamton, MA, 66; Case Western Reserve Univ, PhD(chem), 70. Prof Exp: Fel sch med, Johns Hopkins Univ, 70-71, NIH fel, 72, fel, 72-73; Fulbright-Hays sr fel & vis prof biochem, Univ Cayetano Heredia, Peru, 73; instr physiol chem sch med, Johns Hopkins Univ, 74; RES INVESTR PHYS BIOCHEM, SEARLE LABS, G D SEARLE & CO, 75- Mem: Am Chem Soc; Am Asn Univ Prof; NY Acad Sci. Res: Molecular associations of biological importance; uses of nuclear magnetic resonance in biochemistry and biophysics. Mailing Add: Searle Labs G D Searle & Co Box 5110 Chicago IL 60680

CZEPIEL, THOMAS P, b Deep River, Conn, May 2, 32; m 55; c 3. PAPER CHEMISTRY. Educ: Wesleyan Univ, BA, 54; Inst Paper Chem, Lawrence Univ, MS, 56, PhD(chem), 59. Prof Exp: Res group leader, 59-61, mgr tech servs paper mill, Maine, 61-64, tech dir, Detroit, Mich, 64-68, Pa, 68-72, PAPER MILL SUPT, SCOTT PAPER CO, CHESTER, PA, 72- Mem: Am Chem Soc; Tech Asn Pulp & Paper Indust. Res: Influence of metal traces on color stability of cellulose; preparation and utilization of synthetic fibers for papermaking; pulp and paper technology. Mailing Add: Scott Paper Co Front & Market Sts Chester PA 19013

CZEPYHA, CHESTER GEORGE REINHOLD, b New Brunswick, NJ, Sept 14, 27; m 50; c 5. ATMOSPHERIC PHYSICS. Educ: Lehigh Univ, BSEE, 50; Air Force Inst Technol, MSEE, 56. Prof Exp: Navigator, US Air Force, 50-56, proj officer, Missile Develop, Air Force Missile Test Ctr, 56-60, electronics engr, Air Force Cambridge Res Labs, 62-69, master navigator, Southeast Asia, 69-70, wing chief of plans, Repub of Philippines, 71, chief eng systs prog off, Electronic Systs Div, LG Hascom Field, US Air Force, 71-73; dir metrol, Directorate Metrol, 73, V COMDR, AEROSPACE GUIDANCE AND METROL CTR, 73- Concurrent Pos: US Air Force, 50- Mem: Am Geophys Union; Inst Navigation. Res: Sterics as a means of severe weather identification; development of altitude balloon systems for upper air research. Mailing Add: 634 Kennedy St Newark OH 43055

CZEREPINSKI, RALPH G, synthetic chemistry, see 12th edition

CZERLINSKI, GEORGE HEINRICH, b Cuxhaven, Ger, Dec 31, 24; US citizen; c 1. BIOPHYSICS, BIOCHEMISTRY. Educ: Univ Hamburg, BA, 52; Northwestern Univ, MS, 55; Univ Göttingen, PhD(phys chem), 58. Prof Exp: Res assoc biophys, Univ Pa, 60-64, asst prof, 64-67; ASSOC PROF BIOCHEM, NORTHWESTERN UNIV, CHICAGO, 67- Concurrent Pos: E R Johns Found fel, Univ Pa, 58-60; vis asst prof biochem, Cornell Univ, 66-67. Mem: AAAS; Am Phys Soc; Ger Chem Soc; Biophys Soc; Asn Comput Mach. Res: Instrumentation, theory and application of chemical relaxation for the investigation of the mechanism of enzyme function; rapid kinetics of enzyme reactions; allosteric behavior; temperature-jump method. Mailing Add: Dept of Biochem Northwestern Univ Chicago IL 60611

CZERNIAKIEWICZ, ANASTASIA JUANA, b Buenos Aires, Arg, July 23, 43. MATHEMATICS. Educ: Univ Buenos Aires, Lic, 66; NY Univ, PhD(math), 70. Prof Exp: J F Ritts asst prof math, Columbia Univ, 71-74; ASST PROF MATH, QUEENS COL, CITY UNIV NEW YORK, 74- Mem: Am Math Soc. Res: Theory of graphs; problems involving coloring of graphs, including ramsey type questions, groups of graphs and chromatic numbers. Mailing Add: Dept of Math Queens Col City Univ of New York Flushing NY 11367

CZERNOBILSKY, BERNARD, b Königsberg, Ger, Jan 1, 28; US citizen; m 54; c 2. PATHOLOGY. Educ: Univ Lausanne, MD, 53. Prof Exp: Intern, Jewish Hosp Asn, Cincinnati, Ohio, 53-54, asst resident path, 54-55; asst resident, Israel Beth Hosp, Boston, 57-58, resident, 58-59; chief resident, Mallory Inst, Boston City Hosp, 59-60; assoc pathologist, Women's Med Col Pa, 60-61, from asst prof to assoc prof, 61-65; asst prof surg path, Lab Path Anat, Hosp Univ Pa, 65-69, assoc prof path, 69-70; CHIEF DEPT PATH, KAPLAN HOSP, ISRAEL, 70-; ASSOC PROF PATH, HEBREW UNIV, JERUSALEM, 71- Concurrent Pos: Teaching fel path, Harvard Med Sch, 58-59; instr, Sch Med, Tufts Univ, 59-60. Mem: NY Acad Sci; Am Asn Pathologists & Bacteriologists; Inst Acad Path. Res: Spleen; pancreas; kidneys; breast; cardiovascular system; tissue culture; bilirubin metabolism; gynecologic pathology; ovarian tumors. Mailing Add: Dept of Path Kaplan Hosp Rehovot Israel

CZERWINSKI, ANN LANGLEY, b Omaha, Nebr, Mar 29, 14; m 37; c 1. PHARMACOLOGY. Educ: Creighton Univ, BA, 35, BS, 48, MS, 51; Univ Nebr, PhD, 67. Prof Exp: Teacher & prin, Pub Sch, Nebr, 35-37; from instr to prof, 48-67, actg dean, Sch Pharm, 71-72, PROF PHARMACOL, SCH PHARM, CREIGHTON UNIV, 67-, PROF PHARMACOL, SCH MED, 70- Concurrent Pos: Dir, Omaha Drug Abuse Educ & Info Ctr, 70- Mem: AAAS. Res: Bromides; relationship of estrogens to personality; Tolbutamide and insulin. Mailing Add: Sch of Pharm Creighton Univ Omaha NE 68178

CZERWINSKI, ANTHONY WILLIAM, b St Louis, Mo, Feb 10, 34; m 64; c 3. NEPHROLOGY, PHARMACOLOGY. Educ: St Louis Univ, BS, 55, MD, 59. Prof Exp: Fel clin nephrology, Vet Admin Hosp, Boston, Mass, 63-64; fel metab, Univ NC, 64-66; asst prof med, 69-72, ASSOC PROF MED, HEALTH SCI CTR, UNIV OKLA, 72- Concurrent Pos: Chief spec diag & treatment unit, Oklahoma City Vet Admin Hosp, 73-; assoc chief renal sect & chief clin nephrology, Univ Okla Health Sci Ctr & Oklahoma City Vet Admin Hosp, 75- Mem: Sigma Xi; Am Soc Nephrology; Am Col Physicians; Int Soc Nephrology; Am Soc Pharmacol & Exp Therapeut. Mailing Add: Univ of Okla Health Sci Ctr PO Box 26901 Oklahoma City OK 73190

CZIFFRA, PETER, theoretical physics, see 12th edition

CZUBA, LEONARD J, b East Chicago, Ind, Feb 28, 37; m 60; c 3. MEDICINAL CHEMISTRY. Educ: Ind Univ, AB, 61; Univ Minn, PhD(org chem), 67. Prof Exp: Chemist, Sinclair Res, Inc, 61-63; NIH fel org chem, Mass Inst Technol, 67-68; res chemist, 68-72, PROJECT LEADER, PFIZER, INC, 72- Mem: Am Chem Soc; NY Acad Sci. Res: Synthetic organic chemistry. Mailing Add: Cent Res Pfizer Inc Groton CT 06340

CZYZAK, STANLEY JOACHIM, b Cleveland, Ohio, Aug 21, 16; m 42; c 4. PHYSICS. Educ: Fenn Col, BS(chem eng), 35, BS(civil eng), 36; John Carroll Univ, MS, 39; Univ Cincinnati, DSc(physics), 48. Prof Exp: Res metallurgist, Aluminum Co of Am, 36; chief metallurgist, Master Metals, Inc, 36-38; res engr, Una Welding, Inc, 38-40; assoc res physicist, Argonne Nat Lab, 48-49; res physicist, Battelle Mem Inst, 49-50; asst chief physics sect, Univ Detroit, 50-51; chief basic physics sect, Aeronaut Res Lab, Wright Air Develop Ctr, Wright-Patterson AFB, 51-53; from assoc prof to prof physics, Univ Detroit, 53-61; dir gen physics lab, Aerospace Res Labs, 61-66, PROF ASTRON, OHIO STATE UNIV, 66- Mem: Am Astron Soc; Royal Astron Soc. Res: Atomic structure calculations; transition probabilities; collision cross-sections; gaseous nebulae. Mailing Add: Dept of Astron Ohio State Univ Columbus OH 43210